BIBLIOGRAPHY OF GEOSCIENCE THESES OF THE UNITED STATES AND CANADA

AMERICAN GEOLOGICAL INSTITUTE
Alexandria, Virginia
1993

BIBLIOGRAPHY OF GEOSCIENCE THESES OF THE UNITED STATES AND CANADA

Volume I

Geoscience Theses

Degree Recipients at Each Institution

Thesis Subject Distribution by Decade and by Institution

AMERICAN GEOLOGICAL INSTITUTE
Alexandria, Virginia
1993

Library of Congress Cataloging-in-Publication Data

Bibliography of geoscience theses of the United States and Canada.
    v, 3779 p.   cm.
    Contents: v. 1. Geoscience theses, degree recipients at each institution, thesis subject distribution by decade and by institution — v. 2. Subject and geographic index, A-L — v. 3. Subject and geographic index, M-Z.
    ISBN 0-922152-18-7 (set : alk. paper) : $495.00
    1. Earth sciences--United States--Bibliography. 2. Earth sciences--Canada--Bibliography. 3. Dissertations, Academic--United States--Bibliography. 4. Dissertations, Academic--Canada--Bibliography.
I. American Geological Institute.
Z6034.U5B54 1993
[QE76]
016.557--dc20   92-44740
                                     CIP

ISBN 0-922152-18-7

INTRODUCTION

This bibliography identifies 20,748 doctoral dissertations and 41,464 masters theses in the geosciences issued by academic institutions in the United States and Canada from 1867 through 1988. The subjects covered are described in the beginning of the Subject Distribution section.

Compilation Process

In compiling this publication, we sent listings of theses to each geoscience degree-granting institution for which there were theses in the GeoRef database. Most of these institutions responded by returning the lists with additions and corrections. The individuals who participated are listed in the Acknowledgements section which follows.

It should be pointed out that the librarians and others who supplied or checked much of the theses data included herein worked from the source documents and/or from library catalogues. They devoted hours to varifying this data. Often it was they who supplied the full names of authors and the page numbers. Their efforts are much appreciated and certainly enhanced the quality of this bibliography.

For over fifteen years, AGI has sent letters annually to all geoscience degree-granting institutions in the United States and Canada requesting information on new theses. Responses have been added to GeoRef and are included in this publication. Also, since 1969, relevant sections of *Dissertation Abstracts International* have been scanned for geoscience-related dissertations and these have been added to GeoRef. We also combed over 20 geoscience theses bibliographies for additional theses. These bibliographies are listed in the Acknowledgements section.

Although this theses compilation is by far the most comprehensive in the geosciences, coverage is partial for institutions which did not return the lists sent them and for related sciences and technologies such as agriculture.

The Sections of this Bibliography

Section 1. Geoscience Theses

In this section, all theses are listed alphabetically by degree recipient. Each listing gives the degree recipient's name, thesis title, degree-granting institution, degree level, year, and number of pages. When a thesis has multiple authors, all are listed, and an entry occurs for each author. Names of degree-granting institutions have been standardized, and we have used current institution names. If our thesis source showed the "University of Wisconsin" as the degree-granting institution, we have determined which campus of the university the student attended and used the current name of that campus as the degree-granting institution. In the few cases where we did not determine this, we have put brackets around the degree-granting institution. The brackets indicate some uncertainty on our part, and are shown for about 300 theses. Where number of pages is not shown, our sources did not include it.

Section 2. Degree Recipients at Each Institution

This section gives for each institution its location and postal code, the total doctoral dissertations and masters theses in this bibliography, and lists the degree recipients for each decade, from the earliest we have for each institution through 1988.

Section 3. Theses Subject Distribution

In this section, totals on doctoral dissertations and masters theses in each decade are given for each of 30 subject categories as well as for each institution.

Section 4. Subject and Geographic Index

This index contains, on average, 13.6 entry points per thesis, of which 4.8 are subject headings and 8.8 are cross references. The cross references are based on hierarchical, synonymous, and other relationships between terms, as laid out in the GeoRef Thesaurus. The subject index is the same sort as that to be found in issues of the *Bibliography and Index of Geology*, from January 1993 on.

The GeoRef staff indexed many of the theses from author abstracts. But for some of the theses selected from bibliographies or cited in lists we received from institutions, we had no abstracts and indexing was based on titles only.

The index terms have been selected from the controlled vocabulary in the *GeoRef Thesaurus*. The *Thesaurus*, now in its 6th edition, contains over 27,000 geographic, systematic geoscience, and general terms. The terminology in the *Thesaurus* came from the geoscience literature indexed in the GeoRef database since 1967. As terms were encountered and needed, they were added to the *Thesaurus*. The *GeoRef Thesaurus* is an information retrieval thesaurus, not a Roget's type thesaurus.

Availability of Theses and Dissertations

To obtain copies of masters theses found in this bibliography, the primary source is the degree-granting institution. The best approach is to contact the library in the institution. Seldom will a library lend a thesis, but usually it will supply a photocopy of the thesis for a fee. However, the library may first need to obtain permission from the author to copy the thesis.

University Microfilms can provide copies of many of the doctoral dissertations either on microfilm or paper. To find out if they have a particular dissertation and to place an order call 1-800-521-0600, ext. 2533 or 2534. If University Microfilms cannot provide the dissertation, contact the library of the degree-granting institution.

Acknowledgements

In compiling this bibliography, AGI sent lists of theses already in GeoRef to the degree-granting institutions for checking. Most of the lists were addressed to geology or science librarians while some went to geology departments in the institutions.

Individuals in many of these institutions took the time to edit the data we had sent them and to supplement it with additional thesis and/or dissertation references. Without the generous assistance of the following individuals, this compilation would be much less comprehensive and authoritative.

University of Alaska, Fairbanks, Julia H. Triplehorn, Geophysical Institute

Alfred University, Carla C. Freeman

American University, Kay Schmidt, Bender Library

Amherst College, John T. Cheney, Department of Geology

Arizona State University, Jonathan Fink, Geology Department

Auburn University, Alabama, Robert B. Cook, Department of Geology

Bates College, John W. Creasy, Department of Geology

Baylor University, Judy Brink

Boston University, Christopher T. Baldwin, Department of Geology

Brigham Young University, Richard E. Soares, Library

Brock University, Ian Gordon, Library

Brown University, Linda Worden and Jean Waage

Bryn Mawr College, William A. Crawford

University of Calgary, Marjorie (Midge) King, Gallagher Library

University of California, Berkeley, Philip Hoehn, Earth Sciences Library

University of California, Davis, Carol J. La Russa, Physical Sciences Library

University of California, Los Angeles, Michael M. Noga, Geology Library

University of California, Riverside, Barbara Haner and Lizbeth Langston, Physical Sciences Library

University of California, San Diego, J. Freeman Gilbert, Scripps Institute of Oceanography

University of California, Santa Barbara, Michael Fuller, Department of Geological Sciences

University of California, Santa Cruz, Suzanne Harris, Earth Sciences

Carleton College, Edward Buchwald

California Institute of Technology, Jim O'Donnell, Geology Library

California State University, Chico, Howard Stensrud, Department of Geology and Physical Science

California State University, Hayward, Alexis N. Moiseyev

California State University, Long Beach, Stanley C. Finney, Department of Geological Sciences

California State University, Northridge, Terry Dunn

Carleton University, F. A. Michel, Department of Earth Sciences

Catholic University of America, Harriet Nelson, Library

University of Chicago, Marilyn Bowie, Department of Geophysical Sciences

Clarkson University, Gayle Berry, Schuler Educational Resources Center

Clemson University, Villard S. Griffin, Jr., Department of Earth Sciences

Colorado School of Mines, Lisa G. Dunn and Ann A. Lerew, Arthur Lakes Library

Colorado State University, Donald O. Doehring, Department of Earth Resources

University of Colorado, Colorado Springs, Paul K. Grogger

Columbia University, Susan Klimley, Lamont-Doherty Geological Observatory

Columbia University, Prof. P. Somasundaran, H. Krumb School of Mines

University of Connecticut, Norman Gray, Department of Geology & Geophysics

Cornell University, Donald L. Turcotte, Department of Geological Sciences

Dalhousie University, Patrick J. C. Ryall, Department of Earth Sciences

Dartmouth College, Barbara DeFelice, Kresge Physical Science Library

University of Delaware, Billy P. Glass, Department of Geology

DePauw University, Frederick M. Soster, Department of Geology & Geography

Duke University, Bruce Corliss, Department of Geology

East Carolina University, Scott W. Snyder, Department of Geology

Eastern Kentucky University, Fawn Tribble

Eastern Washington University, Candy Oswald

Ecole Polytechnique, Rene DuFour, Department de Genie Mineral

Emory University, William B. Size

University of Florida, Pam Cenzer

Fordham University, Joseph Loschiavo, Library

Franklin and Marshall College, Edward C. Beutner, Department of Geology

George Washington University, George C. Stephens, Department of Geology

University of Georgia, James A. Whitney, Department of Geology

Georgia Institute of Technology, William L. Chameides

Georgia State University, Vernon J. Henry, Jr., Department of Geology

University of Georgia, Beatrice R. Stephens, Department of Geology

Harvard University, Connie Wick, Kummel Library

University of Houston, Arch M. Reid, Department of Geosciences

University of Idaho, John D. Kawula, Library

University of Illinois at Urbana-Champaign, Lois M. Pausch, Geology Library

Indiana State University, Susan Thompson, Science Library

Indiana University of Pennsylvania, Frank W. Hall, Department of Geoscience

Indiana University, Lois Heiser, Geology Library

University of Iowa, Louise S. Zipp, Geology Library

The Johns Hopkins University, Bruce D. Marsh, Department of Earth & Planetary Sciences

University of Kentucky, Mary R. Spencer

Lakehead University, Graham J. Borradaile, Department of Geology

Lamar University, Donald E. Owen, Department of Geology

Laurentian University, Anthony E. Beswick, Department of Geology

Universite Laval, Michel Rocheleau, Department of Geology

Lehigh University, Bobb Carson, Department of Geological Sciences

Loyola University, Tara Fulton, Library

University of Maine, Muriel A. Sanford

Marshall University, Richard B. Bonnett, Department of Geology

Massachusetts Institute of Technology, Karen Campbell, Lindgren Library

University of Massachusetts, Amherst, Laurence M. Feldman, Biology Science Library

McGill University, S. M. Jackson, Department of Geological Sciences

McMaster University, J. Allen, Department of Geology

Memorial University of Newfoundland, Dianne E. Taylor-Harding, Library

University of Miami, Peter K. Swart, Division of Marine Geology & Geophysics

University of Miami, Coral Gables, Cesare Emiliani, Department of Geology

University of Michigan, Patricia B. Yocum, Library

Michigan State University, Diane Baclawski, Geology Library

Michigan Technological University, William I. Rose, Jr.

University of Minnesota, Kathy Ohler

Mississippi State University, Charles L. Wax, Department of Geology & Geography

University of Missouri, Kansas City, Edwin D. Goebel, Department of Geosciences

Montana College of Mineral Science & Technology, Jean Bishop

Montana State University, Stephan G. Custer, Department of Earth Science

University of Montana, Irene Evers, Mansfield Library

Mount Holyoke College, Mark A. S. McMenamin, Department of Geology and Geography

Murray State University, Neil V. Weber, Department of Geosciences

University of Nevada, Reno, Nancy L. Martineau, Mines Library

University of New Brunswick, Diane Tabor, Department of Geology

University of New Hampshire, S. Lawrence Dingman, Department of Earth Sciences

University of New Mexico, Diane K. Sparago, Department of Geology

New Mexico Institute of Mining and Technology, John Schlue, Department of Geoscience

New Mexico State University, Russell E. Clemons, Department of Earth Science

New Mexico Tech, Laura Wolf, Department of Geoscience

University of North Carolina, Chapel Hill, Rachel Harlan, Geology Library

University of North Carolina, Wilmington, William Harris, Department of Earth Science

University of North Dakota, Frank R. Karner

Northeast Louisiana University, Mervin Kontrovitz, Department of Geosciences

Northern Arizona University, Ronald C. Blakey, Department of Geology

Northern Illinois University, Amy Polzin, Department of Geology

University of Notre Dame, James A. Rigert, Department of Earth Science

University of Oklahoma, Claren M. Kidd, Youngblood Geology Library

Oregon State University, Allen Agnew, Department of Geosciences

University of Oregon, Elizabeth Orr and Bradley Wycoff, Library

University of Ottawa, Suzanne L. Meunier

Pennsylvania State University, Linda Musser

University of Pittsburgh, Jennifer Mayo-Deman

Pomona College, Donald Zenger, Geology Department

Portland State University, Scott Burns and Tom Taylor, Department of Geology

Princeton University, Patricia Gaspari-Bridges, Geology Library

Purdue University, Carolyn J. Laffoon

Universite du Quebec a Chicoutimi, Adam Nagy, Module des sciences de la Terre

Queen's University, Hanne Sherboneau

Queens College (CUNY), Allan Ludman and Bernadette Gatto, Department of Geology

Rensselaer Polytechnic Institute, Nancy Rivers

Rice University, John Stormer, Jr., Department of Geology & Geophysics

University of Rochester, Isabel Kaplan, Carlson Library

Rutgers University, Susan Goodman, Library of Science & Medicine

Saint Louis University, Laurie Hausmann

Slippery Rock University, Kent Bushnell, Department of Geology

Smith College, Rocco Piccinino, Science Library

University of South Carolina, Ann M. Gillon, Earth Sciences and Resources Institute

University of South Dakota, Valentine J. Ansfield

University of South Florida, Mark Stewart, Department of Geology

Southeast Missouri State University, Nicholas H. Tibbs, Department of Earth Science

Southern Illinois University at Carbondale, Harry O. Davis

Southern Methodist University, Mary Ellen Batchelor, Science/Engineering Library

University of Southwestern Louisiana, Daniel Tucker, Department of Geology

Stanford University, Charlotte Derksen, Branner Earth Science Library

State University of New York at Albany, W. S. F. Kidd, Department of Geological Sciences

State University of New York at Binghamton, Ina C. Brownridge

State University of New York at Buffalo, John S. King, Department of Geology

State University of New York College at Cortland, John L. Fauth, Department of Geology

State University of New York College at Fredonia, Walther M. Barnard, Department of Geosciences

State University of New York College at Oneonta, P. Jay Fleisher, Department of Earth Sciences

University of Tennessee, Knoxville, Harry Y. McSween, Jr., Department of Geological Sciences

Texas A&I University, William F. Thomann, Department of Geoscience

University of Texas at Austin, Dennis Trombatore and Jim McCulloch, Walter Geology Library

University of Texas at Dallas, Ellen Safley, University Library
University of Texas of the Permian Basin, Emilio MutisDuplat, Department of Geology
Texas Tech University, Alonzo Jacka, Department of Geoscience
Tulane University, Emily H. Vokes, Department of Geology
Union College, Betty Allen, Schaffer Library
University of Utah, Barbara R. Ritzma, Marriott Library
Utah State University, Donald Fiesinger, Department of Geology
Vanderbilt University, Leonard Alberstadt, Department of Geology
Virginia State University, Constance M. Hill and Christopher P. Egan, Department of Geological Science
University of Virginia, H. H. Shugart, Department of Environmental Science
Washington and Lee University, Samuel J. Kozak, Department of Geology
Washington State University, Eileen E. Brady, Science & Engineering Library
Washington University, St. Louis, Clara McLeod, Earth & Planetary Science Library
University of Waterloo, Johanna Cooper, Davis Centre Library
Wayne State University, Robert B. Furlong, Geology Department
Wesleyan University, Jelle Z. De Boer, Department of Earth & Environmental Science
West Texas State University, Jean Rick, Cornette Library
Western Michigan University, W. Thomas Straw, Department of Geology
University of Western Ontario, Alfred Lenz, Department of Geology
University of Windsor, Elizabeth Chandler, Department of Geology
University of Wisconsin - Madison, Marie Dvorzak, Geology-Geophysics Library
University of Wisconsin - Milwaukee, Linda Brothen, The Golda Meir Library
Woods Hole Oceanographic Institution, Carolyn P. Winn
University of Wyoming, Linda R. Zellmer, Geology Library
Yeshiva University, Pearl Berger, Library

In addition to the assistance received from these people within academic institutions, the following individuals from State agencies and private organizations assisted in completing this project:

Dona Mary Dirlam, Gemological Institute of America
Merrianne Hackathorne, Ohio Department of Natural Resources, Division of Geological Survey
Connie Manson, Washington Department of Natural Resources, Division of Geology and Earth Resources
Klaus Neuendorf, Oregon Department of Geology and Mineral Industries
Janice H. Sorenson, Kansas Geological Survey

In compiling this Bibliography, AGI checked over 20 published bibliographies of geoscience theses. The most productive among these were the Chronic bibliographies. The sources checked are listed below:

A bibliography of theses, dissertations and honors papers on the geology of eastern Massachusetts, by Brewer, T., in *Geology of southeastern New England*, edited by Cameron, B., p. 64-70, 1976.

Abstracts of theses on Georgia geology through 1974, by Moye, Falma J., in *Bulletin*, Earth and Water Division, Department of Natural Resources 89, 1976.

Alphabetical listing of theses and dissertations on North Carolina geology, by Walton, Robert O., III in Open File Report, North Carolina Geological Survey, 1990.

Bibliography of graduate theses and dissertations on Nevada geology to 1976, by Lutsey, I. A. in Nevada Bureau of Mines, Report 31, 1978.

Bibliography of theses and dissertations on Colorado, 1968-1980, by Kirk, Cheryl L., et al. in *The Mountain Geologist* 18(4), p. 96-113, 1981

Bibliography of theses and dissertations on Idaho and North Dakota, by Hanson, M. W., et al. in *The Mountain Geologist* 17(4), p. 108-124, 1980.

Bibliography and index of theses and dissertations on the geology of Louisiana, by Nault, M. J. in Resources Information Series, Louisiana Geological Survey 2, 1980.

Bibliography of theses and dissertations on Utah, 1968-1979, by Hansen, M. W., et al., in *The Mountain Geologist* 17(3), p. 71-87, 1980.

Bibliography of theses in geology, by Chronic, John and Chronic, Halka, Pruett Press, Inc., Boulder, CO, 1958.

Bibliography of theses in geology 1958-1963, by Chronic, John and Chronic, Halka, American Geological Institute, 1964.

Canadian university graduate theses in the geological sciences; abstracts published in the *Canadian Mining Journal*, 1953-1963, by Henderson, J. F., Geological Survey of Canada Paper 64-46, 1965

Compilation and index of theses on Montana geology, 1899-1982, by Daniel, Faith and Berg, Richard B., Montana Bureau of Mines and Geology Special Publication 88, 1983.

Compilation and index of theses on Montana geology, 1983-1988, by Dunn, Lisa G., Montana Bureau of Mines and Geology Special Publication 97, 1990

Geology and geophysics in Hawaii; a bibliography of theses and dissertations, 1909-1977, by Rowell, Unni Havem, Hawaii Institute of Geophysics, University of Hawaii, Data Report 34, 1978

Graduate theses on the geology of Idaho, 1900-1977, by Gaston, M. P., in Idaho Bureau of Mines and Geology, Information Circular 32, 1979.

Index to graduate theses and dissertations on California geology, 1962 through 1972, by Taylor, G. C., California Division of Mines and Geology, Special Report 115, 1974.

Index to graduate theses and dissertations on California geology, 1973 and 1974, by Peterson, D. and Saucedo, G. J., California Division of Mines and Geology, California Geology 31(2), p. 33-40, 1978.

Index to graduate theses and dissertations on California geology, 1975 and 1976, by Peterson, D. and Saucedo, G. J., *California Geology* 31(4), p. 90-96, 1978.

Index to graduate theses on California geology to December 31, 1961, by Jennings, Charles W. and Strand, Rudolph G., California Division of Mines and Geology, Special Report 74, p. 1-39, 1963.

Index to publications, open file reports and theses, 1862-1980, by Lyttle, Norman A. and Gillespie-Wood, Janet in Report of the Nova Scotia Department of Mines 81-6, 1981.

Theses on Washington geology; a comprehensive bibliography, 1901-1979, by Manson, Connie in Information Circular, State of Washington, Department of Natural Resources, Division of Geology and Earth Resources 70, 1980.

Theses on the geology of Washington, 1986 through 1990, by Manson, Connie, State of Washington, Department of Natural Resources, Division of Geology and Earth Resources, 1990.

The Geoscience Information Society (GIS) has long been concerned about bibliographic control of geoscience theses. Many of the individuals acknowledged above are members of GIS. Some years ago GIS undertook its own compilation of geoscience theses. Their work was suspended, partly because of AGI's intent to proceed with this compilation. The information GIS members had gathered was sent to AGI and has been included in this compilation.

At AGI, thousands of new thesis references were added as part of this compilation. The GeoRef editor/indexers who prepared and indexed these were Hal Johnson and George Korvah. As we worked on this, all theses were put in a separate database. This database was maintained by Lesa Read, bibliographer, with help from Beth Shultz and Liza Mallard. They made data corrections, ran consistency checks, and compared the printed bibliographies with the thesis file. Cecile Lethem, George Korvah, and Joe Stables edited the Subject and Geographic Index. Mike Cacic did the programming for page production. Lawrence Berg did the programming needed to produce the subject index. Sharon Tahirkheli provided editorial oversight for the thesis references and participated in subject index development. John Mulvihill worked on design of the bibliography and managed the project.

GEOSCIENCE THESES

A., Jorge Ferretti *see* Ferretti A., Jorge

A., Omar J. Perez *see* Perez A., Omar J.

Aadland, Arne Johannes. The mapping and field stratigraphy of the Monroe Creek Formation (Miocene) in the Wildcat Ridge in Morrill, Scotts Bluff, and Banner counties, Nebraska. M, 1960, University of Nebraska, Lincoln.

Aadland, Rolf. Stratigraphy and petrography of Thaynes Formation (Lower Triassic), western Wyoming and eastern Idaho. M, 1966, University of Nebraska, Lincoln.

Aadland, Rolf Konrad. Cambrian stratigraphy of the Libby Trough, Montana. D, 1980, University of Idaho. 323 p.

Aagaard, Knut. The East Greenland Current north of Denmark Strait. D, 1966, University of Washington. 82 p.

Aagaard, Per. Thermodynamic and kinetic analysis of silicate hydrolysis at low temperatures and pressures. D, 1979, University of California, Berkeley.

Aalto, Kenneth Rolf. Glacial marine sedimentation and stratigraphy of the Toby conglomerate (upper Proterozoic), southeastern British Columbia, northwestern Idaho and northeastern Washington. D, 1970, University of Wisconsin-Madison. 158 p.

Aalto, Kenneth Rolf. The sedimentology of the Late Jurassic (Portlandian) Otter Point Formation of southwestern Oregon. M, 1968, University of Wisconsin-Madison. 60 p.

Aaquist, B. E. Host rock to native copper, Kingston and Centennial mines, Michigan. M, 1977, University of Western Ontario. 90 p.

Aardema, James A. Petrography and geochemistry of three drill cores which penetrate the Suwannee Limestone in Pinellas, Desoto, and Charlotte counties, Florida. M, 1987, University of Florida. 127 p.

Aarden, Henrikus Marinus. Hiortdahlite from Kipawa river, Quebec, Canada; its chemistry and mineralogy. D, 1969, University of Toronto.

Aaron, John Marshal, III. Geology of the Nazareth Quadrangle, Northampton County, Pennsylvania. D, 1971, Pennsylvania State University, University Park. 353 p.

Aarons, Bernard Louis. Geology of a portion of the Las Trampas Ridge and Hayward quadrangles, California. M, 1958, University of California, Berkeley. 78 p.

Aarons, Irwin Isaac. A petrographic study of Canadian and Saint Croixan rocks of a deep test well in Wood County, West Virginia. M, 1957, University of Pittsburgh.

Aaronson, Donald Bruce. A geologic map of a part of Marion County, Missouri. M, 1966, University of Iowa. 163 p.

Aba-Husayn, Mansur Mohammed. Hydrolysis and decomposition of some clays and soils. D, 1972, University of California, Riverside. 125 p.

Abad, Leopoldo F. Methods of evaluation of placer ground. M, 1924, University of California, Berkeley.

Abadie, Victor Hugo, III. Geology of part of the Sierra de Moradillas, Sonora, Mexico. M, 1981, Stanford University. 87 p.

Abashian, Mark S. The Eldora-Bryan Mountain Stock as a natural analog to buried wastes; geochemistry and geochronology. M, 1984, University of New Mexico. 436 p.

Abass, Hazim H. A study of water-coning problems in petroleum production. D, 1987, Colorado School of Mines. 215 p.

Abbas, Fadhil Migbel. A simplified model for salt and boron transport in soils. D, 1984, Utah State University. 152 p.

Abbaszadeh-Dehghani, Maghsood. Analysis of unit mobility ratio well-to-well tracer flow to determine reservoir heterogeneity. D, 1982, Stanford University. 296 p.

Abbe, Cleveland. A general report on the physiography of Maryland. D, 1898, The Johns Hopkins University.

Abbey, Dale A. Gravity study of several Maine coastal plutons (Precambrian). M, 1972, SUNY at Buffalo. 79 p.

Abbey, Marjorie Best. Limestones of Chester Valley, MA. M, 1934, Bryn Mawr College. 30 p.

Abbott, A. C. The silver-tin ores of Oruro, Bolivia. M, 1929, Massachusetts Institute of Technology. 56 p.

Abbott, Agatin Townsend. Geology of northwest portion of Mt. Aix Quadrangle, Washington. D, 1953, University of Washington. 256 p.

Abbott, Caroline L. Bedrock aquifer geometry in the Panther Junction area of Big Bend National Park, Texas. M, 1983, Texas A&M University. 113 p.

Abbott, D. E. The origin of sulphate and the isotope geochemistry of sulphate-rich shallow groundwater in the St. Clair clay plain, southwestern Ontario. M, 1987, University of Waterloo. 134 p.

Abbott, Dallas Helen. Two processes affecting sediment stability and physical properties; hydrothermal circulation and rapid deposition. D, 1982, Columbia University, Teachers College. 218 p.

Abbott, David M., Jr. The Precambrian geology of the East Inlet area, southwestern Rocky Mountain National Park, Colorado. M, 1975, Colorado School of Mines. 96 p.

Abbott, Earl William. Stratigraphy and petrology of the Mesozoic volcanic rocks of southeastern California. D, 1972, Rice University. 196 p.

Abbott, Earl William. Structural geology of the southern Silurian hills, San Bernardino County, California. M, 1971, Rice University. 48 p.

Abbott, James Grant. Structure and stratigraphy of the Mt. Hundere area, southeastern Yukon. M, 1977, Queen's University. 111 p.

Abbott, Jeffrey Tarbell. Geology of Precambrian rocks and isotope geochemistry of shear zones in the Big Narrows area, northern Front Range, Colorado. D, 1970, University of Colorado. 261 p.

Abbott, Jesse Walter. A flood analysis of Paradise Creek, an ungaged stream near Moscow, Idaho. M, 1968, University of Idaho. 97 p.

Abbott, Marvin Milton. A basic evaluation of the uranium potential of the Morrison Formation of northwestern Cimarron County, Oklahoma, and adjoining areas of New Mexico and Colorado. M, 1979, Oklahoma State University. 92 p.

Abbott, Maxine Langford. The American species of Asterophyllites, Annularia, and Sphenophyllum. D, 1954, University of Cincinnati. 153 p.

Abbott, Patrick Leon. The Edwards Limestone in the Balcones fault zone, south-central Texas. D, 1973, University of Texas, Austin.

Abbott, Patrick Leon. The Glen Rose Section in the Canyon Reservoir area, Comal County, Texas. M, 1966, University of Texas, Austin.

Abbott, Ralph E. Insoluble residues of certain formations of West Texas. M, 1939, Texas Tech University. 82 p.

Abbott, Richard N., Jr. Retrograde parageneses in staurolite-kyanite grade metamorphic schists of the Goshen Formation in western Massachusetts. M, 1973, University of Maine. 84 p.

Abbott, Richard Newton, Jr. Petrology of the Red Beach Granite near Calais, Maine. D, 1977, Harvard University.

Abbott, Tom Austin. The distribution of radioactivity in the surficial deposits of Levy, Marion, and Citrus counties, Florida. M, 1988, University of Florida. 122 p.

Abbott, Ward Owen. Cambrian diabase flow in central Utah. M, 1951, Brigham Young University. 70 p.

Abbott, William Harold, Jr. Micropaleontology and paleoecology of Miocene non-marine diatoms from the Harper District, Malheur County, Oregon. M, 1971, Northeast Louisiana University.

Abbott, William Harold, Jr. Vertical and lateral patterns of diatomaceous ooze found between Australia and Antarctica. D, 1972, University of South Carolina. 153 p.

Abboud, Salim A. Chemical forms of cadmium in municipal sewage sludge amended-soils. D, 1987, University of Guelph.

Abby, Darwin G. An investigation of short line triangulation accuracies combined with the field testing of a Kern DKM-3 modified with a 5-wire reticule. M, 1965, Ohio State University.

Abdassah, Doddy. Triple porosity models for representing naturally fractured reservoirs. D, 1984, University of Southern California.

Abdel Rahman, Abdel Fattah Mostafa. Geochemistry of two alkaline granites. M, 1982, University of Toronto.

Abdel Rahman, Mostafa A. Numerical and experimental study of shaft resistance of piles in granular soils. D, 1988, University of Wisconsin-Madison. 181 p.

Abdel Wahab, Mahmoud M. Stratigraphy of the Strawn (Pennsylvanian), Colorado River valley, north-central Texas. M, 1980, Baylor University.

Abdel Wahid, Ibrahim. Area P_n and P_g velocity variation in central United States. M, 1963, University of Kansas.

Abdel Warith, Mostafa Mohamed. Migration of leachate solution through Clay soil. D, 1987, McGill University.

Abdel-aal, Farouk Mostafa. Extension of bed-load formulas to high sediment rates. D, 1969, University of California, Berkeley. 130 p.

Abdel-Aal, Osama Youssef. Alteration of opaque minerals and the magnetization and magnetic properties of volcanic rocks in a drill hole from an active geothermal area in the Azores. D, 1978, Dalhousie University. 215 p.

Abdel-Gawad, A. M. Alteration features associated with some basal Chinle uranium deposits; Colorado Plateau. D, 1960, Columbia University, Teachers College. 174 p.

Abdel-Gawad, Abdel-Moneim M. On the mineralogy of uranium association with magnetite and apatite. M, 1958, Columbia University, Teachers College.

Abdel-Kader, Adel. Landsat analysis of the Nile Delta, Egypt. M, 1982, University of Delaware, College of Marine Studies. 260 p.

Abdel-Monem, Abdalla A. A study of the paleogeography and the source of sediments in the New Jersey Triassic Basin by K-Ar dating. M, 1966, Columbia University. 47 p.

Abdel-Monem, Abdalla A. K-Ar geochronology of volcanism on the Canary islands. D, 1969, Columbia University. 190 p.

Abdel-Rahman, Abdel-Fattah Mostafa. Plutonism and tectonic evolution of the Ras Gharib segment of the northern Nubian Shield. D, 1986, McGill University. 326 p.

Abdelbary, Mohamed Rafeek. Alluvial fan hydraulics and flood hazard mitigation measures. D, 1982, Colorado State University. 166 p.

Abdelhamid, M. S. At rest earth pressure of clays during one-dimensional consolidation. D, 1975, Northwestern University. 412 p.

Abdelmalik, Mohamed B. A. Vertical resolution capability study of direct-current, magnetotelluric, and time-domain electromagnetic methods. D, 1988, Colorado School of Mines. 255 p.

Abdelsaheb, Ibrahim Z. Till paleomagnetism in northeastern Kansas. M, 1988, Emporia State University. 36 p.

Abdelzahir, Mohmed Abdelzahir. The geology of the Carolina slate belt, northern Moore County, North Carolina. M, 1978, North Carolina State University. 67 p.

Abder-Ruhman, Mohammed. Mineralogical characteristics and weathering environments in Texas lignite overburdens. D, 1985, Texas A&M University. 311 p.

Abdon, Abdol-Reza. Field geophysical studies in the Pierrefonds-Ile Bizard region, western Montreal. M, 1985, McGill University. 233 p.

Abdul, Abdul Shaheed. Experimental and numerical studies of the effect of the capillary fringe on streamflow generation. D, 1985, University of Waterloo. 210 p.

Abdul-Latif, Numan A. R. Synthesis of clay minerals. D, 1969, Georgia Institute of Technology. 118 p.

Abdul-Malik, Muhammed M. A geophysical investigation of the Silver Star area of Madison County, southwestern Montana. M, 1977, Montana College of Mineral Science & Technology. 96 p.

Abdul-Rahman, Mogda M. T. A eutrophication model for Lake Ray Hubbard. D, 1979, University of Texas at Dallas. 637 p.

Abdul-Razzaq, Sabeekah. Evolution of middle Devonian species of Euglyphella as indicated by cladistic analysis. M, 1973, University of Michigan.

Abdul-Razzaq, Sabeekah. Study of some Cretaceous Ostracoda of Kuwait. D, 1977, University of Michigan.

Abdulkareem, Talal Faisel. Subsurface stratigraphy and depositional environments of Everton Dolomite, Ancell Group, and Black River Group (Middle and Upper Ordovician) in Indiana. D, 1982, Indiana University, Bloomington. 209 p.

Abdulla, Khalifa Ahmed. Relation of oil entrapment to facies changes in central Sirte Basin (Libya). M, 198?, Syracuse University. 69 p.

Abdullah, S. K. M. Geology of the southern part of the Sand Springs Range, Churchill County, Nevada. M, 1966, University of Nevada. 53 p.

Abdullah, Talat Y. Depositional environments and porosity development in the Funston Formation (Lower Permian), southwestern Kansas. M, 1983, University of Kansas. 140 p.

Abdullatif, A. A. Geologic causes of foundation problems in the city of Amman, Jordan. D, 1975, University of Arizona. 243 p.

Abdullatif, Abdullatif A. Physical testing of engineering properties of collapsing soils in the city of Tucson, Arizona. M, 1969, University of Arizona.

Abdulmumini, Salisu. Determination of regional evapotranspiration using energy balance, planetary boundary layer transfer coefficients and regularly recorded data. D, 1980, University of California, Davis. 137 p.

Abdulrazzak, M. J. Ground water system evaluation for Wadi Nisah, Saudi Arabia. M, 1976, University of Arizona.

Abdulrazzak, Mohamed Jamil. Aquifer recharge from an ephemeral stream and the resulting evolution of water table. D, 1982, Colorado State University. 184 p.

Abe, Joseph Michael. Economic analysis of artificial recharge and recovery of water in Butler Valley, Arizona. M, 1986, University of Arizona. 170 p.

Abed, Fawzi M. A. H. The utilization of potassium from the surface layers of eight Nebraska soils; a mineralogical approach. D, 1965, University of Nebraska, Lincoln.

Abedin, K. Z. Fixed-layer ammonium ion as an indicator of organic transformation. M, 1977, University of Texas, Arlington. 41 p.

Abedrabboh, Walid. Multi-objective decision making applied for watershed development planning of Zarqa River basin in Jordan. D, 1988, University of Arizona. 162 p.

Abegg, Frederick E. Carbonate petrology, paleoecology, and depositional environments of the Clore Formation (upper Chesterian) in southern Illinois. M, 1986, Southern Illinois University, Carbondale. 222 p.

Abegglen, Donn E. The effects of drain wells on the ground water quality of the Snake River plain. M, 1970, University of Idaho. 103 p.

Abel, Cole D. Petrology and sedimentology of the Jacobsville Sandstone (northern Michigan) and Bayfield Group (northern Wisconsin). M, 1985, University of Wisconsin-Madison.

Abel, Kathleen D. Plagioclases of the Dufek Intrusion, Antarctica. M, 1978, University of Missouri, Columbia.

Abel, Vernon G. Stratigraphy of the Medina 7 1/2 minute Quadrangle, New York. M, 1961, SUNY at Buffalo.

Abele, Ralph Warren, Jr. Short-term changes in beach morphology and concurrent dynamic processes, summer and winter periods, 1971-1972, Plum Island, Massachusetts. M, 1973, University of Massachusetts. 166 p.

Abell, Philip Webster. Colorimetric determination of trace quantities of aluminum with aurintricarboxylate. M, 1959, Stanford University.

Abell, Philip Webster. Paleogeology of a portion of Oklahoma and Texas. M, 1959, Stanford University.

Abels, Thomas Allen. A subsurface lithofacies study of the Morrowan Series in the northern Anadarko Basin. M, 1958, University of Oklahoma. 73 p.

Abelson, Robert Stephen. Monte Carlo calculations of the melting point of iron at core pressures. D, 1981, University of California, Los Angeles. 188 p.

Abendshein, Mark. Facies and oil and gas analysis of the Torrey Member of the Moenkopi Formation, south-central Utah. M, 1978, Northern Arizona University. 158 p.

Aber, James S. Erratic-rich drift in the Appalachian Plateau; its nature, origin, and conglomerates. D, 1978, University of Kansas. 115 p.

Aber, James S. Upland glacial stratigraphy in the Binghamton-Montrose region of New York and Pennsylvania. M, 1976, University of Kansas. 58 p.

Abercrombie, Hugh James. Partitioning of fluorine between tremolite, talc and phlogopite. M, 1983, Carleton University. 153 p.

Abercrombie, Hugh James. Water-rock interaction during diagenesis and thermal recovery, Cold Lake, Alberta. D, 1988, University of Calgary. 183 p.

Aberdeen, Esther. The Radiolaria of the San Diego chart of the Caballos Formation. D, 1937, University of Chicago. 96 p.

Aberdeen, Esther J. The location of the break between the Galena and the Platteville limestones. M, 1931, Northwestern University.

Abernathy, Gary Lance. Delineation of the bedrock topography of Montgomery County, Ohio, by the use of the gravity-geologic method. M, 1983, Wright State University. 96 p.

Abernathy, George Elmer. Fault areas of south eastern Kansas coal beds. M, 1926, University of Missouri, Rolla.

Abernathy, George Elmer. Some structural features of the Weir-Pittsburg Coal Bed. M, 1925, University of Kansas. 30 p.

Abernathy, George Elmer. The Cherokee of southeastern Kansas. D, 1936, University of Kansas. 108 p.

Abernathy, Sarah Allison. An X-ray powder diffraction study of the lazulite-scorzalite series. M, 1980, University of Florida. 49 p.

Abernethy, Robert Morris. Brachiopod community evolution across the Silurian-Devonian boundary, Henryhouse and Haragan formations, Arbuckle Mountains, Oklahoma. M, 1987, Stephen F. Austin State University. 96 p.

Aberra, G. B. In situ apparent thermal conductivity measurement at depth in the unsaturated zone. M, 1974, University of Arizona.

Abet Yao, Marcel. Recherche sur les causes des déformations de l'aménagement hydro-électrique de Beauharnois. M, 1984, Ecole Polytechnique. 173 p.

Abeyesekera, R. A. Stress deformation and strength characteristics of a compacted shale. D, 1978, Purdue University. 445 p.

Abid, Iftikhar A. Mineral diagenesis and porosity evolution in the Hibernia oil field, Jurassic-Cretaceous Jeanne d'Arc rift graben, eastern Grand Banks of Newfoundland, Canada. M, 1988, McGill University. 280 p.

Abington, Oscar D. Changing meander morphology and hydraulics, Red River, Arkansas and Louisiana. D, 1972, Louisiana State University.

Abiodun, Adigun A. Analysis of seepage into and the management of a groundwater system. D, 1971, University of Washington. 83 p.

Abitz, Richard. Volcanic geology and geochemistry of the northeastern Black Range Primitive Area and vicinity, Sierra County, New Mexico. M, 1984, University of New Mexico. 174 p.

Able, Kenneth. Life history, ecology and behavior of two new Liparis (Pisces; Cyclopteridae) from the western North Atlantic. D, 1974, College of William and Mary.

Ablordeppey, Victor. Ultrasonic velocities in molten bismuth, tin and 50 atomic % alloy under pressures up to 8.5 kilobars. D, 1970, University of Chicago. 78 p.

Abner, H. Geochemical analysis of carbonate rocks in the upper member of the Callville Formation (Pennsylvanian), Clark County, Nevada. M, 1975, Memphis State University.

Abo-Elela, R. M. Pore water pressure coefficients in organic soils. D, 1975, Wayne State University. 237 p.

Aboaziza, Abdelaziz Hassan. Characterization of sulfur-asphalt-dune sand paving mixtures. D, 1981, University of Arizona. 350 p.

Aboim-Costa, Carlos Alfredo Ferreira. A study of the self-boring pressuremeter on stiff clay. D, 1981, Stanford University. 136 p.

Abolkhair, Yahya Mohammed Sheikh. Sand encroachment by wind in Al-Hasa of Saudi Arabia. D, 1981, Indiana University, Bloomington. 196 p.

Abou-Kassem, Jamal Hussein. Determination of bottom-hole pressure in flowing gas wells. M, 1975, University of Alberta. 164 p.

Abou-seida, Mohamed Mokhles. Bed load function due to wave action. D, 1965, University of California, Berkeley. 83 p.

Abou-Zied, Mohamed Saleh. Geology and mineralogy of the Milford Flat Quadrangle and the Old Moscow Mine, Star District, Beaver County, Utah. D, 1968, University of Utah. 260 p.

Abouchard, Sylvain S. Primary ore shoots in the Pachuca District. M, 1923, Columbia University, Teachers College.

Aboud, Jesus M. Reflected surface waves from a vertical discontinuity. M, 1976, University of Missouri, Rolla.

Aboud, Nelson. A comparison between direct waves and head waves in a seismic refraction profile. M, 1976, University of Missouri, Rolla.

Abracosa, Ramon Panoy. The Philippine environmental impact statement system; an institutional analysis of implementation. D, 1987, Stanford University. 300 p.

Abraham, Amenti. Mineralization in the Purcell Supergroup, southwestern Alberta, southeastern British Columbia. M, 1973, University of Alberta. 94 p.

Abraham, Andrew Peter. The geology of the Otter Lake area and its relation to the southwestern complex of western Churchill Province, Saskatchewan. M, 1986, University of Regina. 237 p.

Abraham • Acker

Abraham, Dwaine G. The mineralogy, petrology and diagenetic history of the Late Cambrian Lamotte Sandstone in southeastern Missouri. M, 1978, Northern Illinois University.

Abraham, E. M. The general geology of Hearst Township, Ontario. M, 1946, Queen's University. 77 p.

Abraham, Mazeeh Younis. Petrology and depositional history of the Lower Cretaceous Bear River and Aspen formations, southern Teton and northern Lincoln counties, western Wyoming. M, 1977, University of Michigan.

Abraham, Nazeen Younis. Petrology and depositional history of the Lower Cretaceous Bear River and Aspen formations, southern Teton and northern Lincoln counties, western Wyoming. M, 1977, University of Michigan.

Abrahams, Jennifer Anne. Isotopic investigation of coastal Paleozoic carbonate diagenesis adjacent to crystalline rock highlands. M, 1986, Arizona State University. 92 p.

Abrahamson, Norman Alan. Estimation of seismic wave coherency and rupture velocity using the SMART 1 strong motion array recordings. D, 1985, University of California, Berkeley. 135 p.

Abrahao, Dirceu. Lacustrine and associated deposits in a rifted coninental margin; the Lagoa Feia Formation, Lower Cretaceous, Campos Basin, offshore Brazil. M, 1987, Colorado School of Mines. 193 p.

Abrahao, Dirceu. Well-log signatures of alluvio-lacustrine reservoirs and source rocks of the Lagoa Feia Formation, Lower Cretaceous, Campos Basin, offshore Brazil. D, 1988, Colorado School of Mines. 279 p.

Abrajano, Teofilo A., Jr. Petrogenesis of the McCartney Mountain Stock, southwest Montana. M, 1982, University of Akron. 90 p.

Abrajano, Teofilo Aniag, Jr. The petrology and low-temperature geochemistry of the sulfide-bearing mafic-ultramafic units of the Acoje Massif, Zambales Ophiolite, Philippines; characterization of mineral and fluid equilibria. D, 1984, Washington University. 474 p.

Abrams, Charlotte E. Geology of the Austell-Frolona Antiform, northwestern Georgia Piedmont. M, 1982, University of Georgia.

Abrams, Gary S. The relationship between electric logs and sidewall core analyses, Wilcox Group (Eocene) in Southwest Mississippi. M, 1986, University of Texas at El Paso.

Abrams, Gerard Joseph. Geology and ore deposits of the Union District, southern Shoshone Mountains, Nye County, Nevada. M, 1979, University of Nevada. 93 p.

Abrams, Mark J. Geology of part of the Inchelium Quadrangle, Stevens County, Washington. M, 1980, Eastern Washington University. 30 p.

Abrams, Michael Allan. Quantitative analysis of the benthic foraminifera in Santa Monica Basin, California. M, 1979, University of Southern California.

Abrams, Nelson Jay. Paleoceanographic and paleoclimatic study of the Pacific sector of the Southern Ocean using Pliocene marine diatoms. M, 1986, Rutgers, The State University, Newark. 215 p.

Abramson, Beth S. The mineralizing fluids responsible for skarn and ore formation at the Continental Mine, Fierro, New Mexico, in light of REE analyses and fluid inclusion studies. M, 1981, New Mexico Institute of Mining and Technology. 143 p.

Abrassart, Chester P. Stratigraphy and sedimentation of the Juniper Mountain area; Moffat County, Colorado. M, 1951, University of Colorado.

Abriel, William Lee. A ground-based study of the Everett-Bedford lineament of Pennsylvania. M, 1978, Pennsylvania State University, University Park. 84 p.

Abriola, Linda Marie. Mathematical modeling of the multiphase migration of organic compounds in a porous medium. D, 1983, Princeton University. 236 p.

Abrishamchi, Ahmad. Dynamic planning in wastewater treatment management. D, 1979, University of California, Davis. 200 p.

Abry, Claude Georges. Quantitative estimation of oil-exploration outcome probabilities in the Tatum Basin, New Mexico. D, 1973, Stanford University. 151 p.

Absher, Bobby Steven. Petrogenesis of granulite facies rocks at Winding Stair Gap, North Carolina Blue Ridge. M, 1984, University of Tennessee, Knoxville. 131 p.

Absi, George A. A study of the volcanic geology of the Starlight, Neeley, and Walcott formations, in Rockland Valley and the American Falls area, southeastern Idaho. M, 1984, Wayne State University. 141 p.

Abt, Steven Roman. Scour at culvert outlets in cohesive bed material. D, 1980, Colorado State University. 170 p.

Abu Ajamieh, M. M. The structure of the Pantano Beds in the Northern Tucson Basin, Arizona (lower Miocene). M, 1966, University of Arizona.

Abu Bakar, M. Y. Assessment of uranium using gamma radiations from thorium-234. M, 1976, University of Western Ontario.

Abu Bakar, Othmanbin. A comparison of selected Entisols and Spodosols occurring in Peninsular Malaysia and Peninsular Florida. D, 1985, University of Florida. 204 p.

Abu Sharar, Taleb M. Stability of soil structure and reduction in hydraulic conductivity as affected by electrolyte concentration and composition. D, 1985, University of California, Riverside. 156 p.

Abu-Agwa, Fawzy El-Shazly. Paleosols in the buried stream valleys of the loessial region of Mississippi. D, 1982, Mississippi State University. 161 p.

Abu-Eid, Ratab Muhmood. Absorption spectra and phase transformations of minerals at pressures up to 200 kilobars. D, 1975, Massachusetts Institute of Technology. 474 p.

Abu-gheida, Othman Mohammad. Effect of stress on ultrasonic wave velocities in rock salt. D, 1964, Michigan State University. 187 p.

Abu-Jaber, Nizar Shabib. Lower Cretaceous sedimentation in the southeastern Mediterranean area and its regional tectonic implications. M, 1987, North Carolina State University. 194 p.

Abu-Moustafa, Adel H. Petrography and geochemistry of the Nashoba Formation (Carboniferous) from the Wachasett-Marlboro tunnel, Massachusetts. D, 1969, Boston University. 237 p.

Abu-Obeid, Hamid A. Dynamic testing of desert soils (Recent). M, 1970, University of Arizona.

Abu-Rizaiza, Omar Seraj. Municipal, irrigational and industrial future water requirements in Saudi Arabia. D, 1982, University of Oklahoma. 212 p.

Abu-Taha, Mohammad F. Efficiency of two water wells (Tucson, Arizona). M, 1970, University of Arizona.

Abuannaja, Abdullah S. General geology of Wadi As-Salile area, Asir Quadrangle, Saudi Arabia. M, 1970, South Dakota School of Mines & Technology.

Abudelgawad, Gilani M. Morphological, mineralogical, and chemical properties of coastal Libyan soils and chemical weathering of glauconite. D, 1973, University of California, Riverside. 160 p.

Abufila, Taher M. A three-dimensional model to evaluate the water resources of the Kufra and Sarir basins, Libya. M, 1984, Ohio University, Athens. 866 p.

Abugares, Youssef Issa. Pecos Slope Abo gas field, Chaves and DeBaca counties, New Mexico. M, 1983, University of Texas at El Paso. 162 p.

Abul-Husn, Adnan Asid. Tectonic analysis of Mere Humorum on the lunar surface. M, 1966, University of Missouri, Rolla.

Abul-Nasr, Radwan Abdel-Aziz. Biostratigraphy and facies analysis of Paleogene rocks in west-central Sinai (Egypt). D, 1987, University of South Carolina. 250 p.

Aburawi, Ramadan. Sedimentary facies of the Holocene Santee River delta. M, 1972, University of South Carolina. 96 p.

Abushagur, Sulaiman Ahmed. Microfacies analysis of the Farrud lithofacies in the Ghani Field, Sirte Basin, Libya. M, 1983, University of Texas at El Paso. 185 p.

Abuzekri, Sadegh Khalifa. Microfacies analysis of the upper Beda Formation in the Balat Field, Sirte Basin, Libya. M, 1984, University of Texas at El Paso.

Abuzied, Hassan T. H. Geology of the Wadi Hamrawin area, Red Sea Hills, Eastern Desert, Egypt. D, 1984, University of South Carolina. 234 p.

Abuzkhar, Ahmed A. The potassium status of eastern and central Oregon soils. M, 1975, Oregon State University. 72 p.

Academia, Imelda Garcia. Stratigraphy and paleontology of the Mountain Springs Formation (Lower to Upper Ordovician), southeastern California. M, 1987, San Diego State University. 250 p.

Acar, Kazim Zafer. Creep properties of Saskatchewan potash. M, 1969, University of Saskatchewan. 48 p.

Acaroglu, Ertan Riza. Sediment transport in conveyance systems. D, 1968, Cornell University. 169 p.

Accame, Guillermo M. Mineralogical, trace-element and Landsat multispectral evaluation of gossans in the Alma mining district, Colorado. M, 1983, Purdue University. 147 p.

Ach, Jay A. The petrochemistry of the Ankara volcanics, central Turkey. M, 1982, SUNY at Albany. 146 p.

Achaibar, Jaikisan. A petrologic and geochemical study of some amphiobolite bodies in the Smith River Allochthon, Virginia. M, 1983, University of Tennessee, Knoxville. 93 p.

Achalabhuti, Charan. Pleistocene depositional systems of central Texas coastal zone. D, 1973, University of Texas, Austin.

Acham, Patrick A. The Mannville Group (lower Cretaceous) of the Hussar area, southern Alberta. M, 1971, University of Alberta. 152 p.

Acharya, Arvind B. Delineation of the bedrock topography and potential aquifers by gravity and seismic refraction techniques. M, 1987, Wright State University. 120 p.

Acharya, Ashutosh. Geology of the Grape Creek area, Fremont County, Colorado. M, 1949, Colorado School of Mines. 61 p.

Acharya, Hemendra Kumar. Wave propagation in inhomogeneous media with Antarctic ice cap as model. D, 1969, University of Wisconsin-Madison. 98 p.

Achauer, Charles Woodrow. Stratigraphy and microfossil studies of the Sappington Formation, southwestern Montana. M, 1957, University of Montana. 50 p.

Achmad, Grufron. An application of residual relaxation to mathematical simulation of petroleum reservoirs. D, 1973, University of Missouri, Rolla.

Achtermann, Roger D. Geology of the Meherrin, Virginia area; tracing formations across the staurolite isograd from the Carolina Slate Belt into the Charlotte Belt. M, 1985, Virginia Polytechnic Institute and State University.

Achtman, Malcolm. Stratigraphy and lithofacies variations of the Upper Cretaceous Smoky River Group, Grande Prairie region, Alberta. M, 1972, University of Alberta. 79 p.

Achuff, Jonathan M. Folding and faulting in the northern Blacktail Range, Beaverhead County, Montana. M, 1981, University of Montana. 69 p.

Ackenheil, Alfred Curtis. A soil mechanics and engineering geology analysis of landslides in the area of Pittsburgh, Pennsylvania. D, 1954, University of Pittsburgh. 243 p.

Acker, Clement J. Geologic interpretations of a siliceous breccia in the Colossal Cave area, Pima County, Arizona. M, 1958, University of Arizona.

Acker, James Gardner. Factors influencing the dissolution of aragonite in the oceanic water column. D,

Geoscience Theses

3

1988, University of South Florida, St. Petersburg. 242 p.

Acker, Kelly L. The carbonate-siliciclastic facies transition in the modern sediments off the northeast coast of Barbados, W.I. M, 1987, McGill University. 230 p.

Acker, Louis L. Geology of the Shining Rock Quadrangle, North Carolina. M, 1982, University of North Carolina, Chapel Hill. 110 p.

Acker, Richard C. The bedrock geology of the northern half of the Clayville Quadrangle, Rhode Island. M, 1950, Brown University.

Ackerbloom, Donald R. An investigation of the effect of temperature on the structure of certain sulfo-salts. M, 1956, Syracuse University.

Ackerman, Dawn Ramsey. Tetrahedrite compositions in the Trixie Mine, East Tintic District, Utah. M, 1987, University of Utah. 76 p.

Ackerman, Walter. Louis beryl pegmatite, Custer County, South Dakota. M, 1953, South Dakota School of Mines & Technology.

Ackermann, D. Physical and chemical response of a shallow groundwater zone to recharge processes. M, 1973, University of Waterloo.

Ackermann, Hans D. Gravity investigations in northeastern and central Pennsylvania. M, 1962, Pennsylvania State University, University Park. 87 p.

Ackman, Brad C. The stratigraphy and sedimentology of the middle Proterozoic Snowslip Formation in Lewis, Whitefish and Flathead ranges, Northwest Montana. M, 1988, University of Montana. 185 p.

Ackroyd, Earl Arthur. Artesian aquifers of the upper James River valley, Brown County, South Dakota. M, 1956, South Dakota School of Mines & Technology.

Acomb, Barry W. The petrology, stratigraphy and origin of phosphatic nodules in Upper Devonian and Lower Mississippian rocks of the Eastern Interior. M, 1979, University of Cincinnati. 88 p.

Acomb, Lawrence Joseph. Stratigraphic relations and extent of Wisconsin's Lake Michigan lobe red tills. M, 1978, University of Wisconsin-Madison.

Acomb, T. J. A watershed analysis; the Mendon Brook Basin, Mendon, Vermont. M, 1977, University of Vermont.

Acosta, Alvaro. Upper Cretaceous paleobotany of northeastern Utah. M, 1961, University of Utah. 98 p.

Acosta, Marcial G. Geology of the Bahia Soledad Embayment, Baja California, Mexico. M, 1966, [University of California, San Diego].

Acosta, Ramon. Geology of Mesa Poleo area, Rio Arriba County, New Mexico. M, 1973, Colorado School of Mines. 80 p.

Acquaviva, Daniel Joseph. Stratigraphy and depositional environments of a part of the lower Breathitt Formation (Penn.) near Williamsburg, Ky. M, 1978, University of Kentucky. 156 p.

Acton, Clifford John. Micropedology and electron probe analysis in a genesis-study of lacustrine soils of Brant County, Ontario. D, 1970, University of Guelph.

Acuff, Hoyt Nealy. Late Cenozoic sedimentation in the Allia Bay area, East Rudolf (Turkana) Basin, Kenya. D, 1976, Iowa State University of Science and Technology. 111 p.

Acunzo, Antonio Carlos. Simulator for the two-dimensional flow of a slightly compressible fluid in a homogeneous isotropic reservoir of variable thickness. M, 1979, Stanford University.

Acurero Salas, Luis Armando. Optimal policies of well completion and production with respect to coning effects for different types of reservoirs. D, 1980, University of Texas, Austin. 284 p.

Adabi, Mohammad H. Sedimentology of the Potomac Group (Cretaceous) and the Aquia Formation (Tertiary) in Prince William County, Va. M, 1978, George Washington University.

Adachi, Kakuichiro. Influence of pore water pressure on the engineering properties of intact rock. D, 1974, University of Illinois, Urbana. 290 p.

Adachi, Toshihisa. Construction of continuum theories for rock by tensor testing. D, 1969, University of California, Berkeley. 344 p.

Adair, Donald H. Geology of the Cherry Creek District, Nevada. M, 1961, University of Utah. 122 p.

Adair, Donald L. Anomalous structures of the Baraboo Basin (Wisconsin). M, 1956, University of Wisconsin-Madison.

Adair, Marcia B. A geophysical study of the Ste. Genevieve fault zone. M, 1975, Southern Illinois University, Carbondale. 101 p.

Adair, Richard Glen. Microseisms in the deep ocean; observations and theory. D, 1985, University of California, San Diego. 189 p.

Adair, Robin N. The pyroclastic rocks of the Crowsnest Formation, Alberta. M, 1986, University of Alberta. 213 p.

Adam, Adam Ibrahim. Mineralogy and micronutrient status of the major soils in the Gezira scheme (Sudan, Africa). D, 1982, Texas A&M University. 110 p.

Adam, David P. Exploratory palynology in the Sierra Nevada, 38°-39° N 119°-121° W, California (Wisconsin-Modern). M, 1965, University of Arizona.

Adam, David Peter. Some palynological applications of multivariate statistics (Osgood swamp, California and southern Arizona). D, 1970, University of Arizona. 149 p.

Adam, M. E. Ground control evaluation of the Sahara No. 20 Mine, Marion, Illinois. M, 1975, University of Missouri, Rolla.

Adam, William Louis. Geology of the Lima Peaks area, Beaverhead County, Montana, and Clark County, Idaho. M, 1948, University of Michigan. 61 p.

Adame, Javier. The effects of a vertical pipe on induced polarization readings. M, 1984, University of Texas at El Paso.

Adamek, James Conrad. Lift and drag forces on a cube on a boundary in a finite, three-dimensional flow field with free-surface effects. D, 1968, Utah State University. 153 p.

Adamek, Scott Harper. Earthquake studies in the Panama-Costa Rica region. M, 1986, University of Texas, Austin. 211 p.

Adamick, John Alton. Depositional and diagenetic analysis of the Hill Sand of the Rodessa Formation (Lower Cretaceous) in the North Shongaloo-Red Rock Field, Webster Parish, Louisiana. M, 1987, Stephen F. Austin State University. 196 p.

Adams, Adam R. Geology of early acid volcanics in Lower Sunlight Basin, Park County, Wyoming. M, 1961, Wayne State University.

Adams, Benjamin Nickolas. The depositional environment, petrography, and tectonics implications of informally named middle to late Eocene marine strata, western Olympic Peninsula, Washington. M, 1988, Western Washington University. 164 p.

Adams, Bradford C. The stratigraphy of the northwestern spur of the Santa Ana Mountains. M, 1932, Stanford University. 57 p.

Adams, Bruce E., Jr. Development and utilization of a resource unit on weathering and erosion for eighth grade earth science. M, 1978, Pennsylvania State University, University Park. 284 p.

Adams, Budd Berwyn. Regional gravity and geologic structure in east-central Minnesota. D, 1957, University of Wisconsin-Madison. 199 p.

Adams, Charles E. An investigation of the X-ray crystallography of bornite (Cu_5FeS_4). M, 1949, University of California, Los Angeles.

Adams, Charles E. Geology of Norwegian Sea cores. M, 1967, University of Washington. 60 p.

Adams, Clifford. Accelerated sedimentation in the Galena River valley, Illinois and Wisconsin. D, 1942, University of Iowa. 67 p.

Adams, Clifford. Modern sedimentation in the Galena River valley, Illinois and Wisconsin. M, 1940, University of Iowa. 71 p.

Adams, Craig W. Geometry, origin, and deformational environment of minor folds in the southern Quitman Mountains, western Trans-Pecos Texas. M, 1983, University of Texas, Arlington. 213 p.

Adams, D. G. A management plan for treatment and disposal of barge cleaning wastes. D, 1974, Texas A&M University. 142 p.

Adams, David D. Geology of the northern contact area of Arrigetch Peaks Pluton, Brooks Range, Alaska. M, 1983, University of Alaska, Fairbanks. 86 p.

Adams, Donald J. Relationships of physical characteristics of the Bluejacket Sandstone (Middle Pennsylvanian) to petroleum production. M, 1959, University of Kansas. 80 p.

Adams, Frank C. An investigation of zonal arrangement of mineral deposits in New Mexico. M, 1924, University of Minnesota, Minneapolis.

Adams, G. W. Precious metal veins of the Berens River Mine, northwestern Ontario. M, 1976, University of Western Ontario. 114 p.

Adams, George Baxter. Stratigraphy of southern Indio Mountains, Hudspeth County, Texas. M, 1953, University of Texas, Austin.

Adams, George F. Glacial waters in the Wallkill Valley. M, 1934, Columbia University. 39 p.

Adams, Gerald Edwin, Jr. An experimental investigation of Ca in olivine in $CaO-MgO-SiO_2$ and $CaO-FeO-MgO-SiO_2$ and its application to geobarometry of spinel and garnet lherzolites. D, 1986, Northwestern University. 119 p.

Adams, Gordon Edward. The geology of Chama area, Rio Arriba County, New Mexico. M, 1957, University of Texas, Austin.

Adams, Gregory T. Laminar-layered calcium carbonate deposits on the Red Rock Canyon alluvial fan, Clark County, Nevada. M, 1974, University of Cincinnati. 86 p.

Adams, Halbert Eden. A crustal structure study of Kentucky and adjacent areas. M, 1975, University of Kentucky. 44 p.

Adams, Henry Clay, Jr. A lithologic analysis of the Galveston beach sand with special emphasis on heavy minerals. M, 1958, Rice University. 51 p.

Adams, Henry James. A study of freshwater limestones in the Conemaugh and Monongahela series, Pittsburgh area. M, 1954, University of Pittsburgh.

Adams, Herbert Gaston. Solid inclusion piezothermometry; experimental calibration of quartz-almandine and sillimanite-almandine. D, 1971, University of California, Los Angeles. 170 p.

Adams, James Bethel, Jr. The petrology and origin of the Simsboro Sand, Bastrop County, Texas. M, 1957, University of Texas, Austin.

Adams, James Warren. Rocks of the Precambrian basement and those immediately overlying it in Missouri and parts of adjacent states. M, 1959, University of Missouri, Columbia.

Adams, Jerald M. An analysis of the gravity-geologic method of mapping buried bedrock topography. M, 1975, Purdue University. 101 p.

Adams, John A. S. Some experiments with the diffusion of cations through geologically significant solids. D, 1951, University of Chicago. 49 p.

Adams, John Bright. Discontinuous outcrops of the Calera Limestone in San Mateo County, California. M, 1956, Stanford University.

Adams, John Bright. Petrology and structure of Stehekin-Twisp Pass area, northern Cascades, Washington. D, 1961, University of Washington. 172 p.

Adams, John Bright. Petrology of isochemically metamorphosed rocks, McGregor Mountain area, Chelan County, Washington. M, 1958, University of Washington. 48 p.

Adams, John Edward. The kinetics of precipitation of amorphous aluminosilicates in near neutral aqueous solutions. M, 1982, University of Rochester. 85 p.

Adams, John Emery. Iron Dike; a pyritized graphite slate of the northern Black Hills of South Dakota. M, 1923, University of Iowa. 78 p.

Adams, John Kendal. Environmental studies of the lower Tertiary formations in New Jersey. D, 1959, Rutgers, The State University, New Brunswick. 194 p.

Adams, John Rodger. Dispersion in anisotropic porous media. D, 1966, Michigan State University. 88 p.

Adams, Keith. Lithospheric structure in northern Pakistan, as inferred from gravity data. M, 1986, Southern Illinois University, Carbondale. 64 p.

Adams, Kenneth D. Taxonomy and paleoecology of the gigantoproductids of Nova Scotia. M, 1978, Acadia University.

Adams, Larry W. Geologic reconnaissance of the proposed Clinton Dam site (Douglas County, Kansas). M, 1961, University of Kansas. 92 p.

Adams, Linn Frank. Stratigraphy of the Mancos Shale of Black Mesa, Arizona. M, 1948, Pennsylvania State University, University Park. 69 p.

Adams, M. Ian and Rice, Emery van Daell. High level silts and gravels. M, 1960, Bryn Mawr College. 27 p.

Adams, Mark Alan. Fault produced fabrics adjacent to the Newberry Mountains detachment fault, Clark County, southern Nevada. M, 1985, San Diego State University. 146 p.

Adams, Mark Allen. Stratigraphy and petrography of the Santiago Peak Volcanics east of Rancho Santa Fe, California. M, 1979, San Diego State University.

Adams, Mark David. A microcomputer assisted study of the relationship of chloride and uranium concentrations in groundwater to geology in the El Dorado 1 x 2 degree Quadrangle, Arkansas. M, 1986, University of Arkansas, Fayetteville.

Adams, Michael Anthony. The origin and distribution of porosity within the Knox Supergroup (Cambrian-Ordovician) in lithofacies subjacent to the regional Knox unconformity, subsurface Indiana. M, 1984, Duke University. 168 p.

Adams, Michael C. Stratigraphy and structure of the McCoy geothermal prospect, Churchill and Lander counties, Nevada. M, 1984, University of Utah. 81 p.

Adams, Nigel Bruce. The Holden Mine; from discovery to production, 1896-1938. D, 1976, University of Washington. 235 p.

Adams, O. Clair. Geology of the Summer Ranch and North Promontory Mountains. M, 1962, Utah State University. 57 p.

Adams, Opal Fay. Geology and ore deposits of the Thunder Mountain mining district, Valley County, Idaho. M, 1985, University of Nevada. 104 p.

Adams, Paul Michael. The mineralogy and petrology of monticellite-clintonite and associated skarns at Clark Mountain, northeastern San Bernardino County, California. M, 1979, University of Southern California.

Adams, Penny R. The foraminiferal paleoecology and biostratigraphy of the Hilliard Formation type area (Upper Cretaceous), Lincoln County, Wyoming. M, 1972, University of Wyoming. 56 p.

Adams, Peter J. A depositional and diagenetic model for a carbonate ramp; Iroquois Formation (Early Jurassic), Scotian Shelf, Canada. M, 1986, Dalhousie University. 150 p.

Adams, R. L. Stratigraphy, petrography, and diagenesis of the lower Oneota Dolomite (Ordovician) of South-central Wisconsin. M, 1975, University of Wisconsin-Madison.

Adams, R. S. The effects of drying on the products of the low-temperature carbonization of North Dakota lignite. M, 1931, University of Minnesota, Minneapolis.

Adams, Robert W. The Greenbrier Limestone of Pennsylvania, Maryland, and West Virginia. D, 1964, The Johns Hopkins University.

Adams, Robert William. Geology of the Cayuse Mountain-Horse Springs-Coulee area, Okanogan County, Washington. M, 1962, University of Washington. 41 p.

Adams, Roy Donald. Late Paleozoic tectonic and sedimentologic history of the Penasco Uplift, north-central New Mexico. M, 1980, Rice University. 79 p.

Adams, Russell Stanley, Jr. Ammonium sorption and release by soil-forming rocks and minerals. D, 1962, University of Illinois, Urbana. 161 p.

Adams, Samuel S. Minor element content of pegmatitic sulfides. M, 1961, Dartmouth College. 81 p.

Adams, Samuel Sherman. Bromine in the Salado Formation (Permian), Carlsbad potash district (Eddy County), New Mexico. D, 1968, Harvard University.

Adams, Scot Crawford. A numerical and statistical study of the Pennsylvanian geology of Sullivan County, Indiana. D, 1980, Purdue University. 159 p.

Adams, Scott A. A biogeochemical investigation of dispersion patterns of the Ste. Genevieve County, Missouri, copper deposits. M, 1974, Southern Illinois University, Carbondale. 72 p.

Adams, Scott Randall. Geochemistry of the Wichita Granite Group in the Wichita Mountains, Oklahoma. M, 1977, Oklahoma State University. 74 p.

Adams, Sidney F. Microscopic study of vein quartz. M, 1920, Stanford University. 59 p.

Adams, Vincent. Oil-brine contamination of Marion Township, Hocking County. M, 1980, Ohio University, Athens. 90 p.

Adams, Wayne C. Analysis of landslide occurrences in Idaho. M, 1988, University of Idaho. 146 p.

Adams, William David. Geologic history of Crescent Lake, Florida. M, 1976, University of Florida. 63 p.

Adams, William L. Geology of the Dry Canyon area, northeastern Ventura County, Southern California. M, 1956, University of California, Los Angeles.

Adams, William Mansfield. Calibration of the Reeff horizontal seismograph. M, 1955, St. Louis University.

Adams, William Mansfield. Some geometrical characteristics of earthquake mechanism as indicated by an analysis of the S-wave. D, 1957, St. Louis University.

Adams, William Russell, Jr. Landsliding in Allegheny County, Pennsylvania; characteristics, causes, and cures. D, 1986, University of Pittsburgh. 306 p.

Adamsen, Floyd James. Nitrogen transformations in soils amended with anaerobically digested sewage sludge. D, 1983, Colorado State University. 163 p.

Adamski, James Clifford. The effect of agriculture on the quality of ground water in a karstified carbonate terrain, Northwest Arkansas. M, 1987, University of Arkansas, Fayetteville.

Adamson, David William. Uranium and thorium abundances in high grade rocks from the the western Saskatchewan shield. M, 1984, University of Regina. 246 p.

Adamson, Richard Floyd. Stratigraphy and conodont biostratigraphy of the Lower and Middle Ordovician carbonates of the Draper Belt, Pulaski and Wythe counties, Virginia. M, 1982, Virginia Polytechnic Institute and State University. 98 p.

Adamson, Robert Clarence. Magnetic investigation in the Kassler area, Jefferson County, Colorado. M, 1956, Colorado School of Mines. 55 p.

Adamson, Robert D. The Salt Lake Group in Cache Valley, Utah and Idaho. M, 1955, Utah State University. 59 p.

Adar, Eilon. Quantification of aquifer recharge distribution using environmental isotopes and regional hydrochemistry. D, 1984, University of Arizona. 269 p.

Adcock, Stephen William. The stability of NA-K aluminosilicates in supercritical water; an experimental and thermodynamic study. D, 1986, Carleton University. 337 p.

Adcock, T. D. Sedimentary character of the Lower Capistrano Formation, upper Miocene age; Dana

Point Harbor to San Clemente, California. M, 1978, University of Southern California.

Addicott, Warren Oliver. Miocene stratigraphy northeast of Bakersfield, California. D, 1956, University of California, Berkeley. 219 p.

Addington, Archie Rombaugh. Topography, physiography, and geology of the Bloomington, Indiana, Quadrangle. M, 1925, Indiana University, Bloomington.

Addington, James W. Petrographic analysis of a sandstone body in the Fayetteville Formation in the SE 1/4, Sec. 29, T. 16 N., R. 24 W. M, 1969, University of Arkansas, Fayetteville.

Addison, Carl C. Stratigraphy and correlation of some Carboniferous sections of Nevada and adjacent states. M, 1929, Stanford University. 173 p.

Addison, Michael Earl. Recent history and future of petroleum production in selected African countries. M, 1963, University of Colorado.

Addo-Abedi, Frederick Yaw. The behaviour of soils in repeated triaxial compression. D, 1980, Queen's University.

Addy, Richard V. Stratigraphy of the Sawatch Sandstone (Cambrian) of the southern Front Range, Colorado. M, 1949, University of Colorado.

Addy, Sunit Kumar. The problem of ore genesis at Ducktown, Tennessee; interpretation of stable isotopes (O_{18}/O_{16}, C_{13}/C_{12} and D/H), microprobe and textural data. D, 1973, University of Colorado. 235 p.

Adediran, Sulleiman Adebayo. Adsorption of copper on particulates along a salinity gradient. D, 1985, McMaster University. 175 p.

Adedokun, Oluwatele Alabi. Genesis and environmental analysis of the reservoir sandstones of Ossu-Izombe oil field, Imo State, Nigeria. M, 1979, University of Houston.

Adedotun, Adekoya Adedayo. Conodont analysis of a limestone lens in the black shale belt of Northeast Washington. M, 1983, Washington State University. 65 p.

Adeff, Sergio Everardo. A sediment-laden three-dimensional-flow numerical model. D, 1988, University of Mississippi. 549 p.

Adegoke, Oluwole Johnson. Analysis of groundwater hydrographs. M, 1982, University of Waterloo. 152 p.

Adegoke, Sylvester. Stratigraphy and paleontology of the Neogene formations of the Coalinga region, California. D, 1966, University of California, Berkeley. 557 p.

Adekeye, Jacob Ishola Dele. Optical investigation of Sn(II) in cassiterite (SnO_2). D, 1985, University of Pittsburgh. 51 p.

Adekoya, Adeodotun Adedayo. Foraminiferal biostratigraphy, depositional environments and hydrocarbon source rock potential of sediments from five wells in the western offshore region of the Niger Delta, Nigeria. D, 1987, Washington State University. 422 p.

Adelman, David. Geochemistry of tributyltin in coastal waters; an experiment in a MERL mesocosm. M, 1988, University of Rhode Island.

Adelseck, Charles G., Jr. Recent and late Pleistocene sediments from the eastern equatorial Pacific Ocean; sedimentation and dissolution. D, 1977, University of California, San Diego. 219 p.

Adem, Adem Osman. Geochemical study of an abandoned strip mine site, Williamson County, Illinois. M, 1985, Southern Illinois University, Carbondale. 99 p.

Aden, Gary David. Quantitative energy dispersive analysis of small particles. D, 1981, Arizona State University. 476 p.

Aden, Leon John. Clay mineralogy and depositional environments of upper Cherokee (Desmoinesian) mudrocks, eastern Kansas, western Missouri, and northeastern Oklahoma. M, 1982, University of Iowa. 126 p.

Adeniji, Francis A. Some aspects of the distribution and mechanics of soil erosion under furrow irrigation in the Bow Island area, Alberta. M, 1970, University of Calgary.

Adeniyi, Joshua O. Geophysical investigations of the central part of Niger State, Nigeria. D, 1984, University of Wisconsin-Madison. 181 p.

Adenle, O. A. Geologic controls on the relationship between surface water and groundwater resources in southwestern Nigeria. D, 1977, George Washington University. 174 p.

Adenle, Oladepo Adeoye. The geomorphic significance of the Lakeland soil in the peninsula Florida. M, 1971, University of Florida. 104 p.

Adenuga, Oladipo S. The petrology of the Brassfield Formation (Lower Silurian) between Richmond, Indiana, and Dayton, Ohio. M, 1985, Miami University (Ohio). 148 p.

Adeyeri, Joseph Babalola. Multiple integral description of the viscoelastic response of cohesive soils. D, 1969, Northwestern University. 211 p.

Adger, John B., Jr. Studies in mass transport in contact metamorphism. M, 1968, Massachusetts Institute of Technology. 60 p.

Adhidjaja, Jopie Iskandar. Finite-difference solutions for transient electromagnetics. D, 1988, University of Utah. 214 p.

Adhidjaja, Jopie Iskandar. Study of major geologic structures indicated by gravity data in the Richfield one-degree by two-degree quadrangle, Utah. M, 1981, University of Utah. 78 p.

Adib, Mazen Elias. Internal lateral earth pressure in earth walls. D, 1988, University of California, Berkeley. 394 p.

Adidas, Eric O. Geophysical characteristics of the Pennsylvanian shales in the Staunton and Linton formations in Clay County, Indiana using gamma-ray logs and X-ray diffractograms. M, 1980, Indiana State University. 118 p.

Adilman, Daivd. Factors affecting the future petroleum development of the North Slope-Beaufort Sea, Alaska. M, 1986, University of Texas, Austin.

Adjali, Salim. Investigation of the static and dynamic interaction behavior of a soil-cantilever wall system. D, 1988, University of Washington. 214 p.

Adkins, John Nathaniel. The Alaskan earthquake of July 22, 1937. D, 1940, University of California, Berkeley. 46 p.

Adkinson, Burton Wilbur. The Alpine glacial history and post glacial adjustments of a section of the Cabinet Mountains, Montana. D, 1942, Clark University. 115 p.

Adler, Alan A. Flood plain sedimentation of Halfmoon Creek. M, 1960, Pennsylvania State University, University Park. 23 p.

Adler, Dennis Marvin. Tracer studies in marine microcosms; transport processes near the sediment-water-interface. D, 1982, Columbia University, Teachers College. 364 p.

Adler, Hans H. Thermal study of the jarosite group. M, 1949, Columbia University, Teachers College.

Adler, Hans Henry. Infrared spectroscopy of carbonate sulfate minerals. D, 1962, Columbia University, Teachers College. 118 p.

Adler, Joseph L. Geologic and petrographic relations of the Coal Creek Quartzite and contiguous crystalline formations in Jefferson, Boulder, and Gilpin counties, Colorado. D, 1930, University of Chicago. 208 p.

Adler, Kevin R. Source rock analysis of oil shale in the Green River Formation, Bridger Basin, Wyoming. M, 1982, University of Wyoming. 71 p.

Adler, Lori Lynn. Adjustment of the Yuba River, California; to the influx of hydraulic mining debris; 1849-1979. M, 1980, University of California, Los Angeles. 180 p.

Adler, Robert C. Intervention analysis of the forest harvesting on streamflow at Coweeta Hydrologic Laboratory, North Carolina. M, 1988, Boston University. 209 p.

Adler, Ron K. H. The application of aero-leveling strip triangulation to the measurement of differential glacier surface movement in areas of limited geodetic control. M, 1963, Ohio State University.

Adlis, David Scott. Stable isotope variations in Late Pennsylvanian brachiopods from cyclic sedimentary deposits; paleoenvironmental and diagenetic implications. M, 1986, Texas A&M University.

Admiraal, Peter. A study of the proposed mechanisms for drumlin formation with reference to drumlins in the Palmyra, New York Quadrangle. M, 1970, Virginia State University. 61 p.

Adotevi-Akue, George Modesto. On the numerical interpretation of gravity and other potential field anomalies caused by layers of varying thickness. D, 1972, Oregon State University. 117 p.

Adshead, John Douglas. Mineralogical studies of bottom sediments from western Hudson and James bays. D, 1973, University of Missouri, Columbia.

Adshead, John Douglas. Petrology of the carbonate rocks of the Siyeh Formation, southwestern Alberta. M, 1963, University of Alberta. 117 p.

Aduamoah-Larbi, Joseph. A study of alkali-aggregate reactions in concrete; measurement and prevention. M, 1987, University of Windsor. 182 p.

Adutwum, John B. New geodetic datum in Ghana. M, 1963, Ohio State University.

Advocate, David Michael. Depositional environments of the Eocene Maniobra Formation, northeastern Orocopia Mountains, Riverside County, Southern California. M, 1983, California State University, Northridge. 112 p.

Aelion, Claire Marjorie. Adaptation of microbial communities from an uncontaminated aquifer to degrade organic pollutants. D, 1988, University of North Carolina, Chapel Hill. 191 p.

Affholter, Kathleen Ann. Petrogenesis of orbicular rock, Tijeras Canyon, Sandia Mountains, New Mexico. M, 1979, University of New Mexico. 124 p.

Affholter, Kathleen Ann. Synthesis and crystal chemistry of lanthanide allanites. D, 1987, Virginia Polytechnic Institute and State University. 218 p.

Affolter, R. H. Geochemistry of mineral matter and selected trace elements for the Herrin (No. 6) and Springfield-Harrisburg (No. 5) coal in northwestern Illinois. M, 1977, Northeastern Illinois University.

Afiattalab, Firooz. Fe, Co and Ni in the metal phases and the classification of ordinary chondrites. M, 1979, University of California, Los Angeles.

Afifi, Abdulkader M. Precambrian geology of the Iris area, Gunnison and Saguache counties, Colorado. M, 1981, Colorado School of Mines. 197 p.

Afifi, Sherif El-sayed Ahmed. Effects of stress history on the shear modulus of soils. D, 1970, University of Michigan. 234 p.

Afify, Abdel-Azeem Hendy. Ground movement and roof control in mining stratified deposits. M, 1963, University of Utah. 254 p.

Afrasiabi, Hedayat. A study of mineralization in the Dugway Range, Tooele County, Utah. M, 1981, University of Utah. 75 p.

Afshar, Abas. An integrated study of a wastewater disposal plan. D, 1979, University of California, Davis. 179 p.

Afshar, Freydown A. Taxonomic revision of the Cretaceous and Cenozoic Tellinidae. D, 1950, The Johns Hopkins University.

Aftabi, Alijan. Geochemical dispersion of Hg, Au, Ag, As, Zn, Cu, Ni, and Zr in relation to gold mineralization at Sigma gold mine, Val d'Or, Quebec. D, 1985, McGill University. 253 p.

Aftabi, Alijan. Polymetamorphism, textural relations and mineralogical changes in Archean massive sulfide deposits at the Garon Lake Mine, Matagami, Quebec. M, 1980, McGill University. 253 p.

Afzali, Behzad. Heavy metal patterns in Big and Aux Vases river drainage, Rolla $1° \times 2°$ Quadrangle, Missouri. M, 1979, University of Missouri, Rolla.

Agagu, Olusegun K. Depositional characteristics of the Frio Formation, subsurface South Texas. M, 1975, University of Texas, Austin.

Agar, William MacDonough. Contact metamorphism in the western Adirondacks. D, 1922, Princeton University.

Agard, Sherry Sue. Investigation of recent mass movements near Telluride, Colorado, using the growth and form of trees. M, 1979, University of Colorado.

Agarwal, A. S. Immobilization and mineralization of nitrogen in Hawaiian soils. D, 1967, University of Hawaii. 159 p.

Agarwal, Ram Gopal. Two-dimensional harmonic analysis of potential fields. D, 1968, University of Alberta. 75 p.

Agashe, Shripad Narayanrao. The extra-xylary tissues in certain Calamites from the American Carboniferous. D, 1964, Washington University. 99 p.

Agasie, John M. Upper Cretaceous palynomorphs from Coal Canyon, Coconino County, Arizona. M, 1967, University of Arizona.

Agatston, Robert Stephen. Pennsylvanian and Lower Permian of northern and eastern Wyoming. D, 1953, Columbia University, Teachers College.

Agatston, Robert Stephen. The structure and lithology of the Inwood Limestone. M, 1947, Columbia University, Teachers College.

Agbe-Davies, Victor F. The geology of the Hartshorne coals in the Spiro and Hackett quadrangles, Le Flore County, Oklahoma. M, 1978, University of Oklahoma. 132 p.

Agee, Carl Bernard. Experimental phase density and mass balance constraints on early differentiation of chondritic mantle. D, 1988, Columbia University, Teachers College. 157 p.

Agee, George Ray. The geology of the Sedalia West Quadrangle. M, 1954, University of Missouri, Columbia.

Agee, Jeffrey J. Petrogenesis of opaque minerals in kimberlite; Elliott County, Kentucky. M, 1980, University of Tennessee, Knoxville. 53 p.

Agegian, Catherine Rose. The biogeochemical ecology of Porolithon gardineri (Foslie). D, 1985, University of Hawaii. 199 p.

Ageli, Hadi S. Atmospheric and deep seated sources of unsupported lead-210 in soil profiles and the use of lead-210 and polonium-210 as indicators of uranium mineralization at depth. M, 1983, Rice University. 52 p.

Agenbroad, Larry D. The geology of the Atlas Mines area, Pima County, Arizona. M, 1962, University of Arizona.

Agenbroad, Larry Delmar. Cenozoic stratigraphy and paleohydrology of the Redington-San Manuel area, San Pedro Valley, Arizona. D, 1967, University of Arizona. 182 p.

Ager, C. A. The three dimensional structure of batholiths as deduced from gravity data. D, 1975, University of British Columbia.

Ager, Charles Arthur. A gravity model for the Guichon Creek Batholith (British Columbia). M, 1972, University of British Columbia.

Ager, T. A. Late Quaternary environmental history of the Tanana Valley, Alaska. D, 1975, Ohio State University. 184 p.

Ager, Thomas A. Surficial geology and Quaternary history of the Healy Lake area, Alaska. M, 1972, University of Alaska, Fairbanks. 127 p.

Ageton, Robert William. Sampling methods for impounded tailings. M, 1946, University of Arizona.

Agey, Charles. The Pavilion gas field, New York. M, 1934, University of Rochester. 94 p.

Aggarwal, Pradeep Kumar. Geochemistry of the Chu Chua massive sulfide deposit, British Columbia. M, 1982, University of Alberta. 81 p.

Aggarwal, Pradeep Kumar. Oxygen-isotope geochemistry and metamorphism of massive sulfide deposits of the Flin Flon-Snow Lake Belt, Manitoba. D, 1986, University of Alberta. 160 p.

Aggarwal, Yash P. Premonitory changes in seismic velocities; observations, causal mechanisms and application to earthquake prediction. D, 1975, Columbia University. 60 p.

Aghajanian, John Gregory. Light and electron microscope and cytochemical studies of morphogenesis in the fresh water red alga Batrachospermum sirodotii Skuja. D, 1977, University of North Carolina, Chapel Hill. 222 p.

Agioutantis, Zacharias George. An investigation into the modeling of ground deformations induced by underground mining. D, 1987, Virginia Polytechnic Institute and State University. 223 p.

Agne, Russell Maynard. Stratigraphy of the Eaton Hill area, Vernon, New York, 7-1/2 minute Quadrangle. M, 1963, Syracuse University.

Agnew, Allen Francis. Bibliographic index of new genera and families of Paleozoic Ostracoda since 1934. M, 1942, University of Illinois, Urbana. 26 p.

Agnew, Allen Francis. The Middle and Upper Ordovician strata of the Upper Mississippi Valley, a restudy. D, 1949, Stanford University. 166 p.

Agnew, Duncan C. Strain tides at Pinon Flat; analysis and interpretation. D, 1979, University of California, San Diego. 189 p.

Agnew, Haddon W. The geology of a part of the Ravenna Quadrangle, California. M, 1948, California Institute of Technology. 18 p.

Agnew, James D. Seismicity of the Central Alaska Range, Alaska, 1904 - 1978. M, 1980, University of Alaska, Fairbanks. 88 p.

Agnew, P. C. Reinterpretation of the Hinesburg Thrust in northwestern Vermont. M, 1977, University of Vermont.

Agocs, William B. A method of determining the time break on deep sea seismic records, and the slope of the sea bottom and its direction from the water sound arrivals. D, 1946, Lehigh University.

Agostino, Patrick N. Theoretical and experimental investigations on ptygmatic structures. M, 1970, Brooklyn College (CUNY).

Agostino, Patrick Noel. An analysis of finite-amplitude single-layer asymmetrical folding. D, 1975, Rensselaer Polytechnic Institute. 198 p.

Agosto, William N. Two Antarctic achondrites; a petrologic and chemical comparison with evidence for phosphorus reduction of iron in non-terrestrial pyroxenes. M, 1981, Rutgers, The State University, New Brunswick. 77 p.

Agra, Jefferson de Mello. Depositional and diagenetic aspects of the Namorado Sandstone, Namorado Field, Campos Basin, Brazil. M, 1984, University of Texas, Austin. 72 p.

Agreda, Francisco Moreno *see* Moreno Agreda, Francisco

Agron, Sam L. Structures and petrology of the Peach Bottom Slate, and its environment. D, 1949, The Johns Hopkins University.

Aguado, E. A. Optimization techniques and numerical methods for aquifer management. D, 1979, Stanford University. 189 p.

Aguado, Edward. The effect of a groin on beach morphology. M, 1977, University of California, Los Angeles. 49 p.

Aguayo, Camargo Joaquim E. Sedimentary environments and diagenetic implications of the El Abra Limestone at its type locality, East Mexico. D, 1975, University of Texas at Dallas. 158 p.

Aguayo, Eduardo. Lithofacies and sedimentary environments of the Glen Rose Limestone (Cretaceous) in north central Texas. M, 1971, Baylor University. 91 p.

Ague, Daria Monica. Natural deformation of plagioclase; a microstructural investigation. D, 1988, University of California, Berkeley. 100 p.

Ague, Jay J. The structure and metamorphism fo the Mullernest Formation, St. Jonsfjorden, Svalbard. M, 1983, Wayne State University. 173 p.

Ague, Jay James. Geochemical modelling of fluid flow and chemical reaction during supergene enrichment of porphyry copper deposits. D, 1987, University of California, Berkeley. 67 p.

Aguerrevere, Pedro I. A study for the development of a field magnetometer based on the principle of the Earth inductor. M, 1929, Colorado School of Mines. 42 p.

Aguilar, Eduardo. Areal/environmental geology of the Luquillo area, northeastern Puerto Rico. M, 1971, Cornell University.

Aguilar, Oscar. Study of the copper-uranium deposits at Vilcabamba, Department of Cuzco, Peru. M, 1962, University of Missouri, Rolla.

Aguilar-Maldonado, A. Methodology for long-term water supply planning; Mexico City case. D, 1979, University of Arizona. 152 p.

Aguilar-Tuñón, Nicholás A. Beach profile changes onshore-offshore sand transport on the Oregon coast. M, 1977, Oregon State University. 58 p.

Aguilera, Raymundo. A technique for the design of digital recursive filters. M, 1969, Rice University. 76 p.

Aguilera, Roberto. Evaluation of fine-grained laminated systems from logs, Wasatch Formation, Utah. D, 1973, Colorado School of Mines. 279 p.

Aguirre-Diaz, Gerardo de Jesus. Eocene and younger volcanism on the eastern flank of the Sierra Madre Occidental, Nazas, Durango, Mexico. M, 1988, University of Texas, Austin. 179 p.

Agunloye, Alfred Olusegun. A nonlinear study of electrode impedance in sulphide minerals. M, 1977, Massachusetts Institute of Technology. 70 p.

Agurkis, Edward N. Depositional history of the Piankasha Sequence (Upper Devonian), southern Arizona, and southwestern New Mexico. M, 1977, Arizona State University. 179 p.

Ahamed, Aziz U. Hydrologic significance of the lithofacies of the late Cenozoic deposits of the northern Tucson Basin, Arizona. M, 1970, University of Idaho. 78 p.

Ahbe, J. B. Southeastern Illinois magnetic anomaly map and its regional geologic implications. M, 1978, Purdue University. 115 p.

Ahdoot, Hooshang. Theoretical and experimental investigation of the phenomenon of sandstone disaggregation. M, 1969, University of Oklahoma. 62 p.

Ahern, J. W. Magma migration. D, 1980, Cornell University. 104 p.

Ahern, Judson Lewis. Aeromagnetic reconnaissance survey of Lake Erie. M, 1975, Ohio State University.

Ahern, Kevin Edward. Ecology and biogeography of Llandovery beyrichiid ostracodes. D, 1978, University of California, Berkeley. 121 p.

Ahern, Kevin Edward. Patterns of Paleozoic gastropod diversity. M, 1974, University of California, Berkeley. 67 p.

Ahern, T. K. An O^{18}/O^{16} study of water flow in natural snow. M, 1975, University of British Columbia.

Ahern, Timothy Keith. The development of a completely automated oxygen isotope mass spectrometer. D, 1980, University of British Columbia.

Ahlborn, Robert C. Mesozoic-Cenozoic structural development of the Kern Mountains, eastern Nevada-western Utah. M, 1976, Brigham Young University. 131 p.

Ahlbrandt, Thomas Stuart. Sand dunes, geomorphology and geology, Killpecker Creek area, northern Sweetwater County, Wyoming. D, 1973, University of Wyoming. 212 p.

Ahlen, Jack L. Regional stratigraphy of the Jordan Sandstone in west central Wisconsin. M, 1952, University of Wisconsin-Madison.

Ahler, Bruce Allen. The effect of faulting and other factors on sharpness ratios of the clay mineral illite. M, 1974, University of Tennessee, Knoxville. 45 p.

Ahlfeld, David Philip. Designing contaminated groundwater remediation systems using numerical simulation and nonlinear optimization. D, 1987, Princeton University. 152 p.

Ahlrichs, John Sigurd. Movement of selenite and selenate by saturated and unsaturated flow in East Texas overburden. D, 1983, Texas A&M University. 136 p.

Ahlschwede, Kelly. Sources and littoral transport of sand in San Diego and southern Orange counties, Southern California; Fourier grain-shape analysis. M, 1988, University of Southern California.

Ahlstrand, Dennis Carl. Permian carbonate facies, Wind River Mountains and western Wind River basin. M, 1972, University of Montana. 104 p.

Ahmad, F. I. Geological interpretation of gravity and aeromagnetic surveys in the Bronson Hill Anticlinorium, southwestern New Hampshire. D, 1975, University of Massachusetts. 274 p.

Ahmad, Fachri. Effect of clay minerals and clay-humic acid complexes on availability and flexation of phosphates. D, 1988, University of Georgia. 235 p.

Ahmad, Khwaja Gulzar. A study of Medicine Hat sandstone (upper Cretaceous), (Alberta, Canada). M, 1969, University of Alberta. 178 p.

Ahmad, Mahmood U. Study of genetic relationship between the Ralston intrusive bodies and the Table Mountain lava flow (Paleocene) near Golden, Jefferson County, Colorado. M, 1971, Colorado School of Mines. 124 p.

Ahmad, N. Evaluation of in-situ testing methods in soils. D, 1975, Louisiana State University. 400 p.

Ahmad, Raisuddin. Stratigraphy, structure and petrology of the Lookingglass and Roseburg formations, Agness-Illahe area, southwestern Oregon. M, 1981, University of Oregon. 150 p.

Ahmed, A. A. The dynamics of land use and woody vegetation changes in Jebel Marra, Darfur, Sudan. D, 1979, University of California, Los Angeles. 181 p.

Ahmed, Abdelazim I. Geology of Kurmuk Sheet, Sudan. M, 1983, Ohio University, Athens. 54 p.

Ahmed, Adnan A. Time-domain electromagnetic sounding curves for short-line source and loop received by means of digital linear filter. M, 1973, Colorado School of Mines. 70 p.

Ahmed, Ahmed El-Sayed. Simulation of simultaneous heat and moisture transfer in soils heated by buried pipes. D, 1980, Ohio State University. 130 p.

Ahmed, Ahmed Zakaria. A new method for orienting the horizontal components in three-component vertical seismic profiling data. M, 1987, University of Utah. 51 p.

Ahmed, Ajaz. The sorption of aqueous nickel (II) species on alumina and quartz. D, 1971, Stanford University. 91 p.

Ahmed, Anees Uddin. Digitized well log evaluation of Mesaverde Formation in Rulison Field, Colorado. M, 1978, University of Oklahoma. 111 p.

Ahmed, Elhag Elhadi Ali. Gravity survey in Madera County, California. M, 1965, University of California, Los Angeles.

Ahmed, Farazi Kamaluddin. The processing and interpretation of Deer Lake seismic data. M, 1984, Memorial University of Newfoundland. 102 p.

Ahmed, Gaafar Abbashar. Petrographic, fluid inclusion and stable isotopic study of the concordant zinc-lead ores of N.E. Washington State. M, 1984, University of Wisconsin-Madison. 57 p.

Ahmed, Gulfaraz. An experimental study of recovery from a 2-D layered sand model. M, 1984, Stanford University.

Ahmed, M. Diagenesis and porosity development of Mission Canyon reservoir interval in 07 43 B pool, Pierson Field, Manitoba. M, 1985, University of Manitoba.

Ahmed, Mesbahuddin. The Brewster magnetite deposit. M, 1948, Columbia University, Teachers College.

Ahmed, Mushtaque. Development of a two-dimensional finite element soil-moisture flow model. D,

1988, Iowa State University of Science and Technology. 141 p.

Ahmed, Riaz. Surface-groundwater interactions and the conjuctive use of the water resources of the Mullica River basin, New Jersey. D, 1973, Rutgers, The State University, New Brunswick. 222 p.

Ahmed, S. Effects of adsorbed cations on the physical properties of soils under arid conditions. D, 1965, University of Hawaii.

Ahmed, Shemsudin. Geology of the Bone Basin area, Madison and Jefferson counties, Montana. M, 1987, Montana College of Mineral Science & Technology. 116 p.

Ahmed, Syed Sirtajuddin. Tertiary geology of part of south Makran, Baluchistan, West Pakistan. M, 1968, University of Kansas. 73 p.

Ahmeduddin, Mir. Subsurface geology of Wheatland area, Cleveland, McClain, Grady, Canadian and Oklahoma counties, Oklahoma. M, 1968, University of Oklahoma. 88 p.

Ahn, Joonhong. Mass transfer and transport of radionuclides in fractured porous rock. D, 1988, University of California, Berkeley. 183 p.

Ahn, Jung-Ho. Mineralogy and diagenesis of phyllosilicates in argillaceous sediments; a TEM study. D, 1986, University of Michigan. 201 p.

Ahn, Jung-Ho. The TEM and AEM characterization of chlorite diagenesis in Gulf Coast argillaceous sediments. M, 1984, University of Michigan.

Ahn, Kyu-Hong. A national program for environmental control in Korea; a developing country; water supply, wastewater management, pollution control. D, 1983, Cornell University. 145 p.

Aho, Aaro Emil. Geology and ore deposits of the property of Pacific Nickel Mines, near Hope, British Columbia. D, 1954, University of California, Berkeley. 151 p.

Aho, Gary D. A reflection seismic investigation of thickness and structure of the Jacobsville sandstone (Precambrian or Cambrian), Keeweenaw peninsula, Michigan. M, 1969, Michigan Technological University. 104 p.

Aho, John E. The fauna and paleoenvironment of the "White River" beds, Rockerville Quadrangle, Pennington County, South Dakota. M, 1974, South Dakota School of Mines & Technology.

Ahr, Wayne M. Petrology and petrography of the Campeche Calcilutite, Yucatan, Mexico. M, 1965, Texas A&M University.

Ahr, Wayne Merrill. Origin and paleoenvironment of some Cambrian algal reefs, Mason County area, Texas. D, 1967, Rice University. 118 p.

Ahrens, Gary Louis. An analysis and interpretation of gravity and magnetic anomalies of the Butte District, Montana. M, 1976, University of Arizona. 48 p.

Ahrens, Thomas J. An ultrasonic interferometer for high-pressure research with results for selected solids. D, 1962, Rensselaer Polytechnic Institute. 67 p.

Ahrens, Thomas Patrick. The use of aerial photographs in geologic studies. M, 1940, George Washington University. 46 p.

Ahrnsbrak, William Frederick. A diffusion model for Green Bay, Lake Michigan. D, 1971, University of Wisconsin-Madison. 121 p.

Ahsan, Abdus Salam. Pre-stack multiple suppression in F-K domain. M, 1983, Ohio University, Athens. 102 p.

Ahuja, H. S. Kimberlites; a review. M, 1977, Northeastern Illinois University.

Ahuja, Suraj Prakash. Structural analysis of the Amisk and Missi rocks (Archaean) north of Amisk lake, Saskatchewan. M, 1970, University of Saskatchewan. 89 p.

Aide, Michael Thomas. The thermodynamics of cation exchange involving smectites and soil clay fractions. D, 1982, Mississippi State University. 92 p.

Aigen, A. A. Early Mississippian gastropods of the Burlington Limestone. M, 1974, University of Illinois, Urbana. 137 p.

Aiinehsazian. Magnetic properties of diabase sills of Agardhdalen, east central Spitsbergen, Svalbard Archipelago. M, 1981, St. Louis University.

Aiken, C. L. V. The analysis of the gravity anomalies of Arizona. D, 1976, University of Arizona. 162 p.

Aiken, Carlos Lynn Virgil. Gravimetric profiles across northern Cascades employing minimum assumptions as to subsurface density distribution. M, 1970, University of Washington. 134 p.

Aiken, Lewis J. Geology of the Adwolf-Thomas Bridge area, Virginia. M, 1967, Virginia Polytechnic Institute and State University.

Aiken, Mary J. Mineralogy and geochemistry of a lacustrine uranium occurrence, Andersen Ranch, Brewster County, Texas. M, 1981, University of Texas at El Paso.

Aiken, Olaf W. Geological significance of surface gravity measurements in the vicinity of the Illinois deep drill hole. M, 1982, University of Texas at El Paso.

Aikin, Andrea R. An assessment of the long-term hydrologic effects of artificial recharge on the Denver ground-water basin using computer simulation methods. M, 1988, Colorado School of Mines. 75 p.

Aikins, Charles W., III. Hydrocarbon production potential of the Lansing-Kansas City E-Zone in southwestern Nebraska. M, 1983, Wichita State University. 106 p.

Ailin-Pyzik, I. B. Micro-scale chemical effects of low temperature weathering of DSDP basaltic glasses. D, 1979, University of Maryland. 186 p.

Aimo, Nino J. Preliminary numerical study of the effects of fluctuating river level on solute transport within riverbank aquifers. M, 1987, Washington State University. 66 p.

Aimone, Catherine Taylor. Three-dimensional wave propagation model of full-scale rock fragmentation. D, 1982, Northwestern University. 304 p.

Aines, Roger Deane. Trace hydrogen in minerals. D, 1984, California Institute of Technology. 295 p.

Ainscough, Harlen R. The geochemical modes of occurrence of cadmium, copper, zink, nickel, and iron in Coal V of Pike and Warrick counties, Indiana. M, 1974, Indiana State University. 76 p.

Ainsworth, B. D. Mineralogical and grain size data on selected samples in the Forest Hills Formation (lower Oligocene) in western Mississippi. M, 1967, University of Mississippi.

Ainsworth, Calvin Carney. Phosphate sorption on goethites. D, 1983, University of Georgia. 92 p.

Ainsworth, Michael R. Geomorphological analysis of the boulder deposit at the mouth of Little Elk Canyon, Black Hills, South Dakota. M, 1981, South Dakota School of Mines & Technology. 112 p.

Airhart, Tom Patterson. The response of a pile-soil system in a cohesive soil as a function of the excess pore water pressure and the engineering properties of the soil. D, 1967, Texas A&M University. 137 p.

Airola, T. M. An assessment of the utility of forest cover type information in land use planning in the Southern Appalachians. D, 1977, Duke University. 267 p.

Aisner, Jonathan Alan. Genesis of a skeletal mound, Arsenic Bank, Florida. M, 1981, University of South Florida, Tampa. 89 p.

Ait-Laoussine, Nordine. Petroleum geology of Algeria. M, 1963, University of Michigan.

Aitken, Alec Edison. The ecology of a subarctic intertidal flat, Pangnirtung Fiord, Baffin Island, N.W.T. and the paleoecology of Quaternary molluscan assemblages. D, 1987, McMaster University. 247 p.

Aitken, B. G. Stability relations of Ca-rich scapolite. D, 1979, Stanford University. 121 p.

Aitken, Frances Kenneth. An x-ray powder diffraction study of potassium feldspar from six possible meteorite impact sites. D, 1970, Pennsylvania State University, University Park. 162 p.

Aitken, James Drynan. Greenstones and associated ultramafic rocks of the Atlin map-area, British Co-lumbia. D, 1953, University of California, Los Angeles.

Aitken, Janet M. Geology of the Mineral Hill area, Millers Falls, Massachusetts. M, 1941, The Johns Hopkins University.

Aitken, Janet M. Petrology and structure of a section of the Hebron Gneiss of eastern Connecticut. D, 1948, The Johns Hopkins University.

Aitken, William Ernest. The loss ratio method as an analyzer of oil well decline curves and its value in their extrapolation. M, 1933, University of Pittsburgh.

Aitkens, D. F. The serpentine rocks of Keith Township. M, 1948, Queen's University. 75 p.

Aivano, John Peter. Revision of the genus Bollia (Ostracoda). M, 1973, Pennsylvania State University, University Park. 85 p.

Aiyer, Arunachalam Kulathu. An analytical study of the time-dependent behavior of underground openings. D, 1969, University of Illinois, Urbana. 259 p.

Aiyesimoju, Kolawole Oluyomi. Numerical prediction of transient water quality in estuarine/river networks. D, 1986, University of California, Berkeley. 156 p.

Ajamieh, M. M. Abu *see* Abu Ajamieh, M. M.

Ajayi, Clement Olatunde. A method for estimating the depth to a gravitating body from its equivalent source. M, 1975, Colorado School of Mines. 67 p.

Ajayi, O. A generalized equation for the shape of the water table between two base levels. M, 1976, University of Arizona.

Ajayi, Owolabi. Institutional models for water resources administration in developing countries; case example, Nigeria. D, 1982, University of Arizona. 390 p.

Ajlani, Mohammad Ghiath. Modeling and migration with an explicit finite difference scheme. M, 1984, University of Utah. 73 p.

Akande, S. O. A comparison and genesis of three vein systems of the Ross Mine, Hislop Township, Ontario. M, 1977, University of Western Ontario. 183 p.

Akande, Samuel Olusegun. Genesis of the lead-zinc mineralization at Gays River, Nova Scotia, Canada; a geologic fluid inclusion and stable isotope study. D, 1982, Dalhousie University. 348 p.

Akbar, Ali Mohammed. Numerical simulation of individual wells in a field simulation model. D, 1973, University of Missouri, Rolla.

Akbari, Gholamlali Estahbanati. The origin and evolution of the Earth's atmosphere and hydrosphere. D, 1984, University of Georgia. 169 p.

Akbarpour, Abbas. Optimal control theory for seismic excited structures. D, 1987, George Washington University. 291 p.

Akehurst, Alfred. The Jasper Formation, Jasper, Alberta. M, 1964, University of Alberta. 30 p.

Akers, D. J. A geochemical study of headwater streams of West Run, near Easton, West Virginia. M, 1976, West Virginia University. 125 p.

Akers, J. P. The Chinle Formation of the Paria Plateau area, Arizona and Utah. M, 1960, Arizona State University.

Akers, Richard H. Conodont biostratigraphy of the lower Oquirrh Group, Samaria Mountain, southeastern Idaho. M, 1984, Washington State University. 113 p.

Akers, Ronald H. Clay minerals of glacial deposits in west central Wisconsin. M, 1961, University of Wisconsin-Madison.

Akers, Ronald Hugh. Unusual surficial deposits in the Driftless Area of Wisconsin. D, 1965, University of Wisconsin-Madison. 198 p.

Akers, Wilburn Holt. A faunal study and tentative correlation of a subsurface Miocene section of South Texas. M, 1947, University of Oklahoma. 43 p.

Akers, Wilburn Holt. Planktonic foraminifera and biostratigraphy of some Neogene formations, northern Florida and Atlantic Coastal Plain. D, 1971, Tulane University. 268 p.

Akersten, William Andrew. Evolution of geomyine rodents with rooted cheek teeth. D, 1973, University of Michigan.

Akersten, William Andrew. Red Light local fauna (Blancan), southeastern Hudspeth County, Texas. M, 1967, University of Texas, Austin.

Akeson, R. S. Comparison of salt marsh corings taken from selected sites along the Massachussetts coast and their biological implications. M, 1969, Boston University. 96 p.

Akgunduz, Recep. Selective shape sorting of the William River delta front sand, Saskatchewan, Canada. M, 1988, University of Pittsburgh.

Akhavi, Manouchehr Sadat. Hydrogeologic investigations in the vicinity of Loveland (Larimer County), Colorado. M, 1967, Colorado State University. 94 p.

Akhavi, Manouchehr Sadat. Occurrence, movement and evaluation of shallow groundwater in the Ames, Iowa area. D, 1970, Iowa State University of Science and Technology. 172 p.

Akhionbare, Monday. A thermodynamic study of phase equilibria involving coal and other geologic material, in the main bituminous field of the Appalachian Plateaus Province, Pennsylvania. M, 1980, Brooklyn College (CUNY).

Akhtar, Salim. Symmetry dependence of magnetic moment of transition metal ions in solids. D, 1968, Stanford University. 175 p.

Akhurst, Maxine Carole. Pore geometry model of the Mississippian Frobisher Beds, Innes Field, southeastern Saskatchewan. M, 1984, University of Regina. 246 p.

Akin, Ralph. Depositional history of the Reynolds Oolite (Upper Jurassic) of southern Arkansas. M, 1966, University of Tulsa. 101 p.

Akinbola, Johnson A. Valley-side slope variation in relation to the control factors of denudation. D, 1971, University of Iowa. 101 p.

Akindunni, Festus Funso Folorunso. Effect of the capillary fringe on unconfined aquifer response to pumping; numerical simulations. D, 1987, University of Waterloo. 164 p.

Akinmade, Olufemi Ernest. Abnormal fluid pressures in the subsurface of the Niger Delta. M, 1974, University of Tulsa. 129 p.

Akiti, T. T. Hydrogeology and groundwater resources of the Fort Saskatchewan area; Alberta, Canada. M, 1975, University of Waterloo.

Akiyama, Juichiro. Two-layer stratified flow analysis of gravity currents and turbidity currents in lakes, reservoirs, and the ocean. D, 1987, University of Minnesota, Minneapolis. 302 p.

Aklstrand, Dennis Carl. Permian carbonate facies, Wind River Mountains and western Wind River basin, (Fremont County) Wyoming. M, 1972, University of Montana. 102 p.

Akmal, Mohammed Gawid. Subsurface geology of Northeast Lincoln and Southeast Payne counties, Oklahoma. M, 1950, University of Oklahoma. 44 p.

Akman, Hulya Hayriye. Resistivity and induced-polarization responses over two different Earth geometries. M, 1988, University of Arizona. 109 p.

Akman, Mustapha S. A map area south of Spadra, two and one-half miles south of Pomona, California. M, 1943, California Institute of Technology. 20 p.

Akol, Halim. The reduction of magnetite occurring in copper converter slag. M, 1952, University of Arizona.

Akpati, Benjamin Nwaka. Micropaleontology of the Jalama (upper Cretaceous) and Anita (middle Eocene) formations, western Santa Ynez Mountains, Santa Barbara County, California. M, 1966, University of California, Los Angeles.

Akpati, Benjamin Nwaka. Sedimentation and foraminiferal ecology in eastern Long Island Sound, New York. D, 1970, University of Pittsburgh. 119 p.

Aksell, Allan Carl. The distribution of uranium in the equigranular monzonite of the Bingham Stock,

Bingham mining district, Utah. M, 1982, University of Utah. 59 p.

Aksoy, M. Zihni. Surface geology of portions of the Sulligent SW Quadrangle and the Vernon Quadrangle, Alabama. M, 1984, Mississippi State University. 102 p.

Aksoy, Rahmi. Geological and geophysical investigations along the Helendale fault zone in the southern Mojave Desert, California. M, 1986, University of California, Riverside. 86 p.

Aksu, Ali Engin. Late Quaternary stratigraphy, paleoenvironmentology and sedimentation history of Baffin Bay and Davis Strait. D, 1980, Dalhousie University. 771 p.

Aksu, Ali Engin. The late Quaternary stratigraphy and sedimentation history of Baffin Bay. M, 1977, Dalhousie University.

Aktan, M. Tunc. Experimental study of temperature dependence of rock elastic moduli, and finite element analysis of thermal stresses induced by hot water injection. D, 1975, Pennsylvania State University, University Park. 259 p.

Al Rawi, Yeha Tawfeq. Cenozoic history of the northern part of Mono Basin (Mono county), California and Nevada. D, 1969, University of California, Berkeley. 188 p.

Al-Aasm, Ihsan Shakir. Stabilization of aragonite to low-Mg calcite; trace elements and stable isotopes in rudists. D, 1985, University of Ottawa. 267 p.

Al-Abdul-Razzaq, Sabeekah K. Study of some Cretaceous Ostracoda of Kuwait. D, 1977, University of Michigan. 436 p.

Al-Abed, Souhail Radhi Ali. Relationship of morphological indicators of soil wetness and frequency and duration of saturation. D, 1986, University of Nebraska, Lincoln. 215 p.

Al-Agidi, Waleed Khalid Hassan. A chronolithosequence of soils on Pleistocene terraces along Maple river in northeastern Clinton County, Michigan, U.S.A. (their morphologic, genetic and geomorphic interrelationship). D, 1970, Michigan State University. 249 p.

Al-Alawi, Jomaah Abd-Ulraheem Awad. Petrography, sulfide mineralogy and distribution, mass transfer, and chemical evolution of the Babbitt Cu-Ni deposit, Duluth Complex, Minnesota. D, 1985, Indiana University, Bloomington. 348 p.

Al-Amri, M. Static correction for shallow seismic reflection. M, 1986, Ohio University, Athens. 226 p.

Al-Ansari, Jasem Mohammad. High temperature thermal energy storage in aquifers with a solar power plant application. D, 1980, University of Alabama. 205 p.

Al-Arabi, Nizar A. Mineralogy and geochemistry of the bentonite marker bed of Graneros Shale (upper Cretaceous) in Lincoln and Russell counties, Kansas. M, 1971, Wichita State University. 49 p.

Al-Aswad, Ahmad Abdullah. Subsurface lithostratigraphy and depositional environments of the Borden Group (Valmeyeran Series, Mississippian System) in Indiana. D, 1986, Indiana University, Bloomington. 405 p.

Al-Awkati, Z. A. On problems of soil bearing capacity at depth. D, 1975, Duke University. 217 p.

Al-Azzaby, Fathi Ayoub. Stratigraphy and sedimentation of the Spencer Formation in Yamhill and Washington counties, Oregon. M, 1980, Portland State University. 104 p.

Al-Bassam, Abdulaziz M. A quantitative study of Haradh Wellfield, Umm-Er-Radhuma Aquifer (Saudi Arabia). M, 1983, Ohio University, Athens. 409 p.

Al-Basso, Khaled M. S. A goal programming approach to alternative decision making rationales for multi-objective public forest land management. D, 1980, University of Missouri, Columbia. 280 p.

Al-Dabagh, Abdulsattar Rashied. Stochastic modeling of the surface flow of the upper Rio Grande system. D, 1986, New Mexico State University, Las Cruces. 222 p.

Al-Dabbagh, Ahmed Assim. Variations and boundary location of static and dynamic soil properties along a desert catena in southern New Mexico. D, 1987, New Mexico State University, Las Cruces. 244 p.

Al-Daghastani, Nabil Subhi. The application of remote sensing to geomorphological mapping and mass movement study in the vicinity of Prove, Utah. D, 1987, Purdue University. 326 p.

Al-Eisa, Abdul-Rahman Mohammed. The structure and stratigraphy of the Columbia River Basalt in the Chehalem Mountains, Oregon. M, 1980, Portland State University. 67 p.

Al-Eryani, Mohamed Lotf. Hydrology and ground water potential of the Tihama, Yemen Arab Republic. M, 1979, University of Arizona. 191 p.

Al-Faraj, Mohammed. Use of post-critical seismic reflection amplitudes, for exploration purposes, including depth estimation. M, 1987, Colorado School of Mines. 113 p.

Al-Fares, Mohammed H. Improvements on conventional VSP processing and the use of spatial median filters. M, 1987, Colorado School of Mines. 170 p.

Al-Geroushi, Rajab A. Probabilistic micromechanics of clay compression and consolidation. D, 1988, McGill University.

Al-Ghamisi, Hezam Hazzaa. Optimal groundwater management model of Saq Aquifer in Gassim region (Saudi Arabia). D, 1988, Colorado State University. 195 p.

Al-Habash, Muyyed. A reconnaissance analysis of organic material, uranium, thorium, and other trace elements in black shale of western New York. M, 1981, SUNY at Buffalo. 75 p.

Al-Hadithi, A. H. Estimated water balance for the proposed Haditha Reservoir on the Euphrates River in Iraq. M, 1976, University of Arizona.

Al-Hadithi, A. H. Optimal utilization of the water resources of the Euphrates River in Iraq. D, 1979, University of Arizona. 275 p.

Al-Hajji, Yacoub Y. A quantitative hydrological study of Field "A", South Kuwait. M, 1976, Ohio University, Athens. 126 p.

Al-Harari, Zaki Y. Three-dimensional seismic physical scale-modeling with applications to the detection of underground cavities. D, 1985, Columbia University, Teachers College. 141 p.

Al-Hashimi, A. R. K. Geochemical prospecting for copper in the northern part of Yuma County, Arizona. M, 1965, University of Missouri, Rolla.

Al-Hashimi, Abdul Razak K. Study of copper dispersion in the Boulder Batholith (late Cretaceous), (southwestern) Montana. D, 1969, Boston University. 145 p.

Al-Hassan, Sumani. Optimal well location in a multi-aquifer groundwater system. D, 1986, Utah State University. 186 p.

Al-Hinai, Khalifa. Subsurface structural geology of Loving County, Texas. M, 1977, University of Texas, Austin.

Al-Hurban, Adeeba. A magnetic survey of the Howell Structure, Lincoln County, Tennessee. M, 1983, Indiana State University. 55 p.

Al-Jallal, Ibrahim Abdulla M. Hydrogeology of the Minjur Aquifer system in Riyadh region, Saudi Arabia. M, 1979, Western Michigan University.

Al-Jassar, Tariq Jamil. A study of radiometric determination and interpretation of uranium and thorium in a section of New Albany Shale. M, 1981, Indiana University, Bloomington. 105 p.

Al-Jassar, Tariq Jamil. Oxygen isotope systematics of the Babbitt Cu-Ni deposit, Duluth Complex, Minnesota. D, 1985, Indiana University, Bloomington. 131 p.

Al-Khafaji, Abdul-Amir Wadi Nasif. Decomposition effects on engineering properties of fibrous organic soils. D, 1979, Michigan State University. 501 p.

Al-Khafaji, S. A. A method for computing the terrain correction for the near zone by the use of triangular prisms. M, 1965, Colorado School of Mines. 52 p.

Al-Khafaji, Saadi Abbas. Studies of the magnetization of the Cambrian Lamotte Formation in Missouri. D, 1968, St. Louis University. 204 p.

Al-Khafif, Soud Mostafa. A study of open channels degradation and corresponding bed roughness. D, 1965, University of California, Davis. 73 p.

Al-Khalifah, Abdul-Jaleel Abdullah. Determination of absolute and relative permeability using well test analysis. D, 1988, Stanford University. 200 p.

Al-Khersan, Hashim Fadil. Carbonate petrography of the Red Eagle limestone (lower Permian), southern Kansas and north-central Oklahoma. D, 1969, University of Oklahoma. 187 p.

Al-Khersan, Hashim Fadil. Regional factors that control oil and gas accumulation. M, 1966, University of Texas, Austin.

Al-Khirbash, Salah. Geology and mineral deposits of the Emery mining district, Powell County, Montana. M, 1982, University of Montana. 60 p.

Al-Khirbash, Salah A. Geology, geochemistry, and copper occurrences of the mid Proterozoic Gateway Formation (member A of Kintla Formation), Purcell Supergroup, SW Alberta and SE British Columbia, Canada. D, 1987, University of Nebraska, Lincoln. 150 p.

Al-Laboun, Abdullah Aziz. Stratigraphic analysis Burgan Sandstone, Arabian Persian Gulf area. M, 1977, [University of Tulsa].

Al-Layla, Mohamad Tayeb Hussain. Study of certain geotechnical properties of Beaumont clay. D, 1970, Texas A&M University. 200 p.

Al-Mashouq, Khalid. Relationship between shear wave birefringence and fracture spacing; a laboratory study. M, 1988, Colorado School of Mines. 36 p.

Al-Mishwt, A. T. and Theyab, A. Geology, mineralogy, and petrochemistry of Al-Halgah Pluton, At-Taif, Saudi Arabia. D, 1977, University of Wisconsin-Madison. 312 p.

Al-Mishwt, Ali Theyab. Contact metamorphism and polymetamorphism, north-western Oquossoc Quadrangle, Maine. M, 1972, University of Wisconsin-Madison.

Al-Momani, Ayman Hassan. Modeling of future water demand and supply in Jordan; a tool for planning. D, 1988, Oklahoma State University. 204 p.

Al-Mooji, Y. Saline groundwaters in the bedrocks of the North Saanich Peninsula, Vancouver Island, British Columbia. M, 1982, University of Western Ontario.

Al-Moussawi, Hassan M. Thermal contraction and crack formation in frozen soil. D, 1988, Michigan State University. 216 p.

Al-Mudaiheem, Khalid Nasser. Water resources and provision problems of Riyadh, Saudi Arabia; an analytical study. D, 1985, University of Oregon. 277 p.

Al-Muneef, Nasser Saad. Productivity and characteristics of the Niagaran pinnacle reefs in northern Michigan. M, 1974, University of Michigan.

Al-Muttair, Fouad Fahad. Design of dynamic equilibrium sand-bed canals. D, 1983, Colorado State University. 193 p.

Al-Nemer, Jaafar Muhammad. P-wave one-dimensional seismic response of a linear velocity transition zone using wave theory. M, 1987, Colorado School of Mines. 132 p.

Al-Omari, Farouk Sunallah. Upper Cretaceous and lower Cenozoic foraminifera of three oil wells in northwestern Iraq. D, 1970, University of Missouri, Rolla. 245 p.

Al-Omari, Farouk Sunallah. Upper Cretaceous-lower Cenozoic Foraminifera from an oil well in northwestern Iraq. M, 1967, University of Missouri, Rolla.

Al-Qayim, Basim A. Facies analysis and depositional environment of the Ames Marine Member (Virgilian) of the Conemaugh Group (Pennsylvanian) in the Appalachian Basin. D, 1983, University of Pittsburgh. 344 p.

Al-Rashid, Y. An elasto-plastic constitutive theory for a quartz sand and its application to anchor problems. D, 1975, Colorado State University. 212 p.

Al-Rawahi, Khalid Hilal. Field and petrographic study of igneous rocks and sulfide deposits in Lasail and Assayab areas, Samail ophiolite complex, northern Oman. M, 1983, University of California, Santa Barbara. 219 p.

Al-Rawi, Amin Hamad. Quantitative mineralogical analysis of some soils in Iraq and of antigo silt loam catena in Menominee County, Wisconsin. D, 1969, University of Wisconsin-Madison. 155 p.

Al-Refai, Badir Hashim. A heavy mineral analysis of contemporary sand derived from the Boone limestone formation in Benton County, Arkansas. M, 1956, University of Arkansas, Fayetteville.

Al-Sanabani, Gaber Ali. Investigation of hydraulic equivalence in beach and dune sands. M, 1987, University of Georgia. 74 p.

Al-Sanad, Hasan Abdul Aziz. Effect of random loading on modulus and damping of sands. D, 1982, University of Maryland. 271 p.

Al-Sarawi, Mohammad. Geomorphology and general geology of Wadi-Al-Batin, Kuwait. M, 1978, Ohio University, Athens. 88 p.

Al-Sarawi, Mohammad A. Morphology of Jal-Az-Zor escarpment and Holocene-Pleistocene sedimentation along the northern Kuwait Bay. D, 1980, University of South Carolina. 184 p.

Al-Shaieb, Zuhair. Geochemical anomalies in the igneous wall rock at Mayflower Mine, Park City District, Utah. D, 1972, University of Missouri, Rolla. 113 p.

Al-Shaieb, Zuhair. Trace element anomalies in the igneous wall rocks of hydrothermal veins in the Searchlight District, Nevada. M, 1969, University of Missouri, Rolla.

Al-Shakhis, Amir A. The relationship between F-K filtering and deconvolution in seismic data processing. M, 1987, Colorado School of Mines. 100 p.

Al-Shamlin, Ali Abdula. Microfacies analysis of lower Wolfcamp (lower Permian) carbonates (Hueco Limestone), Tom Mays Park, El Paso, Texas. M, 1971, University of Texas at El Paso.

Al-Shammari, Lateef T. Seismic interpretation of Bahrah Oilfield, Kuwait. M, 1983, Ohio University, Athens. 176 p.

Al-Shawaf, Taha Daud. Variable modulus model for inelastic finite element analysis. D, 1979, University of California, Berkeley. 136 p.

Al-Sinawi, Sahil Abdulla. An investigation of body wave velocities, attenuation and elastic parameters of rocks subjected to pressure at room temperature. D, 1968, St. Louis University. 159 p.

Al-Sumait, Abdulaziz Jasem. A computer groundwater model for the Tillman alluvium in Tillman County, Oklahoma. M, 1978, Oklahoma State University. 152 p.

Al-Taweel, Bashir Hashim. Soil genesis in relation to groundwater regimes in a hummocky ground moraine area near Hamiota, Manitoba. D, 1982, University of Manitoba.

Al-Temeemi, A. Y. Mineralogical, geochemical and sedimentary parameters of New Caledonia bottom sediments. D, 1977, George Washington University. 268 p.

Al-Temeeni, Ali Yousuf. Carbonate bottom sediments of the Arabian Gulf in relation to environmental parameters. M, 1972, Pennsylvania State University, University Park. 152 p.

Al-Wail, Tahir A. Geoelectric signatures and hydrostratigraphy of reefal limestones, Southwest Florida. M, 1984, University of South Florida, Tampa. 73 p.

Al-Yacoub, Tassier. Application of reflection and refraction to estimate residual static corrections caused by near surface heterogeneity in Saudi Arabia. M, 1988, Colorado School of Mines. 53 p.

Al-Yahya, Kamal Mansour. Velocity analysis by iterative profile migration. D, 1987, Stanford University. 86 p.

Al-Yousef, Hasan Yousef. An improved method to calculate pseudorelative permeabilities to model gravity and capillary effects in reservoir simulation. D, 1985, Stanford University.

Alabert, François Georges. Stochastic imaging of spatial distributions using hard and soft information. M, 1987, Stanford University. 208 p.

Alabi, Adeniyi Oluremi. A study of the North American Central Plains conductivity anomaly. D, 1974, University of Alberta. 251 p.

Alabi, Kolade Ebenezer. Evaluation of chemical methods in assessing lime requirements of sandy soils in northeastern Nebraska. D, 1983, University of Nebraska, Lincoln. 79 p.

Alam el Din, Ibrahim Osman. Water in Ojai Valley, Ventura County, Southern California. M, 1964, University of California, Los Angeles. 104 p.

Alam, Mahmood. Quaternary paleoclimates and sedimentation southwest of the Grand Banks. M, 1976, Dalhousie University. 251 p.

Alam, Mahmood. The effect of Pleistocene climatic changes on the sediments around the Grand Banks. D, 1979, Dalhousie University. 295 p.

Alam, Mohamed Nour. Lateral seismic velocity gradients on Georges Bank, U.S. Atlantic continental margin; a comparison of seismic common depth point (CDP) moveout velocity with sonic and checkshot velocities from ten exploratory wells. M, 1988, Colorado School of Mines. 117 p.

Alam, Muhammad Waqi Ul. Relative permeability measurement at simulated reservoir conditions. D, 1988, University of Oklahoma. 117 p.

Alammawi Alsayed, Alsayed Mouhamed. Some aspects of hydration and interaction energies of montmorillonite clay. D, 1988, McGill University.

Alamos Ovejero, Julio. Subsurface distribution of sandstone in the Petersburg and Dugger formations of southern Knox County, Indiana. M, 1973, Indiana University, Bloomington. 35 p.

Alamri, Abdullah M. Geophysical and hydrogeological effects of a storm water retention pond on the Floridan Aquifer, Hillsborough County, Florida. M, 1985, University of South Florida, Tampa. 248 p.

Alaniz, Roberto Trevino. A study of Recent sediments and the biofacies of Cayo del Grullo, Baffin Bay, Texas. M, 1974, Texas Christian University.

Alao, David Afolayan. Properties of laterites from Ilorin, Nigeria. M, 1982, Michigan Technological University. 134 p.

Alarcon, Alcocer Carolos Felipe. A laboratory study with a light crude oil to determine the effect of high-pressure nitrogen injection on enhanced oil recovery. D, 1982, University of Oklahoma. 329 p.

Alarcon-Guzman, Adolfo. Cyclic stress-strain and liquefaction characteristics of sands; (volumes I and II). D, 1986, Purdue University. 616 p.

Alaric, Saw. General geology and ore deposits of North Star Mountain and Ling Mine, Summit and Park counties, Colorado. M, 1952, Colorado School of Mines. 49 p.

Alatorre, Armando E. Stratigraphy and depositional environments of the Monterrey Formation at San Hilario and San Jose de la Costa areas, Baja California Sur, Mexico. M, 1982, Colorado School of Mines. 102 p.

Alavi, M. Geology of the Bedford Complex and surrounding rocks, southeastern New York. D, 1976, University of Massachusetts. 173 p.

Alawi, Adnan Jassim. Some contributions to the study of scour in long contractions. D, 1985, University of Arizona. 168 p.

Alawi, Jomaah A. Mineralogy, textural relationships, and metamorphic reactions in the North Doherty skarn, Montana. M, 1981, Indiana University, Bloomington.

Alawi, Mohamed Mohamed. Experimental and analytical modeling of sand behavior under nonconventional loading. D, 1988, University of Colorado. 236 p.

Alazar, Tesfalul. Seismic investigation of the Ogaden Basin, onshore Ethiopia. M, 1987, Colorado School of Mines. 95 p.

Alba, Christopher Anthony. Stratigraphy and depositional environments of the Ordovician through Devonian Mountain Springs Formation, southern Nevada. M, 1981, San Diego State University.

Albach, Douglas C. The depositional history of the uppermost Wilcox (lower Eocene) of west-central Beauregard Parish, Louisiana. M, 1979, Louisiana State University.

Albanese, James R. Quantitative paleoecology of selected Cenozoic microfossil assemblages from the Pacific Northwest. M, 1973, Rensselaer Polytechnic Institute. 62 p.

Albanese, John P. A petrographic and structural study of a peridotite body in the Laramie Mountains, Wyoming. M, 1949, University of Wyoming. 64 p.

Albanese, Mary. The geology of three extrusive bodies in the Central Alaska Range. M, 1980, University of Alaska, Fairbanks. 104 p.

Albano, Lorenzo Luis. Relation of physical and geochemical factors to porosities in sandstones of the Bromide (Simpson Group, Ordovician) of Oklahoma. M, 1965, University of Oklahoma. 46 p.

Albano, Michael Anthony. Subsurface stratigraphic analysis, Cherokee Group (Pennsylvanian), Northeast Cleveland County, Oklahoma. M, 1973, University of Oklahoma. 61 p.

Albarracin, Jose. Geophysical exploration in the Orinoco oil belt of Venezuela. M, 1974, University of Tulsa. 81 p.

Albaugh, Frederick W. Study of the wetting of calcite. D, 1941, University of Michigan.

Albayrak, A. Feridun. Environmental impact of mining and preventive control techniques. M, 1978, Stanford University. 144 p.

Albee, Arden Leroy. Geology of the Hyde Park Quadrangle, Vermont. D, 1957, Harvard University.

Albehbehani, Abdulsamee S. K. Carbonate lithofacies, diagenesis, porosity relationships and crude oil geochemistry, Ratawi Formation, Kuwait-Saudi Arabia divided neutral zone. D, 1988, Texas Tech University. 192 p.

Alberding, Herbert. The geology of the northern Empire Mountains, Arizona. D, 1938, University of Arizona.

Albers, Doyle Francis. Geology of the Deadwood-Sawmill Creek area, Custer County, Idaho. M, 1981, University of Idaho. 106 p.

Albers, John Patrick. Geology and ore deposits of the East Shasta copper-zinc district, Shasta County, California. D, 1958, Stanford University. 414 p.

Albers, Sherly Hammond. Paleoenvironment of the upper Triassic-lower Jurassic (?) Nugget (?) Sandstone near Heber, Utah. M, 1975, University of Utah. 94 p.

Alberstadt, Leonard P. Paleontology of the Devils Kitchen Member of the Deese Formation. M, 1962, Tulane University. 66 p.

Alberstadt, Leonard Philip. Brachiopod biostratigraphy of the Viola and "Fernvale" formations (Ordovician), Arbuckle Mountains, south-central Oklahoma. D, 1967, University of Oklahoma. 308 p.

Albert, Daniel Bruce. Sulfate reduction and iron sulfide formation in sediments of the Pamlico River estuary, North Carolina. D, 1986, University of North Carolina, Chapel Hill. 182 p.

Albert, Donald G. Seismic investigations on the Ross Ice Shelf, Antarctica. M, 1978, University of Texas, Austin.

Albert, Nairn R. D. Application of computer-enhanced Landsat imagery to mineral resource assessment in South-central Alaska. M, 1978, San Jose State University. 141 p.

Albert, Nairn Randolph. Dinoflagellate cysts from the Early Cretaceous of the Yukon-Koyukuk Basin and from the Upper Jurassic Naknek Formation, Alaska. D, 1988, Stanford University. 497 p.

Albert, Robert L. A gravity study of earthquake-related structures in the St. Lawrence River valley. M, 1977, University of Rhode Island.

Albert, Tom. Three Crook County (Wyoming) wildcat wells. M, 1958, [University of South Dakota].

Alberta, Patricia Lynne. Depositional facies analysis and porosity development of the (Pennsylvanian) upper Morrow chert conglomerate "Puryear" Member, Roger Mills and Beckham counties, Oklahoma. M, 1987, Oklahoma State University. 135 p.

Alberts, Robert Kirk. Geology of the Truman Group of mineral claims, Salmo area, British Columbia. M, 1952, University of Idaho. 34 p.

Albigese, Muriel and Bell, Jane. Heavy minerals of the Darby Creek region-Chester Quadrangle, Pennsylvania. M, 1940, Bryn Mawr College. 18 p.

Albin, Arthur G. The analysis and recommended design of a high-resolution digital data acquisition system for the in situ measurement of various physical and chemical parameters of sea water. D, 1968, Oregon State University. 123 p.

Albin, Edward Francis, Jr. Mars; volcanic features in the cratered uplands and possible tectonic associations. M, 1986, Arizona State University. 88 p.

Albino, George V. Petrology, geochemistry and mineralization of the Boundary ultramafic complex, Quebec, Canada. M, 1984, Colorado State University. 236 p.

Albino, Katharine Chase. Relative dating and soils of late Quaternary deposits, Devil's Thumb Lake valley, Colorado Front Range. M, 1984, University of Colorado. 170 p.

Albinson, Tawn. Fluid inclusion studies of the Tayoltita Mine and related areas, Durango, Mexico. M, 1978, University of Minnesota, Minneapolis. 91 p.

Albrecht, Jean Irene. Relationship between refractive index and density of certain garnets. M, 1948, Mount Holyoke College. 39 p.

Albrecht, Richard K. The effect of Lake Michigan on precipitation along the western shoreline. M, 1980, University of Wisconsin-Milwaukee. 57 p.

Albright, James. X-ray diffraction studies of aqueous alkaline-earth chloride solutions and aqueous potassium and sodium carbonate solution. D, 1969, University of Chicago. 39 p.

Albritton, Claude Carrol. Geology of the Malone Mountain area, Hudspeth County, Texas. D, 1936, Harvard University.

Albritton, John Allen. Sedimentary features of the Sewanee Conglomerate (Pennsylvanian, Tennessee). M, 1955, Emory University. 52 p.

Albrizzio, Carlos. Geology of the NW of the Mindego Hill and NE of the La Honda quadrangles, California, U.S.A. M, 1957, Stanford University.

Albuquerque, John Stephen de see de Albuquerque, John Stephen

Alcock, Charles William. Stratigraphy of the Pawnee-Lenapah formations of Nowata and Craig counties, Oklahoma. M, 1942, University of Iowa. 112 p.

Alcock, Fred G. Some geochemical aspects of Precambrian iron formation. M, 1971, McMaster University. 122 p.

Alcock, Frederick J. The geology of the Lake Athabaska region. D, 1915, Yale University.

Alcock, R. A. A petrographic analysis of the Tazin-like metasediments (Precambrian) of the Fort Fitzgerald area, northeastern Alberta. M, 1961, University of Toronto.

Alcordo, Isabelo Suelo. The flow properties of montmorillonite, Fe-montmorillonite, Al-montmorillonite, and soil clay suspensions and their changes with anaerobic reduction. D, 1968, University of Illinois, Urbana. 222 p.

Alcorn, Stephen Richard. Mineralogy, petrology, and evolution of a calc-alkaline igneous sequence, Cerros de Tilaran, Puntarenas, Costa Rica. D, 1981, University of Georgia. 189 p.

Alcorn, Steve. Serpentinization of the Holcombe Branch Dunite. M, 1975, University of South Carolina.

Alden, Kathleen A. Automatic detection and recognition of the first arrival phase of seismic event signals contaminated by noise. M, 1986, Southern Methodist University. 301 p.

Alden, William Clinton. The Delavan Lobe of the Lake Michigan glacier of the Wisconsin Stage of glaciation and associated phenomena. D, 1904, University of Chicago. 106 p.

Alderks, David F. The paleoenvironment of the Guttenberg Limestone (Middle Ordovician) in Iowa. M, 1979, University of Wisconsin-Milwaukee. 128 p.

Aldinger, Paul Bruce. Groundwater flow simulation by a stochastic representation of soil. D, 1983, University of Rhode Island. 309 p.

Aldovino, Lino Pineda. Computerized water distribution management for the upper Pampanga River Project, Philippines. D, 1977, University of Arizona.

Aldredge, Robert F. The effect of dipping strata on earth-resistivity determinations. M, 1933, Colorado School of Mines. 24 p.

Aldrich, Arthur G., Jr. Textural analysis of octahedrite meteorites by Fourier series shape approximation. M, 1970, Michigan State University. 50 p.

Aldrich, Charles Allen. Study of near surface elastic-anelastic layers. D, 1965, University of Missouri, Rolla. 119 p.

Aldrich, Henry R. The enrichment of nickel ores. M, 1917, University of Minnesota, Minneapolis. 38 p.

Aldrich, Henry R. The geology of the Gogebic iron range of Wisconsin. D, 1933, University of Wisconsin-Madison.

Aldrich, Jeffrey B. Alteration of volcanic debris and basalts in a submarine remnant arc; Deep Sea Drilling Site 448, Palau-Kyushu Ridge, Philippine Sea. M, 1983, Texas A&M University. 101 p.

Aldrich, John Kenneth. Gravity of the Santa Barbara Channel Islands, California. M, 1969, University of California, Santa Barbara.

Aldrich, Merritt J. Tracing a subsurface structure by joint analysis; Santa Rita-Hanove Axis, southwestern New Mexico. D, 1972, University of New Mexico. 106 p.

Alegre, Julio Cesar. Effect of land clearing and land preparation methods on soil physical and chemical properties and crop performance of an Ultisol in the Amazon Basin. D, 1985, North Carolina State University. 168 p.

Aleinikoff, John N. Petrology and petrochemistry of Cambro-Ordovician volcanics of New England. M, 1975, Dartmouth College. 73 p.

Aleinikoff, John N. Structure, petrology, and uranium-thorium-lead geochronology in the Milford (15') Quadrangle, New Hampshire. D, 1978, Dartmouth College. 247 p.

Aleman, Antenor M. Tectonic and sedimentation of Cretaceous rocks in the Paita Basin-Peru. M, 1977, Northern Illinois University. 126 p.

Alemi, Mohammad A. A gravity survey of Coos Bay, Oregon. M, 1978, University of Oregon. 72 p.

Aleshin, Eugene. The crystal chemistry of pyrochlore. M, 1959, Pennsylvania State University, University Park. 50 p.

Alevizos, Anastasios. The chemical composition of chromite from Skoumtsa-Xerolivado mines of the Vourinos ophiolitic complex, Greece as a petrogenetic indicator. M, 1985, Lehigh University.

Alewine, James W. Investigation of the sources of quartz grains of the Bliss Formation (Cambro-Ordovician), Silver City area, (Grant County) New Mexico. M, 1966, [University of Houston].

Alewine, Ralph Wilson, III. Application of linear inversion theory toward the estimation of seismic source parameters. D, 1974, California Institute of Technology. 303 p.

Alewine, Ralph Wilson, III. Generation and propagation of seismic sea waves using acoustic-gravity wave theory. M, 1970, Brown University.

Alexander, Alexander Emil. A petrographic and petrologic study of some continental shelf sediments. D, 1933, Cornell University.

Alexander, Amanda Joyce. Sedimentology and taphonomy of a middle Clarkforkian (early Eocene) fossil vertebrate locality, Fort Union Formation, Bighorn Basin, Wyoming. M, 1982, University of Michigan.

Alexander, C. Shafe. High frequency enrichment in the P-wave coda of earthquakes. M, 1986, Georgia Institute of Technology. 80 p.

Alexander, Charles Ivan. A preliminary study of the stratigraphic importance of the Ostracoda in the Fredericksburg and Washita formations of North Texas. M, 1926, Texas Christian University. 60 p.

Alexander, Charles Ivan. The Ostracoda of the Cretaceous of North Texas. D, 1928, Princeton University.

Alexander, Charles S. The marine and stream terraces of the Capitola-Watsonville area, California. M, 1950, University of California, Berkeley. 149 p.

Alexander, David William. Petrology and petrography of the Bridal Veil limestone member of the Oquirr Formation at Cascade Mountain, Utah. M, 1978, Brigham Young University.

Alexander, Donald Henry. Geology, mineralogy, and geochemistry of the McClure Mountain alkalic complex, Fremont County, Colorado. D, 1981, University of Michigan. 352 p.

Alexander, Donald Henry. Petrography and origin of an orbicular lamprophyre dike, Fremont County, Colorado. M, 1974, University of Michigan.

Alexander, Earl Betson, Jr. A chronosequence of soils on terraces along the Cauca river, Colombia. D, 1970, Ohio State University. 172 p.

Alexander, Emmit Calvin, Jr. Noble gases; a record of the early solar system. D, 1970, University of Missouri, Rolla. 163 p.

Alexander, Frank. Stratigraphic and structural geology of Blewett-Swauk area, Washington. M, 1956, University of Washington. 64 p.

Alexander, Frank McEwen. Middle Devonian residues of central Tennessee. M, 1942, Vanderbilt University.

Alexander, Frederick John. Structural geology of the Luscar-Sterco Mynheer A Zone, Coal Valley, Alberta. M, 1977, University of Alberta. 72 p.

Alexander, Gilbert R. Areal geology of a portion of Northeast Hunt County, Texas. M, 1967, East Texas State University.

Alexander, Hugh S. Pothole Erosion. D, 1931, University of Minnesota, Minneapolis. 51 p.

Alexander, James Edward. The ecology of iron in tropical waters. D, 1964, University of Miami. 159 p.

Alexander, James Iwan David. Folds and folding in single- and multi-layered rocks; mathematical models and field observations. D, 1981, Washington State University. 145 p.

Alexander, John B. Stratigraphy and structure of part of the Fish Lake Plateau, Sevier County, Utah. M, 1965, Oregon State University. 105 p.

Alexander, Joseph B. Heavy minerals of certain quartzites from Malaya; a study in differentiation and correlation. M, 1950, California Institute of Technology. 77 p.

Alexander, Joseph Watrous. Pennsylvanian geology of eastern Vermilion County, Illinois, northern Vermilion County, Indiana, and Warren and Fountain counties, Indiana. M, 1942, University of Illinois, Urbana. 57 p.

Alexander, Nancy Jo. A study of the depositional environment and diagenesis of the Hill Member of the Rodessa Formation in Caddo Parish, Louisiana. M, 1988, Northeast Louisiana University. 57 p.

Alexander, Nancy Sue. Robert Thomas Hill (1858-1941), father of Texas geology; an account of his life and an appraisal of his contributions to the geological sciences. D, 1973, Southern Methodist University. 324 p.

Alexander, Richard Dolphin. The Desmoinesian fusulinids of northeastern Oklahoma. M, 1953, University of Oklahoma. 74 p.

Alexander, Richard Raymond. Autecological studies of the brachiopod Rafinesquina (Upper Ordovician), the bivalve Anadara (Pliocene), and the echinoid Dendraster (Pliocene). D, 1972, Indiana University, Bloomington. 199 p.

Alexander, Robert Harwood. Geology of the Leon Creek area, Mason County, Texas; paleontology of the lower Wilberns Formation. M, 1956, University of Texas, Austin.

Alexander, Robert Houston. Adaptation of land use to surficial geology in metropolitan Washington, D.C. D, 1981, University of Washington. 200 p.

Alexander, Robert John. A disscussion of the various methods of precisely positioning a ship for the purpose of hydrographic surveying. M, 1955, Ohio State University.

Alexander, Roger Gordon, Jr. The geology of the Whitehall area, Montana. D, 1951, Princeton University. 251 p.

Alexander, Russell James. An objective study of Pennsylvanian Fusulinidae from the Ardmore-Ada areas, Oklahoma. M, 1952, University of Oklahoma. 74 p.

Alexander, Shelton S. Surface wave propagation in the Western United States. D, 1963, California Institute of Technology. 238 p.

Alexander, Shelton Setzer. I, Crustal structure in the Western United States from multimode surface wave dispersion; II, The effects of the continental margin in Southern California on Rayleigh wave propagation. D, 1963, California Institute of Technology. 242 p.

Alexander, Steven C. Diagenesis and secondary porosity development in the late Paleocene Lobo sandstones, Webb County, Texas. M, 1984, Texas A&M University. 88 p.

Alexander, Theodor W. The petrography of the Hannibal and Chouteau Formations in western Illinois. M, 1947, Texas Tech University. 22 p.

Alexander, Velinda D. H. The distribution of iron in staurolite at room and liquid nitrogen temperatures. M, 1988, Brigham Young University. 44 p.

Alexander, Walter Herbert. The physiographic history of the Little Kanawha Valley, West Virginia. M, 1938, University of Cincinnati. 48 p.

Alexander, Wayland B. Areal geology of southern Dewey County, Oklahoma. M, 1965, University of Oklahoma. 42 p.

Alexander, William Albert. The sub-surface structure of Bossier Parish, Louisiana. M, 1928, Louisiana State University.

Alexander, William G. Sedimentology of the lower Wasatch deposits at the Hoe Creek (UCG) Site, Powder River Basin, Wyoming. M, 1982, Colorado State University. 158 p.

Alexander, William J. Reconnaissance geology and geomagnetics of western Flagstaff, Coconino County, Arizona. M, 1974, Northern Arizona University. 67 p.

Alexander, William L. Geology of South Mason area, Texas. M, 1955, Texas A&M University.

Alexandri-Rionda, Rafael. Massive sulfide mineralization at the Teziultan mining district, Puebla, Mexico. M, 1980, Colorado State University. 122 p.

Alexandrov, Eugene A. The origin of Precambrian banded iron ores. M, 1956, Columbia University, Teachers College.

Alexandrov, Eugene Alexander. Problems of genesis of sedimentary manganese deposits. D, 1965, Columbia University. 392 p.

Alexanian, Daniel Albert. Palynology and biostratigraphy of the Midway and Wilcox groups of East Texas. M, 1981, University of Wisconsin-Madison. 62 p.

Alexis, Carl O. The geology of the Lead Mountain area, Pima County, Arizona. M, 1939, University of Arizona.

Alexis, Carl O. The geology of the northern part of the Huachuca Mountains, Arizona. D, 1949, University of Arizona.

Alf, Raymond M. Ecology of the Chadron Formation (Oligocene) of northeast Colorado. M, 1939, University of Colorado.

Alfano, John Joseph. A two-layer ground hydrology model interactive with an atmospheric general circulation model. D, 1981, University of Connecticut. 641 p.

Alfano, Joseph Michael. The geology of Columbia Township, Meigs County, Ohio. M, 1973, Ohio University, Athens. 117 p.

Alfaro, Jorge Daniel Benavides *see* Benavides Alfaro, Jorge Daniel

Alfaro, Luis Domingo. Reliability of soil slopes. D, 1980, Purdue University. 231 p.

Alfaro, Ruben Alberto Mazariegos *see* Mazariegos Alfaro, Ruben Alberto

Alfi, Abdulaziz Adnan Sharif. Mechanical and electron optical properties of a stabilized collapsible soil in Tucson, Arizona. D, 1984, University of Arizona. 331 p.

Alfonsi, Pedro P. Lithostratigraphy and areal geology of east-central Choctaw County, Oklahoma. M, 1968, University of Oklahoma. 57 p.

Alfonso-Roche, J. A study of the electromagnetic fields generated by a finite grounded wire source. M, 1973, University of Toronto.

Alford, David Dorman. Modelling by Markov-chain analysis of river systems of Late Triassic to Middle Jurassic age in the Hartford Basin, Connecticut and Massachusetts. M, 1984, University of Massachusetts. 274 p.

Alford, Dean E. Geology and geochemistry of the Hembrillo Canyon succession, San Andres Mountains, Sierra and Dona Ana counties, NM. M, 1987, New Mexico Institute of Mining and Technology. 180 p.

Alford, Donald Leslie. Cirque glaciers of the Colorado Front Range; mesoscale aspects of a glacier environment. D, 1973, University of Colorado.

Alford, Elizabeth V. Compositional variations of authigenic chlorites in the Tuscaloosa Formation, Upper Cretaceous, of the Gulf Coast basin. M, 1983, University of New Orleans. 66 p.

Alford, Gary W. Petrology and provenance of the Greta Sandstone, Frio Formation (Oligocene), McFaddin Field, Victoria County, Texas. M, 1988, Stephen F. Austin State University. 91 p.

Alford, Robert L. Environmental delineation through petrographic analysis. M, 1971, University of Missouri, Columbia.

Alfors, John Theodore. A structural and petrographic investigation of an area of glaucophane-bearing rocks in Panoche Valley Quadrangle, California. D, 1959, University of California, Berkeley. 126 p.

Algan, Ugur. Numerical simulation of heat transfer in free surface aquifers. D, 1984, University of Missouri, Rolla. 162 p.

Alger, Dean Wesley. Depositional environments of sandstones in the Renault and Aux Vases formations in parts of Posey, Gibson, and Vanderburgh counties, Indiana. M, 1984, Indiana University, Bloomington. 88 p.

Alger, L. H. Near ultraviolet spectral reflectance capabilities for delineating surficial geologic particle size groups. M, 1977, Indiana State University. 108 p.

Alger, Leonard Hugh, Jr. Degradation of an arid environment; earth fissures in central Arizona. D, 1982, Indiana State University. 115 p.

Algeria, Steven Alan. Relative transportability of sand grains as a function of size and density, determined for various bed states in a laboratory flume. M, 1983, Syracuse University. 136 p.

Algermissen, Sylvester Theodore. Gravity survey of the North Leadwood Mine area, Leadwood, Missouri. D, 1957, Washington University. 73 p.

Algermissen, Sylvester Theodore. Resistivity measurements over flint clay deposits. M, 1955, Washington University.

Alhassoun, Saleh Abdullah. SWATCH; a physically-based distributed hydrologic simulation watershed model. D, 1987, Colorado State University. 333 p.

Ali, E. M. Stochastic analysis of seepage in heterogeneous soils. D, 1979, Ohio State University. 203 p.

Ali, Hamzah. Geology of the Carter Lake region from Lyons to north of Hygiene; Boulder County, Colorado. M, 1950, Colorado School of Mines. 71 p.

Ali, Hassan Mohamed. Geology of eastern Thermopolis and Copper Mountain areas, eastern Owl Creek Mountains, Wyoming. M, 1978, University of Wisconsin-Madison.

Ali, Hassan Mohamed. Keweenawan volcanic rocks of the Grandview-Minong area, northwestern Wisconsin. D, 1982, University of Wisconsin-Madison. 376 p.

Ali, Jaafar. Geophysical surveys of groundwater contamination at three sanitary landfills in Essex County, Ontario. M, 1983, University of Windsor. 247 p.

Ali, Khadim S. Sheikh *see* Sheikh Ali, Khadim S.

Ali, M. A. Self burial of offshore pipelines in fine grained cohesive sediment. D, 1977, Texas A&M University. 155 p.

Ali, Mohammad. A seismic reflection investigation in the west-central South Park, Park County, Colorado. M, 1968, Colorado School of Mines. 58 p.

Ali, Mohammad Sanwar. In situ EPR studies of coal pyrolysis. D, 1987, Texas Christian University. 97 p.

Ali, Molla Mohammad. A probabilistic analysis of the distribution of collapsing soil in Tucson using kriging method. D, 1987, University of Arizona. 271 p.

Ali, Odeh Said. Stratigraphy of the lower Triassic sandstone of the northwest Algerian Sahara, Algeria. D, 1969, University of Iowa. 110 p.

Ali, Odeh Said. Subsurface stratigraphy of Van Buren County, Iowa. M, 1966, University of Iowa. 59 p.

Ali, Sayed I. Study of Recent sediments of the beach and delta at the mouth of Alma River (Bay of Fundy). M, 1964, University of New Brunswick.

Ali, Syed A. Chemical composition of pore waters from sediments of subtropical and Arctic environments; a comparison. D, 1974, Rensselaer Polytechnic Institute. 180 p.

Ali, Syed A. Kope and Fairview sedimentary structures, Brown and Adams counties, Ohio. M, 1967, Ohio State University.

Ali, Syed Ikramuddin *see* Ikramuddin Ali, Syed

Aliberti, Elaine Angela. A structural, petrographic, and isotopic study of the Rapid River area and selected mafic complexes in the northwestern United States; implications for the evolution of an abrupt island arc-continent boundary. D, 1988, Harvard University. 219 p.

Alief, Mohammed Hassan. Variation in SiO_2, Al_2O_3, Fe_2O_3, CaO, MgO, Na_2O, K_2O and H_2O in some ignimbrites in southern Twin Falls County, Idaho. M, 1962, University of Idaho. 60 p.

Alioha, Iheakachuku George. Limestone quarrying industry in Southwest Missouri; its problems and future prospects. M, 1981, Southwest Missouri State University. 51 p.

Alipouraghtapeh, Samad. Geochemistry of major and trace elements of the "Raggedy Mountain Gabbro Group," Wichita Mountains, southwestern Oklahoma. M, 1979, Oklahoma State University. 116 p.

Alison, Jamie Richard. Static fatigue and fractographic analysis of the Hamstead Granite. M, 1981, University of New Brunswick.

Alizadeh, Amin. Amount and type of clay and pore fluid influences on the critical shear stress and swelling of cohesive soils. D, 1974, University of California, Davis. 108 p.

Alkaseh, Ahmidi Ali. Hydrogeological and hydrogeochemical aspects of the Jalo area, Libya. M, 1979, University of Nevada. 73 p.

Alkazmi, Rajab Abdussalam. Structural analysis of the Precambrian rocks of the Park Dome area, Custer County, South Dakota. M, 1973, South Dakota School of Mines & Technology.

Alker, Julius. Review of the North American fossil cricetine rodents (Muridae: Mammalia). D, 1967, University of Nebraska, Lincoln. 314 p.

Alkhazmi, Rajab Abdussalam. Geology, geochemistry and petrogenesis of the alkaline syenite series and phonolite of Jabal Arkeno ring complex, Southeast Libya. D, 1981, South Dakota School of Mines & Technology.

Allahiari, Morteza. Morphometric analysis of the Mud Brook basin of northeastern Summit County, Ohio. M, 1983, University of Akron. 85 p.

Allain, Olivier. An artificial intelligence approach to well test interpretation. M, 1987, Stanford University.

Allam, Awad Moustafa. Numerical simulation of heterogeneous fractured gas reservoir systems with turbulence and closure stress effects. D, 1982, University of Oklahoma. 160 p.

Allan, D. W. The formation and development of the earth, with special reference to its thermal history. D, 1957, University of Toronto.

Allan, J. The distribution of the trace ferrides in the magnetites of the Mount Hope Mine and the New Jersey Highlands. D, 1954, Massachusetts Institute of Technology.

Allan, J. A. Geology of the Ice River District, British Columbia. D, 1912, Massachusetts Institute of Technology. 330 p.

Allan, J. D. Geological studies of the Lynn Lake area, North Manitoba. D, 1948, Massachusetts Institute of Technology. 167 p.

Allan, James Frederick. Geological studies in the Colima Graben, SW Mexico. D, 1984, University of California, Berkeley. 144 p.

Allan, James Frederick. Structural control of the 01 and 09 orebodies, Ace and Fay mines, Beaverlodge, Saskatchewan. M, 1963, Queen's University. 82 p.

Allan, John R. Petrographic, stable isotopic, and electron microprobe studies of diagenesis in limestones. D, 1978, Brown University. 257 p.

Allan, John R. Point bar deposition in a tidal creek, Stone Harbor, N.J. M, 1972, Lehigh University.

Allan, Roderick James. Clay mineralogy and geochemistry of soils and sediments with permafrost in interior Alaska. D, 1969, Dartmouth College. 289 p.

Allan, Urban S., Jr. Late Quaternary stream history of western Louisiana and eastern Texas. M, 1956, Columbia University, Teachers College.

Allard, David. Stable isotopic analysis of Narragansett Bay bivalves; environmental records and constraints on growth history. M, 1988, University of Rhode Island.

Allard, Gilles. Comparative study of sulphide ores or the Chibougamau area, Quebec. M, 1953, Queen's University. 126 p.

Allard, Gilles O. The geology of a portion of McKenzie Township, Chibougamau District, Quebec. D, 1956, The Johns Hopkins University.

Allard, Jean-Louis. The electric potential in the neighbourhood of a thin slit. M, 1961, University of British Columbia.

Allard, M. Le Rôle de la géomorphologie dans les inventaires bio-physiques; l'exemple de la région Gatineau-Lievre. D, 1977, McGill University.

Allard, Margaret Jane. Analytical electron microscopy of diagenetic clays in Precambrian stromatolites. M, 1984, University of Michigan.

Allard, Michel. A magnetometer array study of the Kapuskasing structural zone area. M, 1986, Queen's University. 176 p.

Allaway, William. Sedimentology and petrography of the Shelikof Formation, Alaska Peninsula. M, 1982, San Jose State University. 81 p.

Allcock, J. B. Gaspe copper; a Devonian porphyry copper/skarn complex. D, 1978, Yale University. 233 p.

Allday, Edwin. Structure of southern Indio Mountains, Hudspeth County, Texas. M, 1953, University of Texas, Austin.

Alldredge, A. William. Wildlife inventory and analysis of some energy development impacts. D, 1977, Colorado State University. 114 p.

Allegro, Gayle L. The Gilmore Dome tungsten mineralization, Fairbanks mining district. M, 1987, University of Alaska, Fairbanks. 150 p.

Alleman, David G. Stratigraphy and sedimentation of the Precambrian Revett Formation, Northwest Montana and northern Idaho. M, 1983, University of Montana. 103 p.

Allen Dick, Beryl J. Distribution of copper in an Atlantic coastal-plain estuary, Prince Edward Island. M, 1980, Queen's University. 171 p.

Allen, A. R. Wall-rock alteration at Zeballos, British Columbia. M, 1941, University of British Columbia.

Allen, Albert E., Jr. The subsurface geology of Woods and Alfalfa counties, northwestern Oklahoma. M, 1953, University of Oklahoma. 86 p.

Allen, Alice S. Gradation of grain in sediments. M, 1931, University of Wisconsin-Madison.

Allen, Arthur Thomas, Jr. Geology of the Ringgold, Georgia area. D, 1950, University of Colorado. 203 p.

Allen, Arthur Thomas, Jr. The Longview Member of the Kingsport Formation. M, 1947, University of Tennessee, Knoxville. 102 p.

Allen, Ashley V. Geochemistry of the Mecca Quarry Shale and Colchester Coal, Linton Formation (Middle Pennsylvanian), in Parke and Vermillion counties, Indiana. M, 1986, University of Missouri, Kansas City. 79 p.

Allen, Billy Dean. Foraminifera from the Paynes Hammock Sand of Wayne County, Mississippi. M, 1958, University of Missouri, Columbia.

Allen, Bonnie L. A mineralogical study of soils developed on Tertiary and Recent lava flows in northeastern New Mexico. D, 1959, Michigan State University. 185 p.

Allen, Boyd, III. Glaciochemically derived net mass balance for the Rennick Glacier area, Antarctica. M, 1983, University of New Hampshire. 96 p.

Allen, Brian L. Applied geology for land-use planning in the B.A.C.O.G. area. M, 1975, Northern Illinois University.

Allen, Brian L. Surficial geology and land use planning for the Barrington, Illinois area. M, 1975, Northern Illinois University. 67 p.

Allen, Bruce R. Van *see* Van Allen, Bruce R.

Allen, C. C. Volcano-ice interactions on the Earth and Mars. D, 1979, University of Arizona. 145 p.

Allen, C. M. Geology of the Gooderham nepheline pegmatites, Gooderham, Ontario. D, 1953, University of Toronto.

Allen, C. M. Geology of the Wilberforce radioactive pegmatites. M, 1951, University of Toronto.

Allen, C. U. Paleozoic subsurface stratigraphy of east central New York. D, 1961, [University of Michigan].

Allen, Charles C. Geology of north central part of Reindeer Lake South, Saskatchewan. M, 1938, University of Minnesota, Minneapolis. 50 p.

Allen, Charles C. Geology of Poohbah Lake, Ontario. D, 1940, University of Minnesota, Minneapolis. 74 p.

Allen, Charles W. Structure of the northwestern Puente Hills, Los Angeles County, California. M, 1949, California Institute of Technology. 57 p.

Allen, Charlotte Mary. Intrusive relations and petrology of the Slinkard Pluton, central Klamath Mountains, California. M, 1981, University of Oregon. 120 p.

Allen, Clarence Roderick. The San Andreas fault zone in San Gorgonio Pass, California. D, 1954, California Institute of Technology. 147 p.

Allen, Clifford Marsden. A study of the effect of soil formation processes on serpentine rock at Clangula lake, Manitoba. M, 1950, University of Manitoba.

Allen, Corbett U., Jr. Paleozoic subsurface stratigraphy of east central New York. M, 1961, University of Kentucky. 124 p.

Allen, D. M. The permafrost regime in the Mackenzie Delta, Beaufort Sea region, N.W.T. and its paleoclimatic implications. M, 1988, Carleton University. 154 p.

Allen, David Peter Beddome. Diagenesis and thermal maturation of the Eureka Sound Formation, Strand Fiord, Axel Heiberg Island, Arctic Canada. M, 1986, University of British Columbia. 145 p.

Allen, David William. Clay minerals of Tar-Pamlico River sediments (North Carolina). M, 1965, University of North Carolina, Chapel Hill. 34 p.

Allen, David William. Sedimentary texture; a key to interpret deep-marine dynamics. D, 1970, Oregon State University. 167 p.

Allen, Donald Bruce. Petrology of the granitic rocks, Swans Island, Maine. M, 1966, University of Illinois, Urbana. 60 p.

Allen, Donald Bruce. Structure and petrology of the North Sullivan Pluton, Hancock County, Maine. D, 1971, University of Illinois, Urbana. 89 p.

Allen, Donald G. Mineralogy of Stikine Copper's Galore Creek deposits (Stikine River area, N.W. British Columbia). M, 1966, University of British Columbia.

Allen, Douglas Ray. Geology and geochemistry of the Escalante silver veins, Iron County, Utah. M, 1979, University of Utah. 71 p.

Allen, Elan Anderson. The effects of urbanization on stream flow, sediment transport, and sediment deposition in the White Rock Prairie, central Texas. M, 1988, Baylor University. 172 p.

Allen, Elizabeth A. Identification of Gramineae fragments in salt marsh peats and their use in late Holocene paleoenvironmental reconstructions. M, 1974, University of Delaware.

Allen, Elizabeth A. Petrology and stratigraphy of Holocene coastal-marsh deposits along the western shore of Delaware Bay. D, 1978, University of Delaware. 303 p.

Allen, Erlece Paree Green Kovar. Ground water protection from the Federal level, an analysis of the development and implementation of the Safe Drinking Water Act. D, 1981, University of Texas at Dallas. 336 p.

Allen, Fred Mitchell. Structural and chemical variations in vesuvianite. D, 1985, Harvard University. 459 p.

Allen, Gary Curtis. The determination and geochemistry of cesium. M, 1963, Rice University. 33 p.

Allen, Gary Curtiss. Chemical and mineralogic variations during prograde metamorphism in the Great Smoky Mountains of North Carolina-Tennessee. D, 1968, University of North Carolina, Chapel Hill. 79 p.

Allen, George P. An experimental study of imbrication of tabular grains under various conditions of sorting, flow velocity, and initial dip. M, 1965, Rensselaer Polytechnic Institute. 84 p.

Allen, Harris Hughes. Geology of the Browns Creek area, Nashville Quadrangle, Tennessee, with special reference to the Bigby Limestone. M, 1937, Vanderbilt University.

Allen, Harry. A gravity and aeromagnetic interpretation of the Decaturville dome area, Missouri (Precambrian). M, 1969, Washington University. 79 p.

Allen, Harry B. and Van Couvering, Martin. Geology of the Devils Den District, northwestern Kern County, California. M, 1941, University of California, Los Angeles.

Allen, Henry Whitney. The North Jay Granite in the upper quarry area, North Jay, Maine. M, 1949, University of Missouri, Columbia.

Allen, J. D. Metamorphism of the Yellowknife sediments. M, 1940, Queen's University. 40 p.

Allen, J. S. Carbonatization of Red Lake, Ontario. M, 1939, University of Toronto.

Allen, Jack Christopher, Jr. Structure and petrology of the Royal Stock and the Mt. Powell Batholith, Flint Creek Range, western Montana. D, 1962, Princeton University. 112 p.

Allen, James M. The Lyons Falls Pegmatite. M, 1956, Syracuse University.

Allen, James Michael. The chemical and isotopic compositions of fluid inclusions. D, 1959, Yale University.

Allen, James R. Beach dynamics along Sandy Hook spit, New Jersey. D, 1973, Rutgers, The State University, New Brunswick. 127 p.

Allen, Jimmy E. Wall rock alteration, Ontario Mine, Keetley, Utah. M, 1961, University of Utah. 331 p.

Allen, Joan Park. Stratigraphic analysis and syndepositional structural influence on Cambrian units of the Rome Trough in West Virginia. M, 1988, West Virginia University. 123 p.

Allen, John Eliot. 1, Geology of the San Juan Bautista Quadrangle, California; 2, Structures in the chromite deposits of the West Coast. D, 1945, University of California, Berkeley. 92 p.

Allen, John Eliot. Contributions to the structure, stratigraphy and petrography of the lower Columbia River gorge. M, 1932, University of Oregon. 91 p.

Allen, John Francis, Jr. Paleocurrent and facies analysis of the Triassic Stockton Formation in western New Jersey. M, 1979, Rutgers, The State University, New Brunswick. 84 p.

Allen, John Murray. Silicate-carbonate equilibria in calcareous metasediments of the Tudor Township area, Ontario; a test of the P-T-X CO_2-X H_2O model of metamorphism. D, 1976, Queen's University. 227 p.

Allen, John Murray. The genesis of Precambrian uranium deposits in eastern Canada, and the uraniferous pegmatites at Mont Laurier, Quebec. M, 1971, Queen's University. 84 p.

Allen, John O. The geology of the Southwest Wayne Pool, McClain County, Oklahoma. M, 1949, University of Oklahoma. 39 p.

Allen, Julia Coan. Deforestation, soil degradation, and wood energy in developing countries. D, 1983, The Johns Hopkins University. 346 p.

Allen, Marian Ruth. The origin of dolomites in the phosphatic sediments of the Miocene Pungo River Formation, North Carolina. M, 1985, Duke University. 58 p.

Allen, Martin. Geology of Thurber area, Earth County, Texas. M, 1951, University of Texas, Austin.

Allen, Mary E. Theobald. The geochemistry of heavy mineral concentrates from rocks associated with the South Bay massive sulphide deposit, Ontario; an exploration technique. M, 1982, Queen's University. 222 p.

Allen, Merrill Peter. Dissolution and cation exchange studies of asbestos minerals in aqueous media. M, 1972, University of Nevada - Mackay School of Mines. 87 p.

Allen, Michael Lee. A palynological definition of the Devonian-Mississippian boundary in the Midcontinent of the United States. M, 1974, University of Texas at Dallas. 107 p.

Allen, Michael Steven. An evaluation of stream sediment and heavy mineral concentrate geochemistry as aids in geologic mapping and mineral exploration in northern Colorado. M, 1982, Queen's University. 244 p.

Allen, Milton. An evaluation of computerized systems for acceleration of interpretations from soil surveys and other resource information. D, 1983, University of Tennessee, Knoxville. 216 p.

Allen, Myron Bartlett, III. Collocation techniques for modeling compositional flows in porous media. D, 1983, Princeton University. 224 p.

Allen, Nancy Edwards. The geology of base precious metal-bearing quartz veins in Hall and Gwinnett counties, Georgia. M, 1987, Auburn University. 113 p.

Allen, Paula E. The petrology and paleoecology of a Silurian (Niagaran) coral-stromatoporoid association on the northwest margin of the Michigan Basin, Door County, Wisconsin. M, 1986, University of Wisconsin-Green Bay. 87 p.

Allen, Peter. Urban geology along I-35 growth corridor from Hillsboro through Ellis County, Texas. M, 1972, Baylor University. 182 p.

Allen, Peter Martin. Analysis of an escarpment landform for land-use planning, Mountain Creek basin, Dallas-Ellis counties, Texas. D, 1977, Southern Methodist University. 391 p.

Allen, Philip. The geochemistry of the amphibolite-granulite facies transition in central South India. D, 1985, New Mexico Institute of Mining and Technology. 273 p.

Allen, R. Petrology of the Bird River Sill, Manitoba. D, 1966, University of Minnesota, Minneapolis.

Allen, Ralph Orville, Jr. Multi-element neutron activation analysis; development and application to a trace element study of the Bruderheim chondrite. D, 1970, University of Wisconsin-Madison. 243 p.

Allen, Reid. Stratigraphy, carbonate petrology, paleoenvironmental analysis, geologic control on mineralization of the Jones Camp and Dike region, Socorro County, NM. M, 1988, New Mexico Institute of Mining and Technology. 133 p.

Allen, Rhesa McCoy, Jr. Geology and ore deposits of the Volcano District, Elmore County, Idaho. M, 1940, University of Idaho. 42 p.

Allen, Rhesa McCoy, Jr. The geology and mineralization of the Morning Mine and adjacent region, Grant County, Oregon. D, 1947, Cornell University. 78 p.

Allen, Richardson Beardsell. Geologic studies of the Scotia Arc region and Agulhas Plateau. D, 1983, Columbia University, Teachers College. 274 p.

Allen, Risden Tyler. The Triassic rocks of the Wadesboro area. M, 1908, University of North Carolina, Chapel Hill. 19 p.

Allen, Robert. Structure of Sierra de los Fresnos, Chihuahua, Mexico. M, 1957, University of Texas, Austin.

Allen, Robert Dorchester. Variations in chemical and physical properties of fluorite. D, 1950, Harvard University.

Allen, Robert Gordon Hamilton. Granitic rocks of the Halifax harbour–Saint Margaret's bay area (Nova Scotia). M, 1963, Dalhousie University. 35 p.

Allen, Robert S. A computer program for use in computing a first order latitude by the Horrebow-Talcott method. M, 1966, Ohio State University.

Allen, Robert Stanton. Petrology, paleontology and origin of the Clear Creek Chert (Lower Devonian) in Union and Jackson counties, southwestern Illinois. M, 1985, Southern Illinois University, Carbondale. 86 p.

Allen, Rodger F. Stratigraphy of the Sanilac Group (Silurian) of the Michigan Basin. M, 1974, Northwestern University.

Allen, Rodney L. Petrology and chemistry of a komatiite sill and Fe-Ni-Cu sulfide mineralization, Munro and Beatty townships, Ontario. M, 1986, Queen's University. 108 p.

Allen, Ron R. Variations of zircon crop measurements due to weathering of igneous rocks. M, 1968, Texas Tech University. 52 p.

Allen, Roy Frank. Uranium potential of the Cement District, southwestern Oklahoma. M, 1980, Oklahoma State University. 85 p.

Allen, Russell Warren. The geology of South Putnum Mountain, Bannock and Caribou counties, Idaho. M, 1976, Idaho State University. 63 p.

Allen, Samuel W. Seismic interpretation, Riverslea Field, Surat Basin, Southeast Queensland, Australia. M, 1986, Colorado School of Mines. 67 p.

Allen, Stanley Randolph. The Breckenridge oil field, Stephens County, Texas. M, 1930, University of Texas, Austin.

Allen, Stephen H. Stratigraphy and diagenesis of a lower Cretaceous carbonate sand complex, central Texas. M, 1970, Louisiana State University.

Allen, Stephen Jeffrey. Digital hydrologic modeling methods for water resources engineering with application to the Broad Brook watershed. D, 1987, University of Connecticut. 239 p.

Allen, Timothy J. The subsurface stratigraphy of the northern Avra Valley, Pima County, Arizona. M, 1981, Kent State University, Kent. 63 p.

Allen, Victor Thomas. Mineral composition of certain Minnesota sands and its relation to suitability for use in concrete. M, 1922, University of Minnesota, Minneapolis. 43 p.

Allen, Victor Thomas. The Ione Formation of California. D, 1928, University of California, Berkeley. 83 p.

Allen, Walter C. Mineralogy of the brick clays at Kingston, New York. M, 1947, Columbia University, Teachers College.

Allen, Walter Carl. Solid solution in the refractory magnesium spinels. D, 1965, Rutgers, The State University, New Brunswick. 196 p.

Allen, William Burrows. Geology of the Browns Station Anticline, Boone County, Missouri. M, 1941, University of Missouri, Columbia.

Allen, William Hammond. The permeability of certain Illinois oil sands in the direction of the bedding planes and in the transverse direction. M, 1937, University of Illinois, Urbana. 24 p.

Allen, William Henry, Jr. A quantitative study of the drainage net and its related phenomena in the Pigeon Roost Watershed (Mississippi). M, 1963, University of Mississippi.

Allen, William Odis; Tipsword, Howard Lee. Some Lower Pennsylvanian foraminifera from southern Oklahoma. M, 1938, Indiana University, Bloomington. 21 p.

Allen, William Turner. A computer oriented approach to environmental geologic data base construction (with application to the Athens Quadrangle, Tennessee). M, 1980, University of Tennessee, Knoxville. 108 p.

Allen, Woods Wilkinson, Jr. Petrology of the Middle Jurassic (?) La Joya Formation, Sierra Madre Oriental, southwestern Tamaulipas, Mexico. M, 1976, Texas A&M University. 189 p.

Allenby, R. J. Determination by mass spectrometer of isotopic ratios of silicon in rocks. D, 1952, University of Toronto.

Allenson, Sherman. Geology of the southern part of the Camillus, New York 7 1/2 minute quadrangle. M, 1955, Syracuse University.

Aller, R. C. The influence of macrobenthos on chemical diagenesis of marine sediments. D, 1977, Yale University. 628 p.

Allex, Mark. Stratigraphy and paleoenvironments Sawatch Formation (Cambrian) Cement Creek area, Gunnison County, Colorado. M, 1987, University of Kentucky.

Allexan, John Stephen. Geology of the Jenkins Canyon area, Portneuf Range, Bannock County, Idaho. M, 1979, Idaho State University. 45 p.

Alley, Douglas Wayne. Drift prospecting and glacial geology in the Sheffield Lake-Indian Pond area; north-central Newfoundland. M, 1975, Memorial University of Newfoundland. 215 p.

Alley, Hugh A. Diagenesis and porosity evolution within the upper Shunda and Turner Valley formations, Moose Dome area, Alberta. M, 1982, University of Manitoba.

Alley, Lonnie Bruce. Aeromagnetic survey of northeastern Utah. M, 1973, University of Utah. 58 p.

Alley, Neville F. The Quaternary history of part of the Rocky Mountains, foothills, plains and western Porcupine Hills, Alberta. D, 1972, University of Calgary.

Alley, Richard Blaine. Transformations in polar firn. D, 1987, University of Wisconsin-Madison. 430 p.

Alley, Sharon L. The late Quaternary South Atlantic; a nannoplankton study. M, 1981, University of Utah. 181 p.

Alley, William McKinley. Numerical simulation of transient and steady state optimal pumping from artesian aquifers. M, 1975, Stanford University.

Alley, William McKinley. Use of regional water balance models in characterizing hydrologic drought. D, 1984, The Johns Hopkins University. 163 p.

Allie, Adrienne Dee. Stratigraphic controls of carbonate-hosted sulfide mineralization, Iron Rock Creek area, Blanco and Gillespie counties, Texas. M, 1981, University of Texas, Austin. 101 p.

Alling, Harold L. Glacial lakes and other glacial features of the central Adirondacks. M, 1917, Columbia University, Teachers College.

Alling, Harold L. The mineralography of the feldspars. D, 1921, Columbia University, Teachers College.

Alling, Merle K. A study of the physiographic history of Allen's Creek at Corbett's Glen. M, 1924, University of Rochester.

Allinger, Richard Jack. Geology, petrology, and geochemistry of the Viejas Mountain gabbro pluton, San Diego County, California. M, 1979, San Diego State University.

Allingham, John W. Metamorphism at the contact of the Cable Stock, Montana. M, 1954, California Institute of Technology. 58 p.

Allinton, John Richard. Certain phonolitic and related intrusions in the Spearfish Canyon area, Black Hills, South Dakota. M, 1962, University of Nebraska, Lincoln.

Allis, R. G. Constraints on crustal structure from heat flow measurements in lakes of N.W. Ontario. D, 1977, University of Toronto.

Allis, R. G. Geothermal measurements in five small lakes of northwestern Ontario, Canada. M, 1975, University of Toronto.

Allison, David Bryan. A petrographic study of the Haymond Formation, West Texas. M, 1959, University of Houston.

Allison, Dennis A. A pattern change along Drury Creek in southern Illinois and its relation to the threshold concept. M, 1978, Southern Illinois University, Carbondale. 67 p.

Allison, Edwin Chester. Middle Cretaceous faunules of Puerto Santo Tomas, Baja California. M, 1953, University of California, Berkeley. 138 p.

Allison, Edwin Chester. The Bivalvia of the Alsitos Formation, northwestern Baja California, Mexico. D, 1964, University of California, Berkeley. 464 p.

Allison, Eric T. Timing of bottom-water scour recorded by sedimentological parameters in the South Australian Basin. M, 1980, University of Georgia.

Allison, Ira S. The Giants Range Batholith of Minnesota. D, 1924, University of Minnesota, Minneapolis. 96 p.

Allison, Jerry Dewell. Seismicity of the Central Georgia seismic zone. M, 1980, Georgia Institute of Technology. 204 p.

Allison, John P. Petrography of the upper border zone of the Kiglapait Intrusion, Labrador. M, 1984, University of Massachusetts. 220 p.

Allison, M. L. Geophysical studies along the southern portion of the Elsinore Fault. M, 1974, San Diego State University.

Allison, Marvin Dale. Geology of a portion of San Miguel County, New Mexico. M, 1950, University of Iowa. 125 p.

Allison, Mead A. Mineralogy and sedimentology of the clay-sized fraction, Miocene Pungo River Formation, North Carolina continental margin. M, 1988, East Carolina University. 112 p.

Allison, Merle Lee. Structural analysis of the Tensleep Fault, Bighorn Basin, Wyoming. D, 1986, University of Massachusetts. 393 p.

Allison, Michael C. Trace element distribution in pyrite and pyrrhotite from the Ducktown mining district, Tennessee. M, 1984, University of Tennessee, Knoxville. 150 p.

Allison, Michael Duane. Petrology and depositional environments of the Mississippian Chappel bioherms; Hardeman County, Texas. M, 1979, West Texas State University. 55 p.

Allison, Richard Case. The Cenozoic stratigraphy of Chiapas, Mexico, with discussions of the classification of the Turritellidae and selected Mexican representatives. D, 1967, University of California, Berkeley. 450 p.

Allison, Richard Chase. Geology and Eocene megafaunal peleontology of Quimper Peninsula area, Washington. M, 1959, University of Washington. 121 p.

Allman, David William. Analysis of base flow recession slopes of Panther Creek (Woodford and Livingston County) (north-central) Illinois. M, 1968, University of Illinois, Urbana. 169 p.

Allman, David William. Ground-water flow under an irrigated alfalfa field. D, 1973, University of Idaho. 328 p.

Allmaras, J. M. Stratigraphy and sedimentology of the Late Cretaceous Nanaimo Group, Denman Island, British Columbia. M, 1979, Oregon State University. 178 p.

Allmedinger, Roger J. A model for ore-genesis in the Hansonburg mining district, New Mexico. D, 1975, New Mexico Institute of Mining and Technology. 190 p.

Allmendinger, R. W. Structural evolution of the northern Blackfoot Mountains, southeastern Idaho. D, 1979, Stanford University. 296 p.

Allmendinger, Roger J. Hydrologic control over the origin of gypsum at Lake Lucero, White Sands National Monument, New Mexico. M, 1972, New Mexico Institute of Mining and Technology. 82 p.

Allmon, Warren Douglas. Evolution and environment in turritelline gastropods (Mesogastropoda, Turritellidae), lower Tertiary of the U.S. Gulf and Atlantic coastal plains. D, 1988, Harvard University. 826 p.

Allong, Albert F. Sedimentation and stratigraphy of the Saskatchewan Gravels and Sands (Pliocene or early Pleistocene) in central and southern Alberta (Canada). M, 1967, University of Wisconsin-Madison.

Allong, Albert Francis. Hydrogeology of the Scioto drainage basin. D, 1971, Ohio State University. 228 p.

Allotta, Thomas Lawrence. A shallow seismic investigation in the volcanic stratigraphy of the Jemez Mountains, New Mexico. M, 1985, Purdue University. 185 p.

Allouani, Rabah Nadir. Stratigraphy and microfacies analysis of the Berino Formation (Atkon-Desmoines), Vinton Canyon, El Paso County, Texas. M, 1976, University of Texas at El Paso.

Allred, David Hammond. Soil conservation districts; a democratic approach to erosion control. M, 1939, American University. 39 p.

Allred, Ronald Dean. The application of shallow seismic reflection techniques in the detection of mined and unmined coal. M, 1980, Wright State University.

Allshouse, Sharon Dale. Petrographic variation due to depositional setting of the lower Kittanning seam, western Pennsylvania. M, 1984, Pennsylvania State University, University Park. 96 p.

Allsman, Paul Lewis. Oxidation and enrichment of the manganese deposits of Butte, Montana. M, 1955, University of Illinois, Urbana. 92 p.

Allsop, Charles Mark. A study of the basement rock in central Kentucky using refraction techniques. M, 1986, University of Kentucky. 102 p.

Allsop, Heather Allyne. Pore structure, wettability, and two-phase flow behaviour in Pembina Cardium Sandstone. D, 1988, University of Waterloo. 364 p.

Allspach, H. G. Geology of a part of the south flank of the Seminoe Mountains, Carbon County, Wyoming. M, 1955, University of Wyoming. 65 p.

Ally-Gregoire, Raymonde. Anomalie de spectro-réflectivité des feuilles de Populus canadensis Eugenei et Salix pentandra croissant en laboratoire sur un sol contaminé en cuivre. M, 1976, Ecole Polytechnique.

Almalik, Mansour Saleh. An investigation of parameters effecting oil recovery efficiency of carbon dioxide flooding in cross-sectional reservoirs. D, 1988, Texas A&M University. 201 p.

Almand, Charles William. The geology of the Lumpkin Quadrangle, Stewart County, Georgia. M, 1961, Emory University. 86 p.

Almashoor, Syed Sheikh. The petrology of a quartz monzonite pluton in the southern Greenwater Range, Inyo County, California. D, 1983, Pennsylvania State University, University Park. 186 p.

Almasi, Mohammad. Ecology and color variation of benthic foraminifera in Barnes Sound, Northeast Florida Bay. M, 1978, University of Miami. 144 p.

Almasi, Mohammad. Holocene sediments and evolution of the Indian River (Atlantic Coast of Florida). D, 1984, University of Miami. 238 p.

Almasi, Mohammad Naghash. Holocene sediments and evolution of the Indian River (Atlantic coast of Florida). D, 1983, University of Miami. 266 p.

Almeida, Erasto Boretti de *see* de Almeida, Erasto Boretti

Almen, William F. Von *see* Von Almen, William F.

Almen, William Frederick Von *see* Von Almen, William Frederick

Almendinger, James Edward. Lake and groundwater paleohydrology; a groundwater model to explain past lake levels in west-central Minnesota. D, 1988, University of Minnesota, Minneapolis. 133 p.

Almendinger, John Curtis. The late-Holocene development of jack pine forests on outwash plains, north-central Minnesota. D, 1985, University of Minnesota, Minneapolis. 234 p.

Almoghrabi, Hamzah Abdulgader. Numerical evaluation of bright spots and thin layer reflectivity. D, 1986, Oklahoma State University. 131 p.

Almon, W. R. Petroleum-forming reactions; clay catalyzed fatty acid decarboxylation. D, 1974, University of Missouri, Columbia. 135 p.

Almon, William Robert. Ammonia complexes in the interlayers of montmorillonite. M, 1971, Washington University. 72 p.

Almond, P. Petrology of the laminated quartz gneisses and the quartz diorites, Charlebois Lake area, northern Saskatchewan. M, 1953, University of Saskatchewan. 46 p.

Almouslli, Mohamad O. The application of the gamma ray spectralog to the analysis of the Chase Group (Gearyan Stage, Lower Permian Series) in Stevens County, Kansas. M, 1987, Wichita State University. 180 p.

Almy, Charles Coit, Jr. Geological and geophysical studies of a portion of the Little Llano River valley, Llano and San Saba counties, Texas. M, 1960, Rice University. 60 p.

Almy, Charles Coit, Jr. Parguera Limestone, Upper Cretaceous, Mayaguez Group, Southwest Puerto Rico. D, 1965, Rice University. 203 p.

Almy, Robert, III. Petrology and major element geochemistry of albite granite near Sparta, Oregon. M, 1977, Western Washington University. 100 p.

Alnes, Joel R. Joint sets of the Llano Uplift, Texas. M, 1984, Texas Tech University. 57 p.

AlNouri, Ilham. Time dependent strength behavior of two soil types at lowered temperatures. D, 1969, Michigan State University. 98 p.

Alonso, Manuel. Petrology of the metamorphic aureole around McCartney Mountain Stock, southwestern Montana. M, 1985, University of Akron. 97 p.

Alor, Jerjes Pantoja *see* Pantoja Alor, Jerjes

Aloui, Tahar. Calcium/sodium exchange models for soil and reference clays. D, 1984, University of Missouri, Columbia. 262 p.

Aloysius, David L. The distribution of uranium, thorium, and rare earth elements in selected post metamorphic granites of the Southern Appalachian Piedmont; studies from drill cores. M, 1982, SUNY at Buffalo. 94 p.

Alpaslan, Tümer. Spectral behaviour of short-period body waves and the synthesis of crustal structure in Western Canada. M, 1969, University of Alberta. 221 p.

Alpaslan, Tümer. Transfer function analysis. D, 1974, University of Alberta. 213 p.

Alpay, Okan. A study of physical and textural heterogeneity in sedimentary rocks. D, 1963, Purdue University. 236 p.

Alpdogan, Sami. Application of aerial triangulation for cadastral purposes in Turkey. M, 1957, Ohio State University.

Alper, Allen M. Geology of Walnut Wells Quadrangle, Hidalgo County, New Mexico. D, 1961, Columbia University, Teachers College. 212 p.

Alperin, Marc Jon. The carbon cycle in an anoxic marine sediment; concentrations, rates, isotope ratios, and diagenetic models. D, 1988, University of Alaska, Fairbanks. 258 p.

Alpers, Charles N. Geochemical and geomorphological dynamics of supergene copper sulfide ore formation and preservation at La Escondida, Antofagasta, Chile. D, 1986, University of California, Berkeley. 197 p.

Alpert, Stephen P. Trace fossils of the Precambrian-Cambrian succession, White-Inyo mountains, California. D, 1974, University of California, Los Angeles. 175 p.

Alpha, Andrew G. Geology and ground water resources, Burke, Divide, Mountrail and Williams counties, North Dakota. M, 1935, University of North Dakota. 63 p.

Alpha, James W. Behavior of cation saturated Wyoming bentonite under low temperature and pressure. M, 1967, University of Cincinnati. 70 p.

Alptekin, Omer. Focal mechanisms of earthquakes in western Turkey and their tectonic implications. D, 1973, New Mexico Institute of Mining and Technology. 190 p.

Alsayed, Alsayed Mouhamed Alammawi *see* Alammawi Alsayed, Alsayed Mouhamed

Alsayegh, Abdul Hadi Y. Seismic and resistivity studies in Alkali Creek drainage five miles southwest of Sturgis, South Dakota. M, 1966, South Dakota School of Mines & Technology.

Alsharhan, Abdulrahman Sultan. Petrography, sedimentology, diagenesis and reservoir characteristics of the Shuaiba and Kharaib formations (Barremian-mid Aptian) carbonate sediments of Abu Dhabi, United Arab Emirates. D, 1985, University of South Carolina. 189 p.

Alsobrook, Albert Francis. The origin of the granitic gneisses in the northwest quarter of the Woodbury Quadrangle, Connecticut. M, 1965, University of Wisconsin-Madison.

Alsop, Janice L. Shelf carbonates of the Madera Formation (Desmoinesian) of the late Paleozoic Taos Trough, northern New Mexico. M, 1982, University of Texas, Austin.

Alstine, David Ralph Van *see* Van Alstine, David Ralph

Alstine, James Bruce Van *see* Van Alstine, James Bruce

Alstine, Ralph Erkstine Van *see* Van Alstine, Ralph Erkstine

Alston, Mickey Wayne. Depositional environment and diagenesis of the Hill Sand Member of the Rodessa Formation (Lower Cretaceous), Saint Mary Field, Lafayette County, Arkansas. M, 1985, Northeast Louisiana University. 94 p.

Alsup, Stephen Alex. Estimation of Upper Mantle Q beneath the United States from Pn amplitudes. D, 1972, George Washington University.

Alt, David. A review of the geology of Rainy Lake, Minnesota. M, 1958, University of Minnesota, Minneapolis. 73 p.

Alt, David Dolton. Isotope geochemistry of boron. D, 1961, University of Texas, Austin. 104 p.

Alt, Jeffrey C. Alteration of the upper oceanic crust; DSDP Site 417. M, 1983, University of Miami. 116 p.

Alt, Jeffrey C. The structure, chemistry, and evolution of a submarine hydrothermal system, DSDP, Site 504. D, 1984, University of Miami. 303 p.

Altamura, Robert James. Geology of the ultramafic rocks near Westfield, Massachusetts, with integrated geochemistry and geophysics. M, 1983, Wesleyan University. 226 p.

Altan, Ozer. Calculation of the gravimetric effects of topography and isostatic compensation with an electronic digital computer. M, 1961, Colorado School of Mines. 63 p.

Altaner, Stephen Paul. Potassium metasomatism and diffusion in Cretaceous K-bentonites from the Disturbed Belt, northwestern Montana and in the Middle Devonian Tioga K-bentonite, Eastern U.S.A. D, 1985, University of Illinois, Urbana. 192 p.

Altany, Robert M. Facies of the Hurricane Cliffs tongue of the Toroweap Formation, Northwest Arizona. M, 1983, Northern Arizona University. 147 p.

Altena, Peter James Van *see* Van Altena, Peter James

Alter, B. E. K. The propagation of strain waves in lead. D, 1955, Lehigh University.

Alter, Benjamin; Brown, Larry. Seismic velocities of near-surface Precambrian rocks in the Adirondack Mountains, New York, and the Laramie Range, Wyoming. M, 1983, Cornell University.

Alteris, Joseph Thomas de *see* de Alteris, Joseph Thomas

Alterman, Ina Brown. Structure and tectonic history of the Taconic allochthon and surrounding autochthon, east-central Pennsylvania. D, 1972, Columbia University. 287 p.

Alther, George R. Distribution of elements downstream from a sewage treatment plant. M, 1976, University of Toledo. 66 p.

Althoff, Penelope L. Structural refinements of dolomite and a magnesian calcite. D, 1976, Rutgers, The State University, New Brunswick. 24 p.

Altic, Mark Alan. Seismic reflection line, Lake County, Michigan. M, 1988, Wright State University. 96 p.

Altintas, Sabri. Deformation behavior of an idealized crystal. M, 1975, University of California, Berkeley.

Altman, Amy Bentley. Hydrogeology of a proposed development site in Suffolk County, New York. M, 1981, SUNY at Binghamton. 74 p.

Altman, L. W. Distribution of stress in the oceanic lithosphere beneath the Lau-Havre Basin. M, 1978, Texas A&M University.

Altman, Thomas DeWitt. The petrologic study of a microsyenite intrusion in Brewster County, Texas. M, 1970, University of Houston.

Alto, Bruno Raymond. Geology of part of Boylston Quadrangle and adjacent areas, central Washington. M, 1955, University of Washington. 38 p.

Altobelli, Randall Wilson. Geology of Three Bayou Bay Field and Manilla Village Field, Jefferson Parish, Louisiana. M, 1981, University of New Orleans. 88 p.

Altschaeffl, Adolph G. Effect of soil moisture and other natural variables on aerial photo gray tones. M, 1955, Purdue University.

Altschuld, Kenneth R. Stratigraphy and depositional environments of the Dakota Group, Canon City area, Colorado. M, 1980, Colorado School of Mines. 180 p.

Altschuler, Zalman Samuel. The joint pattern in Osage County, Oklahoma, as interpreted from aerial photographs. M, 1947, University of Cincinnati. 20 p.

Aluka, Innocent J. Pleistocene stratigraphy of southern Milwaukee County, Wisconsin. M, 1981, University of Wisconsin-Milwaukee. 71 p.

Aluka, Maduegboaka Innocent Jude. Stratigraphy of Early Ordovician El Paso Group of the southern Hueco Mountains, Hudspeth County, Texas. D, 1984, University of Texas at El Paso. 520 p.

Alusow, Edward W. A study of ore and breccia relationships at Magmont Mine, Iron County, Missouri. M, 1976, University of Missouri, Rolla.

Alvarado, Alfredo. Phosphate retention in Andepts from Guatemala and Costa Rica as related to other soil properties. D, 1982, North Carolina State University. 89 p.

Alvarado, Pacifico M. The engineering significance of the airphoto patterns of northern Indiana soils. M, 1946, Purdue University.

Alvarado, Rodrigo B. Photogeomorphic analysis of northern basin, Trinidad. M, 1972, Pennsylvania State University, University Park.

Alvarado, Salvador. Nonlinear effects near an explosion. M, 1969, Rice University. 75 p.

Alvarez, Leonardo Schultz. Geology of the Isabella area, Kern County, California. M, 1962, University of Utah. 20 p.

Alvarez, Walter. Geology of the Simarua and Carpintero areas, Guajira Peninsula, Colombia. D, 1967, Princeton University. 205 p.

Alvarez-Ayesta, Jose Alfredo. A probabilistic analysis of the Emerald Bay rock slide, El Dorado County, California. M, 1987, University of Nevada. 224 p.

Alvarez-Bejar, Ramon. Electrical conduction phenomena in rocks. D, 1972, University of California, Berkeley. 172 p.

Alvarez-Borrego, Saúl. Oxygen-carbon dioxide-nutrients relationships in the northeastern Pacific Ocean and southeastern Bering Sea. D, 1973, Oregon State University. 171 p.

Alvarez-Osejo, J. Alberto. A compilation of formulae for computing the flow of water under several types of boundary conditions. M, 1963, Columbia University, Teachers College.

Alves, Carlos A. Rock mechanics instrumentation applied to longwall coal mining. M, 1977, Colorado School of Mines. 224 p.

Alves, David J. Aspects of regional short-period wave propagation; a study of the December 1967 Koyna earthquakes, Maharashtra, India. M, 1987, Pennsylvania State University, University Park. 55 p.

Alvi, Javaid. Effect of rate of strain on isotrapically anisotrop and consolidated cohesive plate. D, 1967, University of Pittsburgh. 179 p.

Alvir, Antonio Delgado. A geological study of the Angat-Novaliches region. D, 1930, University of Chicago. 60 p.

Alvis, Mark R. Metamorphic petrology, structural and economic geology of a portion of the central Mazatzal Mountains, Gila and Maricopa counties, Arizona. M, 1984, Northern Arizona University. 129 p.

Alwahhab, Riyadh Mostafa. Bond and slip of steel bars in frozen sand. D, 1983, Michigan State University. 280 p.

Alwin, B. W. Sedimentation of the middle Precambrian Tyler Formation of Northcentral Wisconsin and Northwestern Michigan. M, 1976, University of Minnesota, Duluth.

Alwin, John Arnold. Clastic dikes of the Touchet beds, southeastern Washington. M, 1970, Washington State University. 87 p.

Aly, Saleh Mohamed. Polymer and water quality effects on soil strength and flocculation of montmoril-

lonite. D, 1988, University of California, Riverside. 100 p.

Alyanak, Nancy. "Framboidal" chalcocite from White Pine, Michigan. M, 1973, Michigan State University. 37 p.

Alzaydi, A. A. Flow of gases through porous media. D, 1975, Ohio State University. 191 p.

am Ende, Barbara Ann. Depositional environments and paleontology of the Upper Cretaceous Dakota Formation, Kane County, Utah. M, 1987, Northern Arizona University. 190 p.

Amabeoku, Maclean Oluka. Temperature effects on low tension surfactant flooding efficiency in consolidated sands. D, 1981, University of Southern California.

Amadi, Philip Uchenna Mbanu. Geohydrology and ground water chemistry of the Flatwoods area, Owen and Monroe counties, Indiana. M, 1979, Indiana University, Bloomington. 95 p.

Amadi, Philip Uchenna Mbanu. Ground-water chemistry and hydrochemical facies distribution as related to flow in the Mississippian carbonates, Harrison County, Indiana. D, 1981, Indiana University, Bloomington. 338 p.

Amaefule, Jude Ogbonnah. The effect of interfacial tensions on relative oil-water permeabilities of consolidated porous media. D, 1981, University of Southern California.

Amajor, Levi Chukwuemeka. A regional subsurface correlation of some bentonite beds in the Lower Cretaceous Viking Formation of south-central Alberta, Canada. M, 1978, University of Alberta. 104 p.

Amajor, Levi Chukwuemeka. Chronostratigraphy, depositional patterns and environmental analysis of subsurface Lower Cretaceous (Albian) Viking reservoir sandstones in central Alberta and part of southwestern Saskatchewan. D, 1980, University of Alberta. 596 p.

Amalfi, Frederick Anthony. Urban lake sediment chemistry; lake design, runoff, and watershed impact. D, 1988, Arizona State University. 241 p.

Amanat, Jamshid. Analysis and prediction of subsidence caused by longwall mining using numerical techniques. D, 1987, West Virginia University. 300 p.

Amankwah, Samuel Asare. Comparison of three separation techniques for arsenic (III) and arsenic (V) in sea water. D, 1987, University of Rhode Island. 145 p.

Amant, Marcel M. Y. St. *see* St. Amant, Marcel M. Y.

Amante, A. Gravity investigation of Blue Mounds, Wisconsin, system of caves. M, 1954, University of Wisconsin-Madison.

Amara, Mark Steven. Mechanical analyses of stratigraphic layers from the Lind Coulee archaeological site in central Washington. M, 1975, Washington State University. 129 p.

Amaral, Eugene Jordan. Textural analysis of the St. Peter Sandstone, southwestern Wisconsin. M, 1975, University of Cincinnati. 428 p.

Amarasiriwardena, Dulasiri Dayananda. Qualitative and quantitative analysis of iron oxides and oxyhydroxides by Mössbauer spectroscopy. D, 1987, [University of North Carolina, Chapel Hill]. 278 p.

Amateis, Larry Joe. The geology of the Permian Garden Valley Formation, Eureka County, Nevada. M, 1981, University of Nevada. 99 p.

Amato, Francis L. The oscillation chart as a method of correlation. M, 1949, University of Nebraska, Lincoln.

Amato, Roger V. Structural geology of the Salem area, Virginia. M, 1968, Virginia Polytechnic Institute and State University.

Amba, Etim Anwanna. Effects of rainfall characteristics, tillage systems and soil physicochemical properties on sediment and runoff losses from micro-erosion plots. D, 1983, Ohio State University. 201 p.

Ambers, Clifford P. Metasomatism associated with ductile deformation in a granodioritic gneiss, central

Connecticut. M, 1988, Indiana University, Bloomington. 87 p.

Ambjah, Rachmadi. An apparatus to determine water dispersion through sand. M, 1972, University of Utah. 53 p.

Ambler, J. S. The stratigraphy and structure of the Lloydminster oil and gas area. M, 1951, University of Saskatchewan. 111 p.

Ambos, Elizabeth Luke. Applications of ocean bottom seismometer data to the study of forearc and transform fault systems. D, 1984, University of Hawaii. 275 p.

Ambrason, Ellen P. Amplitude analysis of the synthetic seismic response of a thinly bedded carbonate zone. M, 1984, Stanford University.

Ambrose, John W. Geology of the northeast portion of the Flinflon map area, Manitoba. D, 1935, Yale University.

Ambrose, William Anthony. Aggraded fluvial and tide-dominated deltaic deposits of the Spoon Formation (Pennsylvanian) in the southern Illinois Basin; a shallow embayment. M, 1983, University of Texas, Austin. 105 p.

Ambs, Loran D. A shallow geothermal survey of Durkee oil field and Woodgate Fault, Harris County, Texas. M, 1980, [University of Houston].

Ambuehl, Alan W. Surficial authigenic silica in Gulf Coast Tertiary formation. M, 1979, Louisiana State University.

Amdurer, Michael. Chemical speciation and cycling of trace elements in estuaries; radiotracer studies in marine microcosms. D, 1983, Columbia University, Teachers College. 500 p.

Amdurer, Michael. Geochemistry, hydrology, and mineralogy of the Laguna Madre Flats, South Texas. M, 1978, University of Texas, Austin.

Amegee, Kodjo Yahwondu. Application of geostatistics to regional evapotranspiration. D, 1985, Oregon State University. 176 p.

Amell, Alexander Renton. Solution mining of sedimentary uranium deposits; factors influencing the solution rate of uranium dioxide under conditions applicable to in situ leaching. M, 1979, Pennsylvania State University, University Park. 124 p.

Amendolagine, Emanuel. Uranium mineralization and mapping of Cayzor Athabaska Mines Ltd., Uranium City, Saskatchewan, Canada. M, 1956, Columbia University, Teachers College.

Amenta, Roddy V. Geology of the Northwest Branch of the Anacostia River, Montgomery County, Maryland. M, 1966, George Washington University.

Amenta, Roddy Vincent. The structural and metamorphic history of the Wissahickon Formation (lower Paleozoic?) Montgomery and Philadelphia counties, Pennsylvania. D, 1970, Bryn Mawr College. 42 p.

Amer, Hamed H. A study of the change in effective permeability of oil sands to water of different chemical compositions. M, 1948, Stanford University. 122 p.

Amer, Mohamed Ibrahim Mohamed. Sample size effect on dynamic properties of sand in the simple shear test. D, 1984, University of Maryland. 772 p.

Amerigian, C. Factors controlling the acquisition of primary and secondary magnetizations in sediments. D, 1977, University of Rhode Island. 209 p.

Amerman, Roger E. Geology of a portion of the Jeff Cabin Creek 7 1/2' Quadrangle, Caribou County, Idaho. M, 1987, Colorado School of Mines. 138 p.

Amerson, Michael Daniel. Depositional and diagenetic interpretation of the Sligo (Pettet) Formation, Pine Island Field, Caddo Parish, Louisiana. M, 1988, Northeast Louisiana University. 89 p.

Ames, D. E. Stratigraphy and alteration of gabbroic rocks near the San Antonio gold mine in the Rice Lake area, southeastern Manitoba. M, 1988, Carleton University. 200 p.

Ames, Edward W. Notes on the geology of the Forest of Dean Mine. M, 1918, Columbia University, Teachers College.

Ames, H. G. Certain fragmental rocks in the Porcupine area. M, 1940, Queen's University. 51 p.

Ames, Herbert Tate. Plant microfossils in a Colorado Cretaceous coal. M, 1951, University of Massachusetts. 81 p.

Ames, John Alfred. The geology of the Dawson Butte area, Castle Rock Quadrangle, Colorado. M, 1950, University of Illinois, Urbana. 225 p.

Ames, Lloyd Leroy, Jr. Analysis of fluid inclusion in minerals and their significance. D, 1956, University of Utah. 66 p.

Ames, Lloyd Leroy, Jr. Methods for chemical analyses of fluid inclusions in quartz. M, 1955, University of Utah. 62 p.

Ames, Martha H. Las Trampas, New Mexico; dendrochronology of a Spanish colonial church. M, 1972, University of Arizona.

Ames, Richard M. Geochemistry of the Grenville basement rocks from the Roseland District, Virginia. M, 1981, University of Georgia.

Ames, Roger Lyman. Sulfur isotopic study of the Tintic mining districts, Utah. D, 1962, Yale University. 179 p.

Ames, Vincent Eugene. Geology of McCarty Park and vicinity, Costilla and Huerfano counties, Colorado. M, 1957, Colorado School of Mines. 122 p.

Amick, David Carroll. Crustal structure studies in the South Carolina Coastal Plain. M, 1979, University of South Carolina.

Amick, Harold Clyde. Geology of Booker Creek Quadrangle. M, 1923, University of North Carolina, Chapel Hill. 19 p.

Amig, Bruce Clement. Lithofacies and paleoenvironment, Lower Pennsylvanian rocks, Kentucky State Route Eighty near the Rockcastle River. M, 1988, University of Kentucky. 99 p.

Amin, Isam Eldin. A general mathematical model for the interpretation of tracer data and calculation of transit times in hydrologic systems. D, 1987, University of Nevada. 90 p.

Amin, Magdy Ibrahim. The compartmention bed load trap and its use in comparing the bed load formulae. D, 1980, Cornell University. 362 p.

Amin, Mohammad. A nonstationary stochastic model for strong-motion earthquakes. D, 1966, University of Illinois. 125 p.

Amin, Surendra R. Heavy mineral study of the intrusive rocks (Tertiary) of the Antelope Range, Piute County, Utah. M, 1950, University of Utah. 40 p.

Amini, Farshad. Dynamic soil behavior under random excitation conditions. D, 1986, University of Maryland. 255 p.

Amini, Mohamed Karim. Structural cross sections of the Arkoma Basin, Arkansas. M, 1980, University of Arkansas, Fayetteville.

Amini, Mohammad-Hassan. Geochronology, paleomagnetism and petrology of the upper Cenozoic Bruneau Formation in the western Snake River plain, Idaho. D, 1983, University of Colorado. 236 p.

Aminian, Khashayar. The synergetic study of Silurian-Niagaran pinnacle reef belt around the Michigan Basin for exploration and production of oil and gas. D, 1982, University of Michigan. 568 p.

Amisial, Roger A. Gabriel. Analog computer solution of the unsteady flow equations and its use in modeling the surface runoff process. D, 1969, University of Utah. 189 p.

Ammentorp, Alan David. Depositional systems, petrography and petroleum geology of a Caddo Conglomerate (Atokan) wave-reworked braid delta in north-central Texas. M, 1988, Oklahoma State University. 141 p.

Ammentorp, Willis Fay. Thermoluminescence of anhydrite. M, 1957, University of Wisconsin-Madison. 57 p.

Ammer, Bobby R. Geology of the Hilda-Southwest area, Mason County, Texas. M, 1959, Texas A&M University.

Ammerman, Michael Lee. A gravity and tectonic study of the Rome Trough. M, 1976, University of Kentucky. 70 p.

Ammon, Charles James. Time domain teleseismic P waveform modeling and the crust and upper mantle structure beneath Berkeley, California. M, 1985, SUNY at Binghamton. 97 p.

Ammon, Walter L. Geology and plate tectonic history of the Marfa Basin, Presidio County, Texas. M, 1977, Texas Christian University. 44 p.

Amna, Kei. An aeromagnetic survey in the Valley of Ten Thousand Smokes, Alaska. M, 1971, University of Alaska, Fairbanks. 97 p.

Amonette, James Edward. The role of structural iron oxidation in the weathering of trioctahedral micas by aqueous solutions. D, 1988, Iowa State University of Science and Technology. 260 p.

Amor, Stephen Donald. Application of discriminant analysis to a study of geochemical dispersion around massive sulphide deposits in Superior Province. D, 1983, Queen's University. 404 p.

Amoruso, John Joseph. A structural study of the Northville oil field, Wayne, Washtenaw, and Oakland counties, Michigan. M, 1957, University of Michigan.

Amos, Dewey Harold. Diking and its structural control in the Littleton Formation in New Hampshire. M, 1950, University of Illinois, Urbana. 37 p.

Amos, Dewey Harold. Geology and petrology of the Calais and Robbinston quadrangles, Maine. D, 1958, University of Illinois, Urbana. 251 p.

Amr, Amr M. Sedimentological activity in the drainage basin of Reelfoot Lake, Tennessee. M, 1987, Eastern Kentucky University. 66 p.

Amrhein, Christopher. The effect of an exchanger phase, carbon dioxide, and mineralogy on the rate of geochemical weathering. D, 1984, Utah State University. 139 p.

Amringe, John Howard Van see Van Amringe, John Howard

Amrstrong, H. S. The genus Stigmatella as represented in the Ordovician of the central Ontario Basin. M, 1939, University of Toronto.

Amsbury, David Leonard. Geology of Pinto Canyon area, Presidio County, Texas. D, 1957, University of Texas, Austin.

Amsden, Thomas W. Stratigraphy and palaeontology of the Brownsport Formation (Silurian) of western Tennessee. D, 1947, Yale University.

Amsden, Thomas William. Ordovician conodonts from the Bighorn Mountains of Wyoming. M, 1941, University of Iowa. 17 p.

Amster, Andrew L. Subsurface stratigraphy of carbonate bioherms and biostromes within the lower Fort Payne Formation (Lower Mississippian) of South-central Kentucky. M, 1982, University of Kentucky. 143 p.

Amstutz, Platte T., Jr. An investigation in the control of water production from the Arbuckle Dolomite wells in western Kansas. M, 1943, University of Kansas. 36 p.

Amukan, Samuel E. Petrography of the gold-bearing vein rocks (Archean) from Bissett area, south eastern Manitoba. M, 1970, University of Manitoba.

Amundson, Burton. Hamilton stratigraphy of southern Indiana. M, 1957, University of Wisconsin-Madison. 37 p.

Amundson, Ronald Gene. A chronosequential evaluation of the effects of reclamation on a saline-sodic soil. D, 1984, University of California, Riverside. 159 p.

Amy, Gary Lee. Contamination of groundwater by organic pollutants leached from in situ spent shale. D, 1978, University of California, Berkeley. 311 p.

Amy, Vincent Peter. Effect of strain rate on the strength of brittle rocks. M, 1963, Columbia University, Teachers College.

Anagnos, Thalia. A stochastic earthquake recurrence model with temporal and spatial dependence. D, 1985, [Stanford University]. 149 p.

Anagnostos, Nicholas. The comparison of crude oil levels between Newark Bay and Great Bay. M, 1984, Montclair State College. 19 p.

Anan, Fayez Shaban. Petrology and paleocurrent study of the Dagger Flat (Cambrian), Marathon Basin, Texas. M, 1965, University of Texas, Austin.

Anan, Fayez Shaban. Provenance and statistical parameters of sediments of the Merrimack Embayment, Gulf of Maine. D, 1971, University of Massachusetts. 391 p.

Anan-Yorke, Rowland. A microfaunal study of the Bearpaw Formation (upper Cretaceous), Lethbridge area, Alberta (Canada). M, 1969, University of Alberta. 129 p.

Anan-Yorke, Rowland. Devonian Chitinozoa and Acritarcha from exploratory oil wells on the shelf and coastal region of Ghana, West Africa. D, 1973, University of Alberta. 324 p.

Ananaba, Simon Enyinnah. An integrated profile interpretation of the gravity anomalies over the Benue Trough and the Younger granite province in Nigeria. M, 1976, Pennsylvania State University, University Park. 67 p.

Anandarajah, Annalingham. In situ prediction of stress-strain relationships of clays using a bounding surface plasticity model and electrical methods. D, 1982, University of California, Davis. 127 p.

Anastasi, Frank S. An analysis of excursions and hydrogeologic testing methods at four in-situ uranium mines. M, 1984, University of Idaho. 191 p.

Anastasio, David John. Thrusting, halotectonics and sedimentation in the external Sierra, southern Pyrenees, Spain. D, 1987, The Johns Hopkins University. 247 p.

Anastasio, John. Stratigraphic paleontology within the Jamesville limestone formation in central New York. M, 1954, Syracuse University.

Anazalone, Salvatore A. Geology of the Inspiration and Silver-Ore Mine, Coeur d'Alene District, Idaho. M, 1956, Columbia University, Teachers College.

Ancieta, Hugo Alfonso. Observations on the stratigraphy through Cave Point, California. M, 1959, Stanford University.

Ander, Holly Dockery. Rotation of late Cenozoic extensional stresses, Yucca Flat region, Nevada Test Site, Nevada. D, 1984, Rice University. 145 p.

Ander, Mark Embree. Geophysical study of the crust and upper mantle beneath the central Rio Grande Rift and adjacent Great Plains and Colorado Plateau. D, 1980, University of New Mexico. 205 p.

Anderegg, Charles. Petrology and cementation of the Tuscarora Sandstone. M, 1955, West Virginia University.

Anderegg, Fred. The stratigraphy and paleontology of the Yorktown Formation of Virginia. M, 1930, University of Virginia. 107 p.

Anderhalt, Robert W. Stratigraphy of the Crowder Formation and the Santa Ana Sandstone, San Bernardino Mountains, California. M, 1976, University of California, Los Angeles.

Anderhalt, Robert Walter. Beach foreshore sedimentation by organic and inorganic processes. D, 1981, University of California, Los Angeles. 208 p.

Anderhalt, Robert Walter. Sediment textural relationships, Southern California borderland. M, 1964, University of California, Los Angeles. 64 p.

Anderman, George Gibbs. Geology of a portion of the north flank of the Uinta Mountains in the vicinity of Manila, Summit and Doggett counties, Utah, and Sweetwater County, Wyoming. D, 1955, Princeton University. 658 p.

Anders, Fred John. Sand deposits as related to interactions of wind and topography in the Mojave Desert, near Barstow, California. M, 1974, University of Virginia. 66 p.

Anders, Mark Hill. Nature of microfracturing near naturally occurring faults. M, 1982, University of Michigan.

Anders, Nuni-Lyn E. Sawyer. Relation between clay extraction techniques and experimentally determined chemistry and mineralogy of clays; implications for interpretation of clay chemistry. M, 1988, California State University, Long Beach. 161 p.

Andersen, Allen E., Jr. Bedrock geology of part of the Goshen Quadrangle, Massachusetts. M, 1959, University of Massachusetts. 78 p.

Andersen, C. Brannon. Sedimentary gradients and depositional evolution of a high-energy lagoon; Snow Bay, San Salvador, Bahamas. M, 1988, Miami University (Ohio). 148 p.

Andersen, Christine L. Stratigraphy and petrology of Silurian rocks from the subsurface at Genoa, Ohio. M, 1980, Bowling Green State University. 87 p.

Andersen, D. and Spring, S. A geologic investigation of the Ferguson Pegmatite (South Dakota). M, 1957, South Dakota School of Mines & Technology.

Andersen, David John. Internally consistent solution models for Fe-Mg-Mn-Ti oxides. D, 1988, SUNY at Stony Brook. 214 p.

Andersen, David William. Sedimentology of the Duchesne River Formation (Eocene-Oligocene), northern Uinta Basin, northeastern Utah. D, 1973, University of Utah. 84 p.

Andersen, David William. Stratigraphy of the Duchesne River Formation (Eocene-Oligocene), northern Uinta Basin, northeastern Utah. M, 1973, University of Utah. 94 p.

Andersen, Harold V. Recent foraminiferal faunules from the Louisiana Gulf Coast. D, 1950, Louisiana State University.

Andersen, Henrik T. Analytic continuation in the interpretation of transient electromagnetic data. D, 1987, Colorado School of Mines. 116 p.

Andersen, Kristine Louise. Reflectance spectroscopy as a remote sensing technique for the identification of porphyry copper deposits. D, 1978, Massachusetts Institute of Technology. 240 p.

Andersen, Marvin John. Subsurface stratigraphy and sedimentation of the West Franklin-Cutler Limestone zones of southern Illinois. M, 1956, University of Illinois, Urbana. 32 p.

Andersen, R. L. Geology of the Playa San Felipe Quadrangle, Baja California, Mexico. M, 1973, [University of California, San Diego].

Andersen, Richard L. The geology and slope stability analysis of the Mitchell Canyon landslide, Mount Diablo State Park, California. M, 1974, [University of New Mexico].

Andersen, Rodney E. The taxonomy, biostratigraphy, and paleoecology of a group of Pennsylvanian eurypterids from Kansas. M, 1974, Wichita State University. 111 p.

Andersen, Sarah Beale. Long-term changes (1930-32 to 1984) in the acid-base status of forest soils in the Adirondacks of New York. D, 1988, University of Pennsylvania. 534 p.

Andersland, Orlando B. The clay-water system and the shearing resistance of clays. D, 1960, Purdue University.

Anderson, Alan. The geochemistry, mineralogy and petrology of the Cross Lake pegmatite field, central Manitoba. M, 1984, University of Manitoba.

Anderson, Alfred L. Geology and mineral resources of eastern Cassia County, Idaho. D, 1931, University of Chicago. 169 p.

Anderson, Alfred Leonard. Geology and ore deposits of the Silver Hill District, Spokane County, Washington. M, 1923, University of Idaho. 100 p.

Anderson, Alfred Titus, Jr. A contribution to the mineralogy and petrology of the Brule Lake anorthosite massif, Quebec. D, 1963, Princeton University. 183 p.

Anderson, Amy Louise. Conodont biostratigraphy of the Lower Triassic Dinwoody Formation in the Meade Plate, southeastern Idaho. M, 1984, University of Wisconsin-Milwaukee. 86 p.

Anderson, Arthur Edward. Geology of the Coldspring area and clay mineralogy of the Fleming Formation, San Jacinto County, Texas. M, 1958, University of Texas, Austin.

Anderson, Arthur Erick. X-ray and electron microscope methods of studying the clay minerals in clay and shale formations. M, 1952, University of Nebraska, Lincoln.

Anderson, Arthur H. Laboratory formation of gypsum compared with natural occurrences. M, 1963, American University. 59 p.

Anderson, Arthur Taylor. Development of petroleum reservoirs in fractured rocks of the Monterey Formation, California. D, 1954, Stanford University. 148 p.

Anderson, Barbara. Nickel-zinc mineralization in the southern Oquirrh Mountains, Utah. M, 1959, University of Utah. 64 p.

Anderson, Barbara Jean. Methods of detecting compositional differences in some common zeolites. D, 1966, University of Utah. 192 p.

Anderson, Brent C. Use of temperature logs to determine intra-well-bore flows, Yucca Mountain, Nevada. M, 1987, Colorado School of Mines. 59 p.

Anderson, Brian D. Basin margin facies variations of the Trinity Group, central Texas. M, 1982, Baylor University. 117 p.

Anderson, Brooks D., II. Geology of the western half of the San Juan de Guadalupe Quadrangle, southeastern Durango, Mexico. M, 1965, Bowling Green State University. 70 p.

Anderson, Burton R. A study of American petrified Calamites. D, 1954, Washington University.

Anderson, C. E. A water balance model for agricultural watersheds on deep loess soils. D, 1975, Kansas State University. 166 p.

Anderson, Carl W. Large scale velocity anisotropy and the Q-ellipsoid method. M, 1978, Michigan State University. 65 p.

Anderson, Charles Alfred. Surficial geology of the Fall City area, Washington. M, 1965, University of Washington. 70 p.

Anderson, Charles Alfred. The geology and ore deposits of the Engles and Superior mines, Plumas County, California. D, 1928, University of California, Berkeley. 85 p.

Anderson, Charles Peter. A study of the crystal chemistry of the rutile-tapiolite family. M, 1970, University of California, Riverside. 135 p.

Anderson, Chester Washington, III. Geomagnetic depth sounding and the upper mantle in the western United States. D, 1970, University of Alberta. 257 p.

Anderson, Christian A. Quantitative theory of the ion microprobe analyzer and its application to geology. D, 1970, University of California, Santa Barbara.

Anderson, Christian Donald. Telluric current surveys in Utah. D, 1966, University of Utah. 118 p.

Anderson, Christine Alexis. The crystal structure of weeksite. M, 1980, Pennsylvania State University, University Park. 82 p.

Anderson, Curtis A. Pleistocene geology of the Comstock-Sebeka area, west-central Minnesota. M, 1976, University of North Dakota. 111 p.

Anderson, Cynthia S. A new cation ordering pattern in amesite-2H$_2$. M, 1980, University of Wisconsin-Madison.

Anderson, D. H. An empirical estimate of the relative mobilities of the common rock-forming elements. M, 1956, Massachusetts Institute of Technology. 67 p.

Anderson, D. K. Recovering oil from an unconsolidated sand by means of combustion. M, 1951, University of Missouri, Rolla.

Anderson, D. V. Model seismology; a study of the reflection of acoustic pulses. D, 1951, University of Toronto.

Anderson, Dale Lewis. Timing and mechanism of formation of selected talc deposits in the Ruby Range, southwestern Montana. M, 1987, Montana State University. 90 p.

Anderson, Daniel Harvie. Uranium-thorium-lead ages of zircons and model lead ages of feldspars from the Saganaga, Snowbank, and Giants Range granites (Precambrian) of northeastern Minnesota. D, 1965, University of Minnesota, Minneapolis. 140 p.

Anderson, Darrell John. Marls in Salt Lake County (Utah) and low-fired marl brick. M, 1966, University of Utah. 229 p.

Anderson, David Lawrence. A search for solutions to conflicting demands of outdoor recreation in the Oregon dunes coastal environment. D, 1974, Oregon State University. 191 p.

Anderson, David Lee. Phosphate rock dissolution in soil. D, 1981, University of Wisconsin-Madison. 114 p.

Anderson, David V. Model seismology; a study of reflection of acoustic pulse. D, 1950, University of Toronto.

Anderson, Don Lynn. Surface wave dispersion in layered anisotropic media. D, 1962, California Institute of Technology. 95 p.

Anderson, Don Randolph. Moodys Branch Formation (Eocene) in Gulf Coastal Plain; a model for transgressive marine sedimentation. D, 1971, Texas A&M University. 171 p.

Anderson, Don Randolph. Some middle Tertiary foraminifera from the subsurface of Toledo Bend, Louisiana. M, 1965, Louisiana State University.

Anderson, Donald G. Dynamic modulus of cohesive soils. D, 1974, University of Michigan. 332 p.

Anderson, Donald Thomas. The distribution of copper and nickel in magmatic sulphides. M, 1959, University of Manitoba.

Anderson, Donald Thomas. The distribution of nickel, copper, and cobalt in igneous rocks and in synthetic sulphide-silicate metals. D, 1963, University of Manitoba.

Anderson, Donna Schmidt. Provenance and tectonic implications of mid-Tertiary nonmarine deposits, Santa Maria Basin and vicinity, California. M, 1980, University of California, Los Angeles.

Anderson, Douglas B. Stratigraphy and depositional history of the Deadwood Formation (Upper Cambrian and Lower Ordovician), Williston Basin, North Dakota. M, 1988, University of North Dakota. 330 p.

Anderson, E. H. An investigation of the Martian atmospheric haze. D, 1977, University of Washington. 207 p.

Anderson, E. R. Propagation and interaction of pressurized cracks in photoelastic gelatin. M, 1978, Stanford University. 40 p.

Anderson, E. R. Stratigraphy and petrology of Rome-Shady (Lower Cambrian) transition zone in southwestern Virginia. M, 1968, Virginia Polytechnic Institute and State University.

Anderson, Edwin E. Petrography and petrology of some bed-rock types in Becker and Ottertail counties, Minnesota. M, 1957, University of Minnesota, Minneapolis. 47 p.

Anderson, Edwin Joseph. Paleoenvironments of the Coeymans Formation (Lower Devonian) of New York State. D, 1967, Brown University. 192 p.

Anderson, F. D. Areal geology of the Woodstock map-area, 1 inch to 1 mile, and Millville areas, New Brunswick. D, 1956, McGill University.

Anderson, F. D. The McDougall-Segur Conglomerate (Lower Cretaceous, Alberta). M, 1951, McGill University.

Anderson, Frank J. Quantitative factors and distribution of low density soils in the Tucson, Arizona vicinity. M, 1968, University of Arizona.

Anderson, Frank Marion. Lower Cretaceous deposits in California and Oregon. D, 1930, Stanford University. 339 p.

Anderson, Frank Marion. The geology of Point Reyes Peninsula. M, 1897, University of California, Berkeley.

Anderson, Franklin Joseph. Geology of the Buffalo Fork Mountain area, Wyoming. M, 1940, University of Iowa. 58 p.

Anderson, Franz E. Changes in textural and compositional properties of a beach sediment near the mouth of the Cape Fear River, North Carolina. M, 1962, Northwestern University.

Anderson, Franz Elmer. Stratigraphy of late Pleistocene and Holocene sediments from the Strait of Juan de Fuca (Between Washington and Canada). D, 1967, University of Washington. 168 p.

Anderson, G. M. A study of lead sulfide solubility in aqueous solutions of hydrogen sulfide and its geological significance. D, 1962, University of Toronto.

Anderson, G. M. Regional structure in the Precambrian of Manitoba. M, 1956, University of Toronto.

Anderson, G. Witt. Northwest Power Planning Council's Fish and Wildlife Program and the water budget; analysis and implications. M, 1983, University of Idaho. 97 p.

Anderson, Garry E. Geology of the southern two-thirds of Volcano Butte and Hoover Springs quadrangles, Meagher County, Montana. M, 1986, Montana College of Mineral Science & Technology. 69 p.

Anderson, Garry Gayle. A quantitative analysis of selected ostracode species from the Des Moines Series (Pennsylvanian) of Iowa. D, 1971, Indiana University, Bloomington. 177 p.

Anderson, Garth S. Surface geology of land use planning, Minot, North Dakota area. M, 1980, University of North Dakota. 83 p.

Anderson, Gary Dale. Distribution of copper and zinc in ground water in a selected area of known mineralization, Cabarrus and Stanley counties, North Carolina. M, 1972, University of North Carolina, Chapel Hill. 43 p.

Anderson, Gary Swen. Environment and distribution of thermal relief features in the northern foothills section, Alaska. M, 1963, Iowa State University of Science and Technology.

Anderson, George A., III. An investigation of the possibility of cryptic layering and the implications of garnet coronas in the Jay Mountain layered metagabbro. M, 1971, Rensselaer Polytechnic Institute. 82 p.

Anderson, George B. The observation of gravity at sea. M, 1962, Ohio State University.

Anderson, George H. Geology of the north half of the White Mountain Quadrangle; Vol. 1, General geology; Vol. 2, Petrography. D, 1933, California Institute of Technology. 414 p.

Anderson, George H. Stratigraphy and faunal relationships of Pliocene beds of San Diego age in the vicinity of Las Llajas Canyon, Simi Valley. D, 1933, California Institute of Technology. 58 p.

Anderson, Gerald D. Geophysical investigation in southern Pope and Massac counties, Illinois. M, 1970, Southern Illinois University, Carbondale. 157 p.

Anderson, Gerald Edward. Copper-nickel mineralization at the base of the Duluth Gabbro. M, 1956, University of Minnesota, Minneapolis. 74 p.

Anderson, Gerald J. Mineralogy and occurrence of some commercial clays near Mayfield, Kentucky. M, 1951, University of Wisconsin-Madison.

Anderson, Gerald K. The micropaleontology of the Walnut Formation in West Texas. M, 1950, Texas Tech University. 59 p.

Anderson, Gerald Kenneth. Geology of Crolund Island, Lake of the Woods, Ontario. M, 1958, University of Iowa. 129 p.

Anderson, Gery F. An earthquake hazard study of a dune sand site in San Francisco (San Francisco County), California. M, 1969, San Jose State University. 78 p.

Anderson, Glenn Richard. A stratigraphic and tectonic analysis and synthesis of a portion of the non-marine Neogene of the Northwestern United States. M, 1980, University of Washington. 157 p.

Anderson, Gordon John. Distribution patterns of Recent foraminifera of the Arctic and Bering seas. M, 1961, University of Southern California.

Anderson, Gregg R. Petrology and occurrence of oolite bodies in the Ste. Genevieve Formation (Mississippian) Massac, Union and Marion counties, Illinois. M, 1976, Southern Illinois University, Carbondale. 88 p.

Anderson, Gustavus E. An experimental study of the mechanical wear of sand grains. D, 1925, University of Chicago. 57 p.

Anderson, Gustavus E. Studies in certain Paleozoic corals; on the origin and development of the inner wall. M, 1906, Columbia University, Teachers College.

Anderson, Harold Dean. The geochemistry and mineralogy of the Pennsylvanian clays of Mahaska County, Iowa. D, 1970, University of Iowa. 73 p.

Anderson, Henry Robert. The ground-water geology of the Rahway, New Jersey area. M, 1961, University of Iowa. 78 p.

Anderson, Henry W., Jr. The conodont fauna of the lower part of the Decorah Shale in the upper Mississippi River valley. M, 1959, University of Minnesota, Minneapolis. 32 p.

Anderson, Howard T. Head and neck musculature of phytosaurs. M, 1935, University of California, Berkeley. 43 p.

Anderson, I. Carl. Geology and petrology of the ultramafics of Lake Victor, Horseshoe Lake Quadrangle, Bridger Wilderness, Wind River Mountains, western Wyoming. M, 1986, Idaho State University. 129 p.

Anderson, Irvin J. The geology of an area along Shoal Creek north of Hancock Drive, Austin, Texas, to State Highway 29. M, 1946, University of Texas, Austin.

Anderson, J. D. Igneous and metamorphic rocks from the Andes of central Peru. D, 1928, The Johns Hopkins University.

Anderson, J. L. Petrology and geochemistry of the Wolf River Batholith. D, 1975, University of Wisconsin-Madison. 317 p.

Anderson, J. L. Water movement through some apedal and pedal soil materials. D, 1976, University of Wisconsin-Madison. 282 p.

Anderson, J. Lawford. Petrologic study of a migmatite-gneiss terrain in central Wisconsin and the effect of biotite-magnetite equilibria on partial melts in the granite system. M, 1972, University of Wisconsin-Madison.

Anderson, James Arthur. Geochemistry of manganese oxides, a guide to sulfide ore deposits. D, 1965, Harvard University.

Anderson, James E. Stratigraphy and depositional environments of part of the Cretaceous Mesaverde Formation, north of Sinclair, Wyoming. M, 1967, University of Wyoming. 89 p.

Anderson, James Howard. Depositional facies and carbonate diagenesis of the downslope reefs in the Nisku Formation (U. Devonian), central Alberta, Canada. D, 1985, University of Texas, Austin. 412 p.

Anderson, James J. Identification of tracing of water masses with an application near the Galapagos Island. D, 1977, University of Washington. 144 p.

Anderson, James Lee. The stratigraphy and structure of the Columbia River Basalt in the Clackamas River drainage. M, 1978, Portland State University. 136 p.

Anderson, James Lee. The structural geology and ages of deformation of a portion of the Southwest Columbia Plateau, Washington and Oregon. D, 1987, University of Southern California. 283 p.

Anderson, James M. The origin of rare earth, thorium, and uranium mineralization in the northern Tendoy Mountains, Beaverhead County, Montana. M, 1981, Western Washington University. 101 p.

Anderson, James P. A geological and geochemical study of the southwest part of the Black Rock Desert and its geothermal area; Washoe, Pershing, and Humboldt counties, Nevada. M, 1977, Colorado School of Mines. 86 p.

Anderson, James Rodney. The polymetamorphic sequence in the Paleozoic rocks of northern Vermont; a new approach using metamorphic veins as petrologic and structural markers. D, 1977, California Institute of Technology. 679 p.

Anderson, James W. Radiometric study of selected middle Precambrian quartzite-conglomerate units in northern Wisconsin. M, 1979, University of Wisconsin-Milwaukee.

Anderson, Jane L. Palynology of the Lower Kittanning (No. 5) Coal (Pennsylvanian) of eastern Ohio. M, 1985, Kent State University, Kent. 130 p.

Anderson, Jay Earl, Jr. Geology of the Green Mountain area, Llano County, Texas. M, 1960, University of Texas, Austin.

Anderson, Jay Earl, Jr. Igneous geology of the central Davis Mountains, Jeff Davis County, Texas. D, 1965, University of Texas, Austin. 212 p.

Anderson, Jerome E. Geology and geomorphology of the Santo Domingo basin, Sandoval and Santa Fe counties, New Mexico. M, 1960, University of New Mexico. 110 p.

Anderson, Jerry Myron. A study of the possible correlation of strain, tilt and earthquakes. M, 1972, University of Utah. 89 p.

Anderson, John B. Marine geology of the Weddell Sea (Antarctica). D, 1972, Florida State University.

Anderson, John B. Structure and stratigraphy of the western margin of the Nacimiento uplift, New Mexico. M, 1970, University of New Mexico. 44 p.

Anderson, John G. Studies of strong ground motion and the rupture process of earthquakes. D, 1975, Columbia University. 136 p.

Anderson, John H. Hydrogeology of reservoir sites in the Meramec River basin, Missouri. M, 1963, University of Missouri, Rolla.

Anderson, John J. Bedrock geology of Antarctica. M, 1962, University of Minnesota, Minneapolis. 234 p.

Anderson, John Jerome. Geology of northern Markagunt Plateau, Utah. D, 1965, University of Texas, Austin. 218 p.

Anderson, John R. The gastropod genera Donaldina, Orthonema, and Streptacis in the Pennsylvanian System of the Appalachian Basin. M, 1981, Bowling Green State University. 102 p.

Anderson, John Robert, II. Gastropod biostratigraphy and biofacies of the Upper Carboniferous (Pennsylvanian) System in the Appalachian (Dunkard) Basin. D, 1986, University of Pittsburgh. 366 p.

Anderson, Keith Elliott. Copper mineralization near Oxbow, Jefferson County, New York. M, 1947, University of Rochester. 85 p.

Anderson, Kenneth Clyde. The occurrence and orgin of pegmatites and miarolitic cavities in the Wichita Mountain system, Oklahoma. M, 1946, University of Oklahoma. 66 p.

Anderson, Kenneth Robert. Automatic processing of local earthquake data. D, 1978, Massachusetts Institute of Technology. 162 p.

Anderson, Kenneth Robert. The shallow structure of the Moon. M, 1972, Massachusetts Institute of Technology. 76 p.

Anderson, Kurt Soe. Sedimentology, sedimentary petrology, and tectonic setting of the lower Miocene Clallam Formation, northern Olympic Peninsula, Washington. M, 1985, Western Washington University. 135 p.

Anderson, L. Marlow. The stratigraphy of the Fredericksburg Group, East Texas Basin. M, 1987, Baylor University. 284 p.

Anderson, L. W. Rates of cirque glacier erosion and source of glacial debris, Pangnirtung Fiord area, Baffin Island, N.W.T., Canada. M, 1976, University of Colorado.

Anderson, Lance Christopher. The Pilot Knob hematite deposit, Pilot Knob, Southeast Missouri. M, 1976, University of Wisconsin-Madison.

Anderson, Laurie C. Ostracoda of the Pinecrest Sand Member, upper Tamiami Formation, Macasphalt pit mine, Sarasota, Florida. M, 1987, Bowling Green State University. 168 p.

Anderson, Linda Davis. Transport by soil waters of chromium and nitrate in solid tannery wastes. M, 1984, University of California, Santa Cruz.

Anderson, Lynn G. Partitioning of trace metals in glacial till, southern Michigan. M, 1986, Michigan State University. 44 p.

Anderson, Marilyn Lea. Santa Maria, Azores; lava petrogenesis and upper mantle geochemistry. M, 1983, University of Illinois, Chicago. 74 p.

Anderson, Max Leroy. Degradation of polychlorinated biphenyls in sediments of the Great Lakes. D, 1980, University of Michigan. 271 p.

Anderson, Michael B. The hydrogeology of the Neuse wastewater treatment site. M, 1975, North Carolina State University. 185 p.

Anderson, Nancy Louise. A geological and geophysical investigation of the "DW" Valley massive sulfide deposits, Delta District, east-central Alaska. M, 1982, University of Alaska, Fairbanks. 126 p.

Anderson, Neil L. An expanding spread seismic reflection survey across the Snake Bay-Kakagi Lake greenstone belt, Northwest Ontario. M, 1979, University of Manitoba.

Anderson, Neil L. An integrated geophysical/geological analysis of the seismic signatures of some Western Canadian Devonian reefs. D, 1986, University of Calgary. 333 p.

Anderson, Norman K. Recent sediments of Deep Inlet, Alaska; a fjord estuary. D, 1962, University of Washington. 152 p.

Anderson, Norman N. Geology of the Bob Smith Creek area, northern Portneuf Range, Bannock and Caribou counties, Idaho. M, 1978, Idaho State University. 44 p.

Anderson, Norman Roderick. Glacial geology, Mud Mountain, King County, Washington. M, 1954, University of Washington. 48 p.

Anderson, Norman Roderick. Upper Cenozoic stratigraphy of the Oreana, Idaho 15′ Quadrangle. D, 1965, University of Utah. 261 p.

Anderson, Orin J. Pleistocene geology and archeological problems of the Seistan Basin, Southwest Afghanistan. M, 1973, University of New Mexico. 105 p.

Anderson, Patricia Ann. A quantitative study of the groundwater flow in the Upper East Coast Planning Area of Florida. M, 1985, Ohio University, Athens. 472 p.

Anderson, Paul. Determination of ocean borehole horizontal seismic sensor orientation at DSDP Site 581. M, 1987, University of Hawaii. 81 p.

Anderson, Paul Bradley. Geology and coal resources of the Pine Canyon Quadrangle, Carbon County, Utah. M, 1978, University of Utah. 143 p.

Anderson, Paul Gordon. Structural and lithological controls on the formation of gold bearing veins at the Erickson gold mine, north-central British Columbia. M, 1986, Queen's University. 146 p.

Anderson, Paul Leon. Bloating clays, shales, and slates for lightweight aggregate, Salt Lake City and vicinity, Utah. M, 1960, University of Utah. 20 p.

Anderson, Paul Ralph. Bulk and surface characteristics in mixed oxide suspensions. D, 1988, University of Washington. 209 p.

Anderson, Paul Victor. Pre-batholithic stratigraphy of the San Felipe area, Baja California, Mexico. M, 1982, San Diego State University. 100 p.

Anderson, Peter. Geology of the Quien Sabe Valley 7 1/2 minute Quadrangle, San Benito County, California. M, 1984, San Jose State University. 89 p.

Anderson, Peter Ascroft MacKenzie. Phase relations of the hydroxyl feldspathoids in the system NaAlSiO$_4$-NaOH-H$_{20}$O. D, 1968, McMaster University. 78 p.

Anderson, Peter Ascroft MacKenzie. The system NaAlSiO$_4$-NaOH-H$_2$O at 1 kilobar and 440°C to 540°C. M, 1963, McMaster University. 55 p.

Anderson, Peter Wilfred William. Deterministic stream-quality model of oxygen resources in the Manasquan River basin, New Jersey. D, 1978, Rutgers, The State University, New Brunswick. 259 p.

Anderson, Philip J. A study of some of the geochemical processes of rock weathering. M, 1961, Massachusetts Institute of Technology. 72 p.

Anderson, Phillip. The Proterozoic tectonic evolution of Arizona. D, 1986, University of Arizona. 445 p.

Anderson, R. S., Jr. Paleo-oceanography of the Mediterranean sea; some consequences of the Wurm glaciation. M, 1965, United States Naval Academy.

Anderson, Randall L. Hydrogeologic study of the Deer Park, Washington, aquifer system. M, 1986, Eastern Washington University. 80 7 plates p.

Anderson, Randall L. The origin and significance of photolineaments in southeastern Nebraska. M, 1981, University of Nebraska, Lincoln.

Anderson, Raymond R. Environmental geology and land-use analysis, Lake MacBride area, northern Johnson County, Iowa. M, 1975, University of Iowa.

Anderson, Richard. Pebble and sand lithology of the major Wisconsin glacial lobes of the central lowland. D, 1955, University of Chicago. 34 p.

Anderson, Richard C. A gravity survey of the Rio Grande Valley near Socorro, New Mexico. M, 1953, New Mexico Institute of Mining and Technology. 32 p.

Anderson, Richard E. Geology and micropaleontology of Point Loma, San Diego County, California. M, 1962, University of Southern California.

Anderson, Richard G. Sand budget for Capitola Beach. M, 1971, Naval Postgraduate School.

Anderson, Richard Garland. The bore hole compensated amplitude log. M, 1983, University of Texas, Austin. 220 p.

Anderson, Richard J. A study of the silver ores of the Sunshine Mine, Coeur d'Alene District, Idaho. M, 1938, Columbia University, Teachers College.

Anderson, Richard Kent. Organic geochemistry of an oil and gas seep in northern Gulf of Mexico sediments. D, 1984, University of Texas, Austin. 147 p.

Anderson, Richard Kime. Rubidium-strontium age determinations from the Churchill Province of northern Manitoba. M, 1974, University of Manitoba.

Anderson, Richard L. Reef structures in the Louisville Limestone (Silurian) in Bullitt County, Kentucky. M, 1980, Eastern Kentucky University. 48 p.

Anderson, Richard Lee. The Denali Fault (Hines Creek Strand) in the Wood River area, central Alaska Range. M, 1973, University of Wisconsin-Madison.

Anderson, Richard Mark. Cordierite-mullite composites; a study of their mechanical, thermal and dielectric properties. D, 1987, Rutgers, The State University, New Brunswick. 175 p.

Anderson, Robert. Nonlinear induced polarization spectra. M, 1981, University of Utah. 159 p.

Anderson, Robert C. Lineament analysis and tectonic interpretation for the central Tharsis region, Mars. M, 1985, Old Dominion University. 132 p.

Anderson, Robert C. Northern termination of the Massabesic Gneiss, New Hampshire. M, 1978, Dartmouth College. 111 p.

Anderson, Robert Frederick. The marine geochemistry of thorium and protactinium. D, 1981, Massachusetts Institute of Technology. 287 p.

Anderson, Robert G. Identification of some of the chlorite minerals. M, 1949, University of Utah. 18 p.

Anderson, Robert Gordon. Geology of the Hotailuh Batholith and surrounding volcanic and sedimentary rocks, north-central British Columbia. D, 1983, Carleton University. 669 p.

Anderson, Robert H. Geophysical determination and computer modelling of ground water flux through lake sediments; Lake St. Clair, Michigan/Ontario. M, 1987, University of Wisconsin-Milwaukee. 151 p.

Anderson, Robert John. Upper Mississippian and Lower Pennsylvanian formations of Bridger Moun-

tains, Montana. M, 1957, University of Wisconsin-Madison. 53 p.

Anderson, Robert Matthews. Geology of part of the lower Piru Creek area, Ventura and Los Angeles counties, California. M, 1960, University of California, Los Angeles.

Anderson, Robert Neil. Thermodynamics of nitride reactions in molten uranium-tin alloys and applications to nuclear fuel reprocessing. D, 1969, Stanford University. 379 p.

Anderson, Robert Stewart. A biography of Clarence Edward Dutton (1841-1912); nineteenth century geologist and geographer. M, 1978, Stanford University. 126 p.

Anderson, Robert Stewart. Sediment transport by wind; saltation, suspension, erosion and ripples. D, 1986, University of Washington. 174 p.

Anderson, Robert W. Sedimentologic and stratigraphic relationships of the Pennington Formation (Upper Mississippian) and the lower tongue of the Breathitt Formation (Lower Pennsylvanian) in Rowan and Menifee counties, Kentucky. M, 1977, Eastern Kentucky University. 97 p.

Anderson, Robert Warner. Geology of the northwest quarter of the Brownsville Quadrangle, Oregon. M, 1963, University of Oregon. 62 p.

Anderson, Rodney Scott. Late-Quaternary environments of the Sierra Nevada, California. D, 1987, University of Arizona. 304 p.

Anderson, Roger N. A geophysical study of the eastern equatorial Pacific. M, 1971, University of Oklahoma. 119 p.

Anderson, Roger Neeson. The implications of topography, gravity, and heat flow on midocean ridges. D, 1973, University of California, San Diego. 122 p.

Anderson, Roger Y. Uranium accumulation in plants. M, 1955, University of Arizona.

Anderson, Roger Yates. Cretaceous-Tertiary palynology of the eastern side of the San Juan Basin, New Mexico. D, 1960, Stanford University. 167 p.

Anderson, Ronald E. Northern Minnesota crustal refraction profile. M, 1978, University of Minnesota, Duluth.

Anderson, Roy Arnold. Fusulinids of the Granite Falls Limestone and their stratigraphic significance. M, 1936, Washington State University. 24 p.

Anderson, Roy Arnold. Structural control of the ore deposits in the Coeur d'Alene District, Idaho. D, 1971, University of Idaho. 61 p.

Anderson, Roy Arnold, Jr. Stability of slopes in clay shales interbedded with Columbia River Basalt. M, 1971, University of Idaho. 297 p.

Anderson, Roy Ernest. Geology of lower Bass Creek Canyon, Bitterroot Range, Montana. M, 1959, University of Montana. 70 p.

Anderson, Roy Ernest. Igneous petrology of the Taum Sauk area, Missouri. D, 1962, Washington University. 116 p.

Anderson, Rudolph F. A subsurface study of the Hunton Formation in central Oklahoma. M, 1939, University of Oklahoma. 30 p.

Anderson, Ruth Anne. Sedimentology of Lower Cretaceous strata near Redding, California. M, 1986, University of California, Davis. 201 p.

Anderson, S. R. Earth fissures in the Stewart area of the Willcox Basin, Cochise County, Arizona. M, 1978, University of Arizona.

Anderson, Scott A. Landslide evolution, Green Mountain reservoir, Colorado. M, 1988, Colorado State University. 111 p.

Anderson, Stanley Wayne. Structural analysis of the McCarthy Mountain Stock, Madison County, Montana. M, 1973, Indiana University, Bloomington. 136 p.

Anderson, Stephen Robert. Stratigraphy and structure of the Sunrise Peak area south of Brighton, Utah. M, 1974, Brigham Young University. 150 p.

Anderson, Susan Jean. Lithology and lithofacies of the Fort Payne Formation (Lower Mississippian) in central Tennessee. M, 1981, Memphis State University. 83 p.

Anderson, Susan Leslie. Investigation of the Mesa earth crack, Arizona, attributed to differential subsidence due to groundwater withdrawal. M, 1973, Arizona State University. 111 p.

Anderson, T. C. I. Geology of part of Northwest San Juan County, New Mexico. M, 1951, University of California, Los Angeles.

Anderson, Thane Wesley. Postglacial vegetative changes in the Lake Huron-Lake Simcoe District, Ontario with special reference to Glacial Lake Algonquin. D, 1971, University of Waterloo.

Anderson, Thomas. A fission track investigation of portions of the Pine Mt. Belt, and inner Piedmont in W. central Georgia. M, 1981, Florida State University.

Anderson, Thomas B. Stratigraphy of the Cambrian and Ordovician rocks of the southern Mosquito Range (Lake, Park and Chaffee counties), Colorado. M, 1965, University of Colorado.

Anderson, Thomas Bertram, III. Mineralogy and geochemistry of Recent carbonate sediments, Timor sea, Australia. D, 1969, University of Colorado. 164 p.

Anderson, Thomas C. Compound faceted spurs and recurrent movement in the Wasatch fault zone, north central Utah. M, 1977, Brigham Young University. 101 p.

Anderson, Thomas F. The measurement of surface area and of the self-diffusion rates of carbon and oxygen in calcite by isotopic exchange with carbon dioxide. D, 1967, Columbia University. 170 p.

Anderson, Thomas Howard. Geology of the middle third of La Democracia Quadrangle, Guatemala. M, 1967, University of Texas, Austin.

Anderson, Thomas Howard. Geology of the San Sebastian Huehuetenango Quadrangle, Guatemala, Central America. D, 1969, University of Texas, Austin. 277 p.

Anderson, Thomas P. A geochemical study of the Miocene age Conejo Volcanics, Santa Monica Mountains, Los Angeles County, California. M, 1980, California State University, Northridge.

Anderson, Thomas P. Geology of the Turkey Creek-Strains Gulch area, Jefferson County, Colorado. M, 1949, Colorado School of Mines. 71 p.

Anderson, Thornton Earl. Geology of the Boggs Field, Barber County, Kansas. M, 1949, University of Wisconsin-Madison. 37 p.

Anderson, Timothy Allan. Geology of the lower Tertiary Gualanday group, upper Magdalena valley, Colombia. D, 1970, Princeton University. 127 p.

Anderson, Timothy D. Magnetic investigations of the Baraga County diabase, Baraga County, Michigan. M, 1980, Michigan Technological University. 160 p.

Anderson, V. H. Glaciological observations in Marie Byrd Land, Antarctica. M, 1959, University of Wyoming. 269 p.

Anderson, Virginia Ruth. Calcareous surface sediments of the U. S. Virgin platform. M, 1981, Duke University. 79 p.

Anderson, Walter A. Sedimentation study in the Pinnacle Range, Monroe County, New York. M, 1956, University of Rochester. 86 p.

Anderson, Warren LeGrande. Geology of the northern Silver Island Mountains, Box Elder and Tooele counties, Utah. M, 1957, University of Utah. 131 p.

Anderson, Wayne Irvin. Stratigraphy of the Phosphoria Formation, north-central Wyoming. M, 1961, University of Iowa. 199 p.

Anderson, Wayne Irvin. Upper Devonian and Lower Mississippian conodonts from north central Iowa. D, 1964, University of Iowa. 215 p.

Anderson, Wells Foster. Calcium sulphate in western New York and the Ontario Peninsula. D, 1930, University of Wisconsin-Madison.

Anderson, Wells Foster. The gypsum and anhydrite deposits at Oakfield, New York. M, 1929, University of Wisconsin-Madison.

Anderson, William B. Cooling history and uranium mineralization of the Buckshot Ignimbrite, Presidio and Jeff Davis counties, Texas. M, 1975, University of Texas, Austin.

Andersson, Kent Albert. Early lithification of limestones in the Redwater Shale Member of the Sundance Formation (Jurassic) of southeastern Wyoming. M, 1978, University of Wyoming. 74 p.

Andersson, Knut Albert. Permian trace fossils of western Wyoming and southwestern Montana; systematics, paleoenvironments and diagenesis. D, 1982, University of Wyoming. 263 p.

Anderton, Arlo Jo Payne. Geology of the Goldmine Creek area, Llano County, Texas. M, 1971, University of Texas, Austin.

Anderton, Lesley Jean. Quaternary stratigraphy and geomorphology of the lower Thompson Valley. M, 1970, University of British Columbia.

Anderton, Peter Wightman. Structural glaciology of a glacier confluence, Kaskawulsh Glacier, Yukon Territory, Canada. D, 1967, Ohio State University. 205 p.

Andes, Jerry Philip, Jr. Mineralogic and fluid inclusion study of ore-mineralized fractures in drillhole State 2-14, Salton Sea Scientific Drilling Project, California, U.S.A. M, 1987, University of California, Riverside. 125 p.

Ando, Clifford Joseph. Structural and petrologic analysis of the North Fork terrane, central Klamath Mountains, California. D, 1979, University of Southern California.

Andors, Allison Victor. Giant groundbirds of North America (Aves, Diatrymidae). D, 1988, Columbia University, Teachers College. 599 p.

Andrade Nery Leao, Zelinda Margarida de *see* de Andrade Nery Leao, Zelinda Margarida

Andrau, W. E. Geology of the West Pine Creek area, Bonneville County, Idaho. M, 1958, University of Wyoming. 64 p.

André, Richard A. Geochemical characterization of kerogens, shales and coal refuse associated with Allegheny Formation coals in northern West Virginia. M, 1987, West Virginia University. 332 p.

Andreae, Meinrat O. The marine biogeochemistry of arsenic. D, 1978, University of California, San Diego. 92 p.

Andren, Anders Wikar. The geochemistry of mercury in three estuaries from the Gulf of Mexico. D, 1973, Florida State University. 176 p.

Andres, Alan Scott. Geology and ground-water flow in the Potomac-Raritan-Magothy aquifer system, Logan Township, New Jersey. M, 1984, Lehigh University. 166 p.

Andres, Robert J. Sulfur dioxide and particle emissions from Mount Etna, Italy. M, 1988, New Mexico Institute of Mining and Technology. 159 p.

Andresen, Arild. Stratigraphy and structural history of the lower Paleozoic metasediments on Hardangervidda, South Norway. D, 1982, University of California, Davis. 260 p.

Andresen, Brian Dean. Identification of components in the neutral fraction of Green River Shale. M, 1972, Massachusetts Institute of Technology. 101 p.

Andresen, Marvin John. Geology and petrology of the Trivoli Sandstone (Pennsylvanian) in the Illinois Basin. D, 1960, University of Missouri, Columbia. 194 p.

Andress, Edward C. Neve studies of the Juneau icefield, Alaska, 1961 with special reference to glaciohydrology on the Lemon Glacier. M, 1962, Michigan State University. 174 p.

Andress, Noel Eugene. The distribution of foraminifera in the southeastern Gulf of Mexico. M, 1970, Florida State University.

Andretta, Daniel B. Geology of the Moose Creek Stock, Highland Mountains, Montana. M, 1961, Montana State University. 67 p.

Andrew, Anne. Lead and strontium isotope study of five volcanic and intrusive rock suites and related

mineral deposits, Vancouver Island, British Columbia. D, 1987, University of British Columbia.

Andrew, James Alexander. The structure and geology of Djebel Chambi, a Triassic evaporite cored anticline in central Tunisia. M, 1979, University of South Carolina.

Andrew, John Alexander. Sediment distribution in deep areas of the northern Kara Sea. D, 1973, University of Wisconsin-Madison.

Andrew, John Alexander. Size distribution of the sand and heavy minerals in the Ironton Sandstone (Upper Cambrian) of western Wisconsin. M, 1965, University of Wisconsin-Madison.

Andrew, Kathryn Pauline Elizabeth. Geology and genesis of the Wolf precious metal epithermal prospect and the Capoose base and precious metal porphyry-style prospect, Capoose Lake area, central British Columbia. M, 1988, University of British Columbia. 300 p.

Andrews, Alday Bishop. Frequency dependence of seismic wave attenuation. D, 1957, Pennsylvania State University, University Park. 105 p.

Andrews, Alday Bishop. Photomechanical wave analyzer for seismic wave analysis. M, 1954, Pennsylvania State University, University Park. 67 p.

Andrews, Anthony James. Petrology and geochemistry of alteration in Layer 2 basalts, DSDP Leg 37. D, 1978, University of Western Ontario. 327 p.

Andrews, Barbara Ann. A petrologic study of Weddell Sea sediments; implications for provenance and glacial history. M, 1984, Rice University. 203 p.

Andrews, C. B. M. The structure of southeastern portion of the Island of Oahu, Hawaiian Islands. M, 1909, Rose-Hulman Institute of Technology. 29 p.

Andrews, Charles Bryce. An analysis of the impact of a coal fired power plant on the groundwater supply of a wetland in central Wisconsin. M, 1976, University of Wisconsin-Madison.

Andrews, Charles Bryce. The simulation of groundwater temperature in shallow aquifers. D, 1978, University of Wisconsin-Madison. 306 p.

Andrews, Charles Hubert. Application of the Wenner resistivity method to the detection of buried shallow faults in the Gulf Coast Province. M, 1961, Texas A&M University. 64 p.

Andrews, D. A. The trilobites of the Kinderhookian of Missouri. M, 1928, University of Missouri, Columbia.

Andrews, Douglas James. The petrology, chemistry and sulphide mineralogy of the Dixie 150—17B, 18 and 19 pyritic copper-zinc-silver prospects, Red Lake, northwestern Ontario. M, 1979, University of Manitoba. 185 p.

Andrews, Edmund Daniel. Hydraulic adjustment of an alluvial stream channel to the supply of sediment. D, 1977, University of California, Berkeley. 161 p.

Andrews, Edward J. Profiles of gravity and magnetics across the Baraga Basin, in Michigan's Upper Peninsula. M, 1975, Bowling Green State University. 63 p.

Andrews, Edwin Eads, III. The source of channel sands in tidal creeks, North Island Quadrangle, South Carolina. M, 1973, University of South Carolina.

Andrews, Franklin. Structure and areal extent of the Colchester (No. 2) Coal in Adams, Brown, Schuyler and McDonough counties, Illinois. M, 1957, Florida State University.

Andrews, George W. The Windrow Formation of the Upper Mississippi Valley region; a sedimentary and stratigraphic study. D, 1955, University of Wisconsin-Madison.

Andrews, George William. Morphologic studies of the brachiopod genus Composita. M, 1953, University of Wisconsin-Madison. 63 p.

Andrews, Harold Edward. Middle Ordovician conodonts from the Joachim Formation of Eastern Missouri. M, 1966, University of Missouri, Rolla.

Andrews, Harold Edward, III. Turritella mortoni (Gastropoda) and biostratigraphy of the Aquia Formation (Paleocene) of Maryland and Virginia. D, 1972, Harvard University.

Andrews, James Einar. The Bahama Canyon system. D, 1967, University of Miami. 114 p.

Andrews, Joe A. The geometry of Government Creek, Union County, Illinois. M, 1974, Southern Illinois University, Carbondale. 77 p.

Andrews, L. A. Structure of the area north of Roanoke, Virginia. D, 1952, The Johns Hopkins University.

Andrews, Leslie W. Uraninite occurrence at Hazelton, British Columbia. M, 1951, University of Washington. 55 p.

Andrews, Mark Stephen. Contact metamorphism of the Virginia Formation at Dunka Road, Minnesota; chemical modifications and implications for ore genesis in the Duluth Complex. M, 1987, Indiana University, Bloomington. 93 p.

Andrews, Mary Catherine. The relocation of microearthquakes in the Mississippi Embayment. M, 1985, University of Wisconsin-Madison. 189 p.

Andrews, Peter Bruce. Facies and genesis of a hurricane washover fan, Saint Joseph Island, central Texas coast. D, 1967, University of Texas, Austin. 290 p.

Andrews, Peter W. Chemical characteristic of some volcanic rocks of the Superior Province of the Canadian Shield. M, 1964, University of Manitoba.

Andrews, Peter William. The effect of import quotas on the United States zinc industry. D, 1969, Pennsylvania State University, University Park. 202 p.

Andrews, Philip. Geology of the Pinnacles National Monument. M, 1933, University of California, Berkeley. 85 p.

Andrews, R. W. The digital simulation of areal salt transport to evaluate water management proposals in a coastal aquifer. D, 1979, University of Illinois, Urbana. 226 p.

Andrews, Ralf E., Jr. The surface geology of the Raiford area, McIntosh County, Oklahoma. M, 1957, University of Oklahoma. 52 p.

Andrews, Richard Duane, Jr. The geology and geochemistry of plutonic bodies in central Marathon County, Wisconsin. M, 1976, University of Wisconsin-Milwaukee.

Andrews, Robert Sanborn. Synthetic seismograms from marine sediment models. D, 1970, Texas A&M University. 176 p.

Andrews, Sarah. Physical modeling of controls on primary uranium ore deposition in alluvial fans. M, 1981, Colorado State University. 96 p.

Andrews, Thomas G. A study of the carbon ratios of the coals of northern Tennessee and their relation to oil and gas. M, 1927, Vanderbilt University.

Andrews, Thomas G. Insoluble residues as an aid in stratigraphic studies of limestones of central Tennessee. D, 1932, University of Minnesota, Minneapolis. 90 p.

Andrews, Wayne. Seismic velocity and facies analyses from multichannel seismic reflection data Exuma Sound, Bahamas. M, 1986, University of Delaware.

Andrews, William J. Nitrate occurrence in ground and surface waters, DeKalb County, Illinois. M, 1988, Northern Illinois University. 211 p.

Andriashek, Laurence Douglas. Quaternary stratigraphy of the Sand River area, NTS 73L. M, 1985, University of Alberta. 387 p.

Andricevic, Roko. Control of water resources systems under uncertainty. D, 1988, University of Minnesota, Minneapolis. 142 p.

Andrichuk, John Michael. Regional stratigraphic analysis of the Devonian System in Wyoming, Montana, southern Saskatchewan, and Alberta. D, 1951, Northwestern University. 341 p.

Andrichuk, John Michael. Stratigraphy of the area including Majeau Lake No. 1 Well, Edmonton area. M, 1949, University of Alberta. 129 p.

Andrievich, Ellen. Slope failure characteristics in two facies of the Jurong Formation, Singapore. M, 1986, Colorado State University. 66 p.

Andrisin, Mary E. A geologic field trip in Mercer County, Ohio. M, 1970, Virginia State University. 33 p.

Andronaco, Margaret. Hurricane effects and post-storm recovery, Pinellas County, Florida (1985-1986). M, 1987, University of South Florida, Tampa. 118 p.

Anepohl, Jane K. Seasonal distribution of living benthonic foraminifera of the South Texas outer continental shelf. M, 1976, University of Texas, Austin.

Anessi, Thomas Joseph. Strength and consolidation properties of raw and stabilized Oklahoma shales. D, 1970, University of Oklahoma. 199 p.

Anestad-Fruth, Elizabeth. Uranium series disequilibrium in Recent volcanic rocks. M, 1963, Columbia University, Teachers College.

Angel, Jose M. Mejia *see* Mejia Angel, Jose M.

Angel, Loren Henry. Geology of a portion of the St. Helena Quadrangle, California. M, 1948, University of California, Berkeley. 25 p.

Angelakis, Andreas Nikolaos. Transient movement and transformations of insoluble, soluble, and gaseous carbon in soil. D, 1981, University of California, Davis. 245 p.

Angelich, Michael Terry. Inverse spectral modeling of geopotential fields over sedimentary basins. D, 1985, Louisiana State University. 126 p.

Angelo, Michael V. De *see* De Angelo, Michael V.

Angeloni, Linda Marie. The role of water in the formation of granulite and amphibolite facies rocks, Tobacco Root Mountains, Montana. M, 1988, University of Montana. 51 p.

Angerman, Thomas Westley. A study of the Pennsylvanian Vanport Limestone in a portion of the Foxburg Quadrangle, Pennsylvania. M, 1955, University of Pittsburgh.

Angevine, Charles Leon. Thermal histories of sedimentary basins and compaction of sediments. D, 1983, Cornell University. 150 p.

Angino, Ernest Edward. Pressure-induced thermoluminescence as a geologic age determination method. M, 1959, University of Kansas.

Angino, Ernest Edward. The effects of nonhydrostatic pressures on radiation-damage thermoluminescence. D, 1961, University of Kansas. 80 p.

Angle, David G. Organic carbon in Amazon continental shelf sediments; an isotopic analysis. M, 1985, North Carolina State University. 54 p.

Angle, Michael Paul. Quaternary stratigraphy of part of Richfield Township, Summit County, Ohio. M, 1982, University of Akron. 155 p.

Angley, Joseph Timothy. An evaluation of the attenuation mechanisms for dissolved aromatic hydrocarbons from gasoline sources in a sandy surficial Florida aquifer. D, 1987, University of Florida. 338 p.

Anglin, Carolyn Diane. Geology, structure and geochemistry of gold mineralization in the Geraldton area, northwestern Ontario. M, 1987, Memorial University of Newfoundland. 283 p.

Anglin, F. M. Subsurface temperatures in Western Canada. M, 1964, University of Western Ontario.

Anglin, Marion Edward. The petrography of the bioherms of the St. Joe Limestone of northeastern Oklahoma. M, 1964, University of Tulsa. 109 p.

Angoran, Yed Esaie. Induced polarization of metallic minerals; a study of its chemical basis. D, 1976, Massachusetts Institute of Technology. 186 p.

Angstadt, David Moris. Late Cretaceous - Recent seismic stratigraphy and geologic history of the southeastern Gulf of Mexico/southwestern Straits of Florida. M, 1983, University of Texas, Austin. 206 p.

Angulo, Raul E. Major strike-slip faults of northern South America and their role in the tectonics of the region. M, 1972, Stanford University.

Anikouchine, William Alexander. A model of the distribution of dissolved species in interstitial water in clayey marine sediments. D, 1966, University of Washington. 52 p.

Anikouchine, William Alexander. Bottom sediments of Rongelap Lagoon, Marshall Islands. M, 1961, University of Washington. 137 p.

Aniku, Jacob Robert Francis. Trends of pedogenic iron oxides in a marine terrace chronosequence. D, 1986, University of California, Davis. 227 p.

Anisgard, Harry W. Ostracoda of the Vienna Limestone. M, 1939, Columbia University, Teachers College.

Aniya, M. Numerical analyses of glacial valleys and cirques in the Victoria Valley system, Antarctica, from photogrammetrically derived terrain data. D, 1975, University of Georgia. 313 p.

Anjos, Sylvia Maria Couto dos *see* Couto dos Anjos, Sylvia Maria

Anjos, Sylvia Maria Couto dos *see* dos Anjos, Sylvia Maria Couto

Ankenbauer, Gilbert A., Jr. Early Precambrian tonalites, Morton area, Minnesota. M, 1975, Northern Illinois University. 41 p.

Ankeny, Lee Andrew. A seismic study of the subsurface structure of the Jemez volcanic field, New Mexico. M, 1985, Purdue University. 61 p.

Ankeny, Mark Dwight. Characterization of soil macropores by infiltration measurements. D, 1988, Iowa State University of Science and Technology. 72 p.

Anna, Lawrence O. Geology of the Kirk Hill area, Lawrence-Meade counties, South Dakota. M, 1973, South Dakota School of Mines & Technology.

Annaki, M. Liquefaction of sand in triaxial tests using uniform and irregular cyclic loading. D, 1975, University of California, Los Angeles. 361 p.

Annambhotla, Venkata Subramanya Shastri. Statistical properties of bed forms in alluvial channels in relation to flow resistance. D, 1969, University of Iowa. 137 p.

Annan, A. P. Radio interferometry depth sounding. D, 1970, University of Toronto.

Annan, Alexander Peter. The equivalent source method for electromagnetic scattering analysis and its geophysical application. D, 1974, Memorial University of Newfoundland. 242 p.

Annas, R. M. Boreal ecosystems of the Fort Nelson area of northeastern British Columbia. D, 1977, University of British Columbia.

Annesley, Irvine R. A field petrographic and chemical investigation of the Amer Lake mafic and ultramafic komatiites and associated mafic volcanic rocks. M, 1981, University of Windsor. 159 p.

Annis, David Robert. Petrochemical variations in post-Laramide igneous rocks in Arizona and adjacent regions; geotectonic and metallogenic implications. M, 1986, Arizona State University. 693 p.

Annis, Malcolm Paul. The geology and metallogeny of the upper Numedal region, Norway. D, 1980, University of Georgia. 307 p.

Anno, Phil Dean. An application of the compressional and shear-wave synergism to the detection of changes in rock properties in Kingfisher County, Oklahoma. M, 1980, University of Oklahoma. 113 p.

Anooshehpoor, Abdolrasool. Foam rubber model studies of problems in seismology and earthquake engineering. D, 1988, [University of California, San Diego]. 144 p.

Anovitz, Lawrence M. Pressure-temperature-time constraints on the metamorphism of the Grenville Province, Ontario. D, 1987, University of Michigan. 493 p.

Anovitz, Lawrence Michael. Phase equilibria in the system $CaCO_3$-$MgCO_3$-$FeCO_3$. M, 1982, University of Michigan.

Ansal, A. M. An endochronic constitutive law for normally consolidated cohesive soils. D, 1977, Northwestern University. 182 p.

Ansari, A. M. Azheruddin. Resistivity investigations for the location of ground water in the Barrens area of central Pennsylvania. M, 1959, Pennsylvania State University, University Park. 84 p.

Ansari, Gholam Reza. A laboratory study of submerged multi-body systems in earthquakes. D, 1983, University of California, Berkeley. 370 p.

Ansari, Homayon Jaberi. A study of the ceramic properties of the Peorian and Loveland loess deposits in southeastern Nebraska. M, 1956, University of Nebraska, Lincoln.

Ansari, Noorul Wase. Subsurface geology of Smith, Jewell, Mitchell and Osborne counties, Kansas, related to petroleum accumulation. M, 1965, Kansas State University. 59 p.

Ansdell, Kevin Michael. Fluid inclusion and stable isotope study of the Tom Ba-Pb-Zn deposit, Yukon Territory, Canada. M, 1985, University of Alberta. 134 p.

Ansell, Mark Willis. Carbonate petrology and depositional environments within the Cambrian Emigrant Formation, northern Last Chance Range, California. M, 1987, Eastern Washington University. 70 p.

Ansell, Valerie. Mineralogy and geochemistry of thorium. M, 1985, Carleton University. 112 p.

Ansfield, Valentine J. The geology of northwestern Rib Mountain Township (Marathon County, Wisconsin). M, 1967, University of Wisconsin-Madison.

Ansfield, Valentine Joseph. The stratigraphy and sedimentology of the Lyre Formation, northwestern Olympic Peninsula, Washington. D, 1972, University of Washington. 131 p.

Anson, Gwendolyn L. The effects on the post-depositional remanent magnetization of synthetic sediment. M, 1985, Lehigh University. 70 p.

Anspach, David Harold. Geology of the Normangee Lake area, Leon County, Texas. M, 1972, Texas A&M University.

Anstett, Terrance F. Distribution of burrows in Upper Cambrian sandstones, Baraboo area, Wisconsin. M, 1977, University of Wisconsin-Madison. 83 p.

Anstett, Terrance F. Grade-tonnage models of silver-copper-lead-zinc vein deposits of the Coeur d'Alene mining district, Idaho. D, 1986, University of Wisconsin-Madison. 175 p.

Anstey, Robert Leland. The trepostome bryozoan fauna of the Eden shale (Ordovician) in southeastern Indiana and adjacent areas in Kentucky and Ohio. D, 1970, Indiana University, Bloomington. 171 p.

Antai, A. E. Groundwater hydrology of the Benin Formation in Mid-Western State of Nigeria. M, 1976, Purdue University. 78 p.

Anthony, Clyde C., Jr. The Bluffport Marl Member of the Demopolis Formation in Clay County, Mississippi. M, 1959, Mississippi State University. 48 p.

Anthony, Elizabeth Youngblood. Geochemical evidence for crustal melting in the origin of the igneous suite at the Sierrita porphyry copper deposit, southeastern Arizona. D, 1986, University of Arizona. 98 p.

Anthony, Gaylord Dean. Studies of the thermal decomposition of kaolin and sediments. D, 1969, University of Akron. 164 p.

Anthony, Helen V. Fossil foraminifera from the Island of St. Croix, American Virgin Islands. M, 1927, Smith College. 53 p.

Anthony, James Michael. Structural analysis of asymmetrical folds using the finite element method. M, 1976, University of Oklahoma. 66 p.

Anthony, John W. An investigation of physical property variations related to thorium content of synthetic monazite. D, 1964, Harvard University.

Anthony, John W. Geology of the Montosa-Cottonwood Canyons area, Santa Cruz County, Arizona. M, 1951, University of Arizona.

Anthony, Stephen S. Hydrogeochemistry of a small limestone island; Laura, Majuro Atoll, Marshall Islands. M, 1987, University of Hawaii. 114 p.

Antine, Helen M. The origin, occurrence, interrelationships and economic importance of the ultra-basic igneous rocks. M, 1933, Columbia University, Teachers College.

Antoine, John W. An hypothesis concerning the distribution of salt and salt structures in the Gulf of Mexico. M, 1970, Texas A&M University.

Anton, Ann. Paleolimnology of an equatorial lake in the Inter-Andean Plateau of Ecuador. D, 1987, Ohio State University. 172 p.

Antoniuk, Stephen Alexander. A mineralogic study of the Saskatchewan sands and gravels. M, 1954, University of Alberta. 61 p.

Antonuk, Caroline-Nathalie. Geology, structure and microstructure of Farmer's Island and adjacent islands, north-central Newfoundland; with a discussion of the Dunnage Melange. M, 1986, University of New Brunswick. 194 p.

Antony, John J. Geology and copper deposits of the Acimiento Mine Nacimiento-Eureka Mesa area, Sandoval and Rio Arriba counties, New Mexico. M, 1972, Colorado School of Mines. 63 p.

Antosch, Larry Michael. Management of a gravel-pit lake system to optimize future water quality. D, 1982, Iowa State University of Science and Technology. 199 p.

Antrim, Lisa Kay. Fine scale study of a small overlapping spreading center at 12°54′N on the East Pacific Rise. M, 1986, University of California, Santa Barbara. 79 p.

Antrobus, Edmund S. A. A study of lime-rich metamorphic rocks from Cree Lake, Manitoba. M, 1949, McGill University.

Antrobus, Edmund S. A. A study of the Witwatersrand System. D, 1955, McGill University.

Anttonen, Gary Jacob. Hydrothermal ore deposition simulation model. M, 1969, Stanford University.

Anttonen, Gary Jacob. Mechanical analysis of Miocene sandstones from the Slac site, Stanford University, California. M, 1966, Stanford University.

Anttonen, Gary Jacob. Trace elements in high cascade volcanic rocks, Three Sisters area, Oregon. D, 1972, Stanford University. 101 p.

Antweiler, Ronald Chisholm. The chemistry of weathering of a Pliocene volcanic ash; field and laboratory studies. D, 1981, University of Wyoming. 166 p.

Anzalone, Salvatore A. Geology of the Inspiration and Silver ore mine, Coeur d'Arlene District, Idaho. M, 1956, Columbia University, Teachers College.

Anzoleaga, Rodolfo. Crustal structure in western Lake Superior from the integration of seismic and gravity data. D, 1971, University of Wisconsin-Madison. 64 p.

Anzoleaga, Rodolfo. Shallow seismic refraction studies, western Lake Superior. M, 1969, University of Wisconsin-Madison.

Apak, Sukru Nail. Subsurface stratigraphy and sedimentologic control on the productive Middle Devonian age Richfield Member of the Lucas Formation in the Michigan Basin. M, 1985, Western Michigan University.

Aparicio, Agustin. Petrography of the Fort Union Formation in uranium mineralized areas, Cave Hills, Harding County, South Dakota. M, 1977, South Dakota School of Mines & Technology.

Aparicio, Miguel Pablo. Seismic model studies of a transition zone between two media. M, 1967, St. Louis University.

Aparisi, Michelle. Stratigraphy and structure of the Ganson Hill area; northern Taconic Allochthon. M, 1984, SUNY at Albany. 128 p.

Apel, Robert A. The geology and geochemistry of the Chicken Creek dike and greisen, Kougarok Mountain, Alaska. M, 1984, University of Wisconsin-Madison. 91 p.

Apfel, Earl Taylor. The post-Cretaceous, pre-Pleistocene history of Iowa. M, 1925, University of Iowa. 45 p.

Apfel, Earl Taylor. The pre-Illinoian Pleistocene (geology) of Iowa. D, 1926, University of Iowa. 251 p.

Apfel, John B. The stratigraphy and structure of the west half of the Oneida 7 1/2 minute Quadrangle (New York). M, 1958, Syracuse University.

Apgar, Julie L. Stratigraphy, sedimentology, and diagenesis of the Flathead Sandstone, Libby Trough, Montana. M, 1986, University of Idaho. 95 p.

Apgar, Michael A. Groundwater pollution potential of a sanitary landfill above the water table. M, 1971, Pennsylvania State University, University Park.

Apmann, Robert Proctor. The diffusion of sediment in a non-uniform flow field. D, 1968, SUNY at Buffalo. 190 p.

Apodaca, Lori E. Geochemical study of the Cochiti mining district, Sandoval County, NM. M, 1987, New Mexico Institute of Mining and Technology. 99 p.

Apon, Johan Frederik. Upper Devonian conodonts from Hay River-Fort Simpson area. M, 1980, University of Alberta. 116 p.

Apostolou, Charalampos. Growth and longevity of Crassostrea carolinensis in different biofacies of the Choptank Formation (Miocene, Maryland). M, 1986, Queens College (CUNY). 136 p.

Apotria, Theodore G. The stability and evolution of triple junctions. M, 1985, University of Connecticut. 155 p.

Appelbaum, B. S. Surface microtextures of deep water quartz sands from Colombia and Sigsbee basins. D, 1974, University of California, Los Angeles. 311 p.

Appelbaum, Bruce S. Geological investigation of a portion of upper continental slope; northern Alaminos Canyon region (northwestern Gulf of Mexico). M, 1971, Texas A&M University.

Appelt-Rossi, Herbert. Interactions between organic compounds, minerals and ions in volcanic ash derived soils. D, 1974, University of California, Riverside. 133 p.

Apple, Olive F. The relation of rock alteration to mineralization at East Tintic, Utah. M, 1929, Northwestern University.

Applebaum, S. Geology of the Palo Verde Ranch area, Owl Head mining district, Pinal County, Arizona. M, 1975, University of Arizona.

Appleby, Alfred Noel. A study of joint patterns in highly folded and crystalline rocks, with particular reference to northern New Jersey. D, 1942, New York University.

Appledorn, Conrad R. Some volcanic structures in the Chuska Mountains, Navajo Reservation, Arizona-New Mexico. M, 1954, University of Minnesota, Minneapolis. 45 p.

Applegate, James K. A seismic investigation of the offshore Santa Maria area, California. M, 1968, Colorado School of Mines. 33 p.

Applegate, Robert Lewis. Geology of the Sour Dough Mountain area, southern Alaska. M, 1966, University of California, Berkeley. 65 p.

Applegate, Shelton Pleasants. Additions and review of the paleobiology of the Triassic of Virginia. M, 1956, University of Virginia.

Applegate, Shelton Pleasants. The vertebrate fauna of the (Upper Cretaceous) Selma Formation of Alabama; Part VII, Fish. D, 1961, University of Chicago. 48 p.

Appleman, Daniel E. The crystal structures of liebigite and johannite. D, 1956, The Johns Hopkins University.

Appleyard, Edward Clair. Metasomatic or magmatic origin of nepheline-bearing gneisses at Wolfe, Lyndoch Township, Ontario. M, 1959, Queen's University. 190 p.

Applin, Kenneth Richard. Ground water geochemistry as a prospecting tool for uranium deposits in Pennsylvania. M, 1978, Pennsylvania State University, University Park. 145 p.

Applin, Kenneth Richard. Theoretical, experimental, and field studies concerning the diffusion of aqueous oxidized sulfur species and the diagenesis of anoxic coastal sediments. D, 1982, Pennsylvania State University, University Park. 207 p.

Appling, Richard N. Economic geology of the Brattain mining area, Paisley, Oregon. M, 1950, University of Oregon. 74 p.

Apps, John Anthony. The stability field of analcime. D, 1970, Harvard University.

Appuhn, Richard A. A shaking experiment to determine the seismic response of soft ground. M, 1964, University of California, Berkeley. 123 p.

April, Richard H. Clay mineralogy and geochemistry of the Triassic-Jurassic sedimentary rocks of the Connecticut Valley. D, 1978, University of Massachusetts. 221 p.

April, Richard H. Trace element distributions in the sediments of Lake Champlain (Vermont). M, 1972, University of Vermont.

Apsouri, Constantin N. Contributions to the study of the Oriskany Sandstone, basal Onondaga, and the contact thereof. M, 1934, Syracuse University.

Apsouri, Constantin Nicolas. The pegmatites of the Keystone area, South Dakota. D, 1940, University of Minnesota, Minneapolis. 172 p.

Apted, Michael John. Rare earth element partitioning between garnet and andesitic melt; implications for the genesis of orogenic andesites. D, 1980, University of California, Los Angeles. 123 p.

Aquilar, Jannette. Geochemistry of mafic rocks units of the southern Oklahoma aulacogen, southwestern Oklahoma. M, 1988, University of Oklahoma. 167 p.

Arabasz, Walter Joseph, Jr. Geological and geophysical studies of the Atacama fault zone, in northern Chile. D, 1971, California Institute of Technology. 275 p.

Aragon, R. Chemical equilibria and kinetics associated with reactions in the magnetite-ulvospinel system. D, 1979, Purdue University. 363 p.

Araktingi, Udo Gaetan. Viscous fingering in heterogeneous porous media. D, 1988, Stanford University. 265 p.

Aram, Richard Bruce. Cenozoic geomorphic history relating to Lewis and Clark Caverns, Montana. M, 1979, Montana State University. 150 p.

Arama, Rachelle Brooker. Structural geology of Tumbledown Mountain, Culberson County, Texas. M, 1987, University of Texas of the Permian Basin. 76 p.

Arancibia Ramos, Olga Nanet. Mineralogy and chemistry of two nickeliferous laterite soil profiles, Soroako, Sulawesi, Indonesia. M, 1975, Queen's University. 199 p.

Aranda-Gomez, Jose. Metamorphism, mineral zoning, and paragenesis in the San Martin Mine, Zacatecas, Mexico. M, 1978, Colorado School of Mines. 90 p.

Aranda-Gomez, Jose Jorge. Ultramafic and high grade metamorphic xenoliths from central Mexico. D, 1982, University of Oregon. 254 p.

Aranz, William B. Petrographic analysis of the Monteagle Limestone (Mississippian) of south central Tennessee and North Alabama. M, 1972, University of Georgia.

Araujo Filho, Jose Oswaldo de *see* de Araujo Filho, Jose Oswaldo

Arauz, Alejandro J. Evaluation of stream sediments in areas of known mineralization, San Jose and Talamanca quadrangles, Costa Rica; an orientation survey. M, 1986, Colorado School of Mines. 133 p.

Aravena, R. O. ^{18}O, ^{1}H and ^{13}C in tree rings and their relation to the environment. M, 1982, University of Western Ontario.

Araya Montoya, Rodrigo. Seismic hazard analysis; improved models, uncertainties and sensitivities. D, 1988, University of California, Berkeley. 158 p.

Araya, Kidane. Seismic investigation of the southern Red Sea, offshore Ethiopia. M, 1986, Colorado School of Mines. 70 p.

Arbab, Mahmood. Simulation of soil moisture, evapotranspiration and deep percolation in agricultural watersheds. D, 1980, University of Nebraska, Lincoln. 218 p.

Arbogast, Jeffrey Scott. Fluvial deposition of Triassic red beds, Durham Basin, North Carolina. M, 1976, Duke University. 117 p.

Arbour, Guy. L'utilisation du modèle Cole-Cole dans le domaine du temps en polarisation provoquée. M, 1982, Ecole Polytechnique. 189 p.

Arbour, Roger. Géologie le long du contact Macquereau-Mictaw (Cambrian?, Ordovician; Quebec). M, 1962, Universite Laval. 70 p.

Arbucci, R. P. Some aspects of the geochemistry of Sr, Mg and U in Pleistocene corals. M, 1974, Queens College (CUNY). 64 p.

Arca, M. Specificity in monovalent-divalent cation adsorption by clay minerals. D, 1966, University of California, Riverside. 132 p.

Arcaro, Nick P. The control of lutite mineralogy by selective transport, late Pleistocene and Holocene sediments of northern Cascadian Basin-Juan de Fuca abyssal plain (northeastern Pacific Ocean); a test of clay mineral size dependency. M, 1978, Lehigh University. 91 p.

Arce, Carlos. Cordillera Oriental (Eastern Cordillera) Tertiary sediments in the upper Magdalena valley, Colombia, South America. M, 1969, Lehigh University.

Arce, Gary N. Volcanic hazards from Makushin Volcano, northern Unalaska Island, Alaska. M, 1983, University of Alaska, Fairbanks. 142 p.

Arce, Rodolfo Gagarin. Water chemistry of Alabama ponds. D, 1980, Auburn University. 63 p.

Archambault, Marthe. Geology and mineralography of the Silver Creek Deposit, Midway Property, north-central British Columbia. M, 1985, University of British Columbia. 93 p.

Archambeau, Charles Bruce. Elastodynamic source theory. D, 1965, California Institute of Technology. 396 p.

Archbold, Norbert Lee. Late Precambrian diabase dikes in eastern Ontario and western Quebec. D, 1962, University of Michigan. 155 p.

Archbold, Norbert Lee. Relationships of calcium carbonate to lithology and vanadium-uranium deposits in the Salt Wash Sandstone. M, 1956, University of Michigan.

Archer, Allen W. Analysis of Upper Cretaceous trace-fossil assemblages, U.S. Western Interior. D, 1983, Indiana University, Bloomington. 499 p.

Archer, Allen W. Growth banding in Ordovician and Silurian tabulate corals. M, 1979, Indiana University, Bloomington. 427 p.

Archer, Hugh Victor. Water quality and national development; an economic assessment model for water quality management planning. D, 1982, University of Pittsburgh. 258 p.

Archer, Jerry Alan. A hydrogeological evaluation of alluvial fans in northern Big Bend National Park, Texas, using geophysical methods. M, 1982, Texas A&M University. 106 p.

Archer, Katherine. Some Cephalopoda from the Buda Limestone. M, 1936, University of Texas, Austin.

Archer, Paul. Interprétation de l'environnement volcano-sédimentaire de la Formation de Blondeau dans la section stratigraphique du lac Barlow, Chibougamau. M, 1983, Universite du Quebec a Chicoutimi. 160 p.

Archer, Paul Lawrence. An experimental investigation of seawater/basalt interactions; the role of water/rock ratios and temperature gradients. M, 1978, Texas A&M University. 79 p.

Archer, Rex Donald. A photogeological study of the regional structural geology of the southern one-half of the Junction City Quadrangle. M, 1951, Kansas State University. 36 p.

Archer, Robert Edward. Seismic stratigraphy of the Northwest Gulf of Mexico. M, 1985, Rice University. 179 p.

Archibald, Douglas Arthur. Geochronology and tectonic implications of magmatism and metamorphism; southern Kootenay Arc and neighbouring regions, southeastern British Columbia. D, 1983, Queen's University. 154 p.

Archibald, Douglas Arthur. The K-Ar geochronology of the Bay of Islands Complex, western Newfoundland. M, 1975, Queen's University. 107 p.

Archibald, Gary Mervyn. General geology and mineral deposits of the Chibougamau District of Quebec. M, 1960, University of Michigan.

Archibald, James David. Fossil Mammalia and testudines of the Hell Creek Formation, and the geology of the Tullock and Hell Creek formations, Garfield County, Montana. D, 1977, University of California, Berkeley. 705 p.

Archibald, John C., Jr. Studies of antimony occurrence with special relation to its effective extraction, San Jose antimony mines, Wadley, San Luis Potosi, Mexico. M, 1951, Montana College of Mineral Science & Technology. 41 p.

Archibald, Lawrence Eben. Stratigraphy and sedimentology of the Bisbee Group in the Whetstone Mountains, Pima and Cochise counties, southeastern Arizona. M, 1982, University of Arizona. 195 p.

Archie, Andrea. Orthophosphate adsorption by iron oxide complexes in Lakeland soil profiles of Lexington County. M, 1974, University of South Carolina.

Archinal, Brent Allen. Determination of Earth rotation by the combination of data from different space geodetic systems. D, 1987, Ohio State University. 198 p.

Archinal, Bruce Edward. The lithostratigraphy of the Atoka Formation; (Lower Pennsylvanian) along the southwestern margin of the Arkoma Basin, Oklahoma. M, 1977, University of Oklahoma. 172 p.

Archuleta, Ralph J. Experimental and numerical three-dimensional simulations of strike-slip earthquakes. D, 1976, University of California, San Diego. 144 p.

Arcilise, Casper and Vigoren, LaVerne. Uraniferous siltstone of the Lonesome Pete No. 2 Claim, South Cave Hills, Harding County, South Dakota. M, 1957, South Dakota School of Mines & Technology.

Arco, Eugenio Nunez del *see* Nunez del Arco, Eugenio

Arcuri, J. Design and testing of seismic detectors. M, 1974, McGill University. 84 p.

Ardell, Robert J. A geologic study of Little Gobi Desert, Pottawatomie County, Kansas. M, 1965, Kansas State University. 48 p.

Arden, Daniel Douglas. Sediments from borings along the east side of San Francisco Bay, California. D, 1961, University of California, Berkeley.

Arden, Daniel Douglas, Jr. The microstratigraphy of a Carolina bay. M, 1949, Emory University. 86 p.

Ardila, Luis Ernesto. Subsurface study of the "Devonian limestone" of the Permian Basin. M, 1968, University of Texas, Austin.

Ardila, V. F. Cordero *see* Cordero Ardila, V. F.

Ardito, Cynthia Paula. The relationship between mineralogy, groundwater chemistry and groundwater flow in the Moreno Hill Formation. M, 1987, New Mexico Institute of Mining and Technology. 183 p.

Arditty, Patricia C. The Earth tide effects on petroleum reservoirs; preliminary study. M, 1978, Stanford University. 140 p.

Ardrey, Robert Holt. Diagenesis of the Middle Ordovician Trenton Formation in southern Michigan. M, 1974, University of Michigan.

Arehart, Gregory B. Geology and geochemistry of the Black Cloud #3 Zn-Pb-Ag replacement orebody, Leadville District, Lake County, Colorado. M, 1978, Colorado State University. 100 p.

Arem, Joel Edward. Crystal chemistry and structure of idocrase. D, 1970, Harvard University.

Arenas, Mario J. Observations on petrography and hydrothermal alteration at Toquepala and related copper deposits, Peru. M, 1965, University of Cincinnati. 72 p.

Arendt, Ward W. The geology of La Joyita Hills, Socorro County, New Mexico. M, 1971, University of New Mexico. 75 p.

Arengi, Joseph. Sedimentary evolution of the Sudbury Basin, Ontario. M, 1977, University of Toronto.

Arens, Nan Crystal. Salona-Coburn bryozoans, systematics, paleoecology and sedimentologic interpretation of a Middle Ordovician fauna from central Pennsylvania. M, 1988, Pennsylvania State University, University Park. 135 p.

Arenson, John Dean. Downward continuation of Bouguer gravity anomalies and residual aeromagnetic anomalies by means of finite differences. M, 1974, University of Arizona.

Arestad, John F. Resistivity studies in the upper Arkansas Valley and northern San Luis Valley, Colorado. M, 1977, Colorado School of Mines. 129 p.

Arfa, Hossein. Surface water and groundwater interactions in a surficial aquifer in Northwest Iowa. D, 1980, Iowa State University of Science and Technology. 247 p.

Arfele, Anthony Thomas, Jr. Uranium evaluation of the Whitsett Formation, East Texas (Polk, Tyler, Jasper, and Angelina counties). M, 1980, Stephen F. Austin State University.

Arfman, John Frederick, Jr. A software package for analysis of geophysical measurements. M, 1971, United States Naval Academy.

Argast, Scott Frederick. Sepiolite and palygorskite from Ninetyeast Ridge and Wharton Basin. M, 1981, SUNY at Binghamton. 73 p.

Argast, Scott Frederick. The hydraulic differentiation of sedimentary components and the composition and sources of Archean siliciclastic rocks from the Sargur, Javanahalli and Dharwar sequences in Karnataka, South India. D, 1986, SUNY at Binghamton. 256 p.

Argenal, R. Covariance structure of aeromagnetic data from Slave Lake and Churchill structural provinces, Canada. D, 1975, University of Texas, Austin. 87 p.

Argenbright, Dean Nelson. Geology and structure of Proterozoic rocks of the Yavapai Supergroup in Crazy Basin, central Arizona. M, 1986, North Carolina State University. 81 p.

Arghin, Salem Saleh. Hydrogeology of the Al-Marj Basin, Libya. M, 1980, University of Nevada. 73 p.

Arguelles, Victor. Trend analysis study of ore grades in the Carlos Francisco, Consuelo, and Aguas Calientes veins, Casapalca Mine, Peru. M, 1969, Stanford University.

Arguello, Ottoniel. Discharge model of the Mississippi River; evaluation of the impact of diversion of water to Texas. D, 1972, Louisiana State University.

Arias, Rojo Hector Manuel. Modeling the movement of tebuthiuron in runoff and soil water. D, 1986, University of Arizona. 1013 p.

Ariathurai, Chita Ranjan. A finite element model for sediment transport in estuaries. D, 1974, University of California, Davis. 192 p.

Arick, Millard Boston. The Eagle Ford Formation. M, 1928, University of Texas, Austin.

Ariey, Catherine A. Molluscan biostratigraphy of the upper Poul Creek and lower Yakataga formations, Yakataga District, Gulf of Alaska. M, 1978, University of Alaska, Fairbanks. 250 p.

Arif, Abdul H. Analysis of selected tests of aquifer characteristics, West Pakistan. M, 1964, University of Arizona.

Arihara, Norio. A study of non-isothermal single and two-phase flow through consolidated sandstones. D, 1974, Stanford University. 206 p.

Arikan, Ender. Surface geology of portions of the Hightogy and Millport NW quadrangles, Lamar County, Alabama. M, 1984, Mississippi State University. 108 p.

Arikan, Fehmi. Spalled quartz overgrowths as a potential source of quartz silt. M, 1988, University of Pittsburgh.

Aristarain, Lorenzo Francisco. Caliche deposits of New Mexico. D, 1963, Harvard University.

Arjmand, Olya. Computer simulation of the irrigation potential of selected low water holding capacity soils. D, 1982, Iowa State University of Science and Technology. 151 p.

Arkani Hamed, Jafar Gholi. Lateral variations of density in the Earth's mantle. D, 1969, Massachusetts Institute of Technology. 160 p.

Arkell, Brian W. Geology and coal resources of the Cub Mountain area, Sierra Blanca coal field, New Mexico. M, 1983, New Mexico Institute of Mining and Technology. 104 p.

Arkle, Thomas, Jr. Economic geology and stratigraphy of Switzerland Township and immediate environs. M, 1950, Ohio State University.

Arledge, Edward Abner. The geology of the Hub oil field, Marion County, Mississippi. M, 1957, Mississippi State University. 49 p.

Arleth, K. F. Marine structural geology and geologic evolution south of Santa Rosa and San Miguel Islands, California. M, 1977, San Diego State University.

Armbrust, George Aimé Wall rock alteration and paragenesis of the Tribag Mine (Precambrian), Batchawana Bay, Ontario (Canada). D, 1967, University of Colorado. 108 p.

Armbruster, John David. The effects of bedrock lithology on sediment production in small drainage basins in south-central Colorado. M, 1983, Wichita State University. 147 p.

Armbrustmacher, Theodore J. Petrography and petrology of ten square miles in northern Fremont County, Colorado. M, 1963, Miami University (Ohio). 100 p.

Armbrustmacher, Theodore Joseph. Mafic dikes of the Clear Creek drainage area, eastern Bighorn Mountains, Wyoming. D, 1966, University of Iowa. 170 p.

Armentrout, John M. Molluscan biostratigraphy and paleontology of the Lincoln Creek Formation (late Eocene-Oligocene), southwestern Washington; 2 volumes. D, 1973, University of Washington. 479 p.

Armentrout, John Myers. The Tarheel (Miocene) and Empire (Pliocene) formations, geology and paleontology of the type sections, Coos Bay (Coos County), Oregon. M, 1967, University of Oregon. 155 p.

Armin, Richard A. Geology of the southeastern Stansbury Mountains and southern Onaqui Mountains, Tooele County, Utah, with a paleoenvironmental study of part of the Oquirrh Group. M, 1979, San Jose State University. 105 p.

Armin, Richard Alan. Red chert-clast conglomerate in the Earp Formation (Pennsylvanian-Permian), southeastern Arizona; stratigraphy, sedimentology, and tectonic significance. D, 1986, University of Arizona. 459 p.

Armistead, Gary Anthony. The occurrence and geological implications of carbon dioxide clathrate hydrate on Mars. M, 1979, University of Houston.

Armor, Mildred Virginia. The Bois d'Arc Formation of the Arbuckle Mountains. M, 1931, University of Oklahoma. 84 p.

Armour, Michael D. Geochemical and petrographic investigation of selected Tertiary igneous intrusions of the northern Black Hills, South Dakota. M, 1975, University of Toledo. 67 p.

Armstrong, Augustus Keathly. Mississippian System of west-central New Mexico. M, 1957, University of Cincinnati. 134 p.

Armstrong, Augustus Keathly. The paleontology and stratigraphy of the Mississippian System of southwestern New Mexico and southeastern Arizona. D, 1960, University of Cincinnati. 337 p.

Armstrong, Bobby D. Areal geology of southeastern Canadian County, Oklahoma. M, 1958, University of Oklahoma. 64 p.

Armstrong, Calvert William. Role of replacement processes in the formation of complex pegmatites (Canada). D, 1969, University of Western Ontario. 109 p.

Armstrong, Charles F. The moisture balance concept as applied to the American river watershed (Placer and El Dorado counties, California). M, 1967, University of Nevada. 38 p.

Armstrong, D. K. Trace element geochemistry and petrology of the Kettle Point Formation (Upper Devonian) a black shale unit of southwestern Ontario. M, 1986, University of Waterloo. 234 p.

Armstrong, Elizabeth Jeanne. Mylonization of hybrid rocks near Philadelphia. D, 1939, Bryn Mawr College. 43 p.

Armstrong, Ernest Elwood. The use of Fourier transforms in the identification of interstratified clay minerals. M, 1980, Wright State University. 119 p.

Armstrong, Escar Weldon. Structural geology of an area in northeastern Fremont County, Colorado. M, 1951, Iowa State University of Science and Technology.

Armstrong, Frank C. Preliminary report on geology of Atlantic City-South Pass mining district. M, 1948, University of Washington. 65 p.

Armstrong, Frank Clarkson. The Bannock thrust zone, southeastern Idaho. D, 1963, Stanford University. 118 p.

Armstrong, H. S. The gold ores of Little Long Lac area, Ontario. D, 1942, University of Chicago. 74 p.

Armstrong, Jane Crozier and Dedman, Kathryn K. Geology of the Foxcroft area. M, 1939, Bryn Mawr College. 23 p.

Armstrong, Jeffrey A. Correlations of the Devonian formations of California and Nevada. M, 1978, California State University, Fresno.

Armstrong, Jenifer Ann. Hydrocarbon trapping mechanisms in the Miller Creek area of the Powder River basin, Wyoming. M, 1975, Texas A&M University. 93 p.

Armstrong, Jim Richard. Geochemical and petrologic characteristics of selected freshwater limestones. M, 1978, [Oklahoma State University].

Armstrong, John. The geology of the Florida Aquifer System in E. Martin and St. Lucie counties, Florida. M, 1980, Florida State University.

Armstrong, John Edward. The stratigraphy, structure and ore deposits of the southern Yukon. M, 1935, University of British Columbia.

Armstrong, Lee C. The geologic conditions favorable for the accumulation of marl, with special reference to east central Minnesota. M, 1927, University of Minnesota, Minneapolis. 38 p.

Armstrong, Lee Charles. Decomposition and alteration of microline, albite, and kunzite (spodumene) by water. D, 1937, University of Minnesota, Minneapolis. 49 p.

Armstrong, Lisa Fellows. Metamorphic mineral parageneses in Mesozoic and Paleogene rocks, southern East-West Cross-Island Highway, Taiwan. M, 1982, University of California, Los Angeles. 150 p.

Armstrong, Paul F. The geology and ore deposits of the Elbow Lake mining area, northern Manitoba, Canada. D, 1923, Yale University.

Armstrong, Peter B. Copper, zinc, chromium, lead, and cadmium in the unconsolidated sediments of Great Bay estuary, New Hampshire. M, 1974, University of New Hampshire. 85 p.

Armstrong, Richard L. Relationship between solar radiation and ablation on the Blue Glacier, Washington. M, 1976, University of Colorado. 69 p.

Armstrong, Richard Lee. Geochronology and geology of the eastern Great Basin in Nevada and Utah (with) Section 2; illustrations. D, 1964, Yale University. 290 p.

Armstrong, Robert Clarke. Studies on phase equilibria in the system Cu-Hg-S between 400° and 100°C. M, 1971, Queen's University. 137 p.

Armstrong, Robert Clarke. The dispersion of mercury and other metals related to mineral deposits in the Canadian Cordillera. D, 1975, Queen's University. 298 p.

Armstrong, Robert Morgan. Environmental geology of the Provo-Orem area. M, 1974, Brigham Young University.

Armstrong, Scott C. Engineering geologic analysis of reclaimed spoil at a Southwest Texas Gulf Coast surface lignite mine. M, 1987, Texas A&M University. 179 p.

Armstrong, W. D. The geology of the central portion of the Iron Blossom ore run, Tintic District, Utah. M, 1969, Dartmouth College. 123 p.

Armstrong, W. P. Geology of the Ajax-Monte Carlo property. M, 1973, University of British Columbia.

Armstrong, William B. Photogeologic investigation of bedrock fractures along the Bowling Green Fault-Lucas County Monocline, northwest Ohio. M, 1976, University of Toledo. 52 p.

Armstrong, William M. Analysis and modeling of gravity anomalies related to the Knox unconformity. M, 1987, Wright State University. 119 p.

Arnal, Robert E. Limnology, sedimentation and microorganisms of the Salton Sea, California. D, 1957, University of Southern California.

Arnaud, Elaine P. Earth history as an historical science. M, 1933, George Washington University. 25 p.

Arndorfer, David James. Process and parameter interaction in Rattlesnake Crevasse, Mississippi River Delta. D, 1970, Louisiana State University. 88 p.

Arndt, B. Michael. An X-ray and mechanical analysis of Pleistocene Paleosols. M, 1971, University of South Dakota. 82 p.

Arndt, Michael B. Stratigraphy of offshore sediment of Lake Agassiz, North Dakota. D, 1975, University of North Dakota. 97 p.

Arndt, N. T. Ultramafic rocks of Munro Township and their volcanic setting. D, 1975, University of Toronto.

Arndt, Richard E. Mineralogy and petrology of a diabase dyke and adjacent Sokoman Iron Formation (Precambrian), Howells River area, Labrador. M, 1971, University of Vermont.

Arne, D. C. A study of zonation of the Nanisivik Zn-Pb-Ag mine, Baffin Island, Canada. M, 1985, Lakehead University.

Arneill, Lynn. Paleogeography of the Ordovician-Silurian Ely Springs Dolomite, Mazourka Canyon, Inyo Mountains, California. M, 1981, San Diego State University.

Arnestad, Kenneth H. The geology of a portion of the Lompoc Quadrangle, Santa Barbara County, California. M, 1950, University of California, Los Angeles.

Arnett, Bruce A. Subsurface geology of northwestern Pawnee County, Kansas. M, 1974, Wichita State University. 66 p.

Arney, Barbara Holota. Geochemistry of Eyjafjoll, a volcano in southern Iceland. M, 1978, Massachusetts Institute of Technology. 133 p.

Arnold, Anthony J. Distribution of benthic foraminifera in the surface sediments of the Georgia-South Carolina continental slope. M, 1977, University of Georgia. 228 p.

Arnold, Anthony Jay. Hierarchical structure in evolutionary theory; applications in the Foraminiferida. D, 1982, Harvard University. 255 p.

Arnold, B. Petrogenesis of the Spanish Peaks igneous complex, Colo.; major element, rare earth element, and strontium isotopic data. M, 1977, Kansas State University.

Arnold, Billy M., Sr. The subsurface geology of Kingfisher County, Oklahoma. M, 1956, University of Oklahoma. 84 p.

Arnold, Chester Arthur. The genus Callixylon from the Upper Devonian of central and western New York. D, 1929, Cornell University.

Arnold, Dwight Ellsworth. Geology of the northern Stansbury Range, Tooele County, Utah. M, 1956, University of Utah. 57 p.

Arnold, Eldon Drewes. Temperature correlation of the reverse shift reaction. M, 1951, West Virginia University.

Arnold, Eve Maureen. Seafloor echo character and geologic history of survey Site E₂, the type locality for deep sea red clay deposition, Northwest Pacific. M, 1987, SUNY at Buffalo. 80 p.

Arnold, George Edward. A petrographic study of sandstone weathering. M, 1978, West Virginia University.

Arnold, Harold B. Geology of part of the Muddy Mountains, Clark County, Nevada. M, 1977, Eastern Washington University. 61 p.

Arnold, Herbert Julius. The selection, organization, and evaluation of localities available for unspecialized field work in earth science in the New York City region. D, 1936, Columbia University, Teachers College.

Arnold, Joseph Jenk, Jr. Prairie mounds and their climatic implications. M, 1959, University of Arkansas, Fayetteville.

Arnold, K. C. Ice ablation measured by stakes and by terrestrial photogrammetry; a comparison on the lower part of the White Glacier, Axel Heiberg Island, Canada. D, 1978, McGill University.

Arnold, L. D. The climatic response in the partitioning of the stable isotopes of carbon in juniper trees from Arizona. D, 1979, University of Arizona. 209 p.

Arnold, Leavitt Clark. Supergene mineralogy and processes in the San Xavier Mine area, Pima County, Arizona. M, 1964, University of Arizona.

Arnold, Leavitt Clark, Jr. Structural geology along the southeastern margin of the Tucson Basin, Pima County, Arizona. D, 1971, University of Arizona. 158 p.

Arnold, Lynne J. A scale model study of the effects of meteorological, soil, and house parameters on radon entry. D, 1988, Rutgers, The State University, New Brunswick. 250 p.

Arnold, Ralph. The geology of San Pedro Bay (California). M, 1900, Stanford University.

Arnold, Ralph. The paleontology and stratigraphy of the marine Pliocene and Pleistocene of San Pedro, California. D, 1903, Stanford University. 420 p.

Arnold, Ralph Gunther. A preliminary account of the mineralogy and genesis of the uraniferous conglomerate of Blind River, Ontario. M, 1954, University of Toronto.

Arnold, Ralph Gunther. Pyrrhotite-pyrite equilibrium relations between 325°C and 743°C. D, 1958, Princeton University. 108 p.

Arnold, Randal Irad. Geology and mineral deposits of Little Whitewater Canyon, Holt Gulch, Catron County, New Mexico. M, 1974, University of Texas at El Paso.

Arnold, Scott. Discovering Mount St. Helens; a guide to Mount St. Helens National Volcanic Monument. M, 1984, University of Washington. 252 p.

Arnold, Walter Allen. Dipping slab effects on seismic source mechanisms at subduction zones. M, 1982, Pennsylvania State University, University Park. 68 p.

Arnon, Boas. Recognizing terrigenous depositional environments with the aid of the computer. M, 1978, Rensselaer Polytechnic Institute. 268 p.

Arnonne, Robert. Floccule characterization of the Satilla Estuary. M, 1972, Georgia Institute of Technology. 56 p.

Arnott, Robert William Charles. Sedimentology of an ancient clastic nearshore sequence, Lower Cretaceous Bootlegger Member, north-central Montana. D, 1987, University of Alberta. 282 p.

Arnott, Ronald James. A study of some radio-active minerals by x-ray diffraction; with special reference to the oxides containing columbium, tantalum and titanium. M, 1949, University of Manitoba.

Arnott, Ronald James. Particle sizes of clay minerals by small-angle X-ray scattering. D, 1954, Columbia University, Teachers College. 63 p.

Arnott, William Charles. Proximal channel deposits of the Hadrynian Hector Formation, Lake Louise, Alberta. M, 1984, University of Alberta. 164 p.

Arnow, Jill Ann. Deep structure from continental collision; COCORP deep seismic reflection profiles across the Southern Appalachians. M, 1986, Cornell University. 82 p.

Arnow, Theodore. The groundwater resources of Albany County, New York. M, 1950, Columbia University, Teachers College.

Arnseth, Richard Wayne. Carbonate and clay mineral reactions in a modern mixing zone environment, Salt River estuary, St. Croix, U.S. Virgin Islands. D, 1983, Northwestern University. 132 p.

Arnstein Breuer, Roberto John. Some foraminifera from the Heron Island Reef. M, 1978, Pennsylvania State University, University Park. 287 p.

Arntson, Ronald Hughes. Solubility of metallic sulfides in certain aqueous ore-forming solutions. M, 1958, University of California, Los Angeles.

Arola, John L. Origin of aluminous laterite and bauxite. M, 1974, Northern Illinois University. 69 p.

Aron, Gert. Optimization of conjunctively managed surface and ground water resources by dynamic programming. D, 1969, University of California, Davis. 160 p.

Aronoff, Steven Martin. A reexamination of the type material of some Upper Ordovician nautiloid cephalopods (Orthocerida and Actinocerida) of the Cincinnati, Ohio area. M, 1973, University of Cincinnati. 79 p.

Aronoff, Steven Martin. Phenon and interspecific variation in selected Paleozoic orthoconic nautiloids (Mollusca, Cephalopoda). D, 1981, University of Iowa. 361 p.

Aronovici, Vladimir S. Geo-pedological reconnaissance survey of the Placerville Project. M, 1937, Columbia University, Teachers College.

Aronow, Saul. Problems in late Pleistocene and Recent history of the Devils Lake region, North Dakota. D, 1955, University of Wisconsin-Madison.

Aronow, Saul. The geomorphic system of W. M. Davis. M, 1946, University of Iowa. 106 p.

Aronson, David. Petrology of the Shady Dolomite (Lower Cambrian) in Virginia. D, 1967, Virginia Polytechnic Institute and State University.

Aronson, David Allen. The stratigraphy, petrology, and origin of the Copper Ridge Dolomite (Upper Cambrian), Blacksburg, Virginia. M, 1966, Virginia Polytechnic Institute and State University.

Aronson, James Louis. The geochronology of the plutonic and metamorphic rocks of New Zealand. D, 1966, California Institute of Technology. 243 p.

Arora, C. R. Environmental resource maps for land-use planning in Carlisle, Massachusetts. D, 1975, Boston University. 167 p.

Arora, Harpal Singh. Clay dispersion as influenced by chemical environment and mineralogical properties. D, 1969, University of California, Riverside. 96 p.

Arora, O. P. Semiquantitative methods for determining cerium, lanthanum, and thorium in monazite sands. M, 1954, Colorado School of Mines. 44 p.

Arora, Sushil Kumar. Water resources management in relation to groundwater salinity for a hydroagronomic system. D, 1982, University of California, Davis. 314 p.

Arper, William Burnside, Jr. A three dimensional study of the Wilcox Group in Mississippi. M, 1942, University of Oklahoma. 36 p.

Arper, William Burnside, Jr. The Smackover Formation in southern Arkansas and northern Louisiana, and adjacent areas of northeastern Texas and west-central Mississippi. D, 1953, University of Kansas. 74 p.

Arpin, Marc. Etude petrographique et petrologique du massif de St. Nazaire-de-Chicoutimi. M, 1985, Universite de Montreal.

Arrigo, John A. The hydrostratigraphy of selected formations in Southeast Idaho. M, 1983, University of Idaho. 111 p.

Arrington, Jimmy L. The geology of the Rodgers Quadrangle, Benton County, Arkansas. M, 1962, University of Arkansas, Fayetteville.

Arrington, Robert Newton. Geology of Berry Creek Quadrangle, Williamson County, Texas. M, 1954, University of Texas, Austin.

Arriola Torres, Alfredo. An experimental study of the effects of viscous and capillary forces on the trapping and mobilization of oil drops in capillary constrictions. D, 1980, University of Kansas. 281 p.

Arro, Eric. Morrowan Sandstones (Pennsylvanian) in the subsurface of the Hough area, Texas County, Oklahoma. M, 1965, University of Oklahoma. 50 p.

Arroyo, Patricio Goyes. Seismic stratigraphy and structure of the Progreso Basin, Ecuador. M, 1987, Texas A&M University.

Arruda, Edward Charles. U-Pb systematics and ages of rocks from the zone of cataclasis, Bitterroot Dome, southwestern Montana. M, 1981, Western Michigan University. 91 p.

Arsalan, Ahmad. Photogeology of the Loveland region (Front range, Colorado). M, 1968, Wesleyan University.

Arsdale, B. E. Van see Van Arsdale, B. E., Jr.

Arseneau, Gildar Joseph. Stockwork molybdenum and porphyry copper potential, North Dome, Rocher Deboule Stock, Hazelton, British Columbia. M, 1984, University of Western Ontario. 89 p.

Arth, Joseph George, Jr. Geochemistry of Early Precambrian igneous rocks, Minnesota-Ontario. D, 1973, SUNY at Stony Brook. 164 p.

Arth, Joseph George, Jr. Rb-Sr whole rock and mineral isochron ages of the early Precambrian Northern Lights gneiss, District of Thunder Bay, Ontario. M, 1970, SUNY at Stony Brook.

Arthur, Andrew John. Mesozoic stratigraphy and paleontology of the west side of Harrison Lake, southwestern British Columbia. M, 1987, University of British Columbia. 139 p.

Arthur, Michael A. Sedimentologic and geochemical studies of Cretaceous and Paleogene pelagic sedimentary rocks; Part I, The Gubbio sequence; Part II,...and some global paleoceanographic trends and events. D, 1979, Princeton University. 508 p.

Arthur, Michael A. Stratigraphy and sedimentation of lower Miocene non-marine strata of the Orocopia Mountains; constraints from late Tertiary slip on the San Andreas fault system in Southern California. M, 1974, University of California, Riverside. 200 p.

Arthur, Randolph Clyde. The fluid- and chemical-dynamics of base-metal sulfide recovery from geothermal systems. D, 1983, Pennsylvania State University, University Park. 186 p.

Arthur, Robert S. Effect of islands on surface waves; and evaluation of the influence of refraction, diffraction and variability in direction on the wave pattern. D, 1949, University of California, Los Angeles.

Artioli, Gilberto. Structural studies of the water molecules and hydrogen bonding in zeolites. D, 1985, University of Chicago. 149 p.

Artusy, Ray. A survey of the Pliocene microfauna in the Pico Formation of Ventura County, California. M, 1938, University of Southern California.

Artusy, Raymond Longino. Ostracods of the Stone City Beds at Stone City Bluff, Texas. D, 1960, Louisiana State University. 156 p.

Artzner, D. G. Palynology of a volcanic ash in the Fox Hills Formation (Maestrichtian) of Emmons, Morton, and Sioux counties, North Dakota. M, 1974, Kent State University, Kent. 122 p.

Arulanandan, Kandiah. Electrical response characteristics of clays and their relationships to structure. D, 1966, University of California, Berkeley. 203 p.

Arulmoli, Kandiah. Electrical characterization of sands for in situ prediction of liquefaction potential. D, 1982, University of California, Davis. 124 p.

Aruna, Muhammadu. The effects of temperature and pressure on absolute permeability of sandstones. D, 1976, Stanford University. 102 p.

Arunapuram, Sundararajan. Computer simulation of mining and reclamation operations of a sub-arctic surface coal mine. M, 1985, University of Alaska, Fairbanks. 232 p.

Arur, Manohar. Analysis of latitude observations for detection of crustal movement. M, 1970, Ohio State University.

Arvidson, Raymond Ernst. An analysis of aeolian processes on Mars. M, 1971, Brown University.

Arvidson, Raymond Ernst. Five studies related to sedimentary processes on Mars. D, 1974, Brown University. 148 p.

Ary, M. D. Geology of the eastern part of the Thermopolis and Lucerne Anticline, Hot Springs County, Wyoming. M, 1959, University of Wyoming. 64 p.

Aryani, Cyrus. Analysis of two-dimensional autocorrelation functions for shear strength of tailings dam material. D, 1984, Utah State University. 241 p.

Arzaghi, Mohamad Mehdi. Conodont biostratigraphy of the Valmont Formation (Upper Ordovician) in the Sacramento Mountains, Otero County, New Mexico. M, 1981, West Texas State University. 66 p.

Arzi, Avner A. Partial melting in rocks; rheology, kinetics and water diffusion. D, 1974, Harvard University.

Asad, Syed Ali. The significance of hydraulic equivalence in transportation and deposition of heavy minerals in beach sands. M, 1970, McGill University. 99 p.

Asamoa, Godfried Kofi. Discontinuity location and micromorphological analysis in a genesis study of the soils of the Honeywood catena (southern Ontario). D, 1969, University of Guelph.

Asbury, Larry Marshall. Geology of the South Carbon area, Eastland County, Texas. M, 1961, University of Texas, Austin.

Ascencious, Alejandro C. The San Alberto lead-zinc ore body at Cerro de Pasco Mine, Cerro de Pasco, Peru. M, 1966, University of Arizona.

Aschemeyer, Esther Louise. The urban geography of the clay products industries of metropolitan St. Louis. M, 1943, Washington University.

Aschenbrenner, Bert Claus. Tridimensional shape analysis of sand grains with a photogrammetric method. M, 1954, Southern Methodist University. 47 p.

Asemota, Isaac. Sulfide mineral distribution of northern New Jersey rock formations and their surface water induced acid generating capacity. M, 1987, Rutgers, The State University, Newark. 64 p.

Asgian, Margaret Isabelle. A numerical study of fluid flow in deformable, naturally fractured reservoirs. D, 1988, University of Minnesota, Minneapolis. 229 p.

Ash, D. W. Geomorphology and littoral processes between Jones Beach and Montauk Point, Long Island, New York. D, 1979, SUNY at Binghamton. 296 p.

Ash, Henry O. The Jurassic Todilto Formation of New Mexico. M, 1958, University of New Mexico. 63 p.

Ash, Nadim F. Application of numerical simulation using a progressive failure approach to underground coal mine stability analysis. D, 1987, University of Alabama. 156 p.

Ash, P. O. Improvement of oil sands tailings sludge disposal behaviour with overburden material. M, 1986, University of Waterloo. 107 p.

Ash, Sidney R. Geology and ground-water resources of northern Lea County, New Mexico. M, 1961, University of New Mexico. 66 p.

Ash, Simon Harry. Mining operations at the New Black Diamond Mine of the Pacific Coast Coal Company. M, 1929, University of Washington. 67 p.

Ashabranner, Donald. A 2D phase-shift migration algorithm for laterally varying velocity fields and an analysis of ophiolite-derived models of accretion. M, 1986, University of Houston.

Ashbaugh, Alexander Cleveland. An experimental study for the selection of geological concepts for intermediate grades. D, 1964, University of Georgia. 138 p.

Ashbaugh, James Graham. The historical geography of Cripple Creek, Colorado. M, 1953, University of Colorado.

Ashby, William H. The solubility of natural gas in oil field brines. M, 1948, Louisiana State University.

Ashenden, David D. Stratigraphy and structure, northern portion of the Pelham Dome, north-central Massachusetts. M, 1973, University of Massachusetts. 132 p.

Asher, Bruce Robert. Electrical resistivity as a means to monitor groundwater contamination. M, 1976, University of Tennessee, Knoxville. 45 p.

Asher, Roderick R. Geology and mineral resources of a portion of the Silver City region, Owyhee County, Idaho. M, 1968, University of Idaho. 218 p.

Ashleman, James C. The geology of the western part of the Kachess Lake Quadrangle, Washington. M, 1979, University of Washington. 88 p.

Ashley, Burton Edward. A study of certain rocks of the Lake Tanganyika District, northern Rhodesia, Africa. M, 1936, University of Minnesota, Minneapolis. 44 p.

Ashley, G. M. Sedimentology of a freshwater tidal system, Pitt River - Pitt Lake, British Columbia. D, 1977, University of British Columbia.

Ashley, Gail M. Rhythmic sedimentation in Glacial Lake Hitchcock, Massachusetts/ Connecticut. M, 1972, University of Massachusetts. 148 p.

Ashley, George H. Geology of the Paleozoic area of Arkansas south of the novaculite region. M, 1897, [Stanford University].

Ashley, George H. Neocene of the Santa Cruz Mountains. D, 1894, [Stanford University].

Ashley, Randal Jack. Offshore geology and sediment distribution of the El Capitan-Gaviota continental shelf, northern Santa Barbara Channel, California. M, 1974, San Diego State University.

Ashley, Roger Parkmand. Metamorphic petrology and structure of the Burnt River Canyon area, northeastern Oregon. D, 1967, Stanford University. 236 p.

Ashley, William H. Geology of the Kennady Peak-Pennock Mountain area, Carbon County, Wyoming. M, 1948, University of Wyoming. 71 p.

Ashman, S. H. Geology of the Peshastin Creek area, Washington. M, 1974, University of Washington. 29 p.

Ashmead, Lawrence Peel. Geology of the Flatwater ultramafic pluton, Baie Verte, Newfoundland. M, 1958, University of Rochester. 66 p.

Ashour, Abdurrahim Mohammed. Petrographic and geochemical analysis of middle Permian dolomites; a comparison of subsurface and outcrop data. D, 1973, Texas Tech University. 69 p.

Ashour, Abdurrahim Mohammed. Stratigraphic and environmental study of Chispa Summit Formation, upper Cretaceous (southwestern Texas). M, 1969, Texas Tech University. 50 p.

Ashouri, Ali Reza. Stratigraphical analysis of Triassic and Lower Jurassic rocks in northeastern Arizona. M, 1980, University of Arizona. 72 p.

Ashraf, Abdul Aziz. The earthquake phase SKS. M, 1972, Texas Tech University. 72 p.

Ashraf, Saied. Conodont biostratigraphy of the Montoya Formation (Ordovician) in the Sacramento Mountians, Otero County, New Mexico. M, 1981, West Texas State University. 149 p.

Ashton, Bessie L. The geonomic aspects of the Illinois Waterway. D, 1926, University of Wisconsin-Milwaukee.

Ashton, Donald Alan. Eocene lithofacies and depositional history in a portion of the Del Mar Quadrangle, San Diego County, California. M, 1987, San Diego State University. 212 p.

Ashton, Jean Hadley. Distribution of intraspecific variation of nonagnostid trilobites of the pterocephaliid biomere of the Great Basin. M, 1974, University of Kansas. 61 p.

Ashton, Kenneth Earl. Precambrian geology of the southeastern Amer Lake area (66H/1), near Baker Lake, N.W.T.; a study of the Woodburn Lake Group, an Archean, orthoquartzite-bearing sequence in the Churchill structural province. D, 1988, Queen's University. 335 p.

Ashton, Kenneth Earl. The geology of the Milton Island map sheet (east half), Saskatchewan (64D/10E). M, 1979, University of Saskatchewan. 131 p.

Ashton, L. W. Influences on the geochemistry of ground water in Welby, Colorado. M, 1978, University of Colorado.

Ashwal, Lewis D. Metamorphic hydration of augite-orthopyroxene monzodiorite to hornblende granodiorite gneiss, Belchertown Batholith, west-central Massachusetts. M, 1974, University of Massachusetts. 117 p.

Ashwal, Lewis D. Petrogenesis of massif-type anorthosites; crystallization history and liquid line of descent of the Adirondack and Morin complexes. D, 1979, Princeton University. 143 p.

Ashwall, L. Primary pyroxene-bearing rocks from the Belchertown intrusive complex and their metamorphic reconstitution. M, 1974, University of Massachusetts.

Ashwill, Walter R. The geology of the Winberry Creek area, Lane County, Oregon. M, 1951, University of Oregon. 63 p.

Ashworth, E. T. Foraminifera from the Cretaceous of the Central Cordillera of Guatemala. D, 1974, Ohio State University. 128 p.

Ashworth, Edwin Thomas. Strawn Fusulinidae of north-central Texas. M, 1954, University of Texas, Austin.

Ashworth, Kathryn King. Genesis of gold deposits at the Little Squaw Mines, Chandalar mining district, Alaska. M, 1983, Western Washington University. 98 p.

Ashworth, Richard Allan. Methodology delineating groundwater contamination by household wastewater. D, 1984, University of Arkansas, Fayetteville. 228 p.

Asihene, Edmund Buahin. Geology of the Buena Vista mercury mine, Klau Mining District, San Luis Obispo County. M, 1962, University of California, Los Angeles.

Asihene, Kwame ANane Buahin. The Texada Formation (upper Triassic) of British Columbia and its associated magnetite concentrations. D, 1970, University of California, Los Angeles. 217 p.

Askar, Hasan Ghuloom. Development and application of infinite elements to ground freezing problems. D, 1982, University of Colorado. 217 p.

Askeland, James Philip. A gravity interpretation of subsurface structures in northeast Washington County, Arkansas. M, 1978, University of Arkansas, Fayetteville.

Askenaizer, Daniel Jay. Public policy issues in developing a drinking water standards bill for California; a case-study. D, 1988, University of California, Los Angeles. 181 p.

Askren, David R. Holocene stratigraphic framework; southeastern Bering Sea continental shelf. M, 1972, University of Washington. 104 p.

Asmerom, Yemane. Geochemistry of biotite from a part of the Loon Lake Batholith and its relationship to uranium mineralization at the Midnite uranium mine, Stevens County, Washington. M, 1981, Eastern Washington University. 110 p.

Asmerom, Yemane. Mesozoic igneous activity in the southern Cordillera of North America; implications for tectonics and magma genesis. D, 1988, University of Arizona. 232 p.

Asmussen, Loris E. Clay mineralogy of some Permian shales and limestones. M, 1958, Kansas State University. 73 p.

Asnake, Mesfin. Biostratigraphic and evolutionary relationships of cylindrodontid rodents of the Chadronian (early Oligocene) of Montana. M, 1984, University of Washington. 97 p.

Aso, Kazuo. Phenomena involved in pre-splitting by blasting. D, 1966, Stanford University. 177 p.

Asper, Vernon L. Accelerated settling of particulate matter by "marine snow" aggregates. D, 1985, Woods Hole Oceanographic Institution. 189 p.

Asper, Vernon L. Accelerated settling of particulate matter by marine aggregates. D, 1986, Massachusetts Institute of Technology. 186 p.

Aspiroz, Rogilio. Automatic detection of seismic reflection. M, 1969, Rice University. 76 p.

Aspler, Lawrence B. Geology of Nonacho Basin (early Proterozoic), NWT. D, 1985, Carleton University. 385 p.

Asquith, Donald O. Geology of a portion of the Sangre de Cristo Range northwest of Cuchara Camps, Colorado. M, 1952, University of Kansas. 98 p.

Asquith, Donald Owen. Depositional topography and major marine environments, Late Cretaceous, Wyoming. D, 1972, University of California, Los Angeles.

Asquith, George B. Origin of the Precambrian Wisconsin rhyolites. M, 1963, University of Wisconsin-Madison.

Asquith, George B. The marine dolomitization of the Mifflin Member, Platteville Limestone, (Middle Ordovician) in southwest Wisconsin. D, 1966, University of Wisconsin-Madison.

Asrar, Ghassem. Short duration evapotranspiration estimated by Class A pan and meteorological parameters. D, 1981, Michigan State University. 139 p.

Asreen, Robert C., Jr. Stratigraphy, petrology and depositional environments of the Middle Ordovician upper Lenoir through lower "Bays" formations near Calhoun, Tennessee. M, 1985, University of Tennessee, Knoxville. 201 p.

Assad, Jamal M. Geology of the upper Elkhorn No. 3 coal in the eastern Kentucky coal field. M, 1988, University of Kentucky. 114 p.

Assad, Robert Joseph. The formation of certain granite-like rocks in the foot-wall of the Sudbury Norite northwest of Sudbury Basin, Ontario. M, 1954, McGill University.

Assad, Robert Joseph. The geology of the East Sullivan Deposit, Val d'Or, Quebec. D, 1958, McGill University.

Assadi, Seid Mohamad. Structure of the Golden Gate Mountain, Pima County, Arizona. M, 1964, University of Arizona.

Assaf, K. K. Digital simulation of aquifer response to artificial groundwater recharge. D, 1976, University of Texas Health Science Center at Houston School of Public Health. 289 p.

Assaf, Karen Klare. Mathematics in the earth science curriculum project textbook. M, 1968, Iowa State University of Science and Technology.

Asseez, Laidiyu Olayinka. A chemical and petrographic analysis of Wooden No. 6 Well; Middle Ordovician from Hillsdale County, Michigan. M, 1964, Michigan State University. 60 p.

Asseez, Liadiyu Olayinka. Stratigraphy and paleogeography of the Lower Mississippian sediments of the Michigan Basin. D, 1967, Michigan State University. 318 p.

Assefa, Getaneh. Mineralogy and petrology of selected rocks from the Hawthorne Formation, Marion and Alachua counties, Florida. M, 1969, University of Florida. 79 p.

Assefa, Getaneh. Stratigraphy and sedimentology of the Mesozoic sequence in the upper Abbay (Blue Nile) River valley region, Ethiopia. D, 1975, University of Minnesota, Minneapolis. 321 p.

Asselin, Esther. Les Chitinozoaires du Silurien inférieur au Synclinorium de la Baie des Chaleurs, Gaspésie; biostratigraphie et systématique. M, 1988, Universite Laval. 161 p.

Astarita, Arthur Michael. Depositional trends and environments of "Cherokee" sandstones, east-central Payne County, Oklahoma. M, 1975, Oklahoma State University. 54 p.

Aster, Richard C. Hypocenter locations and velocity structure in the Phlegraean Fields, Italy. M, 1986, University of Wisconsin-Madison. 124 p.

Asthana, Virendra. Summary of the marine Permian formations of India and adjacent countries. M, 1963, University of Wisconsin-Madison.

Asthana, Virendra. The Mount Simon Formation (Dresbachian stage, upper Cambrian) of (southwestern) Wisconsin. D, 1969, University of Wisconsin-Madison. 172 p.

Astin, Gary Kent. Structure and stratigraphy of the Co-op Creek Quadrangle, Wasatch County, Utah. M, 1977, Brigham Young University. 35 p.

Astiz Delgado, Luciana Maria de los Angeles. I, Source analysis of large earthquakes in Mexico; II, Study of intermediate-depth earthquakes and interplate seismic coupling. D, 1987, California Institute of Technology. 295 p.

Aston, H. F. Spectrographic and microscopic investigation of some Saskatchewan and Manitoba pyrrhotites. M, 1949, University of Saskatchewan. 32 p.

Aston, Robert Earl, Jr. Stratigraphy, petrology, and depositional environment of the Crystal Mountain Sandstone and lowermost Mazarn Shale, Montgomery County, Arkansas. M, 1987, University of New Orleans. 144 p.

Astwood, Phillip M. A petrographic study of two alpine-type ultramafic bodies. M, 1970, University of South Carolina.

Astwood, Phillip M. An examination of the effects of contrasting teaching strategies in a college earth science course. D, 1979, University of South Carolina. 87 p.

Aszklar, Stanley Joseph. A contribution to a revision of Platteville Middle Ordovician Gastropoda. M, 1939, University of Wisconsin-Madison. 58 p.

Atakol, Kenan. The effect of rate of shearing deformation on the shearing resistance of a cohesionless soil. D, 1967, University of Virginia. 120 p.

Atalan, Namik K. A petrographic study of the Yeniliman serpentinized ultramafic body, Karaburun, Turkey. M, 1970, Miami University (Ohio). 47 p.

Atallah, Nicolas Jamil. The hydrologic balance to a two-aquifer system under integrated use of surface and ground water. D, 1966, University of California, Davis. 111 p.

Atallah, Raja Hanna. New solvent extraction and column techniques for preconcentration and speciation; I, Flow injection and closed loop solvent extraction of copper and uranium; II, Metal speciation using silica and C18 bonded silica columns. D, 1987, University of Washington. 148 p.

Atchison, D. W. Garnet gneiss near Bancroft, Ontario. M, 1937, Queen's University. 57 p.

Atchison, Dick Eric. Geology of Brushy Creek Quadrangle, Williamson County, Texas. M, 1954, University of Texas, Austin.

Atchison, Michael E. Stratigraphy and depositional environments of the Comox Formation (Upper Cretaceous), Vancouver Island, British Columbia. M, 1968, Northwestern University.

Atchley, Frank William. Geology of the Marcona iron deposits, Peru. D, 1956, Stanford University. 209 p.

Atchley, Stacy C. The pre-Cretaceous surface in central, North, and West Texas; the study of an unconformity. M, 1986, Baylor University. 233 p.

Ateiga, Abdalla A. Diagenesis and depositional environment of Noodle Creek Limestone eastern shelf, Fisher County, Texas. M, 1985, Texas Tech University. 88 p.

Aten, Robert Eugene. Geomorphology of the east flank of the Crazy Mountains, Montana. D, 1974, Purdue University. 136 p.

Ater, Patricia C. Petrology and geochemistry of eclogite xenoliths from Colorado-Wyoming kimberlites. M, 1983, Colorado State University. 251 p.

Atha, Thomas M. A subsurface study of the Cambrian-Ordovician Rose Run Sandstone in eastern Ohio. M, 1981, Ohio University, Athens. 81 p.

Athaide, D. J. A. A model for the evolution of the chemical systems of the Earth's crust and mantle defined by radiogenic strontium distribution, and the rubidium-strontium geochemistry of the Shulaps Range and other ultramafic bodies in and near southwestern British Columbia. M, 1975, University of British Columbia.

Athanas, Louis Chris. The effects of wastewater enrichment on a coastal plain forest ecosystem. D, 1982, University of Maryland. 159 p.

Athanasiou-Grivas, D. Reliability of slopes of particulate materials. D, 1976, Purdue University. 174 p.

Athanasopoulos, George Andreas. Time effects of low-amplitude shear modulus of cohesive soils. D, 1981, University of Michigan. 300 p.

Athaullah, Muhammad. Prediction of bed forms in erodible channels. D, 1968, Colorado State University. 161 p.

Atherton, Charles C. Planktonic foraminiferal biostratigraphy of the Niobrara Formation (upper Cretaceous), Centennial Valley, Albany County, Wyoming. M, 1971, University of Wyoming. 81 p.

Atherton, Elwood. An investigation of thrust faulting. D, 1937, University of Chicago. 44 p.

Athy, Lawrence F. Geology and mineral resources of the Herscher Quadrangle. D, 1925, University of Chicago. 185 p.

Atia, Abdel-Kadir Mohamed H. Differential thermal analysis and high temperature X-ray study of uraninite. D, 1964, Columbia University, Teachers College.

Atik, Ertugrul A. Tectonic and geologic relationships of various mineral deposits in eastern North America. M, 1951, University of Tennessee, Knoxville. 87 p.

Atkin, Kenneth Thomas James. Foraminifers and other microfossils from the Early Cretaceous Mannville Group in Saskatchewan. M, 1986, University of Saskatchewan. 89 p.

Atkin, Steven Allen. Magmatic history, alteration, and mineralization of the Clayton Peak Stock, Utah. D, 1982, The Johns Hopkins University. 282 p.

Atkin, Steven Allen. Submarine volcanic rocks on the west coast of San Juan Island, Washington. M, 1972, University of Washington. 21 p.

Atkins, Elizabeth Dale. Calcareous microfossils in the Pliocene central Arctic Ocean. M, 1988, University of Wisconsin-Madison. 122 p.

Atkins, Frank Pearce, Jr. The geology of the Higginsport Quadrangle (north of the Ohio River). M, 1940, University of Cincinnati. 69 p.

Atkinson, Dorothy. Catheart Mountain, an Ordivician porphyry copper molybdenum occurrence in Northern Appalachia. D, 1978, University of Western Ontario. 192 p.

Atkinson, Elizabeth Allen. Metal separations by means of hydrobromic acid gas; indium in tungsten minerals. D, 1898, [Pennsylvania State University, University Park].

Atkinson, Gerald. Planetary effects of magnetic activity. M, 1964, University of British Columbia.

Atkinson, Ian Athol Edward. Rates of ecosystem development on some Hawaiian lava flows. D, 1969, University of Hawaii. 208 p.

Atkinson, James Ernest. Geology of the Guadalupe quicksilver mine and vicinity (New Almaden Quadrangle, California). M, 1942, University of California, Berkeley. 50 p.

Atkinson, Jon Charles. Chemical quality of the groundwater system in Hall County, Nebraska. M, 1973, University of Nebraska, Lincoln.

Atkinson, K. D. A compilation of the total intensity aeromagnetic and geological data for Nova Scotia. M, 1971, Technical University of Nova Scotia.

Atkinson, Lee Chaflin. A laboratory and numerical investigation of steady-state, two-regime, radial flow to a well from rough, horizontal, deformable fractures. D, 1987, Memorial University of Newfoundland. 321 p.

Atkinson, Marlin J. Phosphate metabolism of coral reef flats. D, 1981, University of Hawaii. 98 p.

Atkinson, Robert F. Conodonts from the middle Ordovician Platteville Formation in (southwestern) Wisconsin. M, 1969, University of Wisconsin-Madison.

Atkinson, Ross David. Geology of the Pony Gulch area near Mystic, South Dakota. M, 1976, South Dakota School of Mines & Technology.

Atkinson, Susan J. The relation between fabric, jointing and flowage in Recent Icelandic lava flows. M, 1973, Acadia University.

Atkinson, Walter Edward. A subsurface study of the South Palacine oil field, Stephens County, Oklahoma. M, 1952, University of Oklahoma. 42 p.

Atkinson, William W., Jr. Experiments on the synthesis of high-purity quartz. D, 1972, Harvard University.

Atkinson, William W., Jr. Geology of the San Pedro Mountains, Santa Fe County, New Mexico. M, 1960, University of New Mexico. 81 p.

Atlas, Elliot L. Phosphate equilibria in seawater and interstitial waters. D, 1976, Oregon State University. 154 p.

Atlas, Leon Morris. Solid state equilibria in the system $MgSiO_3$-$CaMgSi_2O_6$, the polymorphism of $MgSiO_3$. D, 1950, University of Chicago. 53 p.

Atlee, William Augustus. The geology of the Paluxy Sand in the central Texas area. M, 1960, Baylor University. 56 p.

Atluri, Chandrasekera Rao. The possible antimatter content of the Tunguska Meteor of 1908. M, 1965, University of California, Los Angeles. 35 p.

Atobrah, K. A digitally recording, shallow-well geophysical logging system. M, 1977, University of Waterloo.

Atobrah, Kobina. Groundwater flow in the crystalline rocks of the Accra Plains of Ghana, West Africa. D, 1983, Princeton University. 252 p.

Atta, Robert Otis Van *see* Van Atta, Robert Otis

Attanayake, Premadasa M. A field investigation of the groundwater ridging hypothesis. M, 1983, University of Windsor. 203 p.

Attaway, Dorothy Claire. The boundary element method for the diffusion equation. D, 1988, Boston University. 115 p.

Attaya, James S. The geology and mineral resources of Lafayette County, Mississippi. M, 1951, University of Mississippi.

Attig, J. W., Jr. Quaternary stratigraphy and history of the central Androscoggin River valley, Maine. M, 1975, University of Maine. 56 p.

Attig, John William, Jr. The Pleistocene geology of Vilas County, Wisconsin. D, 1984, University of Wisconsin-Madison. 310 p.

Attiga, M. A. Freshwater losses to the sea and water level decline in northwest Libya. M, 1971, University of Missouri, Rolla.

Attoh, Kodjopa. Metamorphic reactions in the Michigamme Formation, Iron County, Michigan. D, 1973, Northwestern University.

Attoh, Kodjopa. Structure and petrology of the iron-rich metasediments in the Hell Roaring Lakes area, Beartooth Mountains, Montana. M, 1970, University of Cincinnati. 69 p.

Attridge, John. On a complete therocephalian skeleton from the Cistecephalus Zone of South Africa. M, 1954, University of California, Berkeley. 73 p.

Attwood, Paul J. Origin and diagenesis of lime mud; scanning electron microscope observations from upper Pleistocene lagoonal limestones, northeastern Tucatan Peninsula, Mexico. M, 1981, University of Kansas. 81 p.

Atuanya, Udemezue Obidigwe. Paleomagnetism of the Gunflint Iron Formation of northwestern Ontario. M, 1982, University of Windsor. 63 p.

Atwater, B. F. Serpentinite diapir in Woodside and Redwood City, California. M, 1975, Stanford University. 118 p.

Atwater, Brian Franklin. Attemps to correlate late Quaternary climatic records between San Francisco Bay, the Sacramento-San Joaquin Delta, and the Mokelumne River, California. D, 1980, University of Delaware. 227 p.

Atwater, David E. Ontogeny and valve microstructure of *Helminthochiton Simplex* (Raymond) (Polyplacophora) from the Allegheny Group (Pennsylvan-

ian) of Ohio. M, 1979, Bowling Green State University. 215 p.

Atwater, Gordon I. Geology of the Tyler and Copps Formation of the Gogebic iron district, Michigan and Wisconsin. D, 1936, University of Wisconsin-Madison.

Atwater, Gordon I. The geology of a portion of the Stoddard Quadrangle, southwestern Wisconsin. M, 1930, University of Iowa. 155 p.

Atwater, Marshall A. Comparison of numerical methods for computing radiative temperature changes in the atmospheric boundary layer. M, 1965, New York University.

Atwater, Tanya Maria. A detailed near-bottom geophysical study of the Gorda Rise and implications of plate tectonics for the Cenozoic tectonic evolution of Western North America. D, 1972, University of California, San Diego. 104 p.

Atwell, Buddy H. Elastic wave propagation in highway pavements. D, 1967, Texas A&M University.

Atwill, Edward Robert, IV. Stratigraphic nomenclature in Sierra Pilares, Chihuahua, Mexico. M, 1960, University of Texas, Austin.

Atwood, Alfred R. The glacial geology of Quincy, Massachusetts. M, 1911, University of Vermont.

Atwood, Dorothy Fisher. Management of contaminated groundwater with aquifer simulation and linear programming; the development of a hydraulic gradient control procedure. M, 1984, Stanford University. 89 p.

Atwood, Glen William. Geology of the San Juan Peak area, San Mateo Mountains, Socorro County, New Mexico; with special reference to the geochemistry, mineralogy, and petrogenesis of an occurrence of riebeckite-bearing rhyolite. M, 1982, University of New Mexico. 156 p.

Atwood, Rollin S. Physiography of the Southern Rocky Mountains. M, 1925, Clark University.

Atwood, Wallace W. Physiographic relationships between the Causses and Cevennes of southern France. D, 1930, Clark University.

Atwood, Wallace W., Jr. Physical features of the Sudbury Basin. M, 1927, Clark University.

Atwood, Wallace Walter. Glaciation of the Uinta and Wasatch mountains. D, 1903, University of Chicago. 72 p.

Atzet, T. Description and classification of the forests of the upper Illinois River drainage of southwestern Oregon. D, 1979, Oregon State University. 211 p.

Atzmon, Gil. Platinum group metals. M, 1987, University of Texas, Austin.

Au, Andrew Yu-Chung. Theoretical modelling of the elastic properties of mantle silicates. D, 1984, SUNY at Stony Brook. 146 p.

Au, Chong Ying Daniel. Crustal structure from an ocean bottom seismometer survey of the Nootka fault zone. D, 1981, University of British Columbia.

Au, Chong Ying Daniel. Lower mantle P wave travel time under Asia. M, 1977, University of Alberta. 69 p.

Au, G. H. C. X-ray mineralogy and paleoecology of the Caroline Basin. M, 1977, University of Hawaii. 166 p.

Au, Wing-Cheong. Dynamic shear modulus and damping in additive-treated expansive soils. D, 1979, Rutgers, The State University, New Brunswick. 195 p.

Au-Ngoc-Ho. Geology and ore deposits of the Norwich Mine, Silver Bow County, Montana. M, 1957, Montana College of Mineral Science & Technology. 70 p.

Aubele, Jayne Christine. Geology of the Cerros del Rio volcanic field, Santa Fe, Sandoval, and Los Alamos counties, New Mexico. M, 1978, University of New Mexico. 136 p.

Aubert, Winton G. Fate and transport of low-pH hazardous materials after deep well disposal. D, 1986, Louisiana State University. 261 p.

Aubin, Thomas E. St. *see* St. Aubin, Thomas E.

Auble, Gregor Thomas. Biogeochemistry of Okefenokee Swamp; litterfall, litter decomposition, and surface water dissolved cation concentrations. D, 1982, University of Georgia. 324 p.

Aubrey, David G. Statistical and dynamical prediction of changes in natural sand beaches. D, 1978, University of California, San Diego. 213 p.

Aubrey, W. M., III. The structure and stratigraphy of the northern ridges of Camels Hump Mountain, Camels Hump Quadrangle, North central Vermont. M, 1977, University of Vermont.

Aubry, Brain Francis. Runoff, sediment transport, and rill formation by sheetflow. M, 1984, University of Washington. 106 p.

Aubut, Alan James. The geology and mineralogy of a Tertiary buried placer deposit, southern British Columbia. M, 1979, University of Alberta. 98 p.

Auch, Timothy W. Paleoecology of middle Pleistocene estuarine deposits in southern Beaufort County, North Carolina. M, 1987, East Carolina University. 188 p.

Auclair, François. Une etude des variations geochimiques dans les laves du complexe ophiolitique de Troodos, Chypre. M, 1987, Universite de Montreal.

Aucremann, Leslie J. Diagenesis and porosity evolution of the Jurassic Nugget Sandstone, Anschutz Ranch East Field, Summit County, Utah. M, 1984, University of Missouri, Columbia. 116 p.

Aucremann, Leslie J. Diagenetic controls of porosity and permeability in the Nugget Sandstone in a Western Overthrust Belt oil reservoir. M, 1983, University of Missouri, Columbia.

Audemard, Felipe. Geology and copper mineralization of the La Quinta Formation, Sierra de Perija, western Zulia, Venezuela. M, 1982, Colorado School of Mines. 75 p.

Audesey, Joseph Louis, Jr. Petrographic analysis of the beach sands of Chatham County, Ga. M, 1954, University of Alabama. 17 p.

Audet-Lapointe, Martine. Aménagement de la nappe d'eau souterraine du Cap-de-la-Madeleine. M, 1984, Universite Laval. 117 p.

Audibert, Jean M. E. Prediction and measurement of strain fields in soils. D, 1972, Duke University.

Auerbach, Pauline Dorothy and Benedict, Dorothy K. The geology of Ithan Creek valley and vicinity (Pennsylvania). M, 1939, Bryn Mawr College. 25 p.

Aufdemberge, Theodore Paul. Energy balance studies over glacier and tundra surfaces, summer 1969 (Capps Glacier and Chitistone Pass, Alaska). D, 1971, University of Michigan.

Auffenberg, Walter. A study of the fossil snakes of Florida. D, 1956, University of Florida. 279 p.

Aufmuth, Raymond E. The Lansing, Kansas City, and Marmaton groups (Pennsylvanian) of the Lemon-Victory oil and gas field, Haskell County, Kansas. M, 1966, Miami University (Ohio). 45 p.

Auger, P. E. Spectrographic study of minor elements in sulphides. D, 1940, Massachusetts Institute of Technology. 145 p.

Aughenbaugh, Nolan B. Degradation of base course aggregates during compaction. D, 1963, Purdue University.

Aughenbaugh, Nolan Blaine. Geology of the Sangre de Cristo Mountains at Mosca and Morris creeks (Colorado). M, 1959, University of Michigan.

August, Lisa Layne. Model for aquifer deterioration during an injection/withdrawal cycle of fresh water in a brackish water aquifer using laboratory data from column experiments. M, 1986, George Washington University. 123 p.

Augustin, B. D. Soils; a multi-media teaching unit for the middle grades. D, 1976, University of Northern Colorado. 94 p.

Aukeman, Frederick Neil. Pleistocene molluscan faunas of the Oakhurst Deposit, Franklin County, Ohio. M, 1960, Ohio State University.

Aukland, Merrill Forrest. Geology of Waterloo and Lee townships, Athens, Ohio. M, 1952, Rutgers, The State University, New Brunswick.

Auld, Bruce Charles. Seismicity off the coast of northern California determined from ocean bottom seismic measurements. M, 1969, Columbia University. 37 p.

Auld, Thomas W. Facies analysis of the Virgin Limestone Member, Moenkopi Formation, Northwest Arizona and Southwest Utah. M, 1976, Northern Arizona University. 83 p.

Aulia, Karsani. Stratigraphy of the Codell Sandstone and Juana Lopez members of the Carlile Formation (Upper Cretaceous), El Paso and Fremont counties, Colorado. M, 1982, Colorado School of Mines. 145 p.

Aulstead, Kathy Loree. Origin and diagenesis of the Manetoe Facies, southern Yukon and Northwest Territories, Canada. M, 1987, University of Calgary. 143 p.

Ault, Curtis Henry. Petrology of the upper Cloverly Formation along the eastern flank of the Big Horn Mountains. M, 1956, University of Illinois, Urbana.

Ault, Ruey A. The geology of portions of Dutch Mills and Cane Hill townships, Washington County, Arkansas. M, 1958, University of Arkansas, Fayetteville.

Ault, Wayne U. Isotopic geochemistry of sulfur. D, 1957, Columbia University, Teachers College. 285 p.

Ault, Wayne U. Variations of sulfur isotope abundances in sulfide minerals. M, 1956, Columbia University, Teachers College.

Aultman, W. L. The subsurface Jurassic Bay Springs Sand. M, 1975, University of Southern Mississippi.

Aumento, Fabrizio. An X-ray study of some Nova Scotia zeolites. M, 1962, Dalhousie University. 104 p.

Aumento, Fabrizio. Thermal transformation of selected zeolites and related hydrated silicates. D, 1965, Dalhousie University. 281 p.

Aurelia, Michael Anthony, III. The distribution of foraminifera in a moderately polluted New England estuary and adjacent salt marsh. M, 1972, Brown University.

Aurin, Fritz. Asphalt of Oklahoma. M, 1915, University of Oklahoma. 109 p.

Aurisano, R. Upper Cretaceous dinoflagellate zonation of the subsurface Toms River section near Toms River, New Jersey. M, 1975, Queens College (CUNY).

Aurisano, Richard Warren. Upper Cretaceous subsurface dinoflagellate stratigraphy and paleoecology of the Atlantic Coastal Plain of New Jersey. D, 1980, Rutgers, The State University, New Brunswick. 204 p.

Ausburn, Brian Edwin. Subsurface disposal of natural brines in western Oklahoma and northern Texas. M, 1961, University of Oklahoma. 105 p.

Ausburn, Kent. A geochemical and petrographic study of the Hansen sedimentary uranium orebody N.W. of Canon City, Col. M, 1981, Florida State University.

Ausburn, Mark P. Stratigraphy, structure and petrology of the northern half of the Topton Quadrangle, North Carolina. M, 1983, Florida State University.

Ausich, William Irl. Community organization, paleontology, and sedimentology of the Lower Mississippian Borden delta platform (Edwardsville Formation, southern Indiana). D, 1978, Indiana University, Bloomington. 433 p.

Ausich, William Irl. The functional morphology and evolution of Pisocrinus (Crinoidea; Silurian). M, 1976, Indiana University, Bloomington. 48 p.

Ausmus, Judith Erlene. The hydrogeochemistry of an area strip-mined for coal, near McCurtain, Haskell County, Oklahoma. M, 1987, Oklahoma State University. 105 p.

Austin, C. Bradford. A crustal structure study of the Mississippi Embayment. M, 1978, University of Texas at El Paso.

Austin, C. Thomas. Analysis of an aeromagnetic profile across the Mill Creek Syncline, Anadarko Basin,

southern Oklahoma. M, 1969, University of Tulsa. 34 p.

Austin, Carl Fulton. Geochemical exploration in silicated limestone at Darwin, California. D, 1958, University of Utah. 166 p.

Austin, Carl Fulton. Geochemical prospecting as applied to replacement ores in limestone. M, 1955, University of Utah. 77 p.

Austin, Earl Bowen, Jr. Geology of the Wylie area, Collin County, Texas. M, 1948, Southern Methodist University. 29 p.

Austin, George S. Photogeology of Cook County, Minnesota. M, 1961, University of Minnesota, Minneapolis.

Austin, George Stephen. The stratigraphy and petrology of the Shakopee Formation, Minnesota. D, 1971, University of Iowa. 216 p.

Austin, James Albert, Jr. Geology of the passive margin of New England. D, 1979, Massachusetts Institute of Technology. 201 p.

Austin, James Albert, Jr. Geology of the passive margin off New England. D, 1978, Woods Hole Oceanographic Institution. 201 p.

Austin, Jesse William. Plastic injection of limestone to reproduce porosity. M, 1948, University of Tulsa. 29 p.

Austin, John Michael. Subsurface stratigraphy and petroleum geology of the "Clinton" sandstone, northern Portage County, northeastern Ohio. M, 1979, Kent State University, Kent. 90 p.

Austin, Michael Neal. Seismic stratigraphy of the Upper Pennsylvanian Swope Limestone, Comanche County, Kansas, and Woods County, Oklahoma. M, 1988, University of Oklahoma. 158 p.

Austin, Robert Burton. The Chelsea Sandstone and associated strata east and northeast of Claremore, Oklahoma. M, 1947, University of Oklahoma. 35 p.

Austin, Roger Seth. The geology of Southeast Elbert County, Georgia. M, 1965, University of Georgia. 68 p.

Austin, Roger Seth. The origin of the kaolin and bauxite deposits of Twiggs, Wilkinson, and Washington counties, Georgia. D, 1972, University of Georgia. 185 p.

Austin, Steven Arthur. Critique of uniformitarianism. M, 1971, San Jose State University. 121 p.

Austin, Steven Arthur. Depositional environment of the Kentucky No. 12 coal bed (Middle Pennsylvanian) of western Kentucky, with special reference to the origin of coal lithotypes. D, 1979, Pennsylvania State University, University Park. 411 p.

Austin, Ward Hunting and Stoever, Edward Carl, Jr. Reconnaissance geology of the south flank of Cinnamon Mountain, Gallatin County, Montana. M, 1950, [University of Michigan]. 102 p.

Austria, Benjamin Suarez. Geochemical implications of iron in sphalerite. D, 1975, Harvard University.

Austria, V. B., Jr. Primary and secondary dispersion of metals near molybdenum and tungsten mineralization in granite; Welsford, Queens County, New Brunswick. M, 1972, University of New Brunswick.

Autels, David des *see* des Autels, David

Autin, Leonard J. Subsurface study of DeQuincy and Perkins fields, Calcasieu Parish, Louisiana. M, 1952, Louisiana State University.

Autin, Whitney J., Jr. Geomorphic, sedimentologic, and stratigraphic analysis of two Northwest Mississippi watersheds. M, 1978, University of Mississippi. 98 p.

Autio, Laurie Knapp. Compositional diversity of mid-ocean ridge basalts; an experimental and geochemical study with emphasis on the depleted Costa Rica rift zone basalts. D, 1984, University of Massachusetts. 368 p.

Autra, Katherine Balshaw. A syngenetic model for the origin of the copper mineralization in the Precambrian Nonesuch Shale, White Pine, Michigan. M, 1977, Michigan State University. 45 p.

Autra, Marshall D. A regional study of the Niagaran and lower Salina of the Michigan Basin. M, 1977, Michigan State University. 96 p.

Auxt, Tara Lou. Effectiveness of four soil amendments in controlling toxic levels of aluminum and manganese in orchard subsoils. D, 1981, West Virginia University. 147 p.

Avadisian, Antoine Mehran. Subsurface Paleozoic geology of Upton County, Texas. M, 1963, University of Texas, Austin.

Avakian, Robert W. Determination of the subsurface geology of Mt. Hillers, Utah, by use of geophysical techniques; gravity and magnetics. M, 1970, Stanford University.

Avanessian, Vahe. Global-local finite element analysis of steady soil-structure interaction. D, 1984, University of California, Los Angeles. 114 p.

Avary, Katherine Lee. Stratigraphy and sedimentology of the Back Creek Siltstone (Devonian) in Virginia and West Virginia. M, 1978, University of North Carolina, Chapel Hill. 89 p.

Avasthi, Jitendra Mohan. Hydrofracturing in inhomogeneous, anisotropic and fractured rocks. D, 1981, University of Wisconsin-Madison. 253 p.

Avcin, Matthew J., Jr. Des Moinesian conodont assemblages from the Illinois Basin. D, 1974, University of Illinois, Urbana. 157 p.

Avcin, Matthew John, Jr. Stratigraphic and environmental study of the Summum (No. 4) and the Springfield (No. 5) coals, and the intervening strata (Pennsylvanian-Desmoinesian series) in south central Illinois. M, 1969, University of Illinois, Urbana.

Avedisian, Gary Edward. Geology of the western half of Phoenix South Mountain Park, Arizona. M, 1966, Arizona State University. 52 p.

Avedissian, Yeghise Murad. Correlation of creep of rock with its dynamic properties. D, 1968, Purdue University. 257 p.

Aven, Russell Edward. Movement of water in porous materials under artesian conditions. D, 1963, University of Tennessee, Knoxville. 262 p.

Avenius, Christopher Gerald. Tectonics of the Monterrey Salient, Sierra Madre Oriental, northeastern Mexico. M, 1982, University of New Orleans. 84 p.

Avenius, Rodney. Petrology of the Cheney Pond area, Adirondacks, New York. M, 1948, Syracuse University.

Avent, Jon Carlton. Cenozoic geology of the Pueblo Mountains region, Oregon-Nevada. D, 1965, University of Washington. 119 p.

Avent, Jon Carlton. Structure and stratigraphy of the Antelope Range, NE White Pine Co., Nev. M, 1962, University of Washington. 67 p.

Avera, William Edgar. Iterative techniques in linearized free surface flow. M, 1981, Oregon State University. 83 p.

Averill, E. L. Areal geology of the Campbell Chibougamau Mine and adjacent areas, Quebec. D, 1955, McGill University.

Averill, E. L. Some aspects of stress-strain theories and their use in the interpretation of fracture patterns in rocks. M, 1953, McGill University.

Averill, Sally Ann. A sedimentologic study of the Mississippi River valley fill at Alton, Illinois. M, 1984, Southern Illinois University, Carbondale. 125 p.

Averitt, Paul. A sedimentary study of the Brannon Formation. M, 1931, University of Kentucky. 39 p.

Averitt, S. D. Bituminous sandstone of Kentucky. M, 1900, University of Kentucky.

Avers, Darrell D. Stratigraphy of the lower part of the Council Grove Group (Early Permian) in southeastern Nebraska and eastern Kansas. M, 1968, University of Nebraska, Lincoln.

Avery, C. Groundwater geology of Johnson County, Nebraska. M, 1978, University of Nebraska, Lincoln.

Avery, Donald C. Flow structures and zonation of ring fracture domes and their relation to ore control in the Mogollon mining district, Catron County, New Mex-

ico. M, 1988, New Mexico State University, Las Cruces. 63 p.

Avignone, Joseph. A palynological analysis of the Big Mary Coal, Windrock, Tennessee. M, 1964, University of Tennessee, Knoxville. 61 p.

Avila, Fred Angelo. Middle Tertiary stratigraphy of Santa Rosa Island, California. M, 1968, University of California, Santa Barbara.

Avis, Loren E. Connate water resistivity, Mississippian System, western Kansas. M, 1973, Wichita State University. 56 p.

Avison, A. T. Fracture patterns in heterogeneous rocks. M, 1954, McGill University.

Avolio, G. W. Granulometric analysis of Recent sediments of Tillamook Bay, Oregon. M, 1973, Portland State University. 67 p.

Avon, Lizanne. A contaminant transport model of trichloroethylene movement in groundwater at Castle Air Force Base, California. M, 1988, Stanford University. 72 p.

Avotins, Peter. Adsorption and coprecipitation studies of mercury on hydrous iron oxide. D, 1975, Stanford University.

Avramenko, Walter. Volcanism and structure in the vicinity of Echo Mountain, central Oregon Cascade Range. M, 1981, University of Oregon. 156 p.

Awai-Thorne, B. V. Palynology of the Bearpaw Formation (Campanian) and contiguous Upper Cretaceous strata from the CPOG strathmore EV Well, southern Alberta, Canada. M, 1972, University of Toronto.

Awak, Collins T. Origin of the Lake Bonaparte wollastonite deposit, Remington Corners, Lewis County, New York. M, 1985, Lehigh University.

Awald, John Theodore. A systems approach to mineral exploration. D, 1970, Stanford University. 184 p.

Awan, Muhammed Amjad. Petrology and geochemistry of the Dargai ultramafic complex, Pakistan. D, 1987, Purdue University. 213 p.

Awbrey, Elizabeth. A comparative study of species of the ostracode genus Cytheresis of the Washita Group in North Texas. M, 1939, University of Oklahoma. 55 p.

Awosika, Olakunle A. Geophysical interpretation of magnetic data from the northeastern United States and adjoining part of Canada. M, 1983, SUNY at Buffalo. 106 p.

Awramik, Stanley M. The origin and evolution of stromatolites with special reference to the Gunflint Iron Formation. D, 1973, Harvard University.

Axelrod, Daniel I. A Miocene flora from the western border of the Mohave Desert. D, 1938, University of California, Berkeley. 192 p.

Axelrod, Russell B. Tertiary sedimentary facies, depositional environments, and structure, Jefferson Basin, Southwest Montana. M, 1984, University of Montana. 64 p.

Axelsen, Claus. Pennsylvanian stratigraphy in South-central Idaho and adjacent areas. M, 1973, Oregon State University. 129 p.

Axelsson, Gudni. Hydrology and thermomechanics of liquid-dominated hydrothermal systems in Iceland. D, 1986, Oregon State University. 291 p.

Axelsson, Gudni. Tidal tilt observations in the Krafla geothermal area in North Iceland. M, 1981, Oregon State University. 152 p.

Axen, Gary James. Geology of the La Madre Mountain area, Spring Mountains; southern Nevada. M, 1980, Massachusetts Institute of Technology. 170 p.

Axenfeld, Sheldon. Geology of the North Scranton area, Tooele County, Utah. M, 1952, Indiana University, Bloomington. 32 p.

Axtell, Drew Cunningham. Geology of the northern part of the Malad Range (Oneida County), Idaho. M, 1967, Utah State University. 65 p.

Axtmann, Tyrrell Charles. Structural mechanisms and oil accumulation along the Mountain View-Wayne

Fault, south-central Oklahoma. M, 1983, University of Oklahoma. 70 p.

Ayala, Luis. Viscosity-controlled sediment transport in the absence of suspended load. D, 1980, University of California, Berkeley. 248 p.

Ayan, Cosan. Multiphase pressure buildup analysis; a history matching approach. D, 1988, Texas A&M University. 192 p.

Ayatollahi, M. S. Stress and flow in fractured porous media. D, 1978, University of California, Berkeley. 154 p.

Aycock, Lester C. The Claiborne-Jackson contact in Sabine and San Augustine counties, Texas. M, 1942, Louisiana State University.

Aycox, Tracie Leigh. Relationship between morphological variation and environment in Holocene North Atlantic benthic foraminifera. M, 1985, University of New Orleans. 59 p.

Aydin, A. Faulting in sandstone. D, 1978, Stanford University. 282 p.

Aydin, A. F. A numerical example in block adjustment. M, 1957, Ohio State University.

Aydin, Fehmi Numan. Mathematical simulation of unsteady flows and the mechanics of dispersion in estuaries. D, 1976, Rutgers, The State University, New Brunswick. 376 p.

Aydinoglu, Mustafa Ali. Energy in earthquakes. D, 1949, University of California, Berkeley. 146 p.

Aye, Tin. Deposits of the Des Moines Lobe margin, Marshall County, Iowa. M, 1955, Iowa State University of Science and Technology.

Aye, Tin. X-ray study of biotite from the Scott Mine, Sterling Lake, New York. D, 1958, University of Illinois, Urbana. 60 p.

Ayeni, O. O. Considerations for automated digital terrain models with applications in differential photo mapping. D, 1976, Ohio State University. 208 p.

Ayer, China O. Depositional environments of the Ordovician Hillier Limestone (Trenton Group) of northwestern New York. M, 1980, Boston University. 131 p.

Ayer, John Albert. The Mazinaw Lake metavolcanic complex, Grenville Province, southeastern Ontario. M, 1979, Carleton University. 113 p.

Ayer, Nathan John, Jr. Possible relationship between color loss in hyacinth zircons and meteoritic impact (Arizona). M, 1965, [University of California, San Diego].

Ayer, Nathan John, Jr. Statistical and petrographic comparison of artificially and naturally compacted carbonate sediments. D, 1971, University of Illinois, Urbana. 92 p.

Ayer, Robert Mitchell. Wave propagation over a submarine trench of arbitrary shape. D, 1983, University of Southern California.

Ayers, Jerry Floyd. A hydrologic study of an alpine karst, Flathead County, Montana. M, 1976, Washington State University. 86 p.

Ayers, Jerry Floyd. Unsteady behavior of fresh-water lenses in Bermuda with applications of a numerical model. D, 1980, Washington State University. 286 p.

Ayers, John C. Partitioning of elements between silicate melt and H_2O-NcCl fluids at mantle conditions. M, 1988, Pennsylvania State University, University Park. 132 p.

Ayers, Mark W. Depositional processes; their influence on the mineralogy of the Hatteras abyssal plain sediments. M, 1979, East Carolina University. 126 p.

Ayers, Robert L. A comparison of macroseismic and microseismic ground motion in Virginia. M, 1974, Virginia Polytechnic Institute and State University.

Ayers, Vincent Leonard. An olivine-nodule basalt and other recent lavas from Rice, Arizona. D, 1927, [University of Michigan].

Ayers, Vincent Leonard. The geologic structure of the lower Nittany Valley. M, 1920, Pennsylvania State University, University Park. 26 p.

Ayers, Walter Barton, Jr. Depositional systems and coal occurrence in the Fort Union Formation (Paleocene), Powder River basin, Wyoming and Montana. D, 1984, University of Texas, Austin. 226 p.

Aylor, Joseph Garnett, Jr. The geology of Mummy Mountain, Phoenix, Arizona. M, 1973, Arizona State University. 86 p.

Aylor, Richard Burns. A detailed study of the Triassic-Jurassic contact in central Wyoming. M, 1947, University of Missouri, Columbia.

Ayoub, Wafic Tawfic. Soil set-up due to pile driving. D, 1981, [Tulane University]. 480 p.

Ayres, Dean Esmond. Iron oxide genesis in the Brockman iron formation (Precambrian) and associated ore deposits, Western Australia. D, 1970, University of Wisconsin-Madison. 134 p.

Ayres, Emma. The insoluble residues of the Holston Marble. M, 1933, University of Tennessee, Knoxville. 58 p.

Ayres, Lorne Dale. Early Precambrian stratigraphy of part of Lake Superior Provincial Park, Ontario, Canada, and its implications for the origin of the Superior Province. D, 1969, Princeton University. 439 p.

Ayrton, William G. A study of the York River Formation in the Rimouski-Matapedia area, Quebec. M, 1962, Northwestern University.

Ayrton, William Grey. A structural study of the Chandler-Port Daniel area, Gaspe Peninsula, Quebec. D, 1964, Northwestern University. 263 p.

Aytuna, Sezgin. Forms of large folds in the Central Appalachians, Pennsylvania. D, 1984, University of Cincinnati. 248 p.

Ayuso, R. A. Electron microprobe major element determination and comparison of beads prepared by the flux and flux-free methods. M, 1975, SUNY at Stony Brook.

Ayuso, Robert Armando. Geology of the Bottle Lake Complex, Maine. D, 1982, Virginia Polytechnic Institute and State University. 264 p.

Azari, Mehdi. Unsteady state pressure behavior of a well in the presence of a slanted flow barrier. D, 1983, University of Southern California.

Azaroff, L. F. A one-dimensional Fourier analogue computer and its application to the refinement of the structure of cubanite. D, 1954, Massachusetts Institute of Technology. 15 p.

Azevedo, Antonio Expedito Gomes de *see* de Azevedo, Antonio Expedito Gomes

Azevedo, Joao Jose Rio Tinto de. Characterization of structural response to earthquake motion. D, 1984, Stanford University. 239 p.

Azevedo, Luiz Otavio Roffe. Infra-red spectrophotometry and X-ray diffractometry as tools in the study of nickel laterites. M, 1985, University of Arizona. 118 p.

Azevedo, Roberto Francisco. Centrifugal and analytical modeling of excavations in sand. D, 1983, University of Colorado. 239 p.

Azih, O. D. The carbothermic solvent-metal reduction process. M, 1979, Stanford University. 49 p.

Azim, Bilqees. Elemental analysis of zircon samples from Pacific Northwest beaches by INAA. M, 1988, Oregon State University. 111 p.

Azim, Syed A. The low-temperature Fischer assay distillation tests of certain coals and a correlation of the products of chronohorization with proximate and ultimate analysis. M, 1948, Colorado School of Mines. 60 p.

Azimi, Esmaeil. Use of remote sensing for fracture discrimination and assessment of pollution susceptibility of a limestone-chert aquifer in northeastern Oklahoma. M, 1978, Oklahoma State University. 98 p.

Aziz, Nadim Mahmoud. Sediment transport; a continuum mechanics approach. D, 1984, University of Mississippi. 117 p.

Azmon, Emanuel. Heavy minerals in sediments of Southern California. D, 1960, University of California, Los Angeles. 166 p.

Azmon, Emanuel. The geology of Point Mugu Quadrangle. M, 1956, University of California, Los Angeles.

Azrag, Elfadil Abd Elrhaman. Ground water aquifers management model for hydraulics and solute transport. D, 1987, University of California, Davis. 159 p.

Azuola Valls, Hannia. Occurrences and origin of heavy mineral placers in braided stream facies of the Agujas River, Osa Peninsula, Costa Rica. M, 1985, Pennsylvania State University, University Park.

Azzaria, L. M. Distribution of copper, lead, and zinc in the minerals of a granite. D, 1960, University of Toronto.

Baab, Patricia. The Pawnee Limestone Formation, Novinger Field, Meade County, Kansas. M, 1981, Wichita State University. 40 p.

Baafi, Ernest Yaw. Application of mathematical programming models to coal quality control. D, 1983, University of Arizona. 117 p.

Baag, Chang-Eob. Computation of the shear-coupled PL wave. D, 1983, Pennsylvania State University, University Park. 180 p.

Baag, Czang-go. Contributions to paleomagnetism; A, Geomagnetic secular variation Model E; B, Remanent magnetization of a 50m core from the Moenkopi Formation, western Colorado; C, Evidence for penecontemporaneous magnetization of the Moenkopi Formation. D, 1973, University of Texas at Dallas. 189 p.

Baag, Czang-Go. Magnetic investigations of a Cretaceous andesite ring dike in South Korea. M, 1968, Washington University. 76 p.

Baar, Hein J. W. de *see* de Baar, Hein J. W.

Baars, Donald Lee. Pre-Pennsylvanian paleotectonics of southwestern Colorado (San Juan County and vicinity) and east-central Utah. D, 1965, University of Colorado. 207 p.

Baba, Jumpei. Experiments on the settling behavior of irregular grains in a fluid. M, 1981, Oregon State University. 47 p.

Baba, Jumpei. Terrigenous sediments in two continental margin environments; western South America and the Gulf of California. D, 1986, Oregon State University. 200 p.

Baba, Nobuyoshi. A numerical investigation of Lake Ontario; dynamics and thermodynamics. D, 1974, Princeton University.

Babaei, Abdolali. Structural geometry and deformational history of western Benton Uplift, Arkansas. D, 1984, University of Missouri, Columbia. 189 p.

Babaei, Abdolali. The structural geology of part of the limestone hills in the Wichita Mountains, Caddo and Comanche counties, Oklahoma. M, 1980, Oklahoma State University. 63 p.

Babaie, Hassan Ali. Structural and tectonic history of the Golconda Allochthon, southern Toiyabe Range, Nevada. D, 1984, Northwestern University. 275 p.

Babalola, Olufemi O. Seismic stratigraphic analysis and tectono-depositional evolution of the unimbricated Ellesmerian Sequence (Mississippian to Lower Cretaceous), National Petroleum Reserve, North Slope, Alaska. M, 1984, University of Texas, Austin.

Babalola, Stephen Oladele. Modes of occurrence and textures of pyrite in Iowa coal. M, 1986, Iowa State University of Science and Technology. 139 p.

Babashoff, George, Jr. Temporal and spatial distribution of benthic foraminifera in the northeastern Gulf of Mexico. M, 1982, University of Miami. 103 p.

Babb, Carlton Scott. Geologic interpretive study and data resource evaluation of the Saint Clair and Macomb counties, Michigan subsurface 1969. M, 1969, Michigan State University. 105 p.

Babb, Janet L. Tertiary ash-flow tuffs in the southern Quitman Mountains and their relationship to calderas in the Eagle and northern Quitman mountains, Hudspeth County, Texas. M, 1988, West Texas State University. 187 p.

Babb, Josiah Smith. The conglomerates of the Jones-Ford Quadrangle. M, 1921, University of North Carolina, Chapel Hill. 19 p.

Babb, Robert Frederick, II. A study of compositional zoning in garnets from the Ducktown mining district, Tennessee; implications for mathematical modelling of garnet growth. D, 1981, University of Illinois, Urbana. 475 p.

Babcock, Burt A. Ground water occurrence and quality, San Diego County. M, 1958, University of Southern California.

Babcock, Dorothy Fern. The qualitative and quantitative analysis of a silicate mineral. M, 1933, University of Missouri, Columbia.

Babcock, Douglas Lee. The effects of weathering on the petrographic and fluorescent properties of sub-bituminous coal from the Fort Union Formation, Colorado. M, 1981, Southern Illinois University, Carbondale. 140 p.

Babcock, Elkanah Andrew. Geology and geophysics of the Durmid area, Imperial valley, California. D, 1969, University of California, Riverside. 145 p.

Babcock, Elkanah Andrew. Structural aspects of the folded belt area near Leeds, New York. M, 1966, Syracuse University.

Babcock, Gary B. Petrography and sedimentology of the Spearfish Formation (Permian and Triassic), Black Hills region (western South Dakota). M, 1967, University of Nebraska, Lincoln.

Babcock, Harold Earl. The historical geology of Devil's Lake, North Dakota. M, 1952, University of Washington.

Babcock, J. W. The late Cenozoic Coso volcanic field, Inyo County, California. D, 1977, University of California, Santa Barbara. 257 p.

Babcock, Jack Arthur. Geological significance of amino acids in selected Cenozoic and Cretaceous fossils. M, 1969, University of Wisconsin-Madison.

Babcock, Jack Arthur. The role of algae in the formation of the Capitan Limestone (Permian, Guadalupian), Guadalupe Mountains, West Texas-New Mexico. D, 1974, University of Wisconsin-Madison. 256 p.

Babcock, James B. Structural geology of the Batchelder Springs area, Inyo County, California. M, 1971, University of California, Santa Barbara.

Babcock, James Nissen. Geology of a portion of the Pinyon Well Quadrangle, Riverside County, California. M, 1961, University of California, Los Angeles.

Babcock, L. L. Mineralogy, geochemistry, and genesis of the magnetite-jacobsite mineral series and manganese-ferrite-bearing iron-formation from Champion Mine, Champion, Michigan. D, 1974, Michigan Technological University. 195 p.

Babcock, Laurel Clarke. Comparison of six late Paleozoic siliceous shales, Ouachita mountains, Oklahoma. M, 1969, University of Wisconsin-Madison.

Babcock, Laurel Clarke. Statistical approaches to the conodont paleoecology of the Lamar Limestone, Permian reef complex, West Texas. D, 1974, University of Wisconsin-Madison. 187 p.

Babcock, Loren Edward. Devonian and Mississippian conulariids of North America. M, 1986, Kent State University, Kent. 279 p.

Babcock, Randall Scott. Geochemistry of the main-state migmatitic gneisses in the Skagit gneiss complex (Mesozoic, central northern Washington). D, 1970, University of Washington. 147 p.

Babcock, Robert Earl. Longitudinal dispersion of thermal energy in unconsolidated packed beds. D, 1964, University of Oklahoma. 139 p.

Babcock, Russel C. Petrogeny of the granophyre and intermediate rock in Duluth Gabbro of northern Cook County, Minnesota. M, 1959, University of Wisconsin-Madison.

Babenroth, Donald L. Axinite from the Mother Lode, California. M, 1938, Columbia University, Teachers College.

Babiker, Hashim Musa, II. Geology of the Clegg copper mine and vicinity, Lee and Chatham counties, North Carolina. M, 1978, North Carolina State University. 67 p.

Babineau, Jacques. Evolution géochimique et pétrologique des series volcaniques de la région de Cadillac-Malartic, Abitibi. M, 1982, Universite de Montreal. 112 p.

Babisak, Julius. The geology of the southeastern portion of the Gunnison Plateau, Utah. M, 1949, Ohio State University.

Babits, Steven J. Gravity anomalies and lithospheric flexure beneath the Denver Basin; evidence for a buoyant subcrustal load. M, 1987, University of Wyoming. 49 p.

Babley, Ralph E. Eocene stratigraphy, Sekiu River area, Olympic Peninsula. M, 1959, University of Washington.

Babu, Peethambaram. Certain features of alteration in the Michigan lavas and their genetic interpretation. M, 1962, University of Missouri, Rolla.

Babuin, Michael L. Fluvial depositional processes of a tropical river, Colombia, South America. M, 1985, Old Dominion University. 139 p.

Baca, Brian R. Depositional environments of the Vaqueros Formation in the Lake Casitas - Red Mountain area, Ventura County, California. M, 1981, University of California, Santa Barbara. 130 p.

Bacelar de Oliveira, Ruy Bruno. Exploration for buried channels by shallow seismic reflection and resistivity and determination of elastic properties at Rocky Flats, Jefferson County, Colorado. M, 1975, Colorado School of Mines. 131 p.

Bach, William Earl. A petrographic and sedimentary study of the dolomites of Ordovician and Lower Silurian formations in Kentucky. M, 1938, University of Kentucky. 76 p.

Bachechi, Fiorella. Crystal structure of the naturally occurring Au telluride-montbrayite, Au_2Te_3. D, 1970, University of Toronto.

Bacheller, John, III. Quaternary geology of the Mojave Desert-eastern Transverse Ranges boundary in the vicinity of Twentynine Palms, California. M, 1978, University of California, Los Angeles.

Bachhuber, Frederick W. Paleolimnology of Lake Estancia and the Quaternary history of the Estancia Valley, central New Mexico. D, 1971, University of New Mexico. 238 p.

Bachhuber, Frederick Willard. Pollen analysis from Hansen Marsh, an upland site, south-central Wisconsin. M, 1966, University of Wisconsin-Madison.

Bachinski, Donald John. Geology, geochemistry and genesis of selected skarn deposits, northern New Brunswick, Canada. M, 1965, University of Missouri, Rolla.

Bachinski, Donald John. Metamorphism of cupriferous iron-sulfide-rich rocks in ophiolitic terrains. D, 1973, Yale University.

Bachinski, Sharon L. W. Alkali feldspars; their subsolidus behavior. D, 1972, Yale University.

Bachman, Leon Joseph. Geology for development planning in the Moshannon Valley region, Centre and Clearfield counties, Pennsylvania. M, 1980, Pennsylvania State University, University Park. 163 p.

Bachman, Mattias Edgar. Geology of the Water Hollow fault zone, Sevier and Sanpete counties, Utah. M, 1959, Ohio State University.

Bachman, Steven Bruce. Depositional and structural history of the Waucobi Lake bed deposits, Owens Valley, California. M, 1974, University of California, Los Angeles.

Bachman, Steven Bruce. Sedimentation and margin tectonics of the coastal belt Franciscan, Mendocino coast, northern California. D, 1979, University of California, Davis. 166 p.

Bachman, Wayne R. Paleogene conglomerates from Point Arena, California, through Valle de las Palmas, Baja California. M, 1984, San Diego State University. 253 p.

Bacho, Andrew Benjamin. A geologic and preliminary hydrologic comparison of McDonalds Branch and Middle Branch watersheds, Lebanon State Forest, New Jersey. M, 1955, University of Iowa. 58 p.

Bachrach, Ruth Esther. Upper Mississippian and Pennsylvanian stratigraphy in Teton, Lincoln, and Sublette counties, Wyoming. M, 1946, University of Michigan.

Bachtel, Steven L. Stratigraphy, depositional environments, and porosity development of the Mission Canyon Formation, Southeast Idaho and western Wyoming. M, 1984, University of Idaho. 113 p.

Bachus, Robert Charles. An investigation of the strength deformation response of naturally occurring lightly cemented sands. D, 1983, Stanford University. 350 p.

Back, David Bishop. Hydrogeology, digital solute-transport simulation of nitrate and geochemistry of fluoride in ground water of Comanche County, Oklahoma. M, 1985, Oklahoma State University. 134 p.

Back, Judith Mae. Petrologic characteristics of igneous rocks in the Van Horn Peak cauldron complex, Salmon River Mountains, central Idaho. M, 1982, University of Oregon. 203 p.

Back, William. Chemical hydrogeology of the carbonate peninsulas of Florida and Yucatan. D, 1969, University of Nevada. 81 p.

Back, William. Reconnaissance of geology and groundwater of Smith River plain, Del Norte County, California. M, 1955, University of California, Berkeley. 115 p.

Backlund, Alvin Lorenzo, Jr. A Pennsylvanian gastropod fauna from the Snyderville Quarry, Cass County, Nebraska. M, 1953, University of Nebraska, Lincoln.

Backman, Olen L. Makamik area, Abitibi District, Quebec. D, 1933, University of Minnesota, Minneapolis. 72 p.

Backsen, L. B. Geohydrology of the aquifer supplying Ames, Iowa. M, 1963, Iowa State University of Science and Technology.

Backus, M. M. Mass spectrometric determination of the relative isotope abundances of calcium and the determination of geologic age. D, 1956, Massachusetts Institute of Technology. 186 p.

Backush, Ibrahim M. A quantitative model of Gefara Plain-central part, NW Libya. M, 1983, Ohio University, Athens. 476 p.

Baclawski, Paul. Seismic stratigraphic study of six Pleistocene submarine canyons of the Northwest Gulf of Mexico. M, 1980, Brigham Young University.

Bacon, Charles Robert. High-temperature heat capacity of silicate glasses. D, 1975, University of California, Berkeley. 62 p.

Bacon, Charles Sumner, Jr. Geologic history of the Spokane region, Washington. M, 1923, University of Chicago. 43 p.

Bacon, Douglas J. Chert genesis in a Mississippian sabkha environment (saline plain), (Bayport Limestone, Huron County, Michigan). M, 1971, Michigan State University. 47 p.

Bacon, Joan Irene. A geochemical study of some manganese nodules. M, 1967, University of Tulsa. 53 p.

Bacon, Lloyal Orrin. A study of the potentials measured on an electrolyte-clay-electrolyte system. M, 1948, Pennsylvania State University, University Park. 73 p.

Bacon, Michael Putnam. Applications of Pb-210/Ra-226 and Pb-210/Pb-210 disequilibria in the study of marine geochemical processes. D, 1976, Massachusetts Institute of Technology. 165 p.

Bacon, Randall W. Geology of the northern Sierra de Catorce, San Luis Potosi, Mexico. M, 1978, University of Texas, Arlington. 124 p.

Bacon, W. R. Economic geology of the Polaris-Taku Mine, Tulsequah, British Columbia. M, 1952, University of British Columbia.

Bacon, Walter S. Character of the Franconia Sandstone at Taylors Falls, Minnesota. M, 1938, University of Minnesota, Minneapolis. 38 p.

Bacon, William R. The geology and mineral deposits of the Sechelt Peninsula-Jervis Inlet area, British Columbia. D, 1952, University of Toronto.

Bacon, William Russell. The economic geology of the Polaris Taku Mine, Talsequah, BC. M, 1942, University of British Columbia.

Bacopoulos, Ioannis. Solution-generated collapse (SGC) structures and formation-fluid hydrodynamics in southern Manitoba. M, 1988, University of Windsor. 233 p.

Bada, Jeffrey Lee. The reversible deamination of aspartic acid and its geochemical implications. D, 1968, [University of California, San Diego]. 109 p.

Badachhape, Abhaya Ramachandra. Mid-Cretaceous unconformities in the East Texas Basin and the Sabine Uplift. M, 1988, University of Texas, Austin. 76 p.

Badawy, Assem M. Sulfide-silicate metamorphism at the Blue Hill copper-zinc mine, Maine. D, 1978, Boston University. 186 p.

Badayos, Rodrigo Briones. Soils of the Oregon coastal fog belt in relation to the proposed "Andisol" order. D, 1983, Oregon State University. 145 p.

Baddley, Elmer R. A study of the Tintic standard ore deposit, Dividend, Utah. M, 1924, [Stanford University].

Baddour, Frederick R. Petroleum hydrocarbons in the canals that drain the Miami International Airport. M, 1983, University of Miami. 134 p.

Bader, Jeffrey W. Surface and subsurface structural relations of the Cherokee Ridge arch, south-central Wyoming. M, 1987, San Jose State University. 68 p.

Bader, Richard G. A quantitative study of some physical, chemical and biological variants of modern marine sediments. D, 1952, University of Chicago. 75 p.

Bader, Robert S. Variation and rates of evolution in the oreodonts. D, 1954, University of Chicago. 21 p.

Bader, Roger H. Stratigraphy, paleontology and palynology of Holocene deposits in southwestern Lake Michigan. M, 1981, Northeastern Illinois University. 165 p.

Badger, Robyn Lucas. Geology of the western Lion Canyon Quadrangle, Ventura County, California. M, 1957, University of California, Los Angeles.

Badger, William Barton. Phase relations in the system $Li_2O-Na_2O-Al_2O_3-SiO_2$ with special emphasis on the silica polymorphs. D, 1968, Pennsylvania State University, University Park. 311 p.

Badger, William Wiley. Structure of friable (Sioux City) Iowa loess. D, 1972, Iowa State University of Science and Technology.

Badgley, Catherine Elizabeth. Community reconstruction of a Siwalik mammalian assemblage. D, 1982, Yale University. 384 p.

Badgley, Edmund Kirk, Jr. Correlation of some Late Cretaceous-early Tertiary sediments in eastern Montana. M, 1953, University of Wyoming. 105 p.

Badgley, Peter Coles. Stratigraphy, sedimentology and oil and gas geology of the lower Cretaceous in central Alberta. D, 1952, Princeton University. 293 p.

Badham, John Patrick Nicholas. Volcanogenesis, orogenesis and metallogenesis, Camsell River area, Northwest Territories. D, 1973, University of Alberta. 334 p.

Badie, Ahmad. Stability of spread footing supported by clay soil with an underground void. D, 1983, Pennsylvania State University, University Park. 232 p.

Badiei, Jalil. Correlation and comparison of the Lost River Chert of the Ste. Genevieve Formation in South-central Indiana and North-central Kentucky. M, 1981, Indiana State University. 81 p.

Badiey, Mohsen. Acoustic normal mode propagation in shallow waters over inhomogenous anisotropic porous sediments. D, 1988, University of Miami. 87 p.

Badiozamani, Khosrow. The Dorag dolomitization model-application to the Middle Ordovician of Wisconsin. D, 1972, Northwestern University.

Badir, Hashim Al-Refai. A heavy mineral analysis of contemporary sand derived from the Boone Limestone Formation in Benton County, Northwest Arkansas. M, 1956, University of Arkansas, Fayetteville.

Badley, Ruth Hall. Petrography and chemistry of the East Fork dike swarm, Ravalli County, Montana. M, 1978, University of Montana. 54 p.

Badon, Calvin L. Tectonic history of a portion of the Hurricane fault zone, northwestern Arizona. M, 1963, University of Kansas. 45 p.

Badon, Calvin Lee. Petrology of the Norphlet and Smackover formations (Jurassic), Clarke County, Mississippi. D, 1973, Louisiana State University.

Badri, Mohammed. Seismic wave attenuation in uncondolidated sediments. D, 1985, University of Minnesota, Minneapolis. 537 p.

Baechler, F. E. The effect of Metro Kitchener, as an urban area on the sediment load and regime of the Grand River. M, 1974, University of Waterloo.

Baedecker, Mary Jo. Organic material in sediments of Great Salt Lake, Utah; influence of changing depositional environments. D, 1985, George Washington University. 44 p.

Baedecker, Philip A. The distribution of gold and iridium in meteoritic and terrestrial materials. D, 1967, [University of Kentucky]. 110 p.

Baegi, Mohamed Bashir. Turbidite deposition on Bellingshausen abyssal plain; sedimentologic implications. M, 1985, Rice University. 189 p.

Baehr, William M. An investigation of the relationship between rock structure and drainage in the southern half of the Junction City, Kansas, Quadrangle. M, 1954, Kansas State University. 68 p.

Baer, David Walter. A reflection seismic study in the upper crust in the Precambrian Shield near Kenora, (Ontario). M, 1972, University of Manitoba.

Baer, James Logan. Geology of the Star Range, Beaver County, Utah. M, 1962, Brigham Young University. 52 p.

Baer, James Logan. Paleoecology of cyclic sediments of the lower Green River Formation, central Utah. D, 1968, Brigham Young University.

Baer, Norbert Sebastian. Topochemical studies on calcite. D, 1969, New York University. 184 p.

Baer, Roger Lawrence. Petrology of quaternary lavas and geomorphology of lava tubes, south flank of Medicine Lake, Highland, California. M, 1973, University of New Mexico. 120 p.

Baerg, James R. Analysis of COCRUST seismic refraction and wide angle reflection experiments from southern Saskatchewan and Manitoba. M, 1985, University of Western Ontario. 155 p.

Baerns, Rudolph. The urbanization of the lower Fox River valley. M, 1950, Washington University. 70 p.

Baesemann, John F. Missourian conodonts (Pennsylvanian) of northeastern Kansas. D, 1971, University of Iowa. 112 p.

Baesemann, John Frederick. Silicified Spiriferidina of the Middle Devonian Lingle Limestone of southwestern Illinois. M, 1966, Southern Illinois University, Carbondale. 57 p.

Baetcke, Gustav Berndt. Stratigraphy of the Star range (about 10 miles S.W.W. of Milford, Utah, Beaver County, Utah) and reconnaissance study of three selected mines. D, 1969, University of Utah. 359 p.

Baetcke, Gustav Berndt and Erickson, Lance. Structure of the Lobate thrust sheet, Huerfano Park, Colorado. M, 1957, University of Michigan.

Baffaut, Claire. Knowledge techniques for the analysis of urban runoff using SWMM. D, 1988, Purdue University. 321 p.

Bagan, Richard John. The Greenhorn Formation of western South Dakota. M, 1955, South Dakota School of Mines & Technology.

Baganz, Bruce P. Carboniferous and Recent Mississippi lower delta plain sedimentation. M, 1975, University of South Carolina.

Baganz, Bruce P. Paleoenvironmental model for the Pocahontas No. 1, No. 2, and No. 3 coal seams, southern West Virginia. D, 1979, University of South Carolina. 115 p.

Bagby, W. C. Geology, geochemistry, and geochronology of the Batopilas Quadrangle, Sierra Madre Occidental, Chihuahua, Mexico. D, 1979, University of California, Santa Cruz. 271 p.

Bagchi, S. Mathematical model for groundwater reservoir operation. D, 1977, Polytechnic University. 372 p.

Bagdadi, Khaled Ali. Application of the gravity and the seismic refraction methods for mapping bedrock topography in East-central Minnesota; a feasibility study. M, 1977, University of Minnesota, Minneapolis. 237 p.

Bagdadi, Khaled Ali. Tectonic studies of the Midcontinent gravity high in East-central Minnesota and western Wisconsin; Part I, Gravity field and crustal structure; Part II, Application of phase transition models to the Precambrian vertical motion history. D, 1981, University of Minnesota, Minneapolis. 196 p.

Bagg, Rufus Mather. The Cretaceous foraminifera of New Jersey. D, 1895, The Johns Hopkins University. 89 p.

Baggett, Stephen Myles. Ground-water resources of the Bryan area, Williams County, Ohio. M, 1987, University of Toledo. 145 p.

Baggs, Charles Chaplin. Bed load dynamics. D, 1984, Colorado State University. 101 p.

Baghai, Nina Lucille. The Miocene Clarkia flora of Idaho; Liriodendron megascopic and microscopic statistical analysis of the fossil and living species. M, 1983, University of Idaho. 147 p.

Baghanem, Ali M. Geology of the lake beds near Turabah, Saudi Arabia. M, 1972, South Dakota School of Mines & Technology.

Bagheri, Saeed. Water resources in the northeast part of Sedgwick County and the southeast part of Harvey County, Kansas. M, 1980, Wichita State University. 77 p.

Baghoomian, Ovaness. Excess pore pressure and in situ measurement of shear strength gain in clays. D, 1972, University of Utah. 230 p.

Bagley, Ralph Eugene. Eocene stratigraphy of Sekiu River area, Olympic Peninsula, Washington. M, 1959, University of Washington. 130 p.

Bagley, Roy Louis, Jr. The boulder beds of the Marathon Formation (Lower Ordovician), Marathon Basin, West Texas. M, 1972, Texas Christian University. 68 p.

Baglio, Joseph V., Jr. Hydrogeologic investigations of geothermal systems in the vicinity of the Bear River Range, southeastern Idaho. M, 1983, University of Idaho. 83 p.

Bagshaw, Lawrence H. Paleoecology of the lower Carmel Formation of the San Rafael Swell, Emery County, Utah. M, 1977, Brigham Young University. 62 p.

Bagshaw, Rebecca Lillywhite. Foraminiferal abundance related to bentonitic ash beds in the Tununk Member of the Mancos Shale (Cretaceous) in southeastern Utah. M, 1977, Brigham Young University. 49 p.

Bagstad, David Peter. Structural analysis of folding of Paleozoic sequence, Solitario Uplift, Trans-Pecos, Texas. M, 1981, Texas Tech University. 42 p.

Bahabri, Mohammad Sultan. A study of amphibolite rocks in different metamorphic environments, Black Hills, South Dakota. M, 1975, South Dakota School of Mines & Technology.

Bahadoran, Behzad. Water resources in the Nemo area, South Dakota. M, 1976, South Dakota School of Mines & Technology.

Bahadur, Sher. In situ stress measurements, Black Hills, S. Dakota. D, 1972, South Dakota School of Mines & Technology.

Bahan, Walter George. Microfauna of the Joli Fou Formation in north central Alberta. M, 1951, University of Alberta. 66 p.

Baharaloui, Abdolhossein. Quantitative study of electrical parameters, Prue Sand (Pennsylvanian), southern Creek County, Oklahoma. M, 1964, University of Oklahoma. 173 p.

Baharloui, Abdolhossein. A comparison of the chemical composition of interstitial waters of shales and associated brines. D, 1973, University of Tulsa. 83 p.

Bahavar, Manouchehr. The reflection of channel waves from obstructions in a coal seam with dissimilar roof and floor; numerical modeling. D, 1984, Colorado School of Mines. 197 p.

Bahia-Guimaraes, Paulo Fernando. The genesis of the antimony-mercury deposits of the Stayton District, California. D, 1972, Stanford University. 185 p.

Bahjat, Abdullah Mohammed. Development of a methodology to determine optimal investment strategies for water storage projects. D, 1988, Colorado State University. 171 p.

Bahjat, Dhari Saaid. Decoupling and source function for explorations in a three-dimensional model. D, 1968, St. Louis University. 205 p.

Bahjat, Dhari Saaid. Seismic model study of reflections in media containing fluids. M, 1964, St. Louis University.

Bahlburg, William C. Depositional environments of the Tamaroa Sequence (Mississippian) of southeastern Arizona, southwestern New Mexico, and northern Mexico. M, 1977, Arizona State University. 214 p.

Bahmanyar, Gholam Hossein. A study of buried-pipe failures in cold climates. D, 1982, University of Wisconsin-Madison. 205 p.

Bahr, Charles H. Precambrian geology of the Aero Lakes area, Beartooth Mountains, Montana. M, 1980, Northern Illinois University. 138 p.

Bahr, Jean Marie. Applicability of the local equilibrium assumption in mathematical models for groundwater transport of reacting solutes. D, 1987, Stanford University. 252 p.

Bahr, Jean Marie. Hydrostratigraphic interpretation and ground water flow modelling of sedimentary deposits in the vicinity of Gloucester Landfill, Ontario, Canada. M, 1984, Stanford University. 90 p.

Bahr, John Robert. Sulphur isotopic fractionation between H_2S, $S°$ and SO_4^{2-} in aqueous solutions in possible mechanisms controlling isotopic equilibrium in natural systems. M, 1976, Pennsylvania State University, University Park. 75 p.

Bahr, Tim J. Chemical evolution of ground waters within upper Pottsville aquifers, eastern Columbiana County, Ohio. M, 1985, University of Akron. 143 p.

Bahyrcz, G. S. Geology of the Grey River area, Newfoundland, with special reference to metamorphism. M, 1957, McGill University.

Baiamonte, Matthew J. Rb-Sr isotopic study of the Rosetown Complex, New York, and regional implications. M, 1976, Brooklyn College (CUNY).

Baichtal, James Fay. The geology of Waldron, Bare, and Skipjack islands, San Juan County, Washington. M, 1982, Washington State University. 222 p.

Baie, Lyle Frederick. Post-Cretaceous structures and sediment of the northeast Campeche platform, Gulf of Mexico. D, 1970, Texas A&M University. 145 p.

Baie, Lyle Frederick. Submarine topography and sediments of the lower continental slope off east-central Mexico. M, 1967, Texas A&M University.

Baier, David. Mechanical properties of polycrystalline freshwater ice as a function of optic axis orientation. M, 1977, University of Wisconsin-Milwaukee.

Baier, Roger W. Lead distribution in coastal waters. D, 1971, University of Washington. 189 p.

Baik, Ho Yeal. Paleozoic stratigraphy and petrology along the Plum River fault zone in Jackson and Clinton counties, eastern Iowa. M, 1980, University of Iowa. 185 p.

Bail, Pierre. Problèmes géomorphologiques de l'englacement et de la transgression marine pleistocènes en Gaspésie sud-orientale. D, 1983, McGill University.

Bailar, Elizabeth F. Geochemistry and petrogenesis of the Cranberry magnetite deposit, Blue Ridge Province, North Carolina. M, 1984, University of North Carolina, Chapel Hill. 78 p.

Bailes, Alan H. Sedimentology and metamorphism of a Proterozoic volcaniclastic turbidite suite that crosses the boundary between the Flin Flon and Kisseynew belts, File Lake, Manitoba, Canada. D, 1979, University of Manitoba.

Bailes, Alan Harvey. The geology and geochemistry of the Pilot-Smuggler shear zone, Rice Lake, Manitoba. M, 1969, University of Manitoba.

Bailes, Richard James. The Cariboo-Bell alkaline stock, British Columbia. M, 1977, University of Manitoba.

Bailey, Alan. Comparison of low-temperature with high-temperature diffusion of sodium in albite. D, 1970, Michigan State University. 89 p.

Bailey, Alan. Course and extent of alteration of selected ferromagnesian silicates by aqueous solutions of oxalic acid. M, 1967, Michigan State University. 51 p.

Bailey, Alan Clarke. Solute and particle gradients in the benthic boundary layer. D, 1988, Clemson University. 152 p.

Bailey, Alvin Cornell. Yielding of unsaturated soils. M, 1967, Auburn University. 157 p.

Bailey, Arthur C., Jr. Geology of the Smith's Crossroads area, Troup County, Georgia. M, 1969, University of Georgia. 52 p.

Bailey, Clarence G. The geology along the Green Mountain front in the region of Burlington, Vermont. M, 1936, University of Vermont.

Bailey, David G. Stratigraphy and geochemistry of the Troodos Ophiolite extrusive sequence in the Margi area, Cyprus. M, 1984, Dalhousie University. 215 p.

Bailey, David Gerard. The geology of the Morehead Lake area, South central British Columbia. D, 1978, Queen's University. 198 p.

Bailey, Edgar H. Mineralogy, petrology, and geology of Santa Catalina Island, California. D, 1941, Stanford University. 193 p.

Bailey, G. B. The occurrence, origin, and economic significance of gold-bearing jasperoids in the central Drum Mountains, Utah. D, 1975, Stanford University. 354 p.

Bailey, George B. Wall rock alteration as a quide to lode deposit ore in the Park City District, Utah. M, 1971, University of Iowa. 161 p.

Bailey, Gordon C. Stratigraphic relationships of the Banting Group, Yellowknife Supergroup, and a reappraisal of post-Archean faults near Yellowknife, N.W.T. M, 1987, Queen's University. 131 p.

Bailey, Harry P. Physical geography of the San Gabriel Mountains, California. D, 1950, University of California, Los Angeles. 246 p.

Bailey, Henry H. Sand-shale ratio and isopach maps as aids in determining the depositional environment of the Tertiary formations of the Velasquez Field area, Colombia, South America. M, 1956, University of Wisconsin-Madison.

Bailey, J. B. Systematics, functional morphology, and ecology of middle Devonian bivalves from the Solsville Member (Marcellus Formation), Chenango Valley, New York. D, 1975, University of Illinois, Urbana. 303 p.

Bailey, James Peter. Seismicity and contemporary tectonics of the Hebgen Lake-Centennial Valley, Montana area. M, 1977, University of Utah. 115 p.

Bailey, James Stuart. A stratigraphic analysis of Rico strata in the Four Corners region. D, 1955, University of Arizona. 116 p.

Bailey, James Stuart. Hydrogeochemical investigation of ground water in the northern Wallowa Mountains, Oregon. M, 1984, Washington State University. 85 p.

Bailey, James Stuart. Permian system of the southern Middle Rocky Mountains region. M, 1953, Northwestern University.

Bailey, Janet Allen. Structural chemistry and possible origin of illite and smectite in Lake Turkana, Kenya. M, 1986, University of Massachusetts. 105 p.

Bailey, Janet E. Occurrence of dissolved methane resources in formation waters of the hydropressure zone in southern Louisiana. M, 1978, Louisiana State University.

Bailey, John Richard. A petrofabric investigation for preferred lattice orientation of quartz aggregate in concrete from Hale Bar Dam (Tennessee) and its possible relation to concrete movement (creep). D, 1964, University of Mississippi.

Bailey, Jonas William. Stratigraphy, environments of deposition, and petrography of the Cotton Valley Terryville Formation in eastern Texas. M, 1983, University of Texas, Austin. 229 p.

Bailey, Lee Eldon. Geology of Scott Summit, Klamath Mountains, Northern California. M, 1980, University of Oregon. 111 p.

Bailey, Lester. The geology of the Ozora District of Ste. Genevieve County, Missouri. M, 1921, University of Missouri, Columbia.

Bailey, Loren T. Lower Pennsylvanian (upper Caseyville-lower Abbott) depositional environments in southwestern Jackson County, Illinois. M, 1975, Southern Illinois University, Carbondale. 131 p.

Bailey, M. E. The magnetic properties of pseudo-single domain grains. M, 1975, University of Toronto.

Bailey, Marcia Lynn. Hazardous waste ground-water task force evaluation of Envirosafe Services, Inc. Site B, Grand View, Idaho. D, 1988, University of California, Los Angeles. 239 p.

Bailey, Mark H. Bedrock geology of western Taylor Park, Gunnison County, Colorado. M, 1980, Oregon State University. 123 p.

Bailey, Marshall W. A sedimentary study of Ogallala Formation. M, 1949, Texas Tech University. 25 p.

Bailey, Michael Mathewson. Revisions to stratigraphic nomenclature and petrogenesis of the Picture Gorge Basalt subgroup, Columbia River Basalt Group. D, 1988, Washington State University. 238 p.

Bailey, Monika Ella. Magnetic properties of deep-ocean basalts. D, 1980, University of Toronto.

Bailey, Palmer K. Periglacial landforms and processes in the southern Kenai Mountains, Alaska. M, 1980, University of North Dakota. 159 p.

Bailey, Paul T. The development of and history of the Garza Field, Garza County, Texas. M, 1953, Texas Tech University. 62 p.

Bailey, R. A. Volcanism, structure, and petrology of Long Valley Caldera, California. D, 1978, The Johns Hopkins University. 154 p.

Bailey, Ralph Fraser. A summary of photogeologic techniques with some applications to the Cypress Hills, Saskatchewan (Canada). M, 1953, Pennsylvania State University, University Park. 101 p.

Bailey, Reed Warner. A contribution to the geology of the Bear River Range, Utah. M, 1927, University of Chicago. 63 p.

Bailey, Richard Hendricks. Paleoenvironment, paleoecology, and stratigraphy of molluscan assemblages from the Yorktown Formation (upper Miocene-lower Pliocene) of North Carolina. D, 1973, University of North Carolina, Chapel Hill. 110 p.

Bailey, Richard Hendricks. Relationships between pelecypod assemblages and sediment types in the Yorktown Formation (upper Miocene) along the Chowan River, northeastern North Carolina. M, 1971, University of North Carolina, Chapel Hill. 56 p.

Bailey, Robert. Recent marine sediments off the Texas coast. M, 1940, Louisiana State University.

Bailey, Robert G. Soil slips on the San Dimas Experimental Forest, Northridge, California. M, 1967, San Fernando Valley State University.

Bailey, Robert Gale. Landslides and related hazards in Teton National Forest, northwest Wyoming. D, 1971, University of California, Los Angeles. 323 p.

Bailey, Robert L. Geology and ore deposits of the Alder Gulch area, Little Rocky Mountains, Montana. M, 1974, Montana State University. 81 p.

Bailey, Roy Alden. Fusion of arkosic sandstone by intrusive andesite, Valles Mountain, New Mexico. M, 1954, Cornell University.

Bailey, Stephen Milton. Paleocurrent analysis of the Cretaceous Rosario Formation, Baja California, Mexico. M, 1966, [University of California, San Diego].

Bailey, Sturges Williams. Liquid inclusions in granite thermometry. M, 1948, University of Wisconsin-Madison. 39 p.

Bailey, T. P. A hydrogeological and subsurface study of Imperial Valley geothermal anomalies, Imperial Valley, California. M, 1977, University of Colorado.

Bailey, Thomas Corwin, III. A geochemical and petrologic study of core MQ-63-1 penetrating the Columbus Limestone (Middle Devonian) and Detroit River Group (Middle Devonian) at Marblehead, Ohio. M, 1968, Bowling Green State University. 140 p.

Bailey, Thomas Laval. The Gueydan Tuff, a new middle Tertiary formation from the south-western coastal plain of Texas. D, 1926, University of California, Berkeley. 213 p.

Bailey, Thomas Laval. The petrography and origin of the Recent sediments along the east side of San Francisco Peninsula. M, 1921, University of California, Berkeley. 175 p.

Bailey, Willard Francis. A study of the micropaleontology and stratigraphy of the Lower Pennsylvanian of central Missouri. D, 1934, University of Missouri, Columbia.

Bailey, Willard Francis. Invertebrates from the Triassic of the mid-western United States. M, 1926, University of Missouri, Columbia.

Bailey, William. Petrology of selected granites in northern Wisconsin. M, 1983, University of Wisconsin-Milwaukee. 115 p.

Bailey, William C. Economic evaluation of Idaho's water supply. M, 1975, University of Idaho. 96 p.

Bailey, William M. Geology of the northern half of the Horseshoe Mountain Quadrangle, Nelson County, Virginia. M, 1983, University of Georgia.

Baillie, Andrew D. Devonian stratigraphy of Lake Manitoba-Lake Winnipegosis area, Manitoba. M, 1950, University of Manitoba.

Baillie, Andrew Dollar. Devonian system of the Williston Basin area. D, 1953, Northwestern University. 265 p.

Baillie, Priscilla Woods. Microalgal - microhabitat interactions; the structure and function of the epipelic diatom community in intertidal estuarine sediments. D, 1983, University of Connecticut. 160 p.

Baillie, William Norman. Geology of the Fogg Hill area, Bonneville and Teton counties, Idaho. M, 1960, University of Idaho. 45 p.

Baillieul, Thomas A. The Cascade slide; a mineralogical investigation of a calc-silicate body on Cascade Mountain, Town of Keene, Essex County, New York. M, 1976, University of Massachusetts. 126 p.

Baillio, Robert H. Stratigraphy, petrography and clay mineralogy of Oceana Ridge, Virginia Beach, Virginia. M, 1964, Virginia Polytechnic Institute and State University.

Bailly, Paul Alain. Geology of the southeastern part of Mineral Ridge, Esmeralda County, Nevada. D, 1952, Stanford University. 186 p.

Bain, George L. Geology of the Nokesville Quadrangle, Virginia. M, 1959, West Virginia University.

Bain, George W. Geology and problems of the Webb-wood area, Canada. D, 1927, Columbia University, Teachers College.

Bain, George W. The specular haematite deposits and Huronian formations of Deroche Township, Ontario. M, 1923, Columbia University, Teachers College.

Bain, H. Foster. Relations of the Wisconsin and Kansas drift sheets in central Iowa and related phenomena. D, 1897, University of Chicago. 43 p.

Bain, Ian. Structure and stratigraphy of the Grenville in the vicinity of the Elzevir Batholith (Quebec). M, 1953, University of Toronto.

Bain, Ian. The geology of the Grenville Belt through actinolite, Ontario. D, 1960, University of Toronto.

Bain, James S. The nature of the Cretaceous, pre-Cretaceous contact, north-central Texas. M, 1973, Baylor University. 112 p.

Bain, Leslie Gay. The nature, distribution, and origin of terraces in the Cuyahoga River valley between Akron and Peninsula, Ohio. M, 1975, University of Akron. 60 p.

Bain, Roger J. Paleoecology of some Leonardian patch reefs in the Glass Mountains, Texas. D, 1967, Brigham Young University.

Bain, Roger John. The stratigraphy of the Permian Taku Group of the Tagish Lake area, Yukon Territory. M, 1964, University of Wisconsin-Madison.

Bain, Roland John. Geology of the Eureka Canyon area, Ventura County, California. M, 1954, University of California, Los Angeles.

Bainbridge, C. W., III. The Carnegie Plateau. M, 1973, University of Hawaii. 73 p.

Bainbridge, Russell Benjamin, Jr. Stratigraphy of the upper member, Koobi Fora Formation, southern Karari Escarpment, East Turkana Basin, Kenya. M, 1976, Iowa State University of Science and Technology.

Bainer, Robert W. The effect of seasonal variations on geophysical measurements along the Hayward Fault. M, 1982, California State University, Hayward. 175 p.

Bains, Trilochan Singh. A study of non-clay minerals in Bovill clay deposits. M, 1961, University of Idaho. 48 p.

Bair, Edwin Scott. Numerical simulation of the hydrogeologic effects of open-pit anthracite mining. D, 1980, Pennsylvania State University, University Park. 308 p.

Bair, Edwin Scott. Permeability distribution in surficial glacial outwash revealed by a shallow geothermal prospecting technique. M, 1976, Pennsylvania State University, University Park. 91 p.

Bair, John H. Palynology of some Lower Crectaceous sediments from the Falkland Plateau. M, 1980, Louisiana State University.

Baird, Alex K. Geology of a portion of San Antonio Canyon, San Gabriel Mountains, California. M, 1956, Pomona College.

Baird, Alexander Kennedy. Superposed deformations in the central Sierra Nevada Foothills, east of the Mother Lode. D, 1960, University of California, Berkeley. 102 p.

Baird, David McCurdy. Geology and mineral deposits of the Burlington Peninsula, Newfoundland. D, 1947, McGill University.

Baird, David McCurdy. Titaniferous magnetites in anorthosite east of St. Georges, Newfoundland. M, 1943, University of Rochester. 100 p.

Baird, Donald. Modern exhibition of vertebrate paleontology. M, 1949, University of Colorado.

Baird, Donald. Three reptilian ichnite faunules from the Newark Triassic of Milford, New Jersey. D, 1955, Harvard University. 96 p.

Baird, Donald Wallace. Pennsylvanian fusulinids from the Piedra River valley, Archuleta County, Colorado. M, 1956, University of Illinois, Urbana. 133 p.

Baird, Gordon C. Paleoecology of the Beil Limestone (Upper Pennsylvanian) in the northern midcontinent region. M, 1971, University of Nebraska, Lincoln.

Baird, Gordon Cardwell. Paleoecology and taphonomy associated with submarine discontinuities in the geologic record. D, 1975, University of Rochester. 107 p.

Baird, James Kaye. Geology of the Alcester Quadrangle, South Dakota-Iowa. M, 1957, University of South Dakota. 136 p.

Baird, John. Potential drop ratios in the case of a stratified medium. D, 1940, Colorado School of Mines. 56 p.

Baird, Lucille Bailey. Museum presentation of vertebrate paleontology. M, 1949, University of Colorado.

Baird, Mary Rebecca. Conodont biostratigraphy of the Kaibab Formation, eastern Nevada and west-central Utah. M, 1975, Ohio State University.

Baird, Robert Alan. A geochemical survey of the top of the Knox Dolomite in central Kentucky. M, 1981, University of Kentucky. 85 p.

Baitis, Hartmut W. Geology, petrography, and petrology of Pinzon and Santiago islands, Galapagos Archipelago. D, 1976, University of Oregon. 271 p.

Bajabaa, Saleh A. S. Hydrogeology of Wadi Turabah, Saudi Arabia. M, 1984, South Dakota School of Mines & Technology.

Bajak, Doris M. An investigation of strain patterns and mesoscopic structures associated with ramp-induced folds. M, 1983, Virginia Polytechnic Institute and State University. 120 p.

Bajc, A. F. Molluscan paleoecology and Superior Basin water levels, Marathon, Ontario. M, 1986, University of Waterloo. 271 p.

Bajsarowicz, Caroline J. Diagenesis of the Bromide, Simpson Group, Oklahoma. M, 1983, University of Missouri, Columbia.

Bajwa, Rajinder Singh. Irrigation potentials in humid regions of eastern United States based on drought and market conditions. D, 1980, University of Michigan. 215 p.

Bajza, Charles Carl. A special case of circumvallation at Hindostan Falls region, Martin County, Indiana. M, 1944, Indiana University, Bloomington. 33 p.

Bajza, Ester Ruth. A study of the Chester Series of central Indiana. M, 1944, Indiana University, Bloomington.

Bakar, M. Y. Abu *see* Abu Bakar, M. Y.

Bakar, Othmanbin Abu *see* Abu Bakar, Othmanbin

Bakbak, Mohamed Rida. Three-dimensional numerical modelling in resistivity and IP prospecting. D, 1977, University of California, Berkeley. 132 p.

Bakel, Allen J. Sulfur and nitrogen compounds in oils, asphaltenes, kerogens, and coals. M, 1987, University of Oklahoma. 93 p.

Baken, Jeffrey Frank. The structural geology and tectonic history of the northern Flint Creek Range, western Montana. M, 1984, Montana State University. 125 p.

Baker, Alfred H., Jr. Foraminiferal paleoecology of the Fredericksburg and Washita groups (Lower Cretaceous) from North Texas. M, 1976, Louisiana State University.

Baker, Arthur A. Geology of the Moab District, Grand and San Juan counties, Utah. D, 1931, Yale University.

Baker, Arthur, III. Pyrometasomatic ore deposits at Johnson Camp, Arizona. D, 1953, Stanford University. 93 p.

Baker, B. A. Stratigraphy and depositional environment of lower Pliocene strata from the Ventura Avenue Field, California. M, 1975, University of Houston.

Baker, Bernard Boyd. A lithofacies and current direction study of the Dunkard Group in southwestern West Virginia. M, 1964, Miami University (Ohio). 171 p.

Baker, Bruce H. The effects of solid solution on the unit cell dimensions and index of refraction of sphalerite. M, 1963, Miami University (Ohio). 88 p.

Baker, Bruce W. Geology and depositional environments of Upper Cretaceous rocks, Sevilleta Grant, Socorro County, New Mexico. M, 1981, New Mexico Institute of Mining and Technology. 159 p.

Baker, C. L. A microfabric study of the Maryhill, Mornington and Stirton tills near Waterloo, Ontario. M, 1978, University of Waterloo.

Baker, Cathy. Intraspecific variation in the Morrowan ammonoid Arkanites relictus (Quinn, McCaleb, and Webb). M, 1978, University of Arkansas, Fayetteville.

Baker, Cathy. Selected studies in Permian ammonoids. D, 1986, University of Iowa. 220 p.

Baker, Charles Allen. A study of estuarine sedimentation in South Slough, Coos Bay, Oregon. M, 1978, Portland State University. 104 p.

Baker, Charles Hays. Slope stability investigation of the Snake Hill Shale, Hudson Valley region, New York. M, 1981, Purdue University. 126 p.

Baker, Charles Laurence. Geology and underground waters of the northern Llano Estacado of Texas. M, 1916, University of California, Berkeley.

Baker, Christopher Thomas. Forecasting Kentucky coal production. D, 1987, University of Kentucky. 263 p.

Baker, Colin Woods. Evolution and hybridization in the radiolarian genera Theocorythium and Lamprocyclas. M, 1982, University of Michigan.

Baker, D. J. Significance of differences between $^{40}Ar/^{39}Ar$ and K-Ar uplift ages of portions of the northwestern Reading Prong; New York-New Jersey. M, 1974, Ohio State University.

Baker, Dan M. Balanced structural cross section of the central Salt Range and Potwar Plateau of Pakistan; shortening and overthrust deformation. M, 1988, Oregon State University. 120 p.

Baker, David Alan. Subsurface geology of southwestern Pawnee County, Oklahoma. M, 1958, University of Oklahoma. 46 p.

Baker, David W. Application of the geometric inequality to the solution of systems of nonlinear equations. D, 1980, Colorado School of Mines. 54 p.

Baker, David Warren. X-ray analysis and representation of preferred orientations in crystal aggregates. D, 1969, University of California, Los Angeles.

Baker, Dewey Allen. Geology of the Stockton Quadrangle, Cedar County, Missouri. M, 1962, University of Missouri, Columbia.

Baker, Don Read. The compositions of melts coexisting with plagioclase, olivine, augite, orthopyroxene, and pigeonite at pressures from one atmosphere to 20 kbar and application to petrogenesis in intraoceanic island arcs. D, 1985, Pennsylvania State University, University Park. 256 p.

Baker, Donald Frederick. Geology and geochemistry of an alkali volcanic suite (Skinner Cove Formation) in the Humber Arm Allochthon, Newfoundland. M, 1979, Memorial University of Newfoundland. 314 p.

Baker, Donald John. The metamorphic and structural history of the Grenville Front near Chibougamau, Quebec. D, 1980, University of Georgia. 464 p.

Baker, Donald Roy. Geology of the Edison Area, Sussex County, New Jersey. D, 1955, Princeton University. 275 p.

Baker, Edward Daniel. Geology of the Phoebe Tip-Trapper Peak area, Boundary County, Idaho. M, 1979, University of Idaho. 132 p.

Baker, Edward Michael. A finite element model of the Earth's anomalous gravitational potential. D, 1988, Ohio State University. 116 p.

Baker, Edward Thomas, Jr. Nephelometry and mineralogy of suspended particulate matter in the waters over the Washington continental slope and Nitinat Deep-Sea Fan. D, 1973, University of Washington. 142 p.

Baker, Eldon R. Geology of north-central Burleson and south-central Milam counties, Texas. M, 1956, Texas A&M University. 90 p.

Baker, Elizabeth. Tentative evaluation of some of the factors influencing topographic development in central New York State. M, 1930, Cornell University.

Baker, Ethel. Distribution of Middle Devonic Brachiopoda. M, 1913, Columbia University, Teachers College.

Baker, Forrest Grant. The geology of a portion of the Younts Peak, Wyoming, Quadrangle. M, 1951, University of Iowa. 72 p.

Baker, Frank Elmore. A mechanical and petrographic analysis of the Garber Sandstone in Cleveland and Pottawatomie counties. M, 1951, University of Oklahoma. 68 p.

Baker, Fred G. Development of a pore interaction model for hydrodynamic dispersion during flow through porous media. D, 1985, University of Colorado. 161 p.

Baker, George Oliver. Paleoecology of the fauna and paleoenvironmental analysis of the Portersville Shale in Morgan, Perry, and Muskingum counties, Ohio. M, 1975, Ohio University, Athens. 143 p.

Baker, Gerald Lawrence. Investigation into the intrusive and extrusive origin of a small section of Second Watchung Mountain, North Caldwell, New Jersey. M, 1972, Montclair State College. 98 p.

Baker, Glen E. Source parameters of the magnitude 7.1, 1949, South Puget Sound, Washington earthquake determined from long-period body waves. M, 1985, University of Washington. 68 p.

Baker, Glenn J. A study in base exchange in the zeolites. M, 1928, University of Wisconsin-Madison.

Baker, Gordon K. An environmental study of the Morrison Formation (Jurassic), Freezeout Hills-Como Bluff area, Carbon County, Wyoming. M, 1965, University of Wyoming. 142 p.

Baker, Grant Cody. Salt redistribution during freezing of saline sand columns with applications to subsea permafrost. D, 1987, University of Alaska, Fairbanks. 248 p.

Baker, Gus Bowman. Structure of northeastern Chispa Quadrangle, Culberson and Jeff Davis counties, Texas. M, 1952, University of Texas, Austin.

Baker, Harold Wellington, Jr. Environmental sensitivity of submicroscopic surface textures on quartz sand grains; a statistical evaluation. M, 1974, University of Wisconsin-Madison. 62 p.

Baker, Herbert Arney. The geology of the Gaysport and Skelley limestones in Athens, Meigs, Morgan, and Perry counties, Ohio. M, 1967, Ohio University, Athens. 147 p.

Baker, Jack. Glacial geology of Geauga County, Ohio. D, 1957, University of Illinois, Urbana. 118 p.

Baker, James John. Microstructural defects and their importance in highly crystalline pyrolytic graphites. D, 1970, Pennsylvania State University, University Park. 128 p.

Baker, James Morgan. Investigation of mineralization of the ore minerals at the Taylor-Windfall Mine, B.C. M, 1936, University of Toronto.

Baker, Joel Eric. The particle-mediated geochemistry of hydrophobic organic contaminants in large lakes. D, 1988, University of Minnesota, Minneapolis. 234 p.

Baker, Joel F. Log interpretation of shaly sandstones. M, 1987, Texas A&M University.

Baker, John Hudson. Aeromagnetic investigations in southeast Missouri (western margin of the Saint Francois Mountains). M, 1967, Washington University. 86 p.

Baker, Joseph M. A stratigraphic study of the Marmaton Group in Cowley, Elk, and Chautauqua counties in Kansas. M, 1984, Wichita State University. 63 p.

Baker, K. H. Heat flow studies in Colorado and Wyoming. M, 1976, University of Wyoming. 142 p.

Baker, Kevin L. An investigation of ^{2}H and ^{13}C abundance in a sequence of related oils of increasing maturity. M, 1987, Indiana University, Bloomington. 106 p.

Baker, Lawrence Alan. Mineral and nutrient cycles and their effect on the proton balance of a softwater, acidic lake. D, 1984, University of Florida. 159 p.

Baker, Linda J. The stratigraphy and depositional setting of the Spencer Formation, west-central Willamette Valley, Oregon; a surface-subsurface analysis. M, 1988, Oregon State University. 171 p.

Baker, Lisa M. Sedimentology and diagenesis of the basinal facies of the Dimple Limestone, Marathon Basin, West Texas. M, 1985, Texas Tech University. 97 p.

Baker, Lora May. A physiographic study of a selected area in Brown County, Indiana. M, 1923, Indiana University, Bloomington.

Baker, Lynn Edward. Dinoflagellates and Hystrichosphaerids from the Tithonian (Upper Jurassic), Grenoble, France. M, 1967, Texas Tech University. 79 p.

Baker, Mark Richard. Application of time series analysis to the enhancement of seismic refraction data interpretation. M, 1979, Purdue University. 73 p.

Baker, Mark Richard. Quantitative interpretation of geological and geophysical well data. D, 1988, University of Texas at El Paso. 146 p.

Baker, Merle V. The geology of a portion of Harrison and Bloom Townships, Scioto County, Ohio. M, 1931, Ohio State University.

Baker, Michael Baldwin. Evolution of lavas at Mt. Shasta Volcano, N. California; an experimental and petrologic study. D, 1988, Massachusetts Institute of Technology. 430 p.

Baker, Nancy Tucker. Use of a rainfall runoff model to simulate the effects of urbanization on the hydrologic regimen of Ward Creek drainage basin, East Baton Rouge Parish, Louisiana. D, 1987, Louisiana State University. 118 p.

Baker, P. A re-interpretation of the Eocene Capay Formation, Yolo County, California. M, 1975, Stanford University. 112 p.

Baker, Paul Arthur. Coral growth rate; variation with depth. M, 1975, Pennsylvania State University, University Park. 11 p.

Baker, Paul Arthur. The diagenesis of marine carbonate sediments; experimental and natural geochemical observations. D, 1981, University of California, San Diego. 153 p.

Baker, Philip Craig. Trace element analysis of limestones by secondary fluorescence X-ray spectroscopy. M, 1981, University of Florida. 86 p.

Baker, Ralph N. A two-level weighted factor evaluation of the metallic mineralization potential of central Baja California using satellite data and computer-assisted enhancement techniques. D, 1979, University of Delaware. 339 p.

Baker, Ralph N. The foraminiferal paleoecology of the Upper Cretaceous Exogyra ponderosa and Exogyra costata zones of the western Georgia Coastal Plain. M, 1968, Rutgers, The State University, New Brunswick. 84 p.

Baker, Randall Keith. The depositional environment of the Pennsylvanian Upper Marchand sandstones, Northeast and East Binger fields, Oklahoma. M, 1978, University of Oklahoma. 122 p.

Baker, Ray Gordon. A study of early S motion. M, 1956, St. Louis University.

Baker, Reginald Anthony. A study of ore and rock specimens from the Nkana Mine, northern Rhodesia. M, 1951, University of British Columbia.

Baker, Richard C. Geology of the Swiftwater area, Moosilauke Quadrangle, New Hampshire. M, 1954, Dartmouth College. 47 p.

Baker, Richard Calvin. The geology and ore deposits of the southeast portion of the Patagonia Mountains, Santa Cruz County, Arizona. D, 1962, University of Michigan. 325 p.

Baker, Richard Graves. Late-Wisconsin geology and vegetation history of the Alborn area, St. Louis County, Minnesota. M, 1964, University of Minnesota, Minneapolis. 44 p.

Baker, Richard Graves. Pollen and plant macrofossils from late Pinedale (later Pleistocene) and Recent sediments in an abandoned lagoon of Yellowstone lake, Wyoming. D, 1969, University of Colorado. 147 p.

Baker, Richard K. The ground water capacity of the Newcastle-Dakota sandstones in central and western South Dakota. M, 1973, South Dakota School of Mines & Technology.

Baker, Robert Allison, III. Experimental studies in ripple mark formation. M, 1965, University of Florida. 72 p.

Baker, Robert Allison, III. Stratigraphy and sedimentology of the Cañon del Tule Formation (upper Cretaceous), Parras Basin, northeastern Mexico. D, 1970, University of Texas, Austin. 355 p.

Baker, Robert J. Geology of the Big Clifty Formation in the Wheatonville Consolidated oil field in Gibson County, Indiana. M, 1980, Ball State University. 85 p.

Baker, Robert Jethro, Jr. Phase equilibria studies in the system MgO-Al$_2$O$_3$-ZrO$_2$. D, 1964, University of Illinois, Urbana. 209 p.

Baker, Robert W. Quaternary geology of parts of the Blue Hill, Ellsworth, Mount Desert, and Orland quadrangles, Maine. M, 1971, University of Maine. 81 p.

Baker, Robert W. The influence of ice-crystal size and dispersed-solid inclusions on the creep of polycrystalline ice. D, 1977, University of Minnesota, Minneapolis. 111 p.

Baker, Roger Crane. The age and fauna of the Olentangy Shale of central Ohio. M, 1938, University of Iowa. 16 p.

Baker, Seymour R. A non-destructive core analyses technique using x-rays. M, 1969, Rensselaer Polytechnic Institute. 41 p.

Baker, Seymour R. Sedimentation in an Arctic marine environment; Baffin Bay between Greenland and the Canadian Arctic Archipelago. D, 1971, Rensselaer Polytechnic Institute. 118 p.

Baker, Sherry Lynn. Depositional environment of the "Springvale" Sandstone of central New York and its relationship to the Oriskany Sandstone. M, 1983, Syracuse University. 121 p.

Baker, Stephanie Ashburn. Holocene palynology and reconstruction of paleoclimates in north-central Alaska. M, 1984, University of Colorado. 204 p.

Baker, T. N. Analysis of six years of pelagic tidal heights from the eastern North Pacific Ocean. M, 1975, Columbia University. 27 p.

Baker, Vernon R. Geology of the foothills of the Front Range between Masonville and Bellvue (Larimer County), Colorado. M, 1948, University of Colorado.

Baker, Victor Richard. Paleohydrology and sedimentology of Lake Missoula flooding in eastern Washington. D, 1971, University of Colorado. 152 p.

Baker, W. F. A study of the erosive action of the Red River, the materials carried in solution and in suspension, with a chapter on the state of silica in natural waters. M, 1925, University of Manitoba.

Baker, Walker Holcombe. Geologic setting and origin of the Grouse Creek Pluton, Box Elder County, Utah. D, 1959, University of Utah. 214 p.

Baker, Wallace Hayward. Mohr-Coulomb strength theory for anisotropic soils. D, 1968, Northwestern University. 229 p.

Baker, Walter Edwin. The Spraberry Sandstone of West Texas. M, 1952, University of Michigan.

Baker, William Laird. Characteristics of the P phase. D, 1960, University of California, Berkeley. 158 p.

Baker, William Samuel. Mineral occurrences in the District of Columbia and vicinity. M, 1941, George Washington University. 48 p.

Bakheit, Abdalla Kodi. Petrography of Cu-Ni mineralization in Mineral Lake area, Ashland County, Wisconsin. M, 1981, University of Wisconsin-Madison. 104 p.

Bakhshandeh, Farhad. A study of the aromatic fraction of oil shales and other carbonaceous deposits from the Green River Formation in Utah. M, 1977, Colorado School of Mines. 68 p.

Bakhtar, Khosrow. Large scale behavior of rock joints. D, 1985, University of Utah. 221 p.

Bakhtiar, Steven Norouz. Uranium fallout from nuclear-powered satellites and volcanic eruptions. D, 1987, University of Arkansas, Fayetteville. 203 p.

Bakhtiari, Raiani Hamid. Hydrogeologic and digital model studies of a shallow unconfined aquifer in the Souris River basin, Manitoba. M, 1971, University of Manitoba.

Bakke, Arne A. Investigation of selected skarns, central Alaska Range, Alaska. M, 1987, University of Alaska, Fairbanks. 124 p.

Bakken, Barbara M. Petrology and chemistry of the Elk Creek and Bozeman corundum deposits, Madison and Gallatin counties, Montana. M, 1980, University of Washington. 69 p.

Bakken, Wallace E. The surficial geology of north-central Kidder County, North Dakota. M, 1960, University of North Dakota. 93 p.

Bakker, Allen. A geochemical and petrographic study of the Cortlandt Complex, Peekskill, New York. M, 1980, Adelphi University.

Bakker, Daniel. Origin of the lenticular sand bodies in the Raton Formation (Cretaceous-Tertiary). M, 1954, University of Wisconsin-Madison.

Bakker, Robert Thomas. Progressivism and non competitive extinction. D, 1976, Harvard University.

Bakos, Nancy A. Ultrapetrography of a Miocene chalky limestone, Kingshill Marl, U. S. Virgin Islands. M, 1975, Northern Illinois University. 81 p.

Bakr, Muhammed Abu. Structure of the western Ras Koh Range, Chagai and Kharan districts, Kalat Division, West Pakistan. M, 1962, University of Michigan.

Baksi, A. K. Whole rock dating of extusives by the potassium-argon method. D, 1970, University of Toronto.

Bakun, William Henry. Body wave spectra and crustal structure; an application to the San Francisco Bay region: 1 volume. D, 1970, University of California, Berkeley.

Bakush, Sadeg H. Carbonate microfacies, depositional environments and diagenesis of the Galena Group (Middle Ordovician) along the Mississippi River (Iowa, Wisconsin, Illinois and Missouri), United States. D, 1985, University of Illinois, Urbana. 223 p.

Bakush, Sadeg Hasan. The geology of the Clinton Sandstone of east-central Ohio. M, 1975, Ohio University, Athens. 69 p.

Balacek, Kenneth Joseph. Radioactivity in tektites. M, 1966, Rice University. 53 p.

Balachandran, K. A theory of the solar wind interaction with the geomagnetic field. D, 1967, Colorado School of Mines. 69 p.

Balachandran, Nambath K. Propagation of acoustic-gravity waves in the atmosphere. D, 1968, Columbia University. 101 p.

Baladi, George Youssef. Distribution of stresses and displacements within and under long elastic and viscoelastic embankments. D, 1968, Purdue University. 196 p.

Balakrishna, Thirumale. Petrography of some Silurian rocks from northern Michigan. M, 1972, Wayne State University.

Balasubramanian, V. Adsorption, denitrification, and movement of applied ammonium and nitrate in Hawaiian soils. D, 1974, University of Hawaii. 167 p.

Balazs, Rodney J. Hydraulic factors controlling the migration of dunes and sand waves in a tide dominated environment. M, 1971, University of Illinois, Urbana. 59 p.

Balbach, Margaret Kain. Morphology of Paleozoic lycopsid fructifications and spore correlations. D, 1964, University of Illinois, Urbana. 160 p.

Balcells, Roberto. Stratigraphy, age, and paleoecology of benthonic foraminifera from the type Kevin Shale Member, Marias River Formation, northwestern Montana. M, 1978, University of Wyoming. 114 p.

Balcer, Richard Allen. Stratigraphy and depositional history of the Pantano Formation (Oligocene-early Miocene), Pima County, Arizona. M, 1984, University of Arizona. 107 p.

Balch, Duane Clark. Sedimentology of the Santiago Peak volcaniclastic rocks, San Diego County, California. M, 1981, San Diego State University.

Bald, Roberta C. Petrogenesis of early Archean, gneissic tonalite-granodiorite from the English River Subprovince, Gundy Lake area, northwestern Ontario. M, 1981, University of Manitoba. 120 p.

Baldar, Nouri Amin. Occurrence and formation of soil zeolites. D, 1968, University of California, Davis. 150 p.

Baldasari, Arthur. Iron-titanium oxides in the Elberton Granite. M, 1981, University of Georgia.

Baldauf, Jack Gerald. Diatom biostratigraphic and paleoceanographic studies of Neogene material recovered from the North Atlantic and Equatorial Pacific oceans. D, 1985, University of California, Berkeley. 470 p.

Baldauf, Phillip Dayle. Ostracoda of the Brightseat Paleocene formation of Maryland, and a review of the acceptance of the Paleocene as an epoch. M, 1960, George Washington University.

Balderas, Jack Moreno, Jr. Stone City foraminifera in eastern Burleson County, Texas. M, 1953, Texas A&M University. 70 p.

Balderman, Morris Aaron. Relationship of yield stress and strain-rate in hydrolytically weakened synthetic quartz. M, 1972, University of California, Los Angeles.

Baldridge, Warren Scott. Petrology and petrogenesis of basaltic rocks and their inclusions; studies from the Rio Grande Rift, the Roman comagmatic province, and Oceanus Procellarum. D, 1979, California Institute of Technology. 336 p.

Baldwin, Andrew B. The nature and genesis of the iron ores of the Bayot Lake area, Labrador–New-Quebec in comparison with those of New Brunswick. M, 1954, University of New Brunswick.

Baldwin, Andrew Bennett. A study of the Precambrian hematite ore deposits at Fort Gouraud, Mauritania (West Africa). D, 1965, University of Toronto. 236 p.

Baldwin, Arthur D., Jr. Ice-push deformation of lithified and unlithified sediments in Northeast Kansas. M, 1963, University of Kansas. 62 p.

Baldwin, Arthur Dwight, Jr. Geologic and geographic controls upon the rate of solute eroison from selected coastal river basins between Half Moon Bay and Davenport, California. D, 1967, Stanford University. 164 p.

Baldwin, David Arthur. Garnet-cordierite-anthophyllite rocks at Rat Lake, Manitoba. M, 1971, University of Manitoba.

Baldwin, David Arthur. Physical volcanology of the northwest segment of the Karsakawigamak Block, Proterozoic Rusty Lake metavolcanic belt, northern Manitoba. D, 1987, University of Manitoba.

Baldwin, Donald Carl. Late Pleistocene clays of the Sault Ste. Marie area and vicinity. M, 1951, University of Illinois, Urbana. 60 p.

Baldwin, Dorothy Esther. The geology of the eastern half of the Joyabaj Quadrangle, west-central Guatemala. M, 1972, University of Florida. 98 p.

Baldwin, E. A. Paleoecology of benthonic foraminifera from the Fort Hays Member, Niobrara Formation, West-central Kansas. M, 1976, University of Wyoming. 99 p.

Baldwin, E. Joan. Pliocene turbidity current deposits in Ventura Basin, California. M, 1959, University of Southern California.

Baldwin, Ellwood E. Urban geology of the growth corridor along Interstate 35; Belton, Texas to Hillsboro, Texas. M, 1972, Baylor University. 142 p.

Baldwin, Evelyn Joan. Environments of deposition of the Moenkopi Formation in north-central Arizona. D, 1971, University of Arizona. 268 p.

Baldwin, Ewart Merlin. Late Cenozoic diastrophism along the Olympia coast. M, 1939, Washington State University. 44 p.

Baldwin, Ewart Merlin. Structure and stratigraphy of the northern half of Lost River Range, Idaho. D, 1943, Cornell University.

Baldwin, H. T. The genesis of the sediments of Wind River age in the Laramie Basin. M, 1935, University of Wyoming. 35 p.

Baldwin, Harry L., Jr. Gravity survey of part of the Snake River plain, Idaho. M, 1960, Colorado School of Mines. 90 p.

Baldwin, James L. A crustal seismic refraction study in southwestern Indiana and southern Illinois. M, 1980, Purdue University. 99 p.

Baldwin, Jeffrey. The vertical and lateral distribution of trace elements in the Fire Clay Coal Seam near Hazard, Kentucky. M, 1975, University of South Carolina.

Baldwin, Joe Allen. Analysis of water resource problems related to mining in the Blackbird mining district, Idaho. M, 1977, University of Idaho. 232 p.

Baldwin, John Schuyler. Grain size distribution and clay occurrence; five sand dominant environments. M, 1983, University of New Orleans. 155 p.

Baldwin, Kenneth Charles. An analysis of 3.5 kHz normal incidence acoustic reflections and sediment physical properties. D, 1982, University of Rhode Island. 137 p.

Baldwin, Otha Don. Igneous rocks of the Texas Gulf Coast; a temporal and petrochemical evaluation. D, 1971, Rice University.

Baldwin, Randall Wayne. Depositional environments and diagenesis of Cisco carbonates. M, 1980, University of Texas at Dallas. 87 p.

Baldwin, Richard Taylor. A test of the simple lithospheric stretching model; application to the Delaware and Albuquerque basins in New Mexico. M, 1986, Purdue University. 113 p.

Baldwin, Suzanne Louise. Fission track dating of detrital zircons from the Scotland Sandstones, Barbados, West Indies. M, 1984, SUNY at Albany. 97 p.

Baldwin, Suzanne Louise. Thermochronology of a subduction complex in western Baja California. D, 1988, SUNY at Albany. 261 p.

Baldwin, Thomas Ashley. A geological, geophysical investigation of the Depew area, Creek County, Oklahoma. M, 1985, Wright State University. 75 p.

Baldwin, W. Brewster. Gray iron ores of Tallaseehatchee District, Talladega County, Alabama. M, 1944, Columbia University, Teachers College.

Baldwin, W. Brewster. The geology of the Sioux Formation. D, 1951, Columbia University, Teachers College.

Baldwin, William Felbert Jackson, Jr. The geology of the Kirkwood Quadrangle. M, 1953, Washington University.

Bales, James. Environmental geology of the Tempe Quadrangle, Maricopa County, Arizona; Part II. M, 1985, Arizona State University. 103 p.

Bales, Roger Curtis. Surface chemical and physical behavior of chrysotile asbestos in natural waters and water treatment. D, 1985, [University of California, San Diego]. 274 p.

Bales, William E. Geology of the lower Brice Creek area, Lane County, Oregon. M, 1951, University of Oregon. 53 p.

Balgord, William D. Dissolved hydrolysis products of artificially pulverized selected silicate minerals. M, 1961, University of Missouri, Columbia.

Balgord, William Dwyer. Crystal chemical relationships in the analcite family and a study of cation-H_2O coordination in certain synthetic and natural zeolites.

D, 1966, Pennsylvania State University, University Park. 208 p.

Balise, Michael John. The relation between surface and basal velocity variations in glaciers, with application to the mini-surges of Variegated Glacier. D, 1988, University of Washington. 205 p.

Balistrieri, Laurie S. The basic surface characteristics of goethite. M, 1977, University of Washington. 65 p.

Balke, Bennie Kuno. Structural geology along Mount Bonnell Fault, south central Travis County, Texas. M, 1958, University of Texas, Austin.

Balke, Scott Carter. The petrology, diagenesis, stratigraphy, depositional environment, and clay mineralogy of the Red Fork Sandstone in north-central Oklahoma. M, 1984, Oklahoma State University. 118 p.

Balkwill, Hugh R. Geology of the central New York Mountains, California (eastern Mojave Desert). M, 1965, University of Southern California.

Balkwill, Hugh Robert. Structural analysis of the Western Ranges, Rocky Mountains near Golden, British Columbia (Canada). D, 1969, University of Texas, Austin. 216 p.

Ball, Alexander Ross. Geology of the northern and western parts of the San Clemente Quadrangle, Orange County, California. M, 1961, University of Southern California.

Ball, Andrew David. Ground-water supply as a criterion for subdivision approval; the administration of Arizona revised statute 45-513. M, 1977, University of Arizona.

Ball, Clayton Garrett. The mineral constituents of Coal Number 6 in the New Orient Mine, Franklin County, Illinois. D, 1935, Harvard University.

Ball, David S. The Pennsylvanian Haskell-Cass section, a perspective on controls of Midcontinent cyclothem deposition. M, 1985, University of Kansas. 147 p.

Ball, F. D. Sulphide transformations and magnetic expression of a thermally metamorphosed contact aureole in the Goldenville Formation, Nova Scotia. M, 1974, University of New Brunswick.

Ball, John R. The geology of the Pike River area, Kenosha, Wisconsin. M, 1918, Northwestern University.

Ball, John R. The stratigraphy and areal distribution of the "Hunton Formation", Oklahoma. D, 1936, Northwestern University.

Ball, John Rice. The faunas of the Brassfield and Bainbridge limestones of southeastern Missouri. D, 1927, University of Chicago. 425 p.

Ball, Mahlon Marsh. Geology of southwestern Franklin County, Kansas. M, 1957, University of Kansas. 85 p.

Ball, Mahlon Marsh. Gravity and magnetic measurements in eastern Kansas. D, 1960, University of Kansas. 125 p.

Ball, Nancy Beatrice. Colloidal properties of coagulated calcium-montmorillonite suspensions. D, 1981, University of California, Riverside. 145 p.

Ball, Stanton Mock. Geology of eastern Franklin County, Kansas. M, 1958, University of Kansas. 75 p.

Ball, Stanton Mock. Stratigraphy of the Douglas Group (Pennsylvanian, Virgilian) in the northern Midcontinent region. D, 1964, University of Kansas. 674 p.

Ball, Stephen. The Partridge Point area of the Kiglapait layered intrusion. M, 1984, University of Massachusetts. 98 p.

Ball, Sydney H. General geology of the Georgetown Quadrangle, Colorado. D, 1910, University of Wisconsin-Madison.

Ball, Theodore T. Geology and fluid inclusion investigation of the Crown King breccia pipe, Yavapai County, Arizona. M, 1983, Colorado School of Mines. 141 p.

Balla, John C. The relationship of Laramide stocks to regional structure in central Arizona. D, 1972, University of Arizona.

Balla, John Coleman. The geology and geochemistry of beryllium in southern Arizona. M, 1962, University of Arizona.

Ballantyne, Geoffrey Hugh. Chemical and mineralogical variations in propylitic zones surrounding porphyry copper deposits. D, 1981, University of Utah. 227 p.

Ballantyne, J. C. The political economy of Peruvian Gran Mineria. D, 1976, Cornell University. 336 p.

Ballantyne, Judith Mary. Geochemistry of hydrothermal sericite and chlorite. D, 1981, University of Utah. 146 p.

Ballard, David Wayne. A compressional and shear-wave seismic refraction study for near-surface earth layers in northwestern Oklahoma. M, 1980, University of Oklahoma. 139 p.

Ballard, Eva Oakley. Limestone in the Heterostegina Zone (Oligocene-Miocene) on Damon Mound salt dome, Brazoria County, Texas. M, 1961, University of Houston.

Ballard, Frederick V. The structural and stratigraphic relationships in the Paleozoic rocks of eastern North Dakota. M, 1963, University of North Dakota. 175 p.

Ballard, Geoffrey Edwin Hall. The plastic limit of the clay-silt-water system. D, 1963, Washington University. 117 p.

Ballard, J. A. Geology of a stable intraplate region; the Cape Verde/Canary Basin. D, 1980, University of North Carolina, Chapel Hill. 248 p.

Ballard, James Alan. The geology of the Merry Oaks section of the Durham Triassic basin of North Carolina. M, 1958, University of North Carolina, Chapel Hill. 44 p.

Ballard, James H. Seismic and gravity investigation of the crust and upper mantle in southwestern Montana. M, 1980, University of Montana. 98 p.

Ballard, Martha M. Remagnetizations in Late Permian and Early Triassic rocks from southern Africa and their implications for Pangea reconstructions. M, 1985, University of Michigan. 24 p.

Ballard, R. D. The nature of Triassic continental rift structures in the Gulf of Maine. D, 1974, University of Rhode Island. 132 p.

Ballard, Ronald Lee. Distribution of beach sediment near the Columbia River. M, 1964, University of Washington. 82 p.

Ballard, Sanford. Terrestrial heat flow and thermal structure of the lithosphere in Southern Africa. D, 1987, University of Michigan. 151 p.

Ballard, Sanford, III. The development of the Powell Valley Anticline, Pine Mountain Block, Southwest Virginia. M, 1984, University of Michigan.

Ballard, Stanton Neal. Structural geologic controls at the San Luis Mines, Tayoltita, Durango, Mexico. M, 1980, University of Arizona. 121 p.

Ballard, William Norval. An isopachous map of the Hunton Formation in the Seminole District (Oklahoma). M, 1930, University of Oklahoma. 33 p.

Ballard, William Turpin. Sedimentology and controls on drainage development in the glaciofluvial White River system, northeast-central Indiana. M, 1985, Indiana University, Bloomington. 70 p.

Ballard, William Wayne. A subsurface study of the Morrow and Atoka series in a portion of the Arkansas Valley of western Arkansas. M, 1956, University of Oklahoma. 95 p.

Ballard, William Wayne. Sedimentary petrology of post-Madison-pre-Kootenai rock, north flank of Little Belt Mountains, Montana. D, 1961, University of Texas, Austin. 379 p.

Ballaron, Paula Balcom. A stratigraphic and sedimentologic study of NW-directed paleoflow in the Alton-Lagrange esker, Maine. M, 1979, University of Maine. 122 p.

Ballenger, Benjamin David. Subsurface study of selected Mesozoic strata of northern Columbia County, Arkansas. M, 1985, University of Arkansas, Fayetteville.

Ballew, Gary I. Quantitative geologic analysis of multiband photography from the Mono Craters area, California. M, 1968, Stanford University.

Ballew, William Harold. The geology of the Jarre Canyon area, Douglas County, Colorado. M, 1957, Colorado School of Mines. 86 p.

Ballinger, Philip. An analysis of the energy and nonfuel minerals of the U.S.S.R. M, 1983, University of Texas, Austin.

Ballivy, Gérard. Contribution à l'étude des caractéristiques géologiques et géotechniques des dépôts d'argile du nord-ouest du Québec (région de Matagami-Fort Rupert). M, 1970, Ecole Polytechnique. 222 p.

Ballmann, Donald Lawrence. The geology of the Knight Range, Grant County, New Mexico. D, 1959, University of Illinois, Urbana. 73 p.

Ballmann, Donald Lawrence. The geology of the southeast quarter of the Red Rock Quadrangle, New Mexico. M, 1956, University of Illinois, Urbana. 29 p.

Ballon, Wilfredo. Sedimentary petrology of the Lagonda Formation (Desmoinesian) of Saint Louis County (Missouri) and vicinity. M, 1968, Washington University.

Ballotti, Dean M. Permeability of young oceanic crust. M, 1988, Purdue University. 98 p.

Ballou, Robert L. Geology of the Co-op Spring Quadrangle, Oneida County, Idaho. M, 1979, Brigham Young University.

Ballou, S. W. Ecology and economics of strip-mine reclamation in Mahaska County, Iowa. D, 1975, University of Wisconsin-Milwaukee. 179 p.

Ballou, William D. and Sturgeon, David A. The formation and detailed description of a portion of Wind Cave, South Dakota. M, 1958, South Dakota School of Mines & Technology.

Balls, Scott Nelson. A simple shear apparatus for study of microstructural deformation of rock analogs. M, 1985, Michigan Technological University. 120 p.

Balogh, Randy J. A study of the Middle Silurian Belle River Mills, Peters, and Ray pinnacle reefs from the Michigan Basin. M, 1980, Bowling Green State University. 202 p.

Balogh, Richard Stephen. Subglacial pluvial erosion in the vicinity of Tuolumne Meadows, Yosemite National Park, California. M, 1976, University of California, Los Angeles.

Balombin, Michael T. The St. Peter Sandstone in Michigan. M, 1974, Michigan State University. 59 p.

Balsam, Robert C., Jr. Geology and geophysics of the South Pass area, Fremont County, Wyoming. M, 1986, University of Wyoming. 100 p.

Balsam, William Lando. Ecological interactions in an early Cambrian archaeocyathid reef community. D, 1973, Brown University.

Balsam, William Lando. Reinterpretation of a lower Cambrian archaeocyathid reef; Shady Formation, southwestern Virginia. M, 1969, Brown University.

Balsdon, Jason Thomas. Environmental isotope and geochemical study of landfill leachate migration. M, 1988, University of Windsor. 229 p.

Balseiro, Lina M. Upper Cretaceous foraminifera of southern Colombia, South America. M, 1954, Miami University (Ohio). 37 p.

Balshaw-Biddle, Katherine M. Antarctic glacial chronology reflected in the Oligocene through Pliocene sedimentary section in the Ross Sea. D, 1981, Rice University. 150 p.

Balsinger, Daniel Francis. Mississippian and Pennsylvanian cross-stratification and paleocurrents in a portion of Somerset County, Pennsylvania. M, 1960, University of Pittsburgh.

Balsley, James Robinson, Jr. Remanent magnetization and anisotropic susceptibility of Adirondack rocks. D, 1960, Harvard University.

Balsley, John Kimball. Origin of fossiliferous concretions in the Ferron sandstone (upper Cretaceous), southeastern Utah. M, 1969, University of Utah. 73 p.

Balsley, Steven Devry. The petrology and geochemistry of the Tshirege Member of the Bandelier Tuff, Jemez Mountains volcanic field, New Mexico, U.S.A. M, 1988, University of Texas, Arlington. 188 p.

Balster, Clifford A. Surface geology of Calhoun County. M, 1950, Iowa State University of Science and Technology.

Baltensperger, Paul. Stratigraphic framework and depositional environments of the lower Graham Formation (Pennsylvanian) of north-central Texas. M, 1985, University of Texas, Arlington. 266 p.

Balter, Howard and Heusser, Calvin J. Forest-soil relations on limestone and gneiss in southeastern New York and northern New Jersey. D, 1980, New York University. 177 p.

Balthaser, Lawrence Harold. Petrology and paleoecology of middle Chester (Mississippian) rocks of the southwestern Indiana outcrop. D, 1969, Indiana University, Bloomington. 292 p.

Baltuck, Miriam. Mesozoic pelagic sedimentation in a Tethyan continental margin basin, Pindos Mountains, Greece. D, 1982, University of California, San Diego. 265 p.

Baltz, Elmer Harold, Jr. Stratigraphic relationships of Cretaceous and Early Tertiary rocks of a part of northwestern San Juan Basin, New Mexico and Colorado. M, 1953, University of New Mexico. 101 p.

Baltz, Elmer Harold, Jr. Stratigraphy and geologic structure of uppermost Cretaceous and Tertiary rocks of the east-central part of the San Juan Basin, New Mexico. D, 1962, University of New Mexico. 294 p.

Baltz, Rachel May. Geology of the Arica Mountains. M, 1982, San Diego State University. 92 p.

Balzarini, Maria Anne. Paleoecology of Everson-age glaciomarine drifts in northwestern Washington and southwestern British Columbia. M, 1981, University of Washington. 109 p.

Bam, Swagat Arvind. Late Miocene paleoecology and biostratigraphy of southeastern Maryland. M, 1982, Rutgers, The State University, New Brunswick. 66 p.

Bambach, Richard Karl. Bivalvia of the Siluro-Devonian Arisaig group, Nova Scotia. D, 1969, Yale University. 524 p.

Bamber, Edward Wayne. Mississippian corals from northeastern British Columbia, Canada. D, 1961, Princeton University. 199 p.

Bamberger, Mark J. A quantitative examination of alluvial fans on the piedmont of the San Juan Mountains, southcentral Colorado. M, 1986, Wichita State University. 120 p.

Bambrick, James. Spectral analysis and filtering techniques applied to a lithologic interpretation of high resolution aeromagnetic data from the Timmins area, Ontario, Canada. D, 1984, University of Toronto.

Bambrick, James, Jr. Gravity investigation of the Triassic Newark Basin and adjacent Precambrian highlands in the vicinity of the Watchung Mountains. M, 1976, Rutgers, The State University, New Brunswick. 36 p.

Bambrick, Thomas C. Seismic determination of the vertical and areal extent of the Oak Openings sand belt in Lucas, Fulton and Henry counties, Ohio. M, 1986, Bowling Green State University. 99 p.

Bamburak, James David. The Upper Cretaceous and Paleocene stratigraphy of Turtle Mountain, Manitoba. M, 1973, University of Manitoba.

Bame, Dorthe Ann. A source model for low frequency volcanic earthquakes. M, 1984, University of Washington. 95 p.

Bamford, John Ross. Paleoecology of the Albrights Reef, Onondaga Limestone (Devonian), eastern New York. M, 1966, University of Nebraska, Lincoln.

Bamford, Robert Wendell. Genesis of the magnesite deposits of Stevens County, Washington. D, 1970, Stanford University. 96 p.

Bamford, Thomas Sayers. The physical properties of two limestones at permafrost temperatures. M, 1973, University of Saskatchewan. 69 p.

Bammel, B. H. Stratigraphy of the Simsboro Formation, East-central Texas. M, 1977, Baylor University. 179 p.

Bamwoya, J. J. Exploration geochemistry in the Burnt Hill area, New Brunswick; distribution of elements in bedrock and in heavy and light fractions of stream sediments. D, 1978, University of New Brunswick.

Banaee, Jila. Microfacies and depositional environment of the Bailey Limestone (Lower Devonian) southwestern Illinois, U.S.A., a carbonate turbidite. M, 1981, University of Illinois, Urbana. 61 p.

Banar, Frank J. Petrography of the early basalts sheets, Absaroka Mountains, Wyoming. M, 1963, Wayne State University.

Banas, P. J. An investigation of the circulation dynamics of a Louisiana bar-built estuary. M, 1978, Louisiana State University.

Banaszak, K. J. Genesis of the Mississippi Valley-type lead-zinc ores. D, 1975, Northwestern University. 146 p.

Banaszak, Konrad J. Geology of the southern Sand Springs range, Mineral and Churchill counties, Nevada. M, 1969, Northwestern University.

Bancroft, Genevieve R. Calyx variations in a new species of Dorycrinus from the Mississippian Saint Joe Formation, northeastern Oklahoma. M, 1965, University of Wisconsin-Madison.

Band, Lawrence Ephram. Environmental constraints on the development of small, headwater stream networks. M, 1979, University of California, Los Angeles. 68 p.

Band, Lawrence Ephram. Measurement and simulation of hillslope development. D, 1983, University of California, Los Angeles. 149 p.

Bandoian, Charles Asa. Geomorphology of the Animas river valley, San Juan County, New Mexico. M, 1969, University of New Mexico. 88 p.

Bandoian, Charles Asa. The Pliocene and Pleistocene rocks of Bonaire, Netherlands Antilles. D, 1973, Rutgers, The State University, New Brunswick. 54 p.

Bandurski, E. L. Analysis of insoluble organic material in carbonaceous meteorites by combined vacuum pyrolysis-gas chromatography-mass spectrometry. D, 1975, University of Arizona. 91 p.

Bandy, James Chapman. The geology of the McKinley Pool, Washington County, Illinois. M, 1950, University of Illinois, Urbana. 27 p.

Bandy, Mark Chance. Corundum gems. M, 1924, Columbia University, Teachers College.

Bandy, Mark Chance. I, Geology and petrology of Easter Island; II, Mineralogy of three sulphate deposits of northern Chile. D, 1938, Harvard University.

Bandy, Orville Lee. Eocene and Oligocene foraminifera from Little Stave Creek, Clarke County, Alabama. D, 1948, Indiana University, Bloomington. 270 p.

Bandy, Orville Lee. Invertebrate paleontology of Cape Blanco. M, 1941, Oregon State University. 138 p.

Bandy, William F., Jr. Paleontology and stratigraphy of the Mattoon Formation (Pennsylvanian), Coles County, Illinois. M, 1981, Indiana University, Bloomington. 123 p.

Bandy, William Lee. Structure and seismic stratigraphy of the Bonin trench-arc system. M, 1982, Texas A&M University. 155 p.

Bandyopadhyay, Pinaki. Release of nitrogen compounds from dredged sediment. D, 1981, University of Texas at Dallas. 244 p.

Bandyopadhyay, Sunirmal. Consolidation characteristics of compressible soils determined in-situ by electro-osmosis. D, 1978, University of California, Berkeley. 149 p.

Banerdt, William Bruce. The rheology of single crystal sodium chloride at high temperatures and low stresses and strains. D, 1983, University of Southern California.

Banerjee, Ajit Kumar. A study of the morphogenesis of gray-brown podzolic soils in southern Ontario. D, 1969, University of Toronto.

Banerjee, Anil K. Structure and petrology of the Oracle granite, Pinal County, Arizona. D, 1957, University of Arizona. 155 p.

Banerjee, Bakul. Interpretation of the Earth's gravity field. D, 1981, The Johns Hopkins University. 219 p.

Banerjee, Nani Gopal. Cyclic behaviour of dense coarse-grained materials in relation to the seismic stability of dams. D, 1979, University of California, Berkeley. 252 p.

Banerjee, Syamadas. Two possible sources of tin in the base metal sulphide deposits (Precambrian) at Geco, Manitouwadge, Ontario. M, 1972, University of Western Ontario. 95 p.

Banet, A. C., Jr. Kinetics of a clay catalyzed cracking reaction. M, 1976, University of Missouri, Columbia.

Banfield, Armine F. The geology of Beattie gold mines, Duparquet, Quebec, Canada. D, 1940, Northwestern University.

Banfield, Armine F. The micrography of the lead and zinc deposits of the Upper Mississippi Valley and its bearing on their origin. M, 1933, Northwestern University.

Banfield, Oscar M. Analysis of and proposed mine design for coal property in Raleigh County, West Virginia. M, 1950, Ohio State University.

Bangert, James C. Geologic report and selected geotechnical aspects of the city of Larkspur, California. M, 1974, [University of New Mexico].

Banghar, Amru Ram. Focal mechanisms of earthquakes in the Indian Ocean and adjacent regions. M, 1968, Columbia University. 55 p.

Banghart, Roger Clinton. Geology and ore deposits of the northeast portion of Breckenridge District (Summit County), Colorado. M, 1957, Colorado School of Mines. 78 p.

Bangs, Carol Lynn. Analysis of muddy siliciclastic rocks and provenance determination. M, 1988, Indiana University, Bloomington. 83 p.

Bangsainoi, S. Estimating runoff from rainfall and basin characteristics in northeastern Thailand. M, 1973, University of Arizona.

Bangsund, William J. Hydrogeology of Upper Cretaceous shales and surficial deposits, Igloo area, Fall River County, South Dakota. M, 1985, South Dakota School of Mines & Technology.

Banherjee, Anil K. Structural and petrological study of the Oracle Granite. D, 1957, University of Arizona. 155 p.

Banholzer, Gordon S., Jr. Distribution and shock metamorphism of crystalline components in the Bunte Breccia, Ries Crater, Germany. M, 1979, University of Houston.

Banikowski, Jeffrey E. Prospect location; a comparison of theoretical and practical limits using subsurface well data. M, 1984, Syracuse University.

Banino, George M. Stratigraphy and structure of the Paleozoic rocks of the Musconetcong valley, Hackettstown, New Jersey. M, 1969, Brooklyn College (CUNY).

Bank, Evelyn Ruth Jastram. A study of the transformation of sphalerite to wurtzite in atmospheres of zinc and sulfur. D, 1967, University of Colorado. 58 p.

Bank, Walter. The ore deposits of the Taxco District, Guerrero, Mexico. M, 1947, Columbia University, Teachers College.

Banka, Eleanor C. A petrographic study of some porphyry intrusives in the Beartooth Mountains, near Red Lodge, Montana. M, 1960, Smith College. 78 p.

Bankley, Erik Stefan. The relationship between land subsidence and subsurface barrier island sands in metropolitan New Orleans. M, 1980, University of New Orleans. 85 p.

Bankole, B. O. The oil shale of the Green River Formation, Colorado and the Athabasca Tar Sands, Alberta, Canada; potential fossil fuel energy resources. M, 1977, Northeastern Illinois University.

Banks, Carlie Elisabeth. Precambrian gneiss at Sheephead mountain, Carbon County, Wyoming, and its relationship to Laramide structure. M, 1970, University of Wyoming. 36 p.

Banks, Craig Stewart. Geochronology, general geology and structure of Hill Island Lake-Tazin Lake areas. M, 1980, University of Alberta. 109 p.

Banks, D. C. A reanalysis of the East Culebra Slide, Panama Canal. D, 1978, University of Illinois, Urbana. 264 p.

Banks, Elizabeth Young. Petrographic characteristics and provenance of fluvial sandstone, Sunnyside oil-impregnated sandstone deposit, Carbon County, Utah. M, 1981, University of Utah. 12 p.

Banks, Eric W. Liquefaction potential at the Naval Weapons Center, China Lake, California. M, 1982, University of Nevada. 102 p.

Banks, Ernest Robey. Isostatic and Bouguer gravity anomalies along the Inside Passage of Alaska and British Columbia. M, 1969, Oregon State University. 56 p.

Banks, Harlan Parker. Devonian plants from southeastern New York. D, 1940, Cornell University.

Banks, John L., Jr. The upper Santa Ana Watershed (California); water resources. M, 1949, University of California, Los Angeles. 233 p.

Banks, Joseph Edwin. The Kent fenster, Virginia. M, 1937, University of Iowa. 31 p.

Banks, Luis Maria. Mineralization of Sheridan copper prospect, Burnet County, Texas. M, 1949, University of Texas, Austin.

Banks, Norman Guy. Geology and geochemistry of the Leadville Limestone (Mississippian, Colorado) and its diagenetic, supergene, hydrothermal and metamorphic derivatives. D, 1967, University of California, San Diego. 321 p.

Banks, P. T., Jr. A geologic analysis of the side looking airborne radar imagery of southern New England. M, 1975, Boston College.

Banks, Philip Oren. Systematics of the distribution of uranium and lead in relation to the petrology of the Mt. Rubidoux Granites, Riverside County, California. D, 1963, California Institute of Technology. 197 p.

Banks, Roland Stewart. Stratigraphy of the Eocene Santee Limestone in three quarries of the Coastal Plain of South Carolina. M, 1977, University of North Carolina, Chapel Hill. 97 p.

Banks, Thomas H. Geology of the Culebra Island, Puerto Rico. M, 1962, Rice University. 75 p.

Bannahan, Annabelle Richardson. Foraminifera and paleoecology of a portion of the Marquez Member, Reklaw Formation (middle Eocene), in Bastrop County, Texas. M, 1950, University of Texas, Austin.

Bannan, David Bruce. Stratigraphy and sedimentology of late Quaternary sediments in the high depressions of Yuma and Kit Carson counties, Colorado. M, 1980, University of California, Davis. 162 p.

Bannatyne, Barry B. The geology of the Rankin Inlet area and North Rankin Nickel Mines, Limited, Northwest Territories. M, 1958, University of Manitoba.

Banner, Jay Lawrence. Petrologic and geochemical constraints on the origin of regionally extensive dolomites of the Mississippian Burlington and Keokuk fms., Iowa, Illinois and Missouri. D, 1986, SUNY at Stony Brook. 382 p.

Bannerman, Harold MacCall. The Nickel Lake iron range, Ontario. D, 1927, Princeton University.

Banning, Davey Lee. Variations of certain transition elements in the oxides in marine manganese nodules. M, 1979, Washington State University. 113 p.

Bannister, E. N. Topographic potential energy and channel development on engineered slopes in unconsolidated backfill. D, 1976, University of Pittsburgh. 268 p.

Bannister, John Richard. A magnetometer array study of polar magnetic substorms. D, 1977, University of Alberta. 235 p.

Bannister, Timothy Allen. Engineering geology assessment for the reclamation of an abandoned strip mine. M, 1979, Purdue University. 127 p.

Banowsky, Bill Raymond. Structure and hydrocarbon potential of the western part of the Snedaker Basin Quadrangle, part of the Disturbed Belt, west-central Montana. M, 1984, University of New Mexico. 110 p.

Bansbach, Louis Philip, III. Property evaluation of the Tensleep Reservoir, Quealy Dome Field, Albany County, Wyoming. M, 1964, Stanford University. 99 p.

Banta, Edward R. Groundwater flow patterns in the Dakota Group Aquifer in an area near Pueblo, Colorado. M, 1983, Colorado State University. 55 p.

Banta, Howard E. Faunal studies of the Sappington Sandstone, southwestern Montana. M, 1951, Montana College of Mineral Science & Technology. 45 p.

Banta, John Elliott. Seismic studies in the Great Salt Lake Desert of northwestern Utah. M, 1987, University of Oklahoma. 87 p.

Banting, Douglas Ralph. Characterization of Arctic soils; interrelationships with site and vegetation. D, 1982, University of Western Ontario.

Banwell, Gail Marie. Helium-4 and radon-222 concentrations in groundwater and soil gas as indicators of the extent and depth of fracture concentration in rock. M, 1986, Pennsylvania State University, University Park. 136 p.

Baptista, Braulio M. Petrology of the limestones of the Arnheim Formation (Cincinnatian Series, Ordovician) from western Butler County, Ohio. M, 1969, Miami University (Ohio). 93 p.

Bapuji, Soli Jehangir. The Cenozoic geology of Wylie Mountains, Culberson County, Texas. M, 1951, University of Texas, Austin.

Bar, Jules R. Du *see* Du Bar, Jules R.

Barabas, A. H. Petrologic and geochemical investigations of porphyry copper mineralization in West central Puerto Rico. D, 1977, Yale University. 491 p.

Barackman, Milan A. A study of the mineral glauconite in Apalachicola Bay, Florida; its distribution, mode of occurrence and source. M, 1964, Florida State University.

Baragar, W. R. A. Nepheline gneisses of York River, Ontario. M, 1952, Queen's University. 113 p.

Baraka, Moustafa Ahmed. A combined GPS-photogrammetric solution. D, 1988, Ohio State University. 143 p.

Barakos, Peter A. On the theory of acoustic wave scattering and refraction by internal waves. M, 1965, University of Washington. 164 p.

Baranoski, Mark T. Joint analysis in the Valley and Ridge and Allegheny Plateau in portions of Allegany County, Maryland, Mineral and Hampshire counties, West Virginia. M, 1982, University of Toledo. 126 p.

Baranovic, Michael Joseph. Persistence of brine contamination in central Ohio. M, 1975, Ohio State University.

Baranowski, James. $^{238}U/^{230}Th$ isotope systematics of rhyolites from Long Valley, California. M, 1977, Michigan State University. 122 p.

Baranowski, Jean M. Determination of seismic attenuation using observed phase shift in sedimentary rocks. M, 1982, Massachusetts Institute of Technology. 74 p.

Barany, Istvan, Jr. Basaltic breccias of the Clipperton fracture zone; sedimentation and tectonics in a fast-slipping oceanic transform. M, 1987, Duke University. 171 p.

Baranyai, Paul D. The crystallization of amorphous silica using salt catalysts. M, 1964, Michigan State University. 52 p.

Baranyi, Elizabeth. Geophysical electromagnetics at very-high and ultra-high frequencies. M, 1985, McGill University. 185 p.

Barari, Rachel A. Computer analyses of geologic data in South Dakota. M, 1976, University of South Dakota. 33 p.

Barazangi, Muawia. Geophysical investigation of the Hudson-Afton Horst, Dakota County, Minnesota. M, 1967, University of Minnesota, Minneapolis. 75 p.

Barazangi, Muawia. Three studies of the structure and dynamics of the upper mantle adjacent to a descending lithospheric slab (Tonga island arc, southwest Pacific and Great Basin, western U.S.). D, 1971, Columbia University. 147 p.

Barba Pingarron, Luis Alberto. The ordered application of geophysical, chemical and sedimentological techniques for the study of archaeological sites; the case of San Jose Ixtapa, Mexico. M, 1984, University of Georgia. 162 p.

Barbaro, Jeffrey Ralph. Early diagenesis of particulate organic matter in bioadvective sediments, Lowes Cove, Maine. M, 1985, SUNY at Binghamton. 78 p.

Barbaro, Ralph Wesley. Uncertainty and risks of reserve estimation for coal quantity and quality. D, 1987, Pennsylvania State University, University Park. 267 p.

Barbash, J. E. Distribution and behaviour of volatile organic compounds in groundwater at a landfill site in Gloucester, Ontario. M, 1983, University of Waterloo. 107 p.

Barbato, Lucia Sabina. Effect of beach replenishment at Coronado, California 1985-1986. M, 1988, University of California, Los Angeles. 118 p.

Barber, B. G. Origin of folding in Paleozoic rocks near Theresa, New York. M, 1977, Syracuse University.

Barber, David Williams. A photoelastic study of the effects of surface geometry on fault movements. M, 1973, Texas A&M University.

Barber, David Williams. Analytical study of displacements along faults with irregular fault-plane geometry. D, 1976, Texas A&M University. 65 p.

Barber, Dean Austin. The origin of sheet joints and exfoliation in the upper San Joaquin Basin, California. M, 1962, University of Idaho. 70 p.

Barber, Irene E. Hardness determinations of various opaque ore minerals. M, 1956, Columbia University, Teachers College.

Barber, Larry Billingsley, II. Geochemistry of organic and inorganic compounds in a sewage contaminated aquifer, Cape Cod, Massachusetts. M, 1985, University of Colorado.

Barber, Thomas David. A geochemical study of phosphates and glauconite found in the Eagle Fort Group of the Cretaceous. M, 1942, Texas Christian University. 64 p.

Barber, William Bruce. A geologic and geophysical survey of glacial drift in Streetsboro, Portage County, Ohio. M, 1981, Kent State University, Kent. 181 p.

Barberio, Stephen John. Sedimentation patterns on the lagoonal side of a barrier island along the North Carolina coast. M, 1971, North Carolina State University. 180 p.

Barbis, Frederic C. Rb-Sr geochronology of the Precambrian basement of Ohio. M, 1978, Ohio State University.

Barbosa Levy, Miguel Rudy. The safety and reliability of rockfill embankments. D, 1987, Texas A&M University. 190 p.

Barbosa, Aluzio Liciniode Miranda *see* Miranda Barbosa, Aluzio Liciniode

Barbour, Erwin Hinckley. The osteology of the Heloderma. D, 1887, Yale University.

Barbour, George B. The geology of the Kalgan area (China). D, 1930, Columbia University, Teachers College.

Barby, Boardman Gene. Subsurface geology of the Pennsylvanian and Upper Mississippian of Beaver County, Oklahoma. M, 1956, University of Oklahoma. 56 p.

Barca, Richard Albert. Geology of the northern portion of Old Dad Mountain Quadrangle, San Bernardino County, California. M, 1960, University of Southern California.

Barcas, Kestutis. Examination of Lower Silurian clastic deposits along the Niagara Escarpment, Bruce Peninsula, Ontario. M, 1980, Wayne State University.

Barcellona, Bruce. Application of photo-optical and Landsat MSS data to monitor surface aggregate mining; Lowndes and Monroe counties, Mississippi. M, 1981, Mississippi State University. 145 p.

Barcelo-Duarte, Jaime. Lower Cretaceous stratigraphy and depositional systems in northwestern Coahuila, Mexico. D, 1983, University of Texas, Austin. 583 p.

Barclay, C. S. Geology of the Gore canyon, Kremmling area, Grand County, Colorado. M, 1968, University of Arizona.

Barclay, Craig C. Sedimentary structures and depositional history of the coarse clastics of the Cuyahoga Formation in northern Ohio. M, 1969, Kent State University, Kent. 65 p.

Barclay, Joseph Ellis. The Cherokee-Marmaton contact in northeastern Randolph County, Missouri. M, 1949, University of Missouri, Columbia.

Bard, Catherine Sundstrom. Mineralogy and chemistry of pyroxenes from the Imnaha and lower Yakima basalts of west-central Idaho. M, 1978, Washington State University. 75 p.

Bard, Gary G. Petrology and diagenetic features of the Fort Dodge gypsum beds. D, 1982, Iowa State University of Science and Technology. 103 p.

Bard, Thomas R. Hydrothermal alteration in two deep exploratory wells in Dixie Valley, Nevada. M, 1980, University of Nevada. 92 p.

Barday, Robert J. Structure of the Panama Basin from marine gravity data. M, 1974, Oregon State University. 99 p.

Bardecki, Michal James. Wetlands in southern Ontario; a policy science approach. D, 1981, York University.

Bardet, Jean Pierre. Application of plasticity theory to soil behavior; a new sand model. D, 1984, California Institute of Technology. 207 p.

Bardoux, Marc-Victor. Stratigraphy and structural interpretation in the Aklavik Range, eastern slope of the northern Richardson Mountains, District of Mackenzie, N.W.T. M, 1984, Queen's University. 135 p.

Bardsley, Stanford Ronald. Evaluating oil shale (Tertiary, Green River Formation) by log analysis. M, 1962, University of Utah. 47 p.

Bardwell, Jennifer. The paleoecological and social significance of the zooarchaeological remains from Central Plains tradition earthlodges of the the Glenwood locality, Mills County, Iowa. M, 1981, University of Iowa. 80 p.

Barendregt, R. W. A detailed geomorphological survey of the Pakowki-Pinhorn area of southeastern Alberta. D, 1977, Queen's University.

Barfield, Billy Joe. Studies of turbulence in shallow sediment laden flow with superimposed rainfall. D, 1968, Texas A&M University. 115 p.

Barfus, Brian Lawrence. Lithologic and seismic determination and correlation of Recent depositional environments in western Mississippi Sound. M, 1984, University of Mississippi. 109 p.

Bargar, Keith E. Effects of water storage regulation upon sedimentation in Lexington and Coyote reservoirs, Santa Clara County, California. M, 1975, San Jose State University. 151 p.

Bargen, David J. Von see Von Bargen, David J.

Bargen, Nikolaus von see von Bargen, Nikolaus

Barger, Edith M. The continental shelf on the western side of North America. M, 1925, Columbia University, Teachers College.

Bargeron, Dorothy L. An economic, mineralogic and chemical investigation of the offshore phosphate deposit, People's Republic of the Congo, West Africa. M, 1984, University of Mississippi. 76 p.

Barghoorn, Steven Frederick. Magnetic polarity stratigraphy of the Tesuque Formation, Santa Fe Group, in the Española Valley, New Mexico, with a taxonomic review of the fossil camels. D, 1985, Columbia University, Teachers College. 489 p.

Barghouthi, Amjad Fawzi. Pile response to seismic waves. D, 1984, University of Wisconsin-Madison. 183 p.

Barghusen, Herbert R. Functional anatomical changes in the chelonian temporal region. D, 1960, University of Chicago. 67 p.

Barham, Bruce A. The geology of the New Fox alteration zone. M, 1987, Carleton University. 110 p.

Barham, Samuel D., III. A geologic field trip in the Richmond, Virginia area. M, 1967, Virginia State University. 32 p.

Bari, Shah Fazoul. Air-photo lineament analysis in Gundy and Broaderick townships (Kenora area, Ontario, Canada). M, 1967, University of Manitoba.

Baria, Lawrence Robert. The geology of the Red Dirt game management area (Louisiana), and a petrologic study of the upper Catahoula Formation (Miocene). M, 1971, Northeast Louisiana University.

Barifaijo, Erasmus. Petrography, chemistry and origin of magnetite in the rocks of the Nutbush Creek fault zone and adjacent area, North Carolina and Virginia. M, 1986, University of North Carolina, Chapel Hill. 60 p.

Baril, Roger Wilfrid. A study of some of the morphological, physical and chemical characteristics of a virgin and cultivated Lordstown gravelly silt loam. M, 1939, Cornell University.

Barinaga, Charles Joe. Sulfur-containing gases; I, Improved gas chromatographic analysis; II, Applied to geochemical exploration. D, 1987, University of Idaho. 156 p.

Bariss, Nicholas. Geomorphological characteristics of loess terrain; a comparative study of five sample areas in the Midwestern United States. D, 1967, Clark University. 300 p.

Barkdull, James Edwin. Geology of a portion of the Grandin Quadrangle, Missouri. M, 1957, University of Missouri, Columbia.

Barkeley, Susan Jeanette. Lower to Middle Pennsylvanian conodonts from the Klawak Formation and Ladrones Limestone, S. E. Alaska. M, 1981, University of Oregon. 116 p.

Barkemeyer, Eric. Rill sinuosity and watercourse meandering as a function of slope as developed in clay pits in the Perth Amboy area, N.J. M, 1984, Rutgers, The State University, Newark. 45 p.

Barker, Benjamin Joseph. Transient flow to finite conductivity vertical fractures. D, 1977, Stanford University. 146 p.

Barker, Charles. Tidal hydraulics and morphodynamics of Scarboro River inlet, Maine. M, 1988, Boston University. 128 p.

Barker, Charles Edward. Vitrinite reflectance geothermometry in the Cerro Prieto geothermal system, Baja California, Mexico. M, 1979, University of California, Riverside. 127 p.

Barker, Craig Alan. Upper-crustal structure of the Milford Valley and Roosevelt Hot Springs, Utah region, by modeling of seismic refraction and reflection data. M, 1986, University of Utah. 101 p.

Barker, Daniel Stephen. Hallowell granite and associated rocks, south central Maine. D, 1961, Princeton University. 240 p.

Barker, David L. The geology and gold deposits of selected areas of the Jicarilla District, Lincoln County, NM. M, 1986, New Mexico Institute of Mining and Technology. 73 p.

Barker, Fred. Pre-Cambrian and Tertiary geology of the Las Tablas Quadrangle, New Mexico. D, 1954, California Institute of Technology. 230 p.

Barker, Fred. The Coast Range Batholith between Haines, Alaska and Bennett Lake, British Columbia. M, 1952, California Institute of Technology. 45 p.

Barker, G. S. An investigation of the ore zone and other basic intrusions at Populus Lake, Kenora, Ontario. M, 1961, University of Manitoba.

Barker, Gary Wayne. Palynology of the Admire Group ("Lower Permian") of eastern Kansas. D, 1983, Kent State University, Kent. 329 p.

Barker, Gregg S. A rock magnetic study of subglacial volcanoes in Iceland. M, 1982, University of Georgia.

Barker, J. F. Methane in groundwaters; a carbon isotope geochemical study. D, 1979, University of Waterloo.

Barker, James Charles. The geology of a portion of the Lawrence Uplift, Pontotoc County, Oklahoma. M, 1950, University of Oklahoma. 63 p.

Barker, James Franklin. Rare earth elements in the Whitestone Anorthosite. M, 1972, McMaster University. 93 p.

Barker, James Michael. Geology and petrology of the Toad Spring Breccia, Abel Mountain, California, and its relation to the San Andreas Fault. M, 1972, University of California, Santa Barbara.

Barker, Jeffrey Scott. A seismological analysis of the May 1980 Mammoth Lakes, California, earthquakes. D, 1984, Pennsylvania State University, University Park. 300 p.

Barker, Jeffrey Scott. Moment tensor inversion of complex earthquakes; the 1979 Thessalonika, Greece, earthquake. M, 1981, Pennsylvania State University, University Park.

Barker, K. Scott. Stratigraphy, petrography and paleoenvironmental interpretation of the Mississippian-Pennsylvanian Amsden Formation, south of Labarge Guard Station, Salt River Range, western Wyoming. M, 1985, Idaho State University. 81 p.

Barker, M. E. The clay resources of South Carolina. M, 1924, University of South Carolina. 27 p.

Barker, M. S. Formation and dispersal of shells in a shore-line system of sedimentation. M, 1963, Pennsylvania State University, University Park.

Barker, Norman Kay. The stratigraphy and structure of Eldorado Springs North Quadrangle, Missouri. M, 1953, University of Iowa. 94 p.

Barker, Peter Arnold. Glaciation of the Chelan Trough. M, 1968, Washington State University. 52 p.

Barker, Rachel M. An areal study and petrographic description of the Triassic igneous and sedimentary rocks of the Black Rock area, South Hadley, Massachusetts. M, 1949, Smith College. 73 p.

Barker, Richard M. Constituency and origins of cyclic growth layers in pelecypod shells. D, 1970, University of California, Berkeley. 265 p.

Barker, Richard M. The formation and dispersal of shells in a shoreline system of sedimentation. M, 1963, Pennsylvania State University, University Park. 163 p.

Barker, Robert E. An environmental analysis of Recent sediments and microfauna of the continental shelf of the Southeastern United States. M, 1958, University of Oklahoma. 63 p.

Barker, Robert Wadhams. The formation of sulfides in the basal zone of the Stillwater intrusion (Precambrian), Montana. D, 1971, University of California, Berkeley. 270 p.

Barker, Sally. Magmatic system geometry at Mt. St. Helens inferred from the stress field associated with post-eruptive earthquakes. D, 1988, University of Washington. 76 p.

Barker, Steven A. Experimental apparatus for the continuous liquefaction of coal. M, 1976, Colorado School of Mines. 91 p.

Barker, Steven E. Mineral chemistry and crystallization history of basalts from holes 483, 483B, and 485A, DSDP Leg 65, East Pacific Rise, Gulf of California. M, 1981, University of New Mexico. 90 p.

Barker, Terrance G. Static strain in a non-uniform solid; with applications to California. D, 1974, University of California, San Diego. 104 p.

Barker, Terrance Gordon. Response of an elastic layer over an elastic half-space to a point source. M, 1970, Massachusetts Institute of Technology. 83 p.

Barker, William Samuel. Mineral occurrences in the District of Columbia and vicinity. M, 1941, George Washington University.

Barker, William Wayne. An electron microscopic study of extralamellar organoclay complexes. D, 1988, University of Georgia. 188 p.

Barkley, Richard A. A one-dimensional model of the vertical distribution of dissolved phosphorus in the oceans. D, 1961, University of Washington. 42 p.

Barkman, Leilya K. The crystalline rocks of the Scunemunk Quadrangle in New York State. M, 1930, Columbia University, Teachers College.

Barkmann, Peter E. A reconnaissance investigation of active tectonism in the Bitterroot Valley, western Montana. M, 1984, University of Montana. 85 p.

Barks, Ronald E. Flux growth of single crystal R_2O_3 oxides with the corundum structure. D, 1966, Pennsylvania State University, University Park. 334 p.

Barks, Ronald E. The effect of temperature and other parameters upon the uranium content of natural carbonates. M, 1962, Rice University. 72 p.

Barksdale, Joe M. The geology of the Mason Bend Quadrangle, Alabama. M, 1958, University of Alabama.

Barksdale, Julian Deverau. The Shonkin Sag Laccolith. D, 1936, Yale University. 38 p.

Barlaz, Dora. Sedimentation in the Duluth-Superior Harbor, Lake Superior. M, 1983, University of Minnesota, Minneapolis. 122 p.

Barlow, Charles A. Radar geology and tectonic implications of the Choco Basin, Colombia, South America. M, 1981, University of Arkansas, Fayetteville. 102 p.

Barlow, David A. Petrology of the Mount Shields Formation (Belt Supergroup), western Montana, northern Idaho. M, 1983, University of Montana. 140 p.

Barlow, James A., Jr. Geology of the Laprele Creek-Boxelder Creek area, Converse County, Wyoming. M, 1950, University of Wyoming. 49 p.

Barlow, James A., Jr. The geology of the Rawlins Uplift, Carbon County, Wyo. D, 1953, University of Wyoming. 179 p.

Barlow, James Arthur. Geology of the Fairmont East 7 1/2 Minute Quadrangle (West Virginia). M, 1965, West Virginia University.

Barlow, James Arthur. Stratigraphy and paleobotany of the youngest Pennsylvanian strata in the Caryville (Campbell county), Tennessee area. D, 1969, University of Tennessee, Knoxville. 342 p.

Barlow, James L. Geology of the central third of the Lyon's Quadrangle, Oregon. M, 1955, University of Oregon. 79 p.

Barlow, Lisa Katharine. Event stratigraphy, paleoenvironments, and petroleum source rock potential of the lower Niobrara Formation (Cretaceous), northern Front Range, Colorado. M, 1986, University of Colorado.

Barlow, Rodney A. Crystalline inclusions from Muong Nong tektites and their implication concerning tektite genesis. M, 1977, University of Delaware.

Barlow, Roger Brock. Major and minor element geochemistry of the Archean volcanic rocks of the southern Knee Lake area, Manitoba. M, 1973, Michigan Technological University. 124 p.

Barlow, Steven Gus. Geology and slope stability of Point Delgada, California. M, 1980, California State University, Long Beach. 85 p.

Barlow, Wallace Dudley. Geology of the Salem area. M, 1936, Virginia Polytechnic Institute and State University.

Barnard, Frederick L. Structural geology of the Sierra de Los Cucapas, northeastern Baja California, Mexico and Imperial Valley, California. D, 1968, University of Colorado. 189 p.

Barnard, Leo Allen. Marine manganese accretions; nodule-micronodule comparisons among major ocean basins. D, 1982, Texas A&M University. 291 p.

Barnard, Leo Allen. The kinetics and thermodynamics of silica sorption in marine sediments. M, 1977, University of South Florida, St. Petersburg. 90 p.

Barnard, Pamela Bright. A geochemical and mineralogical analysis of weathered andesitic basalt in the tropics. M, 1987, University of Maryland.

Barnard, R. S. Morphologic and morphometric response to channelization in Big Pine Creek ditch, Benton County, Indiana. M, 1976, Purdue University. 86 p.

Barnard, Ralph M. Geology of the Ricardo Beds in the western portion of Saltdale Quadrangle, Kern County. M, 1950, University of Southern California.

Barnard, Walther M. Optical properties of selected plagioclases. M, 1961, Dartmouth College. 70 p.

Barnard, Walther M. Solubilities of selected chalcophile elements in hydrothermally synthesized β-ZnS (sphalerite). D, 1965, Pennsylvania State University, University Park. 138 p.

Barnard, William Dana. Late Cenozoic sedimentation on the Washington continental slope. D, 1973, University of Washington. 255 p.

Barnard, William Rives. Determination of fluoride in precipitation samples and its implications for the geochemical cycling of fluorine. M, 1980, University of Virginia. 51 p.

Barndt, Jeffrey K. The magnetic polarity stratigraphy of the type locality of the Dhok Pathan faunal stage, Potwar Plateau, Pakistan. M, 1977, Dartmouth College. 105 p.

Barner, Wendell. Investigation of the geology and hydrology of Stone County, Missouri to determine the extent of groundwater contamination. M, 1988, Southwest Missouri State University.

Barnes, C. R. Conodont biofacial analysis of some Wilderness (Middle Ordovician) Limestone, Ottawa Valley, Ontario. D, 1964, University of Ottawa. 369 p.

Barnes, Calvin Glenn. Geology and petrology of the Wooley Creek Batholith, Klamath Mountains, northern California. D, 1982, University of Oregon. 287 p.

Barnes, Calvin Glenn. The geology of the Mount Bailey area, Oregon. M, 1978, University of Oregon. 123 p.

Barnes, Charles. Model studies of subsurface gravity gradients in vicinity of salt domes. M, 1977, Texas A&M University.

Barnes, Charles Winfred. Geology of the Salmon River headwaters area, Blaine and Camas counties, Idaho. M, 1962, University of Idaho. 47 p.

Barnes, Charles Winfred. Reconnaissance geology of the Priest River area, Idaho. D, 1965, University of Wisconsin-Madison. 145 p.

Barnes, Charles Wynn. Geology of the manganese deposits, Walnut Grove, Alabama. M, 1946, University of Virginia. 34 p.

Barnes, David A. Basin analysis of volcanic arc-derived Jura-Cretaceous sedimentary rocks, Vizcaino Peninsula, Baja California Sur, Mexico. D, 1982, University of California, Santa Barbara. 313 p.

Barnes, Farrell Francis and Butler, John W., Jr. The Pre-Cambrian rocks of the Sawatch Range, Colorado. D, 1935, Northwestern University.

Barnes, Farrell Francis and Butler, John W., Jr. The structure and stratigraphy of the Columbia River gorge and Cascade Mountains in the vicinity of Mount Hood. M, 1930, University of Oregon. 73 p.

Barnes, Frederick Q. Correlation of granitic rocks, Aylmer Lake area, Northwest Territories. M, 1949, University of Toronto.

Barnes, Frederick Q. The Snowdrift and McLeod Bay map-areas, Great Slave Lake, Northwest Territories. D, 1953, University of Toronto.

Barnes, Gordon L. A determination of the effect of bubble length and position on the sensitivity of the Wild T-4 suspension level. M, 1966, Ohio State University.

Barnes, Harley, Jr. Stratigraphy and structure of the Cretaceous rocks on the north flank of the San Juan Basin, Colorado and New Mexico. D, 1954, The Johns Hopkins University.

Barnes, Harley, Jr. The geology of Kittatinny and Little Mountains, north of Harrisburg, Pennsylvania. M, 1939, Northwestern University.

Barnes, Hubert Lloyd. The source of base metal deposits. D, 1958, Columbia University, Teachers College. 123 p.

Barnes, I. L. An investigation of a new method for potassium-argon age determination. D, 1963, University of Hawaii.

Barnes, Ivan Keiler. The system Pb-U-O-H, water saturated at 250 degrees centigrade. D, 1961, Harvard University.

Barnes, James Irvin. Measurement of mercury in soil gas; an aid to minerals exploration in Nevada. M, 1971, University of Nevada. 60 p.

Barnes, James Virgil. Geology of the northern one-half of the Montgomery City Quadrangle (Loutre River area) Callaway and Montgomery counties, Missouri. M, 1944, University of Missouri, Columbia.

Barnes, James Virgil. Structural analysis of the northern end of the Tobacco Root Mountains, Madison County, Montana. D, 1954, Indiana University, Bloomington. 178 p.

Barnes, Jerry D. A gravity and magnetic investigation of the southern Colombian Andes. M, 1971, University of Missouri, Columbia.

Barnes, John H. Sedimentological studies of (Pleistocene) tills of the Gowanda area, New York. M, 1972, SUNY at Buffalo. 81 p.

Barnes, John McGregor, Jr. A regional study of the Denver-Julesburg Basin. M, 1953, University of Michigan.

Barnes, Kenneth Burton. The extraction of petroleum from reservoirs by liquid drive. M, 1933, Pennsylvania State University, University Park. 48 p.

Barnes, Laverne Ellsworth. Heavy minerals in the Pennsylvanian Sewanee, Bon Air and Rockcastle conglomerates of the Mayland Quadrangle, Tennessee. M, 1954, Vanderbilt University.

Barnes, Lawrence Gayle. Late Tertiary cetacea of the Northeast Pacific Ocean. D, 1972, University of California, Berkeley. 494 p.

Barnes, Lawrence Gayle. Miocene Desmatophocinae from California. M, 1969, University of California, Berkeley. 113 p.

Barnes, Marvin Peterson. Porphyry copper deposits; a computer analysis of significance of geological parameters (Basin and Range Province). D, 1970, University of Utah. 200 p.

Barnes, Mary Elizabeth. Subsurface Pennsylvanian studies in Hamilton County, Illinois. M, 1948, University of Illinois, Urbana. 18 p.

Barnes, Melanie Ames Weed. The geology of Cascade Head, an Eocene volcanic center. M, 1981, University of Oregon. 94 p.

Barnes, Neal E. The areal geology and Holocene history of the eastern half of Mahone Bay, Nova Scotia. M, 1976, Dalhousie University. 125 p.

Barnes, Nora L. Local study of the Coalmont Aquifer in the North Park Basin, Colorado. M, 1983, Texas Tech University. 53 p.

Barnes, Peter William. Marine geology and oceanography of Santa Cruz Basin off southern California. D, 1970, University of Southern California. 192 p.

Barnes, Philips Jeffrey. Burial diagenesis of shales in the lower Tuscaloosa Formation (Cretaceous), Louisiana and Mississippi. M, 1986, University of New Orleans. 187 p.

Barnes, Robert Clay. Geological and geophysical investigation of sulfide deposits near New Canton, Virginia. M, 1963, University of Virginia. 88 p.

Barnes, Robert Howell. Pre-Catheys geology of the Elkton Quadrangle, Giles County, Tennessee. M, 1958, Vanderbilt University.

Barnes, Robert P. The Upper Mississippian and Lower Pennsylvanian strata of northern Arkansas. M, 1960, Northwestern University.

Barnes, Robert Stith. A trace element survey of selected waters, sediments, and biota of the Lake Washington drainage. M, 1976, University of Washington. 169 p.

Barnes, Ross Owen. Noble gas concentrations in the pore fluids of marine sediments and the construction of an in situ pore water sampler. D, 1973, University of California, San Diego.

Barnes, Sarah-Jane. The origin of the fractionation of platinum group elements in Archean komatiites of the Abitibi greenstone belt, northern Ontario, Canada. D, 1983, University of Toronto.

Barnes, Stephen J. Petrology and geochemistry of a portion of the Howland (J. M.) Reef of the Stillwater Complex, Montana. D, 1983, University of Toronto. 213 p.

Barnes, Stephen John. Petrology and geochemistry of the Katiniq nickel deposit and related rocks, Ungava, northern Quebec. M, 1979, University of Toronto.

Barnes, Steven Solomon. The formation of oceanic ferromanganese nodules. D, 1967, [University of California, San Diego]. 68 p.

Barnes, Sydney U. The foraminifera of the Times Point Formation at San Pedro, Los Angeles County, California. M, 1938, University of Southern California.

Barnes, Thomas John. Geology of the Sand Mountain area, western Wheeler County, Oregon. M, 1978, Oregon State University. 119 p.

Barnes, Virgil Everett. Changes in hornblende at about 800° C. D, 1930, University of Wisconsin-Milwaukee.

Barnes, Virgil Everett. The geology of the Oakesdale quadrangle. M, 1927, Washington State University. 51 p.

Barnes, W. C. Geology of the North Park syncline area, Jackson County, Colorado. M, 1958, University of Wyoming. 55 p.

Barnes, William Charles. Geology of the northeast Whitefish Range, northwest Montana. D, 1963, Princeton University. 102 p.

Barnes, William Charles, III. An investigation of the subsurface Bouguer anomaly in the vicinity of shallow salt domes. M, 1977, Texas A&M University.

Barness, Donald Lawrence. Guembelinids of the Upper Cretaceous Taylor Formation, Texas. M, 1960, University of Wisconsin-Madison.

Barnett, Albert Prinnon. Residential flood irrigation and microclimate in Phoenix, Arizona. D, 1980, Arizona State University. 124 p.

Barnett, Colin T. Theoretical modelling of induced polarization effects due to arbitrarily shaped bodies. D, 1972, Colorado School of Mines. 239 p.

Barnett, D. E. Chemical and crystallographic properties of phases in the ZnS-In_2S_3 system; 600-1080°C. M, 1972, University of New Brunswick.

Barnett, D. M. Glacial geomorphology in a sub-polar proglacial lake basin; a process-response model. D, 1977, University of Western Ontario.

Barnett, Daniel. A petrological study, Kramer Island macrodike, East Greenland. M, 1987, Dartmouth College. 109 p.

Barnett, Donald W. The tradeoff between energy and the environment; the case of crude oil supplies for California. D, 1975, Pennsylvania State University, University Park. 319 p.

Barnett, Douglas B. A paleoenvironmental reconstruction of the upper Roslyn Formation, central Washington, with implications for coal exploration. M, 1985, Eastern Washington University. 167 p.

Barnett, E. S. Geology of the Agnico-Eagle gold mine, Quebec. M, 1979, University of Western Ontario. 147 p.

Barnett, Jack Arnold. Ground-water hydrology of Emigration Canyon, Salt Lake County, Utah. M, 1966, University of Utah. 101 p.

Barnett, James Matthew. Sedimentation rate of salt determined by micrometeorite analysis. M, 1983, Western Michigan University. 98 p.

Barnett, Jarrall. Late Tertiary foraminifera from the Macasphalt shell pit, Sarasota, Florida. M, 1986, Bowling Green State University. 95 p.

Barnett, Lloyd Oris. Storm runoff characteristics of three small watersheds in western Oregon. M, 1963, Colorado State University. 91 p.

Barnett, Peter James. Quaternary stratigraphy and sedimentology, north-central shore, Lake Erie, Ontario, Canada. D, 1987, University of Waterloo. 335 p.

Barnett, Peter James. Till matrix characteristics of the upper and lower till of the Niagara Peninsula, Ontario. M, 1975, University of Waterloo.

Barnett, R. L. The contact aureole at Copper Flat, Grant County, New Mexico. M, 1973, University of Western Ontario. 132 p.

Barnett, Richard S. A quantitative study of late Eocene Nummulites (Foraminiferida), Jackson stage, southeastern United States. M, 1969, University of Houston.

Barnett, Robert G. Metamorphism of the Pittsburgh Coal and its relationship to volatile matter. M, 1983, University of Akron. 82 p.

Barnett, Stanley David. A paleoenvironmental analysis based on Ostracoda of the Cane River Formation (Eocene) of north central Louisiana. M, 1987, Northeast Louisiana University. 112 p.

Barnett, Stephen F. Palynology and age of the Alvord Creek Formation, Steens Mountain, southeastern Oregon. M, 1984, Loma Linda University. 71 p.

Barnett, Stockton Gordon, III. Conodonts from the Jacksonburg Limestone (Middle Ordovician) of northwestern New Jersey and eastern Pennsylvania. M, 1964, University of Iowa. 142 p.

Barnett, Stockton Gordon, III. Late Cayugan and Helderbergian stratigraphy of southeastern New York and northern New Jersey. D, 1966, Ohio State University. 240 p.

Barnett, T. R. Hydrogeology for urban planning in Boone County, Illinois. M, 1975, Northeastern Illinois University.

Barnett, Thomas MacDonough. A method for determining an optimal solid waste/recreational system for rural regions. M, 1978, University of Virginia. 76 p.

Barnett, Wayne Stephen. A morphometric, stratigraphic and paleoecologic analysis of the genera Voluticorbis and Athleta (Gastropoda, Voluta) from the Paleocene and Eocene epochs of Alabama. M, 1981, Auburn University. 222 p.

Barnette, Carr Howard. Carbonate petrology of the zone of Orbitolina, Hudspeth, Brewster, and Presidio counties, Texas. M, 1961, Texas Tech University. 108 p.

Barney, Emmet C. Petrology of the chert in the Boone Group and its stratigraphic relation in the Southwest Ozark area. M, 1959, University of Kansas. 80 p.

Barnhardt, M. L. Late Quaternary glacial geomorphology of the Kearn Creek-South Piney Creek area, Bighorn Mountains, Wyoming. D, 1979, University of Illinois, Urbana. 241 p.

Barnhart, John T. Distribution of Recent sediments and foraminifera across the Bermuda Platform. M, 1963, University of Houston.

Barnhart, Stephen F. A possible extension of the Oregon-Nevada Lineament to the Devils Gate area, Eu-

reka County, Nevada. M, 1978, Ohio University, Athens. 50 p.

Barnhill, Susan Jean. Geology and genesis of tungsten-molybdenum mineralization at Mt. Reed-Mt. Haskin, northern British Columbia. M, 1982, Queen's University. 170 p.

Barnhill, William Burrough. Jeff Conglomerate, northern Davis Mountains, Texas (B 266). M, 1950, University of Texas, Austin.

Barnhisel, Richard Irven. The formation and stability of aluminum interlayers in clays. D, 1965, Virginia Polytechnic Institute and State University. 186 p.

Barnhouse, John Douglas, Jr. A classification system for microfractures and other microscopic zones of fluid transmissibility in clastic rocks. M, 1978, Ohio University, Athens. 82 p.

Barnick, Sandra K. Feasibility of the proposed Garden Lane recharge basin in Portage, Michigan. M, 1987, Western Michigan University.

Barning, Kwasi. Petrology of the northwestern part of the Holyrood Granite Batholith, Avalon Peninsula, Newfoundland. M, 1965, Memorial University of Newfoundland. 88 p.

Barnola, Alberto. Pre-Cretaceous paleogeology of Arizona and Utah. M, 1957, Stanford University.

Barnola, Alberto. Stratigraphy of the Zayante and Lompico creeks area, Santa Cruz County, California. M, 1958, Stanford University.

Barnosky, Anthony David. A skeleton of mesoscalops from the Miocene Deep River Formation, Montana; evolutionary, functional, and stratigraphic implications. M, 1980, University of Washington. 181 p.

Barnosky, Anthony David. Geology and mammalian paleontology of the Miocene Colter Formation of Jackson Hole, Teton County, Wyoming. D, 1983, University of Washington. 332 p.

Barnosky, Cathy Lynn. Late Quaternary vegetational history of the southern Puget Lowland; a long record from Davis Lake, Washington. M, 1979, University of Washington. 47 p.

Barnosky, Cathy Whitlock. Late-Quaternary vegetational and climatic history of southwestern Washington. D, 1983, University of Washington. 201 p.

Barnsley, John Anthony. A sedimentological study of the glacigenic deposits at Mohawk Bay, near Dunnville, Ontario. M, 1985, Brock University. 219 p.

Barnum, Bruce E. Petrography and petrology of the lower Stillwater igneous complex (Precambrian), northwest Iron Mountain area, Montana. M, 1971, Colorado State University. 92 p.

Barnum, Dean C. Modern coal lens analogs, Wakulla County, Florida. M, 1966, Florida State University.

Barnum, Dean Charles. A petrologic, stratigraphic, and crushing properties study of the Pahasapa Limestone (Madison) of the northern Black Hills, South Dakota. D, 1973, South Dakota School of Mines & Technology.

Barnum, Harry P. 1-D synthetic seismogram analysis of seismic anomalies in the Lockport Formation, Alleghany County, New York. M, 1982, SUNY at Buffalo. 78 p.

Barnum, James Bradford. Ecological structure and function of small agricultural streams in central Iowa. D, 1984, Iowa State University of Science and Technology. 192 p.

Barnwell, George F. The alteration of a diabase dike, Isabella Mine, Marquette iron district, Michigan. M, 1922, University of Wisconsin-Milwaukee.

Barnwell, William W. Geology of the South Hahns Peak District. M, 1955, University of Wyoming. 91 p.

Baroffio, James Richard. Environmental mapping of the Lower Allegheny Series of the Appalachian Basin. D, 1964, University of Illinois, Urbana. 101 p.

Baroffio, James Richard. Structure and stratigraphy of Sandy and eastern Pike townships, Stark County, Ohio. M, 1958, Ohio State University.

Baron, Donald M. The petrology of a Precambrian schist from Bristol, northeastern South Dakota. M, 1971, University of South Dakota. 60 p.

Baron, J. A geotechnical survey of the five red mud "tailings" storage sites in Jamaica. M, 1977, University of Waterloo.

Baron, William. The response of cohesive soils to shock loadings. D, 1966, Purdue University. 228 p.

Barondeau, Bernard. Les Phénomènes de fracture avec la mise en place du dome anorthositique de Morin. M, 1974, Universite de Montreal.

Barondeau, Bernard. Mise en place du Massif de Morin (southwestern Quebec, Precambrian). M, 1971, Universite de Montreal.

Barone, John. Field relationships and emplacement of the Baltimore gabbro complex along the Susquehanna River, Cecil County, Maryland. M, 1978, University of Delaware.

Barone, William Edward. Depositional environments and diagenesis of the lower San Andres Formation. M, 1976, Texas Tech University. 93 p.

Barongo, Justus Obiko. Magnetic model theory in the analysis of vertical gradient anomalies. M, 1977, Queen's University. 119 p.

Barosh, Patrick James. Geology of the Beaver Lake Mountains, Utah. M, 1959, University of California, Los Angeles.

Barosh, Patrick James. Lower Permian stratigraphy of east-central Nevada and adjacent Utah. D, 1964, University of Colorado. 272 p.

Barovich, Karin Marie. Age constraints on early Proterozoic deformation in the northern Front Range, Colorado. M, 1986, University of Colorado. 49 p.

Baroyant, V. Numerical models for the mechanical properties of saturated clay-water systems. D, 1974, University of Utah. 113 p.

Barozzi, Rolando. Sonic resonance and related engineering properties of selected soils and rock. M, 1965, University of Arizona.

Barr, David John. Use of side-looking airborne radar imagery for engineering soils studies. D, 1968, Purdue University. 206 p.

Barr, Frank Theodore. Paleontology and stratigraphy of the Pennsylvanian and Permian rocks of Ward Mountain, White Pine County, Nevada. M, 1957, University of California, Berkeley. 97 p.

Barr, Joe William. The geography of the water supply of metropolitan Saint Louis. M, 1937, Washington University. 121 p.

Barr, John George. The geologic structure of Point Township, Posey County, Indiana. M, 1951, Indiana University, Bloomington. 13 p.

Barr, Kelton. Estimation of groundwater and lake chemistry parameters and their inter-relationships in glaciated terranes. M, 1978, University of Minnesota, Duluth.

Barr, Richard Kevin. Economic geology of the Zacatecas mining district. M, 1976, University of New Orleans. 100 p.

Barr, Sandra Marie. Geology of the northern end of Juan de Fuca Ridge and adjacent continental slope. D, 1972, University of British Columbia. 268 p.

Barragy, Edward J. The geology of the Deadwood Formation of the Lead Quadrangle of the Black Hills of South Dakota. M, 1929, University of Iowa. 109 p.

Barranco, Frank Thomas, Jr. The distribution of brick red lutite in the western North Atlantic; evidence from visible light spectra. M, 1988, University of Texas, Arlington. 102 p.

Barrash, Warren. Geology of the NE 1/4 Twin Peaks Quadrangle, Custer County, Idaho. M, 1978, University of Idaho. 99 p.

Barrash, Warren. Hydrostratigraphy and hydraulic behavior of fractured Brule Formation in Sidney Draw, Cheyenne County, Nebraska. D, 1986, University of Idaho. 205 p.

Barratt, John C. Fales Member (Upper Cretaceous) deltaic and shelf bar complex, central Wyoming. M, 1982, University of Texas, Austin.

Barratt, Michael William. Regional study of the Permo–Pennsylvanian strata of South Dakota. M, 1969, Michigan State University. 56 p.

Barraud, Claude. Etude géologique des domes de granulites au sud du Massif de Morin (S.W. Quebec). D, 1971, Universite de Montreal.

Barraud, Claude. Pétrographie et étude métaleucotroctolites géochimique des anorthosites, des paratroctolites, des métaleucotroctolites et des métagabbros de la région de Saint-Calixte—New Glasgow, P.Q., Canada. M, 1971, Universite de Montreal.

Barreda, Willy Z. Rodriguez. Strain release in the central Aleutians from 1966 to 1970; observed seismicity vs estimated creep. M, 1972, Colorado School of Mines. 139 p.

Barreiro, Barbara Anne. Lead isotope evidence for crust-mantle interaction during magmagenesis in the South Sandwich island arc and in the Andes of South America. D, 1982, University of California, Santa Barbara. 185 p.

Barrell, Joseph. The geology of the Elkhorn District, Montana. D, 1900, Yale University.

Barrell, Kirk Arthur. Structure and distribution of abnormal pressures in the Vicksburg Formation (Oligocene), Hinde Field, Starr County, Texas. M, 1988, Texas A&M University. 123 p.

Barrell, Sharon. Stratigraphic and depositional environments of Upper Devonian rocks in east-central West Virginia and adjacent Virginia. M, 1986, University of North Carolina, Chapel Hill. 115 p.

Barrell, Steve. Analysis of the effect on graded beds of changing gravitational acceleration with time; a protype paleogravimeter. M, 1976, University of South Carolina.

Barrera, Enriqueta. Isotopic paleotemperatures; 1, Effect of diagenesis; II, Late Cretaceous temperatures. D, 1986, Case Western Reserve University. 331 p.

Barrera, Enriquetta. Middle and late Miocene oceanography of the northeastern Pacific Ocean. M, 1983, Case Western Reserve University.

Barrero, Dario. Geology of the area west of Payande Tolima, Colombia. D, 1968, Indiana University, Bloomington. 65 p.

Barrero, Dario. Geology of the central Western Cordillera, west of Buga and Roldanillo, Colombia. D, 1977, Colorado School of Mines. 154 p.

Barreto, Laura L. The surficial geology of northern Washington County, Wisconsin. M, 1988, University of Wisconsin-Milwaukee. 76 p.

Barrett, Christopher Mathias. A gravity and magnetic study of the Kingfisher anomaly, North-central Oklahoma. M, 1980, University of Oklahoma. 45 p.

Barrett, Daniel Patrick. A hydrogeologic assessment of the Ozan Formation, central Texas. M, 1988, Baylor University. 141 p.

Barrett, David Wilburn. Micropaleontology of the Evanston Formation, southwestern Wyoming. M, 1953, University of Utah. 60 p.

Barrett, Donald Larry. Seismic crustal studies in eastern Canada; Atlantic coast of Nova Scotia. M, 1963, Dalhousie University.

Barrett, Elizabeth E. Paleosalinity studies of the Kiamichi, Denton, and Pawpaw formations (Lower Cretaceous), north-central Texas. M, 1986, University of Texas, Arlington. 90 p.

Barrett, Gary Edward. Infiltration in water repellent soil. D, 1988, University of British Columbia.

Barrett, Harold Elliott, Jr. Paleoenvironmentally determined maceral, lithotype, and elemental relationships in two coals of the Cumberland Plateau, Tennessee. M, 1979, University of Tennessee, Knoxville. 78 p.

Barrett, Henry Haldred. Geology of the Calwood area, Callaway County, Missouri. M, 1940, University of Missouri, Columbia.

Barrett, James Edward. A guide to the highway geology of Washington and Sullivan counties, Tennessee. M, 1974, East Tennessee State University.

Barrett, Jonathan R. Physiographic interpretation of the stratigraphy of Block Island, Rhode Island. M, 1977, University of Rhode Island.

Barrett, Karen W. Silicoflagellates from Tertiary deposits along the Rappahannock River in Virginia. M, 1977, University of Rhode Island.

Barrett, Kent R. The volcanology and sedimentology of the Kaminaki Group, Carr Lake District of Keewatin, Nortwest Territories. M, 1981, University of Manitoba. 63 p.

Barrett, Larry Frank. Igneous intrusions and associated mineralization in the Saddle Mountain mining district, Pinal County, Arizona. M, 1972, University of Utah. 90 p.

Barrett, Lynn Wandell, II. Subsurface study of Morrowan rocks in central and southern Beaver County, Oklahoma. M, 1961, University of Oklahoma. 60 p.

Barrett, Mary L. Stratigraphic bank and off-bank facies of the Winchell Limestone (Pennsylvanian), Possum Kingdom area, North-central Texas. M, 1980, Stephen F. Austin State University.

Barrett, Mary Louise. The dolomitization and diagenesis of the Jurassic Smackover Formation, Southwest Alabama. D, 1987, The Johns Hopkins University. 383 p.

Barrett, Michael E. Depositional systems in the Rinconada Formation, Taos County, New Mexico. M, 1979, University of Texas, Austin.

Barrett, Peter John. The post-glacial Permian and Triassic Beacon Rocks in the Beardmore Glacier area, central Transantarctic Mountains, Antarctica. D, 1968, Ohio State University.

Barrett, Philip M. Petrology and structure, Bayonne Batholith, southeastern British Columbia. M, 1982, University of Montana. 87 p.

Barrett, Ramsay A. The maturation of the Mississippian Chainman Shale in Railroad Valley, Nye County, Nevada. M, 1987, University of Wyoming. 82 p.

Barrett, Richard B. Geology of the Spring Valley Quadrangle, Washington, Benton and Madison counties, Arkansas. M, 1965, University of Arkansas, Fayetteville.

Barrett, Robert E. An optimization algorithm for the regional identification of aquifer hydraulic conductivity. M, 1984, Colorado State University. 125 p.

Barrett, Ruth Anne. The geology, mineralization, and geochemistry of the Milestone Hot-Spring silver-gold deposit near the Delamar silver mine, Owyhee County, Idaho. M, 1985, University of Idaho. 237 p.

Barrett, Stephen A. A three-dimensional magnetic model of Augustine Volcano. M, 1978, University of Alaska, Fairbanks. 175 p.

Barrett, Stephen Francis. Paleoecology and stratigraphy of Devonian sediments in the Northern Andes, Colombia; paleogeographic implications. D, 1986, University of Chicago. 329 p.

Barrett, Vernon A. Mineralogical study and crystal structure refinement of Canadian sodalite. M, 1968, University of Manitoba.

Barrett, William J. Geology of an area in southwestern Riley County, Kansas. M, 1958, Kansas State University. 69 p.

Barrette, Paul Dominique. Structure and deformational history of the Hawasina Complex in the Sufrat and Dawh Range, western foothills of the Oman Mountains, eastern Arabian Peninsula. M, 1985, Memorial University of Newfoundland. 139 p.

Barretto, Paulo Marcos de Campos. Emanation characteristics of terrestrial and lunar materials and the ^{222}Rn loss effect on the U-Pb system discordance. D, 1973, Rice University. 179 p.

Barretto, Paulo Marcos de Campos. Radon-222 emanation from rocks, soils, and lunar dust. M, 1971, Rice University. 100 p.

Barrick, James Edward. Silurian conodonts of the Clarita Formation, Arbuckle Mountains, Oklahoma. M, 1975, University of Iowa. 207 p.

Barrick, James Edward. Wenlockian (Silurian) depositional environments and conodont biofacies, South-central United States. D, 1978, University of Iowa. 273 p.

Barrick, Paula Jean. The petrogenesis of the alkaline rocks of the Judith Mountains, central Montana. M, 1982, Montana State University. 106 p.

Barrie, Charles Prescott. The geology of the Khayyam and Stumble-On deposits, Prince of Wales Island, Alaska. M, 1984, University of Texas, Austin. 171 p.

Barrie, Charles Q. Late Quaternary geological history of Makkovik Bay, Labrador. M, 1980, Dalhousie University. 265 p.

Barrie, Donald Show. Engineering geology of northern Estancia Valley, north-central New Mexico. M, 1987, New Mexico Institute of Mining and Technology. 110 p.

Barrie, F. J. A gravimetric survey of South-central New Mexico and West Texas; Volume I & II. M, 1975, University of Texas at El Paso.

Barrie, K. A. The conodont biostratigraphy of the Black Prince Limestone (Pennsylvanian) of southeastern Arizona. M, 1975, University of Arizona.

Barrientos, Sergio Eduardo. Seismic sources from geodetic observations. D, 1987, University of California, Santa Cruz. 110 p.

Barriga, Fernando Jose Arraiano de Sousa. Hydrothermal metamorphism and ore genesis at Aljustrel, Portugal. D, 1983, University of Western Ontario. 368 p.

Barrilleaux, Janell. The geomorphology and Quaternary history of the Houston barrier segment of the Ingleside strandplain, Calcasieu Parish, Louisiana. M, 1986, University of Southwestern Louisiana.

Barringer, Julia L. Handy *see* Handy Barringer, Julia L.

Barringer, Richard Alan. The stratigraphy of late Quaternary deposits in the Payan mining district, Narino, Colombia, South America. M, 1987, Old Dominion University. 172 p.

Barrington, Jonathan. Piercement features near Cameron, Arizona; Part 1, A breccia pipe at Black Peak; Part 2, Collapse features and silica plugs, Part 3, Alteration of Tuba dike. D, 1961, Columbia University, Teachers College. 193 p.

Barrington, Jonathan. Uranium mineralization at the Midnite Mine, Spokane, Washington. M, 1959, Columbia University, Teachers College. 41 p.

Barron, Andrew Morrow. The effects of thermal cycling on the magnetic properties of lunar analogs. M, 1982, University of Wyoming. 76 p.

Barron, Barbara Rae. Diffusion rate estimates from pyroxene, Garnet Ridge, Arizona. M, 1985, University of Texas, Austin. 112 p.

Barron, Eric James. Paleogeography and climate, 180 million years to the present. D, 1980, University of Miami.

Barron, Eric James. Suspended sedimentation processes, Marco Island, Florida. M, 1976, University of Miami. 183 p.

Barron, John Arthur. The late Miocene-early Pliocene marine diatom assemblage of southern California; biostratigraphy and paleoecology. D, 1974, University of California, Los Angeles.

Barron, Lance S. The paleoecology of the Upper Devonian-Lower Mississippian black shale sequence in eastern Kentucky. M, 1982, University of Kentucky. 309 p.

Barron, Lawrence Murray. Thermodynamic multicomponent silicate equilibrium phase calculations. D, 1970, McGill University. 156 p.

Barron, Terry Jay. Pre-exploration economic feasibility study of the southern Mexico massive sulfide province, Mexico and Guerrero, Mexico. M, 1980, University of Texas, Austin. 120 p.

Barros, Jose Antonio. Stratigraphy, structure and paleogeography of the Jurassic-Cretaceous passive margin in western and central Cuba. M, 1987, University of Miami. 139 p.

Barrou, S. S. Chelcenham lead works. M, 1876, Washington University.

Barrow, Kenneth Thomson. Trout Creek Formation, southeastern Oregon; stratigraphy and diatom paleoecology. M, 1983, Stanford University. 121 p.

Barrow, Leonidas Theodore. The geology and the building stone of Cedar Park and vicinity. M, 1923, University of Texas, Austin.

Barrow, Thomas Davies. Mesozoic stratigraphy and structural history of the East Texas Basin. D, 1953, Stanford University. 222 p.

Barrow, Thomas Davies. The geology of the Backbone Ridge area, Llano and Burnet counties, Texas. M, 1948, University of Texas, Austin.

Barrow-Hurlbert, Sarah A. Geology and neotectonics of the upper Nevis Basin, South Island, New Zealand. M, 1986, Oregon State University. 161 p.

Barrows, Allan Geer, Jr. Geology of the Hamburg-McGuffy creek area, Siskiyou County, California, and petrology of the Tom Martin ultramafic complex. D, 1969, University of California, Los Angeles. 354 p.

Barrows, Katherine Jadwiga. Geology of the southern Desatoya Mountains, Churchill and Lander counties, Nevada. D, 1971, University of California, Los Angeles.

Barrows, Lawrence J. Earth strain; episodic noise due to soil moisture variations. M, 1973, Colorado School of Mines. 51 p.

Barrows, Lawrence J. Gravitational tectonic and static seismic modeling with finite elements. D, 1978, Colorado School of Mines. 141 p.

Barrows, Peter Scott. The mineralogy and petrology of the Crystal Peak skarn, Gunnison County, Colorado. M, 1982, University of Massachusetts. 172 p.

Barrows, Walter L. A fulgurite from the Raritan Sands of New Jersey with historical sketch and bibliography. M, 1910, Columbia University, Teachers College.

Barrus, Robert Bruce. The bedrock geology of southeastern Alpine Lake Quadrangle, Wind River Mountains, Wyoming. M, 1968, University of Washington. 45 p.

Barrus, Robert Bruce. The petrology of the Precambrian rocks of the High Peaks area, Wind River mountains, Fremont County, Wyoming. D, 1970, University of Washington. 86 p.

Barry, Dereck Michael. A regional subsurface study of the upper Wilcox in East Texas. M, 1979, Texas Christian University. 50 p.

Barry, J. C. The pelvic anatomy and adaptations of extant fissiped carnivores. D, 1976, Yale University. 292 p.

Barry, Jeffrey M. Hydrogeochemistry of Golden Valley, Nevada, and the chemical interactions during artificial recharge. M, 1985, University of Nevada. 190 p.

Barry, John O'Keefe. Louisiana Midway Eocene Pelecypoda. M, 1941, Louisiana State University.

Barry, John P. Fracture analysis of the northern end of the Leinster Granite. M, 1987, Pennsylvania State University, University Park. 86 p.

Barry, Thomas Leo. Kinetics, phase equilibria, and crystal chemical studies in rare earth oxide-alkaline earth oxide systems. D, 1964, Pennsylvania State University, University Park. 129 p.

Barry, William Leo. A comparative study of geologic information derived from visually interpreted SLAR and Landsat imagery. M, 1983, Indiana State University. 40 p.

Barsdate, Robert John. Contribution to the geochemistry of molybdenum in the aquatic environment. D, 1963, University of Pittsburgh.

Barsky, C. K. Geochemistry of basalts and andesites from the Medicine Lake highland, California. D, 1975, Washington University. 373 p.

Barsotti, A. F. Resource and energy constraints on regional and global availability of aluminum, copper, and iron, 1975-2000; a computer study. D, 1979, Case Western Reserve University. 482 p.

Barsotti, A. T. Structural and paleomagnetic analysis of eastern Umtanum Ridge, south-central Washington. M, 1986, Washington State University. 204 p.

Bart, Henry A. A sedimentologic and petrographic study of the Miocene Arikaree Group of south-eastern Wyoming and west-central Nebraska. D, 1974, University of Nebraska, Lincoln. 140 p.

Barta, Leslie Ann. Thermodynamic investigations; 1, Thermodynamic properties of aqueous aluminum ion; 2, Kinetics and energetics of low temperature oxidation of Athabasca bitumen. D, 1987, University of Alberta. 131 p.

Bartberger, C. E. Stability of slipface bedforms. D, 1976, Syracuse University. 151 p.

Bartel, David Clark. Interpretation of Crone pulse electromagnetic data. M, 1984, University of Utah. 47 p.

Bartel, David Clark. Spectral analysis in airborne electromagnetics. D, 1988, University of California, Berkeley. 237 p.

Bartel, Douglas J. Structure and stratigraphy of the western Red Hills (east-central Nevada White Pine County), Nevada. M, 1968, University of Nebraska, Lincoln.

Bartel, James Robert. Petrology of fluvial and shoreline sands in a modern arc-trench gap, Guatemala. M, 1981, Western Michigan University. 66 p.

Bartelle, John Clemente. Water pollution analysis of the Broad River, Columbia, South Carolina. M, 1974, Virginia State University. 23 p.

Bartelmehs, Kurt Lane. Bond length and bonded radii variations in sulfide and crystals containing main group elements. M, 1987, Virginia Polytechnic Institute and State University.

Bartels, Richard L. An experimental study in the oxidation of triphylite series minerals. M, 1965, South Dakota School of Mines & Technology.

Bartels, William Stephen. Fossil reptile assemblages and depositional environments of selected early Tertiary vertebrate bone concentrations, Bighorn Basin, Wyoming. D, 1987, University of Michigan. 645 p.

Bartels, William Stephen. Reptilian fauna of the Clarkforkian land-mammal age (Paleocene-Eocene) in western North America. M, 1981, University of Michigan.

Barten, Paul Kevin. Modelling streamflow from headwater catchments in the northern lake states. D, 1988, University of Minnesota, Minneapolis. 322 p.

Barter, Charles F. Geology of the Owl Head mining district, Pinal County, Arizona. M, 1962, University of Arizona.

Barthelemy, Ernst. Utilisation du dilatomètre CSM dans la roche et dans la glace. M, 1980, Ecole Polytechnique.

Barthelemy, Ernst Nels. Utilisation du dilatomètre CSM dans la roche et dans la glace. M, 1979, Universite de Montreal.

Barthelman, William Bruce. Upper Arbuckle (Ordovician) outcrops in the Unap Mountain–Saddle Mountain area, northeastern Wichita Mountains, Oklahoma. M, 1969, University of Oklahoma. 67 p.

Barthelmy, D. A. Geology of the El Arco-Calmalli area, Baja California, Mexico. M, 1975, San Diego State University.

Bartholome, Paul Marie. Structural and petrological studies in Hamilton County, New York. D, 1956, Princeton University. 211 p.

Bartholomew, Mervin J. Geology of the southern portion of the Fonts Point and the southwest portion of the Seventeen Palms Quadrangle, San Diego County, California. M, 1968, University of Southern California.

Bartholomew, Mervin Jerome. Geology of the Humpback Mountain area of the Blue Ridge in Nelson and

Augusta counties, Virginia. D, 1971, Virginia Polytechnic Institute and State University. 211 p.

Bartholomew, Paul Richard. Geology and metamorphism of the Yale Creek area, British Columbia. M, 1979, University of British Columbia.

Bartholomew, Paul Richard. The Fe-Mg solution properties of olivine, enstatite, anthophyllite and talc, from ion-exchange experiments with aqueous chloride solutions. D, 1985, University of British Columbia. 212 p.

Bartle, Glenn Gardner. A geologic and physiographic study of the region in the vicinity of Raccoon Creek, and the Wabash River, lying principally in Parke County, Indiana. M, 1922, Indiana University, Bloomington. 70 p.

Bartle, Glenn Gardner. The geology of the Blue Springs gas field, Jackson County, Missouri. D, 1932, Indiana University, Bloomington. 237 p.

Bartlein, P. J. The influence of short-period climatic variations on streamflow in the United States and southern Canada 1951-1970. D, 1978, University of Wisconsin-Madison. 288 p.

Bartleson, Bruce Landon. Stratigraphy and petrology of the Gothic Formation (Pennsylvanian, Desmoinesian), Elk Mountains, (western) Colorado. D, 1968, University of Colorado. 220 p.

Bartleson, Bruce Landon. Studies of the Rosiclare Member in Coles and Douglas counties, Illinois. M, 1958, University of Illinois, Urbana. 37 p.

Bartlett, Charles Samuel, Jr. Anatomy of the lower Mississippian delta in southwestern Virginia. D, 1974, University of Tennessee, Knoxville. 373 p.

Bartlett, Charles Samuel, Jr. Geology of the Southern Pines (North Carolina) Quadrangle. M, 1967, University of North Carolina, Chapel Hill. 101 p.

Bartlett, D. S. Spectral reflectance of tidal wetland plant canopies and implications for remote sensing. D, 1979, University of Delaware, College of Marine Studies. 253 p.

Bartlett, Elizabeth Hancock. Lithostratigraphy and environments of deposition of the Upper Dornick Hills Group (Middle Pennsylvanian) of the southern part of the Ardmore Basin, Oklahoma. M, 1981, University of Oklahoma. 207 p.

Bartlett, Grant A. Foraminifera and their relations to bottom sediments on the Southeast Scotian Shelf. M, 1962, Carleton University. 262 p.

Bartlett, Grant Aulden. Benthonic foraminiferal ecology in Saint Margarets Bay and Mahone Bay, Southeast Nova Scotia. D, 1964, New York University. 239 p.

Bartlett, James Rodney. Stratigraphy, physical volcanology and geochemistry of the Belmont Lake metavolcanic complex, southeastern Ontario. M, 1983, Carleton University. 218 p.

Bartlett, Jeremy John. An analysis of sequence boundaries of the event stratigraphy of the Cardium Formation, Alberta. M, 1987, McMaster University. 185 p.

Bartlett, John David. The geology of Union Township, Morgan County, Ohio. M, 1950, Ohio State University.

Bartlett, Mark William. Petrology and genesis of carbonate-hosted Pb-Zn-Agoyes, San Cristobal District, Department of Junin, Peru. D, 1984, Oregon State University. 287 p.

Bartlett, Peter McIntyre. Global potential of marine nonfuel minerals. D, 1987, University of Texas, Austin. 480 p.

Bartlett, Ralph L. A study of the peneplains of the Southeast Appalachian District, with special reference to the section from Blue Mountain to Quakertown, Pennsylvania. M, 1917, Lehigh University.

Bartlett, Robert D. Geology of an Oligocene-age acid hot spring, San Luis Hills, Conejos and Costilla counties, Colorado. M, 1984, Colorado State University. 120 p.

Bartlett, Timothy R. Synthetic dolomitization; rate effects of variable mineralogy, surface area, external CO_3^{2-} and crystal seeding. M, 1984, Michigan State University. 69 p.

Bartlett, Wade A. Peru fore-arc sedimentation; SeaMarc II side-scan interpretation of an active continental margin. M, 1987, University of Hawaii. 87 p.

Bartlett, Wayne A. Distribution of sulfur in the West Kiernan Sill, Iron County, Michigan. M, 1975, Bowling Green State University. 93 p.

Bartlett, Wendy Louise. Experimental wrench faulting of confining pressure. M, 1980, Texas A&M University. 98 p.

Bartley, John Michael. Structural geology, metamorphism, and Rb/Sr geochronology of East Hinnoy, North Norway. D, 1981, Massachusetts Institute of Technology. 263 p.

Bartley, John William. A numerical analysis of Fistulipora (Bryozoa) from the Henryhouse Formation (Silurian) in south-central Oklahoma. M, 1979, University of Oklahoma. 78 p.

Bartley, John William. Wall calcification and rhythmic growth in the Permian stenolaemate bryozoan Tabulipora carbonaria. D, 1988, Michigan State University. 131 p.

Bartley, M. W. The geology and iron deposits at Steeprock Lake, Ontario. D, 1940, University of Toronto.

Bartley, M. W. The relationship of lamprophyres to gold deposits, particularly those of Precambrian age. M, 1937, University of Toronto.

Bartley, Ronald Clark. Geology of the East Evans Creek area, Trail Quadrangle, Oregon. M, 1955, Oregon State University. 81 p.

Bartling, Brian T. Rayleigh wave propagation in fractured sedimentary rock. M, 1983, University of Wisconsin-Milwaukee. 113 p.

Bartling, William Allen. Sedimentology of Upper Cretaceous strata, San Miguel Island, California, and its relation to the tectonics of the southern California borderland. M, 1981, San Diego State University.

Bartolini, Claudio. Regional structure and stratigraphy of Sierra El Aliso, central Sonora, Mexico. M, 1988, University of Arizona. 189 p.

Bartolino, James R. Cenozoic geology of the eastern half of the La Flor Quadrangle, Durango and Chihuahua, Mexico. M, 1988, West Texas State University. 182 p.

Bartolomucci, Henry A. Sedimentological study of the Niagara Falls moraine (Pleistocene), (Genesee, Erie and Niagara counties, New York). M, 1969, SUNY at Buffalo. 53 p.

Bartolucci-Castedo, Luis A. Computer-aided processing of remotely sensed data for temperature mapping of surface water from aircraft altitudes. M, 1973, Purdue University. 143 p.

Bartolucci-Castedo, Luis A. Digital processing of satellite multispectral scanner data for hydrologic applications. D, 1976, Purdue University. 232 p.

Barton, Anna Louise. The correlation of the Rogers Gap-Fulton formations in the Bluegrass. M, 1934, University of Kentucky.

Barton, C. L. Carbonate surface sediments of Tanner and Cortes banks, California continental borderland. M, 1976, University of Southern California. 73 p.

Barton, Charles Addison. The sediments of Biloxi Bay, Mississippi. M, 1952, University of Illinois, Urbana. 23 p.

Barton, Christopher Cramer. Systematic jointing in the Cardium Sandstone along the Bow River, Alberta, Canada. D, 1983, Yale University. 337 p.

Barton, Churchill J. Geophysical investigation of basement structure in southeast peninsular Florida. M, 1984, University of South Florida, Tampa. 50 p,

Barton, Colleen. Paleomagnetic studies of intrusive rocks in the Southern Appalachian Piedmont. M, 1980, Amherst College. 104 p.

Barton, Colleen A. Development of in-situ stress measurement techniques for deep drillholes. D, 1988, Stanford University. 211 p.

Barton, Donald Clinton. Arkose; its definition, classification, and geologic significance. D, 1914, Harvard University.

Barton, Eric Watson. Samah Formation; carbonate accumulation on a fault block, Sirte Basin, Libya. M, 1979, University of New Orleans.

Barton, Erika S. The significance of Rb-Sr and K-Ar ages of selected sedimentary rock units, Eastern Townships, Quebec. M, 1973, McGill University. 86 p.

Barton, Gary James. Land use effects on shallow earth resistivity. M, 1984, Western Michigan University. 125 p.

Barton, Gerald Stanley. Fortran program for the generation of mean gravity anomalies by least squares. M, 1971, University of Texas, Austin.

Barton, Jackson M. A geochronologic and stratigraphic study of the Precambrian rocks north of Montreal. D, 1971, McGill University. 117 p.

Barton, Jackson M. Geochronology of Chatham-Grenville type rocks, Quebec, (Canada). M, 1968, McGill University.

Barton, James Wesley. Stratigraphic study of the lower Miocene Planulina sediments in Cameron Parish, Louisiana. M, 1984, University of Nebraska, Lincoln.

Barton, M. D. The Ag-Au-S system. M, 1978, Virginia Polytechnic Institute and State University.

Barton, Mark David. The thermodynamic properties of topaz and some minerals in the $BeO\text{-}Al_2O_3\text{-}SiO_2\text{-}H_2O$ system, with petrologic applications. D, 1981, University of Chicago. 152 p.

Barton, Otis. The phylogeny and geologic history of the Mustelidae. M, 1929, Columbia University, Teachers College.

Barton, Paul B., Jr. Interpretation and evaluation of the uranium occurrences near Goodsprings, Nevada. D, 1955, Columbia University, Teachers College.

Barton, Paul B., Jr. Preliminary report on interpretation and evaluation of uranium occurrences in the Bird Spring and adjacent mining districts, Nevada. M, 1954, Columbia University, Teachers College.

Barton, Raymond. Geology of the Kennaday Peak-Pennock Mountain area, Carbon County, Wyoming. M, 1974, University of Wyoming. 74 p.

Barton, Robert A. Contrasting depositional processes of sub-Clarksville and Woodbine reservoir sandstones, Grimes County, Texas. M, 1982, Texas A&M University.

Barton, Robert H. Periglacial features in Lake City area, southeastern Minnesota. M, 1957, University of Minnesota, Minneapolis. 86 p.

Barton, William Thomas. A test of the seismic reflection profiling technique in the Adirondacks, New York. M, 1977, Cornell University.

Bartos, Frances Maribel. Pollen in fecal pellets as an environmental indicator. M, 1972, University of Arizona.

Bartos, Paul Joseph. Mineralization, alteration, and zoning of the Cu-Pb-Zn-Ag lodes at Quiruvilca, Peru. M, 1984, Stanford University. 219 p.

Bartow, J. A. Sedimentology of the Simmler and Vaqueros formations in the Caliente Range-Carrizo Plain area, California. D, 1974, Stanford University. 163 p.

Bartow, James Alan. Stratigraphy and sedimentation of the Capistrano Formation. M, 1964, University of California, Los Angeles.

Bartsch-Winkler, Susan. Geology of the Oak Flat Ranch area, Santa Clara County, California. M, 1976, San Jose State University. 52 p.

Bartz, Gerald L. Distribution of selected rare-earth elements in fluvial-deltaic sediment, Rio Grande de Anasco, western Puerto Rico. M, 1973, University of Wisconsin-Milwaukee.

Barua Remy, Victor Felix. Gravimetric separation of accessory minerals of the rocks of the Ground Hog Mine, Vanadium, New Mexico. M, 1950, Indiana University, Bloomington. 26 p.

Barua, Mridul C. Geology of uranium-molybdenum-bearing rocks of the Aillik-Makkovik Bay area, Labrador. M, 1969, Queen's University. 76 p.

Barwick, Arthur R. A study of the phosphorus content of certain interbedded iron ores. D, 1926, New York University.

Barwick, Arthur Richardson. Studies in oil shale. M, 1922, New York University.

Barwin, J. R. Stratigraphy of the Mesaverde Formation in the southeastern part of the Wind River basin, Fremont and Natrona counties, Wyoming. M, 1961, University of Wyoming. 78 p.

Barwis, John H. The sedimentologic and stratigraphic characteristics of beach ridge and tidal channel deposits in mesotidal barrier systems. D, 1979, University of South Carolina. 143 p.

Bary, David O. Some chemical constituents of Central American lavas. M, 1965, Dartmouth College. 111 p.

Basan, Paul B. Actual paleontology of a modern salt marsh near Sapelo Island, Georgia. D, 1975, University of Georgia. 272 p.

Basan, Paul B. Aspects of sedimentation and development of a carbonate bank in the Barracuda keys, Florida. M, 1970, SUNY at Binghamton. 71 p.

Basaran, A. K. T. A mathematical model for simulating water quality under non-steady state conditions. D, 1976, North Carolina State University. 175 p.

Bascle, Barbara. Geological study of the Admire 650′ Sandstone, micellar-polymer flood project, El Dorado Field, Butler County, Kansas. M, 1976, University of South Carolina.

Bascle, Robert. Pattern correlation and well-log interpretation of eastern North Carolina subsurface geology. M, 1976, University of South Carolina.

Bascom, Florence. A contribution to the geology of South Mountain, PA. D, 1893, The Johns Hopkins University.

Bascom, Florence. The sheet gabbros of Lake Superior. M, 1884, University of Wisconsin-Milwaukee.

Basden, Wayne A. Inversion of dispersed Love wave phase for velocity data multiple continuous parameters. M, 1985, University of Texas at El Paso.

Baseler, Thomas W. A study of the structure of the basement rock of Rhode Island Sound. M, 1965, University of Rhode Island.

Basham, Hal J. Lower Frio and upper Vicksburg formations of the La Reforma Field area of South Texas; relating growth faulting and structure to hydrocarbon accumulation. M, 1988, Texas A&I University. 60 p.

Basham, P. W. Time domain studies of short period teleseismic P phases. M, 1967, University of British Columbia.

Basham, Sandra Lynne. Analyses of shoreline springs in the Mono Basin, California, with applications to the groundwater system. M, 1988, University of California, Santa Cruz.

Basham, William L. Structure and metamorphism of the Pre-Cambrian rocks of the South Manzano Mountains, New Mexico. M, 1951, Northwestern University.

Basham, William Lassiter. Paleomagnetic evidence for structural rotation in the Pacific Northwest. D, 1978, University of Oklahoma. 170 p.

Basharkhah, Mohammad Ali. Reliability aspects of structural control. D, 1983, Purdue University. 243 p.

Basharmal, M. Isotopic composition of aqueous sulfur at landfill sites. M, 1985, University of Waterloo. 172 p.

Basher-Riani, Mustafa. Clay mineralogy and thermal maturity of the Tagrifet Formation in Concession 47, Sirte Basin, Libya. M, 1981, Michigan Technological University. 79 p.

Bashford, Howard H. A study of the effect of soluble minerals in hydraulic structures constructed from soil. D, 1984, Brigham Young University. 88 p.

Bashore, William McClellan, Jr. Upper crustal structure of the Salt Lake Valley and the Wasatch Fault from seismic modeling. M, 1982, University of Utah. 95 p.

Basick, James T. The characterization of organic pollutants in the surface waters of Bloomington, Indiana. M, 1980, Indiana University, Bloomington. 79 p.

Basile, Laura Lorraine. Sclerosponges-comparative generalities, modern species, Enewetak reef-dwellers, and Turkish fossils. M, 1978, Pennsylvania State University, University Park. 174 p.

Basilone, Tim. The diagenesis and epigenesis of the Oriskany Sandstone, south-central Somerset County, Pennsylvania. M, 1984, University of Pittsburgh.

Basinski, Paul. The mineralogy and uranium potential of bedded zeolites in the northern Reese River valley, Lander County, Nevada. M, 1979, University of Nevada. 98 p.

Baskerville, Charles Alexander. A micropaleontological study of Cretaceous sediments on Staten Island, New York. D, 1965, New York University. 99 p.

Baskerville, Charles Alexander. Geological aspect of engineering foundation design. M, 1958, New York University.

Baskette, Harry Buchanan. Geology of Blacktail Butte, Jackson Hole, Wyoming. M, 1946, University of Iowa. 86 p.

Baskin, David A. Elecrical and thermal measurements for D-C grounding electrode design. M, 1983, California State University, Hayward. 49 p.

Baskin, George D. The petrology and chemistry of a portion of the north limb of the Dore Lake Complex Chibougamau, Quebec, Canada. M, 1975, University of Georgia.

Baskin, J. A. Small vertebrates of the Bidahochi Formation, White Cone, northeastern Arizona. M, 1975, University of Arizona.

Baskin, Jon Alan. Carnivora from the late Clarendonian Love Bone Bed, Alachua County, Florida. D, 1980, University of Florida. 229 p.

Baskin, Yehuda. A study of authigenic feldspars. D, 1955, University of Chicago. 5 p.

Basko, Donald B. Stratigraphy of the Phosphoria Formation of northern Carbon County, Wyoming. M, 1954, University of Wyoming. 85 p.

Basler, Albert L. Geology of the Emigrant Peak Intrusive Complex, Park County, Montana. M, 1965, Montana State University. 52 p.

Basmaci, Y. Groundwater flow in double porosity media; carbonate rocks. D, 1977, Iowa State University of Science and Technology. 141 p.

Basocak, Cihat. Land use and land cover classification with remote sensor data in the Tuz Golu (Lake) Basin of Turkey. M, 1981, University of Kansas. 54 p.

Bass, Charles E. Geology and petrography of the Draper Quadrangle, Virginia. M, 1928, University of Virginia. 44 p.

Bass, Irvin. The Rafinesquinae of the Eden Formation (Upper Ordovician), Cincinnati, Ohio, and vicinity. M, 1957, University of Cincinnati. 99 p.

Bass, Jay David. An experimental determination of the stability and thermochemical properties of the mineral trolleite. M, 1977, Lehigh University. 109 p.

Bass, Jay David. The relationship between elasticity and crystal chemistry for some mantle silicates and aluminates. D, 1982, SUNY at Stony Brook. 193 p.

Bass, John H. The Dakota Formation in San Miguel and Mora counties, New Mexico. M, 1951, Texas Tech University. 41 p.

Bass, Manuel Nathan. A vertebrate fauna from late Tertiary beds near Frazier Mountain, California. M, 1951, California Institute of Technology. 37 p.

Bass, Manuel Nathan. An interpretation of the geologic history of part of the Timiskaming Subprovince, Canada. D, 1956, Princeton University. 340 p.

Bass, Ralph Oswald. Geology of the western portion of the Point Dume Quadrangle, Los Angeles County, California. M, 1960, University of California, Los Angeles.

Bass-Laszlo, Sarah Luann. Ordovician and Silurian conodonts, Nopah Range, Inyo County, California. M, 1984, San Diego State University. 171 p.

Bassarab, Dennis Rudyard. Clay mineralogical study of the Kope and Fairview formations (Cincinnatian) in the Cincinnati region. M, 1965, University of Cincinnati. 50 p.

Basse, R. A. Stratigraphy, sedimentology, and depositional setting of the late Precambrian Pahrump Group, Silurian Hills, California. M, 1978, Stanford University. 79 p.

Bassett, Allen M. Ore deposition at Naica, Chihuahua, Mexico; a preliminary report. M, 1950, Columbia University, Teachers College.

Bassett, Allen Mordorf. Geology and mineralization of the Naica mining district, Chihuahua, Mexico. D, 1955, Columbia University, Teachers College. 66 p.

Bassett, Ann B. Staining reactions of some clay minerals and their relation to pH values. M, 1949, Columbia University, Teachers College.

Bassett, Charles Fernando. Stratigraphy and paleontology of Dundee Limestone of southeastern Michigan. D, 1933, University of Michigan.

Bassett, Charles Fernando. The stratigraphy of the Devonian formations of the Alto Pass Quadrangle, Illinois. M, 1924, [University of Illinois, Urbana].

Bassett, Henry G. The Hay River Limestone (Upper Devonian), Northwest Territories. M, 1950, McGill University.

Bassett, Henry Gordon. Correlation of Devonian sections in northern Alberta and Northwest Territories. D, 1952, Princeton University. 313 p.

Bassett, Herbert. Geology and physiography of the Macomb and Colchester quadrangles, Illinois. M, 1916, University of Wisconsin-Milwaukee.

Bassett, John L. Hydrology and geochemistry of karst terrain, upper Lost River drainage basin, Indiana. M, 1974, Indiana University, Bloomington. 102 p.

Bassett, R. L. The geochemistry of boron in thermal waters. D, 1977, Stanford University. 305 p.

Bassett, Randy Lynn. A geochemical investigation of silica and fluoride in the unsaturated zone of a semi-arid environment. M, 1973, Texas Tech University. 93 p.

Bassett, William A. Studies relating to the nature and genesis of vermiculite with special emphasis on copper vermiculite. D, 1960, Columbia University, Teachers College.

Bassett, William S. The association of uranium minerals with fluorite. M, 1956, Columbia University, Teachers College.

Bassin, N. J. Analysis of total suspended matter in the Caribbean Sea. D, 1975, Texas A&M University. 118 p.

Bassir, Franklin. Marine geophysical examination of the Southeast Indian Ocean. M, 1978, Brooklyn College (CUNY).

Bassler, Harvey. Filicales and Pteridospermae of the Monongahela Formation of Maryland. D, 1913, The Johns Hopkins University.

Basson, Philip Walter. The fossil flora of the Drywood Formation of southwestern Missouri. D, 1965, University of Missouri, Rolla. 221 p.

Bastanchury, Ruth Frances. Geology of the Bradley Mountain area, Lincoln County, Wyoming. M, 1947, University of Michigan.

Bastedo, Jerold C. Pelecypods of the Upper Devonian Chadakoin Formation in Cattaraugus and Chautauqua counties, New York. M, 1980, SUNY at Buffalo.

Bastidas, Ramon O. An economic analysis of offshore oil field development. M, 1980, University of Oklahoma. 82 p.

Bastien, Thomas W. Geology of the Tertiary volcanic rocks of the Jones Mountains, Antarctica. M, 1963, University of Minnesota, Minneapolis. 99 p.

Bastin, Edson Sunderland. Chemical composition as a criterion in identifying metamorphosed sediment. D, 1909, University of Chicago. 28 p.

Bastug, Mustafa Cengiz. Subsurface geology of Bronte field, Cook County, Texas. M, 1970, University of Texas, Austin.

Basu, Abhijit. Petrology of Holocene fluvial sand derived from plutonic source rocks; implications to provenance interpretation. D, 1975, Indiana University, Bloomington. 138 p.

Basu, Asish Ranjam. Petrogenesis of the ultramafic xenoliths from San Quintin volcanic field, Baja California. D, 1975, University of California, Davis. 229 p.

Basu, Debabrata. Genesis of the Grace Mine magnetite deposit, Morgantown, Berks County, southeastern Pennsylvania. D, 1974, Lehigh University. 317 p.

Bat, David Thomas. A subsurface facies analysis of the distribution, depositional environments, and diagenetic overprint of the Evans and Lewis Sandstone units in northern Mississippi and northwestern Alabama. M, 1987, Oklahoma State University. 225 p.

Bata, Mazin Y. Strain determination techniques and finite strain in rocks from Prins Karls Forland. M, 1982, Wayne State University. 137 p.

Bataille, Klaus Dieter. Inhomogeneities near the coremantle boundary inferred from short-period scattered waves recorded at the GDSN. D, 1987, University of California, Santa Cruz. 117 p.

Batchelder, Eric C. Lithofacies, depositional environments, and diagenesis of the Knox Group exposed along Alligator Creek, Bibb County, Alabama. M, 1984, University of Alabama. 238 p.

Batchelder, Gail. Hydrogeochemical cycling in the headwaters region of the Fort River watershed. M, 1984, University of Massachusetts. 128 p.

Batchelder, George Lewis. Postglacial ecology at Black Lake, Mono County, California. D, 1970, Arizona State University. 198 p.

Batchelder, John N. A study of stable isotopes and fluid inclusions at Copper Canyon, Lander County, Nevada. M, 1973, San Jose State University. 92 p.

Batchelor, Carl F. Subsidence over abandoned coal mines; Bellingham, Washington. M, 1982, Western Washington University. 112 p.

Batchelor, James W. The paleontology and stratigraphy of the Paoli Limestone, a basal Chester Formation from southern Indiana. M, 1948, Miami University (Ohio). 60 p.

Bate, George L. Variations in the isotopic composition of common lead and the history of the crust of the Earth. D, 1955, Columbia University, Teachers College.

Bate, Matthew Adam. Temblor and Big Blue formations; interpretation of depositional environment sequence on Coalinga Anticline, Fresno County, California. M, 1984, Stanford University. 123 p.

Bateman, Alan M. Geology and ore deposits of Bridge River District, British Columbia. D, 1913, Yale University.

Bateman, Barry Lynn. Generation of representative geologic strata suitable for simulation of fluid flow. D, 1970, Texas A&M University. 71 p.

Bateman, John D. Geology and gold deposits of Uchi-Slate lakes area, Ontario. D, 1939, Yale University.

Bateman, Marcus Kelden. Petrology and clay mineralogy of the Tertiary Poul Creek (Oligocene-Miocene) and Yakataga (Miocene-Pliocene) formations, Cape Yakataga, Alaska. M, 1967, University of Iowa. 131 p.

Bateman, Paul Charles. The geology of the Bishop 15-minute Quadrangle, California. D, 1957, University of California, Los Angeles.

Bateman, Philip Walker. Geological and geochemical characteristics of selected stratiform mercury deposits in Nevada and their relationship to precious metal deposits. D, 1988, Colorado School of Mines. 275 p.

Bateman, Philip Walker. Rock alteration at the Bousquet gold mine, Quebec. M, 1984, University of Western Ontario. 159 p.

Bateman, Richard L. Environmental controls on occurrence and chemistry of ground water in a basic volcanic terrane, eastern Sierra Nevada (Nevada county and Placer County, California). M, 1970, University of Nevada. 115 p.

Bateman, Richard L. Influence of watershed characteristics on water quality behavior of streams in the eastern Sierra Nevada. D, 1976, University of Nevada - Mackay School of Mines. 159 p.

Bateman, Richard L. The geology of the south-central part of the Sawtooth Creek Quadrangle, Oregon. M, 1961, University of Oregon. 97 p.

Bateridge, Thomas Earl. Effects of clearcutting on water and nutrient discharge, Bitterroot National Forest, Montana. M, 1974, University of Montana. 68 p.

Bates, Allan Clifford. Slowness-azimuth measurements and P wave velocity distribution. D, 1976, University of Alberta. 207 p.

Bates, Allan Clifford. Upper mantle structure deduced from seismic records acquired during Project Edzoe in southern Saskatchewan and western Manitoba between distances of about 790 kilometers and 1285 kilometers. M, 1971, University of Manitoba.

Bates, Allen Neal. Zonation of the Comanchean echinoids of central Texas. M, 1958, Baylor University. 155 p.

Bates, Carl Hobart. Phase equilibrium studies germanium and silicon at high pressures. D, 1966, Pennsylvania State University, University Park. 70 p.

Bates, Charles. Sources and distribution of fine quartz sand in the northeastern Gulf of Mexico; a paleogeographic reconstruction using Fourier grain shape analysis. M, 1985, Texas A&M University.

Bates, Charles C. Physical and geological processes of delta formation. D, 1953, Texas A&M University.

Bates, Clair Ellen. Some foraminifera of the Niobrara Formation (Upper Cretaceous) of Colorado. M, 1949, University of Colorado.

Bates, Edmon Elkins, Jr. Stratigraphic analysis of the Cambrian Carrara Formation, Death Valley region, California-Nevada. M, 1965, University of California, Los Angeles.

Bates, Edward R. The Niagaran reefs and overlying carbonate evaporite sequence in southeastern Michigan. M, 1970, Michigan State University. 86 p.

Bates, John Davis. Geology of a portion of Orange County, New York, in the vicinity of Bear Mountain. M, 1941, University of Virginia. 110 p.

Bates, Leonard Gordon, Jr. Interpretation of deep seismic reflections in the Blake-Bahama Basin using synthetic seismograms. M, 1986, University of Delaware. 139 p.

Bates, Peter Paul. Base level study of select organochloride pesticides and polychlorinated biphenyls in Quohog clams (Mercenaria mercenaria), sediments and soil taken from Brevard County, Florida. M, 1977, Florida Institute of Technology.

Bates, Robert Ellery. Geomorphic history of the Kickapoo region, Wisconsin. D, 1939, Columbia University, Teachers College.

Bates, Robert Ellery. Physiography of the Sinking Creek area, Washington County, Indiana. M, 1932, Indiana University, Bloomington. 33 p.

Bates, Robert G. The Upper Cambrian, Ordovician, and Devonian stratigraphy along the northeast flank of the Wind River Mountains, Wyoming. M, 1951, Dartmouth College. 86 p.

Bates, Robert Latimer. Geology of the Powell Valley in northeastern Lee County, Virginia. D, 1938, University of Iowa. 140 p.

Bates, Robert Latimer. Stratigraphy and structure of the Big A Mountain area, Virginia. M, 1936, University of Iowa. 78 p.

Bates, Roger G. Tectonic rotations in the Cascade Mountains of southern Washington. M, 1980, Western Washington University. 86 p.

Bates, Thomas F. Diffraction variations in the spinel group. M, 1940, Columbia University, Teachers College.

Bates, Thomas F. Origin of the Edwin Clay, Ione, California. D, 1944, Columbia University, Teachers College.

Bates, Thomas Robert. Late Pleistocene geology of pluvial Lake Mound, Lynn and Terry counties, Texas. M, 1968, Texas Tech University. 45 p.

Bates, Tim Frank. The stereochemistry of silenes and alpha-lithio silanes. D, 1987, University of North Texas. 113 p.

Bates, Wayne E. Geology of the Red Dirt Creek area, Grand County, Colorado. M, 1957, University of Kansas. 47 p.

Bateson, J. T. Lateral variations of clast sorting in the Eureka Valley Tuff, East-central Sierra Nevada. M, 1979, Stanford University. 56 p.

Bath, Thomas Patrick. Igneous lamination and layering in the nepheline syenite quarry, Sec. 36, T1S, R14W, Saline County, Arkansas. M, 1983, University of Arkansas, Fayetteville. 111 p.

Bath, William W. Geomorphic processes at Palo Duro Canyon, Texas Panhandle. M, 1980, University of Texas, Austin.

Bathen, K. H. A descriptive study of the physical oceanography of Kanehoe Bay, Oahu, Hawaii. M, 1968, University of Hawaii. 353 p.

Bathke, Susan Ann. Stratigraphy, petrology, and environment of deposition of the Blakely Sandstone, Ouachita Mountains, western Garland County, Arkansas. M, 1984, University of New Orleans. 148 p.

Bathurst, B. W. Chemical activities in magmas. M, 1975, Dartmouth College. 147 p.

Bathurst, Bruce Warren. Some relations among the directions of curves at an invariant point on a thermodynamic phase diagram. D, 1985, Princeton University. 191 p.

Batiza, Rodey. Oceanic crustal evolution; evidence from the petrology and geochemistry of isolated oceanic central volcanoes. D, 1977, University of California, San Diego. 312 p.

Batllo-Ortiz, Josep. Studies of seismicity in the Cerdanya region of the eastern Pyrenees. M, 1968, St. Louis University.

Batory, Bruce L. Analysis of the lacustrine sediments of the Creede Formation, Mineral County, Colorado. M, 1981, New Mexico Institute of Mining and Technology. 120 p.

Batra, Ravi. Travel-times and crustal structure in the Nevada region from earthquakes, nuclear explosions, and mine blasts. M, 1970, University of Nevada. 78 p.

Batsche, Ralph W. Field study and geological interpretation of a gravity anomaly located in the Fayette County, Ohio area. M, 1963, Ohio State University.

Batt, Paul. Electron microprobe analysis of Fe_2SiO_4 and $(Fe,Mg)_2$ SiO_4 olivines at pressures up to 250 kilobars and temperatures up to 900°C. M, 1974, University of Rochester.

Batt, Richard J. Foraminiferal biostratigraphy and marine paleoecology of the Frontier Formation, southwestern Wyoming. M, 1981, University of Wyoming. 84 p.

Batt, Richard James. Pelagic biofacies of the Western Interior Greenhorn Sea (Cretaceous); evidence from ammonites and planktonic foraminifera. D, 1987, University of Colorado. 797 p.

Battelle, Sarah Jane. A study of thermal waters in Valle Trinidad, Baja California, Mexico. M, 1980, San Diego State University.

Batten, Roger L. A study of certain Permian pleurotomarian gastropods. D, 1955, Columbia University, Teachers College.

Batten, Roger L. Permian Gastropoda of the Southwestern United States; 2, Pluerotomariacea, Portlockiellidae, Phymatopleuridae, and Eotomariidae. D, 1958, Columbia University, Teachers College.

Batten, Roland Wesley. The sediments of the Beaufort Inlet area, North Carolina. M, 1959, University of North Carolina, Chapel Hill. 38 p.

Batterson, Ted Randall. Arsenic in Lake Lansing, Michigan. D, 1980, Michigan State University. 79 p.

Battié, John E. Seismic stratigraphy of the Baltimore Canyon Trough area. M, 1981, University of Houston.

Battis, James Craig. Zones of anomalous seismic propagation in the vicinity of Mount Rainier. M, 1973, University of Washington. 88 p.

Battle, J. M., Jr. Stratigraphy and mineralogy of Pleistocene loesses in West Tennessee. M, 1977, Memphis State University.

Batuk, Hamzer. Occurrence and mining methods of lignite. M, 1944, University of Pittsburgh.

Baty, Joseph Bruce. Fission track age dates from three granitic plutons in the Flint Creek Range, (Granite County), western Montana. M, 1973, University of Montana. 37 p.

Batzle, Michael Lee. Fracturing and sealing in geothermal systems. D, 1978, Massachusetts Institute of Technology. 289 p.

Batzner, Jay C. The hydrogeology of Lomerio de Peyotes, Coahuila, Mexico. M, 1976, University of New Orleans. 64 p.

Bau, A. F. S. History of regional deformation of Archean rocks in the Kashabowie-Lac des Mille Lacs area, N.W. Ontario. D, 1979, University of Toronto.

Bauch, John H. A. Geology of the central area of the Lomas de las Canas Quadrangle, Socorro County, NM. M, 1982, New Mexico Institute of Mining and Technology. 115 p.

Bauchman, James Bell. The geology of the southeast corner of Travis County, Texas. M, 1950, University of Texas, Austin.

Bauchman, John Allen. Geology of the Josephine Quadrangle, Collin and Hunt counties, Texas. M, 1949, Southern Methodist University. 27 p.

Baud, Richie Darren. Calculating crack density values from compressional and shear wave velocities. M, 1988, Purdue University. 100 p.

Bauder, J. M. Geology of the Cedar Mountain region, northern Diablo Range, California. M, 1975, Stanford University. 93 p.

Baudino, Frank Joseph. The geology of the Glendale Quadrangle, Los Angeles County, California. M, 1934, University of Southern California.

Bauer, Bernard Oswald. Nearshore morphodynamics and the relative role of low frequency wave motion. D, 1988, The Johns Hopkins University. 284 p.

Bauer, C. Max. Geology of the southeastern part of the Wind River basin, Wyoming. D, 1932, University of Colorado.

Bauer, Charles Bruce. Electric log study of the "Trivoli" Sandstone, Louden Pool, Fayette County, Illinois. M, 1946, University of Illinois, Urbana. 30 p.

Bauer, David Thomas. The structure of the Pine Mountain Belt, Upson, Meriwether, and Talbot counties, Georgia. M, 1976, Florida State University.

Bauer, Dennis A. Trilobites, biostratigraphy, lithostratigraphy, and environments of deposition of the Reagan Formation and Honey Creek Limestone, Kindblade Ranch area, Wichita Mountains, southern Oklahoma. M, 1985, University of Missouri, Columbia. 109 p.

Bauer, E. J. Geology of the Wagonhound Creek area, Carbon County, Wyoming. M, 1952, University of Wyoming. 58 p.

Bauer, Francis Harry. Marine terraces between Terrace Creek and Stewarts Point, Sonoma County, California. M, 1952, University of California, Berkeley. 273 p.

Bauer, Glenn R. The geology of Tofua island, Tonga. M, 1969, University of Hawaii.

Bauer, Herman Louis, Jr. Fluorspar deposits north end of Spor Mountain, Thomas Range, Juab County, Utah. M, 1952, University of Utah. 47 p.

Bauer, Hubert A. Tides of the Puget Sound and adjacent inland waters. M, 1928, University of Washington. 114 p.

Bauer, Janet. Late Quaternary planktonic foraminiferal paleoecology of the Central Pacific Ocean. M, 1981, Rutgers, The State University, Newark. 89 p.

Bauer, Jeffrey A. Conodont biostratigraphy, correlation, and depositional environments of Middle Ordovician rocks in Oklahoma. D, 1987, Ohio State University. 366 p.

Bauer, John W. Base and total stream flow variability in carbonate basins. M, 1969, Pennsylvania State University, University Park. 107 p.

Bauer, Jon F. Crystal chemistry and stability of a high-pressure hydrous $10A^o$ phyllosilicate in the system MgO-SiO_2-H_2O. D, 1977, Lehigh University. 260 p.

Bauer, Larry Paul. The alteration of sedimentary rocks due to burning coal in the Powder River Basin, Campbell County, Wyoming. M, 1972, Texas Tech University. 65 p.

Bauer, Linda Rose. Gas phase migration of C-14 through barrier materials applicable for use in a high-level nuclear waste repository located in tuff. D, 1988, Purdue University. 125 p.

Bauer, Mary Alvina. Ecology and distribution of living planktonic foraminifera of the South Texas outer continental shelf. M, 1976, Rice University. 125 p.

Bauer, Michael Steven. Heat flow at the upper Stillwater dam site, Uinta Mountains, Utah. M, 1985, University of Utah. 94 p.

Bauer, Paul G. Geology of the Redwood Retreat-Croy Ridge area of Santa Clara County, California. M, 1971, San Jose State University. 74 p.

Bauer, Paul Winston. Geology of the Precambrian rocks of the southern Manzano Mountains, New Mexico. M, 1983, University of New Mexico. 133 p.

Bauer, Paul Winston. Precambrian geology of the Picuris Range, north-central New Mexico. D, 1987, New Mexico Institute of Mining and Technology. 81 p.

Bauer, Robert Alan. Prediction of compressive strength from point-load and moisture content indices of highly anisotropic, coal bearing strata of the Illinois Basin. M, 1983, University of Illinois, Urbana. 62 p.

Bauer, Robert L. Structural geology and petrology of the granite gneiss near Granite Falls and Montevideo, Minnesota. M, 1974, University of Missouri, Columbia.

Bauer, Robert Louis. The petrology and structural geology of the western Lake Vermilion area, northeastern Minnesota. D, 1981, University of Minnesota, Minneapolis. 207 p.

Bauer, Stephen Joseph. Semibrittle deformation of granite at upper crustal conditions. D, 1984, Texas A&M University. 203 p.

Bauer, Stephen Joseph. The effects of unconfined slow uniform heating on the mechanical and transport properties of the Westerly and Charcoal granites. M, 1980, Texas A&M University. 106 p.

Bauer, W. Geology of a portion of the Briones Valley Quadrangle. M, 19??, [University of California, Berkeley].

Bauer, William Henry. Geology of the Jarre Canyon-Dawson Butte region, Douglas County, Colorado. M, 1959, Colorado School of Mines. 163 p.

Bauerlein, Henry J. Stratigraphy and structure of the Cave Mountain area of Virginia. M, 1966, Virginia Polytechnic Institute and State University.

Bauernfeind, Paul Edward. The Misener Sandstone in portions of Lincoln and Creek counties, Oklahoma. M, 1980, University of Oklahoma. 72 p.

Bauernschmidt, August John. The Saganaga Granite of northeastern Minnesota and Canada. M, 1926, University of Minnesota, Minneapolis. 44 p.

Baughman, Richard Lee. Sedimentology of an Eocene volcanic mudflow deposit, Beaver Rim, Fremont County, Wyoming. M, 1988, Northern Arizona University. 159 p.

Baughman, Russell Leroy. The geology of the Musinia Graben area, Sevier and Sanpete Counties, Utah. M, 1959, Ohio State University.

Baughn, Milton H. Echinoids from the Weches Member of the Mount Selman Formation (middle Eocene) of Texas. M, 1932, Texas A&M University.

Baulsir, George Edward. The geological history of the Macksburg oil field. M, 1939, George Washington University. 15 p.

Baum, Bruce R. A petrographic study of the carbonates of the west central Taconic region (Vermont). M, 1960, Rensselaer Polytechnic Institute. 89 p.

Baum, Gerald Robert. Cetacean paleontology and biostratigraphy of the Yorktown and St. Mary's formations, Surry County, Virginia. M, 1974, University of North Carolina, Chapel Hill. 103 p.

Baum, Gerald Robert. Stratigraphic framework of the middle Eocene to lower Miocene formations of North Carolina. D, 1977, University of North Carolina, Chapel Hill. 139 p.

Baum, Lawrence Frederick. Geology and Cu-Zn-Ni-Fe geochemistry of some eugeosynclinal rocks, Stevens County, Washington. D, 1975, University of Idaho. 256 p.

Baum, Lawrence Frederick. Geology and mineral deposits, Vesper Peak stock area, Snohomish County, Washington. M, 1968, University of Washington. 75 p.

Baum, Rex Lee. Engineering geology and relative slope stability in the area of the Fay Apartments and in part of Mount Airy Forest, Cincinnati, Ohio. M, 1983, University of Cincinnati. 73 p.

Baum, Rex Lee. The Aspen Grove landslide, Ephraim Canyon, central Utah. D, 1988, University of Cincinnati. 460 p.

Bauman, Bruce John. Soil suitability for on-site sewage treatment in the Flathead Valley, Montana; soil permeability, variability, and ground-water contamination. D, 1985, Montana State University. 221 p.

Bauman, Carl F., Jr. Cretaceous rudistids from near Cuernavaca, Morelos, Mexico. M, 1950, Columbia University, Teachers College.

Bauman, Jeanette M. The geology and mineralogy of the Colchester underclays and shales. M, 1978, Indiana University, Bloomington. 77 p.

Bauman, Paul Thomas. External morphogenetic relationships among articulate brachiopods. M, 1954, Washington University.

Baumann, D. K. Chemical diagenesis during burial of Tertiary Gulf Coast pelitic sediments. M, 1975, University of Missouri, Columbia.

Baumann, Dean R. The occurrence and distribution of mineral matter in coal lithotypes in the Herrin (No. 6) Coal seam under marine and nonmarine influences. M, 1982, Southern Illinois University, Carbondale. 151 p.

Baumann, Rodney M. Metamorphism of metasediments in the Precambrian English River Subprovince, western Ontario. M, 1985, University of North Dakota. 144 p.

Baumann, Warren A. In situ stress measurements in a freshwater ice sheet. M, 1979, University of Wisconsin-Milwaukee. 115 p.

Baumeister, Dorothy. Surface and subsurface studies of some Pennsylvanian formations in Hughes, Seminole, and Pottawatomie counties, Oklahoma. M, 1942, University of Oklahoma. 41 p.

Baumer, Otto Weichsel. The governing partial differential equation for one-dimensional flow of water in a deforming body of soil. D, 1986, University of Nebraska, Lincoln. 100 p.

Baumgaertner, I. von see von Baumgaertner, I.

Baumgardner, John Rudolph. A three-dimensional finite element model for mantle convection. D, 1983, University of California, Los Angeles. 284 p.

Baumgardner, Robert Welcome, Jr. A quantitative geomorphic study of the Riberao do Mandaguari, Sao Paulo, Brazil. M, 1979, Texas A&M University. 135 p.

Baumgardner, Thomas F. A comparison of regional and individual parameters in National Weather Service River Forecast System. M, 1984, Colorado State University. 73 p.

Baumgardt, Douglas Reid. Seismic body-wave study of vertical and lateral heterogeneity in the Earth's interior. D, 1981, Pennsylvania State University, University Park. 558 p.

Baumgarten, Diane M. A comparison of linears and curvilinears mapped from digitally processed Landsat thematic mapper data to faults depicted on geologic maps; Wells Creek Structure, Tennessee. M, 1984, Murray State University. 72 p.

Baumgarten, Hortense. A study of potash in the United States. M, 1933, Columbia University, Teachers College.

Baumgartner, Eric Paul. Some magnetic properties of archeomagnetic materials. M, 1973, University of Oklahoma. 91 p.

Baumgartner, T. R. Tectonic evolution of the Walvis Ridge and West African margin, South Atlantic Ocean. M, 1974, Oregon State University. 79 p.

Baumgartner, Timothy Robert. High resolution paleoclimatology from the varved sediments of the Gulf of California. D, 1988, Oregon State University. 287 p.

Baumgartner, Warren Francis. Ontogeny of the Pennsylvanian ostracod Cavellina sp. M, 1955, Washington University. 40 p.

Baumgras, Lynne M. Liquefaction potential evaluation of surficial deposits in the City of Buffalo, New York. M, 1988, SUNY at Buffalo. 105 p.

Baumiller, George N. Surficial geology of the East Liberty, Ohio, Quadrangle. M, 1917, Ohio State University.

Bausch, Walter Charles. Development of a soil moisture model for use with passive microwave remote sensors. D, 1980, Texas A&M University. 185 p.

Baver, Leonard D., Jr. Studies in the lateral and vertical distribution of Virgilian calcareous algae in Dry and Beeman canyons, Sacramento Mountains, New Mexico. M, 1960, University of Wisconsin-Madison.

Baveye, Philippe. The nonequilibrium statistical mechanics of simultaneous steady-state transport of water and solutes in soils. D, 1985, University of California, Riverside. 145 p.

Bawden, William Frederick. Two-phase flow through rock fractures. D, 1981, University of Toronto.

Baxendale, R. W. A study of the morphology and anatomy of cordaitean organ genera from Middle Pennsylvanian Kansas and Iowa coal balls. D, 1977, University of Kansas. 297 p.

Baxter, David A. Geology and geochemistry of hydrothermal alteration associated with precious metal mineralization in the Clark Creek region, Marquette County, Michigan. M, 1988, Michigan Technological University. 197 p.

Baxter, Franklin Paul. Sedimentology of Woodfordian glacial materials and subsequent biocycling in derived soils in a mixed forest of northeastern Wisconsin; ant pedoturbation in a prairie soil of southwestern Wisconsin. D, 1970, University of Wisconsin-Madison. 196 p.

Baxter, James Edward. The Quaternary geology of the West Branch of the Susquehanna River valley near Lock Haven, Pennsylvania. M, 1983, Pennsylvania State University, University Park. 135 p.

Baxter, James W. Distribution of radioactivity in uranium-bearing rocks from the Potash Sulphur Springs igneous complex, Fayetteville. M, 1952, University of Arkansas, Fayetteville.

Baxter, James Watson. Stratigraphy and texture of the Salem Limestone in southwestern Illinois. D, 1958, University of Illinois, Urbana. 103 p.

Baxter, Sonny. Conodont biostratigraphy of the Mississippian of western Alberta and adjacent British Columbia, Canada. D, 1972, Ohio State University. 274 p.

Baxter, Sonny. Mississippian conodonts from Alberta (Canada). M, 1969, University of Calgary. 87 p.

Bay, Annell Russell. Deposition of prograding carbonate sand shoals and their subsequent diagenesis; lower Glen Rose (Cretaceous), South Texas. M, 1980, University of Texas, Austin.

Bay, Harry X. A sedimentary study of certain Pennsylvanian conglomerates of Texas. D, 1931, University of Iowa. 211 p.

Bay, Harry X. The geology of a portion of Benton County, Missouri. M, 1927, University of Iowa. 116 p.

Bay, Kirby Whitmarsh. Stratigraphy of Eocene sedimentary rocks in the Lysite mountain area, Hot Springs, Fremont and Washakie counties, Wyoming. D, 1969, University of Wyoming. 181 p.

Bay, Roger Rudolph. Interpretation of fundamental hydrologic relationships of forested bogs in northern Minnesota. D, 1967, University of Minnesota, Minneapolis. 123 p.

Bay, Thomas A., Jr. Mississippian System of Andrews and Gaines counties, Texas. M, 1954, University of Texas, Austin.

Bayan, Mohammad Reza. A comparison of methods for quantifying the commonly occurring minerals in soils and sediments. D, 1984, University of Kentucky. 268 p.

Baybrook, Thomas G. Investigation of Earth modeling from geophysical and geopotential data. M, 1972, Ohio State University.

Bayer, James Lawrence. Maceral segregation in commercially prepared coal products. M, 1960, Pennsylvania State University, University Park. 76 p.

Bayer, Marvin Benno. Contributions to water resources management; optimization of river basin water quality models using nonlinear programming. D, 1972, University of Toronto.

Bayer, Robert J. Trace metal distribution in bottom sediment from the Tennessee River-Fort Loudon lake system, Knoxville, Knox County, Tennessee. M, 1974, University of Tennessee, Knoxville. 45 p.

Bayer, Thomas Norton. The Maquoketa Formation (Upper Ordovician) in Minnesota (Fillmore, Mower, and Olmstead counties) and an analysis of its benthonic communities. D, 1965, University of Minnesota, Minneapolis. 224 p.

Bayer, Thomas Norton. The subsurface bedrock stratigraphy of northwestern Minnesota. M, 1959, University of Minnesota, Minneapolis. 77 p.

Bayha, David C. The petrography of selected limestones of the McMillan Formation from the Trenton Quadrangle, Butler County, Ohio. M, 1965, Miami University (Ohio). 128 p.

Baykal, Gokhan I. The effect of micromorphological development on the elastic moduli of fly ash-lime stabilized bentonite. D, 1987, Louisiana State University. 283 p.

Baykal, Orhan. Dispersion of Rayleigh waves across the Atlantic Ocean basin. M, 1947, University of Michigan.

Baykal, Turan S. Bituminous coal stripping in the United States. M, 1944, University of Pittsburgh.

Bayless, Edward Randall. A sensitivity analysis of the parameters affecting time-dependent glacier flow in a Pleistocene ice lobe. M, 1987, Indiana University, Bloomington. 91 p.

Bayless, John Clinton. Geology of the Snake River Range near Alpine, Idaho. M, 1947, University of Michigan.

Bayless, Michael Lynn. A sedimentologic and stratigraphic study of the Merom Sandstone near Merom, Indiana and Marshall, Illinois. M, 1988, Indiana State University. 68 p.

Bayless, Richard C. A feasibility study of artificially recharging the Bassano-Gem gravel and sand aquifer, Alberta, Canada. M, 1984, Ohio University, Athens. 125 p.

Bayley, Emery Perham, Jr. Bedrock geology of the Twin Peaks area, an intrusive complex near Wenatchee, Washington. M, 1965, University of Washington. 47 p.

Bayley, Richard William. A heavy mineral study of the Morrison Formation of southcentral Utah. M, 1950, Ohio State University.

Bayley, Richard William. Geology of the Lake Mary Quadrangle, Iron County, Michigan. D, 1956, Ohio State University.

Bayley, W. S. Contact metamorphism of the slates and sandstones of Pigeon Point. D, 1886, The Johns Hopkins University.

Bayliss, W. Z. Applications of geophysical prospecting. M, 1929, University of Minnesota, Minneapolis. 65 p.

Bayly, Maurice. A comparison of rock viscosities by measurements on similar folds and a determination of the sampling error in modal analysis. D, 1962, University of Chicago. 141 p.

Baynard, David Nicoll. Depositional settings in the coal bearing, upper Tradewater Formation in western Kentucky with emphasis on the Mannington (No. 4) Coal zone. M, 1983, University of Kentucky. 91 p.

Baynas, Christopher H. The petrology and stratigraphy of the Cambrian strata in northeastern Illinois. M, 1976, Northern Illinois University. 148 p.

Bayne, Bradley J. Depositional analysis of conglomerates in the Mazatzal Group and related strata, central Arizona. M, 1987, Northern Arizona University. 186 p.

Bayne, George Wallace. An auxiliary method of geologic mapping in the Galice Quadrangle, Oregon. M, 1950, University of Washington. 63 p.

Bayo, Eduardo. Numerical techniques for the evaluation of soil structure interaction effects in the time domain. D, 1983, University of California, Berkeley. 158 p.

Bayoumi, Abdel Rehim Imam. Gravity and magnetic study in the San Joaquin Valley. D, 1961, Stanford University. 82 p.

Bayri, Halis Muhtesem. On the bearing estimation of radiating sources. D, 1987, SUNY at Stony Brook. 161 p.

Bayrock, Luboslaw Antin. Glacial geology of an area in east central Alberta. M, 1954, University of Alberta. 59 p.

Bayrock, Luboslaw Antin. Glacial geology of the Alliance-Galahad-Hardisty-Brownfield area, Alberta, Canada. D, 1960, University of Wisconsin-Madison. 161 p.

Bays, Alan R. Development of an electromagnetic modeling apparatus. M, 1982, University of Calgary. 148 p.

Bays, Carl A. Stratigraphy of the Platteville Formation. D, 1938, University of Wisconsin-Madison.

Baysal, Edip. Modeling and migration by the Fourier transform method. D, 1982, University of Houston. 139 p.

Baysinger, Billy Lynn. A magnetic investigation of the Nemaha Anticline in Wabaunsee, Geary and Riley counties, Kansas. M, 1963, Kansas State University. 47 p.

Baz, Farouk El *see* El Baz, Farouk

Baz, Farouk el Sayed El *see* El Baz, Farouk el Sayed

Bazakas, Peter C. The bedrock geology of Easthampton Quadrangle, Massachusetts. M, 1960, University of Massachusetts. 132 p.

Bazaraa, A. S. Experimental/analytical investigation of the recharge rates to a groundwater table. D, 1979, Colorado State University. 164 p.

Bazard, David R. Paleomagnetism of the Late Cretaceous Ventura Member of the Midnight Peak Formation, Methow-Pasayten Belt, north-central Washington. M, 1987, Western Washington University. 168 p.

Bazinet, J. Paul. Geology of the Gander Group, Gander River ultramafic belt and Davidsville Group in the Jonathan's Pond-Weir's Pond area, Northeast Newfoundland. M, 1980, Brock University. 154 p.

Bazinet, Robert. Measure inductive de la conductivité des roches. M, 1975, Ecole Polytechnique.

Bazinet, Robert. Prospection telluric en sondage. D, 1979, Ecole Polytechnique. 209 p.

Bazrafshan, Khosrow. Geology and economic geology of the Upper Cunningham Gulch, Southeast Silverton, Colorado. M, 1981, Wichita State University. 128 p.

Be, A. W. H. Planktonic foraminifera from several western South Atlantic submarine cores. M, 1954, Columbia University, Teachers College.

Be, Allan W. Ecology of Recent planktonic foraminifera in the North Atlantic. D, 1958, Columbia University, Teachers College.

Be, Kenneth. Geological and thermal aspects of the southern San Joaquin Basin, California; application of the 40Ar/30Ar stepwise heating technique to detrital microclines. M, 1983, SUNY at Albany. 97 p.

Beach, A. R. An investigation of the magnetic character of Kilauea lava. M, 1939, University of Hawaii. 30 p.

Beach, David Kent. Depositional and diagenetic history of Pliocene-Pleistocene carbonates of northwestern Great Bahama Bank; evolution of a carbonate platform. D, 1982, University of Miami. 600 p.

Beach, Gary A. Late Devonian and Early Mississippian biostratigraphy of central Utah. M, 1961, Brigham Young University. 54 p.

Beach, H. Hamilton. The geology of the coal seams of Edmonton and district. M, 1934, University of Alberta. 157 p.

Beach, Hugh H. The geology of Moose Mountain area, Alberta. D, 1940, Yale University.

Beach, J. S. The carbonaceous sediments of the Narragansett Basin, their extent, correlation and content. M, 1932, Brown University.

Beach, Paul Ronald. Ostracodes of the family Bairdiidae from the Middle Permian Schwagerina grassectoria Zone of West Texas. M, 1952, Washington University. 46 p.

Beach, Terry Lee. The palynology of Albian strata from North-central and East Texas, U.S.A. M, 1981, University of Texas at Dallas. 236 p.

Beakhouse, Gary Philip. A structural analysis of the Nelson Lake area, Manitoba. M, 1974, University of Manitoba.

Beakhouse, Gary Philip. Geological, geochemical and Rb-Sr and U-Pb zircon geochronological investigations of granitoid rocks from the Winnipeg River Belt, northwestern Ontario and southeastern Manitoba. D, 1983, McMaster University. 376 p.

Beal, Carl H. The geology of the Monterey Quadrangle, California. M, 1916, Stanford University. 50 p.

Beal, Kenton L. The opening of a back-arc basin; the northern Mariana Trough. M, 1987, University of Hawaii. 56 p.

Beal, Laurence Hastings. Wall-rock alteration in the western portion of the Robinson mining district, Kimberly, Nevada. M, 1957, University of California, Berkeley. 91 p.

Beal, Miah Allan. Bathymetry and structure of the Arctic Ocean. D, 1969, Oregon State University. 205 p.

Beal, William A. Contaminant migration of oil-and-gas drilling fluids within the glaciated sediments of north-central North Dakota. M, 1986, University of North Dakota. 242 p.

Beale, Richard A. Beach-nearshore sediment dispersal, Matunuck Point, Rhode Island. M, 1975, University of Rhode Island.

Beales, Francis W. The sedimentation and diagenesis of certain late Paleozoic rocks of Southwest Alberta. D, 1952, University of Toronto.

Beall, Arthur Oren, Jr. Fabric and mineralogy of Woodbine sediments, east-central Texas. D, 1964, Stanford University. 188 p.

Beall, Arthur Oren, Jr. Stratigraphy of the Taylor Marl, central Texas. M, 1963, Baylor University. 64 p.

Beall, George H. Differentiation controls in siliceous gabbros. D, 1962, Massachusetts Institute of Technology. 271 p.

Beall, George H. Some aspects of atmosphere-Earth energy relationships. M, 1958, McGill University.

Beall, Joseph John. Mechanics of intrusion and petrochemical evolution of the Adel Mountain Volcanics. D, 1973, University of Montana. 104 p.

Beals, Harold Oliver. A study of the anatomy and degradation in mid-western Pleistocene woods and some of their implications. D, 1960, Purdue University. 96 p.

Beaman, Brian Roy. Natural hazards and geologic resources in the Coeur d'Alene area, Kootenai County, Idaho. M, 1982, University of Idaho. 79 p.

Beamish, Peter C. Reflection of seismic waves at oblique angles from deep sea sediments. M, 1964, Massachusetts Institute of Technology. 34 p.

Bean, Beryl Kenneth. Ostracods of the Gimlet cyclothem (Pennsylvanian) near Peoria. M, 1938, University of Illinois, Urbana. 36 p.

Bean, Clarke Lee. A geochemical study of the North Doherty intrusive complex, Jefferson County, Montana. M, 1981, Indiana University, Bloomington. 134 p.

Bean, Daniel Joseph. Seasonal variations in free gas ebullition from lake sediments. D, 1969, University of Rhode Island. 62 p.

Bean, Daniel W. Pulsating flow in alluvial channels. M, 1977, Colorado State University. 132 p.

Bean, David. Some diagenetic changes and potassium-argon relationships of a Cambrian limestone as a function of burial depth. M, 1981, Georgia Institute of Technology. 77 p.

Bean, Ernest F. Fiords of Prince William Sound, Alaska. M, 1911, University of Wisconsin-Madison.

Bean, Lawrence Edward. A geologic and hydrologic investigation of Fisher Ridge Cave system. M, 1987, Wayne State University. 112 p.

Bean, Merit W., Jr. The Alleghanian granites of Trumbull (SW CT); an integrated tectonic, geophysical, and geochemical study of the Pinewood Adamellite and associated igneous rocks. M, 1982, Wesleyan University. 145 p.

Bean, Robert Jay. The relation of gravity anomalies to the geology of central Vermont and New Hampshire. D, 1951, Harvard University.

Bean, Robert Taylor. The geology of the Mississippian System of Bledsoe and southern Rhea counties, Tennessee. M, 1942, Ohio State University.

Bean, Susan M. Volcanotectonics and geothermal potential in the Big Jack Lake area, Lassen County, California. M, 1980, Colorado School of Mines. 103 p.

Beane, John E. Petrology of ophiolitic and granodioritic rocks from Mine Ridge and the Bullrun Creek valley, Baker County, Oregon. M, 1984, Washington State University. 123 p.

Beane, John Edward. Flow stratigraphy, chemical variation and petrogenesis of Deccan flood basalts from the Western Ghats, India. D, 1988, Washington State University. 576 p.

Beane, Richard E. An investigation of the manner and time of formation of malachite. M, 1968, University of Arizona.

Beane, Richard E. The thermodynamic analysis of the effect of solid solution on the hydrothermal stability of biotite. D, 1972, Northwestern University.

Bearce, Denny N. The Sacajawea Formation and the Darwin Sandstone in the Southeast Wind River Mountains, Wyoming. M, 1963, University of Missouri, Rolla.

Bearce, Denny Neil. Geology of the Chilhowee Group (Lower Cambrian) and the Ocoee Series (Precambrian) in the southwestern Bald Mountains, Greene and Cocke counties (Tennessee). D, 1966, University of Tennessee, Knoxville. 147 p.

Bearce, Steven Craig. Characterization of fracture patterns in the area in and between the Barletts Ferry and Goat Rock fault zones, Lee County, Alabama. M, 1988, Auburn University. 140 p.

Beard, Charles Noble. Drainage development in the vicinity of Monterey Bay, California. D, 1941, University of Illinois, Urbana. 123 p.

Beard, Charles Noble. The physiography of the Brownstown Hills. M, 1936, Indiana University, Bloomington. 85 p.

Beard, Donald Chamberlin. The geology of the Deerfield Anticline at the southeast end of Walker Mountain, Augusta and Bath counties, Virginia. M, 1954, University of Virginia. 103 p.

Beard, James Sudler. The geology and petrology of the Smartville intrusive complex, northern Sierra Nevada foothills, California. D, 1985, University of California, Davis. 356 p.

Beard, John H. Microfauna of Upper Cretaceous-lower Tertiary rocks of the western Book Cliffs, Carbon County, Utah. M, 1959, University of Utah. 113 p.

Beard, Les Paul. Assessment of 2D resistivity structures using 1D inversions. M, 1987, Texas A&M University. 109 p.

Beard, Linda Sue. Precambrian geology of the Cottonwood Cliffs area, Mohave County, Arizona. M, 1985, University of Arizona. 115 p.

Beard, Richard C. A photogeological study of a part of the Arvon area of northeastern Baraga County, Michigan. M, 1964, Michigan Technological University. 59 p.

Beard, Robert D. Geology and geochemistry of the central part of the Gold Hill District, Hidalgo and Grant counties, New Mexico. M, 1987, University of New Mexico. 157 p.

Beard, Thomas. Geology of part of the Blacktail Range, Beaverhead County, Montana. M, 1949, University of Michigan. 46 p.

Beard, Thomas Christopher. Photogeology of the Spike "S" Ranch, southern Hueco Mountains, Hudspeth County, Texas. M, 1983, University of Texas at El Paso. 65 p.

Beard, Thomas Noble. Outwash in the Nuna ramp-twin glacier areas, Nunatarssuak, Greenland. M, 1954, Ohio State University.

Beard, William Clarence. Phase relations in the systems titania and titania - boric oxide. D, 1965, Ohio State University. 119 p.

Beardall, Geoffrey Bonser, Jr. Depositional environment, diagenesis and dolomitization of the Henryhouse Formation, in the western Anadarko Basin and northern shelf, Oklahoma. M, 1983, Oklahoma State University. 128 p.

Bearden, Bennett L. Hydrocarbon trapping mechanisms in the Carter Sandstone (Upper Mississippian) in the Black Warrior Basin, Fayette and Lamar counties, Alabama. M, 1984, University of Alabama. 203 p.

Beardsley, Donald W. Paleoecology of the Choctawhatchee deposits (late Miocene) at Alum Bluff, Florida. M, 1961, University of Houston.

Beardsley, Henry S., Jr. A lithologic and structural study of certain Cambro-Ordovician formations in the Redfield gas storage area, Dallas County, Iowa. M, 1956, University of Nebraska, Lincoln.

Beardsley, Reginald Huse. Modal analysis of the Granite Mountain Pulaskite, Pulaski County, Arkansas. M, 1982, University of Arkansas, Fayetteville. 60 p.

Bearzi, James Paul. Soil development, morphometry, and scarp morphology of fluvial terraces at Jack Creek, southwestern Montana. M, 1987, Montana State University. 131 p.

Beasley, Charles Alfred. The conversion of mineral resources to mineral reserves; general and specific process models and a comparative case-study. D, 1969, University of Minnesota, Minneapolis. 255 p.

Beasley, Elizabeth L. Change in the diatom assemblage of Rehoboth Bay, Delaware and the environmental implications. M, 1987, University of Delaware. 209 p.

Beasley, Thomas M. Lead-210 in selected marine organisms. D, 1969, Oregon State University. 82 p.

Beathard, R. Michael. Effects of tributaries on the Upper Mississippi River. M, 1976, Colorado State University. 109 p.

Beatie, Robert Lee. The geology of the Sunland-Tujunga area, Los Angeles County, California. M, 1958, University of California, Los Angeles.

Beaton, N. S. Geology of the area between the N'Kana and Roan Antelope mines. M, 1931, Queen's University. 23 p.

Beaton, N. S. Wall rock alteration accompanying Canadian Pre-Cambrian gold mineralization. D, 1935, Massachusetts Institute of Technology. 86 p.

Beaton, William Douglas. Trace element partition in sulphides, Noranda, Quebec. D, 1970, McGill University.

Beattie, Donald Andrew. Geology of part of southeastern Moffat County, Colorado. M, 1958, Colorado School of Mines. 176 p.

Beattie, Edward T. Structural style affected by Devonian facies change, Ram Range, Alberta. M, 1984, University of Calgary. 146 p.

Beatty, Charles A. Foraminiferal evidence for the development of anoxic conditions, Niobrara Formation (Upper Cretaceous), Boulder, Colorado, with paleoceanographic implications for the Western Interior Seaway. M, 1985, University of Wyoming. 198 p.

Beatty, Gwendolyn Faye. The study of lineaments and fracture traces and correlation to springs along the Suwannee River from Mayo, Florida, to Branford, Florida. M, 1977, University of Florida. 90 p.

Beatty, William A. The Phosphoria Formation of the northeastern Big Horn Basin, Wyoming. M, 1957, University of Missouri, Rolla.

Beaty, Chester B. Gradational processes in the White Mountains of California and Nevada. D, 1960, University of California, Berkeley. 261 p.

Beaty, David Wayne. Part I, Comparative petrology of the Apollo 11 mare basalts; Part II, The oxygen isotope geochemistry of the Abitibi greenstone belt. D, 1980, California Institute of Technology. 475 p.

Beauchamp, Benoit. Stratigraphy and facies analysis of the Upper Carboniferous to Lower Permian Canyon Fiord, Belcher Channel and Nansen formations, southwestern Ellesmere Island. D, 1987, University of Calgary. 370 p.

Beauchamp, Robert G. Stratigraphy and depositional environments of Paleocene sediments in the lower Potomac river valley. M, 1969, George Washington University.

Beauchamp, Robert George. Glauconite diagenesis and analysis of Eocene coastal plain sediments in Maryland and Virginia. D, 1988, University of Maryland. 259 p.

Beauchamp, Weldon Harold. The structural geology of the southern Slick Hills, Oklahoma. M, 1983, Oklahoma State University. 119 p.

Beauchemin, Yves. Une chaîne de programmes statistiques pour le chercheur en sciences de le terre. M, 1976, Ecole Polytechnique.

Beauclair, William Alfred. Paleontology and stratigraphy of two well cores in the Middle Devonian rocks of Claire County, Michigan. M, 1951, [University of Michigan].

Beaudin, Jean. Analyse structurale du Groupe des Shickshock et de la peridotite alpine du Mont Albert, Gaspesie. D, 19??, Universite Laval.

Beaudoin, Alain. Pétrographie et géochimie de l'altération reliée au gîte aurifère Dest-Or, Abitibi, Québec. M, 1986, Ecole Polytechnique. 287 p.

Beaudoin, Alwynne Bowyer. Holocene environmental change in the Sunwapta Pass area, Jasper National Park. D, 1984, University of Western Ontario.

Beaudoin, Georges. La Mine Roy-Ross, St.-Fabien, Québec; structure, pétrographie, terres rares et isotopie du plomb de la minéralisation à barite-galène-sphalérite. M, 1987, Universite Laval. 128 p.

Beaudry, Charles. The geology and geochemistry of Archean volcanic rocks in Daniel Township, Matagami, Quebec. M, 1984, McGill University. 129 p.

Beaudry, Desiree. Depositional history and structural evolution of a sedimentary basin in a modern forearc setting, western Sunda Arc, Indonesia. D, 1983, University of California, San Diego. 168 p.

Beaudry, Donald Arthur. Pore space reduction in some deeply buried sandstones. M, 1950, University of Cincinnati. 45 p.

Beaudry, Frederick H. Calcareous nannofossils recovered from some Pleistocene cores taken on leg fifteen of the Deep Sea Drilling Project, (Caribbean Sea). M, 1972, University of Illinois, Urbana. 108 p.

Beauheim, Richard Louis. The effects of dredging on groundwater-lake interactions at Lilly Lake, Wisconsin. M, 1980, University of Wisconsin-Madison.

Beaujon, James Sherman. The relationship of height to geomorphic degradation rates of terrace scarps in Idaho and Wyoming. M, 1987, University of Cincinnati. 240 p.

Beaulieu, Giselle M. Oxygen isotopic geochemistry, major element chemistry, and petrography of the Devonian New Hampshire magma series of northeastern Vermont. M, 1981, University of Georgia.

Beaulieu, Jean. Pétrographie du flysch Ordovician (UTICA) domaines autochtone et parautochtone de Saint Antoine de Tilly, Québec. M, 1976, Universite de Montreal.

Beaulieu, John David. San Andreas fault and the Cenozoic stratigraphy of the Santa Cruz mountains, California. D, 1970, Stanford University. 323 p.

Beaulieu, Patrick Leo. Determination of aluminum, chromium, magnesium, and silicon in trace amounts in iron meteorites. D, 1970, Arizona State University. 210 p.

Beaumont, Christopher. Tilts and tides; a study of the deformation of the Earth by ocean tide loading. D, 1973, Dalhousie University. 300 p.

Beaumont, Donald F. Significance of clays in Southeast Missouri lead ores. D, 1953, Columbia University, Teachers College.

Beaumont, Donald F. X-ray data on certain hydrous uranium minerals. M, 1951, Columbia University, Teachers College.

Beaumont, Edward A. Depositional environments of Fort Union sediments (Tertiary, Northwest Colorado) and their relations to the occurrence of coal. M, 1977, University of Kansas. 109 p.

Beaumont, Edward C. Correlation between electric and lithologic characteristics in four wells in San Juan County, New Mexico. M, 1948, University of New Mexico. 41 p.

Beaupre, Gary Scott. SH wave propagation in a laminated composite. D, 1983, Stanford University. 167 p.

Beaupre, Michel. Stratigraphie et structure du complexe de St. Germain et de la partie frontale des Appalaches, de Drummondville au Lac Champlain, Québec. M, 1975, Universite de Montreal.

Beausoleil, Yvan Joseph. A ground-water management model for the Enid Isolated Terrace Aquifer in Garfield County, Oklahoma. M, 1981, Oklahoma State University. 66 p.

Beavan, A. P. Petrology of the Mayville iron ore of Wisconsin. M, 1934, Queen's University. 88 p.

Beaven, Arthur P. The geology and gold deposits of Goldfields, Lake Athabaska, Saskatchewan. D, 1938, Princeton University. 170 p.

Beaven, Lee Wilson. Anomalous longitudinal stream profiles on a selected section of the Tipton Till plain of Indiana; a case study of knickpoint migration in Warren and Benton counties. D, 1982, Indiana State University. 527 p.

Beaver, Albert John. Characteristics and genesis of some bisequal soils in eastern Wisconsin. D, 1966, University of Wisconsin-Madison. 233 p.

Beaver, Dennis Earl. Metal zonation and fluid characteristics in the vein and skarn system, South Mountain mining district, Owyhee County, Idaho. M, 1986, Washington State University. 153 p.

Beaver, Donald W. Two-fluid interface motion in porous media. D, 1973, New Mexico Institute of Mining and Technology. 300 p.

Beaver, Frank W., Jr. The effects of fly ash and flue-gas desulfurization wastes on groundwater quality in a reclaimed lignite strip mine disposal site. D, 1986, University of North Dakota. 467 p.

Beaver, Frank W., Jr. The effects of seismic blasting on shallow water wells and aquifers in western North Dakota. M, 1984, University of North Dakota. 403 p.

Beaver, George. The geology of the Medicine Lake area and it's relationship to the lake water quality. M, 1973, University of South Dakota. 77 p.

Beaver, Harold H. Morphology and stratigraphic occurrence of the blastoid genus Pentremites. D, 1954, University of Wisconsin-Madison.

Beaver, James L. Deposition and diagenesis of Abo and Wichita carbonates, northern Midland Basin, Texas. M, 1982, Texas Tech University. 127 p.

Beavers, Alvin H. The mineral relationship between recently deposited alluvium and the loessial soils on the bluffs of the Missouri and Mississippi rivers. M, 1948, University of Missouri, Columbia.

Beavers, W. M. The depositional environments of the Hartselle Sandstone in Colbert and Franklin counties, Alabama. M, 1977, University of Alabama.

Beazley, Robert W. Study of the distribution of cultivable bacteria in lagoonal waters and sediments. M, 1973, Florida Institute of Technology.

Bebel, Dennis. Depositional environments of the lower Pottsville Group (Pennsylvanian) in Jackson County, Ohio. M, 1981, Ohio University, Athens. 83 p.

Bebout, Don Gray. Conemaugh corals from Ohio. M, 1954, University of Wisconsin-Madison.

Bebout, Don Gray. Desmoinesian fusulinids of Missouri. D, 1961, University of Kansas. 150 p.

Bebout, Gray Edward. Fluid evolution and transport during metamorphism; evidence from the Llano Uplift, Texas. M, 1984, University of Texas, Austin. 134 p.

Bebout, John Wardell. Palynology of the Paleocene-Eocene Golden Valley Formation of western North Dakota. D, 1977, Pennsylvania State University, University Park. 402 p.

Becher, Jack W. Deposition, diagenesis, and pore space evolution of the upper Smackover Formation (Jurassic), Walker Creek Field, southern Arkansas. M, 1975, Louisiana State University.

Becher, Max A. A study and interpretation of the fossils of the limestone from Spring Creek, Weber County, Utah. M, 1910, University of Wisconsin-Madison.

Bechtel, Michael Joseph. Subsurface geology of Breton Sound, offshore Louisiana. M, 1974, University of New Orleans.

Bechtel, William Lott. The geology of Snyder County, Pennsylvania. M, 1974, Pennsylvania State University, University Park. 37 p.

Beck, Barry Frederick. Erection and calibration of the Rice University radiocarbon dating laboratory and its application to some carbonate samples from Northeast Yucatan, Mexico. D, 1972, Rice University. 103 p.

Beck, Barry Frederick. Speleogenesis in Comal County, Texas. M, 1968, Rice University. 32 p.

Beck, Brian. Geologic and gravity studies of the structures of the northern Bullfrog Hills, Nye County, Nevada. M, 1984, California State University, Long Beach. 86 p.

Beck, Candyce Lynne. A re-examination of the use of glauconite for age dating. M, 1981, Rice University. 64 p.

Beck, Carl Wellington. An improved method of differential thermal analysis and its use in the study of natural carbonates. D, 1946, Harvard University.

Beck, Chris C. Geological engineering assessment of the Boise Foothills, Ada County, Idaho. M, 1988, University of Idaho. 75 p.

Beck, Fredrick M. Geology of the Sphinx Mountain area, Madison and Gallatin counties, Montana. M, 1959, University of Wyoming. 80 p.

Beck, George Frederick. Pine and pine-like woods of the West American Tertiary. M, 1948, University of Washington. 69 p.

Beck, Henry Voorhees. Geology and ground water of the Kansas River valley between Kiro and the Vermillion River. D, 1955, University of Kansas. 147 p.

Beck, Henry Voorhees. The Quaternary geology of Riley County, Kansas. M, 1949, Kansas State University. 67 p.

Beck, James Aubrey, Jr. Geology of the Lexington Hill-Pillar Peak area, Lawrence County, South Dakota. M, 1976, South Dakota School of Mines & Technology.

Beck, James Aubrey, Jr. Mechanisms of intraformational folding in the Minnekahta Limestone and relation to major structures, Black Hills, South Dakota-Wyoming. D, 1980, South Dakota School of Mines & Technology.

Beck, John H. Subsurface structural analysis of the Southeast Hoover Field and vicinity, northern Arbuckle Mountain region, southern Oklahoma. M, 1987, Baylor University. 147 p.

Beck, John Walter. Sulfide ores within the Quill ore body, Bunker Hill Mine, Kellogg, Idaho. M, 1980, Washington State University. 129 p.

Beck, John Warren. Implications for early Proterozoic tectonics and the origin of continental flood basalts, based on combined trace element and neodymium/strontium isotopic studies of mafic igneous rocks of the Penokean Lake Superior Belt, Minnesota, Wisconsin and Michigan. D, 1988, University of Minnesota, Minneapolis. 273 p.

Beck, Kevin Charles. Clay mineral diagenesis in a marine cove. D, 1971, Harvard University.

Beck, Lawrence D. Petrologic investigation of the plutonic and related rocks (Precambrian) of Comanche peak (Front range), Colorado. M, 1969, SUNY at Buffalo. 52 p.

Beck, Myrl Emil. Paleomagnetism of the Table Mountain Latite, Alpine, Tuolumne, and Stanislaus counties, California. M, 1960, Stanford University.

Beck, Myrl Emil, Jr. Paleomagnetism and magnetic intensities of Keweenawan intrusive rocks (Precambrian) from northeastern Minnesota. D, 1969, University of California, Riverside. 206 p.

Beck, Paul J. The southern Nevada-Utah border earthquakes, August to December 1966. M, 1970, University of Utah. 62 p.

Beck, Ray Hall. Depositional mechanics of the Cherry Canyon Formation (Lower Permian) Delaware Basin, Texas. M, 1967, Texas Tech University. 107 p.

Beck, Robert Lynn. The impacts of Indiana's reclamation laws on land use. D, 1982, Indiana State University. 115 p.

Beck, Robert W. Lower Mississippian formations of North America. M, 1934, University of Chicago. 74 p.

Beck, Stuart Murray. Computer-simulated deformation of meandering river patterns. D, 1988, University of Minnesota, Minneapolis. 218 p.

Beck, Susan L. Deformation in the Willard thrust plate in northern Utah and its regional implications. M, 1982, University of Utah. 79 p.

Beck, Susan Lynn. Rupture process of subduction zone earthquakes. D, 1987, University of Michigan. 221 p.

Beck, Warren R. A study of high titania-baria glasses. M, 1948, University of Minnesota, Minneapolis. 26 p.

Beck, William C. Joint analyses of the Casper Mountain/Emigrant Gap Anticline juncture, Natrona County, Wyoming. M, 1984, University of Akron. 101 p.

Beck, William W., Jr. Correlation of Pleistocene barrier islands in the lower coastal plain of South Carolina as inferred by heavy minerals. M, 1972, SUNY at Buffalo. 43 p.

Beck-von-Peccoz, Charles Morse, Jr. The characterization of a wind-dominated, tidal sand flat, southeastern Cape Cod Bay, Massachusetts. M, 1985, University of California, Los Angeles. 163 p.

Becker, Bruce D. Reciprocity of clastic and carbonate sediments, Pennsylvanian, Missourian series, Wheeler County, Texas. M, 1977, University of Texas, Austin.

Becker, Dale A. Late Wisconsin stratigraphy, upper St. John River, northwestern Maine. M, 1983, SUNY at Buffalo. 95 p.

Becker, David G. Structural implications of thermal maturity data; Sawtooth Mountains; Montana. M, 1988, University of Oklahoma. 136 p.

Becker, David Joseph. Modelling of the hydrothermal system, Mineral Mountains vicinity, Utah. M, 1985, Southern Methodist University. 136 p.

Becker, Dennis Eldon. Settlement analysis of intermittently-loaded structures founded on clay subsoils. D, 1981, University of Western Ontario.

Becker, Dennis Lee. Quantifying the environmental impact of particulate deposition from dry unpaved roadways. M, 1978, Iowa State University of Science and Technology.

Becker, Ervin S. Stratigraphic studies of the Lower Cretaceous rocks of Brewster County, Texas. M, 1956, University of Wisconsin-Madison.

Becker, Herman Frederick. An Oligocene flora from the Ruby River basin in southwestern Montana. D, 1957, University of Michigan. 218 p.

Becker, Jacques. The freshwater Ostracoda of the lower Humbolt Formation (Miocene) in northern Nevada, northern Utah and southeastern Idaho. M, 1969, University of Minnesota, Minneapolis. 55 p.

Becker, John E. Sedimentology of the Upper Triassic Shinarump Member of the Chinle Formation, southern Nevada. M, 1986, Southern Illinois University, Carbondale. 122 p.

Becker, Jonathan, J. The fossil birds of the late Miocene and early Pliocene of Florida. D, 1985, University of Florida. 245 p.

Becker, Joseph. Structural analysis of the western Llano Uplift with emphasis on the Mason Fault. M, 1985, Texas A&M University.

Becker, Joseph Henry. Upper Devonian conodonts from the eastern Great Basin. M, 1959, Southern Methodist University. 91 p.

Becker, Keir. Heat flow studies of spreading center hydrothermal processes. D, 1981, University of California, San Diego.

Becker, L. R. A survey of a sandy beach and bay, Appletree Bay, Lake Champlain, Vermont. M, 1978, University of Vermont.

Becker, Leroy Everett. Ostracodes from the Snyder Creek Shale of Callaway and Montgomery counties, Missouri. M, 1940, University of Missouri, Columbia.

Becker, Michael James. Depositional and diagenetic history of the Lower Mississippian Stobo carbonate mound, Monroe County, Indiana. M, 1988, Indiana University, Bloomington. 81 p.

Becker, Paul John. Middle Pennsylvanian cyclical sedimentation in the Minturn Formation of South-central Colorado. M, 1978, University of Iowa. 79 p.

Becker, R. M. Geophysical investigation of Iron Mountain titaniferous magnetite deposit, Fremont County, Colorado. M, 1960, Colorado School of Mines. 28 p.

Becker, Richard Charles. Oligocene foraminiferal zonation of a well in the Main Pass area (Gulf of Mexico). M, 1982, Washington University. 47 p.

Becker, Robert William. Geology of a part of the Tendoy Mountains, west of Lima, Beaverhead County, Montana. M, 1948, University of Michigan. 43 p.

Becker, Russell Dail. Stratigraphy and depositional characteristics of the Bromide "Dense" unit and Viola Limestone (Middle and Upper Ordovician), subsurface, central Oklahoma. M, 1988, University of Tulsa. 126 p.

Becker, Susan Ward. Field relations and petrology of the Burro Mesa "Riebeckite" Rhyolite, Big Bend National Park, Texas. M, 1976, University of California, Santa Cruz.

Becker, Thomas Edward. Correlations for drill-cuttings transport in directional-well drilling. D, 1987, University of Tulsa. 298 p.

Beckerman, Joseph H. The formational environment and characterization of the Lower Kittanning coal seam, northeastern Ohio. M, 1982, Miami University (Ohio). 100 p.

Beckett, John Randall. The origin of calcium-, aluminum-rich inclusions from carbonaceous chondrites; an experimental study. D, 1986, University of Chicago. 373 p.

Beckett, K. L. Water characteristics and circulation in the passages of the northern Channel Islands, California, 1968. M, 1974, University of Southern California.

Beckett, M. Patricia. Hydrocarbons associated with particulate matter in the northern Gulf of Mexico. M, 1979, Florida State University.

Beckett, Martyn Frank. Phase relations in synthetic alkali-bearing dolomite carbonatites and the effect of alkalinity and fluorine content on the solubility of pyrochlore and the formation of niobium deposits in carbonates. M, 1987, University of Toronto.

Beckett, Robert L. Geology of the Red Canyon area, Eagle County, Colorado. M, 1955, University of Colorado.

Beckham, Elizabeth. Diagenesis of the Muddy Sandstone (Cretaceous), Peoria Field, Denver Basin, Colorado. M, 1979, University of Texas, Austin.

Beckham, Wallace Edgar, Jr. Electrical polarization in carbonate rocks. D, 1968, Washington University. 162 p.

Beckham, Wallace Edgar, Jr. The gravity and magnetic fields of the Iron Mountain ore body. M, 1964, Washington University. 76 p.

Beckman, Charles Allan. Geology of the Wimsattville Basin central mining district, New Mexico. M, 1957, University of Minnesota, Minneapolis. 37 p.

Beckman, Gary Lee. A comparison of calcium silicate bricks manufactured from various combinations of Florida sands and environmental waste materials. M, 1972, University of Florida. 104 p.

Beckman, Marian C. Saline facies of Recent sediments in Great Salt Lake, Utah. M, 1960, Columbia University, Teachers College.

Beckman, Michael A. Subsurface stratigraphy of the Lower Atoka sandstones of central Franklin County, Arkansas. M, 1974, University of Arkansas, Fayetteville.

Beckman, Michael W. The stratigraphy and age of the Seguin Formation of central Texas. M, 1941, Texas A&M University.

Beckman, Richard John. Petrology of the Brassfield Limestone in southern Indiana. M, 1961, Indiana University, Bloomington. 38 p.

Beckman, Scott Warren. Paleoenvironmental reconstructions and organic matter characterizations of peats and associated sediments from cores in a portion of the LaFourche Delta. D, 1985, Louisiana State University. 313 p.

Beckman, Steven C. Petrology and depositional environments of the Mississippian Newman Limestone in Hancock and Claiborne counties, Tennessee. M, 1985, University of North Carolina, Chapel Hill. 88 p.

Beckmann, Douglas D. The Frontier Formation (Late Cretaceous) of the Uinta Mountains area, Utah and Colorado. M, 1967, University of Nebraska, Lincoln.

Beckner, Jeffrey S. Organic sedimentation processes of Buck Basin, Brewers Bar, and upper Blue Basin within Reelfoot Lake, Tennessee. M, 1986, Eastern Kentucky University. 90 p.

Beckwith, Clyde Grosvenor, Jr. The geology of the Fort Gibson area, Cherokee and Muskogee counties, Oklahoma. M, 1950, University of Oklahoma. 74 p.

Beckwith, Radcliffe H. The geology and ore deposits of the Buffalo Hump District. D, 1928, Columbia University, Teachers College.

Becraft, George Earle. Definition of the Tieton Andesite on lithology and structure. M, 1950, Washington State University. 26 p.

Becraft, George Earle. Geology of the southern part of Turtle Lake quadrangle, northeastern Washington. D, 1959, University of Washington. 101 p.

Bedaiwy, Mohamed Naguib A. Densification of soil surface under simulated high intensity rainfall. D, 1988, University of California, Davis. 206 p.

Bedard, Jean H. J. Fractional crystallization and intrusion mechanisms, Spur Slice (Block 4), Cape Smith, New Quebec. M, 1981, McGill University. 260 p.

Bédard, L. Paul. Pétrographie et géochimie du Stock de Dolodau; syénite et carbonatite associée. M, 1988, Universite du Quebec a Chicoutimi. 186 p.

Beddoes, Leslie R., Jr. Foraminiferal populations of the Goodland Formation, Tarrant County, Texas. M, 1956, Southern Methodist University. 35 p.

Bedell, Frank G. The Sudbury mining district. D, 1906, University of Kansas.

Bedell, Frank G. The surveying methods used in the location of a claim and in the opening and development of an ore body as practiced at La Cananea, Sonora, Mexico. M, 1904, University of Kansas.

Bedell, Theodore E., III. Petrology, depositional environments and diagenesis of the Ste. Genevieve Limestone (Upper Mississippian) in Washington County, Virginia. M, 1986, East Carolina University. 121 p.

Bedford, Betty. A depositional model for the Cedar Grove and adjoining coal seams, Boone County, West Virginia. M, 1980, University of South Carolina.

Bedford, John Phillips. Pleistocene deposits at Brook Hill Farm, Pike County, Missouri. M, 1952, Washington University. 89 p.

Bedford, John William. Geology of the Horse Mountain area, Mitchell Quadrangle, Oregon. M, 1954, Oregon State University. 90 p.

Bedient, Philip Bruce. Hydrologic-land use interactions in a Florida river basin. D, 1975, University of Florida. 262 p.

Bediz, Pertev I. Depth determination of semi-infinite horizontal conductive layer by electromagnetic galvanic methods (T593). M, 1942, Colorado School of Mines. 18 p.

Bednarski, Sheila Palmer. Paleoecology of the San Carlos Formation type section, Presidio County, Texas. M, 1987, University of Texas of the Permian Basin. 136 p.

Bedosky, Stephen J. Recent sediment history of Apalachicola Bay, Florida. M, 1987, Florida State University. 247 p.

Bedrossian, Trinda Lee. Paleoecological evolution of the "Merced" formation (Pliocene, upper and Pleistocene ?) Sonoma County, California. M, 1970, University of California, Davis. 92 p.

Bedwell, John Lewis. Geophysical investigation of basement relief conditions in the southwest quarter of McClain County, Oklahoma. M, 1967, University of Oklahoma. 52 p.

Bée, Michel. A comparison of seismic properties of young and mature oceanic crust. D, 1984, Oregon State University. 183 p.

Bée, Michel. Marine seismic refraction study between Cape Simpson and Prudhoe Bay, Alaska. M, 1979, Oregon State University. 83 p.

Beebe, Matthew A. Conodont biofacies analysis of the Desmoinesian limestones (Middle Pennsylvanian) of the Robledo Mountains, Dona Ana County, New Mexico. M, 1986, New Mexico State University, Las Cruces. 113 p.

Beebe, Morris Wilson. A study of Kentucky oil shale. M, 1928, [University of Kentucky].

Beebe, Ward Joseph. A paleomagnetic study of the southern Coast Range Ophiolite, California, and tectonic implications. M, 1986, University of California, Santa Barbara. 148 p.

Beecham, Arthur W. A structural study of the ABC Fault, Beaverlodge, Saskatchewan. M, 1969, Queen's University. 124 p.

Beecher, Charles E. Brachiospongiidae; a memoir on a group of Silurian sponges. D, 1889, Yale University.

Beechler, Theodore W. Petrology of the Pooleville Limestone Member of the Bromide Formation (middle Ordovician), Arbuckle area, Oklahoma. M, 1974, Tulane University.

Beede, Joshua W. Carboniferous invertebrates. D, 1899, University of Kansas. 189 p.

Beeden, David Robert. Sedimentology of some turbidites and related rocks from the Cloridorme Group, Ordovician, Quebec. M, 1983, McMaster University. 256 p.

Beeder, John Ralph. The bedrock geology and placer deposits of the Dixie District, Idaho County, Idaho. M, 1958, University of Idaho. 50 p.

Beeghly, Sallie. Paleoenvironment of the Cretaceous Mowry Shale in Colorado. M, 1977, University of Wisconsin-Madison.

Beegle, Douglas Brian. Potassium buffering behavior of three Pennsylvania soils. D, 1983, Pennsylvania State University, University Park. 89 p.

Beek, Johannes Laurens Van *see* Van Beek, Johannes Laurens

Beeker, Ralph Edward. Chemical variations in Dundee brine and their relation to structure. M, 1940, University of Michigan.

Beekly, Emerson Keagy. Geology of the Wolf Creek area, Bighorn Mountains, Wyoming. M, 1948, University of Iowa. 104 p.

Beeler, Nick M. A preliminary investigation of pressure solution deformation in theory and experiment. M, 1986, Arizona State University. 172 p.

Beem, Kenneth Alan. Benthonic foraminiferal paleoecology of the Choctawhatchee deposits (Neogene) of northwestern Florida. D, 1973, University of Cincinnati. 215 p.

Beem, Kenneth Alan. Paleoecology of the Choctawhatchee Formation (middle and upper Miocene) in Calhoun and western Liberty counties, Florida. M, 1967, Tulane University. 39 p.

Beem, Leigh Ivan. Investigations into fractured granite rock by detailed geologic and geophysical methods. M, 1988, Joint program, Idaho State Univ. and Boise State Univ. 75 p.

Beer, Johannes Hendrik de *see* de Beer, Johannes Hendrik

Beer, Lawrence P. Geology of the Thompson Lakes NW 15-minute quadrangle, Northwest Montana. M, 1960, University of Massachusetts. 96 p.

Beer, Lawrence Peter. Ground-water hydrogeology of southern Cache Valley, Utah. D, 1967, University of Utah. 251 p.

Beer, Robert M. Suspended sediment over Redondo submarine canyon and vicinity, southern California. M, 1969, University of Southern California.

Beerbower, D. C. Relationship between clay and dolomitization in the Pipe Creek Junior Reef (Silurian), Grant County, Indiana. M, 1977, Ball State University. 57 p.

Beers, Armand Henry. Trace element analysis and trend surface analysis of the Oquirrh mountains (Oquirrh mountains, Utah). D, 1970, University of Utah. 210 p.

Beers, Charles A. Geology of the Precambrian rocks of the southern Los Pinos Mountains, Socorro County, New Mexico. M, 197?, New Mexico Institute of Mining and Technology. 228 p.

Beers, L. D. The Earth's gravity field from a combination of satellite and terrestrial gravity data. D, 1971, University of Hawaii. 82 p.

Beers, R. F. The distribution of radioactivity in ancient sediments. D, 1943, Massachusetts Institute of Technology. 293 p.

Beery, John Arlington. Depositional history and paleoenvironments of the lower and middle Miocene Temblor, northern San Joaquin Basin, California. M, 1987, Stanford University. 132 p.

Beeson, Dale Clayton. The relative significance of tectonics, sea level fluctuations, and paleoclimate to Cretaceous coal distribution in North America. M, 1984, University of Colorado. 202 p.

Beeson, David L. Large scale bedforms in pyroclastic flow deposits; Mount Saint Helens, Washington. M, 1988, University of Texas, Arlington. 216 p.

Beeson, John H. The geology of the southern half of the Huntington Quadrangle, Oregon. M, 1955, University of Oregon. 79 p.

Beeson, Kenneth Crees. The relation of soils to the cobalt requirements of ruminants. D, 1948, Cornell University.

Beeson, Marvin H. The geology of the north-central part of the Sawtooth Creek Quadrangle, Oregon. M, 1962, University of Oregon. 92 p.

Beeson, Marvin Howard. A trace element study of silicic volcanic rocks. D, 1969, University of California, San Diego. 144 p.

Beeson, Melvin H. Geology of the Lostine Valley, Wallowa County, Oregon. M, 1963, University of Oregon. 105 p.

Beeson, Melvin Harry. Petrology, mineralogy and geochemistry of the lavas of East Molokai Volcano, Hawaii. D, 1973, University of California, Santa Cruz. 160 p.

Beeunas, Mark Anthony. Preserved stable isotopic signature of subaerial diagenesis in the 1.2 b.y. Mescal Limestone, central Arizona; implications for the timing and development of a terrestrial plant cover. M, 1984, Arizona State University. 91 p.

Beever, Frances Kay. The depositional environment of the Lucas Formation, Amherst Quarry, Amherstburg, Ontario, Canada. M, 1982, Wayne State University. 100 p.

Beever, Hank G. Van *see* Van Beever, Hank G.

Beffort, Joseph D. Investigation of possible groundwater contamination in the vicinity of a sanitary landfill, Vermillion, South Dakota. M, 1972, University of South Dakota. 81 p.

Beger, Richard M. Geology of the Pater Mine, Blind River area, Ontario. M, 1963, Michigan Technological University. 91 p.

Beger, Richard M. The structure of pollucite. M, 1967, Massachusetts Institute of Technology. 54 p.

Beger, Richard Myron. Aluminum-silicon ordering in sillimanite and mullite. D, 1979, Harvard University.

Beget, James Earl. Postglacial eruption history and volcanic hazards at Glacier Peak, Washington. D, 1981, University of Washington. 192 p.

Beget, James Earl. Stratigraphy and sedimentology of three permafrost cores from New Harbor, South Victoria Land, Antarctica. M, 1977, University of Washington. 52 p.

Beggs, George. Interpretation of seismic reflection data from the Hartsel and Sulphur Mountain quadrangles, Park County, Colorado. M, 1976, Colorado School of Mines. 91 p.

Beggs, John McIntyre. Stratigraphy, petrology, and tectonic setting of the Alisitos Group, Baja California, Mexico. D, 1983, University of California, Santa Barbara. 236 p.

Beghtel, Floyd Woodrow. Ammonoid fauna of the Pennsylvanian Wewoka Formation of Oklahoma. M, 1959, University of Iowa. 128 p.

Beghtel, Floyd Woodrow. Desmoinesian ammonoids of Oklahoma. D, 1962, University of Iowa. 402 p.

Begin, Ze've B. Aspects of degradation of alluvial streams in response to base-level lowering. D, 1979, Colorado State University. 257 p.

Begle, Elsie A. The weathering of granite, Llano region, central Texas. M, 1978, University of Texas, Austin.

Begley, Alisa L. Experimental evidence for the existence of hydropyrope. M, 1985, Lehigh University. 63 p.

Behairy, Abdel-Kader Ali. Mineralogy and geochemistry of bentonite beds of the middle and lower Graneros Shale (Upper Cretaceous), Ellsworth and Russell counties, Kansas. M, 1972, Wichita State University. 76 p.

Behbehani, Abdulsamee S. K. Structural and petrographic analysis of the Mishrif Member of the Magwa Formation in the Hout, Dorra, and Khafji oil fields, offshore Kuwait-Saudi Arabia divided neutral zone. M, 1980, University of Toledo. 44 p.

Beheiry, Salah El Deen Abdalla. Epigene sedimentary forms in northeastern Coachella Valley, Southern California. D, 1964, University of California, Los Angeles. 315 p.

Behensky, J. F., Jr. Reassessment of the distributin of benthic foraminifera of the shelf and slope of the Atlantic margin and Gulf of Mexico of the U.S. M, 1977, University of Miami.

Behken, Fred Henry. Late Permian conodonts from Wyoming and Nevada. M, 1969, University of Wisconsin-Madison.

Behling, Richard W. The paleoenvironment of the Quimbys Mill Member, Platteville Formation (Middle Ordovician) of southwestern Wisconsin. M, 1972, University of Wisconsin-Milwaukee.

Behling, Robert E. A detailed study of the Wisconsin stratigraphic sections of the upper Lamoille Valley, north-central Vermont. M, 1965, Miami University (Ohio). 125 p.

Behling, Robert Edward. Pedological development on moraines of the Meserve Glacier, Antarctica. D, 1971, Ohio State University.

Behm, Juan Joaquin. The petrology of the serpentinite of Richmond County (Staten Island), New York. M, 1954, New York University.

Behnami, Farhad. Subsurface geology of the West Baden Group within the Elliot oil field and surrounding area in Vanderburgh County, Indiana. M, 1981, Ball State University. 50 p.

Behnke, Jerold J. Geology of the southern portion of Martin's Ridge Monitor Range, Nye County, Nevada. M, 1961, University of Nevada - Mackay School of Mines. 42 p.

Behnke, Jerold J. Soil clogging in recharge and irrigation operations. D, 1967, University of Nevada. 61 p.

Behnken, Fred Henry. Leonardian and Guadalupian (Permian) conodont biostratigraphy and evolution in Western and Southwestern United States. D, 1972, University of Wisconsin-Madison. 198 p.

Behr, Christina B. Geochemical analyses of ore fluid from the St. Cloud-U.S. Treasury vein system, Chloride mining district, NM. M, 1988, New Mexico Institute of Mining and Technology.

Behre, Charles H. Slate in Northampton County, Pennsylvania. D, 1925, University of Chicago. 488 p.

Behre, Charles H., Jr. Slate in Northampton County, Pennsylvania. M, 1926, New York University.

Behrendt, John Charles. A regional magnetic study of the Uinta Mountains, Utah. M, 1957, University of Wisconsin-Madison.

Behrendt, John Charles. Geophysical studies in the Filchner Ice Shelf area of Antarctica. D, 1961, University of Wisconsin-Madison. 191 p.

Behrens, Earl William. An extension and reinterpretation of a regional magnetometer survey of part of southeastern Michigan. M, 1958, University of Michigan.

Behrens, Earl William. Environment reconstruction for a part of the Glen Rose Limestone, central Texas. D, 1963, Rice University. 262 p.

Behrens, Gene K. Stratigraphy, sedimentology, and paleoecology of a Pliocene reef tract; St. Croix, U.S. Virgin Islands. M, 1976, Northern Illinois University. 93 p.

Behrensmeyer, Anna K. The taphonomy and paleoecology of Plio-Pleistocene vertebrate assemblages east of Lake Rudolf, Kenya. D, 1973, Harvard University.

Behrman, Philip George. Paleogeography and structural evolution of a middle Mesozoic volcanic arc-continental margin; Sierra Nevada foothills, California. D, 1978, University of California, Berkeley. 301 p.

Behum, Paul. Geology and mineral resources of a portion of the Cheat Mountain coal field, Randolph and Pocahontas counties, West Virginia. M, 1984, University of Pittsburgh.

Behymer, Thomas David. Photolysis of polycyclic aromatic hydrocarbons adsorbed on fly ash. D, 1987, Indiana University, Bloomington. 163 p.

Beier, Joy Ann. Petrologic analysis of beachrock, San Salvador Island, Bahamas. M, 1984, Indiana University, Bloomington. 118 p.

Beier, Joy Ann. Sulfide, phosphate, and minor element enrichment in the New Albany Shale (Devonian-Mississippian) of southern Indiana. D, 1988, Indiana University, Bloomington. 141 p.

Beiersdorfer, Raymond Emil. Metamorphic petrology of the Smartville Complex, northern Sierra Nevada foothills. M, 1982, University of California, Davis. 152 p.

Beike, Dieter. Engineering economic evaluation of mining in Antarctica; a case study of platinum. M, 1988, University of Texas, Austin.

Beikirch, Dale W. Decapod crustaceans from the Pflugerville Member, Austin Formation (Campanian; Late Cretaceous) of Travis County, Austin, Texas. M, 1979, Kent State University, Kent. 40 p.

Beimfohr, Oliver Wendell. Battle Creek Coal, Marion County, Tennessee. M, 1941, Vanderbilt University.

Bein, Margaret G. A quantitative photogrammetric analysis of Narragansett Bay, Rhode Island shoreline changes. M, 1981, University of Rhode Island.

Beinert, Richard James. Conodonts (Devonian) of the Maple Mill Shale, (Washington County), southeast Iowa. M, 1968, University of Iowa. 100 p.

Beinert, Richard James. Thalassoceratidae, upper Paleozoic "ceratitic" ammonoids. D, 1971, University of Iowa. 249 p.

Beinkafner, Katherine Jorgensen. Deformation of the subsurface Silurian and Devonian rocks of the southern tier of New York State. D, 1983, Syracuse University. 481 p.

Beiser, Erna. Rubidium-strontium age determination of the carbonaceous chondrite "Murray". M, 1964, Massachusetts Institute of Technology. 33 p.

Beissel, Dennis R. Geophysical studies of fractured rock. M, 1971, Colorado State University. 111 p.

Beiswenger, Jane Miller. Late Quaternary vegetational history of Grays Lake, Idaho and the Ice Slough, Wyoming. D, 1987, University of Wyoming. 240 p.

Beiter, David P. Calcite saturation in an eastern Kentucky karst stream. M, 1970, University of Kentucky. 74 p.

Beitzel, John Edward. Geophysical investigations in Marie Byrd Land, Antarctica. D, 1972, University of Wisconsin-Madison. 123 p.

Bejnar, Craig Russel. Paleocurrents and depositional environments of the Dakota Group (Cretaceous), San Miguel County, New Mexico. M, 1975, University of Arizona.

Bejnar, Waldemere. Lithologic control of ore deposits in the San Juan Mountains, Colorado. M, 1947, University of Michigan.

Bejnar, Waldomere. Geology of the Ruin Basin area, Gila County, Arizona. D, 1952, University of Arizona.

Beka, Francis Thomas. Geology of the Squaw Creek Prospect, Northport, Washington. M, 1978, Washington State University. 90 p.

Beka, Francis Thomas. Upper Palaeozoic metasedimentary and metavolcanic rocks and associated mineral deposits between Glasgo Lakes and Little Sheep Creek, Stevens County, Washington. D, 1980, Washington State University. 172 p.

Bekkar, Hamed. Laramide structural elements and relationship to Precambrian basement, southeastern Wyoming. M, 1973, University of Wyoming. 70 p.

Bela, Jim. Engineering geology of the Rapley trail loop, Portola Valley, California. M, 1971, Stanford University.

Belak, Ronald. Stratigraphy and sedimentology of the Cobleskill Formation (Upper Silurian), New York State. M, 1978, Indiana University, Bloomington. 192 p.

Belan, Ricky Allen. Hydrogeology of a portion of the Santa Catalina Mountains (Tucson, Arizona). M, 1972, University of Arizona.

Beland, A. Management of the Greenbrook well field. M, 1977, University of Waterloo.

Beland, Jacques. Geology of the Shawinigan map area, Champlain and Saint Maurice counties, Quebec. D, 1953, Princeton University. 198 p.

Béland, Jacques Robert. The relationships of mafic minerals to feldspar in anthrosite gabbro near Bourget, Chicoutimi County. M, 1949, Universite Laval.

Beland, Rene. Petrography of the Ruth, Josephine, and Goulais River iron ranges, District of Algoma, Ontario. M, 1944, University of Toronto.

Beland, Rene. Synthesis of the silver-sulpho-minerals in alkali sulphide solutions. D, 1946, University of Toronto.

Bélanger, Jacques. Caractérisation pétrographique et géochimique de la zone cupro-zincifère "B-5" et de ses roches encaissantes, secteur Cooke, mines Opémisca, Chapais. M, 1979, Universite du Quebec a Chicoutimi. 145 p.

Bélanger, Jean. Etude de la zone de transition entre la Formation de Waconichi et la Formation de Gilman, Groupe de Roy, Chibougamau, Quebec. M, 1979, Universite du Quebec a Chicoutimi. 83 p.

Belanger, Paul Edward. Late Cenozoic benthic foraminifera of the Norwegian-Greenland Sea. D, 1981, Brown University. 300 p.

Bélanger-Davis, Colette E. Mineralogical and petrophysical changes after steam testing in carbonate rocks of the Grosmont Formation, Alberta. M, 1985, University of Calgary. 142 p.

Belcher, R. C. The geomorphic evolution of the Rio Grande. M, 1975, Baylor University. 210 p.

Belcher, Robert C. Depositional environments, paleomagnetism, and tectonic significance of Huizachal red beds (lower Mesozoic); northeastern Mexico. D, 1979, University of Texas, Austin. 292 p.

Belcher, Steven W. Spatial resolution by single sweep recording and processing. M, 1984, Virginia Polytechnic Institute and State University. 50 p.

Belcher, Wayne R. Assessment of aquifer heterogeneities at the Hanford Nuclear Reservation, Washington using inverse contaminant plume analysis. M, 1988, Colorado School of Mines. 54 p.

Belchic, George. Plasticity of the Dakota clays of Kansas. M, 1915, University of Kansas. 31 p.

Belchic, George, Jr. The potentiometric model and its application to the problems of secondary recovery. M, 1948, Louisiana State University.

Belden, William A. The stratigraphy of the Toroweap Formation, Aubbey Cliffs, Coconino County, Arizona. M, 1954, University of Arizona.

Belding, Herbert Frederick. Ordovician section in the vicinity of Rohallion. M, 1941, University of Toronto.

Belforte, Alberto Santiago. Pre-canyon structural geology of the southern end of Fort Worth Basin, central Texas. M, 1971, University of Texas, Austin.

Belgea, Paul. Grinding and impact type abrasion of selected molluscan shell microstructures. M, 1980, University of North Carolina, Chapel Hill.

Belik, G. D. Geology of the Harper Creek copper deposit. M, 1973, University of British Columbia.

Belinger, Eric Vance. Mineralogy and oxygen isotope ratios of hydrothermal and low-grade metamorphic argillaceous rocks. D, 1971, Case Western Reserve University.

Bélisle, Jacqueline. The origin of Bt horizons in some Luvisols of southern Ontario. M, 1981, University of Guelph.

Belisle, Jean-Marc. Méthodologie et programmathèque pour l'étude géostatistèque des gisements de type porphyrique. M, 1981, Universite de Montreal. 230 p.

Belitz, Kenneth. Hydrodynamics of the Denver Basin; an explanation of subnormal fluid pressures. D, 1985, Stanford University. 214 p.

Beljin, Milovan. A three-dimensional hydrogeological model of the Sarir well fields, Libya. M, 1981, Ohio University, Athens. 237 p.

Beljin, Milovan Slavko. Testing and validation of groundwater solute transport models. D, 1987, Ohio University, Athens. 253 p.

Belk, Jerrel Keith. A petrographic analysis of the volcaniclastic Woodbine Formation (Upper Cretaceous) of southwestern Arkansas. M, 1985, Stephen F. Austin State University. 101 p.

Belkacemi, Smain. Laboratory study in a calibration chamber of a pressuremeter test on silt. D, 1988, Tufts University. 325 p.

Belknap, Barton Austin. TXL Devonian field, Ector County, Texas. M, 1951, Texas Tech University. 29 p.

Belknap, Daniel F. Application of amino acid geochronology to stratigraphy of late Cenozoic marine units of the Atlantic Coastal Plain. D, 1979, University of Delaware. 567 p.

Belknap, Daniel F. Dating of late Pleistocene and Holocene relative sea levels in coastal Delaware. M, 1975, University of Delaware.

Belknap, Ralph Leroy. Physiographic studies in the Holstenborg District of southwestern Greenland. D, 1929, University of Michigan.

Bell, Alan George Ridley. Application of two multivariate classification techniques to the problem of seismic discrimination. M, 1978, Pennsylvania State University, University Park. 163 p.

Bell, Alfred H. The geology of Whitehorse District, Yukon Territory. D, 1926, University of Chicago. 122 p.

Bell, Archibald M. Major structural patterns in parts of the Canadian Shield. D, 1935, University of Wisconsin-Madison.

Bell, Archibald M. The relation of silver ores to diabase. M, 1929, University of Toronto.

Bell, Brian H. Slope maintenance, Big Mountain, Ventura County, California. M, 1969, San Fernando Valley State University.

Bell, Bruce McConnell. A study of North American Edrioasteroidea. D, 1972, University of Cincinnati. 1022 p.

Bell, Bruce McConnell. An introduction to the Cincinnatian edrioasteroids (Echinodermata). M, 1966, University of Cincinnati. 168 p.

Bell, Christopher K. Some aspects of the geochemistry of gallium. D, 1953, Massachusetts Institute of Technology. 200 p.

Bell, Christy Anne. Regional uranium and thorium anomalies associated with sedimentary uranium deposits in Pennsylvania and Colorado. M, 1980, Pennsylvania State University, University Park. 123 p.

Bell, David Allan. Structural and age relationships in the Embudo Granite, Picuris Mountains, New Mexico. M, 1985, University of Texas at Dallas. 175 p.

Bell, David I. A geologic investigation of landslides of a northern portion of the Santa Monica Mountains, California. M, 1966, University of Southern California.

Bell, David L. A comparative study of glauconite and the associated clay fraction in modern marine sediments. M, 1966, Florida State University.

Bell, Douglas Alan. Characterization of a Lower Mississippian oil producing sandstone reservoir in central West Virginia. M, 1987, University of Maryland.

Bell, Elaine J. Origin of the auriferous clays in the Fairbanks area, Alaska. M, 1974, Arizona State University. 63 p.

Bell, Frank James. A phylogenetic study on the Oligocene rabbits of Nebraska. M, 1941, University of Nebraska, Lincoln.

Bell, Frank W. The stratigraphy and foraminiferal fauna of the Santa Susana Formation. M, 1933, California Institute of Technology. 39 p.

Bell, Gerald Laverne. The geology of Salem Township, Washington County (Ohio). M, 1950, Ohio State University.

Bell, Gordon Knox, Jr. The disputed structures of the Mesonacidae and their significance. M, 1930, Columbia University, Teachers College.

Bell, Gordon Lennox. Devonian stratigraphy and paleontology of the Ram River area, Alberta. M, 1951, University of British Columbia.

Bell, Gordon Leon. A geologic section of the Santa Lucia Mountains, Coast Range, California. M, 1940, University of California, Berkeley. 54 p.

Bell, Gordon Leon. Geology of the Precambrian metamorphic terrain, Farmington Mountains, Utah. D, 1951, University of Utah. 101 p.

Bell, Hillis F. The geologic and subsurface features of a part of the Electra oil field, Texas. M, 1926, University of Oklahoma. 16 p.

Bell, J. A. Benthonic foraminifera of Kaneohe Bay, Oahu, Hawaii. M, 1976, University of Hawaii. 109 p.

Bell, James John. Geology of the foothills of Sierra de los Pinos, northern Chihuahua, near Indian Hot Springs, Hudspeth County, Texas. M, 1963, University of Texas, Austin.

Bell, James M. The petrology of a copper rich conglomerate and its bearing on in situ leaching, Centennial Mine, Houghton County, Michigan. M, 1974, Michigan Technological University. 64 p.

Bell, James Mackintosh. Report on the Michipicoten iron range. D, 1904, Harvard University.

Bell, Jane. A study of zones of metamorphism in the Wissahickon Schist by means of heavy mineral analysis. M, 1940, Bryn Mawr College. 21 p.

Bell, Jane and Albigese, Muriel. Heavy minerals of the Darby Creek region-Chester Quadrangle, Pennsylvania. M, 1940, Bryn Mawr College. 18 p.

Bell, Jean Louise. Data structures for scientific simulation programs. D, 1983, University of Colorado. 268 p.

Bell, John Sebastian. Geology of the Camatagua area, Estado Aragua, Venezuela. D, 1967, Princeton University. 327 p.

Bell, John W. Environmental geology of the Fairbanks area, Alaska. M, 1974, Arizona State University. 113 p.

Bell, John William. Refractories from Pacific Northwest olivines. M, 1940, University of Washington. 62 p.

Bell, K. C. Geology of the Balachey Lake area, Northwest Territories. M, 1950, McGill University.

Bell, K. G. Geology of the Boston metropolitan area. D, 1948, Massachusetts Institute of Technology. 435 p.

Bell, K. G. Preliminary investigation into the radioactive properties of crude oils and associated elements. M, 1940, Massachusetts Institute of Technology. 31 p.

Bell, Kenneth Robert. Incorporation of remotely sensed soil moisture data into a hydrologic runoff model. D, 1983, University of Maryland. 244 p.

Bell, Kenneth William. Pebble deformation in the San Antonio Formation (Archaean), Rice Lake area, (southeastern) Manitoba (Canada). M, 1968, University of Manitoba.

Bell, L. M. Factors influencing the sedimentary environents of the Squamish River delta in southwestern British Columbia. M, 1975, University of British Columbia.

Bell, L. V. Geology of the Boston-Skead area with special reference to rock alteration (carbonation) in this and related Canadian Precambrian areas. D, 1930, University of Toronto.

Bell, L. V. The lithology and alteration of the Timiskaming sediments in the Kirkland Lake area. M, 1928, University of Toronto.

Bell, Leslie. Impact of sediment clay mineralogy and organic matter on lead in estuarine water. M, 1977, University of South Florida, Tampa. 98 p.

Bell, Lyndon H. Stratigraphy and depositional history of the Cambrian Flathead Sandstone, northern Park County, Wyoming. M, 1968, University of Wyoming. 164 p.

Bell, M. L. Models for dilatancy, creep, and flow in rocks. D, 1977, Stanford University. 122 p.

Bell, Mary K. Sedimentology of the Planulina palmerae in portions of Vermilion and Iberia parishes, Louisiana. M, 1986, University of Southwestern Louisiana. 87 p.

Bell, Michael Stewart. Geology of the Chatham Fault, central Taconic region. M, 1978, Cornell University.

Bell, Pamela Elizabeth. The role of anaerobic bacteria in the neutralization of acid mine drainage. D, 1988, University of Virginia. 232 p.

Bell, Patricia J. Environments of deposition, Pliocene Imperial Formation, Southeast Coyote Mountains, Imperial County, California. M, 1980, San Diego State University.

Bell, Peter M. Triple point and phase relations in the system Al_2SiO_5. M, 1963, Harvard University.

Bell, Peter Mayo. Gibbsite deposition at Rikanau Hill, Suriname. M, 1959, University of Cincinnati. 48 p.

Bell, R. T. Photoelastic studies of geologic structure. M, 1962, University of Toronto.

Bell, Raymond T. Ground water reservoir response to Earth tides. M, 1970, Virginia Polytechnic Institute and State University.

Bell, Richard Arthur. Comparison of terrestrial rock glaciers and other flow types with similar appearing lunar features. M, 1972, South Dakota School of Mines & Technology.

Bell, Richard C. Geology of central Sierra Pena Blanca. M, 1981, University of Texas at El Paso.

Bell, Richard Thomas. Precambrian rocks of the Tuchodi Lakes map area, northeastern British Columbia, Canada. D, 1966, Princeton University. 235 p.

Bell, Robert A. Environmental significance of submicroscopic imperfections in quartz. D, 1953, University of Wisconsin-Madison.

Bell, Robert E. Geology and ore deposits of Shadow Mountains, San Bernardino County, California (Precambrian, Paleozoic, Tertiary). M, 1971, University of San Diego.

Bell, Robert E. Geology and stratigraphy of the Fort Peck fossil field, northwest McCone County, Montana. M, 1965, University of Minnesota, Minneapolis. 97 p.

Bell, Robert Joe. Pre-Pennsylvanian subsurface geology of the East Lindsay area, Garvin County, Oklahoma. M, 1959, University of Oklahoma. 46 p.

Bell, Stephen Craig. Sedimentary history and early diagenesis of Holocene reef limestone on Rota (Mariana Islands). M, 1988, University of Maryland.

Bell, Steven R. Post-Camerina structure and stratigraphy of Kaplan Field, Vermilion Parish, Louisiana. M, 1981, University of Southwestern Louisiana. 48 p.

Bell, Stuart A. Sediments of Nhatrang Bay, South Vietnam. M, 1966, University of Southern California.

Bell, T. C. Ground-water quality of the Abilene area, Kansas. M, 1974, Kansas State University. 87 p.

Bell, Thomas Edward. Deposition and diagenesis of the Brushy Basin and upper Westwater Canyon members of the Morrison Formation in Northwest New Mexico and its relationship to uranium mineralization. D, 1983, University of California, Berkeley. 102 p.

Bell, Thomas Howard. Stratigraphic problems and sulfide mineralization in a section of the Granite Mountain Quadrangle, Pershing County, Nevada. M, 1983, University of Western Ontario. 144 p.

Bell, Trevor J. Quaternary geomorphology, glacial history and relative sea level change in outer Nachvak Fiord, northern Labrador. M, 1987, Memorial University of Newfoundland. 267 p.

Bell, Vernon Lynn. Petrography and paleoecology of the Otter Creek Coral Bed, Upper Ordovician, east-central Kentucky. M, 1978, Eastern Kentucky University. 75 p.

Bell, W. G. Geology of the southeastern flank of the Wind River Mountains, Fremont County, Wyoming. D, 1955, University of Wyoming. 204 p.

Bell, Walter A. Stratigraphy of the Horton-Windsor District, Nova Scotia. D, 1920, Yale University.

Bell, Walton. Surface geology of the Muskogee area, Muskogee County, Oklahoma. M, 1959, University of Oklahoma. 113 p.

Bell, William Charles. Montana Middle Cambrian Brachiopoda. M, 1936, University of Montana. 105 p.

Bell, William Charles. Revision of Cambrian Brachiopoda from Montana. D, 1939, University of Michigan. 42 p.

Belland, Rene J. The disjunct bryophyte element of the Gulf of St. Lawrence region; glacial and postglacial dispersal and migrational histories. D, 1984, Memorial University of Newfoundland. 269 p.

Bellatti, John T. A field investigation comparing conventional compressional-wave, converted-wave, and horizontally-polarized shear-wave reflections. M, 1981, Colorado School of Mines. 161 p.

Belle, Eddie R. A geochemical study of the organic matter within the Lower Cretaceous Mesilla Valley Shale, Cerro de Cristo Rey Uplift, Dona Ana County, New Mexico. M, 1987, University of Texas at El Paso.

Bellehumeur, Claude. Lithogéochimie des calcaires supérieurs de Gaspé M, 1988, Ecole Polytechnique. 186 p.

Bellemin, George Jean. A petrologic study of the Whittier conglomerates (Southern California). M, 1938, Pomona College.

Bellemore, Barbara A. Gulf of Alaska; an example of cold-water carbonate deposition. M, 1981, Rensselaer Polytechnic Institute. 71 p.

Bellerjeau, Orwyn Tilton. Geology of Pinetop and Camels Hump, Bethlehem, Pennsylvania. M, 1952, Lehigh University. 30 p.

Belling, A. J. Postglacial migration of Chamaecyparis thyoides (L.) B.S.P. (southern white cedar) in the northeastern United States. D, 1977, New York University. 220 p.

Bellis, Brian James. Inorganic analysis of polluted groundwater in the Piedmont of North Carolina; a case study. M, 1985, North Carolina State University. 92 p.

Bellis, Caroline Johnson. Shoreline changes at Kill Devil Hills and Kitty Hawk, North Carolina. M, 1985, North Carolina State University. 110 p.

Bellis, William Henry. Distribution and thickness of pre-Desmoinesian rock units in south-central Oklahoma. M, 1961, University of Oklahoma. 39 p.

Belliston, William Hilton. Materials for teaching geological concepts in junior high school science. M, 1964, University of Utah. 181 p.

Bellizzia, Alirio Antonio. Sedimentary study of the limestones and shales of the Simpson Group in the Anderson-Prichard No. 1 Chipman, Murray County, Oklahoma. M, 1950, University of Oklahoma. 61 p.

Bellizzia, Cecilia Martin. Sedimentary study of the sandstones of the Simpson Group in the Anderson-Prichard No. 1 Chipman, Murray County, Oklahoma. M, 1950, University of Oklahoma. 54 p.

Bello, Anthony Eugene Dal see Dal Bello, Anthony Eugene

Bello, Donald M. Pillow lavas and other volcanic structures of Jurassic age; upper flow unit of the Orange Mountain Basalt, Newark Basin. M, 1982, Rutgers, The State University, Newark. 154 p.

Bellotti, M. J. Patterns of eutrophication for the Fulton Chain of Lakes, Herkimer and Hamilton counties, New York. M, 1976, Syracuse University.

Belmont, Ronald A. A regional depositional-systems analysis of the south shore and southwestern Narragansett Bay, Rhode Island. M, 1978, University of Rhode Island.

Belnap, Dennis Wayne. Petrology and geochemistry of shoal water carbonates of the Virgin limestone member, Triassic Moenkopi Formation, Clark County, Nevada. M, 1971, Brigham Young University. 184 p.

Belsky, Theodore. Chemical evolution and organic geochemistry. D, 1966, University of California, Berkeley. 131 p.

Belt, Charles B., Jr. A petrographic and alteration study of the Hanover-Fierro intrusive, New Mexico. M, 1955, Columbia University, Teachers College.

Belt, Charles Banks, Jr. Intrusions and ore depositions in three New Mexico mining districts. D, 1959, Columbia University, Teachers College. 197 p.

Belt, Edward Scudder. Stratigraphy and sedimentology of the Mabou group (middle Carboniferous), Nova Scotia, Canada. D, 1963, Yale University. 486 p.

Beltagy, Ali Ibrahim. The geochemistry of some Recent marine sediments from the Gulf of Saint Lawrence; a study of the less than 63 fraction. D, 1974, University of British Columbia.

Beltrame, Robert J. Petrography and chemistry of metasedimentary rocks at the base of the Stillwater Complex, Montana. M, 1972, University of Cincinnati. 166 p.

Belvedere, Paul Gerard. Depositional environment and diagenesis of the upper Wilcox Sandstone (Eocene), South Harmony Church Field, Allen Parish, Louisiana. M, 1988, University of Tulsa. 205 p.

Belvin, William Mark. Sedimentary environments of the basal Chickamauga Group in a portion of Raccoon Valley, Anderson County, Tennessee. M, 1975, University of Tennessee, Knoxville. 124 p.

Belyea, Helen R. The geology of the Musquash area, New Brunswick. D, 1939, Northwestern University.

Belyea, Richard R. Stratigraphy and depositional environments of the Sespe Formation, northern Peninsular Ranges, California. M, 1984, San Diego State University. 206 p.

Bemben, Stanley Michael. The influence of controlled strain restraints on the strength and behavior during shear of a sand tested with a constant volume. D, 1966, Cornell University. 330 p.

Bembia, Paul J. Bioadvective sediment mixing and beryllium-7 diagenesis in intertidal sediments of Lowes Cove, Maine. M, 1985, SUNY at Binghamton. 60 p.

Bement, Kenneth Arthur. Pacific Plate-North American Plate relative motion, 70-0 myBP. M, 1982, San Diego State University. 110 p.

BeMent, W. Owen. Sedimentological aspects of middle Carboniferous sandstones on the Cumberland overthrust sheet. D, 1976, University of Cincinnati. 182 p.

Bement, W. Owen. The sedimentologic and paleogeographic history of the lower Kicking Horse River valley, Southeast British Columbia, during deglaciation. M, 1972, University of Illinois, Chicago.

Ben Omran, Abdelmoneim. A geological study of the oil fields in Rawlins County, Kansas. M, 1972, Wichita State University. 56 p.

Ben-Avraham, Zvi. Structural framework of the Sunda Shelf and vicinity. D, 1973, Massachusetts Institute of Technology. 269 p.

Ben-Menahem, Ari. Radiation of seismic surface-waves from finite moving sources. D, 1961, California Institute of Technology. 124 p.

Ben-Miloud, K. Geophysical mapping of the subsurface at a landfill site near North Bay, Ontario. M, 1986, University of Waterloo. 69 p.

Ben-Saleh, Faraj F. Stratigraphic analysis of Pennsylvanian rocks in southeastern Anadarko Basin, Oklahoma. M, 1970, University of Tulsa. 76 p.

Benaissa, Saddok. Geologic factors influencing oil and gas accumulation, southeast flank, Rock Springs Uplift, Sweetwater County, Wyoming. M, 1977, University of Wyoming. 81 p.

Benak, Joseph Vincent. Engineering properties of the late Pleistocene loess in the Omaha-Council Bluffs area. D, 1967, University of Illinois, Urbana. 333 p.

Benamy, Elana. Physical alteration of the structure of the hard-shelled clam, Mercenaria. M, 1984, University of Delaware. 226 p.

Benante, Joanne M. Studies of natural and artificial radionuclides in Recent Lake Michigan sediments. M, 1984, University of Wisconsin-Milwaukee. 108 p.

Benavides Alfaro, Jorge Daniel. Wall-rock alteration and mineralogical zoning in a section of the Julcani mining district, Peru. M, 1983, Stanford University. 199 p.

Benavides, Victor. Notes concerning the improvement of reflection seismograms. M, 1962, Stanford University.

Benavides-Caceres, Victor E. Cretaceous System in northern Peru. D, 1956, Columbia University, Teachers College.

Benavidez, Alberto. The constitutive equations for an electrochemically polarizable medium. M, 1982, Texas A&M University. 102 p.

Benavidez, Alberto. The pattern recognition method applied to the forecast of strong earthquakes in South American seismic phone areas. D, 1986, Texas A&M University. 217 p.

Bence, Alfred E. Differentiation history of the Earth by rubidium-strontium isotopic relationship. D, 1966, Massachusetts Institute of Technology. 254 p.

Bence, Alfred Edward. Geothermometric study of quartz deposits in the Ouachita Mountains, Arkansas. M, 1964, University of Texas, Austin. 68 p.

Bench, Bernard M. Reservoir technology. M, 1946, Colorado School of Mines. 120 p.

Benda, Lee E. Debris flows in the Tyee sandstones of the Oregon Coast Range. M, 1988, University of Washington. 134 p.

Benda, M. N. Trace element distribution in wallrocks from the Jefferson City Mine, Tennessee. M, 1977, Queens College (CUNY). 108 p.

Benda, William K. The distribution of foraminifera and Ostracoda off part of the Gulf Coast of Peninsular Florida in the vicinity of Cape Romano. M, 1962, Florida State University.

Bender, Gary. The distribution of snow accumulation on the Greenland ice sheet. M, 1984, University of Alaska, Fairbanks. 110 p.

Bender, Gretchen L. Silurian through Devonian biostratigraphy and depositional environments, Inyo Mountains, California. M, 1978, San Diego State University.

Bender, Hallock John. Structure and stratigraphy of hidden anticline, Fremont County, Wyoming. M, 1944, University of Missouri, Columbia.

Bender, Joel R. Metal releases during the incineration of municipal refuse. D, 1974, Drexel University. 156 p.

Bender, John F. Regional metamorphism of pelites in southeastern Pennsylvania. M, 1972, Pennsylvania State University, University Park.

Bender, John Francis. Petrogenesis of the Cortlandt Complex. D, 1980, SUNY at Stony Brook. 321 p.

Bender, Martin S. The carnivores of the New Paris Sinkholes, Bedford County, Pennsylvania. M, 1955, University of Pittsburgh.

Bender, Marvin J. A micro-faunal study of the upper Neva Limestone (Permian) in Nebraska and Kansas. M, 1951, University of Nebraska, Lincoln.

Bender, Michael L. Helium-uranium dating of fossil corals. D, 1971, Columbia University. 149 p.

Bender, Russell B., Jr. Statistical analysis of pebbles from Pleistocene gravel deposits of the Monroe area, Louisiana. M, 1971, Northeast Louisiana University.

Bender, Russell Berryman, Jr. Petrology and geochemistry of the Silver Plume-age plutons of the southern and central Wet Mountains, Colorado. D, 1983, Louisiana State University. 200 p.

Bendig, Daniel J. Finite element solution for the stresses and displacements associated with igneous intrusions. M, 1978, SUNY at Buffalo. 84 p.

Bendimerad, Mohamed Fouad. Modeling of recorded three-dimensional earthquake motion in the frequency-time domain. D, 1985, Stanford University. 183 p.

Bending, David Alexander Glen. A reconaissance study of the stratigraphic and structural setting, timing and geochemistry of mineralization in the Metaline District, northeastern Washington, U.S.A. M, 1983, University of Toronto. 324 p.

Bendixen, Roald L. An analysis of the bottom sediments of the southern basin of Eagle Lake, California. M, 1971, California State University, Chico. 42 p.

Bendula, Richard A. A geotechnical evaluation of existing and potential landfill sites in Portage County, northeastern Ohio. M, 1985, Kent State University, Kent. 193 p.

Benecke, Daniel M. A geothermal gradient analysis of the Paradox Basin, Colorado and Utah. M, 1983, Memphis State University.

Benedetti, Steven Jess. Paleozoic conodonts from the Placer de Guadalupe area, East-central Chihuahua, Mexico. M, 1976, Texas Christian University.

Benedict, Barry Arden. Hydrodynamic lift in sediment transport. D, 1968, University of Florida. 206 p.

Benedict, Dorothy K. and Auerbach, Pauline Dorothy. The geology of Ithan Creek valley and vicinity (Pennsylvania). M, 1939, Bryn Mawr College. 25 p.

Benedict, Ellis Neil. A subsurface study of the pre-Knox unconformity and related rock units in the State of Ohio. M, 1967, Michigan State University. 109 p.

Benedict, Frank Christopher. Geology and trace element geochemistry, east-central Alpine County, California. M, 1984, Colorado School of Mines. 164 p.

Benedict, G. L., III. Lithofacies and depositional environments of the Lebanon Limestone (Ordovician) in central and East-central Tennessee. M, 1974, Vanderbilt University.

Benedict, James B., Jr. Neoglacial history of the Colorado Front Range. D, 1968, University of Wisconsin-Madison. 84 p.

Benedict, Jonathan F. The geology and mineral potential of the Schiestler Peak area, Temple Peak Quadrangle, Wyoming. M, 1982, University of Wyoming. 119 p.

Benedict, Louis G. Maceral and mineral concentrations in chance cone products. M, 1962, Pennsylvania State University, University Park. 119 p.

Benedict, Nathan Blair. The vegetation and ecology of subalpine meadows of the southern Sierra Nevada, California. D, 1981, University of California, Davis. 128 p.

Benedict, Platt C. Geology of Deception Gulch and the Verde Central Mine (Arizona). M, 1923, Massachusetts Institute of Technology. 90 p.

Benedict, Reba Ward. Aplite-pegmatite dikes of Tenaya Canyon, Yosemite; a study of their structures, textures, and mineralogy. D, 1959, University of California, Berkeley. 230 p.

Benekas, Sandy L. Detection of "serpentine plugs" using Landsat (MMS) and magnetic data. M, 1986, Texas Christian University.

Benelmouffok, Djamel E. Two-dimensional numerical modeling of a wet detention pond. D, 1988, University of Virginia. 123 p.

Benenati, Francis E. An assessment of the effects of zinc, lead, cadmium, and arsenic in soil, vegetation, and water resources surrounding a zinc smelter. D, 1974, University of Oklahoma. 152 p.

Benes, Paula S. Relationship between physical condition of the carbonate fraction and sediment environments; northern shelf of Puerto Rico. M, 1988, Duke University. 178 p.

Benfer, Jon Alan. The petrology of the eastern phase of the Toroweap Formation (Permian), Walnut Canyon (Coconino County), Arizona. M, 1971, Northern Arizona University. 85 p.

Bengert, Stephen R. Pediment development in the Sierrita Mountains, Pima County, Arizona. M, 1981, Indiana State University. 46 p.

Bengochea, Jose Ignacio R. Garcia see Garcia Bengochea, Jose Ignacio R.

Bengston, Kermit Bernard. Further studies of olivine chlorination at high temperatures. M, 1955, University of Washington. 72 p.

Bengston, Nels A. Meander of the Missouri River; progress and consequences. M, 1908, University of Nebraska, Lincoln.

Bengtson, Carl Aners. Geology of the Terrace Mountain area, Wyoming. M, 1940, University of Iowa. 54 p.

Bengtson, Kermit Bernard. Further studies of olivine chlorination at high temperatures. M, 1955, University of Washington. 72 p.

Bengtson, Mark Eric. Geophysical field study of the Slate Islands cryptoexplosion site. M, 1984, University of Wisconsin-Madison.

Bengtson, Richard Lee. Predicting storm runoff from small grassland watersheds with the USDAHL hydrologic model. D, 1980, Oklahoma State University. 168 p.

Bengtsson, Terrance. Flow-net analysis of strip-island lenses and calculations of time-related parameters. M, 1987, University of South Florida, Tampa. 130 p.

Benham, John R. The Meade Peak phosphatic shale member of the Permian Phosphoria Formation in the Snake River Range, Bonneville and Teton counties, Idaho, Lincoln and Teton counties, Wyoming. M, 1984, Eastern Washington University. 81 p.

Benham, Julia Anne. Geology and uranium content of middle Tertiary ash-flow tuffs in the southen Nightingale Mountains and northern Truckee Range, Washoe County, Nevada. M, 1982, University of Nevada. 112 p.

Benham, Steven R. Calcareous green algal distribution and sediment composition and texture, Bahia Honda Key, Florida. D, 1979, Indiana University, Bloomington. 283 p.

Benimoff, Alan Irwin. A comparative statistical study of the chemistry of the lunar and terrestrial rocks. M, 1976, Brooklyn College (CUNY).

Benimoff, Alan Irwin. Characterization and stability of phase "A"; a high pressure phase in the system MgO-SiO₂-H₂O. D, 1984, Lehigh University. 97 p.

Benioff, Hugo. (1) A linear strain seismograph; (2) The physical evaluation of seismic destructiveness; (3) A method for the instrumental determination of the extent of faulting. D, 1935, California Institute of Technology. 60 p.

Benites, Lois A. Study of the mechanical properties of soils affected by piping near the Benson area, Cochise County, Arizona. M, 1967, University of Arizona.

Benito, Hugo Oscar. Use of multiple correlation for forecasting streamflow. M, 1970, University of Nevada. 58 p.

Benitt, Theodore G. Evaluation of the high-low phase inversion in quartz, using a new D.T.A. procedural technique. M, 1971, Brooklyn College (CUNY).

Beniwal, R. The effect of grain characteristics on the shear modulus of gravels. D, 1977, University of Kentucky. 192 p.

Benjamin, Michael T. Fission track ages on some Bolivian plutonic rocks; implications for the Tertiary uplift and erosion history of the Altiplano-Cordillera Real. M, 1986, Dartmouth College. 58 p.

Benjamin, Timothy Miller. Experimental actinide element partitioning between whitlockite, apatite, diopsidic clinopyroxene, and anhydrous melt at one atmosphere and 20 kilobars pressure. D, 1980, California Institute of Technology. 243 p.

Benjamins, Janet. An investigation of the Squantum Formation (Devonian or Carboniferous). M, 1968, Boston University. 56 p.

Benke, Mary Lee. Integrated subsurface geological mapping in the presence of a velocity gradient in the North Lansing Field, Harrison and Gregg counties, Texas. M, 1957, University of Houston.

Benmore, William C. Stratigraphy and paleoecology of the lower Johnnie Formation, southern Nopah Range, eastern California. M, 1974, University of California, Santa Barbara.

Benmore, William C. Stratigraphy, sedimentology, and paleoecology of the late Paleophytic or earliest Phanerozoic Johnnie Formation, eastern California and southwestern Nevada. D, 1978, University of California, Santa Barbara. 263 p.

Benn, Keith. Petrology of the Troodos plutonic complex in the Caledonian Falls area, Cyprus. M, 1986, Universite Laval. 226 p.

Benne, Robert R. The stratigraphy of the lower Gobbler Formation, Sacramento Mountains, New Mexico. M, 1975, University of Oklahoma. 141 p.

Benner, Richard Walter. Geology of the Lima Peak's area of the Tendoy Mountains, Beaverhead County, Montana and Clark County, Idaho. M, 1948, University of Michigan. 55 p.

Benner, Velma. Webster Groves, Missouri; residential satellite of Saint Louis. M, 1950, Washington University. 116 p.

Bennet, Robert Edwin, Jr. Fluid inclusion study of sphalerite from the northern Arkansas zinc-lead district. M, 1974, University of Michigan.

Bennett, B. G. Environmental aspects of americium. D, 1979, New York University. 216 p.

Bennett, Billie. The principal igneous rock forming minerals and their metamorphic products. M, 1937, Colorado College.

Bennett, Brian R. A long-period magnetotelluric study in California. M, 1985, Massachusetts Institute of Technology. 155 p.

Bennett, Bruce A. Structural traps in the Lower and Middle Devonian sediments of Erie County, Pennsylvania; a general survey and tectonic synthesis. M, 1987, SUNY, College at Fredonia. 58 p.

Bennett, Catheryn MacDonald. Radius effect of the alkaline earths on the rate of inversion of aragonite to calcite. M, 1972, University of Arizona.

Bennett, Curtis Owen. Analysis of fractured wells. D, 1982, University of Tulsa. 335 p.

Bennett, Debra Kim. Cenozoic rocks and faunas of north-central Kansas; with an appendix concerning taxonomy and evolution of the genus "Equus". D, 1984, University of Kansas. 242 p.

Bennett, Debra Kim. Paleontology, paleoecology, sedimentology, and biostratigraphy of the Rhinoceras Hill fauna (Hemphillian; latest Miocene), Wallace County, Kansas. M, 1977, University of Kansas. 209 p.

Bennett, Earl Healen, II. Sedimentation patterns on the Outer Banks of North Carolina between Nags Head and Ocracoke (North Carolina). M, 1970, North Carolina State University. 108 p.

Bennett, Earl Healen, II. The petrology and trace element distribution of part of the Idaho Batholith compared to the White Cloud Stock, in Custer County, Idaho. D, 1973, University of Idaho. 172 p.

Bennett, Ethel Evans. Stratigraphic and faunal studies of the Grayson Formation in North Texas. M, 1939, Texas Christian University. 129 p.

Bennett, G. T. A seismic refraction survey along the southern Rocky Mountain Trench. M, 1973, University of British Columbia.

Bennett, G. V. Late Wisconsinan geology of the Annsville Creek and Sprout Brook valleys, northwestern Westchester County, New York. M, 1978, Queens College (CUNY). 99 p.

Bennett, George H. Sedimentology of the Antrim Shale from five drill sites in the Michigan Basin. M, 1978, Michigan Technological University. 75 p.

Bennett, Gerald. The petrology of the Stewarton Igneous Complex, Queens County, New Brunswick. M, 1965, University of New Brunswick.

Bennett, Gordon D. Determination of specific capacities in a multiaquifer well. M, 1961, Pennsylvania State University, University Park. 104 p.

Bennett, Hugh Frederick. An investigation into velocity anisotropy through measurements of ultrasonic wave velocities in snow and ice cores from Greenland and Antarctica. D, 1968, University of Wisconsin-Madison. 68 p.

Bennett, John N., Jr. Paleocurrent analysis of the upper Miocene formations, Los Angeles Basin, California. M, 1967, University of Arizona.

Bennett, Jon Lewis. The aerobic biodegradation of quinoline and analogs in soils from a creosote-contaminated site in Pensacola, Florida. M, 1988, Colorado School of Mines. 127 p.

Bennett, Joseph E. The determination of the astro-geodetic deflections of the vertical. M, 1962, Ohio State University.

Bennett, Joseph Thomas. The biogeochemical significance of zooplankton fecal material in a biologically productive temperate fjord. D, 1980, University of Washington. 271 p.

Bennett, K. C. Geology and origin of the breccias in the Morenci-Metcalf District, Greenlee County, Arizona. M, 1975, University of Arizona.

Bennett, Kathleen C. A detailed magnetic survey of Alachua County, Florida. M, 1978, University of Florida. 75 p.

Bennett, Lee C., Jr. In situ measurements of acoustic absorption in unconsolidated marine sediments. D, 1966, Bryn Mawr College. 107 p.

Bennett, Marvin Edward, III. Geology and petrography of pre-Mesozoic and Mesozoic rocks of the northern Rio Nima area, Departamento Del Valle, Colombia, S.A. M, 1986, Stephen F. Austin State University. 80 p.

Bennett, Michael J. Depositional environments and geotechnical properties of Quaternary sediment from South San Francisco Bay, San Mateo County, California. M, 1979, San Jose State University. 133 p.

Bennett, Nathan Paul. The mineralogy and physical and chemical properties of the Porters Creek Clay. M, 1976, Indiana University, Bloomington. 70 p.

Bennett, Paul J. The geology and mineralization of the Sedimentary Hills area, Pima County, Arizona. M, 1957, University of Arizona.

Bennett, Paul Joseph. The economic geology of some Virginia kyanite deposits. D, 1961, University of Arizona. 163 p.

Bennett, R. H. Clay fabric and geotechnical properties of selected submarine sediment cores from the Mississippi Delta. D, 1976, Texas A&M University. 286 p.

Bennett, Reb. E. Geology of the Dexter Canyon area, Santa Clara County, California. M, 1972, San Jose State University. 67 p.

Bennett, Richard Edwin. Geology of East Bourland and Simpson Springs mountains, Brewster County, Texas. M, 1959, University of Texas, Austin.

Bennett, Robert. A petrographic study of a Pitkin reef complex (Mississippian) located near Wesley, Arkansas. M, 1965, University of Arkansas, Fayetteville.

Bennett, Robert B. Geology and its relation to the occurrence of ground-water in the Big Springs area, Texas. M, 1939, University of Nebraska, Lincoln.

Bennett, Robert Turner. Geology of the northern portion of Manitou Park, Colorado. M, 1940, University of Iowa. 82 p.

Bennett, Sara L. Where three oceans meet; the Agulhas retroflection region. D, 1988, Massachusetts Institute of Technology. 367 p.

Bennett, Sean Joseph. Temporal variations in channel morphology and hydrology of the eastern Susquehanna River in New York State; causes and implications. M, 1987, SUNY at Binghamton. 221 p.

Bennett, Theodore W. X-ray diffraction in minerals. M, 1931, University of Minnesota, Minneapolis. 34 p.

Bennett, Theron Joseph. Determination of source characteristics of underground nuclear explosions from analysis of teleseismic body waves. D, 1972, St. Louis University.

Bennett, Timothy John. Geochemistry and genesis of calc-alkaline volcanics from the islands of St. Lucia and St. Vincent, Lesser Antilles. M, 1978, Wright State University.

Bennett, Truman W. Shape and optic properties of quartz grains of mature sandstone. M, 1962, Ohio State University.

Bennett, W. R. Determination of geologic time by radioactive disintegration; volumetric determination of uranium in radioactive minerals. D, 1933, Purdue University.

Bennett, William Alfred Glenn. A petrographic study of greenstones near Blue Creek, Stevens County, Washington. M, 1928, Washington State University. 30 p.

Bennett, William Alfred Glenn. Stratigraphic and structural studies in the Colville Quadrangle, Washington. D, 1937, University of Chicago. 62 p.

Bennetts, Kimberly Robert Winter. Characteristics of three individual turbidites from the Hispaniola-Caicos Basin. M, 1974, Duke University. 131 p.

Benninger, L. K. The uranium-series radionuclides as tracers of geochemical processes in Long Island Sound. D, 1976, Yale University. 161 p.

Bennington, Kenneth Oliver. Role of shearing stress and pressure in differentiation as illustrated by some mineral reactions in the system MgO-SiO2-H2O. D, 1960, University of Chicago. 19 p.

Bennington, Kenneth Oliver. The paragenesis of the Chewelah District metalliferous deposits. M, 1951, Washington State University. 36 p.

Bennion, Douglas Wilford. A stochastic model for predicting variations in reservoir rock properties. D, 1965, Pennsylvania State University, University Park. 117 p.

Benniran, M. M. Casper Formation limestones, southwestern Laramie Mountains, Albany County, Wyoming. M, 1970, University of Wyoming. 116 p.

Benoit, E. G. Application of quantitative mapping techniques to the geologic evaluation of sand-gravel distribution in the Kitchener-Waterloo-Cambridge area. M, 1975, University of Waterloo.

Benoit, Edward L. The Desmoinesian Series, Edmond area, central Oklahoma. M, 1957, University of Oklahoma. 40 p.

Benoit, Fernand Wilbrod. Geology of the St. Sylvestre and St. Joseph west half areas (Quebec). D, 1958, Universite Laval.

Benoit, Fernand Wilbrod. Investigation into the chemical composition of the upper part of the Cap Bon Ami Formation and the lower part of the Grande Greve Formation. M, 1955, McGill University.

Benoit, Gaboury. The biogeochemistry of ^{210}Pb and ^{210}Po in fresh waters and sediments. D, 1988, Woods Hole Oceanographic Institution. 304 p.

Benoit, Jean. Analysis of self-boring pressuremeter tests in soft clay. D, 1984, Stanford University. 376 p.

Benoit, Jeffrey R. A shoreline erosion study of the Atlantic Intracoastal Waterway of Georgia, classification and methods of erosion control. M, 1978, Georgia Institute of Technology. 80 p.

Benoit, Paul Harland. An experimental and field evaluation of the oxonium alunite-potassium alunite solid-solution series as a potential geothermometer. M, 1987, Lehigh University. 114 p.

Benoit, Walter Richard. Vertical zoning and differentiation in granitic rocks; central Flint Creek Range, Montana. M, 1971, University of Montana. 53 p.

Benomran, Omran. A change of facies of the Sirte Basin, Libya. M, 1978, University of South Carolina.

Bensinger, Herbert Schatzlein. Joint patterns in quarries of southeastern Michigan. M, 1961, University of Michigan.

Bensley, David F. Petrographic and fluorescent properties of liptinite macerals from cutinite-rich "paper coals" from Indiana. M, 1981, Southern Illinois University, Carbondale. 134 p.

Benson, A. Raeburn. The physiography of New York State. M, 1934, Union College. 81 p.

Benson, Anthony Lane. The Devonian system in western Wyoming and surrounding area. D, 1965, Ohio State University. 141 p.

Benson, Carl Sidney. A review of some problems associated with ice formation in geological environments. M, 1955, University of Minnesota, Minneapolis. 149 p.

Benson, Carl Sidney. Stratigraphic studies in the snow and firn of the Greenland ice sheet. D, 1960, California Institute of Technology. 213 p.

Benson, Christopher Joseph. Geology of the Kilgore Prospect area, Clark County, Idaho. M, 1986, Arizona State University. 111 p.

Benson, D. G., Jr. Dinoflagellate biostratigraphy of the Cretaceous-Tertiary boundary, Round Bay, Maryland. M, 1975, Virginia Polytechnic Institute and State University.

Benson, Dale L. Kansas oil fields. M, 1923, University of Kansas. 79 p.

Benson, David G. Genesis and variation of the Hampstead granitic stock (New Brunswick). M, 1953, University of New Brunswick.

Benson, David G. The mineralogy of the New Brunswick sulphide deposits. D, 1959, McGill University.

Benson, Donald Joe. Bottom sediments of western Lake Erie, Ohio. M, 1971, University of Cincinnati. 77 p.

Benson, Donald Joe. Lithofacies and depositional environments of Osagean-Meramecian platform carbonates, southern Indiana, central and eastern Kentucky. D, 1976, University of Cincinnati. 223 p.

Benson, Gilbert Thomas. Structural geology of the Stonewall area, Colorado. D, 1963, Yale University.

Benson, Gregory Scott. Structural investigations of the Italian Trap Allochthon, Redington Pass, Pima County, Arizona. M, 1981, University of Arizona. 245 p.

Benson, James C. A petrographic study of the Mississippian Heath Formation, Sumatra oil field, central Montana. M, 1956, University of Wisconsin-Madison. 56 p.

Benson, Larry V. Electron microprobe studies of carbonates. D, 1974, Brown University. 184 p.

Benson, Lawrence I. The geology of the Texas Valley area, Knox and Union counties, Tennessee. M, 1963, University of Tennessee, Knoxville. 26 p.

Benson, Paul Harrison, III. A comparison of the clay mineralogy of marsh and adjacent non-marsh environments along the North Carolina coast. M, 1965, University of North Carolina, Chapel Hill. 31 p.

Benson, Paul Harrison, III. The depositional environment of the Upper Cretaceous Black Creek Formation in North and South Carolina. D, 1968, University of North Carolina, Chapel Hill. 149 p.

Benson, Richard Hall. Ostracoda from the type section of the Fern Glen Formation. M, 1953, University of Illinois, Urbana.

Benson, Richard Hall. The ecology of the Recent ostracods of the Todos Santos Bay region, Baja, California, Mexico. D, 1953, University of Illinois, Urbana.

Benson, Richard Norman. Recent Radiolaria from the Gulf of California. D, 1966, University of Minnesota, Minneapolis. 622 p.

Benson, Robert G. Hydrothermal alteration and geothermometry in the area of the Lucky Boy Mine, Yankee Fork mining district, Custer County, Idaho. M, 1985, University of Idaho. 89 p.

Benson, Sally Merrick. Characterization of the hydrologic and transport properties of the shallow aquifer under Kesterson Reservoir, Merced County, California. D, 1988, University of California, Berkeley. 341 p.

Benson, William E. B. Geology of the Knife River basin, North Dakota. D, 1952, Yale University.

Bent, James VanEtten. Petrographic reconnaissance of upper Oligocene-middle Miocene sandstones of the San Joaquin Basin, California. M, 1986, Stanford University.

Benthack, Louis. High melting products from petroleum asphalts. M, 1933, University of Pittsburgh.

Benthien, Ross Howard. Origin of magnetization in the Phosphoria Formation, Wyoming; a possible relationship with hydrocarbons. M, 1987, University of Oklahoma. 76 p.

Bentkowski, James Edward. The Morrow Formation in eastern Dewey County, Oklahoma. M, 1985, Oklahoma State University. 84 p.

Bentley (Pyzanowski), Barbara. Carbonate lithofacies and diagenetic features of the Guelph Formation (Middle Silurian) in the Amoco production, Berg-Brege 1-21 unit well, Presque Isle County, Michigan. M, 1979, Rensselaer Polytechnic Institute. 147 p.

Bentley, Charles R. Seismic measurements on the Greenland ice cap. D, 1959, Columbia University, Teachers College.

Bentley, Craig B. Upper Cambrian stratigraphy of western Utah. M, 1958, Brigham Young University. 78 p.

Bentley, Eugene Macke, III. The effect of marshes on water quality. D, 1969, University of Wisconsin-Madison. 228 p.

Bentley, L. R. Crustal structure of the Carnegie Ridge, Panama Basin and Cocos Ridge. M, 1974, University of Hawaii. 49 p.

Bentley, Michael Emmons. Hydrogeology of the Beaumont Formation (Pleistocene), Brazoria County, Texas. M, 1980, University of Texas, Austin.

Bentley, R. H. Candidate siting area for nuclear power facilities within the State of Oklahoma. D, 1976, University of Oklahoma. 147 p.

Bentley, Robert Donald. Geologic evolution of the Beartooth Mountains, Montana and Wyoming; Part 9, Cloverleaf Lakes area. D, 1969, Columbia University. 148 p.

Benton, Douglas Chamberlin. Petrology of fine clastic, terrigenous rocks of the Dunkard Group (Pennsylvanian-Permian) within the central portion of the Dunkard Basin, western West Virginia and southeastern Ohio. M, 1983, Miami University (Ohio). 139 p.

Benton, Edward Raymond. The Richmond boulder trains. D, 1878, Harvard University.

Benton, John William. Subsurface stratigraphic analysis, Morrow (Pennsylvanian), north central Texas County, Oklahoma. M, 1971, University of Oklahoma. 60 p.

Benton, Lynda M. Ammonium geochemistry of sedimentary exhalative Pb-Zn-Ag deposits; a possible exploration tool. M, 1984, Dartmouth College. 114 p.

Benton, Stephen B. Holocene evolution of a nanotidal brackish marsh-protected bay system, Roanoke Island, North Carolina. M, 1980, East Carolina University. 175 p.

Bentson, Herdis. A monographic study of the fossil gastropod Exilia. M, 1937, University of California, Berkeley. 155 p.

Bentson, Herdis. The stratigraphy and faunas of the Capay Eocene of the Sacramento Valley. D, 1941, University of California, Berkeley.

Bentz, Mark G. Progressive structural and stratigraphic events affecting the Roberts Mountains Allochthon in the Devil's Gate area, Nevada. M, 1984, Ohio University, Athens. 239 p.

Bentzel, Ruby H. Stratigraphic study of the fossil flora of the Dunkard (Washington and Greene) strata. M, 1952, West Virginia University.

Bentzin, David Allan. Geology of the Weasel Creek area, northern Whitefish Range, Flathead and Lincoln counties, Montana. M, 1960, University of Montana. 34 p.

Benvegnu, Carl Jerome. Stratigraphy and structure of the Croydon-Henefer Grass Valley area, Morgan and Summit counties, Utah. M, 1963, University of Utah. 33 p.

Benvenuto, Gary Louis. Structural evolution of the Hosmer Thrust Sheet, southeastern British Columbia. D, 1978, Queen's University. 184 p.

Benyamin, Ninos B. Digital restoration of strain steps from strain-meter transients. M, 1968, Colorado School of Mines. 34 p.

Benz, Harley Mitchell. Kinematic source modeling of elastic waves using the finite element method. D, 1986, University of Utah. 134 p.

Benz, Harley Mitchell. Simultaneous inversion for lateral velocity variations and hypocenters in the Yellowstone region using earthquake and refraction data. M, 1982, University of Utah. 105 p.

Benz, Robert. Origin and occurrence of chert and other forms of silica in the limestones near Syracuse. M, 1952, Syracuse University.

Benz, Sandra. The stratigraphy and paleoenvironment of the Triassic Moenkopi Formation at Radar Mesa, Arizona. M, 1980, Northern Arizona University. 45 p.

Benzel, William Marc. Cation exchange properties of montmorillonite at temperatures simulating subsurface environments. M, 1978, University of Illinois, Urbana. 45 p.

Benzel, William Marc. The efect of smectite layer thickness and fabric upon salt filtration at room temperature. D, 1982, University of Illinois, Urbana. 114 p.

Benzing, William Martin, III. Experimental stress wave propagation in media containing liquid inclusions. D, 1974, University of Chicago. 163 p.

Beraithen, Mohammed I. A quantitative study of the Minjur Aquifer (Saudi Arabia). M, 1982, Ohio University, Athens. 507 p.

Bérard, Jean. Géologie de la région du Lac aux Feuilles, Nouveau-Québec. D, 1959, Universite Laval. 368 p.

Berard, Jean. Geology of the Leaf Bay area, New Quebec. M, 1959, Universite Laval.

Beras, Manuel E. Determining the optimum sinking path in open pits. D, 1988, West Virginia University. 147 p.

Berberian, George Assadour. Infrared absorption as a guide to the crystallographic analysis of certain inorganic salts, particularly those of uranium and vanadium. M, 1967, American University. 48 p.

Berc, Jeri Lynne. A pedologically based agro-economic soil rating system; a computer assisted approach to soil resource appraisal. D, 1988, University of California, Berkeley. 256 p.

Bercaw, Louise B. Geology and gold deposits of central Mineral Ridge, Esmeralda County, Nevada. M, 1986, University of Colorado. 156 p.

Berchenbriter, Dean Kenneth. The geology of La Caridad Fault, Sonora, Mexico. M, 1976, University of Iowa. 127 p.

Bercutt, Henry. Isopachous and paleogeologic studies in eastern Oklahoma north of the Choctaw Fault. M, 1958, University of Oklahoma. 65 p.

Berdan, Jean M. Brachiopoda and Ostracoda of the Manlius Group of New York State. D, 1949, Yale University.

Berdanier, Charles Reese, Jr. Genesis of some calimorphic soils in the New Jersey coastal plain. D, 1967, Rutgers, The State University, New Brunswick. 92 p.

Berelson, William M. Barrier island evolution and its effect on lagoonal sedimentation; Shackleford Banks, Back Sound, and Harkers Island; Cape Lookout National Seashore. M, 1979, Duke University. 227 p.

Berelson, William Max. Studies of water column mixing and benthic exchange of nutrients, carbon and radon in the Southern California Borderland. D, 1985, University of Southern California.

Berendsen, Pieter. The solubility of calcite in CO_2-H_2O solutions, from 100° to 300°C, 100 to 1000 bars, and 0 to 10 weight per cent CO_2, and geologic applications. D, 1971, University of California, Riverside. 228 p.

Berent, Louis J. An investigation of the relationships between the thermoluminescence of quartz and its temperature of formation. M, 1974, Brooklyn College (CUNY).

Bereskin, S. Robert. Miocene biostratigraphy of southwestern Santa Cruz Island, California. M, 1966, University of California, Santa Barbara.

Bereskin, Stanley Robert. Carbonate petrology and biostratigraphy of the Sultan Limestone (Devonian), southeastern California and southern Nevada. D, 1969, University of California, Santa Barbara.

Berg van Saparoea, C. M. G. Van den see Van den Berg van Saparoea, C. M. G.

Berg, Alexander Nicolaas Van Den see Van Den Berg, Alexander Nicolaas

Berg, Arthur Brede. The geology of the northwestern corner of the Tobacco Root Mountains, Madison County, Montana. M, 1959, University of Minnesota, Minneapolis. 75 p.

Berg, Clifden A. Mississippian stratigraphy of the Kindersley area, Saskatchewan. M, 1953, University of Saskatchewan. 30 p.

Berg, D. A. The Minnelusa Formation of the eastern Black Hills area. M, 1951, South Dakota School of Mines & Technology.

Berg, Edgar Lowndes. Geology of the Sierra de Samalayuca, Chihuahua, Mexico. M, 1971, University of Texas, Austin.

Berg, Eric L. Evaluation of a method of crustal exploration based on converted waves from microearthquakes. M, 1968, New Mexico Institute of Mining and Technology. 83 p.

Berg, Ernest Lyle. A study of the basic border facies of intrusive igneous rocks. M, 1935, University of Minnesota, Minneapolis.

Berg, Gilman A. The solubility of silica from silicates at elevated temperatures. M, 1932, University of Minnesota, Minneapolis. 10 p.

Berg, Jacob Van Den see Van Den Berg, Jacob

Berg, James A. Petrography and bromine geochemistry of the Hutchinson Salt Member of the Wellington Formation in Ellsworth County, Kansas; a paleoenvironmental analysis. M, 1981, University of Kansas. 74 p.

Berg, James Donald. A mathematical simulation model of ecosystem response to natural stresses. D, 1983, University of Tennessee, Knoxville. 228 p.

Berg, John Robert. Petrography of the Tertiary igneous rocks, Nigger Hill District, Wyoming-South Dakota. M, 1940, University of Iowa. 66 p.

Berg, John Robert. Pre-Cambrian geology of the Galena-Roubaix District, Black Hills, South Dakota. D, 1942, University of Iowa. 79 p.

Berg, John Stoddard. Direct shear testing of marine sediment. M, 1971, United States Naval Academy.

Berg, Jonathan Henry. Mineralogy and petrology of the contact aureoles of the anorthositic Nain Complex, Labrador. D, 1976, University of Massachusetts. 143 p.

Berg, Jonathan Henry. The petrology of the outer and inner border of the (Precambrian) Kiglapait layered intrusion (Labrador). M, 1971, Franklin and Marshall College. 133 p.

Berg, Joseph Wilbur, Jr. Conductivity study of aqueous kaolin-NaCl mixtures. M, 1952, Pennsylvania State University, University Park. 35 p.

Berg, Joseph Wilbur, Jr. Effect of stemming on the energy-content of explosion-generated seismic pulses. D, 1954, Pennsylvania State University, University Park. 105 p.

Berg, Orville Roger. Quantitative study of the Cherokee-Marmaton groups (Pennsylvanian) west flank of the Nemaha Ridge, north-central Oklahoma. D, 1968, University of Oklahoma. 73 p.

Berg, Orville Roger. The depositional environment of a portion of the Bluejacket Sandstone. M, 1963, University of Tulsa. 87 p.

Berg, Richard Blake. Petrology of anorthosite bodies, Bitterroot Range, Ravalli County, Montana. D, 1965, University of Montana. 158 p.

Berg, Richard M., Jr. Processing and interpretation of a seismic line across the resurgent dome of Long Valley Caldera, California. M, 1988, University of Wyoming. 115 p.

Berg, Robert R. The Franconia Formation of Minnesota and Wisconsin. D, 1951, University of Minnesota, Minneapolis.

Berg, S. A. Theoretical determination of cleavage direction. M, 1974, SUNY at Buffalo. 86 p.

Berg, Thomas Miles. Pennsylvanian biohermal limestones of Marble Mountain (Saguache and Custer counties), south-central Colorado. M, 1967, University of Colorado.

Berg, W. W., Jr. Chlorine chemistry in the marine atmosphere. D, 1976, Florida State University. 263 p.

Berg, William R. Subsidence due to underground mining in the Hanna coal field. M, 1980, University of Wyoming. 106 p.

Bergado, Dennis Taganajan. Probabilistic assessment of the safety of earth slopes including pore-pressure uncertainty. D, 1982, Utah State University. 227 p.

Bergan, Gail Renae. Shoreline depositional environments of the Glen Rose Formation (Lower Cretaceous) in the type area, Somervell and Hood counties, Texas. M, 1987, University of Texas, Arlington. 139 p.

Bergantz, George. Double-diffusive boundary layer convection in a porous medium; implications for fractionation in magma chambers. M, 1985, Georgia Institute of Technology. 91 p.

Bergantz, George Walter. Convection and solidification in tall magma chambers. D, 1988, The Johns Hopkins University. 183 p.

Berge, Charles William. Heavy minerals study of the intrusive bodies of the central Wasatch Range, Utah. M, 1960, Brigham Young University. 31 p.

Berge, Charles William. Sedimentation of Arklow Bank, Irish Sea. D, 1972, University of Wisconsin-Madison.

Berge, Johannes Christian Van den see Van den Berge, Johannes Christian

Berge, John Stuart. Stratigraphy of the Ferguson Mountain area, Elko County, Nevada. M, 1960, Brigham Young University. 63 p.

Berge, Olaf T. Petrography as applied to determining coarse aggregate rock suitable for concrete. M, 1932, University of Minnesota, Minneapolis. 31 p.

Berge, Timothy Bryan Swearingen. Structural evolution of the Malone Mountains, Hudspeth County, Texas. M, 1981, University of Texas, Austin. 95 p.

Berge, William Victor. Some magnetotelluric modeling techniques. M, 1968, University of Wisconsin-Madison.

Bergen, Christopher L. Petrology and depositional environment of the Cove Creek and Girkin limestones (Mississippian) in Washington County, Virginia. M, 1985, East Carolina University. 71 p.

Bergen, Donald Von see Von Bergen, Donald

Bergen, Frederick Winfield. A restudy of the upper Mohnian-lower Delmontian boundary near Calabasas, California. M, 1955, University of California, Los Angeles.

Bergen, James A. Jurassic nannofossils from Portugal. D, 1987, Florida State University. 583 p.

Bergen, James Alan. Calcareous nannoplankton from Deep Sea Drilling Project Leg 78A; evidence for imbricate underthrusting at the Lesser Antillian active margin. M, 1982, Florida State University.

Bergenback, Richard E. A petrographic study of the Bald Eagle, Juniata, and Tuscarora formations in central and eastern Pennsylvania. M, 1950, Lehigh University.

Bergenback, Richard Edward. The geochemistry and petrology of the Vanport Limestone, western Pennsylvania. D, 1964, Pennsylvania State University, University Park. 197 p.

Bergeon, Thomas C. Seismic character study, Ismay Cycle, Paradox Formation, Paradox Basin, Southwest Colorado. M, 1986, Colorado School of Mines. 179 p.

Berger, A. R. Recent volcanic ash deposit, Yukon Territory. M, 1958, Dalhousie University.

Berger, Ben R. Stratigraphy of the western Lake St. Joseph greenstone terrain, northwestern Ontario. M, 1981, Lakehead University. 152 p.

Berger, Byron Roland. Petrogenesis of the Green Acres Gabbro, Riverside County, California. M, 1975, University of California, Los Angeles.

Berger, Deborah J. The Blancan equid, Equus shoshonensis, from Hagerman, Idaho. M, 1987, Bowling Green State University. 85 p.

Berger, Ernst. Seismic response of axisymmetric soil-structure systems. D, 1976, University of California, Berkeley. 190 p.

Berger, Glenn W. $^{40}Ar/^{39}Ar$ step heating of biotite, hornblende and potassium feldspar from a zone of contact metamorphism, Eldora, Colorado. D, 1973, University of Toronto.

Berger, Jonathan. Investigation of earth strain using a laser strain meter. D, 1970, University of California, San Diego. 147 p.

Berger, Jonathan Joseph. Explosion crustal studies on the continental shelf, the Sable island area (Nova Scotia). M, 1964, Dalhousie University.

Berger, Kent H. Beach and chenier sediments of western Cameron Parish, Louisiana; a textural and mineralogical analysis. M, 1981, Stephen F. Austin State University. 69 p.

Berger, Louis. Effects of static loading and dynamic forces on the density of cohesionless soils. M, 1940, Massachusetts Institute of Technology. 77 p.

Berger, P. S. Extension structures in the Central Appalachians. M, 1978, West Virginia University.

Berger, Paula Marie. Structural geology of northern Dome and southern McDonough townships, Red Lake, northwestern Ontario. M, 1984, Queen's University. 99 p.

Berger, Philip S. Analysis of bed-duplication folding. D, 1986, University of Cincinnati. 133 p.

Berger, Richard J. Compilation and interpretation of the geologic factors affecting land use planning, Chassell Quadrangle, Houghton County, Michigan. M, 1973, Michigan Technological University. 51 p.

Berger, Richard Lee. Relation of petrographic variations in mine run and tipple prepared Illinois Herrin (No. 6) coal samples to mining and preparation procedures. M, 1960, University of Illinois, Urbana. 54 p.

Berger, Richard Lee. The effect of adsorbed ions on the structural derangement temperature of illite and phlogopite. D, 1965, University of Illinois, Urbana. 109 p.

Berger, Roger John. Skeletal morphology, variability, and ecology of the bryozoan species Idmonea atlantica in the modern reefs of Bermuda. M, 1974, Pennsylvania State University, University Park.

Berger, Thomas J. A simple numerical model for the study of baroclinic estuarine shelf interactions. D, 1987, Old Dominion University. 105 p.

Berger, Wolfgang H. Pennsylvanian biohermal limestones of Marble Mountain, south-central Colorado; Saguache and Custer counties. M, 1963, University of Colorado.

Berger, Wolfgang Helmut. Areal geology of the central part of Minturn Quadrangle, Colorado. M, 1963, University of Colorado.

Berger, Wolfgang Helmut. Planktonic Foraminifera; shell production and preservation. D, 1968, University of California, San Diego. 258 p.

Berger, Z. Stream adjustment to drop in base level tested through dynamic equilibrium and geomorphic threshold concepts; a case study of some of the Allegheny's tributaries. D, 1978, University of Pittsburgh. 191 p.

Bergeron, Alain. Pétrographie et géochimie du complexe igné alcalin de Crevier et de son encaissant minéralisé M, 1980, Universite du Quebec a Chicoutimi. 129 p.

Bergeron, Brian P. Gasoline contamination of the alluvial aquifer in east-central Rapid City, South Dakota. M, 1986, South Dakota School of Mines & Technology.

Bergeron, Dalton J. Stratigraphy and sedimentation of the Anahuac Discorbis Zone, Southwest Louisiana. M, 1977, University of Southwestern Louisiana. 83 p.

Bergeron, Marcel P. The hydrogeology of the Milligan Canyon area, Montana. M, 1979, Indiana University, Bloomington. 120 p.

Bergeron, Mario. La distribution et le comportement du bore dans la lithosphere oceanique. D, 1985, McMaster University. 272 p.

Bergeron, Michel. Etude minéralogique du minerai titanifère du Lac Tio, Québec. M, 1973, Ecole Polytechnique. 144 p.

Bergeron, Michel. Minéralogie et géochimie de la suite anorthositique de la région du Lac Evolution des membres mafiques et origine des gîtes massif d'ilménite. D, 1986, Ecole Polytechnique. 487 p.

Bergeron, Robert. A study of the Quebec-Labrador iron belt between Derby Lake and Larch River. D, 1958, Universite Laval.

Bergeron, Robert. Etude de la réflectivité spectrale volcaniques de la région de l'Abitibi dans l'intervale 350 nm.-2600 nm. M, 1976, Ecole Polytechnique.

Bergeron, Robert. Geology of Forbes Lake area, Ungava. M, 1952, Columbia University, Teachers College.

Bergeron, Thomas Joseph. Stratigraphy of the (Pennsylvanian) Beeman Formation in Dry Canyon, Alamogordo, New Mexico. M, 1957, University of Wisconsin-Madison.

Bergeron, William Joseph. Finite element analysis of salt pillar models. D, 1968, Louisiana State University. 178 p.

Bergeron, William M. Stream table simulation of physical sedimentation processes in alluvial channels. M, 1974, Northeast Louisiana University.

Bergeson, Jerry R. Development of a facility for studying absorption phenomena. M, 1961, University of Oklahoma. 31 p.

Bergey, W. R. Pressure and temperature measurements of vein minerals in gold deposits. M, 1951, University of Toronto.

Bergfelder, W. The origin of the thermal water at Hot Springs, Arkansas. M, 1976, University of Missouri, Columbia.

Bergford, Paul M. On the determination of the dimensions of the Earth ellipsoid. M, 1960, Ohio State University.

Berggreen, R. G. Petrography, structure, and metamorphic history of a metasedimentary roof pendant in the Peninsular Ranges Batholith, San Diego County, California. M, 1976, San Diego State University.

Berggren, C. F. A regional gravity study of Hunt County, Texas. M, 1977, East Texas State University.

Berggren, William Alfred. Opisthobranch gastropods from the type locality of the Stone City Beds (middle Eocene) of Texas. M, 1957, [University of Houston].

Bergh, Hugh W. Paleomagnetism of the Stillwater Complex (Precambrian), Montana. D, 1968, Princeton University. 207 p.

Berghorn, Claude E. Benthonic foraminifera of the Niobrara Formation (Upper Cretaceous) at Pueblo, Colorado. M, 1973, Colorado School of Mines. 122 p.

Berglof, William Randall. Absolute age relationships in selected Colorado plateau uranium ores. D, 1970, Columbia University. 150 p.

Berglund, Pete. Tectonically and experimentally induced phyllosilicate preferred orientations in deep-sea sediments. M, 1978, Lehigh University.

Bergman, Denzil W. The Greenhorn Limestone in Kansas. M, 1950, Kansas State University.

Bergman, Eric Allen. Intraplate earthquakes and the state of stress in oceanic lithoshere. D, 1984, Massachusetts Institute of Technology. 438 p.

Bergman, Katherine Mary. Erosion surfaces and gravel shoreface deposits; the influence of tectonics on the sedimentology of the Carrot Creek Member, Cardium Formation (Turonian, Upper Cretaceous), Alberta, Canada. D, 1987, McMaster University. 404 p.

Bergman, Katherine Mary. The distribution and ecological significance of the boring sponge Cliona Viridis on the Great Barrier Reef, Australia. M, 1983, McMaster University. 69 p.

Bergman, Sheldon Cornelius. Geology of the Horse Creek Paleozoics, Fremont County, Wyoming. M, 1950, Miami University (Ohio). 36 p.

Bergman, Steven Clark. Petrogenetic aspects of the alkali basaltic lavas and included megacrysts and nodules from the Lunar Crater volcanic field, Nevada, USA. D, 1982, Princeton University. 447 p.

Bergmann, Peter C. Comparison between sieving and settling tube determinations of environments of deposition. M, 1982, Florida State University.

Bergmann, Robert J. A quantitative analysis of a microfaunule of the Ordovician lower Whitewater Member, Butler County, Ohio. M, 1959, Miami University (Ohio). 80 p.

Bergquist, Harlan Richard. A petrographic study of the crystalline rocks from the Opelike Quadrangle, Alabama. M, 1935, University of Minnesota, Minneapolis. 55 p.

Bergquist, Harlan Richard. The Cretaceous of the Mesabi Range. D, 1938, University of Minnesota, Minneapolis. 96 p.

Bergquist, J. R. Depositional history and fault-related studies, Bolinas Lagoon, California. D, 1978, Stanford University. 248 p.

Bergquist, Stanard Gustaf. Pleistocene history of the Tahquameon and Manistaque drainage region of the Northern Peninsula of Michigan. D, 1933, University of Michigan.

Bergquist, Stanard Gustaf. Report on marl deposits of Oceana County, Michigan. M, 1927, University of Michigan.

Bergren, Arthur Learoyde, Jr. Geology of the Jackson area, Teton County, Wyoming. M, 1947, University of Michigan.

Bergstresser, Thomas James. Foraminiferal biostratigraphy and paleobathymetry of the Pierre Shale, Colorado, Kansas and Wyoming. D, 1981, University of Wyoming. 351 p.

Bergstresser, Thomas James. Planktonic foraminifera from the lower part of the Niobrara Formation, Laporte, Colorado. M, 1978, University of Wyoming. 75 p.

Bergstrom, Frank W. Episodic behavior in badlands; its effects on channel morphology and sediment routing. M, 1980, Colorado State University. 224 p.

Bergstrom, John R. Geology of the east portion of Casper Mountain and vicinity. M, 1950, University of Wyoming. 55 p.

Bergstrom, John R. The stratigraphy of the Mesaverde "Formation" of southeastern Wyoming. D, 1954, University of Wyoming. 221 p.

Bergstrom, Robert E. Correlation of some Pennsylvanian limestones of the mid-continent by thermoluminescence. D, 1953, University of Wisconsin-Madison.

Bergstrom, Robert Edward. Stratigraphy of the Munterville, Seahorne, and Wiley cyclothems in Warren and Marion counties, Iowa. M, 1950, University of Wisconsin-Madison. 62 p.

Beriault, André Analyse tectonique et stratigraphique des groupes d'Armagh et de Rosaire, (Cambrian), région de Saint Malachie, Appalaches du Québec. M, 1975, Universite de Montreal.

Beriault, Andre. Etudes tectoniques du Groupe de Rosaire (Cambrian) dans la région de St-Malachie, Québec. M, 1971, Universite de Montreal.

Beringer, Robert O. Bentonite beds as key beds; a correlation study of several (Lower Cretaceous) Mowry bentonite beds from Alkali Anticline to Kane in the northeast area of Bighorn Basin in Wyoming. M, 1958, University of Wisconsin-Madison.

Berk, Jeffrey A. Development of groundwater resources for the Kent well field aquifer, Portage County, Ohio. M, 1983, Kent State University, Kent. 199 p.

Berkebile, Charles Alan. Growth and properties of transition metal monoxide crystals. D, 1965, Boston University. 206 p.

Berkel, Gary Joseph Van see Van Berkel, Gary Joseph

Berkelhamer, Louis Harry. Properties and uses of olivine from deposits of the Pacific Northwest. M, 1936, University of Washington. 126 p.

Berkey, Charles Peter. Geology of the St. Croix dalles. D, 1897, University of Minnesota, Minneapolis.

Berkey, Edgar. Terrestrial modification of trace elements in iron meteorites. D, 1967, Cornell University. 131 p.

Berkheiser, Samuel W., Jr. Petrographic analysis of the Boyle Dolomite (Devonian) of eastern Kentucky. M, 1971, Eastern Kentucky University. 74 p.

Berkhouse, Gregory A. Sedimentology and diagenesis of the Lower Cretaceous Kootenai Formation in the Sun River canyon area, northwestern Montana. M, 1985, Indiana University, Bloomington. 151 p.

Berkhout, Aart Wouter Jan. Gravity in the Prince of Wales, Somerset, and northern Baffin islands region, District of Franklin, Northwest Territories. D, 1968, Queen's University. 149 p.

Berkland, James. Geology of the Novato Quadrangle (Marin county), California. M, 1969, San Jose State University. 33 p.

Berkley, John L. The geology of the Deer Lake gabbro peridotite complex (Precambrian), Itassa County, Minnesota. M, 1972, University of Missouri, Columbia.

Berkley, John Lee. A petrochemical characterization of certain DSDP subaqueous basalts and andesites from the Indian Ocean. D, 1977, University of New Mexico. 310 p.

Berkley, Richard J. Modification of a magnetic airborne detector (AN/ASQ-1A) for use in geophysical prospecting. M, 1955, New Mexico Institute of Mining and Technology. 46 p.

Berkman, Frederick Eugene. Interpretation of gravity and magnetic data from the northern Appalachian Basin. D, 1988, Colorado School of Mines. 145 p.

Berkson, Jonathan Milton. A gravity survey in the vicinity of Michipicoten island, Lake Superior. M, 1969, University of Wisconsin-Madison.

Berkson, Jonathan Milton. Microrelief of western Lake Superior. D, 1972, University of Wisconsin-Madison.

Berlanga, J. M. Oil exploration outcome probabilities in the Tabasco Basin, Mexico as estimated by use of seismic information. D, 1979, Stanford University. 300 p.

Berlanga-Galindo, E. R. Exploration geology of the Aurora area, South central Sonora, Mexico. M, 1975, University of Arizona.

Berlau, Charles E. Lithostratigraphy and depositional history of the Brentwood Member, Bloyd Formation (Morrowan), in northern Arkansas. M, 1981, University of Arkansas, Fayetteville. 131 p.

Berler, Daniel H. The formation and preservation of submariliths on the South Florida Shelf. M, 1988, University of Miami. 226 p.

Berlin, L. A. Petrology and mineralogy of the alkaline rocks of the Stettin area, Wisconsin. M, 1977, Northeastern Illinois University.

Berman, Arthur E. Permian stratigraphy and paleotectonics, Bellvue-Livermore area, Larimer County, Colorado; relation to petroleum in the Lyons Formation. M, 1978, Colorado School of Mines. 86 p.

Berman, Brigitte Helene. Biostratigraphy of the Cozy Dell Formation in Ventura and Santa Barbara counties, California. M, 1979, California State University, Long Beach. 119 p.

Berman, Byrd Louis. A petrographic study of Devonian, Mississippian and Pennsylvanian sandstones of the Eastern Interior Basin. M, 1953, University of Illinois, Urbana. 57 p.

Berman, David S. Vertebrate fossils from the Lueders Formation, lower Permian of north-central Texas. D, 1969, University of California, Los Angeles. 172 p.

Berman, Harry. Constitution and classification of the natural silicates. D, 1936, Harvard University.

Berman, Jack E. Geology of Elk Mountain and Tabernacle Butte area, Sublette County, Wyoming. M, 1955, University of Wyoming. 63 p.

Berman, Joel. A comparison of two models used to predict atmospheric refraction in VLBI. M, 1979, Massachusetts Institute of Technology. 27 p.

Berman, Joseph. A detailed petrological and structural study of Springfield Quarry, Delaware County, Pennsylvania. M, 1937, University of Pennsylvania.

Berman, Joseph Harold. Geology of the Upper Tick Canyon area, California. M, 1950, California Institute of Technology. 72 p.

Berman, Robert G. The Coquihalla volcanic complex, southwestern British Columbia. M, 1979, University of British Columbia.

Berman, Robert Glenn. A thermodynamic model for multicomponent melts, with application to the systm $CaO-MgO-Al_2O_3-SiO_2$. D, 1983, University of British Columbia.

Berman, Robert Morris. The role of lead and excess oxygen in uraninite. D, 1957, Harvard University.

Berman, Roslyn. Rb-Sr age determinations of lepidolites by X-ray fluorescence and isotope dilution. M, 1961, Pennsylvania State University, University Park. 105 p.

Bermes, Boris John. Correlation of precipitation and ground water levels on Long Island, New York. M, 1953, University of Utah. 62 p.

Bermudez, Vilma Isabel Perez. Geohydrology of the Big Creek alluvial aquifer of Hays and vicinity, Ellis County, Kansas. M, 1986, Fort Hays State University.

Bernabo, J. Christopher. Sensing climatically and culturally induced environmental changes using palynological data. D, 1977, Brown University. 223 p.

Bernal, Juan Bautista. A method for computing the axial capacity of drilled shafts in sand. D, 1984, University of Texas, Austin. 201 p.

Bernard, B. B. Light hydrocarbons in marine sediments. D, 1978, Texas A&M University. 154 p.

Bernard, Hugh Allen. Pleistocene Ostracoda and foraminifera of southwestern Louisiana. M, 1940, Louisiana State University.

Bernard, Hugh Allen. Quaternary geology of Southeast Texas. D, 1950, Louisiana State University.

Bernardi, Mitchell L. Petrology of the crystalline rocks of Vedder Mountain, British Columbia. M, 1977, Western Washington University. 137 p.

Bernardini, Gian P. Differential thermal analysis of some copper sulfides. M, 1960, Columbia University, Teachers College.

Bernardon, Milo A. A mechanical and statistical analysis of the Middle Devonian Rogers City-Dundee formations in Michigan. M, 1957, Michigan State University. 49 p.

Bernasek, Rodney A. Stratigraphy and rhythmic sedimentation of the basal Big Blue Series (Permian) in southeastern Nebraska. M, 1967, University of Nebraska, Lincoln.

Bernaski, Greg E. Laramide deformation in the Southeast Uinta Mountains, northwestern Colorado, and northeastern Utah. M, 1985, University of Wyoming. 157 p.

Bernat, Phoebe E. Heavy mineral distribution in sediments in the vicinity of the Citronelle escarpment between Orangeburg and St. Matthews, South Carolina. M, 1963, University of South Carolina. 49 p.

Bernath, Hans Jakob. Integration of remote sensing and photogrammetry; a unified digital approach to interpretation and mapping of multispectral aerial photography. D, 1974, University of Washington. 138 p.

Bernatowicz, Thomas James. Noble gases in ultramafic xenoliths from San Carlos, Arizona. D, 1980, Washington University. 322 p.

Berndt, Kathleen A. Petrology and tectonic setting of the satellitic Tertiary Coryell Intrusives of southeastern British Columbia. M, 1983, University of Alberta. 126 p.

Berndt, Marian Patricia. Metal partitioning in a sand and gravel aquifer contaminated by crude petroleum, Bemidji, Minnesota. M, 1987, Syracuse University. 64 p.

Berndt, Michael Eugene. Experimental and theoretical constraints on the origin of mid-ocean ridge geothermal fluids. D, 1987, University of Minnesota, Minneapolis. 168 p.

Berndt, Michael Eugene. Experimental brine-mud interaction at 250°C and 500 bars pressure. M, 1983, University of Wisconsin-Madison. 121 p.

Berner, Paul C. The petrography and structure of "Bald Knob", Huron Mountains, Michigan. M, 1952, Wayne State University.

Berner, Robert Arbuckle. Continental Tertiary sediments of Huerfano Park, Colorado. M, 1958, University of Michigan.

Berner, Robert Arbuckle. Experimental studies of the formation of sedimentary iron sulfides. D, 1962, Harvard University.

Berner, Ruth Eva. New species of Favosites from the Niagaran Series of Michigan. M, 1930, University of Michigan.

Bernero, Clare Ann. The Tertiary bathymetry of the Norwegian-Greenland Sea and Eurasia Basin and the Cenozoic environment. M, 1983, University of Oklahoma. 193 p.

Bernhagen, Ralph John. Stratigraphy and micropaleontology of a deep well in Calhoun County, Florida. M, 1939, Ohio State University.

Bernhard, Joan M. Characteristic benthic foraminiferal assemblages of anoxic deposits, Jurassic to Recent. M, 1982, University of California, Davis. 137 p.

Bernhard, Ronald P. Rheological properties of high-temperature drilling fluids. M, 1981, Texas Tech University. 98 p.

Bernhardt, Carl A. Gas analyses of thermal waters in New Mexico. M, 1982, New Mexico Institute of Mining and Technology. 121 p.

Bernier, Louis. Géologie, minéralogie et pétrographie de la zone aurifère nord du gisement métamorphisé de Zn-Pb-Au-Ag-Cu de Montauban-les-Mines, Qué. M, 1985, Ecole Polytechnique. 283 p.

Bernier, Pierre Yves. VSAS2; a revised source area simulator for small forested basins. D, 1982, University of Georgia. 152 p.

Berninghausen, William Henry. Upper Mississippian ostracods from the Pella Beds of south-central Iowa. M, 1949, University of Iowa. 46 p.

Bernitsas, Nikolaos. Determination of velocity heterogeneities in the weathered zone by inversion of seismic refraction traveltime residuals. M, 1985, Ohio University, Athens. 136 p.

Bernitz, John Alexander. Isochemical ductile deformation in the Honey Hill fault zone, Connecticut; an oxygen isotopic and geochemical study. M, 1987, Indiana University, Bloomington. 88 p.

Bernold, Stanley. The bedrock geology of the Guilford 7 1/2-minute quadrangle, Connecticut. D, 1962, Yale University. 190 p.

Bernstein, Howard Alan. Modern sediments of White Oak River estuary, North Carolina. M, 1977, University of North Carolina, Chapel Hill. 78 p.

Bernstein, Lawrence Mark. Stratigraphy and sedimentology of the Espanola Formation (Huronian) in the Whitefish Falls area, Ontario. M, 1985, University of Western Ontario. 183 p.

Bernstein, Lawrence Richard. Aspects of germanium mineralogy and geochemistry. D, 1985, Stanford University. 193 p.

Bernstein, Robert L. Neutrally buoyant float measurements of internal tides off the Gulf of Maine. M, 1968, Columbia University. 27 p.

Bernt, John Dodson. Geology of the southern J-P Desert, Owyhee County, Idaho. M, 1982, University of Idaho. 73 p.

Bernthal, Mike. Analytical solution to a dispersion, advection, reaction equation. M, 1982, University of Missouri, Columbia.

Bero, David Alex. Petrology of the Heather Lake Pluton, Klamath Mountains, California. M, 1980, California State University, Fresno.

Berquist, Carl R., Jr. Analysis of structure contour mapping of the Bellwood Quadrangle in central Tennessee. M, 1970, Vanderbilt University.

Berquist, Carl Richard, Jr. Stratigraphy and heavy mineral analysis in the lower Chesapeake Bay, Virginia. D, 1986, College of William and Mary. 125 p.

Berrange, Jevan Pierre. Dispersion of certain metals from mineralized zones in a glaciated Precambrian terrain, as indicated by humus and moss. M, 1958, McGill University.

Berrera, Andreas Ramirez *see* Ramirez Berrera, Andreas

Berri, Dulcy Annette. Geology and hydrothermal alteration, Glass Buttes, Southeast Oregon. M, 1982, Portland State University. 125 p.

Berrill, John Beauchamp. A study of high-frequency strong ground motion from the San Fernando earthquake. D, 1975, California Institute of Technology. 277 p.

Berry, Andrew David. A study of the Aldridge Formation, St. Mary Lake area, British Columbia. M, 1951, University of Alberta. 107 p.

Berry, Anne L. Thermoluminescence of Hawaiian basalts. M, 1972, Stanford University.

Berry, Archie William, Jr. Precambrian volcanic rocks associated with the Taum Sauk Caldera, Saint Francois mountains, Missouri. D, 1970, University of Kansas. 147 p.

Berry, Bernard Richard. Early Pennsylvanian foraminifera from southern Indiana and northern Kentucky. M, 1961, Indiana University, Bloomington. 21 p.

Berry, Cameron George. Stratigraphy of the Cherokee Group, eastern Osage County, Oklahoma. M, 1963, University of Tulsa. 84 p.

Berry, Catherine A. Bedload transport processes in a cobble bed channel. M, 1985, Colorado State University. 129 p.

Berry, Charles T. Miocene and Recent Ophiura skeletons. D, 1934, The Johns Hopkins University.

Berry, David R. Paleoecology and biostratigraphy of the foraminiferal sequence at Grimes Canyon, Ventura County, California. M, 1970, University of California, Berkeley. 147 p.

Berry, David T. Geology of the Portola and Reconnaissance Peak 7.5' quadrangles, Plumas County, California. M, 1979, University of California, Davis.

Berry, Dean Harold. Subsurface geology of the western half of Texas County, Oklahoma. M, 1957, University of Oklahoma. 84 p.

Berry, Donald W. The stratigraphy of the (Permian) Phosphoria Formation along the northwestern margin of the Wind River basin, Wyoming. M, 1961, University of Wisconsin-Madison.

Berry, E. Willard. Lepidocyclina from the Verdun Formation of northwestern Peru. D, 1929, The Johns Hopkins University.

Berry, Edward Clark. Fracture fabric in the Piedmont and Blue Ridge of North and South Carolina. D, 1953, University of North Carolina, Chapel Hill. 45 p.

Berry, Frederick Almet Fulghum. Hydrodynamics and geochemistry of the Jurassic and Cretaceous systems in the San Juan Basin, northwestern New Mexico and southwestern Colorado. D, 1959, Stanford University. 466 p.

Berry, G. W. Glacial gravels of the upper Chenango River valley. M, 1936, Colgate University.

Berry, George Willard. Stratigraphy and structure at Three Forks, Montana. D, 1941, Cornell University. 49 p.

Berry, George Willard. The relation of fracture cleavage to faults. M, 1938, Cornell University.

Berry, Hally L. The Ogishke Conglomerate of northwestern Minnesota. M, 1926, Northwestern University.

Berry, Henry N., IV. Bedrock geology of the Camden Hills, Maine. M, 1986, University of Maine. 138 p.

Berry, J. E. and Hollingsworth, J. A. C. A seismic investigation in the Deer Creek area, Colorado (T 763). M, 1952, Colorado School of Mines. 48 p.

Berry, J. R. Malcolm. Study of the variation in Stromatocerium in the upper Black River Limestone (Ordovician) at Ouareau River, Quebec. M, 1966, McGill University.

Berry, John L. The evolution of the southern North Sea Basin. M, 1966, Columbia University. 109 p.

Berry, Joyce Kempner. Environmental values and regional planning. M, 1976, Colorado State University. 98 p.

Berry, Keith David. Micropaleontological zonation of the Niobrara Formation, El Paso County, Colorado. M, 1951, Iowa State University of Science and Technology.

Berry, Leonard Gascoigne. A study of copiapite. M, 1938, University of Toronto.

Berry, Leonard Gascoigne. Studies of mineral sulphosalts. D, 1941, University of Toronto.

Berry, Michael John. Some interpretation techniques in crustal seismology with application to the Lake Superior experiment. D, 1965, University of Toronto.

Berry, Norman J. The geology of the northeast quarter of the Huntington Quadrangle, Oregon. M, 1956, University of Oregon. 78 p.

Berry, Richard A. Stratigraphy and petrology of the Middle Bloyd Sandstone; Morrowan, Southeast Madison County, Arkansas. M, 1978, University of Arkansas, Fayetteville.

Berry, Richard Harry. Precambrian geology of the Putnam-Whitehall quadrangles, New York. D, 1961, Yale University.

Berry, Richard M. Tertiary stratigraphy of the Bates Hole-Alcova region, central Wyoming. M, 1950, University of Wyoming. 61 p.

Berry, Richard Warren. Geochemistry and mineralogy of the clay-sized fractions of some North Atlantic and Arctic ocean bottom sediments. D, 1963, Washington University. 110 p.

Berry, Richard Warren. Location of lithologic contacts obscured by mantle. M, 1957, Washington University. 33 p.

Berry, Robert Chapman. Mid-Tertiary volcanic history and petrology of the White Mountain volcanic province, southeastern Arizona. D, 1976, Princeton University. 385 p.

Berry, W. F. Descriptive study of crinoid columnals from Hamilton Group of New York. M, 1952, University of Massachusetts. 119 p.

Berry, William B. N. Graptolite faunas of the Marathon region, West Texas. D, 1957, Yale University.

Berry, William Francis. The thermal behavior of coal constitutents. D, 1963, Pennsylvania State University, University Park. 286 p.

Berry-Spark, K. L. Gasoline contaminants in groundwater; a field experiment. M, 1987, University of Waterloo. 196 p.

Berryhill, Alan Walter. Structural analysis of progressive deformation within a complex strike-slip fault system; southern Narragansett Basin, Rhode Island. M, 1984, University of Texas, Austin. 79 p.

Berryhill, Henry Lee, Jr. Stratigraphy and structure of the Wyndale Quadrangle, Washington County, Virginia. M, 1949, University of North Carolina, Chapel Hill. 62 p.

Berryhill, Louise Russell. Stratigraphy and foraminifera of the Standard Oil of New Jersey Hatteras Light Well No. 1. M, 1948, University of North Carolina, Chapel Hill. 41 p.

Berryhill, Richard A. Subsurface geology of south-central Pawnee County, Oklahoma. M, 1960, University of Oklahoma. 71 p.

Berryhill, Walter. The micropaleontology and the sedimentology of the Coon Creek Tongue. M, 1955, Mississippi State University. 86 p.

Berryman, R. J. Geology of the Deer Creek-Little Deer Creek area, Converse County, Wyoming. M, 1942, University of Wyoming. 55 p.

Berryman, William M. Lithology and environment of deposition of Pliocene rocks cored at Elk Hills, Kern County, California. M, 1973, University of Wyoming. 172 p.

Bersch, Michael G. Petrology and geology of the southern West Potrillo basalt field, Dona Ana County, New Mexico. M, 1977, University of Texas at El Paso.

Bersticker, Albert C. Selected petrologic studies in the Precambrian complex of the Wind River Range, Fremont County, Wyoming. M, 1958, Miami University (Ohio). 150 p.

Berta, Ann Alisa. Quaternary evolution and biogeography of the larger South American Canidae (Mammalia; Carnivora). D, 1979, University of California, Berkeley. 267 p.

Bertagne, Allen John. Seismic stratigraphic investigation, western Gulf of Mexico. M, 1980, University of Texas, Austin.

Bertagnolli, Alex J., Jr. Geology of the southern part of the La Barge Ridge, Lincoln County, Wyoming. M, 1940, University of Wyoming. 19 p.

Bertani, Renato Tadeu. Microfacies, depositional models and diagenesis of Lago Feia Formation (Lower Cretaceous), Campos Basin, offshore Brazil. D, 1984, University of Illinois, Urbana. 214 p.

Berthiaume, Sheridan Alba. Stratigraphy and foraminiferal fauna of the Meganos and Vacaville formations (Eocene) of California. D, 1938, Cornell University.

Bertholf, Harold Wyman. Geology and oil resources of the Timber Canyon area, Ventura County, California. M, 1967, University of California, Los Angeles.

Bertholf, W. E., Jr. Graded unconformity, Washeibemaga Lake area, Ontario. M, 1946, University of Chicago. 45 p.

Berthoud, Charles E., Jr. Soil variability over short distances. M, 1978, Rutgers, The State University, New Brunswick. 28 p.

Berti, Albert A. Palynology and stratigraphy of the mid-Wisconsin in the eastern Great Lakes region, North America. D, 1971, University of Western Ontario. 178 p.

Bertine, Kathe Karlyn. The marine geochemical cycle of chromium and molybdenum. D, 1970, Yale University. 125 p.

Bertka, Constance M. Martian mantle primary melts; an experimental study of iron rich garnet lherzolite minimum melt composition. M, 1987, Arizona State University. 61 p.

Bertl, Jeffery D. Comparative osteology of Graptemys caglei, Haynes and McKnown and Graptemys versa, Stejneger (Testudines, Emydidae). M, 1981, West Texas State University. 44 p.

Bertoli, Lou. A fracture analysis in portions of Hardy and Grant counties, West Virginia. M, 1982, University of Toledo. 94 p.

Bertoni, R. S. Geological and geophysical investigation of Block Island Sound between Fishers and Gardiners islands. D, 1975, University of Connecticut. 127 p.

Bertrand, Aimee. A PKP study of the Earth's core using the Warramunga seismic array. M, 1972, University of British Columbia.

Bertrand, Claude. L'hypersthène alumineux du Lac St. Jean. M, 1963, Ecole Polytechnique. 64 p.

Bertrand, Claude E. Metamorphism at the Normetal Mine (Precambrian), northwestern Quebec (Canada). D, 1969, University of Western Ontario.

Bertrand, Rudolf. Biométrie de Cryptolithus (Trilobite ordovicien). M, 1971, Universite de Montreal.

Bertrand, Rudolf. Etude des Trilobites du "Trenton" (Ordovician, Ontario and Quebec). D, 1971, Universite de Montreal.

Bertrand, Walley E. Geology and petrofabric analysis of the Bear Mountain-Medicine Mountain area, southwestern Pennington County, South Dakota. M, 1965, South Dakota School of Mines & Technology.

Bertrand, Wayne Gerrard. A geological reconnaissance of the Dellwood Seamount area, Northeast Pacific Ocean, and its relationship to plate tectonics. M, 1972, University of British Columbia.

Bertsch, Paul Michael. The behavior of aluminum in complex solutions and its role in the exchange equilibria of soils. D, 1983, University of Kentucky. 274 p.

Bertucci, Paul Frederick. Petrology and provenance of Upper Jurassic-Lower Cretaceous Great Valley sequence conglomerate, northwestern Sacramento Valley, California. M, 1980, University of California, Davis. 143 p.

Berumen, Manuel, Jr. Fort Chadbourne fault system, eastern Coke County, Texas. M, 1979, University of Texas, Austin.

Berven, Robert James. Lower Cardium (Upper Cretaceous) shoestring sands, Crossfield-Garrington area, Alberta. M, 1965, University of Saskatchewan. 65 p.

Berwanger, D. J. Aerobic biorestoration of groundwater contaminated with chlorinated hydrocarbons facilitated by hydrogen peroxide addition. M, 1988, University of Waterloo. 130 p.

Besancon, James Robert. Disordering kinetics in orthopyroxenes. D, 1975, Massachusetts Institute of Technology. 128 p.

Beschta, R. L. Streamflow hydrology and simulation of the Salt River basin in central Arizona. D, 1974, University of Arizona. 152 p.

Beshears, Glenn T. Geology of the Northeast Wakefield area, Clay County, Kansas. M, 1958, Kansas State University. 35 p.

Beshish, Ghaith K. An investigation into the influence of rock discontinuities on slope stability of an open pit mine. M, 1984, Michigan Technological University. 149 p.

Besien, Alphonse Camille Van see Van Besien, Alphonse Camille

Beske, Suzanne J. Paleomagnetism of the Snoqualmie Batholith, central Cascades, Washington. M, 1972, Western Washington University. 67 p.

Beskid, Nicholas J., Jr. Behavior of Cu, Ni, Mn, Zn, Ca, Mg, and Fe in two New Zealand mafic-ultramafic complexes. M, 1971, Miami University (Ohio). 52 p.

Besse, Linda. Geochemistry of Tl and selected elements in four carlin-type gold deposits. M, 1985, Eastern Washington University. 56 p.

Bessey, Larry Eugene. Beneficiation of low-grade Idaho phosphate rock. M, 1958, University of Idaho. 25 p.

Best, Bryan M. Stratigraphy and petrology of the Upper Cretaceous Selma Chalk, West-central Alabama to Southwest Louisiana. M, 1978, University of New Orleans.

Best, David Malcolm. Determining best-fit two-dimenstional gravity anomaly models by total derivatives. M, 1970, University of North Carolina, Chapel Hill. 27 p.

Best, David Malcolm. Spatial and temporal relationships of earthquake activity in the Circum-Caribbean and adjacent regions. D, 1977, University of North Carolina, Chapel Hill. 149 p.

Best, Edward W. Pre-Hamilton Devonian stratigraphy southwestern Ontario, Canada. D, 1953, University of Wisconsin-Madison.

Best, James R. Stratigraphy, sedimentology and uranium geology of the Francis Mesa area, McKinley County, New Mexico. M, 1973, Colorado School of Mines. 107 p.

Best, M. A. Detailed stratigraphy of part of the upper Morien Group, (Upper Carboniferous), North Sydney, Cape Breton Island, Nova Scotia. M, 1984, University of Ottawa. 214 p.

Best, Myron Gene. Metamorphic and igneous rocks in the Cathay area, western Mariposa County, California. D, 1961, University of California, Berkeley. 154 p.

Best, Raymond Victor. A Lower Cambrian trilobite fauna from near Cranbrook, British Columbia. M, 1952, University of British Columbia.

Best, Raymond Victor. Taxonomic revision of North American olenellid trilobites. D, 1959, Princeton University. 231 p.

Best, William Allen. A sedimentologic and stratigraphic study of the Paleocene Fort Union Formation in the South Cave Hills of Harding County, South Dakota. M, 1987, South Dakota School of Mines & Technology.

Bestland, Erick Anthony. Stratigraphy and sedimentology of the Oligocene Colestin Formation, Siskiyou Pass area, southern Oregon. M, 1985, University of Oregon. 150 p.

Beswick, Barry Thomas. A multi-component hydrogeologic evaluation of a shallow groundwater flow system in glacial drift. M, 1971, University of Manitoba.

Betcher, R. N. Temperature distributions in deep groundwater flow systems; a finite element model. M, 1977, University of Waterloo.

Bethel, Horace Lloyd. Geology of the southeastern part of the Sultan Quadrangle, King County, Washington. D, 1951, University of Washington. 244 p.

Bethel, John Patterson. An investigation of the primary and secondary mineralogy of a sequence of glacial outwash terraces along the Cowlitz River, Lewis County, Washington. M, 1982, University of Washington. 90 p.

Bethke, Craig Martin. Compaction-driven groundwater flow and heat transfer in intracratonic sedimentary basins and genesis of the upper Mississippi Valley mineral district. D, 1985, University of Illinois, Urbana. 125 p.

Bethke, Karl J. Moss bluff caprock thickness. M, 1973, University of Wisconsin-Madison. 84 p.

Bethke, P. M. X-ray methods applied to the study of the teimannite-metacinnabar group. M, 1954, Columbia University, Teachers College.

Bethke, Philip Martin. The sulfo-selenides of mercury and their occurrence at Marysvale, Utah. D, 1957, Columbia University, Teachers College. 181 p.

Bethke, William Martin. Structural reconnaissance of the High Ridge area, Jefferson County, Missouri. M, 1956, Washington University. 42 p.

Bethune, Pierre F. Considerations on the symmetry of mountain chains. M, 1933, University of Wisconsin-Madison.

Bettinger, Charles Edward. The geology of portions of Beartrap Canyon and Quail quadrangles, California. M, 1948, University of Southern California.

Bettini, Cláudio. Forecasting populations of undiscovered oil fields with the log-Pareto distribution. D, 1987, Stanford University. 237 p.

Bettison, Lori Ann. Authigenic phyllosilicate mineralogy of the Point Sal remnant, California Coast Range ophiolite. M, 1986, University of California, Davis. 126 p.

Betz, Christopher E. Facies analysis and depositional environments of the upper part of the Richmond Group (Upper Ordovician) Richmond, Indiana, to Xenia, Ohio. M, 1984, Miami University (Ohio). 159 p.

Betz, Frederick, Jr. Geology and mineral deposits of the Canada Bay area, northern Nfld. D, 1938, Princeton University.

Betz, J. E. Design and construction of apparatus for measuring the electrical conductivity and magnetic permeability of drill core specimens of metallic ore. M, 1954, University of Toronto.

Betz, J. E. Measurement of some electrical and magnetic properties of a suite of ores and rocks. M, 1953, University of Toronto.

Betzer, Peter R. The concentration and distribution of particulate iron in waters of the northwestern Atlantic Ocean and Caribbean Sea. D, 1971, University of Rhode Island.

Beu, Robert D. The geology of the Cuchara Pass area, Colorado. M, 1952, University of Kansas. 95 p.

Beullac, Raymond. Etude pétrologique du complexe ophiolitique du Lac Nicolet, Québec. M, 1983, Universite Laval. 80 p.

Beuren, Victor Vignot Van see Van Beuren, Victor Vignot

Beus, Stanley S. Geology of the northern part of Wellsville Mountain, northern Wasatch Range, Utah. M, 1958, Utah State University. 84 p.

Beus, Stanley Spencer. Geology of the central Blue Spring Hills, Utah-Idaho. D, 1963, University of California, Los Angeles. 282 p.

Beuthin, John D. Facies interpretation of the lower Pocono Formation (Devonian-Mississippian); Garrett County, Maryland, and vicinity. M, 1986, University of North Carolina, Chapel Hill. 91 p.

Beutner, Edward Chandler. Structure and tectonics of the southern Lemhi Range, Idaho. D, 1968, Pennsylvania State University, University Park. 181 p.

Beutner, Edward L. Intrusive rocks of the Marenisco Range, Michigan and Wisconsin. M, 1932, Northwestern University.

Bevan, Arthur Charles. Geology of the Beartooth Mountains, Montana. D, 1921, University of Chicago. 41 p.

Bevan, Stewart. The stratigraphy, composition, and origin of the Tesnus Formation. M, 1938, Texas A&M University. 82 p.

Bever, James Edward. Geology of the Guffy area, Colorado. D, 1954, University of Michigan.

Beveren, Oscar F. Van see Van Beveren, Oscar F.

Beveridge, Alexander James. Heavy minerals in lower Tertiary formations in the Santa Cruz Mountains, California. D, 1959, Stanford University. 119 p.

Beveridge, Alexander James. Relation of some British Columbia intrusives to the Alberta sedimentary basin. M, 1956, University of Alberta. 90 p.

Beveridge, Richard Clark. The subsurface geology of Major County, Oklahoma. M, 1954, University of Oklahoma. 74 p.

Beveridge, Thomas Robinson. Subdrift valleys of southeastern Iowa. M, 1947, University of Iowa. 57 p.

Beveridge, Thomas Robinson. The geology of the Weaubleau Quadrangle, Missouri. D, 1949, University of Iowa. 199 p.

Beverly, Burt, Jr. Some graphite deposits of Los Angeles, California. M, 1933, Cornell University.

Beverly, Charles E. The petrology and structure of Precambrian crystalline and Tertiary igneous rocks in the northeast quarter of the Rustic Quadrangle, Larimer County, Colorado. M, 1969, Colorado State University. 111 p.

BeVier, Laura M. Depositional environments of the Upper Ordovician Fremont Formation, northern Canon City Embayment, Colorado. M, 1987, Colorado School of Mines. 85 p.

Bevier, M. L. Field relations and petrology of the Rainbow Range shield volcano, West-central British Columbia. M, 1978, University of British Columbia.

Bevier, Mary Lou. Geology and petrogenesis of Mio-Pliocene Chilcotin Group basalts, British Columbia. D, 1982, University of California, Santa Barbara. 124 p.

Bevill, James Cecil, Jr. Uranium groundwater orientation survey for central Arkansas. M, 1981, University of Arkansas, Fayetteville. 108 p.

Bevis, Michael Graeme. Hypocentral trend surface analysis; regional and fine structure of selected intermediate depth Benioff zones. D, 1982, Cornell University. 127 p.

Bexon, Roger. A petroleum engineering study of the upper Miocene oil reservoirs, Main Quarry-West Coora Field, Trinidad, B.W.I. M, 1952, University of Tulsa. 72 p.

Bey, Ahmad. Meso climatonomy modeling of four restoration stages following Krakataus 1883 destruction (pumice ash desert-savanna-young forest-rainforest). D, 1981, University of Wisconsin-Madison. 130 p.

Beydoun, Wafik Bulind. Asymptotic wave methods in heterogeneous media. D, 1985, Massachusetts Institute of Technology. 260 p.

Beydoun, Wafik Bulind. Sources of seismic noise in boreholes. M, 1982, Massachusetts Institute of Technology. 95 p.

Beyer, Betsy Jo. Petrology and geochemistry of ophiolite fragments in a tectonic melange, Kodiak Islands, Alaska. D, 1980, University of California, Santa Cruz. 231 p.

Beyer, Harry. Basement structure on the western flank of the Ozark uplift, Missouri (Paleozoic). M, 1969, Washington University. 50 p.

Beyer, Joan Brown. Petrology of a mafic facies in the Organ Needle Pluton, Dona Ana County, New Mexico. M, 1986, New Mexico State University, Las Cruces. 89 p.

Beyer, John Henry, Jr. Telluric and D.C. resistivity techniques applied to the geophysical investigation of Basin and Range geothermal systems. D, 1977, University of California, Berkeley. 477 p.

Beyer, Larry Albert. Measuring gravity on the sea floor in deep water. M, 1964, University of California, Riverside. 66 p.

Beyer, Larry Albert. The vertical gradient of gravity in vertical and near-vertical boreholes. D, 1971, Stanford University. 230 p.

Beyer, Paul Joseph. Alteration of slate belt xenoliths in a granodioritic stock, Orange County, North Carolina. D, 1973, University of North Carolina, Chapel Hill. 85 p.

Beyer, Robert F. Variation of Rm, Rmc, and Rmf for specific mud additives under differing temperature and concentration. M, 1964, Tulane University. 99 p.

Beyer, Robert Lee. Magma differentiation at Newberry Crater in central Oregon. D, 1973, University of Oregon. 93 p.

Beyer, Samuel W. The Sioux Quartzite, near Sioux Falls in South Dakota, with special reference to an intrusive diabase. D, 1895, The Johns Hopkins University.

Beyer, William C. Petrology and genesis of a uranium-bearing system of pegmatite dikes, Nancy Creek area, northeastern Washington. M, 1981, University of Montana. 82 p.

Beyke, Robert J. Urea decomposition and environmental implications at an airport site, northeast Ohio. M, 1986, University of Akron. 60 p.

Bez, Lauri. Mineralogy and geochemistry of the Morro Agudo zinc-lead deposit, Brazil. M, 1979, University of Missouri, Columbia.

Bezan, Abduelhafid Mohamed. The geology and the hydrology of Rock Creek basin. M, 1974, University of Idaho. 118 p.

Bezara, Miguel. Computer optimization of open pit design. M, 1967, Stanford University. variously paginated p.

Bezore, Stephen Patrick. Petrology of ultramafic and associated mafic rocks near Middletown, California. M, 1972, University of California, Davis. 81 p.

Bezys, Ruth Krista Angela. An ichnological ans sedimentological study of Devonian black shales from the Long Rapids Formation, Moose River basin, northern Ontario. M, 1987, McMaster University. 244 p.

Bezzerides, Theodore L. Triassic stratigraphy and geology of the O'Neil Pass area, Elko County, Nevada. M, 1967, University of Oregon. 74 p.

Bhagat, Snehal. Changes in rock fabric, trace-element content, and stable-isotope composition accompanying cleavage development during pressure-solution deformation of limestone; implications for volume-loss strain. M, 1988, University of Illinois, Urbana. 88 p.

Bhagavathula, Rao V. Petrogenesis of sulfides in the Dunka Road Cu-Ni deposit, Duluth Complex, Minnesota, with special reference to the role of contamination by country rock. D, 1981, Indiana University, Bloomington. 304 p.

Bhambhani, Deepak J. High resolution methods for treating marine seismic data. M, 1973, Texas A&M University.

Bhambri, Inder Jit. Shear strength of clay as function of degree of consolidation. D, 1971, Catholic University of America. 121 p.

Bhardwag, M. C. Effect of high pressure on crystalline solubility in the system NaCl-KCl. M, 1970, Pennsylvania State University, University Park.

Bhasavanija, Khajohn. A finite difference model of an acoustic logging tool; the borehole in a horizontally layered geologic medium. D, 1983, Colorado School of Mines. 247 p.

Bhasin, R. N. Pore size distribution of compacted soils after critical region drying. D, 1975, Purdue University. 242 p.

Bhasker, Rao Kidiyoor. A water flow-salt transport model to simulate salt efflorescence. D, 1982, Utah State University. 165 p.

Bhate, Uday Ramesh. Trace metal distributions in natural salt marsh sediments. M, 1972, Georgia Institute of Technology. 83 p.

Bhatia, D. M. S. Geochemistry, paragenetic relationship, spatial distribution and factor analysis of selected trace elements at the Magmont Mine, Southeast Missouri. D, 1976, University of Missouri, Rolla. 331 p.

Bhatia, Dil Mohan Singh. Facies change in iron formation at Brunswick Number Twelve Mine, Bathurst (New Brunswick, Canada). M, 1970, University of New Brunswick.

Bhatia, Shobha Krisna. The verification of relationships for effective stress method to evaluate liquefaction potential of saturated sands. D, 1983, University of British Columbia.

Bhatrakarn, Tanakarn. Correlation of gravity observations with the geology of parts of Burnet, Blanco, and Llano counties, Texas. M, 1961, University of Texas, Austin.

Bhatt, Bharat K. Petrology and stratigraphy of Swift and Morrison formations near Drummond, Montana. M, 1967, University of Montana. 135 p.

Bhatt, Bipinkumar J. Geology and hydrothermal alteration of the West Fork, Yankee Fork, Custer County, Idaho. M, 1978, University of Idaho. 85 p.

Bhatt, Jagdish J. Cretaceous history of India and adjoining countries. M, 1963, University of Wisconsin-Madison.

Bhatt, Kireet Jivanram. Geology of Mount Bohemia, Michigan. M, 1952, Michigan Technological University. 53 p.

Bhattacharjee, Shyama Bijoy. Ripple-drift cross-lamination in turbidites of the Ordovician Cloridorme Formation, Gaspe, Quebec. M, 1970, McMaster University. 167 p.

Bhattacharji, Somdev. Theoretical and experimental investigations of cross-folding. D, 1959, University of Chicago. 42 p.

Bhattacharya, Bireswar. A frequency domain study of Love waves from the Illinois earthquake of November 9, 1968. M, 1974, St. Louis University.

Bhattacharya, Nityananda. Weathering products of tills in Indiana. D, 1960, Indiana University, Bloomington. 90 p.

Bhattacharya, Prabhat Kumar. An investigation of changes in the magnetic field of the Earth (magnetic anisotropy of sedimentary rocks). D, 1950, California Institute of Technology. 100 p.

Bhattacharyya, Debaprasad. Sedimentology of the Late Cretaceous Nubia Formation at Aswan, southeast Egypt, and origin of the associated ironstones. D, 1981, Princeton University. 229 p.

Bhattacharyya, P. J. The Mid-Atlantic Ridge near 45°N; analysis of bathymetric and magnetic data. D, 1972, Dalhousie University.

Bhattacharyya, Tapan Kanti. Seismic model investigations of energy partitioning in multilayered media. D, 1961, Texas A&M University. 115 p.

Bhattacharyya, Tapas. Tectonic evolution of the Shoo Fly Formation and the Calaveras Complex, central Sierra Nevada, California. D, 1986, University of California, Santa Cruz. 351 p.

Bhatti, Nasir Ali. Depositional environment of phosphate deposits of the Hazara District, southern Himalaya, Pakistan. D, 1981, University of Idaho. 150 p.

Bhatti, Nasir Ali. Geology and phosphate resources of a part of the Red Ridge Quadrangle, Bonneville County, Idaho. M, 1971, University of Idaho. 70 p.

Bhatti, Sabir A. Analysis of pumping well near a stream. M, 1967, University of Arizona.

Bhoojedhur, S. Adsorption and heavy metal partitioning in soils and sediments of the Salmon River area, British Columbia. D, 1975, University of British Columbia.

Bhutta, Mohammed Afhar. Geology of the Salida area, Chaffee and Fremont counties, Colorado. D, 1954, Colorado School of Mines. 173 p.

Bhuyan, Ganesh Ch. An analysis of some regional gravity data in Arizona. M, 1965, University of Arizona.

Biadgelgne, Abraham. Third derivative analysis of gravity anomalies from prismatic sources. M, 1979, Pennsylvania State University, University Park. 75 p.

Biaglow, Joseph Anthony. Study to determine the effectiveness of the hydraulic mine seal at Big Four Hollow Creek near Lake Hope State Park in southeastern Ohio. M, 1988, Wright State University. 169 p.

Biancardi, John M. Tin in granites; a review and contribution. M, 1973, Stanford University.

Bianchetti, Susan Fullam. Variable dissolution rates of deformed and undeformed calcite. M, 1986, SUNY at Stony Brook. 162 p.

Bianchi, Luiz. Geology of the Manitou Springs-Cascade area, El Paso County, Colorado, with a study of the permeability of its crystalline rocks. M, 1968, Colorado School of Mines. 197 p.

Bianchi-Mosquera, Gino Cesar. Adsorption of Ag, Co, Cu, Ni, Pb, and Zn on goethite. M, 1986, Pennsylvania State University, University Park. 192 p.

Bianchini, Gary Francis. The effects of anisotropy and strain on the dynamic properties of clay soils. D, 1982, Case Western Reserve University. 327 p.

Biasi, Glenn Paul. The streaming potential method applied to a low gradient hydrologic environment along the Mojave River, San Bernardino County, California. M, 1988, University of California, Riverside. 91 p.

Bibbins, Arthur. Notes of the Mesozoic and Cenozoic geology of Chesapeake Basin. M, 1891, The Johns Hopkins University.

Bibby, R. Flow between the confined aquifer of the Fox Hills Sandstone and the alluvial aquifer in the north Kiowa-Bijou District, Colorado. M, 1969, Colorado State University. 84 p.

Bibee, Leonard D. Crustal structure in areas of active crustal accretion. D, 1979, University of California, San Diego. 169 p.

Bible, Gary G. Landslide phenomena in Shell and Tensleep canyons, Bighorn Mountains, Wyoming. D, 1978, Iowa State University of Science and Technology. 565 p.

Bible, Gary Gill. Geologic and engineering properties of Quaternary age materials at the Scranton II Roadcut, Scranton, Iowa. M, 1975, Iowa State University of Science and Technology.

Bibler, Carol Jean. Stratigraphy and depositional environment of the Upper Cretaceous Horsethief Formation northwest of Augusta, Montana. M, 1985, Montana State University. 57 p.

Bice, David Clifford. Tephra stratigraphy and physical aspects of Recent volcanism near Managua, Nicaragua. D, 1980, University of California, Berkeley. 422 p.

Bick, Kenneth Fletcher. Geology of the Deep Creek Quadrangle, western Utah. D, 1958, Yale University.

Bickart, K. Jeffrey. The birds of the late Miocene-early Pliocene Big Sandy Formation, Mohave County, Arizona. M, 1986, University of Michigan. 158 p.

Bickel, Charles Eliot. Bedrock geology of the Belfast Quadrangle, Maine. D, 1972, Harvard University.

Bickel, Edwin David. Pleistocene non-marine Mollusca of the Gatineau valley and Ottawa area of Quebec and Ontario, Canada. D, 1970, Ohio State University. 193 p.

Bickel, James. Diagenetic and hydrothermal reactions in calcareous ooze; an experimental study. M, 1982, University of Minnesota, Minneapolis. 79 p.

Bickford, Barbara J. Analysis of contaminant movement from a manure storage facility. M, 1983, University of Wisconsin-Madison.

Bickford, David A. Economic geology of the Jones Camp iron deposit, Socorro County, NM. M, 1980, New Mexico Institute of Mining and Technology. 226 p.

Bickford, Fred E. Petrology and structure of the Barlow Gap area, Wyoming. M, 1977, University of Wyoming. 76 p.

Bickford, Hugh L. Petrology of the upper Sawatch Formation (White River Plateau, Colorado). M, 1974, Colorado State University. 114 p.

Bickford, Kenneth F. A sedimentary and petrographic analysis of the Spearfish and Sundance formations of the Black Hills, South Dakota. M, 1939, University of Minnesota, Minneapolis. 23 p.

Bickford, Marion Eugene, Jr. A textural and mineralogical study of some diorites from Mt. Desert Island, Maine. M, 1958, University of Illinois, Urbana. 39 p.

Bickford, Marion Eugene, Jr. Petrology and structure of the gabbro in the Jonesport-Milbridge area, Maine. D, 1960, University of Illinois, Urbana. 107 p.

Bicki, Thomas James. Geomorphology, stratigraphy, and soil development in the Iowa and Cedar River valleys in southeastern Iowa. D, 1981, Iowa State University of Science and Technology. 348 p.

Bickley, John A. Petrology and stratigraphy of the upper Sewickley, Benwood and Arnoldsburg members of the middle Monongahela Group (Pennsylvanian) in Morgan County, Ohio. M, 1976, University of Akron. 91 p.

Bickley, William B., Jr. Paleoenvironmental reconstruction of the late Quaternary lacustrine sediments (Seibold site) in southeastern North Dakota. M, 1970, University of North Dakota. 88 p.

Bickley, William B., Jr. Stratigraphy and history of the Sakakawea Sequence, south-central North Dakota. D, 1972, University of North Dakota. 218 p.

Bicknell, James Scott. Depositional environment and hydrodynamic flow in Lower Cretaceous J Sandstone, Lonetree Field, Denver Basin, Colorado. M, 1986, Texas A&M University.

Bicknell, John Dee Rogers. Tectonics of mid-ocean ridges; deep-tow and Sea beam studies at fast and slow spreading centers at 19°30'S on the East Pacific Rise and in the Nereus Deep, Red Sea. M, 1986, University of California, Santa Barbara. 105 p.

Biddle, John H. The geology and mineralization of part of the Bird Creek Quadrangle, Lemhi County, Idaho. M, 1985, University of Idaho. 119 p.

Biddle, Kevin Thomas. Characteristics of Triassic carbonate buildups of the Dolomite Alps, Italy; evidence from the margin-to-basin depositional system. D, 1979, Rice University. 228 p.

Biddle, Kevin Thomas. Physical and biogenic sedimentary structures of a recent coastal lagoon. M, 1976, Rice University. 115 p.

Bidgood, Thomas Warren. Alteration associated with sandstone-type uranium mineralization in the Black Hills, South Dakota and Wyoming. M, 1973, South Dakota School of Mines & Technology.

Bidgood, Thomas Warren. Petrography and trace element distribution across a gold ore body in the Homestake Mine, Lead, South Dakota. D, 1977, South Dakota School of Mines & Technology.

Bidin, Abdul-Aziz. Phosphate in Malaysian Ultisols and Oxisols as evaluated by a mechanistic model. D, 1982, Purdue University. 151 p.

Bidwell, Matthew E. Structural analysis of shear zones in the central Vermilion District, NE Minnesota. M, 1988, University of Missouri, Columbia. 174 p.

Bieber, Charles L. The stratigraphy and paleontology of the Trenton rocks of the Upper Mississippi Valley. D, 1942, Northwestern University.

Bieber, David William. Geology and reservoir characteristics of the Cherokee Group in the Start oil field of western Kansas. M, 1984, University of Colorado. 154 p.

Bieberman, Robert Arthur. The Mississippian-Pennsylvanian unconformity in middle western Indiana. M, 1950, Indiana University, Bloomington. 42 p.

Bieda, George E. Measurement of the viscoelastic and related mass-physical properties of some continental terrace sediments. M, 1970, United States Naval Academy.

Biederman, Donald D. Recent sea-level change in the Pacific Northwest. M, 1967, University of Washington. 24 p.

Biederman, Edwin Williams, Jr. Shoreline sedimentation in New Jersey. D, 1958, Pennsylvania State University, University Park. 292 p.

Biederman, John L. Petrology of the Chain Lakes Massif along Route 27 in central-western Maine; a Precambrian high-grade terrane. M, 1984, University of Maine. 125 p.

Biegel, Ronald L. The fractal structure of fault gouge and its implications for fault stability. D, 1988, University of Southern California.

Biegler, Norman W. Percolation characteristics of various sediments. M, 1952, Kansas State University.

Biehle, Alfred A. Comparison of petrophysical characteristics of selected carbonate rocks to their well log response; Smackover Formation (Upper Jurassic). M, 1986, Stephen F. Austin State University. 94 p.

Biehler, Shawn. A geophysical study of the Salton Trough, Southern California. D, 1964, California Institute of Technology. 145 p.

Biel, Ralph. Foraminifera from the Upper Cretaceous of Delaware. M, 1955, New York University.

Bielak, James Walter. The origin of Cherry Creek amphibolites from the Winnipeg Creek area of the Ruby Range, southwestern Montana. M, 1978, University of Montana. 46 p.

Bielak, LeRon E. Pleistocene biostratigraphy, chronostratigraphy and paleocirculation of the southeast Pacific central water core RC11-220. M, 1975, University of Cincinnati. 110 p.

Bielenstein, Hans U. The Rundle thrust sheet, Banff, Alberta. D, 1969, Queen's University. 149 p.

Bielenstein, Hans Uwe. The Miette Formation, Jasper, Alberta. M, 1964, University of Alberta. 65 p.

Bieler, Barrie Hill. Primary uranium mineralization in some hydrothermal vein deposits in Boulder Batholith, Montana. D, 1955, Pennsylvania State University, University Park. 155 p.

Bieler, Barrie Hill. The design, construction, and testing of a low magnification camera to photograph polished ore specimens. M, 1952, California Institute of Technology. 64 p.

Bieler, David B. A study of some lamprophyric dikes associated with the White Mountain Magma Series. M, 1973, Dartmouth College. 67 p.

Bieler, David Bruce. Melanges and glaucophanic schists of the Mona Complex (Precambrian), Southeast Anglesey, North Wales. D, 1983, University of Illinois, Urbana. 210 p.

Biemesderfer, George K. Structural observations on the Peach Bottom slate belt, Lancaster County, Pennsylvania, east of the Susquehanna River. M, 1948, Columbia University, Teachers College.

Bieneman, Paul Martin. Dunes on the Navajo Uplands of northeastern Arizona; their relationship to selected environmental variables. D, 1982, University of Oklahoma. 149 p.

Bienkowski, Henry G. A finite-difference model of ground-water flow in a portion of Kent County, Maryland, and New Castle County, Delaware. M, 1978, University of Toledo. 84 p.

Bienkowski, Lisa Sophia. The paleoecology of echinoderms of the Middle Ordovician Benbolt Formation of East Tennessee. M, 1985, University of Tennessee, Knoxville. 131 p.

Bienvenu, Léo R. The chemical composition of some metavolcanic Archaean rocks. M, 1955, Universite Laval.

Bierei, Mark Alan. Hydrocarbon maturation, source rock potential, and thermal evolution of the Late Cretaceous and early Tertiary rocks of the Hanna Basin, Southeast Wyoming. M, 1987, University of Wyoming. 151 p.

Bierley, Janice. Upper Cretaceous foraminifera from the Gas Point Member, Budden Canyon Formation, northwestern Sacramento Valley, California. M, 1977, University of Washington. 131 p.

Bierley, Robert. Physical properties of Lake Maracaibo recent sediments. M, 1977, University of Tulsa. 113 p.

Bierman, Victor Joseph, Jr. Dynamic mathematical model of algal growth in eutrophic freshwater lakes. D, 1974, University of Notre Dame. 155 p.

Bierschenk, William Henry. Groundwater resources of southern Crawford County, Illinois. M, 1953, University of Illinois, Urbana.

Biersel, Thomas P. V. Van *see* Van Biersel, Thomas P. V.

Bierwagen, Elmer Emanuel. Geology of the Black Mountain area, Lewis and Clark and Powell counties, Montana. D, 1964, Princeton University. 166 p.

Biery, Jerry N. Geology of a part of the footwall of the Great Smoky fault, Monroe County, Tennessee. M, 1968, University of Tennessee, Knoxville. 52 p.

Biesinger, James C. Mineral and chemical content of the deep-water sediment sequences of Bear Lake, Utah-Idaho. M, 1973, Utah State University. 105 p.

Biesiot, Peter Gerard. Conodonts and fish remains in the Missouri Series (Pennsylvanian) of Nebraska. M, 1950, University of Nebraska, Lincoln.

Bifani, Ronald. A Triassic sequence of southern Tunisia; a subsurface study. M, 1975, University of South Carolina.

Bifano, Francis Vincent. Ostracode paleoecology from shales of the Wreford Megacyclothem (Lower Permian), Kansas and Oklahoma. M, 1974, Pennsylvania State University, University Park. 124 p.

Bigarella, Laertes P. Potential methane drainage using hydro fracturing, Carbondale coal field, Pitkin County, Colorado. M, 1981, Colorado School of Mines. 181 p.

Bigelow, Eric A. Techniques of volatile analysis in volcanic glass by quadrupole mass spectrometry and application to Mount Erebus, Antarctica. M, 1985, New Mexico Institute of Mining and Technology. 149 p.

Bigelow, G. E. Distribution and characteristics of dikes in the southeast part of the Koolau Range. M, 1968, University of Hawaii. 27 p.

Bigelow, G. E. Geology of the Black Mountain-Greens Peak area near Eagle Lake, Lassen County, California. D, 1972, University of Hawaii. 106 p.

Bigelow, Nelson, Jr. Environment of deposition of the Funston cyclothem (Lower Permian) of Riley and Geary counties, Kansas. M, 1954, University of Kansas. 102 p.

Bigelow, Phillip Kenneth. The petrology, stratigraphy and basin history of the Montesano Formation, southwestern Washington and southern Olympic Peninsula. M, 1987, Western Washington University. 263 p.

Biggane, John Howard. The low-temperature geothermal resource and stratigraphy of portions of Yakima County, Washington. M, 1982, Washington State University. 149 p.

Biggar, Norma E. A geological and geophysical study of Chena Hot Springs, Alaska. M, 1973, University of Alaska, Fairbanks.

Biggers, Barbara. Reef to back-reef facies and diagenesis of the Permian Tansill/Capitan formations in Dark Canyon, Guadalupe Mountains, New Mexico. M, 1984, University of Texas at Dallas. 113 p.

Biggers, James Virgil. Experimental determination of some thermodynamic properties of oxide solid solutions containing manganese oxide and cobalt oxide. D, 1966, Pennsylvania State University, University Park. 113 p.

Biggerstaff, Brad P. Geology and ore deposits of the Steeple Rock-Twin Peaks area, Grant County, New Mexico. M, 1974, University of Texas at El Paso.

Biggi, Robert J. Gravity evidence for fault controlled intrusion, southeast coast, Maine. M, 1973, SUNY at Buffalo.

Biggs, Charles A. Stratigraphy of the Amsden Formation of the Wind River Range and adjacent areas in northwestern Wyoming. M, 1951, University of Wyoming. 121 p.

Biggs, Donald Lee. Chert nodules in carbonate rocks of Illinois. D, 1957, University of Illinois, Urbana.

Biggs, Donald Lee. The infrared spectra of some pulverized rock-forming minerals. M, 1950, University of Missouri, Columbia.

Biggs, Maurice Earl. Interrelationships of seismic characteristics and physical parameters of Indiana stratigraphy. D, 1973, Indiana University, Bloomington. 263 p.

Biggs, Maurice Earl. Structure of an area near Fourteenmile Creek, Clark County, Indiana. M, 1950, Indiana University, Bloomington.

Biggs, Robert Bruce. Deposition and early diagenesis of modern Chesapeake Bay muds. D, 1963, Lehigh University. 129 p.

Biggs, Thomas H. The geology of the McCormick area, McCormick County, South Carolina. M, 1982, University of Georgia.

Bigham, Gary Neil. Clay mineral transport on the inner continental shelf of Georgia. M, 1972, Georgia Institute of Technology. 49 p.

Biglow, Crague C. Stratigraphy and sedimentology of the Cretaceous Newark Canyon Formation in the Cortez Mountains, north-central Nevada. M, 1986, Eastern Washington University. 123 p.

Bigman, Nathan. Formation of boudins by stress-driven diffusion. M, 1985, Rensselaer Polytechnic Institute. 97 p.

Bihl, Gerhard. Palynological correlation of Late Cretaceous beds, Sheerness, Alberta (Canada). M, 1968, University of Alberta. 76 p.

Bihl, Gerhard. Palynostratigraphic investigation of upper Maastrichtian and Paleocene strata near Tate Lake, N.W.T. D, 1973, University of British Columbia.

Bijak, Martin Kenneth. A fluid inclusion study of the Bunker Hill Mine, Coeur d'Alene District, Idaho. M, 1985, New Mexico Institute of Mining and Technology. 189 p.

Bikerman, Michael. A geologic-geochemical study of the Cat Mountain Rhyolite. M, 1962, University of Arizona.

Bikerman, Michael. Geological and geochemical studies of the Roskruge Range, Pima County, Arizona. D, 1965, University of Arizona. 135 p.

Bikun, James V. Fluorine and lithophile element mineralization at Spor Mountain, Utah. M, 1980, Arizona State University. 195 p.

Bilan, Larry Joseph. North American Tentaculitidae. M, 1961, University of Alberta. 103 p.

Bilbey, Sue Ann. Petrology and geochemistry of the Morrison Formation, Dinosaur Quarry Quadrangle, Utah. M, 1973, Utah State University. 102 p.

Bilbrey, Don Gene. Economic geology of Rim Rock Country, Trans-Pecos, Texas. M, 1957, University of Texas, Austin.

Bilby, R. E. The function and distribution of organic debris dams in forest stream ecosystems. D, 1979, Cornell University. 148 p.

Bild, R. W. A study of primitive and unusual meteorites. D, 1976, University of California, Los Angeles. 254 p.

Bilelo, Maria A. Marques. The Fusulinidae of the Winchell Limestone (Pennsylvanian), Brazos and Trinity river valleys, Texas. M, 1967, Southern Methodist University. 55 p.

Biles, Norman. A study of vertical ground motion showing the free surface effect. M, 1967, New Mexico Institute of Mining and Technology. 48 p.

Bilgesu, Huseyin Ilkin. A multi-purpose numerical model for petroleum reservoirs. D, 1984, Pennsylvania State University, University Park. 280 p.

Bilinski, Peter Walter. Subsurface geology of the Cockfield Formation, Allen and Beauregard parishes, Louisiana. M, 1980, University of New Orleans. 65 p.

Bill, Steven David. Paleoceanography of the eastern Indian Ocean during late Pleistocene time. D, 1982, Case Western Reserve University. 481 p.

Biller, Edward J. Stratigraphy and petroleum possibilities of lower Upper Devonian (Frasnian and lower Famennian) strata, southwestern Utah. M, 1976, Colorado School of Mines. 105 p.

Biller, Martin J., III. Seismic refraction studies in Lake Michigan sediments. M, 1984, University of Wisconsin-Milwaukee. 91 p.

Billings, Gale Killmer. A geochemical investigation of the Valley Spring Gneiss and Packsdale Schist, Llano Uplift, Texas. M, 1962, Rice University. 40 p.

Billings, Gale Killmer. Major and trace element relationships within coexisting minerals of the Enchanted Rock Batholith, Llano Uplift, Texas. D, 1963, Rice University. 79 p.

Billings, Gladys D. Pennsylvanian Ostracoda of the Wayland Shale of Texas. M, 1931, Columbia University, Teachers College.

Billings, M. Hewitt. Origin and distribution of the thick Freeport Coal of Allegheny County, Pennsylvania. M, 1931, University of Pittsburgh.

Billings, M. Hewitt. The Nocana oil field, Montague County, Texas. D, 1934, University of Illinois, Urbana. 85 p.

Billings, Marland P. Geology of the North Conway Quadrangle, New Hampshire. D, 1927, Harvard University.

Billings, Patty. Fission track annealing related to vein mineralization and hydrothermal alteration, Ouray County, Colorado. M, 1980, Colorado School of Mines. 42 p.

Billings, Roger Lewis. Geology of eastern Noble County, Oklahoma. M, 1956, University of Oklahoma. 56 p.

Billingsley, Arthur Lee. Unconventional energy resources in Tuscaloosa sediments of the "Tuscaloosa Trend", South Louisiana. M, 1980, University of New Orleans. 63 p.

Billingsley, George H., Jr. General geology of Tuckup canyon, central Grand Canyon, Mohave County, Arizona. M, 1970, Northern Arizona University. 115 p.

Billingsley, Harold Ray. A subsurface study of the Sholem Alechem oil field, Stephens and Carter counties, Oklahoma. M, 1949, University of Oklahoma. 61 p.

Billingsley, Lee T. Stratigraphy and clay mineralogy of the Trinidad Sandstone and associated formations (Upper Cretaceous), Walsenburg area, Colorado. M, 1977, Colorado School of Mines. 105 p.

Billingsley, Lee Travis. Geometry and mechanism of folding related to growth faulting, Wilcox Formation, De Witt County, Texas. D, 1983, Texas A&M University. 133 p.

Billingsley, Paul. The Shawangunk Grit, its structure, origin and stratigraphic significance. M, 1911, Columbia University, Teachers College.

Billingsley, Randal Lee. The Cantwell Formation (Paleocene) as a possible offset feature along the Denali Fault, central Alaska Range, Alaska. M, 1977, University of Wisconsin-Madison.

Billington, Edward. Spatial and temporal seismicity variations in the South Sandwich and northwestern South American subduction zones. M, 1986, Georgia Institute of Technology. 154 p.

Billington, Selena. The morphology and tectonics of the subducted lithosphere in the Tonga-Fiji-Kermadec region from seismicity and focal mechanism solutions. D, 1980, Cornell University. 228 p.

Billman, Robert B., III. Conodonts of the Cuivre Shale Member of the Hannibal Formation (Kinderhookian) from northeastern Missouri. M, 1984, Bowling Green State University. 131 p.

Billman, Thomas A., Jr. Interpretation and geochemical analysis of "Clinton" brines from northern Summit and Portage counties, Ohio. M, 1986, University of Akron. 108 p.

Billo, Saleh Mohammad. Petrology and geochemistry of varved Halite II Member, Permian Castile Formation, Delaware Basin, Texas and New Mexico. D, 1973, University of New Mexico. 100 p.

Bills, Bruce Gordon. A harmonic and statistical analysis of the topography of the Earth, Moon and Mars. D, 1978, California Institute of Technology. 273 p.

Bills, Terry Vance, Jr. Geology of Waco Springs Quadrangle, Comal County, Texas. M, 1957, University of Texas, Austin.

Bilodeau, Bruce Joseph. Structure and emplacement of the Sage Hen Flat Plateau, White Mountains, California. M, 1981, University of California, Los Angeles. 133 p.

Bilodeau, W. L. Early Cretaceous tectonics and deposition of the Glance Conglomerate, southeastern Arizona. D, 1979, Stanford University. 176 p.

Bils, Julie. Morphometric trends in tabulate corals, Coralville Member, Cedar Valley Limestone (Devonian), Johnson County, Iowa. M, 1983, University of Iowa. 67 p.

Bilzi, Paul. Study of geology and landsliding in the Monte Bello slopes area near Cupertino, California. M, 1973, Stanford University.

Bin Mohamad, Ramli. Evaluation of seismically-induced liquefaction flow failure of earth dams. D, 1985, Rensselaer Polytechnic Institute. 318 p.

Bina, Craig Richard. Mineralogic transformations and seismic velocity variations in the upper mantle of the Earth. D, 1987, Northwestern University. 87 p.

Binda, Louis S. Petrology of the Archaean sediments in the West Hawk Lake area, Manitoba. M, 1954, University of Manitoba.

Binda, Pier Luigi. Sedimentology and vegetal micropaleontology of the rocks associated with the Cretaceous Kneehills tuff of Alberta (Canada). D, 1970, University of Alberta. 294 p.

Binder, Alan B. Stratigraphy and structure of the Cleomedes Quadrangle of the Moon. D, 1967, University of Arizona.

Binder, Charles Regis. Electron spin resonance; its application to the study of thermal and natural histories of organic sediments. D, 1965, Pennsylvania State University, University Park. 136 p.

Binder, Frank Hewson. A Mississippian conodont fauna from Grundy County, Iowa. M, 1960, Iowa State University of Science and Technology.

Bindschadler, R. A. A time-dependent model of temperate glacier flow and its application to predict changes in the surge-type Variegated Glacier during its quiescent phase. D, 1978, University of Washington. 245 p.

Binford, M. W. Holocene paleolimnology of Lake Valencia, Venezuela; evidence from animal microfossils and some chemical, physical and geological features. D, 1979, Indiana University, Bloomington. 134 p.

Bingaman, Paul T. Late Paleozoic stratigraphy, east central Sublett Range, Power and Oneida counties, Idaho. M, 1979, Idaho State University. 50 p.

Bingert, Neil J. Geology of the northeast one-quarter of the Prineville Quadrangle, north-central Oregon. M, 1984, Oregon State University. 141 p.

Bingham, Clair C. Recent sedimentation trends in Utah Lake. M, 1974, Brigham Young University.

Bingham, Douglas K. Ice motion and heat flow studies on Mount Wrangell, Alaska. M, 1967, University of Alaska, Fairbanks. 142 p.

Bingham, Douglas K. Paleosecular variation of the geomagnetic field in Alaska. D, 1971, University of Alaska, Fairbanks. 155 p.

Bingham, Edgar. Land use, Watauga River basin, North Carolina. M, 1948, University of Tennessee, Knoxville.

Bingham, Henry Todd. A study of three to nine period microseisms recorded at St. Louis. M, 1951, St. Louis University.

Bingham, Michael W. An investigation of local site amplification of ground motion in the Upper Peninsula of Michigan. M, 1978, Michigan Technological University. 188 p.

Bingler, Edward C. The investigation and interpretation of the niobium-bearing Sanostee heavy mineral deposit of the San Juan Basin, northwestern New Mexico. M, 1961, New Mexico Institute of Mining and Technology. 112 p.

Bingler, Edward Charles. Precambrian geology of the Madera Quadrangle, Rio Arriba County, New Mexico. D, 1964, University of Texas, Austin. 363 p.

Binney, Edwin, Jr. The Oregon Basin gas fields, Park County, Wyoming. D, 1953, Yale University.

Binney, William Paul. Lower Carboniferous stratigraphy and base-metal mineralization, Lake Enon, N.S. M, 1975, Queen's University. 93 p.

Binsariti, Abdalla. Newton's iteration in unsaturated flow. M, 1974, New Mexico Institute of Mining and Technology.

Binsariti, Abdalla Abdurazig. Statistical analyses and stochastic modeling of the Cortaro Aquifer in southern Arizona. D, 1980, University of Arizona. 258 p.

Binstock, Jutta Lore Hager. Petrology and sedimentation of Cambrian manganese-rich sediments of the Harlech Dome, North Wales. D, 1978, Harvard University.

Bint, Anthony Neil. Mid-Cretaceous dinoflagellates from the Western Interior, U. S. A. D, 1984, Stanford University. 287 p.

Bintasan, L. The analysis and design of filters for pulse reception. D, 1953, University of Missouri, Columbia.

Bippus, William John. Geochemical prospecting study of the Einstein vein, Madison County, Missouri. M, 1979, University of Missouri, Rolla.

Birak, Donald J. Mineralogy and petrology of the middle Precambrian rocks, Groveland iron mine, Dickinson County, Michigan. M, 1978, Bowling Green State University. 149 p.

Birch, Douglass Wanell. Dominance in marine ecosystems; a new perspective. D, 1982, University of Southern California.

Birch, Francis S. Some heat flow measurements in the Atlantic Ocean. M, 1964, University of Wisconsin-Madison.

Birch, Francis Sylvanus. Geological and geophysical studies of the Barracuda fault zone in the western north Atlantic Ocean. D, 1969, Princeton University. 219 p.

Birch, Michael Joseph. Relationship of the basal "Big Lime" gas producing zone to the lower Newman Limestone (Mississippian) of the Hyden West Pool area, Leslie County, Kentucky. M, 1980, University of Kentucky. 68 p.

Birch, Peter Barrett. Sedimentation in lakes of the Lake Washington drainage basin. M, 1974, University of Washington. 163 p.

Birch, Peter Barrett. The relationship of sedimentation and nutrient cycling to the trophic status of four lakes in the Lake Washington drainage basin. D, 1976, University of Washington. 200 p.

Birch, Rondo O. The geology of the Little Willow District, Wasatch Mountains, Utah. M, 1940, University of Utah. 44 p.

Birchard, George Franklin. Soil radon monitoring for earthquake research; a study of radon concentrations using alpha-particle sensitive films in shallow soil holes along the San Jacinto fault zone in southern California. D, 1978, University of California, Los Angeles. 115 p.

Birchfield, Gene E. A systematic method for introduction of quasi-hydrostatic and quasi-geostrophic approximations into general dynamic equations. D, 1962, University of Chicago. 50 p.

Birchum, Jack Roy. The foraminifera of the Owl Creek Formation in Tippah County, Miss. M, 1955, Mississippi State University. 78 p.

Birchum, Joe Michael. Areal geology of northwestern Dewey County, Oklahoma. M, 1963, University of Oklahoma. 33 p.

Bird, Allan G. Petrology and ore deposits of Schwartzwalder uranium mine, Jefferson County, Colorado. M, 1958, University of Colorado.

Bird, David Neil. Time - term analysis using linear programming and its application to refraction data from the Queen Charlotte Islands. M, 1981, University of British Columbia.

Bird, Debra L. Paleontology of the Meadville Member of the Cuyahoga Formation (Mississippian) in Medina County, Ohio. M, 1988, West Virginia University. 185 p.

Bird, Dennis K. Geology and geochemistry of the Dunes hydrothermal system, Imperial Valley of California. M, 1975, University of California, Riverside. 123 p.

Bird, Dennis Keith. Chemical interaction of aqueous solutions with epidote-feldspar mineral assemblages in geologic systems. D, 1978, University of California, Berkeley. 121 p.

Bird, Donna J. The depositional environments of the late Carboniferous, coal-bearing Sydney Mines Formation at Point Aconi, Cape Breton Island, Nova Scotia. M, 1987, Dalhousie University. 343 p.

Bird, George Peter. Thermal and mechanical evolution of continental convergence zones; Zagros and Himalayas. D, 1976, Massachusetts Institute of Technology. 423 p.

Bird, Gordon W. Metamorphic reactions in the system $K_2O-MgO-Al_2O_3-SiO_2-H_2O$ at water pressures to 10 kilobars. D, 1971, University of Toronto.

Bird, Gordon Winslow. Petrology, geochemistry and geochronology of a pegmatite-gneiss complex (Precambrian) near Viking Lake, Saskatchewan (Canada). M, 1967, University of Alberta. 60 p.

Bird, John M. The Precambrian geology of the area northwest of Brasie Corners, St. Lawrence County, New York. M, 1959, Rensselaer Polytechnic Institute. 10 p.

Bird, John Malcolm. Geology of the Nassau Quadrangle, Rensselaer County, New York. D, 1962, Rensselaer Polytechnic Institute. 204 p.

Bird, John Wilbur. Operational hydrology and reservoir size determination. D, 1970, University of Nevada. 131 p.

Bird, Kenneth John. Biostratigraphy of the Tyee Formation (Eocene), southwest Oregon. D, 1967, University of Wisconsin-Madison. 237 p.

Bird, Kenneth John. Lower Cretaceous stratigraphy and paleontology of the Kukpuk-Kukpowruk region, Northwest Alaska. M, 1964, University of Wisconsin-Madison.

Bird, Melvin Leroy. A stratigraphic study of the Des Moines Series of north central Texas. M, 1956, Northwestern University.

Bird, Melvin Leroy. Distribution of trace elements in olivines and pyroxenes; an experimental study. D, 1971, University of Missouri, Rolla.

Bird, Samuel Oscar. A Pennsylvanian pelecypod fauna from Gaptank Formation, West Texas. M, 1958, University of Wisconsin-Madison.

Bird, Samuel Oscar, II. Upper Tertiary Arcacea of the Midatlantic Coastal Plain. D, 1962, University of North Carolina, Chapel Hill. 197 p.

Bird, Shane R. Lithologic and thickness variations of the Brereton Limestone in southwestern Illinois. M, 1981, Southern Illinois University, Carbondale. 111 p.

Bird, William H. Development of the southern portion of the Platoro Caldera complex and its related mineral deposits, Southeast San Juan Mountains, Colorado. D, 1973, Colorado School of Mines. 186 p.

Birdsall, Barton C. Eastern Gulf of Mexico, continental shelf phosphorite deposits. M, 1979, University of South Florida, Tampa. 87 p.

Birdsall, Edwin F. A seismic and gravimetric study of the Golden Fault near Golden, Colorado. D, 1956, Colorado School of Mines. 51 p.

Birdsall, John Manning. Crossley Clays (New Jersey). M, 1932, Columbia University, Teachers College.

Birdseye, Richard Underwood. Glacial and environmental geology of east-central Jefferson County, Washington. M, 1976, North Carolina State University. 96 p.

Birdseye, Richard Underwood. Quaternary geology of the Five Springs area, northeastern Bighorn Basin, Wyoming. D, 1985, Iowa State University of Science and Technology. 155 p.

Birdwell, Bobby Thomas. The relationship between farmers' soil conservation ethics and soil erosion. D, 1982, Oklahoma State University. 102 p.

Birdwell, Maurice Nixon. The trilobite fauna of the St. Clair Limestone (Silurian) of Independence County, Arkansas. M, 1969, Northeast Louisiana University.

Biren, Helen A. An electron microscope and differential thermal analysis study of the serpentine minerals. D, 1954, Columbia University, Teachers College.

Biren, Helen Antine. An electron microscope and a differential thermal analysis study of the serpentine minerals. D, 1958, Columbia University, Teachers College. 131 p.

Birgisson, Gunnar Ingi. Investigation of the dry and wet engineering behavior of Icelandic lava-gravels. D, 1983, University of Missouri, Rolla. 260 p.

Birk, Robert H. The petrology, petrography, and geochemistry of the Black Jack breccia pipe, Silver Star plutonic complex, Skamania County, Washington. M, 1980, Western Washington University. 107 p.

Birk, W. Dieter. The nature and timing of granitoid plutonism in the Wabigoon volcanic-plutonic belt, northwestern Ontario; geochemistry, Rb/Sr geochronology, petrography and field investigation. D, 1978, McMaster University. 496 p.

Birkeland, Peter Wessel. Pleistocene history of the Truckee area, north of Lake Tahoe, California. D, 1962, Stanford University. 171 p.

Birkenhauer, Henry Francis. A study of house vibrations from quarry blasts. D, 1945, St. Louis University.

Birkenhauer, Henry Francis. The Illinois earthquake of November 23, 1939 and crustal structure east of St. Louis. M, 1941, St. Louis University.

Birkett, Tyson C. Metamorphism of a Cambro-Ordovician sequence in south-eastern Quebec. D, 1981, Universite de Montreal. 268 p.

Birkett, Tyson Clifford. Petrology and chemistry of Archean metabasalts from Thackeray Township, Ontario. M, 1974, Queen's University. 176 p.

Birkhahn, P. C. Neutron capture cross section of small samples. M, 1973, [University of California, San Diego].

Birkhead, Paul Kenneth. A study of some of the Bryozoa and Stromatoporoidea from the Devonian of Missouri. M, 1960, University of Missouri, Columbia.

Birkhead, Paul Kenneth. Stromatoporoidea of Missouri. D, 1964, University of North Carolina, Chapel Hill. 172 p.

Birkhimer, Cheryl Patricia. Mineralogy of pre-Illinoian (Pleistocene) glacial outwash terraces in the Hocking River valley, Ohio. M, 1971, Ohio University, Athens. 65 p.

Birkholz, Donald O. Geology of the Camas Creek area, Meagher County, Montana. M, 1967, Montana College of Mineral Science & Technology. 68 p.

Birle, John David. An investigation of the high pressure, subsolidus phase equilibrium relations in the system $Al_2O_3-GeO_2$ as a model for the ultra high pressure behavior of kyanite. D, 1967, Ohio State University. 81 p.

Birman, Joseph H. Geology of the upper Tick Canyon area, Los Angeles County, California. M, 1950, California Institute of Technology.

Birman, Joseph Harold. Glacial geology of the upper San Joaquin drainage, Sierra Nevada, California. D, 1957, University of California, Los Angeles.

Birmingham, Scott Daniel. The Cripple Creek volcanic field, central Colorado. M, 1987, University of Texas, Austin. 295 p.

Birmingham, Thomas John. The stratigraphy and petroleum geology of the Mission Canyon Formation in Richland County, Montana. M, 1979, University of Colorado. 37 p.

Birney, Carol C. Sedimentology and petrology of the Scots Bay Formation (Lower Jurassic), Nova Scotia, Canada. M, 1985, University of Massachusetts. 325 p.

Birnie, D. J. Geochronology of the Clachnacudainn Gneiss, located near Revelstoke, B.C. M, 1976, University of British Columbia.

Birnie, Richard W. Infrared radiation thermometry of Central American volcanoes. M, 1971, Dartmouth College. 120 p.

Birnie, Richard Williams. The crystal chemistry and paragenetic association of geocronite. D, 1975, Harvard University.

Birnie, Robert I. Late Quaternary environments and archaeology of the Snake Range, east central Nevada. M, 1986, University of Maine. 377 p.

Birnie, Thomas Alexander. Normal faulting in the Indianhead Creek area of the Alberta Rocky Mountains. M, 1960, University of British Columbia.

Biron, Serge. Pétrographie et pétrochimie d'un gîte de pépérites spiritiques des environs du Poste-de-la-Baleine, Nouveau-Québec. M, 1972, Universite Laval. 85 p.

Biros, Daniel J. Geochemistry of a tufa deposit on a highway embankment, Cuyahoga County, Ohio. M, 1981, Kent State University, Kent. 83 p.

Birsa, D. S. The North Horn Formation, central Utah; sedimentary facies and petrography. M, 1974, Ohio State University. 189 p.

Birsa, David S. Subsurface geology of the Palo Duro Basin, Texas Panhandle. D, 1977, University of Texas, Austin. 379 p.

Birse, Donald John. Dolomitization processes in the Paleozoic horizons of Manitoba. M, 1928, University of Manitoba.

Birsoy, Rezan. Coloring of fluorites and problems related to their origin. D, 1977, New Mexico Institute of Mining and Technology. 115 p.

Birsoy, Yuksel K. Groundwater recharge in the plant, soil and atmosphere continuum. D, 1977, New Mexico Institute of Mining and Technology. 145 p.

Birsoy, Yuksel K. Laboratory measurements of tritium in natural waters, Socorro, New Mexico. M, 1972, New Mexico Institute of Mining and Technology.

Bisbee, Wallace A. The paleontology and stratigraphy of the Magdalena Group of northern and central New Mexico. M, 1932, University of New Mexico. 99 p.

Bisby, Curtis G. Depositional environments of the Wood Siding Formation and the Onaga Shale (Pennsylvanian-Permian) in Northeast Kansas. M, 1986, Kansas State University. 73 p.

Biscaye, Pierre Eginton. Mineralogy and sedimentation of the deep-sea sediment fine fraction in the Atlantic Ocean and adjacent seas and oceans. D, 1964, Yale University. 189 p.

Bischke, Richard E. The structure and metamorphism of the Cervandone-Alp Buscagna area, Lepontine Alps, northern Italy. M, 1968, University of Wisconsin-Milwaukee.

Bischke, Richard Edward. A viscoelastic model of convergent plate margins based on the recent tectonics of Shikoku, Japan. D, 1973, University of Colorado. 127 p.

Bischoff, James Louden. Kinetics of crystallization of calcite and aragonite. D, 1966, University of California, Berkeley. 150 p.

Bischoff, William David. Magnesian calcites; physical and chemical properties and stabilities in aqueous solution of synthetic and biogenic phases. D, 1985, Northwestern University. 144 p.

Bish, David Lee. The occurrence and crystal chemistry of nickel in silicate and hydroxide minerals. D, 1977, Pennsylvania State University, University Park. 161 p.

Bishea, Douglas M. Induced polarization measurements and their relationship to hydraulic parameters in Lake Michigan. M, 1983, University of Wisconsin-Milwaukee. 114 p.

Bishko, Donald. Planning for common mineral resources in the urban-suburban environment; a computer simulation approach. D, 1969, Rensselaer Polytechnic Institute. 237 p.

Bishop, A. G. The geology and genesis of the Calabogie iron deposits, Renfrew County, Ontario. M, 1978, Carleton University.

Bishop, Allen David, Jr. The polymerization of silicic acid in dilute aqueous solutions. D, 1967, [University of Houston]. 234 p.

Bishop, Barbara Parks. Correlation of hydrothermal sericite composition with temperature and permeability, Coso Hot Springs geothermal field, Inyo County, California. M, 1985, Stanford University. 45 p.

Bishop, Bobby A. A stratigraphic study of the Kiamichi Formation of the Lower Cretaceous of Texas. M, 1957, Texas Christian University. 171 p.

Bishop, Bobby Arnold. Stratigraphy and carbonate petrography of the Sierra de Pichachos and vicinity, Nuevo Leon, Mexico. D, 1966, University of Texas, Austin. 515 p.

Bishop, D. T. Retrogressive metamorphism in the Seminoe Mountains, Carbon County, Wyoming. M, 1964, University of Wyoming. 49 p.

Bishop, Donald Thomas. Petrology and geochemistry of the Purcell sills, Boundary County, Idaho and adjacent areas. D, 1974, University of Idaho. 167 p.

Bishop, Finley C. Partitioning of Fe^{2+} and Mg between ilmenite and some ferromagnesian silicates. D, 1976, University of Chicago. 137 p.

Bishop, Gale A. Biostratigraphic mapping in the upper Pierre Shale (Upper Cretaceous) utilizing the cephalopod genus Baculites, Cedar Creek Anticline, Montana. M, 1967, South Dakota School of Mines & Technology. 18 p.

Bishop, Gale A. Fossil decapod crustacea from the Pierre Shale (Upper Cretaceous) of South Dakota. D, 1971, University of Texas, Austin.

Bishop, James Corwith, Jr. Petrology of the Blasingame Gabbro, Fresno County, California. M, 1979, California State University, Fresno.

Bishop, Joseph Michael. Formulation, solution and application of a non-linear Miles-Phillips spectral component growth mechanism. D, 1973, New York University.

Bishop, Mary Augusta. Engineering geology study of highwall stability for a proposed lignite mine in Grimes County, Texas. M, 1977, Texas A&M University. 142 p.

Bishop, Michele Gregg. Clastic depositional processes in response to rift tectonics in the Malawi Rift, Malawi, Africa. M, 1988, Duke University. 112 p.

Bishop, Ottey M. Geology and ore deposits of the Richmond Basin area, Gila County, Arizona. M, 1935, University of Arizona.

Bishop, R. S. Shale diapirism and compaction of abnormally pressured shales in South Texas. D, 1977, Stanford University. 180 p.

Bishop, Richard S. Fractures in the Columbia, Missouri 7.5 minute quadrangle. M, 1969, University of Missouri, Columbia.

Bishop, Robert Eugene. Eocene carbonates of Phulji #1, West Pakistan. M, 1959, University of Illinois, Urbana. 24 p.

Bishop, Samuel Wills. Geology and petrography of the Mount Mestas area, Colorado. M, 1952, University of Kansas. 89 p.

Bishop, Sue Lynn. Stratigraphy and areal distribution of depositional facies in the Woodbine Formation, central Texas. M, 1982, Baylor University.

Bishop, Thomas Norton. Identification and deconvolution of deep sea sound waves. M, 1974, University of Washington. 57 p.

Bishop, W. E. The distribution and effects of selected heavy metals in a contaminated lake. D, 1976, Purdue University. 160 p.

Bishop, William C. Geology of southern flank of Santa Susana Mountains, county line to Limekiln Canyon, Los Angeles County, California. M, 1950, University of California, Los Angeles.

Bishop, William Clifton. Study of the Albion-Scipio field of Michigan. M, 1967, Michigan State University. 99 p.

Bishop, William F. The petrology of the Darwin Sandstone Member of the Carboniferous Amsden Formation of west central Wyoming. M, 1957, Miami University (Ohio). 56 p.

Bishop, William H. The geology of the Brentwood-Sulphur area, Washington County, Arkansas. M, 1961, University of Arkansas, Fayetteville.

Bispham, Robert G. Minerals of Brook Hollow, New York. M, 1940, Columbia University, Teachers College.

Bisque, Ramon Edward. Limestone aggregate as a possible source of chemically reactive substances in concrete. M, 1957, Iowa State University of Science and Technology.

Bisque, Ramon Edward. Silicification of argillaceous carbonate rocks. D, 1959, Iowa State University of Science and Technology. 66 p.

Bissada, Kadry Kaddis. Cation-dipole interactions in clay-organic complexes. D, 1968, Washington University. 89 p.

Bissada, Kadry Kaddis. Interpretation of aeromagnetic data from Southwest Missouri. M, 1965, Washington University. 100 p.

Bissada, Mona. Foraminiferal zonation and paleoecologic study of subsurface Neogene sediments, offshore, Louisiana. M, 1968, Washington University. 29 p.

Bissell, Bradford. The Devonian section at Roche Miette, Alberta. M, 1930, Cornell University.

Bissell, Clinton Randall. Stratigraphy of the McAlester Formation (Booch Sandstones) in the Eufaula Reservoir area, east-central Oklahoma. M, 1984, Oklahoma State University. 120 p.

Bissell, Harold Joseph. Pennsylvanian and Lower Permian stratigraphy in the southern Wasatch Mountains, Utah. M, 1936, Iowa State University of Science and Technology.

Bissell, Harold Joseph. Pleistocene sedimentation in southern Utah Valley, Utah. D, 1948, Iowa State University of Science and Technology.

Bissell, Joseph Harold. Pennsylvanian and Lower Permian stratigraphy in the southern Wasatch Mountains, Utah. M, 1936, University of Iowa. 29 p.

Bissell, Joseph Harold. Pleistocene sedimentation in southern Utah Valley, Utah. D, 1948, University of Iowa. 364 p.

Bissell, Malcolm H. Glacial geology of a portion of the area between the Connecticut River and Lake Pocotopaug, Connecticut. M, 1918, Yale University.

Bissell, Malcolm H. The Triassic area of the New Cumberland Quadrangle, Pennsylvania. D, 1921, Yale University.

Bissett, David H. A survey of hydrothermal uranium occurrences in southeastern Arizona. M, 1958, University of Arizona.

Bisson, M. A. Turgor regulation in the marine alga Codium decorticatum (Woodw.) Howe. D, 1976, Duke University. 154 p.

Bissonnette, R. Pétrologie et structure de l'anorthosite de Borgia. M, 1978, University of Ottawa. 226 p.

Biswas, Nirendra Nath. The upper mantle structure of the United States from the dispersion of surface waves. D, 1971, University of California, Los Angeles. 190 p.

Bitar, Richard F. Heavy mineral analysis of beach sands and its use in the interpretation of sedimentary provenance. M, 1977, Wright State University. 149 p.

Bitgood, Charles D. A gravity and magnetic investigation of the Matador Uplift. M, 1968, Texas Tech University. 50 p.

Bither, Katherine M. Quaternary stratigraphy and hydrologic significance of paleodrainage patterns within the lower Androscoggin River valley, Durham, Maine. M, 1988, University of New Hampshire. 129 p.

Bitten, Bernard I. Age of the Potato Hill volcanic rocks near Deary, Latah County, Idaho. M, 1951, University of Idaho. 65 p.

Bittencourt-Netto, Otto. Some aspects of side-looking airborne radar for mapping purposes. M, 1971, Stanford University. 82 p.

Bitter, Mark R. Sedimentology and petrology of the Chicontepec Formation, Tampico-Misantla Basin, eastern Mexico. M, 1986, University of Kansas. 174 p.

Bitter, Peter H. Von *see* Von Bitter, Peter H.

Bitter, Peter Hans von *see* von Bitter, Peter Hans

Bittinger, Morton Wayne. Simulation and analysis of stream-aquifer systems. D, 1968, Utah State University. 118 p.

Bittner, M. Structural analysis and geologic map of Godfrey Ridge. M, 1980, SUNY at Binghamton. 67 p.

Bittner-Gaber, Enid. Geology of selected migmatite zones within the Bitterroot Lobe of the Idaho Batholith, Idaho and Montana. D, 1983, University of Idaho. 187 p.

Bittson, Andrew George. Analysis of gravity data from the Cienega Creek area, Pima and Santa Cruz counties, Arizona. M, 1976, University of Arizona.

Bivens, Wilmer E., Jr. The geology of northwestern Blanco County. M, 1939, University of Texas, Austin.

Bix, Cecil Charles. Geology of Chinato Peak Quadrangle, Trans-Pecos, Texas. D, 1953, University of Texas, Austin.

Biyikoglu, Yusuf. Geoseismic modelling study of the Morrow Formation in eastern Dewey County, Oklahoma. M, 1988, Oklahoma State University. 122 p.

Bizzio, Renato R. Mineral exploration in the St. Francois Mountains, southeastern Missouri, a case study. M, 1981, Memphis State University. 147 p.

Bjarnason, Ingi Thorleifur. Contemporary tectonics of the Wasatch Front region, Utah, from earthquake focal mechanisms. M, 1987, University of Utah. 79 p.

Bjelke, William. Recharge capacity of Purdue gravel pit (Lafayette, Indiana). M, 1960, Purdue University.

Bjerklie, David M. The use of dissolved organic carbon (DOC) as an indicator of ground water contamination. M, 1977, University of New Hampshire. 62 p.

Bjerstedt, Thomas W. Stromatoporoid paleosynecology in the Lucas Dolostone (Middle Devonian) on Kelleys Island, Ohio. M, 1983, Kent State University, Kent. 152 p.

Bjerstedt, Thomas William. Stratigraphy and deltaic depositional systems of the Price Formation (Upper Devonian-Lower Mississippian) in West Virginia; (Volume I and II). D, 1986, West Virginia University. 754 p.

Bjoraker, Robert Wayne. Upper Devonian conodonts from the Lime Creek Formation of northern Iowa. M, 1955, Iowa State University of Science and Technology.

Bjorck, Frederick Richard. Engineering and economic analysis of a west Texas limestone oil field. M, 1965, Stanford University. 139 p.

Bjork, Philip R. Stratigraphy and paleontology of the Slim Buttes Formation in Harding County, South Dakota. M, 1964, South Dakota School of Mines & Technology.

Bjork, Philip Reese. The Carnivora of the Hagerman local fauna (late Pliocene) of southwestern Idaho. D, 1968, University of Michigan. 201 p.

Bjorklund, Thomas Keith. Structure of Horse Mountain Anticline (southwest extension), Brewster County, Texas. M, 1962, University of Texas, Austin.

Bjorlie, Peter F. Stratigraphy and depositional setting of the Carrington shale facies (Mississippian), of the Williston Basin. M, 1978, University of North Dakota. 114 p.

Bjornerud, Marcia G. Structural evolution of a Proterozoic metasedimentary terrane, Wedel Jarlsberg Land, SW Spitsbergen. D, 1987, University of Wisconsin-Madison. 228 p.

Bjornerud, Marcia G. The structure and stratigraphy of Proterozoic rocks in the Thiisfjellet area, Wedel Jarlsberg Land, West Spitsbergen. M, 1985, University of Wisconsin-Madison. 135 p.

Bjornsson, Bjorn Johann. Engineering geology of the Woodrat Mountain area, north-central Idaho. M, 1977, University of Idaho. 86 p.

Bjornstad, Bruce N. Sedimentology and depositional environment of the Touchet Beds, Walla Walla River basin, Washington. M, 1980, Eastern Washington University. 137 p.

Bjurstrom, Stanley Theodore. The petrology of the Ames Limestone in a portion of southeastern Ohio. M, 1960, Ohio University, Athens. 139 p.

Blacet, Philip Merrell. Precambrian geology of the SE 1/4 Mount Union Quadrangle, Bradshaw Mountains, central Arizona. D, 1968, Stanford University. 244 p.

Blacic, James Donald. Hydrolytic weakening of quartz and olivine. D, 1971, University of California, Los Angeles. 222 p.

Blacic, Jan Marie. New microorganisms from the Bitter Springs Formation, late Precambrian of the north-central Amadeus Basin, Australia. M, 1971, University of California, Los Angeles.

Black, Bruce Allan. Nebo Overthrust, southern Wasatch Mountains, Utah. M, 1975, Brigham Young University.

Black, Bruce Allen. Geology of the northern and eastern parts of the Otero Platform, Otero and Chaves counties, New Mexico. D, 1973, University of New Mexico. 158 p.

Black, Bruce Allen. Origin of isolated sandstone masses in shales of late Paleozoic flysch, Ouachita mountains, southeastern Oklahoma. D, 1969, University of Wisconsin-Madison. 265 p.

Black, Bruce Allen. The geology of the northern and eastern parts of the Ladron Mountains, Socorro County, New Mexico. M, 1964, University of New Mexico. 117 p.

Black, Clarence J. The preparation of certain inorganic and organic compounds which may be of use in mineralogical separations. M, 1928, University of Missouri, Rolla.

Black, Curtis Wendell. Hydrogeology of the Hickory Sandstone Aquifer, Upper Cambrian, Riley Formation, Mason and McCulloch counties, Texas. M, 1988, University of Texas, Austin. 194 p.

Black, Cynthia E. Environment of deposition and reservoir facies of the Taylor "B" Sandstone, Cotton Valley Group (Upper Jurassic), Kildare Field, Cass County, Texas. M, 1983, Texas A&M University. 137 p.

Black, David Charles. Trapped helium, neon, and argon in meteorites; boundary conditions on the formation and evolution of the solar system. D, 1970, University of Minnesota, Minneapolis. 112 p.

Black, Douglas F. B. Geology of the Bridger area, west central South Dakota. M, 1962, South Dakota School of Mines & Technology.

Black, Ernest D. A petrographic study of the metamorphic rocks of Little Manicouagan Lake area (Quebec). M, 1958, McGill University.

Black, Frederick Michael. The geology of the Turtle-Flambeau area; Iron and Ashland counties, Wisconsin. M, 1977, University of Wisconsin-Madison. 150 p.

Black, Geoffrey Alan. Geology and geothermal system near Jackson, Beaverhead County, Montana. M, 1983, Montana State University. 111 p.

Black, Gerald Lee. Structural geology of the southeast quarter of the Dutchman Butte Quadrangle, Oregon. M, 1979, Portland State University. 108 p.

Black, Grant Eugene. Geology of the Upper Cretaceous Nacatoch Sand of South Arkansas. M, 1980, Oklahoma State University. 74 p.

Black, J. P. The geology of the Silsbee oil and gas field, Hardin County, Texas. M, 1947, University of Houston.

Black, James M. The Bell River igneous complex. D, 1942, McGill University.

Black, James Murray. Geology and mineral deposits of the eastern contact of the Coast Range Batholith. M, 1936, University of British Columbia.

Black, Janette Louise Young. Soil mineralogy used to distinguish solifluction deposits formed under a periglacial environment on the Boulder Batholith, Jefferson County, Montana. M, 1984, Montana State University. 111 p.

Black, John Ernest. Mineralization and wallrock alteration at the Rawhide gold-silver deposit, Mineral County, Nevada. M, 1988, Stanford University. 100 p.

Black, Kenneth C. A microcomputer groundwater data analysis program. M, 1988, Southern Illinois University, Carbondale. 298 p.

Black, Kenneth D. Petrology, alteration, and mineralization of two Tertiary intrusives, Sierra Blanca igneous complex, New Mexico. M, 1977, Colorado State University. 137 p.

Black, Nancy R. Petrography and diagenesis of the Galena (Middle Ordovician) - Maquoketa (Late Ordovician) contact and the basal Maquoketa phosphorites in eastern Missouri and eastern Iowa, U.S.A. M, 1985, University of Illinois, Urbana. 133 p.

Black, Paul R. A quantitative test of a subduction model for observed heat-flow distributions in the western United States. M, 1974, University of Wyoming. 136 p.

Black, Paul R. Seismic and thermal constraints on the physical properties of the continental crust. D, 1978, Purdue University. 223 p.

Black, Paula Jo. Urban geology of Dallas County, Texas. M, 1974, University of Texas, Arlington. 43 p.

Black, Peter Elliot. Timber and water resource management; a physical and economic approach to multiple use on Denver's municipal watershed. D, 1961, Colorado State University. 146 p.

Black, Philip T. A study of Archean sediments of the Canadian Shield. M, 1949, McGill University.

Black, Philip T. The geology of Malartic gold mine, Halet, Quebec. D, 1954, McGill University.

Black, Robert B. Anomalous sandstone bodies of Morrowan age in Northwest Arkansas. M, 1975, University of Arkansas, Fayetteville.

Black, Robert Bernard. Petrology, sedimentology and depositional environments of the Prairie Grove Member of the Hale Formation (Morrowan) in northwestern Arkansas. D, 1987, University of Tulsa. 231 p.

Black, Robert F. A preliminary geological report of the Huntington Forest (New York). M, 1942, Syracuse University.

Black, Robert Foster. Fabrics of ice wedges. D, 1953, The Johns Hopkins University.

Black, Ross Allen. Geophysical processing and interpretation of Magsat satellite magnetic anomaly data over the U. S. Midcontinent. M, 1981, University of Iowa. 116 p.

Black, Rudolph Allan. A magnetic investigation of the Round Mountain Stock, Castle Valley, Utah. M, 1955, Washington University. 31 p.

Black, Thomas Cummins. Concentration uncertainty for stochastic analysis of solute transport in a bounded, heterogeneous aquifer. D, 1988, Stanford University. 294 p.

Black, Thomas John. A test apparatus for frozen soil in complex stress. D, 1967, Dartmouth College. 224 p.

Black, Tyrone James. Bedrock topography and overburden thickness of northeastern Alpena and eastern Presque Isle counties, Michigan. M, 1977, Kansas State University.

Black, William A. A study of the marked positive gravity anomaly in the northern mid-continent region of the United States. M, 1954, University of Wisconsin-Madison.

Black, William E. A geophysical investigation of the Yuha Desert, Imperial County, California. M, 1974, University of California, Riverside. 71 p.

Black, William W. Geochemistry of the Triassic Watchung basalts (New York). M, 1972, Rutgers, The State University, New Brunswick. 52 p.

Black, William W. The geochronology and geochemistry of the Carolina slate belt of North-central North Carolina. D, 1977, University of North Carolina, Chapel Hill. 118 p.

Blackadar, Donald William. The Aristifats Diatreme; a Proterozoic copper-lead-cobalt-nickel-silver deposit, Northwest Territories. M, 1981, University of Alberta. 327 p.

Blackadar, R. G. Differentiation and assimilation in the Logan Sills, Port Arthur, Ontario. D, 1954, University of Toronto.

Blackadar, R. G. The greenstone intrusions of the Mayo District, Yukon. M, 1951, University of Toronto.

Blackbeer, Lawrence E. Trace fossils; their classification and use in paleoenvironmental interpretation. M, 1973, Brooklyn College (CUNY).

Blackburn, Charles. Structure and metamorphism of the McKim Formation (Precambrian) at Espanola, Ontario. M, 1967, University of Western Ontario. 123 p.

Blackburn, Wilbert Howard. Simulated rainfall studies of selected plant communities and soils in five rangeland watersheds of Nevada. D, 1973, University of Nevada - Mackay School of Mines. 152 p.

Blackburn, William H. The spatial degree of chemical equilibrium in some high grade chemical rocks. D, 1967, Massachusetts Institute of Technology. 230 p.

Blackee, Benson D. Manganese deposits associated with the sedimentary rocks of the Ungava Trough, Labrador-Quebec. M, 1950, Ohio State University.

Blackerby, Bruce Alfred. The Conejo Volcanics (Miocene) in the Malibu Lake area of the western Santa Monica Mountains, Los Angeles County, California. D, 1965, University of California, Los Angeles. 194 p.

Blackett, Robert Earl. Landslide hazards in the Weber River delta near Ogden, Utah. M, 1979, University of Utah. 72 p.

Blackey, Mark E. Detection of subsurface cavities by electrical resistivity, with a field study at Miller Cave, Pa. M, 1984, Pennsylvania State University, University Park. 93 p.

Blackhall, Raymond N. A comparative evaluation of seismic refraction and earth resistivity techniques for the study of Pleistocene deposits near Fairhaven, Ohio. M, 1974, Miami University (Ohio). 101 p.

Blackie, Gary William. A mathematical and digital model of the Athens Water Well Field, Athens, Ohio. M, 1973, Ohio University, Athens. 113 p.

Blackman, Abner. Pleistocene Ostracoda from Bayou Manchac, Ascension Parish, Louisiana. M, 1960, Louisiana State University.

Blackman, Abner. Pleistocene stratigraphy of cores from the southeast Pacific Ocean. D, 1966, University of California, San Diego.

Blackman, Donna Kay. Axial structure of fast spreading mid-ocean ridges; implications for overlapping spreading centers. M, 1986, Massachusetts Institute of Technology. 91 p.

Blackman, John Tristan. Geology of a vent of the Mount Dutton Formation (Miocene), southwest Tushar Mountains, Utah. M, 1985, Kent State University, Kent. 80 p.

Blackman, M. J. The trace element geochemistry of selected serpentinized alpine-type ultramafic intrusions in Vermont. D, 1975, Ohio State University. 133 p.

Blackman, Myron James. A detailed study of the Pleistocene history of a portion of Preble County, Ohio. M, 1970, Miami University (Ohio). 160 p.

Blackman, Thomas Donald. Geophysical investigations of the Hayfield dry lake area in the western Chuckwalla Valley, Southern California. M, 1988, University of California, Riverside. 157 p.

Blackmer, Andrew John. The occurrence, hydrogeology and geochemistry of two salt-water springs in northern Ontario. M, 1984, University of Waterloo. 112 p.

Blackmer, Gale Corless. Engineering geologic parameters and their relationship to roof falls in a coal mine on the Appalachian Plateau, Pennsylvania. M, 1987, Pennsylvania State University, University Park. 87 p.

Blackmer, Joanne. Geology of the Steamboat Springs area, Routt County, Colorado, with special emphasis on thermal springs. M, 1939, University of Colorado.

Blackmon, Paul D. Pleistocene geology of the East Aurora (New York) and vicinity. M, 1955, SUNY at Buffalo.

Blackmur, Robert. Late Cretaceous sedimentation in the La Panza Range, California. M, 1978, University of California, Santa Barbara.

Blackport, Raymond J. A hydrogeologic response model for excavations in large multiple aquifer systems. M, 1980, University of Waterloo.

Blackstone, Donald LeRoy, Jr. Brachiopoda from the Madison Limestone in Montana. M, 1934, University of Montana. 110 p.

Blackstone, Donald LeRoy, Jr. Structure and stratigraphy of the Pryor Mountains, Montana. D, 1936, Princeton University. 87 p.

Blackstone, James P. A study of the so-called subsurface Navarro and Taylor groups of a part of South Texas. M, 1958, Texas A&M University. 42 p.

Blackstone, Robert E. Contact relationships of Laramie Anorthosite and associated rocks, Poe Mountain area, Albany County, Wyoming. M, 1976, University of Wyoming. 187 p.

Blackwelder, Blake W. Cluster analysis of the molluscan assemblages from the Duplin (upper Miocene) and Waccamaw Formation (lower Pliocene), North Carolina and South Carolina. M, 1971, George Washington University.

Blackwelder, Blake Winfield. Morphometry, evolution, and phylozones of the molluscan genus cavilinga (Bivalvia: Lucinidae) in the late Miocene to Holocene of the southern Atlantic Coastal Plain. D, 1972, George Washington University. 115 p.

Blackwelder, Eliot. Post-Cretaceous history of the mountains of central western Wyoming. D, 1914, University of Chicago. 77 p.

Blackwelder, Patricia L. Temperature relationships in coccolith morphology and dimension in fossil and living Emiliania huxleyi (Chrysophyta; Haptophyceae). D, 1976, University of South Carolina. 117 p.

Blackwelder, Patricia Lurie. Electron miscroscopy of quartz sand grains from the Eastern United States and Scotian continental margin. M, 1970, Duke University. 173 p.

Blackwell, Bonnie. Archeometry of five Pleistocene sites as inferred from uranium and thorium isotopic abundances in travertine. M, 1980, McMaster University. 538 p.

Blackwell, Bonnie. Problems in amino acid racemization dating analyses; bones and teeth from the archeological sites Lachaise and Montgaudier (Charente, France). D, 1987, University of Alberta. 654 p.

Blackwell, David Douglas. Terrestrial heat flow determinations in the northwestern United States. D, 1967, Harvard University. 190 p.

Blackwell, David L. Geology of the Park Butte-Loomis Mountain area, Washington (eastern margin of the Twin Sisters Dunite). M, 1983, Western Washington University. 253 4 plates p.

Blackwell, John Michael. Surficial geology and geomorphology of the Harding Lake area, Big Delta Quadrangle, Alaska. M, 1965, University of Alaska, Fairbanks. 91 p.

Blackwell, Michael Lloyd. The influence of pore fluids on the frictional properties of quartzose sandstone. M, 1973, Texas A&M University.

Blackwell, Richard Joseph. The analysis of gravity data over Lake Superior type iron formations. M, 1964, [University of Michigan].

Blackwell, Thomas Sanford. Geology of Big Sunday Creek area, Erath County, Texas. M, 1952, University of Texas, Austin.

Blackwood, Charles F. Dakota Group of the northeast flank of the Canon City Embayment, Colorado. M, 1960, University of Oklahoma. 90 p.

Blackwood, Reginald Frank. The relationship between the Gander and Avalon zones in the Bonavista Bay region, Newfoundland. M, 1976, Memorial University of Newfoundland. 155 p.

Blade, Lawrence V. The petrology of the reddish-brown sediments of the Bonavista Bay area, eastern Newfoundland. M, 1949, Michigan Technological University. 62 p.

Blades-Zeller, Elizabeth L. Ash-flow tuff and interelated volcaniclastic sedimentary rocks exposed at Johnson Shut-ins, Reynolds County, Missouri. M, 1980, University of Kansas. 45 p.

Bladh, Katherine Laing. Petrology of O'Leary Peak volcanics, Coconino County, Arizona. M, 1972, University of Arizona.

Bladh, Katherine Liana. Rapakivi formation of O'Leary Peak porphyry. D, 1976, University of Arizona. 141 p.

Bladh, Kenneth Walter. The clay mineralogy of selected fault gouges. M, 1973, University of Arizona.

Bladh, Kennith Walter. The weathering of sulfide-bearing rocks associated with porphyry-type copper deposits. D, 1978, University of Arizona. 110 p.

Blaeser, Christopher. Recognition of bottom water activity in the Vema Channel on the basis of sediment size distributions. M, 1981, University of Georgia.

Blagbrough, John. The red clay deposits of Otisco Valley. M, 1951, Syracuse University.

Blagbrough, John Wilkinson. Quaternary geology of the northern Chuska Mountains and Red Rock Valley, northeastern Arizona and northwestern New Mexico. D, 1965, University of New Mexico. 138 p.

Blaik, Maurice. An empirical investigation of microseism ground motion at Palisades, New York and Weston, Massachusetts. M, 1953, New York University.

Blain, Christopher F. Regional geochemistry in the Superior Province of the Canadian Shield. D, 1972, Queen's University. 242 p.

Blain, Paul Guy. Infiltration and recharge rates in fractured crystalline rock terrain. M, 1983, San Diego State University. 359 p.

Blair, Alexander Marshall. Surface extraction of non-metallic minerals in Ontario southwest of the Frontenac axis. D, 1965, University of Illinois, Chicago.

Blair, Arthur John, II. Geomagnetic delineation of the basement surface southeast McClain County and southern Cleveland County, Oklahoma. M, 1968, University of Oklahoma. 45 p.

Blair, Barry. Basin analysis of the Topanga Formation, Orange County, California. M, 1978, University of California, Santa Barbara.

Blair, Donald. The distribution of planktonic foraminifera in deep-sea cores from the Southern Ocean, Antarctica. M, 1965, Florida State University.

Blair, Helen Mae. Equilibrium studies in the lithia-alumina-silica system. D, 1931, Ohio State University.

Blair, John Anthony. Surficial geology of the Cimarron Valley from Interstate 35 to Perkins, north-central Oklahoma. M, 1975, Oklahoma State University. 62 p.

Blair, Kevin P. Structural analysis of the Paired Ridges Syncline, Montgomery County, Arkansas. M, 1984, University of Wyoming. 73 p.

Blair, Michael L. Geology of the lower Grider Creek-Grider Ridge area, Siskiyou County, California. M, 1983, University of Nevada. 128 p.

Blair, Michael Reed. Depositional history of the Des Moinesian Series (Pennsylvanian), type region in central Iowa. M, 1978, Iowa State University of Science and Technology.

Blair, R. D. Hydrogeochemistry of an inactive pyritic, uranium tailings basin, Nordic Mine, Elliot Lake, Ontario. M, 1981, University of Waterloo.

Blair, Robert G. and Crawford, James J., Jr. Geology of the magnetite deposits near Russell, Costilla County, Colorado. M, 1960, University of Michigan.

Blair, Robert W., Jr. Weathering and geomorphology of the Pikes Peak Granite in the southern Rampart Range, El Paso County, Colorado. D, 1975, Colorado School of Mines. 200 p.

Blair, Terence C. Alluvial fan deposits of the Todos Santos Formation of central Chiapas, Mexico. M, 1981, University of Texas, Arlington. 134 p.

Blair, Terence C. Paleoenvironments, tectonic and eustatic controls of sedimentation, regional stratigraphic correlation, and plate tectonic significance of the Jurassic-lowermost Cretaceous Todos Santos and San Ricardo formations, Chiapas, Mexico. D, 1986, University of Colorado. 265 p.

Blais, Alan G. Variations of Charlestown Beach, Rhode Island. M, 1986, University of Rhode Island.

Blais, Roger A. A petrologic and decrepitometric study of the gold mineralization at the O'Brien Mine, northwestern Quebec. D, 1954, University of Toronto.

Blais, Roger A. La pétrologie de la région de Lauzon. M, 1950, Universite Laval.

Blaisdell, George L. Influence of the weak bedding plane in Michigan Antrim Shale on laboratory hydraulic fracture orientation. M, 1979, Michigan Technological University. 98 p.

Blaisdell, Robert Clark. Stratigraphy and foraminifera of the Matilija, Cozy Dell, and "Coldwater" formations near Ojai, California. M, 1955, University of California, Berkeley. 93 p.

Blaisdell, Thomas. Computer simulation of sedimentation. M, 1972, Stanford University.

Blake, Alan Brian. Engineering geology of a selected area in Baldwin County, Alabama. M, 1978, University of Alabama.

Blake, Bonnie Janine. Geochemistry of the epigenetic uranium-bearing Cretaceous Lakota Formation, southern Black Hills, South Dakota. M, 1988, South Dakota School of Mines & Technology.

Blake, Brenda Jean. An experimental investigation of dislocation glide in olivine. M, 1976, Massachusetts Institute of Technology. 40 p.

Blake, Daniel B. Gosport Eocene Ostracoda from Little Stave Creek, Alabama. M, 1949, Louisiana State University.

Blake, Daniel B. The stratigraphy of the Au Train Formation at Au Train Falls and Wagner Falls, Alger County, northern Michigan. M, 1962, Michigan State University. 49 p.

Blake, Daniel Bryan. Skeletal structures on selected asteroids of the order Phanerozonia. D, 1966, University of California, Berkeley. 386 p.

Blake, David Edward. The geology of the Grissom area, Franklin, Granville, and Wake counties, North Carolina; a structural and metamorphic analysis. M, 1986, North Carolina State University. 300 p.

Blake, David Frederick. Crystallochemistry and diagenesis of inorganic and biogenic magnesian calcite. D, 1983, University of Michigan. 266 p.

Blake, David Frederick. The sequence and mechanism of low-termperture dolomite formation; calcian dolomites from a Pennsylvanian echinoderm. M, 1980, University of Michigan.

Blake, David William. Geology, alteration, and mineralization of the San Juan Mine area, Graham County, Arizona. M, 1971, University of Arizona.

Blake, Donald A. W. The geology of the Forget Lake and Nevins Lake map-areas, North Saskatchewan. D, 1952, McGill University.

Blake, G. H. The distribution of benthic foraminifera in the outer borderland and its relationship to Pleistocene marl biofacies. M, 1976, University of Southern California. 143 p.

Blake, Gregory Howard. The faunal response of California continental margin benthic foraminifera to the oceanographic and depositional events of the Neogene. D, 1985, University of Southern California.

Blake, J. Roger. The influence of coastal and sea floor geometry on natural electromagnetic variations of 10^{-4} to 10^{3} Hz. D, 1968, University of Alaska, Fairbanks. 440 p.

Blake, Jerry Wayne. Hydrothermal shale and clay-iron compound studies. D, 1968, Ohio State University. 111 p.

Blake, John W. Geology of the Bald Mountains intrusive, Ruby Mountains, Nevada. M, 1964, Brigham Young University. 35 p.

Blake, Mabel Louise. The Bushberg Sandstone in the vicinity of St. Louis. M, 1948, Washington University. 97 p.

Blake, Milton Clark, Jr. Structure and petrology of low-grade metamorphic rocks, Blueschist facies, Yolla Bolly area, Northern California. D, 1965, Stanford University. 123 p.

Blake, Natalie Ruth. Partial oxidation of C_4 hydrocarbons. D, 1987, University of Minnesota, Minneapolis. 168 p.

Blake, Oliver Duncan. The geology of Gallia County, Ohio. D, 1952, Ohio State University. 142 p.

Blake, Robert Whitney. The structural geology of the tectonized ultramafic suite of the Table Mountain Massif, Bay of Islands Complex, Newfoundland. M, 1982, SUNY at Albany. 188 p.

Blake, Rolland Laws. A study of iron silicate minerals in iron-formations of the Lake Superior region, with emphasis on the Cuyuna District, Minnesota. D, 1958, University of Minnesota, Minneapolis. 147 p.

Blake, Thomas Ford. Depositional environments of the Simmler Formation in southern Cuyama Valley, Santa Barbara and Ventura counties, California. M, 1981, California State University, Northridge. 151 p.

Blake, Vachel. A paper on the Chattanooga black shale. M, 1910, Vanderbilt University.

Blake, Weston, Jr. Geomorphology and glacial geology in Nordaustlandet, Spitsbergen. D, 1962, Ohio State University. 492 p.

Blakeley, David C. Subsurface geology of north-central Pawnee County, Oklahoma. M, 1959, University of Oklahoma. 75 p.

Blakely, Richard J. Short geomagnetic polarity intervals in marine magnetic profiles (Pacific Ocean). D, 1972, Stanford University. 59 p.

Blakely, Robert Fraser. Some aspects of 4-10 second microseisms recorded at Bloomington, Indiana. D, 1974, Indiana University, Bloomington. 194 p.

Blakeman, William B. A study of the mineralogic and magnetic characteristics of metamorphosed iron formation from the Julian deposit, Wabush Lake area, Labrador. M, 1968, University of Vermont.

Blakeney, Beverly A. Origin and paleoenvironmental significance of soft-sediment deformation structures in the Upper Cretaceous Parkman Sandstone, Northwest Wyoming, and Neogene sediments, Mal Pais, Costa Rica. M, 1986, Pennsylvania State University, University Park. 170 p.

Blakeney, R. S. The geochemistry and stratigraphy of the Herbert River Limestone Member of the Mississippian Windsor Group in Atlantic Canada. M, 1974, Acadia University.

Blakestad, Robert Byron, Jr. Geology of the Kelly mining district, Socorro County, New Mexico. M, 1978, University of Colorado.

Blakey, Ronald Clyde. Geology of the Paria northwest quadrangle, Kane County, Utah. D, 1970, University of Utah. 171 p.

Blakey, Ronald Clyde. Stratigraphy, depositional environments, and economic geology of the Moenkopi Formation, southeastern Utah. D, 1973, University of Iowa. 269 p.

Blakney, William Gilbert G. Application of stereometer instruments to cadastral mapping. M, 1959, Ohio State University.

Blanc, Robert Parmelee. Geology of the Deep Spring Valley area, White-Inyo Mountains, California. M, 1958, University of California, Los Angeles.

Blanchard, Allan Marc. Relationships between grain shape characteristics and source areas or depositional histories in Jackson Hole, Wyoming; multivariate rotation method of quantitative grain shape analysis. M, 1986, Lehigh University.

Blanchard, Chrystian. La carte géotechnique de l'Ile Jésus (Laval), Québec. M, 1988, Ecole Polytechnique. 76 p.

Blanchard, Frank Nelson. Drift analysis as a guide to the character of the bedrock in Oneida County, Wisconsin. D, 1960, University of Michigan. 198 p.

Blanchard, Frank Nelson. Thermoluminescent properties of fluorite in relation to geologic occurrence. M, 1954, University of Michigan.

Blanchard, J. E. Resistivity methods of diamond drill-hole surveying. D, 1952, University of Toronto.

Blanchard, Jonathan E. Resistivity methods of diamond drill hole surveying. D, 1951, University of Toronto.

Blanchard, Kenneth Stephen. Quaternary alluvium of the Washita River valley in western Caddo County, Oklahoma. M, 1951, University of Oklahoma. 49 p.

Blanchard, Margaret C. Investigation of the shallow fractured dolomite aquifer in Door County, Wisconsin. M, 1988, University of Wisconsin-Madison. 186 p.

Blanchard, Marie-Claude. Geochemistry and petrogenesis of the Fisset Brook Formation, western Cape Breton Island, Nova Scotia. M, 1982, Dalhousie University. 225 p.

Blanchard, Maxwell B. A contribution to the analysis of near Earth interplanetary dust particles. M, 1968, San Jose State University. 55 p.

Blanchard, Melbourne Kenneth. Properties of refractories from Pacific Northwest olivines. M, 1936, University of Washington. 83 p.

Blanchard, Paul Edward. Fluid flow in compacting sedimentary basins. D, 1987, University of Texas, Austin. 206 p.

Blanchard, Ralph C. The geology of the western Buckskin Mountains, Yuma County, Arizona. D, 1913, Columbia University, Teachers College.

Blanchard, Richard Lee. Uranium decay series disequilibrium in age determination of marine calcium carbonates. D, 1963, Washington University. 175 p.

Blanchard, William O. The geography of southwestern Wisconsin. D, 1921, University of Wisconsin-Milwaukee.

Blanchard, William O. The geography of the Sparta-Tomah quadrangles. M, 1917, University of Wisconsin-Milwaukee.

Blancher, Donald W. Origin of the Hazel Patch Sandstone Member of the Lee Formation of Southeast Kentucky. M, 1970, University of Kentucky. 57 p.

Blancher, Donald W., Jr. Sediments and depositional environments of the Maccrady Formation and the Greenbrier Group (Mississippian) of the Hurricane Ridge Syncline of southwestern Virginia and West Virginia. D, 1974, Virginia Polytechnic Institute and State University.

Blanco, Julius M. Geology, fracture studies and lineament analysis in the Tracy, Fountain Run and Gamaliel 7.5 minute quadrangles, south-central Kentucky. M, 1985, University of Toledo. 257 p.

Blanco, Stephen R. Geology of the Rye-Colorado City area, Pueblo and Huerfano counties, Colorado. M, 1971, Colorado School of Mines. 48 p.

Bland, Alan Edward. Geochemistry of the Meadow Flats Complex (Precambrian), Orange County, North Carolina. M, 1972, University of North Carolina, Chapel Hill. 49 p.

Bland, Alan Edward. Trace element geochemistry of volcanic sequences of Maryland, Virginia, and North Carolina and its bearing on the tectonic evolution of the Central Appalachians. D, 1978, University of Kentucky. 350 p.

Bland, Douglas M. Geology and structure of the Pocatello Range, Bannock County, Idaho. M, 1982, University of Wyoming. 73 p.

Bland, Michael J. Holocene geologic history of Little Sarasota Bay, Florida. M, 1985, University of South Florida, Tampa. 101 p.

Blane, John P. An historical geography of the Amur-Ussuri region of the USSR. M, 1953, University of Cincinnati. 264 p.

Blaney, Geoffrey W. Lateral response of a single pile in overconsolidated clay to relatively low frequency harmonic pile-head loads and harmonic ground surface loads. D, 1983, [University of Houston]. 505 p.

Blank, Efrom. Petrology of the First Watchung Mountain at Prospect Park Quarry, Patterson, New Jersey. M, 1960, New York University.

Blank, Horace Richard, Jr. Geology of Bull Valley District, Washington Coounty, Utah. D, 1959, University of Washington. 177 p.

Blank, Horace Richard, Jr. Rate of sedimentation in the Skagit Bay region by radium analysis. M, 1950, University of Washington. 28 p.

Blank, Marsha Ann. Mineralogy of aerosols and sediments from the western North Pacific; evidence of major eolian inputs from Asia. M, 1984, University of Miami. 12 p.

Blank, Richard. Paleoceanography of the southeastern Indian Ocean and paleoglacial history of Antarctica as revealed by Late Cenozoic deep-sea sediments. M, 1973, University of Hawaii at Manoa. 61 p.

Blank, Richard G. Correlation of Cenozoic deep sea sediments of the equatorial Pacific Ocean; an example of a new chronostratigraphic system of measurement. D, 1977, University of Washington. 138 p.

Blank, Robert Raymond. Genesis of silica and calcite cemented soils in Idaho. D, 1987, University of Idaho. 308 p.

Blank, Robert Raymond. Physical, chemical, micromorphological, and mineralogical comparisons of long-term cultivated and adjacent virgin soils in north-central South Dakota. M, 1984, University of Idaho. 239 p.

Blankenship, Asa Lee, Jr. Areal geology of Painthorse Quadrangle, Culberson County, Texas. M, 1952, University of Texas, Austin.

Blankenship, Dana G. Net shore-drift of Mason County, Washington. M, 1983, Western Washington University. 172 p.

Blankenship, Donald D. P-wave anisotropy in the high polar ice of East Antarctica. M, 1982, University of Wisconsin-Madison. 143 p.

Blankenship, John C., Jr. Stratigraphy and petrology of the El Paso Formation (Lower Ordovician) in the Silver City Range, New Mexico. M, 1972, [University of Houston].

Blankenship, Joseph Croxton. Geology of central-northeastern Burleson County, Texas. M, 1955, Texas A&M University. 76 p.

Blankenship, O. The effects of sedimentation and other geologic processes on a small lake in Texas. M, 1977, East Texas State University.

Blankenship, Robert William. The stratigraphy and petrography of the Hale Formation from Hale Mountain, Arkansas, to Fort Gibson, Oklahoma. M, 1962, University of Tulsa. 153 p.

Blankenship, William Dave. Sedimentology of the outer Texas coast. M, 1953, University of Texas, Austin.

Blanton, Carol Lynn. Quantitative analyses of Tournaisian, Visean, and Namurian (Carboniferous) brachiopod biogeography; comparison of paleobiogeography with paleogeography. M, 1984, University of Mississippi. 130 p.

Blanton, Jackson O. The subsurface frontal zone beneath the subtropical convergence in the northeast Pacific Ocean. D, 1968, Oregon State University. 93 p.

Blanton, T. L., III. The Cavern Gulch faults and the Fountain Creek Flexure, Manitou Spur, Colorado. M, 1973, Syracuse University.

Blanton, Thomas Lindsay, III. Effect of strain rates from 10^{-2} to 10 S^{-1} in triaxial compression tests on three rocks. D, 1976, Texas A&M University. 77 p.

Blanz, R. E. The compilation of diagrammatic transects and their application to preliminary environmental assessment of water resources development in Texas. D, 1976, Texas A&M University. 234 p.

Blaricom, Richard Van see Van Blaricom, Richard

Blasch, Sheila R. A fluid inclusion study of the Jumbo lead mine, Linn County, Kansas. M, 1986, University of Missouri, Kansas City. 118 p.

Blasdel, Eugene Sherwood. Geology of the Granite Creek area, Gros Ventre mountains, Wyoming. M, 1969, University of Michigan.

Blasius, Karl Richard. Topical studies of the geology of the Tharsis region of Mars. D, 1976, California Institute of Technology. 96 p.

Blatt, Harvey. Sedimentation in New Jersey beaches. M, 1958, University of Texas, Austin.

Blatt, Harvey. The character of quartz grains in sedimentary rocks and source rocks. D, 1963, University of California, Los Angeles. 212 p.

Blatter, C. L. The interaction of clay minerals with distilled water and saline solutions at elevated temperatures. D, 1974, SUNY at Binghamton. 120 p.

Blau, Barbara J. Sources of mineral aggregates in Cape Cod glacial deposits (Massachusetts). M, 1957, Columbia University, Teachers College.

Blau, Jan G. Geology of southern part of the James Peak Quadrangle, Utah. M, 1975, Utah State University. 55 p.

Blau, Peter E. Petrology of the Goodland Limestone (Lower Cretaceous), southeastern Oklahoma. M, 1961, University of Oklahoma. 148 p.

Blauch, Matthew E. Dolomitization, diagenesis and paleoenvironments of a Middle to Upper Silurian shallowing upward sequence exposed near Peebles, Ohio. M, 1988, University of Akron. 162 p.

Blauser, William H. Geology of the southern Sierra de Catorce and stratigraphy of the Taraises Formation in north central Mexico. M, 1979, University of Texas, Arlington. 80 p.

Blaustein, Morton Katz. Relation of petrology and structure to productivity in a stratigraphic trap, Lindemann (McMillan Sand) oil field, Runnels County, Texas. D, 1955, Stanford University. 215 p.

Blauvelt, Bessie. The continental shelf of Western Europe. M, 1925, Columbia University, Teachers College.

Blauvelt, Robert P. Fracture patterns in a part of the Gettysburg Intrusive; Devil's Den, Gettysburg National Military Park, Pennsylvania. M, 1978, Rutgers, The State University, Newark. 83 p.

Blaxland, Alan. Occurrence of zinc in granitic biotite. M, 1970, Washington University. 66 p.

Blay, Oliver T. Delineation of uraniferous granitoids using computing and statistical analysis of stream sediment geochemical data. M, 1982, University of Georgia.

Blazenko, Eugene J. Geology of the South Erick gas area, Beckham and Greer counties, Oklahoma. M, 1964, University of Oklahoma. 88 p.

Blazey, Edward Brice. Fossil flora of the Mogollon rim. D, 1971, Arizona State University. 207 p.

Blazey, Philip Thomas. The areal geology of the Mount Peale 2NW Quadrangle, San Juan County, Utah. M, 1967, Texas Tech University. 131 p.

Blazquez, R. Endochronic model for liquefaction of sand deposits as inelastic two-phase media. D, 1978, Northwestern University. 226 p.

Blecha, Matthew. Origin of mineralized breccias in Batachewana area, Ontario (Canada). M, 1967, McGill University.

Blecha, Matthew. Origin of mineralized breccias in Batchawana area, Ontario (Canada). D, 1969, McGill University.

Blechschmidt, Gretchen Louise. Paleoenvironmental characteristics of Paleocene sedimentation in the deep sea as evidenced by calcareous nannofossils. D, 1983, University of Washington. 235 p.

Bleem, Jeanice C. Quaternary geology of the Darling Creek drainage, Williams Fork Mountains, Colorado. M, 1982, University of Wisconsin-Milwaukee. 123 p.

Blegen, Ronald Paul. Field comparison of groundwater sampling methods. M, 1988, University of Nevada, Las Vegas. 115 p.

Bleifuss, Rodney L. Correlation of magnetite content with magnetic susceptibility measurements of Minnesota Pre-cambrian rocks. M, 1952, University of Minnesota, Minneapolis. 46 p.

Bleifuss, Rodney L. The origin of the iron ores of southeastern Minnesota. D, 1966, University of Minnesota, Minneapolis. 142 p.

Bleil, C. E. Streaming potentials in spherical-grain sands. D, 1953, University of Oklahoma. 73 p.

Bleiwas, Donald I. The McNamara-Garnet Range transition (Precambrian Missoula Group). M, 1977, University of Montana. 91 p.

Blencoe, James G. An experimental study of muscovite-paragonite stability relations. D, 1974, Stanford University. 216 p.

Blenkinsop, John. An integrated geophysical study of the Zuni lineament in New Mexico. M, 1966, University of British Columbia.

Blenkinsop, John. Computer-assisted mass spectrometry and its application to rubidium-strontium geochronology. D, 1972, University of British Columbia.

Bless, Stephen J. Production of high pressures with a magnetic pinch. M, 1968, Massachusetts Institute of Technology. 48 p.

Blessing, Dennis R. Computer methods for the prediction of the coking and blending potential of selected Ohio coals. M, 1986, University of Toledo. 442 p.

Bleuer, Ned Kermit. Geology of the southeast quarter of the Shelbyville, Illinois, Quadrangle. M, 1967, University of Illinois, Urbana.

Bleuer, Ned Kermit. Glacial stratigraphy of south-central Wisconsin. D, 1971, University of Wisconsin-Madison. 186 p.

Blevens, Dale M., Jr. Computer assisted structural analysis of the western termination of the Flat Top Anticline, Carbon County, Wyoming. M, 1984, University of Wyoming. 119 p.

Blewett, Kenneth W. Sydney Lake gneiss and sulphide deposit, northwestern Ontario. M, 1976, University of Manitoba.

Blexrud, Owen Hefte. The areal geology and stratigraphy of the Prairie du Chien Quadrangle, Wisconsin. M, 1947, University of Wisconsin-Madison. 84 p.

Blick, Nicholas Hammond. Stratigraphic, structural and paleogeographic interpretation of upper Proterozoic glaciogenic rocks in the Sevier orogenic belt, northwestern Utah. D, 1979, University of California, Santa Barbara. 708 p.

Blickle, Arthur H. Ohio psaronii. D, 1940, University of Cincinnati. 62 p.

Blickle, Arthur H. On the genus Callixylon from the Ohio Black Shale. M, 1936, University of Cincinnati. 41 p.

Blickwede, Jon Frederic. Stratigraphy and petrology of Triassic(?) "Nazas Formation," Sierra de San Julian, Zacatecas, Mexico. M, 1981, University of New Orleans. 100 p.

Blickwedel, Roy. Rapid-quench experiments using dilute chloride solutions applied to the system MgO-SiO2-Al2O3-H2O-HC1. M, 1983, Indiana University, Bloomington. 50 p.

Bliefnick, Deborah Marie. Sedimentology and diagenesis of bryozoan- and sponge-rich carbonate buildups, Great Bahama Bank. D, 1980, University of California, Santa Cruz. 299 p.

Blij, Harm Jan de. The physiographic provinces and cyclic erosion surfaces of Swaziland. D, 1959, Syracuse University. 219 p.

Blinman, Eric. Pollen analysis of Glacier Peak and Mazama volcanic ashes. M, 1978, Washington State University. 49 p.

Blinn, L. J. A study of seismic surface waves on the Reykjanes Ridge. M, 1975, Dalhousie University.

Bliss, Douglas Allen. Design, construction, and field monitoring of an experimental tailings impoundment,

Coeur d'Alene mining district, Idaho. M, 1982, University of Idaho. 165 p.

Bliss, Eleanora Frances and Jonas, Anna Isabel. Relation of the Wissahickon micagneiss to the Shenandoah Limestone and to the Octoraro mica-schist of the Doe Run-Avondale District, Coatsville Quadrangle, Pa. D, 1912, Bryn Mawr College. 64 p.

Bliss, Eleanora Frances. Structural relations of the rock formations in the Doe Run region, Pa. and their bearing upon the stratigraphy of the Piedmont Plateau. M, 1904, Bryn Mawr College.

Bliss, Franklin E. The "concrete layer" (a unit resembling the Hitz Bed) of the Saluda Formation (Upper Ordovician) in Frankling County, Indiana. M, 1984, Miami University (Ohio). 121 p.

Bliss, James D. Selected topics in the geochemistry of mercury. M, 1974, Arizona State University. 109 p.

Bliss, Miranda C. Hoch. Strontium isotopic composition and abundances of strontium, REE, and other trace elements as indicators of fluid mixing during the crystallization of nonsulfide minerals from the Elmwood Mine, Tennessee. M, 1983, Miami University (Ohio). 93 p.

Bliss, Neil Welbourne. A comparative study of two ultramafic bodies at the SW end of the Manitoba Nickel Belt; with special reference to the chromite mineralogy. D, 1973, McGill University.

Bliss, Neil Welbourne. The Deweras Formation (upper Precambrian) south of the Umfuli River, Rhodesia. M, 1965, McGill University.

Blissenback, Erich. The geology of alluvial fans in Arizona. M, 1951, University of Arizona.

Bliven, F. L., Jr. Sediment transport phenomena generated by a combined wave and steady flow condition. D, 1977, North Carolina State University. 144 p.

Blixt, John Elmer. Geology of the North Moccasin Mountains, Fergus County, Montana. M, 1932, Montana College of Mineral Science & Technology. 40 p.

Bloch, John Daniel. Diagenesis of the Upper Cambrian Lamotte Sandstone in the southwest quadrant of the Rolla 1° × 2° Quadrangle, Missouri. M, 1985, Washington University. 80 p.

Bloch, S. Mineralogy and geochemistry of metalliferous sediments from the Line Islands Oceanic Formation, equatorial East Pacific. D, 1978, George Washington University. 332 p.

Block, Douglas A. Glacial geology of the north half of Barnes County, North Dakota. D, 1965, University of North Dakota. 162 p.

Block, Douglas Alfred. The geology of the Deadwood Formation between Bear Butte and Spring Creeks, Black Hills, South Dakota. M, 1952, University of Iowa. 40 p.

Block, Fred. A multivariate chemical characteristic of rocks from the Monteregian petrographic province Quebec, Canada. D, 1972, Pennsylvania State University, University Park. 184 p.

Block, Fred. Zircons in some pegmatites and associated country rocks of the New Jersey Highlands. M, 1964, Rutgers, The State University, New Brunswick. 46 p.

Blodget, H. W. Lithology mapping of crystalline shield test sites in western Saudi Arabia using computer-manipulated multispectral satellite data. D, 1977, George Washington University. 223 p.

Blodget, Herbert. Geological and geomorphological comparison of space craft and aircraft imagery in selected areas of the Arabian Peninsula. M, 1971, George Washington University.

Blodgett, Daniel D. Hydrogeology of the San Augustin Plains, New Mexico. M, 1972, New Mexico Institute of Mining and Technology.

Blodgett, Jack W. Some depositional features along the east coast of Florida between St. Augustine and Fernandine Beach. M, 1956, Emory University. 45 p.

Blodgett, Robert B. Biostratigraphy of the Ogilvie Formation and limestone and shale member of the McCann Hill Chert (Devonian), East-central Alaska and adjacent Yukon Territory. M, 1978, University of Alaska, Fairbanks. 142 p.

Blodgett, Robert B. Taxonomy and paleobiogeographic affinities of an early Middle Devonian (Eifelian) gastropod faunule from the Livengood Quadrangle, east-central Alaska. D, 1987, Oregon State University. 139 p.

Blodgett, Robert Hugh. Comparison of Oligocene and modern braided stream sedimentation on the High Plains. M, 1974, University of Nebraska, Lincoln.

Bloemker, J. Mark. Subsurface stratigraphy and paleoecology of the Saluda Formation (Upper Ordovician) of Indiana. M, 1981, Ball State University. 105 p.

Bloeser, Bonnie. Structurally complex microfossils from shales of the late Precambrian Kwagunt Formation (Walcott Member, Chuar Group) at the eastern Grand Canyon, Arizona. M, 1980, University of California, Los Angeles.

Blohm, Susan Jeanne. Geologists and land use; community communication. M, 1974, University of California, Davis. 173 p.

Blois, Roland de. Petrography and petrology of rocks of the Shickshock Series [Quebec]. M, 1949, Universite Laval.

Blok, Jack H. Man's intervention in the evolution of the northern coastal area of the Netherlands. M, 1967, University of California, Los Angeles. 136 p.

Blom, Ronald George. Geologic mapping from thematic mapper simulator images in the Ubehebe Peak and Dry Mountain quadrangles, eastern California. D, 1987, University of California, Santa Barbara. 372 p.

Blom, Ronald George. Spectral reflectance studies of plutonic rocks in the 0.45 to 2.45 micron region. M, 1978, California State University, Northridge. 341 p.

Blomberg, N. E. The carbon-nitrogen ratios in Hawaiian soils. M, 1958, University of Hawaii. 22 p.

Blome, Charles David. Upper Triassic Radiolaria from eastern Oregon and British Columbia. D, 1981, University of Texas at Dallas. 327 p.

Blomquist, John T. Current directions in the Diamond Peak Formation, an upper Mississippian-lower Pennsylvanian clastic wedge, east-central Nevada. M, 1971, University of Nevada. 79 p.

Blomshield, Richard J. Superposed deformations on the Isles of Shoals, Maine-New Hampshire. M, 1975, University of New Hampshire. 57 p.

Blondeau, Kenneth M. Sedimentation and stratigraphy of the Mount Rogers Formation, Virginia. M, 1975, Louisiana State University.

Blood, Elizabeth Reid. Surface water hydrology and biogeochemistry of the Okefenokee Swamp watershed. D, 1981, University of Georgia. 206 p.

Blood, Pearl. A profile study of erosion surfaces in Pennsylvania and Maryland. M, 1924, Columbia University, Teachers College.

Blood, W. Alexander. Geology, history and economics of the Sunnyside Mine, Eureka mining district, San Juan County, Colorado. D, 1968, Colorado School of Mines.

Blood, W. H. A simulation method for water supply firm yield evaluation. D, 1977, Utah State University. 114 p.

Bloodgood, Mary Anne. Deformational history, stratigraphic correlations and geochemistry of eastern Quesnel terrane rocks in the Crooked Lake area, east central British Columbia. M, 1987, University of British Columbia. 154 p.

Bloodworth, Billy Lloyd. Subsurface correlations of Trinity and Upper Jurassic formations of parts of Arkansas, Louisiana, and Texas. M, 1941, University of Texas, Austin.

Bloom, Arthur L. Late Pleistocene changes of sea level in southwestern Maine. D, 1959, Yale University.

Bloom, Duane N. Geology of the Horseshoe District and ore deposits of the Hilltop Mine, Park County, Colorado. D, 1965, Colorado School of Mines. 211 p.

Bloom, Harold. Field methods for the determination of nickel using dimethylglyoxime. M, 1961, Colorado School of Mines. 28 p.

Bloom, James Clifford. A study of synthetic seismograms. M, 1961, Indiana University, Bloomington. 38 p.

Bloom, John R. Correlating Ordovician strata by sedimentary petrography. M, 1927, University of Minnesota, Minneapolis. 25 p.

Bloom, Jonathan I. A mineralogical model of the Floridan Aquifer in the Southwest Florida Water Management District. M, 1982, University of Florida. 140 p.

Bloom, Laurie. The relationships among river, beach and sands in the southern Santa Barbara littoral cell, Ventura County, California, Fourier grain shape analysis. M, 1979, University of Southern California.

Bloom, Lynda B. Dispersion and mode of occurrence of uranium in stream sediments. M, 1980, Queen's University. 308 p.

Bloom, M. A. A subsurface geologic study of northern Brewster County, Texas. M, 1988, Sul Ross State University.

Bloom, Margaret Louise Haroldson. Cretaceous intrusive rocks and the deformational fabric of thrusts in the Sawtooth Mountains, Disturbed Belt, northwestern Montana, U.S.A. M, 1986, University of Nebraska, Lincoln. 69 p.

Bloom, Mark Alan. Open-system studies of water-feldspathic sand interactions at 200°C and 1 kilobar; an experimental investigation. M, 1986, Texas A&M University.

Bloom, Mark S. Mineral paragenesis and contact metamorphism in the Jarilla Mountains, Orogrande, NM. M, 1975, New Mexico Institute of Mining and Technology. 107 p.

Bloom, Mark Stephen. Geochemistry of fluid inclusions and hydrothermal alteration in vein- and fracture-controlled mineralization, stockwork molybdenum deposits. D, 1983, University of British Columbia.

Bloomberg, Diane. Cone-penetrometer exploration of sinkholes; stratigraphy and soil properties. M, 1987, University of South Florida, Tampa. 132 p.

Bloomer, Alfred Travers. A regional study of the middle Devonian Dundee dolomite in the Michigan Basin. M, 1969, Michigan State University. 76 p.

Bloomer, Daniel R. A hydrographic investigation of Winyah Bay, South Carolina and the adjacent coastal waters. M, 1973, Georgia Institute of Technology. 57 p.

Bloomer, Gail. Petrology and sedimentation of the Schenectady-Frankfort Formation (Ordovician), New York State. D, 1972, Harvard University.

Bloomer, Gail Elizabeth. Geology, mineralogy, and geochemistry of the Iron Crown calcic iron skarn deposit, Vancouver Island, British Columbia. M, 1986, Washington State University. 115 p.

Bloomer, Philip A., Jr. Subsurface study of the Delhi area, Franklin and Richland parishes, Louisiana. M, 1947, Louisiana State University.

Bloomer, Richard Rodier. Basal Cambrian volcanics in the central Blue Ridge Mountains of Virginia. M, 1941, University of Virginia. 73 p.

Bloomer, Richard Rodier. Geology of the Christmas and Rosillos mountains, Brewster County, Texas. D, 1949, University of Texas, Austin.

Bloomer, Robert Oliver. Geology of the Blue Ridge in the Buena Vista Quadrangle, Virginia. D, 1941, University of North Carolina, Chapel Hill. 68 p.

Bloomer, Robert Oliver. The geology of the Piedmont in Chesterfield and Henrico counties, Virginia. M, 1938, University of Virginia. 53 p.

Bloomer, Sherman Harrison. Mariana Trench; petrologic and geochemical studies; implications for the structure and evolution of the inner slope. D, 1982, University of California, San Diego. 286 p.

Bloomfield, G. D. Geology of a portion of the western Santa Monica Mountains, Los Angeles County, California. M, 1952, University of California, Los Angeles.

Bloomfield, James Miller. Volcanic ash in the White Cloud Peaks-Boulder Mountain region, south-central Idaho. M, 1983, Lehigh University.

Bloomfield, Susan L. The Proterozoic Greyson-Spokane transition sequence; a stratigraphic and gravity study, west-central Montana. M, 1983, University of Montana. 106 p.

Bloomquist, Marvin Gaines. Propagation characteristics of micropulsations. D, 1967, University of Texas, Austin. 166 p.

Bloor, David Trent. Geologic planiform maps of Southern Peninsula of Michigan. M, 1960, University of Michigan.

Bloss, Fred Donald. Relationship between light absorption and composition in the solid solutional series between $Ni(NH_4)_2 \cdot 6H_2O$ and $Mg(NH_4)_2 \cdot 6H_2O$. D, 1951, University of Chicago. 63 p.

Bloss, Pamela. Geochemistry and petrogenesis of the Mitchell Dam Amphibolite; Chilton and Coosa counties, Alabama. M, 1979, University of Alabama.

Blouin, Cecil F. The chemical analysis of oil well waters and their relation to the structure of the Lockport oil field. M, 1933, Louisiana State University.

Blount, Alice McDaniel. The crystal structure of a secondary hydrated calcium aluminum phosphate mineral. D, 1970, University of Wisconsin-Madison. 77 p.

Blount, Ann E. Two years after the Metula oil spill, Strait of Magellan, Chile; oil interaction with coastal environments. M, 1978, University of South Carolina.

Blount, Charles Werner. The solubility of anhydrite in the systems $CaSO_4$-H_2O and $CaSO_4$-NaCl-H_2O and its geologic significance. D, 1965, University of California, Los Angeles. 184 p.

Blount, Donald Neal. Geology of the Chiantla Quadrangle, Guatemala (central). D, 1967, Louisiana State University. 168 p.

Blount, Donald Neal. Geology of the Honey Creek area, Llano County, Texas. M, 1962, University of Texas, Austin.

Blount, Gerald C. Stratigraphy, depositional environment, and diagenesis related to porosity formation and destruction in the Devonian Jefferson Formation, Sawtooth Range, northwestern Montana. M, 1986, University of Idaho. 100 p.

Blount, Howard Grady, II. Regional aeolian dynamics from remote sensing; origin of the Gran Desierto, Sonora, Mexico. D, 1988, Arizona State University. 227 p.

Blount, Jonathan G. The geology of the Rancho los Filtros area, Chihuahua, Mexico. M, 1982, East Carolina University. 83 p.

Blount, Scott Brian. Depositional and diagenetic history of the Hosston Formation (Travis Peak), Nuevo Leon Group, Trawick Field, Nacogdoches County, Texas. M, 1987, Stephen F. Austin State University. 177 p.

Blount, William. Paleoenvironmental analysis of the Lower Mississippian Caballero Formation and the Andrecito Member of the Lake Valley Formation in the northern Sacramento Mountains, Otero County, New Mexico. M, 1985, Texas A&M University. 192 p.

Blouse, R. S. An analysis of one degree by one degree mean free-air anomalies and associated tectonic features in ocean areas. M, 1975, Washington University. 117 p.

Blowes, D. B. The influence of the capillary fringe on the quantity and quality of runoff in an inactive uranium mill tailings impoundment. M, 1983, University of Waterloo. 57 p.

Bloxsom, Walter Eden. A Lower Cretaceous (Comanchean) prograding shelf and associated environments of deposition, Northern Coahuila, Mexico. M, 1972, University of Texas, Austin.

Bloy, Graeme Richard. U-Pb geochronology of uranium mineralization in the East Arm of Great Slave

Lake, Northwest Territories. M, 1979, University of Alberta. 64 p.

Blue, Donald M. Geology and ore deposits of the Lucin mining district, Box Elder County, Utah, and Elko County, Nevada. M, 1960, University of Utah. 121 p.

Blueford, Joyce Raia. Miocene spumellarian Radiolaria from the equatorial Pacific. D, 1980, University of California, Santa Cruz. 144 p.

Bluemle, John P. Erosional surfaces and glacial geology along the southwest flank of the Crazy Mountains, Montana. M, 1962, Montana State University. 151 p.

Bluemle, John P. Geology of McLean County, North Dakota. D, 1971, University of North Dakota. 82 p.

Bluemle, Mary E. Natural science of the Great Plains as it relates to the American Indian; a syllabus and sourcebook. D, 1975, University of North Dakota. 234 p.

Bluhm, Christopher T. The structure of the Frank, alpine-type dunite-lherzolite complex, Avery County, North Carolina. M, 1976, Southern Illinois University, Carbondale. 153 p.

Blum, Alex E. Chemical weathering and controls on the chemistry of infiltrating solutions in a forested watershed, Medicine Bow Mountains, Wyoming. M, 1984, University of Wyoming. 89 p.

Blum, Brian Allen. A hydrogeologic investigation of the East Branch, Swift River basin, north-central Massachusetts. M, 1986, University of Massachusetts. 132 p.

Blum, Cynthia Elizabeth. Depositional environments of the Mississippian-Pennsylvanian Amsden and Pennsylvanian lower Wells formations, west-central Wyoming and southeastern Idaho. M, 1982, University of Kansas.

Blum, Joel David. The petrology, geochemistry and isotope geochronology of the Gilmore Dome and Pedro Dome plutons, Fairbanks District, Alaska. M, 1982, University of Alaska, Fairbanks. 107 p.

Blum, Justin Lawrence. Geologic and gravimetric investigation of the South Lake Tahoe groundwater basin, California. M, 1979, University of California, Davis. 96 p.

Blum, Michael David. Late Quaternary sedimentation in the upper Pedernales River, Texas. M, 1987, University of Texas, Austin.

Blum, Norbert. Geochemical studies of Archean iron formations and associated volcanic rocks. D, 1986, McMaster University. 410 p.

Blum, Victor Joseph. Seismometric study of the moderately deep earthquake of June 24, 1935. M, 1936, St. Louis University.

Blum, Victor Joseph. The magnetic field and the geology of the Canon City area. D, 1944, St. Louis University.

Bluman, Dean Edward. An experimental study of convective heat transfer to a solid-in-gas-suspension. D, 1966, West Virginia University.

Blumberg, A. F. A numerical investigation into the dynamics of estuarine circulation. D, 1975, The Johns Hopkins University.

Blumberg, George Micah Connor. A refraction study of the median ridge of the Kane fracture zone. M, 1987, Massachusetts Institute of Technology. 51 p.

Blumberg, Randolph. Analysis of ocean waves and wave forces by a filtering technique. D, 1955, Texas A&M University.

Blumberg, Roland Krezdorn. A new displacement seismograph. D, 1948, Harvard University.

Blumenthal, Morris B. Subsurface geology of the Prague-Paden area, Lincoln, and Okfuskee counties, Oklahoma. M, 1956, University of Oklahoma. 64 p.

Blumentritt, Russell A. A theoretical earth resistivity study with applications on the Llano Estacado, Texas. M, 1969, Texas Tech University. 108 p.

Blumer, John W. Geology of the Deadman Canyon-Copperopolis area, Meagher County, Montana. M,

1971, Montana College of Mineral Science & Technology. 68 p.

Blumreich, William, III. Petroleum geology and the environment of the Glade Sandstone (Devonian) in (southwestern) New York. M, 1968, SUNY at Buffalo. 45 p.

Blumstengel, Wayne. Studies of an active rock glacier, east side Slims River valley, Yukon Territory. M, 1988, University of Calgary.

Blumthal, James E. Selective trace element distribution in pyrite in coal seams of western Kentucky. M, 1977, Southern Illinois University, Carbondale. 84 p.

Blundell, J. Stuart. Structural trends of Precambrian rocks, Sheep Ridge Anticline, western Owl Creek Mountains, Wyoming. M, 1988, University of Wyoming. 81 p.

Blundell, Lane Cameron. Depositional environments of the Calabasas and Modelo formations, Big Mountain area, Ventura County, California. M, 1980, San Diego State University.

Blundell, Michael Craig. Depositional environments of the Vaqueros Formation in the Big Mountain Area, Ventura County, California. M, 1981, California State University, Northridge. 102 p.

Blundon, Sandra J. An attempt to obtain deep crustal reflections using a low level source. M, 1969, Dalhousie University. 92 p.

Blunt, David J. Geochemistry of amino acids in mollusks and wood, Pacific Northwest, United States. M, 1982, California State University, Hayward. 183 p.

Blusson, Stewart Lynn. Geology and tungsten deposits near the headwaters of Flat River, Yukon, and southwest District of Mackenzie, Canada. D, 1965, University of California, Berkeley. 170 p.

Bluth, Gregg J. Sulfur isotope study of sulfide-sulfate chimneys on the East Pacific Rise, 11 and 13°N latitudes. M, 1987, Pennsylvania State University, University Park. 50 p.

Blyskun, George James. The hydrogeology of the Bellbrook, Ohio, area. M, 1983, Wright State University. 99 p.

Blythe, Ernest W., Jr. Environmental study of a portion of the middle Ordovician in Sequatchie Valley, eastern Tennessee. D, 1974, University of Tennessee, Knoxville. 197 p.

Blythe, Ernest W., Jr. Lithostratigraphy of the Chickamauga Group (Ordovician) in Campbell and Claiborne counties, Tennessee. M, 1967, University of Tennessee, Knoxville. 74 p.

Blythe, Jack Gordon. Areal variation of the sedimentary characteristics of the Chattanooga Shale in eastern Kansas. M, 1950, Northwestern University.

Blythe, Jack Gordon. The Atoka Formation on the north side of the McAlester Basin. D, 1957, University of Oklahoma. 142 p.

Boa Hora, Marco Polo Pereira da *see* da Boa Hora, Marco Polo Pereira

Boadi, Issac Opoku. Gold mineralization and Precambrian geology of the Hopewell area, Rio Arriba County, NM. M, 1986, New Mexico Institute of Mining and Technology. 107 p.

Boadu, Fred Kofi. Constrained minimum entropy deconvolution. M, 1987, University of Calgary. 129 p.

Boak, Jeremy Lawrence. Geology and petrology of the Mount Chaval area, North Cascades, Washington. M, 1977, University of Washington. 87 p.

Boak, Jeremy Lawrence. Petrology and geochemistry of clastic metasedimentary rocks of the Isua supracrustal belt, West Greenland. D, 1983, Harvard University. 241 p.

Boakye, Samuel Yamoah. Hydraulic fracturing in earth dams. D, 1984, University of Illinois, Urbana. 358 p.

Boardman, Darwin Rice, II. A new model for the depth related community succession of Pennsylvanian Midcontinent cyclothems with implications on the black shale problem. M, 1983, Ohio University, Athens. 100 p.

Boardman, Donald C. Sedimentation and stratigraphy of the Jordan and Madison sandstones in central Wisconsin. D, 1952, University of Wisconsin-Madison.

Boardman, Donald Chapin. The Minnelusa Formation in the Rapid Canyon area, Black Hills, South Dakota. M, 1942, University of Iowa. 108 p.

Boardman, James Joseph. Petrology of the Salmon Mountain Stock, Klamath Mountains, California. M, 1985, Texas Tech University. 78 p.

Boardman, M. R. Holocene deposition in Northwest Providence Channel, Bahamas; a geochemical approach. D, 1978, University of North Carolina, Chapel Hill. 164 p.

Boardman, Richard Stanton. The trepostomatous Bryozoa of the Hamilton Group of New York State. D, 1955, University of Illinois, Urbana. 176 p.

Boardman, Shelby Jett. Precambrian geology and mineral deposits of the Salida area, Chaffee County, Colorado. D, 1971, University of Michigan.

Boardman, Shelby Jett. The world-Wide occurrence of native bismuth. M, 1969, University of Michigan.

Boast, John. Structural evolution of Djebel Chambi, central Tunisia. M, 1979, University of South Carolina.

Boatright, Byron B. Fluid phenomena in porous subsurface strata. D, 1936, University of Colorado.

Boatright, Daniel Thomas. Development of a national hazardous waste management system model. D, 1981, University of Oklahoma. 304 p.

Boatwright, J. L. The nature of the earthquake focus as inferred from body-wave observations. D, 1979, Columbia University, Teachers College. 220 p.

Bob, Matthew Regis. Significance of earth movements along Lake Austell in Village Creek State Park, Arkansas. M, 1981, Memphis State University. 90 p.

Bobba, Arabinda Ghosh. Temperature survey of coal mines producing acid water (S.E. Ohio). M, 1971, Ohio University, Athens. 178 p.

Bobbitt, John Bailey. Petrology, structure, and contact relations of part of the Yuba Rivers Pluton, northwestern Sierra Nevada foothills, California. M, 1982, University of California, Davis. 160 p.

Bobeck, Patricia Ann. Igneous petrology and structural geology of Nine Point Mesa, Brewster County, Texas. M, 1985, University of Texas, Austin. 101 p.

Boben, Carolyn L. Geological comparison of three precious metal prospects in Marquette County, Michigan. M, 1986, Michigan Technological University. 77 p.

Bober, Danny R. Study of hydrocarbon production and potential production of the Ireland Sandstone (Douglas Group, Upper Pennsylvanian) in northern Barber County, Kansas. M, 1985, Wichita State University. 169 p.

Boberg, Walter W. Transportation and precipitation of uranium in the South Platte river, Colorado. M, 1970, University of Colorado.

Boble, John D. The relationship between permeability and gravity-drainage in oil-producing formations. M, 1951, Ohio State University.

Bobola, John M. An analysis of weathering products from Brazil. M, 1969, SUNY at Buffalo. 47 p.

Bobrow, Danny J. Geochemistry and petrology of Miocene silicic lavas in the Socorro-Magdalena area of New Mexico. M, 1984, New Mexico Institute of Mining and Technology. 145 p.

Bobyarchick, A. R. Tectogenesis of the Hylas Zone and eastern Piedmont near Richmond, Virginia. M, 1976, Virginia Polytechnic Institute and State University.

Bobyarchick, Andy Russell. Structure of the Brevard Zone and Blue Ridge near Lenoir, North Carolina, with observations on oblique crenulation cleavage and a preliminary theory for irrotational structures in shear zones. D, 1983, SUNY at Albany. 360 p.

Bochensky, Paul. Contact metamorphism effects of the Laramie anorthosite complex at Morton Pass, Wyoming. M, 1982, University of Wyoming. 58 p.

Bochneak, Diane Lynn. A subsurface study of the Doyle Field, Stephens County, Oklahoma. M, 1982, Baylor University. 157 p.

Bock, Charles Mitchell. The distribution of some selected alkaline metals and alkaline earths in the Stronghold Granite, Cochise County, Arizona. D, 1962, University of Arizona.

Bock, Charles Mitchell. The petrography of a section of Eocene undifferentiated tuff, and general geology of the upper Warm Spring Creek area, Fremont County, Wyoming. M, 1958, Miami University (Ohio). 49 p.

Bock, Wayne D. The benthonic foraminifera of southwestern Florida Bay. M, 1961, University of Wisconsin-Madison.

Bock, Wayne Dean. Monthly variation in the foraminiferal biofacies on thalassia and sediment in the Big Pine Key area, Florida. D, 1967, University of Miami. 304 p.

Bock, Yehuda. The use of baseline measurements and geophysical models for the estimation of crustal deformations and the terrestrial reference system. D, 1982, Ohio State University. 220 p.

Bockheim, James Gregory. Effects of alpine and subalpine vegetation on soil development, Mount Baker, Washington. D, 1972, University of Washington. 171 p.

Bockius, Samuel Harrison. Geophysical mapping of the extent of basaltic rocks in the Moscow groundwater basin, Latah County, Idaho. M, 1985, University of Idaho. 83 p.

Bockoven, Frances Dart. Source, transport, and deposition of the (Eocene) Yegua sediments of the middle Texas Gulf Coast. M, 1985, University of Texas, Austin. 127 p.

Bockoven, Neil T. Petrology and volcanic stratigraphy of the El Sueco area, Chihuahua, Mexico. M, 1976, University of Texas, Austin.

Bockoven, Neil Thomas. Reconnaissance geology of the Yecora-Ocampo area, Sonora and Chihuahua, Mexico. D, 1980, University of Texas, Austin. 241 p.

Boctor, Nabil Zaki. The mercury-selenium-sulfur system and its geological implications; Part I, Phase relations in the mercury-selenium-sulfur system; Part II, The sulfoselenides and sulfides of mercury; mineralogy and geochemistry. D, 1976, Purdue University. 237 p.

Bodden, Wilfred Rupert, III. Depositional environments of the Eocene Domengine Formation in the Mount Diablo coal field, Contra Costa County, California. M, 1981, Stanford University. 111 p.

Bode, Francis D. Characters useful in determining the position of individual teeth in the permanent cheektooth series of Merychippine horses. M, 1931, California Institute of Technology. 11 p.

Bode, Francis D. Fauna of the Merychippus Zone, North Coalinga District, California. D, 1934, California Institute of Technology. 68 p.

Bode, Francis D. The structural geology of the San Joaquin Hills, Orange County, California. D, 1934, California Institute of Technology. 29 p.

Bodell, John Michael. Heat flow in the North-central Colorado Plateau. M, 1981, University of Utah. 134 p.

Boden, Brian P. The crustaceans of the order Euphausiacea from the temperate North-east Pacific, with notes of their biology. D, 1950, University of California, Los Angeles. 114 p.

Boden, David R. Stratigraphy of the Tshirege Member of the Bandelier Tuff and structural analysis of the Pajarito fault zone, Bandelier National Monument area, Jemez Mountains, New Mexico. M, 1980, Colorado School of Mines. 187 p.

Boden, David Rendall. Geology, structure, petrology, and mineralization of the Toquima caldera complex, central Nevada. D, 1987, Stanford University. 285 p.

Boden, James R. Lithology, hydrothermal petrology, and stable isotope geochemistry of three geothermal exploration drill holes, upper Clackamas River area. M, 1985, University of California, Riverside. 137 p.

Boden, Linda. Sound speed profile inversion in the ocean. M, 1985, Colorado School of Mines. 83 p.

Bodenlos, Alfred J. A study of metamorphic structures, Plains, New York, and vicinity. M, 1942, Columbia University, Teachers College.

Bodenlos, Alfred J. Geology of the Red Mountain magnesite district, Santa Clara and Stanislaus counties, California. D, 1950, Columbia University, Teachers College.

Bodge, Kevin Robert. Short term impoundment of longshore sediment transport. D, 1986, University of Florida. 346 p.

Bodholt, Frederick Brunson. Relation of shale petrology to variations in physical properties of clastic sediment across the Montana disturbed belt. D, 1976, University of Montana. 94 p.

Bodie, Jeffrey G. Formation and development of beach cusps on Del Monte Beach, Monterey, California. M, 1974, Naval Postgraduate School.

Bodily, Norman Mark. An armored dinosaur from the Lower Cretaceous of Utah. M, 1968, Brigham Young University.

Bodine, John Howard. The thermo-mechanical properties of the oceanic lithosphere. D, 1981, Columbia University, Teachers College. 338 p.

Bodine, Marc W., Jr. Geology of the Capitan coal fields, Lincoln County, New Mexico. M, 1953, Columbia University, Teachers College.

Bodman, Geoffrey Baldwin. Some characteristics and criteria of the soils of the Red Drift. D, 1927, University of Minnesota, Minneapolis.

Bodmer, Rene. Induced electrical polarization and groundwater. M, 1967, University of California, Berkeley. 102 p.

Bodnar, Dirk A. Stratigraphy, age, depositional environment and hydrocarbon source-rock potential of the Otuk Formation, northcentral Brooks Range, Alaska. M, 1984, University of Alaska, Fairbanks. 232 p.

Bodnar, R. J. Fluid inclusion study of the porphyry copper prospect at Red Mountain, Arizona. M, 1978, University of Arizona.

Bodnar, Robert John. Pressure-volume-temperature-composition (PVTX) properties of the system H$_2$O-NaCl at elevated temperatures and pressures. D, 1985, Pennsylvania State University, University Park. 199 p.

Bodnar, Theodor. Petrography of the Green Creek Complex, central and southern Albion Range, Cassia County, Idaho. M, 1983, Idaho State University. 82 p.

Bodnar, Theodore. Crustal structure of the Great Plains of North America from Rayleigh wave analysis. M, 1982, University of Texas at El Paso.

Bodner, Daniel Paul. Heat variations caused by groundwater flow in growth faults of the South Texas Gulf Coast Basin. M, 1985, University of Texas, Austin. 188 p.

Bodvarsson, Gudmundur Svavar. Mathematical modeling of the behavior of geothermal systems under exploitation. D, 1982, University of California, Berkeley. 353 p.

Bodvarsson, Gudrun M. Ocean wave-generated microseism at the Oregon coast. D, 1975, University of Oregon. 83 p.

Bodvarsson, Gunnar. Thermal activity and related phenomena in Iceland. D, 1957, California Institute of Technology. 156 p.

Bodwell, Willard Arthur. Geologic compilation and nonferrous minerals potential, Precambrian section, northern Michigan. M, 1972, Michigan Technological University.

Boebel, Richard Wallin. Sedimentology of St. Lawrence dolomite and relationship to Wisconsin Arch. M, 1950, University of Wisconsin-Madison. 108 p.

Boeck, Robert V. Geology and stratigraphy of the Boaz Quadrangle, Wisconsin. M, 1959, University of Wisconsin-Madison.

Boeckerman, Ruth Bastanchury. Geology of the southwest quarter of the Jackson Quadrangle, Wyoming. D, 1950, University of Michigan.

Boeckman, Charles H. A subsurface study of the Lower Pennsylvanian sediments of northern Grady and Caddo counties, Oklahoma. M, 1955, University of Oklahoma. 52 p.

Boeckman, G. O. An investigation as to the uses of topographic map and improvements desired by the general public. M, 1952, University of Missouri, Rolla.

Boehl, J. E. An application of the Hilbert transform to the magnetotelluric method. D, 1977, University of Texas, Austin. 107 p.

Boehm, Mark Charles Francis. Biostratigraphy, lithostratigraphy, and paleoenvironments of the Miocene-Pliocene San Felipe marine sequence, Baja California Norte, Mexico. M, 1982, Stanford University. 326 p.

Boehm, P. D. The transport and fate of hydrocarbons in benthic environments. D, 1977, University of Rhode Island. 193 p.

Boehme, Richard William. A litho-biostratigraphic study of the Hamilton Group (Devonian) in the Buffalo area, New York. M, 1964, SUNY at Buffalo.

Boehmer, Walter Richard. Erasure of sediment surface features by Mellita quinquiesperforata (Leske). M, 1970, Old Dominion University. 58 p.

Boehner, R. C. The Lower Carboniferous stratigraphy of the Musquodoboit Valley, central Nova Scotia. M, 1977, Acadia University.

Boekenkamp, Richard Paul. Stratigraphy of the (Devonian) Sherburne Member (Genesee Formation) of the Cayuga Trough, central New York. M, 1963, Cornell University.

Boeker, Ralph. Chinese energy and mineral resource development; implications for world supply. M, 1983, University of Texas, Austin.

Boellstorff, John David. Geology of the Tasersiaq Peninsula, southwest Greenland. M, 1968, University of Nebraska, Lincoln.

Boellstorff, John David. Tephrochronology, petrology, and stratigraphy of some Pleistocene deposits in the Central Plains, U.S.A. D, 1973, Louisiana State University.

Boenig, Charles Martin. Deltaic and coastal interdeltaic environments of the Carrizo Formation (Eocene), Milam County, Texas. M, 1970, Texas A&M University. 85 p.

Boerboom, Terrence John. Tourmalinites, nelsonites, and related rocks (early Proterozoic) near Philbrook, Todd County, Minnesota. M, 1987, University of Minnesota, Duluth. 212 p.

Boerner, David E. The calculation of electromagnetic fields from an arbitrary source in a horizontally layered Earth. M, 1983, University of Toronto. 130 p.

Boerner, David Eugene. A generalized approach to the interpretation of controlled source electromagnetic data collected in sedimentary basins. D, 1987, University of Toronto.

Boerner, Ralph E. J. Post-fire mineral cycling and ecosystem stability in the New Jersey Pine Barrens. D, 1980, Rutgers, The State University, New Brunswick. 262 p.

Boersma, Anne. The evolution, geographic and stratigraphic distribution of the genus Uvigerina; part I, The Eocene. M, 1970, Brown University.

Boersma, Anne. Time-space distribution of Uvigerina; a Tertiary benthonic foraminiferal genus. D, 1976, Brown University. 254 p.

Boettcher, Arthur Lee. Geology and petrology of the Rainy Creek Intrusive near Libby, Montana. M, 1963, Pennsylvania State University, University Park. 70 p.

Boettcher, Arthur Lee. The Rainy Creek igneous complex near Libby, Montana. D, 1966, Pennsylvania State University, University Park. 155 p.

Boettcher, Richard Scott. Foraminiferal trends of the central Oregon Shelf. M, 1967, Oregon State University. 134 p.

Boettger, William M. Origin and stratigraphy of Holocene sediments, Souris and Des Lacs glacial-lake spillways, north-central North Dakota. M, 1986, University of North Dakota. 186 p.

Boetzkes, Peter C. A spinner magnetometer for susceptibility anisotropy in rocks. D, 1973, University of Alberta. 426 p.

Bogaert, Barbara Mary. An aftershock study of the Santa Barbara earthquake of August 13, 1978. M, 1984, University of California, Santa Barbara. 55 p.

Bogard, Donald Dale. Krypton anomalies in achondritic meteorites. D, 1966, University of Arkansas, Fayetteville. 70 p.

Bogardus, Egbert Hall. Lower Pennsylvanian of the Richland Springs area, San Saba County, Texas. M, 1957, University of Texas, Austin.

Bogardus, James W. The effect of aqueous chemical environments on the shock failure and rubblization of Green River oil shale. M, 1988, Indiana University, Bloomington. 132 p.

Bogart, Lowell Eldon. The Hueco (Gym) Limestone, Luna County, New Mexico. M, 1953, University of New Mexico. 91 p.

Bogart, Vera Jo. The paleoecology of the Ostracoda of the Corsicana Marl at Onion Creek, Travis County, Texas. M, 1952, University of Houston.

Bogdanski, John K. Significance of compositional and textural modifications of sand sized sediments in the south fork Salmon River, west-central Idaho. M, 1987, Idaho State University. 114 p.

Bogen, Nicholas Louis. Studies of the Jurassic geology of the west-central Sierra Nevada of California. D, 1983, Columbia University, Teachers College. 337 p.

Boger, J. B. Conodont biostratigraphy of the Upper Beekmantown Group and the St. Paul Group (Middle Ordovician) of Maryland and West Virginia. M, 1976, Ohio State University. 180 p.

Boger, P. D. A study of the history of sedimentation in the Red Sea by means of isotopic and geochemical methods. D, 1976, Ohio State University. 182 p.

Bogert, Bernard O. Rock fans of Cedar Mountain, Wyoming. M, 1935, Columbia University, Teachers College.

Bogert, Elizabeth A. The evolution of Connecticut river drainage. M, 1947, Smith College. 131 p.

Boggs, Ann S. Petrology of lower Eocene sandstones in south central Colorado compared to their time equivalents in Texas. M, 1978, University of Texas, Austin.

Boggs, Russell Calvin. Mineralogy and geochemistry of the Golden Horn Batholith, northern Cascades, Washington. D, 1984, University of California, Santa Barbara. 211 p.

Boggs, Russell Calvin. Okanoganite, a new rare-earth borofluoro-silicate from the Golden Horn Batholith, Okanogan County, Washington. M, 1980, University of California, Santa Barbara. 24 p.

Boggs, Sam, Jr. Stratigraphy and petrology of the Upper Minturn Formation (Pennsylvanian), east-central Eagle County, Colorado. D, 1964, University of Colorado. 265 p.

Boggs, Samuel W. The physiography of Tibet. M, 1924, Columbia University, Teachers College.

Boghrat, Alireza. The design and construction of a piezoblade and an evaluation of the Marchetti dilatometer in some Florida soils. D, 1982, University of Florida. 312 p.

Bogle, Edward Warren. Factors affecting lake sediment geochemistry in the southern Grenville Province. D, 1980, Queen's University. 659 p.

Bogle, Edward Warren. The primary geochemical dispersion associated with the Lac Frotet volcanic cycle, Quebec. M, 1977, Queen's University. 349 p.

Bognar, Edwin J. The geology of Pike and Sandy townships, Stark County, Ohio. M, 1926, Ohio State University.

Bogner, Jean A. Regional relations of the Lemont Drift (Pleistocene, Northern Illinois). M, 1973, University of Illinois, Chicago.

Bogner, Kurt A. An isoseismal study of northwest Ohio and southeast Michigan based on data from the January 31, 1986 Geauga County, Ohio earthquake. M, 1988, Bowling Green State University. 75 p.

Bogrett, J. W. Geology of the northwestern end of the Rattlesnake Hills, Natrona County, Wyoming. M, 1951, University of Wyoming. 70 p.

Boguchwal, Lawrence Allen. Dynamic scale modeling of bed configurations. D, 1978, Massachusetts Institute of Technology. 149 p.

Bogucki, Donald Joseph. Debris slides and related flood damage associated with the September 1, 1951, cloudburst in the Mount Le Conte-Sugarland Mountain area, Great Smoky Mountains National Park. D, 1970, University of Tennessee, Knoxville. 246 p.

Bogue, Richard G. A petrographic study of the Mount Hood and Columbia River basalt formations. M, 1932, University of Oregon. 88 p.

Bogue, Scott Weatherly. Behavior of the geomagnetic field during successive reversals recorded in basalts on Kauai, Hawaii. D, 1982, University of California, Santa Cruz. 243 p.

Bohacs, Kevin Michael. Flume studies on the kinematics and dynamics of large-scale bedforms. D, 1981, Massachusetts Institute of Technology. 179 p.

Bohanan, Earl Roger, Jr. A petrographic analysis of several soapstone artifacts from Tennessee and soapstone deposits in North Carolina and South Carolina in an attempt to determine the source area of the artifacts. M, 1975, University of Tennessee, Knoxville. 110 p.

Bohannon, Robert G. Mid-Tertiary nonmarine rocks along the San Andreas Fault in southern California. D, 1976, University of California, Santa Barbara. 327 p.

Bohanon, James Paul. Sedimentology and petrography of deltaic facies in the Aguja Formation, Brewster County, Texas. M, 1987, Texas Tech University. 149 p.

Bohart, Morris Fielding. The distribution and life histories of the Grypheas of the Cretaceous System of Texas. M, 1927, Texas Christian University.

Bohart, Philip H., Jr. Subsurface geology of the Purdy oil field, Garvin County, Oklahoma. M, 1958, University of Oklahoma. 70 p.

Bohidar, Naikananda K. Stress distribution around a cylindrical opening under uniaxial compression and lateral confinement. D, 1968, University of Utah. 205 p.

Bohlen, Steven Ralph. Feldspar and oxide thermometry of granulites in the Adirondack highlands. M, 1977, University of Michigan.

Bohlen, Steven Ralph. Pressure, temperature, and fluid composition of Adirondack metamorphism as determined in orthogneisses, Adirondack Mountains, New York. D, 1979, University of Michigan. 315 p.

Bohlin, Howard G. Genetic studies of the Cretaceous and associated sands and clays encountered along the route of the proposed intracoastal canal across New Jersey. D, 1935, New York University.

Bohlin, Howard G. Notes on the petrography of the Grenville series, as represented at Wanaque, New Jersey. M, 1925, Columbia University, Teachers College.

Bohling, Geoffrey C. A ground penetrating radar study of water table elevation in a portion of Wisconsin's Central Sand Plain. M, 1988, University of Wisconsin-Madison. 119 p.

Bohlke, Brenda. Mississippi prodelta "crusts"; a clay fabric and geotechnical analysis. M, 1978, University of Miami. 95 p.

Bohlke, John Karl. Alteration of deep-sea basalt from Site 396B, DSDP. M, 1978, University of Miami. 226 p.

Böhlke, John Karl Friedrich Paul. Local wall rock control of alteration and mineralization reactions along discordant gold quartz veins, Alleghany, California. D, 1986, University of California, Berkeley. 308 p.

Bohlken, Bruce Arthur. Distribution and interpretation of late Wisconsinan glacial landforms and materials in Boone County, Iowa. M, 1980, Iowa State University of Science and Technology.

Bohm, Burkhard W. Controls on ground water chemistry in central and western Nevada. D, 1982, University of Nevada. 230 p.

Bohm, Steven M. Regional facies of the Caseyville Formation (Lower Pennsylvanian) in Johnson and Pope counties, Illinois. M, 1981, Southern Illinois University, Carbondale. 110 p.

Bohman, R. P. A geochemical study of the Portage Lake Basalts in the Winona Quadrangle, Houghton County, Michigan. M, 1976, Wayne State University.

Bohmer, Harold, Jr. Geology of a rhyolite plug, Pinal County, Arizona. M, 1960, University of Cincinnati. 76 p.

Bohmer, Harold, Jr. Mineralogy of the tetrahedrite series. D, 1964, University of Cincinnati. 176 p.

Bohn, Ralph T. A subsurface correlation of Permian-Triassic strata in Lisbon Valley, Utah. M, 1977, Brigham Young University. 116 p.

Bohnert, John E. The effect of urban land use on total runoff; a case study for Salt Creek basin, Cook and Dupage counties, Illinois. D, 1971, Southern Illinois University, Carbondale. 144 p.

Bohnsack, Richard Lee. The relation of porosity and permeability to the textural properties of arenites. M, 1958, University of Oklahoma. 91 p.

Bohon, William Oliver, Jr. Geologic field trip in the Princeton area of Mercer County, West Virginia. M, 1968, Virginia State University. 80 p.

Bohor, Bruce Forbes. Characterization of illite and its associated mixed layers. D, 1959, University of Illinois, Urbana. 165 p.

Bohor, Bruce Forbes. Origin of the Allegheny underclays of western Indiana. M, 1955, Indiana University, Bloomington. 64 p.

Bohrer, Vorsila Laurene. Paleoecology of an archaeological site near Snowflake, Arizona. D, 1968, University of Arizona. 109 p.

Bohrn, Marie T. A study of selected Bryozoa from Jeptha Knob, Shelby County, Kentucky. M, 1945, Columbia University, Teachers College.

Boicourt, W. C. The circulation of water on the continental shelf from the Chesapeake Bay to Cape Hatteras. D, 1973, The Johns Hopkins University.

Boies, Robert B. Structure and stratigraphy of a part of the upper Horse Creek valley near Dubois, Wyoming. M, 1949, Miami University (Ohio). 16 p.

Boily, Michel. Evolution géochimique et isotopique du magmatisme Andin dans la sud du Pérou. D, 1988, Universite de Montreal.

Boily, Michel. Les Basaltes de la ceinture métavolcanique archéenne de l'Abitibi et les basaltes modernes; une étude géochimique comparative. M, 1981, Universite de Montreal.

Boison, Paul Joseph. Late Pleistocene and Holocene alluvial stratigraphy of three tributaries in the Escalante River basin, Utah. M, 1983, University of California, Santa Cruz.

Boissonnade, Auguste Claude. Earthquake damage and insurance risk. D, 1984, Stanford University. 312 p.

Boissonnault, Jean. La minéralogie des intrusions alcalins du Mont St-Hilaire. M, 1966, Ecole Polytechnique. 100 p.

Boisture, Timothy A. Carbonate petrology and environments of deposition of the Fredonia Member of the Ste. Genevieve Formation (Mississippian) near Fredonia, Kentucky. M, 1983, Southern Illinois University, Carbondale. 99 p.

Boisvert, Denis. La Banque de donnees gites-estrie; collecte et traitement informatique de donnees gitologiques et metallogeniques. M, 1988, Universite du Quebec a Montreal. 183 p.

Boker, Thomas A. Sand dunes on northern Padre Island, Texas. M, 1956, University of Kansas. 100 p.

Boker, Thomas Dominic Nmah. Resilient characteristics of Michigan cohesionless roadbed soils in correlation to the soil support values. D, 1978, Michigan State University. 232 p.

Bokhari, S. M. H. Management of water resources under different socio-economic conditions. D, 1975, University of Arizona. 306 p.

Bokman, John W. Petrology and genesis of the Stanley and Jackfork formations, Ouachita Mountains, Arkansas and Oklahoma. D, 1951, University of Chicago. 140 p.

Bokuniewicz, H. J. Estuarine sediment flux evaluated in Long Island Sound. D, 1976, Yale University. 176 p.

Bolakas, John Frank. Investigation of flood plain development, Brandywine Creek, Chadds Ford, Pennsylvania. M, 1985, University of Delaware. 125 p.

Boland, David Craig. Upper Jurassic (Portlandian) sedimentology and palynofacies of Cabo Espichel, Portugal. M, 1987, Memorial University of Newfoundland. 261 p.

Boland, Gary D. Petrographic analysis of the Pettet porosity, Sligo Formation (Lower Cretaceous), in the Kerlin oil field, Columbia County, Arkansas. M, 1980, Stephen F. Austin State University.

Boland, John J. A framework for evaluating power plant siting decisions in the Chesapeake Bay region. D, 1973, The Johns Hopkins University.

Bolander, Richard John. A gravimetric study of a portion of the Polochic fault zone near Huehuetenango, Guatemala, C. A. M, 1969, Louisiana State University.

Bolduc, Andree-Monique. The Quaternary history of Churchill Falls, Labrador. M, 19??, Carleton University. 95 p.

Bolduc, Pierre-Michel. A feasibility study for a portable long period seismograph. M, 1972, University of British Columbia.

Bole, Clifton Eugene. Potassium-argon ages, argon diffusion studies and petrography in the northern Front Range, Manhatten (Rusting quadrangle), Colorado. M, 1971, Ohio State University.

Bole, George Robert. The geology of a part of the Manhattan Quadrangle, Gallatin County, Montana. M, 1962, University of California, Berkeley. 84 p.

Bolen, Michael M. A geophysical and structural investigation of the northern talc orebody Winterboro, Alabama. M, 1987, Rutgers, The State University, Newark. 80 p.

Boleneus, David E. Diagenesis of some Holocene, intertidal carbonate sands, Saint Croix, U.S. Virgin Islands. M, 1972, Louisiana State University.

Boler, Frances Michele. Aeromagnetic measurements, magnetic source depths, and the Curie point isotherm in the Vale-Owyhee, Oregon geothermal area. M, 1979, Oregon State University. 104 p.

Boler, Frances Michele. Brittle dynamic fracture experiments; a comparison of radiated seismic energy and fracture energy. D, 1985, University of Colorado. 248 p.

Boler, Milton E. Pre-Desmoinesian isopach and paleogeologic study of northwestern Oklahoma. M, 1959, University of Oklahoma. 53 p.

Boles, James R. Zeolites and authigenic feldspar along a part of the Beaver Rim, Fremont County, Wyoming. M, 1968, University of Wyoming. 64 p.

Boles, Jennifer Lee Snow see Snow Boles, Jennifer Lee

Bolger, George Walton. Chemical evidence that the Mid-Atlantic Ridge is a source of hydrothermally derived suspended particulate material to the deep Atlantic. M, 1976, University of South Florida, St. Petersburg. 145 p.

Bolger, Robert Courtney. An investigation of the Mercer Clay, Clearfield County, Pennsylvania. M, 1949, Washington University. 93 p.

Bolha, J., Jr. A sedimentological investigation of a progradational foreshore sequence; C.F.B. Borden. M, 1986, University of Waterloo. 207 p.

Bolich, Leonard C. Petrography of the western portion of the Juriquipa stock, Nacozari, Sonaro (Sutton county, Texas). M, 1969, University of Texas at El Paso.

Bolich, Richard E. Geophysical and geochemical investigation of diabase from the western portion of the Durham Mesozoic basin, North Carolina. M, 1986, North Carolina State University. 40 p.

Bolin, David Samuel. A geochemical comparison of some barren and mineralized igneous complexes of southern Arizona. M, 1976, University of Arizona.

Bolin, Edward J. Some foraminifera, Radiolaria, and Ostracoda from the Cretaceous of Minnesota. M, 1954, University of Minnesota, Minneapolis. 62 p.

Bolitho, Mason R. A subsurface study of the "A" Zone (Pennsylvanian System) in Red Willow County, Nebraska. M, 1982, University of Nebraska, Lincoln. 121 p.

Bolivar, Stephen L. Geochemistry of the Prairie Creek, Arkansas and Elliott County, Kentucky intrusions. D, 1977, University of New Mexico. 286 p.

Bolivar, Stephen L. Kimberlite of Elliott County, Kentucky. M, 1972, Eastern Kentucky University. 61 p.

Bolka, David F. The marine geochemistry of argon. M, 1964, Massachusetts Institute of Technology. 76 p.

Bolla, William Owen. Gravity investigation of a Silurian reef in north-central Ohio. M, 1984, Wright State University. 93 p.

Bolland, Robert Finley. Paleoecological interpretation of the diatom succession in the recent sediments of Utah Lake, Utah. D, 1974, University of Utah. 100 p.

Bolle, Doris J. Lithofacies of Guelph-Lockport Group, Lake Erie area. M, 1978, SUNY, College at Fredonia. 69 p.

Bolles, Lawrence W. Geology of the Las Flores and Dry Canyon quadrangles, Los Angeles County, California. M, 1932, California Institute of Technology. 57 p.

Bolles, Myrick Nathanial. The concentration of gold and silver in iron bottoms reduced from highly ferruginous copper mattes. D, 1903, University of Chicago. 30 p.

Bollin, Edgar M. The geology of the Kaibab Formation, Marble Platform, Coconino County, Arizona. M, 1955, University of Arizona.

Bollin, Edgar Marshall. Differential thermal pyrosynthesis and analysis. D, 1961, Columbia University, Teachers College. 117 p.

Bolling, Sharon J. Neogene stratigraphy of Gulf County, Florida. M, 1982, Florida State University.

Bollinger, Becky. United States federal lands as a source of critical and strategic minerals to the United States. M, 1984, University of Texas, Austin.

Bollinger, Gilbert Arthur. Determination of earthquake fault parameters from long period P waves. D, 1967, St. Louis University. 136 p.

Bollinger, Marsha Spencer. Radium isotopes in salt marsh and estuarine environments. D, 1986, University of South Carolina. 158 p.

Bollman, D. D. Geology of the east part of Mackay 3 SE and the west part of Mackay 4 SW Quadrangle, Blaine, Butte, and Custer counties, Idaho. M, 1971, University of Wisconsin-Milwaukee.

Bollman, James Franklin. Geology of the Murphy area, Mayes County, Oklahoma. M, 1950, University of Oklahoma. 61 p.

Bollmann, Dennis Dean. Geology of the east part of Mackay 3 SE and the west part of Mackay 4 SW Quadrangle, Blaine, Butte, and Custer counties, Idaho. M, 1971, University of Wisconsin-Milwaukee.

Bollow, George Edward. Economic effects of deep ocean minerals exploitation. M, 1971, United States Naval Academy.

Bollwinkel, Donald. Facies study of the Coeymans Limestone (Devonian) in the Hudson Valley, East central New York State. M, 1963, New York University.

Bolm, John Gary. Geology of the Lone mountain area, southwestern Montana. M, 1969, University of Idaho. 38 p.

Bolm, John Gary. Structural and petrographic studies in the Shuswap Terrane. D, 1975, University of Idaho. 93 p.

Bolsenga, Stanley Joseph. Spectral reflectances of freshwater ice and snow from 340 through 1100 nm. D, 1981, University of Michigan. 159 p.

Bolsover, Leslie Ruth. Petrogenesis of the Sybille iron-titanium oxide deposit, Laramie anorthosite complex, Laramie Mountains, Wyoming. M, 1986, SUNY at Stony Brook. 78 p.

Bolt, Charles T. An investigation of the structure in the South Atlantic Ocean west of the Mid-Atlantic Ridge from Rayleigh wave dispersion. M, 1975, Pennsylvania State University, University Park. 61 p.

Bolt, Larry R. Petrology of the San Martin de Porres area, Nacozari, Sonora, Mexico. M, 1969, University of Tennessee, Knoxville. 53 p.

Boltin, William Randolph. Geology of the Hollister 7 1/2-minute quadrangle, Warren and Halifax counties, North Carolina; metamorphic transition in the eastern slate belt. M, 1985, North Carolina State University. 87 p.

Bolton, Beatrice. Structure of the coral reef in the Hamilton Shale of Onondaga Valley (New York). M, 1927, Syracuse University.

Bolton, Beatrice E. Gastropoda of the Barton (upper Eocene). D, 1931, Cornell University.

Bolton, David W. Water chemistry of springs and streams in the St. Louis Limestone, Lawrence County, Indiana. M, 1980, Indiana University, Bloomington. 101 p.

Bolton, Edward Warren. Problems in nonlinear convection in planar and spherical geometries. D, 1985, University of California, Los Angeles. 205 p.

Bolton, James Christopher. Structure and stratigraphy of the Rich Mountain area, North Carolina. M, 1985, University of Kentucky. 111 p.

Bolton, Thomas E. Silurian stratigraphy and palaeontology of the Niagara Escarpment in Ontario. D, 1955, University of Toronto.

Bolton, Thomas E. Stratigraphy and palaeontology of the Silurian section at De Cew Falls, Ontario. M, 1949, University of Toronto.

Bolton, W. R. Precambrian geochronology of the Sevillita Metarhyolite and the Los Pinos, Sepultura, and Priest plutons of the southern Sandia Uplift, central New Mexico. D, 1976, New Mexico Institute of Mining and Technology. 57 p.

Bolyard, Dudley W. Pennsylvanian and Permian stratigraphy in the Sangre de Cristo Mountains between La Veta Pass and Westcliffe (Custer and Saguache counties), Colorado. M, 1956, University of Colorado.

Bolyard, Garrett L. Pleistocene features of the Palos Park region (Chicago). M, 1923, University of Chicago. 51 p.

Bolyard, Thomas Harner. Empirical relationships between barrier island hydrology and physiography. M, 1978, University of Virginia. 38 p.

Bolze, Claude E. Concentration of some chemical elements in volcanic rocks from St. Lucia and St. Vincent, Lesser Antilles. M, 1974, Wright State University. 82 p.

Bomah, Andrew Kelvinson. An analysis of the physical and landuse variables affecting soil erosion in the Njala area of Sierra Leone, Southern Province. D, 1982, Clark University. 193 p.

Bomback, Ronald L. The geology of the "Clinton" Formation in parts of Portage, Trumbull, and Mahoning counties, Ohio. M, 1986, University of Akron. 105 p.

Bomber, Brenda Jean. Uranium mineralization along a fault plane Tertiary sedimentary rocks in the McLean 5 Mine, Live Oak County, Texas. M, 1980, Texas A&M University. 121 p.

Bombolakis, Emanuel G. Geology of the Hot Sulphur Springs-Parshall area of Middle Park, Grand County, Colorado. M, 1959, Colorado School of Mines. 146 p.

Bombolakis, Emanuel G. Photoelastic stress analysis of crack propagation within a compressive stress. D, 1963, Massachusetts Institute of Technology. 96 p.

Bommer, P. M. A streamline-concentration balance model for in situ uranium leaching and site restoration. D, 1979, University of Texas, Austin. 271 p.

Bona, Pietro Alphonse di *see* di Bona, Pietro Alphonse

Bonaparte, Rudolph. A time-dependent constitutive model for cohesive soils. D, 1982, University of California, Berkeley. 356 p.

Bonar, Chester Milton. Geology of Ephraim area, Utah. M, 1948, Ohio State University.

Bonar, Kermit Mark. A study of marcasite equilibrium. M, 1963, Brown University.

Bonavia, Franco Ferdinand. The geology and geochemistry of Radiore 2 Mine, Matagami, Quebec. M, 1981, McGill University. 120 p.

Bonchonsky, Andrew P. A subsurface geological and geophysical investigation in Barton County, Kansas. M, 1957, Kansas State University. 51 p.

Bond, Frederick William. A parametric model calibrated with a physically based model for runoff prediction from ungaged streams. M, 1977, University of Arizona.

Bond, Gerad C. Bedrock geology of the Gulkana Glacier area, Alaska Range, Alaska. M, 1965, University of Alaska, Fairbanks. 45 p.

Bond, Gerard Clark. Permian volcanics, volcaniclastics, and limestones in the Cordilleran eugeosyncline, east-central Alaska range, Alaska. D, 1970, University of Wisconsin-Madison. 163 p.

Bond, Grego Benton. Sedimentology and Holocene stratigraphy of a carbonate mangrove buildup, Twin Cays, Belize, Central America. M, 1988, Texas A&M University.

Bond, Ivor John. Lithostratigraphy and conodont biostratigraphy of the Lower Ordovician of the Ottawa-St. Lawrence Lowlands, Ontario in New York. D, 1974, Queen's University. 467 p.

Bond, James F. Geology of the tin granite and associated skarn mineralization at Ear Mountain, Seward Peninsula, Alaska. M, 1983, University of Alaska, Fairbanks. 141 p.

Bond, John Gilbert. Geology of the Clearwater Embayment in Idaho. D, 1962, University of Washington. 193 p.

Bond, John Gilbert. Sedimentary analysis of Kummer Formation within Green River Canyon, King County. M, 1959, University of Washington. 113 p.

Bond, Linda Darlene. Origins of seawater intrusion in a coastal aquifer; a case study of the Pajaro Valley. M, 1986, Stanford University. 40 p.

Bond, Marc A. An integrated geophysical study of the Shaw Warm Spring area, San Luis Valley, South-central Colorado. M, 1981, Colorado School of Mines. 162 p.

Bond, Paul Norman. Sublethal predation of Upper Mississippian (Chesterian) ammonoids. M, 1984, Bryn Mawr College. 80 p.

Bond, Paulette Alice. A sequence of development for the Henderson augen gneiss and its adjacent cataclastic rocks. M, 1974, University of North Carolina, Chapel Hill. 53 p.

Bond, Ralph Hurd. A study of some conodonts from the lower part of the Ohio Shale in central Franklin County, Ohio. M, 1937, Ohio State University.

Bond, Robert M. Conditions of quartz mineralization in the Martinsburg Formation, eastern Pennsylvania and New Jersey. M, 1985, Lehigh University. 49 p.

Bond, Steven Craig. Origin and distribution of platinum-enriched heavy mineral accumulations in a beach placer near Platinum, Alaska. M, 1982, University of Texas, Austin. 64 p.

Bond, Thomas Alden. Palynology of Quaternary terraces and flood plains of the Washita and Red rivers, central and southeastern Oklahoma. D, 1966, University of Oklahoma. 100 p.

Bond, Thomas Alden. Palynology of the Weir-Pittsburgh Coal (Pennsylvanian) of Oklahoma and Kansas. M, 1963, University of Oklahoma. 103 p.

Bond, Wendell A. Stratigraphy and sedimentation of upper Paleozoic volcanics, volcaniclastics and limestones in the eastern Alaska Range, Alaska. M, 1973, University of Colorado.

Bond, William D. Preferred orientations of C-axes in quartz grains and their relationship to the tectonite structure of the Homestake gold mine at Lead, South Dakota. M, 1982, South Dakota School of Mines & Technology. 77 p.

Bond, William Douglas. The ovoid anorthositic gabbro at Bernic Lake, Manitoba. M, 1973, University of Manitoba.

Bond, William Earl. A study of the engineering characteristics of the 1971 San Fernando earthquake records using time domain techniques. D, 1980, Rensselaer Polytechnic Institute. 179 p.

Bond, William Howard. Characterization of a low-level radioactive waste site, applied to the former Weldon Spring feed materials plant, Weldon Spring, Missouri. M, 1985, University of Missouri, Rolla. 173 p.

Bonds, James A. Sedimentology and environments of deposition of the Wildcat Group, Northern California. M, 1983, Texas A&M University. 130 p.

Bondurant, Charles E. Structural geology of a portion of the Hurricane fault zone, northern Mohave County, Arizona. M, 1963, University of Kansas. 49 p.

Bondurant, William Stewart. A trace element study of sphalerite in the Central Kentucky mineral district. M, 1978, University of Kentucky. 52 p.

Boneham, Roger F. Occurrence of tasmanites, plant spores, and Chitinozoa in Middle and Upper Devonian formations of Michigan, Ohio, and Ontario. M, 1965, Wayne State University.

Boneham, Roger Frederick. Palynology of three Tertiary coal basins in south-central British Columbia. D, 1968, University of Michigan. 114 p.

Bonelli, Douglas T. The geology and skarn mineralization at the Ataspaca Prospect, Tacna, Peru. M, 1983, Oregon State University. 101 p.

Bonem, Rena Mae. Comparison of ecology and sedimentation in Pennsylvanian (Morrowan) bioherms of northeastern Oklahoma with modern patch reefs in Jamaica and the Florida Keys. D, 1975, University of Oklahoma. 194 p.

Bonem, Rena Mae. Upper Cambrian (Dresbachian) faunas of the Pilgrim Formation in southwestern Montana. M, 1971, New Mexico Institute of Mining and Technology. 234 p.

Bones, Dennis George. Seismicity of the Intermountain seismic belt in southeastern Idaho and western Wyoming, and tectonic implications. M, 1978, University of Utah. 130 p.

Boness, David Arno. The electronic thermodynamics of iron under earth core conditions. M, 1985, University of Washington. 80 p.

Bonham, Harold F. Areal geology of the northern half of Washoe County, Nevada. M, 1963, University of Nevada - Mackay School of Mines. 83 p.

Bonham, Lawrence Cook. A geochemical investigation of certain crude oils. D, 1950, Washington University. 92 p.

Bonham, Lawrence Cook. The geology of the southwest part of the Ironton, Missouri, Quadrangle. M, 1948, Washington University. 181 p.

Bonham, Lawrence D. Structural geology of the Hoosick Falls area, New York-Vermont, in relation to the theory of Taconic overthrust. D, 1950, University of Chicago. 111 p.

Bonham, Oliver J. H. Mineralization controls at the Yava lead deposit, Salmon River, Cape Breton County, Nova Scotia. M, 1983, Dalhousie University. 251 p.

Bonham-Carter, Graeme Francis. The geology of the Pennsylvanian sequence of the Blue Mountains, northern Ellesmere Island, Canada. D, 1966, University of Toronto.

Bonilla Franco, Jose Vicente. Laboratory simulation of crude oil migration; possible implications for oil exploration. D, 1986, [University of Oklahoma]. 109 p.

Bonilla, Manuel George. Landslides in the San Francisco South Quadrangle, California. M, 1960, Stanford University.

Bonillas, Ygnacio, III. A study of Miocene vulcanism in Southern California. M, 1935, California Institute of Technology. 29 p.

Bonini, William Emory. Subsurface geology in the area of the Cape Fear Arch as determined by seismic-refraction measurements. D, 1957, University of Wisconsin-Madison. 218 p.

Bonis, Samuel B. Geologic reconnaissance of the Alta Verapaz Fold Belt, Guatemala (east central). D, 1967, Louisiana State University. 180 p.

Bonkowski, Michael Steven. Tectonic geomorphology and neotectonics of the San Andreas fault zone Indio, Hills, Coachella Valley, California. M, 1981, University of California, Santa Barbara. 120 p.

Bonnar, R. U. Measurements of some nitrogenous components of three deep Pacific sediment cores. M, 1975, University of Hawaii. 61 p.

Bonné, Jochanan. Stochastic simulation of monthly streamflow by a multiple regression model utilizing precipitation data. D, 1970, University of Nevada. 114 p.

Bonner, Brian P. The effect of water content on elastic velocities and moduli in porous materials. M, 1971, Rensselaer Polytechnic Institute. 72 p.

Bonner, William Paul. Iodination of public water supplies. D, 1967, University of Florida. 155 p.

Bonnet, Adrienne Thornley. Lithostratigraphy and depositional setting of the limestone-rich interval of the LaHood Formation (Belt Supergroup), southwestern Montana. M, 1979, Montana State University. 87 p.

Bonnet, Daniel J. Experiments on the foundation of water escape structures. M, 1979, Louisiana State University.

Bonnett, Richard Brian. Glacial sequence of the upper Boulder creek drainage basin in the Colorado Front range (Colorado). D, 1970, Ohio State University.

Bonnett, Richard Brian. The Pleistocene and early post Pleistocene history of portions of the Orland and Bucksport quadrangles, Maine. M, 1963, University of Maine. 123 p.

Bonneville, John W. Surficial geology of southern Logan County, North Dakota. M, 1961, University of North Dakota. 87 p.

Bonnichsen, Bill. General geology and petrology of the metamorphosed Biwabik Iron Formation (Precambrian), Dunka River area, Minnesota. D, 1968, University of Minnesota, Minneapolis. 269 p.

Bonora, P. F. Geology of the Flower Mountain area, Nacogdoches County, Texas. M, 1977, Stephen F. Austin State University.

Bonsib, Ray Myron. The genesis of the Red soil of the oolitic limestone of southern Indiana. M, 1911, Indiana University, Bloomington.

Bonsteel, Jay A. The soils of St. Mary's County, Maryland, showing the relationships of the geology of the soils. D, 1901, The Johns Hopkins University.

Bonzo, Kevin M. Chemical and optical zoning in metamorphic garnets from the Black Hills, South Dakota. M, 1985, University of Akron. 139 p.

Boodoo, W. A sand development study of the lower member of the late Miocene, Morne l'Enfer Formation, south of the Los Bajos Fault, Trinidad, W. I. M, 1976, University of Wyoming. 138 p.

Book, Gerald Wayne. Structural and metamorphic history of Precambrian rocks, upper Gallinas Creek area, San Miguel County, New Mexico. M, 1978, Texas Tech University. 49 p.

Book, James Burgess, IV. A survey of shallow oil in the United States including estimated reserves. M, 1964, University of Michigan.

Book, Patricia O'Donnell. A study of vein minerals in Deep Sea Drilling Project Hole 462A. M, 1983, Iowa State University of Science and Technology. 82 p.

Booker, John Ratcliffe. Geomagnetic secular variations and the kinematics of the Earth's core. D, 1968, University of California, San Diego. 150 p.

Bookman, Marcia. Transgressive sedimentary facies of lower New York Bay. M, 1978, Brooklyn College (CUNY).

Bookout, John Frank, Jr. The geology of a part of the Cushing Quadrangle, northwestern Nacagdoches and southwestern Rush counties, Texas (B644). M, 1950, University of Texas, Austin.

Bookstrom, A. A. Magnetite deposits of El Romeral, Chile; physical geology, sequence of events, and processes of formation. D, 1975, Stanford University. 431 p.

Bookstrom, Arthur Albin. Geology of the Yellowjacket Anticline, Rio Blanco County, northwestern Colorado. M, 1964, University of Colorado.

Boon, John D., 3rd. Quantitative analysis of beach sand movement, Virginia Beach, Virginia. M, 1968, College of William and Mary.

Boon, John Daniel, III. Sediment transport processes in a salt marsh drainage system. D, 1974, College of William and Mary.

Boone, Charles Glenn. An evaluation of a stable isotope of dysprosium for labeling and tracing sedimentary particles. M, 1972, Old Dominion University. 56 p.

Boone, Erica Rechnitzer. Petrology and tectonic implications of the Hellancourt Volcanics, northern Labrador Trough, Quebec. M, 1987, McGill University. 95 p.

Boone, Gary McGregor. Geology of the Fish River Lake District, northern Maine. D, 1959, Yale University.

Boone, Gary McGregor. Petrology of the metamorphic and igneous rocks in the vicinity of Farmington, Maine. M, 1954, Brown University.

Boone, John L. Lake margin depositional systems of the Dockum Group (Upper Triassic) in Tule Canyon, Texas Panhandle. M, 1979, University of Texas, Austin.

Boone, Peter A. Stratigraphy of the basal Trinity Sands (Lower Cretaceous) of central Texas. M, 1966, Baylor University. 181 p.

Boone, Peter Augustine. Facies patterns during transgressive sedimentation; Antlers Formation, Lower Cretaceous; north-central Texas. D, 1972, Texas A&M University. 183 p.

Boone, Richard L. Groundwater recharge and subsurface flow processes on a hillslope in the Clear Creek watershed, eastern Sierra Nevada. M, 1983, University of Nevada. 164 p.

Boone, Richard Lee. Areal geology of Bruno and Lane quadrangles, Atoka County, Oklahoma. M, 1961, University of Oklahoma. 99 p.

Boone, William J. Microseismicity near Long Valley Caldera, California, June 29 to August 12, 1982. M, 1985, University of Wisconsin-Madison.

Boonlayangoor, C. The nature and characteristics of colloids in Lake Michigan. D, 1979, Illinois Institute of Technology. 224 p.

Boore, David M. Finite difference solutions to the equations of elastic wave-propagation, with application to Love waves over dipping interfaces. D, 1970, Massachusetts Institute of Technology. 240 p.

Boorman, Roy Slater. Subsolvus studies in the system FeS-ZnS: 0-400°C. D, 1967, University of Toronto.

Boos, C. M. The geology of the Big Thompson River valley in Colorado from the continental divide to the foothills area. M, 1924, University of Chicago.

Boos, Maynard. Geology of the Dongola Quadrangle, southern Illinois. M, 1922, University of Chicago. 71 p.

Boosman, Jaap W. Geology of the southeast quarter of the Vernon, New York, 7-1/2 minute quadrangle. M, 1959, Syracuse University.

Boosman, Jaap Wim. Sedimentation processes on the Argentine Continental Shelf. D, 1973, George Washington University.

Booth, Arthur L. The subsurface structure and stratigraphy as related to petroleum accumulation in Cowley County, Kansas. M, 1962, Kansas State University. 52 p.

Booth, Charles Clinton. Geology of Chalk Mountain Quadrangle, Bosque, Erath, Hamilton, and Somervell counties, Texas. M, 1956, University of Texas, Austin.

Booth, Charles Vincent. Geology of the west central portion of the Orestimba Quadrangle, California. M, 1950, University of California, Berkeley. 58 p.

Booth, Colin John. A numerical model of groundwater flow associated with an underground coal mine in the Appalachian Plateau, Pennsylvania. D, 1984, Pennsylvania State University, University Park. 490 p.

Booth, Derek Blake. Deformation of freezing and thawing ground. M, 198?, Stanford University. 59 p.

Booth, Derek Blake. Glacier dynamics and the development of glacial landforms in the eastern Puget Lowland, Washington. D, 1984, University of Washington. 217 p.

Booth, Franklin O., III. Geology of the Galisteo Creek area, Lamy to Canoncito, Santa Fe County, New Mexico. M, 1976, Colorado School of Mines. 122 p.

Booth, G. Martin, III. Heavy mineral study of the Coal Valley sediments in the lower Truckee River Canyon, Washoe County, Nevada. M, 1965, University of Nevada. 69 p.

Booth, Geoffrey Warren. The petrology and geochemistry of the Pamiutuq Lake Batholith, Northwest Territories. M, 1983, University of Toronto. 132 p.

Booth, Gregory Seeley. The ecology and distribution of rock-boring pelecypods off Del Monte Beach, Monterey, California. M, 1972, United States Naval Academy.

Booth, J. S. Early diagenesis in southern California continental borderland sediments. D, 1973, University of Southern California.

Booth, James S. Sediment dispersion in the northern Channel Island passages, California. M, 1971, University of Southern California.

Booth, Michael Cameron. Carbonate formation in Mars-like environments. M, 1976, University of California, Los Angeles. 151 p.

Booth, Michael Cameron. Carbonate formation on Mars. D, 1980, University of California, Los Angeles. 157 p.

Booth, Robert T. Ostracoda of the Wayland Shale Member of the Graham Formation, Pennsylvanian, Graham, Texas. M, 1932, Columbia University, Teachers College.

Booth, Sherry Linette. Structural analysis of portions of the Washita Valley fault zone, Arbuckle Mountains, Oklahoma. M, 1978, University of Oklahoma. 50 p.

Booth, Verne H. Stratigraphy and structure of the Oak Hill succession in Vermont. D, 1948, Columbia University, Teachers College.

Booth, Verne H. The petrology of the conglomeratic members of the Pinnacle Formation. M, 1940, Columbia University, Teachers College.

Boothby, Donald Roy. Coastal sedimentation of volcanogenic sands, Guatemala. M, 1978, Texas Tech University. 100 p.

Boothroyd, Jon C. Outwash fan sedimentation, northeast Gulf of Alaska. D, 1974, University of South Carolina.

Booty, William Gordon. Lithology and groundwater geochemistry of Beulah and Marshay townships, Ontario. M, 1977, McMaster University. 78 p.

Booty, William Gordon. Watershed acidification model and the soil acid neutralization capacity concept. D, 1983, McMaster University. 194 p.

Booy, Emmy C. Minerals of the Nile River sediments in the region of the Second Cataract, Sudan. M, 1963, Columbia University, Teachers College.

Booy, Emmy Catherine. Mineralogy and physical properties of clays involved in certain Cordilleran landslides. D, 1968, Columbia University. 163 p.

Bopp, Frederick, III. Trace metal environments near Shell Banks in Delaware Bay. M, 1973, University of Delaware.

Bopp, Frederick, III. Trace metal geochemistry of upper Delaware Bay. D, 1980, University of Delaware. 387 p.

Bopp, R. F. The geochemistry of polychlorinated biphenyls in the Hudson River. D, 1979, Columbia University, Teachers College. 207 p.

Bor, Sheng-Sheang. Scaling for seismic source spectra and energy attenuation in the Chelan region, eastern Washington. M, 1977, University of Washington. 76 p.

Borah, D. K. The dynamic simulation of water and sediment in watersheds. D, 1979, University of Mississippi. 199 p.

Borahay, Abd El Aziz El Hady A. Saint Louis Limestone (upper Mississippian), stratigraphy and petrography, near its type locality. M, 1970, University of Missouri, Rolla.

Borahay, Abd El-Aziz El-Hady Ahmad. Petrography, diagenesis and environment of deposition of the Gasconade Formation, lower Ordovician, southern Missouri. D, 1973, University of Missouri, Rolla.

Borak, Barry. Progressive and deep burial diagenesis in the Hunton (Late Ordovician to Early Devonian) and Simpson (Early to Middle Ordovician) groups of the deep Anadarko Basin of southwestern Oklahoma. M, 1978, Rensselaer Polytechnic Institute. 182 p.

Boras, Jaime Buitrago. Genetic study of the Northwest Butterfly Field, Garvin County, Oklahoma. M, 1978, University of Oklahoma. 53 p.

Borawski, Teddy W., Jr. Analysis of a magnetic anomaly present in Bedford County, Virginia. M, 1978, University of Pittsburgh.

Borax, Eugene. Mineralogy of the upper Miocene sandstones of the northernmost San Joaquin Valley. M, 1942, University of California, Los Angeles.

Borbas, Steve. The geology of radioactive mineral occurrences near Marietta, Nevada. M, 1977, University of Nevada. 73 p.

Borbely, Evelyn Susanna. Development and chemical quality of groundwater system in cast overburden at the Gibbons Creek lignite mine. M, 1988, Texas A&M University. 199 p.

Borch, Mary Ann. A quantitative study of injection wells in southeastern Ohio. M, 1988, Ohio University, Athens. 201 p.

Borchardt, Glenn Arnold. Neutron activation analysis for correlating volcanic ash soils. D, 1970, Oregon State University. 219 p.

Borchelt, Thomas R. The petrology of an andesitic intrusion in the Absaroka volcanic field, northwestern Wyoming. M, 1983, Miami University (Ohio). 50 p.

Borcherdt, Roger D. Inhomogeneous body and surface plane waves in a generalized viscoelastic half-space. D, 1971, University of California, Berkeley. 308 p.

Bordeau, Kenneth Vernon. Micropaleontology of the Fernvale Formation (Upper Ordovician) of Tennessee and Oklahoma. D, 1967, University of Oklahoma. 303 p.

Bordeau, Kenneth Vernon. Palynology of the Dry-wood Coal (Pennsylvanian) of Oklahoma. M, 1964, University of Oklahoma. 207 p.

Bordelon, Irion, Jr. Geologic controls affecting hydrocarbon accumulation in the Plio-Pleistocene of the East Cameron Block 271 Complex, offshore Louisiana. M, 1988, University of New Orleans.

Borden, Eugene William. The Hystrichosphaeridae; a group of microfossils of uncertain affinities. M, 1956, University of Illinois, Urbana. 90 p.

Borden, Richard K. Structural geology and stratigraphy of the southwestern margin of the Meade Thrust, Harrington Peak, Georgetown and Meade Peak quadrangles, southeastern Idaho. M, 1986, South Dakota School of Mines & Technology.

Borden, Robert Leslie. An Upper Ordovician coral fauna from the lower Mackenzie River area, Northwest Territories. M, 1956, University of Alberta. 92 p.

Bordine, Burton W. Paleoecologic implications of strontium, calcium, and magnesium in Jurassic rocks near Thistle, Utah. M, 1965, Brigham Young University.

Bordine, Burton William. Neogene biostratigraphy and paleoenvironments, lower Magdalena Basin, Colombia. D, 1974, Louisiana State University.

Boreck, Donna L. Geologic factors affecting mine development at the Hawk's Nest Mines, Somerset, Colorado. M, 1983, Colorado School of Mines. 105 p.

Borella, L. G. Prediction of uniaxial and triaxial strength of rocks from other measured physical properties. M, 1976, [Texas Tech University].

Borella, P. E. Petrologic and stratigraphic relationships among middle Ordovician limestones from central Kentucky to central Tennessee. D, 1975, University of Southern California.

Borella, Peter Edward. Sediment transport, Fire island, New York. M, 1969, George Washington University.

Borengasser, Marcus X. Remote sensor linear analysis as an aid for petroleum exploration, Arkoma Basin, Arkansas. M, 1980, University of Arkansas, Fayetteville. 126 p.

Borenstein, Herbert. Statistical comparisons among subjective analyses and an objective analysis. M, 1979, Purdue University. 98 p.

Boreske, John R., Jr. The Brachiopoda indigenous to the Mahantango coral reef (middle Devonian) of Monroe County, Penna. M, 1969, Brooklyn College (CUNY).

Boreske, John Robert, Jr. A review of the North American fossil amiid fishes. D, 1972, Boston University. 277 p.

Boreski, Charles V. Sedimentation and diagenesis of the Mississippian carbonates of the Twining oil field, Alberta. M, 1978, University of Calgary. 163 p.

Borg, Heinz. Estimating soil hydraulic properties from texture data. D, 1982, Washington State University. 87 p.

Borg, Iris Parnell. Studies in metamorphism; I, Planar structures within calcite of artificially deformed Yule Marble; II, Glaucophane and related schists near Healdsburg, California. D, 1954, University of California, Berkeley. 124 p.

Borg, Scott Gerald. Granitoids of northern Victoria Land, Antarctica. D, 1984, Arizona State University. 356 p.

Borg, Scott Gerald. Petrology and geochemistry of the Wyatt Formation and the Queen Maud Batholith, Upper Scott Glacier area, Antarctica. M, 1980, Arizona State University. 100 p.

Borgeld, Jeffrey Calvert. Holocene stratigraphy and sedimentation on the Northern California continental shelf. D, 1985, University of Washington. 177 p.

Borger, Harvey Daniel. Ostracodes of the Macoupin Cyclothem, Upper Pennsylvanian of southeastern Illinois. M, 1939, University of Illinois, Urbana. 39 p.

Borger, John Godfrey, II. Bedrock geology of a portion of extinct glacial Lake Calvin (Johnson and Louisa counties), eastern Iowa. M, 1965, University of Iowa. 60 p.

Borgerding, Janet Lee. Evolution of the diductor muscle system in articulate brachiopods with respect to mechanical efficiency of hingement. M, 1976, Indiana University, Bloomington. 45 p.

Borges Olivieri, Cesar A. Relationship between alteration and strength in South Table Mountain Lavas and Pikes Peak Granite, Jefferson County, Colorado. M, 1982, Colorado School of Mines. 147 p.

Borges, Rafael E. Decision-making in the preliminary stages of mineral development, using the Monte Carlo simulation. M, 1976, Colorado School of Mines. 161 p.

Borgia, Andrea. Physical aspects of eruptions at Arenal and Poas volcanoes, Costa Rica. D, 1988, Princeton University. 236 p.

Borglin, Edward Kenneth. The geology of part of the Morgan Valley Quadrangle, California. M, 1949, University of California, Berkeley. 86 p.

Boris, C. M. An integrated methodology for analysis of water for energy development with special application to the Yellowstone River basin. D, 1979, University of Michigan. 404 p.

Borja, Ronaldo Israel. Finite element analysis of the time-dependent behavior of soft clays. D, 1984, Stanford University. 239 p.

Bork, Jonathan. A gravity survey in the vicinity of Mellen, Wisconsin. M, 1967, Michigan State University. 91 p.

Bork, Kennard Baker. Bryozoa (Ectoprocta) of Champlainian age (Middle Ordovician) from northwestern Illinois, northeastern Iowa, and southwestern Wisconsin. D, 1967, Indiana University, Bloomington. 240 p.

Bork, Kenneth R. Post Miocene history of Antarctic bottom water paleospeed and polar front zone migrations in the Argentine Basin. M, 1988, San Jose State University. 96 p.

Borkland, Jay A. Mechanisms of emplacement of the Middle Mountain laccolithic complex, La Sal Mountains, Utah. M, 1986, SUNY at Buffalo. 185 p.

Borkowski, Annette Hottman. Testing and implementation of a phase-difference polarization filter and application of cepstral analysis to enhance regional seismic phases. M, 1987, Pennsylvania State University, University Park. 177 p.

Borkowski, Richard M. Geochemical and optical characterization of diagenetic and hydrothermal dolomite from the Bonneterre Formation within the southeastern Missouri lead zinc district. M, 1983, Texas A&M University. 111 p.

Borland, Gerald, C. and Lee, Charles A. The geology and ore deposits of the Cuprite mining district (Arizona). M, 1935, University of Arizona.

Born, Kendall Eugene. The brown iron ores of the western Highland Rim of Tennessee. M, 1931, Vanderbilt University.

Born, Peter. Geology of the East Bull Lake layered complex, District of Algoma, Ontario. M, 1979, Laurentian University, Sudbury.

Born, Stephen M. Geology of the southeastern third of the Powers Quadrangle, Oregon. M, 1963, University of Oregon. 83 p.

Born, Stephen Michael. Deltaic sedimentation at Pyramid lake (Reno Ana), Nevada. D, 1970, University of Wisconsin-Madison. 256 p.

Bornemann, E. Well-log analysis as a tool for lithofacies determination in the Viola Limestone (Ordovician) of South-central Kansas. D, 1979, Syracuse University. 202 p.

Bornhold, Brian Douglas. Carbonate turbidites in Columbus Basin, Bahamas. M, 1970, Duke University. 120 p.

Bornhold, Brian Douglas. Late Quaternary sedimentation in the eastern Angola Basin. D, 1973, Massachusetts Institute of Technology. 212 p.

Bornhorst, Laurie E. Mineralogy and engineering properties of Pleistocene silts and clays from selected sites in the Upper Peninsula, Michigan. M, 1985, Michigan Technological University. 192 p.

Bornhorst, Theodore Joseph. Major- and trace-element geochemistry and mineralogy of upper Eocene to Quaternary volcanic rocks of the Mogollon-Datil volcanic field, southwestern New Mexico. D, 1980, University of New Mexico. 1108 p.

Bornhorst, Theodore Joseph. Volcanic geology of the Crosby Mountains and vicinity, Catron County, New Mexico. M, 1976, University of New Mexico. 113 p.

Borns, David James. Blueschist metamorphism of the Yreka-Fort Jones area, Klamath Mountains, northern California. D, 1980, University of Washington. 155 p.

Borns, Harold William, Jr. The geology of the Skowhegan Quadrangle, Maine. D, 1959, Boston University. 183 p.

Boronow, Thomas Carlton. Petrography and diagenesis of upper Smackover Formation (Oxfordian), Atlanta and Pine Tree fields, southwestern Arkansas. M, 1984, University of New Orleans. 84 p.

Borovicka, Thomas G. The pre-Tertiary geology of the Lower Salmon River canyon between Packers Creek and Billy Creek, west-central Idaho. M, 1988, University of Idaho. 178 p.

Borowski, Robert D. Gravity survey in the Mount Eisenhower area, Banff-Kootenay national parks. M, 1975, University of Calgary. 125 p.

Borowski, Walter S. Petrology, depositional environments, and stratigraphic analysis of a part of the Middle Ordovician Chickamauga Group limestones near Clinton, Tennessee. M, 1982, University of Tennessee, Knoxville. 239 p.

Borquaye, Samuel A. A parametric study of basin thickness determined from Rayleigh wave inversion. M, 1988, University of Wisconsin-Milwaukee. 176 p.

Borras, Jaime Buitrago. Genetic study of the Northwest Butterly Field, Garvin County, Oklahoma. M, 1978, University of Oklahoma. 53 p.

Borrowman, George. The clays of Nebraska. D, 1916, University of Nebraska, Lincoln.

Borry, Barrett E. A geophysical investigation of the Bunker Hills, Jonestown, Pennsylvania. M, 1973, Lehigh University. 44 p.

Borst, Roger Lee. A mineralogical study of some Helderberg rocks of the central Hudson Valley, New York. D, 1965, Rensselaer Polytechnic Institute. 197 p.

Borst, Roger Lee. The granites of Big Falls, Wisconsin. M, 1958, University of Wisconsin-Madison.

Borstal, B. E. von see von Borstal, B. E.

Borthwick, Alastair Andrew. The geology and geochemistry of the Big Trout Lake Complex, N.W. Ontario. M, 1984, University of Toronto.

Borton, Robert L. Jade; its occurrences, mineralogy, geology, uses, and technology. M, 1952, University of New Mexico. 91 p.

Bortz, Louis C. Geology of the Copenhagen Canyon area, Monitor Range, Eureka County, Nevada. M, 1960, University of Nevada - Mackay School of Mines. 56 p.

Borucki, Mark K. A re-examination of southeastern Wisconsin bluff stratigraphy and correlation to offshore Pleistocene deposits. M, 1988, University of Wisconsin-Milwaukee. 128 p.

Borum, Jeffrey. Mass-wasting, geomorphology, and geology headwaters of the south fork of the Trinity River in Northern California. M, 1985, San Jose State University. 119 p.

Borys, Edmund. A horizontal intensity magnetic survey across the Rosamond Fault (California). M, 1936, California Institute of Technology. 16 p.

Boryta, Mark D. Geochemistry and origin of igneous rocks from the Archean Belt Bridge Complex, Limpopo Belt, South Africa. M, 1988, New Mexico Institute of Mining and Technology. 111 p.

Bosc, Eric Antoine. Geology of the San Agustîn Acasaguastlán Quadrangle and northeastern part of El Progreso Quadrangle. D, 1971, Rice University. 161 p.

Bosch, Herman F. Ecology of Poden polyphemoides (Crustacea, Branchiopoda) in Chesapeake bay (Maryland and Virginia). D, 1970, The Johns Hopkins University.

Bosch, Silverio C. Depositional pattern in the growth-faulted Frio Formation (Oligocene), McAllen-Pharr Field, South Texas. M, 1975, University of Texas, Austin.

Boschetto, Harold Bradley. Geology of the Lothidok Range, northern Kenya. M, 1988, University of Utah. 203 p.

Bosco, Francis N. Petroleum resources and their development in the Denver-Cheyenne Basin (T 589). M, 1941, Colorado School of Mines. 141 p.

Boslough, Mark Bruce. Shock-wave properties and high-pressure equations of state of geophysically important materials. D, 1984, California Institute of Technology. 179 p.

Bosma-Douglas, Julia. Geology and hydrothermal alteration of a portion of the eastern Excelsior Mountains, Mineral County, Nevada. M, 1987, University of Nevada. 159 p.

Boss, Reuel Lee. A general descriptive key to the Tetrabranchiate Cephalopoda, with special reference to those species of Acanthoscaphites, Desmoscaphites, Discoscaphites, Scaphites, and Baculites represented in the University of Colorado collection. M, 1928, University of Colorado.

Boss, Stephen K. Parameters controlling sediment composition of modern and Pleistocene Jamaican reefs. M, 1985, Utah State University. 101 p.

Boss, Theodore Robert. Vegetation ecology and net primary productivity of selected freshwater wetlands in Oregon. D, 1983, Oregon State University. 236 p.

Bosscher, Peter Jay. Soil arching in sandy slopes. D, 1981, University of Michigan. 134 p.

Bosse, Jacqueline Y. van see van Bosse, Jacqueline Y.

Bosse, Mark K. Diagenesis and petrology of the Cypress Sandstone in hydrocarbon producing areas of the central Illinois Basin. M, 1986, Southern Illinois University, Carbondale. 196 p.

Bossler, John David. Bayesian inference in geodesy. D, 1972, Ohio State University. 84 p.

Bossong, Clifford Robert. Teays deposits and drainage modification in southern Athens County, Ohio. M, 1975, Ohio University, Athens. 115 p.

Bossort, Dallas Overton. The geology of adjacent portions of the Adair and Smithfield quadrangles, Fulton County, Illinois. M, 1950, University of Iowa. 142 p.

Bossy, K. V. H. Morphology, paleoecology, and evolutionary relationships of the Pennsylvanian urocordylid nectrideans (Subclass Lepospondyli, Class Amphibia). D, 1976, Yale University. 493 p.

Boster, Mark Alan. Colorado River trips within the Grand Canyon National Park and Monument; a socioeconomic analysis. M, 1972, University of Arizona.

Boster, Mark Alan. Evaluation of agricultural adjustment to irrigation water salinity; a case study for Pinal County, Arizona. D, 1976, University of Arizona.

Boster, Ronald S. A study of ground-water contamination of the oil field brines in Morrow and Delaware counties, Ohio, with emphasis on the direct utilization of electrical resistivity techniques. M, 1967, Ohio State University.

Boster, Ronald Stephen. The value of primary versus secondary data in interindustry analysis; an Arizona case study emphasizing water resources. D, 1971, University of Arizona.

Bostick, Kent Anthony. Application of stable isotopes to the study of Santa Cruz River recharge to the Tucson Basin Aquifer. M, 1978, University of Arizona.

Bostick, Neely Hickman. Thermal alteration of clastic organic particles (phytoclasts) as an indicator of contact and burial metamorphism in sedimentary rocks. D, 1970, Stanford University. 220 p.

Bostik, Wayne Charles. Micropaleontology of the upper Eagle Ford and lower Austin groups, Big Bend National Park, Texas. M, 1960, Texas Tech University. 131 p.

Bostock, C. A. Minimizing costs in well field design in relation to aquifer models. D, 1975, University of Arizona. 157 p.

Bostock, Charles Alexander. Groundwater study, Rivers area, Manitoba. M, 1965, University of Saskatchewan. 75 p.

Bostock, Hewitt Hamilton. Petrology of the Shingle Creek porphyry near Penticton. M, 1956, University of British Columbia.

Bostock, Hewitt Hamilton. The petrology of the Shingle Creek porphyry. D, 1961, University of Wisconsin-Madison. 180 p.

Bostock, Hugh Samuel. A petrological study of the dyke rocks of the Ayox District, British Columbia. M, 1925, McGill University.

Bostock, Hugh Samuel. The geology and ore deposits of the Nickel Plate Mountain, Hedley, British Columbia. D, 1929, University of Wisconsin-Milwaukee.

Bostock, Michael Gerhard. Seismic detection of collapse structures; case study at Rocanville Mine, Saskatchewan. M, 1988, Queen's University. 126 p.

Boston, William Bryan. The surficial geology, paleontology, and paleoecology of the Finis Shale, Lower Virgilian, Pennsylvanian, of Jack County, Texas. M, 1988, Ohio University, Athens. 293 p.

Bostwick, David Arthur. Stratigraphy of the Gaptank Formation, Glass Mountains, Texas. D, 1958, University of Wisconsin-Madison. 108 p.

Bostwick, Douglas Leland. Structural geology of southern Indio Mountains, Hudspeth County, Texas. M, 1953, University of Texas, Austin.

Bostwick, T. R. The effect of Mn in the stability and phase relations of iron-rich pyroxenes. M, 1976, [SUNY at Albany].

Boswell, Ernest Harrison. Geology and mineral resources of the Ground Hog area, western Jefferson County, Alabama. M, 1949, University of Alabama.

Boswell, Ray Marcellus. Basin analysis of the Acadian clastic wedge in northern West Virginia and adjacent areas. D, 1988, West Virginia University. 373 p.

Boswell, Ray, Jr. Stratigraphy and sedimentation of the Acadian clastic wedge in northern West Virginia. M, 1985, West Virginia University. 179 p.

Bosworth, William P. Structural petrology of a mixed phyllite-marble terrane, Pownal, Vermont. M, 1977, Rensselaer Polytechnic Institute. 152 p.

Bosworth, William Paul. Structural geology of the Fort Miller, Schuylerville and portions of the Schaghticoke 7-1/2' quadrangles, eastern New York, and its implications in Taconic geology and experimental and theoretical studies of solution transfer in deforming heterogeneous systems. D, 1980, SUNY at Albany. 346 p.

Botbol, Joseph Moses. Characteristic analysis of base metal deposits in the continental United States. D, 1968, University of Utah. 257 p.

Botbol, Joseph Moses. Geochemical exploration for gilsonite. M, 1961, University of Utah. 42 p.

Botero, Gilberto. The coal-bearing Cali Series, Cauca System, Tertiary of Colombia, South America. M, 1942, University of Chicago. 123 p.

Botha, Willem J. Basis of the theory and interpretation of the dual frequency method. D, 1980, Colorado School of Mines. 191 p.

Bothner, Bryan R. Conodonts of the Plattin Limestone (Middle Ordovician), northern Arkansas. M, 1988, University of New Orleans.

Bothner, M. H. The radiation dosimetry for the natural thermoluminescence of carbonate deep sea cores. M, 1968, Dartmouth College. 60 p.

Bothner, Michael Henry. Mercury; some aspects of its marine geochemistry in Puget Sound, Washington. D, 1973, University of Washington. 126 p.

Bothner, W. A., Jr. Petrology, structural geometry, and economic geology of the Cooney Hills, Platte County, Wyoming. D, 1967, University of Wyoming. 164 p.

Botoman, George G. Precambrian and Paleozoic stratigraphy and potential mineral deposits along the Cincinnati Arch of Ohio. M, 1975, Ohio State University.

Botros, Effat S. Diagenetic alteration in tuffaceous sediments of the Duff Formation, Trans-Pecos, Texas. M, 1976, University of Texas, Arlington. 104 p.

Botsford, Jack William. Depositional history of Middle Cambrian to Lower Ordovician deep water sediments, Bay of Islands, western Newfoundland. D, 1988, Memorial University of Newfoundland. 509 p.

Bott, Winston. Mineralogy and diagenesis of Gulf Coast Tertiary shales Ann-Mag Field, Brooks County, Texas. M, 1985, Texas A&M University. 138 p.

Bottaro, Joseph L. Geology of the Middle Buttes volcanic complex, Mojave District, Kern County, California. M, 1987, San Jose State University. 94 p.

Bottinga, Jan. Oxygen isotopes in geology. M, 1963, University of British Columbia.

Bottinga, Yan. Isotopic fractionation in the system; calcite-graphite-carbon dioxide-methane-hydrogen-water. D, 1968, University of California, San Diego. 142 p.

Bottjer, David J. Paleoecology, ichnology, and depositional environments of Upper Cretaceous chalks (Annona Formation; Chalk Member of Saratoga Formation), southwestern Arkansas. D, 1978, Indiana University, Bloomington. 424 p.

Bottjer, David J. Stromatoporoid beds in the Jamesville Member, Manlius Formation (Lower Devonian), New York State. M, 1975, SUNY at Binghamton. 43 p.

Bottjer, Richard J. Depositional environments and tectonic significance of the South Pass Formation, southern Wind River Range, Wyoming. M, 1984, University of Wyoming. 229 p.

Bottner, R. W. Wallrock alteration associated with the Mesozoic pyrrhotite mineralization at Cuttingsville, Vermont. M, 1977, University of Vermont.

Botto, R. I. Josephinite; a unique nickel-iron. D, 1976, Cornell University. 278 p.

Bottomley, D. J. Sources of stream flow and dissolved constituents in a small Precambrian Shield watershed. M, 1974, University of Waterloo.

Bottomley, Dennis James. The isotope geochemistry and origins of fracture calcites in the Grenville gneisses, Chalk River, Ontario. D, 1988, University of Ottawa. 264 p.

Bottomley, R. J. Argon-40/argon-39 dating of basalts from the Columbia River Plateau. M, 1975, University of Toronto.

Bottomley, Richard John. ^{40}Ar-^{39}Ar dating of melt rock from impact craters. D, 1982, University of Toronto.

Bottoms, Kenneth P. A study of the middle Devonian strata based on a core from Grosse Ile, Michigan. M, 1959, University of Michigan.

Botts, Michael Edward. The effects of slaking on the engineering behavior of clay shales. D, 1986, University of Colorado. 412 p.

Botts, Michael Edward. The stratigraphic sequence of volcanic and sedimentary units in the north polar region of Mars. M, 1979, Washington University. 63 p.

Bou-Rabee, Firyal. Quantification of pore complex geometry via image analysis and its relationship to reservoir quality as related to the Wafra Field in Kuwait. M, 1983, Wichita State University. 171 p.

Bou-Rabee, Firyal Ahmed. The geology and geophysics of Kuwait. D, 1986, University of South Carolina. 246 p.

Bouchard, A. B. Natural resources analysis of a section of the Gros Morne National Park, in Newfoundland, Canada. D, 1975, Cornell University. 163 p.

Bouchard, Gilles. Environment géologique du gisement aurifère de Gwillim, Chibougamau, Québec. M, 1987, Universite du Quebec a Chicoutimi. 83 p.

Bouchard, Karl. Correction de l'effect d'une topographie de surface bi-dimensionnelle sur les levés magnétotelluriques. M, 1988, Ecole Polytechnique. 318 p.

Bouchard, Michel. Etude de la géomorphologie et des conditions périglaciaires de l'Ile Herschel, Territoire du Yukon. M, 1971, Universite de Montreal.

Bouchard, Michel. Géologie des depots meubles de l'Ile Herschel, Territoire du Yukon. M, 1974, Universite de Montreal.

Bouchard, Michel A. Late Quaternary geology of the Temiscamie area, central Quebec. M, 1981, McGill University. 284 p.

Bouchard, Stephen M. Structural changes in margarite with increasing temperature. M, 1988, University of Illinois, Chicago.

Boucher, Gary Wynn. The studies of microearthquakes; I, The microearthquake seismicity of the Denali fault; II, Microearthquakes investigations in Nevada; III, Earthquakes associated with underground nuclear explosions. D, 1969, Columbia University. 125 p.

Boucher, Michael Lee. The crystal structure of alamosite PbSiO3. M, 1967, University of Michigan.

Boucherle, Mary M. An ecological history of Elk Lake, Clearwater Co., Minnesota, based on Cladocera remains. D, 1982, Indiana University, Bloomington. 128 p.

Bouchon, Michel Paul. Discrete wavenumber representaion of seismic wave fields with application to various scattering problems. D, 1976, Massachusetts Institute of Technology. 180 p.

Boucot, Arthur J. The Lower Devonian rocks of west central Maine. D, 1953, Harvard University.

Boudette, Eugene L. Stratigraphy and structure of the Kennebago Lake Quadrangle, central western Maine. D, 1978, Dartmouth College. 342 p.

Boudette, Eugene L. The thermal history of selected plagioclases. M, 1959, Dartmouth College. 130 p.

Boudouris, James. A lithofacies analysis of the Bass Island and Salina formations in the Michigan Basin. M, 1955, University of Michigan.

Boudra, Robert A. Refraction seismic and direct current resistivity surveying at a proposed sanitary landfill site in western Washington County, Arkansas. M, 1988, University of Arkansas, Fayetteville.

Boudreau, Alan Ernest. Fine-scale layering in the Stillwater Complex, Montana. M, 1982, University of Oregon. 66 p.

Boudreau, Alan Ernest. The role of fluids in the petrogenesis of platinum-group element deposits in the Stillwater Complex, Montana. D, 1986, University of Washington. 247 p.

Boudreau, Bernard Paul. Diagenetic models of biological processes in aquatic sediments. D, 1985, Yale University. 541 p.

Boudreau, Cynthia Elizabeth. A reconnaissance of the structure and petrology of the Octoraro Phyllite. M, 1948, Bryn Mawr College. 19 p.

Boudreau, Francis C. Paleozoic geology of the southwest portion of the Des Arc Quadrangle, Mo. M, 1960, St. Louis University.

Boudreau, Lawrence Joseph. Depositional environment of sands associated with the Planulina palmerae biozone, offshore Louisiana. M, 1982, University of Massachusetts. 201 p.

Boudreau, M. Diagenetic and deformational structures in cherts of the Havallah Sequence, Nevada. M, 1988, Queens College (CUNY).

Boudreault, Alain P. Pétrographie et géochimie des laves et des filons-couches mafiques et ultramafiques du Canton de Richardson, Chibougamau, Québec. M, 1977, Universite du Quebec a Chicoutimi. 117 p.

Boudreaux, Joseph Edes. Biostratigraphy and calcareous nannoplankton of the Submarex section (Caribbean Sea, Nicaragua Rise). M, 1967, University of Illinois, Urbana.

Bouett, Lawrence W. The effect of transverse mixing on tracer dispersion in a fracture. M, 1986, Stanford University.

Bougan, Shelley June. Aeolian bedforms on Venus; an investigation of roles of particle size and wind speed. M, 1985, Arizona State University. 85 p.

Boughton, Carol Jean. Integrated geochemical and hydraulic analyses of Nevada Test Site ground water systems. M, 1986, University of Nevada. 135 p.

Bouillon, Jean Jacques. Etude pétrographique et structurale des granulites de la bordure nord. M, 1973, Universite de Montreal.

Bouknight, Dan L. Paleogeography and environmental interpretation of the Brassfield Formation (Silurian) across the Cincinnati Arch in south-central Kentucky. M, 1980, Eastern Kentucky University. 47 p.

Boule, Mark E. Geomorphic interpretation of vegetation on Fisherman Island, Virginia. M, 1976, College of William and Mary.

Bouley, Bruce Albert. Volcanic stratigraphy, stratabound sulfide deposits, and relative age relationships in the East Penobscot Bay area, Maine. D, 1978, University of Western Ontario. 168 p.

Boulger, Martha Lillian. The Kings River fan; a geographical interpretation. M, 1938, University of California, Berkeley. 130 p.

Boulware, Joe Wood. The relationship of the soils to the geology of Hillsborough County, Florida. M, 1963, University of Florida. 32 p.

Bounds, Jon Dudley. The geology of the Floras Creek area, Curry County, Oregon. M, 1982, Portland State University. 75 p.

Bounk, Michael Joseph. The petrology and depositional environment of the Bainbridge Limestone (middle and upper Silurian) of Southeast Missouri and Southwest Illinois. M, 1975, University of Iowa. 85 p.

Bourbie, Thierry. Effects of attenuation on reflections. D, 1982, Stanford University. 223 p.

Bourbon, William Bruce. The origin and occurrences of talc in the Allamoore District, Culberson and Hudspeth counties, Texas. M, 1981, West Texas State University. 65 p.

Bourbonniere, R. A. Geochemistry of humic matter in Holocene Great Lakes sediments. D, 1979, University of Michigan. 387 p.

Bourcier, William Louis. Stabilities of chloride and bi-sulfide complexes of zinc in hydrothermal solutions. D, 1983, Pennsylvania State University, University Park. 194 p.

Bourgault, Gilles. Pétrographie et géochimie del l'indice minéralisé "Swanson", une syénite aurifère. M, 1988, Ecole Polytechnique. 312 p.

Bourgeois, Jason. Middle Miocene foraminiferal paleoecology, Gibson Field, Terrebonne Parish, Louisiana. M, 1988, University of New Orleans.

Bourgeois, Joanne. Sedimentology and tectonics of Upper Cretaceous rocks, Southwest Oregon. D, 1980, University of Wisconsin-Madison. 309 p.

Bourget, André Pétrographie et distribution de l'or autour du gîte S-50 de la Mine Kiena, Val d'Or, Québec. M, 1986, Ecole Polytechnique. 119 p.

Bourgoin, Bernard Patrick. Mytilus edulis shells as environmental recorders for lead contamination. D, 1987, McMaster University. 135 p.

Bourke, John Francis. Differential transport of sand grains; ripples and dunes. M, 1979, Syracuse University.

Bourke, Robert H. Monitoring coastal upwelling by measuring its effects within an estuary. M, 1969, Oregon State University. 54 p.

Bourke, Robert Hathaway. A study of the seasonal variation in temperature and salinity along the Oregon-northern California coast. D, 1972, Oregon State University. 107 p.

Bourland, William C. Tectogenesis and metamorphism of the Piedmont from Columbia to Westview, Virginia along the James River. M, 1976, Virginia Polytechnic Institute and State University.

Bourn, Oscar B. Regional stratigraphic analysis of the Dakota and Colorado groups, Julesburg Basin area. M, 1952, Northwestern University.

Bourne, Donald Alleyne. Wall rock alteration in the Nipissing diabase sill, Cobalt, Ontario. M, 1951, McMaster University. 81 p.

Bourne, Douglas Randal. Electric analog simulation of groundwater flow patterns at a potash waste disposal pond located near Esterhazy, Saskatchewan. M, 1976, University of British Columbia.

Bourne, Harold L. Techniques and results of an evaluation of the Port Deposit granodiorite Stoltzfus quarry, northeastern Maryland as a potential source of alumina for the glass industry. M, 1969, University of Toledo. 46 p.

Bourne, James H. A tectonic reconstruction along the Grenville Front (Ontario). M, 1971, Queen's University. 163 p.

Bourne, James Hillary. The petrogenesis of the Humit Group minerals regionally metamorphosed marbles of the Grenville Supergroup. D, 1974, Queen's University. 159 p.

Bourque, Michael W. Stratigraphy of the Upper Pennsylvanian - Lower Permian portion of the Bird Spring Group, Battleship Wash, Arrow Canyon Range, Clark County, Nevada. M, 1978, University of Illinois, Urbana. 117 p.

Bourque, Paul Daniel. A metallogenic study of the Antigonish area, Nova Scotia, with special reference to the copper occurrences of the Ohio-Sylvan Glen Belt. M, 1980, Dalhousie University. 117 p.

Bourque, Pierre Andre. Stratigraphie du Silurien et du Devonien basal du Nord-Est de la Gaspésie. D, 1973, Universite de Montreal.

Bourque, Pierre-André Stratigraphie du Silurien et du Dévonien Inférieur du nord-est de la Gaspésie. M, 1969, Universite de Montreal.

Bouse, Robin. Isotopic, chemical and petrographic characteristics of stratiform manganese oxide deposits in the Lake Mead region, southeastern Nevada; implications for the origin of the deposits. M, 1988, University of Rhode Island.

Bousfield, John Channing. The geology of the Manchester Quadrangle, Missouri. M, 1949, Washington University. 61 p.

Boutilier, Robert Francis. The transformation of a volcanic sequence under special conditions. D, 1963, Boston College. 218 p.

Boutlier, R. F. The transformation of a volcanic sequence under special conditions. D, 1963, Boston University. 162 p.

Bouton, Katherine Alice. A spectral reflectance study of the sedimentary Paleozoic rocks around Racetrack Valley in Death Valley National Monument. M, 1984, California State University, Northridge. 152 p.

Boutte, Andre L. Facies analysis; Fredericksburg Limestone; Callahan Divide, Texas. M, 1969, Louisiana State University.

Boutte, Brian. Prediction before drilling and detection after drilling of abnormal subsurface pressures in wildcat wells in the Gulf of Mexico. M, 1979, Tulane University.

Boutwell, Gordon Powers, Jr. On the yield behavior of cohesionless materials. D, 1968, Duke University. 278 p.

Boutwell, Jerry. Geology of the Spring Valley Quadrangle. M, 1966, University of Arkansas, Fayetteville.

Bova, John A. Peritidal cyclic and incipiently drowned platform sequences; Lower Ordovician Chepultepec Formation, Virginia. M, 1982, Virginia Polytechnic Institute and State University. 175 p.

Bove, Dana Joseph. Evolution of the Red Mountain alunite deposit, Lake City caldera, San Juan Mountains, Colorado. M, 1988, University of Colorado. 179 p.

Bovell, George R. L. Sedimentation and diagenesis of the Nordegg Member in central Alberta. M, 1979, Queen's University. 149 p.

Bow, Craig Sherwood. The geology and petrogeneses of the lavas of Floreana and Santa Cruz islands; Galapagos Archipelago. D, 1979, University of Oregon. 322 p.

Bowden, Douglas Richard. Volcanic rocks of the Pelican River massive sulfide deposit, Rhinelander, Wisconsin; a study in wallrock alteration. M, 1978, Michigan Technological University. 62 p.

Bowden, Kenneth Lester. The geology of the Lorado Taft field campus and vicinity. M, 1957, Northern Illinois University. 64 p.

Bowden, Thomas D. Depositional processes and environments within the Revett Formation, Precambrian Belt Supergroup, northwestern Montana and northern Idaho. M, 1977, University of California, Riverside. 161 p.

Bowden, William Breckenridge. Nitrogen cycling in the sediments of a tidal freshwater marsh. D, 1982, North Carolina State University. 158 p.

Bowditch, Samuel I. The geology and ore deposits of Cerro de Pasco, Peru. D, 1935, Harvard University.

Bowdler, Jay Laurence. Mid-Cretaceous calcareous nannoplankton paleobiogeography and paleo-oceanography of the Atlantic Ocean. M, 1978, University of Utah. 182 p.

Bowdon, M. M. A petrologic and petrographic investigation of the lower Tuscaloosa in the Feliciana parishes of Louisiana. M, 1987, University of Southwestern Louisiana. 118 p.

Bowe, Richard J. Depositional history of the Dakota Formation (Cretaceous) in eastern Nebraska. M, 1972, University of Nebraska, Lincoln.

Bowen, Anita Schenck. The stratigraphy and micropaleontology of the Mississippian strata above the Berea Formation in northern Geauga and southern Lake counties, Ohio. M, 1951, Ohio State University.

Bowen, Boone Moss, Jr. The structural geology of a portion of southeastern Dawson County, Georgia. M, 1961, Emory University. 45 p.

Bowen, Bruce E. The geology of the upper Cenozoic sediments in the East Rudolf Embayment of the Lake Rudolf Basin, Kenya. D, 1974, Iowa State University of Science and Technology.

Bowen, Bruce Eugene. Stratigraphy of the Jebel el Qatrani Formation (Oligocene), Fayum depression (Egypt, U.A.R.). M, 1970, Iowa State University of Science and Technology.

Bowen, Charles F. Pleistocene geology of the Uinta Mountains. M, 1903, University of Wisconsin-Madison.

Bowen, Charles Henry. Geology of Clayton and Pike townships, Perry County, Ohio. M, 1947, Ohio State University.

Bowen, Charles Henry. The petrology and economic geology of the Sharon Conglomerate in Geauga and Portage counties, Ohio. D, 1952, Ohio State University.

Bowen, Corey Scott. Structure and stratigraphy of the Marietta District, Excelsior Mountain, west-central Nevada. M, 1982, Rice University.

Bowen, David G. Subsurface study of the Lee Formation in Buchanan County, Virginia. M, 1963, Virginia Polytechnic Institute and State University.

Bowen, David Wayne. Solid waste disposal site suitability evaluation in Montana. M, 1980, Montana State University. 87 p.

Bowen, James Howland. Regional finite strain analysis of basal Devonian orthoquartzites; Valley and Ridge Province, Pennsylvania. M, 1986, University of Connecticut. 45 p.

Bowen, John E. The sediments of the Washita Arm of Lake Texoma. M, 1959, University of Oklahoma. 59 p.

Bowen, Oliver Earle, Jr. Geology of the Sidewinder and Granite Mountains, California. M, 1950, University of California, Berkeley. 99 p.

Bowen, Richard Gordon. The geology of the Beulah area, Malheur County, Oregon. M, 1956, University of Oregon. 80 p.

Bowen, Richard Lee. A geological report on the Guffie area, McLean and Daviess counties, Kentucky. M, 1951, Indiana University, Bloomington. 26 p.

Bowen, Timothy Dana. Estimation of subsurface temperature gradients and heat flow variation by linear least squares analysis. M, 1982, University of Michigan.

Bowen, Ward Culver. A review of theories of origin of the zinc ores of Sussex County, New Jersey. D, 1935, Cornell University.

Bowen, Zeddie Paul. Brachiopoda of the Keyser Limestone (Siluro-Devonian), Maryland and adjacent areas. D, 1963, Harvard University.

Bower, Guy Joseph. Origin of the basaltic clays near Troy, Idaho. M, 1940, University of Idaho. 35 p.

Bower, P. M. Burdens of industrial cadmium and nickel in the sediments of Foundry Cove, Cold Spring, New York. M, 1976, Queens College (CUNY). 162 p.

Bower, Peter Michael. Addition of radiocarbon to the mixed-layers of two small lakes; primary production, gas exchange, sedimentation and carbon budget. D, 1981, Columbia University, Teachers College. 259 p.

Bower, Richard R. Dispersal centers of sandstones in the Douglas Group (Pennsylvanian) of Kansas. M, 1961, University of Kansas. 19 p.

Bower, Richard Raymond. Stratigraphy of Red Peak Formation, Alcova Limestone, and Crow Mountain Member of Popo Agie Formation (Triassic) of central Wyoming. D, 1964, University of Oklahoma. 174 p.

Bower, Scott M. An analysis of data sources for predicting runoff using the Soil Conservation Service curve number method; East Fork, Massac Creek watershed, Kentucky. M, 1985, Murray State University. 58 p.

Bowerman, James Nelson. Geology and mineral resources of the Pasour Mountain area, Gaston County, North Carolina. M, 1954, North Carolina State University. 56 p.

Bowers, Dale. Potassium-argon age dating and petrology of the Mineral Mountains Pluton, Utah. M, 1978, University of Utah. 76 p.

Bowers, Fred Howard. Effects of windthrow on soil properties and spatial variability in Southeast Alaska. D, 1987, University of Washington. 185 p.

Bowers, Gerald Frank. Geology of the Kissick Canyon area, Beaverhead County, Montana. M, 1949, University of Michigan. 34 p.

Bowers, Glenn Lee. The influence of pore fluid on the stability of a rock mass with a weakened zone. D, 1982, University of Illinois, Urbana. 107 p.

Bowers, Howard E. Geology of the Tony Butte area and vicinity, Mitchell Quadrangle, Oregon. M, 1953, Oregon State University. 152 p.

Bowers, John Richard. Areal geology of the Cheyenne area, Roger Mills County, Oklahoma. M, 1967, University of Oklahoma. 62 p.

Bowers, Keith Douglas. Mixed water phreatic dolomitization of Jurassic oolites in the upper Smackover Member, East Texas basin; petrologic and isotopic evidence. M, 1986, Texas A&M University. 138 p.

Bowers, Mark Thomas. In situ soil testing system for collapsible soils. D, 1986, Arizona State University. 265 p.

Bowers, Richard F. Effects of oil field operations in ground water on parts of Pennsylvania. M, 1969, West Virginia University.

Bowers, Robert Alwyn. Gravity in Northern California. D, 1958, University of California, Berkeley. 67 p.

Bowers, Roger Lee. Petrography and petrogenesis of the Alibates dolomite and chert (Permian) northern Panhandle of Texas. M, 1975, University of Texas, Arlington. 155 p.

Bowers, Teresa Suter. Calculation of the thermodynamic and geochemical consequences of nonideal mixing in the system H_2O-CO_2-NaCl on phase relations in geologic systems. D, 1982, University of California, Berkeley. 136 p.

Bowers, Timothy L. Upper Niagara-lower Salina (Mid-Silurian) sedimentology, and conodont-based biostratigraphy and thermal maturity of the Southeast Michigan Basin. M, 1987, Wayne State University. 120 p.

Bowers, William E. Geology of the East Portrillo Hills, Dona Ana County, New Mexico. M, 1960, University of New Mexico. 67 p.

Bowersox, J. R. Nearshore environments of the late Pleistocene Nestor Terrace, Point Loma, California. M, 1974, San Diego State University.

Bowes, William A. Distribution and thickness of phosphate ore zones in the Phosphoria Formation (Permian). M, 1952, University of Utah. 60 p.

Bowie, Mark R. Geology of the Dripping Springs Valley chabazite and related zeolite deposits, Southeast Arizona and Southwest New Mexico. M, 1985, New Mexico Institute of Mining and Technology. 139 p.

Bowin, Carl Otto. Geology of central Dominican Republic. D, 1960, Princeton University. 249 p.

Bowin, Carl Otto. Geology of the Moxie Mountain-Moosehead Lake area, Maine. M, 1957, Northwestern University.

Bowker, David E. Remanent magnetization of Eastern U.S. Triassic rocks. D, 1960, Massachusetts Institute of Technology. 166 p.

Bowker, Kent Alan. Stratigraphy, sedimentology, and uranium potential of Virgilian through Leonardian strata in western Marietta Basin and central Muenster-Waurika Arch, Oklahoma and Texas. M, 1980, Oklahoma State University. 100 p.

Bowland, Christopher Lee. Seismic stratigraphy and structure of the western Colombian Basin, Caribbean Sea. M, 1984, University of Texas, Austin. 247 p.

Bowlby, David C. Lithostratigraphy of the Morrow Formation (Lower Pennsylvanian), Tenkiller Ferry Reservoir area, northeastern Oklahoma. M, 1968, University of Oklahoma. 151 p.

Bowlby, J. R. Late Glacial ice wedge casts in the Kingston Basin of Lake Ontario. M, 1975, Queen's University. 167 p.

Bowler, E. L. The geologic and economic aspects of copper. M, 1932, McMaster University.

Bowles, Edgar. The family Turritellidae in the Eocene of the Atlantic and Gulf Coastal Plain of North America. D, 1939, The Johns Hopkins University.

Bowles, Frederick A. Solid state reaction study in the system PbO-Al_2O_3-SiO_2. M, 1965, Miami University (Ohio). 93 p.

Bowles, Frederick Albert. Electron microscopy investigation of the microstructure in sediment samples from the Gulf of Mexico. D, 1969, Texas A&M University. 145 p.

Bowles, J. H. Copper deposits of Shannon County, Missouri (T 426). M, 1921, University of Missouri, Rolla.

Bowles, Jack Paul Fletcher, Jr. Subsurface geology of Woods County, Oklahoma. M, 1959, University of Oklahoma. 79 p.

Bowles, Jane Margaret. Effects of human disturbance on the sand dunes at Pinery Provincial Park. D, 1980, University of Western Ontario.

Bowles, John Hyer. Investigations to determine a possible source of the Carboniferous sandstone of the Ozark region. M, 1921, University of Missouri, Rolla.

Bowles, Kelly L. Carbonate progradational-submergence sequences in the Lower Cretaceous (upper Albian) of Southwest Texas. M, 1986, University of Texas, Arlington. 182 p.

Bowles, Oliver. Economic geology of slate. D, 1922, George Washington University.

Bowlin, Benjamin. Descriptive, interpretive and quantitative study of coarse clastics in the Middle Ordovician Tellico Formation in the vicinity of South

Holston Dam, Tennessee. M, 1979, University of Tennessee, Knoxville. 256 p.

Bowling, David Lynn. The geology and genesis of the Apex gallium-germanium deposit, Washington County, Utah. M, 1988, University of Utah. 81 p.

Bowling, Edward C., III. A petrologic study of the Middle Silurian Laurel Dolomite in north central Kentucky. M, 1980, Eastern Kentucky University. 58 p.

Bowling, G. Patrick. Geology and geochemistry of early Proterozoic supracrustal rocks from the western Dos Cabeza Mountains, Cochise County, Arizona. M, 1987, New Mexico Institute of Mining and Technology. 126 p.

Bowman, Anthony Frank. An investigation of Al$_2$SiO$_5$ phase equilibrium utilizing the scanning electron microscope. M, 1975, University of Oregon. 80 p.

Bowman, Delbert A. Cambrian and Ordovician limestones of Paulinskill Valley, New Jersey. M, 1951, Lehigh University.

Bowman, Edgar Cornell. Stratigraphy and structure of the Orient area, Washington. D, 1950, Harvard University. 161 p.

Bowman, Eugene A. The subsurface geology of southeastern Noble County, Oklahoma. M, 1956, University of Oklahoma. 42 p.

Bowman, Flora Jean. The geology of the north half of Hampton Quadrangle, Oregon. M, 1940, Oregon State University. 72 map p.

Bowman, Frank Otto, Jr. The Carolina slate belt near Albemarle, North Carolina. D, 1954, University of North Carolina, Chapel Hill. 85 p.

Bowman, Isaiah. The geography of the Central Andes. D, 1909, Yale University.

Bowman, J. Roger. Anisotropy and attenuation in the upper mantle in the Fiji and Tonga region. D, 1985, University of Colorado. 187 p.

Bowman, J. Roger. Occurrence of foreshocks in the central Aleutian Island arc. M, 1982, University of Colorado. 96 p.

Bowman, James Floyd, II. Petrology of the Pensauken Formation (Pleistocene: New Jersey and northern Delaware). D, 1966, Rutgers, The State University, New Brunswick. 155 p.

Bowman, Jimmy Dean. The geology of the Centerview, Missouri, Quadrangle. M, 1963, University of Iowa. 96 p.

Bowman, John K. Geology and mineralization of the Meadow Lake mining district, Nevada County, California. M, 1983, California State University, Hayward. 268 p.

Bowman, John Randall. Contact metamorphism, skarn formation, and origin of C-O-H skarn fluids in the Black Butte aureole, Elkhorn, Montana. D, 1978, University of Michigan. 485 p.

Bowman, John Randall. Use of the isotopic composition of strontium and SiO$_2$ content in determining the origin of Mesozoic basalt from Antarctica. M, 1971, Ohio State University.

Bowman, John Thomas. Compositional variation in granite of eastern Wake and northwestern Johnston counties, North Carolina. M, 1970, North Carolina State University. 33 p.

Bowman, Kenneth Charles, Jr. Sedimentation, economic enrichment and evaluation of heavy mineral concentrations on the southern Oregon continental margin. D, 1972, Oregon State University. 136 p.

Bowman, Kenneth Paul. Effects of ice sheets on the sensitivity of climate. D, 1984, Princeton University. 146 p.

Bowman, Linda Gail. Analysis of the risk of deep groundwater pollution due to surface mining for lignite in Grimes County, Texas. M, 1978, Texas A&M University. 99 p.

Bowman, Michael J. Implementation and application of an interactive seismic data processing package on the IBM 3083 mainframe. M, 1986, Purdue University. 85 p.

Bowman, Patricia A. The post-Pleistocene foraminifera of Little Lake, San Salvador, Bahamas. M, 1982, University of Akron. 118 p.

Bowman, Paul L. Correlation of gravity and magnetic data over central North America. M, 1978, Purdue University. 176 p.

Bowman, Paul W. Study of a peat bog near the Matamek River, Quebec, Canada, by the method of pollen analysis. D, 1930, University of Virginia. 34 p.

Bowman, Phillip Robert. Depositional and diagenetic interpretations of the Lower Silurian Waucoma Limestone in Northeast Iowa. M, 1985, University of Iowa. 141 p.

Bowman, R. Scattering of seismic waves by small inhomogeneities. D, 1955, Massachusetts Institute of Technology. 86 p.

Bowman, Richard Spencer. Stratigraphy and paleontology of the Niagaran Series in Highland County, Ohio. D, 1956, Ohio State University. 266 p.

Bowman, Robert Stephen. Prediction of nickel sorption and mobility in soils. D, 1982, New Mexico State University, Las Cruces. 135 p.

Bowman, Rudolf A. Microbial denitrification in sandy soils. D, 1973, University of California, Riverside.

Bowman, Scott A. Miocene extension and volcanism in the Caliente Caldera Complex, Lincoln County, Nevada. M, 1985, Colorado School of Mines. 142 p.

Bown, John S. The geology of the Searchlight mining district, Clark County, Nevada. M, 1977, University of Missouri, Rolla.

Bown, T. M. Geology and mammalian paleontology of the Sand Creek Facies, lower Willwood Formation (early Eocene), Washakie County, Wyoming. D, 1977, University of Wyoming. 567 p.

Bownocker, John A. The paleontology and stratigraphy of the Carboniferous rocks of Ohio. D, 1897, Ohio State University.

Bowring, Samuel Anthony. U-Pb zircon geochronology of early Proterozoic Wopmay Orogen, N.W.T. Canada; an example of rapid crustal evolution. D, 1985, University of Kansas. 158 p.

Bowser, Carl James. Geochemistry and petrology of the sodium borates in the non-marine evaporite environment. D, 1965, University of California, Los Angeles. 307 p.

Bowser, William Franklin. Subsidence of Cretaceous rocks at Big Spring, Texas. M, 1927, Texas Christian University. 61 p.

Bowyer, Robert C. Landsat and airborne MSS reconnaissance of glacial drift-derived materials in northern Summit County, Ohio. M, 1978, University of Akron. 65 p.

Box, Jerry W. X-ray studies of iron deposits (Eocene) in Lincoln Parish (Louisiana). M, 1968, Louisiana Tech University.

Box, Michael R. Petrology and depositional environments of the Smackover and Buckner formations (Upper Jurassic), South Texas. M, 1973, University of Texas, Arlington. 146 p.

Box, Stephen Edward. Mesozoic tectonic evolution of the northern Bristol Bay region, southwestern Alaska. D, 1985, University of California, Santa Cruz. 199 p.

Boxwell, Mimi A. Metamorphic history of the Standing Pond and Putney volcanics in the Claremont, Bellows Falls and Saxtons River quadrangles in southeastern Vermont. M, 1986, University of New Hampshire. 245 p.

Boyce, Barry James Simon. Four Devonian pinnacles in the Windfall Reef complex, central Alberta; facies analysis, paleoecology and reef development. M, 1974, University of Manitoba.

Boyce, Glenn Markland. A study of stress determination in rock salt by the method of hydraulic fracturing. D, 1988, University of California, Berkeley. 272 p.

Boyce, Helen. The mines of the upper Harz from 1514 to 1589. D, 1917, University of Chicago. 122 p.

Boyce, Henry Spurgeon. Geology of the Burhenglen Knob Quadrangle. M, 1922, University of North Carolina, Chapel Hill. 23 p.

Boyce, Joseph Michael. The structure and petrology of older Precambrian crystalline rocks, Bright Angel Canyon, Grand Canyon, Arizona. M, 1972, Northern Arizona University. 88 p.

Boyce, Malcom Walter. The macropaleontology of the Osgood Formation (Niagaran) in Ripley County, Indiana. M, 1956, Indiana University, Bloomington. 140 p.

Boyce, Robert E. Electrical resistivity of modern marine sediments from the Bering Sea. M, 1967, San Diego State University.

Boyce, Robert L. Depositional systems in the LaHood Formation, Belt Supergroup, Precambrian, southwestern Montana. D, 1975, University of Texas, Austin. 247 p.

Boyce, Steven Craig. Laboratory determination of horizontal stress in cohesionless soil. D, 1983, Cornell University. 329 p.

Boyce, William Douglas. Early Ordovician trilobite faunas of the Goat Harbour and Catache formations (St. George Group) in the Goat Harbour-Cape Norman area, Great Northern Peninsula, western Newfoundland. M, 1984, Memorial University of Newfoundland. 272 p.

Boyd, Alston. Geology of the western third of La Democracia Quadrangle, Guatemala. M, 1966, University of Texas, Austin.

Boyd, Austin E., III. A Miocene flora from the Oviatt Creek basin, Clearwater County, Idaho. M, 1985, University of Idaho. 208 p.

Boyd, Cynthia Stiles. Variation in genus Phacops (Trilobita) from the Silica Formation (Devonian, near Sylvania, Ohio). M, 1967, Bowling Green State University. 40 p.

Boyd, Daniel T. Stratigraphic relations within the Devonian Martin, Swisshelm and Portal formations of Cochise County, Arizona. M, 1978, University of Arizona.

Boyd, Donald Ray. Stratigraphy of the Difunta Group in area north of Saltillo, Coahuila, Mexico. M, 1959, Louisiana State University.

Boyd, Donald W. Permian sedimentary facies, central Guadalupe Mountains, New Mexico. D, 1959, Columbia University, Teachers College.

Boyd, Donald William. Simulation via time-partitioned linear programming; a ground and surface water allocation model for the Gallatin Valley of Montana. D, 1968, Montana State University. 227 p.

Boyd, Felicia Michelle. Hydrogeology of the northern Salt Basin of West Texas and New Mexico. M, 1982, University of Texas, Austin. 135 p.

Boyd, Francis Raymond, Jr. Geology of the Yellowstone rhyolite plateau. D, 1957, Harvard University. 134 p.

Boyd, Gilbert H. A geologic study of the Chickamauga Formations of Raccoon Valley, Roane County, Tennessee. M, 1955, University of Tennessee, Knoxville. 34 p.

Boyd, Harold Alfred, Jr. Eocene foraminifera from the "Vacaville Shale". M, 1949, University of California, Berkeley. 76 p.

Boyd, Harold Alfred, Jr. Geology of the Capay Quadrangle, California. D, 1956, University of California, Berkeley. 203 p.

Boyd, Harry R. Geology and geochemical exploration of the southern Sonoma and northern Tobin ranges, Pershing County, Nevada. M, 1972, Stanford University.

Boyd, Harry William. A multi-purpose remedial/resource room for the teaching of college geology. D, 1979, University of South Carolina. 91 p.

Boyd, Harry William. Depositional sub-environments and erosion on the Outer Banks at Pea Island, North Carolina. M, 1971, North Carolina State University. 81 p.

Boyd, Homer Joe. Formation resistivity factors in carbonate rocks. D, 1988, University of Texas at Dallas. 393 p.

Boyd, James. A preliminary study of Pre-Cambrian ore deposits of Colorado. M, 1932, Colorado School of Mines. 69 p.

Boyd, James. Pre-Cambrian mineral deposits of Colorado. D, 1934, Colorado School of Mines. 115 p.

Boyd, John. A study of relative dating techniques and their application to glacial chronologies in the central Brooks Range, Alaska. M, 1987, SUNY at Buffalo. 110 p.

Boyd, John A. Geology of a portion of the Reading Prong in southeastern Berks County. M, 1956, University of Pittsburgh.

Boyd, John Malcolm. Transition from brittle to ductile behaviour in alabaster under confining pressures to five kilobars. M, 1967, University of Toronto.

Boyd, John Ritchie. Domain observations on naturally occurring magnetite. M, 1986, University of California, Santa Barbara. 284 p.

Boyd, John Whitney, IV. The geology and alteration of the Slate Creek breccia pipe, Whatcom County, Washington. M, 1983, University of Arizona. 89 p.

Boyd, Kenneth A. A theoretical approach to the adsorption of water on mineral surfaces. M, 1971, Rensselaer Polytechnic Institute. 80 p.

Boyd, Oliver Ray. The petrology and diagenesis of the upper Smackover Member, Boyd Hill Field, Miller County, Arkansas. M, 1987, Northeast Louisiana University. 62 p.

Boyd, R. Michael. The petrology and field occurrence of some pegmatite dikes (Precambrian) in Conger Township, Parry Sound District, Ontario, Canada. M, 1970, Bowling Green State University. 41 p.

Boyd, R. W. An investigation of the mineral deposits of the Sierra de Pinta, Baja California, Mexico. M, 1976, San Diego State University.

Boyd, Richard G. Geology of the Lee's Summit and Grandview quadrangles. M, 1951, University of Missouri, Columbia.

Boyd, Robin Francis. Comparison of late Pleistocene Arctic Ocean climates with the O^{18} record using spectral analysis. M, 1981, University of Wisconsin-Madison. 120 p.

Boyd, Thomas Lee. Geology and joint pattern study of the Turkey Mountains, Mora County, New Mexico. M, 1983, West Texas State University. 120 p.

Boyd, Thomas M. Bedrock geology of the Whitewood Peak area, Lawrence County, South Dakota. M, 1975, University of Toledo. 89 p.

Boyd, Thomas M. High resolution determination of the Benioff zone geometry beneath southern Peru. M, 1983, Virginia Polytechnic Institute and State University. 91 p.

Boyd, Thomas Muryl. Seismic studies of the Aleutian Arc. D, 1988, Columbia University, Teachers College. 260 p.

Boydell, Anthony N. Multiple glaciation in the foothills, Rocky Mountain House area, Alberta. D, 1972, University of Calgary.

Boydell, Anthony N. Relationship of Late Wisconsin Rocky Mountain and Laurentide Ice in the vicinity of Sundre, Alberta. M, 1970, University of Calgary.

Boydell, H. C. Role of colloidal solutions in the formation of mineral deposits. D, 1924, Massachusetts Institute of Technology. 207 p.

Boyden, Ernest Duree. Geology of the Steptoe Warm Springs Pluton, White Pine County, Nevada. M, 1972, University of Nebraska, Lincoln.

Boyden, Thomas A. Ground-water geology of Payson-Benjamin, Utah. M, 1951, Brigham Young University. 50 p.

Boydston, Donald; Hamil, Brenton M. and Santos, Elmer S. Geology of the east central portion of the Huerfano Quadrangle, Huerfano County, Colorado. M, 1954, University of Michigan.

Boyer, Bruce W. Grain size parameters and constituent composition of Recent carbonate sediments from south Floria. M, 1969, University of Texas, Austin.

Boyer, D. G. Prediction of hydraulic conductivity changes using soil characteristics. M, 1978, University of Arizona.

Boyer, David Layne. Total sulfur content and the iron-sulfide morphotype distribution in the Parachute Creek Member of the Green River Formation (Eocene), Piceance Creek basin, Colorado. M, 1979, Southern Illinois University, Carbondale. 95 p.

Boyer, Douglas Gene. Frequency analysis and prediction of minimum streamflow rates. D, 1980, West Virginia University. 156 p.

Boyer, James Francis, Jr. Study of the structural geology of the southeastern section of Grand Bay oil field, Plaquemines Parish, Louisiana. M, 1958, University of Pittsburgh.

Boyer, Jannette Elaine. Sedimentary facies and trace fossils in the Eocene Delmar Formation and Torrey Sandstone, California. M, 1974, Rice University. 176 p.

Boyer, Jeffrey Alan. Hydrogeology of the Carefree Ranch area, Maricopa County, Arizona. M, 1974, University of Arizona.

Boyer, Jeffrey T. Potential erosion from montane zone roads. M, 1981, Colorado State University. 114 p.

Boyer, Larry Fred. Production and preservation of surface traces in the intertidal zone. D, 1980, University of Chicago. 434 p.

Boyer, Luc. A geological investigation of potential dam sites, Yukon Territory. M, 1960, Washington University. 62 p.

Boyer, Paul Rice. Patterns of convection in the Earth's mantle. M, 1965, University of Illinois, Urbana.

Boyer, Paul Rice. Structure of the continental margin of Brazil, Natal to Rio de Janeiro (South America). D, 1969, University of Illinois, Urbana.

Boyer, Paul Slayton. Actuopaleontology of the larger invertebrates of the coast of Louisiana (Recent). D, 1970, Rice University. 299 p.

Boyer, Renee C. Depositional environments and diagenetic history of the Springer Formation; Ardmore Basin, Oklahoma. M, 1983, University of Texas at Dallas. 189 p.

Boyer, Robert Eernst. Geology of the southern Wet Mountains, Colorado. D, 1959, University of Michigan. 246 p.

Boyer, Robert Ernst. The geology of the structural anomaly near Kentland, Indiana. M, 1953, Indiana University, Bloomington. 54 p.

Boyer, S. E. Structure and origin of Grandfather Mountain Window, North Carolina. D, 1978, The Johns Hopkins University. 306 p.

Boyer, Stephen Joseph. Pre-Wisconsin and neoglacial ice limits in Maktak Fiord, Baffin Island (Canada); a statistical analysis. M, 1972, University of Colorado.

Boyke, Waldimer Paul. Mineralization and wall rock alteration, Flin Flon area, Saskatchewan. M, 1953, University of Saskatchewan. 52 p.

Boylan, David Michael. The hydrogeologic resources of North Padre Island; coastal South Texas. M, 1986, Baylor University. 156 p.

Boylan, John C. Eastmanosteus, a placoderm from the Devonian of North America. D, 1973, Columbia University. 523 p.

Boyle, Albert C. Geology and ore deposits of Bully Hill mining district, California. D, 1914, Columbia University, Teachers College.

Boyle, Albert C. The geology in the vicinity of Winthrop, Shasta County, California. M, 1910, Columbia University, Teachers College.

Boyle, Edward Allen. The marine geochemistry of trace metals. D, 1976, Massachusetts Institute of Technology. 156 p.

Boyle, Huron Lee. Stratigraphy and structure in the Chippewa River valley between Nelson and Durand, Wisconsin. M, 1932, University of Iowa. 106 p.

Boyle, John, Jr. Ore dressing. M, 1887, Washington University.

Boyle, Joseph. Determination of recharge rates using temperature-depth profiles in wells. M, 1978, University of Illinois, Chicago.

Boyle, Michael William. Stratigraphy, sedimentation, and structure of an area near Point Arena, California; 1 volume. M, 1967, University of California, Berkeley. 71 p.

Boyle, Ray E. A rapid electrical resistivity prospecting technique. M, 1982, Colorado State University. 70 p.

Boyle, Robert W. The mineralization of the Yellowknife gold belt with special reference to the factors which controlled its localization. D, 1954, University of Toronto.

Boyle, Steven T. A Landsat satellite lineament study of Rhode Island. M, 1981, University of Rhode Island.

Boyle, Walter Victor. Simpson Group, Pecos County, Texas. M, 1955, University of Texas, Austin.

Boyle, William John. The influence of initial stresses on the movement of rock blocks around underground openings. D, 1987, University of California, Berkeley. 147 p.

Boyles, James McGregor. Zoogeography of the herpetofauna of central Florida. D, 1966, University of Alabama. 220 p.

Boyles, Joseph Michael. Depositional history and sedimentology of Upper Cretaceous Mancos Shale and lower Mesaverde Group, northwestern Colorado; migrating shelf-bar and wave-dominated shoreline deposits. D, 1983, University of Texas, Austin. 321 p.

Boynton, William Vandegrift. An investigation of the thermodynamics of calcite-rhodochrosite solid solutions. D, 1971, Carnegie-Mellon University. 156 p.

Bozanic, Dan. A subsurface study of the Muddy Sandstone in the northern part of the Julesburg Basin, Wyoming, Colorado and Nebraska. M, 1952, University of Wyoming. 55 p.

Bozanich, Richard G. The Bell Canyon and Cherry Canyon formations, southern Delaware Basin, Texas. M, 1978, University of Texas, Austin.

Bozbag, Hamdi A. Pre-Cambrian geology of an area near the mouth of the Golden Gate Canyon, Jefferson County, Colorado. M, 1943, Colorado School of Mines. 57 p.

Bozorgnia, Yousef. Linearization methods in earthquake analysis and design of hysteretic structural systems. D, 1981, University of California, Berkeley. 139 p.

Bozovich, Slobodan. Pre-Mississippian carbonate rocks in the Hollis Basin of Oklahoma. M, 1963, University of Oklahoma. 85 p.

Bozza, A. W. Application of groundwater flow modelling techniques to the nitrogen contamination problem at Olean, New York. M, 1986, SUNY at Binghamton. 60 p.

Braatz, Barbara Vanston. Recent relative sea-level change in eastern North America. M, 1987, Massachusetts Institute of Technology. 61 p.

Brabb, Earl Edward. Description of Whittlesey beach sediments, Ann Arbor area, Michigan. M, 1952, University of Michigan.

Brabb, Earl Edward. Geology of the Big Basin area, Santa Cruz Mountains, California. D, 1960, Stanford University. 197 p.

Brabb, Earl Edward. Paleogeologic study of the Silurian and Devonian systems in the eastern interior basin. M, 1956, Stanford University.

Brabec, Dragan. A geochemical study of the Guichon Creek Batholith, British Columbia. D, 1971, University of British Columbia.

Brabston, William Newell. Deformation characteristics of compacted subgrade soils and their influence in flexible pavement structures. D, 1982, Texas A&M University. 198 p.

Brace, Benjamin R. Pleistocene stratigraphy of the Hamilton Quadrangle. M, 1968, Miami University (Ohio). 138 p.

Brace, William F. Rock deformation of the Rutland, Vermont area. D, 1953, Massachusetts Institute of Technology. 104 p.

Bracewell, L. W. The contribution of wastewater discharges to surface films and other floatables on the ocean surface. D, 1977, University of California, Berkeley. 229 p.

Bracey, Dewey R. Geophysics and tectonic development of the Caroline Basin. M, 1981, University of Alaska, Fairbanks. 81 p.

Brachman, Steven H. Controls of base-metal deposits, the Little Eightmile and northern Gilmore mining districts, Lemhi County, Idaho. M, 1983, Pennsylvania State University, University Park. 214 p.

Brackebusch, Fred W. Economic geology of the Queen mountain area, a part of the Purcell Range, Boundary county, Idaho. M, 1969, University of Idaho. 76 p.

Bracken, Barth W. The geology of the Sivell's Bend oil field, Cooke County, Texas and Love County, Oklahoma. M, 1958, University of Oklahoma. 52 p.

Bracken, Bryan Reed. Environment of deposition and diagenesis of sandstones, La Joya Formation, Huizachal Group, northeastern Mexico. M, 1982, University of Texas, Austin. 178 p.

Bracken, Bryan Reed. Sedimentology of channel and floodplain rocks, lower Eocene Wind River Formation, Wyoming, and Late Cretaceous/Tertiary North Horn Formation, central Utah. D, 1987, University of Utah. 190 p.

Brackett, Michael Howard. Alderwood soil series. M, 1966, Washington State University. 100 p.

Brackett, Robert Stevens. Characteristics of a modern storm-generated sedimentary sequence; north insular shelf, Puerto Rico. M, 1985, Duke University. 140 p.

Brackmann, Anne Jordan. Diffusion of Zn and Fe in sphalerite. M, 1984, University of Toronto.

Brackmier, Gladys Helen. The ostracod genus Glyptopleura. M, 1930, Columbia University, Teachers College.

Braco, Paulo, Jr. Isothermal behaviour of gassy soils. D, 1988, University of Alberta. 724 p.

Bradbeer, G. E. Hydrogeologic evaluation of the Sonoita Creek Aquifer. M, 1978, University of Arizona.

Bradbrook, Christopher James. The nature and origin of the auriferous Hill-Sloan-Tivey quartz horizon, Uchi Lake, northwestern Ontario. M, 1982, University of Western Ontario. 237 p.

Bradburn, Frederick R. Experimental study of basin evolution as a function of slope. M, 1983, Southern Illinois University, Carbondale. 94 p.

Bradbury, Albert E. Geology of part of the Parkfield Syncline, Monterey County, California. M, 1941, Stanford University. 75 p.

Bradbury, J. W. A reconnaissance study of fluid inclusions from Tertiary intrusives in Colorado. M, 1976, University of Colorado.

Bradbury, James Clifford. Mineralogy and mineralization controls in the northwestern Illinois zinc-lead district. D, 1958, Harvard University.

Bradbury, John. Reconnaisance study of fluid inclusions in Tertiary intrusives of Colorado. D, 1983, Pennsylvania State University, University Park.

Bradbury, John Platt. Origin, paleolimnology, and limnology of Zuni Salt Lake Maar, west central New Mexico. D, 1967, University of New Mexico. 247 p.

Bradbury, John William. Pyrrhotite solubility in hydrous albite melts. D, 1983, Pennsylvania State University, University Park. 149 p.

Bradbury, Kenneth Rhoads. Hydrogeologic relationships between Green Bay of Lake Michigan and onshore aquifers in Door County, Wisconsin. D, 1982, University of Wisconsin-Madison. 324 p.

Bradbury, Kenneth Rhoads. Sedimentation and soil alteration, Monroe County, Indiana. M, 1977, Indiana University, Bloomington. 83 p.

Braddock, J. A. A gravity survey applied to water and sand and gravel resources in the Binghamton, New York area. M, 1982, SUNY at Binghamton. 45 p.

Braddock, William A. The geology of the Jewel Cave SW Quadrangle, South Dakota and its bearing on the origin of the uranium deposits in the southern Black Hills. D, 1959, Princeton University. 197 p.

Braden, William Joseph. The Oread Megacyclothem (Upper Pennsylvanian) in the Forest City Basin. M, 1958, University of Nebraska, Lincoln.

Bradfield, Herbert Henry. Pennsylvanian Ostracoda of the Ardmore Basin, Oklahoma. D, 1933, Indiana University, Bloomington. 228 p.

Bradfish, Larry James. Petrogenesis of the Tea Cup Granodiorite, Pinal County, Arizona. M, 1979, University of Arizona. 160 p.

Bradford, C. E. Structure of the southern part of Sheep Mountain, Albany County, Wyoming. M, 1934, University of Wyoming. 15 p.

Bradford, Cynthia A. Depositional environments and diagenesis of the Jurassic Smackover Formation, Escambia County, Alabama. M, 1982, University of Texas, Austin. 153 p.

Bradford, Donald Comnick. A study of the occurrence of microseisms at Sitka, Alaska, during the period January 1, 1929, to December 1931, inclusive. M, 1934, St. Louis University.

Bradford, John Allan. Geology and genesis of the Midway silver-lead-zinc deposit, north-central British Columbia. M, 1988, University of British Columbia. 265 p.

Bradford, Martin Ronald. The distribution and paleoecology of the dinoflagellate cysts in the Persian Gulf and adjacent regions. D, 1977, University of Saskatchewan. 555 p.

Bradford, Wesley L. Calcium analysis in seawater by an iron sensitive electrode. M, 1968, Oregon State University. 48 p.

Bradford, William Fay. An investigation and comparison of various techniques of disaggregation as applied to shales. M, 1958, Michigan State University. 56 p.

Bradford, William T. Marine scour effects on the sea floor off Southern California. M, 1966, University of Southern California.

Bradhurst, Stephen Thomas. Cartographic description of Nevada population. M, 1972, University of Nevada. 51 p.

Bradley, Charles C. Geology of the eastern half of the New Glarus Quadrangle (Wisconsin). D, 1936, University of Wisconsin-Madison.

Bradley, Charles C. Petrogenesis of the granite of Mount Whitney, California, and its relationship to depth of erosion. D, 1950, University of Wisconsin-Madison.

Bradley, Cheryl. Modified meandering river regimes; effects on Plains cottonwood regeneration, Milk River valley, S.E. Alberta and North Montana. M, 1982, University of Calgary.

Bradley, Christopher H. The structure and stratigraphy of the Anahuac Discorbis Zone of central Southwest Louisiana. M, 1978, University of Southwestern Louisiana. 106 p.

Bradley, Daniel Albert. Geology of the Fortune Bay area, Newfoundland. M, 1948, University of Michigan.

Bradley, Daniel Albert. Geology of the Gisburn Lake-Terrenceville area, Fortune Bay region, southeastern Newfoundland. D, 1954, University of Michigan. 220 p.

Bradley, Dwight C. Late Paleozoic strike slip tectonics of the Northern Appalachians. D, 1984, SUNY at Albany. 356 p.

Bradley, Edward. The physical and mineralogical properties of several groups of Minnesota clays. M, 1949, University of Minnesota, Minneapolis. 53 p.

Bradley, James E. A study of the Cambridge and Ames limestones in Muskingum County, Ohio. M, 1964, Ohio State University.

Bradley, James W. Magnetic measurements of Saginaw Lobe glacial tills in the Southern Peninsula of Michigan. M, 1968, Michigan State University. 158 p.

Bradley, Jeffrey Brent. Hydraulics and bed material transport at high fine suspended sediment concentrations. D, 1986, Colorado State University. 155 p.

Bradley, John H. The fauna and stratigraphy of the Kimmswick Limestone of Missouri and Illinois. D, 1924, University of Chicago. 215 p.

Bradley, John S. Permian Capitan Formation of Guadaloupe Mts. of Texas and New Mexico. D, 1952, University of Washington. 109 p.

Bradley, Lauren Magin. Structural analysis of deformed Carboniferous strata, Mispec Beach, southern New Brunswick. M, 1984, SUNY at Albany. 134 p.

Bradley, M. F. Evaluation and application of dye tracers in karst terrain for determining aquifer characteristics. M, 1970, University of Missouri, Rolla.

Bradley, Michael Dennis. Geology of northeastern Selway Bitterroot Wilderness area, Idaho and Montana. M, 1981, University of Idaho. 51 p.

Bradley, Michael Dennis. Structural evolution of the Uinta Mountains, Utah, and their interaction with the Utah-Wyoming salient of the Sevier overthrust belt. D, 1988, University of Utah. 178 p.

Bradley, Michael T. Gully development and valley stability in northeastern Colorado. M, 1980, Colorado State University. 150 p.

Bradley, Scott D. Granulite facies and related xenoliths from Colorado-Wyoming kimberlite. M, 1984, Colorado State University. 179 p.

Bradley, Walter W. Quicksilver resources of California, with a section of metallurgy and ore dressing. M, 1918, University of California, Berkeley.

Bradley, Wayles B. The geology of the region between Little Thompson River and Carter Lake, Larimer County, Colorado. M, 1951, Colorado School of Mines. 69 p.

Bradley, Whitney A. Jurassic and Pre-Mancos Cretaceous stratigraphy of the eastern Uinta Mountains, Colorado (Moffat and Rio Blanco counties); Utah. M, 1952, University of Colorado.

Bradley, Wilbur C. A subsurface study of the Simpson Group in Harper County, Kansas. M, 1961, University of Oklahoma. 60 p.

Bradley, William Crane. Marine terraces and sedimentation in the Santa Cruz area, California. D, 1956, Stanford University. 110 p.

Bradley, Wilmot H. Origin and microfossils of the oil shale in the Green River Formation of Colorado and Utah. D, 1927, Yale University.

Bradof, Kristine Lynn. Environmental impacts of drainage-ditch and road construction on Red Lake Peatland, northern Minnesota; drainage history, hydrology, water chemistry, and tree growth. M, 1988, University of Minnesota, Minneapolis. 174 p.

Bradshaw, B. A. Petrological comparison of Lake Superior iron formations [Minnesota-Ontario]. D, 1956, University of Toronto.

Bradshaw, David Curtis. An experimental investigation of magnetic susceptibility in weak fields of different frequencies. M, 1975, University of Oklahoma. 92 p.

Bradshaw, Donald Dean. Areal geology of the Stanley Group (Mississippian) in northeastern Pushmataha County, Oklahoma. M, 1962, University of Oklahoma. 56 p.

Bradshaw, Donald T. Geologic variables influencing production in the Eastburn Field, Vernon County, Missouri. M, 1985, Wichita State University. 125 p.

Bradshaw, Emily C. Structure in the Mazatzal Quartzite, Del Rio, Arizona. M, 1975, Northern Arizona University. 67 p.

Bradshaw, Herbert E. Geology of the Palmer Volcanics. M, 1964, University of Oregon. 109 p.

Bradshaw, John Stratlii. Ecology of living planktonic foraminifera in the North and Equatorial Pacific

Ocean. D, 1957, University of California, Los Angeles. 256 p.

Bradshaw, John Yates. Petrology and mineralogy of interlayered eclogite and high-grade blueschist from the Franciscan Formation, California. M, 1978, University of Calgary. 183 p.

Bradshaw, K. L. An assessment of the Standard Percolation Test. M, 1986, University of Waterloo. 215 p.

Bradshaw, Lael Marguerite Ely. Conodonts from the Fort Pena Formation, Marathon Basin, Brewster County, Texas. D, 1966, University of Texas, Austin. 270 p.

Bradshaw, Robert Donald. The Autwine Field, Kay County, Oklahoma. M, 1959, University of Oklahoma. 81 p.

Bradstreet, Theodore E. Pollen influx diagram and associated quaternary geology from Moulton Pond, Maine. M, 1973, University of Maine. 75 p.

Brady, Arthur Gerald. Studies of reponse to earthquake ground motion. D, 1966, California Institute of Technology. 161 p.

Brady, Brian K. Porosity and diagenetic sequence of the Whirlpool Sandstone. M, 1987, SUNY at Buffalo. 74 p.

Brady, Frank Howard. Some problems of the Minnelusa Formation near Beulah, Wyoming. M, 1930, University of Iowa. 100 p.

Brady, Howard Thomas. Late Neogene diatom biostratigraphy and paleoecology of the Dry Valleys and McMurdo Sound, Antarctica. M, 1977, Northern Illinois University. 72 p.

Brady, John Ballard. Intergranular diffusion in metamorphic rocks. D, 1975, Harvard University.

Brady, John Francis, Jr. Qualitative reconnaissance methods of geochemical exploration. M, 1960, University of New Mexico. 20 p.

Brady, John M. Ore and sedimentation of the lower sandstone at the White Pine Mine, Michigan. M, 1960, Michigan Technological University. 101 p.

Brady, Karen S. Iron precipitates from acid coal mine drainage in southeastern Ohio; origin, occurrence and regional significance. D, 1983, [University of Oklahoma]. 198 p.

Brady, Lawrence Lee. An experimental study in a large flume of heavy mineral segregation under alluvial flow conditions. D, 1971, University of Kansas. 115 p.

Brady, Lawrence Lee. Stratigraphy and petrology of the Morrison Formation (Jurassic) in the vicinity of Canon City, Colorado. M, 1968, University of Kansas. 111 p.

Brady, Michael B. A petrographic and structural study of an area in the Huron Mountains, Michigan. M, 1952, Wayne State University.

Brady, Michael J. Thrusting in the southern Wasatch Mountains, Utah. M, 1965, Brigham Young University.

Brady, Michael John. Sedimentology and diagenesis of carbonate muds in coastal lagoons of NE Yucatan. D, 1972, Rice University. 288 p.

Brady, R. T. Geology of the east flank of the Laramie Range in the vicinity of Federal and Hecia, Laramie County, Wyoming. M, 1949, University of Wyoming. 41 p.

Brady, Roland Hamilton, III. Cenozoic geology of the northern Avawatz Mountains in relation to the intersection of the Garlock and Death Valley fault zones, San Bernardino County, California. D, 1986, University of California, Davis. 292 p.

Brady, Steven C. Diagenetic changes in Texas Gulf Coast sediments. M, 1988, Arizona State University. 80 p.

Brady, Thomas James. Geology of part of the central Santa Monica Mountains east of Topanga Canyon, Los Angeles County, California. M, 1957, University of California, Los Angeles.

Brady, Timothy Brian. Early Proterozoic structure and deformational history of the Sheep Basin Mountain

area, northern Sierra Anchas, Gila County, Arizona. M, 1987, Northern Arizona University. 122 p.

Braga, B. P. F., Jr. An evaluation of streamflow forecasting models for short-range multi-objective reservoir operation. D, 1979, Stanford University. 123 p.

Bragdon, Frederick F. Geology of the Fish Spring Creek and adjacent area, Lincoln and Sublette counties, Wyoming. M, 1965, University of Wyoming. 64 p.

Bragg, Susan Lynn. Characteristics of Martian soil at Chryse Planitia as inferred by reflectance properties, magnetic properties, and dust accumulation on Viking Lander 1. M, 1977, Washington University. 112 p.

Bragonier, William Atwood. Genesis and geologic relations of the high-alumina Mercer fireclay (Pennsylvanian), western Pennsylvania. M, 1970, Pennsylvania State University, University Park. 212 p.

Brahana, John Van. Geology and tectonic analysis of the northwestern Illinois zinc-lead district. M, 1968, University of Missouri, Columbia.

Brahana, John Van. Systematic jointing in slightly deformed rocks. D, 1973, University of Missouri, Columbia.

Brahma, Chandra Sekhar. Analysis of granular soil deformation as a stochastic process. D, 1969, Ohio State University. 191 p.

Braide, J. O. The effects of soil physical parameters on the diffusion of phosphorus in Hawaiian soils. M, 1971, University of Hawaii. 68 p.

Braide, Sokari Percival. Clay mineral burial diagenesis in Tertiary sediments from the eastern flank of the Niger Delta. D, 1982, University of Cincinnati. 180 p.

Braile, Lawrence Wendell. Seismic interpretation of crustal structure across the Wasatch Front and applications of geophysical data inversion. D, 1973, University of Utah. 143 p.

Braile, Lawrence Wendell. The isostatic condition and crustal structure of Mount Saint Helens, as determined from gravity data. M, 1970, University of Washington. 37 p.

Brailey, David Elton. Structural analysis of a Mesozoic sequence in the Kluane Ranges, Yukon Territory; evidence for terrane accretion and offset. M, 1986, University of Wisconsin-Madison. 150 p.

Brainard, Ray Carter. Geology of the Colville area, Washington, township 35 north, east half of range 38 east, west half of range 39 east. M, 1982, Washington State University. 60 p.

Brainard, Richard H. Pennsylvanian (Virgilian) scolecodonts from Shawnee Group limestones in eastern Kansas and southeastern Nebraska. M, 1978, Wichita State University. 140 p.

Brainerd, A. E. Pleistocene deposits of a section of the Larger Cicero Swamp (near Syracuse, New York). M, 1912, Syracuse University.

Brainerd, William F. Dana glacial lake terrace and the Great Delta of Onondaga Valley. M, 1922, Syracuse University.

Braithwaite, Lee Fred. Graptolites from the Pogonip group (lower Ordovician) of western Utah. D, 1970, Brigham Young University.

Braithwaite, Philip. Cretaceous stratigraphy of northern Rim Rock Country, Trans-Pecos, Texas. M, 1958, University of Texas, Austin.

Braithwaite, R. J. Air temperature and glacier ablation; a parametric approach. D, 1977, McGill University.

Brake, Christopher French. Sedimentology of the North Victoria Land continental margin, Antarctica. M, 1982, Rice University. 175 p.

Brake, James Frank. Analysis of historic and pre-historic slip on the Elsinore Fault at Glen Ivy Marsh, Temescal Valley, Southern California. M, 1987, San Diego State University. 103 p.

Brakel, W. H. The ecology of coral shape; microhabitat variation in the colony form and corallite

structure of Porites on a Jamaican reef. D, 1976, Yale University. 256 p.

Brakenridge, George Robert. Alluvial stratigraphy and geochronology along the Duck River, central Tennessee; a history of changing floodplain sedimentary regimes. D, 1982, University of Arizona. 109 p.

Bralower, Timothy James. Part A; An integrated Mesozoic biochronology and magnetochronology; Part B, Studies of Cretaceous black shales. D, 1986, University of California, San Diego. 443 p.

Bramadat, Kelvin. Preliminary investigations on activation analysis. M, 1954, University of Manitoba.

Braman, David E. Interpretation of gravity anomalies observed in the Cascade Mountain Province of northern Oregon. M, 1981, Oregon State University. 144 p.

Braman, Dennis Richard. Palynology and paleoecology of the Mattson Formation, Northwest Canada. M, 1976, University of Calgary. 169 p.

Braman, Dennis Richard. Upper Devonian-Lower Devonian Carboniferous miospore biostratigraphy of the Imperial Formation, District of Mackenzie and Yukon. D, 1981, University of Calgary. 377 p.

Bramble, Dennis M. Afrolamia, a hornless meiolaniid from the east African Miocene. M, 1968, University of California, Berkeley. 123 p.

Bramble, Dennis Marley. Functional morphology and evolution of gopher tortoises with an introduction to paleoecology of North American land tortoises. D, 1971, University of California, Berkeley. 341 p.

Brame, Jeffrey W. The stratigraphy and geologic history of Caladesi Island, Pinellas County, Florida. M, 1976, University of South Florida, Tampa. 109 p.

Brame, Simon. Mineralisation near the northeast margin of the Nelson Batholith, Southeast British Columbia. M, 1979, University of Alberta. 146 p.

Bramkamp, Richard A. Stratigraphy and molluscan fauna of the Imperial Formation of San Gorgonio Pass, California. D, 1934, University of California, Berkeley. 312 p.

Bramlett, Kenneth. Geology of the Johnston area S.C. and its regional implications. M, 1980, University of South Carolina.

Bramlett, Richard Randall. The relationship of hydrocarbon production fracturing in the Woodford Formation of southern Oklahoma. M, 1981, Oklahoma State University. 106 p.

Bramlette, Milton N. Geology of the Arkansas bauxite region. D, 1936, Yale University.

Bramlette, William Allen. The fauna of the Nostoceras Zone of the Taylor Formation. M, 1934, University of Texas, Austin.

Bramlette, William Allen. The Pennsylvanian-Permian boundary in north-central Texas. D, 1943, University of Kansas. 206 p.

Bramson, Emil. Trace-element study of Tertiary volcanic rocks from the Sierra Madre Occidental, Mexico, to Trans-Pecos, Texas. M, 1984, University of Texas, Austin. 243 p.

Branca, John. Petrology and structure of the Glenarm Series and associated rocks in the Mill Creek area, Delaware. M, 1979, University of Delaware.

Branch, Charles N. Re-analysis and modelling of COCORP's Sevier Desert Line 1; implications for models of Cenozoic extension in west-central Utah. M, 1985, University of Wyoming. 112 p.

Branch, Colby Lloyd. Deposition and diagenesis of the Upper Devonian Nisku Formation, Toole County, Montana. M, 1988, Texas Tech University. 138 p.

Branch, Jerry D. Stratigraphy of the Hale Mountain area (Pennsylvanian), Washington County, Arkansas. M, 1972, University of Arkansas, Fayetteville.

Branch, Jill N. Haywa. Conodonts from the Welden Limestone (Osagean, Mississippian), south-central Oklahoma. M, 1988, Texas Tech University. 113 p.

Branch, John Russell. The geomorphology and Pleistocene geology of the Big Bone Lick area of Kentucky. M, 1946, University of Cincinnati. 49 p.

Branckstone, Hugh R. Determination and interpretation of bulk and grain densities of rock samples. M, 1932, University of Pittsburgh.

Brand, John Frederick. Low grade metamorphic rocks of the Ruppert and Hobbs coasts of Marie Byrd Land, Antarctica. M, 1979, Texas Tech University. 49 p.

Brand, John P. The Ordovician limestones of the Cumberland River valley, Russell County, Kentucky. M, 1947, Miami University (Ohio). 35 p.

Brand, John Paul. Cretaceous of Llano Estacada of Texas. D, 1952, University of Texas, Austin.

Brand, S. R. Geology, petrology and geochemistry of the lower Kingurutik River area, Labrador, Canada. D, 1976, Purdue University. 265 p.

Brand, U. Lower Ordovician conodonts from the Kindblade Formation, Arbuckle Mountains, Oklahoma. M, 1976, University of Missouri, Columbia.

Brand, Uwe. Geochemistry of Paleozoic corals, crinoids and associated carbonate rocks from Arctic Canada, Iowa and Missouri. D, 1979, University of Ottawa. 176 p.

Brande, S. Biometric analysis and evolution of two species of Mulinia (Bivalvia; Mactridae) from the late Cenozoic of the Atlantic Coastal Plain. D, 1979, SUNY at Stony Brook. 242 p.

Brandenbure, F. Merrill. The pre-Cambrian geology of the upper Iron Creek area of the Black Hills of South Dakota. M, 1932, University of Iowa. 93 p.

Brandenburg, John L. Late Cretaceous vertebrate faunas from the upper Hell Creek Formation of Montana and North Dakota. M, 1983, University of Minnesota, Minneapolis. 117 p.

Brandenburg, Norman R. Pozzolanic properties of Oregon volcanic ash. M, 1951, Oregon State University. 74 p.

Brandes, William Frederick. A method for identifying water resources research needs and setting priorities among them. D, 1986, [Vanderbilt University]. 410 p.

Brandley, Richard T. Depositional environments of the Triassic Ankareh Formation, Spanish Fork Canyon, Utah. M, 1988, Brigham Young University. 233 p.

Brandon, Alan D. Geochemical features of the Bear Creek lavas, Deschutes and Crook counties, Oregon. M, 1987, University of Oregon. 122 p.

Brandon, Dale Edward. Oceanography of the Great Barrier Reef and the Queensland continental shelf, Australia. D, 1970, University of Michigan. 217 p.

Brandon, David Earl. Analysis of potential seismic precursors by Wiener predictive filtering. M, 1980, SUNY at Binghamton. 56 p.

Brandon, James P. Geology of the Ridgefield area, southwestern Connecticut and southeastern New York. M, 1981, University of Massachusetts. 125 p.

Brandon, Mark Thomas. Deformational processes affecting unlithified sediments at active margins; a field study and a structural model. D, 1984, University of Washington. 159 p.

Brandon, Mark Thomas. Structural geology of Middle Cretaceous thrust faulting on southern San Juan Island, Washington. M, 1980, University of Washington. 130 p.

Brandon, Steven Howard. A study on probable geologic influences controlling shear strength of the Fort Hayes Member, Niobrara Chalk Formation, Gregory County, S.D. M, 1986, University of Missouri, Rolla. 108 p.

Brandon, Thomas Lyle. Thermal conductivity and thermal instability of sand. D, 1985, University of California, Berkeley. 252 p.

Brandon, William Campbell. An origin for the McCartney's Mountain salient of the southwestern Montana fold and thrust belt. M, 1984, University of Montana. 128 p.

Brandorff, William A. Bentonites of the Niobrara Formation in northeastern Nebraska and southeastern South Dakota. M, 1954, University of Nebraska, Lincoln.

Brandriss, Mark Elliott. Effects of glass transition temperature on the structures of silicate glasses, and implications for the effects of temperature on the structures of silicate liquids; silicon-29 NMR results. M, 1988, Stanford University. 39 p.

Brandsdóttir, Bryndis. Precise measurements of coda buildup and decay rates of western Pacific P, P_o and S_o phases and their relevance to lithospheric scattering. M, 1987, Oregon State University. 153 p.

Brandstrom, Gary Wayne. A morphologic analysis of Granicus Valles, Mars. M, 1986, Texas A&M University. 66 p.

Brandt Velbel, Danita. Stratigraphic resolution in the Nicolet Formation (Upper Ordovician), Quebec, Canada. D, 1985, Yale University. 401 p.

Brandt, Danita Sue. Phenotypic variations and paleoecology of Flexicalymene (Arthropoda; Trilobita) in the Cincinnatian Series (Upper Ordovician) near Cincinnati, Ohio. M, 1980, University of Cincinnati. 148 p.

Brandt, John Lawrence. Stratigraphic clay-mineral distribution in the Cretaceous Colorado Group near Saskatoon (Saskatchewan). M, 1965, University of Saskatchewan. 60 p.

Brandt, William Otis. Forsterite refractories from Washington materials. M, 1936, University of Washington. 46 p.

Brandwein, Sidney S. Selected shallow electrical resistivity surveys at Volcano Cliffs, Bernalillo County, and Warm Springs, Sandoval County, New Mexico. M, 1974, University of New Mexico. 139 p.

Branham, Alan David. Gold mineralization in low angle faults American Girl Valley, Cargo Muchacho Mountains, California. M, 1988, Washington State University. 144 p.

Branham, Keith L. Cavity detection using high-resolution seismic reflection methods. M, 1986, University of Kansas. 75 p.

Branham, Thomas B. A lithofacies study of the Frio Formation in the upper Gulf Coast region of Texas. M, 1958, University of Houston.

Brann, Bethany Celia. Microfossils of the Trinity river delta, Texas. M, 1969, Rice University. 36 p.

Brannon, Charles Andrew. Hydrothermal alteration of volcanic cover rocks, Tintic District, Utah. M, 1982, University of Arizona. 91 p.

Brannon, James M. Seasonal variations of nutrients and physicochemical properties in the salt marsh soils of Barataria Bay, Louisiana. M, 1973, Louisiana State University.

Brannon, James Milton. The transformation, fixation, and mobilization of arsenic and antimony in contaminated sediments. D, 1983, Louisiana State University. 206 p.

Brannon, Joyce C. Geochemistry of successive lava flows of the Keweenawan North Shore Volcanic Group. D, 1984, Washington University. 312 p.

Branson, Carl C. The paleontology and stratigraphy of the Phosphoria Formation. D, 1929, University of Chicago. 126 p.

Branson, Carl Colton. Paleontology and stratigraphy of the Phosphoria Formation. D, 1930, University of Missouri, Columbia.

Branson, Carl Colton. The stratigraphy and paleontology of the upper part of the Phosphoria Formation on the east side of the Wind River Mountains, Montana. M, 1927, University of Missouri, Columbia.

Branson, Edwin Bayer. The structure and relationships of the American Labyrinthodontidae. D, 1905, University of Chicago. 43 p.

Branson, John Wallace. The correlation of the Gosport and Lisbon formations of western Alabama with formations to the west. M, 1947, University of Oklahoma. 53 p.

Branson, Robert Burns. The geology of the Vinita-Ketchum area, Craig County, Oklahoma. M, 1952, University of Oklahoma. 65 p.

Branstrator, Jon Wayne. A redescription of Hudsonasterid (Asterozoa) types. M, 1969, University of Cincinnati. 115 p.

Branstrator, Jon Wayne. Paleobiology and revision of the Ordovician Asteriadina (Echinodermata; Asteroidea) of the Cincinnati area. D, 1975, University of Cincinnati. 269 p.

Brant, David Mann. Paleogeological study of Trans-Pecos Texas. M, 1960, Stanford University.

Brant, Lynn Alvin. A palynological investigation of post-glacial sediments at two locations along the Continental Divide near Helena, Montana. D, 1980, Pennsylvania State University, University Park. 162 p.

Brant, Lynn Alvin. A study of the faunal succession in the Beach Creek Shale (Pennsylvanian), near Shelocta, Pennsylvania. M, 1971, Pennsylvania State University, University Park. 86 p.

Brant, Ralph A. Stratigraphy of the Meramac and Chester series of Mayes County, Oklahoma. M, 1941, University of Tulsa. 44 p.

Brant, Russell Alan; Elmer, Nixon; Gillespie, W. A. and Peterson, John Robert. Geology of the Armstead area, Beaverhead County, Montana. M, 1949, University of Michigan. 118 p.

Brantley, Mims McGehee. A study of the possibilities of the occurrence of petroleum in the Union area of Jones County, Texas. M, 1954, University of Virginia. 96 p.

Brantley, Susan Louise. The chemistry and thermodynamics of natural brines and the kinetics of dissolution precipitation reactions of quartz and water. D, 1987, Princeton University. 237 p.

Brantly, John E., Jr. Geology of a portion of the Birmingham Valley in the vicinity of Village Springs, Alabama. M, 1948, University of Virginia. 49 p.

Brar, N. S. Mechanisms and kinetics of the polymorphic transitions in $CaCO_3$ and GeO_2/SiO_2 systems. D, 1979, University of Western Ontario.

Bras, Ronan J. Le *see* Le Bras, Ronan J.

Brasaemle, Joan E. Chemical character of ground water in and adjacent to the Cuyahoga Valley National Recreation Area, Ohio. M, 1978, University of Akron. 164 p.

Brasaemle, Karla Anne. A refinement of a heat-flow based model for the evolution of the continental crust. M, 1988, University of Minnesota, Minneapolis. 31 p.

Braschayko, Thomas. Japan's strategic mineral dependence. M, 1983, University of Texas, Austin.

Brasher, George Kirt. Geology of part of the Snowcrest Range, Beaverhead County, Montana. M, 1950, University of Michigan.

Brasher, James Everett. Diagenesis of Miocene arkoses of the southern San Joaquin Valley. M, 1982, Texas A&M University. 124 p.

Brasino, John Sheldon. A simple stochastic model predicting conservative mass transport through the unsaturated zone into ground water. D, 1986, University of Wisconsin-Madison. 267 p.

Braslau, David. On an earthquake and aftershock mechanism relating to a model of the crust and mantle. D, 1966, University of California, Berkeley. 292 p.

Brass, Garret William. The sources of marine strontium and the Sr^{87}/Sr^{86} ratio in the sea throughout Phanerozoic time. D, 1973, Yale University. 118 p.

Bratberg, David. Hydrogeology of Dodge Flat and its relation to flow and quality changes in the Truckee River. M, 1980, University of Nevada. 100 p.

Brathovde, James Edgar. Stratigraphy of the Grand Wash Dolomite (Upper ? Cambrian), western Grand Canyon, Mohave County, Arizona. M, 1986, Northern Arizona University. 140 p.

Brathwaite, Samuel L. A. An investigation of hydraulic connection of sand lenses in silty clay till deposits of Essex County, Ontario. M, 1988, University of Windsor. 252 p.

Bratney, William A. Mineralogical and elemental trends in the Stibnite Hill Mine, Sanders County, Montana. M, 1977, University of Montana. 120 p.

Bratt, Steven Richard. The structure and thermal tectonics of planetary lithospheres; mid-ocean ridges and lunar impact basins. D, 1984, Massachusetts Institute of Technology. 386 p.

Bratton, Samuel T. Some geographical influences in the development of type railroads in Missouri. M, 1917, University of Missouri, Columbia.

Brauer, Clemens P. The geology of the Cookson Hills area, Cherokee County, Oklahoma. M, 1952, University of Oklahoma. 103 p.

Brauer, Constance J. Genetic mapping and erosional history of the surface sediments of Shackleford Banks, North Carolina. M, 1974, Duke University. 101 p.

Brauer, D. F. Barinophytacean plants from the Upper Devonian Catskill Formation of northern Pennsylvania. D, 1980, SUNY at Binghamton. 112 p.

Brauer, Julie Fay. Experimental hydrothermal dedolomitization. M, 1985, Duke University. 61 p.

Braumiller, Allen Spooner. Some late Tertiary foraminifera from the Makron Coast, Pakistan. M, 1957, University of Illinois, Urbana. 94 p.

Braun, D. D. The relation of rock resistance to incised meander form in the Appalachian Valley and Ridge Province. D, 1976, The Johns Hopkins University. 248 p.

Braun, Don. The stratigraphy of the Niland Tongue of the Wasatch Formation in the western Washakie Basin, southwestern Wyoming. M, 1982, San Jose State University. 124 p.

Braun, Emma Lucy. The Cincinnatian Series and its brachiopods in the vicinity of Cincinnati. M, 1912, University of Cincinnati. 48 p.

Braun, Eric R. Geology and ore deposits of the Marble peak area, Santa Catalina mountains, Pima County, Arizona. M, 1969, University of Arizona.

Braun, Gerald E. Atomic substitution in the tetrahedrite-tennantite series. M, 1969, University of Arizona.

Braun, Gerald E. Chemographic relationships in multicomponent systems. D, 1976, University of Minnesota, Minneapolis. 134 p.

Braun, Jordan C. A stratigraphic study of the Sycamore and related formations in the southeastern Anadarko Basin. M, 1958, University of Oklahoma. 80 p.

Braun, Roger Elmer. The transition from the Judith River Formation to the Bearpaw Shale (Campanian), north-central Montana. M, 1983, Montana State University. 66 p.

Braun-Adams, Karla A. Petrologic and stratigraphic analysis of the Upper Cretaceous Cotuí Formation and the unnamed limestone member of the Sabana Grande Formation, Southwest Puerto Rico. M, 1984, University of North Carolina, Chapel Hill. 81 p.

Braund, Robert William. Magnetometric and geologic observations on a part of the Canadian Shield. M, 1939, Michigan Technological University. 47 p.

Brauneck, Dorothy A. A study of the variation in radiographs produced by the radioactive minerals. M, 1938, Columbia University, Teachers College.

Brauner, J. F. Morphology and taxonomy of some southern South American Ceramiaceae (Rhodophyta). D, 1977, Duke University. 347 p.

Brauner, Michael R. A petrological study of the Housatonic Highlands gneiss (Precambrian) and the Dalton Formation (lower Cambrian) of the South Canaan Quadrangle, Connecticut. M, 1969, University of Wisconsin-Madison.

Braunsdorf, Neil Robert. Isotopic trends in gangue carbonates from the Viburnum Trend; implications for mississippi valley-type mineralization. M, 1983, University of Michigan.

Braunstein, Jules. A study of a collection of fossils from the Traverse Group, Alpena County, Michigan. M, 1936, Columbia University, Teachers College.

Brautigam, Gerald L. Faulting in the Silver City Range, Grant County, New Mexico. M, 1979, University of Houston.

Bravinder, Kenneth Mason. Stratigraphy and paleontology of the Oligocene in the eastern portion of the Puget Sound basin. M, 1932, University of Washington. 38 p.

Bravo, Jorge Oswaldo Munoz *see* Munoz Bravo, Jorge Oswaldo

Brawley, Tommy R. Micropaleontology of the Del Rio Formation of Brewster County, Texas. M, 1961, Texas Tech University. 142 p.

Brawner, J. D., Jr. The neutron bombardment of solutions similar to oil field brines. M, 1943, Massachusetts Institute of Technology. 61 p.

Bray, A. C. A petrographic study of certain cordierite-bearing rocks. M, 1929, McGill University.

Bray, Alexander G. The structure and stratigraphy of the Proterozoic Hecla Hoek succession near Floyfjellet, Wedel Jarlsberg Land, Spitsbergen. M, 1985, University of Wisconsin-Madison. 135 p.

Bray, Cynthia Jean. Deformation and material properties of sediments at convergent plate margins. D, 1986, Cornell University. 193 p.

Bray, Edward Arthur du *see* du Bray, Edward Arthur

Bray, John Thomas. The behavior of phosphate in the interstitial waters of Chesapeake Bay sediments. D, 1973, The Johns Hopkins University.

Bray, Joseph M. Distribution and relationships of the minor chemical elements in some igneous rocks from the Front Range, Colorado. D, 1940, Massachusetts Institute of Technology. 114 p.

Bray, R. C. E. A comparison of the non-opaque minerals of certain parts of the Waite-Amulet area, Quebec. M, 1940, McGill University.

Bray, R. G. Ecology and "life history" of mid-Devonian brachiopod clusters, Erie County, New York. D, 1971, McMaster University. 162 p.

Bray, R. G. The paleoecology of some middle Devonian fossil clusters, Erie County, New York. M, 1969, McMaster University. 68 p.

Bray, Richard A. Petrography and petrology of the Oligocene intrusives of western Lane County, Oregon. M, 1958, University of Oregon. 116 p.

Bray, Robert Eldon. Igneous rocks and alterations in the Carr Fork area of Bingham Canyon, Utah. M, 1967, University of Utah. 116 p.

Bray, Thomas F., Jr. The sedimentology and stratigraphy of a transgressive barrier at Sheldon's Marsh State Nature Preserve, Erie County, Ohio. M, 1988, University of Akron. 125 p.

Bray, Timothy D. Stratabound zinc-lead deposits in the Monte Cristo Limestone, Goodsprings, Nevada. M, 1983, Dartmouth College. 235 p.

Brayer, Roger C. Some Ostracoda from the Spergen (Salem) Limestone (Mississippian) of Missouri. M, 1950, St. Louis University.

Brayton, Darryl M. The physical stratigraphy of the Rundle Formations as determined by petrographic study. D, 1963, University of Washington. 322 p.

Brayton, Darryl Merritt. Cretaceous flora of Genesee, Alberta. M, 1953, University of Alberta. 129 p.

Brazeau, André L'Inventaire des granults au Québec. M, 1988, Universite Laval. 115 p.

Brazell, Thomas Nathan. The geology and economic potential of the western half of the Lake Wylie 7.5' Quadrangle, South Carolina. M, 1984, University of Georgia. 91 p.

Brazie, Mike E. An assessment of runoff and erosion and the possible effects of mining operations in the Nogal Creek drainage basin, Lincoln County, New Mexico. M, 1979, Colorado School of Mines. 126 p.

Brazil, Larry E. Multilevel calibration strategy for complex hydrologic simulation models. D, 1988, Colorado State University. 237 p.

Breakey, A. R. A mineralogical study of the gold-quartz lenses in the Campbell Shear, Con Mine, Yellowknife, N.W.T. M, 1976, McGill University. 117 p.

Breakey, B. The concentrically banded siliceous nodules of the (Silurian) Eramosa Dolomite in the vicinity of Dundas (Ontario). M, 1936, McMaster University.

Breakey, E. C. Sedimentology of the lower Paleozoic shelf-slope transition, Levis, Quebec. M, 1976, McGill University. 190 p.

Breaks, Frederick W. Origin of the Silent Lake Pluton and its quartz-sillimanite nodules, near Bancroft, Ontario. M, 1971, McMaster University. 119 p.

Bream, Susan Elaine. Depositional environment, provenance, and tectonic setting of the upper Oligocene Sooke Formation, Vancouver Island, B.C. M, 1986, Western Washington University. 228 p.

Breard, Sylvester Q. Macrofaunal ecology, climate and biogeography of the Jackson Group in Louisiana and Mississippi. M, 1978, Northeast Louisiana University.

Brearley, Mark. Ultramafic xenoliths from British Columbia, Canada; petrological and dissolution studies. D, 1986, University of Alberta. 182 p.

Breaux, James E. Neogene foraminifera from a well in Timbalier Bay Field, Louisiana. M, 1961, Louisiana State University.

Brecher, Aviva. Part I, Vapor condensation of Ni-Fe phases and related problems; Part II, The paleomagnetic record in carbonaceous chondrites. D, 1972, University of California, San Diego.

Brecher, Henry H. Surface velocity measurements on the Kaskawulsh Glacier, Yukon Territory, Canada. M, 1966, Ohio State University.

Breckenridge, Roy M. Quaternary and environmental geomorphology of the upper Wood River area, Absaroka Range, Wyoming. D, 1974, University of Wyoming. 139 p.

Breckenridge, Roy Melvin. Neoglacial geology of upper Fall creek basin, Mummy range (Larimer County) Colorado. M, 1969, University of Wyoming. 59 p.

Breckling, Robert. Gravity and magnetic modelling of the Thermopolis, Wyoming, Anticline. M, 1987, Bowling Green State University. 93 p.

Breckon, Curtis Eugene. Sedimentology and facies of the Pennsylvanian Jackfork Group in the Caddo Valley and Degray quadrangles, Clark County, Arkansas. M, 1988, University of Tulsa. 134 p.

Bredall, S. Early Eocene fanglomerate, northwestern Big Horn Basin, Wyoming. M, 1971, Iowa State University of Science and Technology.

Bredehoeft, John Dallas. Hydrogeology of the lower Humboldt River basin, Nevada. D, 1962, University of Illinois, Urbana. 128 p.

Bredehoeft, John Dallas. Refraction seismic studies in the Havana Lowland area, Mason County, Illinois. M, 1957, University of Illinois, Urbana. 33 p.

Breed, Charles E. The Dakota Group (Lower Cretaceous) in northwestern Colorado; Moffat and Rio Blanco counties and vicinity. M, 1956, University of Colorado.

Breed, William J. River terraces and other geomorphic features of the Castle Hill Basin, Canterbury, New Zealand. M, 1960, University of Arizona.

Breeding, J. Ernest, Jr. Refraction of gravity water waves (Panama City, Florida). D, 1972, Columbia University. 161 p.

Breeding, N. Kelly. Geochemical soil analysis in the central (Kentucky) fluorspar district. M, 1972, University of Kentucky. 41 p.

Breedlove, Robert Leeroy. The Balcones fault zone north of Buda. M, 1935, University of Texas, Austin.

Breedon, David H. Stratigraphy and sedimentation of the Mississippian Ste. Genevieve Limestone and Cedar Bluff Group in Daviess County, Indiana. M, 1983, Indiana University, Bloomington. 57 p.

Breemen, Otto Van *see* Van Breemen, Otto

Breemer, Dale A. A lunar terrain analysis and slope frequency study in the lunar equatorial belt. M, 1964, Emporia State University.

Breen, David John. Foraminiferal biostratigraphy and paleoenvironmental reconstruction of the Oligocene Marianna Limestone in Mississippi and Alabama. M, 1987, Northeast Louisiana University. 173 p.

Breen, Kevin John. Control of Eh and pH to evaluate the rate of pyrite oxidation by ferric ion. M, 1982, Pennsylvania State University, University Park. 52 p.

Breen, Nancy Ann. Three investigations of accretionary wedge deformation. D, 1987, University of California, Santa Cruz. 132 p.

Breene, Victor Martin. Preliminary study of the heavy minerals in the Wisconsin and Illinoian tills near Cincinnati, Ohio. M, 1957, University of Cincinnati. 30 p.

Breese, T. E. Depositional environment of the middle Pennsylvanian (Desmoinesian) Fort Scott Limestone in the Putnam area, Dewey County, Oklahoma. M, 1973, Washington University. 114 p.

Breeser, Patsy J. General geology and highway realignment considerations in the Pinehurst-New Meadows-Tamarack area, west central Idaho. M, 1972, University of Idaho. 103 p.

Breeze, Arthur F. Abnormal–subnormal pressure relationships in the Morrow Sands (lower Pennslyvanian) of northwestern Oklahoma. M, 1970, University of Oklahoma. 122 p.

Breger, Irving A. The role of lignin in coal genesis. D, 1950, Massachusetts Institute of Technology. 171 p.

Bregman, Martin L. Geology of the New Castle area, Craig County, Virginia. M, 1967, Virginia Polytechnic Institute and State University.

Bregman, Martin Louis. Structural geology of the Sheep Creek and Rattlesnake Mountain quadrangles, Lewis and Clark County, Montana. D, 1971, University of New Mexico. 100 p.

Bregman, Nina Diane. Tomographic inversion of crosshole seismic data. D, 1987, University of Toronto.

Breidenstein, J. F. A study of water circulation in Monterey harbour (California) using Rhodamine B dye. M, 1965, United States Naval Academy.

Breiner, Sheldon. The piezomagnetic effect in seismically active areas. D, 1967, Stanford University. 210 p.

Breit, George Nicholas. Geochemical exploration study of the Polaris mining district and vicinity, Beaverhead County, Montana. M, 1980, Colorado School of Mines. 265 p.

Breit, George Nicholas. Geochemical study of authigenic minerals in the Salt Wash Member of the Morrison Formation, Slick Rock District, San Miguel County, Colorado. D, 1986, Colorado School of Mines. 267 p.

Breitenbach, Eugene Allen. A computer simulation of gravity drainage in oil reservoirs. D, 1964, Stanford University. 169 p.

Breitenwischer, Robert and Richter, James B. Geology of the northern half of the Marquand Quadrangle, Madison and Bollinger counties, Missouri. M, 1953, University of Michigan.

Breithart, Mark S. The significance of the distribution of clastic lenses within the Negaunee Iron Formation at the eastern end of the Palmer Basin, Marquette Synclinorium, northern Michigan. M, 1983, Michigan State University. 112 p.

Breithaupt, Brent H. Paleontology and paleoecology of Lance equivalent strata, east flank of Rock Springs Uplift, Sweetwater County, Wyoming. M, 1981, University of Wyoming. 113 p.

Breitling, William F. Crustal structure and attenuation derived from the Boston earthquake of October 16, 1963. M, 1965, [Boston University].

Breitsprecher, Charles Hepner. Correlation of foraminifera and megafossils from the Upper Cretaceous, Sucia Island, Washington. M, 1962, University of Washington. 82 p.

Breitzman, Lynne L. Fission track ages of intrusives of the Chagai District, Baluchistan, Pakistan. M, 1979, Dartmouth College. 69 p.

Brekke, David W. Mineralogy and chemistry of clay-rich sediments in the contact zone of the Bullion Creek and Sentinel Butte formations (Paleocene), Billings County, North Dakota. M, 1979, University of North Dakota. 94 p.

Breland, Fritz Clayton, Jr. A petrologic and paleoocurrent analysis of the Lower Pennsylvanian Pottsville Formation of the Warrior Basin in Alabama. M, 1972, University of Mississippi.

Breland, Fritz Clayton, Jr. Stratigraphic, paleoenvironmental, and echinoderm distributional patterns in an unfaulted Middle Ordovician (Chickamauga Group) shelf-edge to on-shelf transect in East Tennessee. D, 1980, University of Tennessee, Knoxville. 217 p.

Breland, Jabe A., II. Chemical and physical characteristics of a saline geothermal submarine spring off Florida's southwestern coast. M, 1980, University of South Florida, St. Petersburg. 105 p.

Brem, Gerald F. Petrogenesis of late Tertiary potassic volcanic rocks in Sierra Nevada and western Great Basin. D, 1977, University of California, Riverside. 361 p.

Bremaecker, Jean-Claude De *see* De Bremaecker, Jean-Claude

Bremer, Mary Lee. Abyssal benthonic foraminifera and the carbonate saturation of sea water and benthonic foraminiferal carbonate saturation history for the Cape Verde Basin for the last 550,000 years. D, 1982, Woods Hole Oceanographic Institution. 174 p.

Bremer, Mary Lee. The quantitative paleobathymetry and paleoecology of the late Pliocene-early Pleistocene foraminifera of La Castella (Calabria, Italy). M, 1977, University of Cincinnati. 77 p.

Bremner Cramer, Jane Alison. Micro-facies analysis of depositional and diagenetic history of the Redwall Limestone in the Chino and Verde valleys. M, 1986, Northern Arizona University. 161 p.

Bremner, James M. The geology of Wreck Bay, Vancouver Island (late Jurassic–Recent). M, 1971, University of British Columbia.

Bremner, Trevor John. Metamorphism in the Fraser Canyon, BC. M, 1972, University of British Columbia.

Brenchley, Gayle Anne. On the regulation of marine infaunal assemblages at the morphological level; a study of the interactions between sediment stabilizers, destabilizers and their sedimentary environment. D, 1978, The Johns Hopkins University.

Brenckle, Paul Louis. Smaller Mississippian and lower Pennsylvanian calcareous foraminifers from Nevada. D, 1970, University of Colorado. 291 p.

Brennan, Brian Daniel. Silurian and Devonian rocks along the upper reaches of the north fork of the Big Lost River, Custer County, Idaho. M, 1987, University of Wisconsin-Milwaukee. 72 p.

Brennan, Charles Victor. Geology and proposed methods of mining and milling for the Eldorado Prospect, British Columbia. M, 1941, University of Washington. 108 p.

Brennan, Daniel J. Geological reconnaissance of Cienega Gap, Pima County, Arizona. D, 1957, University of Arizona.

Brennan, Daniel J. Sedimentation in the Pennsylvanian of the Black Hills and vicinity. M, 1953, South Dakota School of Mines & Technology.

Brennan, J. F. and Meaux, R. P. Observation of the nearshore water circulation off a sand beach. M, 1964, United States Naval Academy.

Brennan, Jeanne L. Interpretation of Vibroseis reflections within the Catoctin Formation of central Virginia. D, 1985, Virginia Polytechnic Institute and State University.

Brennan, Peter Anderson. Geology of the NASA Arizona sedimentary test site, Mohave County, Arizona. M, 1968, University of Nevada. 93 p.

Brennan, Sandra F. Analysis of bluff erosion along the southern coastline of Lake Ontario, New York. M, 1979, SUNY at Buffalo. 98 p.

Brennan, Terrance Patrick. Use of synthetic seismograms to determine subsurface structure in the Linton Gas storage field, Greene County, Indiana. M, 1979, Wright State University. 59 p.

Brennan, William Joseph. Little Paddy Creek dam site. M, 1964, Washington University. 59 p.

Brennan, William Joseph. Structural and surficial geology of the west flank of the Gore range, Colorado. D, 1969, University of Colorado. 140 p.

Brenneman, Lionel. Preliminary sedimentary study of certain bodies in the Apalachicola Delta. M, 1957, Florida State University.

Brenner, Gilbert J. A zoogeographic analysis of some shallow-water foraminifera in the Gulf of Mexico. M, 1958, New York University.

Brenner, Gilbert Jay. The spores and pollen of the Potomac Group of Maryland. D, 1962, Pennsylvania State University, University Park. 302 p.

Brenner, Mark. Paleolimnology of the Maya region. D, 1983, University of Florida. 249 p.

Brenner, Robert Lawrence. Geology of the Lubrecht experimental forest, Missoula County, Montana. M, 1964, University of Montana. 90 p.

Brenner, Robert Lawrence. Oxfordian sedimentation in the Western Interior U.S.A. D, 1973, University of Missouri, Columbia. 189 p.

Brenninkmeyer, Benno Max S.J. Synoptic surf zone sedimentation patterns. D, 1973, University of Southern California.

Brent, William Bonney. The geology of the Harrisonburg Quadrangle, Virginia. D, 1955, Cornell University. 104 p.

Brent, William Bonney. Toccoa Quartzite and adjacent rocks in Stephens County, Georgia. M, 1952, Cornell University. 42 p.

Brera, A. M. Application of Landsat imagery to monitor sand dunes movement in the Sahara Desert. D, 1979, University of Tennessee, Knoxville. 225 p.

Brereton, Roy G. A magnetometer survey of the Weldon Spring Quadrangle, Missouri. M, 1954, St. Louis University.

Brereton, Roy George. Magnetic study of Meteor Crater, Arizona. M, 1957, St. Louis University.

Breshears, Terry L. Regional gravity effects of igneous intrusions, central Arkansas. M, 1983, University of Missouri, Columbia.

Breslin, Patricia Anne. Geology and geochemistry of a young cinder cone in the Cima volcanic field, eastern Mojave Desert, California. M, 1982, University of California, Los Angeles. 121 p.

Bressler, Calder Tupper. The petrology of the Roslyn Arkose, central Washington. D, 1951, Pennsylvania State University, University Park. 175 p.

Bressler, Jason Robert. Structural geology of China Bend, Stevens County, Washington. M, 1979, Washington State University. 105 p.

Bressler, S. L. Paleomagnetism of the Pliocene Verde Formation, Yavapai County, Arizona. M, 1977, University of Arizona. 95 p.

Bretches, John E. A geologic study of East Butte, a rhyolitic volcanic dome on the eastern Snake River plain, Idaho. M, 1984, SUNY at Buffalo. 159 p.

Breteler, Ronald Johannes. The cycling of mercury in Spartina marshes and its availability to selected biota. D, 1980, McGill University.

Bretnall, Robert Edward, Jr. Electromagnetic signature of the salt-water/fresh-water transition zone, west-central Florida. M, 1988, University of South Florida, Tampa. 178 p.

Bretsky, Peter William. The geology of the upper Canyon Group, Stephens and Palo Pinto counties, Texas. M, 1963, Southern Methodist University. 171 p.

Bretsky, Peter William, Jr. Upper Ordovician ecology of the central Appalachians. D, 1968, Yale University. 432 p.

Bretsky, Sara Stewart. Phenetic and phylogenetic classifications of the Lucinidae (Mollusca, Bivalvia). D, 1969, Yale University. 564 p.

Brett, B.D. An interpretation of recorded information relating to the (Precambrian) Grenville "Front" (Ontario and New York). M, 1960, McGill University.

Brett, C. E. Studies on the paleontology and stratigraphy of the western New York area. M, 1975, SUNY at Buffalo. 97 p.

Brett, C. E. Systematics and paleoecology of Late Silurian (Wenlockian) pelmatozoan echinoderms from western New York and Ontario. D, 1978, [University of Michigan]. 595 p.

Brett, Charles Everett. Paleoecologic and faunal analyses of some fossil assemblages of the Cretaceous Black Creek and Peedee Formation. M, 1961, University of North Carolina, Chapel Hill. 120 p.

Brett, Charles Everett. Relationships between marine invertebrate in fauna distribution and sediment type distribution in Bogue Sound, North Carolina. D, 1963, University of North Carolina, Chapel Hill. 202 p.

Brett, James Walter. A vertical magnetic intensity survey of a portion of south central Michigan. M, 1960, Michigan State University. 48 p.

Brett, Peter R. Experimental data from the system Cu-Fe-S and the bearings on exsolution textures and reaction rates in ores. D, 1963, Harvard University.

Brett-Surman, Michael Keith. The appendicular anatomy of hadrosaurian dinosaurs. M, 1975, University of California, Berkeley. 70 p.

Bretz, J. Harlen. The glaciation of the Puget Sound basin. D, 1913, University of Chicago. 237 p.

Bretz, Richard F. Stratigraphy, mineralogy, paleontology and paleoecology of the Crow Creek Member, Pierre Shale (Late Cretaceous), South central South Dakota. M, 1979, Fort Hays State University.

Breuer, Joseph W. Mineralogy and distribution of clay sediments in the Great Miami River basin. M, 1965, University of Cincinnati. 57 p.

Breuer, Roberto John Arnstein *see* Arnstein Breuer, Roberto John

Breunig, Peter A. A crustal model for northern New England. M, 1980, Boston College.

Breuninger, Ray Hubert. Late Pennsylvanian and early Permian carbonate mounds in the Arco Hills and southern Lemhi Range, Idaho. D, 1971, University of Montana. 216 p.

Brew, David Alan. Synorogenic sedimentation of Mississippian age, Eureka Quadrangle, Nevada. D, 1964, Stanford University. 312 p.

Brew, David Scott. Seismic modelling at Deep Sea Drilling Project Site 603; the Lower Continental Rise Hills. M, 1986, Dalhousie University. 219 p.

Brew, Douglas Crocker. Stratigraphy of the Naco Formation (Pennsylvanian) in central Arizona. D, 1965, Cornell University. 230 p.

Brew, Douglas Crocker. Stratigraphy of the Upper Devonian Fall Creek Conglomerate in northern Pennsylvania and southern New York. M, 1963, Cornell University.

Brewer, Bobby Lee. Holocene Ostracoda from Barnegat Bay, New Jersey. M, 1987, Northeast Louisiana University. 86 p.

Brewer, Charles. Origin and preservation of sandstone oil reservoirs. M, 1927, University of Pittsburgh.

Brewer, George F. Sulfur, heavy metal, and major element chemistry of sediments from four eastern Maine ponds. M, 1986, University of Maine. 140 p.

Brewer, Jonathan Andrew. COCORP traverses in Wyoming and Oklahoma. D, 1981, Cornell University. 153 p.

Brewer, Quenton L. Geology and ore deposits of a section of the Chalk Creek mining district, Chaffee County, Colorado. M, 1931, Colorado School of Mines. 39 p.

Brewer, R. L. Applications of radon distribution and radon flux for the determination of oceanic mixing and air-sea gas exchange. M, 1977, Texas A&M University.

Brewer, Roger Clay. Structure and petrography of the basement complexes east of the Hot Springs Window,

Blue Ridge Province, western North Carolina. M, 1986, University of Tennessee, Knoxville. 196 p.

Brewer, Thomas. A mass balance study of Fox glacier, Yukon Territory, Canada. M, 1969, Boston University. 43 p.

Brewer, Thomas. Morphological characteristics of Hadley Rille, the Moon. D, 1975, Boston University. 329 p.

Brewer, W. S., Jr. Characterization of organic substances found in lake sediments. D, 1975, University of North Carolina, Chapel Hill. 184 p.

Brewer, Wayne Martin. Possibly offset plutons along the Denali Fault (McKinley Strand), central Alaska Range, Alaska. M, 1977, University of Wisconsin-Madison.

Brewer, Wayne Martin. Stratigraphy, structure, and metamorphism of the Mount Deborah area, central Alaska Range, Alaska. D, 1982, University of Wisconsin-Madison. 528 p.

Brewer, William August, III. A method of remote sensing based on laser-induced fluorescence. D, 1965, University of California, Berkeley. 100 p.

Brewer, William August, III. The geology of a portion of the China Mountain Quadrangle, California. M, 1955, University of California, Berkeley. 47 p.

Brewster, Arthur V. Subsurface geology within the Panhandle Field and its relation to ground magnetometer survey, Carson and Hutchinson counties, Texas. M, 1985, West Texas State University. 57 p.

Brewster, Charles L. The geology and origin of the Sawyer uranium prospect, Live Oak County, Texas. M, 1982, University of Tennessee, Knoxville. 68 p.

Brewster, Christine. Oxygen isotope ratios in amphibolites from southwestern Montana. M, 1980, Purdue University.

Brewster, David P. Stratigraphy, biostratigraphy, and depositional history of the Big Snowy and Amsden formations (Carboniferous) of southwestern Montana. M, 1984, Indiana University, Bloomington. 164 p.

Brewster, G. R. Genesis of volcanic ash-charged soils having Podzolic morphologies, Banff and Jasper national parks, Alberta. D, 1979, University of Western Ontario.

Brewster, James B. The geology of the Hidalgo Bluff area, Washington County, Texas. M, 1965, Texas A&M University. 109 p.

Brewster, Nancy A. Cenozoic biogenic silica sedimentation in the Antarctic Ocean, based on two Deep Sea Drilling Project sites. M, 1977, Oregon State University. 98 p.

Brewster, Renee Harrison. The distribution and chemistry of rare-earth minerals in the South Platte pegmatite district, Colorado, and their genetic implications. M, 1986, University of New Orleans. 139 p.

Brewton, Joseph Lawrence. Heavy mineral distribution in the Carrizo Formation (Eocene), east Texas. M, 1970, University of Texas, Austin.

Breyer, John Albert. The Plio-Pleistocene boundary in western Nebraska with a contribution to the taxonomy of the Pleistocene camelids. D, 1974, University of Nebraska, Lincoln.

Breymann, Marta T. von *see* von Breymann, Marta T.

Brezina, James Lewis. Stratigraphy and petrology of the Pecan Gap Formation (Taylor Group, Upper Cretaceous) in its type area. M, 1974, University of Texas, Arlington. 63 p.

Brezinski, David Kevin. Dynamic lithostratigraphy and paleoecology of Upper Mississippian (Chesterian) strata of the northcentral Appalachian Basin. D, 1984, University of Pittsburgh. 132 p.

Brezonik, Patrick Lee. The dynamics of the nitrogen cycle in natural waters. D, 1968, University of Wisconsin-Madison. 277 p.

Briar, David W. Water resource analysis of the Sullivan Flats area near Niarada, Flathead Indian Reservation, Montana. M, 1987, University of Montana. 184 p.

Briard, Vernon Eugene. A study of faulting in the Luling oil field, Texas, by the gravimetric method. D, 1952, University of Iowa. 387 p.

Briard, Vernon Eugene. A study of staining methods applied to well cuttings. M, 1932, University of Iowa. 48 p.

Brice, Donald A. Comparison of textural and hydrogeologic properties of Carboniferous and modern fluvial point bar deposits. M, 1985, University of Kentucky. 122 p.

Brice, Glyn Alan. Sequence of measurement investigation for slow learners. M, 1968, University of Utah. 461 p.

Brice, James Coble. Geology of a portion of the Lower Lake Quadrangle, California. M, 1948, University of California, Berkeley. 75 p.

Brice, James Coble. Geology of the Lower Lake Quadrangle, California. D, 1950, University of California, Berkeley. 72 p.

Brice, William Charles. An analysis technique for mineral resource planning. D, 1981, University of Minnesota, Minneapolis. 774 p.

Brice, William Riley. Subsolidus phase relations in the system magnesium carbonate-calcium carbonate-strontium carbonate-barium carbonate. D, 1971, Cornell University. 145 p.

Briceño M., Henry O. Application of remote sensing to diamond placer exploration in a tropical jungle environment, Caroni River, Venezuela. D, 1982, Colorado School of Mines. 176 p.

Briceño M., Henry O. Geology and geochemistry of the Magmont Mine, Iron County, Missouri. M, 1975, Colorado School of Mines. 153 p.

Briceño-Guarupe, Luis Alberto. The crustal structure and tectonic framework of the Gulf of Panama. M, 1979, Oregon State University. 71 p.

Bricker, Owen. Stability relations in the system Mn-O_2-H_2O at 25°C and 1 atmosphere total pressure. D, 1964, Harvard University.

Brickey, David Wayne. A geochemical and structural study of the South Mountain Batholith of southern Nova Scotia for tin and uranium mineralization. M, 1987, Pennsylvania State University, University Park. 158 p.

Brickman, Eugene. Cs^{137} chronology in marsh and lake samples from Delaware. M, 1978, University of Delaware.

Bridge, Dane Alexander. The geology and geochemistry of cassiterite as a minor constituent of the zinc-copper massive sulphide deposit at South Bay mines, Confederation Lake, Ontario. M, 1972, University of Manitoba.

Bridge, J. and Ingerson, M. J. The Middle Ordovician section in east central Missouri. M, 1922, University of Missouri, Rolla.

Bridge, Josiah. Geology of the Eminence and Cardareva quadrangles. D, 1929, Princeton University.

Bridge, Thomas E. Contact metamorphism in siliceous limestone and dolomite in Marble Canyon and geology of related intrusion, Culberson County, Trans-Pecos Texas. D, 1966, University of Texas, Austin. 165 p.

Bridge, Thomas E. The petrology and petrography of the igneous rocks of Riley County, Kansas. M, 1953, Kansas State University. 57 p.

Bridger, J. R. Variations in the granite-syenite masses occurring near Kingston, Ontario. M, 1933, Queen's University. 65 p.

Bridges, Katherine H. Assessment of tectonic uplift through analysis of consequent stream channel profiles, central Columbia Basin, Washington. D, 1985, Indiana State University. 97 p.

Bridges, Kenneth Francis. The areal geology and Cretaceous stratigraphy of southern Marshall County, Oklahoma. M, 1979, University of Oklahoma. 126 p.

Bridges, Lindell C. Determination of optimum SAR sensor configurations from analysis of geologic terrain models. M, 1982, University of Arkansas, Fayetteville.

Bridges, Luther Wadsworth, II. Geology of Mina Plomosas area, Chihuahua, Mexico. D, 1962, University of Texas, Austin.

Bridges, Luther Wadsworth, II. Revised Cenozoic history of Rim Rock Country, Trans-Pecos, Texas. M, 1958, University of Texas, Austin.

Bridges, Samuel Rutt. Evaluation of stress drop of the August 2, 1974 Georgia-South Carolina earthquake and aftershock sequence. M, 1975, Georgia Institute of Technology. 103 p.

Bridges, Steven Dwayne. Mapping, stratigraphy and tectonic implications of lower Permian strata, eastern Wichita Mountains, Oklahoma. M, 1985, Oklahoma State University. 125 p.

Bridges, William Clayton. Megafauna of the Pawnee Formation, Adair County, Missouri. M, 1957, University of Missouri, Columbia.

Bridges, William Elmer. Beach sediments of Galveston, Chambers, and Jefferson counties, Texas. M, 1959, University of Texas, Austin.

Bridgett, L. S. Fission track dating of the Tar Springs Formation, a Mississippian Period sandstone of the Illinois Basin. M, 1978, Washington University. 48 p.

Bridwell, Richard Joseph. Finite element applications to mechanical problems in structural geology. D, 1973, University of Utah. 162 p.

Bridwell, Richard Joseph. Geology of the Kerber creek area, Saguache County, Colorado. M, 1968, Colorado School of Mines. 104 p.

Briedis, John. Effects of stress of microfabric of oil shale. M, 1966, University of Arizona.

Briedis, N. A. Geology of the sulfur-sulfate deposits at the Apache Mines, Baja California, Mexico. M, 1976, San Diego State University.

Briel, Lawrence I. An investigation of the U^{234}/U^{238} disequilibrium in the natural waters of the Santa Fe River basin of North-central Florida. D, 1976, Florida State University. 241 p.

Brienzo, Richard K. Velocity and attenuation profiles in the Monterey deep-sea fan. D, 1987, [University of California, San Diego]. 123 p.

Brier, Ervin Ernest. A study of the petrographic and certain textural characteristics of the "Dakota Group" in northeastern Nebraska. M, 1940, University of Nebraska, Lincoln.

Brieva, Jorge A. Petrology of the Mazarn and Blakely formations, Montgomery County, Arkansas. M, 1963, University of Tulsa. 89 p.

Briggeman, Homer W. Subsurface geology of the Iuka-Carmi Pool area, Pratt County, Kansas, in relation to petroleum accumulation. M, 1959, Kansas State University. 67 p.

Briggs, David F. Petrology of the Roxboro Metagranite, North Carolina. M, 1974, Virginia Polytechnic Institute and State University.

Briggs, Garrett. A paleocurrent study of Upper Mississippian and Lower Pennsylvanian rocks of the Ouachita Mountains and Arkoma Basin, southeastern Oklahoma. D, 1963, University of Wisconsin-Madison. 138 p.

Briggs, Garrett. Paleocurrent study of the Brazos River sandstone member of the Garner Formation, Palo Pinto County, Texas. M, 1960, Southern Methodist University. 20 p.

Briggs, John Peter. A sedimentation study of loess and loess-like deposits in the Big Badlands of South Dakota. M, 1974, South Dakota School of Mines & Technology.

Briggs, Louis Isaac. Geology of the Ortigalita Peak Quadrangle, California. D, 1950, University of California, Berkeley. 157 p.

Briggs, Otis E. The relation between the faults and igneous flows of the Sierra Nevada mountains. M, 1920, [Stanford University].

Briggs, Peter B. Effects of a sanitary landfill on ground water quality in Ashland, New Hampshire. M, 1974, University of New Hampshire. 55 p.

Briggs, Peter Laurence. Pattern recognition applied to uranium exploration. D, 1978, Massachusetts Institute of Technology. 233 p.

Briggs, Thomas D. The influence of fracture aperture on the propagation of acoustic waves along a fluid-filled borehole. M, 1986, Washington State University. 60 p.

Briggs, William Melrose, Jr. Late Tertiary Ostracoda from Palau Islands, western Caroline Islands. M, 1963, George Washington University.

Brigham, Julie K. Stratigraphy, amino acid geochronology, and genesis of Quaternary sediments, Broughton Island, East Baffin Island, Canada. M, 1980, University of Colorado.

Brigham, Julie Kay. Marine stratigraphy and amino acid geochronology of the Gubik Formation, western Arctic Coastal Plain, Alaska. D, 1985, University of Colorado. 379 p.

Brigham, Robert Hoover. K-feldspar genesis and stable isotope relations of the Papoose Flat Pluton, Inyo Mountains, California. D, 1984, Stanford University. 194 p.

Brigham, Robert John. An investigation of total and exchangable manganese as a means of correlating argillaceous sediments. M, 1958, Michigan State University. 39 p.

Brigham, Robert John. Structural geology of southwestern Ontario and southeastern Michigan (Cambrian-Mississippian). D, 1972, University of Western Ontario. 219 p.

Bright, Edward G. Geology of the Topley Intrusives (Jurassic and/or Lower Cretaceous) in the Endako area, British Columbia. M, 1967, University of British Columbia.

Bright, Mont Jackson, Jr. Some metasediments in the Grandfather Mountain Fenster. M, 1956, University of North Carolina, Chapel Hill. 41 p.

Bright, Robert C. Geology of the Cleveland area, southeastern Idaho. M, 1960, University of Utah. 262 p.

Bright, Robert C. Pleistocene lakes Thatcher and Bonneville, south-eastern Idaho. D, 1963, University of Utah. 292 p.

Bright, William E. Geology of the Irmo NE, South Carolina quadrangle. M, 1962, University of South Carolina. 40 p.

Brightman, George Forsha. The Tom Sauk Limestone of south-eastern Missouri. M, 1937, Washington University. 157 p.

Brikowski, Tom Harry. A quantitative analysis of hydrothermal circulation around mid-ocean ridge magma chambers. D, 1987, University of Arizona. 65 p.

Brikowski, Tom Harry. Geology and petrology of Gearhart Mountain; a study of calc-alkaline volcanism east of the Cascades in Oregon. M, 1983, University of Oregon. 157 p.

Brill, Edward Tobias. The significance and the use of optimal parameters in defining the Genesee River plume in the Rochester (New York) Embayment of Lake Ontario. M, 1973, University of Rochester. 26 p.

Brill, Kenneth Gray. Brachiopods from the Permian red beds of Texas and Oklahoma. M, 1938, University of Michigan.

Brill, Kenneth Gray. Pennsylvanian rocks of the Gore area, Colorado. D, 1939, University of Michigan.

Brill, R. C., Jr. The geology of the lower southwest rift of Haleakala, Hawaii. M, 1975, University of Hawaii at Manoa. 65 p.

Brill, Russell Martin. Differential geomagnetic field measurements at the edge of the Denver Basin. D, 1974, University of Colorado.

Brill, Virgil August. The Travis Peak Formation. M, 1928, University of Texas, Austin.

Brilliant, Renee M. Dispersion of the Rayleigh waves under the continental United States. M, 1951, Columbia University, Teachers College.

Brillinger, Allan. Investigation of Trinity River paleochannels in the central business district, Dallas, Texas. M, 1985, Texas A&M University. 92 p.

Brim, Raymundo Jose Portella. Geochemical investigations on the Lakeview mining district, Bonner County, Idaho. M, 1968, University of Idaho. 82 p.

Brimberry, David L. Depositional environments, diagenesis and conodont biostratigraphy of the Montoya Group (Late Ordovician), Sacramento Mountains, New Mexico. M, 1984, Texas Tech University. 164 p.

Brimhall, George H., Jr. Early hydrothermal wall rock alteration at Butte, Montana. D, 1972, University of California, Berkeley. 103 p.

Brimhall, Ronald M. Digital analysis of borehole-measured aquifer resistivity to determine water quality. M, 1969, New Mexico Institute of Mining and Technology. 147 p.

Brimhall, Willis H. Stratigraphy and structural geology of the northern Deer Creek Reservoir area, Provo Canyon, Utah. M, 1951, University of Arizona.

Brimhall, Willis Hone. Concentration distribution of uranium, thorium and some alkali, alkaline earth, and transition metals in three cores of Conway Granite, New Hampshire. D, 1966, Rice University. 187 p.

Brindle, Wendy D. Carboniferous annulate orthoconic nautiloids from the Mid-Continent. M, 1985, Bowling Green State University. 224 p.

Briner, William D. Geology of part of the Cucomunga area, Esmeralda County, Nevada. M, 1980, University of Nevada. 78 p.

Bringhurst, Kelly Norman. Major element chemistry and mineralogy in well Fee #5, in the Salton Sea geothermal field, California. M, 1987, University of California, Riverside. 135 p.

Brink, Marilyn Rae Buchholtz ten *see* ten Brink, Marilyn Rae Buchholtz

Brink, Norman Wayne Ten *see* Ten Brink, Norman Wayne

Brink, Ronald. A microstructural investigation of the deformational conditions and kinematics of flow in peridotite tectonics from the North Arm Mountain Massif, Bay of Islands ophiolite complex, Newfoundland. M, 1987, University of Houston.

Brink, Terry L. Geology of a portion of the Wells Tannery and Everett East 7 1/2-minute quadrangles, Bedford and Fulton counties, Pennsylvania. M, 1982, Indiana University of Pennsylvania. 95 p.

Brink, Uri S. Ten *see* Ten Brink, Uri S.

Brinker, Willard Franklin. Placer tin deposits north of Basin, Montana. M, 1944, Montana College of Mineral Science & Technology. 33 p.

Brinkley, Charles Alexander. Glacial geology of northern Montour and Northumberland counties, Pennsylvania between north and west branches of Susquehanna River. M, 1960, Pennsylvania State University, University Park. 103 p.

Brinkman, D. B. The structural and functional evolution of the diapsid tarsus. D, 1979, McGill University.

Brinkmann, Robert. Surficial geology of the Eau Galle drainage basin, west-central Wisconsin. M, 1987, University of Wisconsin-Milwaukee. 67 p.

Brinkmann, Robert Terry. The photodissociation of water vapor, evolution of oxygen and escape of hydrogen in the Earth's atmosphere. D, 1969, California Institute of Technology. 162 p.

Brinster, Kenneth F. Molluscan paleontology of the Pierre shale (upper Cretaceous), Bowman County, North Dakota. M, 1970, University of North Dakota. 136 p.

Brinton, Edward. Distribution, faunistics, and evolution of Pacific euphausiids. D, 1957, University of California, Los Angeles. 498 p.

Brinton, Lise. Deposition and diagenesis of Middle Pennsylvanian (Desmoinesian) phylloid algal banks,

Paradox Formation, Ismay Zone, Ismay Field and San Juan Canyon, Paradox Basin, Utah and Colorado. M, 1986, Colorado School of Mines. 314 p.

Briones, Angelina Mariano. Nature and distribution of organic nitrogen in tropical soils. D, 1969, University of Hawaii. 152 p.

Brisbin, William Corbett. Regional Bouguer anomalies and crustal structure in southern California. D, 1970, University of California, Los Angeles. 246 p.

Briscoe, Harry J., Jr. Stratigraphy of the Fox Hills Sandstone with some comments on its suitability as an aquifer. M, 1972, Colorado School of Mines. 79 p.

Briscoe, Howard. A statistical evaluation of mineral deposit sampling. M, 1954, Massachusetts Institute of Technology. 46 p.

Briscoe, James A. The general geology of the Picacho Peak area (Pinal County, Arizona). M, 1967, University of Arizona.

Briscoe, Melanie. An investigation of the invertebrate paleontology of the Tsegi and Naha formations in Tsegi Canyon, Arizona. M, 1974, Northern Arizona University. 77 p.

Brisebois, Daniel. Géométrie d'une partie de l'assemblage de conglomerat et de grès chenalisés du quai de l'Islet. M, 1972, Universite de Montreal.

Briskey, Joseph A., Jr. Geology, petrology, and geochemistry of the Jersey, East Jersey, Huestis, and Iona porphyry copper-molybdenum deposits, Highland Valley, British Columbia. D, 1980, Oregon State University. 399 p.

Briskin, Madeleine. Pleistocene stratigraphy and quantitative paleooceanography of tropical North Atlantic core V16-205. D, 1973, Brown University.

Bristol, Calvert C. An investigation of ore textures using the vacuum heating stage. M, 1962, University of Manitoba.

Bristol, Calvert C. The quantitative X-ray powder diffraction determination of minerals in some metamorphosed volcanic rocks. D, 1965, University of Manitoba.

Bristol, David Arthur, Jr. Structural evolution and metamorphism of mid-Proterozoic basement in the Northwest Van Horn Mountains, Trans-Pecos, Texas. M, 1987, University of Texas, Austin. 104 p.

Bristol, Hubert M. New ostracodes from the Menard Formation. M, 1939, University of Chicago. 50 p.

Bristow, Joseph Dalton. Oil and gas prospecting in northeast Mississippi. M, 1952, Mississippi State University. 97 p.

Bristow, Keith Leslie. Simulation of heat and moisture transfer through a surface residue-soil system. D, 1983, Washington State University. 74 p.

Bristow, Milton M. The geology of the northwestern third of the Marcola Quadrangle, Oregon. M, 1959, University of Oregon. 70 p.

Brite, S. E. A. Seasat orbital radar imagery applied to lineament analysis and relationships with hydrocarbon production in the Wartburg Basin area, Tennessee. M, 1982, University of Tennessee, Knoxville. 93 p.

Brito, Ignacio Autrliano Machado *see* Machado Brito, Ignacio Autrliano

Brito, Ignacio Machado. Silurian and Devonian Acritarcha from Maranhao Basin, Brazil. M, 1966, Stanford University.

Brito, Sergio N. A. De *see* De Brito, Sergio N. A.

Britsch, Louis D. Migration of Isles Dernieres; past and future. M, 1984, Tulane University. 89 p.

Britt, Claude J. Archaic occupation of west-central Ohio. M, 1967, Bowling Green State University. 122 p.

Britt, Terence L. The geology of the Twin Peaks area, Pima County, Arizona. M, 1955, University of Arizona.

Brittain, Alan Lee. Geometry, stratigraphy, and depositional environment of the Springfield Coal and the Dugger Formation (Pennsylvanian) north of Winslow,

Pike County, Indiana. M, 1975, Indiana University, Bloomington. 64 p.

Brittain, Richard L. Geology and ore deposits of the western portion of the Hilltop Mine area, Cochise County, Arizona. M, 1954, University of Arizona.

Brittain, William H. The use of aerial photographs in locating and evaluating gravel deposits in New Brunswick with special reference to forested areas. M, 1958, University of New Brunswick.

Brittenham, Marvin Del. Permian Phosphoria bioherms and related facies, southeastern Idaho. M, 1973, University of Montana. 213 p.

Brittingham, Peter Lane. Structural geology of a portion of the White Tank Mountains, central Arizona. M, 1985, Arizona State University. 107 p.

Britton, Douglas R. The occurrence of fish remains in modern lake systems; a test of the stratified-lake model. M, 1988, Loma Linda University. 39 p.

Britton, J. W. M. The iron formation of the Albany Basin with special reference to the associated gold deposits. M, 1938, University of Toronto.

Britton, Joe M. Petrology and petrography of the Pedro Dome plutons (Cretaceous), (about fifteen miles northeast of Fairbanks, Alaska). M, 1969, University of Alaska, Fairbanks. 52 p.

Britton, Kathy Booth. Effects of experimental conditions on the sorption of organic compounds to soils and sediments. D, 1988, University of New Hampshire. 162 p.

Britton, Nathaniel L. The geology of Staten Island; the examination of American tellurium minerals. D, 1881, Columbia University, Teachers College.

Britton, Thomas Abbot, Jr. Depositional environment of the Tombigbee Sand Member-Mooreville Formation (Cretaceous) contact at selected outcrops in Alabama. M, 1968, University of Alabama.

Brixey, Austin Day, Jr. An Eocene ostracode fauna from Parris Island, South Carolina. M, 1951, Columbia University, Teachers College.

Brizzolara, Donald William. Geology of the northern halves of the Crocker Mountain and Dixie Mountain 7.5' quadrangles, Plumas County, California. M, 1979, University of California, Davis. 110 p.

Broad, D. Lower Devonian Heterostraci from the Peel Sound Formation, Prince of Wales island, Northwest Territories. M, 1968, University of Ottawa. 138 p.

Broadbent, Robert C. Numerical modeling of the effects of artificial recharge in Las Vegas Valley, Nevada. M, 1980, University of Nevada. 124 p.

Broadhead, Ronald Frigon. Gas-bearing Ohio Shale (Devonian) along Lake Erie. M, 1979, University of Cincinnati. 177 p.

Broadhead, Roxane. Hydrologic summarization and management classification of tributary basins in the upper Snake River basin. M, 1987, University of Idaho. 117 p.

Broadhead, Sean. A Fourier grain shape analysis of the source and littoral transport of sand in the Black's Beach area, San Diego County, California. M, 1988, University of Southern California.

Broadhead, T. W. Biostratigraphy and paleoecology of the Floyd Shale, upper Mississippian, Northwest Georgia. M, 1975, University of Texas, Austin.

Broadhead, Thomas Webb. Carboniferous camerate crinoid subfamily Dichocrininae. D, 1978, University of Iowa. 247 p.

Broatch, Jane Catherine. Palynologic zonation and correlation of the Peace River Coalfield, northeastern British Columbia. M, 1987, University of British Columbia. 67 p.

Broberg, Steven Kent. Lithofacies and environment of deposition of the upper Gasconade and Roubidoux formations (Lower Ordovician) of south central Missouri. M, 1985, University of Missouri, Columbia. 171 p.

Brobst, Donald Albert. Geology of the Plumtree area, Spruce Pine District, North Carolina. D, 1953, University of Minnesota, Minneapolis. 117 p.

Brocculeri, Thomas. The depositional environments and early diagenetic controls on the mineralogy of the Kittanning Members, Allegheny Group, eastern Ohio. M, 1988, University of Akron. 166 p.

Broch, Michael John. Igneous and metamorphic petrology, structure, and mineral deposits of the Mineral Ridge area (Moses mining district), Colville Indian Reservation, Washington. M, 1979, Washington State University. 204 p.

Brocher, Thomas Mark. Reflectivity analysis of multichannel seismic reflection data with applications to COCORP Rio Grande Rift survey. D, 1980, Princeton University. 176 p.

Brock, Byron B. The so-called load metamorphism of the Shuswap Terrane of British Columbia. D, 1934, University of Wisconsin-Madison.

Brock, Dennis. An engineering and petrographic study of fissile and nonfissile shales in northeast Ohio. M, 1988, Kent State University, Kent. 100 p.

Brock, Frank C., Jr. Walker Creek revisited; a reinterpretation of the diagenesis of the Smackover Formation of Walker Creek Field, Arkansas. M, 1980, Louisiana State University.

Brock, Jerome A. Shallow and brackish-water foraminifera of Tampa Bay, Florida. M, 1958, New York University.

Brock, John Campbell. Petrology of the Mobley Mountain Granite, Amherst Co., Virginia. M, 1981, University of Georgia. 130 p.

Brock, Kenneth Jack. Mineralogy of the Garnet Hill skarns (Calaveras county, California). D, 1970, Stanford University. 91 p.

Brock, Lafayette Breckinridge. The geology of Jessamine County (Kentucky). M, 1899, [University of Kentucky].

Brock, M. Michael. A two-dimensional computer model of the Clayton Aquifer in southwestern Georgia. M, 1987, University of Georgia. 76 p.

Brock, Martha Morgan. Granitization of amphibolite, Tallulah Falls Formation, North Carolina. M, 1984, University of Kentucky. 141 p.

Brock, Michael D. De *see* De Brock, Michael D.

Brock, William G. Characterization of the Muddy Mountain-Keystone thrust contact and related deformation. M, 1973, Texas A&M University.

Brockhouse, Robert Burton. Ecology of marine Ostracoda. M, 1951, University of Illinois, Urbana. 64 p.

Brockhouse, Thomas E. The geology and paleontology of a portion of the Santa Monica Mountains. M, 1932, University of Southern California.

Brockman, Allen R. The underclays and overshales of the Brazil Formation (Pennsylvanian) in western Indiana. M, 1986, Indiana University, Bloomington. 105 p.

Brockman, Charles Scott. The engineering geology, relative stability and Pleistocene history of the Dry Run Creek area, Hamilton County, Ohio. M, 1983, University of Cincinnati. 147 p.

Brockman, George Frederic. Late Cretaceous stratigraphy in western Alabama. M, 1962, University of Alabama.

Brockmann, Mark E. The geology of the southeastern half of Whiskey Mountain along the northeast flank of the Wind River Range in northwestern Fremont County, Wyoming. M, 1985, Miami University (Ohio). 112 p.

Brockunier, Sawyer R. Geology of the Little Rocky Mountains, Montana. D, 1936, Yale University. 130 p.

Brockway, C. E. Investigation of natural sealing processes in irrigation canals and reservoirs. D, 1977, Utah State University.

Brocoum, Alice V. Paragenesis and fluid inclusions of the Tayoltita silver-gold bearing quartz vein deposit, Durango, Mexico. M, 1971, Columbia University. 50 p.

Brocoum, Stephan John. Structural and metamorphic history of the major Precambrian gneiss belt in the

Hailesboro-West Fowler-Balmat area, Adirondack lowlands, New York. D, 1971, Columbia University. 194 p.

Broderick, Alan T. Geology of the southern part of the San Antonio Mountains, Nevada. D, 1949, Yale University.

Broderick, Gregory Philip. Stabilization of compacted clay against attack by concentrated organic chemicals. D, 1987, University of Texas, Austin. 259 p.

Broderick, John C. The geology of Granite Hill, Luna County, New Mexico. M, 1984, University of Texas at El Paso.

Broderick, Jon P. Structure and petrography of the Piety Hill area, Pima County, Arizona. M, 1967, University of Arizona.

Broderick, Thomas M. The locus and conditions of the cementation of the shales. M, 1914, University of Wisconsin-Madison.

Broderick, Thomas M. The relation of the iron ores in northeastern Minnesota to the Duluth gabbro. D, 1917, University of Minnesota, Minneapolis. 63 p.

Brodersen, Ray Arlyn. The petrology, structure, and age relationships of the Cathedral Peak porphyritic quartz monzonite, central Sierra Nevada, California. D, 1962, University of California, Berkeley. 219 p.

Brodie, David R. The Deville (detrital) formation of the Kindersley area, Saskatchewan. M, 1955, University of Alberta. 59 p.

Brodie, Gregory A. Engineering and environmental geology for land use planning in Hamilton County, Indiana. M, 1979, Purdue University.

Brodsky, Harold. The Mesaverde Group at Sunnyside, Utah. M, 1960, University of Colorado.

Brodsky, Nancy S. An investigation of the fracture of granite under triaxial stress. D, 1985, University of Colorado. 273 p.

Brodsky, Nancy S. Strain rate and moisture dependent deformation of Ralston intrusive. M, 1980, University of Colorado.

Brodylo, Leslie A. Sedimentology of the Sparky Formation, Lower Cretaceous (Albian), Wainwright heavy oil pool, east-central Alberta. M, 1988, University of Calgary. 212 p.

Broecker, W. S. Present status of the lead method of age determination. M, 1954, Columbia University, Teachers College.

Broedel, Carl Huntington. The structures of the gneiss domes near Baltimore, MD. D, 1935, The Johns Hopkins University.

Broekstra, Bradley R. Diagenetic changes in a Cambrian shale as a function of burial depth. M, 1978, Georgia Institute of Technology. 150 p.

Broekstra, Scott Douglas. The genesis of the Onverdacht bauxite deposit at Moengo, Suriname. M, 1986, Indiana University, Bloomington. 142 p.

Broersma, K. Dark soils of the Victoria area, British Columbia. M, 1974, University of British Columbia.

Brogan, George E. Geology of the Pancake range near Duckwater, Nye County, Nevada. M, 1969, University of San Diego.

Brogan, John P. Geology of the Suplee area, Dayville Quadrangle, Oregon. M, 1952, Oregon State University. 139 p.

Brogden, William B., Jr. Modification of hydrocarbons and fatty acids in sediments by marine bacteria. M, 1968, Florida State University.

Brogdon, Dewey Robert. Beach sands of the Gulf Coast, northern Tamaulipas, Mexico. M, 1954, University of Texas, Austin.

Broida, Saul. Interpretation of geostrophy in the Straits of Florida. D, 1966, University of Miami. 119 p.

Broili, Christopher J. Geology and alteration of the southeastern part of Yankee Fork mining district, Custer County, Idaho. M, 1974, University of Idaho. 114 p.

Broin, Irene J. Historical geography of the Leadville mining district, Colorado. M, 1953, University of Colorado.

Broin, Thayne Leo. Geology of the Owl Canyon-Bellvue area, Larimer County, Colorado. M, 1952, University of Colorado.

Broin, Thayne Leo. Stratigraphy of the Lykins Formation (Permian, Triassic) of eastern Colorado. D, 1957, University of Colorado. 256 p.

Brojanigo, Antonio. Keweenaw Fault; structures and sedimentology. M, 1984, Michigan Technological University. 124 p.

Brokau, Arnold Leslie. Spores from Coal No. 5 (Springfield-Harrisburg) in Illinois. M, 1942, University of Illinois, Urbana.

Brokaw, Albert D. The solution and precipitation of gold in secondary enrichment of ore deposits. D, 1913, University of Chicago. 16 p.

Brokaw, Mark Alan. Upper crustal interpretation of Yellowstone determined from ray-trace modeling of seismic refraction data. M, 1985, University of Utah. 179 p.

Broman, Moritz L. Some experiments on replacement of minerals. M, 1928, University of Minnesota, Minneapolis. 40 p.

Broman, William H. Geophysical investigation of the Bear Lake area, Houghton County, Michigan. M, 1953, Michigan Technological University. 46 p.

Bromberg, Janet. Seismic risk in the central United States. M, 1979, University of Illinois, Chicago.

Bromberger, Samuel H. Basal Maquoketa phosphatic beds. D, 1968, University of Iowa. 209 p.

Bromberger, Samuel H. Mineralogy and petrology of basal Maquoketa (Ordovician) phosphatic beds, Iowa. M, 1965, University of Iowa. 159 p.

Bromble, Sandra L. Thermal metamorphic history of the Digdeguash Formation within the inner portion of the Pocomoonshine gabbro-diorite contact aureole, Big Lake Quadrangle, southeastern Maine. M, 1983, Queens College (CUNY). 113 p.

Bromery, Randolph W. Geological interpretation of aeromagnetic and gravity surveys of the northeastern end of the Baltimore-Washington anticlinorium, Harford, Baltimore, and part of Carroll counties, Maryland. D, 1968, The Johns Hopkins University. 156 p.

Bromery, Randolph Wilson. Aeromagnetic and gravimetric interpretation of the geology of the Malone, Rochester, Pe Ell, and Adna quadrangles, Pacific Lewis, Grays Harbor, and Thruston counties, Washington. M, 1962, American University. 44 p.

Bromfield, Calvin S. Geology of the Maudina Mine area, northern Santa Catalina Mountains, Pinal County, Arizona. M, 1950, University of Arizona.

Bromfield, Calvin Stanton. Geology of the Wilson Peak Stock; San Miguel Mountains, Colorado. D, 1962, University of Illinois, Urbana. 182 p.

Bromley, Bruce Warren. Intramolecular stable carbon isotopic distributions in Krebs cycle intermediates. D, 1981, Indiana University, Bloomington. 127 p.

Bromley, Karl Sydney. Stable isotope study of the Corral Canyon and Wasatch Fault shear zones, Utah. M, 1986, University of Utah. 111 p.

Bronaugh, Richmond Lee. Geology of Brazos River terraces in McLennan County, Texas. M, 1950, University of Texas, Austin.

Bronder, Joseph Bertram. Downward continuation applied to the theoretical magnetic data. M, 1957, Washington University. 30 p.

Brondos, Michael David. Distribution and paleoecology of ostracodes and foraminifers in phylloid-algal mound and continuous terrigenous deltaic facies of the Stanton Formation (Upper Pennsylvanian), southeastern Kansas. D, 1983, University of Kansas. 192 p.

Brondos, Michael David. Diversity of assemblages of late Paleozoic Ostracoda. M, 1974, University of Kansas. 46 p.

Bronner, Raymond L. Lithofacies and environmental interpretation of the Jeffersonville Limestone (Middle Devonian) in Jefferson County, Kentucky. M, 1984, Eastern Kentucky University. 88 p.

Bronson, Beth Joy. Geology of the Mono Syncline, southwestern San Rafael Mountains, Santa Barbara Co., California. M, 1986, University of California, Santa Barbara. 172 p.

Bronson, Brent R. An engineering analysis of the stability of Slide Mountain, Yosemite National Park, California. M, 1987, University of Nevada. 175 p.

Bronson, Roy DeBolt. Weathering sequence of micaceous minerals. D, 1959, Purdue University. 83 p.

Brook, Charles A. Saddlebag Lake roof pendant; Sierra Nevada; California. M, 1973, California State University, Fresno.

Brook, Doyle Kenneth, Jr. Relative ages of stratiform sulfide ore and wall rock determined by a structural analysis of the Copper Queen Mine, Yavapai County, Arizona. M, 1974, University of Arizona.

Brook, Edward J. Particle-size and chemical control of metals in Clark Fork River bed sediment. M, 1988, University of Montana. 126 p.

Brook, Hilary James. Evaluation of potential flood water detention sites for artificial recharge, Black Hills, South Dakota. M, 1981, South Dakota School of Mines & Technology. 56 p.

Brookby, Harry England. Upper Arbuckle (Ordovician) outcrops in the Richards Spur–Kindblade Ranch area, northeastern Wichita Mountains, Oklahoma. M, 1969, University of Oklahoma. 73 p.

Brooke, Gerald L. Sub-Huronian (Timiskaming) sediments of the Lake Enchantment area, Marquette County, Michigan. M, 1951, Michigan State University. 44 p.

Brooke, Jefferson Packard. Interpretation of cross-beds in the Bigby Facies of the Bigby-Cannon Formation (Ordovician) in the central basin of Tennessee. M, 1971, Vanderbilt University.

Brooke, John Percival. Alteration and trace elements in the San Francisco Mountains. D, 1964, University of Utah. 172 p.

Brooke, John Percival. Trace elements in pyrite from Bingham, Utah. M, 1959, University of Utah. 43 p.

Brooke, Margaret Martha. Jurassic microfaunas and biostratigraphy of Saskatchewan and north-central Montana. D, 1971, University of Saskatchewan. 413 p.

Brooke, Robert Clymer, Jr. Chert breccia in Pfeiffer-Big Sur State Park and vicinity, Monterey County, California. M, 1957, Stanford University.

Brooke, Robert Clymer, Jr. Pre-Mississippian paleogeology of Oklahoma and Arkansas. M, 1956, Stanford University.

Brooke, Robert Clymer, Jr. Stratigraphy and structure of the Point Sur area, Monterey County, California. M, 1957, Stanford University.

Brooker, Donald Duane. Pyrite porphyroblast paragenesis at the Cherokee Mine, Ducktown, Tennessee. M, 1984, Virginia Polytechnic Institute and State University. 113 p.

Brooker, E. J. The metamict state and a study of artificial and naturally occurring uranium oxides. M, 1951, University of Toronto.

Brooker, Edward James. The interpretation of the reciprocal lattice in randomly oriented single crystals. D, 1964, University of Toronto.

Brookes, Ian Alfred. The glaciation of southwestern Newfoundland. D, 1970, McGill University.

Brookins, D.G. Rubidium-strontium age investigations in the middle Haddam and Glastonbury quadrangles, Connecticut. D, 1963, Massachusetts Institute of Technology. 213 p.

Brookley, Arthur Clifford. The paleontology of the Beech Creek (Chester) Limestone of Indiana. M, 1955, Indiana University, Bloomington. 105 p.

Brookner, Paul L. Lithofacies produced by rapid retreat of the Casement Glacier, Glacier Bay National Park, Alaska. M, 1986, University of Nebraska, Lincoln. 142 p.

Brooks, A. James. Stratigraphy and sedimentology of the Bryon Formation, Silurian, east-central Wisconsin. M, 1978, University of Wisconsin-Madison. 193 p.

Brooks, B. G. The geology of the Wheatland Reservoir area, Albany County, Wyoming. M, 1957, University of Wyoming. 42 p.

Brooks, Betty Watt. Fossil plants from the Payette Formation, Snake River basin, Idaho, with reference to their distribution. D, 1934, University of Pittsburgh.

Brooks, D. A. Wind-forced continental shelf waves in the Florida Current. D, 1975, University of Miami. 281 p.

Brooks, Elwood Ralph. Geology of the Bridalveil Creek area, Yosemite National Park, California. M, 1958, University of California, Berkeley. 115 p.

Brooks, Elwood Ralph. Nature and origin of the Grenville Front north of Georgian Bay, Ontario. D, 1964, University of Wisconsin-Madison. 330 p.

Brooks, Frank Leroy. A study of the substratum of a portion of the continental shelf of the northeastern Gulf of Mexico. M, 1962, Mississippi State University. 65 p.

Brooks, Gregg R. Recent carbonate sediments of the Florida middle ground reef system; northeastern Gulf of Mexico. M, 1981, University of South Florida, St. Petersburg. 164 p.

Brooks, Gregg R. Recent continental slope sediments and sedimentary processes bordering a non-rimmed carbonate platform; southwestern Florida continental margin. D, 1986, University of South Florida, St. Petersburg. 166 p.

Brooks, Howard. Geology of a uranium deposit in Virginia Mountains, Washoe County, Nevada. M, 1956, University of Nevada - Mackay School of Mines. 50 p.

Brooks, Irving Harvey. Lake currents associated with the thermal bar. D, 1971, Case Western Reserve University.

Brooks, J. M. Sources, sinks, concentrations and sublethal effects of light aliphatic and aromatic hydrocarbons in the Gulf of Mexico. D, 1975, Texas A&M University. 362 p.

Brooks, Jack Alexander. Interfacial tension and its relations to the underground movement of oil. M, 1940, Texas Christian University. 86 p.

Brooks, James E. Facies of the Salem Limestone in the Eastern Interior basin. M, 1950, Northwestern University.

Brooks, James Elwood. Regional Devonian stratigraphy in central and W. Utah. D, 1954, University of Washington. 193 p.

Brooks, James Mark. The distribution of organic carbon in the Brazos river basin (Texas). M, 1970, Texas A&M University.

Brooks, James Rolland. Contamination of wallrock adjoining granite pegmatites. M, 1956, South Dakota School of Mines & Technology.

Brooks, Jeffrey W. A mineralogical, fluid inclusion, and oxygen isotope study of the Mammoth Revenue Vein, Platoro Caldera, San Juan Mountains, Colorado. M, 1986, Washington State University. 206 p.

Brooks, John A. Geology of New Hampshire's inner continental shelf. M, 1985, University of New Hampshire. 137 p.

Brooks, John E. A study in seismicity and structural geology. M, 1959, Boston College.

Brooks, Kenneth J. Petrology and depositional environments of the Hanna Formation in the Carbon Basin, Carbon County, Wyoming. M, 1977, University of Wyoming. 105 p.

Brooks, Lon Clyde. Biostratigraphy of the Purgatoire Formation, west central Quay County, New Mexico. M, 1959, Texas Tech University. 123 p.

Brooks, Mark W. Reconnaissance geochemical survey of the west flank of the Gravelly Range, Madison County, Montana. M, 1984, Wright State University. 132 p.

Brooks, Paul L., Jr. Geologic photography. M, 1950, University of Houston.

Brooks, Rebekah. Distribution and concentration of metals in sediments and water of the Clark Fork River floodplain, Montana. M, 1988, University of Montana. 105 p.

Brooks, Richard O. The effect of photo-lineaments and season on water chemistry of the Boone-St. Joe Aquifer of Benton County, Arkansas. M, 1979, University of Arkansas, Fayetteville.

Brooks, Robert Alexander. Geochemical and mineralogical relationships in the sediments on the Louisiana continental shelf. D, 1970, Louisiana State University. 135 p.

Brooks, Robert Alexander. Sedimentary geochemistry and clay mineralogy of Lake Pontchartrain and Lake Maurepas, Louisiana. M, 1969, Louisiana State University.

Brooks, Robert Andrew, Sr. A bottom gravity survey of the shallow water regions of southern Monterey Bay and its geological interpretation. M, 1973, United States Naval Academy.

Brooks, Thomas David. Hydrology of the proposed North Henry Mine, southeastern Idaho. M, 1982, University of Idaho. 46 p.

Brooks, Warren W. Recent foraminifera from the southern coast of Puerto Rico. M, 1971, University of Houston.

Brooks, William Earl, Jr. Stratigraphy and structure of the Columbia River Basalt in the vicinity of Gable Mountain, Benton County, Washington. M, 1974, University of Washington. 39 p.

Brooks, William Earl, Jr. Volcanic stratigraphy of part of McLendon Volcano, Yavapai County, Arizona. D, 1982, University of Washington. 139 p.

Broom, B. B. Mineralogy and geochemistry of Upper Jurassic evaporites, Southwest Alabama. M, 198?, University of Missouri, Columbia.

Broome, James Richard. Geology of the Denison Dam, Oklahoma-Texas Quadrangle, Bryan and Marshall counties, Oklahoma and Grayson County, Texas. M, 1983, University of Texas, Arlington. 142 p.

Broomfield, Robin E. Structural geology and igneous petrology of the Canones area, Rio Arriba County, New Mexico. M, 1977, University of New Mexico. 75 p.

Broomhall, Robert W. Geology of Dry Creek-West Fork area, central Arizona. M, 1978, Northern Arizona University. 99 p.

Brophy, Gerald P. Geology and hydrothermal alteration, Papsy's Hope Prospect, Marysvale, Utah. M, 1953, Columbia University, Teachers College.

Brophy, Gerald Patrick. Hydrothermal alteration and uranium mineralization in the Silica Hills area, Marysvale District, Utah. D, 1954, Columbia University, Teachers College. 210 p.

Brophy, James G. Geology of the Virgin Valley-Rock Springs Table area, Humboldt County, Nevada. M, 1980, Colorado School of Mines. 147 p.

Brophy, James Gerald. The chemistry and physics of Aleutian Arc volcanism; the Cold Bay volcanic center, southwestern Alaska. D, 1984, The Johns Hopkins University. 438 p.

Brophy, John Allen. Mineralogy of Sangamon weathering profiles. D, 1958, University of Illinois, Urbana. unpaginated p.

Brophy, John Allen. The ostracode genus Hollinella. M, 1949, University of Illinois, Urbana. unpaginated p.

Broscoe, Andy J. Quantitative analysis of longitudinal stream profiles of small watersheds. D, 1960, Columbia University, Teachers College.

Brosnahan, David Ramsey. High temperature single-crystal X-ray diffraction study of monoclinic chlorapatite. M, 1977, University of Michigan.

Bross, Gerald L. Distribution of Layton Sandstone (Pennsylvanian), Logan County, Oklahoma. M, 1960, University of Oklahoma. 40 p.

Brossard, L. Geology of the Beaufort Mine, Pascalis and Louvicourt townships, Quebec. M, 1940, McGill University.

Brosseau, Hubert N. The geology of the Glen Almond area, Papineau County, Quebec. M, 1961, University of Kansas. 39 p.

Brossman, James J. Surficial deposits of Kendall County, Illinois. M, 1982, Northern Illinois University. 66 p.

Brost, Roger L. A mineralogical study of some Helderberg rocks (Devonian) of the central Hudson Valley, New York. D, 1965, Rensselaer Polytechnic Institute.

Broster, Bruce Elwood. Compositional variations in the St. Joseph Till units in the Goderich area. D, 1982, University of Western Ontario. 307 p.

Brothers, Jack Anthony. The physical, chemical and ion-exchange properties of laumontite from the Osceola No. 6 mine, Houghton County, Michigan. M, 1967, Michigan Technological University. 55 p.

Brothers, James. The Antlers Sand in North Texas. M, 1984, Baylor University. 163 p.

Brothers, Sara C. Theoretical phase relations in the assemblage; rutile, ilmenite (hematite), aluminum silicate, quartz (RISQ); implications for geothermometry and oxygen barometry. M, 1987, University of New Mexico. 77 p.

Brotherton, Mark Allison. Geology of aggregate resources in the Brazos and Colorado floodplains on the coastal plains of Texas. M, 1982, Texas A&M University.

Brott, Charles Arthur. Heat flow and tectonics of the Snake River plain, Idaho. D, 1976, Southern Methodist University. 197 p.

Broucker, Gilles De see De Broucker, Gilles

Broughan, Fionnuala M. Palaeoenvironments of the Eastend, Whitemud (Maestrichtian), and Ravenscrag (Palaeocene) formations, eastern Cypress Hills, Saskatchewan. M, 1984, University of Saskatchewan. 178 p.

Broughton, D. W. Deformation, alteration and vein emplacement at the Davidson-Tisdale Mine, Timmins. M, 1987, University of Waterloo. 202 p.

Broughton, John Gerard. Comparison of Precambrian and Paleozoic structures in northwestern New Jersey. D, 1940, The Johns Hopkins University. 119 p.

Broughton, John Gerard. Petrology of the Sugarloaf-St. Kevin mining district, Lake County, Colorado. M, 1938, University of Rochester. 78 p.

Broughton, Martin Napoleon. The relation of crushing strength to mineral content in rocks at Hamilton dam site, Burnet County, Texas. M, 1931, University of Texas, Austin.

Broughton, Paul L. Structural geology of the Pulaski-Salem thrust sheet and the eastern end of the Christiansburg Window, southwestern Virginia. D, 1971, Virginia Polytechnic Institute and State University.

Broughton, Paul L. Structural geology of the Pulaski-Salem thrust sheet and the eastern end of the Christiansburg window, southwestern Virginia. M, 1971, Virginia Polytechnic Institute and State University.

Brouillard, Lee A. Geology of the northeastern Gallinas Mountains, Socorro County, NM. M, 1984, New Mexico Institute of Mining and Technology. 161 p.

Broussard, Marion U. Vanderpoolia, a new genus of Ostracoda from the Gulf Coast of North America. M, 1932, Louisiana State University.

Broussard, Matthew C. Chester depositional systems (Upper Mississippian) of the Black Warrior Basin. M, 1979, University of Mississippi. 164 p.

Browder, George Thomas. The geology of Shaws Park, Fremont County, Colorado. M, 1958, University of Oklahoma. 110 p.

Brower, Gilbert K. The influence of differential porosity on oil sand erosion. M, 1930, University of Pittsburgh.

Brower, James Clinton. Evolution and classification of selected actinocrinitids. D, 1964, University of Wisconsin-Madison. 160 p.

Brower, James Clinton. Sedimentation and sedimentary tectonics of the stratigraphic and structural development of Little North Mountain, Virginia. M, 1961, American University. 107 p.

Brower, John Charles. Geology of the East Fork Mine and vicinity, Sevier County, Tennessee. M, 1973, University of Tennessee, Knoxville. 112 p.

Brower, Ross Dean. The hard part morphology of Oreastir reticulatis (living starfish from Bahamas). M, 1969, University of Illinois, Urbana.

Brown, A. S. Detailed structural studies of Wells-Round Top Mountain (Cariboo), 1 inch to 1000 feet (Canada). D, 1952, [Princeton University].

Brown, Aaron Donald. Chemical weathering of pyrite in soils. D, 1985, Utah State University. 191 p.

Brown, Albert Anthony. The southern margin of the Springhill coal basin, Nova Scotia. M, 1950, University of New Brunswick.

Brown, Alexander Cyril. Mineralogy at the top of the cupriferous zone, White Pine Mine, Ontonagon County, Michigan. M, 1965, University of Michigan.

Brown, Alexander Cyril. Zoning in the White Pine copper deposit (late Precambrian), Ontonagon County, Michigan. D, 1968, University of Michigan. 234 p.

Brown, Alfred Louis. Geology of the Round Mountain Intrusive, Gunnison County, Colorado. M, 1950, University of Kentucky. 27 p.

Brown, Alison Y. Reprocessing and interpretation of USGS Tennessee-North Carolina seismic lines 3 and 5. M, 1986, University of Wyoming. 161 p.

Brown, Alton Arthur. Studies of deposition and diagenesis of the Canyon Group (Pennsylvanian, Missourian) of Texas. D, 1980, Brown University. 259 p.

Brown, Alton R. Geology of Wenter-Unger pool area, Marion County, Kansas. M, 1959, Kansas State University. 52 p.

Brown, Amalie Jo. Channel profiles in arid badlands near Borrego Springs, California. M, 1978, University of California, Los Angeles. 56 p.

Brown, Amalie Jo. Space and time relationships on Ventura County beaches, California. D, 1983, University of California, Los Angeles. 179 p.

Brown, Amos Peaslee. A comparative study of the chemical behavior of pyrite and marcasite. D, 1893, University of Pennsylvania.

Brown, Anton. fracture analysis in Opemiska Mine area (northwestern Quebec). D, 1971, Queen's University. 415 p.

Brown, Arthur R. Geology of a portion of the southeastern San Jacinto Mountains, Riverside County, California. M, 1968, University of California, Riverside. 95 p.

Brown, Bahngrell Walter. A study of the northern Black Hills Tertiary petrogenic province with notes on the geomorphology involved. D, 1954, University of Nebraska, Lincoln. 131 p.

Brown, Bahngrell Walter. A study of the southern Bear Lodge Mountains intrusive. M, 1952, University of Nebraska, Lincoln. unpaginated.

Brown, Bert N. Geology of the Sedillo-Cedro Canyon area, Bernalillo County, New Mexico. M, 1962, University of New Mexico. 64 p.

Brown, Billy H. Recent sedimentation in Rockefeller Wildlife Refuge, Cameron Parish. M, 1964, University of Southwestern Louisiana.

Brown, Bobby Joe. Sediment transport in hyperconcentrated flows in sand-bed streams of volcanic origin. D, 1987, Colorado State University. 188 p.

Brown, Bruce A. Geology of the Hanis Lake-Cochrane River area, Saskatchewan-Manitoba. M, 1971, University of Oregon. 111 p.

Brown, Bruce A. The role of granite diapirism in the deformational history of Archean greenstones of the central Lake of the Woods area, northwestern Ontario. D, 1984, University of Manitoba.

Brown, Bruce E. The crystal structure of chromium chlorite. M, 1959, University of Wisconsin-Madison.

Brown, Bruce Elliot. The crystal structure of maximum microcline. D, 1962, University of Wisconsin-Madison. 120 p.

Brown, Calvin C. Experimental blasting in the Cananea open pit mine. M, 1972, University of Arizona.

Brown, Calvin Smith. Contributions to the coal flora of Tracy City, Tennessee. D, 1892, Vanderbilt University.

Brown, Carl B. A preliminary geological survey of the Wadesboro Triassic area. M, 1931, University of Cincinnati. 132 p.

Brown, Carolyn Ruth. Analysis and interpretation of Magsat anomalies over North Africa. M, 1985, Southern Methodist University. 96 p.

Brown, Carroll Parker. Analysis of magnetic anomalies of the Lahore Quadrangle, Virginia. M, 1971, Virginia State University. 46 p.

Brown, Charles Douglas. Depositional systems and provenance of the Mississippian middle Stanley Group, southern Ouachita Mountains, Arkansas. M, 1986, Southern Methodist University. 127 p.

Brown, Charles Edward. Environmental geology for planning in the Branford Quadrangle, Connecticut. M, 1974, Pennsylvania State University, University Park.

Brown, Charles Edward. Multivariate analysis of petrographic and chemical properties influencing porosity and permeability in selected carbonate aquifers in central Pennsylvania. D, 1977, Pennsylvania State University, University Park. 220 p.

Brown, Charles K. An acid-etching study of the Mascot Member of the Knox Dolomite at Thorn Hill and Lee Valley, Tennessee. M, 1956, University of Tennessee, Knoxville. 52 p.

Brown, Charles Michael. The lithostratigraphy and depositional history of the Marble Falls Formation (Pennsylvanian) in the subsurface immediately north of the Llano Uplift, central Texas. M, 1983, University of Oklahoma. 215 p.

Brown, Charles N. The origin of caliche on the northeastern Llano Estacado, Texas. D, 1953, University of Chicago. 15 p.

Brown, Charles Quentin. Clay mineralogy of sediments and source materials in the York River tributary basin. D, 1959, Virginia Polytechnic Institute and State University.

Brown, Charles Quentin. The clay minerals of the Neuse River sediments. M, 1953, University of North Carolina, Chapel Hill. 16 p.

Brown, Charles William. Stratigraphic and structural geology of north-central-northeast Yellowstone National Park, Wyoming and Montana. D, 1957, Princeton University. 222 p.

Brown, Clifford L., III. Geology of the northwest Wortham area, Navarro, Limestone, and Freestone counties, Texas. M, 1966, Texas A&M University. 66 p.

Brown, Cyril Benjamin. Influence of layer charge on X-ray properties of muscovite. D, 1961, Purdue University. 90 p.

Brown, D. Studies in matrix methods modelling microseismic excitation of a layered oil sand. M, 1988, University of Alberta.

Brown, D. A. Toxicology of trace metals; metallothionen production and carcinogenesis. D, 1978, University of British Columbia.

Brown, Daniel R. Hydrogeology and water resources of the Aztec Quadrangle, San Juan County, New Mexico. M, 1976, New Mexico Institute of Mining and Technology. 174 p.

Brown, Darren Leo. The stratigraphy and petrography of the Wapanucka Formation along the northeastern flank of the Arbuckle Mountains in southern Oklahoma. M, 1987, University of Oklahoma. 127 p.

Brown, David Edward. A study of the petrology and structure of a portion of the Boyds Creek mafic-ultramafic complex, Chambers County, Alabama. M, 1982, Auburn University. 155 p.

Brown, David McKendree. Geology of the southern Bear Mountains, Socorro County, New Mexico. M, 1972, New Mexico Institute of Mining and Technology. 110 p.

Brown, David P. Marine geology of Cay Sal Bank, Bahamas. M, 1972, University of South Florida, Tampa. 102 p.

Brown, Delmer. Geology of the Mount Bross-Mineral Park area, Park County, Colorado. M, 1962, Colorado School of Mines. 83 p.

Brown, Dennis Lewis. A structural analysis of the Grenville Front zone, northeast Gagnon terrane, western Labrador. M, 1981, Memorial University of Newfoundland. 190 p.

Brown, Derek Anthony. Geological setting of the volcanic-hosted Silbak Premier Mine, northwestern British Columbia (104A/4,B/1). M, 1987, University of British Columbia. 203 p.

Brown, Don M. Sedimentology of the Upper Jurassic-Lower Cretaceous Hibernia Member of the Missisauga Formation in the Hibernia oil field, Jeanne d'Arc Basin. M, 1986, Carleton University. 157 p.

Brown, Donald D. The Canoe-Desert Lake wrench fault and associated structure. M, 1961, Queen's University. 49 p.

Brown, Donald Dawson. Hydrogeology of Taylor Island, New Brunswick (Mississippian-Pennsylvanian). D, 1971, University of Western Ontario. 215 p.

Brown, Donald L. Investigation and development of ground water of Chalone Creek (Pinnacles National Monument, Calif. M, 1962, University of Missouri, Rolla.

Brown, Donald Marvin. The Pleistocene geology of Clark County, Ohio. M, 1948, Ohio State University.

Brown, Dorothy Claire. Fold geometry in the Belt Sequence, Salmon, Idaho. M, 1973, Franklin and Marshall College. 71 p.

Brown, Douglas F. An analysis of the Saint George "Black Granite" (New Brunswick). M, 1938, University of New Brunswick.

Brown, Dwight Alan. Erosion and morphometry of small drainage basins in eastern Puerto Rico. D, 1969, University of Kansas. 139 p.

Brown, Dwight Delon. Palynology of the Hannibal Formation (lower Mississippian) of northeast Missouri and western Illinois. D, 1969, University of Missouri, Columbia. 293 p.

Brown, Earl D. The geography of the Chippewa Valley. M, 1922, University of Wisconsin-Madison.

Brown, Edward Charles. Tiltmeter analysis of Mount St. Helens, Skamania County, Washington. M, 1984, Portland State University. 153 p.

Brown, Edwin Hacker. Metamorphic facies and structure of a region in eastern Otago, New Zealand. D, 1966, University of California, Berkeley. 160 p.

Brown, Ethan Douglas. Frequency dependent Q coda in the Livermore Valley region, California. M, 1985, University of California, Santa Cruz.

Brown, Ethan Douglas. Seismicity patterns associated with the 1973 Colima and 1981 Playa Azul, Mexico earthquakes. M, 198?, University of California, Santa Cruz.

Brown, Ethan Douglas. The May 2, 1983 Coalinga, California earthquake (M_1 = 6.7) sequence; evidence for a complex aftershock zone. M, 198?, University of California, Santa Cruz.

Brown, Eugene. Study of the lithofacies, thickness and pre-Cretaceous geology of a portion of the Mississippi Embayment. M, 1957, Stanford University.

Brown, Eugene. The geochemistry of the ground waters of northeastern Florida and southeastern Georgia. D, 1952, University of Florida. 151 p.

Brown, Eugene H. Geology of the Rancho San Carlos area, Monterey County, California. M, 1962, Stanford University.

Brown, Francis Harold. Volcanic petrology of the Toro-Ankale region, western Uganda. D, 1971, University of California, Berkeley. 152 p.

Brown, G. H. A structural and stratigraphic study of the keewatin-type and shebandowan-type rocks, west of Thunder Bay. M, 1985, Lakehead University.

Brown, George Donald. Trepostomatous Bryozoa from the Logana and Jessamine limestones (Mohawkian) of the Kentucky Bluegrass region. D, 1963, Indiana University, Bloomington. 104 p.

Brown, George Donald, Jr. Some Upper Cretaceous foraminifera of northern Mississippi and southern Tennessee. M, 1955, University of Illinois, Urbana.

Brown, George Earl. Geology of parts of the Calabasas and Thousand Oaks quadrangles, Los Angeles and Ventura counties, California. M, 1957, University of California, Los Angeles.

Brown, George W. Some physical and chemical soil properties as possible causes to piping erosion. M, 1961, Colorado State University. 67 p.

Brown, Gilbert Daleth. Subsurface study of the Puryear Formation of the upper Morrow Group in the eastern Texas Panhandle. M, 1979, West Texas State University. 77 p.

Brown, Glen Arthur. Geology of the Mahoney Mine, Gym Peak area, Florida Mountains, Luna County, New Mexico. M, 1982, New Mexico State University, Las Cruces. 82 p.

Brown, Glen F. The geology and ground water of Al Kharj District, Nejd, Saudi Arabia. D, 1949, Northwestern University.

Brown, Glen Francis. Late Tertiary sediments in South Park, Colorado. M, 1940, Northwestern University.

Brown, Gordon Edgar, Jr. Crystal chemistry of the olivines. D, 1970, Virginia Polytechnic Institute and State University. 142 p.

Brown, Gordon Edgar, Jr. Crystal structure of osumilite. M, 1968, Virginia Polytechnic Institute and State University.

Brown, H. Gassaway, IV. Petrology and geochemistry of shonkinite, syendiorite, and nepheline monzonite of the Island of Kauai, Hawaii. M, 1977, University of New Mexico. 173 p.

Brown, Harry L. Loess deposits of northern Arkansas. M, 1958, University of Arkansas, Fayetteville.

Brown, Henry Seawell. Anorthite variations around titaniferous iron deposits, Laramie Range, Wyoming. M, 1954, University of Illinois, Urbana.

Brown, Henry Seawell. Geology of the Ore Knob and Elk Knob copper deposits, North Carolina. D, 1952, University of Illinois, Urbana. 196 p.

Brown, Hugh Matiland. A theoretical and experimental study of germanium isotope fractionation. M, 1962, University of Alberta. 94 p.

Brown, I. C. Brucite deposits in Grenville limestone. M, 1942, Queen's University. 41 p.

Brown, Ian Roderick. Behaviour of the Bay Area Rapid Transit tunnels through the Hayward Fault. D, 1981, University of California, Berkeley. 208 p.

Brown, Ira Charles. Structure of the Yellowknife gold belt, Northwest Territories. D, 1949, Harvard University.

Brown, Ira Otho. Limestone sinks in the Bloomington Indiana, Quadrangle. M, 1920, University of Chicago. 64 p.

Brown, Irvin Cecil. A sedimentary study of some Mississippi Valley soils. D, 1929, University of Iowa. 127 p.

Brown, Irving Foster. The nitrogen cycle and heat budget of a subtropical lagoon, Devil's Hole, Harrington Sound, Bermuda; implications for nitrous oxide production and consumption in marine environments. D, 1981, Northwestern University. 338 p.

Brown, Isobel Julia. Gold-bismuth-copper skarn mineralization in the Marn Skarn, Yukon. M, 1985, University of Alberta. 158 p.

Brown, J. R. Adsorption of metal ions by calcite and iron sulphides; a quantitative X-ray photoelectron

spectroscopy study. D, 1978, University of Western Ontario. 220 p.

Brown, J. W. Stratigraphy, petrology, and depositional history of the Kaibab Formation (Permian) between Cameron and Desert View, Coconino County, Arizona. M, 1969, University of Arizona.

Brown, Jack Ralston. The geology of the Kent area, Culberson and Jeff Davis counties. M, 1948, University of Texas, Austin.

Brown, James Alexander, Jr. Thrust contact between Franciscan Group (Jurassic-Cretaceous) and Great Valley sequence northeast of Santa Maria, California. D, 1968, University of Southern California. 273 p.

Brown, James Eugene. Relationship of structure to reservoir petrology in some middle Miocene sands of a producing South Louisiana oil field (Jefferson Parish). M, 1973, North Carolina State University. 61 p.

Brown, James Harrison, Jr. The mineralogy of a radioactive shale from near Ste. Genevieve, Missouri. M, 1952, University of Missouri, Columbia.

Brown, James Lee. Comparison of heavy minerals in the Hillsboro, Sylvania, and Oriskany sandstones and the clastic zone at the base of the Ohio Shale. M, 1951, University of Cincinnati. 31 p.

Brown, James Lee, Jr. Paleoenvironment and diagenetic history of the Moffat Mound, Edwards Formation, central Texas. M, 1975, Louisiana State University.

Brown, James O. Search for a western North American stratotype for the Lower-Middle Devonian boundary in Eureka County, Nevada. M, 1982, Oregon State University. 75 p.

Brown, James Oliver, Jr. The paleoecology of the Palliseria facies of the Antelope Valley Limestone. M, 1981, University of California, Santa Barbara. 97 p.

Brown, James Peter, Jr. Geophysical investigations of the Empire cave, north of Santa Cruz, California. M, 1970, San Jose State University. 59 p.

Brown, Janet L. Sedimentology of a late Pleistocene and Holocene continental and estuarine deposit, Mountain View, CA. M, 1978, San Jose State University. 74 p.

Brown, Jean C. The relation between lattice constants and composition in the scheelite-powellite series. M, 1945, Columbia University, Teachers College.

Brown, Jeffrey. Origin and hydrodynamic history of quartz sand on the Southeastern United States continental shore; Fourier grain shape analysis. D, 1978, University of South Carolina.

Brown, Jeffrey. Variations in South Carolina coastal morphology. M, 1975, University of South Carolina.

Brown, Jeffrey C. Paleoecology of a Lower Permian crinoid locality, Battleship Wash, Clark County, Nevada. M, 1973, Washington State University. 92 p.

Brown, Jeffrey Scott. A structural investigation of a basement-involved thrust system in southern Sphinx Mountain Quadrangle, (Madison Range) southwestern Montana. M, 1986, Western Michigan University.

Brown, Jennifer Ruth. Variations in volcanic processes along the median valley of the Mid-Atlantic Ridge; 22°50'N to 23°40'N. M, 1987, Duke University. 118 p.

Brown, Jim McCaslin. Bedrock geology and ore deposits of the Pedro Dome area, Fairbanks mining district, Alaska. M, 1963, University of Alaska, Fairbanks. 137 p.

Brown, Jim McCaslin. Structure and origin of the Grenville front south of Coniston, Ontario (Canada). D, 1968, University of Wisconsin-Madison. 199 p.

Brown, John Michael. Computed tomography and magnetic resonance imaging of water distribution in plants and soils. D, 1987, North Carolina State University. 111 p.

Brown, John Michael. Thermodynamic properties for iron at very high pressures; implications for the Earth's core. D, 1980, University of Minnesota, Minneapolis. 149 p.

Brown, John S. Graphite deposits of Ashland, Alabama. D, 1925, Columbia University, Teachers College.

Brown, Johnnie Boyd. Geology, hydrology, and soil conservation on Flat Top Ranch, Bosque County, Texas. M, 1964, Baylor University. 30 p.

Brown, Judith Barbara Moody. A chemical and mineralogical study of some synthetic potassium-hydronium jarosites. M, 1967, University of Michigan.

Brown, K. Elizabeth. Stratigraphy and deposition of the Price Formation coals in Montgomery and Pulaski counties, Virginia. M, 1983, Virginia Polytechnic Institute and State University. 76 p.

Brown, Karen B. Geology of the southern Canoncito de la Uva area, Socorro County, NM. M, 1987, New Mexico Institute of Mining and Technology. 116 p.

Brown, Kenneth Alan. A palynostratigraphic and morphologic study of striate spores (Schizaeaceae) from the Potomac Group of the Atlantic Coastal Plain. M, 1976, University of Michigan.

Brown, L. S. Age of the Gulf border salt deposits. D, 1933, Massachusetts Institute of Technology. 145 p.

Brown, Larry Douglas. Recent vertical crustal movements from geodetic measurements; Alaska and the eastern United States. D, 1976, Cornell University. 194 p.

Brown, Lauren Shelley. Structure of the northern Cedar Mountains, west-central Nevada; a study utilizing balanced cross-sections and surface data. M, 1986, Rice University. 124 p.

Brown, Lawrence Edward. Genesis of the ores of the Jardine-Crevasse Mountain area, Park County, Montana. M, 1965, Kansas State University. 43 p.

Brown, Lawrence Gregory. Recent fault scarps along the eastern escarpments of the Sierra San Pedro Martir, Baja California. M, 1978, San Diego State University.

Brown, Leonard F., Jr. A paleoecologic study of foraminifera in selected Pennsylvanian cyclothems. D, 1955, University of Wisconsin-Madison.

Brown, Leonard Franklin, Jr. Description and stratigraphic relationships of a Carboniferous cephalopod fauna, Confusion Range, Utah. M, 1953, University of Wisconsin-Madison. 88 p.

Brown, Levi S. A petrographic description of a section of the Garber, Wellington and Stillwater sandstones in Cleveland and Pottawatamie counties, Oklahoma. M, 1928, University of Oklahoma. 53 p.

Brown, Lewis M. The stratigraphy and structure of the Herculaneum Quadrangle (Jefferson County) Missouri. M, 1967, University of Iowa. 86 p.

Brown, Lionel F. A gravity survey of central Dona Ana County, New Mexico. M, 1977, New Mexico State University, Las Cruces. 56 p.

Brown, Lynn A. Engineering geology; divide tunnel, Lake and Pitkin counties, Colorado. D, 1969, Colorado School of Mines. 205 p.

Brown, Lynn A. Properties and foundation problems of loess in the St. Louis area. M, 1954, Washington University. 32 p.

Brown, Mark A. Depositional and diagenetic history of a Middle Mississippian shoal and related facies, Salem Limestone, Lawrence County, Indiana. M, 1987, Indiana University, Bloomington. 165 p.

Brown, Mark Theodore. Energy basis for hierarchies in urban and regional landscapes. D, 1980, University of Florida. 359 p.

Brown, Michael Jonathan. Applications of hydrologic modeling to water resource assessment. D, 1983, Harvard University. 1759 p.

Brown, Michael L. Geology of Sierra de Los Chinos-Cerro La Cueva area, Northwest Chihuahua, Mexico. M, 1985, University of Texas at El Paso.

Brown, Michael P. An integrated geophysical survey over a water-filled sink system. M, 1976, University of South Florida, Tampa. 99 p.

Brown, Nicholas Arthur. Tracking down turbidites with trace elements. M, 1979, Duke University. 72 p.

Brown, Noel King, Jr. Upper Cretaceous foraminifera from Seco Creek, Medina County, Texas. M, 1952, University of Texas, Austin.

Brown, Norman Elwood. The geology of Buchans Junction area of Newfoundland. M, 1952, McGill University.

Brown, Norman James. The geology of the Stillwater, Minnesota Quadrangle. M, 1956, University of Minnesota, Minneapolis. 161 p.

Brown, Orman Presley. The geology of the Difficulty-Little Shirley Basin area, Carbon County, Wyoming. M, 1939, University of Wyoming. 22 p.

Brown, Owen Cleveland, Jr. Geology of the Verona area, Collin County, Texas. M, 1949, Southern Methodist University. 21 p.

Brown, P. J. Origin and hydrodynamic history of quartz sand on the southeastern United States continental shelf; Fourier grain shape analysis. D, 1978, University of South Carolina. 65 p.

Brown, Paul Henry. The geology of the Lower Silurian Albion Group of Mahoning County, Ohio. M, 1979, Kent State University, Kent. 90 p.

Brown, Paul Jeffrey. Coastal morphology of South Carolina. M, 1975, University of South Carolina.

Brown, Paul M. Gas-liquid chromatography of some alkyl phenols. M, 1957, West Virginia University.

Brown, Perry L. Stratigraphy and depositional environment of Takoma Bluff rocks, east-central Alaska. M, 1983, University of Alaska, Fairbanks. 81 p.

Brown, Peter Alan. Basement-cover relationships in Southwest Newfoundland. D, 1975, Memorial University of Newfoundland. 220 p.

Brown, Peter Alan. Structural and metamorphic history of the gneisses of the Port aux Basques Region, Newfoundland. M, 1973, Memorial University of Newfoundland. 102 p.

Brown, Peter McKay. A paleomagnetic study of Piedmont metamorphic rocks from northern Delaware. M, 1982, University of Michigan.

Brown, Peter McKay. Paleomagnetism of the latest Precambrian-Cambrian Unicoi Basalts from the Blue Ridge, Northeast Tennessee and Southwest Virginia; evidence for Taconic deformation. M, 1982, University of Michigan.

Brown, Philip Edward. A petrologic and stable isotopic study of skarn formation and mineralization at the Pine Creek, California tungsten mine. D, 1980, University of Michigan. 250 p.

Brown, Philip Edward. Sphalerite geobarometry in the Balmat-Edwards District, New York. M, 1976, University of Michigan.

Brown, Prescott L. Occurrence and origin of the known trona deposits in Sweetwater and Uinta counties, Wyoming. M, 1950, University of Wyoming. 82 p.

Brown, R. A. C. Preliminary study of Acanthoscaphites nodosus in the Upper Cretaceous of Western Canada. M, 1943, University of Toronto.

Brown, R. A. C. Upper Paleozoic stratigraphy and paleontology in the Mt. Greenock area (Alberta). D, 1950, University of Toronto.

Brown, R. W. A computer study of lower Manville channel sands in SE Alberta. M, 1975, [University of Alberta].

Brown, Ralph Edward. A multi-layered finite element model for predicting mine subsidence. D, 1968, Carnegie-Mellon University. 182 p.

Brown, Ralph H. The economic geography of the middle Connecticut Valley of Massachusetts. D, 1925, University of Wisconsin-Madison.

Brown, Ralph Sherman. Geology of the Payson Canyon-Picayune Canyon area, southern Wasatch Mountains, Utah. M, 1950, Brigham Young University.

Brown, Randall B. Genesis of soils in the central western Cascades of Oregon. M, 1975, Oregon State University. 172 p.

Brown, Randall E. Geology and petrography of the Mount Washington area, Oregon. M, 1941, Yale University. 139 p.

Brown, Raymon Lee. A dislocation approach to plate interaction. D, 1975, Massachusetts Institute of Technology. 449 p.

Brown, Richard L., Jr. The geology of the Fire Tower Mountain area, Middlebury, Vermont. M, 1954, Syracuse University.

Brown, Richard Shaw. The geology of the Grimes Canyon area, Moorpark and Fillmore quadrangles, Ventura County, California. M, 1959, University of California, Los Angeles.

Brown, Robert A. The geology of the north shore of Gaspe Bay, Quebec. D, 1939, McGill University.

Brown, Robert Bruce. Depositional history and diagenesis of the Noodle Creek Limestone (Wolfcampian), Wallace Ranch and North Rough Draw fields, Kent County, Texas. M, 1988, Stephen F. Austin State University. 120 p.

Brown, Robert D. The geology of the McMinnville Quadrangle, Oregon. M, 1951, University of Oregon. 54 p.

Brown, Robert Ernest. Bedrock topography, lithology, and glacial drift thickness of Lapeer and Saint Clair counties, Michigan. M, 1963, Michigan State University. 54 p.

Brown, Robert J. Petrography, microfacies, and reservoir potential of the Ely Limestone, in east-central Nevada and west-central Utah. M, 1986, University of Idaho. 120 p.

Brown, Robert James. Isostasy and crustal structure in the English river gneissic belt (northwestern Ontario, Canada). M, 1968, University of Manitoba.

Brown, Robert Lewis. Shallow ground-water systems beneath strip and deep coal mines at two sites, Clearfield County, Pennsylvania. M, 1971, Pennsylvania State University, University Park. 216 p.

Brown, Robert Ludger. Steady state crystallization from solutions of metal salts under hydrogen atmospheres at elevated temperatures and pressures. M, 1965, University of Oklahoma. 76 p.

Brown, Robert Parker. A regional gravity survey of the Sanpete-Sevier valley and adjacent areas in Utah. M, 1975, University of Utah. 73 p.

Brown, Robert Wesley. Folds of the Osage type, Osage County, Oklahoma. D, 1927, University of Chicago. 93 p.

Brown, Robert Wesley. The geology and occurrence of the lead and zinc ores of the Kennedy Mine near Hazel Green, Wisconsin. M, 1913, University of Illinois, Chicago.

Brown, Rodney Stuart. Computer analysis of sand and petroleum distribution in the Mannville Group, Turin area, southern Alberta. M, 1976, University of Alberta. 104 p.

Brown, Roger James Evan. Permafrost in Canada; its effects on development in a region of marginal human activity. D, 1961, Clark University. 450 p.

Brown, Roland Wilbur. Composition and environment of the Green River flora (Wyoming). D, 1926, The Johns Hopkins University.

Brown, Ronald G. Geochemical survey of the Oracle vicinity (Pinal County) of Arizona. M, 1970, Arizona State University. 50 p.

Brown, Ronald LaBern. Geology and ore deposits of the Twin Buttes District. M, 1926, University of Arizona.

Brown, Roy Harold. The Argana Basin of Morocco; a Triassic model for early rifting. M, 1974, University of South Carolina.

Brown, Sally Sue Liggett. General stratigraphy and depositional environment of the Elgin Sandstone (Pennsylvanian) in south-central Kansas. M, 1966, University of Kansas. 52 p.

Brown, Seward Ralph. Diagenesis of chlorophylls in lacustrine sediments. D, 1962, Yale University. 60 p.

Brown, Sharon-Dale. Environmental characteristics and sediment diatoms of 51 lakes on southern Vancouver Island and Saltspring Island. D, 1980, University of Victoria. 485 p.

Brown, Stephen H. Geology of the Mill Creek Mountain area, Garland County, Arkansas. M, 1982, University of Missouri, Columbia.

Brown, Stephen Phillip. Geometry and mechanical relationship of folds to a thrust fault using a minor thrust in the Front Ranges of the Canadian Rocky Mountains. M, 1976, University of Calgary. 176 p.

Brown, Stephen Ray. Fundamental study of the closure property of joints. D, 1984, Columbia University, Teachers College. 242 p.

Brown, Stephen W. The effect of algal photosynthesis on the rate of precipitation and crystal structure of calcium carbonate. M, 1974, Miami University (Ohio). 25 p.

Brown, T. C., Jr. An economic analysis of chaparral conversion on national forest lands in the Salt-Verde Basin, Arizona. M, 1973, University of Arizona.

Brown, T. J. Potash. M, 19??, Colgate University.

Brown, Thomas C. A new lower Tertiary fauna from Chappaquiddick Island, Martha's Vineyard. M, 1905, Columbia University, Teachers College.

Brown, Thomas D. Geology of the Brush Hollow area, Fremont County, Colorado. M, 1964, University of Kansas. 30 p.

Brown, Thomas E. Relationship between basin parameters and landform configuration, Lampasas cut plain, central Texas. M, 1988, Baylor University. 166 p.

Brown, Thomas E. Stratigraphy of the Washita group (lower Cretaceous) in central Texas. M, 1970, Baylor University. 133 p.

Brown, Thomas Edwards. Cation exchange reactions on size-fractionated montmorillonites. D, 1963, University of Texas, Austin. 127 p.

Brown, Thomas Edwards. Mineralogy and crystal chemistry of glauconite. M, 1958, University of Texas, Austin.

Brown, Thomas Howard. Theoretical predictions of equilibria and mass transfer in the system $CaO-MgO-SiO_2-H_2O-CO_2-NaCl-HCl$. D, 1970, Northwestern University.

Brown, Thomas William. The formation of pedogenic calcrete; its stratigraphic and diagenetic significance in the Quaternary limestones on San Salvador Island, Bahamas. M, 1986, Indiana University, Bloomington. 226 p.

Brown, Timothy Reed. Eskers and heavy mineral prospecting, northeastern Minnesota. M, 1988, University of Minnesota, Duluth. 103 p.

Brown, Timothy S. Geology of the Owdoms Quadrangle, South Carolina. M, 1968, University of South Carolina.

Brown, Vernon Max. Metamorphic history of the older granites, central Rhode Island. M, 1976, University of Oklahoma. 110 p.

Brown, Vernon Max. The Precambrian volcanic stratigraphy and petrology of the Des Arc NE 7 1/2 minute Quadrangle, south central St. Francois Mountains, Missouri. D, 1983, University of Missouri, Rolla. 235 p.

Brown, W. R. An analysis of seismic noise in conjunction with the noise profile performed in Creek County, Oklahoma. M, 1984, Wright State University. 66 p.

Brown, Walter Emerson. Digital data acquisition from tidal recording gravimeters for investigation of secular changes in gravity and free oscillations of the Earth. D, 1986, Columbia University, Teachers College. 212 p.

Brown, Walter William. Geochemistry of selected soil profiles developed on ultramafic rocks, Del Norte County, California. D, 1965, Stanford University. 185 p.

Brown, Warren L. Land measurements in Louisiana. M, 1950, [University of Houston].

Brown, William Donald. Stratigraphy and petrology of Cenozoic volcanics, Zacatecas, Mexico. M, 1976, University of New Orleans. 119 p.

Brown, William G. Stratigraphy of the Beil Limestone, Virgilian of eastern Kansas. M, 1958, University of Kansas. 189 p.

Brown, William Gregor. Structural style of the Laramide Orogeny, Wyoming foreland (Volumes I and II). D, 1987, University of Alaska, Fairbanks. 619 p.

Brown, William Horatio. The mineral zones of the Whitecross District and neighboring deposits in Hinsdale County, Colorado. D, 1924, University of Minnesota, Minneapolis. 69 p.

Brown, William J. Geology and mineral deposits of the Bauer Mine and vicinity, Jo Davies County, Illinois. M, 1970, Colorado School of Mines. 82 p.

Brown, William Lee. Conodonts from the basal sandstone of the Noel Shale, southwestern Missouri. M, 1959, University of Missouri, Columbia.

Brown, William Lester. A study of rocks associated with radium deposits. M, 1932, University of Toronto.

Brown, William Lester. Luminescence in minerals. D, 1937, University of Toronto.

Brown, William Lindop. Pleistocene and ground-water geology of Surrey Municipality and the western part of Delta Municipality, British Columbia, Canada. M, 1953, University of Kansas.

Brown, William Randall. Coastal-plain geology of the Richmond area, Virginia. M, 1939, University of Virginia. 111 p.

Brown, William Randall. Geology of the Lynchburg-Rustburg area, Virginia. D, 1942, Cornell University.

Brown, William T., Jr. Igneous geology of the Rio Puerco necks (Mio-Pliocene?) Sandoval and Valencia counties, New Mexico. M, 1969, University of New Mexico. 89 p.

Brown, William W. Flare triangulation as a rapid survey technique. M, 1962, Ohio State University.

Brown, Willis Reider. Geology of Artist Point and vicinity south side of Mt. Diablo, Contra Costa County, California. M, 1957, Stanford University.

Brown, Wilton J. Geology of a part of the Cushing Quadrangle, Nacogdoches County, Texas. M, 1953, University of Texas, Austin.

Brownawell, Bruce J. The role of colloidal organic matter in the marine geochemistry of PCB's. D, 1986, Massachusetts Institute of Technology. 318 p.

Browne, Bryant Alan. Transformations of aluminum in waters draining podzolic forest soils. D, 1988, Syracuse University. 364 p.

Browne, David Richard. Study of the transport of water through the Haulover Canal. M, 1974, Florida Institute of Technology.

Browne, James L. and Haynes, Charles W. Geology of the Mount Gratiot area, Keweenaw County, Michigan. M, 1956, Michigan Technological University. 44 p.

Browne, Jonathan F. The geology of the Cuprite Mine area, Pima County, Arizona. M, 1958, University of Arizona.

Browne, Kathleen M. Thrombolites of Lower Devonian Manlius Formation of central New York. M, 1986, SUNY at Binghamton. 581 p.

Brownell, George McLeod. Stability relationships of the silicates of alumina in metamorphosed argillaceous rocks. M, 1925, University of Manitoba.

Brownell, George McLeod. The geology of the Lyndhurst area, Ontario. D, 1928, University of Minnesota, Minneapolis. 147 p.

Brownell, James R. Geobiological influences on natural redistribution of rubidium relative to potassium in plants and soils. D, 1969, University of California, Davis.

Brownell, James R. Stratigraphy of unlithified deposits in the central sand plain of Wisconsin. M, 1986, University of Wisconsin-Madison.

Brownfield, Michael E. Geology of Floras Creek drainage, Langlois Quadrangle, Oregon. M, 1972, University of Oregon. 104 p.

Brownfield, Robert Lee. The structural history of the Centralia area, Illinois. M, 1955, University of Illinois, Urbana. 49 p.

Browning, James S. Methods of concentrating the black sands of the Idaho placer deposits. M, 1948, University of Idaho. 38 p.

Browning, John Leverett. Foraminifera from the upper Santa Susana Shale. M, 1952, University of California, Berkeley. 118 p.

Browning, Lawrence A. Source of nitrate in water of the Edwards Aquifer, south-central Texas. M, 1977, University of Texas, Austin.

Browning, Timothy D. Age of copper mineralization in the flow tops of the Portage Lake Lava Series determined by paleomagnetic methods. M, 1986, Michigan Technological University. 62 p.

Browning, William Fleming. Significance of airphoto patterns to geologic mapping. M, 1951, University of Virginia. 103 p.

Brownlee, Diane Elizabeth. Stratigraphic and structural investigation of the Eola Klippe, Garvin County, Oklahoma. M, 1981, University of Oklahoma. 133 p.

Brownlee, M. E. Stable carbon and oxygen isotopes of carbonate coal balls and associated carbonates of the Illinois Basin. M, 1975, University of Illinois, Urbana. 102 p.

Brownlie, William Robert. Prediction of flow depth and sediment discharge in open channels. D, 1982, California Institute of Technology. 425 p.

Brownlow, Arthur H. Serpentine and associated rocks and contact minerals near Westfield, Massachusetts. D, 1960, Massachusetts Institute of Technology. 177 p.

Brownson, Allyn R. The stratigraphy of the Triassic in eastern North Carolina. M, 1915, University of North Carolina, Chapel Hill. 9 p.

Brownson, Ernest Maitland. A deep water facies of the Mount Laurel Formation. M, 1956, Rutgers, The State University, New Brunswick. 114 p.

Brownstein, Jack M. Analysis of the non-clay fractions of the mudstones from a lower Cincinnatian Series core, Butler County, Ohio. M, 1960, Miami University (Ohio). 53 p.

Brox, George Stanley. The geology (Wasatch Formation, Eocene) and erosional development of northern Bryce Canyon National Park, (Garfield County, Utah). M, 1961, University of Utah. 70 p.

Broxton, David E. The petrology and structural geology of the Crandall Creek area, Park County, Wyoming. M, 1977, University of New Mexico. 79 p.

Broyles, M.L. The structure of the East Rift Zone of Kilauea, Hawaii from seismic refraction, gravity and magnetic surveys. M, 1977, University of Hawaii. 85 p.

Bruan, Gerald E. Atomic substitution in the tetrahedrite tennantite series. M, 1969, University of Arizona.

Bruce, Clemont Hughes. Utica oil field, Daviess County, Kentucky. M, 1949, University of Kentucky. 24 p.

Bruce, Everend L. Geology and ore deposits of Rossland. M, 1912, Columbia University, Teachers College.

Bruce, Everend L. Geology and ore deposits of Rossland, Victoria, British Columbia. D, 1915, Columbia University, Teachers College.

Bruce, G. S. W. Economic mineral potentialities of Newfoundland. M, 1953, University of Toronto.

Bruce, J. R. G. Investigation of the nature of the fractured till at the Lambton generating station. M, 1987, University of Waterloo. 197 p.

Bruce, James L. Geology of the eastern portion of the Indian Wells Canyon mining district, Kern County, California. M, 1981, University of Nevada. 133 p.

Bruce, L. G. Economic geology of the Maxville Limestone, Newton Township, Muskingum County, Ohio. M, 1974, Ohio State University.

Bruce, Lorraine. Solute transport in a groundwater system with reversing regional gradients. M, 1984, University of Nevada. 139 p.

Bruce, R. C. A study of the relationship between soil and quantitative terrain factors. D, 1971, University of Hawaii. 202 p.

Bruce, Richard L. An evaluation of some factors affecting slope stability of the Dobbin Landslide. M, 1966, University of South Dakota. 53 p.

Bruce, Robert MacAllister. Petrochemical evolution and physical construction of an Andean Arc; evidence from the southern Patagonian Batholith at 53°S. D, 1988, Colorado School of Mines. 114 p.

Bruce, Wayne Royal. Geology, mineral deposits, and alteration of parts of the Cuddy Mountain District, western Idaho. D, 1971, Oregon State University. 165 p.

Bruck, Glenn Ralph. Engineering and environmental geology of Guadalupe Quadrangle, Maricopa County, Arizona; Part I. M, 1983, Arizona State University. 76 p.

Bruckenthal, Eileen A. The dehydration of phyllosilicates and palagonites; reflectance spectroscopy and differential scanning caliometry. D, 1987, University of Hawaii.

Bruder, Karl Fritz and Wheeler, Charles Thomas, Jr. Geology of the Greaser Creek area, west central Huerfano Park, Huerfano County, Colorado. M, 1955, University of Michigan.

Bruderer, Barry E. A preliminary investigation of the sediments of the Great South bay, Long Island, New York. M, 1970, Long Island University, C. W. Post Campus.

Brueckmann, John Edward. Areal geology of the Gosport Quadrangle, Indiana. M, 1958, Indiana University, Bloomington. 48 p.

Brueckner, Hannes Kurt. Relations of anorthosite, eclogite, and ultramafic rock to the country rock, Tafjord, Norway. D, 1968, Yale University.

Brueggeman, John Lyle. Beach cusps of Monterey Bay, California. M, 1971, United States Naval Academy.

Bruehl, Donald H. Petrology and structure of an area north of Cooke city, Montana. M, 1961, Wayne State University.

Bruemmer, Jerry L. Gravity models for peridotite injection and local metamorphism along north-northeast trending zones of tectonism, Southwest Oregon. M, 1964, University of Oregon. 124 p.

Bruen, James N. Radiation pattern of Rayleigh waves from the southeastern Alaska earthquake of July 10, 1958. D, 1961, Columbia University, Teachers College.

Bruen, Michael P. Morphology and process of a cirque glacier and rock glaciers at Atigun Pass, Brooks Range, Alaska. M, 1980, SUNY at Buffalo. 95 p.

Bruff, Stephan Cartland. The Pleistocene history of the Newport Bay area, Southern California. M, 1939, University of California, Berkeley. 30 p.

Brugger, Keith A. Till provenance studies and glacier reconstructions in the Wildhorse Canyon area, Idaho. M, 1985, Lehigh University.

Brugman, Melinda Mary. Water flow at the base of a surging glacier. D, 1987, California Institute of Technology. 280 p.

Bruha, Douglas James. Paragenetic and fluid-inclusion study of Pb-Zn-Ag mineralization at Mina Teresita, Huachocolpa District, Peru. M, 1983, University of Nevada. 55 p.

Bruhl, Elliot J. Ground water management in eight western states under the prior appropriation doctrine. M, 1988, University of Idaho. 174 p.

Bruhn, Henry H. Some notes on the Apache Group of the Santa Catalina Mountains and other sections in southeastern Arizona. M, 1927, University of Arizona.

Bruhn, R. L. Middle Cretaceous deformation in the Andes of Tierra del Fuego; an example of aborted obduction. D, 1976, Columbia University, Teachers College. 109 p.

Bruin, James H. De see De Bruin, James H.

Bruin, Rodney H. De see De Bruin, Rodney H.

Brukardt, Susan A. Gravity survey of Waukesha County, Wisconsin. M, 1983, University of Wisconsin-Milwaukee. 131 p.

Bruland, Kenneth W. Pb-210 geochronology in the coastal marine environment. D, 1974, University of California, San Diego. 118 p.

Brule, D. G. An ESCA investigation of the adsorption of metals on hydrous manganese dioxide. M, 1977, University of Western Ontario. 120 p.

Brumbaugh, David Scott. Structural analysis of the complexly deformed Big Hole River area, Beaverhead, Madison, and Silver Bow counties, Montana. D, 1972, Indiana University, Bloomington. 96 p.

Brumbaugh, Mark Virgil. A study of Werner deconvolution and its application to the interpretation of aeromagnetic data. M, 1985, Purdue University. 143 p.

Brumbaugh, Richard L. Reconnaissance geology of the Sebec Lake area, Maine. M, 1964, Brown University.

Brumbaugh, Robert Wayne. Hillslope gullying and related changes, Santa Cruz Island, California. D, 1983, University of California, Los Angeles. 210 p.

Brumbaugh, William Donald. Gravity survey of the Cove Fort-Sulphurdale KGRA and the North Mineral Mountains area, Millard and Beaver counties, Utah. M, 1978, University of Utah. 131 p.

Brumfield, Kevin E. A geophysical survey of the tectonic setting of the 1980 Sharpsburg, Kentucky earthquake. M, 1986, University of Kentucky. 64 p.

Brummer, Johannes J. Areal geology of Holland Township (northwest quarter), Gaspe North County, Quebec. D, 1955, McGill University.

Brumund, William Frank. Subsidence of sand due to surface vibration. D, 1969, Purdue University. 143 p.

Brundage, Walter L., Jr. Recent sediments of the Nisqually River delta, Puget Sound, Washington. M, 1961, University of Washington. 69 p.

Brundrett, Jesse Lee. Cretaceous stratigraphy of the northeastern front of Davis Mountains. M, 1955, University of Texas, Austin.

Brune, William Arthur. Application of dimensional analysis to the control of strata surrounding a mine opening. D, 1952, Pennsylvania State University, University Park. 131 p.

Bruneau, Jeffrey Allan. Geologic map of a portion of the Salt River Canyon area and the geochemistry of the Tomato Juice uranium mine, Gila County, Arizona. M, 1981, New Mexico Institute of Mining and Technology. 84 p.

Brunelle, Thomas M. Non-steady state sulfur diagenesis in softwater lakes; an initial investigation. M, 1984, University of Rochester. 61 p.

Brunengo, M. J. Environmental geology of the Kellogg Creek - Mt. Scott Creek and lower Clackamas River drainage areas, northwestern Clackamas County, Oregon. M, 1978, Stanford University. 184 p.

Bruner, Diane Hyslop. The influence of cap rock on the development of slopes; a quantitative analysis of slopes developed on two sequences of flat-lying carbonate rocks in the presence of and in the absence of resistant sandstone cap rocks (southwestern Kentucky). M, 1972, Indiana State University. 75 p.

Bruner, Frank Henry. Contributions to the exact age of a Canadian uraninite. D, 1936, University of Missouri, Columbia.

Bruner, William Michael. Effects of time-dependent crack growth on the unroofing and unloading behavior of rocks. D, 1980, University of California, Los Angeles. 220 p.

Brunger, Allan G. The Drummond, Hector and Peyto glaciers; their wastage and deposits. M, 1966, University of Calgary.

Bruning, Curtis J. Analytical determination of Wood Fordian-Valeran boundary in a part of southeastern Wisconsin. M, 1970, University of Wisconsin-Milwaukee.

Bruning, James Earl. Origin of the Popotosa Formation, North-central Socorro County, New Mexico. D, 1973, New Mexico Institute of Mining and Technology. 131 p.

Bruning, James Earl. Petrochemical investigation of the mafic intrusive bodies in the Bergland and Thomaston quadrangles, Michigan. M, 1969, University of Toledo. 96 p.

Brunner, C. A. Late Neogene and Quaternary paleoceanography and biostratigraphy of the Gulf of Mexico. D, 1978, University of Rhode Island. 356 p.

Bruno, Anthony C. Derivation of synorogenic clastic in the Mississippian and Pennsylvanian strata of southern West Virginia. M, 1982, University of Kentucky.

Bruno, Lawrence. Depositional environments and diagenesis of the Cable Canyon Sandstone and Upham Dolomite, Ordovician of southern New Mexico. M, 1987, University of Houston.

Bruno, Pierre W. Lithofacies and depositional environments of the Ashtabula Till, Lake and Ashtabula counties, Ohio. M, 1988, University of Akron. 207 p.

Bruno, Sherrie L. Diagenetic characteristics of the Vanport Limestone interval (Pennsylvanian) of eastern Ohio and Lawrence County, Pennsylvania. M, 1988, University of Akron. 116 p.

Brunotte, D. The origin of ocean floor ferromanganese nodule ores by alteration of oceanic basalts. M, 1977, Iowa State University of Science and Technology.

Bruns, Dennis L. Geology of the Lake Mountain Northeast Quadrangle, Saguache County, Colorado. M, 1971, Colorado School of Mines. 79 p.

Bruns, Joan Marie Burnet. A coal and ash blend analysis of a bituminous and a subbituminous coal. M, 1981, Iowa State University of Science and Technology. 236 p.

Bruns, John J. Petrology of the Tijeras Greenstone, Bernalillo County, New Mexico. M, 1959, University of New Mexico. 119 p.

Bruns, Joseph John. A study of Midwestern bituminous coal fly ash. M, 1980, Iowa State University of Science and Technology.

Bruns, Richard Harte. The role of storm winnowing in producing the characteristic stratification of the Cincinnatian rocks in the Cincinnati region. M, 1953, University of Cincinnati. 45 p.

Bruns, Terry Ronald. Gravity study of the Southern California continental borderland. M, 1969, Massachusetts Institute of Technology. 57 p.

Bruns, Terry Ronald. Tectonics of an allochthonous terrane, the Yakutat Block, northern Gulf of Alaska. D, 1984, University of California, Santa Cruz. 256 p.

Brunsing, Thomas Peter. The effect of a tensile prestress on the efficiency of drag bit cutting in hard rock. D, 1980, University of California, Berkeley. 205 p.

Brunskill, Gregg John. Fayetteville Green Lake, New York; I, Physical and chemical limnology; II, Precipitation and sedimentation of calcite in a meromictic lake with laminated sediments. D, 1968, Cornell University. 183 p.

Brunsman, Mark J. Three-dimensional fluorescence spectroscopy of the sporinite and resinite macerals from coals of different rank. M, 1986, University of Toledo. 142 p.

Brunson, Burlie Allen. Shear wave attenuation in unconsolidated laboratory sediments. D, 1984, Oregon State University. 242 p.

Brunson, Karen L. Clay mineralogy of some loess-derived soils and Sangamon Paleosols in southwestern Indiana. M, 1976, Indiana University, Bloomington. 37 p.

Brunson, Wallace Edward. Type sections of the Cox and Finlay Formation, Hudspeth County, Trans-Pecos, Texas. M, 1954, University of Texas, Austin.

Brunton, Frank R. Silurian (Llandovery-Wenlock) patch reef complexes of the Chicotte Formation, Anticosti Island, Quebec. M, 1988, Laurentian University, Sudbury. 190 p.

Brunton, George Delbert. The crystal structure of callaghanite. D, 1957, Indiana University, Bloomington. 47 p.

Brunton, George Delbert. The study of an unusual augite near Cabezon Peak, Sandoval County, New Mexico. M, 1952, University of New Mexico. 40 p.

Brunton, James S. L. A geological reconnaissance on the west coast of Newfoundland between St. Georges and Bonne Bay. M, 1914, Columbia University, Teachers College.

Brush, E. R. The relationship of bottom sediments to the current-generated topography on Crumps Bank and Willoughby Bank, southern Chesapeake Bay, Virginia. M, 1978, Old Dominion University. 147 p.

Brush, Laurence Henry. The solubility of some phases in the system UO_3-Na_2O-H_2O in aqueous solutions at 60 and 90°C. D, 1980, Harvard University. 195 p.

Brush, Lucien Munson, Jr. Drainage basins, channels, and flow characteristics of streams in central Pennsylvania. D, 1956, Harvard University.

Brush, Randal Moorman. Bias in engineering estimation; a case study. M, 1980, Stanford University. 186 p.

Brusky, Eugene S. A comparison study of two mesoscale convective complexes evolving within similar large-scale environments. M, 1987, University of Wisconsin-Milwaukee. 98 p.

Brusseau, Mark Lewis. A short-term analysis of the USLE-SDR sediment yield model; South Lake MacBride watershed. M, 1984, University of Iowa. 76 p.

Bruton, Roger L. The geology of a fault area in southeastern Riley County, Kansas. M, 1958, Kansas State University. 44 p.

Brutsaert, Willem Frans. Immiscible multiphase flow in ground water hydrology; a computer analysis of the well flow problem. D, 1970, Colorado State University. 129 p.

Brutvan, William J., Jr. The economic feasibility of limestone production in Washington and Weston townships, Wood County, Ohio. M, 1964, Bowling Green State University. 30 p.

Bryan, Benjamin K., Jr. The Greenwood Anomaly of the Cumberland Plateau, a test application of GRAVMAP, a system of computer programs for reducing and automating contouring land gravity data. M, 1975, University of Tennessee, Knoxville. 118 p.

Bryan, Carol J. An analysis of pressure-temperature-time histories of eroding orogenic belts. M, 1987, Massachusetts Institute of Technology. 128 p.

Bryan, Charles R. Geology and geochemistry of mid-Tertiary volcanic rocks in the eastern Chiricahua Mountains, southeastern Arizona. M, 1988, University of New Mexico. 137 p.

Bryan, Dennis Paul. The geology and mineralization of the Chalk Mountain and Westgate Mining districts, Churchill County, Nevada. M, 1972, University of Nevada. 78 p.

Bryan, G. Gregory. Mineragraphy and paragenesis of the ore of the Park City consolidated mine, Park City, Utah. D, 1935, University of Utah. 54 p.

Bryan, Jack Howard. A petrographic study of the Pottsville Formation at Hold Dam Site, Alabama. M, 1963, University of Alabama.

Bryan, Jonathan R. Macrofaunal changes across the Cretaceous-Tertiary boundary, Braggs, Lowndes County, Alabama. M, 1987, University of Florida. 101 p.

Bryan, Joseph Jefferson. Hydrothermal alteration of granite in Wayne County, Missouri. M, 1931, University of Missouri, Columbia.

Bryan, Kirk. Geology, physiography, and water resources of the Papago country, Arizona. D, 1920, Yale University.

Bryan, Lynn Claire. Magnetization in the sediments of a modern fan delta, Baja California, Mexico. M, 1984, University of Oklahoma. 126 p.

Bryan, Michael R. The geology of the Lake Three quartz monzonite area (Precambrian), Marquette County, Michigan. M, 1970, Bowling Green State University. 72 p.

Bryan, Robert A. Thin-bed resolution from cepstrum analysis. M, 1985, Virginia Polytechnic Institute and State University.

Bryan, Robert Calvin. The subsurface geology of the Deer Creek, Webb, and North Webb oil pools, Grant and Kay counties, Oklahoma. M, 1950, University of Oklahoma. 46 p.

Bryan, Thomas Scott. Mineralogical and elemental trends in the Lucky Friday Mine, Mullan, Idaho. M, 1974, University of Montana. 94 p.

Bryan, Timothy Michael. Sedimentology of the Farmers Member, Borden Formation (Mississippian) in northeastern Kentucky, and equivalent units in adjacent Ohio and West Virginia. M, 1982, University of Cincinnati. 90 p.

Bryan, Tolbert Wilson. Geology of the Oak Cliff Quadrangle, Dallas County, Texas. M, 1952, Southern Methodist University. 43 p.

Bryan, Wilfred Bottrill, Jr. High-silica alkaline lavas of Clarion and Socorro Islands, Mexico; their genesis and regional significance. D, 1959, University of Wisconsin-Madison. 180 p.

Bryan, Wilfred Bottrill, Jr. Structural evolution of the Black Hills sandstone dikes. M, 1956, University of Wisconsin-Madison.

Bryant, Boyd LaVerl. Concentration of Latah County beryl. M, 1939, University of Idaho. 15 p.

Bryant, Brian Alan. Geology of the Sierra Santa Rosa Basin, Baja California, Mexico. M, 1986, San Diego State University. 75 p.

Bryant, Bruce Hazelton. Petrology and reconnaissance geology, Snowking area, northern Cascades, Washington. D, 1955, University of Washington. 321 p.

Bryant, David Gerald. Geology of the Gray Horse area, Osage County, Oklahoma. M, 1957, University of Oklahoma. 118 p.

Bryant, Deborah Jean Allen. Analysis of Des Moines Lobe glacial sediments and landforms, Martin-Marietta's "Cook's" quarry site, Ames, Iowa. M, 1987, Iowa State University of Science and Technology. 58 p.

Bryant, Donald Glassell. Intrusive breccias and associated ore of the Warren (Bisbee) mining district, Cochise County, Arizona. D, 1964, Stanford University. 347 p.

Bryant, Donald L. Stratigraphy of the Permian System in southern Arizona. D, 1955, University of Arizona.

Bryant, Donald L. The geology of the Mustang Mountains, Santa Cruz County, Arizona. M, 1951, University of Arizona.

Bryant, George F. Geology of the Schep-Panther Creek area, Mason County, Texas. M, 1959, Texas A&M University.

Bryant, George T. The general geology of the northernmost part of the Pine Forest Mountains, Humboldt County, Nevada. M, 1970, Oregon State University. 75 p.

Bryant, J. G. Geochemistry, geological environment and genesis, Weiss Deposit, Ergani Maden, Turkey. M, 1977, University of Western Ontario. 128 p.

Bryant, James Elwood. The areal geology of the Buda Manchaca area, Austin, Texas (B841). M, 1948, University of Texas, Austin.

Bryant, Jay Clark. A refinement of the upland glacial drift border in southern Cattaraugus County, New York. M, 1955, Cornell University.

Bryant, Jeffrey W. Origin and stratigraphic relations of Cambrian quartzites in Arizona. M, 1978, University of Arizona.

Bryant, Laurie J. Early Miocene lagomorphs and rodents from the Wounded Knee area (Shannon

county), South Dakota. M, 1969, South Dakota School of Mines & Technology.

Bryant, Laurie Jean. Non-dinosaurian lower vertebrates across the Cretaceous-Tertiary boundary, northeastern Montana. D, 1985, University of California, Berkeley. 287 p.

Bryant, Luther Frye. The mechanical analysis of sand and gravel deposits from selected localities of Mississippi. M, 1955, Mississippi State University. 69 p.

Bryant, Nancy Lee. Hydrothermal alteration at Roosevelt Hot Springs KGRA; DDH 1976-1. M, 1977, University of Utah. 87 p.

Bryant, Peter Franklin. Bioturbation of Recent abyssal sediments in the eastern equatorial Pacific. M, 1977, University of Utah. 84 p.

Bryant, Ray Baldwin. Genesis of the Ryker and associated soils in southern Indiana. D, 1981, Purdue University. 234 p.

Bryant, Samuel Morris. Applications of tailings flow analyses to field conditions. D, 1983, University of California, Berkeley. 312 p.

Bryant, Vaughn Motley, Jr. Late full-glacial and postglacial pollen analysis of Texas sediments. D, 1969, University of Texas, Austin. 179 p.

Bryant, Vicki Yolanda. A study of the occurrence of garnet in siliceous igneous rocks of the Mount Pilchuck area, Snohomish County, Washington. M, 1975, University of Washington. 31 p.

Bryce, Robert William. Phosphorus transfer between the liquid and solid phase in Lahontan Reservoir, Nevada. M, 1981, University of Nevada. 122 p.

Bryda, Anthony P. Nitrate-nitrogen profiles documenting land use practice effects on ground water in and around Sidney, Nebraska. M, 1988, University of Nebraska, Lincoln. 119 p.

Bryden, Elmer Louis. Geology of an area north of Gardiner, Montana. M, 1950, Wayne State University. 47 p.

Bryers, Wesley E. The Mesa Central of Mexico as a mineralogenic province. D, 1964, Columbia University, Teachers College. 244 p.

Bryn, Sean M. Determination and distribution of the hydraulic conductivity and specific yield of the Ogallala Aquifer in the Northern High Plains of Colorado. M, 1984, Colorado School of Mines. 123 p.

Bryndzia, L. Taras. The composition of chlorite as a function of sulfur and oxygen fugacity; an experimental study with application to metamorphosed sulfide ores, Snow Lake area, Manitoba. D, 1985, University of Toronto.

Bryndzia, Taras Lubomyr. Mineralogy, geochemistry and mineral chemistry of siliceous ore and altered footwall rocks in the Uwamuki No. 4 deposit. M, 1980, University of Toronto.

Bryndzia, Taras Lubomyr. The composition of chlorite as a function of sulfur and oxygen fugacity; an experimental study with application to the metamorphosed sulfide ores, Snow Lake Area, Manitoba. D, 1985, University of Toronto.

Bryner, Leonid. Geology of the South Comobabi Mountains and Ko Vaya Hills, Pima County, Arizona. D, 1959, University of Arizona. 181 p.

Bryson, Herman Jennings. Stratigraphy of Deep River coals [North Carolina]. M, 1924, University of North Carolina, Chapel Hill. 41 p.

Bryson, Robert P. Faulted fanglomerates at the mouth of Perry Aiken Creek, northern Inyo Range, California-Nevada. M, 1937, California Institute of Technology. 51 p.

Bryson, William R. Clay mineralogy and clay paragenesis of the upper Chase and lower Sumner limestones and shales in north and east central Kansas. M, 1959, Kansas State University. 126 p.

Brzozowy, Carl P. Magnetic and seismic reflection surveys of Lake Superior. M, 1973, University of Wisconsin-Milwaukee.

Buapang, Somkid. A nonlinear hydrologic cascade. M, 1977, New Mexico Institute of Mining and Technology.

Bubb, John Neal. Diagenesis of the Liston Creek Member of the Wabash Formation, Grant County, Indiana. D, 1963, Indiana University, Bloomington. 101 p.

Bubb, John Neal. Geology of part of the Greenhorn Range and vicinity, Madison County, Montana. M, 1961, Oregon State University. 165 p.

Bubeck, Robert Clayton. A kinetic study of the evaluation of carbon dioxide from an aqueous solution. M, 1965, Pennsylvania State University, University Park. 69 p.

Bubeck, Robert Clayton. Some factors influencing the physical and chemical limnology of Irondequoit Bay, Rochester, N.Y. D, 1972, University of Rochester. 291 p.

Bublitz, Richard F. A comparison of four contemporary seismic methods, T 35 N, R 93 W., Big Horn County, Wyoming. M, 1962, Colorado School of Mines. 56 p.

Bubnick, Steven C. Net shore-drift along the Strait of Juan de Fuca coast of Clallam County, Washington. M, 1986, Western Washington University. 69 p.

Bucci, S. A. A modeling approach for determining key chemical variables in natural and man-made lakes. D, 1976, University of Massachusetts. 89 p.

Bucek, Milena F. Pleistocene geology and history of the west branch of Susquehanna River valley near Williamsport, Pennsylvania. D, 1975, Pennsylvania State University, University Park. 238 p.

Buch, I. Philip. Upper Permian (?) and Lower Triassic stratigraphy near El Marmol, Baja California, Mexico. M, 1984, San Diego State University. 137 p.

Buchan, K. Paleomagnetism of Precambrian Grenville intrusive rocks from Haliburton County, Ontario. M, 1973, University of Toronto.

Buchan, K. L. Paleomagnetic and rock magnetic studies of multi-component remanences in metamorphosed rocks of the Grenville Province of the Canadian Precambrian shield. D, 1977, University of Toronto.

Buchanan, Cathy McGhee. Precambrian structural and metamorphic history of the Rociada area, New Mexico. M, 1976, Texas Tech University. 57 p.

Buchanan, Douglas Mitchell. Carbonate petrology of the Negli Creek Limestone Member, Kinkaid Formation (Chesterian) in southern Illinois. M, 1985, Southern Illinois University, Carbondale. 61 p.

Buchanan, Hugh. Environmental stratigraphy of Holocene carbonate sediments near Frazers Hog cay, British West Indies. D, 1970, Columbia University. 229 p.

Buchanan, J. J. Skeletal morphology and budding pattern in the bryozoan genus Melicerites. M, 1975, Wayne State University.

Buchanan, John Petrella. Annual behavior of sand-bed channels. D, 1985, Colorado State University. 237 p.

Buchanan, John Petrella. Channel morphology and sedimentary facies of the Niobrara River, north-central Nebraska. M, 1981, Colorado State University. 137 p.

Buchanan, John Wilson. Geology of the Double Lakes area, Lynn County, Texas. M, 1973, Texas Tech University. 121 p.

Buchanan, Kenneth J. An investigation of the causes of differences between on-hand coal resources and calculated coal reserves at the Toledo Edison Bay Shore power generation station. M, 1979, University of Toledo. 132 p.

Buchanan, Larry J. The Las Torres Mines, Guanajuato, Mexico; ore controls of a fossil geothermal system. D, 1979, Colorado School of Mines. 156 p.

Buchanan, Peter. Debris avalanche and debris torrent initiation, Whatcom County, Washington, U.S.A. M, 1988, University of British Columbia. 218 p.

Buchanan, Peter H. Volcanic geology of the Guffey area, Park County, Colorado. M, 1967, Colorado School of Mines. 95 p.

Buchanan, R. M. The paragenesis of accessory minerals in intermediate igneous rocks. M, 1949, University of Toronto.

Buchbinder, Goetz Gustav Rudolf. Properties of the core-mantle boundary and observations of PcP. D, 1968, Columbia University. 77 p.

Buchbinder, Goetz Gustav Rudolph. The possibility of measuring absolute stresses in the Earth's crust. M, 1962, Dalhousie University. 67 p.

Bucher, Edward J. Subsurface geology of the Middle Ordovician Ottawa Limestone Supergroup of eastern Ohio. M, 1979, Kent State University, Kent. 142 p.

Bucher, Gerald Joseph. Heat flow and radioactivity studies in the Ross Island-Dry Valley area, Antarctica and their tectonic implications. D, 1980, University of Wyoming. 204 p.

Bucher, Robert Louis, II. Crustal and upper mantle structure beneath the Colorado Plateau and Basin and Range provinces as determined from Rayleigh wave dispersion. D, 1970, University of Utah. 119 p.

Buchheim, H. P. Paleolimnology of the Laney Member of the Eocene Green River Formation. D, 1978, University of Wyoming. 113 p.

Buchheim, Martin Paul. Non-traditional approaches to classification of high resolution satellite imagery in a microcomputer environment. D, 1988, University of Wisconsin-Madison. 503 p.

Buchheit, Richard L. The opaque minerals of the basic igneous rocks of Minnesota. M, 1959, University of Minnesota, Minneapolis. 109 p.

Buchholtz, Herbert F. The fracture pattern in the Crestmore Mine, Southern California. M, 1960, Pomona College.

Buchi, Sylvia Duncan Hall. Studies on the natural history of beryl. D, 1962, University of Michigan. 179 p.

Buchwald, Caryl Edward. Sedimentary microstructures of the Fiddlers Green Dolomite Member, Bertie Formation, Upper Silurian (New York). M, 1963, Syracuse University.

Buchwald, Caryl Edward. Types and distribution of sandstones in the Belly River and Edmonton formations (uppermost Cretaceous) of the north Saskatchewan River area, west-central Alberta. D, 1966, University of Kansas. 76 p.

Buck, Arvo Viktor. The shallow geologic features of the upper continental slope, northern Gulf of Mexico. M, 1981, Texas A&M University.

Buck, Brian Willima. Physico-chemical properties of sensitive soils in the lower Jordan Valley, Utah. M, 1976, University of Utah. 82 p.

Buck, Caroline Jo. Monitoring acoustic emissions during water flow through granular soils. M, 1985, University of Nevada. 128 p.

Buck, D. J. Petrology and paleoclimatology of claystones near Corona, Riverside County, California. M, 1977, San Diego State University.

Buck, Don E., Jr. Subsurface geology of West Gueydan Field. M, 1976, University of Southwestern Louisiana. 55 p.

Buck, Peter Stanley. A caldera sequence in the early Precambrian, Favourable Lake volcanic complex, northwestern Ontario. M, 1978, University of Manitoba. 140 p.

Buck, Shane. Structural studies and gabbro mylonitization within the Barton Bay deformation zone, Geraldton, Ontario. M, 1986, Brock University. 172 p.

Buck, Stanley P. Geology of a portion of southern Campbell County, Wyoming. M, 1977, University of Colorado.

Buck, W. K. The geology of the Lake Wasa property, Beauchastel Township, Quebec. M, 1951, McGill University.

Buck, Walter Roger. Small-scale convection and the evolution of the lithosphere. D, 1984, Massachusetts Institute of Technology. 256 p.

Bucke, David Perry, Jr. Effect of diagenesis upon clay mineral content of interlaminated Desmoinesian sand-stones and shales in Oklahoma. D, 1969, University of Oklahoma. 123 p.

Bucke, David Perry, Jr. Petrology of the Watonga Dolomite (Permian) in Blaine County, Oklahoma. M, 1968, University of Oklahoma. 103 p.

Buckham, Alex Fraser. A petrographical study of gabbro-like Precambrian rock types in the Francois River District, Great Slave Lake, Northwest Territories. M, 1936, University of Alberta. 137 p.

Buckingham, James Gordon and Callender, Edward. An investigation of the folding in Forgie Township, Ontario (Canada). M, 1968, University of Manitoba.

Buckingham, Martin L. Fluvio-deltaic sedimentation patterns of the Upper Cretaceous to lower Tertiary Sabbath Creek section, Arctic National Wildlife Refuge (ANWR), northeastern Alaska. M, 1985, University of Alaska, Fairbanks. 165 p.

Buckingham, W. R. A heat budget analysis of Pendrell Sound. M, 1976, University of British Columbia.

Buckingham, William Forrest. A mineralogical characterization of rock surfaces formed by hydrothermal alteration and weathering; application to remote sensing. D, 1981, University of Maryland. 206 p.

Buckland, Francis Channing. The dolomitic magnesite deposits of Grenville Township, Argenteuil County, Quebec. D, 1937, McGill University.

Buckland, Francis Channing. The geology and petrography of a section along the tramway, Mount Royal, Montreal, Quebec. M, 1932, McGill University.

Buckler, William Roger. Rates and implications of bluff recession along the Lake Michigan shorezone of Michigan and Wisconsin. D, 1981, Michigan State University. 208 p.

Buckley, Christopher P. Structure and stratigraphy of the Kingsley Mountains, Elko and White Pine counties, Nevada. M, 1967, San Jose State University. 50 p.

Buckley, Christopher Paul. The structural position and stratigraphy of the Palmetto complex in the northern Silver Peak mountains, Nevada. D, 1971, Rice University. 89 p.

Buckley, D. W. Petrology of Ordovician ignimbrites and spilites, East Murrelrea Syncline, Ireland. M, 1977, Acadia University.

Buckley, Dale E. Recent marine sediments of Lancaster Sound (Northwest Territories). M, 1963, University of Western Ontario. 113 p.

Buckley, Ernest R. The building and monumental stone of Wisconsin. D, 1898, University of Wisconsin-Madison.

Buckley, Glenn R. Genesis of the amphiboles of Iron Hill, Delaware. M, 1970, Wayne State University.

Buckley, Glenn Robert. The effect of diffusion on garnet zoning. D, 1973, University of Illinois, Urbana. 113 p.

Buckley, J. R. Currents, winds, and tides in Howe Sound. D, 1977, University of British Columbia.

Buckley, R. A. The geology of the Weedon Pyrite and Copper Corporation Ltd. Mine (Quebec). M, 1959, McGill University.

Buckley, S. B. Study of post-Pleistocene ostracod distribution in the soft sediments of southern Lake Michigan. D, 1975, University of Illinois, Urbana. 299 p.

Bucknam, Robert Campbell. Structure and petrology of Precambrian rocks in part of the Glen Haven Quadrangle, Larimer County, Colorado. D, 1969, University of Colorado. 118 p.

Buckner, Dean Alan. A portion of the system; $CaOSiO_2$-$SrO \cdot SiO_2$-H_2O. M, 1955, Pennsylvania State University, University Park. 67 p.

Buckner, Dean Alan. Synthesis of epistilbite and its phase relationships to other calcium zeolites. D, 1958, University of Utah. 54 p.

Buckner, Duane Herbert. Salinity gradient study of Oklahoma. M, 1972, Oklahoma State University. 71 p.

Buckovic, William Alan. The Cenozoic stratigraphy and structure of a portion of the west Mount Rainier area, Pierce County, Washington. M, 1974, University of Washington. 123 p.

Buckstaff, Sherwood. A study of the distribution of magnetite in one of the Keweenawan lava flows. M, 1923, University of Wisconsin-Madison.

Buckwalter, Tracy Vere. Geology of a part of the Dillon Quadrangle, Colorado. M, 1964, University of Michigan.

Bucurel, Hildred Gail. Stratigraphy and coal deposits of the Upper Cretaceous, Campanian, Mesaverde Group in the southern Wasatch Plateau. M, 1977, University of Utah. 113 p.

Buczynski, Chris. The sedimentology and diagenesis of the Lueders Formation (Permian) of Baylor County, Texas. M, 1985, University of Houston.

Budai, Christine M. Depositional model of the Antelope coal field, Wyoming. M, 1983, Portland State University.

Budai, Joyce Margaret. A diagenetic study of the lower Coralline Limestone (Oligocene), the Maltese Islands. M, 1980, Rice University. 105 p.

Budai, Joyce Margaret. Diagenesis and deformation of the Madison Group, Wyoming and Utah overthrust belt. D, 1984, University of Michigan. 216 p.

Budd, David A. Bathymetric zonation and paleoecological significance of microboring organisms in Puerto Rican shelf and slope sediments. M, 1978, Duke University. 140 p.

Budd, David A. Freshwater diagenesis of Holocene ooid sands, Schooner Cays, Bahamas. D, 1984, University of Texas, Austin. 517 p.

Budd, Harrell J., Jr. Geology of the McKinney area, Collin County, Texas. M, 1950, Southern Methodist University. 20 p.

Buddemeier, Robert Worth. A radiocarbon study of the varved marine sediments of Saanich inlet, British Columbia. D, 1969, University of Washington. 136 p.

Budden, R. T. The Tortolita-Santa Catalina Mountain Complex. M, 1975, University of Arizona.

Budding, Karin Elisabeth. Eruptive history, petrology, and petrogenesis of the Joe Lott Tuff Member of the Mount Belknap Volcanics, Marysvale volcanic field, west-central Utah. M, 1982, University of Colorado. 104 p.

Buddington, Arthur Francis. The Precambrian rocks of southeastern Newfoundland. D, 1916, Princeton University.

Budenstein, David. Attenuation of seismic energy near an explosion. M, 1954, Pennsylvania State University, University Park. 88 p.

Buder, Theodore A. The geology of the northwestern portion of the Lesterville Quadrangle, Missouri. M, 1961, University of Missouri, Columbia.

Budge, David R. Paleontology and stratigraphic significance of Late Ordovician-Silurian corals from the eastern Great Basin. D, 1972, University of California, Berkeley. 572 p.

Budge, David Rush. Stratigraphy of the Laketown Dolomite (Silurian, north central Utah). M, 1966, Utah State University. 86 p.

Budge, Suzanne. Trace elements distribution around precious and base metal veins, Idaho Springs District, Colorado. M, 1982, Colorado School of Mines. 152 p.

Budge, Wallace Don. A method of identification and volume change prediction of a stiff, fissured clay shale. D, 1964, University of Colorado. 212 p.

Budinger, Thomas F. Geology, biology, and hydrography of Cobb Seamount. M, 1957, University of Washington. 54 p.

Budke, Earl Hugo, Jr. Electrical transients in sandstones. M, 1964, Washington University. 87 p.

Budlong, Gerald Michael. Processes of beach change at Tahoe Keys, California; an example of man and nature as geomorphological agent. M, 1971, California State University, Chico.

Budnick, Dorene M. Analysis of structural fabric data on the unit sphere. M, 1978, Indiana University, Bloomington. 219 p.

Budnik, Roy Theodore. The geologic history of the Valdez Group, Kenai Peninsula, Alaska; deposition and deformation at a late Cretaceous consumptive plate margin. D, 1974, University of California, Los Angeles.

Budo, Shoro. Hydrologic-system analysis of the Wind River Formation (Eocene) with special reference to underground mining in the Shirley Basin area, Wyoming. M, 1965, University of Arizona.

Budrevics, Valdis. Detailed stratigraphy of the Slave Point Formation in the Nabesche River area, northeastern British Columbia. M, 1974, University of Manitoba.

Budros, Ronald Charles. The stratigraphy and petrogenesis of the Ruff Formation, Salina Group in Southeast Michigan. M, 1974, University of Michigan.

Buehler, Edward J. The morphology and taxonomy of the Halysitidae. D, 1953, Yale University.

Buehler, Edward J. The Tetracorallia of an Onondaga Limestone coral reef. M, 1942, SUNY at Buffalo.

Buehner, J. H. Geology of an area north of the Sierra Madre, Carbon County, Wyoming. M, 1936, University of Wyoming. 37 p.

Buehrer, David W. Environmental geology of Sioux Falls, South Dakota, and its relation to urban and industrial development. M, 1971, University of South Dakota. 73 p.

Buehrle, Paul Michael. Luminescence petrography of pressure solution in silica-cemented sandstones. M, 1976, University of Texas, Arlington.

Buelow, Kenneth L. Geothermal studies in Wyoming and northern Colorado, with a geophysical model of the Southern Rocky Mountains near the Colorado-Wyoming border. M, 1980, University of Wyoming. 150 p.

Buelter, Donald Paul. Geochemical investigation of epigenetic dolomite associated with lead-zinc mineralization of the Viburnum Trend, Southeast Missouri. M, 1985, Southern Illinois University, Carbondale. 105 p.

Buenaventura, Alfredo Capistrano. Geologic feasibility of dam and reservoir sites, Blacksmith Fork Canyon, Utah. M, 1968, Utah State University. 50 p.

Buer, Kill Yngvar. Stratigraphy, structure and petrology of a portion of the Smartville Ophiolite, Yuba County, California. M, 1979, University of California, Davis. 120 p.

Buerger, Martin J. The deformations of ore minerals; a preliminary investigation. M, 1927, Massachusetts Institute of Technology. 67 p.

Buerger, Martin J. Translation-gliding in crystals. D, 1929, Massachusetts Institute of Technology. 203 p.

Buerger, Newton W. An X-ray investigation of the solid phase of the system Cu_2S-CuS. D, 1940, Massachusetts Institute of Technology. 74 p.

Buerger, Newton W. Chalcopyrite-sphalerite relations. M, 1934, Massachusetts Institute of Technology. 18 p.

Buermann, John M. Aluminum. M, 1942, Colgate University.

Bues, Diane Jean. The hydrologic relationship between Lake Michigan and a shallow dolomite aquifer at Mequon, Wisconsin. M, 1983, University of Wisconsin-Milwaukee. 211 p.

Buettner, Peter Erhard. The petrography of the Moorefield Formation of the western flanks of the Ozarks. M, 1963, University of Tulsa. 93 p.

Bufe, Charles Glenn. An estimate of the configuration of the surface of the Earth's core from the consideration of surface focus PcP travel times. D, 1969, University of Michigan. 107 p.

Buff, Paul Jeffrey. The foraminifera and paleoecology of the Vacaville Shale, Vacaville, California. M, 1976, University of Nevada. 68 p.

Buffa, John Warren. Geology of the SE1/4 Twin Peaks Quadrangle, Custer County, Idaho. M, 1976, University of Idaho. 74 p.

Buffam, Basil Scott Whyte. Destor area, Abitibi County, Quebec. D, 1927, Princeton University.

Buffett, R. N. Geology of the Little Canyon Creek area, Washakie County, Wyoming. M, 1958, University of Wyoming. 85 p.

Buffington, E. C. The Aleutian-Kamchatka trench convergence; an investigation of lithospheric plate interaction in light of modern geotectonic theory. D, 1973, University of Southern California.

Buffington, Edwin Conger. An invertebrate fauna from the "Modelo" of Dry Canyon, Los Angeles County, California. M, 1947, California Institute of Technology. 40 p.

Buffington, John William. Tertiary geology of the Gangplank area, Colorado and Wyoming. M, 1961, University of Nebraska, Lincoln.

Buffler, Richard Thurman. The Browns Park Formation and its relationship to the late Tertiary geologic history of the Elkhead region, northwestern Colorado-south central Wyoming. D, 1967, University of California, Berkeley. 215 p.

Buffo, Lynn Karen. Extraction of zinc from sea water. M, 1967, Oregon State University. 49 p.

Buffum, James Ted. Correlation of well cuttings by insoluble residues. M, 1931, University of Missouri, Columbia.

Bufo, David. Detailed record of high amplitude secular variation in Pichileufu-age glaciolacustrine sediments, northwestern Patagonia, Argentina. M, 1988, Lehigh University.

Buford, Selwyn Oliver. Characteristics of the Taylor Marl of Travis County, Texas. M, 1928, University of Texas, Austin.

Bugden, G. Ice movement and modification in the Gulf of St. Lawrence. M, 197?, Dalhousie University.

Bugenig, Dale C. Well loss in fractured rock; a comparison of radial-flow and linear-flow analyses. M, 1985, University of Nevada. 48 p.

Bugh, James Edwin. Geomorphic evolution of the southeastern Sangre de Cristo Mountains, New Mexico. D, 1968, Case Western Reserve University. 166 p.

Bugh, James Edwin. Glacial geology of north-central Hancock County, Ohio. M, 1962, Bowling Green State University. 28 p.

Bugliosi, Edward F. Meander activity and flood plain development of Cattaraugus Creek, New York, a gravel-bearing, Lake Erie, tributary. M, 1977, SUNY, College at Fredonia. 112 p.

Bugosh, Nicholas. Field verification of predictive bedload formulas in a coarse bedload mountain stream. M, 1988, Montana State University. 98 p.

Bugry, Raymond. Geochemistry of boron in pelitic sediments. M, 1964, McMaster University. 146 p.

Bugs, Donna M. The persistence of copper in the Robinson Reservoir, South Carolina. M, 1988, University of Wisconsin-Milwaukee. 198 p.

Buhay, William M. A theoretical study on ESR dating of geological faults in Southern California. M, 1987, McMaster University. 136 p.

Buhl, P. H. An investigation of organic compounds in the Mighei meteorite. D, 1975, University of Maryland. 152 p.

Buhl, Peter. Extraction and enhancement of Earth structure and velocity information from marine seismic data. D, 1984, Columbia University, Teachers College.

Buhle, Merlyn Boyd. Ground water supplies in the vicinity of the tri-cities; Davenport, Iowa, Rock Island and Moline, Illinois. M, 1935, University of Iowa. 79 p.

Buhn, William Kenneth. The geology and ore deposit of the Rianda Mine, Santa Clara County, California. M, 1951, University of Idaho. 32 p.

Bui, Elisabeth Nathalie. Relationships between pedology, geomorphology and stratigraphy in the Dallol Bosso of Niger, West Africa. D, 1986, Texas A&M University. 247 p.

Buie, Bennett Francis. Dikes and related intrusives of the Highwood Mountains area, Montana. D, 1939, Harvard University. 126 p.

Buie, Bennett Frank. A report of investigations of the Allentown limestone formation. M, 1932, Lehigh University.

Buika, James Alexander. A seismicity study for portions of the Los Angeles Basin, Santa Monica Basin, and Santa Monica Mountains, California. M, 1979, University of Southern California.

Buika, Paul H. Preliminary investigation of the sedimentary zonation of the Ogallala Aquifer southern High Plains. M, 1979, Texas Tech University. 83 p.

Buis, Otto J. The structural geology of a part of the Potrero de Padilla, Coahuila, Mexico. M, 1958, Louisiana State University.

Buis, Patricia. Geochemistry of fluorite from the ore body of the Sterling Hill Mine in Ogdensburg, New Jersey. M, 198?, Queens College (CUNY). 116 p.

Buis, Patricia Frances. A geochemical study of minor and trace elements in the carbonate country rock surrounding Sterling Hill, a zinc-iron-manganese ore deposit in Ogdensburg, New Jersey. D, 1987, University of Pittsburgh. 392 p.

Buising, Anna Valetta. Depositional and tectonic evolution of the northern proto-Gulf of California and lower Colorado River, as documented in the Mio-Pliocene Bouse Formation and bracketing units, southeastern California and western Arizona. D, 1988, University of California, Santa Barbara. 269 p.

Bujak, Catherine A. Recent palynology of the Goat Lake and Lost Lake, Waterton Lakes National Park. M, 1974, University of Calgary.

Bukhari, Mohamed. Sedimentary environments and stratigraphic framework of the medial Ordovician Denley Limestone (Trenton Group) in New York State. M, 1977, Boston University. 157 p.

Bukhari, Syed Amir. A linear programming approach to the interpretation of Earth resistivity data. M, 1961, University of Alberta. 52 p.

Bukowski, Charles T., Jr. Depositional environment of Woodbine and Eagleford sandstones at OSR-Halliday Field, Leon and Madison counties, Texas. M, 1984, Texas A&M University. 109 p.

Bukry, John, David. Cretaceous (Santonian-Campanian) nannofossils of Texas. D, 1967, Princeton University. 342 p.

Buland, Raymond P. Retrieving the seismic moment tensor. D, 1976, University of California, San Diego. 75 p.

Bulau, James Ronald. Intergranular fluid distribution in olivine-liquid basalt systems. D, 1982, Yale University. 129 p.

Bulfinch, Douglas L. Western boundary undercurrent delineated by sediment texture at base of North American continental rise. M, 1981, University of Georgia.

Bulger, Paul R. Evaluation of the interaction between seepage from a municipal waste stabilization lagoon, McVille, North Dakota, and a shallow unconfined aquifer. M, 1987, University of North Dakota. 225 p.

Bulin, George Vincent, Jr. First-motion studies and directional analysis of Lake Superior seismic record. D, 1966, University of Minnesota, Duluth.

Bulin, George Vincent, Jr. The role of sericite in hydrothermal alteration. M, 1958, University of Minnesota, Minneapolis. 76 p.

Bull, Louis. Facies analysis of the Schulz Ranch Sandstone Member, Williams Formation (Upper Cretaceous), Santa Ana Mountains, Southern California. M, 1986, California State University, Long Beach. 116 p.

Bull, Stratton Hemptead. Geology of the Clover Divide area, Snowcrest Range, Montana. M, 1949, University of Michigan. 55 p.

Bull, William Benham. Alluvial fans and near-surface subsidence, western Fresno County, California. D, 1961, Stanford University. 308 p.

Bulla, Edward William. A mineragraphic study of ore from the Tamarack Mine, Burke, Idaho. M, 1951, University of Idaho. 37 p.

Bullard, Fred M. Preliminary report on the geology of Love County, Oklahoma. M, 1922, University of Oklahoma. 51 p.

Bullard, Fred Mason. Lower Cretaceous of western Oklahoma, a study of the outlying areas of Lower Cretaceous in Oklahoma and adjacent states. D, 1928, University of Michigan.

Bullard, Fredda Jean. Microfauna of the Del Rio Formation (Lower Cretaceous) of central Texas. M, 1951, University of Texas, Austin.

Bullard, Reuben George. Geology of the Tell Gezer, Israel archeologic site and specific applications of the geologic investigations to archeologic interpretation. D, 1969, University of Cincinnati. 277 p.

Bullard, Reuben George. The philosophical basis of geology. M, 1964, University of Cincinnati. 183 p.

Bullard, Thomas Fitts. Influence of bedrock geology on complex geomorphic responses and late Quaternary geomorphic evolution of Kim-Me-Ni-Oli Wash drainage basin, northwestern New Mexico. M, 1985, University of New Mexico. 408 p.

Bullen, Susan Brook. Synthetic dolomite textures. M, 1983, Michigan State University. 182 p.

Bullen, Thomas D. Structure and stratigraphy of the eastern Cardigan Quadrangle, central New Hampshire. M, 1977, Dartmouth College. 81 p.

Bullen, Thomas Darwin. Magmagenesis in the Devils Garden lava field; implications for the nature of the subcontinental lithosphere at an active continental margin. D, 1986, University of California, Santa Cruz. 354 p.

Buller, Robert David. The chemical geohydrology of Twin Lakes, Ohio. M, 1974, Kent State University, Kent. 83 p.

Bulley, Enid Joan. Areal distribution of sulfate reduction products in lake sediments receiving acid mine drainage. M, 1988, University of North Carolina, Chapel Hill. 68 p.

Bulling, Thomas Peter. Exploring for subtle traps with high resolution paleogeographic maps; the Reklaw 1 Interval in South Texas. M, 1987, Texas Christian University. 43 p.

Bullion, David Nelson. Geology and tectonics of Mendana fracture zone between latitudes 10°S-12°S and longitudes 80°W-83°W. M, 1988, Texas A&M University. 84 p.

Bullock, D. B. A microfaunal study of the basal Lloydminster Shale. M, 1950, University of Alberta. 76 p.

Bullock, Jack M. A correlation of core analysis and micro-logs of the West Short Junction Field. M, 1958, University of Oklahoma. 54 p.

Bullock, James M. Gypsum deposits in the northwestern part of the Bighorn Basin, Park County, Wyoming. M, 1964, University of Wyoming. 91 p.

Bullock, James S. Biostratigraphic study of foraminifera from the Smeltertown Formation (Lower Cretaceous, Albian) of southeastern New Mexico. M, 1985, University of Texas at El Paso.

Bullock, Kenneth C. A study of the geology of the Timpanogos Caves, Utah. M, 1942, Brigham Young University. 66 p.

Bullock, Kenneth C. Geomorphology of Lake Mountain, Utah. D, 1949, University of Wisconsin-Madison.

Bullock, Ladell R. Paleoecology of the Twin Creek Limestone (Upper Jurassic) in the Thistle, Utah area. M, 1965, Brigham Young University.

Bullock, Nedra D. Summary of the Pennsylvanian sedimentation of Utah. M, 1940, University of Utah. 45 p.

Bullock, Paul A. The feasibility of developing forecast systems to predict changes in beach sand volume on ocean beaches during storms. M, 1971, College of William and Mary.

Bullock, Peter. The zone of degradation at the eluvial-illuvial interface of some New York soils. D, 1968, Cornell University. 197 p.

Bullock, Reuben M. [Lynn]. The geology of the Lehi Quadrangle, Utah. M, 1958, Brigham Young University. 59 p.

Bullock, Walter Richard. A geochronological study of the Grenville Front near Coniston, Ontario, Canada. M, 1987, University of North Carolina, Chapel Hill. 50 p.

Bullwinkel, Henry J. Age investigation of syenites from Coldwell, Ontario. M, 1958, Massachusetts Institute of Technology. 50 p.

Bullwinkel, Paul E. Caustic aided parameter estimation; an application to geophysical travel time inversion. D, 1987, University of Miami. 224 p.

Bultman, Mark William. An analysis of oil field size distributions in consanguineous basins. D, 1986, University of Wisconsin-Madison. 262 p.

Bultman, T. R. Geology and tectonic history of the Whitehorse Trough west of Atlin, British Columbia. D, 1979, Yale University. 296 p.

Bultman, Thomas R. The Denali Fault (Hines Creek Strand) near the Nenana River, Alaska. M, 1972, University of Wisconsin-Madison.

Bultz, Deanna Jean. The Precambrian-Cambrian unconformity in Wisconsin. M, 1981, University of Wisconsin-Madison.

Buma, Grant. Trace element distributions and the origins of some New England granites. M, 1970, Massachusetts Institute of Technology. 141 p.

Bumgardner, John E. Geology of the gabbro sill surrounding Mariscal Mountain, Big Bend National Park, Texas. M, 1976, University of Texas, Austin.

Bumgardner, Louis Samuel. The geology of an area near Hannibal, Missouri. M, 1928, University of Missouri, Columbia.

Bumgarner, Edward L. The geology of the Portage Lake Volcanics in the M.T.U. Mining Laboratory, Hancock, Michigan. M, 1980, Michigan Technological University. 138 p.

Bumgarner, James G. Stratigraphy and structure of the Knox Dolomite in the Fair Garden area, Sevier and Jefferson counties, Tennessee. M, 1956, University of Tennessee, Knoxville. 46 p.

Bunch, James W. A fluorescent tracer study at a tidal inlet (Rudee inlet, Virginia). M, 1969, Old Dominion University. 55 p.

Bunch, Rosella L. The stratigraphy and paleontology of the Silurian System of parts of Highland County, Virginia. M, 1940, Oberlin College.

Bunch, Theodore Eugene. A petrologic study of the granodiorite-tonalite complex of a portion of the Warm Spring Mountain Quadrangle, Fremont County, Wyoming. M, 1962, Miami University (Ohio). 87 p.

Bunch, Theodore Eugene. A study of shock-induced microstructures and solid state transformations of several minerals from explosion craters. D, 1966, University of Pittsburgh. 175 p.

Bundtzen, Thomas K. Geology and mineral deposits of the Kantishna Hills, Mt. McKinley Quadrangle, Alaska. M, 1981, University of Alaska, Fairbanks. 238 p.

Bundy, Jerry Lowell. Contact metamorphism in Lancaster County, South Carolina. M, 1965, University of South Carolina.

Bundy, Wayne Miley. The geology and mineralization of the Cochiti mining district, New Mexico. M, 1954, Indiana University, Bloomington. 48 p.

Bundy, Wayne Miley. Wall rock alteration in the Cochiti mining district, New Mexico. D, 1957, Indiana University, Bloomington. 106 p.

Bunk, Gregory L. Structural and stratigraphic analysis of an area in south-central Kay County, Oklahoma using mini-sosie and geophysical well log data. M, 1982, Wright State University. 59 p.

Bunker, Rachel. A micro-faunule group from Puerto Armuelles, Panama. M, 1938, Columbia University, Teachers College.

Bunker, Russell Craig. Catastrophic flooding in the Badger Coulee area, South-central Washington. M, 1980, University of Texas, Austin.

Bunn, R. L. Action of aqueous alkali on pyrite. D, 1976, Iowa State University of Science and Technology. 145 p.

Bunnell, Mark D. Roof geology and coal seam characteristics of the No. 3 Mine, Hardscrabble Canyon, Carbon County, Utah. M, 1985, Brigham Young University.

Bunnell, Victoria D. The water structure of the San Pedro Basin, California borderland. M, 1969, University of Southern California.

Bunner, W. D. A paleoecological investigation of palynomorphs in the lower Gull River Formation (Middle Ordovician) of southwestern Ontario. M, 1987, University of Waterloo. 267 p.

Buongiorno, Benny. Handbook of Tierra Vieja Mountains, Presidio and Jeff Davis counties, Trans-Pecos, Texas. M, 1955, University of Texas, Austin.

Burack, Anna Camille. Geology along the Pinnell Mountain Trail, Circle Quadrangle, Alaska. M, 1983, University of New Hampshire. 98 p.

Burak, Roman W. Fluctuations in late Quaternary diatom abundances; stratigraphic and paleoclimatic implications from subantarctic deep-sea cores. M, 1987, Rutgers, The State University, Newark. 183 p.

Buratti, Bonnie Jean. Photometric properties of Europa and the icy satellites of Saturn. D, 1983, Cornell University. 226 p.

Burbach, George VanNess. Intermediate and deep seismicity and lateral structure of subducted lithosphere in the Circum-Pacific region. D, 1985, University of Texas, Austin. 127 p.

Burbach, George VanNess. Seismicity and tectonics of the subducted Cocos Plate. M, 1983, University of Texas, Austin. 93 p.

Burbach, S. P. Depositional environments of the Kodiak Shelf, Alaska. M, 1977, Texas A&M University.

Burbanck, George Palmer. Sediment and macro-faunal trends in the Altamaha Estuary, Georgia. M, 1972, Emory University. 124 p.

Burbank, Douglas West. Late Holocene glacier fluctuations on Mount Rainier and their relationship to the historical climate record. M, 1979, University of Washington. 84 p.

Burbank, Douglas West. The chronologic and stratigraphic evolution of the Kashmir and Peshawar intermontane basins, northwestern Himalaya. D, 1982, Dartmouth College. 291 p.

Burbank, W. S. Geology of the region about Melrose, Massachusetts. M, 1920, Massachusetts Institute of Technology. 46 p.

Burbey, Thomas J. Three dimensional numerical simulation of tritium and chloride-36 migration. M, 1984, University of Nevada. 124 p.

Burbidge, Geoffrey Harrison. A late Quaternary submarine outwash fan at St. Lazare, Quebec. M, 1985, University of Ottawa. 209 p.

Burbridge, Clarence E. The geology of a portion of the State of Anzoategui, Venezuela. M, 1930, Columbia University, Teachers College.

Burch, Alvin L. Petrology and U-Th potential of the eastern portion of the Precambrian Rawah Batholith, Larimer County, Colorado. M, 1983, Colorado State University. 400 p.

Burch, Charles Ivan. Enhancement and detection of linear features with application to the Appalachian region. M, 1985, Pennsylvania State University, University Park. 65 p.

Burch, Ellen. Tensleep Sandstone; diagenetic relation to tilted oil-water contacts in oil fields of the Southeast Big Horn Basin. M, 1982, University of Wyoming. 59 p.

Burch, Gary K. Provenance and diagenesis of the Ivishak Sandstone, northern Alaska. M, 1984, Texas A&M University. 178 p.

Burch, S. L. An evaluation of the Mississippian bedrock as a potential aquifer for Ames, Iowa. M, 1977, Iowa State University of Science and Technology.

Burch, Stephen Howell. Tectonic emplacement of the Burro Mountain ultramafic body, southern Santa Lucia Range, California. D, 1965, Stanford University. 193 p.

Burch, Terrill Lee. The morphology and hydrodynamics of potholes. M, 1971, University of Minnesota, Minneapolis. 18 p.

Burcham, Donald Preston. Radium analysis of submarine cores from Skagit Bay area. D, 1942, University of Washington. 30 p.

Burchard, Ernest F. The lignites of Dakota County, Nebraska. M, 1903, Northwestern University.

Burchett, Raymond Richard. Stratigraphy of the upper part of Kansas City Group (Pennsylvanian) in southeastern Nebraska and adjacent regions. M, 1959, University of Nebraska, Lincoln.

Burchfiel, Burrell Clark. Geology of the Spector Range Quadrangle, Nevada. D, 1961, Yale University.

Burchfiel, Burrell Clark. Geology of the Two Bar Creek area, Boulder Creek, California. M, 1958, Stanford University. 57 p.

Burchfield, Gail Robert. The geology of a portion of the Bridger Range, Montana. M, 1951, University of Iowa. 100 p.

Burchfield, George Edward. The geology of the Brookhaven oil field, Lincoln County, Mississippi; production from Tuscaloosa to Upper Cretaceous. M, 1950, Mississippi State University. 48 p.

Burchfield, Margaret R. Petrology of Tertiary igneous intrusives near Windy Mountain, Beartooth Butte Quadrangle, Absaroka Range, Northwest Wyoming. M, 1983, Southern Illinois University, Carbondale. 70 p.

Burchfield, William W., Jr. The Unaka Mountains of Tennessee and North Carolina. M, 1941, University of Tennessee, Knoxville.

Burck, Martin Stuart. The stratigraphy and structure of the Columbia River Basalt Group in the Salmon River area, Oregon. M, 1986, Portland State University. 108 p.

Burckle, Lloyd H. Late-Cenozoic planktonic diatoms from the eastern equatorial Pacific. D, 1971, New York University. 297 p.

Burckle, Lloyd H. Some Mississippian fenestrate Bryozoa from central Utah. M, 1964, Brigham Young University.

Burd, James R. A study of the clays of the Southeast Missouri Bariet District, Washington County, Missouri. M, 1980, Southern Illinois University, Carbondale. 84 p.

Burdelik, William Joseph. Crustal model beneath McMurdo Sound from seismic refraction and gravity data. M, 1981, Northern Illinois University. 114 p.

Burden, Cecil Ronald. A quantitative evaluation of zircon-crop measurements. M, 1973, Texas Tech University. 39 p.

Burden, Elliott T. Lower Cretaceous terrestrial palynomorph biostratigraphy of the McMurray Formation, northeastern Alberta. D, 1982, University of Calgary. 422 p.

Burden, Elliott Thomas. Pollen and algal assemblages in cored sediments from Gignac Lake and Second Lake (Simcoe Co., Ontario); relationship with lacustrine facies, geochemistry and vegetation. M, 1978, University of Toronto.

Burdette, David James. Sedimentology of Moenave-Kayenta equivalent strata in the northern Black Mountains, southern Nevada; early Mesozoic depositional environments and implications for Cenozoic tectonics. M, 1986, Southern Illinois University, Carbondale. 106 p.

Burdick, Dennis W. Upper Mississippian crinoids of Alabama. M, 1971, University of Iowa. 364 p.

Burdick, Donald G. Mixed clastic to carbonate deposition, Falmouth, Jamaica. M, 1984, University of Oklahoma. 134 p.

Burdick, Joseph Matthew. A gravity survey of Greene County, Ohio. M, 1985, Wright State University. 62 p.

Burdick, Lawrence James. Broad-band seismic studies of body waves. D, 1977, California Institute of Technology. 157 p.

Burdige, David Jay. The biogeochemistry of manganese redox reactions; rates and mechanisms. D, 1983, University of California, San Diego. 271 p.

Burditt, Marvin Reece. Surface and subsurface observations on the Sundance Formation of the southwestern Black Hills area. M, 1948, University of Oklahoma. 41 p.

Bureau, Serge. Zones de brèches associées à des gîtes de porphyres cuprifères dans le région de Chibougamau, Chibougamau, Quebec. M, 1981, Universite du Quebec a Chicoutimi. 103 p.

Buren, Mark Van see Van Buren, Mark

Buren, Wayne Martin Van see Van Buren, Wayne Martin

Burfeind, Walter John. A gravity investigation of the Tobacco Root Mountains, Jefferson Basin, Boulder Batholith, and adjacent areas of southwestern Montana. D, 1967, Indiana University, Bloomington. 90 p.

Burfoot, James D., Jr. The barite deposits of the Southern Appalachian states. M, 1925, University of Virginia. 70 p.

Burfoot, James Dabney, Jr. Talc and soapstone deposits of Virginia. D, 1929, Cornell University.

Burford, Arthur Edgar. Geology of the Medano Peak area, Sangre de Cristo Mountains, Colorado. D, 1960, University of Michigan. 230 p.

Burford, Arthur Edgar. Geology of the Sheep Springs Quadrangle, San Juan County, New Mexico. M, 1954, University of Tulsa. 54 p.

Burford, Robert Oliver. Strain analysis across the San Andreas Fault and Coast Ranges of California. D, 1967, Stanford University. 83 p.

Burg, John Parker. Maximum entropy spectral analysis. D, 1975, Stanford University. 123 p.

Burg, Robert Stanley. Study of the subsurface conditions in the north-central Texas oil fields. M, 1923, University of Missouri, Rolla.

Burgchardt, Carl Robert. Geology of the northwest portion of the Richwoods Quadrangle, Missouri. M, 1952, University of Iowa. 83 p.

Burgdorf, Gregory John. Geophysical exploration of the buried valley systems in Greene County, Ohio, for ground-water resources. M, 1983, Wright State University. 127 p.

Burge, Donald Lockwood. Intrusive and metamorphic rocks of the Silver Lake Flat area, American Fork Canyon, Utah. M, 1959, Brigham Young University. 46 p.

Burge, Furman Horace, Jr. Annotated bibliography of carbonate rock genesis. M, 1958, University of Michigan.

Burgel, William D. Structural geology of the Rapid Creek area, Pocatello Range, north of Inkom, Southeast Idaho. M, 1986, Idaho State University. 72 p.

Burgener, J. D., IV. Petrography of the Queen Maud Batholith, central Transantarctic Mountains, Ross Dependency, Antarctica. M, 1975, University of Wisconsin-Madison.

Burgener, John Albert. The stratigraphy and sedimentation of the Pictured Cliffs Sandstone and Fruitland Formation, Upper Cretaceous of the San Juan Basin. M, 1953, University of Illinois, Urbana.

Burger, Henry Robert. Structure, petrology, and economic geology of the Sheridan District, Madison County, Montana. D, 1966, Indiana University, Bloomington. 156 p.

Burger, John Allan. Geology of central Uinta County, Wyoming. M, 1955, University of Utah. 62 p.

Burger, John Allan. Mesa Verde Group in adjoining areas of Utah, Colorado and Wyoming. D, 1959, Yale University. 298 p.

Burger, John Robert. The structure and petrography of the Humboldt Mine area (Marquette, Michigan). M, 1970, Michigan State University. 56 p.

Burger, Roy W. Source mechanism of the May 18, 1983 St. Helens eruption from regional surface waves. M, 1984, Pennsylvania State University, University Park. 72 p.

Burger, Theodore Bernhard. A stochastic model to predict the effect of random waste inputs upon the variance of dissolved oxygen in a tidal river. D, 1983, Polytechnic University. 239 p.

Burger, William Hunt. A study of measurements made with the Model MRA-1 tellurometer. M, 1965, Ohio State University.

Burgert, Barrett L. Petrology of the Cambrian Tapeats Sandstone, Grand Canyon, Arizona. M, 1972, Northern Arizona University. 156 p.

Burgess, Anne E. The Tertiary stratigraphy of the North American coastal plain and the Caribbean region. M, 1929, Smith College. 272 p.

Burgess, Carl Foulds. The structural and stratigraphic evolution of Lake Tanganyika; a case study of continental rifting. M, 1985, Duke University. 42 p.

Burgess, Charles Harry. Stocks of the Highwood Mountains, Montana. D, 1936, Harvard University. 169 p.

Burgess, Christopher. Geology of the Sais Basin. M, 1975, University of South Carolina.

Burgess, Curtis William, Jr. Devonian stratigraphy and faunas of two cores from Antrim and Crawford counties, Michigan. M, 1950, University of Michigan.

Burgess, Diane Eleanor. Small scale deformation adjacent to the Darby Thrust, western Wyoming. M, 1974, University of Michigan.

Burgess, Frances C. Major economic geographic regions of West Virginia. M, 1927, Columbia University, Teachers College.

Burgess, Jack Donald. Cretaceous deposits of part of the Mesabi Range. M, 1955, University of Missouri, Columbia.

Burgess, Lawrence C. N. A study of colloid-mineral relationships with special reference to the Wisconsin lead-zinc ores. M, 1950, Northwestern University.

Burgess, Lawrence Charles Norman. The application of airphoto interpretation to watershed planning and development with special reference to flood susceptibility and frequency determinations. D, 1970, Cornell University. 783 p.

Burgess, Margaret V. A morphometric study of Bulimina aculeata d'Orbigny and B. marginata d'Orbigny, from 7,000 years B.P. to Recent, in the Gulf of Maine. M, 1988, University of Maine. 88 p.

Burgess, William J. Petrography and origin of the West Spring Creek Formation, Lower Ordovician, in Oklahoma. M, 1964, Columbia University, Teachers College.

Burgess, William Joseph. Carbonate paleoenvironments in the Arbuckle Group, West Spring Creek Formation, Lower Ordovician, in Oklahoma. D, 1968, Columbia University. 165 p.

Burgett, Thomas L. Air permeability study of the Kokomo Limestone of northwest Indiana. M, 1974, University of Toledo. 45 p.

Burggraf, Daniel Robert, Jr. Stratigraphy of the upper member, Koobi Fora Formation, southern Karari Escarpment, East Turkana Basin, Kenya. M, 1976, Iowa State University of Science and Technology.

Burggraf, Gloria Butson. Glacial geology of the West Tensleep Creek drainage basin, Bighorn Mountains, Wyoming. M, 1978, Iowa State University of Science and Technology.

Burgis, Winifred Ann. Late-Wisconsin history of northeastern Lower Michigan. D, 1977, University of Michigan. 444 p.

Burgis, Winifred Ann. The Imlay outlet of glacial lake Maumee, Imlay City (Lapeer County), Michigan. M, 1970, University of Michigan.

Burgoyne, Alfred Alexander. Mineralogy of coesite and petrographic association and distribution at Meteor Crater, Arizona. M, 1966, University of New Mexico. 127 p.

Burgy, Jacob H. The influence of topography on the construction of certain railroads in the United States. M, 1925, University of Wisconsin-Madison.

Burianyk, Michael J. A. A two-dimensional field study of the 1985 ice island reflection experiment. M, 1988, University of Saskatchewan. 203 p.

Burk, Cornelius Franklin, Jr. A regional study of the Silurian stratigraphy of the Gaspe Peninsula, Quebec. D, 1959, Northwestern University. 100 p.

Burk, Creighton A. Geology and structural history of the Alaska Peninsula. D, 1964, Princeton University. 189 p.

Burk, Creighton Alvin. Stratigraphy of the Frontier Formation along the southern margin of the Bighorn Basin, Wyoming. M, 1953, University of Wyoming. 115 p.

Burk, Mitchell Keith. Facies and depositional environments of the Energy Shale (Pennsylvanian) in southwestern Jefferson County, Illinois. M, 1982, Southern Illinois University, Carbondale. 138 p.

Burk, Raymond Ronald. Geological setting of the Teck-Corona gold-molybdenum deposit, Hemlo, Ontario. M, 1987, Queen's University. 241 p.

Burk, Robert L. Factors affecting $^{18}O/^{16}O$ ratios in cellulose. D, 1979, University of Washington. 125 p.

Burk, Robert L. Geology of a portion of the Shadow Mtns., Mojave Desert, Calif. M, 1971, University of Washington.

Burk, Roger. Palynological dating of sediment of E. central Florida. M, 1973, Florida State University.

Burkalow, Anastasia Van *see* Van Burkalow, Anastasia

Burkard, Richard Killiam. Changes in the formula of normal gravity resulting from changes in the reference ellipsoid of the Earth. M, 1959, Ohio State University.

Burkart, Burke. Geology of the Esquipulas, Chanmagua and Cerro Montecristo quadrangles, southeastern Guatemala. D, 1965, Rice University. 148 p.

Burkart, Burke. Thermoluminescence of calcite and aragonite. M, 1960, University of Texas, Austin.

Burkart, Michael R. Pollen biostratigraphy and late Quaternary vegetation history of the Bighorn Mountains, Wyoming. D, 1976, University of Iowa. 70 p.

Burkart, Michael R. Stratigraphy, structure, and petrography of Carboniferous rocks from Crystal mountain region, Arkansas. M, 1969, Northern Illinois University. 75 p.

Burke, Christopher Brian. Numerical simulation of the hillslope runoff processes. D, 1983, Purdue University. 164 p.

Burke, Collette Dick. Ostracod shape analysis and deep basin paleoecology of the Lake Michigan basin. D, 1983, University of Wisconsin-Milwaukee. 268 p.

Burke, David Alan. Oxygen isotope and thorium values of Arkansas bauxite and bauxitic kaolins. M, 1985, Indiana University, Bloomington. 77 p.

Burke, Dennis Bernan. An aerial photograph survey of Dixie Valley (Churchill and Pershing counties, west central Nevada). M, 1967, Stanford University. 51 p.

Burke, Dennis Bernan. Reinterpretation of the Tobin Thrust; pre-Tertiary geology of the southern Tobin Range, Pershing County, Nevada. D, 1973, Stanford University. 144 p.

Burke, Harold W. The petrography of the Mariana Limestone, Tinian, Mariana Islands. D, 1953, Stanford University. 144 p.

Burke, James Charles. A resistivity study of groundwater on Sandy Point, San Salvador Island. M, 1985, University of Akron. 139 p.

Burke, James M. Geological examination of northern Chester County, Pennsylvania. M, 1921, Lehigh University.

Burke, Jenie Lee, III. Sedimentology and paleohydraulics of the terraces of South Canadian River. M, 1959, University of Oklahoma. 103 p.

Burke, John James. The fauna of the Ames Limestone from Painter Hollow, Wellsburg, West Virginia. M, 1930, University of Pittsburgh.

Burke, John Joseph. Magnetic survey of Cape Cod (Massachusetts) region. M, 1951, Boston College.

Burke, Karl D. The differential extraction of radiogenic lead for use as a uranium pathfinder. M, 1976, Eastern Washington University. 19 p.

Burke, Margaret M. Compressional wave velocities in rocks from the Ivrea-Verbano and Strona-Ceneri zones, Southern Alps, northern Italy; implications for models of crustal structure. M, 1987, University of Wyoming. 78 p.

Burke, Michael R. Stratigraphic analysis of the Oak Openings sand, Lucas County, Ohio. M, 1973, University of Toledo. 108 p.

Burke, Raymond. Neoglaciation of Boulder Valley, Mount Baker, Washington. M, 1972, Western Washington University. 47 p.

Burke, Raymond Merle. Multiparameter relative dating (RD) techniques applied to morainal sequences along the eastern Sierra Nevada, California, and Wallowa Lake area, Oregon. D, 1979, University of Colorado.

Burke, Robert Francis. Multi-band aerial photography in geological analysis simulating orbital satellite photography of planetary bodies. D, 1967, Boston University. 179 p.

Burke, Roger Allen, Jr. Stable hydrogen and carbon isotopic compositions of biogenic methanes. D, 1985, University of South Florida, St. Petersburg. 118 p.

Burke, S. K. Recent benthic foraminifera of the Ontong Java Plateau. M, 1977, University of Hawaii.

Burke, Todd M. The petrography and chemistry of detrital ultramafic material at the Mineral Hill Mine, Sykesville mining district, Maryland and the role of accessory chromite in determining the origin of the body and associated sulfide ores. M, 1987, University of Maryland.

Burke, Willard F. The second derivative of the Earth's magnetic field in central United States. M, 1951, University of California, Berkeley. 21 p.

Burke-Griffin, Barbara Mary. Geology, petrology, and geochemistry of Black Butte volcanic neck, Gravelly Range, Montana. M, 1978, Wright State University. 102 p.

Burket, John Maxwell. The geology of the city of Waco and environs. M, 1960, Baylor University.

Burkett, David. Experimental alteration of basaltic glass under submarine conditions. M, 1970, Florida State University.

Burkett, Gerald G. A subsurface study of the Middle Pennsylvanian rocks of western McClain County, Oklahoma. M, 1957, University of Oklahoma. 49 p.

Burkett, Gerald R. A detailed grain size analysis of DSDP Leg 86 Northwest Pacific ash layers to determine the history of explosive volcanism on Japan and the Kuril Islands. M, 1987, SUNY at Buffalo. 118 p.

Burkett, Patti Jo. Significance of the microstructure of Pacific red clays to nuclear waste disposal. M, 1987, Texas A&M University.

Burkett, William C. Morphology and taxonomy of the eichwaldiid brachiopods. M, 1969, Miami University (Ohio). 72 p.

Burkhard, N. Upper mantle structure of the Pacific Basin from inversion of gravity and seismic data. D, 1977, University of California, Los Angeles. 179 p.

Burkhart, Patrick A. A stream-sediment reconnaissance survey of the northern Snowcrest Range, Madi-

son County, Montana. M, 1987, Wright State University. 80 p.

Burkholder, James Franklin. A subsurface and petrologic study of the Glen Rose Limestones (Lower Cretaceous) of South Texas. M, 1972, Memphis State University.

Burkholder, Paul. A study of the physical characteristics of the Arikaree Group of Wildcat Ridge in western Nebraska. M, 1941, University of Nebraska, Lincoln.

Burklew, Richard Hill, Jr. The hydrogeologic system of Trail Ridge above the basal clays near Folkston, Georgia. M, 1988, University of Florida. 191 p.

Burkley, Lewis A. Geochronology of the central Venezuelan Andes. D, 1976, Case Western Reserve University. 160 p.

Burks, Rachel Jane. Alleghenian deformation and metamorphism in southwestern Narragansett Basin, Rhode Island. M, 1981, University of Texas, Austin. 93 p.

Burks, Rachel Jane. Incremental and finite strains within ductile shear zones, Narragansett Basin, Rhode Island. D, 1985, University of Texas, Austin. 177 p.

Burley, Brian John. A study of some volcanic rocks from Harrison Mills, British Columbia. M, 1954, University of British Columbia.

Burley, Brian John. The physical stability of natrolite. D, 1956, McGill University.

Burling, Robert Jeffrey. Petrography of the McClosky oolitic limestone (Mississippian) on the Clay City Anticline of southeastern Illinois. M, 1974, Ohio University, Athens. 133 p.

Burma, Benjamin H. Some aspects of the theory and practice of quantitative invertebrate paleontology. D, 1947, University of Wisconsin-Madison.

Burman, Howard Richard, Jr. Grain orientation, paleocurrents, and reservoir trends. M, 1973, Oklahoma State University. 84 p.

Burmester, Russell Frederick. Llanite, a hypabyssal rhyolite porphyry from Llano County, Texas. M, 1966, University of Texas, Austin.

Burmester, Russell Frederick. Utility of granitic rocks for paleomagnetic research with an example from the Sierra Nevada of California. D, 1974, Princeton University. 314 p.

Burn, Christopher Robert. On the origin of aggradational ice in permafrost. D, 1986, Carleton University. 233 p.

Burnaman, M. D. Interpretation of regional gravity anomalies on the margin of the Northwest Gulf of Mexico. M, 1974, [University of Houston].

Burne, Eleanor. Economic versus pure science in American geology. M, 1916, Columbia University, Teachers College.

Burnel, Ralph Sherman. Geology of the Ahngayakasrakuvik Creek area, Romanzof Mountains, Alaska. M, 1959, University of Michigan.

Burnell, J. R., Jr. Petrology and structural relations of the Brule Lake intrusions, Cook County, Minnesota. M, 1976, University of Minnesota, Duluth.

Burnell, James Russell, Jr. Experimental and analytical studies in pelitic metamorphism. D, 1983, Brown University. 133 p.

Burnet, Frederick William. Felsic volcanic rocks and mineral deposits in the Buck Mountain Formation andesites, Okanogan County, Washington. M, 1976, University of Washington. 24 p.

Burnett, Adam S. Alluvial stream response to neotectonics in the Lower Mississippi Valley. M, 1982, Colorado State University. 194 p.

Burnett, Andrew Isaac. A new method for strain measurement, with a test case using conglomerates of the Missi Group in the Flin Flon Basin, Flin Flon, Manitoba. M, 1975, University of Saskatchewan. 90 p.

Burnett, Harold Morris. Applications of time-resolved spectroscopy to spectrochemical analysis. D, 1966, University of Texas, Austin. 100 p.

Burnett, Jerome B. A geological study of northeastern Coahuila, Mexico. D, 1918, University of Nebraska, Lincoln.

Burnett, John L. The geology of the southern portion of Frazier Mountain near Gorman, California. M, 1960, University of California, Berkeley.

Burnett, Linda S. Biostratigraphic identification of the Pliocene/Pleistocene boundary at Deep Sea Drilling Project Site 502, Colombian Basin, Caribbean Sea. M, 1981, Boston University. 48 p.

Burnett, Michael Welch. The occurrence and distribution of Ca, Sr, Ba, and Pb in marine ecosystems. D, 1980, California Institute of Technology. 250 p.

Burnett, Neill C. A biological evaluation of the effect of a flood pain sanitary landfill site (Lawrence, Kansas) on ground water quality. D, 1972, University of Kansas.

Burnett, R. D. An alternating direction Galerkin technique for simulation of groundwater contaminant transport in three dimensions. M, 1985, University of Waterloo. 133 p.

Burnett, Ronald Gordon. An evaluation of a shallow groundwater flow regime near Taber, Alberta. M, 1981, University of British Columbia. 275 p.

Burnett, Thomas L. Sedimentology of the Piankatank Estuary. M, 1966, University of South Carolina.

Burnett, Thomas Lawrence, Jr. Petrology of southeastern piedmont river sands, Georgia, South Carolina and North Carolina. D, 1971, Texas A&M University. 210 p.

Burnett, W. C. Phosphorite deposits from the sea floor off Peru and Chile; radiochemical and geochemical investigations concerning their origin. D, 1974, University of Hawaii. 164 p.

Burnett, W. C. Trace-element variations in some Central Pacific and Hawaiian sediments. M, 1971, University of Hawaii. 112 p.

Burnette, Charles R. Geology of the Middle Canyon, Whetstone Mountains, Cochise County, Arizona. M, 1957, University of Arizona.

Burnette, John Paul, III. Framework, processes, and evolution of the New River Inlet Complex. M, 1977, North Carolina State University. 187 p.

Burney, David Allen. Late Quaternary environmental dynamics of Madagascar. D, 1986, Duke University. 285 p.

Burnham, C. Wayne. Metallogenic provinces of the Southwestern United States and northern Mexico. D, 1955, California Institute of Technology.

Burnham, Charles W. The structures and crystal chemistry of the aluminum-silicate minerals. D, 1961, Massachusetts Institute of Technology. 505 p.

Burnham, Robert Lawrence. Mylonite zones in the crystalline basement rocks of Sixmile Creek and Yankee Jim Canyon, Park County, Montana. M, 1982, University of Montana. 94 p.

Burnham, Robyn Jeanette. Foliar morphological analysis of the Ulmoideae (Ulmaceae) from the early Tertiary of North America. M, 1983, University of Washington. 109 p.

Burnham, Robyn Jeanette. Inferring vegetation from plant-fossil assemblages; effects of depositional environment and heterogeneity in the source vegetation on assemblages from modern and ancient fluvial-deltaic environments. D, 1987, University of Washington. 235 p.

Burnham, Rollins. The geology of the southern part of the Pueblo Mountains, Humboldt County, Nevada. M, 1971, Oregon State University. 114 p.

Burnham, Willis Lee. The geology and ground water conditions of the Etiwanda-Fontana area (California). M, 1953, Pomona College.

Burnie, S. W. Sulphur isotopes in the White Pine Mine, Ontonagon County, Michigan. M, 1971, McMaster University. 117 p.

Burnie, Stephen Wilbur. A sulphur and carbon isotope study of hydrocarbons from the Devonian of Alberta, Canada. D, 1979, University of Alberta. 339 p.

Burnitt, Seth Charles. Geology of the Red Mountain area, Llano, Gillespie, and Blanco counties, Texas. M, 1961, University of Texas, Austin.

Burnley, Gertrude I. The conodonts of the shale overlying the Lexington coal bed of Lafayette County and Jackson County, Missouri. M, 1938, University of Missouri, Columbia.

Burnley, Pamela Carol. Metamorphic petrology, structure and stratigraphy of the Chloride Cliff area, Funeral Mountains, Death Valley, California. M, 1986, University of California, Davis. 200 p.

Burns, Allan Fielding. The role of interlayer cations in micaceous minerals as revealed by X-ray diffraction and infrared absorption studies. D, 1963, Purdue University. 81 p.

Burns, Beverly Ann. The sedimentology and significance of a middle Proterozoic braidplain; Chediski Sandstone Member of the Troy Quartzite, central Arizona. M, 1987, Northern Arizona University. 143 p.

Burns, C. A. The Clare River area of southeastern Ontario. M, 1951, Queen's University. 90 p.

Burns, Christopher. A study of North American microtektites from Barbados, West Indies. M, 1986, University of Delaware.

Burns, Danny E. Age and origin of the sedimentary dikes of the Pipe Creek Junior (Silurian) Reef, Grant County, Indiana. M, 1984, Ball State University. 141 p.

Burns, David Bruce. A synthesis, stability, and X-ray study of some of the mercury minerals from Terlingua, Brewster County, Texas. M, 1980, Kent State University, Kent. 70 p.

Burns, Donna Jane. A trace element study of carbonate sediments from the Lord Howe Rise, Southwest Pacific Ocean. M, 1985, Duke University. 110 p.

Burns, Douglas A. Speciation and equilibrium modelling of soluble aluminum in a small headwater stream, Shenandoah National Park, Virginia. M, 1982, University of Virginia. 105 p.

Burns, Gerald Ray. Foraminiferal biostratigraphy of the Upper Cretaceous rocks in the subsurface of Lakhra, West Pakistan. M, 1965, University of Illinois, Urbana.

Burns, Gregory K. Middle Precambrian black slate of the Baraga Basin, Baraga County, Michigan. M, 1975, Bowling Green State University. 129 p.

Burns, James Arthur. Late Quaternary palaeoecology and zoogeography of southwestern Alberta; vertebrate and palynological evidence from two Rocky Mountain caves. D, 1984, University of Toronto.

Burns, James Richard. The geology of Fredericksburg, Virginia, and vicinity. M, 1950, University of Virginia. 77 p.

Burns, James William. Regional study of the Upper Silurian Salina evaporites in the Michigan Basin. M, 1962, Michigan State University. 90 p.

Burns, Lary Kent. Petrogenesis and chemistry of cordierite and associated minerals in contact-metamorphic rocks. D, 1970, University of California, Berkeley. 147 p.

Burns, Lary Kent. Sedimentary petrography of the Umpqua Formation in the axial part of the southern Coast Range, Oregon. M, 1964, University of Oregon. 154 p.

Burns, Laurel E. The Border Ranges ultramafic and mafic complex; plutonic core of an intraoceanic island arc. D, 1983, Stanford University. 203 p.

Burns, Phillip E. The regime of Grizzly Glacier, central Brooks Range, Alaska. M, 1984, SUNY at Buffalo. 94 p.

Burns, R. G. An improved sediment delivery model for Piedmont forests. D, 1978, University of Georgia. 81 p.

Burns, Richard L., Jr. Controls on observed variations in surface geochemical exploration; a case study, the Albion Trend oilfield. M, 1986, Wayne State University.

Burns, Robert Donald. The geology of the Jardun Mine, Sault Ste. Marie, Ontario. M, 1956, Michigan Technological University. 58 p.

Burns, Robert Earle. A model of sedimentation in a small, sill-less embayed estuary of the Pacific Northwest. D, 1962, University of Washington. 117 p.

Burns, Robert Earle. Geology along Monocacy Creek at Bethlehem, Pennsylvania. M, 1950, Lehigh University.

Burns, Robert Obed. Geiger counter characteristics. D, 1937, University of Illinois, Urbana.

Burns, Robert Parker. New facts concerning the stratigraphic position of the Independence Shale of Iowa. M, 1954, University of Wisconsin-Madison.

Burns, Robert R. Effects of manganese and iron on algal growth in an Adirondack lake. D, 1977, Clarkson University. 229 p.

Burns, Roger George. Electronic spectra of silicate minerals; application of crystal-field theory to aspects of geochemistry; 1 volume. D, 1965, University of California, Berkeley.

Burns, Scott D. Comparison of statically, seismically, and ultrasonically determined Young's modulus for polycrystalline freshwater-ice. M, 1979, University of Wisconsin-Milwaukee. 93 p.

Burns, Scott Frimoth. Alpine soil distribution and development, Indian Peaks, Colorado Front Range. D, 1980, University of Colorado. 425 p.

Burns, Stephen James. Sedimentary processes of a deep-water carbonate slope; southern Little Bahama Bank, Bahamas. M, 1983, University of North Carolina, Chapel Hill. 145 p.

Burns, Stephen James. Three studies of the origin and geochemistry of dolomite. D, 1987, Duke University. 269 p.

Burns, Thomas Daniel. Petrology of the Red Bluff Clay-Bumpnose Limestone; Oligocene; southwestern Alabama. M, 1974, University of New Orleans.

Burns, William A. Collecting and analyzing deep-sea sediments from central Gulf of Mexico. M, 1970, Texas A&M University.

Burnside, Michael James. Alteration petrology in the Potosi mining district, Tobacco Root Mountains, Montana. M, 1975, University of Montana. 70 p.

Burnson, Terry Quentin. Sedimentological study of Matanzas Inlet, Florida, and adjacent areas. M, 1972, University of Florida. 104 p.

Buros, Oscar Krisen. Wastewater reclamation at St. Croix, U. S. Virgin Islands. D, 1975, University of Florida. 348 p.

Burr, Cynthia D. Paleomagnetism and tectonic significance of the Goble Volcanics of southern Washington. M, 1978, Western Washington University. 235 p.

Burr, Freeman F. Some geologic factors in the distribution of the red cedar (Juniperus virginians). M, 1913, Columbia University, Teachers College.

Burr, John H., Jr. Ostracoda of the Dubuque and Maquoketa formations of Minnesota and northern Iowa. M, 1958, University of Minnesota, Minneapolis. 60 p.

Burr, Jonathan L. Bedrock geology of the Ellsworth and eastern part of the Amenia quadrangles, Connecticut and New York. M, 1986, University of Massachusetts. 155 p.

Burr, Norman Charles. The relationship of source parameters of oceanic transform earthquakes to plate velocity and transform length. M, 1977, Massachusetts Institute of Technology. 87 p.

Burr, S. V. Local variations in the intrusive near Kingston, Ontario. M, 1940, Queen's University. 65 p.

Burrell, Herbert Cayford. A statistical and laboratory investigation of ore types at Broken Hill, Australia. D, 1946, Harvard University.

Burrell, Jennifer Ann. Distribution, ecology, and taxonomy of recent freshwater Ostracoda of Lake Mendota, Wisconsin. M, 1971, University of Wisconsin-Madison.

Burrell, Stephen D. Geology of an area southwest of Silverton, San Juan County, Colorado. M, 1967, University of Colorado.

Burrell, Steve C. Estimating groundwater recharge in irrigated areas of an agricultural basin. M, 1987, University of Idaho. 172 p.

Burress, George Thomas. Spectrochemical correlation of the Clay Creek salt dome formations. M, 1951, Texas Tech University. 73 p.

Burridge, Paul Brian. Failure of slopes. D, 1987, California Institute of Technology. 257 p.

Burrier, Dale. The paleoecology of the Chadakoin Formation of Chautauqua County. M, 1977, SUNY, College at Fredonia. 129 p.

Burris, Robert Leroy. Effect of tension cutoff between the soil and foundation on structural response. D, 1981, University of Maryland. 154 p.

Burritt, E. C. A ground water study of part of the southern Laramie Basin, Albany County, Wyoming. M, 1962, University of Wyoming. 167 p.

Burrough, Herman C. Surface geology of the Jumbo Quadrangle, Pushmataha County, Oklahoma. M, 1960, University of Oklahoma. 131 p.

Burroughs, R. H., III. The structural and sedimentological evolution of the Somali Basin; paleooceanographic interpretations. D, 1974, Woods Hole Oceanographic Institution. 285 p.

Burroughs, Richard Hansford. The structural and sedimentological evolution of the Somali Basin; paleooceanographic interpretation. D, 1975, Massachusetts Institute of Technology. 285 p.

Burroughs, Richard K. Structural geology of the Enola earthquake swarm area, Faulkner County, Arkansas. M, 1987, University of Arkansas, Fayetteville.

Burroughs, Richard L. Geology of the San Luis Hills, south central Colorado. D, 1972, University of New Mexico. 140 p.

Burroughs, Richard Lee. The structural geology of the Foy Ridge area, Twin Buttes, Arizona. M, 1960, University of Arizona.

Burroughs, Wilbur Greeley. Geography and stratigraphy of the Susquehanna Basin, in the Towanda region, Pennsylvania. D, 1932, Cornell University.

Burroughs, William Alfred. Direct determination of hydralic equivalence using fluorescent sand tracers in the beach nearshore zone. D, 1982, Syracuse University. 148 p.

Burroughs, William Alfred. Structure of the Dead River Formation in the Forks Quadrangle, west-central Maine. M, 1979, Syracuse University.

Burrows, David Robert. Geology and geochemistry of molybdenite mineralization within an Archean granodiorite intrusion, Mink Lake, N.W. Ontario. M, 1984, University of Toronto.

Burrows, Lees Joslyn, Jr. Earthquake magnitude evaluation at Florissant. M, 1954, St. Louis University.

Burrows, Lloyd A., III. A gravity and magnetic survey in the Hueco Mountains, western Diablo Plateau, Texas. M, 1984, University of Texas at El Paso.

Burrows, Steven Mark. Oxygen isotope evidence for the conditions of diagenesis, Muddy Formation, east flank of the Powder River basin, Montana and Wyoming. M, 1985, Case Western Reserve University. 226 p.

Burrows, Vernon C. Subsurface stratigraphy and paleoenvironmental interpretation of the Mississippian Berea Sandstone and Bedford Formation of Medina County, Ohio. M, 1988, Kent State University, Kent. 121 p.

Burruss, Robert Carlton. Analysis of fluid inclusions in graphitic metamorphic rocks from Bryant Pond, Maine, and Khtada Lake, British Columbia; thermodynamic basis and geologic interpretation of observed fluid compositions and molar volumes. D, 1977, Princeton University. 167 p.

Bursaw, Richard B. The influence of hydrogeologic variables on well yields on Swans Island, Maine. M, 1978, University of New Hampshire. 129 p.

Burshears, C. A. The geology and petrology of a portion of the Elkahatchee Quartz Diorite Gneiss, Coosa and Tallapoosa counties, Alabama. M, 1978, Memphis State University.

Burst, John Frederick. The clay mineralogy of two typical Missouri fireclays. D, 1950, University of Missouri, Columbia.

Burston, Michael R. Tectonics and sedimentation during the deposition of the Breathitt Formation (Pennsylvanian) in part of the Eastern Kentucky coal field. M, 1983, Eastern Kentucky University. 66 p.

Burt, Donald McLain. Mineralogy and geochemistry of Ca-Fe-Si skarn deposits. D, 1972, Harvard University.

Burt, Edward Ramsey, III. Geology of the northwest eighth of the Troy, North Carolina, Quadrangle. M, 1967, University of North Carolina, Chapel Hill. 34 p.

Burt, Edward Ramsey, III. Petrology of the Mitchell Mesa rhyolite (Oligocene and Younger(?)), Trans-Pecos Texas. D, 1970, University of Texas, Austin. 117 p.

Burt, Robert John. Springs in the Pigeon Point area, San Mateo County, California. M, 1966, Stanford University.

Burt, Ronald Allen. Ground-water chemical evolution and diagenetic processes in the upper Floridan Aquifer, southern South Carolina and northeastern Georgia. D, 1988, University of South Carolina. 270 p.

Burt, William D. The geology of the Collier Butte area, Southwest Oregon. M, 1963, University of Wisconsin-Madison. 93 p.

Burtch, Steven Douglas. Diagenesis of fine-grained Upper Cretaceous and Tertiary clastic rocks from the Nova Scotia shelf and slope. M, 1986, University of Alberta. 132 p.

Burtner, Don Reed. Analysis of coal by neutron activation. D, 1983, University of California, Irvine. 151 p.

Burtner, Roger L. Paleocurrent and petrographic analysis of the Catskill facies of southeastern New York and northeastern Pennsylvania. D, 1965, Harvard University.

Burtner, Roger Lee. Geology of the upper Crystal Springs Reservoir, Cahill Ridge area, San Mateo County, California. M, 1959, Stanford University.

Burtner, Roger Lee. Paleogeology of the central Panhandle of Texas. M, 1959, Stanford University.

Burton, Benjamin Paul. Thermodynamic analysis of the systems $CaCO_3MgCO_3$, $\propto$-Fe_2O_3, and Fe_2O_3-$FeTiO_3$. D, 1982, SUNY at Stony Brook. 187 p.

Burton, Bradford Robert. Stratigraphy of the Wood River Formation in the eastern Boulder Mountains, Blaine and Custer counties, south-central Idaho. M, 1988, Idaho State University. 165 p.

Burton, Bruce H. Paragenetic study of the San Martin Mine, near Sombrerete, Mexico. M, 1975, University of Minnesota, Minneapolis. 98 p.

Burton, Dale M. Petrology of metadiabase intrusions in the White Rock Quadrangle, western North Carolina. M, 1979, Eastern Kentucky University. 77 p.

Burton, Donald MacLaren. The geology of the Cam uranium deposit, Cardiff Township, Ontario, Canada. M, 1984, University of New Brunswick. 220 p.

Burton, Elizabeth Ann. Laboratory investigation of the effects of seawater chemistry on carbonate mineralogy. D, 1988, Washington University. 306 p.

Burton, Elizabeth Ann. X-ray diffraction of natural high and low Mg calcites. M, 1984, University of Miami. 148 p.

Burton, F. R. Geology of the district about Lake Aylmer, Eastern Township, Quebec. D, 1933, McGill University.

Burton, Guy C., Jr. Geology of the Squaw Creek area, Wind River Mountains, Fremont County, Wyoming. M, 1952, University of Missouri, Columbia.

Burton, Jacqueline C. Experimental and mineralogical studies of skarn silicates. D, 1978, University of Tennessee, Knoxville. 155 p.

Burton, Jacqueline C. Lithologic control of temperature and CO_2 pressure during metamorphism of siliceous carbonates at Wind Mountain, Otero County, New Mexico. M, 1974, Northeast Louisiana University.

Burton, James Hutson. Selected petrologic applications of back-scattered electron imaging. D, 1986, Arizona State University. 114 p.

Burton, James Hutson, III. Wyrdite; petrogenesis of an extraordinary mining assemblage. M, 1981, Cornell University.

Burton, Jeffrey P. Radioactive mineral occurrences, Mt. Prindle area, Yukon-Tanana Uplands, Alaska. M, 1981, University of Alaska, Fairbanks. 72 p.

Burton, Michael D. Chemical and mineralogical characteristics of suspended particulate matter and surface sediments near Cape Hatteras, North Carolina. M, 1977, University of South Florida, St. Petersburg. 151 p.

Burton, Robert Clyde. Conodonts from the Kinkaid Formation in the Illinois Basin. M, 1959, Texas Tech University. 77 p.

Burton, Robert Clyde. Conodonts of the Mississippian System in the Sacramento Mountains, New Mexico. D, 1965, University of New Mexico. 215 p.

Burton, Steven Mark. Structural geology of the northern part of Clarkston Mountain, Malad Range, Utah and Idaho. M, 1973, Utah State University. 54 p.

Burton, Vinston. Effect of earth movements on slope and drainage characteristics, western San Gabriel Mountains, California. D, 1974, University of California, Los Angeles.

Burton, W. D. Ore deposits at Premier, British Columbia. M, 1925, Massachusetts Institute of Technology.

Burton, William Chapin. Geology of the Scott Bar Mountains, Northern California. M, 1982, University of Oregon. 120 p.

Burton, William Dunn. Geology of the western part of the La Madre Mountain area, Clark County, Nevada. M, 1962, University of California, Los Angeles.

Burwash, E. M. J. The geology of Michipicoten Island. D, 1914, University of Toronto.

Burwash, Edward M. J. The geology of Vancouver and vicinity. D, 1915, University of Chicago. 103 p.

Burwash, Elizabeth Jean. Diagenesis of Tertiary clastic rocks of the Carmanah Group, Vancouver Island, Canada. M, 1986, University of Alberta. 127 p.

Burwash, Ronald Allan. The Precambrian under the Central Plains of Alberta. M, 1951, University of Alberta. 121 p.

Burwash, Ronald Allen McLean. A reconnaissance of the subsurface Precambrian of the Province of Alberta, Canada. D, 1955, University of Minnesota, Minneapolis. 74 p.

Burwell, Howard Beirne. The stratigraphy and paleontology of the Lobelville Formation of the central Tennessee Basin. M, 1930, Vanderbilt University.

Bury, Curtis A. The geology of the Cedar Mountain Complex, Minnesota River valley. M, 1958, University of Minnesota, Minneapolis. 31 p.

Busacca, Alan James. Geologic history and soil development, northeastern Sacramento Valley, California. D, 1982, University of California, Davis. 372 p.

Busanus, James William. Paleontology and paleoecology of the Mauch Chunk Group in northwestern West Virginia. M, 1974, Bowling Green State University. 388 p.

Busbey, Arthur B., III. Functional morphology of the head of Pristichampsus vorax (Crocodilia, Eusuchia) from the Eocene of North America. M, 1977, University of Texas, Austin.

Busby, Clarence Edward. A detailed study of the Table Rock Anticline and the Humboldt Fault in southeastern Nebraska. M, 1931, University of Nebraska, Lincoln.

Busby, John Cifford. Sedimentation characteristics of Recent sands of the Middle Loup and Dismal rivers. M, 1950, University of Nebraska, Lincoln.

Busby, Linda Lucille. Studies in the functional morphology of some fenestellid bryozoas. M, 1974, University of California, Davis. 77 p.

Busby, Roswell F. Sediments and reef corals of Cayo Arenas, Campeche Bank, Yucatan, Mexico. M, 1965, Texas A&M University.

Busby-Spera, Cathy Jeanne. Paleogeographic reconstruction of a submarine volcanic center; geochronology, volcanology and sedimentology of the Mineral King roof pendant, Sierra Nevada, California. D, 1983, Princeton University. 317 p.

Busch, Danial Adolph. Tetraseptate corals of the Hamilton of western New York. M, 1936, Ohio State University.

Busch, Daniel Adolph. The stratigraphy and paleontology of the Niagaran strata (Silurian) of west-central Ohio and adjacent northern Indiana. D, 1939, Ohio State University.

Busch, John Daniel. Geometry, depositional environments, diagenesis, and economic potential of the Codell Sandstone (Upper Cretaceous) in western Nebraska and eastern Wyoming. M, 1976, University of Nebraska, Lincoln.

Busch, Karl M. Structure and stratigraphy of the B Zone (Lansing-Kansas City groups) in Red Willow County, Nebraska. M, 1977, University of Nebraska, Lincoln.

Busch, Richard Munroe. Stratigraphic analysis of Pennsylvanian rocks using a hierarchy of transgressive-regressive units. D, 1984, University of Pittsburgh. 449 p.

Busch, Robert Edward, Jr. Techniques for the recognition of articulate brachiopods in thin section, with examples from the Mankomen Group (Permian), Alaska. D, 1983, University of California, Davis. 428 p.

Busch, William Henry. Dish structures in some ancient subaqueous sandy debris flow deposits. M, 1977, University of Illinois, Urbana.

Busch, William Henry. The physical properties, consolidation behavior, and stability of the sediments of the Peru-Chile continental margin. D, 1981, Oregon State University. 149 p.

Buschbach, Thomas Charles. Lithology and distribution of the Chouteau Limestone in Illinois. M, 1951, University of Illinois, Urbana.

Buschbach, Thomas Charles. Stratigraphy of Cambrian and Ordovician formations of northeastern Illinois. D, 1959, University of Illinois, Urbana. 132 p.

Busche, F. D. Mineralogy of the pegmatoids of Moiliili Quarry. M, 1968, University of Hawaii. 38 p.

Busche, Frederick D. Major and minor element contents of coexisting olivine, orthopyroxene, and clinopyroxene in ordinary chondritic meteorites. D, 1975, University of New Mexico. 75 p.

Buscheck, Thomas Alan. The hydrothermal analysis of aquifer thermal energy storage. D, 1984, University of California, Berkeley. 216 p.

Buseck, Peter Robert. Contact metasomatic deposits at Concepcion del Oro, Mexico; Tem Piute, Nevada; and Silver Bell, Arizona. D, 1962, Columbia University, Teachers College. 254 p.

Busen, Karen E. Silicoflagellate stratigraphy, Leg 36, Deep Sea Drilling Project. M, 1978, Florida State University.

Busen, Ken. Historical and sedimentary analysis of the parallel beach ridges at Hammond Bay, Mich. M, 1978, Florida State University.

Busenberg, Eurybiades. Investigation of blistering phenomena in serpentinite (verde antique) from Rochester, Vermont. M, 1967, New York University.

Busenberg, Eurybiades. The kinetics of dissolution of potassium feldspars and plagioclases at 25 C and 1 atmosphere P_{CO2}. D, 1975, SUNY at Buffalo. 138 p.

Bush, Alfred Lerner. Petrology of zinc ores at Edwards, St. Lawrence County, New York. M, 1946, University of Rochester. 65 p.

Bush, Asahel. Economic geology of the Billy Goat copper prospect, Okanogan County, Washington. M, 1970, University of Washington. 36 p.

Bush, Bruce Allen. Investigation of groundwater quality in unincorporated Greene County, Missouri. M, 1980, Southwest Missouri State University. 60 p.

Bush, Charles Vincent. Dundee Fields in the central Michigan Basin. M, 1983, Michigan State University. 95 p.

Bush, Daniel A. The stratigraphy and paleontology of the Niagaran strata of west-central Ohio and adjacent northern Indiana. D, 1938, Ohio State University.

Bush, David M. Equilibrium sedimentation; north insular shelf of Puerto Rico. M, 1977, Duke University. 120 p.

Bush, Edward Allen, Jr. Drift thickness and bedrock topography of the Toledo area, Ohio. M, 1966, Bowling Green State University. 24 p.

Bush, Edward Calvin. A paleocurrent study of the Conemaugh Series in southeastern Ohio. M, 1965, Ohio University, Athens. 59 p.

Bush, Gordon L. Geology of upper Ojai Valley. M, 1956, University of California, Los Angeles.

Bush, James. A preliminary report on the foraminifera of Biscayne Bay, Florida, and their ecological relations. M, 1949, Indiana University, Bloomington. 50 p.

Bush, James. Foraminifera and sediments of Biscayne Bay, Florida, and their ecology. D, 1958, University of Washington. 158 p.

Bush, James Gilbert. Geology of the northeast part of the Nemo Quadrangle, Black Hills, South Dakota. M, 1982, South Dakota School of Mines & Technology. 165 p.

Bush, John Harold, Jr. The basalts of Yellowstone Valley, southwestern Montana. M, 1967, Montana State University. 66 p.

Bush, John Harold, Jr. The Upper Cambrian stratigraphy of central Colorado. D, 1973, Washington State University. 231 p.

Bush, Louise Altha. Mine examinations, valuations, and reports. M, 1929, University of Michigan.

Bush, Mark M. The geology of Round Mountain, a bimodal volcanic field in Northwest Arizona. M, 1986, SUNY at Buffalo. 110 p.

Bush, Richard R. Gravity survey of Tyhee area, Bannock County, Idaho. M, 1980, Idaho State University. 33 p.

Bush, Robert Nelson. Small Salinian ultramafic near Jamesburg, California. M, 1981, Stanford University. 123 p.

Bush, Thomas A. Hydrothermal alteration and copper mineralization at Copper Glance, Okanogan County, Washington. M, 1983, Washington State University. 155 p.

Bush, William Robert, Jr. An econometric model of the world zinc industry in the twentieth century. D, 1979, Stanford University. 93 p.

Bushara, Mohammed N. Tectonic evolution of the Lake Van area, relation of the North Anatolia and Zagros deformation belts; application of the large format camera LFC imagery. M, 1987, University of Washington. 84 p.

Bushee, Jonathan. Petrology and potassium-argon age dating of several alkaline igneous rock bodies, southeastern Brazil. D, 1971, University of California, Berkeley. 145 p.

Bushman, Arthur Vern. Pre-Needles Range silicic volcanism; Cowboy Pass Tuff of west central Utah. M, 1973, Brigham Young University. 190 p.

Bushman, James Richard. Geology of the Barquisimeto area, Venezuela. D, 1958, Princeton University. 218 p.

Bushnell, David. Continental shelf sediments in the vicinity of Newport, Oregon. M, 1964, Oregon State University. 107 p.

Bushnell, Hugh Pearce. Geology of the McRae Canyon area, Sierra County, New Mexico. M, 1953, University of New Mexico. 106 p.

Bushnell, Kent Orpha. The geology of the Rowland Quadrangle, Nevada. D, 1955, Yale University. 188 p.

Bushnell, Steven Ensign. Paragenesis and zoning of the Cananea-Duluth breccia pipe, Sonora, Mexico. D, 1982, Harvard University. 476 p.

Buskirk, Donald Robert Van see Van Buskirk, Donald Robert

Buskirk, Steven C. Van see Van Buskirk, Steven C.

Buss, Barbara Ann. Suspended sediments in continental shelf waters off Cape Hatteras, North Carolina. M, 1972, University of Illinois, Chicago.

Buss, David R. An investigation of some physicochemical properties of a pozzolanic material to determine its suitability as a sanitary landfill liner. M, 1977, University of Toledo. 110 p.

Buss, David Roger. Evaluation of the connector well roof dewatering method in the abatement of acidic mine drainage from the Arnot No. 2 Mine, Tioga County, Pennsylvania. D, 1986, Pennsylvania State University, University Park. 741 p.

Buss, Fred Earle. The physiography of the southern Wasatch Mountains and the adjacent valley lands with especial reference to the origin of topographic forms. M, 1924, Stanford University. 87 p.

Buss, L. W. Competition of marine hard-substrata; pattern, process and mechanism. D, 1980, The Johns Hopkins University. 123 p.

Buss, Walter R. A preliminary study of the physiographic types of Utah. M, 1933, Brigham Young University. 188 p.

Buss, Walter Richard. A model study, implemented by slow motion photography, of rapid mass movements in granular materials. D, 1964, Stanford University. 202 p.

Bussa, Kathleen Louise. Sulfide-silicate equilibria in rocks of staurolite grade in NW Maine. M, 1973, University of Wisconsin-Madison. 79 p.

Bussey, Floyd Robert. Analysis of gravity measurements in the Great Lakes region. M, 1949, University of Michigan.

Bussey, Steven D. Geology of the iron dyke copper-gold massive sulfide deposit and its relationship to facies of the Permian Hunsaker Creek Formation, Snake River Canyon, Oregon and Idaho. D, 1988, Colorado School of Mines. 202 p.

Bussières, Louise. La Série flyschoide de l'Ordovicien moyen tardif du Comté de Charlevoix, Québec. M, 1978, Universite Laval.

Bussod, Gilles Yves Albert. Thermal and kinematic history of mantle xenoliths from Kilbourne Hole, New Mexico. M, 1981, University of Washington. 74 p.

Bust, Vivian Kay. Constitutional supercooling, a mechanism for oscillatory zoning in plagioclase. M, 1980, Michigan State University. 74 p.

Bustin, R. Marc. The Eureka Sound and Beaufort formations, Axel Heiberg and West Central Ellesmere Islands, District of Franklin. M, 1977, University of Calgary. 208 p.

Bustin, Robert Marc. Structural features of coal measures of the Kootenay Formation, southeastern Canadian Rocky Mountains. D, 1980, University of British Columbia.

Buswell, Karyn. A depositional model for the lower Freeport coal seam in central western Pennsylvania. M, 1980, University of South Carolina.

Buswell, Michael Douglas. Subsurface geology of the Oshoto uranium deposit, Crook County, Wyoming. M, 1982, South Dakota School of Mines & Technology. 84 p.

Buszka, Paul Mark. Hydrogeochemistry of contaminated groundwater in a sand aquifer at a landfill near North Bay, Ontario. M, 1982, University of Waterloo. 92 p.

Butcher, Robert H. Comparison of portable geophysical exploration instruments. M, 1962, University of Colorado.

Butcher, Seldon D. A study of the Morrison and Ingalls districts, Oklahoma. M, 1923, University of Oklahoma. 19 p.

Butcher, William S. Part I, Lithology of the offshore San Diego area; Part II, Foraminifera, Coronado Bank, California. D, 1951, University of California, Los Angeles.

Butera, Joseph G. Depositional environments and petrology of Mesaverde Formation (upper Cretaceous), central Wyoming. M, 1971, University of Missouri, Columbia.

Buterbaugh, Gary Jay. Petrology of the lower Middle Cambrian Langston Formation, north-central Utah and southeastern Idaho. M, 1982, Utah State University. 166 p.

Buteux, Christopher Blaine. Variations in magnitude and direction of longshore currents along the central New Jersey coast. M, 1982, Rutgers, The State University, New Brunswick. 132 p.

Butherus, D. L. Heavy metal contamination in soil around a lead smelter in Southeast Missouri. M, 1975, University of Missouri, Rolla.

Buthman, B. David. Correlation of zones within the "Hunton" Formation in southeastern Nebraska. M, 1949, University of Nebraska, Lincoln.

Butkus, Timothy Anton. Sedimentology and depositional environments on the Great Blue Limestone (late Mississippian), Northcentral Utah. M, 1975, University of Utah. 143 p.

Butler, Arthur Pierce, Jr. Tertiary and Quaternary geology of the Tusas-Tres Piedras area, New Mexico. D, 1946, Harvard University.

Butler, B. F. An examination of the application of the draft convention on the Law of the Sea to prospective polymetallic sulfide mining at ocean rift zones. M, 1982, University of Washington. 78 p.

Butler, Barbara A. Petrology and geochemistry of the Ringing Rocks Pluton, Jefferson County, Montana. M, 1983, University of Montana. 70 p.

Butler, Bert S. Petrographic study of rocks from Yakutat Bay region, Alaska. M, 1907, Cornell University.

Butler, Bertram Theodore. The geomorphology of the Triassic basin in New Jersey. D, 1933, New York University. 220 p.

Butler, Brian Faraday. Tops of epithermal veins in the Axell District, Paltoro Caldera, San Juan Mountains, Conejos County, southeastern Colorado. M, 1985, University of Washington. 67 p.

Butler, Charles R. Structure of the Post-Cambrian formations in the vicinity of Coal Creek, Colorado. M, 1950, University of Colorado.

Butler, D. M. The ionosphere of Venus. D, 1975, Rice University. 263 p.

Butler, David. Potential-field-data analysis. D, 1969, Colorado School of Mines. 157 p.

Butler, David Ray. "Cambrian" oil horizons in a portion of west-central Texas. M, 1957, University of Oklahoma. 71 p.

Butler, David Ray. Late Quaternary glaciation and paleoenvironmental changes in adjacent valleys, east-central Lemhi Mountains, Idaho. D, 1982, University of Kansas. 436 p.

Butler, Denise M. Planktic and benthic foraminifera of the Manchioneal formations near Port Maria, north coast Jamaica. M, 1984, Tulane University. 196 p.

Butler, Dwain Kent. Microgravimetry and the theory, measurement and application of gravity gradients. D, 1983, Texas A&M University. 275 p.

Butler, Edward Taylor. Methods of determining pyroelectricity in tourmaline. M, 1962, American University. 40 p.

Butler, Edwin Farnham, Jr. The geology and geochemical case history of the Juniper Canyon copper-molybdenum prospect, Pershing County, Nevada. M, 1981, University of Arizona. 82 p.

Butler, Elizabeth M. Cretaceous Ostracoda from Rayburn's Salt Dome, Louisiana. M, 1957, Louisiana State University.

Butler, Godfrey Phillip. Holocene supratidal evaporites; an analogue of ancient evaporites. D, 1969, University of California, Riverside. 176 p.

Butler, Howard Putnam. The study of beneficiation of Georgia talc. M, 1949, Emory University. 82 p.

Butler, James. Geology of the Charcas mineral district, San Luis Potosi, Mexico. D, 1972, Colorado School of Mines. 170 p.

Butler, James Hall. Cycling of reduced trace gases and hydroxylamine in coastal waters. D, 1986, Oregon State University. 207 p.

Butler, James Johnson, Jr. Pumping tests in nonuniform aquifers; a deterministic/stochastic analysis. D, 1986, Stanford University. 220 p.

Butler, James Johnson, Jr. The automated hydrogeologic system; description and application. M, 1982, Stanford University. 101 p.

Butler, James M. A study of seismic road noise. M, 1975, Georgia Institute of Technology. 46 p.

Butler, James Robert. Geology of an area north of Cotopaxi, Fremont County, Colorado. M, 1954, University of Colorado.

Butler, James Robert. Geology of the Cathedral Peak area, Beartooth Mountains, Montana. D, 1962, Columbia University, Teachers College. 108 p.

Butler, Jeannette Stier. The evaluation of the elutriate test as a means of predicting the behavior of selected chlorinated hydrocarbons in dredged sediments during open-water disposal. D, 1981, [University of Texas at Dallas]. 212 p.

Butler, John B. Pennsylvanian stratigraphy of the northern shelf and eastern part of the Anadarko Basin of Oklahoma and Kansas. M, 1960, Kansas State University. 55 p.

Butler, John Charles. A petrologic and X-ray spectrochemical investigation of the Trinity Lake amphibolites, Poundridge, New York. M, 1965, Miami University (Ohio). 133 p.

Butler, John Charles. Selected aspects of the crystal chemistry of $BaSO_4$, $SrSO_4$, and $PbSO_4$. D, 1968, Miami University (Ohio). 108 p.

Butler, John W. Origin of the Emery Deposits near Peekskill, New York. D, 1936, Columbia University, Teachers College.

Butler, John W., Jr. and Barnes, Farrell Francis. The Pre-Cambrian rocks of the Sawatch Range, Colorado. D, 1935, Northwestern University.

Butler, John W., Jr. and Barnes, Farrell Francis. The structure and stratigraphy of the Columbia River gorge and Cascade Mountains in the vicinity of Mount Hood. M, 1930, University of Oregon. 73 p.

Butler, Kenneth Bryan. Acoustic probing of salt using sonar. M, 1977, Texas A&M University.

Butler, Kim R. A structural analysis of Cambrian-Ordovician strata on the north flank of the Wichita Mountains, Oklahoma. M, 1980, Wichita State University. 80 p.

Butler, Kim Robert. Andean-type foreland deformation; structural development of the Neiva Basin, Upper Magdalena Valley, Colombia. D, 1953, University of South Carolina. 543 p.

Butler, Louis Winters. Geology of a submarine valley on the continental slope off Baja California, Mexico. M, 1964, San Diego State University.

Butler, Louis Winters, II. Shallow structure of the continental margin, southern Brazil and Uruguay (South America). D, 1969, University of Illinois, Urbana. 58 p.

Butler, Mark L. A theoretical mechanical analysis applied to Santa Susana-San Fernando-type reverse faults, Ventura Basin, California. M, 1977, Ohio University, Athens. 69 p.

Butler, Patrick, Jr. Magnetite from intrusives and associated contact deposits, Lincoln County, New Mexico. M, 1964, New Mexico Institute of Mining and Technology. 63 p.

Butler, Patrick, Jr. Mineral compositions and equilibria in the metamorphosed (Proterozoic) iron formation

of the Gagnon region, Quebec, Canada. D, 1968, Harvard University.

Butler, Paul Ray. Geology, structural history, and fluvial geomorphology of the southern Death Valley fault zone, Inyo and San Bernardino counties, California. D, 1984, University of California, Davis. 122 p.

Butler, Paul Ray. Movement of cobbles in a gravel-bed stream during a flood season. M, 1976, University of California, Berkeley. 50 p.

Butler, Paula Jean. Upper Cretaceous turbidites, Southwest Monterey County, California. M, 1984, Stanford University. 67 p.

Butler, Phillip Edward. Morphologic classification of sponge spicules, with descriptions of siliceous spicules from the Lower Ordovician Bellefonte Dolomite in central Pennsylvania. M, 1964, Pennsylvania State University, University Park. 36 p.

Butler, R. B. Terrestrial heat flow in the Saint Lawrence Plain (Quebec). M, 1961, McGill University.

Butler, Raymond Darrell. Hydrogeology of a sanitary landfill, Mandan, North Dakota. M, 1973, University of North Dakota. 124 p.

Butler, Raymond Darrell. Stratigraphy, sedimentology, and depositional environments of the Hell Creek Formation (Late Cretaceous) and adjacent strata, Glendive area, Montana. D, 1980, University of North Dakota. 538 p.

Butler, Rhett Giffen. Seismological studies using observed and synthetic waveforms. D, 1979, California Institute of Technology. 289 p.

Butler, Robert. Hydrogeology of the upper drainage, Middle Fork, South Platte River, Park County, Colorado. M, 1976, Colorado School of Mines. 156 p.

Butler, Robert D. Geology and zonal mineralization of Horseshoe-Sacramento region, Mosquito Range, Colorado. D, 1937, Massachusetts Institute of Technology. 285 p.

Butler, Robert E. Paleontology and stratigraphy of the Cynthiana Formation. M, 1954, Miami University (Ohio). 59 p.

Butler, Robert F. The effect of neutron irradiation on remanent magnetization in iron and kamacite (Apollo 11 and 12 lunar samples). D, 1972, Stanford University. 95 p.

Butler, Robert Grant, Jr. Sedimentology of the Upper Cambrian Danby Formation of western Vermont; an example of mixed siliciclastic and carbonate platform sedimentation. M, 1986, University of Vermont. 137 p.

Butler, Robert Scott. Geology of La Plata Canyon, Stillwater Range, Nevada. M, 1979, University of Nevada. 102 p.

Butler, Roy. A study of Pennsylvanian-Permian arkoses in north-central New Mexico. M, 1950, Texas Tech University. 33 p.

Butler, Roy Elbert. Comparative subsurface structure of Pennsylvanian and Upper Mississippian rocks in Posey County, Indiana. M, 1967, Indiana University, Bloomington. 25 p.

Butler, Roy Leslie. The geology of Madsen Red Lake gold mine. M, 1955, University of Manitoba.

Butler, Theresa Meade. Seismic attenuation studies using frequency domain synthetic seismograms. M, 19??, Texas A&M University.

Butler, Thomas Abraham. Geology of the Irma-Republic mines and vicinity; New World mining district, Montana and Wyoming. M, 1965, University of Idaho. 48 p.

Butler, Thomas Allen. Quartz and feldspar types; key indicators of provenance and diagenesis in the Hennessey Shale, southwestern Oklahoma. M, 1985, University of Oklahoma. 119 p.

Butler, Thomas Harry. A structural interpretation of a portion of the eastern Laramie Mountain flank, Albany, Platte, and Laramie counties, Wyoming. M, 1982, University of Oklahoma. 69 p.

Butler, Todd. Stratigraphy, petrography, and hydrothermal alteration in the Stedman mining district, Mojave Desert, California. M, 1979, University of California, Santa Barbara.

Butler, William C. The upper Paleozoic stratigraphy of Total Wreck ridge, Pima County, Arizona. M, 1969, University of Arizona.

Butler, William Charles. Permian conodonts from southeastern Arizona. D, 1972, University of Arizona. 155 p.

Butlien, Lawrence J. Diagenesis and porosity relationships in the Muddy Sandstone (Cretaceous), Powder River basin, Wyoming. M, 1983, Bowling Green State University. 114 p.

Butram, Glen N. The geology of the northeast corner of the Cuyapaipe Quadrangle, California. M, 1961, University of Southern California.

Butrenchuk, Stephen. Metamorphic petrology (Archean) of the Bird lake area, southeastern Manitoba. M, 1970, University of Manitoba.

Butt, Khurshid Alam. Genesis of granitic stocks in southwestern New Brunswick. D, 1976, University of New Brunswick.

Butt, William Horace. Igneous rocks of the Jones Ford Quadrangle. M, 1921, University of North Carolina, Chapel Hill. 18 p.

Butterfield, Gale Eugene. A geological study of the Irvine Ranch area, Johnson and Campbell counties, Wyoming. M, 1957, University of Missouri, Rolla.

Butterman, William Charles. Equilibrium phase relations among oxides in the systems GeO_2-B_2O_3, HfO_2-B_2O_3, ZrO_2-SiO_2-B_2O_3, and ZrO_2-SiO_2. D, 1965, Ohio State University.

Butterman, William Charles. Insoluble residues of the Silurian section in western Ohio. M, 1961, Ohio State University.

Butters, Greg Lee. Field scale transport of bromide in unsaturated soil. D, 1987, University of California, Riverside. 265 p.

Butters, Roy M. Permian or Perm-Carboniferous of the eastern foothills of the Rocky Mountains in Colorado. M, 1912, University of Colorado.

Butterworth, Joseph E. Paleomagnetic investigation of mid-Tertiary volcanic rocks in the Castle Dome and southern Kofa Mountains, Yuma County, southwestern Arizona. M, 1984, San Diego State University. 195 p.

Butterworth, Nancy Ann. Controlled-source audio-frequency magnetotelluric responses of three-dimensional bodies. M, 1988, University of Utah. 60 p.

Butterworth, Ronald Arthur. Sedimentology of the Maxon Formation (Cretaceous), west Texas. M, 1970, University of Texas, Austin.

Buttgereit, Charles D. A quantitative study of the least-squares method of residual potential field determination. M, 1968, University of Utah. 89 p.

Buttleman, Kim Parker. The evolution of the federal role in coastal zone management. M, 1981, University of Virginia. 183 p.

Buttner, Peter J. R. On systems analysis and environmental modeling. D, 1973, Rensselaer Polytechnic Institute. 295 p.

Button, R. M. Petrography of the Encrucijada Pluton, Estado Bolivar, Venezuela. M, 1970, University of Pennsylvania.

Buttram, George Franklin. Glass sands of Oklahoma. M, 1912, University of Oklahoma. 141 p.

Buttram, Glen Neil. The geology of the Agua Caliente Quadrangle, California. M, 1962, University of Southern California.

Buttrick, S. C. The alpine vegetation ecology and remote sensing of Teresa Island, British Columbia. D, 1978, University of British Columbia.

Butts, Christopher Lloyd. Modeling the evaporation and temperature distribution of a soil profile. D, 1988, University of Florida. 241 p.

Butts, JoLynn. Sediment to water radium exchange processes in the Pee Dee River-Winyah Bay estuary, South Carolina. M, 1986, University of South Carolina. 55 p.

Butts, Rayburn L. The geology of a portion of the Bartletts Ferry mylonite zone at Bartletts Ferry Dam, Georgia. M, 1986, Auburn University. 97 p.

Butz, Todd R. Mobilization of heavy metals by naturally occurring organic acids. D, 1976, University of Missouri, Rolla. 127 p.

Butz, Todd Randall. The hydrogeology of a sandstone knob overlain by a glacial till in Geauga County, Ohio. M, 1973, Kent State University, Kent. 61 p.

Buwalda, John Peter. New mammalian faunas from Miocene sediments near Tehachapi Pass in the southern Sierra Nevada. D, 1915, University of California, Berkeley. 10 p.

Buxton, Bruce Edward. Geostatistical determination of the precision of global recoverable reserve estimates. D, 1985, Stanford University. 314 p.

Buxton, Donna S. Desorption and leachability of sorbed DBCP residues in Hawaii soils. M, 1987, University of Hawaii. 185 p.

Buxton, Herbert T. Contribution of western New York streams to the Lake Erie sediment budget. M, 1977, SUNY, College at Fredonia. 105 p.

Buxton, Rebecca E. Transport mechanisms of shell fragments. M, 1980, Boston College.

Buxton, Timothy Montrose. Lithologic control of pressure solution; Alpena Limestone, Alpena, Michigan. M, 1981, Michigan State University. 70 p.

Buyannanonth, V. Biostratigraphic correlation in the area of the Ontong Java Plateau. M, 1971, University of Hawaii. 58 p.

Buyce, M. Raymond. Significance of authigenic K-feldspar in Cambrian-Ordovician carbonate rocks of the Proto-Atlantic Shelf in North America. D, 1975, Rensselaer Polytechnic Institute. 104 p.

Buyce, Michael R. The geology of the Guilford Quadrangle Mine. M, 1964, Brown University.

Buza, John W. Dispersal patterns of lower and middle Tertiary sedimentary rocks in portions of the Chiwaukum graben, east-central Cascade Range, Washington. M, 1977, University of Washington. 40 p.

Buzarde, Laverne Ernest. A study of the Upper Ordovician Bryozoa by zones. M, 1956, Emory University. 103 p.

Buzas, Alfons. The interpretation of the aeromagnetics of southeastern Marquette County, Michigan. M, 1960, Michigan State University. 45 p.

Buzas, Martin Alexander. Benthonic foraminifera from late Paleozoic clays of Waterville, Maine. M, 1960, Brown University.

Buzas, Martin Alexander. Ecology of the foraminifera in Long Island Sound. D, 1963, Yale University.

Buzzalini, Arnold. Study of geologic and economic aspects of the (Devonian) Onondaga Limestone, Onondaga County, New York. M, 1957, Syracuse University.

Bwerinofa, Obadiah K. Geology and mineralization in the David Mine area, Rowe, Massachusetts. M, 1972, University of Massachusetts. 152 p.

Bybee, Halbert Pleasant. The Aviculidae of the Permian. M, 1913, Indiana University, Bloomington.

Bybee, Halbert Pleasant. The flood of 1913 in the lower White River region of Indiana. D, 1915, Indiana University, Bloomington. 223 p.

Bybell, Laurel Mary. Middle Eocene calcareous nannofossils at Litte Stave Creek, Alabama. M, 1974, University of Miami. 175 p.

Bye, Bethany Ann. Volcanic stratigraphy and red beds of the Olorgesailie Formation, southern Kenya. M, 1984, University of Utah. 112 p.

Bye, Doris Lippincott. Cyrtospirifer disjunctus species in Pennsylvania and southwestern New York. M, 1949, Pennsylvania State University, University Park. 62 p.

Byer, Gregory B. The geology and geochemistry of the Cotter Basin stratabound+vein copper-silver deposit, Helena Formation, Lincoln, Lewis and Clark County, Montana. M, 1987, University of Montana. 174 p.

Byer, John W. Geology and engineering properties of the Portuguese tuff (Miocene, middle), Palos Verdes Hills, California. M, 1969, University of Southern California.

Byerlee, James D. The frictional characteristics of Westerly Granite (Pennsylvanian or younger) (SW Rhode Island and SE Connecticut). D, 1966, Massachusetts Institute of Technology. 179 p.

Byerley, Keith Alan. A field test for borehole gravity meter precision in shallow wells. M, 1977, University of Colorado.

Byerly, Benjamin Edward. Effects of clay minerals on petrophysical properties in a tight-gas sand, Cotton Valley Group, East Texas. M, 1987, University of Alabama. 135 p.

Byerly, Don Wayne. Structural geology along a segment of the Pulaski Fault, Greene County, Tennessee. D, 1966, University of Tennessee, Knoxville. 94 p.

Byerly, Don Wayne. The geology of the northern portion of Dutch Valley, Anderson County, Tennessee. M, 1957, University of Tennessee, Knoxville. 43 p.

Byerly, Gary Ray. A model for surface area, mineralogy, and metamorphic grade in carbonates. D, 1974, Michigan State University. 23 p.

Byerly, Gary Ray. Grain boundary processes and development of metamorphic plagioclase. M, 1972, Michigan State University. 20 p.

Byerly, John Robert. The relationship between watershed geology and beach radioactivity. M, 1963, University of California, Berkeley. 52 p.

Byerly, Perry E. Dispersion of energy with dispersion of frequency in transverse elastic waves in earthquakes. D, 1924, University of California, Berkeley. 78 p.

Byerly, Perry Edward. Regional gravity in the central Coast Ranges and San Joaquin Valley, California. D, 1954, Harvard University.

Byers, Alfred R. The geology and mineral deposits of the Night Hawk Lake area, Ontario. D, 1936, McGill University.

Byers, Charles D. A geochemical investigation of volatiles in abyssal glasses from the Galapagos spreading center at 85°W and 95°W, East Pacific Rise at 21°N, and Loihi Seamount, Hawaii, using high temperature mass spectrometry. D, 1984, University of Hawaii. 232 p.

Byers, Charles, Wesley, II. Biogenic structures of black shale paleoenvironments. D, 1973, Yale University. 296 p.

Byers, Frank M. The petrology of Umnak and Bogoslof islands, Alaska. D, 1955, University of Chicago. 189 p.

Byers, Gerald Eugene. The effects of acid rain on the movement of ions in a typic quartzipsamment soil under natural vegetation in Florida. D, 1984, University of Florida. 310 p.

Byers, Jay Morgan. An evaluation of Archie's equation when applied to unconsolidated sediment. M, 1985, University of Wisconsin-Milwaukee. 75 p.

Byers, Peter N. Mineralogy and origin of the Eastend and Whitemud formations of south-central and southwestern Saskatchewan and southeastern Alberta. M, 1966, Queen's University. 134 p.

Byers, Philip C. Facies of the Todilto Formation (Upper Jurassic) in north-central New Mexico. M, 1959, University of Kansas. 51 p.

Byers, R. A. Stratigraphy and paleoenvironments of the St. Lawrence Formation, western Wisconsin. M, 1979, University of Wisconsin-Madison.

Bykerk-Kauffman, Ann. Kinematic analysis of deformation at the margin of a regional shear zone, Buehman Canyon area, Santa Catalina Mountains, Arizona. M, 1983, University of Arizona. 79 p.

Byle, Chris S. Conodonts of the Moscow Formation (Middle Devonian) of central New York. M, 1980, Duke University. 36 p.

Bynum, Fred J., Jr. The geology of the south one-half of the Ware Shoals West Quadrangle, S.C. M, 1982, University of Georgia.

Byram, Kelly Gene. Petrographic analysis of the Gilmer Limestone, Louark Group (Upper Jurassic), Box Church Field, Limestone County, Texas. M, 1988, Stephen F. Austin State University. 150 p.

Byrd, C. Leon. Origin and history of the Uvalde gravel (Cenozoic) of central Texas. M, 1970, Baylor University.

Byrd, James Tillman. The marine geochemistry of tin. D, 1984, Florida State University. 219 p.

Byrd, John Odard Dutton. Geology of the Alisal Ranch area, south of Solvang, Santa Barbara County, California. M, 1983, University of California, Santa Barbara. 169 p.

Byrd, Phillip E. The effects of two water tracing agents on passive cotton dye detectors. M, 1981, University of Kentucky. 51 p.

Byrd, Richard E. A study of limestones from the Glass Mountains, Texas, that contain silicified fossils. M, 1951, Columbia University, Teachers College.

Byrd, Thomas Wayne. A statistical study of grain-size variation around Indian Peninsula, Gulf County, Florida. M, 1959, Mississippi State University. 70 p.

Byrd, William. Potential applications of magnetic gradients to marine geophysics. M, 1967, Massachusetts Institute of Technology. 133 p.

Byrd, William David, II. Geology of the bituminous sandstone deposits (Cretaceous; Eocene), southeastern Uinta Basin, Uinta and Grand counties, Utah. M, 1967, University of Utah. 44 p.

Byrd, William J. Geology of the Ely Springs range, Lincoln County, Nevada. M, 1970, University of Illinois, Urbana. 41 p.

Byrd, William John. Petrology of the Cambrian Shady Dolomite in North Carolina, Northeast Tennessee, and Southwest Virginia. D, 1973, University of North Carolina, Chapel Hill. 152 p.

Byrd, William Martin. The geology of a portion of the Combs Ranch, Brewster County, Texas. M, 1958, University of Texas, Austin.

Byrkit, James W. Operations at New Cornelia copper smelter of Phelps Dodge Corporation. M, 1956, University of Nevada - Mackay School of Mines. 10 p.

Byrne, A. W. The stratigraphy and paleontology of the Beekmantown Group in the Saint Lawrence Lowlands, Quebec. D, 1958, McGill University.

Byrne, Christian Jean. The geochemical cycling of hydrocarbons in Lake Jackson, Florida. D, 1980, Florida State University. 180 p.

Byrne, Frank E. Moschoides romeri; a new dinocephalian from the Karroo of South Africa. D, 1940, University of Chicago. 77 p.

Byrne, James R. Holocene depositional history of Lavaca Bay, central Texas Gulf Coast. D, 1975, University of Texas, Austin. 163 p.

Byrne, James Richard. Sedimentation of selected abyssal plain sands. M, 1972, SUNY, College at Oneonta. 84 p.

Byrne, John V. Oolites of the Great Bahama Bank. M, 1953, Columbia University, Teachers College.

Byrne, John Vincent. The marine geology of the Gulf of California. D, 1957, University of Southern California.

Byrne, Patrick James Sherwood. Sediments associated with the Kneehills Tuff in the Edmonton area. M, 1951, University of Alberta. 67 p.

Byrne, Patrick James Sherwood. The effect of variations in montmorillonite upon the nature of montmorillonite-organic complexes. D, 1953, University of Illinois, Urbana.

Byrne, Peter Michael. Elastic-viscoplastic response of earth structures to earthquake motion. D, 1969, University of British Columbia.

Byrne, R. H. Iron speciation and solubility in sea water. D, 1974, University of Rhode Island. 217 p.

Byrne, Richard Michael. The Ordovician-Silurian contact and related formations in Clark, Montgomery, and Bath counties. M, 1961, University of Kentucky. 65 p.

Byrne, Robert J. Some effects of particle-bed geometry in selective sorting. D, 1964, University of Chicago. 69 p.

Byrne, Timothy Briggs. Tectonic and structural evolution of the Ghost Rocks Formation, Kodiak Island, Alaska. D, 1981, University of California, Santa Cruz. 184 p.

Byrnes, M. E. Provenance study of late Eocene arkosic sandstones in southwest and central Washington. M, 1985, Portland State University. 65 p.

Byrnes, Mark Richard. Holocene geology and migration of a low-profile barrier island system, Metompkin Island, Virginia. D, 1988, Old Dominion University. 422 p.

Byron, G. G. Mineragraphy and paragenesis of the ore of the Park City consolidated mine, Park City, Utah. D, 1935, University of Utah. 54 p.

Byrum, Scott R. Foraminiferal biostratigraphy and paleoecology of upper Pleistocene sediments from the outer banks of North Carolina. M, 1978, East Carolina University. 65 p.

Byun, B. S. A linearly constrained adaptive algorithm for seismic array processing. M, 1973, Texas A&M University.

Byun, B. S. The corrective gradient projection method, and some adaptive algorithms for linearly constrained array processing. D, 1975, Texas A&M University. 121 p.

C., Enrique Morales *see* Morales C., Enrique

C., Ignacio A. Reyes *see* Reyes C., Ignacio A.

C., Jose A. Furnaguera *see* Furnaguera C., Jose A.

Caamano, Edward. A geochemical investigation of the northern portion of the Palisades diabase intrusion and New City Park Dike, Rockland County, New York. M, 1981, Rutgers, The State University, Newark. 66 p.

Cabaniss, Gerry Henderson. Crustal tilt in coastal New England; an experimental study. D, 1975, Boston University. 107 p.

Cabaup, Joseph John. Origin and differentiation of the gabbro in the Concord ring dike, North Carolina Piedmont. M, 1969, University of North Carolina, Chapel Hill. 42 p.

Cabe, Suellen. Cretaceous and Cenozoic stratigraphy of the upper and middle Coastal Plain, Harnett County area, North Carolina. D, 1984, University of North Carolina, Chapel Hill. 101 p.

Cabe, Suellen. Post-Eocene stratigraphy of the Carthage and Southern Pines 71/2′ quadrangles, North Carolina. M, 1980, University of North Carolina, Chapel Hill. 97 p.

Cabeen, Charles K. The petrography of Randolph, Vermont. M, 1922, Syracuse University.

Cabeen, William Ross. Geology of the Aliso and Browns canyons area, Santa Susana Mountains (California). M, 1939, California Institute of Technology. 36 p.

Cable, Emmett James. Some phases of the Pleistocene of Iowa, with special reference to the Peorian interglacial epoch. D, 1917, University of Iowa. 65 p.

Cable, Gregory. Paleoenvironments and hydrocarbon potential of the Chicla and Cabao formations, Northwest Libya. M, 1978, University of South Carolina.

Cable, Louis Walter. A petrographic study of the Edwards Limestone core from the Lone Star No. 1-A Tom Well, Atascosa County, Texas. M, 1961, Texas Christian University. 64 p.

Cable, Mark Stephan. Aspects of subsurface Cambrian and Early Ordovician lithostratigraphy and structure, eastern Kentucky, western West Virginia and Ohio. D, 1984, University of South Carolina. 347 p.

Cable, Mark Stephan. Depositional environments of the Rhode Island Formation, Carboniferous Narragansett Basin, southeastern New England. M, 1980, University of South Florida, St. Petersburg.

Cable, Steven W. A characterization of the Oregon Basin Thrust, Big Horn Basin, Wyoming. M, 1986, Colorado School of Mines. 57 p.

Cabral de Farias, Luiz Carlos. Preliminary study of the effects of counterflow on relative permeability. M, 1963, Stanford University.

Cabrera, John George. Geological and engineering properties of basaltic flows and interbeds throughout the upper Paraná Basin, Brazil. D, 1971, Cornell University. 248 p.

Cabri, Louis Jean Pierre. Phase relations in the Au-Ag-Te. D, 1965, McGill University. 152 p.

Caccaviello, Vivian M. Sedimentary analysis of sediments from Kaaterskill Creek, Catskill Mountains. M, 1939, Columbia University, Teachers College.

Cacchione, David A. Experimental study of internal gravity waves over a slope. D, 1970, Woods Hole Oceanographic Institution. 239 p.

Cacek, Terrance L. An ecological interpretation of north central Colorado. D, 1974, Colorado State University. 263 p.

Caceres, Victor B. Cretaceous System in northern Peru. D, 1955, Columbia University, Teachers College.

Caddey, Eric Lee. Geology of the Castle Peak area, Black Hills, South Dakota. M, 1986, University of Idaho. 51 p.

Caddey, Stanton William. Structural geometry of the "J" vein, the Bunker Hill Mine, Kellogg, Idaho. D, 1974, University of Idaho. 352 p.

Cade, Cassius M., III. The geology of the Marmaton Group of northeastern Nowata and northwestern Craig counties, Oklahoma. M, 1952, University of Oklahoma. 49 p.

Cade, Perry A. Magnetic and gravity survey over a buried Precambrian mafic body, Todd Co., Minnesota. M, 1987, Bowling Green State University. 115 p.

Cadieux, Bernard. La dispersion glaciaire des fragments des roches dans la region du Mistassini, Quebec. M, 1986, Universite de Montreal.

Cadigan, Robert Allen. The correlation of the Jurassic Bluff and Junction Creek sandstones in southeastern Utah and southwestern Colorado. M, 1952, Pennsylvania State University, University Park. 163 p.

Cadman, John Denys. The origin of exfoliation joints in granitic rocks. D, 1970, University of California, Santa Barbara.

Cadot, H. Meade, Jr. Magnesium content of calcite in carapaces of benthic marine Ostracoda. D, 1975, University of Kansas. 111 p.

Cadot, H. Meade, Jr. Statistical analysis of intraspecific variation in carapace morphology of selected Recent Ostracoda from Bermuda. M, 1970, University of Kansas. 44 p.

Cadwell, Donald Herbert. Geomorphic study of several flood plains in drainage basins in Lancaster County, Pennsylvania. M, 1969, Franklin and Marshall College. 40 p.

Cadwell, Donald Herbert. Late Wisconsinan deglaciation chronology of the Chenango River valley and vicinity, New York. D, 1972, SUNY at Binghamton. 102 p.

Cadwgan, Richard Morgan. Stratigraphy and sedimentology of the Gaptank Formation (middle and late Pennsylvanian), Pecos County, Texas. M, 1970, University of Texas, Austin.

Cady, Candace Clark. Hydrothermal alteration in the Corral Canyon shear zone, Mineral Mountains, Utah. M, 1983, University of Utah.

Cady, Francis H. Thermal metamorphism of sedimentary rocks of Hanover, New Mexico. M, 1938, Northwestern University.

Cady, Gilbert H. A preliminary study of the geology of the West Frankfort Quadrangle, Illinois. M, 1911, Northwestern University.

Cady, Gilbert H. The structure of the LaSalle Anticline. D, 1917, University of Chicago. 94 p.

Cady, Gilbert Victor. Model studies of geothermal fluid production. D, 1969, Stanford University. 82 p.

Cady, James Richard. Effect of rate of deformation on intergranular failure of aluminum near the melting temperature. D, 1955, Stanford University. 141 p.

Cady, John Gilbert. Some relationships between forest type and certain chemical and mineralogical properties of Podzol and brown Podzolic forest soil profiles. D, 1941, Cornell University.

Cady, John W. Magnetic and gravity anomalies in the California Great Valley and western Sierra Nevada metamorphic belt (Mesozoic). D, 1972, Stanford University. 104 p.

Cady, Wallace M. Areal and structural geology of the north end of the Taconic Syncline. M, 1936, Northwestern University.

Cady, Wallace M. Stratigraphy and structure of west-central Vermont. D, 1944, Columbia University, Teachers College.

Caelles, Juan Carlos. The geological evolution of the Sierras Pampeanas Massif, La Rioja and Catamarca provinces, Argentina. D, 1979, Queen's University. 514 p.

Caffall, Nancy M. Paleomagnetism and tectonic interpretations of the Taos Plateau volcanic field, Rio Grande Rift, New Mexico. M, 1986, University of Massachusetts. 58 p.

Caffee, Marc William. Pre-compaction irradiation of meteorites. D, 1986, Washington University. 191 p.

Caffery, Stephen. The sedimentology of the gray sands of Cotton Valley and Ivan Fields, Webster and Bossier parishes, Louisiana. M, 1986, University of Southwestern Louisiana.

Caffey, Kyle C. Depositional environments of the Olmos, San Miguel, and Upson formations (Upper Cretaceous), Rio Escondido Basin, Coahuila, Mexico. M, 1978, University of Texas, Austin.

Caffrey, Gregory Michael. Petrology and alteration of ultramafic rocks of Bullrun Mountain, Baker County, Oregon. M, 1982, Washington State University. 130 p.

Caffrey, Jane Marie. The influence of physical factors on water column nutrients and suspended sediments in Fourleague Bay, Louisiana. M, 1983, Louisiana State University. 85 p.

Caggiano, Joseph A., Jr. Glacial drainage history and drift petrography of the Vernon, New York Quadrangle. M, 1966, Syracuse University.

Caggiano, Joseph Anthony, Jr. Surficial and applied surficial geology of the Belchertown Quadrangle, Massachusetts. D, 1978, University of Massachusetts. 238 p.

Cagle, Clinton D. Seismic stratigraphy of the central Tonga Arc, SW Pacific. M, 1986, University of Texas, Austin. 158 p.

Cagle, David Anthony. Foraminifera and paleobathymetry of the Bluffport Marl Member of the Demopolis Chalk (Upper Cretaceous) in eastern Mississippi and west-central Alabama. M, 1985, University of New Orleans. 219 p.

Cagle, Fred R., Jr. Evaporite deposits of the central Namib Desert, Namibia. M, 1974, University of New Mexico. 155 p.

Cagle, Joseph W., Jr. The geology of the Knoxville Quadrangle, Knox and Blount counties, Tennessee. M, 1948, University of Tennessee, Knoxville. 68 p.

Cahall, Leavitt P. The Oriskany Fauna of Yawgers Woods, New York. M, 1956, University of Cincinnati. 118 p.

Cahill, Kevin Edwin. Geology of the Burtville Quadrangle, Johnson County, Missouri. M, 1962, University of Iowa. 218 p.

Cahill, Richard Allen. Geochemistry of Recent Lake Michigan surficial sediments. M, 1980, University of Illinois, Urbana. 143 p.

Cahn, Joe Harold. The origin of the Venango Second Sand Formation. M, 1940, University of Pittsburgh.

Cahn, Lorie S. Development of guidelines for design of sampling programs to predict groundwater discharge. M, 1987, University of British Columbia. 152 p.

Cahoon, Bobby G. Mineralogical and geochemical studies of wallrock alteration at the Temperly-Thompson deposit, New Diggings, Wisconsin. M, 1967, University of Wisconsin-Madison.

Cahoon, Bobby Glenn. Geochemical and mineralogical studies of the system Cu-As. D, 1968, Harvard University.

Cahoon, Elizabeth Jerabek. Lower Cretaceous pollen and spores from the southern Black Hills. D, 1964, University of Minnesota, Minneapolis. 153 p.

Cahow, A. C. Glacial geomorphology of the southwest segment of the Chippewa Lobe moraine complex, Wisconsin. D, 1976, Michigan State University. 218 p.

Caicedo, Nelson Oswaldo Luna. Identification aquifer parameters using transient observations and the discrete kernel approach. D, 1983, Colorado State University. 102 p.

Caillouet, Howard J. Geophysical (velocity) study of Ellenburger of West Texas. M, 1957, [University of Houston].

Cain, Bruce. Carboniferous paleomagnetics of the Appalachian Basin; stratigraphy, depositional environments, and some paleomagnetic relationships of Pennsylvanian strata in western Pennsylvania. M, 1977, University of Pittsburgh.

Cain, Douglas L. The determination and interpretation of rare earth abundances in hybrid granitoid rocks of the southern Snake Range, Nevada. M, 1974, Colorado School of Mines. 94 p.

Cain, James A. Ordovician volcanism in Great Britain and Ireland; nature, distribution, and thickness of the volcanic rocks with special reference to the Cross Fell Inlier. M, 1960, Northwestern University.

Cain, James Allan. Precambrian granitic complex of northeastern Wisconsin. D, 1962, Northwestern University. 187 p.

Cain, Parham Mikell. Deformation of the Leoville Chondrite and implications for its asteroidal parent body. M, 1985, University of Tennessee, Knoxville. 78 p.

Cain, W. F. Carbon-14 in tree rings of twentieth-century America. D, 1975, [University of California, San Diego]. 120 p.

Caine, Jennifer M. Sources of dissolved humic substances of a subalpine bog in the Boulder watershed, Colorado. M, 1982, University of Colorado. 85 p.

Caines, Gary L. Florida land-pebble phosphorite; the mineralogy and an evaluation of electrostatic beneficiation. M, 1981, Georgia Institute of Technology. 68 p.

Cains, William Tyson. Geochemical investigation of stream sediments in mineralized areas of central Arkansas. M, 1980, University of Arkansas, Fayetteville.

Cairnes, Clive Elmore. Coquihalla area, British Columbia. D, 1920, Princeton University.

Cairns, James Lowell. Internal wave measurements from a midwater float. D, 1974, University of California, San Diego.

Cairns, Janet Lorraine. Diagenetic aureoles indiced by hydrocarbon migration in the Permian redbeds of south-central Oklahoma. M, 1985, Oklahoma State University. 151 p.

Caithamer, Celine E. Mineralogical sources for barium in Cambro-Ordovician aquifers of northeastern Illinois. M, 1983, Northern Illinois University. 89 p.

Caixeiro, E. Correlations of reservoir oil properties, Miranga Field, Brazil. M, 1976, Stanford University. 40 p.

Calabro, Camille Elinore. Chemical and petrological investigations of spinel-silicate intergrowths in xenoliths from African kimberlite pipes. M, 1978, University of California, Davis. 78 p.

Calacal, Elias L. Sintering characteristics of diatomites. D, 1980, University of Washington. 218 p.

Calagari, Ali Ashghar. The geology and sulfide mineralization of the Sand Creek drainage, Spokane In-

dian Reservation, Stevens County, Washington. M, 1981, Washington State University. 127 p.

Calandra, John D. A ground magnetic survey of kimberlite intrusives in Elliott County, KY. M, 1986, Marshall University. 161 p.

Calavan, Charles W. Depositional environments and basinal setting of the Cretaceous Woodbine Sandstone, Northeast Texas. M, 1985, Baylor University. 225 p.

Calaway, Edward Lee. Availability of non-fuels materials; the strategic importance of Southern Africa. M, 1983, University of Texas, Austin.

Calbeck, James Morgan. Geology of the central Wise River Valley, Pioneer Mountains, Beaverhead County, Montana. M, 1975, University of Montana. 89 p.

Calcagno, Frank, Jr. The fracture pattern of the gas producing Devonian (Ohio) shale in outcrop in northeastern Ohio. M, 1979, Case Western Reserve University. 120 p.

Calcutt, Michael. The stratigraphy and sedimentology of the Cinq Isles Formation, Fortune Bay, Newfoundland. M, 1974, Memorial University of Newfoundland. 104 p.

Calder, Craig Paul. Geochemistry of the Eureka-Excelsior gold-lode deposit and associated greenstones and metasedimentary rocks, Cracker Creek District, Baker County, Oregon. M, 1986, Eastern Washington University. 83 p.

Calder, Dale R. Hydrozoa of southern Chesapeake Bay. D, 1968, College of William and Mary.

Calder, John Archer. Carbon isotope effects in biochemical and geochemical systems. D, 1969, University of Texas, Austin. 138 p.

Calder, Lynn M. Field and laboratory determinations of Cr(VI) mobility in an unconfined sand aquifer. M, 1984, University of Waterloo. 62 p.

Calderón Riveroll, Gustavo. A marine geophysical study of Vizcaino Bay and the continental margins of western Mexico between 27° and 30° north latitude. D, 1979, Oregon State University. 178 p.

Calderon, Gonzalo Cruz see Cruz Calderon, Gonzalo

Calderon-Riveroll, Gustavo. Circulation models and oceanographic parameters of the northern Gulf of California from Earth Resources Technology Satellite-1. M, 1974, University of Arizona.

Calderone, Gary Jude. Paleomagnetism of Miocene volcanic rocks in the Mojave-Sonora Desert region, Arizona and California. D, 1988, University of Arizona. 175 p.

Calderwood, Keith W. Geology of the Cedar Valley Hills area, Utah. M, 1951, Brigham Young University. 116 p.

Caldwell, Charles Keith. A study of the Sickle Conglomerate with special reference to the zircon content (P∈). M, 1951, University of Manitoba.

Caldwell, Craig D. Depositional environments of the Swasey Limestone, Middle Cambrian shelf carbonate of the Cordilleran miogeocline. M, 1980, University of Kansas. 84 p.

Caldwell, Dabney Withers. Channel characteristics and bed materials of streams in the Mt. Katahdin area, Maine. D, 1959, Harvard University.

Caldwell, Dabney Withers. The glacial geology of parts of the Farmington and Livermore, Maine, quadrangles. M, 1953, Brown University.

Caldwell, Donald Martin. A sedimentological study of an active part of a modern tidal delta, Moriches Inlet, Long Island, New York. M, 1971, Columbia University. 70 p.

Caldwell, Eleanor. The Tertiary larger foraminifera of Puerto Rico. M, 1941, Smith College. 69 p.

Caldwell, Gary. A study of the volcanic rocks on the southern half of Carriacou, Grenadines, West Indies. M, 1983, University of Windsor. 156 p.

Caldwell, James Phaon. Geology of the Red Canyon-Buffalo Basin area, Owl Creek Mountains, Wyoming. M, 1977, University of Iowa. 99 p.

Caldwell, John Gerrit. The mechanical behavior of the oceanic lithosphere near subduction zones. D, 1978, Cornell University. 161 p.

Caldwell, John William. Stratigraphic analysis of the Upper Cretaceous Pictured Cliff Sandstone in northeastern San Juan County, New Mexico. M, 1953, University of New Mexico. 59 p.

Caldwell, Katherine G. Petrographic and chemical analysis of rocks (Cambrian-Ordovician) from the Newtown area, Connecticut. M, 1972, University of Vermont.

Caldwell, Kenneth Robert. Depositional history and paleotectonic implications of the Pennsylvanian-Permian Sangre de Cristo Formation, northern New Mexico. M, 1987, University of Texas at Dallas. 149 p.

Caldwell, Lorin T. A study of the stratigraphy and the pre-glacial topography of the DeKalb and Sycamore quadrangles (Illinois). M, 1936, University of Chicago. 22 p.

Caldwell, William Stone. Stratigraphic control of oil accumulation in the Brown's East area, Wabash and Edwards counties, Illinois. M, 1950, University of Illinois, Urbana.

Calengas, Peter Leonard. Mineral resources as an element in land use planning for East central Indiana. D, 1977, Indiana University, Bloomington. 142 p.

Caless, Jonathan R. Geology, paragenesis, and geochemistry of sphalerite mineralization at the Young Mine, Mascot-Jefferson City zinc district, East Tennessee. M, 1983, Virginia Polytechnic Institute and State University. 161 p.

Caley, J. F. The Ordovician of Manitoulin Island, Ontario. D, 1934, University of Toronto.

Caley, W. F. Distribution of iron, nickel, and magnesium between olivine, metal and basaltic liquid. M, 1970, Queen's University. 78 p.

Calhoun, Frank Gilbert. Micromorphology and genetic interpretations of selected Columbian andosoils. D, 1971, University of Florida. 198 p.

Calhoun, Fred Harvey Hall. The relations of the Keewatin ice sheet to the mountains of Montana. D, 1902, University of Chicago. 62 p.

Calhoun, Jeannette A. Structural geology of the Morales Canyon and Taylor Canyon region of the Cuyama Basin, southern Coast Ranges, California. M, 1986, Oregon State University. 81 p.

Calhoun, John C., Jr. An investigation of the flow of homogeneous fluids through porous media. D, 1946, Pennsylvania State University, University Park. 84 p.

Calhoun, Steven H. Crinoids of the Devonian Cedar Valley Limestone in Iowa. M, 1983, University of Iowa. 151 p.

Calhoun, Thomas Addison, II. A study of a fragmental ore occurrence, Clementine prospect, Buchans, Newfoundland. M, 1979, University of Western Ontario. 95 p.

Calich, Rade. A study of Precambrian banded iron formation. M, 1954, University of Manitoba.

Caliga, Charles F. Petrography and geology of the rocks in the Ruckersville, Virginia area. M, 1948, University of Virginia. 94 p.

Calk, Lewis Clifton. The effects of a basalt intrusion on the fission track ages of accessory minerals in the Cathedral Peak Granite, California. M, 1972, San Jose State University. 37 p.

Calkin, Parker Emerson. Geomorphology and glacial geology of the Victoria Valley system, southern Victoria Land, Antarctica. D, 1963, Ohio State University.

Calkin, Parker Emerson. The geology of the Lummi and Eliza islands, Whatcom County, Washington. M, 1959, University of British Columbia. 140 p.

Calkin, William S. Geology, alteration and mineralization of the Alum Creek area, San Juan volcanic field, Colorado. D, 1967, Colorado School of Mines. 177 p.

Calkin, William S. The pre-Wisconsin drainage in the Orono and Bangor quadrangles (Maine). M, 1960, University of Maine. 62 p.

Calkins, Cary Peter Howard. Stratigraphic and structural geology of the northern half of The Solitario, Presidio and Brewster counties, Texas. M, 1980, Sul Ross State University.

Calkins, James A. The geology of the northeastern half of the Jamieson Quadrangle, Oregon. M, 1954, University of Oregon. 72 p.

Calkins, James A. The geology of the western limb of the Hazara-Kashmir syntaxis, West Pakistan, and Kashmir. D, 1966, Pennsylvania State University, University Park. 142 p.

Calkins, R. D. Geology and geography of the Prairie du Sac region, Wisconsin. M, 1911, University of Chicago. 42 p.

Calkins, William G. Magnetic and gravity study of Desert Mountain, Juab County, Utah. M, 1970, University of Utah. 64 p.

Call, Richard D. Analysis of geologic structure for open pit slope design. D, 1972, University of Arizona.

Call, Richard D. Geology of Cerro del Mercado, Coahuila, Mexico. M, 1960, Columbia University, Teachers College.

Call, Richard Ellsworth. Geology of Crowley's Ridge, Arkansas. M, 1891, Indiana University, Bloomington.

Call, Terry D. Application of filter methods to gravity data in southern Rhode Island. M, 1982, University of Rhode Island.

Callaghan, Eugene. A contribution to the structural geology of central Massachusetts. D, 1931, Columbia University, Teachers College.

Callaghan, Eugene. Geology of the Heceta Head District. M, 1927, University of Oregon. 72 p.

Callahan, Chester James, Jr. Aden basalt volcanic depressions, Dona Ana County, New Mexico. M, 1973, University of Texas at El Paso.

Callahan, Debra. The Paradise Sill and bordering migmatite found within the El Capitan Quadrangle, Idaho and Montana. M, 1983, University of Northern Colorado.

Callahan, Elaine J. Uranium and thorium concentrations in Precambrian gneissic terrain in the western Dharwar Craton in southern India, Hassan, Karnataka State. M, 1985, University of North Carolina, Chapel Hill. 60 p.

Callahan, Gary Delmar. A plasticity approach for rock containing planes of weakness. D, 1982, University of Minnesota, Minneapolis. 230 p.

Callahan, James Emmett. Structure of Houston Valley, Georgia. M, 1956, Emory University. 32 p.

Callahan, Jeffrey Edwin. The structure and circulation of deep and bottom waters in the Antarctic Ocean. D, 1971, The Johns Hopkins University. 149 p.

Callahan, John Edward. A regional heavy mineral petrographic and stream sediment geochemical survey applied to mineral exploration, Churchill Falls area, Labrador. D, 1973, Queen's University. 257 p.

Callahan, John Edward. Source rock determination on the basis of the heavy minerals in the saprolite in Piedmont, North Carolina. M, 1968, University of North Carolina, Chapel Hill. 63 p.

Callahan, Joseph Thomas. The geology of the Glen Canyon Group of the Echo Cliffs region, Arizona. M, 1951, University of Arizona.

Callahan, Richard. Stratigraphy, sedimentation and petrology of the Cambro-Ordovician Cow Head Group at Broom Point and Martin Point, western Newfoundland. M, 1974, University of Massachusetts. 119 p.

Callahan, Robert L. Lateral distribution of clay minerals in the Harlem Coal Underclay (Ohio). M, 1966, West Virginia University.

Callahan, W. H. The magnetite deposit at Cornwall. M, 1927, Massachusetts Institute of Technology. 49 p.

Callander, P. F. The nature and properties of highly sensitive soils. M, 1983, University of Waterloo. 175 p.

Callard, J. G. Improving coal dewatering and drying efficiencies by chemically lowering interfacial energies. M, 1977, Stanford University. 54 p.

Callaway, Jack M. Systematics, phylogeny, and ancestry of Triassic ichthyosaurs; a review. M, 1987, University of Rochester.

Callaway, Richard Joseph. Selenium content of pyrites of the central and Western United States. M, 1965, University of Minnesota, Minneapolis. 26 p.

Callaway, T. M. Correlation of formations in the Kettleman City oil field and vicinity, San Joaquin Valley, California. M, 1978, San Diego State University.

Callen, Jon M. The origin of surface lineaments in Atchison, Jefferson, and Leavenworth counties, Kansas. M, 1983, Wichita State University. 69 p.

Callender, A. D., Jr. Middle and upper Eocene biostratigraphy (foraminifera) of the Cascade Head area, Lincoln and Tillamook counties, Oregon. M, 1977, Portland State University. 212 p.

Callender, Alistaire Blyden Bruce. An investigation of the presence and mobility of pollutants at sites used for land treatment of oily residues. D, 1983, University of Oklahoma. 200 p.

Callender, Dean Lynn. Petrology of the Queen City Formation, Bastrop County, Texas. M, 1958, University of Texas, Austin.

Callender, Edward and Buckingham, James Gordon. An investigation of the folding in Forgie Township, Ontario (Canada). M, 1968, University of Manitoba.

Callender, Edward. Physical limnology and sedimentology of Miller Lake, Martin River glacier, south-central Alaska. M, 1964, University of North Dakota. 139 p.

Callender, Edward. The postglacial sedimentology of Devils Lake (Ramsey County), North Dakota. D, 1968, University of North Dakota. 336 p.

Callender, Jonathan Ferris. Geology of the York Mountain area, southern Coast Ranges, California. D, 1975, Harvard University.

Callender, William Russell. Paleoecology of the Mountain Lake Member, Bromide Formation (Middle Ordovician), Arbuckle Mountains, Oklahoma. M, 1985, Stephen F. Austin State University. 136 p.

Callewaert, David L. Hydrocarbon production potential of the Iola Limestone (Kansas City Group, Upper Pennsylvanian) in parts of Stafford and Pawnee counties, Kansas. M, 1987, Wichita State University. 155 p.

Callian, James Thomas. A paleomagnetic study of Miocene volcanics from the Colorado River and mainland Mexico regions. M, 1984, San Diego State University. 114 p.

Callier, Douglas Rean. Upper Paleozoic stratigraphy of the Masonville-Lyons area, Colorado. M, 1947, University of Colorado.

Callihan, John Brent. Quaternary and environmental geology of northwestern Iberia Parish, Louisiana. M, 1988, University of Southwestern Louisiana. 116 p.

Callmeyer, Thomas J. The structural, volcanic, and hydrothermal geology of the Warm Springs Creek area, eastern Garnet Range, Powell County, Montana. M, 1984, Montana State University. 84 p.

Calmes, Grady Allen. The average zero time lag cross-correlation of LASA earth noise signals. M, 1972, Texas A&M University. 80 p.

Calogero, Frank. Development of stylolites in sandstones of the Gulf Coast. M, 1979, Texas A&M University. 83 p.

Calton, Robert. Possibilities of concentrating ilmenite from the beach sands of the north shore of Lake Superior. M, 1936, University of Minnesota, Minneapolis. 31 p.

Caluda, Carol Ann. Analysis of fine-grain turbidites in intraslope basins of the Louisiana-Texas continental slope. M, 1987, University of New Orleans. 119 p.

Calver, James Lewis. Glacial and post-glacial history of the Platte and Crystal lake depressions, Benzie County, Michigan. D, 1942, University of Michigan.

Calverley, Anne. Sedimentology and geomorphology of the modern epsilon cross-stratified point bar deposits on the Athabasca Delta. M, 1984, University of Calgary.

Calvert, Craig Steven. Chemistry and mineralogy of iron-substituted kaolinite in natural and synthetic systems. D, 1981, Texas A&M University. 138 p.

Calvert, Gary C. Subsurface structure of the northern portion of the Burning Springs Anticline. M, 1987, West Virginia University. 128 p.

Calvert, H. Thomas. The complex resistivity response of mineralized rocks at sub-freezing temperatures. M, 1988, University of Calgary. 208 p.

Calvert, Ronald H. Geology of the Milner Mountain area, Larimer County, Colorado. M, 1963, University of Colorado.

Calvert, Ronald Harold. Stratigraphy of a Chester cycle (Upper Mississippian) in the Kentucky part of the eastern interior basin. D, 1968, Washington University. 58 p.

Calvert, Stephen Edward. The diatomaceous sediments of the Gulf of California. D, 1964, University of California, San Diego. 265 p.

Calvete, Francisco Javier Samper see Samper Calvete, Francisco Javier

Calvin, Don G. Incidence of oil and gas in the Cottage Grove Sandstone. M, 1963, Wichita State University. 63 p.

Calvin, James S. Preliminary mine feasibility study of disseminated gold deposits located in the Fairbanks mining district, Alaska. M, 1985, University of Alaska, Fairbanks. 99 p.

Calzada, Carlos Edgardo Ruiz see Ruiz Calzada, Carlos Edgardo

Calzia, James P. Petrography of a part of the Owlshead Pluton, San Bernardino County, California. M, 1973, University of Southern California.

Camacho, Ricardo. Stratigraphy of the upper Pierre Shale, Fox Hills Sandstone, and lower Laramie Formation (Upper Cretaceous), Leyden Gulch area, Jefferson County, Jefferson County, Colorado. M, 1969, Colorado School of Mines. 84 p.

Camara, Michael. Glacial stratigraphy of southeastern North Dakota. M, 1977, University of North Dakota. 66 p.

Camara, Richard P. De see De Camara, Richard P.

Camargo, Antonio. Anomalies in P-wave velocities in the Tampico Embayment, Gulf Coast of Mexico. M, 1968, Rice University. 79 p.

Camargo, Jorge M. T. Rapid lateral velocity variations and variations in seismic reflection amplitudes in the Ubarana field area; offshore northeastern Brazil. M, 1982, University of Texas, Austin. 156 p.

Cambareri, Greg. Inverse modelling of apparent resistivity over a thin inclined dike. M, 1988, North Carolina State University. 50 p.

Cambareri, Thomas Christian. Hydrogeology and hydrochemistry of a sewage effluent plume in the Barnstable Outwash of the Cape Cod Aquifer, Hyannis, Massachusetts. M, 1986, University of Massachusetts. 130 p.

Cambridge, Thomas Ross. Stratigraphy and sedimentation of the upper Shawnee Group (Upper Pennsylvanian) in southeastern Nebraska and adjacent regions. M, 1959, University of Nebraska, Lincoln.

Cameron, Alexander Rankin. Lateral variations in the lower Cedar Grove coal seam, Logan and Mingo counties, West Virginia. M, 1954, Pennsylvania State University, University Park. 179 p.

Cameron, Alexander Rankin. Some petrological aspects of the Harbour Seam, Sydney Coalfield, Nova Scotia. D, 1961, Pennsylvania State University, University Park. 450 p.

Cameron, Aubrey T., Jr. Physical conditions of low-grade metamorphism of the Jacksonburg Formation, Northampton County, Pennsylvania. M, 1977, Lehigh University. 86 p.

Cameron, Barry Winston. Displaced and mixed foraminiferal assemblages in hemipelagic sediments and turbidites from a deep-sea fan, northeastern Bahamas. M, 1965, Columbia University. 45 p.

Cameron, Barry Winston. Stratigraphy and sedimentary environments of lower Trentonian Series (Middle Ordovician) in northwestern New York and southeastern Ontario (Canada). D, 1968, Columbia University. 272 p.

Cameron, Bruce. Permian fauna of the Yukon Territory. M, 1962, University of Alberta. 109 p.

Cameron, Christopher Paul. Paleomagnetism of Tertiary volcanics from Shemya and Adak islands, Aleutian islands, Alaska. D, 1970, University of Alaska, Fairbanks. 177 p.

Cameron, Christopher Scott. Geology of the Sugarloaf and Delamar Mountain areas, San Bernardino Mountains, California. D, 1981, Massachusetts Institute of Technology. 399 p.

Cameron, Christopher Scott. Structure and stratigraphy of the Potosi Mountain area, southern Spring Mountains, Nevada. M, 1978, Rice University. 83 p.

Cameron, Cornelia C. Comparative study of fossils of two loess sections at North Liberty and Iowa City, Iowa. M, 1935, University of Iowa. 46 p.

Cameron, Cornelia C. The fossils of the Peorian loess of Iowa. D, 1940, University of Iowa. 79 p.

Cameron, Diana. A study of glauconitic pellets from the Navesink and Red Bank formations (Upper Cretaceous) in New Jersey. M, 1985, Montclair State College. 39 p.

Cameron, Donald Edward, Jr. Depositional and diagenetic histories and trapping mechanism of the Devonian Misener Sandstone in Garfield and Grant counties, Oklahoma. M, 1986, West Texas State University. 55 p.

Cameron, Donald Eugene. Contact metamorphism of metapelites in the Front Range, Colorado; a study of disequilibrium reactions. M, 1976, University of Arizona.

Cameron, Donald Kenzie. Bryozoan fauna of the Jefferson Lake Section. M, 1954, Indiana University, Bloomington. 35 p.

Cameron, Douglas R. Modeling silver transport in the soil. D, 1973, Colorado State University. 128 p.

Cameron, Eugene N. Geology and mineralization of the northeastern Humboldt Range, Nevada. D, 1939, Columbia University, Teachers College.

Cameron, Gregory Joseph. Geology and mineralization history of the Jasay-los Gavilanes tungsten district, Baja California Norte, Mexico. M, 1984, San Diego State University. 255 p.

Cameron, H. L. The gold deposits of Fifteen Mile Stream, Nova Scotia. M, 1941, McGill University.

Cameron, Harriet V. The Winnfield, Louisiana salt dome. M, 1949, Louisiana State University.

Cameron, Jeri Lynn. The Lucky Five Pluton in the Southern California Batholith; a history of emplacement and solidification under stress. M, 1980, University of California, Los Angeles.

Cameron, John Baades. Earthquakes in Northern California coastal region. D, 1959, University of California, Berkeley. 82 p.

Cameron, John Ian. Investment in the nickel industry since 1959; theory and practice. D, 1977, Pennsylvania State University, University Park. 523 p.

Cameron, Kenneth Allan. Geology of the southcentral margin of the Tillamook Highlands; southwest quarter of the Enright Quadrangle, Tillamook County, Oregon. M, 1980, Portland State University. 87 p.

Cameron, Kenneth L. A study of the modification of stream sediments during transportation, Elk Creek, Black Hills, South Dakota. M, 1968, University of Houston.

Cameron, Kenneth L. An experimental study of coexisting amphiboles; phase relations along the join, $Mg_{3.5}Fe_{3.5}Si_8O_{22}(OH)_2$–$Ca_2Mg_{2.5}Fe_{2.5}Si_8O_{22}(OH)_2$. D,

1971, Virginia Polytechnic Institute and State University.

Cameron, Kevin J. Petrochemistry of mafic rocks from the Harbour Main Group (Western Block), Conception Bay, Avalon Peninsula, Newfoundland. M, 1986, Memorial University of Newfoundland. 209 p.

Cameron, Maryellen. Geology of the Rattlesnake Mountain intrusion, Big Bend National Park, Texas. M, 1969, University of Houston.

Cameron, Maryellen. The crystal chemistry of tremolite and richterite; a study of selected anion and cation substitutions. D, 1971, Virginia Polytechnic Institute and State University.

Cameron, Narcissa S. Lower and Middle Devonian formations near Everett, Bedford Co., Pennsylvania. M, 1941, Bryn Mawr College.

Cameron, Peter Forrest. Allochthonous carbonate debris deposits adjacent to the Southesk-Cairn carbonate complex (Upper Devonian), Wapiabi Gap, Alberta. M, 1975, University of Manitoba.

Cameron, Peter John. Structural analysis of the Thierry copper-nickel deposit in Northwest Ontario, Canada. M, 1980, Pennsylvania State University, University Park. 108 p.

Cameron, R. A. An experimental study of the effects of heat, pressure, and fluids on sedimentary materials. D, 1956, McGill University.

Cameron, R. A. Facies change in the Sokoman iron formation, John Lake area, Quebec-Labrador. M, 1951, University of Toronto.

Cameron, Richard Leo. Glaciological studies at Wilkes Station, Budd Coast, Antarctica. D, 1963, Ohio State University. 234 p.

Cameron, Richard William. An interpretation of the heavy mineral suites of the Argentine continental shelf. M, 1968, Columbia University. 59 p.

Cameron, Robert E. Development of an integrated computer package for oil shale information retrieval and feasibility analysis of public lands (Volumes I and II). D, 1985, University of Utah. 907 p.

Cameron, Suzanne P. Ostracodes of pluvial Lake Cochise (late Pleistocene), Cochise County, southeastern Arizona. M, 1971, Arizona State University. 65 p.

Cameron, William Maxwell. On the dynamics of inlet circulation. D, 1951, University of California, Los Angeles.

Cameron-Schimann, Monique. Electron microprobe study of uranium minerals and its application to some Canadian deposits. D, 1978, University of Alberta. 342 p.

Camfield, P. Adrian. Studies with a two-dimensional magnetometer array in Northwestern United States and southwestern Canada. D, 1973, University of Alberta. 203 p.

Camfield, Paul Adrian. Measurement of Curie temperatures of modern red sedimentary rocks. M, 1966, Massachusetts Institute of Technology. 87 p.

Cammack, C. H. Triaxial shear of soil with stress path control by performance feedback. D, 1978, Iowa State University of Science and Technology. 156 p.

Cammack, James. The subsurface geology of the Stamps area, Lafayette County, Arkansas. M, 1958, University of Arkansas, Fayetteville.

Cammarata, Thomas Joseph. The use of Tl in a pedogeochemical survey for gold mineralization, Howard mining district, Oregon. M, 1987, Eastern Washington University. 73 p.

Camp, Mark J. Pleistocene lacustrine deposits and molluscan paleontology of western Ohio, eastern Indiana, and southern Michigan. M, 1972, University of Toledo. 161 p.

Camp, Mark J. Pleistocene Mollusca of three southeastern Michigan marl deposits. D, 1974, Ohio State University. 147 p.

Camp, Quentin Walter Van see Van Camp, Quentin Walter

Camp, S. H. Terranes of Northfield, Vermont. M, 1917, Syracuse University.

Camp, Scott Gregory van *see* van Camp, Scott Gregory

Camp, Thomas M. A lithofacies and paleocurrent analysis of the Dunkard Group (Pennsylvanian-Permian) in the central Dunkard Basin of West Virginia and Ohio. M, 1968, Miami University (Ohio). 108 p.

Camp, Victor E. The mineralogy and petrology of a contact aureole, Jefferson County, Montana. M, 1972, Miami University (Ohio). 66 p.

Camp, Victor Eric. Petrochemical stratigraphy and structure of the Columbia River Basalt, Lewiston Basin area, Idaho-Washington. D, 1976, Washington State University. 201 p.

Camp, Wayne K. Structure and stratigraphy of the Laramie River valley, Larimer County, Colorado. M, 1979, Colorado State University. 235 p.

Campagna, David John. Regional geology and tectonic interpretation of southern Jalisco and Colima states, Mexico. M, 1982, University of Kentucky. 68 p.

Campana, M. E. Determination of hydraulic parameters in a fractured rock aquifer. M, 1973, University of Arizona.

Campana, M. E. Finite-state models of transport phenomena in hydrologic systems. D, 1975, University of Arizona. 270 p.

Campau, Donald Edmund. The stratigraphic distribution of Ostracoda within the Traverse Formation of the Michigan Basin. M, 1950, Michigan State University. 55 p.

Campbell, A. Richard. Some ostracodes from the basal Ludlowville (Hamilton), Genesee County, New York. M, 1952, University of Rochester. 102 p.

Campbell, Alan Neil. Search and discovery; an investigation of traditional assumptions about the economics of finding new ore deposits and the proposal of a new hypothesis. D, 1976, Stanford University. 120 p.

Campbell, Alice M. Hydrology and hydrochemistry of the Verdugo Basin, Los Angeles Co., Cal. M, 1978, California State University, Los Angeles.

Campbell, Allan G. Sediment storage trends in several channels along the San Gabriel mountain front, Southern California. M, 1986, Colorado State University. 143 p.

Campbell, Andrew Clare. Geodesy at sea. M, 1965, Ohio State University.

Campbell, Andrew Craig. Geochemistry of hydrothermal clouds in the Guaymas Basin, Gulf of California. D, 1985, University of California, San Diego. 281 p.

Campbell, Andrew Robert. Genesis of the tungsten-base metal ores at San Cristobal, Peru. D, 1983, Harvard University. 186 p.

Campbell, Archibald R., III. Volcanic rocks of the La Perla area, Chihuahua, Mexico. M, 1977, University of Texas, Austin.

Campbell, Arthur Byron. The paragenesis of the lead-zinc ores, Pine Creek District, Shoshone County, Idaho. M, 1949, Washington University. 64 p.

Campbell, Bruce G. Geology and development constraints of the North Augusta, South Carolina, riverfront. M, 1982, University of Missouri, Kansas City. 77 p.

Campbell, Bruce S. Stratigraphy and metamorphism in the Balmat-Edwards zinc mining district, Northwest Adirondacks, New York. M, 1976, Cornell University.

Campbell, Bruce Samuel. Geology and electrical and electromagnetic modeling of volcanogenic sulfide bodies near Savant Lake, Ontario. D, 1980, Pennsylvania State University, University Park. 261 p.

Campbell, Carl Earl. A seismic refraction and electrical resistivity investigation of a karst environment in Lawrence County, Indiana. M, 1967, Indiana University, Bloomington. 44 p.

Campbell, Catherine Chase. Post glacial changes in bottom conditions of the southeastern New England coast as indicated by foraminifera. D, 1933, Harvard University.

Campbell, Charles D. The Kruger alkaline syenites of southern British Columbia. D, 1934, Stanford University. 84 p.

Campbell, Charles Duncan. Titanium; a summary of the literature. M, 1931, University of Michigan.

Campbell, Charles L. Upper Permian evaporites of western Kansas between the Blaine Formation and the Stone Corral Formation. M, 1963, University of Kansas. 73 p.

Campbell, Charles Lillie. A study of the gravel deposits (Pleistocene) of Saint Helena and Tangipahoa Parishes, Louisiana. D, 1971, Tulane University. 372 p.

Campbell, Charles Lyman. The washing of Iowa coals. D, 1937, University of Iowa. 79 p.

Campbell, Charles Virgil. A study of the paleogeology of the Paleozoic Era in Wyoming. M, 1949, Stanford University. 27 p.

Campbell, Charles Virgil. The Phosphoria Formation in the southeastern Bighorn Basin, Wyoming. D, 1956, Stanford University. 203 p.

Campbell, Colin Robert. Tertiary Dasyuridae and Peramelidae (Marsupialia) from the Tirari Desert, South Australia. D, 1976, University of California, Berkeley. 220 p.

Campbell, D. C. Geology and ore control of Eldorado Mine, Port Radium uranium deposits, Northwest Territories. D, 1954, [California Institute of Technology].

Campbell, D. E. Suspended matter and light attenuation in a turbid estuary. M, 1977, Old Dominion University. 143 p.

Campbell, David Gwynne. Geology of the Jamesville area, Muskogee and Okmulgee counties, Oklahoma. M, 1957, University of Oklahoma. 91 p.

Campbell, David Lowell. The loading problem for a viscoelastic Earth. D, 1969, University of California, Berkeley. 182 p.

Campbell, Dennis R. Stratigraphy of pre-Needles Range Formation ash-flow tuffs in the northern Needle Range and southern Wah Wah Mountains, Beaver County, Utah. M, 1976, Brigham Young University.

Campbell, Donald F. Geology of the Colquiri tin mine, Bolivia. D, 1946, University of Arizona.

Campbell, Donald Fergus. Geology of the Bonanza King Mine, Humboldt Range, Pershing County, Nevada. M, 1938, Cornell University.

Campbell, Donald H. The transport of road derived sediment as a function of slope characteristics and time. M, 1984, Colorado State University. 54 p.

Campbell, Donald Harvey. Petrography of the Cretaceous Hensell Sandstone, central Texas. M, 1962, University of Texas, Austin.

Campbell, Donald Harvey. Petrology of the Early Cretaceous Yucca Formation in the southern Quitman Mountains and vicinity, Trans-Pecos, Texas. D, 1968, Texas A&M University. 339 p.

Campbell, Donald Lorne. Cretaceous microfauna from Cameron Hills, Northwest Territories. M, 1956, University of Alberta. 85 p.

Campbell, Douglas Dean. Geology of the pitchblende deposits of Port Radium, Great Bear Lake, Northwest Territories. D, 1955, California Institute of Technology. 323 p.

Campbell, Douglas Patrick. Biostratigraphy of the Albertella and Glossopleura zones (lower middle Cambrian) of northern Utah and southern Idaho. M, 1974, University of Utah. 295 p.

Campbell, E. E. The geology of Baltic Mines, Ltd., Missanable, Ontario. M, 1948, Queen's University.

Campbell, Edith Ciora Allison. Palynology and paleoecology of the Miocene lignites of the Goose Creek basin, Idaho, Nevada, and Utah. M, 1979, University of Utah. 66 p.

Campbell, F. A. Nickeliferous sulphide deposits and associated basic rocks at Quill Creek, and White River, Yukon Territory. M, 1956, Queen's University. 155 p.

Campbell, F. H. A. Paleocurrents and sedimentation of part of the Meguma Group (Lower Paleozoic) of

Nova Scotia, Canada. M, 1966, Dalhousie University. 91 p.

Campbell, Finley Alexander. The geology of the Torbrit silver mine (British Columbia). D, 1958, Princeton University. 124 p.

Campbell, Frank Howard, III. Dikes and recrystalline carbonates in the central Shenandoah Valley of Virginia. M, 1962, University of Virginia. 82 p.

Campbell, Frank W. A stratigraphic study of dental variations in the Oligocene rodent Ischyromys from Slim Buttes. M, 1975, South Dakota School of Mines & Technology.

Campbell, Frederick E. Stability of nickel-magnesium olivine. M, 1966, Queen's University. 71 p.

Campbell, Frederick H. A. Sedimentation and stratigraphy of part of the Rice Lake Group (Precambrian), Manitoba. D, 1971, University of Manitoba.

Campbell, Frederick Kedney. Geology of the upper Precambrian Flambeau Quartzite, Chippewa County, North-central Wisconsin. M, 1981, University of Minnesota, Duluth.

Campbell, G. J. Petrography and stratigraphy of the north band of Iron Formation, Sherman Mine, Temagami, Ontario. M, 1978, Laurentian University, Sudbury.

Campbell, George E. The effect of organic fluids on the fabric and mechanical behavior of clays. M, 1986, SUNY at Buffalo. 64 p.

Campbell, Gregg Tyler. Depositional environments, biofacies, and paleoecology of the Howell's Ridge Member of the Aptian-Albian U-Bar Formation, Little Hatchet Mountains, New Mexico. M, 1988, University of Colorado. 175 p.

Campbell, Ian. A geological reconnaissance of the McKenzie River section of the Oregon Cascades with petrographic descriptions of some of the more important rock types. M, 1925, University of Oregon. 56 p.

Campbell, Ian. The petrography of the Tonopah mining district, Nevada; including some observations on the ore deposits and the rock alteration. D, 1931, Harvard University.

Campbell, Ian D. Pollen-sedimentary environment relations and late Holocene palynostratigraphy of the Ruby Range, Yukon Territory, Canada. M, 1987, University of Ottawa. 108 p.

Campbell, J. A. Nutrient losses and related processes in a seasonally-operated septic bed soil under favourable conditions. D, 1979, McMaster University.

Campbell, J. B., Jr. Geographic analysis of variation across a soil boundary and within soil mapping units. D, 1976, University of Kansas. 267 p.

Campbell, J. K. Beekmantown Formation; middle Ordovician limestone unconformity on the northwest limb of the Green Ridge Anticline near Fincastle, Virginia. M, 1975, Virginia Polytechnic Institute and State University.

Campbell, Jack Albert. Cenozoic structural and geomorphic evolution of the Canyon Range, central Utah. D, 1978, University of Utah. 158 p.

Campbell, Janet E. The sedimentology and stratigraphy of a gravelly meander lobe in the Saskatchewan River, near Nipawin, Saskatchewan. M, 1988, University of Saskatchewan. 131 p.

Campbell, Jeffrey Erle. Dielectric properties of moist soils at RF and microwave frequencies. D, 1988, Dartmouth College. 178 p.

Campbell, Jeffrey Erle. Gravity and magnetic analysis of the Mascoma Dome and surrounding structures. M, 1986, Dartmouth College. 41 p.

Campbell, Jock Albert. Geology and structure of a portion of the Rio Puerco Fault Belt, western Bernalillo County, New Mexico. M, 1967, University of New Mexico. 89 p.

Campbell, John Arthur. Petrology of the Permo-Triassic red beds of the northern Colorado Front Range and south-central Wyoming. M, 1957, University of Colorado.

Campbell, John Arthur. The Devonian system of west-central Colorado. D, 1966, University of Colorado. 207 p.

Campbell, John S. Pleistocene turbidites of the Canada Abyssal Plain, Arctic Ocean. M, 1973, University of Wisconsin-Madison.

Campbell, Josephine Kay. The change in molar volumes at the B1-B2 phase transformation for KCl, KBr, and KI. M, 1969, University of Rochester. 14 p.

Campbell, K. W. The use of seismotectonics in the Bayesian estimation of seismic risk. D, 1977, University of California, Los Angeles. 177 p.

Campbell, Kenneth Eugene. The Pleistocene avifauna of the Talara Tar Seeps, northwestern Peru. D, 1973, University of Florida. 227 p.

Campbell, Kenneth Eugene, Jr. A comparison of the postcranial skeletons of Hypolagus and Pratilepus (Lagomorpha). M, 1967, University of Michigan.

Campbell, Kenneth M. The sedimentology of simple and reticulated transverse bars in a low wave energy environment. M, 1979, Florida State University.

Campbell, Kenneth Vincent. Metamorphic petrology and structural geology of the Crooked Lake area, Cariboo Mountains, British Columbia. D, 1971, University of Washington. 192 p.

Campbell, Kenneth Vincent. The bedrock geology of the Crooked lake region of the Cariboo mountains, British Columbia (Canada). M, 1969, University of Washington. 56 p.

Campbell, Kerry Jacquith. Surficial geology of the Northfield Quadrangle, Massachusetts, New Hampshire, and Vermont. M, 1975, University of Massachusetts. 126 p.

Campbell, Kevin. Slope failure along the urbanizing White Rock Escarpment, central Texas. M, 1988, Baylor University. 294 p.

Campbell, Kevin Todd. A study of Landsat lineament data observed in Michigan. M, 1981, Michigan State University. 132 p.

Campbell, L. B. Organic and inorganic phosphorus content, movement, and mineralization of phosphorus in soil beneath a feedlot. M, 1974, University of Manitoba.

Campbell, L. F., Jr. An analysis of the Colorado experience in the National Topographic Program. D, 1974, University of Colorado. 393 p.

Campbell, Lois Jeannette. The late glacial and lacustrine deposits of Erie and Huron counties, Ohio. D, 1955, Ohio State University. 219 p.

Campbell, Louis F. The North Sea area; a comparative analysis of petroleum exploration by the peripheral states. M, 1966, Catholic University of America. 69 p.

Campbell, Lyle D. Paleoecology of the Lone Star Industries pit, Yorktown Formation (Pliocene), Chuckatuck, Virginia. D, 1976, University of South Carolina. 198 p.

Campbell, Lyle David. Stratigraphy and paleontology of the Kinzers Formation, southeastern Pennsylvania. M, 1969, Franklin and Marshall College. 228 p.

Campbell, Marie A. Derricks and ditches; cultural-historical study of an oil boom area. D, 1971, Bowling Green State University. 250 p.

Campbell, Michael D. Geology of the northern Buck Mountain area, White Pine County, Nevada. M, 1981, Ohio University, Athens. 71 p.

Campbell, Michael David. Paleoenvironmental and diagenetic implications of selected siderite zones and associated sediments in the upper Atoka Formation, Arkoma Basin, Oklahoma-Arkansas. M, 1976, Rice University. 124 p.

Campbell, Michael James. Combined interpretation of gravity, magnetic, and seismic reflection data as related to the northeast extension of the New Madrid fault zone. M, 1983, Purdue University. 78 p.

Campbell, Michael John. Prebatholithic stratigraphy of the northeastern Sierra la Asamblea, Baja California, Mexico. M, 1985, San Diego State University. 133 p.

Campbell, Michael P. The geology of the southern Sierra Santa Rita, Chihuahua, Mexico. M, 1984, University of Texas at El Paso.

Campbell, Nancy E. Subsurface stratigraphic analysis of the Skiatook and "Cherokee" groups in the southwestern part of Noble County, northern part of Logan County, and the southeastern part of Garfield County, Oklahoma. M, 1984, Oklahoma State University. 44 p.

Campbell, Neil. The geology of the Con-Rycon mines. D, 1943, Massachusetts Institute of Technology. 161 p.

Campbell, Newell Paul. Stratigraphy and petrology of the Jefferson Formation (Upper Devonian), Little Belt Mountains, Montana. M, 1966, University of Colorado. 90 p.

Campbell, R. C. The interaction of fatty acids with calcite; adsorption, reaction and decomposition to hydrocarbons. M, 1977, SUNY at Binghamton.

Campbell, Richard Allan. Stratigraphy of the Barrachera Anticline, Municipio de Ojinaga, Chihuahua, Mexico. M, 1959, University of Texas, Austin.

Campbell, Richard B. Reconnaissance glacial geology of the Hoboe Valley, Atlin Provincial Wilderness Park, British Columbia, Canada. M, 1988, University of Idaho. 164 p.

Campbell, Richard Bradford. Continental glaciation in the Glenlyon area, Pelly River district, Yukon, Canada. M, 1951, California Institute of Technology. 37 p.

Campbell, Richard Bradford. The texture, origin, and emplacement of the granitic rocks of Glenlyon Range, Yukon Territory, Canada. D, 1959, California Institute of Technology. 299 p.

Campbell, Richard N. Geochemical prospecting. M, 1949, Miami University (Ohio).

Campbell, Robert A. A new approach for estimating fluid saturations from neutron-density logs. M, 1974, University of Oklahoma. 53 p.

Campbell, Robert Anderson. Petrography and spectral IP response of the Sturgeon Lake massive sulfide and Temagami iron formation deposits, Ontario. M, 1985, University of Western Ontario. 176 p.

Campbell, Robert Arthur. Refractory and other uses of Northwest silica. M, 1949, University of Washington. 53 p.

Campbell, Robert Emerson. Geophysical investigation of the Silver Mountain area, Houghton County, Michigan. M, 1952, Michigan Technological University. 26 p.

Campbell, Robert W. Ground-truth interpretation of thermal infrared imagery, Carrizo Plains, California and Goldfield, Nevada. M, 1969, Stanford University.

Campbell, Russell Boone. The economic geology of Des Moines County, Southeast Iowa. M, 1966, University of Iowa. 183 p.

Campbell, Russell Boone. The feasibility of exploration and development of the coal resources of Iowa. D, 1973, University of Iowa. 266 p.

Campbell, Sarah L. C. Holocene and Pliocene molluscan biogeography of the western North Atlantic. M, 1976, University of South Carolina. 55 p.

Campbell, Steven K. Geology, petrography, and geochemistry of the Lemon Springs Pluton and associated rocks, Lee County, North Carolina. M, 1985, East Carolina University. 132 p.

Campbell, Susan Wendy. Geology and genesis of copper deposits and associated host rocks in and near the Quill Creek area, southwestern Yukon. D, 1981, University of British Columbia.

Campbell, Susan Wendy. Mineralization in the Canoe Lake Stock, Lake of the Woods-Shoal Lake area, northwestern Ontario. M, 1973, University of Manitoba.

Campbell, Thomas J. Phosphate mineralogy of the Tip Top Pegmatite, Custer, South Dakota. M, 1984, South Dakota School of Mines & Technology.

Campbell, William G. Changes in the hydrology of the Pack River following the Sundance fire. M, 1987, University of Idaho. 101 p.

Campiglio, Carlo. La géochimie et la pétrologie du batholite de Bourlamaque. D, 1974, Ecole Polytechnique. 294 p.

Campion, K. M. Diagenetic alteration and formation of authigenic minerals in the Miocene "Rome Beds", Southeast Oregon. D, 1979, Ohio State University. 197 p.

Campion, Kirt Michael. Burrow morphologies of two modern decapod crustaceans from coastal Mississippi and their potential as environmental indicators. M, 1974, University of Nebraska, Lincoln.

Campisano, Cynthia Dane. Geochemical, hydrologic factors affecting radon-222 and radium-226 concentrations in ground waters at Deerfield, New Hampshire. M, 1987, University of New Hampshire. 132 p.

Campisi, Joseph S., Jr. Evaluation of hydrogeology and geochemistry at two different fly ash landfills. M, 1988, SUNY at Binghamton. 114 p.

Campo, Arthur M. Geology and structural analysis of the SE 1/4 of the Burke 15′ Quadrangle, Shoshone County, Idaho. M, 1984, University of Idaho. 190 p.

Campuzano, Jorge. The structure of the Cretaceous rocks in the southeastern part of Sierra de Juarez, Chihuahua, Mexico. M, 1973, University of Texas at El Paso.

Camur, Mehmet Zeki. Thermodynamic and chemical analyses of fracture fillings within the granodiorite of Lake Edison, east-central Sierra Nevada, California. M, 1986, University of Cincinnati. 124 p.

Canaan, Morris. A study of the Ranger Limestone in McCulloch County, Texas. M, 1937, University of Texas, Austin.

Canales-L., L. Inversion of realistic fault models. D, 1975, Stanford University. 131 p.

Canard, Carlos. Spectral analysis of well logs for lithologic subsurface correlation. M, 1966, University of Kansas. 54 p.

Canas, Jose Antonio. Rayleigh wave propagation and attenuation across the Atlantic Ocean. D, 1980, St. Louis University. 254 p.

Canas, Jose Antonio. Surface wave attenuation along several paths in the Pacific Ocean. M, 1978, St. Louis University.

Canavello, Douglas A. Geology of some Paleocene coal-bearing strata of the Powder River basin, Wyoming and Montana. M, 1980, North Carolina State University. 63 p.

Cancienne, Gary Peter. Topographic corrections to gravity data in northeastern Mexico. M, 1987, University of New Orleans. 163 p.

Cancienne, Joseph A. Sedimentary facies relationships and depositional environment of a Mississippian (St. Genevieve Formation) crossbedded-grainstone sequence in the Greendale Syncline, Washington County, southwestern Virginia. M, 1984, University of Southwestern Louisiana. 107 p.

Cande, S. C. The shapes of marine magnetic anomalies; paleomagnetic field behavior, paleomagnetic poles and ridge crest processes. D, 1977, Columbia University, Teachers College. 93 p.

Candee, Christopher R. The geology of the Lincoln Gulch Stock (Tertiary), Pitkin County, Colorado. M, 1972, Colorado School of Mines. 86 p.

Candela, Philip Anthony. Copper and molybdenum in silicate melt-aqueous fluid systems. D, 1982, Harvard University. 151 p.

Candelaria, Magell Phillip. Sedimentology and depositional environment of upper Yates Formation siliciclastics (Permian, Guadalupian), Guadalupe Mountains, Southeast New Mexico. M, 1982, University of Wisconsin-Madison.

Candido, Aladino. Stratigraphy and reservoir geology of the Carmopolis oil field, Brazil. M, 1984, University of Calgary. 132 p.

Candito, Robert J. A geochemical baseline study of surficial materials in the vicinity of oil shale tract C-a, Rio Blanco County, Colorado. M, 1977, Colorado School of Mines. 101 p.

Candler, Charlotte Evans. Subsurface stratigraphic analysis of the Prue, Skinner and Red Fork sandstones, southern Noble County, Oklahoma. M, 1976, Oklahoma State University. 49 p.

Candler, Rudolph John, II. Characterization of metalorganic complexes in aspen and birch forest soils in interior Alaska. D, 1987, University of Alaska, Fairbanks. 123 p.

Candy, Graham John. The geochemistry of some Ordovician and Silurian shales from S.W. Ontario. M, 1963, McMaster University. 175 p.

Caneer, William T. The stratigraphic and clay mineral relationships in fire clay pit number 6D of the A. P. Green Firebrick Company. M, 1956, University of Missouri, Columbia.

Canepa, Julie Ann. The behavior of fluorine, chlorine and sulfur in basalts. D, 1985, Arizona State University. 226 p.

Caner, Bernard. Electrical conductivity structure of the lower crust and upper mantle in western Canada. D, 1969, University of British Columbia.

Canfield, Beverly Anne. Deposition, diagenesis and porosity evaluation of Silurian carbonates in the Permian Basin. M, 1985, Texas Tech University. 138 p.

Canfield, Charles R. A contribution to the petrology of the graywacke series of Mongolia. M, 1932, Columbia University, Teachers College.

Canfield, D. E., Jr. Prediction of total phosphorus concentrations and trophic states in natural and artificial lakes; the importance of phosphorus sedimentation. D, 1979, Iowa State University of Science and Technology. 95 p.

Canfield, Donald Eugene. Sulfate reduction and the diagenesis of iron in anoxic marine sediments. D, 1988, Yale University. 269 p.

Canfield, Douglas John. Quantification of the cephalopod suture pattern. M, 1977, Michigan State University. 66 p.

Canfield, Howard Evan. Thermoluminescence dating and the chronology of loess deposition in the central United States. M, 1985, University of Wisconsin-Madison.

Canfield, Joyce. A depositional model for the middle and lower Newman Formation near Olive Hill, Kentucky. M, 1975, University of South Carolina.

Canich, M. R. and Gold, D. P. A study of the Tyrone-Mount Union lineament by remote sensing techniques and field methods. M, 1977, Pennsylvania State University, University Park. 92 p.

Canis, Wayne Francis. Arenaceous foraminifera from the Chouteau Limestone, Boone County, Missouri. M, 1963, University of Missouri, Columbia.

Canis, Wayne Francis. Conodonts and biostratigraphy of the Lower Mississippiian of Missouri. D, 1967, University of Missouri, Columbia. 186 p.

Cann, Robert Michael. Geochemistry of magnetite and the genesis of magnetite-apatite lodes in the Iron Mask Batholith, British Columbia. M, 1979, University of British Columbia.

Cann, Ross S. Recent calcium carbonate facies of the north-central Campeche Bank, Yucatan, Mexico. D, 1963, Columbia University, Teachers College.

Cannaday, Francis X. The OH Vein and its relation to the Amethyst Fault; Mineral County, Colorado. M, 1950, Colorado School of Mines. 57 p.

Cannata, Stan Lee. Hydrogeology of a portion of the Washita Prairie Edwards Aquifer; central Texas. M, 1988, Baylor University. 205 p.

Canney, Frank C. Some aspects of the geochemistry of potassium, rubidium, cesium, and thallium in sediments. D, 1952, Massachusetts Institute of Technology. 256 p.

Cannizzaro, Carl R. Depositional analysis of the upper Ismay and lower Desert Creek intervals (Pennsylvanian) of the southern Paradox Basin. M, 1985, Rensselaer Polytechnic Institute. 175 p.

Cannon Suva, Melinda S. The investigation of a hydrothermal convective system along the Cincinnati-Findlay Arch in northwestern Ohio. M, 1982, Kent State University, Kent. 142 p.

Cannon, Debra May. The stratigraphy, geochemistry and mineralogy of two ash-flow tuffs in the Deschutes Formation, central Oregon. M, 1985, Oregon State University. 142 p.

Cannon, M. R. Conceptual models of interactions of mining and water resource systems in the southwestern Idaho phosphate field. M, 1979, University of Idaho.

Cannon, Philip Jan. Application of radar and infrared imagery to a quantitative geomorphological investigation of the Mill Creek drainage basin, south-central Oklahoma. D, 1973, University of Arizona.

Cannon, Philip Jan. Pleistocene geology of the North Fork and Salt Fork of the Red River in Oklahoma. M, 1967, University of Oklahoma. 38 p.

Cannon, Ralph S. Geology of the Piseco Lake Quadrangle, New York. D, 1935, Princeton University. 173 p.

Cannon, Ralph S., Jr. Genesis of the pre-Cambrian rocks of Tenmile Valley, Colorado. M, 1933, Northwestern University.

Cannon, Robert Lee. The fauna of the Escondido Formation. M, 1922, University of Texas, Austin.

Cannon, Robert P. Petrology and geochemistry of the Palisades Sill, New Mexico. M, 1976, University of North Carolina, Chapel Hill. 76 p.

Cannon, Susan Hilary. The lag rate and the travel-distance potential of debris flows. M, 1985, University of Colorado.

Cannon, Wayne Harry. Geomagnetic depth-sounding in southern British Columbia and Alberta (Canada). M, 1967, University of British Columbia.

Cannon, Wayne Harry. The free air laser strain meter (a feasibility study of a new method of observing tectonic motions). D, 1971, University of British Columbia.

Cannon, William Francis. Plutonic evolution of the Cutler area (Precambrian) (Lake Huron, north shore), Ontario (Canada). D, 1968, Syracuse University. 116 p.

Cannon, William Francis. The petrography of the tills of northern Vermont. M, 1964, Miami University (Ohio). 79 p.

Cano, Octavio. A solution to the matrix absorption problem in quantitative elemental X-ray fluorescence analysis. D, 1966, Pennsylvania State University, University Park. 145 p.

Canova, Judy Lynn. Late Holocene isotopic and sedimentologic records contained in carbonate lagoonal cores, northern Little Bahama Bank. M, 1988, Texas A&M University.

Canstein, Ruth Marilyn. The stratigraphy and faunas of the Dakota and Colorado groups (upper Lower and Lower Cretaceous) of the Muddy Creek Anticline, Huerfano Park, Colorado. M, 1964, University of Michigan.

Cant, Douglas J. Braided stream deposits of the Battery Point Formation, Gaspe, Quebec. M, 1974, McMaster University. 125 p.

Cant, Douglas J. Braided stream sedimentation of the South Saskatchewan River. D, 1977, McMaster University. 246 p.

Canter, Karen Lyn. Stratigraphy, lithologic descriptions, depositional environment and ecological successional stages of the Juniper Gulch Member of the Snaky Canyon Formation, east-central Idaho. M, 1984, University of Idaho. 108 p.

Canter, Neil W. Paleogeology and paleogeography of the Big Mountain area, Santa Susana, Moorpark, and Simi Quadrangles, Ventura County, California. M, 1974, Ohio University, Athens. 58 p.

Cantrell, Carlton L. The geology of the Irvine Formation, Irvine, Kentucky. M, 1973, Eastern Kentucky University. 65 p.

Cantrell, Dave Lee. Anatomy, diagenesis, and geochemistry of a Middle Ordovician oolite shoal, East Tennessee. M, 1982, University of Tennessee, Knoxville. 153 p.

Cantrell, Jason Kirk. Rare earth element speciation in sea water. M, 1986, University of South Florida, St. Petersburg.

Cantrell, Kirk Jason. Rare element speciation in sea water. M, 1986, University of South Florida, St. Petersburg.

Cantrell, Peggy Francis Parthenia. A faunal study of the Pumpkin Creek Limestone Member of the Dornick Hills Formation in the Ardmore Basin. M, 1949, University of Oklahoma. 79 p.

Cantrell, Ralph B. Subsurface geology of a portion of Freestone County, Texas. M, 1937, Texas Tech University. 28 p.

Cantwell, Jonathan. A gravity study of the Blackfoot-Nevada Valley area, northwestern Montana. M, 1980, University of Montana. 39 p.

Cantwell, Richard J. Fossil Sigmodon from southeastern Arizona (Blancan; post-Blancan; Irvingtonian; Recent?), San Pedro and San Simon valleys near Benson (Cochise County) and Safford (Graham County). M, 1967, University of Arizona.

Cantwell, Thomas. The detection and analysis of low frequency magnetotelluric signals. D, 1960, Massachusetts Institute of Technology. 172 p.

Cao, Song. Sensitivity analysis of 1-D dynamical model for basin analysis. D, 1987, University of South Carolina. 325 p.

Cao, Tianqing. Simulation of seismicity pattern and recurrence behavior on a heterogeneous fault using laboratory friction laws. D, 1986, Massachusetts Institute of Technology. 174 p.

Caoili, Abraham Albano. Performance of rainfall simulator and ponding infiltrometer at different soil conditions. D, 1982, Utah State University. 321 p.

Capaccioli, Deborah Ann. Biostratigraphy, petrology, and paleodepositional environments of the Ross Formation, Upper Silurian-Lower Devonian, west-central Tennessee. M, 1987, University of Tennessee, Knoxville. 125 p.

Capaldo, Paul S. Environmental factors influencing the zonation of three semi-terrestrial decapods; Uca pugnax, Uca minax, and Sesarma reticulatum. M, 1976, Florida Institute of Technology.

Caparis, Petros P. Stable isotope geochemistry of the reservoir rocks of the active geothermal system in Valles Caldera, New Mexico. M, 1987, University of California, Davis. 88 p.

Capaul, William A. Volcanoes of the Chiapas volcanic belt, Mexico. M, 1987, Michigan Technological University. 92 p.

Cape, David Frank. An investigation of the radioactivity of the Pilot Lake area, Northwest Territories. M, 1977, University of Alberta. 156 p.

Capel, Paul David. Distributions and diagenesis of chlorinated hydrocarbons in sediments. D, 1986, University of Minnesota, Minneapolis. 256 p.

Capen, Robert C. Some aspects of the tectonic history of Lancaster County, Pennsylvania, recorded in the Miller quarry, Lancaster township. M, 1969, Millersville University.

Capers, William A., Jr. Contact metamorphism of the Mancos Shale near Crested Butte, Colorado. M, 1974, Rice University. 59 p.

Capichioni, Maria Luisa. A sedimentological examination of the Berea Sandstone at Windsor Mills, northeastern Ohio. M, 1986, Case Western Reserve University. unpaginated p.

Capitani, Christian Emile De *see* De Capitani, Christian Emile

Capo, Rosemary Clare. Petrology and geochemistry of a Cambrian Paleosol developed on Precambrian granite, Llano Uplift, Texas. M, 1984, University of Texas, Austin. 110 p.

Capobianco, Christopher. Partitioning of germanium between forsterite and silicate melts in Di-Fo-An system. M, 1980, Rensselaer Polytechnic Institute. 56 p.

Capobianco, Christopher John. Thermodynamic relations of several carbonate solid solutions. D, 1986, Arizona State University. 223 p.

Caponi, E. A. A three-dimensional model for the numerical simulation of estuaries. D, 1974, University of Maryland. 234 p.

Caporuscio, Florie Andre. Petrogenesis of mantle eclogites from South Africa. D, 1988, University of Colorado. 164 p.

Caporuscio, Florie Andre. Petrology and geochemistry of the San Carlos Arizona alkalic lavas. M, 1980, Arizona State University. 105 p.

Capozza, Frank Cataldo. The geology of the Bone Basin and Perry Canyon intrusions (Late Cretaceous and early Tertiary), Madison County, Montana. M, 1967, Indiana University, Bloomington. 48 p.

Cappa, James A. The depositional environment, paleocurrents, provenance and dispersal patterns of the Abo Formation in part of the Cerros de Amado region, Socorro County, New Mexico. M, 1975, New Mexico Institute of Mining and Technology. 153 p.

Cappallo, Roger James. The rotation of the Moon. D, 1980, Massachusetts Institute of Technology. 104 p.

Cappel, Howard Noble, Jr. A study of Lower Ordovician Bryozoa in Northwest Georgia. M, 1957, Emory University. 74 p.

Capps, Richard C. The geology of the Rancho El Papalote area, Chihuahua, Mexico. M, 1981, East Carolina University. 73 p.

Capps, Stephen Reid, Jr. The Pleistocene geology of the Leadville Quadrangle, Colorado. D, 1907, University of Chicago. 99 p.

Capps, William M. Stratigraphic analysis of the Missourian and lower Virgilian series in northwestern Oklahoma. M, 1959, University of Oklahoma. 48 p.

Caprara, John Robert. Feldspar dispersal patterns in mudrocks of the Vanoss Group, south-central Oklahoma. M, 1978, University of Oklahoma. 64 p.

Caprez, Lionel Preston. Sedimentary analysis of alluvial fill (sand and gravel pits) along the Saline and Smoky Hill rivers, Ellis County, Kansas. M, 1974, Fort Hays State University. 45 p.

Caprio, Eugene R. Water resources of the western slopes of the Sandia Mountains, Bernalillo and Sandoval counties, New Mexico. M, 1960, University of New Mexico. 176 p.

Caprio, R. C., Jr. Quantitative appraisal of till in South-central New York. D, 1980, SUNY at Binghamton. 212 p.

Capuano, Regina M. Chemical mass transfer and solution flow in Wyoming roll-type uranium deposits. M, 1977, University of Arizona. 81 p.

Capuano, Regina Marie. Chemical equilibria and fluid flow during compaction diagenesis of organic-rich geopressured sediments. D, 1988, University of Arizona. 135 p.

Caputo, Mario Vicente. Stratigraphy, tectonics, paleoclimatology and paleogeography of northern basins of Brazil. D, 1984, University of California, Santa Barbara. 605 p.

Caputo, Mario Vincent. Depositional history of Middle Jurassic clastic shoreline sequences in southwestern Utah. M, 1980, Northern Arizona University. 203 p.

Caputo, Mario Vincent. Origin of sedimentary facies in the upper San Rafael Group (Middle Jurassic), east-central Utah. D, 1988, University of Cincinnati. 462 p.

Car, Dwayne Peter. A volcaniclastic sequence on the flank of an early Precambrian stratovolcano, Lake of the Woods, northwestern Ontario. M, 1980, University of Manitoba. 111 p.

Caraco, Nina Marie. Phosphorous, iron, and carbon cycling in a salt stratified coastal pond. D, 1986, Boston University. 224 p.

Caram, Hector L. Depositional environment and reservoir characteristics of the upper Frio sandstones, Willamar Field, Willacy County, Texas. M, 1988, Texas A&M University. 210 p.

Caramanica, Frank P. Coral paleontology and paleoecology of the "Centerfield" biostromes of northeastern Pennsylvania. M, 1968, SUNY at Binghamton. 251 p.

Caramanica, Frank Phillip. Ordovician corals of the Williston Basin periphery. D, 1973, University of North Dakota. 570 p.

Caran, Samuel Christopher. Environmental geology of abandoned lignitic- and bituminous-coal mines of Texas. M, 1984, University of Texas, Austin. 255 p.

Carapetian, Ara G. Frequency analysis of short-period microseisms generated by trains. M, 1966, New Mexico Institute of Mining and Technology. 42 p.

Caravella, Frank. A study of Poisson's ratio in the Socorro area. M, 1977, New Mexico Institute of Mining and Technology.

Caravella, Joseph Christopher. Lacustrine deposits in northeastern Yahualica de Gonzales Gallo Quadrangle, Jalisco, Mexico. M, 1985, University of New Orleans. 118 p.

Carballo, Jose Domingo, Jr. Holocene dolomitization of supratidal sediments, Sugarloaf Key, Florida. M, 1985, University of Texas, Austin. 130 p.

Carboneau, Come. Geology of the Big Berry Mountains map area, Gaspe Peninsula, Quebec. D, 1953, McGill University.

Carboni, Salvatore. Photointerprétation et analyse d'image des linéaments photogéologiques de l'Ile d'Anticosti, Québec. M, 1988, Ecole Polytechnique. 126 p.

Carbonneau, C. Geology of the Big Berry Mountains map-area, Gaspé Peninsula, Quebec. D, 1953, McGill University.

Carbonneau, Come. Stratigraphy and structure of a portion of the Schick Schok Mountains, Gaspe Peninsula. M, 1949, University of British Columbia.

Carbotte, Suzanne Marie. Geological and geophysical characteristics of the Tuzo Wilson Knolls and vicinity; triple junction tectonics in the NE Pacific. M, 1986, Queen's University. 124 p.

Card, J. W. Spontaneous deposition of radon decay products from air. M, 1978, Carleton University. 108 p.

Card, Kenneth Darius Huycke. Geology of the Agnew Lake area, Ontario; a study in Precambrian stratigraphy, structure, and metamorphism. D, 1963, Princeton University. 266 p.

Carda, Dan D. A study of the radium content of the ground waters in western South Dakota with emphasis on the Madison (Pahasapa) Limestone. D, 1975, South Dakota School of Mines & Technology. 64 p.

Cardea, Harry S. The Terra Alta gas field, Preston County, West Virginia. M, 1959, West Virginia University.

Carden, John R. Chemical and petrographic variations in McCartys basalt flow (late Cenozoic), Valencia County, New Mexico. M, 1972, Kent State University, Kent. 77 p.

Carden, John R. The comparative petrology of blueschists and greenschists in the Brooks Range and Kodiak-Seldovia schist belts. D, 1978, University of Alaska, Fairbanks. 258 p.

Cardenas, Jack Roger Palomino *see* Palomino Cardenas, Jack Roger

Cardenas-Cala, Rafael. Interactions between zinc and cadmium in clays. D, 1976, University of California, Riverside. 117 p.

Carder, Dean Samuel. A comparative study of the seismic surface waves and the crustal structure of the Pacific region. D, 1933, University of California, Berkeley. 69 p.

Carder, Dean Samuel. Origin of the Palouse soil in the vicinity of Moscow, Idaho. M, 1925, University of Idaho. 43 p.

Cardinal, D. F. Geology of the Crystal Creek-Balk Mountain area, Big Horn and Sheridan counties, Wyoming. M, 1958, University of Wyoming. 114 p.

Cardinell, Alex Phillip. An investigation of mass movement structures on the Mississippi River delta front off southwest pass. M, 1983, University of Illinois, Urbana. 253 p.

Cardneaux, Christopher A. Depositional systems of upper Bloyd and lower Atoka sandstones (Pennsylvanian), eastern Crawford and western Franklin counties, Arkansas. M, 1978, University of Arkansas, Fayetteville.

Cardona, Alberto. Effect of the Delaware River in the contamination of the Raritan and Magothy aquifers in Burlington County, New Jersey. M, 1988, Montclair State College. 60 p.

Cardone, Anthony Thomas. A statistical forecasting of engineering properties and compression index of soils, Salt Lake City, Utah. M, 1966, University of Utah. 67 p.

Cardott, Brian Joseph. A comparative study on the occurrence and distribution of fluorescent macerals coals from three major coal basins of the United States. M, 1981, Southern Illinois University, Carbondale. 159 p.

Cardwell, Aubrey L. The petroleum source-rock potential of the Arbuckle and Ellenburger groups (Cambro-Ordovician and Ordovician), southern Mid-Continent United States. M, 1977, Colorado School of Mines. 172 p.

Cardwell, Dudley H. Geology of the Melrose Granite and associated rocks of Campbell, Halifax, and Pittsylvania counties, Virginia. M, 1925, University of Virginia. 48 p.

Cardwell, Richard Kenneth. Geometry of the lithosphere beneath the eastern Indonesian and Philippine islands as determined from the spatial distribution of earthquakes and focal mechanism solutions. D, 1980, Cornell University. 157 p.

Careaga, Richard Oliver. Paleomagnetism of part of the Jonathan Creek Formation (upper Mississippian) of east-central Ohio. M, 1971, Ohio State University.

Caress, Mary Elizabeth. Volcanology of the youngest Toba Tuff, Sumatra. M, 1985, University of Hawaii. 175 p.

Carew, James Leslie. Faunal analysis of Permo-Carboniferous shales, North-central Texas. D, 1978, University of Texas, Austin. 321 p.

Carew, James Leslie. Ostracod species distribution, Harbor island, Texas. M, 1969, University of Texas, Austin.

Carey, Byrl D., Jr. Geology of the eastern part of Flat Top Anticline, Albany and Carbon counties, Wyoming. M, 1950, University of Wyoming.

Carey, Byrl D., Jr. Geology of the Rattlesnake Hills, Tertiary volcanic field, Natrona County, Wyoming. D, 1959, University of Wyoming. 247 p.

Carey, Daniel Irvin. Hydrologic and economic models in reservoir design. D, 1975, [University of Kentucky]. 189 p.

Carey, Dwight L. Form and processes in the pseudokarstic topography, Arroyo Tapiado, Anza Borrego Desert State Park, California. M, 1976, University of California, Los Angeles.

Carey, Dwight Lee. Environmental activities for three geothermal resources exploration well projects in the Geysers-Calistoga Known Geothermal Resource Area, California. D, 1982, University of California, Los Angeles. 174 p.

Carey, J. Anne. Geology of late Proterozoic Miette Group, southern Main Ranges, Cushing Creek area, B.C. M, 1984, University of Calgary. 119 p.

Carey, James William. Petrology and metamorphic history of metapelites in the Boehls Butte Quadrangle, northern Idaho. M, 1985, University of Oregon. 171 p.

Carey, Mary Alice. Chesterian-Morrowan conodont biostratigraphy from northeastern Utah. M, 1973, University of Utah. 83 p.

Carey, Max Raymond. Quantitative determination of chlorophyll in the Indian River lagoon. M, 1973, Florida Institute of Technology.

Carey, Neal J. An evaluation of the effectiveness of bentonite as a soil additive for the cover material on the KL Avenue sanitary landfill, Kalamazoo, Michigan. M, 1986, Western Michigan University.

Carey, Stephen Paul. Post-Sonoman conodont biofacies of the Triassic of northwestern Nevada. D, 1984, University of Wisconsin-Madison. 335 p.

Carey, Steven Norman. Studies on the generation, dispersal and deposition of tephra in the marine and terrestrial environment. D, 1982, University of Rhode Island. 383 p.

Cargill, D. G. Geology of the "Island Copper" Mine, Port Hardy, British Columbia. D, 1975, University of British Columbia.

Cargill, Donald G. Radioactive deposits in Donaldson Lake area, Beaverlodge, Saskatchewan. M, 1969, Queen's University. 71 p.

Cargill, Simon M. Strain analysis of three rock types in South mountain, in the vicinity of Myersville, Maryland. M, 1968, George Washington University.

Cargo, David N. Mineral deposits of the Granite Gap area, Hidalgo County, New Mexico. M, 1959, University of New Mexico. 70 p.

Cargo, David Niels. Polygonal strain systems in the Cordilleran Foreland, Western United States. D, 1966, University of Utah. 140 p.

Carhart, Grace M. A study of the Peekskill Granite. M, 1925, Columbia University, Teachers College.

Cariani, Anthony Robert. The geology of the Anson Quadrangle, Maine. D, 1958, Boston University. 200 p.

Carignan, Jacques. Géochimie et géostatistique appliquées à l'exploration des gisements volcanogènes; le Gisement de Millenbach. D, 1979, Universite de Montreal.

Carignan, Jacques. Géochimie et géostatistique appliquées a l'éxploration des gisements volcanogènes; le gisement de Millenbach. D, 1980, Ecole Polytechnique. 351 p.

Carignan, Jacques. Pétrographie et géochimie des roches volcaniques des Cantons Destor et Dufresnoy, Abitibi. M, 1975, Ecole Polytechnique.

Carini, George Francis. Inferences of the petrography and depositional environment of the Tiawah and Seville limestones of west-central Missouri. M, 1959, University of Missouri, Columbia.

Carini, George Francis. Regional petrographic and paleontologic analysis of the Triassic Alcova Limestone Member in central Wyoming. D, 1964, University of Missouri, Columbia. 180 p.

Cariolou, Marios Andreou. Extraction, partial purification, and characterization of the abalone shell peptides. D, 1985, University of California, Santa Barbara. 171 p.

Caristan, Yves Denis. High temperature mechanical behaviour of Maryland diabase. D, 1981, Massachusetts Institute of Technology. 155 p.

Carius, Terry L. Variation of the geotechnical properties of the Gulf of Maine and San Diego Trough and their geologic implications. M, 1973, Lehigh University.

Carkin, Brad A. The geology and petrology of the Fifes Peak Formation in the Cliffdell area, central Cascades, Washington. M, 1988, Western Washington University. 157 p.

Carl, James Dudley. An investigation of the minor element content of potash feldspar from pegmatites, Haystack Range, Wyoming. D, 1961, University of Illinois, Urbana. 83 p.

Carl, James Dudley. X-ray fluorescent analysis of iron content of host rock, Sterling Lake, New York. M, 1960, University of Illinois, Urbana.

Carl, Joseph Buford. Geology of the Black Dog area, Osage County, Oklahoma. M, 1957, University of Oklahoma. 105 p.

Carlat, James Eugene. The geology of the southwestern portion of the Jamieson Quadrangle, Malheur County, Oregon. M, 1954, University of Oregon. 86 p.

Carlberg, R. G. A model of a magnetic star. M, 1975, University of British Columbia.

Carle, Henry Mark. Modeling oceanic crustal magnetization using MAGSAT derived scalar anomalous field data. M, 1983, University of Miami. 58 p.

Carleton, Alfred Townes, Jr. Structural history of part of Kent Station Quadrangle, Culberson and Jeff Davis counties, Texas. M, 1952, University of Texas, Austin.

Carleton, Natalie E. The petrography and petrology of the rocks of the northern Champlain Lowlands. M, 1936, University of Vermont.

Carlin, Joseph T. The dependence of measured charge separation upon the nature and polarization of the base during freezing of dilute aqueous solutions. M, 1956, New Mexico Institute of Mining and Technology. 36 p.

Carlin, Rachel Ann. A geochemical study of the Eagle Creek Formation in the Columbia River gorge, Oregon. M, 1988, Portland State University. 90 1 plate n.

Carlisle, Craig L. The subsurface structure of the Ivanpah Valley, California, as determined by geophysical measurements. M, 1982, University of California, Santa Barbara. 90 p.

Carlisle, Donald. Conservation of mineral resources. D, 1950, University of Wisconsin-Madison.

Carlisle, Donald. Vanadium in an interlava sediment, Quadra Island, B.C. M, 1944, University of British Columbia.

Carlisle, Donald Hugh. A continuous seismic profiling survey off the coast of Lebanon. M, 1965, Massachusetts Institute of Technology. 138 p.

Carlisle, Frank Jefferson, Jr. Characteristics of soils with fragipans in a Podzol region. D, 1954, Cornell University.

Carlisle, Joel Christie. Economic geology of Candelaria area, Presidio County, Trans-Pecos Texas. M, 1955, University of Texas, Austin.

Carlisle, Joseph T. A laboratory study of the Beech Granite plagioclase. M, 1956, University of Tennessee, Knoxville. 28 p.

Carlisle, Scott P. Velocity and energy studies of microseismic data. M, 1986, University of Idaho. 97 p.

Carlisle, W. Joseph. Upper Cretaceous stratigraphy, Lincoln and Sublette counties, western Wyoming thrust belt. M, 1979, University of Wyoming. 103 p.

Carlmark, Jon William. Penetration of free-falling objects into deep-sea sediments. M, 1971, United States Naval Academy.

Carloss, James C. Depositional environments from borehole measurements, lower Cretaceous, Peoria fields and adjacent areas, Arapahoe County, Colorado. M, 1974, Colorado School of Mines. 335 p.

Carlsen, Greg M. Variations of gravity field intensity in the Mississippi Embayment region. M, 1978, Northern Illinois University. 121 p.

Carlson, Alane R. Surface geology of the Beaverton Quadrangle, Alabama. M, 1986, Mississippi State University. 92 p.

Carlson, Allan Eugene. The influence of the laminar flow boundary layer on crystals growing from solution. D, 1958, University of Utah. 172 p.

Carlson, Barry. Biostratigraphy of the Keokuk and Warsaw formations (Mississippian) of western Illinois. M, 1961, University of Wisconsin-Madison.

Carlson, Barry Albin. A gravity study of the geology of northeastern Wisconsin. D, 1974, Michigan State University. 120 p.

Carlson, Carl A. A statistical study of the geochemical evolution of a platinum-bearing magma from near Goodnews Bay, Alaska. M, 1983, California State University, Hayward. 55 p.

Carlson, Charles Edward. Areal geology and stratigraphy of the Red Fork Powder River area, Johnson County, Wyoming. M, 1949, University of Wyoming. 52 p.

Carlson, Charles G. A test of the feldspar method for the determination of the origin of metamorphic rocks. M, 1920, University of Wisconsin-Madison.

Carlson, Christine. Sedimentary serpentinites of the Wilbur Springs area; a possible Early Cretaceous structural and stratigraphic link between the Franciscan Complex and the Great Valley Sequence. M, 1981, Stanford University. 105 p.

Carlson, Clarence G. Stratigraphy of the Winnipeg and Deadwood Formations in North Dakota. M, 1960, University of North Dakota. 149 p.

Carlson, Craig Iver. Electron probe microanalysis; empirical data reduction and an application to a carbonate concretion. M, 1970, University of Oregon. 89 p.

Carlson, David John. Chemistry and microbiology of surface microlayers of estuarine and coastal waters of the Gulf of Maine. D, 1981, University of Maine. 205 p.

Carlson, David Roy. Gravity and bedrock geology study of east-central Minnesota. M, 1971, Northern Illinois University. 31 p.

Carlson, Diane Helen. Geology and petrochemistry of the Keller Butte pluton and associated intrusive rocks in the south half of the Nespelem and northern half of the Grand Coulee Dam quadrangles, and the development of cataclasites and fault lenses along the Manila Pass Fault, northeastern Washington. D, 1984, Washington State University. 181 p.

Carlson, Diane Helen. Petrology of the migmatite complex along the South Fork of the Clearwater River, Idaho. M, 1981, University of Minnesota, Duluth.

Carlson, E. N. Notes on the origin, weathering, and secondary enrichment of manganese ores. M, 1921, University of Minnesota, Duluth.

Carlson, Earl Reinhold. The metallogeny of Finno-Scandia. M, 1923, University of Minnesota, Minneapolis. 121 p.

Carlson, Ernest H. Experiments in the solid-vapor systems HgS-H$_2$S. D, 1966, McGill University.

Carlson, Ernest H. Geology of the Big English Gulch area, Lake County, Colorado. M, 1960, University of Colorado.

Carlson, F. R. Measurement of cobble abrasion in natural streams. M, 1974, University of Arizona.

Carlson, Fred Albert. Some relations of organic matter in soils. D, 1922, Cornell University.

Carlson, Garry J. Crustal structure within southwestern Montana and adjacent northeastern Idaho; a seismic refraction study. M, 1985, University of Montana. 101 p.

Carlson, Gerald G. Geology of the Bailadores massive sulfide deposit. M, 1974, Michigan Technological University. 53 p.

Carlson, Gerald G. Mapping ultramafic rocks by computer analysis of digital LANDSAT data. D, 1978, Dartmouth College. 225 p.

Carlson, Gorden Anders, Jr. Translocation and attenuation of wastewater phosphorus in streams. D, 1977, Rensselaer Polytechnic Institute. 260 p.

Carlson, Gregory D. Depositional modelling of Carboniferous rocks applied to coal exploration, northern Cumberland Plateau, Tennessee. M, 1978, University of South Carolina.

Carlson, Gustaf Magnus. Applications of the microscope to ore testing with special reference to a Butte ore. M, 1932, University of Minnesota, Minneapolis. 64 p.

Carlson, H. D. Origin of the corundum deposits of Renfrew County (Ontario). D, 1953, Queen's University. 146 p.

Carlson, Harry W. Geology of the Elysian Park-Silver Lake District, California. M, 1945, California Institute of Technology. 20 p.

Carlson, Hugh Douglas. The influence of temperature on the fundamental strength of rocks. D, 1951, University of Toronto.

Carlson, Hugh Douglas. The intrusive rocks of the northeastern portion of the Timagami Lake area, Ontario. M, 1950, University of Toronto.

Carlson, John E. An investigation into ground-water conditions in the vicinity of Dayton, Nevada. M, 1958, University of Nevada - Mackay School of Mines.

Carlson, John Edward. A lithologic study of some medium-grained Upper Cretaceous sedimentary rocks of southeastern Idaho. M, 1951, University of Idaho. 34 p.

Carlson, Jon Andrew. Exploration for kimberlite and geophysical delineation of diatremes, W. State Line District, Colo./Wyo. M, 1983, Colorado State University. 245 p.

Carlson, K. J. Corals of the Gilmore City Limestone, Iowa. M, 1962, Iowa State University of Science and Technology.

Carlson, Kelley Elaine. A combined analysis of gravity and magnetic anomalies in east-central Minnesota. M, 1985, University of Minnesota, Minneapolis. 138 p.

Carlson, Kenneth W. Seismic noise and microseismicity in a Nevada geothermal prospect. M, 1975, Colorado School of Mines.

Carlson, Marvin Paul. Lithostratigraphy and correlation of the Mississippian System in Nebraska. M, 1963, University of Nebraska, Lincoln.

Carlson, Marvin Paul. Stratigraphic framework of Precambrian and lower and middle Paleozoic rocks in the subsurface of Nebraska. D, 1969, University of Nebraska, Lincoln. 92 p.

Carlson, Mikel Carl. A petrologic analysis of surface and subsurface Atoka Formation (Lower Pennsylvanian) sandstone, western margin of the Arkoma Basin, Oklahoma. M, 1988, University of Tulsa. 238 p.

Carlson, Paul Richard. Geology and engineering properties of Cenozoic sediments near Point Barrow, Alaska. M, 1957, Iowa State University of Science and Technology.

Carlson, Paul Roland. Marine geology of Astoria submarine canyon. D, 1968, Oregon State University. 259 p.

Carlson, R. E. Phosphorus cycling in a shallow eutrophic lake in southwestern Minnesota. D, 1975, University of Minnesota, Minneapolis. 183 p.

Carlson, Randall. Depositional environments, cyclicity and diagenetic history of the Wahoo Limestone, eastern Sadlerochit Mountains, northeastern Alaska. M, 1988, University of Alaska, Fairbanks. 189 p.

Carlson, Richard L. A gravity study of the Cypress Island peridotite, Washington. M, 1972, University of Washington. 39 p.

Carlson, Richard L. Cenozoic plate convergence in the vicinity of the Pacific Northwest; a synthesis and assessment of plate tectonics in the northeastern Pacific. D, 1976, University of Washington. 129 p.

Carlson, Richard Walter. Crust-mantle differentiation on the Earth and Moon; evidence for isotopic studies for contrasting mechanisms and duration. D, 1980, University of California, San Diego. 234 p.

Carlson, Robert John. Correlation of the Pennsylvanian stratigraphy across the Illinois Basin. M, 1954, Miami University (Ohio). 33 p.

Carlson, Roger Allan. Geology and petrology of the volcanic rocks south of Hawkins Basin, S. E. Idaho. M, 1968, Idaho State University. 66 p.

Carlson, Roseann J. Foraminifera from the upper Twin River Formation, Hoko river area, Olympic peninsula, Washington. M, 1969, University of Washington. 27 p.

Carlson, Sandra Jean. Functional analysis of a carpoid aulacophore. M, 1982, University of Michigan.

Carlson, Sandra Jean. Ontogenetic and evolutionary trends in the articulate brachiopod hinge mechanism. D, 1986, University of Michigan. 282 p.

Carlson, Steven Michael. Investigations of Recent and historical seismicity in East Texas. M, 1984, University of Texas, Austin. 197 p.

Carlson, Steven R. Fluid inclusion studies of the Tourmalina copper-molybdenum bearing breccia pipe, northern Peru. M, 1979, University of Minnesota, Minneapolis. 61 p.

Carlson, Steven Ray. Microcrack porosity and in situ stress in Illinois borehole UPH-3. M, 1985, University of Wisconsin-Madison. 222 p.

Carlson, Thomas R. Surficial deposits and land utilization in Athens Township, Ohio. M, 1977, Ohio University, Athens. 209 p.

Carlson, Thomas Warren. Deep-water currents and their effect on sedimentation in Lake Superior. D, 1982, University of Minnesota, Minneapolis. 190 p.

Carlson, Thomas Warren. Opaline claystones in the Paleocene descriptions of southern North Carolina. M, 1976, University of Minnesota, Minneapolis. 73 p.

Carlson, William A. Quaternary geology and groundwater resources of the Kansas River valley between Newman and Lawrence, Kansas. M, 1952, University of Kansas. 94 p.

Carlson, William Douglas. Experimental studies of metamorphic petrogenesis; Part II, The calcite-aragonite equilibrium; Part II, Aragonite-calcite transformation kinetics; Part III, One-atmosphere subsolidus equilibria in model peridotite. D, 1980, University of California, Los Angeles. 178 p.

Carlson, William S. Scientific report of the 4th University of Michigan Greenland Expedition (1930-31). D, 1938, University of Michigan.

Carlston, Charles W. Appalachian drainage and the Highland border sediments of the Newark Series. D, 1946, Columbia University, Teachers College.

Carlston, Karen Jean. Electric field ratio telluric survey of the Roosevelt Hot Springs, Utah. M, 1982, University of Utah. 103 p.

Carlton, Cleet Francis. Tectonic significance of late Miocene deposits in the southern Fry Mountains, Mojave Desert, California. M, 1988, University of California, Riverside. 130 p.

Carlton, Dennis R. The paleoecology of the Deer Creek Formation, Upper Pennsylvanian of eastern Kansas. M, 1975, Wichita State University. 314 p.

Carlton, James L. The geology of the Bartelso oil field, Clinton County, Illinois. M, 1940, University of Chicago. 58 p.

Carlton, Keith H. Depositional environment of the Rodessa Formation, west central protion of Anderson County, Texas. M, 1981, Texas Christian University. 101 p.

Carlton, Richard W. Stratigraphy, petrology, and mineralogy of the Colestin Formation in Southwest Oregon and Northern California. D, 1972, Oregon State University. 208 p.

Carlton, Richard W. The structure and stratigraphy of a portion of the Trout Creek Mountains, Harney County, Oregon. M, 1969, Oregon State University. 116 p.

Carlton, Ronald Ray. Models of dolomitization from petrographic and selected trace element data within the Middle Devonian carbonates of the Reed City storage field, Lake and Osceola counties, Michigan. M, 1982, Michigan State University. 162 p.

Carlton, Stephen M. Fish Springs multibasin flow system, Nevada and Utah. M, 1985, University of Nevada. 103 p.

Carluccio, Leeds Mario. Contributions to the morphology and anatomy of the Devonian progymnosperm Archaeopteris. D, 1966, Cornell University. 162 p.

Carlyle, George Alva. Metasomatism of a basic intrusive near Bedford, New York. M, 1955, Miami University (Ohio). 42 p.

Carmack, Eddy. On the hydrography of the Greenland Sea. D, 1972, University of Washington. 185 p.

Carmack, Ray P. The history of secondary recovery of oil in the United States. M, 1953, Texas Tech University. 128 p.

Carman, Erick P. The water balance, hydrogeology and chemical loading from ground water in Beaver Lake, Waukesha County, Wisconsin. M, 1988, University of Wisconsin-Milwaukee. 240 p.

Carman, Joel Ernest. The Pleistocene geology of northwestern Iowa. D, 1915, University of Chicago. 210 p.

Carman, John H. Petrographic study and composition analysis of olivine phenocrysts, Bernalillo County, New Mexico. M, 1960, New Mexico Institute of Mining and Technology. 87 p.

Carman, John Homer. The study of the system $NaAlSiO_4$-$Mg_2SiO_4SiO_2$-H_2O from 200 to 5000 bars and 800°C to 1100°C and its petrologic applications. D, 1969, Pennsylvania State University, University Park. 302 p.

Carman, K. W. Shallow-water foraminifera of Bermuda. D, 1933, Massachusetts Institute of Technology. 181 p.

Carman, Mary Ruth Cote. Conodonts of the Lake Valley Formation (Lower Mississippian), Sacramento Mountains, New Mexico. M, 1984, University of Iowa. 77 p.

Carman, Max F., Jr. Geology of the Lockwood Valley area, Kern and Ventura counties, California. D, 1954, University of California, Los Angeles.

Carmichael, A. D. Geology of the Norbeau Mine. M, 1940, Queen's University. 41 p.

Carmichael, Alan Barnett. Mineralogy and geochemistry of Upper Cretaceous clay mineral assemblages from the Star Lake-Torreon coal fields, San Juan Basin, New Mexico. M, 1982, New Mexico Institute of Mining and Technology. 87 p.

Carmichael, Dugald Macaulay. Structure and progressive metamorphism in the Whetstone Lake area, Ontario (Canada), with emphasis on the mechanism of prograde metamorphic reactions. D, 1967, University of California, Berkeley. 112 p.

Carmichael, Mary F. Catalytic action in the oxidation of sulphides and arsenides. M, 1928, University of Toronto.

Carmichael, Robert Stewart. Pressure magnetization of ferromagnetics as applied to rock magnetism. D, 1967, University of Pittsburgh. 226 p.

Carmichael, Robert Stewart. Synthetic seismograms and reflection seismology. M, 1964, University of Pittsburgh.

Carmichael, Scott Matthew McKenzie. Sedimentology of the Lower Cretaceous Gates and Moosebar formations, Northeast Coalfields, British Columbia. D, 1983, University of British Columbia.

Carmichael, Thomas J. An investigation into the origin of some Adirondack (New York) pegmatites. M, 1970, SUNY at Buffalo. 58 p.

Carmichael, Vernon Owen. The Ripley Formation and the Bluffport Member of the Demopolis Chalk in the Buena Vista, Mississippi, Quadrangle. M, 1960, Mississippi State University. 56 p.

Carmichael, Virgil Wesly. The Pumpkin Creek lignite deposit, Powder River County, Montana. D, 1967, University of Idaho. 83 p.

Carmichael, Virgil Wesly. The relationship of the "soils" of the Palouse to the Columbia River Basalt. M, 1956, University of Idaho.

Carmony, John Rodman. Stratigraphy and geochemistry of Mount Belknap Series, Tushar Mountains, Utah. M, 1977, Pennsylvania State University, University Park. 153 p.

Carnaghi, Gary Louis. The Kankakee Flood and the origin of the Parkland Sand, northeastern Illinois. M, 1979, University of Iowa. 93 p.

Carnahan, Chalon L. Non-equilibrium thermodynamic treatment of transport processes in groundwater flow. D, 1975, University of Nevada. 91 p.

Carnahan, Gary L. Geology of the southwestern part of Eagle Cap Quadrangle, Wallowa Mountains, Oregon. M, 1962, Oregon State University. 98 p.

Carnahan, George Gilbert. Secondary recovery from a Bartlesville sandstone rock. M, 1955, University of Oklahoma. 89 p.

Carne, Robert Clifton. Upper Devonian stratiform barite-lead-zinc-silver mineralization at Tom Claims, Macmillan Pass, Yukon Territory. M, 1979, University of British Columbia.

Carnein, C. R. Geology of the Suncook 15-minute quadrangle, New Hampshire. D, 1976, Ohio State University. 222 p.

Carnein, Carl Robert. Mass balance of the Meserve Glacier, Wright Valley, Antarctica. M, 1967, Ohio State University.

Carnes, J. B. Conodont biostratigraphy in the lower middle Ordovician of the western Appalachian thrustbelts in northeastern Tennessee. D, 1975, Ohio State University. 300 p.

Carnes, Lynne H. Geology of the south half, Old Fields Quadrangle, West Virginia. M, 1980, University of Akron. 34 p.

Carnes, Susan Fraker. Mollusks from southern Nichupte Lagoon, Quintana Roo, Mexico. M, 1975, Ohio State University.

Carnese, Michael J. Gravity study of intrusive rocks in West-central Maine. M, 1981, University of New Hampshire. 97 p.

Carnevali, J. Structural analysis of the Timmins-South Porcupine area, Ontario. M, 1976, University of Waterloo.

Carney, Cindy Kay. Petrology and diagenesis of the Upper Mississippian Greenbrier Limestone in the central Appalachian Basin and of the Lower Carboniferous Great Limestone in northern England. D, 1987, West Virginia University. 420 p.

Carney, Craig A. A study of vertical leachate transport mechanisms using analogous two dimensional dispersion models. M, 1979, University of Akron. 119 p.

Carney, Frank. Pleistocene geology of the Moravia Quadrangle (New York, west of Cortland). D, 1909, Cornell University.

Carney, George. Geology of the Osage Quadrangle, Carroll County, Arkansas. M, 1970, University of Arkansas, Fayetteville.

Carney, John L. Paleoenvironments of sediments of the shallow subsurface of offshore Louisiana near the Mississippi River delta. M, 1975, Louisiana State University. 84 p.

Carney, Keith F. The nature and importance of fine-grained sediment aggregation processes in the coastal lagoon complex at Stone Harbor, N.J. M, 1982, Lehigh University. 121 p.

Carney, Kevin Michael. A study of controlled point bar deposition on the Kentucky River. M, 1975, University of Kentucky. 119 p.

Carney, Michael Joseph. The relation of permeability to the porosity, grain size, and packing geometry of aggregates of artificial and natural grains. M, 1988, University of Wisconsin-Milwaukee. 195 p.

Caron, Alain. Analyse structurale des tectonites de la région de Ham-Nord, Québec. M, 1985, Universite Laval. 73 p.

Carothers, Marshall C. Depositional environments of the Cambridge Limestone (Missourian, Pennsylvanian) of the Appalachian Basin. D, 1976, University of Pittsburgh. 226 p.

Carothers, Thomas Arthur. Geology and hydrology of the Fish Creek Basin near Kent, Ohio. M, 1973, Kent State University, Kent. 90 p.

Carothers, William W. Aliphatic acids and stable carbon isotopes of oil field waters in the San Joaquin Valley, California. M, 1976, San Jose State University.

Carpenter, Alden Bliss. Mineralogy of the system $CaO-MgO-CO_2-H_2O$ at Crestmore, California. D, 1963, Harvard University.

Carpenter, Carey C. Preliminary investigation of the extent of volatile organic compounds contamination of ground water near Massillon, Ohio. M, 1987, Ohio University, Athens. 189 p.

Carpenter, David. Stratigraphy and sedimentation of the Glenshaw Formation (Conemaugh group) (Pennsylvanian) in central Preston County, West Virginia. M, 1969, West Virginia University.

Carpenter, David W. Hydrothermal alteration at Santa Rita, New Mexico. M, 1958, Pennsylvania State University, University Park. 136 p.

Carpenter, David W., Jr. The effect of CO_2 on the occurrence and distribution of epidote in basic dikes in the Republic metamorphic node, Marquette County, Michigan. M, 1974, Miami University (Ohio). 57 p.

Carpenter, Donald J. Elemental, isotopic, and mineralogic distributions within a tabular-type uranium-vanadium deposit, Henry Mountains mineral belt, Garfield County, Utah. M, 1980, Colorado School of Mines. 156 p.

Carpenter, F. Owen. New constraints on transient creep of mantle materials. M, 1986, Arizona State University. 120 p.

Carpenter, Gene Charles. Insoluble residue techniques in a portion of the Cynthiana Formation. M, 1960, University of Cincinnati. 62 p.

Carpenter, Glenn F. Geology of the Sutherland Creek area, Manitou Embayment, Colorado (El Paso County) with emphasis on petrography as evidence for the Sawatch Sandstone (Upper Cambrian) source of the sandstone dikes. M, 1967, Louisiana State University.

Carpenter, Gregory Wallace. Assessment of the triaxial falling head permeability testing technique. D, 1982, University of Missouri, Rolla. 173 p.

Carpenter, John Martin. The effects of acid mine drainage on the decomposition of leaf litter in a reservoir. M, 1981, University of Virginia. 92 p.

Carpenter, John Richard. A study of the uranium series disequilibria in Pleistocene shallow-water carbonates as a possible basis for absolute age determination. M, 1962, Florida State University.

Carpenter, John Richard. Influence of structural deformation on some aspects of metamorphic differentiation. D, 1964, Florida State University. 120 p.

Carpenter, John Tyer. A tentative correlation of northwestern Tertiary strata. M, 1932, University of Idaho. 27 p.

Carpenter, Lee B. Cretaceous and Tertiary nautiloids from Angola. M, 1951, University of Iowa. 55 p.

Carpenter, Leo C. Geology of the eastern end of the Ferris Mountains, Carbon County, Wyoming. M, 1951, University of Wyoming. 65 p.

Carpenter, Margaret. Frondicularia of the Upper Cretaceous formations in Texas. M, 1926, Texas Christian University. 43 p.

Carpenter, Martha Alice. Metamorphic trends as shown by underclay mineralogy in Pennsylvania. M, 1986, Lehigh University.

Carpenter, Paul Kenneth. Origin of comb layering in the Willow Lake Intrusion, N.E. Oregon. M, 1983, University of Oregon. 119 p.

Carpenter, Perry Albert, III. Geology of the Wilton area, Granville County, North Carolina. M, 1970, North Carolina State University. 106 p.

Carpenter, Philip John. Apparent Q for upper-crustal rocks in the Rio Grande Rift of central New Mexico from the analysis of microearthquake spectra. D, 1984, New Mexico Institute of Mining and Technology. 298 p.

Carpenter, Robert D. Mineral beneficiation of gravity concentration. M, 1948, University of Idaho. 25 p.

Carpenter, Robert H. A preliminary report on the Campbell Orebody Mines Division, Copper Queen Division, Phelps Dodge Corporation, Bisbee, Arizona. M, 1941, Stanford University. 39 p.

Carpenter, Robert H. Some vein-wall rock relations in the White Pine Mine, Ontonagon County, Michigan. M, 1962, University of Wisconsin-Madison.

Carpenter, Robert H. The geology and ore deposits of the Vekol Mountains, Pinal County, Arizona. D, 1947, Stanford University. 108 p.

Carpenter, Robert Heron. A study of the ore minerals in cupriferous pyrrhotite deposits in the Southern Appalachians. D, 1965, University of Wisconsin-Madison. 81 p.

Carpenter, Roger M. Lithofacies and depositional environments of the Upper Ordovician-Lower Silurian carbonates of the eastern Great Basin, Utah and Nevada. M, 1981, University of Wisconsin-Milwaukee. 139 p.

Carpenter, Roy. The marine geochemistry of fluorine. D, 1968, [University of California, San Diego]. 149 p.

Carpenter, S. R. Submersed aquatic vegetation and the process of eutrophication. D, 1979, University of Wisconsin-Madison. 178 p.

Carpenter, T. W. Stratigraphy and sedimentation of Middle Mississippian rocks of Gilmer and Braxton counties, West Virginia. M, 1976, West Virginia University.

Carpenter, W. David, Jr. Reexamination of the regional unconformity in the Carboniferous strata of Buchanan and Tazewell counties, Virginia. D, 1974, University of South Carolina.

Carpenter, William H. The standard baseline and an investigation of reading accuracy in baseline measurement with in-var wires. M, 1963, Ohio State University.

Carr, Cynthia. Hydrogeology of Zio and Yoto prefectures, Togo, West Africa. M, 1988, Carleton University. 165 p.

Carr, David Alan. Facies and depositional environments of the coal-bearing upper carbonaceous member of the Wepo Formation (Upper Cretaceous) northeastern Black Mesa, Arizona. M, 1987, Northern Arizona University. 238 p.

Carr, David L. Late Paleozoic siliciclastic shelf-bars, Sacramento Mountains, New Mexico. M, 1983, University of Texas, Austin. 127 p.

Carr, Donald D. Subsurface geology of Ellis County, Kansas. M, 1958, Kansas State University. 39 p.

Carr, Donald Dean. Geometry and origin of oolite bodies in the Ste. Genevieve limestone (Mississippian) in the Illinois Basin. D, 1969, Indiana University, Bloomington. 169 p.

Carr, Donald R. Potassium-argon method of geochronometry. D, 1958, Columbia University, Teachers College.

Carr, Donald R. Surface area of deep sea sediments. M, 1950, Columbia University, Teachers College.

Carr, George T. A study of methods for transporting natural gas from off-shore wells. M, 1950, University of Pittsburgh.

Carr, Gerald L. Marine manganese nodules; identification and occurrence of minerals. M, 1970, Washington State University. 101 p.

Carr, James E. Sedimentary tectonics and the Cenozoic history of the Verde Valley near Camp Verde, Yavapai County, Arizona. M, 1986, Northern Arizona University. 197 p.

Carr, James Russell. Application of the theory of regionalized variables to earthquake parametric estimation and simulation. D, 1983, University of Arizona. 278 p.

Carr, James Russell. Inventory of Arizona mined lands through classification of satellite remote sensing data. M, 1981, University of Arizona. 90 p.

Carr, John Lawrence. The geology of the Highwood-Elbow area, Alberta. M, 1946, University of Alberta. 198 p.

Carr, John Loften, III. The thermal maturity and clay mineralogy of the Chattanooga Formation along a transect from the Ozark Uplift to the Arkoma Basin. M, 1986, University of Arkansas, Fayetteville.

Carr, John William. The sedimentology and micropaleontology of the Tombigbee Sands (Tombigbee; Upper Cretaceous). M, 1954, Mississippi State University. 77 p.

Carr, Leo C. Geology of the Big Spring area, Washington County, Arkansas. M, 1963, University of Arkansas, Fayetteville.

Carr, Mary Margaret. Facies and depositional environments of the Lower Cretaceous McKnight Formation evaporites, Maverick Basin, Southwest Texas. M, 1987, University of Texas, Arlington. 144 p.

Carr, Michael David. Structure and stratigraphy of the Goodsprings District, southern Spring Mountains, Nevada. D, 1978, Rice University. 179 p.

Carr, Michael David. The geology of the Oued Sedjenane area, northern Tunisia and its bearing on the Numidian flysch problem. M, 1976, Rice University. 66 p.

Carr, Michael H. The geochemistry of cobalt. D, 1960, Yale University.

Carr, Michael J. Seismicity and upper mantle structure under the Japanese arcs. M, 1971, Dartmouth College. 46 p.

Carr, Michael J. Tectonics of the Pacific margin of northern Central America. D, 1974, Dartmouth College. 159 p.

Carr, Peter Alexander. Geology and hydrogeology of the Moncton map area, New Brunswick, Canada. D, 1964, University of Illinois, Urbana. 119 p.

Carr, Richard S. Geology and petrology of the Ossipee Ring-complex, Carroll County, New Hampshire. M, 1980, Dartmouth College. 174 p.

Carr, Robert Sidney. A finite element stream-aquifer model. D, 1985, Iowa State University of Science and Technology. 102 p.

Carr, Sharon D. The Valkyr shear zone and the Slocan Lake fault; Eocene structures that bound the Valhalla Complex, southeastern British Columbia. M, 1986, Carleton University. 106 p.

Carr, Theodore George. Two-dimensional seismic model studies of dispersion in layered media. M, 1961, St. Louis University.

Carr, Timothy R. Conodont biostratigraphy of the Skinner Ranch and Hess formations (Permian) Glass Mountains, West Texas. M, 1977, Texas Tech University. 43 p.

Carr, Timothy Robert. Paleogeography, depositional history and conodont paleoecology of the Lower Triassic Thaynes Formation in the Cordilleran miogeosyncline. D, 1981, University of Wisconsin-Madison. 227 p.

Carr, William Lester. A study of the pebbles and heavy minerals of the Olean, Salamanca, and Wolf Creek conglomerates of southwestern New York. M, 1947, Cornell University.

Carragan, William Dillard. Tsunamis; theory and observation. M, 1961, Rensselaer Polytechnic Institute. 28 p.

Carraher-Muto, Ruth. Structural geology and related mineralization of the Antelope Springs District, Pershing County, Nevada. M, 1979, University of Nevada. 72 p.

Carranza, Carlos. Surficial geology of a portion of south Panamint Valley, Inyo County, California. M, 1965, University of Massachusetts. 225 p.

Carrara, Alberto. Structural geology of lower Paleozoic rocks, Mt. Albert area, Gaspe Peninsula, Quebec. D, 1972, University of Ottawa. 206 p.

Carrara, Paul Edward. Late and neoglacial history in Smirling and Sulung valleys, eastern Baffin Island, Northwest Territories, Canada. M, 1972, University of Colorado.

Carrasco-Velazquez, Baldomero. The upper Austin Group (Cretaceous) in Jiminez, Coahuila (Mexico). M, 1968, University of Texas, Austin.

Carrell, Charles Howard. The insoluble residues of the Texas Cretaceous. M, 1932, Texas Christian University. 93 p.

Carrell, Olleon. A restudy of the type localities of the middle Washita. M, 1928, Texas Christian University.

Carrera-Ramirez, Jesus. Estimation of aquifer parameters under transient and steady-state conditions. D, 1984, University of Arizona. 276 p.

Carrick, Stanley J. Streamflow and sediment transport in the Whatcom Creek basin, Bellingham, Washington. M, 1981, Western Washington University. 105 p.

Carriel, James Turner. Jones zone water flood study. M, 1955, University of Southern California.

Carrier, John Baldwin. Ordovician cephalopods from the Bighorn Mountains of Wyoming. M, 1941, University of Iowa. 53 p.

Carriere, G. E. The geology of Suffield Mine, Sherbrooke, Quebec. M, 1954, McGill University.

Carriere, Patrick Edwige. A hydrologic and statistical evaluation of storage real-location for multipurpose reservoir system operations. D, 1988, Texas A&M University. 253 p.

Carrigan, Francis J. A geologic investigation of contact metamorphic deposits in the Coyote Mountains, Pima County, Arizona. M, 1971, University of Arizona.

Carrigan, Francis John. Computer-assisted decision aid for the estimation of mineral endowment; uranium in the San Juan Basin, New Mexico, a case study. D, 1983, University of Arizona. 911 p.

Carrigan, John A. Geology of the Rye Formation; New Castle Island and adjacent areas of Portsmouth Harbor, New Hampshire and Maine. M, 1984, University of New Hampshire. 128 p.

Carrigan, P. H. Regional flood maxima. D, 1975, Colorado State University. 81 p.

Carriker, Neil Edward. Heavy metal interactions with natural organics in aquatic environments. D, 1977, University of Florida. 155 p.

Carrilho, Cid. Geology of the northeast Burlington Quadrangle (Alamance County, North Carolina). M, 1973, North Carolina State University. 58 p.

Carrillo, Victor. A geological and geophysical assessment of the Hammond Field area, Zavala County, Texas. M, 1988, Baylor University. 126 p.

Carrington, Thomas J. Claryville Clays and associated deposits, Campbell County, Kentucky. M, 1960, University of Kentucky. 45 p.

Carrington, Thomas Jack. Stratigraphy and petrography of the Iron Mountain Formation (Precambrian) (basal part of the Unicoi Group) in southwestern Virginia. D, 1965, Virginia Polytechnic Institute and State University. 222 p.

Carrol, Michael J. Seismic models of porosity variations in the Smackover Formation, Jay Field, Florida-Alabama. M, 1978, University of New Orleans.

Carroll, Alan Robert. Lawrence Lake, Michigan; Holocene carbonate deposition in a temperate-region lacustrine system. M, 1983, University of Michigan.

Carroll, B. J. The stratigraphy and conodont biostratigraphy of the Montoya Group (Middle-Upper Ordovician) in southeastern Arizona. M, 1977, University of Arizona. 68 p.

Carroll, Beverly Mildred. Thermal environment of the East Pacific Rise at 39°S. M, 1973, Massachusetts Institute of Technology. 61 p.

Carroll, Cynthia J. Petrographically-sited stable isotopes of the Bonneterre Formation; insights into diagenesis. M, 1983, University of Missouri, Columbia.

Carroll, Forrest Arthur. A numerical model for potential conjunctive use in San Dieguito Basin, San Diego County, California. M, 1985, San Diego State University. 196 p.

Carroll, Gerald V. Geology of the Dover Plains Quadrangle, New York. D, 1952, Yale University.

Carroll, James C. Mining economics of copper-zinc deposits in the foothill copper belt of the Sierra Nevada in California. M, 1971, University of Nevada. 122 p.

Carroll, James F. A resistivity survey of the Saginaw Formation in the Lansing, Michigan, area. M, 1963, Michigan State University. 59 p.

Carroll, John Edward. Protection and preservation of coastal saltmarsh on the northeast Atlantic coast. D, 1974, Michigan State University. 268 p.

Carroll, Kipp W. Depositional and paragenetic controls on porosity development, upper Red River Formation, North Dakota. M, 1978, University of North Dakota. 153 p.

Carroll, Michael Robert. Sulfur in evolved magmas; experimental data and geochemical implications. D, 1986, Brown University. 130 p.

Carroll, Michael Timothy. A model for karst-like development in calcareous outwash deposits. M, 1979, University of Illinois, Urbana. 130 p.

Carroll, Neil Patrick. Upper Eocene and lower Oligocene biostratigraphy, Hoko River area, northern Olympic Peninsula, Washington. M, 1959, University of Washington. 101 p.

Carroll, Paul Richard. Petrology and structure of the pre-Tertiary rocks of Lummi and Eliza islands, Washington. M, 1980, University of Washington. 78 p.

Carroll, Richard E. Geology of the Standardville 71/2' Quadrangle, Carbon County, Utah. M, 1984, Brigham Young University. 31 p.

Carroll, Richard M. Spectral analysis of the gravity effect due to finite mass distribution. M, 1973, Pennsylvania State University, University Park.

Carron, Michael Joseph. The Virginia Chesapeake Bay; Recent sedimentation and paleodrainage. D, 1979, College of William and Mary. 333 p.

Carroon, C. Petrochemistry and petrography of the Newfoundland Stock, northwestern Utah. D, 1977, Stanford University. 262 p.

Carsola, A. J. Marine geology of the Arctic Ocean and adjacent seas off Alaska and Northwest Canada. D, 1953, University of California, Los Angeles.

Carsola, Alfred James. Depth distribution of Recent Ostracoda collected off Baja California. M, 1947, University of Southern California.

Carson, Bobb. Stratigraphy and depositional history of Quaternary sediments in northern Cascadia Basin and Juan de Fuca abyssal plain, northeast Pacific Ocean. D, 1971, University of Washington. 249 p.

Carson, Carlton M. The paleontology and stratigraphy of the marine Pliocene of Southern California. M, 1924, Stanford University. 44 p.

Carson, Charles Edward. The limestone "cap rock" of the Kimball Formation in western Nebraska. M, 1957, University of Nebraska, Lincoln.

Carson, Charles Edward. The oriented lakes of Arctic Alaska. D, 1962, Iowa State University of Science and Technology. 83 p.

Carson, Dana Woodruff. A regional analysis of calcite twinning strain in the Bighorn Mountains, northern Wyoming. M, 1988, Iowa State University of Science and Technology. 72 p.

Carson, David B. The geology of Branch Quadrangle, Franklin and Logan counties, Arkansas, with special emphasis on the coal resources. M, 1977, Northeast Louisiana University.

Carson, David John Temple. Geology of Mount Washington, Vancouver Island, British Columbia. M, 1960, University of British Columbia.

Carson, David John Temple. Metallogenic study of Vancouver Island with emphasis on the relationships of mineral deposits to plutonic rocks. D, 1968, Carleton University. 238 p.

Carson, David Marshall. Canadian-Whiterockian (Ordovician) conodont biostratigraphy of the Arctic Platform, southern Devon Island, eastern Canadian Arctic Archipelago. M, 1980, University of Waterloo.

Carson, J. M. In situ and laboratory gamma-ray spectrometer studies of the Bancroft area (Ontario). M, 1970, University of Western Ontario.

Carson, John Richard. Stratigraphy and structure of the east flank of the Bighorn Mountains, west-southwest of Buffalo, Wyoming. M, 1956, University of Iowa. 81 p.

Carson, Matt Wayne. Quality evaluation of fine limestone used in asphalt mixtures. M, 1980, University of Florida. 108 p.

Carson, Matthew V., III. Geology of the Santiago Peak Quadrangle, Orange and Riverside counties, California. M, 1966, University of Southern California.

Carson, Robert James, III. Physical stratigraphy of the post-Beekmantown–pre-Liberty Hall limestones (Ordovician) of central Rockbridge County, Virginia. M, 1967, Tulane University. 121 p.

Carson, Robert James, III. Quaternary geology of the south-central Olympic Peninsula, Washington. D, 1970, University of Washington. 67 p.

Carson, Scott E. Liquefaction susceptibility in the San Bernardino Valley and vicinity, California. M, 1987, California State University, Hayward. 101 p.

Carson, Thomas Gordon. A sedimentary study of the Demopolis chalk in the Artesia, Mississippi, Quadrangle. M, 1961, Mississippi State University. 45 p.

Carson, Thomas L. Radiolarian responses to the 1982-83 California El Nino and their implications. M, 1985, Rice University. 241 p.

Carson, Thomas M. U/Pb zircon geochronology of the Island Lake greenstone belt, eastern Manitoba. M, 1984, University of Windsor. 106 p.

Carson, William Pierce. Computerized lineament tectonics and porphyry copper deposits in S.E. Arizona and S.W. New Mexico. D, 1970, Stanford University. 310 p.

Carss, Brian Williams. A lithological and environmental study of the Ely Springs Dolomite, Arrow Canyon Range, Nevada. M, 1962, University of Illinois, Urbana.

Carss, Brian Williams. Microfacies study of Upper Devonian rocks, Arrow Canyon Range, Clark County, Nevada. D, 1964, University of Illinois, Urbana.

Carstea, Dumitru Dumitru. Formation and stability of Al, Fe, and Mg interlayers in montmorillonite and vermiculite. D, 1967, Oregon State University. 117 p.

Carsten, Forrest Paul. Petrology of the Upper Cretaceous strata of Orcas Island, San Juan County, Washington. M, 1982, Washington State University. 120 p.

Carstensen, Andrew B. Geology and ore genesis of the Slate Creek area, Custer County, Idaho. M, 1983, University of Montana. 90 p.

Carswell, Allan. A multioffset vertical seismic profiling experiment for the mapping of fracture zones. M, 1985, University of Manitoba.

Carswell, Henry Thomas. Geology and ore deposits of the Summit Camp, Boundary District, British Columbia. M, 1957, University of British Columbia.

Carswell, Henry Thomas. Origin of the Sullivan lead-zinc-silver deposit, British Columbia. D, 1961, Queen's University. 148 p.

Cartaya, Rafael A. Four-variables trend analysis study of porosity in the Mercedes M-2 and M-210 fault blocks, Las Mercedes oil field, Venezuela. M, 1968, Stanford University.

Carten, Richard Bell. Sodium-calcium metasomatism and its time-space relationship to potassium metasomatism in the Yerington porphyry copper deposit, Lyon County, Nevada. D, 1981, Stanford University. 306 p.

Carten, Thomas L. Pennsylvanian spores from the Sandia Formation, Santa Fe County, New Mexico. M, 1959, University of New Mexico. 54 p.

Carter, A. L. An age determination by carbon 14 analyses of wood from Avonport, Nova Scotia. M, 1954, Dalhousie University.

Carter, Alan Scott. Gold in stratabound ankerite rock, Shenandoah Prospect, Plumas County, California. M, 1984, University of Western Ontario. 148 p.

Carter, Anna Dombrowski. Conodont biostratigraphy of the Maury Shale, Northcentral Alabama. M, 1975, University of Florida. 288 p.

Carter, Berkeley Roger. Comparison of compression characteristics with some physical properties for a residual soil from Wake County, North Carolina. M, 1967, North Carolina State University. 60 p.

Carter, Brian John. Soil genesis and classification studies in Pennsylvania; Part I, Soil temperature regimes of the Northern Appalachian Mountains; Part II, Genesis of soils developed in pre-Wisconsinan glacial till. M, 1979, Pennsylvania State University, University Park. 136 p.

Carter, Brian John. The effect of slope gradient and aspect on the genesis of soils formed on a sandstone ridge in central Pennsylvania. D, 1983, Pennsylvania State University, University Park. 245 p.

Carter, Bruce Alan. Structure and petrology of the San Gabriel anorthosite-syenite body, Los Angeles County, California. D, 1980, California Institute of Technology. 393 p.

Carter, Bruce Applegate. Geology of the Eocene volcanic sequence, Mt. Baldy-Union Peak area, central Garnet Range, Montana. M, 1982, University of Montana. 55 p.

Carter, Burchard D., III. Paleoenvironmental aspects of Middle Ordovician (Black River and Trenton) carbonates; Germany Valley, Pendleton County, West Virginia. D, 1981, West Virginia University. 231 p.

Carter, C. H. Geology of the Palassou ridge area, California. M, 1970, San Jose State University. 23 p.

Carter, Carl Mitchell. Nuclear magnetic resonance studies in organic single crystals; chemical shift anisotropy of aromatic molecules. D, 1987, University of Utah. 111 p.

Carter, Carol S. Morphometry and evolution of a rapidly expanding drainage basin at Poquonock, Connecticut. M, 1957, Brown University.

Carter, Charles Henry. Miocene-Pliocene beach and tidal flat sedimentation, southern New Jersey. D, 1972, The Johns Hopkins University.

Carter, Charles William. The Upper Cretaceous deposits of the Chesapeake and Delaware Canal of Maryland and Delaware. D, 1938, The Johns Hopkins University.

Carter, Claire. Ontogeny of the graptolite Phyllographus. M, 1966, Stanford University.

Carter, Cole H. The geology of part of the Yellowjacket mining district, Lemhi County, Idaho. M, 1981, University of Idaho. 131 p.

Carter, Daniel Bradley. Amelioration and revegetation of smelter contaminated soils in the Coeur d'Alene mining district of northern Idaho. M, 1977, University of Idaho. 57 p.

Carter, Darryl Wayne. A study of strike-slip movement along the Washita Valley Fault, Arbuckle Mountains, Oklahoma. M, 1979, University of Oklahoma. 96 p.

Carter, David C. The Late Devonian-Early Carboniferous Albert Formation; a sedimentological approach to depositional history and facies relationships in a fluvial/deltaic and lacustrine basin. M, 1981, University of New Brunswick.

Carter, David Powell. Liquefaction potential of sand deposits under low levels of excitation. D, 1988, University of California, Berkeley. 365 p.

Carter, Elizabeth Sibbald. Early and Middle Jurassic radiolarian biostratigraphy, Queen Charlotte Island, B.C. M, 1985, University of British Columbia. 291 p.

Carter, George F. E. Ordovician ostracoda from the Saint Lawrence Lowlands of Quebec. D, 1958, McGill University.

Carter, George F. E. The Dunham Dolomite in St. Armand Township, Quebec, Canada. M, 1954, McGill University.

Carter, J. A. Pacific stress patterns and possible stress transmission. M, 1978, University of Hawaii.

Carter, J. G. Ecology and evolution of the Gastrochaenacea (Mollusca, Bivalvia) with notes on the status of the Myoida. D, 1976, Yale University. 660 p.

Carter, Jack Bryan. Depositional environments and tectonic history of the type Temblor Formation, Chico Martinez Creek, Kern County, California. M, 1985, Stanford University. 191 p.

Carter, Jack M. The stratigraphy, structure, and sedimentology of the Cretaceous Nanaimo Group, Galiano Island, British Columbia. M, 1977, Oregon State University. 203 p.

Carter, James A., Jr. The geology of the Pearsonia area, Osage County, Oklahoma. M, 1954, University of Oklahoma. 114 p.

Carter, James Allen. Regional gravity and aeromagnetic surveys of the Mineral Mountains and vicinity, Millard and Beaver counties, Utah. M, 1978, University of Utah. 178 p.

Carter, James Franklin. Spores of the Pennsylvanian Toronto Limestone in Kansas and Oklahoma. M, 1960, University of New Mexico. 121 p.

Carter, James Lee. The origin of olivine bombs and related inclusions in basalts. D, 1965, Rice University. 264 p.

Carter, James Neville. Paleomagnetism of the eastern Transverse Ranges and tectonic implications. M, 1985, University of California, Santa Barbara. 121 p.

Carter, James W. Environmental and engineering geology of the Astoria Peninsula area, Clatsop County, Oregon. M, 1976, Oregon State University. 138 p.

Carter, Jesse Louis, Jr. A pedology-geochemistry-botany study of gob piles, spoil banks, reclaimed land, and unmined areas, associated with coal mining in Franklin and Logan counties, Arkansas. M, 1984, Northeast Louisiana University. 178 p.

Carter, John Lyman. Mississippian brachiopods from the Chappel Limestone of central Texas. D, 1966, University of Cincinnati. 407 p.

Carter, John Swain, Jr. The origin of some granites and gneisses from a portion of the Teton Range, Wyoming. M, 1964, Texas Tech University. 48 p.

Carter, Karen Eileen. Deformation and metamorphism of the Red Mountain area, Llano County, Texas. M, 1985, University of Texas, Austin. 63 p.

Carter, Kent. The geochemistry and petrology of plagiogranites from North Arm Mountain, Bay of Islands Ophiolite, Newfoundland, Canada. M, 1985, University of Houston.

Carter, Kristine D. Middle and late Wisconsinan (Pleistocene) insect assemblages from Illinois. M, 1985, University of South Dakota. 124 p.

Carter, L. M. The effect of human activity on the middle course of the Tualatin River, Oregon. D, 1975, Portland State University. 180 p.

Carter, Lee Scott. Recent marine sediments of Carmel Bay, California. M, 1971, United States Naval Academy.

Carter, Lee Steven. The relationship between electrical resistivity and brine saturation in reservoir rocks. M, 1951, University of Texas, Austin.

Carter, Lionel. Surficial sediments of Barkley Sound and adjacent continental shelf, Vancouver Island, B. C. D, 1971, University of British Columbia.

Carter, Louis D. Late Quaternary erosional and depositional history of Sierra del Mayor, Baja California, Mexico. D, 1977, University of Southern California.

Carter, Matthew. Experimental investigations of a recent fluxgate theory. M, 1988, University of British Columbia. 92 p.

Carter, Maurice Wylde. A preconcentration-spectrographic method for the determination of trace elements in plant materials and the application of the biogeochemical method at the Silver Mine lead deposit, Cape Breton, Nova Scotia. M, 1966, Carleton University. 116 p.

Carter, Michael Howard. Paleoenvironmental analysis of the Ralston Creek Formation within the Canon

City Embayment, Canon City, Colorado. M, 1984, University of Oklahoma. 90 p.

Carter, N. C. Geology and geochronology of porphyry copper and molybdenum deposits in west-central British Columbia. D, 1974, University of British Columbia.

Carter, Neal Allen. Geology of Kinkaid Township, Campbell Hill Quadrangle, Jackson County, Illinois. M, 1964, Southern Illinois University, Carbondale. 46 p.

Carter, Neville Louis. Experimental deformation and recrystallization of quartz. D, 1963, University of California, Los Angeles. 225 p.

Carter, Neville Louis. Geology of the Fernwood-Topanga Park area, Santa Monica Mountains, California. M, 1958, University of California, Los Angeles.

Carter, Nicholas C. Geology of a part of the Republic Trough, Marquette County, Michigan. M, 1962, Michigan Technological University. 55 p.

Carter, O. F. The structural features of the gold deposits of Ontario. M, 1936, University of Toronto.

Carter, Pamela Hobart. Wisconsinan sedimentation in a glacial drainage trough, Deer Creek, Carroll County, north-central Indiana. M, 1987, Indiana University, Bloomington. 263 p.

Carter, Paul Henry, Jr. The microstructure and mineral composition of the shell of Recent muricid gastropods. M, 1964, Texas Christian University.

Carter, Paul W. Aspects of the biogeochemistry of organic matter in Recent carbonate sediments. M, 1976, University of Texas at Dallas. 95 p.

Carter, Phillip K. A speculative critique of the Absaroka Sequence in the Rocky Mountains and Colorado Plateau. M, 1973, University of Washington.

Carter, Richard Michael. An experimental study of inertial wave propagation in a rotating liquid cone. M, 1969, Massachusetts Institute of Technology. 85 p.

Carter, Robert Daniel. Geology of Travis County between latitude 30° and 20'N and Colorado River east of Austin, Texas. M, 1948, University of Texas, Austin.

Carter, Terry Robert. Copper-antimony-gold-silver deposits of the Lavant-Darling area, southeastern Ontario; geology, genesis, and metallogenetic significance. M, 1981, University of Toronto.

Carter, Thomas E. Pediment development along Book Cliffs, Utah. M, 1981, Colorado State University. 99 p.

Carter, Thomas R. A geologic study of the Ropes Reef Reservoir, Hockley County, Texas. M, 1956, Texas A&M University. 31 p.

Carter, Virginia Perkins. Relation of hydrogeology, soils and vegetation on the wetland-to-upland transition zone of the Great Dismal Swamp, Virginia and North Carolina. D, 1988, George Washington University. 323 p.

Carter, William Horace. Geology of the northeast corner of the Calistoga Quadrangle, California. M, 1948, University of California, Berkeley. 73 p.

Carter, William W., Jr. Geology of Cedar and Talladega mountains, Clay County, Alabama. M, 1985, University of Alabama. 132 p.

Carthew, John Arthur. Mono County, California; a geographical interpretation of California's Leedside Sierra high plateau country. D, 1970, University of California, Los Angeles.

Cartier, Gerard Leon Marcel. Geology of the Bellshill Lake oil field, eastern Alberta. M, 1976, University of Alberta. 117 p.

Cartier, Kenn D. W. Sediment, channel morphology, and streamflow characteristics of the Bitterroot River drainage basin, southwestern Montana. M, 1984, University of Montana. 191 p.

Cartmill, John Craig. Some properties and flow characteristics of fine dispersed oil-in-water emulsions in porous media. M, 1968, University of Tulsa. 69 p.

Cartwright, George C. The Precambrian geology of the Lake Abundance area, Beartooth Mountains, Montana and Wyoming. M, 1984, Northern Illinois University. 78 p.

Cartwright, Jack Cleveland. Simpson Group; Midland, Ector, and eastern Winkler counties, Texas. M, 1955, University of Texas, Austin.

Cartwright, Keros. A study of the Lake Lahontan sediments in the Winnemucca area, Nevada. M, 1961, University of Nevada. 52 p.

Cartwright, Keros. The effect of shallow groundwater flow systems on rock and soil temperatures. D, 1973, University of Illinois, Urbana. 117 p.

Cartwright, Lon D., Jr. Sedimentation of the Pico Formation in the Ventura Quadrangle, California. M, 1927, Stanford University. 107 p.

Cartwright, Richard. Provenance and sedimentology of carbonate turbidites from two deep-sea fans, Bahamas. M, 1985, University of Miami. 166 p.

Cartwright, Weldon Emerson. A study of the Cretaceous Inocerami of Texas. M, 1933, University of Texas, Austin.

Carty, James Michael. Elastic anisotropy in the Baraboo Quartzite; Baraboo, Wisconsin. M, 1985, Michigan State University. 79 p.

Caruccio, Frank Thomas. An evaluation of factors influencing acid mine drainage production from various strata of the Allegheny Group (Mississippian) and the ground water interactions in selected areas of western Pennsylvania (Clearfield County). D, 1967, Pennsylvania State University, University Park. 248 p.

Caruccio, Frank Thomas. Hydrogeology of the sewage disposal experiment area, northwest of State College, Pennsylvania. M, 1963, Pennsylvania State University, University Park. 132 p.

Caruso, Joel W. Burrowing and bioturbation by fiddler crabs in the Florida Keys. M, 1978, Bowling Green State University. 52 p.

Caruso, Louis J. The stability of lizardite. M, 1979, University of Maine. 46 p.

Caruso, Nancy E. Depositional environments and age of the Cambrian Clarks Spring Member of the Secret Canyon Shale and the lower Hamburg Dolomite of the Eureka mining district, Nevada. M, 1984, University of Kansas. 41 p.

Carvalho Martins, Verónica E. de Sousa see de Sousa Carvalho Martins, Verónica E.

Carvalho, Antone V., III. Gahnite-franklinite intergrowths at the Sterling Hill zinc deposit, Sussex County, New Jersey; an analytical and experimental study. M, 1978, Lehigh University. 131 p.

Carvalho, I. G. Geology of the Western Mines District, Vancouver Island, British Columbia. D, 1979, University of Western Ontario. 294 p.

Carver, Gary Alen. Glacial geology of the Mountain Lakes Wilderness and adjacent parts of the Cascade Range, Oregon. D, 1972, University of Washington. 75 p.

Carver, Gary Alen. Quaternary tectonism and surface faulting in the Owens lake basin, California. M, 1969, University of Nevada. 105 p.

Carver, George Evans, Jr. A subsurface study of the Northeast Arcadia-Coon Creek area, Oklahoma and Logan counties, Oklahoma. M, 1947, University of Oklahoma. 59 p.

Carver, John Arthur. Sedimentation of the Sespe and Alegria formations, Santa Barbara County, California. M, 1960, University of California, Los Angeles.

Carver, Robert E. Geology of the Clinton North Quadrangle, Henry County, Missouri. M, 1959, University of Missouri, Columbia.

Carver, Robert Elliott. Petrology and paleogeography of the Roubidoux Formation (Ordovician) of Missouri. D, 1961, University of Missouri, Columbia. 178 p.

Carwile, Roy H. A study of strontium isotope ratios in a core from the Atlantis II deep, Red Sea hot brine area. M, 1970, Ohio State University.

Cary, Logan W. The subsurface geology of the Garber area, Garfield County, Oklahoma. M, 1954, University of Oklahoma. 51 p.

Cary, Russell S., Jr. Structure and stratigraphy of the Horse Mountain Quadrangle, Garfield County, Colorado. M, 1960, University of Colorado.

Carye, Jeffrey Alyn. Structural geology of part of the Crooked Lake area, Quesnel Highlands, British Columbia. M, 1986, University of British Columbia. 111 p.

Casaceli, Robert J. The geology and mineral potential of the Hahn Peak intrusive porphyry, Routt County, Colorado. M, 1984, Oregon State University. 223 p.

Casadevall, Tom. Gold mineralization in the Sunnyside Mine, Eureka mining district, San Juan County, Colorado. M, 1974, Pennsylvania State University, University Park. 78 p.

Casadevall, Tom. Sunnyside Mine, Eureka mining district, San Juan County, Colorado; geochemistry of gold and base metal ore formation in the volcanic environment. D, 1976, Pennsylvania State University, University Park. 157 p.

Casagrande, Daniel Joseph. Geochemistry of amino acids in selected Florida peats. D, 1970, Pennsylvania State University, University Park. 267 p.

Casals, Javier Fernandez see Fernandez Casals, Javier

Casarta, Lawrence Joseph. The effects of interlayer slip on the folding of layered rocks. M, 1980, Texas A&M University.

Casas, Enrique. Diagenesis of salt halite. M, 1987, SUNY at Binghamton. 82 p.

Casas, Federico Pardo. Application of the finite difference method in the study of wave propagation in a borehole. M, 1984, Massachusetts Institute of Technology. 77 p.

Casas, Jamie Lopez. Geology and preliminary groundwater investigations of the Codazzi area, northeastern Colombia, South America. M, 1960, University of Wisconsin-Madison.

Casavant, Deborah Ilgenfritz. Sedimentation history and sedimentary environment of upper Mississippi River Pool 19. M, 1985, University of Illinois, Urbana. 100 p.

Cascaddan, Ann E. Petrology and petrofabrics of the Jacobsville sandstone in the vicinity of the Keweenaw fault (northern Michigan). M, 1969, Michigan State University. 62 p.

Cascia, Malvin Charles. A petrographic study of coals from the Trinidad coal field, Colorado, including a comparison of fluorescence spectra with rank parameters. M, 1980, Southern Illinois University, Carbondale. 101 p.

Case, Ermine C. On the osteology and relationships of Protostega. D, 1896, University of Chicago.

Case, Ermine Cowles. The drumlin region of western New York together with some experiments on ice action. M, 1895, Cornell University.

Case, Harvey Lee, III. An approach for the evaluation of environmental impact. M, 1973, Oklahoma State University. 97 p.

Case, James Boyce. A comparison of photogrammetric methods in glacier mapping. D, 1959, Ohio State University. 87 p.

Case, James Boyce. Photogrammetric mapping of the Lemon Creek Glacier, Alaska. D, 1957, Ohio State University.

Case, James E. Maximum felt intensities of earthquakes as a factor of the seismicity of the Western United States. M, 1955, University of Arkansas, Fayetteville.

Case, James Edward. Geology of a portion of the Berkeley and San Leandro hills, California. D, 1963, University of California, Berkeley. 319 p.

Case, Leslie C. The physical history of the Appalachian Mountains. M, 1925, University of Missouri, Columbia.

Case, Lyle Eldon. General geology of the Horn Fault region, Bighorn Mountains, Wyoming. M, 1957, University of Iowa. 87 p.

Case, Robert William. A regional gravity survey of the Sevier Lake area, Millard County, Utah. M, 1977, University of Utah. 95 p.

Casella, Clarence Joseph. Geologic evolution of the Beartooth Mountains, Montana and Wyoming; Part 5, The Line Creek area. D, 1963, Columbia University, Teachers College. 96 p.

Caserotti, Phillip Mark. Magnetic properties of the Precambrian rocks of a small area near Pony, Madison County, Montana. M, 1975, Indiana University, Bloomington. 67 p.

Casey, Brian J. The geology of a portion of the northeastern Bristol Mountains, Mojave Desert, California. M, 1981, University of California, Riverside. 128 p.

Casey, Donald Joseph. Relationship of geology to stream base flow in Tennessee. M, 1962, American University. 25 p.

Casey, James Michael. Depositional systems and basin evolution of the late Paleozoic Taos Trough, northern New Mexico. D, 1980, University of Texas, Austin. 253 p.

Casey, John Eagle. Morrow Group in southern Boston Mountains. M, 1953, University of Arkansas, Fayetteville.

Casey, John Francis. The geology of the southern part of the North Arm mountain massif, Bay of Islands ophiolite complex, western Newfoundland, with application to ophiolite obduction and the genesis of the plutonic portions of oceanic crust and upper mantle. D, 1980, SUNY at Albany. 680 p.

Casey, John Joseph. Geology of the Heart Peaks Volcanic Centre, northwestern British Columbia. M, 1980, University of Alberta. 120 p.

Casey, Tom Ann L. Lithology and facies relationships of the Big Blue Formation near Cantua Creek, California. M, 1974, Stanford University.

Casey, William Howard. Geology and geochemistry of mineralization and alteration in the central portion of the West Shasta Cu-Zn district, Shasta County, California. M, 1980, University of California, Davis. 149 p.

Casey, William Howard. Solute transport and models for sulfate reduction and radionuclide migration in marsh sediment. D, 1985, Pennsylvania State University, University Park. 297 p.

Cash, Leon D. Petrographic analysis of selected Arkansas terrace gravels from Crowley's Ridge related to their engineering properties. M, 1982, Memphis State University.

Cashby, Susan Margaret. Geologic and geophysical investigation of a portion of the upper continental slope, Northwest Gulf of Mexico. M, 1978, Rice University. 91 p.

Cashion, W. W. Relation of seven springs and Pulaski thrusts in Virginia. M, 1968, Virginia Polytechnic Institute and State University.

Cashman, Katharine Venable. Crystal size distributions in igneous and metamorphic rocks. D, 1986, The Johns Hopkins University. 359 p.

Cashman, Patricia Hughes. Geology of the Forks of Salmon area, Klamath Mountains, California. D, 1979, University of Southern California.

Cashman, Susan Moran. Geology of the Peshastin Creek area, Washington. M, 1974, University of Washington. 29 p.

Cashman, Susan Moran. Structure and petrology of part of the Duzel Formation and related rocks in the Klamath Mountains southwest of Yreka, California. D, 1977, University of Washington. 93 p.

Cashore, Jac. A new Monte Carlo model of the development of the lunar megaregolith. M, 1987, University of Houston.

Caskey, Charles Frederick, Jr. The Needles Range Formation in southwestern Utah; paleomagnetism and stratigraphic correlation. D, 1975, University of Utah. 117 p.

Caskey, Deborah Jane. Geology and hydrothermal alteration of the Iron Beds area, Hinsdale County, Colorado. M, 1979, University of Texas, Austin.

Caskey, M. C. The recharge of the Waikapu Aquifer, Maui. M, 1968, University of Hawaii. 75 p.

Caskey, Thomas Lee. Geology of western Coryell County, Texas. M, 1961, University of Houston.

Caskie, Dennis Raymond Mac see Mac Caskie, Dennis Raymond

Caskie, Robert A. Geology of Staunton and vicinity, Augusta County, Virginia. M, 1951, University of Virginia. 106 p.

Caskie, Robert Alden. A mineralogical study of the Cold Spring kaolin deposit and correlation with kaolin exposures along the western flank of the Blue Ridge Mountains. M, 1957, University of Virginia. 61 p.

Caslavsky, Jaroslav Ladislav. Dislocation structures and the mechanical properties of mica. D, 1969, Pennsylvania State University, University Park. 109 p.

Caslick, James Frederick. Municipal water supply source protection in New York State; an evaluation of the public water supply rules and regulations program. D, 1982, State University of New York, College of Environmental Science and Forestry. 367 p.

Cason, Cynthia Lynn. Engineering geologic feasibility of lignite mining in alluvial valleys by hydraulic dredging methods. M, 1982, Texas A&M University.

Cason, James Hubert. The geochemistry of Lake Izabal, Guatemala. M, 1972, University of Florida. 47 p.

Cason, Russell R. Paleoenvironments of the Denton, Weno and Pawpaw formations (Lower Cretaceous), central Texas. M, 1986, Stephen F. Austin State University. 146 p.

Caspall, Frederick Charles. The spatial and temporal variations in loess deposition in northeastern Kansas. D, 1970, University of Kansas. 306 p.

Caspar, Barry Christman. Deposition of a Late Permian mud rich sabkha in northern Caddo County, Oklahoma. M, 1987, University of Oklahoma. 107 p.

Casquino Rey, Walter T. The development of conical stress forms by explosive charges. M, 1966, University of Missouri, Rolla.

Cass, Hilton K. Geology and alteration of the Jordan Creek area, Yankee Fork mining district, Custer County, Idaho. M, 1974, University of Idaho. 92 p.

Cass, John I. Paleoenvironmental interpretation of the Beekmantown Group within the Ottawa Basin. M, 1979, University of Ottawa. 90 p.

Cass, John Tufts. Geology of the northern part of the Highland Mountains, Montana. M, 1953, Indiana University, Bloomington. 32 p.

Cassa, Mary Rose. Stratigraphy and petrology of the Onondaga Limestone (Middle Devonian), eastern Lake Erie region of New York, Pennsylvania, and Ontario. M, 1980, SUNY at Binghamton. 82 p.

Casse, F. J. The effect of temperature and confining pressure on fluid flow properties of consolidated rocks. D, 1975, Stanford University. 132 p.

Cassell, David Terrance. Neogene foraminifera of the Limon Basin of Costa Rica. D, 1986, Louisiana State University. 338 p.

Cassell, David Terrance. Sedimentary petrography and depositional models for the non-marine carbonate rocks of the Verde Formation, northern Verde Valley, Arizona. M, 1980, Northern Arizona University. 153 p.

Cassell, Dwight Eugene. Geology of the Coldspring area and petrology of the Fleming Formation, San Jacinto County, Texas. M, 1958, University of Texas, Austin.

Cassell, John K. Variation of the type Monterey Formation, California, near the type locality. M, 1949, Stanford University. 62 p.

Casselman, M. J. Petrology and alteration of the Bathurst Norsemines central area deposits, Northwest Territories. M, 1977, Carleton University. 227 p.

Cassetta, Dominick M. The inversion of gravity anomalies in portions of Henry, Lucas, and Wood counties, Ohio, as a method of locating a portion of the Bowling Green Fault. M, 1980, Bowling Green State University. 57 p.

Casshyap, Satyendra M. Sedimentary petrology of the Huronian rocks (Precambrian), Espanola-Willisville area, Ontario. D, 1967, University of Western Ontario. 285 p.

Cassidy, Martin M. Stratigraphy, petrology, and partial geochemistry of the Excello Shale, Pennsylvanian (Desmoinesian), of northeastern Oklahoma. M, 1962, University of Oklahoma. 107 p.

Cassidy, William Arthur. Phase equilibrium studies on tektite and meteorite systems. D, 1961, Pennsylvania State University, University Park. 139 p.

Cassie, Robert MacGregor. The evolution of a domal granitic gneiss and its relation to the geology of the Thomaston Quadrangle, Connecticut. D, 1965, University of Wisconsin-Madison. 152 p.

Cassiliano, M. L. Stratigraphy and paleontology of the Horse Creek-Trail Creek area, Laramie County, Wyoming. M, 1976, University of Wyoming. 183 p.

Cassin, Richard Joel. A geological study of the Johnsonville oil field, Wayne County, Illinois. M, 1949, University of Illinois, Urbana.

Cassity, Paul Edward. Pennsylvanian and Permian fusulinids of the Bird Spring Group from Arrow Canyon, Clark County, Nevada. M, 1965, University of Illinois, Urbana.

Cassol, Elemar Antonino. Sediment transport and deposition of various textured soils in shallow flow. D, 1988, Purdue University. 331 p.

Casson, Robert N. The geohydrology of Manhasset Neck, Long Island, New York. M, 1986, Queens College (CUNY). 88 p.

Cast, Martha E. Petrography and provenance of the Eocene Simsboro Formation, central Texas. M, 1986, University of Texas, Austin. 314 p.

Cast, Mary Elizabeth. An investigation of stratabound chromium concentrations in the Tecoyas Formation, Palo Duro Canyon, Texas. M, 1986, University of Texas at Dallas. 131 p.

Castagna, John Patrick. Energetics of clinopyroxene equilibrium reactions. M, 1981, Brooklyn College (CUNY).

Castagna, John Patrick. Methods for the analysis of sonic log waveforms. D, 1983, University of Texas, Austin. 334 p.

Castano, John R. The mechanism of replacement of the Clinton iron ores. M, 1950, Northwestern University.

Castano, Juan Carlos. The determination of crustal thickness in Central America from the spectrum of dilatational body waves. M, 1967, St. Louis University.

Casteel, Mitch. Geology of a portion of the Northwest Last Chance Range, south of Hanging Rock Canyon, Inyo County, California. M, 1986, California State University, Fresno. 115 p.

Casteleiro, M. Mathematical model of one-dimensional consolidation and desiccation of dredged materials. D, 1975, Northwestern University. 125 p.

Castellano, Rocco Horatio. A study in clay mineralogy and the relationship of the clays to soils and texture in selected exposures of the Loveland and Peorian formations in eastern Nebraska and western Iowa. D, 1961, University of Nebraska, Lincoln. 391 p.

Castellano, Rocco Horatio. A study of the Kirkwood Formation (Miocene) of New Jersey. M, 1952, University of Nebraska, Lincoln.

Castellanos, Mario Ruiz. Rubidium-strontium geochronology of the Oaxaca and Acatlan metamorphic areas of southern Mexico. D, 1979, University of Texas at Dallas. 178 p.

Castellanos, Mario Ruiz see Ruiz Castellanos, Mario

Castello, R. R. Bearing capacity of driven piles in sand. D, 1979, Texas A&M University. 151 p.

Castens, Pamela G. Morphology and sedimentation in Catalina Harbor, Santa Catalina Island, California. M, 1988, University of California, Los Angeles. 125 p.

Caster, Kenneth Edward. Microscopic and macroscopic paleontology of Angola. M, 1931, Cornell University.

Caster, Kenneth Edward. Pre-Pennsylvanian stratigraphy of northwestern Pennsylvania. D, 1933, Cornell University.

Caster, W. A. Near-bottom currents in Monterey (California) submarine canyon and on the adjacent shelf. M, 1969, United States Naval Academy.

Castillo Ron, Enrique. Dispersion of a contaminant in jointed rock (a two-dimensional mathematical model). D, 1972, Northwestern University.

Castillo S., Jesus M. Analyses of well flow tests considering pressure-dependent rock and fluid properties for a compressible reservoir. M, 1973, Stanford University. 154 p.

Castillo, David Andrew. A near-bottom geophysical investigation of the Vema fracture zone and its intersection with the Mid-Atlantic Ridge. M, 1984, University of California, Santa Barbara. 182 p.

Castillo, G. Luis Del see Del Castillo, G. Luis

Castillo, Paterno R. Petrology of the Precambrian granite and granite-related rocks of Casper Mountain, Wyoming. M, 1983, University of Akron. 106 p.

Castillo, Paterno Reyes. Geology and geochemistry of Cocos Island, Costa Rica; implications for the evolution of the aseismic Cocos Ridge. D, 1987, Washington University. 300 p.

Castillon, David Alan. The relationships between morphostratigraphy, rock stratigraphy, and aspects of till fabric in central Illinois. D, 1972, Michigan State University. 151 p.

Castle, Bruce. Pedogeochemical and biogeochemical trends at the Heddleston porphyry copper-molybdenum deposit, Lewis and Clark County, Montana. M, 1978, University of Montana. 195 p.

Castle, J. W. Comparative sedimentology of some modern Pacific trenches and the Caples Group (Permo-Triassic?), New Zealand. D, 1978, University of Illinois, Urbana. 188 p.

Castle, James W. The deposition and metamorphism of the Polarstar Formation (Permian), Ellsworth Mountains, Antarctica. M, 1974, University of Wisconsin-Madison. 102 p.

Castle, Margaret. Gravity survey of the Simi Valley, California, and structural implications. M, 1986, California State University, Long Beach. 114 p.

Castle, R. O. Paleogeological studies in the Maritime Provinces. M, 1949, McGill University.

Castle, Richard A. Stratigraphy and sedimentology of basal Cretaceous sediments, north-central Texas. M, 1969, Louisiana State University.

Castle, Robert Oliver. Geology of the Andover Granite and surrounding rocks, Massachusetts. D, 1964, University of California, Los Angeles. 627 p.

Castleberry, Joe P., II. An engineering geology analysis of home foundations on expansive clays. a, 1974, Texas A&M University.

Castor, Stephen Baird. Geology of the central Pine Nut and northern Buckskin ranges, Nevada; a study of Mesozoic intrusive activity. D, 1972, University of Nevada. 334 p.

Castro, Antonio Amilcar Ubiera see Ubiera Castro, Antonio Amilcar

Castro, Celso Filho de see de Castro, Celso Filho

Castro, E. J. A subsurface study of the Tipton Member of the Green River Formation west of the Rock Springs Uplift. M, 1962, University of Wyoming. 66 p.

Castro, Gerardo. Analysis of underground pipelines for seismic wave propagation. D, 1986, Rensselaer Polytechnic Institute. 281 p.

Castro, Louis Reyes. Computation of synthetic seismograms by the method of characteristics. D, 1979, Texas Tech University. 59 p.

Castro, Louis Reyes. Origin and seismicity of the Gulf of California. M, 1972, Texas Tech University. 42 p.

Caswell, William Bradford, Jr. The hydrogeology of thin-limestone layers in east-central Ohio. D, 1969, Ohio State University. 228 p.

Catacosinos, Paul A. Stratigraphy and paleontology of the Pennsylvanian rocks of the rim of the Sandia Mountains, Sandoval and Bernalillo counties, New Mexico. M, 1962, University of New Mexico. 76 p.

Catacosinos, Paul Anthony. Cambrian stratigraphy of the lower Peninsula of Michigan. D, 1972, Michigan State University. 62 p.

Catalano, Lee Edward. Geology of the Hartshorne Coal, McCurtain and Lafayette quadrangles, Haskell and Le Flore counties, Oklahoma. M, 1978, Oklahoma State University. 61 p.

Cataldo, Robert Mario. Sediment transport along the coast of New Jersey. M, 1980, Syracuse University.

Catanzaro, E. J. A preliminary petrographic study of the Lake Owens mafic complex, Albany County, Wyoming. M, 1956, University of Wyoming. 39 p.

Catanzaro, Edward John. A study of discordant zircons from the Little Belt (Montana), Beartooth (Montana), and Santa Catalina (Arizona) Mountains. D, 1962, Columbia University, Teachers College. 149 p.

Catchings, Rufus. Crustal structure from seismic refraction in the Medicine Lake area of the Cascade Range and Modoc Plateau, Northern California. M, 1983, University of Wisconsin-Madison. 97 p.

Catchings, Rufus Douglas. Crustal structure of the northwestern United States. D, 1987, Stanford University. 194 p.

Cate, Alta. Population dynamics of selected species of Brachiopoda and Gastropoda from the Strawn, Canyon, and Cisco groups (Upper Pennsylvanian) of north-central Texas. M, 1987, University of Houston.

Cate, Paul David. The geology of the Fayetteville Quadrangle, Washington County, Arkansas. M, 1962, University of Arkansas, Fayetteville.

Cathcart, Stanley Holman. Origin of concretions in certain shales of the Bellefonte Quadrangle. M, 1916, Pennsylvania State University, University Park.

Cather, Steven Martin. Petrology, diagenesis, and genetic stratigraphy of the Eocene Baca Formation, Alamo Navajo Reservation and vicinity, Socorro County, New Mexico. M, 1980, University of Texas, Austin.

Cather, Steven Martin. Volcano-sedimentary evolution and tectonic implications of the Datil Group (latest Eocene-early Oligocene), west-central New Mexico. D, 1986, University of Texas, Austin. 534 p.

Catherinet, Jules. The geology of the ore deposits of Copper Mountain (British Columbia). M, 1904, Columbia University, Teachers College.

Cathey, Carl A. Precambrian geology of the Montezuma, New Mexico Quadrangle. M, 1973, Texas Tech University. 70 p.

Cathey, Joseph B., Jr. The geology of the eastern half of the Lake City Quadrangle, Anderson and Campbell counties, Tennessee. M, 1950, University of Tennessee, Knoxville. 74 p.

Cathey, William B. Implications of the geology and geochemistry of the Maclean Five uranium deposit, Three Rivers, Texas. M, 1980, University of Tennessee, Knoxville. 126 p.

Cathles, Lawrence M., III. The viscosity of the Earth's mantle. D, 1971, Princeton University. 475 p.

Catlin, James E. Intrapopulational variation in Miogypsina. M, 1972, Northern Illinois University.

Catlin, Steven Allen. Mineralogy, zoning, and paragenesis of sulfide ores at the Ground Hog Mine, Central District, New Mexico. M, 1981, University of Arizona. 52 p.

Cato, Kerry D. Variation in physical rock properties determined from sonic logs at a South Texas lignite mine. M, 1985, Texas A&M University.

Caton, Paul William. The Surtsey Pond problem; determination of mean sea level. D, 1973, University of Tulsa. 194 p.

Catt, Diane M. Depositional environments and diagenesis of the Ratcliffe Interval, Madison Group (Mississippian), Williston Basin, North Dakota. M, 1982, University of North Dakota. 180 p.

Cattafe, Joseph S., Jr. Effects of deep burial on diagenetic textures in carbonate rocks from a depth of 20,000 feet, 6 kilometers. M, 1981, Rensselaer Polytechnic Institute. 49 p.

Cattalani, Sergio. A fluid inclusion and isotope study of the St. Robert W-Ag-Bi vein deposit, Eastern Townships, Quebec. M, 1987, McGill University. 112 p.

Cattany, Ronald W. Colorado energy analysis; the economy and the state energy industry with methods for calculating the energy intensity of economic sectors and the resources of oil and gas in the state. M, 1977, Colorado School of Mines. 205 p.

Catto, Antonio J. Fluid content effect on acoustic impedance and limits of direct detection capability; illustrated on an offshore prospect. M, 1980, University of Texas, Austin.

Catto, Norman Rhoderick. Quaternary geology of the western Cypress Hills region, Alberta and Saskatchewan. M, 1981, University of Alberta. 385 p.

Catto, Norman Rhoderick. Quaternary sedimentology and stratigraphy, Peel Plateau and Richardson Mountains, Yukon and Northwest Territories. D, 1986, University of Alberta. 751 p.

Catts, John G. Adsorption of Cu, Pb, and Zn onto birnessite. D, 1982, Colorado School of Mines. 227 p.

Caty, Jean Louis. Pétrographie et pétrologie du flanc sud-est du complexe du Lac Doré (Précambrien), (Québec, Canada). M, 1970, Universite de Montreal.

Caty, Jean Louis. Stratigraphie des roches protérozoiques du Bassin Papaskwasati et du Bassin des Monts Otish (Québec). D, 1971, Universite de Montreal.

Caudill, Samuel Jefferson. The Irvine, Kentucky oil and gas field. M, 1918, Pennsylvania State University, University Park.

Caudle, Karen L. Effects of off-road vehicles on vegetation and soils, Lake Meredith recreation area, Texas. M, 1983, West Texas State University. 65 p.

Cauffman, Lewis B. A study of the mineralogy and geochemistry of the Melville Island Group (Middle-Upper Devonian) and Imperial Formation (Upper Devonian), N.W.T. M, 1974, University of Calgary. 63 p.

Cauffman, Toya L. Determination of transport parameters of coincide inorganic and organic plumes in the Savannah River Plant M-area, Aiken, South Carolina. M, 1987, Texas A&M University.

Caughey, Charles A. Paluxy Formation (Lower Cretaceous) of Northeast Texas; depositional systems and distribution of groundwater, oil, and gas resources. M, 1973, University of Texas, Austin.

Caughey, Michael E. A study of the dissolved organic matter in the pore waters of carbonate-rich sediment cores from Florida Bay. M, 1982, University of Texas at Dallas. 69 p.

Caughey, Michael Eugene. Biogeochemistry of organic matter deposition and diagenesis in Bering-Chukchi and Gulf of Mexico sediments. D, 1988, University of Texas, Austin. 162 p.

Cauller, Stephen J. Quantitative shape analysis of benthic foraminifera from southern Maryland; a new approach to the paleoecology of the middle Miocene. M, 1987, Lehigh University. 93 p.

Causey, James D. Geology, geochemistry and lava tubes in Quaternary basalts, northeastern part of Zuni Lava Field, Valencia County, New Mexico. M, 1971, University of New Mexico. 57 p.

Causey, Marion E., Jr. The Chickamauga rocks of a portion of Raccoon Valley, western Knox County, Tennessee. M, 1956, University of Tennessee, Knoxville. 45 p.

Causey, Miles Andrew. A sedimentary study of the Cretaceous Tertiary contact in Mississippi. M, 1959, University of Mississippi.

Cavaleri, Mark Eugene. Volume and entropy systematics of materials at high pressures and temperatures by heated diamond-anvil energy-dispersive techniques. D, 1984, University of Minnesota, Minneapolis. 151 p.

Cavalero, Richard. Geology of the Clifton Township area in the Orono and Great Pond quadrangles, Maine. M, 1965, University of Maine. 117 p.

Cavaliere, Antonio. A magnetotelluric investigation of the Kapuskasing structural zone in the Superior Provinces; implications for the nature of the lower crust. M, 1987, University of Toronto.

Cavallaro, Nancy. Sorption and fixation of Cu, Zn and phosphate by soil clays as influenced by the oxide fraction. D, 1982, Cornell University. 153 p.

Cavallas, Joe F. Geology of the Natural Dam area, Crawford County, Arkansas. M, 1959, University of Arkansas, Fayetteville.

Cavallero, Lillian. Scanning electron microscopic examination of morphotypic variation in Globorotalia inflata from north Atlantic deep-sea cores. M, 1971, Queens College (CUNY). 92 p.

Cavanaugh, Eugene T. Stratigraphy of the Frontier Formation, Emigrant Gap Anticline, Natrona County, Wyoming. M, 1976, Colorado School of Mines. 173 p.

Cavanaugh, James F. Diagenesis in the supratidal facies of the Lower Cretaceous Cupido Formation, northern Mexico. M, 1980, Louisiana State University.

Cavanaugh, Lorraine Marie Monnier. Biostratigraphy of the Cache Creek Group, Horsefeed Formation, Tagish and Tutshi lakes area, South-central Yukon Territory and Northwest British Columbia, Canada. D, 1980, University of Wisconsin-Madison. 382 p.

Cavanaugh, M. D. The paleomagnetism of the Otto Stock near Kirkland Lake, Ontario, and its application to Precambrian tectonics. M, 1977, SUNY at Buffalo.

Cavanaugh, Martin James. Comparison of radar probing and nonlinear sonar probing through rock salt. D, 1984, Texas A&M University. 112 p.

Cavanaugh, Michael Dennis. A paleomagnetic study of post-Penokean intrusions from the Minnesota River valley and apparent polar wander for the interval 2800-1400 Ma for North America. D, 1983, University of South Carolina. 81 p.

Cavanaugh, Patrick Charles. The geology of the Little Boulder Creek molybdenum deposit, Custer County, Idaho. M, 1979, University of Montana. 100 p.

Cavanaugh, Thom. Three-dimensional ray tracing through heterogeneous media. M, 1975, University of North Carolina, Chapel Hill. 82 p.

Cavanaugh, Thomas Daniel. Finite difference wave models and the detection of caves. D, 1977, University of North Carolina, Chapel Hill. 163 p.

Cavaroc, Carolyn Wynn. A new species of Orbitolina from upper Albian rocks of Texas. M, 1968, Louisiana State University.

Cavaroc, Victor Viosca, Jr. "Allegheny" stratigraphy of southern West Virginia. M, 1963, Louisiana State University.

Cavaroc, Victor Viosca, Jr. Geology of some Carboniferous outer deltaic plain sediments of western Pennsylvania. D, 1969, Louisiana State University. 211 p.

Cavazza, William. Sedimentology, petrology, and basin analysis of the Tesuque Formation (Miocene) in the central Espanola Basin, Rio Grande Rift, New Mexico. M, 1985, University of California, Los Angeles. 125 p.

Cave, Deborah L. Geochemical reactions between primary-treated sewage and volcanic phase assemblages near Tahoe City, California. M, 1987, University of Nevada. 165 p.

Cave, W. R. The biological availability of metals in Juan de Fuca Strait. M, 1977, University of British Columbia.

Cavell, Patricia Anne. The geochronology and petrogenesis of the Big Spruce Lake alkaline complex. D, 1986, University of Alberta. 471 p.

Cavender, Wayne S. Structural geology of the Livermore area, Larimer County, Colorado. M, 1951, University of Colorado.

Cavender, Wayne Sherrell. Integrated mineral exploration in the Osgood Mountains, Humboldt County, Nevada. D, 1963, University of California, Berkeley. 231 p.

Cavendor, Philip N. Seasonal fluctuations of the hydrogeochemistry of selected springs and wells in Ba, Mn, and Hg mineralized areas, Ouachita Mountains, Arkansas. M, 1980, University of Arkansas, Fayetteville.

Cavin, Richard E. Significance of the interbasalt sediments in the Moscow Basin, Idaho. M, 1964, Washington State University. 97 p.

Cavin, William J. Precambrian geology of the northern Manzanita Mountains, Bernalillo County, New Mexico. M, 1985, University of New Mexico. 144 p.

Cavit, Douglas S. Statistical study of seismicity associated with geothermal reservoirs. M, 1981, University of California, Riverside. 132 p.

Cavounidis, S. Effective stress-strain analysis of earth dams during construction. D, 1975, Stanford University. 174 p.

Cawlfield, Jeff D. A first-order reliability approach to stochastic analysis of groundwater flow. D, 1987, University of California, Berkeley. 195 p.

Cawthorne, Nigel George. Geology and petrology of the Troitsa Lake property, Whitesail Lake map area, B.C. M, 1973, University of British Columbia.

Cayce, William Powers, Jr. The use of trace elements in correlating rhyolitic lava flows. M, 1963, Texas Tech University. 76 p.

Caylor, Floyd Martin. Secondary recovery from a consolidated rock. M, 1954, University of Oklahoma. 83 p.

Caylor, James Warren. Subsurface geology of western Garfield County, Oklahoma. M, 1957, University of Oklahoma. 78 p.

Cazavant, Alain. Etablissement d'un modèle géostatistique et graphique pour le calcul des réserves d'un gisement volcanogenique de sulfures massifs. M, 1987, Ecole Polytechnique. 279 p.

Cazeau, Charles Jay. An analysis of some Chattahoochee River sediments. M, 1955, Florida State University.

Cazeau, Charles Jay. Upper Triassic deposits of West Texas and northeastern New Mexico. D, 1962, University of North Carolina, Chapel Hill. 94 p.

Cazes, Debra Kay. Geochemical analyses of stream sediment and heavy mineral concentrates collected near a stratabound copper/silver occurrence in the Cabinet Mountains Wilderness area, Montana. M, 1981, University of Idaho. 104 p.

Cazier, Edward Coin, III. Late Paleozoic tectonic evolution of the Norfolk Basin, southeastern Massachusetts. M, 1984, University of Texas, Austin. 147 p.

Ceazan, Marnie L. Migration and transformations of ammonium and nitrate in a sewage-contaminated aquifer at Cape Cod, Massachusetts. M, 1987, Colorado School of Mines. 145 p.

Cebeci, Ahmet. A study of quartz deposits near highway highlands, Los Angeles County, California. M, 1944, California Institute of Technology. 37 p.

Cebull, Stanley Edward. Bedrock geology of the southern Grant Range, Nye County, Nevada. D, 1967, University of Washington. 130 p.

Cebull, Stanley Edward. The structure and stratigraphy of portions of the Mare Island, Sears Point and Richmond quadrangles, California. M, 1958, University of California, Berkeley. 79 p.

Cebulski, Donald E. Distribution of foraminifera in the barrier reef and lagoon of British Honduras. M, 1961, Texas A&M University.

Cechova, Irina Maclin. Projection of water resources availability in relation to future requirements of the Houston Gulf Coast area. D, 1973, University of Texas Health Science Center at Houston School of Public Health. 307 p.

Ceci, Vincent. The petrology and geochemistry of Precambrian basement. M, 1985, University of Pittsburgh.

Cecil, Charles Blaine. Effects of grain coatings on quartz growth. M, 1966, West Virginia University.

Cecil, Charles Blaine. Silica in experimental diagenesis of sandstones. D, 1969, West Virginia University. 102 p.

Cecil, Thomas Martin. Variability of bottom nepheloid layers near the Flower Garden Banks, Gulf of Mexico. M, 1983, Texas A&M University. 139 p.

Cecile, Michael P. Lithofacies analysis of the Proterozoic Thelon Formation, Northwest Territories (including computer analysis of field data). M, 1973, Carleton University. 119 p.

Cecile, Michael P. Stratigraphy and depositional history of the upper Goulburn Group, Kilohigok Basin, Bathurst Inlet, N.W.T. D, 1976, Carleton University. 200 p.

Cedarleaf, Darwin C. Petrography of basal Tertiary sandstone, Brewster and Presidio counties, Texas. M, 1951, University of Minnesota, Minneapolis. 63 p.

Cederberg, Gail Anne. TRANQL; a ground-water mass-transport and equilibrium chemistry model for multicomponent systems. D, 1985, Stanford University. 130 p.

Cederstrom, D. John. Geology of the central Dragoon Mountains, Arizona. D, 1946, University of Arizona.

Cedillo, Hector Enrique Febres *see* Febres Cedillo, Hector Enrique

Cedraro, Rodulfo Prieto *see* Prieto Cedraro, Rodulfo

Cefola, David Paul. Two-dimensional elastic wave propagation in a duraluminum sheet. M, 1982, Texas A&M University.

Cehrs, D. Petrology and stratigraphy of the Monte Cristo Limestone (Mississippian); southeastern California and southern Nevada. M, 1975, California State University, Fresno.

Celasun, Merih. Some planning problems in mineral exploration. D, 1964, Columbia University, Teachers College. 209 p.

Celaya, Michael Augustine. Time series analysis of paleomagnetic polarity transition zones. M, 1988, Texas A&M University. 141 p.

Celerier, Bernard. Models for the evolution of the Carolina trough and their limitations. D, 1986, Massachusetts Institute of Technology. 206 p.

Celestian, Susan Myers. A faunal analysis of the Devonian Martin Formation in the Verde Valley, Arizona. M, 1979, Northern Arizona University. 68 p.

Celestino, Tarcisio Barreto. Deformation modes of filled joints in rock. D, 1981, University of Calgary. 121 p.

Celles, Peter G. De *see* De Celles, Peter G.

Cember, Richard Paul. Two oceanographic studies in the Red Sea. D, 1988, Columbia University, Teachers College. 163 p.

Cemen, Ibrahim. Geology of the Sespe-Piru Creek area, Ventura County, California. M, 1977, Ohio University, Athens. 69 p.

Cemen, Ibrahim. Stratigraphy, geochronology and structure of the selected areas of the northern Death Valley region, eastern California-western Nevada, and implications concerning Cenozoic tectonics of the region. D, 1983, Pennsylvania State University, University Park. 321 p.

Cenedella, Louise G. Petrogenesis of the Leggett peridotite (Jurassic?), Mendocino County, California. M, 1969, University of Wisconsin-Madison.

Centanni, James Patrick. The composition, synthesis and thermal stability of celadonitic muscovite with regards to its occurrence in plutonic rocks. M, 1983, University of California, Santa Barbara. 67 p.

Centini, Barry Austin. Structural geology of the Hanging Rock area, Stokes County, North Carolina. D, 1968, University of North Carolina, Chapel Hill. 57 p.

Centini, Barry Austin. Structural geology of the Hanging Rock State Park area, North Carolina. M, 1964, North Carolina State University. 38 p.

Centorino, J. R. Geothermal energy in New England. M, 1975, Boston College. 150 p.

Century, Jack Remo. The Animas Formation in the northern part of the San Juan Basin, Colorado. M, 1952, University of Illinois, Urbana.

Cepeda Díaz, Abel Fernando. An experimental investigation of the engineering behavior of natural shales. D, 1987, University of Illinois, Urbana. 708 p.

Cepeda, Joseph C. Geology and geochemistry of the igneous rocks of the Chinati Mountains, Presidio County, Texas. D, 1977, University of Texas, Austin. 195 p.

Cepeda, Joseph Cherubini. Geology of Precambrian rocks of the El Oro Mountains and vicinity, Mora County, New Mexico. M, 1972, New Mexico Institute of Mining and Technology. 63 p.

Cepek, Robert Joseph. Acoustical and mass physical properties of deep ocean Recent marine sediments. M, 1972, United States Naval Academy.

Cercone, Karen Rose. Diagenesis of Niagaran (Middle Silurian) pinnacle reefs, Northwest Michigan. D, 1984, University of Michigan. 382 p.

Cerling, Bettina. Diagenesis of Miocene bedded cherts, Berkeley Hills, California. M, 1976, University of California, Berkeley. 64 p.

Cerling, Thure Edward. Correlation of Plio-Pleistocene tuffs utilizing O-18/O-16 isotope ratios, East Rudolf Basin, Kenya. M, 1973, Iowa State University of Science and Technology.

Cerling, Thure Edward. Paleochemistry of Plio-Pleistocene Lake Turkana and diagenesis of its sediments. D, 1977, University of California, Berkeley. 180 p.

Cermignani, C. Metamorphic reactions in the system albite-anorthite-nepheline-Na$_2$CO$_3$-CaCO$_3$-H$_2$O, with application to the Haliburton-Bancroft alkaline rocks. D, 1979, University of Toronto.

Cerna, Ivanka. Mineralogy and paragenesis of amblygonite-montebrasite with special reference to the Tanco (Chemalloy) pegmatite, Bernic lake, Manitoba. M, 1970, University of Manitoba.

Cernock, Paul John. Consolidation characteristics and related physical properties of selected sediments from the Gulf of Mexico. M, 1967, Texas A&M University.

Cernock, Paul John. Sound velocities in Gulf of Mexico sediments as related to physical properties and simulated overburden pressures. D, 1970, Texas A&M University. 124 p.

Cerny, John H. Compaction and porosity evolution in the Smackover; Oaks Field, Louisiana, and Walker Creek Field, Arkansas. M, 1985, Southern Methodist University. 77 p.

Ceroici, W. J. Groundwater hydrology of the Edmonton area (southwest quarter, Alberta, Canada). M, 1979, University of Waterloo.

Cerri, S. T. Correlation of strong ground-motion and local site geology in Los Angeles during the San Fernando earthquake, February 1971. M, 1976, University of Southern California. 106 p.

Cerrillo, Lawrence A. The hydrogeology of the Beaver Creek drainage basin, Larimer County, Colorado. M, 1967, Colorado State University. 55 p.

Cervantes R., J. Alfredo. Geological reconnaissance of the Sierra del Kilo, Villa Ahumada, Chihuahua, Mexico. M, 1983, Colorado State University. 126 p.

Cervantes Silva, Juan Jose. Impact of Mexico's recent mining legislation on national mineral production goals. M, 1980, University of Arizona. 86 p.

Cervantes, Jose Enrique. A trigger type cluster model for flood analysis. D, 1981, Purdue University. 192 p.

Cervantes, Michael Arthur. Foraminiferal biofacies of middle to late Paleogene rocks in the western San Emigdio Mountains, California. M, 1988, University of Texas, Austin. 121 p.

Cervantes-Montoya, Jesus Alberto. Use of geostatistics in developing drilling programs at the Cananea copper mine. M, 1981, University of Arizona. 86 p.

Cerven, James F. Refinement of the crystal structure of clinozoisite. M, 1968, Southern Illinois University, Carbondale. 46 p.

Cervenka, Robert E. The meteorology and provenance of central Texas dust storms, 1975-78. M, 1983, Baylor University. 92 p.

Cerveny, Philip F., III. Uplift and erosion of the Himalaya over the past 18 million years; evidence from fission track dating of detrital zircons and heavy mineral analysis. M, 1986, Dartmouth College. 198 p.

Cervik, Joseph. Comparison of X-ray diffraction and k-factor studies of clay content in artificial cores. M, 1955, Pennsylvania State University, University Park. 47 p.

Ceryak, Ronald Joseph. A mineralogical and textural analysis of Recent beach sands of the Gulf Coast of Florida from Anclote Key to Marco Is. M, 1974, University of Florida. 41 p.

Cessaro, Robert K. The study of P and S phases from regional earthquakes recorded by a borehole seismometer in the Northwest Pacific. D, 1987, University of Hawaii.

Cetinay, Huseyin Turgut. The geology of the eastern end of the Canelo Hills, Santa Cruz County, Arizona. M, 1967, University of Arizona.

Cetrone, Ronald and Paschal, Lawrence W., Jr. Correlation between a well in Fallon County, Montana, and a well in Harding County, South Dakota. M, 1957, South Dakota School of Mines & Technology. 35 p.

Ceylan, Rasit. Geology and ground water resources of Saltdale Quadrangle, California. M, 1952, University of Southern California.

Ch'ih, Chi Shang. Preliminary microscopic study of serpentine and its associated rocks. M, 1947, Bryn Mawr College. 31 p.

Ch'ih, Chi Shang. Structural petrology of the Wissahickon Schist near Philadelphia, with special reference to granitization. D, 1949, Bryn Mawr College. 87 p.

Chabot, Nathalie. Etude de la provenance de la Formation de Cabana (Caradocien-Llandoverien), comte de Rimouski, Quebec. M, 1986, Universite de Montreal.

Chaboudy, Louis R., Jr. Long-term phosphorus flux to Neogene Antarctic and Pacific deep sea sediments. M, 1984, Ohio University, Athens. 112 p.

Chacartegui, Fernando J. Geochemistry and hydrography of Laguna de Tacarigua, Venezuela. M, 1981, University of South Florida, Tampa. 96 p.

Chace, Frederick A. The granites of southeastern Rhode Island. M, 1932, Brown University.

Chace, Frederick Mason. Tin-silver veins of Oruro, Bolivia. D, 1947, Harvard University.

Chacko, Soman. A finite-element model of lunar thermal evolution. D, 1980, Rice University. 167 p.

Chacko, Thomas. Petrologic, geochemical and isotopic studies in the charnockite-khondalite terrain of southern Kerala, India; the deposition and granulite-facies metamorphism of a Precambrian sedimentary sequence. D, 1987, University of North Carolina, Chapel Hill. 191 p.

Chacko, Thomas. The nature of Acadian plutonism in northern New England. M, 1983, Pennsylvania State University, University Park. 113 p.

Chackowsky, L. E. Mineralogy, geochemistry and petrology of pegmatitic granites and pegmatites at Red Sucker and Gods Lake, northeastern Manitoba. M, 1987, University of Manitoba.

Chacon, Roberto. Geology of the San Carlos Dome, Manuel Benairdes, Chihuahua, Mexico. M, 1972, University of Texas at El Paso.

Chadbourn, Charles H. The Trojan Mines, Trojan, South Dakota. M, 1921, University of Minnesota, Minneapolis. 32 p.

Chadeayne, Dennis. The geology of the Pass Creek Ridge, Saint Mary's and Cedar Ridge anticlines, Carbon County, Wyoming. M, 1966, University of Wyoming. 87 p.

Chadima, Sarah Anne. Trace element abundances in mineral separates from Adirondack anorthosites. M, 1982, Iowa State University of Science and Technology. 78 p.

Chadwick, George H. Brachiopod fauna of the basal Port Ewen beds (New York). M, 1907, University of Rochester. 50 p.

Chadwick, Michael L. Identification and geological significance of petrified wood from the Oligocene Catahoula Formation, Jasper County, Texas. M, 1988, Stephen F. Austin State University. 77 p.

Chadwick, Oliver Austin. Incipient silica cementation in central Nevada alluvial soils influenced by tephra. D, 1985, University of Arizona. 191 p.

Chadwick, Robert Aull. Mechanisms of pegmatite emplacement. D, 1956, University of Wisconsin-Madison.

Chadwick, Russell John. Depositional environments, diagenesis, and porosity/permeability relationships of the Ellenburger Group in the Pegasus Field, Midland/Upton counties, Texas. M, 1987, Texas Tech University. 136 p.

Chadwick, William Ward, Jr. Ground deformation precursory to extrusions at Mount St. Helens, 1981-1982; tectonic analogues and mechanical modeling. D, 1988, University of California, Santa Barbara. 108 p.

Chael, Eric Paul. Constraints on the Earth's anelastic and aspherical structure from antipodal surface waves. D, 1983, California Institute of Technology. 147 p.

Chaemsaithong, Kanchit. Design of water resources systems in developing countries; the Lower Mekong Basin. D, 1973, University of Arizona. 244 p.

Chafetz, Henry S. Petrography and stratigraphy of the strata of the Lower-Middle Ordovician contact, central Pennsylvania. M, 1967, Michigan State University. 86 p.

Chafetz, Henry Simon. Petrology and stratigraphy of the lower part of the Wilberns Formation, upper Cambrian of central Texas. D, 1970, University of Texas, Austin. 287 p.

Chaffe, Robert Gibson. Indications of Cretaceous New Jersey shore lines; Part I, A possible Mount Laurel-Wenonoh barrier beach. M, 1941, University of Pennsylvania.

Chaffe, Robert Gibson. The Deseadan fauna of the Scarritt Pocket, Patagonia. D, 1952, Columbia University, Teachers College.

Chaffee, Maurice A. Dispersion patterns as a possible guide to ore deposits in the Cerro Colorado District, Pima County, Arizona. M, 1964, University of Arizona.

Chaffee, Maurice Ahlborn. A study of the geology and hydrothermal alteration north of the Creede mining district, Mineral, Hinsdale, and Saguache counties, Colorado. D, 1967, University of Arizona. 247 p.

Chaffin, David Leland. Implications of regional gravity and magnetic data for structure beneath western Pennsylvania. M, 1981, Pennsylvania State University, University Park. 72 p.

Chaffin, Herbert Scott, Jr. Sedimentary history of a middle Oligocene valley in northwestern Nebraska. M, 1966, University of Nebraska, Lincoln.

Chagarlamudi, Pakiraiah. Mapping rock outcrops from Landsat digital data. D, 1980, University of Manitoba.

Chagarlamudi, Pakiriah. Resistivity and seismic refraction surveys over Pleistocene deposits in southern Manitoba. M, 1971, University of Manitoba.

Chagnon, Andre. Sédimentologie des claystones rouges et verts du facies flysch des Appalaches du Québec (Cambro-Ordovicien). M, 1970, Universite de Montreal.

Chagnon, Jean Yves. Experimental studies on the growth of minerals in sediments. M, 1961, McGill University.

Chagnon, Jean Yves. The geology of the Des Quinze Lake area, Temiscamingue County, Quebec. D, 1965, McGill University.

Chai, B. H.-T. The kinetics and mass transfer of calcite during hydrothermal recrystallization process. D, 1975, Yale University. 231 p.

Chaichanavong, Thira. Dynamic properties of ice and frozen clay under cyclic triaxial loading conditions. D, 1976, Michigan State University. 460 p.

Chaiffetz, Michael. Lateral variations in relative frequencies of chitinozoan taxa from the Red Mountain Formation of Northeast Alabama. M, 1973, Florida State University.

Chaille, John Lee. Statistical analysis of the cricetid rondent Eumys from the Brule Formation, Slim Buttes, Harding County, South Dakota. M, 1980, South Dakota School of Mines & Technology.

Chainey, Daniel. Paramètres pétrographiques et géochimiques du gisement d'or de la Mine Camflo, Québec, Canada. M, 1983, Universite de Montreal. 215 p.

Chainey, Michel. Paramètres pétrographiques et géochimiques du gisements d'or de la Mine Camflo, Québec, Canada. M, 1983, Ecole Polytechnique. 215 p.

Chaipayungpun, W. Geophysical studies in the High Plains of northeastern Colorado. D, 1977, University of Texas at Dallas. 81 p.

Chairat, Trakarn. Mathematical modeling of a petroleum refinery for optimization by linear programming techniques. M, 1971, Colorado School of Mines. 68 p.

Chaisrakeo, Meechai. Photogrammetric determination of single points. M, 1960, Ohio State University.

Chaivre, Kenneth R. The feasibility of determining joint orientations by sampling all sizes of joints. M, 1972, Wayne State University.

Chaiyadhuma, Wirote. Rainfall infiltration for different soil surface conditions. D, 1981, Utah State University. 240 p.

Chakarun, John Douglas. Geology, mineralization, and alteration of the Jhus Canyon, Cochise County, Arizona. M, 1973, University of Arizona.

Chakel, John Anthony. Collision processes in triple quadrupole mass spectrometry; characteristics and applications. D, 1982, Michigan State University. 160 p.

Chaki, Susan J. A sedimentary investigation of the beach ridge sets composing Cape Canaveral, Florida. M, 1974, Florida State University.

Chakoumakos, Bryan Charles. Systematics of the pyrochlore structure type and theoretical molecular orbital study of silanol--water interactions. D, 1984, Virginia Polytechnic Institute and State University. 212 p.

Chakrabarti, Chinmoy. Diffusion of sediment in the lee of dune-like bedforms. D, 1976, Louisiana State University. 415 p.

Chakravorty, Sailendra K. Late Mississippian Bryozoa from Missouri. D, 1951, University of Kansas. 180 p.

Chakridi, Rachid. Conception de modèles reduits en simulation électromagnétique. M, 1986, Ecole Polytechnique. 59 p.

Chalcraft, Richard George. A petrographic study of Mesozoic dolerites from eastern North America. D, 1973, University of North Carolina, Chapel Hill. 175 p.

Chalcraft, Richard George. Petrography and geophysics of the Rock Hill gabbro pluton, York County, South Carolina. M, 1969, University of North Carolina, Chapel Hill. 40 p.

Chalhoub, Michel Soto. Theoretical and experimental studies of earthquake isolation and fluid containers. D, 1987, University of California, Berkeley. 279 p.

Chalmer, John Roy. The origin and significance of the chert in the lead-zinc district of Missouri-Kansas-Oklahoma. D, 1936, Harvard University.

Chalmers, A. G. Pools of nitrogen in a Georgia salt marsh. D, 1977, University of Georgia. 178 p.

Chalmers, Ann L. Stradley. Quaternary glacial geology and geomorphology of the Teton River drainage area, Montana. M, 1968, Montana State University. 83 p.

Chalmers, John A. Spring deposits of the Clear Lake area of Manitoba. M, 1939, McMaster University.

Chalokwu, Christopher I. Petrology and geochemistry of ultramafic (komatiitic) rocks of Karnataka, India. M, 1980, Northeastern Illinois University.

Chalokwu, Christopher Iloba. A geochemical, petrological, and compositional study of the Partridge River Intrusion, Duluth Complex, Minnesota. D, 1985, Miami University (Ohio). 232 p.

Chamberlain, Alan K. Biostratigraphy of the Great Blue Formation. M, 1977, Brigham Young University.

Chamberlain, Barry N. A gravimetric survey across the Kowaliga lithotectonic block, Tallapoosa County, Alabama. M, 1986, Memphis State University. 57 p.

Chamberlain, C. Page. Metamorphic zonation in south-central New Hampshire. M, 1981, Dartmouth College. 191 p.

Chamberlain, C. Page. Tectonic and metamorphic history of a high-grade terrane, southwestern New Hampshire. D, 1985, Harvard University. 295 p.

Chamberlain, Charles Kent. Trace fossils and paleoecology of the Ouachita Mountains of southeast Oklahoma. D, 1970, University of Wisconsin-Madison. 148 p.

Chamberlain, John Andrew, Jr. Fluid mechanics of the ectocochliate cephalopod shell; an experimental study. D, 1971, University of Rochester. 311 p.

Chamberlain, John David. The Vernon Shale in central and western New York. M, 1948, University of Rochester. 73 p.

Chamberlain, John Mark. Microfacies analysis of the Ames Limestone (Pennsylvanian-Conemaugh Group) in east-central Ohio. M, 1985, Wright State University. 110 p.

Chamberlain, Joseph Annandale. Structural control of pitchblende orebodies, Eldorado, Saskatchewan. D, 1959, Harvard University.

Chamberlain, Kent C. Carboniferous trilobites of central Utah. M, 1966, Brigham Young University.

Chamberlain, Randy C. Structure and stratigraphy of the Rex Peak Quadrangle, Rich County, Utah. M, 1980, Brigham Young University.

Chamberlain, Rick Earl. Geology and the Forest City Basin in Iowa; tectonic activity and influence on Pennsylvanian sedimentation. M, 1980, Iowa State University of Science and Technology.

Chamberlain, Theodore K. Mechanics of mass sediment transport in Scripps submarine canyon, California. D, 1960, University of California, Los Angeles.

Chamberlin, Dale S. Study of the formation of dolomite in Beekmantown Limestone. M, 1920, Lehigh University.

Chamberlin, David C. Gully erosion in submontane basins near Trinidad, Colorado. M, 1979, Colorado State University. 93 p.

Chamberlin, Peter. The Middle Fork of Stony Creek, California; a study in mountain anthropogeomorphology. M, 1971, California State University, Chico.

Chamberlin, Richard M. Cenozoic stratigraphy and structure of the Socorro Peak volcanic center, central New Mexico. D, 1980, Colorado School of Mines. 462 p.

Chamberlin, Richard Martin. Geology of the Council Rock District, Socorro County, New Mexico. M, 1974, New Mexico Institute of Mining and Technology. 134 p.

Chamberlin, Rollin Thomas. The gases occluded in rocks. D, 1907, University of Chicago. 80 p.

Chamberlin, Thomas L. Stratigraphy of the Ordovician Ely Springs Dolomite in the southeastern Great Basin, Utah and Nevada. D, 1975, University of Illinois, Urbana. 209 p.

Chamberlin, Thomas L. The Ordovician-Silurian boundary in the Arrow Canyon Range, Clark County, Nevada. M, 1971, University of Illinois, Urbana. 36 p.

Chambers, Arthur Edwin. Geology and mineral deposits of part of the Bayhorse mining district, Custer County, Idaho. D, 1966, University of Arizona. 280 p.

Chambers, David Marshall. Holocene sedimentation and potential placer deposits on the continental shelf off the Rogue river, Oregon. M, 1969, Oregon State University. 103 p.

Chambers, Donald D. Cirque characteristics and their influence on the maintenance of glaciers; Sierra Nevada, California. M, 1979, University of California, Los Angeles.

Chambers, E. F. Geology of portions of the Whitaker Peak and Beartrap Canyon quadrangles, California. M, 1947, University of Southern California.

Chambers, Henry Peyton. A regional ground motion model for historical seismicity along the Rock Creek Fault, western Wyoming. M, 1988, University of Wyoming. 102 p.

Chambers, J. F. A.C. demagnetization as a method of palaeomagnetic cleaning. M, 1957, University of Toronto.

Chambers, Jack. Ostracoda of the lower Jackson Eocene of Louisiana. M, 1935, Louisiana State University.

Chambers, Jefferson K. Resistivity modelling of three-dimensional structures, using the finite element technique. M, 1985, University of California, Riverside. 157 p.

Chambers, John Mark. The geology and structural petrology of ultramafic and associated rocks in the Northeast Marble Mountain wilderness, Klamath Mountains, California. M, 1983, University of Oregon. 149 p.

Chambers, Richard Lee. Information content of grain-size frequency distributions in coastal deposits on the eastern shore of Lake Michigan. D, 1975, Michigan State University. 130 p.

Chambers, Richard Lee. Sedimentation in glacial Lake Missoula. M, 1971, University of Montana. 100 p.

Chambers, T. M. Late Wisconsinan events of the Ontario ice lobe in the southern and western Tug Hill region, New York. M, 1978, Syracuse University.

Chamblin, William Jack. Pennsylvanian conodont faunas from Iowa. M, 1956, University of Illinois, Urbana.

Chameau, Jean-Lou. Probabilities and hazard analysis for pore pressure increase in soils due to seismic loading. D, 1981, Stanford University. 244 p.

Chamness, Ralph S. Stratigraphy of the Eagle Ford Group, McLennan County, Texas. M, 1963, Baylor University. 85 p.

Chamney, T. P. Investigation of the Colorado Group of central Colorado. M, 1954, Colorado School of Mines. 86 p.

Chamon, Nagib. Structural and magnetic characteristics of the Belton anomaly, Cass County, Missouri. M, 1965, University of Missouri, Rolla.

Champ, John G. Geology of the northern third of the Dixonville Quadrangle, Oregon. M, 1969, University of Oregon. 86 p.

Champagne, Michel. Paléo-milieux sédimentaires, géochimie d'éléments-traces et carbone organique total, dans quatre coupes stratigraphiques des Basses-Terres du Saint-Laurent. M, 1983, Universite Laval. 63 p.

Champeny, Jon Duckett. Paleocene and Upper Cretaceous stratigraphy of Santa Ynez Canyon. M, 1962, University of California, Los Angeles.

Champigny, Normand. A geological evaluation of the Cinola (Specogna) gold deposit, Queen Charlotte Islands, B.C. M, 1981, University of British Columbia. 199 p.

Champion, Duane E. The relationship of large scale surface morphology to lava flow direction, Wapi Lava Field, southeastern Idaho. M, 1973, SUNY at Buffalo.

Champion, Duane Edwin. Holocene geomagnetic secular variation in the western United States; implications for the global geomagnetic field. D, 1980, California Institute of Technology. 325 p.

Champion, J. W., Jr. A detailed gravity study of the Charleston, South Carolina, epicentral zone. M, 1975, Georgia Institute of Technology. 97 p.

Champlin, Maurice Anthony. Geology of the Mertens Quadrangle, Ellis, Hill, and Navarro counties, Texas. M, 1976, University of Texas, Arlington. 90 p.

Champlin, Stephen C. A stratigraphic study of the Sycamore and related formations in the eastern Arbuckle Mountains. M, 1959, University of Oklahoma. 66 p.

Champney, Richard D. Structural geology of a rhyolite flow in the Tucson Mountains. M, 1962, University of Arizona.

Champney, Richard Daniel. Studies of geologic structures by paleomagnetic methods. D, 1971, University of Arizona. 109 p.

Chan, Chien-Lu. Experimental study of diopside-enstatite-jadeite equilibrium at 800 to 1400°C, 15 and 20 kbar. M, 1983, Purdue University. 52 p.

Chan, Chun Young. A study of the effects of water incursion in the Coalinga oil field of California. D, 1922, University of California, Berkeley. 110 p.

Chan, Gee Hung. Seismic diffraction from wedges. D, 1987, University of Alberta. 233 p.

Chan, King Nam. Modeling of West Texas crustal structure from earthquake data. M, 1977, University of Texas, Austin.

Chan, Kwok Chun. Studies in modeling, reduction and filtering of potential field data. D, 1978, University of California, Berkeley. 84 p.

Chan, Lap-yan. Application of block theory and simulation techniques to optimum design of rock excavations. D, 1987, University of California, Berkeley. 248 p.

Chan, Lung Sang. Paleomagnetism and tectonic studies of the Scaglia Rossa pelagic limestones (Upper Cretaceous-Eocene) from the Umbria-Marches Apennines, Italy. D, 1984, University of California, Berkeley. 293 p.

Chan, Mankin Kenneth. Interfacial activity in alkaline flooding enhanced oil recovery. D, 1981, University of Southern California.

Chan, Marjorie Ann. Comparison of sedimentology and diagenesis of Eocene rocks, Southwest Oregon. D, 1982, University of Wisconsin-Madison. 322 p.

Chan, Paul Chi-Keung. A study of dynamic load-deformation and damping properties of sands concerned with a pile-soil system. D, 1968, Texas A&M University. 219 p.

Chan, Samuel S. M. Syntheses and X-ray investigations within the system FeS₂-CoS₂. M, 1962, University of Missouri, Rolla.

Chan, Samuel Shu Mou. Mineralogical and geochemical studies of the vein materials from the Galena Mine, Shoshone County, Idaho. D, 1966, University of Idaho. 251 p.

Chan, Siew Hung. A theoretical study on the interpretation of resistivity sounding data measured by the Wenner electrode system. D, 1969, University of Missouri, Rolla. 210 p.

Chan, Wing-Wah Winston. Crustal structure of Spitsbergen; waveform synthesis and surface wave studies. M, 1981, St. Louis University.

Chan, Wing-Wah Winston. Structure and tectonics of the Barents Shelf. D, 1983, St. Louis University. 255 p.

Chance, John Matthew. The geology of the eastern half Hebron Quadrangle. M, 1971, Virginia State University. 27 p.

Chance, Patrick Neville. Petrogenesis of a low-Ti, potassic suite; Kuh-e-Lar Caldera subsidence complex, East Iran. M, 1981, University of Western Ontario. 166 p.

Chancellor, Robert E. Nature and possible source beds of the fragments in the Deepskill breccias near Pleasant Valley, New York. M, 1948, Columbia University, Teachers College.

Chander, Ramesh. On the synthesis of sheer-coupled PL waves. D, 1970, Columbia University. 78 p.

Chandiok, Kailash Chandra. A detailed gravimetric survey of an area immediately south and east of St. Charles, Missouri. M, 1950, St. Louis University.

Chandiok, Kailash Chandra. Geology of the area south of Golden, Jefferson County, Colorado. M, 1948, Colorado School of Mines. 87 p.

Chandlee, George O. Patterns of morphological clines (spatial and temporal) in species populations of flexicalymene senaria. M, 1978, University of Rochester.

Chandlee, George Oliver. Anatomical variation in Paucicrura rogata (Brachiopoda; Dalmanellidae); systematics, geographic variation, and evolution along an Ordovician depth gradient. D, 1982, Cornell University. 251 p.

Chandler, Carol Reisen. Outdoor laboratory; Lamkin area, north-central Texas. M, 1979, Baylor University. 150 p.

Chandler, Charles. A petrographic study of sedimentary rocks of Peregrina Canyon, State of Tamaulipas, Mexico. M, 1957, Louisiana State University.

Chandler, Frederick William. Geology of the Huronian rocks (Precambrian) of Harrow township and surrounding areas, north shore of Lake Huron, Ontario (Canada). D, 1969, University of Western Ontario. 328 p.

Chandler, Gary Wayne. An experimental investigation of high-temperature interactions between seawater and rhyolite, andesite, basalt and peridotite. M, 1979, Texas A&M University. 101 p.

Chandler, John Frederick. Determination of the dynamical properties of the Jovian system by numerical analysis. D, 1979, Massachusetts Institute of Technology. 206 p.

Chandler, Mark Arnold. Depositional controls on permeability in an eolian sandstone sequence, Page Sandstone, northern Arizona. M, 1986, University of Texas, Austin. 132 p.

Chandler, P. B. Geologic elements for urban planning, Estero Bay, California. M, 1977, University of Southern California. 179 p.

Chandler, Philip Prescott. The geology of the McBride area, Cherokee and Wagoner counties, Oklahoma. M, 1950, University of Oklahoma. 69 p.

Chandler, R. L. Water and chemical transport in layered porous media. D, 1979, Colorado State University. 165 p.

Chandler, V. W. Correlation of gravity and magnetic data over the Great Lakes region, North America. D, 1977, Purdue University. 188 p.

Chandler, Val William. A gravity and magnetic investigation of the McCarthy Mountain area, Beaverhead and Madison counties, Montana. M, 1973, Indiana University, Bloomington. 47 p.

Chandler, William Alton. A subsurface study of the Katie Field, Garvin County, Oklahoma. M, 1949, University of Oklahoma. 47 p.

Chandler, William E. Graben mechanics at the junction of the Hartford and Deerfield basins of the Connecticut Valley, Massachusetts. M, 1979, University of Massachusetts. 151 p.

Chandra, B. Thermal convection under an axially symmetric force field. M, 1969, University of Western Ontario.

Chandra, Bhuvanesh. A laboratory study of thermal convection under a central force field. D, 1972, University of British Columbia.

Chandra, Deb Kumar. Geology of the Colfax and Foresthill quadrangles, California. D, 1954, University of California, Berkeley. 148 p.

Chandra, Jagdesh James. Ground investigation of airborne gamma-ray radiometric anomalies in New Brunswick by truck mounted and hand held gamma-ray sensors. M, 1981, University of New Brunswick.

Chandra, Nellutla Naveena. Converted-wave method for determining crustal structure in southern Alberta. M, 1966, University of Alberta. 48 p.

Chandra, Nellutla Naveena. Microseisms in Canada. D, 1970, University of Alberta. 170 p.

Chandra, Umesh. Analysis of body wave spectra for earthquake energy. D, 1969, St. Louis University. 203 p.

Chaney, Dan Scott. The Jacona microfauna (late Barstovian), Pojoaque Member, Tesuque Formation; north central New Mexico; its geology, taphonomy, paleoecology, Insectivora. M, 1988, University of California, Riverside. 158 p.

Chaney, Gregory Paul. Coastal geomorphology of Puget Bay, Alaska. M, 1987, University of Idaho. 190 p.

Chaney, J. B. The barite industry in the United States. M, 1949, University of Missouri, Rolla.

Chaney, Mark Anthony. A gravity model of the Williston Basin. M, 1987, University of Oklahoma. 113 p.

Chaney, R. C. Deformations of earthdams under earthquake loading. D, 1978, University of California, Los Angeles. 480 p.

Chaney, Ralph Works. The ecological significance of the Eagle Creek flora of the Columbia River gorge. D, 1919, University of Chicago. 15 p.

Chang, Andre Chi-Chao. Love waves and the crust-upper mantle structure of the southwestern United States. D, 1968, Rice University. 92 p.

Chang, Chen Ping. The constitution of chlorite. M, 1924, University of Wisconsin-Madison.

Chang, Cheng-Jung. Seismic safety analysis of slopes. D, 1981, Purdue University. 137 p.

Chang, Chi-Chung. Hydrogeologic approach to the characterization of aquifer contamination and restoration using mathematical models. M, 1985, Oklahoma State University. 179 p.

Chang, Chia-Yu. Three-dimensional velocity structure and precise earthquake locations on the Calaveras Fault in the Morgan Hill area, California. M, 1988, Indiana University, Bloomington. 183 p.

Chang, Chin-yung. Finite element analyses of soil movements caused by deep excavation and dewatering. D, 1969, University of California, Berkeley. 317 p.

Chang, Ching Shung. Analysis of consolidation of earth and rockfill dams. D, 1976, University of California, Berkeley. 243 p.

Chang, Chung Chin. A gravity study of the Triassic valley in southern Connecticut. M, 1968, Wesleyan University. 108 p.

Chang, David. Design of an indirect sounding system. M, 1964, New York University.

Chang, Eric Yea-Yuan. Sulfation reactions of limestone/dolomite and nahcolite/trona. D, 1988, Northwestern University. 156 p.

Chang, Feng-Keng. A vertical magnetometer survey in a seismic zone. M, 1957, St. Louis University.

Chang, Fi-John. Refinement of hydrogeochemical models of the ecological impact of acid deposition. D, 1988, Purdue University. 227 p.

Chang, Fong-shun. The structure of the upper mantle of Eurasia and tectonic interpretation from a study of Rayleigh wave dispersion. D, 1979, University of California, Los Angeles. 225 p.

Chang, Heu-Cheng. Effect of displacement rate and gouge composition on the sliding behavior of simulated gouge. M, 1982, Texas A&M University. 69 p.

Chang, Hsing Chi. X-ray diffraction studies of test boring samples from the glacial lake plain in Wayne County, Michigan. M, 1968, Wayne State University.

Chang, Hui Sing. A time domain finite element-boundary element method for two dimensional non-linear soil-structure interaction. D, 1988, Northwestern University. 217 p.

Chang, Hung Kiang. Diagenesis and mass transfer in Cretaceous sandstone-shale sequences, offshore Brazil. D, 1983, Northwestern University. 360 p.

Chang, Ja-Shian. A comparison of several methods for analyzing earthquake resistant soil-building interaction systems. D, 1986, Stevens Institute of Technology. 196 p.

Chang, Jui-Yuan. Emergent angle dependent deconvolution. M, 1983, University of Texas, Austin. 165 p.

Chang, K.-R. Determination of soil subsidence due to well pumping by numerical analysis. D, 1976, North Carolina State University. 236 p.

Chang, Ker-Chi. Co-disposal of low-level radioactive waste within sanitary landfills. D, 1982, Georgia Institute of Technology. 203 p.

Chang, Ki Hong. A study of stratigraphic classification in Korea with special reference to synthems. D, 1976, Princeton University. 459 p.

Chang, Luke. Subsolidus phase relations in the systems BaCO$_3$-SrCO$_3$, CaCO$_3$-SrCO$_3$, and CaCO$_3$-BaCO$_3$. D, 1963, University of Chicago. 22 p.

Chang, Ming-Fang. Static and seismic lateral earth pressures on rigid retaining structures. D, 1981, Purdue University. 488 p.

Chang, N.-Y. Probabilistic approach to the consolidation of varved clay. D, 1976, Ohio State University. 232 p.

Chang, Peter Yun Tsin. The Orange Phyllite and its metamorphism. M, 1928, Yale University.

Chang, Ping-Hsi. Structure and metamorphism of the Bridgewater-Woodstock area, Vermont. D, 1950, Harvard University.

Chang, Pingsheng. Inversion of seismic and gravity data for southern Nevada. M, 1986, Colorado School of Mines. 127 p.

Chang, Pyoung Wuck. Parametric study of vibratory densification of granular soils. D, 1982, Rutgers, The State University, New Brunswick. 223 p.

Chang, Shih-Bin Robin. Bacterial magnetite in sedimentary deposits and its geophysical and paleoecological implication. D, 1988, California Institute of Technology. 263 p.

Chang, Susie Yung. The surficial geology of the Mount Vernon, Illinois area. M, 1974, Southern Illinois University, Carbondale. 52 p.

Chang, Syhhong. Complete wavefield modeling and seismic inversion for lossy-elastic layered half-space due to surface force. D, 1988, University of Southern California.

Chang, Tien-Po. Landslide investigation techniques. M, 1971, [Colorado State University].

Chang, Tien-Po. Watershed model study using Landsat image data. D, 1981, Colorado State University. 331 p.

Chang, Tien-Show. An X-ray study of heat-treated chlorites. M, 1974, Pennsylvania State University, University Park. 50 p.

Chang, Ting Pao. Statistical analysis of meandering river geometry. D, 1969, Purdue University. 177 p.

Chang, Ting-Chieh. Crack growth in an elastic-primary creeping material. D, 1985, Ohio State University. 198 p.

Chang, Toshi. Fourier transform window migration applied to data of COCORP, Socorro, New Mexico, survey. M, 1981, University of Texas at Dallas. 108 p.

Chang, Wen-Fong. Reverse time migration of offset VSP data using the excitation time imaging condition. M, 1985, University of Texas at Dallas. 71 p.

Chang, Yi-Maw. Morphological variation of the foraminifera Ammonia beccarii (Linne) from the At-lantic Coast of the United States. D, 1973, University of Kansas. 84 p.

Chang, Yi-Maw. Paleoecologic study of Pleistocene foraminifera from an offshore well, Ship Shoal area, Louisiana. M, 1965, Washington University. 28 p.

Chang, Yung-Kang Mark. Diamond synthesis. M, 1972, University of Rochester.

Changkakoti, Amarendra. Mineralogic, fluid inclusions and isotopic investigations of the Great Bear Lake silver deposits, Northwest Territories. D, 1985, University of Alberta. 240 p.

Channabasappa, Kenkere C. An integrated study of petrographic carbonization and chemical properties of some bituminous coal seams from the Appalachian region. D, 1954, Pennsylvania State University, University Park. 367 p.

Chanpong, R. R. Depositional environment of the Ardath Shale, San Diego, California. M, 1975, San Diego State University.

Chanton, Jeffrey Paul. Sulfur mass balance and isotopic fractionation in an anoxic marine sediment. D, 1985, University of North Carolina, Chapel Hill. 406 p.

Chao, Benjamin Feng. Symmetry, excitation and estimation in terrestrial spectroscopy. D, 1981, University of California, San Diego. 123 p.

Chao, Chung-Huei. Surface wave profiling for 3-dimensional shear wave structure of the Pacific Ocean basin. D, 1984, University of Southern California.

Chao, Edward C. T. Granitization and related geochemical changes due to cation diffusion. D, 1948, University of Chicago. 111 p.

Chao, George Yien-Chi. The crystal structure of orthorhombic D K Al Ge$_3$O$_8$ and its relation to feldspar structure. D, 1958, University of Chicago. 27 p.

Chao, Jih-Yuh. Study of the Plum River fault zone in Fairfield Township, Jackson County, Iowa. M, 1980, University of Iowa. 101 p.

Chao, Pao-Chin. Effect of supersaturation on the kinetics of the gypsum-anhydrite transition. M, 1969, University of New Mexico. 80 p.

Chao, Tsu Ko. Quantitative analysis of near fields in a halfspace. D, 1969, Colorado School of Mines. 93 p.

Chao, Tsu Ko. Source study for two dimensional modeling. M, 1967, Colorado School of Mines. 44 p.

Chao, Tze. Sulfatization of alumina with gaseous sulfur trioxide and its kinetics. D, 1966, Pennsylvania State University, University Park. 165 p.

Chaoka, Thebeyame R. The geology, petrology and geochemistry of the Kanye volcanics and Nnywane Formation; Ramotswa area, Southwest Botswana. M, 1988, Memorial University of Newfoundland. 137 p.

Chaouai, Nour-Eddine. Etude géostatistique de Mount Leyshon Mine (Autralie). M, 1988, Ecole Polytechnique. 200 p.

Chapel, J. D. Petrology and depositional history of Devonian carbonates in Ohio. D, 1975, Ohio State University. 290 p.

Chapelle, Francis H. The behaviour of selected trace metals in trench-disposed sewage sludge; a case study in Montgomery County, Maryland. M, 1979, George Washington University.

Chapelle, Francis Hughes. Hydrogeology, digital solute-transport simulation, and geochemistry of the Lower Cretaceous aquifer system near Baltimore, Maryland. D, 1984, George Washington University. 234 p.

Chapietta, Richard. Palynologic correlation of two subsurface sections of Ghana, West Africa. M, 1960, New York University.

Chapin, Charles E. Geologic and petrologic features of the Thirty-Nine Mile volcanic field (Fremont County), central Colorado; Part 1, General geology; Part 2, Correlation of ash-flow tuffs by the use of uptic axial angles of alkali feldspar crystals. D, 1965, Colorado School of Mines. 177 p.

Chapin, Daivd A. Geological interpretations of a detailed Bouguer gravity survey of the Chattolanee Dome, near Baltimore, Maryland. M, 1981, Lehigh University.

Chapin, Douglas William. Source identification of tar deposits from the Santa Barbara Channel coastline by trace metal analysis. M, 1972, University of California, Santa Barbara.

Chapin, Mark A. Analysis of Laramide basement-involved deformation, Fra Cristobal Range, New Mexico. M, 1986, Colorado School of Mines. 92 p.

Chapin, Robert Ira. Short-term variations, sampling techniques, and accuracy of analyses of the concentration of nitrate in produced municipal ground waters, North Texas. M, 1981, University of Texas, Austin. 105 p.

Chapin, Theodore. Geology of the foothills of the Rincon Mountains, Tucson Quadrangle, Pima County, Arizona. M, 1913, [Stanford University].

Chapin, Thomas Scott. Geology of Fort Burgwin Ridge, Taos County, New Mexico. M, 1981, University of Texas, Austin. 151 p.

Chaplin, Catherine Elizabeth. Isotope geology of the Gloserheia granite pegmatite, South Norway. M, 1981, University of Alberta. 111 p.

Chaplin, James R. Paleoecology of the Pamlico Formation in the Nixonville Quadrangle, Horry County, South Carolina. M, 1962, University of Houston.

Chaplinsky, Peter P., Jr. Deformational aspects and tectonic setting of the Fish Creek-Soda Creek mylonite zone. M, 1987, University of Iowa. 127 p.

Chapman, Bradley. Soil gas geochemistry over the Mount Taylor uranium deposit. M, 1986, University of Pittsburgh.

Chapman, Carleton Abramson. Geology of the Mascoma Quadrangle, New Hampshire. D, 1937, Harvard University.

Chapman, Clark Russell. Surface properties of asteroids. D, 1972, Massachusetts Institute of Technology. 392 p.

Chapman, Curtis R. A biometrical study of five species of enteletacean brachiopods from the Upper Ordovician. M, 1979, Miami University (Ohio). 139 p.

Chapman, David Spencer. Heat flow and heat production in Zambia. D, 1976, University of Michigan. 100 p.

Chapman, Diana Ferguson. Petrology and structure of the Byram Cove synform Precambrian highlands, New Jersey. M, 1966, Rutgers, The State University, New Brunswick. 118 p.

Chapman, Diana Ferguson. Petrology, structure and metamorphism of a concordant granodiorite gneiss in the Grenville Province of southeastern Ontario. D, 1968, Rutgers, The State University, New Brunswick. 107 p.

Chapman, Diane. Distribution of benthonic foraminiferal thanatocoenoses, Arlington Reef complex, Great Barrier Reef, Australia. M, 1974, Duke University. 193 p.

Chapman, Donald Harding. Late-glacial and post-glacial history of the Champlain region. D, 1931, University of Michigan.

Chapman, Edward Dewey. Structure and tectonics of the Arctic region. M, 1973, Massachusetts Institute of Technology. 119 p.

Chapman, Jeannette Burgen. Hydrogeochemistry of the unsaturated zone of a salt flat in Hudspeth County, Texas. M, 1984, University of Texas, Austin.

Chapman, Jimmy Lee. A model for evaluating the economics of offshore oil field development. M, 1974, University of Oklahoma. 84 p.

Chapman, John Gary. Stratigraphy of the basal Atoka Formation, Washington and Crawford counties, Arkansas. M, 1978, University of Arkansas, Fayetteville.

Chapman, John J. The geology of the eastern Clear Creek-Golden Gate Canyon area, Jefferson County, Colorado. M, 1948, Colorado School of Mines. 74 p.

Chapman, John Judson. Sand distribution in the Cypress Formation, Clay County and vicinity, Illinois. D, 1953, University of Illinois, Urbana. 74 p.

Chapman, John S. Conodonts from the Manticoceras Zone, Confusion Range, Millard County, Utah. M, 1958, University of Kansas. 53 p.

Chapman, Kenneth R. Conodonts from the Lower Ordovician of central Mo. M, 1984, University of Missouri, Columbia. 74 p.

Chapman, M. E. D. Shape of the ocean surface and implications for the Earth's interior. D, 1979, Columbia University, Teachers College. 210 p.

Chapman, Mary G. Depositional and compositional aspects of volcanogenic clasts in the upper member of the Carmel Formation, southern Utah. M, 1987, Northern Arizona University. 93 p.

Chapman, R. P. Evaluation and comparison of different statistical and computerized methods of interpreting multi-element geochemical drainage data. M, 1973, University of New Brunswick.

Chapman, R. S. G. Some geological exploration criteria for lode-gold deposits in British Columbia, Canada. M, 1981, Queen's University. 74 p.

Chapman, Ralph Ebener. An ecological explanation for growth rules in fossil and Recent arthropods. M, 1977, University of Rochester.

Chapman, Randolph Wallace. Geology of the Percy Quadrangle in New Hampshire. D, 1934, Harvard University.

Chapman, Raymond Scott. A numerical simulation of two-dimensional separated flow in a symmetric open-channel expansion using the depth-integrated two-equation (k-ε) turbulence closure model. D, 1982, Virginia Polytechnic Institute and State University. 111 p.

Chapman, Rodger Hale. Gravity methods in iron ore prospecting. D, 1956, University of Wisconsin-Madison. 138 p.

Chapman, Ruthven Hoyt. Geology of the Lake Guija District, El Salvador, Central America. M, 1957, University of Texas, Austin.

Chapman, Stanley Lane. Quartz, vermiculite, and montmorillonite determination, and clay mineralogy of selected soils from the north central United States. D, 1970, University of Wisconsin-Madison. 178 p.

Chapman, William Brewer and Fitzpatrick, Robert Charles. Terrestrial tides as determined from deviations of gravity at Ann Arbor, Michigan. M, 1950, University of Michigan.

Chapman, William Frank. Glacial and postglacial drainage changes in the Mascoma River system, west-central New Hampshire. M, 1968, University of Michigan.

Chapman, William Frank. Glacial history of the northern part of the Hoback River basin, west-central Wyoming. D, 1972, University of Michigan.

Chapman, Willie Eugene. Ore occurrence and controls of ore deposition in the Jeff Price Mine, Cave-In-Rock, Illinois. M, 1971, University of Alabama.

Chapman, Wilson A. A study of the pressure-volume relationship of an underground gas storage reservoir. M, 1948, University of Wisconsin-Madison.

Chapman, Winifred M. A study of feldspar twinning in a differentiated sill. M, 1936, University of Wisconsin-Madison.

Chapnick, Susan D. Manganese oxidation; geochemical and microbiological studies in Oneida Lake, New York. M, 1980, University of South Carolina. 45 p.

Chappars, Michael S. The geology of the Corinth region of Grant and Scott counties, Kentucky. M, 1930, Ohio State University.

Chappell, D. F. Comparative cytology of Cephaleuros, Dunaliella, and Helicodictyon; three green algae of uncertain taxonomic affinities. D, 1977, Miami University (Ohio). 130 p.

Chappell, George A., Sr. Identification and correlation of the fluvial terraces in the Ohio River valley near Proctorville, Ohio and Cox Landing, West Virginia. M, 1988, Marshall University. 148 p.

Chappell, J. F. The Clare River structure and its tectonic setting. M, 1978, Carleton University. 184 p.

Chappell, Walter Miller. Endogenetic alteration of the Camas Land irruptive, Chelan County, Washington. M, 1931, University of Washington. 37 p.

Chappell, Walter Miller. Geology of the Wenatchee quadrangle, Washington. D, 1936, University of Washington. 249 p.

Chappelle, John C. Mineralization in the Bear River Range, Utah-Idaho. M, 1975, Utah State University. 63 p.

Chapple, William Massee. Mathematical study of finite-amplitude rock folding. D, 1964, California Institute of Technology. 176 p.

Chapra, Steven Christopher. Long-term models of interactions between solids and contaminants in lakes. D, 1982, University of Michigan. 215 p.

Chapuis, Ralph A. Sedimentary response to climatic changes as recorded in deep sea piston cores from the Southern Ocean. M, 1974, Florida State University.

Chapusa, Frank Winthrope Peter. Geology and structure of Stansbury Island (S.W. Great Salt Lake, Utah). M, 1969, University of Utah. 83 p.

Charania, Eqbalali Hassanali. Study of pile uplift characteristics in swelling clays using a newly developed test. D, 1983, Arizona State University. 367 p.

Charbeneau, Randall J. Finite element modeling of groundwater injection-extraction systems. D, 1978, Stanford University. 128 p.

Charbonneau, Jean-Marc. Analyse structurale des tectonites métamorphiques du Groupe de Oak Hill dans la région de Saint-Sylvestre, Appalaches du Québec. M, 1975, Universite Laval.

Charbonneau, Robert. Géochronologie Rb-Sr du domaine granulitique de Pikwitonei, Manitoba. M, 1981, Universite de Montreal.

Chareonsri, Prachon. A geochemical and microscopic study of a recently discovered deposit in southern Illinois fluorspar district. M, 1975, University of Missouri, Rolla.

Charest, Marc H. Petrology, geochemistry and mineralization of the New Ross area, Lunenburg County, Nova Scotia. M, 1976, Dalhousie University. 292 p.

Charland, Anne. Sedimentologie d'une sequence pyroclastique de fond Canonville, Montagne Pelee, Martinique. M, 1986, Universite de Montreal.

Charles, Donald Franklin. Studies of Adirondack Mountain (N.Y.) lakes; limnological characteristics and sediment diatom-water chemistry relationships. D, 1982, Indiana University, Bloomington. 428 p.

Charles, Homer H. The oil and gas resources of Anderson County, Kansas. M, 1927, University of Kansas. 109 p.

Charles, John Roy. Underground water supply of the City of Paris. M, 1937, University of Pittsburgh.

Charles, Maureen E. The maximum post-Pleistocene marine level of Maine. M, 1949, Smith College. 109 p.

Charles, Michel. Velocity and stress distributions in the Earth's mantle due to secular variation of the geomagnetic field. D, 1969, New Mexico State University, Las Cruces. 86 p.

Charles, Raymond Grover. The petrology of the solid silica plus feldspar fraction of Recent interbedded muds and sands. M, 1977, University of Oklahoma. 117 p.

Charles, Robert John. Hydrogeology of a proposed surface lignite mine, southwestern Harrison County, Texas. M, 1979, Texas A&M University. 126 p.

Charles, Robert W. The physical properties and phase equilibria of the hydrous Mg-Fe richterites. D, 1972, Massachusetts Institute of Technology. 157 p.

Charles, William Curtis. The east border of the Durham Triassic basin of North Carolina. M, 1959, University of North Carolina, Chapel Hill. 47 p.

Charleston-Aviles, Santiago. Some Aptian cephalopods from the La Pena Formation of Serrania del Burro, Coahuila and the Nazas Valley, Durango, Mexico (Lower Cretaceous). M, 1966, University of Michigan.

Charleston-Aviles, Santiago. Stratigraphy, tectonics, and hydrocarbon potential of the lower Cretaceous, Coahuila Series, Coahuila, Mexico. D, 1974, University of Michigan.

Charlesworth, Lloyd James, Jr. Bay, inlet and nearshore marine sedimentation; Beach Haven-Little Egg Inlet region, New Jersey (coast). D, 1968, University of Michigan. 614 p.

Charlesworth, Lloyd James, Jr. Case-hardening of the (Upper Cretaceous) Hygiene Sandstone. M, 1958, University of Michigan.

Charlet, Laurent. Anion effects on Na-Ca and Na-Mg exchange on Wyoming bentonite at 298.15°K. M, 1982, University of California, Riverside. 59 p.

Charleton, F. Studies of the genus Amphistegina. M, 1927, Columbia University, Teachers College.

Charletta, Anthony Charles. Dinoflagellate biostratigraphy of the Upper Cretaceous Navesink Formation, New Jersey coastal plain. M, 1976, Rutgers, The State University, Newark. 33 p.

Charletta, Anthony Charles. Eocene benthic foraminiferal paleoecology and paleobathymetry of the New Jersey continental margin. D, 1980, Rutgers, The State University, New Brunswick. 84 p.

Charlewood, G. H. The nature of carbonates in altered rocks and in veins. D, 1933, University of Toronto.

Charlie, W. A. Two cut slopes in fibrous organic soils; behavior and analysis. D, 1975, Michigan State University. 259 p.

Charlton, David S. The ecological and sedimentological history of a lake in northern Wisconsin. M, 1969, University of Wisconsin-Madison.

Charlton, David Samuel. The characterization and evolution of carbonate tidal deltas, upper Florida Keys. D, 1981, University of Wisconsin-Madison. 421 p.

Charlton, Douglas W. Structures of the protrusive rhyodacite domes of the Julcani volcanic center, Peru. M, 1974, University of Nevada. 61 p.

Charmatz, Richard. Aquitanian ostracodes from Victoria, Australia. D, 1967, New York University. 109 p.

Charmatz, Richard. Recent foraminifera of Long Island Sound. M, 1961, New York University.

Charoen-Pakdi, Dawaduen. Petrographic and petrophysical characteristics of the Upper Cretaceous Turner Sandy Member of the Carlile Shale, Todd oil field, Weston County, Wyoming. M, 1988, South Dakota School of Mines & Technology.

Charpentier, Ronald Russell. Chitinozoans of the Middle Ordovician Decorah Subgroup in Iowa and Minnesota. M, 1976, University of Wisconsin-Madison.

Charpentier, Ronald Russell. Conodonts through time and space; studies in conodont provincialism. D, 1983, University of Wisconsin-Madison. 167 p.

Chartier, Torrie A. Detailed record of SO_2 emissions from PU'u 'O'o between episodes 33 and 34 of the 1983-86 East Rift Zone eruption of Kilauea, Hawaii. M, 1986, Michigan Technological University. 119 p.

Chartrand, Francis. Evolution diagénétique des dépôts stratiformes de cuivre du Lac Coates, ceinture cuprifère de Redstone, T. du N.O., Canada. M, 1982, Ecole Polytechnique. 137 p.

Chartrand, Francis. Evolution diagénétique des dépôts stratiformes de cuivre du Lac Coates, ceinture cuprifère de Redstone, T. du N.O., Canada. M, 1981, Universite de Montreal. 137 p.

Chartrand, Francis. Evolution diagénétique des dépôts stratiformes de la ceinture de cuivre de Redstone, Territoires du Nord-Ouest, Canada et de Kamoto, Shaba, Zaire. D, 1987, Ecole Polytechnique. 326 p.

Charukalas, Banhan. Hydraulic conductivity of sandstone under different confining pressures. M, 1976, New Mexico Institute of Mining and Technology.

Charukalas • Cheang

Charukalas, Benjamin. Strength variation of limestone. M, 1976, New Mexico Institute of Mining and Technology.

Charusiri, Boonsiri. Ore mineralogy of manganese deposits in Thailand. M, 1988, Queen's University. 143 p.

Charvat, William A. The nature and origin of the bentonite-rich Eagle Ford rocks, central Texas. M, 1985, Baylor University. 220 p.

Chary, K. Narasimha *see* Narasimha Chary, K.

Chase, Carol. The Holocene geologic history of the Maurice River Cove and its marshes, eastern Delaware Bay, New Jersey. M, 1979, University of Delaware.

Chase, Clement Grasham. Tectonic history of the Fiji plateau. D, 1970, University of California, San Diego. 95 p.

Chase, Duane D. Radon distribution controls and possible sources in the Gilbertown oil field area, Choctaw County, Alabama. M, 1984, University of Alabama. 80 p.

Chase, Gerald Warren. The igneous rocks of the Roosevelt area, Oklahoma. M, 1950, University of Oklahoma. 108 p.

Chase, H. D. The mafic associates of the Belchertown Tonalite (Massachusetts, Connecticut). M, 1932, Massachusetts Institute of Technology. 34 p.

Chase, Jack S. Operation UP-SAILS; sub-bottom profiling in Lake Champlain (Vermont). M, 1972, University of Vermont.

Chase, Leonard Richard. Secondary manganese deposits of Pennsylvania. M, 1955, Pennsylvania State University, University Park. 119 p.

Chase, Livingston. Appalachian magnetite deposits in relation to tectonics and metamorphism. M, 1961, University of Illinois, Urbana.

Chase, Richard Lionel St. Lucien. The Itamaca Complex, the Panamo Amphibolite, and the Guri Trondhjemite; Precambrian rocks of the Adjuntos-Panamo Quadrangle, State of Bolivar, Venezuela. D, 1963, Princeton University. 224 p.

Chase, Robert Gordon. A subsurface irrigated, controlled traffic, no-tillage system. D, 1982, University of Hawaii.

Chase, Robert Perkins. Theoretical wave competence in suspension of sea floor sediments. M, 1972, Millersville University.

Chase, Robert Perkins. Transport dynamics and kinematic behavior of aquatic particulates. D, 1977, University of Chicago. 174 p.

Chase, Ronald Buell. Geology of lower Sweathouse Creek Canyon, Bitterroot Range, Ravalli County, Montana. M, 1961, University of Montana. 83 p.

Chase, Ronald Buell. Petrology of the northeastern border zone of the Idaho Batholith, Bitterroot Range, Montana. D, 1968, University of Montana. 189 p.

Chase, Terry Lee. The variation in growth habits and ecology of the Stony corals from Don Quixote Bank, Florida Bay. M, 1973, University of Michigan.

Chasen, Edith A. Crystal habit as related to seed posture during growth. M, 1970, Boston University. 63 p.

Chasey, Kenneth LeMay. Preliminary report on the geology and structure of the Sierra Nevada west of Bishop, California. M, 1933, University of Rochester. 44 p.

Chastain, Charlette Elizabeth. Hydrodynamic processes and sea cliff/platform erosion, Favorite Channel, Alaska. D, 1976, University of California, Santa Cruz. 78 p.

Chatarpaul, L. Laboratory studies on nitrogen transformations in stream sediments. D, 1978, University of Guelph.

Chatelain, Edward Ellis. Ammonoids of the Marmaton Group Middle Pennsylvanian (Desmoinesian), Arkoma Basin, Oklahoma. D, 1984, University of Iowa. 296 p.

Chatelain, Edward Ellis. Autecology of selected genera of Mississippian, Permian, and Triassic ammonoids; analysis of coiling geometries. M, 1977, Utah State University. 168 p.

Chatellier, Jean-Yves. Sedimentology of the Mississippian Banff Formation in southwestern Alberta mountains and plains. M, 1983, University of Calgary. 159 p.

Chatfield, Ernest J. Stratigraphy of western Ferrisburg, Addison County, Vermont. M, 1959, Columbia University, Teachers College.

Chatham, Ernest Walter, Jr. Upper Cretaceous foraminifer genus Globotruncana and its stratigraphic distribution. M, 1950, University of Texas, Austin.

Chatham, James Randall. The applications of solution-mineral equilibria concepts in prospecting for sandstone-type uranium deposits. D, 1981, Colorado School of Mines. 177 p.

Chatham, Randall James. A study of uranium distribution in an upper Jackson lignite-sandstone ore body, South Texas. M, 1979, Texas A&M University.

Chatterjee, A. K. Mineralization and associated wall rock alteration in the George River Group, Cape Breton Island, Nova Scotia. D, 1980, Dalhousie University. 197 p.

Chatterjee, Biswanath. Paleomagnetism of the Crater Creek basalts, Unmak Island, Aleutian Islands, Alaska. M, 1971, University of Alaska, Fairbanks. 85 p.

Chatterjee, Subir K. Geochemistry of hornblende gneiss and associated rocks of the Ralston Butte, Colorado. M, 1981, SUNY at Buffalo. 91 p.

Chaturvedi, Lokeshwa Nathr N. Geological structure and its effect on the geothermal hydrology in southwestern Hreppar, Iceland. D, 1969, Cornell University. 180 p.

Chatwin, Stephen C. Permafrost aggradation and degradation in a sub-Arctic peatland. M, 1981, University of Alberta. 163 p.

Chatzis, Ioannis. A network approach to analyze and model capillary and transport phenomena in porous media. D, 1980, University of Waterloo.

Chaudet, Roy Edward. The petrology and geochemistry of precaldera magmas, Long Valley Caldera, eastern California. M, 1986, Virginia Polytechnic Institute and State University.

Chaudhri, Ata-ur-Rehman. Analysis of hydrologic performance tests in unconfined aquifers. M, 1966, University of Arizona.

Chaudhry, Mohammad Naveed Hayat. The effects of moisture, cation concentration, temperature, density and composition of soils on their electrical resistivity. M, 1984, Brock University. 151 p.

Chaudhuri, Sanbhudas. The geochronology of the Keweenawan rocks of Michigan and the origin of the copper deposits. D, 1966, Ohio State University. 145 p.

Chauff, K. M. Multielement conodont species from the Osagean (Early Mississippian) Burlington carbonate shelf, Midcontinent North America, and the Chappel Limestone of Texas. D, 1978, University of Iowa. 182 p.

Chauff, K. M. The micropaleontology of the Ordovician-Devonian unconformity of east central Missouri. M, 1973, Washington University. 155 p.

Chauhan, Ehsanul Haque. Application of emission spectrographic analysis of soils in geochemical exploration of the Lakeview District, Bonner County, Idaho. M, 1968, University of Idaho. 116 p.

Chauvel, Jean Paul. The geology of the Arroyo Parida Fault, Santa Barbara and Ventura counties, California. M, 1958, University of California, Los Angeles.

Chauvin, Aaron L. Geology of the Pontotoc Northwest area, San Saba County, Texas. M, 1962, Texas A&M University.

Chauvin, E. Noel. The geology of the East River Mountain area, Giles County, Virginia. M, 1958, Virginia Polytechnic Institute and State University.

Chauvin, Edward N. The geology of the East River Mountain area, Giles County, Virginia. M, 1957, Virginia Polytechnic Institute and State University.

Chauvin, Luc. Etude sédimentologique des dépôts glacio-estuariens de la vallée de la rivière Sainte Anne, Gaspésie. M, 1977, Universite de Montreal.

Chauvot, Isabelle P. Study of the gold deposits at the War Eagle Mine, Idaho County, Idaho. M, 1986, Oregon State University. 143 p.

Chave, Alan Dana. Applications of time series analysis to geophysical data. D, 1980, Woods Hole Oceanographic Institution. 179 p.

Chave, Keith Ernest. Some aspects of the biogeochemistry of magnesium. D, 1952, University of Chicago. 73 p.

Chavez D., Eduardo E. Evaluation of Landsat digital data for a hierarchical structure of soil survey in Venezuela. D, 1984, Colorado State University. 116 p.

Chavez Martinez, Mario Luis. A potential supply system for uranium based upon a crustal abundance model. D, 1982, University of Arizona. 509 p.

Chavez Rodriguez, Adolfo. Modeling mountain-front recharge to regional aquifers. D, 1987, University of Arizona. 177 p.

Chavez, David Eliseo. Surface wave dispersion in the Great Basin and northern Central America. M, 1980, University of Nevada. 66 p.

Chavez, Joel Edmund. The relationship of sulfur gases desorbed from surface soils to massive sulfide and elemental sulfur mineralization at depth. M, 1986, University of Idaho. 117 p.

Chavez, Raul. Mexican oil fields. M, 1921, University of Missouri, Rolla.

Chavez, William Xavier, Jr. Geologic setting and the nature and distribution of disseminated copper mineralization of the Mantos Blancos District, Antofagasta Province, Chile. D, 1985, University of California, Berkeley. 148 p.

Chavez-Martinez, Marlo Luis. The unit regional value of the mineral resources of Mexico. M, 1978, Pennsylvania State University, University Park. 288 p.

Chavez-Quirarte, Ramon. Stratigraphy and structural geology of Sierra de Sapello, northern Chihuahua, Mexico. M, 1986, University of Texas at El Paso.

Chavez-Sequra, Rene Efrain. Simultaneous inversion of gravity and magnetic data using the linear programming method. D, 1985, University of Toronto.

Chawner, William D. The geology of Catahoula and Concordia parishes. D, 1937, Louisiana State University.

Chawner, William D. The Montrose-La Crescenta flood of January 1, 1934, and its sedimentary aspects. M, 1934, California Institute of Technology. 42 p.

Chayes, Felix. Alkaline and carbonate intrusives near Bancroft, Ontario. D, 1942, Columbia University, Teachers College.

Chayes, Felix. Geology of the alkaline intrusives in the immediate vicinity of Bancroft, Hastings County, Ontario. M, 1939, Columbia University, Teachers College.

Chazal, Suzanne Marie de *see* de Chazal, Suzanne Marie

Chazen, Stephen I. Distribution of titanium and phosphorus in oceanic basalts as a test of origin. D, 1973, Michigan State University. 34 p.

Cheadle, Burns Alexander. Stratigraphy and sedimentation of the middle Proterozoic Sibley Group, Thunder Bay District, Ontario. D, 1986, University of Western Ontario. 435 p.

Cheadle, S. A gravity study of an Archean crustal segment near Thunder Bay, Ontario. M, 1982, Lakehead University.

Cheadle, Scott Philip. Applications of physical modeling and localized slant slacking to a seismic study of subsea permafrost. D, 1988, University of Calgary. 237 p.

Cheang, K. K. Structure and polytypism in synchysite and parisite from Mont St. Hilaire, Quebec. M, 1977, Carleton University. 82 p.

138

Bibliography of Geoscience Theses

Cheang, Kok Keong. Oxygen isotope, fluid inclusion, microprobe and petrographic studies of the Precambrian granites from the southern Wind River Range and the Granite Mountains, central Wyoming, U.S.A.; constraints on origin, hydrothermal alteration and uranium genesis. D, 1982, University of Georgia. 154 p.

Cheatham, Bruce Ned. Geology of the Mountain Creek Lake area, Dallas County, Texas. M, 1955, University of Oklahoma. 49 p.

Cheatham, Michael M. An isotopic study of the Chain Lakes Massif, Maine. M, 1985, University of New Hampshire. 163 p.

Cheatham, Terri L. The enrichment in the Conway Granite, Mad River Pluton, New Hampshire. M, 1985, University of New Hampshire. 119 p.

Cheatham, Thomas L. Paleoenvironments of the upper Albian Stage (Cretaceous) of eastern Trans-Pecos, Texas. M, 1983, Stephen F. Austin State University. 153 p.

Cheatum, Craig E. A microearthquake study of the San Jacinto Valley, Riverside County, California. M, 1974, University of California, Riverside. 59 p.

Chebaane, Mohamed. Stochastic modeling of seasonal intermittent streamflow in arid and semi-arid areas. D, 1988, Colorado State University. 352 p.

Cheek, Catherine A. A remote sensing study of the volcanoes of the Central Andes between 21° and 24°S. M, 1988, University of Texas, Arlington. 245 p.

Cheek, Robert B. Sulfur facies of the upper Freeport coal (Pennsylvanian) of northwestern Preston County, West Virginia. M, 1969, West Virginia University.

Cheek, William M., Jr. Geology of the southern half of the Romney Quadrangle, West Virginia. M, 1980, University of Akron. 42 p.

Cheel, Richard James. Late Quaternary glacio-marine deposits of the Stittsville area, near Ottawa, Canada. M, 1979, University of Ottawa. 154 p.

Cheel, Richard James. Sediment transport and deposition under upper flow regime plane bed conditions. D, 1984, McMaster University. 183 p.

Cheema, M. R. Sedimentation and gas production of the Upper Devonian Benson sand in North-central West Virginia; a model for exogeosynclinal mid-fan turbidites off a delta complex. D, 1977, West Virginia University. 127 p.

Cheeseman, Raymond J. The geology of the Webb Mountain District, Gila Bend Mountains, Maricopa County, southwestern Arizona. M, 1974, Northern Arizona University. 69 p.

Cheesman, William C. The geology and copper mineralization at Poison Mountain, British Columbia. M, 1957, University of Oregon. 72 p.

Cheetham, Alan. Late Eocene zoogeography of the eastern Gulf Coast region. D, 1959, Columbia University, Teachers College. 232 p.

Cheetham, Robert Nelson. General geology of the Del Rio area, Texas. M, 1949, University of Iowa. 169 p.

Cheevers, Craig W. Stratigraphic analysis of the Kaibab Formation in northern Arizona, southern Utah and southern Nevada. M, 1980, Northern Arizona University. 144 p.

Cheevers, Craig Wallace. Weathering, genesis and classification of selected basaltic soils of the San Francisco volcanic field, northern Arizona. D, 1982, University of California, Riverside. 233 p.

Chege, A. M. Effect of organic material concentration and placement on soil aggregation. M, 1979, University of Guelph.

Chehata, Mondher. Spectral domain decomposition method for the solution of the groundwater flow equation. D, 1988, Colorado State University. 207 p.

Chelikowsky, Joseph R. Geologic distribution of fire clays in the United States. D, 1935, Cornell University. 7 p.

Chelikowsky, Joseph R. Investigation of unsupported fragments in veins. M, 1932, Cornell University.

Chelini, J. M. Market study and compendium of data on industrial minerals and rocks of Montana. M, 1966, Montana College of Mineral Science & Technology. 190 p.

Chelius, Carl Robert. Vertical velocities in continental cumuli. M, 1973, Pennsylvania State University, University Park. 51 p.

Chellman, Kathryn King. Provenance of some nearshore marine sandstones in south-central New York. M, 1983, University of Rochester. 55 p.

Chelton, Dudley Boyd, Jr. Low frequency sea level variability along the west coast of North America. D, 1980, University of California, San Diego. 231 p.

Chen Mao Ge *see* Mao Chen Ge

Chen, Albert T. F. Plane strain and axisymmetric primary consolidation of saturated clays. D, 1967, Rensselaer Polytechnic Institute. 180 p.

Chen, C. Y. Investigation and statistical analysis of the geotechnical properties of coal mine refuse. D, 1976, University of Pittsburgh. 210 p.

Chen, Cary Ching-chi. A gravity survey of a portion of the Cuyahoga River valley, Summit County, Ohio. M, 1982, University of Akron. 137 p.

Chen, Chang. Structural comparison between the Santa Monica and Santa Ana mountains, Southern California; a strain evaluation approach. M, 1983, University of California, Los Angeles. 58 p.

Chen, Chang-Sen. An investigation of geological and geochemical characteristics of late Quaternary sediments in the Georgian Bay region, southern Ontario. M, 1981, Brock University. 180 p.

Chen, Chao-Hsia. Stochastic estimation of standard free energies of formation of silicate minerals. M, 1973, University of California, Berkeley. 64 p.

Chen, Chen-Hong. Petrology, geochemistry, and petrogenesis of ultramafic xenoliths from 1800-1801 Kaupulehu flow, Hualalai Volcano, Hawaii. D, 1986, University of Texas at Dallas. 103 p.

Chen, Cheng-Hsing. Seismic modelling of deep foundations. D, 1984, University of California, Berkeley. 127 p.

Chen, Chi-Chieu. Application of thermoluminescence (TL) to mineral exploration based on studies of replacement ore deposits in carbonate host rocks at Charcas, Mexico, Toggenburg, S. W. Africa, and Bisbee, Arizona. D, 1976, Columbia University. 224 p.

Chen, Chia-Shyun. Temperature distribution around a well during thermal injection and a graphical determination of thermal properties of the aquifer. D, 1980, Texas A&M University. 128 p.

Chen, Chih Shan. The petrology of Lower Pennsylvanian Suwannee Sandstone, Lookout Mountain, Alabama and Georgia. M, 1960, Florida State University. 101 p.

Chen, Chin. Ecological studies of shellbearing pteropods from the western part of the North Atlantic Ocean. D, 1962, Boston University. 177 p.

Chen, Chin. Fine scale seismic structure of upper crust on the East Pacific Rise. M, 1985, University of California, Davis. 148 p.

Chen, Chin. The subsurface geology and oil and gas fields of southeastern Michigan. M, 1957, Wayne State University. 100 p.

Chen, Chin Shan. Regional lithostratigraphic analysis of Paleocene and Eocene rocks of Florida. D, 1964, Northwestern University.

Chen, Ching-Rua. Temperature disturbance in a geothermal borehole system. M, 1975, Georgia Institute of Technology. 49 p.

Chen, Christina M. Holocene benthonic foraminiferal distributions and the deep Norwegian Sea overflow. M, 1984, Duke University. 78 p.

Chen, Chu-Yung. Geochemical and petrologic systematics in lavas from Haleakala Volcano, East Maui; implications for the evolution of Hawaiian mantle. D, 1982, Massachusetts Institute of Technology. 344 p.

Chen, D. D.-S. A probabilistic representation for drained creep in clays. D, 1976, McGill University.

Chen, Deng-Bo. Stress effects on some petrophysical properties and fluid pressure generation in sealed reservoir rocks. D, 1982, Colorado School of Mines. 271 p.

Chen, Gianming James. A study of seismicity and spectral source characteristics of small earthquakes; Hansel Valley, Utah, and Pocatello Valley, Idaho areas. M, 1988, University of Utah. 119 p.

Chen, Harold Hwai. Genesis of certain secondary vanadates, molybdates, and associated minerals. D, 1927, Harvard University.

Chen, Hsien Su. The response of sulfide-affiliated copper, nickel, and zinc in shale overburdens associated with coals to simulated leaching. D, 1982, University of South Carolina. 114 p.

Chen, Hsien Wu. Stress-strain and volume change characteristics of tailings materials. D, 1984, University of Arizona. 166 p.

Chen, Hsien-su. Study on apophyllite from Poona, India. M, 1968, Dalhousie University. 64 p.

Chen, Hsiu-hsiung. A hydrologic and economic model for the optimization of irrigation water utilization on the Humboldt River system. D, 1978, University of Nevada. 270 p.

Chen, Hsiu-Kuo. Measurement of water content in porous media under geothermal fluid flow conditions. D, 1977, Stanford University. 162 p.

Chen, Huei-Tsyr Jeremy. Dynamic stiffness of non-uniformly embedded foundations. D, 1984, University of Texas, Austin. 253 p.

Chen, James Chian-Tung. Application of modeling techniques to the study and forecasting of energy needs, energy supply, environmental impacts, the assessment of new technology, and alternatives. D, 1973, University of Oklahoma. 227 p.

Chen, James H-Young. Zircon geochronology of the Sierra Nevada Batholith. D, 1977, University of California, Santa Barbara. 210 p.

Chen, James Huei-Young. U/Th/Pb radiometric investigations of the Allende carbonaceous chondrite. M, 1974, University of California, Santa Barbara.

Chen, Jen-Hwa. A probabilistic approach to seismic soil-structure interaction analysis. D, 1982, University of California, Berkeley. 161 p.

Chen, Jian-Chu. Analysis of local variations in free field seismic ground motion. D, 1980, University of California, Berkeley. 263 p.

Chen, Jin-Song. An exploration for rheological properties of montmorillonite at low electrolyte concentrations and assessment of irrigation practices in northwestern Indiana. D, 1988, Purdue University. 93 p.

Chen, Jing-Jong. Lateral variation of surface wave velocity and Q structure beneath North America. D, 1985, St. Louis University. 249 p.

Chen, Jing-Wen. Stress path effect on static and cyclic behavior of Monterey No. 0/30 sand. D, 1988, University of Colorado. 531 p.

Chen, Jinxing. Depositional environments and diagenesis of the San Andres Formation in Howard-Glasscock Field, Howard County, Texas. M, 1986, Texas Tech University. 95 p.

Chen, John Teh-Jen. A rheological stress-strain-time relationship for Seattle soils. D, 1969, University of Washington. 214 p.

Chen, Joy C. The hydrothermal component in ferro-manganese nodules from the Southeast Pacific Ocean. M, 1985, University of Michigan. 22 p.

Chen, Ju-Chin. Petrofabric studies of some fine grained rocks, by means of X-ray diffraction. M, 1965, Rice University. 130 p.

Chen, Ju-Chin. Petrological and chemical studies of Utuado Pluton (post-Lower Cretaceous pre-lower Miocene), Puerto Rico. D, 1967, Rice University. 188 p.

Chen, K. H. Transient problems in a layered elastic medium due to a finite line and a cylindrical source. D, 1978, Columbia University, Teachers College. 171 p.

Chen, Kuang-Chian. Stochastic model study of diffusion of solid particles. D, 1968, University of Pittsburgh. 109 p.

Chen, Kuang-Hsiang. Dynamic analysis of embedded 3-D foundations by boundary element method. D, 1987, SUNY at Buffalo. 245 p.

Chen, Kung-Yung. The effects of aging at elevated temperatures on certain physical properties of drilling fluids. M, 1957, New Mexico Institute of Mining and Technology. 52 p.

Chen, Li-King. Surface energy determinations in plexiglas. M, 1970, University of Missouri, Rolla.

Chen, Ly Kwong. Genesis of the tungsten deposits of the United States. M, 1935, University of Michigan.

Chen, M. P. Calcareous nannoplankton biostratigraphy and paleoclimatic history of the late Neogene sediments of the Northwest Florida continental shelf. D, 1978, Texas A&M University. 478 p.

Chen, Pei-Hsin. An X-ray diffraction study of Silurian carbonate rocks from the subsurface of Grand Traverse County, Michigan. M, 1969, Wayne State University.

Chen, Pei-Hsin. Post-Paleocene Antarctic Radiolaria; their taxonomy, biostratigraphy and phylogeny, and the development of late Neogene cold-water faunas. D, 1975, Columbia University. 405 p.

Chen, Pei-Yuan. Clay deposits in northwestern Taiwan (Formosa), China. M, 1957, Indiana University, Bloomington. 70 p.

Chen, Pei-Yuan. Geology and mineralogy of the white bentonite beds of Gonzales County, Texas. D, 1968, University of Texas, Austin. 233 p.

Chen, Ping-Fan. Geology and mineral resources of the Goose Creek area near Roanoke, Virginia. D, 1959, Virginia Polytechnic Institute and State University.

Chen, Ping-fan. The geology and oil possibilities of Taiwan, China. M, 1957, University of Cincinnati. 74 p.

Chen, Ran-Jay. Finite element modelling of interface behavior in geologic media. D, 1986, West Virginia University. 254 p.

Chen, Roland Lee-Ping. Ground magnetic study of the Anderson Ridge Quadrangle, Fremont County, Wyoming. M, 1964, University of Missouri, Rolla.

Chen, Rong-Her Jimmy. Three-dimensional stability analysis. D, 1981, Purdue University. 323 p.

Chen, Shiahn-Jauh. Structural geology of the Eureka Complex in the Seven Devils Terrane, eastern Oregon and western Idaho. M, 1985, Rice University. 94 p.

Chen, Shih-Fang. Determination of Fourier site response spectra during earthquakes. M, 1968, University of Washington. 69 p.

Chen, Shih-Tsu. Regional geological engineering mapping by ERTS-1 imagery; a case study on Taiwan. M, 1974, University of Missouri, Rolla.

Chen, Shu-Meei. A chemical, thermogravimetric and X-ray study of cancrinite. M, 1970, McMaster University. 65 p.

Chen, Tai-Shan. Application of partial cross-correlation to extend the depth coverage of Vibroseis seismic reflection data. M, 1986, Purdue University. 96 p.

Chen, Thomas Chui-Tung. Reflection and transmission of obliquely incident Rayleigh waves at vertical discontinuity. D, 1980, Columbia University, Teachers College. 148 p.

Chen, Tse Pu. Sulfide film formation on the surface of chrysocolla under flotation conditions. M, 1961, New Mexico Institute of Mining and Technology. 68 p.

Chen, Tu Kao. Computer solutions to depletion drive hydrocarbon systems. M, 1967, University of Missouri, Rolla.

Chen, Tzann-Hwang. Structural analysis of the Hillabee Schist, Clay County, Alabama. M, 1974, University of Southern Mississippi.

Chen, Tzann-Hwang. The seismicity of forearc marginal wedges (accretionary prisms) and seismotectonics of convergent margins. D, 1981, University of Texas at Dallas. 111 p.

Chen, Tzerhong. Strain analysis of Monterey Formation rocks in the Los Prietos Syncline, Santa Maria Basin, California. M, 1986, University of California, Los Angeles. 101 p.

Chen, Tzong-Tzyy. Cell parameters, optical properties and chemical composition in some natural sodic amphiboles. M, 1969, Carleton University. 98 p.

Chen, Tzong-Tzyy. Compositional and thermal study on natural and synthetic phases in the system Ag_2S-Cu_2S-Sb_2S_3-Bi_2S_3. D, 1971, Cornell University.

Chen, Victor J. A synthetic unit sedimentgraph for ungaged watersheds. D, 1984, Virginia Polytechnic Institute and State University. 132 p.

Chen, Wang-Ping. Seismic studies of central Asia. D, 1979, Massachusetts Institute of Technology. 241 p.

Chen, Wen-Lan. Continuous generation and tyndallmetric measurement of dust clouds, an investigation of a possible method to evaluate mine dust. D, 1950, Pennsylvania State University, University Park. 143 p.

Chen, Wu-Shong. Two-dimensional seismic modelling. M, 1976, University of Tulsa. 102 p.

Chen, Xunhong. Depositional environments and paleogeography of the lower Saugus Formation, northern San Fernando Valley, Los Angeles County, California. M, 1988, California State University, Northridge. 79 p.

Chen, Y. Jurassic and Cretaceous palynostratigraphy of a Madagascar well. D, 1978, University of Arizona. 278 p.

Chen, Yao-Tang. Sedimentation on the Balearic abyssal plain, western Mediterranean Sea. M, 1980, Duke University. 108 p.

Chen, Yohchia. Dynamic soil-structure interaction of reinforced concrete lifelines under earthquake effects. D, 1988, University of Minnesota, Minneapolis. 313 p.

Chen, Yong-Qi. Analysis of deformation surveys; a generalized method. D, 1983, University of New Brunswick.

Cheney, Eric S. Stable isotopic geology of the Gas Hills uranium district, Wyoming. D, 1964, Yale University. 342 p.

Cheney, J. T. Mineralogy and petrology of lower sillimanite through sillimanite + K-feldspar zone pelitic schists, Puzzle Mountain area, N.W. Maine. D, 1975, University of Wisconsin-Madison. 306 p.

Cheney, John Thomas. Petrologic relationships of layered meta-anorthosites and associated rocks, Bass Creek, western Montana. M, 1972, University of Montana. 112 p.

Cheney, Monroe G., Jr. An annotated bibliography of secondary recovery of oil. M, 1952, Columbia University, Teachers College.

Cheney, Richard Stephen. Geophysical investigation of the Raton Basin. M, 1982, Texas Tech University. 75 p.

Cheng, Albert. Cosmic ray interaction with iron meteorites and high energy production of ^{36}Cl. D, 1972, SUNY at Stony Brook.

Cheng, Amy I-Mei. Satellite magnetic survey in southern high latitudes. D, 1988, University of Wisconsin-Madison. 269 p.

Cheng, Amy I-Mei. Structure of the Tjörnsletta area, Wedel Jarlsberg Land, Spitsbergen. M, 1984, University of Wisconsin-Madison.

Cheng, Bih-Ling Monica. A study of geomorphologic instantaneous unit hydrograph. D, 1982, University of Illinois, Urbana. 238 p.

Cheng, Chang-Chi. Hydrologic regionalization based on flow-duration relationships. D, 1988, Syracuse University. 97 p.

Cheng, Chao-nang. Determination of amino acid-n in deepsea sediments and the relationship between amino acid levels in three cores from the Argentine Basin and past climatic changes. D, 1969, University of Illinois, Urbana. 104 p.

Cheng, Ching-Chau Abe. Deformation of the San Andreas fault system inferred from trilateration measurements. D, 1985, University of California, Los Angeles. 386 p.

Cheng, Chiung-Chuan. Determination of crustal Q structure from multimode surface waves. D, 1980, St. Louis University.

Cheng, Chuen Hon Arthur. Seismic velocities in porous rocks; direct and inverse problems. D, 1978, Massachusetts Institute of Technology. 255 p.

Cheng, Chunyuen Raymond. Boundary element analysis of single and multidomain problems in acoustics. D, 1988, University of Kentucky. 209 p.

Cheng, Hung-Darh Alexander. Boundary integral equation formulation for porous-elasticity with applications in soil mechanics and geophysics. D, 1981, Cornell University. 168 p.

Cheng, J.-D. A study of the stormflow hydrology of small forested watersheds in the Coast Mountains of southwestern British Columbia. D, 1976, University of British Columbia.

Cheng, Jauh-Tai. Crystal growth kinetics in K_2O-CaO-SiO_2. D, 1987, Case Western Reserve University. 242 p.

Cheng, Kuei-Yu Yeh. Taxonomic studies of Lower Jurassic Radiolaria from the Nicely Formation, the Hyde Formation, and the Snowshoe Formation, east-central Oregon. D, 1985, University of Texas at Dallas. 335 p.

Cheng, Kuei-Yu Yeh. Upper Triassic Radiolaria from the Fields Creek Formation, John Day Inlier, east-central Oregon. M, 1982, University of Texas at Dallas. 107 p.

Cheng, Mary Mei-Ling Huang. Isotopic thorium and isotopic uranium compositions of modern and fossil marine molluscan shells. D, 1966, Washington University. 92 p.

Cheng, Minkang. Time variations in the Earth's gravity field from Starlette orbit analysis. D, 1988, University of Texas, Austin. 209 p.

Cheng, Mo Chun. Two-dimensional simulation of heat and fluid flow in geothermal systems. D, 1980, University of Oklahoma. 106 p.

Cheng, Shiang-ho. Study on the Earth constants in the central North American continent. D, 1977, St. Louis University. 233 p.

Cheng, Shih-Cheang. Zinc and copper content of stream sediment in south-central Missouri. M, 1965, University of Missouri, Rolla.

Cheng, Shui-Tuang. Overtopping risk evaluation for an existing dam. D, 1982, University of Illinois, Urbana. 214 p.

Cheng, Song-Lin. Application of stable isotopes of oxygen, hydrogen, and carbon to hydrogeochemical studies, with special reference to Canada Del Oro Valley and the Tucson Basin. D, 1984, University of Arizona. 222 p.

Cheng, Song-Lin. The metamorphism and wall rock alteration of the Dixie Prospect, Red Lake, Ontario. M, 1979, Wright State University. 124 p.

Cheng, Wen-Lon. Effects of soil properties on liquefaction potential during earthquakes. D, 1980, University of Washington. 206 p.

Cheng, Yen-Nien. Upper Paleozoic (Upper Devonian and Carboniferous) Radiolaria from the Ouachita Mountains, Oklahoma and Arkansas. D, 1985, University of Texas at Dallas. 540 p.

Cheng, Ying-Yu. Estimation of the depth of focus of the May 10, 1963 Ecuador earthquake. M, 1966, Pennsylvania State University, University Park. 85 p.

Cheng, Yung-Yu. The complex torsional oscillations of an anelastic Earth model. M, 1969, Brown University.

Chenoweth, Philip A. The pegmatite dikes of Manhattan Island, New York. M, 1947, Columbia University, Teachers College.

Chenoweth, William Charles. The sedimentary and igneous rocks, structure, and mineral deposits of the southeastern Bear Lodge Mountains, Crook County, Wyoming. M, 1955, University of Iowa. 220 p.

Chenoweth, William Lyman. The variegated member of the Morrison Formation in the southeastern part of the San Juan Basin, Valencia County, New Mexico. M, 1953, University of New Mexico. 85 p.

Cheriton, Camon Glenn. Correlation of the area including Kimberley, Metaline, and Coeur d' Alene. M, 1949, University of British Columbia. 82 p.

Cheriton, Camon Glenn. Disorder in chalcopyrite. D, 1953, Harvard University.

Cherkauer, Douglas Stuart. An analysis of Holocene valley alluviation and Modern streamflow mechanics, Treia River, Italy. D, 1972, Princeton University. 150 p.

Cherkener, Douglas S. Longitudinal profiles of ephemeral streams in southeastern Arizona. M, 1969, University of Arizona.

Chermak, Andrew. A structural study of the Romanche fracture zone based on geophysical data. M, 1979, University of Miami. 223 p.

Chermak, John Alan. The rates of oxidation of galena and sphalerite in acidic ferric chloride solutions. M, 1986, Virginia Polytechnic Institute and State University.

Chern, How-Hueir. 3-D modeling of transmitted seismic wave energy with geometrical synthetic seismograms. M, 1987, University of Texas at Dallas. 57 p.

Chern, Jin-Ching. Undrained response of saturated sands with emphasis on liquefaction and cyclic mobility. D, 1985, University of British Columbia.

Chern, Laura Allison. Late Pliocene-early Pleistocene paleoclimatology of the Arctic Ocean. M, 1984, University of Wisconsin-Madison.

Cherng, Juling-Chaun. Depositional environment and diagenesis of the Chico Ridge limestone bank (Upper Pennsylvanian), north-central Texas. M, 1980, University of Texas at Dallas. 119 p.

Chernicoff, Stanley Edward. The Superior-lobe and Grantsburg-sublobe tills; their compositional variability in East-central Minnesota. D, 1980, University of Minnesota, Minneapolis. 268 p.

Chernis, Peter J. Microcrack structures in plutonic rocks from the Whiteshell Nuclear Establishment, eastern Manitoba, and Atikokan, northwestern Ontario. M, 1985, Carleton University. 145 p.

Chernoff, C. N. Lithology of the Interlake Group in Saskatchewan. M, 1960, University of Saskatchewan. 65 p.

Chernosky, Joseph V. An experimental investigation of the serpentine and chlorite group minerals in the system $MgO-Al_2O_3-SiO_2-H_2O$. D, 1973, Massachusetts Institute of Technology. 106 p.

Chernosky, Joseph V. Metasomatic zoning at Tamerack lake, Trinity County (California). M, 1969, University of Wisconsin-Madison.

Chernow, Robert Michael. Biogenic controls on the development of magnetic fabric in coastal sediments. M, 1984, University of Georgia. 128 p.

Cherry, Janet G. A water balance and hydrologic analysis on Crumarine Creek watershed. M, 1986, University of Idaho. 120 p.

Cherry, Jesse Theodore, Jr. A line source of SH-waves overlying a plane boundary separating two semi-infinite elastic media. D, 1961, St. Louis University.

Cherry, Jesse Theodore, Jr. The energy contained in a record of ground displacement as determined by the autocorrelation function. M, 1956, St. Louis University.

Cherry, John A. Sand movement along a portion of the Northern California coast. M, 1964, University of California, Berkeley. 150 p.

Cherry, John Anthony. Geology of the Yorkton area, Saskatchewan. D, 1966, University of Illinois, Urbana. 169 p.

Cherry, M. E. The petrogenesis of granites in the St. George Batholith, southwestern New Brunswick, Canada. D, 1976, University of New Brunswick.

Cherry, Philip John. Hydrogeology of the Smyrna-Clayton area, Delaware. M, 1984, University of Delaware. 173 p.

Cherukupalli, Nehru E. Petrology of the olivine-dolerite sill, Sierra Ancha Mountains, Arizona. M, 1960, Columbia University, Teachers College.

Cherven, Victor B. High- and low-sinuosity stream deposits of the Sentinel Butte Formation (Paleocene) McKenzie County, North Dakota. M, 1973, University of North Dakota. 73 p.

Cherven, Victor Bruce. A delta-slope-submarine fan model for the Maastrichtian part of the Great Valley Sequence, southern Sacramento and northern San Joaquin basins, California. D, 1982, Stanford University. 153 p.

Chery, D. L., Jr. An approach to the simplification of watershed models for applications purposes. D, 1976, Utah State University. 207 p.

Cheshier, Roby Albert. Geological features of a mineralized shear zone, Kettle Falls, Stevens County, Washington. M, 1982, Washington State University. 67 p.

Chesley, John Theodore. A combined $^{18}O/^{16}O$ and D/H isotopic study of molybdenite mineralization at Pear Lake and related areas in the Pioneer Batholith, Southwest Montana. M, 1986, Oregon State University. 91 p.

Chesner, Craig Alan. Geochemistry and evolution of the Fuego volcanic complex, Guatemala; constraints on magma chambers at Fuego and other nearby volcanos. M, 1982, Michigan Technological University. 69 p.

Chesner, Craig Alan. The Toba tuffs and caldera complex, Sumatra, Indonesia; insights into magma bodies and eruptions. D, 1988, Michigan Technological University. 428 p.

Chesney, Claybourne. Petrology of Mississippian bioherms in the Pitkin limestone in northern Arkansas. M, 1969, University of Missouri, Columbia.

Chesser, Kevin C. Recognition of sedimentary structures in the Skinner Sandstone of Strauss Field by discriminant analysis. M, 1987, University of Kansas. 67 p.

Chesser, William La Grand. The nature and development of the Esplanade, in the Grand Canyon, Arizona. M, 1971, Brigham Young University. 48 p.

Chester, Frederick Dixon. Crystalline rocks of Delaware. M, 1886, Cornell University.

Chester, Frederick Dixon. The gabbros and associated rocks in Delaware. M, 1887, Cornell University.

Chester, Frederick M. Mechanical properties and fabric of the Punchbowl fault zone, California. M, 1983, Texas A&M University. 143 p.

Chester, Frederick Michael. The transition from cataclasis to intracrystalline plasticity in experimental shear zones. D, 1987, [University of Michigan]. 151 p.

Chester, Judith. Deformation of layered rocks in the ramp regions of thrust faults; a study with rock models. M, 1985, Texas A&M University.

Chester, Mik E. The stratigraphy of a modern oyster reef in the Ten Thousand Islands area of Southwest Florida. M, 1979, University of Toledo. 66 p.

Chesterman, Charles Wesley. Contact metamorphism of the Twin Lakes region, Fresno County, California. M, 1940, University of California, Berkeley. 61 p.

Chestnut, Donald R. Echinoderms from the lower part of the Pennington Formation (Chesterian) in south-central Kentucky. M, 1980, University of Kentucky. 178 p.

Chestnut, Donald Rader, Jr. Stratigraphic analysis of the Carboniferous rocks of the central Appalachian Basin. D, 1988, University of Kentucky. 363 p.

Chesworth, Ward. The origin of certain granitic rocks occurring in Glamorgan Township, southeastern Ontario. D, 1967, McMaster University. 161 p.

Cheung, C. H. Influence of salt on the unfrozen water in frozen clays. D, 1979, McGill University.

Cheung, Henry P. Y. Crustal structure near Explorer Ridge; ocean-bottom seismometer results parallel to Revere-Dellwood fracture zone. M, 1978, University of British Columbia.

Cheung, Paul Chi-Tak. The mineralogy, geochemistry and petrogenesis of the Mount Poser gabbroic pluton, Southern California. M, 1981, University of Windsor. 136 p.

Cheung, Paul Kwon-Shun. The geothermal gradient in sedimentary rocks in Oklahoma. M, 1978, Oklahoma State University. 55 p.

Cheung, Sha-Pak. Rubidium-strontium whole-rock ages from the Oxford Lake-Knee Lake greenstone belt, northern Manitoba. M, 1978, University of Manitoba. 41 p.

Chevalier, Jean. Diagénèse de grès Cambro-Ordovicien, St. Fabien, Québec. M, 1976, Universite de Montreal.

Chevillon, Chas. Victor. Petrology and structural geology of the Mount Cindy-Bearpaws Peaks area, Jackson County, Colorado. M, 1973, SUNY at Buffalo. 107 p.

Chew, Chye Heng. Acoustical properties of coal. D, 1980, Georgia Institute of Technology. 201 p.

Chew, Randall T. Geology of the Mineta Ridge area, Pima and Cochise counties, Arizona. M, 1952, University of Arizona.

Chewaka, S. The petrochemistry and structure of Elsie Mountain metabasalts; possible connection with some Sudbury sulfide mineralization, Sudbury, Ontario. M, 1975, University of Western Ontario. 104 p.

Chewning, John R. Wavenumber domain analysis of aeromagnetic data. M, 1973, Pennsylvania State University, University Park.

Chi, Byung I. A petrologic comparison of the Frenchman and upper Edmonton formations (upper Cretaceous, S. Saskatchewan and S.E. Alberta, Canada). M, 1966, University of Alberta. 125 p.

Chi, Byung I. Devonian megaspores and their stratigraphic significance in the Canadian Arctic. D, 1974, University of Calgary. 368 p.

Chi, Wen-Wei. The sampling problem in the petrographic study of granite plutons. M, 1966, University of Toronto.

Chia, Yee Ping. Digital simulation of compaction in sedimentary sequences. D, 1980, University of Illinois, Urbana. 119 p.

Chia, Yee-Ho. Rubidium, strontium, yttrium, zirconium, and titanium in island arc rocks. M, 1973, Columbia University.

Chiang, Chao-Sheng. Synthetic seismogram studies of laterally inhomogeneous earth structure. D, 1984, Purdue University. 180 p.

Chiang, Chen Yu. Numerical simulation of ground water contaminant transport on a supercomputer with injection-pumping networks using the modified MOC and MFE method. D, 1986, Rice University. 158 p.

Chiang, E. The implication of sediment dynamics and heavy metal dispersion to ocean disposal at the "NB" buoy, south of Fire Island, New York. M, 1974, Adelphi University.

Chiang, Hsien-Hsiang. A probabilistic updating model for assessing landslide hazards. M, 1984, University of Idaho. 56 p.

Chiang, J. C. Determination of abnormal pressures by the interpretation of well logs, seismic data, and drilling data. M, 1978, Stanford University. 16 p.

Chiang, J. H. Relation of radon anomalies in groundwater to seismicity; field and laboratory studies at Lake Jocassee, South Carolina. M, 1977, University of South Carolina. 77 p.

Chiang, Kam Kuen. Silurian (middle) brachiopods from the Fossil Hill and Amabel Formation, Ontario. M, 1970, University of Western Ontario. 137 p.

Chiang, Liann. Geometric/geomorphologic measures of ocean volcanic islands. D, 1988, University of Pittsburgh. 108 p.

Chiang, Ming Chen. Element partition between hornblende and biotite in the rocks from Loon Lake Aureole, Chandos Township, Ontario. M, 1966, McMaster University. 83 p.

Chiang, Robert Huai. Quantitative geomorphology of the Hangman Creek drainage basin, Washington and Idaho. M, 1982, Eastern Washington University. 63 p.

Chiang, Wei-Ling. Models for uncertainty propagation; applications to structural and earthquake engineering. D, 1988, Stanford University. 168 p.

Chiang, Yu-Jen. Two kinds of vacillation in rotating laboratory experiments. M, 1968, Columbia University. 28 p.

Chiarenzelli, Jeffrey Robert. Mid-Proterozoic chemical weathering, regolith, and silcrete in the Thelon Basin, Northwest Territories. M, 1983, Carleton University. 205 p.

Chiburis, Edward F. Reliability of the LaCoste-Romberg surface ship gravity meter S-9 during Cruise 60-H-13 of the Texas A & M research vessel "Hidalgo". M, 1962, Texas A&M University. 63 p.

Chiburis, Edward Frank. Crustal structures in the Pacific Northwest states from phase-velocity dispersion of seismic surface waves. D, 1966, Oregon State University. 170 p.

Chichester-Constable, David John. A study of the chemical composition, fate and movement of landfill leachate. D, 1986, University of Connecticut. 427 p.

Chico, Eduardo. Mineralogy, paragenesis, and fluid inclusion studies in three veins of the Fresnillo mining district, Zacatecas, Mexico. M, 1986, Dartmouth College. 115 p.

Chico, Raymundo L. The geology of the uranium-vanadium deposit of the Diamond No. 2 Mine, near Gallup, New Mexico. M, 1959, University of Missouri, Rolla.

Chicoine, Stephen Duane. A physical model of a geothermal system; its design and construction and its application to reservoir engineering. M, 1975, Stanford University. 97 p.

Chidester, Alfred Hermann. Steatization of serpentine bodies in north-central Vermont. D, 1959, University of Chicago. 70 p.

Chidsey, Thomas C., Jr. Intrusions, alteration, and economic implications in the northern House Range, Utah. M, 1977, Brigham Young University.

Chien, Tammy Chin-Hsia. The effect of fly-ash on groundwater at Port Washington, Wisconsin. M, 1976, University of Wisconsin-Milwaukee.

Chierton, Camon G. Correlation of the area including Kimberley, Metaline and Coeur d'Alene. M, 1949, University of British Columbia.

Chieruzzi, Gianni Oswaldo. Geochemical flow modeling of the supergene alteration of porphyry copper deposits. M, 1988, University of Texas, Austin. 222 p.

Chilakos, Peter. [14]C, Th/U, amino acid and stable isotope measurements in late Quaternary deposits of Fuerteventura, Canary Islands; a critical approach. M, 1988, Universite du Quebec a Montreal. 123 p.

Chilcoat, Steven R. Applications of the computer analysis of dispersed waves. M, 1977, Colorado School of Mines. 137 p.

Child, Charles Joseph. The reliability of paleotemperatures from Cretaceous DSDP Inoceramus. M, 1988, University of Miami. 107 p.

Childerhose, Allan Jerome. Genetic relationships in the Flin Flon, Mandy, and Oiseau River ore deposits. M, 1928, University of Manitoba.

Childers, David Wayne. Diagenesis of feldspars in the Minturn Formation (Pennsylvanian) of Colorado. M, 1979, University of Oklahoma. 129 p.

Childers, M. O. Geology of the French Creek area, Albany and Carbon counties, Wyoming. M, 1957, University of Wyoming. 58 p.

Childers, Milton Orville. Structure and stratigraphy of the Southwest Marias Pass area, Flathead County, Montana. D, 1960, Princeton University. 181 p.

Childs, Constance Smythe. Some mineralogic changes through a roll-type uranium deposit, South Texas. M, 1981, University of Texas, Austin.

Childs, Gerald Dewitt. The petrology of the Myrna Lake and Fraser Lake basic intrusive bodies. M, 1950, University of Manitoba.

Childs, John Frazer. Contact relationships of Mount Carlyle stock, Slocan, British Columbia. M, 1969, University of British Columbia.

Childs, John Frazer. Geology of the Precambrian Belt Supergroup and the northern margin of the Idaho Batholith, Clearwater County, Idaho. D, 1982, University of California, Santa Cruz. 510 p.

Childs, Lewis. A study of a karst area in Orange and Lawrence counties, Indiana. M, 1940, Indiana University, Bloomington. 111 p.

Childs, Michael S. Thermal alteration of central Utah coals. M, 1986, Southern Illinois University, Carbondale. 173 p.

Childs, Orlo E. A physiographic study of Morgan Valley, Utah. M, 1938, University of Utah. 42 p.

Childs, Orlo Eckersley. Geomorphology of the valley of the Little Colorado River, Arizona. D, 1945, University of Michigan.

Childs, Philip. Stratigraphy and petrology of the Silurian Brassfield and Noland formations in eastern Kentucky and southern Ohio. M, 1969, SUNY at Binghamton. 100 p.

Childs, Susan Marie Stell. The petrographic characterization, coking potential and factors affecting coalification of coals along the eastern and southern margins of the Piceance Creek basin in Colorado. M, 1980, Southern Illinois University, Carbondale. 177 p.

Childs, Theodore S. Stratigraphic significance of Lepidocyclina in the Vaqueros Formation, San Luis Obispo County, California. M, 1941, Stanford University. 61 p.

Childs, Vicki L. Geology of the Granite Wash Formation; eastern Hartley County of the Texas Panhandle. M, 19??, West Texas State University. 92 p.

Chilingar, George V. Use of Ca/Mg ratio of limestones and dolomites as a geologic tool. D, 1956, University of Southern California.

Chilton, Rosalie. Analysis of aircraft multipolarization and spacecraft radar imagery implications for the Venus Radar mapping mission and characteristics of multipolarization radar date in vegetated terrain. M, 1985, University of Arkansas, Fayetteville.

Chimahusky, J. S. Dolomite nonstoichiometry; its effect on X-ray diffraction estimates of dolomite in carbonates and its use as a facies parameter. M, 1978, Memphis State University.

Chimene, Calvin Alphonse. A study of the subsurface Tertiary stratigraphy of the central Texas Gulf Coast, with cross-sections from Northwest Bastrop and Lee counties to the Gulf of Mexico. M, 1952, University of Houston.

Chimene, Julius, III. Ostracoda biostratigraphy and depositional environments of the upper Taylor Group (Campanian, Upper Cretaceous) in central Texas. M, 1983, University of Houston.

Chimney, Peter J. Paleoecology of trepostome ectoprocts of the Richmond Group (Upper Ordovician), southwestern Ohio-southeastern Indiana. M, 1977, Miami University (Ohio). 147 p.

Chin, Chen. Ecological studies of shell-bearing pteropods from the western part of the North Atlantic Ocean. D, 1962, Boston College. 225 p.

Chin, Danton J. A crystal chemical study of garnet, biotite, cordierite, and cummingtonite from the metamorphic rocks of Copper Mountain in Wyoming. M, 1977, Brooklyn College (CUNY).

Chin, David Arthur. A mathematical model of dispersion in coastal waters. D, 1982, Georgia Institute of Technology. 313 p.

Chin, John. Late Quaternary coastal sedimentation and depositional history, south-central Monterey Bay, California. M, 1984, San Jose State University. 130 p.

Chin, Lennard Hilton. An experimental investigation on the brightness improvement of Bovill clays. M, 1962, University of Idaho. 69 p.

Chin, Maureen. Provenance of abyssal silts from the southern Iceland margin and its paleoclimatic implications, Fourier grain shape analysis. M, 1980, University of South Carolina.

Chin, S-Len Richard. The selection of best management practices for controlling the non-point source pollution in Oklahoma. D, 1984, University of Oklahoma. 254 p.

Chin, Ti-hau. On-stream particle-size analysis by a dynamic equilibrium sieving method. D, 1977, Stanford University. 105 p.

Chin, Wai Suey. Studies of the characteristics of limestone porosity. M, 1951, University of Texas, Austin.

Chin, Yu-Tung. System design and implementation of a computerized Rhode Island water resources information system. D, 1987, University of Rhode Island. 194 p.

Chinburg, Susan. Sediment dynamics in Monterey Canyon, central California. M, 1985, San Jose State University. 89 p.

Ching, Paul D. An investigation of the Cambrian Deadwood Sandstone in central and southern Black Hills, South Dakota, as potential industrial silica sands. M, 1973, South Dakota School of Mines & Technology.

Ching, Paul W. Economic gravel deposits of the lower Cache La Poudre River (Colorado). M, 1972, Colorado State University. 103 p.

Chinn, Alan F. A study in clonal populations of the foraminifera Rosalina Floridana (Cushman) (Recent, Florida). M, 1972, Dalhousie University. 52 p.

Chinn, Alvin A. Dynamic interpretation of calcite twin lamellae in limestone of Northwest Arkansas. M, 1967, University of Arkansas, Fayetteville.

Chinn, Douglas Samuel. Accurate source depths and focal mechanisms of shallow earthquakes in western South America and in the New Hebrides island arc. D, 1982, Cornell University. 238 p.

Chinn, E. Y. H. Soil profiles along Kipapa Gulch, Oahu, Hawaii, as modified by altitude and climate. M, 1936, University of Hawaii. 113 p.

Chinn, William. Structural and mineralogical studies at the Homestake Mine, Lead, South Dakota. D, 1969, University of California, Berkeley. 191 p.

Chinsomboon, Vichol. Surficial geology along the Arkansas Valley from Ponca City northward to Kirk's Hill top, north-central Oklahoma. M, 1976, Oklahoma State University. 52 p.

Chintakovid, Vanit. Evaluation of aggregate needs and problems along the Oregon coast. M, 1979, Oregon State University. 107 p.

Chiou, Jyh-Dong. Mathematical modeling and simulation of three dimensional flow systems. D, 1988, University of Pittsburgh. 99 p.

Chiou, Shyh-Jeng. Regional variation of crust and upper mantle structure beneath northern Canada and the Arctic Ocean from surface wave inversion. M, 1986, St. Louis University.

Chiou, Wen-An. A study of clay minerals in surface and suspended sediments from the Anclote River and estuary, central Florida. M, 1976, University of South Florida, St. Petersburg. 62 p.

Chiou, Wen-An. Clay fabric of gassy submarine sediments. D, 1981, Texas A&M University. 166 p.

Chipera, Steve J. Metamorphism in the eastern Lac Seul region of the English River Subprovince, Ontario. M, 1985, University of North Dakota. 168 p.

Chipman, David Walter. Partial melting of a spinel lherzolite at 20 kilobars pressure. D, 1972, Harvard University.

Chipman, David Walter. Preferential ionic substitution in the crystal lattice of the chromite spinel series. M, 1966, Michigan State University. 37 p.

Chipp, Eddie Ray. The geology of the Klondike mining district, Esmeralda County, Nevada. M, 1969, University of Nevada. 52 p.

Chipping, David Hugh. The petrology and paleogeography of Cretaceous and lower Tertiary strata in the vicinity of Cuyama valley, California. D, 1970, Stanford University. 184 p.

Chiquito, Freddy Jesus. Yarbrough and Allen field, Ector County, Texas. M, 1970, University of Texas, Austin.

Chirlin, Gary Richard. Flow and convective transport in a porous medium with periodic barriers. D, 1982, Princeton University. 339 p.

Chisholm, David B. The paragenesis of some ferromagnesian minerals in certain granitoid rocks of the Marblehead, Massachusetts area. M, 1925, Syracuse University.

Chisholm, Duncan M. Gas potential south of the Bearpaw Mountains, north-central Montana. M, 1975, Stanford University. 74 p.

Chisholm, Earl J. Sedimentary petrology of the San Andres Formation of central New Mexico. M, 1950, Texas Tech University. 40 p.

Chisholm, Edward Owen. The geology of Balmer Township, District of Kenora. M, 1947, University of Toronto.

Chisholm, Sallie W. Studies on daily rhythms of phosphate uptake in *Euglena* and their potential ecological significance. D, 1974, SUNY at Albany. 150 p.

Chisnell, Thomas Cutter. A correlation test of knickpunkte. M, 1941, Cornell University.

Chisnell, Thomas Cutter. Recognition and interpretation of proglacial strand lines in the Cayuga Basin. D, 1951, Cornell University.

Chisolm, Stoney P. The effect of nonlinear soil response on the behavior of single piles and pile groups in clay. D, 1978, University of Virginia. 264 p.

Chitale, Dattatraya V. Laterites, bauxites and associated clays from western India. D, 1986, Texas Tech University. 149 p.

Chitale, Jayashree Dattatraya. Study of petrography and internal structures in calcretes of West Texas and New Mexico. D, 1986, Texas Tech University. 107 p.

Chitsazan, Manouchehr. Hydrogeologic evaluation of the Boone-St. Joe carbonate aquifer. M, 1980, University of Arkansas, Fayetteville.

Chittenden, David Morse, II. Noble gases in granites and accessory minerals. D, 1966, University of Arkansas, Fayetteville. 122 p.

Chittrayanont, Sumeth. A geohydrologic study of the Garfield Township coal basin area, Bay County, Michigan. M, 1978, Michigan Technological University. 183 p.

Chitwood, Lawrence Allan. Stratigraphy, structure, and petrology of the Snoqualmie Pass area, Washington. M, 1976, Portland State University. 68 p.

Chiu, Hung-Chie. The effects of topography and subsurface structure on the seismic amplification. D, 1986, University of Southern California.

Chiu, Jer-Ming. Structural features of subduction zone determined by detailed analysis of short period seismic waves from earthquakes recorded in the New Hebrides island arc. D, 1982, Cornell University. 218 p.

Chivalak, S. Pore-pressure function for preconsolidated saturated clay. D, 1975, Wayne State University. 292 p.

Chmelik, Frank B. The structural geology of the Green Acres Ranch area, Larimer County, Colorado. M, 1957, University of Colorado.

Chmelik, Frank Bernard. An investigation of changes induced in macrostructures of pelitic sediments during primary consolidation. D, 1970, Texas A&M University. 143 p.

Chmelik, James. Pleistocene geology of northern Kidder County, North Dakota. M, 1960, University of North Dakota. 63 p.

Cho, Byung C. Significance of bacteria in biogeochemical fluxes in the pelagic ocean. D, 1988, University of California, San Diego.

Cho, D. Variational principles for transient electromagnetics and their applications to the exploration of deep-seated mineral deposits. D, 1978, Columbia University, Teachers College. 75 p.

Cho, Moonsup. A kinetic study of the clinochlore composition at 2 kbar water pressure. M, 1982, University of Toronto.

Cho, Moonsup. Petrochemical and experimental studies of subgreenschist facies metamorphism of basaltic rocks. D, 1986, Stanford University. 194 p.

Cho, Taechin F. Continuum and discrete modelings of porous and jointed rock; application to the design of near surface annular excavations. D, 1988, University of Wisconsin-Madison. 286 p.

Cho, Y.-Y. Small strain shear modulus and disturbance in sand sample. D, 1975, University of South Carolina. 66 p.

Choate, M. L. Energy budget study, lower Colorado River, Arizona. M, 1973, University of Arizona.

Choate, Raoul and Hilles, Robert. The geology of the foothills west of Loveland, Larimer County, Colorado. M, 1954, University of Michigan.

Chobthum, Worapot. Zinc availability and mobility in West Virginia soils. D, 1988, West Virginia University. 156 p.

Chockalingam, Solayappa. Reduced heatflow studies in North America. M, 1979, University of Texas at Dallas. 112 p.

Choi, A. P. Rays and caustics in vertically inhomogeneous elastic media. M, 1978, University of Alberta. 109 p.

Choi, Duck Keun. Paleopalynology of the Upper Cretaceous-Paleogene Eureka Sound Formation of Ellesmere and Axel Heiberg islands, Canadian Arctic Archipelago. D, 1983, Pennsylvania State University, University Park. 593 p.

Choi, Jin Beom. Mossbauer spectroscopy and crystal chemistry of aenigmatites. M, 1983, Massachusetts Institute of Technology. 59 p.

Choi, Ling-Kit. Consolidation behavior of natural clays. D, 1982, University of Illinois, Urbana. 446 p.

Cholach, Michael S. A geochemical survey and the nature of lead-silver ores in Sixty Mile river area (pre-Mesozoic), Yukon Territory (Canada). M, 1969, University of Alberta. 85 p.

Chollett, D. Analysis of the stresses induced in rock by rock splitters. M, 1975, University of Missouri, Rolla.

Chomyn, Beverley A. The relation between magnetic properties, density, opaque mineralogy and chemistry in drill core from the Lac du Bonnet Batholith, Manitoba. M, 1987, Carleton University. 78 p.

Chon, C. S. Dynamic response of friction piles. D, 1977, University of Michigan. 232 p.

Chonglakmani, Chongpan. The fauna of the (upper Jurassic) Sundance Formation (western Great Plains, U.S.). M, 1971, South Dakota School of Mines & Technology.

Choo, Kangsoo. Celestite mineralization at Enon Lake, Cape Breton County, Nova Scotia. M, 1972, Dalhousie University.

Chopra, Anil Kumar. Earthquake effects on dams. D, 1966, University of California, Berkeley. 104 p.

Chopra, Kailash C. Performance of tellurometer over short distances. M, 1963, Ohio State University.

Choquette, Marc. La Stabilisation à la chaux des sols argileux du Québec. D, 1988, Universite Laval. 184 p.

Choquette, Philip W. Petrology and structure of the Cockeysville Formation near Baltimore, Maryland. D, 1957, The Johns Hopkins University.

Chork, C. Y. The application of some statistical and computer techniques to the interpretation of soil and stream sediment geochemical data. D, 1977, University of New Brunswick.

Chorley, D. W. A finite element model for analysis of multiaquifer systems. M, 1976, University of Waterloo.

Chorlton, Lesley B. The effect of boron on phase relations in the granite-water system. M, 1973, McGill University. 95 p.

Chorlton, Lesley Bronwyn. Geological development of the southern Long Range Mountains, Southwest Newfoundland; a regional synthesis. D, 1984, Memorial University of Newfoundland. 581 p.

Chormann, Frederick H., Jr. The occurrence of arsenic in soils and stream sediments, Town of Hudson, NH. M, 1985, University of New Hampshire. 222 p.

Chormann, Jeffrey. The origin of reverse deltas in Maine. M, 1983, Boston University. 72 p.

Chorn, John Douglas. A trimerorhachid amphibian from the Upper Pennsylvanian of Kansas. D, 1984, University of Kansas. 123 p.

Chornesky, Elizabeth Ann. The consequences of direct competition between scleractinian reef corals; development and use of sweeper tentacles. D, 1984, University of Texas, Austin. 122 p.

Chorney, Raymond. Experiment station scheelite and mineral association at the Reaper Mines, Gold Hill, Utah. M, 1943, University of Utah. 48 p.

Chotimon, A. The properties and genesis of four soils in southwestern Kauai, Hawaii. M, 1969, University of Hawaii. 65 p.

Chou Hsi T'an *see* Hsi Chou T'an

Chou, Chen-Lin. Gallium and germanium in the metal and silicate phases of L- and LL- chondrites. D, 1971, University of Pittsburgh. 129 p.

Chou, Chung-chi. The catalytic activity of montmorillonite by heating-oscillating X-ray diffraction and differential thermal analysis. D, 1968, Baylor University. 169 p.

Chou, Gin. Geochemistry of the Togo Metapelites from the Midnite uranium mine, northeastern Washington. M, 1983, Eastern Washington University. 84 p.

Chou, I-M. Geochemical studies; Precambrian banded iron-formations and acid-base buffers. D, 1974, The Johns Hopkins University. 99 p.

Chou, Lei. Study of the kinetics and mechanisms of dissolution of albite at room temperature and pressure. D, 1985, Northwestern University. 320 p.

Chou, Lynn. Seismic imaging beneath volcanics; an experiment in the Poplar area, southeastern Idaho. M, 1983, University of Houston.

Chou, Shuh-Dar Frank. Geology of Lake Hermitage Field, Plaquemines Parish, Louisiana. M, 1976, University of New Orleans. 14 p.

Chou, Teh-An George. Continental and oceanic mantle structure from dispersion of higher modes of surface wave. D, 1979, Harvard University.

Choudary, Narendra. An investigation of failure patterns of layered specimens loaded in uniaxial compression. M, 1967, University of Missouri, Rolla.

Choudary, Narendra Y. B. An investigation of the influence of grain size on stress wave attenuation and dispersion in rock like materials. D, 1973, University of Missouri, Rolla.

Choudhary, Muhammad Rafiq. Optimal conjunctive use of surface and groundwater in a watercourse command area of the Indus Basin. D, 1987, Colorado State University. 223 p.

Choudhry, Abdul G. The petrology and geochemistry of the Gamitagama Lake igneous complex, near Wawa, North central Ontario. M, 1981, University of Windsor. 245 p.

Chouet, Bernard A. Photoballistic analysis of volcanic jet dynamics at Stromboli, Italy. M, 1973, Massachusetts Institute of Technology. 66 p.

Chouet, Bernard Alfred. Source, scattering and attenuation effects on high frequency seismic waves. D, 1976, Massachusetts Institute of Technology. 183 p.

Chough, S. K. Morphology, sedimentary facies and processes of the Northwest Atlantic Mid-Ocean Channel between 61° and 52°N, Labrador Sea. D, 1978, McGill University. 167 p.

Chouinard, Paul Norman. Reflection seismology; synthetics and inversion. D, 1987, University of Saskatchewan.

Choung, Haeung. Paleoecology, stratigraphy, and taxonomy of the foraminifera of the Weches Formation of East Texas and the Cane River Formation of Louisiana. M, 1975, Louisiana State University. 300 p.

Chouteau, Michel C. Acquisition et traitement des signaux à large bande de fréquence pour sondage magneto-telluriques miniers et pétroliers. M, 1976, Ecole Polytechnique.

Chouteau, Michel C. Prospection magnétotellurique sur des structures conductrices à trois-dimensions. D, 1982, Ecole Polytechnique. 281 p.

Chow, Andre M. C. Sedimentology and paleontology of the Attawapiskat Formation (Silurian) in the type area, northern Ontario. M, 1986, McGill University. 239 p.

Chow, Jinder. Biostratigraphy and marine paleoenvironments of the Gulf of Mexico, the western Caribbean, and the eastern Equatorial Pacific. D, 1985, Texas A&M University. 309 p.

Chow, Minchen Ming. The Pennsylvanian Mill Creek Limestone in Pennsylvania. D, 1950, Lehigh University.

Chow, Nancy. Sedimentology and diagenesis of Middle and Upper Cambrian platform carbonates and siliciclastics, Port au Port Peninsula, western Newfoundland. D, 1987, Memorial University of Newfoundland. 458 p.

Chow, Tsaijwa J. The determination and distribution of copper in sea water. D, 1953, University of Washington. 100 p.

Chowdhury, Dipak Kumar. Seismic model investigation of wave propagation along layers in multilayered media. D, 1961, Texas A&M University. 127 p.

Chowdhury, Yusuf. Mineral interactions in natural systems as shown by the physical chemistry of the aqueous phase. D, 1970, University of Missouri, Columbia. 326 p.

Chowdiah, Attru Mallikarjuniah. Stress and strain distribution around openings in underground salt formations. D, 1963, Michigan State University. 159 p.

Chown, Edward H. M. Amphibolites of the Papachouesati River area, Mistassini Territory, Quebec. D, 1963, The Johns Hopkins University.

Chown, Edward Holton MacPhail. The geology of the Willroy Property, Manitouwadge Lake, Ontario. M, 1957, University of British Columbia.

Choy, G. L. Theoretical seismograms of core phases calculated by a frequency-dependent full wave theory, and their interpretation. D, 1977, Columbia University, Teachers College. 146 p.

Choy-Manzanilla, Jose Enrique. Use of body waves amplitudes in constraining nodal planes of small local earthquakes. M, 1981, University of California, Riverside. 114 p.

Chrisinger, Danny L. Economic geology of the Tameapa area, Badiraguato Municipality, Sinaloa, Mexico. M, 1975, University of Iowa. 178 p.

Chrisman, Louis Paul. The geology of the Big Cabin Creek area, Craig County, Oklahoma. M, 1951, University of Oklahoma. 66 p.

Chrismas, Lawrence Philip. A geologic, economic and isotopic evaluation of Craigmont mines (Mesozoic) (Highland Valley area, near Merritt), British Columbia. M, 1968, University of Alberta. 141 p.

Chriss, Terry Michael. Glacial marine sedimentation, Ross Sea, Antarctica. M, 1971, University of California, Los Angeles.

Christ, Charles Milton. Study of horizontal oil wells. M, 1947, University of Pittsburgh.

Christ, Janice M. A preliminary palynological analysis of the upper Canadaway and basal Chadakoin (Cuba) formations in southwestern New York State. M, 1979, SUNY at Buffalo.

Christe, Geoff. The geology and petrology of the eastern Mesozoic belt, northern Sierra, California. M, 1987, University of Vermont. 350 p.

Christen, Randolph Frederick. Effects of mine openings on seismic waves; a two-dimensional model study. M, 1978, Pennsylvania State University, University Park. 72 p.

Christensen, Andrew Dougan. Part of the geology of the Coyote Mountain area, Imperial County, California. M, 1957, University of California, Los Angeles.

Christensen, Andrew L. Geology and physiography of Deer Creek and Silver Fork tributaries of American Fork Canyon, Wasatch Mountains, Utah. M, 1928, Stanford University. 103 p.

Christensen, Andrew Lee. Igneous geology of the Elkhead Mountains, Colorado. D, 1942, University of California, Berkeley. 180 p.

Christensen, Arthur Francis. Stratigraphy and sedimentation of the subsurface Cockfield Formation (middle Eocene) in southern Beauregard Parish, Louisiana. M, 1967, University of Southwestern Louisiana.

Christensen, Cleo M. Pleistocene stratigraphy of Clay County, South Dakota. M, 1966, University of South Dakota. 102 p.

Christensen, D. F. Regional stratigraphic analysis of the Triassic of the Northern Great Plains of Canada. M, 1953, Northwestern University.

Christensen, Douglas H. Intraplate earthquakes and seismic coupling. D, 1987, University of Michigan. 190 p.

Christensen, E. R. Trace metals in urban runoff and their influence on phytoplankton growth in the receiving waters. D, 1977, University of California, Irvine. 206 p.

Christensen, Eric Joseph. Digital change detection; a quantitative evaluation of image registration and wetland phenological characteristics using high resolution multispectral scanner data. D, 1987, University of South Carolina. 203 p.

Christensen, Evart Wayne. Petrography of plumasite, altered enstatite, and metamorphosed quartz-iron rocks from the Gilliam Prospect, Madison County, Montana. M, 1956, Indiana University, Bloomington. 41 p.

Christensen, Frank Deon. A paleomagnetic investigation of the Permo-Carboniferous Maroon and Upper Permian-Lower Triassic State Bridge formations in north-central Colorado. M, 1974, University of Texas at Dallas. 141 p.

Christensen, John Edward. Petrology of selected Mississippian cherts, Marshall and Keokuk counties, Iowa. M, 1964, Iowa State University of Science and Technology.

Christensen, John Neil. A strontium isotopic study of processes in a silicic magma chamber; the Bishop Tuff, Long Valley, California. M, 1987, University of California, Los Angeles. 67 p.

Christensen, Kim C. The stratigraphy and petrography of a light-colored siliceous horizon within the Fort Union Formation (Paleocene), southeastern Montana. M, 1984, Montana College of Mineral Science & Technology. 183 p.

Christensen, Mark Newell. Geologic structures of the Mineral King area, California. D, 1959, University of California, Berkeley. 102 p.

Christensen, Michael W. A strip chart recording method for determining grain size in sedimentary analysis. M, 1971, East Texas State University.

Christensen, N., IV. Numerical simulation of free oscillations of enclosed basins on a rotating Earth. D, 1972, University of Hawaii. 181 p.

Christensen, Nikolas I. Amphibolites and related rocks in the West Torrington Quadrangle, Connecticut. M, 1961, University of Wisconsin-Madison.

Christensen, Nikolas Ivan. The Hodges mafic complex; a study of the structural control of basic intrusives. D, 1963, University of Wisconsin-Madison. 128 p.

Christensen, O. D. Metamorphism of the Manning Canyon and Chainman formations. D, 1975, Stanford University. 174 p.

Christensen, Paul K. The hydrogeology of Sunset Crater and Wupatki national monuments, Coconino County, Arizona. M, 1982, Northern Arizona University. 138 p.

Christensen, Philip Russel. The nature of the Martian surface as derived from thermophysical properties. D, 1981, University of California, Los Angeles.

Christensen, R. J. Effect of state of stress on frictional properties of rock joints. D, 1975, University of Utah. 140 p.

Christensen, R. J. Petrographic and textural analysis of a barchan dune south-west of the Salton Sea, Imperial County, California. M, 1973, [University of California, San Diego].

Christensen, Ralph Warren. North Florence Dunal Aquifer study. M, 1982, University of Oregon. 189 p.

Christensen, Richard M. Stratigraphy of the (Ordovician) Pogonip and Eureka groups in south-central Nevada. M, 1957, University of Nebraska, Lincoln.

Christensen, Richard Martin. Geology of the Paria-Araya Peninsula, northeastern Venezuela. D, 1961, University of Nebraska, Lincoln. 158 p.

Christensen, Roberta Smith. Forminiferal studies in the lower Tertiary Soquel Creek, Santa Cruz County, California. M, 1960, University of California, Berkeley. 175 p.

Christensen, Thomas Hoejlund. Cadmium sorption onto two mineral soils. D, 1980, University of Washington. 307 p.

Christenson, Bruce W. Structure, petrology and geochemistry of the Chilliwack Group near Sauk Mountain, Washington. M, 1981, Western Washington University. 181 p.

Christenson, Donald Robert. Magnesium release and uptake from selected Michigan soils. D, 1968, Michigan State University. 95 p.

Christenson, Gary E. Environmental geology of the McDowell Mountains area, Maricopa County, Arizona; Part I. M, 1976, Arizona State University. 69 p.

Christenson, Lief G. Genesis of gold mineralization in the Lone Jack Mine area, Mount Baker mining district, Washington. M, 1986, Western Washington University. 87 p.

Christenson, Tod D. Organic geochemistry and petroleum source potential of the Cretaceous Aspen Formation, Idaho-Wyoming thrust belt. M, 1983, Idaho State University. 65 p.

Christian, Barbara S. McBirneyite, $CU_3(VO_4)_2$, a new sublimate mineral from the fumaroles of Izalco Volcano, El Salvador. M, 1985, Miami University (Ohio). 32 p.

Christian, Harry E. Geology of the Marble City area, Sequoyah County, Oklahoma. M, 1953, University of Oklahoma. 160 p.

Christian, James Terry. A preliminary study of planktonic Foraminifera from the Mooreville Formation (Upper Cretaceous), Lowndes County, Mississippi. M, 1968, Mississippi State University. 135 p.

Christian, Louis. Cretaceous history of France. M, 1952, Stanford University.

Christiansen, C. R. Marcasite from the Upper Mississippi Valley zinc area as a source of sulphur. M, 1952, University of Missouri, Rolla.

Christiansen, David James, Jr. Petrology and biostratigraphy of Middle, Lower Devonian strata, southern Cortez Mountains, Nevada. M, 1980, University of California, Riverside. 160 p.

Christiansen, Earl Alfred. Glacial geology of the Moose Mountain area, Saskatchewan. M, 1956, University of Saskatchewan. 61 p.

Christiansen, Earl Alfred. Glacial geology of the Swift Current area, Saskatchewan. D, 1969, University of Illinois, Urbana.

Christiansen, Eric H. Geology and geochemistry of topaz rhyolites from the western United States. D, 1981, Arizona State University. 351 p.

Christiansen, Francis Wyman. Geology of economic possibilities of the alunite deposits in Sevier and Piute counties, Utah. M, 1937, University of Utah. 97 p.

Christiansen, Francis Wyman. Geology of the Canyon Range, Utah. D, 1948, Princeton University. 140 p.

Christiansen, Jack Hilbert. Gravity, magnetic and electrical resistivity investigation of selected fault and stratigraphic contacts in southern Jefferson and northern Madison counties, Montana. M, 1970, Indiana University, Bloomington. 63 p.

Christiansen, Robert Lorenz. Structure, metamorphism, and plutonism in the El Paso Mountains, Mojave Desert, California. D, 1961, Stanford University. 208 p.

Christiansen, William D. Stratigraphy and structure of the Precambrian metamorphic rocks in the Grace Coolidge Creek area, Custer State Park, South Dakota. M, 1984, South Dakota School of Mines & Technology.

Christiansen, William Joseph. Geology of the Fish Springs mining district, Juab County, Utah. M, 1977, University of Utah. 66 p.

Christianson, Carlyle Bruce. Chemical denitrification in frozen soils. D, 1981, University of Manitoba.

Christianson, Linda J. A geophysical interpretation of the Skalafell magnetic anomaly in Southwest Iceland. M, 1988, University of Iowa. 189 p.

Christie, Archibald. The geology of the Goldfields area, Saskatchewan. D, 1948, McGill University.

Christie, Brian James. Alteration and gold mineralization associated with a sheeted veinlet zone at the Campbell Red Lake Mine, Balmertown, Ontario. M, 1986, Queen's University. 334 p.

Christie, David L. The Otto Fiord Formation; a Carboniferous submarine evaporite unit of the Canadian Arctic Islands. M, 1975, University of Calgary. 117 p.

Christie, David Mark. Petrologic effects of mid-oceanic rift propagation; the Galapgos spreading center at 95.5°W. D, 1984, University of Hawaii. 235 p.

Christie, Fritz Jay. Analysis of gravity data from the Picacho Basin, Pinal County, Arizona. M, 1978, University of Arizona. 105 p.

Christie, Harold Hans. Geology of the southern part of the Gravelly Range, southwestern Montana. M, 1961, Oregon State University. 159 p.

Christie, James Stanley. Geology of Vaseaux Lake area. D, 1973, University of British Columbia.

Christie, Kyle. The geology of the Sunol regional wilderness. M, 1985, San Jose State University. 112 p.

Christie, Michael Alexander. Delineation of the Seven Mile Creek buried valley between the cities of Somerville and Collinsville, Ohio, using gravity and an analysis of the area's groundwater resources. M, 1986, Wright State University. 106 p.

Christie, N. T. Matthews Mine magnetic deposit. M, 1950, Queen's University.

Christie, R. L. Andesites and their relation to plutonic intrusives. M, 1951, University of Toronto.

Christie, R. L. Plutonic rocks of the Coast Range Batholith in the Bennett area, British Columbia. D, 1954, University of Toronto.

Christie, Robert Loring. Andesites and their relations to plutonic rocks. M, 1952, University of Toronto.

Christie, Robert Loring. Geology of the plutonic rocks of the Coast Mountains in the vicinity of Bennett, British Columbia. D, 1958, University of Toronto.

Christman, Jerry L. Geology and alteration of the Copper Basin porphyry copper deposit, Yavapai County, AZ. M, 1978, University of Arizona.

Christman, Joseph Robert. A provenance and environmental analysis of the upper Eocene and Oligocene sediments of the Fayum Depression, Egypt. M, 1980, Iowa State University of Science and Technology.

Christman, Robert Adam. Age of the high erosion surface in the Wyoming ranges. M, 1947, University of Michigan.

Christman, Robert Adam. Geology of Saint Bartholomew, Saint Martin, and Anguilla, West Indies. D, 1950, Princeton University. 172 p.

Christner, James Blaine. The Comanche Peak Formation. M, 1929, University of Texas, Austin.

Christoffersen, Roy Gray. Studies of feldspar kinetics; I, Disordering kinetics and stability of low-andesine; II, Na-K interdiffusion in the alkali feldspars. D, 1982, Brown University. 168 p.

Christofferson, Eric. Linear magnetic anomalies in the Colombia Basin, central Caribbean Sea. D, 1973, University of Rhode Island. 76 p.

Christofferson, Keith A. Subsurface geology of the Mississippian of the Virginia Hills area, Alberta. M, 1959, Michigan State University. 271 p.

Christopher, Cranston C. Pennsylvanian kirkbyacean and hollinacean ostracodes of the Appalachian Basin. M, 1986, Bowling Green State University. 106 p.

Christopher, James Ellis. An investigation of Lake Erie shore erosion between Fairport Harbor and the Mentor Yacht Club, Lake County, Ohio. M, 1956, Ohio State University.

Christopher, James Ellis. Geology of the Ohio shore of Lake Erie between Fairport and the Pennsylvania border. D, 1959, Ohio State University. 296 p.

Christopher, Michael T. Structure and petrology of the Fountain Quarry Granite. M, 1979, East Carolina University. 74 p.

Christopher, P. A. Fission track ages of Younger intrusions in southern Maine. M, 1968, Dartmouth College. 45 p.

Christopher, Peter Allen. Application of K-Ar and fission-track dating to the metallogeny of porphyry and related mineral deposits in the Canadian Cordillera. D, 1973, University of British Columbia.

Christopher, Raymond A. Lower Upper Cretaceous plant microfossils from Block Island, Rhode Island. M, 1967, University of Rhode Island.

Christopher, Raymond Anthony. The application of statistical techniques to the palynofloral analysis of the Coker Formation (upper Cretaceous), western Alabama (Volumes I and II). D, 1971, Louisiana State University. 485 p.

Christopherson, Karen Rae. A geophysical study of the Steamboat Springs, Colorado geothermal systems. M, 1979, University of Colorado.

Christy, Joseph J. A comparative study of the "miner" and least squares location techniques as used for the seismic location of trapped coal miners. M, 1982, Pennsylvania State University, University Park. 191 p.

Christy, Michael Scott. Investigations into the dynamics of water movement on a low relief landscape on the Eastern Shore of Maryland. D, 1980, University of Maryland. 169 p.

Christy, Robert Brandt. Some Permian fusulinid faunas near Lee Canyon, Clark County, Nevada. M, 1958, University of Illinois, Urbana.

Chroback, David Allen. Evaluation of various weathering and sedimentologic analyses for the differentiation of river terraces in the Nenana Valley, Alaska. M, 1980, Southern Illinois University, Carbondale. 116 p.

Chronic, Byron John, Jr. Paleozoic stratigraphy along the Alaska Highway in northeastern British Columbia, Canada. M, 1947, University of Kansas. 89 p.

Chronic, Byron John, Jr. Upper Paleozoic of Peru, invertebrate paleontology (excepting fusulinids and corals). D, 1949, Columbia University, Teachers College.

Chronic, Felicie Jane. Geology of the Guano-Guayes rare-earth element bearing skarn property, Pelly Mountains, Yukon Territory. M, 1979, University of British Columbia.

Chronic, Halka. Molluscan fauna from the Permian Kaibab Formation, Walnut Canyon, Arizona. D, 1953, Columbia University, Teachers College.

Chronic, Lucy M. The interrelation of fauna and lithology across a Late Cambrian biomere boundary in Wyoming. M, 1988, University of Wyoming. 70 p.

Chrow, James Kenneth. A paleogeologic study of the Michigan Basin. M, 1958, University of Michigan.

Chruscicki, Jean B. Selected trace element distribution in water and sediment in three mine-impacted regions in southern Illinois. M, 1980, Southern Illinois University, Carbondale. 154 p.

Chryssafopoulos, Hanka Wanda Sobczak. Identification of young tills and study of some of their engineering properties in the Greater Chicago area. D, 1964, University of Illinois, Urbana. 471 p.

Chrzastowski, Michael J. Net shore-drift of King County, Washington. M, 1982, Western Washington University. 153 p.

Chrzastowski, Michael J. Stratigraphy and geologic history of a Holocene lagoon; Rehoboth Bay and Indian River Bay, Delaware. D, 1986, University of Delaware. 365 p.

Chu, A. E. C. Differential fixation of phosphate by the Hawaiian soils. M, 1951, University of Hawaii. 42 p.

Chu, Chaw-Long. Constitutive relations of clays and clayey fault gouge at high pressures. D, 1984, University of California, Berkeley. 394 p.

Chu, Ching-Jui. Another method of determining the tilt of an aerial photograph and some suggestions on improving the stereocomparagraph method of mapping. D, 1945, Cornell University.

Chu, Gordon Robert. The behavior and transport of anthropogenic radionuclides in the Peconic River, New York. M, 1981, SUNY at Stony Brook.

Chu, H. J. Zonal arrangements of ore deposits of South China. D, 1944, University of Minnesota, Minneapolis. 29 p.

Chu, Jean Juming. Induced polarization data at Roosevelt Hot Springs geothermal area, Utah. M, 1980, University of Utah. 85 p.

Chu, Jia-Bao. Oxygen and carbon isotopes and mineral chemistry of metamorphic rocks from the Nanao District, eastern Taiwan. M, 1980, Purdue University. 113 p.

Chu, Jiaw. Windowed filtering of gravity data. M, 1976, Southern Methodist University. 35 p.

Chu, Kai-Dee. The kinetics of neutralization of Po-218. D, 1987, University of Illinois, Urbana. 94 p.

Chu, Peter H. T. Metamorphism of the Meguma Group in the Shelburne area, Nova Scotia. M, 1978, Acadia University.

Chu, Sen. Geomorphologic history of the Nanking Hills. M, 1937, Columbia University, Teachers College.

Chu, Shi-Chih. Geotechnical properties and disposal considerations for flue gas desulfurization sludges. D, 1983, Northwestern University. 264 p.

Chu, Show-Chuyan. Development of water resources in Tachia River basin. M, 1968, [Colorado State University].

Chu, Ting Oo. The study of hydrothermal talc. M, 1922, University of Wisconsin-Madison.

Chu, Tyan-Ming. Investigation of the thermal regime in a river-aquifer system near Ashland, Nebraska. M, 1988, University of Nebraska, Lincoln. 113 p.

Chu, Wei Chun. Well test analysis for two-phase flow. D, 1981, University of Tulsa. 191 p.

Chu, Wen-kuan. The adsorption of zinc by soil minerals. D, 1968, University of California, Berkeley. 66 p.

Chuamthaisong, Charoen. Geology and groundwater resources of the Chiangmi Basin, Thailand. M, 1970, University of Alabama.

Chuang, W. S. Propagation and generation of internal tides on the continental margin. D, 1980, The Johns Hopkins University. 126 p.

Chubb, P. A. Genesis, nature, and occurrence of the ore deposits at the Ross Mine. M, 1937, University of Toronto.

Chuber, Stewart. Late Mesozoic stratigraphy of the Elk Creek-Fruto area, Glenn County, California. D, 1961, Stanford University. 115 p.

Chuchla, Richard Julian. Reconnaissance geology of the Sierra Rica area, Chihuahua, Mexico. M, 1981, University of Texas, Austin. 199 p.

Chuilli, Allan T. The geology and stratigraphy of the northeast quarter of White Cross Quadrangle, Orange County, North Carolina. M, 1987, University of North Carolina, Chapel Hill. 74 p.

Chulick, John Alexander. A comparison of M_L to m_{Lg} for large California earthquakes with scaling considerations for strong ground motion. M, 1985, St. Louis University.

Chuman, Richard Wayne. The stratigraphy of the Mesaverde Formation and Lewis Shale north of Rawlins, Wyoming. M, 1954, University of Nebraska, Lincoln.

Chun, E. H. L. Determination of radioactivity of Hawaiian lavas by means of nuclear emulsions. M, 1954, University of Hawaii. 32 p.

Chun, Insik. The determination of local scour potential around circular piles subjected to waves oblique to current. D, 1987, University of Texas, Austin. 326 p.

Chun, Joong Hee. The average dissipation curve of an attenuative layered Earth medium. D, 1972, University of Missouri, Rolla. 90 p.

Chun, Joong Hee. Two-dimensional seismic model studies. M, 1966, University of Missouri, Rolla.

Chun, Kin-Yip. Crustal shear velocity modelling in Nevada; a study of broadband multi-mode surface waves. D, 1983, University of California, Berkeley. 157 p.

Chung, Donald Ta-Lung. Real-time expert system for fault detection and failure prevention; and application. D, 1988, University of Maryland. 227 p.

Chung, Gong Soo. Application of nuclear fission track mapping of uranium to the study of diagenesis in carbonate rocks. D, 1988, University of Miami. 375 p.

Chung, H. M. Isotope fractionation during the maturation of organic matter. D, 1976, Texas A&M University. 174 p.

Chung, Hung Tan. The interpretation of massive sandstones in turbidite sequences; an example from the Miocene Stevens Sandstone in the Paloma oil field, Kern County, California. M, 1988, University of California, Riverside. 221 p.

Chung, Ming-Ping. Seismic analysis of buried pipeline networks. D, 1984, Arizona State University. 406 p.

Chung, Pham Kim. Mississippian Coldwater Formation of the Michigan Basin. D, 1973, Michigan State University. 159 p.

Chung, R. M. Directional variation of compressibility and permeability in an anisotropic kaolin clay. D, 1977, Northwestern University. 381 p.

Chung, Sang-Ok. Stochastic modeling of water movement in the saturated-unsaturated zone. D, 1985, Iowa State University of Science and Technology. 173 p.

Chung, Wai-Ying. I, Variation of seismic source parameters and stress drop within a descending slab as revealed from body-wave pulse-width and amplitude analysis; II, A seismological investigation of the subduction mechanism of aseismic ridges. D, 1979, California Institute of Technology. 179 p.

Chung, Yu-Chia. Pacific deep and bottom water studies based on temperature, radium, and excess random measurements. D, 1971, University of California, San Diego.

Chunkao, Kasem. A comparison of methods for evaluating aggregate stability of mountain soils. M, 1965, Colorado State University. 65 p.

Church, Barry Neil. Geology of the White Lake area (south-central British Columbia). D, 1968, University of British Columbia.

Church, Barry Neil. Petrology of some early Tertiary lavas of the Kettle River region, British Columbia. M, 1963, McMaster University. 161 p.

Church, Clifford C. A laboratory study of certain Tertiary formations of the northern Wasatch Mountains, Utah. M, 1925, Stanford University. 41 p.

Church, Earl. Effect of deflections of the vertical upon measurements of horizontal angles. M, 1916, Syracuse University.

Church, H. Victor. The structural geology of the central portion of the Gros Ventre Range, Wyoming. D, 1940, University of Chicago. 23 p.

Church, John A. The Comstock Lode, its formation and history. D, 1879, Columbia University, Teachers College.

Church, Joseph F. A study of the amphibolites from Sulphide, Ontario. M, 1954, University of New Brunswick.

Church, M. A. Baffin Island sandar; a study of Arctic fluvial environments. D, 1975, University of British Columbia.

Church, Mary S. A quantitative petrographic study of the Black Mountain Intrusion at West Dummerston, Vermont. M, 1935, Smith College. 98 p.

Church, Norman K. Lithostratigraphy and carbonate-evaporite petrology of the Middle Devonian Wapsipinicon Formation in (eastern) Iowa. M, 1967, University of Iowa. 120 p.

Church, Richard E. and Durfee, M. Charles. Geology of the Fossil Creek area, White Mountains, Alaska. M, 1961, University of Alaska, Fairbanks. 96 p.

Church, Richard R. A geological survey of the Warm Spring Mountain, Fremont County, Wyoming. M, 1950, Miami University (Ohio). 80 p.

Church, S. B. The paleoenvironment of the Lower Triassic Thaynes Formation near Cascade Springs, Wasatch County, Utah. M, 1974, Brigham Young University.

Church, Stanley Eugene. Lead and strontium isotope geochemistry of the Cascade mountains (Washington, Oregon and northern California). D, 1970, University of California, Santa Barbara. 124 p.

Church, Stanley Eugene. Nuclear fission-track ages of minerals from granitic rocks (Precambrian) from the Sawatch Range, Colorado. M, 1967, University of Kansas. 33 p.

Church, Stephen B. Lower Ordovician patch reefs in western Utah. M, 1974, Brigham Young University. 62 p.

Church, Victor. The stucture and stratigraphy of the Upper Cretaceous near Redding, California. M, 1937, California Institute of Technology. 23 p.

Churcher, P. L. Clay distribution in carbonate reservoirs; examples from the Silurian of southwestern Ontario. M, 1987, University of Waterloo. 69 p.

Churchill, R. R. Topoclimate as a controlling factor in badlands hillslope development. D, 1979, University of Iowa. 173 p.

Churchill, Ronald. A geochemical and petrological investigation of the Cu-Ni sulfide genesis in the Duluth Complex, Minnesota. M, 1978, University of Minnesota, Duluth.

Churchill, Ronald Keith. Meteoric water leaching and ore genesis at the Tayoltita silver-gold mine, Durango, Mexico. D, 1980, University of Minnesota, Minneapolis. 173 p.

Churkin, Michael, Jr. Middle Paleozoic stratigraphy of central Idaho. D, 1961, Northwestern University. 142 p.

Churkin, Michael, Jr. Silurian stratigraphy of part of the Yreka and China Mountain quadrangles, Siskiyou County, California. M, 1958, University of California, Berkeley. 86 p.

Churnet, Habte Giorgis. The relationships between dolomitization and sphalerite mineralization in the Lower Ordovician upper Knox carbonate rocks of the Copper Ridge District, East Tennessee. D, 1979, University of Tennessee, Knoxville. 237 p.

Churney, Garth. Computation of synthetic seismograms by ray methods. M, 1977, University of Alberta. 92 p.

Chute, John Lawrence, Jr. Polarization studies of geomagnetic rapid variations. D, 1969, Columbia University. 100 p.

Chute, Michael Earl. Geochemistry of the Aulneau granitic batholith, District of Kenora, Ontario, Canada. M, 1977, University of Manitoba.

Chute, Newton E. The orientation of veins to batholiths. M, 1931, University of Minnesota, Duluth.

Chute, Newton Earl. The upper contact of the Sudbury nickel intrusive. D, 1937, Harvard University.

Chyenoweth, Clyde E. The petrography of some Huronian rocks in Baraga County, Michigan. M, 1951, Wayne State University.

Chyi, Kwo-Ling. Hydrothermal stability study of supercalcine. M, 1980, University of Toledo. 95 p.

Chyi, Kwo-Ling. Relationships between crystal structure, bonding and thermal stability of amphiboles. D, 1987, Virginia Polytechnic Institute and State University. 110 p.

Chyi, Lindgren Lin. Distribution of some noble metals in sulphide and oxide minerals in Strathcona Mine, Sudbury. D, 1972, McMaster University. 183 p.

Chyi, Lindgren Lin. The geochemistry of Pd, Os, Ir and Au in the Mount Albert ultramafic pluton Quebec. M, 1969, McMaster University. 90 p.

Chyi, Michael So. The geochemical study of the Jurassic submarine hydrothermal manganese deposits of the Franciscan assemblage, California. D, 1984, Princeton University. 242 p.

Chytalo, Karen Nadia. PCBs in dredged materials and benthic organisms in Long Island Sound. M, 1979, SUNY at Stony Brook.

Ciampa, John David. Microcracks, residual strain, velocity, and elastic properties of igneous rocks from a geothermal test-hole at Fenton Hill, New Mexico. M, 19??, Texas A&M University.

Ciancanelli, Eugene V. Structural geology of the western edge of the Granite Wash Mountains, Yuma County, Arizona. M, 1965, University of Arizona.

Ciaramilla, Philip Stephen. The geology of the Tannersville area, Tazewell County, Virginia. M, 1959, Virginia Polytechnic Institute and State University.

Cibula, Duane Allen. The geology and ore deposits of the Cosala mining district, Cosala Municipality, Sinaloa, Mexico. M, 1975, University of Iowa. 145 p.

Ciccone, Anthony Donato. Flow visualization/digital image analysis of saltating particle motions in a wind-generated boundary-layer. D, 1988, University of Toronto.

Cicerone, Robert D. The attenuation of crustal seismic waves in New England. M, 1980, Boston College.

Cichan, Michael Anthony. Vascular cambium and wood development in selected Carboniferous plants. D, 1984, Ohio State University. 256 p.

Cichetti, Maureen J. Serpentinites of the New York City area; a study of the origin and petrology. M, 1977, Rutgers, The State University, Newark. 55 p.

Cichowicz, Nancy Lee. Development and application of a linked unsaturated-saturated flow model. M, 1979, University of Wisconsin-Madison.

Ciciarelli, John Anthony. Geomorphic evolution of anticlinal valleys in central Pennsylvania. D, 1971, Pennsylvania State University, University Park. 98 p.

Ciciarelli, John Anthony. Some factors affecting the geologic interpretability of aerial photographs (Centre County, Pennsylvania). M, 1967, Pennsylvania State University, University Park. 78 p.

Cid-Dresdner, Hilda. Crystal structure of potassium hexatitanate $K_2Ti_6O_{13}$. M, 1962, Massachusetts Institute of Technology. 58 p.

Ciesielski, Paul F. Silicoflagellate biostratigraphy and paleoecology of Neogene and Oligocene recovered from piston and drill cores off of East Antarctica. M, 1974, Florida State University.

Ciesielski, Paul F. The Maurice Ewing Bank of the Malvinas (Falkland) Plateau; depositional and erosional history and its paleoenvironmental implications. D, 1978, Florida State University. 292 p.

Cieslewicz, Walter John. Recent Soviet research on the organic origin of petroleum. M, 1959, University of Colorado. 172 p.

Cifelli, Richard. Bathonian (Jurassic) foraminifera of England. D, 1959, Harvard University.

Cifelli, Richard. Eocene foraminifera from the Point of Rocks area, California. M, 1951, University of California, Berkeley. 95 p.

Cifelli, Richard Lawrence. The origin and phylogeny of the South American Condylarthra and early Tertiary Litopterna (Mammalia). D, 1983, Columbia University, Teachers College. 424 p.

Cifuentes, Ines Lucia. Source process of the great 1960 Chilean earthquake. D, 1988, Columbia University, Teachers College. 202 p.

Cifuentes, Luis Arturo. Sources and biogeochemistry of organic matter in the Delaware Estuary. D, 1987, University of Delaware, College of Marine Studies. 243 p.

Cifuentes, Luis Arturo. The character and behavior of organic nitrogen in the Delaware Estuary salinity gradient. M, 1982, University of Delaware, College of Marine Studies. 81 p.

Cihlar, Josef. Soil moisture and temperature regimes and their importance to microwave remote sensing of soil water. D, 1975, University of Kansas. 167 p.

Ciliberti, Vito A., Jr. The Libby Dam Project; an ex-post facto analysis of selected environmental impacts, mitigation commitments, recreation usage and hydroelectric power production. D, 1979, University of Montana. 138 p.

Cimon, Jules. Etude de la kaolinisation d'une anorthosite à Château-Richer, Comté de Montmorency, Québec (Canada). M, 1969, Universite Laval.

Cinco Ley, Heber. Unsteady-state pressure distributions created by a slanted well, or a well with an inclined fracture. D, 1974, Stanford University. 173 p.

Cindrich, Richard B. Application of stable isotope and geochemical techniques to problems in geothermal recharge; the Reykjanes Peninsula. M, 1984, Stanford University.

Ciner, Attila T. Stratigraphy and depositional environment of the Bayport Limestone of the southern Michigan Basin. M, 1988, University of Toledo. 133 p.

Cinnamon, Charles Gerald. Parameters and structural variations of the intermediate composition, low plagioclases. D, 1969, University of Wisconsin-Madison. 218 p.

Cinnamon, Charles Gerald. X-ray and optical parameters of low plagioclases; An$_{15-40}$ and An$_{80-90}$. M, 1965, University of Wisconsin-Madison.

Cinque, Mark J. Geology and uranium mineralization of the Hallelujah Junction, Red Rock Canyon area, Lassen County, California, and Washoe County, Nevada. M, 1979, University of Nevada. 117 p.

Cinquemani, L. J. Late Quaternary sea levels in the lower Hudson Estuary. M, 1977, Queens College (CUNY). 34 p.

Cintala, Mark John. The role of planetary variables in impact cratering processes. D, 1980, Brown University. 209 p.

Cintron, John, Jr. The geology of Garfield Quadrangle, Benton County, Arkansas. M, 1962, University of Arkansas, Fayetteville.

Ciolkosz, Edward John. I, The mineralogy and genesis of the Dodge catena of southeastern Wisconsin; II, Rhizosphere weathering and synthesis of soil minerals. D, 1967, University of Wisconsin-Madison. 143 p.

Cipar, John J. Rayleigh wave dispersion and geological provinces of North America. M, 1973, SUNY at Binghamton. 166 p.

Cipar, John Joseph. Seismic source processes and tectonics; observations of four intracontinental earth-

quakes. D, 1981, California Institute of Technology. 156 p.

Cipolletti, Debbie L. Morphometry of Central American composite cones. M, 1988, Rutgers, The State University, New Brunswick. 56 p.

Cipolletti, Robert M. Age determination and paleoecology of marine Paleogene strata, Baja California Norte. M, 1986, Rutgers, The State University, New Brunswick. 148 p.

Cirbus, Lisa. Investigating the accuracy of atmospheric circulation simulated for the last glacial maximum with a general circulation model. M, 1986, Kent State University, Kent. 144 p.

Cirello, John. Transfer of NH_4-N from benthal deposits and NO_3-N losses of overlying waters of the upper Passaic River. D, 1975, Rutgers, The State University, New Brunswick. 280 p.

Ciriacks, Kenneth Wilmer. Permian and Eotriassic bivalves of the Middle Rockies. D, 1962, Columbia University, Teachers College. 316 p.

Cirilli, Jerome P. Late Quaternary diatom paleoecology in late Quaternary sediment from the Pacific sector of the Southern Ocean. M, 1982, Rutgers, The State University, Newark. 111 p.

Cisar, Marilyn Taggi. Chemical trends in the Marcy Anorthosite, Adirondacks, New York. M, 1978, Iowa State University of Science and Technology.

Cisne, John Luther. The anatomy of triarthrus eatoni and its bearing on the phylogeny of Trilobita and Arthropoda. D, 1973, University of Chicago. 288 p.

Cisneros, Lee Anne. Regional paleoenvironmental analysis of coeval Upper Ordovician epeiric sea carbonates in the North American Midcontinent. M, 1988, University of Illinois, Urbana. 86 p.

Cisowski, Stanley M. The effect of shock on the magnetic properties of natural material. D, 1977, University of Pittsburgh. 182 p.

Cissell, Milton Charles. Chemical features of the Columbia River plume off Oregon. M, 1969, Oregon State University. 45 p.

Cisternas, Armando. I, The radiation of elastic waves from a spherical cavity in a half space. II. Precision determination of focal depths and epicenters of earthquakes. D, 1965, California Institute of Technology. 179 p.

Citron, Gary P. Idavada ash-flows in the Three Creek area, southwestern Idaho, and their regional significance. M, 1976, Cornell University.

Citron, Gary P. The Hidden Bay Pluton, Alaska; geochemistry, origin and tectonic significance of Oligocene magmatic activity in the Aleutian island arc. D, 1980, Cornell University. 250 p.

Citrone, Jeffrey. A petrographic study of the Challis Volcanics, White Knob Mountains, Custer County, Idaho. M, 1978, Lehigh University. 142 p.

Civco, Daniel Louis. Knowledge-based classification of Landsat thematic mapper digital imagery. D, 1987, University of Connecticut. 214 p.

Claassen, Daniel R. Mineralogy and diagenesis of a uranium prospect in the Goliad Sandstone, Duval County, Texas. M, 1981, University of Missouri, Columbia.

Clabaugh, Charles Donald. Statistical identification of salinity sources in a shallow water-table aquifer. M, 1987, University of Akron. 137 p.

Clabaugh, Patricia Sutton. Petrofabric analysis of salt crystals from Grand Saline salt dome. M, 1962, University of Texas, Austin.

Clabaugh, Stephen Edmund. Corundum deposits of Montana. D, 1950, Harvard University. 100 p.

Clabaugh, Stephen Edmund. Geology of the northwestern portion of the Cornudas Mountains, New Mexico. M, 1941, University of Texas, Austin.

Clack, W. J. F. Recent carbonate reef sedimentation off the east coast of Carriacou, West Indies. D, 1976, McGill University. 218 p.

Clack, William J. F. Sedimentology of the Mannville group (Cretaceous), in the Cold Lake area, Alberta (Canada). M, 1967, University of Calgary. 95 p.

Clague, David A. The Hawaiian-Emperor seamount chain; its origin, petrology, and implications for plate tectonics. D, 1974, University of California, San Diego. 331 p.

Clague, John Joseph. Landslides in the southeastern part of Point Reyes National Seashore, California. M, 1969, University of California, Berkeley. 107 p.

Clague, John Joseph. Late Cenozoic geology of the southern Rocky Mountain Trench, British Columbia. D, 1973, University of British Columbia.

Clair, Ann Elizabeth St. *see* St. Clair, Ann Elizabeth

Clair, Charles S. Saint *see* Saint Clair, Charles S.

Clair, Charles Spencer Saint *see* Saint Clair, Charles Spencer

Clair, Donald W. Saint *see* Saint Clair, Donald W.

Clair, Gregory M. Saint *see* Saint Clair, Gregory M.

Clair, J. R. Oil and gas resources, Jackson County, Missouri. M, 1938, University of Missouri, Rolla.

Clair, James H. St. *see* St. Clair, James H.

Clair, Joseph Robinson, Jr. Stratigraphy, structure, and economic geology of the Belton area, Cass County, Missouri. D, 1941, University of North Carolina, Chapel Hill. 132 p.

Clair, Stuart St. *see* St. Clair, Stuart

Clair, Virginia. Some Fusulinidae from San Miguel County, New Mexico. M, 1950, Texas Tech University. 39 p.

Clancy, J. Cation exchange capacity of fine-grained sediments. M, 1973, University of Southern California. 85 p.

Clancy, Robert Todd. A finite element model of the San Andreas Fault. M, 1978, Cornell University.

Clanton, Jerry S. Paragenesis of the ores at the San Antonio Mine, Chihuahua, Mexico. M, 1975, Texas Tech University. 63 p.

Clanton, Quinton David. The geology of the Pea Ridge Quadrangle, Benton County, Arkansas. M, 1963, University of Arkansas, Fayetteville.

Clanton, Uel S., Jr. High-temperature X-ray diffraction. M, 1960, University of Texas, Austin.

Clanton, Uel S., Jr. Sorption and release of strontium-89 and cesium-137 by Recent sediments of the Guadalupe River of Texas. D, 1968, University of Texas, Austin. 96 p.

Clapp, C. H. Igneous rocks of Essex County, Massachusetts. D, 1910, Massachusetts Institute of Technology. 241 p.

Clapp, Michael M. Beltian stratigraphy and structure in southern part of Ovando Quadrangle, Montana. M, 1936, University of Montana. 45 p.

Clapp, Roger Burnham. A wetting-front model of soil water dynamics. D, 1982, University of Virginia. 317 p.

Clappin, Philip F. The depositional and diagenetic history of the upper Sinnipegosis Formation (Middle Devonian) in the Williston Basin of North Dakota. M, 1987, Queens College (CUNY). 234 p.

Clardy, Arthur L. A petrotectonic study of the Tesnus Formation in the Marathon Basin, West Texas. M, 1954, University of Houston.

Clardy, Benjamin F. Subsidence structure of the Price Mountain Syncline (Washington and Benton counties, Arkansas). M, 1964, University of Arkansas, Fayetteville.

Clardy, Bruce. Origin of the lower and middle Tertiary Wishbone and Tsadaka formations, Matanuska Valley, Alaska. M, 1974, University of Alaska, Fairbanks. 74 p.

Clare, Patrick Henry. Subsurface geology of Pawnee County, Oklahoma. M, 1961, University of Oklahoma. 139 p.

Clarey, Timothy L. Geologic investigation of the Shirley Fault and vicinity, Carbon County, Wyoming. M, 1984, University of Wyoming. 63 p.

Clark, Alan Raymond. Petrofabric analysis of potash ore beds, Esterhazy, Saskatchewan. M, 1964, University of Saskatchewan. 41 p.

Clark, Albert F. Utah hydrocarbons. M, 1924, University of Utah. 59 p.

Clark, Alex. The Cool-Water Timm's Point Pleistocene horizon at San Pedro, California. M, 1932, California Institute of Technology. 17 p.

Clark, Allen LeRoy. Geology of the Clarkia area, Idaho. M, 1963, University of Idaho. 57 p.

Clark, Allen LeRoy. Wall rock alteration and trace element distribution in the Galena Mine, Shoshone County, Idaho. D, 1968, University of Idaho. 140 p.

Clark, Anthony Miles Stapleton. A reinterpretation of the stratigraphy and deformation of the Aillik Group, Makkovik, Labrador. D, 1974, Memorial University of Newfoundland. 346 p.

Clark, Anthony Miles Stapleton. A structural reinterpretation of the Aillik Series, Labrador. M, 1971, Memorial University of Newfoundland. 79 p.

Clark, Armin Lee. Petrology of the Eocene sediments in Henry, Weakley, and Carroll counties, Tennessee. D, 1973, University of Tennessee, Knoxville. 130 p.

Clark, Armin Lee. Pleistocene molluscan faunas of the Castalia Deposit, Erie County, Ohio. M, 1958, Ohio State University.

Clark, Arthur Roy. The use of drill holes in geophysical investigations. D, 1940, University of Toronto.

Clark, Arthur Watts. Geology of a portion of the St. Helena Quadrangle, California. M, 1948, University of California, Berkeley. 91 p.

Clark, B. Christopher. Comparison of the capabilities of shuttle imaging radar imagery and Landsat MSS/RBV in detecting ancient fluvial system in hyperarid environments. M, 1986, University of Arkansas, Fayetteville.

Clark, Betty A. Stratigraphy of the J Sandstone (Lower Cretaceous), Boulder County and Southwest Weld County, Colorado. M, 1978, Colorado School of Mines. 190 p.

Clark, Bruce Lawrence. Fauna of the San Pablo Group of Middle California. D, 1914, University of California, Berkeley. 187 p.

Clark, Bruce Robert. The origin of slaty cleavage in the Coeur d'Alene district, Idaho. D, 1968, Stanford University. 124 p.

Clark, Bryan Malcolm. Styles of late Pleistocene deltaic sedimentation in a large glaciated lacustrine basin. M, 1986, University of Toronto.

Clark, Burton W. The peridotite dikes of Onondaga County, New York. M, 1908, Syracuse University.

Clark, Burton W. The Trenton Limestone at Rathbone Brook, Herkimer County, New York; its stratigraphy, fauna, and age. D, 1912, The Johns Hopkins University.

Clark, Charles Roosevelt. A study of some connate oil field waters. M, 1932, University of Illinois, Chicago.

Clark, Charles W. The stratigraphy and structure of the Paleozoic and Mesozoic rocks of the West Flank of the Armstead Anticline area, Beaverhead County, Montana. M, 1986, Oregon State University. 209 p.

Clark, Christopher N. Trace elements in Missouri barite. M, 1970, University of Missouri, Columbia.

Clark, Clare M. A preliminary petrologic study of the Dakota Group in southeastern Nebraska. M, 1933, University of Nebraska, Lincoln.

Clark, Clifford Charles. Geophysical studies of permafrost in the Dry Valleys. M, 1973, Northern Illinois University. 97 p.

Clark, Clifton Wirt. Geology and ore deposits of the Sante Fe District, Mineral County, Nevada. D, 1917, University of California, Berkeley. 74 p.

Clark, Clifton Wirt. The process of oil and gas accumulation and its relation to structure. M, 1915, University of Illinois, Chicago.

Clark, Connie M. A Fortran IV program for calculating X-ray powder diffraction patterns; version 5". M, 1972, Pennsylvania State University, University Park.

Clark, D. F. Quaternary nannofossils from the Mid-Atlantic Ridge at 45°N. D, 1974, Dalhousie University. 111 p.

Clark, Daryl Darnell. Textural and compositional analyses of modern Holocene carbonate beach sediment, San Salvador Island, Bahamas. M, 1988, Mississippi State University. 151 p.

Clark, David L. Stratigraphy and sedimentation of the Gardner Formation in central Utah. M, 1954, Brigham Young University. 60 p.

Clark, David Leigh. Marine Triassic stratigraphy in the eastern Great Basin. D, 1957, University of Iowa. 121 p.

Clark, David Raymond. Sedimentology and depositional environments of a hazardous waste site in the sand and gravel aquifer of Escambia County, Florida. M, 1987, Florida State University. 178 p.

Clark, David W. Inorganic chemistry of urban runoff and hydrologic relationships; Bloomington, Indiana. M, 1979, Indiana University, Bloomington. 93 p.

Clark, Dean Stanley. Reconnaissance geology of a portion of Powell and Deer Lodge counties, Montana. M, 1953, Indiana University, Bloomington. 42 p.

Clark, Dean Stanley. Stratigraphy, genesis, and economic potential of the southern part of the Florida land-pebble phosphate field. D, 1972, University of Missouri, Rolla.

Clark, Donald L. The geology of the Juab Quadrangle, Juab County, central Utah. M, 1987, Northern Illinois University. 324 p.

Clark, Douglas B. Trace element chemistry of the Kirkpatrick Basalt, Storm Peak, central Transantarctic Mountains. M, 1974, Ohio State University.

Clark, Douglas H. Tectonic evolution of the Red Hills-southwestern Kern Mountains area, east-central Nevada. M, 1985, Stanford University. 51 p.

Clark, Douglas Robison. Regional hydrologic models for climatology; an application to three Wisconsin watersheds. D, 1981, University of Wisconsin-Madison. 105 p.

Clark, E., Jr. The physical environment of Greater Dallas. M, 1976, Baylor University.

Clark, Edward Lee. Geology of the Cassville Quadrangle, Missouri. D, 1941, University of Missouri, Columbia.

Clark, Edward Lee. The geology of a portion of the Halltown Quadrangle, Missouri. M, 1931, University of Iowa. 19 p.

Clark, Eugene E. Late Cenozoic volcanic and tectonic activity along the eastern margin of the Great Basin, in the proximity of Cove Fort, Utah. M, 1976, Brigham Young University. 114 p.

Clark, Frederick. Upper Cretaceous foraminifera and biostratigraphy of C-Y Creek, Western Australia. M, 1979, Carleton University. 317 p.

Clark, Frederick Emory. Stratigraphic analysis of the middle Miocene Modelo Formation at lower Piru Creek, Ventura County, California. M, 1988, California State University, Northridge. 110 p.

Clark, Genevieve. Glacial physiography of the Quinsigamond River valley, Grafton, Massachusetts. M, 1948, Clark University.

Clark, George Hollinger. The geology of the eastern flank of the San Joaquin Hills (Orange County, California). M, 1952, Pomona College.

Clark, George Michael. Structural geomorphology of a portion of the Wills Mountain Anticlinorium, Mineral and Grant counties, West Virginia. D, 1967, Pennsylvania State University, University Park. 201 p.

Clark, George Richmond, II. Shell characteristics of the family Pectinidae as environmental indicators. D, 1969, California Institute of Technology. 108 p.

Clark, George S. Isotopic age study of metamorphism and intrusion in western Connecticut and southeastern New York. D, 1967, Columbia University. 95 p.

Clark, George Sydney. Feldspars in the Saint Stephen mafic igneous complex. M, 1962, University of New Brunswick.

Clark, Horace B., III. Preliminary seismic reflection studies of the Brevard Zone near Rosman, North Car-

olina. M, 1974, Virginia Polytechnic Institute and State University.

Clark, Howard Charles, Jr. The remanent magnetization, cooling history, and paleomagnetic record of the Marys Peak Sill, Oregon. D, 1967, Stanford University. 67 p.

Clark, Ian D. Isotope hydrogeology and geothermometry of the Mount Meager geothermal area. M, 1980, University of Waterloo.

Clark, J. A. Global sea level changes since the last glacial maximum and sea level constraints on the ice sheet disintegration history. D, 1977, University of Colorado. 164 p.

Clark, J. Robert. The geology and trace element distributions of the sulfide bodies at Orchan Mine, Matagami, Quebec. D, 1983, Colorado School of Mines. 446 p.

Clark, J. W. Runoff and soil moisture. M, 1953, University of Missouri, Rolla.

Clark, Jackson L. Structure and petrology pertaining to a beryl deposit, Baboquivari Mountains, Arizona. M, 1956, University of Arizona.

Clark, James Alan. The petrology and depositional environment of the Grayhorse Limestone in north-central Oklahoma. M, 1980, Oklahoma State University. 167 p.

Clark, James G. Age, chemistry, and tectonic significance of Easter and Sala y Gomez islands. M, 1975, Oregon State University. 131 p.

Clark, James Gregory. Geology and petrology of South Sister Volcano, High Cascade Range, Oregon. D, 1983, University of Oregon. 235 p.

Clark, James Samuel. Climate change, fire occurrence, and forest influences during the last 750 years in northwestern Minnesota. D, 1988, University of Minnesota, Minneapolis. 183 p.

Clark, Jeffrey L. Sulfide mineralization in the Kona Dolomite, Marquette County, Michigan. M, 1974, Michigan Technological University. 49 p.

Clark, Jerry H. Geology of the East Lake Creek area, Eagle County, Colorado. M, 1960, University of Colorado.

Clark, John. New turtle from the Duchesne Oligocene of the Uinta Basin, northeastern Utah. M, 1932, University of Pittsburgh.

Clark, John. Stratigraphy and paleontology of the Chadron Formation in the Big Badlands of South Dakota. D, 1935, Princeton University. 96 p.

Clark, John H. Geology of the northeast quarter of the Mt. Wright Quadrangle, Montana. M, 1964, Washington State University.

Clark, John Harris. The geology of the Ordovician carbonate formations in the State College, Pennsylvania area and their relationships to the general occurrence and movement of ground water. M, 1965, Pennsylvania State University, University Park. 114 p.

Clark, John Robert. The petrology and geochemistry of the Seroyer Branch mafic-ultramafic complex, Chambers County, Alabama. M, 1973, University of Alabama.

Clark, John Thaddeus. Geology of the Bartlett Springs Trend, northern Coast Ranges, California. M, 1983, University of Texas, Austin. 65 p.

Clark, Jon Peter. Biomarker characterization of evaporite-carbonate source rocks and crude oils, and oil-source correlations, in the Black Creek basin, Alberta, Canada. M, 1988, University of Oklahoma. 192 p.

Clark, Joseph Clyde. Subsurface Pennsylvanian geology, western Coke County, Texas. M, 1959, University of Texas, Austin.

Clark, Joseph Clyde. Tertiary stratigraphy of the Felton-Santa Cruz area, Santa Cruz Mountains, California. D, 1966, Stanford University. 240 p.

Clark, Joseph Marsh. The geology of an area near Mineola, Missouri. M, 1926, University of Missouri, Columbia.

Clark, Joyce Lucas. Stratigraphic palynology of the Paleocene and Eocene of the western Santa Ynez

Mountains, Santa Barbara County, California. M, 1981, University of California, Santa Barbara. 238 p.

Clark, Karen A. Bedrock geology and karst development in Grant and Mineral counties, West Virginia. M, 1976, University of Toledo. 100 p.

Clark, Kenneth Frederick. Geology and ore deposits of the Eagle Nest Quadrangle, New Mexico. D, 1966, University of New Mexico. 363 p.

Clark, Kenneth Frederick. Hypogene zoning in the Lordsburg mining district, Hidalgo County, New Mexico. M, 1962, University of New Mexico. 136 p.

Clark, Kenneth W. Interactions between the Clark Fork River and Missoula Aquifer, Missoula County, Montana. M, 1986, University of Montana. 157 p.

Clark, Lindsey D. A contrast in the chemistry of phosphogenesis in selected areas of southwestern and southeastern North America. M, 1985, North Carolina State University. 202 p.

Clark, Lloyd A. Phase relations in the Fe-As-S system. D, 1959, McGill University.

Clark, Lloyd A. Sulphide deposits of the Hanson Lake area. M, 1956, University of Saskatchewan. 53 p.

Clark, Lorin D. Precambrian geology of the Norway Lake area, Dickinson County, Michigan. D, 1955, University of Chicago. 72 p.

Clark, M. E. Variations in skeletal characters of cystiphyllid corals from the Middle Devonian Hamilton Group of New York. M, 1977, SUNY at Binghamton. 104 p.

Clark, Malcolm Mallory. Pleistocene glaciation of the drainage of the West Walker River, Sierra Nevada, California. D, 1967, Stanford University. 170 p.

Clark, Martha Ann. Surface to subsurface study of the Desmoinesian conglomerates of the Ardmore Basin, south-central Oklahoma. M, 1983, Baylor University. 177 p.

Clark, Megan Elizabeth. The geology of the Victory gold mine, Kambalda, Western Australia. D, 1987, Queen's University. 199 p.

Clark, Michael B. Stratigraphy of the Sperati Point Quadrangle, McKenzie County, North Dakota. M, 1966, University of North Dakota. 108 p.

Clark, Michael Lee. Protolith and tectonic setting of an Archean quartzofeldspathic gneiss sequence in the Blacktail Mountains, Beaverhead County, Montana. M, 1987, Montana State University. 86 p.

Clark, Michael Neil. Tectonic geomorphology and neotectonics of the Ojai Valley and upper Ventura River. M, 1982, University of California, Santa Barbara. 77 p.

Clark, Michael S. Planktonic foraminifera of the Ripley Formation (Maestrichtian), Mississippi. M, 1980, Mississippi State University. 203 p.

Clark, Michael Sidney. Clay mineralogy of the Upper Jurassic to Cretaceous section of the Great Valley Sequence exposed at Putah Creek in California. M, 1979, University of California, Davis. 97 p.

Clark, Murlene Wiggs. Paleogene abyssal environments of the eastern South Atlantic; a foraminiferal study. D, 1983, Florida State University. 295 p.

Clark, O. S. Geology of Henderson County, North Carolina. M, 1942, University of North Carolina, Chapel Hill. 32 p.

Clark, P. U. Late Quaternary history of the Malone area, New York. M, 1980, University of Waterloo.

Clark, Pamela Elizabeth. Correction, correlation, and theoretical consideration of lunar X-ray fluorescence intensity ratios. D, 1979, University of Maryland. 184 p.

Clark, Patricia J. An investigation of the crystal structure of thomsenolite $NaCaAlF_6 \cdot H_2O$. M, 1951, University of Manitoba.

Clark, Peter Underwood. Glacial geology of the Kangalaksiorvik-Abloviak region, northern Labrador, Canada. M, 1984, University of Colorado. 261 p.

Clark, Redmond R. The role of climate-associated force and resistance functions in regional erosion rate variations. D, 1978, Southern Illinois University, Carbondale. 118 p.

Clark, Reino. Stratigraphy, sedimentology and copper-uranium occurrences of the upper part of the Middle Pennsylvanian Minturn Formation, Sangre de Cristo Mountains, Colorado. M, 1982, Colorado School of Mines. 151 p.

Clark, Richard C. The structural geology of the Thomsom Formation; Cloquet and Esko quadrangles, east-central Minnesota. M, 1985, University of Minnesota, Duluth. 114 p.

Clark, Richard D. Industrial development for southeastern New Mexico, a case study. M, 1969, New Mexico Institute of Mining and Technology. 107 p.

Clark, Robert C., Jr. Saturated hydrocarbons in marine plants and sediments. M, 1966, Massachusetts Institute of Technology. 96 p.

Clark, Robert Charles, Jr. The biogeochemistry of aromatic and saturated hydrocarbons in a rocky intertidal marine community in the Strait of Juan de Fuca. D, 1983, University of Washington. 268 p.

Clark, Robert E. Prediction of trafficability in the Big Delta, Alaska area from aerial photographs. M, 1950, Purdue University. 113 p.

Clark, Robert Owen. The morphology of slope profiles; a statistical method to critically analyze the slope hypotheses of William M. Davis, Walther Penck, and Lester C. King. D, 1970, University of Denver. 458 p.

Clark, Robert S. The structure and stratigraphy of Government Ridge, Utah. M, 1953, Brigham Young University. 75 p.

Clark, Robert Scarth. A fauna from the Pierre concretions of southwestern South Dakota. M, 1934, University of Missouri, Columbia.

Clark, Robert Stephen. Iodine, uranium and tellurium in meteorites. D, 1969, University of Arkansas, Fayetteville. 59 p.

Clark, Robert Watson. Geology and oil and gas development in Okmulgee, Oklahoma and a study of the origin of the oil-bearing domes of the Okmulgee, Oklahoma district. D, 1924, University of Michigan.

Clark, Robey Harned. A petrologic study of some sandstones of the Des Moines Series of south-central Iowa. M, 1949, University of Wisconsin-Madison. 29 p.

Clark, Roger Alan. Stratigraphy, depositional environments, and carbonate petrology of the Toroweap and Kaibab formations (Lower Permian), Grand Canyon region, Arizona. D, 1981, SUNY at Binghamton. 454 p.

Clark, Roger Nelson. Spectroscopic studies of water and water/regolith mixtures on planetary surfaces at low temperatures. D, 1980, Massachusetts Institute of Technology. 337 p.

Clark, Russell Gould, Jr. Petrography and petrology of the Middlefield granite (late Devonian), Chester Quadrangle (west central), Massachusetts. M, 1968, Michigan State University. 52 p.

Clark, Russell Gould, Jr. Petrologic and structural aspects of the Kinsman Quartz monzonite (south-central New Hampshire) and some related rocks. D, 1972, Dartmouth College. 225 p.

Clark, S. B. Subsurface geology of the Mississippian System of western Kiowa County, Kansas. M, 1974, Wichita State University. 80 p.

Clark, S. R. Deliverability of gas wells with turbulence and pressure dependent gas properties. M, 1977, Stanford University.

Clark, Samuel Gilbert. The Milton Formation of the Sierra Nevada of California. D, 1930, University of California, Berkeley. 206 p.

Clark, Samuel Harvey, Jr. The Eocene Point of Rocks Sandstone; provenance, mode of deposition, and implications for the history of offset along the San Andreas Fault in Central California. D, 1973, University of California, Berkeley. 302 p.

Clark, Sandra Helen Becker. Geology of the Priest River area, Idaho. M, 1964, University of Idaho. 83 p.

Clark, Sandra Helen Becker. Structure and petrology of the Priest River–Hoodoo Valley area, Bonner County, Idaho. D, 1967, University of Idaho. 137 p.

Clark, Stacy Lon. Structural and petrologic comparison of the southern Sapphire Range, Montana, with the northeast border zone of the Idaho Batholith. M, 1979, Western Michigan University. 88 p.

Clark, Stephen Christopher Lane. Evolution of a multicyclic caldera system and magma chamber; the Tejeda Caldera, Gran Canaria, Spain. D, 1988, Princeton University. 342 p.

Clark, Stephen Rex. Petrology of the Bear Poplar and Barber mafic intrusions, North Carolina. M, 1980, University of Tennessee, Knoxville. 77 p.

Clark, Sydney Procter, Jr. Terrestrial heat flow in the Swiss Alps. D, 1955, Harvard University.

Clark, Sylvia Robb. A mineralogical study of the Blister Pegmatite near Gilsum, New Hampshire. M, 1958, Smith College. 97 p.

Clark, Terry E. Zeolites from the Kings Valley and Coffin Butte areas, Benton County, Oregon. M, 1964, University of Oregon. 99 p.

Clark, Thomas. Equilibria between iron-nickel olivines and iron-nickel sulfides. M, 1970, University of Toronto.

Clark, Thomas. Geology of an ultramafic complex on the Turnagain River, northwestern British Columbia. D, 1975, Queen's University. 454 p.

Clark, Thomas Henry. Studies in the Beekmantown series of Levis, Quebec. D, 1923, Harvard University.

Clark, Thomas Phillips. Hydrogeology, geochemistry, and public health aspects of environmental impairment at an abandoned landfill near Austin, Texas. M, 1972, University of Texas, Austin.

Clark, Tony Franklin. A geological model of the Lesser Antilles subduction zone complex. D, 1974, University of North Carolina, Chapel Hill. 81 p.

Clark, Virginia Ann. Effect of volatiles on seismic attenuation and velocity in sedimentary rocks. D, 1980, Texas A&M University. 189 p.

Clark, W. S. Radium transport through a granular geologic porous medium. M, 1982, University of Waterloo. 116 p.

Clark, Walter Thomas, Jr. Petrology and stratigraphy of Kiamichi Formation, West Texas, eastern New Mexico. M, 1948, Texas Tech University. 45 p.

Clark, Wesley Inman. X-ray fluorescent determination of trace elements on ion exchange membranes. M, 1962, University of Nevada. 78 p.

Clark, William D. Geology of the White Rock-Chapin area, South Carolina. M, 1969, University of South Carolina. 54 p.

Clark, William K. Ostracod microfaunas of the Lower Permian of Riley County, Kansas. M, 1950, Kansas State University. 65 p.

Clark, William O. Ground water resources of the Niles Cone and adjacent areas, California. M, 1915, [Stanford University].

Clarke, Alexander C. Study of some sulphide ores. M, 1912, Columbia University, Teachers College.

Clarke, Anthony David. Light absorption characteristics of the remote background aerosol; methods and measurements. D, 1983, University of Washington. 144 p.

Clarke, Anthony Orr. Quaternary surficial deposits and their relationship to landforms in the San Bernardino Valley. D, 1977, University of California, Los Angeles. 186 p.

Clarke, Barbara G. Quantitative geomorphology of Furnace Run basin, Summit County, Ohio. M, 1983, University of Akron. 96 p.

Clarke, Charles Edward. Conodonts from the Glen Dean Formation of Kentucky and equivalent formations of Virginia and West Virginia. M, 1959, Texas Tech University. 62 p.

Clarke, Craig W. The geology of the El Tiro Hills, West Silverbell Mountains, Pima County, Arizona. M, 1965, University of Arizona.

Clarke, Denis Edmund. A geochemical and statistical study of the Okanogan Range, Washington. D, 1974, Stanford University. 261 p.

Clarke, Donald Shane. Cu-Zn mineralization and alteration in the Waite area, Noranda District, Quebec. M, 1983, University of Western Ontario. 174 p.

Clarke, Gerald K. C. Geophysical measurements on the Kaskawulsh and Hubbard glaciers, Yukon Territory, Canada. M, 1964, University of Toronto.

Clarke, Gerald K. C. Optimum time-variable digital filters for seismic signal enhancement. D, 1967, University of Toronto.

Clarke, Harris G. The geology of the Howard Quarter area, Claiborne County, Tennessee. M, 1958, Mississippi State University. 35 p.

Clarke, James Wood. The geology of the Thomaston Quadrangle, Georgia. D, 1950, Yale University. 165 p.

Clarke, John Edward Hughes. The geological record of the 1929 "Grand Banks" earthquake and its relevance to deep-sea clastic sedimentation. D, 1988, Dalhousie University.

Clarke, K. E. Ecology of a subarctic coastal system, North Point, James Bay, Ontario. M, 1980, University of Guelph.

Clarke, Michael. Hydrothermal geochemistry of silver-gold vein formation in the Tayoltita Mine and San Dimas mining district, Durango and Sinaloa, Mexico. D, 1986, University of Arizona. 166 p.

Clarke, Murray K. The Farmers siltstone (lower Mississippian); a flysch-like deposit in northeastern Kentucky. M, 1969, University of Kentucky. 69 p.

Clarke, Pamela R. Effects of hydrostatic pressure on remanent magnetization of magnetite. M, 1979, University of Iowa. 92 p.

Clarke, Peter J. Geology in the vicinity of the Grenville front, Mount Wright District, Quebec. D, 1964, University of Manitoba.

Clarke, Peter John. A petrographic and chemical study of the Bevier coal seam. M, 1953, University of Missouri, Rolla.

Clarke, Peter Johnston. A reconnaissance study of joints in the Precambrian and Paleozoic rocks near Kingston (Ontario). M, 1959, Queen's University. 118 p.

Clarke, R. J. A reconnaissance study of joints in Precambrian and Palaeozoic rocks near Kingston, Ontario. M, 1959, Queen's University.

Clarke, R. S., Jr. Schreibersite growth and its influence on the metallography of coarse structured iron meteorites. D, 1976, George Washington University. 207 p.

Clarke, Robert Travis. Palynology of the Secor Coal (Pennsylvanian) of Oklahoma. M, 1961, University of Oklahoma. 131 p.

Clarke, Robert Travis. Palynology of the Vermejo Formation coals (Upper Cretaceous) in the Canon City coal field, Fremont County, Colorado. D, 1963, University of Oklahoma. 180 p.

Clarke, Theodore S. Glacier runoff, balance and dynamics in the upper Susitna Basin, Alaska. M, 1986, University of Alaska, Fairbanks.

Clarke, Thomas Graham. The geology of the crystalline rocks of the southern half of the Chapel Hill, North Carolina, Quadrangle. M, 1957, University of North Carolina, Chapel Hill. 59 p.

Clarke, W. J. G. Some comparisons between the Sudbury Basin and Bushveld igneous complex. M, 1940, University of Toronto.

Clarkson, Gerry W. Implications for thermal histories of the San Juan Basin and San Juan Mountains since Late Cretaceous time. D, 1984, New Mexico Institute of Mining and Technology. 107 p.

Clary, James Heath. Engineering geology of the Centerville, Texas landslides. M, 1975, Texas A&M University. 153 p.

Clary, Michael R. Geology of the eastern part of the Clark Mountain Range, San Bernardino County, California. M, 1959, University of Southern California.

Clary, Thomas A. Geology of the Schultze granite (Tertiary(?)), (central Arizona) and related copper mineralization. M, 1970, Arizona State University. 37 p.

Classen, David Farley. Thermal drilling and deep ice-temperature measurements on the Fox glacier, Yukon. M, 1970, University of British Columbia.

Classen, James Stark. The geology of a portion of the Half Moon Bay Quadrangle, San Mateo County, California. M, 1959, Stanford University. 57 p.

Classen, Willard John, Jr. Eocene foraminifera from the vicinity of Las Cruces, Santa Barbara County, California. M, 1953, University of California, Berkeley. 86 p.

Claughton, James L. Geology of the lower part of the Wilcox Group of the South Laredo area, Webb and Zapata counties, Texas. M, 1977, West Texas State University. 72 p.

Claus, Richard John. Rotation properties of certain manganese oxide minerals. M, 1954, University of Wisconsin-Madison.

Clausen, Benjamin L. Stratigraphy and structure of the Miocene "Esmeralda" Formation in Stewart Valley, Mineral County, Nevada. M, 1983, Loma Linda University. 77 p.

Clausen, Eric N. Piping, slope development and field sampling studies of Wyoming badlands. D, 1969, University of Wyoming. 132 p.

Clausen, George Samuel. Optimal operation of water-supply systems. D, 1970, University of Arizona. 153 p.

Clausen, John Campbell. Methods for assessing the quality of runoff from Minnesota peatlands. D, 1981, University of Minnesota, Minneapolis. 200 p.

Clausen, John Eric. Paleoecology and stratigraphy of Upper Devonian (Frasnian - Famennian) clastic facies units in Randolph County, West Virginia. M, 1981, Rutgers, The State University, New Brunswick. 115 p.

Clausner, Edward, Jr. Characteristic features of the Florida current. M, 1967, [University of Miami].

Clauson, Victor. Geology of the Sudbury Basin area, Ontario, Canada. D, 1947, University of Washington. 135 p.

Claussen, John P. Direct detection of Niobrara gas using seismic techniques; Yuma County, Colorado. M, 1982, Colorado School of Mines. 72 p.

Clauter, Dean A. An experimental investigation of models of thermoremanence for magnetite powders. D, 1979, University of Pittsburgh. 322 p.

Clautice, Karen H. The geology and sampling of a tungsten occurrence, Hodzana Highlands, Alaska. M, 1988, University of Alaska, Fairbanks. 120 p.

Clavan, Walter. Some hypersthenes from southeastern Pennsylvania and Delaware. D, 1954, Bryn Mawr College. 82 p.

Claveau, Jacques. Geology of the Wakeham-Forget lakes region, north shore, Gulf of St. Lawrence. D, 1944, University of Toronto.

Claveau, Jacques. Origin of the zinc deposits and associated rocks at Long Lake, Ontario. M, 1942, University of Toronto.

Clawsey, Patrick J. The glacial epoch theory refuted. M, 1925, Catholic University of America. 32 p.

Clawson, Paul Norman. Geology of the Greenough barite mine, Missoula County, Montana. M, 1957, University of Montana. 42 p.

Clawson, Steven Ralph. The lateral inverse Q-structure of the upper crust in Yellowstone National Park from seismic refraction data. M, 1981, University of Utah. 183 p.

Claxton, Charles Dale. The geology of the Welch area, Craig County, Oklahoma. M, 1952, University of Oklahoma. 43 p.

Clay, Donald W. Late Cenozoic stratigraphy in the Dry Mountain area, Graham County, Arizona. M, 1960, University of Arizona.

Clay, Donald Wayne. Stratigraphy and petrology of the Mineta Formation (Miocene?, lower) in Pima and eastern Cochise counties, Arizona. D, 1970, University of Arizona. 234 p.

Clay, J. Withers. Relations of metallic ore minerals to the igneous rocks in New England and Eastern Canada. M, 1923, University of Minnesota, Minneapolis. 24 p.

Clay, John Otis. Electric resistivity study of the Pennsylvanian underclays of the Illinois Basin. M, 1948, University of Illinois, Urbana.

Clay, Julia S. The petrological evolution of the gabbroic rocks in the Palisades Intrusion, Haverstraw Quarry, Rockland County, New York. M, 1988, Lehigh University. 97 p.

Claypool, Chester Burns. Limestone oil reservoirs of Winkler County, Texas. M, 1929, University of Illinois, Chicago.

Claypool, Chester Burns. The Wilcox of central Texas. D, 1933, University of Illinois, Urbana.

Claypool, George Edwin. Anoxic Diagenesis and Bacterial Methane Production in Deep Sea Sediments. D, 1974, University of California, Los Angeles. 276 p.

Clayton, C. M. Geology of the Breitenbush Hot Springs area, Cascade Range, Oregon. M, 1975, Portland State University. 80 p.

Clayton, Carol Aurelia. The geology of the West Walker Canyon area, Mono County, California, with emphasis on late magmatic stage veining. M, 1981, University of Nevada. 88 p.

Clayton, Daniel Noble. Volcanic history of the Teanaway Basalt, east-central Cascade Mountains, Washington. M, 1973, University of Washington. 55 p.

Clayton, Deborah A. Geochemical and biogeochemical prospecting for copper at Bear Lake and Mt. Bohemia in the Keweenaw Peninsula of Michigan. M, 1980, Bowling Green State University. 105 p.

Clayton, Geoffrey Alden. Geology of the White Pass area, south-central Cascade Range, Washington. M, 1983, University of Washington. 212 p.

Clayton, James Lindou. Weathering rate, nutrient supply, and denudation in forested watersheds, southwestern Idaho Batholith. D, 1985, Oregon State University. 107 p.

Clayton, Janine. Geomorphology of selected alluvial fans of southeastern Idaho. M, 1981, Idaho State University. 66 p.

Clayton, Jerry L. Organic diagenesis in Gulf Coast pelitic sediments. M, 1974, University of Missouri, Columbia.

Clayton, John Mason. Paleodepositional environments of the "Cherokee" (Pennsylvanian) sands of central Payne County, Oklahoma. M, 1965, University of Oklahoma. 46 p.

Clayton, Lee. Glacial geology of northern Logan County, south-central North Dakota. M, 1962, University of North Dakota. 116 p.

Clayton, Lee Stephen. Late Pleistocene geology of the Waiau valleys, North Canterbury, New Zealand. D, 1965, University of Illinois, Urbana.

Clayton, Nancy Ann. Land use planning; Dinosaur Mountain area, Pinal County, Arizona. M, 1975, Arizona State University. 57 p.

Clayton, Robert N. Variations in oxygen isotope abundances in rock minerals. D, 1955, California Institute of Technology. 85 p.

Clayton, Robert W. Fault kinematics and paleostress determined from slickenlines in an area of unusual fault patterns, southwestern Utah. M, 1987, Brigham Young University. 79 p.

Clayton, Robert W. The deconvolution of teleseismic recordings. M, 1975, University of British Columbia.

Clayton, Robert Webster. Wavefield inversion methods for refraction and reflection data. D, 1981, Stanford University. 109 p.

Clayton, Timothy J. The evolution of a temperate hardwater lake; Otsego Lake, New York. M, 1985, University of Massachusetts. 148 p.

Claytor, Gale Catherine. An evaluation of geochemical parameters of the overburden and the water quality associated with strip mining of the Sewanee Seam in the southern coalfield of East Tennessee. M, 1978, University of Tennessee, Knoxville. 111 p.

Clear, Harry C. Detrital quartz within the lower division of the Arkansas Novaculite. M, 1978, Louisiana State University.

Clearbout, Jon. Digital filters and applications to seismic detection and discrimination. M, 1963, Massachusetts Institute of Technology. 89 p.

Cleary, John Gladden. Geothermal investigation of the Alvord Valley, Southeast Oregon. M, 1976, University of Montana. 71 p.

Cleary, M. P. Fundamental solutions for fluid-saturated porous media and application to localised rupture phenomena. D, 1976, Brown University. 320 p.

Cleary, Michael Duane. Description and interpretation of chemical geothermometry as applied to Utah spring and well waters. M, 1978, University of Utah. 73 p.

Cleary, William J. Genesis and distribution of sand; a profile study from Carolina Piedmont soils to the Hatteras abyssal plain. D, 1971, University of South Carolina.

Cleary, William James, Jr. Marine geology of Onslow Bay (North Carolina). M, 1968, Duke University. 73 p.

Cleath, Richard Allen. Geology of the Precambrian rocks in the Silver Creek area, near Rochford, Black Hills, South Dakota. M, 1986, South Dakota School of Mines & Technology.

Cleath, Timothy S. Ground-water geology of the San Luis Obispo area, California. M, 1978, California State University, Los Angeles.

Cleaveland, Malcolm Kent. X-ray densitometric measurement of climatic influence on the intra-annual characteristics of Southwestern semiarid conifer tree rings. D, 1983, University of Arizona. 206 p.

Cleaveland, Thomas H. Geostatistical analysis of the facies and thickness variation in the Brereton Limestone and the Anna Shale in part of southwestern Illinois. M, 1983, Southern Illinois University, Carbondale. 140 p.

Cleaver, Jeffrey S. Distribution, facies, and reservoir potential of the Elgin sandstones (Upper Pennsylvanian) Ochiltree and Roberts counties, Texas. M, 1988, West Texas State University. 55 p.

Cleaves, Arthur B. Geology of the Wamsutta red beds and their relation to associated sediments and igneous rocks in the Narragansett Basin. M, 1927, Brown University.

Cleaves, Arthur Bailey. The geology of the New Bloomfield Quadrangle, Pennsylvania. D, 1933, Harvard University.

Cleaves, Arthur W., II. Upper Desmoinesian-lower Missourian depositional systems (Pennsylvanian), North-central Texas. D, 1975, University of Texas, Austin. 466 p.

Cleaves, Arthur Wordsworth, II. Depositional environments in the middle part of the Glen Rose Limestone (lower Cretaceous), Blanco and Hays counties, Texas. M, 1971, University of Texas, Austin.

Cleaves, Emery Taylor. Chemical weathering and landforms in a portion of Baltimore County, Maryland. D, 1973, The Johns Hopkins University.

Cleavinger, Howard Ben, II. Paleoenvironments of deposition of the Upper Cretaceous Ferron Sandstone near Emery, Emery County, Utah. M, 1974, Brigham Young University. 224 p.

Clebnik, S. M. Surficial geology of the Willimantic Quadrangle, Connecticut. D, 1975, University of Massachusetts. 253 p.

Clee, T. Edward. Seismic wave attenuation in glacial ice. M, 1968, University of Toronto.

Clee, T. Edward. Simulation of short-period seismic wave propagation in layered media with curved interfaces. D, 1973, University of Toronto.

Clegg, Kenneth Edward. Metamorphism of coal by peridotite dikes in southern Illinois. M, 1953, University of Illinois, Urbana.

Clegg, Robert Henry. Martian channel source and origin; a morphometric examination of the fluvial hypothesis. D, 1982, University of Maryland. 507 p.

Cleland, Herdman F. A study of fossil faunas in the Hamilton, State of New York. D, 1900, Yale University.

Cleland, Jane M. Petrology of the contact metamorphic zone of the Calumet Mine area, Chaffee County, Colorado. M, 1988, West Texas State University. 117 p.

Clemenceau, George Robert. The environmental significance of detrital muscovite grain shape. M, 1981, Duke University. 107 p.

Clemency, Charles. Contacts of the Peekskill Granite (New York). M, 1958, New York University.

Clemency, Charles Valentine. Variation in iron content of pyroxene amphibolite and constituent minerals at the Scott Mine, Sterling Lake, New York. D, 1961, University of Illinois, Urbana. 82 p.

Clemens, Karen E. Along-coast variations of Oregon beach-sand compositions produced by the mixing of sediments from multiple sources under a transgressing sea. M, 1987, Oregon State University. 75 p.

Clemens, Robert. Selected environmental criteria for the designed artificial structures on the Southeast Shore. M, 1975, University of South Carolina.

Clemens, William Alvin, Jr. Fossil mammals of the Type Lance Formation (Cretaceous), Wyoming. D, 1960, University of California, Berkeley. 529 p.

Clement, Bradford Mark. Details of geomagnetic polarity transitions as recorded in deep-sea sediments. D, 1985, Columbia University, Teachers College. 262 p.

Clement, Craig Robert. The occurrence and paleoecology of echinoderm assemblages in the Ludlowville Shales (Middle Devonian) of western Erie County, New York. M, 1981, University of Cincinnati. 97 p.

Clement, George Muller. Paleozoic stratigraphy and structure on the St. Croix River. D, 1933, University of Iowa. 47 p.

Clement, George Muller. Some Pleistocene mammals of Iowa. M, 1932, University of Iowa. 120 p.

Clement, James Hallowell. Correlation of Paleozoic formations and pre-Jurassic structure in central and north-central Montana. M, 1951, Montana College of Mineral Science & Technology. 37 p.

Clement, Jean F. The geology of the northeastern Baca Grant area, Saguache County, Colorado. M, 1952, Colorado School of Mines. 129 p.

Clement, Joseph Frederick. The geology of the Garland-Richardson area (Texas). M, 1952, Southern Methodist University. 34 p.

Clement, Mark Anthony, Jr. Stratigraphy of the Taylor and Navarro groups in the Royse City area, Rockwall and Collin counties, Texas. M, 1950, Southern Methodist University. 27 p.

Clement, Maurice James Young. The reflection and transmission of Rayleigh waves. M, 1961, University of British Columbia.

Clement, Monica Diane. Heat flow and geothermal assessment of the Escalante Desert, part of the Oligocene to Miocene volcanic belt in southwestern Utah. M, 1981, University of Utah. 118 p.

Clement, Richard F. A study of the sediment size distribution of Malletts Bay (Vermont). M, 1967, University of Vermont.

Clement, Sara J. The Sherburne siltstone; late Devonian sedimentation in south central New York. M, 1969, Syracuse University.

Clement, Stephen C. Structure and mineralogy of the Copper Canyon Pluton, Battle Mountain, Nevada. M, 1961, University of Utah. 32 p.

Clement, Stephen Caldwell. Mineralogy and petrology of the Copper Canyon quartz monzonite porphyry, Battle Mountain, Nevada. D, 1964, Cornell University. 127 p.

Clement, William Gilbert. Pre-Pennsylvanian stratigraphy of the West half of the Durham Quadrangle (Georgia). M, 1952, Emory University. 69 p.

Clement, William Glenn. An investigation of the Earth's gravitational field in the northern San Francisco Bay area, California. D, 1965, Stanford University. 114 p.

Clement, William P. Crustal structure of northwestern Montana using seismic refraction techniques. M, 1986, University of Montana. 101 p.

Clements, Donald Harry. Later S phases of the Michigan earthquake of August 9, 1947. M, 1952, University of Michigan.

Clements, Jake E., Jr. Pliocene and Aftonian planation surfaces of the Paleozoic region of Arkansas. M, 1958, University of Arkansas, Fayetteville.

Clements, Kenneth Paul. Subsurface study of the Skinner and Red Fork sand zones (Pennsylvanian) in portions of Noble and Kay counties, Oklahoma. M, 1961, University of Oklahoma. 62 p.

Clements, Thomas. Geology of a portion of the southeast quarter of the Tejon Quadrangle, Los Angeles County, California. M, 1929, California Institute of Technology. 45 p.

Clements, Thomas. The geology of the southeastern portion of the Tejon Quadrangle (California). D, 1932, California Institute of Technology. 165 p.

Clemons, Robert Rickard. Remote sensing detection and geological interpretation of the Idaho Primitive Area. M, 1980, Texas Tech University. 62 p.

Clemons, Robert Rickard. The remote sensor exploration of the Ardmore and Marietta basins of Oklahoma. D, 1984, Texas Tech University. 134 p.

Clemons, Russell E. and McLeroy, Donald F. Geology of Torreon and Pedricenas quadrangles, Coahuila and Durango, Mexico. M, 1962, University of New Mexico. 182 p.

Clemons, Russell Edward. Geology of the Chiquimula Quadrangle, Guatemala, Central America. D, 1966, University of Texas, Austin. 156 p.

Clendenen, William Sterling. Finite strain study of the Sudbury Basin, Ontario. M, 1986, University of Colorado. 253 p.

Clendenin, Charles William, Jr. Stibnite-bearing veins at Stibnite Hill, Burns mining district, Sanders County, Montana. M, 1973, Montana College of Mineral Science & Technology. 53 p.

Clendenin, Thomas P. A summary of four weeks field work in southeastern New York. M, 1921, Columbia University, Teachers College.

Clendening, John A. Spores of the middle Dunkard coals (Permian) and their application to correlation. M, 1960, West Virginia University.

Clendening, John Albert. Sporological evidence on the geological age of the Dunkard Strata in the Appalachian Basin. D, 1970, West Virginia University. 478 p.

Clendining, John Albert. Sporological evidence of the geological age of the Dunkard strata (Pennsylvanian and Permian) in the Appalachian basin. D, 1970, West Virginia University. 478 p.

Cleneay, Charles A. Modern sediments and sedimentary structures of the Bogue Inlet - White Oak Estuary area, North Carolina. M, 1974, Bowling Green State University. 87 p.

Clere, David Russell. The effects of Pleistocene glaciation and drainage-basin characteristics upon streamflow in Indiana and Illinois. D, 1975, Indiana University, Bloomington. 201 p.

Clerk, R. V. Le *see* Le Clerk, R. V., II

Clerman, Robert Joseph. Shrub growth patterns on Assateague Island; effects of dune stabilization. M, 1978, University of Virginia. 38 p.

Cleveland, Courtney E. The geology of the Empire Mine, Bralorne, British Columbia. D, 1940, McGill University.

Cleveland, Courtney E. The geology of the vicinity of Bralorne Mines, British Columbia. M, 1938, McGill University.

Cleveland, Gaylord. Geology and mineralization of the Port Antonio-Berridale area, Portland Parish, Jamaica. M, 1979, University of Arizona. 179 p.

Cleveland, George B. Some controls in biogeochemical prospecting. M, 1955, University of California, Los Angeles.

Cleveland, John Herbert. Paragenesis of the magnesite deposit at Gabbs, Nye County, Nevada. D, 1963, Indiana University, Bloomington. 88 p.

Cleveland, John Herbert. Variation of iron sulfide content in sphalerites of Mississippi Valley deposits. M, 1961, University of Wisconsin-Madison.

Cleveland, Michael N. Radiolarian densities, diversities, and taxonomic composition in Recent sediment and plankton of the Southern California continental borderland; relationship to water circulation and depositional environments. M, 1985, Rice University. 90 p.

Cleveland, Michael N. Study of Moss Beach. M, 1976, Stanford University.

Cleveland, Scott R. Deposition and diagenesis of the Middle Member of the Metaline Formation, Clugston Creek, Stevens County, Washington. M, 1982, University of Idaho. 98 p.

Cleven, G. W. A statistical analysis of erosion surfaces in the vicinity of Laramie, Wyoming. M, 1956, University of Wyoming. 64 p.

Cliadakis, William C. Structural implications of six vertical magnetic profiles in the northern section of the Palisades (New York). M, 1959, New York University.

Click, David L. Depositional and diagenetic history of an oolitic grainstone reservoir in the Lower Cretaceous Pettet (Sligo) Formation, northeast Nacogdoches County, Texas. M, 1987, Bowling Green State University. 152 p.

Clifford, Edward. Trace element concentrations as indicators of fresh-brackish-marine depositional environments; Hamilton Group (Middle Devonian) of New York State. M, 1980, Brooklyn College (CUNY).

Clifford, Michael James. Feasibility of deep-well injection industrial liquid wastes in Ohio. M, 1972, Ohio State University.

Clift, William Orrin. Sedimentary history of the Ogaden District, Ethiopia. D, 1956, Columbia University, Teachers College. 43 p.

Clifton, Amy E. Tectonic analysis of the western border fault zone of the Mesozoic Hartford Basin, Connecticut and Massachusetts. M, 1987, Wesleyan University. 201 p.

Clifton, Billy Dean. A sedimentary study of the Ogallala Group, Crosby County, Texas. M, 1951, Texas Tech University. 40 p.

Clifton, Clarence Cathcart. The volcanic tufas of the Pacific Coast and their utilization. M, 1921, University of Washington. 30 p.

Clifton, H. Edward. The Pembroke Breccia of Nova Scotia. D, 1963, The Johns Hopkins University.

Clifton, Roland LeRoy. Areal extent and stratigraphy of the Whitehorse Sandstone. M, 1925, University of Oklahoma. 38 p.

Clikeman, Paul W. Power-generation potential and geologic feasibility of a low-head hydroelectric complex at Prattsville, New York. M, 1980, SUNY, College at Oneonta. 122 p.

Clinch, J. Michael. The relative importance of glacial and glaciofluvial transport processes at lobate glacier margins in Alaska, Svalbard and Antarctica. D, 1988, Lehigh University. 510 p.

Cline, Charles W. Stratigraphy of Douglas Creek Member, Green River Formation, Piceance Creek basin, Colorado. M, 1957, Brigham Young University. 49 p.

Cline, Denzel Riste. The geology and groundwater resources of upper Black Earth Creek basin, Wisconsin. M, 1961, University of Michigan.

Cline, George Douglas. Geologic factors influencing well yields in a folded sandstone-siltstone-shale terrane within the east Mahantango creek watershed, Pennsylvania. M, 1968, Pennsylvania State University, University Park. 180 p.

Cline, Jeffrey Thomas. Pathways and interactions of copper with aquatic sediments. D, 1974, Michigan State University. 150 p.

Cline, Joel Dudley. Denitrification and isotopic fractionation in two contrasting marine environments; the eastern tropical North Pacific Ocean and the Cariaco Trench. D, 1973, University of California, Los Angeles. 288 p.

Cline, Justis H. The geology of the Fauquier-Culpeper slate district, Virginia. M, 1910, Northwestern University.

Cline, K. Michael. Geology of the Mackay Reservoir area, White Knob Mountains, Mackay 2NE and 1NW quadrangles, Custer County, Idaho. M, 1979, Idaho State University. 128 p.

Cline, Lawrence B. The origin of talc at Johnson, Vermont and the geology of its occurrence. M, 1960, University of Vermont.

Cline, Lewis Manning. Osage blastoids; Part I, The genera Schizoblastus and Cryptoblastus. D, 1935, University of Iowa. 164 p.

Cline, Lewis Manning. Osage formations of the southern Ozark region. M, 1934, University of Iowa. 48 p.

Cline, Marlin George. The significance of differences among mapping units and a method of generalization of the detailed soil map of western Livingston County, New York. D, 1942, Cornell University.

Cline, Patricia V. Behavior of partially miscible organic compounds in simulated ground water systems. D, 1988, University of Florida. 194 p.

Cline, Robert Bruce. Fusulinid paleontology and paleoecology of eastern Nevada. M, 1967, University of Nevada. 83 p.

Cline, Robert William. Properties and uses of olivine from the Pacific Northwest, Part II. M, 1937, University of Washington. 119 p.

Cline, Royce L. Geomorphology of the lower Saline River valley in north-central Kansas. M, 1974, Kansas State University. 49 p.

Cline, Wilford La Verne. Certain crystallographic structures from Permian sediments of Oklahoma. M, 1934, University of Missouri, Columbia.

Clingman, William Warren. A magnetotelluric investigation of the Lookout Mountain area in the central Oregon High Cascades. M, 1988, University of Oregon. 81 p.

Clinkenbeard, John Patrick. The mineralogy, geochemistry, and geochronology of the La Posta Pluton, San Diego and Imperial counties, California. M, 1987, San Diego State University. 215 p.

Clissold, Roger J. Mapping of naturally occurring surficial phenomena to determine groundwater conditions in two areas near Red Deer, Alberta (Canada), Quaternary. M, 1967, University of Alberta. 126 p.

Clister, William Eugene. Hydrogeologic factors in radioactive waste management at the Whiteshell Nuclear Research Establishment, Manitoba. M, 1973, University of Manitoba.

Clodfelter, Rebecca A. Petrography of the Providence Limestone Member (Upper Pennsylvanian) of the Sturgis Formation in western Kentucky. M, 1987, Eastern Kentucky University. 75 p.

Cloft, Harriet S. The sedimentology and stratigraphy of the Ledge Sandstone Member, Ivishak Formation, in the Arctic National Wildlife Refuge, northeastern Alaska. M, 1984, University of Texas at Dallas. 193 p.

Cloke, P. L. Quartz solubility in potassium hydroxide solutions under elevated pressures and temperatures, with some geological applications. D, 1954, Massachusetts Institute of Technology. 93 p.

Cloos, Mark Peter. Studies in Franciscan geology; Part 1, Flow melanges, numerical modeling and geologic constraints on their origin in the Franciscan subduction complex, California; Part 2, Metamorphism and deformation of the shale matrix of the Franciscan melange belt, California; Part 3, Controls and mechanisms for dewatering subducted sediments with emphasis on the Franciscan melange belt, California. D, 1981, University of California, Los Angeles. 230 p.

Clopine, Gordon A. Pleistocene sediments of the Houston area, Texas. M, 1960, University of Houston.

Clopine, William Walter. The lithostratigraphy, biostratigraphy and depositional history of the Atokan Series (Middle Pennsylvanian) in the Ardmore Basin, Oklahoma. M, 1986, University of Oklahoma. 161 p.

Cloran, Courtenay A. Geophysics, hydrology, and geothermal potential of the Tonopah Basin, Maricopa County, Arizona. M, 1977, Arizona State University. 109 p.

Close, Edward Joseph. An application of ray theory to the investigation of seismic head wave post-Pn phases. D, 1981, University of Connecticut. 174 p.

Close, Jay C. Lithofacies and environments of the Pennington (Mississippian), Breathitt and Lee formations (Pennsylvanian); a stratigraphic section in southcentral Kentucky. M, 1985, Miami University (Ohio). 153 p.

Close, Jay Charles. Coalbed methane potential of the Raton Basin, Colorado and New Mexico. D, 1988, Southern Illinois University, Carbondale. 432 p.

Closs, L. Graham. Geology of the Waite lake mines property (Duprat township, Quebec, Canada). M, 1970, University of Vermont.

Closs, Lloyd Graham. An evaluation of selected multivariate mathematical techniques as aids in interpretation of the reconnaissance geochemical stream sediment data of the Halls Bay Concession, Newfoundland. D, 1973, Queen's University. 270 p.

Cloud, Kelton Wayne. The diagenesis of the Austin Chalk. M, 1975, University of Texas at Dallas. 70 p.

Cloud, Preston E. Systematic revision of Silurian and Devonian terebratuloid brachiopods. D, 1940, Yale University.

Cloud, Robert A. Geology of the Munds Park-Oak Creek Canyon area, central Arizona. M, 1983, Northern Arizona University. 159 p.

Cloud, Thomas Arthur. Temporal and spatial distribution of landslides, Gros Ventre River valley, Gros Ventre Mountains, Wyoming. M, 1981, Iowa State University of Science and Technology. 98 p.

Cloud, Wilbur Frank. The subsurface structure and stratigraphy of the Davenport oil field, Lincoln County, Oklahoma. M, 1926, University of Oklahoma. 22 p.

Clough, Albert Hughes. Geology of the NE 1/4 of the Calder Quadrangle, Shoshone County, Idaho. M, 1981, University of Idaho. 105 p.

Clough, George A. Structure of Juniper Mountain (Moffat County), Colorado. M, 1951, University of Colorado.

Clough, James G. Depositional environments and petrology of the Ogilvie Formation and the limestone and shale member of the McCann Hill Chert, east-central Alaska and adjacent Yukon Territory. M, 1981, University of Alaska, Fairbanks. 157 p.

Clough, John W. Measurements of electromagnetic wave velocity in the East Antarctic Ice sheet. M, 1970, University of Wisconsin-Madison.

Clough, John W. Propagation of radio waves in the Antarctic ice sheet. D, 1974, University of Wisconsin-Madison. 130 p.

Clough, Stephen Ronald. Facies development and evolution of fluvial channels in the Cumberland marshes of Saskatchewan, Canada. M, 1983, University of Illinois, Chicago.

Clough, William Allen. Geology of the northern portion of the Bolar Valley Anticline, Highland County, Virginia. M, 1949, University of Virginia. 94 p.

Clouser, Robert H. Upper mantle P-wave velocity structure beneath Southern Africa from P_{nl} waves. M, 1988, Pennsylvania State University, University Park. 48 p.

Cloutier, Marc-André Lithogéochimie et pédogéochimie comme outils d'évaluation du potentiel minéral d'un grand axe conducteur dans le Canton de Richardson, région de Chibougamau. M, 1986, Ecole Polytechnique. 302 p.

Cloutis, Edward Anthony. Interpretive techniques for reflectance spectra of mafic silicates. M, 1985, University of Hawaii. 192 p.

Clow, William Henry Arthur. Structure of the Carbondale area, Alberta. M, 1950, University of Alberta. 87 p.

Clowe, Celia A. The relation of Fo₂, ferric-ferrous ratio (R), and physical properties of four natural clinoamphiboles. M, 1987, Texas A&M University. 71 p.

Clowers, Stanley. Pleistocene Mollusca of the Box Marsh deposit, Admaston Township, Renfrew County, Ontario, Canada. M, 1966, Ohio State University.

Clowes, Ronald Martin. Deep crustal seismic reflections at near-vertical incidence. M, 1966, University of Alberta. 152 p.

Clowes, Ronald Martin. Seismic reflection investigations of crustal structure in southern Alberta (Canada). D, 1969, University of Alberta. 190 p.

Cluer, Brian L. A gravity model of basement geometry and resulting hydrogeologic implications of the Camas Prairie, south-central Idaho. M, 1987, Northern Arizona University. 99 p.

Cluff, Garnet Robert. Mineralogy and alteration of the Murex Breccia, Mount Washington, Vancouver Island, Canada. M, 1981, University of Saskatchewan. 109 p.

Cluff, Robert Murri. Paleoecology and depositional environment of the Mowry Shale (Albian), Black Hills region. M, 1976, University of Wisconsin-Madison.

Clukey, Edward Charles. Laboratory and field investigation of wave-sediment interaction. D, 1984, Cornell University. 411 p.

Clupper, David Richie. The lithostratigraphy and depositional environments of the Pitkin Formation (Mississippian) in Adair County, northeastern Oklahoma. M, 1978, University of Oklahoma. 115 p.

Clute, Peter R. Chautauqua Lake sediments. M, 1973, SUNY, College at Fredonia. 127 p.

Clutterbuck, Donald Booth. Structure in northern Sierra Pilares, Chihuahua, Mexico. M, 1958, University of Texas, Austin.

Clyma, Wayne. Analysis of soil heat transfer for the evapotranspiration system. D, 1971, University of Arizona.

Clymer, Richard W. Rayleigh wave phase velocities and the upper shear velocity structure of Fennoscandia. M, 1973, University of California, Los Angeles.

Clymer, Richard Wayne. High-precision travel-time monitoring with seismic reflection techniques. D, 1980, University of California, Berkeley. 200 p.

Clynne, Michael. Stratigraphy and major element geochemistry of the Lassen volcanic center, California. M, 1983, San Jose State University. 168 p.

Coachman, Lawrence K. On the water masses of the Arctic Ocean. D, 1962, University of Washington. 94 p.

Coakley, J. The history and bottom sediments of Stanwell-Fletcher Lake, Somerset Island, Northwest Territories. M, 1966, University of Ottawa. 83 p.

Coakley, John Phillip. Evolution of Lake Erie based on the postglacial sedimentary record below the Long Point, Point Pelee, and Pointe-aux-Pins forelands. D, 1985, University of Waterloo. 362 p.

Coale, Kenneth Hamilton. Copper complexation in the North Pacific Ocean. D, 1988, University of California, Santa Cruz. 262 p.

Coalson, Edward B. Geology of the New Rambler Mine area, Albany County, Wyoming. M, 1971, University of Wyoming. 51 p.

Coan, Eugene Victor. Taxonomic studies on the tellinacean bivalve mollusks of the northeastern Pacific. D, 1969, Stanford University. 263 p.

Coash, John Russell. A sedimentary study of the Pennsylvanian-Permian section in the Howard-Wellsville area, Fremont County, Colorado. M, 1949, University of Colorado.

Coash, John Russell. Geology of part of the Mt. Velma Quadrangle, Elko County, Nevada. D, 1954, Yale University.

Coates, Anna L. Hydrologic performance of multilayer landfill covers; a field verification and modeling assessment of HELP and SOILINER. M, 1987, University of Kentucky. 325 p.

Coates, Colin. Copper-nickel sulphide deposits in amphibolite, Uchi Lake District, Ontario. M, 1959, Queen's University.

Coates, D. F. A new hypothesis for determination of pillar loads. D, 1965, McGill University.

Coates, Donald Allen. Fill deposits in small coastal stream valleys in the Santa Cruz area, California. M, 1964, University of Colorado.

Coates, Donald Allen. Origin of the Sauce Grande Formation (lower Carboniferous), southern Buenos Aires Province, Argentina. D, 1969, University of California, Los Angeles. 182 p.

Coates, Donald R. Quantitative geomorphology of small drainage basins of southern Indiana. D, 1956, Columbia University, Teachers College.

Coates, Donald R. Structure of gneiss hills near Scunemunk Mountain, New York. M, 1948, Columbia University, Teachers College.

Coates, Eugene Joseph. The occurrence of Wilcox lignite in west-central Louisiana. M, 1979, Louisiana State University.

Coates, Howard James. The structural and metamorphic history of the Pacquet Harbour–Grand cove area of the Burlington Peninsula, Newfoundland. M, 1970, Memorial University of Newfoundland. 79 p.

Coates, James A. The redstone bedded copper deposit and a discussion on the origin of red bed copper deposits, Northwest Territories, Canada. M, 1965, University of British Columbia.

Coates, Mary H. Application of a Markov Chain to the Mazomanie-Reno sandstones (Cambrian) in southcentral Wisconsin. M, 1967, Northwestern University.

Coates, Maurice E. Geology of the Killala Lake Igneous Complex (middle Proterozoic), District of Thunder Bay, Ontario (Canada). M, 1967, McGill University. 128 p.

Coatney, Richard Lee. Modal analysis of the granitic rocks of the Northern Sierra Nevada between Yosemite and Lake Tahoe, California. D, 1965, Pennsylvania State University, University Park. 164 p.

Coats, Colin J. A. Sepentinized ultramafic rocks of the Manitoba nickel belt. D, 1966, University of Manitoba.

Coats, Colin John Alastair. Copper-nickel sulfide deposits in amphibolite, Uchi Lake district, Ontario. M, 1959, Queen's University. 174 p.

Coats, Robert Roy. A contribution to the geology of the Comstock Lode. D, 1938, University of California, Berkeley. 128 p.

Coats, Robert Roy. The ore deposits of the Apex gold mine, Money Creek, King County, Washington. M, 1932, University of Washington. 48 p.

Cobb, Craig Carroll. A gravity and magnetic study of the West Jefferson area, Madison and Franklin counties, Ohio. M, 1979, Wright State University. 105 p.

Cobb, Edward H. Geology of the Gibson Lake region, Montana. M, 1941, Yale University. 25 p.

Cobb, James C. Sedimentology of an outwash fan deposit in the Woodfordian (late Wisconsinan) of northeastern Illinois. M, 1974, Eastern Kentucky University. 77 p.

Cobb, James Collins. Geology and geochemistry of sphalerite in coal. D, 1981, University of Illinois, Urbana. 220 p.

Cobb, James Curtis. Isotopic geochemistry of uranium and lead in the Swedish Kolm and its associated shale. D, 1959, Columbia University, Teachers College. 139 p.

Cobb, LaVerne Burkhart. A study of the distribution of trace elements (iron and manganese) and cathodoluminescent zonation in mineralized fracture fillings. M, 1974, University of Tennessee, Knoxville. 44 p.

Cobb, Margaret Cameron. The origin of corundum associated with dunite in western North Carolina. D, 1922, Bryn Mawr College. 43 p.

Cobb, Robert Charles. Structural geology of the Santiago Mountains between Pine Mountain and Persimmon Gap, Trans-Pecos Texas. M, 1980, University of Texas, Austin.

Cobb, Robert Eugene. Upper Devonian sediments of the Grand Canyon of Pennsylvania. M, 1953, Pennsylvania State University, University Park. 171 p.

Cobb, Steven Lloyd. A laboratory facility for testing the performance of borehole plugs in rock subjected to polyaxial loading. M, 1981, University of Arizona. 157 p.

Cobb, William Battle. A comparison of the development of soils from the acid and basic crystalline rocks of Piedmont North Carolina. D, 1927, University of North Carolina, Chapel Hill. 41 p.

Cobb, William Battle. Geology in relation to soil types and soil fertility in Chapel Hill, North Carolina region. M, 1913, University of North Carolina, Chapel Hill. 12 p.

Cobb, William Marshall. Transient flow in two-layer reservoirs with commingled fluid production. D, 1970, Stanford University. 128 p.

Cobban, William Aubrey. Stratigraphy of the Colorado and Montana groups (Upper Cretaceous) of the central and northern Great Plains with descriptions of the Colorado scaphites. D, 1949, The Johns Hopkins University. 408 p.

Coble, James F. Geology of the Limestone Cove Window, Tennessee and North Carolina. M, 1976, East Carolina University. 62 p.

Coble, Ronald Winner. Electrolyte groundwater tracer studies in a cone-of-depression, Big Bend area, Hamilton County, Ohio. M, 1962, University of Cincinnati. 153 p.

Coblentz, Alex. Taxonomy, biostratigraphy, and paleoecology of Upper Cretaceous dinoflagellates of North America. M, 1986, University of Delaware.

Cobucci, Dolores Ann. Guide to the geology of selected field sites in southwestern Pennsylvania. D, 1974, University of Pittsburgh.

Coburn, Jo Ann. Depositional environments of the upper Caloosahatchee Marl (Pleistocene) based on the ostracod fauna. M, 1973, University of Akron. 114 p.

Coch, Nicholas K. Textural and mineralogical variations in some Lake Ontario beach sands. M, 1961, University of Rochester. 57 p.

Coch, Nicholas Kyros. Post-Miocene stratigraphy and morphology of the inner coastal plain, southeastern Virginia. D, 1965, Yale University. 205 p.

Cochran, Ann. Fluid inclusion populations in quartz-rich gold ores from the Barberton greenstone belt, eastern Transvaal, South Africa. M, 1982, University of Arizona. 208 p.

Cochran, Bruce Duane. Late Quaternary stratigraphy and chronology in Johnson Canyon, central Washington. M, 1978, Washington State University. 81 p.

Cochran, Bruce Duane. Significance of Holocene alluvial cycles in the Pacific Northwest Interior. D, 1988, University of Idaho. 255 p.

Cochran, Daniel John. Geology of an area east of Tincup Mountain, Caribou Range, Southeast Idaho. M, 1983, Idaho State University. 57 p.

Cochran, George Raymond. The in situ determination of the dynamic elastic constants of soils by an impact method. M, 1962, University of California, Berkeley. 103 p.

Cochran, Gilbert Francis. Policy development and evaluation in conjunctive ground and surface water management. D, 1973, University of Nevada. 251 p.

Cochran, H. Merle. The petrography of the Twin Hills, Pima County, Arizona. M, 1914, University of Arizona.

Cochran, J. K. The geochemistry of ^{226}Ra and ^{228}Ra in marine deposits. D, 1979, Yale University. 276 p.

Cochran, J. R. A study of the intermediate and long wavelength gravity field of the world's oceans. D, 1977, Columbia University, Teachers College. 226 p.

Cochran, John A. The sampling problem; sedimentary petrography; a contribution. M, 1960, Pennsylvania State University, University Park. 94 p.

Cochran, Karen Anne. The relationship between hydrocarbon migration and authigenic magnetite; testing the hypothesis. M, 1986, University of Oklahoma. 90 p.

Cochran, Michael D. A re-evaluation of the Hager rhyolite porphyry (Precambrian), (northeastern Wisconsin). M, 1966, Bowling Green State University. 85 p.

Cochran, Michael David. Long period leaking modes. D, 1969, Rice University. 91 p.

Cochran, Michael Patrick. Geophysical investigation of eastern Yakima Ridge, south-central Washington. M, 1982, Washington State University. 99 p.

Cochran, Walter House. Subsurface geology of Bayou Jean La Croix Field, Terrebonne Parish, Louisiana. M, 1987, University of New Orleans. 32 p.

Cochran, Wendell. Silicified Brachiopoda of the Callaway Limestone. M, 1956, University of Missouri, Columbia.

Cochrane, Donald R. A copper deposit near Denbigh, Ontario; a study of the petrology, mineralogy, structure, hydrothermal alteration, geothermometry, trace elements and possible origin of the copper deposits. M, 1964, Queen's University. 139 p.

Cochrane, Judith Christian. Petrogenesis of the Willis Mountain and East Ridge kyanite quartzite, Buckingham County, Virginia. M, 1986, Virginia Polytechnic Institute and State University.

Cochrane, N. A. Geomagnetic and geoelectric variations in Atlantic Canada. D, 1873, Dalhousie University.

Cochrane, Norman. A new approach to geomagnetic depth sounding in western Canada. M, 1969, Dalhousie University.

Cochrane, Peter John. The structure and stratigraphy of the Hecla Hoek Sequence in the Turrsjödalen-Brevassdalen area, Wedel Jarlsberg Land, Spitsbergen. M, 1984, University of Wisconsin-Madison.

Cochrum, Arthur Leroy. Structural history of northern Hurd Draw and southern Foster quadrangles, Culberson County, Texas. M, 1952, University of Texas, Austin.

Cockburn, Daniel. La Cartographie géotechnique de la région de Québec; essai méthodologique. M, 1983, Universite Laval. 93 p.

Cocke, Elton Cromwell. A further study of Dismal Swamp peat. D, 1934, University of Virginia. 395 p.

Cocke, Julius Marion. Dissepimental rugose corals of Pennsylvanian-Missourian rocks of Kansas. D, 1970, University of Iowa. 242 p.

Cocke, Julius Marion. Stratigraphy and coral fauna of the Dewey Formation (Missourian), Washington and Nowata counties, Oklahoma. M, 1962, University of Oklahoma. 115 p.

Cocke, Robert Robinson, III. Subsurface geology, north half of Nacogdoches County, Texas. M, 1951, University of Texas, Austin.

Cocker, Mark David. A new method for the determination of refractive indices using the interference microscope. M, 1974, Ohio State University.

Cocker, Mark David. Multiple intrusion, hydrothermal alteration, and related mineralization in the northern Breckenridge mining district, Summit County, Colorado. D, 1978, Ohio State University. 296 p.

Cockerham, Kirby L., Jr. A lithologic and mineralogic study of a water well at Kinder, Louisiana. M, 1952, Louisiana State University.

Cockerham, Robert S. Stratigraphy of the Columbia River Group, north-central Oregon. M, 1975, Western Washington University. 56 p.

Cockfield, James E. Effects of permeability, pressure gradient, and temperature on the flow of a single-phase fluid through heterogeneous media. M, 1953, Ohio State University.

Cockfield, William Egbert. Sixtymile and Ladue rivers area, Yukon. D, 1918, Princeton University. 156 p.

Cockran, Wendell A., Jr. The silicified Brachiopoda of the Callaway Limestone. M, 1956, University of Missouri, Columbia.

Cockrell, Dale Reed. Stratigraphy, distribution, and structural geology of Lower and Middle Pennsylvanian sandstones in adjacent portions of Okfuskee and Seminole counties, Oklahoma. M, 1985, Oklahoma State University. 55 p.

Cockrum, Amil Blake, Jr. Some laboratory procedures and a preliminary study of the Rio Grande River sediments. M, 1940, University of Texas, Austin.

Cockrum, Dave A. Lithostratigraphy of the Prichard Formation, and geology of the Marble Mountain Quadrangle, Shoshone County, Idaho. M, 1986, University of Idaho. 216 p.

Coco, Arlene M. A study of pore concentration variations in the test walls of Neogloboquadrina dutertrei eggeri (Rhumbler). M, 1977, Rutgers, The State University, Newark. 67 p.

Cocovinis, Dimitri Basil. Areal geologic map of southeastern Williamson County, Texas. M, 1949, University of Texas, Austin.

Cocroft, John E. Hydrogeochemical study of a mined region in the Carbondale Group (Pennsylvanian), southwestern Indiana. M, 1984, Indiana University, Bloomington. 198 p.

Codding, David B. Precambrian geology of the Rio Mora area, New Mexico; structural and stratigraphic relations. M, 1983, University of New Mexico. 128 p.

Coddington, Wayne J. Relationships between dynamic nearshore processes and beach changes at Napatree Beach, Rhode Island. M, 1976, University of Rhode Island.

Codispoti, Louis A. Denitrification in the eastern tropical North Pacific Ocean. D, 1973, University of Washington. 118 p.

Cody, Clyde A. Sedimentary petrology and depositional environments of the Beaver Bend Limestone (lower Chesterian) near French Lick, Indiana. M, 1978, Indiana University, Bloomington. 134 p.

Cody, Martha P. Microfauna of the Stones River Group of central Tennessee. M, 1953, Miami University (Ohio). 31 p.

Cody, R. D. Foraminifera from one section of the Late Cretaceous Steele Shale, Mesaverde Group, and Lewis Shale, Carbon County, Wyoming. M, 1962, University of Wyoming. 89 p.

Cody, Robert Dow. Boron content of late Paleozoic shale and mudstone from flanks of ancestral Front Range, Colorado. D, 1968, University of Colorado. 65 p.

Coe, Curtis. The economic geology of the Citronelle Formation in Escambia and Santa Rosa counties, FL. M, 1979, Florida State University.

Coe, Robert Stephen. Paleo-intensities of the geomagnetic field determined from Tertiary and Quaternary rocks. D, 1966, University of California, Berkeley. 408 p.

Coen, Gerald Marvin. Clay mineral genesis in some New York spodosols. D, 1970, Cornell University. 215 p.

Coen, Larry P. Structure and chemical composition of the otoliths of Cyprinus carpio. M, 1973, University of Missouri, Rolla.

Coenraads, Robert Raymond. Patterns of induced microearthquakes at the Sullivan Mine, Kimberley, B.C. M, 1982, University of British Columbia. 154 p.

Coetzee, Gerrard Louis. The origin of the Sangu carbonate rock complex and associated rocks, Karema Depression, Tanganyika Territory, East Africa. D, 1963, University of Wisconsin-Madison. 131 p.

Cofer, Harland Elbert, Jr. Petrology, petrography, mineralogy, and structure of the Arabia Mountain Gneiss, De Kalb County, Georgia. M, 1948, Emory University. 64 p.

Cofer, Harland Elbert, Jr. Structural relations of the granites and the associated rocks of South Fulton County, Georgia. D, 1958, University of Illinois, Urbana. 139 p.

Cofer, Richard S., III. Geology of the Shelly Cauldron Complex, Pinto Canyon area, Presidio County, Texas. M, 1980, University of Texas at El Paso.

Cofer-Shabica, Nancy B. The late Pleistocene deep water history of the Venezuela Basin in the eastern Caribbean Sea. M, 1988, University of Miami. 120 p.

Coffelt, Richard M. The origin of primitive basaltic magmas. M, 1973, San Francisco State University.

Coffey, Michael Lynn. Paleoecology and biostratigraphy of the Fetzer Formation, Middle Ordovician, of eastern Tennessee. M, 1986, University of Tennessee, Knoxville. 140 p.

Coffey, William S. The mineralogy and geochemistry of the Batesville manganese district, Arkansas. M, 1981, University of Arkansas, Fayetteville.

Coffield, Dana Quentin. Structural styles associated with accommodation zone terminations within the western Gulf of Suez Rift, Egypt. D, 1988, University of South Carolina. 228 p.

Coffin, Brian D. Conodont biostratigraphy of the upper Aspen Range Formation and lower Wells Canyon Formation, Aspen Range, Caribou County, Idaho. M, 1982, Brigham Young University.

Coffin, Catherine. Origin of deltaic deposits and long sand shoal in eastern Long Island Sound as related to late-Wisconsinan deglaciation of the lower Connecticut River valley. M, 1988, Boston University. 135 p.

Coffin, D. Todd. The stability of discontinuously jointed rock slopes. M, 1986, Purdue University. 140 p.

Coffin, Greg C. Geology of the northwestern Gallinas Mountains, Socorro County, New Mexico. M, 1981, New Mexico Institute of Mining and Technology. 202 p.

Coffin, Jeffrey Hart. Analysis of water flow problems in the Hells Canyon reach of the Snake River. M, 1977, University of Idaho. 166 p.

Coffin, Millard Filmore, III. Evolution of the conjugate East African-Madagascan margins and the western Somali Basin. D, 1985, Columbia University, Teachers College. 338 p.

Coffin, Peter E. Geology of the Slate Creek Quadrangle, Idaho County, Idaho. M, 1967, University of Idaho. 57 p.

Coffman, Bruce P. Geology of Sand Rock Creek Beds, Crosby County, Texas. M, 1979, Texas Tech University. 54 p.

Coffman, Bryan Keith. Origin of gaseous hydrocarbons in east-central Texas groundwaters. M, 1988, Texas A&M University.

Coffman, Daniel M. Advances in the theory of classification and modeling of stream channel networks. D, 1972, Purdue University. 214 p.

Coffman, James F. Relict valley asymmetry in the Pleistocene periglacial zone of southeastern Ohio. D, 1981, University of Wisconsin-Madison. 195 p.

Coffman, Jeffery Dale. Seismic stratigraphy and structure of the Barter Island sector of the western Beaufort Sea. M, 1988, Texas A&M University. 88 p.

Coffman, Paul Eugene, Jr. Optimal production from gas and oil reservoirs. D, 1982, University of Texas, Austin. 192 p.

Coffman, Richard L. Mineralogy and geochemistry from zeolitized tuffs from the Barstow Formation in the Mud Hills, San Bernardino County, California. M, 1983, University of California, Riverside. 102 p.

Coffman, W. Elmo. A progress report of metals in Utah. M, 1932, Brigham Young University. 125 p.

Cofield, William H. Geology of the Western Graphite Belt of Clay County, Alabama. M, 1956, University of Tennessee, Knoxville. 42 p.

Coflin, Kevin C. The effects of absorption on seismic signals propagated in a one-dimensional medium. M, 1985, University of Calgary. 126 p.

Cogan, J. The creep of weak rocks in the Burgin Mine. D, 1975, Stanford University. 328 p.

Cogbill, A. H., Jr. The relationship between seismicity and crustal structure in the western Great Basin. D, 1979, Northwestern University. 294 p.

Cogen, William M. A study of the heavy minerals of the Modelo Formation in the eastern portion of the Santa Monica Mountains. M, 1933, California Institute of Technology. 36 p.

Cogen, William M. Mechanics of landslides. D, 1937, California Institute of Technology. 73 p.

Coggins, Vernon. Hydrology of Willow Creek, Lassen County, California. M, 1970, California State University, Chico.

Coggon, John Henry. Induced polarization anomalies. D, 1972, University of California, Berkeley.

Cogo, Neroli Pedro. Effect of residue cover, tillage-induced roughness, and slope length on erosion and related parameters. D, 1981, Purdue University. 373 p.

Cohee, George Vincent. A regional lithologic study of the Hanover and Brereton limestone (Pennsylvanian). M, 1932, University of Illinois, Chicago.

Cohee, George Vincent. Ocean bottom sediments of the Mid-Atlantic Coast. D, 1937, University of Illinois, Urbana.

Cohen, Andrew Scott. Ecological and paleoecological aspects of the rift valley lakes of East Africa. D, 1982, University of California, Davis. 314 p.

Cohen, Arthur David. The petrology of some peats of southern Florida (with special reference to the origin of coal); vol. 1, Part I, General introduction, petrography, and modern environments; Vol. 2, Part II, The alteration of plant materials. D, 1968, Pennsylvania State University, University Park. 382 p.

Cohen, C. J. Structure of the gabbro complex at Baltimore, MD. D, 1934, The Johns Hopkins University.

Cohen, Carel Lodewijk David. Preliminary study of coccoliths and discoasters from Mancos Shale of eastern Utah and western Colorado. M, 1959, University of Utah. 158 p.

Cohen, Charles. An experimental investigation of the strength and failure characteristics of anisotropic rocks under confining pressure. M, 1961, Columbia University, Teachers College.

Cohen, David Ronald. Biogeochemistry; a geochemical method for gold exploration in the Canadian Shield. M, 1986, Queen's University. 498 p.

Cohen, Donald K. The geology and petrography of the Millet Ranch plutons; a mixed magma. M, 1980, University of Nevada. 62 p.

Cohen, Gary B. Dispersion and sorption of dissolved hydrocarbons in aquifer materials. M, 1982, University of Minnesota, Minneapolis. 84 p.

Cohen, Howard Melvin. Densification of glass at very high pressures. D, 1962, Pennsylvania State University, University Park. 153 p.

Cohen, J. P. An X-ray and neutron diffraction study of hydrous low cordierite. M, 1975, Virginia Polytechnic Institute and State University.

Cohen, Joseph Charles, Jr. Investigation of greisen mineralization near Howard, Colorado. M, 1974, University of Wisconsin-Milwaukee.

Cohen, Karen Kluger. Utilization of the paleomagnetism of Mesozoic miogeosynclinal and magmatic arc deposition from the North American Western Cordillera as a test of the sinistral displacement along the Mojave-Sonora megashear. D, 1981, University of Pittsburgh. 712 p.

Cohen, Lawrence Kenneth. An investigation of tritium in the aquatic environment. D, 1977, New York University. 167 p.

Cohen, Lewis Hart. Melting and phase relations in an anhydrous basalt to 45 kilobars. D, 1965, University of California, San Diego. 179 p.

Cohen, Lynne. A uranium disequilibrium study of the submarine springs of Spring Creek, Florida. M, 1977, Florida State University.

Cohen, Mitchell L. The petrography of variously ranked Puget Lowlands coals. M, 1983, Southern Illinois University, Carbondale. 120 p.

Cohen, Philip. Geologic studies in the vicinity of Black Mountain, West Dummerston, Vermont. M, 1956, University of Rochester. 135 p.

Cohen, Philip Leon. Reconnaissance study of the "Russell" Basalt Aquifer in the Lewiston Basin of Idaho and Washington. M, 1979, University of Idaho. 163 p.

Cohen, Robert Mark. Pollution abatement of landfill leachate by spray irrigation in a northeastern forested karst terrane. M, 1982, Pennsylvania State University, University Park. 314 p.

Cohen, Robert Stuart. An ion-selective electrode method for the determination of alkali feldspars. M, 1975, Rutgers, The State University, Newark. 60 p.

Cohen, Ronald Elliott. Thermodynamics of aluminous pyroxenes; effects of short-range order. D, 1985, Harvard University. 362 p.

Cohen, Stephen. Ostracods of the West Franklin Limestone (Middle Pennsylvanian) of southern Indiana. M, 1962, Indiana University, Bloomington. 55 p.

Cohen, Theodore Jerome. Explosion seismic studies of the Mid-Continent gravity high. D, 1966, University of Wisconsin-Madison. 346 p.

Cohen, William J. An investigation of the absorption spectra of some minerals with the Cary spectrometer. M, 1952, Columbia University, Teachers College.

Cohen, Yuval. Studies on the marine chemistry of nitrous oxide. D, 1978, Oregon State University. 112 p.

Cohenour, Bernard C. Study of petroleum degradation by bacteria in the waters encompassing Kennedy Space Center. M, 1976, Florida Institute of Technology.

Cohenour, Robert Eugene. Geology of the Sheep Rock Mountains, Tooele and Juab counties. D, 1957, University of Utah. 201 p.

Cohenour, Robert Eugene. Some techniques for sampling bottom sediments of Great Salt Lake, Utah. M, 1952, University of Utah. 44 p.

Cohn, B. P. A forecast model for Great Lakes water levels. D, 1975, Syracuse University. 247 p.

Cohn, B. P. Accretion and erosion of a Lake Ontario beach, Selkirk Shores, New York. M, 1974, Syracuse University.

Cohn, Stephen Norfleet. Holographic in-situ stress measurement in geophysics. D, 1983, California Institute of Technology. 168 p.

Coholich, Philip A. Longitudinal conductance derived from direct inversion. M, 1987, University of Wisconsin-Milwaukee.

Cohoon, Richard Roy. Geology of the Mount Ida area, Ouachita Mountains, Arkansas. M, 1959, University of Oklahoma. 69 p.

Cohoon, Richard Roy. The environmental geology of the Russellville, Arkansas area. D, 1974, Oklahoma State University. 179 p.

Coil, Fay. The character of the leucoxene in the Permian of Oklahoma. M, 1932, University of Oklahoma. 36 p.

Coish, R. A. Igneous and metamorphic petrology of the mafic units of the Betts Cove and Blow-Me-Down ophiolites, Newfoundland. D, 1977, University of Western Ontario. 228 p.

Cok, Anthony E. Morphology and surficial sediments of the eastern half of the Nova Scotian shelf. D, 1970, Dalhousie University.

Coke, John McBrien. Foothills structure of the Box Elder and Sandcreek region, Larimer County, Colorado. M, 1934, University of Colorado.

Coker, Alfred Eugene. A mineralogical study of an Ordovician metabentonite near Clinton, Anderson County, Tennessee. M, 1962, University of Tennessee, Knoxville. 49 p.

Coker, C. Eugene. Geology of the Coleman Junction Limestone in Shackelford County, Texas. M, 1960, Rice University. 83 p.

Coker, Diane. Shallow ground water resources of a portion of Rapid Valley, Pennington County, South Dakota. M, 1981, South Dakota School of Mines & Technology. 96 p.

Coker, William Bernard. Lake sediment geochemistry in the Superior Province of the Canadian Shield. D, 1974, Queen's University. 297 p.

Colazas, Zenophon C. Subsidence, compaction of sediments and efforts of water injection, Wilmington and Long Beach offshore oil fields, Los Angeles County, California. M, 1971, University of Southern California.

Colbath, George Kent. Organic walled microplancton from the Eden Shale (Ordovician), Indiana, U.S.A. M, 1978, University of California, Los Angeles.

Colbath, George Kent. Paleoecology of palynomorphs from the Upper Ordovician-Lower Silurian of the Southern Appalachians, U.S.A. D, 1983, University of Oregon. 314 p.

Colbath, Sharon Larson. Gastropod predation, Molluscan community paleoecology, and depositional environment of the Miocene Astoria Formation at Beverly Beach State Park, Oregon. M, 1981, University of Oregon. 68 p.

Colbeck, Samuel C. The flow law for temperate glacier ice. D, 1970, University of Washington. 149 p.

Colberg, Mark Robert. The origin of the Auburn Formation migmatites, Lee County, Alabama. M, 1988, Auburn University.

Colbert, Edwin H. A new fossil peccary, Prosthennops niobrarensis, from Brown County, Nebraska. M, 1930, Columbia University, Teachers College.

Colbert, Edwin H. Distributional and phylogenetic studies of Indian fossil mammals. D, 1935, Columbia University, Teachers College.

Colborne, Gerald Laverne. Sedimentary iron in the Cretaceous of the Clear Hills area, Alberta. M, 1958, University of Alberta. 102 p.

Colburn, Ivan P. The tectonic history of Mount Diablo, California. D, 1961, Stanford University. 276 p.

Colburn, Ivan Paul. Stratigraphy and structure along the Whittier Fault near Yerba Linda, California. M, 1953, Pomona College.

Colburn, James A. Clay minerals in stylolites of limestones. M, 1965, Dartmouth College. 18 p.

Colby, Robert E. The stratigraphy and structure of the Recreation Red Beds, Tucson Mountain Park, Arizona. M, 1958, University of Arizona.

Colchin, Michael P. Problems evaluating the composition of ground water in Catahoula Formation aquifers, South Texas. M, 1978, University of Texas, Austin.

Colcleugh, V. D. Geology of the Brookbank gold deposit, (Irwin Township, northwestern Ontario, near Beardmore). M, 1946, University of Manitoba.

Coldiron, Ronn William. The phylogenetic relationships of edopoid amphibians with a description of Orvillerpeton bonneri, a new genus from Kansas. D, 1982, Columbia University, Teachers College. 356 p.

Cole Hoerster, Mary Lou. Detailed texture and heavy mineralogy of Recent sands along the northeastern Texas Gulf Coast and a resulting model for barrier formation. D, 1982, Rice University. 164 p.

Cole, Charles Andrew. Wetland ecosystem development on a reclaimed surface coal mine in southern Illinois. D, 1988, Southern Illinois University, Carbondale. 307 p.

Cole, Charles Taylor. The black shale basin of West Texas. M, 1939, University of Texas, Austin.

Cole, Clarence A. Sedimentation of the Colorado River in Coleman and Runnels counties, Texas. M, 1937, Texas Tech University. 31 p.

Cole, D. N. Man's impact on wilderness vegetation; an example from Eagle Cap Wilderness, northeastern Oregon. D, 1977, University of Oregon. 320 p.

Cole, David Andrew. Free radical and mineral chemistry in the low temperature oxidation of coal. D, 1988, [Lehigh University]. 227 p.

Cole, David Lee. A preliminary investigation of the lithological characteristics of the Troutdale Formation in portions of the Camas, Sandy, Washougal, and Bridal Veil quadrangles. M, 1982, Portland State University. 69 p.

Cole, David Lee. Petrology of the Mesaverde Sandstone, Big Horn Basin, Wyoming. M, 1960, University of Missouri, Columbia.

Cole, David M. Causes of color differences in Landsat color ratio composite images of limonitic areas in Southeast Utah. M, 1984, Colorado School of Mines. 100 p.

Cole, David Martin. A numerical boundary integral equation method for transient motions. D, 1980, California Institute of Technology. 234 p.

Cole, David Robert. Mechanisms and rates of stable isotopic exchange in hydrothermal rock-water systems. D, 1980, Pennsylvania State University, University Park. 319 p.

Cole, David Robert. The geology and the geochemistry of lead and zinc in soils in the Thurman area, southeastern Adirondack Mountains, New York. M, 1976, Pennsylvania State University, University Park. 199 p.

Cole, Gary A. The ontogeny and population structure of an Isorophida edrioasteroid from the Kinkaid Formation of southern Illinois (Upper Mississippian). M, 1977, Southern Illinois University, Carbondale. 30 p.

Cole, George Paul. Permian carbonate facies, southern Big Horn Basin–Owl Creek Mountains area, (Hot Springs County), north central Wyoming. M, 1970, University of Montana. 71 p.

Cole, Gregory Lawrence. Cenozoic 'plate' interactions in the Antarctic region. M, 1976, Cornell University.

Cole, J. C. Geology of East-central Rocky Mountain National Park and vicinity, with emphasis on the emplacement of the Precambrian Silver Plume Granite in the Longs Peak-St. Vrain Batholith. D, 1977, University of Colorado. 393 p.

Cole, James Henderson. Fall line belt studies in central South Carolina. M, 1948, University of South Carolina. 47 p.

Cole, James Morgan. Geology of the Roaring Springs oil field, Motley County, Texas. M, 1958, University of Oklahoma. 69 p.

Cole, Jay Timothy. Late Tertiary paleomagnetic data from Leyte, Philippines; implications for Philippine fault zone motion. M, 1988, Texas A&M University. 129 p.

Cole, Jeffrey E. Techniques and limitations in developing a detailed Q-model on the outer continental shelf of Texas. M, 1984, Stanford University.

Cole, Jimmy Dale. Design of longwall systems. D, 1980, University of Missouri, Rolla. 259 p.

Cole, John Albert. Subsurface geology of east central Lincoln County, Oklahoma. M, 1955, University of Oklahoma. 59 p.

Cole, John Frederick. Mineralogy and physical properties of clays underlying US-45 Highway relocation, Ontonagon, Michigan. M, 1970, Michigan Technological University. 69 p.

Cole, John M. Study of part of the Lucerne Granite contact aureole in eastern Penobscot County, Maine. M, 1961, University of Virginia. 76 p.

Cole, Joseph Glenn. Regional surface and subsurface study of the Marmaton Group, Pennsylvanian (Desmoinesian) of northeastern Oklahoma. M, 1965, University of Oklahoma. 51 p.

Cole, Joseph Glenn. Stratigraphic study of the Cherokee and Marmaton sequences, Pennsylvanian (Desmoinesian), east flank of the Nemaha Ridge, north-central Oklahoma. D, 1968, University of Oklahoma. 334 p.

Cole, Kathleen Patricia Hicks. Uranium and thorium series isotopes as indicators of geochemical processes in Recent Venezuela Basin sediments. D, 1988, Texas A&M University. 278 p.

Cole, Kenneth Arthur. Geology for planning in the Martis Valley, California. M, 1975, University of California, Davis. 43 p.

Cole, Kenneth Lee. Late Quaternary environments in the eastern Grand Canyon; vegetational gradients over the last 25,000 years. D, 1981, University of Arizona. 206 p.

Cole, Mark R. Paleocurrent and basin analysis (Tertiary) on San Nicolas island, California. M, 1970, Ohio University, Athens. 110 p.

Cole, Mark Rolland. Petrology and dispersal patterns of Jurassic and Cretaceous sedimentary rocks in the Methow River area, North Cascades, Washington. D, 1973, University of Washington. 110 p.

Cole, Marshall F. Geology of the Twin Peaks area, Benton and Washington counties, Arkansas. M, 1958, University of Arkansas, Fayetteville.

Cole, Marshall Morris. Nature, age, and genesis of quartz-sulfide-precious-metal vein systems in the Virginia City mining district, Madison County, Montana. M, 1983, Montana State University. 76 p.

Cole, Mary Jane. Geology of northern part of the Fort Worth Basin as shown in the Clingingsmith area, Montague County, Texas. M, 1951, University of Oklahoma. 49 p.

Cole, Mary Lou. Nearshore glacial marine sedimentation, based on late Pleistocene deposits of the Puget Lowlands, Washington and British Columbia. M, 1979, Rice University. 176 p.

Cole, Melanie Ruth Will. Rare earth element abundances in some mafic igneous rocks. M, 1980, Iowa State University of Science and Technology.

Cole, R. C., Jr. Management information systems for the mineral industry. M, 1972, University of Arizona.

Cole, Rex Don. Sedimentology and sulfur isotope geochemistry of Green River Formation (Eocene), Uinta Basin, Utah, Piceance Creek Basin, Colorado. D, 1975, University of Utah. 290 p.

Cole, Robert Dennis. Geology of the Butte Valley and Warm Springs formations, southern Panamint Range, Inyo County, California. M, 1986, California State University, Fresno. 126 p.

Cole, Sally Ann. The effect of thermal stress conditions on benthic foraminifera in Biscayne Bay, Florida. M, 1974, University of Illinois, Urbana.

Cole, Stuart Loren. Implementation aspects of an ecosystems approach; the Conservation of Natural Resources Program, University of California, Berkeley, 1969-1972. D, 1975, University of California, Berkeley. 367 p.

Cole, Tony. A surface to subsurface study of the Sycamore Limestone (Mississippian) along the north flank of the Arbuckle Anticline. M, 1988, University of Oklahoma. 140 p.

Cole, Virgil Bedford. The relations between the Paleozoic and Mesozoic of New Mexico and Arizona. M, 1923, University of Missouri, Rolla.

Cole, W. Storrs. A foraminiferal fauna from the Guayabal Formation in Mexico. M, 1928, Cornell University.

Cole, W. Storrs. Contributions to the Tertiary paleontology of the Tampico Embayment area. D, 1930, Cornell University.

Cole, William Fletcher. Geology and engineering geology of the Wilcox lignite deposit, northeastern Rusk County, Texas. M, 1980, Texas A&M University. 186 p.

Coleman, Carl R. A sedimentary study of the Pennsylvanian outcrops in Abo Canyon, New Mexico. M, 1950, Texas Tech University. 24 p.

Coleman, Craig J. The stranded bar; a study of a bar formation during storm surges at Carrabelle Beach, Florida. M, 1978, Florida State University.

Coleman, D. D. Isotopic characterization of Illinois natural gas. D, 1976, University of Illinois, Urbana. 185 p.

Coleman, Dennis D. Investigation of a method for determining the rate of strontium diffusion in a potassium feldspar. M, 1970, University of Arizona.

Coleman, Derrick Job. An examination of bankfull discharge frequency in relation to floodplain formation. D, 1982, The Johns Hopkins University. 153 p.

Coleman, Donald A. Shelf to basin transition of Silurian-Devonian rocks, Porcupine River area, east-central Alaska. M, 1985, University of Alaska, Fairbanks. 162 p.

Coleman, George L., II. Recent marine ostracodes from the eastern Gulf of Mexico. M, 1960, University of Kansas. 110 p.

Coleman, James Lawrence, Jr. Stratigraphy and depositional environments of the carbonates of the Vicksburg Group; Oligocene in Mississippi and adjacent areas. M, 1978, Mississippi State University. 81 p.

Coleman, James Malcolm. Recent coastal sediments and late Recent rise of sea level in Vermillion, Iberia, and St. Mary parishes, Louisiana. D, 1966, Louisiana State University. 123 p.

Coleman, James Malcom. Recent sedimentation and processes in central coastal Louisiana. M, 1962, Louisiana State University.

Coleman, James Mark. Present and past nutrient dynamics of a small pond in Southwest Florida. D, 1979, University of Florida. 158 p.

Coleman, John M. Industrial development in Clay, Lowndes, and Monroe counties, Mississippi, along the proposed route of the Tennessee-Tembigbee Waterway. M, 1962, Mississippi State University. 104 p.

Coleman, K. Fred. Interpretation of magnetic anomalies in the Dillsburg area, York County, Pennsylvania. M, 1966, Pennsylvania State University, University Park. 77 p.

Coleman, L. C. Mineralogy of the Yellowknife Bay area, N.W.T. M, 1952, Queen's University.

Coleman, Leslie Charles. Mineralogy of the Giant Yellowknife gold mine, Yellowknife, N.W.T. D, 1955, Princeton University. 54 p.

Coleman, Neil Lloyd. Laboratory experiments in selective grain transport. D, 1960, University of Chicago. 49 p.

Coleman, Neil M. Gravity investigation of basement structure in northwest peninsular Florida. M, 1979, University of South Florida, Tampa. 59 p.

Coleman, Robert Griffin. John Day Formation in the Picture Gorge Quadrangle, Oregon. M, 1949, Oregon State University. 211 p.

Coleman, Robert Griffin. Mineralogy and petrology of the New Idria District, California. D, 1957, Stanford University. 191 p.

Coleman, Teresa Ann. Nannoplankton biostratigraphy of the Tepetate Formation, Baja California Del Sur. M, 1979, University of Southern California.

Coleman, Walter F. Surface geology of the Rentiesville area, Muskogee and McIntosh counties, Oklahoma. M, 1958, University of Oklahoma. 100 p.

Coles, Joan Link. A study of some synthetic apatites. D, 1963, University of Utah. 86 p.

Coles, Kenneth Spencer. Stratigraphy and structure of the Pinecone sequence, Roberts Mountains Allochthon, Nevada, and aspects of mid-Paleozoic sedimentation and tectonics in the Cordilleran geosyncline. D, 1988, Columbia University, Teachers College. 327 p.

Coles, Richard Leslie. A thermomagnetic analysis of rocks 20-700°C; the design and construction of measuring apparatus, and its application to rocks from the Whiteshell area, Manitoba. M, 1970, University of Manitoba.

Coles, Richard Leslie. Relationships between measured rock magnetizations and interpretations of longer wavelength anomalies in the Superior Province of the Canadian Shield. D, 1973, University of Manitoba.

Coley, Katharine Lancaster. Structural evolution of the Warwick Hills, Marathon Basin, West Texas. M, 1987, University of Texas, Austin. 140 p.

Coley, Michael J. Intraplate seismicity in central Alaska and Chukotka. M, 1983, Michigan State University. 97 p.

Coley, Tyrol B. The stratigraphic distribution and correlation of some Middle Devonian Ostracoda. M, 1950, Wayne State University.

Colgrove, G. L. FeS-NiS system. M, 1940, Queen's University. 90 p.

Colgrove, Gordon L. The system Fe-Ni-S. D, 1942, University of Wisconsin-Madison.

Colinvaux, Paul Alfred. The environment of the Bering Land Bridge. D, 1962, Duke University. 297 p.

Collagan, Robert Bruce. Stratigraphic differentiation of selected Nevada black shales by inorganic analyses. M, 1958, University of Nevada. 78 p.

Collamer, Stephen Vaughn. Aspects of deposition and diagenesis in the Lower Ordovician Rockview Member (Axemann Limestone) in the vicinity of Bellefonte and State College, Pennsylvania. M, 1985, SUNY at Binghamton. 230 p.

Collar, Paul David. Petrography and geochemistry of quartz veins from the abandoned Housely Mine and vicinity, Hot Springs County, Arkansas. M, 1986, University of Arkansas, Fayetteville.

Collender, Jack. Gravity survey across the Simi Fault and engineering geology implications. M, 1988, California State University, Long Beach. 98 p.

Collett, Raymond T. Interpretation of the California landscape. D, 1967, University of California, Berkeley.

Collett, Timothy S. Detection and evaluation of natural gas hydrates from well logs, Prudhoe Bay, Alska. M, 1983, University of Alaska, Fairbanks. 78 p.

Collette, Ronald. A three-dimensional model for simulating flow and mass transport in groundwater systems. M, 1985, University of Alberta. 116 p.

Coley, Ralph S. A geological history of the McConnells Mills area, Pennsylvania. M, 1935, University of Pittsburgh.

Collie, Carolyn. Abnormal formation pressures. M, 1978, Tulane University.

Collie, George Lucius. The igneous and metamorphic rocks of Conanicut Islands, Rhode Island. D, 1893, Harvard University.

Collier, Frederick J. A taxonomic study of two bryozoan genera from the Hamilton Group (Middle Devonian) of New York. M, 1965, George Washington University.

Collier, J. Maurice. Geology of upper Ruby Basin, Beaverhead County, Montana. M, 1965, University of Arizona. 56 p.

Collier, James D. Geology and uranium mineralization of the Florida Mountain area, Needle Mountains, southwestern Colorado. D, 1982, Colorado School of Mines. 218 p.

Collier, James E. Topographic and soil factors in Kentucky's agriculture. M, 1938, University of Cincinnati. 231 p.

Collier, Kenneth. The fauna and paleoecology of the Santa Margarita Formation at San Juan Creek, San Luis Obispo County, California. M, 1978, University of California, Santa Barbara.

Collier, M. P. Structural analysis of folds in Cambrian rocks adjacent to walls of the Grand Canyon. M, 1978, Stanford University. 100 p.

Collier, Robert L. Subsurface geology and sand distribution patterns of the Marginulina ascensionensis Zone-Hester area, South Louisiana. M, 1988, University of New Orleans.

Collier, Robert William. Trace element geochemistry of marine biogenic particulate matter. D, 1981, Massachusetts Institute of Technology. 200 p.

Collier, William Wayne. The phylogeny of the Devonian ostracod genus *Ctenoloculina* Bassler. D, 1972, [University of Michigan].

Colligan, Richard Vincent. Structural geology of Lake Clear and vicinity (Ontario, Canada). M, 1940, Cornell University.

Colligan, Thomas Henry. Mineralization, deformation, and host rock diagenesis of the upper Yellowhead zinc-lead deposit. M, 1984, University of Washington. 65 p.

Collin, Martin L. Weathering of the Port Deposit Granodiorite (post-Lower Ordovician) and the formation of clays in derived soils at Newark, Delaware. M, 1967, West Virginia University.

Collings, Gay Madsen. Geology and geochemistry of the Colt Mesa copper deposit, Circle Cliffs area, Utah. M, 1975, University of Utah. 52 p.

Collings, Stephen P. Geology and geochemistry of the Omai Mine area, Omai, southwestern Guyana. M, 1969, Colorado School of Mines. 223 p.

Collins, A. Gene. Geochemical classification of formation waters of use in hydrocarbon exploration and production. M, 1972, University of Tulsa. 63 p.

Collins, Amy Hutsinpiller. Geological applications of remote sensing in the Virginia Range, Nevada. D, 1988, University of Nevada. 254 p.

Collins, Ann M. Petrology of the Eocene Marquez Shale Member of the Reklaw Formation, Bastrop County, Texas. M, 1982, University of Texas, Austin.

Collins, B. I. The copper geochemistry of a small mine tailings stream in central eastern Vermont. M, 1971, Dartmouth College. 36 p.

Collins, B. P. Geophysical investigation of the structural framework of the southern portion of the Carboniferous Narragansett Basin, Rhode Island. D, 1978, University of Rhode Island. 182 p.

Collins, Barbara Jane Schenck. Textural and morphological studies of some clay minerals. D, 1955, University of Illinois, Urbana. 125 p.

Collins, Benjamin I. A petrochemical model for the tungsten-bearing skarns in the Pioneer Mountains, Montana. D, 1975, University of Montana. 140 p.

Collins, Brian David. Erosion of tephra from the 1980 eruption of Mount St. Helens. M, 1984, University of Washington. 181 p.

Collins, Bruce A. Geology of the coal deposits of the Carbondale, Grand Hogback, and southern Danforth Hills coal fields, southeastern Piceance Basin, Colorado. D, 1975, Colorado School of Mines. 218 p.

Collins, Bruce A. Geology of the coal-bearing Mesaverde Formation (Cretaceous), Coal Basin area, Pitkin County, Colorado. M, 1970, Colorado School of Mines. 116 p.

Collins, C. B. Application of mass spectrometry to geological age determinations. D, 1951, University of Toronto.

Collins, Catherine M. Quantitative analysis of the faunal and sediment associations on Roberts Hill Reef, Onondaga Formation, in eastern New York. M, 1978, Rensselaer Polytechnic Institute. 77 p.

Collins, Curtis A. Description of measurements of current velocity and temperature over the Oregon continental shelf, July 1965–February 1966. D, 1968, Oregon State University. 154 p.

Collins, Daniel E. Metamorphic geology of a portion of the Bagdad mining district, Yavapai County, Arizona. M, 1977, University of Nevada - Mackay School of Mines. 113 p.

Collins, Daniel John. Morphology, hydrodynamics and subsurface stratigraphy of an ebb-tidal delta; Indian River Inlet, Delaware. M, 1982, University of Delaware. 222 p.

Collins, David G. The multivariate rotation method of quantitative grain shape analysis. M, 1984, Lehigh University. 77 p.

Collins, David J. Hydraulic interpretation of grain size distributions of suspended sands collected over large

intertidal bedforms on Selma Bar, Cobequid Bay, Nova Scotia. M, 1985, McMaster University. 240 p.

Collins, Desmond Harold. Upper Devonian nautiloids from the Fitzroy and Carnarvon basins, Western Australia. D, 1966, University of Iowa. 389 p.

Collins, Donald Francher. The geology of the southern third of the Orestimba Quadrangle, Stanislaus and Merced counties, California. M, 1950, University of California, Berkeley. 54 p.

Collins, Donald N. A study in sediment transportation by the Arkansas River in Colorado and Kansas. M, 1959, University of Kansas. 67 p.

Collins, Donald W. Stratigraphy of late glacial and post-glacial sediments in Gosport Harbor, Isles of Shoals, New Hampshire. M, 1988, University of New Hampshire. 102 p.

Collins, Earl M. Metamorphic petrology and structural geology of the Huehuetenango Quadrangle, Guatemala. M, 1978, Louisiana State University.

Collins, Eric S. Recent benthic foraminifera of Breton and Stake islands, northern Gulf of Mexico. M, 1988, Old Dominion University. 208 p.

Collins, Gary Brent. Implications of diatom succession in postglacial sediments from two sites in northern Iowa. D, 1968, Iowa State University of Science and Technology. 197 p.

Collins, George Alexander. Differential thermal analysis of certain sulphide minerals. M, 1952, Washington University. 18 p.

Collins, George Alexander. Pyrite, a natural semiconductor. D, 1961, Washington University. 72 p.

Collins, George M., Jr. The joint pattern on the limbs of a syncline of Paleozoic rocks infaulted and infolded in the Pre-Cambrian. M, 1953, Columbia University, Teachers College.

Collins, Glendon Elmer. The bedrock geology of a portion of the Ellington, Connecticut, Quadrangle. M, 1953, Brown University.

Collins, H. L. The Tethyan aspect of Neogene deepsea Krithe (Ostracoda) of the Indian and South Atlantic oceans. M, 1975, Washington University. 50 p.

Collins, Harold P. Effects of mechanical site preparation on selected physical properties of four volcanic ash influenced forest soils in northern Idaho. M, 1982, University of Idaho. 86 p.

Collins, Henry B. The system CrO-SiO$_2$ at low oxygen partial pressures. M, 1978, Pennsylvania State University, University Park.

Collins, Horace R. Small spores assemblages of the Harlem Coal (Pennsylvanian, West Virginia). M, 1959, West Virginia University.

Collins, Isabelle D. A micro method for the determination of silica. M, 1927, University of California, Berkeley.

Collins, James Finnbarr. The Eh-pH environment and iron-manganese equilibria in soils. D, 1968, North Carolina State University. 250 p.

Collins, John Anthony. A search for layering in the oceanic crust. D, 1988, Woods Hole Oceanographic Institution. 195 p.

Collins, John Bartlett. The kinetics of Co(II) and Zn(II) adsorption to goethite. D, 1988, University of California, Riverside. 167 p.

Collins, Jon A. Carbonate lithofacies and diagenesis related to sphalerite mineralization near Daniel's Harbour, western Newfoundland. M, 1971, Queen's University. 184 p.

Collins, Jon Alexander. The sedimentary copper universal; from sedimentologic and stratigraphic syntheses of the Proterozoic of Icon, Quebec; Grinnell Formation, Alberta; Nonesuch Shale, Michigan; and the Mississippian Horton-Windsor formations of Nova Scotia. D, 1974, Queen's University. 259 p.

Collins, Kathryn Hope. Depositional and diagenetic history of the Permian rocks in the Meers Valley, southwestern Oklahoma. M, 1985, Oklahoma State University. 110 p.

Collins, Kenneth Alan. Phase distributions of the rare-earth elements in sediments. D, 1967, University of Wisconsin-Madison. 192 p.

Collins, Kenneth Robert. Kinetics of oxidation of aqueous sulfur (IV) as catalyzed by manganous ion. D, 1987, Georgia Institute of Technology. 205 p.

Collins, Laurel Smith. Environmental gradients and morphologic variation in Bulimina aculeata and B. marginata. D, 1988, Yale University. 254 p.

Collins, Lorence Gene. Geology of the magnetite deposits and associated gneisses near Ausable Forks, New York. D, 1959, University of Illinois, Urbana. 165 p.

Collins, Lorence Gene. Refractive index studies of magnetite-bearing pyroxene amphibolites, Scott Mine, Sterling Lake, New York. M, 1955, University of Illinois, Urbana. 147 p.

Collins, Margaret R. Surficial geology (Quaternary) of the Sandwich morainal complex, Cape Cod, Massachusetts. M, 1971, Franklin and Marshall College. 116 p.

Collins, Mary Elizabeth. A model for evaluating differentiating characteristics of the Colo soils as mapped in the North Central Region. D, 1980, Iowa State University of Science and Technology. 439 p.

Collins, Melvin J. Some anticlines of Moffat County, Colorado. M, 1921, University of Colorado.

Collins, R. E. L. Miocene Pelecypoda and Scaphopoda from the isthmus of Tehuantepec, Mexico. D, 1928, The Johns Hopkins University.

Collins, R. L. Para- and diamagnetic susceptibilities in fields of 100 to 1000 oersteds. M, 1950, University of Oklahoma. 43 p.

Collins, Robert F. Pickeringite from the Cleveland Shale (northern Ohio). M, 1924, Columbia University, Teachers College.

Collins, Robert Joseph, Sr. The stratigraphy and Ostracoda of the Ozan, Annona, and Marlbrook formations of southwestern Arkansas. D, 1960, Louisiana State University. 244 p.

Collins, Sam G. Geology of the Winner Quadrangle, South Dakota. M, 1958, University of South Dakota. 67 p.

Collins, Stephen E. Some observed textural and petrographic variations of the basal Knox Sandstone in East Tennessee. M, 1956, University of Tennessee, Knoxville. 41 p.

Collins, Steve Allan. Development and implementation of a hypoelastic constitutive theory to model the behavior of sand. D, 1988, Georgia Institute of Technology. 464 p.

Collins, Steven Lee. Submarine fan deposits in the Shields Formation (Precambrian, Ocoee Series) of East Tennessee. M, 1976, University of Tennessee, Knoxville. 114 p.

Collins, Susan J. B. A biogeochemical study of a borate deposit in a desert environment. M, 1977, University of Missouri, Rolla.

Collins, Terry Moore. Geology of the Stateline District, Utah-Nevada. M, 1977, University of Missouri, Rolla.

Collins, W. Spectroradiometric detection and mapping of areas enriched in ferric iron minerals using airborne and orbiting instruments. D, 1976, Columbia University, Teachers College. 129 p.

Collins, William E. Corals of the St. Louis Limestone in the area of Calhoun County, Illinois and St. Louis County, Missouri. M, 1973, University of Missouri, Rolla.

Collins, William H. The geology of Gowganda mining division. D, 1911, University of Wisconsin-Madison.

Collinson, Charles William. Lower Mississippian ammonoids of Missouri. M, 1950, University of Iowa. 119 p.

Collinson, Charles William. The Upper Ordovician cephalopod fauna of Baffin Island. D, 1952, University of Iowa. 439 p.

Collinson, James Waller. Permian and Triassic biostratigraphy of the Medicine Range, Elko County, Nevada. D, 1966, Stanford University. 156 p.

Collinson, Thomas Barnes. Hydrothermal mineralization and basalt alteration in stockwork zones of the Bayda and Lasail massive sulfide deposits, Oman Ophiolite. M, 1986, University of California, Santa Barbara. 164 p.

Collister, Morton C. Terranes of Albany, Vermont. M, 1912, Syracuse University.

Collord, E. J. Geology and ore deposits of the Sand Springs mining district, Churchill County, Nevada. M, 1980, University of Nevada. 63 p.

Collyer, P. L. A cluster analysis using nonquantitative data of tungsten districts in North America. M, 1972, Syracuse University.

Colman, J. A. Measurement of eddy diffusivity and solute transport through the near-sediment boundary layer of ice-covered lakes, using radon-222. D, 1979, University of Wisconsin-Madison. 222 p.

Colman, Royce Luther. The carbonate petrology and conodont biostratigraphy of the Old Whalen Member of the Lone Mountain Dolomite (Lower Devonian), Sulphur Springs Range, Nevada. M, 1979, University of California, Riverside. 76 p.

Colman, Steven Michael. The development of weathering rinds on basalts and andesites and their use as a Quaternary dating method, Western United States. D, 1977, University of Colorado. 249 p.

Colman, Steven Michael. The history of mass movement processes in the Redwood Creek Basin, Humbolt County, California. M, 1974, Pennsylvania State University, University Park. 180 p.

Colman-Sadd, Stephen Peter. Geology of the iron deposits near Stephenville, Newfoundland. M, 1969, Memorial University of Newfoundland. 92 p.

Colman-Sadd, Stephen Peter. The geologic development of the Bay d'Espoir area, southeast Newfoundland. D, 1974, Memorial University of Newfoundland. 271 p.

Colmenares, Omar A. A palynological study of the south-east region of the Boscan Field, Venezuela. M, 1986, Michigan State University. 141 p.

Colomb, Herbert Palfrey, Jr. Recent marine sediments of the central California continental shelf between Point Lobos and Point Sur. M, 1972, United States Naval Academy.

Colombini, Victor Domenic. The geology of the Rangeley Quadrangle, Maine. D, 1961, Boston College. 203 p.

Colony, Wayne Edward. Prehnite dissociation curve below 3 kilobars water pressure. M, 1970, University of Arizona.

Colopietro, Margaret R. Compositional zoning of tourmalines; petrogenetic history and usefulness as a geothermometer, Black Hills, South Dakota. M, 1987, University of Akron. 125 p.

Colpitts, Robert Moore, Jr. Geology of the Sierra de la Cruz area, Socorro County, New Mexico. M, 1986, New Mexico Institute of Mining and Technology. 166 p.

Colquhoun, Donald John. The stratigraphy and paleontology of the Nipissing and Mattawa areas. M, 1956, University of Toronto.

Colquhoun, Donald John. Triassic stratigraphy of western central Canada. D, 1960, University of Illinois, Urbana. 171 p.

Colson, C. M. Rock bursts. M, 1950, University of Missouri, Rolla.

Colson, Calvin Thomas. Pottsville and Allegheny corals of Ohio. M, 1963, Ohio University, Athens. 103 p.

Colson, Russell Owen. Temperature and compositional dependence of trace element partitioning in silicate systems. D, 1986, University of Tennessee, Knoxville. 131 p.

Colson, William Edward. A new Mississippian conodont fauna from Kentucky. M, 1943, University of Missouri, Columbia.

Colten, Virginia Ann. Experimental determination of smectite hydration states under simulated diagenetic conditions. D, 1985, University of Illinois, Urbana. 154 p.

Colten, Virginia Ann. The behavior of lithium in the Mississippi River estuary. M, 1980, Louisiana State University.

Colton, Clark Roper. Igneous geology of Rim Rock country, Trans-Pecos, Texas. M, 1957, University of Texas, Austin.

Colton, Earl Glenn, Jr. Simpson Group, Lea County, New Mexico. M, 1957, University of Texas, Austin.

Colton, Ilsley Daniel. The equilibrium oxygen fugacities associated with seven mid-ocean ridge basalts. M, 1986, Queen's University. 103 p.

Colton, Richard C. The subsurface geology of Hamilton Co. Florida with emphasis on the Oligocene age Suwannee Limestone. M, 1978, Florida State University.

Coltrin, Donald George, Jr. Seismic reflection imaging problems resulting from a rough surface at the top of the accretionary prism at convergent margins. M, 1987, University of Texas, Austin. 116 p.

Colucci, Michael T. Ductile deformation mechanisms controlled by impurities in a shear zone. M, 1984, Rutgers, The State University, New Brunswick. 129 p.

Columbus, Nathan. Viscous model study of sea-water intrusion. M, 1964, New Mexico Institute of Mining and Technology. 82 p.

Colvard, Elizabeth Monroe. Petrogenesis of the Sunlight Basin Intrusions, Park County, Wyoming. M, 1986, University of New Mexico. 121 p.

Colville, Alan Andrew. Paleomagnetic investigation in the vicinity of the northern Tobacco Root Mountains, Madison County, Montana. D, 1961, Indiana University, Bloomington. 148 p.

Colville, Patricia Ann. A study of the relationships between cell parameters and chemical compositions of monoclinic amphiboles. M, 1966, University of California, Los Angeles.

Colville, Valerie R. Ordovician-Silurian eustatic sea level changes at Limestone Mountain, Houghton County, Michigan. M, 1983, University of Wisconsin-Milwaukee. 143 p.

Colvin, John McRae. Geology of the Centerville Quadrangle, Hickman County, Tennessee. M, 1958, Vanderbilt University.

Colvin, Michael Dale. Petrology of the Madison Limestone (Mississippian), middle Warm Spring Canyon, Dubois, Wyoming. M, 1980, Miami University (Ohio). 112 p.

Colvine, A. C. The petrology, geochemistry and genesis of sulphide-related alteration at the Temagami Mine, Ontario. D, 1974, University of Western Ontario. 205 p.

Colwell, Jane. A new notoungulate of the family Leontiniidae from the Miocene of Colombia. M, 1965, University of California, Berkeley. 74 p.

Colwell, John Allison. Geochemistry and petrology of the Nipissing Diabase in Ontario. D, 1967, Michigan State University. 106 p.

Colwell, John Allison. Trace elements in sphalerite, galena and associated sulfides from Eastern Canadian deposits. M, 1963, Queen's University. 149 p.

Comba, C. D. A. Copper-zinc zonation in tuffaceous exhalites, Millenbach Mine, Noranda, Quebec. M, 1975, Queen's University. 107 p.

Combellick, Rodney Alan. Petrology and economic value of beach sand from southern Monterey Bay, California. M, 1976, University of Southern California.

Combs, Douglas W. The mineralogy, petrology and bromine geochemistry of selected samples of the Salado Salt, Lea and Eddy counties, New Mexico; a potential horizon for the disposal of radioactive waste. M, 1975, University of Tennessee, Knoxville. 76 p.

Combs, James B. Terrestrial heat flow in north central United States. D, 1970, Massachusetts Institute of Technology. 317 p.

Combs, M. J. The use of probability plots in sedimentology. M, 1976, University of Missouri, Columbia.

Combs, Samuel Theodore. The Froude number as a boundary condition for sediment transport in sand-bed streams. D, 1988, Colorado State University. 204 p.

Comeau, Reg L. Transported slices of the Coastal Complex Bay of Islands, western Newfoundland. M, 1972, Memorial University of Newfoundland. 105 p.

Comella, Joseph Robert. Infrared spectra of iron-rich chamasite and chrysotile. M, 1964, Northern Illinois University. 14 p.

Comer, C. Drew. Upper Tertiary stratigraphy of the lower Coastal Plain of South Carolina. M, 1973, University of South Carolina. 19 p.

Comer, John Bennett. Genesis of Jamaican bauxite (Tertiary). D, 1972, University of Texas, Austin. 142 p.

Comer, John Bennett. Sedimentology and physical-chemical environment of a tropical lagoon, western Puerto Rico. M, 1969, University of Wisconsin-Milwaukee.

Comer, Robert Pfahler. Tsunami generation by earthquakes. D, 1982, Massachusetts Institute of Technology. 232 p.

Cominguez, Alberto Horacio. Estimation of the seismic wavelet using a multi-channel procedure. D, 1983, University of Houston. 102 p.

Comline, Stuart Robert. A study of the Piche Group and vein systems at Darius Mine, Cadillac, Quebec. M, 1979, University of Western Ontario. 136 p.

Commerford, Janine. Comparative stratigraphy of the lower part of the Carboniferous-Permian Bird Spring Formation, Spring Mountains, Clark County, Nevada. M, 1984, Massachusetts Institute of Technology. 51 p.

Comparan, Jose L. An electromagnetic sounding profiling method for quasi-stratified terrain. M, 1973, University of Toronto.

Compton, Joe Larry. Diatoms of the Lubbock Lake Site, Lubbock County, Texas. M, 1975, Texas Tech University. 119 p.

Compton, John Sternbergh. Early diagenesis and dolomitization of the Monterey Formation, California. D, 1986, Harvard University. 194 p.

Compton, Nancy H. A field and laboratory study of iron ores specimens from the Ticonderoga, New York Quadrangle. M, 1952, Smith College. 99 p.

Compton, Robert R. Petrology of the southwestern part of the Bidwell Bar (30′) Quadrangle, California. D, 1949, Stanford University. 92 p.

Comstock, S. C. Stratigraphy and structure along the western Big Pine Fault, Santa Barbara County, California. M, 1976, University of California, Santa Barbara.

Comstock, Sherman S. Geology of the Monarch and Michael Breen mining properties, Ouray County, Colorado. M, 1950, Colorado School of Mines. 38 p.

Comstock, Theodore B. The geology and vein-structure of southwestern Colorado. D, 1886, Cornell University.

Comstock, W. O. Leadville Colorado ores and their treatment. M, 1880, Washington University.

Conant, Louis Cowles. Contributions to the geology of the Ossipee Mountains of New Hampshire. M, 1929, Cornell University.

Conant, Louis Cowles. Geology of the New Hampshire garnet deposits. D, 1934, Cornell University.

Conard, JoAnn B. Geology of the Castle Rock 7.5 minute Quadrangle, Fremont County, Wyoming. M, 1981, University of Wyoming. 106 p.

Conard, Nicholas John. Measurement of ^{36}Cl in glacial ice from Greenland and Antarctica. M, 1986, University of Rochester. 59 p.

Conatore, Paul D. Mineralogy and petrology of the transition from the garnet zone to the lower silliman-

ite zone, north-central Rumford area, Maine. M, 1974, University of Wisconsin-Madison. 139 p.

Conatser, Willis E. The contorted strata of the Cynthiana Limestone. M, 1958, University of Cincinnati. 108 p.

Conatser, Willis Eugene. The Grand Isle barrier island complex. D, 1969, Tulane University. 262 p.

Conaway, J. G. Application of geophysical exploration techniques to the detection of solution caverns. M, 1973, University of Western Ontario.

Conaway, J. G. Continuous logging of borehole temperature gradients. D, 1976, University of Western Ontario.

Conca, James Louis. Differential weathering effects and mechanisms. D, 1985, California Institute of Technology. 268 p.

Condie, Kent C. Petrogenesis of the Mineral Range Pluton, Southwest Utah. M, 1960, University of Utah. 92 p.

Condie, Kent Carl. Petrology and geochemistry of the late Precambrian rocks of the northeastern Great Basin. D, 1965, University of California, San Diego. 263 p.

Condit, Carlton. Late Tertiary floras of Central California. D, 1939, University of California, Berkeley. 198 p.

Condit, Christopher D. The geology of Shadow Mountain, Coconino County, Arizona. M, 1973, Northern Arizona University. 71 p.

Condit, Christopher Dana. The geology of the western part of the Springerville volcanic field, east-central Arizona. M, 1984, University of New Mexico. 453 p.

Condit, Daniel D. The Conemaugh Formation in Ohio. M, 1910, Columbia University, Teachers College.

Condit, William S. Some chemical and physical properties of Lake Powell sediments. M, 1977, Dartmouth College. 75 p.

Condon, David Delancey. A preliminary study of the social and economic geography of Utah with special emphasis on the Tintic mining district. M, 1935, Brigham Young University. 155 p.

Condon, James Carl. Mississippian limestones of Tazewell County, Virginia. M, 1942, University of Iowa. 44 p.

Condra, George E. The Bryozoa of the Upper Carboniferous of Nebraska. D, 1902, University of Nebraska, Lincoln.

Conel, James Ekstedt. Studies on the development of fabrics in some naturally deformed limestones. D, 1962, California Institute of Technology. 257 p.

Coney, Peter J. A petrofabric and insoluble residue analysis of the dolomite marble at Warren, Maine. M, 1953, University of Maine. 69 p.

Coney, Peter James. Geology and geography of the Cordillera Huayhuash, Peru. D, 1964, University of New Mexico. 256 p.

Congdon, Roger Duane. Petrology and geochemistry of the Honeycomb Hills Rhyolite, Utah. M, 1987, University of Utah. 139 p.

Conger, Susan Jane Deutsch. Wall ultrastructure of Rotaliina. M, 1974, University of California, Davis. 84 p.

Congram, Angela M. A crustal refraction study in the Williston Basin of southern Saskatchewan. M, 1984, University of Saskatchewan. 273 p.

Conhaim, Howard J. A study of two problems in secondary enrichment of copper ores. M, 1923, University of Minnesota, Minneapolis. 15 p.

Coniglio, Mario. Origin and diagenesis of fine-grained slope sediments; Cow Head Group (Cambro-Ordovician), western Newfoundland. D, 1985, Memorial University of Newfoundland. 684 p.

Coniglio, Mario. Sedimentology of Pleistocene carbonates from Big Pine Key, Florida. M, 1981, University of Manitoba. 261 p.

Conkey, Laura Elizabeth. Dendroclimatology in the northeastern United States. M, 1979, University of Arizona. 78 p.

Conkey, Laura Elizabeth. Eastern U.S. tree-ring widths and densities as indicators of past climate. D, 1982, University of Arizona. 219 p.

Conkhite, George. The analytical composition of the calcareous oolite of the Great Salt Lake. M, 1936, University of Utah. 18 p.

Conkin, Barbara Moyer. Microfossils of the Deer Creek Limestone of Kansas and northern Oklahoma. M, 1954, University of Kansas. 89 p.

Conkin, James E. Stratigraphy and paleontology of lower New Providence Beds in Jefferson and Bullitt counties, Kentucky. M, 1953, University of Kansas. 102 p.

Conkin, James Elvin. Mississippian smaller foraminifera of southern Indiana, Kentucky, northern Tennessee, and southcentral Ohio. D, 1960, University of Cincinnati. 424 p.

Conklin, Bonnie Jean. The structural geology of Cambrian rocks in the Flat Creek area, Stevens County, Washington. M, 1981, Washington State University. 94 p.

Conklin, Carleton Veith. The foraminifera fauna sequence in the Plio-Pleistocene strata of South Florida. M, 1967, University of Florida. 103 p.

Conklin, Jack S. Paleoenvironmental analysis of the Cupido Formation, N.E. New Mexico. M, 1978, Louisiana State University.

Conklin, Michael W. Depositional environments and hydrocarbon potential of the Copper Ridge Dolomite in Union County, Tennessee. M, 1987, Bowling Green State University. 158 p.

Conley, C. D. Petrology of the Leadville Limestone (Mississippian) White River Plateau, Colorado. D, 1964, University of Wyoming. 124 p.

Conley, Curtis D. Geology of the Chromo Anticline, Archuleta County, Colorado, and Rio Arriba County, New Mexico. M, 1959, Colorado School of Mines. 114 p.

Conley, Daniel Joseph. Mechanisms controlling silica flux from sediments and implications for the biogeochemical cycling of silica in Lake Michigan. D, 1987, University of Michigan. 113 p.

Conley, David E. Stratigraphy and depositional history of the Owens Valley Formation at the type locality, Inyo County, California. M, 1978, San Jose State University. 62 p.

Conley, Glenn. Surficial geology and stratigraphy of Killarney-Holmfield area, southwestern Manitoba. M, 1986, University of Manitoba.

Conley, Jack Francis. Structure, stratigraphy, and gas storage potential in Cass, Fulton and Pulaski counties surrounding Royal Center, Indiana. M, 1961, Indiana University, Bloomington. 72 p.

Conley, James Franklin. Selected geologic studies in the central and southwestern Virginia Piedmont. D, 1982, University of South Carolina. 196 p.

Conley, James Franklin. The glacial geology of Fairfield County, Ohio. M, 1956, Ohio State University.

Conley, R. H. A petrographic investigation of the Indian Creek amphibolite, White Pass, Washington. M, 1980, Pacific Lutheran University. 77 p.

Conley, Steven J. Stratigraphy and depositional environment of the Buckhorn Conglomerate Member of the Cedar Mountain Formation (Lower Cretaceous), central Utah. M, 1986, Fort Hays State University. 127 p.

Conley, Steven Meril. The distribution of diatoms in the inside passage of Alaska. M, 1973, University of Oregon. 186 p.

Conlin, Richard Renault. A geologic study of the coal mine roof strata of the Herrin (No. 6) coal bed near Virden, Illinois. M, 1954, University of Illinois, Urbana. 46 p.

Conlon, Paul Joseph. The influence of geochemical variables on selected engineering and electrical properties of re-sedimented Leda Clay. D, 1983, Carleton University.

Conlon, Sean Thomas. Volcanic geology of the general Trias-Tutuaca area, Chihuahua, Mexico. M, 1985, University of Texas, Austin. 187 p.

Connair, Dennis P. Structural geology of the Darwin Peak-upper Gros Ventre River area, Teton County, Wyoming. M, 1985, Miami University (Ohio). 70 p.

Connally, George Gordon. The Almond Moraine of the western Finger Lakes region, New York. D, 1964, Michigan State University. 102 p.

Connally, Gordon. Heavy minerals in glacial drift of western New York. M, 1959, University of Rochester.

Connally, Thomas Chambless, Jr. A facies analysis of the modern Colorado Delta, Matagorda County, Texas. M, 1981, University of Texas, Austin. 106 p.

Connard, Gerald George. Analysis of aeromagnetic measurements from the central Oregon Cascades. M, 1980, Oregon State University. 101 p.

Connare, Kevin M. Rb-Sr geochronology, geochemistry and petrography of two high grade gneisses found near Parry Sound, Ontario. M, 1986, McMaster University. 164 p.

Connary, Stephen Dodd. Investigations of the Walvis Ridge and environs (South Atlantic). D, 1972, Columbia University. 228 p.

Connary, Stephen Dodd. Nepheloid layer and bottom circulation in the eastern basins of the south Atlantic. M, 1970, Columbia University. 44 p.

Connaughton, Charles Robert. A subsurface study of the Jackson Formation in Vanderburgh County, Indiana. M, 1953, Indiana University, Bloomington. 34 p.

Connell, Douglas Edward. Distribution, characteristics, and genesis of joints in fine-grained till and lacustrine sediment, eastern and northwestern Wisconsin. M, 1984, University of Wisconsin-Madison.

Connell, James Frederick Louis. Additions to the brachiopod fauna of the Henryhouse Formation (Silurian) of Oklahoma. M, 1951, University of Oklahoma. 62 p.

Connell, James Frederick Louis. Stratigraphy and paleontology of the Jackson Group of Georgia. D, 1955, University of Oklahoma. 348 p.

Connell, Joseph Francis, Jr. An areal finite element hydrothermal model for a semi-confined aquifer. D, 1984, Virginia Polytechnic Institute and State University. 174 p.

Connell, Larry E. The geology of the northern Butte Mountains, White Pine County, Nevada. M, 1985, University of Nevada. 191 p.

Connelly, Francis B. A heavy mineral study of the Atoka sandstone formation from the Hamm Well in Sec. 25, T. 9N., R. 29W., Franklin County, Arkansas. M, 1955, University of Arkansas, Fayetteville.

Connelly, James Leslie. A study of the Kinderhookian and Osagean microcrinoids. M, 1950, University of Missouri, Columbia.

Connelly, James N. The emplacement history of the Elzevir Batholith relative to the regional deformation of the supracrustal rocks in the Central metasedimentary belt, Grenville Province, southeastern Ontario. M, 1986, Queen's University. 153 p.

Connelly, Jeffrey B. Intragranular strain within early Mesozoic limestone conglomerates of the Gettysburg Basin. M, 1986, Bowling Green State University. 51 p.

Connelly, John R. Study of oolitic textures in sedimentary rocks. M, 1970, Lehigh University.

Connelly, Johnston P., II. Ground-water heat pumps; are they feasible for use in Wisconsin homes?. M, 1980, University of Wisconsin-Madison.

Connelly, Joseph Peter. The Chouteau Formation of east central Missouri. M, 1915, University of Missouri, Columbia.

Connelly, Michael C. Origin and petrography of the James Formation in the Fairway Field of East Texas. M, 1963, University of Tulsa. 69 p.

Connelly, Michael Peter. Petrologic and fluid inclusion studies of the Bawana and Maria copper skarns, Rocky Range, SW Utah. M, 1984, University of Utah. 80 p.

Connelly, William Ronald. Mesozoic geology of the Kodiak islands and its bearing on the tectonics of southern Alaska. D, 1976, University of California, Santa Cruz. 234 p.

Conner, Joanne Marie. Analysis of Cycloclypeus. M, 1964, Cornell University.

Conner, John. Geology of the Sage Valley 7 1/2′ Quadrangle, Caribou County, Idaho, and Lincoln County, Wyoming. M, 1980, Brigham Young University.

Conner, Jon J. Precambrian petrology and structural geology of the Gray Rock-Livermore Mountain area, Larimer County, Colorado. D, 1962, University of Colorado.

Conner, Steven P. Diagenesis of the Upper Cretaceous Teapot Sandstone, Well Draw Field, Converse County, Wyoming. M, 1983, Texas A&M University. 154 p.

Conner, Trent. The mineralogy and water content of Paradox Basin evaporite deposits. M, 1983, Georgia Institute of Technology. 112 p.

Conner, W. G. Response of a soft-bottom ecosystem to physical perturbation. D, 1977, University of South Florida, St. Petersburg. 86 p.

Connerney, J. E. P. Deep crustal electrical conductivity in the Adirondacks. D, 1979, Cornell University. 232 p.

Conners, John Anthony. Quaternary history of northern Idaho and adjacent areas. D, 1976, University of Idaho. 504 p.

Connolly, F. Thomas. The geology of the Passport oil pool, Clay County, Illinois. M, 1949, University of Cincinnati. 33 p.

Connolly, Garrett Morgan. Environmental significance of salt-cavern storage of hydrocarbons in Saskatchewan. M, 1982, University of Windsor. 120 p.

Connolly, James Alexander. Hydrothermal alteration in the Climax Granite Stock at the Nevada Test Site. M, 1981, Arizona State University. 110 p.

Connolly, James Alexander Denis. Calculation of multivariable phase diagrams; a computer strategy based on generalized thermodynamics. D, 1988, Pennsylvania State University, University Park. 300 p.

Connolly, James R. Geology of the Precambrian rocks of Tijeras Canyon, Bernalillo County, New Mexico. M, 1981, University of New Mexico. 147 p.

Connolly, John Patrick. The effect of sediment suspension of adsorption and fate of kepone. D, 1980, [University of Texas, Austin]. 223 p.

Connolly, Joseph P. The Tertiary mineralization of the northern Black Hills. D, 1927, Harvard University.

Connolly, Marc Robert. The geology of the middle Precambrian Thomson Formation in southern Carlton County, East-central Minnesota. M, 1981, University of Minnesota, Duluth.

Connolly, William Marc. Microfacies analysis, paleoecology, and environment of deposition of Morrowan shelf carbonates, Magdalena Limestone (lower division), Hueco Mountains, El Paso County, West Texas. M, 1985, Texas A&M University. 437 p.

Connor, C. L. Holocene sedimentation history of Richardson Bay, California. M, 1975, Stanford University. 112 p.

Connor, Cathy Lynn. Late Quaternary glaciolacustrine and vegetational history of the Copper River basin, south-central Alaska. D, 1984, University of Montana. 115 p.

Connor, Charles Benjamin. Cinder cone distribution described using cluster analysis and two-dimensional Fourier analysis in the central TransMexican volcanic belt, Mexico, and in SE Guatemala and NW El Salvador. D, 1987, Dartmouth College. 234 p.

Connor, Charles Benjamin. Structure of the Michoacan volcanic field, Mexico. M, 1984, Dartmouth College. 189 p.

Connor, James J., Jr. Characteristics of explosion-generated waves in limestone at small distances. M, 1957, St. Louis University.

Connor, Jon James. Precambrian petrology and structural geology of the Gray Rock-Livermore Mountain Area, Larimer County, Colorado. D, 1962, University of Colorado.

Connor, Mike and Kelly, Herbert A. The geology of a portion of Rapid Canyon (South Dakota). M, 1957, South Dakota School of Mines & Technology.

Connors, Donald Nason. The partial equivalent conductances of salts in seawater. D, 1967, Oregon State University. 48 p.

Connors, Harry E., III. Structure of the Heart Mountain-Tinaja Spring area, Brewster County, Texas. M, 1977, University of Texas, Austin.

Connors, Katherine A. The petrology of Tertiary intrusions associated with epithermal veins in the Georgetown-Silver Plume District; Clear Creek County, Colorado. M, 1985, Colorado School of Mines. 103 p.

Connors, Kathryn Francis. Simulated SPOT imagery for the investigation of geomorphic features and hydrologic processes. M, 1985, Pennsylvania State University, University Park. 147 p.

Connors, Roland A. Geology and alteration of the Crater Creek area, southern San Juan Mountains, Colorado. M, 1975, Colorado School of Mines. 57 p.

Connors, Stephen Dennis. Some engineering properties of soils in the town of Arcade, Wyoming County, New York and their relation to land use. M, 1976, Brooklyn College (CUNY).

Conomos, John T. Geologic aspects of the Recent sediments of San Francisco Bay. M, 1963, San Jose State University. 118 p.

Conomos, Tasso John. Processes affecting suspended particulate matter in the Columbia River-effluent system. D, 1968, University of Washington. 141 p.

Conover, Dale. Preliminary site studies for critical facilities using geotechnical units derived from engineering geologic analyses. M, 1985, Texas A&M University. 152 p.

Conover, William V. Depositional systems and genetic stratigraphy of the lower Miocene Planulina Trend in south central Louisiana. M, 1987, University of Texas, Austin. 62 p.

Conrad, Clarence F. Geology of the Ana River section, Summer Lake, Oregon. M, 1953, Oregon State University. 92 p.

Conrad, Curtis Paul. Uranium in the Oatman Creek Granite of central Texas and its economic potential. M, 1982, Texas A&M University.

Conrad, Eric Hale. Dissolution of carbonate minerals and rocks in dilute organic acids and its geochemical implications to carbonate precipitation and diagenesis. M, 1971, University of South Florida, Tampa. 98 p.

Conrad, George J. Possibilities of petrographic correlation of Pennsylvanian sandstones in Monongalia County, West Virginia. M, 1949, West Virginia University.

Conrad, Gregory Stevens. The biostratigraphy and mammalian paleontology of the Glenns Ferry Formation from Hammett to Oreana, Idaho. D, 1980, Idaho State University. 351 p.

Conrad, James Matthew. Stratigraphy, structure, and hydrocarbon production of the Silurian Lockport Dolomite in eastern Kentucky and Wayne County, West Virginia. M, 1987, West Virginia University. 129 p.

Conrad, John Adrian. Shelf sedimentation above storm wave base in the Upper Ordovician Reedsville Formation in central Pennsylvania. M, 1985, University of Delaware.

Conrad, Keith T. Petrology of the Arumbera Sandstone, late Proterozoic(?)-Early Cambrian, northeastern Amadeus Basin, central Australia. M, 1981, Utah State University. 289 p.

Conrad, Malcolm A. Contact metamorphic effects of three pegmatitic dikes in the Upper Peninsula of Michigan. M, 1952, Michigan Technological University. 31 p.

Conrad, Malcolm Alvin. Ultraviolet piezobirefringence of diamond. D, 1960, University of Michigan. 56 p.

Conrad, Mark E. Variations within the layering of the Skaergaard Intrusion, East Greenland. M, 1982, Dartmouth College. 219 p.

Conrad, Omar G. The sub-till chert gravels of northeastern Kansas. M, 1963, University of Kansas. 87 p.

Conrad, Rae L. Rb-Sr geochemistry of cataclastic rocks of the Vincent Thrust, San Gabriel Mountains, Southern California. M, 1976, California State University, Los Angeles.

Conrad, Robert D. Pre-Devonian unconformity, Gila County, Arizona. M, 1964, University of Arizona.

Conrad, Robert E., II. Microscopic feather fractures in the faulting process. M, 1974, Texas A&M University. 34 p.

Conrad, Stanley D. Geology of the eastern portion of the Simi Hills, Los Angeles, and Ventura counties, California. M, 1949, University of California, Los Angeles.

Conrad, Walter Karr. Petrology and geochemistry of igneous rocks from the McDermitt caldera complex, Nevada-Oregon, and Adak Island, Alaska; evidence for crustal development. D, 1983, Cornell University. 343 p.

Conrey, Bert Louis. Sedimentary history of the early Pliocene in the Los Angeles Basin, California. D, 1959, University of Southern California. 331 p.

Conrey, Bert Louis, Jr. Geology of the southern portion of the Morgan Valley Quadrangle, California. M, 1948, University of California, Berkeley. 88 p.

Conrey, Guy Woolard. Geology of Wayne County, Ohio. D, 1921, Ohio State University.

Conrey, Richard M. Volcanic stratigraphy of the Deschutes Formation, Green Ridge to Fly Creek, north-central Oregon. M, 1985, Oregon State University. 349 p.

Conroy, A. Richard. Geology of several thorium deposits, Wet Mountains (Custer County), Colorado. M, 1960, University of Colorado.

Conroy, Elizabeth A. Composition and structure of fossil assemblages in the Elvinia Zone trilobite fauna, of the central Great Basin, Utah and Nevada. M, 1987, University of Kansas. 57 p.

Conroy, Nels. Classification of Precambrian Shield lakes based on factors controlling biological activity. M, 1971, McMaster University. 142 p.

Conroy, Peter J. Investigation of roof shales in Illinois coal mines. D, 1973, University of Missouri, Rolla.

Conselman, Frank Buckley. The geology and stratigraphy petrography of the Auxvasse Creek Quadrangle, Callaway County, Missouri. D, 1934, University of Missouri, Columbia.

Consolmagno, Guy Joseph. Thermal history models of icy satellites. M, 1975, Massachusetts Institute of Technology. 202 p.

Consort, James Jeremiah. The biostratigraphy and depositional environments of the Bonanza King Formation near Pahrump, Nevada. M, 1979, San Diego State University.

Constable, Catherine Gwen. Some statistical aspects of the geomagnetic field. D, 1987, University of California, San Diego. 126 p.

Constans, Richard E. A study of fluctuations in the carbonate compensation depth in the Southern Ocean south of Australia using calcareous nannofossils. M, 1975, Florida State University.

Constant, Warren LeRoy. The microfauna of the Forth (Fort) Worth Formation of southern Oklahoma and northern Texas. M, 1937, University of Oklahoma. 71 p.

Constantino-Herrera, Sergio E. Geology of the Cleveland Gabbro, Rowan County, North Carolina. M, 1971, University of North Carolina, Chapel Hill. 47 p.

Constantopoulos, James Theodore. Fluid inclusions and geochemistry of fluorite from the Challis 1° × 2° Quadrangle, Idaho. M, 1985, University of Idaho. 127 p.

Constantz, Brent Richard. The skeletal ultrastructure of scleractinian corals. D, 1986, University of California, Santa Cruz. 185 p.

Constenius, Kurt N. Stratigraphy, sedimentation, and tectonic history of the Kishenehn Basin, northwestern Montana. M, 1981, University of Wyoming. 118 p.

Contant, Cheryl Katherine. Cumulative impact assessment; design and evaluation of an approach for the Corps of Engineers permit program at the San Francisco District. D, 1984, Stanford University. 388 p.

Conte, Jonathan A. Geochemistry and tectonic significance of amphibolites within the Precambrian Ashe Formation, northwestern North Carolina. M, 1986, University of Tennessee, Knoxville. 122 p.

Contessa, Joseph V. The comparative influence of plant moisture stress and grazing on erosion in two small waterways in New Mexico. D, 1975, Southern Illinois University, Carbondale. 151 p.

Conti, Edward P. The evolution of a carbonate bank in the Caicos Islands, British West Indies. M, 1987, Duke University. 185 p.

Conti, Louis Joseph, Jr. Stratigraphy of the lower Monongahela (Pennsylvanian) in north-central West Virginia. M, 1961, West Virginia University.

Conti, Mario A. Chromatographic characterization of carbonate rocks. M, 1951, Northwestern University.

Conti, Robert D. Stratigraphic distribution of hydrocarbons with differing API gravities in the East Texas Basin. M, 1982, University of Texas, Austin.

Contreras, Carlos E. An inquiry into modern geology and orogenic history of South America. M, 1954, Louisiana State University.

Contrino, Charles Thomas. A study of a buried valley in west-central Ohio using seismic refraction and two-dimensional gravity model studies. M, 1973, Wright State University. 74 p.

Converse, David Rhys. Flow rates in the East Pacific Rise (21°N) hot springs, and numerical investigations of two regimes of hydrothermal circulation. D, 1984, Harvard University. 473 p.

Converse, Glenn Leland. Some wave propagation problems in a bounded, heterogeneous, elastic medium. D, 1965, Stanford University. 122 p.

Convert, Jean. Petrology, structure and petrogenesis of the Mount Arabia Migmatite, Lithonia District, Georgia. M, 1986, Emory University. 137 p.

Convery, Michael P. The behavior and movement of petroleum products in unconsolidated surficial deposits. M, 1979, University of Minnesota, Minneapolis. 175 p.

Conway, Charles Daniel. The development of slaty cleavage and its relationship to mesoscopic and macroscopic structure in the Pulaski thrust sheet near Bristol, Tennessee. M, 1983, University of Tennessee, Knoxville. 129 p.

Conway, Clay Michael. Petrology, structure, and evolution of a Precambrian volcanic and plutonic complex, Tonto Basin, Gila County, Arizona. D, 1976, California Institute of Technology. 523 p.

Conway, Edward Spurgeon. The geology of the Calf Creek area, McCulloch County, Texas. M, 1939, University of Texas, Austin.

Conway, Ernest F. Terrains of Irasburg, Vermont. M, 1911, Syracuse University.

Conway, Michael Francis. Chromium valency response and environmental fate in three-phase aquatic microcosms. M, 1984, University of Connecticut. 82 p.

Conway, Richard Dean. Geochemistry of the Cypress Island ultramafic body, Washington. D, 1971, University of Washington. 108 p.

Conway, Richard Dean. Structure and stratigraphy of a portion of the South Schell Creek Range, White Pine County, Nevada. M, 1965, University of Washington. 64 p.

Conway, Stephen W. Depositional environments and diagenesis of a sequence in a borehole; the Gailor Formation (Lower Ordovician) of the Mohawk Valley-Saratoga region, New York. M, 1977, Rensselaer Polytechnic Institute. 125 p.

Conybeare, Charles Eric Bruce. Structure and metamorphism in the Goldfields area, Saskatchewan, with special reference to the pitchblende deposits. D, 1950, Washington State University. 183 p.

Conybeare, Charles Eric Bruce. The geology of Fraser Valley, Lytton to Lillooet, B.C. M, 1947, University of Alberta. 202 p.

Conyers, Lawrence B. Depositional environments of the Supai Formation in central Arizona. M, 1975, Arizona State University. 85 p.

Conyers, William Patrick. The geology of Upper Spring Creek and a petrographic analysis of the igneous intrusives of the area. M, 1957, University of Kentucky. 35 p.

Conzelman, William E. Notes upon the mines and mills of Colorado. M, 1885, Washington University.

Coode, Alan Melvill. Electric and magnetic fields associated with a vertical fault. M, 1963, University of British Columbia.

Coody, Gilbert L. Geology of the Durst Mountains, Huntsville area, Morgan and Weber counties, Utah. M, 1957, University of Utah. 63 p.

Coogan, Alan Hall. Early Pennsylvanian stratigraphy, biostratigraphy, and sedimentation of the Ely Basin, Nevada. D, 1962, University of Illinois, Urbana. 129 p.

Coogan, Alan Hall. The stratigraphy and paleontology of the Nosoni and Dekkas formations (Permian), Shasta County, California. M, 1957, University of California, Berkeley. 124 p.

Cook, Ariadne Helen Olga. Diagenesis of the Maryville and upper Honaker formations (Cambrian), Tennessee and Virginia, with emphasis on dolomitization and silicification. M, 1983, Duke University. 164 p.

Cook, Beverly Kay Gatlin. Detailed petrology of the Buchanan Massif (Precambrian), Llano and Burnet counties, Texas. M, 1967, Rice University. 23 p.

Cook, Beverly Kay Gatlin. Petrochemistry of the Buck Hill volcanic series (Tertiary), Cathedral mountain quadrangle, Brewster County, Texas. D, 1970, Rice University. 66 p.

Cook, Billy C. Depositional environment of the Stringer Sands Member, lower Tuscaloosa Formation (Cretaceous), Mallalieu Field, Mississippi. M, 1968, Texas A&M University. 97 p.

Cook, Brad D. The effect of stone content, size, and shape on the engineering properties of a compacted silty clay. M, 1988, Kent State University, Kent. 128 p.

Cook, C. W. A mechanical well log study of the Poplar Interval of the Mississippian Madison Formation in North Dakota. M, 1974, University of North Dakota. 154 p.

Cook, Carroll Edwin. Areal geology of the Catahoula Formation in Gonzales and Karnes counties, Texas, with notes on outcrops in adjacent counties. M, 1932, University of Texas, Austin.

Cook, Charles Wilford and Staples, Lloyd William. Microscopic investigation of molybdenite ore from Climax, Colorado. D, 1913, University of Michigan.

Cook, Charles Wilford. The brine and salt deposits of Michigan, their origin and exploitation. D, 1913, University of Michigan.

Cook, Clarence W. Fluid inclusions and petrogenesis of the Harding Pegmatite, Taos County, New Mexico. M, 1979, University of New Mexico. 143 p.

Cook, David Bruce. Seismology and tectonics of the North American Plate in the Arctic; Northeast Siberia and Alaska. D, 1988, Michigan State University. 250 p.

Cook, David Olney. Sand transport by shoaling waves. D, 1969, University of Southern California. 157 p.

Cook, Donald G. Structural analysis of the Buck Lake Syncline (Frontenac County, Ontario). M, 1965, Queen's University. 135 p.

Cook, Donald Jean. The magnetic susceptibilities of copper minerals and cassiterite. D, 1961, Pennsylvania State University, University Park. 121 p.

Cook, Donald Jean. The magnetic susceptibilities of rutile and sphene. M, 1958, Pennsylvania State University, University Park. 71 p.

Cook, Douglas R. Palynological investigation of the Hazard No. 7 coal of eastern Kentucky; a possible indicator of paleoecological conditions in the ancient peat-forming swamp. M, 1986, Eastern Kentucky University. 117 p.

Cook, Douglas R. The geology of the Pride of the West Vein System, San Juan County, Colorado (two volumes). D, 1952, Colorado School of Mines. 137 p.

Cook, E. R. A tree ring analysis of four tree species growing in southeastern New York State. M, 1977, University of Arizona. 121 p.

Cook, Earl Ferguson. Geology of area north of Bacon Creek on Skagit River, Washington. M, 1947, University of Washington. 58 p.

Cook, Earl Ferguson. Geology of Pine Valley Mts., Utah. D, 1954, University of Washington. 236 p.

Cook, F. A. Transient heat flow models and gravity models in the Rio Grande Rift of southern New Mexico. M, 1975, University of Wyoming. 114 p.

Cook, Frances Govean. Neogene foraminifera from the Philippines. M, 1965, University of California, Berkeley. 122 p.

Cook, Frederic S. The geology of the Seven Dash Ranch area, Cochise County, Arizona. M, 1938, University of Arizona.

Cook, Frederick Ahrens. COCORP seismic reflection traverse across the Southern Appalachian Orogen. D, 1981, Cornell University. 122 p.

Cook, Gregory Allan. Heat flow in the Central Plateau of northern Mexico. M, 1978, University of Florida. 101 p.

Cook, Gregory Lee. Land-resource capability units of the Wagoner County area, northeastern Oklahoma. M, 1973, Oklahoma State University. 55 p.

Cook, Harry Edgar, III. Geology of the southern part of the Hot Creek Range, Nevada. D, 1966, University of California, Berkeley. 116 p.

Cook, Harwin T. The hydro-capacitor; a new laboratory modelling technique. M, 1966, University of Oklahoma. 89 p.

Cook, J. G. The late Precambrian structural history of Bear and West mountains, Orange and Rockland counties, New York. M, 1978, SUNY at Stony Brook.

Cook, J. R. Comparative surface water chemistry of three lakes in Susquehanna County, Pennsylvania. M, 1977, SUNY at Binghamton. 77 p.

Cook, Janet Arlene. Geology of the Fall Creek area, Snake River Range, Idaho. M, 1948, University of Michigan.

Cook, Jeffrey A. Sonobuoy refraction study of the crust in the Gorda Basin. M, 1981, Oregon State University. 120 p.

Cook, Jerry Robert. Petrology of the Pennsylvanian-Permian Wells Formation; Southeast Idaho and Southwest Wyoming. M, 1983, Idaho State University. 63 p.

Cook, John Call. Analysis of airborne surveying for surface radioactivity. D, 1951, Pennsylvania State University, University Park. 123 p.

Cook, John Dee. Geology of the Shefoot Mountain area, Shoshone County, Idaho. M, 1964, University of Idaho. 64 p.

Cook, John Thomas. The geology of the southeast quarter (7 1/2 min.) of the Oxford area, Granville and Vance Counties, North Carolina. M, 1968, North Carolina State University. 117 p.

Cook, Kenneth L. Relative abundance of the isotopes of potassium in Pacific kelps and in rock of different geologic age. D, 1943, University of Chicago. 15 p.

Cook, Kerry Brian. Landslide hazard assessment along a portion of U.S. Highway 95, Idaho County, Idaho. M, 1984, Washington State University. 85 p.

Cook, Kevin H. Conodont color alteration; a possible exploration tool for ore deposits. M, 1986, New Mexico Institute of Mining and Technology. 125 p.

Cook, Kevin V. Water resources as a possible growth-limiting factor in a world futures model. M, 1982, University of Minnesota, Minneapolis. 102 p.

Cook, Lawrence Paul. Metamorphosed carbonate rocks of the Nashoba Formation, eastern Massachusetts. D, 1974, Harvard University.

Cook, Lawrence Paul. Phase relations in the system $NaCl-SiO_2$. M, 1969, University of Texas, Austin.

Cook, N. W. A preliminary investigation of the distribution of phosphorus in North Atlantic Ocean deep sea and continental slope sediments. M, 1977, Florida State University.

Cook, P. L., Jr. Petrography of an upper Cretaceous volcanic sequence in Humphreys County, Mississippi. M, 1975, University of Southern Mississippi.

Cook, Peter John. The petrology and geochemistry of the Meade Peak Member of the Phosphoria Formation. D, 1968, University of Colorado. 220 p.

Cook, Philip G. The optical and X-ray properties of the ferromagnesian olivine minerals with charts to aid identification. M, 1950, California Institute of Technology. 36 p.

Cook, Philip Ray. Sedimentary structures as possible indicators of depositional environment in the McShan Formation (Upper Cretaceous) in Mississippi and Alabama. M, 1986, Mississippi State University. 105 p.

Cook, R. J. B. Heavy detritals and glacial stratigraphy in southern Ontario. M, 1953, University of Western Ontario.

Cook, Raymond Arnold. Composition and stratigraphy of late Quaternary sediments from the northern end of Juan de Fuca Ridge. M, 1981, University of British Columbia. 107 p.

Cook, Robert A. A computer simulation of the Lancaster water well field. M, 1979, Ohio University, Athens.

Cook, Robert Annan. Mechanisms of sandstone deformation; a study of the drape folded Weber Sandstone in Dinosaur National Monument, Colorado and Utah. D, 1976, Texas A&M University. 145 p.

Cook, Robert Annan. Scaled dependencies in structural analysis as illustrated by chevron folds along the Beartooth Front, Wyoming. M, 1972, Texas A&M University. 98 p.

Cook, Robert B. Diabase dykes in the Kingston-Brockville area of the Frontenac Axis. M, 1964, Queen's University. 94 p.

Cook, Robert Bigham, Jr. The geologic history of massive sulfide bodies in west-central Georgia. D, 1970, University of Georgia. 163 p.

Cook, Robert Bigham, Jr. The geology of a part of west-central Wilkes County (Georgia). M, 1967, University of Georgia. 53 p.

Cook, Robert Bradley. The biogeochemistry of sulfur in two small lakes. D, 1981, Columbia University, Teachers College. 262 p.

Cook, Robert Davis. Geology of the Moth Bay pluton, Ketchikan area, southeastern Alaska. M, 1986, Bryn Mawr College. 77 p.

Cook, Robert H. Some effects of hypoxia on three marine Crustacea. M, 1965, Dalhousie University.

Cook, Rufus E. The Upper Cretaceous Chico Formation at the type locality (California). M, 1949, Stanford University. 69 p.

Cook, Stephen James. The physical-chemical conditions of contact skarn formation at Alta, Utah. M, 1982, University of Utah. 169 p.

Cook, Sterling S. Supergene copper mineralization at at the Lakeshore Mine, Pinal County, Arizona. M, 1985, University of Oregon. 122 p.

Cook, Terry Allen. Stratigraphy and structure of the central Talladega slate belt, Alabama Appalachians. M, 1983, University of Alabama. 117 p.

Cook, Theodore Davis. Eocene foraminifera from the Devil's Den District, San Joaquin Valley, California. M, 1950, University of California, Berkeley. 116 p.

Cook, Timothy Ralph. The dispersion of uranium and thorium during weathering of a rapakivi granite near Wausau, Wisconsin. M, 1980, University of Wisconsin-Milwaukee.

Cook, Vance Oliver. Geology of the Brentwood Limestone in Madison County, Northwest Arkansas. M, 1953, University of Arkansas, Fayetteville.

Cook, W. R., III. Structural geology of the Aspen Tunnel area, Uinta County, Wyoming. M, 1976, University of Wyoming. 58 p.

Cook, William R., Jr. A study of serpentine by electron microscopy and thermal analysis. M, 1950, Columbia University, Teachers College.

Cook, William Riley, Jr. The copper-sulfur phase diagram. D, 1971, Case Western Reserve University. 138 p.

Cook, William T. Geology of the Minden salt dome (Webster county, Louisiana). M, 1969, Northeast Louisiana University.

Cooke, Alison. Sedimentary environment of Gulf Coast barrier island, Horn Island, Mississippi. M, 1981, Tulane University.

Cooke, Charles W. The Greenbrier Formation in Maryland; a contribution to Mississippian paleontology. D, 1912, The Johns Hopkins University.

Cooke, D. W. Variations in the seasonal extent of sea ice in the Antarctic during the last 140,000 years. D, 1978, Columbia University, Teachers College. 302 p.

Cooke, David Lawrence. The petrography and paragenesis of some base metal veins. M, 1961, University of Toronto.

Cooke, David Lawrence. The Timiskaming volcanics and associated sediments of the Kirkland Lake area. D, 1966, University of Toronto.

Cooke, Dennis. Two-dimensional instantaneous seismic attributes. D, 1987, Colorado School of Mines. 188 p.

Cooke, Dennis A. Generalized linear inversion of reflection seismic data. M, 1981, Colorado School of Mines. 130 p.

Cooke, Gary. Pore water constituents of deeply buried Gulf Coast muds as indicators of diagenetic alteration. M, 1977, Georgia Institute of Technology. 188 p.

Cooke, Harold Caswell. The secondary enrichment of silver ores. D, 1912, University of Chicago. 28 p.

Cooke, Hermon Richard, Jr. The distribution of gold in the original Sixteen to One Vein, Alleghany, California. D, 1945, Harvard University.

Cooke, Horace B., Jr. The structure and petrography of the Rockfish Conglomerate, Virginia. M, 1952, University of Virginia. 97 p.

Cooke, James Crawford. The Belle Glade coral fauna of the Pleistocene Bermont Formation of South Florida. M, 1975, Tulane University. 92 p.

Cooke, Laurence S., Jr. A quantitative sedimentary analysis of the Ordovician deposits in Michigan Basin. M, 1956, Michigan State University. 53 p.

Cooke, Selman C. Caliche of the Lubbock region. M, 1951, Texas Tech University. 36 p.

Cooke, Strathmore Ridley Barnott. The microscopic structure and concentratability of the more important iron ores of the United States. D, 1933, University of Missouri, Columbia.

Cookey, Melanie. The application of linear predictive coding to deconvolution in seismic processing. M, 1987, North Carolina State University. 72 p.

Cookman, Charles Willard. Petrology of the Rockport Quarry Limestone (Middle Devonian Traverse Group), Alpena, Presque Isle and Montmorency counties, Michigan. M, 1976, Western Michigan University.

Cookman, Richard G. Structural analysis of Precambrian metasedimentary rocks of the Swede Gulch

Formation, Nahant, South Dakota. M, 1981, Western Michigan University. 120 p.

Cookro, T. M. Petrology of Precambrian granitic rocks from the Ladron Mountains, Socorro County, New Mexico. M, 1978, New Mexico Institute of Mining and Technology. 86 p.

Cooksey, Calvin Leavelle. The micropaleontology of the Boggy Formation of Oklahoma. M, 1933, University of Oklahoma. 63 p.

Cooksey, Charlton D., Jr. The geology of portions of the Humphreys, Sylmar, Newhall, and Saugus quadrangles, Los Angeles County, California. M, 1934, California Institute of Technology. 42 p.

Cool, Colin Anthony. Effects of aquifer boiling and CO_2 volatilization on the chemical composition of hydrothermal aquifers and hot springs of the Idaho Batholith, central Idaho. M, 1984, University of Washington. 83 p.

Cool, T. E. Cretaceous calcareous nannoplankton biostratigraphy, sedimentation history, and paleoceanography of western North Atlantic Ocean. D, 1979, Texas A&M University. 361 p.

Cool, Thomas Edward. Dredged-material disposal and suspended matter offshore from Galveston, Texas. M, 1976, Texas A&M University.

Coolbaugh, David F. The mechanism of induced polarization and its decay. D, 1961, Colorado School of Mines. 109 p.

Cooley, Beaumont Brewer, Jr. Areal geology of Dry Cimarron Canyon, Union County, New Mexico. M, 1956, University of Texas, Austin.

Cooley, Douglas R. Facies change in the Oread Limestone in southern Kansas and northern Oklahoma. M, 1952, University of Kansas. 61 p.

Cooley, Gerald Allison. Insoluble residues of Detroit River and Dundee formations. M, 1947, University of Michigan.

Cooley, Keith R. Rainfall and runoff relationships along the central highlands of Arizona and western New Mexico. M, 1966, University of Arizona.

Cooley, Lavell I. The Devonian of the Bear River Range, Utah. M, 1928, Utah State University.

Cooley, Maurice E. Geology of the Chinle Formation in the upper Little Colorado drainage area, Arizona and New Mexico. M, 1957, University of Arizona.

Cooley, Richard Lewis. The geohydrologic variables which control seepage from a stream crossing an alluvial fan. D, 1968, Pennsylvania State University, University Park. 357 p.

Cooley, Scott Alfred. Depositional environments of the lower and middle Miocene Temblor Formation of Reef Ridge and vicinity, Fresno and Kings counties, California. M, 1982, Stanford University. 89 p.

Cooley, Tillman Webb, Jr. Post-Cretaceous stratigraphy of the central Sandhills region, North and South Carolina. D, 1970, University of North Carolina, Chapel Hill. 137 p.

Coolidge, C. N. Relations between source rock and Recent stream mineralogy in Horse Creek basin, western Wyoming. M, 1978, University of Wyoming. 86 p.

Coombs, Howard Abbott. The geology of Mount Rainier National Park. D, 1935, University of Washington. 141 p.

Coombs, Howard Abbott. The geology of the southern slope of Mt. Rainier. M, 1932, University of Washington. 47 p.

Coombs, Margery Chalifoux. The Schizotheriinae (Mammalia, Perissodactyla, Chalicotheriidae), with emphasis on the genus Moropus. D, 1973, University of Colorado. 463 p.

Coombs, Stanley L. Footwall features as related to ore occurrence in the Kingston conglomerate, Precambrian, Keeweenaw County, Michigan. M, 1969, University of Arizona.

Coombs, Vincent B. The structural geology of the Cottonwood Limestone in Riley County, Kansas. M, 1948, Kansas State University. 26 p.

Coombs, Walter Preston, Jr. The Ankylosauria. D, 1971, University of Colorado. 496 p.

Cooms, Howard Abbott. The geology of the southern slope of Mount Rainier. M, 1932, University of Washington. 42 p.

Coon, Cathy A. Occurrence and distribution of trace elements in some Permian coals from the Parana Basin, Brazil. M, 1980, University of Toledo. 64 p.

Coon, Lester A. Sedimentation of the "Upper Salt Series" of the Delaware Basin, Texas and New Mexico. M, 1940, Texas Tech University. 33 p.

Coon, Richard F. Surficial geology of the Syracuse East and Manlius, New York 7-1/2 minute quadrangles. M, 1960, Syracuse University.

Coon, Richard Floyd. Correlation of engineering behavior with the classification of in-situ rock. D, 1968, University of Illinois, Urbana. 249 p.

Cooney, Robert Lawrence. The mineralogy of the Jensen and Henshaw quarries near Riverside, California. M, 1956, University of California, Los Angeles.

Cooney, Thomas Francis. Spectroscopic and calorimetric studies of olivine glass and crystal and the origin of ultramafic liquid immiscibility. D, 1988, University of Oregon. 192 p.

Coonrod, David L. Dispersion of dissolved metals from the Webster County, Missouri, sanitary landfill. M, 1985, Southwest Missouri State University. 61 p.

Coons, Lawrence M. Simulated rainfall runoff and infiltration characteristics of selected rangelands in New Mexico. M, 1984, New Mexico State University, Las Cruces. 165 p.

Coons, Richard L. Precambrian basement geology and Paleozoic structure of the Mid-Continent gravity high. D, 1966, University of Wisconsin-Madison. 406 p.

Coons, William E., III. Cobalt as an analogue for iron in high temperature experiments on basaltic compositions. D, 1978, Arizona State University. 181 p.

Coons, William Ellsworth, III. Effects of thermal assimilation and low grade metamorphism on a xenolith-bearing basalt dike in Marquette, Michigan. M, 1975, Western Michigan University.

Cooper, A. J. Pre-Catfish Creek tills of the Waterloo area. M, 1975, University of Waterloo.

Cooper, A. K. The origin of the Bering Sea marginal basin; implications from marine magnetic data. D, 1974, Stanford University. 108 p.

Cooper, Alan. Structure of the continental shelf west of San Francisco, California. M, 1971, San Jose State University. 65 p.

Cooper, B. J. Studies of multielement Silurian conodonts. D, 1974, Ohio State University. 222 p.

Cooper, Brian J. The effects of heating and dehydration on the crystal structure of hemimorphite up to 600°C. M, 1978, Virginia Polytechnic Institute and State University.

Cooper, Brian Jay. The investigation of the optical properties of polytypic minerals. D, 1988, Virginia Polytechnic Institute and State University. 136 p.

Cooper, Budoin-Brutus J. Sedimentary environment and provenances of the Snowbird Group, Ocoee Supergroup in Tennessee and North Carolina. M, 1987, Wright State University. 196 p.

Cooper, Byron Nelson. Geology of the Draper Mountain area, Virginia. D, 1937, University of Iowa. 189 p.

Cooper, Byron Nelson. Stratigraphy and structure of the Marion area, Virginia. M, 1935, University of Iowa. 76 p.

Cooper, Chalmer L. Conodonts from a Bushberg-Hannibal horizon in Oklahoma. D, 1945, University of Chicago. 43 p.

Cooper, Chalmer Lewis. The Sycamore Limestone of the Arbuckle Mountains of Oklahoma. M, 1926, University of Oklahoma. 37 p.

Cooper, Daniel Howell. The theory and geological application of ternary statistics. D, 1981, University of Mississippi. 106 p.

Cooper, David Michael. Radiation induced growth, modification, and decay of thermoluminescence in synthetic fluorite. M, 1967, University of Florida. 69 p.

Cooper, David Michael. Sedimentation, stratigraphy and facies variation of the lower to middle Miocene, Astoria Formation in Oregon. D, 1981, Oregon State University. 524 p.

Cooper, Gerald E. Geology of the Johan Beetz area, Saguenay County, Quebec. D, 1953, McGill University.

Cooper, Gordon Evans. The geology of the Tulks Hill area, central Newfoundland. M, 1968, Memorial University of Newfoundland. 111 p.

Cooper, Gustav Arthur. Stratigraphy of the Hamilton Group of New York. D, 1929, Yale University. 650 p.

Cooper, Herman William. Attenuation in igneous rocks at seismic frequencies. D, 1979, Massachusetts Institute of Technology. 139 p.

Cooper, Herman William. The use of geophysical methods to determine the lateral distribution of permafrost. M, 1975, Queen's University. 90 p.

Cooper, Herschel H. Origin and occurrence of the iron ores of Washington; 1 volume. M, 1922, Washington State University.

Cooper, Herschel T. Geology of the western Seminoe Mountains. M, 1951, University of Wyoming. 88 p.

Cooper, J. Calvin. Geology of the Fondo Negro region, Dominican Republic. M, 1983, SUNY at Albany. 145 p.

Cooper, J. Calvin. The geology of the central Apennines and foreland basin, Italy. D, 1988, Rice University. 401 p.

Cooper, J. F. Critical relations along an early Paleozoic carbonate-slate contact between Pleasant Valley and Upton Lake, Dutchess County, New York. M, 1955, Columbia University, Teachers College.

Cooper, Jack Charles and Kelly, Robert Bowen. Geology of a portion of the Santa Susana Quadrangle, Los Angeles and Ventura counties, California. M, 1941, University of California, Los Angeles.

Cooper, Janice Hatchell. Function and evolution in the eye of Phacops rana, a Middle Devonian trilobite of North America. M, 1986, University of California, Davis. 68 p.

Cooper, Jeanne L. Chemical and isotopic variations within late Cenozoic tholeiitic and alkalic basalts in relation to the Colorado Plateau and Basin and Range provinces, east-central Arizona. M, 1986, Miami University (Ohio). 127 p.

Cooper, John Doyne. Geology of Spring Branch area, Comal and Kendall counties, Texas. M, 1964, University of Texas, Austin.

Cooper, John Doyne. Stratigraphy and paleontology of Escondido Formation (upper Cretaceous), Maverick County, Texas, and northern Mexico. D, 1970, University of Texas, Austin. 352 p.

Cooper, John Edmond, Jr. Petrography of the Moroni Formation, southern Cedar Hills, Utah. M, 1956, Ohio State University.

Cooper, John Roberts. Geology of the southern half of the Bay of Islands igneous complex. D, 1935, Princeton University. 123 p.

Cooper, Laurence C. Geochemical approach to the study of rocks and minerals; a syllabus. M, 1958, University of Utah. 111 p.

Cooper, M. R. Lunar structure and seismicity at the Apollo 17 landing site. D, 1975, Stanford University. 248 p.

Cooper, Margaret. The peneplanes of the Ouachita Mountains. M, 1937, Columbia University, Teachers College.

Cooper, Michelle. Replacement textures in the Silurian Clinton-type iron ores from the Birmingham District, Alabama. M, 1981, University of Missouri, Rolla. 78 p.

Cooper, Milton L. Systematic paleontology of the Pennsylvanian scaphopods of North America. M, 1972, Kent State University, Kent. 66 p.

Cooper, Murray F. J. Geology of the Rose property porphyry copper occurrence, northwestern British Columbia. M, 1978, Queen's University. 220 p.

Cooper, Neil F. Trace element geochemistry and origin of the Andover iron deposit, Andover, New Jersey. M, 1978, University of Delaware.

Cooper, Patricia Ann. Seismicity, focal mechanisms and morphology of subducted lithosphere in the Papua New Guinea-Solomon Islands region. D, 1985, University of Hawaii.

Cooper, Reid Franklin. The structure and rheology of partially molten olivine-basalt aggregates. D, 1983, Cornell University. 172 p.

Cooper, Robert Peyton. The foraminifera of the Non-ionella cockfieldensis Zone in Bee County, Texas. M, 1939, University of Texas, Austin.

Cooper, Roger Brian. Copper, lead and zinc sorption on Wyoming montmorillonite. M, 1976, University of British Columbia.

Cooper, Roger Wayne. Lineament and structural analysis of the Duluth Complex, Hoyt Lakes-Kawishiwi area, northeastern Minnesota. D, 1978, University of Minnesota, Duluth. 295 p.

Cooper, Roger Wayne. Middle Precambrian and Keweenawan rocks north of the Gogebic Range in Wisconsin. M, 1973, University of Wisconsin-Madison.

Cooper, Warren W. Dinoflagellates and acritarchs from the Kiowa Shale (Cretaceous) of western Kansas; a biostratigraphic study. M, 1967, Wichita State University. 90 p.

Cooper, William Amos, Jr. Fresh-water deltaic sedimentation in Lake Maracaibo, Venezuela. M, 1976, University of Tulsa. 124 p.

Cooper, William Clinton. Intertidal foraminifera of the California and Oregon coast. M, 1961, University of Southern California.

Cooper, William James. Short term variability of hydrogen peroxide in surface oceans. D, 1987, University of Miami. 149 p.

Coorough, Patricia Jo. Brachiopod provinciality in the Late Ordovician-Early Silurian. M, 1986, University of Wisconsin-Milwaukee. 129 p.

Cope, Edward L. Geology of the Alice E. Breccia Pipe and vicinity, New World mining district, Park County, Montana. M, 1984, Colorado State University. 132 p.

Copeland, Charles Wesley, Jr. Eocene and Miocene foraminifera from two localities in Duplin County, North Carolina. M, 1961, University of North Carolina, Chapel Hill. 204 p.

Copeland, David Ashley. Chemical variation in hornblende syenite from the Red mountain area, southern Laramie mountains, Wyoming. M, 1970, University of Wyoming. 86 p.

Copeland, J. G. Alteration associated with sulphide mineralization at the Waite-Amulet Mine, northwestern Quebec. M, 1951, University of Toronto.

Copeland, J. G. Structure and stratigraphy of the Waite-Amulet area, Quebec. D, 1953, University of Toronto.

Copeland, Jerry Harold. Geology of the upper Middle Creek area, Huerfano County, Colorado. M, 1959, University of Nebraska, Lincoln.

Copeland, M. J. Stratigraphy and paleontology of the Joggins coal measures, Joggins, Nova Scotia. M, 1951, University of Toronto.

Copeland, Murray John. The Pennsylvanian Arthropoda of the Maritime Provinces of Canada. M, 1953, [University of Michigan].

Copeland, Murray John. The Upper Carboniferous arthropods from the Maritime Provinces of Canada. D, 1955, University of Michigan. 181 p.

Copeland, Peter. Geochemistry and geology of the Pinal Schist, Cochise, and Pima counties, Arizona. M, 1986, New Mexico Institute of Mining and Technology. 176 p.

Copeland, Richard A. Trace element distribution and sediments in the Mid-Atlantic ridge. D, 1970, Massachusetts Institute of Technology. 103 p.

Copeland, Richard Evan. The geology of the northern portion of the Wadesboro Triassic basin, North Carolina. M, 1974, University of Florida. 99 p.

Copeland, W. A. A mineragraphic study of ores from the Parry Sound District, Ontario. M, 1920, University of Minnesota, Minneapolis. 50 p.

Copeland, William Barton. Structural and metamorphic constraints on fault displacement between coherent blueschist terranes near Ball Mountain, eastern belt, Franciscan Complex, Northern California. M, 1988, University of Texas, Austin. 245 p.

Copenhaver, George C. Geochemical prospecting for nickel in the Julian-Cuyamaca area (San Diego County), California. M, 1970, University of San Diego.

Coperude, Shane Patrick. Geologic structure of the western continental margin of south central Baja California based on seismic and potential field data. M, 1978, Oregon State University. 66 p.

Copes, Joe Lenon. A pedagogical model for studying the geomorphology of drainage basins. D, 1973, Clark University.

Copland, John Robin. Laramide structural deformation at the interface between the Laramie Range and the Denver-Julesburg Basin, southeastern Wyoming. M, 1984, University of Wyoming. 49 p.

Coplen, Tyler B., II. Isotopic fractionation of water ultrafiltration. D, 1970, University of Chicago. 96 p.

Copley, Albert J. Areal geology of the Duke area, Oklahoma. M, 1961, University of Oklahoma. 100 p.

Copley, David L. Preliminary high sensitivity measurement of the alpha-beta quartz inversion temperature in chert. M, 1971, University of Toledo. 63 p.

Copp, John Frederick. Soil mercury geochemistry in the Coso volcanic field, Inyo County, California. M, 1981, Arizona State University. 179 p.

Copper, Lon M. An evaluation of surface geophysics as applied to a hydrogeologic study in Routt and Jackson counties, Colorado. M, 1983, Colorado School of Mines. 194 p.

Copper, Paul. Middle Devonian atrypids from northwestern Canada. M, 1962, University of Saskatchewan. 137 p.

Coppersmith, Kevin Joseph. Activity assessment of the Zayante-Vergeles Fault, central San Andreas fault system, California. D, 1979, University of California, Santa Cruz. 333 p.

Coppinger, Walter W. Stratigraphic and structural study of Belt Supergroup and associated rocks in a portion of the Beaverhead Mountains, southwest Montana and east-central Idaho. D, 1974, Miami University (Ohio). 224 p.

Coppinger, Walter W. The distribution and petrology of the surficial Wisconsin loess in portions of Preble and Butler counties, southwestern Ohio. M, 1968, Miami University (Ohio). 117 p.

Coppold, Murray. Stratigraphy and lithofacies of the southwest margin of the Ancient Wall carbonate complex, Chetamon thrust sheet, Jasper National Park, Alberta. M, 1973, McGill University. 122 p.

Corbato, Charles Edward. Gravity investigation of the San Fernando Valley, California. D, 1960, University of California, Los Angeles.

Corbeil, Paul. Géologie du Quaternaire de la région de Rigaud/Rivière-Beaudette (Québec); quelques applications à l'environnement. M, 1984, Universite du Quebec a Montreal. 106 p.

Corbet, T. F. Interaction between open-pit mining and groundwater flow systems at the east area of the Gay Mine, Idaho. M, 1979, University of Idaho.

Corbett, Clifton S. Geology of the Fogarty and Baltic mines and vicinity, Iron River District, Michigan. M, 1914, University of Wisconsin-Madison.

Corbett, Clifton S. Levierrite as a schist-forming mineral. D, 1921, University of Wisconsin-Milwaukee.

Corbett, Cynthia Rena Ruth. The Champion Lake Fault; a Tertiary, east-side-down normal fault in southeastern British Columbia. M, 1985, University of Calgary. 103 p.

Corbett, Edward John. Seismicity and crustal structure studies of Southern California; tectonic implications from improved earthquake locations. D, 1984, California Institute of Technology. 247 p.

Corbett, Edward Sisk. Hydrologic evaluation of the stormflow generation process on a forested watershed. D, 1979, Pennsylvania State University, University Park. 149 p.

Corbett, John Dennen. A geophysical investigation in Richmond Basin and upper Camp Bird Valley of San Juan Mountains, Colorado. M, 1958, Colorado School of Mines. 63 p.

Corbett, Kevin Patrick. Structural stratigraphy of the Austin Chalk. M, 1982, Texas A&M University.

Corbett, Marshall Keene. Groundwater surveys; an application of aerial photographic studies. M, 1956, Cornell University.

Corbett, Marshall Keene. Tertiary igneous petrology of the Mt. Richthofen-Iron Mt. area, north-central Colorado. D, 1964, University of Colorado. 137 p.

Corbett, Mary Diane. Stratigraphy, structure and skarn deposits within the Toroweap, Kaibab and Moenkopi section, Lincoln mining district, southern Mineral Range, Beaver County, Utah. M, 1984, Colorado School of Mines. 180 p.

Corbett, Robert Guy. The formation of hydroxyapatite in the oceans at 25 degrees centigrade. M, 1958, University of Michigan.

Corbett, Robert Guy. The geology and mineralogy of Section 22 Mine, Ambrosia Lake uranium district, New Mexico. D, 1964, University of Michigan. 220 p.

Corbett, Ronald K. A method of obtaining x-ray powder photographs from single crystals. M, 1972, University of Arizona.

Corbin, M. W. The structure of the Fitzwilliam, Troy and Marlboro granite. D, 1935, The Johns Hopkins University.

Corbin, Samuel W. The thermal regime of a stream in central Alaska. M, 1977, University of Alaska, Fairbanks. 144 p.

Corbitt, Christopher L. Petrography and geochemistry of the metasedimentary/metavolcanic host rocks and associated gold deposits of the Portis gold mine, Franklin County, North Carolina. M, 1987, East Carolina University. 99 p.

Corbitt, Lisa B. The petrology and stratigraphy of the Reynolds Limestone Member of the Bluefield Formation (Mississippian) in southeastern West Virginia. M, 1986, East Carolina University. 106 p.

Corbitt, Lonnie L. Structure and stratigraphy of the Florida Mountains, Luna County, New Mexico. D, 1971, University of New Mexico. 115 p.

Corbitt, Lonnie Leroy. Structural and modal analysis of the Butler Hill Granite, southeastern Missouri. M, 1966, Southern Illinois University, Carbondale. 29 p.

Corbo, Salvatore. Production, preservation and facies significance of trace fossils in epicratonic submarine-fan deposits, Upper Devonian, New York. M, 1980, SUNY at Binghamton. 175 p.

Corby, Grant W. The geology and paleontology of the San Joaquin and Miguel hills, Orange County, California. M, 1922, Stanford University.

Corchary, George Sutter. The unconsolidated sediments of a part of the Tanana River valley, Alaska. M, 1959, University of Illinois, Urbana. 33 p.

Corcoran, Eugene Francis. Quatitative studies of organic matter and associated biochromes in marine sediments. D, 1957, University of California, Los Angeles. 142 p.

Corcoran, Raymond Ervin. The geology of the east-central portion of the Mitchell Butte Quadrangle, Oregon. M, 1953, University of Oregon. 80 p.

Cordalis, Charles. Mineralogy and petrology of the Copper King platinum prospect, Park County, Montana. M, 1984, Montana College of Mineral Science & Technology. 62 p.

Cordas, John J. Determination of paleotemperatures associated with the post-Permian thermal events in the Parana Basin, Brazil, by reflectance studies of coals. M, 1978, University of Toledo. 116 p.

Cordell, Donald Allen. Geology of the Pedro Mountain tonalite and associated rocks, northeastern Oregon. M, 1973, University of Oregon. 92 p.

Cordell, Lindreth E. Pennsylvanian paleotectonics in central northwestern New Mexico. M, 1962, University of New Mexico. 62 p.

Cordell, Robert James. Ostracodes from the Trivoli cyclothem. M, 1940, University of Illinois, Urbana. 87 p.

Cordell, Robert James. Ostracodes from the Upper Pennsylvanian of Missouri. D, 1949, University of Missouri, Columbia.

Cordero Ardila, V. F. Gravity and seismic reflection studies over the Ferguson Crossing salt dome, Grimes and Brazos counties, Texas. M, 1977, Texas A&M University.

Cordero, Vladimir. Gravity, magnetic and reflection seismic studies over the Ferguson Crossing salt dome. M, 1977, Texas A&M University.

Cordes, Edwin H. The temperature and moisture distribution in an unsaturated soil column subjected to surface evaporation. M, 1965, University of Arizona.

Cordes, R. E. Measurements of the velocity field of the Fraser River plume. M, 1977, University of British Columbia.

Cording, Edward James. The stability during construction of three large underground openings in rock. D, 1967, University of Illinois, Urbana. 272 p.

Cordiviola, Steven. Hillslope processes in southwestern Arizona. M, 1974, University of Cincinnati. 99 p.

Cordoba-Mendez, Diego Arturo. Geology of Apizolaya Quadrangle (east half), northern Zacatecas, Mexico. M, 1964, University of Texas, Austin.

Cordova, Ernesto Villaescusa *see* Villaescusa Cordova, Ernesto

Cordova, Robert Murray. The general geology and petrology of the greenstone of the Southwestern Mountains, Virginia. M, 1955, University of Virginia. 119 p.

Cordova, Simon. Geology of the Piru area, Ventura County, California. M, 1956, University of California, Los Angeles.

Cordova, Tommy. Active faults in Quaternary alluvium, and seismic regionalization, in a portion of the Mount Rose Quadrangle, Nevada. M, 1969, University of Nevada. 53 p.

Cordry, Cletus D. Heavy minerals in the Roubidoux and other sandstones of the Ozark region. M, 1928, University of Missouri, Rolla.

Cordsen, Andreas. Ocean bottom seismometer refraction results from the continental margin off Nova Scotia. M, 1980, Dalhousie University. 147 p.

Cordsen, Andreas. The crystal structure of libethenite; a structure refinement based on data collected with a four-circle diffractometer. M, 1976, Queen's University. 45 p.

Cordua, William Sinclair. Precambrian geology of the southern Tobacco Root Mountains, Madison County, Montana. D, 1973, Indiana University, Bloomington. 247 p.

Cordy, Gail E. Environmental geology of the Paradise Valley Quadrangle, Maricopa County, Arizona; Part II. M, 1978, Arizona State University. 89 p.

Core, Eugene Howard. Wall rock alteration in the Tonopah mining district, Nevada. M, 1959, Indiana University, Bloomington. 34 p.

Corea, William Charles. A method for synthesizing sedimentary structures generated by migrating bedforms. M, 1978, Massachusetts Institute of Technology. 58 p.

Corea, William Charles. Flume studies of large-scale cross-stratification produced by migrating bed forms.

D, 1981, Massachusetts Institute of Technology. 181 p.

Corey, Allen Frank. Kyanite deposits of the Petaca District, Rio Arriba County, New Mexico. D, 1954, University of Michigan. 169 p.

Corey, Ronald Stewart and Ryan, Richard C. Geology of the Slide Mountain area, Huerfano and Costilla counties, Colorado. M, 1960, University of Michigan.

Corey, William Henry. Paleontology and stratigraphy of the Vaqueros Formation (lower Miocene) of Oak Ridge and South Mountain, Ventura County, California, with notes on Vaqueros sections adjoining and in the western Santa Ynez Mountains of Santa Barbara County. M, 1928, University of California, Berkeley. 64 p.

Corgan, James X. Some quantitative characteristics of poikilohaline molluscan association. M, 1958, Columbia University, Teachers College.

Corgan, James Xavier. Quaternary micromolluscan fauna of the Mudlump Province, Mississippi River Delta. D, 1967, Louisiana State University. 325 p.

Corken, R. James. Stratigraphic analysis of the Morrison Formation in the Shiprock-Beautiful Mountain area, northwestern New Mexico. M, 1979, Northern Arizona University. 189 p.

Corkery, Maurice Timothy. A study of the geology of the sulphide ore bodies at South Bay Mines, northwestern Ontario. M, 1977, University of Manitoba.

Corking, W. P. Genesis of the Sherritt Gordon ore deposits. M, 1938, University of Toronto.

Corlett, Mabel. Mineralogy of pyrrhotite. D, 1964, University of Chicago. 64 p.

Corley, Daniel I. Prudhoe Bay gas, can we afford it?. M, 1983, Stanford University. 214 p.

Corley, J. B. A model for the prediction of elevation changes beneath a slab barrier on expansive clay. D, 1979, [University of Texas, Arlington]. 192 p.

Corley, Robert Andrew. A petrographic analysis and diagenetic history for Brazos River Formation in north-central Texas. M, 1985, Stephen F. Austin State University. 137 p.

Corliss, B. H. Studies of Cenozoic deep-sea benthonic foraminifera in the Southern Ocean. D, 1978, University of Rhode Island. 276 p.

Corliss, Bruce Clyde. Southeastern Michigan geology; a guidebook for teachers. M, 1960, University of Michigan.

Corliss, John Burt. Structure and morphology of the Mid-Atlantic ridge at the Vema fracture zone. D, 1970, University of California, San Diego. 162 p.

Corman, David. Formation of a mature siliceous chert conglomerate; the Cadomin Conglomerate (Lower Cretaceous) of the Rocky Mountain foothills, Alberta. M, 1972, University of Delaware.

Cormie, J. M. Geology and ore deposits of the central Patricia Gold Mines, Ltd. M, 1935, Queen's University. 29 p.

Cormier, Randall F. Rubidium-strontium ages of glauconite and their application to the construction of an absolute post-Precambrian time scale. D, 1957, Massachusetts Institute of Technology. 133 p.

Cormier, V. F. Full wave theory applied to a discontinuous velocity increase; the inner core boundary. D, 1976, Columbia University, Teachers College. 133 p.

Corn, Russell Morrison. The geology of the Mount Bross-Buckskin Creek area, Lake County, Colorado. M, 1957, Colorado School of Mines. 128 p.

Corneil, Barry David. A paleoecological study of the Ante Creek reef, middle to late Devonian in age, west central Alberta (Canada). M, 1969, University of Alberta. 122 p.

Corneille, E. S., Jr. Bostonite and lamprophyre dikes from the Champlain Valley, Vermont. M, 1975, University of Vermont.

Cornejo, John. Pleistocene molluscan faunas of the Souder Lake Deposit, Franklin County, Ohio. M, 1959, Ohio State University.

Cornelius, Howard E. Synthetic seismograph modeling and its application to data from the Rio Grande Rift. M, 1980, University of Texas at El Paso.

Cornelius, Jeffrey Bernard. Electron spin echo and millimeter wave EPR studies of disordered solids. D, 1987, University of Illinois, Urbana. 260 p.

Cornelius, Kenneth D. Geology of the footwall formations of the Veta Madre, Guanajuato mining district, Guanajuato, Mexico. M, 1964, University of Arizona.

Cornelius, Reinold R. Geology and tectonic setting of Rock Creek Butte, Klamath Mountains, California. M, 1984, University of Texas, Austin.

Cornelius, Scott Bedford. The geology of upper Val Masino; 1 volume. M, 1972, University of California, Berkeley.

Cornelius, Searle H., Jr. The structure and inclusions of the Bonsall Tonalite near Fallbrook, California. D, 1933, Harvard University.

Corneliussen, Eric Frantz. Ectoproct (Bryozoan) genera Monotrypa, Hallopora, Amplexopora, and Hennigopora from the Brownsport Formation (Niagaran-Silurian) in western Tennessee. D, 1970, Indiana University, Bloomington. 213 p.

Cornell, James Richard. The ostracode zones of the Decorah Shale. M, 1956, University of Minnesota, Minneapolis. 91 p.

Cornell, Josiah H., III. Geology of the northwest quarter of the Canyonville Quadrangle, Oregon. M, 1971, University of Oregon. 64 p.

Cornell, William C. Archaeomonadaceae from the preliminary Project Mohole drilling. M, 1967, University of Rhode Island.

Cornell, William Crowninshield. Chrysomonad cysts and silicoflagellates from the Marca Shale Member, Moreno Formation (Maestrichtian), Fresno County, California. D, 1972, University of California, Los Angeles.

Cornell, Winton C. Study of the mineralogy of plutonic nodules from three eruptions of Somma-Vesuvius volcano, Italy, evidence for subvolcanic processes. M, 1981, University of Rhode Island.

Cornell, Winton Charles. Volcanologic and petrologic studies of tephra from the Neapolitan volcanic area, Italy. D, 1988, University of Rhode Island. 264 p.

Corner, Richard George. An early Valentinian vertebrate local fauna from southern Webster County, Nebraska. M, 1976, University of Nebraska, Lincoln.

Cornet, F. H. Analysis of the deformation of saturated porous rocks in compression. D, 1975, University of Minnesota, Minneapolis. 283 p.

Cornet, Walter Bruce. The palynostratigraphy and age of the Newark Supergroup. D, 1977, Pennsylvania State University, University Park. 527 p.

Cornett, Duane Charles. An evaluation of goal programming for multiple land use planning at Mineral King, California. D, 1987, University of California, Davis. 266 p.

Cornett, Shawn. Fluvial coal in the Crownest River. M, 1979, University of Calgary.

Cornford, E. H. G. Geology and ore deposits of the Howey Mine, Red Lake, Ontario. M, 1940, Queen's University. 40 p.

Cornish, Bruce E. A magnetic survey of the Herod Quadrangle, Illinois. M, 1975, Southern Illinois University, Carbondale. 87 p.

Cornish, Cornelia Baker. The geography and history of Cortland County. M, 1929, Cornell University.

Cornish, Frank G. Tidally influenced deposits of the Hickory Sandstone, Cambrian, central Texas. M, 1975, University of Texas, Austin.

Cornuelle, Bruce Douglas. Inverse methods and results from the 1981 ocean acoustic tomography experiment. D, 1983, Woods Hole Oceanographic Institution. 361 p.

Cornwall, Dean Torrey. The geology of a portion of the Richland Center, Wisconsin, Quadrangle. M, 1926, University of Iowa. 83 p.

Cornwall, Diane Elinor. Paleoecology of Upper Triassic bioherms in the Pilot Mountains, Mineral County, West-Central Nevada. M, 1979, University of Nevada. 138 p.

Cornwall, F. W. A detailed study of the distribution of calcium, sodium, and potassium in metasomatic zones bordering sulphide mineralization. M, 1953, McGill University.

Cornwall, F. W. Rock alteration and primary base-metal dispersion at Barvue, Golden Manitou, and New Calumet mines, Quebec. D, 1956, McGill University.

Cornwall, Henry B. Manual of blowpipe analysis, qualitative and quantitative, with a complete system of determinative mineralogy. D, 1888, Columbia University, Teachers College.

Cornwall, Henry Rowland. Differentiation in doleritic lavas of the Michigan Keweenawan and the origin of the copper deposits. D, 1947, Princeton University.

Cornwell, Jeffrey Clayton. The geochemistry of manganese, iron and phosphorus in an Arctic lake. D, 1983, University of Alaska, Fairbanks. 249 p.

Cornwell, Kevin J. Evaluation of volcanic ash as a stratigraphic marker in playa basins, western Texas. M, 1984, Texas Tech University. 47 p.

Cornyn, Michael Robert. Stream structure and late glacial climate in Greene County, Ohio. M, 1977, Wright State University. 114 p.

Coron, Cynthia R. Algal origin and diagenesis of a Silurian carbonate mud mound, Wabash, Indiana. M, 1971, University of North Carolina, Chapel Hill. 38 p.

Coron, Cynthia Rose. Facies relations and ore genesis of the Newfoundland zinc mines deposit, Daniel's Harbour, western Newfoundland. D, 1982, University of Toronto.

Corona, Charles Jude. Geology of the Midland-Easterwood area, Acadia Parish, Louisiana. M, 1970, Tulane University.

Corona, Francesco. Reconnaissance geology of Sierra La Gloria and Cerro Basura, northwestern Sonora, Mexico. M, 1980, University of Pittsburgh.

Coronato, James Allen. Dynamics of earth-retaining gravity walls. D, 1988, Carnegie-Mellon University. 148 p.

Correa, A. C. A geological and statistical approach to the exploration for alpine-type chromite in California. D, 1974, Stanford University. 316 p.

Correa, Aberbal Caetano. Geology of the southern half of Casamero lake (7.5 min.) quadrangle, McKinley County, New Mexico. M, 1970, Colorado School of Mines. 187 p.

Correa, Brian Paul. The Taylor Creek Rhyolite and associated tin deposits, southwestern New Mexico. M, 1981, Arizona State University. 105 p.

Correa, Orlando Gonzales. Significance of statistical parameters in the environmental interpretation of beach sediments. M, 1970, University of California, Los Angeles.

Correa, Victor Julio Lopez. Electromagnetic soundings in California, New Mexico and Wisconsin. D, 1981, University of Texas at Dallas. 157 p.

Correa, Victor Julio Lopez Lopez see Lopez Correa, Victor Julio Lopez

Correia, Francisco Carlos da Graca Nunes see Nunes Correia, Francisco Carlos da Graca

Correig, Antoni M. Rayleigh wave amplitudes and regional variations of anelasticity in the Eastern Pacific. M, 1977, St. Louis University.

Corrigan, Anthony F. The evolution of a cratonic basin from carbonate to evaporite deposition and the resulting stratigraphic and diagenetic changes, Upper Elk Point Subgroup, north-eastern Alberta. D, 1975, University of Calgary. 328 p.

Corrigan, Donald. The paleomagnetism and magnetic mineralogy of the Medford diabase dike (Medford area of Boston, Massachusetts). M, 1973, University of Rhode Island.

Corrigan, Jeffrey Delon. Geology of the Burica Peninsula, Panama-Costa Rica; neotectonic implications for the southern Middle America convergent margin. M, 1986, University of Texas, Austin. 152 p.

Corriveau, Louise. Physical conditions of the regional and the retrograde metamorphism in the pelitic gneiss of the Chicoutimi area, Quebec. M, 1982, Queen's University. 264 p.

Corry, Andrew Vincent. A study of the continental fault near Butte, Montana. M, 1931, Montana College of Mineral Science & Technology. 46 p.

Corry, C. E. The emplacement and growth of laccoliths. D, 1976, Texas A&M University. 321 p.

Corry, Charles Elmo. The origin of the Solitario Formation (Upper Ordovician (?)), Trans-Pecos, Texas. M, 1972, University of Utah. 151 p.

Corso, William. Sedimentology of rocks dredged from Bahamian Platform slopes. M, 1983, University of Miami. 200 p.

Corso, William P. Development of the Early Cretaceous Northwest Florida carbonate platform. D, 1987, University of Texas, Austin. 145 p.

Corson, Donald L. The western Washington coal industry, 1875-1935; a study of an ephemeral industry. M, 1974, California State University, Los Angeles. 63 p.

Cort, John J., Jr. Some contact metamorphic effects of the Duluth Gabbro south of Kekequabic Lake, Minnesota. M, 1931, Northwestern University.

Cort, Thierry Michael de see de Cort, Thierry Michael

Corta, Hugues de see de Corta, Hugues

Cortellini, Edmund A. Geology of some intrusions of the Guffey volcanic center. M, 1979, Miami University (Ohio). 47 p.

Cortes Lombana, Abdon. Climosequence of ash-derived soils in the Central Cordillera of Colombia. D, 1971, Purdue University. 260 p.

Cortes, Jorge N. The coral reef at Cahuita, Costa Rica; a reef under stress. M, 1981, McMaster University. 176 p.

Corvalan, Jose Idamor. Early Mesozoic biostratigraphy of the Westgate area, Churchill County, Nevada. D, 1962, Stanford University. 262 p.

Corvinus, Dorothy Anne. Micropetrographic characteristics of peats from modern coal-forming environments in Okefenokee Swamp, Georgia and Albemarle-Pamlico penninsular swamps, North Carolina. D, 1982, University of South Carolina. 129 p.

Corvinus, Dorothy Anne. Pre-maceral characteristics of carbonaceous sediments from Snuggedy Swamp, S.C. M, 1978, University of South Carolina.

Corwin, Bert N. Paleoenvironmental history of Holocene Ostracoda, Storr's Lake, San Salvador, Bahamas. M, 1985, University of Akron. 115 p.

Corwin, Gilbert. The petrology and structure of the Palau volcanic islands. D, 1952, University of Minnesota, Minneapolis. 298 p.

Corwin, Robert F. Offshore applications of self-potential prospecting. D, 1973, University of California, Berkeley.

Corwine, John W. Measurement of the surface energy of rock by the three-point bending of notched beams. M, 1971, University of Missouri, Rolla.

Coryell, George Fossas. Stratigraphy, sedimentation, and petrology of the Tertiary rocks in the Bear Creek-Wickiup Mountain-Big Creek area, Clatsop County, Oregon. M, 1978, Oregon State University. 178 p.

Coryell, Horace Noble. Bryozoan faunas of the Stone River Group of central Tennessee. D, 1919, University of Chicago. 79 p.

Coryell, Horace Noble. The bryozoan faunas of the Black River and Trenton limestones of the Trenton Falls and the Mohawk Valley. M, 1915, Indiana University, Bloomington.

Coryell, Jeffrey J. Structural geology of the Henneberry Ridge area, Beaverhead County, Montana. M, 1983, Texas A&M University. 126 p.

Coryell, Lawrence Ritchie Brooke and King, C. W. A gravity study of the northern boundary of the Boston Basin. M, 1958, Massachusetts Institute of Technology. 44 p.

Coryell, Lewis S. The stratigraphy of Texas. M, 1917, University of Minnesota, Minneapolis. 32 p.

Cos, Raymond Lee. Age and correlation of the Sooke Formation (Oligocene-Miocene) with a section of its palynology (British Columbia). M, 1962, University of British Columbia.

Cosens, Barbara Anne. Initiation and collapse of active circulation in hydrothermal system at Mid-Atlantic Ridge, 23 degrees N. M, 1982, University of Washington. 77 p.

Cosgrave, T. M. An investigation of shallow stratigraphic reflections from ground penetrating radar. M, 1987, University of Waterloo. 230 p.

Cosgriff, John William. A new form of capitosaur from the Arizona Triassic. M, 1959, University of California, Berkeley. 90 p.

Cosgriff, John William. Triassic vertebrates from western Australia. D, 1963, University of California, Berkeley. 288 p.

Coskey, Robert J. Stratigraphy, petrography and structural analysis of the Schunnemunk Mountain region, Orange County, New York. M, 1978, SUNY, College at Oneonta. 192 p.

Coskren, Thomas Dennis. Equilibration between aluminous clinopyroxene and plagioclase; a geobarometer-geothermometer. M, 1972, University of Kentucky. 45 p.

Coskren, Thomas Dennis. Mathematical analysis of distribution curves in high-grade metamorphic carbonates; a new approach to time in metamorphism. D, 1983, University of Kentucky. 115 p.

Coskun, Sefer B. Multivariate data analysis for map comparison for petroleum exploration, Pratt County, Kansas. M, 1988, Wichita State University. 113 p.

Cosmos, George J. Wave erosion rates along the abandoned ocean shore railroad, San Mateo County, California. M, 1974, San Francisco State University.

Coss, James R. Geology of the eastern Santa Ynez Valley, central Santa Barbara County, California. M, 1980, University of California, Santa Barbara.

Coss, John Michael. Paleoenvironments of the upper Fort Union Formation at Pine Ridge, western Powder River basin, Wyoming. M, 1985, University of Colorado.

Cossaboom, C. Carey. Alteration petrology and mineralization of the Flathead Mine, Hog Heaven mining district, Montana. M, 1981, University of Montana. 104 p.

Cossette, Danielle. Variabilité de certaines espèces de trinucleidae (Trilobita) de l'Ordovicien Moyen et Supérieur. M, 1987, Universite de Montreal.

Cossette, Denis. Etablissement et validation d'un modèle géostatistique des reserves de la Mine Niobec. M, 1982, Ecole Polytechnique. 139 p.

Cossey, Steve. An analysis of Devono-Carboniferous sedimentary facies in the Eastern Meseta of Morocco. M, 1975, University of South Carolina.

Cossey, Steve P. J. Jurassic mass movement and turbidite sequences associated with growth faulting, northern Tunisia. D, 1978, University of South Carolina. 68 p.

Costa, John Emil. Geomorphic evolution and environmental geology of Western Run Watershed, Baltimore County, Maryland. D, 1973, The Johns Hopkins University.

Costa, Steven L. Sediment transport dynamics in tidal inlets. D, 1978, University of California, San Diego. 243 p.

Costa, Umberto Raimundo. Hydrothermal footwall alteration and ore formation at Mattagami Lake Mine, Matagami, Quebec. D, 1980, University of Western Ontario. 289 p.

Costa-Ribeiro, Carlos Antonio De Leers. Design and development of a detector to measure integrated radon concentrations. D, 1974, New York University.

Costain, John Kendall. Geology of the Gilson Mountains and vicinity, Juab County, Utah. D, 1960, University of Utah. 139 p.

Costanza, Suzanne Helene. Morphology and systematics of Cordaites of Pennsylvanian coal swamps of Euramerica. D, 1984, University of Illinois, Urbana. 198 p.

Costanzo, Patricia M. Synthesis and some properties of synthetic hydrated halloysite. M, 1982, SUNY at Buffalo. 47 p.

Costanzo, Patricia Marie Vogt. Synthesis and characterization of hydrated kaolinites and the chemical and physical properties of the interlayer water. D, 1984, SUNY at Buffalo. 113 p.

Costanzo-Alvarez, Vincenzo Francesco. A paleomagnetic study of two traverses in the Kapuskasing structural zone, Chapleau-Foleyet region, northwestern Ontario. M, 1986, University of Toronto.

Costaschuk, Suzanne Mickey. Uraniferous pegmatites in the Grease River area, northern Saskatchewan. M, 1979, University of Alberta. 174 p.

Costello, David K. Multiple-sample sediment trap with holographic particle imaging capability design and function of the ADIOS laser trap. M, 1988, University of South Florida, St. Petersburg.

Costello, J. Patrick. Electrical resistivity survey of the upper Boundary Creek thermal area, Yellowstone National Park, Wyoming. M, 1983, University of Missouri, Kansas City. 76 p.

Costello, Oliver P., Jr. The geology of the crystalline rocks of the south shore of Lake Murray, S. C. M, 1979, University of South Carolina.

Costello, Stephen Charles. A paleomagnetic investigation of mid-Tertiary volcanic rocks in the lower Colorado River area, Arizona and California. M, 1985, San Diego State University. 109 p.

Costello, Warren Russell. Braided river deposits and their relationship to the Pleistocene history of the Credit Valley, Ontario. M, 1970, McMaster University. 164 p.

Costello, Warren Russell. Development of bed configurations in coarse sands. D, 1974, Massachusetts Institute of Technology. 206 p.

Costley, Wayne J. A magnetic reconnaissance of the northern part of the Connecticut River valley lowland in Connecticut. M, 1956, Boston College.

Costolnick, David E. Sedimentology of the Devonian-Mississippian Spechty Kopf Formation in northeastern Pennsylvania. M, 1987, SUNY, College at Oneonta. 117 p.

Coston, Wendell Ray. Clay mineralogy and engineering properties of a shale-derived soil. D, 1969, University of Arkansas, Fayetteville. 49 p.

Cota, T. F. Stratigraphy and structural geology of the Yanert Glacier area, East-central Alaska Range, Alaska. M, 1975, University of Wisconsin-Madison.

Côté, Denis. Pétrographie, pétrologie et étude géochimique du dyke de diorite, de l'intrusion troctolitique et des deux petits massifs anorthositiques de Canton Taché M, 1986, Universite du Quebec a Chicoutimi. 181 p.

Cote, Philip Richard. Lower Carboniferous sedimentary rocks of the Horton Group in parts of Cape Breton Island, and their relation to similar strata of the Anguille Group in the southwestern Newfoundland. D, 1964, University of Ottawa. 184 p.

Cote, Pierre E. Geology and petrology of the anorthosite and associated rocks of the Chertsey map area. D, 1949, McGill University.

Cote, R. P. The structure and sedimentation of the Carboniferous sediments in the Craigmore-Long Point area, Cape Breton Island. M, 1958, Acadia University.

Cote, William Emerson. Grain-size analysis of sediments, southeastern Lake Michigan. M, 1967, University of Illinois, Urbana.

Cotera, Augustus S., Jr. Petrology and petrography of Mississippian-Pennsylvanian Tesnus Formation, Marathon Basin, Trans-Pecos, Texas. D, 1962, University of Texas, Austin.

Cotera, Augustus S., Jr. Petrology of the Cretaceous Woodbine Sand in Northeast Texas. M, 1956, University of Texas, Austin.

Cotey, Bradford James and Oliver, Donald McCreery. The geology of the southern part of Sheep Mountain, Freemont County, Wyoming. M, 1935, University of Missouri, Columbia.

Cotkin, Spencer Jerome. Petrology and geochemistry of the Russian Peak Pluton, Klamath Mountains, Northern California. D, 1987, University of Wisconsin-Madison. 306 p.

Cotkin, Spencer Jerome. The petrogenesis and structural geology of the Feragen Peridotite and associated rocks, Sor-Trondelag, east-central Norway. M, 1983, University of Wisconsin-Madison. 180 p.

Cotman, Richard Matthew. Potassium-argon evidence for shifting of the axial rift zone in northern Iceland. M, 1979, Case Western Reserve University.

Cott, Harrison C. Van see Van Cott, Harrison C.

Cotter, Edward Joseph. Mississippian carbonate banks in central Montana. D, 1963, Princeton University. 56 p.

Cotter, James F. P. The glacial geology of the north fork of the Big Lost River, Custer County, Idaho; a pedologic approach. M, 1981, Lehigh University.

Cotter, James F. P. The minimum age of the Woodfordian deglaciation of northeastern Pennsylvania and northwestern New Jersey. D, 1984, Lehigh University. 180 p.

Cotter, Mark Patrick. Paleoecology of the foraminifers of the Presumpscot Formation, Penobscot Valley, Maine. M, 1985, University of Maine. 319 p.

Cotter, R. D. A study of immiscibility in the carbonates from the metamorphic rocks of southeastern Vermont. M, 1953, University of Missouri, Rolla.

Cotton, Mary Lou. Seasonality in benthic foraminiferal populations of the Southern California borderland. M, 1979, University of Southern California.

Cotton, William R. Geology of the Halls Valley-Mount Hamilton area, California. M, 1967, San Jose State University. 152 p.

Cotton, William Robert, Jr. Hydrochemistry of ground water near Pullman, Washington. M, 1982, Washington State University. 89 p.

Cottrell, Daniel J. The sedimentary petrology and geomorphological aspects of the Marismas Nacionales, Mexico. M, 1973, SUNY at Buffalo. 72 p.

Cottrell, John. The paleoecological significance of inclined cephalopods, Cherry Valley Limestone (Devonian) N.Y. M, 1973, University of Rochester.

Cottrell, John Francis. Taphonomy and population dynamics of Recent bivalve populations and their implications for paleopopulation analyses. D, 1978, University of Rochester. 190 p.

Cottrill, Nancy Elaine. Whitewater River valley sediment distribution patterns; significance for land planning. M, 1972, University of Cincinnati. 123 p.

Couch, Albert Harris. Silica sands of Washington. M, 1935, University of Washington. 98 p.

Couch, Amber W. A drastic evaluation for part of the inner Bluegrass karst region, Kentucky; the Centerville, Georgetown, Lexington East and Lexington West Quadrangle. M, 1988, University of Kentucky. 94 p.

Couch, David L. Seismicity and crustal structure of the east continent gravity high. M, 1986, University of Kentucky. 158 p.

Couch, Elton Leroy. Boron sorption and fixation by illites. D, 1967, University of Illinois, Urbana. 102 p.

Couch, Elton Leroy. Description of some sand bodies and related strata of the Pennsylvanian Lazy Bend Formation of Parker County, Texas. M, 1961, Texas Christian University. 100 p.

Couch, Herbert E. Sedimentary study of the Edwards Formation in Borden, Garza, and Scurry counties, Texas. M, 1950, Texas Tech University. 58 p.

Couch, Nathan Pierce, Jr. Metamorphism and reconnaissance geology of the eastern McDowell Moun-

tains, Maricopa County, Arizona. M, 1981, Arizona State University. 57 p.

Couch, Richard William. Gravity and structures of the crust and subcrust in the Northeast Pacific Ocean west of Washington and British Columbia. D, 1969, Oregon State University. 179 p.

Couch, Robert F. An environmental interpretation of the lower Freeport Limestone of eastern Ohio. M, 1971, Ohio State University.

Couchot, Michael Lee. Paleodrainage and lithofacies relationships of the Sharon Conglomerate of southern Ohio. M, 1972, Ohio University, Athens. 64 p.

Coughanowr, Christine. The effects of wind waves on sedimentation in northeastern Delaware Bay. M, 1986, University of Delaware.

Coughlan, James Patrick. Depositional environment and diagenesis of the Teapot Sandstone (Upper Cretaceous) Converse and Natrona counties, Wyoming. M, 1983, University of Wyoming. 116 p.

Coughlin, Mary St. Patrick. The development and relative importance of the Amherst, Ohio, sandstone quarries. M, 1965, Catholic University of America. 73 p.

Coughlin, Matthew Kent. The potential for recovery of residual oil by microbial enhanced oil recovery from the Southeast Vassar Vertz Sand Unit. M, 1988, University of Oklahoma. 97 p.

Coughlin, Robert Michael. Reefs and associated facies of the Onondaga Limestone (Middle Devonian), West-central New York. M, 1980, SUNY at Binghamton. 194 p.

Coughlin, Terry Lee. Geologic and environmental factors affecting groundwater in the Boon Limestone of north central Washington County, Arkansas. M, 1975, University of Arkansas, Fayetteville.

Coughlon, John Patrick. Stratigraphy and microfacies analysis of the Pennsylvanian System, Spike-S Ranch, El Paso and Hudspeth counties, Texas. M, 1983, University of Texas at El Paso. 125 p.

Coughran, Theodore. The geology of the Doss North area, Mason and Gillespie counties, Texas. M, 1959, Texas A&M University.

Coulbourn, W. T. Sedimentology of Kahana Bay, Oahu, Hawaii. M, 1971, University of Hawaii. 141 p.

Coulbourn, W. T. Tectonics and sediments of the Peru-Chile Trench and continental margin at the Arica Bight. D, 1977, University of Hawaii. 243 p.

Coulomb, Jean-Jacques. Paragenèse des assemblages quartzo-feldspathiques dans les porphyres de l'Intrusion de Big Ben, Montana, U.S.A. M, 1972, Universite de Montreal.

Coulson, Francis M. and Rosenberg, Louis J. The formation and detailed description of a portion of Wind Cave (South Dakota). M, 1958, South Dakota School of Mines & Technology.

Coulson, Otis B. Geology of the Sweetwater Drive area, and correlation of the Santa Cruz Valley gravels. M, 1950, University of Arizona.

Coulter, David H. Petrology and stratigraphy of the Wadi Natash Volcanics, Eastern Desert, Egypt. M, 1981, Bryn Mawr College. 73 p.

Coulter, Henry W., Jr. Geology of the southeast portion of the Preston Quadrangle, Idaho-Utah. D, 1954, Yale University.

Coulter, Lynne E. Core and well log study; Medina Formation, Chautauqua Co., N.Y. M, 1981, SUNY, College at Fredonia.

Coulter, R. E. An analysis of the thermal structure within the Gulf Stream's shoreward boundary and inshore waters. M, 1977, Texas A&M University.

Coulthard, Dale Eugene. Nearshore sediments off Galveston Island and jetty system, Texas. M, 1977, Texas A&M University.

Coultrip, Robert. Regional gravity anomalies of the Four Corners states. M, 1982, University of Texas at El Paso.

Council, Edward Augustus, III. Provenance of sands within the lower Chesapeake Bay based on ilmenite composition. M, 1987, Old Dominion University. 116 p.

Council, Konrad Koert. Structure and petrography of the Precambrian rocks of the Yearling Head Mountain area, southern Llano County, Texas. M, 1972, Texas Christian University.

Council, Konrad Koert. The geology and petrography of Paisano Peak, Brewster County, Texas. M, 1972, Texas Christian University. 72 p.

Councill, Richard Jefferson. Mineralogy and particle attributes of Neuse River (North Carolina) sands. M, 1956, University of North Carolina, Chapel Hill. 40 p.

Countryman, Robert Loren. The subsurface geology, structure, and mineralogy of the Billie borate deposit, Death Valley, California. M, 1978, University of California, Los Angeles.

Counts, C. David. Design of transmitted light simple shear deformation apparatus with three dimensional boundary conditions. M, 1987, Michigan Technological University. 106 p.

Coupal, Frank Edward. Stratigraphy and paleontology of a well core from Sharon Township, Washtenaw County, Michigan. M, 1954, University of Michigan.

Couples, Gary Douglas. Analytical solutions to selected boundary value problems and their application to Rocky Mountain foreland deformation. M, 1977, Rice University. 49 p.

Couples, Gary Douglas. Kinematic and dynamic considerations in the forced folding process as studied in the laboratory (experimental models) and in the field (Rattlesnake Mountain, Wyoming). D, 1986, Texas A&M University. 335 p.

Courchesne, François. Mechanisms regulating sulfate movement in some Podzols from Quebec. D, 1988, McGill University.

Courdin, James L. Petrology of the Fort Union Formation (Upper Cretaceous and Paleocene), Wind river basin, Wyoming. M, 1969, University of Missouri, Columbia.

Couri, Clay C. Temperature related shell variation in Holocene Gastrocopta procera (Gastropoda; Pulmonata) from the continental United States; a biometric analysis. M, 1976, Kent State University, Kent. 51 p.

Court, James Edward. Geology of the southern part of the Guinda Quadrangle, Yolo County, California. M, 1966, University of California, Davis. 80 p.

Courten, Frank L. De *see* De Courten, Frank L.

Courtier, William H. Physiography and geology of south-central Kansas. D, 1934, University of Kansas. 94 p.

Courtier, William Henry. Magnetometric investigation of gold placer deposits on a portion of Clear Creek basin (T512). M, 1928, Colorado School of Mines. 16 p.

Courtillot, Vincent E. Inverse filtering of marine magnetic anomalies. M, 1972, Stanford University.

Courtis, David Michael. Geology of the Cutoff Mountain area, Park County, Montana. M, 1965, University of Michigan. 67 p.

Courtney, Cecil. A geologic study of the Krotz Springs oil field, Louisiana. M, 1950, University of Houston.

Courtney, P. S. Geology of the Culberson Quadrangle, Georgia-North Carolina. M, 1979, University of Georgia.

Courtright, Kelly Dean. A proposed property evaluation format to rapidly compare surface coal properties. M, 1982, University of Idaho. 73 p.

Courtright, T. R. Geology of the upper Boxelder Creek area, Larimer County, Colorado. M, 1974, University of Colorado.

Cousens, Ellis E. Jack tests and free-cylinder tests on anisotropic rocks, expected relationships among results. M, 1981, Rensselaer Polytechnic Institute. 46 p.

Couser, Chester Wendall. Paleozoic stratigraphy and structure in the Minnesota River valley. D, 1934, University of Iowa. 32 p.

Couser, Chester Wendall. The geology of the Wauzeka Quadrangle north of the Wisconsin River, Wisconsin. M, 1931, University of Iowa. 79 p.

Cousineau, Pierre. Stratigraphie et faciés des andésites Amulet près de la mine Norbec, Rouyn-Noranda, Québec. M, 1980, Universite du Quebec a Chicoutimi. 104 p.

Cousineau, Pierre A. Analyse tectonostratigraphique d'une partie de la Zone de Dunnage à l'est de la Rivière Chaudière, Québec. D, 1988, Universite Laval. 271 p.

Cousineau, Yvon. Newboro and Eagle Lake titaniferous magnetic deposits. M, 1940, Queen's University. 43 p.

Cousino, Matthew A. Pore geometry and reservoir characteristics of the Kirkwood Sandstone (Mississippian) in southern Illinois. M, 1986, University of Toledo. 163 p.

Cousins, Noel Boyd. Relationship of black calcite to gold and silver mineralization in the Sheep Tanks mining district, Yuma County, Arizona. M, 1972, Arizona State University. 43 p.

Cousminer, Harold Leopold. Devonian Chitinozoa and other palynomorphs of medial South America and their biostratigraphic value. D, 1964, New York University. 304 p.

Cousminer, Harold Leopold. Polymorphism in an Operculina population from the Paleocene of Saudi Arabia. M, 1956, New York University.

Cousser, Kurt H. de *see* de Cousser, Kurt H.

Couto dos Anjos, Sylvia Maria. Depositional environment, diagenetic history, and reservoir geology of the Santiago Member sandstones of the Pojuca Formation (Lower Cretaceous) in the Araçás oil field, Recôncavo Basin, Brazil. D, 1987, University of Illinois, Urbana. 265 p.

Coutren, Lewis Anderson Van *see* Van Coutren, Lewis Anderson

Couture, Diane. Interprétation de diagraphies électromagnétiques T.B.F. M, 1983, Ecole Polytechnique. 127 p.

Couture, Jean-François. Géologie de la Formation de Gilman dans la partie centrale du Canton de Roy, Chibougamau, Québec. M, 1987, Universite du Quebec a Chicoutimi. 138 p.

Couture, Rex A. Synthesis of some clay minerals at 25°C; palygorskite and sepiolite in the oceans. D, 1977, University of California, San Diego. 259 p.

Couvering, John Anthony Van *see* Van Couvering, John Anthony

Couvering, Martin Van *see* Van Couvering, Martin

Covaleski, Ann M. Serpentinization of olivine by diffusion of water. M, 1981, University of Cincinnati. 38 p.

Coveney, Raymond Martin. Quartz as a geobarometer. M, 1968, University of Michigan.

Coveney, Raymond Martin, Jr. Gold mineralization at the Oriental Mine, Alleghany, California. D, 1972, University of Michigan.

Covert, John Joseph. Origin of skarn-forming fluids and wallrock/skarn interaction at CanTung, Northwest Territories, Canada. M, 1983, University of Utah. 84 p.

Covert, Linda Lee. A gravity and tectonic study of Trans-Pecos Texas. M, 1976, University of Kentucky. 64 p.

Covey, Curtis E. A geologic study of the Zenith-Peace Creek Mississippian gas field, Reno County, Kansas. M, 1985, Wichita State University. 72 p.

Covey, Michael Conrad. Sedimentary and tectonic evolution of the western Taiwan Foredeep. D, 1984, Princeton University. 152 p.

Covey, W. P., III. Late Cretaceous to Quaternary sedimentary development of the Lord Howe Rise and the Dampier Ridge. M, 1973, University of Hawaii. 67 p.

Covington, Daniel Joseph. Mangrove peats of Belize, Central America. D, 1988, Texas A&M University. 89 p.

Covington, George H., III. Geology of Powder Wash oil and gas field, Moffat County, northwestern Colorado. M, 1967, Colorado School of Mines. 124 p.

Covington, Morton Douglas. A gravity study of Hill County, Texas. M, 1984, Southern Methodist University. 46 p.

Covington, Sidney L., Jr. Hydrothermal alteration of the Iron Queen igneous body, Blawn Mountain, Utah. M, 1973, Florida State University.

Covington, Thomas. Diagenetic history and evolution of porosity of the Cotton Valley Limestone, southeastern Smith County, Texas. M, 1985, Texas A&M University.

Cowan, A. Gordon and Pontius, David C. The geology of a portion of Crescent Valley and Hilltop quadrangles, Nevada. M, 1950, University of California, Los Angeles.

Cowan, C. A. "Coontail" fluorite rhythmites of the Cave-in-Rock District, southern Illinois; evidence for spelean hydrothermal mineralization. M, 1985, University of Michigan. 32 p.

Cowan, Darrel Sidney. Petrology and structure of the Franciscan Assemblage (Jurassic and Cretaceous) northwest of Pacheco Pass, California. D, 1972, Stanford University. 74 p.

Cowan, Ellen Anne. Deposition of bottom sediment in Lake Ellyn, Glen Ellyn, Illinois. M, 1982, Northern Illinois University. 109 p.

Cowan, Ellen Anne. Sediment transport and deposition in a temperate glacial fjord, Glacier Bay, Alaska. D, 1988, Northern Illinois University. 432 p.

Cowan, Jack Vincent. The effect of a free gas saturation on recovery efficiency of an LPG flood. M, 1958, University of Oklahoma. 76 p.

Cowan, John Christopher. Nickeliferous peridotite in Preissac and La Motte townships, northwestern Quebec. M, 1960, University of Western Ontario. 195 p.

Cowan, John R. Ordovician and Silurian stratigraphy in the Interlake area, Manitoba. M, 1977, University of Manitoba. 73 p.

Cowan, Kenneth Lee. The fractal dimension as a petrophysical parameter. M, 1987, University of Texas, Austin. 147 p.

Cowan, Michael F. Elemental patterns revealed by analyses of ores; a basic geochemical approach to the study of ore deposits. M, 1966, Queen's University. 139 p.

Cowan, Patricia. The gold content of interflow metasedimentary rocks in the Red Lake area. M, 1979, McMaster University. 121 p.

Cowan, W. R. Stratigraphy and quantitative analysis of Wisconsinan tills, Brantford-Woodstock area, Ontario, Canada. D, 1975, University of Colorado. 253 p.

Coward, Julian Michael Henry. Paleohydrology and streamflow simulation of three karst basins in southeastern West Virginia, U.S.A. D, 1975, McMaster University. 394 p.

Coward, Robert Irvin. Geology of the Buist Quadrangle, southeastern Idaho. M, 1979, Bryn Mawr College. 30 p.

Coward, Robert Irvin. Structural geology, stratigraphy and petrology of the Elkhorn Ridge Argillite, in the Sumpter area, northeastern Oregon. D, 1983, Rice University. 195 p.

Cowart, Jack H. Ichnology and depositional environments of the Athabasca Oil Sands, Steepbank River area (McMurray Formation). M, 1983, University of Georgia.

Cowart, James B. ^{234}U and ^{238}U in the Carrizo Sandstone aquifer of South Texas. D, 1974, Florida State University.

Cowart, James Bryant. Thermoluminescence, radioactivity, and time since crystallization of fluorite. M, 1962, University of Florida. 39 p.

Cowell, Peter F. Structure and stratigraphy of part of the North Fish Creek Range, Eureka County, Nevada. M, 1987, Oregon State University. 96 p.

Cowell, R. C. Sedimentology and scanning electron microscope study of the loess and related sediments along Nonconnah Creek, Memphis, Tennessee. M, 1977, Memphis State University.

Cowen, James Prather. Fe and Mn depositing bacteria in open ocean pelagic environments. D, 1983, University of California, Santa Cruz. 160 p.

Cowen, Michael Terence. Petrography of the volcanic rocks of the Jefferson Island Quadrangle, Montana. M, 1958, Indiana University, Bloomington. 57 p.

Cowen, Rachel. Characterization and genesis of palagonite and authigenic mineralization, Hanaoma Bay and Koko Crater, Oahu, Hawaii. M, 1988, University of New Mexico. 237 p.

Cowen, William F. Available phosphorus in urban runoff and Lake Ontario tributary waters. D, 1974, University of Wisconsin-Madison. 309 p.

Cowie, Roger H. Geology of the Zumbro Valley region. D, 1941, University of Minnesota, Minneapolis. 125 p.

Cox, Allan Verne. The remanent magnetization of some Cenozoic volcanic rocks. D, 1959, University of California, Berkeley. 193 p.

Cox, Benjamin B. Stratigraphy and structure in the area about Princeton, Kentucky. M, 1922, University of Chicago. 131 p.

Cox, Billie Lea. Stress modeling of the Nazca Plate; advances in modeling ridge-push and slab-pull forces. M, 1983, University of Arizona. 108 p.

Cox, Brett Forrest. Stratigraphy, sedimentology, and structure of the Goler Formation (Paleocene), El Paso Mountains, California; implications for Paleogene tectonism on the Garlock fault zone. D, 1982, University of California, Riverside. 296 p.

Cox, Bruce Ellis. The petrogenesis of orbicular migmatites bordering the Idaho Batholith, Shoup, Idaho. M, 1973, University of Montana. 58 p.

Cox, Charles L., Jr. Pleistocene terraces of the lower Brazos River, Texas. M, 1950, Louisiana State University.

Cox, Dennis Purver. Geology of the Helena Quadrangle, Trinity County, California. D, 1956, Stanford University. 147 p.

Cox, Doak. Studies of tsunamis. D, 1965, Harvard University.

Cox, Edwin Payne. A contribution to the technique of studying and making inorganic correlations of sedimentary subsurface formations. D, 1926, University of Oregon. variously paginated p.

Cox, Eugene W. Geology and gold-silver deposits in the San Jose mining district, southern San Mateo Mountains, Socorro County, NM. M, 1985, New Mexico Institute of Mining and Technology. 82 p.

Cox, G. L. The effects of smelter emissions on the soils of the Sudbury area. M, 1975, University of Guelph.

Cox, Gary Wriston. Some Carboniferous Paleosol characteristics in a modeled fluvio-deltaic depositional framework and similar Recent fluvial soil processes. M, 1978, North Carolina State University. 106 p.

Cox, Gordon F. N. Brine drainage in sodium chloride ice. D, 1974, Dartmouth College. 179 p.

Cox, Gordon, F. N. Variation of salinity in the multi-year sea ice at the AIDJEX '72 camp. M, 1972, Dartmouth College. 37 p.

Cox, Gregory M. Bedrock geology of the Connor Creek area, Baker County, Oregon. M, 1977, Oregon State University. 116 p.

Cox, Guy H. The origin of the lead and zinc ores of the Upper Mississippi Valley. D, 1911, University of Wisconsin-Madison.

Cox, Herbert M. The combined effect of crystal structure and environment on crystal growth. M, 1971, Ohio State University.

Cox, Herbert Michael. The measurement of crystal-solution-vapor contact angles and the growth rates of corresponding crystal facies. D, 1973, Ohio State University.

Cox, Hollace Lawton. An electron diffraction study of helium, neon, argon, krypton, and xenon. D, 1967, Indiana University, Bloomington. 111 p.

Cox, J. Michael. Investigation of late Tertiary to Recent movement along the Kentucky River fault system in Northwest Madison and Southeast Jessamine counties, Kentucky. M, 1983, Eastern Kentucky University. 87 p.

Cox, Jason Charles. Stratigraphic, economic, and environmental analysis of the "Clinton" sandstone, southern Portage County, Ohio. M, 1980, Kent State University, Kent. 119 p.

Cox, Jeffrey Martin. The isotopic composition of strontium in sediment from cores 1474P and 1445P and in a water sample from the Black Sea. M, 1971, Ohio State University.

Cox, John. Crustal stretching and subsidence in the northeastern Gulf of Mexico. M, 1986, University of Houston.

Cox, John. The geology of the northwest margin of the Nelson Batholith, British Columbia. M, 1979, University of Alberta. 97 p.

Cox, John Waldron. Geology and mineralization of the Atlanta District, Lincoln County, Nevada. M, 1981, University of Nevada. 83 p.

Cox, Kathleen L. Oil and gas potential of portions of Larimer and Weld counties, Colorado. M, 1973, Colorado School of Mines. 87 p.

Cox, Larry Paul. Numerical simulation of the final stages of terrestrial planet formation. D, 1978, Massachusetts Institute of Technology. 174 p.

Cox, Martin L. A study of the massive-bedded tuffs of the Padre Miguel Group, southeastern Guatemala, C.A. M, 1981, University of Texas, Arlington. 105 p.

Cox, Norman J. The geology of a portion of the Well Spring Quadrangle, Campbell and Claiborne counties, Tennessee. M, 1962, University of Tennessee, Knoxville. 34 p.

Cox, Randel T. Style and timing of movement along the Washita Valley fault, Murray County, Oklahoma. M, 1987, University of Arkansas, Fayetteville.

Cox, Raymond Lee. Age and correlation of the Sooke Formation with a section on its palynology. M, 1962, University of British Columbia.

Cox, Ricky G. Use of remote sensing techniques and statistics in petroleum exploration; Permian Basin, Texas. M, 1982, Texas Tech University. 81 p.

Cox, Robert Gerald. Sub-surface stratigraphy and general ground-water resources of Kossuth County, Iowa. M, 1956, University of Iowa. 143 p.

Cox, Robert Sayre. Distributions and ecology of encrusting-cheilostome bryozoans of Enewetak Atoll. M, 1983, Pennsylvania State University, University Park. 256 p.

Cox, Roy Edwin. Subsurface geochemical exploration of stratabound copper in Lower Permian redbeds in north-central Oklahoma. M, 1978, Oklahoma State University. 117 p.

Cox, Rulon Walter. High school teachers' guide to the geology of Washington County, Utah. M, 1967, University of Utah. 89 p.

Cox, Simon Jonathan David. An experimental study of shear fracture in rocks. D, 1987, Columbia University, Teachers College. 210 p.

Cox, Thomas Philip. Evaluation of various processing techniques applied to seismic data recorded from an ice island in the Arctic Ocean, Canada. M, 1987, University of Western Ontario. 220 p.

Cox, Timothy L. A structural study of the serpentinites of the Benton Uplift, Saline County, Arkansas. M, 1986, University of Arkansas, Fayetteville.

Cox, W. E. Ground water management in Virginia; a comparative evaluation of the institutional frame-work. D, 1976, Virginia Polytechnic Institute and State University. 402 p.

Cox, Walter M. Volumetric relations in stratigraphic units. M, 1960, Northwestern University.

Cox, William B. Survey of the Ogallala Formation in eastern New Mexico. M, 1950, Texas Tech University. 35 p.

Cox, William Edgerton. The geology of an area of approximately five square miles west of Austin, Texas. M, 1934, University of Texas, Austin.

Cox, Williard E. A subsurface study of certain Ordovician formations in the Redfield gas storage area, Dallas County, Iowa. M, 1956, University of Nebraska, Lincoln.

Coxe, Berton Woodward. The Virginius vein ore deposit, northwestern San Juan Mountains, Colorado; a study of the mineralogy, structure, and fluid inclusions of an epithermal base-metal and silver vein in a volcanic environment. M, 1985, University of New Mexico. 126 p.

Coxe, Cynthia Louise. Pennsylvanian (lower Desmoinesian) fluvial-deltaic depositional systems in the Taos Trough, northern New Mexico. M, 1981, University of Texas, Austin. 108 p.

Coxon, Donald Allan. A study of the use of natural gas storage reservoirs in the United States and their possible use in the Pacific Northwest. M, 1956, University of Washington. 116 p.

Coyle, David Alexander. Glaciotectonic folding of a chalk-diamicton sequence from Hvideklint, Mon, Denmark. M, 1986, University of Windsor. 116 p.

Coyle, James L. A geochemical investigation of Cretaceous volcanics of western Ecuador. M, 1982, University of Kentucky. 155 p.

Coyle, John M. Bedrock and surficial geology of the San Jose-Milpitas foothills area, Santa Clara County, California. M, 1985, San Jose State University. 234 p.

Coyle, Lynn A. The application of borehole gravimetry to remote sensing of anomalous masses. M, 1976, Purdue University. 89 p.

Coyne, Carmel Anne. Late Albian dinoflagellate cysts and acritarchs from Saskatchewan, Canada. M, 1986, University of Saskatchewan. 274 p.

Coyne, John C. A photoelastic study of slip on an artificial fault in a multilayered medium. M, 1979, Texas A&M University.

Coyner, Karl B. Effects of stress, pore pressure, and pore fluids on bulk strain, velocity, and permeability in rocks. D, 1984, Massachusetts Institute of Technology. 361 p.

Coyuran, Vedat. Ground-water resources of Ogden Valley. D, 1972, University of Utah. 134 p.

Cozza, Leonard Martin. Low temperature ashing of sub-bituminous coals from the Little Dirty coal seam, Centralia, Washington. M, 1984, Washington State University. 31 p.

Cozzens, Arthur B. Stratigraphy and petrology of the Kinderhook formations near St. Albans. M, 1929, Washington University. 75 p.

Cozzens, Arthur B. The natural regions of the Ozarks. D, 1937, Washington University.

Cozzens, James Robert. A shear vertical seismic profile in Dawson County, Texas. M, 1988, University of Wyoming. 113 p.

Crabaugh, Jeff Patrick. Depositional history of the Lower Permian Abo Formation and lower Meseta Blanca Member of the Yeso Formation, north central New Mexico. M, 1988, University of Texas, Arlington. 214 p.

Crabb, John Johnson. Stratigraphy and structure of the flatland area of southeastern BC. M, 1951, University of British Columbia.

Crabtree, Billy J. Lithostratigraphy and depositional environments in the upper Bloyd and lower Atoka Formation in SE Johnson, W. Pope, E. Logan and N. Yell counties, Arkansas. M, 1980, University of Arkansas, Fayetteville.

Crabtree, David Rockwell. The early Campanian flora of the Two Medicine Formation, northcentral Montana. D, 1987, University of Montana. 357 p.

Crabtree, Harry Thomas. A geological model of the Innes Oilfield; Mississippian. M, 1982, University of Regina. 166 p.

Crabtree, Sterling James, Jr. Algorithmic development of a petrographic image analysis system. D, 1983, University of South Carolina. 129 p.

Craddock, Greg F. Comparative sedimentology, calcarenites of the Wapanucka Formation and Recent carbonate sands of the Florida Reef Tract. M, 1983, Baylor University. 208 p.

Craddock, J. Campbell. Preliminary report on structural and stratigraphic studies in the Kinderhook Quadrangle, New York. M, 1953, Columbia University, Teachers College.

Craddock, John Paul. Deformation history of the Prospect thrust plate, Overthrust Belt, Northwest Wyoming. M, 1983, University of Michigan.

Craddock, John Paul. The evolution of the Idaho-Wyoming fold-and-thrust belt along latitude 42°45′. D, 1988, University of Michigan. 288 p.

Craddock, Robert Anthony. High resolution thermal infrared mapping of Martian outflow and fretted channels. M, 1987, Arizona State University. 100 p.

Craddock, William Percival. Areal geology of the Carrizo Sandstone at Bastrop, Bastrop County, Texas. M, 1947, University of Texas, Austin.

Crafford, T. C. SO₂ emission of the October-November 1974 eruption of Volcan Fuego, Guatemala. M, 1975, Dartmouth College. 56 p.

Craft, Helen Faith. Geological field study of North Bull Pasture Mountain, Virginia. M, 1975, Virginia State University. 35 p.

Craft, J. L. Pleistocene local glaciation in the Adirondack Mountains, New York. D, 1976, University of Western Ontario. 226 p.

Craft, James Homer. Wave-dominated marine shelf deposition in the Upper Devonian of south-central New York. M, 1985, SUNY at Binghamton. 136 p.

Craft, Jesse Leo, Jr. The Falkirk-Fiddlers Green Member of the Bertie Formation (Silurian) in central New York. M, 1963, Syracuse University.

Crafts, Anne S. Geochemical and mineralogical characterization of surficial residuum formed from lower Knox Group dolostone, West Chestnut Ridge, Oak Ridge, Tennessee. M, 1987, University of Tennessee, Knoxville. 121 p.

Crafts, Frederick S. A torsion microbalance for weighing living ostracods. M, 1960, University of Michigan.

Crafts, Shirley. Origin of the feldspar pegmatites near Batchellerville, Saratoga County, New York. M, 1947, Cornell University.

Cragin, Francis W. The paleontology of the Malone Jurassic Formation of Texas. D, 1899, The Johns Hopkins University.

Craig, Alan S. The geology and geochemistry of the Proterozoic metavolcanic rocks of the Green Mountain Formation, Sierra Madre Range, Wyoming. M, 1982, New Mexico Institute of Mining and Technology.

Craig, Bruce Gordon. Pleistocene deposits of Mariposa Township, Victoria County, Ontario. M, 1950, University of Michigan.

Craig, Bruce Gordon. Surficial geology of the Drumheller area, Alberta, Canada. D, 1956, University of Michigan. 165 p.

Craig, Christy. Geology and Eocene sedimentology of the southwestern portion of the Camp Pendleton Marine Base, San Diego County, California. M, 1984, San Diego State University. 155 p.

Craig, Daniel J. Predictive models for assessing the effects of land disposal wastewater systems on groundwater. M, 1987, University of Wisconsin-Madison. 141 p.

Craig, Dexter Hildreth. Stanolind stratigraphic test well, Gozar Quadrangle, Reeves County, Texas. M, 1952, University of Texas, Austin.

Craig, Douglas Bennell. Structure and petrology within the Shuswap Complex (Precambrian) Revelstoke, British Columbia. D, 1966, University of Wisconsin-Madison. 179 p.

Craig, Douglas E. The paleomagnetism of a thick middle Tertiary volcanic sequence in northern California. M, 1981, Western Washington University. 131 p.

Craig, Genevieve Susan. Holocene - sedimentation in a Pleistocene depression, South Florida. M, 1983, University of Miami. 120 p.

Craig, Gerald N., II. Depositional environments and petrology of the Hanna Formation, Hanna UCG Site, Wyoming. M, 1982, Colorado State University. 189 p.

Craig, Harmon B. The geochemistry of the stable carbon isotopes. D, 1951, University of Chicago. 135 p.

Craig, Harold Douglas. Intertidal animal and trace zonations in a macrotidal environment, Minas Basin, Bay of Fundy, Nova Scotia. M, 1977, McMaster University. 205 p.

Craig, Howard Reid. The geology of the Red Mountain area, Sharon, Connecticut. M, 1963, University of Cincinnati. 106 p.

Craig, J. D. Geological and geotechnical investigation of sediment redistribution on the central equatorial Pacific seafloor. D, 1979, University of Hawaii. 267 p.

Craig, J. D. The distribution of ferromanganese nodule deposits in the North Equatorial Pacific. M, 1975, University of Hawaii. 104 p.

Craig, James Roland. A systematic study of phase equilibria in the Ag-Bi-S system and exploration of the geologically significant portion of the Ag-Bi-Pb-S system. D, 1965, Lehigh University.

Craig, Lawrence C. Lower Middle Ordovician of south central Pennsylvania. D, 1949, Columbia University, Teachers College.

Craig, Lawrence C. Lower Mohawkian stratigraphy of central New York State. M, 1941, Columbia University, Teachers College.

Craig, Lisa Ellen. Silicification and porosity development in Lower Ordovician Knox Group carbonates, Burkesville, Kentucky. M, 1982, University of Texas, Austin. 104 p.

Craig, R. M. Morphology and genesis of soils formed in various parent materials on the degradation sideslope landforms, upper coastal plain, NC. M, 1963, North Carolina State University.

Craig, Richard Gary. A simulation model of landform erosion. D, 1979, Pennsylvania State University, University Park. 555 p.

Craig, Richard Gary. Comparison of patterns from Earth Resources Technology Satellite multispectral scanner and glacial drift, northwestern Pennsylvania. M, 1976, Pennsylvania State University, University Park. 378 p.

Craig, Richard Michael. Use of finite difference computer model to estimate dewatering needs and effects of a proposed underground uranium mine, Gas Hills, Wyoming. M, 1978, University of Arizona.

Craig, Robert R. Contact metamorphism of the Gilmore Stock, Lemhi Range, Lemhi County, Idaho. M, 1978, Idaho State University. 70 p.

Craig, Steven D. The geology, alteration, and mineralization of the Turquoise Lake area, Lake County, Colorado. M, 1980, Colorado State University. 188 p.

Craig, Ted William. Groundwater of the Uncompahgre Valley, Montrose County, Colorado. M, 1971, University of Missouri, Rolla.

Craig, Thomas Council. The mineralogy of the channel sands north of Columbia, Missouri. M, 1928, University of Missouri, Columbia.

Craig, William Warren. Aptian nonmarine ostracods of the subfamily Cyprideinae from the northern

Rocky Mountain area. M, 1961, University of Missouri, Columbia.

Craig, William Warren. The stratigraphy and conodont paleontology of Ordovician and Silurian strata, Batesville District, Independence and Izard counties, Arkansas. D, 1968, University of Texas, Austin. 422 p.

Craiglow, Carol Jean. Tectonic significance of the Pass Fault, central Bridger Range, Southwest Montana. M, 1986, Montana State University. 40 p.

Crain, Hugh F. Stratigraphy and geology of the Port Barre salt dome, St. Landry Parish, Louisiana. M, 1947, University of Kansas.

Crain, John Russell. Study of some X-ray properties of unheated and heated plagioclase feldspars. M, 1961, University of Nevada - Mackay School of Mines. 40 p.

Crain, William E. The areal geology of the Red Wing Quadrangle. M, 1957, University of Minnesota, Minneapolis. 105 p.

Cram, Ira H. A crustal structure refraction survey in South Texas. D, 1961, Rice University. 59 p.

Cram, Ira H. The Grassy Island Granite. M, 1924, University of Minnesota, Minneapolis. 45 p.

Cramer, C. H. Teleseismic travel time residuals applied to the question of P-velocity changes preceding earthquakes in central and northern California. D, 1976, Stanford University. 173 p.

Cramer, George H., II. The structural geology and igneous petrology of the Paleozoic rocks of the May Mountains, Belize (British Honduras). M, 1976, Louisiana State University.

Cramer, Howard Ross. Coral zones in the Mississippian of the Great Basin area. D, 1954, Northwestern University. 193 p.

Cramer, Howard Ross. Tertiary fresh water Ostracoda from the Uinta Basin, Utah. M, 1950, University of Illinois, Urbana. 31 p.

Cramer, Jane Alison Bremner *see* Bremner Cramer, Jane Alison

Cramer, John A., Jr. The Jurassic Ralston Formation in the southern Colorado Front Range. M, 1962, University of Kansas. 117 p.

Cramer, L. W. A statistical analysis of jointing in the Sheep Mountain-Jelm Mountain area, Albany County, Wyoming. M, 1962, University of Wyoming. 66 p.

Cramer, Richard Stanley. Petrographic, fluid inclusion, and light stable isotope study of the Gray Eagle Cu-Au massive sulfide deposit, Siskiyou County, California. M, 1982, University of California, Davis. 101 p.

Cramer, Ronald Thomas. Mineralogy of Wisconsin loess in southeastern South Dakota and northwestern Iowa. M, 1975, University of South Dakota. 58 p.

Cramer, Scott L. Paleocurrent analysis of the Upper Triassic sandstones of the Texas high plains. M, 1973, West Texas State University. 31 p.

Crampton, Janet Wert. Some late glacial sediments near South Hadley, Massachusetts. M, 1964, Mount Holyoke College. 69 p.

Crandall, Bradford G. Stratigraphy of the Buckhorn Sandstone (Upper Cretaceous), Santa Barbara and San Luis Obispo counties, California. M, 1961, University of California, Los Angeles.

Crandall, Gregory John. A marine seismic refraction study of the Santa Barbara Channel. M, 1982, University of California, Santa Barbara. 132 p.

Crandall, Hector. Geology of the La Jolla Quadrangle, San Diego County, California. M, 1916, [Stanford University].

Crandall, Robert. Diatoms and magnetic anisotropy as means of distinguishing glacial till from glaciomarine drift. M, 1979, Western Washington University. 62 p.

Crandall, Roderic. The geology of the San Francisco Peninsula. M, 1907, Stanford University. 53 p.

Crandall, Thomas M. Sedimentology and depositional environments of the Upper Cretaceous Fox Hills Formation in Eastcentral Wyoming. M, 1978, Rutgers, The State University, New Brunswick. 79 p.

Crandell, Dwight R. Geology of parts of Hughes and Stanley counties, South Dakota. D, 1951, Yale University.

Crane, Calista. Fossil Bryozoa. M, 1927, Boston University.

Crane, D. J. Structural reconnaissance of the San Diego mainland shelf. M, 1977, San Diego State University. 67 p.

Crane, David Clinton. Geology of the Jocotan and Timushan quadrangles, southeastern Guatemala. D, 1965, Rice University. 85 p.

Crane, James John. An investigation of the geology, hydrogeology, and hydrochemistry of the Lower Suwannee River basin. D, 1983, Florida State University. 398 p.

Crane, Kathleen. Hydrothermal activity and near-axis structure at mid-ocean spreading centers. D, 1977, University of California, San Diego. 297 p.

Crane, Marilyn Joyce. A new occurrence of Mississippian Ostracoda in Michigan. M, 1955, Michigan State University. 27 p.

Crane, Robert Lee. A study of the crystal structure of langbeinite. D, 1969, Ohio State University. 67 p.

Crane, Robert Lee. Phase relations in the system MgF_2-MgO-H_2O at one kilobar. M, 1966, Ohio State University.

Crane, Ronald Clinton. The influence of clay mineralogy on ceramic properties. D, 1960, Indiana University, Bloomington. 116 p.

Crane, Stephen Ernest. Structural chemistry of the marine manganate minerals. D, 1981, University of California, San Diego. 311 p.

Craney, Dana Leon. Distribution, structure, origin, and resources of the Hartshorne coals in the Panama Quadrangle, Le Flore County, Oklahoma. M, 1978, University of Oklahoma. 126 p.

Cranor, John Ira. Geochemistry of the Skarn Zones in the Sunrise Peak area, south of Brighton, Utah. M, 1973, Brigham Young University.

Cranor, John Ira. Petrology and geochemistry of the calc-silicate zone adjacent to the Alta and Clayton Peak stocks near Brighton, Utah. M, 1974, Brigham Young University. 176 p.

Cranston, Raymond Earle. Chromium species in natural waters. D, 1979, University of Washington. 312 p.

Cranstone, Donald A. Petrology of the Granville Lake Gabbro (Precambrian, Manitoba). M, 1964, University of Manitoba.

Cranstone, Donald Alfred. An analysis of ore discovery cost and rates of ore discovery in Canada over the period 1946 to 1977. D, 1982, Harvard University. 611 p.

Cranstone, John R. Quartz-sillimanite knots and metamorphism at Southern Indian Lake, Manitoba (Precambrian). M, 1971, University of Manitoba.

Cranswick, J. S. The coral fauna of the Abitibi River Limestone. M, 1953, University of Toronto.

Cranswick, Mark S. The stratigraphy, structure, and petrography of Keyes Mountain, Tertiary upper Clarno Formation, Wheeler County, Oregon. M, 1980, Oregon State University. 122 p.

Crass, David B. The stratigraphy, petrography, and diagenesis of the Frobisher-Alida interval, Mission Canyon Formation (Mississippian), northern Bottineau and Renville Counties, North Dakota. M, 1987, Baylor University. 93 p.

Cratsley, D. W. Recent deltaic sedimentation, Atchafalaya Bay, Louisiana. M, 1975, Louisiana State University.

Crattie, Thomas Bradford. A petrographic and geochemical study of coexisting limestones and dolostones from the Mascot-Jefferson City zinc district, East Tennessee, and the Right Fork area, central Tennessee. M, 1985, University of Tennessee, Knoxville. 140 p.

Cravens, Daniel Lester. Subsurface investigations for petroleum on the Emery Uplift, Southeast Utah. M, 1987, New Mexico State University, Las Cruces. 60 p.

Cravens, Stuart James. The hydrogeology and hydrochemistry of glacial till in Hand and Hyde counties, South Dakota. M, 1985, University of Toledo. 146 p.

Cravotta, Charles Angelo, III. Spatial and temporal variations of groundwater chemistry in the vicinity of carbonate-hosted zinc-lead occurrences, Sinking Valley, Blair County, Pennsylvania. M, 1986, Pennsylvania State University, University Park. 405 p.

Craw, David. Metamorphism, structure and stratigraphy of the Park Ranges (western Rocky Mountains), British Columbia. M, 1977, University of Calgary. 140 p.

Crawford, Adrian Mercer. Rate-dependent behaviour of rock joints. D, 1980, University of Toronto.

Crawford, Arthur L. A petrographic study of certain Pre-Cambrian rocks of Medicine Bow Mountains, Wyoming. M, 1926, Stanford University. 179 p.

Crawford, Charles Mark. Correlation of Pleistocene deposits along the Wisconsinan border in the Greenhills and Glendale quadrangles of southwestern Ohio. M, 1977, Miami University (Ohio). 148 p.

Crawford, D. Steven. Stratigraphy and faunal aspects of Upper Mississippian formations, eastern Idaho. M, 1976, University of Iowa.

Crawford, Frank C. The subsurface correlation and lithology of the James Limestone. M, 1950, Louisiana State University.

Crawford, Frank D. Facies analysis and depositional environments in the Middle Devonian Fort Vermillion and Slave Point formations of northern Alberta. M, 1972, University of Calgary. 99 p.

Crawford, Gayle Posey. Some ostracode species from the lower Glen Rose Formation of central Texas. M, 1942, University of Texas, Austin.

Crawford, George Allan. Depositional history and diagenesis of the Goat Seep Dolomite (Permian, Guadalupian), Guadalupe Mountains, West Texas - New Mexico. D, 1981, University of Wisconsin-Madison. 348 p.

Crawford, George Allan. Post-inundation depositional history of the Weaver Bottoms, Upper Mississippi River, based on cesium-137 data. M, 1976, University of Wisconsin-Madison.

Crawford, Harry Michael. Heavy minerals of Chatham Rise, New Zealand. M, 1976, University of Southern California.

Crawford, Ian Douglas. The geology and structure of the Lethbridge coal field. M, 1947, University of Alberta. 135 p.

Crawford, Jack W. Stratigraphy and sedimentology of the Tongue River Formation (Paleocene), southeast Golden Valley County, North Dakota. M, 1967, University of North Dakota. 73 p.

Crawford, James J., Jr. and Blair, Robert G. Geology of the magnetite deposits near Russell, Costilla County, Colorado. M, 1960, University of Michigan.

Crawford, John P. Alberta foothills structures. M, 1955, Columbia University, Teachers College.

Crawford, John P. Paleozoic facies from the miogeosynclinal to the eugeosynclinal belt in thrust slices, central Nevada. D, 1964, Columbia University, Teachers College.

Crawford, Joseph Edward. Kinetics of phase transformation phenomena in the system Fe-S-H_2 between 350°C and 400°C. M, 1981, Auburn University. 91 p.

Crawford, K. A. Sedimentology and tectonic significance of the Late Cretaceous-Paleocene Echo Canyon and Evanston synorogenic conglomerates of the North-central Utah thrust belt. M, 1979, University of Wisconsin-Madison.

Crawford, K. E. The geology of the Franciscan tectonic assemblage near Mount Hamilton, California. D, 1975, University of California, Los Angeles. 182 p.

Crawford, Kenneth Edgar. Petrology of the Hope Valley Roof Pendants, California. M, 1969, University of California, Davis. 81 p.

Crawford, Lisa Doris. Paleomagnetic dating of calcite speleothems in Arbuckle Group limestones, southern Oklahoma; a possible relationship between hydrocarbons and authigenic magnetite. M, 1987, University of Oklahoma. 75 p.

Crawford, María Luisa Busé Plagioclase feldspar equilibria in some semi-pelitic schists. D, 1965, University of California, Berkeley. 95 p.

Crawford, Mark Justin. Geology and lead-zinc mineralization of the sandstone-hosted Shawangunk Mine, New York, as compared to other lead-zinc sandstone deposits. M, 1982, University of Toronto.

Crawford, Michael F. The simultaneous use of magnetics and Landsat in mineral exploration. M, 1979, Stanford University. 41 p.

Crawford, Michael Francis. The simultaneous use of Landsat and geophysical data in exploration for non-renewable resources. D, 1981, Stanford University. 355 p.

Crawford, N. C. Subterranean stream invasion, conduit cavern development and slope retreat; a surface-subsurface erosion model for areas of carbonate rock overlain by less soluble and less permeable caprock. D, 1978, Clark University. 505 p.

Crawford, Ralph D. Geology and ore deposits of the Monarch and Tomichi districts, Colorado. D, 1913, Yale University.

Crawford, Ralph D. Geology and petrography of the Sugarloaf District, Boulder County, Colorado. M, 1907, University of Colorado.

Crawford, Samuel W. Distribution of certain elements in four areas of hydrothermal alteration. M, 1963, Pennsylvania State University, University Park. 148 p.

Crawford, Thomas Jones. The geology of the Indian Mountain area, Polk County, Georgia. M, 1957, Emory University. 57 p.

Crawford, William A. Effect of gamma radiation from black shales on the thermoluminescence of adjacent limestones. M, 1960, University of Kansas. 38 p.

Crawford, William Arthur. Studies in Franciscan metamorphism near Jenner, California. D, 1965, University of California, Berkeley. 142 p.

Crawford, William James Page. Geology of the Canada tungsten mines, SW District of Mackenzie, Canada. M, 1963, University of Washington. 81 p.

Crawford, William James Page. Metamorphic iron formations of Eqe Bay and adjacent parts of northern Baffin Island. D, 1973, University of Washington. 101 p.

Crawley, Mark Edward. A geochemical model for lithium and boron. M, 1977, Texas Tech University. 110 p.

Crawley, Richard Alvin. Flood effects on Sanderson creek, Trans-Pecos, Texas, June 11, 1965. M, 1969, University of Texas, Austin.

Crawley, Richard Alvin. Stratigraphy and sedimentology of the Cerro Grande Formation (upper Cretaceous), Parras Basin, northeastern Mexico. D, 1975, University of Texas, Austin. 312 p.

Cray, Edward Joseph, Jr. The geochemistry and petrology of the Morgan Creek Pluton, Inyo County, California. D, 1981, University of California, Davis. 454 p.

Creager, Kenneth Clark. Geometry, velocity structure, and penetration depths of descending slabs in the western Pacific. D, 1984, University of California, San Diego. 229 p.

Creager, Marcus Orange. Radium content of a core sample taken from East Sound as a function of particle size. M, 1948, University of Washington. 28 p.

Creagor, Joe Scott. Geomorphic interpretation of the bathymetry of the Bay of Campeche seaward of the continental shelf. M, 1953, Texas A&M University.

Creagor, Joe Scott. Marine geology of the Bay of Campeche. D, 1958, Texas A&M University.

Creamer, Frederick. A method to predict a large earthquake in an aftershock sequence. M, 1987, Georgia Institute of Technology. 55 p.

Creasey, Carol La Vopa. A laboratory and field study of sample bias introduced by soil water samplers. M, 1985, University of California, Santa Cruz.

Creasey, Cyrus. Geology of the Humboldt region and the Iron King mining district, Yavapai County, Arizona. D, 1949, University of California, Los Angeles.

Creasy, David Edward Jack. On-site evaluation of contaminant attenuation and remobilization properties of organic sediments. D, 1981, Queen's University. 346 p.

Creasy, John W. Mineralogy and petrology of the White Mountain Batholith, Franconia and Crawford Notch quadrangles, New Hampshire. D, 1974, Harvard University.

Crebs, Terry Joseph. Gravity and ground magnetic surveys of the central Mineral Mountains, Utah. M, 1976, University of Utah. 129 p.

Crecca, Arthur Joseph. DeQueen Limestone; petrology, ostracode, paleontology and depositional environments. M, 1977, University of New Orleans.

Crecelius, Eric A. The geochemistry of arsenic and antimony in Puget Sound and Lake Washington, Washington. D, 1974, University of Washington. 146 p.

Crecraft, Harrison Ruffin. Silicic volcanism at Twin Peaks, west-central Utah; geology and petrology, chemical and physical evolution, oxygen and hydrogen isotope studies. D, 1984, University of Utah. 240 p.

Cree, Allan. Tertiary intrusives in the Hessie-Tolland area, Boulder and Gilpin counties, Colorado. D, 1948, University of Colorado.

Cree, Susan Bentley. A biostratigraphic study of the asphalt-bearing limestones of Pennsylvanian age in the Arbuckle Mountains, Oklahoma. M, 1984, University of Texas, Arlington. 82 p.

Creed, Robert M. Barite-fluorite mineralization at Lake Ainslie, Inverness County, Nova Scotia (Canada). M, 1968, Dalhousie University. 158 p.

Creel, David Versal. Stratigraphy and petrography of the Hindsville Formation of Mayes County, Oklahoma. M, 1963, University of Tulsa. 103 p.

Creely, Robert Scott. Geology of the Oroville Quadrangle, California. D, 1955, University of California, Berkeley. 269 p.

Creger, Robert B. Stratigraphic analysis of the New Bielau-7500 foot Wilcox Field, western Colorado County, Texas. M, 1977, University of Houston.

Cregg, Allen Kent. Paleoenvironment of an upper Cotton Valley (Knowles Limestone) patch reef, Milam County, Texas. M, 1982, Texas A&M University. 144 p.

Creighton, Ann. Taphonomy and occurrence of early Eocene terrestrial gastropods from the Willwood Formation, Bighorn Basin, Wyoming. M, 1988, University of Colorado. 82 p.

Crelling, John Crawford. A petrologic study of a thermally altered coal from the Purgatoire River Valley of Colorado. M, 1967, Pennsylvania State University, University Park. 66 p.

Crelling, John Crawford. Petrology of the Bear Creek lamprophyre dike in the Spanish Peaks igneous complex, Colorado. D, 1973, Pennsylvania State University, University Park. 133 p.

Cremer, Edward A., III. Gravity determination of basement configuration, southern Deer Lodge Valley, Montana. M, 1966, University of Montana. 23 p.

Crempien, Jorge Laborie. A time-frequency evolutionary model for earthquake ground motion and structural response. D, 1988, University of California, Berkeley. 136 p.

Crepeau, Pierre M. Géologie de l'emplacement du barrage de Carillon. M, 1963, Ecole Polytechnique. 145 p.

Crerar, David Alexander. Solvation and deposition of chalcopyrite and chalcocite assemblages in hydrother-

mal solutions. D, 1974, Pennsylvania State University, University Park. 175 p.

Crerar, David Alexander. The solubility of quartz in dilute sodium hydroxide solutions at elevated temperatures and pressures. M, 1969, University of Toronto.

Crespi, Jean M. Strain studies in metamorphic core complexes of the Southwestern United States. D, 1985, University of Colorado. 320 p.

Crespi, Jean Marie. The relationship of cleavage in carbonate rocks to folding and faulting near Agua Verde Wash, Arizona; implications of volume loss. M, 1982, University of Arizona. 123 p.

Crespo, Esteban. Slope stability of the Cangahua Formation, a volcaniclastic deposit from the Interandean Depression of Ecuador. M, 1987, Cornell University. 146 p.

Cress, Leland D. Geology of the Carlin Window area, Eureka County, Nevada. M, 1972, San Jose State University. 103 p.

Cress, Leland D. Stratigraphy and structure of the south half of the Deep Creek Mountains, Oneida and Power counties, Idaho. D, 1981, Colorado School of Mines. 251 p.

Cressey, George Babcock. Some notes on the sand dunes of Northwest Indiana. M, 1921, University of Chicago. 63 p.

Cressey, George Babcock. The Indiana shore lines and sand dunes of Lake Michigan and its predecessors. D, 1923, University of Chicago.

Cresswell, Richard George. Radiocarbon dating of iron using accelerator mass spectrometry. M, 1987, University of Toronto.

Cressy, Frank B., Jr. Stratigraphy and sedimentation of the Neahkahnie Mountain- Angora Peak area, Tillamook and Clatsop counties, Oregon. M, 1974, Oregon State University. 148 p.

Creveling, J. B. Factors affecting the strength of subcoal materials and the prediction of strength from index properties. M, 1976, University of Missouri, Rolla.

Creveling, J. G. Silver-tin ores of Potosi, Bolivia. M, 1926, Massachusetts Institute of Technology. 36 p.

Creveling, Louis. Research report on air spring seismometer. M, 1958, Stanford University.

Crevello, Paul. Debris flow deposits and turbidites of a modern carbonate basin, Exuma Sound, Bahamas. M, 1978, University of Miami. 139 p.

Crevier, Michel. Pétrographie et géochimie de granitoïdes du socle du bassin Otish et estimation de leur préconcentration en uranium. M, 1981, Université du Quebec a Chicoutimi. 109 p.

Crewdson, Robert A. Geophysical studies in the Black Rock Desert geothermal prospect, Nevada. D, 1976, Colorado School of Mines. 292 p.

Crewdson, Robert A. The geology and mineral deposits of part of the Selenite Range, Pershing County, Nevada. M, 1974, Colorado School of Mines. 65 p.

Crews, A. L. Petrology of the Queen City Sandstone (Eocene); Leon County, Texas. M, 1975, [University of Houston].

Crews, Anita Lucille. Sedimentology of a Lower Cambrian marine shelf sequence; Zabriskie Quartzite, Saline Valley Formation, and related strata, southern Great Basin, U.S.A. D, 1980, University of California, Los Angeles. 163 p.

Crews, Gary Alan. The feasibility of detecting lithologic variations using magnetotelluric methods. M, 1972, University of Oklahoma. 48 p.

Crews, George. Geology of a part of northeast Moffat County, Colorado. M, 1963, Colorado School of Mines. 124 p.

Crews, Patricia Ann. Petrology and stratigraphic relations of a mafic volcanic complex and associated sediments of an ophiolite in southern Haiti. M, 1978, University of Florida. 82 p.

Cribb, Robert Eugene. Areal geology of the northern half of Calhoun Quadrangle, Georgia. M, 1953, Emory University. 47 p.

Crichlow, Henry Brent. Heat transfer in hot fluid injection in porous media. D, 1972, Stanford University. 223 p.

Crick, Rex E. Morphologic variations in the ammonite genus Scaphites of the Blue Hill Member, Carlile Shale, Upper Cretaceous, Kansas. M, 1976, University of Kansas. 87 p.

Crick, Rex Edward. Ordovician nautiloid biogeography; a probabilistic and multivariate analysis. D, 1978, University of Rochester. 166 p.

Crick, Richard Wayne. Stratigraphy and paleontology of a well core from Saybrook Township, Ashtabula County, Ohio. M, 1953, University of Michigan.

Crickmay, Colin H. The geology and paleontology of the Harrison Lake District, British Columbia, together with a general review of the Jurassic faunas and stratigraphy of western North America. D, 1925, Stanford University. 140 p.

Crickmay, Geoffrey W. The geology of the Matapedia River map-area, Quebec. D, 1930, Yale University.

Crider, Don. Structural and stratigraphic effects on geomorphology and groundwater movement in Knox Dolostone terrane. M, 1981, University of Tennessee, Knoxville. 162 p.

Crider, Richard L. Synthetic seismograms and character modeling; an aid to the determination of the Earth's structure from Rayleigh waves. M, 1980, Texas Tech University. 80 p.

Crider, Steven Snowden. Groundwater solute transport modeling using a three-dimensional scaled model. D, 1987, Clemson University. 249 p.

Crider, Val Joe. Petrographic analysis of Southern California sands and their relationship to plate tectonics. M, 1986, Bowling Green State University. 51 p.

Cridland, Arthur Albert. Amyelon iowense (Pierce and Hall) comb. nov., with notes on other cordaitean roots. D, 1961, University of Kansas. 174 p.

Crierson, James Douglas, Jr. Devonian lycopods of New York State. D, 1962, Cornell University.

Crifasi, Robert R. Statigraphy and alluvial architecture of Laramide orogenic sediments; Denver Basin, Colorado. M, 1988, University of Colorado. 89 p.

Crill, Patrick Michael. Methane production and sulfate reduction in an anoxic marine sediment. D, 1984, University of North Carolina, Chapel Hill. 200 p.

Cripe, Jerry Dale. The total sulfur content of lunar samples and terrestrial basalts. M, 1972, Arizona State University. 70 p.

Cripe, Jerry Dale. Total sulfur content and distribution in the lunar samples. D, 1976, Arizona State University. 170 p.

Crippen, Robert E. Landforms of the Frazier Mountain region, Southern California. M, 1979, University of California, Santa Barbara.

Criscenti, Louise Jacqueline. The origin of macrorhythmic units in the Stillwater Complex. M, 1984, University of Washington. 109 p.

Crisi, Peter A. An expert system to assist in processing vertical seismic profiles. M, 1988, Colorado School of Mines. 53 p.

Crisler, R. M. The detailed paleontology and stratigraphy of the Mississippian System of Lookout Mountain in Tennessee. M, 1954, Emory University. 69 p.

Crisman, David P. Groundwater geochemistry in the Portage Lake Volcanics, Hancock, Michigan, and implications for native copper stability. M, 1982, Michigan Technological University. 138 p.

Crisman, T. L. North Pond, Massachusetts; postglacial variations in lacustrine productivity as a reflection of changing watershed-lake interactions. D, 1977, Indiana University, Bloomington. 112 p.

Crisp, David. Radiative forcing of the Venus mesosphere. D, 1984, Princeton University. 208 p.

Crisp, Edward Lee. The skeletal trace element chemistry of freshwater bivalves. D, 1975, Indiana University, Bloomington. 185 p.

Crisp, Edward Lee. Variations in molluscan shell chemistry with environment, species, and time of burial. M, 1972, University of Kentucky. 139 p.

Crisp, Jeffery. Sources and distribution of coarse silt on the South Texas continental shelf. M, 1985, Texas A&M University.

Crisp, Joy Anne. The Mogan and Fataga formations of Gran Canaria (Canary Islands); geochemistry, petrology, and compositional zonation of the pyroclastic and lava flows; intensive thermodynamic variables within the magma chamber; and the depositional history of pyroclastic flow E/ET. D, 1984, Princeton University. 317 p.

Criss, Robert Everett. An $^{18}O/^{16}O$, D/H and K-Ar study of the southern half of the Idaho Batholith. D, 1981, California Institute of Technology. 415 p.

Crissman, Stephen C. Neuse River flood plain development, Johnston and Wayne counties, North Carolina. M, 1973, Duke University. 61 p.

Crissman, Susan Elizabeth. Controls on element distributions; Negaunee Iron-Formation, Empire Mine, Palmer, Michigan. M, 1988, Michigan State University. 82 p.

Crist, Claude Walker, Jr. A petrographic study of garnet outcrops in the Virginia Piedmont. M, 1959, University of Virginia. 109 p.

Crist, Elliott McDonald. Geology of the upper Lightning Creek and upper Jordan Creek area, Custer County, Idaho. M, 1978, University of Idaho. 91 p.

Crist, W. Konrad. Shorevol; a computer wave refraction-shoreline evolution model for coastal erosion studies. M, 1981, Rensselaer Polytechnic Institute. 189 p.

Cristil, Anita Ione. Analysis of the upper Lenoir and Holston formations deposited along a Middle Ordovician shelf reentrant, Rutledge Pike locality, Knoxville, Tennessee. M, 1985, University of Tennessee, Knoxville. 151 p.

Criswell, C. William. Chronology and pyroclastic stratigraphy of the May 18, 1980, eruption of Mount St. Helen's, Washington. M, 1987, University of New Mexico. 76 p.

Criswell, James Richard. A comparison of refraction and gravity data for determining bedrock elevations, Clare County, Michigan. M, 1986, Wright State University. 60 p.

Crites, William Henry. The subsurface geology of the New Harmony-Poseyville area, Posey County, Indiana. M, 1952, Indiana University, Bloomington. 29 p.

Crittenden, Max D., Jr. Geology of the San Jose-Mount Hamilton area, California. D, 1949, University of California, Berkeley. 155 p.

Crocetti, Charles Alfred. Isotopic and chemical studies of the Viburnum Trend lead ores of Southeast Missouri. D, 1985, Harvard University. 592 p.

Crock, James Gerard. Determination of trace amounts of lead in geological materials. M, 1975, Pennsylvania State University, University Park. 85 p.

Crock, James Gerard. The determination of selected rare earth elements in geological materials by inductively coupled argon plasma-optical emission spectroscopy. D, 1981, Colorado School of Mines. 169 p.

Crocker, James R. Prebatholithic geology of the Bahia Calamajue area, Baja California, Mexico. M, 1987, San Diego State University. 126 p.

Crocker, Jonathan A. Sediment deposition in Lake Pontchartrain from the 1973 Bonnet Carre Spillway operation. M, 1988, University of New Orleans.

Crocker, Marvin Carey, Jr. A comparison of the textural and mineralogical properties of river and beach sands in Southeast Texas. M, 1963, Texas A&M University. 98 p.

Crocker, Marvin Carey, Jr. The relationship of cation-adsorption capacity to exchange adsorption in selected feldspar minerals. D, 1966, Texas A&M University. 81 p.

Crocket, James H. A radiochemical measurement of the distribution coefficient of zinc ion between crystalline $(Ca,Zn)CO_3$ and its saturated aqueous solution. D, 1961, Massachusetts Institute of Technology. 174 p.

Crockett, Frederick James. A preliminary study of the use of oxygen isotope ratios as an exploration tool in the Park City District, Utah. M, 1971, University of Utah. 58 p.

Crockett, J. A. Carbonate diagenesis in holes 400A and 401, Deep Sea Drilling Project, The bay of Biscay. M, 1977, Memphis State University.

Crockett, N. E. Foraminifera of the type section of the Archusa Marl of Mississippi (Eocene, Claiborne Group, Wautubbee Formation). M, 1954, University of Missouri, Rolla.

Crockford, Michael Bertram Bray. A study of the Cache Creek Series, B.C. M, 1935, University of Alberta. 58 p.

Croes, Marc Kalman. Magnetostratigraphy of the Ankareh and Chinle formations. M, 1978, University of Utah. 226 p.

Croft, John S. Upper Permian conodonts and other microfossils from the Pinery and Lamar Limestone members of the Bell Canyon Formation and from the Rustler Formation, West Texas. M, 1978, Ohio State University.

Croft, Mack G. Geology of the northern Onaqui Mountains, Tooele County, Utah. M, 1956, Brigham Young University. 44 p.

Croft, Melvin. Ecology and stratigraphy of the echinoids of the Ocala Limestone. M, 1980, Florida State University.

Croft, Steven Kent. Impact craters from centimeters to megameters. D, 1979, University of California, Los Angeles. 282 p.

Croft, Wayne S. Deposition and diagenesis of the Jurassic, upper Smackover Formation, North Haynesville Field, northern Louisiana. M, 1980, Louisiana State University.

Croft, William J. An X-ray line study of uraninite. D, 1954, Columbia University, Teachers College.

Croley, Clifford W. A high resolution gravity analysis using low altitude spacecraft tracking data. M, 1985, University of Akron. 138 p.

Croll, Timothy Caryl. Stratigraphy and depositional history of the Deming sand in northwestern Washington. M, 1980, University of Washington. 57 p.

Crombie, G. P. A study of the insoluble residues of the Paleozoic rocks of southwestern Ontario. D, 1943, University of Toronto.

Crombie, G. P. The carbonated rocks of the Larder Lake area and their relation to the occurrence of gold. M, 1939, University of Toronto.

Crombie, Richard Howard. The stratigraphy of Newark, Madison, and part of Hanover townships, Licking County, Ohio. M, 1952, Ohio State University.

Crompton, James S. An active seismic reconnaissance survey of the Mount Princeton area, Chaffee County, Colorado. M, 1976, Colorado School of Mines. 61 p.

Cromwell, David Williams. The stratigraphy and environment of deposition of the lower Dornick Hills Group (lower Pennsylvanian), Ardmore Basin, Oklahoma. M, 1974, University of Oklahoma. 138 p.

Cromwell, John E. Processes, sediments, and history of Laguna Superior, Oaxaca, Mexico. D, 1975, University of California, San Diego. 177 p.

Cronce, Richard Charles. The genesis of soils overlying dolomite in the Nittany Valley of central Pennsylvania. D, 1988, Pennsylvania State University, University Park. 413 p.

Crone, A. J., Jr. Laboratory and field studies of mechanically infiltrated matrix clay in arid fluvial sediments. D, 1975, University of Colorado. 174 p.

Crone, Walter Richard. The use of Fe/Mn oxide-rich fracture coatings in the geochemical exploration for precious metal deposits; a comparison with standard rock and soil geochemistry. M, 1982, University of Nevada. 93 p.

Croneis, Carey G. Notes on the geology of Giles County, Virginia. M, 1923, University of Kansas.

Croneis, Carey G. The Fayetteville Formation; its fauna and stratigraphy (Arkansas). D, 1928, Harvard University.

Cronenwett, Charles Emanuel. A subsurface study of the Simpson Group in east central Oklahoma. M, 1956, University of Oklahoma. 63 p.

Cronin, Christopher. Stratigraphy and sedimentation of the Ravalli Group (middle Proterozoic Belt Supergroup) in the Mission, Swan, and Flathead Ranges, Northwest Montana. M, 1988, University of Montana. 246 p.

Cronin, Kenneth Stewart. An Edwards Georgetown erosional interval. M, 1932, University of Texas, Austin.

Cronin, Thomas Crawford. A study of the Silurian systems and a Silurian reef in west Texas and southern New Mexico. M, 1971, Texas Tech University. 121 p.

Cronin, Thomas Paul. Petrogenesis of the Webster-Addie ultramafic body, Jackson County, North Carolina. M, 1983, University of Tennessee, Knoxville. 112 p.

Cronin, Vincent S. The physical and magnetic polarity stratigraphy of the Skardu Basin, Baltistan, northern Pakistan. M, 1982, Dartmouth College. 226 p.

Cronin, Vincent Sean. Cycloid tectonics; a kinematic model of finite relative plate motion. D, 1988, Texas A&M University. 202 p.

Croninger, Adele Bullen. A geophysical analysis of the Southwestern Michigan peach landscape. M, 1948, Washington University. 208 p.

Cronk, Caspar. Glaciological investigations near the ice sheet margin, Wilkes Station, Antarctica. D, 1968, Ohio State University.

Cronk, R. J. Geology and mineralization of the Jefferson Mine, Hardin County, Illinois. M, 1951, University of Missouri, Rolla.

Cronkhite, George. The analytical composition of the calcareous oolite of the Great Salt Lake. M, 1936, University of Utah. 18 p.

Cronoble, James M. Geology of South Baggs-West Side Canal gas field, Carbon County, Wyoming and Moffatt County, Colorado. M, 1969, Colorado School of Mines. 46 p.

Cronoble, James M. Stratigraphy and petroleum potential of Dakota Group, North Park, Laramie, and Northwest Denver basins, Wyoming and Colorado. D, 1978, Colorado School of Mines. 503 p.

Cronoble, William R. Petrology of the Hogshooter Formation (Missourian), Washington and Nowata counties, Oklahoma. M, 1961, University of Oklahoma. 253 p.

Cronyn, Brian Sullivan. Underwater gravity survey of northern Monterey Bay. M, 1973, United States Naval Academy.

Crook, Maurice C. Formation and development of the Gulf of Mexico and the Louann Salt province. D, 1984, SUNY at Buffalo. 216 p.

Crook, Maurice Clifford. Formation and development of the Gulf of Mexico and the Louann salt province. D, 1985, University of Kansas. 229 p.

Crook, Stephen R. Structural geology of the northern part of Elkhorn Mountain, Bannock Range, Idaho. M, 1985, Utah State University. 79 p.

Crook, Theo Helsel. The geology of a portion of the Santa Ynez Mountains south of the town of Solvang, California. M, 1921, University of California, Berkeley. 37 p.

Crook, Wilson Walter, III. The geology, mineralogy and geochemistry of the rare-earth pegmatites, Llano and Burnet counties, Texas. M, 1977, University of Michigan.

Crooks, Deborah Marie. Petrology and stratigraphy of the Morrison (Upper Jurassic) and Cedar Mountain (Lower Cretaceous) formations, Emery County, Utah. M, 1986, Fort Hays State University. 133 p.

Crooks, Harold Fordyce. On the lower Paleozoic stratigraphy of the Northern Peninsula of Michigan. M, 1918, University of Illinois, Urbana.

Crooks, Thomas J. Water losses and gains across the Pahasapa Limestone (Lower Mississippian), Box

Elder Creek, Black Hills, South Dakota. M, 1968, South Dakota School of Mines & Technology.

Croonenberghs, Robert Emile. Organic toxic substances monitoring in Virginia. D, 1983, College of William and Mary. 344 p.

Croose, Daivd. Structure and stratigraphy of part of the French Mesa Quadrangle, Rio Arriba County, New Mexico. M, 1985, University of New Mexico. 91 p.

Cropp, Frederick William III. Pennsylvanian spore floras from the Warrior Basin, Mississippi and Alabama. M, 1956, University of Illinois, Urbana. 30 p.

Cropp, Frederick William, III. Pennsylvanian spore succession in Tennessee. D, 1958, University of Illinois, Urbana. 102 p.

Cropper, A. G. Mass physical properties of deep-sea sediments in the Hawaiian area. M, 1968, University of Hawaii. 74 p.

Cropper, John Philip. Tree-ring response functions; an evaluation by means of simulations. D, 1985, University of Arizona. 132 p.

Cropper, Wallace John. Survey of the epithermal precious metal deposits with particular reference to mineralogic and structural changes with depth. M, 1951, University of Michigan.

Crosbie, James Morton. Some foraminifera from the Greenhorn Formation and adjacent beds in south central Nebraska. M, 1941, University of Nebraska, Lincoln.

Crosby, Donald Gladstone, Jr. The Wolfville map-area, Kings and Hants counties, Nova Scotia. D, 1951, Stanford University. 187 p.

Crosby, E. C. Simulation of the effects of contour coal strip mining on stormwater response and suspended sediment yield in the New River watershed, Tennessee. D, 1979, University of Tennessee, Knoxville. 208 p.

Crosby, Eleanor J. A study of Pennsylvanian microorganisms. M, 1939, Northwestern University.

Crosby, Garth M. The mineral belts north of Osburn Fault, Coeur d'Alene District, Shoshone County, Idaho. M, 1959, University of Minnesota, Minneapolis. 39 p.

Crosby, Gary Wayne. Geology of the South Pavant Range, Millard and Sevier counties, Utah. M, 1959, Brigham Young University. 59 p.

Crosby, Gary Wayne. Structural evolution of the Middlebury Synclinorium, West-central Vermont. D, 1963, Columbia University, Teachers College. 136 p.

Crosby, George. Geology of Hancock County, Indiana for engineering and environmental purposes. M, 1982, Purdue University. 104 p.

Crosby, George S. Applications of electrical methods to groundwater development. M, 1952, Boston College.

Crosby, James Thompson. Stratigraphy beneath and geologic origin of the northern Florida Straits from recent multichannel seismic reflection data. M, 1980, University of Delaware.

Crosby, James Winfeld, III. Limestones in the Anarchist Series, Okanogan County, Washington. M, 1949, Washington State University. 15 p.

Crosby, Mary Francis. The urbanization of the River Des Peres. M, 1933, Washington University.

Crosby, Percy. Structure and petrology of the central Kootenay Lake area, British Columbia. D, 1960, Harvard University.

Crosby, R. M. Distribution of some trace metals in the secondary environment in southwestern New Brunswick. M, 1973, University of New Brunswick.

Crosby, Spurgeon C., II. The basement of North Dakota. M, 1958, University of North Dakota. 46 p.

Cross, Bradley D. Application of remote sensing to the groundwater hydrology of Big Bend National Park, Texas. M, 1984, Texas A&M University. 80 p.

Cross, C. H. The relationship of metals and soil organic components with particular reference to copper. D, 1975, University of British Columbia.

Cross, H. Hydrogeology, hydrochemistry and environmental impact of urbanization of the Birch Cove and Sackville areas, Halifax County, Nova Scotia. M, 1975, Dalhousie University. 231 p.

Cross, J. H. Elastic properties of rocks and minerals under high pressures and temperatures (C884). M, 1951, University of Texas, Austin.

Cross, Joseph William. Fracture detection in a volcanic oil reservoir using discriminant analysis of well log data. M, 1986, Old Dominion University. 113 p.

Cross, Ralph Herbert, III. Water surge forces on coastal structures. D, 1966, University of California, Berkeley. 109 p.

Cross, Scott Lewis. Submarine diagenesis in Lower Cretaceous coral-rudist reefs, Mural Limestone, southeastern Arizona. M, 1987, Texas A&M University.

Cross, T. A. Changing patterns of Cenozoic igneous activity in the western United States; relation to "absolute" North American Plate motion. D, 1976, University of Southern California.

Cross, Timothy Aureal. The Mississippian Lake Valley Formation of the Sacramento mountains, New Mexico; an environmental interpretation. M, 1970, University of Michigan.

Cross, Wayne A. Megascopic and microscopic petrography of the Francis (Hazard No. 8) coal seam in central eastern Kentucky. M, 1982, Eastern Kentucky University. 112 p.

Cross, Whitman, II. Groundwater resources of the western half of Albemarle County, Virginia. M, 1959, University of Virginia. 120 p.

Crossan, Arthur Brook, III. The Raritan River 1972; a study of the effect of the American Cyanamid Company on the river ecosystem. D, 1973, Rutgers, The State University, New Brunswick. 269 p.

Crossen, Kristine J. Glaciomarine deltas in southwestern Maine; formation and vertical movements. M, 1985, University of Maine. 124 p.

Crossey, Laura J. The origin and role of water soluble organic compounds in clastic diagenetic systems. D, 1985, University of Wyoming. 146 p.

Crossey, Laura Jones. Intraflow chemical variations in Servilleta basalts; Taos Plateau, New Mexico. M, 1979, Washington University. 105 p.

Crossley, D. J. Gravity and temperature measurement on the Fox glacier, Yukon. M, 1969, University of British Columbia.

Crossley, David John. Magnetoelastic interactions in the Earth's core. D, 1973, University of British Columbia.

Crossley, Robert W. A geological investigation of foundation failures in small buildings in Tucson, Arizona. M, 1969, University of Arizona.

Crosson, L. S. Stereorthophotos; soil-landform relationships-soil classification. D, 1972, University of Guelph.

Crosson, Robert Scott. Regional gravity survey of parts of western Millard and Juab counties, Utah. M, 1964, University of Utah. 34 p.

Crosson, Robert Scott. Seismic wave radiation from sources moving over a finite distance. D, 1966, Stanford University. 126 p.

Croteau, Paul. Dynamic interactions between floating ice and offshore structures. D, 1983, University of California, Berkeley. 355 p.

Crotty, Kevin J. Paleoenvironmental interpretation of ostracod assemblages from Watling's Blue Hole, San Salvador Island, Bahamas. M, 1982, University of Akron. 79 p.

Crouch, James H. Piezometric level and shallow aquifers, Dillon and Meramec Spring quadrangles, Missouri. M, 1965, University of Missouri, Rolla.

Crouch, James K. Marine geology and tectonic evolution of the northwestern margin of the California continental borderland. D, 1979, University of California, San Diego. 116 p.

Crouch, Michael Leonard. A clay mineral and detrital quartz assessment of the Heebner Shale (Pennsylvanian, Virgilian) of eastern Kansas. M, 1971, Wichita State University. 90 p.

Crouch, Robert Wheeler. Pliocene Ostracoda from Southern California. M, 1948, University of Southern California.

Crouch, Smith A. A study of selected megafaunas of the Santa Monica Mountains. M, 1930, University of Southern California.

Crouch, Steven L. The influence of failed rock on the mechanical behavior of underground excavations. D, 1970, University of Minnesota, Minneapolis.

Crouch, Walter H., Jr. A photometric determination of trace quantities of iodine in meteorites and rocks. M, 1962, University of Arkansas, Fayetteville. 46 p.

Crough, S. T. Thermal and mechanical studies of the Earth's lithosphere and asthenosphere. D, 1976, Stanford University. 153 p.

Crough, Sherman Thomas. A method for relocating seismic events using surface waves. M, 1973, University of Michigan.

Crous, Christiaan Mauritz. Computer-assisted interpretation of electrical soundings. M, 1971, Colorado School of Mines. 108 p.

Crouse, George W. Deformation of the Haley Creek Terrane, southern Alaska; Mesozoic transcurrent movement along the southern Alaska margin. M, 1985, Lehigh University.

Crouse, Harry Lynn. Chemical and mineralogical interaction of acid mine drainage with stream sediment, Babb Creek, Tioga County, Pennsylvania. M, 1977, Pennsylvania State University, University Park. 64 p.

Crout, R. L. Momentum balance on a shallow shelf; Moskito Bank, Nicaragua. M, 1978, Louisiana State University.

Crow, Billy B. Relation of gravity anomalies to the subsurface geology in northern Douglas County, Kansas. M, 1960, University of Kansas. 55 p.

Crow, Henry Clay. Geochemistry and origin of the late Archean-early Proterozoic volcanics of the Kaapvall Craton, South Africa. D, 1988, New Mexico Institute of Mining and Technology. 363 p.

Crow, Henry Clay, III. Geochemistry of shonkinites, syenites, and granites associated with the Sulfide Queen carbonatite body, Mountain Pass, California. M, 1984, University of Nevada, Las Vegas. 56 p.

Crow, Louis Milton. The geology of the western third of the Coldwater Quadrangle, Missouri. M, 1930, Washington University. 41 p.

Crow, Roxane J. Geotechnical characterization and environmental significance of expansive soils in Johnson County, Kansas. M, 1985, University of Missouri, Kansas City. 93 p.

Crow, Sidney Alfred, Jr. Microbiological aspects of oil intrusion in the estuarine environment. D, 1974, Louisiana State University.

Crowder, Rowley Keith. Anatomy of a Pennsylvanian fluvial sheet sandstone, Northwest Arkansas. M, 1982, University of Arkansas, Fayetteville.

Crowder, Rowley Keith. Latest Cretaceous sedimentation and paleogeography, Western Interior, U.S.A. D, 1983, University of Iowa. 427 p.

Crowder, W. T., Jr. Depositional environments of the Red Bluff Clay (Oligocene) in eastern Mississippi and southwestern Alabama. M, 1977, Memphis State University.

Crowe, Allan. Chemical and hydrological simulation of prairie lake-watershed systems. M, 1979, University of Alberta. 205 p.

Crowe, Allan S. A numerical model for simulating mass transport and reactions during diagenesis in clastic sedimentary basins. D, 1988, University of Alberta. 289 p.

Crowe, Bruce. The Ubehebe Craters, Northern Death Valley, (Inyo County) California. M, 1972, University of California, Santa Barbara.

Crowe, Bruce Mansfield. Cenozoic volcanic geology of the southeastern Chocolate Mountains, California. D, 1973, University of California, Santa Barbara.

Crowe, C. Transient heat flow methods for determining thermal constants. D, 1956, University of Western Ontario.

Crowe, Clifford T. The Vacherie salt dome, Louisiana and its development from a fossil salt high. M, 1975, Louisiana State University.

Crowe, Douglas E. Stratigraphy and geologic history, Bunces Key, Pinellas County, Florida. M, 1983, University of South Florida, Tampa. 75 p.

Crowe, Gregory George. Structural evolution of the Mackie plutonic complex, southern British Columbia. M, 1981, University of Calgary. 154 p.

Crowe, John. Mechanisms of heat transport through the floor of the Equatorial Pacific Ocean. D, 1981, Woods Hole Oceanographic Institution. 232 p.

Crowell, Catherine Shafer. The mineralogic analysis of the underclays of the lower Allegheny Formation in Jackson, Vinton, Hocking and Perry counties, Ohio. M, 1979, Miami University (Ohio). 130 p.

Crowell, Douglas L. Phase relations in the systems $A_2O-WO_3-SiO_2$ (where A = Li, Na and K). M, 1977, Miami University (Ohio). 66 p.

Crowell, Gordon D. A study of upper Windsor Limestones (Mississippian) as exposed on the Meander River, Hants County, Nova Scotia. M, 1968, Acadia University.

Crowell, John Chambers. Geology of the Tejon Pass region, California. D, 1947, University of California, Los Angeles.

Crowell, Mark. Faunal variation and its potential for sampling bias in the Morgarts Beach Member of the Yorktown Formation (Pliocene). M, 1988, Virginia Polytechnic Institute and State University.

Crowell, William R. The iron ores of Lake Superior. D, 1917, Columbia University, Teachers College.

Crowl, George Henry. Erosion surfaces in the Adirondacks. D, 1950, Princeton University. 48 p.

Crowl, William James. Geology of the central Dome Rock Mountains, Yuma County, Arizona. M, 1979, University of Arizona. 76 p.

Crowley, Appleton J. The relationship of the Hinckley Sandstone to the St. Croix Series. M, 1939, University of Minnesota, Minneapolis. 28 p.

Crowley, Donald Joe. Paleoecology of an algal carbonate bank complex in the late Pennsylvanian Wyandotte Formation of eastern Kansas. D, 1966, Brown University. 193 p.

Crowley, Donald Joe. The benthic fauna and sediment relationships of eastern Narragansett Bay, Rhode Island. M, 1962, Brown University.

Crowley, Francis A., Jr. Regional seismograms. M, 1951, Boston College.

Crowley, Francis Allen. Niobium-rare earth deposits in southern Ravalli County, Montana. M, 1958, Montana College of Mineral Science & Technology. 67 p.

Crowley, Jack Arthur. Geology and ore deposits of the Cosumnes Copper Mine skarn deposit, El Dorado County, California. M, 1974, California State University, Fresno.

Crowley, Julia Coolidge. Strontium isotope and rare earth element analyses of Rio Grande Rift basalts; implications for magmagenesis in continental rifts. D, 1984, Brown University. 129 p.

Crowley, Karl C. Geology of the Seneca-Silvies area, Grant County, Oregon. M, 1960, University of Oregon. 44 p.

Crowley, Kevin David. Origin, structure, and internal stratification of three hierarchical classes of bedforms in unidirectional flows; examples from laboratory rivers and the channels of the Platte River basin in Colorado and Nebraska. D, 1982, Princeton University. 248 p.

Crowley, Mark Thomas. Evolution and palaeontology. D, 1930, [Fordham University].

Crowley, Michael Summers. The effect of solid solubility on synthesis, stability polytypism of the micas. D, 1959, Pennsylvania State University, University Park. 105 p.

Crowley, Peter Duncan. The structural and metamorphic evolution of the Sitas area, Northern Norway and Sweden. D, 1985, Massachusetts Institute of Technology. 53 p.

Crowley, Sharon S. The effects of volcanic ash partings on maceral composition and chemistry of the C coal bed, Ferron Sandstone Member of the Mancos Shale, Utah. M, 1987, George Washington University. 105 p.

Crowley, Thomas John. Fluctuations of the eastern North Atlantic gyre during the last 150,000 years. D, 1976, Brown University. 288 p.

Crowley, Thomas John. Quantitative paleoclimatic analysis of two cores from the central north Atlantic. M, 1971, Brown University.

Crowley, William Patrick. Stratigraphic interpretation of meta-igneous rocks in south central Connecticut; bedrock geology of Long Hill and Bridgeport 7 1/2' quadrangles. D, 1967, Yale University. 144 p.

Crown, Robert W. Time-trends of chemical constituents in Illinois streams. M, 1978, Northern Illinois University. 255 p.

Crowson, Ronald A. Nearshore rock exposures and their relationship to modern shelf sedimentation, Onslow Bay, North Carolina. M, 1980, East Carolina University. 128 p.

Croxell, Thomas Ray. A geothermal probe; an instrument for the measurement of groundwater flow rates. M, 1973, New Mexico Institute of Mining and Technology.

Crozier, Robert N. Analysis of a San Joaquin Valley oil pool. M, 1957, Stanford University.

Cruciani, Cynthia L. W. The prediction of nitrate contamination potential using known hydrogeologic properties. M, 1987, University of Wisconsin-Milwaukee. 181 p.

Cruden, David Milne. Structural analysis of part of the Brazeau Range Anticline, near Nordegg, Alberta, (Canada). M, 1966, University of Alberta. 107 p.

Cruft, Edgar Frank. The geochemistry of apatite. D, 1962, McMaster University. 215 p.

Cruickshank, M. J. Technological and environmental considerations in the exploration and exploitation of marine minerals. D, 1978, University of Wisconsin-Madison. 256 p.

Cruickshank, Roy Douglas. Chloritoid-bearing pelitic rocks of the Horsethief Creek Group, southeastern British Columbia. M, 1976, University of Calgary. 141 p.

Cruikshank, Dale Paul. Infrared colorimetry of the Moon. D, 1968, University of Arizona. 154 p.

Cruikshank, Dale Paul. The origin of certain classes of lunar maria ridges. M, 1965, University of Arizona.

Cruikshank, Kenneth M. Numerical solutions for flow of debris in rectangular channels and computer simulation of growth of duplex structures. M, 1987, University of Cincinnati. 226 p.

Crum, James Robert. Soils and till stratigraphy of west-central Minnesota. D, 1984, University of Minnesota, Duluth. 250 p.

Crum, Steven V. The mechanics of subduction zone topography; a study of age effect in the Pacific Northwest. M, 1987, Pennsylvania State University, University Park. 58 p.

Crume, Robert W. Avenal Sandstone (middle Eocene) of Reef Ridge, California. M, 1940, Stanford University. 56 p.

Crump, James O., Jr. Lithofacies analysis and paleoenvironmental history of the upper Kindblade Formation (Lower Ordovician), Arbuckle Mountains, south-central Oklahoma. M, 1985, University of Texas at Dallas. 185 p.

Crump, Robert M. Origin of hard iron ores of the Marquette District. D, 1948, University of Wisconsin-Madison.

Crump, Terry Richard. The relationship of scheelite mineralization to late-stage changes in selected tact-

ites in southwestern Montana. M, 1976, University of Arizona. 108 p.

Crumpler, Larry S. Alkali basalt-trachyte suite and volcanism, northern part of the Mount Taylor volcanic field, New Mexico. M, 1977, University of New Mexico. 131 p.

Crumpler, Larry Steven. Io; models of volcanism and interior structure. D, 1983, University of Arizona. 162 p.

Crumpley, Bobby Kelly. A field reconnaissance of the geology of southeastern Sequoyah County, Oklahoma. M, 1949, University of Oklahoma. 34 p.

Crumpton, Carl F. The Pleistocene loesses of a part of the Junction City Quadrangle. M, 1951, Kansas State University. 51 p.

Cruson, Michael G. Geology and ore deposits of the Grizzly Peak Cauldron Complex, Sawatch Range, Colorado. D, 1973, Colorado School of Mines. 181 p.

Crutcher, John Fulton. The geologic setting of the Pillikin chromite mines, El Dorado County, California. M, 1959, University of California, Berkeley. 65 p.

Crutcher, Thomas Dent. Nomenclature of Cretaceous rocks, St. Johns vicinity, Apache County, Arizona, and Catron County, New Mexico. M, 1958, University of Texas, Austin.

Crutchfield, William Henry, Jr. The geology and silver mineralization of the Calistoga District, Napa County, California. M, 1953, University of California, Berkeley. 71 p.

Cruver, Jack Richard. Petrology and geochemistry of the Haystack Mountain Unit, Lower Skagit Valley, Washington. M, 1983, Western Washington University. 149 p.

Cruver, Susan Kinder. The geology and mineralogy of bentonites and associated rocks of the Chuckanut Formation, Mt. Higgins area, North Cascades, Washington. M, 1981, Western Washington University. 105 p.

Cruz Calderon, Gonzalo. The relationship between large earthquakes, tsunamis, and mean sea level along the Pacific Coast of middle America and a source model for the 1962 Acapulco (M_s = 7.0) earthquake. D, 1983, University of Colorado. 141 p.

Cruz, Felipe Ortigoza see Ortigoza Cruz, Felipe

Cruz, Jaime A. Geometry and origin of the Burbank Sandstone and Mississippian "chat" in T.25.N., R.6.E., and T.26.N., R.6.E., Osage County, Oklahoma. M, 1963, University of Tulsa. 43 p.

Cruz, Jaime Yap. Disturbance vector in space from surface gravity anomalies using complementary models. D, 1985, Ohio State University. 155 p.

Cruz, Luis A. De la *see* De la Cruz, Luis A.

Cruz, Nga de la *see* de la Cruz, Nga

Cruz, Servando De la *see* De la Cruz, Servando

Cruz, V. H. Santillan *see* Santillan Cruz, V. H.

Cruz-Orozco, R. Morphodynamics and sedimentation of the Rio Guayas Delta, Ecuador. D, 1974, Louisiana State University. 115 p.

Cruz-Orozco, Rodolfo. Suspended solids concentrations and their relations to other environmental factors in selected water bodies in the Barataria Bay region of southern Louisiana. M, 1971, Louisiana State University.

Crysdale, Bonnie L. Fluid inclusion evidence for the origin, diagenesis and thermal history of sparry calcite cement in the Capitan Limestone, McKittrick Canyon, West Texas. M, 1987, University of Colorado. 78 p.

Csejtey, Bela, Jr. Geology of the southeast flank of the Flint Creek Range, western Montana. D, 1962, Princeton University. 208 p.

Cserna, Eugene. A study of the rock fractures in the Cortlandt Complex. M, 1950, Columbia University, Teachers College.

Cserna, Eugene George. Structural geology and stratigraphy of the Fra Cristobal Quadrangle, Sierra

County, New Mexico. D, 1955, Columbia University, Teachers College. 119 p.

Cserna, Gloria A. Lower Cretaceous pelecypods and gastropods from the San Juan Raya-Zapotitlan region, State of Puebla, Mexico. M, 1956, Columbia University, Teachers College.

Cserna, Zoltan de *see* de Cserna, Zoltan

Cubberley, Alan J. Metapodial and first phalange analysis of Equus (Hippotigris) simplicidens Cope, from the late Pliocene-early Pleistocene Broadwater Formation, Broadwater, Nebraska. M, 1987, Bowling Green State University. 57 p.

Cubins, Arnis Gunnars. Magnetic and audio frequency magnetotelluric (AMT) investigations of varied astroblemes in the Williston Basin. M, 1979, University of Toronto.

Cucci, M. A. Chara-production of sediment, Preble Green Lake, Cortland County, New York. M, 1974, Syracuse University.

Cuddihee, John Lee. The organic geochemistry and water-rock system across a contact metamorphic profile in the Mancos Shale near Crested Butte, Colorado. M, 1982, Rice University. 129 p.

Cuddy, Robert Graham. Structural study of Kakagi Lake area, northwestern Ontario. M, 1971, McMaster University. 128 p.

Cudjoe, John E. A programme for geophysical exploration of Ghana. M, 1961, Massachusetts Institute of Technology. 114 p.

Cudrak, Constance Frances. Shallow crustal structure of the Endeavour Ridge segment, Juan de Fuca Ridge, from a detailed seismic refraction survey. M, 1988, University of British Columbia. 157 p.

Cudzill, Mary R. Fluvial, tidal and storm sedimentation in the Chilhowee Group (Lower Cambrian), northeastern Tennessee. M, 1985, University of Tennessee, Knoxville. 164 p.

Cudzilo, Thomas Frederick. Geochemistry of early Proterozoic igneous rocks, northeastern Wisconsin and upper Michigan. D, 1978, University of Kansas. 202 p.

Cudzilo, Thomas Frederick. Reconnaissance geochronology of Precambrian rocks in the central Uncompaghre plateau, west-central Colorado. M, 1971, University of Kansas.

Cuellar, V. Rearrangement measure theory applied to dynamic behavior of sand. D, 1974, Northwestern University. 110 p.

Cuellar-Chavez, R. An application of a regional methodology to estimate point and non-point pollutional loads discharged into a stream. D, 1976, University of Texas, Austin. 119 p.

Cuevas Leree, Juan Antonio. Analysis of subsidence and thermal history in the Sabinas Basin, northeastern Mexico. M, 1985, University of Arizona. 81 p.

Cuffey, Roger James. The bryozoan species Tabulipora carbonaria in the Wreford Megacyclothem (Lower Permian) in Kansas. D, 1966, Indiana University, Bloomington. 489 p.

Cuffney, Francie Lou Smith. Ecological interactions in the moss habitat of streams draining a clearcut and a reference watershed. D, 1987, University of Georgia. 174 p.

Cuffney, Robert G. Geology of the White Ledges area, Gila County, Arizona. M, 1976, Colorado School of Mines. 141 p.

Culberson, Charles H. Pressure dependence of the apparent dissociation constants of carbonic and boric acids in seawater. M, 1968, Oregon State University. 85 p.

Culberson, Charles Henry. Processes affecting the oceanic distribution of carbon dioxide. D, 1972, Oregon State University. 178 p.

Culbert, Llewellyn Borlaug. A gravity survey in the northern part of the Forest City Basin in Southwest Iowa and Southeast Nebraska. M, 1976, University of Iowa. 101 p.

Culbert, Richard Revis. A study of tectonic processes and certain geochemical abnormalities in the Coast Mountains of British Columbia. D, 1971, University of British Columbia.

Culbertson, Alex E. The oolites of Kansas City and their fauna. M, 1915, University of Kansas. 52 p.

Culbertson, John A. The Brassfield Formation of Jefferson County, Indiana. M, 1924, University of Chicago. 73 p.

Culbertson, John Archer. The paleontology and stratigraphy of the Pennsylvanian strata between Caseyville, Kentucky and Vincennes, Indiana. D, 1932, University of Illinois, Urbana.

Culbertson, Thomas Milton. Areal geology of the Jollyville Plateau and the regional ground water. M, 1948, University of Texas, Austin.

Culbreth, Mark A. Geophysical investigation of lineaments in South Florida. M, 1988, University of South Florida, Tampa. 97 p.

Culek, Thomas E. A multiple linear regression model of a gravity survey over a buried valley, Brimfield Township, Portage County, Ohio. M, 1985, Kent State University, Kent. 119 p.

Cull, Frances Anne. Oxidation of diamond at high temperature and 1 atm total pressure with controlled oxygen fugacity. M, 1985, Purdue University. 62 p.

Cullen, Andrew. The geology and petrology of Isla Pinta, Galapagos Archipelago. M, 1985, University of Oregon. 77 p.

Cullen, James J., IV. Digital computer simulation of karst ground water flow. M, 1979, Rensselaer Polytechnic Institute. 417 p.

Cullen, James Leo. Climatic variation in the northern Indian Ocean; analysis of the distribution, ecology, and preservation of planktonic foraminifera in late Quaternary sediments. D, 1984, Brown University. 319 p.

Cullen, Janet M. Impact of a major eruption of Mount Rainier on public service delivery systems in the Puyallup Valley, Washington. M, 1978, University of Washington. 203 p.

Cullen, John. Lime resources and industry in Oklahoma. M, 1917, [University of Oklahoma].

Cullen, John D. Metamorphic petrology and geochemistry of the Goldenville Formation metasediments, Yarmouth, Nova Scotia. M, 1983, Dalhousie University. 241 p.

Cullen, Michael Paul. Geology of the Bass River Complex, Cobequid Highlands, Nova Scotia. M, 1984, Dalhousie University. 183 p.

Cullen, Terry R. The petrology of the Reagan Sandstone (Late Cambrian), Wichita Mountains, Oklahoma. M, 1981, Wichita State University. 84 p.

Cullen, Timothy R. The bedrock geology of the western edge of the Berkshire Massif, Pittsfield East and Cheshire quadrangles, Massachusetts. M, 1979, City College (CUNY).

Cullers, Robert Lee. The partitioning of the rare-earth elements among rock-forming silicate phases and water. D, 1971, University of Wisconsin-Madison. 150 p.

Culligan, Leland B. Geology of the foothills structures west of Loveland, Colorado. M, 1947, University of Colorado.

Culligan, Leland B. Geology of the Loveland Fold area, Loveland (Larimer County), Colorado. M, 1948, University of Colorado.

Cullinan, Thomas Anthony. Contributions to the geology of Washington and St. Tammany parishes, Louisiana. D, 1969, Tulane University. 391 p.

Cullinan, Thomas Anthony. Preliminary study of the movement of silt and clay in a water-bearing formation. M, 1959, Texas Tech University. 63 p.

Cullins, Henry Long, Jr. The subsurface study of the Morrowan Sandstones (Pennsylvanian), Ellis County, Oklahoma. M, 1959, University of Oklahoma. 68 p.

Cullison, J. S. Revision of the Jefferson City Formation in the Rolla Quadrangle. M, 1930, University of Missouri, Rolla.

Cullison, J. S. The stratigraphy of some Lower Ordovician formations of the Ozarks. D, 1942, Yale University.

Cullough, Dale Alan. A systems classification of watersheds and streams. D, 1987, Oregon State University. 230 p.

Culp, B. R. Aqueous complexation of copper with sewage and naturally occurring organics. D, 1975, University of Michigan. 128 p.

Culp, Chesley Key, Jr. Stratigraphic relations of the Sycamore Limestone (Mississippian) in southwestern Oklahoma. M, 1960, University of Oklahoma. 46 p.

Culp, Eugene Forrest. The sedimentary petrography of the Devil's Kitchen Member of the Deese Formation in the Ardmore Basin. M, 1950, University of Oklahoma. 73 p.

Culp, Randolph Alan. Ground truth evaluation of the continuous surficial sediment sampling system for marine pollution assessment. M, 1988, University of Georgia. 112 p.

Culp, S. K. Recent benthic foraminifera of the Ontong Java Plateau. M, 1977, University of Hawaii. 68 p.

Culp, Stuart L. Geology and mineralization of the PP-LZ massive sulfide horizon, east-central Alaska Range, Alaska. M, 1982, Colorado State University. 133 p.

Culshaw, Nicholas G. Geology, structure and microfabric of Grenville rocks, Cardiff area, Ontario. D, 1983, University of Ottawa. 387 p.

Culver, Harold E. Geology and mineral resources of the Morris Quadrangle (Illinois). D, 1923, University of Chicago. 114 p.

Culver, Harold E. The formation of laterite. M, 1911, University of Wisconsin-Madison.

Culver, Stephen Eric. A gravity investigation of the ancestral Teays and Hamilton drainage systems in southwestern Ohio. M, 1988, Wright State University. 148 p.

Cumberlidge, John T. Some experiments on surface and strain energy in minerals. D, 1959, McGill University.

Cumbest, Randolph J. Crystal-plastic deformation and chemical evolution of clinoamphibole. D, 1988, Virginia Polytechnic Institute and State University.

Cumbest, Randolph Josh. Tectonothermal overprinting of the Western Gneiss Terrane, Senja, Troms. M, 1987, University of Georgia. 179 p.

Cumela, Stephen Paul. Sedimentary history and diagenesis of the Pictured Cliffs Sandstone, San Juan Basin, New Mexico and Colorado. M, 1981, University of Texas, Austin.

Cumella, Ronald. A lithofacies study of the San Rafael Group (Jurassic) in the San Juan Basin area. M, 1957, University of Oklahoma. 115 p.

Cumerlato, Calvin Lee. A shallow-seismic study of the Plum River fault zone at the Pleasant Hill Devonian outlier in east-central Iowa. M, 1983, University of Iowa. 92 p.

Cumings, E. R. and Galloway, Jesse James. The stratigraphy and paleontology of the Tanner's Creek Section of Cincinnati Series of Indiana. D, 1913, Indiana University, Bloomington. 128 p.

Cumings, Edgar R. The morphogenesis of Platystrophia; a study of the evolution of a Paleozoic brachiopod. D, 1902, Yale University.

Cumings, Edgar Roscoe and Schrock, Robert Rakes. The geology of the Silurian rocks of northern Indiana. D, 1928, Indiana University, Bloomington. 293 p.

Cumley, Russell Walters. A geologic section across Caldwell County, Texas. M, 1931, University of Texas, Austin.

Cumming, Bradley R. Petrography, petrochemistry and petrogenesis of Huronian volcanic rocks of the Elliot Lake region, Ontario. M, 1986, Brock University. 97 p.

Cumming, George Leslie. A petrographic and radiometric study of the Tazin meta-sediments of the Charlebois Lake area, northeastern Saskatchewan. M, 1952, University of Saskatchewan. 59 p.

Cumming, George Leslie. The correlation of age determinations with arcuate discontinuities in the structure of North America. D, 1955, University of Toronto.

Cumming, Harry John Karns. Gravity survey of southern Marsh Valley, Bannock County, Idaho. M, 1980, Idaho State University. 66 p.

Cumming, Leslie M. Silurian and Lower Devonian sedimentary rocks of eastern Gaspe, Quebec. D, 1955, University of Wisconsin-Madison.

Cumming, William B. Crustal structure from a seismic refraction profile across southern British Columbia. M, 1977, University of British Columbia.

Cummings, David. A study of the basal Chepultepec Sandstone (Cambrian-Ordovician boundary) in the Ridge and Valley Province of Tennessee. M, 1959, University of Tennessee, Knoxville. 47 p.

Cummings, David. Geology of the Bays Mountain synclinorium, Northeast Tennessee. D, 1962, Michigan State University. 152 p.

Cummings, David O. Deformational and metamorphic history of a Precambrian terrane in Gunnison and Saguache counties, Colorado. M, 1987, University of Kansas. 49 p.

Cummings, George Howard, III. Reefs and related sediments of the Cap Haitien area, Haiti. M, 1973, University of Florida. 93 p.

Cummings, J. S. Zircon in pegmatite and country rock at Greenwood, Maine. M, 1955, Columbia University, Teachers College.

Cummings, Jan A. The geology of the southern third of the Courtrock Quadrangle, Oregon. M, 1958, University of Oregon. 54 p.

Cummings, John Moss. The weathered rocks of Hong Kong. M, 1935, University of British Columbia.

Cummings, Jon Clark. Geology of the Langley Hill-Waterman Gap area, Santa Cruz Mountains, California. D, 1960, Stanford University. 334 p.

Cummings, Jon Clark. Reconnaissance geology of the Mindego Hill area, California. M, 1956, Stanford University. 38 p.

Cummings, Leslie M. A heavy mineral study of the Pennsylvanian sedimentary rocks of the Minto-Chipman District, New Brunswick. M, 1951, University of New Brunswick.

Cummings, Michael L. Structure and petrology of Precambrian amphibolite, Big Falls County Park, Eau Claire County, Wisconsin. M, 1975, University of Minnesota, Minneapolis.

Cummings, Michael Levi. Metamorphism and mineralization of the Quinnesec Formation, northeastern Wisconsin. D, 1978, University of Wisconsin-Madison. 203 p.

Cummings, Robert Adams. Methods and environmental factors in the production of thorium from vein deposits. M, 1979, University of Arizona. 139 p.

Cummings, Warren LeRoy. Properties of carbonate rock aggregate effecting the skid resistance of bituminus concrete pavement. M, 1976, Rutgers, The State University, New Brunswick. 47 p.

Cummins, Catharine. Scale modelling of the parallel-line mode of operation of the Turam prospecting system and a comparison with the conventional mode. M, 1986, University of Calgary. 523 p.

Cummins, Dean Lewis. Geology of the Tendoy-Medicine Lodge area, Beaverhead County, Montana. M, 1948, University of Michigan.

Cummins, Gloria. Paleoenvironmental factors affecting mine planning of the Pocahontas No. 3 coal seams in southeastern Virginia. M, 1979, University of South Carolina.

Cummins, James Walter. The petroleum geology of Muskingum County, Ohio. M, 1931, Ohio State University.

Cummins, Laura E. The development of zoned joint-blocks in a basalt flow near Wagon Mound, New Mexico. M, 1981, Bowling Green State University. 132 p.

Cummins, Laura Elaine. Geochemistry, mineralogy and origin of Mesozoic diabase dikes of Virginia. D, 1987, Florida State University. 500 p.

Cummins, Phillip. Seismic body waves and the Earth's inner core. D, 1988, University of California, Berkeley. 136 p.

Cummins, Robert Hays. Taphonomic processes in modern estuarine death assemblages along the Texas coast; rate of taphonomic loss, covariance of species, and size frequency analysis. D, 1984, Texas A&M University. 206 p.

Cunderla, Brent Joseph. Stratigraphic and petrologic analysis of trends within the Spencer Formation sandstones; from Corvallis, Benton County, to Henry Hagg Lake, Yamhill and Washington counties, Oregon. M, 1986, Portland State University. 135 p.

Cundiff, Jerry Allen. Petrography of the Clinch Sandstone of Northeast Tennessee. M, 1951, University of Tennessee, Knoxville. 37 p.

Cundy, Terrance William. An analysis of the effects of spatial variability of point infiltration rates on the comparison of small and large plot rainfall-runoff. D, 1982, Utah State University. 125 p.

Cunha, Roberto Pereira da *see* Pereira da Cunha, Roberto

Cunion, Edward Joseph, Jr. Analysis of gravity data from the southeastern Chino Valley area, Yavapai County, Arizona. M, 1985, Northern Arizona University. 110 p.

Cunliffe, James E. Description, interpretation, and preservation of growth increment patterns in shells or Cenozoic bivalves. D, 1974, Rutgers, The State University, New Brunswick. 171 p.

Cunliffe, James Edwin. Petrology of the Cretaceous Peedee Formation and Eocene Castle Hayne Limestone in northern New Hanover County, North Carolina. M, 1968, University of North Carolina, Chapel Hill. 128 p.

Cunniff, Robert Thomas. The tectonic style of the Keweenawan deformation. M, 1988, Michigan State University. 60 p.

Cunningham, Alan E. Lithofacies, depositional environments, and diagenetic alteration of the Smackover Formation in the Manila Embayment of Southwest Alabama. M, 1984, University of Alabama. 205 p.

Cunningham, Alfred B. Modeling and analysis of hydraulic interchange of surface and ground water. D, 1977, University of Nevada - Mackay School of Mines. 174 p.

Cunningham, Arnold Bryce. The sedimentology and provenance of the Upper Cretaceous Rosario Formation between Punta Banda and Punta San Jose, Baja California, Mexico. M, 1985, San Diego State University. 291 p.

Cunningham, Charles Godvin, Jr. Multiple intrusion and venting of the Italian Mountain intrusive complex, Gunnison County, Colorado. D, 1973, Stanford University. 168 p.

Cunningham, Chris P. Analysis of sand-sized sediments from Lemon Bay, Florida, using a visual-accumulation settling tube. M, 1974, Bowling Green State University. 38 p.

Cunningham, Cindy Carolyn. Spatial and temporal monitoring of the Jovian atmosphere. D, 1987, University of Arizona. 223 p.

Cunningham, Cynthia Taylor. Geology and geochemistry of a massive sulfide deposit and associated volcanic rocks, Blue Creek District, southwestern Oregon. M, 1979, Oregon State University. 165 p.

Cunningham, David. Textural criteria for the discrimination of water-laid and wind-laid barrier island sands; a North Padre Island, Texas example. M, 1985, Texas A&M University.

Cunningham, David A. Computer and physical VSP modeling of Yucca Mountain, Nevada. M, 1988, Colorado School of Mines. 313 p.

Cunningham, Frederick Franklin, Jr. Stratigraphy and petrology of the Columbus Limestone (Devonian) and Detroit River Group (Devonian) in north-central

Ohio. M, 1972, Bowling Green State University. 85 p.

Cunningham, Gary G. Ablation studies of an artificial meteor of olivine composition. M, 1973, San Jose State University. 59 p.

Cunningham, Gregory D. The Plio-Pleistocene Dipodomyinae and geology of the Palm Spring Formation, Anza-Borrego Desert, California. M, 1984, Idaho State University. 193 p.

Cunningham, James P. Downward continuation of potential fields. M, 1983, Georgia Institute of Technology. 104 p.

Cunningham, John Edward. Geology of the North Tumacacori Foothills, Santa Cruz County, Arizona. D, 1964, University of Arizona. 197 p.

Cunningham, K. D. Petrology and petrography of Permian volcanogenic and carbonate rocks near Las Delicias, Coahuila, Mexico. M, 1975, Texas Christian University.

Cunningham, Leslie L. A laser theodolite. M, 1965, Ohio State University.

Cunningham, Paul S. Earthquake locations and three-dimensional seismic structure of southern Peru. M, 1984, Massachusetts Institute of Technology. 176 p.

Cunningham, R. C. Taconites in the vicinity of Lake Superior. M, 1951, University of Toronto.

Cunningham, Richard Carson, Jr. Investigation of littoral transport between Virginia Beach and Sandbridge, Virginia. M, 1974, Old Dominion University. 63 p.

Cunningham, Robert Lester. Genesis of the soils along a traverse in Asotin County, Washington. D, 1964, Washington State University. 158 p.

Cunningham, Robert, Jr. Organic-inorganic interactions of marine humic substances from carbonate sediments; metal binding and adsorption studies. D, 1980, University of Texas at Dallas. 127 p.

Cunningham, Russ DeWitt. Genetic analysis of Rayleigh-Taylor perturbations (RTPs) in the Calvin Sandstone (Pennsylvanian) near Okmulgee, Oklahoma. M, 1978, University of Tulsa. 60 p.

Cunningham, S. E. Surface sediment analysis of five carbonate banks on the Texas continental shelf. M, 1977, Texas A&M University.

Cunningham, Thomas. Correlation of geophysical data in North America. M, 1972, University of Pittsburgh.

Cunningham-Dunlop, P. K. Structural geology of Ontario pyrites deposits, Sudbury, Ontario. M, 1954, University of Toronto.

Cunningham-Dunlop, Peter K. Geology of economic uraniferous pegmatites in the Bancroft area, Ontario (Canada). D, 1967, Princeton University. 290 p.

Cunnington, F. A. Porcupine-Beattie gold belt (Quebec and Ontario). M, 1947, McGill University.

Cuomo, Marie Carmela. The ecological and paleoecological significance of sulphides in marine sediments. D, 1984, Yale University. 198 p.

Cuong, Pham Giem. Thermal convection and magnetic field generation in rotating spherical shells. D, 1979, University of California, Los Angeles. 153 p.

Cupal, Jerry J. A study of the remanent magnetization in the basal plane of natural hematite. M, 1967, Michigan Technological University. 63 p.

Cuppels, Norman Paul. Desilication of some igneous magmas in a part of the northwestern Adirondack Mountains. M, 1952, Rutgers, The State University, New Brunswick. 111 p.

Curatolo, Joel Charles. The petrography and petrology of granitic and related rocks near Griffin, Georgia. M, 1986, Pennsylvania State University, University Park. 234 p.

Curchin, John Montgomery. Tidal triggering of intermediate and deep focus earthquakes. M, 1985, University of Texas, Austin. 75 p.

Curet, A. F. Geology of the Cretaceous-Tertiary(?) rocks of the southwest quarter of the Monte Guilarte Quadrangle, West-central Puerto Rico. M, 1976, University of Minnesota, Duluth.

Curi, Nilton. Lithosequence and toposequence of Ox-isols from Goias and Minas Gerais states, Brazil. D, 1983, Purdue University. 174 p.

Curiale, Joseph Anthony. Source rock geochemistry and liquid and solid petroleum occurrences of the Ouachita Mountains, Oklahoma. D, 1981, University of Oklahoma. 305 p.

Curiel-Mitchell, Helen. Geology and mineralization as related to detachment faulting in Copper Basin, south-eastern Whipple Mountains, San Bernardino County, California. M, 1987, San Diego State University. 234 p.

Curl, J. E. A glacial history of the South Shetland Islands, Antarctica. M, 1976, Ohio State University. 176 p.

Curl, Mary W. Peritidal carbonate lithofacies, diagenetic sabkha overprinting, and paleoenvironment correlation of the subsurface Tribes Hill Formation (Lower Ordovician) Mohawk Valley, New York. M, 1983, Rensselaer Polytechnic Institute. 220 p.

Curlin, Kimberly B. Upper Eocene Archaeomonadaceae from Ottenthal, Austria. M, 1982, University of Rhode Island.

Curran, Claude Warren. The Mendocino Chaparral; a problem in resource management. D, 1973, University of Oklahoma. 139 p.

Curran, Donald Walter. Geology of the Siguatepeque Quadrangle, Honduras, Central America. M, 1980, SUNY at Binghamton. 194 p.

Curran, Harold Allen. Conodonts from the Whistle Creek Limestone (Middle Ordovician) of Virginia. M, 1965, University of North Carolina, Chapel Hill. 105 p.

Curran, Harold Allen. Upper Cretaceous foraminifera and subsurface stratigraphy of the S. E. North Carolina coastal plain. D, 1968, University of North Carolina, Chapel Hill. 245 p.

Curran, John F. Eocene stratigraphy of the Chico Martinez Creek area, Kern County, California. M, 1942, Stanford University. 54 p.

Curran, Theodore Allan. Surficial geology of the Issaquah area, Washington. M, 1965, University of Washington. 57 p.

Curray, Joseph Ross. An analysis of sphericity and roundness of quartz grains. M, 1951, Pennsylvania State University, University Park. 77 p.

Curray, Joseph Ross. Sediments and history of the Holocene transgression continental shelf, Northwest Gulf of Mexico. D, 1959, University of California, Los Angeles. 164 p.

Curren, Robert F. An investigation of some methods of chemical precipitation in the artificial growth of calcite. M, 1953, Michigan State University. 27 p.

Currens, James Calvin. Correlation and depositional environments of the Upper Elthorn Number 3 coal zone in southeastern Kentucky. M, 1978, Eastern Kentucky University. 66 p.

Current, George Thomas. An investigation of the hydrology and geology of the Hocking River valley fill in Athens and Hocking counties. M, 1967, Ohio University, Athens. 97 p.

Currey, D. R. Geology of the Keystone area, Albany County, Wyoming. M, 1959, University of Wyoming. 74 p.

Currey, Donald Rusk. Neoglaciation in the mountains of the Southwestern United States. D, 1969, University of Kansas. 187 p.

Currie, A. L. Analysis of structure in a part of the Clare River synform, Ontario. M, 1972, University of Toronto.

Currie, Donald Varcoe. Hydrogeology of Tri-Creek Basin (Recent), Alberta (Canada). M, 1969, University of Alberta. 187 p.

Currie, John B. The occurrence and relationships of some mica and apatite deposits in southeastern Ontario. D, 1950, University of Toronto.

Currie, John B. Zones of metamorphism in the greenstones of the Morton Lake area, Manitoba. M, 1947, University of Toronto.

Currie, John Morgan. Structural interpretation by field and laboratory methods. M, 1949, Miami University (Ohio). 31 p.

Currie, Kenneth Lyell. Mechanics of metasomatism. D, 1959, University of Chicago. 42 p.

Currie, Lester J. E. Petroleum potential of Trenton Group (Ordovician), Quebec (Canada). M, 1968, McGill University.

Currie, Lisel D. Geology of the Allan Creek area, Cariboo Mountains, British Columbia. M, 1988, University of Calgary. 152 p.

Currie, Michael Thomas. Subsurface stratigraphy and depositional environments of the "Corniferous" (Silurian-Devonian) of eastern Kentucky. M, 1981, University of Kentucky. 108 p.

Currie, Philip John. The osteology and relationships of aquatic eosuchians from the Upper Permian of Africa and Madagascar. D, 1981, McGill University. 526 p.

Currie, Ralph Gordon. A comparison of Long Shot and earthquakes. M, 1967, University of British Columbia.

Currier, Debra Ann. Structures and microfabrics of a zone of superposed deformation, Foothills fault zone, east flank of the Huachuca Mountains, Southeast Arizona. M, 1985, University of Arizona. 167 p.

Currier, John D., Jr. Stratigraphy and areal geology of southwestern Bryan County, Oklahoma. M, 1968, University of Oklahoma. 76 p.

Currier, Louis W. Geology of the La Salle and Streater quadrangles, Illinois. D, 1930, Syracuse University.

Currier, Louis W. The fluorspar deposits of Illinois. M, 1920, Northwestern University.

Curry, Ben Brandon. Age of High Rock and Summit Lake landslides, and overflow history of their associated basins, Humboldt County, Nevada. M, 1984, Purdue University. 263 p.

Curry, David James. The organic geochemistry of kerogen and humic acids in Recent sediments from the Gulf of Mexico. D, 1981, University of Texas, Austin. 221 p.

Curry, Donald Lee. The geology of the Cordero quicksilver mine area, Humboldt County, Nevada. M, 1960, University of Oregon. 60 p.

Curry, H. Donald. The geology of the southern part of the Randolph Quadrangle, New York. M, 1930, University of Iowa. 153 p.

Curry, Richard Porter. Upper Devonian miospores from the Greenland Gap Group, Allegheny Front, eastern United States. D, 1972, University of Connecticut. 173 p.

Curry, Robert R. Geobotanical correlations in the Alpine and Subalpine regions of the Tenmile Range, Summit County, Colorado. M, 1962, University of Colorado.

Curry, Robert Rodney. Quaternary climatic and glacial history of the Sierra Nevada, California. D, 1968, University of California, Berkeley. 238 p.

Curry, Sharon G. Petrofabrics of carbonate rocks determined by X-ray diffraction. M, 1962, Iowa State University of Science and Technology.

Curry, William Baetzel. Isotopic fractionation patterns in planktonic foraminifera from Holocene sediments in the Indian Ocean. D, 1980, Brown University. 210 p.

Curry, William Hirst, III. Stratigraphy and paleogeography of Upper Jurassic and Lower Cretaceous rocks of central Wyoming. D, 1959, Princeton University. 216 p.

Curth, Patrick J. A preliminary reconnaissance of the Kingshill Marl Formation, St. Croix, U. S. Virgin Islands. M, 1976, University of Wisconsin-Milwaukee.

Curtin, Gary C. Structural geology of the Precambrian rocks in the Bear Gulch area, Larimer County, Colorado. M, 1965, University of Colorado.

Curtin, George. Hydrogeology of the Sutter Basin, Sacramento Valley, California. M, 1971, University of Arizona.

Curtin, Margaret H. The Trap dikes of the Adirondack region. M, 1904, Columbia University, Teachers College.

Curtis, Alan Deane and Dott, Robert Henry, Jr. Geology of the northern half of the Coldwater Quadrangle, Madison County, Missouri. M, 1950, University of Michigan.

Curtis, Bruce F. Petrographic Studies on the Post-Laramie sediments of the Denver Basin, Colorado. M, 1942, University of Colorado.

Curtis, Bruce Franklin. Structure and stratigraphy of the Linwood Spring Creek area, Utah-Wyoming. D, 1949, Harvard University.

Curtis, C. S. An analysis of the environmental impact statement of the Warm Springs dam project. M, 1976, University of Arizona.

Curtis, Carl Edward. Enamel-making properties of the flints and feldspars of the State of Washington. M, 1925, University of Washington. 87 p.

Curtis, Charles M. Sedimentology of the northern half of the Laguna Salada, Baja California (Mexico). M, 1966, University of Southern California.

Curtis, Dwight Kenneth. The geological interpretation of geography in the intermediate grades of elementary school. M, 1936, University of Iowa. 158 p.

Curtis, Garniss Hearfield. The geology of the Topaz Lake Quadrangle, and the eastern half of the Ebbetts Pass Quadrangle. D, 1951, University of California, Berkeley. 316 p.

Curtis, George. The stratigraphy and structure of portions of the Bonanza, King, and Schell Mountain quadrangles, Trinity County, California. M, 1980, San Jose State University. 132 p.

Curtis, Janet. Natural abundance deuterium nuclear magnetic resonance spectroscopy in structure and mechanism of selected organic systems. D, 1987, University of Utah. 139 p.

Curtis, John B., Jr. The behavior of strontium isotopes in surface waters of the Scioto River drainage basin, Ohio. M, 1972, Miami University (Ohio). 41 p.

Curtis, L. W. Petrology of the Red Wine Complex, central Labrador. D, 1975, University of Toronto.

Curtis, Lawrence W. Geology in the vicinity of Dry Run, Pennsylvania. M, 1942, Columbia University, Teachers College.

Curtis, Patchin Crandall. The structure and petrology of Central Island, Lake Turkana, Kenya. M, 1987, Duke University. 79 p.

Curtis, Rene Virginia. Sedimentology of the Holocene ooid shoals, Eleuthera Bank, Bahamas. M, 1985, University of Texas, Austin. 137 p.

Curtis, Robert. Depositional environment of the Upper Jurassic Norphlet and Smackover formations, Hatters Pond Field, Mobile County, Alabama. M, 1982, Texas A&M University. 165 p.

Curtis, Robert E., Jr. A hydrogeochemical reconnaissance for uranium in Trans-Pecos, Texas. M, 1978, University of Texas at El Paso.

Curtis, Thomas Gray, Jr. Simulation of salt water intrusion by analytic elements. D, 1983, University of Minnesota, Minneapolis. 77 p.

Curtiss, Brian. Evaluation of the physical properties of naturally occurring iron(III) oxyhydroxides on rock surfaces in arid and semi-arid regions using visible and near infrared reflectance spectroscopy. D, 1985, University of Washington. 106 p.

Curtiss, Robert Eugene. The geology of the Isabel Quadrangle, South Dakota. M, 1953, Washington State University. 145 p.

Cushing, Edward John. Late-Wisconsin pollen stratigraphy in east-central Minnesota. D, 1963, University of Minnesota, Minneapolis. 188 p.

Cushing, Grant W. Tectonic evolution of the eastern Yukon-Tanana Upland, east-central Alaska. M, 1984, SUNY at Albany. 255 p.

Cushing, Henry Platt. Geology of the Thousand Island region. D, 1909, Cornell University.

Cushing, Henry Platt. The serpentine areas of Staten Island, Hoboken, New Jersey, etc. M, 1884, Cornell University.

Cushman, Curtis Dean. Geology and sedimentary petrology of the Taneum/Manastash creeks area, Kittitas County, Washington. M, 1984, Eastern Washington University. 132 p.

Cushman, Robert A., Jr. Palynology and paleoecology of the Fossil Butte Member of the Eocene Green River Formation in Fossil Basin, Lincoln County, Wyoming. M, 1983, Loma Linda University. 88 p.

Cushman, Robert V. Geology of the central part of the Lake Champlain Lowland. M, 1941, Northwestern University.

Cushman, Solomon Frederick. The Ohio standard baseline. D, 1972, Ohio State University. 186 p.

Cushman-Roisin, Mary. The hydrological inverse problem; reconsideration and application to the microcomputer. D, 1986, Florida State University. 113 p.

Cusimans, George. An investigation of the geochemical behavior of beryllium in coastal marine and continental natural water systems. M, 1988, University of Southern California.

Cuskley, Virginia A. Some new ostracodes from the "White Mound" section, Haragan Shale, Murray County, Oklahoma. M, 1932, Columbia University, Teachers College.

Custer, E. S., Jr. Influence of sedimentary processes on grain size distribution curves of bottom sediments in the sounds and estuaries of North Carolina. M, 1974, University of North Carolina, Chapel Hill. 88 p.

Custer, Edward Scheid, Jr. Depositional environments of the subsurface Cretaceous deposits of southeastern North Carolina. D, 1981, University of North Carolina, Chapel Hill. 116 p.

Custer, Richard Lewis Payzant. Paleocurrents of the Triassic Durham Basin, North Carolina. M, 1966, North Carolina State University. 34 p.

Custer, Stephan G. Shallow ground-water salinization in dryland-farm areas of Montana. D, 1976, University of Montana. 215 p.

Custer, Stephan Gregory. Stratigraphy and sedimentation of Black Point Volcano. M, 1973, University of California, Berkeley. 114 p.

Custis, Kit H. Geology and dike swarms of the Homer Mountain area, San Bernardino County, California. M, 1984, California State University, Northridge. 168 p.

Custodi, George L. A survey of mercury in the Gulf of Mexico. M, 1971, Texas A&M University.

Cutcliffe, William E. A study of bedrock geology in part of the Tomhannock and North Troy quadrangles, New York. M, 1961, Rensselaer Polytechnic Institute. 66 p.

Cuthbert, Frederick L. A geological study of Cattaraugus Creek and vicinity with special reference to Pleistocene sediments. M, 1937, SUNY at Buffalo.

Cuthbert, Frederick L. Petrography of two Iowa loess materials. D, 1940, Iowa State University of Science and Technology.

Cuthbert, Margaret Elizabeth. Formation of chalcopyrite and bornite at atmospheric temperature and pressure. M, 1961, University of Michigan.

Cutler, Alan Hughes. Functional morphology of the jaw in canids. M, 1977, University of Rochester.

Cutler, Elizabeth Reinen. Geology of the upper Green River area between the Gros Ventre and Wind River mountains, Sublette County, Wyoming. M, 1984, University of Wyoming. 103 p.

Cutler, Jodi Lyn. Hydrogeochemical ground-water reconnaissance in Monroe County, Indiana. M, 1987, Indiana University, Bloomington. 139 p.

Cutler, John Fredrick. Morphology, taxonomy and evolution of the bryozoan Constellaria from the Cincinnati Arch. D, 1968, Columbia University. 297 p.

Cutler, Jonathan Mitchell. Inhomogeneous deformation; the theory and geological application of finite strain compatibility. D, 1984, The Johns Hopkins University. 182 p.

Cutler, Mark A. The Middle Carboniferous-Permian stratigraphy of Midterhuken Peninsula, Spitsbergen. M, 1981, University of Wisconsin-Madison.

Cutler, Robert M. Laboratory studies of the effects of compressed air energy storage on selected reservoir rock and cap rock. M, 1979, University of Wisconsin-Milwaukee. 78 p.

Cutler, S. Geophysical investigation of the Nazca Ridge. M, 1977, University of Hawaii.

Cutler, Sandra Kay. New Upper Jurassic foraminifera from the Knoxville Formation, Elk Creek, California. M, 1979, University of Nevada. 46 p.

Cutler, William Gerald. Stratigraphy and sedimentology of the Upper Devonian Grosmont Formation, Alberta, Canada. M, 1982, University of Calgary. 191 p.

Cutolo-Lozano, Francisco José Subsurface stratigraphic analysis of northern Seminole County, Oklahoma and portions of Pottawatomie and Okfuskee counties, Oklahoma. M, 1966, University of Oklahoma. 61 p.

Cutright, Bruce Lee. Hydrogeology of a cypress swamp; north central Alachua County, Florida. M, 1974, University of Florida. 83 p.

Cutsforth, David H. The geology of a portion of the San Jose Hills. M, 1949, California Institute of Technology. 25 p.

Cutshall, Norman Hollis. Chromium-51 in the Columbia River and adjacent Pacific Ocean. D, 1967, Oregon State University. 64 p.

Cutten, William. The economic development of the Rocky Mountain oil fields. M, 1941, Massachusetts Institute of Technology. 133 p.

Cutter, Gregory Allan. Processes affecting the distribution and speciation of selenium in seawater. D, 1982, University of California, Santa Cruz. 173 p.

Cutter, Paul Frank. The economics of the mining of Colorado oil shale. M, 1930, Case Western Reserve University.

Cutter, Russell C. Geology of the Big Thompson Valley, west of Loveland (Larimer County), Colorado. M, 1949, Colorado School of Mines. 90 p.

Cutts, James Alfred John. Martian spectral reflectivity properties from Mariner 7 observations. D, 1971, California Institute of Technology. 92 p.

Cuvelier, Gaëtan Jean Francois Joseph. Radio source mapping for precision geodesy. M, 1982, Massachusetts Institute of Technology. 139 p.

Cuyler, Robert Hamilton. The Georgetown Formation of central Texas and its North Texas equivalents. M, 1927, University of Texas, Austin.

Cuyler, Robert Hamilton. The Travis Peak Formation of central Texas. D, 1931, University of Texas, Austin.

Cuzella, Jerome J. Stratigraphic mineralogy, origin, and paragenesis of the Negaunee Iron Formation (Precambrian), section 18, T 47 N, R 26 W, Marquette County, Michigan. M, 1973, Bowling Green State University. 136 p.

Cuzzi, Jeffrey Nicholas. The subsurface nature of Mercury and Mars from thermal microwave emission. D, 1973, California Institute of Technology. 174 p.

Cvancara, Alan M. Gastropoda from Pierre Shale (Upper Cretaceous) of Emmons County, south-central North Dakota. M, 1957, University of North Dakota. 76 p.

Cvancara, Alan Milton. Bivalves and biostratigraphy of the Cannonball Formation (Paleocene) in North Dakota. D, 1965, University of Michigan. 481 p.

Cwiak, Ronald Alvin. A review of the theory of continental drift. M, 1969, Northern Illinois University. 95 p.

Cwick, Gary J. A geobotanical assessment of the silver mine area of southeastern Missouri using Landsat thematic mapper. D, 1987, Indiana State University. 164 p.

Cybriwsky, Zenon Alexander. Spectral excitation of Lg within the south-central Appalachian region. D, 1979, Pennsylvania State University, University Park.

Cygan, Gary L. The effect of Na_2O and MgO on the liquid immiscibility gap in the system K_2O-FeO-Al_2O_3-SiO_2. M, 1979, University of Illinois, Chicago.

Cygan, Norbert Everett. Cambrian and Ordovician conodonts from the Big Horn Mountains, Wyoming. D, 1962, University of Illinois, Urbana. 105 p.

Cygan, Norbert Everett. The stratigraphy and paleontology of the Ordovician Bighorn Dolomite of north central Wyoming. M, 1956, University of Illinois, Urbana. 47 p.

Cygan, Randall Timothy. Chemical diffusion and dielectric polarization processes in silicate minerals. D, 1983, Pennsylvania State University, University Park. 185 p.

Cygan, Randall Timothy. Crystal growth and the formation of chemical zoning in natural garnets. M, 1980, Pennsylvania State University, University Park. 194 p.

Cynn, Hyunchae. Geology and geochemistry of Precambrian metamorphic rocks and Late Cretaceous igneous rocks in the Transverse Ranges, California. M, 1987, University of California, Los Angeles. 196 p.

Cys, John McKnight. Pre-Curtis Jurassic stratigraphy of northwest and west-central Colorado. M, 1965, University of Colorado.

Czajkowski, Jaroslaw. Cosmo and geochemistry of the Jurassic hardgrounds. D, 1987, [University of California, San Diego]. 443 p.

Czamanske, Gerald Kent. A study of the solubilities of iron, lead, and manganese sulfides at elevated temperatures under conditions of geologic interest. D, 1961, Stanford University. 114 p.

Czaplewski, Nicholas Jay. Pliocene vertebrates of the upper Verde Formation, Arizona. D, 1987, [Northern Arizona University]. 334 p.

Czarnecki, John Brian. Characterization of the subregional ground-water flow system of a potential site for a high-level nuclear waste repository. D, 1988, University of Minnesota, Minneapolis. 359 p.

Czechowski, Douglas A. Petrologic comparison of Holocene stream sands and Triassic sandstones in the central Piedmont of Virginia; evidence for Triassic paleoclimate. M, 1982, Southern Illinois University, Carbondale. 96 p.

Czekalski, Steve James. A geophysical survey using gravity and seismic refraction data to approximate the bedrock surface in eastern Clark County, Ohio. M, 1985, Wright State University. 73 p.

Czerniakowski, Lana Ann. Marine burial diagenesis of the Austin Chalk; a stable isotope investigation. M, 1982, University of Michigan.

Czimer, Marilyn A. The description and origin of the Big Slide, southern Montana. M, 1975, Southern Illinois University, Carbondale. 62 p.

Czoer, Kenneth E. Paleoecology of the upper Pliocene portion of the Tamiami Formation, southwestern Florida. M, 1982, University of Cincinnati. 120 p.

Czyscinski, Kenneth. The development of acid sulfate soils ("cat clays") on the Annandale Plantation, Georgetown County, South Carolina. D, 1975, University of South Carolina. 153 p.

Czyscinski, Kenneth. The origin of phillipsite in marine sediments. M, 1971, University of South Carolina.

d'Acierno, Carlos Enrique Reijenstein see Reijenstein d'Acierno, Carlos Enrique

d'Agnese, Susanne L. The engineering geology of the Fountain landslide, Hood River County, Oregon. M, 1986, Portland State University. 174 p.

D'Agostino, Anthony. Foraminiferal biostratigraphy, paleoecology, and systematics of DSDP Site 273, Ross Sea, Antarctica. M, 1980, Northern Illinois University. 124 p.

D'Allura, Jad Alan. Stratigraphy, structure, petrology, and regional correlations of metamorphosed upper Paleozoic volcanic rocks in portions of Plumas, Sierra and Nevada counties, California. D, 1977, University of California, Davis.

D'Aluisio-Guerrieri, Gary M. Holocene lagoonal sedimentation, Vieques, Puerto Rico. M, 1982, University of South Florida, Tampa. 123 p.

D'Amico, Angela. Development of a tidal prism model and its application to the Pagan River, Virginia. M, 1976, College of William and Mary.

D'Amore, Denis W. Hydrogeology and geomorphology of the Great Sanford outwash plain, York County, Maine with particular emphasis on the Branch Brook watershed. D, 1983, Boston University. 164 p.

D'Amore, Denis W. Hydrogeology of the Monatiquot River watershed. M, 1982, Boston University. 105 p.

D'Andrea, Julie. The geochemistry of the Socorro K_2 anomaly, N. Mex. M, 1981, Florida State University.

D'Andrea, R. A. Probabilistic partial safety factor design techniques for undrained soil stability problems. D, 1980, Cornell University. 202 p.

D'Andrea, Ralph F. Replacement of marble by ZnS in chloride solutions. M, 1976, Pennsylvania State University, University Park. 32 p.

D'Angelo, Louellen. Effects of clear-water discharge on a small gravel-bed stream in central New Jersey. M, 1979, Rutgers, The State University, Newark. 102 p.

d'Angelo, Richard M. Correlation of seismic reflection data with seismicity over the Ramapo, New Jersey, fault zone. M, 1985, Virginia Polytechnic Institute and State University.

d'Anglejan-Chatillon, Bruno F. The marine phosphorite deposit of Baja California, present environment and recent history. D, 1965, University of California, San Diego. 214 p.

d'Astous, Jacques. Etude de l'infiltration et de l'écoulement souterrain au Lac Laflamme au moyen des méthodes thermiques. M, 1986, Universite Laval. 160 p.

D'Iorio, Marc A. Quantitative biostratigraphic analysis of the Cenozoic of the Labrador Shelf and Grand Banks. D, 1988, University of Ottawa. 404 p.

D'Lugosz, Joseph J. The description and origin of intraformational folds near Tres Ritos and Holman Hill, New Mexico. M, 1971, Texas Tech University. 88 p.

D'Orazio, Timothy Bruno. The behavior of steel oil storage tanks on compressible foundations. D, 1982, University of California, Berkeley. 165 p.

d'Orsay, Albert Murray. Stratigraphy and sedimentology of Carboniferous rocks in the northwestern Minas Basin and Channel region of Nova Scotia. M, 1986, University of New Brunswick. 301 p.

D'Urso, Gary John. An investigation of the Precambrian rocks of the Point of Rocks Quadrangle, Frederick County, Maryland and Loudoun County, Virginia. M, 1981, University of Pittsburgh.

D., Eduardo E. Chavez see Chavez D., Eduardo E.

D., Francisco H. Vargas see Vargas D., Francisco H.

D., Ignacio Martinez see Martinez D., Ignacio

da Boa Hora, Marco Polo Pereira. Maximum entropy spectral analysis of two-dimensional magnetic and gravity anomalies. D, 1980, St. Louis University. 213 p.

da Cunha, Roberto Pereira see Pereira da Cunha, Roberto

Da Prat, Giovanni C. Well test analysis for naturally-fractured reservoirs. D, 1981, Stanford University. 214 p.

da Silva, Joao Batista Correa. Three-dimensional magnetic inversion. D, 1982, University of Utah. 176 p.

da Silva, Maria Augusta Martins. The Araripe Basin, northeastern Brazil; regional geology and facies analysis of a Lower Cretaceous evaporitic depositional complex. D, 1983, Columbia University, Teachers College. 308 p.

da Silva, Zenaide Carvalho Goncales. Studies on jadeites and albatites from Guatemala. M, 1967, Rice University. 21 p.

Dabbagh, Abdullah E. Geology of the Skyland and Dunsmore Mountain quadrangles, western North Carolina. D, 1975, University of North Carolina, Chapel Hill. 228 p.

Dabbagh, Ali A. Study of thermal stress effects on hydraulic fracturing of rock. M, 1979, University of Nevada. 95 p.

Dabbagh, Mohammad Eesa. Environmental interpretation and tectonic significance of the Wajid Sandstone, Saudi Arabia. M, 1981, University of North Carolina, Chapel Hill. 84 p.

Dabbagh, Mohammad Eesa. Tertiary and associated rocks of Yanbu, Dhaylan and Aznam areas, northwestern Saudi Arabia, and their relationship to the opening of the Red Sea. D, 1988, University of North Carolina, Chapel Hill. 156 p.

Dabitzias, Spyros Georgiou. Petrology and genetic model of the Vavdos cryptocrystalline magnesite deposit, Chalkidiki Peninsula, northern Greece. M, 1977, Queen's University. 102 p.

Dablain, Mark Albert. Low velocity underwater explosions; a non-L model. D, 1980, St. Louis University.

Dablow, John F. Late stage, semi-brittle deformation in the Lake Vermilion Formation of Northeast Minn. M, 1975, University of Minnesota, Minneapolis.

DaBoll, Joan. Holocene sediments of the Parker River estuary (Massachusetts). M, 1969, University of Massachusetts. 138 p.

Dabous, Adel Ahmed. Mineralogy, geochemistry and radioactivity of some Egyptian phosphorite deposits. D, 1981, Florida State University. 217 p.

Dachille, Frank. High pressure studies of the systems Mg_2GeO_4-Mg_2SiO_4 and GeO_2 with special reference to the olivine-spinel transition. D, 1959, Pennsylvania State University, University Park. 83 p.

Dacre, George J. Petrology, mineral chemistry, and geothermometry of some lower Precambrian granulite facies rocks, East-central Minnesota. M, 1981, University of Missouri, Columbia.

Dade, William Brian. Bryozoans of the modern Wallops-Chincoteague coast, Va. M, 1983, Pennsylvania State University, University Park. 217 p.

Dadgari, Farzad. Pedogenesis of sodium ion- and magnesium ion-affected Sedgefield soils (fine, mixed, thermic, Aquultic Hapludalfs) in the North Carolina Piedmont. D, 1983, North Carolina State University. 250 p.

Dadkhah, Arsalan. Heat transfer from a solar pond through saturated groundwater flow. D, 1985, Utah State University. 156 p.

Dadkhah, Manouchehr. The influence of rock cover, vegetal cover, grass species, and simulated trampling on infiltration rates and sediment production. D, 1979, Utah State University. 187 p.

Dadoly, John Peter. Gold mineralization in a regional metamorphic terrane; a wall-rock alteration study of the Chichagof and Hirst-Chichagof gold mines, Southeast Alaska. M, 1987, South Dakota School of Mines & Technology.

Dadourian, P. Polyphase deformation and metamorphism in the northern part of the Manhattan Prong in northern Westchester and southern Putnam counties, N.Y. M, 1978, Queens College (CUNY). 108 p.

Dadson, A. S. A study of some Canadian apatites. M, 1933, University of Toronto.

Dadson, A. S. The role of electrical potential in ore deposition in the Timiskaming District, Ontario. D, 1938, University of Toronto.

Daemen, J. J. K. Tunnel support loading caused by rock failure. D, 1975, University of Minnesota, Minneapolis. 431 p.

Daetwyler, Calvin Crowell, Jr. Marine geology of Tomales Bay, Central California. D, 1965, University of California, San Diego. 219 p.

Daftary, Homayoun. A quantitative study of the Ghazvin Plain. M, 1979, Ohio University, Athens.

Dagel, Mark A. Stratigraphy and chronology of Stage 6 and 2 glacial deposits, Marshall Valley, Antarctica. M, 1985, University of Maine. 102 p.

Dagenais, Georges Roman. The oxygen isotope geochemistry of granitoid rocks from the southern and central Yukon. M, 1984, University of Alberta. 182 p.

Dagenais, Solange. Pétrographie et stratigraphie de la séquence des paragneiss de St. Fulgence, région du Haut-Saguenay, Québec. M, 1983, Universite du Quebec a Chicoutimi. 165 p.

Dagenhart, Thomas Vernon, Jr. The acid mine drainage of Contrary Creek, Louisa County, Virginia; factors causing variations in stream water chemistry. M, 1980, University of Virginia. 215 p.

Daggett, Joyce M. Sediments of the Indian River and the impounded waters near Kennedy Space Center. M, 1973, Florida Institute of Technology.

Daggett, Larry Leon. An optimization technique for the development of a ground water management program. D, 1969, Arizona State University. 162 p.

Daggett, Maxcy DeWitt, III. The structure and petrology of the Cedar Mountain Complex, Redwood County, Minnesota. M, 1980, University of Minnesota, Minneapolis. 85 p.

Daggett, Paul H. A geophysical investigation of the crust using active and passive seismic methods with examples from southern New Mexico and eastern Egypt. M, 1977, New Mexico State University, Las Cruces. 53 p.

Daggett, Paul Henry. An integrated geophysical study of the crustal structure of the southern Rio Grande Rift. D, 1982, New Mexico State University, Las Cruces. 208 p.

Daghlian, C. P. Coryphoid palms from the lower and middle Eocene of southeastern North America. D, 1977, University of Texas, Austin. 189 p.

Dahl, A. R. Petrography of till and loess, south-central Iowa. M, 1958, Iowa State University of Science and Technology.

Dahl, Arthur Richard. Missouri River studies; alluvial morphology and Quaternary history. D, 1961, Iowa State University of Science and Technology. 281 p.

Dahl, Charles Laurence. Trace ferrides in iron ores from the Iron Springs District, Cedar City (Iron County), Utah. M, 1959, University of Utah. 62 p.

Dahl, David Alvin. Petrology and geochemistry of nepheline trachytes and phonolites in the Black Hills area, Brewster County, Trans-Pecos, Texas. M, 1984, Texas Christian University. 103 p.

Dahl, Gardar G., Jr. General geology of the area drained by the north fork of the Smith River, Meagher County, Montana. M, 1971, Montana College of Mineral Science & Technology. 58 p.

Dahl, Harry M. Fluorescence analysis of columbium and tantalum. M, 1951, Columbia University, Teachers College.

Dahl, Harry Martin. Alteration in the central uranium area, Marysvale, Utah. D, 1954, Columbia University, Teachers College. 160 p.

Dahl, Hilbert Douglas. A finite element model for anisotropic yielding in gravity loaded rock. D, 1969, Pennsylvania State University, University Park. 167 p.

Dahl, Jeremy Eliot. A study of the carbon cycle. M, 1982, Rice University. 88 p.

Dahl, John. Surficial geology of the Mechanicville and Schaghticoke quadrangles, NY. M, 1978, Rensselaer Polytechnic Institute. 54 p.

Dahl, Mary Katherine. Structural and stratigraphic control of ore through the Crescent, Sunshine, Silver Summit, Coeur and Galena mines, Coeur d'Alene mining district, Idaho. M, 1981, University of Idaho. 97 p.

Dahl, Peter Steffen. The mineralogy and petrology of Precambrian metamorphic rocks from the Ruby

Mountains, southwestern Montana. D, 1977, Indiana University, Bloomington. 280 p.

Dahl, William Martin. Progressive-burial diagenesis in lower Tuscaloosa sandstones, Louisiana and Mississippi. M, 1984, University of New Orleans. 145 p.

Dahlberg, Eric Charles. A multivariate study of some aspects of trace metals in stream sediments as guides to locating mineral deposits. D, 1967, Pennsylvania State University, University Park. 163 p.

Dahlberg, Eric Charles. Statistical analysis of petrographic variability in a graded bed. M, 1964, Pennsylvania State University, University Park. 156 p.

Dahleen, William. Physical stratigraphy of the lower and middle Miocene rocks of the San Joaquin Hills, California. M, 1971, University of California, Riverside. 82 p.

Dahlem, David Harrison. Geology of the Lookout Mountain area, Fremont County, Colorado. D, 1965, University of Michigan. 204 p.

Dahlem, David Harrison. Geology of the Yaak River - Kootenai River confluence. M, 1959, Montana College of Mineral Science & Technology. 131 p.

Dahlem, Robert Dale. Stratigraphy and paleontology of the Hibbard No. 1 well core, Barry County, Michigan. M, 1959, University of Michigan.

Dahlen, Francis Anthony, Jr. The normal modes of a rotating elliptical Earth. D, 1969, University of California, San Diego. 201 p.

Dahlen, Margaret. Seismic stratigraphy of the Ventura mainland shelf, California; late-Quaternary history of sedimentation and tectonics. M, 1988, University of Southern California.

Dahlgren, Paul B. Petrology of late Triassic lacustrine carbonates in the Newark Basin, New Jersey. M, 1975, Rutgers, The State University, New Brunswick. 80 p.

Dahlin, Brian B. A contribution to the study of meandering. M, 1974, University of Minnesota, Minneapolis. 70 p.

Dahlstrom, C. D. A. Petrological studies in the Tazin Group of Lake Athabasca. M, 1949, University of Saskatchewan. 37 p.

Dahlstrom, Clinton D. A. Statistical analysis of folds. M, 1953, University of Saskatchewan.

Dahlstrom, Clinton Dennis Augustine. Statistical analyses of folds. D, 1952, Princeton University. 86 p.

Dahlstrom, David James. Fluvial architecture of the Lower Cretaceous Lakota Formation, southwestern flank of the Black Hills Uplift, South Dakota. M, 1986, South Dakota School of Mines & Technology.

Dahm, Clifford Neal. Studies on the distribution and fates of dissolved organic carbon. D, 1980, Oregon State University. 160 p.

Dahm, Cornelius G. A study of dilatational wave velocity in the Earth as a function of depth, based on a comparison, of P, P' and PcP phases. D, 1934, St. Louis University.

Dahm, Cornelius G. Location of the epicenter of Hawke Bay (New Zealand) earthquake of February 2, 1931. M, 1932, St. Louis University.

Dahm, Cornelius G. The geology of the Gulf Coast salt domes and the geophysical methods used in locating them. M, 1930, St. Louis University.

Dahm, Jerry B. Geometry and timing of Tertiary deformation of the Dome Rock Mountains, Yuma County, Arizona. M, 1983, San Diego State University. 176 p.

Dahmani, Mohamed Amine. The effects of alcohols and crude oil composition on the performance and mechanisms of alkaline flooding of oil reservoirs. D, 1986, Louisiana State University. 254 p.

Dahy, James P. The geology and igneous rocks of the Yogo sapphire deposit and the surrounding area, Little Belt Mountains, Judith Basin County, Montana. M, 1988, Montana College of Mineral Science & Technology. 92 p.

Dai, Ting-fang. Kirchoff elastic wave migration. D, 1988, Columbia University, Teachers College. 78 p.

Daigle, Deborah M. Origin of deep marine dolomite in the Gulf of Mexico and the Caribbean Sea. M, 1986, Memphis State University. 118 p.

Daigneault, Réal. Géochimie et géologique du gisement d'or de la mine Lamaque, Val d'Or, Québec. M, 1984, Ecole Polytechnique. 174 p.

Dail, John Hugh. The geology of the Maryville Quadrangle, Blount and Knox counties, Tennessee. M, 1950, University of Tennessee, Knoxville. 59 p.

Dail, Rhea A. The geology of the Louisville Quadrangle, Blount and Knox counties, Tennessee. M, 1950, University of Tennessee, Knoxville. 60 p.

Dailey, Dale V. A comparative analysis of three soil compactors for use with the Mini-Sosie recording system. M, 1985, Wright State University. 80 p.

Dailey, Donald Howard. Early Cretaceous foraminifera from the Budden Canyon Formation, northwestern Sacramento valley, California. D, 1969, University of California, Berkeley. 299 p.

Dailey, Donald Howard. Stratigraphic paleontology of the Jalama Formation, western Santa Ynez Mountains, Santa Barbara County, California. M, 1960, University of California, Los Angeles.

Daily, Micheal Irvin. Applications of imaging radar to geology. D, 1984, University of California, Santa Barbara. 339 p.

Dainty, Anton Michael. Crustal studies in eastern (maritime) Canada. D, 1967, Dalhousie University. 144 p.

Dainty, Norman Dale. Petrology of the strata of Patos Island, San Juan County, Washington. M, 1981, Washington State University. 107 p.

Dake, Charles L. The formation of iron sediments at the present time in its relation to the origin of sedimentary iron ores. M, 1912, University of Wisconsin-Madison.

Dake, Charles L. The problem of the St. Peter Sandstone. D, 1922, Columbia University, Teachers College.

Dakessian, Suren. Strength characteristics of root-reinforced soils. D, 1980, University of California, Berkeley. 203 p.

Dakin, Francis. Local mineralogical variations within a limited gabbro of Cape Neddick, Maine. M, 1968, Massachusetts Institute of Technology. 74 p.

Dakin, R. A. The origin of salts in groundwater, Mayne Island, British Columbia. M, 1974, University of Waterloo.

Dakoski, Andrea Marie. Recent microbial mats, stromatolites, and related sediments of Granny Lake, San Salvador, Bahamas. M, 1986, University of Akron. 100 p.

Dal Bello, Anthony Eugene. Stratigraphic position and petrochemistry of the Love Cove Group, Glovertown-Traytown map area, Bonavista Bay, Newfoundland, Canada. M, 1977, Memorial University of Newfoundland. 159 p.

Dalby, Charles E. Evaluation of the potential effects of land use on the hydrology and fluvial geomorphology of the North Fork of the Flathead River. M, 1986, University of Montana. 223 p.

Dale, Christopher Thomas. A study of high resolution seismology and sedimentology on the offshore late Quaternary sediments northeast of Newfoundland. M, 1979, Dalhousie University. 181 p.

Dale, Michael W. Hydrology of the southern part of the Peavine Mountain-Silver Lake sub-basin. M, 1987, University of Nevada. 124 p.

Dale, Nelson C. The Cambrian manganese deposits of Conception and Trinity bays, Newfoundland. D, 1914, Princeton University.

Dale, R. H. Relationship of ground-water tides to ocean tides; a digital simulation model. D, 1974, University of Hawaii. 150 p.

Dale, Robert H. The geology of the southwest quarter of the Dale Quadrangle, Oregon. M, 1957, University of Oregon. 44 p.

Dale, Ruth. Petrology of the Cretaceous sandstones in the Pleasanton area, Alameda and Contra Costa counties, California. M, 1969, Stanford University.

Dale-Bannister, Mary Ann. On the types of rocks at the Viking Lander sites. M, 1986, Washington University. 82 p.

Dalen, Stephen Craig Van see Van Dalen, Stephen Craig

Daleon, Benjamin. Some Philippine upper Tertiary foraminifera. M, 1943, California Institute of Technology. 12 p.

Dales, Benton. Contributions to the chemistry of the rare earths of the yttrium group. D, 1901, Cornell University.

Dales, R. Graeme. Deformation of massive sulphide lenses and their wall rocks, Benny Belt, Ontario. M, 1978, University of Toronto.

Daley, A. Cowles and Poole, David M. A geologic section in east-central California eastward from Donner Pass. M, 1949, University of California, Los Angeles.

Daley, E. Ellen. Petrology, geochemistry, and the evolution magmas from Augustine Volcano, Alaska. M, 1985, University of Alaska, Fairbanks. 106 p.

Daley, Roberta L. Patterns and controls of skeletal silicification in a Mississippian fauna, northwestern Wyoming. M, 1987, University of Wyoming. 140 p.

Dalgleish, Janet Blair. Sinkhole distribution in Winona County, Minnesota. M, 1985, University of Minnesota, Minneapolis. 95 p.

Dalheim, Peggy Ann. Calculation of empirical correlation coefficients in multicomponent oxide systems for the reduction of electron microprobe data. M, 1977, University of Oregon. 88 p.

Dali, Ayad H. Depositional environment of the upper Silurian of the Michigan Basin. M, 1975, Michigan State University. 44 p.

Dalke, Roger A. Nonlinear effects on wave propagation in a cylindrical elastic body. D, 1986, Colorado School of Mines. 76 p.

Dallemagne, Pierre G. Determination of trace inorganic and organometallic compounds of mercury in sea water. M, 1974, Florida Institute of Technology.

Dallmeyer, Mary D. Gilmore. Quaternary biostratigraphy of deep-sea benthic formainifera from the eastern Caribbean Sea. M, 1977, University of Georgia.

Dallmeyer, Ray David. Structural and metamorphic history of the northern Reading Prong, southeastern New York and northern New Jersey. D, 1972, SUNY at Stony Brook. 295 p.

Dally, David J. Depositional environment of Canyon (Cisco) sandstones, North Jameson Field, Mitchell County, Texas. M, 1983, Texas A&M University. 105 p.

Dally, Jesse LeRoy. The stratigraphy and paleontology of the Pocono Group in West Virginia. D, 1956, Columbia University, Teachers College. 280 p.

Dalness, William Michael. The Parunuweap Formation (Pliocene?) in the vicinity of Zion National Park, Utah. M, 1969, University of Utah. 77 p.

Dalphin, Richard James. A Markov-Weibull model of hydrologic drought in the Farmington River basin of Connecticut and Massachusetts. D, 1983, University of Connecticut. 879 p.

Dalrymple, Don W. The geology and petrography of the Parlin area, Gunnison County, Colorado. M, 1962, Wichita State University. 81 p.

Dalrymple, Don Wayne. Recent sedimentary facies of Baffin Bay, Texas. D, 1964, Rice University. 250 p.

Dalrymple, Gary Brent. Potassium-argon dates and the Cenozoic chronology of the Sierra Nevada, California. D, 1963, University of California, Berkeley. 109 p.

Dalrymple, Margaret Ann. Enhancement of copper anomalies in stream sediments of the Big Delta Quadrangle, Alaska. M, 1981, Kent State University, Kent. 110 p.

Dalrymple, Robert Walker. Sediment dynamics of macrotidal sand bars, Bay of Fundy. D, 1977, McMaster University. 635 p.

Dalton, Dale V. The subsurface geology of Northeast Payne County, Oklahoma. M, 1960, University of Oklahoma. 69 p.

Dalton, David C. Long-period background earth noise as measured in shallow, hand excavated holes. M, 1988, Virginia Polytechnic Institute and State University. 36 p.

Dalton, Edward. Sedimentary facies and diagenesis of the Lower Devonian Temiscouata and Fortin formations, Northern Appalachians, Quebec and New Brunswick. M, 1987, McGill University. 228 p.

Dalton, Mary Chalk. Characteristics of the oil fields producing from Mesozoic and Cenozoic Formations of Texas. M, 1938, University of Texas, Austin.

Dalton, Matthew G. Geochemistry of the contact between bicarbonate and upwelling sulfate waters in the Floridan Aquifer. M, 1978, University of South Florida, Tampa. 101 p.

Dalton, Richard Clyde. Stratigraphy and areal geology of central Choctaw County, Oklahoma. M, 1966, University of Oklahoma. 68 p.

Dalton, Russell O., Jr. Stratigraphy of the Bass Formation (Late Precambrian, Grand Canyon, Arizona). M, 1972, Northern Arizona University. 140 p.

Daly, Alan Ronald. Processes of sedimentation and depositional environments in a temperate barrier bar-lagoonal complex. M, 1976, Queen's University. 99 p.

Daly, C. J. Analytical/numerical methods for groundwater flow and quality problems. D, 1979, Colorado State University. 178 p.

Daly, Cathryn Hayes. An evaluation of an area of potential molybdenum mineralization, Chicago Park, Gunnison County, Colorado. M, 1983, South Dakota School of Mines & Technology.

Daly, Edward Joseph. The trepostomatous bryozoan genus Hallopora Bassler in the Dillsboro Formation (Cincinnatian) of southeastern Indiana. M, 1979, Boston College.

Daly, Eleanor. The unarmored dissorophids (Amphibia, Labyrinthodontia), with a description of a new genus from the Pennsylvanian of Kansas. D, 1981, University of Kansas. 172 p.

Daly, John W. The geology and mineralogy of the limestone deposits at Crestmore, Riverside County, California. M, 1931, California Institute of Technology. 72 p.

Daly, Paul J. Comparison between the electrical resistivity and seismic methods for depth determination. M, 1951, Boston College.

Daly, Philip J. Sorbent selection criteria for Ohio carbonates. M, 1983, University of Akron. 76 p.

Daly, Reginald Aldworth. Studies on the so-called porphyritic gneiss of New Hampshire. D, 1896, Harvard University.

Daly, Stephen F. Convection with decaying heat sources and the thermal evolution of the mantle. D, 1978, University of Chicago. 170 p.

Daly, William E. Basement control of the deposition of the Cambrian Deadwood Formation in the eastern Black Hills, South Dakota. M, 1981, South Dakota School of Mines & Technology. 89 p.

Dalziel, Mary Catherine. Geochemistry of Antarctic sediments, with emphasis on the distribution of barium. M, 1975, University of Miami. 58 p.

Dam, Dale A. Van *see* Van Dam, Dale A.

Dam, George Henry Van *see* Van Dam, George Henry

Dam, William L. The geochemistry of metals in groundwaters after in-situ uranium mining. M, 1984, University of Wyoming. 60 p.

Damassa, Sarah Pierce. Early Tertiary dinoflagellates from the Coastal Belt of the Franciscan Complex, northern California. D, 1979, University of California, Los Angeles. 270 p.

Dameron, Wyllie Frank. An investigation of geological correlation by specific gravity. M, 1950, Texas Christian University. 31 p.

Damiata, Brian Neal. Geothermal exploration for direct use applications in the vicinity of Lake Elsinore, Southern California. M, 1986, University of California, Riverside. 196 p.

Damle, Mayurika. Velocity analysis based on pre-stack imaging. M, 1987, University of Houston.

Damm, Karen Louise Von *see* Von Damm, Karen Louise

Dammann, Arthur. The preliminary study of the properties and uses of Pacific Northwest diatomites. M, 1939, University of Washington. 137 p.

Damon, Henry Gordon. Cretaceous conglomerates on the east side of the Llano Uplift, Texas. D, 1940, University of Iowa. 92 p.

Damon, Henry Gordon. The vertical displacement in the main fault of the Balcones fault system at a point west of the City of Austin, Texas. M, 1924, University of Texas, Austin.

Damp, Jeffery N. Controls of Precambrian dike emplacement in the Hanging Canyon area, Grand Teton National Park, Wyoming. M, 1976, Idaho State University. 74 p.

Dampney, C. N. G. From wide angle reflection to leaking mode seismograms, a theoretical and experimental study. D, 1970, University of Toronto.

Damron, Larry A. Physical modeling of lateral variations of resistivity in transient electromagnetics. M, 1986, Colorado School of Mines. 85 p.

Damuth, John Douglas. The evaluation of the degree of community structure preserved in assemblages of fossil mammals. D, 1982, University of Chicago. 248 p.

Damuth, John Erwin. Arkosic sands of the last glacial stage in the western equatorial Atlantic off northeast South America. M, 1968, Columbia University. 49 p.

Damuth, John Erwin. The western equatorial Atlantic; morphology, Quaternary sediments, and climatic cycles. D, 1973, University of Colorado. 629 p.

Dana, Edward S. Trap rocks of the Connecticut Valley. D, 1876, Yale University.

Dana, Frederick F. Oil pay shapes in the Hundred Foot sand. M, 1931, University of Pittsburgh.

Dana, George F. The subsurface geology of Grant County, Oklahoma. M, 1954, University of Oklahoma. 61 p.

Dana, John Kenneth. Alteration of sedimentary rocks related to uranium mineralization in the North Cave Hills of Harding County, South Dakota. M, 1978, South Dakota School of Mines & Technology.

Dana, Richard H., Jr. Stratigraphy and structural geology of the Lake Warmaug area, western Connecticut. M, 1977, University of Massachusetts. 108 p.

Dana, Robert W. Measurements of 8-14 micron emissivity of igneous rock and mineral surfaces. M, 1969, University of Washington. 78 p.

Dana, Stephen W. A sedimentary study of bottom mud samples dredged by the Velero III. M, 1942, University of Southern California.

Dana, Stephen W. Amplitudes of seismic waves reflected and refracted at the Earth's core. D, 1944, California Institute of Technology. 161 p.

Danahy, Thomas V. Hillsdale Limestone (Mississippian, Meramecian), Washington County, VA. M, 1986, East Carolina University. 107 p.

Danbom, S. H. Sediment classification by seismic reflectivity in eastern Block Island Sound. D, 1975, University of Connecticut. 158 p.

Danbom, Stephen H. A gravity and magnetic investigation of the Amarillo Uplift. M, 1969, Texas Tech University. 60 p.

Dance, John Thomas. Evaluation of reactive solute transport in a shallow unconfined sandy aquifer. M, 1980, University of Waterloo.

Dandavati, Kumar S. Sedimentology and uranium potential of the Inyan Kara Group near Buffalo Gap, South Dakota. M, 1980, South Dakota School of Mines & Technology.

Dandekar, Dattataya. Variation in elastic constants of calcite with pressure. D, 1967, University of Chicago. 49 p.

Dando, Kathryn Hickman. Stratigraphy and conodont paleontology of the Mazarn Shale south of Lake Ouachita, Garland County, Arkansas. M, 1980, University of New Orleans. 69 p.

Dando, Mark. Structural geology of the northern portion of the Jessieville Quadrangle, Ouachita Mtns., Arkansas. M, 1978, University of Missouri, Columbia.

Dane, Carle H. Geology of the Salt Valley Anticline and the northwest flank of the Uncompahgre Plateau, Utah. D, 1932, Yale University.

Dane, Ernest B., Jr. Origin of the openings occupied by veins. D, 1936, Harvard University.

Daneker, Thomas M. Sedimentology of the Precambrian Shinumo Sandstone, Grand Canyon, Arizona. M, 1975, Northern Arizona University. 195 p.

Daneshvar, Mohammad Reza. Easterly extension of the Flambeau resistivity anomaly in northern Wisconsin. M, 1977, University of Wisconsin-Madison.

Daneshvar, Mohammad Reza. Imaging of rough surfaces and planar boundaries using passive seismic signals. D, 1977, University of Wisconsin-Madison. 193 p.

Daneshy, Abbas Ali. Numerical inversion of the Laplace transformation and the solution of the viscoelastic wave equations. D, 1969, University of Missouri, Rolla.

Danforth, Carroll F. Oil shale; a competitor of natural petroleum. M, 1947, University of Pittsburgh.

Danforth, Isabel Levin. Sedimentation in Pollet Bay (north end of Lake Saint Clair, north of Detroit, Michigan, near the town of Algonac). M, 1967, University of Michigan.

Danforth, William W. Petrology of the western Scituate Granite. M, 1986, University of Rhode Island.

Daniel, Barbara J. Pliocene-Pleistocene paleoceanography of the Southeast Indian Ocean based on magnetostratigraphic-micropaleontologic studies of piston cores. M, 1983, University of Georgia.

Daniel, Charles Camp, III. Metamorphism of sheet silicates in response to temperature and pressure gradients. D, 1974, University of North Carolina, Chapel Hill. 136 p.

Daniel, David Edwin, Jr. Moisture movement in soils in the vicinity of waste disposal sites. D, 1980, University of Texas, Austin. 253 p.

Daniel, Debra L. Experimental studies in the system $CaAl_2Si_2O_8$-$MgSiO_3$-$FeSiO_3$ and their application to achondrite petrogenesis. M, 1978, Arizona State University. 36 p.

Daniel, Herbert R. Geology of the Log Cabin area, near Questa molybdenum mine, Taos County, New Mexico. M, 1967, University of Arizona.

Daniel, Joseph Hawkins. Engineering geology and geohydrology of the Burkeville confining system northeast of Conroe, Texas. M, 1988, Texas A&M University. 124 p.

Daniel, Maria M. Depositional environment and diagenesis of Upper Jurassic-Lower Cretaceous limestone and dolomite, Galicia margin, offshore Spain. M, 1988, University of Tulsa. 154 p.

Daniel, Marshall Edward, IV. Alteration of shale adjacent to a fluorspar ore body in the southern Illinois fluorspar district. M, 1972, Southern Illinois University, Carbondale. 68 p.

Daniel, P. E. Longshore currents in the vicinity of a breakwater. M, 1978, University of British Columbia.

Daniel, R. G. Evaluation of intermediate-period seismic waves as a geothermal exploration tool. D, 1979, Stanford University. 143 p.

Daniel, T. H. Cruise AP-4; acoustic investigations in the eastern tropical Pacific. M, 1973, University of Hawaii. 76 p.

Daniel, T. H. Tri-axial electric field measurements for determining deep ocean water movements; techniques and a preliminary application. D, 1978, University of Hawaii. 185 p.

Danielito, Tan Franco. Operations policy for the Upper Pampanga River project reservoir system in the Philippines. D, 1977, University of Arizona.

Daniels, David Lee. Igneous intrusive relationships in Patterson Mountain Quadrangle, California. M, 1963, University of California, Berkeley. 82 p.

Daniels, Eric Joseph. Origin and distribution of minerals in shale and coal from the anthracite region, eastern Pennsylvania. M, 1988, University of Illinois, Urbana. 97 p.

Daniels, Jeffrey Irwin. Environmental, health, safety, and socioeconomic impacts associated with oil recovery from tar-sand deposits in the United States. D, 1981, University of California, Los Angeles. 181 p.

Daniels, Jeffrey J. Two dimensionality in magnetic interpretation. M, 1970, Michigan State University. 40 p.

Daniels, Lawrence David. Diagenesis and paleokarst of the Burlington-Keokuk Formation (Mississippian), central and southwestern Missouri. M, 1986, SUNY at Stony Brook. 401 p.

Daniels, Paul A., Jr. Heavy-mineral distribution in the White Oak Estuary-Bogue Inlet area, North Carolina. M, 1968, Bowling Green State University. 66 p.

Daniels, Robert P. Pennsylvanian (Desmoinesian) stratigraphy and petroleum potential, Southeast Colorado. M, 1985, Colorado School of Mines. 129 p.

Daniels, Shirland Augustus. Sediment contaminant assessment techniques; Toronto Harbour dredging/spoils disposal bio-assessment applications. D, 1988, University of Waterloo. 242 p.

Danielson, Daryl Arthur, Jr. Stratigraphy, petrology, and structure of the Blakely Sandstone, Mountain Pine Quadrangle, Garland County, Arkansas. M, 1987, University of New Orleans. 244 p.

Danielson, Joanne. Lithology and geochemistry of the French Gulch Inlier, Klamath Mountains, Northern California. M, 1988, California State University, Chico. 230 p.

Danielson, Richard Earl. Geology of a portion of the Augusta Quadrangle. M, 1956, Washington University. 29 p.

Danielson, Stephen Eric. Provenance of the lower Jackfork Sandstone, Ouachita Mountains, Arkansas and eastern Oklahoma. M, 1987, University of New Orleans. 187 p.

Daniyan, Muhammad Abdullahi. Theoretical magnetotelluric response for simple three dimensional models. D, 1983, Southern Methodist University. 144 p.

Danker, Jeanne A. A micro-environmental study of the hydrocarbons of the Long Island Sound (New York; Connecticut) sediments. M, 1964, New York University.

Danko, Jeffrey H. Stratigraphy and microfacies analysis of the Canutillo Formation (late Middle Devonian), Franklin Mountains, Texas and New Mexico, and Bishop Cap Hills, New Mexico. M, 1981, University of Texas at El Paso.

Danley, William M. Textural, mineralogical, and chemical aspects of dunite body, Bushveld complex (Precambrian), South Africa. M, 1969, University of Wisconsin-Madison.

Dann, Jesse C. Major-element variation within the Emperor igneous complex and the Hemlock and Badwater volcanic formations. M, 1978, Michigan Technological University. 198 p.

Danna, James G. Experimental study of clast orientation in gravels deposited by unidirectional flow. M, 1985, Massachusetts Institute of Technology. 84 p.

Dannemiller, Gary Thomas. Water quality changes associated with fall overturn at Lake Lynn, West Virginia. M, 1972, University of Akron. 77 p.

Dannemiller, George D. Geology of the central part of the James River valley, Mason County, Texas. M, 1957, Texas A&M University.

Dannenberg, Roy Berry. The subsurface geology of Coal County, Oklahoma. M, 1952, University of Oklahoma. 63 p.

Danner, David Lee. Planning criteria for urban water pollution control. D, 1982, Catholic University of America. 162 p.

Danner, Wilbert Roosevelt. A contribution to the geology of the Olympic Mountains, Washington. M, 1948, University of Washington. 67 p.

Danner, Wilbert Roosevelt. A stratigraphic reconnaissance in northwestern Cascades and San Juan Islands of Washington State. D, 1957, University of Washington. 562 p.

Dansart, William Joseph. A petrographic study of the Josephine Breccia in the Metaline District of northeastern Washington using cathodoluminescence. M, 1982, University of Idaho. 72 p.

Danser, James Weart. The geology of township 50 North, range 13 West, Boone County, Missouri. M, 1950, University of Missouri, Columbia.

Danskin, Wesley Robert. Hydrologic optimization techniques used to evaluate management policies for a surface-water/ground-water system near Livermore, California. M, 1985, Stanford University. 146 p.

Danusawad, Thawisak. An investigation of electrical transient shapes. M, 1960, Washington University. 211 p.

Danzl, Ralph. Depositional and diagenetic environment of the Salt Lake Group at Oneida Narrows, southeastern Idaho. M, 1985, Idaho State University. 177 p.

Dapples, Edward C., Jr. The weathering of Illinois coal. M, 1934, Northwestern University.

Dapples, Edward Charles. The geology and coal deposits of the Anthracite-Crested Butte quadrangles, Colorado. D, 1938, University of Wisconsin-Madison.

Dar, Ikram-ul-Hag. An investigation of the creep characteristics of Portland cement mortar under static and dynamic strain. D, 1967, University of Missouri, Rolla.

Darakos, William Efstratos. Free-swelling property of the Chilton Coal of Logan County, West Virginia. M, 1955, University of Pittsburgh.

Darbha, D. M. Thermal conductivity of Earth materials at high pressures and temperatures. M, 1977, University of Western Ontario.

Darby, David Grant. A revision of the Ordovician trilobite Asaphus platycephalus. M, 1961, University of Michigan.

Darby, David Grant. Ecology and taxonomy of Ostracoda in the vicinity of Sapelo Island, Georgia. D, 1964, University of Michigan. 202 p.

Darby, Dennis Arnold. Carbonate cycles and clay mineralogy of Arctic Ocean sediment cores. D, 1971, University of Wisconsin-Madison. 129 p.

Darby, Dennis Arnold. Coralgal deposits from Abaco Island, Bahamas. M, 1968, University of Pittsburgh.

Darby, W. P. Use of indirect indicators of physical activities and land use planning for the management of urban watersheds. D, 1975, Carnegie-Mellon University. 262 p.

Darden, Larry B. Carbonate petrology and microfacies analysis of the El Abra reef complex (Cretaceous), Mexico. M, 1968, Texas Tech University. 92 p.

Dargush, Gary Franklyn. Boundary element methods for the analogous problems of thermomechanics and soil consolidation. D, 1987, SUNY at Buffalo. 289 p.

Dark, William M. Hydrocarbon potential of the eastern Arkoma Basin. M, 1985, University of Arkansas, Fayetteville.

Darken, William H. A finite difference model of channel waves in a coal seam. M, 1975, Colorado School of Mines. 78 p.

Darling, Bruce K. Depositional environments of the DeQueen Formation (Trinity Group) in the Nathan Quadrangle of southwestern Arkansas. M, 1984, University of Southwestern Louisiana. 139 p.

Darling, R. M. Relation of particle size and mineral distribution to physical properties of Gulf Coast drilling muds (C249). M, 1944, University of Texas, Austin.

Darling, Richard Graydon. The distribution of fluorine and other elements in quartz monzonite near ore, Pine Creek contact-metasomatic tungsten deposits, Inyo County, California. D, 1967, Stanford University. 173 p.

Darling, Robert S. The geology and ore deposits of the Carrietown silver-lead-zinc district, Blaine and Camas counties, Idaho. M, 1987, Idaho State University. 168 p.

Darling, Robert William. Geology of eastern half of the Durham Quadrangle, Northwest Georgia. M, 1952, Emory University. 107 p.

Darlington, Julian Trueheart. The Turbellaria of two granite outcrops in Georgia. D, 1953, University of Florida. 86 p.

Darmody, Robert George. Geomorphic and pedogenic relationships of Elioak and associated soils in the Piedmont of Maryland. D, 1980, University of Maryland. 212 p.

Darnell, Nancy Rebecca. A comparison of surficial, in situ sediments overlying plutonic rocks of the Boulder Batholith and gneissic rocks of the southern Tobacco Root Mountains in Montana. M, 1974, Indiana University, Bloomington. 126 p.

Darnell, Richard Douglas. Mississippian rocks of the Claremore-Wagoner area, Oklahoma. M, 1957, University of Oklahoma. 87 p.

Darnell, William I. The Imperial Valley; its physical and cultural geography. M, 1959, San Diego State University.

Darr, James M. Geochemistry of ground water, southwestern Missouri. M, 1978, University of Missouri, Columbia.

Darrach, Mark Edward. A kinematic and geometric structural analysis on an early Proterozoic crustal-scale shear zone; the evolution of the Shylock fault zone, central Arizona. M, 1988, Northern Arizona University. 78 p.

Darrag, Ahmad Amr. Capacity of driven piles in cohesionless soils including residual stresses. D, 1987, Purdue University. 418 p.

Darragh, Robert Bernard. Analysis of near-source waves; separation of wave types using strong motion array recordings. D, 1987, University of California, Berkeley. 156 p.

Darrell, James H., Jr. The palynology of a lignite in Northwest Georgia. M, 1966, University of Tennessee, Knoxville. 83 p.

Darrell, James Harris, II. Statistical evaluation of palynomorph distribution in the sedimentary environments of the modern Mississippi River delta (Volumes I and II). D, 1973, Louisiana State University.

Darrow, Arthur Charles. Origin of the basalts of the Big Pine volcanic field, California. M, 1972, University of California, Santa Barbara.

Darrow, Douglas D. Drainage problems and correction measures in Sedgwick County, Kansas. M, 1974, Wichita State University. 174 p.

Darrow, Richard Lee. The geology of the northwest part of the Montara Mountain Quadrangle, California. M, 1951, University of California, Berkeley. 58 p.

Dart, Robert Henry. An analysis of potential utilization of medium temperature geothermal fluid for the production of electricity. M, 1976, University of Idaho. 110 p.

Dart, Stephen W., Jr. The Sirenian shoulder and forelimb; a study of variation and function. M, 1974, University of Kansas. 253 p.

Darwin, Helen. The geochemistry of banded sands in the vicinity of the University of Mississippi. M, 1948, University of Mississippi.

Darwin, Robert Louis. Geoelectric stratigraphy and subsurface evaluation of Quaternary deposits at Cooper Basin, Northeast Texas. M, 1988, University of Texas, Arlington. 140 p.

Das Gupta, Samir Kumar. Paragenesis and composition of crustified sulphide ores. D, 1955, University of Toronto.

Das Gupta, U. A study of fractured reservoir rocks, with special reference to Mississippian carbonate rocks of Southwest Alberta. D, 1978, University of Toronto.

Das, Braja M. Erosion of compacted cohesive soils. D, 1972, University of Wisconsin-Madison.

Das, Shamita. A numerical study of rupture propagation and earthquake source mechanism. D, 1976, Massachusetts Institute of Technology. 217 p.

Dasch, E. Julius, Jr. Strontium isotopes in deep-sea sediments, weathering profiles, and sedimentary rocks. D, 1969, Yale University. 101 p.

Dasch, Ernest Julius, Jr. Dike swarm of northern Rim Rock country, Trans-Pecos, Texas. M, 1959, University of Texas, Austin.

Dasch, L. E. U/Pb geochronology of the Sierra de Perija, Venezuela. M, 1982, Case Western Reserve University.

Dasgupta, Biswajit. Vibration isolation of structures in a homogeneous elastic soil medium. D, 1987, University of Minnesota, Minneapolis. 294 p.

Dasgupta, Shivaji N. Paleomagnetism of a Paleocene pluton of Jamaica. M, 1974, St. Louis University.

Dash, Umakant. Erosive behavior of cohesive soils. D, 1968, Purdue University. 240 p.

Dasog, Ghulappa. Properties, genesis and classification of clay soils in Saskatchewan. D, 1987, University of Saskatchewan.

Dass, A. S. Wall rock alteration of the silver deposits, Cobalt, Ontario. D, 1970, Carleton University. 149 p.

Dass, P. Water and sediment routing in nonuniform channels. D, 1975, Colorado State University. 171 p.

Dastanpour, Mohammad. An investigation of the carbonate rocks in the Reynolds oil field, Montcalm County, Michigan. M, 1977, Michigan State University. 59 p.

Dastidar, Priyabrata Ghosh. A study of trace elements in selected Appalachian sulphide deposits (New Brunswick and Newfoundland), Canada. D, 1969, University of New Brunswick.

Datondji, Apollinaire. The mechanical evolution of the Williston Basin. M, 1981, SUNY at Buffalo. 136 p.

Datta, Pabindranath. Location of hydroxyl hydrogen atoms in the structures of minerals. D, 1972, SUNY at Buffalo. 95 p.

Datta, Ranajit Kumar. Order-disorder in spinels. D, 1961, Pennsylvania State University, University Park. 167 p.

Dattilo, Benjamin F. Depositional environments of the Fillmore Formation (Lower Ordovician) of western Utah. M, 1988, Brigham Young University. 94 p.

Daub, Gerald J. Stratigraphic interpretation of a possible paleostream channel of the ancient Nueces River, South Texas. M, 1979, University of Rhode Island.

Daud, Badruddin Haroon. Calculating variability of evaluation estimates of grades of ore deposits from techniques of variogram analyses and data simulation; 1 volume. M, 1971, University of California, Berkeley.

Daudt, Carl Ransford. Finite difference synthetic seismogram calculations utilizing acoustic models and explicit and implicit formulations. M, 1983, Purdue University. 79 p.

Daughdrill, William E. Diagnostic foraminifera from the Archusa Marl of Mississippi. M, 1956, University of Missouri, Columbia.

Daughdrill, William Eugene. Distribution and accumulation of mercury pollutants in the Pearl River, its delta and the flanking estuaries. D, 1974, Tulane University.

Daugherty, Clarence Gordon, Jr. A preliminary report on the Jackson Formation of South Texas. M, 1939, University of Oklahoma. 38 p.

Daugherty, Colleen Mae. A gravity survey of Darke County, Ohio. M, 1987, Wright State University. 64 p.

Daugherty, David R. Characteristics and origins of joints and sedimentary dikes of the Bahama Islands. M, 1986, Miami University (Ohio). 110 p.

Daugherty, Franklin Wallace. Geology of the Pico Etero area, Municipio de Acuna, Coahuila, Mexico. D, 1962, University of Texas, Austin. 170 p.

Daugherty, Franklin Wallace. Structure of Sierra Pilares, Municipio de Ojinaga, Chihuahua, Mexico. M, 1959, University of Texas, Austin.

Daugherty, Gregory Lynn. Geobotanical interpretation of the geology and phosphate occurrences of north-central Florida as determined by Landsat imagery. M, 1977, Florida State University.

Daugherty, Helen Roberta. Geophysical well log investigation and environment of deposition of upper Fort Union coals, Powder River basin, Wyoming. M, 1981, Wright State University. 71 p.

Daugherty, Kenneth Ivan. An investigation of the normal vertical gradient of gravity. M, 1964, Ohio State University.

Daugherty, Lloyd F. The Mollusca and foraminifera of Depoe Bay, Oregon. M, 1951, University of Oregon. 77 p.

Daugherty, Thomas D. A petrologic and mechanical analysis of the Lion Mountain and Welge sandstones of southern Mason County, Texas. M, 1960, Texas A&M University. 88 p.

Dauphin, Joseph Paul. Eolian quartz granulometry as a paleowind indicator in the Northeast Equatorial Atlantic, North Pacific and Southeast Equatorial Pacific. D, 1983, University of Rhode Island. 355 p.

Dauphin, Joseph Paul. Size distribution of chemically extracted quartz used to characterize fine-grained sediments. M, 1972, Oregon State University. 63 p.

Daus, A. D., III. An alternating direction Galerkin technique for simulation of contaminant transport in complex groundwater systems. M, 1983, University of Waterloo. 62 p.

Daus, Steven Jean. A user sensitive clustering algorithm for use with Landsat multispectral scanner (MSS) digital data in remote sensing aided inventories of natural resources. D, 1979, University of California, Davis. 328 p.

Daut, Steven William. Gravity study of the Thurman-Redfield structural zone in Southwest Iowa. M, 1980, University of Iowa. 118 p.

Dauzacker, Modesto Victor. Basin analyses of evaporitic and post-evaporitic depositional systems, Espirito Santo Basin, Brazil, South America. D, 1981, University of Texas, Austin. 157 p.

Davenport, Peter H. The application of geochemistry to base-metal exploration in the Birch-Uchi Lakes volcano-sedimentary belt, northwestern Ontario. D, 1972, Queen's University. 411 p.

Davenport, R. R. An investigation of the downward continuation of gravity. M, 1974, University of Hawaii. 43 p.

Davenport, Ronald E. Geophysical investigation of the Sedimentary Hills area, Pima County, Arizona. M, 1963, University of Arizona.

Davenport, Ronald Edmond. Geology of the Rattlesnake and older ignimbrites in the Paulina Basin and adjacent area, central Oregon. D, 1971, Oregon State University. 132 p.

Davenport, Thomas Edward. Sediment and nutrients transport during storm runoff, Hoquat Creek. M, 1981, University of Washington. 63 p.

Davey, John Raymond. Energy flow dynamics of a desert grassland ecosystem. D, 1980, [Northern Arizona University]. 268 p.

David, Briant LeRoy. Petrology and petrography of the igneous rocks of the Stansbury Mountains, Tooele County, Utah. M, 1959, Brigham Young University. 56 p.

David, Edwin Philip. Some studies of gold and its associated minerals. M, 1939, University of British Columbia.

David, Jimmy Leon. Pendleton-Many Field Sabine Parish, Louisiana. M, 1964, University of Missouri, Columbia.

David, Mark Barnett. Organic and inorganic sulfur cycling in forested and aquatic ecosystems in Adirondack region of New York State. D, 1983, State University of New York, College of Environmental Science and Forestry. 286 p.

David, Newton Fraser Gordon. The petrography of the rocks of Hong Kong. M, 1926, University of British Columbia.

David, Peter Pascal. A study of the roundness of wind-blown sands from Hungary and the Canadian plains. M, 1961, McGill University.

David, Peter Pascal. Surficial geology and ground water resources of the Prelate area (72-K), Saskatchewan. D, 1965, McGill University.

David, Pierre. A pedological study of clayey soils of Savane Diane, Central Plateau, Haïti. D, 1984, Rutgers, The State University, New Brunswick. 169 p.

David, Richard S. Geology and groundwater resources of the Mosul Liwa and northern Jezira Desert, Iraq. M, 1956, University of Southern California.

David, Timothy L. Regional variations of cleavage and associated strain in Martinsburg Graywackes. M, 1986, Bowling Green State University. 101 p.

Davidheiser, Carolyn Elizabeth. Bryozoans and bryozoan-like corals; affinities and variability of Diplotrypa, Monotrypa, Labyrinthites, and Cladopora (Trepostomata and Tabulata) from selected Ordovician-Silurian reefs in eastern North America (Newfoundland, Pennsylvania, Michigan). D, 1980, Pennsylvania State University, University Park. 694 p.

Davidian, Beth Eileen Kramer. Evaluation of quantitative analysis of geologic materials with an energy dispersive system on a scanning electron microscope. M, 1978, University of Minnesota, Minneapolis. 21 p.

Davids, Robert Norman. A paleoecologic and paleobiogeographic study of Maestrichtian planktonic foraminifera. D, 1966, Rutgers, The State University, New Brunswick. 241 p.

Davidsen, Erik Kennedy. Petrology and structure of greenstone blocks encased in Franciscan mud-matrix melange near San Simeon, California. M, 1986, University of Texas, Austin. 256 p.

Davidson, Alexander, Jr. Lower Cretaceous foraminifera from Mount Goodenough, Northwest Territories. M, 1960, University of Alberta. 74 p.

Davidson, Anthony. A study of okaite and associated rocks near Oka, Quebec. M, 1963, University of British Columbia.

Davidson, Anthony. Metamorphism and intrusion in the Benjamin Lake map area, Northwest Teritories. D, 1967, University of British Columbia.

Davidson, D. D. The Carrot River ultramafic complex, Manitoba. M, 1974, Acadia University.

Davidson, Dean Frederick. Some aspects of geochemistry and mineralogy of Bear Lake sediments, Utah-Idaho. M, 1969, Utah State University. 67 p.

Davidson, Donald Alexander. Surface geology at the Granduc Mine. M, 1960, University of British Columbia.

Davidson, Donald M. Nature and origin of silica occurrences in the Mount Peal Quadrangle, Grand and San Juan counties, Utah. M, 1963, Columbia University, Teachers College.

Davidson, Donald Miner. Geology and petrology of the Mineral Hill mining district, Lemhi County, Idaho. D, 1928, University of Minnesota, Minneapolis. 65 p.

Davidson, Donald Miner. The Animikie slate of northeastern Minnesota. M, 1926, University of Minnesota, Minneapolis. 29 p.

Davidson, Donald Miner, Jr. Ore emplacement and associated features, Kane Creek, Utah. D, 1965, Columbia University, Teachers College. 240 p.

Davidson, E. L. The relationships between the petrology, mineralogy, and sulfide mineralization in a portion of the Duluth Complex, Minnesota. M, 1979, University of Wisconsin-Madison.

Davidson, Edward Sheldon. The geological relationship and petrography of a nepheline syenite near Beemerville, Sussex County, New Jersey. M, 1948, Rutgers, The State University, New Brunswick. 140 p.

Davidson, James M. Surficial geology and Quaternary history of the Central Plains Experimental Range, Colorado. M, 1988, Colorado State University. 106 p.

Davidson, Jeanne R. Geology and mammalian paleontology of the Wind River Formation, Laramie Basin, southeastern Wyoming. M, 1987, University of Wyoming. 110 p.

Davidson, Joe Dwain. The physical stratigraphy of the Avant limestone member of the Iola Formation, southern Osage County and parts of nearby counties, Oklahoma. M, 1978, Oklahoma State University. 123 p.

Davidson, John Matthew. The structure and petrology of some ultramafic bodies near Proctorsville, Vermont. M, 1981, Harvard University. 68 p.

Davidson, John W. The geology, petrology, geochemistry and economic mineral resources of east-central Oglethorpe County, Georgia. M, 1981, University of Georgia.

Davidson, L. W. On the physical oceanography of Burrard Inlet and Indian Arm, British Columbia. M, 1979, University of British Columbia.

Davidson, Lloyd Arthur. The biology, morphology and relationships of Pilosiphonia dubia (Gruber), n. gen. (Foraminiferida). M, 1969, University of California, Berkeley. 39 p.

Davidson, Lynn Blair. A theoretical analysis of reservoir behavior during the intermittent injection of steam. D, 1966, Stanford University. 192 p.

Davidson, Maurice James. The spatial coherence of geomagnetic rapid variations. D, 1966, Columbia University. 144 p.

Davidson, Oscar B. Paleoenvironmental investigation of the Hazard Coal, Breathitt Formation (Pennsylvanian) eastern Kentucky coal field. M, 1986, Eastern Kentucky University. 71 p.

Davidson, Paula Marie. Thermodynamic analysis of quadrilateral pyroxenes and olivines. D, 1983, SUNY at Stony Brook. 149 p.

Davidson, Peter S. Stratigraphy of the Upper Cretaceous Lewis Formation of the Laramie and Carbon basins, Albany and Carbon counties, Wyoming. M, 1966, University of Wyoming.

Davidson, R. N. Favorability of dolomites as host rocks for ore deposition. M, 1950, Colorado School of Mines. 79 p.

Davidson, Robert H. Petrology of pinnacle reefs in the Guelph Formation (Niagaran), northern Michigan. M, 1981, Stephen F. Austin State University. 93 p.

Davidson, S. C. A petrographical study of specimens of rocks and ores from the Sullivan Mine, Kimberley, British Columbia. M, 1925, McGill University.

Davidson, Stanley C. Geology and ore deposits of Tayoltita, District of San Dimas, Durango, Mexico. D, 1933, Harvard University.

Davidson, William R. Petrology and environmental interpretation of the synorogenic Bacon Ridge Formation, Southeast Idaho and Northwest Wyoming. M, 1986, University of Idaho. 121 p.

Davidson, William T. Deposition and diagenesis of the Glorieta Formation (Permian), northern Midland Basin, Texas. M, 1986, Baylor University. 176 p.

Davidson-Arnott, R. Form, movement and sedimentological characteristics of wave formed bars; a study of their role in the nearshore equilibrium, Kouchibouguac Bay, New Brunswick. D, 1975, University of Toronto.

Davie, Ellen Ingraham, II. The geology and petrology of Three Fingered Jack, a High Cascade volcano in central Oregon. M, 1981, University of Oregon. 138 p.

Davies, Ben. A correlation between the clay mineral content and sulfuric extraction of alumina from some Pacific Northwest clays. M, 1943, University of Washington. 74 p.

Davies, David John. Taphonomic signature as a function of environmental process; sedimentation and taphofacies of shell concentration layers and event beds, Holocene of Texas. D, 1988, Texas A&M University. 259 p.

Davies, David K. Toward a measurement of sedimentary structures. M, 1964, Louisiana State University.

Davies, E. H. Jurassic and Lower Cretaceous dinoflagellate cysts of the Sverdrup Basin, Arctic Canada; taxonomy, biostratigraphy, chronostratigraphy. D, 1979, University of Toronto.

Davies, Edward Jullian Llewelyn. Ordovician and Silurian of the Northern Rocky Mountains between Peace and Muskwa rivers, British Columbia. D, 1966, University of Alberta. 139 p.

Davies, Geoffrey Frederick. Elasticity of solids at high pressures and temperatures; theory, measurement and geophysical application. D, 1973, California Institute of Technology. 276 p.

Davies, H. F. Examination of the ores of the Geyser and Bassick mines, Custer County, Colorado. M, 1922, Massachusetts Institute of Technology. 41 p.

Davies, Herman F. Some silver deposits of Mexico. M, 1921, University of Minnesota, Minneapolis. 50 p.

Davies, Hope M. Apatite and volatiles in the Kiglapait layered intrusion, Labrador. M, 1974, University of Massachusetts. 50 p.

Davies, Hugh Lucius. Peridotite-gabbro-basalt complex in eastern Papua; an overthrust plate of oceanic mantle and crust. D, 1970, Stanford University. 120 p.

Davies, Ian Charles. Transport of conglomerate into deep water; a study of the Cambro-Ordovician Cap Enrage Conglomerate at St. Simon de Rimouski, Quebec. M, 1973, McMaster University. 91 p.

Davies, J. F. The geology and the gold deposits of Rice Lake, Manitoba. D, 1951, University of Toronto.

Davies, J. F. The origin of the Cree Lake "Intrusives" and basic gneisses of the Kisseynew Series, Sherridon area, Manitoba. M, 1948, University of Manitoba.

Davies, James Dudley. Stratigraphy and ostracode distribution of the Cherokee and Henrietta formations of Randolph, Chariton, and Boone counties. M, 1936, University of Missouri, Columbia.

Davies, James Frederick. Geology and gold deposits of the Rice Lake-Wanipigow River area, Manitoba. D, 1963, University of Toronto. 184 p.

Davies, John. Paleomagnetism of the Dohkan Volcanics and Hammamat Sediments, Eastern Desert, Egypt. M, 1979, University of South Carolina.

Davies, John C. Maximum entropy spectral analysis of free oscillations of the Earth; the 1964 Alaska event. M, 1976, University of British Columbia.

Davies, John C. The petrology and geochemistry of basic intrusive rocks, Kakagi Lake-Wabigoon Lake area, District of Kenora, Ontario. D, 1966, University of Manitoba.

Davies, John Clifford. The petrology of the Fisher Lake area District of Kenora, Ontario. M, 1957, University of Manitoba.

Davies, John Leslie. The geology and geochemistry of the Austin Brook area, Gloucester County, New Brunswick, with special emphasis on the Austin Brook iron formation. D, 1972, Carleton University. 253 p.

Davies, John Leslie. Trace elements in the rocks and ores of the Bathurst-Newcastle District, New Brunswick. M, 1960, University of New Brunswick.

Davies, John Norman. Crustal morphology of central Alaska. M, 1970, University of Alaska, Fairbanks. 57 p.

Davies, John Norman. Seismological investigations of plate tectonics in south central Alaska. D, 1975, University of Alaska, Fairbanks. 193 p.

Davies, Kyle Linton. Hadrosaurian dinosaurs of Big Bend National Park, Brewster County, Texas. M, 1983, University of Texas, Austin. 235 p.

Davies, Linda M. Inversion of travel time data. M, 1973, University of Alberta. 121 p.

Davies, Mark Allen. Raman spectroscopy and Raman optical activity of organic molecules. D, 1987, City College (CUNY). 160 p.

Davies, Nathan C. "Flow by head" and recovery in oil wells. M, 1929, University of Minnesota, Minneapolis. 33 p.

Davies, Peter Bowen. Deep-seated dissolution and subsidence in bedded salt deposits. D, 1984, Stanford University. 391 p.

Davies, Raymond. Geology of ore Mutton Bay Intrusion and surrounding area, north shore, Gulf of Saint Lawrence, Quebec (Canada). D, 1968, McGill University.

Davies, Rhys J. A correlation study of the Cynthiana, Gallatin County, Kentucky. M, 1958, Miami University (Ohio). 74 p.

Davies, Russell King. Relations between left-lateral strike-slip faults and right-lateral monoclinal kink bands in granodiorite, Mt. Abbot Quadrangle, Sierra Nevada, California. M, 1985, University of Cincinnati. 110 p.

Davies, Scott L. A parametric study of Rayleigh energy radiated from forging sources. M, 1977, University of Wisconsin-Milwaukee.

Davies, Simon Henry Richard. Mn(II) oxidation in the presence of metal oxides. D, 1985, California Institute of Technology. 188 p.

Davies, Stephen. Examination of selected Eastern United States lakes for organic protolytes, chemical species, and interrelated variables. M, 1987, McMaster University. 163 p.

Davies, Stephen Farrel. Geology of the Grayback Hills, North-central Tooele County, Utah. M, 1980, University of Utah. 206 p.

Davies, Tarin Smith. A ground water model for analysis of the Amador Subbasin of the Livermore Valley, California. M, 1981, Stanford University. 140 p.

Davies, Thomas Daniel. Peat formation in Florida Bay and its significance in interpreting the Recent vegetational and geological history of the Bay area. D, 1980, Pennsylvania State University, University Park. 338 p.

Davies, William Edward. Stratigraphic significance of crinoidal columns of the Traverse of Michigan. M, 1941, Michigan State University. 55 p.

Davies, William James. Pre-loessial "Lafayette type" gravel in the vicinity of St. Louis. M, 1953, Washington University. 80 p.

Davies-Colley, Robert James. Estuarine sediment controls on trace metal distributions. D, 1982, Oregon State University. 224 p.

Daviess, S. N.; Kellum, L. B. and Swinney, C. M. Geology and oil possibilities of the southwestern part of the Wide Bay Anticline, Alaska. M, 1945, University of Michigan.

Daviess, Steven Norman. Contact relationship between Mint Canyon Formation and upper Mesozoic marine beds in eastern Ventura Basin, Los Angeles County, California. M, 1942, University of California, Los Angeles.

Davila-Alcocer, Victor M. Biostratigraphic studies of Jurassic, Cretaceous Radiolaria of the Eugenia and Asuncion formations, Vizcaino Peninsula, Baja California Sur, Mexico. M, 1986, University of Texas at Dallas. 157 p.

Davin, Christopher Gerard. A seismic stratigraphic interpretation and geohistory analysis of the Sable Island area of the Scotian Shelf, Eastern Canada. M, 1985, University of Houston. 106 p.

Davis, A. M. Reconstructions of local and regional Holocene environments from the pollen and peat stratigraphies of some Driftless Area peat deposits. D, 1975, University of Wisconsin-Madison. 240 p.

Davis, A. M. The cosmochemical history of the palasites. D, 1977, Yale University. 298 p.

Davis, Alan. The artificial coalification of wood samples of Taxodium distichum (L.) Rich. M, 1961, Pennsylvania State University, University Park. 89 p.

Davis, Alice S. Artificial meteor ablation studies; iron and nickel-iron. M, 1975, San Jose State University. 67 p.

Davis, Andrew. Factors controlling lead accumulation in the sediments of two remote Adirondack lakes. M, 1979, University of Virginia. 215 p.

Davis, Andrew Owen. Geochemical interactions between uranium tailings fluids and subjacent bedrock, Canon City, Colorado; use of the computer model MINTEQ. D, 1985, University of Colorado. 233 p.

Davis, Ann H. Post fluvial climates of the Southwest United States. M, 1949, University of Michigan.

Davis, Annin-Gray. Field notes; area southeast of Gladwyne. M, 1958, Bryn Mawr College.

Davis, Arden D. Digital models of ground-water flow, solute transport, and dispersion for part of the Spearfish Valley Aquifer, Lawrence County, South Dakota. D, 1983, South Dakota School of Mines & Technology.

Davis, Arden D. Hydrogeology of the Belle Fourche water infiltration gallery area, Lawrence County, South Dakota. M, 1979, South Dakota School of Mines & Technology.

Davis, B. M. Some tests of hypothesis concerning variograms. D, 1978, University of Wyoming. 162 p.

Davis, Brent E. Oxygen isotope paleotemperature geothermometry of fossil carbonates in the Purisima Formation. M, 1976, Stanford University.

Davis, Brian Thomas Canning. Geology of the Saint Regis Quadrangle, New York. D, 1963, Princeton University. 196 p.

Davis, Briant Leroy. High-pressure X-ray investigation of CaCO3-11 and CaCO3-111 at 25°C, and of the 300-500°C temperature interval. D, 1964, University of California, Los Angeles. 147 p.

Davis, C. W. The petrography of a section of Westmount Mountain (Montreal, Quebec). M, 1937, McGill University.

Davis, Carl G. An investigation of the clay mineral suite in a type section of the Jackfork Group, Ouachita Mountains, Arkansas. M, 1968, Northern Illinois University. 91 p.

Davis, Carol Waite. Linear and areal measurements as estimates of grain volume. M, 1963, Louisiana State University.

Davis, Charles George. Stratigraphy of the Capitola-Sunset Beach area, southern Santa Cruz County, California. M, 1957, Stanford University.

Davis, Charles H. The geology of the Santa Lucia Mountains, California. M, 1912, [Stanford University].

Davis, Charles L. Petrography of the Brentwood Limestone in Washington County, Arkansas. M, 1961, University of Arkansas, Fayetteville.

Davis, Charles Moler. The high plains of Michigan. D, 1935, University of Michigan.

Davis, Charles Wayne. Topographic expression of the carbonate rocks of the Bedford County portion of Morrison Cove in central Pennsylvania. M, 1973, Millersville University.

Davis, Clarence Jackson. The geology of the Lavergne Quadrangle, Tennessee. M, 1959, Vanderbilt University.

Davis, Clarence King. Correlation of photogeology with airborne magnetometer survey at Marmora, Ontario. M, 1959, Cornell University.

Davis, Clinton L. Structural geology of southeastern margin of Bear River Range, Idaho. M, 1969, Utah State University. 68 p.

Davis, Craig B. Geology of the Quinnesec Formation in southeastern Florence County, Wisconsin. M, 1977, University of Wisconsin-Milwaukee.

Davis, Curtiss Owen. Effects of changes in light intensity and photoperiod on the silicate-limited continuous culture of the marine diatom Skeletonema costatum (Grev.) Cleve. D, 1973, University of Washington. 122 p.

Davis, Dan Arthur. Geology of the south portion of Childress County, Texas. M, 1940, University of Iowa. 52 p.

Davis, Daniel Michael. Thin-skinned deformation and plate driving forces associated with convergent margins. D, 1983, Massachusetts Institute of Technology. 306 p.

Davis, Danny Ray. Geology of Aragon Hill area, Sierra and Socorro counties, New Mexico. M, 1988, University of Texas of the Permian Basin. 62 p.

Davis, Darrell E. Geology of southeastern and south-central Jefferson County, Kansas. M, 1959, University of Kansas. 81 p.

Davis, David Chandler. The geology of the southern portion of the Lovell Quadrangle, Knox and Loudon counties, Tennessee. M, 1950, University of Tennessee, Knoxville. 49 p.

Davis, Deborah Ann. Gravity survey of Utah and Goshen valleys and adjacent areas, Utah. M, 1983, University of Utah. 141 p.

Davis, Del E. A taxonomic study of the Mississippian corals of central Utah. M, 1956, Brigham Young University. 49 p.

Davis, Donald Ray. The measurement and evaluation of certain trace metal concentrations in the nearshore environment of the northwest Gulf of Mexico and Galveston Bay. D, 1968, Texas A&M University. 80 p.

Davis, Donald Wayne. Determination of the ^{87}Rb decay constant; an Rb/Sr and Pb/Pb study of the Labrador Archean Complex. D, 1978, University of Alberta. 122 p.

Davis, Dorothy Taylor. The Popo Agie Formation of Wyoming. M, 1937, University of Missouri, Columbia.

Davis, Doug. Stratigraphy and hydrothermal alteration of the Gagne Lake Project; an occurrence of volcanogenic-type massive sulfides near Mine Centre, northwestern Ontario, Canada. M, 1987, University of Minnesota, Duluth. 110 p.

Davis, Dudley L. Anaconda's operation at Darwin Mines, Inyo County, California. M, 1948, University of Nevada - Mackay School of Mines. 11 p.

Davis, Earl Edwin. The northern Juan de Fuca Ridge; a geophysical investigation of an active sea floor spreading center. D, 1975, University of Washington. 184 p.

Davis, Edward Louis. An analysis of the joint systems in the Paleozoic rocks of the northern Tobacco Root Mountains (Montana). M, 1967, Indiana University, Bloomington. 100 p.

Davis, Elizabeth R. Paleoecology and distribution of Albian rudists of north-central Texas with special emphasis on the Edwards Formation in Bell, Bosque, McLennan and Coryell counties. M, 1977, Baylor University. 127 p.

Davis, Elmer Fred. A study of the cherts and associated shales of the Franciscan Formation. M, 1912, University of California, Berkeley. 36 p.

Davis, Elmer Fred. The rocks of the Franciscan Group, with special reference to the origin of the radiolarian cherts. D, 1917, University of California, Berkeley.

Davis, Ethel M. Glacial geology of the Palmyra and Pultneyville quadrangles, New York. M, 1941, University of Rochester. 114 p.

Davis, Flavy Eugene. A study of some species of Textularia from the Tertiary of Texas. M, 1937, University of Texas, Austin.

Davis, Frank Willard. The history of submerged aquatic vegetation at the head of Chesapeake Bay; a stratigraphic study. D, 1983, The Johns Hopkins University. 140 p.

Davis, Franklin Theodore. The distribution of phosphorus in Washington coals. M, 1942, University of Washington. 43 p.

Davis, Gary J. The southwestern extension of the Middletown-Lowndesville cataclastic zone in the Greensboro, Georgia area and its regional implications. M, 1980, University of Georgia.

Davis, Gene H. A gravity study of the San Luis Basin, Colorado. M, 1978, University of Texas at El Paso.

Davis, Gene Michael. Geology of the southern Plomosa Mountains. M, 1985, Arizona State University. 159 p.

Davis, George H. Geology of the eastern third of La Democracia Quadrangle, Guatemala. M, 1966, University of Texas, Austin.

Davis, George Hardy, III. Contact between the Manlius Limestone and Coeymans Limestone in upper New York State. M, 1952, University of Michigan.

Davis, George Herbert. Structural analysis of the Caribou sulfide deposit Bathurst, New Brunswick, Canada. D, 1971, University of Michigan. 173 p.

Davis, Gregory Arlen. Metamorphic and igneous geology of pre-Cretaceous rocks, upper Coffee Creek area, northeastern Trinity Alps, Klamath Mountains, California. D, 1961, University of California, Berkeley. 202 p.

Davis, H. Clyde. Geology of the Culmer gilsonite vein of Duchesne, Utah. M, 1952, Brigham Young University. 90 p.

Davis, Harold G. The Precambrian geology of the Sheep Mountain area, Beartooth Mountains, Montana. M, 1979, Northern Illinois University. 120 p.

Davis, Harry Osmond. Canadian Coal Geology; an annotated, toponymic bibliography of Geological Survey of Canada publications, 1845-1962. M, 1964, University of Western Ontario. 279 p.

Davis, Harry Towles. Notes to accompany map of the Triassic contacts of East Chapel Hill, North Carolina. M, 1920, University of North Carolina, Chapel Hill. 6 p.

Davis, Henry G. The Utica Mine of the Slocan District, British Columbia, Canada. M, 1929, Columbia University, Teachers College.

Davis, Hugh R. Deposition of the Lower Cretaceous Mowry Shale. D, 1987, University of Wisconsin-Madison. 225 p.

Davis, J. D. The petrography and petrochemistry of mafic rocks in Chedabucto-St Peter Bay area, Nova Scotia. M, 1972, Technical University of Nova Scotia.

Davis, J. H. Short term study in beach sand movement adjacent to Monterey Canyon. M, 1966, Naval Postgraduate School.

Davis, James A. Adsorption of trace metals and complexing ligands at the oxide water interface. D, 1978, Stanford University. 305 p.

Davis, James Edward. Potential pollution loading to natural waters as a function of land use; a methodology applied to a coastal plain in Spain. M, 1986, Stanford University. 65 p.

Davis, James Frazier. A petrologic examination of the Iron-Formation and associated pyroclastic breccias and graywackes of the Negaunee Formation (Precambrian), Palmer area, Marquette District, Michigan. D, 1965, University of Wisconsin-Madison. 205 p.

Davis, James Frazier. An examination of the effects of polishing on the optical properties of anisotropic ore minerals in reflected light. M, 1956, University of Wisconsin-Madison.

Davis, James Harrison. Geology of El Vado area, Rio Arriba County, New Mexico. M, 1960, University of Texas, Austin.

Davis, James Howell. Mineralization in the Southeast Missouri lead district. D, 1960, University of Wisconsin-Madison. 130 p.

Davis, James L. Buried river channels in the Ft. Berthold Indian Reservation, North Dakota. M, 1953, University of Arkansas, Fayetteville.

Davis, James Peter. Large scale heterogeneity of the upper mantle inferred from measurements of free oscillations. D, 1986, Princeton University. 136 p.

Davis, James R. Stratigraphy and depositional history of upper Mesaverde Formation (Cretaceous) of southeastern Wyoming, volume 1-text, vol. 2-measured stratigraphic sections. D, 1967, University of Wyoming. 124 p.

Davis, James R. The "Mesaverde" Formation of the Kindt Basin, Carbon County, Wyoming. M, 1963, University of Wyoming. 134 p.

Davis, James W. Structure and petrogenesis of the Knoblick Granite, Missouri. M, 1969, St. Louis University.

Davis, James William. Stratigraphy of the Flagstaff Formation, southeastern Utah County, Utah. M, 1967, Ohio State University.

Davis, Jeffery A. Major and trace element distributions in the Glens Falls and Albany quadrangles. M, 1981, Rensselaer Polytechnic Institute. 114 p.

Davis, Jerry Dean. Geothermometry, geochemistry, and alteration at the San Manuel Porphyry copper orebody, San Manuel, Arizona. D, 1974, University of Arizona.

Davis, Jerry Dean. The distribution and zoning of the radioelements potassium, uranium, and thorium in selected porphyry copper deposits. M, 1971, University of Arizona.

Davis, Jerry Douglas. The origin of arcuate sand ridges in the Okefenokee Swamp. D, 1987, University of Georgia. 319 p.

Davis, Jerry Lee. The scattering of elastic waves by void cavities. D, 1972, University of Missouri, Rolla. 170 p.

Davis, John C. Geology of the Clay Spur bentonite district, Crook and Weston counties. M, 1963, University of Wyoming. 79 p.

Davis, John C. Petrology of the Mowry Shale (Lower Cretaceous) (Wyoming, Colorado, Montana, west South Dakota, and Utah). D, 1967, University of Wyoming. 141 p.

Davis, John Daniel. Paleoenvironments of the Moenave Formation, St. George, Utah. M, 1977, Brigham Young University. 31 p.

Davis, John Roland. A field study of the Checkerboard Limestone of northeastern Oklahoma. M, 1939, University of Oklahoma. 32 p.

Davis, John Steven. A microstructural perspective of orogenesis in the Pioneer Mountains, central Idaho. M, 1984, University of Montana. 55 p.

Davis, John Wesley. Paleomagnetic assessment of basement rotation along the eastern flank of the Front Range near Boulder, Colorado. M, 1987, University of Colorado. 58 p.

Davis, Jonathan O. Computer-assisted three-dimensional modelling in geology. M, 1974, University of Idaho. 119 p.

Davis, Jonathan O. Quaternary tephrochronology of the Lake Lahontan area, Nevada and California. D, 1977, University of Idaho. 150 p.

Davis, Jondahl. Clay mineralogy of Lower Cretaceous sediments of the eastern Powder River basin, Wyoming, Montana, and western South Dakota. M, 1981, Texas Tech University. 113 p.

Davis, Joseph Redmon. Late Cenozoic geology of Clayton Valley, Nevada, and the genesis of a lithium-enriched brine. D, 1981, University of Texas, Austin. 273 p.

Davis, Joseph Redmond. Late Quaternary sedimentation in the San Pedro Valley, southeastern Arizona. M, 1975, Southern Methodist University. 101 p.

Davis, Karleen Ethel. Lead isotope ratios in the Bayhorse mining district, Custer County, Idaho. M, 1977, Massachusetts Institute of Technology. 58 p.

Davis, Katherine Renee. Heavy mineral criteria for delineating environments of deposition and source of the Muddy Sandstone (lower Cretaceous), Wind River Basin, Wyoming. M, 1975, Miami University (Ohio). 80 p.

Davis, Keith William. Stratigraphy and depositional environments of the Glen Rose Formation, north-central Texas. M, 1973, Baylor University. 217 p.

Davis, Kenneth E. The foraminifera and lithologic characteristics of the Reklaw Formation between Bastrop County, Texas, and Leon County, Texas. M, 1961, Texas A&M University. 98 p.

Davis, Kenneth R. Petrology of carbonate rocks in the lower portion of the Floyd Shale, Rome area, Georgia. M, 1982, University of Georgia.

Davis, L. C. Late Pleistocene geology and paleoecology of the Spring Valley basin, Meade County, Kansas. D, 1975, University of Iowa. 179 p.

Davis, Larry Eugene. Conodont biostratigraphy and carbonate petrology of the West Canyon Limestone and equivalents, southeastern Idaho and northern Utah. D, 1987, Washington State University. 200 p.

Davis, Larry Eugene. Conodont biostratigraphy of the Alaska Bench Formation, central Montana. M, 1983, Washington State University. 110 p.

Davis, Laura E. Pegmatitic muscovites; effect of composition on optical and lattice parameters. M, 1985, Virginia Polytechnic Institute and State University.

Davis, Leland J. Characteristics, occurrence and uses of the solid bitumens of the Uinta Basin, Utah. M, 1951, Brigham Young University. 93 p.

Davis, Leo Carson. The herpetofauna of Peccary Cave, Newton County, Arkansas. M, 1973, University of Arkansas, Fayetteville.

Davis, Lisa M. A geochemical survey of the Loma Plata area, Presidio County, Texas. M, 1987, University of Texas at El Paso.

Davis, Louis Lloyd, Jr. Petrology of the Claiborne Group and part of the Wilcox Group, Southwest Georgia and Southeast Alabama. M, 1974, University of Texas, Austin.

Davis, Martin Jones. Geology as related to plant growth, with special reference to the Chapel Hill area. M, 1916, University of North Carolina, Chapel Hill. 14 p.

Davis, Mary. Petrology, structure and geochemistry of a portion of the Carthage Colton mylonite zone, South Edwards 7 1/2 minute Quadrangle, northwestern Adirondacks, New York. M, 1981, SUNY at Binghamton.

Davis, Mary Walter. Late Paleozoic crustal composition and dynamics in the Southeastern United States. D, 1972, Michigan State University. 86 p.

Davis, Matthew C. Evaluation of low-temperature geothermal potential in north-central Box Elder County, Utah. M, 1984, Utah State University. 130 p.

Davis, Melvin K. Topographic problems in the Cumberland Gap area. M, 1915, University of Wisconsin-Madison.

Davis, Michael E. Ostracoda of the limestones of the Permian System Wolfcamp series in Kansas. M, 1951, Kansas State University. 98 p.

Davis, Michael Paul. Investigation of uranium and thorium variation in selected intrusive rocks of the southeastern Piedmont. M, 1977, University of Florida. 107 p.

Davis, Michael W. Interpretation of the depositional environment of sandstone channels above the Kentucky No. 9 coal in the Western Kentucky Coal Field. M, 1979, Eastern Kentucky University. 51 p.

Davis, Morgan Jefferson, Jr. Structure of eastern Apache Mountains, Culberson County, Trans-Pecos, Texas; Part I, Regional structure. M, 1953, University of Texas, Austin.

Davis, Newton Fraser Gordon. Geology of the Clearwater Lake map-area, British Columbia. D, 1929, Princeton University. 112 p.

Davis, Nicholas Falconer. Experimental studies in the system $NaAlSi_3O_8$-H_2O; Part I, the apparent solubility of albite in supercritical water; Part II, the partial specific volume of H_2O in $NaAlSi_3O_8$ melts with petrologic implications. D, 1972, Pennsylvania State University, University Park. 323 p.

Davis, Norman Bruce. A review of the theories of plasticity of clays and its relations to their mode of origin. M, 1914, Cornell University.

Davis, Owen Kent. Vegetation migration in southern Idaho during the late-Quaternary and Holocene. D, 1982, University of Minnesota, Minneapolis. 262 p.

Davis, P. T. Quaternary glacial history of Mt. Katahdin, Maine. M, 1976, University of Maine. 155 p.

Davis, Philip A. Development and application of resistivity sounding inversion for several field arrays. M, 1979, University of Minnesota, Minneapolis. 152 p.

Davis, Philip A., Jr. K, Rb, Sr and Sr isotopes in the Preacher Creek ultramafic intrusion, Wyoming. M, 1974, Miami University (Ohio). 45 p.

Davis, Philip A., Jr. Trace element model studies of late Precambrian-early Paleozoic greenstones of Virginia, Maryland and Pennsylvania. D, 1977, University of Kentucky. 149 p.

Davis, Philip Thompson. Late Holocene glacial, vegetational, and climatic history of Pangnirtung and Kingnait Fiord area, Baffin Island, N.W.T., Canada. D, 1980, University of Colorado. 378 p.

Davis, Phillip Burton. An ecological approach for highway routing in Michigan. D, 1979, Michigan State University. 101 p.

Davis, Phillip Nixon. Oklahoma palynology and stratigraphy of the Lower Cretaceous rocks of northern Wyoming. D, 1963, University of Oklahoma. 239 p.

Davis, Phillip Nixon. Palynology of the Rowe Coal (Pennsylvanian) of Oklahoma. M, 1961, University of Oklahoma. 153 p.

Davis, R. A., Jr. Paleocurrent analysis of the Upper Cretaceous, Paleocene and Eocene strata, Santa Ana Mountains, California. M, 1978, University of Southern California.

Davis, R. L. Finite element stress analysis of the role of thermal expansion in small scale elastic crustal deformation. M, 1974, Ball State University.

Davis, R. W. Stratigraphy of the Wasatch Formation, east flank of the Rock Springs Uplift, Sweetwater County, Wyoming. M, 1958, University of Wyoming. 76 p.

Davis, Ralph A. The origin, age, and correlation of the Ingleside Formation of north central Colorado. M, 1947, University of Colorado.

Davis, Ralph K. Hydrogeologic interrelations of the Platte River basin and the upper Big Blue River basin, in the Polk County area of Nebraska. M, 1986, University of Nebraska, Lincoln. 75 p.

Davis, Rhonda Lynn. Sedimentology and petrology of lacustrine carbonates; the mid-Tertiary Camp Davis Formation of northwestern Wyoming. M, 1982, University of Michigan.

Davis, Richard Albert, Jr. Paleoecology of the hurricane lentil, Cook Mountain Formation, East Texas. M, 1961, University of Texas, Austin.

Davis, Richard Albert, Jr. Sedimentation in the nearshore environment, southeastern Lake Michigan. D, 1964, University of Illinois, Urbana. 131 p.

Davis, Richard Arnold. A conodont fauna from the Edgewood Limestone (Lower Silurian) of (Jackson County) Iowa. M, 1965, University of Iowa. 75 p.

Davis, Richard Arnold. Mature modifications and dimorphism in selected late Paleozoic ammonoids (Carboniferous) (Permian). D, 1968, University of Iowa. 179 p.

Davis, Richard D. The geology along a portion of the Saltville fault, Knox County, Tennessee. M, 1970, University of Tennessee, Knoxville. 30 p.

Davis, Richard Laurence. Some physical characteristics of a small agricultural watershed in east-central Illinois. M, 1972, Washington University. 59 p.

Davis, Richard Laurence. The effects of urbanization on small watersheds and ground water in Penfield, New York. D, 1980, University of Rochester. 355 p.

Davis, Richard Spencer. Geology and groundwater resources of the Mosul Liwa and northern Jezira Desert, Iraq. M, 1956, University of Southern California.

Davis, Richard Warren. A geophysical investigation of hydrologic boundaries in the Tucson Basin, Pima County, Arizona. D, 1967, University of Arizona. 99 p.

Davis, Robert Alvin. A bandlimited magnetotelluric study of an area in Harvard, Massachusetts. M, 1979, Massachusetts Institute of Technology. 203 p.

Davis, Robert E. Geology of the Mary G Mine area, Pima County, Arizona. M, 1955, University of Arizona.

Davis, Robert G. Pre-Grenville ages of basement rocks in central Virginia; a model for the interpretation of zircon ages. M, 1974, Virginia Polytechnic Institute and State University.

Davis, Robert Irving and Planck, Robert F. Geology of the McKenzie Canyon area, Beaverhead County, Montana. M, 1949, University of Michigan. 46 p.

Davis, Robert Irving. The geology and ore deposits of the Santa María del Oro gold-copper district, Durango, Mexico. D, 1954, University of Michigan. 291 p.

Davis, Robert L. Geology of the Dog Valley-Granddad Peak area, southern Pavant Mountains, Millard County, Utah. M, 1982, Brigham Young University.

Davis, Robert Martin. Geomorphic base for recent urban expansion in the Sacramento (California) urban area. M, 1972, University of Nevada. 49 p.

Davis, Robin Alane. The influence of phylloid algae on carbonate sedimentation in the Winchell Limestone (Canyon Group), north-central Texas Ranger Quarry, Ranger, Texas. M, 1988, Stephen F. Austin State University. 201 p.

Davis, Ronald. Strip mining effects on water quality. M, 1972, [Colorado State University].

Davis, Ronnie McConnell. Sedimentary structures in river-estuary transition zones of North Carolina. M, 1981, University of North Carolina, Chapel Hill. 60 p.

Davis, S. E. A study of vegetation in relation to coastal formation and stability in the Indiana Dunes National Lakeshore. M, 1976, Purdue University. 112 p.

Davis, Scott Daniel. Investigations of natural and induced seismicity in the Texas Panhandle. M, 1985, University of Texas, Austin. 230 p.

Davis, Stanley Graham. An exploration gravity survey of the south basin of Vekol Valley, Pinal and Maricopa counties, Arizona. M, 1984, University of Arizona. 99 p.

Davis, Stanley N. Pleistocene geology of Platte County, Missouri. D, 1955, Yale University.

Davis, Stanley N. Quaternary geology and groundwater resources in the vicinity of Topeka, Kansas. M, 1951, University of Kansas. 74 p.

Davis, Stanton Hoffman. The stratigraphy and structure of the Sheep Mountain area, Converse County, Wyoming. M, 1952, University of Iowa. 166 p.

Davis, Steven Allen. An analysis of the eastern margin of the Portland Basin using gravity surveys. M, 1988, Portland State University. 135 p.

Davis, Terry E. Strontium isotope and trace element geochemistry of igneous and metamorphic rocks in the Franciscan assemblage, California. D, 1969, University of California, Santa Barbara.

Davis, Terry E. The optical properties of the natural plagioclase feldspars and their heat treated modifications. M, 1962, University of Nevada. 32 p.

Davis, Thomas C., Jr. Structures in the upper snow layers of the southern dome of the Greenland ice

sheet. M, 1962, University of Alaska, Fairbanks. 107 p.

Davis, Thomas Edward. Controls on ore deposition, Polaris mining district, Pioneer Mountains, Beaverhead County, Montana. M, 1980, Montana State University. 88 p.

Davis, Thomas L. Velocity variations around Leduc reefs (upper Devonian, Alberta). M, 1971, University of Calgary. 89 p.

Davis, Thomas Lealand. Late Cenozoic structure and tectonic history of the western "Big Bend" of the San Andreas Fault and adjacent San Emigdio Mountains. D, 1983, University of California, Santa Barbara. 938 p.

Davis, Thomas Mooney. Theory and practice of geophysical survey design. D, 1974, Pennsylvania State University, University Park. 147 p.

Davis, Thomas Weldon. An analysis of coastal recession models with application to the coast of North Carolina. M, 1985, Duke University. 108 p.

Davis, Thornton. Review of the geology and production of the oil fields of South America. M, 1923, Columbia University, Teachers College.

Davis, V. H.; Harper, J. N. and Neish, J. F. A short-term study of beach sand migration adjacent to Monterey canyon (California). M, 1966, United States Naval Academy.

Davis, Willard Frew. Two methods of automatic depth determination applied to a study of the magnetic basement of western PA. M, 1980, Pennsylvania State University, University Park. 72 p.

Davis, William Edwin, Jr. Conodont fauna of the Tully Limestone, Middle Devonian, New York State. D, 1966, Boston University. 163 p.

Davis, William Edwin, Jr. Geology of Lime Kiln Quadrangle, Hays County, Texas. M, 1962, University of Texas, Austin.

Davison, C. C. A hydrogeochemical investigation of a brine disposal lagoon-aquifer system, Esterhazy, Saskatchewan. M, 1976, University of Waterloo.

Davison, Charles H., Jr. Simulation by finite difference methods of water levels in the Horizon City area, Texas. M, 1984, University of Texas at El Paso.

Davison, Fred C., Jr. Classification of thermomagnetic characteristics of southeastern granites and correlation with previously reported magnetic results. M, 1982, University of Georgia.

Davison, Frederick Corbet, Jr. Stress tensor estimates derived from focal mechanism solutions of sparse data sets; applications to seismic zones in Virginia and eastern Tennessee. D, 1988, Virginia Polytechnic Institute and State University. 202 p.

Davison, Gordon E. A geophysical model of the gravity magnetic high, Virginia coastal plain. M, 1985, Old Dominion University. 147 p.

Davison, J. Lynne. Sources, mode of transport and dispersal of Recent sediment to Canada abyssal plain; Fourier grain-shape analysis. M, 1987, Wichita State University. 153 p.

Davison, James Gregory. Physical volcanology, sedimentology, stratigraphy and petrochemistry of the Berry Creek metavolcanics; an Archean calc-alkaline complex, Lake of the Woods, Ontario. M, 1984, Brock University. 303 p.

Davison, Wilbert Lloyd. The geology of upper Frobisher bay (Baffin Island, Northwest Territories. M, 1950, Dalhousie University.

Davison, William D., Jr. Petrology of the Ashe Formation, Shortt's Knob area, Floyd County, Virginia. M, 1973, Virginia Polytechnic Institute and State University.

Davoren, Anthony. Hydrologic characteristics of the Pokaiwhenua River basin, North Island, New Zealand. D, 1982, Washington State University. 176 p.

Davy, W. M. Ore deposition in the Bolivian tin-silver deposits. D, 1920, Massachusetts Institute of Technology. 59 p.

Daw, John Laurence. Seismic detection of subsurface cavities; a feasibility study. M, 19??, North Carolina State University. 63 p.

Daw, Robert Norman. Origin and occurrence of the soapstone deposits, Saline County, Arkansas. M, 1956, University of Oklahoma. 45 p.

Dawans, Jean-Michel L. Distribution and petrography of late Cenozoic dolomites beneath San Salvador and New Providence islands, the Bahamas. M, 1988, University of Miami. 106 p.

Dawe, Steven E. The geology of the Mountain Pine Ridge area and the relation of the Mountain Pine Ridge Granite to the late Paleozoic and early Mesozoic geological history, Belize, Central America. M, 1984, SUNY at Binghamton. 52 p.

Dawers, Nancye Helen. The basement geology of the Catlin Lake-Goodnow Pond area, Adirondack Mountains, New York, and its relationship to the October 7, 1983 Goodnow earthquake. M, 1987, University of Illinois, Urbana. 79 p.

Dawes, Bruce Cameron. Trilobite faunas of the upper Cambrian, Bison Creek Formation, southwestern Alberta. M, 1973, Queen's University. 224 p.

Dawson, Charles A., Jr. Petrology of the igneous complex near Lang, California. M, 1937, California Institute of Technology. 105 p.

Dawson, James Clifford. The geology of the Bluff Cove area, Falkland Islands. M, 1967, University of California, Los Angeles.

Dawson, James Clifford. The sedimentology and stratigraphy of the Morrison Formation (upper Jurassic) in northwestern Colorado and northeastern Utah. D, 1970, University of Wisconsin-Madison. 142 p.

Dawson, James M. Regional geology of the Topsail-Foxtrap area (Newfoundland). M, 1963, Memorial University of Newfoundland. 132 p.

Dawson, James W. Determination of fractured aquifer characteristics from evaluation of pump tests of wells in the crystalline rocks of the Blue Ridge Allocthon. M, 1988, Virginia Polytechnic Institute and State University. 152 p.

Dawson, John W. Geology of the Birch Creek area, Dayville Quadrangle, Oregon. M, 1951, Oregon State University. 98 p.

Dawson, Kenneth Murray. Geology of the Endako Mine, British Columbia. D, 1972, University of British Columbia.

Dawson, Kenneth R. A petrographic description of wall rocks and alteration products associated with pitchblende-bearing veins in the Goldfields region, Saskatchewan. D, 1952, University of Toronto.

Dawson, Kenneth R. Differentiation in an altered diabase dyke, Dassarat Township, Quebec. M, 1949, University of Toronto.

Dawson, M. K. The paleontology of the Cabrillo Formation. M, 1978, San Diego State University.

Dawson, Malcolm Robert, II. Geochemistry and origin of mafic schists from the Pelona, Orocopia and Rand schists; structure and metamorphism of the Orocopia Schist, southern California. D, 1987, Iowa State University of Science and Technology. 161 p.

Dawson, Malcom Robert, II. Trace element distributions and mineral associations in Iowa coals. M, 1982, Iowa State University of Science and Technology. 161 p.

Dawson, Mary L. Comparison of contact metamorphism to burial metamorphism in argillaceous sediments. M, 1979, University of Texas, Arlington. 64 p.

Dawson, Robert Scott. Strontium isotope geology of a section of the Flagstaff Limestone, central Utah. M, 1978, Ohio State University.

Dawson, Robert William. Ecologic and faunal analyses of the Miocene fossil invertebrate faunas at Grimesland and Magnolia, North Carolina. M, 1958, University of North Carolina, Chapel Hill. 85 p.

Dawson, Robin Humphrey. Petrography and stratigraphy of the late Paleozoic rocks in the Wildhay River-Rock Lake area, Alberta. M, 1966, University of British Columbia.

Dawson, Ross Elmo, Jr. A study of the non-opaque heavy minerals in the Lissie Sandstone. M, 1958, Rice University. 39 p.

Dawson, Thomas Albert. Devonian of southeastern Indiana. M, 1940, Indiana University, Bloomington. 66 p.

Dawson, Trevor William. Three-dimensional electromagnetic induction in thin sheets. D, 1979, University of Victoria. 409 p.

Dawson, William Craig. Petrography and sedimentation of the Upper Cretaceous Farmersville Member, Pecan Gap Formation, Taylor Group of Northeast Texas. M, 1976, University of Texas, Arlington. 158 p.

Dawson, William Craig. Petrography, sedimentology, diagenesis, and reservoir characteristics of some Pennsylvanian phylloid algal limestones; Kansas and Utah, U.S.A. D, 1984, University of Illinois, Urbana. 269 p.

Dawud, Awni Yaqub. Variation of soil properties within the soil mapping unit; a statistical study. D, 1979, Oklahoma State University. 302 p.

Day, Blaine Spencer. Stratigraphy of the Upper Triassic(?) Moenave Formation of southwestern Utah (Washington County). M, 1967, University of Utah. 58 p.

Day, Damon P. Petrography, origin and environment of deposition of the Horsethief Sandstone (Upper Cretaceous), Montana. M, 1965, Michigan Technological University. 95 p.

Day, David L. The glacial geomorphology of the Trout Creek areas, Porcupine Hills, Alberta. M, 1971, University of Calgary.

Day, David William, Jr. The geology of a part of the Mount Vaca Quadrangle, California. M, 1951, University of California, Berkeley. 58 p.

Day, Garrett Arthur. Source of recharge to the Beowawe Geothermal System, north-central Nevada. M, 1987, University of Nevada. 82 p.

Day, Harry Clinton, Jr. Stratigraphic/sedimentological parameters of the Westwater Canyon Member of the Morrison Formation; relationship to uranium deposits, east half of Grants uranium region, New Mexico. M, 1984, University of Colorado. 106 p.

Day, Howard Wilman. A theoretical and experimental study of some equilibria in the system K-Fe-Al-Si-O-H. D, 1971, Brown University.

Day, Howard Wilman. Rb-Sr ages of some granites from western Rhode Island. M, 1968, Brown University.

Day, James E., II. Stratigraphy, carbonate petrology, and paleoecology of the Jerome Member of the Martin Formation in east central Arizona. M, 1983, Northern Arizona University. 147 p.

Day, James Edgar, II. Stratigraphy, biostratigraphy, and depositional history of the Givetian and Frasnian strata in the San Andres and Sacramento Mountains of southern New Mexico. D, 1988, University of Iowa. 238 p.

Day, Katherine Walton. Determination of the processes controlling vertical trace element distribution in Filson Creek bog, Lake County, Minnesota. M, 1986, Colorado School of Mines. 236 p.

Day, L. Wayne. Zircon geochronology of northeastern Alberta. M, 1975, University of Alberta. 72 p.

Day, M. J. Movement and hydrochemistry of groundwater in fractured clayey deposits in the Winnipeg area. M, 1978, University of Waterloo.

Day, Peter Conrad. The nature and distribution of surficial sediments on the submerged northern Channel Islands platform, Southern California borderland. M, 1979, University of Southern California.

Day, Ronald. The effect of grain size on the magnetic properties of the magnetite-ulvospinel solid solution series. D, 1974, University of Pittsburgh.

Day, Stephen John. Sampling stream sediments for gold in mineral exploration, southern British Columbia. M, 1988, University of British Columbia. 213 p.

Day, Steven M. Finite element analysis of seismic scattering problems. D, 1977, University of California, San Diego. 165 p.

Day, Sumner Daniel. The petrology of a mafic dike complex near Smartville, Yuba County, California. M, 1977, University of California, Davis. 113 p.

Day, Theodore James. Geothermal temperature study in and around Beowawe (Nevada). M, 1975, Stanford University.

Day, Warren C. Petrology and geochemistry of the Vermilion granitic complex, northern Minnesota. D, 1983, University of Minnesota, Minneapolis. 210 p.

Day-Lewis, Robert Ernest. A hydrological survey of the Town of Tyngsborough, Massachusetts. M, 1980, University of Massachusetts. 263 p.

Dayal, R. Clay-seawater interactions at elevated pressures. D, 1975, Dalhousie University. 234 p.

Dayal, Umesh. Instrumented impact cone penetrometer. D, 1974, Memorial University of Newfoundland. 222 p.

Dayley, Richard D. Seismic investigation of flint clay deposits in east-central Missouri. M, 1962, University of Missouri, Rolla.

Dayton, Frank Herbert, Jr. Quantitative drainage basin studies and possible application. M, 1968, University of Alabama.

Dayvault, Richard D. The geology of lower Santa Clara Canyon, Chihuahua, Mexico. M, 1979, East Carolina University. 124 p.

de Albuquerque, John Stephen. Stratigraphy and depositional environments of the Middle Jurassic (Callovian) Ralston Creek Formation, Beulah-Wetmore area, south-central Colorado. M, 1988, Fort Hays State University. 114 p.

de Almeida, Erasto Boretti. Correlation between some ore deposit belts in Brazil and Africa. M, 1973, Stanford University.

de Almeida, Erasto Boretti. Geology of the bauxite deposits of the Pocos de Caldas District, State of Minas Gerais, Brazil. D, 1977, Stanford University. 273 p.

de Alteris, Joseph Thomas. The sedimentary processes and geomorphic history of Wreck Shoal, an oyster reef of the James River, Virginia. D, 1986, College of William and Mary. 243 p.

de Andrade Nery Leao, Zelinda Margarida. Morphology, geology and developmental history of the southernmost coral reefs of western Atlantic, Abrolhos Bank, Brazil. D, 1982, University of Miami. 239 p.

De Angelo, Michael V. Geophysical anomalies in southwestern New Mexico and adjacent areas. M, 1988, University of Texas at El Paso.

de Araujo Filho, Jose Oswaldo. Geology of the Aiken NW Quadrangle, South Carolina. M, 1975, University of Georgia.

de Azevedo, Antonio Expedito Gomes. Atmospheric distribution of carbon dioxide and its exchange with the biosphere and the oceans. D, 1982, Columbia University, Teachers College. 185 p.

de Baar, Hein J. W. The marine geochemistry of the rare earth elements. D, 1983, Woods Hole Oceanographic Institution. 278 p.

de Beer, Johannes Hendrik. Magnetometer array studies and electrical conductivity in Southern Africa. D, 1976, University of Alberta. 260 p.

De Bremaecker, Jean-Claude. Amplitude of Pn from 3 degrees to 23 degrees. D, 1952, University of California, Berkeley. 105 p.

De Brito, Sergio N. A. The influence of geology in the construction of four tunnels. M, 1970, University of Illinois, Urbana. 115 p.

De Brock, Michael D. Pennsylvanian (Desmoinesian) Polyplacophora from Texas. M, 1982, Bowling Green State University. 56 p.

De Broucker, Gilles. Evolution tectonostratigraphique de la boutonnière Maquereau-Mictaw (Cambro-Ordovicien), Gaspésie, Québec. D, 1986, Universite Laval. 322 p.

De Bruin, James H. The structure and stratigraphy of the southern half of the Syracuse West Quadrangle (Pennsylvanian). M, 1956, Syracuse University.

De Bruin, Rodney H. Precambrian geology near Bruce Mountain, northern Bighorn Mountains, Wyoming. M, 1975, Iowa State University of Science and Technology.

De Camara, Richard P. Trace-element investigation of northern Arkansas sphalerite. M, 1969, University of Missouri, Columbia.

De Capitani, Christian Emile. The computation of chemical equilibrium and the distribution of Fe, Mn and Mg among sites and phases in olivines and garnets. D, 1987, University of British Columbia.

de Castro, Celso Filho. Effects of liming on characteristics of a Brazilian Oxisol at three levels of organic matter as related to erosion. D, 1988, Ohio State University. 276 p.

De Celles, Peter G. Sedimentation and diagenesis in a tectonically partitioned, nonmarine foreland basin; the Lower Cretaceous Kootenai Formation, southwestern Montana. D, 1984, Indiana University, Bloomington. 422 p.

de Chazal, Suzanne Marie. Uranium, plutonium and the rare earth elements in the phosphates of ordinary chondrites and the potential of the 244Pu chronometer. M, 1985, Washington University. 118 p.

de Cort, Thierry Michael. Petrology and diagenesis of upper Minnelusa sandstones, carbonates and evaporites, Northeast Powder River basin, Wyoming. M, 1982, Southern Illinois University, Carbondale. 155 p.

de Corta, Hugues. Les Dépôts quaternaires de la région Lac Rohault-Lac Boisvert (sud de Chibougamau); aspect de la dispersion glaciaire clastique. M, 1988, Universite du Quebec a Montreal. 112 p.

De Courten, Frank L. Trace fossils of the Kaibab Formation (Permian) of northern Arizona. M, 1976, University of California, Riverside. 72 p.

de Cousser, Kurt H. Evidence to the production of commercial oil from the Emerson-Texon fold. M, 1925, University of Missouri, Rolla.

de Cserna, Zoltan. Geology of the Sierra de Santa Rita, Zacatecas, Mexico. M, 1952, University of New Mexico. 71 p.

de Cserna, Zoltan. Structural geology of southeastern Coahuila, and adjacent parts of Nuevo León, Mexico. D, 1955, Columbia University, Teachers College.

De Den, F. Michael. Stratigraphy, depositional environments, and diagenesis of the Devonian Simaison and Guilmette formations in the White Pine, Egan, and Schell Creek ranges, White Pine County, Nevada. M, 1988, Oregon State University. 143 p.

De Eston, S. M. Syllabus on rock mechanics. M, 1975, Stanford University. 238 p.

de Farias, Luiz Carlos Cabral *see* Cabral de Farias, Luiz Carlos

De Fauw, Sherri Lynn. The appendicular skeleton of African dicynodonts. D, 1986, Wayne State University. 292 p.

de Figueiredo Filho, Paulo Miranda. The effect of grinding on kaolinite and attapulgite as revealed by X-ray studies. M, 1963, University of Illinois, Urbana. 49 p.

de Figueiredo, Alberto Garcia, Jr. Submarine sand ridges; geology and development, New Jersey, U.S.A. D, 1984, University of Miami. 459 p.

de Figueiredo, Antonio Manuel Ferreira. Depositional systems in the Lower Cretaceous Morro do Chaves and Coqueiro Seco formations, and their relationship to petroleum accumulations, Middle Rift Sequence, Sergipe-Alagoas Basin, Brazil. D, 1981, University of Texas, Austin. 302 p.

de Figueiredo, Joao Neiva. Field of application of diamond drilling in prospecting, exploring, and appraisal of mineral properties. M, 1943, University of Minnesota, Minneapolis. 74 p.

de Figueiredo, Mario Cesar Heredia. Geochemistry of high-grade metamorphic rocks, northeastern Bahia,

Brazil. D, 1980, University of Western Ontario. 221 p.

de Freitas, Timothy A. The Silurian carbonate-platform margin and contiguous sponge biostromes of east-central Cornwallis Island, Canadian Arctic. M, 1986, University of Western Ontario. 260 p.

De Gasparis, Aurelio Alfonso Amedeo. Magnetic properties of tektites and impact glasses. D, 1973, University of Pittsburgh.

de Gasparis, Silvana. Dinoflagellate stratigraphy and paleoecology at the Cretaceous-Tertiary boundary Alabama. D, 1985, University of Toronto.

de Graaff, Fredric R. van *see* van de Graaff, Fredric R.

de Graaff, Fredric Ray Van *see* Van de Graaff, Fredric Ray

de Graff, Jaye Ellen Up *see* Up de Graff, Jaye Ellen

De Groff, Edward. The joints in Silurian and Devonian limestones and shales of Onondaga County. M, 1954, Syracuse University.

de Groot, Philip Henry. Simulation of variably saturated subsurface flow by sectional partitioning. D, 1987, Utah State University. 148 p.

de Gruyter, D. A. Glacial geology of the Sierra Nevada de Santa Marta, Colombia. M, 1976, Ohio State University. 132 p.

de Gruyter, Philip Clarence. The petrogenesis and tectonic setting of the Egyptian alkaline complexes. M, 1983, Michigan State University. 284 p.

De Hon, René Aurel. A maar origin for Hunts Hole, Dona Ana County, New Mexico. M, 1965, Texas Tech University. 70 p.

De Hon, René Aurel. Photogeologic study of the Agrippa region of the Moon. D, 1970, Texas Tech University. 109 p.

de Irisarri, A. M. A resume of the geology of Colombia, South America. M, 1929, University of Missouri, Columbia.

de Jong, Bernardus Hermenigildus W. S. A spectroscopic and molecular orbital study on the polymerization of silicate and aluminate tetrahedra in aluminosilicate melts, glasses, and aqueous solutions. D, 1981, Stanford University. 179 p.

de Jong, Sybren Hendrik. The skew Mercator projection in rectangular plane coordinate systems. D, 1968, Ohio State University. 159 p.

de Jonge, E. J. Coen Kiewiet *see* Kiewiet de Jonge, E. J. Coen

de Kamp, Peter Cornelius Van *see* Van de Kamp, Peter Cornelius

de Kehoe, J. R., Jr. Geology of the Atascadero area, San Luis Obispo County, California. M, 1973, [University of California, San Diego].

De Keyser, Thomas Lee. The Early Mississippian of the Sacramento Mountains, New Mexico-an ecofacies model for carbonate shelf margin deposition. D, 1979, Oregon State University. 304 p.

De Kimpe, Nancy. Dune recession on the Outer Banks of North Carolina. M, 1987, Virginia State University.

de Kruyter, Mark. The Quaternary geology of Plum Creek valley, Wisconsin. M, 1985, University of Wisconsin-Milwaukee. 104 p.

de l'Etoile, Robert. La Géostatistique des gisements de charbon. M, 1982, Ecole Polytechnique. 144 p.

De la Cruz, Luis A. Stratigraphy of Jurassic and Cretaceous rocks, Hanna area, Duchesne County, and Bridger Lake field, Summit County, Utah. M, 1971, Colorado School of Mines. 103 p.

de la Cruz, Nga. Chemical and isotopic variations within the Ordovician Kinnekulle A₁ Bentonite, Kinnekulle, Sweden. M, 1968, University of Alberta. 51 p.

De la Cruz, Servando. Thermal convection and its geophysical significance. M, 1973, University of Toronto.

de la Fuente Duch, M. F. F. Evaluation of integrated exploration programs for revitalization of old mining districts. D, 1979, University of Arizona. 133 p.

De la Fuente Duch, Mauricio Fernando. Aeromagnetic study of the Colorado River delta area, Mexico. M, 1973, University of Arizona.

de la Garza, Fernando Javier Rodriguez *see* Rodriguez de la Garza, Fernando Javier

de la Montagne, John. Cenozoic history of the Saratoga Valley area, Wyoming and Colorado. D, 1955, University of Wyoming. 140 p.

de la Montagne, John. Geomorphology of the north end of Centennial Valley-Table Mountain area, Wyoming. M, 1951, University of Wyoming. 121 p.

de la Pena Horcasitas, Gerardo Ruiz *see* Ruiz de la Pena Horcasitas, Gerardo

De La Pena, Edward C. Reconnaissance geology of the Corralitos area, Sierra Madre Occidental, Chihuahua, Mexico. M, 1978, Wayne State University.

de la Torre Robles, Jorge. Clay mineral distribution in Recent sediments from the north Pacific coast of Mexico. D, 1965, University of Pittsburgh. 167 p.

De Laca, Ted Edwin. Distribution of benthic foraminifera and the habitat ecology of marine benthic algae of Antarctica. D, 1976, University of California, Davis. 208 p.

de Lacroix, Pierre. The mineralogy, petrography, and diagenesis of the Whetstone Gulf Formation (Upper Ordovician) of North-Central New York State. M, 1980, Syracuse University.

de Laguna, Wallace. Geology of the Atlantic City District, Wyoming. D, 1938, Harvard University.

de Landro, Wanda-Lee. Deep seismic sounding in southern Manitoba and Saskatchewan. M, 1981, University of Manitoba. 126 p.

De Leon, Alfredo Aniano. Hydraulics of debris flow. D, 1982, Utah State University. 144 p.

de Leon, Jose G. Ponce *see* Ponce de Leon, Jose G.

de Lima, Edmilson Santos. Metamorphism and tectonic evolution in the Serido region, northeastern Brazil. D, 1986, University of California, Los Angeles. 228 p.

De Long, James Henry. The paleontology and stratigraphy of the Pleistocene at Signal Hill, California. M, 1939, California Institute of Technology. 29 p.

de Long, Richard F. Geology of the Hall Gulch plutonic complex, Elmore and Camas counties, Idaho. M, 1986, University of Idaho. 90 p.

de Long, Richard F. United States mining policy; hardrock minerals and Federal lands. M, 1984, University of Idaho. 55 p.

de los Reyes, A. Guitron *see* Guitron de los Reyes, A.

de Matos, Milton Martins. Mobility of soil and rock avalanches. D, 1988, University of Alberta. 360 p.

de Mauret, Kevin. Paleo-oceanographic study of late Quaternary sediment from the Atlantic sector of the Southern Ocean using marine diatoms. M, 1985, Rutgers, The State University, Newark. 215 p.

de Medina, Diana Magdalena Diez *see* Diez de Medina, Diana Magdalena

De Melas, John P. The geochemistry, petrology, and provenance of the Pinal Schist. M, 1983, New Mexico Institute of Mining and Technology. 128 p.

de Miranda, Antonio Nunes. Behavior of small earth dams during initial filling. D, 1988, Colorado State University. 249 p.

de Mohrenschildt, George Sergius. A review of the principles of reservoir performance for petroleum geologists. M, 1945, University of Texas, Austin.

de Mora, Stephen John. Manganese chemistry in the Fraser Estuary. D, 1981, University of British Columbia.

De Moraes, Joao A. General conclusions concerning the hydrogeology of Major Valley (Missouri River Valley) alluvium in Central United States. D, 1972, University of Missouri, Columbia.

de Moraes, Joao A. P. Hydrogeology of the McBaine area, central Missouri. M, 1969, University of Missouri, Columbia.

De Nault, Kenneth James. Origin of sandstone type uranium deposits in Wyoming. D, 1974, Stanford University. 352 p.

De Noyer, John Milford. The energy in seismic waves. D, 1958, University of California, Berkeley. 72 p.

de Oliveira Morais, Francisco Ilton. Charge characteristics and ion exchange equilibria in soils from the humid tropics of Brazil. D, 1975, University of California, Riverside. 100 p.

de Oliveira, Ruy Bruno Bacelar *see* Bacelar de Oliveira, Ruy Bruno

de Poll, Henk Wouter van *see* van de Poll, Henk Wouter

de Prado, Connie A. High-latitude Paleocene diatoms of the Southern Oceans. M, 1981, Northern Illinois University. 256 p.

De Ratmiroff, Gregor N. Origin and metamorphism of the major paragenesis of the Precambrian Imataca Complex; Upata Quadrangle, State of Bolivar, Venezuela. D, 1964, Rutgers, The State University, New Brunswick. 199 p.

de Reep, Thomas W. Van *see* Van de Reep, Thomas W.

De Renne, Paul. Geochemistry of the Idaho Batholith. M, 1972, University of Idaho. 128 p.

De Renne, Paul. Rb-Sr isotope geology of the Idaho Batholith and some associated rocks. D, 1976, University of Idaho. 131 p.

de Rezende, Servulo Batista. Geomorphology, mineralogy and genesis of four soils on gneiss in southeastern Brazil. D, 1980, Purdue University. 158 p.

De Rito, Robert F. Geothermal influences on the long-term thickness of the mechanical lithosphere during flexure. D, 1984, SUNY at Buffalo. 103 p.

de Rojas, Isabel. Stratigraphy of the Mowry Shale (Cretaceous), western Denver Basin, Colorado. M, 1980, Colorado School of Mines. 148 p.

De Romer, Henry Severyn. The geology of the eastern border of the "Labrador Trough", east of Thevenet Lake, New Quebec. M, 1956, McGill University.

De Rose, Nicholas. Bouguer gravity crustal models of the Curacao Ridge, Venezuelan borderland. M, 1981, Rutgers, The State University, Newark. 69 p.

De Rosen-Spence, A. F. Stratigraphy, development and petrogenesis of the central Noranda volcanic pile, Noranda, Quebec. D, 1976, University of Toronto. 439 p.

De Saboia, L. A. The West Manasan and the Pipe Mine 2 ultramafic bodies of the Thompson nickel belt. M, 1978, University of Western Ontario. 168 p.

De Santis, J. E. The petrology of the ultramafic rocks in the Wissahickon Formation, Philadelphia. M, 1978, Temple University.

De Silva, G. L. R. Slope stability problems induced by human modification of the soil covered hill slopes of Oahu, Hawaii. D, 1974, University of Hawaii. 451 p.

De Silva, G. L. R. The effects of marine attack on the exposed portions of the Honolulu Series tuff cones. M, 1966, University of Hawaii. 171 p.

de Simpson, R. Systematic paleontology and paleoenvironmental analysis of the upper Hueco Formation, Robledo and Dona Ana mountains, Dona Ana County, New Mexico. M, 1976, University of Texas at El Paso.

De Sonneville, Joseph Leonardus Johannes. Development of a digital groundwater model with application to aquifers in Idaho. D, 1974, University of Idaho. 228 p.

de Sousa Carvalho Martins, Verónica E. A microstructural study of S-C mylonites of part of the Tanque Verde Mountains, Tucson, Arizona. M, 1984, University of Arizona. 52 p.

de Souza, Euler Magno. Ground control, brine inflow arrest and backfill designs in the search for new mining strategies for the potash industry. D, 1988, Queen's University.

de Souza, Jairo Marcondes. Transmission of seismic energy through the Brazilian Parana Basin layered basalt stack. M, 1982, University of Texas, Austin. 303 p.

de St. Jorre, Louise. Economic mineralogy of the North T-Zone Deposit, Thor Lake, Northwest Territories. M, 1986, University of Alberta. 248 p.

de Vasconcelos, Jose Aluizio. Planning and executing campaigns in geochemical exploration with particular reference to deeply weathered terrains. M, 1973, Stanford University.

De vay, Joseph C. The petrology and provenance of the Shoo Fly Formation quartz sandstone northern Sierra Nevada, California. M, 1981, University of California, Davis. 114 p.

de Verg, Philip E. Van *see* Van de Verg, Philip E.

de Villiers, Johan Pieter Roos. The crystal structures of aragonite, strontianite, and witherite. D, 1969, University of Illinois, Urbana. 92 p.

De Villiers, Johanne. Application des méthodes géostatistiques à l'évaluation d'un gisement aurifère; la Mine Chadbourne Inc. M, 1987, Ecole Polytechnique. 125 p.

De Vincenzo, Theresa E. A gravitational and stratigraphic analysis of Orange Dome, Orange, Texas. M, 1984, SUNY at Buffalo. 77 p.

De Vito, Steven A. An environmental geology study, Bath Township, Greene County, Ohio. M, 1975, Wright State University. 130 p.

De Voogd, Beatrice. Deep structures and magmatic processes in two continental rifts; studies using COCORP seismic reflection profiling in Death Valley and in the Rio Grande Rift. D, 1986, Cornell University. 187 p.

de Voorde, Barbara Wiley Van *see* Van de Voorde, Barbara Wiley

De Voto, Richard H. Geology of southwestern South Park, Park and Chaffee counties, Colorado. D, 1961, Colorado School of Mines. 323 p.

De Vries, George A. Structural geology of the southern part of Elkhorn Mountain, Bannock Range, Idaho. M, 1977, Utah State University. 91 p.

de Vries, Janet L. Evaluation of low-temperature geothermal potential in Cache Valley, Utah. M, 1982, Utah State University. 96 p.

de Vries, Thomas John. The geology and paleontology of tablazos in Northwest Peru; (Volumes I-III). D, 1986, Ohio State University. 108 p.

De Wys, Egbert Christiaan. A restudy of the Stones River Group Bryozoa fauna of central Tennessee, with descriptions of new species. M, 1951, Miami University (Ohio). 38 p.

de Zoeten, Ruurdjan. Structure and stratigraphy of the central Cordillera Septentrional, Dominican Republic. M, 1988, University of Texas, Austin. 299 p.

De, Aniruddha. Petrology of dikes emplaced in the ultramafic rocks of southeastern Quebec. D, 1960, Princeton University. 310 p.

Dea, Peter A. Glacial geology of the Ovando Valley, Powell County, Montana. M, 1981, University of Montana. 107 p.

Deacon, Robert J. Cowlitz-Keasey formational boundary in northwestern Oregon. M, 1953, Oregon State University.

Deal, Clyde Stanley. The areal geology and stratigraphy of Boscobel Quadrangle, Wisconsin. M, 1947, University of Wisconsin-Madison. 55 p.

Deal, D. E. Geology of Jewel Cave National Monument, Custer County, South Dakota, with special reference to cavern formation in the Black Hills. M, 1962, University of Wyoming. 183 p.

Deal, Dwight Edward. Quaternary geology of Rolette County, North Dakota. D, 1970, University of North Dakota. 168 p.

Deal, Edmond Graham. Geology of the northern part of the San Mateo Mountains, Socorro County, New Mexico; a study of a rhyolite ash-flow tuff cauldron and the role of laminar flow in ash-flow tuffs. D, 1973, University of New Mexico. 136 p.

Deal, Edmond Graham. Volcanic geology of the Interior valley, San Francisco Mountain (Coconino County), Arizona. M, 1969, Arizona State University. 82 p.

Deal, James E. The geology of the Goshen Quadrangle. M, 1961, University of Arkansas, Fayetteville.

DeAlteris, Joseph T. The Recent history of Wachapreague Inlet, Virginia. M, 1973, College of William and Mary.

Deamer, Gay A. Compaction-induced inclination shallowing in natural and synthetic sediments. M, 1988, Lehigh University. 139 p.

Dean, Bradley W. Focal mechanism solutions and tectonics of the Middle America Arc. M, 1976, Dartmouth College. 81 p.

Dean, Christopher William. Provenance study and environments of deposition of the Pennsylvanian-Permian Wood River Formation, south central Idaho, and the paleotectonic character of the Wood River basin. M, 1982, Texas A&M University. 157 p.

Dean, Claude S. Stratigraphy and structure of the Sunapee Septum, southwestern New Hampshire. D, 1976, Harvard University.

Dean, D. A. Geology, alteration, and mineralization of the El Alacran area, northern Sonora, Mexico. M, 1975, University of Arizona.

Dean, J. R. Recent sedimentary environments and diagenesis of marine carbonates from the Persian Gulf. M, 1977, University of Houston.

Dean, James Scott. Depositional environments and paleogeography of the Lower to Middle Jurassic Cuyan Group, Neuquen Basin, Argentina. D, 1987, Colorado School of Mines. 587 p.

Dean, Jim. Carbonate petrology and depositional environments of the Sinbad limestone member of the Moenkopi Formation in the Teasdale area, Wayne and Garfield counties, Utah. M, 1980, Brigham Young University.

Dean, John R. A petrographic study of argillaceous sandstones. M, 1959, West Virginia University.

Dean, Kenneson. A geologic evaluation of Landsat imagery in interior Alaska. M, 1979, University of Alaska, Fairbanks. 156 p.

Dean, Lewis Shepherd. Petrogenetic relationships of trondhjemitic rocks in the northern Alabama Piedmont. M, 1981, Emory University. 125 p.

Dean, Michael Edward. Diagenesis of the Viking Formation, south-central Alberta. M, 1986, University of Alberta. 210 p.

Dean, Nancy E. Archaeological geology and geochemistry of the Carrara Marble, Carrara, Italy. M, 1984, University of Georgia. 110 p.

Dean, Paul L. The volcanic stratigraphy and metallogeny of Notre Dame Bay, Newfoundland. M, 1978, Memorial University of Newfoundland. 204 p.

Dean, Reginald S. Some chemical problems in geology. M, 1916, University of Missouri, Rolla.

Dean, Richard Lloyd, II. The influence of marine algal succession on the invertebrate community. D, 1983, University of California, Santa Barbara. 245 p.

Dean, Robert M. Improving water well yields in crystalline rocks. M, 1979, [Colorado State University].

Dean, Ronald S. A compositional study of calcareous (Ordovician) Lorraine sedimentary rocks (Quebec). M, 1958, McGill University.

Dean, Ronald S. A study of the St. Lawrence Lowland shales (Pleistocene; Quebec and New York). D, 1963, McGill University.

Dean, Stuart Linden. Geology of the Great Valley of West Virginia. D, 1966, West Virginia University. 396 p.

Dean, Stuart Linden. Geology of the Shepherdstown Quadrangle, West Virginia-Maryland, Quadrangle. M, 1963, West Virginia University.

Dean, Walter Edward, Jr. Petrologic and geochemical variations in the Permian Castile varved anhydrite, Delaware Basin, Texas and New Mexico. D, 1967, University of New Mexico. 326 p.

Dean, Walter Edward, Jr. Slope deposits of the Pennsylvanian Haymond Formation, Marathon region, Texas. M, 1964, University of New Mexico. 75 p.

Deane, Harold Lutz. The geology of Miami County, Indiana. M, 1952, Indiana University, Bloomington. 74 p.

Deane, Roy E. The Pleistocene deposits of the Lake Simcoe area, Ontario. D, 1949, University of Toronto.

Deans, Brian D. Petroleum geology and hydrodynamic analysis of the Stoney Point Field, Trenton-Black River Group, Ordovician, Michigan Basin. M, 1988, Michigan Technological University. 86 p.

Dear, Timothy B. Paleoenvironments of the upper Albian Stage (Cretaceous) of western Trans-Pecos Texas. M, 1981, Stephen F. Austin State University. 128 p.

Dearden, E. Study of certain Ontario radioactive minerals. M, 1955, University of Toronto.

Dearden, Melvin O. Geology of the central Boulter Mountains area, Utah. M, 1954, Brigham Young University. 85 p.

Deardorff, D. L. Stratigraphy and oil shales of the Green River Formation southwest of the Rock Springs Uplift, Wyoming. M, 1959, University of Wyoming. 98 p.

Dearlove, J. P. L. Geology of the alkaline rocks in the Kirkland Lake-Timmins District, Ontario. M, 1984, University of Waterloo. 176 p.

Dearth, Albert E. The geology of the western part of Chase Gulch, Gilpin County, Colorado. M, 1955, University of Colorado.

Deas, A. A crustal study using teleseismic P phases recorded near Port Arthur, Ontario (Canada). M, 1969, University of British Columbia.

Deatherage, Jas. H. The development of the sonic pulse technique and its comparison with the conventional static method for determining the elastic moduli of rock. M, 1966, University of Missouri, Rolla.

Deaton, B. C. Relationship of the Colotenango Conglomerate of Guatemala to the motion of the Polochic Fault during the Tertiary. M, 1982, University of Texas, Arlington. 97 p.

Deaver, B. G. Morphology and systematic position of some anomalinid foraminifera. M, 1955, University of Missouri, Rolla.

Deaver, Boyd Edwin. Deposition and diagenesis along the updip limit of the Jurassic Smackover Formation, Red River County, Texas. M, 1979, West Texas State University. 78 p.

Deaver, Franklin Kennedy, Jr. Faunal utilization at archeological site 45AD2, Adams County, Washington. M, 1973, Washington State University. 40 p.

DeBartolo, Bruce Alan. Geology of the Pachuta Creek-Nancy-East Nancy area, Clarke County, Mississippi. M, 1970, Tulane University. 42 p.

DeBoer, Daniel Alan. Petrology of the Death Canyon Limestone Member (Gros Ventre Formation) and some Middle Cambrian algal buildups, northwestern Wyoming. M, 1981, Fort Hays State University. 99 p.

Deboo, Phili B. Biostratigraphic correlation of the Shubuta and Red Bluff clays in Southeast Mississippi and their equivalents in Alabama. D, 1963, Louisiana State University. 127 p.

DeBord, Philip L. Gallatin Mountain "petrified forests"; a palynological investigation of the in situ model. D, 1977, Loma Linda University. 98 p.

DeBrine, Bruce E. Electrolytic model study for collector wells under river beds. M, 1965, New Mexico Institute of Mining and Technology. 59 p.

DeCamp, Dodd Werner. Structural geology of Mesa de Anguila, Big Bend National Park, Trans-Pecos, Texas. M, 1981, University of Texas, Austin. 186 p.

DeCaprariis, Pascal Peter. Photoelastic stress analyses of the effects of friction on crack growth in a compressive stress field. M, 1967, Boston College.

deCaprariis, Pascal Peter. The effects of non-Newtonian behavior upon the development of single layer folds. D, 1973, Rensselaer Polytechnic Institute. 73 p.

Decatur, Stephen Henley. The geology of southeastern Beaver County, Utah. M, 1979, Kent State University, Kent. 40 p.

DeCesar, Richard T. Natural-gradient tracer tests in a highly fractured soil. D, 1987, Oregon Graduate Institute of Science and Technology. 266 p.

Dechert, Curt Peter. Bedrock geology of the northern Schell Creek Range, White Pine County, Nevada. D, 1967, University of Washington. 266 p.

Dechert, Curt Peter. Structure and stratigraphy of northernmost Shell Creek Range, White Pine County, Nevada. M, 1963, University of Washington. 83 p.

Dechert, Hedy S. Paleoecology and correlation of the ""Aftonian'' (Pleistocene) fauna from Harrison and Monona counties, Iowa. M, 1968, University of Iowa. 85 p.

Dechert, Thomas Van. Land systems inventory of the Palouse District, Idaho. M, 1982, University of Idaho. 352 p.

Dechesne, Roland George. The geology of the southern Cariboo Mountains near Blue River, British Columbia. M, 1986, University of Calgary. 262 p.

Dechow, Ernest W. C. The geology of the Heath Steel Mines, Newcastle, New Brunswick, Canada. D, 1959, Yale University.

Deck, Bruce L. A limnological reconnaissance of Crossman's Pond (Rochester, New York). M, 1972, University of Rochester. 22 p.

Deck, Bruce Linn. Nutrient-element distributions in the Hudson Estuary. D, 1981, Columbia University, Teachers College. 416 p.

Deck, Linda Theresa. Ostracodes of the Piney Point Formation (middle Eocene) of Virginia. M, 1984, Virginia Polytechnic Institute and State University. 120 p.

Deckelman, James A. The petrology of the Early Middle Cambrian Giles Creek and upper Chandler formations, northeastern Amadeus Basin, central Australia. M, 1985, Utah State University. 187 p.

Decker, Billy Louis. Some remarks on the disturbing portion of the Earth's gravity field. M, 1965, Ohio State University.

Decker, Charles Elijah. Studies in minor folds. D, 1917, University of Chicago. 89 p.

Decker, Deborah Ann. Phase relations of the composition $KAlSi_3O_8$. M, 1981, Washington University. 41 p.

Decker, Donald James. Geology and ore deposits of the Arabia District, Pershing County, Nevada. M, 1972, University of Nevada. 44 p.

Decker, Edward Ronald. Terrestrial heat flow in Colorado and New Mexico. D, 1967, Harvard University.

Decker, Gary L. Preliminary report on the geology, geochemistry, and sedimentology of Flathead lake (Pleistocene and Recent), northwestern Montana. M, 1966, University of Montana. 91 p.

Decker, Guy M. A surface and subsurface study along the northwest margin of the Val Verde Basin, Pecos, Terrell, and Brewster counties, Texas. M, 1981, University of Texas at El Paso.

Decker, Jack Minrod. Cretaceous ammonites of the family Vascoceratidae Spath. M, 1950, University of Illinois, Urbana.

Decker, John E. Geology of the Mt. Galen area, Mt. McKinley National Park, Alaska. M, 1975, University of Alaska, Fairbanks. 77 p.

Decker, John Evans, Jr. Geology of a Cretaceous subduction complex, western Chichagof Island, southeastern Alaska. D, 1981, Stanford University. 199 p.

Decker, LaVerne. The Devonian of Boone County, Missouri. M, 1925, University of Missouri, Columbia.

Decker, Paul Lloyd. Geometries and mechanisms of strain in upper Proterozoic conglomerates, western Wedel Jarlsberg Land, Spitsbergen. M, 1986, University of Wisconsin-Madison. 141 p.

Decker, Paul Lloyd. Structural style, deformational mechanics, and stratigraphy of the Absaroka Volcanic Supergroup, east-central Absaroka Range, Wyoming. D, 1988, University of Wisconsin-Madison. 394 p.

Decker, Robert W. Geology of the southern Centennial Range, Elko County, Nevada. D, 1953, Colorado School of Mines. 201 p.

Decker, Robert W. Structural transition across the Appalachian fold front in north central Pennsylvania. M, 1951, Massachusetts Institute of Technology. 108 p.

Decker, Stephen. Geological and geophysical reconnaissance of the shallow subbottom in Raleigh Bay, North Carolina. M, 1986, University of Delaware.

Declercq, Eric P. Hydrogeologic investigation of the Raritan Bay area, Middlesex-Monmouth counties, New Jersey. M, 1986, Lehigh University.

DeCook, Kenneth James. Economic feasibility of selective adjustments in use of salvageable waters in the Tucson region, Arizona. D, 1970, University of Arizona.

DeCook, Kenneth James. Geology of San Marcos Springs Quadrangle, Hays County, Texas. M, 1957, University of Texas, Austin.

DeCoster, George L. Selected mineral deposits of north central Mexico. M, 1948, [University of New Mexico].

DeCoster, Judith M. Landslide deposits, their slope, exposure, and degree of slope in the area of Mt. Hamilton, California. M, 1979, San Jose State University. 43 p.

Decroix, Dominique. Etude préliminaire de la carte géotechnique de l'Île de Montréal. M, 1983, Université du Quebec a Montreal. 173 p.

DeDecker, Kenneth Arnold. The correlation and palynology of some Cherokee (Pennsylvanian) coals from Appanoose, Davis, and Wapello counties, Iowa. M, 1980, Iowa State University of Science and Technology.

Dedes, George C. Baseline estimation from simultaneous satellite laser tracking. D, 1987, Ohio State University. 197 p.

Dedick, Eugene. The densities of KCl and K_2SO_4 solutions as a function of temperature. M, 1987, University of Miami. 93 p.

Dedman, Kathryn K. and Armstrong, Jane Crozier. Geology of the Foxcroft area. M, 1939, Bryn Mawr College. 23 p.

Dedoes, Robert E. Kinetic equations for the precipitation of carbonates within the thermodynamic stability field of dolomite. M, 1987, Michigan State University. 35 p.

Dedolph, Richard Edwin. A thermodynamic model of the hydrolysis of microcline. M, 1977, University of Utah. 63 p.

Dedominic, Joseph Robert. Deposition of the Woodbine-Eagleford sandstones, Aggieland Field, Brazos County, Texas. M, 1988, Texas A&M University. 110 p.

Dee, Mark Philip. Crustal and P-wave velocity study of portions of Southwest New Mexico and Southeast Arizona using open pit mining explosions. M, 1973, New Mexico Institute of Mining and Technology.

Deemer, Sharon J. Seismic reflection profiling in the Long Valley Caldera, California; data acquisition, processing, and interpretation. M, 1985, University of Wyoming. 195 p.

Deen, Arthur Harwood. Some concretion-like forms of the Wilberns Formation of Mason County, Texas. M, 1923, University of Texas, Austin.

Deen, Patricia Ann. Ordovician to Lower (Middle?) Devonian lithology and depositional environments, Nopah Range, Inyo County, California. M, 1984, San Diego State University. 295 p.

Deen, Roy D. Geology and mineralization of the Precambrian rocks of the northern Franklin Mountains, El Paso County, Texas. M, 1978, University of Texas at El Paso.

Deere, Don Uel. Engineering properties of the Pleistocene and Recent sediments of the San Juan Bay Area, Puerto Rico. D, 1955, University of Illinois, Urbana. 155 p.

Deere, Don Uel. Foundation conditions in the University of Colorado Campus area. M, 1949, University of Colorado.

Deere, Raymond Edward. A mineralogical study of the Lower Jurassic in west central Alberta. M, 1968, University of Calgary.

Deering, Lynn Greiner. The uses and limitations of 210Pb and 137Cs geochronology to date recent sediments in Green Bay, Lake Michigan. M, 1985, University of Wisconsin-Milwaukee. 184 p.

Deering, Mark F. Exploration of mineral deposits in the carbonate formations of northwestern Ohio; a hydrogeochemical approach. M, 1981, Kent State University, Kent. 160 p.

Deery, John Richard. Washover fans of the Georgia coast. M, 1976, University of Georgia.

Dees, Jerry Lee. Acoustic impedance synthetic seismograms. M, 1970, Southern Methodist University. 64 p.

Deetae, Suchint. Release and transport of radium during weathering in central and North Florida. D, 1986, Florida State University. 149 p.

Deeter, Jerald D. Quaternary geology and stratigraphy of Kitsap County, Washington. M, 1979, Western Washington University. 175 p.

Deetz, Stephan F. Mineralogy and petrology of the Chester Pluton. M, 1980, University of Tennessee, Knoxville. 55 p.

Defandorf, May. Paleontology and petrography of Carboniferous strata in the Sloan area, San Saba County, Texas. M, 1960, University of Texas, Austin.

Defant, Marc J. A geochemical and petrogenic analysis of the Almond and Blakes Ferry plutons, Randolph County, Alabama. M, 1981, University of Alabama. 118 p.

Defant, Marc J. The potential origin of the potassium depth relationship in the Bataan Orogene, The Philippines. D, 1985, Florida State University. 644 p.

DeFelice, David. Model studies of epiphytic and epipelic diatoms of upper Florida Bay and associated sounds. M, 1975, Duke University. 193 p.

DeFelice, David R. Surface lithofacies, biofacies, and diatom diversity patterns as models for delineation of climatic change in the Southeast Atlantic Ocean. D, 1979, Florida State University. 240 p.

DeFeo, Nancy J. Remote sensing of geobotanical associations in clastic sedimentary terrane. M, 1986, Dartmouth College. 76 p.

Defeu, Edwin Leroy. Environment of deposition of the Redstone Limestone and a description of the ostracods of the Monongahela Formation, Belmont County, Ohio. M, 1956, Ohio State University.

Deffeyes, Kenneth Stover. Late Cenozoic sedimentation and tectonics development of central Nevada. D, 1959, Princeton University. 129 p.

Defieux, R. J. Ground-water resource evaluation in Sudbury, Massachusetts. M, 1977, Boston University. 111 p.

DeFries, Ruth Sarah. Sedimentation patterns in the Potomac Estuary since European settlement; a palynological approach. D, 1981, The Johns Hopkins University. 174 p.

Deganello, Sergio. A study of weathering of clay materials in the Brunswick Formation (Triassic) (New Jersey). M, 1968, Brooklyn College (CUNY).

Deganello, Sergio. A thermal study on a member of the olivine structure type (αNa_2BeF_4). D, 1971, University of Chicago. 177 p.

Degenfelder, George John. Geology of the southwest portion of the Berryman Quadrangle, Missouri. M, 1950, University of Iowa. 97 p.

Degenstein, Joel A. Geology of the Flathead Formation (Middle Cambrian), on the perimeter of the Bighorn Basin, Beartooth Mountains, and Little Belt Mountains in Wyoming and Montana. M, 1979, University of North Dakota. 59 p.

Degnan, Keith Terence. Organic-walled microplankton paleoecology and biostratigraphy of the Upper Cretaceous Ripley Formation, southwestern Georgia. M, 1987, Virginia Polytechnic Institute and State University.

Degner, Dennis A. A chemical investigation of the effect of a flood plain solid waste disposal site on ground water quality. D, 1974, University of Kansas. 385 p.

DeGrace, John Russell. Structural and stratigraphic setting of sulphide deposits in Ordovician volcanics south of King's Point, Newfoundland. M, 1971, Memorial University of Newfoundland. 62 p.

DeGraff, Alejandra Escobar. Lithostratigraphic correlation of the Mount Watson Formation, Uinta Mountains, Utah; test of a method using proportions of grains with inherited overgrowths. M, 1985, Purdue University. 121 p.

DeGraff, James Michael. Mechanics of columnar joint formation in igneous rocks. D, 1987, Purdue University. 221 p.

DeGraff, Jerome Vernon. Quaternary geomorphic features of the Bear River Range, North-central Utah. M, 1976, Utah State University. 215 p.

DeGraffenreid, Norman Bruce. The geology of the Wauhillau area, Cherokee and Adair counties, Oklahoma. M, 1951, University of Oklahoma. 128 p.

DeGraw, Harold M. Subsurface relations of the Cretaceous and Tertiary in western Nebraska. M, 1969, University of Nebraska, Lincoln.

DeGregorio, Vincent B. The influence of grain size, grain shape and sample fabric on the static liquefaction of saturated granular materials. D, 1987, Clarkson University. 195 p.

DeGroodt, James H., Jr. Determination of temperatures of fluorite formation by fluid inclusion thermometry, central Tennessee zinc district. M, 1973, University of Tennessee, Knoxville. 75 p.

deGruchy, James H. B. Water fluctuation as a factor in the life of the higher plants of a 3300 acre lake of the Permian red beds of central Oklahoma. D, 1952, Oklahoma State University. 117 p.

DeHaas, Ronald J. A localized dolomite along the Boonesboro Fault, Clark County, Kentucky. M, 1973, Ohio State University.

DeHaven, Eric C. Integrated study of the freshwater/salt-water interface on a barrier island. M, 1987, University of South Florida, Tampa. 150 p.

Dehen, Timothy. A geochemical investigation of volcanic islands in the eastern Aegean Sea. M, 1987, University of South Florida, Tampa. 126 p.

DeHerrera, Milton Augusto. A time domain analysis of seismic ground motions based on geophysical parameters. D, 1981, Stanford University. 268 p.

Dehlinger, Martin Emery. Geology of northern Hurd Draw and Boracho quadrangles, Culberson County, Texas. M, 1951, University of Texas, Austin.

Dehlinger, Peter. A magnetic survey of Sand Canyon for placer deposits, San Gabriel Mountains, California. M, 1943, California Institute of Technology. 21 p.

Dehlinger, Peter. Shear wave vibrational directions and related fault movements in Southern California earthquakes. D, 1950, California Institute of Technology. 82 p.

Dehlinger, Peter. The relationship of the Modelo and Ridge Route formations in the southern Ridge Basin, California. D, 1950, California Institute of Technology. 31 p.

Deibler, Deborah. Paleobiology and depositional environments of the Middle Ordovician Lenoir Formation, St. Clair, Tennessee. M, 1980, Vanderbilt University.

Deick, Jan F. Ground water resources in a portion of Payette County, Idaho. M, 1986, University of Idaho. 98 p.

Deike, George Herman. Cave development in Burnsville Cove, west-central Virginia, with special reference to Breathing Cave. M, 1961, University of Missouri, Columbia.

Deike, George Herman, III. The development of caverns of the Mammoth Cave region (Kentucky). D, 1967, Pennsylvania State University, University Park. 252 p.

Dein, James Lindall. Effect of transformation on superplastic properties within the olivine-spinel transition zone of the Earth's mantle. D, 1979, Pennsylvania State University, University Park. 200 p.

Deines, Peter. Instrumental factors limiting the precision and accuracy of strontium isotopic composition measurements. M, 1964, Pennsylvania State University, University Park. 140 p.

Deines, Peter. Stable carbon and oxygen isotopes of carbonatite carbonates and their interpretation. D, 1967, Pennsylvania State University, University Park. 239 p.

Deininger, Donald T. An investigation of groundwater in northeastern Florida and southeastern Georgia by analysis of its tritium content. M, 1973, Florida State University.

Deininger, James W. Petrology of the Wrangell Volcanic nNear Nabesna, Alaska. M, 1972, University of Alaska, Fairbanks. 66 p.

Deininger, Robert W. Ferrous iron and uranium concentrations and distribution in 100 selected limestones and dolomites. D, 1964, Rice University. 92 p.

Deino, Alan L. I, Stratigraphy, chemistry, K-Ar dating, and paleomagnetism of the Nine Hill Tuff, California-Nevada; II, Miocene/Oligocene ash-flow tuffs of Seven Lakes Mountain, California-Nevada; III, Improved calibration methods and error estimates for ^{40}K-^{40}Ar dating of young rocks. D, 1985, University of California, Berkeley. 498 p.

Deischl, Dennis George. The postcranial anatomy of Cretaceous multituberculate mammals. M, 1964, University of Minnesota, Minneapolis. 85 p.

Deisner, Richard Herbert. An investigation of the Palos Park partition of the forest preserve district of Cook County, describing the geologic factors which influence its forest environment. M, 1959, Northern Illinois University. 69 p.

Deiss, A. P. Planktonic foraminifera from the type area of the Fort Hays Member of the Niobrara Formation, West-central Kansas. M, 1978, University of Wyoming. 60 p.

Deiss, Charles Fred, Jr. Description and stratigraphic correlation of the Fenestellidæ from the Devonian strata of Michigan. D, 1928, University of Michigan.

Deister, Walter J., Jr. Minor elements distributions in color banded sphalerites. M, 1974, University of Kentucky. 150 p.

DeLure, Anita Mary. The effect of storms on sediments in Halifax Inlet, Nova Scotia. M, 1983, Dalhousie University. 216 p.

deJarnett, Jeffrey G. Stratigraphy, sedimentation, corals, and paleogeographic significance of the Mississippian sequence at Mt. Darby, Overthrust Belt, Wyoming. M, 1984, University of Wyoming. 231 p.

DeJarnett, Presley J. Minor structures of the Boston Mountain Monocline T. 10, 11, and 12 N., R. 29 W. (Arkansas). M, 1954, University of Arkansas, Fayetteville.

DeJarnette, Mark Lynn. The petrology, paleoecology, and depositional environments of the Hulett Sandstone Member and associated facies of the Sundance Formation (Jurassic), northeastern Bighorn Basin and southern Pryor Mountains, Wyoming and Montana. M, 1982, Southern Illinois University, Carbondale. 153 p.

DeJoia, Frank J. Stratigraphy of the Sulphur Spring Range, central Nevada. M, 1952, Columbia University, Teachers College.

DeJong, Gerard. The subsurface geology of Blaine County, Oklahoma. M, 1959, University of Oklahoma. 82 p.

Deju, Raul A. A mathematical and experimental study of the water silicate interface in porous media. D, 1969, New Mexico Institute of Mining and Technology. 123 p.

deKemp, Eric A. Stratigraphy, provenance and geochronology of Archean supracrustal rocks of western Eyapamikama Lake area, northwestern Ontario. M, 1987, Carleton University. 89 p.

DeKeyser, Thomas Lee. Upper Devonian (Frasnian) community patterns in Western Canada and Iowa. M, 1974, University of Wisconsin-Madison. 27 p.

Dekker, Frederik Ernst. Sedimentology of the Upper Cambrian Lion Mountain and Welge sandstones, central Texas. M, 1966, University of Texas, Austin.

Deklerk, Robert Peter. Paleomagnetic investigation of Mesozoic plutons located in the southern Kootenay Arc, southeastern British Columbia. M, 1987, University of Windsor. 121 p.

DeKoster, Gene R. Petrology of the Phosphoria Formation carbonates, eastern Big Horn Basin, Wyoming. M, 1960, Iowa State University of Science and Technology.

del Arco, Eugenio Nunez *see* Nunez del Arco, Eugenio

Del Castillo, G. Luis. Magnetic investigations in the vicinity of Derby, Adams County, Colorado. M, 1968, Colorado School of Mines. 85 p.

Del Mar, Robert. The instability of folded structures. M, 1956, University of New Mexico. 55 p.

Del Mauro, Gene Louis. Geology of Miller Hill and Sage Creek area, Carbon County, Wyoming. M, 1953, University of Wyoming. 144 p.

Del Monte, Lois. Correlation of the Pennsylvanian and Permian rocks, southeastern Wyoming. M, 1949, University of Wyoming. 62 p.

Del Prete, Anthony. Postglacial diatom changes in Lake George, New York. D, 1972, Rensselaer Polytechnic Institute. 110 p.

del Prete, Anthony. The location of underground points by means of a magnetic dipole. M, 1963, University of Missouri, Rolla.

del Rozas Elqueta, Eduardo. Preliminary geologic studies for the construction of a multi-purpose dam and reservoir on the Cache la Poudre River. M, 1970, Colorado School of Mines. 41 p.

Del Signore, Alan G. Uranium and thorium distribution of seven granitoid plutons of different depths of emplacement, Maine. M, 1982, SUNY at Buffalo. 79 p.

Del Solar, Carlos W. Petrology and development of the Kennett Formation reef complex in the Shasta Lake area, Shasta County, California. M, 1964, Stanford University.

del Valle, Raul. Model parameterization in refraction seismology. M, 1986, University of Toronto.

DeLancey, Charles, Jr. A study of the sands of Live Oak Bar from Baffins Bay to Matagorda Bay, Texas. M, 1942, University of Texas, Austin.

Delancey, Peter R. Geology of the Tarefare fluorite deposit, Newfoundland. M, 1970, University of Manitoba.

Deland, André Normand. Areal geology of Surprise Lake area, Abitibi-East County, 1 inch to 1 mile (Quebec). D, 1955, Yale University.

Deland, Andre Normand. The geology of part of the Three Rivers map area, Quebec. M, 1952, McGill University.

DeLand, C. R. Cambrian stratigraphy and Upper Cambrian trilobites of the southwestern flank of the Wind River Mountains, Wyoming. M, 1954, University of Wyoming. 135 p.

Delaney, Garry D. A reconnaissance study of the metamorphic petrology and volcanic geochemistry of a portion of the Island Lake greenstone belt, Manitoba. M, 1976, Brock University. 106 p.

Delaney, Gary Donald. The middle Proterozoic Wernecke Supergroup, Wernecke Mountains, Yukon Territory. D, 1985, University of Western Ontario. 373 p.

Delaney, Holly Johanna. Distribution of foraminifera on the Virginia inner continental-shelf and Chesapeake bay entrance. M, 1969, University of Virginia. 107 p.

Delaney, J. R. Distribution of volatiles in the glassy rims of submarine pillow basalts. D, 1977, University of Arizona. 132 p.

Delaney, Joan. Calculation of the thermodynamics of dehydration in subducting oceanic crust to 100 kb and 800°. D, 1976, University of California, San Diego.

Delaney, John Rutledge. Geology and reconnaissance geochemistry of the Catheart Mountain-Parlin Pond area, Somerset County, Maine. M, 1967, University of Virginia. 100 p.

Delaney, Margaret Lois. Foraminiferal trace elements; uptake, diagenesis, and 100 M.Y. paleochemical history. D, 1983, Woods Hole Oceanographic Institution. 253 p.

Delaney, Patrick. The surface stratigraphy of the Vicksburg equivalent of Louisiana. M, 1957, Louisiana State University.

Delaney, Paul Theodore. Magma flow, heat transport and brecciation of host rocks during dike emplacement near Ship Rock, New Mexico. D, 1980, Stanford University. 262 p.

Delaney, Paul Theodore. Structures related to emplacement of plugs and volcanic necks of basaltic composition. M, 1976, Stanford University.

Delano, J. W. Constraints on the composition and chemical evolution of the Earth's Moon. D, 1977, SUNY at Stony Brook. 182 p.

Delany, Anthony Charles. The search for cosmic dust; solar cosmic ray effects within lunar samples. D, 1970, [University of California, San Diego]. 137 p.

Delany, Joan Marie. A spectral and thermodynamic investigation of synthetic pyrope-grossular garnets. D, 1981, University of California, Los Angeles. 201 p.

Delany, Joan Marie. Calculations of the thermodynamics of dehydration in subducting oceanic crust to 100 kb and 800°C. M, 1976, University of California, Berkeley. 92 p.

Delavan, Gerald. Subsurface lithostratigraphy and stratigraphic framework of the Bloyd Formation (Morrowan) in Franklin, Johnson, and Pope counties, Arkansas. M, 1985, University of Arkansas, Fayetteville.

DeLay, John Milton. Late Paleozoic and early Mesozoic stratigraphy of the western Canon City Embayment, Colorado. M, 1955, University of Oklahoma. 105 p.

Delcore, Manrico. Rate of recharge to a heterogeneous aquifer; an investigation using bomb tritium. M, 1985, Michigan State University. 42 p.

Delcourt, Hazel Marie. Late Quaternary vegetation history of the eastern Highland Rim and adjacent Cumberland Plateau of Tennessee. D, 1978, University of Minnesota, Minneapolis. 218 p.

Delcourt, Hazel Roach. Late Quaternary history of the mixed Mesophytic forest in Mississippi and Louisiana. M, 1974, Louisiana State University. 76 p.

Delcourt, Paul A. Quaternary vegetation history of the Gulf Coastal Plain. D, 1978, University of Minnesota, Minneapolis. 251 p.

Delcourt, Paul Allen. Quaternary geology and paleoecology of West and East Feliciana parishes, Louisiana, and Wilkinson County, Mississippi. M, 1974, Louisiana State University.

DeLeen, John L. Geology and mineral deposits of Calico mining district. M, 1950, University of California, Berkeley.

Deleen, John L. The geology and mineralogy of the Little Billy Mine, Texada Island, BC. M, 1946, University of British Columbia.

Delfel, Deborah Lynn. Palynostratigraphy and paleoecology of the La Ventana Formation, Cretaceous (Maestrichtian); San Juan Basin, New Mexico. M, 1979, Pennsylvania State University, University Park. 106 p.

Delfino, Joseph John. Aqueous environmental chemistry of manganese. D, 1968, University of Wisconsin-Madison. 383 p.

Delgado, D. J. Syntectonic limestone conglomerate lithofacies, Laborcita and Abo formations (Wolfcampian), North central Sacramento Mountains, New Mexico. M, 1975, University of Wisconsin-Madison.

Delgado, Luciana Maria de los Angeles Astiz *see* Astiz Delgado, Luciana Maria de los Angeles

Delgado, Victor Garcia *see* Garcia Delgado, Victor

Delia, Ronald G. The geology of the southern half of the Willington 7 1/2' Quadrangle, South Carolina-Georgia. M, 1982, University of Georgia.

Deliman, Daryl G. The oxidation of sulfide in seawater by nitrate and nitrite. M, 1976, University of Washington.

Delimata, John J. Fort Union (Paleocene) mollusks from southern Golden Valley and southeastern Billings counties, North Dakota. M, 1969, University of North Dakota. 73 p.

Delimata, John J. Petrology and geochemistry of the Killdeer carbonates. D, 1975, University of North Dakota. 271 p.

DeLise, Knoxie Carlton. Foraminifera and Mollusca from the San Emigdio Formation, Kern County, California. M, 1957, University of California, Berkeley. 150 p.

Delisle, Georg. The effect of migrating fluids on heat transfer around intrusions. M, 1973, SUNY at Buffalo.

DeLisle, Mark James. Lead-alpha ages of the clasts in the Poway Formation. M, 1963, San Diego State University.

Delitala, Frank Antony. The mineralogy and geochemistry of the Kettle Point oil shale, southwestern Ontario. M, 1984, University of Western Ontario. 194 p.

Deliz, Michael John. Stratigraphy and petrology of the Plattin Limestone (Middle Ordovician) in Newton and Searcy counties, Arkansas. M, 1984, University of New Orleans. 259 p.

Dell'Agnese, Daniel James. Cretaceous and Eocene diatoms, silicoflagellates, archaeomonads and ebridians from the Arctic Ocean; Core FL-437 and FL-422. M, 1988, University of Wisconsin-Madison. 139 p.

Dell'Angelo, Lisa Nicole. Experimental deformation of quartzo-feldspathic rocks. D, 1987, Brown University. 142 p.

Dell, Carol I. Mineralogical analysis of some Pleistocene deposits in southern Ontario. M, 1958, University of Toronto.

Dell, Carol Irene Green. Late Quaternary sedimentation in Lake Superior. D, 1971, University of Michigan.

Della Valle, R. S. Uranium mineralization in Lemhi County, Idaho. M, 1975, Queens College (CUNY). 125 p.

Della Valle, Richard Saverio. Geochemical studies of the Grants mineral belt, New Mexico; 2 volumes. D, 1981, University of New Mexico. 667 p.

Dellechaie, Frank. Chemical and petrographic variations in the Cerro Colorado and Paxton Springs basalt flows, Valencia County, New Mexico. M, 1973, Kent State University, Kent. 30 p.

Dellen, Kenneth J. Van *see* Van Dellen, Kenneth J.

Dellenback, Charles Richard. A paleotectonic and paleogeologic study of the Mississippian System in the Western Interior of United States. M, 1953, University of Illinois, Urbana.

Dellinger, Philip B. Stratigraphic and facies relationships of the White Cap - Yellow Bed sequence within the upper Albian Santa Elena Formation, Terrell County, Texas. M, 1987, University of Texas, Arlington. 155 p.

Delliquadri, Lawrence Michael. Physiography and history of glacial Lake Arthur. M, 1953, Clark University.

DelloRusso, Vincent. Geology of the eastern part of the Lincoln Massif, central Vermont. M, 1986, University of Vermont. 255 p.

Dellwig, Louis Field. Origin of the Salina Salt of Michigan. D, 1954, University of Michigan.

Dellwig, Louis Field. The origin of chert underlying a bentonite bed. M, 1948, Lehigh University.

Delmastro, G. A. Diagenesis and pore mineralogy of the Clinton Formation in part of northeastern Ohio. M, 1987, University of Akron. 132 p.

Delmet, Dale Aaron. Astogony and intraclonal variability of an Ordovician bryozoan colony; fourier shape analysis. M, 1972, Michigan State University. 107 p.

Delnore, Victor E. Surface layer temperature structure as evidence for Rossby waves southwest of Bermuda. M, 1967, [University of Miami].

Delo, David Marion. A revision of the Phacopidae, a family of trilobites. D, 1935, Harvard University.

Delo, David Marion. Notes on the paleontology of the Oread Limestone of Kansas. M, 1928, University of Kansas. 106 p.

DeLoach, William, Jr. Error analysis of a trilateration network with and without Laplace condition. M, 1963, Ohio State University.

DeLong, James Edward, Jr. Geology of the Chalk Mountain area, Climax, Colorado. M, 1966, University of Michigan.

DeLong, Stephen Edwin. Distribution of Rb, Sr, Ni, and Sr^{87}/Sr^{86} in igneous rocks (Precambrian), central and western Aleutian Islands, Alaska. D, 1971, University of Texas, Austin.

DeLong, Stephen Edwin. Lead isotope model with geologic constraints. M, 1969, University of Texas, Austin.

Delorey, Catherine Marie. Magnetism and paleomagnetism of silicic dike rocks of early Mesozoic age, northeastern North Caroline Piedmont. M, 1983, North Carolina State University. 61 p.

Delorme, Denis Larry. A stratigraphic study of the Regina Basin, Saskatchewan. M, 1962, University of Alberta. 118 p.

Delorme, Denis Larry. Pleistocene and post-Pleistocene ostracods of Saskatchewan. D, 1966, University of Saskatchewan. 286 p.

DeLorme, Donaldson C. The Wheaton River District, Yukon Territory, Canada. D, 1910, Yale University.

deLorraine, William F. Geology of the Fowler Orebody, Balmat Mine, Northwest Adirondacks, N.Y. M, 1979, University of Massachusetts. 159 p.

Delph, Bryan Clifford. Channel morphology of a bedrock stream, Huber Branch, Perry County, Missouri. M, 1982, Southern Illinois University, Carbondale. 77 p.

Delsemme, Jacques Andre. Spectral analysis and migration of major earthquakes; application to South America. D, 1982, University of California, Santa Cruz. 154 p.

Delson, Eric. Fossil colobine monkeys of the circum-Mediterranean region and the evolutionary history of the Cercopithecidae (Primates, Mammalia). D, 1973, University of Colorado. 862 p.

Delu, Jacqueline. Sedimentary processes of Boonton Reservoir. M, 1982, Rutgers, The State University, New Brunswick. 124 p.

DeLuca, Frederick Peter. The development and implementation of structured inquiry methods and materials for an introductory geology laboratory course and their effectiveness as compared with the traditional course. D, 1970, University of Oklahoma. 192 p.

DeLuca, James L. Facies patterns and controls on sedimentation in the Triassic Chinle Formation of Northeast New Mexico. M, 1986, Virginia Polytechnic Institute and State University.

DeLury, Justin Sarsfield. Geology, topography and resources of the Wapawekka and Deschambault lakes area of Saskatchewan. D, 1925, University of Minnesota, Minneapolis. 57 p.

Demaison, G. Geology of Twin Mountain area, northwest of Canon City, Fremont County, Colorado. M, 1955, Colorado School of Mines. 55 p.

DeMar, Robert Eugene. A review of the family Dissorophidae (amphibians) with emphasis on the Upper Wichita and Clear Fork (Permian) genera (Texas, Oklahoma). D, 1961, University of Chicago. 213 p.

Demarcke, Janet A. S. Analysis of Seasat radar imagery for geologic mapping in Louisiana, Arkansas, and Oklahoma. M, 1980, University of Arkansas, Fayetteville.

Demaree, Randall Gene. Geology of Palen Pass, Riverside County, California. M, 1981, San Diego State University.

Demarest, Harold Hunt, Jr. Extrapolation of elastic properties to high pressure in the alkali halides. M, 1971, Columbia University. 30 p.

Demarest, Harold Hunt, Jr. Lattice model calculations and the properties of solids at high pressure and high temperature. D, 1974, University of California, Los Angeles.

Demarest, James Monroe, II. Genesis and preservation of Quaternary paralic deposits on Delmarva Peninsula. D, 1981, University of Delaware. 253 p.

Demarest, James Monroe, II. The shoaling of Breakwater Harbor-Cape Henlopen area, Delaware Bay, 1842 to 1971. M, 1978, University of Delaware.

Demars, K. R. Strength and stress-strain behavior of deep-ocean carbonate soils. D, 1975, University of Rhode Island. 266 p.

Demars, Lorenzo C. Geology of the northern part of Dry Mountain, southern Wasatch Mountains, Utah. M, 1956, Brigham Young University. 49 p.

DeMarte, Domenic L. A spectrographic analysis of the dikes of the Gogebic Range, Michigan. M, 1959, Michigan State University. 95 p.

Demases, Tamrara. Depositional model of the Farmers Member of the Borden Formation in northeastern Kentucky and southern Ohio. M, 1984, University of Texas, Arlington. 236 p.

DeMasi, Amy E. An investigation of the compatibility of zirconium with the structures of zeolites Na-A and faujasite. M, 1988, University of Toledo. 75 p.

DeMaster, D. J. The marine budgets of silica and ^{32}Si. D, 1979, Yale University. 324 p.

DeMatties, Theodore A., Jr. The geology and titaniferous magnetite deposit of the southern Lake Sanford District, New York. M, 1974, SUNY, College at Oneonta. 46 p.

Dembele, Yahaya. Interprétation géochimique de l'environnement volcano-sédimentaire de la Formation de Blondeau dans la section stratigraphique Cuvier-Barlow, Chibougamau. M, 1987, Universite du Quebec a Chicoutimi. 132 p.

Dembicki, Harry. Paleoecological interpretations and primary migration of the organic matter in a Mississippian carbonate sequence. D, 1977, Indiana University, Bloomington. 118 p.

Dembroff, Glenn Rind. Tectonic geomorphology and soil chronology of the Ventura Avenue Anticline, Ventura County, California. M, 1983, University of California, Santa Barbara. 152 p.

DeMent, James Alderson. The morphology and genesis of the subarctic Brown forest soils of central Alaska. D, 1962, Cornell University.

Demers, Denis. Corrélations entre certaines propriétés viscosimétriques et géotechniques d'argiles marines remaniées. M, 1986, Universite Laval. 105 p.

Demeter, Eugene J. Lower Ordovician pliomerid trilobites from western Utah. M, 1973, Brigham Young University. 65 p.

DeMets, Dennis Charles. Four studies using plate motion data to measure distributed deformation of the lithosphere. D, 1988, Northwestern University. 192 p.

Demfange, W. C. Von *see* Von Demfange, W. C., Jr.

Demicco, Peter M. Hydrogeology of the southern half of the Marydel Quadrangle, Delaware. M, 1983, University of Delaware. 243 p.

Demicco, Robert Victor. Comparative sedimentology of an ancient carbonate platform; the Conococheague Limestone of the Central Appalachians. D, 1981, The Johns Hopkins University. 393 p.

Deming, David. Geothermics of the thrust belt in north-central Utah. D, 1988, University of Utah. 197 p.

Deming, H. Michael. Sedimentology of the Hannibal Formation in north-eastern Missouri and western Illinois. M, 1978, University of Missouri, Rolla.

Demir, Ilham. Electrokinetic and chemical aspects of transport of chloride brines through compacted smectite layers at elevated pressures. D, 1984, University of Illinois, Urbana. 209 p.

Demirer, Ali. The Mill Creek Formation; a strike-slip basin filling in the San Andreas fault zone, San Bernardino County, California. M, 1985, University of California, Riverside. 108 p.

Demirmen, Ferruh. Paleogeologic studies in central Oklahoma. M, 1959, Stanford University.

Demirmen, Ferruh. Petrographic and statistical study of part of Pennsylvanian Honaker Trail Formation, southeastern Utah. D, 1969, Stanford University. 545 p.

Demirmen, Ferruh. Sedimentation and diagenesis in the Permian McCloud Limestone, Shasta County, California. M, 1960, Stanford University. 110 p.

DeMis, William Dermot. Geology of the Hell's Half Acre, Marathon Basin, Texas. M, 1983, University of Texas, Austin. 110 p.

Demitrack, Anne. A search for bacterial magnetite in the sediments of Eel Marsh, Woods Hole, Massachusetts. M, 1982, Stanford University. 57 p.

Demitrack, Anne. The late Quaternary geologic history of the Larissa Plain, Thessaly, Greece; tectonic, climatic, and human impact on the landscape. D, 1986, Stanford University. 146 p.

Demmon, Floyd Earl, III. Investigations of the origins and metamorphic history of Pre-Cambrian gneisses, Dowingive $7^{1/2}$ minute quadrangle, southeastern Pennsylvania. M, 1977, Bryn Mawr College. 71 p.

Demorest, Max H. Structural geology of a part of the east front of the Bighorn Mountains, near Buffalo, Wyoming. D, 1938, Princeton University. 69 p.

Demorest, Max Harrison. Physiography and glaciation of the upper Nugssuaq Peninsula, West Greenland. M, 1936, University of Cincinnati. 37 p.

DeMott, Lawrence L. Middle Ordovician trilobites of the Upper Mississippi Valley. D, 1964, Harvard University.

Dempsey, Earle V. Metallurgy of the Blackbird cobalt ore. M, 1956, University of Nevada - Mackay School of Mines. 77 p.

Dempsey, James Edward. A palynological investigation of the lower and upper McAlester Coals (Pennsylvanian) of Oklahoma. D, 1964, University of Oklahoma. 133 p.

Dempsey, James Edward. Tertiary stratigraphy and guide foraminifera of the middle and upper Texas Gulf Coast. M, 1961, University of Oklahoma. 144 p.

Dempsey, John. Gold; a world survey of a precious metal. M, 1988, University of Texas, Austin.

Dempsey, William Joseph. A minerographic study of ores from Lakins Point, Great Bear Lake, Canada. M, 1940, Catholic University of America. 34 p.

Dempster, Kelly. The external anatomy of the Ordovician trilobite Cryptolithus bellulus. M, 1976, Cornell University.

Dempster, R. E. Geology of the northeastern part of the Gonzales Quadrangle, California. M, 1951, University of California, Berkeley. 51 p.

Demshar, Ludwig Stanley. A study of the heavy minerals in the Connellsville Sandstone in the Pittsburgh area. M, 1953, University of Pittsburgh.

den Berg van Saparoea, C. M. G. Van *see* Van den Berg van Saparoea, C. M. G.

Den Berg, Alexander Nicolaas Van *see* Van Den Berg, Alexander Nicolaas

Den Berg, Jacob Van *see* Van Den Berg, Jacob

den Berge, Johannes Christian Van *see* Van den Berge, Johannes Christian

den Heuvel, Peter Van *see* Van den Heuvel, Peter

Den, F. Michael De *see* De Den, F. Michael

Denatale, Douglas Robert. Hydrogeology of Vincent-Brattle Brook site and adjacent areas, Pittsfield and Dalton, Massachusetts. M, 1983, University of Massachusetts. 145 p.

DeNault, Kenneth, Jr. Geology and distribution of copper, lead and zinc in streams and soil, Broadway Mine area, Carbon County, Wyoming. M, 1967, University of Wyoming. 45 p.

Denburgh, Alber Stevens Van *see* Van Denburgh, Alber Stevens

Denby, Gordon Morrison. Self-boring pressuremeter study of the San Francisco Bay mud. D, 1978, Stanford University. 360 p.

Denechaud, E. Barton. Rare-earth activation analysis; development and application to stretishorn dike and Duluth Complex. D, 1969, University of Wisconsin-Madison. 159 p.

Denehie, R. B. Clay mineralogy and engineering properties of surficial deposits in northern Forrest County, Mississippi. M, 1975, University of Southern Mississippi.

Denere, Thomas Ashley. Pliocene to middle Pleistocene siliceous microfossil biostratigraphy of D.S.D.P. sites 173, 183 and 47.2, North Pacific Ocean. M, 1978, University of Southern California.

Denesen, Stephen Louis. Stratigraphy, petrography, and depositional environments of the Banzet Formation (Middle Pennsylvanian) in southeastern Kansas and northeastern Oklahoma. M, 1985, University of Iowa. 141 p.

Dengler, Alfred Theodore. Silicates and diatoms in New England estuaries. M, 1973, Massachusetts Institute of Technology. 89 p.

Dengler, Lorinda Ann. The microstructure of deformed graywacke sandstones. D, 1979, University of California, Berkeley. 273 p.

Dengo, Carlos Arturo. Frictional characteristics of serpentine from the Montagua fault zone in Guatemala; an experimental study. M, 1978, Texas A&M University. 89 p.

Dengo, Carlos Arturo. Structural analysis of the Polochic fault zone in western Guatemala, Central America. D, 1982, Texas A&M University. 326 p.

Dengo, Gabriel. Geology of the bentonite deposits near Casper, Natrona County, Wyoming. M, 1946, University of Wyoming. 28 p.

Dengo, Gabriel. Geology of the Caracas region, Venezuela. D, 1949, Princeton University. 141 p.

Denham, Charles R. Geomagnetic secular variation during the past 30,000 years, as recorded in (Pleistocene) sediments (California, England and Aegean Sea). D, 1972, Stanford University. 70 p.

Denham, Richard L. The igneous rocks at Skrainka, Madison County, Missouri. M, 1934, Washington University. 50 p.

DenHartog, Stephen Ludwig. Gravity profiles of the 1958-59 Victoria Land traverse, Antarctica. M, 1961, Montana College of Mineral Science & Technology. 50 p.

Denholm, J. G. An investigation of induced electrode polarization. M, 1964, University of Western Ontario.

DeNiro, M. J. I, Carbon isotope distribution in food chains; II, Mechanism of carbon isotope fractionation associated with lipid synthesis. D, 1977, California Institute of Technology. 198 p.

Denis, Bertrand T. Guillet Township map area (Quebec). D, 1938, McGill University.

Denis, John R. The location and origin of uranium deposits in the Cutler Formation in the Lisbon Valley Anticline, southeastern Utah. M, 1982, University of Wyoming. 106 p.

Denison, Albert Rodger. The Robberson Field, Garvin County, Oklahoma. M, 1925, University of Oklahoma. 19 p.

Denison, David Floyd. Geology of an area north of Bartlett Mountain, Climax District, Colorado. M, 1963, University of Michigan.

Denison, David Floyd. Geology, sulfide mineralization, and alteration of a central section (N-4400) of the Chuquicamata ore body, (probably Eocene–Miocene, debatable), (Atacama desert, northern Chile). D, 1969, University of Michigan. 259 p.

Denison, Robert H. A review of the fauna and correlation of the Paleocene and lower Eocene. M, 1934, Columbia University, Teachers College.

Denison, Robert H. The broad-skulled Pseudocreodi. D, 1938, Columbia University, Teachers College.

Denison, Rodger E. The basement complex of Comanche, Stephens, Cotton and Jefferson counties, Oklahoma. M, 1958, University of Oklahoma. 81 p.

Denison, Rodger Espy. Basement rocks in adjoining parts of Oklahoma, Kansas, Missouri, and Arkansas. D, 1966, University of Texas, Austin. 328 p.

Denlinger, R. P. Geophysics of The Geysers geothermal field, northern California. D, 1979, Stanford University. 96 p.

Denman, Cedric Eugene. The stratigraphy of western Washington with special reference to the coal measures. M, 1924, Washington State University. 121 p.

Denman, Harry Edward, Jr. Implications of seismic activity at the Clark Hill Reservoir. M, 1974, Georgia Institute of Technology. 103 p.

Denne, Jane Elizabeth. The hydrogeology of the Waubee Lake area, Oconto County, Wisconsin, with implications for land and water management. M, 1976, University of Wisconsin-Madison.

Dennen, Mark M. The brachiopod biostratigraphy and paleoecology of the upper Dornick Hills Group, Middle Pennsylvanian, Ardmore Basin, southern Oklahoma. M, 1987, University of Oklahoma. 227 p.

Dennen, W. H. Spectrographic investigation of chemical variations across igneous contacts. D, 1949, Massachusetts Institute of Technology. 119 p.

Dennen, William Llewellyn. The Yampa vein of Bingham, Utah. M, 1920, Massachusetts Institute of Technology. 40 p.

Denney, Phillip P. Geology of the southeast end of the Paleozoic portion of the Canelo Hills, Santa Cruz County, Arizona. M, 1968, University of Arizona.

Denning, Reynolds McConnell. Directional variation of relative grinding hardness in diamond. D, 1953, University of Michigan.

Denning, Reynolds McConnell. The petrology of the Jacobsville Sandstone, Lake Superior. M, 1949, Michigan Technological University. 72 p.

Dennis, Alan. Bay-fill paleoecology in the Carboniferous of eastern Kentucky. M, 1978, University of South Carolina.

Dennis, Eldon. A preliminary survey of Utah nonmetallic minerals, exclusive of mineral fuels, with special reference to their occurrence and markets. M, 1931, Brigham Young University. 150 p.

Dennis, Frank Anthony Richard. Petrology and mineralization of the Deep Cove Pluton, Gabarus Bay, Cape Breton Island, Nova Scotia. M, 1988, Acadia University.

Dennis, John G. Geology of Lyndonville area, Vermont. D, 1956, Columbia University, Teachers College.

Dennis, John G. Stratigraphy of the Lyndonville area, Vermont. M, 1956, Columbia University, Teachers College.

Dennis, John G. The geology of the Lyndonville area, Vermont. D, 1957, Columbia University, Teachers College.

Dennis, Leonard Stanley. Aeromagnetic survey of Saint Paul Rocks and Saint Peter Rocks (Atlantic Ocean). M, 1967, American University. 26 p.

Dennis, Lyman Clark. Geology of the North Fork Valley, Montana. M, 1936, University of Minnesota, Minneapolis. 45 p.

Dennis, Michael D. Stratigraphy and petrography of pumice deposits near Sugarloaf Mountain, northern Arizona. M, 1981, Northern Arizona University. 79 p.

Dennis, Norman Dale, Jr. Development of correlations to improve the prediction of axial pile capacity. D, 1982, University of Texas, Austin. 352 p.

Dennis, Scott Timothy. The effect of well efficiency on in-situ permeability test results. M, 1987, Western Michigan University.

Dennis, Terence Edwin. The structure of the Horse Heaven Hills of Washington. M, 1938, Washington State University. 35 p.

Dennis, Wilbert C. Igneous rocks of the Valley of Virginia. M, 1934, University of Virginia. 76 p.

Dennison, Douglas I. Rare earth element geochemistry of mineralized intrusive complexes, North and South Moccasin Mountains, Montana. M, 1986, Eastern Washington University. 102 p.

Dennison, John Manley. Stratigraphy of Devonian Onesquethaw Stage in West Virginia, Virginia, and Maryland. D, 1960, University of Wisconsin-Madison. 373 p.

Denny, Charles Storrow. The Cenozoic geology of the San Acacia area, Socorro County, New Mexico. D, 1938, Harvard University.

Denny, James H., Jr. Mesoscopic structural analysis of Precambrian rocks, Lake Buchanan area, northeast Llano Uplift, Llano and Burnet counties, central Texas. M, 1982, Stephen F. Austin State University. 132 p.

Denny, Stuart. Crustal structure of the Juan de Fuca Ridge as determined from deep towed seismic reflection profiles. M, 1988, University of Washington. 95 p.

DeNooyer, LeRoy L. Petrogenesis, classification, and variation in the glacial deposits in southwestern Michigan using multivariate statistical analysis. M, 1971, University of Kansas. 72 p.

Denslow, Lathrop V. B. Instrumentation for model studies of two-horizontal-coplanar-coil method of electromagnetic exploration. M, 1960, Colorado School of Mines. 68 p.

Denson, John Lane, III. Geology of southern half of Salt Draw Quadrangle, Culberson County, Texas. M, 1950, University of Texas, Austin.

Denson, M. Elner, Jr. Longitudinal waves through the Earth's crust. D, 1950, California Institute of Technology.

Denson, Mayette Elner, Jr. The Madison (Mississippian) Limestone of the Big Horn Basin, Wyoming. D, 1950, California Institute of Technology. 177 p.

Denson, Norman Maclaren. Late Middle Cambrian trilobite faunas and stratigraphy of Alberta, Montana, Wyoming and Utah. D, 1942, Princeton University. 195 p.

Denson, Norman Maclaren. Trilobites from the Park Shale of Montana and Yellowstone National Park. M, 1939, University of Montana. 73 p.

Dent, Brian Edward. Studies of large impact craters. D, 1974, Stanford University. 109 p.

Dent, Jimmie Duane. A biviscous modified Bingham model of snow avalanche motion. D, 1982, Montana State University. 192 p.

Dentan, Catherine M. Some applications of principal components analysis to mineral crystal structures. M, 1976, SUNY at Buffalo. 96 p.

Dentler, Patricia L. Geology and ore deposits of the Cross Mine, Boulder County, Colorado. M, 1984, University of Colorado. 108 p.

Denton, Alexander W. S. Tectonics and sediment geochemistry of Tuzo Wilson Seamounts, Northeast Pacific Ocean. M, 1986, University of British Columbia. 183 p.

Denton, George H. Correlation of (Pennsylvanian) lower Allegheny coal beds of Columbia County, Ohio, with coal beds of other areas. M, 1957, West Virginia University.

Denton, George Henry. Late Pleistocene glacial chronology, northeastern Saint Elias Mountains, Canada. D, 1965, Yale University. 100 p.

Denton, Harry Paul. The use of soil classification in prediction of crop response to agronomic management practices. D, 1983, North Carolina State University. 259 p.

Denton, W. E. The metamorphism of the Gordon Lake sediments, Northwest Territories. M, 1940, McGill University.

DePangher, Michael. Quantitative assessment of metasomatic composition-volume changes; techniques for identifying actual protoliths and conserved components. D, 1988, University of Utah. 90 p.

DePangher, Michael. The geology, geochemistry, and petrology of the Quinnesec Group east of Pembine, Marinette County, Wisconsin. M, 1982, University of Utah. 210 p.

DePaolo, D. J. Study of magma sources, mantle structure and the differentiation of the Earth from variations of $^{143}Nd/^{144}Nd$ in igneous rocks. D, 1978, California Institute of Technology. 373 p.

Depatie, Jean. Tuffaceous rocks of the Beauceville area (Quebec). M, 1965, Universite Laval. 65 p.

DePaul, Gilbert John. Depositional environment and reservoir morphology of the upper Wilcox sandstones, Katy gas field, Waller County, Texas. M, 1979, Texas A&M University. 161 p.

Depke, T. J. Surface and subsurface geology, Manchester Quadrangle, Missouri. M, 1973, Washington University. 181 p.

DePree, Lynn Julius. Structure and physiography of the Jackson Hole region, Wyoming. M, 1942, University of Michigan.

Deputy, Edward James. Petrology of the Middle Cambrian Ute Formation, north-central Utah and southeastern Idaho. M, 1984, Utah State University. 124 p.

deQuadros, Antonio Melicio. The distribution and occurrence of copper in the Karmutsen Group, Vancouver Island, British Columbia. M, 1968, University of California, Los Angeles.

Der Flier, Eileen Van *see* Van Der Flier, Eileen

der Hoeven, G. A. Van *see* Van der Hoeven, G. A.

der Hoya, H. Austin von *see* von der Hoya, H. Austin, II

Der Kiureghian, Ahmen. A line-source model for seismic risk analysis. D, 1976, University of Illinois, Urbana. 145 p.

der Laan, Sieger Robbert van *see* van der Laan, Sieger Robbert

der Leeden, Fritz Van *see* Van der Leeden, Fritz

der Leeden, John van *see* van der Leeden, John

der Loop-Avery, Mary Louise van *see* van der Loop-Avery, Mary Louise

der Meyden, Hendrik Jan van *see* van der Meyden, Hendrik Jan

der Osten, Erimar Alfred von *see* von der Osten, Erimar Alfred

der Plank, Adrian Van *see* Van der Plank, Adrian

Der Pluijm, Bernardus Adrianus Van *see* Van Der Pluijm, Bernardus Adrianus

der Poel, Washinton I. Van *see* Van der Poel, Washinton I., III

der Sarkissian, Volga. Clay mineralogy of the Delmar and Friars formations, San Diego County, California. M, 1983, San Diego State University. 261 p.

der Spuy, Peter M. Van *see* Van der Spuy, Peter M.

der Staay, Robert Van *see* Van der Staay, Robert

der Ven, Paulus Hendrikus Van *see* Van der Ven, Paulus Hendrikus

Der, Zoltan Andrew. Studies of the seismic inverse problem and data analysis of vertical arrays. D, 1971, Southern Methodist University. various pagination p.

Deragon, Robert. Etude geothermo-barometrique et geotachymetrique des gneiss associes au complexe anorthositique de Morin et essai de modelisation du role thermique des plutonites, Province de Grenville, Quebec. M, 1987, Universite de Montreal.

Derby, Andrew W. The Algoman intrusives of the Matachewan-Kirkland Lake area, northern Ontario. D, 1935, University of Toronto.

Derby, J. V. Investigation of lunar and terrestrial materials for alkali metal degradation. D, 1970, University of Hawaii. 208 p.

Derby, James R. Geology of the Damascus area. M, 1961, Virginia Polytechnic Institute and State University.

Derby, James Richard. Paleontology and stratigraphy of the Nolichucky Formation in Southwest Virginia and Northeast Tennessee. D, 1966, Virginia Polytechnic Institute and State University. 569 p.

DeReamer, John. An environmental interpretation of the Berea Sandstone Formation (Lower Mississippian) in south-central Ohio, northeastern Kentucky and adjacent West Virginia; an outcrop, core and subsurface study. M, 1984, University of Cincinnati. 145 p.

Deremer, Lori Ann. The geologic and chemical evolution of Volcan Tepetiltic, Nayarit, Mexico. M, 1986, Tulane University. 158 p.

Derewetzky, Aram Noah. Early Cretaceous shallow water foraminifera from Northern California. M, 1987, University of California, Davis. 41 p.

DeRidder, Eduard. The relative accuracy of some magnetic depth-determination techniques. M, 1972, Colorado School of Mines. 78 p.

Derieg, George William. Border relations as a key to the mode of emplacement of the Harney Peak Batholith, Black Hills, South Dakota. M, 1962, University of Nebraska, Lincoln.

Derkey, Pamela Dunlap. Geology of the Eightmile Creek area, Custer County, Idaho. M, 1977, University of Idaho. 100 p.

Derkey, Robert Erwin. Geology of the Blackbutte mercury mine, Lane County, Oregon. M, 1973, University of Montana. 66 p.

Derkey, Robert Erwin. Geology of the Silver Peak Mine, Douglas County, Oregon. D, 1982, University of Idaho. 188 p.

Derksen, Charlotte Meynink. Geology of the eastern end of Elkhorn Ridge (Baker County), Oregon. M, 1968, University of Oregon. 81 p.

Derksen, S. J. Glacial geology of the Brady Glacier region, Alaska. D, 1976, Ohio State University. 143 p.

Derksen, S. J. Raised marine terraces southeast of Lituya Bay, Alaska. M, 1974, Ohio State University.

Derman, A. Sami. Upper Jurassic-Lower Cretaceous non-marine foreland basin sedimentation in the western states. M, 1984, University of Michigan.

Dermengian, John Michael. On the application of teleseismic body wave modeling to study the source characteristics and tectonic implications of the Kalapana, Hawaii foreshock mainshock sequence of 11/29/75. M, 1981, Pennsylvania State University, University Park. 238 p.

Dermer, Michele Suzy. The mineralogy and sedimentology of some Long Island tills, and their correlation with lobes of the late Wisconsinan ice sheet, Long Island, New York. M, 1981, University of Massachusetts. 192 p.

DeRose, William R. A programmed learning unit for the middle school student on plate tectonics. M, 1978, Pennsylvania State University, University Park. 111 p.

deRosset, W. H. M. Petrology of the Mount Airy Granite. M, 1978, Virginia Polytechnic Institute and State University.

Derr, John Sebring. Internal structure of the Earth inferred from free oscillations. D, 1968, University of California, Berkeley. 498 p.

Derr, Michael E. Sedimentary structures and depositional environment of paleochannels in the Jurassic Morrison Formation near Green River, Utah. M, 1974, Brigham Young University. 39 p.

Derstler, Kraig Lawrence. Camptostroma is an edrioasteroid; au grand serieux. M, 1977, University of Rochester.

Derstler, Kraig Lawrence. Studies on the morphological evolution of echinoderms. D, 1985, University of California, Davis. 438 p.

DeRudder, Ronald Dean. Analysis of the glaze peeling of specific Pennsylvanian underclays. M, 1960, Indiana University, Bloomington. 35 p.

DeRudder, Ronald Dean. Mineralogy, petrology, and genesis of the Willsboro Quadrangle, New York. D, 1962, Indiana University, Bloomington. 156 p.

Deruyck, Bruno Guy. Interference well test analysis for a naturally fractured reservoir. M, 1980, Stanford University.

des Autels, David. The paleoecology of an Ordovician marine community in the Galena Limestone, Goodhue County, Minnesota. M, 1978, University of Minnesota, Duluth.

des Rivières, Jean. Etude de l'intrusif et de la minéralisation aurifère des collines Gemini et St.-Éloi, Canton Desboues, Abitibi, Québec. M, 1985, Ecole Polytechnique. 160 p.

Desai, Arvind A. Structure contours on the top and base of the M bed, the relationship of mineralization to structures, and the variation of the M bed thickness in the Tri-state District, Missouri, Kansas and Oklahoma. M, 1966, University of Missouri, Rolla.

Desai, Chandrakant S. Solution of stress-deformation problems in soil and rock mechanics using finite element methods. D, 1969, University of Texas, Austin. 266 p.

Desai, Jyotindra Ishwarbhai. Petrography of the Kaibab and Plymption formations (Permian), near Ferguson mountain, Elko County, Nevada. M, 1970, Brigham Young University. 81 p.

Desai, Kantilal Panachand. Sequential measurement of longitudinal and shear velocities of rock samples under triaxial pressure. D, 1967, University of Tulsa. 168 p.

Desai, Pankaj. Hydrogeochemical analysis of the Dakota Aquifer in Northwest Kansas. M, 1986, Fort Hays State University. 118 p.

Desaulniers, Donald Edouard. Groundwater origin, geochemistry and solute transport in three major glacial clay plains of east-central North America. D, 1986, University of Waterloo. 445 p.

Desaulniers, Donald Edouard. Origin, age and movement of pore water in argillaceous Quaternary deposits at four sites in southwestern Ontario. M, 1980, University of Waterloo.

Desbarats, Alexander Jean. Stochastic modeling of flow in sand-shale sequences. D, 1987, Stanford University. 185 p.

Desbarats, Alexandre. Influence du minage selectif sur le design optimal des carrières. M, 1982, Ecole Polytechnique. 106 p.

Desbiens, Harold. Géochimie des sédiments Cambro-Ordoviciens des Appalaches, Estrie et Beauce. M, 1988, Universite du Quebec a Montreal. 133 p.

Desbiens, Rejean. A magnetotelluric sounding in the Kapuskasing structural zone at Racine Lake. M, 1986, University of Toronto.

Desbiens, Sylvain. Trilobites ordoviciens du Saguenay-Lac St. Jean, Quebec. M, 1988, Universite de Montreal.

Desborough, George Albert. Bedrock geology of the Pomona Quadrangle. M, 1960, Southern Illinois University, Carbondale. 147 p.

Desborough, George Albert. Differentiation of Precambrian olivine diabase in southeastern Missouri. D, 1966, University of Wisconsin-Madison. 66 p.

Desborough, John. Deep crustal reflections in eastern Canada. M, 1969, Dalhousie University. 77 p.

DesCamps, Julius Robert. Gravity survey of a portion of Sanpete Valley, Utah. M, 1971, Ohio State University.

Descarreaux, Jean. Géochimie des roches volcaniques de l'Abitibi, Province de Québec. D, 1972, Universite Laval.

Descarreaux, Jean. Géologie et géostatistique de la Mine Lorraine (Québec, Canada). M, 1967, Universite de Montreal.

Deschamps, Fernand. Interprétation conjointe de données EM-16 et EM-16-R. M, 1981, Ecole Polytechnique. 197 p.

Deschamps, Fernand, Jr. Interpretation conjointe de donnees EM-16 et EM-16-R. M, 1980, Universite de Montreal. 197 p.

Deshler, Richard M. The geology of El Borracho Caldera, Chihuahua, Mexico. M, 1985, University of Texas at El Paso.

DeSimone, David J. Glacial geology of the Schuylerville Quadrangle, NY. M, 1977, Rensselaer Polytechnic Institute. 75 p.

DeSimone, David J. The late Woodfordian history of southern Washington County, New York. D, 1985, Rensselaer Polytechnic Institute. 193 p.

DeSimone, Leslie. Paleoenvironmental interpretation of the Lower Silurian Tuscarora Formation palynomorph suite in central Pennsylvania. M, 1988, Boston University. 136 p.

DeSisto, John A. Calculation of synthetic seismograms in one and two dimensional media. M, 1985, University of Utah. 101 p.

Desjardins, Jean-Pierre. Cartographie des sols par photo-interprétation pour des fins de génie. M, 1971, Ecole Polytechnique. 173 p.

Desjardins, Louis Hosea. The pre-glacial physiography of the Cincinnati region. M, 1934, University of Cincinnati. 43 p.

Deslauriers, Edward Charles. Geophysics and hydrology of the lower Verde River valley, Maricopa County, Arizona. M, 1977, Arizona State University. 61 p.

Desloges, Joseph Robert. Paleohydrology of the Bella Coola River basin; an assessment of environmental reconstruction. D, 1987, University of British Columbia.

Desmarais, David John. The origin and distribution of carbon in lunar soils. D, 1974, Indiana University, Bloomington. 119 p.

Desmarais, Luc. Géologie et géomorphologie quaternaire; secteur du Lac Matapédia et de la Rivière Mitis (Québec). M, 1988, Universite du Quebec a Montreal. 148 p.

Desmarais, Neal Raymond. Geology and geochronology of the Chief Joseph plutonic-metamorphic complex, Idaho-Montana. D, 1983, University of Washington. 150 p.

Desmarais, Neal Raymond. Structural and petrologic study of Precambrian ultramafic rocks, Ruby Range, southwestern Montana. M, 1978, University of Montana. 88 p.

Desmarais, Ralph J. A reinterpretation of regional seismic surveys in northern Ontario and eastern Manitoba between distances of 127 kilometers and 775 kilometers using seismic model studies of the crust and upper mantle. M, 1976, University of Manitoba.

Desmond, Robert J., Jr. Stratigraphy and depositional environments of the middle member of the Minnelusa Formation, central Powder River basin, Wyoming. M, 1985, University of Wyoming. 115 p.

Desonie, Dana. Geology and petrology of Isla San Esteban, Gulf of California, Mexico. M, 1958, University of Oregon. 78 p.

Desormier, William Leo. A section of the northern boundary of the Sapphire tectonic block. M, 1975, University of Montana. 65 p.

Despot, Camille C. Subsurface Sligo Formation of northwestern Louisiana and adjoining East Texas. M, 1956, University of Oklahoma. 52 p.

Despot, Martin. Chacahoula Basin depositional and structural analysis. M, 1981, University of Southwestern Louisiana.

Desrochers, Andre. Etude sédimentolologique de la Formation de La Vieille dans la région de Port-Daniel, Baie-des-Chaleurs. M, 1982, Universite Laval. 49 p.

Desrochers, Andre. The Lower and Middle Ordovician platform carbonates of the Mingan Islands, Quebec; stratigraphy, sedimentology, paleokarst, and limestone diagenesis. D, 1987, Memorial University of Newfoundland. 454 p.

Desrochers, Gary J. Geology of part of the Hilltop District, Lander County, Nevada. M, 1984, University of Nevada. 97 p.

Desselle, Bruce A. Biostratigraphy, paleoecology, and taxonomy of the Marginulina Zone in Cameron and Calcasieu parishes. M, 1988, University of Southwestern Louisiana. 156 p.

Dessouki, Abdelrahim Khalil Mohamed. Stability of soil-steel structures. D, 1985, [University of Windsor].

Destefano, Mark. The diagenetic alteration of volcanic rock fragments in the Stevens Sandstone, San Joaquin Basin, California. D, 1978, Texas A&M University. 94 p.

DeTample, Craig. Taphonomic analysis of the Miocene Flint Hill north locality, Bennett County, South Dakota. M, 1988, South Dakota School of Mines & Technology.

Detenbeck, J. C. Stress and strain analysis of dolomite and quartz lamellae, Colchester Pond, West-central Vermont. M, 1977, University of Vermont.

Dethier, David Putnam. Dissolved constituents in Williamson Creek, Snohomish County, Washington; a preliminary report. M, 1974, University of Washington. 33 p.

Dethier, David Putnam. Geochemistry of Williamson Creek, Snohomish County, Washington. D, 1977, University of Washington. 315 p.

Detournay, Emmanuel Michel. Two-dimensional elastoplastic analysis of a deep cylindrical tunnel under non-hydrostatic loading. D, 1983, University of Minnesota, Minneapolis. 141 p.

Detra, Earl H. Structural investigation of a section through the Seven Devils Mountains, Idaho. M, 1980, University of Montana. 49 p.

Detrick, Robert Sherman. The crustal structure and subsidence history of aseismic ridges and mid-plate island chains. D, 1978, Massachusetts Institute of Technology. 182 p.

Detterman, Mark E. Geology of the Metal Mountain District, In-Ko-Pah Mountains, San Diego County, California. M, 1984, San Diego State University. 216 p.

Deul, M. Origin of the Big Iron deposits near Montezuma, Summit County, Colorado. M, 1947, University of Colorado.

Deurmyer, James Justin. The geology of the South Fork Thrust and subsurface geology of the Belknap Ranch, Park County, Wyoming. M, 1959, University of New Mexico. 79 p.

Deusen, John Ernest Van see Van Deusen, John Ernest, III

Deuser, Werner George. Rubidium-strontium age determinations of muscovites and biotites from pegmatites of the Blue Ridge and Piedmont. M, 1961, Pennsylvania State University, University Park. 55 p.

Deuser, Werner George. The effects of temperature and water pressure on the apparent Rb-Sr age of micas. D, 1963, Pennsylvania State University, University Park. 116 p.

Deuters, Barrie Eugene. A geochemical study of some igneous rocks from the White Mountain magma series using spectrographic methods. M, 1958, McMaster University. 60 p.

Deuth, John Eakle. A study of the Silurian contacts based on subsurface data from Benton and Linn counties, Iowa. M, 1948, University of Iowa. 78 p.

Deutsch, Clayton Vernon. A probabilistic approach to estimate effective absolute permeability. D, 1987, Stanford University. 165 p.

Deutsch, Harvey A. Systematics and evolution of early Eocene Hyaenodontidae (Mammalia, Creodonta) in the Clark's Fork Basin, Wyoming. M, 1979, University of Michigan.

Deutsch, William Louis, Jr. Determination of the preconsolidation pressure in soil using acoustic emission technology; a laboratory and field study (Volumes 1 and II). D, 1985, Drexel University. 495 p.

Devaney, J. R. Sedimentology and stratigraphy of the northern and central metasedimentary belts in the Beardmore-Geraldton area of northern Ontario. M, 1987, Lakehead University.

Devaul, Robert W. Reconnaissance of groundwater resources in Daviess, Hancock, McLean and Ohio counties, Kentucky. M, 1959, Syracuse University.

Devender, Thomas Roger Van see Van Devender, Thomas Roger

Devening, Donald Clayton. The petrology of the sandstone in the Linton Formation. M, 1953, Indiana University, Bloomington. 38 p.

Devenny, D. W. Strength mechanisms and response of highly sensitive soils to simulated earthquake loading. D, 1975, Purdue University. 437 p.

Deventer, Bruce Robert Van see Van Deventer, Bruce Robert

Deventer, James Bartlett Van see Van Deventer, James Bartlett

Dever, Garland Ray., Jr. Stratigraphic relationships in the lower and middle Newman Limestone (Mississippian), east central and northeastern Kentucky. M, 1973, University of Kentucky. 121 p.

Devera, Joseph A. Micropaleontology, carbonate petrography and environments of deposition of the Grand Tower Limestone (Middle Devonian) in southwestern Illinois and southeastern Missouri. M, 1986, Southern Illinois University, Carbondale. 322 p.

Devereaux, Alfred B., Jr. The determination of lateral refraction by the Kukkamaki method. M, 1963, Ohio State University.

Devereaux, Alfred Boyce. Investigations into the feasibility of employing a hypothetical panoramic-frame camera system in aerial triangulation. D, 1973, Ohio State University. 171 p.

Deverse, George D. The geochemistry of iron-rich and silica-rich layering in the Gunflint Iron-Formation, Ontario. M, 1976, University of Wisconsin-Milwaukee.

Devery, Dora Maria. Morphologic changes in synthetic Mg calcites. M, 1979, Texas Christian University. 48 p.

Devery, Hugh Blase. Analysis of the microfauna, facies variation, and stratigraphy of selected outcrops of the Bangor Limestone (Chesterian; Mississippian) in Colbert, Franklin, and Lawrence counties, Northwest Alabama. M, 1987, Mississippi State University. 210 p.

Devery, Justin V. Sedimentary petrology of the upper Paleozoic carbonate near Bavispe, Sonora, Mexico. M, 1979, Texas Christian University. 79 p.

DeVilbiss Munoz, J. W. Wave dispersion and absorption in partially saturated rocks. D, 1980, Stanford University. 140 p.

DeVilbiss, J. W. Wave dispersion and absorption in partially saturated rocks. D, 1979, Stanford University. 128 p.

Devilbiss, Thomas S. An investigation into the movement of an agricultural pesticide within the groundwater system of a karst swallet. M, 1988, Eastern Kentucky University. 84 p.

DeVine, Carolyn S. Sedimentology of the Morrison and related upper Jurassic-lower Cretaceous rocks, central Wyoming. M, 1975, University of Missouri, Columbia.

Devine, Joseph Driscoll, III. Role of volatiles in Lesser Antilles island arc magmas. D, 1987, University of Rhode Island. 628 p.

Devine, Paul Ellis. Depositional patterns in the Point Lookout Sandstone, Northwest San Juan Basin, New Mexico. M, 1980, University of Texas, Austin.

Devine, Stanley Bevan. Mineralogical and geochemical aspects of the surficial sediments of the deep Gulf of Mexico. D, 1971, Louisiana State University. 175 p.

Devine, Steven C. Sedimentary petrology and depositional environment of the Salt Lake Formation in the Raft River geothermal area, Cassia County, Idaho. M, 1980, University of Idaho. 104 p.

Devlin, Barry David. Geology and genesis of the Dolly Varden Silver Camp, Alice Arm area, northwestern British Columbia. M, 1987, University of British Columbia. 122 p.

Devlin, Frank. Observations on shore erosion and shore processes of Lake Michigan adjacent to Northwestern University. M, 1942, Northwestern University.

Devlin, John Frederick. The distribution of volatile organic contaminants in a landfill leachate plume, Gloucester, Ontario. M, 1986, Queen's University. 146 p.

Devlin, William Joseph. Geologic and isotopic studies related to latest Proterozoic-Early Cambrian rifting and initiation of a passive continental margin, southern British Columbia, Canada, and northeastern Washington, U.S.A. D, 1986, Columbia University, Teachers College. 376 p.

Devon, John W. Stratigraphy of the Maryon's Lake-Fleming area, Knob Lake vicinity, Labrador Trough, Canada. M, 1969, SUNY at Buffalo. 83 p.

Devore, Cynthia Helen. The molluscan fauna of the Illinoian Butler Springs sloth locality from Meade County, Kansas. M, 1974, University of Michigan.

DeVore, George W. A structural environment for the formation of actinolite and actinolite-chlorite schist. D, 1952, University of Chicago. 75 p.

Devorkin, Donald B. Representation schemes for investigating non-linear processes. D, 1963, Massachusetts Institute of Technology. 94 p.

Devou, Marie L. The geology and geography of the Evanston Quadrangle. M, 1929, Northwestern University.

DeVries, Christiaan D. S. Metamorphism and structure of the Esplanade Range, British Columbia. M, 1971, University of Calgary. 131 p.

Devries, David A. Paleoecology and paleontology of a Chaetetes biostrome in Madison County, Iowa. D, 1955, University of Wisconsin-Madison.

Devries, Donald Charles. The geology of a suspect "Fourth" Watchung in Towaco, New Jersey. M, 1986, Montclair State College. 49 p.

DeVries, Neal H. Contact between the North Horn and Price River formations, east front of the Gunnison Plateau, Sanpete County, Utah. M, 1959, Michigan State University. 87 p.

DeVries, Robert Charles. Phase equilibria in the system CaO-TiO$_2$-SiO$_2$ and their significance. D, 1953, Pennsylvania State University, University Park. 118 p.

Dew, Elizabeth Ann. Sedimentology and petrology of the Du Noir Limestone (Upper Cambrian) southern Wind River Range, Wyoming. M, 1985, Northern Arizona University. 193 p.

Dew, Mary McClure. Spatial distribution of conodonts in the Decorah Formation near Barnhart, Missouri. M, 1987, University of Missouri, Columbia. 51 p.

Dewall, A. E. Geomorphology and coastal processes along the southeastern Florida coast. M, 1977, George Washington University.

Dewart, Gilbert. Size investigations of ice properties and bedrock topography at the confluence of two glaciers, Kaskawulsh Glacier, Yukon Territory, Canada. D, 1968, Ohio State University. 223 p.

Dewart, Gilbert. The relation of microseisms to meteorological conditions. M, 1954, Massachusetts Institute of Technology. 114 p.

Dewees, Allen H. Lithologic studies of the Tensleep Sandstone. M, 1953, Northwestern University.

Dewees, Donald J. A structural interpretation of deformed metasedimentary rocks in northern Waterbury

Quadrangle, Connecticut. M, 1966, University of Wisconsin-Madison.

Dewey, Arthur Howard. Pleistocene history of Lee County. M, 1917, University of Iowa. 52 p.

Dewey, Christopher Paul. The taxonomy and palaeoecology of Lower Carboniferous ostracodes and peracarids (Crustacea) from southwestern Newfoundland and central Nova Scotia. D, 1983, Memorial University of Newfoundland. 383 p.

Dewey, David E. A mechanical and chemical analysis of the Middle Devonian River Group above the Sylvania in the Michigan Basin. M, 1958, Michigan State University. 52 p.

Dewey, James William. Seismicity studies with the method of joint hypocenter determination. D, 1971, University of California, Berkeley. 164 p.

Dewey, Robert Flanders. The geology of the Southwest Antioch Field, Garvin County, Oklahoma. M, 1948, University of Oklahoma. 28 p.

Dewhurst, JoAnna. Chemical ratios of Laramide igneous rocks and their relation to a paleosubduction zone under Arizona. M, 1976, University of Arizona.

Dewhurst, Warren Taylor. The simultaneous inversion of vector and total field aeromagetic data using Bayesian inference; a case study from the Cobb Offset of the Juan de Fuca Ridge. D, 1988, Colorado School of Mines. 249 p.

DeWilliam, Patrick P. Cost parameters of drilling in mine exploration. M, 1967, University of Arizona.

DeWindt, Justus Thomas. Preliminary studies concerning the paleoecology of agnathous vertebrates. M, 1967, Franklin and Marshall College. 178 p.

Dewing, Keith. Upper Ordovician and Lower Silurian stratigraphy and paleontology of Southampton Island, Northwest Territories. M, 1988, Laurentian University, Sudbury. 177 p.

Dewis, Frederick John. Relationship between mineralogy and trace element chemistry in sediments from two fresh water deltas and one marine delta within the Mackenzie River drainage basin, Canada. M, 1971, University of Calgary. 99 p.

DeWitt, David Bruce. Applications of sphalerite barometry in Grenville marbles. M, 1976, University of Michigan.

DeWitt, Ed. Precambrian geology and ore deposits of the Mayer-Crown King area, Yavapai County, Arizona. M, 1976, University of Arizona.

DeWitt, Ed Howard. Geology and geochronology of the Halloran Hills, southeastern California, and implications concerning Mesozoic tectonics of the southwestern Cordillera. D, 1980, Pennsylvania State University, University Park. 317 p.

DeWitt, Gary R. Geology of the Cedar Mountain area, Llano County, Texas. M, 1966, Texas A&M University.

DeWitt, Grant Whitney. Parametric studies of induced polarization spectra. M, 1979, University of Utah. 168 p.

Dewitt, Henry Gray. Downstream changes in bedload composition; Little South Fork and Cache la Poudre rivers, Colorado. M, 1978, Colorado State University. 141 p.

Dewitt, Nancy. Modern sediments of the Brazos River delta, Brazoria County, Texas. M, 1985, Stephen F. Austin State University. 112 p.

DeWitt, Ronald H. Hydrogeology of Fort Valley, Coconino County, Arizona. M, 1981, Northern Arizona University. 63 p.

DeWitte, Leendert. Experimental studies on the characteristics of the electro-chemical potentials encountered in drill holes. D, 1950, California Institute of Technology. 56 p.

DeWitte, Leendert. Factors governing accumulation of oil and gas in stratigraphic traps. D, 1950, California Institute of Technology. 52 p.

DeWitte, Leendert. Studies of infiltration of porous formations by drilling fluids in relation to the quantitative analysis of electrologs. D, 1950, California Institute of Technology. 128 p.

Dews, Jon Robert. A search for isotopic anomalies in silver and lithium from meteorites. D, 1965, University of California, Berkeley. 85 p.

DeWyk, Bruce H. Bedrock geology of the northern half of the Southbury Quadrangle, Connecticut. M, 1960, University of Massachusetts. 112 p.

DeWyk, Bruce H. The use of punch cards in ore estimation and grade control at the Keystone Mine, Silver City District, Nevada. M, 1963, University of Nevada. 75 p.

Dexter, James J. Uranium-lead zircon ages and crustal contamination of the northeastern Idaho Batholith. M, 1984, Western Michigan University.

Dextraze, Brenda Lynn. The paleoenvironmental effect of high latitudes on the internal growth lines of the late Eocene bivalve Eurhomalea antarctica. M, 1987, Purdue University. 95 p.

Dey, Abhijit. Finite source electromagnetic response of layered and inhomogeneous Earth models. D, 1972, University of California, Berkeley. 397 p.

Dey, Sarmistha. Compositional and textural analysis of beach and shallow-marine carbonate environments in the southern Grenadines, Lesser Antilles volcanic island arc; facies model for ancient island-arc limestones. M, 1985, Queen's University. 281 p.

Dey, Sudhindra Nath. Palaeomagnetism of Memesagamesing Lake and Caribou Lake norites, Parry Sound District, Ontario. M, 1981, University of Windsor. 77 p.

Dey, Thomas Nathanel. Brittle failure of crystalline rock; experiment and analysis. D, 1980, University of California, Berkeley. 179 p.

Dey-Sarkar, S. A Rb-Sr isotopic analysis of the Blake River Group of volcanics in the Superior Province of the Canadian Precambrian Shield. M, 1971, University of Toronto.

Dey-Sarkar, S. K. Upper mantle P-wave velocity distributions beneath western Canada. D, 1974, University of Toronto.

Deyo, Allen Elwin. Salinity investigations of Mancos landforms and springs in the upper Colorado River basin. D, 1984, University of California, Davis. 184 p.

DeYoung, John Hulbert, Jr. Geology of an area south of Los Pilares, Nacozari District, Sonora, Mexico. M, 1969, University of Michigan.

Dézé, Jean-Francois. An electromagnetic loop-loop sounding system at low induction numbers. M, 1981, Colorado School of Mines. 104 p.

Dezfulian, Houshang. Seismic response of soil deposits underlain by inclined boundaries. D, 1969, University of California, Berkeley. 214 p.

DeZoysa, Terence Henry. Geology and base metal mineralization of Lockport area, Notre Dame bay, Newfoundland (Canada). M, 1969, Memorial University of Newfoundland. 99 p.

Dhaliwal, Hardave. A comparison of theoretical rock models with laboratory data, and applications to geophysical hydrocarbon exploration. M, 1987, Pennsylvania State University, University Park. 170 p.

Dhar, B. B. Stresses in depth around an oval opening in an elastic modulus. M, 1966, McGill University.

Dheeradilok, Phisit. A detailed study of the deformational history of part of the Rockerville area, Black Hills, South Dakota. M, 1972, South Dakota School of Mines & Technology.

Dhindsa, Rupinder Singh. Conodont biostratigraphy and palaeontology of the lower Windsor Group (Visean) type area, Nova Scotia, Canada. M, 1984, University of Toronto.

Dhowian, A. W. Consolidation effects on properties of highly compressible soils; peats. D, 1978, University of Wisconsin-Madison. 421 p.

di Bona, Pietro Alphonse. Paleoenvironmental interpretation of the upper part of the Muldoon Canyon Formation, Pioneer Mountains, south-central Idaho. M, 1982, Washington State University. 69 p.

di Giovanni, Marcel, Jr. Stratigraphy and environments of deposition of the lower Pride Mountain Formation (Upper Mississippian) in the Colbert County area, Northwest Alabama. M, 1984, University of Alabama. 152 p.

Di Guiseppi, William Harris. Geomorphic and tectonic significance of Quaternary sediments in southern White River valley, Nevada. M, 1988, University of Utah. 62 p.

Di Paolo, William Dominic. The petrography of the Iron Mountain trachybasalts, Iron Mountain, Montana. M, 1971, Northern Illinois University. 35 p.

di Tullio, Lee Dolores. Fault rocks of the Tanque Verde Mountain decollement zone, Santa Catalina metamorphic core complex, Tucson, Arizona. M, 1983, University of Arizona.

Diaby, Ibrahima. Microfacies of the Salem Limestone (Middle Mississippian), southwestern Illinois. M, 1981, University of Illinois, Urbana. 61 p.

Diaby, Ibrahima. Petrography, diagenesis and depositional models of the St. Louis Limestone, Valmeyeran (Middle Mississippian), Illinois Basin, U.S.A. D, 1984, University of Illinois, Urbana. 198 p.

Dial, Don C. Bedrock geology of the northwest quarter of the Vienna Quadrangle, Illinois. M, 1963, Southern Illinois University, Carbondale. 66 p.

Diallo, Mamadou Mouctar. Morphology, genesis, and classification of selected soils in Klamath Mountains, California. M, 1988, University of California, Riverside. 113 p.

Diamond, Benjamin T. Oligocene foraminifera from Palma Real, Vera Cruz, Mexico. M, 1928, Columbia University, Teachers College.

Diamond, Fiona M. Study of a proposed transcontinental fault using Bouguer anomaly data. M, 1986, University of Western Ontario. 118 p.

Diamond, William Patrick. Differentiation of marine and nonmarine carboniferous environments (Kentucky) based on clay mineralogy. M, 1972, University of Kentucky. 126 p.

Dias Souto, Antonio Pedro. A bottom gravity survey of Carmel Bay, California. M, 1973, United States Naval Academy.

Dias, Carlos Alberto. A non-grounded method for measuring induced electrical polarization and conductivity. D, 1968, University of California, Berkeley. 272 p.

Dias, Manuel de Bettencourt. The geology of Ute Canyon, Clear Creek County, Colorado. M, 1951, Colorado School of Mines. 60 p.

Díaz, Abel Fernando Cepeda see Cepeda Díaz, Abel Fernando

Diaz, Eugenio G. Petrogenesis of a niobium-rich carbonatite dike, Oka Complex, Quebec. M, 1988, Boston University. 129 p.

Diaz, Henry Frank. A comparison of twentieth century climatic anomalies in northern North America with reconstructed patterns of temperature and precipitation based on pollen and tree-ring data. D, 1985, University of Colorado. 255 p.

Diaz, R. J. The effects of pollution on benthic communities of the tidal James River, Virginia. D, 1977, University of Virginia. 158 p.

Diaz, Ricardo Navarro. Variational method in wave propagation problems. M, 1971, Colorado School of Mines. 73 p.

Diaz-Garzon, Alfonso E. Planimetric adjustment of a super-long aerial triangulation performed according to the bz = 0 method. M, 1961, Ohio State University.

Diba, Mahmoud Hossein. Geology and logging study of the Bartlesville (Red Fork) Sand of the Southwest Clearbrook Field in a portion of Cleveland County, Oklahoma. M, 1974, University of Oklahoma. 130 p.

Dibaj, Mostafa. Non-linear seismic response of earth structures. D, 1969, University of California, Berkeley. 154 p.

Dibb, Jack. Hydrogeochemical kinetics of cadmium distribution in the Vestal, New York, well field No.1. M, 1983, SUNY at Binghamton. 60 p.

Dibb, Jack Eaton, Jr. The dynamics of beryllium-7 in Chesapeake Bay. D, 1988, SUNY at Binghamton. 128 p.

Dibble, Edwin Thompson. Stratigraphy and paleontology of a core from Silurian and Devonian strata of Wayne County, Michigan. M, 1956, University of Michigan.

Dibble, Walter Earl, Jr. Non-equilibrium water/rock interactions. D, 1980, Stanford University. 177 p.

DiBiagio, Elmo Lawrence. Stresses and displacements around an unbraced rectangular excavation in an elastic medium. D, 1966, University of Illinois, Urbana. 145 p.

Diblin, Mark C. Description, occurrence and origin of an enigmatic sedimentary structure from the Ocala Limestone, Florida. M, 1987, University of Florida. 116 p.

Dice, Bruce Burton. A quantitative study of composite Devonian lithofacies in the Michigan Basin. M, 1955, Michigan State University. 48 p.

Dice, Mark A. Computer analysis and synthesis of the stratigraphy and petroleum geology of the Upper Cambrian strata of Morrow County, Ohio. M, 1981, Kent State University, Kent. 113 p.

DiCesare, Joseph A. Physiochemical characterization of selected power plant fly ash. M, 1984, University of Toledo. 221 p.

Dicey, Timothy R. Magnetic, gravity, seismic refraction and seismic reflection profiles across Las Alturas geothermal anomaly, New Mexico. M, 1980, New Mexico State University, Las Cruces. 61 p.

Dick, Beryl J. Allen *see* Allen Dick, Beryl J.

Dick, H. J. The origin and emplacement of the Josephine Peridotite of southwestern Oregon. D, 1976, Yale University. 445 p.

Dick, Jay D. Geothermal reservoir temperatures in Chaffee County, Colorado. M, 1976, Northeast Louisiana University.

Dick, Jeffrey C. Evaluation of the ground water resources of Kirtland, Ohio, and adjacent areas. M, 1982, Kent State University, Kent. 180 p.

Dick, Lawrence Allan. A comparative study of the geology, mineralogy, and conditions of formation of contact metasomatic mineral deposits in the northeastern Canadian Cordillera. D, 1980, Queen's University. 471 p.

Dick, Lawrence Allan. Metamorphism and metasomatism at the MacMillan Pass tungsten deposit, Yukon and District of Mackenzie, Canada. M, 1976, Queen's University. 227 p.

Dick, Robert Irving. Coal fields of south-central Iowa. M, 1913, University of Iowa. 45 p.

Dick, Vincent B. Taphonomy and depositional environments of Middle Devonian (Hamilton Group) pyritic fossil beds in western New York. M, 1982, University of Rochester. 118 p.

Dickas, Albert Binkley. A regional stratigraphic study of the Tuscaloosa Group and associated Upper Cretaceous rocks of the central Mississippi Embayment. D, 1962, Michigan State University. 157 p.

Dickas, Albert Binkley. The geology of Lower Jakeys Fork, Fremont County, Wyoming. M, 1956, Miami University (Ohio). 130 p.

Dicke, Craig A. Modeling the feasibility of strontium remobilization from goethite surfaces in a saturated tuff at Yucca Mountain, Nye County, Nevada. M, 1988, University of Texas at El Paso.

Dicke-Burke, Collette. A detailed micropaleontological study and environmental interpretation of the microfauna of the Skelley Member and lower Birmingham Member (Pennsylvanian), Noble County, Ohio. M, 1981, University of Akron. 82 p.

Dickenson, Melville Pierce, III. The structural role and homogeneous redox equilibiria of iron in peralkaline, metaluminous and peraluminous silicate melts. D, 1984, Brown University. 150 p.

Dickerman, Robert W. An ecological survey of the Three-Bar Game Management Unit located near Roosevelt, Arizona. M, 1954, University of Arizona.

Dickerson, Carol Adrienne. The effect of cycled loading on the shear strength along a rock discontinuity. M, 1977, Cornell University.

Dickerson, Eddie Joe. Bolson fill, pediment and terrace deposits of Hot Springs area, Presidio County, Trans-Pecos Texas. M, 1966, University of Texas, Austin.

Dickerson, Robert Paul. A study of the polymetamorphism and mineral chemistry of the Little Bigelow Mountain and Bingham quadrangles, Maine. M, 1984, Southern Methodist University. 115 p.

Dickerson, Roy Ernest. Fauna of the Martinez Eocene of California. D, 1914, University of California, Berkeley. 19 p.

Dickert, Paul Fisher. Neogene phosphatic facies in California. D, 1971, Stanford University. 377 p.

Dickey, B. C. Palynology of the Newcastle Coal of the Pueblo Formation of the Wichita Group in North-central Texas. M, 1975, East Texas State University.

Dickey, Douglas B. Tertiary sedimentary rocks and tectonic implications of the Farewell fault zone, McGrath Quadrangle, Alaska. M, 1982, University of Alaska, Fairbanks. 54 p.

Dickey, Jerry Bland. Geology of the Barton Creek area, northern Wake County, North Carolina. M, 1964, North Carolina State University. 55 p.

Dickey, John Sloan, Jr. Magmatic and tectonic mafic layering in the Serrania de la Ronda (post Triassic–pre Eocene) and other alpine type peridotites (Betic cordillera, southern Spain). D, 1969, Princeton University. 135 p.

Dickey, Leonard Claude. Volume variation in an underground natural gas storage reservoir. M, 1948, University of Pittsburgh.

Dickey, M. L. Upper Cretaceous and Cenozoic history of the Green Mountain-Whiskey Peak area, Fremont County, Wyoming. M, 1962, University of Wyoming. 85 p.

Dickey, Parke Atherton. The igneous rocks and tectonics of the Lesser Antilles and northern South America; a reconnaissance of the igneous rocks of the Peninsula of Paraguana, Venzuela; a reconnaissance of the igneous rocks of the Island of Aruba, Dutch West Indies. D, 1932, The Johns Hopkins University.

Dickey, Robert I. The petrography and petrology of the Devil's Slide, Stark, New Hampshire. M, 1933, Washington University. 65 p.

Dickey, Robert McCullough. Literature and geologic occurrence of beryllium. M, 1931, University of Michigan.

Dickey, Robert McCullough. The granite sequence in the southern complex of Upper Michigan. D, 1936, University of Wisconsin-Madison.

Dickhaut, Lisa A. The strontium isotope and rare-earth element systematics of nonsulfide minerals from the Illinois-Kentucky fluorspar district. M, 1983, Miami University (Ohio). 64 p.

Dickie, Geoffrey. Geology of Cretaceous and Jurassic oil and gas pools in Alberta. D, 1972, University of Alberta. 136 p.

Dickie, George Allan. Subsurface studies of the Clore Formation, White County, Illinois. M, 1955, University of Illinois, Urbana. 33 p.

Dickie, J. R. Geological, mineralogical and fluid inclusion studies at the Dunbrack lead-silver deposit, Musquodoboit Harbour, Halifax County, Nova Scotia. M, 1978, Dalhousie University.

Dickin, R. C. Landfill leachate contamination. M, 1980, University of Waterloo.

Dickins, D. G. Least-squares inverse filtering of seismic reflection data. M, 1973, University of Alberta. 80 p.

Dickinson, Clyde G. The environment of marine limestone deposition. M, 1927, University of Wisconsin-Madison.

Dickinson, James Edward, Jr. Part I, Minor phases in late-stage lunar magmas; Part II, Rutile solubility and titanium coordination in silicate melts. D, 1984, Brown University. 172 p.

Dickinson, Kendall A. The Ostracoda and Cladocera of the Humboldt Formation in northeastern Nevada. M, 1959, University of Minnesota, Minneapolis. 47 p.

Dickinson, Kendall A. The Upper Jurassic stratigraphy of Mississippi and southwestern Alabama. D, 1962, University of Minnesota, Minneapolis. 180 p.

Dickinson, Marc. A search for intrasedimentary aeromagnetic anomalies over a known oil field in the Denver Basin. M, 1986, Colorado School of Mines. 70 p.

Dickinson, Robert G. Correlations of the El Paso Formation in western Texas, southwestern New Mexico, and southeastern Arizona based on insoluble residues. M, 1960, University of Arizona.

Dickinson, Robert Gerald. Geology of the Cerro Summit Quadrangle, Montrose County, Colorado. D, 1966, University of Arizona. 158 p.

Dickinson, Robert Rowland. A unified approach to second moment modeling of probabilistic physical systems. D, 1987, University of Waterloo. 146 p.

Dickinson, Roger G. Ground water study of the Santa Clara River road area, Eugene, Oregon. M, 1972, University of Oregon. 99 p.

Dickinson, Stanley Key, Jr. Investigations of the synthesis of diamonds. D, 1968, Harvard University.

Dickinson, Tamara L. Germanium abundances in lunar basalts; evidence of mantle metasomatism. D, 1988, University of New Mexico. 137 p.

Dickinson, Tamara Lynn. Petrogenesis of Apollo 14 aluminous mare basalts. M, 1984, University of New Mexico. 95 p.

Dickinson, Warren William. Isotope geochemistry of carbonate minerals in nonmarine rocks; northern Green River basin, Wyoming. D, 1984, University of Colorado. 159 p.

Dickinson, William Richard. Geology of the Izee area, Grant County, Oregon. D, 1958, Stanford University. 487 p.

Dickinson, William Richard. Tertiary stratigraphy and structure west of the Arroyo Seco, Monterey County, California. M, 1956, Stanford University. 160 p.

Dickman, Everitt W. Aerial triangulation using statoscope data. M, 1960, Ohio State University.

Dickman, Lynn R. A statistical model of mineral wealth in west-central Montana. M, 1986, University of Montana. 77 p.

Dickman, Steven Richard. Plate motions and polar wandering. D, 1977, University of California, Berkeley. 138 p.

Dicks, Alice L. Regimen and load of a small stream. M, 1975, Slippery Rock University. 72 p.

Dickson, B. A. A chronosequence of soil and vegetation on Recent mud-flow deposits on the lower slopes of Mount Shasta, California. D, 1952, University of California, Berkeley.

Dickson, Beryl Ann. Environmental mapping of the Topeka and Deer Creek megacycles of the Shawnee Group (Upper Pennsylvanian) of the Midcontinent. M, 1965, University of Illinois, Urbana.

Dickson, Charles W. Ore-deposits of Sudbury, Ontario. D, 1903, Columbia University, Teachers College.

Dickson, Elizabeth. Interpretation of glacial geology from aerial photographs. M, 1940, University of Cincinnati. 22 p.

Dickson, Frank W. Geochemical and petrographic aspects of mercury ore deposits. D, 1956, University of California, Los Angeles.

Dickson, Geoffrey Owen. Magnetic anomalies and ocean floor spreading in the south Atlantic Ocean. D, 1968, Columbia University. 86 p.

Dickson, John Richard. A test of the non-uniform distribution of azimuths of overlapping lunar craters. M, 1973, University of New Mexico. 45 p.

Dickson, Peter A. Permo-Carboniferous non-marine Bivalvia of the Appalachian Basin; their adaptive strategies and stratigraphical utility. D, 1977, University of Pittsburgh. 189 p.

Dickson, William Lawson. Geology, geochemistry and petrology of the Precambrian and Carboniferous igneous rocks between Saint John and Beaver Harbour, southern New Brunswick. D, 1985, University of New Brunswick.

Dickson, William Lawson. The general geology and geochemistry of the granitoid rocks of the northern Gander Lake Belt, Newfoundland. M, 1974, Memorial University of Newfoundland. 167 p.

Dicus, Joseph Martin. The geology and stratigraphy of the Cedar Grove Quadrangle of Northwest Georgia. M, 1952, Emory University. 55 p.

Dideriksen, Carrell J. Sedimentology of saline facies, Green River-Uinta formations (Eocene), Uinta Basin, Utah. M, 1968, University of Nebraska, Lincoln.

Didwall, Edna Mary. The electrical conductivity of the Earth's upper mantle as estimated from satellite measured magnetic field variations. D, 1981, The Johns Hopkins University. 178 p.

Didy, Sherif Mohamed Ahmed El see El Didy, Sherif Mohamed Ahmed

Diebel, John Keith. The mineralogy of the Bonanza silver deposit, Great Bear Lake, North West Territories. M, 1948, University of British Columbia.

Diebel, Lyndall J. Historical geography of the City of Columbus, Ohio. M, 1928, University of Wisconsin-Madison.

Diebold, Frank E. Contributions to the solution of the problem of dolomite genesis. D, 1967, Colorado School of Mines. 86 p.

Diebold, Frank Enri. X-ray methods applied to quantitative study of carbonate rocks. M, 1961, Iowa State University of Science and Technology.

Diebold, J. B. The traveltime equation, tau-p mapping and inversion for common midpoint seismic data with applications to the geology of the Venezuela Basin. D, 1980, Columbia University, Teachers College. 171 p.

Diebold, Michael Patrick. Complexes of niobium and tantalum containing metal-metal bonds. D, 1987, Texas A&M University. 126 p.

Diecchio, Richard Joseph. Lower and Middle Silurian ichnofacies and their paleoenvironmental significance, central Appalachian Basin of the Virginias. M, 1973, Duke University. 100 p.

Diecchio, Richard Joseph. Post-Martinsburg Ordovician stratigraphy, central Appalachian Basin. D, 1980, University of North Carolina, Chapel Hill. 227 p.

Diedrich, Robin P. The effect of fractures on the compressional wave velocity of Paleozoic carbonate rock. M, 1981, Bowling Green State University. 129 p.

Diefendorf, Andrew F. The parametrics of small watershed sediment yields. M, 1973, University of Missouri, Columbia.

Diegel, Fredric A. Geometry of imbricate thrusting in the Mountain City Window, Tennessee. D, 1986, The Johns Hopkins University. 203 p.

Diegel, Scott G. Metamorphic and structural geology of the Mount Cheadle area, northern Monashee Mountains, British Columbia. M, 1988, University of Calgary. 160 p.

Diehl, J. F. Paleomagnetic studies of the Casper Formation of southeastern Wyoming and the Ingelside Formation of northern Colorado. D, 1977, University of Wyoming. 116 p.

Diehl, James. Paleomagnetic reconnaissance of the Platoro Caldera, southeastern San Juan Mountains, Colorado. M, 1972, Western Washington University. 64 p.

Diehl, Katharine Benkelman. Middle Devonian paleoecology of the Mahantango Formation of northern Virginia and the Hamilton Group of central New York. M, 1980, University of North Carolina, Chapel Hill. 166 p.

Diehl, Paul Emmett. Structural fabric of the Wyomissing Hills-Laureldale Synclinorium and its relation to the Reading Prong, Pennsylvania. M, 1967, Franklin and Marshall College. 110 p.

Diehl, Paul Emmett. The stratigraphy, depositional environments, and quantitative petrography of the Precambrian-Cambrian Wood Canyon Formation, Death Valley. D, 1979, Pennsylvania State University, University Park. 430 p.

Diehl, S. B. The paleomagnetism of the Mississippian Madison Limestone, North-central Wyoming. D, 1977, University of Wyoming. 81 p.

Diehl, Wesley W. Depositional environments and paleoecology of the Middle and Upper Ordovician Martinsburg Formation of Grainger County, Tennessee. M, 1982, University of Tennessee, Knoxville. 242 p.

Diem, Robert Denton. The geology of the southwestern quarter of the Aurora Quadrangle, Missouri. M, 1953, University of Missouri, Columbia.

Diemer, J. A. Isovels and secondary circulation in a high mountain braided delta. M, 1979, SUNY at Binghamton. 96 p.

Diemer, John Andrew. Sedimentology of the fluvial-marine transition in the Upper Devonian/Lower Carboniferous of Kerry Head, County Kerry, Ireland. D, 1985, SUNY at Binghamton. 383 p.

Diemer, Raymond A. Titaniferous magnetite deposits of the Laramie Range, Wyoming. M, 1940, University of Wyoming. 27 p.

Dienger, Jennifer Lynn. Facies modelling of siliciclastic and carbonate sediments in the Lower Cambrian middle member Deep Spring Formation, White-Inyo Mountains, California. M, 1986, University of California, Davis. 108 p.

Dierberg, Forrest Edward. The effects of secondary sewage effluent on the water quality, nutrient cycles and mass balances, and accumulation of soil organic matter in cypress domes. D, 1980, University of Florida. 287 p.

Diery, Hassan Deeb. Stratigraphy and structure of Yakima Canyon between Roza Gap and Kittitas Valley, central Washington. D, 1967, University of Washington. 117 p.

Diery, Hassen D. Petrography and petrogenetic history of a quartz monzonite intrusive, Swisshelm Mountains, Cochise County, Arizona. M, 1964, University of Arizona.

Diesl, Warren F. A geoelectrical resistivity study of glacial outwash deposits, Chipuxet River valley, Kingston, Rhode Island. M, 1976, University of Rhode Island.

Dieterich, James Herbert. Sequence and mechanics of folding in the area of New Haven, Naugatuck and Westport, Connecticut. D, 1968, Yale University. 194 p.

Dieterle, Gifford Aas. Shallow subsurface investigation through resistivity methods in the Riverdale section of the Bronx (New York). M, 1959, New York University.

Dietrich, Donald R. A study of spinels in the upper zone of the Stillwater Complex, Montana. M, 1986, University of Arizona. 118 p.

Dietrich, Ernest S. A zonation of the Pierre Shale of western Nebraska based on fossil foraminifera. M, 1951, University of Nebraska, Lincoln.

Dietrich, John William. Geology of Presidio area, Presidio County, Texas. D, 1965, University of Texas, Austin. 349 p.

Dietrich, John William. Geology of Presidio-Ocotillo area, Presidio County, Trans-Pecos, Texas. M, 1954, University of Texas, Austin.

Dietrich, Ray Francis, Jr. The Simpson Group along the northern flank of the Anadarko Basin. M, 1954, University of Oklahoma. 54 p.

Dietrich, Richard V. The Fish Creek phacolith and surrounding area, New York. D, 1951, Yale University.

Dietrich, William Eric. Flow, boundary shear stress, and sediment transport in a river meander. D, 1982, University of Washington. 261 p.

Dietrich, William Eric. Sediment production in a mountainous basaltic terrain in central coastal Oregon. M, 1975, University of Washington. 81 p.

Dietsch, Harold A. Paleoecology of the shale in the Eiss Limestone Member of the Bader Limestone Formation (Lower Permian), Chase County, Kansas. M, 1956, University of Kansas. 80 p.

Dietterich, Robert J. Porosity and permeability analysis in Mississippian System, St. Louis Limestone Formation, Damme Field, Finney County, Kansas; utilizing petrographic image analysis. M, 1985, Wichita State University. 223 p.

Dietz, David Delbert. Geophysical investigation of concealed bedrock pediments in central Avra Valley, Pima County, Arizona. M, 1985, University of Arizona. 75 p.

Dietz, Earl Daniel. A study of the superheating of an albite feldspar. D, 1965, Ohio State University. 85 p.

Dietz, Robert Sinclair. Clay minerals in Recent marine sediments. D, 1941, University of Illinois, Urbana. 68 p.

Dietz, Robert Sinclair. Phosphorite deposits on the sea floor off California. M, 1939, University of Illinois, Urbana.

Diez de Medina, Diana Magdalena. Geochemistry of the Sandy Creek Gabbro, Wichita Mountains, Oklahoma. M, 1988, University of Oklahoma. 163 p.

Dieziger-Kim, Donna. Analysis and generation of low-flow sequences for Idaho streams using disaggregation modeling. M, 1985, University of Idaho. 314 p.

Diffenbach, Robert Nevin. The molybdenum deposits of Nedelec and Guerin townships, Timiskaming County, Quebec. M, 1961, Cornell University.

Diffendal, Robert F., Jr. The biostratigraphy of the Delaware Limestone (middle Devonian) of southwestern Ontario, Canada. D, 1971, University of Nebraska, Lincoln.

Diffendal, Robert Francis, Jr. Paleoecology of Pseudozaphrentoides verticillatus (Barbour) in the Plattsmouth Limestone (Pennsylvanian). M, 1964, University of Nebraska, Lincoln.

Difford, Winthrop C. Relation of structure and stratigraphy to gas production in the Devonian brown shales in Wayne Quadrangle, Wayne County, West Virginia. M, 1947, West Virginia University.

Difford, Winthrop Cecil. Development of the lithofacies map and its practical application to earth material; borrow area investigation. D, 1954, Syracuse University. 107 p.

Digel, Mark Richard. Metamorphism and migmatization in the Whitewater-Mojikit lakes area, English River Subprovince, northwest Ontario. M, 1988, University of Ottawa. 161 p.

Digert, Frederick E. The depositional environment of the Navajo Sandstone of the southwestern Colorado Plateau. M, 1955, University of Wisconsin-Madison.

Diggles, Michael. Geologic and mineral-resource evaluation of the White Mountains, California and Nevada. M, 1983, San Jose State University. 120 p.

Diggs, William Edward. Geology of the Otter River area, Bedford County, Virginia. M, 1955, Virginia Polytechnic Institute and State University.

DiGiacomo, Harry Joseph, Jr. Non-specific ion exchange of Zn and Ca in a Ca-Montmorillonite. M, 1976, Pennsylvania State University, University Park. 85 p.

Digman, Ralph E. Geology of the Guilford Quadrangle, Connecticut. D, 1949, Syracuse University.

Dignes, T. W. Late Quaternary abyssal foraminifera of the Gulf of Mexico. D, 1979, University of Maine. 176 p.

Dignes, Thomas W. Latest Quaternary benthic foraminiferal paleoecology and sedimentary history of vibracores from Block Island Sound. M, 1976, University of Rhode Island.

Dihrberg, Edward Ernest. Brachiopod biostratigraphy and biofacies analysis of the Marble Falls Formation (Pennsylvanian) of central Texas. M, 1988, University of Oklahoma. 98 p.

Dijkerman, Joost Christiaan. Properties and genesis of textural subsoil lamellae. D, 1965, Cornell University. 148 p.

Dike, Paul A. Megascopic structures and metamorphism of the Wissahickon Formation of central Delaware County, PA. M, 1950, Bryn Mawr College. 48 p.

Dikeou, Panayes John. A thermal and mechanical model for subsidence and changes in lithospheric thickness in the Michigan Basin. M, 1985, University of Oklahoma. 68 p.

Diker, Salahi. Investigation of stress anomalies associated with intrusive geologic structures. D, 1952, Colorado School of Mines. 128 p.

Dikmen, Seyyit Umit. Seismic response and liquefaction of saturated sands. D, 1981, University of Illinois, Urbana. 207 p.

Dilabio, R. N. Glacial dispersal of rocks and minerals in the Lac Mistassini-Lac Waconichi area, Quebec, with special reference to the Icon dispersal train. D, 1976, University of Western Ontario. 173 p.

Dilamarter, Ronald R. Some relationships of glaciation to land use in Blackberry and northern Sugar Grove townships, Kane County, Illinois. M, 1967, Northern Illinois University.

Dilamarter, Ronald Raymond. Areal variations of selected topographic variables on glacial drift sheets of different ages in Iowa. D, 1972, University of Iowa. 248 p.

Dilday, T. F., III. Hydrogeochemical exploration for Mississippian valley-type mineral deposits, Arkansas. M, 1982, University of Arkansas, Fayetteville. 82 p.

Dilek, Yildirim. Structure and petrology of the Big Bend Fault and association mafic dike complex, northern Sierra Nevada foothills, California. M, 1985, University of California, Davis. 276 p.

DiLeonardo, Christopher G. Structural evolution of the Smith Canyon Fault, northeastern Cascades, Washington. M, 1987, San Jose State University. 85 p.

Diles, Shawn James. Sedimentology of the lower part of the Morien Group in the southeast portion of the Sydney coal basin, Nova Scotia. M, 1984, University of Ottawa. 196 p.

Dill, Charles Edward, Jr. Formation and distribution of glauconite on the North Carolina continental shelf and slope. M, 1968, Duke University. 59 p.

Dill, George Meyer. Structure of northern Sierra de Ventana, Municipio de Ojinaga, Chihuahua, Mexico. M, 1961, University of Texas, Austin.

Dill, Robert Floyd. Contemporary submarine erosion in Scripps Submarine Canyon. D, 1964, University of California, San Diego. 299 p.

Dill, Robert Floyd. Environmental analysis of sediment from the sea floor off Point Arguello, California. M, 1952, University of Southern California.

Dillaha, Theo Alvin, III. Modeling the particle size distribution of eroded sediments during shallow overland flow. D, 1981, Purdue University. 200 p.

Dillard, Taylor W. The geohydrology and water quality of the upper-Buffalo River basin, Newton County, Arkansas. M, 1978, University of Arkansas, Fayetteville.

Dillé, Alan Charles Francis. Paleotopography of the Precambrian surface of northeastern Oklahoma. M, 1956, University of Oklahoma. 66 p.

Dille, Glenn Scott. Geology and mineral resources of the Saddleback Hills Quadrangle, Carbon County, Wyoming. M, 1924, University of Iowa. 88 p.

Dille, Glenn Scott. The stratigraphy and paleontology of the Mississippian of the Black Hills of South Dakota. D, 1929, University of Iowa. 438 p.

Dillenberger, Douglas S. Petrology of the Haney and Glen Dean members of the upper Newman Formation (upper Mississippian) of eastern Kentucky. M, 1976, Eastern Kentucky University. 68 p.

Dilles, John Hook. The petrology and geochemistry of the Yerington Batholith and the Ann-Mason porphyry copper deposit, western Nevada. D, 1984, Stanford University. 462 p.

Dilles, Peter Alden. Skarn formation and mineralization within the Lower Cretaceous Cantarranas Formation, El Mochito Mine, Honduras. M, 1982, University of Alaska, Fairbanks. 97 p.

Dilley, Thomas E. Holocene tephra stratigraphy and pedogenesis in the middle Susitna River valley, Alaska. M, 1988, University of Alaska, Fairbanks. 97 p.

Dillingham, Hervie, Jr. Ostracoda of the Ostrea thirsae Zone of the Nanafalia Formation, Wilcox Eocene of Alabama. M, 1952, Washington University. 37 p.

Dillman, George. Structural investigation and tectonic history of central Parras Basin, Saltito, Coahuila, Mexico. M, 1985, University of Houston.

Dillman, Scott Brian. Subsurface geology of the Upper Devonian-lower Mississippian black-shale sequence in eastern Kentucky. M, 1980, University of Kentucky. 72 p.

Dillon, Andrew Crawford, III. Rill and gully erosion in spoils from surface mining near Searles, Tuscaloosa County, Alabama. M, 1975, University of Alabama.

Dillon, David Lloyd. Geology of the Sand Creek porphyry molybdenum prospect. M, 1983, Brock University. 89 p.

Dillon, Edward Lamblin. Stratigraphy of an area near Lima, Beaverhead County, Montana. M, 1949, University of Illinois, Urbana. 103 p.

Dillon, Edward Patrick. A multi-element geochemical study of the pseudobreccia host rock of the Newfoundland zinc mine. M, 1979, University of Toronto.

Dillon, George R. Subsurface geology of Franklin County, Oklahoma. M, 1956, [University of Oklahoma].

Dillon, John F. Growth line analysis of a recent scallop population and its potential for paleoecology. M, 1974, University of New Mexico. 68 p.

Dillon, John T. Geology of the Chocolate and Cargo Muchacho mountains, southeasternmost California. D, 1976, University of California, Santa Barbara. 575 p.

Dillon, M. J. The integration of bio-physical and socio-economic data in the land use planning process; a case study of the town of Andes, Delaware County, N.Y. D, 1976, Cornell University. 203 p.

Dillon, P. J. The prediction of phosphorus and chlorophyll concentrations in lakes. D, 1974, University of Toronto.

Dillon, Vincent Francis. Variation of velocity of earthquake surface waves with length of path under the sea. M, 1928, St. Louis University.

Dillon, William A. Monogenetic trematodes of some New Zealand fishes. M, 1963, College of William and Mary.

Dillon, William Gregory. One-dimensional iterative inversion of vertical seismic profiles. D, 1985, Texas A&M University. 178 p.

Dillon, William P. A petrographic study of the Niagaran carbonates of an area of western Kentucky. M, 1961, Rensselaer Polytechnic Institute. 139 p.

Dillon, William Patrick. Structural geology of the southern Moroccan continental margin. D, 1969, University of Rhode Island. 92 p.

Dillon-Leitch, Henry C. H. The distribution of platinum-group elements and platinum-group minerals from the Donaldson West and surface deposits, Cape Smith Belt, Quebec; the roles of thermal and dynamic metamorphism. D, 1988, Carleton University. 356 p.

Dillon-Leitch, Henry C. H. Volcanic stratigraphy, structure and metamorphism in the Courageous-Mackay Lake greenstone belt, Slave Province, Northwest Territories. M, 1981, University of Ottawa. 178 p.

Dilts, Roger David. Occurrence and behavior of gold in selected Utah thermal springs. M, 1986, University of Utah. 77 p.

Dilworth, Ottis L. Upper Cretaceous Farmington Sandstone of northeastern San Juan County, New Mexico. M, 1960, University of New Mexico. 96 p.

DiMarco, Michael J. Stratigraphy, sedimentology, and sedimentary petrology of the early Archean Coongan Formation, Warrawoona Group, eastern Pilbara Block, Western Australia. D, 1986, Louisiana State University. 338 p.

Dimelow, Thomas E. Stratigraphy and petroleum, Lyons Sandstone, northeastern Colorado. M, 1972, Colorado School of Mines. 127 p.

Diment, William Horace. A regional gravity survey in Vermont, western Massachusetts and eastern New York. D, 1953, Harvard University.

Dimit, Georgette. An investigation of slope failure along the Maumee River from Grand Rapids, Ohio, to Toledo, Ohio. M, 1988, University of Toledo. 128 p.

Dimitrakopoulos, Roussos-Georgios. Conditional simulation and kriging as an aid to oil sands development; an application in part of the Athabasca Deposit. M, 1985, University of Alberta. 272 p.

Dimitriadis, B. D. Methods in data reduction. D, 1977, SUNY at Buffalo. 243 p.

Dimitroff, Pencho B. Structural geology of Foote Creek Anticline and adjacent area, Carbon and Albany counties, Wyoming. M, 1968, University of Wyoming. 71 p.

Dimmick, Charles William. Hinge and early shell development of representative venerid Bivalvia. D, 1969, Tulane University. 89 p.

Dimmick, Charles William. The age of the blue-green clay outcropping in the Crescent Beach Spring. M, 1964, University of Florida. 105 p.

Dimmick, Ross Alan. Foraminiferal paleoecology of the Plio-Pleistocene Taiwan foredeep basin and its tectonic implications. M, 1987, Rutgers, The State University, New Brunswick. 116 p.

Dimmock, Pamela E. A hydrogeologic investigation of the stamping ground, Kentucky area. M, 1988, University of Kentucky. 74 p.

Dimopoullos, Thomas J. A palynologic study of subsurface samples from Long Island, New York. M, 1961, New York University.

Din, Ibrahim Oslam Alam El *see* El Din, Ibrahim Oslam Alam

Din, Ibrahim Osman Alam el *see* Alam el Din, Ibrahim Osman

Din, Min. Microfossils of the Middle and Upper Ordovician formations in deep wells of Alpena, Ogemaw, Bay, and Ingham counties of Michigan. M, 1950, Michigan State University. 59 p.

Dincel, Mehmet Bedi. Geology of Toyah Lake area, Reeves County, Texas. M, 1952, University of Texas, Austin.

Dincer, Hikmet. The southernmost extension of Grenville Limestone in Peekskill, New York. M, 1951, Columbia University, Teachers College.

Dineen, Michael J. Geology of the Triassic Ellerbe Basin, Richmond County, North Carolina. M, 1983, North Carolina State University. 69 p.

Dineen, R. J. Surficial geology of the Voorheesville Quadrangle. M, 1977, Rensselaer Polytechnic Institute.

Diner, David Joseph. I, Silicon Vidicon imaging of Jupiter 4100-8300 angstroms; spectral reflectivity, limb-darkening, and atmospheric structure; II, Simultaneous ultraviolet (0.36 micron) and infrared (8-20 micron) imaging of Venus; properties of clouds in the upper atmosphere. D, 1978, California Institute of Technology. 233 p.

Diner, Yehuda A. The HY precious metals lode prospect, Mineral County, Nevada. M, 1983, Stanford University. 220 p.

Dinga, Carl F. A quantitative analysis of the effect of Earth rotation on meandering alluvial rivers. M, 1969, Indiana State University. 126 p.

Dinga, Carl F. An analysis of the relationship between geologic structure and the geometric surface form of homoclinal ridges. D, 1971, Indiana State University. 209 p.

Dingee, Brad E. Geology, hydrology and geochemistry of the geothermal area east of Lowman, Idaho. M, 1987, Washington State University. 138 p.

Dinger, James S. Stratigraphic, structural, and petrofabric investigations of lake ice, Lake Champlain, Vermont. M, 1969, University of Vermont.

Dinger, James Sheldon. Relation between surficial geology and near-surface ground water quality, Las Vegas Valley, Nevada. D, 1977, University of Nevada - Mackay School of Mines. 215 p.

Dingler, Craig Mitchell. Reconnaissance glacial geology of the Selway-Bitterroot Wilderness and surrounding lower elevations, Idaho and Montana. M, 1981, University of Idaho. 184 p.

Dingler, John R. A flume study of ripple growth in fine sand. M, 1968, Massachusetts Institute of Technology. 84 p.

Dingler, John R. Wave-formed ripples in nearshore sands. D, 1974, University of California, San Diego. 152 p.

Dingman, Lorraine Elizabeth. The sedimentology and paleohydraulics of the Upper Devonian Catskill facies of the Unadilla-Hancock area, south-central New York State. M, 1982, SUNY at Binghamton. 99 p.

Dingman, Stanley Lawrence. Hydrology of the Glenn Creek watershed, Tanana River drainage, central Alaska. D, 1970, Harvard University.

Dings, McClelland G. The petrology of the (Dry St. Vrain) area, Colorado. M, 1935, Washington University. 62 p.

Dings, McClelland Griffith. Igneous history of the Stony Mt. Stock, Colorado. D, 1937, University of Rochester. 183 p.

Dingus, Delmar D. The nature and properties of amorphous colloids formed from Mazama tephra. D, 1974, Oregon State University. 95 p.

Dingus, Lowell W. The Warm Springs fauna (Mammalia, Hemingfordian) from the western facies of the John Day Formation, Oregon. M, 1979, University of California, Riverside. 178 p.

Dingus, Lowell Wilson. A stratigraphic review and analysis for selected marine and terrestrial sections spanning the Cretaceous-Tertiary boundary. D, 1983, University of California, Berkeley. 200 p.

Dingus, William Frederick. Morphology, paleogeographic setting, and origin of the middle Wilcox Yoakum Canyon, Texas coastal plain. M, 1987, University of Texas, Austin. 78 p.

Dingwell, Donald Bruce. Investigations of the role of fluorine in silicate melts; implications for igneous petrogenesis. D, 1984, University of Alberta. 150 p.

Dinkel, Ted Richard. Vertical and lateral variations in a welded tuff. M, 1969, University of Missouri, Rolla.

Dinkelman, Menno Gustaaf. Late Quaternary radiolarian paleo-oceanography of the Panama Basin, eastern equatorial Pacific. D, 1974, Oregon State University. 123 p.

Dinkins, Theo H., Jr. The Mooreville Formation in the Southeast Artesia and southwest McCrary quadrangles, southern Lowndes County, Mississippi. M, 1960, Mississippi State University. 31 p.

Dinkmeyer, Paul R. Stratigraphy of Upper Cambrian and Lower Ordovician rocks in the Manitou Park-Manitou Springs-Deadman Canyon area, Douglas and El Paso counties, Colorado. M, 1977, Colorado School of Mines. 218 p.

Dinnean, Robert F. Lower Carboniferous Bangor Limestone in Alabama; a multicycle clear water epeiric sea sequence. D, 1974, Louisiana State University. 120 p.

Dinnean, Robert F. The geology of the Chestnut Salt Dome area, Natchitoches Parish, Louisiana. M, 1958, Louisiana State University.

Dinnel, Scott Page. Circulation and sediment dispersal on the Louisiana-Mississippi-Alabama continental shelf. D, 1988, Louisiana State University. 187 p.

Dinnel, Scott Page. Distribution and residence time of freshwater on the West Louisiana and Texas continental shelves. M, 1984, Louisiana State University. 89 p.

Dinsmore, James P. The geology of the Hamma Hamma and North Fork Skokomish River valleys. M, 1953, University of Puget Sound. 69 p.

Dion, Claude. Géologie de la région de Troodhitissa, complexe plutonique du Troodos, Chypre. M, 1987, Universite Laval. 279 p.

Dion, Eric Paul. Trace elements and radionuclides in the Connecticut River and Amazon River estuary. D, 1983, Yale University. 253 p.

Dionisio, Leonard C., Jr. Structural analysis and mapping of the eastern Caddo Anticline, Ardmore Basin, Oklahoma. M, 1975, University of Oklahoma. 83 p.

Dionne, J. V. Geology of the Middle Pennsylvanian Stockton-Lewiston coal seam and associated clastic rocks in West Virginia. D, 1980, George Washington University. 267 p.

Dionne, J. V. Systematics, stratigraphy and paleo-environments of three Cretaceous ahermatypic coral genera from the Sierra del Montsech, Spanish Pyrenees. M, 1974, George Washington University.

DiPiazza, Nicholas John. An aeromagnetic and gravity survey of eastern Virginia. M, 1964, Rensselaer Polytechnic Institute. 33 p.

DiPietro, Joseph Anthony, III. Contact relations in the late Precambrian Pinnacle and Underhill formations, Starksboro, Vermont. M, 1983, University of Vermont. 132 p.

DiPlacido, Arthur J. A petrographic analysis of the Tamiami Formation, Florida,. M, 1980, University of Toledo. 51 p.

Dippon, Duane Roy. The economic consequences of intensifying forest management in Douglas County, Oregon. D, 1982, Oregon State University. 178 p.

Diprime, Leonard Joseph, Jr. A field study of flow through fractured shale and its comparison to two mathematical models. M, 1988, Emory University. 111 p.

Dircksen, Paul Eric. Geology and mineralization of the Pine Grove-Rockland mining districts, Lyon County, Nevada. M, 1975, University of Nevada. 69 p.

DiRenzo, Vincent N., Jr. Petrology and depositional environments of the Little Valley Limestone (Upper Mississippian), Washington County, Virginia. M, 1986, East Carolina University. 154 p.

Dirin, J. Grain size analysis and mineralogy of Galveston Island beach sands. M, 1975, East Texas State University.

Diringer, Michael F. Subsurface configuration of the Bottle Lake Complex and Center Pond Pluton, east-central Maine. M, 1982, SUNY at Buffalo. 64 p.

Dirks, Thomas N. The upper Paleozoic stratigraphy of the Qumby Ranch area, southern Guadalupe Canyon Quadrangle, Cochise County, Arizona. M, 1966, University of Arizona.

Dirlam, D. M. Pollen analysis of mid-Altonian peat in Schelke Bog, Lincoln County, Wisconsin. M, 1979, University of Wisconsin-Madison.

Dirom, G. E. Potassium-argon age determinations on biotites and amphiboles. M, 1965, University of British Columbia.

Disbrow, Alan Eastman. The structural geology of the Cerillos Hills area, New Mexico. M, 1953, University of New Mexico. 35 p.

Dischinger, James B., Jr. Late Mesozoic and Cenozoic stratigraphic and structural framework near Hopewell, Virginia. M, 1979, University of North Carolina, Chapel Hill. 84 p.

Dishman, Bill D. Stratigraphy and cyclic deposition of portions of the Council Grove and Chase groups (Permian) in Nebraska and Kansas. M, 1969, University of Nebraska, Lincoln.

Disney, Ralph Willard. A subsurface study of the pre-Pennsylvanian rocks of Cleveland and McClain Counties, Oklahoma. M, 1950, University of Oklahoma. 39 p.

Disney, Ralph Willard. The subsurface geology of the McAlester Basin, Oklahoma. D, 1960, University of Oklahoma. 116 p.

Distefano, Lee. Correlation of compositional changes in a coal core with density as recorded by a high resolution density log. M, 1980, Southern Illinois University, Carbondale. 106 p.

Distefano, M. P. The diagenetic alteration of volcanic rock fragments in the Stevens Sandstone, San Joaquin Basin, California. D, 1978, Texas A&M University. 107 p.

Distefano, Mark. The mineralogy and petrology of the Eau Claire Formation, west-central Wisconsin. M, 1973, Northern Illinois University. 132 p.

Distelhorst, Carl A. R. Geology of the bedrock formations of Milwaukee County (Wisconsin). M, 1967, University of Wisconsin-Milwaukee.

Distler, George E. The morphology and taxonomy of Zygospira Hall (Brachiopoda). M, 1967, Miami University (Ohio). 89 p.

Distler, George Edward. The morphology, taxonomy, and phylogeny of some Ordovician atrypids (brachiopods). D, 1972, Miami University (Ohio). 235 p.

Ditbanjong, Sandusit. Consolidation behavior of clays under the constant rate of strain test with special emphasis on the secondary compression. D, 1981, North Carolina State University. 221 p.

Ditmars, Richard C. Increase in lithospheric thickness during the subsidence of the Williston Basin. M, 1985, University of Oklahoma. 47 p.

Ditson, G. M. Metallogeny of the Vancouver-Hope area, British Columbia. M, 1978, University of British Columbia.

Ditsworth, Glenn William. The origin of calcium sulfate deposits. M, 1931, University of Iowa. 128 p.

Ditteon, Richard P. Daily temperature variations on Mars. D, 1981, University of California, Los Angeles. 112 p.

Dittman, Fred Melvin, Jr. Petrology of the Precambrian igneous and metamorphic rocks in the vicinity of St. Cloud, Minnesota. M, 1971, Northern Illinois University. 75 p.

Dittmar, Edward I. Environmental interpretation of paleocurrents in the Triassic Durham Basin. M, 1979, University of North Carolina, Chapel Hill. 91 p.

Dittmer, Eric Rheydt. A sediment budget analysis of Monterey Bay. M, 1972, San Jose State University. 132 p.

Dittrich, Harold Steven. Studies on Rhacophyton ceratangium from the Upper Devonian of West Virginia. M, 1982, Southern Illinois University, Carbondale. 67 p.

Ditty, Patrick Scott. Stratigraphy and sedimentation in the Hispaniola-Caicos abyssal basin. M, 1974, Duke University. 126 p.

Ditty, William E. The development of landscapes exemplified by the Bradford Pennsylvania area. M, 1967, Virginia State University. 30 p.

Ditzel, Krista D. Mineralogical and chemical changes related to oxidation and leaching of uranium and associated elements of the Blind River-Elliot Lake District, Ontario. M, 1980, Bowling Green State University. 114 p.

Ditzell, Curtis Leon. Sedimentary geology of the Cambro-Ordovician Signal Mountain Formation as exposed in the Wichita Mountains of southwestern Oklahoma. M, 1984, Oklahoma State University. 165 p.

Ditzell, Leon Sebastian. The geology of the Southwest portion of the Millersburg Quadrangle, Boone County, Missouri. M, 1950, University of Missouri, Columbia.

Divens, Donald F. Geology of the Telegraph Pass area, Diamond Range, White Pine County, Nevada. M, 1957, University of Nevada. 59 p.

Diver, Bradford Babbitt Van *see* Van Diver, Bradford Babbitt

Divi, Sri Ramachandra Rao *see* Rao Divi, Sri Ramachandra

Divine, Douglas Wayne. Guide to Arkansas geology. D, 1972, Northeast Louisiana University.

Divis, Allan F. The geology and geochemistry of the Sierra Madre Mountains, Wyoming. D, 1975, University of California, San Diego. 250 p.

Dix, Fred Andrew, Jr. Foraminifera from the Mobridge Member of the Pierre Shale. M, 1957, Rutgers, The State University, New Brunswick. 101 p.

Dix, George Roger. Shallow-burial diagenesis of deep-water Neogene and Quaternary periplatform carbonates, northern Bahamas. D, 1988, Syracuse University. 292 p.

Dix, George Roger. The Codroy Group (Upper Mississippian) of the Port au Port Peninsula; stratigraphy, palaeontology, sedimentation and diagenesis. M, 1982, Memorial University of Newfoundland. 219 p.

Dix, N. F. A study of weathering at the Pre-Cambrian-Paleozoic contact. M, 1947, Queen's University.

Dixit, Shailaja R. A geologic interpretation of the Vinton geophysical anomaly, in east-central Iowa. M, 1984, University of Iowa. 151 p.

Dixit, Sushil Sharan. Algal microfossils and geochemical reconstruction of Sudbury lakes; a test of paleoindicator potential of diatoms and chrysophytes. D, 1986, Queen's University.

Dixon, Bryan W. A method for activation analysis of ruthenium in sea water. M, 1965, Texas A&M University.

Dixon, Denise Yvonne. A study of dewatering effects at three longwall mines in the Northern Appalachian coal field. M, 1988, West Virginia University. 251 p.

Dixon, Frederik Sigurd. Paleoecology of an Eocene mud-flat deposit (Avon Park Formation, Claibornian) in Florida. M, 1972, University of Florida. 45 p.

Dixon, George H. Records of wells drilled for oil and gas in New Mexico. M, 1954, University of New Mexico. 195 p.

Dixon, Gordon Elliott. Petrography of a biotite granite and granophyre, Mount Desert Island and Schoodic Peninsula, Maine. M, 1960, University of Illinois, Urbana. 31 p.

Dixon, Helen Roberta. The bedrock geology of the Plainfield–Danielson area (Windham County), Connecticut. D, 1968, Harvard University.

Dixon, Helen Roberts. Two petrographic studies; I, Homogeneity of fabric in deformed marble; II, Layered basic plutonic rocks of Seiland, Norway. M, 1956, University of California, Berkeley. 83 p.

Dixon, Howard B. The building and monumental stones of the State of Utah. M, 1938, Brigham Young University. 68 p.

Dixon, Howard Raymond. Ostracods from the Lakota Formation in the northern and eastern Black Hills of South Dakota. M, 1950, University of Iowa. 69 p.

Dixon, J. Stratigraphy and invertebrate paleontology of lower Paleozoic rocks, Somerset and Prince of Wales islands, N.W.T. D, 1973, University of Ottawa. 282 p.

Dixon, J. M. Techniques and tests for measuring joint intensity. D, 1979, West Virginia University. 153 p.

Dixon, James B. Geology of the Wild Horse Canyon area, Fox Range, Washoe County, Nevada. M, 1977, University of Nevada. 63 p.

Dixon, James Robert, Jr. A spinel lherzolite barometer. D, 1980, University of Texas at Dallas. 71 p.

Dixon, James W., Jr. Population studies of the brachiopod Kingena wacoensis occurring in the Lower Cretaceous Georgetown Formation of central Texas. D, 1955, University of Wisconsin-Madison.

Dixon, Jeanette M. Control of the Cuyahoga River drainage system in Summit County, Ohio. M, 1976, University of Akron. 38 p.

Dixon, Joe Boris. Mineralogical analyses of soil clays involving vermiculite-chlorite-kaolinite differentiation. D, 1958, University of Wisconsin-Madison. 86 p.

Dixon, Joe Scott. A statistical study of seven species of the Pennsylvanian-Permian goniatite, Agathiceras. M, 1960, University of Iowa. 58 p.

Dixon, John Charles. Chemical weathering of late Quaternary cirque deposits in the Colorado Front Range. D, 1983, University of Colorado. 190 p.

Dixon, John McConkey. Experimental studies of strain in diapirs, with applications to the mantled gneiss domes of New England. D, 1974, University of Connecticut. 203 p.

Dixon, John Robert. Carbonate petrography of Lower and Middle Paleozoic sediments, west flank, Teton Mountains, Idaho-Wyoming. M, 1964, Texas Tech University. 75 p.

Dixon, Joseph A. Characterization of overburden above coals V and VI in northwestern Greene County, Indiana. M, 1979, Indiana University, Bloomington. 73 p.

Dixon, Kenneth Randall. A model for predicting the effects of sewage effluent on wetland ecosystems. D, 1974, University of Michigan.

Dixon, L. S. A mathematical model of salinity uptake in natural channels traversing Mancos Shale badlands. D, 1978, Utah State University. 213 p.

Dixon, Louis H. Cenozoic cyclic deposition in the subsurface of central Louisiana. D, 1963, Louisiana State University. 163 p.

Dixon, Louis Helprein. Use of aerial photographs in mapping the geology of Gillespie County, Texas. M, 1941, University of Texas, Austin.

Dixon, Mark A. Experimental bias associated with loose grain measurements. D, 1963, Louisiana State University. 46 p.

Dixon, Max Mueller. Geological processes represented in the Cornucopia mining district, Cornucopia, Oregon. M, 1923, Columbia University, Teachers College.

Dixon, Owen Arnold. Middle Devonian tabulate corals from northwestern Canada. M, 1962, University of Saskatchewan. 103 p.

Dixon, Richard A. Lithologic study of Cambro-Ordovician core, Delta County, Michigan. M, 1961, Michigan State University. 99 p.

Dixon, Richard Lee. The geology and ore deposits of the Red Canyon mining district, Douglas County, Nevada. M, 1971, University of Nevada. 88 p.

Dixon, Roy Wilbur. Geology of Isabella-Sweetgrass area, Okanogan County, Washington. M, 1959, University of Washington. 64 p.

Dixon, Timothy H. The evolution of continental crust in the late Precambrian Egyptian Shield. D, 1979, University of California, San Diego. 246 p.

Dixon, Val R. Microstratigraphy of the Leavenworth Limestone, Virgilian of eastern Kansas. M, 1960, University of Kansas. 125 p.

Dixon, William Gordon. Petrology of the Mohawkian (Ordovician) Series in Cass County, Indiana. M, 1966, Indiana University, Bloomington. 23 p.

Dixon, William Hyatt. The application of paper chromatography in the qualitative analysis of the sulfosalt mineral group. M, 1959, University of Michigan.

Dixon, William Ronald. Subsurface geology of an area in Bastrop and Fayette counties, Texas. M, 1958, University of Texas, Austin.

Djamgouz, Okay Tewfik. Relationship between ferromagnetic particles and airborne chrysotile fibres in the asbestos mines and mills of Quebec. D, 19, McGill University.

Djeddi, Rabah. An integrated geophysical and geological study of the deep structure and tectonics of the Permian Basin of West Texas and southeastern New Mexico. M, 1979, University of Texas at El Paso.

Djordjevic, N. Denver air-pollution study with air trajectories. M, 1966, Colorado State University. 113 p.

Djuanda, Handoko. Carbonate depositional facies and porosity development of the Ratcliffe Beds, Charles Formation (Mississippian), Williston Basin, Sheridan County, Montana. M, 1988, Colorado School of Mines. 238 p.

Djuth, Gerald Joseph. An investigation of the use of magnetic susceptibility for stratigraphic correlation of Columbia River basalts. M, 1983, Washington State University. 92 p.

Do Lam Sinh. Détection de directions non-apparentes sur photos aeriennes par filtrage optique de fréquences spatiales de la transformée de Fourier. M, 1973, Ecole Polytechnique.

Do, Lam Sinh. Détection de directions non-apparentes sur photos aériennes par filtrage optique de fréquence spatiale de la transformée de Fourier. M, 1973, Ecole Polytechnique. 104 p.

Doak, Robert Alvin, Jr. The geology of the Double Mountain-Grassy Mountain area, Mitchell Butte Quadrangle, Oregon. M, 1953, University of Oregon. 94 p.

Doak, Roger W. Sedimentology and stratigraphy of late Wisconsin glaciogenic deposits near Kettle Falls, Washington. M, 1986, Eastern Washington University. 146 p.

Doak, William H. Cation retention and solute transport related to porosity of pumiceous soils. D, 1972, Oregon State University. 104 p.

Doan, David Bentley. The method of "Ronov" as applied to Appalachian geology. M, 1949, Pennsylvania State University, University Park. 41 p.

Doane, George H. The distribution of oil shale. M, 1921, Columbia University, Teachers College.

Doane, Virginia L. Petrography of Nonesuch Shale, White Pine, Michigan. M, 1956, Michigan Technological University. 104 p.

Dobak, Paul J. Alteration and paragenesis of the Paradise Peak gold/silver deposit. M, 1988, Colorado State University. 157 p.

Dobar, Walter I. Simulated lunar igneous surface rock. M, 1968, Michigan State University. 76 p.

Dobb, David E. Major metal and fluorine analysis of Fort Union Formation coal. M, 1976, Montana College of Mineral Science & Technology. 108 p.

Dobbin, C. E. The continuity of the lithologic units in the Fox Hills, Lance, and Fort Union formations of eastern Montana, and its bearing on the Laramie problem. D, 1924, The Johns Hopkins University. 105 p.

Dobbins, Charles Nelson. Soil survey of the Chapel Hill Triassic area. M, 1917, University of North Carolina, Chapel Hill. 19 p.

Dobbins, David A. Stratigraphy of Middle and Upper Devonian strata of the northeastern part, Orbisonia Quadrangle, Pennsylvania. M, 1959, University of Minnesota, Minneapolis. 82 p.

Dobbins, David Ashmun. Sedimentary geochemistry and hydrogeochemistry of the Pamlico River-Pamlico Sound estuary, North Carolina. D, 1968, University of North Carolina, Chapel Hill. 79 p.

Dobbins, Lloyd R. Study of the basal Cretaceous clastics of central Texas. M, 1974, Baylor University. 124 p.

Dobbins, Ralph J. Cheilostome Bryozoa of the northern Mosquito Bank (Nicaragua and Honduras). M, 1972, Louisiana State University.

Dobbs, Danny M. A geochemical soil survey for selected elements in Clay County, Alabama. M, 1971, University of Alabama.

Dobbs, Frederick Courtrite. Microbiology of bioturbated sediments; the burrows of Callianassa and the deposit-feeding system of Ptychodera. D, 1987, Florida State University. 109 p.

Dobbs, Phillip Hale. Geology of the central part of the Clark Mountain Range, San Bernardino County, California. M, 1961, University of Southern California.

Dobbs, Steven Lawrence. Linearized inversion of plane-wave seismograms. M, 1987, University of Texas, Austin. 159 p.

Dobecki, Thomas Lee. Seismic modeling and depth migration for two and three dimensional Earth geometrics. D, 1972, Indiana University, Bloomington. 272 p.

Dobell, Joseph Porter. Geology of the Antone District, Wheeler County, Oregon. M, 1948, Oregon State University. 102 p.

Dobell, Joseph Porter. The petrology and general geology, Kettle River-Toroda Creek District of northeastern Washington. D, 1955, University of Washington. 273 p.

Dobell, P. E. R. A study of dinoflagellate cysts from Recent marine sediments of British Columbia. M, 1978, University of British Columbia.

Dobervich, George. Areal geology of the Hulbert area, Cherokee and Wagoner counties, Oklahoma. M, 1952, University of Oklahoma. 93 p.

Dobey, Allen B. Reconnaissance geology of the Meyers Cover Point Quadrangle, Lemhi County, Idaho. M, 1972, University of Idaho. 60 p.

Dobkins, James E., Jr. Pebble shape development on Tahiti-Nui. M, 1968, University of Texas, Austin.

Doblas, Miguel M. S/C deformed rocks; the example of the Sierra de San Vicente sheared granitoids (Sierra de Gredos, Toledo, Spain). M, 1985, Harvard University.

Dobrick, Edward G. Gaseous transfer of compounds in aqueous solutions and its bearing on magmatic differentiation. M, 1938, University of Minnesota, Minneapolis. 28 p.

Dobrick, Edward G. The effect of high pressure on the solubility of quartz. M, 1937, University of Minnesota, Minneapolis. 19 p.

Dobrin, Milton B. A seismic investigation of Bikini Atoll; Part 1, Subsurface constitution of Bikini Atoll as indicated by a seismic refraction survey; Part 2, Submarine geology of Bikini Lagoon as indicated by dispersion of water-borne explosion waves. D, 1949, Columbia University, Teachers College.

Dobrin, Stefanie Z. The gamma ray radioactivity of pegmatites of New York City. M, 1949, Columbia University, Teachers College.

Dobrovolskis, Anthony Robert. The rotation of Venus; I, Atmospheric tides; II, Obliquity and evolution. D, 1979, California Institute of Technology. 250 p.

Dobrzykowski, David B. A homogeneous, mixed-oxide compositional model for the lower mantle and inferred mantle temperatures. M, 1976, Pennsylvania State University, University Park. 96 p.

Dobson, Benjamin Mark. The influence of stratigraphy on the behavior of expansive soils. M, 1978, Texas A&M University. 116 p.

Dobson, David Charles. Geology and geochemical evolution of the Lost River, Alaska, tin deposit. D, 1984, Stanford University. 262 p.

Dobson, Mary Lynn. Petrography, chemistry, and origin of Fe-Ti oxides in the Concord gabbro-syenite, North Carolina. M, 1987, University of Tennessee, Knoxville. 75 p.

Dobson, Patrick Foley. The petrogenesis of boninite; a field, petrologic, and geochemical study of the volcanic rocks of Chichi-Jima, Bonin Islands, Japan. D, 1986, Stanford University. 178 p.

Dobson, Patrick Foley. Volcanic stratigraphy and geochemistry of the Los Azufres geothermal center, Mexico. M, 1984, Stanford University. 67 p.

Docekal, Jerry. Earthquakes of the stable interior, with emphasis on the midcontinent. D, 1970, University of Nebraska, Lincoln.

Docekal, Jerry. Topography and geology of the Pennsylvanian surface in parts of Douglas, Sarpy, Cass, and Washington counties, Nebraska. M, 1959, University of Nebraska, Lincoln.

Docherty, Paul. Ray theoretical modeling, migration and inversion in two-and-one-half-dimensional layered acoustic media. D, 1987, Colorado School of Mines. 174 p.

Docka, Janet Anne. Mineral chemistry and thermometry of metamorphosed mafic rocks in the contact aure-

ole of the Kiglapait Intrusion. M, 1980, Northern Illinois University. 155 p.

Docka, Janet Anne. Petrology and origins of Mn-Fe metasediments in New England. D, 1985, Harvard University. 314 p.

Dockal, James Allan. Bedrock geochemical sampling near lead-zinc mineralization, Dubuque, Iowa. M, 1973, Iowa State University of Science and Technology.

Dockal, James Allan. Petrology and sedimentary facies of Redwall Limestone (Mississippian) of Uinta Mountains, Utah and Colorado. D, 1980, University of Iowa. 423 p.

Dockery, David T., III. Depositional systems in the upper Claiborne and lower Jackson groups (Eocene) of Mississippi. M, 1976, University of Mississippi. 110 p.

Dockery, Holly A. Cenozoic stratigraphy and extensional faulting in the southern Gabbs Valley Range, west-central Nevada. M, 1982, Rice University. 71 p.

Dockery, W. Lyle. Underground water resources of Horse Creek and Bear Creek valleys, southeastern Wyoming. M, 1939, University of Wyoming. 53 p.

Dockstader, David R. Fluid instability as a mechanism for the formation of "fingers" on sill periphery. M, 1973, University of Rochester.

Dockstader, David Roy. The mechanics of fingered sheet intrusions. D, 1978, University of Rochester. 167 p.

Dockter, Roger D. Stratigraphy and fusulinid paleontology of Permian exposures in the Diamond Springs Quadrangle, Nevada. M, 1978, San Jose State University. 128 p.

Dockum, Mark Steven. Greenschist-facies carbonates, eastern Coyote Mountains, western Imperial County, California. M, 1982, San Diego State University. 89 p.

Dod, B. D. Palynostratigraphy of the Rico Formation, southern San Juan County, Utah. M, 1970, East Texas State University.

Dodd, James Robert. Paleoecological implications of the mineralogy, structure, and strontium and magnesium contents of shells of the West Coast of the genus Mytilus. D, 1961, California Institute of Technology. 222 p.

Dodd, Kurt A. Geochemical and mineralogical exploration for galkin-type talc deposits, Vermont. M, 1986, Pennsylvania State University, University Park. 132 p.

Dodd, Philip H. A study of the controls of the crystallization of some minerals. M, 1950, Northwestern University.

Dodd, Robert Taylor, Jr. Precambrian geology of the Popolopen Lake Quadrangle, southeastern New York. D, 1962, Princeton University. 201 p.

Dodd, Stanton P. The mineralogy, petrology and geochemistry of the uranium-bearing vein deposits near Boulder, Montana, and their relationship to faulting and hot spring activity. M, 1981, Western Washington University. 71 p.

Dodds, Christopher James. A study of garnets and host rocks from the central Kootenay Lake area, southeastern British Columbia. M, 1966, University of Alberta. 154 p.

Dodds, R. Kenneth. Geology of the western half of the Svensen Quadrangle, Oregon. M, 1963, University of Oregon. 114 p.

Doden, Arnold Gabriel. Petrology and geochemistry of net-veined complexes in the Austurhorn and Vesturhorn intrusions, S.E. Iceland. M, 1988, Iowa State University of Science and Technology. 270 p.

Dodge, C. H. Artificial control of some interstitial water chemistry, Krause Lagoon, St. Croix, U.S. Virgin Islands. M, 1976, Northwestern University.

Dodge, Charles Fremont. The stratigraphy of the Woodbine Formation in the Arlington area, Tarrant County, Texas. M, 1952, Southern Methodist University. 25 p.

Dodge, Charles Fremont, III. Microstratigraphy of the Channel Sandstones of the Woodbine Formation (Upper Cretaceous, Gulf Series), Tarrant County, Texas. D, 1966, University of New Mexico. 181 p.

Dodge, Constance Nuss. An analysis and comparison of pebbles from the Chinle and Morrison formations, Arizona and New Mexico. M, 1973, University of Arizona.

Dodge, Franklin Charles Walter. A mineralogical study of the intrusive rocks of the Yosemite Valley area, California. D, 1963, Stanford University. 158 p.

Dodge, Harry W., Jr. Fusulinids from the Sublett Range, Idaho. M, 1956, University of Kansas. 61 p.

Dodge, Nelson B. Petrology, structure and age relations of the granites and syenites of the Childwold Quadrangle, New York. M, 1936, University of Rochester. 129 p.

Dodge, R. E. The natural growth records of reef building corals. D, 1978, Yale University. 253 p.

Dodge, Rebecca Lee. Evaluation of Skylab photographs for mapping Quaternary geologic features, West central Smoke Creek Desert, Nevada. M, 1978, Colorado School of Mines. 69 p.

Dodge, Rebecca Lee. Seismic and geomorphic history of the Black Rock fault zone, Northwest Nevada. D, 1982, Colorado School of Mines. 271 p.

Dodge, Sheridan Lee. Soil texture, glacial sediments, and woodlot species composition in Northeast Ingham County, Michigan. D, 1984, Michigan State University. 181 p.

Dodge, Theodore A. Optic angle determination with the universal stage. M, 1933, University of Wisconsin-Madison.

Dodge, Theodore A. The amphibolites of the Lead area, South Dakota. D, 1936, Harvard University.

Dodge, William Stuart. Flow direction studies of mid-Tertiary volcanic rocks, Southwest New Mexico and early Paleozoic volcanic rocks of the Carolina slate belt, North Carolina. M, 1979, North Carolina State University. 34 p.

Dodson, Alexander S. The role of electrical potential in ore deposition in the Timiskaming District, Ontario. D, 1938, University of Toronto.

Dodson, Edward Auld. Geology of Camp San Saba Ranch Branch area, McCulloch and Mason counties, Texas. M, 1941, University of Texas, Austin.

Dodson, Peter. A study of relative growth in some living and fossil reptiles. D, 1974, Yale University.

Dodson, Peter John. Sedimentology and taphonomy of the Oldman Formation (Campanian), Dinosaur Provincial Park, Alberta (Canada). M, 1970, University of Alberta. 115 p.

Dodson, Richard E. Continuous measurements of the natural remanent magnetization of sediments from Lake Michigan and Lake Tahoe. D, 1977, University of Pittsburgh. 228 p.

Dodson, Russell Leslie. Topographic and sedimentary characteristics of the Union Streamlined Plain and surrounding morainic areas. D, 1985, Michigan State University. 169 p.

Dodt, Matthew Edward. Strain analysis in the Bullbreen Group (Early Silurian), central western Spitsbergen, using antitaxial extension veins. M, 1986, University of Windsor. 157 p.

Doe, B. R. Geothermometry at the Balmat No. 2 Mine, New York by the FeS-ZnS system. M, 1956, University of Missouri, Rolla.

Doe, Bruce Roger. The distribution and composition of sulfide minerals at Balmat, New York. D, 1960, California Institute of Technology. 68 p.

Doe, Thomas William. Deformed cross-bedding from the Weber Formation (Pennsylvanian-Permian), northeastern Utah and northwestern Colorado. M, 1973, University of Wisconsin-Madison.

Doe, Thomas William. Geologic factors in siting tunnels for superconductive energy storage magnets. D, 1980, University of Wisconsin-Madison. 261 p.

Doebrich, Jeff L. Geology and hydrothermal alteration at the Mahd Adh Dhahab precious-metal deposit, Kingdom of Saudi Arabia. M, 1984, Colorado School of Mines. 148 p.

Doehler, Robert William. The ostracod Hollinella radlerae. M, 1953, University of Illinois, Urbana.

Doehler, Robert William. Variation in the mineral composition of underclays; its cause and significance. D, 1957, University of Illinois, Urbana. 83 p.

Doehne, Eric Ferguson. Geochemistry, petrology and diagenesis of Cretaceous-Tertiary boundary sediments from Zumaya, Spain. M, 1987, University of California, Davis. 158 p.

Doehring, Donald O. A procedure for discriminating between pediments and alluvial fans from topographic maps. D, 1968, University of Wyoming. 40 p.

Doehring, Donald O. Fire-flood sequences in the San Gabriel Mountains (California). M, 1965, Pomona College.

Doelger, Nancy Micklich. Depositional environments of the Nugget Sandstone, Red Canyon rim, Fremont County, Wyoming. M, 1981, University of Wyoming. 188 p.

Doell, Edward Charles. Correlation of the Campus and Orinda formations in the Berkeley Hills, California. M, 1934, University of California, Berkeley. 78 p.

Doell, Richard Rayman. Remanent magnetism in sediments. D, 1955, University of California, Berkeley. 156 p.

Doelling, Helmut Hans. Geology of the northern Lakeside Mountains and the Grassy Mountains and vicinity, Tooele and Box Elder counties, Utah. D, 1964, University of Utah. 463 p.

Doering, Effie A. The climates of Asia. M, 1926, University of Wisconsin-Madison.

Doermus, Eugene Henry. The sediments of the Manasquan River. M, 1952, University of Missouri, Columbia.

Doerner, David P. Lower Tertiary biostratigraphy of Santa Cruz Island. M, 1968, University of California, Santa Barbara.

Doerr, John E., Jr. An investigation of the characteristics of mud cracks. M, 1926, University of Wisconsin-Madison.

Doerr, John Timothy. The structural controls of the Vale-Rhinehart Buttes Complex, Vale KGRA, Malheur County, Oregon. M, 1986, Portland State University. 130 p.

Doesburg, James M. Effects of channel control structures of flood levels for a selected reach of the Mississippi River. M, 1986, University of Missouri, Columbia. 124 p.

Dogan, Ahmet U. Geology of the Mudurnu Valley area with special emphasis on the northern Anatolian Fault, in Bolu Province, Turkey. M, 1979, Ohio University, Athens.

Dogan, Ahmet Umran. Stratigraphy, petrology, depositional and post-depositional histories and their effects upon reservoir properties of the Parkman Formation of the Mesaverde Group, Powder River basin, Wyoming. D, 1984, University of Iowa. 265 p.

Dogan, Meral. Geology of the Dokurcun area, Mudurnu Valley, Bolu Province, Turkey. M, 1979, Ohio University, Athens.

Dogan, Nevzat. A subsurface study of middle Pennsylvanian rocks in east central Oklahoma. M, 1969, University of Tulsa. 67 p.

Doggett, Michael David. Canadian gold mining trends; an historical perspective. M, 1987, Queen's University. 163 p.

Doggett, Ruth Allen. Origin of the abnormally steep dips in the Niagaran reefs of the Chicago region. M, 1925, University of Chicago.

Doggett, Ruth Allen. The geology and petrology of the Columbia Falls Quadrangle, Maine. D, 1930, Harvard University.

Doh, Seong-Jae. Paleomagnetic record of late Quaternary sediments from Fargher Lake, Washington. M, 1982, Eastern Washington University. 25 p.

Doh, Seong-Jae. Rock-magnetic and paleomagnetic studies of marine sediments from the central and western North Pacific; stratigraphic and paleoceanographic implications. D, 1987, University of Rhode Island. 387 p.

Doheny, Edward John. Petrology and subsurface stratigraphy of the Detroit River Formation (Middle Devonian) in northern Indiana. D, 1967, Indiana University, Bloomington. 120 p.

Doherty, D. J. Ground surge deposits in eastern Idaho. M, 1976, Wayne State University. 114 p.

Doherty, J. T. Apatite and zircon fission track ages of White Mountain plutonic-volcanic series intrusives. M, 1975, Dartmouth College. 106 p.

Doherty, S. M. Structure independent seismic velocity estimation. D, 1975, Stanford University. 110 p.

Dohlen, Edward Lee Von *see* Von Dohlen, Edward Lee

Dohm, Christian F. Mississippi River delta sedimentation. M, 1936, Louisiana State University.

Dohm, Francis Paul. The Lower Mississippian of the northern Paint Creek Uplift, Kentucky. M, 1963, University of Kentucky. 106 p.

Dohrenwend, J. C. Marine geology of continental shelf between Point Lobos and Point Sur; California, a reconnaissance. M, 1971, Stanford University.

Dohrenwend, J. C. Plio-Pleistocene geology of the central Salinas Valley and adjacent uplands, Monterey County, California. D, 1975, Stanford University. 291 p.

Doig, D. J. The environment of deposition, petrography and diagenesis of the basal Belly River Sandstone of central Alberta. M, 1986, University of Calgary. 228 p.

Doig, R. Terrestrial heat flow. M, 1961, McGill University.

Doiron, Linda. Geomorphic processes active in the Southwest Louisiana Canal. M, 1974, Louisiana State University.

Dokka, R. K. Structural geology of a portion of western Newberry Mountains, San Bernardino County, California. M, 1976, University of Southern California. 71 p.

Dokka, Roy Karl. Late Cenozoic tectonics of the central Mojave Desert, California. D, 1980, University of Southern California.

Dokozoglu, Hilmi. A method of mining the Kumtepi lignite deposit of Turkey. M, 1951, University of Missouri, Rolla.

Dolan, James Francis. Paleogene sedimentary basin development in the eastern Greater Antilles; three studies in active-margin sedimentology. D, 1988, University of California, Santa Cruz. 247 p.

Dolan, John D. The petrology of the Whitewood Formation from the Ragged Top Mountain cores, Black Hills, South Dakota. M, 1977, South Dakota School of Mines & Technology.

Dolan, Robert. Relationships between nearshore processes and beach changes along the outer banks of North Carolina. D, 1965, Louisiana State University.

Dolan, William M. Location of geologic features by radio ground wave measurements in the mountainous terrain. M, 1957, University of Utah. 36 p.

Dolberg, David Michael. A duplex beneath a major overthrust plate in the Montana Disturbed Belt; surface and subsurface data. M, 1986, University of Montana. 57 p.

Dolcater, David Lee. Cation exchange selectivity and mineralogy of soil clay materials. D, 1970, University of Wisconsin-Madison. 123 p.

Dolch, William L. Permeability and absorbitivity of Indiana limestone coarse aggregates. D, 1961, Purdue University.

Dolch, William L. Solubility studies of Indiana limestones. M, 1949, Purdue University.

Dole, Hollis M. Petrography of Quaternary lake sediments of northern Lake County, Oregon. M, 1942, Oregon State University.

Dole, Robert Malcolm, Jr. Postglacial vegetation of northern Vermont. D, 1969, University of Michigan. 157 p.

Dolenc, Max Rudolph. Geochemical exploration in a geothermal area using sulfur gases in near-surface soils. M, 1988, University of Idaho. 48 p.

Dolence, Jerry D. The Pokegama quartzite in the Mesabi Range. M, 1961, University of Minnesota, Minneapolis. 72 p.

Doler, Robert Earl. Study and classification of Potlatch Corporation timberlands in Idaho for mineral and related higher-use potentials. M, 1979, University of Idaho. 247 p.

Dolfi, Robert Michael. Geochemical facies analysis and stratigraphic correlation of Miocene phosphorite beds in the Aurora District of North Carolina. M, 1983, North Carolina State University. 88 p.

Dolfi, Ronald U. Lower Ordovician (Canadian) lithofacies of New York and New England and their inferred depositional environments. M, 1982, Rensselaer Polytechnic Institute. 40 p.

Dolgoff, Abraham. Geology of the Ossining Quadrangle, New York. M, 1958, New York University.

Dolgoff, Abraham. The volcanic geology of the Pahranagat Range and certain adjacent areas, Lincoln County, southeastern Nevada. D, 1960, Rice University. 159 p.

Dolgoff, Anatole. The petrology and structure of the Siscowit Granite, Fairfield County, Connecticut, and Westchester County, New York. M, 1960, Miami University (Ohio). 57 p.

Doll, Barry. Ultrabasic body in the Santa Lucia Range, California. M, 1969, Stanford University.

Doll, Charles G. The Memphremagog Quadrangle and the southeastern portion of the Irasburg Quadrangle of Vermont. D, 1950, Columbia University, Teachers College.

Doll, William Eugene. Computer controlled laboratory studies in transient electromagnetic scale modeling. M, 1980, University of Wisconsin-Madison.

Doll, William Eugene. Seismic diffraction processing applied to data from Ashland County, Wisconsin. D, 1983, University of Wisconsin-Madison. 242 p.

Dollar, P. S. Geochemistry of formation waters, southwestern Ontario, Canada and southern Michigan, U.S.A.; implications for origin and evolution. M, 1988; University of Waterloo. 129 p.

Dollar, S. J. Zonation of reef corals off the Kona Coast of Hawaii. M, 1975, University of Hawaii. 183 p.

Dollase, Wayne A. Geology of the Tyson Lake area of Ontario. M, 1962, University of Wisconsin-Madison.

Dollase, Wayne A. The crystal structures of some phases of silica. D, 1966, Massachusetts Institute of Technology. 127 p.

Dolle, Yves. Etudes théoriques et applications des techniques de profilage magnétotellurique. M, 1978, Ecole Polytechnique.

Dollen, Bernard Halloran. Pre-glacial drainage of the northeastern United States. M, 1931, University of Rochester. 81 p.

Dollinger, Gerald Lee. A petrofabric study of olivine from the Seiad ultramafic complex (Jurassic?), Seiad valley, California. M, 1969, University of Wisconsin-Madison.

Dollinger, Gerald Lee. Development of preferred orientation in uniaxially deformed olivine aggregates. D, 1978, University of California, Los Angeles. 208 p.

Dollison, Roberts S. Structural history and stratigraphy of the Emma-Triple "N" oil field, Andrews County, Texas. M, 1959, University of Wisconsin-Madison.

Dolliver, Claire Vincent. Late Tertiary and Quaternary geomorphic history of Kyle canyon, Spring mountains, (southwestern) Nevada. D, 1968, Pennsylvania State University, University Park. 140 p.

Dolliver, Claire Vincent. The geomorphology of the west-central Florida Peninsula. M, 1965, University of Florida. 54 p.

Dolliver, Paul N. Cenozoic evolution of the Canadian River basin. M, 1982, Baylor University. 278 p.

Dollof, John H. Upper Cretaceous stratigraphy in Miller County, Arkansas. M, 1951, University of Minnesota, Minneapolis. 42 p.

Dolloff, Mary Helen. Stratigraphy and paleontology of the Barren Hills, southeastern Cochise County, Arizona. M, 1975, University of Arizona.

Dolloff, Norman H. The pegmatite minerals of Standish, Maine. M, 1936, Columbia University, Teachers College.

Dolly, Edward Dawson. Palynology of the Bevier Coal (Pennsylvanian) of Oklahoma. M, 1965, University of Oklahoma. 115 p.

Dolly, Edward Dawson. Stratigraphic, structural and geomorphological factors controlling oil accumulation in upper Cambrian strata of central Ohio. D, 1969, University of Oklahoma. 203 p.

Dolmage, Victor. The geology of the Telkwa River District, British Columbia. D, 1917, Massachusetts Institute of Technology. 224 p.

Dolman, Jan Dirk. Genesis, morphology, and classification of organic soils in the Tidewater region of North Carolina. D, 1967, North Carolina State University. 147 p.

Dolman, Phil B. Efficient field methods for the petroleum geologist. M, 1920, University of Missouri, Rolla.

Dolphin, Dale Robert. Stratigraphy and paleoecology of an upper Devonian carbonate bank, Saskatchewan river crossing, Alberta (Canada). M, 1969, University of Calgary. 95 p.

Dolsen, Charles Philip. An investigation of some relationships between pyrite and uranium in the Chattanooga Shale. M, 1957, Pennsylvania State University, University Park. 104 p.

Dolson, John. Depositional environments and petrography of the Dakota Group. M, 1981, Colorado State University. 342 p.

Dolton, Gordon Lee. The geology of the southwest portion of the San Joaquin Hills, Orange County, California. M, 1952, Pomona College.

Domack, Eugene Walter. Facies of late Pleistocene glacial marine sediments on Whidbey Island, Washington. D, 1982, Rice University. 393 p.

Domack, Eugene Walter. Glacial marine geology of the George V.-Adelie continental shelf, East Antarctica. M, 1980, Rice University. 142 p.

Domagala, Mark A. Lithofacies and paleoenvironments of a peritidal carbonate setting; Lockport Formation (Middle Silurian), east-central New York. M, 1982, SUNY, College at Fredonia. 94 p.

Domagalski, Joseph Leo. Trace metal and organic geochemistry of closed basin lakes; volumes 1 and 2. D, 1988, The Johns Hopkins University. 588 p.

Doman, Robert Charles. Causes of optical scatter in plagioclase feldspar. D, 1961, University of Wisconsin-Madison. 95 p.

Doman, Robert Charles. Geology of House Springs area. M, 1955, Washington University. 49 p.

Dombkowski, Francis. The Silurian environments and importance of the Maumee Quarry (Ohio) and Capac gas field (Michigan) carbonates. M, 1978, Wayne State University.

Dombroski, Daniel R., Jr. A geological and geophysical investigation of concealed contacts near an abandoned barite mine, Hopewell, New Jersey. M, 1980, Rutgers, The State University, New Brunswick. 33 p.

Dombrouski, Richard Paul, Jr. Engineering geology of the Friar Tuck strip mine area, Greene-Sullivan counties, Indiana, to provide basic information for surface reclamation. M, 1985, Purdue University. 150 p.

Dombrowski, Anna. Carbonate petrology and conodont biostratigraphy of Devonian western and transitional assemblage rocks in the upper plate of the Roberts Mountains Thrust, North-central Nevada. D, 1980, University of Iowa. 243 p.

Dombrowski, John. Variation in grain-size distribution on two Washington beaches. M, 1976, Washington State University. 108 p.

Dombrowski, Thomas. Abundance, distribution, and origin of thorium in the Georgia kaolins. M, 1982, Indiana University, Bloomington. 85 p.

Domenick, Michael A. An isotopic study of crustal evolution; two examples from the Sierra Nevada and Southeast New York. M, 1981, University of Rochester. 60 p.

Domenico, Patrick A. Geology and groundwater hydrology of the (Upper Cretaceous) Edmonton Formation in central Alberta, Canada. M, 1963, Syracuse University.

Domenico, Patrick Anthony. Valuation of a groundwater supply for management and development. D, 1967, University of Nevada. 74 p.

Domenico, Samuel N. Spherical wave motion and dynamic strain measurements. D, 1951, Colorado School of Mines. 155 p.

Dominey, Jerry. Benthonic foraminiferal paleoecology of the Yorktown Formation, Edgecombe County, North Carolina. M, 1983, Tulane University. 200 p.

Dominguez, Jose Maria Landim. Quaternary sealevel changes and the depositional architecture of beach-ridge strandplains along the east coast of Brazil. D, 1987, University of Miami. 327 p.

Dominguez-Vargas, Guillermo Cruz. A simulation study of detergent flooding in a vertical cross section. D, 1983, University of Southern California.

Dominic, D. F. Evaluation of bedload sediment transport models. M, 1983, SUNY at Binghamton. 89 p.

Dominic, David Francis. Quantitative interpretation of Upper Pennsylvanian-Lower Permian sandstones of northern West Virginia using mathematical models of sedimentation. D, 1988, West Virginia University. 162 p.

Dominick, Wayne Paul. Mechanical model analysis of the interaction of soil and framed structures. D, 1969, New Mexico State University, Las Cruces. 156 p.

Domino, Grant A. A marine paleoecological simulation on the CDC 6400 computer. M, 1974, Lehigh University.

Domning, Daryl Paul. Sirenian evolution in the North Pacific Ocean. M, 1970, University of California, Berkeley. 176 p.

Domning, Daryl Paul. Systematics, morphology, and evolution of North Pacific sirenians. D, 1975, University of California, Berkeley. 398 p.

Domoracki, William Joseph. Integrated geophysical survey of the Golden Thrust north of Golden, Colorado. M, 1986, Colorado School of Mines. 134 p.

Domurat, G. W. A deterministic model of the vertical component of sediment motion in a turbulent fluid. M, 1977, Old Dominion University. 178 p.

Donaghy, Thomas James. The petrology of the Thekulthili Lake area, Northwest Territories. M, 1977, University of Alberta. 152 p.

Donahoe, Rona Jean. An experimental investigation of some physicochemical controls on zeolite formation; implications for authigenesis. D, 1984, Stanford University. 174 p.

Donahue, Jack David. Depositional environments of the Salem Limestone (Mississippian, Meramec) of southcentral Indiana. D, 1967, Columbia University. 109 p.

Donahue, Jessie Gilchrist. Diatoms as Quaternary biostratigraphic and paleoclimatic indicators in high latitudes on the Pacific Ocean. D, 1970, Columbia University. 230 p.

Donahue, John C. The geology and petrochemistry of the Patino-Lemoine Township, Quebec. M, 1982, University of Georgia.

Donahue, John Joseph, Jr. An analysis of drainage intensity in western New York State. D, 1970, Syracuse University. 233 p.

Donahue, John Michael. The driving force of plate tectonics evaluated in spherical coordinates. M, 1985, Texas A&M University. 42 p.

Donald, Peter G. Geology of the Fresnal Peak area, Baboquivari Mountains, Arizona. M, 1959, University of Arizona.

Donald, Robert B. Mac *see* Mac Donald, Robert B.

Donald, Roberta L. Sedimentology of the Mist Mountain Formation, in the Fording River area, southeastern Canadian Rocky Mountains. M, 1984, University of British Columbia. 180 p.

Donaldson, A. C. A micropaleontological study of the Vicksburg Formation from five Texas wells. M, 1953, University of Massachusetts.

Donaldson, Alan Chase. Stratigraphy of Lower Ordovician Stonehenge and Larke formations in central Pennsylvania. D, 1959, Pennsylvania State University, University Park. 404 p.

Donaldson, Francis, Jr. Ore deposits of the State of Chihuahua, Mexico. M, 1940, Columbia University, Teachers College.

Donaldson, J. Allen. Geology of the Marion Lake area, Quebec-Labrador. D, 1960, The Johns Hopkins University.

Donaldson, James C. Stratigraphic relationships of Lower Permian rocks, Johnson County, Wyoming. M, 1982, Colorado School of Mines. 116 p.

Donaldson, John William. A paleomagnetic study of the Permian basalts at Las Delicias, Coahuila, Mexico. M, 1970, Rice University. 50 p.

Donaldson, Marybeth. Two models of the influence of surfactants on fracture strength tested in the shale DTAB system. M, 1977, University of North Carolina, Chapel Hill. 67 p.

Donaldson, William Allen, III. Carbonate petrology and diagenesis of the Pennsylvanian Collier Limestone, Hitchland Field, Hansford County, Texas. M, 1983, West Texas State University. 101 p.

Donath, Fred Arthur. Basin-range structure of south-central Oregon. D, 1958, Stanford University. 197 p.

Donato, Mary Margaret. Geology and petrology of a portion of the Ashland Pluton, Jackson County, Oregon. M, 1975, University of Oregon. 89 p.

Donato, Mary Margaret. Metamorphic and structural evolution of an ophiolitic tectonic melange, Marble Mountains, Northern California. D, 1985, Stanford University. 313 p.

Donchin, Jason H. Stratigraphy and sedimentary environments of the Miocene-Pliocene Verde Formation in the southeastern Verde Valley, Yavapai County, Arizona. M, 1983, Northern Arizona University. 182 p.

Dondanville, Richard Fred. The geology of part of northwestern Glenn County, California. M, 1958, University of California, Berkeley. 60 p.

Donegan, David P. Modern and ancient marine rhythmites from the Sea of Cortez and California continental borderland; a sedimentological study. M, 1982, Oregon State University. 123 p.

Donelick, Raymond A. Mesozoic-Cenozoic thermal evolution of the Atlin Terrane, Whitehorse Trough, and Coast plutonic complex from Atlin, British Columbia to Haines, Alaska as revealed by fission track geothermometry techniques. M, 1986, Rensselaer Polytechnic Institute. 167 p.

Donelick, Raymond Allen. Etchable fission track length reduction in apatite; experimental observations, theory and geological applications. D, 1988, Rensselaer Polytechnic Institute. 414 p.

Donellan, Monica Sue. A mineralogical study of the Pleistocene terraces along the Mississippi and Red rivers. M, 1972, Louisiana State University.

Doner, Harvey Ervin. Solubility of calcium carbonate precipitated in clay suspensions and aqueous solutions. D, 1967, University of California, Riverside. 84 p.

Dones, Henry Cling, Jr. Dynamical and photometric studies of Saturn's rings. D, 1987, University of California, Berkeley. 573 p.

Doney, Hugh Holt. Geology of the Cebolla Quadrangle, Rio Arriba County, New Mexico. D, 1966, University of Texas, Austin. 322 p.

Doney, Hugh Holt. Pelecypods of the Conemaugh Series from Pittsburgh, Pennsylvania, and neighboring areas. M, 1954, University of Pittsburgh.

Doney, Kim Elizabeth. The geology of Wilcox Sands (Eocene) of the East Hamel Field, Colorado County, Texas. M, 1984, Northeast Louisiana University. 85 p.

Dong, Allen. Dispersibility of soils and elemental composition of soils, sediments and dust and dirt from the Menominee River basin, Wisconsin. D, 1980, University of Wisconsin-Madison. 158 p.

Donica, David R. The geology of the Hartshorne coals (Desmoinesian) in parts of the Heavener 15' Quadrangle, Le Flore County, Oklahoma. M, 1978, University of Oklahoma. 128 p.

Donk, Jan Van *see* Van Donk, Jan

Donley, Michael William. The pumice landscape of Guatemala; a study of basin morphology and landform in the upland Tropics. D, 1971, University of Oregon. 197 p.

Donlon, Thomas J. The development of terrestrial analogs for proposed Martian fluvial features. M, 1978, University of Rhode Island.

Donn, Thomas Frank. Erosion of a rocky carbonate coastline, Andros Island, Bahamas. M, 1986, Miami University (Ohio). 91 p.

Donn, William L. Studies of frontal, cyclonic and hurricane microseisms generated in the western North Atlantic. D, 1951, Columbia University, Teachers College.

Donnan, Gary Thomas. Stratigraphy and structural geology of the northern part of the northern Animas Mountains, Hidalgo County, New Mexico. M, 1987, New Mexico State University, Las Cruces. 121 p.

Donnay, Joseph D. H. The Engels copper deposit, Plumas County, California. D, 1929, Stanford University. 133 p.

Donnelly, Andrew Charles Alexander. Meteoric water penetration in the Frio Formation, Texas Gulf Coast. M, 1988, University of Texas, Austin. 137 p.

Donnelly, Ann Tipton. The Refugian Stage of the California Tertiary; foraminiferal zonation, geologic history, and correlations with the Pacific Northwest. D, 1975, University of California, Santa Barbara. 316 p.

Donnelly, Brian James. Structural geology of the Nancy Creek area, east flank of the Kettle Dome, Ferry County, Washington. M, 1978, Washington State University. 251 p.

Donnelly, Clarence William. Post Maquoketa ground water resources of the Iowa City, Iowa area. M, 1937, University of Iowa. 95 p.

Donnelly, J. M. Geochronology and evolution of the Clear Lake volcanic field. D, 1977, University of California, Berkeley. 52 p.

Donnelly, M. F. Geology of the Sierra del Pinacate volcanic field, northern Sonora, Mexico, and southern Arizona, U.S.A. D, 1974, Stanford University. 961 p.

Donnelly, Maurice. Geology and mineral deposits of the Julian District, San Diego County, California. D, 1935, California Institute of Technology. 117 p.

Donnelly, Maurice. Preliminary report on the geology of the Julian region, California. M, 1933, California Institute of Technology. 41 p.

Donnelly, Maurice. The lithia pegmatites of Pala and Mesa Grande, San Diego County, California. D, 1935, California Institute of Technology. 117 p.

Donnelly, Michael E. Petrology and structure of a portion of the Precambrian Mullen Creek metaigneous mafic complex, Medicine Bow Mountains, Wyoming. M, 1979, Colorado State University. 476 p.

Donnelly, Thomas Wallace. Geology of Saint Thomas and Saint John, Virgin Islands. D, 1959, Princeton University. 262 p.

Donner, Henry Frederick. Geology of the McCoy area, Eagle and Routt counties, Colorado. D, 1936, University of Michigan.

Donofrio, Chris Joseph. Bedforms of the North Loup River, Nebraska; a braided stream. M, 1982, University of Nebraska, Lincoln. 212 p.

Donofrio, Richard R. The magnetic environment of tektites. D, 1977, University of Oklahoma. 183 p.

Donoghue, Joseph Francis. Estuarine sediment transport and Holocene depositional history, upper Chesapeake Bay, Maryland. D, 1981, University of Southern California.

Donohoe, H. V., Jr. Analysis of structure in the St. George area, Charlotte County, New Brunswick. D, 1978, University of New Brunswick.

Donohue, John Joseph. A laboratory and field investigation of the sediments of Stover Cove, South Harpswell, Maine; a correlation of Mya arenaria Linnaeus with its geological environment; 2 volumes. M, 1949, University of Maine.

Donohue, John Joseph. Littoral sediments of Maine. D, 1951, Rutgers, The State University, New Brunswick. 325 p.

Donovan, Arthur Dean. Stratigraphy and sedimentology of the Upper Cretaceous Providence Formation (western Georgia and eastern Alabama). D, 1985, Colorado School of Mines. 235 p.

Donovan, Jack H. Intertonguing of Green River and Wasatch formations in part of Sublette and Lincoln counties, Wyoming. M, 1950, University of Utah. 47 p.

Donovan, Joe D. Subsurface geology of northern Hardin County, Texas. M, 1957, University of Oklahoma. 41 p.

Donovan, John Francis. Geology of the Woman River iron range, District of Sudbury, Ontario, Canada. D, 1963, Cornell University. 132 p.

Donovan, Kevin. Quaternary stratigraphy of the east side of the Cuyahoga Valley between State Route 303 and State Route 82. M, 1983, University of Akron. 115 p.

Donovan, Peter R. The geology of the Little Falls area, Boise County, Idaho. M, 1962, Colorado School of Mines. 111 p.

Donovan, Robert C. Taphonomy of the Meekoceras beds, Thaynes Formation (Lower Triassic), Idaho, Utah and Nevada. M, 1986, University of Wyoming. 100 p.

Donovan, Sean. A lineament evaluation and structural study of northeastern Washington using Landsat and SLAR imagery and aeromagnetic maps. M, 1983, Eastern Washington University. 188 p.

Donovan, Terrence J. An hypothesis to explain abnormally high porosities of the sedimentary rocks of the Santa Fe Springs Anticline, California. M, 1963, University of California, Riverside. 65 p.

Donovan, Terrence John. Surface mineralogical and chemical evidence for buried hydrocarbons, Cement Field, Oklahoma. D, 1972, University of California, Los Angeles.

Doolan, Barry Lee. The structure and metamorphism of the Santa Marta area, Colombia, South America. D, 1971, SUNY at Binghamton. 200 p.

Dooley, Duane. The geology of the Onion Creek Quadrangle, Ellis County, Texas. M, 1960, Southern Methodist University. 15 p.

Dooley, John J. An investigation of near-bottom currents in the Monterey (California) submarine canyon. M, 1968, United States Naval Academy.

Dooley, Peter Comstock. Laboratory studies of plate motions on a thin fluid layer undergoing Rayleigh-Benard Convection. M, 1972, University of Michigan.

Dooley, Robert E. K-Ar relationships in dolerite dikes of Georgia. M, 1977, Georgia Institute of Technology. 185 p.

Dooley, Robert Ervin. Paleomagnetism and geochronology of igneous rocks in the South Carolina Piedmont. D, 1982, University of South Carolina. 193 p.

Dooley, Robert Lee. Geology and land use considerations in the vicinity of the Green Valley Fault, Solano County, California. M, 1973, University of California, Davis. 47 p.

Doolittle, Russell C. The theory, construction, and field use of a direct current potentiometer for measuring Earth resistivity. M, 1940, California Institute of Technology. 107 p.

Doorn, Stacy Seaman. Theoretical energetics of the formation of iron nickel alloy and magnetite in olivine. M, 1988, Arizona State University. 116 p.

Doose, Paul Robin. The bacterial production of methane in marine sediments. D, 1980, University of California, Los Angeles. 262 p.

Doraibabu, Peethambaram. Trace base metals - petrography - rock alteration of the productive Tres Hermanas Stock, Luna County, New Mexico. D, 1971, University of Missouri, Rolla.

Dorais, Michael John. The geology and petrology of the Shelly Lake Pluton, central Klamath Mountains, California. M, 1983, University of Oregon. 147 p.

Dorais, Michael John. The mafic enclaves of the Dinkey Creek Granodiorite and the Carpenter Ridge Tuff; a mineralogical, textural, and geochemical study of their origins with implications for the generation of silicic batholiths. D, 1987, University of Georgia. 197 p.

Doré, Guy. Microzonage séismique de la région de Québec. M, 1985, Universite Laval. 77 p.

Dore, James Ernest. The use of olivine in earthenware and semiporcelain bodies. M, 1951, University of Washington. 69 p.

Doremus, Dale M. Groundwater circulation and water quality associated with the Madison Aquifer in the northeastern Bighorn Basin, Wyoming. M, 1986, University of Wyoming. 81 p.

Doremus, Eugene Henry. Sediments of the Manasquan River. M, 1952, University of Missouri, Columbia.

Dorf, Erling. Pliocene floras of California. D, 1930, University of Chicago. 246 p.

Dorfman, M. H. Geologic habitats of geothermal energy and methods of exploration. D, 1975, University of Texas, Austin. 244 p.

Dorheim, Fred H. Petrography of selected limestone aggregates. M, 1950, Iowa State University of Science and Technology.

Doria-Medina, Jorge. A study of channel, bank and bottom gravels along Mountain Fork Creek, Crawford County, Arkansas. M, 1962, University of Arkansas, Fayetteville.

Dorich, Roderick Alan. Algal availability of phosphorus in sediments derived from cropland. D, 1981, Purdue University. 230 p.

Dorman, C. E. The southern Monterey bay (California) littoral cell; a preliminary sediment budget study. M, 1968, United States Naval Academy.

Dorman, Henry J. Structure of the Priest Granite, Masano Mountains, New Mexico. M, 1951, Northwestern University.

Dorman, Henry J.; Ewing, M. and Oliver, J. Study of shear-velocity distribution in the upper mantle by mantle Rayleigh waves. D, 1962, Columbia University, Teachers College.

Dorman, James H. The megafauna of the Sulphur Springs Formation of Missouri. M, 1956, University of Missouri, Columbia.

Dorman, James Hubert. Megafauna of the Sulphur Springs Formation of Missouri. M, 1958, University of Missouri, Columbia.

Dorman, Leroy Myron. Anelasticity and the spectra of body waves. M, 1969, University of Wisconsin-Madison.

Dorman, Leroy Myron. The theory of the determination of the Earth's isostatic response to a concentrate load. D, 1970, University of Wisconsin-Madison. 34 p.

Dorn, Geoffrey Alan. Numerical resistivity modeling of a downhole electrode with application to geothermal reservoir mapping. M, 1978, University of New Mexico. 133 p.

Dorn, Geoffrey Alan. Radiation impedance and radiation pattern of torsionally vibrating seismic sources. D, 1980, University of California, Berkeley. 245 p.

Dorn, Mary S. A critical analysis of continental drift. M, 1970, Virginia State University. 34 p.

Dorn, Ronald I. Rock varnish; a key to the Quaternary period in western North American deserts. D, 1985, University of California, Los Angeles. 429 p.

Dorn, Thomas Franz. The development and testing of a portable single-trace seismic recording system for crustal studies. M, 1974, University of Manitoba.

Dorn, Wolfgang Ulrich. Late Miocene hiatuses and related events in the central Equatorial Pacific; their depositional imprint and paleoceanographic implications. D, 1987, University of Hawaii. 179 p.

Dornbach, John Ellis. An experiment in the delineation of types of framing areas in the Big Muddy River basin of southern Illinois. M, 1950, Washington University. 126 p.

Dornian, Nicholas. A study of the sulphides and oxides of the nickel-copper deposits of Lynn lake, Manitoba. M, 1951, University of Manitoba.

Dorobek, Steven Louis. Stratigraphy, sedimentology, and diagenetic history of the Siluro-Devonian Helderberg Group, Central Appalachians. D, 1984, Virginia Polytechnic Institute and State University. 252 p.

Doroshenko, Jerry. Structures and textures of aggrading stream deposits near Cincinnati, Ohio. M, 1948, University of Cincinnati. 33 p.

Dorr, Andre. Magnetite deposits in the northern part of the Dore Lake complex (Pε), Chibougamau District, Quebec (Canada). M, 1969, McGill University. 57 p.

Dorr, James B. The stratigraphic relations of the Trenton and Utica formations. M, 1923, Columbia University, Teachers College.

Dorr, John Adam, Jr. and Wheeler, Walter H. Geology of a part of the Ruby Basin, Madison County, Montana. M, 1948, University of Michigan.

Dorr, John Adam, Jr. Paleocene and early Eocene strata and vertebrate paleontology of the Hoback Basin, central western Wyoming. D, 1951, University of Michigan.

Dorr, Lawrence L. The Upper Devonian and Lower Mississippian of St. Clair and northeastern Macomb counties. M, 1981, Wayne State University. 108 p.

Dorrell, Carter Victor. The Lykins Formation of eastern Colorado. D, 1940, University of Colorado.

Dorrell, Carter Victor. The use of mechanical analyses in the correlation of Cretaceous bentonites and metabentonites in eastern Colorado. M, 1934, University of Colorado.

Dorrler, Richard. Variation of the chemical contaminants for selected water quality changes at induced infiltration sites along the Naugatuck river in Connecticut. M, 1969, University of Connecticut. 67 p.

Dorsett, Russell K. Petrographic and paleoenvironmental analyses of core samples from the Wyodak Coal, Gillette, Wyoming. M, 1984, South Dakota School of Mines & Technology.

Dorsey, Anna Laura. A faunal study of the foraminifera from the Chesapeake Group (Miocene) of southern Maryland. D, 1940, Bryn Mawr College. 154 p.

Dorsey, Anne E. Elemental distributions and relationships between the Pewee Coal and associated sediments, Wartburg Basin, Tennessee. M, 1982, University of Tennessee, Knoxville. 84 p.

Dorsey, George E. The stratigraphy and structure of the Triassic System of Maryland, with a discussion of the origin and climatic significance of red beds. D, 1918, The Johns Hopkins University.

Dorsey, Michael T. Stratigraphy of the Topeka Limestone in eastern Kansas. M, 1978, Wichita State University. 98 p.

Dorsey, Ridgeley E. Geology of the Marble Canyon area, Waucoba Springs Quadrangle, Inyo County, California. M, 1960, University of California, Los Angeles.

Dort, Wakefield, Jr. Glaciation of the Coeur d'Alene District, Idaho. D, 1955, Stanford University. 55 p.

Dort, Wakefield, Jr. The geology of a portion of eastern Ventura Basin, California. M, 1948, California Institute of Technology. 38 p.

dos Anjos, Sylvia Maria Couto *see* Couto dos Anjos, Sylvia Maria

dos Anjos, Sylvia Maria Couto. Diagenetic evolution of marine shales of the Campos Formation (Cretaceous/Tertiary), Campos Basin, offshore Rio de Janeiro, SE Brazil. M, 1984, University of Illinois, Urbana. 265 p.

Dos Santos, Edson R. Newbury volcanics (Massachusetts). M, 1960, Massachusetts Institute of Technology. 59 p.

Dosch, Earl Fuller. The Las Tablas fault zone and the associated rocks, California. M, 1932, University of California, Berkeley. 42 p.

Doser, Diane Irene. Earthquake recurrence rates from seismic moment rates in Utah. M, 1980, University of Utah. 163 p.

Doser, Diane Irene. Source parameters and faulting processes of the August 1959 Hebgen Lake, Montana, earthquake sequence. D, 1984, University of Utah. 165 p.

Doss, Daniel L. Geology of the southern half of the Mt. Pleasant Quadrangle, Izard and Independence counties, Arkansas, with special emphasis on the high-calcium limestones. M, 1976, Northeast Louisiana University.

Dosso, Harry William. Analytical and analogue methods of studying electromagnetic variations at the Earth's surface. D, 1967, University of British Columbia.

Dosso, Laure. The nature of the Precambrian subcontinental mantle; isotopic study (Sr, Pb, Nd) of the Keweenawan volcanism of the north shore of Lake Superior. D, 1984, University of Minnesota, Minneapolis. 221 p.

Dostal, Jaroslav. Geochemistry and petrology of the Loon Lake Pluton, Ontario. D, 1974, McMaster University. 328 p.

Doten, Robert K. Pegmatites of Ohio City, Colorado and the origin of complex pegmatites. D, 1936, Princeton University.

Dotsey, Pete. Sedimentological analysis of some coarse-grained clastic units in the Ouachita Mountains, Arkansas. M, 1983, Stephen F. Austin State University. 85 p.

Dotson, James M. A method for the continuous determination of coal concentrations in a fluidized mixture of coal and air. M, 1949, West Virginia University.

Dotson, Kirk Wayne. Dynamic impedances of soil layers and piles. D, 1988, Rice University. 216 p.

Dott, Robert H., Jr. Pennsylvanian stratigraphy of Elko and northern Diamond ranges, northeastern Nevada. D, 1955, Columbia University, Teachers College.

Dott, Robert Henry, Jr. and Curtis, Alan Deane. Geology of the northern half of the Coldwater Quadrangle, Madison County, Missouri. M, 1950, University of Michigan.

Dotter, Jay Albert. Prairie Mountain Lakes area, Southeast Skagit County, Washington; structural geology, sedimentary petrography, and magnetics. M, 1978, Oregon State University. 105 p.

Doty, Robert L. Structural geology of northwestern Gray Mountain, Coconino County, Arizona. M, 1982, Northern Arizona University. 120 p.

Doty, Robert W. Petrology and origin of the Warrensburg Sandstone of western Missouri. M, 1960, University of Missouri, Columbia.

Doucet, Daniel. Prospection minière par méthode électrique à l'aide de trous de forage. M, 1977, Ecole Polytechnique.

Doucet, J. A. A possible correlation of Mississippian (Windsor) limestones in insoluable residue, heavy mineral and thin section analyses. M, 1960, St. Francis Xavier University.

Doucet, Pierre. The petrology and geochemistry of the Middle River area, Cape Breton Island, Nova Scotia. M, 1983, Dalhousie University. 339 p.

Doucette, John. The geology of the Copper Chief Prospect, Mineral County, Nevada. M, 1981, Oregon State University. 173 p.

Doucette, William Henry, Jr. Soil survey reliability for intensive land management. D, 1983, North Carolina State University. 205 p.

Dougan, Thomas William, Jr. Origin and metamorphism of Imataca and Los Indios gneisses, Precambrian rocks of the Los Indios-El Pilar area, State of Bolivar, Venezuela. D, 1967, Princeton University. 283 p.

Dougherty, David Emery. On equivalent porous medium modeling of transport in fractured porous reservoirs. D, 1985, Princeton University. 261 p.

Dougherty, Dortha Lea. Study of a possible Hawaiian mantle plume model using shallow reflections of earthquake-generated body waves. M, 1979, University of Oklahoma. 149 p.

Dougherty, J. F. A new Miocene mammalian fauna from the Caliente Mountains, California. M, 1939, California Institute of Technology. 50 p.

Dougherty, M. T. The subsurface stratigraphy of the upper Code Shale; southeastern flank, Powder River basin, Wyoming. M, 1961, University of Pittsburgh.

Dougherty, Percy H. Thermogeographic analysis of groundwater diffusion in the Delaware River Raritan-Magothy Formation interface in southern New Jersey. D, 1980, Boston University. 197 p.

Dougherty, Sean M. Optimal design of autocorrelation functions. M, 1986, University of Calgary. 69 p.

Doughri, Abdoolrhman K. Geology and geophysics of the carbonatite area, near Jabal Uwaynat, south-east Libya. M, 1976, Ohio University, Athens. 54 p.

Doughtery, Daniel. The stratigraphy, structure and metamorphic history of the northern half of the Blairsville Quadrangle, Georgia-North Carolina. M, 1977, University of Georgia.

Doughty, John Daniel. Evaluation of borehole acoustic wave logs for the location and characterization of fractures in basalt. M, 1987, Washington State University. 91 p.

Doughty, Roger Keith. The accumulation of helium in natural gas. M, 1972, University of Oklahoma. 64 p.

Douglas, A. V. Past air-sea interactions over the eastern North Pacific Ocean as revealed by tree-ring data. D, 1976, University of Arizona. 209 p.

Douglas, Arthur Gordon. Soil deformation rates and activation energies. D, 1969, Michigan State University. 78 p.

Douglas, Arthur Vern. Past-air-sea interactions off Southern California as revealed by coastal tree-ring chronologies. M, 1973, University of Arizona.

Douglas, Bruce James. Structural and stratigraphic analysis of a metasedimentary inlier within the Coast Plutonic Complex, British Columbia, Canada. D, 1983, Princeton University. 401 p.

Douglas, Charles H. Fine and medium grained alluvial deposits along the Missouri River between Yankton, South Dakota, and Sioux City, Iowa. M, 1959, University of South Dakota. 65 p.

Douglas, Charlton. Geology of part of the Ironside Mountain Quadrangle, Northern California Klamath Mountains. D, 1979, University of California, Santa Barbara.

Douglas, Chris W. Subsurface geology of the Chachaoula area. M, 1978, University of New Orleans.

Douglas, Clyde Lee, Jr. Silicic acid and oxidizable carbon movement in a Walla Walla silt loam as related to long-term residue and nitrogen applications. D, 1984, Oregon State University. 75 p.

Douglas, Dean Alan. Lithogeochemistry as a guide to volcanogenic massive sulfide deposits. M, 1982, University of Arizona. 84 p.

Douglas, Debra Rena. Factors affecting porosity and permeability variation in the Oriskany Sandstone of west-central New York. M, 1984, Syracuse University. 114 p.

Douglas, G. B. Petrography and petrochemistry of scapolite in the Grenville of southern Ontario. M, 1973, McMaster University. 188 p.

Douglas, Gregory Scott. The geochemistry of copper and chromium organic complexes in Narragansett Bay interstitial waters. D, 1986, University of Rhode Island. 220 p.

Douglas, Hugh. Geology of the Namiquipa Mine, Namiquipa, Chihuahua, Mexico. M, 1951, Columbia University, Teachers College.

Douglas, Ian Hedberg. Geology and gold-silver mineralization of the Tecoma District, Elko County, Nevada, and Box Elder County, Utah. M, 1984, Stanford University. 147 p.

Douglas, J. G. Some Miocene mollusks from northwestern Venezuela. D, 1928, The Johns Hopkins University.

Douglas, J. M. Mineralogy of contrasting mineralization at Gaspe, Quebec. M, 1941, McGill University.

Douglas, Jesse King. Geophysical investigation of the Montana Lineament. M, 1973, University of Montana. 75 p.

Douglas, John Leslie. Geochemistry of the Cambrian manganese deposits of eastern Newfoundland. D, 1983, Memorial University of Newfoundland. 305 p.

Douglas, Nancy Browning. Satellite laser ranging and geologic constraints on plate tectonic motion. M, 1988, University of Miami. 186 p.

Douglas, Richard F. A paragenetic study of copper-zinc ore body in Ontario, Canada. M, 1960, Columbia University, Teachers College.

Douglas, Richard Franklin. Microscopy and origin of Mississippi Valley type zinc-lead deposits in the United States. D, 1966, Columbia University, Teachers College. 99 p.

Douglas, Robert Guy. Upper Cretaceous planktonic foraminiferal biostratigraphy of the western Sacramento Valley, California. D, 1966, University of California, Los Angeles. 508 p.

Douglas, Robert J. W. Callum Creek, Langford creek and gap map-areas, Alberta. D, 1948, Columbia University, Teachers College.

Douglas, Thomas Richard. Environments of deposition of the Borden Island gas zone in the subsurface of the Sabine Peninsula area, Melville Island, Arctic Archipelago. M, 1976, University of Calgary. 181 p.

Douglass, David Neil. Geology and paleomagnetics of three Old Red Sandstone basins; Spitsbergen, Norway and Scotland. D, 1987, Dartmouth College. 241 p.

Douglass, David Neil. Stratigraphy and paleomagnetics of the Morrison and Cloverly formations, Big Horn Basin, Wyoming. M, 1984, Dartmouth College. 126 p.

Douglass, Donald P. Depositional control of the Upper Cretaceous Sunnyside coal seam, Blackhawk Formation, of east-central Utah. M, 1988, Southern Illinois University, Carbondale. 99 p.

Douglass, Earl. The Neogene lake beds of western Montana and descriptions of some new vertebrates from the Loup Fork. M, 1899, University of Montana. 52 p.

Douglass, H. Marvin. Geology of the Yonkers area, Wagoner and Cherokee counties, Oklahoma. M, 1951, University of Oklahoma. 149 p.

Douglass, John Liddell. Upper crust inhomogeneities and their effects on telluric currents. M, 1962, University of California, Berkeley. 83 p.

Douglass, Myrl Robert. Geology and geomorphology of the south-central Big Snowy Mountains, Montana. M, 1954, University of Kansas. 105 p.

Douglass, Peter Mack. Anisotropy of granites; a reflection of microscopic fabric. M, 1968, Pennsylvania State University, University Park. 88 p.

Douglass, Raymond C. Preliminary fusulinid zonation of the Pennsylvanian and Permian rocks of northeastern Nevada. M, 1952, University of Nebraska, Lincoln.

Douglass, Raymond Charles. The foraminiferal genus Orbitolina in North America. D, 1957, Stanford University. 236 p.

Douglass, Robert Marshall. Studies in crystal chemistry of silicates; I, Crystal structure of sanbornite; II, High-temperature study of laumonite; III, X-ray examination of dumortierite. D, 1954, University of California, Berkeley. 110 p.

Dougless, Thomas C. Hydrodynamic potential of Upper Cretaceous Mesaverde Group and Dakota Formation, San Juan Basin, northwestern New Mexico and southwestern Colorado. M, 1984, Texas A&M University. 89 p.

Doukas, Michael. Volcanic geology of Big Chico Creek area, Butte County, California. M, 1983, San Jose State University. 157 p.

Doull, Mary Elizabeth. Turbidite sedimentation in the Puerto Rico Trench abyssal plain. M, 1983, Duke University. 124 p.

Douma, Marten. Clay mineral distribution in Mesozoic and Cenozoic strata of the Labrador Shelf. M, 1987, Dalhousie University. 119 p.

Douma, Stephanie Leigh. The mineralogy, petrology and geochemistry of the Port Mouton Pluton, Nova Scotia, Canada. M, 1988, Dalhousie University. 324 p.

Doumani, George Iskandar. Stratigraphy of the San Pablo Group, Contra Costa County, California. M, 1957, University of California, Berkeley. 72 p.

Dourojeanni, Axel Charles. Hydrologic soil study of an alpine watershed. M, 1969, Colorado State University. 118 p.

Douthit, Thomas D. Nathan. Magnetic survey of the Cap au Gres faulted flexure. M, 1959, St. Louis University.

Douthitt, C. B. The boron content of natural graphites. M, 1975, Dartmouth College. 26 p.

Douthitt, Herman. Structure and relationship of Diplocaulus, an American Permian amphibian. D, 1916, University of Chicago. 41 p.

Douze, Eduard Jan. Reflections and refractions of elastic waves from a transition layer. D, 1960, Stanford University. 107 p.

Dove, Floyd Harvey. Groundwater in the Navajo Sandstone; a subset of "simulation of the effects of coal-fired power developments in the Four Corners region". D, 1973, University of Arizona.

Dove, Jane E. Towards a Holocene paleoecology of the Ghost River and Water Valley areas, southwestern Alberta. M, 1981, University of Calgary.

Dove, Patricia Martin. The solubility and stability of scorodite, $FeAsO_4 \cdot 2H_2O$. M, 1984, Virginia Polytechnic Institute and State University. 58 p.

Dover, James Herbert. Bedrock geology of the Pioneer Mountains, central Idaho. D, 1966, University of Washington. 138 p.

Dover, James Herbert. Geology of the northern Palmetto Mtns., Esmeralda County, Nevada. M, 1962, University of Washington. 50 p.

Dover, R. Joseph. Paleoecology of the lower most part of the Jurassic Carmel Formation, San Rafael Swell, Emery County, Utah. M, 1969, Utah State University. 75 p.

Dover, Robert Allen. Geology and geochemistry of silicified rhyolites on Lick Mt., Montgomery County, North Carolina. M, 1985, University of North Carolina, Chapel Hill. 93 p.

Dow, Donald H. The Lebanon Gabbro of Connecticut. M, 1942, Northwestern University.

Dow, Garnett McCormick. Petrography and origin of the Cranberry Island Series, Maine. M, 1962, University of Illinois, Urbana.

Dow, Garnett McCormick. Petrology and structure of North Haven Island and vicinity, Maine. D, 1965, University of Illinois, Urbana. 165 p.

Dow, Gerald R. Bay fill in San Francisco; a history of change. M, 1973, San Francisco State University.

Dow, John Wilson. Lower and Middle Devonian limestones in northeastern Ohio and adjacent areas. M, 1961, Ohio State University.

Dow, Robert Russell. The geology and mineralization of the Tovar mining district, Tepehuanes Municipality, Durango, Mexico. M, 1978, University of Iowa. 139 p.

Dow, Roberta L. Radiolarian distribution and the late Pleistocene history of the Southeast Indian Ocean. M, 1976, University of Rhode Island.

Dow, Verne Eugene. Stratigraphy and ground water resources of Johnson County, Iowa. M, 1959, University of Iowa. 119 p.

Dow, Wallace G. The Spearfish Formation (Permian and Triassic) in Williston Basin of western North Dakota. M, 1964, University of North Dakota. 127 p.

Dowd, John F. Modeling groundwater flow into lakes. D, 1984, Yale University. 338 p.

Dowdall, Wayne Larry. Stratigraphy, depositional environments and petrology of the New Haven Arkose, Newark Supergroup, southern and central Connecticut. M, 1979, University of Massachusetts. 179 p.

Dowden, John E. Numerical simulation of artificial recharge in Cold Spring Valley, Nevada. M, 1981, University of Nevada. 123 p.

Dowding, Lynn G. Sedimentation within the Cocos Gap, Panama Basin. M, 1976, Oregon State University. 69 p.

Dowdney, Jack R. Geology of the central Delaware Mountains of West Texas. M, 1973, University of Texas at El Paso.

Dowdy, James Marshall. Changes in the strength of the relative orientation elements of stereoscopic models caused by variable locations of the orientation points. M, 1954, Ohio State University.

Dowell, Albert Roger. A magnetic investigation of northern Riley County, Kansas. M, 1964, Kansas State University. 84 p.

Dowell, Knneth E. An observational study of mesoscale cellular convection. M, 1974, Purdue University. 74 p.

Dowell, Thomas Perry Laning, Jr. A geological model for land-use suitability evaluation. D, 1973, University of Illinois, Urbana. 97 p.

Dowell, Thomas Perry Laning, Jr. An automated approach to subsurface correlation. M, 1972, University of Illinois, Urbana. 34 p.

Dowie, Paul G. Muscovite deposits of India. M, 1942, University of Chicago. 118 p.

Dowling, Forrest Leroy. A magnetotelluric investigation of the crust and upper mantle across the Wisconsin arch. D, 1968, University of Wisconsin-Madison. 205 p.

Dowling, Helen E. Geology of the Beaver Brook-Soda Creek area, Colorado. M, 1941, Northwestern University.

Dowling, John Joseph. Travel time curves from velocity distributions with applications to the Earth's upper mantle. D, 1964, St. Louis University. 105 p.

Dowling, Michael John. Nonlinear seismic analysis of arch dams. D, 1988, California Institute of Technology. 171 p.

Dowling, Paul L., Jr. Carbonate petrography and geochemistry of the Eiss Limestone (Permian) of Kansas. M, 1967, Kansas State University. 1959 p.

Dowling, Sharron Lea. A stratigraphic study of the Georgetown Formation (Washita Division, Lower Cretaceous) as it approaches the San Marcos Platform, south central Texas. M, 1981, Texas A&M University. 140 p.

Dowling, William M. Depositional environment of the lower Oak Hill Group, southern Quebec; implications for the late Precambrian breakup of North America in the Quebec Reentrant. M, 1988, University of Vermont. 186 p.

Down-Logan, Kathleen. Geological implications obtained from multivariate analysis of lunar geochemical data. M, 1979, Rensselaer Polytechnic Institute. 43 p.

Downer, H. Crystal habit of potash alum. M, 1933, University of Toronto.

Downes, Hilary. The White River Ash, Yukon Territory; a petrologic study. M, 1979, University of Calgary. 137 p.

Downey, Cameron Ingraham. Depositional environments of some upper Morrow sandstones in Kiowa, Bent, and Prowers counties, Colorado. M, 1978, University of Colorado.

Downey, Lewis Marshal. Geochemistry of hydrothermal chlorites, Gold Hill, Utah. M, 1976, University of Utah. 65 p.

Downey, Marlan Wayne. Geology of the Precambrian rocks of the Rochford area, South Dakota. M, 1957, University of Nebraska, Lincoln.

Downie, Elizabeth Anne. Structure and metamorphism in the Cavendish area, north end of the Chester Dome, southeastern Vermont. D, 1982, Harvard University. 349 p.

Downing, G. Geology of Fayette County, Kentucky. M, 1898, [University of Kentucky].

Downing, John Peabody. Field studies of suspended sand transport, Twin Harbors Beach, Washington. D, 1983, University of Washington. 121 p.

Downing, John Peabody. Some aspects of the suspended particulate matter and surficial sediments in the vicinity of Barkley, Nitinat, and Juan de Fuca submarine canyons. M, 1976, University of Washington.

Downing, Karen Pamela. The depositional history of the Lower Cretaceous Viking Formation at Joffre, Alberta, Canada. M, 1986, McMaster University. 138 p.

Downs, A. Fred. Geology of eastern Logan County, Arkansas. M, 1952, University of Arkansas, Fayetteville.

Downs, Dana V. Diagenetic patterns in Westerville Oolite, Kansas City, Missouri. M, 1986, University of Missouri, Columbia. 45 p.

Downs, George Reed. Geology of the Lukasashi rift valley. D, 1935, University of Minnesota, Minneapolis.

Downs, Harry F. Geology of eastern Logan County, Arkansas. M, 1952, University of Arkansas, Fayetteville.

Downs, James P. Some thermal consequences of a theory of continent growth. D, 1960, Massachusetts Institute of Technology. 103 p.

Downs, James Winston. An experimental examination of the electron distribution in bromellite, BeO, and phenacite, Be$_2$SiO$_4$. D, 1983, Virginia Polytechnic Institute and State University. 184 p.

Downs, James Winston. Bonding in beryllosilicates; the charge density of euclase AlBeSiO$_4$(OH) and molecular orbital studies of beryllium oxyanions. M, 1980, Virginia Polytechnic Institute and State University.

Downs, John W. Lithostratigraphy of the Pitkin Formation; Mississippian, Madison, Newton and Searcy counties, Arkansas. M, 1983, University of Arkansas, Fayetteville.

Downs, Robert Harold. Some Pennsylvanian conodonts from Iowa. M, 1947, University of Iowa. 41 p.

Downs, Robert Harold. The ammonoid fauna of the Pennsylvanian Finis Shale of Texas. D, 1949, University of Iowa. 124 p.

Downs, Theodore. A review of the Mascall Miocene fauna and related assemblages from Oregon. D, 1951, University of California, Berkeley. 386 p.

Downs, W. F. Trace element concentrations in pyrite found in and around uranium deposits. M, 1974, University of Colorado.

Downs, William Frederick. Experimental calibration of the quartz-magnetite oxygen isotope geothermometer. D, 1977, Pennsylvania State University, University Park. 155 p.

Dowse, Alice Mary. Geology of the Medfield-Holliston area, Massachusetts. D, 1949, Harvard University.

Dowse, Mary E. The geology of the Pine Mountain Complex, Alton, New Hamphire. M, 1974, University of New Hampshire. 67 p.

Dowse, Mary Elizabeth. The subsurface stratigraphy of the Middle and Upper Devonian clastic sequence in northwestern West Virginia. D, 1980, West Virginia University. 177 p.

Dowsett, Frederick Richard, Jr. Hydrothermal alteration study of the Hahns Peak Stock, Hahns Peak, Colorado. D, 1973, Stanford University. 113 p.

Dowsett, Harry James. A biochronological model for correlation of Pliocene marine sequences; application of the graphic correlation method. D, 1988, Brown University. 325 p.

Dowty, Eric. Crystal chemistry of titanian garnet; site distribution and valence of cations. D, 1969, Stanford University. 143 p.

Doyen, Philippe Marie. Transport and storage properties of inhomogeneous rock systems. D, 1987, Stanford University. 167 p.

Doyle, Alison Beth. Stratigraphy and depositional environments of the Jurassic Gypsum Spring and Sundance formations, Sheep Mountain Anticline area, Big Horn County, Wyoming. M, 1984, University of Wisconsin-Milwaukee. 99 p.

Doyle, Christopher Denis. The solution of ferrous oxide in aluminosilicate melts and its effect on the solubility of sulphur. D, 1983, University of Connecticut. 95 p.

Doyle, Eibhlin. Geology of the Bear River area, Digby and Annapolis counties, Nova Scotia. M, 1979, Acadia University.

Doyle, Frank Larry. The geology of the Freeport Quadrangle, Illinois. D, 1958, University of Illinois, Urbana. 112 p.

Doyle, James C. Geology of the northern Caballo Mountains, Sierra County, New Mexico. M, 1951, New Mexico Institute of Mining and Technology. 51 p.

Doyle, James D. Depositional patterns of the Miocene along the middle Texas coastal plain. M, 1976, University of Texas, Austin.

Doyle, James David. Depositional environments of the Pennsylvanian and Lower Permian of the Hartville Uplift and adjacent areas in eastern Wyoming, western Nebraska, and Southwest South Dakota. D, 1987, Colorado School of Mines. 309 p.

Doyle, James M. Trend surface analysis of the granite porphyry body of the Mackay Stock, Custer County, Idaho. M, 1979, Idaho State University. 53 p.

Doyle, Kevin Michael. Stratigraphy, sedimentary petrology and hydrocarbon occurrence in Lower Ordovician rocks of the deep central Michigan Basin. M, 1983, University of Wisconsin-Milwaukee. 187 p.

Doyle, Larry James. Black shells. M, 1967, Duke University. 69 p.

Doyle, Larry James. Marine geology of the Baja California continental borderland, Mexico. D, 1973, University of Southern California.

Doyle, Marianne C. Gravity investigation of the Valley and Ridge Province of Allegany County, Maryland and Grant and Mineral counties, West Virginia. M, 1984, Wright State University. 76 p.

Doyle, P. J. Regional geochemical reconnaissance and compositional variations in grain and forage crops on the southern Canadian Interior Plain. D, 1977, University of British Columbia.

Doyle, Patrick Joseph. Regional stream sediment reconnaissance and trace element content of rock, soil and plant material in eastern Yukon Territory. M, 1973, University of British Columbia.

Doyle, Polly Ann. Mobility of major and minor elements associated with Indiana Coal V and overlying sediments. M, 1980, Purdue University. 465 p.

Doyle, Robert. Areal and structural geology of Derby Dome area. M, 1938, University of Missouri, Columbia.

Doyle, Robert Emmett, Jr. Petrology of Packsaddle Mountain area, Llano and Burnet counties, Texas. M, 1957, University of Texas, Austin.

Doyle, Roger Whitney Stevens. Eh and thermodynamic equilibrium in environments containing dissolved ferrous iron. D, 1968, Yale University. 130 p.

Doyon, Martin. Synthèse géologique des roches volcaniques du centre-nord de la Gaspésie. M, 1988, Ecole Polytechnique. 247 p.

Doyuran, Vedat. Ground water resources (pre-Tertiary, Tertiary, Quaternary) of Ogden Valley (Utah). D, 1971, University of Utah. 134 p.

Dozier, Malcolm David. Assessment of change in the marshes of southwestern Barataria Basin, Louisiana, using historical aerial photographs and a spatial information system. M, 1983, Louisiana State University. 102 p.

Drabkowski, Robert S. The spectrochemical investigation of a Paleozoic section exposed near Cody, Wyoming. M, 1952, Wayne State University.

Drafall, Larry Edward. Differential thermal analysis and high temperature x-ray diffraction of the lead sulfantimonides and other selected sulfosalts. D, 1972, University of Cincinnati. 205 p.

Drafall, Larry Edward. Differential thermal analysis of selected sulfosalts. M, 1969, University of Cincinnati. 122 p.

Dragert, Herb. A geomagnetic depth-sounding profile across central British Columbia (Canada). M, 1970, University of British Columbia.

Dragert, Herbert. Broad-band geomagnetic depth-sounding along an anomalous profile in the Canadian Cordillera. D, 1973, University of British Columbia.

Dragg, James L. Spherical harmonics and the gravitational field of Earth. M, 1964, Ohio State University.

Dragone, Annette M. Seismic and gravity surveys near a buried Pleistocene glacial valley; Waterville, Ohio. M, 1986, Bowling Green State University. 101 p.

Dragusanu, Juliu Basile. Repressuring oil fields in Eastern United States. M, 1930, University of Pittsburgh.

Drahovzal, James Alan. The geology of Jefferson County, Iowa. M, 1963, University of Iowa. 274 p.

Drahovzal, James Alan. Upper Mississippian (Visean-Namurian) ammonoids of northern Arkansas. D, 1966, University of Iowa. 339 p.

Drain, Vance Keith. Glaciation of the Boulder River area, southcentral Montana. M, 1986, Montana State University. 109 p.

Drake, A. A. A study of the effect of jointing on the blasting of granite, Graniteville, Missouri. M, 1952, University of Missouri, Rolla.

Drake, Benjamin. Structure and origin of the Tomahawk Volcanics (Eocene), central Black Hills, South Dakota. M, 1967, University of Minnesota, Minneapolis. 303 p.

Drake, David E. Distribution and transport of suspended matter, Santa Barbara Channel, California. D, 1972, University of Southern California.

Drake, Dennis Adolph. Microfauna of the Cisco Series, Brown and Coleman counties, Texas. M, 1958, University of Texas, Austin.

Drake, Edwin A. Paleozoic stratigraphy of the Devils Gate - northern Mahogany Hills area, Eureka County, Nevada. M, 1979, Oregon State University. 109 p.

Drake, Ellen Tan. Robert Hooke and the foundation of geology; a comparison of Steno and Hooke and the Hooke imprint on the Huttonian theory. The tectonic evolution of the Oregon continental margin; rotation of segment boundaries and possible spacetime relationships in the central High Cascades. D, 1981, Oregon State University. 177 p.

Drake, F. Lawrence. The propagation of Love and Rayleigh waves in non-horizontally layered media. D, 1971, University of California, Berkeley. 160 p.

Drake, Gerald M. Effect of obstructions on quartz growth. M, 1970, West Virginia University.

Drake, Jerry T. Energy distribution of Rayleigh waves in the crust and upper mantle. M, 1984, Texas Tech University. 62 p.

Drake, Joan Elizabeth. The effects of tropical storm swell on Southern California summer beach profiles. M, 1980, University of California, Los Angeles. 67 p.

Drake, John Craig. Composition of almandine and co-existing minerals in the mica schists of the Errol Quadrangle, New Hampshire. D, 1968, Harvard University.

Drake, John Franklin. Origin and mineralization of sinks and breccia pipes of the Terlingua Uplift, Brewster and Presidio counties, Texas. M, 1980, West Texas University. 97 p.

Drake, Larson Y. Insoluble residues of some Permian and Pennsylvanian limestones. M, 1951, Kansas State University. 100 p.

Drake, Lon David. Experiments on regelation of ice. M, 1965, University of California, Los Angeles. 72 p.

Drake, Lon David. Till studies in New Hampshire. D, 1968, Ohio State University. 130 p.

Drake, Michael Julian. The distribution of major and trace elements between plagioclase feldspar and magmatic silicate liquid; an experimental study. D, 1972, University of Oregon. 205 p.

Drake, Natalie E. R. Geochemistry of DSDP Leg 82 basalts. M, 1985, University of Massachusetts. 102 p.

Drake, Noah F. A geological reconnaissance of the coal fields of Indian territory. D, 1897, [Stanford University].

Drake, Noah F. Relief map of California. M, 1894, [Stanford University].

Drake, Robert E. Surface-subsurface measurements of an anomaly in the vertical gradient of gravity at Loveland Pass, Colorado. M, 1967, University of California, Riverside. 41 p.

Drake, Robert Edward. The chronology of Cenozoic igneous and tectonic events in the central Chilean Andes. D, 1974, University of California, Berkeley. 44 p.

Drake, Robert Tucker. The relationships of the ostracode genus Schmidtella to Eridoconcha, with descriptions of new species of Schmidtella, from the Bromide (Ordovician) of Oklahoma. M, 1939, University of Missouri, Columbia.

Drake, Steve Allen. A subsurface study of the upper part of the Hoxbar Group (Missourian-Virgilian Series) of the Pennsylvanian System, Caddo and Grady counties, Oklahoma. M, 1982, Texas Christian University. 103 p.

Drake, Thomas George. Experimental flows of granular material. D, 1988, University of California, Los Angeles. 159 p.

Drake, William Edward. A study of ore forming fluids at the Mineral Park porphyry copper deposit, Kingman, Arizona. D, 1972, Columbia University. 245 p.

Drake, William Robert. Surface reflected shear waves of the earthquake of May 25, 1944. M, 1949, University of Michigan.

Dramis, Louis Albert. Depositional patterns of lower Vicksburg (Oligocene) sandstones, McAllen Range Field, Hidalgo County, Texas. M, 1980, Texas A&M University. 105 p.

Draney, David. Geotechnical properties of coal strip mined areas in Macon and Randolph counties, Missouri. M, 1982, University of Missouri, Columbia.

Dransfield, Betsy J. The Cow Creek Anticline; an example of the disharmonic along the front of the Bighorn Mountains. M, 1983, Texas A&M University.

Drapeau, Georges. Geology of the Lake Memphremagog Syncline (Vermont). M, 1961, Massachusetts Institute of Technology. 70 p.

Drapeau, Georges. The sedimentology of the surficial sediments of the western portion of the Scotian shelf. D, 1971, Dalhousie University.

Draper, Louise Pierce. Analysis of foraminifers across the Jackson-Claiborne contact (Boone) in Mississippi and Alabama. M, 1976, University of New Orleans.

Draper, Mary B. Maximum initial subaqueous dips of sediments, slopes prohibitive to permanent deposition, and some regions of the present sea bottom where such conditions exist. M, 1930, University of Wisconsin-Madison.

Draper, Richard Brandt. Anatase, brookite and rutile. D, 1938, Washington University.

Draper, Richard Brandt. The synthesis of minerals involving gas reactions. M, 1935, Washington University. 64 p.

Draper, Stephen Elliot. Urban rainfall-runoff modelling using remote sensing imagery. D, 1981, Georgia Institute of Technology. 258 p.

Drashevska, Lubov. Review of recent U.S.S.R. publications in selected fields of engineering soil science. M, 1958, Columbia University, Teachers College.

Dratler, Jay, Jr. Quartz fiber accelerometers and some geophysical applications. D, 1971, [University of California, San Diego]. 528 p.

Dravis, Jeffrey. Holocene sedimentary depositional environments on Eleuthera Bank, Bahamas. M, 1977, University of Miami. 386 p.

Dravis, Jeffrey James. Sedimentology and diagenesis of the Upper Cretaceous Austin Chalk Formation, South Texas and northern Mexico. D, 1980, Rice University. 532 p.

Drean, Thomas Alen. Reduction of sulfate by methane, xylene and iron at temperatures of 175 to 350°C. M, 1978, Pennsylvania State University, University Park. 90 p.

Dredge, L. A. Quaternary geomorphology of the Quebec North Shore, Godbout to Sept-Iles. D, 1977, University of Waterloo.

Drees, Larry Richard. Soil characteristics and genesis of carbonates in the Rolling Plains of Texas. D, 1986, Texas A&M University. 256 p.

Drees, Linus S. Fluorescence of oil well cuttings under the action of ultraviolet light as pertaining to depth of wells investigated. M, 1950, Fort Hays State University.

Drees, Robert Henry. The geology of the Fort Hays Limestone in Ellis County, Kansas. M, 1974, Fort Hays State University. 53 p.

Dreessen, Richard S. Geology of a portion of the Pancake range, Nye County, Nevada. M, 1969, University of San Diego.

Dreeszen, Vincent Harold. A subsurface study of the Pleistocene deposits in Kearney County and adjoining parts of Adams, Franklin, and Webster counties. M, 1950, University of Nebraska, Lincoln.

Dregne, James Michael. Petrography of the Roan Village area, Carter County, Tennessee. M, 1971, University of Tennessee, Knoxville. 36 p.

Drehle, William. A sedimentary investigation of the large linear sand bodies exposed in Gulf County, Florida. M, 1973, Florida State University.

Dreier, Christine de Angelis. Trace metal accumulations in Delaware salt marshes. M, 1982, University of Delaware. 89 p.

Dreier, J. E., Jr. The geochemical environment of ore deposition in the Pachuca-Real del Monte District, Hidalgo, Mexico. D, 1976, University of Arizona. 126 p.

Dreier, John E. Economic geology of the Sunlight mineralized region, Park County, Wyoming. M, 1967, University of Wyoming. 81 p.

Dreier, RaNaye Beth. The Blackstone Series; evidence for Precambrian Avalonian and Permian Alleghanian tectonism in southeastern New England. D, 1985, University of Texas, Austin. 262 p.

Dreifuss, Sophie M. Textural and compositional changes during diagenesis of high-Mg calcite skeletons. M, 1977, University of Illinois, Urbana. 157 p.

Dreiss, S. J. An application of systems analysis to karst aquifers. D, 1980, Stanford University. 209 p.

Dreiss, Shirley Jean. Hydrogeologic controls on solution of carbonate rocks in Christian County, Missouri. M, 1974, University of Missouri, Columbia.

Drennan, William T., III. Sedimentary petrology and depositional environment of the Longarm Quartzite (late Precambrian) in the Blue Ridge Province of North Carolina-Tennessee. M, 1976, Louisiana State University.

Drennen, Charles William. Geology of the Piedmont-Coastal Plain contact in eastern Alabama and western Georgia. M, 1950, University of Alabama. 42 p.

Drennon, Clarence Bartow, III. Stick-slip friction of lightly loaded rock (limestone and basalt). D, 1972, Iowa State University of Science and Technology.

Dresbach, C. H. Organization of geophysical parties for foreign exploration. M, 1939, University of Missouri, Rolla.

Dresbach, Charles Howard. The change in the yield-pressure ratio with time in gas wells. M, 1937, University of Pittsburgh.

Drescher, Arthur B. A new Pliocene badger from Mexico. M, 1939, California Institute of Technology. 10 p.

Drescher, William J. Ground water in Wisconsin. M, 1956, University of Wisconsin-Madison.

Dresel, P. Evan. The geochemistry of oilfield brines from western Pennsylvania. M, 1985, Pennsylvania State University, University Park.

Dresen, Michael D. Geology and slope stability of part of Pleasanton Ridge, Alameda County, California. M, 1979, California State University, Hayward. 130 p.

Dresser, Anita E. Conodonts from the Marble Falls Formation (Pennsylvanian) of central Texas. M, 1975, University of Texas, Austin.

Dresser, Douglas W. Spectral analysis of the gravitational attraction over two-dimensional trapezoidal prisms having asymmetric cross-sections. M, 1982, Montana College of Mineral Science & Technology. 117 p.

Dresser, H. W. A field study of the Jurassic "lower Sundance" beds in southeastern Wyoming. D, 1959, University of Wyoming. 667 p.

Dresser, Hugh. Notes on the fauna of the Carboniferous Itaituba Formation, Rio Tapajos, Estado do Para, Brasil. M, 1951, University of Cincinnati. 90 p.

Dresser, Myron A. The paragenesis of certain nickel ores of Sudbury, Ontario. M, 1917, University of Minnesota, Minneapolis. 34 p.

Dressner, Elliott Francis. Genesis of the Clifton Mine magnetite deposit, DeGrasse, St. Lawrence County, New York. M, 1950, Cornell University.

Drever, James Irving. Electrophoresis and the study of clay minerals in recent sediments. D, 1968, Princeton University. 177 p.

Dreveskracht, Lloyd R. Ionic effects upon the rates of settling of fine-grained sediments. M, 1939, University of Minnesota, Minneapolis. 57 p.

Drew, Alice J. R. Tidal triggering of microearthquakes at lakes Jocassee and Keowee, South Carolina. M, 1978, University of South Carolina.

Drew, C. Wallace. Mathematical laws of slopes and materials of alluvial terraces. M, 1951, University of Wisconsin-Madison.

Drew, Fred Prescott. A geochemical, petrological, and a geophysical case study of Caryn Seamount. M, 1975, Texas A&M University. 75 p.

Drew, Isabella Milling. Properties of the Bootlegger Cove Clay (Pleistocene), Anchorage, Alaska. D, 1966, Columbia University. 105 p.

Drew, Lawrence James. Grid drilling exploration and its application to the search for petroleum. D, 1966,

Pennsylvania State University, University Park. 141 p.

Drew, Lawrence James. Intercorrelations between petrographic and reservoir properties in the Cow Run Sand, Wirt County, West Virginia. M, 1964, Pennsylvania State University, University Park. 92 p.

Drew, Patricia. Geochemistry of iron and clay mineralogy of Playa sediments from Teels marsh, Nevada. M, 1970, University of Wisconsin-Madison.

Drew, Thomas A. Hydrology and waste disposal considerations for Monroe and Ralls counties, Missouri. M, 1984, University of Missouri, Columbia. 71 p.

Drewes, Harald D. Structural geology of the southern Snake Range, Nevada. D, 1954, Yale University.

Drewett, Margaret E. The origin and metamorphism of the Marlboro greenstones of the Providence area. M, 1932, Brown University.

Drexler, Christopher William. Outlet channels for the post-Duluth lakes in the Upper Peninsula of Michigan. D, 1981, University of Michigan. 371 p.

Drexler, Christopher William. Sand dunes along the east shore of Lake Michigan; their composition, structure, and texture, a review of the available literature. M, 1969, University of Michigan.

Drexler, James Michael and Kildal, Edwin. Geology of the Red Peak's area, Beaverhead County, Montana, and Clark County, Idaho. M, 1949, University of Michigan. 54 p.

Drexler, John William. Geochemical correlations of Pleistocene rhyolitic ashes in Guatemala with deep-sea ash layers of the Gulf of Mexico and equatorial Pacific. M, 1978, Michigan Technological University. 84 p.

Drexler, John William. Mineralogy and geochemistry of Miocene volcanic rocks genetically associated with the Julcani Ag-Bi-Pb-Cu-Au-W deposit, Peru; physicochemical conditions of a productive magma body. D, 1982, Michigan Technological University. 259 p.

Drexler, Timothy John. The depositional and diagenetic history of the Oswego Lime, Aline-Lambert field area, north-central Oklahoma. M, 1980, University of Oklahoma. 149 p.

Drexler, Wallace William. Wilmington Jacksonian hornerids; important cyclostome bryozoans from the upper Eocene Castle Hayne Limestone in southeasternmost North Carolina. D, 1976, Pennsylvania State University, University Park. 122 p.

Dreyer, Boyd V. Stratigraphy of the Trinity Group (lower Cretaceous), (north-east side Llano uplift), central Texas. M, 1971, Baylor University. 142 p.

Dreyer, Charles. Some aspects of dissolved and particulate organic carbon in the nearshore environment of the Gulf of Mexico. M, 1973, Florida State University.

Dreyer, Frances Eaton. The geology of a portion of Mt. Pinos Quadrangle, Ventura County, California. M, 1935, University of California, Los Angeles.

Dreyer, Robert M. Magnetometer examination of the Monte Cristo magnetite-ilmenite deposits. D, 1939, California Institute of Technology. 11 p.

Dreyer, Robert M. Mutual interference in the microchemical determination of ore minerals. M, 1937, California Institute of Technology. 27 p.

Dreyer, Robert M. The geochemistry of quicksilver mineralization. D, 1939, California Institute of Technology. 192 p.

Drez, Paul E. Hydrothermal alteration of low-K tholeiitic dikes and intruded clastic sediments. D, 1977, University of North Carolina, Chapel Hill. 290 p.

Dribus, John R. The petrology and geochemistry of the deposit No. 9 dunite, Macon County, North Carolina. M, 1977, Kent State University, Kent. 85 p.

Driel, James Nicholas Van *see* Van Driel, James Nicholas

Driese, Steven George. Paleoenvironments of the Upper Cambrian Mt. Simon Formation in western and West-central Wisconsin. M, 1979, University of Wisconsin-Madison.

Driese, Steven George. Sedimentology, conodont distribution, and carbonate diagenesis of the upper Morgan Formation (Middle Pennsylvanian), northern Utah and Colorado. D, 1982, University of Wisconsin-Madison. 289 p.

Driesner, Douglas Allen. Geotechnical investigation of the Hat Creek landslide, Idaho County, Idaho. M, 1979, University of Idaho. 67 p.

Drifmeyer, Jeffrey Eugene. Pb, Zn, and Mn concentrations in dredge spoil and estuarine organisms. M, 1974, University of Virginia. 85 p.

Driggs, Allan F. The petrology of three Upper Permian bioherms, southern Tunisia. M, 1977, Brigham Young University. 53 p.

Drindak, Joseph T. Insoluble residues of the Oneota Dolomite of western Wisconsin. D, 1933, University of Wisconsin-Madison.

Dring, Nancy Beth. Planktonic foraminifera from the Smoky Hill Member of the Niobrara Formation in West central Kansas. M, 1977, University of Wyoming. 72 p.

Drinker, Philip A. Boundary shear stresses in curved trapezoidal channels. D, 1961, Massachusetts Institute of Technology. 142 p.

Drinkwater, James. Geology of the northeastern part of the Quien Sabe Volcanics, Merced County, California. M, 1983, San Jose State University. 102 p.

Driscoll, C. T. Chemical characterization of some dilute acidified lakes and streams in the Adirondack region of New York State. D, 1980, Cornell University. 329 p.

Driscoll, Egbert G., Jr. An environmental and heavy mineral study of the "Eastern Sandstones" between Marquette and Grand Marais, Michigan. M, 1956, University of Nebraska, Lincoln.

Driscoll, Egbert Gotzian, Jr. Dimyarian lamellibranchs of the Mississippian Marshall Sandstone of Michigan. D, 1962, University of Michigan. 222 p.

Driscoll, Fletcher G. Formation and wastage of neoglacial surge moraines of the Klutlan Glacier, Yukon Territory, Canada. D, 1976, University of Minnesota, Minneapolis. 333 p.

Driscoll, Mavis Lynn. Application of Seasat altimetry to tectonic studies of fracture zones in the southern oceans. D, 1987, Woods Hole Oceanographic Institution. 165 p.

Driscoll, William J. A gravity study of Dallas County, Texas. M, 1962, Southern Methodist University. 35 p.

Driskill, Lorinda Elizabeth. Phaeodarian radiolarians as indicators of Recent and fossil (Monterey) anoxic events in California. M, 1986, Rice University. 92 p.

Driver, Herschel L. Inglewood oil field, Los Angeles County, California. M, 1939, University of Southern California.

Drnevich, Vincent Paul. Effects of strain history on the dynamic properties of sand. D, 1967, University of Michigan. 164 p.

Drobeck, Pete A. Geology and trace element geochemistry of a part of the Gunnison gold belt, Colorado. M, 1979, Colorado School of Mines. 245 p.

Drobny, Gerald Francis. Sedimentology and stratigraphy of a fluvial and lacustrine interbed near Kendrick, Idaho. M, 1981, Washington State University. 228 p.

Drobnyk, John Wendel. Sedimentation and environments of the Aquia Formation. D, 1962, Rutgers, The State University, New Brunswick. 153 p.

Droddy, M. J., Jr. Contact metamorphism of Cretaceous sedimentary rocks near the Bee Mountain Intrusion, Brewster County, Texas. M, 1974, University of Houston.

Droddy, M. Jackson, Jr. Metamorphic rocks of the Fly Gap Quadrangle, Mason County, Texas. D, 1978, University of Texas, Austin. 213 p.

Drolet, Michel. Développement d'une technologie d'investigation (instrumentation, procédures, modèles) pour suivre les fluctuations de teneurs en eau et températures en zone non saturée dans un sol granulaire. M, 1985, Université Laval. 224 p.

Dromgoole, Edward Lee. Petrology and chemistry of sulfides in xenoliths from Kilbourne Hole, N.M. M, 1984, Washington University. 199 p.

Dronyk, Michael P. Stratigraphy, structure and a seismic refraction survey of a portion of the San Felipe Hills, Imperial Valley, California. M, 1977, University of California, Riverside. 141 p.

Drooker, Penelope B. Application of the Stanford watershed model to a small New England watershed. M, 1968, University of New Hampshire. 39 p.

Droser, Mary Louise. Depositional environments of an unusual limestone and associated strata within the Upper Devonian Chemung magnafacies. M, 1984, SUNY at Binghamton.

Droser, Mary Louise. Trends in extent and depth of bioturbation in Great Basin Precambrian-Ordovician strata, California, Nevada and Utah. D, 1987, University of Southern California.

Drost, B. W. Late Quaternary stratigraphy of the southern Argolid (Peloponnese, Greece). M, 1974, University of Pennsylvania.

Droste, John Brown. Clay mineral composition of till in northeastern Ohio. D, 1956, University of Illinois, Urbana. 37 p.

Droste, John Brown. The Nacimiento and San Jose formations in the San Juan Basin, Colorado. M, 1953, University of Illinois, Urbana.

Drouant, Ronald George. Stratigraphy and Ostracoda of the Exogyra costata Zone of southwestern Arkansas. D, 1960, Louisiana State University. 211 p.

Drouin, Ruth. Performance des lieux d'enfouissement sanitaire de Laterrière et de Ste-Sophie, Québec. M, 1986, Université Laval. 47 p.

Droullard, Emerson Keith. Geology of Packwood Quadrangle, California. M, 1951, University of California, Berkeley. 70 p.

Drowley, David. A detailed stratigraphic study of the Battle Formation (Middle Pennsylvanian), north-central Nevada. M, 1973, University of Nevada. 94 p.

Drown, David Birke. The response of Lake Superior periphyton to heat additions. D, 1973, University of Minnesota, Minneapolis.

Droxler, Andre Willy. Late Quaternary glacial cycles in the Bahamian deep basins and in the adjacent Atlantic Ocean. D, 1984, University of Miami. 186 p.

Druce, Edric Charles. Conodont biostratigraphy of the upper Devonian reef complexes of the Canning Basin, Western Australia. D, 1971, University of Michigan. 589 p.

Druckerman, Daniel. The genus Pecten in the New World. M, 1961, University of California, Berkeley. 119 p.

Druecker, Michael D. The geology of the Bladen Volcanic Series, southern Maya Mountains, Belize, Central America. M, 1978, Colorado School of Mines. 73 p.

Druffel, Ellen Mary. Radiocarbon in annual coral rings of the Pacific and Atlantic oceans. D, 1980, [University of California, San Diego]. 228 p.

Druffel, Leroy. Characteristics and prediction of soil erosion on a watershed in the Palouse. M, 1973, University of Idaho. 95 p.

Drugg, Warren Sowle. Eocene stratigraphy of Hoko River area, Olympic Peninsula, Washington. M, 1958, University of Washington. 192 p.

Druham, Robert M. The southwestern Raleigh Belt and the Nutbush Creek Fault in North Carolina. M, 1983, University of North Carolina, Chapel Hill. 69 p.

Druitt, Charles Edward. The subsurface geology of Jefferson County, Oklahoma. M, 1957, University of Oklahoma. 39 p.

Druke, Carmen B. A classification of pene-contemporaneous deformational structures. M, 1982, University of Rochester. 34 p.

Druliner, Allan Douglas. Groundwater uranium chemistry of the Crawford area, northwestern Nebraska. M, 1984, University of Nebraska, Lincoln.

Drumheller, Richard E. The petrochemistry of the allanite-bearing granitoids, Red Rock area, Washoe County, Nevada. M, 1978, University of Nevada. 89 p.

Drumm, Eric Corman. Testing, modelling, and applications of interface behavior in dynamic soil-structure interaction. D, 1983, University of Arizona. 323 p.

Drummond, Arthur Darryl. Geology of the Alice Arm molybdenum prospect, Skeena mining district, Alice Arm, British Columbia. M, 1961, University of British Columbia.

Drummond, Arthur Darryl. Mineralogical and chemical study of Craigmont Mine, Merritt, British Columbia. D, 1966, University of California, Berkeley. 142 p.

Drummond, C. Hanford. Stratigraphy of the Nojoqui and Las Cruces creeks districts, Santa Barbara County, California. M, 1941, Stanford University. 50 p.

Drummond, Keith F. Petrology of deep-sea silts in the Gulf of Mexico. M, 1970, University of Missouri, Columbia.

Drummond, Kenneth McCoy. Zircon studies in the southeastern Piedmont. M, 1962, University of South Carolina. 27 p.

Drummond, Mark Stephen. Igneous, metamorphic, and structural history of the Alabama tin belt, Coosa County, Alabama. D, 1986, Florida State University. 509 p.

Drummond, Paul Linwood. Gouge zone shale near southern Louisiana salt domes. D, 1955, Columbia University, Teachers College. 82 p.

Drummond, S. E. Distribution of heavy minerals; offshore Alabama and Mississippi. M, 1976, University of Alabama. 90 p.

Drummond, Segal Edward, Jr. Boiling and mixing of hydrothermal fluids; chemical effects on mineral precipitation. D, 1981, Pennsylvania State University, University Park. 397 p.

Drury, Malcolm J. The electrical properties of ocean crust and oceanic island basalts and gabbros; results and implications of a laboratory study. D, 1977, Dalhousie University.

Drury, W. H. Cyclic development of bog flats in interior Alaska. D, 1952, Harvard University.

Drwenski, Vernon R. Geology of the Boxelder-Mormon Canyon area, Converse County, Wyoming. M, 1952, University of Wyoming. 67 p.

Dryden, A. L., Jr. Stratigraphy of the Calvert Formation at the Calvert Cliffs, Maryland. D, 1930, The Johns Hopkins University.

Dryden, Donald A. Petrology and petrofabrics of the Randville Dolomite in the Felch Mountain Trough, Dickinson County, Michigan. M, 1962, Michigan State University. 60 p.

Dryden, Jacob Edward. A study of a well core from crystalline rocks near Manson, Iowa. M, 1955, University of Iowa. 89 p.

Dryer, F. E. Geology of a portion of the Mt. Pinos Quadrangle (California). M, 1935, University of California, Los Angeles.

Dryer, S. Reef ecology and sediments, Castle Harbour, Bermuda. M, 1977, University of New Brunswick.

Drysdale, Charles W. The geology of the Franklin mining district, British Columbia. D, 1912, Yale University.

Du Bar, Jules R. Cretaceous faunas from the northern flank of Ochoco Range, Oregon. M, 1950, Oregon State University. 178 p.

du Bray, Edward Arthur. Geology of the igneous and metamorphic rocks in the Evolution-Goddard region of the Sierra Nevada, California. M, 1977, Stanford University. 123 p.

du Toit, Charl. Fluvial geomorphology, sedimentology and paleohydrology of the piedmont east of Glacier National Park, Montana. M, 1988, University of Calgary. 181 p.

Du, Chenggui. Risk assessment and decision analysis in groundwater contamination. D, 1988, Case Western Reserve University. 237 p.

Du, Ming-Ho. Geology of the Germania tungsten deposits, Stevens County, Washington. M, 1979, Eastern Washington University. 58 p.

Du, Raymonde Le *see* Le Du, Raymonde

Duane, David Bierlein. Heavy minerals of the Lower Pennsylvanian sandstones in portions of the Anadarko and Ardmore basins of Oklahoma. M, 1959, University of Kansas. 48 p.

Duane, David Bierlein. Petrology of Recent bottom sediments of the western Pamlico Sound region, North Carolina. D, 1962, University of Kansas. 108 p.

Duane, John William. A study of the genus Mucrospirifer. M, 1958, University of Michigan.

Duarte, Andrew Henry. Geology of eastern Choctaw County, Oklahoma. M, 1968, University of Oklahoma. 70 p.

Duarte, R. Armando. Update and sensitivity analysis of "Export of coking coal from the Checua-Lenguazaque (Colombia) Coalfield". M, 1976, Colorado School of Mines. 88 p.

Duba, Alfred G. The electrical conductivity of olivine as a function of pressure, temperature, composition and crystallographic orientation. D, 1971, University of Chicago. 68 p.

Duba, Daria. The application of illite crystallinity, organic matter reflectance and isotopic techniques to the exploration for sedimentary-hosted hydrothermal ore deposits, southwestern Gaspé M, 1982, McGill University. 142 p.

DuBar, Jules R. Stratigraphy and paleoecology of the late Neogene strata of the Caloosahatchee River area of southern Florida. D, 1956, University of Kansas. 211 p.

Dubé, Benoît. Géologie, pétrographie et métallogénie d'indices aurifères localisés dans le filon-couche de Bourbeau, centre-nord du Canton de Barlow, Chibougamau, Québec. M, 1986, Universite Laval. 221 p.

Dube, Thomas Eugene. Setting and origin of exhalative bedded barite and associated rocks of the Roberts Mountains Allochthon in north-central Nevada. M, 1987, University of Washington. 76 p.

Dubendorff, Bruce H. Changes in seismic velocity and apparent attenuation due to isotropic and anisotropic scattering; results from physical modeling. D, 1987, Oregon State University. 102 p.

Duberger, Reynald. Etude préliminaire de la microséismicité de la Vallée du Saint-Laurent, Québec. M, 1971, Universite Laval.

Dubiel, Russell F. Sedimentology of the Upper Triassic Chinle Formation, southeastern Utah. D, 1987, University of Colorado. 146 p.

Dubin, David Joel. Fusulinid fauna from the type area of the Earp Formation, Permo-Pennsylvanian, Cochise County, Arizona. M, 1964, University of Arizona.

Dubins, M. Ira. The petrography, geochemistry, and economic utilization of the Fort Hays Chalk in Kansas. M, 1947, University of Kansas. 108 p.

Dubiskas, Richard A. Structure and stratigraphy of the Sierra de los Altares area, northeastern Chihuahua and northwestern Coahuila, Mexico. M, 1985, Wichita State University. 88 p.

DuBois, David L. The spatial and temporal relationships of the North American microtektite layer. M, 1984, University of Delaware. 169 p.

DuBois, Dean Paul. Geology and tectonic implications of the Deer Canyon area, Tendoy Range, Montana. M, 1981, Pennsylvania State University, University Park. 87 p.

Dubois, Ernest Paul. The comparative cranial osteology of some members of the Lacertilia. D, 1942, University of Chicago. 53 p.

Dubois, Henry Mathusalem. Fauna of the Winfield and Harrington limestones. M, 1914, Indiana University, Bloomington.

DuBois, James. The basal gneiss of central Wisconsin. M, 1982, University of Kansas.

DuBois, Martin. Factors controlling the development and distribution of porosity in the Lansing-Kansas City "E" Zone, Hitchcock County, Nebraska. M, 1980, University of Kansas. 100 p.

DuBois, Robert Lee. Petrology and genesis of ores, Holden Mine area, Chelan County, Washington. D, 1954, University of Washington. 246 p.

Dubois, Roger Normand. Seasonal variations in beach and nearshore morphology and sedimentology along a profile of Lake Michigan, Wisconsin. D, 1972, University of Wisconsin-Madison.

DuBois, Sarah Barton. Experiments with megascopic description of coal Hazard No. 8, Perry County, Kentucky. M, 1985, University of Kentucky. 64 p.

DuBois, Susan Morrison. The origin of surface lineaments in Nemaha County, Kansas. M, 1978, University of Kansas. 37 p.

Dubord, M. P. The stratigraphy and petrology of the Ogilvie Formation (Devonian), northern Yukon Territory. M, 1986, University of Waterloo. 189 p.

Dubrovsky, Neil Michael. Geochemical evolution of inactive pyritic tailings in the Elliot Lake uranium district. D, 1986, University of Waterloo. 373 p.

Dubuc, F. Map-area west of Timmins Bay, Lake Attikamagen, Labrador. M, 1950, McGill University.

Duc, Aileen Wojtal. Back-barrier stratigraphy of Kiawah Island, South Carolina. D, 1981, University of South Carolina. 265 p.

Duch, M. F. F. de la Fuente *see* de la Fuente Duch, M. F. F.

Duch, Mauricio Fernando De la Fuente *see* De la Fuente Duch, Mauricio Fernando

Duchaine, R. P. Multiple fold styles in Rheems Quarry, Rheems, Pennsylvania. M, 1978, Temple University.

DuChene, Harvey R. Structure and stratigraphy of Guadalupe Box and vicinity, Sandoval County, New Mexico. M, 1973, University of New Mexico. 100 p.

Duchin, Ralph Charles. Pre-Cenozoic stratigraphy of Candelaria area, Presidio County, Trans-Pecos, Texas. M, 1955, University of Texas, Austin.

Duchmann, Peter Rudolf Jaffe *see* Jaffe Duchmann, Peter Rudolf

Duchosal, Yves R. Contribution to the petrology of an andesitic-rhyolitic volcano in the Precambrian of northern Michigan. M, 1981, California State University, Hayward. 291 p.

Duchossois, George Earl. Quaternary geology of the Jamesville, New York, 7-5-Minute Quadrangle. M, 1980, Syracuse University.

Duck, James H. The Northwest Butner Pool, Seminole County, Oklahoma. M, 1958, University of Oklahoma. 62 p.

Duck, John. An investigation of factors controlling the partitioning of trace germanium and gallium between topaz and quartz. M, 1986, University of Pittsburgh.

Duckett, Kathleen Carey. Site characterization and remedial engineering options at an uncontrolled hazardous waste site; Fulbright Landfill, City of Springfield, MO. M, 1986, University of Missouri, Rolla. 121 p.

Duckson, D. W., Jr. Land use intensity and water quality. D, 1979, University of Colorado. 181 p.

Duckwitz, George Herman. Selected mathematical models for predicting potentiometric head and contaminant transport in ground water. M, 1983, Oklahoma State University. 157 p.

Duckwitz, Lester Dean. Groundwater management of the isolated terrace deposit (Gerty Sand) of the Canadian River in Garvin, McClain, and Pontotoc counties, Oklahoma. M, 1987, Oklahoma State University. 210 p.

Duckworth, D. L. Magnesium concentration in the tests of the planktonic foraminifera Globorotalia truncatulinoides. M, 1977, University of Illinois, Chicago.

Duckworth, Diana. Magnesium content of foraminiferal tests. M, 1975, University of Illinois, Chicago.

Duckworth, P. B. Paleocurrent trend in the latest outwash at the western end of the Oak Ridges Moraine, Ontario. D, 1975, University of Toronto.

Duckworth, Robert M. Q estimates from local coda waves. M, 1983, Georgia Institute of Technology. 91 p.

Ducrot, Claude. Les migrations de Saint Colomban (Province de Grenville) P.Q. M, 1974, Universite de Montreal.

Duda, A. M. Simulation and measurement of the forest soil water balance. D, 1977, Duke University. 399 p.

Dudak, Richard M. The paleoecology of the Chadakoin Formation of Cattaraugus County, New York. M, 1980, SUNY at Buffalo. 99 p.

Dudar, John Steven. A study of the Verna ore, Beaverlodge Lake, Saskatchewan. M, 1957, University of Michigan.

Dudar, John Steven. The geology and mineralogy of the Verna uranium deposit, Beaverlodge, Saskatchewan. D, 1960, University of Michigan. 183 p.

Dudas, Marvin Joseph. Mineralogy and trace element chemistry of Mazama ash soils. D, 1973, Oregon State University. 119 p.

Duddy, Kathleen A. Pleistocene sediment characteristics near the City of Toledo waterlines and the Fondessy hazardous waste facility, eastern Lucas County, Ohio. M, 1987, Bowling Green State University. 106 p.

Duddy, Mark Morgan. Characteristics of thrust fault imbrication along the western margin of the Blue Ridge structural province, Buffalo Mountain, Tennessee. M, 1986, University of Tennessee, Knoxville. 151 p.

Dudek, Kathleen B. A petrographic study of fossiliferous Proterozoic cherts. M, 1985, Tulane University. 386 p.

Dudley, John G. Nutrient enrichment of ground water from septic tank disposal systems. M, 1973, University of Wisconsin-Madison.

Dudley, Jon Steven. Zeolitization of the Howson Facies, Telkwa Formation, British Columbia. D, 1983, University of Calgary. 302 p.

Dudley, Julia Lynn. Laramide folding and post-Laramide faulting in the Little Canyon Creek area, southeastern Bighorn Basin, Washakie County, Wyoming. M, 1984, University of Wyoming. 71 p.

Dudley, P. H., Jr. Geology of the Gavilan region (California). M, 1955, Pomona College.

Dudley, Paul H., Jr. Geology of the area adjacent to the Arroyo Seco Parkway, Los Angeles County, California. M, 1955, University of California, Los Angeles.

Dudley, Priscilla Perkins. Glaucophane schists and associated rocks of the Tiburon Peninsula, Marin County, California. D, 1967, University of California, Berkeley. 116 p.

Dudley, Raymond Wesley. Interpretation of the lithologic sequence constituting seismic reflections in northern Seminole County, Oklahoma. M, 1941, University of Oklahoma. 51 p.

Dudley, W. C., Jr. Paleoceanographic applications of oxygen isotope analyses of calcareous nannoplankton grown in culture. D, 1976, University of Hawaii. 168 p.

Dudley, William Wyatt, Jr. Hydrogeology and groundwater flow system of the central Ruby Mountains, Nevada. D, 1967, University of Illinois, Urbana. 117 p.

Dudley, William Wyatt, Jr. Seismic refraction and earth-resistivity investigation of hydrogeologic problems in the Humboldt River basin, Nevada. M, 1962, University of Illinois, Urbana.

Dudziak, Suzanne. Simulating unsaturated flow through a gravel-covered, multilayered lysimeter. M, 1988, University of Idaho. 78 p.

Duebendorfer, Ernest M. Geology of Frazier Park-Cuddy Valley area, California. M, 1979, University of California, Santa Barbara.

Duebendorfer, Ernest Martin. Structure, metamorphism, and kinematic history of the Cheyenne Belt, Medicine Bow Mountains, southeastern Wyoming. D, 1986, University of Wyoming. 414 p.

Duecker, Gregory Thomas. Devonian rocks of the Roberts Mountains Allochthon in the Roberts Mountains, central Nevada. M, 1985, University of California, Riverside. 99 p.

Duecker, John Cecil. Gravity traverse in northeastern Pennsylvania. M, 1952, Pennsylvania State University, University Park. 53 p.

Duedall, Iver Warren. The partial equivalent volumes of salts in seawater. M, 1966, Oregon State University. 47 p.

Duennebier, F. K. Spectral variation of the T phase. M, 1968, University of Hawaii. 49 p.

Duennebier, Frederick K. Moonquakes and meteorites; results from Apollo passive seismic experiment short period data. D, 1972, University of Hawaii at Manoa. 196 p.

Duerr, Michael David. Vicksburg Formation of southeastern Starr County, Texas; depositional systems, structural framework, and hydrocarbon reservoirs. M, 1986, Texas A&I University. 78 p.

Duerring, Nancy. Ion chemical and Tertiary mineralogical transformations and Devonian shales during hydrous pyrolysis, assimilation petroleum catagenesis. M, 1984, University of Pittsburgh.

Duever, Michael James. The distribution of trace elements in a small reservoir as influenced by two types of discharge. D, 1973, University of Georgia.

Duewel, Dennis Brandon. The stratigraphy of the Jefferson City Quadrangle, Missouri. M, 1957, University of Missouri, Columbia.

Duex, Timothy W. K/Ar age dates and U, Th, K geochemistry of the (Catahoula) Gueydan Formation (Oligocene or Miocene, lower) of south Texas. M, 1970, Rice University. 25 p.

Duex, Timothy William. Geology, geochemistry, and geochronology of volcanic rocks between Cuauhtemoc and La Junta, central Chihuahua, Mexico. D, 1983, University of Texas, Austin. 234 p.

Duey, Herbert David. Mineralogical and chemical study of certain shales from the Allegheny Formation from Clearfield County, Pennsylvania. M, 1957, Pennsylvania State University, University Park. 79 p.

Dufek, Debra Ann. Palynological analysis of floral changes caused by repeated volcanic ash burial of a coal-forming Upper Cretaceous peat swamp, Utah, U.S.A. D, 1987, Michigan State University. 232 p.

Duff, Denny Emerson. Some analyses of Pleistocene deposits in the Edmonton area. M, 1951, University of Alberta. 47 p.

Duff, James Kenneth. Structural geology of the Tony area, the Bunker Hill Mine, Kellogg, Idaho. M, 1978, University of Idaho. 101 p.

Duff, Sheila Louise. Some aspects of ecosystem stability in salt marsh soils. M, 1962, Dalhousie University.

Duffell, Stanley. The lode gold deposits of Canada. M, 1932, University of British Columbia.

Duffell, Stanley. The problem of diffusion and its relation to ore deposition. D, 1935, University of Toronto.

Duffet, Walter Nelson. Arroyo-shoreline relationships in northwest Baja California, Mexico. D, 1969, University of Colorado. 273 p.

Duffield, C. Solar shadow maps. M, 1975, University of Arizona.

Duffield, Glenn M. Intervention analysis applied to the quantity and quality of drainage from an abandoned underground coal mine in north-central Pennyslvania. M, 1985, Pennsylvania State University, University Park. 317 p.

Duffield, James A. Depositional environments of the Hermit Formation, central Arizona. M, 1985, Northern Arizona University. 82 p.

Duffield, John Burton, Jr. Investigation of Newcastle Sandstone (Cretaceous) for a water flood project, Black Thunder Field, Weston County, Wyoming. M, 1961, University of Oklahoma. 217 p.

Duffield, Susan Linda. Late Ordovician-Early Silurian acritarch biostratigraphy and taxonomy, Anticosti Island, Quebec. D, 1982, University of Waterloo. 338 p.

Duffield, Wendell Arthur. The petrology and structure of the El Pinal Tonalite (lower Upper Cretaceous), Baja California, Mexico. D, 1967, Stanford University. 130 p.

Duffin, Michael E. Bend Unit (Atokan), Knox and Stonewall counties, north-central Texas; depositional system, petrology, provenance, diagenesis, authigenic clay mineralogy, and reservoir quality. M, 1985, Baylor University. 360 p.

Duffin, W. J. Investigation of a method of estimating depth to shallow bed rock or water table. M, 1954, Massachusetts Institute of Technology. 34 p.

Dufford, Alvin E. Quaternary geology and groundwater resources of Kansas River valley between Bonner Springs and Lawrence, Kansas. M, 1953, University of Kansas. 177 p.

Duffy, C. J. Phase equilibria in the system MgO-MgF_2-SiO_2-H_2O. D, 1977, University of British Columbia.

Duffy, Christopher J. Investigation of recharge and groundwater conditions in the western region of the Roswell artesian basin. M, 1977, New Mexico Institute of Mining and Technology.

Duffy, Christopher J. Stochastic modeling of spatial and temporal water quality variations in groundwater. D, 1982, New Mexico Institute of Mining and Technology. 146 p.

Duffy, D. M. Some factors affecting soil flexibility. D, 1977, University of Arizona. 160 p.

Duffy, Leo Joseph. The chemistry of vitrinitic macerals from coals in the bituminous range. D, 1967, Pennsylvania State University, University Park. 265 p.

Duffy, Robert E., Jr. Seismic coal seam modeling constrained by depositional environment. M, 1980, University of Houston.

Duffy, William Joseph. Soil characterization, headcut erosion and Landsat classification in the San Juan Basin. M, 1983, Pennsylvania State University, University Park. 59 p.

Duford, James Matthew. Late Pleistocene and Holocene cirque glaciations in the Shuswap Highland area, British Columbia. M, 1976, University of Calgary. 100 p.

Dufresne, Cyrille. A study of the Kaniapiskau System in the Burnt Creek-Goodwood area, New Quebec and Labrador, Newfoundland. D, 1952, McGill University.

Dufresne, Cyrille. Faulting in the Saint Lawrence Plain (Quebec). M, 1947, McGill University.

Dufresne, Denis. Design and applications of a diversity stack computer program. M, 1987, University of Calgary. 85 p.

Dufresne, Douglas Paul. Distribution of MgO, palygorskite and other minerals in a North Florida phosphorite. M, 1988, University of Florida. 117 p.

Dugan, J. T. The relationship of the thermal regime of the soil to the extrinsic environment; with special reference to the North central United States. D, 1978, University of Nebraska, Lincoln. 258 p.

Dugan, Joseph P., Jr. Oxygen isotope geochemistry of the Allende carbonaceous chondrite and its implications for the origin of the Solar System. M, 1978, Northern Illinois University. 118 p.

Dugan, Thomas E. Investigation of late Tertiary to Recent movement along faults within the Kentucky River fault system in northern Madison, southern Fayette and southern Clark counties, Kentucky. M, 1983, Eastern Kentucky University. 93 p.

Dugas, Helene. Application de l'appareil de fluorescence-X de terrain, SOQUEM-FRXT, à l'exploration minière. M, 1982, Ecole Polytechnique. 165 p.

Dugas, Jean. Geology of the Perth map-area, Lanark and Leeds counties, Ontario. D, 1952, McGill University.

Duggan, Dannie E. Porosity variations in two deep Pliocene zones of Ventura Avenue Anticline as a function of structural and stratigraphic position. M, 1964, University of California, Riverside. 57 p.

Duggan, Michael D. Geology of a part of the San Joaquin Hills, Orange County, California. M, 1961, University of California, Los Angeles.

Duggan, Peter M. A study of the till deposits in northwestern Vermont. M, 1974, Miami University (Ohio). 79 p.

Duggan, Toni J. Petrography and mineral chemistry of Mauna Loa lavas. M, 1987, University of New Mexico.

Duggan, William LeRoy. Description of a well core from Portage County, Ohio. M, 1952, University of Michigan.

Dugolinsky, B. K. Chemistry and morphology of deep-sea manganese nodules and the significance of associated encrusting protozoans on nodule growth. D, 1976, University of Hawaii. 228 p.

Dugolinsky, B. K. Sedimentation of the Upper Devonian Sonyea Group of South-central New York. M, 1973, Syracuse University.

Duguid, James O. Flow in fractured porous media. D, 1973, Princeton University.

DuHamel, Jonathan E. Volcanic geology of the upper Cottonwood creek area, Park and Fremont counties, Colorado. M, 1968, Colorado School of Mines. 135 p.

Duhling, William H. Oxide facies iron formation in the Owl Creek mountains, northeastern Fremont County, Wyoming. M, 1970, University of Wyoming. 92 p.

Duhon, Michael P. and Dungan, James R. Subsurface and seismic investigation of the geopressure-geothermal potential of the Abbeville area of South Louisiana. M, 1979, University of Southwestern Louisiana.

Duigon, Mark Thomas. Baseline study of groundwater prior to strip mining, Wilson Site, southeastern Knox County, Indiana. M, 1977, Indiana University, Bloomington. 86 p.

Duke, David Allen. Infrared investigation of the crystal chemistry of olivine and humite minerals. D, 1962, University of Utah. 133 p.

Duke, David Allen. Jasperoid and ore deposits in the East Tintic mining district. M, 1959, University of Utah. 54 p.

Duke, Edward F. Part I; Stratigraphy, structure, and petrology of the Peterborough 15-minute Quadrangle, New Hampshire, and Part II; Graphite textural and isotopic variations in plutonic rocks, south-central New Hampshire. D, 1984, Dartmouth College. 168 p.

Duke, Edward F. Petrology of Spaulding Group tonalites from Penacook Quadrangle, New Hampshire. M, 1978, Dartmouth College. 117 p.

Duke, Genet Ide. Structure and petrology of the eastern Concord Quadrangle, New Hampshire. M, 1984, Dartmouth College. 110 p.

Duke, James Alan. The psammophytes of the Carolina fall-line sandhills. D, 1960, University of North Carolina, Chapel Hill. 81 p.

Duke, John M. The effect of variation of oxygen fugacity on the crystallization of an alkali basalt from the Azores. M, 1971, McGill University.

Duke, John Murray. An experimental investigation of the distribution of the period four transition elements among olivine, calcic clinopyroxene and mafic silicate melt. D, 1973, University of Connecticut. 131 p.

Duke, Michael B. Petrology of the basaltic achondrite meteorites. D, 1963, California Institute of Technology. 362 p.

Duke, Michael John Maclachlan. Geochemistry of the Exshaw Shale of Alberta; an application of neutron activation analysis and related techniques. M, 1983, University of Alberta. 186 p.

Duke, N. A. Petrology of the Blue Mountain and Bigwood nepheline syenite gneiss complexes of the Grenville Province. M, 1975, University of Western Ontario. 247 p.

Duke, Norman Albert. A metallogenic study of the central Virginian gold-pyrite belt. D, 1983, University of Virginia.

Duke, Walter. Turonian (Cretaceous) stratigraphy and micropaleontology, Cumberland Gap, Wyoming-Woodside, Utah. M, 1964, University of Colorado.

Duke, William Lewis. Sedimentology of the Upper Cretaceous (Turonian) Cardium Formation in outcrop in southern Alberta. D, 1985, McMaster University. 724 p.

Dukes, Bill Jady. Geology of the Farris Creek area, Gunnison County, Colorado. M, 1953, University of Kentucky. 46 p.

Dukes, George Houston, Jr. Some Tertiary fossil woods of Louisiana and Mississippi. D, 1961, Louisiana State University. 185 p.

Dukozoglu, Hilmi. A method of mining the Kumtepi lignite deposit of Turkey (T 951). M, 1951, University of Missouri, Rolla.

Dula, Philip Charles. The geology of the uppermost strata of Rocky Mountain, Floyd County, Georgia. M, 1982, Emory University. 104 p.

Dula, William F., Jr. Dynamic analysis of quartzite (Sawatch and Parting formations) and limestone (Leadville Limestone), White River Uplift area, Northwest Colorado. M, 1979, University of Colorado.

Dula, William Frederick, Jr. High temperature deformation of wet and dry artificial quartz gouge; Volumes I and II. D, 1985, Texas A&M University. 405 p.

Dulaney, Brenda Sue. The "first Wilcox Sand" of north-central Payne County, Oklahoma. M, 1987, Oklahoma State University. 106 p.

Dulaney, Ernest N. The velocity behavior of a growing crack. M, 1960, Massachusetts Institute of Technology. 36 p.

Dulaney, James Patrick. Diagenesis of the Wilcox Sandstone of the Texas Gulf Coast province. M, 1982, University of Texas at Dallas. 79 p.

Dulanto, Alejandro Euribe *see* Euribe Dulanto, Alejandro

Dulekoz, Erhan. Geochronology and clay mineralogy of the Eskridge shale (Permian) near Manhattan, Kansas. M, 1969, Kansas State University. 55 p.

Duley, Dale Hamilton. Mississippian stratigraphy of the Meadow Valley and Arrow Canyon ranges, southeastern Nevada. M, 1957, University of California, Berkeley. 103 p.

Dulian, James Joseph. Paleoecology of the Brayton local biota, late Wisconsinan of southwestern Iowa. M, 1975, University of Iowa. 50 p.

Dulin, Lise A. Distribution of Mn and Re in sediment from Northwest Providence Channel, Bahamas-a relation to climate?. M, 1984, Miami University (Ohio). 87 p.

Dull-Coleman, Michele M. Evaluation of acid treatments and paraffin control techniques in Bass Island trend oil reservoirs. M, 1986, SUNY at Buffalo. 63 p.

Dumas, David Byron. Seismic structure around salt dome formations. M, 1976, Rice University. 19 p.

Dumas, David Byron. Seismicity of West Texas. D, 1981, University of Texas at Dallas. 94 p.

Dumas, Michael C. Electrical resistivity and dielectric constant of frozen rocks. M, 1962, Colorado School of Mines. 44 p.

Dumbros, Nicholas. Southwest extension of the Boston Basin. M, 1934, Massachusetts Institute of Technology. 80 p.

Dumesnil, Jean-Claude. Analyse chimique instrumentale des roches. M, 1968, Universite de Montreal.

Dumeyer, John McMurray. Geohydrology of the Rio Grande Valley from South Fork to Del Norte, Colorado. M, 1971, University of Arizona.

Dumitru, Trevor Alan. Plate tectonic controls on time-varying geothermal gradients in the Great Valley forearc basin, California; fission track analysis and computer modeling. M, 1986, University of Texas, Austin. 296 p.

Dumonceaux, Gayle M. Stratigraphy and depositional environments of the Three Forks Formation (Upper Devonian), Williston Basin, North Dakota. M, 1984, University of North Dakota. 189 p.

Dumoulin, J. A. Eocene-Oligocene silicoflagellates of the Kreyenhagen Formation, Fresno County, California. M, 1979, University of Wisconsin-Madison.

Dumper, Thomas A. The relationship of the physical environment to land use potential in Muskegon County, Michigan. M, 1970, Virginia Polytechnic Institute and State University.

Dumper, Thomas Apted. A computer-generated model for land-use decisions based on the physical environment. D, 1972, Virginia Polytechnic Institute and State University. 224 p.

Dunagan, Joseph F., Jr. Geology of the Lower Ordovician rocks of the Choctaw Anticlinorium, southeastern Oklahoma. M, 1976, University of Oklahoma. 46 p.

Dunaway, Sabrina G. Petrology and diagenesis of Trinidad Sandstone (Upper Cretaceous), Huerfano and Las Animas counties, Colorado. M, 1987, Wichita State University.

Dunaway, William Edmond. Structure of Cretaceous rocks, central Travis County, Texas. M, 1962, University of Texas, Austin.

Dunay, Robert Edmund. The palynology of the Triassic Dockum Group of Texas, and its application to stratigraphic problems of the Dockum Group. D, 1972, Pennsylvania State University, University Park. 382 p.

Dunbar, Carl O. The stratigraphy and paleontology of the Devonian of western Tennessee. D, 1917, Yale University.

Dunbar, Cecil, Jr. Geologic field trip through the York County portion of the Middletown Quadrangle, Southeast Pennsylvania area. M, 1967, Virginia State University. 34 p.

Dunbar, Clarence P. A geological and soil survey of the Cul-de-Sac Plain of the Republic of Haiti. M, 1935, Louisiana State University.

Dunbar, David M. A seismic velocity model of the Clark Hill Reservoir area. M, 1977, Georgia Institute of Technology. 59 p.

Dunbar, Elizabeth Urquhart. Glacial lake succession in the Genesee Valley. M, 1923, University of Rochester. 60 p.

Dunbar, Gordon Douglas. Multistory Pennsylvanian channel deposits in the vicinity of the Dotiki Mine, Webster County, Kentucky; effects on the thickness, extent, and minability of associated coals. M, 1988, Southern Illinois University, Carbondale. 172 p.

Dunbar, John Andrew, Jr. Kinematics and dynamics of continental breakup. D, 1988, University of Texas, Austin. 179 p.

Dunbar, Nelia W. Pre-eruptive volatile contents and degassing systematics of rhyolitic magmas from the Taupo volcanic zone, New Zealand. D, 1988, New Mexico Institute of Mining and Technology.

Dunbar, R. O. The geology of Como Bluff Anticline, Albany-Carbon counties, Wyoming. M, 1942, University of Wyoming. 40 p.

Dunbar, Robert Bruce. Sedimentation and the history of upwelling and climate in high fertility areas of the northeastern Pacific Ocean. D, 1981, University of California, San Diego.

Dunbar, W. S. The determination of fault models from geodetic data. D, 1977, Stanford University. 233 p.

Dunbar, W. Scott. An extremal approach to some geophysical problems. M, 1973, University of Toronto.

Duncan, Dennis C. Intertidal Ostracoda of the central California coast. M, 1969, Bowling Green State University. 124 p.

Duncan, Donald A. Geology of the southwest quarter of the Eutawville quadrangle, South Carolina. M, 1963, University of South Carolina. 44 p.

Duncan, Donald Cave. Upper Cambrian trilobites from Montana and Yellowstone National Park. M, 1937, University of Montana. 121 p.

Duncan, Douglas Wells. Structural analysis of the central Tobacco Root Mountains, Southwest Montana. M, 1978, Pennsylvania State University, University Park. 79 p.

Duncan, E. J. Scott. Mesoscopic fractures, microscopic fractures and tensile strength characteristics of the Shield Granite; southeastern Manitoba. M, 1987, University of Manitoba.

Duncan, Edward A. Delineation of delta types, Norias Delta system, Frio Formation, South Texas. M, 1987, University of Texas, Austin. 67 p.

Duncan, Gary Alan. Geology of the Bonneville Peak area, Bannock and Caribou counties, Idaho. M, 1978, Idaho State University. 29 p.

Duncan, Georgianna Hawley. Geologic structure between the Catskills and Berkshires in the region of Hudson, New York. M, 1928, Cornell University.

Duncan, Glen A. Sediment characteristics and sedimentary processes of North Beach, St. Catherines Island, Georgia. M, 1983, University of Georgia.

Duncan, Gregory Wade. Structural geology and stratigraphy of the east half of township 37 north, range 38 east, and the southeastern half of township 37 north, range 39 east, Stevens County, Washington. M, 1982, Washington State University. 68 p.

Duncan, Helen M. Trepostomata from the Traverse Group of Michigan. M, 1937, University of Montana. 162 p.

Duncan, Ian James. The evolution of the Thor-Odin gneiss dome and related geochronological studies. D, 1982, University of British Columbia.

Duncan, John Leslie, Jr. Benthonic foraminifera from the Mohnian (upper Miocene) of Newport Bay, Orange County, California. M, 1979, University of California, Los Angeles.

Duncan, John Russell, Jr. Late Pleistocene and post-glacial sedimentation and stratigraphy of deep-sea environments off Oregon. D, 1968, Oregon State University. 222 p.

Duncan, Mardon. Localization of pressure solution and the formation of discrete solution seams. D, 1988, Texas A&M University.

Duncan, Mark Stewart. Structural analysis of the pre-Beltian metamorphic rocks of the southern Highland Mountains, Madison and Silver Bow counties, Montana. D, 1976, Indiana University, Bloomington. 222 p.

Duncan, Mary Anne. Geology of the West Bourland Mountain area, Marathon Basin, West Texas. M, 1987, University of Texas, Austin. 118 p.

Duncan, P. M. A gravity study of the Saguenay-Lac St. Jean area, Quebec. M, 1975, University of Toronto.

Duncan, P. M. Electromagnetic deep crustal sounding with a controlled, pseudo-noise source. D, 1978, University of Toronto.

Duncan, Q. Randolph. Some investigations of oil shale in Colorado; vicinity of Rifle, Garfield County. M, 1919, University of Colorado.

Duncan, Robert A. Chemical heterogeneity in the upper mantle, an isotopic investigation of the plume model. D, 1972, Stanford University.

Duncan, Robert C. Geochemical investigation of the Chattanooga Shale, Northwest Arkansas. M, 1983, University of Arkansas, Fayetteville. 184 p.

Duncan, Robert Louis. The Mesaverde Formation; Upper Cretaceous. M, 1952, University of Colorado.

Duncan, Ronald C. Hydrogeologic reconnaissance study, incorporating remotely sensed data, in a section of Mali, Africa. M, 1986, University of Idaho. 87 p.

Duncan, Truman E., Jr. Paleoecology of the Upper Ordovician Bulls Fork Formation in the Manchester Islands Quadrangle, northeastern Kentucky. M, 1980, Eastern Kentucky University. 63 p.

Duncan, William M. Geology of the Glengary, West Virginia-Virginia, Quadrangle. M, 1967, West Virginia University.

Duncker, Katherine Elizabeth. Trace-element geochemistry and stable isotope constraints on the petrogenesis of Cerros Del Rio Lavas, Jemez Mountains, New Mexico. M, 1988, University of Texas, Arlington. 158 p.

Dunford-Jackson, Carey Stanly. The geomorphic evolution of the Rappahannock River basin. M, 1978, University of Virginia. 92 p.

Dungan, James R. and Duhon, Michael P. Subsurface and seismic investigation of the geopressure-geothermal potential of the Abbeville area of South Louisiana. M, 1979, University of Southwestern Louisiana.

Dungan, Michael Allen. Structural and petrographic reconnaissance on northern portion of Sultan ultramafic-mafic complex, Snohomish County, Washington. M, 1971, University of Washington. 32 p.

Dungan, Michael Allen. The origin, emplacement, and metamorphism of the Sultan mafic-ultramafic complex, North Cascades, Snohomish County, Washington. D, 1974, University of Washington. 227 p.

Dungan, Q. Randolph. Some investigations of oil shales of Colorado. M, 1919, University of Colorado.

Dunham, John B. Depositional environments, paleogeography, and diagenesis of the Upper Ordovician, Lower Silurian carbonate platform of central Nevada. D, 1977, University of California, Riverside. 153 p.

Dunham, Kingsley Charles. The geology of the Organ Mountains; with an account of the geology and mineral resources of Dona Ana County, New Mexico. D, 1935, Harvard University.

Dunham, Montgomery S. The geology of the Blue Jay Mine area, Helvetia, Arizona. M, 1937, University of Arizona.

Dunham, Robert J. Geology of uranium in the Chadron area, Nebraska and South Dakota. D, 1961, Yale University.

Dunham, Robert Jacob. Structure and orogenic history of the Lake Classen area, Arbuckle Mountains, Oklahoma. M, 1951, University of Oklahoma. 108 p.

Dunkelman, Thomas Julian. The structural and stratigrahic evolution of Lake Turkana, Kenya; as deduced from a multichannel seismic survey. M, 1986, Duke University. 64 p.

Dunkerley, Robert. Stylolites in the vicinity of Syracuse, New York. M, 1950, Syracuse University.

Dunkhase, John A. A comparative study of the whole rock geochemistry of the uranium mineralized central intrusive at Marysvale, Utah, to nonmineralized intrusives in Southwest Utah. D, 1980, Colorado School of Mines. 126 p.

Dunkin, Joyce Sattler. Fluorescence of the liptinite macerals in selected Ohio coals. M, 1981, University of Toledo. 106 p.

Dunkin, Ned R., Jr. Influence of coal characteristics on the free-swelling properties of selected Ohio coals. M, 1982, University of Toledo. 223 p.

Dunkle, David H. The cranial osteology of Notelops brama (Agassiz) an elopid fish from the Cretaceous of Brazil. D, 1939, Harvard University.

Dunlap, Dennis Gordon. Tertiary geology of the Muddy Creek Basin, Beaverhead County, Montana. M, 1982, University of Montana. 133 p.

Dunlap, John Bettes. The geology of the area bordering the Brazos River in southeastern Milam and northeastern Burleson counties, Texas. M, 1955, Texas A&M University. 91 p.

Dunlap, Lloyd E. Hydrogeology in the adjacent uplands of the Saline, Smoky Hill and Solomon rivers in Saline and Dickinson county. M, 1977, Kansas State University. 93 p.

Dunlap, Richard E. The geology of the Pinta Dome–Navajo Springs helium fields, Apache County, Arizona. M, 1969, University of Arizona.

Dunlap, Wayne Alan. Deformation characteristics of granular materials subjected to rapid, repetitive loading. D, 1966, Texas A&M University. 223 p.

Dunlap, William Howard. Compaction of an unsaturated soil under a general state of stress. D, 1969, University of Illinois, Urbana. 145 p.

Dunleavy, Jeffrey M. A geophysical investigation of the contact along the northern margin of the Newark Triassic basin, Hosensack, Pennsylvania, to Gladstone, New Jersey. M, 1975, Lehigh University. 41 p.

Dunlop, David John. Magnetic properties in single domain grains; theory and application. M, 1964, University of Toronto.

Dunlop, David John. The remanent magnetism of rocks containing interacting single domain ferromagnetic grains. D, 1968, University of Toronto.

Dunlop, W. B. Pleistocene and Recent deposits of the Church Point area, Nova Scotia. M, 1952, Acadia University.

Dunn, Anthony Charles. The geomorphological base to economic change in Lassen County, California. M, 1973, University of Nevada - Mackay School of Mines. 112 p.

Dunn, Anthony Price. Zinc and lead occurrences in Cambrian dolostone, Gataga River area, British Columbia. M, 1982, University of Western Ontario. 127 p.

Dunn, Darrel Eugene. Hydrogeology of the Stettler area, Alberta, Canada. D, 1967, University of Illinois, Urbana. 392 p.

Dunn, David E. Geology of the Crystal Peak area, Millard County, Utah. M, 1959, Southern Methodist University. 53 p.

Dunn, David Evan. Evolution of the Chama Basin and Archuleta Anticlinorium, eastern Archuleta County, Colorado. D, 1964, University of Texas, Austin. 142 p.

Dunn, David Lawrence. Conodont biostratigraphy of the Mississippian-Pennsylvanian boundary and Morrowan Series in western United States. D, 1967, University of Wisconsin-Madison. 187 p.

Dunn, David Lawrence. Devonian chitinozoans from the Cedar Valley Formation in Iowa. M, 1959, University of Iowa. 61 p.

Dunn, David Lynn. Cadmium-113m as a biogeochemical tracer for cadmium in Lake Michigan. D, 1987, Clemson University. 133 p.

Dunn, Dean Alan. Miocene sediments of the equatorial Pacific Ocean; carbonate stratigraphy and dissolution history. D, 1982, University of Rhode Island. 320 p.

Dunn, Dennis P. Petrology of the San Antonio scheelite skarn, Baviacora, Sonora, Mexico. M, 1980, Arizona State University. 179 p.

Dunn, Gilbert Riley. Low pressure metamorphism in the Orrs Island-Harpswell Neck area, Maine. M, 1988, West Virginia University. 97 p.

Dunn, Harold L., Jr. Subsurface stratigraphy and foraminifera of the sandstones of the Pierre Formation (Cretaceous) in the Denver Basin; vicinity of Weld and Morgan counties, Colorado. M, 1955, University of Colorado.

Dunn, J. L. Stratigraphy of the Hite Bed and the uppermost part of the Chinle Formation in the Red Canyon-White Canyon area, southeastern Utah. M, 1975, University of Arizona.

Dunn, James Robert. Geology of the western Mono Lake area, California. D, 1951, University of California, Berkeley. 133 p.

Dunn, James V. Error analysis of a laser theodolite. M, 1966, Ohio State University.

Dunn, Jeff L. Interests and attitudes of undergraduate students in "coursette" vs. "conventional" approaches to introductory geology. D, 1979, University of South Carolina. 75 p.

Dunn, John Todd. Investigations of the chemistry of silicate melts; kinetics, structure, and redox equilibria. D, 1983, University of Alberta. 148 p.

Dunn, Lisa Gay. Lithology and stratigraphy of the Naqus Formation, Gulf of Suez region, Egypt. M, 1985, Washington University. 138 p.

Dunn, Martha Jean. Depositional history and paleoecology of an Upper Devonian (Frasnian) bioherm, Mount Irish, Nevada. M, 1979, SUNY at Binghamton.

Dunn, Michael D. Paleomagnetism of Eocene volcanics and intrusives from northeastern Washington. M, 1981, Eastern Washington University. 52 p.

Dunn, Mildred Carneal. The relationship of isostasy to the preservation of playas in the Great Basin. M, 1977, University of Nevada. 75 p.

Dunn, Paul H. Silurian foraminifera of the Mississippi Basin. D, 1932, University of Chicago. 25 p.

Dunn, Paul H. The Cynthiana Formation of north-central Kentucky. M, 1924, Ohio State University.

Dunn, Paul M. Paleoecological analysis of coastal Maine's Presumpscot Formation utilizing sediments and fauna. M, 1985, Ohio University, Athens. 79 p.

Dunn, Pete J. Genthelvite, a member of the helvite group. M, 1974, Boston University. 41 p.

Dunn, Pete J. The lead silicate assemblage at Franklin, New Jersey. D, 1984, University of Delaware. 189 p.

Dunn, Peter Ayres. Alluvial chromite deposits of southern Chester and Lancaster counties, Pa. M, 1962, Pennsylvania State University, University Park. 88 p.

Dunn, Robert Jeffrey. Hydraulic conductivity of soils in relation to the subsurface movement of hazardous wastes. D, 1983, University of California, Berkeley. 338 p.

Dunn, Sandra Louise Dimitre. Spatial and sequential relations between the Jackson and Cache Creek thrusts on Teton Pass, Idaho and Wyoming. M, 1983, Idaho State University. 94 p.

Dunn, Steven Robert. Petrology and metamorphism of basic rocks in the Rodskar area, Gurskoy, western Norway. M, 1984, University of Wisconsin-Madison.

Dunn, Thomas Lowell. Mineral reactions in sandstones of the Lysite Mountain area, central Wyoming. M, 1979, University of Wyoming. 81 p.

Dunn, Townsend H. Electrode studies of stability constants of Cd-river water organic matter complexes. M, 1974, Georgia Institute of Technology. 70 p.

Dunn, William John. Paleomagnetic and petrographic investigation of the Taum Sauk Limestone, Southeast Missouri. M, 1984, University of Oklahoma. 66 p.

Dunne, George Charles. Geology of the Devil's Playground area, eastern Mojave Desert, California. D, 1972, Rice University. 79 p.

Dunne, George Charles. Petrology of a portion of the Pat Keyes pluton, Inyo County, California. M, 1970, San Jose State University. 73 p.

Dunne, James A. Tungsten mineralization at the Swisshelm Mountains, Cochise County, Arizona. M, 1957, Columbia University, Teachers College.

Dunne, James Arthur. Thermal studies of sulfide minerals. D, 1961, Columbia University, Teachers College. 116 p.

Dunne, Lorie A. Depositional environments of the Upper Devonian Oneonta Formation of South-central New York. M, 1980, University of Rhode Island.

Dunnewald, J. B. Geology of the Fish Lake Mountain area, Fremont County, Wyoming. M, 1958, University of Wyoming. 71 p.

Dunning, Charles Preston. Provenance and paleotectonic setting of conglomerates in the Virgilian Holder Formation, northern Sacramento Mountains, New Mexico. M, 1978, Rice University. 111 p.

Dunning, Gregory. The geology and platinum group mineralization of the Roby Zone, Lac des Iles Complex, northwestern Ontario. M, 1979, Carleton University. 124 p.

Dunning, Gregory Ralph. The geology, geochemistry, geochronology and regional setting of the Annieopsquotch Complex and related rocks of Southwest Newfoundland. D, 1984, Memorial University of Newfoundland. 403 p.

Dunning, Jeremy D. A microscopic, submicroscopic, and microseismic analysis of stable crack propagation and the effect of chemical environment on stable crack propagation in synthetic quartz. D, 1978, University of North Carolina, Chapel Hill. 73 p.

Dunning, Jeremy David. Self diffusion of lattice vacancies as a possible mechanism for failure in crystals and polycrystalline aggregates. M, 1975, Rutgers, The State University, New Brunswick. 50 p.

Dunnivant, Frank Morris. Congener-specific PCB chemical and physical parameters for evaluation of environmental weathering of Aroclors. D, 1988, Clemson University. 184 p.

Dunphy, J. L. Surface features of the Stetson Bank area and a non-bank area of comparable depth. M, 1975, Texas A&M University.

Dunrud, C. R. Volcanic rocks of the Jack Creek area, southeastern Absaroka Range, Park County, Wyoming. M, 1962, University of Wyoming. 92 p.

Dunscombe, Thomas D. The silicified brachiopod fauna of the Macy Formation. M, 1959, University of Missouri, Rolla.

Dunsmore, Dennis Joseph. A paleomagnetic study of the Lynn Lake and Fraser Lake gabbros, northern Manitoba. M, 1986, University of Windsor. 99 p.

Dunsmore, Hugh E. Diagenetic model for Keg River Formation (Middle Devonian), Rainbow area, northwestern Alberta. M, 1971, University of Calgary. 154 p.

Dunwiddie, Peter William. Holocene forest dynamics on Mount Rainier, Washington. D, 1983, University of Washington. 129 p.

Duplantis, Merle James. Depositional systems in the Midway and Wilcox groups (Paleocene-lower Eocene), North Mississippi. M, 1975, University of Mississippi. 87 p.

Dupler, Philip C. An analysis of carbonate depositional environments in the (Ordovician) Rockland Formation of northern New York State. M, 1970, SUNY at Binghamton. 93 p.

Dupre, David Carl. Geology of the Unionville gas field, Lincoln Parish, Louisiana. M, 1979, Louisiana Tech University.

Dupre, W. R. Quaternary history of the Watsonville Lowlands, north-central Monterey Bay region, California. D, 1975, Stanford University. 232 p.

Dupré, William Roark. Geology of the Zambrano Quadrangle, Honduras, Central America. M, 1970, University of Texas, Austin.

Dupree, Jean A. Stratigraphic control of uranium mineralization at the Pitch Mine, Saguache County, Colorado. M, 1979, Colorado School of Mines. 111 p.

Dupuis, Roy H. Stratigraphy, structure and metamorphic history of the northern half of the Nottely Dam Quadrangle, Georgia-North Carolina. M, 1975, University of Georgia.

Dupuy, John R. The hydrologic significance of geologic structure within the southeastern Hueco Bolson, El Paso County, Texas. M, 1984, University of Texas at El Paso.

Duque, Pablo. Sodic pyroxenes in blueschists from the northern Cascades, Washington. M, 1977, University of Michigan.

Duque, Thomas A. Geology of organic rich mud in Slocum and Hancock creeks, Craven County, North Carolina. M, 1978, East Carolina University. 54 p.

Duquette, Gilles. Geology of the Weedon Lake area and its vicinity, Wolfe and Compton counties, Quebec. D, 1961, Universite Laval. 308 p.

Durall, Rebecca L. Diagenesis and porosity development of the Mission Canyon and Charles formations (Mississippian), Treetop and Whiskey Joe fields, North Dakota. M, 1987, University of North Dakota. 208 p.

Duran, Alexander Paul. Production and release of N_2O from the Assabet River. D, 1984, Harvard University. 224 p.

Duran, Philip B. The use of electromagnetic conductivity techniques in the delineation of groundwater pollution plumes. M, 1982, Boston University. 174 p.

Duran, William Kent. Geology of the Cass and Yale 7.5 Minute quadrangles of Franklin and Johnson counties, Arkansas. M, 1987, University of Arkansas, Fayetteville.

Durand, Benoit. Le rôle des additifs minéraux dans les réactions alcalis-granulat. M, 1985, Ecole Polytechnique. 199 p.

Durand, Harvey S. Geology of the Sahuaro Lake area, Maricopa County, Arizona. M, 1967, University of Arizona.

Durand, Loyal, Jr. The river systems of Wisconsin. M, 1925, University of Wisconsin-Madison.

Durand, Marc. Etude de propriétes physiques et chimiques de calcaires exploités dans la région de Montréal en rapport avec leur utilisation comme agrégats a béton. M, 1969, Ecole Polytechnique. 123 p.

Durand, Thomas Jean-Paul. Wave-induced seepage in sea beds below offshore structures. D, 1980, University of Wisconsin-Madison. 445 p.

Durazzi, J. T. The shell chemistry of ostracods and its paleoecological significance. D, 1975, Case Western Reserve University. 206 p.

Durazzo, Aldo. Exsolution in the systems bornite-chalcopyrite and pyrrhotite-pentlandite; an approach to textural interpretation. D, 1980, University of Tennessee, Knoxville. 101 p.

Durden, Christopher J. Systematics and morphology of Acadian Pennsylvanian blattoid insects (Dictyoptera: Palaeoblattina); a contribution to the classification and phylogeny of Palaeozoic insects. D, 1972, Yale University.

Dureck, Joseph J. Selected aspects of the Upper Mississippi Valley lead-zinc deposits. M, 1963, Columbia University, Teachers College.

Duree, Dana K. Sedimentology of mid-Permian strata of the Sublett Range, south central Idaho. M, 1983, Texas A&M University. 147 p.

Durek, Joseph John. Some characteristics of weathered outcrops of disseminated copper deposits. D, 1964, Columbia University, Teachers College. 158 p.

Duren, Fred Kenneth, Jr. Optimizing flood control allocation for a multipurpose reservoir. M, 1970, University of Nevada. 118 p.

Durfee, Daniel. Nuclear well-logging as a geologic tool. M, 1977, Marshall University. 39 p.

Durfee, George Austin. A regional gravity survey of the Cuyuna iron range, Minnesota. M, 1956, Michigan Technological University. 26 p.

Durfee, M. Charles and Church, Richard E. Geology of the Fossil Creek area, White Mountains, Alaska. M, 1961, University of Alaska, Fairbanks. 96 p.

Durfee, Steven L. Prediction of liquefaction reactivity and structural analysis of coals using pyrolysis mass spectrometry with computerized pattern recognition. D, 1986, Colorado School of Mines. 212 p.

Durfee, Wilda. Field evidence for the presence of primary structures in the Inwood Limestone in the type locality (New York). M, 1934, Columbia University, Teachers College.

Durfuee, Barbara A. Geology and structure of the Carrizo Mountain metarhyolites, Hudspeth and Culberson counties, Texas. M, 1984, University of Texas at El Paso.

Durgin, Dana C. Petrology, contact metamorphism, and hydrothermal alteration of the Brooksville Greenschist, Brooksville, Maine. M, 1972, University of Washington.

Durgin, Philip Bassett. Geology of North Panamint Valley, Inyo County, California. M, 1974, University of Massachusetts. 137 p.

Durgin, Philip Bassett. The influence of forest leachates on erosion of granitic terrane. D, 1983, University of Idaho. 99 p.

Durham, Charles Albert. The stratigraphy of the Eagle Ford Formation from the Red River southward to Austin. M, 1931, University of Texas, Austin.

Durham, Charles Albert, Jr. Subsurface geology of Southeast Lincoln oil field, Kingfisher County, Oklahoma. M, 1962, University of Oklahoma. 35 p.

Durham, Cordelia Louise. A study of the foraminiferal assemblages from a well in South Pass Area Block 30, Louisiana (offshore). M, 1967, Louisiana State University.

Durham, David Peterson. The Fort Payne Formation (Mississippian) of North Alabama. M, 1960, Emory University. 74 p.

Durham, Forrest. Some sporadic minor buckling of strata in south-central New York. M, 1947, Cornell University.

Durham, Forrest. The geomorphology of the Tioughnioga River of central New York. D, 1954, Syracuse University. 101 p.

Durham, John Wyatt. Epitoniidae of the Mesozoic and Tertiary on the west coast of North America. M, 1936, University of California, Berkeley. 89 p.

Durham, John Wyatt. Zones of the Oligocene of northwestern Washington based on megafossils. D, 1941, University of California, Berkeley. 257 p.

Durham, Jon A. Structural geology of the northern part of the East Helena Quadrangle, Lewis and Clark County, Montana. M, 1972, University of New Mexico. 71 p.

Durham, William Bryan. Plastic flow of single-crystal olivine. D, 1975, Massachusetts Institute of Technology. 253 p.

Durisek, E. Jane. A trace-element study of some hypersolvus and subsolvus granites. M, 1964, Pennsylvania State University, University Park. 84 p.

Durkee, Edward F. Cambrian stratigraphy and paleontology of the east flank Bighorn Mountains, Johnson and Sheridan counties, Wyoming. M, 1953, University of Wyoming. 85 p.

Durkee, Steven. Depositional environments of the Lower Cretaceous Smith's Formation within a portion of Idaho-Wyoming thrustbelt. M, 1979, Idaho State University. 70 p.

Durkin, Thomas V. A palynological study of the Petrified Forest area of the early Eocene Wasatch deposits near Buffalo, Wyoming. M, 1986, South Dakota School of Mines & Technology.

Durler, David L. Geology of soil formation, central Texas. M, 1980, Baylor University. 151 p.

Durning, William Perry. Geology and mineralization of Little Hill Mines area, northern Santa Catalina Mountains, Pinal County, Arizona. M, 1972, University of Arizona.

Durocher, A. C. Geochemistry of mafic intrusions and extrusions of Nova Scotia. M, 1974, Acadia University.

Durocher, M. E. Petrology of the Gracefield Pluton. M, 1977, University of Ottawa. 142 p.

Durocher, Marcel Elzear Emery. The geology of Opemiska Township, Quebec, Canada. D, 1984, University of Georgia. 472 p.

Duroy, Yannick. Subsurface densities and lithospheric flexure of the Himalayan Foreland in Pakistan, interpreted from gravity data. M, 1987, Oregon State University. 76 p.

Durrani, Javaid A. Seismic investigation of the tectonic and stratigraphic history, eastern South Park, Park County, Colorado. D, 1980, Colorado School of Mines. 138 p.

Durrant, Richard Lee. A critical view of the role of academic geomorphology and its place in society; a disciplinary assessment. D, 1986, University of Waterloo. 508 p.

Durrell, Cordell. Metamorphism in the southern Sierra Nevada northeast of Visalia, California. D, 1936, University of California, Berkeley. 231 p.

Durrenberger, Sally. An integrated geophysical investigation and comparison of compressional and shear wave seismic reflection data from the San Juan volcanic area; southwestern Colorado. M, 1986, Colorado School of Mines. 96 p.

Durst, Fred M. The identification and correlation of the marine terraces of the California coast between Santa Cruz and Bolinas. M, 1915, University of California, Berkeley. 84 p.

Durst, T. L. The mineralogical and chemical development of the Nickel Mountain (Oregon) nickel laterite deposit. D, 1976, Case Western Reserve University. 266 p.

Durtsche, J. S. Sliding friction and fracture of rocks. D, 1973, New Mexico Institute of Mining and Technology. 298 p.

Duruewuru, Anthony U. Thermodynamic analysis of transient two-phase flow in oil and gas reservoirs. D, 1985, University of Oklahoma. 263 p.

Durupinar, Ahmet T. Determination of control points using the method of independent geodetic control. M, 1962, Ohio State University.

Duschatko, Robert W. Fracture studies in the Lucero Uplift, New Mexico. M, 1956, Columbia University, Teachers College.

Duschenes, Jeremy David. Evolution of the oceanic lithosphere and shear wave travel time residuals from oceanic earthquakes. M, 1976, Massachusetts Institute of Technology. 125 p.

Dusenberry, Arthur N. The foraminifera from O. D. Hanna's type locality in the Eocene near Vacaville, California. M, 1933, Columbia University, Teachers College.

Duskin, Douglas John. Economic geology of the gypsum deposits (Silurian) at Union Springs, New York. M, 1969, Cornell University.

Duskin, Priscilla. The glacial geology of the Rosendale, New York, Quadrangle. M, 1985, Rensselaer Polytechnic Institute. 58 p.

Dussault, Chantal. Minéralogie et paragenèse des veines aurifères de la Mine Ferderber, Val d'Or, Québec. M, 1986, Ecole Polytechnique. 144 p.

Dussell, Eric. Listwanites and their relationship to gold mineralization at Erickson Mine, British Columbia, Canada. M, 1986, Western Washington University. 90 p.

Duster, David W. Evaluation of the hydrogeochemistry of the Yak Tunnel, Leadville, Colorado, through the use of the geochemical computer program, MINTEQ. M, 1988, Colorado School of Mines. 85 p.

Dustin, J. D. Hydrogeology of Utah Lake with emphasis on Goshen Bay. D, 1978, Brigham Young University. 176 p.

Duston, Nina Marie. Water chemistry, sediment chemistry, and carbonate sedimentation in Littlefield Lake, Michigan; implications for production and diagenesis of lacustrine carbonates. D, 1984, University of Michigan. 163 p.

Dutch, S. I. The Creighton Pluton, Ontario, and its significance to the geologic history of the Sudbury region. D, 1976, Columbia University, Teachers College. 141 p.

Dutcher, Russel R. Certain tectonic and petrographic relationships of the Ashley Falls, Massachusetts-Connecticut, Quadrangle. M, 1953, University of Massachusetts. 52 p.

Dutcher, Russell Richardson. Physical, chemical and thermal properties of selected vitrinitic substances. D, 1960, Pennsylvania State University, University Park. 201 p.

Dutro, John Thomas, Jr. Stratigraphy and paleontology of the Noatak and associated formations, Brooks Range, Alaska. D, 1953, Yale University. 233 p.

Dutrow, Barbara Lee. A staurolite trilogy; I, Lithium in staurolite and its petrologic significance; II, An experimental determination of the upper stability of staurolite plus quartz; III, Evidence for multiple metamorphic episodes in the Farmington Quadrangle, Maine. D, 1985, Southern Methodist University. 228 p.

Dutrow, Barbara Lee. Metric analysis of a late Pleistocene mammoth assemblage, Hot Springs, South Dakota. M, 1980, Southern Methodist University. 165 p.

Dutta, Prodip Kumar. The role of climate in the evolution of detrital and authigenic mineralogy in sandstone from the Gondwana Supergroup, India. D, 1983, Indiana University, Bloomington. 167 p.

Dutta, Virendra Kumar. A permeability scaling technique in the numerical simulation of water coning. M, 1970, University of Missouri, Rolla.

Dutton, Alan Robert. Hydrogeochemistry of the unsaturated zone at Big Brown lignite mine, East Texas. D, 1982, University of Texas, Austin. 259 p.

Dutton, Brian Charles. Sedimentology and diagenesis of the Middle Devonian Dundee Formation in South Lambton and West Middlesex counties, southwestern Ontario. M, 1986, University of Toronto.

Dutton, Carl Evans. A sedimentation study of the Cretaceous-Tertiary sands of southern Illinois. M, 1928, University of Illinois, Urbana.

Dutton, Carl Evans. The conglomerates and structure of the Ensign Lake area, Cook County, Minnesota. D, 1931, University of Minnesota, Minneapolis. 41 p.

Dutton, Shirley Peterson. Diagenesis and burial history of the Lower Cretaceous Travis Peak Formation, East Texas. D, 1986, University of Texas, Austin. 183 p.

Dutton, William George. The geology of the Casitas Pass region, Ventura County, California. M, 1962, University of California, Los Angeles.

Dutton-Melendy, Victoria L. The geology and petrology of the Lackner Lake alkaline complex. M, 1987, Wayne State University. 124 p.

Duttweiler, Karen A. Geology and geochemistry of the Broken Ridge area, southern Wah Wah Mountains, Iron County, Utah. M, 1985, University of Nevada. 107 p.

Duvadi, Ashok K. Geology of the Himalayas and southern Tibet. M, 1985, Carleton University. 92 p.

Duvall, Victor M. Geology of the South Mason-Llano River area, Texas. M, 1955, Texas A&M University.

Duvick, Daniel Nelson. An attempt to verify dendroclimatic reconstructions using independent tree-ring chronologies. M, 1979, University of Arizona. 137 p.

Duweluis, John A. Color-infrared photomicrography of opaque and silicate minerals. M, 1982, Indiana State University. 71 p.

Duym, Dirk Peter Van *see* Van Duym, Dirk Peter

Dvoracek, Douglas. Kinematic history of a ductile shear zone in the Wind River Mountains, Sublette County, Wyoming. M, 1988, Idaho State University. 100 p.

Dvorak, John Joseph. Analysis of small scale lunar gravity anomalies; implications for crater formation and crustal history. D, 1979, California Institute of Technology. 267 p.

Dwairi, Ibrahim. Aspects of the uranium geochemistry of selected lakes in northern Saskatchewan; a preliminary study. M, 1977, University of Regina. 170 p.

Dwelley, Peter C. Geology, mineralization, and fluid inclusion analysis of the Ajax vein system, Cripple Creek, Colorado. M, 1984, Colorado State University. 179 p.

dWest, Joseph E. and West, Joseph Edward. Trace element study of wallrocks adjacent to a mineralized breccia body in the New Market zinc mine, New Market, Tennessee. M, 1970, University of Tennessee, Knoxville. 91 p.

Dwibedi, K. Petrology of the English river gneissic belt (Archean), northwestern Ontario and southeastern Manitoba. D, 1966, University of Manitoba.

Dwiggins, George Albert. Variations in the properties of coal dusts, their significance to the health of occupationally exposed workers, and their relationships to parent material characteristics. D, 1981, University of North Carolina, Chapel Hill. 111 p.

Dwight, Marvin Linn. Geology of the northeastern part of the Glasgow Quadrangle, Missouri. M, 1950, University of Missouri, Columbia.

Dwivedi, Rajeev Lochan. Significance of surface and subsurface lineaments in the tectonic analysis of south-central Kansas. M, 1983, Wichita State University. 104 p.

Dworkin, Stephen Irving. Late Wisconsinan ice-flow reconstruction for the central Great Lakes region. M, 1984, Michigan State University. 30 p.

Dwornik, Edward J. A study of the Bertie Formation of western New York. M, 1948, SUNY at Buffalo.

Dwornik, Stephen E. A study of the Bertie Formation of western New York and southern Ontario. M, 1951, SUNY at Buffalo.

Dwyer, John Joseph. A numerical study of the attenuation of high frequency Lg waves in the New Madrid seismic region. M, 1981, St. Louis University.

Dwyer, Mary Kathleen. The geometry and mechanical development of hanging wall structures of the Livingstone thrust fault, Alberta Foothills. M, 1986, University of Calgary. 141 p.

Dwyer, Ruth-Ann. Seismic reflection investigation of the New Madrid rift zone near Caruthersville, Missouri. M, 1985, Colorado School of Mines. 108 p.

Dwyer, Thomas Edward. Development and verification of a two-dimensional, Galerkin finite-element, solute and heat transport model, and its application to aquifer thermal energy storage evaluation. M, 1986, Kent State University, Kent. 359 p.

Dyar, Melinda Darby. Crystal chemistry and statistical analysis of iron in mineral standards, micas, and glasses. D, 1985, Massachusetts Institute of Technology. 350 p.

Dyar, Robert Francis. The (Upper Mississippian) Bluestone Group, Princeton Conglomerate, and Hinton Group-Mauch Chunk Series in southeastern West Virginia. M, 1957, West Virginia University.

Dyck, Alfred Victor. A method for quantitative interpretation of wideband, drill-hole EM surveys in mineral exploration. D, 1982, University of Toronto.

Dyck, John Henry. The detection of subsurface resistive zones; a study of groundwater geophysics in Saskatchewan. D, 1969, University of Saskatchewan. 293 p.

Dyckes, Jan Allan. Geology of the southwest part of Wardensville 15-minute quadrangle (West Virginia). M, 1964, West Virginia University.

Dye, James L. Geology of the Precambrian rocks in the southern Franklin mountains, El Paso County, Texas. M, 1970, University of Texas at El Paso.

Dye, Jane Elizabeth. Petrographic analysis of the Smackover Formation (Jurassic), Ginger Southeast gas field, Rains County, Texas. M, 1985, Stephen F. Austin State University. 150 p.

Dyer, Alison K. A palynological investigation of the late Quaternary vegetational history of the Baie Verte Peninsula, northcentral Newfoundland. M, 1986, Memorial University of Newfoundland. 182 p.

Dyer, Charles. Spectrochemical study of the Leicester pyrite (Devonian; New York). M, 1959, University of Rochester. 46 p.

Dyer, Charles F. Artesian-water studies in south-central South Dakota. M, 1959, South Dakota School of Mines & Technology.

Dyer, Henry Bennett. Glass optics of analysed plagioclase. M, 1954, University of Wisconsin-Madison.

Dyer, James Russell. Jointing in sandstones, Arches National Park, Utah. D, 1983, Stanford University. 275 p.

Dyer, Julie M. Petrology of the Kula volcanic field, western Turkey. M, 1987, SUNY at Albany. 241 p.

Dyer, Kenneth Lee. Effect of CO_2 on the chemical equilibrium of soil solution and ground water. D, 1967, University of Arizona. 140 p.

Dyer, Roger Gregory. Petrology of the Leonardville kimberlite (Cretaceous) (Riley county, Kansas). M, 1970, Kansas State University. 78 p.

Dyer, William S. Stratigraphy and palaeontology of the Credit River Section of the upper Cincinnati Series of Ontario. D, 1923, University of Toronto.

Dyess, James N. Structural geology of the southeast portion of the Fountain Lake Quadrangle, Ouachita Mountains, Arkansas. M, 1979, University of Missouri, Columbia.

Dygert, Harold Paul, III. The dike pattern of the Spanish Peaks area, Colorado, as an analytical tool for determining regional stress values in the Early Tertiary. M, 1973, University of Rochester.

Dyhrman, R. F. Geology of the Bagby Hot Springs area, Clackamas and Marion counties, Oregon. M, 1976, Oregon State University. 78 p.

Dyk, Karl. On the reduction of seismograms obtained in shaking table experiments. D, 1934, University of California, Berkeley. 24 p.

Dyka, Mary Ann K. Environmental conditions associated with modern dolomite in the lower Florida Keys. M, 1984, Bowling Green State University. 182 p.

Dyke, A. S. Quaternary geomorphology, glacial chronology, and climatic and sea-level history of southwestern Cumberland Peninsula, Baffin Island, Northwest Territories, Canada. D, 1977, University of Colorado. 207 p.

Dyke, Gary A. Structure and stratigraphy of the Silver River area, Baraga County, Michigan. M, 1988, Michigan Technological University. 87 p.

Dyke, Lawrence Dana. Experimental deformation of multilithologic specimens simulating sedimentary facies changes. M, 1976, Texas A&M University.

Dyke, Lawrence Dana. Mechanisms of downhill creep in expansive soils. D, 1979, Texas A&M University. 151 p.

Dyke, Lindell Howard Van *see* Van Dyke, Lindell Howard

Dyke, R. J. Van *see* Van Dyke, R. J.

Dykes, Shaun Methuen. The geochemical and petrological study of the South Bay Mine complex, northwestern Ontario. M, 1979, Queen's University. 186 p.

Dykstra, Franz R. Paragenesis of the ore mineralization at the Eagle Mine, Gilman, Colorado. M, 1947, Columbia University, Teachers College.

Dykstra, Jon D. A geologic study of the Chagai Hills, Baluchistan, Pakistan using Landsat digital data. D, 1978, Dartmouth College. 147 p.

Dykstra, Jon D. Detection of porphyry copper alteration by computer processing of ERTS-1 digital data. M, 1975, Dartmouth College. 61 p.

Dyl, Stanley J., II. Engineering geologic factors affecting the stability of slopes in Ontonagon Clay at the Military Hill Slide, U.S. Highway 45, Ontonagon County, Michigan. M, 1979, Michigan Technological University. 92 p.

Dyman, Thaddeus Stanley. A statistical analysis of Eocene Lepidocyclina. M, 1972, Northern Illinois University.

Dyman, Thaddeus Stanley, Jr. Stratigraphic and petrologic analysis of the Lower Cretaceous Blackleaf Formation and the Upper Cretaceous Frontier Formation (lower part), Beaverhead and Madison counties, Montana. D, 1985, Washington State University. 230 p.

Dymek, R. F. Mineralogic and petrologic studies of Archaean metamorphic rocks from West Greenland, lunar samples, and the meteorite Kapoeta. D, 1977, California Institute of Technology. 392 p.

Dymond, Jack Roland. Potassium-argon geochronology of deep-sea sedimentary material. D, 1966, University of California, San Diego.

Dymond, Randel Leo. Adaptation and implementation of a reservoir quality model in the determination of watershed phosphorus effluent standards. D, 1987, Pennsylvania State University, University Park. 255 p.

Dyni, John Richard. Geology of the nahcolite deposits and associated oil shales of the Green River Formation in the Piceance Creek basin, Colorado. D, 1981, University of Colorado. 182 p.

Dyni, John Richard. Post-depositional history of the Cypress Sandstone near Golconda, Illinois. M, 1955, University of Illinois, Urbana. 43 p.

Dyott, Mark Hamilton. Structural geology of the Appalachian Front from Cumberland, Maryland to Keyser, West Virginia. M, 1956, Cornell University.

Dyroff, Terry L. Areal geology of the Lusk Creek area, Pope County, Illinois. M, 1972, Southern Illinois University, Carbondale. 86 p.

Dysart, Paul Stephen. Moment-radius-stress drop relations and temporal changes in the regional stress from the analysis of small earthquakes in the Matsushiro region, Southwest Honshu, Japan. D, 1985, Virginia Polytechnic Institute and State University. 125 p.

Dyson, James Lindsay. Effect of sodium sulphate on the porosity of limestones. M, 1935, Cornell University.

Dyson, James Lindsay. Ruby Gulch gold mining district, Little Rocky Mountains, Montana. D, 1938, Cornell University. 64 p.

Dytrych, William Joseph. A comparison of theoretical and observed bridging bond lengths and angles in condensed phosphates and sulfates. M, 1983, Virginia Polytechnic Institute and State University. 47 p.

Dyvik, Rune. Strain and pore pressure behavior of fine grained soils subjected to cyclic shear loading. D, 1981, Rensselaer Polytechnic Institute. 360 p.

Dzierwa, David John. A qualitative and quantitative study of the induced polarization phenomenon. M, 1964, Michigan Technological University. 56 p.

Dziewa, T. J. Lower Triassic osteichthyans from the Knocklofty Formation of Tasmania with an analysis of Lower Triassic osteichthyan distribution. D, 1977, Wayne State University. 236 p.

Dzilsky, Thomas Edward. Queen City Formation, Nacogdoches County, Texas. M, 1953, University of Texas, Austin.

Dzimian, Raymond. Hydrothermal synthesis of beryllium oxide. M, 1953, SUNY at Buffalo.

Dzis, R. J. Theoretical and experimental study of linear and nonlinear coaxial deformation paths as recorded by folded and extended planar subfabrics. D, 1975, University of Connecticut. 145 p.

Dzou, Iyh-Ping Leon. In-situ and pyrolytic hydrocarbons in carbonate sediments from Florida Bay. M, 1985, University of Texas at Dallas. 109 p.

Dzurisin, D. 1, Scarps, ridges, troughs, and other lineaments on Mercury; 2, Geologic significance of photometric variations on Mercury. D, 1977, California Institute of Technology. 190 p.

E. Rodriguez Rene *see* Rodriguez E. Rene

Eaby, Jacqueline S. Sr-isotopic variations along the Juan de Fuca Ridge. M, 1983, Stanford University. 39 p.

Eaddy, Donald Workman. Relation between drainage class and certain chemical properties of soil organic matter. D, 1968, North Carolina State University. 119 p.

Eade, Kenneth Edgar. Geology and petrology of the gneiss formation, Clyde area, Baffin Island. D, 1953, McGill University.

Eade, Kenneth Edgar. The Huronian (Precambrian) rocks of northeastern Ontario. M, 1949, McGill University.

Eades, James Lynwood. A study of the mineralogy of clays in soils as a product of the parent material. M, 1953, University of Virginia. 99 p.

Eades, James Lynwood. Reaction of $Ca(OH)_2$ with clay minerals in soil stabilization. D, 1962, University of Illinois, Urbana. 92 p.

Eadie, Dorothy Ann. The metamorphic collar in the sediments around Mount Royal (Montreal, Quebec). M, 1953, McGill University.

Eady, Craig. Hyperfiltration as a calcite cementation mechanism in sandstones; an experimental investigation. M, 1985, Texas A&M University. 79 p.

Eagan, James Matthew. Geology of the Hunt Mountain area, Big Horn County, Wyoming. M, 1976, University of Iowa. 117 p.

Eakins, Gilbert Royal. Soil sampling at the Star Mine, Coeur d'Alene District, Idaho. M, 1953, University of Montana. 71 p.

Eakins, Peter R. Geological setting of Malartic gold deposits, Quebec. D, 1952, McGill University.

Eakins, Peter R. Geology of the Jeep Mine, Rice Lake area, southeastern Manitoba. M, 1949, McGill University.

Ealey, Peter John. Marine geology of north Brazil (South America); a reconnaissance study. D, 1969, University of Illinois, Urbana. 144 p.

Ealey, Peter John. Vegetation alignments and their geologic implications, Arrow Canyon Range, Clark County, Nevada. M, 1966, University of Illinois, Urbana. 69 p.

Eames, Gary B. The late Quaternary seismic stratigraphy, lithostratigraphy, and geologic history of a shelf-barrier-estuarine system, Dare County, North Carolina. M, 1983, East Carolina University. 196 p.

Eames, Leonard E. Palynology of the Berea Sandstone and Cuyahoga Group of northeastern Ohio. D, 1974, Michigan State University. 253 p.

Eames, Valerie. Influence of water saturation on oil retention under field and laboratory conditions. M, 1981, University of Minnesota, Duluth.

Eardley, A. J. Structure, stratigraphy and physiography of the southern Wasatch Mountains, Utah. D, 1930, Princeton University.

Eargle, Dolan Hoye. "Pleistocene soils of the Piedmont of South Carolina.". M, 1946, University of South Carolina. 60 p.

Earl, Forrest C. The surficial geology of the Kingston Quadrangle, New Hampshire. M, 1983, University of New Hampshire. 65 p.

Earl, John Leslie. X-ray fluorescence rubidium strontium age determinations of minerals from the Southern California Batholith. M, 1965, [University of California, San Diego].

Earl, Michael W. Structural analysis of a portion of the Precambrian Packsaddle Schist, southeastern Llano County, Texas. M, 1976, Northeast Louisiana University.

Earl, Richard Allen. Paleohydrology and paleoclimatology of the Skunk Creek basin during Holocene time. D, 1983, Arizona State University. 272 p.

Earl, Thomas Alexander. A hydrogeologic study of an unstable open-pit slope, Miami, Gila County, Arizona. D, 1973, University of Arizona.

Earle, D. W., Jr. Land subsidence problems and maintenance costs to homeowners in East New Orleans, Louisiana. D, 1975, Louisiana State University. 387 p.

Earle, F. M. Regional geography of South Carolina. M, 1926, Columbia University, Teachers College.

Earle, Janet L. A study of the tectonics of Wyoming and adjacent areas using photo linear elements mapped from Landsat-1 and Skylab imagery. M, 1977, University of Wyoming. 90 p.

Earley, Charles F. The sediments of Card Sound, Florida. M, 1967, Florida State University.

Earley, Drummond, III. Structural and petrologic studies of a Proterozoic terrain; "Gold Brick District", Gunnison County, Colorado. M, 1987, University of Minnesota, Minneapolis. 148 p.

Earley, J. W. Description and synthesis of the selenide minerals. D, 1950, University of Toronto.

Earley, Mark A. A remanent magnetic test of the color alterations of ichthyoliths due to secondary heating of the Kaibab Limestone, Flagstaff, Arizona. M, 1983, University of Georgia. 66 p.

Earll, Fred Nelson. Geology of the central mineral ridge, Beaver County, Utah. D, 1957, University of Utah. 161 p.

Earlougher, Robert Charles, Jr. Behavior of transient reservoir pressures considering two-phase flow concepts and well interference. D, 1966, Stanford University. 129 p.

Early, Alberta Jenne Kunz. Microfauna of the Brownwood Shale, Brown and McCulloch counties, Texas. M, 1951, University of Texas, Austin.

Early, Thomas Oren. Rare earths in the eclogite inclusions from the Roberts Victor kimberlite, South Africa. D, 1971, Washington University. 129 p.

Early, William P. Correction for the off-level error in surface ship gravity measurements. M, 1970, Columbia University. 44 p.

Earnest, Patricia Miller. Metal price dependence in relation to probabilistic analysis. M, 1981, University of Utah. 99 p.

Eary, L. Edmond, III. Experimental and theoretical study of mass flow and reaction for the leaching of sandstone uranium ores. D, 1983, Pennsylvania State University, University Park. 217 p.

Eary, L. Edmond, III. The kinetics of uranium dioxide dissolution in acidic hydrogen peroxide solutions. M, 1981, Pennsylvania State University, University Park. 88 p.

Easdon, Michael M. A compilation of graphitic occurrences in the Archean of part of northwestern Quebec. M, 1970, McGill University. 80 p.

Easker, David George. Geology of the Tokio Quadrangle, North Dakota. M, 1948, University of Iowa. 76 p.

Easley, Dale H. Ground-water modeling using geostatistical simulation. M, 1987, University of Wyoming. 47 p.

Easley, Donald T. Probabilistic error regulating functions, their properties and design within an inversion framework. M, 1987, University of Calgary. 126 p.

Eason, James. An example of back-barrier sedimentation from the Upper Carboniferous station of the southern Cumberland Plateau of Tennessee. M, 1973, University of South Carolina.

East, Edwin. Geology, San Francisco Mts. of W. Utah. M, 1957, University of Washington. 138 p.

East, Jennifer S. Geothermal investigations at Manley Hot Springs, Alaska. M, 1982, University of Alaska, Fairbanks. 95 p.

Easterbrook, Donald James. Pleistocene geology of the northern part of the Puget Lowland, Washington. D, 1962, University of Washington. 160 p.

Easthouse, Kurt Allen. The paleoenvironmental analysis of the Lower Silurian Rockwood Formation at Green Gap, Whiteoak Mountain, southeastern Tennessee. M, 1987, University of Tennessee, Knoxville. 218 p.

Eastin, Rene. Geochemical aspects of the Olentangy and Scioto rivers at Columbus, Ohio. M, 1967, Ohio State University.

Eastin, Rene. Geochronology of the basement rocks of the central Transantarctic mountains, Antarctica. D, 1970, Ohio State University.

Eastler, Thomas Edward. Silurian geology of Change Islands and easternmost Notre Dame Bay, Newfoundland. D, 1971, Columbia University. 32 p.

Eastler, Thomas Edward. Silurian geology of the Change Islands and eastern Notre Dame Bay, Newfoundland. M, 1968, Columbia University. 145 p.

Eastman, Benjamin Gordon. The geology, petrography, and geochemistry of the Sierra San Pedro Martir Pluton, Baja California, Mexico. M, 1986, San Diego State University. 154 p.

Eastman, Daniel Brian. Multidirectional deformation history of the Hogback Canyon area of the Western Overthrust Belt, Wyoming. M, 1982, University of Michigan.

Eastman, H. S. Skarn genesis and sphalerite-pyrrhotite-pyrite relationships at the Darwin Mine, Inyo

County, California. D, 1980, Stanford University. 315 p.

Easton, Harry Draper, Jr. A geological and engineering study of producing horizons in the Sabine Uplift territory. M, 1938, University of Oklahoma. 29 p.

Easton, J. A. Buried valleys near Limehouse, Ontario. M, 1988, University of Waterloo. 128 p.

Easton, R. M. The stratigraphy and petrology of the Hilina Formation; the oldest exposed lavas of Kilauea Volcano, Hawaii. M, 1978, University of Hawaii.

Easton, Robert Michael. Tectonic significance of the Akaitcho Group, Wopmay Orogen, Northwest Territories, Canada. D, 1982, Memorial University of Newfoundland. 432 p.

Easton, William Heyden. Mineralogy of the Spruce Pine, North Carolina, area. M, 1938, George Washington University. 59 p.

Easton, William Heyden. The Pitkin Limestone (northern Arkansas and eastern Oklahoma). D, 1940, University of Chicago. 187 p.

Easton, William Wonch. Correlation of stratigraphy and vertebrate faunal distribution of the continental Oligocene of North America. M, 1952, University of Michigan.

Eastty, Frederick D., Jr. Subsurface geology of Rooks County, Kansas. M, 1959, Kansas State University. 52 p.

Eastwood, George Edmund Peter. The origin and geologic history of the Snake Rapids Pluton, Saskatchewan. D, 1949, University of Minnesota, Minneapolis. 102 p.

Eastwood, Raymond Lester. A geochemical-petrological study of mid-Tertiary volcanism in parts of Pima and Pinal counties, Arizona. D, 1970, University of Arizona. 233 p.

Eastwood, Raymond Lester. A spectrochemical investigation of the Bala and Stockdale intrusions in Riley County, Kansas. M, 1965, Kansas State University. 41 p.

Eastwood, William Clifford. Trace element correlation of Tertiary volcanic ashes from western Nevada. M, 1969, University of California, Berkeley. 89 p.

Eastwood, William P. Stratigraphy of the Captain Creek Limestone (Missourian) of eastern Kansas. M, 1958, University of Kansas. 159 p.

Eaton, Andrea Drake. Marine geochemistry of cadmium; a baseline study of a toxic metal in the North Atlantic Ocean. D, 1974, Harvard University.

Eaton, David W. S. An integrated geophysical study of Valhalla gneiss complex, southeastern British Columbia. M, 1988, University of Calgary. 131 p.

Eaton, Gary D. Petrology, petrography and geochemistry of some plutons satellite to the Pioneer Batholith, southwestern Montana. M, 1983, University of Montana. 89 p.

Eaton, George F. The prehistoric fauna of Block Island as indicated by its ancient shell heaps. D, 1898, Yale University.

Eaton, George M. Geology of the central portion of Como Bluff Anticline, Albany County, Wyoming. M, 1960, University of Wyoming. 58 p.

Eaton, Gordon Pryor. Miocene volcanic activity in the Los Angeles Basin and vicinity. D, 1957, California Institute of Technology. 388 p.

Eaton, Harry Nelson. The geology of South Mountain and the Reading Hills, Pennsylvania. D, 1912, University of Pittsburgh.

Eaton, Jeffrey G. Paleontology and correlation of Eocene volcanic rocks in the Carter Mountain area, Park County, southeastern Absaroka Range, Wyoming. M, 1982, University of Wyoming. 153 p.

Eaton, Jeffrey Glenn. Stratigraphy, depositional environments, and age of Cretaceous mammal-bearing rocks in Utah, and systematics of the Multituberculata (Mammalia). D, 1987, University of Colorado. 325 p.

Eaton, Jerome F. The high-temperature creep of dunite. D, 1968, Princeton University. 144 p.

Eaton, Jerry Paul. Theory of the electromagnetic seismograph. D, 1953, University of California, Berkeley. 81 p.

Eaton, Larry G. Geology of the Chloride mining district, Mohave County, AZ. M, 1980, New Mexico Institute of Mining and Technology. 204 p.

Eaton, Marilyn Keller. Study of a vegetation anomaly detected by using Landsat digital data in the Pine Nut Mountains of western Nevada. M, 1979, Stanford University. 95 p.

Eaton, Matthew Richard. Origin of insoluble residue in a deep-sea sediment core from Northwest Province Channel, Bahamas. M, 1986, Miami University (Ohio). 86 p.

Eaton, Perry Alan. The influence of a conductive host on two-dimensional borehole transient electromagnetic responses. M, 1984, University of Utah. 57 p.

Eaton, Perry Alan. Three-dimensional electromagnetic inversion. D, 1987, University of Utah. 193 p.

Eaton, Richard G. Strip mine reclamation of the Cordero Mine, Campbell County, Wyoming. M, 1978, Wright State University. 72 p.

Eaton, Wentworth C. A study of the effect of drying on the critical oxidation temperature of North Dakota lignite. M, 1930, University of Minnesota, Minneapolis.

Eaves, Glenn P. Geology of the southern Sierra del Caballo Muerto area, Brewster County, Texas. M, 1965, Texas A&M University.

Ebanks, Gerald Keith. Structural geology of Keechi Salt Dome, Anderson County, Texas. M, 1966, University of Texas, Austin.

Ebanks, William James, Jr. Recent carbonate sedimentation and diagenesis, Ambergris Cay, British Honduras. D, 1967, Rice University. 276 p.

Ebanks, William James, Jr. Structural geology of the Gass Peak area, Las Vegas Range, Nevada. M, 1965, Rice University. 56 p.

Ebaugh, W. F. Landslide factor mapping and slope stability analysis of the Big Horn Quadrangle, Sheridan County, Wyoming. D, 1977, University of Colorado. 260 p.

Ebaugh, Walter Fielding. Temperature patterns in soil above caves and fractures in bedrock. M, 1973, Pennsylvania State University, University Park. 50 p.

Ebbett, Ballard. Structure and petrology of the metasedimentary rocks of the Good Fortune mining area, Platte County, Wyo. M, 1956, University of Wyoming. 151 p.

Ebblin, Claude Paul. Formation and deformation of a Low-angle fault in southeastern Connecticut. D, 1971, University of Rochester. 154 p.

Ebbott, Kendrick Alan. Hydrogeochemical groundwater reconnaissance in Washington County, Indiana. M, 1985, Indiana University, Bloomington. 170 p.

Ebel, Denton Seybold. Argentian zinc-iron tetrahedrite-tennantite thermochemistry. M, 1988, Purdue University. 95 p.

Ebel, John-Edward. Evidence for fault asperities from systematic time-domain modeling of teleseismic waveforms. D, 1981, California Institute of Technology. 149 p.

Ebeling, Lynn Louis. Numerical simulation of thermal energy storage wells and comparison with field experiments. D, 1984, Texas A&M University. 156 p.

Eberiro, Joseph Onukansi. Structure and crustal type of the northwestern Gulf of Mexico derived from very large offset seismic data. D, 1986, University of Texas, Austin. 170 p.

Eberiro, Joseph Onukansi. Surface wave studies in the Gulf Coast area. M, 1981, University of Texas, Austin. 217 p.

Ebens, Richard J. Petrography of the Eocene Tower Sandstone lenses at Green River, Sweetwater County, Wyoming. M, 1964, University of Wyoming. 46 p.

Ebens, Richard J. Stratigraphy and petrography of Miocene volcanic sedimentary rocks in southeastern Wyoming and north-central Colorado. D, 1966, University of Wyoming. 129 p.

Eberele, Robert Francis. The geology of Wheeling and Flushing townships, Belmont County, Ohio. M, 1936, Ohio State University.

Eberhardt, Ellen. Dynamics of intermediate-size stream outlets, northern Oregon coast. M, 1988, Portland State University. 168 p.

Eberhart, Ginger Lea. The geology and tectonic setting of three hydrothermally active sites along the Mid-Atlantic Ridge. M, 1987, University of Miami. 181 p.

Eberl, Dennis Donald. Experimental diagenetic reactions involving clay minerals. D, 1971, Case Western Reserve University. 158 p.

Eberle, Frederick Claude. Status of cartographic information bases for comprehensive planning in Idaho; results of a survey of county land use planning departments. M, 1984, University of Idaho. 122 p.

Eberlin, John E. Chemical weathering and mechanical properties of the Carlile Formation at Igloo, South Dakota. M, 1985, South Dakota School of Mines & Technology.

Eberly, Lyle Dean. The geology of the Baldy Mountain area, La Plata County, Colorado. M, 1953, University of Illinois, Urbana.

Eberly, Paul O. Brittle fracture petrofabric along a west-east traverse from the Connecticut Valley to the Narragansett Basin. M, 1985, University of Massachusetts. 137 p.

Ebers, Michael L. A study of gangue dolomite mineralization in the Mascot-Jefferson City zinc district, Tennessee, using cathodoluminescence. M, 1976, University of Tennessee, Knoxville. 82 p.

Ebersole, W. C. A seismic reflection investigation of sediments on the eastern Murray Zone. M, 1968, University of Hawaii.

Ebert, Charles H. V. Soils and land use correlation of the outer Atlantic Coastal Plain of Virginia, the Carolinas, and North Georgia. D, 1957, University of North Carolina, Chapel Hill. 360 p.

Ebert, James Roger. Stratigraphy and paleoenvironments of the upper Helderberg Group in New York and northeastern Pennsylvania. D, 1983, SUNY at Binghamton. 173 p.

Ebert, Janet. Hydrologic management using the modified Boussinesq equation for aquifer-stream interaction. M, 1970, Stanford University.

Eberth, David Anthony. Stratigraphy, sedimentology, and paleoecology of Cutler Formation redbeds (Permo-Pennsylvanian) in north-central New Mexico. D, 1987, University of Toronto.

Eberz, Noel. Geology of the Badger Flat Limestone (Middle Ordovician) southeastern California. M, 1985, San Jose State University. 91 p.

Ebinger, Cynthia Joan. Tectonic model of the Malawi Rift, Africa. M, 1986, Massachusetts Institute of Technology. 70 p.

Ebinger, Cynthia Joan. Thermal and mechanical development of the East African Rift system. D, 1988, Massachusetts Institute of Technology. 180 p.

Ebinger, Elizabeth Jane. Regional geology and Cu-Zn mineralization of the Lake Megantic area, southeastern Quebec. M, 1985, University of Western Ontario. 99 p.

Ebisch, James F. Geology and geochemistry of the Belt Supergroup within the Big Cedar Gulch area, Sanders County, Montana. M, 1984, Sul Ross State University. 154 p.

Eble, Cortland F. Palynology and paleoecology of a Middle Pennsylvanian coal bed from the Central Appalachian Basin. D, 1988, West Virginia University. 508 p.

Eble, Edward. Precambrian geology of the Hellroaring Mountain area, southwest Beartooth Mountains, Montana and Wyoming. M, 1976, Northern Illinois University. 77 p.

Ebright, John Richard. The westward limitation of the Appalachian folding in the Northern Appalachian Foreland. M, 1948, University of Pittsburgh.

Ebrom, Daniel. Physical modeling of the acoustic absorption characteristics of a particulate suspension. M, 1987, University of Houston.

Ebtehadj, Khosrow. Palynology of the subsurface Frio Formation in Liberty and Chambers counties, Texas. D, 1969, Michigan State University. 274 p.

Eby, David Eugene. Paleoecology and community structure of two bioherms from the Coeymans Formation (Lower Devonian) of central New York. M, 1972, Brown University.

Eby, David Eugene. Sedimentology and early diagenesis within eastern portions of the "Middle Belt Carbonate Interval" (Helena Formation), Belt Supergroup (Precambrian-Y), western Montana. D, 1977, SUNY at Stony Brook. 781 p.

Eby, Elaine T. The minimization of hazardous waste; a study on the desirability and feasibility of selected options to promote waste minimization at the federal level. M, 1986, Washington State University. 149 p.

Eby, George Nelson. Phase relations in the system As-Sb-S. M, 1967, Lehigh University.

Eby, George Nelson. Rare-earth, yttrium and scandium geochemistry of the Oka carbonatite complex, Oka, Quebec. D, 1971, Boston University. 263 p.

Eby, Henry Eckert. Dynamics of chemically enriched, internally heated mantle plumes. M, 1983, Arizona State University. 78 p.

Eby, James B. The geology and coal resources of the coal bearing portions of Wise and Scott counties, Virginia. D, 1922, The Johns Hopkins University.

Eby, James Robert. The geology and water resources for land-use planning, Potter Township, Centre County, Pennsylvania. M, 1975, Pennsylvania State University, University Park. 169 p.

Eby, Raymond K. Copper oxysalts; the Jahn-Teller effect and its structural implications. M, 1988, University of Manitoba. 304 p.

Eby, Richard Kerr. Hydrogeological study of Mount Zircon Spring, Oxford County, Maine. M, 1987, University of New Hampshire. 114 p.

Eby, Robert George. Early Late Cambrian trilobite faunas of the Big Horse Limestone and correlative units in central Utah and Nevada. D, 1981, SUNY at Stony Brook. 631 p.

Eby, Thomas J., Jr. Geology of the Scranton District, Logan County, Arkansas. M, 1952, University of Arkansas, Fayetteville.

Eccker, Sandra. The effect of lithology and climate on the morphology of drainage basins in northeastern Colorado. M, 1984, Colorado State University. 186 p.

Eccles, John Kerby. Textures of lower Paleozoic rocks of northeastern British Columbia. D, 1958, University of Illinois, Urbana. 97 p.

Eccles, John Kerby. Triassic section in the Kvass Flats map area Alberta, with regional interpretations. M, 1954, Washington State University. 63 p.

Eccleston, Chuck. The mechanics of creep movement and its relationship to progressive landslides on Manastash Ridge, near Ellensburg, Washington. M, 1984, Western Washington University. 194 p.

Ece, Omer Isik. Depositional environment, stratigraphy, petrology, paleogeography and organic thermal maturation of the Desmoinesian cyclothemic Excello Black Shale in Oklahoma, Kansas and Missouri. D, 1985, University of Tulsa. 291 p.

Ece, Omer Isik. Uranium mineralization in Northwest Bee County, Oakville Formation, Texas coastal region. M, 1978, University of Texas, Austin.

Ecevitoglu, Berkan Galip. Use of wavelet distortion from supercritical reflections to detect lateral velocity variations. M, 1984, Virginia Polytechnic Institute and State University.

Ecevitoglu, Berkan Galip. Velocity and Q from reflection seismic data. D, 1987, Virginia Polytechnic Institute and State University. 166 p.

Echavez, Joaquin. The Parral District, Chihuahua, Mexico, as related to the silver metallogenic province of northern Mexico. M, 1968, University of Arizona.

Echegaray, Juan A. Masias *see* Masias Echegaray, Juan A.

Echelbarger, Michael J. Glacial geology of Seneca County, Ohio. M, 1978, Bowling Green State University. 76 p.

Echelmeyer, Keith Alan. Response of Blue Glacier to a perturbation in ice thickness; theory and observation. D, 1983, California Institute of Technology. 364 p.

Echeverria R., L. M. Petrogenesis of metamorphosed intrusive gabbros in the Franciscan Complex, California. D, 1978, Stanford University. 217 p.

Echols, Betty Joan. Microfauna and biostratigraphy of upper Strawn Group, Eastland and Palo Pinto counties, Texas. M, 1959, University of Texas, Austin.

Echols, Betty Joan. Reptile faunas and biostratigraphy of the upper Austin and Taylor groups (Upper Cretaceous) of Texas, with special reference to Hunt, Fannin, Lamar, and Delta counties, Texas. D, 1972, University of Oklahoma. 244 p.

Echols, John Bowlus. The geology of the north and south portions of Rankin County. M, 1961, University of Missouri, Columbia.

Echols, Ronald James. Distribution of Foraminifera and Radiolaria in the sediments of the Scotia Sea area, Antarctic Ocean. D, 1967, University of Southern California. 362 p.

Echols, Ronald James. The bryozoan fauna of the Tamiami Formation (upper Miocene) of Florida. M, 1960, University of Florida. 78 p.

Eck, Orville J. Van *see* Van Eck, Orville J.

Eckel, Edwin B. Geology and ore deposits of the Mineral Hill area, Pima County, Arizona. M, 1930, University of Arizona.

Eckelmann, F. Donald. Archean research in the Beartooth Mountains, Montana-Wyoming; origin and structure of granitic gneiss and migmatites in the Quad Creek area. D, 1956, Columbia University, Teachers College. 119 p.

Eckelmann, F. Donald. The sedimentary origin and stratigraphic equivalence of the so-called Cranberry and Henderson granites in western North Carolina. M, 1954, Columbia University, Teachers College.

Eckelmann, W. R. Age determination of uranium minerals by the Pb-210 method. M, 1954, Columbia University, Teachers College.

Eckert, Anne Douglas. The geology and seismology of the Dudley Gulch Graben and related faults, Piceance Creek basin, northwestern Colorado. M, 1982, University of Colorado. 139 p.

Eckert, James Douglas. Early Llandovery crinoids and stelleroids from the Cataract Group (Lower Silurian) southern Ontario, Canada. M, 1981, University of Toronto.

Eckert, James Olin, Jr. Petrology and tectonic implications of the transition from the staurolite-kyanite zone to the Wayah granulite-facies metamorphic core, Southwest North Carolina Blue Ridge; including quantitative analysis of mineral homogeneity. D, 1988, Texas A&M University. 377 p.

Eckert, Jeffrey C. Paleomagnetism of the Middle Ordovician St. Peter Sandstone in southwestern Wisconsin. M, 1984, University of Wisconsin-Milwaukee. 194 p.

Eckert, Raymond J., Jr. Spatial domain filtering of the Bouguer gravity field of northeastern North America. M, 1985, SUNY at Buffalo. 80 p.

Eckert, Thomas Joseph. The faunas of the upper Bromide Formation. M, 1951, University of Oklahoma. 110 p.

Eckert, William Frederic, Jr. Geology of northern Los Altos Hills, Santa Clara County, California. M, 1958, Stanford University.

Eckerty, Donald Gayle. The geology of the Elkhorn Mountains volcanics (Late Cretaceous) in southern Jefferson and northern Madison counties, Montana. M, 1968, Indiana University, Bloomington. 183 p.

Eckhardt, Donald H. Geomagnetic induction and the electrical conductivity of the Earth's mantle. D, 1961, Massachusetts Institute of Technology. 106 p.

Eckhoff, Oscar Bradley. The hill slopes of the Piedmont of North Carolina. M, 1960, University of North Carolina, Chapel Hill. 77 p.

Eckis, Rollin P. Geology of a portion of the Indio Qudrangle (California). M, 1930, California Institute of Technology.

Eckroade, William M. Geology of the Butt Mountain area, Giles County, Virginia. M, 1962, Virginia Polytechnic Institute and State University.

Eckstein, Barbara Ann. Gravity data analysis of western Vilas County, northern Wisconsin. D, 1986, University of Wisconsin-Madison. 182 p.

Eckstrand, Olof Roger. Igneous and metamorphic hornblendes from the Amisk-Wildnest lakes area, Saskatchewan. M, 1957, University of Saskatchewan. 79 p.

Eckstrand, Olof Roger. The crystal chemistry of chlorite. D, 1963, Harvard University.

Eckwright, Terry Alan. Water flow patterns within the Bunker Hill Mine, Idaho. M, 1982, University of Idaho. 56 p.

Econ, George D. Geology for land use; Cedar Lake watershed, Jackson and Union counties, Illinois. M, 1975, Southern Illinois University, Carbondale. 87 p.

Economides, Michael J. Geothermal reservoir evaluation considering fluid adsorption and composition. D, 1983, Stanford University. 114 p.

Economou, Harris. Part 1, A geological study of Inferno Chasm and Papadakis Butte, two volcanic constructs on the eastern Snake River plain, Idaho and Part II, Selected lava channels on the eastern Snake River, Idaho. M, 1983, SUNY at Buffalo. 97 p.

Eddib, Ali Ahmed. A quantitative study of Kufra well field. M, 1973, Ohio University, Athens. 345 p.

Eddings, Arnold Lester. Correlation studies of the Trivoll No. 8 coal bed in Illinois by plant microfossils. M, 1947, University of Illinois, Urbana. 26 p.

Eddington, Paul Kendall. Kinematics of Great Basin intraplate extension from earthquake geodetic and geologic information. M, 1986, University of Utah. 266 p.

Eddy, Carol Ann. Petrology and geochemistry of the Yuba Rivers Pluton, northwestern Sierra Nevada foothills, California. M, 1986, University of California, Davis. 126 p.

Eddy, Gerald Ernest. Study of the insoluble residues of the Lower Triassic Bell and upper Dundee formations of Michigan. M, 1932, University of Michigan.

Eddy, Greg Edward. Geology of the Cresaptown and Patterson Creek quadrangles in West Virginia. M, 1964, West Virginia University.

Eddy, Greg Edward. Sulfur and ash distribution of Pittsburgh Coal (Pennsylvanian), Mathies Mine, Pennsylvania. D, 1971, West Virginia University.

Eddy, Michele Sharon. Gravity survey, computer modeling and tectonic interpretation of the Missouri gravity low. M, 1984, Washington University. 118 p.

Eddy, Paul Southworth, Jr. A tri-potential resistivity study of fractures, caves, and photo-lineaments in the Boone-St. Joe Aquifer, Northwest Arkansas. M, 1980, University of Arkansas, Fayetteville.

Edelman, Delbert Wayne. The Eocene Germer Basin flora of south central Idaho. M, 1975, University of Idaho. 142 p.

Edelman, Steven Harold. Mesozoic tectonic evolution of the San Vicente Reservoir area, western Peninsular Ranges, California. M, 1980, San Diego State University.

Edelman, Steven Harold. Structure across an ocean-continent suture zone in the northern Sierra Nevada, California, and its implications for ocean accretion processes and regional tectonics. D, 1986, University of California, Davis. 385 p.

Edelman, William Dennis. Geophysical investigation of a possible buried channel; Ocracoke Island, North Carolina. M, 1984, North Carolina State University. 68 p.

Edelson, Stuart K., Jr. System to detect and reduce wide-angle seismic reflection at sea. M, 1970, United States Naval Academy.

Edens, Mark. Hydrogeologic study of a landfill site consisting of highly compressed solid waste. M, 1979, University of Minnesota, Minneapolis. 133 p.

Eder, Richard M. Origin of the narrows of the White River basin in Owen County, Indiana. M, 1977, DePauw University.

Edgar, Alan Douglas. Mineral chemistry of some synthetic cancrinites. M, 1961, McMaster University. 144 p.

Edgar, D. E. Geomorphology and hydrology of selected midwestern streams. D, 1976, Purdue University. 389 p.

Edgar, Dorland E. Geomorphic and hydraulic properties of laboratory rivers. M, 1973, Colorado State University. 171 p.

Edgar, N. Terence. A petrographic and geochemical study of an Upper Devonian biotherm in the Canadian Rocky Mountains, Alberta, Canada. M, 1960, Florida State University.

Edgar, Norman Terence. Seismic refraction and reflection in the Caribbean Sea. D, 1968, Columbia University. 159 p.

Edgar, Thomas Viken. Moisture movement in nonisothermal deformable media. D, 1983, Colorado State University. 293 p.

Edgell, Henry Stewart. Stratigraphy and micropaleontology of the Upper Cretaceous and lower Tertiary of the North-west Basin, Australia. D, 1954, Stanford University. 442 p.

Edgerton, G. K. Ground-water quality and alluvial aquifer thickness in the Eaton area, north of Greeley, Colorado. M, 1974, University of Colorado.

Edie, Ralph William. Geologic studies in the Goldfield area, Saskatchewan and the genesis of pitchblende. D, 1952, Massachusetts Institute of Technology. 220 p.

Edie, Ralph William. Petrography and petrology of the Cameron River volcanic belt, District of Mackenzie, Northwest Territories. M, 1949, University of Alberta. 116 p.

Ediger, Volkan S. Paleopalynological biostratigraphy, organic matter deposition, and basin analysis of the Triassic-(?)Jurassic Richmond rift basin, Virginia, U.S.A. D, 1986, Pennsylvania State University, University Park. 572 p.

Edkins, John E. Geochemical assessment of a sulfide-bearing coal waste deposit in southwestern Illinois. M, 1981, Indiana University, Bloomington. 124 p.

Edman, Janell Diane. Diagenetic history of the Phosphoria, Tensleep and Madison formations, Tip Top Field, Wyoming. D, 1982, University of Wyoming. 248 p.

Edmisten, Neil. Micropaleontology of the Salt Lake Group (Paleocene) Jordan Narrows, Utah. M, 1952, University of Utah. 78 p.

Edmiston, Eudora Fern. Some micro-fauna of the so-called Lower Permian of Kansas. M, 1932, University of Oklahoma. 65 p.

Edmiston, Robert. Thermal gradients and sulfide oxidation in the Silver Bell mining district, Pima County, Arizona. M, 1971, University of Arizona.

Edmond, Brian Alexander. The genesis of the Red Devil mercury deposit, Alaska. M, 1964, University of Toronto.

Edmond, Carolyn Lorraine. Magma immiscibility in the Shonkin Sag Laccolith, Highwood Mountains, Montana. M, 1980, University of Montana. 98 p.

Edmond, Katherine Louise. An analysis of structural cross-section methods. M, 1967, University of Toronto.

Edmonds, Robert J. Geology of the Nutria Monocline, McKinley County, New Mexico. M, 1961, University of New Mexico. 100 p.

Edmonds, Robert Lee. Evolution of the oceanic upper mantle and effects on sea floor subsidence, geoid height and heat flow. D, 1984, The Johns Hopkins University. 131 p.

Edmonds, William Joseph. Grouping of soil profiles in three mapping units by conventional and numerical classifications. D, 1983, Virginia Polytechnic Institute and State University. 399 p.

Edmondson, Samuel A. Sediment transport in the Duwamish Estuary. M, 1973, University of Washington. 92 p.

Edmonson, Park Dale. The Moorefield Formation of Benton and Washington counties, Arkansas. M, 1954, University of Arkansas, Fayetteville.

Edmund, Richard Amos. The geology of a portion of the Bridger Range, Montana. M, 1951, University of Iowa. 120 p.

Edmund, Rudolph William. Pleistocene glaciation of the west slope of the Teton Mountains, Wyoming. M, 1938, University of Iowa. 50 p.

Edmund, Rudolph William. Structural geology and physiography of the northern end of the Teton Mountains, Wyoming. D, 1940, University of Iowa. 163 p.

Edmunds, Frederick Robin. Multivariate analysis of petrographic and chemical data from the Aldridge Formation, southern Purcell Mountain Range, British Columbia, Canada. D, 1977, Pennsylvania State University, University Park. 385 p.

Edmunds, Frederick Robin Kitchener. The structure and petrology of the southern magnetite deposit, Sakami Lake area, New Quebec (Canada). M, 1967, University of Toronto.

Edmunds, William Edward. Some features of reflectance and other related fundamental optical properties of vitrain in coal. M, 1956, University of Pittsburgh.

Edmundson, James W. A study of the subsurface conditions prevailing in the Newhall-Potrero oil field. M, 1947, California Institute of Technology. 44 p.

Edmundson, Raymond Smith. Barite deposits of Virginia. D, 1935, Cornell University.

Edmundson, Raymond Smith. Geology of a portion of the Covesville Quadrangle in the vicinity of Red Hill, Virginia. M, 1930, University of Virginia. 101 p.

Edquist, Ronald K. Geophysical investigation of the Baltazor Hot Springs known geothermal resource area and the Painted Hills thermal area, Humboldt County, Nevada. M, 1981, University of Utah. 90 p.

Edsall, Douglas Wayne. Elimination of some variables in scale model tectonic experiments. M, 1964, American University. 121 p.

Edsall, Douglass W. Submarine geology of volcanic ash deposits; stratigraphy, age, and magmatic composition of Hawaiian and Aleutian tephra; Eocene to Recent. D, 1975, Columbia University. 264 p.

Edsall, Robert W. A seismic reflection study over the Bane Anticline in Giles County, Virginia. M, 1974, Virginia Polytechnic Institute and State University.

Edsall, Thomas D., III. Sands of the continental shelf near Maryland. M, 1955, Rutgers, The State University, New Brunswick. 47 p.

Edson, Dwight James, Jr. Comanche Series, Lancaster Hill, Crockett County, Texas. M, 1951, University of Texas, Austin.

Edson, Fanny C. A detailed study of the exploration on the S.E. 1/4 of the N.E. 1/4 of Sec. 19, T. 47, R. 28, Minnesota. M, 1914, University of Wisconsin-Madison.

Edson, G. M. Some bedded zeolites, San Simon Basin, southeastern Arizona. M, 1977, University of Arizona. 65 p.

Edson, James E. Palynology of Bloyd-age rocks of northwestern Arkansas. M, 1971, University of Arkansas, Fayetteville.

Edson, James E., Jr. Palynology of the upper Cretaceous strata of northeastern Texas. D, 1976, Tulane University. 188 p.

Eduardo, Benjamin E. Mineralization and alteration of the Chisay Zone, Casapalca mining district, Peru. M, 1986, University of Montana. 70 p.

Edvalson, Frederick Merlin. Stratigraphy and correlation of the Devonian in the central Wasatch Mountains, Utah. M, 1947, University of Utah. 58 p.

Edward, Albert. The petrography of the Purcell sills. D, 1930, University of Wisconsin-Madison.

Edwards, Acus Rex. Underground water resources of Chugwater Creek, Laramie River, and North Laramie River valleys, Wyo. M, 1940, University of Wyoming. 47 p.

Edwards, Alan Frances. Geology of the Canyon Creek Quicksilver mine area, Grant County, Oregon. M, 1972, University of Oregon. 65 p.

Edwards, Benjamin. Depositional environments of middle Chester (Upper Mississippian) in Indiana. M, 1956, Indiana University, Bloomington. 32 p.

Edwards, Brian Douglas. Animal-sediment relationships in dysaerobic bathyal environments, California continental borderland. D, 1979, University of Southern California.

Edwards, Brian Douglas. Distribution and significance of endolithic microboring organisms in sediments of the Southeastern United States continental margin; a statistical approach. M, 1972, Duke University. 106 p.

Edwards, C. L. A telemetering system for geophysical data, Socorro, New Mexico. M, 1972, New Mexico Institute of Mining and Technology.

Edwards, C. L. Terrestrial heat flow and crustal radioactivity in northeastern New Mexico and southeastern Colorado. D, 1975, New Mexico Institute of Mining and Technology. 100 p.

Edwards, C. M. Seismic wave attenuation and dispersion in thin layer sequences. M, 1975, Texas A&M University.

Edwards, Charles D. Geology of the Del Valle area, Los Angeles, California. M, 1947, California Institute of Technology. 36 p.

Edwards, Dan Cabe. The geology of Township 15 N, Range 28 W, in Washington and Madison counties, Northwest Arkansas. M, 1950, University of Arkansas, Fayetteville.

Edwards, David Arthur. A solute transport model calibration procedure as applied to a tritium plume in the Savannah River Plant F-area, South Carolina. M, 1988, Texas A&M University. 141 p.

Edwards, David P. Controls on deposition of an ancient fluvial-eolian depositional system; the Early Jurassic Moenave Formation of north central Arizona. M, 1985, Northern Arizona University. 243 p.

Edwards, Dennis S. The detrital mineralogy of surface sediments of the ocean floor in the area of the Antarctic Peninsula, Antarctica. M, 1968, Florida State University.

Edwards, Douglas Edwin, Jr. The physiography of Hinds County, Mississippi. M, 1954, Mississippi State University. 156 p.

Edwards, Duncan L. Petrology and structure of Precambrian rocks in the Bosque Peak Quadrangle, North Manzano Mountains, central New Mexico. M, 1978, University of New Mexico. 105 p.

Edwards, Dwayne Ray. Incorporating parametric uncertainty into flood estimation methodologies for ungaged watersheds and for watersheds with short records. D, 1988, Oklahoma State University. 332 p.

Edwards, Everett C. Foraminifera of the Repetto Hills. D, 1933, California Institute of Technology. 27 p.

Edwards, Everett C. Pliocene conglomerates of the Los Angeles Basin and their paleogeographic significance. D, 1933, California Institute of Technology. 80 p.

Edwards, Everett C. The petrography of a portion of Chippewa and Eau Claire counties, Wisconsin. M, 1920, University of Wisconsin-Madison.

Edwards, Garth Richard. Geochemistry and evolution of an Archean bimodal volcanic-plutonic complex, Wabigoon Subprovince, Ontario. D, 1985, University of Western Ontario.

Edwards, George H., III. Geology of the central Little Burrow Mountains, Grant County, New Mexico. M, 1960, University of Kansas. 60 p.

Edwards, Gerald. A stratigraphic, petrographic, and geochemical study of the manganiferous Cason Shale, Batesville District, Arkansas. M, 1975, Northeast Louisiana University.

Edwards, Gerald. Petrography and geochemistry of the Allamoore Formation, Culberson and Hudspeth counties, Texas. D, 1984, University of Texas at El Paso. 381 p.

Edwards, Gerald B. Late Quaternary geology of northeastern Massachusetts and Merrimack Embayment, western Gulf of Maine. M, 1988, Boston University. 337 p.

Edwards, Goldsborough S. Geology of the West Flower Garden Bank (Texas-Louisiana continental shelf). D, 1971, Texas A&M University.

Edwards, Goldsborough Serpell. Distribution of shelf sediments, offshore from Anton Lizardo and the port of Veracruz, Veracruz, Mexico. M, 1969, Texas A&M University.

Edwards, Harold Hermann. Plasma production of cements and pozzolanas. D, 1988, University of Minnesota, Minneapolis. 162 p.

Edwards, Ira. Cambrian atremate and neotremate brachiopods of the Upper Mississippi Valley. D, 1930, George Washington University.

Edwards, Ira. Pleistocene geology of the Albion-Holley District (New York). M, 1923, University of Rochester. 94 p.

Edwards, James Michael. Sedimentological and environmental analysis of a Holocene salt marsh, Sapelo Island, Georgia. M, 1973, University of Georgia. 79 p.

Edwards, James Michael. Tectonic evolution of the western Sierra Nevada Mountains foothills belt; a case for continental accretion. D, 1978, Rice University. 171 p.

Edwards, Jeffrey Craig. Depositional environments and diagenesis of Member A (Lower to lower Middle Ordovician), Mountain Springs Formation, southern Great Basin. M, 1984, San Diego State University. 88 p.

Edwards, Jeffrey Craig. Stratigraphy, petrology, and diagenesis of the Dewey Formation (Missourian, Upper Pennsylvanian), Midcontinent North America. D, 1987, University of Iowa. 288 p.

Edwards, Jeffrey S. Petrology and contact relationships, SW portion, PC Mullen Creek mafic complex, Medicine Bow Mts., WY. M, 1983, Colorado State University. 170 p.

Edwards, John. The effect of pH on the bacterial oxidation of arsenic sulfides. M, 1974, University of Nevada - Mackay School of Mines. 47 p.

Edwards, John D. Areal geology of the northwest Mangum area, Oklahoma. M, 1958, University of Oklahoma. 96 p.

Edwards, John D. Some Tertiary conglomerates of Mexico. M, 1950, Columbia University, Teachers College.

Edwards, John D. Studies of some Tertiary red conglomerates of central Mexico. D, 1953, Columbia University, Teachers College.

Edwards, John Emerson. The geomorphology and hydrogeology of the Taylor Alluvial Fan, Williamson County, Texas. M, 1974, University of Texas, Austin.

Edwards, John L. Geology of the La Caja Anticline, Zacatecas, Mexico. M, 1953, Columbia University, Teachers College.

Edwards, Jonathan. The geology of the upper Roanoke Valley area, Montgomery and Roanoke counties, Virginia. M, 1959, Virginia Polytechnic Institute and State University.

Edwards, Jonathan, Jr. The petrology and structure of the buried Precambrian basement of Colorado. D, 1966, Colorado School of Mines. 522 p.

Edwards, Kenneth Lang. Upper Eocene fauna of the Stanford Foothills. M, 1961, Stanford University.

Edwards, Larry John. Application of geologic interpretation to highway subgrade and surfacing design procedure on the Kaycee-Barnum state secondary highway, Johnson County, Wyoming. D, 1972, University of Arizona.

Edwards, Lloyd Norman. Geology of the Vaqueros and Rincon formations (lower Miocene), Santa Barbara Embayment, California. D, 1971, University of California, Santa Barbara. 421 p.

Edwards, Lloyd Norman. Miocene sedimentary structures of southwestern Santa Cruz Island, California. M, 1967, University of California, Santa Barbara.

Edwards, Lucy E. Range charts as chronostratigraphic hypotheses, with applications to Tertiary dinoflagellates. D, 1977, University of California, Riverside. 203 p.

Edwards, Margaret Helen. Digital image processing and interpretation of local and global bathymetric data. M, 1986, Washington University. 106 p.

Edwards, Merwin G. The occurrence of aluminum hydrates in clay. M, 1913, University of Wisconsin-Madison.

Edwards, Merwin Guy. The geology of a portion of the Yauli Province, Peru. D, 1923, University of California, Berkeley. 86 p.

Edwards, Paul D. The subfamily Leptochoerinae (Mammalia, Artiodactyla). M, 1974, University of Nebraska, Lincoln.

Edwards, R. Lawrence and Essene, Eric J. Pressure, temperature and C-O-H fluid fugacities during the amphibolite facies-granulite facies metamorphism of the major paragneiss, NW Adirondack Mts., N.Y. M, 1986, University of Michigan. 36 p.

Edwards, Richard Archer. A study of the insoluble residues of the Niagaran of southwestern Ohio. M, 1933, University of Cincinnati. 27 p.

Edwards, Richard Archer. Yorktown and Duplin ostracods of North Carolina and Virginia. D, 1938, University of North Carolina, Chapel Hill. 119 p.

Edwards, Richard Lawrence. High precision thorium-230 ages of corals and the timing of sea level fluctuations in the late Quaternary. D, 1988, California Institute of Technology. 369 p.

Edwards, Robert Garry. Cretaceous Spinney Hill Sand in west-central Saskatchewan. M, 1959, University of Saskatchewan. 82 p.

Edwards, Robert Wheless. The recent sediments of the North River estuary, Morehead City, North Carolina. M, 1961, University of North Carolina, Chapel Hill. 68 p.

Edwards, Stephen Walter. Cenozoic history of Alaskan and Port Orford Chamaecyparis cedars. D, 1983, University of California, Berkeley. 271 p.

Edwards, Stephen Walter. Clarendonian mammals from Bolinger Canyon, California. M, 1975, University of California, Berkeley. 132 p.

Edwards, Thomas Kyle. Hydrogeology of the proposed phosphate mining area in the Diamond Creek drainage, Caribou County, Idaho. M, 1977, University of Idaho. 111 p.

Edwards, Thomas W. D. Aspects of sedimentation and postglacial diatom stratigraphy in arctic lakes, District of Keewatin, Northwest Territories. M, 1980, Queen's University. 144 p.

Edwards, Thomas Wellington Deavitt. Postglacial climatic history of southern Ontario from stable isotope studies. D, 1987, University of Waterloo. 216 p.

Edwards, W. A. D. A study of the formation of several glaciofluvial systems in southeastern Ontario by clast analysis. M, 1978, University of Western Ontario. 205 p.

Edwards, William Arthur. Palynology of the Francis Formation (Pennsylvanian) in the Ada, Oklahoma, region. M, 1966, University of Oklahoma. 113 p.

Edwards, William Russell. Petrography and petrology of the Cynthiana Limestone and lower Eden strata of south-western Ohio and northern Kentucky. M, 1957, Ohio State University.

Edwin, John. The relation of mineral deposits to the pyroxenites and peridotites of North America. M, 1922, University of Minnesota, Minneapolis. 61 p.

Edwin, Robert Bruce. Stratigraphy and structure of the Ordovician limestones of the lower and middle Champlain Valley, Vermont. D, 1959, Cornell University. 114 p.

Edzwald, James Kenneth. Coagulation in estuaries. D, 1972, University of North Carolina, Chapel Hill. 204 p.

Eeckhout, Edward M. Van see Van Eeckhout, Edward M.

Eells, John L. The geology of the Sierra de la Berruga (Precambrian-Paleozoic), northwestern Sonora, Mexico. M, 1972, [University of California, San Diego].

Effimoff, Igol. The chemical and morphological variations of zircons from the Boulder Batholith, Montana. D, 1972, University of Cincinnati. 136 p.

Effinger, James A. Systematics and paleobiology of White River Group Entelodontidae (Mammalia, Artiodactyla). M, 1987, South Dakota School of Mines & Technology.

Effinger, William Lloyd. The Gries Ranch fauna of western Washington. M, 1935, University of California, Berkeley. 169 p.

Effler, Michael Edward. Skarn genesis at Morenci, Arizona. M, 1976, University of Cincinnati. 171 p.

Effler, S. W. A study of the Recent paleolimnology of Onondaga Lake. D, 1975, Syracuse University. 162 p.

Eftaxiadis, Thrasos. Numerical taxonomy and phylogeny of Paleozoic fasciculate bryozoans (ectoprocts). M, 1973, Michigan State University. 103 p.

Eftekharzadeh, Shahriar. Sediment bypass system for impounding reservoirs. D, 1987, University of Arizona. 140 p.

Efteland, Jon Norquist. The potential for land application of municipal sewage sludge in Georgia. D, 1983, University of Georgia. 226 p.

Egan, Christopher Paul. Contribution to the late Neoglacial history of the Lynn Canal and Taku Valley sector of the Alaskan Boundary Range. D, 1971, Michigan State University. 200 p.

Egan, Christopher Paul. Firn stratigraphy and Neve regime trends on the Juneau icefield, Alaska, 1925-65. M, 1966, Michigan State University. 130 p.

Egan, David E. An investigation of the Miami County disposal facility near Troy, Ohio, using electrical resistivity techniques. M, 1983, Wright State University. 106 p.

Egan, Mark S. Dispersive and anisotropic aspects of stratigraphic filtering. D, 1988, University of Houston. 125 p.

Egan, Martin D. Mineralogical aspects of hypersaline lakes in southern Saskatchewan. M, 1984, University of Manitoba.

Egan, Roger Thad. Geology of the Badger-Birch Creek area (Disturbed Belt), Glacier and Pondera counties, Montana. M, 1971, University of Montana. 92 p.

Eganhouse, Robert Paul, Jr. Organic matter in municipal wastes and storm runoff; characterization and budget to the coastal waters of Southern California. D, 1982, University of California, Los Angeles. 248 p.

Egar, Joseph M. Analysis of surface waves in a high velocity layer overlying a low velocity half-space. D, 1959, Texas A&M University.

Egbert, Gary David. A multivariate approach to the analysis of geomagnetic array data. D, 1987, University of Washington. 267 p.

Egbert, Robert. Sedimentology and petrography of the Middle and Upper Jurassic rocks in the Tuxedni Bay area, Cook Inlet, Alaska. M, 1982, San Jose State University. 128 p.

Egbert, Robert Lamar. Geology of the East Canyon area, Morgan County, Utah. M, 1955, University of Utah. 34 p.

Egboka, Boniface Chukwura Ezeanyaoha. Bomb tritium as an indicator of dispersion and recharge in shallow sand aquifers. D, 1980, University of Waterloo.

Egboka, Boniface Chukwura Ezeanyaoho. Field investigations of denitrification in groundwater. M, 1978, University of Waterloo. 148 p.

Ege, John R. In situ stress measured at Rainier Mesa, Nevada, and a few geologic implications. D, 1977, Colorado School of Mines. 172 p.

Ege, John Rodda. Interrelation of physical properties and in-place stability of dacite and granite to local geology. M, 1967, University of Montana. 77 p.

Egemeier, Robert Jack. Upper Cretaceous stratigraphy of Matia, Clark, and Barnes islands, San Juan County, Washington. M, 1981, Washington State University. 149 p.

Egemeier, Stephen Jay. Cavern development by thermal waters with a possible bearing on ore deposition. D, 1973, Stanford University. 116 p.

Egemeier, Stephen Jay. Origin of caves in New York as related to ground water movements. M, 1968, University of Rochester.

Eger, James Douglas. Experiments in offshore shallow high resolution seismic refraction profiling. M, 1975, University of Florida. 71 p.

Eger, Martha J. Erickson. Applications of side-looking airborne radar to hydrocarbon exploration. M, 1982, University of Kansas. 111 p.

Eggboro, M. D. Geochemical zoning of the ground water of Montreal Island. M, 19??, McGill University. 227 p.

Eggers, Albert Allyn. Radiation dosimetry from the natural thermoluminescence of feldspars. M, 1968, Dartmouth College. 75 p.

Eggers, Albert Allyn. The geology and petrology of the Amatitlan Quadrangle, south central Guatemala. D, 1972, Dartmouth College. 221 p.

Eggers, Dwight E. Analytical and numerical continuation methods for conductive temperature fields. D, 1976, Oregon State University. 170 p.

Eggers, Dwight Edward. Downward continuation and transformation of potential fields with application to marine magnetic anomalies. M, 1974, Oregon State University. 64 p.

Eggers, Margaret Royall. The nature of size/shape effects of quartz sand from great and small rivers. D, 1987, University of South Carolina. 284 p.

Eggert, Donald L. Palynology and paleoecology of the Waitersburg Formation Chesterian, (upper Mississippian), of southern Illinois. M, 1974, Southern Illinois University, Carbondale. 79 p.

Eggert, Douglas J. Prediction of future oil production of an oil reservoir using time series analysis. M, 1976, Colorado School of Mines. 98 p.

Eggert, Kenneth George. A hydrologic simulation for predicting nonpoint source pollution. D, 1980, Colorado State University. 434 p.

Eggert, Roderick Glenn. Metamorphic equilibria in the siliceous dolomite system; 6 Kb experimental data and geologic implications. M, 1980, Pennsylvania State University, University Park. 34 p.

Eggler, David Hewitt. Structure and petrology of the Virginia Dale ring-dike complex, Colorado-Wyoming Front Range. D, 1967, University of Colorado. 173 p.

Eggleston, Jane. A comparison of algal reefs, Bermuda and Green Lake, New York. M, 1972, Syracuse University.

Eggleston, Julius Wooster. Eruptive rocks at Cuttingsville, Vermont. D, 1924, Harvard University.

Eggleston, Ted Leonard. Geology of the central Chupadera Mountains, Socorro County, New Mexico. M, 1982, New Mexico Institute of Mining and Technology. 161 p.

Eggleston, Ted Leonard. The Taylor Creek District, New Mexico; geology, petrology and tin deposits. D, 1987, New Mexico Institute of Mining and Technology. 161 p.

Eggleton, Richard Anthony. The crystal structure of stilpnomelane. D, 1965, University of Wisconsin-Madison. 48 p.

Eggleton, Richard Elton. Recognition of volcanoes and structural patterns in the Rümker and Montes Riphaeus quadrangles on the Moon. D, 1970, University of Arizona. 266 p.

Eggleton, Richard Elton. Seiches in a small ice-covered lake. M, 1955, University of Michigan.

Eginton, Charles William. Geology of the White Bear Lake West Quadrangle, Minnesota. M, 1975, Oklahoma State University. 56 p.

Egler, Alan P. Seismic refraction exploration for a buried valley near Morning Sun, Ohio. M, 1974, Miami University (Ohio). 48 p.

Eggleston, David Lewis. Relationship of the magnesium/calcium ratio to the structure of the Reynolds and Winfield oil fields, Montcalm County, Michigan. M, 1958, Michigan State University. 53 p.

Egli, P. Cycling behavior of dissolved lithium in the oceans. D, 1979, Northwestern University. 186 p.

Egner, Barbara E. A palynological analysis of fourteen coals from the Cherokee Group (Pennsylvanian), Marion County, Iowa. M, 1981, University of Iowa. 166 p.

Eheart, J. W. Two-dimensional water quality modeling and waste treatment optimization for wide, shallow rivers. D, 1975, University of Wisconsin-Madison. 371 p.

Ehinger, Robert Ferris. Petrochemistry of the western half of the Philipsburg Batholith, Montana. D, 1971, University of Montana. 124 p.

Ehleiter, John Edward. Mid-Jurassic bryozoans from Wyoming and Utah. D, 1981, Pennsylvania State University, University Park. 211 p.

Ehlen, Judy. Geology of a coastal strip near Bandon, Coos County, Oregon. M, 1969, University of Oregon. 83 p.

Ehlenberger, Robert Gordon. Environmental geology and shoreline changes in Hempstead Harbor, Long Island, New York. M, 1980, Rutgers, The State University, Newark. 116 p.

Ehleringer, Bruce Ernest. Evaluation of fracturing and numerical simulation of contaminant transport within crystalline rock, Rock Haven area, San Diego County, California. M, 1986, San Diego State University. 194 p.

Ehlers, Ernest G. An investigation of the stability relations of the Al-Fe members of the epidote group. D, 1952, University of Chicago. 65 p.

Ehlers, Ernest George, Jr. Intensive parameters of a sulfide and aluminosilicate-bearing granite, Hancock County, Maine. M, 1986, Virginia Polytechnic Institute and State University.

Ehlers, George Marion. Stratigraphy of the Niagaran Series of the northern peninsula of Michigan. D, 1930, University of Michigan.

Ehlig, Perry Lawrence. The geology of the Mount Baldy region of the San Gabriel Mountains, California. D, 1958, University of California, Los Angeles.

Ehlmann, Arthur J. Pyrophyllite in shales of northern central Utah. D, 1958, University of Utah. 105 p.

Ehlmann, Arthur, Jr. The petrology of a stratigraphic section of the Sundance Group, Frye's Gulch, Wyoming. M, 1954, University of Missouri, Columbia.

Ehm, Arlen E. The geology of the Lyons West Oil Field (Kinderhookian Series, Mississippian System), Rice County, Kansas. M, 1965, Wichita State University. 98 p.

Ehman, Donald Allen. Stratigraphic analysis of the Detroit River Group (Devonian) in the Michigan Basin. M, 1964, University of Michigan.

Ehman, Kenneth Dean. Paleozoic stratigraphy and tectonics of the Bull Run Mountains, Elko County. D, 1985, University of California, Davis. 174 p.

Ehrenberg, Stephan Neville. Metamorphism and emplacement of the Feather River ultramafic body, Sierra Nevada Range, California. M, 1973, University of California, Davis. 51 p.

Ehrenberg, Stephen Neville. Petrology of potassic volcanic rocks and ultramafic xenoliths from the Navajo volcanic field, New Mexico and Arizona. D, 1978, University of California, Los Angeles. 276 p.

Ehrenfeld, Frederick. Study of the igneous rocks of York Haven and Stonybrook, Pennsylvania, and their accompanying formations. D, 1898, University of Pennsylvania.

Ehrenspeck, Helmut. Geology and miocene volcanism of the eastern Conejo Hills area, Ventura County, California. M, 1972, University of California, Santa Barbara.

Ehrenzeller, Jeffrey L. Geochemical and hydrological analysis of Harrison Spring, Harrison County, Indiana. M, 1978, Indiana State University.

Ehret, Albert L., Jr. Arenaceous foraminifera from the Council Grove Group (Lower Permian) of Kansas. M, 1961, University of Kansas. 45 p.

Ehret, Gayle Ann. Structural analysis of the John Day and Mitchell fault zones, north-central Oregon. M, 1981, Oregon State University. 195 p.

Ehret, K. S. The application of ultrasonic techniques to the detection of fractures in rock blocks. M, 1983, University of Waterloo.

Ehrets, James Russell. The West Falls Group (Upper Devonian) Catskill Delta Complex; stratigraphy, environments and aspects of sedimentation. M, 1981, University of Rochester. 68 p.

Ehrhard, Louis Edward, III. Low activities of H_2O at the amphibolite-granulite transition, Northwest Adirondacks; evidence for pre-Grenville melting and dehydration. M, 1986, SUNY at Stony Brook. 64 p.

Ehrhart, Thomas P. A petrographic study of the Ranger Formation, Canyon Group, north-central Texas. M, 1982, Stephen F. Austin State University. 116 p.

Ehring, Theodore W. The Murry Formation (Permian), Nevada. M, 1957, University of Southern California.

Ehringer, Robert Ferris. Geological and geochemical studies in the Angel District, Calhoun County, Alabama. M, 1964, University of Alabama.

Ehrlich, Marvin Irwin. Paleomagnetic and rock magnetic investigations of subsurface ore, host-rock, and basic dike specimens from the Precambrian Iron Mountain deposit, Southeast Missouri. M, 1966, Washington University. 122 p.

Ehrlich, Robert. Petrography of some Wilcox sediments at Grand Ecore, Louisiana. M, 1961, Louisiana State University.

Ehrlich, Robert. The geologic evolution of the Black Warrior Detrital Basin (Alabama). D, 1965, Louisiana State University. 77 p.

Ehrlich, Walter Arnold. Pedological processes of some Manitoba soils. D, 1954, University of Minnesota, Minneapolis.

Ehrlinger, Henry Phillip, III. Comparison between xanthate and dithiocarbamate for the collection of sulphide minerals. M, 1956, University of Nevada - Mackay School of Mines. 54 p.

Ehrman, Richard L. Origin of "dissipation" structures, Nebraska Sand Hills. M, 1987, University of Nebraska, Lincoln. 88 p.

Ehrreich, Albert LeRoy. Metamorphism, migmatization, and intrusion in the foothills of the Sierra Nevada, Madera, Mariposa, and Merced counties. D, 1965, University of California, Los Angeles. 320 p.

Ehrreich, Albert LeRoy. The geology of the Dalton Quadrangle, Madera County, California. M, 1955, University of California, Los Angeles.

Eiche, Gregory. Petrogenesis of the basalts of the Alligator Lake alkaline complex, Yukon Territory. M, 1986, University of Toronto.

Eichelberger, A. M., Jr. Application of well-logging methods to shot holes. M, 1939, California Institute of Technology. 80 p.

Eichelberger, John. Granites and syenites of the Pliny Range, New Hampshire. M, 1971, Massachusetts Institute of Technology. 64 p.

Eichelberger, John Charles. Origin of andesite and dacite; the petrographic and chemical evidence for volcanic contamination. D, 1974, Stanford University. 70 p.

Eichen, David. The geology and petrology of the volcanic rocks in the Southern Wolverine Basin, Gravelly Range, Montana. M, 1979, Wright State University. 71 p.

Eichenberger, Nancy L. Trace element distribution and diagenesis of the Anastrasia Formation, northeastern Florida. M, 1977, University of North Carolina, Chapel Hill. 75 p.

Eicher, Constance Carolyn. Marine biosphere diversity through time; provinciality, global shelf area, and sediment survivorship. M, 1978, Michigan State University. 75 p.

Eicher, Don Lauren. Stratigraphy and micropaleontology of the Curtis Formation (Jurassic), northwestern Colorado and northeastern Utah. M, 1955, University of Colorado.

Eicher, Don Lauren. Stratigraphy of the (Lower Cretaceous) Thermopolis Formation (Wyoming). D, 1958, Yale University.

Eicher, Donald B. Conodonts from the Elmhurst Quarry fissure fillings (Illinois). M, 1939, University of Chicago. 63 p.

Eicher, Michael Lee. Major aquifer comparison by examination of geochemical characteristic of groundwater in Marshall County, Kentucky. M, 1974, University of Kentucky. 82 p.

Eicher, Randall Neal. The stratigraphy of the Jackson Group, Grimes County, Texas. M, 1985, Texas A&M University. 181 p.

Eichler, Gary Edward. Engineering properties and lime stabilization of tropically weathered soils. M, 1974, University of Florida. 83 p.

Eichler, Rodney J. Geomorphology and engineering geology of the Cottonwood Creek area, Pikeview and Falcon northwest quadrangles, El Paso County, Colorado. M, 1973, Colorado School of Mines. 194 p.

Eichler, Roland. Carbon and oxygen isotope ratios in marine and fresh-water mollusc shells. M, 1961, Pennsylvania State University, University Park. 80 p.

Eid, Walid Khaled. Scaling effect in cone penetration testing in sand. D, 1987, Virginia Polytechnic Institute and State University. 262 p.

Eidel, John James. The paragenesis and geochemistry of the antimony-mercury deposits of the Antelope Springs mining district, Pershing County, Nevada. M, 1963, University of California, Los Angeles.

Eidie, Harold D., Jr. Geology of an area in the northwest corner of Fremont County, Wyoming. M, 1955, Syracuse University.

Eidlin, Michael B. A study of the mineral industry of Brazil with an outlook for foreign investment. D, 1972, Stanford University. 114 p.

Eier, Douglas Dexter. Feasibility of the reuse of treated wastewater for irrigation, fertilization, and ground water recharge in Idaho. M, 1969, University of Idaho. 198 p.

Eiffert, James Howard. A restudy of the Tepee Trail Formation in the Dubois area, Fremont County, Wyoming. M, 1953, Miami University (Ohio). 55 p.

Eifler, Gus K., Jr. Geology of the Santiago Peak Quadrangle, Texas. D, 1941, Yale University.

Eifler, Gus Kearney, Jr. The Edwards Formation in the Balcones fault zone. M, 1930, University of Texas, Austin.

Eilenberg, Sarah. Variations of sulfur, copper and zinc in volcanoes from El Salvador and adjacent regions. M, 1979, Rutgers, The State University, New Brunswick. 39 p.

Eilender, Herbert. Slope profile analysis of the western Highland Rim edge, in the Bellvue Quadrangle, Tennessee. M, 1973, Vanderbilt University.

Eilers, Lawrence John. The flora and phytogeography of the Iowan lobe of the Wisconsin glaciation. D, 1964, Iowa State University of Science and Technology. 420 p.

Eilers, R. G. Relations between hydrogeology and soil characteristics near Deloraine, Manitoba. M, 1973, University of Manitoba.

Eilerts, Toni Lynn. The diagenetic history of the North Vernon Limestone (Middle Devonian), Illinois Basin. M, 1986, Washington University. 112 p.

Einarsen, Jon M. The petrography and tectonic significance of the Blue Mountain Unit, Olympic Peninsula, Washington. M, 1987, Western Washington University. 175 p.

Einarsson, Paul. Detailed studies of the seismicity of Iceland. D, 1975, Columbia University. 142 p.

Einaudi, Marco Tullio. Pyrrhotite-pyrite-sphalerite relations at Cerro de Pasco, Peru. D, 1969, Harvard University.

Einberger, Carl M. Seasonal changes in groundwater inflow to Crystal Lake, Vilas County, Wisconsin. M, 1986, University of Wisconsin-Madison.

Einsohn, Sudi D. The stratigraphy and fauna of a Pleistocene outcrop in Doniphan County, northeastern Kansas. M, 1971, University of Kansas. 83 p.

Eisbacher, Gerhard Heinz. Tectonic analysis in the Cobequid Mountains, Nova Scotia, Canada. D, 1967, Princeton University. 108 p.

Eiseman, Hope H. Ore geology of the Sunshine cobalt deposit, Blackbird mining district, Idaho. M, 1988, Colorado School of Mines. 191 p.

Eisen, Craig E. The groundwater/surface water interaction in the Menomonee River watershed southeastern Wisconsin. M, 1977, University of Wisconsin-Madison.

Eisenbeis, H. Richard. The petrogenesis of the Milltown Dam Sill, Missoula County, Montana. M, 1958, University of Montana. 38 p.

Eisenberg, Alfredo. Observations on body wave amplitudes and their implication concerning the structure of the Earth's mantle. D, 1972, University of California, Berkeley. 189 p.

Eisenberg, Leonard I. Pleistocene marine terrace and Eocene geology, Encinitas and Rancho Santa Fe quadrangles, San Diego County, California. M, 1983, San Diego State University. 386 p.

Eisenberg, Marvin. An investigation of some volcanic rocks from Mount Rainier National Park, Washington. M, 1950, University of Cincinnati. 36 p.

Eisenberg, Richard Alan. Chronostratigraphy and lithogeochemistry of lower Paleozoic rocks from the Boundary Mountains, west central Maine. D, 1982, University of California, Berkeley. 180 p.

Eisenbraun, Paul H. An electron microprobe investigation of chalcopyrite exsolution from sphalerite. M, 1977, University of Texas, Austin.

Eisenbrey, E. H. Petrology of the metamorphic rocks of the Fishtail Lake area, Harcourt Township, Ontario. M, 1954, University of Toronto.

Eisenhard, Robert M. Characteristics of some Paleozoic clastic sediments of the Central Appalachians. M, 1966, American University. 35 p.

Eisenhardt, William Charles. Physical ages of some Franciscan schists as determined by potassium-argon methods. M, 1958, University of California, Berkeley. 26 p.

Eisenmenger, Karl Kenneth. Seismic stratigraphic study of the Lower Cretaceous Dakota Group, Douglas Creek Arch, western Colorado. M, 1986, Colorado School of Mines. 76 p.

Eisert, Janet Lynn. An evaluation and depositional hypothesis for Holocene black clay laminations in Lake Michigan sediment cores. M, 1982, University of Wisconsin-Milwaukee. 171 p.

Eisinger, V. John. Geology of the Prison Hill-Brunswick Canyon area, Ormsby and Douglas counties, Nevada. M, 1960, University of Nevada - Mackay School of Mines. 69 p.

Eisner, Mark Walter. A gravity and magnetic study of the Martic region; Lancaster County, Pennsylvania. M, 1987, University of Delaware. 371 p.

Eisner, Michael H. A geochemical exploration survey of a portion of Southern Illinois Fluorspar District. M, 1981, Southern Illinois University, Carbondale. 193 p.

Eisner, Stephan Max Leopold. The subsurface geology of the Marchand Conglomerate of the east dome, Cement oil pool, Caddo County, Oklahoma. M, 1949, University of Missouri, Columbia.

Eissler, Holly Kathleen. Investigations of earthquakes and other seismic sources in regions of volcanism. D, 1986, California Institute of Technology. 179 p.

Eitani, Ibrahim Mustafa. An application of the finite element method for simulation of underground excavations and support systems. D, 1981, Virginia Polytechnic Institute and State University. 336 p.

Eittreim, Stephen Lawrence. Suspended particulate matter in the deep waters of the northwest Atlantic Ocean. D, 1970, Columbia University. 166 p.

Eiumnoh, A. Influences of organic matter on soil plow layer properties. D, 1977, North Carolina State University. 173 p.

Eivemark, Michael M. The practice of engineering geology during preconstruction investigations in the Montreal area. M, 1971, McGill University. 216 p.

Ejezie, Samuel Uchechukwu. Probabilistic evaluation of predicted soil behavior under cyclic loading. D, 1984, Carnegie-Mellon University. 294 p.

Ejiaku, Sammuel A. An experimental study of the effect of variable ductility of layered rocks in deformation. M, 1976, Brooklyn College (CUNY).

Ekambaram, Vanavan. Stability relations and crystal chemistry of tourmaline in the system Na_2O-Al_2O_3-SiO_2-B_2O_3-H_2O. M, 1978, Washington State University. 58 p.

Ekas, Leslie Marie. The Chama-El Rito Member of the Tesuque Formation, Espanola Basin, northcentral New Mexico. M, 1985, University of California, Los Angeles. 179 p.

Ekblaw, George Elbert. Clastic deposits in playas. D, 1927, Stanford University. 127 p.

Ekblaw, George Elbert. Concerning the stratigraphy of the Paleozoic rocks along the Mississippi River between Alton and Warsaw, Illinois. M, 1923, University of Illinois, Urbana.

Ekblaw, Sidney Everette. The stratigraphic relations of the Pleasantview Sandstone in western Illinois. M, 1930, University of Illinois, Urbana.

Ekblaw, Walter Elmer. Stratigraphy and paleontology of the Devonian System in Rock Island County, Illinois. M, 1912, University of Illinois, Urbana.

Ekdale, Allan Anton. Ecology and paleoecology of marine invertebrate communities in calcareous substrates, Northeast Quintana Roo, Mexico. M, 1973, Rice University. 159 p.

Ekdale, Allan Anton. Geologic history of the abyssal benthos; evidence from trace fossils in Deep Sea Drilling Project cores. D, 1974, Rice University. 156 p.

Ekdale, Susan Faust. Recent foraminifera associations from northeastern Quintana Roo, Mexico. M, 1973, Rice University. 151 p.

Ekebafe, Samson Bandele. Stratigraphic analysis of the interval from the Hogshooter Limestone (Pennsylvanian) to the Checkerboard Limestone (Pennsylvanian); a subsurface study in north central Oklahoma. M, 1973, University of Tulsa. 82 p.

Eklund, W. A. A microprobe study of metalliferous sediment components. M, 1974, Oregon State University. 77 p.

Ekstrom, Carol. Petrography of the Stonehenge Formation (Lower Ordovician), Washington County, Maryland. M, 1972, George Washington University.

Ekstrom, Goran Anders. A broad band method of earthquake analysis. D, 1987, Harvard University. 226 p.

Ekwurzel, Brenda. The sediment dynamics of a mesotidal, mixed sand and gravel, bayside beach; Herring Cove, Cape Cod, MA. M, 1988, Rutgers, The State University, New Brunswick. 194 p.

El Baz, Farouk. Sedimentary features and geochemistry of the sulphide facies in the transition between the LaMotte Sandstone and the Bonneterre Formation in the Fredericktown area, Madison County, Missouri. M, 1961, University of Missouri, Rolla.

El Baz, Farouk el Sayed. Petrology and mineralogy of certain portions of the Fredericktown deposits, Missouri; a case study of ore genesis in a layered sulphide deposit. D, 1964, University of Missouri, Rolla. 311 p.

El Didy, Sherif Mohamed Ahmed. Two-dimensional finite element programs for water flow and water quality in multi-aquifer systems. D, 1986, University of Arizona. 143 p.

El Din, Ibrahim Oslam Alam. The Qoz; a geographical analysis of sandy western Sudan. D, 1968, University of California, Los Angeles. 277 p.

el Din, Ibrahim Osman Alam *see* Alam el Din, Ibrahim Osman

El Emam, Mohamed. Modifications of the Cole Coal and the Davidson Coal permitivity functions for application to an elastic rock media. M, 1983, University of Pittsburgh.

El Fazzani, Ashour. Stratigraphy of north eastern Libya. M, 1977, University of South Carolina.

El Ghonemy, Hamdi Mohamed Riad. Potential ground-water resources and decrease in natural flow of wells in Dakhla oases, Western Desert, Egypt. M, 1988, University of Arizona. 101 p.

El Ibiary, Nabil Yakout. Practical procedure for the inversion of vertical seismic profiles. M, 1987, University of Texas at Dallas.

El Jard, Mustapha R. Diagenesis of the Hickory Sandstone (Cambrian), McCulloch and Mason counties, central Texas. M, 1982, University of Texas, Austin.

El Mahdy, Omar Rasheed. Origin of ore and alteration in the Freedom No. 2 and adjacent mines at Marysvale, Utah. D, 1966, University of Utah. 319 p.

El Makhrouf, Ali A. Geology, petrology, geochemistry and geochronology of Eghei (Nugay) Batholith alkali rich granites, NE Tibesti, Libya. M, 1984, University of North Carolina, Chapel Hill. 308 p.

El Masry, Alaa Eldin Mohsen. Surface wave analysis in San Juan Basin. M, 1974, University of Texas at Dallas. 252 p.

El Shatoury, Hamad Mohamed. Mineralization and alteration studies in the Gold Hill mining district, Tooele County, Utah. D, 1967, University of Utah. 152 p.

El Shazly, Hanssan. Geological lineament enhancement, identification and extraction from Landsat data; a case study, the Gulf of Suez. D, 1988, University of South Carolina. 276 p.

El Tawashi, A. M. H. A study of the depositional environment of the Halfway Formation, British Columbia. M, 1983, Lakehead University.

El Wardani, Sayed A. Geochemistry of germanium. D, 1956, University of California, San Diego.

El Zouki, Ashour Y. Lithologic study and depositional environments of the Lane Shale (Upper Pennsylvanian) in eastern Kansas. M, 1971, University of Kansas. 46 p.

El-Aghel, Asseddigh M. An analysis of the movement of wind blown sand and its relationship to land-use practices in Coachella Valley during the past twenty-five years. D, 1984, University of California, Los Angeles.

El-Aidi, Bahaa M. Nonlinear earthquake response of concrete gravity dam systems. D, 1988, California Institute of Technology. 178 p.

El-Amamy, Mohaded Muftah. Chemical and mineralogical properties of glauconitic soils as related to K-depletion. D, 1980, University of California, Riverside. 110 p.

El-Ansari, Ahmed. Reservoir mechanics of Lidam Dolomite, Masrab oil field, Sirte Basin, Libya. M, 1984, California State University, Long Beach. 191 p.

El-Arabi, Mahgiub Ali. The displacement of oil by carbon-dioxide in physically scaled models. D, 1981, University of Southern California.

El-Ashry, Mohamed Taha. Effects of hurricanes on parts of the U.S. coastline as illustrated by aerial photographs. M, 1963, University of Illinois, Urbana. 73 p.

El-Ashry, Mohamed Taha. Photointerpretation of shoreline changes in selected areas along the Atlantic and Gulf coasts of the United States. D, 1966, University of Illinois, Urbana. 210 p.

El-Atrash, Mohamed Elmahdi. Foraminifera and biostratigraphy of the Kincaid Formation (Paleocene) in northeast Texas. M, 1971, Texas Christian University.

El-Attar, Hatim Abdelwahab Ahmed. Chemical and mineralogical analyses of some soils of the Nile River basin. D, 1970, University of Wisconsin-Madison. 166 p.

El-Bokle, Farouk Mohamed. Techniques for evaluating and improving the performance of the iron oxides used as barite substitutes in oil well drilling fluids. D, 1982, [University of Oklahoma]. 290 p.

El-Dadah, Ghazi. Permian subsurface carbonate facies of the shelf edge Midland Basin, North Terry and South Hockley counties, Texas. M, 1981, University of Texas at El Paso.

El-Damak, Reda Abdu El-Hay Mohamed Ali. Analysis of water, heat, and solute transfers in unsaturated porous media. D, 1983, University of Maryland. 325 p.

El-Domiaty, Awatif Mohammed. Stress-strain characteristics of a saturated clay soil at various strain rates. D, 1968, University of California, Davis. 232 p.

El-Etr, Hasan A. Pegmatites of the Anderson Ridge Quadrangle, Fremont County, Wyoming. M, 1963, University of Missouri, Rolla.

El-Etr, Hasan A. The technique of lineaments and linear analysis and its application in the minerogenic province of southeast Missouri. D, 1967, University of Missouri, Rolla. 272 p.

El-Far, Sherif Ahmed Kamal. Effects of earthquakes on end-bearing piles. D, 1988, Rensselaer Polytechnic Institute. 191 p.

El-Ghoul, Arebi B. Geology of South Baggs-West Side Canal gas field, Carbon County, Wyoming and Moffat County, Colorado. M, 1982, Iowa State University of Science and Technology.

El-Ghoul, Muhktar Taher. Subsurface structure, stratigraphy, and oil occurrence of the upper member of the Minnelusa Formation in northeastern Wyoming. M, 1982, South Dakota School of Mines & Technology. 82 p.

El-Hakim, Ahmed Zaki. Repeated loading of footings on sand overlying bases of different compressibilities. D, 1983, Queen's University.

El-Haris, Mamdouh Khamis. Soil spatial variability; areal interpolations of physical and chemical parameters. D, 1987, University of Arizona. 149 p.

El-Harram, F. A. Stream morphometry in relation to point bar evolution and bank caving. D, 1979, University of Pittsburgh. 184 p.

El-Hassanin, Adel Saad. Physical, chemical, and mineralogical characteristics of soil vs. erodibility. D, 1983, Oklahoma State University. 189 p.

El-Hawat, Ahmed Saleh. Depositional environments of lower Norian (Upper Triassic) sandstones in the Tobin Range and Augusta Range, northwestern Nevada. M, 1970, Stanford University.

El-Hemry, Ismail Ibrahim Mohamed. Potential of remote sensing for evapotranspiration modeling. D, 1983, University of Maryland. 273 p.

El-Hifnawy, Laila Mahmud. Soil-structure interaction under dynamic loads. D, 1984, University of Western Ontario.

El-Hindi, Mohamed A. Geology and mineralization of the southern part of Pinos Altos area, Fort Bayard Quadrangle, Grant County, New Mexico. M, 1977, Colorado School of Mines. 146 p.

El-Hussain, Issa Watban. Magnetic modeling of the buried Precambrian basement rock in Marion County, eastern Kansas. M, 1986, Wichita State University. 139 p.

El-Idrissi, Mirghani Elsayed. Geophysical investigations of copper and chromite mineralization in the East Piedmont of North Carolina. M, 1980, North Carolina State University. 54 p.

El-Kadi, Aly Ibrahim. Aspects of small-scale and large-scale variability in unconfined groundwater flow. D, 1983, Cornell University. 291 p.

El-Khalidi, Hatem Hussein. A field and petrographic study of the sandstones and conglomerates of the Porcupine Mountains, Ontonagon County, Michigan. M, 1950, Michigan State University. 112 p.

El-Khayal, Abd El-Malik Abd Allah. Planktonic and larger foraminiferal biostratigraphy of the Uppermost Cretaceous and lower Tertiary formations of eastern and northwestern Saudi Arabia. D, 1969, Rutgers, The State University, New Brunswick. 152 p.

El-Moslimany, Ann Paxton. History of climate and vegetation in the eastern Mediterranean and the Middle East from the Pleniglacial to the mid-Holocene. D, 1983, University of Washington. 228 p.

El-Moursi, H. E.-D. H. Probabilistic approach to one-dimensional consolidation settlement. D, 1975, Northwestern University. 197 p.

El-Naggar, Mohamed Mamdouh Abdalla. Solid electrolytic-cell studies of the thermodynamics and kinetics of oxygen absorption in liquid copper and copper alloys. D, 1970, Stanford University. 148 p.

El-Nahal, Mohamed Abdelmonem M. H. Chemical behavior of a zeolitic sodic soil and its response to laboratory treatments. D, 1968, University of California, Davis. 156 p.

El-Sabbagh, Dallilah. Depositional environment and diagenesis of a Miocene temperate-water tuffaceous limestone, La Honda Basin, central California. M, 1988, University of California, Santa Cruz.

El-Sabh, Mohammed Ibrahim. Transport and currents in the Gulf of Saint Lawrence. D, 1974, McGill University.

El-Samani, Karimeldin Z. Geology of the Reed gold mine, Cabarrus County, North Carolina. M, 1978, North Carolina State University. 80 p.

El-Sharnouby, Bahaa el Ahmed. Static and dynamic behavior of pile groups. D, 1984, University of Western Ontario.

El-Shazly, Aley El-Din Khaled. High pressure metamorphism in NE Oman and its tectonic implications. M, 1987, Stanford University. 119 p.

El-Shishtawy, Ahmed Moustafa. Diagenesis of the Thebes Formation (Eocene), Quseir area, Red Sea Coast, Egypt. M, 1985, University of California, Santa Cruz.

Elachi, Charles. Subsurface mapping in the Egyptian Western Desert from satellite data. M, 1983, University of California, Los Angeles. 55 p.

Elam, Jack G. Geology of Seminole Quadrangle, Los Angeles County, California. M, 1948, University of California, Los Angeles.

Elam, Jack Gordon. Geology of Troy South and East Greenbush quadrangles, New York. D, 1960, Rensselaer Polytechnic Institute. 200 p.

Elam, Timothy D. Stratigraphy and paleoenvironmental aspects of the Bedford-Berea Sequence and the Sunbury Shale in eastern and South-central Kentucky. M, 1981, University of Kentucky. 155 p.

Elayer, Robert W. Stratigraphy and structure of the southern Inyo Mountains, Inyo County, California. M, 1974, San Jose State University. 121 p.

Elbakhbkhi, Mohamed Abolgasen. Hydrological study of the Hocking River valley between Enterprise and Athens, Ohio. M, 1970, Ohio University, Athens. 121 p.

Elbashir, Mohamed M. Elhassan. Geology of Jebel Dumbeir, central Sudan. M, 1984, North Carolina State University. 69 p.

Elbel, William P. Decomposition of seismograms by ortho-normal expansion and matched filter approximations. M, 1965, Pennsylvania State University, University Park. 108 p.

Elbert, J. A. An analysis of extreme danger problems associated with abandoned coal mine lands in southwestern Indiana. D, 1987, Indiana State University. 183 p.

Elbert, Julie Ann. The feasibility of using remotely sensed Landsat MSS data for identifying abandoned coal mine land features. M, 1985, Indiana State University. 73 p.

Elberty, William T., Jr. The petrology and provenance of T-3. M, 1955, Dartmouth College. 52 p.

Elberty, William Turner. Effect of clay mineralogy on ceramic properties. D, 1960, Indiana University, Bloomington. 120 p.

Elbishlawi, Mohamed Husni Morad. Effect of free gas saturation on oil recovery by water flooding. D, 1953, Stanford University. 74 p.

Elboushi, Ismail Mudathir. Geologic interpretation of recharge through coarse gravel and broken rock. D, 1966, Stanford University. 169 p.

Elbring, Gregory Jay. A method for inversion of two-dimensional seismic refraction data with applications to the Snake River plain region of Idaho. D, 1984, Purdue University. 137 p.

Eld, Terry Johnson Hammeken. The description and origin of holey limestones in the Key Largo Formation of the Florida Keys. M, 1963, University of Rochester. 51 p.

Elder, Barbara L. A study of mine-related surface subsidence features using Landsat Thematic Mapper and Seasat SAR data; the Western Kentucky coal field. M, 1985, Murray State University. 56 p.

Elder, Ben Frank. The geology of the area along the Cumberland Escarpment between Elverton and Oliver Springs, Roan County, Tennessee. M, 1956, University of Tennessee, Knoxville. 46 p.

Elder, Dorian Lizabeth. A critical examination and evaluation of the structure and stratigraphy in the downtown San Diego area, California. M, 1982, San Diego State University. 149 p.

Elder, R. B. Treatment of Lincoln Mine ore. M, 1920, University of Idaho.

Elder, Ruth L. Paleontology and paleoecology of the Dockum Group, Upper Triassic, Howard County, Texas. M, 1978, University of Texas, Austin.

Elder, Ruth Lucinda. Principles of aquatic taphonomy with examples from the fossil record. D, 1985, University of Michigan. 351 p.

Elder, Susan Rachel. Fossil assemblages of a marine transgressive sand, Moodys Branch Formation (upper Eocene), Louisiana and Mississippi. M, 1981, University of Texas, Austin. 140 p.

Elder, William Perdue. Cenomanian-Turonian (Cretaceous) stage boundary extinctions in the Western Interior of the United States. D, 1987, University of Colorado. 660 p.

Eldougdoug, Abdelmonem Abdelfattah. Petrology and geochemistry of the volcano-sedimentary Glen Township Formation, Aitkin County, east-central Minnesota; implications for gold exploration. D, 1984, University of Minnesota, Minneapolis. 214 p.

Eldredge, Frank E. Surficial geology of the vicinity of Syracuse. M, 1911, Syracuse University.

Eldredge, Robert Niles. Geographic variation and evolution on Phacops rana (Green, 1832) and Phacops iowensis (Delo, 1935), in the middle Devonian of North America. D, 1969, Columbia University. 296 p.

Eldredge, Sarah. Paleomagnetic study of thrust sheet rotations in the Helena and Wyoming salients of the Northern Rocky Mountains. M, 1985, University of Michigan. 40 p.

Eldridge, Charles A. Manitou, Harding, Fremont and Leadville formations of northeastern Gunnison County, Colorado. M, 1957, University of Kentucky. 60 p.

Eldridge, Charles Stewart. A sulfur isotopic, ore textural, chemical, and experimental study on the formation of the Kuroko deposits, Hokuroku District, Japan. D, 1984, Pennsylvania State University, University Park. 310 p.

Eldridge, Charles Stewart. Mineral textures and parageneses of kuroko ores from the Uwamuki No. 4 and some other Kuroko deposits, Hokuroku District, Japan. M, 1981, Pennsylvania State University, University Park. 104 p.

Eldridge, William Frederick. The Golconda Formation in the Illinois Basin; an integrated facies study. M, 1961, University of Illinois, Urbana. 43 p.

Eley, Hugh Moore. The invertebrate paleontology of the Big Bend Park, Marathon, Texas. M, 1938, University of Oklahoma. 118 p.

ElFoul, Djamal. Stratigraphy and microfacies analysis of the Bishop Cap Formation (Desmoines) Vinton Canyon, Franklin Mountains, El Paso County, Texas. M, 1976, University of Texas at El Paso.

Elfrink, Neil. The geology of the east central Desolation Butte Quadrangle, Grant County, Oregon. M, 1988, Oregon State University. 123 p.

Elftman, Arthur Hugo. Some points on the structure and composition of igneous rocks of northeastern Minnesota. D, 1898, University of Minnesota, Minneapolis.

Elftman, Herbert Oliver. Pleistocene Mammals of Fossil Lake, Oregon. M, 1925, University of California, Berkeley. 43 p.

Elgamal, Ahmed-Waeil Metwalli. Nonlinear earthquake-response analysis of earth dams. D, 1985, Princeton University. 330 p.

Elger, Jerry Bruce. Stratigraphy and depositional history of the Mayville Dolomite in eastern Wisconsin. M, 1979, University of Wisconsin-Madison.

Elghadamsi, Fawzi E. Site-dependent inelastic response spectra. D, 1983, Southern Methodist University. 297 p.

Elghazali, M. S. Some photogrammetric investigations of scanning and transmission electron micrography and their applications. D, 1978, Ohio State University. 231 p.

Elhadi, N. D. A. Dispersion characteristics in a channel with a rough top cover. D, 1979, University of New Brunswick.

Elhami, Rahmatollah. Analysis of the distribution of petroleum deposits in the eastern part of Williams County, North Dakota, through computer processing of Landsat multispectral scanner data. M, 1978, Indiana State University. 22 p.

Eliagoubi, Bahlul Ali Hameid. Maastrichtian (upper Cretaceous) foraminifera of northcentral and northwestern Libya. D, 1975, University of Idaho. 135 p.

Eliagoubi, Bahlul Ali Hameid. Vindobonian (Miocene) foraminifera from the Faidia Formation, Umm Errzem region, northeastern Libya. M, 1972, University of Idaho. 107 p.

Elias, D. W. Geology of the Spring Creek area, Moffat County, Colorado. M, 1957, University of Wyoming. 114 p.

Elias, Ghanem. Analysis of underground excavations and their support system. D, 1976, University of California, Berkeley. 110 p.

Elias, Gregory Konrad. The sedimentation and geologic history of certain terraces in Frontier County, Nebraska. M, 1949, University of Nebraska, Lincoln.

Elias, Helen V. Nepheline rocks of the USSR and their bearing on Daly's hypothesis. M, 1944, Columbia University, Teachers College.

Elias, M. R. Prediction and statistical analysis of settlements of shallow foundations on sand. D, 1979, University of Pittsburgh. 151 p.

Elias, Maxim K. Tertiary grasses of the High Plains and their relation to the geology of the region. D, 1939, Yale University.

Elias, Maxim Maximavich. The paleoecology of the Sparland Cyclothem (Pennsylvanian). M, 1939, University of Illinois, Urbana. 32 p.

Elias, Peter. Geochemistry and petrology of granitoid rocks of the Gander Zone, Bay d'Espoir area, Newfoundland. M, 1981, Memorial University of Newfoundland. 271 p.

Elias, Peter. Thermal history of the Meguma Terrane; a study based on ^{40}Ar-^{39}Ar and fission track dating. D, 1986, Dalhousie University. 408 p.

Elias, Robert Jacob. Late Upper Ordovician solitary rugose corals of eastern North America. D, 1979, University of Cincinnati. 525 p.

Elias, Robert Jacob. Solitary rugose corals of the Selkirk Member, Red River Formation (late Middle or Upper Ordovician), southern Manitoba. M, 1977, University of Cincinnati. 232 p.

Elias, Scott Armstrong. Paleoenvironmental interpretations of Holocene insect fossil assemblages from three sites in Arctic Canada. D, 1980, University of Colorado. 347 p.

Eliason, James F. The Hyrum and Beirdneau formations (Devonian) of north-central Utah and southeastern Idaho. M, 1969, Utah State University. 86 p.

Eliason, Jay R. A technique for structural geologic analysis of topography. D, 1984, Washington State University. 166 p.

Elifrits, Charles Dale. A study of subsidence over a room and pillar coal mine. D, 1980, University of Missouri, Rolla. 130 p.

Eligman, Don. Volcanic stratigraphy in the Carolina slate belt near Chapel Hill, North Carolina. M, 1987, University of North Carolina, Chapel Hill. 55 p.

Eliopoulos, Demetrios George. Geochemistry and origin of the Dumagami pyritic gold deposits, Bousquet Township, Quebec. M, 1983, University of Western Ontario. 264 p.

Eliopulos, George J. A geological evaluation of mineralization at Mineral Mountain, Washington County, Utah. M, 1974, University of Arizona.

Eliot, Walter G. The water supply of American cities and towns. D, 1882, Columbia University, Teachers College.

Eliseuson, Thomas G. A geophysical investigation of the ground vibrations associated with an urban forging operation. M, 1974, University of Wisconsin-Milwaukee.

Elison, James H. Geology of the Keigley quarries and the Genola Hills area, Utah. M, 1952, Brigham Young University. 76 p.

Elison, Mark W. Structural geology and tectonic implications of the East Range, Nevada. D, 1987, Northwestern University. 321 p.

Eliuk, Leslie Samuel. Correlation of the Entrance conglomerate (upper Cretaceous), Alberta (Canada) by palynology. M, 1969, University of Alberta. 146 p.

Elizalde, Leonardo. Analysis of element distribution in the No. 9 Coal Seam of the Eastern Interior coal field of western Kentucky. M, 1974, University of Kentucky. 133 p.

Elkhoraibi, M. C. E. Volume change of frozen soils. D, 1975, Carleton University.

Elkin, Robert Rich. Sediment dispersal analysis of the Maroon Formation in the Crested Butte Quadrangle, Colorado. M, 1984, Purdue University. 61 p.

Elkington, Robert B. Structure of the Michigamme Slate along the Huron River, Baraga County, Michigan. M, 1952, Wayne State University.

Elkins, E. D. The petrology of a portion of the San Jacinto Gabbro, near Hemet, California. M, 1978, California State University, Fresno.

Elkins, Linda Tarbox. Phase equilibrium investigations of ternary feldspars. M, 1987, Massachusetts Institute of Technology. 85 p.

Elkins, Ned Zane. Potential mediation by desert subterranean termites in infiltration, runoff, and erosional soil loss on a desert watershed. D, 1983, New Mexico State University, Las Cruces. 154 p.

Elkins, Steven R. Recent sediments of the northern North Sea; factors controlling their composition and distribution. M, 1977, University of Minnesota, Minneapolis. 38 p.

Elks, John E. Air photo interpretation of flood plain features; a means of determining former discharges of the Kansas River. M, 1979, University of Kansas. 77 p.

Elleboudy, A. Characterization of chemical compaction aids for fine-grained soils. D, 1977, Iowa State University of Science and Technology. 141 p.

Ellefsen, Karl J. Application of gravity methods to near-surface modeling and detection of small faults in the Hartford Basin, Connecticut. M, 1984, University of Connecticut. 76 p.

Ellen, Stephenson Davis. The development of folds in layered chert of the Franciscan assemblage near San Francisco, California. D, 1971, Stanford University. 353 p.

Ellenberger, John L. A study of clay vein origin and prediction. M, 1976, Indiana University of Pennsylvania. 31 p.

Eller, Eugene R. Chemung fauna from Canoe Camp Creek near Mansfield, Pennsylvania. M, 1932, University of Pittsburgh.

Eller, John August. Petrology of the amphibolites on Casper Mountain, Wyoming. M, 1982, University of Akron. 129 p.

Eller, Lynn Hansack. Petrology and petrography of selected seams of Pittsburg #8 coal of southeastern Ohio. M, 1982, University of Toledo. 129 p.

Ellerby, R. S. Geology of the Dry Creek-Willow Creek area, Fremont County, Wyoming. M, 1962, University of Wyoming. 106 p.

Ellersick, Donald K. An investigation of the shear strength of Palouse clay as a function of moisture content. M, 1967, Washington State University. 41 p.

Ellinger, Scott T. A stream sediment geochemical survey of the eastern half of the Capitan Mountains; Lincoln County, New Mexico. M, 1988, West Texas State University. 108 p.

Ellinghausen, Robert Henry. An atypical fusuline fauna from the Lenox-Hills Formation, Glass Mountains, Texas. M, 1962, Texas Christian University.

Ellingson, Jack Anton. General geology, Cowlitz Pass area, central Cascade Mountains, Washington. M, 1959, University of Washington. 60 p.

Ellingson, Jack Anton. Late Cenozoic volcanic geology of the White Pass-Goat Rocks area, Cascade Mountains, Washington. D, 1968, Washington State University. 112 p.

Ellington, Michael D. Major and trace element composition of phosphorites of the North Carolina continental margin. M, 1984, East Carolina University. 54 p.

Ellingwood, Robert Whitcomb. Geology of the Lyons-Loveland foothill belt, Colorado, as interpreted from aerial photographs (El 56). M, 1948, University of Illinois, Urbana. 25 p.

Ellins, Katherine Kelly. Isotope hydrology of karst drainage basins in Jamaica and Puerto Rico. D, 1988, Columbia University, Teachers College. 224 p.

Ellinwood, Howard L. Late Upper Cambrian and Lower Ordovician faunas of the Wilberns Formation in central Texas. D, 1953, University of Minnesota, Minneapolis. 241 p.

Elliot, A. J. M. Geology and metamorphism of the Mitchell Mountains ultramafite, Fort St. James map area, British Columbia. M, 1975, University of British Columbia.

Elliot, Arthur H. Geology and chemistry of building stones. D, 1883, Columbia University, Teachers College.

Elliot, John G. Evolution of large arroyos, the Rio Puerco of New Mexico. M, 1979, Colorado State University. 155 p.

Elliot, R. John. A population study of the systematics and stratigraphic variation of Hesperocyon (Mammalia, Canidae). M, 1980, South Dakota School of Mines & Technology.

Elliot, Terence Martin. Recognition of hydrothermal alteration zones emphasizing IR and XRD analyses of felsic rocks from the Henderson molybenite deposit, Clear Creek County, Colorado, and ultramafic rocks from the Manitoba nickel belt, Canada. M, 1973, Stanford University.

Elliot, William John. A process based rill erosion model. D, 1988, Iowa State University of Science and Technology. 123 p.

Elliott, Arthur Beverly, Jr. Recent sediments of Corpus Christi and Nueces County, Texas. M, 1958, University of Texas, Austin.

Elliott, Brian. Stylolites as paleostress indicators in the Minekahta Limestone, Black Hills, South Dakota and Wyoming. M, 1979, University of Toledo. 32 p.

Elliott, Charles S., Jr. Hydrologic processes and mid-winter recharge in a Sierra Nevada watershed. M, 1985, University of Nevada. 76 p.

Elliott, Colleen Georgia. The depositional, intrusive, and deformational history of Southwest New World Island, and its bearing on orogenesis in central Newfoundland. D, 1988, University of New Brunswick.

Elliott, Douglas Howard. Photogeologic interpretations using photogrammetric dip calculations. M, 1951, University of California, Berkeley. 66 p.

Elliott, Edward S. Environment of deposition of the Wasatch Formation, Powder River basin, Wyoming. M, 1976, Wright State University. 81 p.

Elliott, Glenn K. The Great Marsh, Lewes, Delaware; the physiography, classification, and geologic history of a coastal marsh. M, 1973, University of Delaware.

Elliott, Harold Charles. Emery Mine, Powell County, Montana. M, 1939, Montana College of Mineral Science & Technology. 45 p.

Elliott, Herbert A., Jr. A planktonic foraminiferal zonation of the Gulf Coast Eocene. M, 1969, Louisiana State University.

Elliott, James Barry. Seasonal variation in major element distribution within a flooded bituminous coal mine shaft, Belmont County, Ohio. M, 1974, University of Akron. 65 p.

Elliott, James Edward. The mineralogy and geochemistry of the tungsten deposits of the Black Rock Mine area, Mono County, California. D, 1971, Stanford University. 166 p.

Elliott, John L. Geology of the eastern Santa Monica Mountains between Laurel Canyon and Beverly Glen boulevards, Los Angeles County, California. M, 1951, University of California, Los Angeles.

Elliott, Kenneth L. Conodonts, biostratigraphy, and lithostratigraphy of the Lower Ordovician McKenzie Hill and Cool Creek formational boundary interval, Wichita and Arbuckle Mountains, Oklahoma. M, 1984, University of Missouri, Columbia. 47 p.

Elliott, Laura Ann. Depositional facies and stratigraphy of the lower San Andres Formation (Permian), Southeast New Mexico. M, 1985, University of Texas, Austin. 206 p.

Elliott, Monty Arthur. Stratigraphy and petrology of the late Cretaceous rocks near Hilt and Hornbrook, Siskiyou County, California and Jackson County, Oregon. D, 1971, Oregon State University. 171 p.

Elliott, Richard Alden. The Seville Limestone; a Pennsylvanian (Desmoinesian) estuarine deposit (Illinois). M, 1967, University of Illinois, Urbana.

Elliott, Robert G. Applications of geology to land use planning in the Clinton Reservoir sanitation zone, Douglas County, Kansas. M, 1973, University of Kansas. 80 p.

Elliott, Roy William. Evaluation of the max/min groundwater monitoring system. M, 1981, University of Nebraska, Lincoln.

Elliott, S. R. Petrology, lithogeochemistry and metasomatic flux of the alteration zone associated with the Lar Deposit, Lynn Lake area, Manitoba. M, 1987, University of Waterloo. 278 p.

Elliott, Thomas L. Deposition and diagenesis of carbonate slope deposits, Lower Cretaceous, northeastern Mexico. D, 1979, University of Texas, Austin. 352 p.

Elliott, William Crawford. Bentonite illitization in two contrasting cases; the Denver Basin and the southern Appalachian Basin. D, 1988, Case Western Reserve University. 250 p.

Elliott, William J. Geological occurrences of columbium and tantalum bearing minerals. M, 1957, University of Toronto.

Elliott, William J. Geology of a portion of the Temblor Range, San Luis Obispo and Kern counties, California. M, 1966, [University of California, San Diego].

Ellis, Arthur Jackson. The stratigraphy of the Ordovico-Silurian boundary line in northeastern Illinois. M, 1911, University of Illinois, Urbana.

Ellis, Bernett Eston. Geology of the northeastern part of Saline County, Missouri. M, 1948, University of Missouri, Columbia.

Ellis, Brooks Fleming. A study of discoidal foraminifera from Cuba. D, 1932, New York University.

Ellis, Charles Allen. An investigation of an ultrasonic, high pressure waterflood in a linear porous medium. M, 1966, University of Oklahoma. 69 p.

Ellis, Charles Howard. Geology and Pennsylvanian paleontology of Perry Park, Colorado. M, 1958, University of Colorado.

Ellis, Charles W. Marine sedimentary environments in the vicinity of the Norwalk Islands, Connecticut. D, 1960, Yale University.

Ellis, Clarence E. Engineering geology of Long Canyon, Boundary County, Idaho. M, 1974, University of Idaho. 81 p.

Ellis, D. E. Mineralogy and petrology of chloride and carbonate bearing scapolites synthesized at 750°C and 4000 bars. D, 1977, Yale University. 147 p.

Ellis, D. H. A study of the genesis of the copper and other ores of the Eastern Townships of Quebec. M, 1926, McGill University.

Ellis, David Burl. Holocene sediments of the South Atlantic Ocean; the calcite compensation depth and concentrations of calcite, opal and quartz. M, 1972, Oregon State University. 77 p.

Ellis, E. G. Surficial and environmental geology of part of the Upper Williams Fork River basin, Colorado. M, 1976, University of Colorado.

Ellis, Glenn W. Modelling earthquake ground motions in seismically active regions using parametric time series methods. D, 1987, Princeton University. 168 p.

Ellis, H. A. and Shotts, T. W. The investigation of Osage City, Kansas, clays. M, 1911, [University of Kansas].

Ellis, James M. Holocene glaciation of the central Brooks Range, Alaska. M, 1978, SUNY at Buffalo. 114 p.

Ellis, James Manning. Holocene glaciation of the central Brooks Range, Alaska. D, 1982, SUNY at Buffalo. 397 p.

Ellis, Jessie Bird S. A statistical analysis of conodonts from the Upper Devonian of Missouri. D, 1959, University of Missouri, Columbia. 121 p.

Ellis, John Hazle. Diffusion of copper, manganese, zinc, and iron in clays. D, 1970, University of Kentucky. 184 p.

Ellis, Lynn Doyle. Geophysical reconstruction of the preglacial Allegheny River valley, western New York. M, 1980, SUNY at Buffalo. 52 p.

Ellis, M. J. Use of chemical analysis for correlation of carbonate rocks. M, 1960, South Dakota School of Mines & Technology.

Ellis, Margaret Jane. Structural analysis and regional significance of complex deformational events in the Big Maria Mountains, Riverside County, California. M, 1981, San Diego State University.

Ellis, Michael Alexander. Structural morphology and associated strain within parts of the U.S. section of the Kootenay Arc, N.E. Washington. D, 1984, Washington State University. 254 p.

Ellis, Patricia Mench. Diagenesis of the Lower Cretaceous Edwards Group in the Balcones fault zone area, south-central Texas. D, 1985, University of Texas, Austin. 358 p.

Ellis, Paul. Holocene and late Pleistocene sedimentation in the eastern South Atlantic. M, 1972, University of South Carolina.

Ellis, R. B. Fourier series reduction of gravity data to a horizontal plane. M, 1975, University of Arizona.

Ellis, Richard Keller. Podiform chromite occurrences in the Josephine Peridotite, Klamath Mountains, northwestern California. M, 1977, University of California, Berkeley. 120 p.

Ellis, Robert M. Optimum prospecting plans. M, 1958, University of Western Ontario.

Ellis, Robert Malcolm. Analysis of natural ultra low frequency electromagnetic fields. D, 1964, University of Alberta. 108 p.

Ellis, Robert W. Glacial and post-glacial phenomena of the Portage region, with special reference to their bearing on the southwestward drainage of Lake Winnebago. M, 1910, University of Wisconsin-Madison.

Ellis, Roger David. Geology and ore deposits of the Winkler Anticline, Hidalgo County, New Mexico. M, 1971, University of Texas at El Paso.

Ellis, Ross Courtland. The geology of the Dutch Miller Gap area, Washington. D, 1959, University of Washington. 113 p.

Ellis, Steven D. Reconnaissance paleomagnetism and geotectonics of the Idaho Batholith (Idaho). M, 1970, Western Washington University. 50 p.

Ellis, Thomas Morgan. The geology and geophysics of a portion of Potosi, Washington County, Missouri. M, 1953, Washington University.

Ellis, Thomas Morgan. The geology and geophysics of a portion of the Potosi Quadrangle, Missouri. M, 1960, Washington University. 49 p.

Ellison, Adam James Gillmar. The solution behavior of highly-charged cations in high-silica liquids. D, 1988, Brown University. 164 p.

Ellison, Albert H. The Hamill group (lower Paleozoic) of the northern Dogtooth mountains, British Columbia, Canada. M, 1967, University of Calgary. 111 p.

Ellison, Andrew Bell. Behavior of introduced solutes and evaluation of a solute transport model for a low-gradient sand bed stream in northwest Ohio. M, 1988, University of Toledo. 177 p.

Ellison, Bruce Edward. Stratigraphy of the Burns Junction-Rome area, Malheur County, Oregon. M, 1968, Oregon State University. 89 p.

Ellison, Charles Ralph, III. 2-D high-order finite-difference reverse-time migration. M, 1988, University of Texas of the Permian Basin. 89 p.

Ellison, Evard Pitts. A size distribution study of coastal sands of northern Florida. M, 1949, University of Missouri, Columbia.

Ellison, Patrick James. Mineralization and alteration of a composite stock in the Gold Hill area, Stevens County, Washington. M, 1982, Washington State University. 67 p.

Ellison, Robert J. The geophysical characterization of the Arkansas seismic zone, the Arkoma Basin, Arkansas. M, 1985, Southern Illinois University, Carbondale. 67 p.

Ellison, Robert Lee. Middle Devonian Mahantango Formation in parts of south-central Pennsylvania. D, 1961, Pennsylvania State University, University Park. 484 p.

Ellison, Samuel P. Revisions of Pennsylvanian conodonts. D, 1940, University of Missouri, Columbia.

Ellison, Samuel P. The conodonts from the Missouri Series (Pennsylvanian) of Jackson County, Missouri. M, 1938, University of Missouri, Columbia.

Ellsworth, Elmer W. Physiographic history of the Afton Basin, San Bernardino County (California). D, 1932, Stanford University. 99 p.

Ellsworth, Elmer W. Varved clays of Wisconsin. M, 1930, University of Wisconsin-Madison.

Ellsworth, George W., Jr. Depositional environments and stratigraphic relationship of some Carboniferous deposits in eastern Kentucky. M, 1977, Eastern Kentucky University. 69 p.

Ellsworth, H. V. A study of certain minerals from Cobalt, Ontario. D, 1916, University of Toronto.

Ellsworth, Kirk Anthony. Numerical models of two-layer convection for an infinite Prandtl number fluid. D, 1986, University of California, Los Angeles. 317 p.

Ellsworth, Ralph Irving. Geology of southeastern part of the Van Horn Mountains, Trans-Pecos, Texas. M, 1949, University of Texas, Austin.

Ellsworth, Richard Gerald. The use of Hyphomycetes as indicators of available nutrient minerals in soils. M, 1934, Catholic University of America. 29 p.

Ellsworth, William L. Geology of the continental shelf, Point Lobos to Point Sur, California. M, 1971, Stanford University.

Ellsworth, William Leslie. Three-dimensional structure of the crust and mantle beneath the Island of Hawaii. D, 1978, Massachusetts Institute of Technology. 327 p.

Ellwood, B. B. Development and utilization of the standardized anisotropy of magnetic susceptibility parameter F_x. D, 1976, University of Rhode Island. 161 p.

Ellwood, Robert Brian. Surficial geology of the Vermilion area, Alberta, Canada. D, 1961, University of Illinois, Urbana. 159 p.

Ellzey, Robert T., Jr. Mississippian rocks on western flank of Oklahoma City Uplift, Oklahoma. M, 1960, University of Oklahoma. 50 p.

Elman, Stanley Harold. A spectrochemical analysis of the insoluble residues of the Dundee Limestone of Presque Isle County, Michigan. M, 1958, Michigan State University. 371 p.

Elmer, Deborah Ann. Bayside shoreline dynamics along natural and stabilized barrier dune sections of Assateague Island, Maryland. M, 1978, University of Virginia. 87 p.

Elmer, Nixon; Brant, Russell Alan; Gillespie, W. A. and Peterson, John Robert. Geology of the Armstead area, Beaverhead County, Montana. M, 1949, University of Michigan. 118 p.

Elmonayeri, Diaa Salah. Mechanisms and analysis of multiphase flow through soil. D, 1983, McGill University.

Elmore, Richard Douglas. The "Black Shell" turbidite, Hatteras abyssal plain. M, 1976, Duke University. 107 p.

Elmore, Richard Douglas. The Copper Harbor Conglomerate and Nonesuch Shale; sedimentation in a Precambrian intracontinental rift, upper Michigan. D, 1981, University of Michigan. 201 p.

Elmore, Robert Thompson. Geology of the Jack's Cabin area of Gunnison County, Colorado. M, 1955, University of Kentucky. 35 p.

Elmoudi, Salem M. The Santee course and the facies of the lower coastal plain of SC. M, 1972, University of South Carolina. 36 p.

Elms, Morris A. Geology of the Buck Hill Quadrangle, Brewster County, Texas. M, 1937, Texas A&M University.

Elnawawy, Osman Ali. The Cell Analytic-Numerical method for solution of the groundwater solute transport equation. D, 1988, University of Illinois, Urbana. 219 p.

Elorrieta, Nimio Juvenal Tristan. Rb-Sr dates of crystalline rocks from southern Israel. M, 1980, University of Texas at Dallas. 38 p.

Elphic, Lance. Geology of the southern third of the Glide Quadrangle, Oregon. M, 1969, University of Oregon. 78 p.

Elphick, Patricia Margaret. Crystal structure analysis of wodginite from Bernic Lake, Manitoba. M, 1972, University of Manitoba.

Elphick, Stephen Conrad. Metamorphic petrology of the Gillam area (Archean), Manitoba. M, 1970, University of Manitoba.

Elqueta, Eduardo del Rozas see del Rozas Elqueta, Eduardo

Elrick, Maya. Depositional and diagenetic history of the Devonian Guilmette Formation, southern Goshute Range, Elko County, Nevada. M, 1986, Oregon State University. 109 p.

Elrod, Dennis Dean. A geochemical and petrographic survey of the Wellington Formation, north-central Oklahoma. M, 1980, Oklahoma State University. 100 p.

Elrod, Mary Melinda. Paleomagnetic analysis of Recent abyssal sediments from the Northeast Argentine Basin. M, 1988, University of Georgia. 87 p.

Elsbree, Hope Carole. Clay alteration in three ore deposits and associated volcanics of the Troodos Ophiolite, Cyprus. M, 1985, University of Illinois, Urbana. 132 p.

Elsby, Darren C. Structure and deformation across the Quesnellia-Omineca Terrane boundary, Mt. Perseus area, east-central British Columbia. M, 1985, University of British Columbia.

Elsenheimer, Donald William. Petrologic and stable isotopic characteristics of graphite and other carbon-bearing minerals in Sri Lankan granulites. M, 1988, University of Wisconsin-Madison. 122 p.

Elsik, William Clinton. Palynology of the lower Eocene Rockdale Formation, Wilcox Group, Milam and Robertson counties, Texas. D, 1965, Texas A&M University. 253 p.

Elsik, William Clinton. Petrological comparison of some Tertiary and Quaternary sands from Brazos and adjoining counties, Texas. M, 1960, Texas A&M University.

Elsinger, Robert John. Estuarine geochemistry of ^{224}Ra, ^{228}Ra, ^{226}Ra, and ^{222}Rn. D, 1982, University of South Carolina. 95 p.

Elsinger, Robert John. Ra-226 behavior in the Pee Dee River-Winyah Bay estuary. M, 1979, University of South Carolina.

Elson, John Albert. Surficial geology of the Tiger Hills region, Manitoba, Canada. D, 1956, Yale University. 471 p.

Elston, Donald P. Geology of the Pool's Brook Limestone of the Manlius Group (New York). M, 1951, Syracuse University.

Elston, Donald Parker. The geologic classification of meteorites. D, 1968, University of Arizona. 305 p.

Elston, Wolfgang. Cenozoic history of the Sherman Quadrangle, Grant, Luna, and Sierra counties, New Mexico. M, 1953, Columbia University, Teachers College.

Elston, Wolfgang. The geology and mineral resources of the Dwyer Quadrangle, Grant, Luna, and Sierra counties, New Mexico. D, 1953, Columbia University, Teachers College.

Elsworth, Derek. Laminar and turbulent flow in rock fissures and fissure networks. D, 1984, University of California, Berkeley. 192 p.

Elterman, Joan. Shape development of folds in experiment. M, 1978, University of Minnesota, Duluth.

Elthon, Donald Lee. The petrology of the Tortuga ophiolite complex, southern Chile; implications for igneous and metamorphic processes at oceanic spreading centers. D, 1980, Columbia University, Teachers College. 357 p.

Elton, William G. Petrology and stratigraphy of the upper Conasauga Group (late Cambrian) in Northeast Tennessee. M, 1974, Eastern Kentucky University. 57 p.

Elvidge, Christopher David. Distribution and formation of desert varnish in Arizona. M, 1979, Arizona State University. 110 p.

Elvidge, Christopher David. Separation of leaf water and mineral absorption in the 2.22 um Thematic mapper band. D, 1985, Stanford University. 201 p.

Elwell, James Halsey. Behavior of carbonate aggregates from the Otis Member of the Wapsipinicon Formation in highway concretes of various ages. M, 1966, Iowa State University of Science and Technology.

Elwell, James Halsey. Deterioration zone petrology of selected highway concretes. D, 1969, Iowa State University of Science and Technology. 190 p.

Elwood, Michael Warren. Geology of the Black Buttes, Crook County, Wyoming. M, 1978, South Dakota School of Mines & Technology.

Elwood, Sompis Chuntamee. Geological education in the museum. M, 1980, South Dakota School of Mines & Technology.

Ely, Lael Marguerite. Microfauna of the Oakville Formation, La Grange area, Fayette County, Texas. M, 1957, University of Texas, Austin.

Ely, Richard Woodman. Paleontology and stratigraphy of the Opf unit (Pogonip group, middle Ordovician) in the Arrow Canyon range, Clark County, Nevada. M, 1969, University of Illinois, Urbana.

Elzaroughi, A. A. Application of endochronic constitutive law to one-dimensional liquefaction of sand. D, 1978, Northwestern University. 176 p.

Emadian, Nazila. The mineralogy and petrology of the Oreville Formation, Black Hills, South Dakota. M, 1981, Kent State University, Kent. 79 p.

Emam, Mohamed El *see* El Emam, Mohamed

Emanuel, Karl M. A geochemical, petrographic and fluid inclusion investigation of the Zuni Mountains fluorspar district, Cibola County, New Mexico. M, 1982, University of New Mexico. 161 p.

Emanuel, Richard Paul. Hydrogeologic aspects of site characterization studies for underground superconductive energy storage facilities. M, 1979, University of Wisconsin-Madison.

Ember, Leon M. Sources of sedimentary organic matter in Spartina-dominated salt marshes. M, 1985, University of South Carolina.

Embich, John Reigle. The Tranquilla Shale and its foraminiferal fauna. M, 1936, Columbia University, Teachers College.

Embley, R. W. Studies of deep-sea sedimentation processes using high frequency seismic data. D, 1976, Columbia University, Teachers College. 350 p.

Embree, Glen F. Lateral and vertical variations in a Quaternary basalt flow; petrography and chemistry of the Gunlock flow, southwestern Utah. M, 1969, Brigham Young University.

Embree, Glenn F. Structural analysis of the Precambrian rocks, Green Lake basin, Teton Range, Wyoming. D, 1976, University of Idaho. 104 p.

Embree, John Marvin. Design of a solar collector and storage system. M, 1976, University of Virginia. 111 p.

Embry, Ashton Fox, III. A Late Devonian reef tract on northeastern Banks Island, Northwest Territories. M, 1970, University of Calgary. 121 p.

Embry, Ashton Fox, III. The Middle-Upper Devonian clastic wedge of the Franklinian Geosyncline. D, 1976, University of Calgary. 282 p.

Embry, Paige A. Petrogenesis of the Yogo Peak stock, Little Belt Mountains, Montana. M, 1987, University of Montana. 94 p.

Emendorfer, Earl. The alteration of rhyolitic pitchstone to montmorillonite at Soap Hill, Nevada. M, 1939, Columbia University, Teachers College.

Emerick, Christina Marie. Age progressive volcanism in the Comores Archipelago and northern Madagascar. D, 1985, Oregon State University. 228 p.

Emerick, John A. Depositional environments of Ordovician and Silurian rocks near Neda, Wisconsin. M, 1984, University of Wisconsin-Milwaukee. 76 p.

Emerick, William L. Geology of the Golden area, Santa Fe County, New Mexico. M, 1950, University of New Mexico. 66 p.

Emerman, Steven Howard. Some creeping flow solutions in geodynamics. D, 1984, Cornell University. 141 p.

Emerson, Donald. The surficial geology of the Cooking Lake Moraine, east central Alberta, Canada. M, 1977, University of Alberta. 116 p.

Emerson, Donald Orville. Granitic rocks of the northern portion of the Inyo Batholith. D, 1959, Pennsylvania State University, University Park. 153 p.

Emerson, Donald Orville. Secondary uranium minerals at the W. Wilson Mine near Clancy, Montana. M, 1955, Pennsylvania State University, University Park. 73 p.

Emerson, John W. A petrographic and environmental study of the Upper Cretaceous Greenhorn Limestone, New Mexico. M, 1961, University of New Mexico. 94 p.

Emerson, John Wilford. Stratigraphy and petrology of upper Chester (Mississippian) rocks in northern Alabama. D, 1967, Florida State University. 142 p.

Emerson, Mark E. An investigation into the origin of the Steelport main seam of the Bushveld Complex. M, 1955, University of Wisconsin-Madison.

Emerson, Matthew S. Geology of the Warrenburg area, Greene and Cocke counties, Tennessee. M, 1963, University of Tennessee, Knoxville. 36 p.

Emerson, Nancy L. Lower Tithonian volcaniclastic rocks above the Llanada Ophiolite, California. M, 1979, University of California, Santa Barbara.

Emerson, S. Radium-226 and radon-222 as limnologic tracers; the carbon dioxide gas exchange rate. D, 1974, Columbia University. 422 p.

Emerson, William Keith. A review of the eastern Pacific scaphopod mollusks. D, 1956, University of California, Berkeley. 312 p.

Emerson, William Stewart. Geological and deformational characteristics of the Little Maria Mountains, Riverside County, California. M, 1981, San Diego State University.

Emery, Alden H. Geology of the Pearisburg (Va.) region. M, 1923, Ohio State University.

Emery, David James. Genesis of copper deposits in Pre-Cambrian iron formation near the Round Lake Batholith, Boston Creek area, Ontario. D, 1959, Harvard University.

Emery, Herbert M. Geology and geomorphology of southeastern Washington County, Rhode Island. M, 1939, Brown University.

Emery, J. A. Geology of the Crystal Mine, Hardin County, Illinois. M, 1950, University of Missouri, Rolla.

Emery, John Rathbone. The application of a discriminant function to a problem in petroleum petrology. M, 1954, Pennsylvania State University, University Park. 120 p.

Emery, Kenneth Orris. A new coring instrument and its relation to problems of sedimentation. M, 1939, University of Illinois, Urbana. 45 p.

Emery, Kenneth Orris. Lithology of the sea-floor off Southern California. D, 1941, University of Illinois, Urbana. 106 p.

Emery, Martin. A detailed investigation of a Morrowan sandstone reservoir, Lexington Field, Clark County, Kansas. M, 1985, Wichita State University. 183 p.

Emery, Philip A. Stratigraphy of the Pleasanton Group in Bourbon, Neosho, Labette, and Montgomery counties, Kansas. M, 1962, University of Kansas. 55 p.

Emery, Wilson B. Geology of the Carrizo Mountains, Arizona. D, 1914, Yale University.

Emhof, John Warren. A geothermal study of southern Virginia, North Carolina and eastern Tennessee. M, 1977, University of Florida. 82 p.

Emhof, Stewart A. A petrographic and geochemical study of the Central Intrusive, Marysvale, Utah. M, 1984, SUNY, College at Oneonta. 50 p.

Emigh, George Donald. A mineralogical and metallurgical investigation of ore from the Center Star Mine, Idaho. M, 1933, University of Idaho. 29 p.

Emigh, George Donald. The petrography, mineralogy, and origin of phosphate pellets in the western Permian formation and other sedimentary formations. D, 1956, University of Arizona.

Emilia, David Arthur. Numerical methods in the direct interpretation of marine magnetic anomalies. D, 1969, Oregon State University. 100 p.

Emiliani, Cesare. The Oligocene microfaunas of the Northern Apennines, Itlay. D, 1950, University of Chicago. 252 p.

Emmanuel, Robert J. The geology and geomorphology of the White Rock Canyon area, New Mexico. M, 1950, University of New Mexico. 65 p.

Emmart, Laura A. Volatile transfer differentiation of the Gordon Butte magma, northern Crazy Mountains, Montana. M, 1985, University of Montana. 83 p.

Emme, David H. Delineation of subsurface flow in the upper Meadow Valley Wash area, southeastern Nevada. M, 1986, University of Nevada. 90 p.

Emme, James J. Tectonic influence on sedimentation, Lower Cretaceous strata, Osage-Newcastle area, Powder River basin, Wyoming. M, 1981, Colorado School of Mines. 173 p.

Emmendorfer, Alan Paul. Diagenesis and pore evolution of the reef plate, Enewetok Atoll, Marshall Islands. M, 1979, University of Oklahoma. 136 p.

Emmer, Rodney E. Crevasses of the Lower Mississippi River delta. M, 1968, Louisiana State University.

Emmerich, Harry Henry. The geologic history of the caverns of Kentucky. M, 1935, George Washington University. 34 p.

Emmerich, William Eugene. Chemical forms of heavy metals in sewage sludge-amended soils as they relate to movement through soils. D, 1980, University of California, Riverside. 176 p.

Emmerling, Michael Dean. The Recent beach sands of Dog Island, Florida. M, 1974, Florida State University.

Emmet, Peter Anthony. Geology of the Agalteca Quadrangle, Honduras, Central America. M, 1983, University of Texas, Austin. 201 p.

Emmet, Robert Temple. Density studies of aqueous solutions and seawater at various temperatures and pressures. D, 1973, University of Miami.

Emmons, N. H. Report of Mine No. 8 at Felming, Missouri. M, 1889, Washington University.

Emmons, Patrick Jay. Relationship between seismic velocity, degree of weathering, and seepage potential; Watershed 12, Beaver Creek watershed, Coconino County, Arizona. M, 1978, Northern Arizona University. 71 p.

Emmons, Richard C. Studies in mineral separation in a finely divided state. D, 1924, University of Wisconsin-Madison.

Emmons, Richard Conrad. The Rainy Day mineral prospect. M, 1920, University of British Columbia.

Emmons, William Harvey. The geology of Haystack Mountains, Montana. D, 1904, University of Chicago.

Emo, Wallace B. The basic intrusives of the Waco Lake area, Saguenay County, Quebec. M, 1955, McGill University.

Emo, Wallace B. The geology of the Wacouno region, Saguenay County, Quebec. D, 1957, McGill University.

Emond, Diane L. S. Geology, mineralogy and petrogenesis of the Oliver Creek Breccia; vein tin occurrence, McQuesten River, Yukon. M, 19??, Carleton University. 196 p.

Emory-Moore, Margot. Sedimentological and mineralogical characteristics of offshore sediments, southeastern Nova Scotia. M, 1985, University of New Brunswick.

Empie, Joel S. Stratigraphy of the Funston Limestone (Wolfcampian) between the Kansas River and Cottonwood River valleys, Kansas. M, 1961, University of Kansas. 131 p.

Emrich, Grover Harry. Geology of the Ironton and Galesville sandstones in the upper Mississippi Valley. D, 1962, University of Illinois, Urbana. 119 p.

Emrick, Harry W. The investigation of determining geoid profiles by gravimetric leveling. M, 1961, Ohio State University.

Emrick, Harry William. Computation techniques for various gravity anomaly correction terms and their effect upon deflection of the vertical computations for mountainous areas. D, 1973, Ohio State University. 158 p.

Emrie, Gail Estelle. Fecal coloform as an indicator of water quality and recreational carrying capacity at Ozark National Scenic Riverways. M, 1986, Southwest Missouri State University.

Emry, Janet Salyer. The hydrogeology of Nags Head Woods, Dare County, North Carolina. M, 1987, Old Dominion University. 117 p.

Emry, Robert John. Stratigraphy and paleontology of the Flagstaff Rim area, Natrona County, Wyoming. D, 1970, Columbia University. 169 p.

Emslie, Ronald Frank. Age determination of accessory zircon from granitic rocks of the Kenora area, Ontario. M, 1957, University of Manitoba.

Emslie, Ronald Frank. The petrology and economic geology of two mafic intrusions in the Lynn Lake area, northern Manitoba. D, 1961, Northwestern University. 187 p.

Emslie, Steven Douglas. The origin, evolution and extinction of condors in the New World. D, 1987, University of Florida. 172 p.

Enbysk, Betty Joyce Blomgren. Additions to the Devonian and Carboniferous faunas of northeastern Washington with summary of Paleozoic fossil localities. M, 1954, Washington State University. 51 p.

Enbysk, Betty Joyce Blomgren. Distribution of foraminifera in the northeast Pacific. D, 1960, University of Washington. 238 p.

Enciso, Gonzalo. Paleoenvironmental analysis of the Morrison Formation (Late Jurassic) in the Canon City, Colorado, area. M, 1982, University of Kansas. 89 p.

Enciso, Salvador. Geologic report on the Cuencame Quadrangle, State of Durango, Mexico. M, 1967, Stanford University.

Ende, Barbara Ann am *see* am Ende, Barbara Ann

Enderlin, Margot Helene. The origin of Feather Falls, Butte County, California. M, 1970, California State University, Chico.

Enders, Dean W. A study of the Cretaceous and Upper Jurassic of the Monticello-Pope Valley region, Napa County, California. M, 1939, University of California, Berkeley. 48 p.

Enders, Merritt Stephen. The geology, mineralization, and exploration characteristics on the Beck Mine and vicinity, Kimball mining district, Hidalgo County, New Mexico, and Cochise County, Arizona. M, 1981, University of Arizona. 109 p.

Endo, Elliot Toru. Focal mechanisms for the May 15-18, 1970 shallow Kilauea (Hawaii) earthquake swarm. M, 1971, San Jose State University. 165 p.

Endo, Elliot Toru. Seismotectonic framework for the southeast flank of Mauna Loa Volcano, Hawaii. D, 1985, University of Washington. 349 p.

Endo, Howard. Mechanical transport in two-dimensional networks of fractures. D, 1984, University of California, Berkeley. 203 p.

Endrodi, Sandra Monroe. Petrology and geochemistry of Isla San Benedicto, Mexico. M, 1975, University of Oregon. 61 p.

Enfield, David B. Prediction of hazardous Columbia River bar conditions. D, 1974, Oregon State University. 204 p.

Eng, Frank Gee. Geology of the Lucky Horseshoe thorium deposit, Lemhi County, Idaho. M, 1960, University of Idaho. 38 p.

Eng, K. J. Sub-bottom reflection profiling of Recent sediments in Lake Powell, Utah-Arizona. M, 1972, [Dartmouth College].

Engdahl, Eric Robert. Core phase and the Earth's core. D, 1968, St. Louis University. 206 p.

Engebretson, David C. Relative motions between oceanic and continental plates in the Pacific Basin. D, 1983, Stanford University. 218 p.

Engel, Albert Edward John. Geology of the House Springs area, Columbia, Missouri. M, 1939, University of Missouri, Columbia.

Engel, Albert Edward John. The quartz crystal deposits of western Arkansas. D, 1945, Princeton University. 118 p.

Engel, Bernard Allen. Knowledge engineering in soil erosion. D, 1988, Purdue University. 213 p.

Engel, Celeste Gilpin. Desert varnish. M, 1957, University of California, Los Angeles.

Engel, Gregory Allen. Measurement of the complex dynamic rigidity of Recent marine sediments. M, 1972, United States Naval Academy.

Engel, Kevin. Faunal succession and lithostratigraphy across the House Limestone-Fillmore Formation boundary, Pogonip Group, Ibex area, Millard County, Utah; a possible biomere boundary. M, 1984, University of Missouri, Columbia. 93 p.

Engel, Michael Harris. Amino acids in ancient (Precambrian) rocks; their occurrence, abundance and degree of racemization. D, 1980, University of Arizona. 186 p.

Engel, Michael Harris. Racemization of amino acids in wood; experimental results, problems, and perspectives. M, 1976, University of Arizona.

Engel, Paul Louis. Ecology of Ostracoda from Mesquite and Aransas bays, Southwest Texas. M, 1956, University of Minnesota, Minneapolis. 48 p.

Engel, Rene. Geochemical properties of the waters of the Elsinore Quadrangle (California). D, 1933, California Institute of Technology.

Engel, Ruth Flora. A new 5C pyrrhotite. M, 1978, University of Michigan.

Engel, Theodore, Jr. Stratigraphy and petrography of the Holyoke meta-sediments of the Dead River basin, Marquette County, Michigan. M, 1954, Michigan State University. 74 p.

Engelder, James Terry. Quartz fault-gouge; its generation and effect on the frictional properties of sandstone. D, 1973, Texas A&M University.

Engelder, P. Richard. Application of the unit regional value concept to a study of the mineral resources of Australia. D, 1979, Pennsylvania State University, University Park. 374 p.

Engelder, P. Richard. The interrelationships between friability and other measured variables in the Chickies Formation. M, 1976, Pennsylvania State University, University Park. 163 p.

Engelhardt, Claus L. The Paleozoic-Triassic contact in the Klamath Mountains, Jackson County, southwestern Oregon. M, 1966, University of Oregon. 98 p.

Engelhardt, Donald Wayne. A palynological study of post-glacial and interglacial deposits in Indiana. D, 1962, Indiana University, Bloomington. 148 p.

Engelhardt, Nancy L. Paleoecologic and biostratigraphic interpretations of late Miocene foraminifera at DSDP Site 265 (Leg 28), Southeast Indian Ocean. M, 1980, Northern Illinois University. 372 p.

Engelhardt, Richard Lee. The petrology of some igneous dikes of western Kentucky. M, 1973, Eastern Kentucky University. 59 p.

Engelmann, G. F. The logic of phylogenetic analysis and the phylogeny of the Xenarthra (Mammalia). D, 1978, Columbia University, Teachers College. 333 p.

Engeln, Joseph Francis. Seismological studies of the tectonics of divergent plate boundaries. D, 1985, Northwestern University. 147 p.

Engeln, Oscar Dierich Von *see* Von Engeln, Oscar Dierich

Engels, Gary G. The occurrence of zinc, copper, and lead in the Decorah Formation from the southwestern Wisconsin zinc and lead district. M, 1959, University of Wisconsin-Madison.

Engels, Joan Carol. Discordances in K-Ar and Rb-Sr isotopic ages. M, 1963, Columbia University, Teachers College.

Engh, Kenneth R. Structural geology of the Rastus Mountain area, east-central Oregon. M, 1984, Washington State University. 78 p.

Engi, Jill Ellen. Structure and metamorphism north of Quesnel Lake and east of Niagara Creek, Cariboo Mountains, British Columbia. M, 1984, University of British Columbia. 137 p.

Engineer, B. B. Iron formation of Shelgrove Lake, Labrador. M, 1950, McGill University.

England, Anthony W. Equations of state of oxides and silicates and new data on the elastic properties of spinel, magnetite, and cadmium oxide. D, 1970, Massachusetts Institute of Technology. 159 p.

England, D. Kent. A physical model experiment to study the monitoring of enhanced oil recovery using the seismic reflection method. M, 1988, University of Houston.

England, Daniel L. Geology in the Modoc Pb-Ag-Zn district, Inyo County, California. M, 1987, University of Minnesota, Duluth. 82 p.

England, Evan J. Experimental studies of the origin and thermal metamorphism of chondrules in chondritic meteorites. M, 1968, University of Vermont.

England, Lindy Alison. Long-term transient regional groundwater flow in a heterogeneous mature basin with large hydraulic conductivity contrasts. M, 1986, University of British Columbia. 156 p.

England, Richard L. A subsurface study of the Hunton Group (Silurian-Devonian) in the Oklahoma portion of the Arkoma Basin. M, 1961, University of Oklahoma. 82 p.

Engle, Edgar Wallace. Observations of the rare earths. D, 1916, University of Illinois, Urbana.

Engle, Kathryn Yvonne. Earthquake focal mechanism studies of the Cook Inlet area, Alaska. M, 1982, University of Alaska, Fairbanks. 81 p.

Engle, M. S. Carbon, nitrogen and microbial colonization of volcanic debris on Mt. St. Helens. M, 1983, Washington State University. 62 p.

Engleman, Mary. Patterns of cementation in selected shallow marine and peritidal carbonates. M, 1979, University of Kansas. 122 p.

English, Brian L. Zircon morphology and U-Pb zircon isotopic study of the Salisbury Pluton, Rowan County, North Carolina. M, 1984, University of North Carolina, Chapel Hill. 123 p.

English, Douglas John. Regional structural analysis of the Santa Rosa Mountains, San Diego and Riverside counties, California; implications for the geologic history of Southern California. M, 1985, San Diego State University. 170 p.

English, H. Duncan. The geology of the San Timoteo Badlands, Riverside County, California. M, 1953, Pomona College.

English, John Richard. Diagenetic processes in the Oligocene-Miocene sediments; B-2 well, Baltimore Canyon trough. M, 1978, Rutgers, The State University, New Brunswick. 21 p.

English, Jordan W. Coarse clastics of the Johns Valley Formation, south-central Arkansas. M, 1984, Memphis State University. 108 p.

English, Leon E. The geology of the Youngstown oil field (Okmulgee County, Oklahoma). M, 1921, University of Oklahoma. 23 p.

English, Paul James. Gold-quartz veins in metasediments of the Yellowknife Supergroup, Northwest Territories; a fluid inclusion study. M, 1981, University of Alberta. 108 p.

English, Robert D. Depositional environments and lignite petrology of the Calvert Bluff Formation (Eocene) in the C area of the Big Brown surface mine near Fairfield, Texas. M, 1988, Southern Illinois University, Carbondale. 95 p.

English, Van Harvey. Cordilleran glaciation in a section of the Cabinet Mountains, Montana. D, 1942, Clark University. 123 p.

English, Walter A. The Fernando Group near Newhall, California. M, 1914, University of California, Berkeley. 218 p.

Englishman, Doanld Ellsworth. Abundance of distribution of Pb, Zn, Cu and Cd in the Downeys Bluff Limestone, Minerva Mine no. 1, Cave-in-Rock District, Illinois. M, 1968, Northern Illinois University. 48 p.

Englund, Evan John. The bedrock geology of the Holderness Quadrangle, New Hampshire. D, 1974, Dartmouth College. 90 p.

Englund, Kenneth John. Stratigraphy and areal geology of the Fairchild Quadrangle, M, 1950, University of Wisconsin-Madison. 48 p.

Engman, Harry Arthur. An insoluble residue study of the Allentown Formation, near Bethlehem, Pennsylvania. M, 1951, University of Pittsburgh.

Engman, Mary Anne. Depositional systems in the lower part of the Pottsville Formation, Black Warrior Basin, Alabama. M, 1985, University of Alabama. 257 p.

Engstrom, Daniel Russell. Chemical stratigraphy of lake sediments as a record of environmental change. D, 1983, University of Minnesota, Minneapolis. 244 p.

Engstrom, David Bert. Geology of part of Centennial Mountain Quadrangle, Bearpaw Mountains, Montana. M, 1953, University of California, Berkeley. 62 p.

Engstrom, James Charles. Invertebrate megafauna of the Microcyclus zone of the Saint Laurent limestone (middle Devonian), in southwestern Illinois and southeastern Missouri. M, 1969, Southern Illinois University, Carbondale. 70 p.

Engstrom, William Scott. Sedimentology of the Inyan Kara Group east of Piedmont, South Dakota. M, 1979, South Dakota School of Mines & Technology.

Enis, Hunter. Relation of fragment orientation to jointing in the lower Weno Formation, Carter Park, Fort Worth, Texas. M, 1963, Texas Christian University.

Enkeboll, Robert Halfdan. Sedimentary petrology of sands from the Middle America Trench and trench slope; Guatemala and southern Mexico. M, 1978, University of California, Santa Cruz.

Enkin, Randolph Jonathan. Micromagnetic study of pseudo-single-domain structure applications to rock magnetism. M, 1986, University of Toronto.

Enlow, Donald Hugh. Methods and observations in vertebrate paleohistology. M, 1950, [University of Houston].

Enlows, Harold E. Geology and ore deposits of the Little Dragoon Mountains. D, 1939, University of Arizona.

Enlows, Harold E. Some factors influencing the physical characteristics of West Texas crude oils. M, 1936, University of Chicago. 15 p.

Enns, Steve Gerhard. A nickel deposit in southern British Columbia. M, 1971, University of Manitoba.

Enos, Paul Portenier. Anatomy of a flysch; Middle Ordovician Cloridorme Formation, northern Gaspe Peninsula. D, 1965, Yale University.

Enos, Paul Portenier. Geologic problems in the western Vallecitos Syncline, San Benito County, California. M, 1961, Stanford University. 90 p.

Enrich, Grover H. Sedimentation studies in the Hawthorn Formation of northwestern Florida. M, 1957, Florida State University.

Enrico, Roy John. Distribution of living shallow water benthonic foraminiferids of Eniwetok Atoll, Marshall Islands. M, 1978, University of California, Davis. 57 p.

Enright, Richard Louis, Jr. The stratigraphy, micropaleontology and paleoenvironmental analysis of the Eocene sediments of the New Jersey coastal plain. D, 1969, Rutgers, The State University, New Brunswick. 242 p.

Ensign, Paul S. An input electromagnetic investigation of the southern margin of the Jacobsville Sandstone, Upper Peninsula of Michigan. M, 1981, Michigan Technological University. 68 p.

Ensley, Michael. The measurement of thermal gradient in shallow boreholes with a portable Wheatstone Bridge. M, 1970, University of Tulsa. 37 p.

Ensminger, Henry R. Sediments and planktonic Foraminifera of tropical North Atlantic cores. M, 1967, Oregon State University. 69 p.

Enterline, Theodore R. Depositional environment of the Pagoda, Pentagon, and Steamboat formations (Middle Cambrian), Northwest Montana. M, 1978, University of Montana. 105 p.

Entrup, Karen. The effect of land use on the water quality of two coves of Table Rock Reservoir. M, 1986, Southwest Missouri State University. 68 p.

Entsminger, Lee. Beach pads and beach cusps, St. Joseph Peninsula. M, 1978, Florida State University.

Entwhistle, Lawson P. The Chloride Flat mining district, New Mexico. M, 1938, University of Arizona.

Entzeroth, Lee Catherine. Particulate matter and organic sedimentation on the continental shelf and slope of the Northwest Gulf of Mexico. D, 1982, University of Texas, Austin. 271 p.

Entzminger, David Jacob. A revision of the Upper Cretaceous stratigraphy in the Big Hole Mountains, Idaho. M, 1979, Idaho State University. 96 p.

Enwall, Robert E. Application of borehole gamma-ray logging in the reserve estimation of a non-sandstone uranium deposit. M, 1982, Idaho State University. 149 p.

Enyert, Richard Lyle. Geology of the Calamity Point area, Snake River Range, Idaho. M, 1947, University of Michigan.

Enyert, Richard Lyle. Middle Devonian sandstones of the Michigan Basin. D, 1950, University of Michigan.

Enz, Robert David. Geochemistry and petrology of the orbicular rocks, Sandia Mountains, New Mexico. M, 1974, University of New Mexico. 73 p.

Epie, Ebenezer E. E. Metamorphic geology of Cooper and Sheep mountains, Fremont County, Colorado. M, 1968, Southern Illinois University, Carbondale. 64 p.

Epis, Rudy Charles. Geology of the Pedragosa Mountains, Cochise County, Arizona. D, 1956, University of California, Berkeley. 261 p.

Epler, Nathan Andrew. Experimental study of Fe-Ti oxide ores from the Sybille Pit in the Laramie Anorthosite, Wyoming. M, 1987, SUNY at Stony Brook. 67 p.

Epp, David. Age and tectonic relationships among volcanic chains on the Pacific Plate. D, 1978, University of Hawaii.

Eppert, Herbert Charles. Stratigraphy of the upper Miocene deposits in Sarasota County, Florida. M, 1963, University of Florida. 66 p.

Eppert, Herbert Charles, Jr. The marine geology and ecology of an area off the west coast of Kauai, Hawaii. D, 1967, Tulane University. 185 p.

Eppich, Gilbert Keith. Aeromagnetic survey of south-central Utah. M, 1972, University of Utah. 78 p.

Eppihimer, Richard M. A geophysical study of the Roseland anorthosite-titanium district, Nelson and Amherst counties, Virginia. M, 1978, University of Georgia.

Eppler, Dean B. The geology of the San Antonio Mountain area, Tres Piedras, Taos and Rio Arriba counties, New Mexico. M, 1976, University of New Mexico. 77 p.

Eppler, Dean Bener. Characteristics of volcanic blasts, mudflows and rock-fall avalanches in Lassen Volcanic National Park, California. D, 1984, Arizona State University. 262 p.

Eppler, Duane. A previously unrecognized early pre-Imbrian multi ringed basin; Fourier shape analysis. D, 1978, University of South Carolina.

Eppler, Duane T. Late Pleistocene geology of Elm Creek, Edwards, and Pitcairn valleys, Saint Lawrence County, New York. M, 1973, Syracuse University.

Eppler, Duane Thomas. Geologic implications of regional scale variation in lunar crater shape; Fourier crater shape analysis. D, 1980, University of South Carolina. 118 p.

Epps, Lawrence W. History of the Brazos River (Texas). M, 1972, Baylor University. 122 p.

Epstein, Anita Fishman. Stratigraphy of uppermost Silurian and lowermost Devonian rocks and the conodont fauna of the Coeymans Formation and its correlatives in northeastern Pennsylvania, New Jersey, and southeasternmost New York. D, 1970, Ohio State University.

Epstein, Bernard. A correlative unit on igneous activity and the development of igneous rocks. M, 1967, University of Utah. 146 p.

Epstein, Claude Murray. Paleoecological analysis of the open-shelf facies in the Helderberg Group (lower Devonian) of New York State. D, 1971, Brown University.

Epstein, J. B. Geology of part of the Fanny Peak Quadrangle, Wyoming-South Dakota. M, 1958, University of Wyoming. 90 p.

Epstein, Jack Burton. Geology of the Stroudsburg Quadrangle and adjacent areas, Pennsylvania - New Jersey. D, 1970, Ohio State University.

Epstein, Mark L. The distribution and abundance of polychlorinated biphenyls in surface waters of Bloomington, Indiana. M, 1979, Indiana University, Bloomington. 76 p.

Epstein, Rachel Sophia. The eastern margin of the Burlington granodiorite, Newfoundland. M, 1983, University of Western Ontario. 189 p.

Epstein, Samuel A. Cementation and inversion in beach rock and reef rock; a process-oriented approach. M, 1979, Rensselaer Polytechnic Institute. 90 p.

Eralp, Atal Enerjin. The role of iron in the phosphorus cycle in lakes. D, 1973, University of North Carolina, Chapel Hill. 144 p.

Erb, David. Geologic evaluation for potential development of underground space in the Black Hills, South Dakota and Wyoming. M, 1983, South Dakota School of Mines & Technology.

Erb, Denise. An investigation of DSDP Leg 86 ash layers bearing on stratigraphy and volcanic history of the Northwest Pacific. M, 1988, SUNY at Buffalo. 57 p.

Erb, Edward Edeburn, Jr. Geology of the volcanic rocks north of Salina Canyon, Salina, Utah. M, 1971, Ohio State University.

Erb, Edward Edeburn, Jr. Petrologic and structural evolution of ash-flow tuff cauldrons and noncauldron-related volcanic rocks in the Animas and southern Peloncillo Mountains, Hildago County, New Mexico. D, 1979, University of New Mexico. 286 p.

Erb, Elizabeth. Crystal habit. M, 1926, Northwestern University.

Ercan, Ahmet. Ground magnetic survey on the serpentine body and the Franciscan Group in Searville Lake Park, Portola Valley, California. M, 1973, Stanford University.

Erchul, Ronald Anton. The use of electrical resistivity to determine porosity of marine sediments. D, 1972, University of Rhode Island.

Ercit, Timothy Scott. The simpsonite paragenesis; the crystal chemistry and geochemistry of extreme Ta fractionation. D, 1986, University of Manitoba.

Erd, Richard Clarkson. The mineralogy of Indiana. M, 1954, Indiana University, Bloomington. 170 p.

Erdahl, William Mitchell. The origin of martite. M, 1932, Michigan Technological University. 20 p.

Erdem, Fahri. Geology of the Marble Canyon plutonic complex. M, 1981, University of California, Los Angeles. 189 p.

Erdlac, Richard John, Jr. A study of the Chixoy-Polochic Fault and its nature in western Guatemala. M, 1979, University of Pittsburgh.

Erdlac, Richard John, Jr. Structural development of the Terlingua Uplift, Brewster and Presidio counties, Texas. D, 1988, University of Texas, Austin. 403 p.

Erdman, Linda Ruth. Chemistry of Neogene basalts of British Columbia and the adjacent Pacific Ocean floor; a test of tectonic discrimination diagrams. M, 1985, University of British Columbia.

Erdman, Mary Cordelia. A review of Bernard's studies of hinge ontogeny in pelecypod phylogeny. M, 1949, Columbia University, Teachers College.

Erdman, Oscar A. Geology of Alexo and Saunders map-area, Alberta, Canada. D, 1946, University of Chicago. 131 p.

Erdman, Oscar Alvin. A study of some Alberta soils in relation to source and reservoir beds and various physical properties. M, 1941, University of Alberta. 116 p.

Erdmann, Anne Lana. A paleomagnetic investigation of the Ankareh Formation, Overthrust Belt, Utah. M, 1988, University of Minnesota, Minneapolis. 118 p.

Erdmann, Charles Edgar. The ore deposits of the Transvaal and their geologic relations. M, 1924, University of Minnesota, Minneapolis. 76 p.

Erdmann, Charles Edgar. The Pre-Cambrian rock of the Keystone region with notes on the geology and ore deposits of the Bullion Mine. M, 1923, University of Minnesota, Minneapolis. 49 p.

Erdmer, Philippe. Metamorphism in the Stanhope Pluton aureole, Quebec Appalachians; an estimate of conditions from mineral reactions in pelite and calcschist. M, 1979, Queen's University. 140 p.

Erdmer, Philippe. Nature and significance of the metamorphic minerals and structures of cataclastic allochthonous rocks in the White Mountains, Last Peak and Fire Lake areas, Yukon Territory. D, 1982, Queen's University. 254 p.

Erdogan, B. Geology, geochemistry, and genesis of the sulphide deposits of the Ergani-Maden region, SE-Turkey. D, 1977, University of New Brunswick.

Erdogan, Solmaz Z. Cenomanian Buda limestone (Comanche Cretaceous), of west and Trans-Pecos, Texas. M, 1969, Louisiana State University.

Erdosh, George. Modal analyses of rocks by instrumental techniques. D, 1967, McGill University. 140 p.

Erdosh, George. Sphalerite and pyrrhotite geothermometry of the New Calumet sulphide deposits, Calumet, (Quebec) Canada. M, 1962, McGill University.

Erdreich, Emil. A study of productivity of the copper mines industry in the United States. M, 1937, American University. 131 p.

Erel, Bilgin. Physico-chemical model analysis for compressibility of pure clay. D, 1970, Carnegie-Mellon University. 153 p.

Eren, Ahmet Aytac. Surface geology of the Millport Quadrangle and western half of the Kennedy Quadrangle, Alabama. M, 1984, Mississippi State University. 111 p.

Erez, Jonathan. The influence of differential production and dissolution on the stable isotope composition of planktonic foraminifera. D, 1978, Woods Hole Oceanographic Institution. 119 p.

Ergas, Raymond Andrew. Local earthquake traveltimes and spatial variations of crustal velocity in Southern California. D, 1981, University of California, Los Angeles. 119 p.

Ergin, Kazim. 1, Energy ratio of the seismic waves reflected and refracted at a rock-water boundary; II, Amplitudes of PcP, PcS, ScS, and ScP in deep focus earthquakes. D, 1950, California Institute of Technology. 32 p.

Ergin, Kazim. Improved epicenters of earthquakes in Turkey. M, 1943, California Institute of Technology. 32 p.

Eric, John Howard. Geology of the Vermont portion of the Littleton Quadrangle. D, 1942, Harvard University.

Ericksen, George Edward. Geology of the Hualgayoc mining district, departamento de Cajamarca, Peru. D, 1954, Columbia University, Teachers College. 246 p.

Ericksen, George Edward. Petrology of Silurian limestones of northern Indiana. M, 1949, Indiana University, Bloomington. 39 p.

Ericksen, Rick L. Rubidium-strontium geochemistry of mafic and ultramafic inclusions, associated volcanic rocks and basement rocks from the Ross Island area, Antarctica. M, 1975, Northern Illinois University. 76 p.

Erickson, Albert J. The measurement and interpretation of heat flow in the Mediterranean and Black seas. D, 1970, Woods Hole Oceanographic Institution. 433 p.

Erickson, Alvin J. Temperature of calcite deposition in the Upper Mississippi Valley lead-zinc deposits. M, 1964, University of Wisconsin-Madison.

Erickson, Barrett H. Marine seismic studies near Newport, Oregon. M, 1967, Oregon State University. 39 p.

Erickson, Denis R. A study of littoral groundwater seepage at Williams Lake, Minnesota using seepage meters and wells. M, 1980, University of Minnesota, Minneapolis.

Erickson, Denis Roger. A study of littoral groundwater seepage at Williams Lake, Minnesota, using seepage meters and wells. M, 1981, University of Minnesota, Duluth.

Erickson, Edwin Sylvester, Jr. Mineralogical, petrographic, and geochemical relationships in some high-alumina and associated claystone from the Clearfield Basin, Pennsylvania. D, 1963, Pennsylvania State University, University Park. 202 p.

Erickson, Einar C. Geology and uranium mineralization in the East Gas Hills, Wyoming. M, 1957, Brigham Young University. 50 p.

Erickson, Glen P. Potassium-argon dating of basaltic systems. D, 1967, Columbia University, Teachers College.

Erickson, Glen P. Potassium-argon measurements on the Pallisades Sill (New York-New Jersey). M, 1960, Columbia University, Teachers College.

Erickson, Harold Dean. Geology of the Willett and Midland No. 1 quadrangles, Harding County, South Dakota. M, 1958, University of California, Los Angeles.

Erickson, James R. Parameter-estimation technique for the analysis of single-well tracer tests. M, 1985, [Colorado State University].

Erickson, John Mark. Gastropoda of the Fox Hills Formation (Maestrichtian-Upper Cretaceous) of North Dakota. D, 1971, University of North Dakota. 248 p.

Erickson, John Mark. The geologic and limnologic history of Glovers Pond, northwestern New Jersey. M, 1968, University of North Dakota. 149 p.

Erickson, John William. Paleoslope and petrographic analysis of the South Point Formation, Santa Rosa Island, California. M, 1972, University of California, Santa Barbara.

Erickson, Kirth. Surficial lineaments and their structural implications in the Williston Basin. M, 1970, University of North Dakota. 59 p.

Erickson, Lance and Baetcke, Gustav Berndt. Structure of the Lobate thrust sheet, Huerfano Park, Colorado. M, 1957, University of Michigan.

Erickson, Laurie Lynn. A three-dimensional dislocation program with applications to faulting in the Earth. M, 1986, Stanford University. 167 p.

Erickson, Martin Richard. Geology of the western portion of the Pleasanton Quadrangle, California. M, 1946, University of California, Berkeley. 57 p.

Erickson, Max Perry. Thermal metamorphism of the ancient tillite (Precambrian) of the Alta region, Utah. M, 1940, University of Utah. 36 p.

Erickson, Norman K. Hystrichosphaerids of the Devonian Onondaga Formation, Welland County, Ontario, Canada. M, 1956, University of Massachusetts.

Erickson, Ralph LeRoy. A petrographical investigation of the longitudinal deposition within the Mason Esker relative to its origin. M, 1948, Michigan State University. 39 p.

Erickson, Ralph LeRoy. Stratigraphy and petrology of the Tascotal Mesa Quadrangle, Texas. D, 1951, University of Minnesota, Minneapolis. 148 p.

Erickson, Richard A. Stratigraphy and depositional environments, Lyons Formation (Permian), Golden-Morrison area, Jefferson County, Colorado. M, 1977, Colorado School of Mines. 125 p.

Erickson, Robert C. Etching quartz grains for purposes of correlation. M, 1948, University of Wisconsin-Madison.

Erickson, Robert H. Cyclic sedimentation in the Hell Creek Formation (Upper Cretaceous) of northwestern South Dakota. M, 1950, Northwestern University.

Erickson, Roland I. Geology of Bomi Hills, Liberia, Africa. M, 1954, University of North Dakota. 68 p.

Erickson, Rolfe C. Petrology and structure of an exposure of the Pinal Schist, Santa Catalina Mountains, Arizona. M, 1962, University of Arizona.

Erickson, Rolfe Craig. Petrology and geochemistry of the Dos Cabezas mountains, Cochise County, Arizona. D, 1969, University of Arizona. 480 p.

Erickson, Thomas David. An east-west structural traverse of western Ohio from Lancaster, Ohio, to the Ohio River. M, 1959, Ohio State University.

Erickson, Thomas David. Equilibrium relations in the CaO-Al_2O_3-ZrO_2-SiO_2 system, especially in the ternary system CaO-ZrO_2-SiO_2. D, 1970, Ohio State University.

Erickson, Wayne Albert. The Mississippian-Pennsylvanian unconformity in Dubois County, Indiana. M, 1952, Indiana University, Bloomington. 30 p.

Ericson, David B. Geology of the Whittier Hills, California. M, 1933, California Institute of Technology. 37 p.

Ericson, Donald Martin. The detailed physical stratigraphy of the Franconia Formation in southwestern Wisconsin. M, 1951, University of Wisconsin-Madison. 28 p.

Ericson, Norman John. The Ilocos Norte manganese deposits and especially the Siec Group of the Ilocos manganese mining company located in the Commonwealth of the Philippines. M, 1939, University of Nevada - Mackay School of Mines. 27 p.

Ericson, Warren Tongo. Bentonite. M, 1931, University of Minnesota, Minneapolis. 44 p.

Erikesson, Yves. Geology of the upper Ogden Canyon, Weber County, Utah. M, 1960, University of Utah. 55 p.

Eriksen, Harold W. Composition of garnets in metamorphic rocks of Twin Island, Pelham Bay Park, Bronx, New York. M, 1971, Brooklyn College (CUNY).

Erikson, Daniel L. Analysis of water movement in an underground lead-zinc mine, Coeur d'Alene mining district, Idaho. M, 1985, University of Idaho. 115 p.

Erikson, Erik Harold, Jr. Petrology of an eastern portion of the Snoqualmie batholith, central Cascades, Washington. M, 1965, University of Washington. 52 p.

Erikson, Erik Harold, Jr. Petrology of the composite Snoqualmie Batholith (Miocene and Pliocene), central Cascade Mountains, Washington. D, 1968, Southern Methodist University. 111 p.

Erikson, Johan P. Structural study of the Paleogene deformation in and around the southern Puerto Rico fault zone of south central Puerto Rico. M, 1988, Stanford University. 74 p.

Erikson, Robert Lawrence. An experimental and theoretical investigation of plagioclase melting relations. M, 1979, Pennsylvania State University, University Park. 62 p.

Erikson, Susan J. Thermometry as a tool for determining the hydrologic properties of the vadose zone. M, 1988, University of Nevada. 244 p.

Eriksson, Carl L. Petrology of the alkalic hypabyssal and volcanic rocks at Cripple Creek, Colorado. M, 1987, Colorado School of Mines. 114 p.

Erinakes, Dennis C. Stratigraphy of the undifferentiated Silurian strata of the Ragged Lake Quadrangle, Maine. M, 1967, University of Maine. 73 p.

Eriyagama, Sarath Chandra. Geology and reservoir rock types, Mississippian; Ratcliffe Beds, Hummingbird oilfield area, TP 2-3, Rge. 18-19 2M, Saskatchewan. M, 1982, University of Regina. 161 p.

Erjavec, James Laurence. The sulfur distribution in the sedimentary rocks of Southeast Arizona. M, 1981, University of Arizona. 114 p.

Erlandson, D. L. Geology and geophysics of the Lyra Basin. M, 1975, University of Hawaii at Manoa. 69 p.

Erlenwein, Susan D. Geochemical aspects of the Lower Permian and the Upper Pennsylvanian in Greenwood County, Kansas. M, 1978, Wichita State University. 118 p.

Erler, Elise L. Petrology and uranium mineralization of the Idaho Batholith near Stanley, Custer County, Idaho. M, 1980, University of Montana. 98 p.

Erler, Yusuf Ayhan. Alteration and trace elements in the Jenney Horizon of the Park City Formation, Park City District, Utah. D, 1974, University of Utah. 121 p.

Erlich, Robert N. Early Holocene to Recent development and sedimentation of the Roanoke River area, North Carolina. M, 1980, University of North Carolina, Chapel Hill. 83 p.

Ermanovics, Ingomar Frank. Evidence bearing on the origin of the Perth Road Pluton (Precambrian, Grenville), southern Ontario (Canada). D, 1967, Queen's University. 177 p.

Ermanovics, Ingomar Frank. Origin of the Belleoram Stock (Devonian), Newfoundland, Canada. M, 1964, University of Western Ontario. 83 p.

Ern, Ernest Henry, Jr. Bedrock geology of the Randolph Quadrangle, Vermont. D, 1959, Lehigh University. 173 p.

Ernissee, John J. Biostratigraphy and siliceous microfossil paleontology of the Coosawhatchie Clay (Miocene, S.C.) and the Pungo River Formation; Miocene, N.C. D, 1978, University of South Carolina. 187 p.

Ernissee, John J. Mineral binding by a species of Thalassiosira. M, 1975, University of South Carolina.

Ernst, David Raymond. Petrography and geochemistry of Boston Peak and Tomichi Dome, and relation to other plutons in Gunnison County, Colorado. M, 1980, Eastern Washington University. 53 p.

Ernst, Richard Everett. Correlation of Precambrian diabase dike swarms across the Kapuskasing structural zone, northern Ontario. M, 1981, University of Toronto.

Ernst, Robert P. Granite and rhyolite relationships of the Lake City Caldera area, Hinsdale County, Colorado. M, 1981, Eastern Washington University. 60 p.

Ernst, Wallace Gary. Phase relations and stabilities of the alkali amphiboles. D, 1959, The Johns Hopkins University.

Ernst, Wallace Gary. The petrology of the Endion Sill, Duluth, Minnesota. M, 1955, University of Minnesota, Minneapolis. 31 p.

Ernster, Omer Francis. The relation of structure to oil accumulation. M, 1920, University of Minnesota, Minneapolis. 50 p.

Erol, O. Clay structure and creep behavior of clays as a rate process. D, 1977, Iowa State University of Science and Technology. 121 p.

Erol, Vasfi. Least-squares estimation of parameters for the interpretation of gravity anomalies caused by two-dimensional structures. M, 1974, Colorado School of Mines. 80 p.

Erpenbeck, Michael Francis. Stratigraphic relationships and depositional environments of the Upper Cretaceous Pictured Cliffs Sandstone and Fruitland Formation, southwestern San Juan Basin, New Mexico. M, 1979, Texas Tech University. 78 p.

Errington, J. C. Natural revegetation of disturbed sites in British Columbia. D, 1975, University of British Columbia.

Ersavci, M. Nedim. Laboratory and field evaluation of a high frequency acoustic emission/microseismic monitoring system. M, 1988, Pennsylvania State University, University Park. 355 p.

Erskian, Malcolm Gregory. Ecology of benthic foraminifera, Russian River, Sonoma County, California. M, 1971, University of California, Davis. 57 p.

Erskian, Malcolm Gregory. Population dynamics of the foraminiferan Glabratella ornatissima. D, 1979, University of California, Davis. 87 p.

Erskine, Bradley G. A paleomagnetic, rock magnetic and magnetic mineralogic investigation of the northern Peninsular Ranges Batholith, Southern California. M, 1982, San Diego State University. 194 p.

Erskine, Bradley Gene. Mylonitic deformation and associated low-angle faulting in the Santa Rosa mylonite zone, Southern California. D, 1986, University of California, Berkeley. 247 p.

Erskine, Daniel W. Geochemistry and petrogenesis of rocks with shoshonitic affinities, Crandall-Sunlight region, Absaroka volcanic field, Wyoming. M, 1988, University of New Mexico.

Erskine, Mellville Cox, Jr. Electromagnetic response of a sphere in a half space. D, 1970, University of California, Berkeley. 119 p.

Erslev, Eric Allan. Petrology and structure of the Precambrian metamorphic rocks of the southern Madison Range, southwesten Montana. D, 1981, Harvard University. 143 p.

Ersoy, Demir. Temperature distributions and heating efficiency of oil recovery by hot water injection. D, 1969, Stanford University. 99 p.

Ertel, John R. The fate of phytol in the sea surface microlayer. M, 1978, Florida State University.

Ertel, John Richard. The lignin geochemistry of sedimentary and aquatic humic substances. D, 1985, University of Washington. 139 p.

Erten, Zeynep Mujde. Toxicity evaluation and biological treatment of lead and zinc mine-mill effluents in Southeast Missouri. D, 1988, University of Missouri, Rolla. 311 p.

Ervin, Clarence P. Interpretation of aeromagnetic data from central Missouri. M, 1968, Washington University. 58 p.

Ervin, Clarence Patrick. Automated analysis of aeromagnetics. D, 1972, University of Wisconsin-Madison. 253 p.

Ervin, E. M. A study of mass movements on Pierce Hill, Vestal, New York. M, 1982, SUNY at Binghamton. 72 p.

Ervin, James Kirk. Geomorphology of the western part of the Ouachita Province, Oklahoma. M, 1964, University of Oklahoma. 104 p.

Ervin, Melanie K. Sedimentary petrology, depositional environment, and diagenesis of the Middle and Upper Cambrian sequence, Copenhagen Canyon, Bear River Range, Southeast Idaho. M, 1982, University of Idaho. 99 p.

Ervin, Sarah Mills. The relationship between the cloud zone and basal zone Cu-Ni sulfides, and the significance of mafic pegmatites, Minnamax Property, Duluth Complex, Minnesota. M, 1988, University of Minnesota, Duluth. 137 p.

Ervine, W. B. A study of the sulphur isotopes in the rocks of the Noranda area, Province of Quebec. M, 1962, University of Toronto.

Ervine, Warren Basil. The geology and mineral zoning of the Spanish Belt mining district, Nye County, Nevada. D, 1973, Stanford University. 295 p.

Erwin, Charles R., Jr. Geology of Mackay 2SE, 2NE, and part of 1NW quadrangles, Custer County, Idaho. M, 1973, University of Wisconsin-Milwaukee.

Erwin, Douglas Hamilton. The Cerithiacea, Subulitacea, Pyramidellacea and Acteonacea of the Permian Basin, West Texas and New Mexico, with a consideration of Permo-Triassic gastropod dynamics. D, 1985, University of California, Santa Barbara. 304 p.

Erwin, Eugene. Heavy minerals of the Gurley and Cruise sands in south-western Nebraska. M, 1958, University of Nebraska, Lincoln.

Erwin, J. W. Deformation of rock foundations under heavy loads. D, 1975, University of Arizona. 131 p.

Erwin, James Walter. Contributions to the paleontology of the northern part of Randolph County, Georgia. M, 1956, St. Louis University. 34 p.

Erwin, John W. Resistivity methods applied to ground water prospecting in the Big Sandy Creek in parts of Lincoln, and Elbert counties, Colorado. M, 1953, Colorado School of Mines. 82 p.

Erwin, Leslie Eugene. Recognition of tin-bearing granites by multivariate statistical analysis, Pikes Peak Batholith, Colorado. M, 1982, Pennsylvania State University, University Park. 217 p.

Erwin, Parrish Nesbitt, Jr. Stratigraphy, depositional environments, and dolomitization of the Maryville and upper Honaker formations (Cambrian), Tennessee and Virginia. M, 1981, Duke University. 232 p.

Erwin, Robert B. Biostromes and bioherms in the lower Middle Ordovician of Isle La Motte, Vermont. M, 1955, Brown University.

Erwin, Robert Bruce. Stratigraphy and structure of the Ordovician limestones of the lower and middle Champlain Valley, Vermont. D, 1959, Cornell University.

Erxleben, Albert Walter. Depositional systems in the Pennsylvanian Canyon Group of north-central Texas. M, 1974, University of Texas, Austin.

Es, Harold Mathijs van see van Es, Harold Mathijs

Esarey, Ralph Emerson. The relation of the Mt. Carmel and Heltonville faults to the Dennison Anticline (Indiana). M, 1923, Indiana University, Bloomington. 25 p.

Esawi, E. K. Geology and geochemistry of mafic Charlotte Belt, Davie County, North Carolina. M, 1985, University of North Carolina, Chapel Hill.

Esbroeck, Guillaume Van see Van Esbroeck, Guillaume

Escalante, Margarito Coballes. Irrigation water management by simulation for a diversion irrigation system and saturated soil conditions. D, 1981, Iowa State University of Science and Technology. 262 p.

Escalera, Saul J. A flotation study of the quartz-calcite-hematite system. M, 1966, New Mexico Institute of Mining and Technology. 39 p.

Eschman, Donald Frazier. Late Cenozoic history of the Michigan River basin, North Park, Colorado. D, 1953, Harvard University.

Eschner, Stanford. Geology of the central part of the Fillmore Quadrangle, Ventura County, California. M, 1957, University of California, Los Angeles.

Eschner, Terence Brent. Marine destruction of eolian sand seas; an example in the Jurassic Entrada and Curtis formations, northeastern Utah. M, 1983, University of Texas, Austin. 83 p.

Eschner, Thomas Richard. Morphologic and hydrologic changes of the Platte River, south-central Nebraska. M, 1981, Colorado State University. 277 p.

Escobar, Leopoldo Lopez see Lopez Escobar, Leopoldo

Escobar, Ricardo Reyes. Statistical evaluation of stream sediment geochemistry for middle and lower Ordovician rocks in North Sequatchie Valley, and a part of the Oostanaula Valley, Tennessee. M, 1974, University of Tennessee, Knoxville. 69 p.

Escorce, Eufredo B. Porphyry-copper type mineralization in dioritic rocks. M, 1977, New Mexico Institute of Mining and Technology.

Esfandiari, Bijan. Geochemistry and geology of helium. D, 1969, University of Oklahoma. 148 p.

Esfandiari, Bijan. Geological engineering study of the Red Fork Sand (Pennsylvanian) in a portion of Oklahoma County, Oklahoma. M, 1965, University of Oklahoma. 112 p.

Esguerra, Orlando Forero see Forero Esguerra, Orlando

Eshelman, Ralph E. The paleoecology (Pleistocene) of Willard Cave, Delaware County, Iowa. M, 1971, University of Iowa. 72 p.

Eshelman, Ralph Ellsworth. Geology and paleontology of the early Pleistocene Belleville Formation of north central Kansas. D, 1974, University of Michigan.

Eshet, Yoram. Palynological aspects of the Permo-Triassic sequence in the subsurface of Israel. D, 1987, City College (CUNY). 227 p.

Eshett, Ali. Ground water system analysis. D, 1970, University of Colorado. 137 p.

Eshlemann, A. M. A preliminary survey of lead and mercury in the Hawaiian environment. D, 1973, University of Hawaii. 154 p.

Eshler, Lynn M. Hydrogeological analysis of Sharon Township, Medina County, Ohio. M, 1988, University of Akron. 118 p.

Eskelson, Quinn Morrison. Geology of the Soapstone Basin and vicinity, Wasatch, Summit, and Duchesne counties, Utah. M, 1953, University of Utah. 45 p.

Eskenasy, Diane M. The origin of the King Ravine rock glacier in the Presidential Range of the White Mountains of New Hampshire. M, 1978, University of Massachusetts. 85 p.

Esker, George C., III. Biostratigraphy of the Cretaceous-Tertiary boundary in the east Texas Embayment based on planktonic Foraminifera. D, 1968, Louisiana State University. 110 p.

Esker, George Cornelius, III. Foraminiferal zonation of a post-Pliocene well section in the Gulf of Mexico. M, 1964, Washington University. 31 p.

Esling, Steven Paul. Quaternary stratigraphy of the lower Iowa and Cedar River valleys, Southeast Iowa. D, 1984, University of Iowa. 451 p.

Eslinger, Eric Vance. Mineralogy and oxygen isotope ratios of hydrothermal and low-grade metamorphic argillaceous rocks. D, 1971, Case Western Reserve University. 223 p.

Esmail, Omar Jubran. Ultrasonic reflection and transmission by submerged composite plates. D, 1969, University of Texas, Austin. 130 p.

Esmaili, Esmail. Geochemistry of oil shale leachates and interaction with geologic substrates, Piceance Creek basin, Colorado. D, 1983, University of Colorado. 369 p.

Esmaili, Houshang. A solution for determination of aquifer characteristics and unsteady flow through injection wells by numerical methods. D, 1966, University of California, Davis. 96 p.

Esmer, Erkan. The freezing and thawing mechanisms in lime stabilized soils. D, 1969, Virginia Polytechnic Institute and State University. 204 p.

Espach, R. H., Jr. Geology of the Mahogany Ridge area, Big Hole Mountains, Teton County, Idaho. M, 1957, University of Wyoming. 104 p.

Espegren, William A. Sedimentology and petrology of the upper Petrified Forest Member of the Chinle Formation, Petrified Forest National Park and vicinity, Arizona. M, 1985, Northern Arizona University. 228 p.

Espejo, Anibal C. Peekskill granitic pluton (Precambrian), New York. M, 1969, Queens College (CUNY). 66 p.

Espenschied, E. K. Stratigraphy of the Cloverly Formation, Thermopolis Shale, and the Mudy Sandstone around Hanna Basin, Carbon County, Wyoming. M, 1957, University of Wyoming. 91 p.

Espenshade, Edward B. An intensive study of the sphericity and roundness of beach and dune sands of the south of Lake Michigan. M, 1932, University of Chicago. 31 p.

Espenshade, Gilbert H. The geology and ore deposits of the Pilleys Island area, Nfld. D, 1937, Princeton University. 114 p.

Esperanca, Sonia. An experimental and geochemical study of the high-K latites and associated nodules from Camp Creek, Arizona. D, 1984, Arizona State University. 205 p.

Espeseth, Robert Lynn. Geology of the Cheniere and Cadeville Field area, Ouachita and Jackson parishes, Louisiana. M, 1972, Louisiana Tech University.

Espindola, Juan M. Finite-difference synthetic seismograms for kinematic models of the earthquake source. D, 1979, Purdue University. 152 p.

Espinosa, Alvaro F. A transient technique for seismograph calibration. M, 1964, Columbia University, Teachers College.

Espinosa, Alvaro Felipe. Structure from near earthquake surface waves. M, 1963, St. Louis University.

Esposito, Richard A., Jr. Stratigraphic analysis and depositional environments of the Smackover Formation in the Conecuh Basin, Escambia County, southwestern Alabama. M, 1987, Auburn University. 124 p.

Espourteille, François A. An assessment of tributyltin contamination in sediments and shellfish in the Chesapeake Bay. M, 1988, College of William and Mary. 78 p.

Essaid, Hedeff Izzudeen. A quasi-three dimensional finite difference model for the simulation of fresh water and salt water flow in a coastal aquifer system. M, 1984, Stanford University. 128 p.

Essaid, Hedeff Izzudeen. Fresh water-salt water flow dynamics in coastal aquifer systems; development and application of a multi-layered sharp interface model. D, 1987, Stanford University. 293 p.

Essed, A. S. A reconnaissance Bouguer gravity anomaly map of Libya. M, 1978, Purdue University. 98 p.

Essene, Eric J. and Edwards, R. Lawrence. Pressure, temperature and C-O-H fluid fugacities during the amphibolite facies-granulite facies metamorphism of the major paragneiss, NW Adirondack Mts., N.Y. M, 1986, University of Michigan. 36 p.

Essene, Eric John. Petrogenesis of Franciscan metamorphic rocks. D, 1967, University of California, Berkeley. 225 p.

Essenfeld, Martin. Mathematical simulation of solution gas drive in a two-dimensional system. D, 1970, Pennsylvania State University, University Park. 212 p.

Esser, Kjell Bjorgen. Distribution of minerals and chemical elements in soils formed in the Indiana Dunes. D, 1987, University of Wisconsin-Madison. 265 p.

Esser, Robert Worth. The reconnaissance geology of a part of the Woodside Quadrangle, northeast of Skyline Boulevard, San Mateo County, California. M, 1958, Stanford University. 78 p.

Essere, Eric J.; Kesler, Stephen E. and Richardson, Stephen Vance. Origin and geochemistry of the Chapada Cu-Au deposit, Goias, Brazil; a metamorphosed wall rock porphyry copper deposit. M, 1984, University of Michigan.

Essien, Isang O. Uranium, plutonium and thorium isotopes in the atmosphere and the lithosphere. D, 1983, University of Arkansas, Fayetteville. 187 p.

Esslinger, Brad A. Facies and depositional environment of the Simpson Group (Middle Ordovician) of central Oklahoma. M, 1983, University of Missouri, Columbia.

Estabrook, Charles Hershey. Seismotectonics of northern Alaska. M, 1985, University of Alaska, Fairbanks. 139 p.

Estasen, Elena. Petrography and petrology of the Heavenly Hills volcanic sequence, Pleasant Valley (Juab county), Utah. M, 1969, University of Nebraska, Lincoln.

Estelle, Duane Kendall. Stratigraphy of the Stump and Preuss formations, Upper Jurassic, in Lincoln County, Wyoming, Sublette County, Wyoming, Bear Lake County, Idaho, Bonneville County, Idaho. M, 1971, University of Michigan.

Estep, Patricia Anne. Infrared spectroscopic studies of hematite. M, 1972, West Virginia University. 122 p.

Esterle, Joan Sharon. The upper Hance coal seam in southeastern Kentucky; a model for petrographic and chemical variation in coal seams. M, 1984, University of Kentucky. 105 p.

Estes, Carol. Chemical and structural analysis of an aluminum hydroxy-interlayered clay from Terra rossa soil, South Florida. M, 1987, University of South Florida, Tampa. 101 p.

Estes, Ernest Lathan. Roundness of the carbonate fraction. M, 1967, Duke University. 67 p.

Estes, Ernest Lathan, III. Diagenetic alteration of Mercenaria mercenaria as determined by laser microprobe analysis. D, 1972, University of North Carolina, Chapel Hill. 95 p.

Estes, Larry D. Geology of the Saddle Mountain Quadrangle and petrology of the layered series, eastern Wichita Mountains, Oklahoma. M, 1980, University of Texas, Arlington. 172 p.

Estes, Richard Dean. Lizards and salamanders from the (Upper Cretaceous) Lance Formation. M, 1957, University of California, Berkeley. 78 p.

Estes, Richard Dean. Lower vertebrates from the (Upper Cretaceous) Lance Formation, eastern Wyoming. D, 1960, University of California, Berkeley. 362 p.

Estes, Steve A. Seismotectonic studies of lower Cook Inlet, Kodiak Island and the Alaska Peninsula areas of Alaska. M, 1978, University of Alaska, Fairbanks. 142 p.

Estes, Wayne Shelton. Fusulinid fauna of the Horquilla Limestone (Middle and Upper Pennsylvanian) in the Gunnison Hills, Cochise County, Arizona. M, 1968, University of Arizona.

Esteves, Ieda R. Forti. Recent bivalves (Palaeotaxodonta & Pteriomorphia) from Brazilian continental shelf. D, 1978, Tulane University. 197 p.

Estevez L., R. J. Wide-angle diffracted multiple reflections. D, 1977, Stanford University. 116 p.

Estey, Louis Howard. Anisotropy and dislocations in the mantle. D, 1988, University of Colorado. 392 p.

Estill, Robert Eugene. Temporal variations of P-wave travel times and lateral velocity structure across the Wasatch Front, Utah. M, 1976, University of Utah. 181 p.

Estoff, Fritz E. Von see Von Estoff, Fritz E.

Eston, S. M. De see De Eston, S. M.

Estrada M., Armando. Geology and plate tectonic history of the Colombian Andes. M, 1972, Stanford University.

Estrada, Jose Andres. Mineralogical and chemical properties of Peruvian acid tropical soils. D, 1971, University of California, Riverside. 199 p.

Esu, Esu Obukho. Petrography and ultrasonic wave velocity in the Admire Formation (Permian), Kansas. D, 1984, University of Wisconsin-Madison. 163 p.

Etayo-Serna, F. Zonation of the Cretaceous of central Colombia by ammonites. D, 1975, University of California, Berkeley. 491 p.

Etheredge, Forest de Royce. Electric log study of the basal Pennsylvanian sandstone, Louden Pool, Fayette County, Illinois. M, 1953, University of Illinois, Urbana.

Etherington, John Robert. Silurian graptolites from Clearwater Creek, south Nahanni River region, Northwest Territories. M, 1968, University of Alberta. 204 p.

Etherington, Thomas John. Stratigraphy and fauna of the Astoria Miocene of Southwest Washington. D, 1930, University of California, Berkeley. 236 p.

Etherington, Thomas John. The stratigraphy and paleontology of the Oligocene and lower Miocene of the Chehalis River valley, Washington. M, 1926, University of Washington. 62 p.

Ethetton, Laura Kay Herrick. Correlation of mineral and mechanical parameters in selected cores for the red clay soil, Douglas County, Wisconsin. M, 1982, University of Wisconsin-Milwaukee. 57 p.

Ethetton, Lee Wayne. A computer automated land magnetic survey system. M, 1980, University of Wisconsin-Milwaukee. 158 p.

Ethier, Valerie Mary Girling. Probability theory applied to sedimentary sequences; a study of the Banff Formation (Mississippian) of the southern Canadian Rockies. M, 1970, University of Calgary. 121 p.

Ethington, Edgar Francis. Resistivity and induced-polarization modeling for a buried resistive dike and buried resistive cylinder. M, 1985, University of Arizona. 105 p.

Ethington, Raymond Lindsay. Conodonts of the Galena Formation. D, 1958, University of Iowa. 153 p.

Ethington, Raymond Lindsay. Interpretation of the drainage pattern, Benton County, Iowa. M, 1955, Iowa State University of Science and Technology.

Ethridge, Frank G. Variation in grain-size distribution in sedimentary environment. M, 1966, Louisiana State University.

Ethridge, Frank Gulde. Quantitative petrographic criteria for recognition of environments of deposition. D, 1970, Texas A&M University. 182 p.

Ethridge, Loch Lee. Applications of computer-enhanced Landsat imagery for uranium exploration in the Colorado Plateau Province. M, 1981, University of Arizona. 131 p.

Etiebet, Donatus O. Continuous seismic reflection profiling in southwestern Ontario. M, 1971, University of Western Ontario.

Etienne, David. Analysis and correlation of granites in the Alabama Piedmont. M, 1978, Memphis State University. 41 p.

Etienne, John E. Geology and mineral resources of the Lightning Mountain-Rattle Creek area, eastern Bonner County, Idaho. M, 1987, Eastern Washington University. 116 p.

Etson, Neil R. Stratigraphy of the Rondout Formation in central New York. M, 1952, Syracuse University.

Ettensohn, F. R. Stratigraphic and paleoenvironmental aspects of upper Mississippian rocks (upper Newman Group), east-central Kentucky. D, 1975, University of Illinois, Urbana. 330 p.

Ettensohn, Frank Robert. The pre-Illinoian lake clays of the Cincinnati region. M, 1970, University of Cincinnati. 151 p.

Etter, Evelyn Mary. Lithostratigraphy and depositional environments of the Devil's Hollow Member of the Lexington Limestone. M, 1976, University of Kentucky. 100 p.

Etter, John. Ostracodes of the family Kirkbyidae from the Middle Permian of West Texas. M, 1951, Washington University. 41 p.

Etter, Stephen D. Geology of the Lake Pillsbury area, northern Coast Ranges, California. D, 1979, University of Texas, Austin. 314 p.

Ettinger, Leonard J. Geology of the Pat Hills, Cochise County, Arizona. M, 1962, University of Arizona.

Ettinger, Morris J. The Geology of the Hartsel area, South Park, Park County, Colorado. M, 1959, University of Colorado.

Ettlinger, Isadore Aaron. Studies in the persistence of ore with depth. D, 1924, Harvard University.

Etzler, Paul. The geology of the northern Cydonia area, Mars. M, 1981, University of Pittsburgh.

Eubank, R. T. Geology of the southwestern end of the Catawba Syncline, Montgomery County, Virginia. M, 1968, Virginia Polytechnic Institute and State University.

Eubanks, Darrell Lynn. Depositional and diagenetic history of Terryville Sandstone, Cotton Valley Group (Upper Jurassic), Carthage Field, Panola County, Texas. M, 1986, Stephen F. Austin State University. 113 p.

Eubanks, Glen E. A field study of tide-induced sand movement on Del Monte beach, California. M, 1968, United States Naval Academy.

Eulich, Artileus V. Economic geology of northern Angola. M, 1924, University of Missouri, Rolla.

Euribe Dulanto, Alejandro. Foraminifera from the type section of the Heath Formation (Upper Oligocene) of northwestern Peru. M, 1960, Stanford University. 70 p.

Eusden, J. Dykstra. The bedrock geology of part of the Alton, NH and Berwick, ME quadrangles. M, 1984, University of New Hampshire. 114 p.

Eusden, John Dykstra, Jr. The bedrock geology of the Gilmanton 15-minute quadrangle, New Hampshire. D, 1988, Dartmouth College. 245 p.

Eusufzai, Hossain Sekandar H. Khan. Post glacial deposits as foundation media in lower alluvial valleys and deltaic areas. D, 1965, Texas A&M University. 111 p.

Eutsler, Robert L. The porous laminated crust of the lower Florida Keys; a possible Recent algal stromatolite. M, 1974, Bowling Green State University. 36 p.

Evander, Robert L. Rocks and faunas of the type section of the Valentine Formation. M, 1978, University of Nebraska, Lincoln.

Evander, Robert Lane. Middle Miocene horses of North America. D, 1985, Columbia University, Teachers College. 440 p.

Evangelou, Vasilios Petros. Chemical and mineralogical composition and behavior of the Mancos Shale as a diffuse source of salts in the upper Colorado River basin. D, 1981, University of California, Davis. 198 p.

Evans, Allen E., Jr. Discrimination of subparallel beach ridges using detrital ilmenite composition and Fourier shape analysis. M, 1987, Old Dominion University. 100 p.

Evans, Andrew Joseph. A study and evaluation of saltwater intrusion in the Floridan Aquifer by means of a Hele-Shaw model. D, 1975, University of Florida. 213 p.

Evans, Anthony Meredith. Geology of the Bicroft uranium mine, Ontario. D, 1962, Queen's University. 325 p.

Evans, Barbara. Statistical techniques for subsurface reservoir management. M, 1982, Stanford University. 119 p.

Evans, Barry Louis. Compressibility of natural gas. M, 1952, University of Southern California.

Evans, Bruce William. Petrology and geochemistry of certain mafic pegmatites in the eastern Bushveld Complex, South Africa. M, 1978, University of Wisconsin-Madison.

Evans, Calvin Ralph. The Precambrian rocks of the Old Fort Point Formation, Jasper, Alberta. M, 1961, University of Alberta. 83 p.

Evans, Carol Anne. Microstructures and sense of shear in the Brevard Zone, Southern Appalachians. M, 1986, University of Texas, Austin. 125 p.

Evans, Carol Susan. The geology, geochemistry, and alteration of Red Butte, Oregon; a precious metal-bearing paleo hot spring system. M, 1986, Portland State University. 133 p.

Evans, Charles. Facies evolution in a Neogene transpressional basin; Cibao Valley, Dominican Republic. D, 1986, University of Miami. 22 p.

Evans, Charles Carroll. Aspects of the depositional and diagenetic history of the Miami Limestone; control of primary sedimentary fabric over early cementation and porosity-development. M, 1982, University of Miami. 233 p.

Evans, Charles Sparling. Geology of Brisco-Dogtooth map-area, British Columbia. D, 1927, Princeton University.

Evans, Christine Victoria. Soil toposequences in Carbon County, Wyoming; pedogenic processes in a semi-arid region. D, 1987, University of Wyoming. 178 p.

Evans, Cynthia Ann. Petrology and geochemistry of the transition from mantle to crust beneath an island arc-backarc pair; implications from the Zambales

Range ophiolite, Luzon, Philippines. D, 1983, University of California, San Diego. 315 p.

Evans, D. J. A. Glacial geomorphology and chronology in the Selamiut Range; Nachvak Fiord area, Torngat Mountains, Labrador. M, 1984, Memorial University of Newfoundland. 138 p.

Evans, Daniel Frederick. Geology and petrochemistry of the Kitts and Michelin uranium deposits and related prospects, Central Mineral Belt, Labrador. D, 1980, Queen's University. 312 p.

Evans, Daniel S. Quality of groundwater in Cretaceous rocks of Williamson and eastern Burnet counties, Texas. M, 1974, University of Texas, Austin.

Evans, David G. Conodont biostratigraphy of the Montoya Group (Upper Ordovician) from the Mud Springs Mountains of south-central New Mexico. M, 1985, Texas Tech University. 69 p.

Evans, David G. Seismicity of the Sleepy Hollow oil field, Red Williow County, Nebraska. M, 1984, University of Kansas. 92 p.

Evans, David John Alexander. Glacial geomorphology and late Quaternary history of Phillips Inlet and the Wootton Peninsula, Northwest Ellesmere Island, Canada. D, 1988, University of Alberta. 281 p.

Evans, David L. A new Tertiary horizon in Ventura County, California. M, 1928, Stanford University. 60 p.

Evans, David William. Effects of ocean water on the soluble-suspended distribution of Columbia River radionucludes. M, 1973, Oregon State University. 57 p.

Evans, David William. Exchange of manganese, iron, copper and zinc between dissolved and particulate forms in the Newport River estuary. D, 1977, Oregon State University. 218 p.

Evans, Diane Louise. Identification of lithologic units using multichannel imaging systems. D, 1981, University of Washington. 118 p.

Evans, Elizabeth Lee. Pleistocene beach ridges of northwestern Pennsylvania. M, 1970, Bowling Green State University. 50 p.

Evans, George Carman. Geology and sedimentation along the lower Rio Salado in New Mexico. M, 1963, New Mexico Institute of Mining and Technology. 69 p.

Evans, Hilton B. Factors influencing permeability and diffusion of radon in synthetic sandstones. D, 1959, University of Utah. 94 p.

Evans, Ian. A palaeoecological analysis of the Whitestone Member of the Walnut Formation, lower Cretaceous, Travis and Williamson counties, Texas. D, 1971, Texas A&M University. 205 p.

Evans, Ian. The post-mortem history of benthis invertebrate fauna in Anasco Bay, Puerto Rico. M, 1968, University of South Carolina.

Evans, J. E. L. Quartz syenite contact action near Kingston, Ontario. M, 1942, Queen's University. 36 p.

Evans, James Brian. Anhysteretic remanence. M, 1975, University of Minnesota, Minneapolis. 67 p.

Evans, James Brian. Indentation hardness and its relation to mechanical yield in quartz and olivine. D, 1978, Massachusetts Institute of Technology. 208 p.

Evans, James E. L. Porphyry of the Porcupine District, Ontario. D, 1944, Columbia University, Teachers College.

Evans, James Erwin. 210 Pb geochronology in Lake Superior sediments; sedimentation rates, organic carbon deposition, sedimentary environments, and post-depositional processes. M, 1980, University of Minnesota, Minneapolis. 130 p.

Evans, James Erwin. Depositional environments, basin evolution and tectonic significance of the Eocene Chumstick Formation, Cascade Range, Washington. D, 1988, University of Washington. 325 p.

Evans, James George. Structural analysis and movements of the San Andreas Fault zone near Palmdale, Southern California. D, 1966, University of California, Los Angeles. 222 p.

Evans, James L. Geology of the upper Marshall Creek drainage basin, Saguache County, Colorado. M, 1973, Colorado School of Mines. 78 p.

Evans, James P. Structural geology of the northern termination of the Crawford Thrust, western Wyoming. M, 1983, Texas A&M University.

Evans, James Parham, III. Geology and fracture pattern analysis of central western Williamson County, Texas. M, 1965, University of Texas, Austin.

Evans, James Paul. Geometry, mechanisms, and mechanics of deformation in a Laramide thrust sheet. D, 1987, Texas A&M University. 295 p.

Evans, James R. Geology of the Mescal Range, San Bernardino County, California. M, 1958, University of Southern California.

Evans, Jeffrey Clinton. Permeant influence on the geotechnical properties of soils. D, 1984, [Lehigh University]. 260 p.

Evans, John Keith. Depositional environment of a Pennsylvanian black shale (Heebner) in Kansas and adjacent states. D, 1967, Rice University. 178 p.

Evans, John Keith. Stratigraphy of the Cretaceous Bearpaw Formation in the South Saskatchewan River valley. M, 1961, University of Saskatchewan. 122 p.

Evans, John P. Geology of a portion of the Inyan Kara Hogback at Rapid City, South Dakota. M, 1962, South Dakota School of Mines & Technology.

Evans, John R. C. The Richmond fauna of northeastern Illinois. D, 1924, University of Chicago. 191 p.

Evans, John Richard. Restricted-array seismic tomography. D, 1988, Princeton University. 181 p.

Evans, Karl Vierling. Geology and geochronology of the eastern Salmon River Mountains, Idaho, and implications for regional Precambrian tectonics. D, 1981, Pennsylvania State University, University Park. 251 p.

Evans, Karl Vierling. Structural geology of the eastern Whipple Mountains, San Bernardino County, California. M, 1979, University of Southern California.

Evans, Kathryn C. Petrography and petrology of Big Aguja Sill, Northeast Davis Mountains, Jeff Davis and Reeves counties, Texas. M, 1972, Texas Tech University. 94 p.

Evans, Kathryn Christina. Geochemical reconnaissance for uranium in middle Tertiary ash-flow tuffs of the Morey Peak Quadrangle, northern Nye County, Nevada. D, 1981, University of Texas at El Paso. 395 p.

Evans, Kenneth. Aeromagnetic study of the Mexicali-Cerro Prieto geothermal area. M, 1972, University of Arizona.

Evans, Lanny L. Geology of the copper deposits in the Eminence region, Shannon County, Missouri. M, 1959, University of Missouri, Rolla.

Evans, Leonard Thomas. Simplified analysis of laterally loaded piles. D, 1982, University of California, Berkeley. 245 p.

Evans, M. Harrison. The geology and ore deposits of the Manzama Quadrangle, Los Angeles County, California. M, 1936, California Institute of Technology. 63 p.

Evans, Mark David. Undrained cyclic triaxial testing of gravels; the effect of membrane compliance. D, 1987, University of California, Berkeley. 403 p.

Evans, Mark W. Barrier island development from lagoonal stratigraphy and sedimentation; northern Pinellas County, Florida. M, 1983, University of South Florida, St. Petersburg. 148 p.

Evans, Max Thomas. Geology and ore deposits of the Great Western Mine, Marysvale region, Utah. M, 1951, Brigham Young University. 46 p.

Evans, Morris De B. and Penneypacker, N. R. The geology, mining, and development of the Tonopah gold and silver district. M, 1906, Lehigh University.

Evans, Nicholas Hartford. Late Precambrian to Ordovician metamorphism and orogenesis in the Blue Ridge and western Piedmont, Virginia Appalachians. D, 1984, Virginia Polytechnic Institute and State University. 322 p.

Evans, Noreen J. Evaluation and application of radio-chemical neutron activation of noble metal analysis. M, 1987, McMaster University. 131 p.

Evans, Oren Frank. Low and ball of the eastern shore of Lake Michigan. D, 1940, University of Michigan.

Evans, Owen Cope. The petrogenesis of the Saganaga Tonalite revisited. M, 1987, SUNY at Stony Brook. 97 p.

Evans, Richard G. A sedimentological study of greater Gullivan Bay, Florida. M, 1962, Florida State University.

Evans, Robert. Studies in the evaporites of the Maritime Provinces of Canada. D, 1972, University of Kansas. 178 p.

Evans, Robert. The structure of salt deposits at Pugwash, Nova Scotia (Canada). M, 1965, Dalhousie University. 69 p.

Evans, Robert Douglas. Measurements of sediment accumulation and phosphorus retention using lead-210 dating. D, 1981, McGill University.

Evans, Robert George. Optimizing salinity control strategies for the upper Colorado River basin. D, 1981, Colorado State University. 310 p.

Evans, Robert James. High temperature simulation of petroleum formation. D, 1982, University of Rhode Island. 190 p.

Evans, Stanley H., Jr. Studies in basin and range volcanism. D, 1978, University of Utah. 131 p.

Evans, Stephen George. Landslides in layered volcanic successions with particular reference to the Tertiary rocks of south central British Columbia. D, 1983, University of Alberta. 350 p.

Evans, Stewart Thompson and Trudell, Laurence G. Geology of the South Veta Creek area, Huerfano Quadrangle, Huerfano County, Colorado. M, 1958, University of Michigan.

Evans, Thomas J. A stratigraphic study of the Toroweap Formation (Permian) between Sycamore and Oak Creek Canyons, Arizona. M, 1971, University of Arizona.

Evans, Thomas Lester. A reconnaissance study of some western Canadian lead-zinc deposits (Pine Point, Northwest Territories). M, 1965, University of Alberta. 69 p.

Evans, W. D. Paleomagnetism of the Table Mountain flows near Golden, Colorado. M, 1962, Colorado School of Mines. 69 p.

Evans, W. Scot. The role of nitrogen in the Green River Formation, Piceance Creek basin, Colorado. M, 1982, University of Texas, Arlington. 50 p.

Evansin, David Paul. Structural style of the Ouachita Core near Caddo Gap, Arkansas. M, 1976, Southern Illinois University, Carbondale. 94 p.

Evarts, R. C. The Del Puerto ophiolite complex, California; a structural and petrologic investigation. D, 1978, Stanford University. 469 p.

Evashko, Anna Helen. Evidence for magma mixing in selected volcanic rocks of northwestern Nevada. M, 1982, University of Nevada. 46 p.

Eveland, Harmon Edwin, Jr. Pleistocene geology of the Danville region, Illinois and Indiana. D, 1950, University of Illinois, Urbana.

Eveland, Harmon Edwin, Jr. Topographic expression of geology in the Lake Huron Basin. M, 1948, University of Illinois, Urbana.

Evenchick, Carol Anne. Stratigraphy, metamorphism, structure, and their tectonic implications in the Sifton and Deserters ranges, Cassiar and Northern Rocky Mountains, northern British Columbia. D, 1986, Queen's University. 197 p.

Evenden, Leonard Jesse. Quantitative characteristics of drainage basins in the delimitation of geomorphic regions. M, 1962, University of Georgia. 111 p.

Evensen, Charles G. A comparison of the Shinarump Conglomerate of Hoskinnine Mesa with that in other selected areas in Arizona and Utah. M, 1953, University of Arizona.

Evensen, James M. Geology of the Copper Hill areas, Winkelman, Arizona. M, 1961, University of Arizona.

Evensen, James Millard. Geology of the central portion of the Agua Fria mining district, Yavapai County, Arizona. D, 1969, University of Arizona. 168 p.

Evensen, N. M. The nature of the exotic component in lunar soils and its implications for lunar differentiation history. D, 1978, University of Minnesota, Minneapolis. 236 p.

Evenson, Edward B. The relationship of macro- and microfabrics of till and the genesis of glacial landforms in Jefferson County, Wisconsin. M, 1970, University of Wisconsin-Milwaukee.

Evenson, Edward Bernard. Late Pleistocene shorelines and stratigraphic relationships in the Lake Michigan Basin. D, 1972, University of Michigan.

Evenson, W. A. Geology of the southern Kilbeck Hills and an adjacent portion of the Old Woman Mountains, eastern Mojave Desert, San Bernardino County, California. M, 1973, University of Southern California. 51 p.

Everdingen, David Allard van *see* van Everdingen, David Allard

Everett, Ardell Gordon. Clay petrology and geochemistry of Blaine Formation (Permian), northern Blaine County, Oklahoma. M, 1962, University of Oklahoma. 108 p.

Everett, Ardell Gordon. Petrology and trace element chemistry of the Carmel Formation (Jurassic), Iron Springs (Iron County) mining district, Utah. D, 1968, University of Texas, Austin. 267 p.

Everett, Charles Jay. Effects of biological weathering on mine soil genesis and fertility. D, 1981, Virginia Polytechnic Institute and State University. 134 p.

Everett, Edward E. Suitability of lacustrine clays in Bay County, Michigan, for land disposal of hazardous wastes. M, 1977, University of Toledo. 117 p.

Everett, James Edward. An analysis of the variation of the Earth's magnetic field in the frequency of 1 to 12 cycles per day. M, 1962, Massachusetts Institute of Technology. 155 p.

Everett, John Raymond. Geology of the Comayagua Quadrangle, Honduras, Central America. D, 1970, University of Texas, Austin. 183 p.

Everett, John Raymond. Post-Cretaceous structural geology near Del Norte Gap, Brewster County, Texas. M, 1964, University of Texas, Austin.

Everett, Kaye R. Geology and ground water of Skull Valley, Tooele County, Utah. M, 1958, University of Utah. 92 p.

Everett, Lorne Gordon. A mathematical model of primary productivity and limnological patterns in Lake Mead. D, 1972, University of Arizona.

Everett, Wayne Leonard. Regional economics; a subset of "simulation of the effects of coal-fired power development in the Four Corners region. D, 1974, University of Arizona.

Everette, Kaye Ronald. Slope movement in contrasting environments. D, 1963, Ohio State University. 266 p.

Everhart, Donald Lough. Geology of the Cuyamaca Peak Quadrangle, San Diego County, California. D, 1953, Harvard University.

Everitt, Richard. Jointing in the Sudbury Basin, Sudbury, Ontario. M, 1979, Laurentian University, Sudbury. 57 p.

Everly, Robert A. The geology of Huntingdon County, Pennsylvania. M, 1978, Pennsylvania State University, University Park. 79 p.

Evernden, Jack Foord. Direction of approach of Rayleigh waves and related problems. D, 1951, University of California, Berkeley. 164 p.

Everse, Douglas Gene. Packing, pressure solution and cementation in quartz-rich arenites. M, 1983, Michigan State University. 54 p.

Eversoll, Duane A. Environmental geology of western Red Willow County, Nebraska. M, 1977, University of Nebraska, Lincoln.

Everson, C. I. Drift lithology in relation to bedrock geology, Long Island Lake Quadrangle, Cook County, Minnesota. M, 1977, University of Minnesota, Duluth.

Everson, Douglas D. A geophysical study of the Tomahawk volcanic area near Brownsville, Black Hills, South Dakota. M, 1985, South Dakota School of Mines & Technology.

Everson, Joel Earl. Regional variations in the lead isotopic characteristics of late Cenozoic basalts from the southwestern United States. D, 1979, California Institute of Technology. 464 p.

Everts, C. H. The evolution of playa waters, Teels Marsh, Mineral County, Nevada. M, 1969, University of Wisconsin-Madison.

Everts, Craig Hamilton. A rational approach to marine placers. D, 1971, University of Wisconsin-Madison. 287 p.

Everts, James Mitchell. Fatty acid analysis of some fossil and recent bones and teeth. D, 1969, University of Arizona. 80 p.

Evertson, D. W. Borehole strainmeters for seismology. D, 1975, University of Texas, Austin. 153 p.

Eves, Robert Leo. The chemistry/mineralogy of the Adaville No. 1 coal seam, southwestern Wyoming. M, 1985, Washington State University. 129 p.

Evetts, Michael J. Biostratigraphy of the Sage Breaks Shale (upper Cretaceous) in northeastern Wyoming. M, 1975, University of Colorado.

Evitt, W. E., II. Trilobites from the lower Lincolnshire Limestone near Strasburg, Shenandoah County, Virginia. D, 1950, The Johns Hopkins University.

Evola, Gena M. Lower Devonian ostracodes from the hesperius eurekaensis, delta, and pesavis conodont zones of central Nevada. M, 1983, University of California, Riverside. 134 p.

Evoy, Barbara Lynn. Complex mass movements of eastern Little Rattlesnake Mountain, Six Rivers National Forest, Del Norte County, California. M, 1982, University of California, Santa Cruz.

Evoy, E. F. The Mistassini Iron Formation, Quebec. M, 1955, Queen's University. 106 p.

Evoy, Ernest Franklin. Geology of the Gunnar uranium deposit, Beaverlodge area, Saskatchewan. D, 1961, University of Wisconsin-Madison. 92 p.

Evoy, Jeffrey Allen. Precision gravity reobservations and simultaneous inversion of gravity and seismic data for subsurface structure of Yellowstone. M, 1978, University of Utah. 212 p.

Ewald, Frederick Charles. Feasibility of completely automated microfossil identification in petroleum exploration. M, 1975, Michigan State University. 68 p.

Ewald, H. K., III. Faunal communities of the Bloyd Formation (Pennsylvanian, NW Arkansas); their paleoecological and biostratigraphic significance. M, 1971, University of Arkansas, Fayetteville.

Ewart, James Alfred, Jr. Elastic models of Krafla Volcano, North Iceland; 1976-1982. M, 1985, Pennsylvania State University, University Park. 103 p.

Eweida, Ahmed Mahmoud Farag. Hydrologeological studies in Ponoka area, Alberta, Canada. M, 1962, University of Saskatchewan. 81 p.

Ewers, Ralph. A model for the development of subsurface drainage along bedding planes. M, 1972, University of Cincinnati. 84 p.

Ewert, W. D. Metamorphism of siliceous carbonate rocks in the Grenville Province of southeastern Ontario. D, 1977, Carleton University. 276 p.

Ewing, Clair Eugene. The parallel radius method of solving the inverse shoran problem. D, 1955, Ohio State University.

Ewing, David Jay. A study of the thinning effect of faults on a geological section near Bedford, Pennsylvania. M, 1952, University of Pittsburgh.

Ewing, Don R. Trilateration network design and error analysis of figure. M, 1962, Ohio State University.

Ewing, Gerald Neil. Structural framework in the Gulf of Saint Lawrence. M, 1965, Dalhousie University. 53 p.

Ewing, Gifford C. Slicks, surface films, and internal waves. D, 1950, University of California, Los Angeles.

Ewing, J. W. A rare earth element study of the 2150 Vein, Sunnyside Mine, Silverton, Colorado, and a 1.7-B.Y.-old pegmatite near Kittredge, Colorado. M, 1977, University of Colorado.

Ewing, John Maclyn. Origin of the shoestring sands. M, 1952, University of Michigan.

Ewing, M.; Dorman, Henry J. and Oliver, J. Study of shear-velocity distribution in the upper mantle by mantle Rayleigh waves. D, 1962, Columbia University, Teachers College.

Ewing, Rodney Charles. Mineralogy of metamict rare earth AB$_2$O$_6$-type niobium-tantalum-titanium oxides. D, 1974, Stanford University. 161 p.

Ewing, Sydney C. Geology and ore-deposits of the Mount Ingalls District, Plumas County, California. M, 1927, Stanford University. 176 p.

Ewing, Thomas. Geochemistry of the Hackett River Volcanics, Northwest Territories. M, 1977, New Mexico Institute of Mining and Technology. 126 p.

Ewing, Thomas Edward. Geology and tectonic setting of the Kamloops Group, South-central British Columbia. D, 1981, University of British Columbia.

Ewoldsen, Hans Martin. Mechanical properties of the Domengine Sandstone; relationships between laboratory and field measurements. M, 1964, University of California, Berkeley. 136 p.

Ewy, Bradford James. Computer enhanced evaluation procedures of Arctic and subarctic soils. M, 1985, University of Alaska, Fairbanks. 111 p.

Exarchos, Constantine Christos. Studies of the fate of cell wall polymers of higher plants in peat; a contribution to the geochemistry of coal. D, 1976, Pennsylvania State University, University Park. 166 p.

Exton, John. Benthonic microfaunal associations from the Liassic (Lower Jurassic) of Zambujal, west central Portugal. M, 1977, Carleton University. 320 p.

Exum, Frank Allen. Geology of a portion of eastern Cuyama Valley, Ventura and Santa Barbara counties, California. M, 1957, University of California, Los Angeles.

Eyck, James R. Ten *see* Ten Eyck, James R.

Eyde, Theodore Henrik. The Potosi tungsten district, Madison County, Montana. M, 1957, Montana College of Mineral Science & Technology. 79 p.

Eye, John David. Aqueous transport of dieldrin residues in soils. D, 1966, University of Cincinnati. 159 p.

Eyer, Andrew. Geomorphology and morphologic development of Bolivar Roads Inlet and Bolivar Peninsula, Texas. M, 1984, University of Houston.

Eyer, Jerome Arlan. New Pennsylvanian, Permian, and Triassic Charophyta of North America. M, 1961, University of Missouri, Columbia.

Eyer, Jerome Arlan. Stratigraphy and micropaleontology of the Gannett Group of western Wyoming and south-eastern Idaho. D, 1964, University of Colorado. 151 p.

Eyerly, George Brown. The properties and uses of Pacific Northwest diatomite. M, 1941, University of Washington. 70 p.

Eyerly, Terma LeClerc. Geology of Hemphill County, Texas. M, 1909, University of Kansas.

Eyisi, A. E. O. Column studies of CO$_2$ diffusion rates in calcareous sandy soil. M, 1978, University of Waterloo.

Eyles, Carolyn Hope. Sedimentation on glacially-influenced continental shelves. D, 1986, University of Toronto.

Eyles, Carolyn Hope. The sedimentology of the early and middle Wisconsin deposits at Scarborough Bluffs, Ontario. M, 1982, University of Toronto.

Eyles, Nicholas. Medial moraines as part of a glacial debris system; their formation and sedimentology. M, 1976, Memorial University of Newfoundland. 196 p.

Eymann, James L. A study of sand dunes in the Colorado and Mojave deserts. M, 1953, University of Southern California.

Eynon, George. The structure of braid bars; facies relationships of Pleistocene braided outwash deposits, Paris, Ontario. M, 1972, McMaster University. 235 p.

Eyrich, Henry Theodore. Economic geology of part of the New World mining district, Park County, Montana. D, 1969, Washington State University. 130 p.

Eysteinsson, Hjalmar. The inversion of two-dimensional magnetotelluric and magnetic variation data. D, 1988, Brown University. 391 p.

Eyzaguirre, Carlos J. Lopez *see* Lopez Eyzaguirre, Carlos J.

Ezeani, Chuba. Petrology of cores from Lake St. John area, Quebec. M, 1957, Universite Laval. 58 p.

Ezekwe, John Nnaemeka. Effect of paraffinic, naphthenic and aromatic distribution in the hydrocarbon mixture and water on the phase equilibria of carbon dioxide-hydrocarbon systems over the temperature range from 333 K to 366 K. D, 1982, University of Kansas. 370 p.

Ezell, Robert L. Geology of the Rendezvous Park area, Cache and Box Elder counties, Utah. M, 1953, Utah State University. 50 p.

Ezerendu, Friday O. Stratigraphic history, deposition and lithologic study of the Topeka Limestone in Greenwood County, Kansas. M, 1987, Emporia State University.

F., Jr. Litchford R. *see* Litchford R. F., Jr.

Faas, Richard W. Micropaleontology of some Quaternary sediments from the Barrow area, northern Alaska. M, 1962, Iowa State University of Science and Technology.

Faas, Richard William. A study of some late Pleistocene estuarine sediments near Barrow, Alaska. D, 1964, Iowa State University of Science and Technology. 196 p.

Fabbi, Brent P. The occurrence of barium and strontium in typical Nevada gypsum. M, 1965, University of Nevada. 54 p.

Fabbri, A. G. Image processing of geological data. D, 1981, University of Ottawa. 385 p.

Faber, James Warren. Geology of the Sand Ridge area. M, 1955, Washington University. 44 p.

Fabian, Robert S. Relation of biofacies to lithofacies in interpreting depositonal environments in the Pitkin Limestone (Mississippian) in northeastern Oklahoma. M, 1984, University of Oklahoma. 126 p.

Fabiyi, Ekundayo E. The geology and computer analysis of mining data of Itakpe Ridge iron ore deposit, Okene, Kwara State, Nigeria. M, 1977, University of Idaho. 155 p.

Fabry, Fredric Carl. Interpretation of the environment of deposition of Grand Tower Limestone in Jackson and Union counties, Illinois. M, 1964, Southern Illinois University, Carbondale. 82 p.

Fabry, Victoria Joan. Aragonite production by pteropod molluscs; implications for the oceanic calcium carbonate cycle. D, 1988, [University of California, Santa Barbara]. 180 p.

Fabryka-Martin, June Taylor. Production of radionuclides in the Earth and their hydrogeologic significance, with emphasis on chlorine-36 and iodine-129. D, 1988, University of Arizona. 423 p.

Faccioli, Ezio. A discrete Eulerian model for spherical wave propagation in compressible media. D, 1968, University of Illinois, Urbana. 128 p.

Fackler, William C. Clastic dikes in the Keweenawan lavas of Lake Superior. M, 1940, University of Cincinnati. 23 p.

Factor, David F. Paleoecology of Malacostracan arthropods in the Bear Gulch Limestone (Namurian) of central Montana. M, 1984, Kent State University, Kent. 77 p.

Facundus, Michael R. Diagenetic aspects of evaporite solution; Kirschberg Evaporite. M, 1968, Louisiana State University.

Fadaie, Kian. Geophysical and isotopic constraints on the lithosphere of the East African Rift system. D, 1988, Carleton University.

Faddies, Thomas Blair. Brecciation in the Ontario Mine, east flank ore bodies, Park City District, Utah. M, 1973, University of Utah. 95 p.

Fadhli, Fathi Ali. Application of the generalized reciprocal refraction method in mapping a refractor in Albany, Ohio. M, 1986, Ohio University, Athens. 176 p.

Faecke, David Charles. An offshore oil and gas development planning model incorporating risk analysis. D, 1982, University of Texas, Austin. 308 p.

Faflak, Richard E. Chert solubility with application to archaeological dating (in Iowa). M, 1971, University of Iowa. 77 p.

Faflak, Richard E. Quaternary paleogeography of the Badlands region, South Dakota. D, 1987, Indiana State University. 142 p.

Fagadau, Sanford Payne. An investigation of the Flagstaff Limestone between Manti and Willow Creek canyons, in the Wasatch Plateau, central Utah. M, 1949, Ohio State University.

Fagadau, Sanford Payne. Paleontology and stratigraphy of the Logan Formation of central and southern Ohio. D, 1952, Ohio State University.

Fagan, George Lawrence, Jr. Analysis of flood hydrographs from wetland areas. D, 1981, Polytechnic University. 297 p.

Fagan, John J. Carboniferous cherts, turbidites and volcanics in northern Independence Range, Nevada. D, 1960, Columbia University, Teachers College. 125 p.

Fagan, Timothy Jay. Metamorphic petrology and structure of a transect from the Berwick Formation into the Massabesic gneiss complex, southeastern New Hampshire. M, 1986, University of New Hampshire. 108 p.

Fagerlin, Stanley Charles. Lower Ordovician conodonts from the Jefferson City Formation of Missouri. D, 1980, University of Missouri, Columbia. 109 p.

Fagerlin, Stanley Charles. Pleistocene and Recent foraminifera from the Chukchi Rise and Canada Basin areas of the Arctic Ocean. M, 1971, University of Wisconsin-Madison. 95 p.

Fagerstrom, John Alfred. The age, stratigraphic relations, and fauna of the Middle Devonian Formosa reef limestone of southwestern Ontario. D, 1960, University of Michigan. 198 p.

Fagerstrom, John Alfred. The geology of Short Creek Township, Harrison County, Ohio. M, 1953, University of Tennessee, Knoxville. 142 p.

Faggart, Billy E. Sm-Nd study of the Sudbury Complex, Ontario, Canada. M, 1984, University of Rochester. 61 p.

Faggioli, Justin M. Tectonic intepretation of a Great Valley sequence allochthon in southern Pope-Chiles Valley, California. M, 1973, Stanford University.

Faggioli, R. E. The geology of the Liebre Fault (California). M, 1952, University of California, Los Angeles.

Fagin, Stuart. The geology of the Culpeper Basin within Culpeper County. M, 1977, George Washington University.

Fagin, Stuart William. Paleogeography and tectonics of the Redding Section, eastern Klamath Belt, Northern California. D, 1983, University of Texas, Austin. 224 p.

Fago, Thomas Arthur. A seismic shear wave study of the Mad River buried valley system near Dayton, Ohio. M, 1988, Wright State University. 87 p.

Fagrelius, Kurt H. Geology of the Cerro del Viboro area, Socorro County, New Mexico. M, 1982, New Mexico Institute of Mining and Technology. 138 p.

Fahey, Barry D. A quantitative analysis of freeze-thaw cycles, frost heave cycles, and frost penetration in the Front Range of the Rocky Mountains, Boulder County, Colorado. D, 1971, University of Colorado.

Fahey, Patrick Louis. The geology of Island copper mine, Vancouver Island, British Columbia. M, 1979, University of Washington. 52 p.

Fahey, Timothy J. A geochemical study of Cedar Butte, and evolved volcanic construct on the eastern Snake River plain, Idaho. M, 1986, SUNY at Buffalo. 57 p.

Fahlquist, Davis A. Seismic refraction measurements in the western Mediterranean. D, 1963, Massachusetts Institute of Technology. 173 p.

Fahlquist, Lynne S. An experimental investigation of the solubility and complexing of nickel in the system NiO-HCl-H$_2$O-NaCl. M, 1987, Texas A&M University. 49 p.

Fahlstrom, Beverly E. Stratigraphy and depositional history of the Cretaceous Nanaimo Group of the Chemainus area, British Columbia. M, 1982, Oregon State University. 115 p.

Fahner, Lewis George. Applications of a finite-difference model for aquifer simulation to an area in east-central Alberta. M, 1982, University of Alberta. 78 p.

Fahnestock, Robert Kendall. Morphology and hydrology of a glacial stream; White River, Mount Rainier, Washington. D, 1960, Cornell University. 70 p.

Fahrig, Walter F. The petrology of the ultramafic rocks of the Labrador Trough. D, 1954, University of Chicago. 124 p.

Fahy, Michael F. Stability analyses pertinent to processes of rock folding. D, 19??, Texas A&M University.

Faick, John N. Geology and ore deposits of the Gold Hill District. M, 1937, University of Idaho. 56 p.

Faick, John Nicholas. Stratigraphy, structure and composition of cement materials in north central California. D, 1959, University of Arizona. 173 p.

Faigle, George A. Glacial geology of western Wells County, North Dakota. M, 1964, University of North Dakota. 85 p.

Failing, Martha S. River and terrace complexes of the Texas Gulf Coast. M, 1968, Rice University. 14 p.

Faill, Rodger T. Deformational modes of behavior of Crown Point Limestone as a function of confining pressure and total strain. M, 1964, Columbia University, Teachers College.

Faill, Rodger Tanner. The effect of loading rate in the experimental deformation of a limestone. D, 1966, Columbia University, Teachers College. 69 p.

Fails, Thomas G., Jr. Permian stratigraphy at Carlin Canyon, Elko County, Nevada. M, 1955, Columbia University, Teachers College.

Fain, Gilbert. Signal theory applied to continuous seismic profiling. D, 1968, University of Rhode Island. 140 p.

Fainstein, Roberto. Geodynamics of the eastern South American continental margin near the Brazilian bulge. D, 1979, Rice University. 193 p.

Fair, Charles Leroy. Geology of the Fresnal Canyon area, Baboquivari Mountains, Pima County, Arizona. D, 1965, University of Arizona. 111 p.

Fairbairn, Harold Williams. The structure and metamorphism of the mountains of Brome County, Quebec. D, 1932, Harvard University.

Fairbairn, Patrick W. Environmental impact evaluation in freshwater impoundments by vegetation analysis of the terrestrial ecosystem. D, 1974, University of Massachusetts. 192 p.

Fairbank, Philip Keith. Heavy metal reconnaissance survey within the Madison River drainage basin, Gravelly Range, Montana. M, 1984, Wright State University. 154 p.

Fairbanks, Harold Wellman. Auriferous conglomerate in California. D, 1896, University of California, Berkeley. 3 p.

Fairbanks, Harold Wellman. Geology and mineralogy of Shasta County. D, 1896, University of California, Berkeley. 29 p.

Fairbanks, Harold Wellman. Geology of northern Ventura, Santa Barbara, San Luis Obispo, Monterey and San Benito counties. D, 1896, University of California, Berkeley. 36 p.

Fairbanks, Harold Wellman. Geology of Point Sal. D, 1896, University of California, Berkeley. 91 p.

Fairbanks, Harold Wellman. Geology of the Mother Lode gold belt. D, 1896, University of California, Berkeley. 13 p.

Fairbanks, Harold Wellman. Geology of the Mother Lode region. D, 1896, University of California, Berkeley. 67 p.

Fairbanks, Harold Wellman. Mineral deposits of eastern California. D, 1896, University of California, Berkeley. 14 p.

Fairbanks, Harold Wellman. Notes on a further study of the pre-Cretaceous rocks of the California Coast Ranges. D, 1896, University of California, Berkeley. 15 p.

Fairbanks, Harold Wellman. Notes on some localities of Mesozoic and Paleozoic rocks in Shasta County, California. D, 1896, University of California, Berkeley. 6 p.

Fairbanks, Harold Wellman. Notes on the characters of the eruptive rocks of the Lake Huron region. D, 1896, University of California, Berkeley. 10 p.

Fairbanks, Harold Wellman. Notes on the geology and mineralogy of Tehama, Colusa, Lake, and Napa counties. D, 1896, University of California, Berkeley. 21 p.

Fairbanks, Harold Wellman. Notes on the geology of eastern California. D, 1896, University of California, Berkeley. 11 p.

Fairbanks, Harold Wellman. Notes on the occurrence of rubellite and lepidolite in Southern California. D, 1896, University of California, Berkeley. 6 p.

Fairbanks, Harold Wellman. On analcite diabase from San Luis Obispo County, California. D, 1896, University of California, Berkeley. 27 p.

Fairbanks, Harold Wellman. Relation between ore deposits and their inclosing walls. D, 1896, University of California, Berkeley. 3 p.

Fairbanks, Harold Wellman. Review of our knowledge of the geology of the California Coast Ranges. D, 1896, University of California, Berkeley. 31 p.

Fairbanks, Harold Wellman. Some remarkable hot springs and associated mineral deposits in Colusa County, California. D, 1896, University of California, Berkeley. 3 p.

Fairbanks, Harold Wellman. Stratigraphy of the California Coast Ranges. D, 1896, University of California, Berkeley. 18 p.

Fairbanks, Harold Wellman. The pre-Cretaceous age of the metamorphic rocks of the California Coast Ranges. D, 1896, University of California, Berkeley. 13 p.

Fairbanks, Harold Wellman. Validity of the so-called Wallala Beds as a division of the California Cretaceous. D, 1896, University of California, Berkeley. 5 p.

Fairbanks, Paul E. Chemical changes in groundwater of northern Utah Valley, Utah. M, 1982, Utah State University. 82 p.

Fairbanks, Richard G. Geochemistry of marine skeletal carbonate for use in paleoenvironmental reconstructions. D, 1977, Brown University. 197 p.

Fairborn, John William. Gravity survey and interpretation of the northern portion of the Salinas Valley. M, 1963, Stanford University.

Fairbrothers, Gregg E. Magmatic trends at Boqueron Volcano, El Salvador. M, 1977, Rutgers, The State University, New Brunswick. 75 p.

Fairchild, Drena K. T. Paleoenvironments of the Chignik Formation, Alaska Peninsula. M, 1977, University of Alaska, Fairbanks. 168 p.

Fairchild, Lee Hamlin. Lahars at Mount St. Helens, Washington. D, 1985, University of Washington. 374 p.

Fairchild, Lee Hamlin. The Leech River unit and Leech River Fault, southern Vancouver Island, BC. M, 1979, University of Washington. 170 p.

Fairchild, Paul W. The geology and petrography of the gypsum deposits near the Blue Rapids, Marshall County, Kansas. M, 1949, University of Kansas. 94 p.

Fairchild, Raymond Eugene. Geology of two wildcat oil wells in Fremont County, Wyoming. M, 1950, University of Missouri, Columbia.

Fairchild, Roy W. Geology of T. 29 S., R. 11 W. of the Sitkum and Coquille quadrangles, Coos County, Oregon. M, 1966, University of Oregon. 68 p.

Fairchild, T. R. The geologic setting and paleobiology of a late Precambrian stromatolitic microflora from South Australia. D, 1975, University of California, Los Angeles. 290 p.

Fairchild, William W. Foraminifera from the Butano, San Lorenzo, and Vaqueros formations, southern Santa Cruz Mountains, California. M, 1957, University of California, Berkeley. 124 p.

Faircloth, Susan Lynne. Precambrian basement rock from North Dakota, South Dakota and western Minnesota; a petrologic and geochemical analysis of 51 drill core samples. D, 1988, South Dakota School of Mines & Technology.

Fairer, George M. Geology of the Granby Quadrangle, Grand County, Colorado. M, 1971, Colorado School of Mines. 54 p.

Fairfax, Vella L. Demonstration and defense of the importance of geology in urban planning in the Carson City, Nevada area. M, 1979, University of Nevada. 105 p.

Fairhurst, William. Stratigraphy, sedimentology, and petrology of the Upper Cretaceous Horsethief, St. Mary River, and Willow Creek formations, west-central Montana. M, 1984, University of Missouri, Columbia. 129 p.

Fairley, William Merle. Littoral sediments and ecology at Great Bar, Jonesport, Maine. M, 1951, University of Maine. 91 p.

Fairley, William Merle. The Murphy Syncline in the Tate Quadrangle, Georgia. D, 1962, The Johns Hopkins University. 71 p.

Faith, Stuart E. An equilibrium distribution of trace elements in a natural stream environment. M, 1974, New Mexico Institute of Mining and Technology. 54 p.

Fajardo, Ivan. A study of the connate waters and clay mineralogy in southern Louisiana. M, 1969, University of Tulsa. 68 p.

Fakundiny, Robert Harry. Birrimian metamorphic and associated granitic rocks (Precambrian), south central Ghana, west Africa. M, 1967, University of Texas, Austin.

Fakundiny, Robert Harry. Geology of the El Rosario Quadrangle, Honduras, Central America. D, 1970, University of Texas, Austin. 237 p.

Falade, Gabriel Kayode. The dynamics and analysis of vertical pulse testing for formation anisotropy. D, 1974, Stanford University. 216 p.

Falatah, Abdulrazag Mohammed. The effect of elemental sulphur on Zn and Fe in two western Kansas soils. D, 1988, Kansas State University. 161 p.

Falchook, Martin G. Benthonic foraminifera from the Navisink Formation (Upper Cretaceous) of New Jersey. M, 1972, Brooklyn College (CUNY).

Falck, Arnold. Exploration and evaluation of mineral deposits and review of origin and occurrence of manganese ore deposits. M, 1960, University of Washington.

Falcone, Sharon K. Glacial stratigraphy of northwestern Cass County, North Dakota. M, 1983, University of North Dakota. 180 p.

Faldetta, Sarah. The effects of a catastrophic flood upon surface water and groundwater in southwestern New Hampshire. M, 1988, Boston University. 199 p.

Falkenhein, Frank Ulrich Helmut. Carbonate microfacies and depositional evolution of the Macae Formation (Albian-Cenomanian), Campos Basin, Brazil. D, 1981, University of Illinois, Urbana. 199 p.

Falkowski, Stephen Kenneth. The geology and ore deposits of the Johnny M Mine, Ambrosia Lake mining district, Grants, NM. M, 1980, New Mexico Institute of Mining and Technology. 154 p.

Fall, Patricia Lynn. Vegetation dynamics in the Southern Rocky Mountains; late Pleistocene and Holocene timberline fluctuations. D, 1988, University of Arizona. 303 p.

Fall, Steven A. Depositional and diagenetic history of some Jurassic carbonates, Indian Rock-Gilmer Field, Upshur County, Texas. M, 1974, Texas A&M University. 132 p.

Falla, William S., Jr. The petrology and geochemistry of the Clarion Flint Clay (Pennsylvanian), western Pennsylvania. M, 1967, Pennsylvania State University, University Park. 84 p.

Fallah-Araghi, M. H. Random models of spilled oil movement. D, 1975, University of Delaware, College of Marine Studies. 216 p.

Fallaw, Wallace Craft. Geology of the Rocky Point, North Carolina, Quadrangle and an adjacent coastal area. M, 1962, University of North Carolina, Chapel Hill. 88 p.

Fallaw, Wallace Craft. The Pleistocene Neuse Formation in southeastern North Carolina. D, 1965, University of North Carolina, Chapel Hill. 174 p.

Falley, R. Thomas. Distribution and transport of mercury within the Nooksack River drainage, Whatcom County, Washington. M, 1974, Western Washington University. 67 p.

Fallis, Jasper N. An insoluble residue analysis of a section of the (Lower Cretaceous) Glen Rose Formation in the vicinity of the type locality, Hood County, Texas. M, 1958, Texas A&M University. 68 p.

Fallis, John F., Jr. Geology of the Pedernal Hills area, Torrance County, New Mexico. M, 1958, University of New Mexico. 50 p.

Fallis, Susan Mary. Mineral sources of water and their influence on the safe disposal of radioactive wastes in bedded salt deposits. M, 1973, University of Tennessee, Knoxville. 61 p.

Fallon, James H. Depositional environments, biostratigraphy, and geologic history of upper Albian (Cretaceous) rocks, West Texas. M, 1981, Sul Ross State University.

Falls, Darryl Lee. Size, grain type and mineralogical relationships in Recent marine calcareous beach sands. M, 1970, University of North Carolina, Chapel Hill. 35 p.

Falls, Robert Meredith. Geology and metamorphism of the Inc 10 sulphide occurrence, Coppermine River area, N.W.T. M, 1979, University of Western Ontario. 144 p.

Falotico, Robert J., Jr. Ground-water modeling of an isolated kame deposit aquifer. M, 1986, University of New Hampshire. 135 p.

Falteisek, Jan D. The water resources potential of the Missouri River in Missouri. M, 1984, University of Missouri, Columbia. 98 p.

Faltyn, Norbert E. Seismic exploration of the Tully Valley overburden (New York). M, 1957, Syracuse University.

Fambrough, James Warren. Isopach and lithofacies study of Virgilian and Missourian Series of north-central Oklahoma. M, 1962, University of Oklahoma. 89 p.

Famy, Syed Mohamad. Relationship between the cross-strike lineaments and the distribution of oil and gas fields in northwestern Pennsylvania. M, 1979, Pennsylvania State University, University Park. 137 p.

Fan, Gary Guoyou. Investigation of local ordering in amorphous materials. D, 1987, Arizona State University. 181 p.

Fan, Jen-Chen. Measurements of erosion on highway slopes and use of the universal soil loss erosion equation. D, 1987, Purdue University. 390 p.

Fan, Jieun-jeou. Organic geochemical study of the upper Cretaceous Pierre shale. D, 1971, Washington University. 99 p.

Fan, Paul Hsiu-Tsu. The subsurface geology of Iowa County. M, 1945, University of Iowa. 57 p.

Fan, Paul Hsui-Tsu. Subsurface Cambrian geology of Iowa. D, 1947, University of Iowa. 140 p.

Fan, Pow-Foong. Mineral assemblages and sedimentation of the Sespe Formation of South Mountain, Ventura County, California. M, 1963, University of California, Los Angeles.

Fan, Pow-foong. Recent silts in the Santa Clara River drainage basin, California; a mineralogical investigation of their origin and evolution. D, 1965, University of California, Los Angeles. 215 p.

Fan, William Reun-Sen. The damping properties and the earthquake response spectrum of steel frames. D, 1968, University of Michigan. 131 p.

Fanaff, Allan S. A study of landslide phenomena in southeastern Ohio. M, 1964, Ohio University, Athens. 88 p.

Fanale, Fraser Partington. Helium in magnetite. D, 1964, Columbia University, Teachers College. 157 p.

Fandriana, Lilian. Numerical simulator for fluid flow in a geothermal well. M, 1979, Stanford University.

Fandrich, Joe W. Studies of bentonites and related rocks in the Eagle Ford group (upper Cretaceous), central Texas. M, 1968, Baylor University. 91 p.

Fanelli, Eileen M. The adsorption of cadmium, copper, and chromium from a synthetic leachate medium by three earth materials. M, 1985, University of New Hampshire. 137 p.

Fang, Changle. Marine heat flow measurement. D, 1985, Memorial University of Newfoundland. 215 p.

Fang, Jen-Ho. X-ray studies of Ca-Cl$_2$-chabazite and dehydrated natrolite. D, 1961, Pennsylvania State University, University Park. 105 p.

Fang, Yung-Show. Dynamic earth pressures against rotating walls. D, 1983, University of Washington. 147 p.

Fankhauser, Robert E. Geology and mineralization of the southern Cuddy Mountains, Washington County, Idaho. M, 1969, Oregon State University. 138 p.

Fannin, Timothy Edward. Various techniques for estimating riverine water quality concentrations, and their application to a loading model for Flaming Gorge Reservoir, Wyoming. D, 1988, University of Wyoming. 228 p.

Fanning, David James. Metamorphism and tungsten mineralization in the Nightingale Range, Pershing County, Nevada. M, 1982, University of Nevada. 110 p.

Fanning, Delvin Seymour. Mineralogy as related to the genesis of some Wisconsin soils developed in loess and in shale-derived till. D, 1964, University of Wisconsin-Madison. 300 p.

Fanning, Kent Abram. Studies on dissolved silica at the boundaries of the ocean system. D, 1973, University of Rhode Island. 234 p.

Fanshawe, John Richardson. Structural geology of the Wind River Canyon area, Wyoming. D, 1939, Princeton University. 54 p.

Fantel, Richard J. Geology of the north half of the Old Fields Quadrangle, West Virginia. M, 1978, University of Akron. 51 p.

Fantone, Kenneth Scott. Geochemistry and petrology of the carbonate iron-formations and ferruginous cherts of the northeastern Black Hills, South Dakota. M, 1983, South Dakota School of Mines & Technology. 14 tables, 9 plate p.

Fantozzi, Joseph H. Paleontology and stratigraphy of the Paleocene sediments in the Simi Hills (California). M, 1956, University of California, Los Angeles.

Fantozzi, Joseph H. The stratigraphy and biostratigraphy of a portion of the Simi Hills on the south side of the Simi Valley, Ventura County, California. M, 1955, University of California, Los Angeles.

Fara, Daniel Ray. Stratigraphy, sedimentology and diagenesis of the Trenton Limestone (U. Ordovician) in three regionally spaced cores in central and northern Indiana. M, 1986, Indiana University, Bloomington. 94 p.

Fara, Mark. The geology of the Parnassus-Mount Solon area of central Shenandoah Valley, Virginia. M, 1957, Virginia Polytechnic Institute and State University.

Farabee, Michael Jay. Systematics of Aquilapolles and selected Upper Cretaceous and Tertiary pollen. D, 1987, University of Oklahoma. 308 p.

Faraguna, John. The dipmeter as a stratigraphic tool; recognizing sedimentary structures on two-dimensional correlation surfaces. M, 1988, University of Houston.

Farah, Osman Mohamed. The bathymetry, oceanography, and bottom sediments of Dongonab Bay (Red Sea), Sudan. D, 1982, University of Delaware, College of Marine Studies. 163 p.

Farahmand, Seyedhassan Hashemi. Uranium potential in the Spar Canyon and Road Creek areas, Custer County, Idaho. M, 1980, Idaho State University. 100 p.

Faramarzpour, Faramarz. A low contamination mass spectrometer for potassium-argon age determination. M, 1966, Massachusetts Institute of Technology. 21 p.

Farber, Steve L. The limestone petrography of the Southgate and McMicken members, Latonia Formation, Cincinnatian Series, in southern Hamilton County, Ohio. M, 1968, Miami University (Ohio). 121 p.

Fardon, Ross Stuart Harpur. The genesis of lateritic nickel deposits. D, 1968, Harvard University.

Fargo, Thomas R. The stratigraphy, structure and tectonic significance of the Campbellton Sequence and associated rock units, north-central Newfoundland. M, 1985, SUNY at Buffalo. 150 p.

Farhat, J. S. Geochemical and geochronological investigation of the early Archaean of the Minnesota River valley, and the effect of metamorphism on Rb-Sr whole rock isochrons. D, 1975, University of California, Los Angeles. 189 p.

Faria Santos, Claudio A. F. Analysis of floor stability in underground coal mines. D, 1988, Pennsylvania State University, University Park. 205 p.

Faria Santos, Claudio A. F. Effect of laboratory-simulated weathering on the properties of Loyalhanna Sandstone. M, 1986, Pennsylvania State University, University Park. 109 p.

Farias, Luiz Carlos Cabral de *see* Cabral de Farias, Luiz Carlos

Farias-Garcia, Ramon. Geophysical exploration of the Elarco-Calmalli mining district, Baja California, Mexico. M, 1978, University of Arizona.

Farinelli, Joseph Augustine. An evaluation of thin-seam, room-and-pillar coal mining potential in the Appalachian region. M, 1986, Pennsylvania State University, University Park. 118 p.

Faris, Craig Duncan. Planktonic foraminiferal biostratigraphy and correlation of the Aquia Formation in the type area, along the Potomac River, Virginia. M, 1982, Virginia Polytechnic Institute and State University. 84 p.

Farkas, Arpad. A trace element study of the Texas Gulf ore body, Timmins, Ontario. M, 1973, University of Alberta. 149 p.

Farkas, Arpad. The distribution of cobalt and nickel between pyrite and pyrrhotite. D, 1980, University of Toronto.

Farkas, Frank S. Infiltration and laboratory permeability studies of spoils from selected coal strip mines, Powder River basin, Wyoming and Montana. M, 1973, South Dakota School of Mines & Technology. 147 p.

Farkas, Steven Eugene. Geology of the southern San Mateo mountains, Socorro and Sierra counties, New Mexico. D, 1969, University of New Mexico. 137 p.

Farlekas, George M. The geology of part of South Mountain of the Blue Ridge Province north of the Pennsylvania-Maryland border. M, 1961, Pennsylvania State University, University Park. 64 p.

Farley, E. Mineralization at the Turner and Walker deposits, South Mountain Batholith. M, 1978, Dalhousie University.

Farley, Martin B. An assessment of the correlation between miospores and depositional environments of the Dakota Formation (Cretaceous), north-central Kansas and adjacent Nebraska. M, 1982, Indiana University, Bloomington. 181 p.

Farley, Martin Birtell. Sedimentologic and paleoecologic importance of palynomorphs in Paleogene nonmarine depositional environments, central Bighorn Basin (Wyoming). D, 1987, Pennsylvania State University, University Park. 231 p.

Farley, William H. A study of some of the physical and chemical properties of the Bangor, Plaisted, and Marlow soil profiles of central Maine. M, 1953, University of Maine. 62 p.

Farley, William Horace. A pedologic study of the Aura soil (New Jersey). D, 1959, Rutgers, The State University, New Brunswick. 65 p.

Farmer, Cathy L. Tectonics and sedimentation, Newcastle Formation (Lower Cretaceous), southwestern flank, Black Hills Uplift, Wyoming and South Dakota. M, 1981, Colorado School of Mines. 195 p.

Farmer, Eugene Edward. The hydrology of a phosphate mine overburden waste embankment. D, 1988, Utah State University. 223 p.

Farmer, Garland Langhorne. Origin of Mesozoic and Tertiary granite in the Western U.S. and implications for pre-Mesozoic crustal structure. D, 1983, University of California, Los Angeles. 278 p.

Farmer, George Thomas. Physiography and paleontology of the Devonian black shales in Highland County, Virginia. M, 1960, University of Virginia.

Farmer, George Thomas, Jr. Bifoliate Cryptostomata of the Simpson Group, Arbuckle Mountains, Oklahoma. D, 1968, University of Cincinnati. 241 p.

Farmer, Jack D. Variation in the Bryozoan species Fistulipora decora (Moore and Dudley) from the Beil Limestone Member of the Lecompton Limestone (Virgilian, Pennsylvanian) of Kansas. M, 1971, University of Kansas. 120 p.

Farmer, Jack Dewayne and Stump, Thomas Edward. Studies in the form, function, development and evolution of the Bryozoa. D, 1978, University of California, Davis. 232 p.

Farmer, Jeanne. Petroleum potential of the Shandong Province, People's Republic of China. M, 1981, Tulane University.

Farmer, M. W. Effect of light intensity on biomass characteristics of a diatom growing in outdoor continuous culture. D, 1978, City College (CUNY). 123 p.

Farmer, Paul. Geology of the Phlox Mountain area, Hot Springs County, Wyoming. M, 1941, University of Missouri, Columbia.

Farmer, Randall Allen. Seismicity, tectonics and focal mechanisms in the Scotia Sea area. M, 1982, Michigan State University. 120 p.

Farmer, Richard Echols. The geology of the southeastern part of Pulaski County, Kentucky. M, 1950, University of Iowa. 77 p.

Farmer, Ronald B. Pillow basalts and plate tectonic setting of the Precambrian Packsaddle Schist, southeastern Llano County, Texas. M, 1977, Northeast Louisiana University.

Farmer, Russell. The conodonts of the Chouteau Limestone. M, 1932, University of Missouri, Columbia.

Farmilo, Alfred William. A lithological and field study of the Boyer Formation and the overlying and underlying formations. M, 1943, University of Oklahoma. 117 p.

Farnham, Jack D. A facies analysis of the Wilcox Group (Eocene), East Texas. M, 1982, Stephen F. Austin State University.

Farnham, Paul Rex. Crustal structure in the Keweenawan Province of east-central Minnesota and western Wisconsin. D, 1967, University of Minnesota, Minneapolis. 495 p.

Farnham, Paul Rex. Geology and groundwater resources of the Tannersville-Tumbling Creek area, Washington, Smyth, and Tazewell counties, Virginia. M, 1960, Virginia Polytechnic Institute and State University.

Farnsworth, Don Willard. Glacial geology of the west side of the Volcano Iztaccihuatl, Mexico. M, 1957, Ohio State University.

Farnsworth, James W. Relationship of gravity anomalies to a drift filled bedrock valley system in Calhoun County, Michigan. M, 1980, Western Michigan University.

Farnsworth, Ray Lothrop. Erosion surfaces of Massachusetts. D, 1961, Boston University. 172 p.

Farooqui, Saleem M. Stratigraphy and petrology of the Port Orford conglomerate (middle Pliocene), Cape Blanco, Oregon. M, 1969, University of Oregon. 57 p.

Farouq, Fadullah Meer. Frequency spectra of some selected quarry blast seismograms. M, 1963, St. Louis University.

Farquhar, Clyde Randolph. Crystalline rocks of north central Wake County, North Carolina. M, 1952, North Carolina State University. 33 p.

Farquhar, Paul Thomas. Petrology and geochemistry of the Precambrian metavolcanics of Ladron Mountain. M, 1976, New Mexico Institute of Mining and Technology.

Farquhar, Roger P. Computer display of a seismic refraction model. M, 1967, Stanford University.

Farquhar, Ronald McCunn. The lead isotope methods of geological age determination. D, 1954, University of Toronto.

Farquhar, Scot Paul. Depositional and diagenetic history of the Mountain Springs Formation Member C (Lower to Middle Devonian), southern Great Basin. M, 1986, San Diego State University. 125 p.

Farquharson, Robin Bruce. The petrology of several late Tertiary gabbroic plugs in the South Cariboo region, BC. M, 1965, University of British Columbia.

Farr, Jocelyn Elizabeth. The geology, mineralogy, and geochemistry of the 070 faults of the Corbet Mine, Noranda, Quebec. M, 1984, University of Toronto.

Farr, John Brent. P-waves between 10 degrees and 30 degrees. D, 1955, University of California, Berkeley. 135 p.

Farr, John Vail. Loading rate effects on the one-dimensional compressibility of four partially saturated soils. D, 1986, University of Michigan. 374 p.

Farr, Mark Randall. Compositional variation of late dolomite cement as a guide to parent fluid flow directions in the Cambrian Bonneterre Formation, Missouri. D, 1988, University of Texas, Austin. 279 p.

Farr, Mark Randall. Compositional variation of the Elliot Lake Group, Sudbury District, Ontario. M, 1983, University of Toronto.

Farr, Thomas Galen. Surface weathering of rocks in semiarid regions and its importance for geologic remote sensing. D, 1981, University of Washington. 161 p.

Farrand, Richard Brownlow. Cretaceous macrofossil assemblages of the Corsicana Formation with implications on the Cretaceous-Tertiary boundary in east-central Texas. M, 1984, University of Texas, Austin. 160 p.

Farrand, William Hoffman. The relation of a fossil foraminifera fauna from Timms Point, San Pedro, California, to a Recent foraminifera fauna from off the coast of San Pedro, California. M, 1929, University of Southern California.

Farrand, William Richard. Former shorelines in western and northern Lake Superior Basin. D, 1960, University of Michigan. 257 p.

Farrand, William Richard. The regimen of a marginal portion of an ice cap in Northwest Greenland. M, 1956, Ohio State University.

Farrar, Christopher D. Ground-water potential in the middle Paleozoic carbonate rocks, Flagstaff area, Coconino County, Arizona. M, 1980, Northern Arizona University. 91 p.

Farrar, Edward. The extraction and ultra-high vacuum mass spectroscopy of argon from rocks. D, 1966, University of Toronto.

Farrar, Michael J. Processes and patterns of sedimentation at Hammonasset Beach, Madison, Connecticut. M, 1977, Wesleyan University.

Farrar, P. D. A numerical study of frictional entrainment in a cyclonic Gulf Stream ring. M, 1977, Texas A&M University.

Farrar, Robert Lynn, Jr. The argon and helium contents of some rock samples. M, 1947, University of Washington. 22 p.

Farrar, S. S. Petrology and structure of the Glen Quadrangle, southeastern Adirondacks, N.Y. D, 1976, SUNY at Binghamton. 177 p.

Farrar, Willard. Fuller's earth in Southeast Missouri. M, 1934, University of Missouri, Rolla.

Farre, John Andrew. The importance of mass wasting processes on the continental slope. D, 1985, Columbia University, Teachers College. 248 p.

Farrell, Clifton William. Strontium isotopes of Kuroko deposits. D, 1979, Harvard University.

Farrell, Michael Thomas. Trace element analysis of the Meigs Creek (No. 9) Coal in Muskingum and Noble counties of Ohio. M, 1985, Wright State University. 120 p.

Farrell, Stewart C. Present coastal processes, recorded changes, and the Post-Pleistocene geologic record of Saco Bay (York and Cumberland counties), Maine. D, 1972, University of Massachusetts.

Farrell, Stewart C. Sediment distribution and hydrodynamics of the Saco and Scarbe estuaries (Maine). M, 1970, University of Massachusetts. 128 p.

Farrell, Thomas G. Paleoecology of the Brickeys Member (Middle Ordovician) of the Mifflin Formation (Plattin Subgroup) of eastern Missouri. M, 1968, St. Louis University.

Farrell, William E. Gravity tides. D, 1970, University of California, San Diego.

Farrelly, John James. Depositional setting and evolution of the Pliocene-basal Pleistocene section of Southeast Trinidad, West Indies. M, 1987, University of Texas, Austin. 141 p.

Farrelly, Peter Joseph. Molt stages of Amphissites fractocarinatus, new species. M, 1953, University of Illinois, Urbana.

Farrens, Christine M. Styles of deformation in the southeastern Narragansett Basin, Rhode Island and Massachusetts. M, 1982, University of Texas, Austin.

Farrington, Oliver C. Crystallized azurite from Arizona. D, 1891, Yale University.

Farrington, Richard Lee. Active-fault and landslide hazards along the San Andreas fault zone, southeast Santa Cruz County, California. D, 1974, University of California, Santa Cruz.

Farrington, William B. Geology and fracturing of the Spraberry Formation in Midland, Glassock, Upton, and Reagan counties, West Texas. D, 1953, Massachusetts Institute of Technology. 175 p.

Farrington, William Benford. Depth of burial, the major cause of changes in carbon ratio in West Virginia. M, 1949, Cornell University.

Farris, Robert A. Heavy minerals of the Cretaceous Hell Creek and Paleocene Ludlow formations of Slope and Bowman counties, North Dakota. M, 1984, University of North Dakota. 109 p.

Farris, Stephen Robert. Geology and mid-Tertiary volcanism of the McKnight Canyon area, Black Range, Grant County, New Mexico. M, 1981, University of New Mexico. 87 p.

Farrow, Hillary. Applications of first breaks to cross-borehole geotomography. M, 1985, University of Houston.

Farsad, Ebrahim. Proposed mine design of Chehel-Koureh ore deposit. M, 1975, University of Nevada. 110 p.

Faruque, M. O. Development of a generalized constitutive model and its implementation in soil-structure interaction. D, 1983, University of Arizona. 320 p.

Farver, John Richard. The diffusion kinetics of oxygen in diopside, oxygen and strontium in apatite, and applications to thermal histories of igneous (Skye, Scotland) and metamorphic rocks. D, 1988, Brown University. 167 p.

Farvolden, Robert N. The chemistry of the clay-size fraction across the Oldman-Bearpaw contact of the St. Mary River section. M, 1958, University of Alberta. 46 p.

Farvolden, Robert Norman. Geologic controls of base-flow of mountain streams in northern Nevada. D, 1963, University of Illinois, Urbana. 85 p.

Farwell, Fred W. A petrographic study of the Henrys Fork dam site, Snake River, Idaho. M, 1936, Columbia University, Teachers College.

Fary, Raymond Wolcott, Jr. An X-ray study of the Bradford Third and Haskell sands of McKean County, Pennsylvania. M, 1948, Washington University. 27 p.

Fasching, George E. An admittance meter for moisture in coal. M, 1967, West Virginia University.

Fashbaugh, Earl F. Geology of igneous extrusive and intrusive rocks in the Sundance area, Crook County, Wyoming. M, 1979, University of North Dakota. 97 p.

Fashola, Ahmed B. Stratigraphic and petrographic investigation of part of the Niger Delta, Nigeria. M, 1987, University of Akron. 257 p.

Fasnacht, Timothy Lee. A seismic reflection study of the Precambrian basement along the Illinois-Wisconsin state line. M, 1982, Northern Illinois University. 103 p.

Fasola, A. Palynological study of Triassic samples from the Beardmore Glacier area in Antarctica. M, 1974, Ohio State University.

Fasola, Armando. Biostratigraphy and paleoecology of dinoflagellate cysts in late Cenomanian to early Campanian deposits in southwestern Manitoba. D, 1982, University of Toronto.

Fass, Fred W. R. Mineral asymmetry and the direction of flow of mineralizing solutions in the Viburnum Trend, Southeast Missouri. M, 1980, University of Missouri, Rolla.

Fassauer, Patti. Lithofacies analysis and environmental interpretation of the Georgetown Formation from south-central to northern Texas. M, 1982, Baylor University. 139 p.

Fassett, Bernard Donald. The petrology of the Connellsville Sandstone of Meigs, Athens, and Morgan counties, Ohio. M, 1964, Ohio University, Athens. 70 p.

Fassett, Jack W. Seismicity near the Chatfield Dam, Denver, Colorado. M, 1974, Colorado School of Mines. 58 p.

Fassett, James E. Subsurface geology of the upper Cretaceous Kirtland and Fruitland formations of the San Juan Basin, New Mexico and Colorado. M, 1964, Wayne State University.

Fassihi, Mohammad Reza. Analysis of fuel oxidation in in-situ combustion oil recovery. D, 1981, Stanford University. 305 p.

Fassio, Thomas D. Structural anlaysis of the Willow Creek thrust fault of the southeastern Uinta Mountains and its regional implications. M, 1984, University of Utah. 57 p.

Fast, Susan Elaine Jeffries. The ontogeny of Steganocrinus pentagonus (Mississippian). M, 1969, University of Michigan.

Fastovsky, David Eliot. Paleoenvironments of vertebrate-bearing strata at the Cretaceous-Paleogene boundary in northeastern Montana and southwestern North Dakota. D, 1986, University of Wisconsin-Madison. 301 p.

Fate, Thomas. Facies distribution and diagenesis as factors effecting reservoir quality of the Gialu Limestone, Sirte Basin, Libya. M, 1978, University of South Carolina.

Fates, Dailey Gilbert. Mesozoic (?) metavolcaniclastic rocks, northern White Mountains, California; structural style, lithology, petrology, depositional setting and paleogeographic significance. M, 1985, University of California, Los Angeles. 225 p.

Fath, Arthur Earl. Origin of the faults, anticlines, and buried granite ridge of the northern part of the Mid-Continent oil and gas field. D, 1922, University of Chicago. 9 p.

Fathulla, Riyadh Najeeb. Factors affecting persistence of aldicarb residues in the sand-and-gravel aquifer of central Wisconsin. D, 1988, University of Wisconsin-Madison. 192 p.

Faucette, James Robert. The geology of the Marmaton Group of southern Nowata County, Oklahoma. M, 1954, University of Oklahoma. 66 p.

Faucette, Robert Christian. Depositional and diagenetic history of the Upper Jurassic Haynesville Formation, Teague Townsite Field, Freestone County, Texas. M, 1981, Texas A&M University. 119 p.

Faul, Cydney L. A paleomagnetic study of the remagnetized Jurassic-Cretaceous boundary limestone section at Nuevo Leon, Mexico. M, 1986, University of Texas, Arlington. 106 p.

Faul, Henry. Neutron logging; experiments on distribution from a point source in pipes and hydrogenous media. D, 1949, Massachusetts Institute of Technology. 127 p.

Faul, Henry. Prismatophyllum in the Traverse Group of north central Michigan. M, 1942, Michigan State University. 72 p.

Faulhaber, John Jacob. Late Mississippian (late Osagian through Chesterian) conodonts from the Peratrovich Formation, southeastern Alaska. M, 1977, University of Oregon. 173 p.

Faulk, Kenneth L. Distribution and controls of Ostracoda within Coupon Bight (Recent), Big Pine Key, Florida. M, 1972, Ohio University, Athens. 51 p.

Faulk, Niles. Green River Formation, Utah. M, 1948, Ohio State University.

Faulkender, DeWayne J. Source of sand for An Nafud sand sea, Kingdom of Saudi Arabia. M, 1986, Kansas State University. 87 p.

Faulkner, Barry M. Chemical quality of the surface water and sediment of the Kentucky River. M, 1976, Eastern Kentucky University. 87 p.

Faulkner, Edward Leslie. The colourimetric determination of trace elements in pyrrhotite. M, 1960, University of Saskatchewan. 71 p.

Faulkner, Edward Leslie. The distribution of cobalt and nickel in some sulphide deposits of the Flin Flon area, Saskatchewan. D, 1964, University of Saskatchewan. 229 p.

Faulkner, Glen L. Geology of the Bessemer Mountain-Oil Mountain area, Natrona County, Wyoming. M, 1950, University of Wyoming. 63 p.

Faulkner, John. The effect on the choices of lapse time window on coda Q. M, 1988, University of Southern California.

Faure, François M. Thermodynamics and vaporization of sulfides of antimony and arsenic. D, 1972, Stanford University.

Faure, Gunter. The Sr^{87}/Sr^{86} ratio in the oceanic and continental basalts and the origin of igneous rocks. D, 1961, Massachusetts Institute of Technology. 176 p.

Fauria, Thomas. Crustal structure of the northern Basin and Range and western Snake River plain; a ray-trace travel-time interpretation of the Eureka, Nevada, to Boise, Idaho, seismic refraction profile. M, 1981, Purdue University.

Fausak, Leland Edward. The beach water table as a response variable of the beach-ocean-atmosphere system. M, 1970, University of Virginia. 53 p.

Fauser, Walter Bernard, Jr. The paleontology and stratigraphy of a well core penetrating Upper, Middle, Lower Sulurian, and Upper Ordovician rocks from Newaygo County, Michigan. M, 1951, University of Michigan.

Fausett, Robert Julian. An interdisciplinary approach to resource inventory. D, 1982, University of Wisconsin-Madison. 419 p.

Fausey, William Robert. The crystal structure and iron content of wurtzite polytypes 4H, 6H 15R. M, 1981, Pennsylvania State University, University Park. 68 p.

Faust, Charles Russell. Numerical simulation of fluid flow and energy transport in liquid- and vapor-dominated hydrothermal systems. D, 1976, Pennsylvania State University, University Park. 181 p.

Faust, George Tobias. The fusion relations of iron-orthoclase, with a discussion of the evidence for the existence of an iron-orthoclase molecule in feldspars. D, 1934, University of Michigan.

Faust, Michael Jess. Seismic stratigraphy of the Middle Cretaceous unconformity (MCU) in the central Gulf of Mexico Basin. M, 1984, University of Texas, Austin. 164 p.

Faust, Nicholas L. Analysis of the usefulness of automatically processed ERTS multispectral data for geologic purposes in Georgia. M, 1976, Georgia Institute of Technology. 83 p.

Faust, Robert J. X-ray study of lecontite. M, 1962, Southern Illinois University, Carbondale. 12 p.

Faustini, John M. Delineation of groundwater flow patterns in a portion of the Central Sand Plain of Wisconsin. M, 1985, University of Wisconsin-Madison. 117 p.

Faustino, Leopoldo Aldo. Recent and fossil corals from the Philippine Islands in the Stanford University collection. M, 1922, [Stanford University].

Faustino, Leopoldo Aldo. Recent Madreporaria of the Philippine Islands. D, 1924, Stanford University. 310 p.

Faustman, Walter Francis. Paleontology of the Scotia and Centerville Beach, California, sections of the Wildcat Group. M, 1962, University of California, Berkeley. 176 p.

Fauth, John Louis. Geology of South Mountain, northwestern Adams County, Pennsylvania. M, 1962, Pennsylvania State University, University Park. 103 p.

Fauth, John Louis. Geology of the Caledonia Park area, South Mountain, Pennsylvania. D, 1967, Pennsylvania State University, University Park. 276 p.

Fauw, Sherri Lynn De *see* De Fauw, Sherri Lynn

Fava, James Archie. A magnetic survey of a portion of the American Bottoms near Mitchell, Illinois. M, 1952, St. Louis University.

Favorite, Felix. Geostrophic and Sverdrup transports as indices of flow in the Gulf of Alaska. D, 1969, Oregon State University. 74 p.

Faw, Dorothea M. Formation evaluation of Springer Britt and lower Cunningham sandstones in parts of Caddo and Grady counties, Oklahoma. M, 1988, Wright State University. 112 p.

Faw, Jeffrey W. Data processing and interpretation of a crooked seismic line in the Michigan Basin; Lake County, Michigan. M, 1988, Wright State University. 116 p.

Fawcett, John Alan. I; Three dimensional ray-tracing and ray-inversion in layered media; II, Inverse scattering and curved ray tomography with applications to seismology. D, 1983, California Institute of Technology. 236 p.

Fawley, A. P. An electrum-ruby silver deposit. M, 1946, Queen's University. 75 p.

Fawley, Allan Priest. Geology and iron-formation of the Petitsikapau area, Labrador. D, 1948, University of California, Los Angeles.

Fawzy, Aly Mahmoud. Elements of fabric-reinforced soil behavior. D, 1979, Oklahoma State University. 100 p.

Faxon, Michael F. Paleomagnetic direction of Sanpoil Volcanics in the southern Republic Graben and along the Spokane River. M, 1982, Western Washington University. 106 p.

Fay, Albert Hill. Copper mining at Cananea, Sonora, Mexico. M, 1905, University of Missouri, Rolla.

Fay, Donald A. A structural analysis of the Crystal Peak Quadrangle, Gros Ventre Mountains, Wyoming. M, 1986, Miami University (Ohio). 74 p.

Fay, Ignatius Charles. Paleoecology of Lower Silurian bioherms, Manitoulin Island, Ontario. D, 1983, University of Saskatchewan. 396 p.

Fay, Jan E. A consideration of mineral cleavage. M, 1975, Boston University. 112 p.

Fay, Leslie Porter. Late Wisconsinan Appalachian herpetofaunas; stability in the midst of change. D, 1984, Michigan State University. 69 p.

Fay, Leslie Porter. Mammals of the Garrett Farm and Pleasant Ridge local biotas (Holocene), Mills County, Iowa. M, 1978, University of Iowa. 54 p.

Fay, Louis Elwyn. An analytical study of the apparent spatial trend in the groundwater dispersion coefficient. D, 1983, Clemson University. 95 p.

Fay, Robert Lawrence. The influence of cleavage on the size and shape of mineral grains during abrasion. M, 1948, University of Rochester. 48 p.

Fay, Robert Oran. The Blaine and related formations of northwestern Oklahoma and Kansas. D, 1961, University of Kansas. 474 p.

Fay, Vincent Kevin. Lower Devonian volcanic and sedimentary rocks of the Eastport Formation, Southwest New Brunswick. M, 1988, University of New Brunswick. 167 p.

Fay, William Martin. Geologic and geochemical studies of portions of the Lincolnton and Plum Branch quadrangles, Georgia and South Carolina. M, 1980, University of Georgia.

Fayer, Michael James. Shallow water table fluctuations and air encapsulation. D, 1984, University of Massachusetts. 113 p.

Fayyaz, Mohammad S. Ground water resources of the Boulder Canyon area, Lawrence and Meade counties, South Dakota. M, 1978, South Dakota School of Mines & Technology.

Faz, Jorge J. Influence of basement paleotectonics on Mesozoic sedimentation patterns in the San Juan Basin, northwestern New Mexico. M, 1984, Purdue University. 162 p.

Fazzani, Ashour El *see* El Fazzani, Ashour

Feakes, Carolyn Ruth. Recognition and chemical characterization of two Ordovician fossil soils and the advent of large organisms on land. M, 1985, University of Oregon. 120 p.

Fears, Fulton K. Determination of pore size of aggregate from Indiana limestones. M, 1950, Purdue University.

Feast, Charles Frederick. The structural history of two klippen on the west flank of the Pavant Mountains, West central Utah. M, 1979, University of Florida. 60 p.

Feather, Ralph Merle, Jr. Geology of the New Enterprises 7/12 minute Quadrangle, Bedford County, Pennsylvania. M, 1974, Indiana University of Pennsylvania.

Featherstone, Raymond Paul. Biostratigraphy of the Middle Devonian Slave Point and adjoining formations, northern Alberta, Canada. M, 1982, University of Saskatchewan. 196 p.

Feazel, C. T. Ecozonation and sediment distribution of three reef areas on St. John, U. S. Virgin Islands. D, 1975, The Johns Hopkins University. 239 p.

Febres Cedillo, Hector Enrique. Prediction of the maximum response of structures subjected to random earthquake excitation. D, 1987, University of Iowa. 100 p.

Febres-Cordero, E. E. Influence of testing conditions on creep behavior of clays. D, 1974, University of Illinois, Urbana. 233 p.

Fechner, Steven A. Petrology and structure of Precambrian Sherman Granite in the Boswell Creek area, Albany County, Wyoming. M, 1979, Colorado State University. 115 p.

Feden, Robert Henry. A petrofabric analysis of some carbonate rocks from the Chester-Whitemarsh Valley. M, 1963, Bryn Mawr College. 24 p.

Feder, Allen M. Radar geology. M, 1960, SUNY at Buffalo.

Feder, Gerald. The geology of the Oak Grove area, Jefferson County, Tennessee. M, 1963, University of Tennessee, Knoxville. 30 p.

Feder, Gerald Leon. A conceptual model of the hydrologic system supplying the large springs in the Ozarks. D, 1973, University of Missouri, Columbia.

Federici, James M. Statistical design of hydrologic data networks. M, 1977, University of Nevada. 77 p.

Federman, Alan Neil. Correlation and petrological interpretation of abyssal and terrestrial tephra layers. D, 1985, Oregon State University. 241 p.

Fedewa, William. Stratigraphy and phosphate resources of the Murdock Mountain area, Elko County, Nevada. M, 1980, San Jose State University. 98 p.

Fedikow, Mark A. F. Geochemical and paleomagnetic studies at the Sullivan Mine, Kimberley, British Columbia. M, 1978, University of Windsor. 180 p.

Fedor, Dennis George. Geology for planning in Bristol and Oswego townships, Kendall County, Illinois. M, 1971, Northern Illinois University.

Fedors, Randall W. Petrology and garnet geochemistry of two calcic copper skarns, central Mexico. M, 1983, University of Minnesota, Minneapolis. 122 p.

Fedoruk, Richard Alexander. Upper Jurassic and Lower Cretaceous stratigraphy along Cache Creek, Northwest Territories. M, 1980, University of Alberta. 123 p.

Fedosh, Michael Stephen. Lower Chesapeake Bay surface turbidity variations as detected from Landsat images. M, 1984, College of William and Mary. 208 p.

Fee, David Wayne. Lithofacies and depositional environments of the Weno and Pawpaw formations (Lower Cretaceous) of north-central Texas and central Oklahoma. M, 1974, University of Texas, Arlington.

Feeley, Mary Hart. Seismic stratigraphic analysis of the Mississippi fan. D, 1984, Texas A&M University. 247 p.

Feeley, Mary Hart. Structural and depositional relationships of intraslope basins, northern Gulf of Mexico. M, 1982, Texas A&M University. 154 p.

Feely, Herbert W. Lamont natural radiocarbon measurements, I. M, 1952, University of Oregon.

Feely, Herbert W. Origin of Gulf Coast salt-dome sulphur deposits. D, 1958, Columbia University, Teachers College.

Feely, Richard Alan. Chemical characterization of the particulate matter in the near bottom nepheloid layer of the Gulf of Mexico. D, 1974, Texas A&M University.

Feely, Richard Alan. The distribution of particulate aluminum in the Gulf of Mexico. M, 1971, Texas A&M University.

Feeney, J. W. Paleomagnetic evidence for sea floor spreading in the Murray fracture zone. M, 1968, University of Hawaii. 65 p.

Feeney, Thomas Aquinas. A deuterium-calibrated groundwater flow model of western Nevada Test Site and vicinity. M, 1987, University of Nevada. 56 p.

Feenstra, B. H. Late Wisconsin stratigraphy in the northern part of the Stratford-Conestogo area, southern Ontario. M, 1975, University of Western Ontario. 233 p.

Feenstra, Roger Ernest. Evolution of folds in the Blaylock Formation (Silurian), Ouachita Mountains, southeastern Oklahoma. M, 1974, University of Oklahoma. 77 p.

Feenstra, Stanley. The isotopic evolution of sulfate in a shallow groundwater flow system on the Canadian Shield. M, 1980, University of Waterloo.

Fehler, Michael. Seismological investigation of the mechanical properties of a hot dry rock geothermal system. D, 1979, Massachusetts Institute of Technology. 344 p.

Feibel, Craig Stratton. Paleoenvironments of the Koobi Fora Formation, Turkana Basin, northern Kenya. D, 1988, University of Utah. 337 p.

Feibel, Craig Stratton. Stratigraphy and paleoenvironments of the Koobi Fora Formation along the western Koobi Fora Ridge, East Turkana, Kenya. M, 1983, Iowa State University of Science and Technology. 104 p.

Feichtinger, John Rudolph. A geographic study of the city of Coos Bay and its hinterland (Oregon). M, 1950, University of Oregon. 198 p.

Feichtinger, Sylvia H. Geology of a portion of the Norris Quadrangle with emphasis on Tertiary sediments, Madison and Gallatin counties, Montana. M, 1970, Montana State University. 85 p.

Feiereisen, Joseph John. Geomorphology, alluvial stratigraphy, and sediments; lower Siuslaw and Alsea river valleys, Oregon. D, 1981, University of Oregon. 303 p.

Feigenson, Mark D. The petrology and geochemistry of the Loma de Cabrera Batholith of the western Dominican Republic. M, 1978, George Washington University.

Feigenson, Mark Daniel. Aspects of the petrology of oceanic basalts; I, Geochemistry of Kohala Volcano, Hawaii; II, Petrochemistry of basalts associated with manganese ores from California; III, Rheology of subliquidus basalts and other silicate melts. D, 1982, Princeton University. 212 p.

Feigl, Frederick J. The northernmost limit of living foraminifera in the Hudson River. M, 1956, New York University.

Feijtel, Tom Cornelis Jan. Biogeochemical cycling of metals in Barataria Basin. D, 1986, Louisiana State University. 303 p.

Fein, C. D. Some trace elements in lavas from the Lau Islands, Tofua, Tonga, and Tutuila, American Samoa. D, 1971, University of Hawaii. 97 p.

Fein, Matthew. Engineering geology investigation of an underground limestone mine, Indianapolis, Indiana. M, 1983, Purdue University. 175 p.

Fein, Michael Neal. Bivens' Arm, Gainesville, Florida; a biogenic source of sulfur to the atmosphere. M, 1975, University of Florida. 45 p.

Feinberg, Herbert. Geology of the central portion of the Sandia granite, Sandia Mountains, (Precambrian), Bernalillo County, New Mexico. M, 1969, University of New Mexico. 127 p.

Feinberg, Paul M. A study of the water quality of the Neponset River, Massachusetts. M, 1987, Boston University. 103 p.

Feininger, Tomas. Petrology of the Ashaway and Voluntown quadrangles, Connecticut-Rhode Island. D, 1964, Brown University. 232 p.

Feininger, Tomas. Surficial geology of the Hope Valley Quadrangle, Rhode Island. M, 1960, Brown University.

Feinstein, Daniel T. A three-dimensional model of flow to the sandstone aquifer in northeastern Wisconsin with discussion of contamination potential. M, 1986, University of Wisconsin-Madison.

Feinstein, Shimon. Subsidence and thermal history of the Southern Oklahoma Aulacogen. M, 1979, University of Oklahoma. 84 p.

Feirn, W. C. The geology of the early Precambrian rocks of the Jasper Lake area, Cook County, northeastern Minnesota. M, 1977, University of Minnesota, Duluth.

Feiss, Julian William. The geology and ore deposits of Hiltano Camp, Arizona. M, 1929, University of Arizona.

Feiss, Paul Geoffrey. Arsenic and antimony distribution between the tennantite-tetrahedrite and enargite-famatinite series. D, 1971, Harvard University.

Feizna, Sadat. Microfacies and depositional environment of the Glen Dean Formation (Upper Mississippian) southwestern Illinois. M, 1981, University of Illinois, Urbana. 68 p.

Feiznia, Sadat. Depositional and diagenetic environments of carbonate-siliciclastic rocks of the Glen Dean Formation (Upper Mississippian), Illinois Basin, U.S.A. D, 1983, University of Illinois, Urbana. 219 p.

Feizpour, Ali A. Energy dissipation of Rayleigh waves due to absorption along the path of the use of finite-element method. D, 1979, Southern Methodist University. 152 p.

Fekete, Steven Ralph. Petrology and diagenesis of Pennsylvanian arkosic sandstones, Taos Trough, northern New Mexico. M, 1988, University of Texas at Dallas. 173 p.

Fekete, Thomas E. The sedimentology and stratigraphy of the Grayburg Formation and its associated erosion surface along the high western escarpment of the Guadalupe Mountains, Texas. M, 1986, University of Wisconsin-Madison. 174 p.

Felber, B. E. The habit and habitat of pyrite. M, 1955, Columbia University, Teachers College.

Felber, Bernard E. Silurian reefs of southeastern Michigan. D, 1964, Northwestern University. 133 p.

Felch, Roger N. A study of the seismic crustal structure in the Valles Caldera region of northern New Mexico. D, 1987, Pennsylvania State University, University Park. 271 p.

Felch, Roger N. A three-dimensional gravity model of basin structure, Yucca Flat, Nevada. M, 1979, Texas Christian University. 92 p.

Felder, F. Trace element dispersion patterns associated with lead-zinc mineralization in northwestern Spain. M, 1972, University of New Brunswick.

Felder, Wilson Norfleat. Barrier island morphology at Rodanthe, North Carolina. M, 1973, University of Virginia. 33 p.

Felder, Wilson Norfleat. Simulation modeling of offshore bars. D, 1978, University of Virginia. 110 p.

Feldhausen, Peter Homer. Ordination of Recent microorganisms from the Cape Hatteras (North Carolina) continental margin. M, 1967, University of Wisconsin-Madison.

Feldman, Arlen D. Chemical quality of water in relation to water use and basin characteristics, Tucson Basin, Arizona. M, 1966, University of Arizona.

Feldman, Howard Randall. Paleontology and paleoecology of the Somerset Shale Member of the Salem Limestone (Mississippian) in central Kentucky. M, 1984, Indiana University, Bloomington. 114 p.

Feldman, Howard Randall. Spatial distribution and taphonomy of fauna and paleoenvironmental parameters of the Waldron Shale (Silurian) in southeastern Indiana. D, 1987, Indiana University, Bloomington. 249 p.

Feldman, Howard Ross. Brachiopods and community ecology of the Onondaga Limestone. D, 1978, Rutgers, The State University, New Brunswick. 203 p.

Feldman, Lawrence. Ground water resources of Chelmsford, Massachusetts. D, 1978, Boston University. 376 p.

Feldman, Mark David. Hydrogen isotope geochemistry of trace water in sedimentary dolomite. M, 1984, Arizona State University. 64 p.

Feldman, Sandra C. Investigation of geology and hydrothermal alteration using detailed field mapping and high-resolution remote sensing data; Hot Creek Range, Nevada. D, 1988, University of Nevada. 362 p.

Feldman, Sandra C. Surface energies in the CaSO₄H₂O system and crystal properties of gypsum as a function of super-saturation. M, 1970, University of New Mexico. 89 p.

Feldmann, Rodney M. Bivalvia and paleoecology of the Fox Hills Formation (Upper Cretaceous) of North Dakota. D, 1967, University of North Dakota. 383 p.

Feldmann, Rodney M. Pelecypoda from lower Fox Hills Formation (Upper Cretaceous) of Emmons County, North Dakota. M, 1963, University of North Dakota. 75 p.

Feldt, Ann Elizabeth von see von Feldt, Ann Elizabeth

Feliciano, Jose Maria. The relation of concretions to coal seams. D, 1923, University of Chicago. 112 p.

Felix, Charles Jeffrey. A study of the arborescent lycopods of southeastern Kansas. M, 1952, Washington University.

Felix, David W. Origin and recent history of Newport Submarine Canyon, California continental borderland. M, 1969, University of Southern California.

Felkey, Jack R. Three analytical chemistry projects; construction and characterization of a gas chromatograph-mass spectrometer interface; preparation of low-level sulfur gas standards; and comparison of respirable volcanic ash measurement data of current threshold limit valves. M, 1982, University of Idaho. 79 p.

Fellers, Thomas J. SEM study of the flocculation, settling behavior and microtextures of phosphatic clay slimes. M, 1978, Florida State University.

Felling, Richard A. Geology and mineralization of the Mount Evans area, Deer Lodge County, Montana. M, 1985, University of Colorado at Colorado Springs. 159 p.

Fellowes, Terrence Leigh. Geology of Hot Springs National Park and vicinity, central Arkansas. M, 1968, Southern Methodist University. 82 p.

Fellows, Jack D. A geographic information system for regional hydrologic modeling. D, 1983, University of Maryland. 298 p.

Fellows, Larry Dean. Geology of the western Windingstair Range, Latimer and LeFlore counties, Oklahoma. D, 1963, University of Wisconsin-Madison. 181 p.

Fellows, Larry Dean. Stratigraphy and paleontology of a core of Upper Silurian strata from Monroe County, Michigan. M, 1957, University of Michigan.

Fellows, Michael Lewis. Composition of epidote from porphyry copper deposits. M, 1976, University of Arizona.

Fellows, Ralph Harold, Jr. Experiments in the formation of dessication cracks in sediments. M, 1951, Southern Methodist University. 67 p.

Fellows, Ralph Sanborn. The influence of climatic factors on the geomorphology of a part of the coastal plain of northern Alaska. D, 1959, Boston University. 180 p.

Fellows, Robert E. Recrystallization and flowage in Appalachian quartzites. D, 1943, The Johns Hopkins University.

Fellows, Robert Ellsworth. Petrology of Pre-cambrian rocks from the east slope of the Ten-Mile Range, Colorado. M, 1941, University of Rochester. 67 p.

Fellows, Steven Neal. Petrology of a portion of a drill core from the Duluth Complex near Babbitt, Minnesota. M, 1976, Cornell University.

Felmlee, Judith K. Geologic structure along the Huronian-Keweenawan contact (Precambrian), Mellen, Wisconsin. M, 1970, University of Wisconsin-Madison.

Felsher, Murray. Beach studies on the outer beaches of Cape Cod, Massachusetts. M, 1963, University of Massachusetts. 270 p.

Felsher, Murray. Physical sedimentology and bathymetry, Santa Cruz submarine-canyon complex, continental borderland, California. D, 1971, University of Texas, Austin.

Felt, Vince. Geology of the Antelope Peak area of the southern San Francisco Mountains, Beaver County, Utah. M, 1980, Brigham Young University.

Felton, Richard M. Lower Ordovician conodonts of the lower West Spring Creek Formation, Arbuckle Mountains, Arkansas. M, 1979, University of Missouri, Columbia.

Felts, Wayne M. The geology of the Lebanon Quadrangle, Oregon. M, 1936, Oregon State University. 83 p.

Felts, Wayne Moore. A granodiorite stock in the Cascade Mountains of southwestern Washington. D, 1938, University of Cincinnati. 40 p.

Felty, Kent K. Stratigraphy of the Silurian and Devonian in western Marion and southern Nelson counties. M, 1962, University of Kentucky. 41 p.

Fenaish, Taher Ali. Numerical modeling of wave up-rush and induced dune erosion. D, 1988, North Carolina State University. 143 p.

Fender, Hollis Blair. The Mt. Carmel Fault of Indiana. M, 1949, Indiana University, Bloomington. 22 p.

Fendinger, Nicholas Joseph. Chemical characterization of organic components in leachates from coal. D, 1987, University of Maryland. 243 p.

Fenelon, Bernard G. Potential uses of induced polarization in groundwater investigations. M, 1987, University of Wisconsin-Milwaukee. 94 p.

Fenelon, Joseph Martin. Glacial geology of the Cramer Quadrangle, northeastern Minnesota. M, 1986, University of Wisconsin-Milwaukee. 76 p.

Feng, Bing-Cheng. Floristic and ecological significance of coal balls from late Middle Pennsylvanian strata of western Pennsylvania, U.S.A. D, 1987, Michigan State University. 205 p.

Feng, Chi-Chin. A surface wave study of crustal and upper mantle structures of Eurasia. D, 1982, University of Southern California.

Feng, Wei-Lin. A model study of three-dimensional slope stability. M, 1988, University of Illinois, Chicago.

Fengler, Timothy A. A petrological investigation of older mafic rocks in the Beartooth Mountains, Wyoming and Montana. M, 1983, Northern Illinois University. 136 p.

Feniak, Michael Walter. Grain sizes of various minerals in igneous rocks. M, 1943, University of Minnesota, Minneapolis. 21 p.

Feniak, Michael Walter. The geology of Dowdell Peninsula, Great Bear Lake, Northwest Territories. D, 1947, University of Minnesota, Minneapolis. 94 p.

Feniak, Oliver William. The upper Franconia of southeastern Minnesota. M, 1948, University of Minnesota, Minneapolis. 162 p.

Fenk, Edward Michael. Sedimentology and stratigraphy of middle and upper Eocene carbonate rocks, Lake, Hernando and Levy counties, Florida. M, 1979, University of Florida. 133 p.

Fenn, Philip Michael. Nucleation and growth of alkali feldspars from melts in the system NaAlSi₃O₈-KalSi₃O₈-H₂O. D, 1973, Stanford University. 167 p.

Fenne, Frank Karl. Geology and mineral deposits of the central part of the Parker Mountain mining district, Lemhi County, Idaho. M, 1977, University of Idaho. 82 p.

Fennell, Edward L. The relation of gravity to structural geology and hydrological features in parts of Gadsden, Leon, and Wakulla counties, Florida. M, 1969, Florida State University.

Fenneman, Nevin Melancthon. Development of the profile of equilibrium of the subaqueous shore terrace. D, 1902, University of Chicago. 178 p.

Fenner, Clarence N. Features indicative of physiographic conditions prevailing at the time of the trap extrusions in New Jersey. M, 1909, Columbia University, Teachers College.

Fenner, Clarence N. The Watchung Basalt and the paragenesis of its zeolites and other secondary minerals. D, 1911, Columbia University, Teachers College.

Fenner, Frederick Donald. The distribution and geochemistry of iridium in river suspended material and marine sediments. M, 1983, Texas A&M University. 63 p.

Fenner, Linda B. The hydrologic factors that control carbonate hardness in Kalamazoo County lakes, southwestern Michigan. M, 1981, Western Michigan University. 76 p.

Fenner, Peter. Stratigraphy and petrology of the lower Sundance Formation (Upper Jurassic) on the flanks of the Big Horn Mountains, Wyoming. M, 1961, University of Illinois, Urbana. 38 p.

Fenner, Peter. Variations in the mineralogy and trace elements, Esopus Formation, Kingston, New York. D, 1963, University of Illinois, Urbana. 93 p.

Fenner, William E. Foraminiferids of the Cannonball Formation (Paleocene, Danian) in western North Dakota. D, 1976, University of North Dakota. 216 p.

Fenner, William E. The foraminiferids of the Cannonball Formation (Paleocene, Danian) and their paleoenvironmental signficance; Grant, Morton and Oliver counties, North Dakota. M, 1974, University of North Dakota. 206 p.

Fenno, Dan P. Chemical characteristics of ground water in a portion of Greene County, Ohio. M, 1987, Wright State University. 109 p.

Fenoglio, Anthony F. Geology of northeastern Payne County, Oklahoma. M, 1957, University of Oklahoma. 84 p.

Fenske, John M., Jr. A depositional model for middle Mesaverde coals, Yampa Field, northwestern Colorado. M, 1985, University of Kentucky.

Fenske, Paul Roderick. Origin of porosity in the oil-producing reefs. M, 1951, University of Michigan.

Fenske, Paul Roderick. The origin and significance of concretions. D, 1963, University of Colorado. 205 p.

Fensome, Robert Allan. Dinoflagellate cysts and acritarchs from the Middle and Upper Jurassic of Jameson Land, East Greenland. M, 1977, University of Saskatchewan. 271 p.

Fensome, Robert Allan. Miospores from the Jurassic-Cretaceous boundary beds, Aklavik Range, Northwest Territories, Canada. D, 1983, University of Saskatchewan. 762 p.

Fenster, D. F. Structural geology and petrology of the Hudson Highlands near Brewster, New York. M, 1975, Queens College (CUNY). 105 p.

Fenster, Eugene Joel. Quantification of microevolutionary patterns; multivariate rates and patterns of evolution in the Neogene diatom lineage. D, 1988, City College (CUNY). 291 p.

Fenton, Carroll L. The Spirifer orestes phylum; its evolution and ecology. D, 1926, University of Chicago. 151 p.

Fenton, M. M. The Quaternary stratigraphy of a portion of southeastern Manitoba, Canada. D, 1974, University of Western Ontario. 286 p.

Fenton, Mark Macdonald. The Quaternary stratigraphy of the Assiniboine River to Lake Manitoba area, Manitoba. M, 1970, University of Manitoba.

Fenton, Michael Dwight. The evolution of the isotope composition of terrestrial strontium. D, 1969, Ohio State University. 180 p.

Fenton, Michael Dwight. The geology of parts of Mad Creek, Clark, Floyd Peak quadrangles of Routt County, Colorado. M, 1965, University of Wyoming. 39 p.

Fenton, Robert Leo. The syntactic log as a tool for correlating basalt stratigraphy. M, 1974, Washington State University. 59 p.

Fenton, Scott Bruce. Geology of the Bonanza King Formation (Cambrian) at the Desert Range, Clark County, Nevada. M, 1980, San Diego State University.

Fenton, Thomas Eugene. Soils, weathering zones, and landscapes in the upland loess of Tama and Grundy counties, Iowa. D, 1966, Iowa State University of Science and Technology. 340 p.

Fenves, Gregory Louis. Earthquake response of concrete gravity dams. D, 1984, University of California, Berkeley. 227 p.

Fenwick, Donald Kenneth Bruce. Geophysical studies of the continental margin northeast of Newfoundland. M, 1967, [Dalhousie University]. 68 p.

Fenwick, Kenneth George. Origin of a stratabound pyrite deposit in predominantly volcanic derived strata in the Finlayson Lake area, District of Rainy River, Ontario. M, 1971, Michigan Technological University. 90 p.

Fenwick, Willis H. Two neglected factors in the interpretation of magnetic anomalies. M, 1938, Colorado School of Mines. 36 p.

Fenzel, Frank Walker. Faulting in the Dixon Springs area, Pope County, Illinois. M, 1962, Southern Illinois University, Carbondale. 31 p.

Feragen, Edward Sebastian. Geology of the southeastern San Felipe Hills, Imperial Valley, California; with emphasis on the geometry of structural fabrics in the Borrego Formation. M, 1986, San Diego State University. 144 p.

Feray, Daniel E. Foraminifera of the Weches Formation on Colorado River, Smithville, Bastrop County, Texas (F372). M, 1940, University of Illinois, Urbana. 66 p.

Feray, Daniel E. Relation of the foraminifera to the sedimentary characteristics of the Weches Formation in Texas. D, 1948, University of Wisconsin-Madison.

Ferber, Charles Thomas. Environmental interpretation of fish deposits in the Eocene Green River Formation of Utah and Wyoming. M, 1987, Kent State University, Kent. 206 p.

Ferber, Daniel. Structural geology of the southwest portion of the Lonsdale Quadrangle, Ouachita Mountains, Arkansas. M, 1979, University of Missouri, Columbia.

Ferber, Robin J. Depositional and diagenetic history of the Sunniland Formation (Lower Cretaceous), Lehigh Park Field, Lee County, Florida. M, 1984, University of Southwestern Louisiana. 351 p.

Ferderer, Robert J. Gravity and magnetic modeling of the southern half of the Duluth Complex, northeastern Minnesota. M, 1982, Indiana University, Bloomington. 99 p.

Ferderer, Robert Joel, Jr. Werner deconvolution and its application to the Penokean Orogen, east-central Minnesota. D, 1988, University of Minnesota, Minneapolis. 313 p.

Ferdock, Gregory Christopher. Geology of the Menan volcanic complex and related volcanic features, northeastern Snake River plain, Idaho. M, 1987, Idaho State University. 171 p.

Ferebee, T. W., Jr. Resistance and spontaneous potential measurements over Heald Bank, Texas. M, 1974, Texas A&M University.

Ferebee, T. W., Jr. Sedimentation in the Mississippi Trough. D, 1978, Texas A&M University. 191 p.

Ferek, Ronald John. A study of aerosol acidity over the northeastern United States. D, 1982, Florida State University. 164 p.

Ferens, Mary C. The impact of mercuric ions on benthos and periphyton of artificial streams. D, 1974, University of Georgia. 106 p.

Fergus, John Howard, Jr. The polymorphism of dicalcium silicate. D, 1978, Lehigh University. 124 p.

Ferguson, Angus J. Late Quaternary geology of the upper Elk Valley, British Columbia. M, 1978, University of Calgary. 118 p.

Ferguson, Carl Council. Stratigraphy of the Upper Mississippian of the Beans Creek area, Tennessee. M, 1965, Vanderbilt University.

Ferguson, Charles A. Geology of the east-central San Mateo Mountains, Socorro County, New Mexico. M, 1985, New Mexico Institute of Mining and Technology. 118 p.

Ferguson, David Bryan. Subsurface geology of northern Lincoln County, Oklahoma. M, 1962, University of Oklahoma. 48 p.

Ferguson, Floyd Jay. Sedimentation analysis of Red River channel sands collected between Fulton, Arkansas, and Montgomery, Louisiana. M, 1962, University of Southwestern Louisiana.

Ferguson, G. Scott. Sedimentology of the Wapiabi Formation and equivalents (Upper Cretaceous), central and northern foothills, Alberta. M, 1984, McMaster University. 255 p.

Ferguson, Glenn C. A correlation of beds in and around San Diego with certain beds in the Los Angeles Basin, California. M, 1933, [University of Southern California].

Ferguson, Henry G. Geology and ore deposits of the Manhattan District, Nevada. D, 1924, Yale University.

Ferguson, Herman H. Middle Trenton insoluble residues. M, 1940, Vanderbilt University.

Ferguson, Hershal C., Jr. Subsurface geology of the Turtle Bayou Complex, Terrebonne Parish, Louisiana. M, 1961, Louisiana State University.

Ferguson, James Ardon. Fission track and K-Ar dates of the northeastern border zone of the Idaho Batholith (Cretaceous, Idaho). M, 1972, University of Montana. 32 p.

Ferguson, Jerry Duane. The subsurface alteration and mineralization of Permian red beds overlying several oil fields in southern Oklahoma. M, 1977, Oklahoma State University. 95 p.

Ferguson, John. A study of metamorphic strata near Fort Chimo, Quebec. M, 1958, McGill University.

Ferguson, John Alexander. Petrology and properties of certain clays from Queensland, Australia. D, 1950, University of Illinois, Urbana.

Ferguson, John David. Structure of Porvenir area, Trans-Pecos, Texas. M, 1959, University of Texas, Austin.

Ferguson, John F. A theoretical investigation of ground motion due to coastal plain earthquakes. M, 1975, University of North Carolina, Chapel Hill. 80 p.

Ferguson, John Franklin. Geophysical investigations of Yucca Flat, Nevada. D, 1981, Southern Methodist University. 102 p.

Ferguson, Kevin William. Human activity and slope contribution of sediments to an estuarine basin; case study, North River, Massachusetts. D, 1982, Clark University. 156 p.

Ferguson, Kurt Mathew. Geochemistry and petrogenesis of the Harrat Kishb volcanic field and comparison to worldwide alkali basalt fields. M, 1984, Arizona State University. 102 p.

Ferguson, L. J. Petrogenesis of the analcime-bearing and associated volcanic rocks of the Crowsnest Pass area, Alberta. M, 1977, University of Western Ontario. 232 p.

Ferguson, Lori J. Groundwater-lake interactions, Wood Lake, Benson County, North Dakota. M, 1984, University of North Dakota. 425 p.

Ferguson, Luther Short. Oil fields of Sullivan County, Indiana. M, 1922, Indiana University, Bloomington.

Ferguson, Pamela. Stratigraphy of the Ohio Shale in the Bellefontaine Outlier. M, 1978, [University of Toledo].

Ferguson, Richard R. An investigation of lake level fluctuations for the Minneapolis chain of lakes. M, 1981, University of Minnesota, Duluth.

Ferguson, Robert B. A contribution to the mineralogy of the aluminum fluorides of Greenland. D, 1948, University of Toronto.

Ferguson, Robert B. Muscovite from Mattawa Township, Ontario. M, 1943, University of Toronto.

Ferguson, Robert Clark. Petrography, depositional environments, and diagenesis of Bisbee Group carbonates, Guadalupe Canyon area, Arizona. M, 1983, University of Arizona. 91 p.

Ferguson, Scott D. A gravity and magnetic investigation of the Snowcrest Range region, Beaverhead and Madison counties, Montana. M, 1988, Montana College of Mineral Science & Technology. 74 p.

Ferguson, Stewart Alexander. Banded iron formation of the Timagami area, Ontario. M, 1942, University of Toronto.

Ferguson, Stewart Alexander. Ore deposits of the Kamiskotia area, Ontario. D, 1945, University of Toronto.

Ferguson, Sue Ann. The role of snowpack structure in avalanching. D, 1984, University of Washington. 150 p.

Ferguson, Tony Lee. Petrographic and trace element characterization of coarse crystalline carbonate minerals in fractures and breccia bodies and associated mineralization in the Right Fork area, central Tennessee. M, 1981, University of Tennessee, Knoxville. 81 p.

Ferguson, Walter Keene Linscott. Geology of parts of Bastrop and Fayette counties, Texas. M, 1958, University of Texas, Austin.

Fergusson, William Blake. Stratigraphic analysis of the Upper Devonian and Mississippian rocks between the La Salle Anticline and Cincinnati Arch (Ohio). D, 1965, University of Arizona. 108 p.

Ferland, Marie Ann. The stratigraphy and evolution of the southern New Jersey backbarrier region. M, 1985, Rutgers, The State University, New Brunswick. 200 p.

Ferm, John Charles. Cyclothems of the upper Allegheny and basal Conemaugh groups near Brookville, Jefferson County, Pennsylvania. M, 1948, Pennsylvania State University, University Park. 47 p.

Ferm, John Charles. Petrology of the Kittanning Formation near Brookville, Pennsylvania. D, 1957, Pennsylvania State University, University Park. 381 p.

Fermor, Peter Robin. Structural and stratigraphic analysis of rocks adjacent to the Lewis thrust fault around the Cate Creek and Haig Brook windows, British Columbia and Alberta. M, 1980, Queen's University. 134 p.

Fernald, Arthur Thomas. Geomorphology of the upper Kuskokwim region, Alaska. D, 1956, Harvard University.

Fernalld, Thomas. Structure and interstratification of glauconite. M, 1971, SUNY at Binghamton. 120 p.

Fernandes, Nuno Machado *see* Machado Fernandes, Nuno

Fernandes, Robert J. Design and construction of a PVT apparatus to study the behavior of complex hydrocarbon mixtures at elevated temperatures and pressures. M, 1949, University of Southern California.

Fernandez Casals, Javier. Analysis of a landslide along Interstate 70 near Vail, Colorado. M, 1986, Colorado School of Mines. 121 p.

Fernandez, Alfred Peter. Geology of a portion of the Capistrano Basin, California. M, 1960, University of California, Los Angeles.

Fernandez, Bonifacio. Stochastic modeling of periodic streamflow series with gamma distribution. D, 1984, Colorado State University. 239 p.

Fernandez, Carlos E. Experiment in two-dimensional model seismology; a report of research conducted for geophysics 400. M, 1962, Stanford University.

Fernandez, Henry E. Geology of Pride Mine and New Standard Mine area, northwestern Yuma County, Arizona. M, 1965, University of Missouri, Rolla.

Fernandez, J. A. Preliminary seismic zoning in the Tucson area, Arizona. M, 1978, University of Arizona.

Fernandez, Louis Anthony. Chemical petrology of the basaltic complex of Nordeste (Pleistocene to Recent vulcanism), Sao Miguel Island, Azores. D, 1969, Syracuse University. 168 p.

Fernandez, Louis Anthony. The petrography of the Faraway Ranch Formation, Chiricahua National Monument, Arizona. M, 1964, University of Tulsa. 117 p.

Fernandez, Louis Osvaldo. Sedimentation on the Cabo Rojo Shelf, southwestern Puerto Rico. M, 1978, North Carolina State University. 112 p.

Fernandez, Luis Maria. Factors affecting the signal-to-noise ratio of short period seismic records. M, 1964, St. Louis University.

Fernandez, Luis Maria. The determination of crustal thickness from the spectrum of P waves. D, 1966, St. Louis University. 216 p.

Fernandez, Manuel Nicolas. Engineering geology of landslides between Punta Salsipuedes and Punta San Miguel, Baja California, Mexico. M, 1977, University of Southern California.

Fernandez, N. C. The soils of the Molokai family. M, 1963, University of Hawaii. 46 p.

Fernandez, Ramon Norberto. Soil color measurements from reflectance spectra; applications to the study of iron oxide-soil color relationships. D, 1988, Purdue University. 173 p.

Fernandez, Réne L. Stratigraphy and clay mineralogy of Wisconsinan tills in the Cuyahoga Valley National Recreation Area, northeastern Ohio. M, 1983, University of Akron. 94 p.

Fernandez, Ricardo. Temporal variations of the electrical resistivity of the Earth's crust. D, 1981, University of California, Berkeley. 103 p.

Fernandez-Concha, Jaime. The geology and mercury deposits of the Terlingua District, Texas. M, 1944, The Johns Hopkins University.

Fernando, Angelo Ransirimal. Trace heavy metal determination in soil samples using differential pulse anodic stripping voltammetry. D, 1988, University of Alberta. 178 p.

Ferng, Yue-Lang. Wetland functional evaluation and management of Onondaga County, New York. D, 1988, SUNY, College at New Paltz. 280 p.

Fernow, Donald Lloyd. The geology and mineral deposits of the San Luis Range and Osos Valley, San Luis Obispo County, California. M, 1960, University of California, Los Angeles.

Fernow, Leonard Reynolds. Correlation of some limestone members of the (Devonian) Hamilton Group by microfacies (New York). M, 1957, Cornell University.

Fernow, Leonard Reynolds. Paleoecology of the Middle Devonian Hamilton Group in the Cayuga Lake region. D, 1961, Cornell University. 285 p.

Ferns, M. L. The petrology and petrography of the Wrangle Gap-Red Mountain ultramafic body, Klamath Mountains, Oregon. M, 1979, University of Oregon. 125 p.

Ferrall, C. C., Jr. Subsurface geology of Waikiki, Moiliili and Kakaako with engineering application. M, 1976, University of Hawaii. 168 p.

Ferrall, Charles C., Jr. Tectonic stress regime of the Cascades region and tectonic classification of large calderas. D, 1986, University of Hawaii. 421 p.

Ferrall, Kim Wallace. Stratigraphic distribution of Pennsylvanian Echinoidea of Ohio. M, 1974, Bowling Green State University. 78 p.

Ferraro, Thomas Edward. The geology of the coastal tract between Tarbat Ness and Rockfield, Cromarty District, Scotland. M, 1983, Oklahoma State University. 289 p.

Ferrate-Felice, L. A. Analisis y diseno ambiental del paisaje Guatemalteco; caso tipico, la evaluacion ecosistematica de la hoja topografica Totonicapan, a escala 1:50,000. D, 1977, University of Oregon. 329 p.

Ferraz, Celso Pinto. Potential market for bauxite deposits in the Amazon region, Brazil. M, 1975, Stanford University.

Ferreira, Alfredo Augusto Cunhal Goncalves *see* Goncalves Ferreira, Alfredo Augusto Cunhal

Ferreira, Justo Camejo. Thermal metamorphism of the organic matter in the Mancos Shale near Crested Butte, Colorado. M, 1973, Rice University. 81 p.

Ferreira, K. J. The mineralogy and geochemistry of the lower Tanco Pegmatite, Bernic Lake, Man., Canada. M, 1984, University of Manitoba.

Ferreira, Maria da Graca de Vasconcelos Xavier. Composition and origin of soil parent materials in Marathon County, north central Wisconsin. D, 1980, University of Wisconsin-Madison. 145 p.

Ferreira, William S. A physical comparison of subaerial and subaqueous eruptive environments in the Proterozoic Anbalysis Group, Sask. Canada. M, 1984, University of Manitoba.

Ferrell, Alton Durane. Stratigraphy of northern Sierra Pilares, Chihuahua, Mexico. M, 1958, University of Texas, Austin.

Ferrell, Max Everett. Areal and structural geology of Conant Creek Anticline, Fremont County, Wyoming. M, 1940, University of Missouri, Columbia.

Ferrell, Ray Edward, Jr. Paleoenvironment significance of clay minerals in the (Lower Cretaceous) Shell Creek Shale, Muddy Sandstone, and Thermopolis Shale on the east flank of the Bighorn Mountains, Wyoming. M, 1965, University of Illinois, Urbana.

Ferrell, Ray Edward, Jr. Structural aspects of clay mineral alterations. D, 1966, University of Illinois, Urbana. 103 p.

Ferren, Jack E. The geology of the southeastern part of the Shirley Mountains, Carbon County, Wyoming. M, 1935, University of Wyoming. 43 p.

Ferrero, Walter. Foraminifera from the Upper Cretaceous Redbank Formation of New Jersey. M, 1972, Brooklyn College (CUNY).

Ferretti A., Jorge. Study on secondary recovery methods in the Magallanes Basin oil fields (Chile). M, 1964, Stanford University.

Ferri, Filippo. Structure of the Blackwater Range, British Columbia. M, 1984, University of Calgary. 143 p.

Ferrians, Oscar J. Geology of a portion of the Mission Creek area, Nez Perce - Lewis counties, Idaho. M, 1958, Washington State University. 54 p.

Ferrigno, Kenneth F. Paleoecology of the reefs of the Holston Limestone. D, 1973, University of Tennessee, Knoxville. 143 p.

Ferrigno, Kenneth Francis. Conodonts of the Dundee Limestone (Middle Devonian) at Saint Mary's, Ontario (Canada). M, 1968, University of Western Ontario. 110 p.

Ferrill, Benjamin Arnold. Frog Mountain Formation, southwestern Appalachian fold and thrust belt, Alabama. M, 1984, University of Alabama. 187 p.

Ferrill, David Alexander. Analysis of shortening across Cacapon Mountain Anticlinorium in the Central Appalachians of West Virginia. M, 1987, West Virginia University. 171 p.

Ferris, Bernard J. An investigation of possible source beds of petroleum. M, 1948, Colorado School of Mines. 119 p.

Ferris, Clinton S., Jr. Petrology and structure of the Precambrian rocks southeast of Encampment, Wyoming. D, 1964, University of Wyoming. 74 p.

Ferris, Clinton S., Jr. Temperature of formation of the Coronation sulphide ore body, Flin Flon area, Saskatchewan. M, 1961, University of Saskatchewan. 51 p.

Ferris, Phebe. The geography of Canada. M, 1927, [Smith College].

Ferriz-Dominguez, Horacio Gerardo. Los Humeros volcanic center, Puebla, Mexico; geology, petrology, geothermal system, and geoarchaeology. D, 1985, Stanford University. 281 p.

Ferrusquia-V., Ismael. Geology of the Tamazulapan-Teposcolula-Yanhuitlan area (Tertiary), Mixteca Alta, State of Oaxaca, Mexico. D, 1971, University of Texas, Austin.

Ferry, Chamberlain. Experiments on the growth of crystals in rock fractures. M, 1932, Cornell University.

Ferry, James Gerard. A model of the near-surface seismic velocity; southern San Joaquin Valley, California. M, 1987, Texas A&M University. 69 p.

Ferry, Joe P. Investigation of possible metals contamination of groundwater supplies in Sloan's Valley and the Sloan's Valley cave system in Pulaski County, Kentucky. M, 1984, Wright State University. 75 p.

Ferry, John Mott. Metamorphism of calcareous sediments in the Waterville-Vassalboro area, south-central Maine. D, 1975, Harvard University.

Ferry, John Mott. Subsolidus phase relations in the system $KAlSiO_4$-$NaAlSiO_4$. M, 1971, Stanford University.

Ferry, Robert Allen. Integration and comparison of continuous electric analog and numerical hydrogeologic modeling techniques using data from northeastern Ohio. M, 1987, Kent State University, Kent. 208 p.

Fertitta, Jay C. Petrographic analysis of the Canyon "Reef" (Upper Pennsylvanian), Salt Creek Field, Kent County, Texas. M, 1985, Stephen F. Austin State University. 122 p.

Fesko, Gregory R. Regional contour and trend surface maps of selected coal quality parameters for the Pittsburgh No. 8 Coal of Ohio. M, 1983, University of Toledo. 214 p.

Fessenden, Franklin W. A lithologic investigation of the Manlius and Coeymans limestones. M, 1959, Rice University. 49 p.

Fessenden, Franklin Wheeler. The geology and hydrology of the Pepperell Springs area, Pepperell, Massachusetts. D, 1971, Boston University. 204 p.

Fessenden, R. Interpretation of Quaternary sediments in the Cortland through valleys from electrical resistivity data. M, 1974, SUNY at Binghamton. 142 p.

Feth, Elle. Sedimentary petrology and depositional environment of the Shinarump sandstone/conglomerate near Lee's Ferry, Arizona. M, 1985, Purdue University. 67 p.

Feth, John H. The geology of the northern Canelo Hills, Santa Cruz County, Arizona. D, 1947, University of Arizona.

Fett, John David. Geophysical investigation of the San Jacinto Valley, Riverside County, California. M, 1968, University of California, Riverside. 87 p.

Fetter, Charles Willard. The hydrogeology of the south fork of Long Island, New York. D, 1971, Indiana University, Bloomington. 236 p.

Fetter, Franklin. Recent deep-sea benthic foraminifera from the Alpha Ridge province of the Arctic Ocean. M, 1973, Florida State University.

Fetterman, James William. Alumina extraction from a Pennsylvania diaspore clay by an ammonium sulfate process. D, 1961, Pennsylvania State University, University Park. 160 p.

Fetterman, James William. The leaching characteristics of uranium from some South Dakota lignites. M, 1958, Pennsylvania State University, University Park. 146 p.

Fetters, Robert Thomas, Jr. An environmental approach to the stratigraphy of a portion of the Middle Ordovician in East Tennessee. M, 1966, University of Tennessee, Knoxville. 75 p.

Fettke, Charles Reinhard. The limonite deposits of Staten Island, New York. M, 1911, Columbia University, Teachers College.

Fettke, Charles Reinhard. The Manhattan Schist of southeastern New York State and its associated igneous rocks. D, 1914, Columbia University, Teachers College.

Fetyani, Ahmad Ali. Petrographic, chemical and Pb-isotopes composition of sphalerite, pyrite and chalcopyrite from a mineralogically zoned Pb-Zn ore body in the Viburnum Trend of Southeast Missouri. M, 1980, Washington University. 94 p.

Fetzer, Joseph A. Biostratigraphic evaluation of some Middle Ordovician bentonite complexes in Eastern North America. M, 1973, Ohio State University.

Fetzer, Kenneth Rolland. Oil pool geology of the Albion Field, Edwards County, Illinois. M, 1954, Miami University (Ohio). 25 p.

Fetzer, Wallace Gordon. Humic acids and other organic acids as mineral solvents. D, 1934, University of Minnesota, Minneapolis. 56 p.

Fetzer, Wallace Gordon. The relation of the heat gradient to crystallographic orientation. M, 1929, University of Minnesota, Minneapolis. 36 p.

Fetzner, Richard W. Frontal Ouachita facies of the Wapanucka Limestone, Oklahoma.1956, University of Wisconsin-Madison. 63 p.

Feucht, Alex T. Geology of the Swift Reservoir area, Sawtooth Range, (Pondera and Teton County), Montana. M, 1971, University of Montana. 59 p.

Feucht, Lynn Janet. The effects of chemical environment on the frictional properties of a quartzose sandstone. M, 1985, Texas A&M University. 52 p.

Feuer, S. M. Pollen morphology and evolution in the Santalales sen. str., a parasitic order of flowering plants. D, 1977, University of Massachusetts. 444 p.

Feuer, Wendy Jo. Nature and significance of the hangingwall fragmental mica schist of the Cofer Zn-Pb-Cu-Ag deposit, Mineral, Virginia. M, 1980, University of Western Ontario. 155 p.

Feuerbach, Daniel Lee. Geology of the Wilson Ridge Pluton; a mid-Miocene quartz monzonite intrusion in the northern Black Mountains, Mohave County, Arizona, and Clark County, Nevada. M, 1986, University of Nevada, Las Vegas. 79 p.

Feuillet, J. P. Control of clay mineral distribution by estuarine circulation, James River estuary, Virginia. M, 1976, Old Dominion University. 113 p.

Feulner, Alvin John. Groundwater resources of Clayton County, Iowa. M, 1952, University of Iowa. 167 p.

Feves, Michael Lawrence. Characterization of stress-induced cracks in rocks. D, 1977, Massachusetts Institute of Technology. 124 p.

Fewkes, Ronald Hubert. The origin of marine manganese nodules as determined by textural and mineralogical analysis. D, 1976, Washington State University. 169 p.

Fezie, Glenn Stephen. The upper mantle velocity and crustal structure in the Basin and Range Province. M, 1980, University of Nevada. 134 p.

Ffolliott, J. H. Replacements in Knife Lake Slates. M, 1929, University of Minnesota, Minneapolis. 34 p.

Fiadeiro, Manuel E. Numerical modeling of tracer distributions in the deep Pacific Ocean. D, 1975, University of California, San Diego. 254 p.

Fiandt, Dallas N. The geology of the Unionville and Allens Creek quadrangles. M, 1950, Indiana University, Bloomington. 14 p.

Fichter, Lynn Stanton. Geographical distribution and osteological variation in fossil and Recent specimens of two species of Kinosternon (turtles). M, 1967, University of Michigan.

Fichter, Lynn Stanton. The North American beavers of the genus Castor post incisor dentition; a multivariate study (Late Pliocene to Recent). D, 1972, University of Michigan.

Fici, Huseyin. Definition of the B Salt Edge of southern limb of the Michigan Basin using seismic techniques. M, 1988, Western Michigan University.

Fickett, Paul V. An investigation into the red to grey till color transition zone, Lancaster, New York and adjacent areas. M, 1972, SUNY at Buffalo. 45 p.

Fico, Cary. Influence of wave refraction on coastal geomorphology; Bull Island to Isle of Palms, South Carolina. M, 1978, University of South Carolina.

Fiddler, Linda Carol. Cyclic sedimentation in the upper Wilcox (lower Eocene) of eastern and central Mississippi and western Alabama. M, 1971, Northeast Louisiana University.

Fidlar, Marion Moore. Alluviation in a portion of the lower Wabash Valley. M, 1936, Indiana University, Bloomington. 106 p.

Fidlar, Marion Moore. The physiography of the lower Wabash Valley. D, 1942, Indiana University, Bloomington. 260 p.

Fiedler, Forest J. Surficial geology of the Mountain Lake area, Giles County, Virginia. M, 1967, Virginia Polytechnic Institute and State University.

Fiedler, William Morris. The geology of the Jamesburg Quadrangle, California. D, 1942, University of California, Berkeley. 144 p.

Field, George W. The Baxter Hollow Granite. M, 1936, University of Wisconsin-Madison.

Field, M. E. Quaternary evolution and sedimentary record of a coastal plain shelf; central Delmarva Peninsula, Mid-Atlantic Bight, U.S.A. D, 1976, George Washington University. 217 p.

Field, Michael Ehrenhart. Sedimentation on North Carolina lower continental rise. M, 1969, Duke University. 98 p.

Field, Michael Timberlake. Bedrock geology of the Ware area, central Massachusetts. D, 1975, University of Massachusetts. 233 p.

Field, Richard Montgomery. The stratigraphy of the Middle Ordovician formations of central and south-central Pennsylvania. D, 1919, Harvard University.

Field, Robert Joseph. A cross section of the Denver Basin based on electric logs. M, 1951, University of Colorado.

Field, Ross. Physiography of the north end of the Humboldt Range, Nevada, together with a discussion of the origin of pediments. M, 1933, Stanford University. 98 p.

Field, Stephen W. Mineralogy and petrology of the Davis Mine, Rowe, Massachusetts. M, 1985, University of Massachusetts. 305 p.

Field, Stephen Walter. Upper mantle peridotites and metasomites from the Jagersfontein Kimberlite in the Kaapvaal Craton. D, 1988, University of Massachusetts. 293 p.

Field, William. The structural dependence of refractive indices. D, 1965, Harvard University.

Fielden, John R., III. Ground water and nitrate in Paradise Valley, Arizona. M, 1975, Arizona State University. 77 p.

Fielder, Gordon W., III. Lateral and vertical variation of depositional facies in the Cambrian Galesville Sandstone, Wisconsin Dells. M, 1985, University of Wisconsin-Madison. 194 p.

Fielder, William Morris. Structure and stratigraphy of a section across the White Mountains, California. M, 1937, California Institute of Technology. 55 p.

Fielding, D. H. Distribution, petrology, and environment of the Saint Louis-Ste. Genevieve transition zone (upper Mississippian) in Missouri. M, 1971, University of Missouri, Rolla.

Fielding, Howard. The stratigraphy and structure of the southern half of the South Onondaga 7 1/2 minute Quadrangle. M, 1957, Syracuse University.

Fielding, Stanley J. Crystal chemistry of the oxonium alunite-potassium alunite series. M, 1981, Lehigh University.

Fields, Edward D. Precambrian rocks of the Halleck Canyon area, Albany County, Wyoming. M, 1963, University of Wyoming. 90 p.

Fields, Harry Basil. The Benton Group of the northeastern Black Hills. M, 1927, University of Iowa. 181 p.

Fields, Mary Leslie. Physical processes and sedimentation in the intra-jetty area, Barnegat Inlet, New Jersey. M, 1984, Rutgers, The State University, New Brunswick. 158 p.

Fields, Noland E., Jr. A paleontological study of the Chickamauga rocks in Raccoon Valley, Anderson and Knox counties, Tennessee. M, 1960, University of Tennessee, Knoxville. 50 p.

Fields, Noland Embry, Jr. The bryozoan Adeonellopsis in the Paleogene of the Southeastern United States. D, 1971, Louisiana State University.

Fields, Perry Merle, III. Subsurface geology of the Morrowan lower Dornick Hills (Cromwell Sandstone) and the Desmoinesian Krebs (Hartshorne and lower Booch sandstones) in northern Pittsburg County, Oklahoma. M, 1987, Oklahoma State University. 170 p.

Fields, R. W. Late Miocene rodents from Colombia, with a detailed description of continental stratigraphy of the La Venta Badlands. D, 1952, University of California, Berkeley. 297 p.

Fields, Suzanne. A Gatun ostracode fauna from Cativa, Panama. M, 1936, Columbia University, Teachers College.

Fiero, G. William, Jr. Geology of the Upheaval Dome, San Juan County, Utah. M, 1958, University of Wyoming. 87 p.

Fiero, G. William, Jr. Ground-water systems of central Nevada. D, 1968, University of Wisconsin-Madison. 243 p.

Fierro, Pedro. Investigation of relationships between a West Virginia mine spoil, the parent overburden, and the seep discharges. M, 1985, University of Kentucky. 78 p.

Fierstein, John. Morphometric analysis of Littorina irrorata from St. Catherine's Island, Georgia. M, 1977, University of Pittsburgh.

Fies, Michael Wayne. Depositional environments and diagenesis of the Tonkawa Formation (Virgilian) in Woods and part of Woodward counties, Oklahoma. M, 1988, Oklahoma State University. 123 p.

Fiesinger, Donald W. A study of post-Ordovician intrusives from the Lake Champlain valley, New York and Vermont. M, 1969, Wayne State University.

Fiesinger, Donald William. Petrology of the Quaternary volcanic centers in the Quesnel Highlands and Garibaldi Provincial Park areas, British Columbia. D, 1975, University of Calgary. 133 p.

Fiess, K. M. The rare-earth element geochemistry of ultramafic nodules from Southern African kimberlites. M, 1979, Dalhousie University. 130 p.

Fifarek, Richard H. Alteration geochemistry, fluid inclusion, and stable isotope study of the Red Ledge volcanogenic massive sulfide deposit, Idaho. D, 1985, Oregon State University. 187 p.

Fife, Donald L. Geology of the Bahía Santa Rosalía Quadrangle (Cretaceous to Recent), Baja California, Mexico. M, 1968, University of San Diego.

Fifer, Henry Clay. Geology of a portion of Jarbidge I Quadrangle, Elko County, Nevada. M, 1960, University of Oregon. 48 p.

Fifield, J. S. Watershed constituent loading analysis utilizing empirical hydrochemical modeling techniques. D, 1979, Utah State University. 195 p.

Figueiredo Filho, Paulo Miranda de *see* de Figueiredo Filho, Paulo Miranda

Figueiredo, Alberto. Submarine sand ridges; geology and development, New Jersey, U.S.A. D, 1984, University of Miami. 408 p.

Figueiredo, Alberto Garcia de *see* de Figueiredo, Alberto Garcia, Jr.

Figueiredo, Antonio M. Depositional systems in the Lower Cretaceous Morro do Chaves and Coqueiro Seco formations and their relationship to petroleum accumulations, middle rift sequence, Sergipe-Alagoas Basin, Brazil. D, 1981, University of Texas, Austin.

Figueiredo, Antonio Manuel Ferreira de *see* de Figueiredo, Antonio Manuel Ferreira

Figueiredo, Joao Neiva de *see* de Figueiredo, Joao Neiva

Figueiredo, Mario Cesar Heredia de *see* de Figueiredo, Mario Cesar Heredia

Figueroa, J. L. Resilient based flexible pavement design procedure for secondary roads. D, 1979, University of Illinois, Urbana. 334 p.

Figueroa, Julio Aguiles Pastor. The mineralization in San Cristobal Mine (central Cordillera, Peru). M, 1970, University of Arizona.

Figueroa-Garcia, Eduardo Anibal. Design of efficient groundwater monitoring networks. D, 1982, Vanderbilt University. 134 p.

Figuers, Sands Hardin. Structural geology and geophysics of the Pipeline Complex, northern Franklin Mountains, El Paso, Texas. D, 1987, University of Texas at El Paso. 331 p.

Figuli, Samuel P. A computer simulation of faulting in the southwestern Great Basin. M, 1986, Kent State University, Kent. 181 p.

Fiki, M. H. One-dimensional seismic modeling of Tertiary uranium deposits in the Powder River basin, Wyoming. M, 1978, University of Wyoming. 99 p.

Fikkan, Philip R. Granitic rocks in the Dry Valley region of south Victoria Land, Antarctica. M, 1968, University of Wyoming. 119 p.

Fiksdal, Allen James. Geology for land use planning in part of Pierce County, Washington. M, 1979, Portland State University. 88 p.

Filaseta, Leonard. Differentiation in a diabase sill based on the study of amphiboles. D, 1936, University of Wisconsin-Madison.

Filby, Royston Herbert. Spectrographic methods for the determination of sodium, potassium and calcium in minerals and their application to some scapolites. M, 1957, McMaster University. 65 p.

Filep, E. J. Geology and origin of the Cortlandt emery deposits. M, 1977, Queens College (CUNY). 92 p.

Filer, Jonathan K. Modern sedimentation through the fluvial to estuarine transition of the Neuse River and estuary, North Carolina. M, 1979, University of North Carolina, Chapel Hill. 142 p.

Files, Edgar James, Jr. Effects of maintenance dredging on sediment distribution in Apalachicola Bay, Florida. M, 1975, University of Alabama.

Files, Frederic G. Geology of the Lefthand Canyon-Nugget Hill Area, Boulder County, Colorado. M, 1964, University of Colorado.

Files, Frederic Grant. Geology and alteration associated with Wyoming uranium deposits. D, 1970, University of California, Berkeley. 189 p.

Files, Nelson. Geology of the Italy Quadrangle, Ellis and Hill counties, Texas. M, 1977, University of Texas, Arlington. 136 p.

Filewicz, Mark V. The Upper and Lower Cretaceous calcareous nannofossils of South Florida. M, 1976, Florida State University.

Filho, J. M. Water resources management for part of the lower Gila Valley. D, 1974, University of Arizona. 143 p.

Filho, Joel Souto-Maior *see* Souto-Maior Filho, Joel

Filho, Jose Oswaldo de Araujo *see* de Araujo Filho, Jose Oswaldo

Filho, Paulo Miranda de Figueiredo *see* de Figueiredo Filho, Paulo Miranda

Filice, Alan Lewis. Geology of a part of the Stillwater Mountains, Nevada. M, 1967, University of California, Los Angeles.

Filion, Gilles. Le contexte géologique des gisements volcanogènes de Marbridge, La Motte, Abitibi. M, 1979, Ecole Polytechnique.

Filion, Gilles. Le Contexte géologique des gisements volcanogènes de Marbridge, La Motte, Abitibi. M, 1978, Ecole Polytechnique.

Filipek, L. H. Factors influencing the phase partitioning of iron and associated elements during early diagenesis in shallow marine sediments. D, 1979, University of Michigan. 188 p.

Filipo, William Anthony San *see* San Filipo, William Anthony

Filipov, Allan James. Sedimentology of debris-flow deposits, west flank of the White Mountains, California. M, 1986, University of Massachusetts. 184 p.

Filippini, Mark G. The impact of urbanization on a flood-plain aquifer; Bloomington, Indiana. M, 1981, Indiana University, Bloomington. 112 p.

Filkins, Jeffrey Elliott. Geochemistry and depositional history of the coals of the Tatman Formation, Bighorn Basin, Wyoming. M, 1986, Iowa State University of Science and Technology. 132 p.

Filley, Thomas Howard. The hydrogeological response to continental glaciation. M, 1985, Pennsylvania State University, University Park. 210 p.

Fillion, Denis. Selected studies in Cambro?-Ordovician ichnology. M, 1984, University of New Brunswick. 507 p.

Fillipone, Jeffrey A. Structure and metamorphism at the western margin of the Omineca Belt near Boss Mountain, east-central British Columbia. M, 1985, University of British Columbia. 150 p.

Filippone, Walter R. Investigation of the Alaskan earthquake of May 4, 1934. M, 1944, California Institute of Technology. 10 p.

Fillman, Louise Anna. Cenozoic history of the northern Black Hills. D, 1924, University of Iowa. 112 p.

Fillman, Louise Anna. Limestone conglomerates of the Deadwood Formation of the northern Black Hills of South Dakota. M, 1921, University of Iowa. 72 p.

Fillmore, Barbara J. Stratigraphy and source-rock potential of the Mowry Shale (Lower Cretaceous), North Park, Colorado. M, 1986, Colorado School of Mines. 153 p.

Fillo, E. J. Watershed urbanization; response by Fuller Hollow Creek, Vestal, New York. M, 1985, SUNY at Binghamton. 65 p.

Fillo, Wayne Joseph. Mineralogy, geochemistry and petrology of the Catherwood and Kinley Saskatchewan meteorites. M, 1973, University of Saskatchewan. 82 p.

Fillon, Richard H. Sedimentation and Recent geologic history of the Missisquoi delta (Lake Champlain, Vermont). M, 1969, University of Vermont.

Fillon, Richard Henry. Late Cenozoic geology, paleooceanography and paleo-climatology of the Ross Sea, Antarctica. D, 1973, University of Rhode Island. 195 p.

Filloux, Jean Henri. Oceanic electric currents, geomagnetic variations and the deep electrical conductivity structure of the ocean-continent transition of Central California. D, 1967, University of California, San Diego. 187 p.

Filmer, Edwin Alfred. The glacial geology of the Binghamton, New York Quadrangle, U.S. Geological Survey. M, 1928, Cornell University.

Filomena, Joseph James. Petrology, diagenesis and depositional environments of Middle Ordovician-Lower Silurian sequence at Pedley Pass, British Columbia, Canada. M, 1985, Queens College (CUNY). 127 p.

Filson, John Roy. Horizontal transverse motion from earthquakes at near distances. D, 1969, University of California, Berkeley. 118 p.

Filson, Robert H. Petrology and structure of Precambrian metamorphic and igneous rocks of the Shipman Mountain area, Larimer and Jackson counties, Colorado. M, 1973, Colorado State University. 144 p.

Filson, Robert Harold. Instruction in college-level, introductory geology; interactions of two teaching methods and selected student characteristics. M, 1979, University of Washington. 171 p.

Filut, Marlene. The crystal structure analysis of anandite-2or. M, 1984, University of Wisconsin-Madison. 67 p.

Finamore, P. F. The Fenelon Falls outlet of glacial Lake Algonquin, south-central Ontario. M, 1985, University of Waterloo. 174 p.

Finch, Christian C. Geology of an area near San Benito, California. M, 1973, Stanford University.

Finch, Christian Charles. Paleoecology and stratigraphy of a Paleocene foraminiferal assemblage from the Simi Valley, Ventura County, California. D, 1980, University of California, Los Angeles. 411 p.

Finch, Michael O. Liquefaction potential of the Sacramento-San Joaquin Delta, California. M, 1987, University of California, Davis. 184 p.

Finch, Richard C. A petrographic study of contact metamorphism at Italian Mountain, Gunnison County, Colorado. M, 1967, Vanderbilt University.

Finch, Richard Carrington. Geology of the San Pedro Zacapa Quadrangle (Mesozoic), Honduras, Central America. D, 1972, University of Texas, Austin. 263 p.

Finch, Warren Irvin. Geology of the Shinarump No. 1 Mine, Grand County, Utah, with a general account of uranium deposits in Triassic rocks of the Colorado Plateau. M, 1954, University of California, Berkeley. 45 p.

Finch, William Anderson, Jr. The karst landscape of Yucatan. D, 1965, University of Illinois, Urbana. 179 p.

Fincham, William J. The Salina Group of the southern part of the Michigan Basin. M, 1975, Michigan State University. 57 p.

Findlay, A. R. The geology of the Decker Creek molybdenite deposit, British Columbia. M, 1975, Carleton University. 86 p.

Findlay, David Christopher. Peridotites from northern Quebec and Ungava. M, 1958, McGill University.

Findlay, David Christopher. Petrology of the Tulameen ultramafic complex, Yale District, British Columbia. D, 1963, Queen's University. 407 p.

Findlay, Donald J. The Lang Lake copper-molybdenum deposit, northwestern Ontario; a possible early Precambrian porphyry deposit. M, 1981, University of Manitoba. 94 p.

Findlay, Marsha G. Tectonically controlled stratigraphy of the Jurassic ranges of Northeast Tunisia. M, 1976, University of South Carolina.

Findlay, Willard A. Geology of a part of the San Joaquin Hills. M, 1932, California Institute of Technology. 47 p.

Findlay, Willard A. Sources of Miocene sediments in southwestern San Joaquin Valley. D, 1940, California Institute of Technology. 126 p.

Findlay, Willard A. and Popenoe, W. P. Transposed hinge structure in lamellibranchs. D, 1940, California Institute of Technology.

Findlay, William F. Geology of a part of Buck Mountain Quadrangle, east-central Nevada. M, 1960, University of Southern California.

Findley, David Paul. Late Cenozoic tectonic deformation along the northern White Mountains, Mono and Inyo counties, California. M, 1984, University of Nevada. 93 p.

Findley, Richard Lee. Facies relationships in the Stevens Sandstone, Kern County, California. M, 1975, Texas A&M University. 217 p.

Fine, Charles D. A petrographic and field study of basalt flows in Skye, Scotland. M, 1968, Brooklyn College (CUNY).

Fine, Gerald Jonathan. Carbon dioxide in synthetic and natural silicate glasses. D, 1986, California Institute of Technology. 214 p.

Fine, R. A. High pressure P-V-T properties of seawater and related liquids. D, 1975, University of Miami. 222 p.

Fine, Spencer F. Geology of part of the western end of Antelope Valley (California). M, 1947, University of California, Los Angeles.

Finfrock, Lawrence J. The stratigraphy of the Madison Group of south-central Montana and Northwest Wyoming (F949). M, 1948, University of Illinois, Urbana. 80 p.

Finger, Kenneth L. Recent benthic foraminifera from pyroclastic substrates; a biofacies analysis of Deception Island, South Shetland Islands, Antarctica. D, 1976, University of California, Davis. 165 p.

Finger, Larry Wayne. The crystal structures and crystal chemistry of ferro-magnesian amphiboles. D, 1967, University of Minnesota, Minneapolis. 81 p.

Fingleton, Walter George. A study of shore erosion at seventeen sites along eastern Lake Michigan. M, 1973, Western Michigan University.

Finiol, Gary Walter. Morphology and sedimentology of the Charlevoix-Antrim drumlin field. M, 1978, University of Michigan.

Fink, Don Roger. The telluric current method; instrumentation and field procedure. M, 1954, Washington University. 19 p.

Fink, Dwayne Harold. Some mineralogical and physical interpretations of the free-swelling characteristics of montmorillonite-water systems. D, 1965, University of Virginia. 156 p.

Fink, Helen Binkley. Pleistocene deposits of the Alton-Bellefontaine area. M, 1954, Washington University. 60 p.

Fink, J. H. Surface structures on obsidian flows. D, 1979, Stanford University. 221 p.

Fink, J. W. Petrology of the Triassic San Hipolito Formation, Vizcaino Peninsula, Baja California Sur, Mexico. M, 1975, San Diego State University.

Fink, K. A. Petrology of the Eocene Friars Formation, El Cajon, Grossmont, and Tierrasanta areas, southwestern San Diego County, California. M, 1976, San Diego State University.

Fink, Kendrick Claude. Geology and ore deposits of the New Pass mines, Lander County, Nevada. M, 1976, Stanford University. 131 p.

Fink, Loyd Kenneth, Jr. The geology of the Guadeloupe region, Lesser Antilles island arc. D, 1969, University of Miami. 144 p.

Fink, Richard Christopher. Late Cenozoic tectonic features in glacial sediments, Owens Valley, California. M, 1979, University of Nevada. 81 p.

Fink, Richard P. Chitinozoa, Paleozoic Problematica from the Middle Devonian of Ohio. M, 1968, Washington University. 86 p.

Fink, Robert Arthur. The geology and igneous petrology of the Comb Creek dike swarm, Crazy Mountains, Montana. M, 1975, University of Cincinnati. 174 p.

Fink, Sidney. The geology of the Otisville Quadrangle, New York. M, 1959, New York University.

Fink, Susan L. Depositional environments of Upper Mississippian and Lower Pennsylvanian strata in south-central Kentucky. M, 1979, Eastern Kentucky University. 53 p.

Finke, Eberhard A. W. Landscape evolution of the Argive Plain, Greece; paleoecology, Holocene depositional history, and coastline changes. D, 1988, Stanford University. 219 p.

Finkel, Elizabeth A. Stylolitization and cementation in the Mississippian Salem Limestone, west-central Indiana. M, 1986, University of Michigan. 36 p.

Finkelman, Robert Barry. Modes of occurrence of trace elements in coal. D, 1980, University of Maryland. 329 p.

Finkelman, Robert Barry. Physical and chemical analyses and suggested origin of the glass beads from Apollo 11 lunar sand. M, 1970, George Washington University.

Finkelnburg, Oscar Carl. Beneficiation of Idaho phosphate rock. M, 1947, University of Idaho. 31 p.

Finkelstein, Kenneth. Morphologic variations and sediment transport in crenulate bay beaches, Kodiak-Island, Alaska. M, 1979, University of South Carolina.

Finkelstein, Kenneth. The late Quaternary evolution of a twin barrier-island complex, Cape Charles, Virginia. D, 1986, College of William and Mary. 297 p.

Finks, R. M. Upper Paleozoic corals from Peru. M, 1954, Columbia University, Teachers College.

Finlay, Corey D. Discussion of RG propagation in terms of ray paths of P and SV. M, 1984, Texas Tech University. 70 p.

Finlay, George I. Geology of the San Jose District (Tamaulipas, Mexico). D, 1903, Columbia University, Teachers College.

Finlayson, Carroll P. The geology of the Max Patch Mountain area, Lemon Gap Quadrangle, Tennessee-North Carolina. M, 1957, University of Tennessee, Knoxville. 41 p.

Finlayson, George Barry. Seismic wave studies of surficial materials in the Winkler area, Manitoba (Canada). M, 1968, University of Saskatchewan. 263 p.

Finlayson, James Bruce. A chemical study of Hawaiian volcanic gases. D, 1967, University of Hawaii. 181 p.

Finlen, James Rendell. Transport investigations in the Northwest Providence Channel. M, 1966, [University of Miami].

Finley, J. A. Nickel; a review of its supply-demand relationships, and its geology. M, 1978, Stanford University. 208 p.

Finley, Jim B. Experimental studies of erosion from slopes protected by rock mulch. M, 1984, Colorado State University. 78 p.

Finley, Judge Dinsmore. Geologic structure of eastern Apache Mountains, Culberson County, Trans-Pecos Texas; Part II, Local structure. M, 1954, University of Texas, Austin.

Finley, Mark Edward. Geology of Black Mountain Quadrangle, Bighorn County, Wyoming. M, 1982, Iowa State University of Science and Technology. 97 p.

Finley, Robert. Sedimentary and shore processes on South beach, Martha's Vineyard, Massachusetts. M, 1969, Syracuse University.

Finley, Robert J. Morphologic development and dynamic processes at a barrier island inlet, North Inlet, South Carolina. D, 1975, University of South Carolina. 372 p.

Finley, Sharon G. Stress-induced metamorphic differentiation in regions of non-uniform viscosity. M, 1981, Rensselaer Polytechnic Institute. 47 p.

Finley, William R. Subsurface geology, Sparta Interval, Evangeline and St. Landry parishes, Louisiana. M, 1975, University of Southwestern Louisiana. 121 p.

Finn, Carol Ann. Gravity evidence for a shallow intrusion under Medicine Lake Volcano, California. M, 1982, University of Colorado. 53 p.

Finn, Carol Ann. Structure of the Washington convergent margin, implications for other subduction zones and for continental growth processes. D, 1988, University of Colorado. 159 p.

Finn, Christopher Jude. Estimation of three dimensional dip and curvature from reflection seismic data. M, 1986, University of Texas, Austin. 192 p.

Finn, Dale Robert. Geology and ore deposits of the Hayden Hill District, Lassen County, California. M, 1987, University of Nevada. 84 p.

Finn, Dennis D. Prospecting for magmatic and hydrothermal deposits of uranium and associated elements by computer evaluation of the chemistry and petrogenesis of igneous source rocks. M, 1979, Eastern Washington University. 336 p.

Finn, Gregory C. Petrogenesis of the Wanapitei gabbronorite intrusion, a Nipissing-type diabase, from northeastern Ontario. M, 1981, University of Western Ontario. 213 p.

Finn, Gregory Clement. Geochemical and isotopic evolution of the Maggo Gneiss component from the Hopedale Block, Labrador; evidence for late-middle Archean crustal reworking. D, 1988, Memorial University of Newfoundland. 471 p.

Finnegan, David Lawrence. The chemistry of trace elements and acidic species in fumarolic emissions. D, 1984, University of Maryland. 242 p.

Finnegan, Stephen Allan. A study of the effects of transient electromagnetic fields on conductive Earth models. M, 1963, Michigan Technological University. 97 p.

Finnell, Tommy L. Structural relationships of the Seminoe-Shirley Syncline, Carbon County, Wyoming. M, 1951, University of Wyoming. 41 p.

Finnemore, Erhardt John. Moisture movement above a varying water table. M, 1970, Stanford University. 145 p.

Finneran, Jane Beeman. Carbonate petrography and depositional environments of the Upper Jurassic Zuloaga Formation, Sierra de Enfrente, Coahuila, Mexico. M, 1986, Stephen F. Austin State University. 221 p.

Finneran, Joseph M. Calcareous nannofossil biostratigraphy and structure of Upper Cretaceous sediments of the North Carolina coastal plain. M, 1980, Ohio University, Athens. 112 p.

Finnerty, A. A. Partition of trace elements between synthetic forsterite and enstatite. D, 1976, University of California, Los Angeles. 250 p.

Finney, B. A. Random differential equations in water quality modeling. D, 1979, Utah State University. 180 p.

Finney, Bruce Preston. Paleoclimatic influence on sedimentation and manganese module growth during the past 400,000 years at MANOP Site H (eastern

Equatorial Pacific). D, 1987, Oregon State University. 195 p.

Finney, Joseph J. The crystal structure of CeNi. M, 1959, University of New Mexico. 39 p.

Finney, Joseph Jessel. The crystal structures of carminite and authigenic maximum microcline. D, 1962, University of Wisconsin-Madison. 143 p.

Finney, S. C. Graptolites of the middle Ordovician Athens Shale, Alabama. D, 1977, Ohio State University. 603 p.

Finney, Stanley C. Lower Devonian lithostratigraphy and graptolite biostratigraphy, Copenhagen Canyon, Nevada. M, 1971, University of California, Riverside. 62 p.

Finney, Vernon Lee. Mineralogy of ferruginous concretions. M, 1967, Florida State University.

Finnie, John Irwin. An application of the finite element method and two equation (k and E) turbulence model to two and three dimensional fluid flow problems governed by the Navier-Stokes equations. D, 1987, Utah State University. 143 p.

Finno, Richard Joseph. Response of cohesive soil to advanced shield tunneling. D, 1983, Stanford University. 334 p.

Finstick, Sue Ann. Hydrogeology of the Victor and Bing quadrangles, Bitterroot Valley, Montana. M, 1986, University of Montana. 150 p.

Finta, Susan F. Hydrogeologic evaluation of a pond in glacial deposits in northeastern Ohio. M, 1986, Kent State University, Kent. 134 p.

Fiore, Richard N. Geology and geomorphology of the Clear Creek drainage basin, western Burnet County, Texas. M, 1976, University of Texas, Austin.

Fiorillo, Anthony R. Taphonomy of Hazard Homestead Quarry (Ogallala Group), Hitchcock County, Nebraska. M, 1987, University of Nebraska, Lincoln. 121 p.

Fiorito, Thomas Francis. Experimental and theoretical study of tremolite and the assemblage tremolite plus calcite in the presence of H_2O-CO_2 aqueous fluids. M, 1975, University of Cincinnati. 90 p.

Firby, James Ronald. Late Cenozoic non-marine Mollusca of western Nevada and adjacent areas. D, 1969, University of California, Berkeley. 146 p.

Firby, James Ronald. Non-marine Mollusca of the "Esmeralda Formation" (Miocene) of Nevada. M, 1963, University of California, Berkeley. 109 p.

Firby, Jean Brower. Revision of the middle Pleistocene Irvington fauna of California. M, 1968, University of California, Berkeley. 134 p.

Firek, F. Heavy mineral distribution in the lower Chesapeake Bay, Virginia. M, 1975, Old Dominion University. 147 p.

Firouzian, Assadolah. Hydrological studies of the Saginaw Formation in the Lansing, Michigan area. M, 1963, Michigan State University. 53 p.

Firth, D. A. A study of the lamprophyres and porphyries of the Bryce area, Ontario. M, 1941, University of Toronto.

Firth, John Victor. Taxonomy and biostratigraphy of Maastrichtian-Danian dinoflagellates of Southwest Georgia. M, 1984, Virginia Polytechnic Institute and State University.

Fischbein, Steven A. Analysis and interpretation of ice-deformed sediments from Harrison Bay, Alaska. M, 1987, California State University, Hayward. 107 p.

Fischbuch, Norman Robert. Stratigraphy of the Devonian Swan Hills Reef complexes of central Alberta (Canada). D, 1968, University of Saskatchewan. 361 p.

Fischbuch, Norman Robert. Stromatoporoids of the Kaybob Reef, Alberta. M, 1959, University of Alberta. 72 p.

Fischer, Alfred G. Petrology of Eocene limestones in and around the Citrus-Levy County area, Florida. D, 1953, Columbia University, Teachers College.

Fischer, Anne Marie. The origin of gold-bearing quartz veins in Precambrian rocks near Wickenburg, Arizona. M, 1984, University of Arizona. 59 p.

Fischer, Arthur Homer. A summary of mining and metalliferous mineral resources in the State of Washington, with bibliography. M, 1918, University of Washington. 124 p.

Fischer, Brian Frederick Gustav. The geology of the area surrounding Bunde and Bukken fiords, northwestern Axel Heiberg Island, Canadian Arctic Archipelago. M, 1985, University of Calgary. 114 p.

Fischer, Conrad G. Sulfur gas emissions from two temperate climate salt marshes. M, 1987, University of New Hampshire. 92 p.

Fischer, David W. Paleoenvironmental analysis of a late Holocene deposit; Stanton site, west-central North Dakota. M, 1980, University of North Dakota. 107 p.

Fischer, Donald. A sedimentary study of the Franconia Sandstone of the Lake Pepin area, with special reference to glauconite. M, 1932, University of Minnesota, Minneapolis. 38 p.

Fischer, Donald E. Depositional environment of the (Upper Cretaceous) Muddy Sandstone in the Southwest Casper area, Wyoming. M, 1961, Northwestern University.

Fischer, Gerard William. Application of the reflectivity method to central United States refraction studies. M, 1977, St. Louis University.

Fischer, Hans J. E. A study of the variation of rate coefficients controlling the chemical reactions in the decaying nocturnal ionosphere between 20 and 240 km. M, 1965, New York University.

Fischer, Howard J. The lithology and diagenesis of the Metaline Formation, northeastern Washington. D, 1981, University of Idaho. 175 p.

Fischer, Howard J. The petrology of the Open Door Limestone Member of the Gallatin Formation (late Cambrian), Dubois area, Wyoming. M, 1975, Miami University (Ohio). 90 p.

Fischer, Ian A. Tidal flat sedimentation in a macrotidal embayment, Bahia de Lomas, Strait of Magellan, Chile. M, 1977, University of South Carolina.

Fischer, J. N. Simulation of hydrologic processes for surface mined lands. D, 1976, University of Arizona. 134 p.

Fischer, Jeffrey Allan. The use of relative travel time residuals of P-phases from teleseismic events to study the crust in the Socorro, New Mexico area. M, 1977, New Mexico Institute of Mining and Technology.

Fischer, Jeffrey M. Streamflow infiltration and groundwater recharge through the unsaturated zone at Vicee Canyon, Nevada. M, 1988, University of Nevada. 114 p.

Fischer, Joseph Fred. The geology of the White River-Carbon Ridge area, Cedar Lake Quadrangle, Cascade Mountains, Washington. D, 1970, University of California, Santa Barbara. 200 p.

Fischer, Karin J. Diagenesis and mass transfer in an upper Miocene source/reservoir system, southern San Joaquin Basin, California. M, 1986, University of Wyoming. 237 p.

Fischer, Kathleen Mary Brigid. Particle fluxes in the eastern tropical Pacific Ocean--sources and processes. D, 1984, Oregon State University. 225 p.

Fischer, Lorraine Eleanor. Geological interpretation of an aeromagnetic anomaly southwest of Boone, in west-central Iowa. M, 1982, University of Iowa. 181 p.

Fischer, Mark W. Sedimentology of the Clinch Sandstone (Lower Silurian) along Clinch Mountain, northeastern Tennessee. M, 1987, University of Tennessee, Knoxville. 156 p.

Fischer, Peter J. Geologic evolution and quaternary geology of Santa Barbara Basin, Southern California. D, 1972, University of Southern California.

Fischer, Richard Phillip. rt I, Sedimentary deposits of copper, vanadium-uranium and silver in Southwestern United States O Part II, Peculiar hydrothermal copper-bearing veins of the northeastern Colorado Plateau;. D, 1936, Princeton University. 162 p.

Fischer, William A. The foraminifera and stratigraphy of the Colorado Group (Cretaceous) in central and eastern Colorado. D, 1953, University of Colorado.

Fischer, William L. A magnetic study of the Wachusett-Marlborough Tunnel area (central Massachusetts). M, 1965, Boston University.

Fisco, Mary Pamala Polite. Sedimentation on the Weddell Sea continental margin and abyssal plain, Antarctica. M, 1982, Rice University. 170 p.

Fish, Craig B. Computer enhanced modelling of tidal velocities and circulation patterns, Buttermilk Bay, Massachusetts; a model study and development of methods for general application. M, 1988, Boston University. 119 p.

Fish, Ferol Fredric. The effect of water content on certain elastic properties of sedimentary rocks. M, 1956, Indiana University, Bloomington. 44 p.

Fish, Ferol Fredric, Jr. Effect of fluid content on the absorption of elastic waves in sandstone. D, 1961, Pennsylvania State University, University Park. 89 p.

Fish, John L. Type San Ramon Formation California. M, 1957, Stanford University.

Fish, Johnnie Edward. Crustal structure of the Texas Gulf Coastal Plain. M, 1970, University of Texas, Austin.

Fish, Johnnie Edward. Karst hydrogeology and geomorphology of the Sierra de El Abra and the Valles-San Luis Potosi region, Mexico. D, 1978, McMaster University. 469 p.

Fish, Thomas William. Geology of the Catoctin Mountain-South Mountain Anticlinorium. M, 1974, Millersville University.

Fishbaugh, David A. Depositional model for the lower Fountain Formation in the Manitou Embayment, Colorado. M, 1980, Indiana University, Bloomington. 136 p.

Fishbein, Alan S. Shell microstructure as a basis for generic distinction of Pecten-Chlamys of the Choptank Formation (Miocene) of Maryland. M, 1970, Brooklyn College (CUNY).

Fishbein, Evan F. Topography on the lithosphere-asthenosphere boundary. D, 1988, University of California, Los Angeles. 272 p.

Fishburn, Maurice D. The geology of southern Anderson County, Kansas. M, 1962, University of Kansas. 125 p.

Fishel, Ken W. Geometric and kinematic analysis of current ripples (in the Mojave desert, California). M, 1970, University of Southern California.

Fishel, Nathan. The technical aspects of the 1959 Melville Bay, West Greenland "Two Range Raydist" hydrographic season. M, 1960, Ohio State University.

Fisher, Bernard. Igneous rocks of the northeastern Bearpaw Mountains, Montana. D, 1946, Harvard University.

Fisher, Cassius Asa. The geology of Gage, Lancaster, and Saunders counties, in Nebraska. M, 1900, University of Nebraska, Lincoln.

Fisher, Cecil Coleman. A study of some outlet glaciers of the Columbia Icefield. M, 1951, University of Iowa. 83 p.

Fisher, Champe Andrews, Jr. Petrographic analysis of the Smackover Formation (Jurassic), Miller County, Arkansas. M, 1984, University of Arkansas, Fayetteville.

Fisher, D. The petrology of the Mt. Edwards nickel sulphide deposit, Widgiemooltha, Western Australia. D, 1979, University of Toronto.

Fisher, D. Ramsey. Regional diagenesis of the Tuscumbia Limestone (Meramecian-Mississippian) in northern Alabama and northeastern Mississippi. M, 1987, University of Alabama. 263 p.

Fisher, Daniel Claude. Evolution and functional morphology of the Xiphosurida. D, 1975, Harvard University.

Fisher, Daniel J. Geology of the Joliet, Illinois Quadrangle. D, 1923, University of Chicago. 160 p.

Fisher, Daniel J. The Devonian of the Mackenzie River basin. M, 1920, University of Chicago. 47 p.

Fisher, Daniel S. Hydrogeochemical processes in glacial deposits in northeastern Ohio. M, 1986, Kent State University, Kent. 280 p.

Fisher, Darrell Reed. Risk evaluation and dosimetry for indoor radon progeny on reclaimed Florida phosphate lands. D, 1978, University of Florida. 129 p.

Fisher, David. Complex seismic trace attributes from theoretical and scaled physical model data. M, 1983, University of Houston.

Fisher, David Frederick. The origin of the Number Five zone, Horne Mine (Precambrian) Noranda, Quebec. M, 1970, University of Western Ontario. 115 p.

Fisher, David M. Subsurface analysis of the Spencer consolidated oil field, Posey County, Indiana. M, 1981, Ball State University. 33 p.

Fisher, Donald Gene. Geology of the Iron King magnetite deposit, Jackson Mountains, Nevada. M, 1962, University of Utah. 50 p.

Fisher, Donald Myron. Structural evolution of a thickly sedimented convergent margin; evidence from macroscopic structures, microstructures, and incremental strain studies. D, 1988, Brown University. 131 p.

Fisher, Donald W. Lower Ordovician stratigraphy and paleontology of the Mohawk Valley. D, 1952, University of Rochester. 254 p.

Fisher, Donald W. Pre-Cambrian and lower Paleozoic geology of a portion of the Mohawk Valley, New York. M, 1948, SUNY at Buffalo.

Fisher, Fred Eugene. A multivariate determination of some Recent morphospecies of the genus Elphidium from Gulf of Mexico. M, 1968, [University of Houston].

Fisher, Frederick S. Tertiary intrusive rocks and mineralization in the Stinking-Water mining region, Park County, Wyoming. D, 1966, University of Wyoming. 140 p.

Fisher, Frederick S. The geology of the Little Sunlight Creek area, Park County, Wyoming. M, 1962, Wayne State University.

Fisher, George W. The petrology and structure of the crystalline rocks along the Potomac River, near Washington, DC. M, 1962, The Johns Hopkins University.

Fisher, Grace Merriam. Geology of Lake Tahoe and Glen Alpine. M, 1892, University of California, Berkeley.

Fisher, Henry Coleman, Jr. Surface geology of the Belford area, Osage County, Oklahoma. M, 1956, University of Oklahoma. 96 p.

Fisher, Henry Hugh. Stratigraphy and correlation of Precambrian volcanic rocks, Eminence (Shannon county), Missouri. M, 1969, University of Missouri, Rolla.

Fisher, Irving Sanborn. Geology of the Bethel area, Maine. D, 1952, Harvard University.

Fisher, J. D., Jr. Mineralization at some Pre-cambrian ore deposits in Colorado. M, 1936, Massachusetts Institute of Technology. 48 p.

Fisher, James C. Remote sensing applied to exploration for vein-type uranium deposits, Front Range, Colorado. D, 1976, Colorado School of Mines. 158 p.

Fisher, James C. The distribution and characteristics of the Traverse Formation of Michigan. M, 1969, Michigan State University. 72 p.

Fisher, James Edward. Textural characteristics of coarse sediments in selected streams of the Niagara Peninsula, Ontario. M, 1978, Brock University. 98 p.

Fisher, James Harold. Granitic pegmatites near Newport, New Hampshire. M, 1949, University of Illinois, Urbana.

Fisher, James Harold. Paleoecology of the Chattanooga-Kinderhook Shale. D, 1953, University of Illinois, Urbana.

Fisher, James Russell. The ion-product constant of water to 350°C. D, 1969, Pennsylvania State University, University Park. 93 p.

Fisher, Jeanne Anne. Fault patterns in southeastern Michigan. M, 1981, Michigan State University. 80 p.

Fisher, John. The hystrichosphaerids of the Clinton (Silurian) Group of New York State. M, 1955, University of Massachusetts.

Fisher, John B. Effects of tubificid oligochaetes on sediment movement and the movement of materials across the sediment-water interface. D, 1979, Case Western Reserve University.

Fisher, John Bailey. Flume development for a study of bedload and suspended sediment in Clear Creek drainage, eastern Sierra Nevada. M, 1978, University of Nevada. 83 p.

Fisher, John Joseph. Development pattern of relict beach ridges, Outer Banks barrier chain, North Carolina. D, 1967, University of North Carolina, Chapel Hill. 255 p.

Fisher, John Joseph. Geomorphic expression of former inlets along the Outer Banks of North Carolina. M, 1962, University of North Carolina, Chapel Hill. 120 p.

Fisher, John K. Geology and structure of the Citadel Rock area, northern Black Hills, South Dakota. M, 1969, South Dakota School of Mines & Technology.

Fisher, Joseph O. Structure and origin of the Old Tungsten Mine near Trumbull, Connecticut. M, 1944, University of Massachusetts.

Fisher, Juanita Prior. An investigation of the extent of certain arithmetical fundamentals used by employees of the Ohio Oil Company of Rusk County, Texas. M, 1949, [University of Houston].

Fisher, Laura Lee. Paleontology, stratigraphy and depositional environment of the Zorritas Formation, Sierra de Almeida, northern Chile. M, 1988, University of Idaho. 106 p.

Fisher, Lloyd Wellington. Chromite; its paragenesis, mineral, and chemical composition. D, 1929, The Johns Hopkins University.

Fisher, Lloyd Wellington. The Helderberg Limestone of Pennsylvania. M, 1923, Pennsylvania State University, University Park. 107 p.

Fisher, Louis A. The influence of electrolyte on the surface forces of a mineral. M, 1972, Rensselaer Polytechnic Institute. 80 p.

Fisher, Lysabeth Ann. On the structure and stratigraphic relations of the limestone conglomerates south and east of Pleasant Valley, New York. M, 1947, Columbia University, Teachers College.

Fisher, Marci A. Transport and fate of organic chemicals in Bandelier Tuff at Los Alamos National Laboratory chemical waste site, New Mexico. M, 1988, University of Texas at El Paso.

Fisher, Mark P. Sedimentology and stratigraphy of the Pennington Formation, Upper Mississippian in south-central Kentucky. M, 1981, University of Cincinnati. 234 p.

Fisher, Michael A. Investigation of the Earth's interior by deformation caused by atmospheric pressure. M, 1971, Stanford University.

Fisher, Mildred. The geology of Kelleys Island. M, 1948, Ohio State University.

Fisher, Mildred. The geology of Kelleys Island (Ohio). M, 1922, Ohio State University.

Fisher, Neil E. Geology of the Hilda-northwest area, Mason County, Texas. M, 1960, Texas A&M University.

Fisher, Nicholas Seth. Effects of chlorinated hydrocarbon pollutants on growth of marine phytoplankton in culture. D, 1974, SUNY at Stony Brook.

Fisher, R. D. The petrology of the Lower Cretaceous Cheyenne Sandstone of South-central Kansas. M, 1977, Fort Hays State University. 260 p.

Fisher, R. Stephen. Strontium isotopes in surface waters of a portion of the Susquehanna River drainage basin, Pennsylvania and Maryland. M, 1975, Miami University (Ohio). 71 p.

Fisher, Richard Forrest. The bedrock topography of the Arlington Heights, Evanston, Highland Park, Park Ridge, and Wheeling, Illinois quadrangles (F536). M, 1941, University of Illinois, Urbana. 10 p.

Fisher, Richard Virgil. Stratigraphy of the Puget Group and Keechelus Group in the Elbe-Packwood area of south western Washington. D, 1957, University of Washington. 157 p.

Fisher, Robert E. The geochemistry and petrography of the Derby Pluton, Memthramagog Quadrangle, Vermont. M, 1984, SUNY at Binghamton. 113 p.

Fisher, Robert Lloyd. Geomorphic and seismic-refraction studies of the Middle American Trench, 1952-56. D, 1957, University of California, Los Angeles. 67 p.

Fisher, Robert Stephen. Diagenetic history of Eocene Wilcox sandstones and associated formation waters, south-central Texas. D, 1982, University of Texas, Austin. 201 p.

Fisher, Robert Wilson. Geographical variation of maximum grain size in the Tar Springs Sandstone. M, 1956, University of Illinois, Urbana.

Fisher, Stanley Parkins, Jr. Geology of Emmons County, North Dakota. D, 1952, Cornell University.

Fisher, Stanley Perkins. A mineral and sedimentation study of the Simpson Group sandstones, Arbuckle Mountains area, Oklahoma. M, 1948, University of Oklahoma. 57 p.

Fisher, Susan Richards. Sedimentology and stratigraphy of the Late Devonian and Early Mississippian Pinyon Peak and Fitchville formations, central Utah. M, 1984, University of Utah. 162 p.

Fisher, Victor A. The Southern Ocean 700,000 years ago. D, 1968, Florida State University.

Fisher, Victor Arthur. The Bear Rock Formation of northeastern British Columbia. M, 1963, University of Montana. 58 p.

Fisher, Walter William. A leaching mechanism for chalcocite. D, 1970, New Mexico Institute of Mining and Technology. 107 p.

Fisher, William L. Geology of the Menard Formation (Mississippian) in southern Illinois. M, 1958, University of Kansas. 158 p.

Fisher, William L. Stratigraphic and geographic variation of population characteristics of fusulinids in the upper Hughes Creek Shale (Permian) of Wabaunsee County, Kansas. M, 1971, University of Kansas. 76 p.

Fisher, William L. Upper Paleozoic and lower Mesozoic stratigraphy of Parashant and Andrus canyons, Mohave County, northwestern Arizona. D, 1961, University of Kansas. 407 p.

Fisher, William Lee. Variation in stratigraphy and petrology of the uppermost Hamlin Shale and Americus Limestone related to the Nemaha structural trend in northeast Kansas. D, 1980, University of Kansas. 175 p.

Fisher, Wilson, Jr. Magnetic studies on soda-lime silicate glasses containing iron equilibrated at various oxygen partial pressures. M, 1973, Pennsylvania State University, University Park. 53 p.

Fishman, Howard Stephan. Geologic structure and regional gravity of a portion of the High Plateaus of Utah. M, 1976, University of Utah. 133 p.

Fishman, Kenneth Lawrence. Constitutive modeling of idealized rock joints under quasi-static and cyclic loading. D, 1988, University of Arizona. 309 p.

Fishman, Neil Steven. Origin of the Mariano Lake uranium deposit, McKinley County, New Mexico. M, 1981, University of Colorado. 97 p.

Fishman, Paul Harold. Mineralogical analysis and uranium distribution of the sediments from the upper Jackson Formation, Karnes County, Texas. M, 1978, Texas A&M University. 103 p.

Fisk, Edward P. Geology and ground-water resources of a portion of the Great Divide Basin, Carbon and Sweetwater counties, Wyoming. M, 1967, University of Southern California.

Fisk, Harold Norman. A microscopic study of basalt flows. D, 1935, University of Cincinnati. 135 p.

Fisk, Harold Norman. The history and petrography of the basalts of Oregon. M, 1931, University of Oregon. 132 p.

Fisk, Henry Grunsky. Solid solution on the spinel minerals and its relation to their use as refractories. D, 1927, Ohio State University.

Fisk, L. H. Palynology of the Amethyst Mountain "fossil forest", Yellowstone National Park, Wyoming. D, 1976, Loma Linda University. 357 p.

Fisk, M. R. Melting relations and mineral chemistry of Iceland and Reykjanes Ridge basalts. D, 1978, University of Rhode Island. 274 p.

Fisk, Peter J. A seismic method of determining various elastic moduli in homogeneous unconsolidated materials. M, 1967, Boston College.

Fiske, Douglas A. Stratigraphy, sedimentology, and structure of the Late Cretaceous Nanaimo Group, Hornby Island, British Columbia. M, 1978, Oregon State University. 164 p.

Fiske, Richard Sewell. Stratigraphy and structure of early and middle Tertiary rocks, Mount Ranier National Park, Washington. D, 1960, The Johns Hopkins University. 163 p.

Fissell, Donald Evan. Geology of the post-Paleozoic section of the Illipah region, Nevada. M, 1955, University of Southern California.

Fitch, David C. Geology and ore deposits of the Comet District, Lincoln County, Nevada. M, 1969, University of New Mexico. 97 p.

Fitch, Frank Williams, III. Petrochemical investigation of the Herefoss Pluton of South Norway. M, 1975, Stanford University.

Fitch, Richard. A study of petroleum occurrences and possibilities in the Northeast Vanderburgh County area, Indiana. M, 1958, Indiana University, Bloomington. 38 p.

Fitch, Thomas Jelstrup. Mechanism of underthrusting in southwest Japan; a model of convergent plate interactions. D, 1971, University of Colorado. 101 p.

Fitchko, R. M. Topography, shallow structure, and sedimentary processes of the Atlantic continental slope off the Carolina coast. M, 1976, Old Dominion University. 114 p.

Fitchko, Y. The distribution, mobility and accumulation of nickel, copper and zinc in a river system draining the eastern part of the metal-polluted Sudbury smeltering area. D, 1978, University of Toronto.

Fitgerald, Michael G. Anthropogenic influence on the sedimentary regime of an urban estuary; Boston Harbor. D, 1980, Woods Hole Oceanographic Institution. 297 p.

Fithian, Patricia A. The family Cytherideidae (Ostracoda) in the Cretaceous of the Atlantic Coastal Plain. M, 1977, University of Delaware.

Fithian, Patricia Ann. Distribution and taxonomy of the Ostracoda of the Paria-Trinidad-Orinoco shelf. D, 1980, Louisiana State University. 557 p.

Fitter, Francis L. Stratigraphy and structure of the French Mesa area, Rio Arriba County, New Mexico. M, 1958, University of New Mexico. 66 p.

Fitter, Jeffrey L. Seismic investigation near Silver Island in the Great Salt Lake Desert, Utah. M, 1985, University of Oklahoma. 75 p.

Fitterman, David Vincent. Electrical resistivity variations and fault creep behavior along strike-slip fault systems. D, 1975, Massachusetts Institute of Technology. 198 p.

Fitts, Charles R. Physical properties of sediments on the Copper River prodelta, Alaska. M, 1980, Cornell University.

Fituri, Hussein Saleh. Study of selected sandstone reservoirs in the Indiana portion of the Griffin consolidated field that lies above New Harmony. M, 1987, Indiana University, Bloomington. 167 p.

Fitzgerald, Cathleen Marie. The Strategic Petroleum Reserve; environmental impacts associated with the leaching of salt caverns. D, 1981, University of California, Los Angeles. 137 p.

Fitzgerald, Colleen E. Deposition and diagenesis of the Hibernia Member, Jeanne d'Arc Basin, offshore Newfoundland. M, 1987, Dalhousie University. 140 p.

Fitzgerald, David J. Petrology of the Pennsylvanian sandstones of Muscatine County, Iowa. M, 1977, University of Iowa. 215 p.

FitzGerald, Duncan M. Hydraulics, morphology and sediment transport at Price Inlet, South Carolina. D, 1977, University of South Carolina. 96 p.

Fitzgerald, Duncan Martin. Consolidation studies of deltaic sediments (Recent, Gulf of Mexico). M, 1972, Texas A&M University.

Fitzgerald, Edward Leo. Geology of the Ghost River map-area, Alberta. M, 1961, University of Alberta. 118 p.

Fitzgerald, Francis Bell, III. Geology of Little North Mountain and adjacent area in west-central Augusta County, Virginia. M, 1966, University of Virginia. 102 p.

Fitzgerald, Haile Vandenburgh, Jr. Correlation of the Eggleston Formation and related beds in southwestern Virginia. M, 1953, Virginia Polytechnic Institute and State University.

Fitzgerald, James Francis, Jr. Geology of the Woods Peak area, Black Mountain Quadrangle, Idaho. M, 1977, University of Idaho. 70 p.

Fitzgerald, Jennifer Kerry. Gabbro pegmatites of Eureka Peak, Plumas County, California. M, 1958, University of California, Los Angeles.

Fitzgerald, John Joseph. The role of cation exchange of glauconite in the aqueous geochemistry of the Aquia Aquifer in the Maryland Coastal Plain. M, 1988, University of Maryland.

Fitzgerald, Kim. Bedrock geology of the western half of the Royalston Quadrangle, Massachusetts. M, 1960, University of Massachusetts. 120 p.

Fitzgerald, Mary. Clast-contact conglomerates in submarine canyons; possible subaqueous sieve deposits. M, 1988, University of Southern California.

Fitzgerald, Michael Gerard. Anthropogenic influence on the sedimentary regime of an urban estuary; Boston Harbor. D, 1980, Massachusetts Institute of Technology. 297 p.

Fitzgerald, Sharon. The crystal chemistry and structure of vesuvianite. D, 1985, University of Delaware. 299 p.

Fitzgerald, Wilfred Harold. Rhodochrosite-rhodonite relationships in the Butte mining district, Montana. M, 1942, Montana College of Mineral Science & Technology. 31 p.

Fitzgerald, William Donald. Post-glacial history of the Minesing Basin, Ontario. M, 1982, University of Waterloo. 94 p.

Fitzgerald, William Francis. A study of certain trace metals in sea water using anodic stripping voltammetry. D, 1970, Massachusetts Institute of Technology. 180 p.

Fitzgibbon, James Lavern. A survey of the Earth science program in Nebraska middle schools. M, 1971, Chadron State College.

Fitzpatrick, Jack Cleo. Geology of Kent Draw, Culberson County, Texas (F582). M, 1950, University of Texas, Austin.

Fitzpatrick, Joan Juliana. Studies in the microstructure and crystal chemistry of minerals; I, Burbankite from the Green River Formation, Wyoming; II, Electron microscopy of staurolite; III, Crystal structure determination of vaugnatite, CaAlSiO$_4$(OH). D, 1976, University of California, Berkeley. 118 p.

Fitzpatrick, John Cole. Interpretation and significance of a major positive gravity anomaly in central Massachusetts. M, 1978, University of Massachusetts. 45 p.

Fitzpatrick, Kathleen. Ultrastructure and minor element composition of the test wall of Profusulinella from the Ely Limestone, east-central Nevada. M, 1982, University of Georgia.

Fitzpatrick, Kenneth. An electron microscopic examination of quartz-size sand particles. M, 1967, Ohio State University.

Fitzpatrick, Kenneth Thomas. Morphology and taxonomy of Platystrophia (Brachiopoda). D, 1972, Miami University (Ohio). 165 p.

Fitzpatrick, M. Gravity in the eastern townships of Quebec. D, 1957, University of Toronto.

Fitzpatrick, Mark. The geology of the Rossland Group in the Beaver Valley area, southeastern British Columbia. M, 1985, University of Calgary. 151 p.

Fitzpatrick, Michael F. Origin of the present course of the Potomac River near Paw Paw, West Virginia. M, 1980, Indiana State University. 73 p.

Fitzpatrick, Michael Morson. Gravity in the Eastern Townships of Quebec. D, 1960, Harvard University.

Fitzpatrick, Robert Charles and Chapman, William Brewer. Terrestrial tides as determined from deviations of gravity at Ann Arbor, Michigan. M, 1950, University of Michigan.

Fitzpatrick, Thomas Frank. Gold-quartz vein mineralization in Stanly County, North Carolina. M, 1983, Virginia Polytechnic Institute and State University. 8 p.

Fitzsimmons, Clifford Lynn. A trial model demonstrating potentials for detailed simulations of benthic macroinvertebrate populations. M, 1984, University of Idaho. 118 p.

Fitzsimmons, J. Paul. Petrology of SW quarter, Pine Quad., Ore. D, 1949, University of Washington. 154 p.

Fiuzat, Abbas-Ali. Subcritical flow analyses of open channel constrictions. D, 1980, Colorado State University. 34 p.

Fix, Carolyn E. Geology of the western Wind River basin, Wyoming. M, 1953, Syracuse University. 29 p.

Fix, James Edward, Sr. The crust and upper mantle of central Mexico and other seismic studies. D, 1974, Southern Methodist University. 102 p.

Fix, M. F. The paleontology and stratigraphy of the Smithville and Blackrock formations of southeastern Missouri. M, 1975, Washington University. 147 p.

Fix, Philip Forsyth. Structure of the Gallatin Valley, Montana. D, 1940, University of Colorado. 68 p.

Fix, Philip Forsyth. The Knobstone Escarpment north of the glacial boundary. M, 1931, Indiana University, Bloomington.

Flaate, Kaare Sigfred. Stresses and movements in connection with braced cuts in sand and clay. D, 1966, University of Illinois, Urbana. 281 p.

Flaccus, Christopher Edward. Stress on the San Andreas Fault; an analysis of shallow stress relief measurements made near Palmdale, California. M, 1988, University of Arizona. 86 p.

Flach, Klaus Werner. The influence of cultivation and different management systems on the organic matter content of soils in central New York. M, 1954, Cornell University.

Flach, Peter Donald. A lithofacies analysis of the McMurray Formation, lower Steepbank River, Alberta. M, 1977, University of Alberta. 139 p.

Flagg, Charles Noel. The kinematics and dynamics of the New England continental shelf and shelf/slope front. D, 1977, Massachusetts Institute of Technology. 207 p.

Flaherty, G. F. Banded ores and their textures. M, 1929, Massachusetts Institute of Technology. 80 p.

Flaherty, G. F. Geology of Chignecto area, New Brunswick. D, 1933, Massachusetts Institute of Technology. 398 p.

Flaherty, Gerard Martin. The Western Cascade-High Cascade transition in the McKenzie Bridge area, central Oregon Cascade Range. M, 1981, University of Oregon. 178 p.

Flahive, Mary E. Triassic period in the United States; a review of the literature. M, 1940, Smith College. 152 p.

Flammer, Gordon H. The use of ultrasonics in the measurement of suspended sediment size distribution and concentration. D, 1958, University of Minnesota, Minneapolis. 67 p.

Flanagan, J. T. Comparative study of mineral occurrences of the South Chibougamau area (N. Quebec). M, 1954, University of Toronto.

Flanagan, James Joseph. Sedimentology of the upper Miocene Republic Sand of the southern San Joaquin Valley, California. M, 1980, University of Massachusetts. 135 p.

Flanagan, Peter William. Analysis of complex resistivity models in spheroidal geometries. M, 1983, University of Arizona. 119 p.

Flanagan, Philip E. Geology of the Mud Creek area, Hot Springs County, Wyoming. M, 1955, University of Wyoming. 57 p.

Flanagan, William Hamilton. Geology of the southern part of the Snowcrest Range, Beaverhead County, Montana. M, 1958, Indiana University, Bloomington. 41 p.

Flanders, Richard William. Geology of the Chiquibul area, Belize, Central America. M, 1978, University of Idaho. 62 p.

Flangas, William G. Underground mining operations at Ruth, Nevada, 1951-1958. M, 1958, University of Nevada - Mackay School of Mines. 139 p.

Flanigan, Donna M. Herring. Facies and paleogeography of the Middle Silurian Bisher and Lilley formations, Adams County, Ohio. M, 1986, University of Cincinnati. 151 p.

Flanigan, Kenneth R. A sequence of field trips dealing with sedimentary deposition in Clinton County from Late Ordovician Period to the end of the Pennsylvanian Period. M, 1976, Pennsylvania State University, University Park. 74 p.

Flanigan, T. Edward, III. Pore fluid pressure regimes, Brazoria County, Texas; a preliminary study. M, 1980, University of Texas, Austin.

Flannery, James William, Jr. The acoustic stratigraphy of Lake Malawi, East Africa. M, 1988, Duke University. 120 p.

Flaschen, Steward S. The petrology and petrography of the northern group of the Rockefeller Mountains, Antarctica, and a critical analysis of the igneous reaction series. M, 1948, Miami University (Ohio). 49 p.

Flaschen, Steward Samuel. A hydrothermal study of the system FeO-SiO$_2$-H$_2$O. D, 1953, Pennsylvania State University, University Park. 55 p.

Flaten, Luvern L. Stratigraphic relations of the (Lower Ordovician) Prairie du Chien Formation in eastern Wisconsin. M, 1959, University of Wisconsin-Madison.

Flatt, C. D. Origin and significance of the oyster banks in the Walnut Clay Formation, central Texas. M, 1975, Baylor University. 149 p.

Flaugher, David Michael. A gravity survey of the Serpent Mound cryptoexplosion structure and surrounding area in southern Ohio. M, 1973, Wright State University. 114 p.

Flawn, Peter Tyrell. Geology of the Mica Mine area, Culberson and Hudspeth counties, Texas. D, 1950, Yale University.

Flax, Philip. Depositional and diagenetic characteristics of the Upper Devonian (Frasnian) Duperow Formation in west central North Dakota. M, 1987, Queens College (CUNY).

Flebbe, Patricia Ann. Biogeochemistry of carbon, nitrogen, and phosphorus in the aquatic subsystem of selected Okefenokee Swamp sites. D, 1982, University of Georgia. 365 p.

Fleck, Kenneth Stewart. Radiometric and petrochemical characteristics of the Dells Granite, Yavapai County, Arizona. M, 1983, University of Missouri, Rolla. 89 p.

Fleck, Robert Joseph. Structural significance of the contact between Franciscan and Cenozoic rocks, southern San Francisco Peninsula, California. M, 1967, Stanford University.

Fleck, Robert Joseph. The magnitude, sequence and style of deformation in southern Nevada and eastern California. D, 1967, University of California, Berkeley. 99 p.

Fleck, William Pyle. Cyclic sedimentation in the lower Allegheny Series of the Kittanning area. M, 1953, University of Pittsburgh.

Fleckenstein, Martin. Precambrian geology of Squaw Mountain, central Laramie Range, Wyoming. M, 1980, Colorado School of Mines. 88 p.

Fleece, James B. The carbonate geochemistry and sedimentology of the Keys of Florida Bay, Florida. M, 1962, Florida State University.

Fleeger, Gary Mark. Pre-Wisconsinan till stratigraphy in the Avon, Canton, Galesburg, and Maquon 15-minute quadrangles, western Illinois. M, 1980, University of Illinois, Urbana. 90 p.

Fleener, Frank Leslie. The Ordovician section and faunal succession in southeastern Minnesota. M, 1914, University of Illinois, Chicago.

Fleer, Varda Nanette. The dissolution kinetics of anorthite (CaAl$_2$Si$_2$O$_8$) and synthetic strontium feldspar (SrAl$_2$Si$_2$O$_8$) in aqueous solutions at temperatures below 100°C; with applications to the geological disposal of radioactive nuclear wastes. D, 1982, Pennsylvania State University, University Park. 212 p.

Fleer, Varda Nanette. The low-temperature, aqueous solution chemistry of SnO$_2$ (cassiterite). M, 1979, Pennsylvania State University, University Park. 60 p.

Fleet, John M. Van see Van Fleet, John M.

Flegal, A. R., Jr. A biogeochemical study of manganese in the estuarine zone. D, 1979, Oregon State University. 124 p.

Flege, Robert Frederick, Jr. Geology of the Lordsburg Quadrangle, Hidalgo County, New Mexico. D, 1956, Washington University. 72 p.

Flege, Robert Frederick, Jr. Minor structural features associated with the Kentucky River fault zone. M, 1952, University of Kentucky. 35 p.

Fleischer, Peter. Mineralogy of hemipelagic sediments, California continental borderland. D, 1970, University of Southern California. 220 p.

Fleischhauer, Henry Louis, Jr. Quaternary geology of Lake Animas, Hidalgo County, New Mexico. M, 1977, New Mexico Institute of Mining and Technology. 149 p.

Fleischmann, Karl H. The orientation and microstructure of crinoid stems, Grainger Formation (Mississippian), southeastern Kentucky and northeastern Tennessee. M, 1985, Miami University (Ohio). 78 p.

Fleischmann, Marianne Lynn. Experimental determination of permeability of intact and fractured rocks. M, 1978, University of South Carolina.

Fleisher, Penrod Jay. Glacial geology of the Big Pine drainage, Sierra Nevada, California. D, 1967, Washington State University. 128 p.

Fleisher, Penrod Jay. Structural control of the igneous intrusions of the Durham Triassic basin, N.C. M, 1963, University of North Carolina, Chapel Hill. 37 p.

Fleisher, R. L. Early Eocene to early Miocene planktonic foraminiferal biostratigraphy of the western Indian Ocean. D, 1975, University of Southern California.

Flemal, Ronald Charles. Sedimentology of the Sespe Formation (upper Eocene and Oligocene), southwestern California. D, 1967, Princeton University. 258 p.

Fleming, Alfred J. The effects of urbanization on channel morphology in southeastern Wisconsin. M, 1978, University of Wisconsin-Milwaukee.

Fleming, Alfred John. Historical stream channel disequilibrium in north-central Illinois; causes and contemporary manifestations. D, 1987, Northern Illinois University. 156 p.

Fleming, Anthony H. The determination of joint system characteristics from azimuthal resistivity surveys. M, 1986, University of Wisconsin-Madison.

Fleming, H. W. W. The major structural features of the Abitibi-Timiskaming area of Ontario and Quebec. M, 1946, University of Toronto.

Fleming, John M. Petrology of the volcanic rocks of the Whalesback area, Springdale peninsula, Newfoundland. M, 1970, Memorial University of Newfoundland. 85 p.

Fleming, Lorraine Nellita. The strength and deformation characteristics of Alaskan offshore silts. D, 1985, University of California, Berkeley. 171 p.

Fleming, Randall J. A study of the Hartselle Sandstone (Mississippian) in Morgan and Lawrence counties, Alabama. M, 1966, University of Alabama.

Fleming, Ray Edward. Surface geology of northeastern Cleveland and southeastern Oklahoma counties, Oklahoma. M, 1957, University of Oklahoma. 47 p.

Fleming, Richard Howell. Oceanographic studies in the Central American Pacific. D, 1935, University of California, Berkeley. 168 p.

Fleming, Robert Eugene, Jr. Crystalline rocks of the northern half of the Farrington Quadrangle, North Carolina. M, 1958, University of North Carolina, Chapel Hill. 28 p.

Fleming, Robert S., Jr. A geophysical study of the southwestern end of Scranton gravity high. M, 1975, Lehigh University. 40 p.

Fleming, Robert William. Soil creep in the vicinity of Stanford University (Palo Alto, California). D, 1972, Stanford University. 148 p.

Fleming, Sheryl Denise. Paleobiology of the Roberts Mountains Formation, Bullfrog Hills, Nevada. M, 1979, San Diego State University.

Fleming, William Jeffrey. Identification of gypsum using near-infrared photography and digital Landsat imagery. M, 1980, Oklahoma State University. 168 p.

Fleming, William L. Glacial geology of central Long Island. M, 1931, Yale University.

Fleming, William M. Geohydrology of a mountain peat wetland, Medicine Bow Mountains, Wyoming. M, 1966, Colorado State University. 124 p.

Flemings, Peter Barry. The paleogeography of the Maastrichtian and Paleocene Wind River basin; interpreting Rocky Mountain foreland deformation. M, 1987, Cornell University. 175 p.

Flesch, Gary A. Stratigraphy, sedimentology, and environments of deposition of the Morrison (Upper Jurassic) Formation, Ojito Spring Quadrangle, Sandoval County, New Mexico. M, 1975, University of New Mexico. 106 p.

Flessa, Karl Walter. Evolutionary pulsations; evidence from Phanerozoic diversity patterns. D, 1973, Brown University.

Fletcher, Charles Henry III. The detail of beach profile fluctuation under a changing wave regime; a case study. M, 1983, University of Delaware. 212 p.

Fletcher, Charles Henry, III. Stratigraphy and reconstruction of the Holocene transgression; a computer aided study of the Delaware Bay and inner Atlantic Shelf. D, 1987, University of Delaware. 515 p.

Fletcher, Charles S. The geology and hydrogeology of the New Lead Belt, Missouri. M, 1974, University of Missouri, Rolla.

Fletcher, Christopher John Nield. Local equilibrium in a two-pyroxene amphibolite. M, 1968, Queen's University. 80 p.

Fletcher, Christopher John Nield. Structure and metamorphism of Penfold Creek area, near Quesnel Lake, central British Columbia. D, 1972, University of British Columbia.

Fletcher, Clark S. A geochemical study of the effects coal mining has had on the Piney Creek watershed, Van Buren County, Tennessee. M, 1977, University of Tennessee, Knoxville. 92 p.

Fletcher, Claude Osborne. Geology of an area one mile south of Austin, Texas. M, 1932, University of Texas, Austin.

Fletcher, Darby Ian. Geology and genesis of the Waterloo and Langtry silver-barite deposits, California. D, 1986, Stanford University. 212 p.

Fletcher, Frank William. Middle and Upper Devonian stratigraphy of southeastern New York. D, 1964, University of Rochester. 197 p.

Fletcher, G. L. and Irvine, T. N. Areal geology of the Emo area, Rainy River District, 1 inch to 1 mile (Ontario). M, 1953, University of Manitoba.

Fletcher, Herbert C. An interpretation and differentiation of glacial deposits by the use of petrographic methods. M, 1935, University of Missouri, Columbia.

Fletcher, Ian Robert. A lead isotope study of some sulphide mineral deposits in the Grenville structural province of Canada. M, 1974, University of Toronto.

Fletcher, Ian Robert. A lead isotopic study of lead-zinc mineralization associated with the Central Metasedimentary Belt of the Grenville Province. D, 1979, University of Toronto.

Fletcher, John P. B. Studies of the seismicity and tectonics in eastern North America with emphasis on western New York, and an analysis of Blue Mountain Lake accelerograms. D, 1976, Columbia University. 179 p.

Fletcher, Jonathan. Chromite from some western North Carolina ultramafites and its petrogenetic implications. M, 1978, University of South Carolina.

Fletcher, M. R. Distribution patterns of Holocene and reworked Paleozoic palynomorphs in sediments from southeastern Lake Michigan. D, 1979, Case Western Reserve University. 331 p.

Fletcher, Raymond Charles. A finite amplitude model for the emplacement of gneiss domes and salt domes. D, 1967, Brown University. 235 p.

Fletcher, Raymond Charles. Petrology of the Hedgehog Volcanics, Aroostook County, Maine. M, 1962, Brown University.

Fletcher, Ruth Reilly. Planktonic foraminifera from the Oligocene and early Miocene of Jones and Onslow counties, North Carolina. M, 1987, University of Delaware. 417 p.

Fletcher, Thomas E. The structural geology of the Caddo Anticline, Carter County, Oklahoma. M, 1986, Baylor University. 83 p.

Fleury, Bruce. The geology, origin, and economics of manganese at Roads End, California. M, 1961, University of Southern California.

Fleury, Mark Gerald Roland. The size distribution of quartz grains in mudrocks. M, 1971, University of Oklahoma. 44 p.

Flewellen, Barbour H. Geology of Burketown Klippe and vicinity, Harrisonburg Quadrangle, Augusta County and Rockingham County, Virginia. M, 1950, University of Virginia. 82 p.

Flexser, Steven. Geology of a portion of the Sonoma Volcanics near Calistoga, Napa County. M, 1980, University of California, Berkeley. 106 p.

Fliegner, J. F. A groundwater model and the hydrogeology of the San Jacinto Valley, Riverside County, California with emphasis on the Soboba Indian Reservation. M, 1978, Stanford University. 189 p.

Flier, Eileen Van Der *see* Van Der Flier, Eileen

Fligelman, Haim. Drawdown and interference test analysis for gas wells with wellbore storage, damage, and nonlaminar flow effects. D, 1981, Stanford University. 241 p.

Flight, Wilson R. The Holocene sedimentary history of Jeffreys Basin. M, 1973, University of New Hampshire. 173 p.

Flih, Baghdad. Osagian bioherms at the Bishop Cap Hills, Dona Ana County, New Mexico. M, 1976, University of Texas at El Paso.

Flinn, Donald J. Mineralogy of the Black Hills, South Dakota and Wyoming bentonites. M, 1959, South Dakota School of Mines & Technology.

Flinn, Douglas Lowell. The geology of the Cerro Macho area, Sonora, Mexico. M, 1977, Northern Arizona University. 73 p.

Flinn, Edward Ambrose, III. Exact transient solution of some problems of elastic wave propagation. D, 1960, California Institute of Technology. 118 p.

Flint, Arthur. Stratigraphic relations of the Shakopee Dolomite and the St. Peter Sandstone in southwestern Wisconsin. D, 1954, University of Chicago. 25 p.

Flint, David C. Geology, trace element geochemistry and hydrothermal alteration at the Buckhorn gold mine, Eureka County, Nevada. M, 1987, University of Nevada. 98 p.

Flint, David Warren. Estimation of coal resource quantities by statistical methods. M, 1978, University of Alberta. 163 p.

Flint, Delos E. The geology of the Beveridge mining district, California. M, 1941, Northwestern University.

Flint, Frederick F. Diagenesis of Tertiary (Miocene) sedimentary rocks of the Old Woman Mountains area, southeastern California. M, 1987, University of Texas at El Paso.

Flint, Jean Jacques. Hydrogeology and geomorphic properties of small basins between Endicott and Elmira, New York. M, 1968, SUNY at Binghamton. 78 p.

Flint, Jean-Jacques. Fluvial systems; a re-evaluation of Horton's laws (law of stream numbers). D, 1972, SUNY at Binghamton. 115 p.

Flint, Norman Keith. Geology of Monday Creek and Salt Lick townships, Perry County, Ohio. M, 1946, Ohio State University.

Flint, Norman Keith. The geology of Perry County, Ohio. D, 1948, Ohio State University.

Flint, Richard F. The geology of parts of Perry and Cape Girardeau counties, Missouri. D, 1925, University of Chicago. 247 p.

Flint, William B. Disposal of radioactive wastes (literature survey). M, 1958, Stanford University.

Flint, William B. Research on radioactive waste disposal in underground formations. M, 1958, Stanford University.

Flippin, Jerel Wayne. An evaluation of the stratigraphy, structure, and economic aspects of the Paleozoic strata in Erath County, North-central Texas. M, 1978, Texas Christian University. 68 p.

Flis, James Edward. K/Pb fractionation trends. M, 1975, Michigan State University. 67 p.

Flis, Marcus F. Induced polarization effects in time-domain electromagnetic measurements. M, 1985, University of Utah. 82 p.

Flock, William Merle. Mineralogy and petrology of the Andersonville, Georgia, bauxite district. D, 1966, Pennsylvania State University, University Park. 228 p.

Flock, William Merle. Soils and soil clay mineral formation in the Virginia Blue Ridge and Piedmont provinces. M, 1963, Virginia Polytechnic Institute and State University.

Flocks, Gerald Walter. Geology of Township 9 North, Range 26 West, Franklin County, Arkansas. M, 1953, University of Arkansas, Fayetteville.

Floess, C. H. L. Direct simple shear behavior of fine grained soils subjected to repeated loads. D, 1979, Rensselaer Polytechnic Institute. 195 p.

Flood, James Ray, Jr. Areal geology of western (Jefferson county), Oklahoma. M, 1969, University of Oklahoma. 61 p.

Flood, Raymond Edward, Jr. Structural geology of the upper Fishtrap Creek area, central Anaconda Range, Deer Lodge County, Montana. M, 1974, University of Montana. 71 p.

Flood, Roger Donald. Studies of deep-sea sedimentary microtopography in the North Atlantic Ocean. D, 1978, Woods Hole Oceanographic Institution. 360 p.

Flood, Timothy P. Cyclic evolution of a magmatic system; the Paintbrush Tuff, SW Nevada volcanic field. D, 1987, Michigan State University. 147 p.

Flood, Timothy P. Geology of the Cypress, Hanson and South Arm of Knife Lake area, Boundary Waters canoe area; eastern Vermilion District, northeastern Minnesota. M, 1981, University of Minnesota, Duluth.

Flora, Larry. Origin and distribution of sediment in the Pennsylvanian-Permian Keeler Canyon Formation, Inyo County, California. M, 1984, San Jose State University. 151 p.

Floran, R. J. Mineralogy and petrology of the sedimentary and contact metamorphosed Gunflint Iron Formation, Ontario-Minnesota. D, 1975, SUNY at Stony Brook. 365 p.

Floran, Robert J. A petrologic investigation of the Loon Bay Batholith in north-central Newfoundland. M, 1971, Columbia University. 62 p.

Florek, Robert J. Measurements of the flux of dissolved inorganic nutrients from marine coastal and shelf sediments. M, 1979, Wayne State University.

Florentino, Eugene. Distribution, petrographic analysis, and origin of the Granny Lake oolite, San Salvador, Bahamas. M, 1985, University of Akron. 99 p.

Florer, Linda E. The Pleistocene of South Bimini (Bahama Islands). M, 1966, Columbia University. 44 p.

Flores, Jorge G. Study of subnormal formation pressures based on geological and electrical log data, Keyes field, Cimarron County, Oklahoma. M, 1967, University of Oklahoma. 89 p.

Flores, Richard J. Sedimentation model for the Crestone Conglomerate Member of the Sangre de Cristo Formation (Pennsylvanian-Permian), south-central Colorado. M, 1984, Indiana University, Bloomington. 145 p.

Flores, Romeo M. The petrology of the Collier Shale and Crystal Mountain Sandstone of Montgomery County, Arkansas. M, 1962, University of Tulsa. 64 p.

Flores, Romeo Marzo. Middle Allegheny paleogeography in eastern Ohio. D, 1966, Louisiana State University. 148 p.

Flores, Victor. The energy and mineral resources of Mexico; strategies for development. M, 1986, University of Texas, Austin.

Flores, W. Adán Emigdio Z. A stochastic management model for the operation of a stream-aquifer system. D, 1976, New Mexico Institute of Mining and Technology. 209 p.

Flores, W. Adán Emigdio Z. Geohydrology study of Rio Sinaloa groundwater reservoir, Mexico. M, 1972, New Mexico Institute of Mining and Technology.

Flores-Espinoza, Emilio. Depositional environments and coal beds of the Bigford Formation (Eocene), northeastern Mexico. M, 1983, University of Texas, Austin. 120 p.

Flores-Luna, Carlos Francisco. Electromagnetic induction studies over the Meager Creek geothermal area, British Columbia. D, 1986, University of Toronto.

Flori, Ralph Emil, Jr. Numerical simulation of a condensate gas cap, oil, solution gas, and water system. D, 1987, University of Missouri, Rolla. 147 p.

Florian, Marc D. Factors controlling deposition of the Arcola Member of the Mooreville Formation (Upper Cretaceous) in east-central Mississippi and west-central Alabama. M, 1984, Michigan State University. 105 p.

Florsheim, Joan Leslie. Channel form and process; a modeling approach. D, 1988, University of California, Santa Barbara. 187 p.

Florstedt, James Edward. Dolomitization phenomena in reef and reef associated carbonates, Permian reef complex, Apache mountains, Texas. M, 1970, Texas Tech University. 87 p.

Flory, Donald Andrew. Stable carbon isotope ratio measurement of meteoritic carbonaceous material. D, 1969, [University of Houston]. 180 p.

Flory, Richard A. Devonian tabulate corals of the Great Basin. D, 1975, Oregon State University. 400 p.

Flory, Richard A. Paleoenvironments of a Coeymans Formation (lower Devonian) reef complex (eastern Pennsylvania). M, 1969, Temple University.

Flower, Rousseau Hayner. A study of the Pseudorthoceratidae (Nautiloidea). D, 1939, University of Cincinnati. 386 p.

Flowers, George Conrad. Equilibrium and mass transfer during progressive metamorphism of siliceous dolomites. D, 1979, University of California, Berkeley. 95 p.

Flowers, Glen Dwight. The depth to bedrock and the nature of the surficial material in the Carbondale, Illinois area. M, 1969, Southern Illinois University, Carbondale. 93 p.

Flowers, Russel R. A regional subsurface study of the Greenbrier Limestone. M, 1955, West Virginia University.

Floyd, Bobby Joe. Geology of the West Harpeth area, Spring Hill Quadrangle, Williamson County, Tennessee. M, 1951, Vanderbilt University.

Floyd, Jack Curtis. The depositional environments of some Silurian rocks, northern Ohio. M, 1971, Bowling Green State University. 231 p.

Floyd, James Gordon. Stratigraphic distribution of upper Eocene larger foraminifera from northeastern Levy County, Florida. M, 1962, University of Florida. 55 p.

Floyd, Larry Wayne. Stratigraphy and paleoecology of the Curlew Limestone Member of the Tradewater Formation (Pennsylvanian, Desmoinesian) in northern Christian County, Kentucky. M, 1985, Indiana University, Bloomington. 70 p.

Flud, Lowell Randle. Field joint study of the Potato Hills structure. M, 1970, University of Oklahoma. 72 p.

Flueckinger, Linda Ann. Geology of a portion of the Allensville 15' Quadrangle (Huntingdon County), Pennsylvania. M, 1967, Pennsylvania State University, University Park. 69 p.

Flueckinger, Linda Ann. Stratigraphy, petrography and origin of Tertiary sediments off the front of the Beartooth Mountains, Montana–Wyoming. D, 1970, Pennsylvania State University, University Park. 382 p.

Fluegeman, Richard H., Jr. The New Point Tongue of the Brainard Shale (Upper Ordovician) in southeastern Indiana; stratigraphy and community succession. M, 1979, Miami University (Ohio). 104 p.

Fluegeman, Richard Herbert. Paleocene benthic foraminiferal biostratigraphy, paleoecology, and paleoceanography of the eastern Gulf Coastal Plain. D, 1987, University of Cincinnati. 203 p.

Fluet, Darrell Wayne. Genesis of the Deer Trail Zn-Pb-Ag vein deposit, Washington, U.S.A. M, 1986, University of Alberta. 142 p.

Flug, M. Optimal energy-water-salinity strategies; Upper Colorado River basin. D, 1977, Colorado State University. 223 p.

Fluharty, D. L. International regulation of access to and use of resources in the Baltic Sea. D, 1977, University of Michigan. 320 p.

Fluhr, Thomas W. Geology of Jamaica, British West Indies. M, 1928, Columbia University, Teachers College.

Fluker, James C., III. Subsurface stratigraphic and structural study of the Smackover Formation in Claiborne Parish, Louisiana. M, 1978, Northeast Louisiana University.

Fluorie, Eric Juan de Dios. Particulate manganese in sea water stressing regimes in marine anoxic basins. D, 1972, University of Washington. 138 p.

Flure, Ernest La *see* La Flure, Ernest

Flurkey, Andrew J. The mineralogy and sedimentology of the Cambrian strata of eastern Iowa. M, 1976, Northern Illinois University. 122 p.

Flurkey, Andrew James. Depositional environment and petrology of the Medicine Peak Quartzite (early Proterozoic), southern Wyoming. D, 1983, University of Wyoming. 152 p.

Fly, Sterling Harper, III. Shallow marine environments of the Laborcita Formation (Wolfcampian), Sacramento Mountains, New Mexico. M, 1985, University of Texas, Austin. 61 p.

Flynn, Clinton J. Geology of the La Gloria–Presa Rodriguez area (Miocene, Eocene, Cretaceous and possibly older), Baja California, Mexico. M, 1969, University of San Diego.

Flynn, Dan Bruce. Geology of part of the Dixie Flats Quadrangle, Elko County, Nevada. M, 1957, University of California, Los Angeles.

Flynn, Elizabeth Chittenden. Effects of source depth on near source seismograms. M, 1986, Southern Methodist University. 107 p.

Flynn, John J. Correlation and geochronology of middle Eocene strata from the Western United States. D, 1983, Columbia University, Teachers College. 502 p.

Flynn, Lawrence John. Biostratigraphy and systematics of Siwalik Rhizomyidae (Rodentia). D, 1981, University of Arizona. 246 p.

Flynn, Lawrence John. Enamel microstructures of cricetid and heteromyid rodent incisors and their importance in rodent phylogeny. M, 1977, University of Arizona.

Flynn, Ronald Thomas. An experimental determination of rare earth partition coefficients between a chloride containing aqueous phase and silicate melts. D, 1977, Pennsylvania State University, University Park. 114 p.

Flynn, T. Filter-pressed partial melts; experimental formation of migmatites. M, 1976, SUNY at Binghamton. 74 p.

Focazio, Michael Joseph. A simulation study of coupling the surface and ground water in the Broad Brook watershed. D, 1988, University of Connecticut. 213 p.

Focht, John Doster. Pennsylvanian brachiopods from the lower Sangre de Cristo Formation of the Huerfano Quadrangle, Colorado. M, 1956, University of Michigan.

Focht, Thomas. Structural geology of the central portion of the Hamilton Quadrangle, Ouachita Mountains, Arkansas. M, 1981, University of Missouri, Columbia.

Focke, Helen M. The pre-glacial drainage of a portion of northeastern Ohio. M, 1928, Case Western Reserve University. 69 p.

Focken, C. M. A mathematical and physical basis of the Sundberg inductive method of electrical prospecting. M, 1935, Colorado School of Mines. 53 p.

Focone, Joseph A. A statistical study of trace element variation in lunar rocks. M, 1979, Brooklyn College (CUNY).

Fodemesi, Stephen Paul. The removal of convective effects induced during needle probe thermal conductivity measurements. M, 1979, University of Western Ontario.

Fodor, Ronald V. Chemistry, mineralogy, petrology of the mafic and intermediate lavas of the Black Range, New Mexico. D, 1971, University of New Mexico. 158 p.

Fodor, Ronald Victor. Petrography and petrology of the volcanic rocks in the Goldfield mountains (Maricopa County), Arizona. M, 1969, Arizona State University. 66 p.

Foell, Christopher J. Stratigraphy, carbonate petrography, paleontology and depositional environments of the Orman Lake Member, Greenhorn Formation (Upper Cretaceous, Cenomanian), eastern Black Hills, South Dakota and Wyoming. M, 1982, Indiana University, Bloomington. 173 p.

Foerste, August Frederic. The igneous and metamorphic rocks of the Narragansett Basin. D, 1890, Harvard University.

Foerster, Bernhard. Skeletal morphology, variability, and ecology of the bryozoan species Crisia eburnea in the modern reefs of Bermuda. M, 1970, Pennsylvania State University, University Park. 212 p.

Foerster, Eugene Paul. The effect of urbanization on watershed runoff (Tucson, Arizona). D, 1972, University of Arizona.

Fofonoff, Nick P. A theoretical study of zonally uniform oceanic flow. D, 1955, Brown University.

Fogarty, Charles F. Subsurface geology of the Denver Basin; Colorado (two volumes). D, 1952, Colorado School of Mines. 197 p.

Fogarty, Mark. The surficial geology of the North Branch Callicoon Creek valley, Sullivan County, New York. M, 1987, Queens College (CUNY). 76 p.

Fogelson, David E. A gravity survey of a portion of the Redwood Falls area, Minnesota. M, 1956, University of Minnesota, Minneapolis.

Fogg, Graham Edwin, Jr. A ground-water modelling study in the Tucson Basin. M, 1978, University of Arizona.

Fogg, Graham Edwin, Jr. Stochastic analysis of aquifer interconnectedness, with a test case in the Wilcox Group, East Texas. D, 1986, University of Texas, Austin. 236 p.

Fogg, James L. Soil respiration and mineral nitrogen in soil water in a subalpine forest. M, 1976, [Colorado State University].

Foggin, G. Thomas, III. The influence of basin geology, morphometry, and discharge upon the solute concentrations of headwater streams in western Montana. D, 1980, University of Montana. 131 p.

Foglia, Mauro Felice. Radionuclide transport in hydrogeologic media. D, 1981, University of California, Berkeley. 175 p.

Fograscher, Arthur Carl. The stratigraphy of the Green River and Crazy Hollow Formation of part of the Cedar Hills, central Utah. M, 1956, Ohio State University.

Foisy, Raymond Deane. Some fossil mollusks of Yakima County, Washington. M, 1967, Central Washington University. 58 p.

Foit, Franklin Frederick, Jr. A high temperature study of the anorthite structure and the x-ray diffraction effects of selected tektosilicates displaying superlattice maxima. D, 1968, University of Michigan. 314 p.

Foit, Franklin Frederick, Jr. The stability of some common minerals in the lunar environments. M, 1965, University of Michigan.

Fok, Henry W. Deep subsurface structure of the western part of the Pecos County, Texas. M, 1972, University of Texas, Austin.

Fok-Pun, Luis. Settling velocities of planktonic foraminifera; density variations and shape effects. M, 1982, Oregon State University. 33 p.

Folami, Samuel Lekan. Paleomagnetism and rock magnetism of Iceland drill core samples. M, 1980, University of Washington. 64 p.

Foland, Kenneth Austin. Cation and Ar^{40} diffusion in orthoclase. D, 1972, Brown University. 153 p.

Foland, Kenneth Austin. Jurassic and Cretaceous K-Ar ages of the White Mountain magma series (New Hampshire, Vermont, and Maine). M, 1969, Brown University.

Foland, Richard L. Pyrite-marcasite nodule petrogenesis in some central New York Devonian limestones. M, 1972, Syracuse University.

Foland, Sara S. Geochemistry, geochronology, and origin of an Archean greenstone-granite terrain, Wabigoon Subprovince, northwestern Ontario. M, 1982, University of Montana. 166 p.

Folchetti, John Robert. Structure and stratigraphy of the Parish magnetite deposit, St. Lawrence County, New York. M, 1962, Brown University.

Foley, Donald Charles. Diagenesis and porosity relationships of lower San Andres Formation, Quay and Roosevelt counties, New Mexico. M, 1978, Texas Tech University. 98 p.

Foley, Duncan. Environmental geology and land-use planning; on the Big Darby Creek, Ohio, watershed. M, 1973, Ohio State University.

Foley, Duncan. Geology of the Stonewall Mountain volcanic center, Nye County, Nevada. D, 1978, Ohio State University. 162 p.

Foley, Francis Daniel, Jr. Neogene seismic stratigraphy and depositional history of the lower Georgia coast and continental shelf. M, 1984, University of Georgia. 80 p.

Foley, Frank C. An X-ray and spectrographic study of lead sulphantimonites. D, 1938, Princeton University.

Foley, Jeffrey Arthur. Hydronium ion and water interactions with SiOSi, SiOAl, and AlOAl tetrahedral linkages. M, 1986, Virginia Polytechnic Institute and State University.

Foley, Jeffrey Young. Petrology, geochemistry, and geochronology of alkaline dikes and associated plutons in the eastern Mount Hayes and western Tanacross quadrangles, Alaska. M, 1984, University of Alaska, Fairbanks. 95 p.

Foley, L. L. The relation of the green schist to the iron ores of the Cuyuna Range. M, 1918, University of Minnesota, Minneapolis. 20 p.

Foley, Lucy J. Quaternary chronology of the Palouse Loess near Washtucna, eastern Washington. M, 1982, Western Washington University. 137 p.

Foley, Lucy Loughlin. Slack water sediments in the Alpowa Creek drainage, Washington. M, 1976, Washington State University. 55 p.

Foley, M. G. Scour and fill in ephemeral streams. D, 1976, California Institute of Technology. 201 p.

Foley, Mary Kathryn. Ecosystem disburbance and forest development in Acadia National Park. D, 1988, Boston University. 238 p.

Foley, Nora Katherine. Mineralogy and geochemistry of the Austinville-Ivanhoe District, Virginia. M, 1980, Virginia Polytechnic Institute and State University.

Foley, Patricia Louise. Depositional setting of the Permo-Carboniferous redbeds around Hillsborough Bay, Prince Edward Island, Canada. M, 1984, University of New Brunswick. 141 p.

Foley, Robert LeRoy. The Moscow Fissure local fauna, late Pleistocene (Woodfordian) vertebrates from the Driftless Area of Southwest Wisconsin. D, 1982, University of Iowa. 128 p.

Foley, Stephen Francis. Mineralogy, geochemistry, petrogenesis and structural relationships of the Aillik Bay alkaline intrusive suite, Labrador, Canada. M, 1982, Memorial University of Newfoundland. 210 p.

Foley, Steven L. Depositional framework and paleoenvironment of the Heiberg Formation, Sverdrup Basin, Canadian Arctic Archipelago. M, 1988, Laurentian University, Sudbury. 246 p.

Foley, William Clark. Instrumental magnitudes and spectral scaling of several significant pre-1960 eastern North American earthquakes. M, 1982, University of Kentucky. 166 p.

Foley, William James. Heavy mineral analyses of sands from the Texas barrier island complex. M, 1974, University of Texas, Arlington. 77 p.

Folger, Charles Lee, Jr. Wall rock alteration surrounding the Silver Hill massive sulfide deposit, North Carolina. M, 1986, Kent State University, Kent. 180 p.

Folger, David W. Geology of the lower Jack Creek area, Elko County, Nevada. M, 1958, Columbia University, Teachers College.

Folger, David Winslow. Trans-Atlantic sediment transport by wind. D, 1968, Columbia University. 189 p.

Folger, Peter F. The geology and mineralization at the Omar copper prospect, Baird Mountains Quadrangle, Alaska. M, 1988, University of Montana. 152 p.

Folinsbee, J. C. Wall-rock alteration at Giant Yellowknife gold mine, Northwest Territories. M, 1954, University of Western Ontario.

Folinsbee, Robert Allin. Heat flow over the equatorial Mid-Atlantic Ridge. M, 1969, Massachusetts Institute of Technology. 68 p.

Folinsbee, Robert Allin. The gravity field and plate boundaries in Venezuela. D, 1972, Massachusetts Institute of Technology. 160 p.

Folinsbee, Robert E. Gem cordierite from northern Canada. M, 1940, University of Minnesota, Minneapolis. 85 p.

Folinsbee, Robert E. Zone-facies of metamorphism in relation to ore deposits of the Yellowknife-Beaulieu region, Northwest Territories. D, 1942, University of Minnesota, Minneapolis. 103 p.

Folk, Robert Louis. Petrology and petrography of the Lower Ordovician Beekmantown carbonate rocks in the vicinity of State College, Pennsylvania. D, 1952, Pennsylvania State University, University Park. 366 p.

Folk, Robert Louis. Petrology of authigenic silica in the Beekmantown Group of central Pennsylvania. M, 1950, Pennsylvania State University, University Park. 109 p.

Folk, Stewart Huntley. The metamorphism and structure of a stratum of marble in the Harney Peak Granite. M, 1938, University of Iowa. 24 p.

Folkman, David N. Hydrogeology including ground water/surface water relationships in the vicinity of Cook, Nebraska. M, 1981, University of Nebraska, Lincoln.

Folkoff, Donald W. A preliminary investigation of the perthitic feldspars and muscovite micas from some selected Black Hills, South Dakota, pegmatites. M, 1977, University of Toledo. 91 p.

Folkoff, Michael Edward. Environmental controls of the geography of secondary clay minerals in the A-horizon of United States soils. D, 1983, University of Georgia. 187 p.

Follis, M. Drumlins; are they stratified glacial features by size and lithology?. M, 1977, Ball State University.

Follmer, Leon Robert. Solid distribution and stratigraphy in the mollic albaqualf region of Illinois. D, 1970, University of Illinois, Urbana. 155 p.

Follo, Michael Ford. Sedimentology of the Wallowa Terrane, northeastern Oregon. D, 1986, Harvard University. 326 p.

Folsom, Cynthia Elizabeth. Geology of the northern half of the Kings Creek Quadrangle, South Carolina. M, 1982, University of North Carolina, Chapel Hill. 88 p.

Folsom, Jerry Robert. The field stratigraphy and mapping of the Marsland Formation in Sioux County, Nebraska. M, 1954, University of Nebraska, Lincoln.

Folsom, Michael MacKay. Glacial geomorphology of the Hastings Quadrangle (southwestern Michigan). D, 1971, Michigan State University. 166 p.

Folsom, Theodore Robert. Apparatus, experiments, and theory concerning the oceanographic reversing thermometer. D, 1952, University of California, Los Angeles. 126 p.

Folwarczny, Joseph James. Molars from Pleistocene wood rats, Neotoma (Rodentia, Cricetidae). M, 1982, Fort Hays State University. 76 p.

Fon, Neal A. La see La Fon, Neal A.

Fonda, Shirley Smith. Cheilostome bryozoans in modern Bermuda reefs, systematics and ecology. D, 1976, Pennsylvania State University, University Park. 507 p.

Fong, Christopher Chung-Kuen. Paleontology of the lower Cambrian Archaeocyatha-bearing Forteau Formation in southern Labrador. M, 1968, Memorial University of Newfoundland. 227 p.

Fong, David G. Structural and petrographic studies of the Thirty Islands Lake synform (Frontenac County, Ontario). M, 1970, Queen's University. 112 p.

Fong, Kin Lan. Systematic prospecting. M, 1922, Columbia University, Teachers College.

Fonner, Robert F. Pleistocene terraces of the lower Little Kanawha River valley. M, 1951, West Virginia University.

Font, Robert G. The engineering geology of the greater Waco area (Texas). M, 1969, Baylor University. 400 p.

Font, Robert Geoseph. Engineering geology study of the instability of the South Bosque Shale and the Del Rio Clay in the Waco area. D, 1973, Texas A&M University. 140 p.

Fontaine, David Alex. The geology and ore genesis of the Bemco rare-earth deposit at Cranberry Lake, New Jersey. M, 1976, Rutgers, The State University, Newark. 40 p.

Fontaine, Paul Jean. Deleterious minerals in building stone. M, 1938, University of Minnesota, Minneapolis. 18 p.

Fontana, Michael R. Holocene tephrochronology of the Matanuska Valley, Alaska. M, 1988, University of Alaska, Fairbanks. 99 p.

Fontana, Ronald Victor. Depositional environment of the Mitchell Tongue of the Gloyd Member, Rodessa Formation (Lower Cretaceous), Duty, St. Mary, and Walnut Hill fields, Lafayette County, Arkansas. M, 1988, Northeast Louisiana University. 199 p.

Fontes, Mauricio Paulo Ferreira. Iron oxide mineralogy in some Brazilian Oxisols. D, 1988, North Carolina State University. 181 p.

Foo, Wayne Kim. Evolution of transverse structures linking the Purcell Anticlinorium to the western Rocky Mountains near Canal Flats, British Columbia. M, 1979, Queen's University. 146 p.

Foolad, Hamid Reza. Effects of surface and interlayer hydroxy-Al polymers on cation exchange properties of montmorillonite. D, 1984, University of California, Davis. 430 p.

Foord, E. E. Mineralogy and petrogenesis of layered pegmatite-aplite dikes in the Mesa Grande District, San Diego County, California. D, 1976, Stanford University. 358 p.

Foord, Eugene E. Compositional and selected trace element variations in mineral of the lower zone of the Kiglapait layered intrusion (Precambrian), (Lat. 57N, Long. 61.5W), (Labrador). M, 1969, Rensselaer Polytechnic Institute. 103 p.

Foos, Anabelle Mary. The mineralogy, petrography and geochemistry of the Eocene Lone Star iron ores, East Texas, and the Ordovician Hooker Ironstone, Northwest Georgia. D, 1984, University of Texas at Dallas. 260 p.

Foose, Michael Peter. The structure, stratigraphy, and metamorphic history of the Bigelow area, northwest Adirondacks, New York. D, 1974, Princeton University. 224 p.

Foose, Richard M. The geology of Kittatinny and Little Mountains, north of Harrisburg, Pennsylvania. M, 1939, Northwestern University.

Foose, Richard M. The manganese minerals of Pennsylvania. D, 1942, The Johns Hopkins University.

Foote, Charles Whittlesey. Notes upon the geological history of Cayuga and Seneca lakes; together with a few general remarks upon the glacial period. D, 1877, Cornell University.

Foote, Gary Ray. Fracture analysis in northeastern Illinois and northern Indiana. M, 1982, University of Illinois, Urbana. 193 p.

Foote, Martin William. Contact metamorphism and skarn development of the precious and base metal deposits at Silver Star, Madison County, Montana. D, 1986, University of Wyoming. 233 p.

Foote, Mary Ann. The spatial and temporal distribution of suspended algae and nutrients in the upper Hackensack River estuary. D, 1983, Rutgers, The State University, New Brunswick. 229 p.

Foote, Priscilla. The structure and petrology of Coltsfoot Mountain and the surrounding region. M, 1931, Columbia University, Teachers College.

Foote, Robert W. Curie-point isotherm mapping and interpretation from aeromagnetic measurements in the northern Oregon Cascades. M, 1986, Oregon State University. 115 p.

Foote, Royal S. The geology of the Houser Canyon Pegmatite with an X-ray analysis of monazite. M, 1951, California Institute of Technology. 22 p.

Forbes, Charles Frank. Pleistocene shoreline morphology of the Fort Rock Basin, Oregon. D, 1973, University of Oregon. 231 p.

Forbes, Donald Lawrence. Babbage River delta and lagoon; hydrology and sedimentology of an Arctic estuarine system. D, 1981, University of British Columbia.

Forbes, Dorothy Ann. Depositional history of the Ramp Creek Formation and Harrodsburg Limestone (Mississippian), Kinser Pike section, Monroe County, Indiana. M, 1975, Indiana University, Bloomington. 100 p.

Forbes, Edwin H. The epidote from Huntington, Massachusetts. D, 1903, Yale University.

Forbes, Gerald Eugene. Ground-water geology of Boone County, Missouri. M, 1958, University of Missouri, Columbia.

Forbes, Jeffrey. Carbon and oxygen isotopic composition of Holocene lake sediments from Okanogan County, Washington. M, 1987, University of Washington. 101 p.

Forbes, Robert Briedwell. The bedrock geology and petrology, Juneau ice field area, southeastern Alaska. D, 1959, University of Washington. 280 p.

Forbes, Robert Lyle. Average density estimation from Fourier transformed gravity data. M, 1982, University of Colorado. 72 p.

Forbes, Ronald Frederick Scott. Modeling of the electromagnetic response of mineral bodies as a function of conductivity. D, 1957, University of California, Los Angeles.

Forbes, Terence Robert. Sedimentation and soil genesis in a small tropical valley in Abak, Nigeria. D, 1981, Cornell University. 330 p.

Forbes, Warren Clarence, Jr. The system talc-minnesotaite. D, 1966, Brown University. 110 p.

Force, Eric Ronald. Sedimentation, petrology, and stratigraphy of the Kaihikuan, upper middle Triassic of South Island, New Zealand. D, 1970, Lehigh University. 301 p.

Force, Lucy McCartan. Atomodesma limestones of the Hokonui Belt, Permian, South Island, New Zealand. D, 1972, Lehigh University. 210 p.

Ford, Arthur Barnes. Geology and petrology, Glacier Peak Quadrangle, northern Cascades. D, 1959, University of Washington. 337 p.

Ford, Arthur Barnes. The petrology of Sulphur Mountain area, Glacier Peak quadrangle, Washington. M, 1957, University of Washington. 103 p.

Ford, Bruce Hicks. Geochemistry of water in a coastal mixing zone, northeastern Yucatan Peninsula. M, 1985, University of New Orleans. 90 p.

Ford, David Wayne. Devonian fauna in the concretionary Picos Member, Pimenteira Formation (Lower Devonian), Piauí, Brazil. M, 1965, University of Cincinnati. 93 p.

Ford, Gary Wayne. Genesis and trend of the lowermost unit of the Vamoosa Formation (Gypsy Sandstone) in parts of northeastern and central Oklahoma. M, 1978, Oklahoma State University. 68 p.

Ford, Glen Melvin. Some Pennsylvanian ostracodes from western Kentucky (F752). M, 1941, University of Illinois, Urbana. 35 p.

Ford, Graham Rudolph. A study of the Platteville Formation in Dakota, Goodhue and Rice counties, Minnesota. M, 1958, University of Minnesota, Minneapolis. 144 p.

Ford, James Timothy. Structural and geophysical analysis of the Athens Plateau-Ouachita core area, Arkansas. M, 1984, Southern Illinois University, Carbondale. 134 p.

Ford, John Philip. Bedrock geology in southwest Hamilton County, Ohio. D, 1965, Ohio State University.

Ford, John W. Biostratigraphic study of Pennsylvanian and Permian inliers in the Edwards Plateau region of west-central Texas. M, 1988, University of Texas, Arlington.

Ford, Kenneth Lloyd. Geology and geophysics of uraniferous pegmatites, Black Creek area, Palmerston Township, southeastern Ontario. M, 1983, Carleton University. 203 p.

Ford, Leonard N. Dinoflagellates, chlorophytes, and acritarchs from the Cooper Formation (Oligocene) of South Carolina. M, 1979, Virginia Polytechnic Institute and State University.

Ford, Leonard Neal, Jr. Palynology of the Grayson Formation (lower Cenomanian) of Texas, U.S.A. D, 1982, University of California, Los Angeles. 165 p.

Ford, Louis McKee. The insoluble residues of the Ste. Genevieve, Renault, and Paint Creek formations (upper Mississippian) at Mount Vernon, Somerset, and Monticello, Kentucky. M, 1956, University of Kentucky. 30 p.

Ford, Margaret Meisburger. A fluid inclusion and petrographic study of barren and gold-mineralized quartz veins in the central and southern Carolina slate belt, North Carolina. M, 1981, University of North Carolina, Chapel Hill. 72 p.

Ford, Mary Spencer. The influence of lithology on ecosystem development in New England; a comparative paleoecological study. D, 1984, [University of Minnesota, Minneapolis]. 213 p.

Ford, Michael E. The stratigraphy of the Trinity rocks (Comanchean) north of the Colorado River. M, 1987, Baylor University. 351 p.

Ford, Michael J. Geology and mineralization in the Zackly Fe-Cu-Au skarn, central Alaska Range, Alaska. M, 1988, University of Alaska, Fairbanks. 157 p.

Ford, Patrick. Geologic history of the Pettet Zone of the Sligo Formation at Lisbon Field, Clairborne Parish, Louisiana. M, 1985, Texas A&M University. 119 p.

Ford, R. Craig. Comparative geology of gold-bearing Archean iron formation, Slave structural province, Northwest Territories. M, 1988, University of Western Ontario. 233 p.

Ford, Ralph Joseph. Geophysical interpretation of an Airborne Field Intensity Survey. M, 1952, St. Louis University.

Ford, Richard Lee, III. The effect of Quaternary climatic change on alluvial-fan sedimentation in the Harquahala Plain, Sonoran Desert, west-central Arizona. M, 1986, University of New Mexico. 188 p.

Ford, Robert B. Occurrence and origin of the graphite deposits near Dillon, Montana. M, 1952, University of Wisconsin-Madison. 17 p.

Ford, Russell James. A study of the West Elk Breccia in the vicinity of Gunnison, Colorado. M, 1950, University of Kentucky. 42 p.

Ford, Theodore Lester. Petrographic study of the intrusive contact of the Fitchburg Granite in the Milford, New Hampshire and Townsend, Massachusetts quadrangles. M, 1974, Ohio University, Athens. 83 p.

Ford, Waldo Emerson. The geology and oil resources of a portion of the Newhall District, Los Angeles County, California. M, 1941, University of California, Los Angeles.

Ford, William E. Investigations in mineralogy. D, 1903, Yale University.

Ford, William Harry. Geology for planning in DeKalb Township, DeKalb County, Illinois. M, 1974, Northern Illinois University. 69 p.

Ford, William Jack. The subsurface geology of southwest Logan County, Oklahoma. M, 1954, University of Oklahoma. 70 p.

Forde, Margaret E. Summary of the geologic literature of the Tarrytown Quadrangle, New York. M, 1934, Columbia University, Teachers College.

Forde, Robert H. Stratigraphy of the Difunta Group along the Sierra Madre Front Range, Monterrey, Nuevo Leon to Saltillo, Coahuila, Mexico. M, 1959, Louisiana State University.

Fordes, George D. Sedimentary processes at Kitts Hummock Beach, Delaware, as influenced by an off-shore breakwater. M, 1981, University of Delaware. 124 p.

Fordham, C. J. Thermogravimetry in engineering geology. M, 1983, University of Waterloo. 92 p.

Fore, James Gary. A laboratory model study of fluid cross-flow. M, 1964, University of Missouri, Rolla.

Forel, David. Dip moveout correction in three and two dimensions. M, 1986, University of Houston.

Foreman, Dennis Walden, Jr. The $(O_4H_4)^{-4}$ tetrahedron in tricalcium aluminate hexadeuterate determined by means of neutron and X-ray diffraction. D, 1966, Ohio State University.

Foreman, Fred. Hydrothermal experiments on solubility, hydrolysis and oxidation of iron and copper sulphides. D, 1929, University of Minnesota, Minneapolis. 35 p.

Foreman, J. Lincoln. Composition and distribution of carbonate phases in the Lower Devonian Ridgeley Sandstone (Oriskany Group), southwestern Pennsylvania. M, 1986, Bowling Green State University. 118 p.

Foreman, Jerome A. A structural and tectonic study of the Lau-Havre--South Fiji Basin region. M, 1973, University of Hawaii. 104 p.

Foreman, Jerome A. The interaction between primary basement structure and secondary crustal extrusion along the Hawaiian Ridge. D, 1978, University of Hawaii.

Foreman, Neil. Refinement of the nepheline structure at several temperatures. M, 1969, University of Michigan.

Foreman, Terry Lee. Determination of hydrogeologic properties of Missouri River alluvium using numerical modelling techniques. M, 1979, University of Missouri, Columbia.

Foreman, Willie Earl. Phosphate mining in Hamilton County, Florida. M, 1972, Virginia State University. 23 p.

Forero Esguerra, Orlando. The Eocene of northwestern South America. M, 1974, University of Tulsa. 81 p.

Foresman, James B. Mud volcanoes and abnormal pressure, Copper River basin, Alaska. M, 1970, University of Tulsa. 44 p.

Forest, Richard C. Structure and metamorphism of Ptarmigan Creek area, Selwyn Range, B.C. M, 1985, McGill University. 163 p.

Forester, Elisabeth Brouwers. Zoogeography, biofacies and systematics of ostracode assemblages from continental shelf sediments, Gulf of Alaska. D, 1985, University of Colorado. 264 p.

Forester, Elizabeth Brouwers. A systematic and biostratigraphic study of the ostracodes from the Monmouth Formation (Upper Cretaceous; Maestrichtian) of the western shore of Maryland. M, 1977, University of Illinois, Urbana.

Forester, R. W. $^{18}O/^{16}O$ and D/H studies on the interactions between heated meteoric ground waters and igneous intrusions; western San Juan Mountains, Colorado, and the Isle of Skye, Scotland. D, 1975, California Institute of Technology. 373 p.

Forester, Richard M. Concepts of fossil and modern ostracode abundance, distribution and diversity patterns portrayed by probabilistic methodology. D, 1975, University of Illinois, Urbana. 180 p.

Forester, Richard M. Isotropic variability of Cheilostome bryozoan skeletons. M, 1972, University of Illinois, Urbana. 33 p.

Forester, Robert Donald. I, Studies on the travel times, periods, and energy of seismic waves SKP and related phases; II, The iron-rich breccia masses at Iron Mountain, Silver Lake District, San Bernardino County, California. D, 1953, California Institute of Technology. 134 p.

Forester, Robert Donald. The magnetite-rich breccia masses at Iron Mountain, Silver Lake District, San Bernardino County, California. D, 1953, California Institute of Technology. 88 p.

Foresti, Robert J. Macrofabrics, microfabrics, and microstructures of till and Pleistocene geology of the Ilion Quadrangle, Mohawk Valley, New York. M, 1984, Syracuse University.

Forge, Laurence La see La Forge, Laurence

Forgeron, Fabian David. Bay of Fundy bottom sediments (New Brunswick; Nova Scotia). M, 1962, Carleton University. 120 p.

Forgie, David John Leslie. Probabilistic cost-effectiveness relationships for sanitary landfill leachate containment and collection systems. D, 1983, University of Toronto.

Forgotson, James Morris, Jr. A correlation and regional stratigraphic analysis of the formations of the Trinity Group of the Comanchean Cretaceous of the Gulf Coastal Plain. D, 1956, Northwestern University.

Forgotson, James Morris, Jr. Regional stratigraphic analysis of the Cotton Valley Group of the upper Gulf Coastal Plain. M, 1954, Northwestern University.

Forkgen, Peter Edward. Study of the pre-Pennsylvanian geology of Nebraska and South Dakota. M, 1955, Stanford University.

Forlenza, Michael Francis. Fluvial sedimentology of the Middle and Upper Triassic Wolfville Redbeds, Hants County, Nova Scotia. M, 1982, University of Massachusetts. 237 p.

Forman, David John F. Palaeotectonics of Precambrian and Palaeozoic rocks of central Australia. D, 1968, Harvard University.

Forman, John Alexander. Geology of the southern Sulphur Spring Mountains (Nevada). M, 1951, Pomona College.

Forman, Robert Douglas. Palynology, thermal maturation, and time temperature history of three oil wells from the Beaufort-Mackenzie Basin. M, 1988, University of British Columbia. 114 p.

Forman, Steven Lawrence. Quaternary glacial, marine, and soil development history of the Forlandsund area, western Spitsbergen, Svalbard. D, 1986, University of Colorado. 353 p.

Forman, Sydney A. Systematic study of the dark micas. D, 1951, University of Toronto.

Forman, Sydney A. The crystal structure of rickardite $Cu_{2-x}Te$. M, 1946, University of Toronto.

Fornari, Daniel John. Micromorphologic and petrologic investigations of submarine volcanic features using submersibles. D, 1978, Columbia University, Teachers College. 186 p.

Forney, L. Bruce. The Willis Formation of the Texas Gulf Coast. M, 1950, [University of Houston].

Forrest, Joseph Turner. Geologic evolution of a portion of the Murphy marble belt in southwestern North Carolina. D, 1975, Rice University. 128 p.

Forrest, Kimball. Geologic and isotopic studies of the Lik Deposit and the surrounding mineral district, Delong Mountains, western Brooks Range, Alaska. D, 1983, University of Minnesota, Minneapolis. 174 p.

Forrest, Michael. Development of a Talwani-Ewing Fortran program for gravity interpretation. M, 1985, California State University, Long Beach. 62 p.

Forrest, R. M. A colation of Precambrian-Paleozoic contacts in the Kingston vicinity. M, 1949, Queen's University.

Forrest, Richard A. Geology and mineral deposits of the Warm Springs-Giltedge District, Fergus County, Montana. M, 1971, Montana College of Mineral Science & Technology. 191 p.

Forrest, Ronald J. Evolution of marshlands and offshore barrier sandbars in East bay, New York. M, 1969, Southern Illinois University, Carbondale. 89 p.

Forrestal, Geraldine. Geographic contrasts; Taos, Santa Fe and Albuquerque. M, 1946, Washington University. 168 p.

Forrester, James Donald. Structure of the Uinta Mountains. D, 1935, Cornell University.

Forrester, James Donald. Structure of the Wasatch Front between Rock Canyon and Slate Canyon, near Provo City, Utah. M, 1929, Cornell University.

Forrester, John Douglas. Skarn Formation and sulfide mineralization at the Continental Mine, Fierro, New Mexico. D, 1972, Cornell University.

Forrester, Macquorn Rankine. The Lake Renzy nickel deposits, Pontiac County, Quebec. M, 1957, University of Western Ontario. 122 p.

Forslev, Alfred William. The composition and origin of the Valparaiso Till in northeastern Illinois. D, 1960, University of Chicago. 44 p.

Forsman, Nels F. Experimental eolian abrasion of very fine grains at different atmospheric pressures. M, 1978, University of Houston.

Forsman, Nels Frank. Petrology of the Sentinel Butte Formation (Paleocene), North Dakota. D, 1985, University of North Dakota. 236 p.

Forster, C. B. A laboratory study of the effects of packer compliance on pressure pulse tests. M, 1979, University of Waterloo.

Forster, Craig Burton. Interaction of groundwater flow systems and thermal regimes in mountainous terrain; a numerical study. D, 1987, University of British Columbia.

Forster, Douglas Burton. Geology, petrology and precious metal mineralization, Toodoggone River area, north-central British Columbia. M, 1984, University of British Columbia. 313 p.

Forster, John R. Diagenesis of the Lower Ordovician Manitou Formation, El Paso County, Colorado. M, 1977, University of Montana. 111 p.

Forster, Stephen W. Pleistocene geology of the Carthage 15-minute Quadrangle, New York State. D, 1971, Syracuse University.

Forsthoff, Gary M. Composition, source and dispersal of the Muertos Trench turbidites; northeastern Caribbean Sea. M, 1981, University of New Orleans. 119 p.

Forsthoff, Harry S. Petrogenesis and porosity development in the Crystal River Formation of the Ocala Group (Eocene) of Southwest Georgia. M, 1980, University of Georgia.

Forsyth, D. A. G. A refraction survey across the Canadian Cordillera. M, 1973, University of British Columbia.

Forsyth, Donald William. Anisotropy and the structural evolution of the oceanic upper mantle. D, 1973, Woods Hole Oceanographic Institution. 253 p.

Forsyth, Jane Louise. Eden and Maysville groups of the Cincinnatian Series at Cincinnati, Ohio. M, 1946, University of Cincinnati. 122 p.

Forsyth, Jane Louise. The glacial geology of Logan and Shelby counties, Ohio. D, 1956, Ohio State University. 236 p.

Forsyth, Prentice Mark. Multicomponent trace analysis and seismic reflection studies in the Great Salt Lake Desert, Utah. M, 1985, University of Oklahoma. 200 p.

Forsyth, William T. Metamorphic facies of the Ellsworth Schist in Blue Hill, Maine. M, 1953, University of Maine. 51 p.

Forsythe, James Thorp. Pre-glacial drainage of central southern Michigan. M, 1952, University of Michigan.

Forsythe, Leander Harold. Trends of anorthite and triclinicity variations in the Archaean Dome Stock Granite, Red Lake area, Ontario. M, 1966, University of Saskatchewan. 52 p.

Forsythe, Randall David. Geological investigations of pre-Late Jurassic terranes in the southernmost Andes. D, 1981, Columbia University, Teachers College. 298 p.

Forsythe, Roger. Depositional and time framework of a Waulsortian mud mound, Noel, McDonald County, Missouri. M, 1985, University of Arkansas, Fayetteville.

Fortescue, John Adrian Claude. Some Devonian brachiopods reported from Western Canada. M, 1954, University of British Columbia.

Fortesque, John. Some Brachiopoda from the Upper Devonian of the Rocky Mountains. M, 1954, University of British Columbia.

Forth, David R. The structure and stratigraphy of the Sacapulas Quadrangle, Guatemala, with particular emphasis on the Paleozoic rocks. M, 1971, Louisiana State University.

Forth, Michael. Geology of the southwest quarter of the Dayville Quadrangle, Oregon. M, 1966, Oregon State University. 75 p.

Fortier, Alfred Joseph, III. Synthetic seismograms by the slant stack method; interpretations in p-tau and x-t. M, 1983, University of Texas, Austin. 240 p.

Fortier, David Harvey. Conceptual study of potential large underground reservoirs in southern Idaho. M, 1975, University of Idaho. 87 p.

Fortier, J. Daniel. Eocene depositional environments, Lake o' the Pines damsite, Marion County, Texas. M, 1975, University of Texas, Austin.

Fortier, Yves O. Geology of chromite. M, 1941, McGill University.

Fortier, Yves O. Geology of the Orford map-area in the Eastern Townships of the Province of Quebec, Canada. D, 1946, Stanford University. 227 p.

Fortin, Danielle. Contact Beekmantown-Chazy dans les Basses-Terres du St-Laurent; évaluation des effets de la discordance (Ordovicien inférieur). M, 1987, Universite Laval. 63 p.

Fortin, Normand. Etude géostatistique de la minéralisation aurifère à la Mine Sullivan, comte d'Abitibi-Est, Province de Québec. M, 1970, Ecole Polytechnique. 94 p.

Fortner, David William. The effects of composition and bedding of log response, Yowlumne Sandstone, Kern County, California. M, 1988, Texas A&M University. 195 p.

Fortsch, David E. A Late Pleistocene vertebrate fauna from the northern Mojave Desert of California. M, 1972, University of Southern California.

Fortson, Charles Wellborn, Jr. Geology of the Crabtree Creek area northwest of Raleigh, North Carolina. M, 1958, North Carolina State University. 101 p.

Fortuna, Mark Allen. Paleomagnetism and shear history of Precambrian X dikes. M, 1979, Michigan State University. 73 p.

Fortuna, Raymond. Muscovite geothermometry in metamorphic rocks from southwestern New Hampshire. M, 1978, Miami University (Ohio). 79 p.

Fortunato, Kathleen Susan. Depositional framework of the La Casita Formation (Upper Jurassic - Lowermost Cretaceous) near Saltillo, Coahuila, Mexico. M, 1982, University of New Orleans. 198 p.

Fortune, Gladys M. The development of earth science field activities along the shore of Lake Michigan in Milwaukee County. M, 1970, University of Wisconsin-Milwaukee.

Fortune, Irene A. La see La Fortune, Irene A.

Fortune, Kim M. Sedimentation at an inlet at Wilson Harbor on Lake Ontario, New York. M, 1980, SUNY at Buffalo. 87 p.

Foruria, Jon. Geology, alteration and precious metal reconnaissance of the Nogal Canyon area, San Mateo Mountains, NM. M, 1984, Colorado State University. 178 p.

Forward, Frederick. The Doctor Bond Sandstone; Jurassic, Colorado. M, 1938, University of Colorado.

Foscal-Mella, Gabriel. Analyse minéralogique des argiles glaciaires. M, 1976, Ecole Polytechnique.

Foshag, William Frederick. The origin of the colemanite deposits of the Western United States. D, 1923, University of California, Berkeley.

Foshee, Ronald R. Lithostratigraphy and depositional systems of the Bloyd and Hale formations (Pennsylvanian), in the western Arkoma Basin of Arkansas. M, 1980, University of Arkansas, Fayetteville.

Fosness, John Leslie. An investigation of the zonal arrangement of mineral deposits in southeastern Arizona. M, 1925, University of Minnesota, Minneapolis. 43 p.

Foss, Adolph L. The geology of the Vipont silver mine. M, 1923, University of Minnesota, Minneapolis. 21 p.

Foss, Deane Campbell. Depositional environment of Woodbine sandstones, Polk, Tyler, and San Jacinto counties, Texas. M, 1978, Texas A&M University. 234 p.

Foss, Donald J. Geology of the Jerusalem Valley with particular attention to the Jerusalem Raft. M, 1980, Boise State University. 39 p.

Foss, Ted Harry. A textural and mineralogical study of a hornblende granite. M, 1958, University of Illinois, Urbana. 32 p.

Foss, Ted Harry. Chemical and mineralogical variations in the radial dikes of Difficulty Creek intrusive center, San Juan Mountains, Colorado. D, 1964, Rice University. 76 p.

Fossen, John Doan Van *see* Van Fossen, John Doan

Fossey, Kenneth Wayne. Structural geology and slope stability of the southeast slopes of Turtle Mountain, Alberta. M, 1986, University of Alberta. 127 p.

Fossum, Martin Peter. Tracer analysis in a fractured geothermal reservoir; field results from Wairakei, New Zealand. M, 1982, Stanford University.

Foster, A. B. Morphologic variation within three species of reef corals (Cnidaria, Anthozoa, Scleractinia). D, 1978, The Johns Hopkins University. 491 p.

Foster, Allan Royal. Marine manganese nodules; nature and origin of internal features. M, 1970, Washington State University. 131 p.

Foster, Brayton P. Study of the kimberlite-alnoite dikes in central New York (Finger Lakes region). M, 1970, SUNY at Buffalo. 59 p.

Foster, C. T., Jr. Diffusion controlled growth of metamorphic segregations in sillimanite grade pelitic rocks near Rangeley, Maine, U.S.A. D, 1975, The Johns Hopkins University. 216 p.

Foster, Christopher Allen. Paleocene Ostracoda from Nigeria. M, 1981, University of Delaware. 186 p.

Foster, David Allen. Synplutonic dikes of the Idaho Batholith, Idaho and western Montana, and their relationship to the generation of the batholith. M, 1986, University of Montana. 79 p.

Foster, David Wayne. Application of quantitative techniques to selected problems in terrigenous sedimentology; sandstone provenance and analysis of sequences of bedding thickness. D, 1987, University of Kansas. 140 p.

Foster, David Wayne. Morphologic similarity of Loxocorniculum postdorsoalata and L. trincornata (Ostracoda) from the Veracruz-Anton Lizardo reef complex, Mexico. M, 1981, University of Kansas. 69 p.

Foster, Donald I. Lower Pennsylvanian stratigraphy of the southern Egan Range, Nevada. M, 1953, Columbia University, Teachers College.

Foster, Douglas John. Finite difference synthetic seismogram with a point source. M, 1980, University of Missouri, Rolla.

Foster, Douglas John. Velocity analysis and extrapolation of seismic data in vertically inhomogeneous media. D, 1985, Columbia University, Teachers College. 133 p.

Foster, F. L. Electrical procedures and interpretation. M, 1930, Massachusetts Institute of Technology.

Foster, Fess. Volcanic geology and mineralization in the Mt. Jordan vicinity, Custer County, Idaho. D, 1983, University of Montana. 56 p.

Foster, Glen Lloyd. The constituents and their structural arrangement in ostracode carapaces. M, 1958, University of Kansas. 88 p.

Foster, Helen Laura. Geology of the Mount Leidy highlands, Teton County, Wyoming. D, 1946, University of Michigan.

Foster, Helen Laura. Geology of the northern Hogback Mountains, Sublette County, Wyoming. M, 1943, University of Michigan.

Foster, Jack D. The W. E. Robbins Bridgeport water flood (Illinois) (F814). M, 1951, University of Illinois, Urbana. 39 p.

Foster, James A. The Pliocene series of Southeast Lubbock County, Texas. M, 1952, Texas Tech University. 133 p.

Foster, Jeffrey S. Ground water modeling in a small loessial watershed. M, 1981, University of New Hampshire. 221 p.

Foster, John David. Glacial morphology of the Cary (Pleistocene) age deposits in a portion of central Iowa. M, 1969, Iowa State University of Science and Technology.

Foster, John Harold. Paleomagnetic stratigraphy of deep sea sediments. D, 1970, Columbia University. 92 p.

Foster, John Hugh. Late Cenozoic tectonic evolution of Cajon Valley, southern California. D, 1980, University of California, Riverside. 315 p.

Foster, John M. Geology of the Bismark Peak area, North Tintic District, Utah County, Utah. M, 1959, Brigham Young University. 95 p.

Foster, John Robert. Efficiency of acid digestion procedures on rocks. M, 1975, Queen's University. 155 p.

Foster, John Webster. Glaciations and ice lake of Paint Creek valley, Ohio. M, 1950, Ohio State University.

Foster, Joseph F. An investigation of the clay content-permeability relationship of oil sands. M, 1948, University of Southern California.

Foster, Joseph M. Geology and mineralization of the Lucky Hill Pit, Candelaria Mine, Mineral County, Nevada. M, 1988, University of Nevada. 156 p.

Foster, Margaret D. Geochemical relations of the ground waters of the Houston-Galveston area, Texas. D, 1935, American University. 89 p.

Foster, Merrill W. The biology of the Recent Antarctic and subantarctic brachiopods. D, 1970, Harvard University.

Foster, Merrill White. Ordovician receptaculitids, trilobites and brachiopods from the Grapevine Mountains, California. M, 1964, University of California, Berkeley. 190 p.

Foster, Michael. Analysis of small-scale structures in the Duluth Complex at Bardon Peak, St. Louis County, Minnesota. M, 1980, University of Minnesota, Minneapolis. 144 p.

Foster, Norman Holland. Faunal zonation and stratigraphy of the Mississippian Madison Group, northeastern Bighorn Mountains, Wyoming. M, 1960, University of Iowa. 108 p.

Foster, Norman Holland. Faunal zonation and stratigraphy of the Mississippian Madison Group, Wyoming and Montana. D, 1963, University of Kansas. 266 p.

Foster, Paul W. The study of Mosby Butte, a volcanic construct of the south-central Snake River plain, Idaho. M, 1978, SUNY at Buffalo. 48 p.

Foster, Paul Woodward. Evidence of rock-plains in southeastern Missouri. M, 1939, University of Oklahoma. 90 p.

Foster, R. Leon. Geology of the Shiloh School-Liberty Church area, West Burleson County, Texas. M, 1956, Texas A&M University. 74 p.

Foster, Robert A. Geology of the north half of the Romney Quadrangle, West Virginia. M, 1978, University of Akron. 38 p.

Foster, Robert John. A study of the Guye Formation, Snoqualmie Pass, King and Kittitas counties, Washington. M, 1955, University of Washington. 57 p.

Foster, Robert John. The Tertiary geology of a portion of central Cascade Mountains, Washington. D, 1957, University of Washington. 186 p.

Foster, Robert Lutz. The petrography and petrology of the Keweenawan rocks in the Cascade River area, Cook County, Minnesota. M, 1962, University of Missouri, Columbia.

Foster, Robert Lutz. The petrology and structure of the Amy Dome area, Tolovana mining district, east-

central Alaska. D, 1966, University of Missouri, Columbia. 254 p.

Foster, Scot Alan. Structural analysis of the NE 1/4 of the Wallace 15' Quadrangle, Shoshone County, Idaho. M, 1983, University of Idaho. 150 p.

Foster, Stephen Eric. Heat flow near a North Atlantic fracture zone and west of Gibraltar. M, 1973, Massachusetts Institute of Technology. 127 p.

Foster, Vellora Meek. A geologic section along the Mississippi River Bluff from President Street, St. Louis, Missouri to the Meramec River. M, 1927, Washington University. 156 p.

Foster, Wilder D. Effect of calcium chloride solutions on the hydration of minerals occurring in Portland cement. M, 1929, Ohio State University.

Foster, Wilfred R. The system albite-wollastonite-nepheline. D, 1940, University of Chicago. 74 p.

Foster, William H. The anticlinal theory of oil and gas. M, 1915, University of Kansas.

Fotouchi, Manuchehre and Saraby, Fereydoon. Geology of the Dunkleberg District, Drummond Quadrangle, Montana. M, 1958, Michigan Technological University. 95 p.

Fou, Joseph T. K. Thermal conductivity and heat flow at Saint Jerome, Quebec (Canada). M, 1969, McGill University.

Fouch, Thomas Dee. Geology of the northwest quarter of the Brogan Quadrangle, Malheur County, Oregon. M, 1968, University of Oregon. 62 p.

Fouch-Flores, Donna Lynn. Regional uranium resource evaluation using Landsat imagery and N.U.R.E. geochemical data, southern Trans-Pecos, Texas. M, 1982, Texas Christian University. 69 p.

Fouda, Ahmed Ali. The upper mantle structure under the stable regions. D, 1973, University of California, Los Angeles.

Fouke, Bruce William. Quaternary geology and depositional history of Providenciales, Turks and Caicos islands, British West Indies. M, 1984, University of Iowa. 202 p.

Fouke, Michael A. The petrology of the Kessler Limestone Member, Bloyd Formation, in southwestern Washington and southwestern Crawford counties, Arkansas. M, 1976, University of Arkansas, Fayetteville.

Foulkes, Thomas G. The formation of zinc ferrite. M, 1923, Colorado School of Mines. 24 p.

Founie, Alan. Sedimentology of a marine-nonmarine-marine rock sequence in central Wyoming. M, 1976, University of Missouri, Columbia.

Fountain, Aubroy W., II. Pelecypods of the Yorktown Formation (Miocene) of Petersburg, Virginia. M, 1972, Virginia State University. 48 p.

Fountain, David Michael. Evidence for transform faulting in the western Aleutian Trench. M, 1971, University of Michigan.

Fountain, David Michael. Seismic velocities in granulite facies rocks and rocks from the Ivrea-Verbano and Strona-Ceneri zones; a study of continental crustal composition. D, 1974, University of Washington. 178 p.

Fountain, John Crothers. The geochemistry of Mt. Tehama, Lassen Volcanic National Park, California. D, 1975, University of California, Santa Barbara.

Fountain, John F. The determination of rare Earth abundances by X-ray fluorescence and the application of rare Earth data to the origin of andesites. M, 1972, University of California, Santa Barbara.

Fountain, Richard Calhoun. The geology of the northwestern portion of Jasper County, Georgia. M, 1961, Emory University. 65 p.

Fouret, James Howard. Geology of the central Culebra Range region, Costilla, Huerfano, and Las Animas counties, Colorado. M, 1955, Colorado School of Mines. 104 p.

Fouret, Kent L. Depositional and diagenetic environments of the Mississippian Leadville Formation at Lisbon Field, Utah. M, 1982, Texas A&M University.

Fournelle, John Harold. The geology and petrology of Shishaldin Volcano, Unimak Island, Aleutian Arc, Alaska. D, 1988, The Johns Hopkins University. 559 p.

Fournier, Benoit. Les Produits de la réaction alcalis-silice dans le béton; étude de cas de la région de Québec. M, 1986, Universite Laval. 51 p.

Fournier, Jorge C. New methods and techniques in the photography of microfossils. M, 1954, New York University.

Fournier, Rene E. The geochemistry of manganiferous-iron concretions in the Pierre Formation (Cretaceous), South Dakota. M, 1967, University of South Dakota. 97 p.

Fournier, Robert Orville. Mineralization of the porphyry copper deposit near Ely, Nevada. D, 1958, University of California, Berkeley. 179 p.

Fourt, Robert. Post-batholithic geology of the Volcanic Hills and vicinity, San Diego County, California. M, 1979, San Diego State University.

Foust, Denny G. An investigation of cation exchange, calcium for potassium, in muscovite and resulting alterations. M, 1972, Bowling Green State University. 38 p.

Fout, James Scott. Mineral and chemical variations within a compositionally zoned ash flow from the upper Pumice Series, Thera, Greece. M, 1977, Indiana University, Bloomington. 63 p.

Fout, James Scott. The mineralogy and petrology of the Willimantic Dome, eastern Connecticut. D, 1981, Indiana University, Bloomington. 244 p.

Fouts, James A. Petrology and chemistry of some diabase sills in central Arizona. D, 1974, University of Arizona. 172 p.

Fouts, James Allen. The geology of the Metasville area, Wilkes and Lincoln counties, Georgia. M, 1966, University of Georgia. 61 p.

Fouts, John Douglas. Sedimentary and structural features across the Pine Mountain Thrust. M, 1983, University of Kentucky.

Fouts, John Martin, Jr. The geology of North Henderson County, Texas (F829). M, 1939, University of Texas, Austin.

Foutz, Dell R. Geology of the Wash Canyon area, southern Wasatch Mountains, Utah. M, 1960, Brigham Young University. 37 p.

Foutz, Dell Riggs. Stratigraphy of the Mississippian System in northeastern Utah and adjacent states. D, 1966, Washington State University. 218 p.

Fowells, Joseph Edward. Supergene alteration products of the Butte primary minerals. M, 1942, Montana College of Mineral Science & Technology. 26 p.

Foweraker, John Charles. Quantitative studies in river sinuosity with special reference to incised meanders of Ozark rivers. D, 1963, Washington University. 124 p.

Fowkes, Elliott Jay. Pegmatites of Granite Peak Mountain, Tooele County, Utah. M, 1964, Brigham Young University. 127 p.

Fowler, Anthony David. Age and origin of the granulite facies rocks west of the Labrador Trough, Labrador-Quebec. M, 1975, McGill University. 64 p.

Fowler, Anthony David. The age, origin, and rare-earth-element distributions of Grenville Province uraniferous granites and pegmatites. D, 1980, McGill University. 130 p.

Fowler, Charles Sidney. Water absorption by a swelling porous medium. D, 1988, University of Nebraska, Lincoln. 111 p.

Fowler, Chris K. Structure and stratigraphy of the normally pressured sands in an offshore giant; Superior Oil Company's West Cameron Block 71 Gas Field. M, 1982, University of Southwestern Louisiana. 147 p.

Fowler, Claude Stewart. The origin of the sulphur deposits of Mount Adams. M, 1936, Washington State University. 23 p.

Fowler, David. Slip along the San Andreas Fault. M, 1971, Stanford University.

Fowler, Donald R. Miocene paleoecology in Calvert County, Maryland. M, 1966, Northwestern University.

Fowler, Gerald Allan. The stratigraphy, foraminifera and paleontology of the Montesano Formation (Miocene-Pliocene), Grays Harbor County, Washington. D, 1965, University of Southern California.

Fowler, Jack. Analysis of fabric-reinforced embankment test section at Pinto Pass, Mobile, Alabama. D, 1979, Oklahoma State University. 224 p.

Fowler, James W. The paleoecology of the Pamlico Formation (Pleistocene) in the vicinity of Myrtle Beach, Horry County, South Carolina. M, 1963, University of Houston.

Fowler, John Henry. Analysis and interpretation of the 5,000 ft. (1,538 m) redbed section encountered in the Sparks Et Al. 1-8 (Michigan Basin deep borehole). M, 1979, Western Michigan University.

Fowler, Julie Ann. Structural analysis of Mesozoic deformation in the central Slate Range, eastern California. M, 1982, California State University, Northridge. 135 p.

Fowler, Katharine S. Glacial drainage changes in northeastern United States. M, 1926, University of Wisconsin-Milwaukee.

Fowler, Katherine Stevens. The anorthosite area of the Laramie Mountains, Wyoming. D, 1930, Columbia University, Teachers College.

Fowler, Kathleen Anne. Depositional environment and diagenesis of the Aux Vases Sandstone (Mississippian), New Harmony Field, White County, Illinois. M, 1987, Southern Illinois University, Carbondale. 173 p.

Fowler, Linda Leigh. Brecciation, alteration, and mineralization at the Copper Flat porphyry copper deposit, Hillsboro, New Mexico. M, 1982, University of Arizona. 133 p.

Fowler, Michael Lee. Distribution of magnesium and strontium in the skeletons of modern regular echinoids from the United States Coast and Caribbean Sea. D, 1972, Indiana University, Bloomington. 292 p.

Fowler, Paul J. Seismic velocity estimation using pre-stack time migration. D, 1988, Stanford University. 186 p.

Fowler, Phillip Teague. Stratigraphy and structure of the Castleton area, Vermont. D, 1949, Harvard University.

Fowler, Rhonda M. Environment of deposition of lower Wilcox "Mula" sandstones, East Washburn and Bid Mule fields, La Salle and McMullen counties, Texas. M, 1984, Texas A&M University. 98 p.

Fowler, Scott K. Environments of deposition as determined by lateral variations in a Gulf Coast lignite seam. M, 1979, Southern Illinois University, Carbondale. 80 p.

Fowler, Sharon Patricia. Lower Cretaceous foraminiferal microfaunas and biostratigraphy of the Richardson Mountains, Yukon and Northwest Territories. D, 1986, University of Saskatchewan. 394 p.

Fowler, Todd A. Local variations in the modern carbonate depositional environment, Bahia, Honda Key, Florida. M, 1977, University of Texas, Arlington. 96 p.

Fowler, Wayne Edward. Geology of the Trusty Lake-Quartz Hill Gulch area, Beaverhead County, Montana. D, 1955, Indiana University, Bloomington. 88 p.

Fowler, Wayne Edward. The Pleistocene geology of Hendricks County, Indiana. M, 1953, Indiana University, Bloomington. 55 p.

Fowler, William A. Geology of the Manitou Park area, Douglas and Teller counties, Colorado. M, 1952, University of Colorado.

Fowler, William Lane. Potential debris flow hazards of the Big Bend Drive drainage basin, Pacifica, California. M, 1984, Stanford University. 112 p.

Fowles, George Richard. Shock-wave compression of quartz. D, 1961, Stanford University. 76 p.

Fox, Adam Jeffrey. An integrated geophysical study of the southeastern extension of the Midcontinent Rift system. M, 1988, Purdue University. 112 p.

Fox, Andrew J. The Lower Cretaceous McMurray Formation in the subsurface of Syncrude Oil Sands Lease 17, Athabasca Oil Sands, northeastern Alberta; a physical sedimentological study in an area of exceptional drill core control. M, 1988, University of Alberta. 483 p.

Fox, Bruce Wendell. The petrography and petrofabrics of the Deer Creek Laccolith, Park County, Wyoming. M, 1956, Washington University. 74 p.

Fox, C. A. The soil micro-morphology of turbic Cryosols from the Mackenzie River valley and Yukon coastal plain. D, 1979, University of Guelph.

Fox, Charles E. Determination of fracture aperture a multi-tracer approach. M, 1988, Stanford University.

Fox, Charles S. The Honey Creek Formation of the Wichita Mountains, Oklahoma. M, 1958, University of Oklahoma. 61 p.

Fox, Christopher Gene. A quantitative method for analyzing the roughness of the seafloor. D, 1985, Columbia University, Teachers College. 233 p.

Fox, Darwin Eugene. Behavioral predictions of compacted embankments on rigid foundations. D, 1981, Iowa State University of Science and Technology. 231 p.

Fox, Dennis. Evaluation of Seasat and SIR-A radar imagery for reconnaissance geologic mapping in the Anza-Borrego desert area. M, 1984, California State University, Long Beach. 84 p.

Fox, Feramorz. General features of the Wasatch Mountains. M, 1906, University of Utah. 53 p.

Fox, Forrest L. Chemical variations of the Truckee River from Lake Tahoe to Truckee, California, during low flow. M, 1982, University of Nevada. 113 p.

Fox, Frederick Glenn. A Cretaceous grit bed horizon in the districts west and south from Turner Valley, Alberta. M, 1942, University of Alberta. 79 p.

Fox, Frederick Glenn. The stratigraphy of the Devonian and Mississippian rocks in the foothills of southern Alberta, Canada. D, 1948, University of Oklahoma. 137 p.

Fox, G. R. Dispersion of a longitudinal strain pulse in an elastic cylindrical bar. D, 1956, Lehigh University.

Fox, Harold Dixon. Structure and origin of two windows exposed on the Nittany Arch at Birmingham, Pennsylvania. M, 1949, Cornell University.

Fox, Harry Bert. Clay investigations. M, 1905, University of Illinois, Urbana.

Fox, Hewitt Bates. The geology of the Kent Draw area, Culberson County, Texas. M, 1948, University of Texas, Austin.

Fox, James E. Invertebrate fossils and environment of the Fox Hills and Medicine Bow formations (upper Cretaceous) in south central Wyoming. D, 1971, University of Wyoming. 163 p.

Fox, James E. Paleoecology of the Oacoma facies, Pierre Formation (Cretaceous of South Dakota). M, 1968, University of South Dakota. 171 p.

Fox, James Henry. "Cryptovolcanic" force fields. D, 1954, St. Louis University.

Fox, James Henry. A magnetic study of a northeastern Ozark Fold. M, 1953, St. Louis University.

Fox, John Martin. Mineralogy and stratigraphy of Illinoian and pre-Illinoian tills across south-central Illinois. M, 1987, University of Illinois, Urbana. 113 p.

Fox, John Thomas. Geology of the Fairmont West, West Virginia Quadrangle. M, 1965, West Virginia University.

Fox, Joseph Peter. Conservation of the principal power resources of the United States; viz. coal, petroleum, natural gas, and water power. M, 1922, Catholic University of America. 92 p.

Fox, Joseph S. Petrologic study of aluminous metapelitic assemblages from below the staurolite isograd near Agnew lake, Ontario. M, 1970, McGill University.

Fox, Julian C.   Glacial geomorphology of the Cataract Brook valley, Yoho National Park, British Columbia. M, 1974, University of Calgary.

Fox, Kenneth Francis.   Geology of Mill Creek basin, Park County, Montana. M, 1960, Montana College of Mineral Science & Technology. 85 p.

Fox, Kenneth Francis, Jr.   The geology of alkalic complexes in north-central Washington. D, 1973, Stanford University. 293 p.

Fox, Laura.   Porosity and permeability reduction in the Nugget Sandstone, Southwest Wyoming. M, 1979, University of Missouri, Columbia.

Fox, Lewis E.   The geochemistry of humic acid and iron during estuarine mixing. D, 1981, University of Delaware, College of Marine Studies. 231 p.

Fox, Mary Ann Strouse.   Studies of the Bethel Sandstone, Wayne County area, Illinois. M, 1953, University of Illinois, Urbana. 37 p.

Fox, Michael.   Factors affecting the development of conventional fossil fuels on federal lands of the Rocky Mountain region. M, 1984, University of Texas, Austin.

Fox, Norman Albert.   Facies relationships and provenance of the Swift Formation (Jurassic), southwestern Montana. M, 1982, Montana State University. 104 p.

Fox, Paul Jeffrey.   The geology of some Atlantic fracture zones, Caribbean Escarpment and the nature of the oceanic basement and crust. D, 1972, Columbia University, Teachers College. 359 p.

Fox, Peter Edward.   Petrology of Adamant Pluton (Proterozoic), British Columbia. D, 1966, Carleton University. 184 p.

Fox, Peter Edward.   Reaction zones in the Adamant Range Batholith, British Columbia. M, 1962, Queen's University. 73 p.

Fox, Richard C.   A regional gravity survey of south-central Utah, including gravity profiles across the Paunsaugunt Fault. M, 1968, University of Utah. 81 p.

Fox, Richard Dale.   Hydrogeology of the Cascade-Ulm area (Cascade County), Montana. M, 1965, University of Montana. 115 p.

Fox, Richard Lyn.   Galactic cosmic ray produced radionuclides in Antarctic meteorites and a lunar core. D, 1987, [University of California, San Diego]. 150 p.

Fox, Robert Eugene.   Subsurface studies of the Aux Vases Sandstone, Wayne County area, Illinois. M, 1953, University of Illinois, Urbana.

Fox, Stephen Edward.   Biostratigraphy and paleogeography of the Three Forks Formation and the basal McGowan Creek Formation, Custer County, Idaho. M, 1985, University of Idaho. 89 p.

Fox, Stephen Knowlton, Jr.   Stratigraphy and micropaleontology of the Cody Shale in southern Montana and northern Wyoming. D, 1939, Princeton University. 242 p.

Fox, Steven W.   The structure and stratigraphy of the Woodchopper and Coal Creek area, Alaska. M, 1988, University of Alaska, Fairbanks. 114 p.

Fox, Terrance.   Structural and stratigraphic analysis of the west-central Schell Creek Range. M, 1984, California State University, Long Beach. 98 p.

Fox, Thomas P.   Geochemistry of the Hemlock Metabasalt and Kiernan Sills, Iron County, Michigan. M, 1983, Michigan State University. 81 p.

Fox, Walter F., Jr.   Relation of natural gas analyses to geology and reservoir parameters in Beaver County, Oklahoma. M, 1966, University of Oklahoma. 90 p.

Fox, William Blake.   Geology and structural history of the Sage Creek-Winkleman Trend, Fremont County, Wyoming. M, 1957, University of Wisconsin-Madison.

Fox, William Joseph.   The geology of the Crawford and Speegleville quadrangles, McLennan County, Texas. M, 1962, Baylor University. 102 p.

Fox, William Templeton.   Paleoecology of the Frontier Formation, Wind River basin, Wyoming. M, 1960, Northwestern University.

Fox, William Templeton.   Stratigraphy and paleontology of the Richmond Group in southeastern Indiana. D, 1961, Northwestern University. 342 p.

Foxall, William.   A geophysical study of the Skagit Valley. M, 1976, University of Washington. 88 p.

Foxhall, Harold B.   Geology of the upper Jalama Valley area, Santa Barbara County, California. M, 1942, Stanford University. 56 p.

Foxworth, Richard Dear.   Heavy minerals of sand from Recent beaches of the Gulf Coast of Mississippi and associated islands. M, 1958, University of Missouri, Columbia.

Foxworth, Wyckliff Riley.   Economic geology of the Las Cuevas fluorspar deposit, Salitrera, San Luis Potosi, Mexico. M, 1960, Texas Tech University. 30 p.

Foxx, Mark Steven.   Slope failures in the Felton Quadrangle 1981-83 and analysis of factors that control slope failure susceptibility of the Monterey Formation. M, 1984, University of California, Santa Cruz.

Foye, Wilbur Garland.   The Glamorgan gabbro body and its associated rocks. D, 1915, Harvard University.

Fracasso, Michael Anthony.   Cranial osteology, functional morphology, systematics and paleoenvironment of Limnoscelis paludis Williston. D, 1983, Yale University. 648 p.

Fragaszy, Richard John.   Drum centrifuge studies of overconsolidated clay slopes. D, 1979, University of California, Davis. 126 p.

Frailey, Carl David.   Studies on the Cenozoic Vertebrata of Bolivia and Peru. D, 1981, University of Kansas. 268 p.

Fraim, Parke Benjamin.   The geology, mining, and ore dressing of the Cornwall ores. M, 1919, Lehigh University.

Frakas, Steven Eugene.   Analysis of cross-lamination in the Upper Cambrian Franconia Formation of southwestern Wisconsin. M, 1958, University of Wisconsin-Madison.

Frakes, Lawrence Austin.   Paleogeography of the Trimmers Rock Member of the Fort Littleton Formation (Devonian) in southern and eastern Pennsylvania. D, 1964, University of California, Los Angeles. 456 p.

Frakes, Lawrence Austin.   The geology of the Quatal Canyon area, Kern, Ventura, and Santa Barbara counties, California. M, 1959, University of California, Los Angeles.

Fraley, C. M.   An aerodynamic sizing and chemical analysis of the volcanic fume particulate matter from Kilauea Volcano, Hawaii. M, 1976, University of Hawaii. 99 p.

Fraley, Peter Allen.   Alluvial sedimentology and cyclicity of the Wolfville Formation (Middle to Upper Triassic), north shore of Minas Basin, Nova Scotia. M, 1988, Miami University (Ohio). 151 p.

Fralick, Philip W.   Tectonic and sedimentological development of a late Paleozoic wrench basin; the eastern Cumberland Basin, Maritime Canada. M, 1980, Dalhousie University. 178 p.

Fralick, Philip William.   Early Proterozoic basin development on a cratonic margin; the lower Huronian Supergroup of central Ontario. D, 1985, University of Toronto.

Fralick, T. N.   A computer-aided spectral classification of saline deposits, gypsiferous diapirs and the Totora Formation, Rio Desaguadero region, Bolivia. M, 1976, Purdue University. 91 p.

Frame, Philip A.   Landslides in the Mt. Sizer area, Santa Clara County, California. M, 1974, San Jose State University. 87 p.

Frames, Donald Wayland.   Stratigraphy and structure of the lower Coyote Creek area, Santa Clara County, California. M, 1955, University of California, Berkeley. 67 p.

Frampton, James Alan.   The association of Co, Ni, Cu, and Zn with Fe and Mn oxides of soils. D, 1983, University of California, Davis. 149 p.

Franca, Almerio Barros.   Stratigraphy, depositional environment, and reservoir analysis of the Itarare Group, Parana Basin, Brazil. D, 1987, University of Cincinnati. 266 p.

France, Lynne June.   Geochronology, stratigraphy, and petrochemistry of the upper Tertiary volcanic arc, southernmost Peru, Central Andes. M, 1985, Queen's University. 182 p.

France, Noelle A.   Petrology and economic implications of King's Mountain belt metaconglomerates. M, 1983, North Carolina State University. 47 p.

Francek, Mark A.   Spatial characteristics of the New York drumlin field. D, 1988, University of Wisconsin-Madison. 177 p.

Franceschini, Timothy.   Incremental strain analysis in the Martinsburg Formation along a section of the Portland Fault near Newton, New Jersey. M, 1978, Rutgers, The State University, New Brunswick. 30 p.

Francheteau, Jean M.   Paleomagnetism and plate tectonics. D, 1970, University of California, San Diego. 342 p.

Francica, Joseph R.   Geology mapping of the Ladakh Himalaya by computer processing of Landsat data. M, 1980, Dartmouth College. 67 p.

Francis, Alfred.   Treatment of low grade argentiferous zinc ores. M, 1894, Washington University.

Francis, Billy Max, II.   Petrology and sedimentology of the Devonian Misener Formation, north-central Oklahoma. M, 1988, University of Tulsa. 176 p.

Francis, Carl A.   A crystallographic study of the $Fe_{1-x}S$-$Ni_{1-x}S$ monosulfide solid solution. M, 1974, Virginia Polytechnic Institute and State University.

Francis, Carl Arthur.   Magnesium-manganese solid solution in the olivine and humite groups. D, 1980, Virginia Polytechnic Institute and State University. 126 p.

Francis, David Roy.   The Jurassic stratigraphy of the Williston Basin area. M, 1956, Northwestern University. 82 p.

Francis, Dennis C.   The geology and geochemistry of the Cahill gold mine and vicinity, Lander County, Nevada. M, 1973, Colorado School of Mines. 112 p.

Francis, Donald Michael.   Electrum tarnish by sulfur; a study of an irreversible process. M, 1971, University of British Columbia.

Francis, Donald Michael.   Xenoliths and the nature of the upper mantle and lower crust; Nunivak Island, Alaska. D, 1974, Massachusetts Institute of Technology. 237 p.

Francis, George Gregory.   Stratigraphy and environmental analysis of the Swan Peak Formation and Eureka Quartzite, northern Utah. M, 1972, Utah State University. 125 p.

Francis, Kevin Albert.   Geology and geochemistry of the Caribou Mine, Boulder County, Colorado. M, 1987, University of Colorado. 120 p.

Francis, Mary Lee.   The Cathey's Formation and its fauna around Nashville, Tennessee. M, 1923, Vanderbilt University.

Francis, R. M.   Hydrogeological properties of a fractured porous aquifer, Winter River basin, Prince Edward Island. M, 1981, University of Waterloo.

Francis, Robert A.   Systematic trends in the ancient fracture features of the Tharsis region, Mars. M, 1988, University of Massachusetts. 38 p.

Francis, Robert Daniel.   On the fractionation of sulfur, copper, and related transition elements in silicate liquids. D, 1980, University of California, San Diego.

Francis, Robert E. L., Jr.   Gravity and magnetic surveys of the New Castle area, Craig County, Virginia. M, 1967, Virginia Polytechnic Institute and State University.

Francisco, German.   Geology and barite potential of the Raymond Widel area, Cooper County, Missouri. M, 1977, University of Missouri, Rolla.

Franck, C. J. A study of the Holocene stratigraphy and the recent coastal processes of Crab Meadow Marsh and barrier system on the North Shore of Long Island, New York. M, 1974, Columbia University. 80 p.

Francka, Benjamin Joseph. The geology of the Gray Eagle Mine area, Lander County, Nevada. M, 1980, Texas Tech University. 68 p.

Francki, Benjamin Joseph. The geology of the Gray Eagle Mine area, Lander County, Nevada. M, 1980, University of Texas at Dallas.

Franco, D. T. Operations policy for the Upper Pampanga River Project reservoir system in the Philippines. D, 1977, University of Arizona. 406 p.

Franco, Jose Vicente Bonilla *see* Bonilla Franco, Jose Vicente

Franco, Lamberto-Augusto. Deposition and diagenesis of the Yates Formation (Permian), Guadalupe Mountains and central basin platform (New Mexico and Texas). M, 1973, Texas Tech University. 68 p.

Franco, Marcia Clara. Modeling of the Koyna, India, aftershock of December 12, 1967. M, 1982, Pennsylvania State University, University Park. 70 p.

Francoeur, D. Petrographic and structural evolution of granites and gneisses of the Grenville Province, Glamorgan and Monmouth townships, Ontario. M, 1975, University of Ottawa. 141 p.

Francois, Darryl K. The tectonics of the Caribbean Plate. M, 1980, Pennsylvania State University, University Park. 121 p.

Francois, Roger. Some aspects of the geochemistry of sulphur and iodine in marine humic substances and transition metal enrichment in anoxic sediments. D, 1987, University of British Columbia.

Franczyk, Karen J. Stratigraphy of the Upper Cretaceous Toreva and Wepo fms., Northeast Black Mesa, Arizona. M, 1983, Colorado School of Mines. 150 p.

Frand, David M. Environment of deposition of the Permian Lyons Sandstone at Black Hollow Field, West County, Colorado. M, 1984, Texas A&M University. 129 p.

Frank, Albert Joseph. Petrology of the Pennsylvanian cycles of the St. Louis area. M, 1940, St. Louis University.

Frank, Albert Joseph. Properties of the Cheltenham Clay as a cause of building cracking. D, 1948, St. Louis University.

Frank, Andrew Jay. Analysis of gravity data from the Picacho Butte area, Yavapai and Coconino counties, Arizona. M, 1984, Northern Arizona University. 91 p.

Frank, Barbara A. Quaternary micropaleontology of the Rub' al Khali (Empty Quarter), Saudi Arabia. M, 1977, University of Delaware.

Frank, Barbara Joyce. The effects of urbanization on the stream flow of the Reedy River, Greenville, South Carolina. D, 1973, University of South Carolina. 135 p.

Frank, Charles Otis. Relationships between coexisting plagioclase and hornblende in the Harrison Gneiss; southeastern New York; southwestern Connecticut. D, 1973, Syracuse University.

Frank, David Gerard. Hydrothermal processes at Mount Rainier, Washington. D, 1985, University of Washington. 195 p.

Frank, David M. Environment of deposition of the Permian Lyons Sandstone at Black Hollow Field, Weld County, Colorado. M, 1984, Texas A&M University. 89 p.

Frank, Donald James. Deuterium variations in the Gulf of Mexico and selected organic materials. D, 1972, Texas A&M University. 128 p.

Frank, Ernest C. Correlations between physiographic factors and annual streamflow. M, 1962, Colorado State University. 46 p.

Frank, Glenn W. A tidal zone investigation of Stover Cove, South Harpswell, Maine; 2 volumes. M, 1953, University of Maine.

Frank, Gregory Bryan. Petrography and petrochemistry at the DD massive sulfide prospect, Delta District, Alaska. M, 1979, University of Missouri, Rolla.

Frank, Hal J. Stratigraphy of the upper member, Koobi Fora Formation, northern Karari Escarpment, East Turkana Basin, Kenya. M, 1976, Iowa State University of Science and Technology.

Frank, James R. Dolomitization in the Taum Sauk Limestone. M, 1979, University of Missouri, Columbia.

Frank, Jeannette. Variation in some Devonian Stropheodonta. M, 1914, Columbia University, Teachers College.

Frank, Kevin James. A diagenetic study of the Toroweap Formation (Permian), in Toroweap Valley, Arizona. M, 1983, Texas Christian University. 140 p.

Frank, Marc Hilary. Textural and chemical alteration of dolomite; interaction of mineralizing fluids and host rock in a mississippi valley-type deposit, Bonneterre Formation, Viburnum Trend. M, 1986, University of Michigan. 16 p.

Frank, Mark Steven. Image reconstruction from projections, with application to geotomography. D, 1988, Arizona State University. 297 p.

Frank, Robin. Depositional environment, paleoecology and diagenesis of a glaciated surface on the Middle Devonian Columbus Limestone, Marblehead, Ohio. M, 1982, University of Cincinnati. 176 p.

Frank, Ruben Milton. Petrologic study of sediments from selected central Texas caves. M, 1965, University of Texas, Austin.

Frank, T. D. Categorization of watershed land cover units with Landsat data. D, 1979, University of Utah. 172 p.

Frank, Thomas Russell. Geology and mineralization of the Siginaw Hill area, Pima County, Arizona. M, 1970, University of Arizona.

Frank, Wendy L. An environmental study of the impact of highway deicing on the ground water Supplies of Buzzards Bay and Onset, Massachusetts. M, 1972, Boston University. 155 p.

Frank, William. Stratigraphy and sedimentation of the upper Knox Group (Ordovician), Oostanaula Valley northeast of Athens (McMinn County), Tennessee. M, 1967, University of Tennessee, Knoxville. 70 p.

Frank, William M. Continental-shelf sediments off New Jersey. D, 1971, Rensselaer Polytechnic Institute. 117 p.

Franke, Milton Romeu. Natural porosity, diagenetic evolution and experimental porosity development in Macae carbonates (Albian-Cenomanian), Campos Basin, offshore Brazil. D, 1981, University of Illinois, Urbana. 141 p.

Frankel, Arthur David. Earthquake source parameters and seismic attenuation in the northeastern Caribbean. D, 1982, Columbia University, Teachers College. 163 p.

Frankel, Larry. Pleistocene geology and paleoecology of parts of Nebraska and adjacent areas. D, 1956, University of Nebraska, Lincoln. 391 p.

Frankel, Leah Shirley. A crystallographic study of natural dimorphite I and dimorphite II. M, 1972, University of Minnesota, Minneapolis. 57 p.

Frankel, Paul. Application of mathematical modeling to the analysis of ground water flow patterns near a deep underground mine. M, 1986, University of Idaho. 85 p.

Frankenberg, Julian Myron. Structure and development of the underground part of arborescent lycopsida from the Pennsylvanian of North America. D, 1968, University of Illinois, Urbana. 302 p.

Frankfort, Donald. Factors affecting basalt talus slopes in central Connecticut. M, 1968, University of Connecticut. 41 p.

Frankforter, Matthew J. A subsurface study of the C Zone, Lansing-Kansas City groups, in Red Willow County, Nebraska. M, 1982, University of Nebraska, Lincoln. 125 p.

Frankforter, Weldon Deloss. The Pleistocene geology of the middle portion of the Elkhorn River valley. M, 1949, University of Nebraska, Lincoln.

Frankie, Kathleen Adams. A new method for determining pyrite size/form/microlithotype distribution in coal with applications for characterizing pyrite in western Kentucky and western Pennsylvania channel samples and western Kentucky. M, 1984, University of Kentucky. 235 p.

Franklin, Adele. A reconnaissance trip to the Palisades; the petrographic study of igneous rocks from Newfoundland. M, 1919, Columbia University, Teachers College.

Franklin, Arley Graves. Energy dissipation and nonlinear mechanical response in a kaolin clay. D, 1968, Northwestern University. 199 p.

Franklin, Donald Wilbert. Lithologic and stratigraphic study of the Lower Pennsylvanian strata, Orange County, Indiana. M, 1939, University of Illinois, Urbana. 49 p.

Franklin, George Joseph. Geology of Licking County, Ohio. D, 1961, Ohio State University. 601 p.

Franklin, George Joseph. The geology of Mary Ann, Perry, and Fallsbury townships, Licking County, Ohio. M, 1953, Ohio State University.

Franklin, James M. Pyrite zone of the Timagami Mine of Copperfields Mining Company, Timagami, Ontario. M, 1967, Carleton University. 118 p.

Franklin, James McWillie. Metallogeny of the Proterozoic rocks of Thunder Bay District, Ontario (Precambrian). D, 1970, University of Western Ontario. 318 p.

Franklin, Jerry Forest. Vegetation and soils in the subalpine forests of the southern Washington Cascade Range. D, 1966, Washington State University. 132 p.

Franklin, Russell J. Geology and mineralization of the Great Excelsior mine, Whatcom County, Washington. M, 1985, Western Washington University. 119 p.

Franklin, Stanley P. Diagenesis of the Dakota Sandstone, West Lindrith Field, Rio Arriba County, New Mexico. M, 1987, Texas A&M University. 126 p.

Franklin, Steven Eric. The significance of geomorphometric variables in Landsat MSS analysis of a high relief environment. D, 1985, University of Waterloo. 237 p.

Franklin, Wesley Earlynne. Structural significance of meta-igneous fragments in the Prairie Mountain area, North Cascade Range, Snohomish County, Washington. M, 1974, Oregon State University. 109 p.

Franklin, Wesley Earlynne. Vegetation and soils in the subalpine forests of the southern Washington Cascade Range. D, 1966, University of Washington. 132 p.

Franklin, William Talbert. Mineralogical and chemical characteristics of western Oregon andic soils. D, 1971, Oregon State University. 199 p.

Franklyn, Michael T. (Bone). Sr isotopic composition of saline waters and host rock in the Eye-Dashwa Lakes Pluton, Atikokan, Ontario. M, 1987, McMaster University. 108 p.

Frankovic, Edward A. An aerial photograph interpretation and physical model study of Lake Michigan shoreline erosion in the villages of Whitefish Bay, Fox Point, and Shorewood, Wisconsin. M, 1975, University of Wisconsin-Milwaukee.

Frankovits, Nicholas D. The diatom succession of Alder Marsh, Portage County, Ohio. M, 1977, University of Akron. 71 p.

Franks, Alvin LeRoy. Environmental geology; land use planning, erosion and sedimentation, West Martis Creek drainage basin, California. D, 1980, University of California, Davis. 371 p.

Franks, Bernard J. Petrology and stratigraphy of the Putnam Hill Limestone (Pennsylvanian), northeastern Ohio. M, 1973, University of Akron. 42 p.

Franks, Bernard Jeffrey. Transport of organic contaminants in a surficial sand aquifer; hydrogeologic and geochemical processes affecting transport movement. D, 1988, Florida State University. 131 p.

Franks, Christopher D. Heavy metal content of coals and associated rocks in the Indian Fork Watershed, Anderson County, Tennessee. M, 1978, University of Tennessee, Knoxville. 93 p.

Franks, James Lee. Land-resource capability units of Payne County, Oklahoma. M, 1974, Oklahoma State University. 44 p.

Franks, Paul C. Geology of the Beulah area, Colorado. M, 1956, University of Kansas.

Franks, Paul C. Petrology and stratigraphy of the Kiowa and Dakota formations (basal Cretaceous), north central Kansas. D, 1967, University of Kansas. 459 p.

Franks, S. G. Stratigraphy, sedimentology, and petrology of early Paleozoic island arc deposits, Newfoundland. D, 1976, Case Western Reserve University. 357 p.

Franks, Stephen G. A middle Eocene depositional cycle in central Mississippi. M, 1971, University of Mississippi.

Franotovic, Davor. Stratabound copper mineralization in the Revett Formation, Belt Supergroup, North Idaho, Northwest Montana. D, 1982, Colorado School of Mines. 383 p.

Franseen, Evan K. Sedimentology of the Grayburg and Queen formations (Guadalupian) and the shelf margin erosion surface, western escarpment, Guadalupe Mountains, West Texas. M, 1985, University of Wisconsin-Madison.

Fransham, Peter Bleadon. Regional geology and groundwater controls of natural slope stability. D, 1980, McGill University. 215 p.

Franson, Oral M. Sedimentation of the basal Oquirrh Formation, Provo Canyon, Utah. M, 1950, Brigham Young University. 55 p.

Frantes, James R. Petrology and sedimentation of the Archean Seine Group; conglomerate and sandstone, western Wabigoon Belt, northern Minnesota and western Ontario. M, 1987, University of Minnesota, Duluth. 148 p.

Frantes, Thomas J. The geology of the Palomas volcanic field, Luna County, New Mexico and Chihuahua, Mexico. M, 1981, University of Texas at El Paso.

Franti, Thomas George. Modeling the mechanical effects of incorporated residue on rill erosion. D, 1987, Purdue University. 186 p.

Frantti, Gordon E. Geophysical investigations in the eastern half of the Upper Peninsula of Michigan. M, 1954, Michigan Technological University. 58 p.

Frantz, Gregory Alan. The fixation of Tl, Rb, and Cs in micaceous minerals and their influence on the release of interlayer K. D, 1983, University of California, Davis. 121 p.

Frantz, J. C. Origin and nature of ilmenite deposits. M, 1947, University of Toronto.

Frantz, John Duncan. Acid buffers; use of Ag + AgCl for measuring mineral-solution equilibria in the system MgO-Si$_2$-H$_2$O-HCl. D, 1973, The Johns Hopkins University.

Frantz, Wendelin Robert. A study of the Lower Pennsylvanian sands underlying the region of Eldorado, Saline County, Illinois. M, 1956, University of Pittsburgh.

Frantz, Wendelin Robert. A subsurface stratigraphic study from the top Wilcox to the top Vicksburg in central Louisiana. D, 1963, University of Pittsburgh. 328 p.

Frantzen, Danie Ray. Oligocene folding rim rock country, Trans-Pecos, Texas. M, 1958, University of Texas, Austin.

Franz, Arthur J. Sedimentology of the Sugarloaf Arkose, Late Triassic-Early Jurassic of the Connecticut Valley; a succession of alluvial-fan, braided-stream and meandering-stream deposits. M, 1978, University of Massachusetts. 174 p.

Franz, Gilbert Wayne. Melting relationships in the system CaO-MgO-SiO$_2$-H$_2$O; a study of synthetic kimberlites. D, 1965, Pennsylvania State University, University Park. 152 p.

Franz, Kristen Elizabeth. Geochemistry of the sandstone and Silurian aquifers in eastern Wisconsin. M, 1985, Syracuse University. 104 p.

Franz, Richard H. Lithofacies, diagenesis, and petrophysical properties of selected sandstones from the Morrowan Kearny Formation of southwestern Kansas. M, 1984, University of Kansas. 177 p.

Franz, Richard Lewis. Clay mineralogy of the Claiborne Group, Sabine Parish, Louisiana. M, 1973, Louisiana Tech University.

Franzi, David A. Glacial geology of the Richmond-New Paris region of Indiana and Ohio. M, 1980, Miami University (Ohio). 134 p.

Franzi, David Alan. The glacial geology of the Remsen-Ohio region, east-central New York. D, 1984, Syracuse University. 187 p.

Franzmann, Frederick K. History of sedimentation of Duncan Point; a point bar on the Mississippi river. M, 1969, Louisiana State University.

Franzone, Joseph G. Geology, geotechnical properties and vesicular rock classification of Lousetown basalts and latites, Truckee area, California. M, 1980, University of Nevada. 170 p.

Frape, Shaun Keith. A geochemical study of Collins Lake, north of Mingston, eastern Ontario. M, 1974, Queen's University. 214 p.

Frape, Shaun Keith. Interstitial waters and bottom sediment geochemistry as indicators of ground water seepage. D, 1979, Queen's University.

Frappa, Richard H. A quantitative contamination model; application to the Virginia Dale ring-dike complex, Colorado-Wyoming. M, 1987, SUNY at Buffalo. 91 p.

Frarey, Murray James. Geology of the Willbob Lake area, northern Quebec and western Labrador. D, 1954, University of Michigan. 166 p.

Frarey, Murray James. Reconnaissance geology of the Ile a La Crosse area (Saskatchewan). M, 1950, University of Michigan.

Frasca, J. W. The effectiveness of two and three dimensional isarithmic surfaces in communicating magnitude, gradient, and pattern information. D, 1979, University of Oklahoma. 276 p.

Frasco, Barry Richard. Plant ecology of the upland-salt marsh transition zone surrounding several forest islands in southern New Jersey. D, 1980, Rutgers, The State University, New Brunswick. 223 p.

Fraser, Donald M. The geology of the San Jacinto Quadrangle south of San Gorgonio Pass (California). D, 1932, Columbia University, Teachers College.

Fraser, Donald M. The petrology of a section of the Oregon Cascades from Oakridge to Crescent. M, 1926, University of Oregon. 96 p.

Fraser, Douglas Culton. Rotary field electromagnetic prospecting. D, 1966, University of California, Berkeley. 242 p.

Fraser, Douglas Culton. The Tantramar copper swamp, New Brunswick. M, 1960, University of New Brunswick.

Fraser, Douglas R. Late Pleistocene fluvial-lacustrine history of the Hutchins Creek, Clear Creek Valley, southern Illinois. M, 1980, Southern Illinois University, Carbondale. 123 p.

Fraser, Gordon Simon. Petrology of the Hall and Pontiac limestone members (upper Pennsylvanian) in Livingston County, Illinois. M, 1970, University of Illinois, Urbana. 69 p.

Fraser, Gordon Simon. Sedimentology of the Saint Peter-Platteville Transition, middle Ordovician Black Riveran Stage, upper Mississippi Valley region. D, 1974, University of Illinois, Urbana. 148 p.

Fraser, Gregory Thomas. Stratigraphy, sedimentology, and structure of the Swauk Formation in the Swauk Pass area, central Cascades, Washington. M, 1985, Washington State University. 219 p.

Fraser, Horace John. An experimental study of permeability with respect to ore deposition. D, 1930, Harvard University.

Fraser, Horace John. Dolomitization processes in the Paleozoic horizons of Manitoba. M, 1927, University of Manitoba.

Fraser, J. A. Hydrothermal synthesis of pyroxene, garnet, and related materials. D, 1953, University of Wisconsin-Madison.

Fraser, J. R. Nephrite in British Columbia. M, 1973, University of British Columbia.

Fraser, James Allan. The hydrothermal synthesis of hydrogarnet, sphene, and related silicates. D, 1955, University of Minnesota, Minneapolis. 96 p.

Fraser, John Keith. The physiography of Boothia Peninsula, Northwest Territories; a study in terrain analysis and air photo interpretation of an Arctic area. D, 1964, Clark University. 321 p.

Fraser, William Brian. Seismic refraction study of post-Pliocene stratigraphy, Dismal Swamp. M, 1982, Old Dominion University. 178 p.

Frasier, Clint Wellington. Discrete time solution of plane P-SV waves in a plane layered medium. D, 1969, Massachusetts Institute of Technology. 210 p.

Frasner, N. H. C. Genesis of titaniferous pegmatite dyke. M, 1938, Queen's University. 33 p.

Frasse, Frederic I. Geology and structure of the western and southern margins of Twin Sisters Mountain, North Cascades, Washington. M, 1981, Western Washington University. 87 p.

Frater, J. B. Geomorphic interpretation of Skylab photography collected over the Nevada portion of the Great Basin. M, 1975, Purdue University. 86 p.

Fraticelli, Luis. Geology of portions of the Project City and Bella Vista quadrangles, Shasta County, California. M, 1984, San Jose State University. 110 p.

Fratt, Walter James. The Big Snowy Group (Mississippian) in the Bridger Range in Montana. M, 1957, University of Wisconsin-Madison. 52 p.

Fraunfelter, George Henry. The paleontology and stratigraphy of the Cedar City Formation (Middle Devonian) of Missouri. D, 1964, University of Missouri, Columbia. 731 p.

Fraunfelter, George Henry. The Rensseladnia Beds (Middle Devonian) of central Missouri. M, 1951, University of Missouri, Columbia.

Frazee, Charles Joseph. Distribution of loess from a source. D, 1969, University of Illinois, Urbana. 87 p.

Frazell, William Davis. The geology along the Balcones fault zone, east of Oak Hill, Travis County, Texas. M, 1935, University of Texas, Austin.

Frazer, Laurie Neil. Synthesis of shear-coupled PL. D, 1978, Princeton University. 62 p.

Frazier, David E. Sedimentary parameters of lower Barataria Bay, Jefferson Bay, Louisiana. M, 1960, Texas A&M University. 93 p.

Frazier, Don W. Paragenesis of silica in silicified woods of the Whitsett Formation (Eocene), in (Brazos, Karnes, Polk, McMullen counties) Texas. M, 1966, University of Houston.

Frazier, James Edward. Analysis of the Bouguer gravity anomalies in the region surrounding the Elberton and Danburg granites in east-central Georgia. M, 1982, Georgia Institute of Technology. 149 p.

Frazier, Melvin. Vanadium. M, 1987, University of Texas, Austin.

Frazier, Michael K. A revision of the fossil Erethizontidae of North America. M, 1978, University of Florida. 126 p.

Frazier, Noah Arthur. A heavy mineral study of the Morrison Formation and the Indianola Group of central Utah. M, 1951, Ohio State University.

Frazier, Robert H. The Ft. Apache Limestone of east central Arizona. M, 1961, University of Arizona.

Frazier, Samuel Bowman. The Morgantown Sandstone unconformity in the vicinity of Pittsburgh, Pennsylvania. M, 1950, University of Pittsburgh.

Frazier, William James. Carbonate petrology of the Mississippian Pennington Formation, central Tennessee. D, 1974, University of North Carolina, Chapel Hill. 102 p.

Freas, Donald Hayes. High temperatures of mineral formation in the Deardorff and Victory mines of the Cave-in-Rock District as indicated by liquid inclusions. M, 1957, University of Wisconsin-Madison. 40 p.

Freas, Donald Hayes. Occurrence, mineralogy and origin of the Lower Golden Valley kaolinitic clay deposits near Dickinson, North Dakota. D, 1959, University of Wisconsin-Madison. 83 p.

Freas, Robert C. The faunal distribution of the Blanchester Member of the Waynesville Formation in Indiana, Ohio, and Kentucky. M, 1968, Miami University (Ohio). 121 p.

Freberg, R. A. Investigation of the dithizone method of detecting traces of metallic elements. M, 1951, University of Toronto.

Frebold, Fred. Study of the Mississippian faunas of the Rocky Mountains of Canada. M, 1954, University of British Columbia.

Frebold, Fridtjof Albert. Corals from the Rundle Formation (Mississippian) of Banff, Alberta. M, 1955, University of British Columbia.

Frech, Richard Eugene. Basic rocks of the Roosevelt-Cold Springs area, southwestern Oklahoma. M, 1962, University of Oklahoma. 46 p.

Frechette, Andre B. Aerial triangulation with independent geodetic controls using the wild steroplotter A-8. M, 1962, Ohio State University.

Fréchette, Ghislain. Etude tridimensionnelle de la dispersion chimique et détritique dans les sédiments glaciaires et glacio-lacustres reposant sur le complexe rhyolitique de la Mine Hunter, région de Palmarolle, Abitibi. M, 1988, Universite Laval. 189 p.

Frechette, William G. An analysis of ground vibrations from Vibroseis TM vibrators. M, 1983, University of Wisconsin-Milwaukee. 200 p.

Freckman, John T. Fluid inclusion and oxygen isotope geothermometry of rock samples from Sinclair and Elmore boreholes, Salton Sea geothermal field, Imperial Valley, California, U.S.A. M, 1978, University of California, Riverside. 66 p.

Frederick, Daniel. Regional of variations of conodont assemblages in the Decorah Subgroup (Middle Ordovician) of the upper Midwest. M, 1987, Northern Illinois University. 104 p.

Frederick, Jan Elizabeth. Perennial snowcover variations during the last 130 years at Mount Rainier, Washington. M, 1980, University of Washington. 67 p.

Frederick, Lawrence Churchill. Dispersion of the Columbia River plume based on radioactivity measurements. D, 1967, Oregon State University. 134 p.

Frederick, Margaret A. An atlas of Secchi disc transparency measurements and Forel-Ule codes for the oceans of the world. M, 1970, United States Naval Academy.

Frederick, V. R., Jr. The environmental significance of the algal floras from three central Ohio sediment profiles. D, 1977, Ohio State University. 99 p.

Fredericks, Alan D. A method for the determination of dissolved organic carbon in sea water by gas chromatography. M, 1965, Texas A&M University.

Fredericks, Carol M. Petrology and chemical differentiation of the Wallace sill at Whitefish Falls, Ontario, Canada. M, 1970, Queens College (CUNY). 152 p.

Fredericks, Kenneth J. A gravity survey of eastern Vilas County, Wisconsin. M, 1974, Wright State University. 90 p.

Fredericks, Paul Edward, Jr. Volcanic lithofacies and massive sulfide mineralization, East Shasta District, California. M, 1980, University of Texas, Austin.

Fredericks, Robert Warren. Scattering of elastic pulses by obstacles of infinite impedance and semi-infinite dimensions on the surface of a half-space. D, 1959, University of California, Los Angeles. 102 p.

Fredericksen, Rick Stewart. The secondary dispersion of tungsten in some southern Arizona tungsten districts. M, 1974, University of Arizona.

Frederickson, A. F. The mineralogy of a drill core of Arkansas bauxite. D, 1947, Massachusetts Institute of Technology.

Frederickson, Edward Arthur. Cambrian stratigraphy of Oklahoma. D, 1942, University of Wisconsin-Madison.

Frederickson, J. A. Petrology and depositional environments of the Fountain and Ingleside formations, Owl Canyon, Colorado. M, 1978, University of Wyoming. 100 p.

Frederickson, Norman Oliver. Stratigraphy and palynology of the Jackson stage (upper Eocene) and adjacent strata of Mississippi and western Alabama. D, 1969, University of Wisconsin-Madison. 417 p.

Frederiksen, Norman Oliver. Sporomorphae of the Brookville Seam near Brookville, Pennsylvania. M, 1961, Pennsylvania State University, University Park. 273 p.

Frederking, Ray Lynn. Spatial variation of the presence and form of earth mounds on a selected Alp surface, Sangre de Cristo Mountains, Colorado. D, 1973, University of Iowa. 201 p.

Fredrikson, Goran. Geology of the Huitis and La Mision quadrangles (Mesozoic and Tertiary), northernmost Sinaloa, Mexico. M, 1971, University of Texas, Austin.

Fredrikson, Goran. Geology of the Mazatlan area, Sinaloa, western Mexico. D, 1974, University of Texas, Austin.

Free, Brickey Rae. Geology of a portion of the Meers Quadrangle, Oklahoma. M, 1980, University of New Orleans. 92 p.

Free, Dwight Allen, Jr. The stratigraphy of the Templeton Member of the Gulfian Cretaceous Woodbine Formation in Denton, Collin, and Grayson counties, Texas. M, 1956, Southern Methodist University. 25 p.

Free, Michael Royce. Evidence for magmatic assimilation in several diorites of the middle Columbia River gorge. M, 1976, University of Utah. 67 p.

Freeborn, W. P. The distribution of iron and magnesium between olivine and calcic clinopyroxene; an experimental study with some comments on natural occurrences. D, 1976, University of California, Los Angeles. 66 p.

Freed, George Richard. Geological structure of northeastern Monroe County, Indiana. M, 1932, Indiana University, Bloomington.

Freed, Robert Lowell and Williams, Richard S., Sr. Geology of the Buck Mountain area, Costill County, Colorado. M, 1963, University of Michigan.

Freed, Robert Lowell and Williams, Richard Sugden, Jr. Geology of the Buck Mountain area, Costilla County, Colorado. M, 1962, University of Michigan.

Freed, Robert Lowell. The structures and crystal chemistry of margarosanite and johannsenite. D, 1966, University of Michigan. 145 p.

Freedenberg, H. Environment of deposition of the Irondequoit Formation (Middle Silurian), western New York and southern Ontario. M, 1976, SUNY at Buffalo. 67 p.

Freedman, Adam Paul. Marine geophysical applications of Seasat altimetry and the lithospheric structure of the South Atlantic Ocean. D, 1987, Massachusetts Institute of Technology. 220 p.

Freedman, B. Effects of smelter pollution near Sudbury, Ontario, Canada on surrounding forested ecosystems. D, 1978, University of Toronto.

Freedman, Jacob. Stratigraphy and structure of the Mount Pawtuckaway Quadrangle, southeastern New Hampshire. D, 1948, Harvard University.

Freeland, George Lockwood. Carbonate sediments in a terrigenous province; the reefs of Veracruz, Mexico. D, 1971, Rice University. 367 p.

Freeman, Bruce C. The Long Lake Diorite and associated rocks, Sudbury District, Ontario. D, 1932, University of Chicago. 51 p.

Freeman, Charles Edward, Jr. A pollen study of some post-Wisconsin alluvial deposits in Dona Ana County, southern New Mexico. D, 1968, New Mexico State University, Las Cruces. 71 p.

Freeman, Curtis J. Geology and mineral occurrences in the Wood River area, North-central Alaska Range, Alaska. M, 1980, University of Alaska, Fairbanks. 172 p.

Freeman, Edward Bicknell. Effect of pressure and temperature on lattice parameters of nepheline. M, 1958, McMaster University. 50 p.

Freeman, Gary W. Stratigraphy of the Cheverie Formation (Lower Carboniferous), Minas Sub-Basin, Nova Scotia. M, 1972, Acadia University.

Freeman, Harvey Albert. The Connor Facies (of Dunham Dolomite, Lower Cambrian) in northwestern Vermont. M, 1958, University of Rochester. 119 p.

Freeman, James C. Strand line accumulation of petroleum. M, 1947, University of Colorado.

Freeman, James Thomas. Modeling regional groundwater flow with environmental isotopes; Ross Creek Basin, Alberta. M, 1986, University of Alberta. 139 p.

Freeman, John C. Geology of the Cabezon Peak area, Sandoval County, New Mexico. M, 1949, Stanford University. 59 p.

Freeman, John H. Fallout plutonium and naturally occurring radionuclides in annual bands from Montastrea annularis, Broward County, Florida. M, 1985, University of North Carolina, Chapel Hill. 59 p.

Freeman, Kevin John. A textural analysis of the strain state of a shear zone. M, 1974, Michigan State University. 21 p.

Freeman, Kimberley June. Hydrothermal metamorphism of Telkwa Formation volcanics near Terrace, British Columbia. M, 1986, University of Calgary. 156 p.

Freeman, Lawrence K. Geology and tactite mineralization of the South Mountain mining district, Owyhee County, Southwest Idaho. M, 1982, Oregon State University. 99 p.

Freeman, Leroy Bradford; Sweet, John M. and Tillman, Chauncey. Geology of the Henrys Lake Mountains, Fremont County, Idaho and Madison and Gallatin counties, Montana. M, 1950, University of Michigan. 83 p.

Freeman, Louise B. Devonian subsurface strata in western Kentucky. D, 1940, University of Chicago. 72 p.

Freeman, M. Lawrence. The Bluffport Member of the Demopolis Formation in Oktibbeha County, Mississippi. M, 1961, Mississippi State University. 38 p.

Freeman, Mimi J. Magnetic survey of several ultramafic bodies in northern Harford County, Maryland. M, 1983, Kent State University, Kent. 59 p.

Freeman, Paul Swift. Clastic diapirism in the Gueydan (Catahoula) Formation (Miocene (?) and Oligocene(?)), Live Oak and McMullen counties, Texas. M, 1966, University of Texas, Austin.

Freeman, Peter Verner. Copper mineralization in the Coppermine series, Northwest Territories. M, 1953, McGill University.

Freeman, Peter Verner. Geology of the Beraud-Mazerac area, Quebec. D, 1957, McGill University.

Freeman, Peter Verner. Petrographic study of the "A" orebody of the Monroe asbestos mine, Matheson, Ontario. M, 1954, McGill University.

Freeman, Ralph Neptune. Geology of the Round Mountain area of Gunnison County, Colorado and petrological study of the Coffman Conglomerate. M, 1950, University of Kentucky. 46 p.

Freeman, Thomas Jewell, Jr. Carboniferous stratigraphy of the Brady area, McCulloch and San Saba counties, Texas. D, 1962, University of Texas, Austin. 280 p.

Freeman, Thomas, Jr. Stratigraphy of the pre-Atokan Carboniferous in the sub-surface of Pope, Cleburne, Van Buren and White counties, Arkansas. M, 1957, University of Arkansas, Fayetteville.

Freeman, Timothy F. The stratigraphic and petrologic controls of a Pottsville Sandstone (Lower and Middle Pennsylvanian) at Caryville, Tennessee. M, 1966, University of Tennessee, Knoxville. 55 p.

Freeman, Val LeRoy. Geology of part of the Johnny Gulch Quadrangle, Montana. M, 1954, University of California, Berkeley. 79 p.

Freeman, William E. Re-evaluation of the depositional environment of the Navajo Sandstone. M, 1973, University of Tulsa. 119 p.

Freeman, Worth Merle. Geology of east half of Foster Quadrangle, Culberson County, Texas. M, 1950, University of Texas, Austin.

Freeman-Lynde, Raymond Paul. The marine geology of the Bahama Escarpment. D, 1981, Columbia University, Teachers College. 292 p.

Freers, F. Theodore. A structural and morphogenetic investigation of the Vaughan Lewis Glacier and adjacent sectors of the Juneau Icefield, Alaska. M, 1966, Michigan State University. 132 p.

Freethey, Geoffrey W. Hydrogeologic evaluation of pollution potential in mountain dwelling sites (Colorado). M, 1969, Colorado State University. 96 p.

Freeze, Arthur C. Geology of the Fredericton Sheet, New Brunswick. M, 1936, University of New Brunswick.

Freeze, Arthur C. Geology Pinchi Lake, B. C. D, 1942, Princeton University. 158 p.

Freeze, Roy Allan. Theoretical analysis of regional groundwater flow. D, 1966, University of California, Berkeley. 304 p.

Fregeau, Elizabeth J. Trace element partitioning between a silicate melt and a super-critical hydrous fluid; implications for mantle metasomatism. M, 1985, Pennsylvania State University, University Park.

Frei, Leah Shimonah. Paleomagnetic constraints on extension in the Basin and Range and in the North Atlantic area. D, 1986, Stanford University. 197 p.

Freidline, Roger Alan. Seismicity and contemporary tectonics of the Helena, Montana area. M, 1974, University of Utah. 84 p.

Freie, Alvin John. A study of the Silurian and Devonian systems of Bremer County, Iowa. M, 1927, University of Iowa. 91 p.

Freie, Alvin John. The geology of the Anadarko Basin of Oklahoma. D, 1929, University of Iowa. 142 p.

Freiholz, Ginette. The stratigraphic and structural setting of a lead-zinc occurrence near Invermere, southeastern B.C. M, 1983, University of Calgary. 203 p.

Freile, Deborah. A sedimentological and ichnostratigraphical paleofacies interpretation of the Upper Ordovician Juniata Formation and the Lower Silurian Tuscarora Formation of central Pennsylvania. M, 1988, Boston University. 250 p.

Freitag, Helen Clare. Gravity fields of eight North Pacific seamounts; implications for density. M, 1987, Texas A&M University. 127 p.

Freitas, Timothy A. de see de Freitas, Timothy A.

Frelier, Andrew P. Sedimentology, fluvial paleohydrology, and paleogeomorphology of the Dockum Formation (Triassic), Garza County, West Texas. M, 1987, Texas Tech University. 198 p.

French, Alice Elizabeth. Primary evidence for and against the existence of the Taconic Overthrust. M, 1956, University of Michigan.

French, B. E. Geology of Marion Dome, Smyth County, Virginia. M, 1967, Virginia Polytechnic Institute and State University.

French, Bevan M. Stability of siderite $FeCO_3$ and the progressive metamorphism of iron formations. D, 1964, The Johns Hopkins University.

French, Don E. Geology and mineralization of the southeastern part of the Black Pine Mountains, Cassia County, Idaho. M, 1975, Utah State University. 69 p.

French, G. B. Precambrian geology of Washington County area, Missouri. M, 1956, University of Missouri, Rolla.

French, Gordon. Coal reserves of Somerset County, Pennsylvania. M, 1959, University of Pittsburgh.

French, Gregory McNaughton. Relationships of lithology and ore deposits in the Lower Cambrian Deadwood Formation and Precambrian basement in the Lead-Deadwood Dome area, Black Hills, South Dakota. M, 1985, South Dakota School of Mines & Technology.

French, J. J. The Goose Egg Formation in the southern portion of the Bighorn Mountains, Wyoming. M, 1959, University of Wyoming. 80 p.

French, Larry B. Local geology around Bandbox Mountain (Little Belt Mountains) with emphasis on a Mississippian age carbonate buildup, Judith Basin County, Montana. M, 1984, University of Montana. 121 p.

French, Lawrence Nelson. Hydrogeologic aspects of lignite strip mines near Fairfield, Texas. M, 1979, University of Texas, Austin.

French, Leanne Sue. Wall structure of selected species of hyaline foraminifera. M, 1979, University of Georgia.

French, Peter Newton. Water quality modeling using interactive computer graphics. D, 1980, Cornell University. 252 p.

French, Robert Rex. Niagaran dolomites of Adams County (mineralogical analysis). M, 1961, University of Cincinnati. 28 p.

French, Rowland Barnes. Lower Paleozoic paleomagnetism of the North American Craton. D, 1976, University of Michigan. 170 p.

French, Tipperton J. The geology of the Castle Rock Butte Quadrangle, South Dakota. M, 1958, University of South Dakota. 129 p.

French, Tracy Alan. A petrographic study of Devonian sediments of a closed basin in northern Scotland and southern Shetland Islands. M, 1985, Oklahoma State University. 146 p.

French, Vernon Edwin. Geology of the Lake Davy Crockett area, Greene County, Tennessee. M, 1966, University of Tennessee, Knoxville. 46 p.

French, William Edwin. The sedimentary environment and geological evolution of the Manitou Passage area of Lake Michigan. D, 1965, University of Michigan.

French, William Edwin. The sedimentary environment of southern Lake Huron. M, 1960, University of Michigan.

French, William Stanley. Earthquake waves following the P_n phase and their indications of focal depth and crustal structures in the Pacific northwest states. D, 1970, Oregon State University. 175 p.

Freniere, Jon Edmund La see La Freniere, Jon Edmund

Frenkel, Oded J. The flow of aqueous solutions through clay. M, 1970, McGill University.

Frerichs, W. E. Significance of lower Burlington conodont assemblages in southeastern Iowa. M, 1963, Iowa State University of Science and Technology.

Frerichs, William E. Distribution and ecology of Foraminifera in the sediments of the Andaman Sea (eastern Indian Ocean). D, 1967, University of Southern California.

Frese, Ralph Robert Benedict von see von Frese, Ralph Robert Benedict

Frese, Ralph Robert Benedikt von see von Frese, Ralph Robert Benedikt

Freshwater, Norman G. Morgan. Geology and ore deposits of Yukon Territory. M, 1930, University of British Columbia.

Frest, Terrence James. Studies of Silurian echinoderms; (Volumes I and II). D, 1983, University of Iowa. 602 p.

Frest, Terrence James. The Pennsylvanian-Permian ammonoid families Maximitidae and Pseudohaloritidae. M, 1978, University of Iowa. 141 p.

Fretwell, Judy D. Resistivity study of the shallow fresh water-salt water interface in selected coastal areas of Citrus County, Florida. M, 1978, University of South Florida, Tampa.

Freudenberg, Connie M. Paleoenvironmental interpretation of the Guilmette Formation near Ely, Nevada. M, 1981, Eastern Washington University. 105 p.

Freudenheim, Priscilla. The Devonian stratigraphy and paleontology of the Mackenzie River valley, Canada. M, 1946, University of Chicago. 105 p.

Freund, Harold H. Petrogenesis of the paragneiss, Quesnel Lake, Manitoba (Canada). M, 1968, University of Manitoba.

Frew, William Michner. Stratigraphy and paleontology of a well core through the Middle Ordovician in Salem Township, Washtenaw County, Michigan. M, 1955, University of Michigan.

Frey, B. E. Effects of micro-nutrients and major nutrients on the growth and species composition of natural phytoplankton populations. D, 1977, Oregon State University. 81 p.

Frey, D. M. Geology of the Hunt Mountain-Red Gulch area, Big Horn and Sheridan counties, Wyoming. M, 1959, University of Wyoming. 66 p.

Frey, David A. Petrographic analysis of the Dracut Diorite and its associated rocks in the northeastern portion of Pepperell Quadrangle, Massachusetts-New Hampshire. M, 1977, Ohio University, Athens. 119 p.

Frey, John H. Mexican tectonics and seismicity. M, 1959, Boston College.

Frey, John W. The geographic background of the coal export trade of the United Kingdom and the United States. D, 1926, University of Wisconsin-Madison.

Frey, Leo Joseph, III. Rock slope stability analysis along selected areas of I-287 in northeastern New Jersey. M, 1983, Purdue University. 107 p.

Frey, Maurice Gordon. Geology of the Red Wing District (Minnesota). M, 1937, University of Minnesota, Minneapolis. 33 p.

Frey, Maurice Gordon. Geology of the region about the west end of Lake Superior. D, 1939, University of Minnesota, Minneapolis. 83 p.

Frey, Richard Paul. Distribution and genesis of dolomite in Kinderhook rocks (Mississippian) of Missouri. D, 1967, University of Missouri, Rolla. 68 p.

Frey, Richard Paul. Studies on some Recent freshwater ostracodes of St. Louis County, Missouri. M, 1963, Washington University. 51 p.

Frey, Robert Charles. The biostratigraphy of the Richmond Group (Upper Ordovician), Franklin County, Indiana. M, 1976, Miami University (Ohio). 87 p.

Frey, Robert Charles. The paleontology and paleoecology of the Treptoceras duseri shale unit (Late Ordovician, Richmondian) of southwestern Ohio. D, 1983, Miami University (Ohio). 719 p.

Frey, Robert Wayne. Stratigraphy, ichnology, and paleoecology of the Fort Hays limestone member of the Niobrara chalk (upper Cretaceous) in Trego County, Kansas. D, 1969, Indiana University, Bloomington. 345 p.

Frey, Susan. Data acquisition, processing, and structural interpretation of 2D and 3D seismic data from the Woodada gas field, Western Australia. M, 1986, University of Houston.

Freyberg, David Lewis. Models of surface-subsurface flow interaction in an ephemeral channel. D, 1981, Stanford University. 228 p.

Freyman, A. Analysis of an industrial sector; the coal mining industry in Nova Scotia. D, 1967, Columbia University, Teachers College.

Freyne, D. M. Geology of the Ruby Lake S.E., 7.5-minute quadrangle, White Pine County, Nevada. M, 1973, [University of California, San Diego].

Freyr Thorarinsson. A program to interpret roving dipole surveys with a conductive plate model. D, 1987, Colorado School of Mines. 181 p.

Frezon, Sherwood E. and Williams, John Stuart. Cambrian stratigraphy and paleontology in the Teton Mountains, Teton County, Wyoming. M, 1963, University of Michigan.

Friberg, James Frederick. Mineralogy and provenance of the Recent alluvial sands of the Ohio river drainage basin. D, 1970, Indiana University, Bloomington. 207 p.

Friberg, LaVerne Marvin. Petrology of a metamorphic sequence of upper-amphibolite facies in the central Tobacco Root Mountains, southwestern Montana. D, 1976, Indiana University, Bloomington. 147 p.

Frick, Elizabeth A. Quantitative analysis of groundwater flow in valley-fill deposits in Steptoe Valley, Nevada. M, 1985, University of Nevada. 192 p.

Fricke, Carl A. P. The Pleistocene geology and geomorphology of a portion of central-southern Wisconsin. M, 1976, University of Wisconsin-Madison.

Fricke, John N. Geologic investigation of Precambrian mafic intrusive rocks near McGee Siding, Pennington County, South Dakota. M, 1982, South Dakota School of Mines & Technology. 64 p.

Fricke, Rodney A. The hydrogeochemistry and aqueous uranium distribution of Petersen Mountain and Red Rock Valley, Washoe County, Nevada. M, 1983, University of Nevada. 92 p.

Fricker, Aubrey. A development of surface wave analysis and interpretation in the Canadian Shield. M, 1971, Dalhousie University.

Friddell, Michael S. A study of the mineralogy of the selected Cretaceous and Tertiary kaolins of central and eastern Georgia. M, 1981, Georgia Institute of Technology. 94 p.

Fridley, Harry Marion. Identification and correlation of erosion surfaces in south central New York. D, 1928, Cornell University.

Fridley, Mark S. Pennsylvanian (Desmoinesian) non-fusulinid foraminifera from central Missouri. M, 1982, University of Missouri, Columbia.

Fridrich, Christopher J. Reverse zoning in the resurgent intrusions of the Grizzly Peak Cauldron, Sawatch Range, Colorado. M, 1983, Stanford University. 34 p.

Fridrich, Christopher John. The Grizzly Peak Cauldron, Colorado; structure and petrology of a deeply dissected resurgent ash-flow caldera. D, 1987, Stanford University. 227 p.

Friedel, George F. Structural and geothermal relationships in the Lake Charles area of southwestern Louisiana. M, 1978, University of Southwestern Louisiana. 62 p.

Friedel, Michael J. A numerical investigation of the amplitude of ground motion radiated by a Vibroseis system vibrator. M, 1986, University of Wisconsin-Milwaukee. 214 p.

Frieders, T. Y. Stratigraphic differences in Knox and middle Ordovician strata as evidence of the structural development of the Polar Hill Anticline, Giles County, Virginia. M, 1975, Virginia Polytechnic Institute and State University.

Friedl, Arthur John. Some placer deposits of North America and Europe and their relation to glaciation. M, 1923, University of Minnesota, Minneapolis. 31 p.

Friedland, Andrew Jay. Trace metal accumulation, distribution and fluxes in forests of the northeastern United States. D, 1985, University of Pennsylvania. 186 p.

Friedlander, Susan J. Spin-down in a rotating stratified fluid. D, 1972, Princeton University.

Friedman, Daniel Bruce. Geophysical evidence for shallow basement folding, central Adirondacks, New York. M, 1978, Cornell University.

Friedman, Gerald M. A study of the emery deposits in the southeastern part of the Cortlandt Complex. D, 1952, Columbia University, Teachers College.

Friedman, Isidore I. The system H_2O-Na_2O-SiO_2-Al_3O at high temperature. D, 1950, University of Chicago. 48 p.

Friedman, Joan. Anion absorption exchange and fixation capacity determinations on clays. M, 1978, SUNY at Buffalo. 75 p.

Friedman, Jule Daniel. Bedrock geology of the Ellenville area, New York. D, 1957, Yale University.

Friedman, Melvin. Miocene orthoquartzite from New Jersey. M, 1953, Rutgers, The State University, New Brunswick. 73 p.

Friedman, Melvin. Petrofabric analysis of experimentally controlled calcite-cemented sandstones. D, 1961, Rice University. 84 p.

Friedman, R. M. The developmental history of a wetland ecosystem; a spatial modeling approach. D, 1978, University of Wisconsin-Madison. 157 p.

Friedman, Richard M. Geology and geochronometry of the Eocene Tatla Lake metamorphic core complex, western edge of the Intermontane Belt, British Columbia. D, 1988, University of British Columbia.

Friedman, Samuel Arthur. Petrography and petrology of the Maxville Limestone from parts of Muskingum and Perry counties. M, 1952, Ohio State University.

Friedmann, Anton R. Pollen and spore sequence of the Thompson's Branch Mammoth Site. M, 1978, Indiana State University. 43 p.

Friedrich, Nancy E. Depositional environments and sediment transport patterns in the Point Judith/Potter Pond complex. M, 1981, University of Rhode Island.

Friel, J. J. The stability of synthetic armalcolite at high pressures and at varying oxygen fugacities. D, 1975, University of Pennsylvania. 127 p.

Frielinghausen, Karl William. A geological report of the Lewis Quadrangle, Vigo County, Indiana. M, 1950, Indiana University, Bloomington. 20 p.

Friend, Joseph E. Insoluble residues and Silurian stratigraphy of two exploratory tests in Michigan. M, 1963, Wayne State University.

Fries, Carl, Jr. Geology of the State of Morelos and contiguous areas in south-central Mexico. D, 1958, University of Arizona. 387 p.

Friesen, George Henry. Development of the processing capability for crustal exploration on the Canadian Shield by the near vertical reflection technique. M, 1974, University of Manitoba.

Friesen, Larry Jay. Radon diffusion and migration at low pressures, in the laboratory and on the Moon. D, 1974, Rice University. 87 p.

Friesen, Menno. Four simple laboratory tests for determining the mechanical properties of rock. M, 1967, University of Saskatchewan. 156 p.

Friess, John Paul. Closed gravity and magnetic anomalies in West Texas. M, 1972, Texas Tech University. 87 p.

Friestad, Harlan K. Upper Red River Formation (Ordovician) in western North Dakota. M, 1969, University of North Dakota. 82 p.

Friis, Karin L. Petrology of the Route 2-Park avenue roadcut, Lexington guadrangle, Massachusetts. M, 1969, Boston University. 52 p.

Frimpter, Michael Howard. Geology of the Thiells Quadrangle (Rockland-Orange counties), New York, with emphasis on the igneous and metamorphic rocks. D, 1967, Boston University. 181 p.

Frinak, Timothy R. The geology of a part of Northeast Coosa County, Alabama. M, 1984, Auburn University. 173 p.

Frink, John W. Subsurface Pleistocene of Louisiana. M, 1939, Louisiana State University.

Frisch, Adam Arthur. Development, test and application of a new method of particle shape analysis based on the concept of the fractal dimension. D, 1988, College of William and Mary. 208 p.

Frisch, Conny Jean. Investigation of metasomatic phase relations in dolomites of the Adamello Alps. M, 1979, University of California, Berkeley. 104 p.

Frisch, Thomas. Metamorphism and plutonism in northernmost Ellesmere Island, Canadian Arctic Archipelago. D, 1967, University of California, Santa Barbara.

Frische, Richard H. A theoretical study of induced electrical polarization. M, 1956, New Mexico Institute of Mining and Technology. 31 p.

Frischhertz, Robert P. Geology of the southern half of the Lake Toxaway Quadrangle, North Carolina. M, 1987, University of New Orleans. 69 p.

Frischknecht, Frank Conrad. Application of resistolog method of the two-layer resistivity case. M, 1953, University of Utah. 56 p.

Frischknecht, Frank Conrad. Electromagnetic scale model study of geophysical methods using a plane wave source. D, 1973, University of Colorado.

Frischmann, Peter S. A paleoenvironmental study of the Middle Ordovician (Black River) interval in central Pennsylvania. M, 1977, Temple University.

Frishman, David. High- and low-temperature mineral assemblages in the Josephine Peridotite, Del Norte County, California; implications for mineral geothermometers and geobarometers applied to alpine-type harzburgites. D, 1980, University of California, Los Angeles. 356 p.

Frishman, Steven Arthur. Geochemistry of oolites, Baffin Bay, Texas. M, 1969, University of Texas, Austin.

Frisillo, Albert Lawrence. The elastic coefficients of bronzite as a function of pressure and temperature. D, 1972, Pennsylvania State University, University Park. 183 p.

Friske, Peter Wilhelm Bruno. Wall-rock alteration and ore genesis at the Lyon Lake deposits, northwestern Ontario. D, 1984, University of New Brunswick.

Frisken, Jim Gilbert. Pleistocene glaciation of the Brinnon area, east-central Olympic Peninsula, Washington. M, 1965, University of Washington. 75 p.

Frison, Eugene Hubert John. Mid-Devonian cerioid rugose corals from Northwest Territories. M, 1961, University of Saskatchewan. 88 p.

Frith, R. Anthony. Rb-Sr geochronological study of rocks of the Bear and Slave provinces, Northwest Territories. M, 1974, McGill University. 65 p.

Frith, R. Anthony. Rb-Sr isotopic studies of the Grenville structural province (Precambrian) in the Chibougamau and Lac Saint Jean area (Quebec). D, 1971, McGill University. 65 p.

Frith, Robert B. A seismic refraction investigation of the Salton Sea geothermal area, Imperial Valley, California. M, 1978, University of California, Riverside. 124 p.

Fritsche, Albert Eugene. Miocene geology of the central Sierra Madre mountains, Santa Barbara County, California. D, 1969, University of California, Los Angeles. 475 p.

Fritsche, Glen D. Subsurface stratigraphy and structure related to petroleum occurrences in the middle Atoka Formation, Arkoma Basin. M, 1980, University of Arkansas, Fayetteville. 114 p.

Fritsche, Kenneth L. Stratigraphy and petrography of the upper Paluxy Sand in Titus and Morris counties, Texas. M, 1984, University of Arkansas, Fayetteville. 92 p.

Fritts, Charles C., Jr. Carbonaceous matter of the Nonesuch Shale (northeastern Wisconsin). M, 1931, University of Wisconsin-Madison.

Fritts, Crawford Ellsworth. A petrologic study of the Mount Houghton Felsite, Keweenaw Peninsula. M, 1952, Michigan Technological University. 42 p.

Fritts, Crawford Ellsworth. Bedrock geology of the Mount Carmel and Southington quadrangles, Connecticut. D, 1962, University of Michigan. 254 p.

Fritts, John Raymond. Fauna, stratigraphy, and paleoecology of the Foraker Limestone; Osage, Pawnee, Payne, and Lincoln counties, Oklahoma. M, 1980, Oklahoma State University. 142 p.

Fritts, Paul Jan. Some late Cretaceous coccoliths of Colorado and eastern Wyoming. D, 1969, University of Colorado. 231 p.

Fritts, Steven Grant. Suitability of Landsat multispectral scanner and return beam vidicon stereo imagery for reconnaissance engineering geologic mapping. M, 1982, University of California, Los Angeles. 171 p.

Fritz, Barrett Robert. Interactions of radionuclides with particulate organic detritus and subsequent transfer to an estuarine detritivore. M, 1978, University of Virginia. 75 p.

Fritz, Dale A. The ground water hydrology of the Loes Lakes, North Bark Lake, and South Bark Lake wetlands; with implications for a new wetland classification scheme. M, 1982, University of Wisconsin-Milwaukee. 220 p.

Fritz, Deborah M. Ophiolite belt west of Paskenta, northern California Coast Ranges. M, 1975, University of Texas, Austin.

Fritz, Jeffrey Lynn. Systematics, biostratigraphy and paleoenvironment of Desmoinesian, Missourian and Virgilian syringoporid corals of the Bird Spring Group, Arrow Canyon Range, Clark County, Nevada. M, 1981, University of Illinois, Urbana. 299 p.

Fritz, Joseph F. Geology of an area between Honey Creek and Bluff Creek, Mason County, Texas. M, 1954, Texas A&M University.

Fritz, Lloyd G. Petrography of the crystalline rocks south of Okanogan, in North central Washington. M, 1978, Eastern Washington University. 36 p.

Fritz, Madeleine A. The stratigraphy and palaeontology of the Workman's Creek section of the Cincinnatian series of Ontario. D, 1926, University of Toronto.

Fritz, Richard Dale. Structural contour map of Oklahoma on the Pennsylvanian Wapanucka Limestone, Oswego Limestone, base of the Hoxbar Group, and Checkerboard Limestone. M, 1978, Oklahoma State University. 47 p.

Fritz, Steven J. Physical and chemical characteristics of weathering rinds from several plutons in the North Carolina Piedmont. D, 1976, University of North Carolina, Chapel Hill. 241 p.

Fritz, Steven James. Provenance in process deposition of sand layers on the North Carolina continental rise. M, 1971, Duke University. 103 p.

Fritz, T. R. The depositional environments of the Jurassic Carmel Formation of northeastern Utah. M, 1977, Fort Hays State University.

Fritz, William Harold. Structure and stratigraphy of northern Egan Range, White Pine County, Nev. D, 1960, University of Washington. 117 p.

Fritz, William Harold. Structure and stratigraphy, Telegraph Canyon area, N. Egan Range, E. central Nevada. M, 1957, University of Washington. 79 p.

Fritz, William Jon. Depositional environment of the Eocene Lamar River Formation in Yellowstone National Park. D, 1980, University of Montana. 113 p.

Fritz-Miller, Molly. Paleoenvironmental analysis of the Tonoloway and lower Keyser Formation (upper Silurian, West Virginia, Maryland, Pennsylvania, Virginia). M, 1971, George Washington University.

Fritzsche, Hans. Geology and ore deposits of the Silver Star mining district, Madison County, Montana. M, 1935, Montana College of Mineral Science & Technology. 89 p.

Fritzsche, Kurt W. The geology of Lake Fortune Mine. M, 1929, University of Wisconsin-Madison.

Friz, David R. Stratigraphy and sedimentology of the Kayenta Formation, Capital Reef National Park and vicinity, Utah. M, 1985, University of Utah. 134 p.

Frizado, Joseph Pacheco. Humic material-cation interactions in Chesapeake Bay. D, 1980, Northwestern University. 153 p.

Frizzel, Donald L. Studies in the molluscan superfamily Veneracea. D, 1935, Stanford University. 262 p.

Frizzell, Larry Glen. Deposition and diagenesis of the Upper Smackover (Jurassic) grainstone at Tubal Field, South Arkansas. M, 1983, Louisiana Tech University. 117 p.

Frizzell, V. A., Jr. Petrology and stratigraphy of Paleogene nonmarine sandstones, Cascade Range, Washington. D, 1979, Stanford University. 169 p.

Frlich, Waldo J. Geology of the southern half of Whitehall Quadrangle of Virginia and West Virginia. M, 1976, University of Akron. 55 p.

Frodesen, Eric Wells. Petrology and correlation of the Chickasaw Creek Formation (Mississippian), Ouachita Mountains, Oklahoma. M, 1971, University of Wisconsin-Madison.

Froehlich, David J. The rock reefs of the Everglades of South Florida. M, 1979, Syracuse University.

Froelich, Albert Joseph. The geology of a part of the Wisconsin granite-Quinnesec greenstone complex, Florence County, Wisconsin. M, 1953, Ohio State University.

Froelich, P. N., Jr. Marine phosphorus geochemistry. D, 1979, University of Rhode Island. 309 p.

Froese, Edgar. Metamorphosed sediments of the middle Foster Lake area, northern Saskatchewan. M, 1956, University of Saskatchewan. 72 p.

Froese, Edgar. Structural geology and metamorphic petrology of the Coronation Mine area, Saskatchewan. D, 1963, Queen's University. 153 p.

Frohlich, C. A. I, Upper mantle structure beneath the Fiji Plateau; seismic observations of second P-arrivals from the olivine spinel phase transition zone; II, Strainmeter and tiltmeter measurements from the Tonga island arc; III, The case for four-component strainmeters. D, 1976, Cornell University. 166 p.

Frohlinger, Thomas Gordon. Structural history, and plutonic and metamorphic geology of the central southern Indian Lake area, Manitoba. M, 1973, University of Manitoba.

Froidevaux, Claude M. Geology of the Hoback Peak area in the Overthrust Belt, Lincoln and Sublette counties, Wyoming. M, 1968, University of Wyoming. 126 p.

Frolking, Tod Alexander. Loess distribution and soil development in relation to hillslope morphology in Grant County, Wisconsin. D, 1985, University of Wisconsin-Madison. 271 p.

Froman, N. L. Petroleum exploration using Landsat imagery. M, 1976, University of Wyoming. 96 p.

Froming, George T. Conodonts from the Upper Ordovician Maquoketa Formation in Wisconsin. M, 1965, University of Wisconsin-Madison.

Fromm, Kurt A. Heat flow and potential low-temperature geothermal resources in western and central New York. M, 1983, SUNY at Buffalo. 96 p.

Fronabarger, Allen Kem. Petrogenesis of the Calhoun Falls, Mt. Carmel, and Greenwood complexes, South Carolina, and modeling of heat transfer in gabbro. D, 1984, University of Tennessee, Knoxville. 200 p.

Fronabarger, Allen Kem. Petrogenesis of the North Harper Creek uranium prospect. M, 1980, University of Tennessee, Knoxville. 93 p.

Fronczek, Daniel V. A hydrogeologic investigation of the Balkema Wetland. M, 1986, Western Michigan University.

Frondel, Clifford. Crystal habit variation in sodium fluoride. D, 1939, Massachusetts Institute of Technology. 94 p.

Frondel, Clifford. Oriented inclusions of tourmaline in muscovite. M, 1936, Columbia University, Teachers College.

Frondorf, A. F. Interdisciplinary approaches to resource planning issues; the National Heritage Program. D, 1979, University of Arizona. 140 p.

Fronjosa, Ernesto. A study of Oklahoma water flood statistics. M, 1965, University of Oklahoma. 87 p.

Froomer, N. L. Geomorphic change in some Western Shore estuaries during historic times. D, 1978, The Johns Hopkins University. 236 p.

Frossard, Michael Louis. Diagenesis of Woodbine and sub-Clarksville sandstones at the Kurten and Iola field areas, Brazos and Grimes counties, Texas. M, 1982, Texas A&M University. 138 p.

Frossard, Robert Louis. The geology of the Nixa area. M, 1942, University of Missouri, Columbia.

Frost, Bryce Ronald. Contact metamorphism of the Ingalls ultramafic complex at Paddy-go-easy Pass, central Cascades, Washington. D, 1973, University of Washington. 176 p.

Frost, Bryce Ronald. Geology of the Double Lake area, Wind River Mountains, Fremont County, Wyoming. M, 1971, University of Washington. 53 p.

Frost, Eric George. Mid-Tertiary, gravity-induced deformation in Happy Valley, Pima and Cochise counties, Arizona. M, 1977, University of Arizona. 86 p.

Frost, Frederick Hazard. The Pleistocene flora of Rancho LaBrea. D, 1927, University of California, Berkeley. 38 p.

Frost, G. P. Data analysis methods and instrumentation sensors for monitoring liquefied natural gas spill experiments. D, 1977, University of California, Los Angeles. 225 p.

Frost, Jack Philip. A geologic study of the Brassfield Formation in portion of Greene and Clark counties, Ohio. M, 1977, Wright State University. 124 p.

Frost, Jackie Glenn. Algal banks of the Dennis Limestone (Pennsylvanian) of eastern Kansas. D, 1968, University of Kansas. 215 p.

Frost, Jackie Glenn. Stratigraphy of the Edwards Limestone of central Texas. M, 1963, Baylor University. 120 p.

Frost, Jay Miles, III. The geologic significance of heaving shale on the Texas coastal plain. M, 1938, University of Texas, Austin.

Frost, John Elliot. Controls of ore deposition for the Larap mineral deposits, Camarines Norte, Philippines. D, 1965, Stanford University. 173 p.

Frost, Karl Albert. Geology and mineralization of the Luning (New Boston) tungsten - molybdenum prospect, Mineral County, Nevada. M, 1983, University of Nevada. 60 p.

Frost, Kenneth Robert. Petrology and chemistry of Precambrian amphibolites in the Black Hills, South Dakota. M, 1979, Kent State University, Kent. 51 p.

Frost, L. H. Single-well injection; withdrawal test for demonstrating vertical connectivity and determining retardation factors in fractured rock. M, 1985, University of Waterloo. 60 p.

Frost, Richard J. Bedrock geology of the south half of the Clayville Quadrangle of Rhode Island. M, 1950, Brown University.

Frost, Richard W. The subsurface stratigraphy of the Wapanucka Formation in the Arkoma Basin of Oklahoma. M, 1983, Baylor University. 118 p.

Frost, Stanley Harold. Mexican Tertiary biostratigraphy and paleontology; larger foraminifera and corals. D, 1966, University of Illinois, Urbana. 371 p.

Frost, Stanley Harold. The stratigraphy and paleontology of the Piute Formation, Arrow Canyon Range, Nevada. M, 1963, University of Illinois, Urbana.

Frost, Thomas Philip. The Lamarck Granodiorite, Sierra Nevada, California; fractionation and interaction of mafic and felsic magmas. D, 1986, Stanford University. 226 p.

Frost, Victor Le Roy. Oligocene Ostracoda from the State of Mississippi. M, 1934, University of Oklahoma. 88 p.

Frouzan, Faramarz. Cyclic sedimentation of the upper Fayetteville Formation. M, 1960, University of Missouri, Rolla.

Fruchey, R. A. Overthrusting in Mt. Thompson and adjacent areas, Sublette and Lincoln counties, Wyoming. M, 1962, University of Wyoming. 82 p.

Frueh, A. J., Jr. A study of disorder in minerals. D, 1949, Massachusetts Institute of Technology. 91 p.

Frueh, A. J., Jr. The confirmation of the symmetry of claudetite (monoclinic As_2O_3) by means of Harker synthesis. M, 1947, Massachusetts Institute of Technology. 62 p.

Fruehling, S. W. Petrofabric analysis of the prophritic rhyolite west of Grand Marais, Minnesota. M, 1941, University of Minnesota, Minneapolis.

Frugoni, James John and Warner, Marvin Eugene. A magnetic study of selected intrusives in Jefferson, Madison and Gallatin counties, Montana. M, 1958, Indiana University, Bloomington. 54 p.

Fruit, David J. Tide and storm dominated bars on a distal muddy shelf; the Pennsylvanian Cottage Grove Sandstone, northwestern Oklahoma. M, 1986, University of Oklahoma. 83 p.

Fruland, Robert M. The numerical model for the calculation of suspended sediment under estuarine conditions. M, 1974, University of South Florida, St. Petersburg.

Fruland, Ruth Marcia. Impact-generated volatile movement and redistribution in the Rose City Meteorite. M, 1975, University of Houston.

Frund, Eugene. Stratigraphy of the Borden Siltstone in Clinton and adjacent counties, Illinois. M, 1953, University of Illinois, Urbana.

Frush, Mary Penelope. Cenomanian and Turonian foraminifera; Big Bend region of Texas and Mexico. M, 1973, University of Colorado.

Fruth, Elisabeth Anestad. Uranium series disequilibrium in Recent volcanic rocks. M, 1963, Columbia University, Teachers College.

Fruth, Lester Sylvester, Jr. Compaction effects and depth-pressure relationships in Bahamian sediments. D, 1967, Columbia University. 108 p.

Fruth, Lester Sylvester, Jr. The 1929 Grand Banks turbidite and the sediments of the Sohm Abyssal Plain, Northwest Atlantic Ocean. M, 1965, Columbia University. 367 p.

Fry, Harold Chester, Jr. Filiramoporina kretaphila; a new genus and species of bifoliate Tubulobryozoan (Ectoproctai from the Lower Permian Wrefore Megacyclothem of Kansas. D, 1975, Pennsylvania State University, University Park. 127 p.

Fry, Joyce Ann. Analysis of joints on monoclines in the Fanny Peak Quadrangle, Wyoming and South Dakota. M, 1982, South Dakota School of Mines & Technology. 82 p.

Fry, Michael F. Ultrasonic crosshole assessment of crystalline rock. M, 1987, Colorado School of Mines. 126 p.

Fry, Steven. A cyclic sequence of Givetian (upper Middle Devonian) biostromal reefs in northern Spain. M, 1978, California State University, Fresno.

Fry, Virginia Ann. Tidal velocity asymmetries and bedload transport in shallow embayments. M, 1987, Massachusetts Institute of Technology. 55 p.

Fry, Wayne L. Studies of the Carboniferous lycopod Asterodendron gen. nov. and of Cordaitanthus. D, 1953, Cornell University.

Fryberger, John S. The geology of Steens Mountain, Oregon. M, 1959, University of Oregon. 65 p.

Fryberger, Steven G. Stratigraphy of the Weber Formation (Pennsylvanian-Permian), Dinosaur National Monument and adjacent area, Utah and Colorado. M, 1978, Colorado School of Mines. 105 p.

Frydenlund, David Dexter. Structural geology of the continental margin off Pt. Ano Nuevo, California. M, 1974, Naval Postgraduate School.

Frye, Charles I. Glacial and post-glacial erosion of Franconia Notch, New Hampshire. M, 1960, University of Massachusetts. 101 p.

Frye, Charles Isaac. The Hell Creek Formation (Upper Cretaceous) in North Dakota. D, 1967, University of North Dakota. 411 p.

Frye, John Chapman. Additional studies on the history of Mississippi Valley drainage. D, 1938, University of Iowa. 55 p.

Frye, John Chapman. Geology of a portion of the lower Muskingum Valley, Ohio. M, 1937, University of Iowa. 92 p.

Frye, John K. The petrography of the ancient granites of the Minnesota, Ontario boundary region. M, 1959, University of Minnesota, Minneapolis. 55 p.

Frye, John Keith. Composition and crystallization history of the Conway Granite of New Hampshire. D, 1965, Pennsylvania State University, University Park. 137 p.

Frye, Kenneth Lee. The geology and mineralization of the Tubutama area, Sonora, Mexico. M, 1975, University of Iowa. 103 p.

Frye, Mark W. The nautiloid cephalopods of the Upper Ordovician Boda Limestone, province of Dalarna, Sweden; Ascocerida, Tarphycerida, and

Apsidoceratidae (Barrandeocerida). M, 1978, Ohio State University.

Frye, Wayne Herschel. Stratigraphy and petrology of late Quaternary terrace deposits around Tillamook Bay, Oregon. M, 1976, University of Oregon. 112 p.

Fryer, Alan Ernest. Determination of transport parameters from coincident chloride and tritium plumes at the Idaho National Engineering Laboratory. M, 1986, Texas A&M University. 69 p.

Fryer, Brian J. Canadian Precambrian iron-formation ages and trace element compositions. D, 1971, Massachusetts Institute of Technology. 175 p.

Fryer, G. J. Detection of weak seismic arrivals in the presence of microseism noise. M, 1973, University of Hawaii. 32 p.

Fryer, Karen Helene. Metamorphism and deformation in early Proterozoic basic dykes near Scourie, Northwest Scotland. D, 1986, University of Illinois, Urbana. 289 p.

Fryer, P. Petrology and geochemistry of some rock samples from the northern Fiji Plateau. M, 1973, University of Hawaii. 49 p.

Frykberg, W. R. Biota and environment of the Muskegon, Michigan, combined industrial and municipal wastewater storage lagoons. D, 1976, Western Michigan University. 196 p.

Fryklund, Verne C., Jr. The titanium ore deposits of Magnet Cove, Hot Spring County, Arkansas. D, 1949, University of Minnesota, Minneapolis. 118 p.

Fryman, Mark David. The Isom Formation on the Markagunt Plateau in southwestern Utah. M, 1987, Kent State University, Kent. 86 p.

Fryxell, Fritiof M. The glacial geology of Jackson Hole, Wyoming. D, 1929, University of Chicago. 188 p.

Fryxell, Jenny Christine. Depositional environments and provenance of arkosic sandstone, Park Shale, Middle Cambrian, Bridger Range, southwestern Montana. M, 1982, Montana State University. 95 p.

Fryxell, Joan Esther. Structural development of the west-central Grant Range, Nye County, Nevada. D, 1984, University of North Carolina, Chapel Hill. 139 p.

Fryxell, Roald. The contribution of interdisciplinary research to geologic investigation of prehistory, eastern Washington. D, 1970, University of Idaho. various pagination p.

Fu, Ch'eng-Yi. Studies on seismic waves. D, 1944, California Institute of Technology. 57 p.

Fu, Cheng-Ping David. Entrainment effects on the distribution of salinity in the Hudson Estuary. D, 1980, City College (CUNY). 131 p.

Fu, Yun-ta. Lithofacies and diagenesis of Spraberry and Dean sediments, Reagan County, Texas. M, 1984, Texas Tech University. 120 p.

Fuchs, James W. Microfacies analysis of the Edwards Limestone (Lower Cretaceous), central Texas. M, 1981, Stephen F. Austin State University. 113 p.

Fuchs, Jens Peter. Evaluation of a rectilinear motion detector. M, 1969, University of British Columbia.

Fuchs, Robert Louis. Upper Triassic cephalopods from northern Peru. M, 1952, University of Illinois, Urbana.

Fuchs, Viveka. A paleomagnetic study of the Morrison and Kootenay formations at three localities in southwestern Montana. M, 1988, Dartmouth College. 183 p.

Fuchs, William Arthur. Geochemical behavior of platinum, palladium, and associated elements in the weathering cycle in the Stillwater Complex, Montana. M, 1972, Pennsylvania State University, University Park. 92 p.

Fuchs, William Arthur. Tertiary tectonic history of the Castle Mountain - Caribou fault system in the Talkeetna Mountains, Alaska. D, 1980, University of Utah. 162 p.

Fudali, Robert F. Experimental studies bearing on the origin of pseudoleucite and associated problems of

alkaline rock systems. D, 1960, Pennsylvania State University, University Park.

Fudge, Melvin Ray. The Upper Pennsylvanian and Lower Permian strata of northwestern Chautauqua County, Kansas. M, 1974, Wichita State University. 205 p.

Fuehrer, David W. Metamorphic petrology and petrofabrics of the eastern part of the McCaslin Range, northeast Wisconsin. M, 1981, Bowling Green State University. 56 p.

Fuenkajorn, Kittitep. Borehole closure in salt. D, 1988, University of Arizona. 517 p.

Fuenning, Paul. A thickness and structural study of certain divisions of the Cretaceous of Nebraska. M, 1941, University of Nebraska, Lincoln.

Fuente Duch, M. F. F. de la see de la Fuente Duch, M. F. F.

Fuente Duch, Mauricio Fernando De la see De la Fuente Duch, Mauricio Fernando

Fuentes, Ruderico Procopio. Stratigraphy of Santa Clara and Sierra Gomas, Nuevo Leon, Mexico. M, 1964, University of Texas, Austin.

Fuenzalida, Ricardo H. Geological correlation between the Patagonian Andes and Antarctic Peninsula and some tectonic implications. M, 1972, Stanford University.

Fuerst, Samuel I. Sediment transport on the insular slope of Puerto Rico off the La Plata River. M, 1979, Duke University. 101 p.

Fueten, Frank. Spaced cleavage development in the metagreywackes of the Goldenville Formation, Meguma Group, Nova Scotia. M, 1985, McMaster University. 145 p.

Fuex, Anthony Nichols. Stable carbon isotopes in igneous rocks. D, 1970, Rice University. 203 p.

Fuex, Anthony Nichols. Thermoluminescence of shocked granodiorite. M, 1967, University of Houston.

Fugate, George W., Jr. Geology of the area of Smith Mills North and Geneva oil pools, Henderson County, Kentucky. M, 1956, University of Kentucky. 49 p.

Fugate, James K. Methodology for determining the parameters of bioturbation using ash layers in areas of frequent volcanism. M, 1978, University of Hawaii. 80 p.

Fugelso, Leif Eric. Transverse seismic wave propagation in an anisotropic, layered earth. D, 1973, University of Chicago. 96 p.

Fugitt, David Spencer. Structural geology of the Buckville area, Lake Ouachita, Arkansas. M, 1978, Texas A&M University. 80 p.

Fuh, G. F. Design analysis of annular tunnels for super conductive energy storage using the finite element method. D, 1978, University of Wisconsin-Madison. 246 p.

Fuh, Tsu-Min. Correlation of rocks across the Grenville Front near Val D'Or, Quebec. D, 1970, Queen's University. 199 p.

Fuhr, Joseph M. Stratigraphy and depositional history of the Pleistocene bedrock underlying Florida Bay. M, 1988, Stephen F. Austin State University. 120 p.

Fuhrman, Miriam Lea. Petrology and mineral chemistry of the Sybille Monzosyenite and the role of ternary feldspars. D, 1987, SUNY at Stony Brook. 240 p.

Fuis, Gary Stephen. The geology and mechanics of formation of the Fort Rock Dome, Yavapai County, Arizona. D, 1973, California Institute of Technology. 386 p.

Fujii, Takashi. Muscovite-paragonite equilibria. D, 1967, Harvard University.

Fujimoto, C. K. The behavior of manganese in the soil and the manganese cycle. M, 1947, University of Hawaii.

Fujishima, K. Y. Hydrothermal mineralogy of Keolu Hills, Oahu, Hawaii. M, 1975, University of Hawaii. 51 p.

Fujita, K. Tectonics of divergent and convergent plate margins; I, Membrane stresses near mid-ocean ridge - transform intersections; II, Teleseismic relocation of central Aleutian earthquakes; III, Tectonics of the margins of the western Bering Sea. D, 1979, Northwestern University. 308 p.

Fujita, Kazuya. Tectonic evolution of the Arctic Ocean margins of North America and northeastern Siberia. M, 1976, Northwestern University.

Fukuda, Michael K. The entrainment of cohesive sediments in freshwater. D, 1978, Case Western Reserve University. 229 p.

Fukui, Larry M. The mineralogy and petrology of the South Kawishiwi Intrusion, Duluth Complex, Minnesota. M, 1976, University of Illinois, Chicago.

Fukuta, Nobuhiko. Application of Shake Program on estimation of ground response at Satsop nuclear reactor site. M, 1977, University of Washington. 74 p.

Fulcher, Richard Alfred, Jr. The effect of the capillary number and its constituents on two-phase relative permeabilities. D, 1982, Pennsylvania State University, University Park. 189 p.

Fulford, James Kenny. Thermal convection in porous media with application to hydrothermal circulation in the oceanic crust. M, 1979, Georgia Institute of Technology. 54 p.

Fulgham, Henry Leroy. Geology of the San Martine Quadrangle, Reeves and Culberson counties, Texas. M, 1950, University of Texas, Austin.

Fulker, Katharine D. The origin of carbonate cements in Bahama Escarpment limestones. M, 1982, Western Michigan University. 121 p.

Fulkerson, Donald H. Geology of a part of the west flank of the Bighorn Mountains north of Bigtrails, Washakie County, Wyo. M, 1951, University of Wyoming. 45 p.

Full, Roy P. Structural relations north of the Osburn Fault, Coeur d'Alene District, Shoshone County, Idaho. D, 1955, University of Idaho. 34 p.

Full, William Edward. Analysis of quartz detritus of complex provenance via analysis of shape. D, 1982, University of South Carolina. 220 p.

Full, William Edward. Processes of lithosphere thinning and crustal rifting in the Salton Trough, Southern California. M, 1980, University of Illinois, Chicago.

Fullagar, Paul David. Host rock gneiss as a possible source of ore mineralization at Ore Knob, North Carolina. D, 1963, University of Illinois, Urbana. 83 p.

Fullagar, Peter Kelsham. Inversion of horizontal loop electromagnetic soundings over a stratified Earth. D, 1981, University of British Columbia.

Fullam, Timothy Jewell. Lower Columbia River sand waves. M, 1967, University of Washington. 34 p.

Fullam, Timothy Jewell. The measurement of bedload from sand wave migration in Bonneville Reservoir on the Columbia River. D, 1969, University of Washington. 135 p.

Fullas, George H. A study of trace metal concentrations in two species of mytilus along the Pacific Coast of the United States. M, 1973, University of New Mexico. 60 p.

Fulle, Richard M. Paleozoic stratigraphy of the Rosendale area, Ulster County, New York. D, 1929, Princeton University.

Fuller, Arthur Orpen. The Witwatersrand System. D, 1957, Princeton University. 186 p.

Fuller, Brent D. Low frequency electromagnetic response of an N-layered sphere of arbitrary electrical parameters. D, 1971, University of California, Berkeley. 224 p.

Fuller, Brent D. Two dimensional frequency analysis and design of grid operators. M, 1966, University of California, Berkeley. 50 p.

Fuller, Brian N. Seismic detection of Upper Cretaceous stratigraphic oil traps in the Powder River basin, Wyoming. D, 1988, University of Wyoming. 143 p.

Fuller, Brian N. Seismic reflection data acquisition problems in the Columbia River Plateau and Snake River plain. M, 1987, University of Wyoming. 87 p.

Fuller, David Richard. Paleoenvironmental analysis of an Early Cambrian archaeocyathid reef in the White-Inyo Mountains, California and Nevada. M, 1976, Idaho State University. 53 p.

Fuller, James O. Geology and mineral deposits of the Fleur-de-Lys area. D, 1941, Columbia University, Teachers College.

Fuller, Jonathan A. Shallow water high-magnesium calcite mud production and dispersal, St. Croix, U.S. Virgin Islands. M, 1978, Western Michigan University.

Fuller, Lynn Roy. General geology of Triassic rocks at Alaska Canyon in the Jackson Mountains, Humboldt County, Nevada. M, 1986, University of Nevada. 110 p.

Fuller, Margaret B. Post glacial sedimentation in the Connecticut Valley in Massachusetts. M, 1919, University of Chicago. 60 p.

Fuller, Margaret B. The geology of the Big Thompson River valley in Colorado from the Continental Divide to the foothills area. D, 1924, University of Chicago. 191 p.

Fuller, Melville Weston. Marginifera and related genera of the Pennsylvanian. M, 1933, University of Illinois, Chicago.

Fuller, Richard Eugene. Petrology and structural relationship of Steens Mountain volcanic series of southeastern Oregon. D, 1930, University of Washington. 282 p.

Fuller, Richard Eugene. The geology of the northeastern part of Cedar Lake quadrangle with special reference to de-roofed Snoqualmie Batholith. M, 1925, University of Washington. 96 p.

Fuller, Richard H. Some aspects of geochemistry of the water and sediment of Bear Lake, Utah-Idaho. M, 1975, Utah State University. 69 p.

Fuller, Robert L. Geology of the Crossville School area, Mason County, Texas. M, 1957, Texas A&M University.

Fuller, Steven R. Neoglaciation of Avalanche Gorge and the Middle Fork Nooksack River valley, Mt. Baker, Washington. M, 1980, Western Washington University. 68 p.

Fuller, Warren Philips. Weathering of the Ryukyu Formation on Okinawa, Ryukyu Islands. M, 1948, University of Texas, Austin.

Fuller, Wayne Ross. Heat flow reconnaissance of Florida. M, 1976, University of Florida. 78 p.

Fuller, Willard P., Jr. Spectographic study of gold-quartz ores from Alleghany, California. M, 1942, California Institute of Technology. 31 p.

Fullerton, David S. Indian Castle glacial readvance in the Mohawk Lowland, N.Y. and its regional implications. D, 1971, Princeton University. 92 p.

Fullerton, Donald S. The geology of the Silurian and Devonian in the subsurface of the Larue County, Kentucky. M, 1961, University of Kentucky. 34 p.

Fullerton, H. D. A petrographic study of the serpentinized peridotites of the Griffis Lake map area, Quebec. M, 1951, McGill University.

Fullerton, Larry Bryant. The petrology of the lower bed of the Utopia Limestone Member of the Howard Formation (Upper Pennsylvanian) of Kansas. M, 1970, Wichita State University. 95 p.

Fullerton, Marilynn. Dismicrite in the Cedar City Formation (Middle Devonian) of central Missouri. M, 1973, University of Missouri, Columbia.

Fullmer, Corey Y. Geology of the SE 1/4 of the Twin Lakes Quadrangle, Ferry County, Washington. M, 1986, Eastern Washington University. 73 p.

Fulmer, Charles Virgil. Sedimentary petrography of Eocene Series on N. side of Mt. Diablo, Calif. M, 1947, University of Washington. 66 p.

Fulmer, Charles Virgil. Stratigrahy and paleontology of the typical Markley and Nortonville formations. D, 1956, University of California, Berkeley. 327 p.

Fulp, Michael S. Precambrian geology and mineralization of the Dalton Canyon volcanic center, Santa Fe County, New Mexico. M, 1982, University of New Mexico. 199 p.

Fulreader, Rufus Everett, Jr. Geologic studies of the Leicester Pyrite. M, 1957, University of Rochester. 112 p.

Fulthorpe, Craig Stephen. Paleoceanographic and tectonic controls on Neogene carbonate deposition in shelf, slope and pelagic settings. D, 1988, Northwestern University. 262 p.

Fulton, Christopher Robert. The chemistry and origin of the ordinary chondrites; implications from refractory-lithophile and siderophile elements. M, 1984, University of Massachusetts. 66 p.

Fulton, Clark. The glacial geology of McKenzie County, North Dakota. M, 1976, University of North Dakota. 100 p.

Fulton, David A. Sedimentology, structure, and thermal maturity of the lower Atoka Formation, Ouachita frontal thrust belt, Yell and Perry counties, Arkansas. M, 1985, University of Missouri, Columbia. 149 p.

Fulton, Dwight David. Electrical resistivity surveying for tar sands in western Vernon County, Missouri. M, 1982, University of Missouri, Rolla.

Fulton, Fred J. The Omega transformation in a Ti-Mn alloy. M, 1956, University of Nevada. 24 p.

Fulton, G. Lyman. The subsurface geology of the Monument Oil Field of New Mexico. M, 1938, Texas Tech University. 38 p.

Fulton, John W. The distribution of explosive residues and nitrate-nitrogen in the groundwater west of Grand Island, Nebraska. M, 1987, University of Nebraska, Lincoln. 59 p.

Fulton, Kenneth James. Preliminary investigations, subsurface stratigraphy and depositional environments, Rio Grande Delta, Texas. M, 1973, University of Cincinnati. 63 p.

Fulton, Kenneth James. Subsurface stratigraphy, depositional environments and aspects of reservoir continuity, Rio Grande Delta, Texas. D, 1976, University of Cincinnati. 314 p.

Fulton, Robert B., III. Prospecting for zinc using semiquantitative chemical analysis of soils (Ducktown, Tennessee). D, 1949, Stanford University. 87 p.

Fulton, Robert John. Deglaciation of the Kamloops region, British Columbia. D, 1963, Northwestern University. 128 p.

Fulton, Sara M. Mauldin. Burial diagenesis of sandstone reservoirs, Lake St. John Field, east-central Louisiana. M, 1985, University of Missouri, Columbia. 209 p.

Fulton-Bennett, Kim Wilbur. An assessment of coastal protection structures between San Francisco and Carmel, California. M, 1984, University of California, Santa Cruz.

Fults, Michelle Ellen. A trace element geochemical analysis of the Lake Ellen Kimberlite, Crystal Falls, Michigan, U.S.A. M, 1987, Western Michigan University.

Fultz, Lawrence Anthony. Sr isotopic correlations with minor elements in Guatemalan basalts. M, 1979, Michigan Technological University. 52 p.

Fulweiler, Robert Edward. The geology of part of the igneous and metamorphic complex of southeastern Florence County, Wisconsin. M, 1957, Ohio State University.

Fulwider, Roy Wesley. Biostratigraphy of the Tepetate Formation, Baja California del Sur. M, 1976, University of Southern California.

Fumal, Thomas Edward. Correlations between seismic wave velocities and physical properties of near-surface geologic materials in the southern San Francisco Bay region. M, 1978, University of California, Santa Cruz.

Fumerton, Stewart Lloyd. Geology of the Reindeer Lake area, Saskatchewan, with emphasis on granitic rocks. D, 1979, University of Saskatchewan. 472 p.

Funderburg, Eddie Ray. An analysis of farmers' soil testing practices in three Louisiana parishes. D, 1984, Louisiana State University. 156 p.

Fundingsland, Ernest L. Harrisburg oil area. M, 1956, University of Colorado.

Fung, Patrick Chuen-Fai. K, Rb and Tl distributions between coexisting natural and synthetic rock-forming minerals. D, 1979, McMaster University. 263 p.

Funk, Alan C. The relationships of engineering properties to geochemistry in the Taylor Group, Travis County, Texas. M, 1975, University of Texas, Austin.

Funk, James M. Climatic and tectonic effects on alluvial fan systems, Birch Creek valley, East central Idaho. M, 1977, University of Kansas. 246 p.

Funk, John L., III. Geology of the Landers fork in Blackfoot River area, Montana. M, 1969, University of Missouri, Columbia. 119 p.

Funk, Thomas J. Subsurface and petroleum geology of the Mississippian System of southern Ford and northern Clark counties, Kansas. M, 1982, Wichita State University. 81 p.

Funkhouser, Harold J. The Ostracoda of the Golconda Formation (Illinois, Indiana). M, 1938, University of Chicago. 41 p.

Funkhouser, John Gray. The determination of a series of ages of a Hawaiian volcano by the potassium-argon method. D, 1966, University of Hawaii. 168 p.

Funkhouser, Lawrence W. The geology of the Arroyo del Valle area, Alameda County, California. M, 1948, Stanford University. 75 p.

Funkhouser, Roy V. Hydrogeological study of the French Lake area, Vigo County, Indiana. M, 1983, Indiana University, Bloomington. 171 p.

Funkhouser-Marolf, Myra J. The mineralogy and distribution of uranium and thorium in the Sheeprock Granite, Utah. M, 1985, University of Iowa. 60 p.

Fuqua, Frank Jones. Comanche stratigraphy of Hurd Draw Quadrangle, Culberson County, Texas. M, 1951, University of Texas, Austin.

Furbush, Malcolm. The stratigraphy and structure of the northern half of the South Onondaga Quadrangle, New York. M, 1952, Syracuse University.

Furcron, Aurelius Sidney. Geology of the James River iron and marble belt in central Virginia. D, 1931, University of Iowa. 241 p.

Furcron, Aurelius Sidney. The gold deposits of the southeastern Atlantic states. M, 1923, University of Virginia. 142 p.

Furer, L. C. Overthrusting in the Thompson Pass area, Lincoln and Sublette counties, Wyoming. M, 1962, University of Wyoming. 97 p.

Furer, Lloyd Carroll. Sedimentary petrology and regional stratigraphy of the non-marine Upper Jurassic-Lower Cretaceous rocks of western Wyoming and southeastern Idaho. D, 1967, University of Wisconsin-Madison. 214 p.

Furgason, David C. Petrology and geochemistry of part of the southern sector, southeastern Bushveld Complex. M, 1977, University of Wisconsin-Madison.

Furgerson, Robert Bernard. A controlled-source telluric current technique and its application to structural investigations. M, 1970, Colorado School of Mines. 123 p.

Furhmann, Mark. Sedimentology of the New York Bight dredged material dumpsite deposit. M, 1980, Adelphi University.

Furlong, Edward Thomas. Sediment geochemistry of photosynthetic pigments in oxic and anoxic marine and lacustrine sediments; Dabob Bay, Saanich Inlet, and Lake Washington. D, 1986, University of Washington. 214 p.

Furlong, Ira Ellsworth. The geology of the Farmington Quadrangle (Maine). D, 1960, Boston University. 199 p.

Furlong, Kevin Patrick. Time dependent thermal modeling of plate tectonic processes. D, 1981, University of Utah. 149 p.

Furlong, Paul O. Paleomagnetism of the Mount Rainer area, western Washington. M, 1982, Western Washington University. 76 p.

Furlong, Robert Burton. An electron diffraction and micrographic study of the high temperature changes in the clay minerals. D, 1967, University of Illinois, Urbana. 134 p.

Furlong, Robert Burton. Significance of the clay minerals in the Gypsum Spring and Lower Sundance formations (Jurassic), eastern flank of the Big Horn Mountains, Wyoming. M, 1965, University of Illinois, Urbana.

Furlong, William J. Depositional environment of the Chipola Formation, Calhoun County, Florida. M, 1980, Tulane University.

Furlow, Bruce. Subsurface geology of the Kellyville District, Creek County, Oklahoma. M, 1956, University of Oklahoma. 61 p.

Furlow, James Warren. Geology of the San Mateo Peak area, Socorro County, New Mexico. M, 1965, University of New Mexico. 83 p.

Furman, Francis Chandler. Origin and economic potential of olivine in alpine-type ultramafic bodies, the Willits Tract, Jackson County, North Carolina. M, 1981, Bowling Green State University. 263 p.

Furman, Marvin J. Authigenic analcite in the Golden Valley Formation (Eocene), southwestern North Dakota. M, 1970, University of North Dakota. 111 p.

Furnaguera C., Jose A. Economic decision making in mineral ventures by means of statistical techniques. M, 1976, Colorado School of Mines. 188 p.

Furnish, Michael David. Investigating the olivine-spinel transition mechanism by static and shock techniques. D, 1986, Cornell University. 222 p.

Furnish, William Madison. A Middle Ordovician fauna from the upper portion of the Deadwood Formation in the Black Hills. M, 1935, University of Iowa. 14 p.

Furnish, William Madison. Conodonts from the Prairie du Chien beds of the Upper Mississippi Valley. D, 1938, University of Iowa. 66 p.

Furnival, G. M. Large quartz veins at Great Bear Lake. M, 1934, Queen's University. 90 p.

Furnival, G. M. Silver-pitchblende deposit at Contact Lake, Great Bear Lake area, Northwest Territories, Canada. D, 1935, Massachusetts Institute of Technology. 170 p.

Furr, James E. A phreatic paleokarst explanation of some dolomite bodies in the Ste. Genevieve and St. Louis Limestone in south-central Kentucky. M, 1985, Eastern Kentucky University. 75 p.

Furry, Robert Edward. An examination of other errors in short line measurements with the MRA-2 tellurometer. M, 1965, Ohio State University.

Furse, George Norman D. The mountain structures of Asia. M, 1928, Columbia University, Teachers College.

Furst, Bruce Wayne. Petrology of the alkalic Hawi Volcanic Series of Kohala Volcano, Hawaii. M, 1982, California State University, Northridge. 147 p.

Furst, George A. Geology and petrology of the Fairbanks Basalts, Alaska. M, 1968, University of Alaska, Fairbanks. 53 p.

Furst, George Arrowsmith. The melting of plagioclase in the system sodium oxide-calcium oxide-aluminum oxide-silicon oxide-water at high pressure and temperature. D, 1978, Pennsylvania State University, University Park. 95 p.

Furst, Marian Judith. The use of boron in fossil materials as a paleosalinity indicator. D, 1979, California Institute of Technology. 179 p.

Furst, Martha Jean. The reconnaissance petrology of andesites from the Mount Wrangell Caldera, Alaska. M, 1968, University of Alaska, Fairbanks. 83 p.

Furu, Edward James. Sedimentary geology of a portion of southern Camp Pendleton. M, 1982, San Diego State University. 136 p.

Furukawa, Toshiharu. Adsorption and oxidation of aromatic amines on clay minerals. M, 1973, Pennsylvania State University, University Park. 76 p.

Furumoto, Augustine Sadamu. The use of ScS wave data in determining the mechanism at the focus of an earthquake. D, 1961, St. Louis University. 117 p.

Fusilier, Wallace Eaton. An opinion derived nine parameter unweighted multiplicative lake water quality index; the LWQI. D, 1982, University of Michigan. 122 p.

Fuste, Luis Alberto. Effects of coal mine drainage on Wilkeson Creek, a stream in western Washington. M, 1978, University of Washington. 103 p.

Füstös, Árpád. Comparative statistical analysis of the uranium and thorium content distribution in the Denison main reef, Elliot Lake Syncline, Elliot Lake, Ontario. M, 1982, Carleton University. 145 p.

Futyma, Richard Paul. Postglacial vegetation of eastern upper Michigan. D, 1982, University of Michigan. 495 p.

Fyffe, Leslie Robert. Petrogenesis of the adamellite-diorite transition, southwestern New Brunswick. M, 1971, University of New Brunswick.

Fyles, J. T. Geology of the Cowichan Lake area, Vancouver, British Columbia. D, 1954, Columbia University, Teachers College.

Fyles, James Thomas. Geology and manganese deposits of the north shore of Cowichan Lake, Vancouver Island, BC. M, 1949, University of British Columbia.

Fyles, John Gladstone. Geology of the northwest quarter of Whitehorse map area, Yukon and studies of weathered granitic rocks near Whitehorse. M, 1950, University of British Columbia.

Fyles, John Gladstone. Surficial geology of the Horne Lake and Parksville map-areas, Vancouver Island, British Columbia. D, 1956, Ohio State University. 283 p.

Fyock, Tad L. The stratigraphy and structure of the Virgin Valley-Thousand Creek area. M, 1963, University of Washington. 50 p.

Fyon, John Andrew. Field and stable isotopic characteristics of carbonate alteration zones, Timmins area. D, 1986, McMaster University. 399 p.

Fyon, John Andrew. Seawater alteration of early Precambrian (Archean) volcanic rock and exploration criteria for stratiform gold deposits, Porcupine Camp, Abitibi greenstone belt, northeastern Ontario. M, 1980, McMaster University. 238 p.

Fyvie, Donald James. Sedimentation and diagenesis of the Upper Devonian Kakisa Formation, Trout River area, Northwest Territories. M, 1988, University of Alberta. 154 p.

G. Padmanabhan *see* Padmanabhan G.

G., Miguel A. Miranda *see* Miranda G., Miguel A.

Gaal, Robert. Geology of the central portion of the Green Springs Quadrangle, Nevada. M, 1958, University of Southern California.

Gaal, Robert Arthur Paul. Marine geology of Santa Catalina Basin area, California. D, 1965, University of Southern California.

Gaba, Robert G. Geology of the Archean turbidite-hosted Gatlan gold occurrence, Wallace Lake greenstone belt, Southeast Manitoba. M, 1987, University of Western Ontario. 181 p.

Gabay, Steven Howard. Velocity refinement within a generalized inverse framework. M, 1985, University of Texas, Austin. 65 p.

Gabbert, Stephen Charles. Gravity survey of parts of Millard, Beaver, and Iron counties, Utah. M, 1980, University of Utah. 107 p.

Gabelman, Joan L. Precambrian geology of the upper Brazos Box area, Rio Arriba County, N.M. M, 1988, New Mexico Institute of Mining and Technology. 171 p.

Gabelman, John W. Geology and ore deposits of the Fulford mining district, Eagle County, Colorado with reconnaissance of the Brush Creek mining district (T652). D, 1949, Colorado School of Mines. 188 p.

Gabelman, John W. The geology of the Golden Gate-Van Bibber Creek area, Jefferson County, Colorado. M, 1948, Colorado School of Mines. 129 p.

Gabert, Gordon Michael. The geology and hydrogeology of the surficial deposits in the Devon area, Alberta (Canada). M, 1968, University of Alberta. 215 p.

Gabisi, Abdul H. Contact metamorphism of the Muskox Intrusion, Coppermine River area, District of Mackenzie, Northwest Territories. M, 1963, Carleton University. 115 p.

Gabites, Janet Elizabeth. Geology and geochronometry of the Cogburn Creek-settler Creek area, northeast of Harrison Lake, B.C. M, 1985, University of British Columbia. 153 p.

Gable, Douglas M. Delineating Pierre Formation fracture reservoirs using compressional and horizontal shear wave seismic data near Florence, Fremont County, Colorado. M, 1986, Colorado School of Mines. 151 p.

Gable, Kristine M. Conodonts and biostratigraphy of the Olentangy Shale (Middle-Upper Devonian), Ohio. M, 1973, Ohio State University.

Gabor, Reka Katalin. The influence of halloysite content on the shear strength of kaolinite. M, 1981, Portland State University. 121 p.

Gaboury, Bernard E. Geochemical elucidation of gneiss-hosted massive sulfide deposit. M, 1986, University of Manitoba.

Gabriel, Alton. Geochemical data of germanium. D, 1930, Cornell University.

Gabriel, Ralph Henry. The evolution of Long Island; a story of land and sea. D, 1919, Yale University.

Gabriel, Vittaly Gavrilovich. The Castle Rock Conglomerate and associated Placer gold deposits; Douglas County, Colorado. D, 1933, Colorado School of Mines. 35 p.

Gabriel, Vittaly Gavrilovich. The use of the graphical methods in the magnetic surveying and the introduction of the simplified diagrams pertaining to the Eotvos torsion-balance work. M, 1930, Colorado School of Mines. 27 p.

Gabriel, Walter J. A bedrock topography survey by the electrical resistivity method. M, 1950, St. Louis University.

Gabrielse, Hubert. A petrographic study of contact facies of granitic rocks with limestone. M, 1950, University of British Columbia.

Gabrielse, Hubert. Petrology and structure of the McDame ultramafic belt, British Columbia. D, 1955, Columbia University, Teachers College. 141 p.

Gaby, Walter E. The petrography of the Mount Morgan Mine, Queensland, Australia. M, 1915, Columbia University, Teachers College.

Gacek, Walter Frank. Mechanical analyses of sediments from Southwest Lake Erie. M, 1951, University of Michigan.

Gachowski, Christin Marianne. The effectiveness of sodium bicarbonate as a neutralizing agent in the treatment of acidic lake water. M, 1987, Syracuse University. 87 p.

Gadd, Nelson Raymond. A study of some contacts in the Laurentian area north of Quebec. M, 1947, Universite Laval.

Gadd, Nelson Raymond. Pleistocene geology of the Becancour map-area, Quebec. D, 1955, University of Illinois, Urbana. 215 p.

Gaddah, Ali Hadi. Damping ratio for dry sands. M, 1976, Indiana University, Bloomington. 52 p.

Gaddis, Lisa R. Radar and spectral reflectance analyses of volcanic landforms on the Earth and the Moon. D, 1987, University of Hawaii.

Gaddis, M. Francis. Siting criteria in hazardous waste disposal; a study of the implemenetation of the Resource, Conservation and Recovery Act (RCRA) mandate to develop physical geographic criteria and standards for the location of new hazardous waste disposal facilities, 1976-1985. D, 1988, Columbia University, Teachers College. 238 p.

Gadou, Georges S. The petrography of faults on Dugout Mountain, Brewster County, Texas. M, 1986, Sul Ross State University. 88 p.

Gaffey, Michael J. A new graphical representation for the chemical evolution of the igneous rocks. M, 1970, University of Iowa. 76 p.

Gaffey, Michael James. A systematic study of the spectral reflectivity characteristics of the meteorite classes with applications to the interpretation of asteroid spectra for mineralogical and petrological information. D, 1974, Massachusetts Institute of Technology. 355 p.

Gaffey, Susan Jenks. Spectral reflectance of carbonate minerals and rocks in the visible and near infrared (0.35 to 2.55μm) and its applications in carbonate petrology. D, 1984, University of Hawaii. 256 p.

Gaffin, Stuart Roger. Variable seafloor spreading rates and global sea-level; introducing a ridge volume inversion technique. D, 1986, New York University. 131 p.

Gaffke, Thresa M. Stratigraphy of the Trinidad Sandstone and Vermejo Formation (Upper Cretaceous), Canon City coal field, Fremont County, Colorado. M, 1982, Colorado School of Mines. 137 p.

Gaffney, Edward S. Distribution of transition elements in the Port Coldwell Complex, Ontario. M, 1966, Dartmouth College. 80 p.

Gaffney, Edward Stowell. Crystal field effects in mantle minerals. D, 1973, California Institute of Technology. 216 p.

Gaffney, Eugene Spencer. The North American baenoid turtles and the cryptodire-pleurodire dichotomy. D, 1969, Columbia University. 338 p.

Gaffney, Joseph Walter. Glacial geology of the Old Mystic Quadrangle, New London County, Connecticut. M, 1966, University of Massachusetts. 107 p.

Gafford, Edward Leighman, Jr. Determination of depositional environment of the Annelly 1; a section by the K/Rb ratio within the clay mineral illite. M, 1965, Wichita State University. 69 p.

Gafford, Edward Leighman, Jr. Experimental determination of partition coefficients for calcium, strontium, and barium in aragonite precipitated from sea water at low temperatures. D, 1969, University of Oklahoma. 81 p.

Gafni, Abraham. Field tracing approach to determine flow velocity and hydraulic conductivity in saturated peat soils. D, 1986, University of Minnesota, Minneapolis. 197 p.

Gage, D. R. Basic and applied projects in analytical chemistry; I, The development of a sulfur specific detector for gas chromatography based on SO_2 fluorescences; II, The application of laser Raman spectrometry for the determination of quartz and cristobalite in Mount St. Helens volcanic ash. D, 1981, University of Idaho. 125 p.

Gage, John E. Sedimentology of the Cretaceous Mesa Rica Sandstone, Tucumcari Basin area, New Mexico. M, 1976, West Texas State University. 45 p.

Gage, Kenneth S. The effect of stable thermal stratification on the stability of viscous paralleled flows. D, 1968, University of Chicago. 61 p.

Gagen, Patrick Michael. The oxidation rates of arsenopyrite and chalcopyrite in acidic ferric chloride solutions at 0 to 60°C. M, 1987, Virginia Polytechnic Institute and State University.

Gager, Barry Robert. Stratigraphy, sedimentology, and tectonic implications of the Tiger Formation, Pend Oreille and Stevens counties, Washington. M, 1982, University of Washington. 176 p.

Gagliano, Sherwood Moneer. Occupation sequence at Avery Island. D, 1967, Louisiana State University. 217 p.

Gagliardi, James S. Subsurface study of the Guelph-Lockport Group in northeastern Ohio. M, 1979, SUNY, College at Fredonia. 106 p.

Gagnard, Philip E. Temporal and spatial changes in the chemistry of carbonate groundwaters in the Chicago area. M, 1979, University of Illinois, Chicago.

Gagne, Michael P. Structural geology of the California Peak/middle Zapata Creek Lake region, south-central Colorado. M, 1987, Wichita State University.

Gagnier, Michelle Annette. Geochemistry of the Viola Limestone (Ordovician), southwestern Arbuckle Mountains, Oklahoma. M, 1986, Southern Methodist University. 151 p.

Gagnon, Denis-Claude. Estimation géostatistique de la minéralisation uranifère à la mine Fay-Ace, Beaverlodge, Saskatchewan. M, 1971, Ecole Polytechnique. 125 p.

Gagnon, Lawrence Gregory. Structural geology of the Luscar-Sterco Mine, Coal Valley, Alberta. M, 1982, University of Alberta. 98 p.

Gagnon, P. Theoretical investigation on step by step procedures for the adjustment of large horizontal geodetic networks. D, 1976, University of New Brunswick.

Gagnon, Pierre. Les Depôts quaternaires sur la bordure du Bouclier Laurentidien (Rawdon-Québec); cartographie, stratigraphie, sédimentologie et mode de retrait glaciaire. M, 1988, Universite du Quebec a Montreal. 125 p.

Gagnon, Robert E. Simultaneous determination of compressional and shear elastic wave velocities in Canadian east coast sedimentary rocks as functions of pressure and temperature. M, 1981, Memorial University of Newfoundland. 82 p.

Gagnon, William P. Geology of Argosy Creek and adjacent areas, Pioneer Mountains, Blaine and Custer counties, south-central Idaho. M, 1980, University of Wisconsin-Milwaukee. 115 p.

Gagnon, Yves. Etude pétrographique et géochimiques des roches volcaniques acides du Canton Clermont, Abitibi-Ouest, Québec. M, 1983, Ecole Polytechnique. 228 p.

Gahagan, Lisa Marie. The mapping of tectonic features in the ocean basins from satellite altimetry data. M, 1988, University of Texas, Austin. 95 p.

Gahé, Emile. Paléontomagnétisme, pétrophysique et certaines caractéristiques chimiques des unités lithologiques des ceintures archéennes de Protet-Evans et de Matagami-Chibougamau, Québec, Canada. M, 1984, Universite Laval. 90 p.

Gahnoog, Abdillahi. Textural and mineralogical investigation of Iroquois and Warren beaches in Erie and Niagara counties, western New York. M, 1968, SUNY at Buffalo. 56 p.

Gahring, W. Ross. A study of redbeds concretions. M, 1924, University of Oklahoma. 48 p.

Gaidusek, Barbara Ursel Marie. Landslide susceptibility mapping for real estate development, tested along I-35, Dallas to San Antonio, Texas. M, 1982, Texas A&M University. 136 p.

Gail, George Joseph. Subsurface geology of Capitan Aquifer northeast of Carlsbad, Eddy and Lea counties, New Mexico. M, 1974, University of New Mexico. 61 p.

Gaillot, Gary. Hydrogeologic analysis and reclamation alternatives for the Jack Waite Mine, Shoshone County, Idaho. M, 1979, University of Idaho.

Gaines, A. G., Jr. Papers on the geomorphology, hydrography and geochemistry of the Pettaquamscutt River estuary. D, 1975, University of Rhode Island. 296 p.

Gaines, Alan M. An experimental investigation of the kinetics and mechanism of the formation of dolomite. D, 1968, University of Chicago. 83 p.

Gaines, Elizabeth. Landsat linear trend analysis; a tool for groundwater exploration in northern Arkansas. M, 1978, University of Arkansas, Fayetteville.

Gaines, Patrick W. Stratigraphy and structure of the Provo Canyon-Rock Canyon area, south central Wasatch Mountains, Utah. M, 1950, Brigham Young University. 62 p.

Gaines, Richard Venable. The mineralogy, synthesis, and genetic significance of luzonite, famatinite, and some related minerals. D, 1952, Harvard University.

Gaines, Robert Byron, Jr. Statistical study of Irvingella, Upper Cambrian trilobite. M, 1951, University of Texas, Austin.

Gaines, Roberta Kay Sampson. A study of the organic geochemistry of the lower part of the Sharon Springs

Member of the Pierre Shale in western South Dakota. D, 1986, South Dakota School of Mines & Technology.

Gair, Jacob E. Some effects of deformation in the Central Appalachians. D, 1949, The Johns Hopkins University.

Gaiser, James E. Origin-corrected travel-time variations measured across the Wasatch Front. M, 1977, University of Utah. 146 p.

Gaisie, J. S. The effect of well spacing on economic recovery of oil. M, 1977, Stanford University. 24 p.

Gait, Robert I. Computer calculated electrostatic charge distributions in the structures of the feldspars, low albite, high albite, and anorthite. D, 1967, University of Manitoba.

Gait, Robert I. The mineralogy of the chrome spinels of the Bird River Sill, Manitoba. M, 1964, University of Manitoba.

Gaitanaros, Alexandros P. Nonlinear dynamic analysis of 3-D soil-structure systems using a direct time domain BEM-FEM. D, 1988, University of South Carolina. 285 p.

Gaither, Alfred. A study of porosity and grain relationships in experimental sands. M, 1951, University of Cincinnati. 47 p.

Gaither, Bruce Edward. The relation of spring discharge behavior to the hydrologic properties of carbonate aquifers. M, 1977, Pennsylvania State University, University Park. 210 p.

Gaito, Richard A. An interpretation of the possible magnetic anomaly due to sedimentation of Pompton Lake, New Jersey. M, 1980, Montclair State College. 50 p.

Gajewski, Konrad J. On the interpretation of climatic change from the fossil record; climatic change in central and eastern United States for the past 2000 years estimated from pollen data. D, 1983, University of Wisconsin-Madison. 233 p.

Gajkowski, Wynn A. Transient creep and stress history for polycrystalline freshwater-ice. M, 1979, University of Wisconsin-Milwaukee. 67 p.

Gakle, Arthur Frederick. Geology of northwest Keddie Ridge, Plumas and Lassen counties, California. M, 1966, University of Oregon. 49 p.

Gal-Chen, Tzvi. Numerical simulation of convection with topography. D, 1973, Columbia University. 170 p.

Galadanchi, Habeeb I. Application of resistivity inversion method to map shallow earth structures. M, 1988, Pennsylvania State University, University Park. 84 p.

Galagoda, Herath Mahinda. Nonlinear analysis of porous soil media and application. D, 1986, University of Arizona. 300 p.

Galanos, D. A. M. The distribution of Cu, Pb, Zn, Co, Mn, Fe in soil, rocks and stream sediments at Kirki area in northern Greece. M, 1973, University of New Brunswick.

Galarraga, Federico Antonio. Organic geochemistry and origin of the heavy oils in the Eastern Venezuelan Basin. D, 1986, University of Maryland. 364 p.

Galas, Christodoulos Alexander. The stratigraphy and paleontology of Core H-1A of Upper Silurian strata from Taylor Township, Wayne County, Michigan. M, 1958, University of Michigan.

Galas, F. Brian. Petrological and petrographical study of a silicated marble of the Grenville Series (Precambrian) displaying mineralogical evidence of metamorphic facies. M, 1965, SUNY at Buffalo. 89 p.

Galavis, Jose A. Study of minerals and rocks from El Callao gold region, State of Bolivar, Venezuela. M, 1947, Columbia University, Teachers College.

Galbiati, Larry Dale. Ammonium ions occupying potassium sites in glauconite. M, 1982, University of Texas, Arlington. 68 p.

Galbraith, Alan Farwell. The soil water regime of a shortgrass prairie ecosystem. D, 1971, Colorado State University. 137 p.

Galbraith, Frederic W., III. The geology of the Silver King area, Superior, Arizona. D, 1935, University of Arizona.

Galbraith, James Herbert. A study of mine tailings and associated plants and ground water in the Coeur d'Alene district, Idaho. M, 1971, University of Idaho. 138 p.

Galbraith, James Herbert. Regional stream-sediment geochemistry and data analysis in Northeast Bahia, Brazil. D, 1975, University of Idaho. 297 p.

Galbraith, James Nelson, Jr. Computer studies of microseism statistics with applications to prediction and detection. D, 1963, Massachusetts Institute of Technology. 283 p.

Galbraith, Lyman Edgar. Geology of an area in the northern Hogback Range, Wyoming. M, 1949, University of Michigan.

Galbraith, Robert Marshall, IV. Peripheral deformation of the Serpent Mound cryptoexplosion structure in Adams County, Ohio. M, 1968, University of Cincinnati. 47 p.

Galbreath, Edwin C. A contribution to the Tertiary geology and paleontology of northeastern Colorado. D, 1951, University of Kansas. 405 p.

Galbreath, Kevin C. Mass transfer during wall-rock alteration; an example from a quartz-graphite vein, Black Hills, South Dakota. M, 1987, South Dakota School of Mines & Technology.

Galceran, Carlos Manuel, Jr. Detailed subsurface mapping in central Kentucky with electrical resistivity methods. M, 1984, University of Kentucky.

Gale, Arthur S., Jr. Common Ostracoda of the Golconda Formation (Illinois, Indiana, Kentucky). M, 1938, University of Chicago. 54 p.

Gale, George Henry. The primary dispersion of Cu, Zn, Ni, Co, Mn and Na adjacent to sulfide deposits, Springdale Peninsula, Newfoundland. M, 1969, Memorial University of Newfoundland. 143 p.

Gale, Hoyt R. and Grant, Ulysses S., IV. Catalogue of the marine Pliocene and Pleistocene Mollusca of California and adjacent regions, with notes on their morphology, classification, and nomenclature and a special treatment of the Pectinidae and the Turridae (including a few Miocene and Recent species). D, 1930, Stanford University. 1036 p.

Gale, J. E. A numerical, field and laboratory study of flow rocks with deformable fractures, Sambro area, Halifax County, Nova Scotia. D, 1975, University of California, Berkeley.

Gale, John Edward. Analysis of bedrock topography mapping (Paleozoic) in southwestern Ontario. M, 1971, University of Western Ontario. 106 p.

Gale, Peter Edward. Diagenesis of the Middle to Upper Devonian Catskill facies sandstones in southeastern New York State. M, 1985, Harvard University.

Gale, Robert Earle. The geology of Kinskuch Lake area, British Columbia. M, 1957, University of British Columbia.

Gale, Robert Earle. The geology of Mission copper mine, Pima mining district, Arizona. D, 1965, Stanford University. 162 p.

Galegor, William Baker. A study of the Moberly Sandstone. M, 1951, University of Missouri, Columbia.

Galehouse, Jon Scott. Provenance and paleocurrents of the Paso Robles Formation. D, 1966, University of California, Berkeley. 123 p.

Galemore, Joe A. Depositional environment and provenance of the Mojado Formation (Lower Cretaceous), southwestern New Mexico. M, 1986, New Mexico State University, Las Cruces. 103 p.

Galen, William Mamoru. A hydrodynamic interpretation of climbing ripple structures. M, 1973, Massachusetts Institute of Technology. 64 p.

Galey, Jimmy L. Paragenetic study of the ores of Santa Barbara, Chihuahua, Mexico. M, 1971, Texas Tech University. 76 p.

Galey, John T. Mineralization in the Eagle Creek. M, 1971, University of Wyoming. 51 p.

Galicki, Alan M. An application of the lead-210 geochronological method in cores from Meginnis Arm, Lake Jackson, Leon County, Florida. M, 1977, Florida State University.

Galicki, Stanley J. Environment of deposition, diagenesis, and porosity of the Terryville Sandstone (U. Jurassic) from East Texas. M, 1982, Memphis State University.

Galiette, Stephen Joseph. The use of the gravity surveying technique in studying soil thicknesses and bedrock topography. M, 1974, Pennsylvania State University, University Park. 51 p.

Galindo-Griffith, Glenn. Electric charges, sorption of phosphate and cation exchange equilibria in Chilean Dystrandepts. D, 1974, University of California, Riverside.

Galinski, Christine Helen. Selected trace metal concentrations in groundwater, stream and soils in Essex County, Ontario. M, 1983, University of Windsor. 112 p.

Galipeau, Joan Mary. Petrochemistry of the Virginia Dale Ring Dike complex, Colorado-Wyoming. M, 1976, University of North Carolina, Chapel Hill. 95 p.

Gall, Daniel G. Geology of a section of ignimbrites near Sacramento, Chihuahua, Mexico. M, 1977, East Carolina University. 81 p.

Gall, Quentin. Petrography and diagenesis of the carboniferous Deer Lake Group and Howley Formation, Deer Lake subbasin, western Newfoundland. M, 1984, Memorial University of Newfoundland. 242 p.

Gallacher, Mark Hayes. Fractures and surface lineaments in northeastern Utah. M, 1975, University of Utah. 75 p.

Gallagher, Alton V. Geology of the Lower Cretaceous Cutbank Conglomerate in Northwest Montana. M, 1957, Michigan State University. 40 p.

Gallagher, B. J. Optimum hydraulic design of dredged material sedimentation basins. D, 1977, University of California, Los Angeles. 202 p.

Gallagher, Dan Jeffrey. Extensional deformation and regional tectonics of the Charleston Allochthon, central Utah. M, 1985, University of Utah. 74 p.

Gallagher, David. Origin of the magnetic deposits at Lyon Mountain, New York. D, 1935, Yale University.

Gallagher, Gerald L. The petrography of ultramafic inclusions from Bandera Crater, New Mexico. M, 1973, Kent State University, Kent. 49 p.

Gallagher, Helen D. The material and structure of the walls of Paleozoic foraminifera. M, 1928, Columbia University, Teachers College.

Gallagher, James. Alteration of acoustical properties in a saturated saline sediment aggregate caused by sampling. M, 1973, University of Rhode Island.

Gallagher, James Frederick. The variability of water masses in the Indian Ocean. M, 1967, American University. 74 p.

Gallagher, James Robert. A study of Laramide igneous intrusions, east of Ward, Boulder County, Colorado. M, 1978, Colorado School of Mines. 92 p.

Gallagher, John. Fold mechanics in the Wind River Basin, Wyoming. M, 1965, University of Missouri, Rolla.

Gallagher, John Joseph, Jr. Photomechanical model studies relating to fracture and residual elastic strain in granular aggregates. D, 1971, Texas A&M University. 141 p.

Gallagher, John Neil. A method of determining the source mechanism in small earthquakes with application to the Pacific Northwest region. D, 1969, Oregon State University. 187 p.

Gallagher, Maureen T. Substrate controlled biofacies; Recent foraminifera from the continental shelf and slope of Vancouver Island, British Columbia. D, 1979, University of Calgary. 237 p.

Gallagher, Michael G. Hydrogeologic and geochemical factors influencing the impact of chromium-arsenic-copper wastewater discharge on a glacial outwash aquifer. M, 1984, Western Michigan University.

Gallagher, Michael Patrick. Structure and petrology of meta-igneous rocks in the western part of the Shuksan metamorphic suite, northwestern Washington, U.S.A. M, 1986, Western Washington University. 59 p.

Gallagher, Peggie R. A model study of a thin plate in free-space for the EM37 transient system. M, 1984, University of Utah. 88 p.

Gallagher, Robert Anthony. Depositional environments of the Loyalhanna Formation, Centre County, Pennsylvania. M, 1984, Lehigh University. 103 p.

Gallagher, Robert Taylor. Mineral content of the Bevier coal seam. M, 1938, University of Missouri, Columbia.

Gallagher, Ronald Eric. Ottawa River (Tenmile Creek) hydrology; streamflow quantity and erosion at Toledo, Ohio. M, 1978, University of Toledo. 187 p.

Gallahan, David Michael. Group selection and neo-Darwinian theory. D, 1988, Cornell University. 98 p.

Gallaher, Bruce Morris. Recharge properties of the Tucson Basin aquifer as reflected by the distribution of a stable isotope. M, 1979, University of Arizona. 92 p.

Gallaher, David W. Stratigraphic geostatistical analysis of the Supai Group of northern Arizona. M, 1984, Northern Arizona University. 87 p.

Gallaher, John Taylor. The ground water resources of the Jackson, Michigan, area. M, 1956, University of Michigan.

Gallais, Christopher J. Le *see* Le Gallais, Christopher J.

Gallant, Raymond Bockles. An analysis of the physical characteristics of Kootenai sandstones in Montana. M, 1941, Montana College of Mineral Science & Technology. 35 p.

Gallant, William A. A zircon study of some Finnish rapakivi granites. M, 1971, University of Cincinnati. 110 p.

Gallaspy, Irvin Lee. Permian and Pennsylvanian (post-Cherokee) geology of Harper County, Oklahoma. M, 1958, University of Oklahoma. 45 p.

Gallemore, Roy Thornhill. A strip reconnaissance of several physiographic regions. M, 1937, University of Missouri, Columbia.

Gallena, Jane. A foraminiferal analysis of a deep-sea core from the east Equatorial Pacific. M, 1962, Smith College. 107 p.

Galley, Alan George. Volcanic stratigraphy and gold-silver occurrences on the Big Missouri claim group, Stewart, British Columbia. M, 1981, University of Western Ontario. 181 p.

Galli, Carlos Alberto. Geology of the Concepcion area, Chile. D, 1968, University of Utah. 217 p.

Galli, Carlos Alberto. Geology of the Juan de Morales Quadrangle, Tarapaca Province, Chile. M, 1964, University of Utah. 48 p.

Galli, Kenneth G. Differentiation between Olean and Kent tills, upper Cattaraugus Creek Basin, Cattaraugus County, New York, by matrix analysis. M, 1981, Rensselaer Polytechnic Institute. 57 p.

Gallick, Cyril M. The geology of a part of the Blanco Mountain Quadrangle, Inyo County, California. M, 1964, University of California, Los Angeles.

Gallie, Thomas Muir, III. Chemical denudation and hydrology near tree limit, Coast Mountains, British Columbia. D, 1983, University of British Columbia.

Galliher, Edgar W. A study of the Monterey Formation, California, at the type locality. M, 1930, Stanford University.

Galliher, Edgar W. Sediments of Monterey Bay, California. D, 1932, Stanford University. 143 p.

Gallivan, Lyle Bradshaw. Sediment transport in a mesotidal inlet; Matanzas Inlet, Florida. M, 1979, University of South Florida, Tampa. 72 p.

Gallo, Benedict James. The stratigraphy and paleontology of the (Cretaceous) Dakota Group and the Benton Shale (Colorado Group in part) south and east of the Wet Mountains from Badito, Huerfano County, to Rye, Pueblo County, Colorado. M, 1962, University of Michigan.

Gallo, David G. The influence of oceanic transform boundaries on the generation and evolution of oceanic lithosphere. M, 1984, SUNY at Albany. 129 p.

Gallop, Roger G. Organic sediment and overlying water column of Crane Creek. M, 1975, Florida Institute of Technology.

Galloway, Cheryl Leora. Structural geology at eastern part of the Smithfield Quadrangle, Utah. M, 1970, Utah State University. 115 p.

Galloway, James N. Man's alteration of the natural geochemical cycle of selected trace metals. D, 1972, [University of California, San Diego].

Galloway, Jesse J. and Kaska, Harold Victor. The genus Pentremites and its species. M, 1952, Indiana University, Bloomington. 224 p.

Galloway, Jesse James. Faunas of Tanner's Creek Section of the Cincinnati Series. M, 1911, Indiana University, Bloomington.

Galloway, Jesse James and Cumings, E. R. The stratigraphy and paleontology of the Tanner's Creek Section of Cincinnati Series of Indiana. D, 1913, Indiana University, Bloomington. 128 p.

Galloway, John Duncan. Petrology of the Keeton Prophyry, El Paso County, Colorado. M, 1956, Iowa State University of Science and Technology.

Galloway, Kathleen Elise. The effects of reclamation methods on groundwater quality and resaturation rates. M, 1983, Iowa State University of Science and Technology. 140 p.

Galloway, Malcolm Charles Bell Bradsworth. Carboniferous deltaic sedimentation, Fayette and Raleigh counties, southeastern West Virginia. D, 1973, University of South Carolina.

Galloway, Michael J. Qualitative hydrogeologic model of thermal springs in fractured crystalline rock. M, 1977, Montana State University. 197 p.

Galloway, Sherman Elsworth. Geology and groundwater resources of the Pontales Valley area, Roosevelt and Curry counties, New Mexico. M, 1956, University of New Mexico. 167 p.

Galloway, Walter Bruce. The Rb-Sr whole rock of the Bulgarmarsh Granite, Rhode Island and its geological implications. M, 1973, Brown University.

Galloway, William Edmond. Depositional history of the Wilcox Group (Eocene, lower) (Gulf Coast). M, 1966, University of Texas, Austin.

Galloway, William Edmond. Depositional systems and shelf-slope relationships in uppermost Pennsylvanian rocks of the eastern shelf, north central Texas. D, 1971, University of Texas, Austin.

Gallucci, Richard Nicholas. Palynology and biostratigraphy of the Upper Cretaceous Adaville Formation (southwestern Wyoming) and biostratigraphic comparison to the Niobrara Formation (Ridgway, Colorado). M, 1986, Idaho State University. 304 p.

Gallup, Marc R. Lower Cretaceous dinosaurs and associated vertebrates from north-central Texas in the Field Museum of Natural History. M, 1975, University of Texas, Austin.

Gallup, Marc Richmond. Osteology, functional morphology, and palaeoecology of Coloraderpeton brilli Vaughn, a Pennsylvanian aistopod amphibian from Colorado. D, 1982, University of California, Los Angeles. 226 p.

Galpin, Sidney Longman. Studies of flint clays and their associates. D, 1912, Cornell University.

Galpin, Sidney Longman. The effect of varying sizes and percentages of quartz grains upon the property and shrinkage of kaolins. M, 1910, Cornell University.

Galson, Daniel Allen. An investigation of the thermal structure in the vicinity of IPOD sites 417 and 418. M, 1979, Massachusetts Institute of Technology. 155 p.

Galster, Richard William. Geology of Miller-Foss River area, King County, Washington. M, 1956, University of Washington. 96 p.

Galton, Julie Hope. Magnitude and frequency of suspended sediment tranport in four tributary streams of Lake Tahoe. M, 1984, Stanford University. 78 p.

Galvan, Geoffrey Scott. Seismic reflection evaluation of detachment-related crustal extension in the transition zone, Yavapai County, Arizona. M, 1986, San Diego State University. 79 p.

Galvez Sinibaldi, Alfredo Salvador. Characterization of possible kerogen precursors by pyrolysis-gas chromatography. M, 1988, University of Oklahoma. 179 p.

Galvez, J. A. Evaluation of the water resources of the central Luzon Basin, Philippines. D, 1976, University of Arizona. 342 p.

Galvin, Cyril J. Derivations of principal strains from deformed brachiopods. M, 1959, Massachusetts Institute of Technology. 107 p.

Galvin, Cyril J. Experimental and theoretical study of longshore currents on a plane beach. D, 1963, Massachusetts Institute of Technology. 185 p.

Galvin, Timothy Joseph. Stratigraphy, faunal assemblages, and depositional environments of the Surrett Canyon Formation (Chesterian), Lost River Range, South-central Idaho. M, 1981, University of Idaho. 116 p.

Galya, Thomas A. Sedimentary petrology and history of the Pocahontas Formation (lower Pennsylvanian) of southern West Virginia. M, 1975, Northeast Louisiana University.

Galya, Thomas Andrew. Coal petrology of some selected Pocahontas No. 3 and No. 6 coal seams of the Pocahontas Formation (Lower Pennsylvanian) of Mercer, McDowell, and Wyoming counties, West Virginia. D, 1983, Miami University (Ohio). 757 p.

Galyen, Robert L. The Cenozoic deposits of a portion of Marsh Valley, Bannock County, Idaho. M, 1978, Idaho State University. 42 p.

Gamache, Mark Thomas. Fusulinid biostratigraphy of the Bird Spring Formation in the Spring Mountains near Mountain Springs Pass, Clark County, Nevada. M, 1986, Washington State University. 215 p.

Gambell, Neil Austin. A heavy mineral reconnaissance of a portion of the Copper Basin mining district, Yavapai County, Arizona, with emphasis on gold. M, 1973, Northern Arizona University. 69 p.

Gamber, James H. Recent sponges, molluscs, and echinoderms in sediment cores from the central Arctic Ocean. M, 1976, University of Wisconsin-Madison.

Gambill, David Thomas. Tertiary alkalic intrusions of East central Lucero Uplift, central New Mexico. M, 1980, Stanford University. 177 p.

Gambill, John A. Geology of Clover Hollow and surrounding area, Giles and Craig counties, Virginia. M, 1974, Virginia Polytechnic Institute and State University.

Gamble, Bruce Martin. Petrography and petrology of the Mount Cumulus Stock, Never Summer Mountains, Colorado. M, 1979, University of Colorado.

Gamble, Erling Edward. Origin and morphogenetic relations of sandy surficial horizons of upper coastal plain soils of North Carolina. D, 1966, North Carolina State University. 269 p.

Gamble, James A. A study of the Jurassic sediments within a wildcat well, Madison Parish, Louisiana. M, 1986, University of Southwestern Louisiana.

Gamble, James Clifton. Durability-plasticity classification of shales and other argillaceous rocks. D, 1971, University of Illinois, Urbana.

Gamble, James Clifton. Paleoenvironmental study of a portion of upper Desmoinesian (Pennsylvanian) strata of the mid-continent region. M, 1967, University of Illinois, Urbana.

Gamble, James Harold. Geology of the Point Mugu and Camarillo quadrangles, Ventura County, California. M, 1957, University of California, Los Angeles.

Gamble, R. P. The sulfidation of andradite and hedenbergite; an experimental study of skarn-ore genesis. D, 1978, Yale University. 243 p.

Gamboa, Luiz Antonio Pierantoni. Marine geology of the Brazilian continental margin and adjacent oceanic basin between the latitudes of 23° and 37°. D, 1981, Columbia University, Teachers College. 211 p.

Gamero, Gonzalo A. Subsurface stratigraphic analysis of northern Pottawatomie County, Oklahoma. M, 1965, University of Oklahoma. 52 p.

Gamero, Maria Lourdes. Foraminifera from the Punta Gavilan Formation (Miocene), northern Falcon State, Venezuela. M, 1965, University of Oklahoma. 132 p.

Games, Larry Martin. Biogeochemical and environmental applications of natural variations in carbon isotope ratios. D, 1975, Indiana University, Bloomington. 200 p.

Gammill, Laura May. K-40/Ar-40 age dating on glauconites and paleontologic interpretation near the Cretaceous-Tertiary transition in Texas. M, 1984, Rice University. 98 p.

Gammons, Christopher Hall. Studies in hydrothermal phenomena; 1, The solubility of silver sulfide in aqueous sulfde solutions to 300°C; 2, A paragenesis and fluid inclusion study of polymetallic vein mineralization in the Big Creek mining district, central Idaho. D, 1988, Pennsylvania State University, University Park. 365 p.

Gamper-Bravo, Martha Alicia. Correlation and paleoclimatic analysis of three Pliocene-Pleistocene deep-sea north Atlantic cores. M, 1971, Brown University.

Gan, Qigao. Geological implications of ophiolites in the Junggar area, Xinjiang, China. M, 1988, Stanford University. 110 p.

Gan, Tjiang-Liong. Heavy minerals in sediments from Owens, China, Searles and Panamint basins, Southern California. D, 1961, Indiana University, Bloomington. 95 p.

Gancarz, Alexander John, Jr. I, Isotopic systematics in Archean rocks, West Greenland; II, Mineralogic and petrologic investigations of lunar rock samples. D, 1976, California Institute of Technology. 378 p.

Gander, Craig Robert. Sources of stormflow from a forested watershed and a carbonate spring in central Pennsylvania determined by natural deuterium variations. M, 1979, Pennsylvania State University, University Park. 153 p.

Gander, Malcolm J. The geology of the northern part of the Rialto Stock, Sierra Blanca igneous complex, New Mexico. M, 1983, Colorado State University. 148 p.

Gandera, William Edward. Stratigraphy of the middle to late Eocene formations of southwestern Willamette Valley, Oregon. M, 1977, University of Oregon. 75 p.

Gandhi, Rajni K. Estimating bench design parameters for open-cut excavation. M, 1964, University of Missouri, Rolla.

Gandhi, S. M. Exploration rock geochemical studies in and around the Caribou sulphide deposit, New Brunswick, Canada. D, 1978, University of New Brunswick.

Gandhi, Sunilkumar S. Igneous petrology of Mount Yamaska, P.Q. (Canada). D, 1967, McGill University. 267 p.

Gandl, Lynnette Anne. The Waulsortian facies of the St. Joe Limestone (Lower Mississippian) in southwestern Missouri. M, 1983, University of Arkansas, Fayetteville. 88 p.

Gane, Henry Stewart. Some Neocene corals of the United States. D, 1895, The Johns Hopkins University.

Ganesan, Sudalaimuthu. Evolution and recovery of trace elements from a coal gasification plant. D, 1988, North Carolina State University. 370 p.

Gang, Michael W. Controls on heavy metals in surface and ground waters affected by coal mine drainage; Clarion River-Redbank Creek Watershed,

Pennsylvania. M, 1974, Pennsylvania State University, University Park.

Gangloff, R. A. The Archaeocyatha of the central and southwestern Great Basin, California and Nevada. D, 1975, University of California, Berkeley.

Gangloff, Roland Anthony. Archaeocyatha from the Westgard Pass area, Inyo-White mountains, California. M, 1963, University of California, Berkeley. 130 p.

Gangopadhyay, Jibimatra. Reconnaissance study of the stabilities of chloritoid and staurolite and some equilibrium relations in the system $FeO-Al_2O_3-SiO_2-H_2O-O_2$. D, 1967, University of Chicago. 107 p.

Ganguli, Ajit Kumar. Geology of the area southwest of Florence, Fremont County, Colorado. M, 1950, Colorado School of Mines. 37 p.

Ganguli, Debkumar. Gravity survey of part of Saint Clair Quadrangle, Missouri. M, 1955, St. Louis University.

Ganis, George R. Geology of the Ordovician clastic rocks of the Bethel Quadrangle, Pennsylvania. M, 1972, Lehigh University. 31 p.

Ganjei, Hossein. The geochemical analyses of some Pliocene plateau basalts. M, 1976, California State University, Fresno.

Ganjidoost, Hossein. Sorption and desorption of volatile organic compounds by soils and sediments. D, 1988, University of Missouri, Rolla. 189 p.

Ganju, Ashutosh M. Development of a portable seismic reflection data processing package and its application to data collected from the Jemez Mountains of New Mexico. M, 1987, Purdue University. 99 p.

Ganley, David Charles. A seismic reflection crustal model near Edmonton, Alberta. M, 1973, University of Alberta. 82 p.

Ganley, David Charles. The seismic measurement of absorption and dispersion. D, 1980, University of Alberta. 237 p.

Gann, D. E. The geology and microscopy of the Prewitt Copper Shale (Permian), Jackson County, Oklahoma. D, 1976, University of Missouri, Rolla. 425 p.

Gann, Delbert Eugene, Jr. Sedimentary analysis of the Filtro Corporation bentonite quarry at Mineral Springs, Lincoln Parish, Louisiana. M, 1971, Northeast Louisiana University.

Gann, Donald P. Changes in ionic concentrations of effluent from compaction of clay. M, 1965, University of Houston.

Gannett, Marshall W. A geochemical study of the Rhododendron and Dalles formations in the area of Mount Hood, Oregon. M, 1982, Portland State University. 64 p.

Gannett, R. W. The origin, weathering and secondary enrichment of tungsten ores. M, 1918, University of Minnesota, Minneapolis. 38 p.

Gannon, Brian Lee. Geology of a volcanic complex on the south flank of Mount Jefferson, Oregon. M, 1981, Portland State University. 181 p.

Ganoe, Steven J. Investigation of P_n wave propagation in Oregon. M, 1983, Oregon State University. 97 p.

Ganow, H. C. A geotechnical study of the squeeze problem associated with the underground mining of coal. D, 1975, University of Illinois, Urbana. 265 p.

Ganow, Harold C. The clay mineralogy and swell pressures of expansive bentonite clays contained within the Cretaceous Benton Formation, north central Colorado. M, 1969, Colorado State University. 159 p.

Gans, Phillip Bruce. Cenozoic extension and magmatism in the eastern Great Basin. D, 1987, Stanford University. 191 p.

Gans, Phillip Bruce. Mid-Tertiary magmatism and extensional faulting in the Hunter District, White Pine County, Nevada. M, 1982, Stanford University. 179 p.

Gans, Roger Frederick. The steady-state response of a contained, rotating, electrically conducting, viscous fluid subject to precessional and Lorentz forces, and

its implications regarding geomagnetism. D, 1969, University of California, Los Angeles.

Gans, William Thomas. The detailed stratigraphy of the Goodsprings dolomite (Cambro-Devonian), southeastern Nevada-California. D, 1970, Rice University. 225 p.

Ganse, Robert Anthony. A study by Fourier integral analysis of the tide gage response to a tsunami at two different stations. M, 1967, St. Louis University.

Ganse, Robert Anthony. A study of earthquake magnitudes and their relation to the law of seismic spectrum scaling. D, 1974, St. Louis University.

Ganser, Robert W. Geology of the Cumberland area, Bryan, Johnston, and Marshall counties, Oklahoma. M, 1968, University of Oklahoma. 60 p.

Ganster, Maurice W., II. Analysis and interpretation of a magnetic anomaly at Gordonsville, Smith County, Tennessee. M, 1969, Vanderbilt University.

Gant, Jonathan L. Mineralogical and geochemical studies of brine-related metalliferous sediments in the Suakin Deep, Red Sea. M, 1980, University of Wisconsin-Madison.

Gant, Orland James, Jr. Theory and application of uranium ore grade determination by gamma ray self absorption in layered media. D, 1975, Southern Methodist University. 153 p.

Gantela, Christopher. Crustal structure of north western Hudson Bay. M, 1969, University of Saskatchewan. various pagination p.

Ganthavee, Somkiat. A study of placer tin materials from southern Thailand. M, 1962, University of Missouri, Rolla.

Gantnier, Robert. Geology of the Chenango Forks, New York, seven and one-half minute quadrangle. M, 1951, Syracuse University.

Gants, Donald G. Geologic and mechanical properties of the Sevier Desert detachment as inferred by seismic and rheologic modeling. M, 1985, University of Utah. 129 p.

Ganus, William J. Lithologic and structural influences on the hydrodynamics of the Tucson Basin, Arizona. M, 1965, University of Arizona.

Ganus, William Joseph. Analysis of factors controlling groundwater flow and prediction of rates of groundwater movement and changes in quality, Atlantic Coastal Plain. D, 1972, University of Arizona.

Gao, Ruixiang. Deformation characteristics of the eastern Cobequid and Hollow fault zones and Stellarton Basin, Nova Scotia. M, 1987, University of New Brunswick. 240 p.

Gao, Zu-Cheng. The theory of fluxgate sensor systems. D, 1986, University of British Columbia.

Gaona, Michael Thomas. Stratigraphy and sedimentology of the Osburger Gulch Sandstone Member of the Upper Cretaceous Hornbrook Formation, Northern California and southern Oregon. M, 1985, San Diego State University. 164 p.

Gaposchkin, Edward Michael. Some results from satellite geodesy. D, 1969, Harvard University.

Garabedian, Stephen P. Factors influencing streambed infiltration for selected streams in the western Middle Anthracite Field, Pennsylvania. M, 1980, Pennsylvania State University, University Park. 154 p.

Garaycochea-Wittke, Isabel. Revision and refinement of the crystal structure of pharmacosiderite. M, 1966, Massachusetts Institute of Technology. 28 p.

Garbarini, George Stephen. Geology of the McLeod area, Beartooth Range, Montana. D, 1957, Princeton University. 230 p.

Garbarini, J. Michael. Feldspar diagenesis in the Hosston Formation, central Mississippi. M, 1979, University of Missouri, Columbia.

Garber, Jonathan Hunt. ^{15}N-tracer and other laboratory studies of nitrogen remineralization in sediments and waters from Narragansett Bay, Rhode Island. D, 1982, University of Rhode Island. 304 p.

Garber, Lowell Wilbur. Influence of volcanic ash on the genesis and classification of two Spodosols in Idaho. M, 1966, University of Idaho.

Garber, Murray S. Stratigraphy of the Foraker Limestone in east-central Kansas. M, 1962, University of Kansas. 109 p.

Garber, Raymond Alan. Diagenetic patterns of Oligocene reefs in Indonesia. M, 1976, Rensselaer Polytechnic Institute. 116 p.

Garber, Raymond Alan. The sedimentology of the Dead Sea. D, 1980, Rensselaer Polytechnic Institute. 169 p.

Garbett, Elizabeth C. Zoogeography of Recent cytheracean ostracodes in the bays of Texas. M, 1978, University of Houston.

Garbisch, Jon Ootek. Facies variation, depositional environments and paleoecology of the interval between and inclusive of the Mary Lee and Pratt coal groups, (upper Pottsville Formation) in Walker County, Alabama. M, 1988, Mississippi State University. 113 p.

Garbrecht, David A. Lineaments in North-central Nevada and their relations to geothermal areas. M, 1978, University of Nevada. 37 p.

Garbrecht, Jurgen D. The physical basis of stream flow hydrology with emphasis on drainage network morphology. D, 1984, Colorado State University. 299 p.

Garcés-Gonzales, Hernán. The White Elephant Pegmatite, Custer, South Dakota. D, 1945, University of Chicago. 41 p.

Garcia Bengochea, Jose Ignacio R. The determination of aquifer constants using water-level observations in the pumping well. D, 1963, University of Florida. 140 p.

Garcia Delgado, Victor. Soil effects on building earthquake response. D, 1986, University of Texas, Austin. 412 p.

Garcia S., Eduardo. DC-resistivity profiling as an aid in the study of facies distribution in residual-type iron deposits. M, 1973, Colorado School of Mines. 94 p.

García, Alfredo R. Paleomagnetic reconnaissance of the "region de Los Lagos", southern Chile, and its tectonic implications. M, 1986, Western Washington University. 127 p.

Garcia, Daniel. Mammalian zoogeography and faunal dynamics in the northwestern region of the United States during the Arikareean (late Oligocene - early Miocene). M, 1987, University of Montana. 105 p.

Garcia, Esmeralda. Petrography and petrogenesis of Neogene andesites, Cordillera Occidental, southernmost Peru. M, 1986, Queen's University. 162 p.

Garcia, Francisco Raul. Late Paleozoic metamorphism in the Paramo de los Torres area, Estado Trujillo, Venezuelan Andes. M, 1972, Michigan Technological University. 66 p.

Garcia, Joseph E. A geostatistical study of fracture traces and their hydrological significance, southern Piedmont Georgia. M, 1987, University of Georgia. 117 p.

Garcia, M. O. Petrology of the Rogue River area, Klamath Mountains, Oregon; problems in the identification of ancient volcanic arcs. D, 1976, University of California, Los Angeles. 211 p.

Garcia, Marcial V. Physical properties of mine rock and their effect on percussive drillability. M, 1959, Colorado School of Mines. 121 p.

Garcia, Rafael A. Geology and petrology of andesitic intrusions in and near El Paso (Texas). M, 1970, University of Texas at El Paso.

García-Banda, Rosalba. Geology, geochemistry and petrology of the Pizarro and Pinto domes and the Tepeyahualco Flows related to the Los Humeros caldera complex, Puebla, Mexico. M, 1984, McGill University. 68 p.

Garcia-Miragaya, J. Sorption and desorption of cadmium by soils and soil materials. D, 1975, University of California, Riverside. 128 p.

Garcia-Ocampo, Alvaro. Cation exchange in some soils of Northern California. D, 1986, University of California, Riverside. 145 p.

Garcia-Solorzano, Roberto. Depositional systems in the Queen City Formation (Eocene), Central and South Texas. M, 1972, University of Texas, Austin.

Gard, Leonard Meade, Jr. Engineering geology as applied to highway construction. M, 1954, University of Colorado.

Gard, Theodore Max. Tectonics of the Badwater uplift area, central Wyoming. D, 1969, Pennsylvania State University, University Park. 194 p.

Garden, Arthur John. Geology of western Payne County, Oklahoma. M, 1973, Oklahoma State University. 70 p.

Gardescu, Ionel Ion. The occurrence and behavior of natural gas in an oil reservoir. D, 1931, University of California, Berkeley. 43 p.

Gardiner, Errol Murray. A three-dimensional photoelastic study of stress distribution following pillar removal. M, 1969, University of Utah. 66 p.

Gardiner, H. C. Genesis of a climosequence of soils in the Kohala regon. M, 1967, University of Hawaii. 82 p.

Gardiner, Janice L. A field spectrometer and Landsat remote sensing study of the Fresnillo mining district, Mexico. M, 1987, Dartmouth College. 88 p.

Gardiner, John J. Structural geology of the Lupin gold mine, Northwest Territories. M, 1986, Acadia University. 206 p.

Gardiner, Larry L. Environmental analysis of the upper Cambrian Nounan Formation, Bear River Range and Wellsville Mountain, north-central Utah. M, 1974, Utah State University. 121 p.

Gardiner, Mary Anne. Cyclic sedimentation patterns, Middle Ordovician carbonates, central Pennsylvania. M, 1985, University of Delaware. 104 p.

Gardiner, McE. C. The gold deposits of the Atikokan area, Ontario. M, 1939, University of Toronto.

Gardiner, Scott. Sedimentology and local basin analysis of the lower Conception Group (Hadrynian), Avalon Zone, Newfoundland. M, 1984, Memorial University of Newfoundland. 230 p.

Gardinier, Clayton Frank. Phase relations in the systems alkali oxide (Li2O,Na2O)-ZrO2-SiO2 and Na2ZrO3-NaAlO2-SiO2. D, 1980, Miami University (Ohio). 156 p.

Gardinier, Clayton Frank. Phase relations in the systems W-Sn-S, Mo-Sn-S, and Mo-W-S. M, 1977, Miami University (Ohio). 85 p.

Gardner, Charles Hardwood. The geology of central Newton County, Georgia. M, 1961, Emory University. 53 p.

Gardner, David A. Foliations in upper Proterozoic metasedimentary rocks of the Chamberlindalen area, Wedel Jarlsberg Land, Spitsbergen. M, 1988, University of Wisconsin-Madison. 121 p.

Gardner, David Allison. Hydrogeologic investigation of the Montecito ground water basin, Santa Barbara County. M, 1974, University of California, Los Angeles.

Gardner, David Ward. Long-term changes in the elemental composition of soil and vegetation in the vicinity of a coal-fired power plant. D, 1988, University of Minnesota, Minneapolis. 411 p.

Gardner, Dion Lowell. Geology of the Newberry and Ord mountains, southeastern California. M, 1933, University of California, Berkeley. 170 p.

Gardner, Douglas A. C. Structure, geology and metamorphism of calcareous lower Paleozoic slates, Blaeberry River-Redburn Creek area, near Golden, British Columbia. D, 1977, Queen's University. 224 p.

Gardner, Douglas Hansen. Structure and stratigraphy of the northern part of the Snake Mountains, Elko County, Nevada. D, 1968, University of Oregon. 222 p.

Gardner, Edgar Jackson. A study of the insoluble residues of the Wilberns Formation of central Texas. M, 1940, University of Texas, Austin.

Gardner, Frank Johnson. A correlation of characteristics of the oil fields of north and northcentral Texas. D, 1942, University of Texas, Austin.

Gardner, Frank Johnson. A correlation of characteristics of the oil fields of South Texas. M, 1938, University of Texas, Austin.

Gardner, George Dennis. A regional study of landsliding (Holocene) in the lower Cuyahoga River valley, Ohio. M, 1972, Kent State University, Kent. 85 p.

Gardner, Henry J. The Alcova Limestone Member (Triassic) of the Chugwater Formation, Freezeout Mountains, Wyoming. M, 1964, University of Wyoming. 118 p.

Gardner, Howard David. Petrologic and geochemical constraints on genesis of the Jason Pb-Zn deposits, Yukon Territory. M, 1983, University of Calgary. 212 p.

Gardner, Jack Winston. The Northview Formation; its character and extent in Webster County, Missouri. M, 1935, University of Missouri, Columbia.

Gardner, James Edward. Trace element and Nd isotope evidence for the origin of the stratifications of the Endion Sill, Duluth, Minnesota. M, 1987, Washington University. 359 p.

Gardner, James Vincent. The eastern equatorial Atlantic; sedimentation, faunal and sea-surface temperature responses to global climatic changes during the past 200,000 years. D, 1973, Columbia University. 387 p.

Gardner, James Vincent. The submarine geology of the western Coral sea. M, 1969, Columbia University. 65 p.

Gardner, Jamie Neal. Tectonic and petrologic evolution of the Keres Group; implications for the development of the Jemez volcanic field, New Mexico. D, 1985, University of California, Davis. 293 p.

Gardner, John Darrell. Concepts of mine evaluation and design optimization. D, 1981, University of Utah. 324 p.

Gardner, John Kelsey. Earthquakes in the Walker Pass region, California, and their relation to the tectonics of the southern Sierra Nevada. D, 1964, California Institute of Technology. 127 p.

Gardner, Joseph R. Geologic study of subsurface Mississippian rocks, Comanche County, Kanss, with special emphasis on Bird South Field. M, 1985, Wichita State University. 127 p.

Gardner, Joseph Vincent. A quantitative analysis of the relationships between valley asymmetry and two possible controlling factors; lithology and geographical location. D, 1973, Indiana State University. 110 p.

Gardner, Julia Anna. On certain families of the Gastropoda from the Miocene and Pliocene of Virginia and North Carolina. D, 1911, The Johns Hopkins University.

Gardner, Leonard Robert. The Pleistocene geology of the Brodheadsville and Pohopoco Mountain (7.5') quadrangles, Pennsylvania. M, 1966, Pennsylvania State University, University Park. 99 p.

Gardner, Leonard Robert. The Quaternary geology of the Moapa valley, Clark County, Nevada. D, 1968, Pennsylvania State University, University Park. 236 p.

Gardner, Maxwell E. Quaternary and engineering geology of the Orchard, Weldona, and Fort Morgan quadrangles, Morgan County, Colorado. D, 1967, Colorado School of Mines. 283 p.

Gardner, Michael C. Till stratigraphy of Tippecanoe County, Indiana. M, 1979, Purdue University. 125 p.

Gardner, Murray Curtis. Cenozoic volcanism in the high Cascade and Modoc Plateau provinces of Northeast California. D, 1964, University of Arizona. 224 p.

Gardner, Peter M. Petrofabric analysis of a section across Piseco Dome (Precambrian) in the southern Adirondacks (New York). M, 1971, Syracuse University.

Gardner, Robert A. The associations of alluvial parent material from granitic and from sedimentary rock sources as indicated by soil characteristics in Los Banos region, California. M, 1940, University of California, Berkeley.

Gardner, Robert Alexander. Sequence of podzolic soils along the coast of Northern California. D, 1967, University of California, Berkeley. 226 p.

Gardner, Stephen P. A seismic investigation of a buried valley near Cuyahoga Falls, Ohio. M, 1981, University of Akron. 169 p.

Gardner, Thomas W. A model study of a river meander incision. M, 1973, Colorado State University. 96 p.

Gardner, Thomas William. Paleohydrology, paleomorphology and depositional environments of some fluvial sandstones of Pennsylvanian age in eastern Kentucky. D, 1978, University of Cincinnati. 216 p.

Gardner, Walter Hale. Flow of soil moisture in the unsaturated state. D, 1950, Utah State University. 31 p.

Gardner, Weston C. Environmental analysis of the middle Devonian of the Michigan Basin. D, 1971, Northwestern University.

Gardner, Weston Clive. Geology of the West Tintic mining district and vicinity, Juab County, Utah. M, 1954, University of Utah. 43 p.

Gardner, Wilford D. Fluxes, dynamics, and chemistry of particulates in the ocean. D, 1977, Woods Hole Oceanographic Institution. 405 p.

Gardner, Wilford Dana. Fluxes, dynamics, and chemistry of particulates in the ocean. D, 1978, Massachusetts Institute of Technology. 405 p.

Gardner, William Edgar. Geology of the Barnsdall area, Osage County, Oklahoma. M, 1957, University of Oklahoma. 102 p.

Gardner, William Irving. Structural study of the Merrimac Batholith, Sierra Nevada, California. D, 1935, University of Minnesota, Minneapolis.

Gardulski, Anne Frances. A structural and petrologic analysis of a quartzite-pegmatite tectonite, Coyote Mountains, southern Arizona. M, 1980, University of Arizona. 69 p.

Gardulski, Anne Frances. Climatic and oceanographic controls on the Neogene sedimentary framework of the outer West Florida carbonate ramp. D, 1987, Syracuse University. 240 p.

Garet, Gerald H. Geology of the Rosendale 7 1/2 minute Quadrangle (New York). M, 1958, New York University.

Garey, Christopher Lee. Radiolaria from the Otter Point Complex (Oregon) and the volcano-pelagic strata above the Coast Range Ophiolite (California). M, 1987, University of Texas at Dallas. 156 p.

Garfias, Valentine Richard. The oil region of northeastern Mexico. M, 1912, [Stanford University].

Garg, Nek R. Spatial and temporal analysis of electromagnetic survey data. D, 1984, Colorado School of Mines. 161 p.

Garg, Nek R. Synthetic electric sounding surveys over known oil fields (southern portion of San Joaquin Valley, Kern County, California; Payne, Logan and Lincoln counties, Oklahoma, and Lance-Creek Field, Niobrara County, Wyoming). M, 1981, Colorado School of Mines. 192 p.

Garg, Own Prakash. Static and dynamic mechanical properties of a sandstone. M, 1970, University of Saskatchewan. 50 p.

Gargi, Satya Parkash. Geometric and kinematic analysis, and geochemical study of the Corbin Gneiss complex and its associated sheared rocks in the Blue Ridge of NW Georgia. D, 1985, Miami University (Ohio). 250 p.

Garihan, Anne Burroughs Lutz. Stratigraphy and brachiopod genus Composita in the Wreford Megacyclothem (lower Permian) in Kansas and Oklahoma. D, 1973, Pennsylvania State University, University Park. 230 p.

Garihan, Anne Burroughs Lutz. The influence of geologic and geomorphic parameters on the flood response of small drainage basins in Pennsylvania. M, 1970, Pennsylvania State University, University Park. 81 p.

Garihan, John Michael. Geology and talc deposits of the central Ruby Range, Madison County, Montana. D, 1973, Pennsylvania State University, University Park. 282 p.

Garland, George David. The relationship between gravity magnetic anomalies as illustrated by observations made in the Maritime Provinces, and Ontario, Canada. D, 1951, St. Louis University.

Garland, Mary Isabelle. Graphite in the Central Gneiss Belt of the Grenville Province of Ontario. M, 1987, University of Toronto.

Garlauskas, Algirdas Benedict. Environmental management in urban geology; case studies of geologic and legal applications. M, 1974, Kent State University, Kent. 150 p.

Garlick, George Donald. Oxygen isotope ratios in co-existing minerals of regionally metamorphosed rocks. D, 1965, California Institute of Technology. 244 p.

Garlow, Richard A. The stratigraphy, petrology and paleogeographic setting of the Middle Pennsylvanian Tihvipah Limestone of southeastern California. M, 1985, San Jose State University. 134 p.

Garman, George Walter, Jr. Structure of the Beryl Mountain, New Hampshire, pegmatite. M, 1954, Miami University (Ohio). 47 p.

Garman, James E. Mechanism of acquisition of remanent magnetization by airfall ash. M, 1987, Eastern Washington University. 47 p.

Garman, Phyllis Metrolis. Effective use of wells in monitoring sanitary landfills. M, 1981, University of Virginia. 69 p.

Garman, Ray Keith. A geological and geochemical investigation of a deep well on Andros Island, Bahamas. M, 1960, Florida State University.

Garmany, Jan D. Methods of seismic travel time calculation and inversion and of synthesizing high frequency seismograms. D, 1978, University of California, San Diego. 110 p.

Garmezy, Lawrence. Geology and geochronology of the southeast border of the Bitterroot Dome; implications for the structural evolution of the mylonitic carapace. D, 1983, Pennsylvania State University, University Park. 276 p.

Garmezy, Lawrence. Geology and tectonic evolution of the southern Beaverhead Range, east-central Idaho. M, 1981, Pennsylvania State University, University Park. 155 p.

Garmoe, Walter James. A preliminary study of the Mississippian and Lower Pennsylvanian formations in the Park City District, Utah. M, 1958, University of Minnesota, Minneapolis. 62 p.

Garnar, Thomas E., Jr. Igneous rocks of Pendleton County, West Virginia. M, 1951, West Virginia University.

Garner, Gary Lee. Areal geology of the Flint area, Delaware County, Oklahoma. M, 1965, University of Oklahoma. 56 p.

Garner, Hessle Filmore. Lower Mississippian cephalopods of Michigan; coiled nautiloids and ammoniids. D, 1953, University of Iowa. 201 p.

Garner, Hessle Filmore. Lower Mississippian orthoconic nautiloid cephalopods of Michigan. M, 1951, University of Iowa. 72 p.

Garner, John C., Jr. Palynomorphs of the Calvert Formation (Miocene) of Maryland. M, 1976, University of Rhode Island.

Garner, L. Edwin. Environmental geology of the Austin area, Texas. M, 1973, University of Texas, Austin.

Garner, Norman Earl. Experimental study of crater formation in rocks at elevated stress states. D, 1963, University of Texas, Austin. 191 p.

Garnett, John Arthur. Structural analysis of part of the Lubec-Belleisle Fault Zone, southwestern New Brunswick. D, 1973, University of New Brunswick.

Garney, Ronald T. Fauna of the Racine Dolomite (Niagaran), Grafton, Illinois. M, 1983, University of Missouri, Columbia.

Garr, John D. Quaternary geology and tectonic geomorphology of the Pocatello Valley area, Idaho-Utah. M, 1988, Utah State University. 115 p.

Garrels, Robert M. Factors influencing deposition of galena and sphalerite in the mississippi type lead-zinc deposits. D, 1941, Northwestern University.

Garrels, Robert M. Iron deposits of the Bay de Verde Peninsula, Newfoundland. M, 1939, Northwestern University.

Garret, Julius B., Jr. The microfauna of the Sabine Group (Wilcox Eocene) of Louisiana. M, 1933, Louisiana State University.

Garrett, Bruce T. A design for a caustic flood, Bell Creek oil field, Unit E, Powder River County, Montana. M, 1981, Colorado School of Mines. 87 p.

Garrett, Donald Maurice. Variation in the sedimentary parameters of the Saratoga Springs sand dunes, Death Valley National Monument, California. M, 1966, University of Southern California.

Garrett, Elizabeth M. A petrographic analysis of ceramics from Apache-Sitgreaves national forests, Arizona; onsite or specialized manufacture?. D, 1982, Western Michigan University. 207 p.

Garrett, George R. Lineament distribution in southwestern Ohio; an integrated study combining digitally enhanced Landsat, aerial photography, and soil radon techniques. M, 1987, Ohio University, Athens. 184 p.

Garrett, Howard L. The geology of Star Basin and Star Mine, Gunnison County, Colorado (670). M, 1950, Colorado School of Mines. 45 p.

Garrett, James Hugh. A subsurface study of the Tar Springs Formation, Henderson County, Kentucky. M, 1958, University of Illinois, Urbana.

Garrett, Jim R. Depositional model for the auriferous gravels in the Payan mining district, Department of Narino, Colombia, South America. M, 1985, Old Dominion University. 142 p.

Garrett, M. W. Depositional environments of lower Pennsylvanian strata in the Hanging Limb area, Cumberland Plateau, Tennessee. M, 1975, Vanderbilt University.

Garrett, Marion E. A regional study of the Eutaw Formation in the subsurface of central and southern Mississippi. M, 1956, Mississippi State University. 39 p.

Garrett, P. M. An evaluation of the U. S. Atomic Energy Commission's environmental impact statements for nuclear power reactors. D, 1977, University of California, Los Angeles. 98 p.

Garrett, Paul Allen. Relationships between benthic communities, land use, chemical dynamics, and trophic state in Georgetown Lake. D, 1983, Montana State University. 164 p.

Garrett, Paul Winslow, Jr. The geology and groundwater in a section of southern High Plains between Lubbock and Silverton, Texas. M, 1953, Texas Tech University. 69 p.

Garrett, Peter. The sedimentary record of life on a modern tropical carbonate tidal flat, Andros Island, Bahamas. D, 1971, The Johns Hopkins University. 259 p.

Garrett, R. The development and use of soil potential ratings in land use planning; a case study within the Niagara region. M, 1983, University of Guelph.

Garrett, Theodore Watrous, Jr. Geology of a portion of the Sequatchie Anticline in southern Marshall County, Alabama. M, 1973, University of Tennessee, Knoxville. 57 p.

Garrey, George H. Glaciation between the Rockies and the Cascades in northwestern Montana, northern Idaho, and eastern Washington. M, 1902, University of Chicago. 93 p.

Garrido, Robert William. Hydrologic transport of uncontrolled hazardous wastes from Sac River-Fulbright landfills, Springfield, Missouri. M, 1986, University of Missouri, Rolla. 69 p.

Garrison, Edwin J. Heavy and alkali metal content of surface and ground waters in the Joplin area, Missouri. M, 1974, University of Missouri, Rolla.

Garrison, Gene. The lithology of the Saluda Formation. M, 1954, Miami University (Ohio). 110 p.

Garrison, James R., Jr. Petrology and geochemistry of the Precambrian Coal Creek Serpentinite mass and associated metamorphosed basaltic and intermediate rocks, Llano Uplift, Texas. D, 1979, University of Texas, Austin. 278 p.

Garrison, Judy West. Mineralogy and petrology of hydrous groundmass minerals and altered crustal clasts in kimberlite, Elliott County, Kentucky. M, 1981, University of Tennessee, Knoxville. 76 p.

Garrison, Louis Eldred. Cretaceous-Cenozoic development of the continental shelf south of New England. D, 1967, University of Rhode Island. 112 p.

Garrison, Martyna. A comparative study of some species of Cytheropteron and Eocytheropteron of the Washita Series in Texas. M, 1939, University of Oklahoma. 38 p.

Garrison, Robert Edward. Jurassic and early Cretaceous sedimentation in the Unken Valley area, Austria. D, 1965, Princeton University. 188 p.

Garrison, Robert Kent. Paleoenvironmental analysis of the Lost Creek Limestone and associated facies; marine and nonmarine. M, 1977, University of Cincinnati. 338 p.

Garrow, Holly C. Shoreline rhythmicity on a natural beach. M, 1985, Oregon State University. 170 p.

Garshasb, Masoud. Quantitative geochemical exploration for lead-zinc deposits in Rossie area, New York. M, 1979, Syracuse University.

Garside, Larry J. Geology of the Bishop Creek area, Nevada (Elko County). M, 1968, University of Nevada. 52 p.

Garske, David Herman. The study of cleavage in ionic crystals. D, 1970, University of Michigan. 94 p.

Gartig, Derry G. Petrography, structure and age relationship of the dikes in an area north of Cooke, Montana. M, 1957, Wayne State University. 47 p.

Gartland, Eugene F. Conodont biostratigraphy of the Wallington Limestone member of the Reynales Limestone and the lower Sodus Shale (Silurian). M, 1973, University of Rochester.

Gartland, Jeffrey Dale. Experimental dissolution-reprecipitation processes with two Florida limestones. M, 1979, University of Florida. 80 p.

Gartmann, Charles W. Structural geology in northeastern Marathon County, Wisconsin. M, 1979, University of Wisconsin-Milwaukee. 111 p.

Gartmann-Siemann, Susan. Microfauna associated with crinoid root systems of the Wenlockian Waldron Shale in southeastern Indiana. M, 1979, University of Wisconsin-Milwaukee. 61 p.

Gartner, Anne E. Geometry and emplacement history of a basaltic intrusive complex, San Rafael Swell and Capitol Reed areas, Utah. M, 1984, San Jose State University. 149 p.

Gartner, Jeffrey. Density determinations of suspended particles using an electronic particle counter. M, 1978, University of South Florida, St. Petersburg.

Gartner, John F. Geology and geochemistry of the intrusive and volcanic rocks on the Norita and Radiore West properties, Matagami, Quebec. M, 1987, McGill University. 124 p.

Gartner, Stefan, Jr. A biostratigraphic study of the calcareous nannofossils of the Upper Cretaceous of the northwestern Gulf Coast. D, 1965, University of Illinois, Urbana.

Gartner, Stefan, Jr. Paleocene (Ilerdian) planktonic foraminifer from the Tremp Basin, Spain, and Mont Cayla near Agel, France. M, 1962, University of Illinois, Urbana.

Gartzos, Eutheme G. The geology and petrology of the Iron and Manitou Islands alkaline carbonatite complexes at Nipissing Lake, Ontario. M, 1977, McMaster University.

Garven, Audrey Curry. Geology of the Stackhouse-Numabin Bays area, Reindeer Lake, Saskatchewan. M, 1978, University of Regina. 100 p.

Garven, Grant. The role of groundwater flow in the genesis of stratabound ore deposits; a quantitative analysis. D, 1982, University of British Columbia.

Garvey, John Thomas. The hydrogeology of eastern Franklin and western Green Townships, Summit County, Ohio. M, 1988, University of Akron. 127 p.

Garvey, Michael Joseph. Uranium, thorium, and potassium abundances in rocks of the Piedmont of Georgia. M, 1975, University of Florida. 95 p.

Garvey, Phillip L. Oil and gas in Essex County, Ontario. M, 1941, University of Michigan.

Garvin, Donald S. Geology of the Mormon Basin, Wind River Mountains, Fremont County, Wyoming. M, 1951, University of Missouri, Columbia.

Garvin, James Brian. Geological analyses of the surfaces of Venus and Mars from Lander spacecraft images and orbital radar observations. D, 1984, Brown University. 279 p.

Garvin, Paul Lawrence. Phase relations in the Pb-Sb-S system. D, 1969, University of Colorado. 94 p.

Garvin, Robert Franklin. Stratigraphy of the Currant Creek Formation (Cretaceous-Tertiary), Wasatch and Duchesne counties, Utah. M, 1967, University of Utah. 85 p.

Garwick, Robert. A Devonian fauna from Peru. M, 1947, Columbia University, Teachers College.

Garwin, Steven Lee. Structure and metamorphism in the Niagara Peak area, western Cariboo Mountains, British Columbia. M, 1987, University of British Columbia. 166 p.

Gary, Anthony C. Fourier series analysis of the relationship of morphologic variation and bathymetry in Recent Bolivina (foraminiferida) from the northwestern Gulf of Mexico. M, 1984, Old Dominion University. 200 p.

Gary, Anthony Cavedo. Habitat and morphology of Recent benthic foraminifera, northwestern Gulf of Mexico. D, 1988, University of South Carolina. 197 p.

Gary, Steven D. Quaternary geology and geophysics of the upper Madison Valley, Madison County, Montana. M, 1980, University of Montana. 76 p.

Garza, Abato John. Depositional model of the upper San Andres Formation (Central Basin Platform), Ector County, Texas. M, 1982, Stephen F. Austin State University. 72 p.

Garza, Fernando Javier Rodriguez de la see Rodriguez de la Garza, Fernando Javier

Garza, Roberto. A geomorphic investigation of possible fracture control of stream course development in the Nashville Basin and adjacent Highland Rim area of Tennessee. M, 1971, Indiana State University. 85 p.

Garza, Roberto. An island in geographic transition; a study of the changing land use patterns of Padre Island, Texas. D, 1980, University of Colorado. 223 p.

Garza, Sergio. Aquifer characteristics from well-field production records, Edwards Limestone, San Antonio, Texas. M, 1968, University of Arizona.

Gascoyne, Melvyn. Pleistocene climates determined from stable isotope and geochronologic studies of speleothem. D, 1979, McMaster University. 467 p.

Gaskarth, Joseph William. Petrogenesis of Precambrian rocks in the Hanson Lake area, east-central Saskatchewan (Canada). D, 1967, University of Saskatchewan. 464 p.

Gaskell, Barbara Ann. Paleoecology of the Eocene Wheelock Member of the Cook Mountain Formation, in western Houston County, Texas. M, 1988, University of Texas, Austin. 130 p.

Gaskill, Charles H., Jr. Planktonic foraminifera from the Niobrara Formation, Fort Randall Dam core, Fort Randall Dam, South Dakota. M, 1977, University of Wyoming. 75 p.

Gaskill, Daniel Wills. Climatic variability around the Great Lakes specific to the problem of ice forecasting. D, 1982, University of Michigan. 245 p.

Gaskill, David L. Geology of the White Rock Mountain area, Gunnison County, Colorado. M, 1956, University of New Mexico. 175 p.

Gaskill, James R. Geological engineering as applied to highway construction. M, 1962, Washington University. 30 p.

Gasparik, Tibor. Thermodynamic properties of pyroxenes in the NCMAS system saturated with silica. D, 1981, SUNY at Stony Brook. 146 p.

Gasparis, Aurelio Alfonso Amedeo De see De Gasparis, Aurelio Alfonso Amedeo

Gasparis, Silvana de see de Gasparis, Silvana

Gass, Harold. A review of the Paleozoic fish of Arizona. M, 1963, University of Arizona.

Gass, Nicholas James. Pegmatites of the Winnipeg River area, Manitoba. M, 1957, Dalhousie University. 69 p.

Gass, T. E. Subsurface geology of the Santa Cruz well field, Pima County, Arizona. M, 1977, University of Arizona. 63 p.

Gassaway, J. S. A reconnaissance study of Cenozoic geology in West-central Arizona. M, 1977, San Diego State University.

Gassaway, John Duncan. Boron content as a criterion for a marine depositional environment. M, 1961, George Washington University.

Gassaway, John Duncan. The mineralogy and geochemistry of the sediments of the Straits of Florida. D, 1969, George Washington University. 115 p.

Gassaway, Mack A., III. Subsurface expression of the Nemaha Anticline in southeastern Riley County and northwestern Wabaunsee County, Kansas. M, 1959, Kansas State University. 74 p.

Gasser, Michael M. The geology of the southeast portion of the Deadman Mountain Quadrangle, Black Hills, South Dakota. M, 1981, South Dakota School of Mines & Technology. 89 p.

Gassett, Roger. Seismic refraction study at Dome C, Antarctica. M, 1982, University of Wisconsin-Madison. 136 p.

Gast, Paul W. Abundance of Sr[87] during geologic time. M, 1955, Columbia University, Teachers College.

Gast, Paul Werner. Absolute age determinations from early Precambrian rocks (southeastern Manitoba, Montana and Wyoming). D, 1959, Columbia University, Teachers College. 14 p.

Gastaldo, Robert A. Studies on a Middle Pennsylvanian compression-impression flora from the overburden of the Herrin (No. 6) Coal at Carterville, Illinois. D, 1978, Southern Illinois University, Carbondale. 193 p.

Gasteiger, Carla Maria. Strain analysis of a low amplitude fold in north-central Oklahoma using calcite twin lamellae. M, 1980, University of Oklahoma. 90 p.

Gastil, Russel Gordon. The geology of the eastern half of the Diamond Butte Quadrangle, Gila County, Arizona. D, 1954, University of California, Berkeley. 158 p.

Gaston, Andy. Atokan stratigraphy of the eastern Arkoma Basin. M, 1985, University of Arkansas, Fayetteville.

Gaston, L. R. Biostratigraphy of the type Yamhill Formation, Polk County, Oregon. M, 1974, Portland State University. 139 p.

Gaston, Lewis Andrew. Effects of sulfuric and nitric acids on cation leaching from an Ultisol. D, 1987, University of Florida. 162 p.

Gaston, Wilbert P. Paleohydrologic analysis of late Pleistocene fluvial sediments, Brazoria and Galveston counties, Texas. M, 1979, University of Houston.

Gastreich, K. D. A geohydrologic study of the St. Francois County regional sanitary landfill. M, 1974, University of Missouri, Rolla.

Gastrich, Mary Downes. The ecology of planktonic foraminifera and their symbiotic algae. D, 1986, Rutgers, The State University, New Brunswick. 247 p.

Gatchell, John H. The conodonts of the lower Joins (Middle Ordovician) of Oklahoma. M, 1948, University of Missouri, Columbia.

Gatehouse, Colin G. Lower and middle Cambrian trilobites from the Pensacola mountains and Mount Spann, Antarctica. M, 1969, SUNY at Stony Brook.

Gatenby, Glen Michael. Subsurface fluid migrations in the Lake Borgne-Valentine area. M, 1979, Louisiana State University.

Gates, Alexander E. The tectonic evolution of the Altavista area, Southwest Virginia Piedmont. D, 1986, Virginia Polytechnic Institute and State University. 292 p.

Gates, Bruce Cameron. Stratigraphic architecture and depositional history of the lower Miocene, Planulina Zone, southern Louisiana. M, 1987, University of Texas, Austin. 88 p.

Gates, Cameron Herschel. Origin and development of flow cleavage in the Martinsburg Formation, Lehigh and Northhampton counties, Pennsylvania. M, 1962, University of Texas, Austin.

Gates, Edward E. The geology of the Carrizalillo Hills, Luna County, New Mexico. M, 1985, University of Texas at El Paso.

Gates, Gary Rickey. Mineral resource inventories for use in economic development planning. D, 1965, Indiana University, Bloomington. 100 p.

Gates, Gary Rickey. Petrologic investigations of Kramer Lake and associated strata at Boron, California. M, 1960, Indiana University, Bloomington. 52 p.

Gates, Joseph Spencer. Hydrology of the Middle Canyon, Oquirrh Mountain, Tooele County, Utah. M, 1960, University of Utah. 63 p.

Gates, Joseph Spencer. Worth of data used in digital-computer models of ground-water basins (south-central Arizona). D, 1972, University of Arizona.

Gates, Olcott. Geology of the west side of the Gore Range near Radium (Grand and Eagle counties), Colorado. M, 1950, University of Colorado.

Gates, Olcott. Tertiary volcanism and brecciation in the Shoshone Range, Colorado. M, 1956, University of Colorado.

Gates, Richard Holt. Inelastic analysis of slopes by the finite element method. D, 1968, University of Illinois, Urbana. 197 p.

Gates, Robert M. The petrogenic significance of perthite. D, 1949, University of Wisconsin-Madison.

Gates, Robert W. Ground-water geology of the Spanish Fork-Springville area. M, 1951, Brigham Young University. 54 p.

Gates, Timothy Kevin. Optimal irrigation and drainage strategies in regions with saline high water tables. D, 1988, University of California, Davis. 183 p.

Gates, Todd M. Improved dating of Canadian Pre-Cambrian dikes and a revised polar wandering curve. D, 1971, Massachusetts Institute of Technology. 252 p.

Gates, William C. B. Source and transport mechanisms of quartzite boulders in the Red Valley area, Black Hills, South Dakota and Wyoming. M, 1985, South Dakota School of Mines & Technology.

Gatien, M. G. A study in the slope water south of Halifax. M, 1975, Dalhousie University.

Gatje, P. H. and Pizinger, D. D. Bottom current measurements in the head of Monterey submarine canyon (California). M, 1965, United States Naval Academy.

Gatlin, Garnett Auman. Calcination rate of limestone as related to textural features. M, 1966, University of Virginia. 56 p.

Gatlin, Leroy. Geology of southcentral portion of the Cushing Quadrangle, Texas. M, 1951, University of Texas, Austin.

Gatto, Henrietta. The effects of various states of stress on the permeability of Berea sandstones. M, 1984, Texas A&M University.

Gatto, Lawrence W. Sediment distribution on the shelf, slope and in two submarine canyons off Gaviota, Santa Barbara County, California. M, 1970, University of Southern California.

Gaucher, Edwin Henri Stanislas. The magnetic anomaly of the magnetic serpentinite at the Montagne du Sorcier, Chibougamau, Prov. Quebec, Canada. D, 1960, Harvard University.

Gaudet, Donald Joseph, Jr. Stratigraphy, petrology, and depositional environment of the Blakely Sandstone, Ouachita Mountains, Montgomery County, Arkansas. M, 1986, University of New Orleans. 171 p.

Gaudet, Philip Arthur, Jr. A development of the theories of salt dome formation with emphasis on the American occurrences. M, 1959, University of Mississippi.

Gaudette, Henri Eugene. Geochemistry of the Twin Sisters ultramafic body, Washington. D, 1963, University of Illinois, Urbana. 104 p.

Gaudette, Henri Eugene. Textural study of the Champaign, Urbana, and West Ridge End moraines in east-central Illinois. M, 1962, University of Illinois, Urbana. 39 p.

Gaudreau, Denise Claire. Late-Quaternary vegetational history of the Northeast; paleoecological implications of topographic patterns in pollen distributions. D, 1986, Yale University. 285 p.

Gaudreau, Roch. Intrusion syn-volcanique et minéralisations aurifères; exemple du Pluton de Mooshla, Canton de Bousquet, Abitibi. M, 1986, Universite Laval. 42 p.

Gauger, David Justin. Microfauna of the Hilliard Formation (Cretaceous) near Evanston (Uinta County), Wyoming. M, 1952, University of Utah. 95 p.

Gaughan, Maryann. The influence of local bedrock on the composition of pre-Woodfordian tills in Huron and Crawford counties, Ohio. M, 1985, University of Akron. 140 p.

Gaughan, Michael K. Breaking waves; a review of theory and measurements. M, 1974, Oregon State University. 145 p.

Gaughan, Michael K. Prediction of breaker type and measurement of surf-bores on an ocean beach. D, 1976, Oregon State University. 81 p.

Gaul, R. F. Central Zeballos Mines. M, 1941, University of British Columbia.

Gaulin, Raymond. Gîtologie de l'or à la Mine Elder, Canton Beauchastel, Abitibi, Québec. M, 1988, Ecole Polytechnique. 181 p.

Gault, Alta Ray. The foraminifera of the Moodys Branch Marl (Mississippi). M, 1937, University of Mississippi.

Gault, H. R. The petrology, structures, and petrofabrics of the Pinckneyville quartz-diorite complex in eastern Alabama. D, 1972, The Johns Hopkins University.

Gault, Hugh Richard. The petrography of the Mansfield Sandstone of Indiana. M, 1938, University of Missouri, Columbia.

Gault, Hugh Richard. The petrography, structures, and petrofabrics of the Pinckneyville quartz-diorite, Alabama. D, 1942, The Johns Hopkins University.

Gaumond, André. Le gite d'or New Pascalis, Canton de Louvicourt, P.Q.; structure, minéralogie et alteration associée aux veines. M, 1986, Ecole Polytechnique. 203 p.

Gauss, Joseph Charles. Areal geology of the Reevesville Quadrangle, southern Illinois. M, 1967, Southern Illinois University, Carbondale. 138 p.

Gausseres, Richard Francis. Generalized time-dependent behavior of clays consolidated under different stress ratios. D, 1988, Illinois Institute of Technology. 222 p.

Gauthier, André. Etude minéralogique, pétrographique et géochimique de la zone à terres rares de la carbonatite de St. Honoré. M, 1979, Universite du Quebec a Chicoutimi. 181 p.

Gauthier, C. Deglaciation d'un secteur des rivieres Chaudiere et Etchemin, Quebec. M, 1975, McGill University. 169 p.

Gauthier, Gilles. Application de la méthode de datation uranium-plomb aux zircons du Massif Duxbury. M, 1981, Universite de Montreal. 149 p.

Gauthier, Jacques Armand. A cladistic analysis of the higher systematic categories of the Diapsida. D, 1984, University of California, Berkeley. 564 p.

Gauthier, Louise. Paléoécologie des algues ordoviciennes et siluriennes de l'Ile d'Anticosti, Québec. M, 1981, Universite de Montreal.

Gauthier, Lysanne. Analyse structurale et stratigraphique de l'anticlinorium d'Aroostook-Perce au nord de Port Daniel. M, 1986, Universite de Montreal.

Gauthier, Marilyn. The stratigraphy, petrology, and depositional environment of the Wildie Member of the Borden Formation (Mississippian) in southeast-central Kentucky. M, 1988, Eastern Kentucky University. 90 p.

Gauthier, Michel. Métallogénie du zinc dans la région de Maniwaki-Gracefield, Québec. D, 1982, Ecole Polytechnique. 210 p.

Gauthier, Michel. Minéralisations zincifères de la région de Maniwaki, Comté Gatineau, Québec. M, 1978, Ecole Polytechnique.

Gauthier, Michel. Minéralisations zincifères de la région de Maniwaki, comté Gatineau, Quebec. M, 1979, Ecole Polytechnique.

Gautie, Stephen C. Geomorphology of the southern Rhode Island shoreline in relation to erosion and accretion characteristics. M, 1977, University of Rhode Island.

Gautier, Donald Lee. Petrology of shallow gas reservoirs in the northern Great Plains; selected examples from the Eagle Sandstone and equivalent rocks. D, 1980, University of Colorado. 268 p.

Gautier, Theodore Gary. Growth, form, and functional morphology of Permian Acanthocladiid Bryozoa from the Glass Mountains, West Texas. D, 1972, University of Kansas. 217 p.

Gautier, Theodore Gary. Taxonomy and morphology of the bryozoan genus Tabulipora. M, 1968, University of Kansas. 86 p.

Gautreaux, John W. A study of a South Louisiana Tuscaloosa gas well, the SLAPCO J. R. Morris Heirs False River Field, West Baton Rouge Parish. M, 1983, University of Southwestern Louisiana. 158 p.

Gavasci, Anna Teresa. Uranium emplacement at Garnet ridge, Arizona. D, 1969, Columbia University. 88 p.

Gavenda, Alan Paul. Paleontology and paleoecology of the Ames Limestone (Conemaugh-Pennsylvania) of Pittsburgh and eastern environment. M, 1966, University of Pittsburgh.

Gavenda, Robert Thomas. A characterization of the soils and landscape at the Manis Mastodon site, Sequim, Washington. M, 1980, Washington State University. 70 p.

Gavett, Kerry Lea. Coal quality and depositional setting of the Mudseam Coal, northeastern Kentucky. M, 1984, University of Kentucky. 91 p.

Gavigan, Catherine Louise. Composition and stratigraphy of the Purismia Formation in the central California Coast Ranges. M, 1984, Stanford University. 122 p.

Gavin, Brian. Soil and stream sediment base and precious metal geochemical exploration, Bearpaw Mountains, Montana. M, 1981, University of Missouri, Rolla. 135 p.

Gavin, William Morris Bauer. A paleoenvironmental reconstruction of the Cretaceous Willow Creek Anticline dinosaur nesting locality; north central Montana. M, 1986, Montana State University. 148 p.

Gavish, Eliezer Kneidel. Progressive diagenesis in Recent, Pleistocene, and Neogene carbonate sediments of the Mediterranean coast of Israel. D, 1968, Rensselaer Polytechnic Institute. 144 p.

Gavlin, Suzanne. Community paleoecology of the Mifflin Submember (Middle Ordovician) in Wisconsin. M, 1976, University of Wisconsin-Madison.

Gawarecki, Stephen Jerome. Geology of the Front Range Foothills in the Palisade Mountain-Masonville area, Larimer County, Colorado. D, 1963, University of Colorado. 226 p.

Gawarecki, Susan L. Geological investigation of the Railroad Ridge diamicton, White Cloud Peaks area, Idaho. M, 1983, Lehigh University.

Gawarecki, Susan L. Neotectonics of the Ras Issaran region, Gulf of Suez, Egypt. D, 1986, University of South Carolina. 271 p.

Gawell, Mark J. Chemical and petrographic variations in the Cerro-Negro-Cerrito Arizona cinder cone chain, Valencia County, New Mexico. M, 1975, Kent State University, Kent. 57 p.

Gawloski, Ted. Stratigraphy and environmental significance of the continental Triassic rocks of Texas. M, 1981, Baylor University. 224 p.

Gawne, Constance Elaine. Faunas and sediments of the Zia Sand, middle Miocene of New Mexico. D, 1973, University of Colorado. 359 p.

Gawthrop, W. H. Seismicity and tectonics of the central California coastal region. M, 1977, University of Colorado.

Gay, Frank Thomas. Hydrology of Furnace Run and Yellow Creek, Summit County, Ohio. M, 1975, Kent State University, Kent. 120 p.

Gay, Michael Charles. Evaluation of a ground transient electromagnetic remote sensing method for the deep detection and monitoring of salt-water interfaces. M, 1983, University of South Florida, Tampa. 148 p.

Gay, Norman Kennedy. Paleoecology of the Yorktown Formation in Edgecombe County, North Carolina. M, 1980, East Carolina University. 141 p.

Gay, Sylvester Parker. Anomalies in magnetic intensity over thin infinite dikes; a research report. M, 1961, Stanford University.

Gay, Thomas Edwards, Jr. Geology of Upper Coffee Creek, Etna Quadrangle, California. M, 1952, University of California, Berkeley. 104 p.

Gayer, Martin Jerome. Quaternary and environmental geology of northeastern Jefferson County, Washington. M, 1977, North Carolina State University. 140 p.

Gayes, Paul Thomas. Buried paleoshorelines in Long Island Sound; evidence for irregularities in the postglacial marine transgression in to Long Island Sound. D, 1987, SUNY at Stony Brook. 210 p.

Gayes, Paul Thomas. Primary consolidation and subsidence in transgressive barrier island systems. M, 1983, Pennsylvania State University, University Park. 126 p.

Gayle, Henry Boyes. Lead-zinc occurrence and distribution in soil in the vicinity of lead deposits of central Texas. M, 1961, University of Texas, Austin.

Gaylor, Robert Marshall. Evaluation of a provenance model for Carboniferous age sedimentary strata of central West Virginia. M, 1980, North Carolina State University. 83 p.

Gaylord, David Russell. Recent eolian activity and paleoclimate fluctuations in the Ferris-Lost Soldier area, south-central Wyoming. D, 1983, University of Wyoming. 303 p.

Gaylord, David Russell. Stratigraphic and sedimentary controls on uranium mineralization in sandstones; with an example from the Highland Mine, Converse County, Wyoming. M, 1981, University of Wyoming. 131 p.

Gaytan Rueda, Jose E. Exploration and development at the La Negra Mine, Maconi, Queretaro, Mexico. M, 1975, University of Arizona.

Gazdar, M. N. Ionic activity relations in the flocculation of saline and sodic soils. M, 1969, University of Hawaii.

Gazdar, Muhammad Nasir. Tertiary and Quaternary drainage of Southern High Plains. D, 1981, Texas Tech University. 110 p.

Gazdik, Gertrude Christie. A reconnaissance geology of the Toklat River area, Mt. McKinley Quadrangle, Alaska. M, 1957, University of Pittsburgh.

Gazi, Md Nazmul Hossain. Development and investigation of a petroleum prediction method from well data. M, 1980, University of Oklahoma. 145 p.

Gazin, C. Lewis. A Miocene mammalian fauna from southeastern Oregon. D, 1930, California Institute of Technology. 111 p.

Gazin, C. Lewis. Geology of the central portion of the Mt. Pinos Quadrangle, Ventura and Kern counties, Southern California. D, 1930, California Institute of Technology. 65 p.

Gazin, C. Lewis. Tertiary mammal-bearing beds in the upper Cuyama drainage basin, California. M, 1928, California Institute of Technology. 50 p.

Gazonas, George Aristotle. An experimental and elastic-plastic finite element analysis of the deformation on Berea Sandstone. M, 1980, Texas A&M University. 73 p.

Gazonas, George Aristotle. The mechanics of a near-surface crack under uniform pressure or shear in transversely isotropic medium; with applications to hydraulic fracture. D, 1985, Texas A&M University. 209 p.

Gazzam, J. P. Treatment of two lead ores containing copper, gold and silver. M, 1884, Washington University.

Gazzier, Conrad A. Holocene stratigraphy of the Bayou Cumbest fluvial system, southeastern Mississippi. M, 1977, University of Mississippi. 71 p.

Ge Mao Chen see Mao Chen Ge

Gealey, William Kelso. Geology of the Healdsburg Quadrangle, Sonoma County, California. D, 1949, Cornell University.

Gealy, Betty Lee. Topography of the Continental Slope in the northwest Gulf of Mexico. D, 1953, Harvard University.

Gealy, John R. Geology of Cape Girardeau and Jonesboro quadrangles, southeastern Missouri. D, 1955, Yale University.

Gealy, Wendell Baum. Determination of the nature of the distribution of grain size in sediments by statistical methods. D, 1934, University of Pittsburgh.

Gealy, William James. Geology of the Antone Peak Quadrangle, southwestern Montana. D, 1953, Harvard University. 143 p.

Gearhart, Harry L. Subsurface geology of northwestern Pawnee County, Oklahoma. M, 1958, University of Oklahoma. 69 p.

Gearing, P. J. Organic carbon stable isotope ratios of continental margin sediments. D, 1975, [University of Texas, Austin]. 165 p.

Geary, Dana Helen. Evolutionary mode in Pleuriocardium (Cretaceous Bivalvia). M, 1981, University of Colorado. 132 p.

Geary, Dana Helen. The evolutionary radiation of melanopsid gastropods in the Pannonian Basin (late Miocene, eastern Europe). D, 1986, Harvard University. 238 p.

Geary, Edward Eugene. Petrological and geochemical documentation of ocean floor metamorphism in the Zambales Ophiolite, Philippines. M, 1982, Cornell University. 87 p.

Geary, Edward Eugene. Tectonic significance of basement complexes and ophiolites in the northern Philippines; results of geological, geochronological and geochemical investigations. D, 1986, Cornell University. 236 p.

Geasan, Dennis L. The geology of a part of the Olinghouse mining district, Washoe County, Nevada. M, 1980, University of Nevada. 118 p.

Gebben, Dennis J. Geology of the central Peloncillo Mountains, the north third of the Pratt Quadrangle, Hidalgo County, New Mexico. M, 1979, Western Michigan University.

Gebel, Dana Carl. Stratigraphy of the Nankoweap Formation, eastern Grand Canyon, Arizona. M, 1978, Northern Arizona University. 129 p.

Gebelein, Conrad Dennis. Sedimentology and ecology of a Recent carbonate facies mosaic, Cape Sable, Florida. D, 1972, Brown University. 244 p.

Geberl, Hilary Ann Plint. Windsor Group (Lower Carboniferous) conodont biostratigraphy, taxonomy, and palaeoecology, les Iles de la Madeleine, Quebec. M, 1982, University of Toronto.

Gebert, James. The metallogeny of Cu-Ni and Zn-Cu-Pb deposits of the Frederickson Lake area, central Labrador Trough. M, 1988, McGill University. 115 p.

Gebhard, Paul. Petrography and sedimentology of Jefferson City Formation, Cole County, Missouri. M, 1973, University of Missouri, Columbia.

Gebhardt, R. C. Hydrothermal alteration of pyroxene. M, 1930, University of Minnesota, Minneapolis. 32 p.

Gebhardt, Robert L. A systematic study of the Pennsylvanian brachiopods from Rainbow Mountain, east central Alaska. M, 1972, University of Alaska, Fairbanks. 88 p.

Gebhardt, Rudolph Carl. The geology and mineral resources of the Quijotoa Mountains. M, 1931, University of Arizona.

Gedde, Roger W. Geophysical investigation of a magnetite deposit, Chester County, Pennsylvania. M, 1965, Pennsylvania State University, University Park. 59 p.

Geddes, A. J. S. Petrochemical studies of the Thorr Pluton, Donegal. D, 1979, Northwestern University. 142 p.

Geddes, Arthur. Distribution and migration of zirconium in metasomatized amphibolites (North Twin Island, Pelham Bay Park, The Bronx, New York). M, 1970, Brooklyn College (CUNY).

Geddes, Francis N. The Spergen Formation in St. Louis and Jefferson counties, Missouri. M, 1928, Washington University. 121 p.

Geddes, Richard W. Geochemical prospecting for mercury in the Terlingua, Texas, mining district. D, 1969, Texas Tech University. 92 p.

Geddes, Richard W. Structural geology of Little San Pasqual Mountain and the adjacent Rio Grande Trough. M, 1963, New Mexico Institute of Mining and Technology. 64 p.

Geddes, Robert Stewart. The Vixen Lake indicator train, northern Saskatchewan. M, 1980, University of Western Ontario. 164 p.

Gedeon, James E. The petrology of Raker Peak, Badger Mountain, and Crescent Crater in Lassen Volcanic National Park, California. M, 1970, Case Western Reserve University.

Gedney, E. K. Geology and mineralogy of the igneous rocks of Catamint Hill, Rhode Island. M, 1928, Brown University.

Gedney, Larry D. A preliminary study of focal mechanisms of small earthquakes in the central Nevada region. M, 1967, University of Nevada. 59 p.

Gee, Carole Terry. Revision of the Early Cretaceous flora from Hope Bay, Antarctica. D, 1987, University of Texas, Austin. 165 p.

Gee, David Easton. Some characteristics of crustal deformation. M, 1949, University of Texas, Austin.

Gee, Herbert Caran. A model study of Cabrillo Beach, Los Angeles, California. M, 1938, University of California, Berkeley. 54 p.

Gee, Kenneth Homer. The system akermanite-gehlenite-anorthite-diopside. D, 1955, Pennsylvania State University, University Park. 49 p.

Gee, Lauren Louise. Experimental petrology of melilitites-nephelinites. M, 1988, Purdue University. 56 p.

Gee, Wing Lin. Lower Eocene foraminifera from the Bashi Marl Member of the Hatchetigbee Formation in eastern Lauderdale County, Mississippi. M, 1960, Mississippi State University. 86 p.

Geehan, Gregory W. Nearshore sand bars in the Gulf of California. D, 1978, University of California, San Diego. 217 p.

Geehan, R. W. A geological map of Montana. M, 1932, University of Minnesota, Minneapolis.

Geen, Alfred Francis. Planktonic foraminiferal thanatocoenoses in two deep-sea piston cores from the southern Indian Ocean. M, 1971, Duke University. 91 p.

Geer, Kristen Anders. Evaluation of stream sediment sample media in geochemical exploration for gold-vein mineralization in an area contaminated by mining activity. M, 1983, Colorado School of Mines. 206 p.

Geer, Kristen Anders. Geochemistry of the stratiform zinc-lead-barite mineralization at the Meggen Mine, Federal Republic of Germany. D, 1988, Pennsylvania State University, University Park. 191 p.

Geer, Lucius C. The theory of incompetent shales; a possible solution to the genesis of regional faulting in the Gulf Coast area of Texas and Louisiana. M, 1957, University of Houston.

Geesaman, Richard Carl. Sedimentary facies of the Carmel Formation, southeastern Utah. M, 1979, Northern Arizona University. 135 p.

Geeslin, Jill H. Petrography of the Aleman Formation, Upper Ordovician Silver City Range, southwestern New Mexico. M, 1980, University of Houston.

Geffert, Michael A. Lithostratigraphic and biostratigraphic relationships within the Duck Creek Limestone, (Lower Cretaceous), in North-central Texas. M, 1980, Stephen F. Austin State University.

Gehlen, William T. The geology and mineralization of the eastern part of the Little Smoky Creek mining district, Camas County, Idaho. M, 1983, University of Idaho. 136 p.

Gehman, Harry Merrill, Jr. Geology of the Notch Peak intrusive, Millard County, Utah. M, 1954, Cornell University.

Gehman, Harry Merrill, Jr. The petrology of the Beaver Bay Complex, Lake County, Minnesota. D, 1957, University of Minnesota, Minneapolis. 92 p.

Gehrels, George Ellery. Geologic and tectonic evolution of Annette, Gravina, Duke, and southern Prince of Wales islands, southeastern Alaska. D, 1986, California Institute of Technology. 441 p.

Gehrels, George Ellery. The geology of the western half of the La Grande Basin, northeastern Oregon. M, 1981, University of Southern California. 97 p.

Gehrig, John Leonard. Middle Pennsylvanian Brachiopoda of New Mexico. M, 1954, University of Wisconsin-Madison.

Gehris, Clarence Winfred. Pollen analysis of the Cranberry Bog Preserve, Tannersville, Monroe County, Pennsylvania. D, 1964, Pennsylvania State University, University Park. 82 p.

Geib, Horace Valentine and Goddard, Ira. Reconnaissance erosion survey of the Brazos River watershed, Texas. D, 1933, Iowa State University of Science and Technology.

Geidel, Gwendelyn. A laboratory study of the effect of Carboniferous shales from the Pocahontus Basin (eastern Kentucky-West Virginia) on acid mine drainage and water quality. M, 1976, University of South Carolina.

Geidel, Gwendelyn. An evaluation of a surface application of limestone for controlling acid mine discharges from abandoned strip mines, Sewellsville, Ohio. D, 1982, University of South Carolina. 207 p.

Geier, Richard J. Glacial stratigraphy and bluff recession along the Lake Erie coast in New York State. M, 1980, SUNY at Buffalo.

Geiger, Beth Carol. Ductile strain in the overlap zone between the Cordilleran thrust belt and the Rocky Mountain Foreland near Melrose, Montana. M, 1986, University of Montana. 47 p.

Geiger, Charles. Thermodynamic mixing properties of almandine garnet solid solutions. D, 1986, University of Chicago. 150 p.

Geiger, Charles Arthur. The crystal structure of cronstedtite-2H₂. M, 1981, University of Wisconsin-Madison.

Geiger, Earl G., Jr. What is the nature of the pegmatite "mother fluid". M, 1968, Rensselaer Polytechnic Institute. 72 p.

Geiger, Fredric J. Geochemistry of the Ladentown, Union Hill, New Germantown and Sand Brook basalts; lithostratigraphic correlations and tectonic implications for the Newark Basin. M, 1985, Rutgers, The State University, Newark. 107 p.

Geiger, Kenneth J. The engineering geology and relative stability of parts of Newport, Bellevue, and Fort Thomas, Kentucky. M, 1983, University of Cincinnati. 66 p.

Geiger, Kenneth Warren. Genetic aspects of the consolidated Denison-Blind River area uranium deposits (Ontario). M, 1959, Cornell University.

Geiger, Kenneth Warren. Reconnaissance geology and mineral deposits of the Wilson Lake-Winokapau Lake area, Labrador. D, 1961, Cornell University. 141 p.

Geijer, Theresa Anna Maria. Provenance of Eocene conglomerates in the Santa Ynez Mountains and their tectonic implications for western Transverse Ranges, California. M, 1986, University of California, Los Angeles. 243 p.

Geil, Donald D. Structure and stratigraphy of Stafford County, Kansas, related to petroleum accumulation. M, 1957, Kansas State University. 67 p.

Geil, Sharon Anne. Significance of the age and identity of a volcanic ash near DeSoto, Kansas, with respect to the enclosing terrace deposits. M, 1987, University of Kansas. 169 p.

Geilissee, P. J. The bedrock geology of the Newton Quadrangle, Massachussetts. M, 1959, Boston College.

Geirsdottir, Aslaug. Sedimentologic analysis of diamictites and implications for late Cenozoic glaciation in western Iceland. D, 1988, University of Colorado. 290 p.

Geis, Harold Lorenz. Some ostracodes from the Salem Limestone, Mississippian of Indiana. M, 1933, University of Illinois, Urbana.

Geiser, Peter A. Deformation of the Bloomsburg Formation (upper Silurian) in the Catoctin Mount Anticline, Hancock, Maryland. D, 1970, The Johns Hopkins University. 223 p.

Geisinger, Karen Leslie. A theoretical and experimental study of bonding in silicates and related materials. D, 1983, Virginia Polytechnic Institute and State University. 275 p.

Geisler, Jean Marie. Studies on the Pennsylvanian of Colorado. M, 1949, University of Illinois, Urbana.

Geisler, Thomas A. A solid state study of the systems CdO-SiO₂, CdO-SiO₂-Al₂O₃, and CdO-Al₂O₃. M, 1965, Miami University (Ohio). 130 p.

Geisse, Elaine. The petrography of the syenites, nepheline syenites and related rocks west of Wausau, Wisconsin. M, 1951, Smith College. 41 p.

Geissler, Edwin L. A heavy mineral investigation of ancestral Arkansas River sediments. M, 1969, Northeast Louisiana University.

Geissman, John William. Paleomagnetism and tectonics of the Yerington (porphyry copper) mining district, Nevada. D, 1980, University of Michigan. 364 p.

Geissman, John William. Paleomagnetism of the Butte District, Montana. M, 1976, University of Michigan.

Geist, Dennis James. Geology and petrology of San Cristobal Island Galapagos Archipelago. D, 1985, University of Oregon. 152 p.

Geist, Eric. The formation and use of synthetic velocity analyses. M, 1984, Stanford University.

Geitgey, Ronald Paul. Mineralogy [barian fergusonite, weinschenkite, and rhabdophane] of a deeply weathered perrierite-bearing pegmatite in Amherst County, Virginia. M, 1967, University of Virginia. 56 p.

Geitzenauer, Kurt R. A study of the chonetid brachiopod Longispina mucronatus (Hall) in the Hamilton Group (Middle Devonian) of western New York. M, 1965, SUNY at Buffalo. 56 p.

Geitzenauer, Kurt R. Nannoplankton of the Subantarctic Pacific Ocean. D, 1970, Florida State University. 123 p.

Gelber, Arthur Winston. The structure and stratigraphy of the Mystery Ridge Formation, west-central Nevada. M, 1985, Rice University. 68 p.

Gelberg, Russ. Geophysical and geological study of the Greeley Arch, Colorado. M, 1986, Purdue University. 80 p.

Geldart, L. P. Periodic variations of the gravitational force. D, 1949, California Institute of Technology. 33 p.

Geldart, Lloyd Philip. A gravity survey in the Monk Hill area. M, 1949, California Institute of Technology. 25 p.

Gelder, Susan M. van see van Gelder, Susan M.

Geldon, Arthur L. Hydrogeology and water resources of the Missoula Basin, Montana. M, 1979, University of Montana. 114 p.

Geldon, Arthur L. Petrology and origin of a lamprophyre pluton, near the Dead River, Saint Louis County, Minnesota. M, 1972, University of Minnesota, Minneapolis. 71 p.

Geldsetzer, Helmut. Cenozoic stratigraphy and structure of the Owyhee Reservoir-Sucker Creek region, east-central Oregon. M, 1966, University of Washington. 93 p.

Geldsetzer, Helmut. Tectonically controlled sedimentation during the Middle Paleozoic of northeastern North America. D, 1971, Queen's University. 274 p.

Gelfenbaum, Guy Richard. Mechanics of steady turbidity currents. D, 1988, University of Washington. 137 p.

Gélinas, Léopold. Géologie de la région de Fort Chimo et des lacs Gabriel et Thevenet, Nouveau Québec. D, 1966, Universite Laval. 212 p.

Gélinas, Léopold. The nodular aplites near Fort Chimo. M, 1956, Universite Laval. 97 p.

Gelinas, Robert L. Mineral alterations as a guide to the age of sediments vented by prehistoric earthquakes in the vicinity of Charleston, South Carolina. M, 1986, University of North Carolina, Chapel Hill. 304 p.

Gelinas, S. Nodular aplites near Fort Chimo, Quebec. M, 1956, Universite Laval.

Gelineau, William J. The Pleistocene geology of the Inver Grove and St. Paul SW quadrangles, Minnesota. M, 1959, University of Minnesota, Minneapolis. 77 p.

Gell, James Walter. Geochemistry of the lower Keweenawan Powder Mill Group, Upper Michigan. M, 1987, Michigan State University. 51 p.

Gell, W. A. Underground ice in permafrost, Mackenzie Delta-Tuktoyaktuk Peninsula, N.W.T. D, 1976, University of British Columbia.

Geller, Bruce Alan. Bulk chemistry studies of Recent sediments in the western equatorial Atlantic. M, 1981, SUNY at Binghamton. 106 p.

Geller, Kris L. Stratigraphic relationships between the Devonian-Mississippian black-shale sequence in the Appalachian and Illinois basins. M, 1985, University of Kentucky. 165 p.

Geller, Robert James. Part I, Earthquake source models, magnitudes and scaling relations; Part II, Amplitudes of rotationally split normal modes for the 1960 Chilean and 1964 Alaskan earthquakes. D, 1977, California Institute of Technology. 218 p.

Gellis, Allen C. Decreasing sediment and salt loads in the Colorado River basin; a response to arroyo evolution. M, 1988, Colorado State University. 185 p.

Gelnett, Ronald H. Geology of the southern part of Wellsville Mountain, Wasatch Range, Utah. M, 1958, Utah State University. 72 p.

Gelphman, Norman Ray. West Sentinel oil field, Washita County, Oklahoma. M, 1959, University of Oklahoma. 94 p.

Gemmell, D. E. Carboniferous volcanic and sedimentary rocks of the Mount Pleasant Caldera and Hoyt Appendage, New Brunswick. M, 1975, University of New Brunswick.

Gemmell, John Bruce. Metallic trace element geochemistry of volcanic gas from selected Central American and Japanese volcanoes. M, 1982, Dartmouth College. 229 p.

Gemmell, John Bruce. The Santo Nino Ag-Pb-Zn vein, Fresnillo District, Mexico; geology, sulphide and sulphosalt mineralogy, and geochemistry. D, 1987, Dartmouth College. 244 p.

Gemperle, Richard J. Analysis of ground water resource conditions and management alternatives for the Bruneau-Grand View area, southwestern Idaho. M, 1988, University of Idaho. 120 p.

Gendi, Mohamed H. Statistical correlation by graphic method of trace metal distribution in rock types in the Black Hills, South Dakota. M, 1972, South Dakota School of Mines & Technology.

Gendzwill, Don John. A gravity study in the Amisk lake area, Saskatchewan (Canada). D, 1969, University of Saskatchewan. 268 p.

Genes, Andrew Nicholas. Glacial geology of Stord Island, Norway. D, 1973, Syracuse University.

Genik, Gerard Julian. A regional study of the Winnipeg Formation (middle or upper Ordovician) (Manitoba). M, 1963, University of Manitoba.

Gennett, Judith Ann. Palynology and paleoecology of sediments from Blacktail Pond, northern Yellowstone Park, Wyoming. M, 1977, University of Iowa. 74 p.

Geno, Kirk R. Coated grains in Recent fluvial carbonate sediments of central Texas. M, 1980, University of Houston.

Genrich, Donald Allen. Isolation and characterization of sand-, silt-, and clay-size fractions of soils. D, 1972, Iowa State University of Science and Technology.

Gensamer, Alan Richard. Sedimentation and stratabound copper of the lower Miller Peak Formation. M, 1973, University of Montana. 84 p.

Gensmer, Richard P. X-ray diffractometry and its use in the study of the Flagstaff Formation, central Utah. M, 1977, Northern Illinois University. 96 p.

Gent, James Albert, Jr. The impact of intensive forest management practices on the physical properties of lower coastal plain and Piedmont soils. D, 1982, North Carolina State University. 53 p.

Gent, Malcolm Richard. The interpretation of satellite images and airphotos for reconnaissance groundwater exploration in coastal Peru. M, 1981, McGill University. 108 p.

Gentet, Robert Eugene. Depositional model of conglomeratic channel facies in the lower Reagan Formation, Wichita Mountains, Oklahoma. M, 1982, Wichita State University. 54 p.

Gentile, Anthony L. Investigation of phase relations in the high alumina portion of the system lime-alumina-silica. D, 1960, Ohio State University. 76 p.

Gentile, Anthony L. Some features of rhyolite petrogenesis. M, 1957, New Mexico Institute of Mining and Technology. 55 p.

Gentile, Francesco. Etude stratigraphique et structurale du Dôme Lemieux, Comté de Gaspé nord, Québec (Paléozoic inférieur, Canada). M, 1970, Universite de Montreal.

Gentile, Francesco. Nature et origine de la minéralisation cupro-zincifère de la formation des schistes de Weedon (Cambrian-Ordovician), Québec. D, 1973, Ecole Polytechnique.

Gentile, Francesco. Nature et origine de la minéralisation cuprozincifère de la formation des schistes de Weedon, Québec. M, 1972, Universite de Montreal.

Gentile, Fransesco. La métallogenie du cuivre dans la région de Disraéli, comté Wolfe, Québec. D, 1973, Ecole Polytechnique.

Gentile, John Richard. The delineation of landslides in the Lincoln County, Oregon, coastal zone. M, 1978, University of Oregon. 24 p.

Gentile, Leo Frederick. Sedimentology and graptolite biostratigraphy of the Viola Group (Ordovician), Arbuckle Mountains and Criner Hills, Oklahoma. M, 1984, Oklahoma State University. 104 p.

Gentile, Richard J. Probable cause for the variation in thickness of the gypsum deposits in the vicinity of Lander, Wyoming. M, 1958, University of Missouri, Columbia.

Gentile, Richard Joseph. Stratigraphy, sedimentation and structure of the upper Cherokee and lower Marmaton (Pennsylvanian) rocks of Bates County, and portions of Henry and Vernon counties, Missouri. D, 1965, University of Missouri, Rolla. 288 p.

Gentry, Dianna J. Solution cleavage in the Twin Creek Formation and its relationships to thrust fault motions in the Idaho-Utah-Wyoming thrust belt. M, 1983, University of Wyoming. 51 p.

Gentry, Donald W. The determination of residual stresses in the vicinity of the 755 Breccia Pipe at Cananea, Sonora, Mexico. D, 1972, University of Arizona.

Gentry, Donald William. Scheduling production from underground mines by linear programming. M, 1967, University of Nevada. 26 p.

Gentry, Herman Raymond. Geomorphology of some selected soil-landscapes in Whitman County, Washington. M, 1974, Washington State University. 130 p.

Gentry, Robert W. The effect of reservoir and fluid properties on production decline curves. M, 1974, University of Oklahoma. 135 p.

Gentry, Stephen Swift. Trace metal distribution in soil overlying sphalerite-bearing carbonate rock. M, 1973, University of Tennessee, Knoxville. 72 p.

Gentzis, Thomas. Organic petrology and depositional environment of the Hat Creek No. 2 Coal Deposit, British Columbia, Canada. M, 1985, University of Alberta. 151 p.

Geoltrain, Sebastien. Induction lateral sounding. M, 1986, Colorado School of Mines. 157 p.

George, Clement Enos, III. The geology of the Boracho area, Culberson and Jeff Davis counties, Texas. M, 1948, University of Texas, Austin.

George, D'arcy Roscoe. Soapstone deposits in Wake County, North Carolina. M, 1939, North Carolina State University. 28 p.

George, Daniel T. Relationships between the physical and chemical properties of crude oils from the Gulf Coast of the United States and the depositional environments of the rocks from which they were produced. M, 1978, Bowling Green State University. 153 p.

George, Gary D. Investigation of spatial and temporal migration of seismic activity in the California/Nevada area. M, 1974, University of Wisconsin-Milwaukee.

George, Gene Richard. Stratigraphy of part of the Crow Indian Reservation, Big Horn County, Montana. M, 1967, Oregon State University. 151 p.

George, Hubert. Late Quaternary history and engineering geology of the Elk River valley, southeastern British Columbia. D, 1984, Queen's University. 253 p.

George, John H. A subsurface study of Pennsylvanian rocks of the South Wetumka area, northeastern Hughes County, Oklahoma. M, 1960, University of Oklahoma. 38 p.

George, John Louis. Glacial water levels in the Nashua River valley, Massachusetts. M, 1956, Clark University.

George, John Samuel. Mineralogic and trace element variation in the carbonate rocks (Middle and Upper Ordovician) associated with faults in central Kentucky. M, 1983, University of Kentucky. 124 p.

George, Joseph Peter, Jr. A study of the emplacement and mineralization of the ferruginous Upper Breccia on Pilot Knob, Iron County, southeastern Missouri. M, 1983, Southern Illinois University, Carbondale. 67 p.

George, L. The influence of constitutive relations on the bending of a multilayer. M, 1978, Texas A&M University.

George, Larry. An investigation of the effects of material response on the analytical solution of a geologic deformation. M, 1978, Texas A&M University.

George, Lawrence. Quartz grain-size and grain shape variation in the surface sediments of Galveston Island, Texas. M, 1986, Texas A&M University.

George, M. Carbonate equilibrium in the Hosston Formation, central Mississippi. M, 1977, University of Missouri, Columbia.

George, Peter G. A paleoenvironmental study of the Lower Mississippian Caballero Formation and Andrecito Member of the Lake Valley Formation in the south-central Sacramento Mountains, Otero County, New Mexico. M, 1985, Texas A&M University. 240 p.

George, R. P., Jr. The internal structure of the Troodos ultramafic complex, Cyprus. D, 1975, SUNY at Stony Brook. 265 p.

George, Steven E. Structural geology the Pavant Mountain front in the Fillmore and Kanosh quadrangles, Millard County, Utah. M, 1985, Brigham Young University. 60 p.

George, Thomas H. Verification of polygonal resource maps using aerial observations. M, 1985, University of Alaska, Fairbanks. 157 p.

George, William Owsley. The relation of the physical properties of the natural glasses to their chemical composition. M, 1920, University of Minnesota, Minneapolis. 52 p.

Georgens, Robert E. Mineralogy of the Gardiners Clay. M, 1979, New York University.

Georges, Danae. A study of waste fluid injection on the Texas Gulf Coast. D, 1978, Rice University. 156 p.

Georgesen, Neils Christian. The stratigraphy of the Colorado Group of northeastern Nebraska and adjacent areas. M, 1931, University of Iowa. 143 p.

Georgi, Daniel T. The spherical harmonic analysis of paleomagnetic inclination data. M, 1973, Columbia University. 41 p.

Georgiou, John C. Gravity anomalies and crustal mass concentrations in north-central Illinois. M, 1970, Northern Illinois University. 31 p.

Gephart, Carol J. Relative importance of iron-oxide, manganese-oxide, and organic material on the adsorption of chromium in natural water sediment systems. M, 1982, Michigan State University. 125 p.

Gephart, Gregory David. Differentiation and correlation of till sheets using 7Å/10Å X-ray diffraction peak height ratios, Allegan County, Michigan. M, 1982, Michigan State University. 71 p.

Gephart, John Wesley. Studies of stress and deformation in the Earth's crust; I, Determining the state of stress in the Earth from earthquake focal mechanism data, II, Deformation around the Creede Caldera, San Juan volcanic field, Southwest Colorado; implications for caldera mechanics. D, 1986, Brown University. 107 p.

Gephart, John Wesley. The structural geology of the lower North Inlet area, Rocky Mountain National Park, Colorado; an evaluation of the Precambrian F_4 deformation. M, 1981, University of Colorado. 120 p.

Gephart, Roy E. An analysis of the groundwater resources available to the city of New Carlisle, Ohio. M, 1974, Wright State University. 111 p.

Geppert, Timothy J. Small mammals of Shield Trap, East Pryor Mountain, Montana. M, 1984, University of Iowa. 45 p.

Gera, A. V. Geophysical study of alpine meadows in San Diego County. M, 1977, San Diego State University.

Geraghty, E. P. Stratigraphy, structure, and petrology of part of the North Creek 15′ Quadrangle, southeastern Adirondack Mountains, New York. M, 1973, Syracuse University.

Geraghty, James J. Geology of the basal gravel zone of the Magothy Formation in western Long Island, New York. M, 1953, New York University.

Gerald, Rosemary Elaine. An ultrasonic determination of the elastic properties of single-crystal manganosite, MnO. M, 1986, Pennsylvania State University, University Park. 86 p.

Geralnick, Alan. Heavy metals distribution in Long Island Sound bottom sediments. M, 1980, Brooklyn College (CUNY).

Gerami, Abbas. Hydrogeology of St. Marks River basin. M, 1984, Florida State University.

Gerard, Matthew G. The effect of external boundaries on a reservoir with a pinchout. M, 1982, Stanford University.

Gerardin, V. An integrated approach to the determination of ecological groups in vegetation studies. D, 1977, University of Connecticut. 236 p.

Gerasimoff, Michael D. The Hobson Lake Pluton, Cariboo Mountains, British Columbia, and its significance to Mesozoic and early Cenozoic Cordilleran tectonics. M, 1988, Queen's University. 196 p.

Gerber, Murry S. Carbonate microfacies of the Burlington crinoidal limestone (Middle Mississippian), western Illinois, southeastern Iowa, and northeastern Missouri. M, 1978, University of Illinois, Urbana. 78 p.

Gercek, Hasan. Stability of intersections in room-and-pillar coal mining. D, 1982, Pennsylvania State University, University Park. 204 p.

Gerdin, R. B. Application of remote sensing to managing the Earth's environment. D, 1976, University of California, Los Angeles. 335 p.

Gere, Milton A., Jr. A study of samples from the Knowlton Amygdaloid, Caledonia Mine, Ontonagon County, Michigan. M, 1970, Michigan State University. 92 p.

Gereby, Clarissa H. Organic matter decomposition in polluted and unpolluted Lake Erie sediments. M, 1986, Case Western Reserve University. 168 p.

Gerencher, Joseph James. Structural relationships, petrography, chemistry, and magnetic properties of some dike swarms in and around the Mutton Bay pluton, Quebec (Canada). M, 1968, Pennsylvania State University, University Park. 131 p.

Gerencher, Joseph James, Jr. Multivariate study of the interrelationships among selected variables of the organic fraction of samples of United States' coals. D, 1982, Pennsylvania State University, University Park. 641 p.

Gerety, Kathleen M. Quantitative model of point-bar deposition applied to Rainbow Beach, Northampton, Massachusetts. M, 1979, SUNY at Binghamton. 159 p.

Gerety, Kathleen Mary. A wind-tunnel study of the saltation of heterogeneous (size, density) sands. D, 1984, Pennsylvania State University, University Park. 189 p.

Gerety, Michael Thomas. Bipole-dipole electrical technique applied to geothermal exploration in New Mexico. M, 1980, University of New Mexico. 86 p.

Gergel, Thomas Joseph. Morphometric analysis of drainage basin characteristics on the Georgia Piedmont. M, 1964, University of Georgia. 178 p.

Gergel, Thomas Joseph. The regionalization of tidal marshes along the eastern coast of the United States. D, 1969, University of Georgia. 336 p.

Gergen, Leslie Dickson. Petrology and provenance of the Deep Sea Drilling Project sand and sandstone from the north and northeastern Pacific margins. M, 1985, University of California, Los Angeles. 150 p.

Gerhard, F. Bruce, Jr. The crystal structure of the mineral pachnolite. D, 1966, Rensselaer Polytechnic Institute. 229 p.

Gerhard, Jacob Esterly. Paleocene miospores from the Slim Buttes area, Harding County, South Dakota. M, 1958, Pennsylvania State University, University Park. 171 p.

Gerhard, Lee C. Geology of the lower Phantom Canyon area, Fremont County, Colorado. M, 1961, University of Kansas. 33 p.

Gerhard, Lee C. Paleozoic paleogeology of the Canon City Embayment, Colorado. D, 1964, University of Kansas. 69 p.

Gerhard, Roberta G. Trend-surface analysis of the structural development of the Pratt Anticline (Kansas). M, 1965, University of Kansas. 42 p.

Gerhardt, Daniel Joseph. The anatomy and history of a Pleistocene strand plain deposit, Grand Bahama Island, Bahamas. M, 1983, University of Miami. 170 p.

Gerhardt, Roger A. Hydrogeology of three solid waste disposal sites in the Iowa River floodplain at Iowa City, Iowa. M, 1974, University of Iowa. 235 p.

Gerhart, James M. Digital simulation of the yield potential of the Elliot Park-Burgoon Aquifer in eastern Clearfield and western Centre counties, Pennsylvania. M, 1977, Pennsylvania State University, University Park. 142 p.

Gerken, Antony N. The type Niobrara Formation (late Cretaceous) in northeastern Nebraska. M, 1971, University of Nebraska, Lincoln.

Gerla, Philip J. The geology of the Middle Road Unit, Union Complex, Union, Knox County, Maine. M, 1977, University of New Hampshire. 114 p.

Gerla, Philip Joseph. Structure and hydrothermal alteration of the Diamond Joe Stock, Mohave County, Arizona. D, 1983, University of Arizona. 134 p.

Gerlach, David C. Petrology and geochemistry of plagiogranite and related basic rocks of the Canyon Mountain ophiolite complex, Oregon. M, 1980, Rice University. 203 p.

Gerlach, David Christian. Geochemistry and petrology of Recent volcanics of the Puyehue-Cordon Caulle area, Chile (40.5°S). D, 1985, Massachusetts Institute of Technology. 400 p.

Gerlach, George Smith. A regional gravity survey of the St. Francois Mountains of Missouri. M, 1959, Washington University. 52 p.

Gerlach, Paul Joseph. The ground-water hydrology of mine spoils at two coal strip mines in western Sheridan County, Wyoming. M, 1976, South Dakota School of Mines & Technology.

Gerlach, Terrence M. The magnetic and textural character of the Macauley Granite (Precambrian, east of Mount Wisconsin, Northeast Wisconsin). M, 1967, University of Wisconsin-Madison.

Gerlach, Terrence Melvin. The C-O-H-S gas system and its applications to terrestrial and extraterrestrial volcanism. D, 1974, University of Arizona.

Gerlitz, Carol Nan. Chemical interaction between major dissolved components in acidic uranium tailings fluids and adjacent bedrock. M, 1982, University of Colorado. 141 p.

Gerlock, Jeffrey Lee. Sedimentary petrology of the Mesaverde Formation (Upper Cretaceous), east flank, Bighorn Mountains, Wyoming. M, 1986, Memphis State University. 112 p.

Germain, Louis Charles St. *see* St. Germain, Louis Charles

Germain, M. S. D. Contaminant migration in fractured porous media; modelling and analysis of advective-diffusive interaction. D, 1988, University of Waterloo. 166 p.

Germain, M. S. D. Quasi-stable concentration distributions in saturated porous media with a constant solute source. M, 1981, University of Waterloo.

German, Kenneth E., Jr. Secondary uranium enrichment of the Precambrian basement rock of Nebraska. M, 1982, University of Nebraska, Lincoln. 141 p.

German, Rebecca. A multivariate morphometric study of prosimian forelimb and hindlimb. M, 1979, University of Rochester.

German, Robert Allen. A gravity and magnetic investigation of New York-Alabama Lineament. M, 1985, Purdue University. 139 p.

Germano, Richard Joseph. Polyhedral distortions in disordered olivine; a geometric analysis. M, 1978, University of Minnesota, Minneapolis. 64 p.

Germanoski, Dru. Tributary entrenchment and knickpoint development in response to local base level lowering below a dam, Osage River, Missouri. M, 1984, Southern Illinois University, Carbondale. 180 p.

Germeroth, Robert. The Miocene foraminifera of a deep test hole in Dover Air Force Base, Dover, Delaware. M, 1957, New York University.

Germiat, Steven John. An assessment of future coastal land loss in Galveston, Chambers, and Jefferson counties, Texas. M, 1988, University of Texas, Austin. 190 p.

Germinario, Mark Philip. The depositional and tectonic environments of the Julian Schist, Julian, California. M, 1982, San Diego State University. 95 p.

Germine, Mark. Asbestiform serpentine and amphibole group minerals in the northern New Jersey area. M, 1981, Rutgers, The State University, Newark. 239 p.

Germundson, Robert Kenneth. Stratigraphy and micropaleontology of some late Cretaceous-Pal continental formations, Western Interior, North America. D, 1965, University of Missouri, Rolla. 224 p.

Germundson, Robert Kenneth. Wapiabi biostratigraphy. M, 1960, University of Alberta. 92 p.

Gernant, Robert Everett. Paleoecology of the Miocene Choptank Formation of Maryland and Virginia. D, 1969, University of Michigan. 570 p.

Gernant, Robert Everett. Paleoecology of the Oligocene middle Frio Formation (Texas). M, 1965, University of Michigan. 570 p.

Geronsin, Rolin Lee. Chemical relationship of the mississippi-valley type ore deposits in Missouri, Oklahoma, and Kansas. M, 1980, University of Missouri, Rolla.

Gerrard, Thomas A. A petrographic study of the Dayton Formation (Niagaran Series), Harrison and Twin townships, Preble County, Ohio. M, 1959, Miami University (Ohio). 77 p.

Gerrard, Thomas A. Environment studies of the Fort Apache Member, Supai Formation (Permian), east-central Arizona. D, 1964, University of Arizona.

Gerrie, William. Molybdenite veins of Lacorne and Malartic townships, Abitibi, Quebec. M, 1927, University of Toronto.

Gerritsen, Steven Scott. Structural analysis of the Silurian-Devonian cover in the Smoke Holes, WV. M, 1988, West Virginia University. 197 p.

Gerry, David L. An engineering geology feasibility study of the Lefthand Creek dam and reservoir site, Boulder County, Colorado. M, 1975, Colorado State University. 185 p.

Gerryts, Egbert. The geology of the Premier (Transvaal) diamond mine (South Africa). M, 1949, McGill University.

Gerryts, Egbert. The petrology of the kimberlites at the Premier (Transvaal) diamond mine, South Africa. D, 1951, McGill University.

Gersic, Joseph. A limited structural and stratigraphic interpretation of Red River Formation, South Dakota. M, 1973, South Dakota School of Mines & Technology.

Gerson, H. S. Investigation of lead-zinc replacements in limestones where there is no apparent connection with igneous activity. M, 1929, McGill University.

Gerstner, Michael Roy. A fluid inclusion and petrologic study of the Mactung scheelite skarn deposit, Yukon-Northwest Territories, Canada. M, 1987, University of Utah. 75 p.

Gertje, Henry. A microstructural study of the eastern Devonian gas shale; P. D. McCartney #1 gas well, Mahoning County, Ohio. M, 1987, Michigan Technological University. 106 p.

Gertman, Richard Leo. Cenozoic Typhinae (Mollusca: Gastropoda) of the western Atlantic region. M, 1968, Tulane University. 96 p.

Gertson, Rodney Curtis. Interpretation of a seismic refraction profile across the Roosevelt Hot Springs, Utah and vicinity. M, 1979, University of Utah. 116 p.

Gerwe, Jeffrey E. Ag-Ni-Co-As-U mineralization in the Black Hawk mining district, Grant County, New Mexico. M, 1986, New Mexico Institute of Mining and Technology. 85 p.

Gerwels, Richard P. A study of the Golconda lead mines, Shoshone County, Idaho. M, 1951, University of Utah. 56 p.

Gesch, D. An analysis of the utility of Landsat thematic mapper data and digital elevation model data for predicting soil erosion; east form Massac Creek watershed, Kentucky. M, 1984, Murray State University. 56 p.

Gesink, Joel A. Fluid inclusion evidence for multi-solutional depositional processes at the Sweetwater and Eve Mills mississippi valley-type districts, Tennessee. M, 1986, University of Michigan. 23 p.

Gesink, Marc L. Preliminary hydrogeologic investigations of the White River alluvial aquifer, Rio Blanco County, Colorado. M, 1983, Colorado School of Mines. 212 p.

Geslin, Jeffrey K. The Permian Dollarhide Formation and Paleozoic Carrietown Sequence in the SW 1/4 of the Buttercup Mountain Quadrangle, Blaine and Camas counties, Idaho. M, 1986, Idaho State University. 116 p.

Gest, Donald Evan. Preliminary petrogenetic study of absarokites, shoshonites, and associated flows, Absaroka Mountains, Wyoming. M, 1977, University of Oregon. 94 p.

Gester, Kenneth Clark. Evidence for a Paleozoic submarine fan, Shoo Fly Complex, northern Sierra Nevada, California. M, 1987, San Diego State University. 232 p.

Gesumaria, Robert Hugh. Industrial wastewater sludge disposal on agricultural soils of Northwest New Jersey. D, 1981, Rutgers, The State University, New Brunswick. 651 p.

Getsinger, Jennifer Suzanne. A structural and petrologic study of the Chiwaukum Schist on Nason Ridge, northeast of Stevens Pass, North Cascades, Washington. M, 1978, University of Washington. 151 p.

Getsinger, Jennifer Suzanne. Geology of the Three Ladies Mountain/Mount Stevenson area, Quesnel Highland, British Columbia. D, 1985, University of British Columbia. 166 p.

Getting, Ivan Craig. Determination of the pressure of the barium I-II transition with single-stage piston-cylinder apparatus; melting of silicates at high pressure. M, 1967, University of California, Los Angeles. 121 p.

Gettings, Mark Edward. Some thermal models of the Skaergaard Intrusion. D, 1976, University of Oregon. 171 p.

Gettrust, J. F. Robust signal averaging. D, 1974, University of Wisconsin-Madison. 100 p.

Getts, T. R. Gravity and tectonics of the Peru-Chile Trench and eastern Nazca Plate 0°-33°30′S. M, 1975, University of Hawaii at Manoa. 104 p.

Getty, Theodore Alexander. Jurassic and basal Cretaceous ammonites from the Kemaboe Valley, West Irian (western New Guinea). M, 1967, McMaster University. 111 p.

Gettys, William R. Dynamics of oyster populations from Hobcaw Barony, Georgetown County, South Carolina. M, 1972, University of South Carolina. 50 p.

Getz, Albert Julius. The origin and occurrence of the brown hematite ores in Lehigh County. M, 1939, Lehigh University.

Getz, Boyd Steven. Benthic foraminiferal biostratigraphy and paleoecology of the lower Luisian Leisure World locality, Orange County, California. M, 1982, University of California, Los Angeles. 133 p.

Getz, Roger C. Jointing and stratigraphy on Elkhorn Peak, Whitewood, South Dakota and Green Mountain, Sundance, Wyoming. M, 1966, South Dakota School of Mines & Technology.

Getzen, Rufus D., Jr. Cretaceous stratigraphy of the upper coastal plain in central South Carolina. M, 1969, University of South Carolina.

Getzen, Rufus T. The Long Island ground-water reservoir; a case study in anisotropic flow. D, 1974, University of Illinois, Urbana. 154 p.

Geurin, Stanley Paul. Subsurface geology of the Frederick area, Tillman County, Oklahoma. M, 1986, University of Oklahoma. 64 p.

Gevaert, D. M. Location and evaluation of induced infiltration sites near the Grand River in the Kitchener-Waterloo area. M, 1979, University of Waterloo.

Gevirtz, Joel L. Nature and origin of deep-water carbonate sediments of the Red Sea. M, 1965, Rensselaer Polytechnic Institute. 52 p.

Gevirtz, Joel L. Paraecology of benthonic foraminifera and associated microorganisms of the continental shelf off Long Island, New York. D, 1969, Rensselaer Polytechnic Institute. 106 p.

Gevirtzman, Debra Ann. Paleoenvironments of an earliest Cambrian (Tommotian) shelly fauna in Esmeralda County, Nevada. M, 1983, University of California, Davis. 137 p.

Gevrak, Ihsan. Clay mineralogy and sedimentary petrography of lower to middle Paleozoic rocks from a single core from Northwest Georgia. M, 1978, Georgia Institute of Technology. 78 p.

Geyer, Alan Raymond. Geology of the vicinity of the Hershey Mine, Hershey, Pennsylvania. M, 1956, University of Michigan.

Geyer, Richard Adam. Interpretation of the submarine topography of the northwestern Gulf of Mexico. D, 1951, Princeton University. 64 p.

Geyer, Richard G. The effect of subsurface geologic structure in electromagnetic induction prospecting. D, 1970, Colorado School of Mines. 101 p.

Geyer, Robert Lee. Secondary sources of seismic noise. D, 1977, University of Tulsa. 294 p.

Ghaeni, Mohammad R. Gravity survey of the Elsinore-Murrieta Valley, California. M, 1967, University of California, Riverside. 45 p.

Ghaffer-Adly, Rahmat. A detailed gravity survey in the Triassic basin, North Chester County, Pennsylvania. M, 1961, Pennsylvania State University, University Park. 76 p.

Ghafory-Ashtiany, Mohsen. Seismic response for multicomponent earthquakes. D, 1984, Virginia Polytechnic Institute and State University. 250 p.

Ghaheri, Abbas. Numerical solution of two-phase flow equation under varied initial and boundary conditions. D, 1983, Colorado State University. 198 p.

Ghahremani, Darioush Tabrizi. Paleoecology of the spiriferid branchiopods of the Silica Shale Formation (Middle Devonian), S.E. Michigan and N.W. Ohio. M, 1978, Western Michigan University.

Ghahremani, Darioush Tabrizi. Radon prospecting for hydrocarbon; potential strategy for Devonian shale gas in N.E. Ohio. D, 1984, Case Western Reserve University. 279 p.

Ghaly, Fatma M. Abd El Rahman. Importance of soil surveys in comprehensive land use planning. D, 1987, University of California, Los Angeles.

Ghanem, Youcef. Réinterprétation des données aeromagnétiques de la région de Timgaouine, Hoggar, Algérie. M, 1988, Ecole Polytechnique. 96 p.

Ghasemi, Amir Mohammad Soltani. Determination of in-situ stresses within rock masses using the acoustic emission technique. M, 1986, University of Nevada. 158 p.

Ghassemi, Ahmad. Rock slope stability analysis for a highway cut near Hill City, South Dakota. M, 1986, South Dakota School of Mines & Technology.

Ghassemi, Farhad. Steam drive; its extension to thin oil sands and reservoirs containing residual saturation

of high gravity crude. D, 1981, University of Southern California.

Ghatge, Suhas Laxman. A geophysical investigation of a possible astrobleme in southwestern Michigan. M, 1984, Western Michigan University. 74 p.

Ghavidel-Syooki, Mohammed. Palynostratigraphy and paleoecology of the Faraghan Formation of southeastern Iran. D, 1988, Michigan State University. 279 p.

Ghazal, Ralphael Louis. Structural analysis and mapping of the western part of the Caddo Anticline, Carter County, Oklahoma. M, 1975, University of Oklahoma. 61 p.

Ghazali, Fouad Muhammed. Soil stabilization by chemical additives. D, 1981, University of Washington. 212 p.

Ghazarian, Ghazar Boulos. Induced polarization measurements through frequency dependence of resistivity. M, 1960, New Mexico Institute of Mining and Technology. 82 p.

Ghazban, Fereydoun. Geological and stable isotope studies of carbonate-hosted lead zinc deposits in Nanisivik, northern Baffin Island, Northwest Territories, Canada. D, 1988, McMaster University. 374 p.

Ghazi, Ali Mohamad. Mineralogy of the epitaxial copper sulfide coating on sphalerite. M, 1983, University of Nebraska, Lincoln.

Ghazi, Samir Abd-el-Rahman. Petrology and provenance of the Eocene Carrizo Sandstone in Cherokee, Nacogdoches and Rusk counties, Northeast Texas. M, 1981, University of Texas, Austin. 78 p.

Ghazizadeh, Mahmood. Heavy mineral evidence for source of Stanley Shale of Ouachita Mountains, Arkansas. M, 1978, Northeast Louisiana University.

Ghazizadeh, Mahmood. Petrology, depositional environments, geochemistry, and diagenetic history of lower and middle Chickamauga Group (Middle Ordovician) along Highway 58, East Tennessee. D, 1987, University of Tennessee, Knoxville. 363 p.

Ghaznavi, Muhammad Ishaq. The petrographic properties of the coals of Pakistan. M, 1988, Southern Illinois University, Carbondale. 247 p.

Gheddida, Mehemed S. Carbonate petrography of Mississippian dolomite near Springdale, Washington. M, 1988, Eastern Washington University. 96 p.

Gheith, Mohamed Ahmed. Stability relations of ferric oxides and their hydrates, lipscombite; a new synthetic "iron latulite". D, 1951, University of Minnesota, Minneapolis.

Ghellali, Salem M. Geology of the vicinity of the Villa Grove Mine, Saguache County, Colorado. M, 1970, Columbia University. 60 p.

Ghent, Edward Dale. Petrology and structure of the Black Butte area, Hull Mountain and Anthony Peak quadrangles, northern Coastal Ranges, California. D, 1964, University of California, Berkeley. 229 p.

Ghidey, Fessehaie. Terrace location, design, and evaluation by computer. D, 1987, University of Missouri, Columbia. 256 p.

Ghiorso, Mark Stefan. Studies in natural solid-liquid equilibrium. D, 1980, University of California, Berkeley. 371 p.

Ghist, J. M. Devonian Tentaculites of Ohio. M, 1976, Ohio State University. 184 p.

Ghohestani-Bojd, Hamid. Seepage in stochastic and spatially correlated permeability fields with an application to soil liners. D, 1988, University of Nebraska, Lincoln. 241 p.

Ghole, Jagannath Rao. An optimization approach to well spacing for gas storage reservoirs. M, 1969, University of Missouri, Rolla.

Ghoneim, Ghoneim Abdel-Azim. Development of river basin operational guidelines for conjunctive use of surface and groundwater. D, 1988, Colorado State University. 274 p.

Ghonemy, Hamdi Mohamed Riad El see El Ghonemy, Hamdi Mohamed Riad

Ghooprasert, W. Salinity effects on soil consolidation. D, 1978, Colorado State University. 152 p.

Ghorashi-Zadeh, Medhi. Development of hypogene and supergene alteration and copper mineralization patterns, Sar Cheshmeh porphyry copper deposits, Iran. M, 1979, Brock University. 223 p.

Ghorbanzadeh-Rendi, Ali. Non-steady, two-dimensional tile drainage of saturated-unsaturated artesian lands analyzed by finite element method. D, 1980, University of California, Davis. 268 p.

Ghose, Subrata. The crystal structure of cummingtonite and Mg-Re ordering in ferromagnesian amphiboles. D, 1959, University of Chicago. 36 p.

Ghosh, Dipak Kumar. Geochemistry of the Nelson-Rossland area, southeastern British Columbia. D, 1986, University of Alberta. 328 p.

Ghosh, Mrinal Kanti. Interpretation of airborne EM measurements based on thin sheet models. D, 1972, University of Toronto.

Ghosh, Protip Kumar. Use of bentonites and glauconites in potassium-40/argon-40 dating in Gulf Coast stratigraphy. D, 1972, Rice University. 136 p.

Ghosh, S. K. Interpretation of coarsening upwards hemicycles in the upper Paleozoic rocks of South-central West Virginia. D, 1977, SUNY at Binghamton. 275 p.

Ghosh, S. K. Origin and geochemistry of ferromanganese nodules in Oneida Lake, New York. D, 1975, Syracuse University. 249 p.

Ghosh, Sanjib Kumar. Investigation into the problems of relative orientation. D, 1964, Ohio State University. 155 p.

Ghosh, Santi Kumar. Theoretical studies on seismic reflections from vertically inhomogeneous media. D, 1983, University of Minnesota, Minneapolis. 200 p.

Ghosh, Sudipta K. Structure, petrology, and strontium content of barites of Minerva No. 1 Mine, Cave-in-Rock District, southern Illinois. M, 1973, Boston University. 78 p.

Ghosh, Swapan K. Extractable iron minerals and coloration in tills of southeastern Wisconsin and offshore lake deposits at Terry Andrae State Park, Wisconsin. M, 1973, University of Wisconsin-Milwaukee.

Ghul, Sharef. Porosity as related to the depositional and diagenetic history of the Gialo Limestone (middle-upper Eocene); Masrab Field-Sirte Basin, Libya. M, 1981, Ohio University, Athens. 62 p.

Ghuma, Mohamed A. Petrology of the gabbro of Electra Lake, Needle Mountains, Southwest Colorado. M, 1971, University of Kansas.

Ghuma, Mohamed Ali. The geology and geochemistry of the Ben Ghnema Batholith, Tibisti Massif, southern Libya, L.A.R. D, 1976, Rice University. 188 p.

Giacomini, David. Petrology of the Middle Ordovician Trenton Limestone, Pine Mountain overthrust sheet of southwestern Virginia and northeastern Tennessee. M, 1986, University of North Carolina, Chapel Hill. 151 p.

Giambelluca, Thomas Warren. Water balance of the Pearl Harbor-Honolulu Basin, 1946-1975. D, 1983, University of Hawaii. 325 p.

Giammarco, Joseph H. Fluoride saturation levels in groundwater, applications for tin exploration. M, 1983, Pennsylvania State University, University Park. 63 p.

Giammona, C. P., Jr. Octocorals in the Gulf of Mexico; their taxonomy and distribution with remarks on their paleontology. D, 1978, Texas A&M University. 272 p.

Gianella, Vincent P. Geology of the Silver City District and the southern portion of the Comstock Lode, Nevada. D, 1936, Columbia University, Teachers College.

Gianella, Vincent P. The sampling of free gold ores. M, 1920, University of Nevada. 25 p.

Giangrande, Peter Anthony. Geology and sulfide mineralization of the Skeleton Lake Prospect, St. Louis County, Minnesota. M, 1981, University of Minnesota, Duluth. 116 p.

Giannini, William Fenwick. A study of the lead-zinc deposit near Faber, Virginia. M, 1959, University of Virginia. 80 p.

Giannone, Ralph John. Geology of Caldwell Knob area, Bastrop County, Texas. M, 1951, University of Texas, Austin.

Gianotti, Frank B., III. A study of self-purification under unsteady flow conditions using dye tracer techniques. M, 1969, University of Tennessee, Knoxville. 64 p.

Giaramita, Mario Joseph. Structural evolution and metamorphic petrology of the Monarch Canyon area, northern Funeral Mountains, Death Valley, California. M, 1984, University of California, Davis. 145 p.

Giardinelli, Anthony. Stratigraphy, sedimentation and depositional history of the late Tertiary Camp Davis Formation, Teton County, Wyoming. M, 1979, Idaho State University. 54 p.

Giardini, Armando Alfonzo. Piezobirefringence in strontium titanate. D, 1957, University of Michigan. 98 p.

Giardino, J. R. Rock glacier mechanics and chronologies; Mount Mestas, Colorado. D, 1979, University of Nebraska, Lincoln. 244 p.

Giarratana, Ann Marie. A subsurface structural analysis of the Rock Creek Trend, Carter and Love counties, Oklahoma. M, 1984, Baylor University. 53 p.

Gibali, Abdalla Sasi. Nickel adsorption by different soil separates and layer silicates. D, 1977, University of California, Riverside. 154 p.

Gibb, Dorothy Margaret. Aluminium distribution in a Southern Appalachian forested watershed. D, 1988, University of Georgia. 201 p.

Gibbins, Walter A. Geology of the Falcon Lake Stock (Precambrian, southeast Manitoba, Canada). M, 1967, Northwestern University.

Gibbins, Walter A. Rubidium-strontium mineral and rock ages at Sudbury, Ontario. D, 1974, McMaster University. 230 p.

Gibbon, Donald Leroy. The origin and development of the Star Mountain Rhyolite. D, 1964, Rice University. 118 p.

Gibboney, Melissa J. Paleoclimatic interpretations of the Permian Cutler Formation in western Colorado from a comparison of size-composition trends of the Cutler Formation and Holocene sands. M, 1980, Indiana University, Bloomington. 58 p.

Gibbons, Helen. Microstructures and metamorphism in an accretionary prism in Prince William Sound, Alaska. M, 1988, University of California, Santa Cruz.

Gibbons, J. F. Tectonics of the eastern Ozarks area, southeastern Missouri. D, 1974, Syracuse University. 164 p.

Gibbons, James Arthur. The geology of part of the Contact mining district, Elko County, Nevada. M, 1973, University of Nevada. 181 p.

Gibbons, John F., III. A systematic study of fracture patterns of Northwest and west central Arkansas. M, 1962, University of Arkansas, Fayetteville.

Gibbons, Kenneth E. The Pennsylvanian of the north flank of the Anadarko Basin. M, 1956, University of Oklahoma.

Gibbons, Kenneth Edward. Pennsylvanian of the north flank of the Anadarko Basin. M, 1960, University of Oklahoma. 46 p.

Gibbons, Rex Vincent. Experimental effects of high shock pressure on materials of geological and geophysical interest. D, 1974, California Institute of Technology. 215 p.

Gibbons, Rex Vincent. Geology of the Moreton's harbour area, Newfoundland (Canada) with emphasis on the environment and mode of formation of the arsenopyrite veins. M, 1969, Memorial University of Newfoundland. 164 p.

Gibbons, Thomas Lynn. Geochemical and petrographic investigation of the Jones Camp magnetite ores and associated intrusives, Socorro County, New Mexico. M, 1981, New Mexico Institute of Mining and Technology. 156 p.

Gibbs, Alan D. Uranium geology of the granitic rocks east of Round Mountain, Nevada. M, 1976, San Diego State University.

Gibbs, Allan Kendrick. Geology of the Barama - Mazaruni Supergroup of Guyana. D, 1979, Harvard University.

Gibbs, Clare H. The geology of the northeastern portion of the Louisville Quadrangle, Knox and Blount counties, Tennessee. M, 1965, University of Tennessee, Knoxville. 51 p.

Gibbs, Clifford J. The influence of subsurface geology upon the propagation of electro-magnetic waves. M, 1939, Michigan State University. 55 p.

Gibbs, Frank Kendall. The Silurian system in eastern Montana. M, 1967, University of Montana. 46 p.

Gibbs, Gerald V. The crystal structure of proto-amphibole. D, 1962, Pennsylvania State University, University Park. 79 p.

Gibbs, Gerald V. The effect of barium and lithium substitution on the optics of synthetic fluorphlogopite $K_2 \cdot Mg_6 \cdot (Si_3AlO_{10})_2F_4$. M, 1958, University of Tennessee, Knoxville. 35 p.

Gibbs, Graham W. The organic geochemistry of chrysotile asbestos especially from the eastern townships, Quebec (Canada). M, 1969, McGill University. 154 p.

Gibbs, Harley S. Coal moisture as a correlate of oil and gas occurrence in Pennsylvania. M, 1932, University of Pittsburgh.

Gibbs, Harry Daniel. A field study of the Goodland Limestone and the Washita Group in southeastern Choctaw County, Oklahoma. M, 1951, University of Oklahoma. 72 p.

Gibbs, James A. Subsurface study of the Molas Formation in the Paradox Basin, Colorado. M, 1961, University of Oklahoma. 80 p.

Gibbs, James F. Gravity survey of the Ruby Mountain area, northeastern Nevada. M, 1967, University of Colorado.

Gibbs, Ronald E. Stratigraphy and paleontology of the Fish Haven Dolomite, south central Idaho. M, 1960, Northwestern University.

Gibbs, Ronald John. The geochemistry of the Amazon River basin (Brazil). D, 1965, University of California, San Diego. 107 p.

Gibbs, William Kirk, Jr. Geology and geochemistry of uranium mineralization in rhyolites of the Nellie Creek area, Hinsdale County, Colorado. M, 1981, Colorado School of Mines. 190 p.

Gibeaut, James C. Beach sedimentation cycles (1962-1985) along a microtidal wave dominated coast; south shore of Rhode Island. M, 1987, University of Rhode Island. 153 p.

Gibler, Pamela R. Bedrock deformation along the Salt Lake segment of Wasatch Fault; implications for principal stress directions, principal stress magnitudes and seismicity. M, 1985, University of Utah. 60 p.

Giblin, Anne Ellen. Uptake and remobilization of heavy metals in salt marshes. D, 1982, Boston University. 299 p.

Giblin, P. E. The geology and mineralogy of the Basin property, Faraday township, Ontario. M, 1956, University of Toronto.

Giblin, Peter Edwin. A study of the magnetite deposits of Mayo Township, Ontario. D, 1960, University of Toronto.

Gibling, Martin R. Sedimentation of Siluro-Devonian clastic wedge of Somerset Island, Arctic Canada. D, 1978, University of Ottawa. 334 p.

Gibson, B. S. A non-linear least squares solution to the parametric travel time equations. M, 1975, University of Hawaii at Manoa. 70 p.

Gibson, Bruce Sanderson. Seismic imaging and wave scattering in zones of random heterogeneity. D, 1988, Rice University. 226 p.

Gibson, C. R. Geology of the northwest part of Dike Mountain, Huerfano County, Colorado. M, 1977, West Texas State University. 96 p.

Gibson, Carleon, Jr. Stratigraphic and depositional studies of the sand-Glen Rose Limestone transition in Parker County, Texas. M, 1962, Texas Christian University.

Gibson, Christy Rae. Post-Pleistocene biostratigraphic and paleoenvironmental analysis of Salt Lake, Brevard County, Florida. M, 1979, University of Florida. 103 p.

Gibson, David Whiteoak. Triassic stratigraphy and petrology between the Athabasca and Smoky rivers of Alberta, Canada. D, 1966, University of Toronto.

Gibson, Donald Thomas. Sedimentation of the Santa Rosa Sandstone in Guadalupe County, New Mexico. M, 1939, Texas Tech University. 39 p.

Gibson, Everett Kay, Jr. Inert carrier-gas fusion determination of total nitrogen abundances in stony and iron meteorites. D, 1969, Arizona State University. 213 p.

Gibson, Gail G. Phanerozoic geology of the Gallina Quadrangle, Rio Arriba County, North-central New Mexico. D, 1975, University of New Mexico. 188 p.

Gibson, Gail G. Pleistocene non-marine Mollusca of the Richardson Lake deposit (Clarendon Township, Pontiac County), Quebec, Canada. M, 1967, Ohio State University.

Gibson, George Randall. The stratigraphy and structure of the Snowbank Lake area. D, 1934, University of Minnesota, Minneapolis. 70 p.

Gibson, Harold Lorne. Geology of the Amulet Rhyolite Formation, Turcott Lake section, Noranda area, Quebec. M, 1979, Carleton University. 154 p.

Gibson, James Bedford, Jr. The geology of the Yellow Creek oil field, Wayne County, Mississippi (production from Eutaw of Upper Cretaceous). M, 1953, Mississippi State University. 47 p.

Gibson, James M. Anita Formation (Paleocene-middle Eocene) and Sierra Blanca Limestone (Paleocene) of the Western Santa Ynez Mountains, Santa Barbara County, California. M, 1972, University of Southern California.

Gibson, Joan Reynolds. The relationship of vegetation to diabase dikes and sills of the Gettysburg Basin, Pennsylvania. D, 1987, University of North Carolina, Chapel Hill. 298 p.

Gibson, John Frank. A study of three Washington ores. M, 1940, University of Washington. 53 p.

Gibson, John L. Ground-water hydrology of the Panther Junction area of Big Bend National Park, Texas. M, 1983, Texas A&M University. 110 p.

Gibson, Joseph Gallagher. General geology of the Farris creek area, Gunnison County, Colorado and Gothic-Maroon (Pennsylvanian and Permian) stratigraphy. M, 1953, University of Kentucky. 45 p.

Gibson, Kenneth Mark. The design and construction of an experiment to observe porous convection at high Rayleigh numbers. M, 1980, University of Washington. 126 p.

Gibson, Lee B. Palynology and paleoecology of the Iron Post Coal (Pennsylvanian) of Oklahoma. D, 1961, University of Oklahoma. 254 p.

Gibson, Lee B. Population study of fossil and Recent ostracodes. D, 1954, University of Chicago.

Gibson, Lee Boring. The Upper Devonian Ostracoda from the Cerro Gordo Formation (Hackberry Group) of Iowa. M, 1952, Washington University. 52 p.

Gibson, Lisa M. The configuration of the Vallecito-Fish Creek Basin, western Imperial Valley, California; as interpreted from gravity data. M, 1983, Dartmouth College. 87 p.

Gibson, Michael Allen. Paleogeography, depositional environments, paleoecology, and biotic interactions of the Rockhouse Limestone and Birdsong Shale members of the Ross Formation (Lower Devonian), western Tennessee. D, 1988, University of Tennessee, Knoxville. 423 p.

Gibson, Michael Allen. The paleontology and paleoecology of the invertebrate megafauna associated with the upper cliff coals, Plateau coalfield, Alabama. M, 1983, Auburn University. 171 p.

Gibson, Peter Craig. Geology of the Buckskin Mine, Douglas County, Nevada. M, 1987, University of Nevada. 93 p.

Gibson, Richard G., III. Structural evolution of the Max Meadows thrust sheet, Southwest Virginia. M, 1983, Virginia Polytechnic Institute and State University. 131 p.

Gibson, Richard G., III. Structural studies in a Proterozoic gneiss complex and adjacent cover rocks, West Needle Mountains, Colorado. D, 1987, Virginia Polytechnic Institute and State University. 186 p.

Gibson, Robert G. Provenance and stratigraphic relations of Cretaceous nonmarine sediments, middle Atlantic Coastal Plain; an application of quantitative grain shape analysis. M, 1985, Lehigh University. 84 p.

Gibson, Ronald C. Geology of a portion of the Mill Creek area, San Bernadino County, California. M, 1964, University of California, Riverside. 50 p.

Gibson, Roy B., Jr. Crustal structure across the West Florida Escarpment from gravity data. M, 1962, Texas A&M University. 26 p.

Gibson, Russell. A discussion of the criteria of deep-seated gold-quartz deposits. D, 1929, Harvard University.

Gibson, Russell. Structural geology and ore deposits of the Red Cliff mining region (Eagle County), Colorado. M, 1922, University of Colorado.

Gibson, Russell. Structural geology and ore deposits of the Red Cliff mining region, Colorado. M, 1929, University of Colorado.

Gibson, Thomas G. Revision of the Turridae of the Miocene St. Mary's Formation of Maryland. M, 1959, University of Wisconsin-Madison.

Gibson, Thomas George. Benthonic foraminifera and paleoecology of the Miocene deposits of the middle Atlantic Coastal Plain. D, 1962, Princeton University. 218 p.

Gibson, Thomas R. Precambrian geology of the Burned Mountain-Hopewell Lake area, Rio Arriba County, New Mexico. M, 1981, New Mexico Institute of Mining and Technology. 105 p.

Gibson, Victor Rutledge. A review of the Jupiter Iron Works at Carondelet, Missouri. M, 1877, Washington University.

Gibson, Victoria R. Vertical distribution of estuarine phytoplankton in the surface microlayer and at one meter, and fluctuations in abundance caused by surface adsorption of monomolecular films. M, 1971, College of William and Mary.

Gibson, Wayne Ross. Paleoecology of Ostracoda of the Miocene Saint Mary Formation of southeastern Maryland. M, 1970, University of Wisconsin-Milwaukee.

Gibson, William Carleton. A study of the Sparks Hill Diatreme, Hardin County, Illinois. M, 1950, University of Cincinnati. 28 p.

Giddens, Leslie Wylie, Jr. Subsurface geology of Wise County, Texas. M, 1957, University of Texas, Austin.

Giddings, Harrison J. The geology of a portion of the Laramie Basin lying north of Como Anticline, Albany County, Wyoming. M, 1935, University of Wyoming. 45 p.

Giddings, Marston Todd. Induced infiltration at the University of Connecticut well field. M, 1966, University of Connecticut. 35 p.

Giddings, Marston Todd, Jr. Hydrologic budget of Spring Creek drainage basin, Pennsylvania. D, 1974, Pennsylvania State University, University Park. 124 p.

Giddings, Steven D. Petrology, mineralogy, and geochemistry of the Goldlund gold deposit, northwestern Ontario. M, 1986, University of North Dakota. 217 p.

Gidley, James William. Paleocene primates of the Fort Union. D, 1922, George Washington University.

Giedt, Norman Ray. A statistical study in the Mississippian goniatite genus, Cravenoceras. M, 1955, University of Iowa. 69 p.

Giegengack, Robert F. Recent volcanism near Dotsero (Eagle County), Colorado. M, 1962, University of Colorado.

Giegengack, Robert, Jr. Late-Pleistocene history of the Nile Valley in Egyptian Nubia. D, 1968, Yale University.

Gieger, Ronald Maney. Quitman Mountains Intrusion, Hudspeth County, Texas. M, 1965, University of Texas, Austin.

Gielisse, Peter Jacob. Investigation of phase equilibria in the system alumina-boron oxide-silica. D, 1961, Ohio State University. 95 p.

Gierke, William Gordon. Structural geology and geothermal investigation of the White Sulphur Springs area, Montana. M, 1987, University of Montana. 225 p.

Gies, Theodore Fredrick. Palynology of sediments bordering some Upper Cretaceous strand lines in northwestern Colorado. D, 1972, Michigan State University. 356 p.

Giesbrecht, Karl Otto. The deformational history of a portion of the Oxford Lake subgroup, Gods Lake Narrows, Manitoba. M, 1972, University of Manitoba.

Gieschen, Paul Allen. Gravimetrically determined depths of fill in the upper Susquehanna River basin; procedures and interpretations. M, 1974, SUNY, College at Oneonta. 90 p.

Giese, Graham S. Beach pebble movements and shape sorting; indices of swash zone mechanics. D, 1966, University of Chicago. 65 p.

Giese, Graham Sherwood. Coastal orientations of Cape Coy. M, 1964, University of Rhode Island.

Giese, Rossman F., Jr. The infrared absorption spectra of some hydrated borates. M, 1959, Columbia University, Teachers College.

Giese, Rossman Frederick, Jr. The crystal structures of ordered and disordered cobaltite. D, 1962, Columbia University, Teachers College. 64 p.

Gietz, Otto. The Whirlpool Sandstone. M, 1952, McMaster University. 44 p.

Giffin, Charles E. Potassium-argon ages in the Precambrian basement of Colorado. M, 1959, Columbia University. 18 p.

Giffin, Jon W. Stratigraphy and petrography of the Livingston Limestone Member, Bond Formation (Mississippian, Pennsylvanian), of east-central Illinois and western Indiana. M, 1978, Indiana University, Bloomington. 190 p.

Gifford, Charles D. Foraminifera from the Monterey Formation of California. M, 1924, [Stanford University].

Gifford, Gregory Paul. Assessment of shear strength loss of a silty sand subjected to frost action. D, 1984, Worcester Polytechnic Institute. 296 p.

Gifford, John A. A description of the geology of the Bimini Islands, Bahamas. M, 1973, University of Miami. 88 p.

Gifford, John Dempster. Glacial drainage in the Kaosag Quadrangle, New York, as indicated by eskers. M, 1941, University of Oklahoma. 83 p.

Giffuni, Genaro F. Miocene-Pliocene boundary problem; correlation of Mediterranean, Atlantic, Caribbean and Venezuelan sections. M, 1983, University of Rochester.

Giger, M. W. Application of limit analysis to certain problems in geotechnical engineering. D, 1974, Northwestern University. 354 p.

Gigl, Paul Donald. Pressure-temperature studies of the alumina-water and the aluminum-water systems. D, 1972, Pennsylvania State University, University Park. 187 p.

Giglierano, James D. Detection and enhancement of circular features on a Landsat image. M, 1984, Purdue University. 135 p.

Giguère, Christine. Caractérisation pétrographique, structurale, minéralogique et géochimique de la Mine Sigma-2, Canton de Louvicourt, Québec. M, 1988, Ecole Polytechnique. 236 p.

Giguere, J. Major-element distribution in coexisting pyroxenes in granulitic rocks of Somerset Island (Northwest Territories). M, 1966, University of Ottawa. 126 p.

Gil, April V. Whiterock (lower Middle Ordovician) cephalopod fauna from the Ibex area, Millard County, western Utah. M, 1987, New Mexico Institute of Mining and Technology. 127 p.

Gilani, Maqsood Ali Shah. Electric analog analysis of the groundwater system in Chaj Doab area, West Pakistan. M, 1964, University of Arizona.

Gilani, Mohammad Ali Sadighi. Distribution of rare "alkalies" and selected metals around the Peerless and Etta pegmatites, Keystone, South Dakota. M, 1979, South Dakota School of Mines & Technology.

Gilardi, John R. Experimental determination of the effective Taylor dispersivity in a fracture. M, 1984, Stanford University.

Gilb, Scot H. The relationship of total sulfur to organic carbon in the Pennsylvanian age Breathitt Formation of eastern Kentucky. M, 1987, University of Cincinnati. 111 p.

Gilbeau, Kevin P. Geology, geochemistry, and petrogenesis of the upper Keres Group, Ruiz Peak area, Jemez Mountains, New Mexico. M, 1982, University of New Mexico. 132 p.

Gilbert, Charles Merwin. The Cenozoic geology of the region southeast of Mono Lake, California. D, 1938, University of California, Berkeley. 180 p.

Gilbert, Deborah. Geology and geochemistry of the Mahogany Hot Spring gold prospect in the Owyhee region of southeastern Oregon. M, 1988, University of Washington. 76 p.

Gilbert, Deidre M. Deltaic lithofacies of an Upper Pennsylvanian coal bearing sequence; the Meigs Creek Coal, the Upper Sewickley sandstone and shale, and the Benwood Limestone in southeastern Ohio. M, 1982, University of Akron. 60 p.

Gilbert, Francis Louisa. Metamorphism in the Lake Alpine area, Alpine County, California. M, 1959, University of California, Berkeley. 70 p.

Gilbert, J. F., Jr. Seismic wave propagation in a two-layer half-space. D, 1956, Massachusetts Institute of Technology. 140 p.

Gilbert, Jean Ann. Determination of the index of refraction and coefficient of absorption under the Microscope; a new method and some of its applications. D, 1972, Columbia University. 214 p.

Gilbert, Jerry L. Sedimentology of the braided alluvial interval of the Dakota Sandstone. M, 1975, West Texas State University. 55 p.

Gilbert, John M. The origin of the chromites below Steelpoort main seam of the Bushveld Complex (South Africa). D, 1962, University of Wisconsin-Madison.

Gilbert, John Robert. Late Pleistocene history of the Caribou Creek Valley, Boulder County, Colorado. M, 1968, University of Colorado.

Gilbert, John Robert, Jr. Stratigraphy and sedimentology of the upper Miocene Williams Sand of the San Joaquin Valley, California, with a note on the neighboring Temblor Sands. M, 1980, University of Massachusetts. 181 p.

Gilbert, Joseph E. J. The acidic intrusives of the Bachelor Lake area, Abitibi-East, Quebec. M, 1947, McGill University.

Gilbert, Joseph E. J. The geology of the Capsisit Lake area, Abitibi, Quebec. D, 1949, McGill University.

Gilbert, Lawrence William. Lateral load analysis of piles in very soft clay. D, 1980, Tulane University. 129 p.

Gilbert, Leonard C. The geology of Dutcher Knob and Richland Creek valley, Rhea County, Tennessee. M, 1957, University of Tennessee, Knoxville. 32 p.

Gilbert, Lewis Edward. Lineament studies of the Southeastern U.S.A. M, 1985, North Carolina State University. 174 p.

Gilbert, M. Charles. The geology of the western Glen Mountains, Oklahoma. M, 1960, University of Oklahoma. 48 p.

Gilbert, Mark W. Plio-Pleistocene calcareous dinoflagellate cysts of the Arctic Ocean and their paleontologic significance. M, 1981, University of Wisconsin-Madison.

Gilbert, Michel. Géologie du groupe volcanique archéen de Blake River dans la région du Lac Pelletier, Ceinture de l'Abitibi. M, 1986, Ecole Polytechnique.

Gilbert, Michel. Géologie du groupe volcanique archéen de Blake River dans la région du lac Pelletier, ceinture de l'Abitibi. M, 1987, Ecole Polytechnique.

Gilbert, Murray Charles. Synthesis and stability relations of the hornblende, ferropargasite. D, 1965, University of California, Los Angeles. 131 p.

Gilbert, Neil Jay. Chronology of post-Tuscan volcanism in the Manton Area Quadrangle, California. M, 1969, University of California, Berkeley. 72 p.

Gilbert, Oscar Edward, Jr. Paleozoic evolution of the Valley and Ridge thrust belt in Alabama. D, 1981, University of Tennessee, Knoxville. 295 p.

Gilbert, Oscar Edward, Jr. Stratigraphic and structural relationships of the Sylacauga unconformity in Alabama. M, 1974, University of Alabama.

Gilbert, Pat Kader. Mechanical characteristics of folds in upper Cretaceous strata in the Disturbed Belt of northwestern Montana. M, 1974, Texas A&M University. 64 p.

Gilbert, Philip Geoffrey Britton. I, The iron oxides in deposits of magmatic derivation; II, Oxidation and enrichment at Ducktown, Tennessee. D, 1924, Harvard University.

Gilbert, Ray Clark. Middle Ordovician limestones in the valley of the north fork of the Roanoke River, Montgomery County, Virginia. M, 1953, Virginia Polytechnic Institute and State University.

Gilbert, Robert. Observations of lacustrine sedimentation at Lillooet Lake, British Columbia. D, 1974, University of British Columbia.

Gilbert, Wyatt G. Petrology of the Sunday Stock, King County, Washington. M, 1967, University of Washington. 20 p.

Gilbert, Wyatt Graves. Sur fault zone, Monterey County, California. D, 1971, Stanford University. 154 p.

Gilberto, Richard Joseph. A statistical study of microfossils by application of moss data-processing and computer techniques. M, 1967, Northern Illinois University. 58 p.

Gilbertson, Roger Lee. Biostratigraphy of the upper Paleozoic rocks in the Gulkana glacier area, Alaska. M, 1969, University of Wisconsin-Madison.

Gilbertson, Roger Lee. Quantitative studies of some actinocrinitid crinoids. D, 1972, University of Michigan. 139 p.

Gilchrist, Carol Mary. The glacial geology of the southeastern area of the District of Keewatin, Northwest Territories, Canada. D, 1982, University of Massachusetts. 326 p.

Gilchrist, James Michael. Sedimentology of the Lower to Middle Jurassic Portland Arkose of central Connecticut. M, 1978, University of Massachusetts. 166 p.

Gilchrist, William B. Trench investigation of late Tertiary to Recent movement along the southwest-bounding fault of the Shearer Graben within the Kentucky River fault system in Southeast Clark County, Kentucky. M, 1986, Eastern Kentucky University. 51 p.

Gilder, Harold R. Van *see* Van Gilder, Harold R.

Gilder, Kerry L. Van *see* Van Gilder, Kerry L.

Gildersleeve, Benjamin. The Eocene of northern Virginia. M, 1931, University of Virginia. 124 p.

Gildersleeve, Benjamin. The Eocene of Virginia. D, 1939, The Johns Hopkins University.

Giles, Albert H. Physical paleoecology of the Francis Formation (Pennsylvanian) near Ada, Oklahoma. M, 1963, University of Oklahoma. 127 p.

Giles, Albert W. The geology and coal resources of the coal bearing portion of Lee County, Virginia. D, 1922, University of Chicago. 212 p.

Giles, Albert William. Eskers of western New York. M, 1910, University of Rochester. 44 p.

Giles, Alfred E. Geology of the Strake and Squire Field areas, Duval County, Texas. M, 1948, University of Oklahoma. 43 p.

Giles, Alice B. The geology of the Cretaceous Green Island Inlier, Hanover, Jamaica, West Indies. M, 1977, George Washington University.

Giles, Billy E. Ground-water geochemistry of an unmined area in the Carbondale Group, southwestern Indiana. M, 1987, Indiana University, Bloomington. 100 p.

Giles, David Lee. A petrochemical study of compositionally zoned ash-flow tuffs. D, 1967, University of New Mexico. 188 p.

Giles, David Lee. Geology of the northern Sierrita Mountains, Soto Peak, and Gunsight Mountain area, Pima County, Arizona. M, 1963, Miami University (Ohio). 110 p.

Giles, Eugene. Multivariate analysis of Pleistocene and Recent coyotes in California. M, 1956, University of California, Berkeley. 79 p.

Giles, P. S. Stratigraphy, petrology and diagenesis of Beekmantown carbonate rocks in eastern Ontario. D, 1976, University of Western Ontario. 329 p.

Giles, Peter S. Cretaceous gypsum deposit, San Luis Potosí, Mexico. M, 1968, Acadia University.

Giles, Robert Talmadge. Petrography and petrology of the Rabun Bald area, Georgia-North Carolina. M, 1966, University of Georgia.

Giletti, B. J. New age determinations by the lead method. M, 1954, Columbia University, Teachers College.

Gilg, Joseph G. Stratigraphy of Skaneateles Formation in Onondaga County, New York. M, 1955, Syracuse University.

Gilhooly, Murray Gordon. Sedimentology and geologic history of the Upper Devonian (Frasnian) uppermost Ireton and Nisku formations, Bashaw area, Alberta. M, 1987, University of Calgary. 210 p.

Gilinsky, Norman Lawrence. Studies of the functional morphology and evolution of marine snails; with implications for macroevolutionary theory. D, 1983, Harvard University. 181 p.

Gilje, Stephen Arne. An evaluation of the effects of selected groins on south shore of Long Island, New York. M, 1974, SUNY at Binghamton. 106 p.

Gilkey, Arthur K. Fracture pattern of the Zuni uplift. D, 1954, Columbia University, Teachers College.

Gilkey, Arthur K. Structural observations on the main camp nunatak, Juneau ice field, Alaska. M, 1951, Columbia University, Teachers College.

Gilkey, Earle Will. Future possibilities of the Columbia lava plateau. M, 1940, George Washington University. 131 p.

Gilkey, Karen Eileen. Sedimentology of the North Fork and South Fork Toutle River mudflows generated during the 1980 eruption of Mount St. Helens. M, 1983, University of California, Santa Barbara. 254 p.

Gill, Dan. Stratigraphy, facies, evolution and diagenesis of productive Niagaran Guelph reefs and Cayugan sabkha deposits, the Belle River Mills Gas Field, Michigan Basin. D, 1973, University of Michigan.

Gill, Frederick David. Petrography of molybdenite-bearing gneisses, Makkovik area, Labrador, Canada. M, 1966, University of Toronto.

Gill, Gary Arthur. On the marine biogeochemistry of mercury. D, 1986, University of Connecticut. 230 p.

Gill, Gerald M. Size distribution of heavy minerals relative to the hydraulic parameters of sediments. M, 1970, Wayne State University.

Gill, Harold Edward. A stratigraphic analysis of a portion of the Matawan Group. M, 1956, Rutgers, The State University, New Brunswick. 167 p.

Gill, Hugh W., Jr. The Morrowan and Chesteran rocks of southwestern Kansas. M, 1961, Wichita State University. 56 p.

Gill, Ivan P. The sedimentological controls on organic carbon in the Virgin Islands Trough, U.S. Virgin Islands. M, 1982, University of Rochester. 70 p.

Gill, J. C. The geology of the Waskaiwaka and Mystery Lake areas, northern Manitoba. M, 1954, Queen's University. 110 p.

Gill, J. W. The Takiyuak metavolcanic belt; geology, geochemistry, and mineralization. D, 1977, Carleton University. 210 p.

Gill, James Burton. Distribution of K, Sr, Rb, and Ba abundances in the Nain, Labrador anorthosite complex. M, 1968, Franklin and Marshall College. 33 p.

Gill, James E. Gunflint iron-bearing formation, Ontario. D, 1925, Princeton University.

Gill, James Rodger. Flagstaff Limestone of the Spring City-Manti area, Sanpete County, Utah. M, 1950, Ohio State University.

Gill, John F. An investigation of natural fractures in the Ricinus Field, Alberta. M, 1981, University of Toronto.

Gill, Roger. Geology and mineral deposits of the southwest quarter of the Tanacross D-1 quadrangle, Alaska. M, 1977, Western Washington University. 129 p.

Gill, W. R. Cation exchange properties of the gray Hydromorphic soils of the Hawaiian Islands. M, 1949, University of Hawaii. 68 p.

Gillam, Mary L. Contact relations of the Ione and Valley Springs formations in the Buena Vista area, Amador County, California. M, 1974, Stanford University.

Gilland, James Kenneth. Paleoenvironment of a carbonate lens in the Navajo Sandstone near Moab, Utah. M, 1978, Brigham Young University. 36 p.

Gillanders, Earle Burdette. An outline of the general geology and physiography of the western part of north eastern Rhodesia, with notes on correlation and rift valleys. D, 1932, Princeton University. 150 p.

Gillanders, Earle Burdette. Slocan ores. M, 1926, University of British Columbia.

Gillen, K. A paleo-geomagnetic study of Pleistocene sediments in northwestern Canada. M, 1988, University of Alberta.

Gillerman, Elliot. Geology of the central Peloncillo Mountains, Hidalgo County, New Mexico, and Cochise County, Arizona. D, 1957, University of Texas, Austin.

Gillerman, Elliott. Early Osage formations. M, 1937, Washington University. 129 p.

Gillerman, Virginia Sue. Tungsten and copper skarns of the Railroad mining district, Nevada. D, 1982, University of California, Berkeley. 196 p.

Gillert, Martin Peter. The geology of the Simpson Group on the northeast flank of the Arbuckle Mountains, Oklahoma. M, 1952, University of Oklahoma. 76 p.

Gillespie, Alan Reed. Quaternary glaciation and tectonism in the southeastern Sierra Nevada, Inyo County, California. D, 1982, California Institute of Technology. 738 p.

Gillespie, Blake W. Temblor Formation; Maricopa to the eastern San Emigdio Mountains, Kern County, California. M, 1986, Stanford University. 161 p.

Gillespie, Clinton D. Geology of the central Bond Creek area, Nabesna, Alaska. M, 1970, Oregon State University. 67 p.

Gillespie, David R. Lithostratigraphy and petrography of the Atoka Formation in the Treat area, Northwest Arkansas. M, 1985, University of Arkansas, Fayetteville.

Gillespie, Dennis Patrick. Hydrogeology of the Austin Cary Control Dome in Alachua County, Florida. M, 1976, University of Florida. 127 p.

Gillespie, Gary L. The Clearwater region (Idaho); an economic base study. M, 1969, University of Idaho. 215 p.

Gillespie, Janice M. Depositional environments and hydrocarbon potential of the uppermost Cretaceous Lance Formation, Wind River basin, Wyoming. M, 1984, South Dakota School of Mines & Technology.

Gillespie, Randall Thomas. Stratigraphic and structural relationships among rock groups at Old Man's Pond, west Newfoundland. M, 1984, Memorial University of Newfoundland. 198 p.

Gillespie, Robert Howard, Jr. Quaternary geology of South-central New York. D, 1980, SUNY at Binghamton. 306 p.

Gillespie, Ruth Frances. Bartonian foraminifera and ostracods (Pennsylvania, New York, Maryland). M, 1930, Cornell University.

Gillespie, Thomas D. The structure and geochemistry of a breccia dike complex at McAfee, New Jersey. M, 1987, Rutgers, The State University, New Brunswick. 90 p.

Gillespie, W. A.; Brant, Russell Alan; Elmer, Nixon and Peterson, John Robert. Geology of the Armstead area, Beaverhead County, Montana. M, 1949, University of Michigan. 118 p.

Gillespie, W. G. The Huronian of Waconichi Lake area, Quebec. M, 1951, University of Toronto.

Gillespie, Walter Lee. A preliminary survey of the Hydracarina and Ostracoda of twelve counties in southern Ohio. M, 1954, Miami University (Ohio).

Gillet, Lawrence Britton. Anorthosites and syenites of the Mealy Mountain area, Labrador. M, 1956, McGill University.

Gilleti, Bruno J. The geochemistry of tritium; II, Tritium tracer in Arctic problems. D, 1960, Columbia University, Teachers College.

Gillett, Lawrence B. Bedrock and Pleistocene geology of the Vienne-Blaiklock area, Quebec, with observations on magnetic diabase dikes in northeast Ontario and northwest Quebec. D, 1962, Princeton University. 222 p.

Gillett, Stephen Lee. Magnetization and remagnetization processes in some early Paleozoic limestones from the Great Basin. D, 1981, SUNY at Stony Brook. 521 p.

Gillette, Christopher B. Lineament tectonics of the Montana mining districts. M, 1965, Montana College of Mineral Science & Technology. 133 p.

Gillette, David D. A review of North American glyptodonts (Edentata, Mammalia); osteology, systematics, and paleobiology. D, 1973, Southern Methodist University. 648 p.

Gillette, Tracy. Geology of the Clyde and Sodus Bay quadrangles, New York. M, 1934, University of Rochester. 153 p.

Gillette, Tracy. The Clinton of New York State from Rochester to Clinton. D, 1936, The Johns Hopkins University.

Gilley, Jerry D. Geology of Kingston area, Madison County, Arkansas. M, 1966, University of Arkansas, Fayetteville.

Gilley, John Edwards. Soil erosion by sheet flow. D, 1982, Colorado State University. 162 p.

Gilliam, W. B. Distribution of gold values in soil samples from an area of Cleburne County, Alabama. M, 1976, University of Alabama.

Gilliam, William W. Determination of environments of deposition through hydraulic equivalence studies. M, 1983, University of Georgia.

Gilliard, T. C. Sedimentary paleomagnetic evidence for counter-clockwise rotation of the Fiji Plateau. M, 1971, University of Hawaii. 28 p.

Gillies, Donald F. Geology of barium and strontium in sediments. M, 1933, University of Wisconsin-Madison.

Gillies, N. B. The geology of the Formaque Property, Bourlamaque township, Quebec. M, 1946, McGill University.

Gillies, Warren Douglas. The geology of a portion of Cottonwood Springs Quadrangle, Riverside County, California. M, 1958, University of California, Los Angeles.

Gilligan, Eileen Dombroski. Computer simulation of the effects of earth materials on the propagation of radar waves. M, 1979, Syracuse University.

Gilligan, Eileen Dombroski. The effect of organic pore fluids on the fabric and geotechnical behavior of clays. D, 1983, Syracuse University. 291 p.

Gilliland, J. A. A proposed ore control at the Coronation Mine (eastern Saskatchewan) derived from examination of the quantitative mineralogy. M, 1964, University of Manitoba.

Gilliland, J. D. Geology of the Whiskey Mountain area, Fremont County, Wyoming. M, 1959, University of Wyoming. 93 p.

Gilliland, John Dale. A mineral and chemical analysis of the Duck Creek and Fort Worth formations, north central Texas. M, 1960, Texas Christian University. 98 p.

Gilliland, John Michael. Mean field electrodynamics and dynamo theories of planetary magnetic fields. D, 1973, University of Alberta. 512 p.

Gilliland, Martha Winters. Man's impact on the phosphorus cycle in Florida. D, 1973, University of Florida. 269 p.

Gilliland, W. Gravel deposits in central southeastern Mississippi; engineering characteristics and correlation. M, 1973, University of Southern Mississippi.

Gilliland, W. J. A ground water study using earth resistivity methods. M, 1973, Kansas State University. 64 p.

Gilliland, William A. Gravel deposits in central southeastern Mississippi; engineering characteristics and correlations. M, 1966, University of Southern Mississippi.

Gilliland, William Nathan. Geology of the Gunnison Quadrangle, Utah. D, 1948, Ohio State University.

Gillin, John A. Magnetic susceptibilities in weak fields. M, 1932, University of Oklahoma. 42 p.

Gillingham, Thomas E. The geology of the California Mine area, Pima County, Arizona. M, 1936, University of Arizona.

Gillingham, Thomas E., Jr. The solubility and transfer of silica and other nonvolatiles in steam. D, 1946, University of Minnesota, Minneapolis.

Gillings, O. J. Nitrate leaching in soil on Rutgers Agricultural Research Center at Adelphia, New Jersey. M, 1972, Rutgers, The State University, New Brunswick. 46 p.

Gillis, John William. Geology of northwestern Pictou County, Nova Scotia, Canada. D, 1964, Pennsylvania State University, University Park. 130 p.

Gillmeister, Norman Maack. Petrology of Precambrian rocks in the central Tobacco Root Mountain, Madison County, Montana. D, 1972, Harvard University. 201 p.

Gillon, Kenneth A. Stratigraphic, structural, and metamorphic geology of portions of the Cowrock and Helen, Georgia 7 1/2' quadrangles. M, 1982, University of Georgia.

Gillou, Robert B. The geology of the Johnston Grade area, San Bernardino Mountains, California. M, 1951, University of California, Los Angeles.

Gillson, J. L. Certain phases of the geology and ore deposits of the Pend Oreille silver mining region, Bonner County, Idaho. D, 1923, Massachusetts Institute of Technology. 235 p.

Gillson, J. L. Some notes on the geology of Shoshone Canyon, Park County, Wyoming. M, 1921, Massachusetts Institute of Technology. 46 p.

Gillson, Joseph L. Recomposed granite and associated rocks of Saganaga Lake. M, 1920, Northwestern University.

Gilluly, James. Geology of a part of the San Rafael Swell, Utah. D, 1926, Yale University.

Gillum, Cecil Conrad. Areal geology of Northeast Canadian County, Oklahoma. M, 1958, University of Oklahoma. 70 p.

Gillum, J. P. Stratigraphy and structure of the Alkali Creek-Willow Creek, Natrona County, Wyoming. M, 1956, University of Wyoming. 43 p.

Gilman, Chandler R. Geology and geohydrology of the Sitgreaves Mountain area, Coconino County, Arizona. M, 1965, University of Arizona.

Gilman, J. A. Hydrogeology of northern Mercer County, North Dakota with emphasis on the Beulah-Zap lignite bed. M, 1975, Stanford University. 187 p.

Gilman, M. N. Trilobites of the Upper Cambrian Gorge Formation of Highgate Falls, Vermont. M, 1955, University of Wyoming. 108 p.

Gilman, Richard Atwood. An X-ray method for quantitative mineralogical analysis of aphanitic rocks from Mt. Desert Island, Maine. M, 1959, University of Illinois, Urbana. 45 p.

Gilman, Richard Atwood. Petrology and structure of the Milbridge-Whitneyville area, Maine. D, 1961, University of Illinois, Urbana. 130 p.

Gilman, W. F. Tectonites of the northwest portion of the Lovingston area, Virginia. M, 1948, University of Toronto.

Gilmartin, Patricia Purcell. The map context as a source of perceptual error in graduated circle maps. D, 1980, University of Kansas. 167 p.

Gilmer, Allen L. The geology and genesis of the Sierra de Santa Maria metalliferous deposits, Velardena, Durango, Mexico. M, 1987, University of Texas at El Paso.

Gilmer, Douglas R. General geology, landsliding, and slope development of a portion of the north flank of the Uinta Mountains, south-central Uinta County, Wyoming. M, 1986, University of Wyoming. 93 p.

Gilmont, Norman L. Geology of the Puerto la Bandera area, Sonora, Mexico. M, 1979, Northern Arizona University. 112 p.

Gilmore, Herman Lee, Jr. The depositional environment and diagenetic history of waulsortian type mounds in Southcentral Tennessee. M, 1980, Vanderbilt University.

Gilmore, John B. Stratigraphy of the Emporia Limestone (Virgilian; Pennsylvanian) in east-central Kansas. M, 1972, University of Kansas.

Gilmore, John W. Petrology of the argillaceous component of the Rosiclare Member of the Ste. Genevieve Limestone in Indiana. M, 1958, Indiana University, Bloomington. 23 p.

Gilmore, Raymond Maurice. Review of Microtus voles of the subgenus Stenocranius (Mammalia; Rodentia; Muridae), with special discussion of the Bering Strait region. D, 1942, Cornell University. 327 p.

Gilmore, Robert Snee. High-pressure, ultrasonic study of elastic properties and equations of state of fifteen materials. D, 1968, Rensselaer Polytechnic Institute. 163 p.

Gilmore, Susan Potbury. The La Porte flora of Plumas County, California. M, 1935, University of California, Berkeley. 149 p.

Gilmore, Tyler J. Stratigraphy and depositional environments of the Lower Mississippian Joana Limestone, east-central Nevada. M, 1987, University of Idaho. 79 p.

Gilmore, Walter E. Structural and stratigraphic relationships of the sedimentary rocks along the east flank of the Front Range in northern Colorado. M, 1966, Texas A&M University. 64 p.

Gilmour, Ernest Henry. Carbonate petrology and paleontology of the Alaska Bench Formation (Mississippian or Pennsylvanian), central Montana. D, 1967, University of Montana. 152 p.

Gilmour, Ernest Henry. The geology of the southwestern part of the Stryker Quadrangle, northwestern Montana. M, 1964, University of Montana. 70 p.

Gilmour, Ralph G. Stratigraphy of the Phillipsburg and Rosenberg thrust sheets, southern Quebec. M, 1971, McGill University.

Gilotti, Jane A. The role of ductile deformation in the emplacement of the Sarv thrust sheet, Swedish Caledonides. D, 1987, The Johns Hopkins University. 239 p.

Gilpin, A. E. An economic stabilization of laterite soils and gravels. D, 1975, Carnegie-Mellon University. 146 p.

Gilpin, Bernard E., IV. A microearthquake study of the Salton Sea geothermal area, Imperial Valley, California. M, 1977, University of California, Riverside. 181 p.

Gilpin, Lawrence Mellick. Tectonic geomorphology of Santo Island, Vanuatu (New Hebrides). M, 1982, Cornell University. 147 p.

Gilson, Edward S., Jr. The glacial history of the lower Whitewater River valley. M, 1953, University of Cincinnati. 26 p.

Gilson, Mary M. A model of the energy and frequency spectrum of a projectile seismic source. M, 1986, University of Kansas. 132 p.

Giltner, John Patrick. Application of extensional models to the northern Viking Graben, North Sea. M, 1987, University of Texas, Austin. 119 p.

Giltner, Robert. Petrology of the Morrison Limestone of central Colorado. M, 1955, Iowa State University of Science and Technology.

Gimbrede, Louis de A. Foraminiferal markers in the Austin Group, Texas. D, 1961, Louisiana State University.

Gimbrede, Louis de Agramonte. Hurricane Lentil, Cook Mountain, Eocene in East Texas. M, 1951, University of Texas, Austin.

Gimlett, James Irwin. The gravimetric method applied to basin exploration, exemplified by a study of Warm Springs Valley, Washoe County, Nevada. D, 1965, Stanford University. 161 p.

Gimmel, J. C. A depositional model for the middle and lower Newman Formation near Olive Hill, Kentucky. M, 1975, University of South Carolina.

Gin, Thon Too. Mineralization in the Pine Creek area, Coeur d'Alene mining region, Idaho. D, 1953, University of Utah. 73 p.

Gin, Thon Too. The application of differential thermal analyses to the determination of the composition of dolomites. M, 1950, University of Utah. 31 p.

Gindlesperger, Gary D. A hydrothermal investigation of the mineral pollucite. M, 1979, University of Toledo. 62 p.

Gingerich, Philip Dean. Cranial anatomy and evolution of early Tertiary Plesiadapidae (Mammalia, Primates). D, 1974, Yale University.

Gingrich, Dean Alan. North Sea surface sediments; a study of the lateral and vertical variation in the magnetic remanence over short distances. M, 1977, University of Minnesota, Minneapolis. 67 p.

Ginn, A. Variations in the Lochalsh ultrabasic stock. M, 1945, Queen's University. 49 p.

Ginn, R. M. The relationship of the Bruce Series to the granites in the Espanola area. D, 1960, University of Toronto.

Ginn, Robert M. A study of granitic rocks in the Sudbury area. M, 1958, Queen's University. 137 p.

Ginn, Timothy Rollins. A continuous-time solution to the inverse problem of groundwater and contaminant transport modelling. D, 1988, Purdue University. 81 p.

Ginsberg, Marilyn H. Water-budget model of the south-central Sand Hills of Nebraska. D, 1987, University of Nebraska, Lincoln. 210 p.

Ginsburg, Marilyn. Hydrology of the Triassic basin, Pennsylvania. M, 1970, SUNY at Binghamton. 94 p.

Ginsburg, Merrill Stuart. A gravity study of the Boston (Massachusetts) Basin region. M, 1960, Massachusetts Institute of Technology. 69 p.

Ginsburg, Merrill Stuart. The application of downward continuation to two infinite parallel horizontal cylinders. D, 1963, University of Utah. 291 p.

Ginsburg, Robert N. Lithification and alteration processes in South Florida carbonate deposits. D, 1953, University of Chicago. 104 p.

Ginsel, Marvin G. A regional structural analysis and gravity study of the eastern Marquette Synclinorium and Republic Trough, Michigan. M, 1973, Northern Illinois University. 75 p.

Gintautas, Peter Alan. Lake sediment geochemistry, Northern Interior Plateau, British Columbia. M, 1984, University of Calgary. 175 p.

Ginther, Paul G. Geologic investigation and mapping of the North Fork Payette River, Banks to Smith Ferry, Idaho. M, 1981, Washington State University. 77 p.

Ginzel, Edwin Charles. Earth waves generated from quarry blasts. M, 1958, St. Louis University.

Ginzler, S. L. A topological analysis of some braided reaches of the North Platte River, Nebraska. M, 1985, SUNY at Binghamton. 69 p.

Giordano, Thomas Henry. Dissolution and precipitation of lead sulfide in hydrothermal solutions, and the point defect chemistry of galena. D, 1978, Pennsylvania State University, University Park. 167 p.

Giosa, Thomas A. Gabbroic xenoliths in Pliocene to Recent basalts, Nayarit, Mexico. M, 1985, Tulane University. 113 p.

Giovanella, Carlo. Petrology of the Precambrian rocks at Tichborne, SE Ontario. M, 1965, University of Washington. 78 p.

Giovannetti, Dennis. Mio-Pliocene volcanic rocks and associated sediments of southeastern Owens Lake, Inyo County, California. M, 1979, University of California, Berkeley. 76 p.

Giovanni, Marcel di see di Giovanni, Marcel, Jr.

Giovinetto, Mario Bartolomé Glacier landforms of the Antarctic coast and the regimen of the inland ice. D, 1968, University of Wisconsin-Madison. 176 p.

Gipson, Mack Jr. A study of relations of depth, porosity, and mineral orientation in Pennsylvania shales. D, 1963, University of Chicago. 104 p.

Gipson, William Earl. Geology of Kimbro area in Travis and Bastrop counties, Texas. M, 1949, University of Texas, Austin.

Giraldez-Cervera, Juan V. The theory of infiltration and drainage in swelling soils. D, 1976, University of California, Riverside. 184 p.

Girard, Cecil M. Geology of a part of the western Santa Ynez Range, Santa Barbara County, California. M, 1949, Stanford University. 57 p.

Girard, Lewis V. Geological engineering study of a proposed tunnel through South Fork Mountain, Trinity County, California. M, 1959, University of California, Berkeley.

Girard, Marie-Josée. Pétrologie de la série volcanique d'Alausi-Tixan, Equateur. M, 1983, Universite Laval. 193 p.

Girard, Oswald Woodrow, Jr. The geology of the Gainesville West Quadrangle, Alachua County, Florida. M, 1968, University of Florida. 56 p.

Girard, Paul. La pétrologie des laves de Doublet, Fosse du Labrador. M, 1965, Ecole Polytechnique. 109 p.

Girard, Paul. The Madeleine copper mine, Gaspe, Quebec; a hydrothermal deposit. D, 1971, McGill University. 243 p.

Girard, Roselle Margaret. The absence of Salado Salt, central Ward County, Texas. M, 1952, University of Texas, Austin.

Girard, William W. Size, shape, and symmetry of the cross profiles of glacial valleys. D, 1976, University of Iowa. 90 p.

Girard, William W. Stereonet analysis of moderately deformed sedimentary strata. M, 1962, University of Oregon. 50 p.

Girardot, Gerald B. Environment of deposition and diagenesis of the Dean Formation in the Ackerly

Field, Dawson County, Texas. M, 1986, Texas Tech University. 88 p.

Girardot, Stephen Lee. Stratigraphy and sedimentation of the Wabaunsee Group (Upper Pennsylvanian) in southeastern Nebraska and adjacent regions. M, 1962, University of Nebraska, Lincoln.

Giraud, A. P. Silurian and Lower Devonian coral formations of New York and vicinity. D, 1918, Princeton University.

Giraud, Joel Robert. Petrology, diagenesis and environment of deposition of the Middle Devonian Wapsipinicon Formation, from a core and reference exposure, southeastern Iowa and western Illinois. M, 1986, Iowa State University of Science and Technology. 193 p.

Giraud, Richard Ernest. Stratigraphy of volcanic sediments in the McDermitt Mine, Humboldt County, Nevada. M, 1986, University of Idaho. 87 p.

Girault, Pablo. A study of the consolidation of Mexico City clay. D, 1960, Purdue University.

Girdley, William Arch. Petrology of Pennsylvanian limestones, San Juan Mountains, Colorado. D, 1967, Washington State University. 204 p.

Girdley, William Arch. Trenton Limestone as a structural marker bed in north-central and east-central Indiana. M, 1961, Indiana University, Bloomington. 40 p.

Girhard, Mary Nancy. A regional study of lineation in western New Hampshire. M, 1948, University of Illinois, Urbana.

Girijavallabhan, Chiyyarath V. Application of the finite element method to problems in soil and rock mechanics. D, 1967, University of Texas, Austin. 124 p.

Girty, G. Mesozoic cataclastic rocks in the Boyden Cave roof pendant and their regional significance. M, 1977, California State University, Fresno.

Girty, Gary H. The Culbertson Lake Allochthon; a newly identified structural unit within the Shoo Fly Complex; sedimentologic, stratigraphic, and structural evidence for extension of the Antler orogenic belt to the northern Sierra Nevada, California. D, 1983, Columbia University, Teachers College. 184 p.

Girty, George H. A partial revision of the fauna of the lower Helderberg Group. D, 1894, Yale University.

Gishler, C. A. Upper Ordovician Chitinozoa from Manitoulin Island and Bruce Peninsula, Ontario. M, 1976, University of Western Ontario.

Gislason, Sigurdur Reynir. Meteoric water-basalt interactions; a field and laboratory study. D, 1986, The Johns Hopkins University. 254 p.

Gisler, Patrick Michael. Identification of some opaque minerals by direct measurement of polarizing angles. M, 1971, University of Arizona.

Gist, J. G. Geology of the North Freezeout Hills and adjacent areas, Carbon and Albany counties, Wyoming. M, 1957, University of Wyoming. 112 p.

Gitelson, Geoffrey A. Examination of a method of data acquisition for a hole-to-hole seam wave study in an Ohio coal. M, 1985, Wright State University. 90 p.

Gitlin, Ellen C. Remobilization of sulfides and low temperature alteration of submarine basaltic rocks from Leg 53 of the Deep Sea Drilling Project. M, 1981, University of Washington. 106 p.

Gitlin, Ellen C. Wall rock geochemistry of the Lucky Friday Mine, Shoshone County, Idaho. D, 1986, University of Washington. 223 p.

Gitschlag, Gregg. Salinity effects on survival and growth of larvae of the Quahog clams Mercenaria mercenaria, M. campechiensis, and their hybrid. M, 1978, Florida State University.

Gittins, John. Nepheline metagabbro and associated hybrid rocks from Monmouth Township, Ontario. M, 1956, McMaster University. 37 p.

Giudice, Philip Michael. Mineralization at the convergence of the Amethyst and Oh fault systems, Creede District, Mineral County, Colorado. M, 1980, University of Arizona. 95 p.

Giuffria, Ruth. The Vanport Horizon in the Allegheny Series (Pennsylvanian) in southern Stark and northern Tuscarawas counties, Ohio. M, 1977, University of Akron. 67 p.

Giulj, Dominique. An interactive system for display and processing of seismic data. D, 1972, Colorado School of Mines. 333 p.

Giuseffi, David Francis. Paleoecology and community analysis of selected shale intervals in the Mt. Hope Member of the Fairview Formation (Cincinnatian Series, Ordocian), with a comparison to a shale unit in the Ft. Ancient Member of the Waynesville Formation. M, 1982, Miami University (Ohio). 145 p.

Giusti, Ennio Vincent. An investigation of Piedmont streams in terms of their geomorphology and runoff characteristics. M, 19??, American University. 60 p.

Giusti, Lorenzino. The distribution, grades and mineralogical composition of gold-bearing placers in Alberta. M, 1983, University of Alberta. 397 p.

Given, Jeffrey Wayne. Inversion of body-wave seismograms for upper mantle structure. D, 1984, California Institute of Technology. 163 p.

Given, Mary Michie. Foraminifera of the Bearpaw Formation (upper Cretaceous), central southeastern Alberta (Canada). M, 1969, University of Alberta. 160 p.

Given, Robert Kevin. Original isotopic composition of Permian marine carbonates. M, 1983, University of Michigan.

Given, Robert Kevin. The use of carbonate marine cements in geochemical studies at global and regional scales. D, 1986, University of Michigan. 201 p.

Givens, Charles Ray. Eocene molluscan biostratigraphy of the Pine mountain region, Ventura County, California. D, 1969, University of California, Riverside. 289 p.

Givens, David B. A new basic copper phosphate mineral (chinoite) from Santa Rita, New Mexico. M, 1951, University of New Mexico. 12 p.

Givens, Terry J. Paleoecology and environment of deposition of parts of the Brereton and Jamestown cyclothems, Williamson County, Illinois. M, 1968, Southern Illinois University, Carbondale. 182 p.

Gize, Andrew Paul. The organic geochemistry of three mississippi valley-type ore deposits. D, 1984, Pennsylvania State University, University Park. 371 p.

Gjelsteen, Thor W. Origin and timing of uranium mineralization in the Chadron Formation, Northwest Nebraska. M, 1988, University of Wyoming. 92 p.

Gjerde, Michael Wolf. Petrography and geochemistry of the Alverson Formation, Imperial County, California. M, 1982, San Diego State University. 85 p.

Gjere, Robert Allen. Glacial geology of the Meadowlark Lake Quadrangle, Bighorn Mountains, Wyoming. M, 1974, South Dakota School of Mines & Technology.

Glaccum, Robert. The mineralogical and elemental composition of mineral aerosols over the tropical North Atlantic; the influence of Saharan dust. M, 1978, University of Miami. 161 p.

Gladden, Scott Charles. Landslide hazardous areas, Bellevue, Washington. M, 1981, San Diego State University.

Gladen, L. W. Geology of lower Precambrian rocks and associated sulfide mineralization in an area south of Indus, Koochiching County, northern Minnesota. M, 1978, University of Minnesota, Duluth.

Glaeser, J. Douglas. Provenance, dispersal and depositional environments of Triassic sediments in the Newark-Gettysburg Basin. D, 1964, Northwestern University. 367 p.

Glaeser, John Douglas. Petrology of the basal Stockton (Triassic) lithofacies of eastern Montgomery County, Pennsylvania. M, 1958, Miami University (Ohio). 73 p.

Glagola, Peter A. Usefulness of clay minerals vs. environmental indicators in the Conemaugh shales (Pennsylvanian). M, 1973, West Virginia University.

Glaister, Roland P. Lower Cretaceous of southern Alberta and adjoining areas. D, 1958, Northwestern University.

Glancy, Patrick A. Cenozoic geology of the southeastern part of the Gallatin Valley, Montana. M, 1964, Montana State University. 66 p.

Glandon, Robert Paul. Phosphate exchange between littoral sediment and lake water. D, 1982, Michigan State University. 87 p.

Glanzman, Richard K. Fluoride in the confined ground water of the San Luis Valley, Colorado. M, 1972, Colorado School of Mines. 93 p.

Glasby, Virginia June. Silurian bedrock geology of the Muncie area. M, 1980, Ball State University. 54 p.

Glaser, Ann H. A petrographic and petrogenetic investigation of the metamorphosed ultramafic bodies in the Winston-Salem Quadrangle, northwestern North Carolina-southwestern Virginia. M, 1978, Miami University (Ohio). 76 p.

Glaser, Gerald C. Biostratigraphy of south central Lafourche Parish, Louisiana. M, 1962, Tulane University. 39 p.

Glaser, Gerald Clement. Lithostratigraphy and carbonate petrology of the Viola Group (Ordovician), Arbuckle Mountains, south-central Oklahoma. D, 1965, University of Oklahoma. 218 p.

Glaser, John Donald. Non-marine Cretaceous sedimentation in the Middle Atlantic Coastal Plain. D, 1967, The Johns Hopkins University. 467 p.

Glaser, Paul H. Recent plant-macrofossils from the Alaska interior and their relation to late-glacial landscapes in Minnesota. D, 1978, University of Minnesota, Minneapolis. 186 p.

Glasheen, Richard Michael. Geology of the Whetstone Ridge area, Meagher County, Montana. M, 1969, Oregon State University. 137 p.

Glasmann, Joseph Reed. Soil solution chemistry, profile development, and mineral authigenesis in several western Oregon soils. D, 1982, Oregon State University. 142 p.

Glaspey, Robin Gail. A sediment budget of the South Fork Rivanna Reservoir. M, 1981, University of Virginia. 83 p.

Glass, B. D. King Midas and old rip; the Gold Hill mining district of North Carolina. D, 1980, University of North Carolina, Chapel Hill.

Glass, Billy Price. Correlation of Pliocene and Pleistocene events in deep-sea sediments by geomagnetic reversals. D, 1968, Columbia University, Teachers College. 262 p.

Glass, Cecil Robertson. A process-response model for the Salt Fork of the Arkansas delta orifice, Great Salt Plains, Oklahoma. M, 1976, University of Tulsa. 65 p.

Glass, David Lawrence. A structural study of some upper Strawn-lower Canyon beds near Mineral Wells, Texas. M, 1958, Texas Christian University. 44 p.

Glass, Douglas Edward. Prediction of the thermal conductivity of marine sediments from some of their physical properties. M, 1977, University of Georgia.

Glass, Frank R. Structural geology of the Christiansburg area, Montgomery County, Virginia. M, 1970, Virginia Polytechnic Institute and State University.

Glass, Herbert D. Clay mineralogy of the coastal plain formations of New Jersey. D, 1956, Columbia University, Teachers College.

Glass, Herbert D. Fluorite-bearing clay from Darwin, California. M, 1947, Columbia University, Teachers College.

Glass, John Patrick. Analysis of freshwater lens formation in saline aquifers. D, 1979, University of Florida. 172 p.

Glass, Jon Lawrence. Depositional environments, reservoir trends and diagenesis of the Red Fork Sandstone in Grant and eastern Kay counties, Oklahoma. M, 1981, Oklahoma State University. 99 p.

Glass, Steven Wilbur. The Peterson Limestone, Early Cretaceous lacustrine carbonate deposition in western

Wyoming and southeastern Idaho. M, 1978, University of Michigan.

Glass, Susan Elizabeth. Amphissitid and kirkbyid ostracods from the Lower and lower Middle Pennsylvanian of Indiana and adjacent Kentucky. M, 1967, Indiana University, Bloomington. 57 p.

Glass, Theodore Gunter. Stratigraphy of certain Devonian beds in Alabama. M, 1935, University of Alabama.

Glassburn, Tracy. Principle components quantative shapes analysis on middle Miocene bivalves from southern Maryland. M, 1986, Lehigh University.

Glassen, Robert Carl. Oregon Inlet, North Carolina; a case history. M, 1971, University of Virginia. 63 p.

Glasser, Frederick Paul. The ternary system MgO-MnO-SiO$_2$. D, 1958, Pennsylvania State University, University Park. 106 p.

Glassey, Richard. Soil-water equilibria of three Virginia soil profiles. M, 1972, University of Virginia. 60 p.

Glassinger, Craig L. A statistical analysis of the Middle Ordovician conodonts of the Kimmswick Limestone of eastern Missouri. M, 1972, University of Missouri, Columbia.

Glassley, William Edward. A preliminary report on the chemistry of the Crescent volcanic sequence, Olympic Peninsula, Washington. M, 1971, University of Washington. 13 p.

Glassley, William Edward. Part 1, Geochemistry, metamorphism, and tectonic history of the Crescent volcanic sequence, Olympic Peninsula, Washington; Part II, Phase equilibria in the prehnite-pumpellyite facies. D, 1973, University of Washington. 137 p.

Glassman, Donald. The stratigraphy of the Chouteau Limestone of Boone County, Missouri. M, 1925, University of Missouri, Columbia.

Glauser, Alfred. Geology of the syenite belts near Tory Hill and Bancroft, Ontario. M, 1937, University of Toronto.

Glavinovich, Paul S. Trace-element copper distribution and areal geology in a portion of the Clearwater Mountains, Alaska. M, 1967, University of Alaska, Fairbanks. 55 p.

Glawe, Lloyd Neil. Pecten perplanus Stock (Oligocene) of Southeastern United States. D, 1966, Louisiana State University. 242 p.

Glaze, Michael V. A hydrologic study of northern Wood County. M, 1972, University of Toledo. 83 p.

Glazier, Robert M., Jr. Trace metal solubility in a South Texas oil field brine with implications for mississippi valley-type ore deposits. M, 1984, Pennsylvania State University, University Park. 103 p.

Glazner, Allen. Cenozoic evolution of the Mojave Block and adjacent areas. D, 1981, University of California, Los Angeles. 189 p.

Glazzard, Charles F. Streamflow characteristics related to channel geometry of selected streams on the Cumberland Plateau, Tennessee. M, 1981, University of Tennessee, Knoxville. 132 p.

Gle, Dennis Ray. The dynamic lateral response of deep foundations. D, 1981, University of Michigan. 293 p.

Gleason, Charles D. Faults of the eastern flank of the Wind River Mountains, Wyoming. M, 1931, University of Missouri, Columbia.

Gleason, Mark Lawrence. Habitat factors influencing vegetational zonation in two Chesapeake Bay marshes. M, 1976, University of Virginia. 94 p.

Gleason, Patrick James. The origin, sedimentation and stratigraphy of a calcitic mud located in the southern fresh-water Everglades. D, 1972, Pennsylvania State University, University Park. 369 p.

Gleason, Richard J. Wall-rock alteration, fluid inclusion, and mine water analyses of the Panteon vein system, Limon, Nicaragua. M, 1977, Dartmouth College. 105 p.

Gleba, Peter. Investigation of a copper prospect at Copper Mine Hill (Pawtucket Quandrangle), Cumberland, Rhode Island. M, 1967, Boston University. 20 p.

Gledhill, T. L. The gold quartz veins and igneous rocks of the Sturgeon Lake gold-field district of Thunder Bay, Ontario. D, 1925, Massachusetts Institute of Technology. 21 p.

Gleece, James B. The carbonate geochemistry and sedimentology of the Keys of Florida Bay, Florida. M, 1962, Florida State University.

Gleeson, Christopher F. The geology and mineralization of the Pegma Lake area in New Quebec. M, 1956, McGill University.

Gleim, David Thomas. Stratigraphy and paleontology of the Lower Pennsylvanian rocks in southeastern Iowa. D, 1955, University of Iowa. 256 p.

Gleim, David Thomas. Stratigraphy and structure of the upper Labarge Creek area, Lincoln County, Wyoming. M, 1952, University of Iowa. 169 p.

Glen, Craig Richard. Stratigraphy, petrology and sedimentology of the Duwi Formation (Late Cretaceous, eastern Egypt). M, 1980, University of California, Santa Cruz.

Glen, William. Pliocene and lower Pleistocene of the western part of the San Francisco Peninsula. M, 1957, University of California, Berkeley. 119 p.

Glenday, Keith Stuart. Deposition and diagenesis of the upper Mount Head Formation (Mississippian), Highwood River, Alberta. M, 1981, University of Manitoba. 141 p.

Glendinning, Gerald R. The glacial geomorphology of an area near Calgary, Alberta, between the Bow River and Fish Creek valleys. M, 1974, University of Calgary. 145 p.

Glendinning, Robert Morton. The Lake St. Jean lowland, Quebec. D, 1933, University of Michigan.

Glenister, Brian Frederick. Upper Devonian ammonoids from the Manticoceras Zone, Fitzroy Basin, Western Australia. D, 1956, University of Iowa. 145 p.

Glenister, Linda Marie. High resolution stratigraphy and interpretation of the depositional environments of the Greenhorn Cyclothem regression (Turonian; Cretaceous), Colorado Front Range. M, 1985, University of Colorado.

Glenn, Cheryl A. Dudo. A quantitative study of glaciated and periglacial hillslopes and drainage basins in the Slippery Rock area. M, 1974, Slippery Rock University. 83 p.

Glenn, Craig Richard. Sedimentologic and geochemical evidence bearing on the origin of Cretaceous marine phosphorites of Egypt; contrasts to the modern upwelling analogue. D, 1987, University of Rhode Island. 445 p.

Glenn, David Hendrix. Mississippian rocks in the subsurface of the North Tulsa area, Oklahoma. M, 1963, University of Oklahoma. 78 p.

Glenn, E. Charlotte. Morrowan conodonts from Garrett Hollow, Washington County, Arkansas. M, 1972, University of Arkansas, Fayetteville.

Glenn, George Rembert. X-ray studies of lime-bentonite reaction products. D, 1963, Iowa State University of Science and Technology. 265 p.

Glenn, Howard E. An investigation of the physical properties of samples of western Kentucky sand, sandstone, and gravel. M, 1927, [University of Kentucky].

Glenn, Jerry Lee. Late Quaternary sedimentation and geologic history of the Willamette Valley, Oregon. D, 1965, Oregon State University. 231 p.

Glenn, Jeryy Lee. Missouri River studies; alluvial morphology and engineering soil classification. M, 1960, Iowa State University of Science and Technology.

Glenn, John M. Stratigraphy and depositional environment of middle Bloyd Sandstone (Pennsylvanian); Madison and Washington counties, Arkansas. M, 1972, University of Arkansas, Fayetteville.

Glenn, L. C. A contribution to the study of the Pelecypoda of the Miocene of Maryland. D, 1899, The Johns Hopkins University.

Glenn, Lawrence Edward. Heavy mineral study of the recent beach and three terraces near Watsonville, California. M, 1974, Stanford University.

Glenn, Richard Allen. Permeability of sandstone cores. D, 1953, Washington University. 136 p.

Glenn, Robert Jerrell. Water resources of Warm Springs Valley, Washoe County, Nevada. D, 1968, University of Nevada. 93 p.

Glenn, Rosemary Thompson. Clay mineralogy and diagenesis of the Upper Permian Rustler Formation, near Carlsbad, New Mexico. M, 1986, University of New Mexico. 113 p.

Glenn, Scott Michael. A continental shelf bottom boundary layer model; the effects of waves, currents, and a moveable bed. D, 1983, Woods Hole Oceanographic Institution. 237 p.

Glenn, Sidney. Geology of the southern half of the Otisco Valley 7 1/2 minute quadrangle, New York. M, 1957, Syracuse University.

Glenn, Sidney E. Deposition and diagenesis of the Shafter Lake San Andres Formation, Andrews County, Texas. M, 1985, Texas Tech University. 108 p.

Glenn, W. H. and Webb, L. E. An investigation of longshore currents at Moss Landing (Monterey County) California. M, 1966, United States Naval Academy.

Glenn, William Edward. A study of electromagnetic and resistivity sounding with an application of generalized linear inversion. D, 1973, University of Utah. 39 p.

Glennie, James Stanley. Precambrian geology of the Piseco Lake area, south-central Adirondack Mountains, New York. D, 1973, Syracuse University.

Glennon, Mary Ann. A paleomagnetic investigation of glacial lacustrine (varved) sediment of the Kinnickinnic Member of the Pierce Formation, north-central Wisconsin. M, 1986, Michigan Technological University. 70 p.

Glerup, Melvin O. and Seefeldt, David R. Stream channels of the Scenic Member of the Brule Formation (Oligocene), western Big Badlands, South Dakota. M, 1958, South Dakota School of Mines & Technology.

Glerup, Melvin Obert. Economic geology of the Lime Point area, Nez Perce County, Idaho. M, 1960, University of Idaho. 40 p.

Gless, Ingrid Maria Verstraeten. Geochemistry and petrology of the Purcell lava flows Alberta, Canada. M, 1987, University of Nebraska, Lincoln. 156 p.

Glibota, Thomas J. Evaluation of thixotropic effects on hydraulic conductivity of compacted clay soils. M, 1988, University of Colorado. 138 p.

Glick, David C. Variability in the inorganic content of United States' coals; a multivariate statistical study. M, 1984, Pennsylvania State University, University Park. 538 p.

Glick, Linda Lee. Structural geology of the northern Toano Range, Elko County, Nevada. M, 1987, San Jose State University. 141 p.

Glicken, Harry. Rockslide-debris avalanche of May 18, 1980, Mount St. Helens Volcano, Washington. D, 1986, University of California, Santa Barbara. 463 p.

Glickstein, Sara. The skeleton of Pleuraspidotherium aumonieri; 1 volume. M, 1967, University of California, Berkeley.

Glidden, Philip Eugene. The geology of the Pittsfield Quadrangle, Maine. D, 1963, Boston University. 256 p.

Glines, Aubrey Leon. The insoluble residues of the Pennsylvanian limestones of Boone County, Missouri. M, 1926, University of Missouri, Columbia.

Gliozzi, James. A petrographic investigation of a Tertiary volcanic vent, Absaroka Mountains, Wyoming. M, 1959, Ohio State University.

Gliozzi, James. Petrology and structure of the Precambrian rocks of the Copper Mountain District, Owl Creek Mountains, Fremont County, Wyoming. D, 1967, University of Wyoming. 141 p.

Glissmeyer, Carl Howard. Microfauna of the Funk Valley Formation (Cretaceous), central Utah. M, 1959, University of Utah. 61 p.

Globensky, Yvon Raoul. Upper Mississippian conodonts from the Windsor Group of the Maritime Provinces. M, 1962, University of New Brunswick.

Globensky, Yvon Raoul. Upper Mississippian non-carbonate microfauna from the Windsor Group of the Atlantic provinces. D, 1965, University of New Brunswick.

Globerman, Brian R. Geology, petrology and paleomagnetism of Eocene basalts from the Black Hills, Washington Coast Range. M, 1980, Western Washington University. 373 p.

Globerman, Brian Rod. A paleomagnetic and geochemical study of Upper Cretaceous to lower Tertiary volcanic rocks from the Bristol Bay region, southwestern Alaska. D, 1985, University of California, Santa Cruz. 398 p.

Glock, Waldo S. Geology of the east-central part of the Spring Mountain Range, Nevada. D, 1925, Yale University.

Glockzin, Albert Richard. Structural geology of the Red Creek area, Colorado. M, 1942, Louisiana State University.

Gloeckler, Emily Frances. Leaching studies applied to Rb/Sr dating of glaucony from the Nanjemoy Formation of Virginia. M, 1988, University of North Carolina, Chapel Hill. 69 p.

Glohi, Boblai. Petrography of Upper Devonian gas bearing sandstones in the Indiana 7 1/2′ quadrangle, Well No. I1237 Ind25084, Indiana. M, 1984, University of Pittsburgh.

Gloor, Edward Alfred. Solubility of a natural rhyolitic glass in alkaline solutions. M, 1969, University of Wyoming. 35 p.

Glore, Charles Richard. A preliminary aquifer study of Pennsylvanian age sandstones, Busseron creek watershed, Sullivan County, Indiana. M, 1970, Indiana University, Bloomington. 47 p.

Glotfelty, Marvin Frank. Hydrogeology of the Camp Verde area, Yavapai County, Arizona. M, 1985, Northern Arizona University. 142 p.

Glover, Albert Douglas. Environmental maps of the Trivoli and Carlinville cyclothems, Pennsylvanian, and their equivalents. M, 1964, University of Illinois, Urbana.

Glover, Bernard K. Analysis of terraces in the Santa Ynez drainage basin, Santa Barbara County, California. D, 1971, University of California, Los Angeles.

Glover, D. W. Crustal structure of the Columbia Basin, Washington from borehole and refraction data. M, 1985, University of Washington. 71 p.

Glover, Dale Prince. A gravity study of the northeastern Piedmont Batholith of North Carolina. M, 1963, University of North Carolina, Chapel Hill. 41 p.

Glover, David Mark. Processes controlling radon-222 and radium-226 on the southeastern Bering Sea shelf. D, 1985, University of Alaska, Fairbanks. 157 p.

Glover, E. D. Cathodoluminescence, iron and manganese content, and the early diagenesis of carbonates. D, 1977, University of Wisconsin-Madison. 465 p.

Glover, J. Keith. Geology of the Summit Creek map area, southern Kootenay Arc, British Columbia. D, 1978, Queen's University. 144 p.

Glover, Joseph John Edmund. Studies in petrology and mineralogy of sedimentary rocks; I, Stratigraphy and structure of the Chico-Martinez-Bitterwater Creek area, Kern County, California; II, Investigation of authigenic albite with the universal stage. D, 1953, University of California, Berkeley. 110 p.

Glover, Lynn, III. Geology of the Coamo area, Puerto Rico; with comments on Greater Antillean volcanic island arc-trench phenomena. D, 1967, Princeton University. 478 p.

Glover, Lynn, III. The stratigraphy of the Devonian-Mississippian boundary in southwestern Virginia. M, 1953, Virginia Polytechnic Institute and State University.

Glover, Peter. Model for the Earth's crust beneath LASA, Montana, based on Rayleigh wave dispersion data. M, 1967, Stanford University.

Glover, Rebecca Marie. Diatom fragmentation in Grand Traverse Bay, Lake Michigan and its implications for silica cycling. D, 1982, University of Michigan. 300 p.

Glover, Sheldon Latta. Clays and shales of Washington. M, 1922, University of Washington. 368 p.

Glover, Thomas J. Geology and ore deposits of the northwestern Organ Mountains, Dona Ana County, New Mexico. M, 1975, University of Texas at El Paso.

Gluck, Leo. The Forest City Claim, Lake County (Yankee Hill) Colorado. M, 1886, Washington University.

Gluskoter, Harold Jay. A stratigraphic study of the Shoal Creek-Millersville interval in south-central Illinois. M, 1958, University of Iowa. 97 p.

Gluskoter, Harold Jay. Geology of a portion of western Marin County, California. D, 1962, University of California, Berkeley. 185 p.

Glynn, Pierre David. Thermodynamic behaviour of solid-solution aqueous-solution systems; a theoretical and experimental investigation. D, 1986, University of Waterloo. 235 p.

Gnabasik, Barbara J. The hydrogeology of the Niagara Escarpment near Neda, Dodge County, Wisconsin. M, 1985, University of Wisconsin-Milwaukee. 390 p.

Gnagy, Jean. The origin of granitic batholiths through granitization and related processes; an annotated bibliography. M, 1956, University of Washington. 90 p.

Gnidovec, D. M. Taphonomy of the Powder Wash vertebrate quarry, Green River Formation; Eocene, Uintah County, Utah. M, 1978, Fort Hays State University. 45 p.

Gnirk, Paul Farrell. An investigation of some aspects of contained explosion phenomena in rock. D, 1966, [University of Minnesota, Minneapolis]. 179 p.

Goad, Bruce E. Pegmatitic granites of the Winnipeg River area, southeastern Manitoba. M, 1984, University of Manitoba.

Goad, C. C. Application of digital filtering to satellite geodesy. D, 1977, Catholic University of America. 76 p.

Goad, Robin E. The geology, primary and secondary chemical dispersion of the Hemlo Au district metal occurrences, northwestern Ontario; Volumes I and II. M, 1987, University of Western Ontario. 578 p.

Goalen, Jeffrey S. The geology of the Elk Mountain/Porter Ridge area, Clatsop County, Oregon. M, 1988, Oregon State University. 356 p.

Gobel, Volker W. Geology and petrology of Mount Evans area, Clear Creek County, Colorado. D, 1972, Colorado School of Mines. 220 p.

Gobelman, Steven. Sublimation from reconstituted frozen silt. M, 1985, University of Alaska, Fairbanks. 106 p.

Goble, Robert S. The pre-Knox sequence and its relation to the Precambrian subsurface topography in eastern Kentucky. M, 1972, Eastern Kentucky University. 59 p.

Goble, Ronald James. The mineralogy, composition, and crystal structure of selected copper sulphides from the Belt-Purcell Supergroup, Southwest Alberta, Canada. D, 1977, Queen's University. 301 p.

Goble, Ronald James. The Yarrow creek, Spionkop creek copper deposit, southwestern Alberta (Proterozoic). M, 1970, University of Alberta. 116 p.

Gobran, Brian David. The effects of confining pressure, pore pressure and temperature on absolute permeability. D, 1981, Stanford University. 117 p.

Gocevski, Vladimir. Elasto-plastic two surface soil model and its finite element formulation and application. D, 1984, Concordia University.

Gockley, Catherine Kristin. Structure and strain analysis of the Big Elk Mountain Anticline, Caribou Mountains, Idaho. M, 1985, University of Colorado.

Godaih, Sulaiman H. A laboratory study of the collapse characteristics of a saline fine sand. M, 1984, Michigan Technological University. 80 p.

Godard, J. D. Wall-rock alteration at the Bridge River gold veins, British Columbia. M, 1953, McGill University.

Godard, Stephen Thomas. Depositional environments and sandstone trends of the Pennsylvanian Morrowan Series, southern Major and Woodward counties, Oklahoma. M, 1981, Oklahoma State University. 52 p.

Godbey, Will E. Detection of fluvial gravel beneath glacial till. M, 1982, Colorado School of Mines. 154 p.

Godchaux, Martha Miller. Petrology of the Greyback igneous complex and contact aureole, Klamath Mountains, southwestern Oregon. D, 1969, University of Oregon. 223 p.

Goddard, Edwin Newell. Geology and ore deposits of the Jamestown District, Boulder County, Colorado. D, 1936, University of Michigan.

Goddard, Ira and Geib, Horace Valentine. Reconnaissance erosion survey of the Brazos River watershed, Texas. D, 1933, Iowa State University of Science and Technology.

Goddard, John G. Th^{230}/U^{234} dating of saline deposits from Searles Lake, California. M, 1970, Queens College (CUNY). 50 p.

Godfrey, Charlie Brown. The geology of the Seagoville Quadrangle, Dallas and Kaufman counties, Texas. M, 1957, Southern Methodist University. 27 p.

Godfrey, Jack Martin. The subsurface geology of the Mannsville-Madill-Aylesworth Anticline. M, 1956, University of Oklahoma. 53 p.

Godfrey, John D. The deuterium content of hydrous rocks and minerals from the east-central Sierra Nevada and Yosemite National Park. D, 1962, University of Chicago. 30 p.

Godfrey, Joseph E., Jr. Hydrogeologic characteristics of the Jonesboro fault zone in Wake, Harnett, and Chatham counties, North Carolina. M, 1980, North Carolina State University. 83 p.

Godfrey, Pamela E. Geology and economic potential of southern half of the Kings Creek Quadrangle, South Carolina. M, 1983, University of Georgia.

Godfrey, Patricia Kathryn. A comparative study of New Jersey stilbites. M, 1982, Montclair State College. 79 p.

Godfrey, R. J. A non-Gaussian stochastic model applied to deconvolution and seismogram inversion. D, 1979, Stanford University. 97 p.

Godfrey, Stephen C. Rock groups, structural slices and deformation in the Humber Arm Allochthon at Serpentine Lake, western Newfoundland. M, 1982, Memorial University of Newfoundland. 182 p.

Godfrey, Warren C. Geology of calcrete and associated terrace deposits in southern Ellis and northern Hill counties, Texas. M, 1984, University of Texas, Arlington. 113 p.

Godkin, Carl B. Travel time inversion of multi-offset vertical seismic profiles. M, 1985, Massachusetts Institute of Technology. 147 p.

Godlewski, David W. Origin and classification of the middle Wallace Breccias. M, 1981, University of Montana. 74 p.

Godowic, Paul Francis. Secondary recovery of oil. M, 1954, Miami University (Ohio). 36 p.

Godoy, Estanislao Pirzio-Biroli. Geology of central eastern Isabella Quadrangle, Sierra Nevada, California. M, 1973, University of California, Berkeley. 27 p.

Godue, Robert. Etude métallogénique et lithogéochimique du Groupe de Magog, Estrie et Beauce, Québec. M, 1988, Universite du Quebec a Montreal. 70 p.

Godwin, A. G. Stable isotope analyses on postglacial fluvial and terrestrial molluscs from the Kincardine area of southern Ontario. M, 1985, University of Waterloo. 233 p.

Godwin, C. I. Geology of Casino porphyry copper-molybdenum deposit, Dawson Range, Y. T. D, 1975, University of British Columbia.

Godwin, David. A statistical study of bright spot reflection parameters. M, 1981, Texas A&M University.

Godwin, Larry H. Geology of the west side of Peavine Mountain, Washoe County, Nevada. M, 1958, University of Nevada. 59 p.

Godwin, Robert Paul. The Mossbauer effect in surface studies, Fe^{57} on Ag. D, 1966, University of Illinois, Urbana. 98 p.

Goebel, Edwin D. Stratigraphy of Mississippian rocks in western Kansas. D, 1966, University of Kansas. 380 p.

Goebel, Edwin DeWayne. Paleozoic formations of the Ironton Quadrangle, Missouri. M, 1951, University of Iowa. 82 p.

Goebel, J. E. The genesis of microknoll topography in Northeast Texas. M, 1971, East Texas State University.

Goebel, Joseph Edward. Glacial drift characteristics of Minnesota as revealed on Landsat imagery. D, 1978, Texas Tech University. 126 p.

Goebel, Katherine A. Stratigraphy, petrology and interpretation of the Iatan Limestone (Pedee Group, Upper Pennsylvanian), of northwestern Missouri and adjacent states. M, 1985, University of Iowa. 171 p.

Goebel, Leo R. Geology of a portion of the Lostine River valley, Wallowa County, Oregon. M, 1963, University of Oregon. 46 p.

Goebel, Vaughn. Modeling of the Peru-Chile Trench from wide-angle reflection profiles. M, 1974, Oregon State University. 88 p.

Goedicke, Thomas R. Geology of the ilmenite deposits of Caldwell and Watauga counties, North Carolina. D, 1953, [University of North Carolina, Chapel Hill].

Goedicke, Thomas Robert. The Boy Scout-Jones molybdenum property, Halifax County, North Carolina. M, 1948, North Carolina State University. 29 p.

Goeger, Donald Ernest. Geology of the Crooked Creek area, Fremont County, Wyoming. M, 1950, University of Missouri, Columbia.

Goeke, James W. The hydrogeology of Black Squirrel creek basin, El Paso County, Colorado. M, 1970, Colorado State University. 87 p.

Goel, Shailendra Kumar. Seismic wave attenuation, intensity and magnitude relation for Rocky Mountains. M, 1974, St. Louis University.

Goel, Subhash Chandra. Inelastic behavior of multistory building frames subjected to earthquake motions. D, 1968, University of Michigan. 170 p.

Goepfert, W. M. Contour and orthophoto mapping by optical matched filter correlation of complex exponentiated stereotransparencies. D, 1976, Iowa State University of Science and Technology. 165 p.

Goerner, Hugh H. A study on the interrelation of permeability, porosity, and density of rocks. M, 1947, University of Tulsa. 59 p.

Goerold, William Thomas. Geology and geochemistry of tin occurrences in southwestern New Mexico. M, 1981, Pennsylvania State University, University Park. 131 p.

Goerold, William Thomas. The effect of the recent gold price rise on gold mine production, exploration and investment in the United States and South Africa. D, 1983, Pennsylvania State University, University Park. 217 p.

Goers, John William. Geology and groundwater resources, Stockett-Smith River area, Montana. M, 1968, University of Montana. 150 p.

Goertz, Michael Joseph. Regression analysis of peak levels of earthquake strong ground motion; a theoretical study. M, 1980, St. Louis University.

Goetsch, Sherree Ann. The metamorphic and structural history of the Quartz Mountain-Lookout Mountain area, Kittitas County, central Cascades, Washington. M, 1978, University of Washington. 86 p.

Goetschius, David W. Preliminary sedimentological and geomorphological study of certain high terrace sands between the Ocklocknee and Apalachicola rivers, Liberty and Gadsden counties, Florida. M, 1971, Florida State University.

Goett, Harry. The geology of the Catoctin Formation (Precambrian) near Front Royal, Virginia. M, 1969, George Washington University.

Goettel, Kenneth Alfred. Partitioning of potassium, magnesium and calcium between silicates and sulfide melts. M, 1972, Massachusetts Institute of Technology. 48 p.

Goettel, Kenneth Alfred. Potassium in the Earth's core. D, 1975, Massachusetts Institute of Technology. 136 p.

Goettel, Mary Jane Westervelt. Aeolian sediment transport on Mars. M, 1972, Massachusetts Institute of Technology. 72 p.

Goettle, Marjorie S. Geological development of the southern portion of Assateague Island, Virginia. M, 1978, University of Delaware.

Goetz, Alexander Franklin Herman. Infrared 8-13µ spectroscopy of the Moon and some cold silicate powders. D, 1967, California Institute of Technology. 88 p.

Goetz, Carole L. A study of humic acid distribution throughout the Tampa Bay estuarine system. M, 1972, University of South Florida, St. Petersburg. 78 p.

Goetz, James E. A petrologic and petrographic investigation of some limestones near Brooksville, Florida (Hernando County). M, 1973, University of South Florida, Tampa. 42 p.

Goetz, Lisa K. Quaternary faulting in Salt Lake basin graben, West Texas. M, 1977, University of Texas, Austin.

Goetz, Michael J. An aerial photogrammetric survey of long-term shoreline changes, Nantucket Island, Massachusetts. M, 1978, University of Rhode Island.

Goetz, Peter Andrew. Depositional environment of the Sherridon Group and related mineral deposits near Sherridon, Manitoba. D, 1980, Carleton University. 248 p.

Goetze, Brigitte Ricarda. Contextual systems description of an Oregon coastal watershed. D, 1988, Oregon State University. 191 p.

Goetze, Christopher. High temperature elasticity and anelasticity of polycrystalline salts. D, 1970, Harvard University.

Goff, Fraser Earl. Vesicle cylinders in vapor-differentiated basalt flows. D, 1977, University of California, Santa Cruz. 183 p.

Goff, Kenneth J. Hydrology and chemistry of the Shoal Lakes basin, Interlake area, Manitoba. M, 1971, University of Manitoba.

Goff, Stephen Patrick. The magmatic and metamorphic history of the East Arm, Great Slave Lake, Northwest Territories. D, 1984, University of Alberta. 504 p.

Goff, William T. Sedimentation in a modern tidal inlet; Bogue Inlet, North Carolina. M, 1977, University of North Carolina, Chapel Hill. 117 p.

Goforth, Gary F. E. Technical feasibility of centrifugal techniques for evaluating hazardous waste migration. D, 1986, University of Florida. 122 p.

Goforth, Thomas Tucker. Theoretical and empirical studies of the response of the Earth to various types of seismic disturbances. D, 1973, Southern Methodist University. various pagination p.

Goforth, Tommy Tucker. Distribution of rubidium in the system KCl-RbCl-H$_2$O with applications to geo-

logic thermometry. M, 1962, University of Texas, Austin.

Gogas, John G. A gravity investigation of the Smith River basin and White Sulphur Springs area, Meagher County, Montana. M, 1984, Montana College of Mineral Science & Technology. 108 p.

Gogel, Anthony J. Ground water resources of northwestern Jefferson County, Kansas. M, 1969, University of Kansas. 142 p.

Goggin, David Jon. Geologically-sensible modelling of the spatial distribution of permeability in eolian deposits; Page Sandstone (Jurassic), northern Arizona. D, 1988, University of Texas, Austin. 436 p.

Gognat, Timothy A. Fracture trace analysis in southern Illinois. M, 1977, Southern Illinois University, Carbondale. 71 p.

Goh, Rocque Tien-Lock. A marine magnetic survey in the Mackenzie Bay/Beaufort Sea area, Arctic Canada. M, 1972, University of British Columbia.

Goh, Yong Soon. Natraqualfs and associated soils characteristics and their interactions with gypsum. D, 1984, Louisiana State University. 325 p.

Goheen, Hunter C. A structural analysis of Roanoke Dome, South Louisiana. M, 1962, [University of Houston].

Gohn, Gregory S. Metasedimentary units of the Marburg Formation, York County, Pennsylvania. M, 1973, University of Delaware.

Gohn, Gregory S. Sedimentology, stratigraphy, and paleogeography of lower Paleozoic carbonate rocks, Conestoga Valley, southeastern Pennsylvania. D, 1976, University of Delaware. 325 p.

Goin, Fred L. The olivine as foundry sand. M, 1938, University of Washington. 59 p.

Goin, John Samuel, Jr. Analysis of facies sequences and paleoenvironmental interpretation of the Kyrock Sandstone (Pennsylvanian), Edmonson County, Kentucky. M, 1985, Purdue University. 76 p.

Goings, David Bruce. Spring flow in a portion of Grand Canyon National Park, Arizona. M, 1985, University of Nevada, Las Vegas. 58 p.

Goins, Neal Rodney. Lunar seismology; the internal structure of the Moon. D, 1978, Massachusetts Institute of Technology. 666 p.

Goitom, Tesfai. Characteristics of Michigan cohesive subgrade soils under cyclic loading. D, 1981, Michigan State University. 325 p.

Gokce, Ali Onder. Engineering geology and relative stability of Main Ridge and part of Pleasanton Ridge, Alameda County, California. M, 1978, Stanford University. 64 p.

Gokturk, Erkin. Geophone array filtering. M, 1971, Colorado School of Mines. 61 p.

Golchin, Jahanshir. Remote sensing of land resources; application of Landsat satellite imagery. D, 1982, Iowa State University of Science and Technology. 197 p.

Gold, Arthur J. Conservation tillage; impact on agricultural hydrology and water quality in the Saginaw Bay drainage basin. D, 1983, Michigan State University. 160 p.

Gold, Christopher Malcolm. Quantitative methods in the evaluation of the Quaternary geology of the Sand River (73L) map sheet, Alberta, Canada. D, 1978, University of Alberta. 462 p.

Gold, D. P. and Canich, M. R. A study of the Tyrone-Mount Union lineament by remote sensing techniques and field methods. M, 1977, Pennsylvania State University, University Park. 92 p.

Gold, Irwin B. Foraminifera from the Marlbrook Formation of southwestern Arkansas. M, 1955, University of Oklahoma. 154 p.

Gold, K. M. Capillary theory of ore deposition. D, 1932, Massachusetts Institute of Technology.

Gold, Paul B. Diagenesis of middle and upper Miocene sandstones, Louisiana Gulf Coast. M, 1984, University of Texas, Austin.

Goldak, George Robert. A new method for the measurement of X-ray diffraction angles and unit cell dimensions. M, 1960, University of Saskatchewan. 29 p.

Goldak, George Robert. A new theoretical method for the calculation of absolute structure amplitudes from the Patterson function. D, 1967, University of Saskatchewan. 301 p.

Goldberg, David S. Sonic attenuation in consolidated and unconsolidated sediments. D, 1985, Columbia University, Teachers College. 146 p.

Goldberg, David Samuel. The physical properties of deep ocean sediments from the northern Atlantic; a comparison of in situ and laboratory methods. M, 1981, Massachusetts Institute of Technology. 143 p.

Goldberg, Irving. Fixation of molybdenum ions by clay colloids. M, 1950, University of California, Berkeley.

Goldberg, James. The geological significance of the coastal terraces of the Santa Monica Mountains. M, 1940, University of California, Los Angeles.

Goldberg, Jerald Melvin. Geology of the Copeville area, Collin County, Texas. M, 1949, Southern Methodist University. 36 p.

Goldberg, M. A. Geology of the northern half of the Mount Vernon Quadrangle, southeastern New York. M, 1977, Queens College (CUNY). 93 p.

Goldberg, Paul. Sediment analysis of two prehistoric rockshelters in Syria. M, 1968, University of Michigan.

Goldberg, Paul S. Sedimentology, stratigraphy and paleoclimatology of et-Tabun cave, Mount Carmel, Israel. D, 1973, University of Michigan.

Goldberg, Richard Henry. Depth-sensitive analysis of fluorine on lunar sample surfaces and in carbonaceous chondrites by a nuclear resonant reaction technique. D, 1976, California Institute of Technology. 132 p.

Goldberg, Sabine Ruth. A chemical model of phosphate adsorption on oxide minerals and soils. D, 1983, University of California, Riverside. 166 p.

Goldberg, Steven Amiel. Geochemical constraints on the origin of Proterozoic anorthosites, western United States. D, 1983, University of Oregon. 206 p.

Goldblatt, Rosann E. Rayome *see* Rayome Goldblatt, Rosann E.

Goldburg, Barbara L. Geometry and styles of displacement transfer, eastern Sun River canyon area, Sawtooth Range, Montana. M, 1984, Texas A&M University. 128 p.

Golden, Bruce L. Late Quaternary sedimentation on the Hatteras outer ridge, North Carolina continental rise. M, 1970, Duke University. 81 p.

Golden, Jerry B. Conodonts of the Everton Formation (Ordovician, Arkansas-Missouri). M, 1969, University of Missouri, Columbia.

Golden, Julia. Paleoecologic interpretations of Tertiary (middle Eocene-middle Oligocene) sediments based on the foraminifera (Clarke county, Alabama). M, 1969, Washington University. 153 p.

Golden, Paul W. Acoustic properties of thin plastic plates with application to two dimensional seismic modeling. M, 1986, Southern Methodist University. 154 p.

Goldenberg, Jeffrey E. An investigation of the effects of the solid earth tides on the triggering of volcanic eruptions in Central America. M, 1978, Rutgers, The State University, New Brunswick. 61 p.

Goldfarb, Marjorie Styrt. Hydrothermal sulfide deposits on the East Pacific Rise, 21°N. D, 1982, Massachusetts Institute of Technology. 336 p.

Goldfarb, Richard Jeffrey. A stream sediment geochemical reconnaissance of the Golden Trout Wilderness, California. M, 1981, University of Nevada. 242 p.

Goldhaber, Martin Bruce. Equilibrium and dynamic aspects of the marine geochemistry of sulfur. D, 1974, University of California, Los Angeles. 399 p.

Goldhaber, Martin M. The Chickamauga rocks of the Beaver Valley belt from Halls Crossroads, Knox County, to the Union County line, Tennessee. M, 1956, University of Tennessee, Knoxville. 38 p.

Goldhammer, Robert Kent. Constructive and destructive paleosoils capping regressive carbonate cycles, Cochise County, Arizona. M, 1982, University of Oklahoma. 65 p.

Goldhammer, Robert Kent. Platform carbonate cycles, Middle Triassic of northern Italy; the interplay of local tectonics and global eustasy. D, 1987, The Johns Hopkins University. 548 p.

Goldich, Samuel S. A study of rock-weathering. D, 1936, University of Minnesota, Minneapolis. 97 p.

Goldie, Raymond J. Petrology and geochemistry of a Monteregian sill near Ste. Dorothee, Quebec. M, 1972, McGill University. 202 p.

Goldie, Raymond James. The Flavrian and Powell plutons, Noranda area, Quebec; a geological investigation of the Flavrian and Powell plutons and their relationships to other rocks and structures of the Noranda area. D, 1976, Queen's University. 355 p.

Goldin, Alan. Effects of historical land use change on soils in the Fraser Lowland of British Columbia and Washington. D, 1986, University of British Columbia.

Goldman, Barbara Ellen. The extent of stream disequilibrium in northern Illinois. M, 1981, Northern Illinois University.

Goldman, Daniel. Morphology and systematics of the Middle Devonian Ambocoeliidae (Brachiopoda) of western New York. M, 1987, SUNY at Buffalo. 88 p.

Goldman, David John. Petrology of the upper Nugget Sandstone in the Wind River basin, Wyoming. M, 1963, University of Missouri, Columbia.

Goldman, David Marc. Estimating runoff prediction uncertainty using a physically-based stochastic watershed model. D, 1987, University of California, Davis. 373 p.

Goldman, Dennis. Application of a mathematical ground-water modeling technique. M, 1974, University of Idaho. 76 p.

Goldman, Dennis. Development of a low-temperature hydrothermal energy resource. D, 1982, University of Idaho. 298 p.

Goldman, Dennis. Refinements of geologic age and geographic locations for apparent polar-wandering. M, 1971, University of Illinois, Urbana. 95 p.

Goldman, Don. Owens Valley and its water. M, 1960, University of California, Los Angeles. 149 p.

Goldman, Don Steven. Crystal-field and Moessbauer applications to the study of site distribution and electronic properties of ferrous iron in minerals with emphasis on calcic amphiboles, orthopyroxene and cordierite. D, 1977, California Institute of Technology. 304 p.

Goldman, Gary C. The use of benthos to describe the long-term, time-averaged motion of the Saint Joseph River as it moves into Lake Michigan. D, 1973, University of Michigan.

Goldman, Harold B. Tertiary fluviatile deposits in the vicinity of Mokelumne Hills, Calaveras County, California. M, 1964, University of California, Los Angeles.

Goldman, Marcus I. Types of sediments of the Upper Cretaceous of Maryland. D, 1913, The Johns Hopkins University.

Goldmintz, Amy Jo. A theoretical study of surface-energy and interface-energy emphasizing applications to metamorphic systems. M, 1981, Rensselaer Polytechnic Institute. 252 p.

Goldrich, Samuel S. The mechanical composition of the till in the Syracuse region. M, 1930, Syracuse University.

Goldsberry, Stephen Lee. Permeability determination of in situ marine floor sediments. M, 1985, Baylor University. 84 p.

Goldsmith, Julian R. The system $CaAl_2Si_2O_8$-$Ca_2Al_2SiO_7$-$NaAlSiO_4$. D, 1947, University of Chicago. 23 p.

Goldsmith, Locke B. A quantitative analysis of some vein-type mineral deposits in southern British Columbia. M, 1984, University of British Columbia. 86 p.

Goldsmith, Louis H. A study of the Breece, New Mexico, Hishikari, Japan, and Mino, Japan, meteorites. M, 1950, University of New Mexico. 55 p.

Goldsmith, Richard. Petrography of Tiffany-Conconully area, Okanogan County, Washington. D, 1952, University of Washington. 356 p.

Goldsmith, Victor. Coastal processes of a Barrier Island Complex and adjacent ocean floor; Monomoy Island, Nauset Spit, Cape Cod, Massachusetts. D, 1972, University of Massachusetts. 469 p.

Goldsmith, Victor. The Recent sedimentary environment of the Choctawhatchee Bay, Florida. M, 1966, Florida State University.

Goldsmith, William Alee. Radionuclide retention from dilute solutions by Panamanian soil clays. D, 1968, University of Florida. 154 p.

Goldstein, Abram. Seismic stratigraphy and subsurface geology of Tongue of the Ocean, Bahamas. M, 1986, University of Delaware. 270 p.

Goldstein, Alan S. Recent motion of the geomagnetic field direction as recorded in sediments of Lake Tahoe, California-Nevada. M, 1970, University of Southern California.

Goldstein, Arthur Gilbert. Brittle fracture history of the Montague Basin, North-central Massachusetts. M, 1976, University of Massachusetts. 108 p.

Goldstein, Arthur Gilbert. Tectonics of ductile faulting in a portion of the Lake Char mylonite zone, Massachusetts and Connecticut. D, 1980, University of Massachusetts. 177 p.

Goldstein, August, Jr. Dakota Group (Lower Cretaceous) of the Colorado Front Range. D, 1948, University of Colorado.

Goldstein, August, Jr. Sedimentary petrologic provinces of the northern Gulf of Mexico. M, 1942, Louisiana State University.

Goldstein, B. A. Computer applications to conodont paleoecology in the Blackjack Creek and Excello (Marmaton Group, Desmoinesian Series) formations. M, 1977, University of Missouri, Columbia.

Goldstein, Barry Samuel. Stratigraphy, sedimentology, and late-Quaternary history of the Wadena drumlin region, central Minnesota. D, 1986, University of Minnesota, Minneapolis. 234 p.

Goldstein, Bruce Leon. Clay mineralogy of the Butano and Whiskey Hill formations in exposures near Stanford, California. M, 1981, Stanford University. 127 p.

Goldstein, Elaine. Comparison of the chemistry and mineralogy with the distribution and physical aspects of marine manganese concretions of the S. Ocean. M, 1981, Florida State University.

Goldstein, Flora J. Hydrogeology and water quality of Michaud Flats, southeastern Idaho. M, 1981, Idaho State University. 80 p.

Goldstein, Fredric R. The Pleistocene geology of a portion of Butler County, southwestern Ohio. M, 1968, Miami University (Ohio). 102 p.

Goldstein, Fredric Robert. Paleoenvironmental analyses of the Kirkwood Formation. D, 1974, Rutgers, The State University, New Brunswick. 70 p.

Goldstein, Gilbert. The geology of the Sweitzer Formation at San Diego, California. M, 1956, University of California, Los Angeles.

Goldstein, Myron. A paleomagnetic study of a Miocene transition in southeastern Oregon. M, 1967, Massachusetts Institute of Technology. 122 p.

Goldstein, Myron A. Magnetotelluric experiments employing an artificial dipolar source. D, 1971, University of Toronto.

Goldstein, Norman Edward. Numerical filtering of potential field signals as applied to geophysical exploration. M, 1962, University of California, Berkeley. 62 p.

Goldstein, Norman Edward. The separation of induced from remanent magnetism of near-surface rocks by means of in situ measurements. D, 1965, University of California, Berkeley. 205 p.

Goldstein, Peter. Array measurements of earthquake rupture. D, 1988, University of California, Santa Barbara. 211 p.

Goldstein, Robert Fritz. Comparison of Silurian chitinozoans from Florida well samples and the Red Mountain Formation in Northeast Alabama and Northwest Georgia. M, 1970, Florida State University. 90 p.

Goldstein, Robert Howell. Integrative carbonate diagenesis studies; fluid inclusions in calcium-carbonate cement; Paleosols and cement stratigraphy of Late Pennsylvanian cyclic strata, New Mexico. D, 1986, University of Wisconsin-Madison. 361 p.

Goldstein, Robert Howell. Stratigraphy and sedimentology of Pleistocene-Holocene fine-grained turbidite and ice-rafted sediments, Canada Basin, Arctic Ocean. M, 1981, University of Wisconsin-Madison. 161 p.

Goldstein, Steven Joel. Isotopic and chemical systematics of river waters. D, 1987, Harvard University. 277 p.

Goldstein, Steven Lloyd. Isotopic studies of continental and marine sediments, and igneous rocks of the Aleutian island arc. D, 1986, Columbia University, Teachers College. 376 p.

Goldstein, Stuart B. Geology of the civil war mine area, Lincoln County, Georgia. M, 1980, University of Georgia.

Goldstein, Susan Twyla. Biology of a Saccammina from San Francisco Bay. D, 1984, University of California, Berkeley. 151 p.

Goldstein, Susan Twyla. The distribution and ecology of benthic foraminifera in South Florida mangrove environment. M, 1976, University of Florida. 111 p.

Goldston, Walter L., Jr. Soils of the University Station area. M, 1916, University of North Carolina, Chapel Hill. 21 p.

Goldstone, Martin. Stereoradiography as applied to problems of petrology. M, 1963, University of Missouri, Rolla.

Goldstone, Selma L. A mechanical analysis of the Oriskany Sandstone of Northwest Virginia. M, 1938, Oberlin College.

Goldstrand, Patrick M. The Mesozoic stratigraphy, depositional environments, and tectonic evolution of the northern portion of the Wallowa Terrane, northeastern Oregon and western Idaho. M, 1987, Western Washington University. 200 p.

Goldthwait, James Walter. The abandoned shore lines of eastern Wisconsin. D, 1906, Harvard University.

Goldthwait, Richard Parker. The glacial geology of the Presidential Range (New Hampshire). D, 1939, Harvard University.

Golia, Ralph. Depositional environments and paleogeography of the upper Miocene Wassuk Group, Lyon and Mineral counties, west-central Nevada. M, 1983, San Jose State University. 105 p.

Golightly, John Paul. The 3d electronic spectrum of some minerals. D, 1968, McGill University. 153 p.

Golik, Abraham. Foraminiferal ecology and Holocene history, Gulf of Panama. D, 1965, University of California, San Diego. 211 p.

Golike, David C. Origin of sandstones in the Reedsville Formation (Ordovician), central Pennsylvania. M, 1981, University of Delaware. 109 p.

Goll, Carroll Leon. The geology of Seward County, Nebraska. M, 1961, University of Nebraska, Lincoln.

Goll, Robert Miles. Classification and phylogeny of Cenozoic Trissocyclidae (Radiolaria) in the Pacific and Caribbean basins. D, 1967, Ohio State University. 157 p.

Gollhofer, Rolla Linz. The Ste. Genevieve outliers of St. Louis County, Missouri. M, 1933, Washington University. 52 p.

Gollnick, Robert L. Phylogeny of the ostracod family Kirkbyidae with special reference to the genus Amphissites (G 582). M, 1941, University of Illinois, Urbana. 50 p.

Golomb, Berl. Paleogeography of the Basin of Mexico. D, 1965, University of California, Los Angeles. 295 p.

Golombek, Matthew Philip. Structural analysis of lunar grabens and the shallow crustal structure of the Moon. M, 1978, University of Massachusetts. 88 p.

Golombek, Matthew Philip. Structural analysis of the Pajarito fault zone in the Espanola Basin of the Rio Grande Rift, New Mexico. D, 1981, University of Massachusetts. 141 p.

Golovchenko, Xenia. Late Cenozoic history of sedimentation in the Blake-Bahama Basin. M, 1975, University of Delaware.

Golz, David Jon. The Eocene Artiodactyla of southern California. D, 1973, University of California, Riverside. 262 p.

Gomah, Aly Hemedah. Gelatin models for photoelastic study of slope stability in open-pit mines. D, 1965, University of Utah. 190 p.

Gomberg, David. Investigations of a deep-sea sediment core. M, 1972, University of Miami. 167 p.

Gomberg, David Norman. Geology of the Pourtales Terrace, Straits of Florida. D, 1976, University of Miami. 371 p.

Gomberg, Joan Susan. The structure of the crust and upper mantle of Mexico as inferred from seismic data. D, 1986, University of California, San Diego. 214 p.

Gombos, Andrew M., Jr. Neogene and Paleogene diatom stratigraphy in the region of the Falkland Plateau. D, 1976, Florida State University. 295 p.

Gombos, Andrew Michael, Jr. A study of some central Equatorial Pacific Neogene diatoms. M, 1973, University of Illinois, Urbana.

Gombos, Frances. The role of vermiculite in the fixation of potassium in soils and the availability of potassium for vine nutrition in some Niagara vineyards. M, 1977, Brock University. 116 p.

Gomer, M. D. Field evaluation of dispersivity and strontium Kd in a sandy aquifer. M, 1981, University of Waterloo. 86 p.

Gomes, Joao Bosco Ponciano. Paleogeologic study in Southwest Oklahoma and Texas Panhandle. M, 1959, Stanford University.

Gomes, Joao Bosco Ponciano. Study of the Upper Cretaceous along Marsh Creek, Contra Costa County, California. M, 1959, Stanford University.

Gomes, Patricia M. Influence of bedrock structure on ground-water occurrence and movement in the Stony Brook and Jacobs Creek drainage basins, Hunterdon and Mercer counties, New Jersey. M, 1988, Rutgers, The State University, Newark. 71 p.

Gomez Hernandez, Jose Jaime. Dynamic network design for estimation of groundwater flow and mass transport in a one-dimensional aquifer using the Kalman filter. M, 1988, Stanford University.

Gomez Reggio, Jose de Jesus. On the origin of the black pyritic slates from the iron mineral deposits of Iron River District of Michigan (Precambrian). M, 1970, University of Michigan.

Gomez, Ernest. Geology of the south central part of the New River Mesa Quadrangle, Cave Creek area, Maricopa County, Arizona. M, 1979, Northern Arizona University. 144 p.

Gomez, Filiberto P. The geology of the Sierra del Aguila, Chihuahua, Mexico. M, 1983, University of Texas at El Paso.

Gomez-Masso, A. J. Soil structure interaction in an arbitrary seismic environment. D, 1978, University of Texas, Austin. 311 p.

Gomez-Moran, Concepcion. Geochemistry and petrology of xenoliths from Xalapasco de la Joya, State of San Luis Potosi, Mexico. M, 1986, University of Houston.

Gomez-Trevino, Enrique. Geoelectrical soundings in the sedimentary basin of southern Ontario using a pseudo-noise source electromagnetic system. D, 1981, University of Toronto.

Gomez-Trevino, Enrique. The magnetometric resistivity response of two-dimensional structures. M, 1977, University of Toronto.

Goncalves Ferreira, Alfredo Augusto Cunhal. Influence of a shallow water layer over the soil in the erosion by raindrop impact. D, 1984, University of California, Davis. 281 p.

Gondouin, Michel. Analysis of the method of acoustic logging. D, 1952, Colorado School of Mines. 202 p.

Gonen, Behram. Building standards and the earthquake hazard for the Puget Sound basin. M, 1974, University of Washington. 139 p.

Gong, Henry. The geochemistry of cadmium. M, 1975, Pennsylvania State University, University Park. 114 p.

Gong, Jingyao. The evolution of the Canon City Embayment, Colorado. M, 1986, Oklahoma State University. 146 p.

Gonguet, Christophe. Mechanism of attenuation of compressional and shear waves for dry, water and benzene saturated rocks. M, 1985, Massachusetts Institute of Technology. 111 p.

Gonsalves, Ronald George. Foraminifera from the type Astoria Formation. M, 1965, University of California, Berkeley. 153 p.

Gonsiewski, James. Characteristics and distribution of large-scale fluvial features on the Martian surface. M, 1975, University of South Carolina.

Gonterman, J. Ronald. Petrographic study of the Precambrian basement rocks of Ohio. M, 1973, Ohio State University.

Gonthier, Joseph Bernard. The bedrock geology of the northern half of the Torrington Quadrangle, Connecticut. M, 1964, University of Massachusetts. 95 p.

Gontko, Robert N. Geology of the Pontotoc area, Mason, Llano, and San Saba counties, Texas. M, 1962, Texas A&M University.

Gonulden, Parisa. Heavy minerals of the Oakville Formation (Miocene; Texas). M, 1952, University of Texas, Austin.

Gonzales, Arsenio Geronimo. Geology and genesis of the Lepanto copper deposit, Mankayan, Mountain Province, Philippines. D, 1959, Stanford University. 122 p.

Gonzales, B. Norman. A contribution to the paleontology of the Paleozoic faunas of central Oregon. M, 1934, Oregon State University. 115 4 pls map p.

Gonzales, Benjamin Ray. Facies analysis of pinnacle reefs of the Guelph Formation (Middle Silurian), northern Michigan. M, 1981, Stephen F. Austin State University. 150 p.

Gonzales, David Alan. A geological investigation of the early Proterozoic Irving Formation, southeastern Needle Mountains, Colorado. M, 1988, Northern Arizona University. 151 p.

Gonzales, Eduardo. An analysis of volcaniclastic sediments, Pacific offshore, Guatemala. M, 1980, Texas Tech University. 81 p.

Gonzales, Orlando Jose. Significance of statistical parameters in the environmental interpretation of beach sediments. M, 1970, University of California, Los Angeles. 200 p.

Gonzales, Serge. Stratigraphy of the Delhi Quadrangle, east-central New York. D, 1963, Cornell University. 217 p.

Gonzales, Serge. The bedrock geology of the Shandon area, Butler County, Ohio. M, 1961, Miami University (Ohio). 327 p.

Gonzalez, Adelso Vera. A study of teleseismic P wave travel time residuals at Oklahoma seismograph stations. M, 1978, University of Oklahoma. 64 p.

Gonzalez, Alberto Rex. The stratigraphy of Intihuasi Cave, Argentina and its relationships to early lithic cultures of South America. D, 1959, Columbia University, Teachers College. 206 p.

Gonzalez, Argenis Rodriguez see Rodriguez Gonzalez, Argenis

Gonzalez, D. D. Hydraulic effects of underground nuclear explosions, Amchitka Island, Alaska. D, 1977, Colorado State University. 155 p.

Gonzalez, E., Jr. Paleo-environmental implications of shells from several Woodland Stage shell middens in northern New York City. M, 1978, Queens College (CUNY). 40 p.

Gonzalez, F. I., Jr. A descriptive study of the physical oceanography of the Ala Wai Canal. M, 1971, University of Hawaii. 176 p.

Gonzalez, James M. Skarn and sulfide replacement mineralization in the Ward District, White Pine County, Nevada. M, 1988, University of California, Los Angeles. 74 p.

Gonzalez, Jose Grover Percy. Subsurface geology of Strawn-Pennsylvanian Series, northwest quarter of Wise County, Texas. M, 1965, University of Texas, Austin.

Gonzalez, José Manuel Souto. Test of time-domain electromagnetic exploration for oil and gas. D, 1980, Colorado School of Mines. 152 p.

Gonzalez, Luis A. Carbon and oxygen isotopic composition of Recent marine carbonates. M, 1983, University of Michigan.

Gonzalez, Ralph Alan. Petrography and structure of the Pedernal Hills, Torrance County, New Mexico. M, 1968, University of New Mexico. 78 p.

Gonzalez-Bonorino, Gustavo. Sedimentology and paleogeography of a Devonian turbidite basin in Argentina. M, 1973, McMaster University. 137 p.

Gonzalez-Bonorino, Gustavo. Sedimentology and stratigraphy of the Curling Group (Humber Arm Supergroup), central western Newfoundland. D, 1979, McMaster University. 294 p.

Gonzalez-P., Gustavo C. A geological engineering study of the Red Fork Sand in the Oakdale Field, Woods County, Oklahoma. M, 1968, University of Oklahoma. 62 p.

González-Ruiz, Jaime Rogelio. Earthquake source mechanics and tectonophysics of the Middle America subduction zone in Mexico. D, 1986, University of California, Santa Cruz. 105 p.

Gonzalez-Serrano, Alfonso. Wave equation velocity analysis. D, 1982, Stanford University. 97 p.

Gooch, Dee David. Some Ostracoda of the genus Cythereis from the Cook Mountain Eocene of Louisiana. M, 1939, Louisiana State University.

Gooch, Edwin O. Geology of the Woodstock area, Shenandoah County, Virginia. M, 1949, University of Virginia. 54 p.

Gooch, Edwin Octavius. Infolded metasediments near the axial zone of the Catoctin Mountain-Blue Ridge Anticlinorium. D, 1954, University of North Carolina, Chapel Hill. 29 p.

Gooch, James L. Stratigraphy and petrography of the Cayuga Series (Upper Silurian and Lower Devonian) at Anthony Creek Gap, Greenbier County, West Virginia. M, 1967, West Virginia University.

Good, Charles Neil. The environment of deposition of glacial Lake Willard inferred from fossil Diatomacea. M, 1980, University of Akron. 70 p.

Good, George A. Picturesque geology. M, 1920, University of New Brunswick.

Good, John Conrad. MM-wavelength measurements of CO in the atmosphere of Mars and SO2 in the atmosphere of Venus. D, 1983, University of Massachusetts. 155 p.

Good, John Maxwell. Geology of the Eureka Quadrangle. M, 1948, Washington University. 62 p.

Good, L. W. Geology of the Baggs area, Carbon County, Wyoming. M, 1960, University of Wyoming. 90 p.

Good, Richard Standish. A chromographic study of nickel in soils and plants at the Lancaster Gap Mine, Pennsylvania. M, 1955, Pennsylvania State University, University Park. 53 p.

Good, William K. Reconnaissance geology and geochemistry of the Newman Peak 7.5 minute quadrangle, Camas County, Idaho. M, 1986, University of Idaho. 124 p.

Goodale, Jonathan L. An experimental and theoretical study of some quartz grain surface features, with applications to sediment gravity flows. M, 1975, University of Rhode Island.

Goodarzi, Nasrin K. Geomorphological and soil analysis of soil mounds in Southwest Louisiana. M, 1978, Louisiana State University.

Gooday, Andrew J. Taxonomy of ostracoda from the Saint Mary's Formation (middle Miocene) of Maryland. M, 1971, University of Wisconsin-Milwaukee.

Goodbody, Quentin Harald. Silurian Radiolaria from the Cape Phillips Formation of the Canadian Arctic Archipelago. M, 1981, University of Alberta. 388 p.

Goodbody, Quentin Harald. Stratigraphy, sedimentology, and paleontology of the Bird Fiord Formation, Canadian Arctic Archipelago. D, 1985, University of Alberta. 405 p.

Goodbread, Drew Robert. The effect of intrastratal solution on heavy minerals in the Minturn Formation (Pennsylvanian), Colorado. M, 1978, University of Oklahoma. 129 p.

Goode, Harry Donald. Surficial deposits, geomorphology and Cenozoic history of the Eureka Quadrangle, Utah. D, 1959, University of Colorado. 186 p.

Goode, Richard Whitfield, III. Vertical distribution of sulfide in the organic soils of the northern Everglades, Palm Beach County, Florida. M, 1982, University of Florida. 89 p.

Goode, Sterling D. The effect of combined triangulation and trilateration observations in the adjustment of a small network. M, 1963, Ohio State University.

Goodell, Horace Grant. The petrology and petrogenesis of the Frontier Sandstone of Wyoming. D, 1958, Northwestern University. 223 p.

Goodell, Michael W. Lake resistivity measurements and their relationship to hydrogeologic parameters. M, 1981, University of Wisconsin-Milwaukee. 164 p.

Goodell, Philip Charles. Zoning and paragenesis in the Julcani mining district, Peru. D, 1971, Harvard University.

Gooden, Don C. Earth science field trips for ninth grade students in the area of Mendota, Illinois. M, 1968, Northern Illinois University. 67 p.

Goodenkauf, O. The groundwater geology of southern Lancaster County, Nebraska. M, 1978, University of Nebraska, Lincoln.

Goodfellow, Robert W. Petrography and provenance of sandstones from the Otter Point Formation, southwestern Oregon. M, 1987, University of Oregon. 158 p.

Goodfellow, W. D. Rock geochemical exploration and ore genesis at Brunswick No. 12 deposit, New Brunswick. D, 1975, University of New Brunswick.

Goodfield, Alan Granger. Pleistocene and surficial geology of the city of Saint Louis and the adjacent Saint Louis County, Missouri. D, 1965, University of Illinois, Urbana. 280 p.

Goodfield, Alan Granger. Pleistocene geology of the Monroe city-center area, northeastern Missouri. M, 1963, University of Illinois, Urbana. 83 p.

Goodfriend, Cecil Thomas. Environmental significance of glauconite compared with that of other sedimentary iron minerals. M, 1952, University of Michigan.

Goodge, John William. Fold reorientation and quartz microfabric in the Okanogan Dome mylonite zone, Washington; kinematic and tectonic implications. M, 1983, University of Montana. 65 p.

Goodge, John William. Polyphase metamorphic evolution of the Stuart Fork Terrane; a Late Triassic subduction complex in the Klamath Mountains, Northern California. D, 1987, University of California, Los Angeles. 270 p.

Goodhue, William V., Jr. Bedrock geology and stratigraphy of the northern parts of the Armstead and Madigan Gulch anticlines, Beaverhead County, Montana. M, 1986, Oregon State University. 310 p.

Goodin, Sarah Elizabeth. Metamorphic country rocks of the southern Sierra Nevada, California; a petrologic and structural study to determine their early Mesozoic paleotectonic and paleogeographic setting. M, 1978, University of California, Berkeley. 75 p.

Gooding, Ansel M. Geology of the southwest portion of the Richwoods Quadrangle, Missouri. M, 1951, Iowa State University of Science and Technology.

Gooding, Ansel Miller. Geology of the southwest portion of the Richwoods Quadrangle, Missouri. M, 1957, University of Iowa. 96 p.

Gooding, Ansel Miller. Pleistocene terraces in the upper Whitewater drainage basin, southeastern Indiana. D, 1957, University of Iowa. 107 p.

Gooding, J. L. A high temperature study on the vaporization of alkalis from molten basalts under vacuum; a model for lunar volcanism. M, 1975, University of Hawaii. 111 p.

Gooding, James Leslie. Petrogenetic properties of chondrules in unequilibrated H-, L-, and LL-group chondritic meteorites. D, 1979, University of New Mexico. 392 p.

Gooding, Patrick J. Study of the unconformity at the top of the Knox Group (Cambrian-Ordovician) in the subsurface, south-central Kentucky. M, 1983, Eastern Kentucky University. 44 p.

Goodings, C. R. The Kapuskasing structure and its relationship to the Proterozoic movements in the Superior Province. M, 1988, University of Waterloo. 99 p.

Goodknight, Craig S. Structure and stratigraphy of the central Cimarron Range, Colfax County, New Mexico. M, 1973, University of New Mexico. 85 p.

Goodlin, Thomas Charles. Stratigraphic and structural relations of the area south of Hot Springs Canyon, Galiuro Mountains, Arizona. M, 1985, University of Arizona. 101 p.

Goodman, C. Studies in terrestrial radioactivity. D, 1940, Massachusetts Institute of Technology. 218 p.

Goodman, David K. Lower Eocene dinoflagellate assemblages from the Maryland Coastal Plain south of Washington, DC. M, 1975, Virginia Polytechnic Institute and State University.

Goodman, David Karns. Morphology, taxonomy and paleoecology of Cretaceous and Tertiary organic-walled dinoflagellate cysts. D, 1983, Stanford University. 515 p.

Goodman, Dean. Seismic refraction survey of crustal and upper mantle structures in the West Philippine Basin. M, 1983, Oregon State University. 122 p.

Goodman, Emery. Aspects of marine and non-marine sand dispersal and provenance, northern Puerto; Fourier grain shape analysis. M, 1980, University of South Carolina.

Goodman, Howard Mark. Comparison of three stratigraphic sections of the Wood River Formation in the vicinity of Bellevue, Idaho. M, 1984, Washington State University. 92 p.

Goodman, Jerome. An investigation of shoreline processes along Lake Erie between the Vermillion and Huron harbors, Erie County, Ohio. M, 1956, Ohio State University.

Goodman, Karen Jeanne. A comparison of a traditional and an independent laboratory approach in introductory college geology. M, 1975, Cornell University.

Goodman, Kathleen Stack. Petrology and mineral chemical studies of the Lawler Peak Granite and associated tungsten mineralization, Yavapai County, Arizona. M, 1986, University of Oklahoma. 99 p.

Goodman, Marjorie Jeanne. Mesozoic-Cenozoic boundary in the Rocky Mountain region. M, 1954, Mississippi State University. 103 p.

Goodman, Richard Edwin. Prospecting by photogeology. M, 1958, Cornell University.

Goodman, Richard Edwin. The stability of slopes in cohesionless materials during earthquakes. D, 1964, University of California, Berkeley. 259 p.

Goodman, Wayne Richard. The conodont fauna of the Floyds Knob Member, Borden Formation. M, 1975, University of Cincinnati. 289 p.

Goodman, Wyndal M. A hydrogeologic evaluation of the Northwest Arkansas waste management landfill, Washington County, Arkansas. M, 1985, University of Arkansas, Fayetteville.

Goodmen, William Walter. Structural evolution of Corbin Hill, a Precambrian massif in southeastern Dutchess County, New York. M, 1980, Syracuse University.

Goodner, D. C. Paleoecology of the late Cretaceous upper Frontier and Henefer formations (Wanship) near Coalville, Utah. M, 1975, University of Minnesota, Duluth.

Goodner, Ernest Francis. Four typical refractory clays of Washington. M, 1921, University of Washington. 70 p.

Goodney, D. E. Non-equilibrium fractionation of the stable isotopes of carbon and oxygen during precipitation of calcium carbonate by marine phytoplankton. D, 1977, University of Hawaii. 156 p.

Goodnow, Warren Hastings. A simulation of the shaft furnace process. D, 1969, Stanford University. 147 p.

Goodoff, L. R. Analysis of gravity data from the Cortaro Basin area, Pima County, Arizona. M, 1975, University of Arizona.

Goodrich, Cyrena Anne. Petrogenesis of native iron-carbon alloys, Disko Island, Greenland. D, 1983, Cornell University. 362 p.

Goodrich, Donald Larry. A sedimentary analysis of the Lower Devonian Bois Blanc Formation in Michigan. M, 1957, Michigan State University. 39 p.

Goodrich, Edward Arrott. The geology of northwestern Warren County. M, 1952, University of Missouri, Columbia.

Goodrich, J. A. Hydrogeology of Lucerne Valley, California. M, 1978, University of Southern California.

Goodrich, James Alan. Predicting toxic waste concentrations in community drinking water supplies from upstream industrial discharges; a vulnerability analysis. D, 1983, University of Cincinnati. 132 p.

Goodrich, Laurel E. A quadrupole lens ion source for mass spectrometers. M, 1963, University of Alberta. 60 p.

Goodrich, Ross Edgell. Application of the Thornthwaite and Blaney-Criddle methods of estimating irrigation water needs; a problem of scale. D, 1981, University of Wisconsin-Milwaukee. 128 p.

Goodrich, Willard Dale. The geology of the Perigo District, Gilpin County, Colorado. M, 1941, University of Kansas. 130 p.

Goodridge, Randolph. A review of Rocky Mountain structure. M, 1928, Yale University.

Goodroad, Lewis Leonard. Nitrous oxide production in natural and agricultural ecosystem soils of Wisconsin. D, 1983, University of Wisconsin-Madison. 160 p.

Goodrum, Christopher K. A paleoenvironmental and stratigraphic study of the Paleocene Fort Union Formation in the Cave Hills area of Harding County, South Dakota. M, 1983, South Dakota School of Mines & Technology.

Goodson, Robert H. Application of EMI conductivity methods and numerical filtering to mapping of the saltwater interface, Cape Coral, Florida. M, 1987, University of South Florida, Tampa. 69 p.

Goodspeed, Robert Marshall. An investigation of the coexisting feldspars from the Precambrian plutonic rocks in the Wanaque area (Passaic County), New Jersey. D, 1968, Rutgers, The State University, New Brunswick. 189 p.

Goodspeed, Robert Marshall. The petrography of the Togus Plutonic Complex, south-central Maine. M, 1962, University of Maine. 108 p.

Goodwill, David. Application of hydro-carbon pressure-volume-temperature laboratory data to reservoir engineering problems; practical consideration. M, 1952, University of Southern California.

Goodwin, Alan M. Metamorphism in rocks of the Evening-South Gabbro lakes area of Labrador. M, 1951, University of Wisconsin-Madison.

Goodwin, Alan M. The stratigraphy of the Gunflint iron-bearing formation of Ontario (Thunder Bay District). D, 1953, University of Wisconsin-Madison.

Goodwin, Bruce Kesseli. Geology of the Island Pond area, Vermont. D, 1959, Lehigh University. 210 p.

Goodwin, C. N. The geomorphology and sedimentology of a sand-bottom ephemeral stream. M, 1976, University of Wyoming. 91 p.

Goodwin, C. W. Process interaction and soil temperature near Point Barrow, Alaska. D, 1976, University of Michigan. 271 p.

Goodwin, Clinton J. Stratigraphy and sedimentation of the Yaquina Formation, Lincoln County, Oregon. M, 1973, Oregon State University. 121 p.

Goodwin, Elisabeth Rayner Krause. Fracture densities in the Rattlesnake Mountain fold, Wyoming. M, 1979, University of Oklahoma. 90 p.

Goodwin, Jeffrey Thomas. Determination of volatile sulfur compounds in aqueous solutions. M, 1982, Massachusetts Institute of Technology. 51 p.

Goodwin, Jonathan H. Authigenesis of silicate minerals in tuffs of the Green River Formation (Eocene), Wyoming. D, 1971, University of Wyoming. 123 p.

Goodwin, Joseph Grant. The geology of the southern half of the Strawberry Valley Quadrangle, California. M, 1952, University of California, Berkeley. 73 p.

Goodwin, Laurel Bernice. Structural studies of two strongly deformed terranes in California and Arizona. D, 1988, University of California, Berkeley. 143 p.

Goodwin, M. J. Design and testing of in situ devices for measuring geochemical retardation factors. M, 1980, University of Waterloo.

Goodwin, Michael Lawrence. Analysis of the Southeast Missouri earthquake of March 3, 1963. M, 1965, St. Louis University.

Goodwin, Michael Lawrence. Love and Rayleigh phase velocities over United States continental paths. D, 1968, St. Louis University. 225 p.

Goodwin, Peter. Sediment transport in unsteady flows. D, 1986, University of California, Berkeley. 227 p.

Goodwin, Peter B. Geomorphic interpretation of digital SPOT imagery; Hanaupah Canyon alluvial fan, Death Valley, California. M, 1988, Texas Tech University. 94 p.

Goodwin, Peter Warren. Fauna and stratigraphy of the Cambrian and Lower Ordovician of the Bighorn Mountains, Wyoming. M, 1961, University of Iowa. 73 p.

Goodwin, Peter Warren. Ordovician formations of Wyoming. D, 1964, University of Iowa. 201 p.

Goodwin, Ralph T. The organic content of oil shales, Colorado. D, 1922, Columbia University, Teachers College.

Goodwin, Robert Glenn. Neoglacial lacustrine sedimentation and ice advance, Glacier Bay, Alaska. D, 1982, Ohio State University. 272 p.

Goodwin, Robert Glenn. Vegetation response to the Two Rivers Till advance based on a pollen diagram from Kellners Lake, Manitowoc County, Wisconsin. M, 1976, University of Wisconsin-Madison.

Goodwin, Steven Dale. Factors influencing the interstitial water pH of sediments from the Potomac River estuary. M, 1980, University of Virginia. 96 p.

Goodwyn, James T., Jr. A study of the North Magnolia City oil field, Jim Wells County, Texas. M, 1951, Texas A&M University. 22 p.

Goodz, Morrie D. Geology and isotope geochemistry of the Beaver-Temiskaming Silver Mine cobalt, Ontario. M, 1985, Carleton University. 246 p.

Gooldy, Penn Lawrence. Geology of the Beaver Creek-South Sheep Mountain area, Fremont County, Wyoming. M, 1947, University of Wyoming. 75 p.

Goolsby, Jay Lee. A study of the Hickory Sandstone. M, 1957, Texas A&M University. 92 p.

Goolsby, Jimmy Earl. Cenozoic stratigraphy and geomorphology of Lynn and Terry counties, Texas. D, 1975, Texas Tech University. 98 p.

Goolsby, Jimmy Earl. East Rock Glacier of Lone Mountain, Madison and Gallatin counties, Montana. M, 1972, Montana State University. 74 p.

Goolsby, Robert Stark. Geology of the Lamy-Canoncito area, Santa Fe County, New Mexico. M, 1965, University of New Mexico. 68 p.

Gopal, Vijender Nath. Numerical simulation of pressures in a radial system. M, 1969, University of Missouri, Rolla.

Gopalapillai, Sivasithamparam. Non-global recovery of gravity anomalies from a combination of terrestrial and satellite altimetry data. D, 1974, Ohio State University. 108 p.

Gopfert, Wolfgang Martin. Contour and orthophot mapping by optical matched filter correlation of complex exponentiated stereotransparencies. D, 1976, Iowa State University of Science and Technology.

Gopinath, Tumkur R. Lithofacies and lebensspuren as criteria for distinguishing depositional environments; the Muddy Sandstone, Lower Cretaceous, Wind River Basin, Wyoming. D, 1976, Miami University (Ohio). 237 p.

Gopinath, Tumkur Raja Rao. Upper Caseyville (Lower Pennsylvanian) deltaic sedimentation; Jackson, Union, and Johnson counties, Illinois. M, 1972, Southern Illinois University, Carbondale. 84 p.

Goranson, Edwin A. I, The behavior of ore minerals in polarized light; II, Pre-Carboniferous geology of the Bras d'Or, Sydney, and Glace Bay map sheets, Cape Breton, Nova Scotia. D, 1933, Harvard University.

Goranson, Roy Walter. I, Thermodynamic relations in multicomponent systems; II, The solubility of water in granite magmas. D, 1931, Harvard University.

Gorczyca, Nancy Elizabeth. Effects of gasoline contamination on hydraulic conductivity and Atterberg limits. M, 1988, University of Wisconsin-Madison. 151 p.

Gord, Clarence E. The geology of the southern Boone County, Missouri. M, 1950, University of Missouri, Columbia.

Gordanier, Wayne Derek. Sedimentology and relationship to volcanology of Formation K, Favorable Lake metavolcanic-metasedimentary belt, northwestern Ontario. M, 1982, University of Manitoba.

Gorday, Lee. Hydrogeology of alluvial aquifers in or bordering dissected till plains. M, 1981, University of Missouri, Columbia.

Gordey, Steven P. Stratigraphy, structure, and tectonic evolution of the southern Pelly Mountains in the Indigo Lake near Yukon Territory. D, 1977, Queen's University. 200 p.

Gordji, Nasser. Microphotographic study of the effect of capillary forces on the interface behavior of liquids. M, 1966, University of Oklahoma. 77 p.

Gordon, Allen Stewart. Mathematical evaluation of mineral potential as an alternative for use in land planning systems. M, 1974, University of Idaho. 90 p.

Gordon, Andrew Hunt. Time series of, and extinction in, Cretaceous ammonites. M, 1969, University of Rochester. 38 p.

Gordon, Arnold L. Quantitative study of the dynamics of the Caribbean Sea. D, 1965, Columbia University. 236 p.

Gordon, Arthur J. Trace element geochemistry of the San Francisco Peaks, Arizona. M, 1977, Arizona State University. 96 p.

Gordon, Charles Henry. Syenite-gneiss (Leopard Rock) from the apatite region of Ottawa County, Canada. D, 1895, University of Chicago.

Gordon, Clarence E. Geology of Poughkeepsie Quadrangle. D, 1911, Columbia University, Teachers College.

Gordon, Clarence E. Studies in the plan of growth of the primary septa in the Rugosa. M, 1905, Columbia University, Teachers College.

Gordon, David Walker. Geological processes along the south shore of Lake Erie between Lakeline and Mentor-on-the-lake, Lake County, Ohio. M, 1956, Ohio State University.

Gordon, David Walker. Revised hypocenters and correlation of seismicity and tectonics in the Central United States. D, 1983, St. Louis University. 212 p.

Gordon, Donivan Lewis. A morphometric analysis of selected Iowa drainage basins. M, 1960, University of Iowa. 193 p.

Gordon, Elizabeth A. Petrology and field relations of Precambrian metasedimentary and metaigneous rocks west of Twin Bridges, southwestern Montana. M, 1979, University of Montana. 90 p.

Gordon, Elizabeth Adams. Sedimentology, paleohydraulics, and paleontology of the early Frasnian Upper Devonian Catskill facies, southeastern New York. D, 1987, SUNY at Binghamton. 187 p.

Gordon, Ellis Davis. Hydraulic rotary test drilling procedure in groundwater investigations. M, 1949, University of Nebraska, Lincoln.

Gordon, Ernest Rollin. The geology of the Twin Buttes mining district. M, 1922, University of Arizona.

Gordon, George E. The geology and paleontology of the Yegua-Jackson contact in Brazos County, Texas. M, 1957, University of Houston.

Gordon, James Eddie. Geology of Hutton Quadrangle, Williamson County, Texas. M, 1951, University of Texas, Austin.

Gordon, Joan Esther. The Upper Devonian stratigraphy and paleontology of the Silverhorn Dolomite, West Range Limestone, and Pilot Shale at Dutch John Mountain, Lincoln County, Nevada. M, 1962, University of Illinois, Urbana.

Gordon, Julia Perry. Channel changes on the lower Cañada del Oro, 1936-1980, and policies of flood plain management. M, 1983, University of Arizona. 112 p.

Gordon, Larry. Stratigraphy and paleoecology of the Brassfield Formation (Llandoverian) in east-central Kentucky. M, 1980, University of Kentucky. 113 p.

Gordon, Lawrence. An albitized aplite-cataclasite dike at Franklin, New Jersey. M, 1956, Rutgers, The State University, New Brunswick. 76 p.

Gordon, Michael. Hydrologic conditions affecting the growth of macrophytes and algae in the upper Tittabawassee River watershed, Michigan. M, 1982, Northern Arizona University. 95 p.

Gordon, Patrick T. Correlation of gravity observations and geology in the San Antonio region, Texas. M, 1968, Texas A&M University. 85 p.

Gordon, R. G. Plate motions relative to the paleomagnetic axis and the lower mantle. D, 1979, Stanford University. 151 p.

Gordon, S. I. Simulating the water quality impacts of new development around Owasco Lake, New York, using geographic based information. D, 1977, Columbia University, Teachers College. 245 p.

Gordon, Sadie C. B. Geographic influences in the history of Haiti. M, 1937, Columbia University, Teachers College.

Gordon, Stuart. Relations among the Santa Ynez, Pine Mountain, Agua Blanca and Cobblestone Mountain faults, Transverse Ranges, California. M, 1978, University of California, Santa Barbara.

Gordon, Susan L. The environmental geochemistry of total uranium and ^{226}Ra at the South March uranium occurrence, Ontario. M, 1986, Carleton University. 131 p.

Gordon, Terence Michael. Some silicate - carbonate phase relations in H_2O-CO_2 mixtures. D, 1969, Princeton University. 117 p.

Gordon, W. Richard. Toroweap and Kaibab formations in a part of Whitmore Wash, Mohave County, Arizona. M, 1961, University of Kansas. 119 p.

Gordon, Yoram. Water management alternatives for the Colorado river below the Imperial dam, Arizona. D, 1970, University of Arizona.

Gordon, Yoram. Water management for the area downstream from the Imperial dam on the Colorado River. M, 1968, University of Arizona.

Gore, Clayton Edwin. The geology of a part of the drainage basins on Spavinaw, Salina, and Spring creeks, northeastern Oklahoma. M, 1952, University of Tulsa. 79 p.

Gore, Dorothy Jean. Differentiation of the Peorian Loess in the Peoria area. M, 1952, University of Illinois, Urbana.

Gore, Dorothy Jean. Potash metasomatism in granitization as illustrated in the rocks of the Niagara and Neillsville areas of Wisconsin. D, 1963, University of Wisconsin-Madison. 157 p.

Gore, Larry D. The sedimentology, paleontology, and depositional environment of the Precambrian Allamoore Formation, Culberson County, Texas. M, 1985, Texas A&M University.

Gore, Pamela Jeanne Wheeless. Sedimentology and invertebrate paleontology of Triassic and Jurassic lacustrine deposits, Culpeper Basin, northern Virginia. D, 1983, George Washington University. 375 p.

Gore, Richard Z. An evaluation of total field magnetometer surveying in a metamorphic terrain (eastern Massachusetts). M, 1967, Boston College.

Gore, Richard Z. Geology of the porphyritic Ayer Quartz Monzonite and associated rocks in portions of the Clinton, and Ayer quadrangles, Massachusetts. D, 1973, Boston University. 388 p.

Gore, Roger C. A mathematical model of geophysical exploration for elongated ore deposits. M, 1958, Massachusetts Institute of Technology. 35 p.

Goreau, Peter David Efran. The tectonic evolution of the north-central Caribbean plate margin. D, 1981, Massachusetts Institute of Technology. 245 p.

Goreau, Thomas Joaquin. Biogeochemistry of nitrous oxide. D, 1981, Harvard University. 150 p.

Gorelick, Steven Marc. Numerical management models of groundwater pollution. D, 1980, Stanford University. 154 p.

Gorham, Julie M. Seismic velocities of Archean metagreywackes and metavolcanic rocks, Wind River Range, Wyoming. M, 1986, University of Wyoming. 89 p.

Gorham, Scott Brady. Digital high resolution shallow-marine seismic data. M, 1981, University of Texas, Austin. 152 p.

Gorham, Timothy W. Geology of the Galisteo Formation, Hagan Basin, New Mexico. M, 1979, University of New Mexico. 136 p.

Goring, Arthur William. Refractory and petrographic properties of Pacific Northwest chromites. M, 1947, University of Washington. 134 p.

Gorini, M. A. The tectonic fabric of the equatorial Atlantic and adjoining continental margins; Gulf of Guinea to northeastern Brazil. D, 1977, Columbia University, Teachers College. 429 p.

Goris, James. Pressure molded high strength concrete. M, 1962, University of Utah. 62 p.

Gorman, Angela K. Calcareous nannofossils as paleoclimatic indicators in the equatorial and tropical Atlantic, last glacial to Recent. M, 1985, University of Utah. 111 p.

Gorman, Barry E. Petrography, chemistry, and mechanism of deposition of the Don Rhyolites, Rouyn-Noranda, Quebec. M, 1975, Queen's University. 216 p.

Gorman, Barry Edward. A model of flow and fracture in plagioclase; examples from shear zones, Fiskenaesset Complex, West Greenland. D, 1980, University of Western Ontario. 293 p.

Gorman, Charles M. Geology, geochemistry and geochronology of the Rattlesnake Creek Terrane, west-central Klamath Mountains, California. M, 1985, University of Utah. 111 p.

Gorman, D. H. The uranium silicate minerals. D, 1957, University of Toronto.

Gorman, Donald Robert. A study of the Amsden Formation. D, 1962, University of Illinois, Urbana. 181 p.

Gorman, Kelly M. Petrographic analysis of syn-sedimentary and diagenetic features of the High Bridge Group (Middle Ordovician) in the subsurface, Boone County, northern Kentucky. M, 1984, Eastern Kentucky University. 88 p.

Gorman, William Alan. Acid intrusives of the Thetford Mines-Black Lake area, Quebec. M, 1952, McGill University.

Gorman, William Alan. Areal geology of Ste. Justine area, Bellechasse and Dorchester counties, 1 inch to 1 mile, Quebec. D, 1956, McGill University.

Gorman, William Albert. The geology of the lower Seine River area, Rainy River District, Ontario. D, 1933, University of Minnesota, Minneapolis. 64 p.

Gormican, Sheila Catherine. Depositional environment of the Yates Formation in Kermit Field, Winkler, Texas. M, 1988, Texas A&M University. 101 p.

Gornitz, Vivien Monisa. A study of halite from the Dead Sea. M, 1965, Columbia University. 33 p.

Gornitz, Vivien Monisa. Mineralization, alteration, and mechanism of emplacement, Orphan ore deposit, Grand Canyon, Arizona. D, 1969, Columbia University. 186 p.

Gorody, Anthony Wagner. The Lower Ordovician Mascot Formation, upper Knox Group, in North central Tennessee; Part I, Paleoenvironmental history; Part II, Dolomitization and paleohydraulic history. D, 1980, Rice University. 263 p.

Gorrell, H. A. Geologic studies of formations exposed in the T.T.C. subway excavations. M, 1952, University of Toronto.

Gorski, Daniel Everett. Geology and trace transition element variation of the Mitre Peak area, Trans-Pecos Texas. M, 1970, University of Texas, Austin.

Gorsline, Donn Sherrin. Marine geology of San Pedro and Santa Monica basins and vicinity, California. D, 1958, University of Southern California.

Gorsline, Donn Sherrin. Sedimentation in Sebastian Viscaino Bay and vicinity, Baja California, Mexico. M, 1954, University of Southern California.

Gortner, Catherine Willis. Hemipelagic rocks at Bissex Hill, Barbados and a survey of silica diagenesis in accreted rocks on Barbados. M, 1984, Stanford University. 96 p.

Gorton, Kenneth Arnold. Geology of the Cameron Pass area, Grand, Jackson, and Larimer counties, Colorado. D, 1941, University of Michigan.

Gorton, Kenneth Arnold. Structural geology of Wayne County and its bearing on possible oil accumulation. D, 1941, University of Michigan.

Goruk, Gerald L. Petrography of Middle Triassic cross-bedded sandstones in northeastern British Columbia. M, 1963, McMaster University. 83 p.

Gorveatt, Arnold Charles. Permo-Pennsylvanian spiriferids from the Yukon Territory. M, 1967, University of Calgary. 92 p.

Gorycki, Michael A. Ultrastructure of living and fossil foraminifera; a microscopical technique for testate micoorganisms. M, 1967, New York University.

Gorycki, Michael Anthony. Pyroxene geothermometry of the Adirondack Lowlands. D, 1981, Rutgers, The State University, New Brunswick. 134 p.

Gorzynski, George Arthur. Geology and lithogeochemistry of the Cirque stratiform sediment-hosted Ba-Zn-Pb-Ag deposit, northeastern British Columbia. M, 1986, University of British Columbia. 129 p.

Goscinski, John S. The petrology and coking behavior of certain Western Canadian coal and their significance in predicting coke strength. M, 1973, Pennsylvania State University, University Park.

Gose, Wulf Achim. Contributions to paleomagnetism and mineralogy; A, Paleomagnetic studies of Miocene ignimbrites from Nevada; B, Paleomagnetic and rock magnetic studies of the Permian Cutler and Pennsylvanian Rico formations, Utah; C, Mössbauer studies and chemical variations in a large phosphate crystal. D, 1970, Southern Methodist University. 102 p.

Goshorn, JoAnn H. Foraminiferal paleoenvironments of the Eastover Formation (upper Miocene, Virginia). M, 1985, Old Dominion University. 255 p.

Gosman, Robert F. Stratigraphic analysis of the Jurassic of the Western Interior. M, 1953, Dartmouth College. 76 p.

Gosnell, Gary Johnston. Three dimensional analysis of hinged end pile groups in a layered soil system. D, 1968, University of Virginia. 127 p.

Gosnell, George Edward. Structure of the Osage Limestone. M, 1940, University of Arkansas, Fayetteville.

Gosney, Terry C. Conodont biostratigraphy of the Pinyon Peak Limestone and the Fitchville Formation, Late Devonian-Early Mississippian, northern Salt Lake City, Utah. M, 1982, Brigham Young University. 38 p.

Gosnold, William David, Jr. A model for uranium and thorium assimilation by intrusive magmas and crystallizing plutons through interaction with crustal fluids. D, 1976, Southern Methodist University. 131 p.

Gospodarec, Judith A. Clay mineralogy and diagenesis in the Hockingport Sandstone lentil, Dunkard Group (Pennsylvanian-Permian) in Ohio and West Virginia. M, 1983, Miami University (Ohio). 73 p.

Goss, Brian Glen. An analysis of fluid-rock interactions at the Ely porphyry copper deposit by utilization of fluid inclusions. M, 1983, Pennsylvania State University, University Park. 79 p.

Goss, Charles Richard. Geology of the southwest corner of the Calistoga Quadrangle, California. M, 1948, University of California, Berkeley. 47 p.

Goss, Don Woodson. Mica weathering as related to mica species and soil parent material in the North Carolina Slate Belt. D, 1968, North Carolina State University. 177 p.

Goss, Ronald. Empirical relationships between thermal conductivity and other physical parameters in rocks. D, 1974, University of California, Riverside. 216 p.

Goss, William A. Uranium mineralization at the Madawaska Mine, Bancroft, Ontario. M, 1981, Bowling Green State University. 54 p.

Gosselin, Charles. Etude paléo-environnementale d'un faciès à Crinoides silurien, Formation de West Point, Port-Daniel, Gaspésie, Québec. M, 1982, Universite Laval. 46 p.

Gosselin, David Charles. Geology, geochemistry, and petrology of Archean rocks from the Black Hills, South Dakota. D, 1987, South Dakota School of Mines & Technology.

Gosser, Charles F. Petrography and metamorphism (of the Precambrian rocks) of the Star Lake area, Montana. M, 1960, Wayne State University.

Gossett, Charles Joseph. Cambrian stratigraphy in the Big Horn Mountains, Wyoming. M, 1957, University of Illinois, Urbana.

Gossett, Lloyd David. Semiquantitative analysis of crude oil in lacustrine sediments. M, 1973, University of Tulsa. 26 p.

Gossiaux, Barbara Marie. The structural geology of poly-deformed low-grade metasediments in a portion of Tudor Township, southeastern Ontario. M, 1986, University of Windsor. 122 p.

Goswami, Dulal Chandra. Brahmaputra River, Assam (India); suspended sediment transport, valley aggradation and basin denudation. D, 1983, The Johns Hopkins University. 199 p.

Goswami, Ram Kishore. A study of wetlands using geochemical, remote sensing, and multivariate analytical techniques. D, 1982, University of Tennessee, Knoxville. 137 p.

Gotautas, Vito A. The Ostracoda of the Saluda Formation. M, 1951, Miami University (Ohio). 50 p.

Goter, Edwin R., Jr. Depositional and diagenetic history of the windward reef of Enewetak Atoll during the mid to late Pleistocene and Holocene. D, 1979, Rensselaer Polytechnic Institute. 239 p.

Goter, Edwin Robert, Jr. Stratigraphy and sedimentary petrology of the Fort Pena Formation, Marathon Basin, Texas. M, 1974, University of Texas, Austin.

Goth, Joseph Herman, Jr. Fracture porosity in relation to oil reservoirs. M, 1949, University of Pittsburgh.

Gottesfeld, Allen S. Paleoecology of the Chinle Formation (upper Triassic), Petrified Forest National Park, Arizona. M, 1969, University of Arizona.

Gottschalk, Marlin Ralph. The effects of sulfur dioxide and acid precipitation on decomposition nutrient cycling processes in a southeastern deciduous forest soil. D, 1981, Emory University. 112 p.

Gottschalk, Richard Robert. Hydrothermal metasomatic banding in alpine-type peridotites. M, 1979, University of Arizona. 109 p.

Gottschalk, Richard Robert, Jr. Structural and petrologic evolution of the southern Brooks Range near Wiseman, Alaska. D, 1987, Rice University. 306 p.

Gottsdanker, Eugene Nathan. Sedimentation in the Sespe Formation north of Simi Valley, Ventura County, California. M, 1939, University of California, Los Angeles.

Goud, Margaret Redding. Prediction of continental shelf sediment transport using a theoretical model of the wave-current boundary layer. D, 1987, Woods Hole Oceanographic Institution. 211 p.

Goudarzi, Hossein. Geology and ore deposits of the Quartz Hill mining area, Beaverhead County, Montana. M, 1941, Montana College of Mineral Science & Technology. 52 p.

Goudreault, Paul Richard. A two-dimensional, finite-difference groundwater flow model of the Minneapolis chain of lakes and surrounding surficial water table aquifer. M, 1985, University of Minnesota, Minneapolis. 139 p.

Goudy, Clyde LeRoy. A structural and petrographic study of the Vaqueros and associated formations in the southeast portion of the Adelaide Quadrangle, California. M, 1936, University of California, Berkeley. 66 p.

Gough, Lia Ann Fong. Fluid compositions buffered in scapolite-bearing skarns in the Adirondacks, N.Y. M, 1984, University of Michigan.

Gough, William R. The geology and water resources of the Milesburg-Sayers Dam area, Pennsylvania. M, 1977, Pennsylvania State University, University Park. 163 p.

Gouin, Leon Oliver. Metamorphism at the Andrew Yellowknife Property, Northwest Territories. M, 1948, University of British Columbia.

Gouin, Pierre. An electronic seismic transducer. M, 1954, Boston College.

Gould, Charles Newton. Geology and water resources of Oklahoma. D, 1906, University of Nebraska, Lincoln.

Gould, Charles Newton. The Dakota Formation of the Great Plains, with descriptions of fossil leaves in the Nebraska State Museum. M, 1900, University of Nebraska, Lincoln.

Gould, David R. Provenance study of the Palmerton Sandstone. M, 1975, Lehigh University.

Gould, Donald B. Stratigraphy and structure of the Pennsylvanian and Permian rocks in the Salt Creek area, Mosquito Range, Colorado. D, 1934, University of Iowa. 88 p.

Gould, Donald B. The stratigraphy and paleontology of the Fort Riley Limestone in northern Oklahoma. M, 1930, University of Iowa. 116 p.

Gould, Franklin David. Geology of the Paskenta District, Tehama County, California. M, 1962, California State University, Chico. 48 p.

Gould, George F. Maturation and alteration of crude oils in the Cherokee Group (Middle Pennsylvanian) of southeastern Kansas. M, 1975, University of Kansas. 43 p.

Gould, Gerald. Numerical simulation of groundwater flow in bedrock aquifers of southwestern New York and northwestern Pennsylvania. M, 1987, Syracuse University. 80 p.

Gould, Howard Ross. Lake Mead sedimentation. D, 1953, University of Southern California.

Gould, J. Lineation analysis from aerial photographs of the Sierrita Mountains, Pima County, Arizona. M, 1973, University of Arizona.

Gould, James G. A subsurface structural trend in western Rice County, Kansas. M, 1952, University of California, Los Angeles.

Gould, John David. The geology of the southeast corner of Travis County, Texas. M, 1949, University of Texas, Austin.

Gould, Joseph Charles. A study of the Ordovician Ostracoda below the green chert horizon in Northwest Georgia. M, 1957, Emory University. 47 p.

Gould, Laurence McKinley. Geology of the La Sal Mountains of Utah. D, 1925, University of Michigan.

Gould, Martin James. The ground roll phenomenon of applied seismology. D, 1941, California Institute of Technology. 114 p.

Gould, Ramon John. A study of olivine as a foundry sand. M, 1953, University of Washington. 84 p.

Gould, Stephen Jay. Pleistocene and Recent history of the subgenus P. (Poecilozonites) (Gastropoda, Pulmonata) in Bermuda; an evolutionary microcosm. D, 1967, Columbia University. 281 p.

Gould, Wilburn James. Geology of the northern Needle Range, Millard County, Utah. M, 1959, Brigham Young University. 47 p.

Goulden, Clyde Edward. The history of the cladoceran fauna of Esthwaite Water (England) and its limnological significance. D, 1962, Indiana University, Bloomington. 94 p.

Goulet, Normand. Stratigraphy and structural relationships across the Cadillac-Larder Lake Fault, Rouyn-Beauchastel area, Quebec. D, 1978, Queen's University. 141 p.

Goulet, William H. The environment of deposition of the Oligocene Burbank Sandstone, Tulare Lake Field, Kings County, California. M, 1986, Texas A&M University. 123 p.

Goullaud, Lee H. Petrology and structure of a gabbroic body in the Trinity ultramafic pluton, Klamath Mountains, California. M, 1973, University of Washington. 54 p.

Goulter, I. C. A technique for watershed land use planning under uncertainty. D, 1979, University of Illinois, Urbana. 185 p.

Gourley, Albert Carlisle. A geological and petrological study of Heath Steele mines, Northumberland County, New Brunswick. M, 1957, Dalhousie University. 36 p.

Gourley, James W., III. Stratigraphy and sedimentation patterns, upper Miocene Puente Formation (Mohnian–Delmontian), northwestern Puente Hills, Los Angeles County, California. M, 1971, University of Southern California.

Goutier, Françoise Melanie. Galena lead isotope study of mineral deposits in the Eagle Bay Formation, southeastern British Columbia. M, 1986, University of British Columbia. 102 p.

Gouty, John J. Ostracodes of the Fern Glen Formation of eastern Missouri. M, 1953, Washington University. 46 p.

Gouzie, Douglas R. Carbonate geochemistry and chemical discrimination of ground water basins in portions of the Inner Bluegrass Karst region, Kentucky. D, 1986, University of Kentucky. 120 p.

Govean, Frances Marie. Some paleoecologic aspects of the Monterey Formation, California. D, 1980, University of California, Santa Cruz. 320 p.

Govett, Raymond Weston. Geology of Wagoner County, Oklahoma. D, 1959, University of Oklahoma. 182 p.

Govett, Raymond Weston. The geology of the Cabaniss-Arpelar area, Pittsburg County, Oklahoma. M, 1957, University of Oklahoma. 60 p.

Govin, Charles T., Jr. Sedimentation survey, Lake Buchanan, Texas 1973. M, 1973, University of Texas, Austin.

Gow, Thomas T. Radioactivity of the basalt of Table Mountain at Golden (Jefferson County), Colorado, and igneous rocks in the region northwest of Gilmore, Conejos County (and Rio Grande County), Colorado. M, 1914, Colorado School of Mines. 70 p.

Gowan, Samuel Ward. Depositional environment of the San Miguel lignite deposit in Atascosa and McMullen counties, Texas. D, 1985, Texas A&M University. 257 p.

Gowan, Samuel Ward. Exploration and analysis of potential aggregate deposits in Southeast Texas. M, 1981, Texas A&M University. 102 p.

Gowen, Linn H. Pleistocene benches along the Missouri in the vicinity of Blair, Nebraska. M, 1982, University of Nebraska, Lincoln. 146 p.

Gowen, Peter J. Composite land use impacts on water quality on a diversely developed watershed. M, 1981, Colorado State University. 123 p.

Gowen, Walter K. A study of leaching effects upon the measured age of a Cambrian glauconite. M, 1958, Massachusetts Institute of Technology. 42 p.

Gower, Charles Frederick. The tectonic and petrogenetic history of Archean rocks from the Kenora area, English River Subprovince, Northwest Ontario. D, 1979, McMaster University. 622 p.

Gower, David Patrick. Geology and genesis of uranium mineralization in subaerial felsic volcanic rocks of the Byers Brook Formation and the comagmatic Hart Lake Granite, Wentworth area, Cobequid Highlands, Nova Scotia. M, 1988, Memorial University of Newfoundland. 358 p.

Gower, John A. Powder X-ray diffractometer study of the phlogopite-biotite series. D, 1955, Massachusetts Institute of Technology. 67 p.

Gower, John Arthur. The Seagull Creek Batholith and its metamorphic aureole. M, 1952, University of British Columbia.

Goydan, Paul Alexander. Hydrochemistry of the alluvium near McBaine, central Missouri. M, 1971, University of Missouri, Columbia.

Goyette, André Etude pétrographique et géochimique d'intrusions mafiques et ultramafiques situées dans la région d'Acton Vale-Roxton Falls. M, 1987, Universite de Montreal.

Graaff, Fredric R. van de *see* van de Graaff, Fredric R.

Graaff, Fredric Ray Van de *see* Van de Graaff, Fredric Ray

Graaskamp, Garret W. Paleomagnetism and tectonic setting of the Red Mountain intrusive complex (Henderson molybdenum deposit); Clear Creek County, Colorado. M, 1983, Colorado School of Mines. 132 p.

Grabau, Amadeus William. Phylogeny of Gastropoda; I, The Fusidae and their allies. D, 1900, Harvard University.

Grabau, Warren Edward. The significance of oriented coral sections. M, 1950, Michigan State University. 53 p.

Grabb, Robert F. Geology for land use planning, Jack Creek basin, Madison County, Montana. M, 1977, Montana State University. 90 p.

Graber, Ellen Ruth. Diagenesis of Eocene Gulf Coast carbonates; paradoxes in the feculent Weches. M, 1984, University of Texas, Austin. 140 p.

Graber, Karen. Stratigraphy and petrography of bedded barite in phosphatic Devonian Slaven Chert, Toquima Range, Nye County, central Nevada. M, 1988, University of Houston.

Graber, Ronald Gene. Gas analysis studies of fluid inclusions, Casapalca Mine, Peru. M, 1978, University of Minnesota, Minneapolis. 118 p.

Graber, Stuart M. Frequency analysis of aeromagnetic data as a method of correlating areas within the Gren-

ville Province. M, 1980, Bowling Green State University. 81 p.

Grabowski, George Joseph, Jr. Origin, distribution and alteration of organic matter and generation and migration of hydrocarbons in Austin Chalk, Upper Cretaceous, southeastern Texas. D, 1981, Rice University. 295 p.

Grabowski, Richard J. Displacement transfer mechanisms in a portion of the Narrows/Copper Creek thrust sheet, southwestern Virginia. M, 1983, Virginia Polytechnic Institute and State University. 102 p.

Grabyan, R. J. Investigations of the geology and mineralization of the Wingate Wash Mine area, Death Valley, California. M, 1974, University of Southern California.

Grace, James. New microtechnique of size frequency characteristics of sand laminae. M, 1974, University of South Carolina.

Grace, John Dale. A technique for determining uranium equilibrium in rock specimens using alpha and fission fragment radiography. M, 1957, Pennsylvania State University, University Park. 72 p.

Grace, Katherine. A study of the differential thermal analysis of minerals. M, 1947, Columbia University, Teachers College.

Grace, Ken A. The gallium and aluminum content and the gallium-aluminum ratio in selected rocks from New Mexico. M, 1965, New Mexico Institute of Mining and Technology. 42 p.

Grace, Marvin. A geologic study of Pledger Field. M, 1954, University of Houston.

Grace, Robert M. Stratigraphy of the Newcastle Formation, Black Hills region, Wyoming and South Dakota. M, 1951, University of Wyoming. 68 p.

Grace, Scott R. Sedimentary phosphorus in the Myakka and Peace River estuaries, Charlotte Harbor, Florida. M, 1977, University of South Florida, Tampa. 74 p.

Gracey, George Dennis, III. Focal mechanism and seismic wave velocity studies of the Arkansas earthquake swarm. M, 1984, Memphis State University. 118 p.

Graczyk, Edward J., Jr. An insoluble residue study of the Comanche Peak and Edwards limestones (Lower Cretaceous) in central Kimble and eastern Sutton counties, Texas. M, 1962, Texas A&M University. 102 p.

Gradijan, Stephen J. Lebensspuren of a barrier island. M, 1972, Louisiana State University.

Grady, J. C. Distribution of acid volcanic rocks in the Superior Province of the Canadian Shield. M, 1952, McGill University.

Grady, Michael. Stratigraphy, sedimentology, and hydrocarbon potential of the Hoh turbidite sequence (Miocene), western Olympic Peninsula, Washington. M, 1985, University of Idaho. 192 p.

Grady, Stephen J. Hydrogeology and water quality at the refuse dump, Madison, New Hampshire. M, 1977, University of Massachusetts. 116 p.

Grady, Thomas Richard. The effects of snow compaction on water release and sediment yield. M, 1982, Montana State University. 67 p.

Grady, William C. Sediments of Florida Bay near Islamorada, Florida. M, 1978, West Virginia University.

Graeber, Charles Karsner. The geology and petrography of the mica peridotite dike at Dixonville, Pennsylvania. M, 1925, Pennsylvania State University, University Park.

Graeber, Clyde P. The association of gold deposits with igneous rocks in the Western United States. M, 1924, University of Minnesota, Minneapolis. 48 p.

Graeber, Edward John. Crystal structures of potassium ferric sulfate minerals. D, 1970, University of New Mexico. 131 p.

Graebner, Mark. Energy flux of waves in elastic and viscoelastic media. M, 1982, Colorado School of Mines. 147 p.

Graebner, Peter. Remanent magnetism in major rock units of the Thirty-Nine Mile volcanic field, central Colorado. M, 1967, Colorado School of Mines. 165 p.

Graeff, Ronald W. Glaciolacustrine sedimentation in southeastern Winnebago County, Iowa. M, 1986, University of Iowa. 208 p.

Graese, Anne M. A petrographic and chemical model for the evolution of the Tradewater Formation coals in western Kentucky. M, 1985, University of Kentucky. 136 p.

Graetzer, Miguel K. Upper Eocene-lower Miocene planktonic foraminiferal biostratigraphy of wells JS 25-1 and JS 52-1, offshore eastern Java, Indonesia. M, 1980, University of Oklahoma. 112 p.

Graf, Alexander N. An integrated geological study of the Great Hill area, Middlesex County, Connecticut. M, 1970, Wesleyan University.

Graf, Claus Heinrich. Quaternary geology of northeast Venezuela; coastal plains of Falcon and Zulia. D, 1968, Rice University. 261 p.

Graf, Claus Heinrich. The late Pleistocene Ingleside Barrier trend, Texas and Louisiana. M, 1966, Rice University. 83 p.

Graf, Donald L. A study of the mineral ferritungstite. M, 1947, Columbia University, Teachers College.

Graf, Donald L. Trace-element studies, Santa Rita, New Mexico. D, 1949, Columbia University, Teachers College.

Graf, Gary C. Carbonate mudmound complexes of the Upper Silurian Douro and Barlow Inlet formations at Gascoyne Inlet, Devon Island, Arctic Canada. M, 1988, University of Ottawa. 152 p.

Graf, J. B. Nearshore sediment distribution, southwestern Lake Michigan. D, 1975, University of Illinois, Urbana. 270 p.

Graf, J. L., Jr. Rare earth elements as hydrothermal tracers during the formation of massive sulfide deposits and associated iron formations in New Brunswick. D, 1975, Yale University. 244 p.

Graf, Robert Bernard. Reactions in the system MgO-Al_2O_3-SiO_2 as examined by continuous high-temperature X-ray diffraction. D, 1961, University of Illinois, Urbana. 84 p.

Graf, Robert Bernard. Some pedologic implications off the serial variations in the mineralogy of the silt and clay fractions of Gumbotil. M, 1958, Washington University. 26 p.

Graff, Harry J. The Ninnescah shale Beds 1 through 7, of south-central Kansas. M, 1970, Wichita State University. 165 p.

Graff, Jaye Ellen Up de *see* Up de Graff, Jaye Ellen

Graff, P. J. Geology of the lower part of the early Proterozoic Snowy Range Supergroup, Sierra Madre, Wyoming; with chapters on Proterozoic regional tectonics and uraniferous quartz-pebble conglomerates. D, 1978, University of Wyoming. 165 p.

Graff, Paul. The areal geology of the West Fork Lake Quadrangle, northern Colorado. M, 1973, Northern Illinois University. 59 p.

Graham, A. D. Mineralogy, internal structure and genesis of beryl pegmatites, Renfrew County, Ontario. M, 1952, Queen's University. 129 p.

Graham, A. R. X-ray study of chalcosiderite and turquoise. M, 1947, Queen's University. 59 p.

Graham, Albert R. Synthesis and X-ray study of compounds in the systems Pb-Bi-S, Ag-Bi-S, and $Ag(Bi,Sb)S_2$. D, 1950, University of Toronto.

Graham, Charles Edward. A chemical and petrographic study of carbonate constituents of sedimentary rocks. D, 1954, University of Iowa. 164 p.

Graham, Charles Edward. Structure of the western portion of the Lewiston downwarp in southeastern Washington. M, 1949, Washington State University. 36 p.

Graham, D. G. A mass spectrometric investigation of the volatile content of deep submarine basalts. D, 1978, University of Hawaii. 187 p.

Graham, Daniel W. The geology of Paradise Valley, Trans-Pecos Texas. M, 1942, Texas A&M University. 54 p.

Graham, David H. Geology and petroleum possibilities of a part of the Huasana District, San Luis Obispo County, California. M, 1938, University of California, Los Angeles.

Graham, David L. Hydrochemistry of selected parameters at the Raft River KGRA, Cassia County, Idaho. M, 1983, University of Idaho. 120 p.

Graham, David W. Pleistocene environment reconstruction in Palos Verdes Hills, California. M, 1964, Northwestern University.

Graham, David William. Helium and lead isotope geochemistry of oceanic volcanic rocks from the East Pacific and South Atlantic. D, 1987, Massachusetts Institute of Technology. 252 p.

Graham, David Wilson. Stratigraphic and environmental analysis of the Hoxbar Group (Pennsylvanian) - Stephens County, Oklahoma. D, 1966, Northwestern University. 156 p.

Graham, Don. Hydrothermal alteration of serpentinite associated with the Devils Mountain fault zone, Skagit County, Washington. M, 1988, Western Washington University. 125 p.

Graham, Donald Steven. Design criteria for pipeline crossings; evaluation of current practice and proposed tractive force numerical model. D, 1982, The Johns Hopkins University. 264 p.

Graham, Earl Kendall, Jr. The elastic coefficients of forsterite as a function of pressure and temperature. D, 1969, Pennsylvania State University, University Park. 171 p.

Graham, George Martin. An integrated geological and geophysical study of the Diana Mills Pluton, Virginia. M, 1975, University of Kentucky. 49 p.

Graham, Gordon Marion. An investigation of comparative costs for treating drilling fluids to obtain water-loss reduction. M, 1948, University of Pittsburgh.

Graham, Graydon Elliott. Petrographic study of the heavy minerals of certain sandstones of the Sundance Formation of western South Dakota and Wyoming. M, 1950, University of Nebraska, Lincoln.

Graham, Ida Ellen. Land utilization in relation to environment; illustrated by type studies in Missouri. M, 1926, University of Missouri, Columbia.

Graham, Jack Bennett. Mammalian fauna and terrace correlation of the Pleistocene of Garden County, Nebraska. M, 1940, University of Iowa. 67 p.

Graham, Jack Bennett. The Illinoian and post-Illinoian Pleistocene of Iowa. D, 1942, University of Iowa. 260 p.

Graham, John Donald. The appraisal of a mineral exploration venture in the Sandon mining camp, British Columbia. M, 1964, University of British Columbia.

Graham, John Paul. Devonian and Mississippian stratigraphy of the southern Hot Creek Range, Nye County, Nevada. M, 1983, Oregon State University. 125 p.

Graham, John R. The Tertiary igneous intrusions of Green Lakes and Isabel Lakes valleys, Boulder County, Colorado. M, 1934, University of Colorado.

Graham, John W. The stability and significance of magnetism in sedimentary rocks. D, 1949, The Johns Hopkins University.

Graham, Joseph J. The foraminifera of the Marianna Limestone of Florida. M, 1939, Northwestern University.

Graham, Joseph John. The foraminifera of the type Meganos (Eocene) of California. D, 1947, University of California, Berkeley. 390 p.

Graham, Michael James. Chemical systems, Monroe Reservoir, Indiana. M, 1977, Indiana University, Bloomington. 131 p.

Graham, Michael James. Hydrogeochemical and mathematical analyses of aquifer intercommunication, Hanford Site, Washington State. D, 1983, Indiana University, Bloomington. 93 p.

Graham, Rhea L. A paleomagnetic study of Recent sediments in the Santa Barbara Basin. M, 1978, Oregon State University. 39 p.

Graham, Richard C. The Quaternary history of the upper Cache River valley, southern Illinois. M, 1985, Southern Illinois University, Carbondale. 236 p.

Graham, Richard H. Geophysical study of Cape Cod and the islands (Massachusetts). M, 1952, Boston College.

Graham, Robert Alexander Fergus. Metamorphism of Willroy sulphide ore minerals (Precambrian) by diabase dykes (Ontario). M, 1967, University of Western Ontario. 107 p.

Graham, Robert Alexander Fergus. The Mogul base metal deposits (lower Carboniferous), County Tipperary, Ireland. D, 1970, University of Western Ontario. 236 p.

Graham, Robert B. A petrographic correlation of the Wetetnagami area with adjacent areas. M, 1941, University of Toronto.

Graham, Robert B. The geology of the Duquesne and Lanaudiere map-areas, Destor and Duparquet township areas, with particular reference to porphyritization. D, 1948, University of Toronto.

Graham, Robert C. Geomorphology, mineral weathering, and pedology in an area of the Blue Ridge front, North Carolina. D, 1986, North Carolina State University. 212 p.

Graham, Robert L. Assessment of the effects construction will have on a commercial spring, Woonsocket, Rhode Island. M, 1982, Boston University. 59 p.

Graham, Robin Spear. Structure and stratigraphy of a portion of the Murphy Marble belt (Lower Cambrian) in Gilmer County, Georgia. M, 1967, Emory University. 59 p.

Graham, Rodney W. The geology and geochemistry of the Blue Moon polymetallic sulfide deposit, Mariposa County, California. M, 1987, Colorado School of Mines. 163 p.

Graham, Roy. Contributions to the Pennsylvanian flora of Illinois as revealed in coal balls. D, 1933, University of Chicago. 23 p.

Graham, Roy. The Mesozoic plant bearing formations of Canada. M, 1931, University of British Columbia.

Graham, Russell W. Biostratigraphy and paleoecological significance of the Conard Fissure (Arkansas, Newton County) local fauna with emphasis on the genus Blarina (Pleistocene). M, 1972, University of Iowa. 90 p.

Graham, Russell W. Pleistocene and Holocene mammals, taphonomy, and paleoecology of the Friesenhahn Cave local fauna, Bexar County, Texas. D, 1976, University of Texas, Austin.

Graham, S. A. Tertiary sedimentary tectonics of the central Salinian Block of California. D, 1976, Stanford University. 541 p.

Graham, T. Gordon. An evaluation of the point load strength test. M, 1977, University of Calgary. 90 p.

Graham, Theodore Kenne. Geology of the Saylorsburg to Stormville, Pennsylvania. M, 1959, Lehigh University.

Graham, W. F. Atmospheric pathways of the phosphorus cycle. D, 1977, University of Rhode Island. 283 p.

Graham, William A. P. A textural and petrographic study of the Cambrian sandstones of Minnesota and Wisconsin. D, 1927, University of Minnesota, Minneapolis. 104 p.

Graham, William A. P. Experiments in the origin of phosphate pebbles. M, 1924, University of Minnesota, Minneapolis. 29 p.

Graichen, Ronald E. Geology of a northern part of the Pine Forest Mountains northwestern Nevada. M, 1972, Oregon State University. 121 p.

Grainger, James R. Geology of the White Oaks mining district, Lincoln County, New Mexico. M, 1974, University of New Mexico. 69 p.

Grajales-Nishimura, Jose Manuel. Geology, geochronology, geochemistry and tectonic implications of the Juchatengo green rock sequence, State of Oaxaca, southern Mexico. M, 1988, University of Arizona. 145 p.

Gram, Oscar E. Tectonic features of Utley and Berlin regions, Wisconsin. M, 1940, Northwestern University.

Gram, Ralph. Mineralogical changes in Antarctic deep-sea sediments and their paleo-climatic significance. D, 1974, Florida State University. 296 p.

Gram, Ralph. The marine geology of the Recent sediments of central San Francisco Bay (California). M, 1966, San Jose State University. 132 p.

Grambling, Jeffrey A. Geology of Precambrian metamorphic rocks of the Truchas Peaks area, north-central New Mexico. D, 1979, Princeton University. 269 p.

Gramlich, J. W. Improvements in the potassium-argon dating method and their application to studies of the Honolulu volcanic series. D, 1970, University of Hawaii. 157 p.

Gramont, Bertrand. Carbonate sedimentology of the Virgilian part of the Horquilla Formation, Big Hatchet Peak area, Hidalgo County, NM. M, 1987, New Mexico Institute of Mining and Technology. 222 p.

Grams, Bryan A. A subsurface temperature study of three areas in western Canada. M, 1973, University of Calgary. 106 p.

Granados, Eduardo. Calcium carbonate deposition in geothermal wellbores, Miravalles geothermal field, Costa Rica. M, 1983, Stanford University.

Granat, M. A. Dynamics and sedimentology of Inner Middle Ground - Nine Foot Shoal, Chesapeake Bay, Virginia. M, 1976, Old Dominion University. 105 p.

Granata, George Edward. Regional sedimentation of the Late Triassic Dockum Group, West Texas and eastern New Mexico. M, 1981, University of Texas, Austin.

Granata, Glenn Walter. High-energy carbonate banks in the Smackover Formation; a target for exploration in East Texas. M, 1986, Baylor University. 140 p.

Granata, Harold Peter. Ostracodes from the Coon Creek Tongue of the Ripley Formation of McNairy County. M, 1960, University of Missouri, Columbia.

Granata, James Samuel, II. Solar-stimulated luminescence of a phosphorite near New Cuyama, Santa Barbara County, California. M, 1987, Northern Arizona University. 141 p.

Granata, Walter Harold, Jr. Cretaceous stratigraphy and structural development of the Sabine Uplift area, Texas and Louisiana. D, 1960, University of Wyoming. 148 p.

Granata, Walter Harold, Jr. The geology of the northwestern portion of the Mokane Quadrangle, Missouri. M, 1952, University of Missouri, Columbia.

Granath, James Wilton. The deformational and metamorphic history of the Precambrian rocks of Wind River Canyon, central Wyoming. M, 1973, University of Illinois, Urbana. 101 p.

Grand, Stephen Pierre. Shear velocity structure of the mantle beneath the North American Plate. D, 1986, California Institute of Technology. 235 p.

Grande, R. Lance. Paleontology of the Green River Formation, with a review of the fish fauna. M, 1980, University of Minnesota, Minneapolis. 333 p.

Grande, Roger Lance. Recent and fossil clupeomorph fishes with materials for revision of the subgroups of clupeoids. D, 1983, City College (CUNY). 358 p.

Grandi, Rolando Barozzi. Sonic resonance and related engineering properties of selected soils and rock. M, 1965, University of Arizona.

Grandstaff, David Eugene. Kinetics of uraninite oxidation; implications for the Precambrian atmosphere. D, 1974, Princeton University. 153 p.

Graney, Joseph Robert. Geology, alteration, and mineralization at Hasbrouck Mountain, Divide District, Esmeralda County, Nevada. M, 1985, University of Nevada. 106 p.

Grange, John La *see* La Grange, John

Granger, Arthur E. Geologic aspects of torrential floods in northern Utah. M, 1939, University of Washington. 57 p.

Granger, Bernard. Analyse tectonique et stratigraphique des schistes des groupes de Bennett et Rosaire, région de Saint-Malachie, Appalaches du Québec. M, 1973, Universite de Montreal.

Granger, Bernard. Tectonique des schistes de Bennet (Cambrian) de la région de St-Malachie-Est, Québec. M, 1971, Universite de Montreal.

Granirer, Julian L. Paleomagnetic evidence for northward transport of the Methow-Pasayten Belt, north-central Washington. M, 1985, Western Washington University. 143 p.

Grannell, Dana Bradford. Geologic section across the Durham Triassic Basin, North Carolina. M, 1961, North Carolina State University. 67 p.

Grannell, Roswitha Barenberg. Geological and geophysical studies of three Franciscan serpentinite bodies in the southern Santa Lucia range, California. D, 1969, University of California, Riverside. 287 p.

Grannemann, N. G. Hydrogeology of the Missouri River flood plain near Glasgow, Missouri. M, 1976, University of Missouri, Columbia.

Grannis, Jonathon L. Sedimentology of the Middle Cambrian Wheeler Formation, Drum Mountains, Utah. M, 1982, University of Kansas. 135 p.

Granquist, Donald Paul. Utilization of titaniferous deposits. M, 1945, University of Washington. 36 p.

Grant, Alan Carson. Distributional trends in the Recent marine sediments of northern Baffin Bay. M, 1965, University of New Brunswick.

Grant, Alan H. Melting relations of high-iron Archean basalts. M, 1978, Queen's University. 80 p.

Grant, Alan Robert. Bedrock geology of the Dome Peak area, Chelan, Skagit and Snohomish counties, northern Cascades, Washington. D, 1966, University of Washington. 270 p.

Grant, Alan Robert. Geology and petrology of a portion of the Dome Peak area, northern Cascades, Washington. M, 1959, University of Washington. 71 p.

Grant, Allan Marshall. An attempt to correlate the Cambro-Ordovician limestones near Birmingham, Pennsylvania, by means of heavy minerals. M, 1936, Cornell University.

Grant, Brian D. Lithofacies associations and reservoir geology of the Cardium Formation (Late Cretaceous) in Edson-Pine Creek region of west-central Alberta. M, 1988, University of Windsor. 270 p.

Grant, D. A. Land use in Pictou County, Nova Scotia. M, 1951, Acadia University.

Grant, David Edward. On the spatial structure of the acoustic signal field near the deep ocean bottom due to a near-surface cw source. D, 1986, University of Texas, Austin. 153 p.

Grant, David J. Sediment textural patterns on the San Pedro Shelf, California. M, 1972, University of Southern California.

Grant, Douglas R. Pebble lithology of the tills of South-east Nova Scotia. M, 1963, Dalhousie University. 235 p.

Grant, Douglas Roderick. Recent coastal submergence of the Maritime provinces, Canada. D, 1970, Cornell University. 121 p.

Grant, Earl Brian. The structure and development of part of the Kingston Uplift, Kings County, New Brunswick. M, 1972, Carleton University. 88 p.

Grant, F. S. A geophysical interpretation of gravitational anomalies. D, 1952, University of Toronto.

Grant, George C. The Chaetognatha of the inner continental shelf waters off Virginia, their taxonomy, abundance and dependence on physical factors of the environment. M, 1962, College of William and Mary.

Grant, Gordon Elliot. Downstream effects of timber harvest activities on the channel and valley floor morphology of western Cascade streams. D, 1986, The Johns Hopkins University. 367 p.

Grant, Hogn Bruce. A comparison of the chemistry and mineralogy with the distribution and physical aspects of marine manganese concretions of the southern oceans. M, 1967, Florida State University.

Grant, I. C. The tinguaite and related dike rocks of Rosemount Quarry, Montreal East (Quebec). M, 1952, McGill University.

Grant, J. A. Seismic spectrometry and magnitude. M, 1975, University of Western Ontario.

Grant, James Alexander. The granitic rocks of the Grenville (Precambrian) Province in southeastern Ontario. M, 1959, Queen's University. 185 p.

Grant, James Alexander. The nature of the Grenville Front near Lake Timagami, Ontario. D, 1964, California Institute of Technology. 168 p.

Grant, John A., III. Evolution of eastern Margaritifer Sinus, Mars. M, 1985, University of Rhode Island.

Grant, Keith. The Dakota Sandstone (Cretaceous) in the southern part of the Chama Basin, New Mexico. M, 1974, Bowling Green State University. 99 p.

Grant, Lori T. Experimental determination of seismic source characteristics for small chemical explosions. M, 1988, Southern Methodist University. 247 p.

Grant, Maureen. Etude du métamorphisme et de la distribution verticale des teneurs en Au, As et Sb à la Mine Sigma, Val d'Or, Québec. M, 1986, Ecole Polytechnique. 121 p.

Grant, Philip Robert. Limestone units within the Triassic Wild Sheep Creek Formation of the Snake River canyon. M, 1980, Washington State University. 103 p.

Grant, R. W. E. Geochemistry of metavolcanic rocks of the Timmins region, northeastern Ontario. M, 1977, Laurentian University, Sudbury.

Grant, Raymond Wallace. The occurrence of silica minerals in meteorites. D, 1968, Harvard University.

Grant, Richard E. Trilobite distribution, upper Franconia Formation, Wabasha County, Minnesota (southeastern Minnesota). M, 1953, University of Minnesota, Minneapolis. 94 p.

Grant, Richard Evans. Cambrian faunas of the Snowy Ridge Formation, southwestern Montana and northwestern Wyoming. D, 1958, University of Texas, Austin. 553 p.

Grant, S. C. Channel deposits of the Wind River Formation in Fremont County, Wyoming as a guide to uranium ore. M, 1954, University of Wyoming. 70 p.

Grant, Sheldon Kerry. Metallization and paragenesis in the Park City District, Utah. D, 1966, University of Utah. 200 p.

Grant, Stacy Kent. Geologic study of the Viking Formation, Harmattan East Field. M, 1985, University of Alberta. 229 p.

Grant, Stanley Cameron. Thirty-five millimeter color oblique aerial photography as a tool for reconnaissance exploration for uranium mineralization in the Tertiary basins of Wyoming. D, 1971, University of Idaho. 170 p.

Grant, Terry A. Minor folding in the Rabbit Hill Formation, Eureka County, Nevada. M, 1974, University of Nevada. 100 p.

Grant, Timothy C. Geology of the Spry Intrusion, Garfield County, Utah. M, 1979, Kent State University, Kent. 59 p.

Grant, Ulyses Simpson, III. The effect of the National Environmental Education Act on secondary and elementary schools in two mid-western states. D, 1981, University of Michigan. 204 p.

Grant, Ulysses S. Catalogue of the marine Pliocene and Pleistocene Mollusca of California and adjacent regions, with notes on their morphology, classification, and nomeclature and a special treatment of the Pectinidae and Turridae (including a few Miocene and Recent species). D, 1931, Stanford University. 1036 p.

Grant, Ulysses S. The geology of Kekequabic Lake in northeastern Minnesota, with especial reference to an augite soda-granite. D, 1893, The Johns Hopkins University.

Grant, Ulysses S., IV and Gale, Hoyt R. Catalogue of the marine Pliocene and Pleistocene Mollusca of California and adjacent regions, with notes on their morphology, classification, and nomenclature and a special treatment of the Pectinidae and the Turridae (including a few Miocene and Recent species). D, 1930, Stanford University. 1036 p.

Grant, Willard Huntington. The geology of Hart County, Georgia. D, 1955, The Johns Hopkins University. 96 p.

Grant, Willard Huntington. The lithology and structure of the Brevard Schist and the hornblende gneiss in the Lawrenceville, Georgia, area. M, 1949, Emory University. 45 p.

Grantham, Jeremy Hummon. The influence of climate and relief on lithic fragment abundance in modern fluvial sands of the southern Blue Ridge Mountains, North Carolina. M, 1986, Michigan State University. 80 p.

Grantz, Arthur. Strike-slip faults in Alaska. D, 1966, Stanford University. 156 p.

Grapes, Kathyrn J. Lithologic and textural study of the Clear Lake Fe-Zn-Pb-Ag massive sulphide deposit, Yukon Territory, Canada. M, 1987, Carleton University. 329 p.

Grasel, Peter Corbin. The reconnaissance geology of the La Salitrera mining district, San Luis Potosi, Mexico. M, 1979, University of Houston.

Grassmück, Gerhard. Gold enrichment in Montana ores by meteoric waters as shown by microscopic study. M, 1934, Montana College of Mineral Science & Technology. 61 p.

Grasso, Anthony Louis. Hydrogeologic parameters for zoning in South Russell Village, Geauga County, Ohio. M, 1986, Kent State University, Kent. 241 p.

Grasso, Santo Vincent. An analysis of the factors affecting the distribution of heavy metals in a tidal estuary. D, 1979, Rutgers, The State University, New Brunswick. 268 p.

Grasso, Thomas Xavier. Faunal zones of the Middle Devonian Hamilton Group in the Tully Valley, central New York. M, 1966, Cornell University.

Grasty, John S. The limestone of Maryland. D, 1908, The Johns Hopkins University.

Graterol, Magaly. The sulphide mineralogy of the Marbridge No.:3 and No.:4 deposits, Malartic mining district, Quebec, Canada. M, 1969, University of Toronto.

Graterol, Victor. A paleomagnetic study across a line in the north range of the Sudbury irruptive, Ontario. M, 1970, University of Toronto.

Graton, Louis Caryl. Hydrothermal origin of the Rand gold deposits. D, 1930, Cornell University.

Gratton, Patrick John Francis. Geology of the Hidden Mountains, Valencia County, New Mexico. M, 1958, University of New Mexico. 56 p.

Gratton, Sara M. Hydrothermal alteration products of the La Bufa rhyolitic tuff near Zacatecas, Mexico. M, 1988, University of New Orleans.

Gratz, Jeffrey F. Paragenesis and evolution of mineralizing fluids, Gordonsville Mine, central Tennessee zinc district. M, 1986, University of Tennessee, Knoxville. 143 p.

Grau, Gérard. Sec. 1, On the reflection of plane waves by stratified systems (normal incidence); Sec. 2, The determination of seismic velocities in layers with non-parallel interfaces. D, 1957, California Institute of Technology. 75 p.

Grauch, Richard Irons. Geology of the Sierra Nevada south of Mucuchies, Venezuelan Andes; an aluminum-silicate-bearing metamorphic terrain. D, 1971, University of Pennsylvania. 203 p.

Grauch, V. J. S. Correcting aeromagnetic data for magnetic terrain effects, with an example from the Lake City Caldera area, Colorado. D, 1986, Colorado School of Mines. 195 p.

Graul, Ronald Wayne. A subsurface study of the sandstone of the Renault Formation (Mississippian System) in west-central Posey County, Indiana. M, 1986, Southern Illinois University, Carbondale. 126 p.

Graumlich, Lisa. Long-term records of temperature and precipitation in the Pacific Northwest derived from tree rings. D, 1985, University of Washington. 198 p.

Graus, Richard Raphael. Latitudinal gradients in the shell morphology of modern marine gastropods. D, 1971, University of Rochester. 81 p.

Graustein, William Chandler. The effects of forest vegetation on solute acquisition and chemical weathering; a study of the Tesuque watersheds near Santa Fe, New Mexico. D, 1981, Yale University. 671 p.

Gravely, Marion Shelor. The geology of southwestern Madison County, Missouri. M, 1960, Washington University. 29 p.

Gravenor, Conrad Percival. A lithologic study of tills in southwestern Ontario. M, 1950, University of Wisconsin-Madison. 26 p.

Gravenor, Conrad Percival. Pleistocene geology of the Peterborough and Rice Lake districts, Ontario. D, 1952, Indiana University, Bloomington. 95 p.

Graves, Barbara Jean. Determination of hydraulic conductivity of glacial tills from the Wedron Formation in central Illinois; a comparative study of small and large scale methods. M, 1984, Southern Illinois University, Carbondale. 73 p.

Graves, Catherine A. A quantitative geomorphic analysis of sediment yield of watersheds in Azusa and Mt. Wilson quadrangles, Los Angeles County, California. M, 1975, University of Massachusetts. 65 p.

Graves, Frank Douglas. A sedimentary analysis of the Queen and Grayburg formations of southeastern New Mexico. M, 1958, Texas Tech University. 250 p.

Graves, Howard B. Joint systems in the St. Francois Mountains (Missouri). M, 1934, Washington University. 75 p.

Graves, Howard B. The Pre-Cambrian structure of Missouri. D, 1936, Washington University. 78 p.

Graves, John Milton. Subsurface geology of portions of Lincoln and Payne counties, Oklahoma. M, 1954, University of Oklahoma. 81 p.

Graves, Lawrence S. A reconstruction of the environmental conditions for human cultural development in the western Lake Erie Basin during late Holocene time. M, 1977, Bowling Green State University. 113 p.

Graves, M. C. The formation of gold-bearing quartz veins in Nova Scotia; hydraulic fracturing under conditions of greenschist regional metamorphism during early stages of deformation. M, 1976, Dalhousie University. 166 p.

Graves, Ramona M. Biaxial acoustic and static measurement of rock elastic properties. D, 1982, Colorado School of Mines. 213 p.

Graves, Roy William. Conodonts from the Marathon Basin, Brewster County, Texas (T710). M, 1941, University of Missouri, Rolla.

Graves, Roy William. Geology of Hood Spring Quadrangle, Brewster County, Texas. D, 1949, University of Texas, Austin.

Graves, Timothy. Predicting the performance of water wells in crystalline rocks, Front Range, Colorado. M, 1984, Colorado State University. 74 p.

Graves, W. H., Jr. The use of models for the study of equipotential prospecting. M, 1931, Massachusetts Institute of Technology. 106 p.

Graveson, David H., Jr. Uranium distribution in granites of the Shirley and Pedro Mountains, Carbon County, Wyoming. M, 1965, University of Wyoming. 66 p.

Grawbarger, David J. Petrology, paleoecology and diagenesis of the Manitowaning Bioherm, Manitoulin Island, Ontario. M, 1977, University of Calgary. 148 p.

Grawe, Oliver R. The St. Louis Formation with special reference to the conglomerate breccias in the vicinity of St. Louis, Missouri. M, 1923, Washington University.

Grawe, Oliver Rudolph. Shales of the Cromwell oil pool, Oklahoma. D, 1927, University of Iowa. 171 p.

Gray, A. B. Sedimentary facies of the Don Member, Toronto Formation [Ontario]. M, 1950, University of Toronto.

Gray, Alice Viola King. Terraces of the Hoback Basin, western Wyoming. M, 1946, University of Michigan.

Gray, Allan W. Field manual of Oregon geology for Oregon schools. M, 1952, Oregon State University. 82 p.

Gray, Brian Erwin. Petrology of the Alderson Formation (Mississippian) in southeastern West Virginia. M, 1985, East Carolina University. 166 p.

Gray, C. B. Geology of Northeast Santa Ana Mountains (California). M, 1951, Pomona College.

Gray, Carlyle. The geology of the Shawangunk Mine and environs. M, 1947, Columbia University, Teachers College.

Gray, Carlyle. The lead-zinc ores of the Shawangunk Mountain District [New York]. D, 1953, Columbia University, Teachers College. 162 p.

Gray, Clifton H. Geologic structure southwest of Corona, California. M, 1953, Pomona College.

Gray, Dale Franklin. Methods for the economic evaluation of petroleum exploration and synthetic fuels production; an application to Brazil. D, 1981, Massachusetts Institute of Technology. 337 p.

Gray, David S. Geology of the Morgan Butte area, Yavapai County, Arizona. M, 1982, Northern Arizona University. 115 p.

Gray, Donald Harford. Coupled flow phenomena in clay-water systems. D, 1966, University of California, Berkeley. 180 p.

Gray, Donald McLeod. Geology of the Porter Springs area, western Houston County, Texas. M, 1953, University of Texas, Austin.

Gray, Eddie V. Geology, ground water, and surface subsidence, Baytown-La Porte area, Harris County, Texas. M, 1958, Texas A&M University. 66 p.

Gray, F. Anton. Ore deposits of the Mineral Hill District, Lemhi County, Idaho. D, 1927, University of Minnesota, Minneapolis. 95 p.

Gray, Floyd. Geology of the igneous complex at Tincup Peak, Kalmiopsis Wilderness Area, southwestern Oregon. M, 1982, University of Massachusetts. 135 p.

Gray, Gary George. Structural, geochronologic, and depositional history of the western Klamath Mountains, California and Oregon; implications for the early to middle Mesozoic tectonic evolution of the western North American Cordillera. D, 1985, University of Texas, Austin. 224 p.

Gray, Gary W. Cobalt, iron, nickel and grain size distribution; Meteor Crater, Arizona. M, 1977, Arizona State University. 128 p.

Gray, Henry Hamilton. Stratigraphy and sedimentation of Pottsville rocks near Beach City, Ohio. D, 1954, Ohio State University. 150 p.

Gray, Henry Hamilton. Triassic and Jurassic stratigraphy of the Camp Davis region, Wyoming. M, 1946, University of Michigan.

Gray, Irving Bernard. Nature and origin of the Moenkopi-Shinarump hiatus in Monument Valley, Arizona and Utah. D, 1961, University of Arizona. 127 p.

Gray, Irving Raymond. Correlation of the upper flows of the Columbia River basalts between Moses Coulee and Yakima, Washington. M, 1955, University of Washington. 52 p.

Gray, J. E. Petrology and geochemistry of the eastern portion of the Ingalls Complex, central Washington Cascades. M, 1982, University of Kansas. 63 p.

Gray, James R. Genesis of a Holocene carbonate mud mound in Florida Bay. M, 1978, University of South Florida, St. Petersburg.

Gray, James T. Processes and rates of development of Talus slopes and Protalus rock glaciers in the Ogilvie and Wernecke Mountains, Yukon Territory. D, 1972, McGill University. 285 p.

Gray, Jane. Plant microfossils from the Miocene of the Columbia Plateau, Oregon. D, 1958, University of California, Berkeley. 329 p.

Gray, John D. Sedimentology of the upper Sewickley Sandstone (Pennsylvanian) in parts of Morgan, Noble, and Muskingum counties, Ohio. M, 1979, University of Akron. 108 p.

Gray, John Gardiner. The geology and ore deposits of the interior plateaux region south of the 54th parallel of latitude. M, 1935, University of British Columbia.

Gray, L. O., Jr. Geology of the east flank of the Laramie Range in the vicinity of Horse Creek, Laramie County, Wyoming. M, 1947, University of Wyoming. 25 p.

Gray, Lee Malcolm. Stratigraphy and depositional environments of the Centerfield and Chenango members (Middle Devonian) in western and central New York State. D, 1985, University of Rochester. 178 p.

Gray, Lee Malcolm. The evolution of form in the Gastropoda. M, 1976, University of Rochester. 55 p.

Gray, Lewis Richard. Palynology of four Allegheny coals, northern Appalachian coal field. D, 1965, University of Illinois, Urbana. 128 p.

Gray, Lynn D. Geology of Mesozoic basement rocks in the Santa Maria Basin, Santa Barbara and San Luis Obispo counties, California. M, 1980, San Diego State University.

Gray, Marion Glover. The preGulfian rocks of the southwestern Georgia coastal plain. M, 1978, Emory University. 72 p.

Gray, Mary Elizabeth. Geology and strain in the footwall of the Meade Thrust, southeastern Idaho. M, 1986, Bryn Mawr College. 75 p.

Gray, Matthew Dean. Gold mineralization in the Black Cloud #3 carbonate replacement orebody, Leadville mining district, Lake County, Colorado. M, 1988, University of Arizona. 65 p.

Gray, Michael G. Lithostratigraphy and conodont biostratigraphy of the Hindsville Formation, Northwest Arkansas. M, 1983, University of Arkansas, Fayetteville.

Gray, Norman Henry. The growth of plagioclase and clinopyroxene crystals in a solidifying magma. M, 1968, McGill University. 60 p.

Gray, Norman Henry. Thermal histories of small intrusions from petrologic information. D, 1971, McGill University. 275 p.

Gray, Ralph Joseph. Microfossils and general stratigraphy of the Waynesburg coal (northern West Virginia). M, 1951, West Virginia University.

Gray, Ralph L. Reservoir study, South Lovington Field, Lea County, New Mexico. M, 1944, New Mexico Institute of Mining and Technology. 21 p.

Gray, Ralph S. A study of the Wasatch Mountains. M, 1925, University of Utah. 81 p.

Gray, Raymond Franklin. Geology of a portion of the Pine Nut Mountains, Nevada. M, 1953, University of California, Berkeley. 75 p.

Gray, Richard H. The Sydney coalfield (Nova Scotia). D, 1940, McGill University.

Gray, Richard H. The Sydney Coalfield, Nova Scotia. M, 1937, McGill University.

Gray, Robert Charles. Crustal structure from the Nevada Test Site to Kansas as determined by a gravity profile. M, 1966, University of Utah. 96 p.

Gray, Robert Hugh. Stratigraphy of the Mowry Bentonite beds (Cretaceous) in the Greybull-Lovell area, Wyoming. M, 1958, University of Wisconsin-Madison.

Gray, Robert S. Cenozoic geology of Hindu Canyon, Mohave County, Arizona. M, 1959, University of Arizona.

Gray, Robert Stephen. Late Cenozoic sediments in the San Pedro Valley near Saint David, Arizona. D, 1965, University of Arizona.

Gray, Russell Dent. The alteration of peridotite to serpentine. M, 1940, University of Cincinnati. 107 p.

Gray, Russell L. Numerical investigation of reflection and transmission of spherical compressional waves. D, 1968, Colorado School of Mines. 56 p.

Gray, Russell L. Theoretical resistivity profiles over hemispherical sinks. M, 1960, University of Utah. 100 p.

Gray, Shapleigh Gardom. The Spiriferidae of the Chouteau Limestone. M, 1927, University of Missouri, Columbia.

Gray, Stephen Ralph. Crustal and upper-mantle studies in New Mexico along east-west gravity profiles. M, 1968, University of Utah. 96 p.

Gray, Terry Lee. Diagenesis of Strawn Limestone, Lovington East Field, Lea County, New Mexico. M, 1973, Texas Tech University. 63 p.

Gray, Thomas Eastman. Origin and evolution of sub-seafloor Au-Zn-Cu-Pb mineralization and related alteration at the Johnson River Prospect, south-central Alaska. M, 1988, University of Oregon. 219 p.

Gray, Timothy J. Environment of deposition of Chappel Mississippian carbonated buildups on the Bend Arch, north central Texas. M, 1987, Texas A&M University.

Gray, Tony Douglas. Depositional systems of the upper cliff coals in a portion of the Plateau coal field, Alabama. M, 1981, Auburn University. 119 p.

Gray, W. C. Variable norm deconvolution. D, 1979, Stanford University. 109 p.

Gray, Warren Wilbur. Geology of a portion of the Livermore Quadrangle, Colorado. M, 1942, University of Iowa. 80 p.

Gray, Wayland E. Structural geology of the southern part of Clarkston Mountain, Malad Range, Utah. M, 1975, Utah State University. 43 p.

Gray, Wilfred Lee. The geology of the Drinking Water Pass area, Harney and Malheur counties, Oregon. M, 1956, University of Oregon. 86 p.

Graybeal, Frederick T. The partition of trace elements among co-existing minerals in some Laramide intrusive rocks in Arizona. D, 1972, University of Arizona.

Graybeal, Frederick Turner. The geology and gypsum deposits of the southern Whetstone Mountains, Cochise County, Arizona. M, 1962, University of Arizona.

Graybeal, G. E. Inventory of surface water using Landsat data. D, 1978, Texas A&M University. 130 p.

Grayson, John Francis. The postglacial history of vegetation and climate in the Labrador-Quebec region as determined by palynology. D, 1957, University of Michigan. 278 p.

Grayson, Robert C., Jr. Lithostratigraphy and conodont biostratigraphy of the Hindsville Formation, Northwest Arkansas. M, 1976, University of Arkansas, Fayetteville.

Grayson, Robert Calvin, Jr. The stratigraphy of the Wapanucka Formation (Lower Pennsylvanian) along the frontal margin of the Ouachita Mountains, Oklahoma. D, 1980, University of Oklahoma. 353 p.

Grazer, Robert Anthony. Experimental study of current ripples using medium silt. M, 1982, Massachusetts Institute of Technology. 131 p.

Greacen, Katherine Fielding. A bryozoan fauna as a criterion for the correlation of the Vincentown Formation with descriptions of some new species of Bryozoa (New Jersey). D, 1938, Rutgers, The State University, New Brunswick. 164 p.

Greaney, Peter H. The magnetic stability and magnetic mineral composition of the Columbus Limestone (Devonian, Ohio). M, 1972, Ohio State University.

Greason, Arthur. Copper ores and deposits of Shannon County, Missouri. M, 1876, University of Missouri, Rolla.

Greathead, Colin. The geology and petrochemistry of the greenstone belt, south of Hurley and Upson, Iron County, Wisconsin. M, 1975, University of Wisconsin-Milwaukee.

Greb, Stephen Francis. Structural control of Carboniferous sedimentation in the Madisonville and Indian Lake paleovalleys, western Kentucky. M, 1985, University of Kentucky. 155 p.

Greb, Wayne S. Relations between textural parameters and cross-bedding in the Navajo Sandstone (Upper Triassic (?) and Jurassic), eastern Uinta Mountains, Utah. M, 1966, Bowling Green State University. 72 p.

Grebe, Steven Walter. Comparative sedimentology of the Chrysler Member of the Rondout Formation (Silurian-Devonian) in east-central New York State. M, 1985, SUNY at Binghamton. 66 p.

Grebmeier, Jacqueline Mary. The ecology of benthic carbon cycling in the northern Bering and Chukchi seas. D, 1988, University of Alaska, Fairbanks. 204 p.

Greco, John Antonio. The geology of the Bolton oil field, Hinds County, Mississippi. M, 1956, Mississippi State University. 41 p.

Greeley, Michael Nolan. Geology of the Esterbrook area, Converse and Albany counties, Wyoming. M, 1962, University of Missouri, Rolla.

Greeley, Ronald. Cenozoic and Recent lunulithiform bryozoans of the Gulf and Atlantic coasts. D, 1966, University of Missouri, Rolla. 223 p.

Greeley, Ronald. Geology of the southwest quarter of the Buena Vista Quadrangle, Mississippi, and a study of the sediment-foraminifera relations in the Prairie Bluff Formation. M, 1963, Mississippi State University. 60 p.

Green, A. H. Evolution of Fe-Ni sulfide ores associated with Archean ultramafic komatiites, Langmuir Township, Ontario. D, 1978, University of Toronto.

Green, Arthur R. Geology of the Crowley area, Malheur County, Oregon. M, 1962, University of Oregon. 149 p.

Green, Carolyn A. A stratigraphic analysis of the Lower Cretaceous Rodessa Formation in the Louisiana Sabine Uplift area. M, 1982, Duke University. 169 p.

Green, Charles Frederic. Eocene and Cretaceous stratigraphy of the Laguna Seca Hills, Merced County, California. M, 1942, Stanford University.

Green, Connie L. Subsurface geology of the Edwards and Georgetown formations, south-central Texas, and its relation to petroleum occurrence. M, 1982, Baylor University. 207 p.

Green, D. R. Aspects of the long-term fate of petroleum hydrocarbons in the marine environment. D, 1976, University of British Columbia.

Green, David Christopher. Precambrian geology and geochronology of the Yellowknife area, Northwest Territories. D, 1968, University of Alberta. 223 p.

Green, David Ely. Measurement of the electrical resistivity of geological formations. M, 1930, Stanford University. 40 p.

Green, David J. Morphologies and fission track ages of zircons from quartzites of the Hellroaring Lakes area, Pine Creek Quarry and Chrome Mountain area, Beartooth Mountains, Montana. M, 1972, University of Cincinnati. 89 p.

Green, Deborah. Evaluation of ground-water quality impacts of lignite waste disposal at a Texas lignite mine. M, 1984, Texas A&M University. 90 p.

Green, Douglas A. Structural geology of the central part of Clarkston Mountain, Malad Range, Utah. M, 1986, Utah State University. 55 p.

Green, Edward J. A redetermination of the solubility of oxygen in sea water and some thermodynamic implications of the solubility relations. D, 1965, Massachusetts Institute of Technology. 137 p.

Green, Edwin Thomas. Secular variation of the geomagnetic field as determined from playa lake sediments. D, 1977, University of Oklahoma. 158 p.

Green, Francis Earl. Geology of sand dunes, Lamb and Hale counties. M, 1951, Texas Tech University. 74 p.

Green, Francis Earl. The Triassic deposits of northwestern Texas. D, 1954, Texas Tech University. 196 p.

Green, George Bruton, Jr. The geology of the Slate Belt rocks of the Goldston and Bear Creek quadrangles, North Carolina. M, 1977, North Carolina State University. 69 p.

Green, Glen Martin. Physical basis for remotely sensed spectral variation in a semi-arid shrubland and an oak-hickory forest; implications for mapping soil types in vegetated terrains. D, 1988, Washington University. 397 p.

Green, Guy Emmett. The Eagle Ford Formation of Travis County, Texas. M, 1925, University of Texas, Austin.

Green, Harry Western, II. Syntectonic and annealing recrystallization of fine-grained quartz aggregates. D, 1968, University of California, Los Angeles. 220 p.

Green, Howard R. A Paleozoic section in the Egan Range near Sunnyside, Nevada. M, 1951, Columbia University, Teachers College.

Green, Jack. The Marysvale Canyon area, Marysvale, Utah. D, 1954, Columbia University, Teachers College. 231 p.

Green, Jack Harlan. Igneous rocks of the Craterville area, Wichita Mountains, Oklahoma. M, 1952, University of Oklahoma. 63 p.

Green, James A. Global analysis of the shallow geology of large-scale ocean slopes. M, 1982, University of New Orleans. 181 p.

Green, James Edward Peter. Synthetic rainfall and its use in hydrologic modeling. D, 1984, Oklahoma State University. 115 p.

Green, Jimmie Logan. Phosphate fractions and their relation to the weathering and genesis of some soils of the Texas high plains. D, 1970, Texas A&M University. 156 p.

Green, John Chandler. The geology of Errol Quadrangle, New Hampshire-Maine. D, 1960, Harvard University.

Green, Jonathan A. A petrologic and quantitative compositional study of four rocks from the Mid-Atlantic Ridge. M, 1974, University of New Mexico. 61 p.

Green, Julian W. The relevance of linear features found on Landsat imagery of New Hampshire. M, 1977, Dartmouth College. 71 p.

Green, Julian Wiley. Microfossils from the Upper Proterozoic limestone-dolomite "series", central East Greenland. D, 1988, Harvard University. 250 p.

Green, Keith W. Ecology of some Arctic foraminifera. M, 1958, University of Southern California.

Green, Kenneth A. Determination of gas reservoir characteristics in the eastern Devonian shale by pressure transient testing. M, 1983, Stanford University.

Green, Kenneth Edward. Geothermal processes at the Galapagos spreading center. D, 1980, Woods Hole Oceanographic Institution. 226 p.

Green, Lewis H. The relationship between polarization colors and rotation properties of anisotropic minerals. M, 1951, University of Wisconsin-Madison.

Green, Lewis H. Wall-rock alteration associated with certain zinc and lead deposits, formed through the replacement of limestone, Salmo map-area, British Columbia. D, 1954, University of Wisconsin-Madison.

Green, Loring K. A detailed study of a portion of the Hutchinson Salt of Kansas. M, 1963, University of Kansas. 43 p.

Green, Malcolm Omand. Low-energy bedload transport by combined wave and current flow on a southern mid-Atlantic bight shoreface. D, 1987, College of William and Mary. 177 p.

Green, Marsha A. Survey of endolithic organisms from the Northeast Bering Sea, Jamaica, and Florida Bay. M, 1975, Duke University. 112 p.

Green, Mary L. Paleoenvironmental and diagenetic analysis of the Wichita Group, Palo Duro Basin,

Texas Panhandle. M, 1985, Stephen F. Austin State University. 194 p.

Green, Morton. A review of the stratigraphy and vertebrate paleontology of the John Day Formation, Oregon. D, 1954, University of California, Berkeley. 188 p.

Green, N. L. Multistage andesite genesis in the Garibaldi Lake area, southwestern British Columbia. D, 1978, University of British Columbia.

Green, Nathan Louis. Multistage andesite genesis in the Garibaldi Lake area, southwestern British Columbia. D, 1977, University of British Columbia.

Green, Nathan Louis. The volcanic stratigraphy and petrochemistry of the God's Lake subgroup (Precambrian), Knee Lake, Manitoba. M, 1973, University of Manitoba.

Green, Paul Reed. Microfauna of the Allen Valley Shale, (Cretaceous) central Utah. M, 1959, University of Utah. 82 p.

Green, Richard. An analysis of the distribution of the major surface characteristics and the thermal anomalies observed on the eclipsed Moon. D, 1968, University of Washington. 177 p.

Green, Robert Otis. Mapping geological structure in an extensional basin using multitemporal analysis of Landsat MSS imagery, Railroad Valley, Nevada. M, 1986, Stanford University. 148 p.

Green, Ronald Thomas. Gravity survey of the southwestern part of the southern Utah geothermal belt. M, 1981, University of Utah. 107 p.

Green, Ronald Thomas. Radionuclide transport as vapor through unsaturated fractured rock. D, 1986, University of Arizona. 227 p.

Green, Sandra Lynn. The behavior of deep ocean sediments in response to thermo-mechanical loading. D, 1984, University of California, Berkeley. 522 p.

Green, Scott Robert. Computer-aided mapping and statistical analysis of the East Canton Oilfield, Rose Township, Carroll County, northeastern Ohio. M, 1988, University of Akron. 150 p.

Green, Stephen H. Structural, stratigraphic and sedimentary history of the Heterostegina Zone in the Anahuac Formation in parts of Acadia, Lafayette and Vermilion parishes, Louisiana. M, 1980, University of Southwestern Louisiana. 118 p.

Green, Stephen N. Lineament analysis and prospective economic deposits employing radar imagery of Darien Province, Panama. M, 1975, University of Arkansas, Fayetteville.

Green, Susan Molly. Seismotectonic study of the San Andreas, Mission Creek, and Banning fault systems. M, 1983, University of California, Los Angeles. 52 p.

Green, Thomas Edgar, Jr. Disturbed Chinle beds near St. Johns, Apache County, Arizona. M, 1956, University of Texas, Austin.

Green, Thomas Kent. The macromolecular structure of coal. D, 1984, University of Tennessee, Knoxville. 171 p.

Green, Timothy Myron. A petrographic examination of primary and secondary pyrite textures in coal. M, 1987, Iowa State University of Science and Technology. 108 p.

Green, Willard Russell. Geology of East Cooper Mountain area, Fremont County, Wyoming. M, 1955, University of Texas, Austin.

Green, William. A study of the Cohansey Formation (Pliocene) of New Jersey. M, 1952, University of Nebraska, Lincoln.

Green, William. Midchannel islands; sedimentology, physiography, and effects on channel morphology in selected streams of the Great Bend region of the Wabash Valley. M, 1982, Purdue University. 128 p.

Green, William D. The location of trace quantities of uranium and thorium associated with magnetite. M, 1962, University of Arizona.

Green, William Randolph. Geology and mineral deposits of the Blacktail Mountain area, Bonner County, Idaho. D, 1976, Washington State University. 108 p.

Green, William Randolph. Structural control of mineralization at the Aurora mining district, Mineral County, Nevada. M, 1964, University of Nevada. 41 p.

Green, William Robert. Some new approaches to gravity modelling. M, 1973, University of British Columbia.

Greenamyer, Randolph. Marginal geology, Academy Pluton. M, 1974, California State University, Fresno.

Greenberg, Dallas W. Sedimentology and diagenesis of the Short Creek Oolite. M, 1981, University of Missouri, Columbia.

Greenberg, Helene. Depositional and diagenetic study of the Lenapah Limestone (Marmaton Group, late Middle Pennsylvanian), in eastern Kansas and western Missouri. M, 1986, University of Iowa. 209 p.

Greenberg, Jeffrey K. Geochemistry, petrology and tectonic origin of Egyptian Younger Granites. D, 1978, University of North Carolina, Chapel Hill. 134 p.

Greenberg, Redge L. Gravity sliding in the emplacement of the Sheep Mountain Block of the Heart Mountain Fault, Wyoming. M, 1973, University of Texas, Austin.

Greenberg, Seymour Samuel. Petrography of the Ste. Genevieve Limestone in Indiana. D, 1959, Indiana University, Bloomington. 77 p.

Greenberg, Seymour Samuel. Zeolites and associated minerals from southern Brazil. M, 1952, Indiana University, Bloomington. 26 p.

Greenburg, Jeffrey King. A tectonic-geophysical investigation of a portion of the Blue Ridge-Piedmont transitional zone in the Martinsville area, Virginia. M, 1975, University of Kentucky. 82 p.

Greenburg, Joseph Gary. Diagenesis of the Lower Cretaceous James Limestone, Fairway Field, East Texas; a petrographic and geochemical study. M, 1986, University of Texas, Austin. 189 p.

Greene, Adrian Vance. Geology of Newman Ridge and Brushy-Indiana Ridge between Sneedville, Hancock County, Tennessee, and Blackwater, Lee County, Virginia. M, 1959, University of Tennessee, Knoxville. 55 p.

Greene, Bryan A. Geology of the Branch-Point Lance area, Newfoundland. M, 1962, Memorial University of Newfoundland. 74 p.

Greene, David Carl. Structural geology of the Quseir area, Red Sea coast, Egypt. M, 1984, University of Massachusetts. 159 p.

Greene, David Terrell. Effects of pressure dependent fluid properties, formation damage, and high velocity flow on constant terminal pressure production of natural gas from a bounded circular reservoir. M, 1979, Stanford University.

Greene, Earl A. Characterization and movement of Mount St. Helens tephra on a high mountain lake watershed in northern Idaho. M, 1982, University of Idaho. 94 p.

Greene, Frank Cook. The stratigraphy of the Huron Formation of Indiana. M, 1909, Indiana University, Bloomington.

Greene, Frank F. Geology of the northeast corner of the Sparta Quadrangle and vicinity, Oregon. M, 1960, Oregon State University. 107 p.

Greene, G. M. The geochemistry of spinel lherzolites from Xalapasco de la Joya, in Luis Potosi, Mexico. M, 1975, University of Houston.

Greene, Glen Stonefield. Prehistoric utilization in the channeled scablands of eastern Washington. D, 1975, University of Washington. 149 p.

Greene, H. G. Geology of the Monterey Bay region. D, 1978, Stanford University. 484 p.

Greene, H. Gary. Morphology, sedimentation, and seismic characteristics of an Arctic beach, Nome, Alaska, with economic significance. M, 1970, San Jose State University. 139 p.

Greene, Jeremy Theodore. Reflection enhancement by bubble pulse and scattered noise attenuation in the

North Atlantic Transect multichannel data set, Nares Basin. M, 1984, University of Texas, Austin. 134 p.

Greene, John Frederick. Stratigraphy and sedimentary tectonics of the Gypsum Spring Formation, middle Jurassic, Fremont County, Wyoming. M, 1970, University of Michigan.

Greene, John M. Paleozoic stratigraphy of Clear Creek Canyon, Monitor Range, Nye County, Nevada. M, 1953, Columbia University, Teachers College.

Greene, Kimberly Richmond. The morphologic response of three streams to damming and the relation of the response to their watershed characteristics; Binghamton, New York. M, 1982, SUNY at Binghamton. 69 p.

Greene, Laurence Robert. Cyclic sedimentation within the upper member of the Deep Spring Formation (Lower Cambrian), eastern California and western Nevada; the anatomy of a grand cycle. M, 1986, University of California, Davis. 198 p.

Greene, Naomi Esther. Some Upper Cretaceous foraminifera of North Carolina. M, 1937, University of North Carolina, Chapel Hill. 27 p.

Greene, Richard Patrick. Metamorphosed McCoy Mountain Formation of Coxcomb Mountains, California. M, 1968, University of California, Santa Barbara.

Greene, Robert C. Geology of English Mountain and vicinity, Cocke, Jefferson and Sevier counties, Tennessee. M, 1959, University of Tennessee, Knoxville. 54 p.

Greene, Robert C. The geology of the Peterborough Quadrangle, New Hampshire. D, 1964, Harvard University.

Greene, Steven E. Stratigraphy and sedimentation of the Revett Formation, Precambrian Belt Supergroup, Shoshone and Bonner counties, northern Idaho. M, 1984, University of Idaho. 177 p.

Greene, William Mordock. Petrography of the Calvin Formation (Pennsylvanian) in Pontotoc and part of Okmulgee counties, Oklahoma. M, 1965, University of Oklahoma. 87 p.

Greenewalt, David. The origin of remanent magnetism in sedimentary rocks. D, 1960, Massachusetts Institute of Technology. 133 p.

Greenfeld, Joshua Shlomo. A stereo vision approach to automatic stereo matching in photogrammetry. D, 1987, Ohio State University. 166 p.

Greenfield, Leslie Lohr. Stratigraphy and micropaleontology of the Cook Mountain Formation, western Fayette County, Texas. M, 1957, University of Texas, Austin.

Greenfield, Roy Emmett. Mississippian subsurface stratigraphy in Clay, Harlan, and Leslie counties, Kentucky. M, 1957, University of Kentucky. 46 p.

Greenfield, Roy J. The response of a point source in a liquid layer overlying a liquid half space. M, 1962, Massachusetts Institute of Technology. 73 p.

Greengold, Gerald. Engineering geology of an operating area strip mine, Pike County, Indiana. M, 1981, Purdue University. 221 p.

Greenhalgh, Brian R. The Fitchville Formation; a study of the biostratigraphy and depositional environments in west-central Utah County, Utah. M, 1980, Brigham Young University.

Greenhalgh, Stewart A. Studies with a small seismic array in East-central Minnesota. D, 1979, University of Minnesota, Minneapolis. 356 p.

Greenhouse, John Phillips. Geomagnetic time variations on the sea floor off Southern California. D, 1972, University of California, San Diego.

Greenhouse, John Phillips. The application of direct-current resistivity prospecting methods to ice masses. M, 1963, University of British Columbia.

Greenland, Cyril Walter. Gel minerals (colloids). M, 1920, Columbia University, Teachers College.

Greenlee, Arthur. Geology of the Turtle Mountains of North Dakota and Manitoba. M, 1942, University of Colorado.

Greenlee, David Walden. Petrography and petrology of the north central Davis Mountains, Jeff Davis County, Texas. M, 1963, Texas Tech University. 72 p.

Greenlee, Mac. A stratigraphic and sedimentologic study of the upper Monongahela and lower Dunkard groups in southeast Athens County, Ohio. M, 1985, Ohio University, Athens. 176 p.

Greenman, Celia. Stratigraphy of the Silurian and Devonian rocks, northwestern Pend Oreille County, Washington. M, 1977, Oregon State University. 104 p.

Greenman, Elizabeth R. Petrology and mineralogy of Tertiary volcanic rocks in the vicinity of the Rozel Hills and Black Mountain, Box Elder County, Utah. M, 1982, Utah State University. 75 p.

Greenman, Lawrence. The petrology of the footwall breccias in the vicinity of the Strathcona Mine, Levack, Ontario. D, 1970, University of Toronto.

Greenman, Norman N. The origin of the Randville Dolomite of Dickinson and Iron counties, Michigan. D, 1951, University of Chicago. 179 p.

Greenough, John David. Petrology and geochemistry of Cambrian volcanic rocks from the Avalon Zone in Newfoundland and New Brunswick. D, 1984, Memorial University of Newfoundland. 487 p.

Greenough, John David. The geochemistry of Hawaiian lavas. M, 1979, Carleton University. 131 p.

Greensfelder, Roger W. The problem of determining depth focus with applications to Nevada earthquakes. M, 1965, University of Nevada. 33 p.

Greensfelder, Roger Weir. Lithospheric structure of the eastern Snake River plain, Idaho. D, 1981, Stanford University. 197 p.

Greenslade, William Murray. A water cost study in Las Vegas valley, Nevada. M, 1967, University of Nevada. 64 p.

Greenslate, Jimmie L. Manganese-biota associations in northeastern equatorial Pacific sediments. D, 1975, University of California, San Diego. 261 p.

Greenstein, Benjamin J. Mass mortality of the West Indian echinoid Diadema antillarum; a natural experiment in taphonomy. M, 1986, University of Cincinnati. 57 p.

Greenstein, Gerald. The structure of the Amole Arkose north of King Canyon, Tucson Mountains, Arizona. M, 1961, University of Arizona.

Greenup, Wilbur. Physiography and climate (past and present) in the Oregon portion of the Great Basin. M, 1941, University of Oregon. 59 p.

Greenwald, Michael T. The lower vertebrates of the Hell Creek Formation (upper Cretaceous), Harding County, South Dakota. M, 1971, South Dakota School of Mines & Technology.

Greenwald, Robert M. A planning model for ground water plume removal with cleanup time as a management variable. M, 1988, Stanford University. 66 p.

Greenwald, Roy Fuld. Volumetric response of porous media to pressure variations. D, 1980, University of California, Berkeley. 174 p.

Greenwood, Bobby M. Geology of the Smoothington Mountain-North area, Llano and San Saba counties, Texas. M, 1963, Texas A&M University.

Greenwood, David E. The depositional environment of the Navajo Sandstone at Zion National Park, Utah. M, 1978, University of Arizona.

Greenwood, E. Allen, Jr. An investigation of the distribution and partitioning of trace metals in the sediments of Lake Chichankanah, Quintana Roo, Mexico. M, 1984, Tulane University. 139 p.

Greenwood, Eugene. The geology of the Blue Mound area. M, 1950, Texas Christian University. 59 p.

Greenwood, Hugh John. A study of plagioclase zoning. M, 1956, University of British Columbia.

Greenwood, Hugh John. The system $NaAlSi_2O_6$-H_2O-argon and osmotic equilibria in metamorphism. D, 1960, Princeton University. 113 p.

Greenwood, Richard. Gravimetric study of Swartout Valley, Los Angeles and San Bernardino counties,

Southern California. M, 1981, California State University, Long Beach. 120 p.

Greenwood, Richard Charles. Petrogenesis of the gabbroic suite, Mt. Saint Hilaire, Quebec. M, 1983, University of Western Ontario. 194 p.

Greenwood, Richard John. The frequency-magnitude relation of seismo-acoustic events observed in two mines in the Upper Peninsula of Michigan. M, 1969, University of Michigan.

Greenwood, Richard, III. Anorthosite gabbros and related rocks of an area southwest of Wayne, Pennsylvania. M, 1959, Bryn Mawr College. 18 p.

Greenwood, Robert. Geology of the Sugar Pine area, Madera County, California. M, 1943, California Institute of Technology. 38 p.

Greenwood, Robert. Younger intrusive rocks of the Plateau Province, Nigeria, compared with the alkalic rocks of New England. D, 1949, Harvard University.

Greenwood, William Rucker. Genesis and history of the Augen Gneiss of Red River, Idaho County, Idaho. D, 1968, University of Idaho. 76 p.

Greenwood, William Rucker. Polymetamorphism in the Red Ives area, Shoshone County, Idaho. M, 1966, University of Idaho. 79 p.

Greer, C. E. Chinese water management strategies in the Yellow River basin. D, 1975, University of Washington. 234 p.

Greer, Catherine Bowman. A stratigraphic and petrographic analysis of the "Abo" Formation in the Santa Rita, New Mexico, area. M, 1987, Stephen F. Austin State University. 95 p.

Greer, Efford Wayne, Jr. Deltaic sedimentation on an active volcanic continental margin, Rio Achiguate, Pacific Coast, Guatemala. M, 1978, Texas Tech University. 127 p.

Greer, Jerry Kenneth. Subsurface geology of east-central Lincoln County, Oklahoma. M, 1961, University of Oklahoma. 67 p.

Greer, John Craig. Dynamics of withdrawal from stratified magma chambers. M, 1986, Arizona State University. 180 p.

Greer, Phillip L. Genesis and paleogeographic significance of evaporites and associated facies of the Goose Egg Formation (Permo-Triassic), southeastern Wyoming. M, 1985, University of Wyoming. 67 p.

Greer, Sharon A. Geology of sand body geometry and sedimentary facies at the estuarine-marine transition zone; Ossabaw Sound, Georgia, U.S.A. D, 1975, University of Georgia. 208 p.

Greer, William L. C. Geology of the Ponask-Stevenson lakes area. M, 1929, University of Wisconsin-Madison.

Greer, William L. C. Mix-crystals of calcium and manganese. D, 1931, University of Wisconsin-Madison.

Greger, Joel G. Stratigraphy and structural geology of the Gilbert District, Esmeralda County, Nevada. M, 1986, University of Wisconsin-Madison. 125 p.

Gregersen, S. Amplitudes of horizontally refracted Love waves. D, 1974, Columbia University. 135 p.

Gregg, Jack H. Coastal cliff erosion at Point Loma, San Diego, California. M, 1984, Texas A&M University.

Gregg, Jay Mason. Coal geology of parts of the Inola, Chouteau N.W., Catoosa S.E., and Neodesha quadrangles, southeastern Rogers and northern Wagoner counties, Oklahoma. M, 1976, Oklahoma State University. 77 p.

Gregg, Jay Mason. The origin of xenotopic dolomite texture. D, 1982, Michigan State University. 152 p.

Gregg, William J. Structural studies in the Moretown and Cram Hill units near Ludlow, Vermont. M, 1975, SUNY at Albany. 118 p.

Gregg, William J. The development of foliations in low, medium and high grade metamorphic tectonites. D, 1979, SUNY at Albany. 278 p.

Gregg, William Nathan, Jr. The depth of the bedrock surface in the lower Embarrass Valley, Illinois indi-

cated by earth resistivity. M, 1949, University of Illinois, Urbana.

Greggs, Robert George. Archaeocyatha from the Colville and Salmo area of Washington and British Columbia. M, 1957, University of British Columbia. 78 p.

Greggs, Robert George. Upper Cambrian biostratigraphy of the southern Rocky Mountains, Alberta. D, 1962, University of British Columbia.

Gregorowicz, Timothy Joseph. A three-dimensional simulation of the valley-fill and bedrock aquifers in the Williamsport area, Lycoming County, Pennsylvania. M, 1987, Ohio University, Athens. 548 p.

Gregory, A. Dewatering potential study for an open pit mine. M, 1976, McGill University. 147 p.

Gregory, Alan Frank. Analysis and interpretation of gamma radiation patterns obtained through low elevation aeroradiometry. D, 1958, University of Wisconsin-Madison. 251 p.

Gregory, Billy Warren. Geology of the Coloma Quadrangle, Missouri. M, 1954, University of Missouri, Columbia.

Gregory, Cecilia Dolores. Geology of the Stinking Water Creek area, Harney County, Oregon. M, 1962, University of Oregon. 59 p.

Gregory, Daniel I. Geomorphic study of the lower Truckee River, Washoe County, Nevada. M, 1982, Colorado State University. 136 p.

Gregory, Elizabeth. (1) Geology of Mt. Werner, (2) The structure and development of the Brachiopoda. M, 1915, Smith College. 26 p.

Gregory, Herbert E. Geology of the Aroostook volcanic area of Maine. D, 1899, Yale University.

Gregory, James Finley. The Pleistocene geology of Crawford County, Ohio. M, 1956, Ohio State University.

Gregory, James L. Volcanic stratigraphy and K-Ar ages of the Manuel Benavides area, northeastern Chihuahua, Mexico, and correlations with the Trans-Pecos, Texas, volcanic province. M, 1981, University of Texas, Austin. 79 p.

Gregory, Janet L. The stratigraphy and depositional environments of the Kalo Formation (Pennsylvanian) of south-central Iowa. M, 1982, University of Iowa. 218 p.

Gregory, Janice Lynne. Stratigraphy, petrology, and environment of deposition of the Crystal Mountain Sandstone (Lower Ordovician), Ouachita Mountains, Garland County, Arkansas. M, 1986, University of New Orleans. 115 p.

Gregory, Joel P. Variations in metamorphic grade in the Kings Mountain Belt of north-central South Carolina. M, 1981, University of North Carolina, Chapel Hill. 63 p.

Gregory, John L. Analysis of variables affecting water-table fluctuations in the Big Blue River Valley below Tuttle Creek Reservoir, (Kansas). M, 1967, Kansas State University. 82 p.

Gregory, John S. Stratigraphy and geologic history of an emerging barrier complex, northern Pinellas County, Florida. M, 1984, University of South Florida, Tampa. 141 p.

Gregory, Joseph Tracy. Pliocene vertebrates from Big Spring Canyon, South Dakota. D, 1938, University of California, Berkeley. 297 p.

Gregory, M. R. Distribution of benthonic foraminifera in Halifax Harbour, Nova Scotia, Canada. D, 1971, Dalhousie University.

Gregory, Phillip Glyde. Petrology and paleoenvironments of the Mississippian System exposed at Jellico, Tennessee. M, 1981, Memphis State University. 82 p.

Gregory, Robert George. A geothermal study of Alabama, Georgia and South Carolina. M, 1978, University of Florida. 108 p.

Gregory, Robert Theodore. Geology and isotope geochemistry of the Samail ophiolite complex, southeastern Oman Mountains. D, 1981, California Institute of Technology. 368 p.

Gregson, Victor Gregory, Jr. A model study of elastic waves in a layered sphere. D, 1966, Stanford University. 76 p.

Gregson, Victor Gregory, Jr. A regional gravity survey of parts of Franklin, Crawford, and Washington counties, Missouri. M, 1958, Washington University. 35 p.

Gregware, William. Surface geology of the McLain area, Muskogee County, Oklahoma. M, 1958, University of Oklahoma. 101 p.

Greif, M. Andrew. Stratigraphic and petrographic analysis of the Moczygemba ore body, Hobson area, Karnes County, Texas. M, 1980, Southern Illinois University, Carbondale. 174 p.

Greife, John Luverne. Stratigraphy and paleontology of the Mazourka Formation, Middle Ordovician, Independence Quadrangle, California. M, 1959, University of California, Berkeley. 75 p.

Greig, Edmund W. The geology of the Matamec Lake map area, Saguenay County, Quebec. D, 1941, Princeton University. 73 p.

Greig, John. A reconnaissance lead isotope study of base metal occurrences (Mississippian) in west-central Ireland. M, 1972, University of Alberta. 89 p.

Greig, Joseph W. Immiscibility in silicate melts. D, 1927, Harvard University.

Greig, Stanley. Petrology of the intrusive at Rigaud, Quebec (Canada). M, 1968, McGill University. 123 p.

Greimel, Thomas C. Depositional systems and paleogeography of Buck Creek Sandstone, Middle Pennsylvanian, north central Texas. M, 1977, Wayne State University. 95 p.

Greiner, Daniel Joseph. Influence of fluorine versus hydroxyl content on the optics of the amblygonite-montebrasite series. M, 1986, Virginia Polytechnic Institute and State University.

Greiner, Gary Oliver George. Environmental factors causing distribution of Recent foraminifera. D, 1969, Case Western Reserve University. 305 p.

Greiner, Gerald F. Geology of a regressive peritidal sequence with evaporitic overprints; the subsurface Dunham Formation (Lower Cambrian), Franklin, Vermont. M, 1982, Rensselaer Polytechnic Institute. 116 p.

Greiner, Gretchen G. Metamorphic mineral compositions related to bulk composition and grade of metamorphic rocks from Hastings County, Ontario. M, 1978, SUNY at Buffalo. 84 p.

Greiner, Hugo R. Spirifer disjunctus; its evolution and paleoecology in the Catskill Delta. D, 1954, Yale University.

Greiner, Hugo Robert. The petrology and paleontology of the Methy Formation. M, 1951, University of Alberta. 88 p.

Greischar, Larry L. Depth sounding with transient electromagnetic reflections. M, 1973, University of Wisconsin-Madison.

Greischar, Lawrence Lee. An analysis of gravity measurements on the Ross Ice Shelf, Antarctica. D, 1982, University of Wisconsin-Madison. 227 p.

Greisemer, Allen D. Brachiopod fauna from Middle Devonian of southeastern Wisconsin. M, 1963, University of Wisconsin-Madison.

Grekulinski, Edmund F. Foraminifera from some of the Coastal Plain sediments. M, 1954, New York University.

Gremell, Paul E. A gravity study of the Purcell Anticlinorium in the Toby Creek area of British Columbia. M, 1986, University of Calgary. 122 p.

Gremillion, Louis Ray. The origin of attapulgite in the Miocene strata of Florida and Georgia. D, 1965, Florida State University. 188 p.

Grenda, James C. Paleontology of two middle Pennsylvanian black shales in the Brereton and Saint David cyclothems of parts of Jackson and Perry counties, Illinois. M, 1969, Southern Illinois University, Carbondale. 109 p.

Grenda, James C. Paleozoology of cores from the Tyler Formation (Pennsylvanian) in North Dakota, U.S.A. D, 1977, University of North Dakota. 372 p.

Grender, Gordon Conrad. Petrology of the Vaqueros Formation near Gaviota, California. D, 1960, Pennsylvania State University, University Park. 152 p.

Grender, Gordon Conrad. The origin and distribution of structures on coals V, VI, and VII in Knox County, Indiana. M, 1952, Indiana University, Bloomington. 23 p.

Grenier, Paul Emile. Géologie de la région du Lac Albanel, Territoire de Mistassini (Quebec). M, 1949, Universite Laval.

Grenier, Paul Emile. Géologie et pétrologie de la région du Lac Beetz, Comté de Saguenay (Quebec). D, 1952, Universite Laval.

Greninger, Alden Buchannon. An investigation of the electrode-position of chromium and the structure of chromium deposits. M, 1931, Stanford University. 120 p.

Grenot, Charles H. Economic geology of the Middle Ordovician dolomites and limestones at the Narehood Quarry, Blair and Huntington counties, central Pennsylvania. M, 1987, Pennsylvania State University, University Park. 107 p.

Gresens, Randall Lee. A geochemical structural study of metasomatic formation of certain pegmatites. D, 1964, Florida State University. 203 p.

Gresh, Roger Theodore. Mercury in stream sediments from Conestoga Creek, southeastern Pennsylvania. M, 1972, Franklin and Marshall College. 110 p.

Gresham, Cyane W. Cretaceous and Paleocene siliceous phytoplankton assemblages from DSDP sites 216, 214 and 208 in the Pacific and Indian oceans. M, 1985, University of Wisconsin-Madison. 233 p.

Gresko, Mark J. The cause and seismotectonic implications of the Laconia linear aeromagnetic anomaly, Laconia, New Hampshire. M, 1980, Ohio University, Athens. 101 p.

Gresko, Mark Joseph. Analysis and interpretation of compressional (P-wave) and shear (Sh-wave) reflection seismic and geologic data over the Bane Dome, Giles County, Virginia. D, 1985, Virginia Polytechnic Institute and State University. 97 p.

Grethen, Bruce L. Synthesis of dolomite at 150 degrees centigrade. M, 1979, University of Missouri, Columbia.

Grether, William John. Conodonts and stratigraphy of the Readstown Member of the St. Peter Sandstone in Wisconsin. M, 1977, University of Wisconsin-Madison.

Grette, J. F. Cache Creek and Nicola groups near Ashcroft, British Columbia. M, 1978, University of British Columbia.

Greubel, Scott P. The stratigraphy of the lower Tertiary Wasatch Formation in the Government Creek area, Garfield County, Colorado. M, 1987, Colorado School of Mines. 128 p.

Greve, Gordon Madsen. An investigation of the Earth's gravitational and magnetic field on the San Francisco Peninsula, California. D, 1962, Stanford University. 209 p.

Grew, Edward Sturgis. Geology of the Pennsylvanian and pre-Pennsylvanian rocks of the Worcester area, Massachusetts. D, 1971, Harvard University.

Grey, Carllett. Conodonts of the reef and inter-reef facies of the Onondaga Limestone (Devonian, New York). M, 1987, Queens College (CUNY).

Grey, Charles E. Cyclic sedimentation, Dessa Dawn and Rundle formations, Banff-Jasper parks, Alberta. D, 1951, University of Wisconsin-Madison.

Grey, Charles Edwin. Mississippian rocks, north of Wapiti Lake, Wapiti Lake area, British Columbia. M, 1948, University of Kansas. 94 p.

Grey, L. Economic and land use evaluation of piedmont alluvial deposits, Windsor area, Colorado. M, 1974, University of Colorado.

Greybeck, James D. Geology of the Dee Mine area, Elko County, Nevada. M, 1985, University of Idaho. 96 p.

Gribble, G. W. Total chemical analyses of rocks, soils, and clay minerals by X-ray fluorescence quantometer. M, 1974, University of Hawaii. 81 p.

Gribble, Robert F. Zonation in the tilted Ashland Pluton, Klamath Mountains, California and Oregon. M, 1987, Texas Tech University. 107 p.

Grice, Joel Denison. Crystal structures of the tantalum-oxide minerals tantalite and wodginite, and of millerite nickel-monosulfide. D, 1973, University of Manitoba.

Grice, Joel Denison. The nature and distribution of the Tantalum minerals in the Tanco (Chemalloy) mine pegmatite at Bernic lake, Manitoba. M, 1970, University of Manitoba.

Grice, Reginald Hugh. Hydrogeology at a hydroelectric installation on Paleozoic dolomites at Grand Rapids, Manitoba. D, 1964, University of Illinois, Urbana. 296 p.

Grieco, Robert Anthony. Petrology and geochemistry of carbonate veins in the Moe-Reindeer Queen mineral belt of the Coeur d'Alene mining district, Idaho-Montana. M, 1981, Washington State University. 110 p.

Grieg, Paul Bennett, Jr. Geology of Pawnee County, Oklahoma. D, 1957, University of Oklahoma. 247 p.

Grieg, Paul Bennett, Jr. Geology of the Hallett area, Pawnee County, Oklahoma. M, 1954, University of Oklahoma. 91 p.

Griep, Jacobus L. Petrochemistry and metamorphism of the Tallan Lake Sill, Bancroft area, Ontario. M, 1975, McMaster University. 166 p.

Griepentrog, Thomas E. Classification and environment of continental shelf placers (general). M, 1970, University of Arizona.

Grier, Albert William. Lower Tertiary foraminifera from the Simi Valley, California. M, 1953, University of California, Berkeley. 108 p.

Grier, Susan Patricia. Alluvial fan and lacustrine carbonate deposits in the Snake Range; a study of Tertiary sedimentation and associated tectonism. M, 1984, Stanford University. 67 p.

Grierson, James Douglas. Devonian lycopods of New York State. D, 1962, Cornell University. 284 p.

Gries, John C. The structure and Cenozoic stratigraphy of the Pass Creek basin area, Carbon County, Wyoming. M, 1964, University of Wyoming. 69 p.

Gries, John Charles. Geology of the Sierra de la Parra area, northeast Chihuahua, Mexico. D, 1970, University of Texas, Austin. 193 p.

Gries, John Paul. Ordovician scolecodonts. D, 1935, University of Chicago. 76 p.

Gries, John Paul. Scolecodonts from the Decorah Formation of Ste. Genevieve County, Missouri. M, 1933, University of Chicago. 48 p.

Gries, Ruth Roberta Rice. Carboniferous biostratigraphy, western San Saba County, Texas. M, 1970, University of Texas, Austin.

Griesbach, Frederick Richard. Preliminary palynology of the lower Frontier Formation (Cretaceous) subwestern Wyoming. M, 1956, University of Utah. 62 p.

Griesemer, Allan David. Paleoecology of the Ervine creek limestone (late Pennsylvanian) in the mid-continent region. D, 1970, University of Nebraska, Lincoln. 364 p.

Griesemer, Jeffrey Crane. A petrographic study of the Mauch Chunk-Pottsville transition zone in Northeastern Pennsylvania. M, 1980, Lehigh University. 237 p.

Grieve, D. A. Behaviour of some trace metals in sediments of the Fraser River delta-front, southwestern British Columbia. M, 1977, University of British Columbia.

Grieve, Richard Andrew Francis. The petrology of a basic-ultrabasic complex in Monmouth Township, Ontario (Canada). M, 1967, University of Toronto.

Grieve, Richard Andrew Francis. The stability of chloritoid below 10Kb PH₂O. D, 1970, University of Toronto.

Grieve, Robert Oliver. Subsurface study of the Detroit River Group in southwestern Ontario. M, 1951, University of Michigan.

Griffen, D. T. Distortions in the tetrahedral oxyanions of crystalline substances, with crystal structure refinements of slawsonite and celsian. D, 1975, Virginia Polytechnic Institute and State University. 154 p.

Griffen, Dana T. The crystal chemistry of staurolite. M, 1972, Virginia Polytechnic Institute and State University.

Griffen, Villard. Geology of the Walhalla Quadrangle, Oconee County, SC. M, 1974, University of South Carolina.

Griffin, Andree French. Depositional and diagenetic history of the West Smyer Field, Hockley County, Texas. M, 1988, Texas Christian University. 37 p.

Griffin, Bert Eldon. Geology of Monte Cristo district with special reference to ore deposits. M, 1948, University of Washington. 60 p.

Griffin, George Martin. Clay mineral facies development in Recent surface sediments of the northeastern Gulf of Mexico. D, 1960, Rice University. 170 p.

Griffin, George Melvin, Jr. Clay minerals of the Neuse River estuary. M, 1954, University of North Carolina, Chapel Hill. 23 p.

Griffin, J. R., Jr. Paleoecologic study of the Oketo Shale (lower Permian) in north central Kansas. M, 1974, Kansas State University. 179 p.

Griffin, John Joseph. Clay mineralogy of the Mississippi River and major tributaries. M, 1958, Washington University. 50 p.

Griffin, John Roy. A structural study of the Silurian metasediments of central Maine. D, 1973, University of California, Riverside. 157 p.

Griffin, Judson Roy. The fauna of the LaSalle Limestone. D, 1931, University of Illinois, Chicago.

Griffin, Judson Roy. The stratigraphy and paleontology of Boulder Valley, Sweetgrass County, Montana. M, 1927, University of Illinois, Chicago. 36 p.

Griffin, Karen M. Sedimentology and paleontology of thrombolites and stromatolites of the Upper Cambrian Nopah Formation and their modern analog on Lee Stocking Island, Bahamas. M, 1988, University of California, Santa Barbara.

Griffin, M. G. A description, environmental interpretation and correlation of a section of the lower Windsor Group exposed in the Dark Quarry, Wentworth Quarry Complex, Hants County, Nova Scotia. M, 1971, Acadia University.

Griffin, Mark E. Precious metal deposits associated with alkaline rocks, North American Cordillera north of 41 degrees N.; computer exercises in pattern recognition and prediction. M, 1984, Eastern Washington University. 85 p.

Griffin, Mitchell Lee. Procedure to generate precipitation data for simulating long term soil erosion from a hillslope area with event oriented models; the design storm methodology. D, 1988, Purdue University. 337 p.

Griffin, Nancy Lindsley. Paleomagnetic properties of the Dufek intrusion (Mesozoic or possibly younger) Pensacola mountains, Antarctica. M, 1969, University of California, Riverside. 93 p.

Griffin, P. A. Heavy mineral investigation of Carmel bay (California) beach sands. M, 1969, United States Naval Academy.

Griffin, Patrick Maesa. Influence of shear and compression interaction on the response of sand to dynamic loading. D, 1980, University of California, Berkeley. 494 p.

Griffin, Robert Hardy. Structure and petrography of the Hillabee Sill and associated metamorphics, Alabama. D, 1947, University of Cincinnati. 92 p.

Griffin, Thomas T. Modeling phosphorus dynamics in reservoirs. M, 1982, Princeton University.

Griffin, Villard Stuart, Jr. A petrofabric investigation of some Cambrian and Precambrian rocks of the central Blue Ridge of Virginia. M, 1961, University of Virginia. 95 p.

Griffin, Villard Stuart, Jr. Mesoscopic and microscopic fabric relationships across the Catoctin Mountain-Blue Ridge anticlinorium of central Virginia. D, 1965, Michigan State University. 232 p.

Griffin, William Lindsay. Geology of the Babbitt-Embarrass area, Saint Louis County, Minnesota. D, 1967, University of Minnesota, Minneapolis. 285 p.

Griffin, William Lindsay. San Benito gravels. M, 1963, Stanford University.

Griffin, William Timothy. Petrology of the Tertiary carbonates exposed at Belgrade, Onslow County, North Carolina. M, 1982, University of North Carolina, Chapel Hill. 146 p.

Griffing, David H. Petrography, stratigraphy, and depositional environments of the Arco Hills Formation (Chesterian) of east-central Idaho. M, 1987, University of Idaho. 116 p.

Griffis, Arthur T. Some structural features of the Timiskaming series at Kirkland Lake, Ontario. M, 1937, University of Toronto.

Griffis, Arthur Thomas. The Timiskaming series and the early Precambrian. D, 1939, Cornell University.

Griffis, Robert Arthur. Kern Knob Pluton and other highly evolved granitoids in east-central California. M, 1987, California State University, Northridge. 305 p.

Griffis, Robert John. Genesis of the Deloro township talc-magnesite deposit, Ontario, Canada. M, 1970, Washington State University. 121 p.

Griffis, Robert John. Igneous petrology, structure, and mineralization in the eastern Sultan Basin, Snohomish County, Washington. D, 1977, Washington State University. 270 p.

Griffith, Alan Fraser. The effects of hydratable material on the permeability of oil field sands. M, 1955, University of Oklahoma. 134 p.

Griffith, Charles E. Palynostratigraphy of the Sundance Formation (Jurassic) of the Black Hills area, South Dakota and Wyoming. M, 1972, East Texas State University.

Griffith, Christine Marie. Stratigraphy and paleoenvironment of the New Albany Shale (Upper Devonian) of North-central Kentucky. M, 1977, University of Wisconsin-Madison.

Griffith, Earl Francis. Environmental geology of the southeast margin of the Gallatin Valley, Gallatin County, Montana. M, 1982, Montana State University. 106 p.

Griffith, Gary Lee. The Tonganoxie Sandstone in portions of Sedgwick, Butler, and Greenwood counties, Kansas. M, 1981, Wichita State University. 54 p.

Griffith, James Hendrie. Foraminifera from the Carlile Shale of the Republican River valley in Nebraska and Kansas. M, 1948, University of Nebraska, Lincoln.

Griffith, L. A. Depositional environment and conglomerate diagenesis of the Cardium Formation, Ferrier Field, Alberta. M, 1981, University of Calgary. 132 p.

Griffith, Lawrence S. The Carboniferous geology of the Pahranagat Range; a study of the Mississippian and Lower Pennsylvanian stratigraphy and paleontology of the Pahranagat Range, southeastern Nevada. M, 1959, Rice University. 75 p.

Griffith, Michael Craig. Experimental evaluation of seismic isolation for medium-rise structures subject to uplift. D, 1988, University of California, Berkeley. 238 p.

Griffith, Robert Fiske. Economic geology of Peacock Mountain, Okanogan County, Washington. M, 1949, University of Washington. 63 p.

Griffith, Roger Clinton. Geology of the southern Sierra Calamajue area; structural and stratigraphic evidence for latest Albian compression along a terrane boundary, Baja California, Mexico. M, 1987, San Diego State University. 115 p.

Griffith, Thomas W. A geological and geophysical investigation of sedimentation and Recent glacial history of the Gerlache Strait region, Graham Land, Antarctica. M, 1988, Rice University. 449 p.

Griffiths, Donald Ward. The effect of Appalachian Mountain topography on seismic waves. M, 1978, Virginia Polytechnic Institute and State University.

Griffiths, Reginald. The geology of the Wind map-area, Jasper, Alberta. M, 1962, University of Alberta. 52 p.

Griffiths, Sally A. Determination of depth extent of tabular dikes by wavenumber-domain magnetic interpretation. M, 1979, Pennsylvania State University, University Park. 87 p.

Griffiths, Scott A. Depositional environments and provenance of the Trinidad Sandstone, and related formations, Vermejo Park, New Mexico. M, 1981, Indiana University, Bloomington. 195 p.

Griffitts, Wallace Rush. Occurrence of mica-bearing pegmatite in Southeastern United States. M, 1948, University of Michigan.

Griffitts, Wallace Rush. Pegmatite geology of the Shelby District, North and South Carolina. D, 1958, University of Michigan. 162 p.

Grigg, Robert P., Jr. Structure of West Lake Verret Field, St. Martin Parish, Louisiana. M, 1950, Louisiana State University.

Griggs, Allan B. Geology and notes on ore deposits of Canyon-Nine Mile creeks area, Shoshone County, Idaho. D, 1952, Stanford University. 141 p.

Griggs, David Gould. The Ferguson and Parkman Sandstone (Upper Cretaceous) in the Deadhorse-Barber Creek area, Powder River basin, Wyoming. M, 1966, University of Colorado.

Griggs, David T. The strain ellipsoid and ecologic structure. M, 1933, Ohio State University.

Griggs, Gary Bruce. Cascadia channel; the anatomy of a deep-sea channel. D, 1969, Oregon State University. 183 p.

Griggs, Peter Humphrey. Stratigraphic significance of fossil pollen and spores of the Chuckanut Formation (Eocene), Northwest Washington. M, 1965, Michigan State University. 116 p.

Griggs, Peter Humphrey. Stratigraphy and palynology of the Frontier Formation (upper Cretaceous), Big Horn Basin, Wyoming. D, 1970, Michigan State University. 233 p.

Griggs, Roy Lee. The geology of the Bachelor Quadrangle, Callaway County, Missouri. M, 1940, University of Missouri, Columbia.

Grigor, Lynne Jones. Occurrence and sediment characteristics of washovers at Fire Island National Seashore, Fire Island, New York. M, 1985, SUNY at Binghamton. 88 p.

Grigsby, F. Bryan. Quaternary tectonics of the Rincon and San Miguelito oil fields area, western Ventura Basin, California. M, 1986, Oregon State University. 110 p.

Grigsby, Jeffrey D. Sandstone diagenesis in the Rincon Valley and Hayner Ranch formations (Miocene), San Diego Mountain, New Mexico. M, 1984, New Mexico State University, Las Cruces. 60 p.

Griley, Horace Longin. A geologic section along the north side of the Missouri River from Providence, Missouri, to North Jefferson, Missouri. M, 1930, University of Missouri, Columbia.

Grill, Edwin Vatro. A model for the biogeochemical cycle of silica in the sea. D, 1965, University of Washington. 90 p.

Grillot, Larry Ray. An analysis of magnetotelluric measurements made in southwest Iceland. M, 1970, Brown University.

Grillot, Larry Ray. Regional electrical structure beneath Iceland as determined from magnetotelluric data. D, 1973, Brown University.

Grills, Richard Barbee, Jr. Deposition and diagenesis of Early Pennsylvanian shelf edge carbonates, Delaware Basin. M, 1977, Texas Tech University. 128 p.

Grim, Paul J. Some new heat flow measurements from the floor of the equatorial eastern Pacific. M, 1963, Columbia University, Teachers College.

Grim, Ralph Early. The Eocene sediments of Mississippi. D, 1931, University of Iowa. 397 p.

Grimes, Douglas James. Depositional models, subaerial facies, and diagenetic histories of the Rosedale and Fletcher reefs, southwestern Ontario. M, 1987, Queen's University. 120 p.

Grimes, John H. The origin and occurrence of chromiferous muscovite in the Crow Formation of the southern Black Hills, Custer County, South Dakota. M, 1972, South Dakota School of Mines & Technology.

Grimes, Wayne Harlan. The subsurface geology of Beaver County, Oklahoma. M, 1952, University of Oklahoma. 60 p.

Grimes-Graeme, Rhoderick C. H. The origin of the intrusive igneous breccias in the vicinity of Montreal, Quebec. D, 1935, McGill University.

Grimes-Graeme, Rhoderick C. H. The petrology of certain igneous rocks of Newton Township, Ontario. M, 1932, McGill University.

Grimestad, Garry R. A geo-hydrologic evaluation of an infiltration disposal system for Kraft process pulp and paper mill liquid effluents. D, 1977, University of Montana. 187 p.

Grimison, Nina Louise. Diffuse zones of deformation in oceanic lithosphere; the Azores-Gibraltar plate boundary and the Davie Ridge-Madagascar region. D, 1987, University of Illinois, Urbana. 129 p.

Grimm, Eric Christopher. An ecological and paleo-ecological study of the vegetation in the Big Woods region of Minnesota. D, 1981, University of Minnesota, Minneapolis. 379 p.

Grimm, Joan P. Cenozoic pisolitic limestone in Pima and Cochise counties, Arizona. M, 1978, University of Arizona.

Grimm, Joel Patrick. The late Cenozoic geomorphic history of the Lobo Canyon area of the Mount Taylor volcanic field, Cibola County, New Mexico. M, 1985, University of New Mexico. 159 p.

Grimm, Kenneth E. The paragenesis of the West Coast Mines ore body, Nevada. M, 1942, Stanford University. 31 p.

Grimm, Kurt Andrew. Sedimentation and diagenesis at a Late Cambrian biomere extinction horizon. M, 1986, University of Wisconsin-Madison.

Grimm, Michael A. A fracture analysis of a sandstone in the upper portion of the Morrison Formation in the Tenorio Ranch area, north-central, New Mexico. M, 1985, Bowling Green State University. 65 p.

Grimmer, John C. Macro-invertebrates of the Middle Devonian shales of southern Illinois. M, 1966, Southern Illinois University, Carbondale. 58 p.

Grimshaw, Thomas W. Environmental geology of urban and urbanizing areas; a case study from the San Marcos area, Texas. D, 1976, University of Texas, Austin.

Grimshaw, Thomas Walter. Geology of the Wimberley area, Hays and Comal counties, Texas. M, 1970, University of Texas, Austin.

Grimsley, George P. Microscopical study of Ohio limestones. M, 1891, Ohio State University.

Grimsley, George P. The granites of Cecil County, in northeastern Maryland. D, 1894, The Johns Hopkins University.

Grimwood, C. Evaluation of methods to predict the effects of dredging on the quality of adjacent waters in South Louisiana. D, 1978, Tulane University. 218 p.

Grine, Donald. Mechanized determination of epicenters by Geiger's method. M, 1954, Massachusetts Institute of Technology. 23 p.

Grine, Donald R. Finite amplitude stress waves in rocks. D, 1959, Massachusetts Institute of Technology. 79 p.

Gringarten, Alain Charles. Unsteady-state pressure distributions created by a well with a single horizontal fracture, partial penetration, or restricted entry. D, 1971, Stanford University. 110 p.

Grinham, David F. Intertidal sedimentation in Akimiski Strait, James Bay, Canada. M, 1980, University of Guelph.

Grinnell, Daniel Voorhis. Physiography of the continental margin of Antarctica from 125°E to 150°E. M, 1973, University of South Carolina.

Grinnell, George B. The osteology of Geococcyx californianus. D, 1880, Yale University.

Grinnell, Philip Collins. The Sunniland Limestone within the Forty Mile Bend area, Monroe and Dade counties, Florida. M, 1976, University of Florida. 136 p.

Grinnell, Robert Newell. Physical characteristics of the Kirtland Formation, Cretaceous, in the northwest portion of the San Juan Basin. M, 1952, University of Illinois, Urbana.

Grinnell, Robert S., Jr. Specific and infraspecific differences of the brachiopod genus Composita in eastern Kansas. M, 1958, University of Kansas. 83 p.

Grinnell, Robert Stone, Jr. Structure and development of oyster reefs on the Suwannee River delta, Florida. D, 1971, SUNY at Binghamton. 199 p.

Grinshpan, Zvi. The effects of the horizontal component of the Earth stress field in geological structures. M, 1976, University of Arizona.

Grinstead, Gary Patrick. Pliocene paleoclimatic history of the Southwest Atlantic; a quantitative micropaleontological approach. M, 1984, University of Georgia. 101 p.

Grippi, Jack. Sedimentologic and tectonic setting of the Hanover Inlier, western Jamaica. M, 1978, SUNY at Albany. 182 p.

Grisafe, David Anthony. Crystal chemistry and color in apatite. D, 1968, Pennsylvania State University, University Park. 208 p.

Grisafe, David Anthony. Phase relations in the system lead oxide-carbon dioxide. M, 1963, Pennsylvania State University, University Park. 59 p.

Grisak, G. E. Hydrogeologic response analysis and radioactive waste management characteristics of WNRE Manitoba. M, 1974, University of Waterloo.

Griscom, Andrew. Bedrock geology of the Harrington Lake area, Maine. D, 1976, Harvard University.

Griscom, Clement A. The environmental impact statement as a contribution to the library of knowledge. D, 1973, New York University.

Griscom, Melinda. Space-time seismicity patterns in the Utah region and an evaluation of local magnitude as the basis of a uniform earthquake catalog. M, 1980, University of Utah. 134 p.

Grisso, Julie Martin. An analysis of some secondary magnetizations of baked clays. M, 1978, University of Oklahoma. 80 p.

Grissom, Holly D. The environment of deposition of the lower nodular zone of the Brule Formation (Oligocene, middle and upper) in Big Badlands, South Dakota. M, 1968, South Dakota School of Mines & Technology.

Griswold, George Bullard. Some aspects of rock mechanics as applied to Project Mohole. D, 1967, University of Arizona.

Griswold, Mark L. Petrology and structure of the southwest portion of the Precambrian Rawah Batholith, northcentral Colorado. M, 1980, Colorado State University. 265 p.

Griswold, Russell E. Geology and ground water resources of Wayne County, New York. M, 1949, Syracuse University.

Griswold, Thomas Baldwin. A study of the effects of a mafic dike on a granite, Front Georgeville, Antigonish County, Nova Scotia. D, 1978, University of Kentucky. 168 p.

Griswold, Thomas Baldwin. The equations of the eutectic line for two component, three phase systems and application to the prediction of the effects of total pressure and water pressure on the system diopside-anorthite. M, 1970, University of Kentucky. 34 p.

Griswold, Willis R. Some suggestions as to the nature of magmatic waters. M, 1923, University of Minnesota, Minneapolis. 51 p.

Grivetti, L. E. Intertidal foraminifera of the Farallon Islands. M, 1962, University of California, Berkeley. 312 p.

Grivetti, Mark Christopher. Aspects of stratigraphy, diagenesis, and deformation in the Monterey Formation near Santa Maria-Lompoc, California. M, 1982, University of California, Santa Barbara. 155 p.

Groat, Charles George. Geology and hydrology of Rosamond and Rodgers Playa, San Bernardino County, California. M, 1967, University of Massachusetts. 133 p.

Groat, Charles George. Geology of Presidio bolson, Presidio County, Texas and adjacent Chihuahua, Mexico. D, 1970, University of Texas, Austin.

Groat, Lee Andrew. The crystal chemistry of vesuvianite. D, 1988, University of Manitoba.

Grobecker, Alan J. Travel time curves at small distances and wave velocities of principal phases in the Southern California ranges. M, 1941, California Institute of Technology. 42 p.

Groce, John A. Selected physical properties of the relict salt marsh deposits of St. Catherines Island, Georgia. M, 1980, University of Georgia.

Grocock, G. R. Stratigraphy and petrography of the upper member of the Murl Limestone in Southeast Cochise County, Arizona. M, 1975, University of Colorado.

Grodi, Ernest D. Subsurface coal mining works delineation using P-, P-SV-, and SH-waves. M, 1988, Wright State University. 111 p.

Groen, John Corwyn. Gold-enriched rims on placer gold grains; an evaluation of formational processes. M, 1987, Virginia Polytechnic Institute and State University.

Groeneveld Meijer, Willem Otto Jan. The geochemistry of the platinum metals with respect to their occurrence in nickeliferous sulphide deposits. D, 1955, Queen's University. 407 p.

Groenewold, Gerald Henry. Applied geology of the Bismarck-Mandan area, North Dakota. D, 1972, University of North Dakota. 259 p.

Groenewold, Gerald Henry. Concretions and nodules in the Hell Creek Formation (upper Cretaceous), southwestern North Dakota. M, 1971, University of North Dakota. 84 p.

Groenewold, Joanne Van Ornum. Lexicon of bedrock stratigraphic names of North Dakota. M, 1979, University of North Dakota. 205 p.

Groenewold, John Carl. Electrical properties of clay-bearing sandstones. M, 1988, University of Utah. 86 p.

Groening, Donald I. Instrument constants and the adjustment of a torsion balance. M, 1961, Ohio State University.

Groenwold, Bernard Cyrus. Subsurface geology of the Mesozoic formations overlying the Uncompahgre Uplift in Grand County, Utah. M, 1961, University of Utah. 70 p.

Groff, Donald William. Spectrochemical analysis of sandstones and its classification implications. D, 1965, University of Pittsburgh. 429 p.

Groff, Edward De *see* De Groff, Edward

Groff, Sidney Lavern. Petrography of the Kootenai Creek area, Bitterroot Range, Montana. M, 1954, University of Montana. 80 p.

Groff, Sidney Lavern. The geology of the West Tintic Range and vicinity, Tooele and Juab counties, Utah. D, 1959, University of Utah. 244 p.

Groffie, Frank Johannes. Geology and Late Cretaceous depositional environments in the northern corner of Camp Pendleton, Southern California. M, 1985, San Diego State University. 117 p.

Groffman, Louis H. Stratigraphy, sedimentology, and structure of the upper Proterozoic Three Sisters Formation and Lower Cambrian Gypsy Quartzite, Northeast Washington. M, 1986, Washington State University. 208 p.

Grogan, Robert Mann. Geology of a part of the Minnesota shore of Lake Superior northeast of Two Harbors, Minnesota. D, 1940, University of Minnesota, Minneapolis. 110 p.

Grogan, Robert Mann. Geology of Northeast Bay, Minnitaki Lake, Ontario. M, 1936, University of Minnesota, Minneapolis. 36 p.

Grogger, Paul Karl. Glaciation of the High Uintas primitive area, Utah, with emphasis on the northern slope. D, 1974, University of Utah. 209 p.

Groh, Douglas. The economic feasibility of marine mining phosphorite offshore North Carolina. M, 1985, University of Texas, Austin.

Grohskopf, John G. Some results of magnetometric surveying in Missouri. M, 1931, University of Missouri, Rolla.

Grolier, Maurice Jean. Geology of the Big Bend area, in the Columbia Plateau, Washington. D, 1965, The Johns Hopkins University. 265 p.

Gromer, James. Stratigraphy and sedimentation of the Baca Formation of SE Socorro County. M, 1975, New Mexico Institute of Mining and Technology.

Gromet, L. Peter. Rare earths abundances and fractionations and their implications for batholithic petrogenesis in the Peninsular Ranges Batholith, California, USA, and Baja California, Mexico. D, 1979, California Institute of Technology. 357 p.

Grommé, Charles Sherman. Remanent magnetization of igneous rocks from the Franciscan and Lovejoy formations, Northern California. D, 1963, University of California, Berkeley. 225 p.

Grommesh, Mark W. A study of a subsurface algal-mound complex (Toronto Limestone, Upper Pennsylvanian), in parts of Barber, Kingman, and Pratt counties, Kansas. M, 1986, Wichita State University. 158 p.

Gron, Shelley A. A study of the igneous horizons in the St. Jonsfjorden and Lovliebreen groups in West Spitsbergen. M, 1984, Wayne State University. 242 p.

Gronberg, Eric C. Biostratigraphy of the Lower and Middle Devonian of Lone Mountain and Table Mountain, central Nevada. M, 1967, University of California, Riverside. 83 p.

Grondin, Gaston Guy. Géologie régionale des Appalaches et des basse-terres du St. Laurent dans le Québec. M, 1959, Universite Laval.

Groneck, John E., III. Stratigraphy, petrology, and depositional environment of the Sulphur Springs Formation (Upper Devonian-Lower Mississippian), east-central Missouri. M, 1987, Fort Hays State University. 76 p.

Gronewold, Robert L. Trace and major element geochemistry of the Gray Eagle volcanogenic massive sulfide deposit, Siskiyou County, California. M, 1983, University of California, Davis. 154 p.

Gronseth, Kenneth Allen. Seismic velocities of an in situ, jointed block under controlled stress conditions. M, 1976, University of Utah. 72 p.

Groot, Philip Henry de *see* de Groot, Philip Henry

Grootenboer, Johan. Equilibrium fractionation of sulfur isotopes between pyrite, sphalerite and galena. D, 1969, McMaster University. 270 p.

Gros, Frederick Christian. Contact metamorphism in red beds from the Tarryall District, Colorado. M, 1940, University of Rochester. 56 p.

Groschel, Henrike. Effect of urbanization of the geologic and hydrologic behavior of streams in College Station, Texas. M, 1985, Texas A&M University.

Groschen, George Earl. Geochemistry of Williams Lake, Hubbard County, Minnesota. M, 1981, University of Minnesota, Minneapolis.

Grose, Lucius Towbridge. A petrogenetic study of Evans Lake granodiorite, Omak, Washington. M, 1949, University of Washington. 90 p.

Grose, Lucius Trowbridge. Rocks and structure of the northeast part of the Soda Mountains, San Bernardino County, California. D, 1956, Stanford University. 156 p.

Grose, Peter. The stratification and circulation of the subsurface waters of the Gulf of Mexico. M, 1966, Florida State University.

Groselle, Francis X. Amphibolites and associated rocks of Buttermilk Point, Maine. M, 1955, University of Maine. 64 p.

Groshong, Richard Hughes, Jr. Geology and fracture patterns of north central Burnet County, Texas. M, 1967, University of Texas, Austin.

Groshong, Richard Hughes, Jr. Strain in minor folds, Valley and Ridge Province, Pennsylvania. D, 1971, Brown University.

Gross, Barry L. The relationship of the Adirondack anorthosite (Precambrian) with surrounding rocks, eastern New York. M, 1970, Iowa State University of Science and Technology.

Gross, C. M. The fauna of the Reedsville Shale at Antes Gap, Pennsylvania. M, 1955, Pennsylvania State University, University Park.

Gross, David James. Geology of the Ortega area, Ventura County, California. M, 1958, University of California, Los Angeles.

Gross, David Lee. Glacial geology of Kane County, Illinois. D, 1969, University of Illinois, Urbana. 253 p.

Gross, David Lee. Mineralogical gradations within Titusville Till and associated tills of northwestern Pennsylvania. M, 1967, University of Illinois, Urbana.

Gross, David Thomson. Depositional and diagenetic history of the Mississippian Gilmore City Limestone, north central Iowa. M, 1982, University of Iowa. 200 p.

Gross, Eugene B. Contact metamorphism of the Idaho Springs Formation (Precambrian; Park County), Colorado. M, 1948, University of Colorado.

Gross, Eugene Bischoff. Alkalic granites and pegmatites of the Mount Rosa area, El Paso and Teller counties, Colorado. D, 1962, University of Michigan. 196 p.

Gross, G. A. A comparative study of three slate formations in the Ferriman Series in the Labrador Trough. M, 1951, Queen's University. 122 p.

Gross, Geraldo Wolfgang. Geoelectrical investigations in the lower Paleozoic of central Pennsylvania. D, 1959, Pennsylvania State University, University Park. 146 p.

Gross, Gordon Arnold. The metamorphic rocks of the Mount Wright and Matonipi Lake areas of Quebec. D, 1955, University of Wisconsin-Madison.

Gross, James T. A hydrogeological investigation of the Las Cruces geothermal field. M, 1988, New Mexico State University, Las Cruces. 212 p.

Gross, Jonathan A. The Lincoln sanitary landfill; an evaluation of ground water quality and movement, methane production, and percolation potential. M, 1986, University of Nebraska, Lincoln. 107 p.

Gross, Larry Thomas. Stratigraphic analysis of the Mesa Verde Group, (Cretaceous) Uinta Basin, Utah. M, 1961, University of Utah. 115 p.

Gross, Laura Blanche. The stratigraphy and lithology of the glaciogenic sediments of the Two Harbors area, northeastern Minnesota. M, 1982, University of Minnesota, Duluth. 151 p.

Gross, Meredith Grant, Jr. Carbonate sedimentation and diagenesis of Pleistocene limestones in the Bermuda Islands. D, 1961, California Institute of Technology.

Gross, Michael P. Mineralization and alteration in the Greaterville District, Pima County, Arizona. M, 1969, University of Arizona.

Gross, Michael Robert. Reclamation plans for abandoned mill tailing impoundments in the South Fork, Coeur d'Alene River basin. M, 1982, University of Idaho. 141 p.

Gross, Oliver. Compressed air energy storage in Cambro-Ordovician sandstones of eastern Iowa and southwestern Wisconsin. M, 1980, University of Wisconsin-Milwaukee.

Gross, Richard Stewart. A determination and analysis of polar motion. D, 1982, University of Colorado. 260 p.

Gross, Robert Erwin. The geology of the southwestern quarter of the Elsberry Quadrangle, Missouri. M, 1949, University of Iowa. 100 p.

Gross, Robert Olvin. Geology of Sierra Tinaja Pinta and Cornudas Station areas, northern Hudspeth County, Texas. M, 1965, University of Texas, Austin.

Gross, Warren William. Geochemistry and origins of dark inclusions and dikes in the granitic rocks of a portion of the Peninsular Ranges Batholith, San Diego County, California. M, 1984, San Diego State University. 103 p.

Gross, William H. An investigation of the control of gold mineralization in northwestern Ontario. D, 1950, University of Toronto.

Gross, William H. Structure and tectonic history of the Red Lake area, Ontario. M, 1947, University of Toronto.

Grosse, Charles W. Miospores associated with the flint clay parting within the fire clay coal of the Breathitt Formation at selected sites in eastern Kentucky. M, 1987, Eastern Kentucky University. 52 p.

Grosser, Paul W. Application of Bayesian decision theory to groundwater monitoring. D, 1984, University of Illinois, Urbana. 202 p.

Grosskopf, F. W. The hydraulic connection between the Republican River and aquifer(s) along the reach between McCook, Nebraska and Cambridge, Nebraska. M, 1978, University of Nebraska, Lincoln.

Grossman, Ethan Lloyd. Stable isotopes in live benthic foraminifera from the southern California borderland. D, 1982, University of Southern California.

Grossman, Irving. The identification of the tetracoral genera. M, 1947, Columbia University, Teachers College.

Grossman, Jeffrey N. A chemical and petrographic study of chondrules from the Chainpur (LL3.4) and Semarkona (LL3.0) chondrites. D, 1983, University of California, Los Angeles. 284 p.

Grossman, Lawrence. Condensation, chondrites and planets. D, 1972, Yale University.

Grossman, Robert Bruce. Studies on the physical properties of fragipans in New York State. M, 1954, Cornell University.

Grossman, Stuart. The ecology of the Rhizopodea and Ostracoda of the southern Pamlico Sound region, North Carolina. D, 1961, University of Kansas. 285 p.

Grossman, Stuart. The subfamily Glyptopleurinae. M, 1953, University of Illinois, Urbana.

Grossman, William L. Stratigraphy of the Genesee Group of New York. D, 1944, Columbia University, Teachers College.

Grossman, William Lewis. Geology of the Caledonia Quadrangle, New York. M, 1938, University of Rochester. 149 p.

Grossman, Zev N. Distribution and dispersal of Manati River sediments; Puerto Rico north insular shelf. M, 1978, Duke University. 72 p.

Grossnickle, Effie Ann. Tempestites and depositional interpretations of the Middle Ordovician Curdsville-Limestone of central Kentucky. M, 1985, University of Kentucky. 107 p.

Grossnickle, William E. Petrology of the Fort Riley Limestone from four Kansas quarries. M, 1961, Kansas State University. 82 p.

Grosso, Stephen T. Paleoenvironmental analysis of spore assemblages from regressive facies of the Upper Cretaceous in New Jersey. M, 1979, Rutgers, The State University, Newark. 100 p.

Grosvenor, Florence Anne. Brachiopoda and stratigraphy of the Rondout Formation (Silurian) in the

Rosendale Quadrangle, southeastern New York. M, 1965, University of Rochester. 185 p.

Grote, Benjamin. A study of the Tar Springs Sandstone in southwestern Illinois. D, 1949, University of Illinois, Urbana.

Grote, Fred Rankin. Structural geology of the central Bluff Creek area, Mason County, Texas. M, 1954, Texas A&M University.

Grotewold, Andreas Peter. Physiographic development of the Lake Quinsigamond Valley (Massachusetts). M, 1951, Clark University.

Groth, F. A. Stratigraphy of the Chinle Formation in the San Rafael Swell, Utah. M, 1955, University of Wyoming. 76 p.

Groth, Linda Williamson. Late-Glacial and Post-glacial vegetational changes in the Conneaut Marsh region, northwestern Pennsylvania. M, 1966, Pennsylvania State University, University Park. 148 p.

Groth, Peter K. H. Palynological delineation of environments in the Columbiana Shale (Pennsylvanian?) of western Pennsylvania. M, 1966, Pennsylvania State University, University Park. 192 p.

Grothaus, Brian. Size frequency analysis from sand microstructures. M, 1976, University of South Carolina.

Grothaus, Brian T. Depositional environment and structural implications of the Hammamat Formation, Egypt. D, 1979, University of South Carolina. 116 p.

Grotte, James Robert. The development and application of the Werner deconvolution and 2-D filtering to gravity and magnetic data in S.E. Virginia. M, 1985, University of Texas at Dallas. 154 p.

Grotts, Tim Douglas. A study of the Pleistocene marine invertebrate fauna of San Diego County, California. M, 1981, San Diego State University.

Grotzinger, John P. The stratigraphy and sedimentation of the Wallace Formation, Northwest Montana and northern Idaho. M, 1981, University of Montana. 153 p.

Grotzinger, John Peter. Evolution of early Proterozoic passive-margin carbonate platform, Rocknest Formation, Wopmay Orogen, N.W.T., Canada. D, 1985, Virginia Polytechnic Institute and State University. 303 p.

Groughnour, Roy Robert. The soil-ice system and the shear strength of frozen soils. D, 1967, Michigan State University. 139 p.

Grout, Frank F. The Duluth Gabbro and its associated formations. D, 1917, Yale University.

Grout, Marilyn Ann. Rockwall erosion in the northern Sangre de Cristo Mountains, Colorado; contributing factors, sequence, and rate. M, 1981, University of Colorado. 165 p.

Grove, Arlen K. Distribution of selected trace elements in the upper part of the Salamonie Dolomite (Silurian), Jay County, Indiana. M, 1981, Indiana University, Bloomington. 121 p.

Grove, Brandon H. Geology of the northwestern part of the Idaho Batholith and adjacent region in Montana. D, 1934, University of Chicago. 42 p.

Grove, E. W. Detailed geological studies in the Stewart Complex, northwestern B.C. D, 1973, McGill University. 434 p.

Grove, Edward Willis. A study of contact metamorphism at Harrison Ridge, Harrison hot springs, BC. M, 1955, University of British Columbia.

Grove, G. D. Consolidation as a delayed yield factor in confined aquifer response. M, 1970, University of Manitoba.

Grove, Ginny R. A study of fine-grained disseminated gold ore of the Windfall Mine, Eureka County, Nevada. M, 1979, University of California, Santa Barbara.

Grove, John Hamman. The ion exchange chemistry of soils and liming. D, 1980, University of Georgia. 82 p.

Grove, Marty. Metamorphism and deformation of pre-batholithic rocks in the Box Canyon area, eastern Pen-

insula Ranges, San Diego County, California. M, 1987, University of California, Los Angeles. 174 p.

Grove, Thurman Lee. Phosphorus cycles in forests, grasslands, and agricultural ecosystems. D, 1983, Cornell University. 154 p.

Grove, Timothy L. Structural characterization of natural calcic plagioclases. D, 1976, Harvard University.

Grover, Anil. Radiation from an explosion in a non-uniformly prestressed medium. M, 1971, St. Louis University.

Grover, Charles H. List of Kansas minerals with brief notes on their crystallographic form, chemical composition, and the principal localities which they have been reported. M, 1895, University of Kansas. 8 p.

Grover, George A. Fenestral and associated diagenetic fabrics, Middle Ordovician New Market Limestone, Virginia. M, 1976, Virginia Polytechnic Institute and State University.

Grover, George Adelbert, Jr. Cement types and cementation patterns of Middle Ordovician ramp-to-basin carbonates, Virginia. D, 1981, Virginia Polytechnic Institute and State University. 299 p.

Grover, Jeffrey Alan. The geology of Teran Basin, Cochise County, Arizona. M, 1982, University of Arizona. 73 p.

Grover, John Emerson. Two problems in pyroxene mineralogy; a theory of partitioning of cations between co-existing single and multi-site phases and a determination of the stability of low-clinoenstatite under hydrostatic pressure. D, 1972, Yale University.

Grover, Timothy Warren. Polymetamorphism and petrology of aluminous schists along the northwest border zone of the Idaho Batholith. D, 1988, University of Oregon. 221 p.

Grover, Timothy Warren. Progressive metamorphism west of the Condrey Mountain dome, north-central Klamath Mountains, Northern California. M, 1984, University of Oregon. 129 p.

Groves, David Alan. Stratigraphy and alteration of the footwall volcanic rocks beneath the Archean Mattabi massive sulfide deposit, Sturgeon Lake, Ontario. M, 1984, University of Minnesota, Duluth. 141 p.

Groves, James Robert. Development of a remote sensing based hydrologic model. D, 1983, University of Maryland. 197 p.

Groves, John R. Calcareous foraminifers and algae from the type Morrowan (Lower Pennsylvanian) region of northwestern Arkansas and northeastern Oklahoma. M, 1981, University of Oklahoma. 158 p.

Groves, John Reid. Calcareous microfossils and biostratigraphy of the Arco Hills, Bluebird Mountain, and lower Snaky Canyon formations (Mid-Carboniferous) of east-central Idaho. D, 1983, University of Iowa. 253 p.

Grow, Jeffrey S. Pedogenic carbonates (caliche) in the lower Newman Limestone (Mississippian) of northeastern and east-central Kentucky. M, 1982, University of Kentucky. 145 p.

Grow, John A. A gravity study of the western Rift Valley of Africa. M, 1964, Columbia University, Teachers College.

Grow, John Allen. A geophysical study of the central Aleutian Arc. D, 1972, University of California, San Diego. 147 p.

Grow, Sheila Roseanne. Water quality in the Forestville Creek karst basin of southeastern Minnesota. M, 1986, University of Minnesota, Minneapolis. 229 p.

Growitz, Douglas. Geochemistry of mine water, northern bituminous coal field, West Virginia. M, 1967, West Virginia University.

Grubb, Carl Frederich. Conodonts from the Pennsylvanian shales of Illinois. M, 1932, University of Illinois, Chicago.

Grubb, James M. Petrology and petrography of the Munro Sill, Matheson, Ontario (Canada). M, 1968, Bowling Green State University. 75 p.

Grubbs, David M. Fauna of the Niagaran nodules of the Chicago area. D, 1939, University of Chicago. 92 p.

Grubbs, David M. The Edwards Limestone of Bell County, Texas, and surrounding territory. M, 1934, University of Oklahoma. 46 p.

Grubbs, Donald Keeble. Ore-bearing magmatic and metamorphic brine from the Salton Sea volcanic domes geothermal area, Imperial County, California. M, 1963, University of Virginia. 50 p.

Grubbs, Donald Keeble. Weathering of diabase. D, 1969, University of Pittsburgh. 106 p.

Grubbs, Edward L. Variations of porosity and permeability in the Wilcox Group in the Texas upper Gulf Coast. M, 1954, Texas A&M University.

Grubbs, Kenneth Lee. Triassic paleomagnetism of the Idaho-Wyoming overthrust belt (U.S.A.); a study of a mesotectonic structural deformation. M, 1975, University of Michigan.

Grubbs, Larry Stanley. The bedrock geology of the Harrison area, Butler-Hamilton counties, Ohio. M, 1963, Miami University (Ohio). 57 p.

Grubbs, Robert Kent. Pennsylvanian stratigraphy and conodont biostratigraphy of the Mill Creek Syncline, central Arbuckle Mountains, Oklahoma. M, 1981, University of Oklahoma. 227 p.

Grubbs, Robert S. Geology and water quality of the Prairie Grove area, Arkansas. M, 1974, University of Arkansas, Fayetteville.

Grubbs, William Howard. The stratigraphy, sedimentation and paleontology of the Trinity division between the Brazos and Red rivers. M, 1930, Texas Christian University. 72 p.

Grube, John P. Dakota Group (Lower Cretaceous) stratigraphy, northern Front Range, Larimer County, Colorado. M, 1984, Colorado School of Mines. 218 p.

Grube, Michael H. The origin and development of the southern portion of the Oak Openings sand belt, Lucas County, Ohio. M, 1980, Bowling Green State University. 144 p.

Gruber, David P. Regional stratigraphic relations of the Milligen Formation, central and southern Idaho. M, 1975, University of Wisconsin-Milwaukee. 171 p.

Gruber, Paul. A hydrologic study of the unsaturated zone adjacent to a radioactive-waste disposal site at the Savannah River Plant, Aiken, South Carolina. M, 1982, University of Georgia. 80 p.

Grudewicz, Eugene. Lunar rille region stratigraphic and relative age determinations for portions of Schroeter's Valley, Rima Trinz II and Hyginus Rille. M, 1972, University of California, Los Angeles.

Gruebel, Marilyn May. Microfacies of the Comanche Peak Limestone (Lower Cretaceous), north-central Texas. M, 1982, Texas A&M University. 103 p.

Gruen, Mary Abbott. The geology of the Chopawamsic Formation and associated exhalites in the Woods Mountain area, Virginia. M, 1982, University of North Carolina, Chapel Hill. 73 p.

Gruenenfelder, Charles R. Hydrogeology, hydrochemistry and reclamation alternatives for an inactive lead-silver mine in northern Idaho. M, 1987, University of Idaho. 96 p.

Gruenenfelder, Jane B. Stratigraphy and source rock potential of the Miocene Monterey Formation, San Joaquin Basin, California. M, 1987, Stanford University. 119 p.

Grueter, Joyce C. Petrology and structure of the Archean amphibolite facies gneiss terrain exposed adjacent to the Yellow River, Chippewa County, Wisconsin. M, 1982, University of Wisconsin-Milwaukee. 172 p.

Gruetzmacher, Jeff C. Surficial geology of southern Washington County, Wisconsin; with an evaluation of the exploitable sand and gravel deposits. M, 1975, University of Wisconsin-Milwaukee.

Gruman, William Paul. Geology of the Foyil area, Rogers and Mayes counties, Oklahoma. M, 1954, University of Oklahoma. 99 p.

Grumbles, George Robert. Stratigraphy and sedimentation of the (Eocene) Wilcox Formation in the Andersonville bauxite district of Georgia. M, 1957, Emory University. 115 p.

Grummon, Mark Longden. Experimental techniques for studying reaction equilibria at elevated temperatures and pressures. M, 1976, University of Michigan.

Grunder, Anita Lizzie. The Calabozos caldera complex; geology, petrology, and geochemistry of a major silicic volcanic center and hydrothermal system in the Southern Andes. D, 1986, Stanford University. 189 p.

Grundl, Timothy J. A stable-isotopic view of the Cambro-Ordovician aquifer system in northern Illinois. M, 1980, Northern Illinois University. 88 p.

Grundy, Allen T. Geology and geochemistry of the granitic and related rocks of the Littleton and Thelma area, eastern Piedmont, North Carolina. M, 1982, East Carolina University. 82 p.

Grundy, Wilbur D. Geology and uranium deposits of the Shinarump Conglomerate of Nokai Mesa, Arizona and Utah. M, 1953, University of Arizona.

Gruner, John W. Geologic reconnaissance of the southern part of Taos Range, New Mexico. M, 1919, University of Minnesota, Minneapolis. 38 p.

Gruner, John W. Organic matter and the origin of the Biwabik iron-bearing formation of the Mesabi Range. D, 1922, University of Minnesota, Minneapolis. 69 p.

Gruner, Thayer Meredith. The heavy accessory minerals of the Devonian sandstones of southeastern Missouri and southwestern Illinois. M, 1941, Washington University. 164 p.

Grunsky, Eric Christopher. A method of determining orientation and shape parameters of porphyroclasts in mylonitic rocks. M, 1978, University of Toronto.

Grunsky, Eric Christopher. Multivariate and spatial analysis of lithogeochemical data from metavolcanics with zones of alteration and mineralization in Ben Nevis Township, Ontario. D, 1988, University of Ottawa. 710 p.

Gruntmeyer, Paul Alexander. Bedload sediment transport pathways in the lower Connecticut River estuary. M, 1984, Wesleyan University. 137 p.

Grunwald, R. R. The distribution of zirconium in Hawaiian sediments. M, 1967, University of Hawaii. 85 p.

Grunwald, Ross Richard. Geology and mineral deposits of the Galena Mining District, Black Hills, South Dakota. D, 1970, South Dakota School of Mines & Technology.

Gruszka, Thomas Peter. Induced polarization and its interaction with electromagnetic coupling in low frequency geophysical exploration. D, 1987, University of Arizona. 360 p.

Grutzeck, Michael William. Microprobe investigation of Portland cement hydrates. M, 1968, Pennsylvania State University, University Park. 112 p.

Grutzeck, Michael William. The role of nickel in geochemically important oxide phases, as deduced from phase equilibria at liquidus temperatures in the system MgO-NiO-iron oxide-SiO_2. D, 1973, Pennsylvania State University, University Park. 233 p.

Gruver, Barbara L. Pleistocene valley train deposition, Whitewater River, southeastern Indiana and southwestern Ohio. M, 1984, Indiana University, Bloomington. 145 p.

Gruyter, D. A. de *see* de Gruyter, D. A.

Gruyter, Philip Clarence de *see* de Gruyter, Philip Clarence

Grybeck, Donald. Geology of the lead-zinc-silver deposits of Silver Plume area, Clear Creek County, Colorado. D, 1969, Colorado School of Mines. 154 p.

Gryc, George. The Keweenawan geology of the Grand Portage Indian Reservation. M, 1942, University of Minnesota.

Grygo, Roland. Some scolecodonts from the Arnheim and Waynesville formations, Butler County, Ohio. M, 1962, Miami University (Ohio). 78 p.

Gryta, Jeffrey J. Sediment budgets on reclaimed coal surface mines in central Pennsylvania. D, 1987, Pennsylvania State University, University Park. 189 p.

Gryta, Jeffrey John. Landslides along the western shore of Hood Canal, northern Mason County, Washington. M, 1975, North Carolina State University. 120 p.

Grzybek, Paul Stanley. A sedimentological study of the Eighteen Mile Creek drainage basin, Erie County, New York. M, 1966, SUNY at Buffalo. 27 p.

Gsell, Ronald N. The rock asphalt deposits at Natural Rock, Kentucky. M, 1933, Washington University. 76 p.

Gu, Hongren. A study of propagation of hydraulically induced fractures. D, 1987, University of Texas, Austin. 139 p.

Guala, John Riddoch, Jr. Bottom topography effects on ocean currents. D, 1972, University of Delaware, College of Marine Studies. 138 p.

Gualtieri, J. L. Geology of the Crab Orchard Quadrangle and related aspects of Mississippian stratigraphy (Washington). M, 1967, University of Washington.

Guarin, Gilberto. Geochemistry of soils in a portion of the Carolina slate belt, Cabarrus and Stanly counties, North Carolina. M, 1974, University of North Carolina, Chapel Hill. 109 p.

Guarino, Joe C. Feasibility of an alternate route for the Northern Tier Pipeline across Idaho. M, 1982, University of Idaho. 75 p.

Guarnera, Bernard John. Geology of the McKernan Lake phase of the Lac des Iles Intrusive, Thunder Bay mining district, Ontario (Canada). M, 1967, Michigan Technological University. 100 p.

Guarraia, Ernest. A study of Rotalia trochidiformis Lamarck. M, 1954, New York University.

Guass, Joseph Charles. Areal geology of the Reevesville Quadrangle. M, 1966, Southern Illinois University, Carbondale. 138 p.

Gubala, Chad Paul. The cycling of iron and trace metals in the sediments of acidic lakes. D, 1988, Indiana University, Bloomington. 180 p.

Gubbels, Timothy Louis. Structural and geomorphic evolution of the north flank, eastern Owl Creek Mountains, Wyoming. M, 1987, University of Wyoming. 181 p.

Guber, Albert Lee. Some Richmond (Ordovician) ostracodes from Indiana and Ohio. D, 1962, University of Illinois, Urbana. 124 p.

Gubitosa, Matthew. Glacial geology of the Hancock area, western Catskills, New York. M, 1984, SUNY at Binghamton. 102 p.

Gubitosa, Richard. Depositional systems of the Moss Back Member, Chinle Formation, (Upper Triassic), Canyonlands, Utah. M, 1981, Northern Arizona University. 98 p.

Guccione, Margaret J. Differentiation of tills of the Miami Lobe in southwestern Ohio. M, 1972, Miami University (Ohio). 60 p.

Guccione, Margaret Josephine Weatherhead. Stratigraphy, soil development and mineral weathering of Quaternary deposits, Midcontinent, U.S.A. D, 1982, University of Colorado. 320 p.

Gucwa, John Henry. Late Cenozoic geomorphic history of Lee canyon, Spring mountains, Nevada. M, 1969, Pennsylvania State University, University Park. 59 p.

Gucwa, Paul Ramon. Geology of the Covelo/Laytonville area, northern California. D, 1974, University of Texas, Austin. 101 p.

Gucwa, Paul Ramon. Gravity sliding (Recent) south of Bearpaw Mountains, Montana. M, 1971, University of Texas, Austin. 60 p.

Gude, Arthur J., III. Geologic correlation by x-ray diffractions. M, 1949, Colorado School of Mines. 40 p.

Gudjurgis, Paul Joseph. Lead isotope analyses on selected Canadian mineral deposits. M, 1971, University of Alberta. 65 p.

Gudmundsson, Olafur. Velocity inversion of cross-hole data in flood basalts. M, 1984, University of Washington. 178 p.

Gudramovics, Robert. A geochemical and hydrological investigation of a modern coastal marine sabkha. M, 1981, Michigan State University. 107 p.

Guebert, Michael Dean. Relationship of remotely sensed SPOT data to infiltration capacity surface mined land in central Pennsylvania. M, 1988, Pennsylvania State University, University Park. 37 p.

Guendel, Federico David. Seismotectonics of Costa Rica; an analytical view of the southern terminus of the Middle America Trench. D, 1986, University of California, Santa Cruz.

Guendel-Umana, Federico D. On the relationship between Earth tides and volcanic activity at Arenal Volcano, Costa Rica. M, 1978, University of Texas, Austin.

Guennel, Gottfried Kurt. A comparative paleobotanical investigation of the Indiana and Russian paper coals. D, 1960, Indiana University, Bloomington. 301 p.

Guensburg, Thomas Edgar. Echinodermata of the Middle Ordovician Lebanon Limestone, central Tennessee. D, 1982, University of Illinois, Urbana. 277 p.

Guensburg, Thomas Edward. Savagella illinoisensis Miller and Gurley, (1895); a cyclocystoid-like edrioasteroid from the Orchard Creek Shale (Upper Ordovician) in southern Illinois. M, 1977, Southern Illinois University, Carbondale. 40 p.

Guentert, James S. Petrochemistry of the Quinn Canyon silicic intrusives, Nye County, Nevada; a comparison with climax-type porphyry Mo systems. M, 1988, Michigan State University. 98 p.

Guenther, Edwin Michael. The geology of the Mercur gold camp, Utah. M, 1973, University of Utah. 79 p.

Guerin, William F. Studies on the speciation and distribution of low-molecular-weight sulfur compounds in sediments. M, 1978, University of South Florida, St. Petersburg. 99 p.

Guernsey, Terrant Dickie. Cordillera Subarctica. M, 1924, University of British Columbia.

Guernsey, Terrant Dickie. The geology of North Mountain, Cape Breton, Nova Scotia, Canada. D, 1929, Columbia University, Teachers College.

Guerrero U., Alberto Lobo *see* Lobo Guerrero U., Alberto

Guerrero, A. Ortega *see* Ortega Guerrero, A.

Guerrero, Benito. Sedimentology of a fluvial sheet sandstone, Eocene Willwood Formation, Bighorn Basin. M, 1988, Colorado School of Mines. 87 p.

Guerrero, David Hipolito Zamora *see* Zamora Guerrero, David Hipolito

Guerrero, Juan A. Geology of the northern White Pine Range, east-central Nevada. M, 1983, California State University, Long Beach. 105 p.

Guerrero, Richard Gonzales. Classification of carbonate rocks by the versenate method of chemical analysis. M, 1952, Southern Methodist University. 25 p.

Guerrero-Garcia, Jose Celestino. Contributions to paleomagnetism and Rb-Sr geochronology. D, 1976, University of Texas at Dallas. 131 p.

Guertin, David Phillip. A land use planning model to predict the effect of fire and timber harvesting on water resources. M, 1983, Colorado State University. 154 p.

Guertin, David Phillip. Modeling streamflow response from Minnesota peatlands. D, 1984, University of Minnesota, Minneapolis. 241 p.

Guertin, Kateri. Méthodes géostatistiques d'estimation des réserves des gisements d'uranium de type trend de la région de Grants, Nouveau Mexique, et applications. M, 1979, Ecole Polytechnique.

Guertin, Kateri Valerie. Correcting conditional bias in ore reserve estimation. D, 1985, Stanford University. 346 p.

Guess, Roy Hayes, Jr. The geology of a portion of northeastern Gillespie County. M, 1940, University of Texas, Austin.

Guess, Sam C. The use of the scanning electron microscope as a predictive tool in determining asphalt pavement durability. M, 1980, Washington State University. 57 p.

Guest, Hardy Grady. The Ripley Formation in east-central Mississippi and west-central Alabama. D, 1935, University of Iowa. 44 p.

Guest, Henry Grady. The geology of south central Mississippi. M, 1933, University of Iowa. 37 p.

Guest, Michael E. Ostracoda from the Brownstone Formation (Cretaceous) in southwestern Arkansas. M, 1964, University of Oklahoma. 88 p.

Guest, Peter R. Case study; an assessment of groundwater containment system at the Rocky Mountain Arsenal, Denver, Colorado. M, 1988, Colorado School of Mines. 235 p.

Guevara-Sanchez, Edgar Humberto. Depositional systems in the Queen City Formation (Eocene), Central and East Texas. M, 1972, University of Texas, Austin.

Guevara-Sanchez, Edgar Humberto. Pleistocene facies in the subsurface of the southeast Texas coastal plain. D, 1974, University of Texas, Austin.

Guggenheim, S. J. Cation ordering in subgroup symmetry in the micas. D, 1976, University of Wisconsin-Madison. 71 p.

Guha, Shyamal Kanti. The effect of focal depth on the spectrum of P waves. D, 1970, St. Louis University.

Guharoy, Prasanta Kumar. Determination of reservoir characteristics from capillary pressure curves and electrical properties of rocks. M, 1970, University of Missouri, Rolla.

Guidotti, Charles Vincent. Geology of the Bryant Pond Quadrangle, Maine. D, 1963, Harvard University.

Guidroz, Ralph Robert. Gravity features along the western edge of the slate belt, North Carolina. M, 1964, University of North Carolina, Chapel Hill. 34 p.

Guilarte, Fernando Anibal. Non-linear regression analysis to estimate reserves in gas reservoirs. M, 1978, University of Oklahoma. 82 p.

Guilbault, Jean-Pierre. Algues ordoviciennes de basses-terres du Saint Laurent et des régions limitrophes. M, 1975, Universite de Montreal.

Guilbeau, Ellis R. Subsurface geology of the Lake Palourde area, St. Mary, St. Martin, and Assumption parishes, Louisiana. M, 1982, University of Southwestern Louisiana. 57 p.

Guilbeau, Kevin P. Geology, geochemistry, and petrogenesis of the upper Keres Group, Ruiz Peak area, Jemez Mountains, New Mexico. M, 1982, University of New Mexico. 132 p.

Guilbert, John M. The autoradiographic determination of alpha-activity distribution in certain Wisconsin granites and related rocks. M, 1955, University of Wisconsin-Madison.

Guild, Frank Nelson. A microscopic study of the silver ores and their associated minerals. D, 1917, Stanford University. 373 p.

Guild, Phillip W. Petrology and structure of the Moa chromite district, Oriente Province, Cuba. D, 1946, The Johns Hopkins University.

Guilinger, David R. Geology and uranium potential of the Tejana Mesa-Hubbell Draw area, Catron County, New Mexico. M, 1982, New Mexico Institute of Mining and Technology. 129 p.

Guillemette, Renald Norman. Geochemical and experimental investigations of andesite-water interactions and their relationship to island-arc geothermal systems. D, 1983, Stanford University. 336 p.

Guillet, G. R. A chemical and inclusion study of nepheline syenite for petrogenetic criteria. M, 1962, University of Toronto.

Guillette, Brian R. Hydrogeology of a lower Taylor Marl flow system, Blackland Prairies, central Texas. M, 1988, Baylor University. 121 p.

Guillou, Robert Barton. The geology of the Johnston Grade area, San Bernardino Mountains, California. M, 1951, University of California, Los Angeles.

Guimarães, Paulo de Tarso Martins. Basin analysis and structural development of the Sergipe-Alagoas Basin, Brazil. D, 1988, University of Texas, Austin. 214 p.

Guimon, Robert Kyle. Annealing studies of the thermoluminescence of meteorites and implications for their metamorphic history. D, 1986, University of Arkansas, Fayetteville. 164 p.

Guindon, David Leslie. The geochemistry of free gold and its application in exploration. M, 1982, Queen's University. 125 p.

Guinn, Stewart A. Earthquake focal mechanisms in the southeastern United States. M, 1977, Georgia Institute of Technology. 150 p.

Guinness, Edward Albert, Jr. Spatial and temporal variations in the spectral properties of soils exposed at the Viking landing sites. D, 1980, Washington University. 191 p.

Guisepppi, William Harris Di *see* Di Guiseppi, William Harris

Guitjens, Johannes Caspar. The effect of soil-moisture hysteresis on the flow of water into a gravity well. D, 1968, University of California, Davis. 172 p.

Guiton, R. S. A groundwater study of the Ojibway Prairie. M, 1978, University of Waterloo.

Guitron de los Reyes, A. Improved stochastic dynamic programming for optimal reservoir operation based on the asymptotic convergence of benefit differences. M, 1974, University of Arizona.

Guja, Nasser Hossain. Paleomagnetic investigations of Jamaican rocks. D, 1970, St. Louis University. 326 p.

Gulbrandsen, Leif Fontaine. Prediction of erosion using wave refraction techniques. M, 1976, University of Virginia. 29 p.

Gulbrandsen, Robert Allen. Petrology of the Meade Peak Member of the Phosphoria Formation at Coal Canyon, Wyoming. D, 1958, Stanford University. 203 p.

Guldenzopf, Emil Charles. Champlainian biostratigraphy in northern Michigan. D, 1968, Iowa State University of Science and Technology. 139 p.

Guldenzopf, Emil Charles. The conodont fauna and stratigraphy of the Pecatonica Member of the Platteville Formation. M, 1964, University of Iowa. 254 p.

Gulen, Levent. Sr, Nd, Pb isotope and trace element geochemistry of calc-alkaline and alkaline volcanics, eastern Turkey. D, 1984, Massachusetts Institute of Technology. 232 p.

Gulick, Charles Wyckoff, III. Bedrock geology of the southwest quarter of the Aeneas Valley 15' Quadrangle, Okanogan County, Washington. M, 1987, Eastern Washington University. 55 p.

Guliov, Paul. Paleoecology of invertebrate fauna from post glacial sediments near Earl Grey, Saskatchewan. M, 1964, University of Saskatchewan. 95 p.

Gullatt, Ennis Murray, Jr. An investigation of the compressibility of natural gas at elevated pressures. M, 1958, University of Oklahoma. 64 p.

Gulley, Gerald Lee, Jr. The petrology of granulite-facies metamorphic rocks on Roan Mountain, western Blue Ridge Province, N.C.-TN. M, 1982, University of North Carolina, Chapel Hill. 165 p.

Gulliver, Frederic Putnam. Stages in the development of shore lines; geographic criteria for the recognition of continental oscillation. D, 1896, Harvard University.

Gulliver, Rachel M. Structural analysis of Paleozoic rocks in the Talc City Hills, Inyo County, California. M, 1976, University of California, Santa Barbara.

Gullixson, Carl Fredrick. The structure, geologic evolution and regional significance of the Bethel Creek - North Fork area, Coos and Curry counties, Oregon. M, 1981, Portland State University. 86 p.

Gulmon, Gordon W. Cambrian stratigraphy and paleontology of the northwestern Wind River Mountains of Wyoming. M, 1939, Texas A&M University. 56 p.

Gultekin, Savci. Structure and tectonics of the Keban metamorphics in the northern margin of the Bitlis suture zone, southeastern Turkey. M, 1983, SUNY at Albany. 201 p.

Guma'a, G. S. Spatial variability of in situ available water. D, 1978, University of Arizona. 154 p.

Gumati, Yousef Daw. Crustal extension, subsidence, and thermal history of the Sirte Basin, Libya. D, 1985, University of South Carolina. 219 p.

Gumble, Gordon E. Structure section through the Tuolumne intrusive complex, Yosemite National Park. M, 1962, University of Arizona.

Gumbo, F. J. Efficacy of tile drainage system at Agriculture Canada, Morden, for reducing soil salinity. M, 1974, University of Manitoba.

Gumert, William Richard. Helicopter gravity measuring system. M, 1968, University of Texas, Austin.

Gumma, William H. An interpretation of the gravity and magnetic anomalies of the Rivera fracture zone, eastern Pacific Ocean. M, 1974, Oregon State University. 50 p.

Gummer, W. K. Border rocks of granite batholiths, Red Lake, Ontario. M, 1939, Queen's University. 59 p.

Gummer, Wilfred K. The system $CaSiO_3$-$CaAl_2Si_2O_8$-$NaAlSiO_4$. D, 1941, University of Chicago. 127 p.

Gumper, Frank J. Seismic wave velocities and Earth structure on the African Continent. M, 1971, Columbia University. 18 p.

Gunal, Asuman. Clay mineralogy, petrography, chemical composition and stratigraphic correlation of some Middle Ordovician K-bentonites in the eastern Mid-Continent. D, 1979, University of Cincinnati. 253 p.

Gundersen, Thomas D. Petroleum potential of the Ordovician Ellenburger Group along the Fort Chadbourne Fault system, west-central Texas. M, 1986, Baylor University. 130 p.

Gundersen, Wayne Campbell. An isopach and lithofacies study of the Price River, North Horn, and Flagstaff formations of central Utah. M, 1961, University of Nebraska, Lincoln.

Gunderson, Brian M. Three-dimensional transient electromagnetic responses for a grounded-wire source. M, 1985, University of Utah. 75 p.

Gunderson, James Ronald Novotny. The stratigraphy and mineralogy of the metamorphosed Biwabik Iron Formation, eastern Mesabi District, Minnesota. D, 1958, University of Minnesota, Duluth. 198 p.

Gunderson, Richard Paul. Geology and geochemistry of the Benavides-Pozos area, eastern Chihuahua, Mexico. M, 1983, University of California, Santa Cruz.

Gundlach, David Lou. An investigation of groundwater flow through a confined nonuniform porous material. D, 1973, University of Southern California.

Gundlach, Erich R. Biological effects of marine emplaced compacted solid waste bales. M, 1974, University of New Hampshire. 110 p.

Gundlach, Erich R. Oil spill impact on temperate shoreline environments, based on study of the Urquiola (May 1976) and Amoco Cadiz (March 1978) oil spills. D, 1979, University of South Carolina. 260 p.

Gundrum, Lois Elizabeth. Chonetid way of life. M, 1977, Kansas State University.

Gundy, Clarence Edgar Van *see* Van Gundy, Clarence Edgar

Gunn, Christopher Bruce. Provenance of diamonds in the glacial drift (Pleistocene) of the Great Lakes region, North America. M, 1967, University of Western Ontario. 134 p.

Gunn, Donald William. Late Devonian regional unconformity in and around the Uinta Mountains, Utah. D, 1965, Washington State University. 233 p.

Gunn, Donald William. The relationship of grain size and morphological characteristics in some Maine eskers as a guide to construction material sources. M, 1961, University of Minnesota, Minneapolis.

Gunn, Robert C. M. Economic geology of the Tenth potash ore zone; Permian Salado Formation, Carlsbad District, New Mexico. M, 1976, University of Texas at El Paso.

Gunn, Susan Helen. The geology, petrography and geochemistry of the Laguna Juarez Pluton, Baja California, Mexico. M, 1984, San Diego State University. 166 p.

Gunn, Vincent C. Subsurface Pennsylvanian geology, Foard County, Texas. M, 1976, University of Texas, Austin.

Gunnarsson, Bjorn. Petrology and petrogenesis of silicic and intermediate lavas on a propagating oceanic rift; the Torfajokull and Hekla central volcanoes south-central Iceland (Volume I). D, 1987, The Johns Hopkins University. 452 p.

Gunnell, Emory Mitchell. A study of the luminescence in minerals with special emphasis on photo-luminescence. M, 1931, Washington University. 131 p.

Gunnell, Francis Hawkes. The Telotremata of the Brazer Limestone of northern Utah. M, 1930, University of Missouri, Columbia.

Gunnell, Gregg Frederick. Evolutionary history of Microsyopoidea (Mammalia, Primates?) and the relationship of Plesiadapiformes to Primates; (Volumes I and II). D, 1986, University of Michigan. 630 p.

Gunner, John Duncan. Age and origin of the Nimrod Group and of the Granite Harbour intrusives, Beardmore Glacier region, Antarctica. D, 1971, Ohio State University. 246 p.

Gunning, H. C. Economic geology of Lardeau map area, British Columbia. D, 1929, Massachusetts Institute of Technology. 230 p.

Gunning, H. C. Syenite porphyry west of Rouyn, Quebec. M, 1926, Massachusetts Institute of Technology. 39 p.

Gunning, S. P. The distribution of uranium in phosphorites of central Florida. M, 1978, University of Florida. 77 p.

Gunow, Alexander James. The geochemistry of fluorine-rich micas at the Henderson molybdenite deposit. M, 1978, University of Colorado.

Gunow, Alexander James. Trace element mineralogy in the porphyry molybdenum environment. D, 1983, University of Colorado. 609 p.

Gunter, Avril E. An experimental study of synthetic cordierites. D, 1977, Carleton University.

Gunter, Avril E. Fe-Mg exchange in synthetic biotites on the join annite-phlogopite and synthetic clinopyroxene on the join hedenbergite-diopside. M, 1973, Carleton University. 62 p.

Gunter, Bobby Dean. Geochemical and isotopic studies of hydrothermal gases and waters. D, 1968, University of Arkansas, Fayetteville. 125 p.

Gunter, Charles Phillip. A lithofacies study of Desmoinesian rocks in East central Eddy and West central Lea counties, New Mexico. M, 1975, Mississippi State University. 59 p.

Gunter, Craig E. Subsurface study of the Deese Group, western Garvin County, Oklahoma. M, 1959, University of Oklahoma. 38 p.

Gunter, Karl D. Correlation of paleomagnetic parameters with magnetic mineralogy in a detailed vertical traverse of the Holyoke Basalt, Holyoke, Massachusetts. M, 1978, University of Massachusetts. 79 p.

Gunter, Mickey. Relationship between the chemical composition, lattice parameters, and optical properties of andalusite and its isostructural analogs. M, 1982, Virginia Polytechnic Institute and State University. 52 p.

Gunter, Mickey E. Refractometry by total reflection. D, 1987, Virginia Polytechnic Institute and State University. 171 p.

Gunter, W. Richard. A preliminary study of the physical and chemical effects of short term weathering on the Qeenston Shale and their applications to the brick industry. M, 1980, University of Windsor. 100 p.

Gunter, William Daniel. An experimental study of mineral-solution equilibria applicable to metamorphic rocks. D, 1974, The Johns Hopkins University.

Gunter, William Daniel. Feldspars from a tholeiite sill, Grand Manan, New Brunswick (Canada). M, 1967, University of New Brunswick.

Gunter, William L. The petrology and petrochemistry of the Lac Bueil area, Quebec, Canada. M, 1975, University of Georgia.

Gunther, Frederick J. Statistical foraminiferal ecology from seasonal samples, central Oregon continental shelf. D, 1972, Oregon State University. 228 p.

Gunther, Fredrick John. Ostracoda of the Gulf of Panama and Bahia San Miguel (Panama). M, 1967, University of Minnesota, Minneapolis. 204 p.

Gunther, Paul R. Megaspore palynology of the Brazeau Formation (upper Cretaceous), Nordegg area, Alberta. M, 1970, University of Calgary. 164 p.

Gunton, John Eric. Geochemical dispersion associated with porphyry-type mineralization in the Canadian Cordillera. D, 1974, Queen's University. 331 p.

Guo, H. Y. Watershed modeling to evaluate the effects of floodwater retarding structures on water yield from Lewisville Reservoir, Trinity River, Texas. D, 1979, University of Texas, Austin. 284 p.

Guppy, K. H. Interference analysis of fractured wells. M, 1976, Stanford University. 318 p.

Gupta, Alok Krishna. The system forsterite-diopside-akermanite-leucite and its significance in the petrogenesis of potassium-rich mafic and ultramafic volcanic rocks. D, 1969, University of Pittsburgh.

Gupta, Ashok K. The reaction-rate studies of asbestos minerals with acids. M, 1972, University of Nevada. 112 p.

Gupta, Avijit. The effect of seasonal flow and high magnitude floods on channel for and stream behavior in eastern Jamaica. D, 1973, The Johns Hopkins University. 308 p.

Gupta, Barun Kumar Sen. Some larger foraminifera from northwestern Kutch, India. M, 1961, Cornell University.

Gupta, Indra Narayan. Model study of explosion-generated Rayleigh waves in a half space. D, 1964, St. Louis University.

Gupta, Indra Narayan. Resonant oscillations of the overburden excited by seismic waves. M, 1962, St. Louis University.

Gupta, Mrinal Kanti Sen *see* Sen Gupta, Mrinal Kanti

Gupta, Pradip Kumar Sen *see* Sen Gupta, Pradip Kumar

Gupta, Ram Swaroop. Groundwater reservoir operation for drought management. D, 1983, Polytechnic University. 326 p.

Gupta, Ramesh Chandra. Determination of the insitu coefficients of consolidation and permeability of submerged soils using electrical piezoprobe soundings. D, 1983, University of Florida. 303 p.

Gupta, Ravindra Nath. Reflection of seismic waves from transition layers. D, 1965, University of Toronto.

Gupta, Samir Kumar Das *see* Das Gupta, Samir Kumar

Gupta, Santosh Kumar. A distributed digital model for estimation of flows and sediment load from large ungauged watersheds. D, 1974, University of Waterloo.

Gupta, Sujoy. Palynology of Grassy Creek and Saverton Shales of Missouri. D, 1965, University of Missouri, Rolla. 257 p.

Gupta, Sumant K. Three-dimensional Galerkin-finite element formulation of flow and mass transport through porous media. D, 1975, University of California, Davis. 162 p.

Gupta, U. Das *see* Das Gupta, U.

Gupta, V. K. An interpretation of aeromagnetic and gravity data of Caledonian area in southern New Brunswick. D, 1975, University of New Brunswick.

Gupta, V. K. Computed versus observed terrain effects in magnetic methods. M, 1969, Queen's University. 136 p.

Gupton, Charles Pernell. An investigation of iron and manganese oxide surfaces in saprolite in North Carolina. M, 1964, North Carolina State University. 80 p.

Gurbuz, Behic. Structure of the Earth's crust and upper mantle under a portion of Canadian shield deduced from travel times and spectral amplitudes of body waves using data from Project Early Rise (Manitoba, Ontario). D, 1969, University of Manitoba.

Gurbuz, Behic M. Calibration of continuous velocity logs using the comparison of synthetic and field records. M, 1966, University of British Columbia.

Gurel, Mehmet. Insoluble residues of Glen Rose Formation, Mt. Barker area, Austin, Texas. M, 1956, University of Texas, Austin.

Gurkan, T. H. A comparative interpretation of a shallow refraction survey by time-term and conventional methods. M, 1970, University of Western Ontario.

Gurney, James Walter. Contacts of the Almo Pluton, Albion Range, Cassia County, Idaho. M, 1971, Idaho State University. 152 p.

Gurr, Theodore M. The geology of a central Florida peat bog, Section 26, Township 30 South, Range 25 East, Polk County, Florida. M, 1972, University of South Florida, Tampa. 86 p.

Gurrieri, Joseph Thomas. Surficial geology of the Saranac Lake Quadrangle, New York. M, 1983, University of Connecticut. 86 p.

Gurriet, Philippe Charles. Geochemistry of Hawaiian dredged lavas. M, 1988, Massachusetts Institute of Technology. 177 p.

Gurrola, Harold. A crustal structure study of the northern margin of the Gulf of Mexico. M, 1987, University of Texas at El Paso.

Gusey, Daryl L. The geology of southwestern Fidalgo Island, Washington. M, 1978, Western Washington University. 85 p.

Gussow, W. C. Mineral deposits of the Robb-Montbray Mines, Ltd. M, 1935, Queen's University. 62 p.

Gussow, W. C. Petrogeny of the major acid intrusives of the Rouyn-Bell River area, Northwest Quebec. D, 1938, Massachusetts Institute of Technology. 175 p.

Gust, David Allen. Petrology and geochemistry of the Mormon Mountain volcanic field, Arizona. M, 1978, Rice University. 84 p.

Gustafson, Axel Ferdinand. The effect of drying soils on the water-soluble constituents. D, 1921, Cornell University.

Gustafson, Catherine. Debris flows along the Slims River valley, Kluane National Park, Yukon Territory. M, 1986, University of Calgary.

Gustafson, Craig Warren. The fluid mechanics of hydraulic fracturing. D, 1987, University of Illinois, Urbana. 168 p.

Gustafson, Donald L. The geology of the Lake Isabelle area, Boulder County, Colorado. M, 1965, University of Colorado.

Gustafson, Edward Paul. Three-dimensional magnetotelluric response of the Rio Grande Rift near Socorro, New Mexico. M, 1987, San Diego State University. 183 p.

Gustafson, Eric P. Carnivorous mammals of the late Eocene and early Oligocene of Trans-Pecos Texas. D, 1977, University of Texas, Austin. 221 p.

Gustafson, Eric Paul. The vertebrate fauna of the late Pliocene Ringold Formation, south-central Washington. M, 1973, University of Washington. 164 p.

Gustafson, Fridolf V. Regional reconnaissance of the Sheep Pass Formation. M, 1977, University of Nevada. 86 p.

Gustafson, John Kyle. The Homestake gold-bearing formation. D, 1930, Harvard University.

Gustafson, Lewis Allan. Structure and stratigraphy of the Traverse Group in the Lansing area, Michigan. M, 1960, Michigan State University. 63 p.

Gustafson, Lewis Brigham. Paragenesis and hypogene zoning at the Magma Mine, Superior, Arizona. D, 1962, Harvard University.

Gustafson, Rayford B. A microscopic analysis of the Moodys Branch Marl. M, 1951, University of Mississippi.

Gustafson, Richard Dale. A land use projection model applied to Emmet County, Michigan. D, 1983, Michigan State University. 161 p.

Gustafson, Thomas K. Geology and structural analysis between the narrows and St. Clair thrust faults in the Narrows Quadrangle, Giles County, Virginia. M, 1982, Eastern Kentucky University. 84 p.

Gustafson, Timothy J. The structural geology of the Boonesboro limestone mine, Madison County, Kentucky. M, 1986, Eastern Kentucky University. 72 p.

Gustafson, Velman Oscar. Geology of a portion of the Bridger Range, Livingston Quadrangle, Montana. M, 1951, University of Iowa. 96 p.

Gustafson, William G. Geology of copper-lead-zinc-turquoise deposits, southern part of the Cerillos mining district, Santa Fe County, New Mexico. M, 1965, University of New Mexico. 58 p.

Gustafson, William Ivor. Stability relations of andradite, hedenbergite and related minerals in the system Ca-Fe-Si-O-H. D, 1970, University of California, Los Angeles.

Gustajtis, K. Andrew. An experimental hydrothermal study of basalt-seawater interaction in the temperature range 2°-180°C and pressures up to 1 kilobar. M, 1977, Dalhousie University. 265 p.

Gustason, Edmund R. Depositional environment of the basal conglomerate of the Dakota Formation, Orderville Gulch, southwestern Utah. M, 1983, Northern Arizona University. 127 p.

Gustav, Spence H. The sedimentology and paleogeography of the Bridge Formation, (Eocene) of southwestern Wyoming. M, 1974, University of Massachusetts. 82 p.

Gustavson, J. B. Exploration for zinc near Wartrace, Bedford County; contribution to the economic geology of the state of Tennessee. M, 1976, Memphis State University.

Gustavson, Thomas C. Glacial geology of eastern Sheridan County, North Dakota. M, 1964, University of North Dakota. 105 p.

Gustavson, Thomas Carl. Fluvial and lacustrine sedimentation in the proglacial environment, Malaspina Glacier Foreland, Alaska. D, 1973, University of Massachusetts. 177 p.

Gustin, Mae Sexauer. A petrographic, geochemical and stable isotope study of the United Verde Orebody and its associated alteration, Jerome, Arizona. D, 1988, University of Arizona. 270 p.

Guswa, John Henry. Numerical simulation of the multi-layered aquifer system in the Coastal Plain area, southeastern Pennsylvania. D, 1976, Pennsylvania State University, University Park. 351 p.

Gutentag, Edwin D. Pleistocene fresh-water ostracodes from Meade County, Kansas, and vicinity. M, 1958, University of Kansas. 98 p.

Guth, Lawrence Roland. Theories and applications of calcite and quartz paleopiezometers. M, 1983, University of Utah. 227 p.

Guth, Peter Lorentz. Geology of the Sheep Range; Clark County, Nevada. D, 1980, Massachusetts Institute of Technology. 189 p.

Guthe, Otto Emmor. The Black Hills of South Dakota and Wyoming. D, 1933, University of Michigan.

Guthrie, Alan Edgar. Volcanic stratigraphy of the Geneva Lake greenstone belt, Ontario. M, 1981, University of Western Ontario. 215 p.

Guthrie, Daniel Albert. The mammals of the Eocene Lysite Member, Wind River Formation of Wyoming. D, 1964, University of Massachusetts. 180 p.

Guthrie, Gary Eich. Stratigraphy and depositional environment of the Upper Mississippian Big Snowy Group in the Bridger Range, Southwest Montana. M, 1984, Montana State University. 105 p.

Guthrie, James O. The geology of the northern portion of the Belchertown intrusive complex, west-central Massachusetts. M, 1972, University of Massachusetts. 110 p.

Guthrie, John M. Clay mineralogy as an indicator of thermal maturity of Carboniferous strata, Ouachita Mountains. M, 1985, University of Missouri, Columbia. 74 p.

Guthrie, Robert S. Geology of the northern Sierra Boca Grande area, Chihuahua, Mexico. M, 1987, University of Texas at El Paso.

Guthrie, Verner Noel. Geology of the Dorn Mine, McCormick, South Carolina. M, 1980, University of Georgia.

Gutiérrez, Alberto Alejandro. Channel and hillslope geomorphology of badlands in the San Juan Basin, northwestern New Mexico. M, 1980, University of New Mexico. 158 p.

Gutierrez, Carlos Angel Q. Mortera see Mortera Gutierrez, Carlos Angel Q.

Gutierrez, Dora. Foraminiferae faunas from the Tertiary of Ica, southern Peru. M, 1958, University of California, Berkeley. 150 p.

Gutierrez, Francisco Javier. Subsurface study of the Bachaquero interfield area, Bolívar coastal fields, Lake Maracaibo, Venezuela. M, 1968, University of Tulsa. 70 p.

Gutierrez, Gay Nell. Controls on ore deposition in the Lamotte Sandstone, Goose Creek Mine, Indian Creek Subdistrict, southeast Missouri. M, 1987, University of Texas, Austin. 119 p.

Gutierrez, J. A. A substructure method for earthquake analysis of structure-soil interaction. D, 1976, University of California, Berkeley. 137 p.

Gutierrez, Julian Castillo. Hydrologic characteristics of the San Tiburcio Watershed, northern Zacatecas, Mexico. M, 1979, Colorado State University. 120 p.

Gutmann, James Trafton. Eruptive history and petrology of Crater Elegante, Sonora, Mexico. D, 1972, Stanford University. 235 p.

Gutowski, Vincent Peter. An experimental study of flume-generated braided streams. D, 1987, University of Pittsburgh. 213 p.

Gutschick, Raymond Charles. A method of determining the polarization angle, using a single-circle goniometer. M, 1939, University of Illinois, Urbana. 59 p.

Gutschick, Raymond Charles. The Redwall Limestone (Mississippian) of north central Arizona. D, 1942, University of Illinois, Urbana. 96 p.

Gutstadt, Alan Morton. Stratigraphy of the Upper Ordovician rocks in Iowa, Illinois, and Indiana. D, 1954, Northwestern University.

Guttery, Thomas H. Depositional features of Pennsylvanian sediments in the Birmingham area (Alabama). M, 1955, Emory University. 77 p.

Guttormsen, Paul Andrew, Jr. Geology of the Swamp Creek-Triangle Gulch area, Beaverhead County, Montana. M, 1952, Montana College of Mineral Science & Technology. 84 p.

Gutub, Saud Abdulaziz. A finite element model of artificial recharge with an impedance layer above the water table. D, 1988, Colorado State University. 373 p.

Gutzler, Robert Quenton. Petrology and depositional environments of Alabama Tertiary lignites. D, 1979, Pennsylvania State University, University Park. 449 p.

Guu, Cindy Kuei-Ding. Catalog of Chinese earthquakes. M, 1981, Colorado School of Mines. 442 p.

Guu, Jeng-Yih. Studies of seismic guided waves; the continuity of coal seams. D, 1975, Colorado School of Mines. 85 p.

Guu, Jeng-yih. Sulfide assemblages in metamorphic rocks. M, 1969, University of Rochester.

Guvenir, Ibrahim M. The development of an automated in-situ combustion assembly to study effects of clay on the dry forward in-situ combustion process. D, 1980, University of Kansas. 270 p.

Guy, Donald E., Jr. Origin and evolution of Bay Point sand spit, Lake Erie, Ohio. M, 1983, Bowling Green State University. 205 p.

Guy, Jerry L. Computer application to exploration and development of bedded kaolin. M, 1972, Florida State University.

Guy, Russell E. The Dinkey Creek intrusive series, Huntington Lake Quadrangle, Fresno County, California. M, 1980, Virginia Polytechnic Institute and State University.

Guy, Samuel Cole. A heavy mineral analysis of North Carolina beach sands. M, 1964, University of North Carolina, Chapel Hill. 30 p.

Guynes, George Eldridge. Geology of the Triunfo Pass area, Los Angeles and Ventura counties, California. M, 1959, University of California, Los Angeles.

Guyton, J. Stephen; Hutton, John R. and Sokolsky, George E. Geology of the Devil's Hole area, Custer and Huerfano counties, Colorado. M, 1960, University of Michigan.

Guyton, J. W. Ancient landslide deposits and gravity slides in Wyoming. D, 1965, University of Wyoming. 126 p.

Guyton, J. W. Geology of the Lost Soldier area, Sweetwater, Fremont and Carbon counties, Wyoming. M, 1960, University of Wyoming. 70 p.

Guza, Robert T. Excitation of edge waves and their role in the formation of beach cusps. D, 1974, University of California, San Diego. 117 p.

Guzan, Michael John. Petrographic investigation and examination of the thickness of the Brereton Limestone (Middle Pennsylvanian) in a portion of the Illinois Basin. M, 1983, Southern Illinois University, Carbondale. 97 p.

Guzel, Nuri. A pedogenic study of Malcolm, Morrill and Sharpsburg soils in southwestern Saunders County, Nebraska. D, 1970, University of Nebraska, Lincoln. 165 p.

Guzikowski, Michael Vincent. Evolution of pore fluid chemistry during the recrystallization of periplatform carbonates, Bahamas. M, 1987, University of Miami. 215 p.

Guzman, Alfredo Eduardo. Carbonate diagenesis of the Cupido Formation, Lower Cretaceous, Coahuila, Mexico. M, 1973, Texas Tech University. 59 p.

Guzman, Ana Maria Perez see Perez Guzman, Ana Maria

Guzman, Armando. Tin industry of Bolivia. M, 1969, Stanford University.

Guzman, Humberto A. The Abo Formation of Otero and Chaves counties, New Mexico; a coastal alluvial fan deposit. M, 1984, Indiana University, Bloomington. 102 p.

Guzman-Speziale, Marco. The triple junction of the North America, Cocos, and Caribbean plates, seismicity and tectonics. M, 1985, University of Texas, Austin. 67 p.

Guzofsky, David Paul. The relationship between geology and man and how they have affected Wyoming Valley. M, 1975, Pennsylvania State University, University Park. 96 p.

Guzowski, R. V. Stratigraphy, structure, and petrology of the Precambrian rocks in the Black Lake region, Northwest Adirondacks, New York. D, 1979, Syracuse University. 214 p.

Gwilliam, William. Geology of the Clarksburg, West Virginia, 7 1/2 minute Quadrangle. M, 1967, West Virginia University.

Gwinn, Billy W. A Stratigraphic study of some Pennsylvanian rocks in south-central Colorado. M, 1951, University of Kansas. 89 p.

Gwinn, James E. Origin and practical implication of the structure of the Beckley Coal, Stanaford Number 2 Mine, Stanaford, West Virginia. M, 1950, West Virginia University.

Gwinn, Vinton E. Cretaceous and Tertiary stratigraphy and structural geology of the Drummond area, Montana. D, 1960, Princeton University. 182 p.

Gwinner, Don. Structural setting of uranium-bearing quartz-pebble conglomerate of Precambrian age, Northwest Sierra Madre, Carbon County, Wyoming. M, 1979, University of Wyoming. 29 p.

Gwyn, Hugh. Heavy mineral assemblages in tills (late Wisconsin) and their use in distinguishing glacial lobes in the Great Lakes region. D, 1971, University of Western Ontario. 194 p.

Gwyn, Quintin H. J. The role of moisture and temperature cycles in soil movement on Mont Saint Hilaire, Quebec (Canada). M, 1968, McGill University. 114 p.

Gwynn, John Wallace. Instrumental analysis of tars and their correlations in oil-impregnated sandstone beds, Uintah and Grand counties, Utah. D, 1970, University of Utah. 130 p.

Gwynn, Thomas Andrew. Geology of the Provo Slate Canyon in the southern Wasatch Mountains, Utah. M, 1948, Brigham Young University. 91 p.

Gwynne, Charles Sumner. A study of the structural features south of the Helderberg Escarpment in the vicinity of Syracuse, New York. M, 1925, Syracuse University.

Gwynne, Charles Sumner. Weathering of the Pre-Cambrian rocks in central Wisconsin. D, 1927, Cornell University.

Gyan-Gorski, S. S. The Cooke City and Silver Gate area; a recreational potential study. M, 1977, Montana State University.

Gyebi, Osei Kwabena. Finite element model for viscoplastic soil under dynamic loads. D, 1987, Columbia University, Teachers College. 89 p.

Gyongyossy, Zoltan. Lithium; a commodity report and an exploration geochemical orientation survey at Rush Lake, Lac du Bonnet Mining Division, southeastern Manitoba. M, 1981, University of Toronto.

Gyss, Emile B. A geological report of field work done in New York, New Jersey, and Pennsylvania. M, 1923, Columbia University, Teachers College.

H., Alfredo Mederos see Mederos H., Alfredo

Ha, Tiong. CPF transforms and convolution. M, 1987, University of Houston.

Haack, David Arno. The geology of the Selma Farm area, Jefferson County, Missouri. M, 1955, Washington University. 81 p.

Haack, Norman Erwin. Devonian agglutinated foraminifera from Illinois. M, 1956, University of Illinois, Urbana. 55 p.

Haack, Richard C. Organism-sediment interrelationships; Pennsylvanian ostracode assemblages from a mixed carbonate-terrigenous mud environment. M, 1979, University of Kansas. 54 p.

Haag, Gary H. The sedimentologic and hydraulic characteristics of the Raritan River in the Bound Brook reach. M, 1982, Rutgers, The State University, New Brunswick. 73 p.

Haag, William George, Jr. Physiographic interpretation of Kentucky topography. M, 1933, University of Kentucky. 46 p.

Haagensen, Robert B. A chemical, X-ray and infrared investigation of some natural forsterite-fayalite series minerals. M, 1963, Rutgers, The State University, New Brunswick. 44 p.

Haar, Stephen P. Vonder see Vonder Haar, Stephen P.

Haardeng-Pedersen, G. P. Studies on the dynamics of the rotating Earth. D, 1975, Memorial University of Newfoundland. 225 p.

Haarr, Doris T. Magnetic investigation of the Catoctin Formation in Maryland. D, 1976, University of Pittsburgh. 201 p.

Haartz, Eric R. Petrology and origin of the Camp Creek corundum deposit, Southwest Ruby Range, Montana. M, 1979, University of Montana. 53 p.

Haas, Bruno J. The areal geology and stratigraphy of the north-western quarter of the Richland Center Quadrangle, Wisconsin. M, 1955, University of Wisconsin-Madison.

Haas, Christopher Allen. Barrier island depositional systems in the Black Warrior Basin, Lower Pennsylvanian Parkwood and Pottsville formations in northwestern Alabama. M, 1988, Auburn University. 172 p.

Haas, Eugene Anthony. Structural analysis of a portion of the Reagan fault zone, Murray County, Oklahoma. M, 1978, University of Oklahoma. 52 p.

Haas, Herbert. Equilibria in the system Al_2O_3-SiO_2-H_2O involving the stability limits of diaspore and pyrophyllite, and thermodynamic data of these minerals. D, 1971, Southern Methodist University. 30 p.

Haas, James J.; Kaarsberg, J. T. and Stephens, J. J. Geology of the Poison Canyon area, Huerfano County, Colorado. M, 1956, University of Michigan.

Haas, John Lewis, Jr. Solubility of iron in solutions coexisting with pyrite from 25° to 250°C, with geologic implications. D, 1966, Pennsylvania State University, University Park. 127 p.

Haas, Nina. A geophysical study of the North Scappoose Creek - Alder Creek - Clatskanie River Lineament, along the trend of the Portland Hills Fault, Columbia County, Oregon. M, 1983, Portland State University. 109 p.

Haase, C. S. Metamorphic petrology of the Negaunee Iron-Formation, Marquette District, northern Michigan. D, 1979, Indiana University, Bloomington. 246 p.

Haase, P. C. Glacial stratigraphy and landscape evolution of the north-central Puget Lowland, Washington. M, 1987, University of Washington. 73 p.

Habeck, Mark Fredrick. Origin of abnormal pressures in the lower Vicksburg, McAllen Ranch Field, Hidalgo County, Texas. M, 1982, Texas A&M University. 101 p.

Haberfeld, Joeseph L. Late Pleistocene sands of southwestern Illinois. M, 1977, Southern Illinois University, Carbondale. 113 p.

Habermann, Gail M. Textural and Mineralogic and variations of the duricrust in southwestern Wisconsin. D, 1978, University of Wisconsin-Madison. 166 p.

Habermann, Ray Edward. The quantitative recognition and evaluation of seismic quiescence; applications to earthquake prediction and subduction zone tectonics. D, 1981, University of Colorado. 276 p.

Haberyan, Kurt August. Phycology, sedimentology, and paleolimnology near Cape Maclear, Lake Malawi, Africa. D, 1988, Duke University. 262 p.

Habib, Antoine Ghali Elia. Geology of the Bearpaw Formation in South central Alberta. M, 1981, University of Alberta. 103 p.

Habib, Daniel. Distribution of spore and pollen assemblages in the lower Kittanning Coal (Pennsylvanian) of western Pennsylvania. D, 1965, Pennsylvania State University, University Park. 310 p.

Habib, Daniel. Palynological correlation of the Bevier and Wheeler coals. M, 1960, University of Kansas. 86 p.

Habib, M. K. Structural fabric and uranium distribution in shear zones near Cardiff, Ontario. M, 1982, University of Ottawa. 108 p.

Habib, Nashat M. Interpretation of some aeromagnetic maps of southeastern Missouri. M, 1972, St. Louis University.

Habibafshar, Azar. Determination of volatile hydrocarbon emissions from land treatment of petroleum oily sludges. D, 1984, University of Oklahoma. 159 p.

Habrukowich, Richard George. Impact of groundwater quality from the surface application of liquid digested sludge on sandy soils. D, 1982, Rutgers, The State University, New Brunswick. 493 p.

Hachey, Philip Osmund. Geology and groundwater of the Fredericton District, New Brunswick. M, 1955, University of New Brunswick.

Hack, John Tilton. Geography and geology of the Hopi country, Arizona. D, 1940, Harvard University.

Hackathorn, Merrianne. Foraminifera and some associated microorganisms of Lemon Bay, Florida. M, 1975, Bowling Green State University. 103 p.

Hackbarth, Claudia Jane. Depositional modeling of tetrahedrite in the Coeur D'Alene District, Idaho. D, 1984, Harvard University. 302 p.

Hackbarth, Douglas A. Hydrogeologic aspects of spent sulfite liquor disposal at Peshtigo (Marinette County), Wisconsin. D, 1971, University of Wisconsin-Madison.

Hackbarth, Douglas A. The hydrogeology of the Saint Germain area, Wisconsin. M, 1968, University of Wisconsin-Madison.

Hackenberg, Robert L. Uranium and thorium mobilization in selected Precambrian metamorphic rocks in northern Michigan. M, 1981, University of Wisconsin-Milwaukee. 91 p.

Hackenmueller, Joseph M. Environmental effects on the magnetic properties of soils and lake sediments across the forest-prairie ecozone in central Minnesota. M, 1983, University of Minnesota, Minneapolis.

Hacker, Bradley Russell. Experimental deformation and metamorphism of amphibolite and basalt. D, 1988, University of California, Los Angeles. 383 p.

Hacker, Bradley Russell. Stratigraphy and structures of the Yuba Rivers area, Central Belt, northern Sierra Nevada, California. M, 1983, University of California, Davis. 125 p.

Hacker, David B. Structural geology of the southwest flank of Gros Ventre Uplift, upper Shoal Creek area, Teton and Sublette counties, Wyoming. M, 1985, Miami University (Ohio). 93 p.

Hacker, Robert D. The geology, mineralogy, and high gradient magnetic beneficiation of Texas bentonites. M, 1984, Indiana University, Bloomington. 75 p.

Hacker, Robert Norris. The geology of the northwest corner of the Orestimba Quadrangle and northeast corner of the Mount Boardman Quadrangle, California. M, 1950, University of California, Berkeley. 42 p.

Hackett, Barbara E. Geomorphology and sedimentary character of the Redondo submarine fan, southern California. M, 1969, University of Southern California.

Hackett, Gary Kenneth. Ultrasonic absorption and dispersion in a limestone. D, 1967, University of Utah. 155 p.

Hackett, James E. The occurrence and movement of ground water in the Jefferson Junction area, Wisconsin. M, 1952, University of Wisconsin-Madison.

Hackett, James Edward. Groundwater geology of Winnebago County, Illinois. D, 1958, University of Illinois, Urbana. 136 p.

Hackett, John P., Jr. Stratigraphy, extent and geologic history of the Red Sea geothermal deposits. M, 1972, University of Southern California.

Hackett, Robert S. The stratigraphy of the Amsden Formation of the Big Horn, Owl Creek, and Wind River mountains of Wyoming. M, 1930, University of Missouri, Columbia.

Hackett, Steve W. Gravity survey of Beluga Basin and adjacent areas, Cook Inlet region, South-central Alaska. M, 1977, University of Alaska, Fairbanks. 50 p.

Hackett, William Robert. The timing of burial metamorphism of argillaceous sediments; mineralogical, chemical, and ^{40}Ar evidence. M, 1977, Case Western Reserve University.

Hackley, Keith Crowell. Sulfur isotope variations in low-sulfur coals from the Rocky Mountain region. M, 1984, University of Illinois, Urbana. 85 p.

Hackman, David B. Origin and environment of mineralization at the Siskon Mine, Siskiyou County, California. M, 1971, University of Arizona.

Hackman, David Brent. The evaluation of supergene copper deposits for in situ leaching. D, 1982, University of Arizona. 530 p.

Haddad, A. The pipeline transportation of complex mixtures of ammonia, urea and crude oil. M, 1977, Stanford University. 53 p.

Haddad, Geoffrey Allen. A study of carbonate dissolution, stable isotope chemistry and minor element composition of pteropods and forams deposited in the Northwest Providence Channel, Bahamas during the past 500,000 years. M, 1986, Duke University. 197 p.

Haddad, Marwan Najeh. Modeling of limestone dissolution in packed-bed contactors treating dilute acidic water. D, 1986, Syracuse University. 267 p.

Haddad, P. A multivariable-statistical approach to the evaluation of the undrained behaviour of clays. D, 1977, University of Toronto.

Haddad, Richard. Minor structures of the Boston Mountain Monocline in T. 9, 10, 11 N., R. 32 W., of Crawford County, Arkansas. M, 1954, University of Arkansas, Fayetteville.

Haddadin, Munir Abdullah. Mineralogical study of the phosphate rock in Jordan. M, 1971, University of Utah. 411 p.

Hadden, David R. Relationships of grain orientation on ripples to current velocity, and grain size parameters to ripple dimensions of modern sands in Bogue Inlet, (coast of) North Carolina. M, 1968, Bowling Green State University. 47 p.

Haddock, David R. Modeling bedload transport in mountain streams of the Colorado Front Range. M, 1978, Colorado State University. 82 p.

Haddock, Gerald Hugh. Geology of the Cougar Peak volcanic area, Lake County, Oregon. M, 1959, Washington State University. 72 p.

Haddock, Gerald Hugh. The Dinner Creek welded ash-flow tuff (Miocene) of the Malheur Gorge area, Malheur County, Oregon. D, 1967, University of Oregon. 111 p.

Haddox, C. A. Sedimentology of the Munising Formation. M, 1982, University of Wisconsin-Madison. 168 p.

Haddox, Jimmy V. A systematic study of the fracture pattern in the Demopolis Chalk in the Artesia, Mississippi, Quadrangle. M, 1963, Mississippi State University. 55 p.

Haden, J. M. Isotopic studies of sulfide and other mineralization from the middle Paleozoic rocks of western Ohio. M, 1977, Ohio State University. 117 p.

Hadiaris, Amy K. Quantitative analysis of groundwater flow in Spanish Springs Valley, Washoe County, Nevada. M, 1988, University of Nevada. 203 p.

Hadidi-Tamjed, Hassan. Statistical response of inelastic SDOF systems subjected in earthquakes. D, 1988, Stanford University. 288 p.

Hadj Hamou, Taric Aly. Probabilistic evaluation of damage potential due to seismically induced pore pressures. D, 1983, Stanford University. 272 p.

Hadji-Sabbagh, Mehdi. Structural geology of the Crook Mountain and Whitewood area, Lawrence-Meade counties, South Dakota. M, 1979, South Dakota School of Mines & Technology.

Hadler, Harry George. A subsurface study of northwestern Cleveland County, Oklahoma, with emphasis on the south and west Moore oil fields. M, 1947, University of Oklahoma. 33 p.

Hadley, David G. Coal petrology of the Big Seam, Centralia Coal Mine, Centralia, Washington. M, 1981, Western Washington University. 61 p.

Hadley, David Milton. Geophysical investigations of the structure and tectonics of Southern California. D, 1978, California Institute of Technology. 173 p.

Hadley, David Milton. Microearthquake distribution and mechanism of faulting in the Fontana-San Bernardino area of Southern California. M, 1973, University of California, Riverside. 58 p.

Hadley, Donald Gene. The sedimentology of the Huronian Lorrain Formation, Ontario and Quebec, Canada. M, 1968, The Johns Hopkins University.

Hadley, Jarvis Bardwell. Geology of the New Hampshire portion of the Mount Cube Quadrangle. D, 1938, Harvard University.

Hadley, Kate Hill. Dilatancy; further studies in crystalline rock. D, 1975, Massachusetts Institute of Technology. 202 p.

Hadley, Linda M. Seismicity of Colorado; vicinity of Cabin Creek pumped-storage hydroelectric plant. M, 1975, Colorado School of Mines. 74 p.

Hadley, Richard Frederick. Recent sedimentation and erosional history of Five Mile Creek, Fremont County, Wyoming. M, 1950, University of Minnesota, Minneapolis. 48 p.

Hadley, Wade H. The foraminifera of the Navarro at Jones' Crossing, Texas. M, 1933, Cornell University.

Hadsell, Frank A. Scattering of an elastic disturbance by a random medium. D, 1961, Colorado School of Mines. 129 p.

Haeck, Gary Dennis. The geologic and tectonic history of the central portion of the southern Sierra Madre, Luzon, Philippines. D, 1987, Cornell University. 311 p.

Haederle, Wolfgagng F. Structure and metamorphism in the southern Sierra Ladrones, Socorro County, New Mexico. M, 1966, New Mexico Institute of Mining and Technology. 56 p.

Haefka, Delbert John. The structural geology of the East River Mountain-Wolf Creek area, Virginia. M, 1972, University of Akron. 39 p.

Haefner, Ralph J. Mississippi valley-type mineralization and dolomitization of the Trenton Limestone, Wyandot County, Ohio. M, 1986, Bowling Green State University. 123 p.

Haefner, Richard Charles. Emplacement and cooling history of a rhyolite lava flow and related tuff at Deadman Pass, near Death Valley, California. M, 1969, Pennsylvania State University, University Park. 82 p.

Haefner, Richard Charles. Igneous history of a rhyolite lava-flow series near Death Valley, California. D, 1972, Pennsylvania State University, University Park. 348 p.

Haeger, Steven D. Applying the FWQA hydrodynamic model to the Indian River. M, 1978, Florida Institute of Technology.

Haeggni, Walter Tiffany. Geology of the Cowboy Pass area, Confusion Range, Millard County, Utah. M, 1957, University of Texas, Austin.

Haeggni, Walter Tiffany. Geology of the El Cuervo area, northeastern Chihuahua, Mexico. D, 1966, University of Texas, Austin. 477 p.

Haehl, Harry L. Miocene diabase of the Santa Cruz Mountains in San Mateo County, California. M, 1902, Stanford University. 53 p.

Haensel, Jose Mariano. Plagioclase compositions in the Dufek Massif anorthosites, Dufek Intrusion, Antarctica. M, 1988, University of Missouri, Columbia.

Hafen, Preston L. Geology of the Sharp Mountain area, southern part of the Bear River Range, Utah. M, 1961, Utah State University. 72 p.

Haff, John Coles. Multiple dikes of Cape Neddick. D, 1939, Columbia University, Teachers College.

Haffner, B. K. Microscopic investigation of the lead-silver ores of the Bunker Hill and Sullivan Mine. M, 1941, Queen's University. 41 p.

Haffner, Robert Louis. Block caving and other mining methods at the Bunker Hill mines, Kellogg, Idaho. M, 1949, Stanford University. 86 p.

Hafi, Zuhair. Digital-computer model for nitrate-transport in the Rapid Valley Aquifer, Pennington County, South Dakota. M, 1983, South Dakota School of Mines & Technology. 1 table p.

Hafley, Daniel James. Geology and paleontology of the Heart Lake area, White River Plateau, Northwest Colorado. M, 1984, Washington State University. 124 p.

Hafner, Robert Otto. Lithofacies and depositional setting of the San Andreas Formation, Cochran County, West Texas. M, 1979, University of New Orleans. 72 p.

Hafner-Douglass, Katrin. Stratigraphic, structural and geochemical analyses of bedrock geology, Woodsville Quadrangle, New Hampshire-Vermont. M, 1986, Dartmouth College. 117 p.

Haga, Hideyo. Distribution of foraminifera in sediments of the Gulf of Thailand. M, 1963, University of Southern California.

Hagan, Teresa A. Ponderosa pine tree rings; variability of response to precipitation in space and time. M, 1988, Washington State University. 74 p.

Hagan, Wallace Woodrow. Geology of the Cub Run Quadrangle, Kentucky. D, 1942, University of Illinois, Urbana. 252 p.

Hagan, Wallace Woodrow. Inadunate crinoids of the Chester Series; Zeacrinus. M, 1936, University of Illinois, Urbana. 37 p.

Hagar, David Jon. Geomorphology of Coyote Basin, San Bernadino, California. D, 1966, University of Massachusetts. 273 p.

Hagar, Richard Allen. A petrographic analysis of the Home Creek Limestone (Upper Pennsylvanian), north-central Texas. M, 1982, Stephen F. Austin State University. 113 p.

Hage, Conrad O. Accessory minerals in igneous rocks of the central Mississippi Valley. M, 1932, University of Wisconsin-Madison.

Hagedon, Dan Newman. The calculation of synthetic thermal conductivity logs from conventional geophysical well logs. M, 1985, Southern Methodist University. 110 p.

Hagegeorge, Charles G. The geology of the Liberty Hill area, Grainger County, Tennessee. M, 1962, University of Tennessee, Knoxville. 30 p.

Hageman, Mark Robert. Interpretations of the kinematic evolution of kink folds in the northern Valley and Ridge Province. M, 1988, University of North Carolina, Chapel Hill. 95 p.

Hageman, Steven James. Concepts and methods for taxonomic analysis of fenestrate Bryozoa. M, 1988, University of Illinois, Urbana. 189 p.

Hagemeyer, R. Todd. Resistivity study of the lower Withlacoochee River-Cross-Florida Barge Canal complex. M, 1988, University of South Florida, Tampa. 101 p.

Hagen, David J. Spatial temporal variability of ground water quality in a shallow aquifer in north-central Oklahoma. M, 1986, Oklahoma State University. 191 p.

Hagen, Donald Wesley. Geology of the Wheeler Springs area. M, 1957, University of California, Los Angeles.

Hagen, Earl Sven. Hydrocarbon maturation in Laramide-style basins; constraints from the northern Bighorn Basin, Wyoming and Montana. D, 1986, University of Wyoming. 215 p.

Hagen, John C. Some aspects of the geochemistry of platinum, palladium, and gold in igneous rocks with special reference to the Bushveld Complex, Transvaal. D, 1954, Massachusetts Institute of Technology. 310 p.

Hagen, John Christopher. The geology of the Green Mountain Mine, San Juan County, Colorado. M, 1951, Colorado School of Mines. 151 p.

Hagen, Kurt Brian. Mapping of surface joints on air photos can help understand waterflood performance problems at North Burbank Unit, Osage and Kay counties, Oklahoma. M, 1972, University of Tulsa. 85 p.

Hagen, Randall Alan. The geology and petrology of the Northcraft Formation, Lewis County, Washington. M, 1987, University of Oregon. 252 p.

Hagen, Ricky A. A seismic refraction study of the crustal structure in the active seismic zone east of Taiwan. M, 1987, University of Hawaii. 73 p.

Hagen-Leveille, Janice. A statistical study of Alaska skarns with application to resource evaluation. M, 1987, University of Alaska, Fairbanks. 164 p.

Hager, Glenn G. The stratigraphy of some Pennsylvanian rocks southwest of La Veta Pass, Colorado. M, 1952, University of Kansas.

Hager, Glenn M. Petrologic and structural relations of crystalline rocks in the Hoopes Reservoir area, Delaware. M, 1976, University of Delaware.

Hager, James Marion. Approximating the bedrock topography of Clark County, Ohio, using the gravity method. M, 1982, Wright State University. 101 p.

Hager, Michael Warring. Late Pliocene and Pleistocene history of the Donnelly Ranch vertebrate site, southeastern Colorado. D, 1973, University of Wyoming. 140 p.

Hager, Richard Charles. Petrology and depositional environment of the Hampton Formation (Kinderhookian) in central Iowa. M, 1981, University of Iowa. 201 p.

Hagermann, Steffen Gerd. The structure, petrology and geochemistry of the gold-bearing Canastra phyllites near Luziania, Goias, Brazil. M, 1988, University of Wisconsin-Milwaukee. 156 p.

Hagerty, Roayl Moncrief. Clay mineralogy of the southwestern Gulf of Mexico and adjacent river outlets. D, 1969, Texas A&M University.

Haggerty, Janet Ann. Environmental and depth variation in Enewetak Favia. M, 1978, Pennsylvania State University, University Park. 172 p.

Haggerty, Janet Ann. The geologic history of the southern Line islands. D, 1982, University of Hawaii. 217 p.

Haggiagi, Musa A. Stratigraphy and environments of deposition of the Chanute shale (Pennsylvanian) in southeastern Kansas. M, 1970, University of Kansas. 31 p.

Haghighi, Rahim Ghorbanzadeh. Investigation of relationship between rock fragmentation and burden stiffness ratio in confined bench blasting. D, 1985, Ohio State University. 293 p.

Hagihara, H. H. Potassium fixation in Hawaiian soils. M, 1953, University of Hawaii. 49 p.

Hagley, Mark T. A comparison of model approaches for evaluating groundwater flow and transport. M, 1988, Michigan Technological University. 88 p.

Haglund, David S. The distribution of uranium in recent carbonate sediments and skeletons of organisms and the effect of diagenesis on uranium redistribution. D, 1968, Rensselaer Polytechnic Institute. 146 p.

Haglund, David Seymour. An investigation of natural surface charge and zeta potential on quadrangles quartz aggregates by the streaming potential method. M, 1965, University of Virginia. 79 p.

Haglund, John Louis. A method for the inversion of seismic travel times for velocity structure in two-dimensional media. M, 1982, University of California, Davis. 189 p.

Haglund, Wayne Milton. The brachiopod genus Enteletes of the Pennsylvanian of Kansas. M, 1965, University of Kansas. 128 p.

Hagmaier, Jonathan Ladd. Groundwater flow, hydrochemistry, and uranium deposition in the Powder River Basin, Wyoming. D, 1971, University of North Dakota. 60 p.

Hagner, Arthur F. Absorptive clays of the Texas Gulf Coast. M, 1939, Columbia University, Teachers College.

Hagni, Richard Davis. Mineral paragenesis and trace element distribution in the tri-state zinc-lead district, Missouri, Kansas, Oklahoma. D, 1962, University of Missouri, Columbia. 272 p.

Hagni, Richard Davis. Petrology and origin of the Kitchi Conglomerate, Marquette County, Michigan. M, 1954, Michigan State University. 45 p.

Hagood, Allen Roland. Geology of the Monument Peak area, Malheur County, Oregon. M, 1963, University of Oregon. 165 p.

Hagood, Michael Curtis. Structure and evolution of the Horse Heaven Hills in south-central Washington. M, 1985, Portland State University. 202 p.

Hagopian, John J. Petrographic analysis of, and distribution of uranium in, the granitic rocks of the Deer Creek area, northern Laramie Range, Wyoming. M, 1982, Bowling Green State University. 77 p.

Hagstrom, Earl L. Granulometric analysis of foreshore sediments, south shore Rhode Island. M, 1978, University of Rhode Island.

Hagstrum, Jonathan Tryon. Part I; Middle-Tertiary paleomagnetism and hydrothermal alteration of the structurally deformed Latir volcanic field, northern New Mexico; Part II; Mesozoic paleomagnetism and northward translation of the Baja California Peninsula. D, 1985, Stanford University. 173 p.

Hague, Nancy Ellen. Alkalic gneisses near Gorman Lake, Brudenell Township, in the Grenville Province of southeastern Ontario, Canada. M, 1976, Cornell University.

Hague, William C. Geology of the northern part of the Slate Mountains, Pinal County, Arizona. M, 1940, University of Arizona.

Hahman, W. Richard. Geology of the Hedgesville Quadrangle, Berkeley County, West Virginia. M, 1963, West Virginia University.

Hahn, Glenn Walter. Clay mineral relationships in the Ochesky fireclay pit. M, 1954, University of Missouri, Columbia.

Hahn, Kenneth R. Petrology and geochemistry of the Mincey Mine dunite, Macon County, North Carolina. M, 1976, Kent State University, Kent. 49 p.

Hahn, Raimund. Lithogeochemistry and fluid inclusions of the Acme mining claims, Stevens County, Washington. M, 1986, Eastern Washington University. 67 p.

Hahn, Robert. Factors affecting the origin and distribution of methane in the Sparta Aquifer, Brazos and Burleson counties, Texas. M, 1985, Texas A&M University. 162 p.

Hahn, Roger K. Upper mantle velocity structure in eastern Kansas from teleseismic P-wave residuals. M, 1980, University of Kansas. 85 p.

Hahnenberg, James J. The petrology and geochemistry of Keweenawan diabase dikes in Ontonagon, Gogebic, Iron and Dickinson counties, Michigan. M, 1981, Western Michigan University. 90 p.

Haible, William Wilson. Holocene profile changes along a California coastal stream. M, 1976, University of California, Berkeley. 78 p.

Haidarian, Mohammad Reza. Geophysical investigation of a gravity minima in northwestern Ohio. M, 1976, Bowling Green State University. 56 p.

Haidl, Frances Margaret. Geology and fluid distribution in the Lower Cretaceous Mannville Group, Celtic-Westhazel area, Saskatchewan. M, 1986, University of Regina. 269 p.

Haiduk, John Paul. Facies analysis, paleoenvironmental interpretation, and diagenetic history of Britt Sandstone (Upper Mississippian), in portions of Caddo and Canadian counties, Oklahoma. M, 1987, Oklahoma State University. 188 p.

Haig, Trevor D. A new paleomagnetic pole position for the interbedded lava flows within the Copper Harbor Conglomerate of Michigan's Keweenaw Peninsula. M, 1986, Michigan Technological University. 46 p.

Hail, William James, Jr. Petrology of an early Tertiary sandstone, Powder River valley, northern Wyoming. M, 1951, Washington University. 25 p.

Haileab, Bereket. Characterization of tephra from the Shungura Formation, southwestern Ethiopia. M, 1988, University of Utah. 130 p.

Hailer, J. G. The metal content of tree rings as an environmental monitoring method. M, 1977, George Washington University.

Haimes, Robert. An analysis of the formation of ultramafic and mafic layered intrusions. M, 1976, Rensselaer Polytechnic Institute. 127 p.

Haimila, Norman Edward. Contact phenomena of the Central Vancouver Island Intrusion. D, 1973, Michigan State University. 120 p.

Haimovitz, Allan. Ostracodes and stratigraphy of the Brownsport Formation of western Tennessee. M, 1968, Arizona State University. 76 p.

Haimson, Bezalel. Hydraulic fracturing in porous and nonporous rock and its potential for determining in-situ stresses at great depth. D, 1968, University of Minnesota, Minneapolis. 246 p.

Haimson, Marshall. Oxygen isotope studies of silica in the Monterey Formation, California. M, 1983, Arizona State University. 77 p.

Haines, Carroll Eugene. A gravity meter survey of a portion of the American Bottoms. M, 1953, St. Louis University.

Haines, David Vincent. A spectrographic and petrographic study of the ore minerals at Climax, Colorado. M, 1953, Pennsylvania State University, University Park. 105 p.

Haines, Evelyn B. Processes affecting production in Georgia coastal waters. D, 1974, Duke University. 128 p.

Haines, Forest E. Lower Mississippian sedimentation in northwestern Montana. D, 1978, University of Missouri, Rolla. 128 p.

Haines, Forest E., Jr. Petrographic analysis of the Ames Limestone in Meigs, Gallia, and Lawrence counties, Ohio. M, 1965, Ohio University, Athens. 93 p.

Haines, Harvey Hartman. Thermal expansion and compressibility of rocks as a function of pressure and temperature. M, 1982, Massachusetts Institute of Technology. 601 p.

Haines, John Beverly, Jr. Source-inherited shape characteristics of coarse quartz-silt on the Northwest Gulf of Mexico continental shelf. M, 1986, Texas A&M University. 92 p.

Haines, John W. An investigation of low-frequency wave motions; a comparison of barred and planar beaches. D, 1987, Dalhousie University.

Haines, Richard Arthur. The geology of the White Oaks-Patos Mountain area, Lincoln County, New Mexico. M, 1968, University of New Mexico. 63 p.

Hainstock, H. N. The coal deposits of the Hay River District (Alberta). M, 1929, McMaster University.

Hair, Gregory L. Stratigraphy and microfacies analysis of Panther Seep Formation (Virgilian), Franklin Mountains, Texas, New Mexico. M, 1977, University of Texas at El Paso.

Hairr, L. M. An investigation of factors influencing radiocesium cycling in estuarine sediments of the Hudson River. D, 1974, New York University. 198 p.

Hait, Mortimer Hall, Jr. Structure of the Gilmore area, Lemhi Range, Idaho. D, 1965, Pennsylvania State University, University Park. 173 p.

Haitjema, Hendrik Marten. Modeling three-dimensional flow in confined aquifers using distributed singularities. D, 1982, University of Minnesota, Minneapolis. 139 p.

Hajali, Paris Andraos. Remedial Action Evaluation System; a methodology for determining the completion point for aquifer restoration programs. D, 1987, University of Oklahoma. 515 p.

Hajash, Andrew, Jr. Hydrothermal processes along mid-ocean ridges; an experimental investigation. D, 1975, Texas A&M University. 70 p.

Hajash, Andrew, Jr. Paleomagnetism of the Comores Islands, Southwest Indian Ocean. M, 1970, Florida State University.

Haji, Mustapha. Correlation study of regional geophysical data across the conterminous United States. M, 1985, Purdue University.

Haji-Djafari, Sirous. Two-dimensional finite element analysis of transient flow and tracer movement in confined and phreatic aquifers. D, 1976, [Michigan State University]. 267 p.

Haji-Vassiliou, Andreas. The association of uranium with naturally occurring organic materials in the Colorado plateau and other areas. D, 1969, Columbia University. 265 p.

Haji-Vassiliou, Andreas. The cell edge shortening in uranite as related to artificial oxidation. M, 1965, Columbia University. 37 p.

Hajic, Edwin Robert. Geology and paleopedology of the Koster archeological site, Greene County, Illinois. M, 1981, University of Iowa. 107 p.

Hajishafie, Manoutcher. Electric log interpretation of the Sparta Formation, east-central Louisiana. M, 1978, Florida State University.

Hajitaheri, Jafar. Fluid inclusion geothermometry of fluorite, Indian Peak Range, Utah. M, 1980, University of Utah. 94 p.

Hajnal, Zoltan. A continuous deep-crustal seismic refraction and near-vertical reflection profile in the Canadian shield, interpreted by digital processing techniques (Archean, southeastern Manitoba). D, 1970, University of Manitoba.

Hajnal, Zoltan. A palaeomagnetic study of the Coronation Mine area. M, 1963, University of Saskatchewan. 96 p.

Hajosy, Roger Alan. The geology of a portion of the White Hollow Quadrangle, Union County, Tennessee. M, 1960, University of Tennessee, Knoxville. 42 p.

Hake, Benjamin F. A study of faulting in the Coast Ranges, California. M, 1930, Stanford University. 60 p.

Hakes, William G. Trace fossils and depositional environment of four clastic units, upper Pennsylvanian megacyclothems, Northeast Kansas. D, 1976, University of Kansas. 247 p.

Hakes, William G. Trace fossils and the depositional environment of the Lawrence Shale (Upper Pennsylvanian) of eastern Kansas. M, 1972, University of Kansas. 55 p.

Hakim, H. D. M. Geology and the genesis of gold-silver base metal sulphide-bearing veins, at Mahd ad Dahab, Kingdom of Saudi Arabia. D, 1978, University of Western Ontario. 182 p.

Hakkinen, Joseph William. Structural geology and metamorphic history of western Hinnoey and adjacent parts of eastern Hinnoey, North Norway. D, 1977, Rice University. 214 p.

Halabura, Stephen Philip. Depositional environments of Upper Devonian Birdbear Formation, Saskatchewan. M, 1983, University of Saskatchewan. 182 p.

Halada, R. S. A seismic study of the crustal structure of the Ontong Java Plateau and Nauru Basin. M, 1978, University of Hawaii.

Halamicek, William Arnold, Jr. Geology of Toyah Field and Burchard quadrangles, Reeves County, Texas. M, 1951, University of Texas, Austin.

Halbig, Joseph B. Solubility studies of selected chalcophile elements in hydrothermally synthesized galena. M, 1965, Pennsylvania State University, University Park. 111 p.

Halbig, Joseph B. Trace element studies in synthetic sulfide systems; the solubility of thallium in sphalerite and the partition of selenium between sphalerite and galena. D, 1969, Pennsylvania State University, University Park. 165 p.

Halbouty, James Jubron. The subsurface geology of the southwestern portion of the Sabine Uplift, with emphasis on electric log correlation. M, 1943, University of Texas, Austin.

Halbouty, Michel T. The geology of Atacosa County, Texas. M, 1931, Texas A&M University.

Halcomb, Robert Allan. A mathematical model for a horizontal in situ oil shale retort. D, 1980, University of Wyoming. 64 p.

Haldar, A. Probabilistic evaluation of liquefaction of sand under earthquake motions. D, 1976, University of Illinois, Urbana. 250 p.

Halderman, Tom Pepin. Asthenosphere Q below the Rio Grande Rift using a predetermined velocity structure and calculated t values. M, 1987, University of California, Los Angeles. 67 p.

Haldorsen, Helge Hove. Reservoir characterization procedures for numerical simulation. D, 1983, University of Texas, Austin. 576 p.

Hale, Alma Phillips. Evidence for pedogenesis in Pleistocene-Holocene carbonates on San Salvador Island, Bahamas. M, 1984, University of Kentucky. 76 p.

Hale, Christopher. The application of microwave heating to lunar paleointensity determination. M, 1978, University of California, Santa Barbara.

Hale, Christopher James. Evidence of the Archean geomagnetic field. D, 1986, University of Toronto.

Hale, David A. Observations of the nearshore currents off Hutchinson Island, east central Florida. M, 1976, Florida Institute of Technology.

Hale, Elbert, Lee. An analysis of shallow refraction seismograms. M, 1955, Indiana University, Bloomington. 42 p.

Hale, George Robert. Reassessment of the Death Valley-Colorado River overflow hypothesis in light of new evidence. D, 1984, University of California, Berkeley. 291 p.

Hale, Ira David. Dip-moveout by Fourier transform. D, 1983, Stanford University. 94 p.

Hale, Lyle A. Stratigraphy of the Upper Cretaceous Montana Group in the Rock Springs Uplift, Sweetwater County, Wyoming. M, 1950, University of Wyoming. 115 p.

Hale, Robin C. Geology of the Peters Mountain area, Giles County, Virginia. M, 1961, Virginia Polytechnic Institute and State University.

Hale, W. E. Geology of the Black Bay map-area, northern Saskatchewan, with reference to structural control of pitchblende deposits. D, 1953, Queen's University. 208 p.

Hale, W. E. Geology of the Uranium City area (west half) with special reference to the pitchblende deposits. M, 1953, Queen's University.

Hale, Walter R. Application of a regression model in the determination of post-mortem transport of foraminiferal tests. M, 1985, East Carolina University. 175 p.

Hale, William Ernest. Variation in the gabbroic rocks of the Saint Stephen area, Charlotte County, New Brunswick. M, 1950, University of New Brunswick.

Hale-Erlich, Wendy Susan. Morphology and morphometry of central structures in craters and basins on the Moon, Mercury and Mars. D, 1983, Brown University. 159 p.

Halepaska, John C. Drawdown distribution around a well partially penetrating a thick leaky aquifer. M, 1966, New Mexico Institute of Mining and Technology. 53 p.

Hales, Peter O. Geology of the Green Ridge area, Whitewater River Quadrangle, Oregon. M, 1975, Oregon State University. 90 p.

Halet, Robert Alfred Frans. A study of the geology in the vicinity of Corporation Quarry, Mount Royal, Montreal, Canada. M, 1932, McGill University.

Halet, Robert Alfred Frans. The geology and mineral deposits of the Beattie-Galatea area (Quebec). D, 1934, McGill University.

Haley, Alva Justice. Geology of Campana, Chile. M, 1938, University of Minnesota, Minneapolis. 15 p.

Haley, Francis L. A sedimentary-stratigraphic study of insoluble residues of the Oligocene limestones of the Florida Panhandle. M, 1956, Florida State University.

Haley, J. Christopher. Upper Cretaceous (Beaverhead) synorogenic sediments of the Montana-Idaho thrust belt and adjacent foreland; relationships between sedimentation and tectonism. D, 1985, The Johns Hopkins University. 542 p.

Haley, Patrick C. An analysis of the channel sediments of the Ochlockonee River. M, 1956, Florida State University.

Halferdahl, Laurence B. Chloritoid; its composition, X-ray and optical properties, stability, and occurrences. D, 1959, The Johns Hopkins University.

Halferdahl, Laurence Bowes. Trace elements in granitic rocks of the Preissac-La Corne area, Quebec. M, 1954, Queen's University. 128 p.

Halfman, Barbara Mary. Suspended solids in the western arm of Lake Superior. M, 1984, University of Minnesota, Duluth. 49 p.

Halfman, John D. Textural analysis of lacustrine contourites. M, 1983, University of Minnesota, Minneapolis.

Halfman, John David. High-resolution sedimentology and paleoclimatology of Lake Turkana, Kenya. D, 1987, Duke University. 241 p.

Halgedahl, Susan Louise. The dependence of magnetic domain structure upon magnetization state in naturally-occurring pyrrhotite and titanomagnetite. D, 1982, University of California, Santa Barbara. 223 p.

Haliburton, John Leo. The sedimentary petrology of the Pennsylvanian System of the upper Pecos Valley, New Mexico. M, 1948, Texas Tech University. 36 p.

Halim-Dihardja, Marjammanda. Diagenesis and sedimentology of the Late Devonian (Famennian) Wabamun Group in the Tangent, Normandville, and Eaglesham fields, north-central Alberta. M, 1986, University of Toronto.

Halimdihardja, Piushadi. Stratigraphy and depositional environments of the Brigham Group in a portion of the northern Portneuf Range. M, 1987, Idaho State University. 206 p.

Halka, Jeffrey P. An objective method for analysis of growth banding patterns. M, 1972, University of Rochester.

Hall, Anne Marie. The clay mineralogy of the lower San Andres Formation, Palo Duro Basin, Texas. M, 1985, Georgia Institute of Technology. 132 p.

Hall, B. V. Geochemistry and mineralogy of the cordierite-anthophyllite rock within the Amulet upper 'A' alteration pipe, Noranda District, Quebec. M, 1978, University of Waterloo.

Hall, Billy P. Cross-bedding in the sandstones and limestones of the Kansas City Group throughout Kansas. M, 1961, University of Kansas. 33 p.

Hall, Blaine R. Collection, reduction, and interpretation of magnetic data from the Newfoundland Basin. M, 1977, Dalhousie University.

Hall, Bradford A. Nature and petrologic significance of variations in abundance of heavy minerals and statistical parameters of zircon populations at different elevations in a granodiorite dike, Bradford, Rhode Island. M, 1959, Brown University.

Hall, Bruce S. Petrography and geochemistry of a part of the Painted Rocks Pluton, Idaho County, Idaho. M, 1980, Eastern Washington University. 94 p.

Hall, Carol Anne. Survey of surficial bottom sediments and the distribution of manganese nodules in Green bay, Wisconsin. M, 1970, University of Wisconsin-Madison.

Hall, Cassandra. The paleoecology and biostratigraphy of the Anahuac Formation in parts of Vermilion and Acadia parishes, Louisiana. M, 1988, University of Southwestern Louisiana. 141 p.

Hall, Charlene R. Precambrian phosphorites of northern Michigan; mode and environment of deposition. M, 1985, Michigan Technological University. 146 p.

Hall, Chris Michael. Potassium-argon dating of extremely young rocks. M, 1975, University of Toronto.

Hall, Chris Michael. The application of K-Ar and ^{40}Ar/^{39}Ar methods to the dating of Recent volcanics and the Laschamp Event. D, 1982, University of Toronto.

Hall, Clarence Albert, Jr. The geology of the Pleasanton area, Alameda County, California. D, 1956, Stanford University. 269 p.

Hall, Dan O. The petrology of a lower tallohatte zeolitic clay (Eocene, Mississippi). M, 1972, University of Mississippi.

Hall, Daniel W. The hydrogeology of the Saukville fly ash landfill. M, 1977, University of Wisconsin-Milwaukee.

Hall, David Joseph. Compositional variations in biotites and garnets from kyanite and sillimanite zone mica schists, Orange area (Massachusetts and New Hampshire). M, 1970, University of Massachusetts. 110 p.

Hall, David Joseph. Geology and geophysics of the Belchertown Batholith, west-central Massachusetts. D, 1973, University of Massachusetts. 110 p.

Hall, David W. Electrical resistivity survey near the "KL" Avenue Landfill, Kalamazoo, Michigan. M, 1983, Western Michigan University.

Hall, Denis Kane. Hydrothermal alteration and mineralization in the East Camp of the Turquoise District, San Bernardino County, California. M, 1972, University of Arizona.

Hall, Donald D. Dalmanellidae of the Cincinnatian. M, 1960, University of Cincinnati. 112 p.

Hall, Donald D. Paleoecology and taxonomy of fossil ostracoda in the vicinity of Sapelo Island, Georgia. D, 1965, University of Michigan. 239 p.

Hall, Donald H. The geophysical analysis of magnetic anomalies. D, 1959, University of British Columbia.

Hall, Donald Lewis, Jr. Contact metamorphism, hydrothermal alteration, and iron-ore deposition in the south-central Marble Mountains, San Bernardino County, California. M, 1985, University of California, Riverside. 240 p.

Hall, Dorothy Kay. Analysis of the origin of water which forms large aufeis fields on the Arctic Slope of Alaska using ground and Landsat data. D, 1980, University of Maryland. 137 p.

Hall, Douglas Charles. The petrology of xenoliths from the Orapa AK1 kimberlite pipe, Botswana. M, 1985, Queen's University. 186 p.

Hall, Durand A. The origin of the magnetite-pyrrhotite ores of the Rush Lake region, Ontario. M, 1915, University of Wisconsin-Madison.

Hall, Dwight Lyman. Stratigraphy and sedimentary petrology of the Mesozoic rocks of the Waterman Mountains, Pima County, Arizona. M, 1985, University of Arizona. 92 p.

Hall, Ellis A. Stratigraphy of the Redcliff District (Eagle County), Colorado; the Groundhog Mine. M, 1922, University of Colorado.

Hall, Francis R. Geology of the southwestern portion of the Las Flores Quadrangle, Los Angeles County, California. M, 1952, University of California, Los Angeles.

Hall, Francis Ramey. Geology and ground water of a portion of eastern Stanislaus County, San Joaquin Valley, California. D, 1961, Stanford University. 312 p.

Hall, Frank Reginald. A study of the effects of consolidation (compaction) on the remanent magnetism of argillaceous sediments. M, 1983, Lehigh University.

Hall, Frank Washington, II. Bedrock geology, north half of Missoula 30' Quadrangle, Montana. D, 1969, University of Montana. 253 p.

Hall, Frank Washington, II. Geology of the Northwest Pleasant Valley Quadrangle, Montana. M, 1962, University of Montana. 60 p.

Hall, Gary L. Sediment transport processes in the nearshore waters adjacent to Galveston Island and Bolivar Peninsula. D, 1976, Texas A&M University. 325 p.

Hall, Gary L. The subenvironments of deposition in San Antonio Bay, Texas. M, 1973, Texas A&M University.

Hall, Gary Owen. Stratigraphy and geologic history of the Cannonball Formation (Paleocene). M, 1958, University of North Dakota. 64 p.

Hall, George Frederick. Geomorphology and soils of the Iowan-Kansan border area, Tama County, Iowa. D, 1965, Iowa State University of Science and Technology. 300 p.

Hall, George Ian. A study of the Precambrian greenstone in northeastern Marinette County, Wisconsin. M, 1971, University of Wisconsin-Milwaukee.

Hall, George Waverly Briggs, Jr. Geology of the Preston Hollow Quadrangle, Dallas, Collin, and Denton counties, Texas. M, 1953, Southern Methodist University. 11 p.

Hall, Henry Thompson. Structural analysis of a polymetamorphic tectonite along the South Fork of the Clearwater River, Idaho. M, 1961, University of Idaho. 38 p.

Hall, Henry Thompson. The systems Ag-Sb-S, Ag-As-S, and Ag-Bi-S; phase relations and mineralogical significance. D, 1966, Brown University. 180 p.

Hall, Hubert H. Mississippian stratigraphy in southwestern Alberta and northwestern Montana. D, 1952, University of Wisconsin-Madison. 110 p.

Hall, J. D. The geology of the lower "A" ore-body, Waite-Amulet (Quebec). M, 1939, McGill University.

Hall, Jack Charles. Conodonts and conodont biostratigraphy of the Middle Ordovician in the western overthrust region and Sequatchie Valley of the Southern Appalachians. D, 1986, Ohio State University. 362 p.

Hall, Jack Charles. Conodonts from the Centerfield and Stone Mill limestone members of the Ludlowville Formation (Hamilton Group, Middle Devonian) of central New York State. M, 1981, University of North Carolina, Chapel Hill. 96 p.

Hall, James Creevey. A numerical model of the dynamics of large ice sheets. D, 1987, University of Massachusetts. 497 p.

Hall, James Monroe, III. Geology of the northwestern part of the Casper Arch. M, 1962, University of Kentucky. 71 p.

Hall, Jeffrey D. Devonian-lowermost Mississippian lithostratigraphy and conodont biostratigraphy, northern Arkansas. M, 1978, University of Arkansas, Fayetteville.

Hall, Jennifer Lynn. Processes leading to the formation of tectonic features on the Moon and Mars. D, 1985, Massachusetts Institute of Technology. 308 p.

Hall, John F., Jr. Paleokarst and other dissolution features of the Devonian Dyer and Mississippian Leadville formations, central Colorado. M, 1987, Colorado School of Mines. 142 p.

Hall, John Frederick. The geology of southern Hocking County, Ohio. D, 1951, Ohio State University. 255 p.

Hall, John Kendrick. Arctic Ocean geophysical studies; the Alpha cordillera and Mendeleyev ridge. D, 1970, Columbia University. 125 p.

Hall, John W. Louisiana survey streams; their antecedents, distribution and characteristics. D, 1970, Louisiana State University.

Hall, John Whitling. Lithology of Missouri south of the Missouri River. M, 1963, Southern Illinois University, Carbondale. 61 p.

Hall, Johnny L. Paleoecology and age of the upper Eocene Basilosaurus octoides beds of Louisiana, Mississippi and southwestern Alabama. M, 1976, Northeast Louisiana University.

Hall, Kenneth McCoy. A mathematical model to describe the consolidation process in fine-grained soils. D, 1969, Arizona State University. 197 p.

Hall, Leo Matthew. Contact metamorphism in the Peacham area, northeastern Vermont. M, 1956, University of Cincinnati. 73 p.

Hall, Leo Matthew. Geology of the St. Johnsbury Quadrangle, Vermont-New Hampshire. D, 1959, Harvard University.

Hall, Louis S., III. Lead and zinc contamination of ground and surface waters at the municipal landfill of Danbury, Connecticut. M, 1982, Western Connecticut State University. 35 p.

Hall, M. M. Geographical analysis of West Virginia motor vehicle accidents; socio-economic and environmental implications. M, 1976, West Virginia University. 155 p.

Hall, Mark H. Structural geology of the Fairbanks mining district, central Alaska. M, 1985, University of Alaska, Fairbanks. 68 p.

Hall, Mary Jo. The distribution of sediments and adsorbed trace metals on the inner continental shelf off

southern New Jersey. D, 1981, Lehigh University. 219 p.

Hall, Michael Scott. Oblique slip faults in the northwestern Picuris Mountains of New Mexico; an expansion of the Embudo transform zone. M, 1988, University of Texas, Austin. 69 p.

Hall, Minard Lane. Chemical relations between coexisting minerals in a progressively metamorphosed pelite. D, 1969, Case Western Reserve University. 292 p.

Hall, Minard Lane. Intrusive truncation of the Precambrian-Cambrian succession of the White Mountains, California. M, 1964, University of California, Berkeley. 93 p.

Hall, Monte R. Cathode luminescence of cassiterite; an electron microprobe study. M, 1968, Virginia Polytechnic Institute and State University.

Hall, Morris D. Gravity geophysical delineation of buried channel aquifers in northern Missouri. M, 1979, University of Missouri, Columbia.

Hall, Nelson Timothy. Late Quaternary history of the eastern Pleito thrust fault, northern Transverse Ranges, California. D, 1984, Stanford University. 314 p.

Hall, Nelson Timothy. Petrology of the type Merced Group (Pliocene, upper and Pleistocene), San Francisco Peninsula, California. M, 1965, University of California, Berkeley. 127 p.

Hall, Pamela S. Deformation and metamorphism of the aluminous schist member of the Setters Formation, Cockeysville, Maryland. M, 1988, West Virginia University. 99 p.

Hall, Peter C. Some aspects of deformation fabrics along the highland/lowland boundary, northwestern Adirondacks, New York State. M, 1984, SUNY at Albany. 124 p.

Hall, R. E. Scattering of Rossby waves by topography in a stratified ocean. D, 1976, [University of California, San Diego]. 126 p.

Hall, Richard Drummond. Metamorphism of sulfide schists, Limerick Township, Ontario. D, 1980, University of Western Ontario. 441 p.

Hall, Richard J. Petrology of diamond drill core from Judith Peak-Red Mountain area, Fergus County, Montana. M, 1976, Eastern Washington University. 38 p.

Hall, Richard P. Distribution of sediments at Cheat Lake, West Virginia. M, 1966, West Virginia University.

Hall, Robert Arthur. Physical considerations for land use planning of an area immediately west of Manhattan, Kansas. M, 1979, Kansas State University. 72 p.

Hall, Robert B. Stibnite deposits of Sevier County, Arkansas. M, 1940, Northwestern University.

Hall, Robert Dean. Heavy minerals in Recent alluvium along the eastern flank of the Front Range, Golden to the Colorado-Wyoming line. M, 1966, University of Colorado.

Hall, Robert Dean. Sedimentation and alteration of loess in southwestern Indiana. D, 1973, Indiana University, Bloomington. 103 p.

Hall, Robert Forrest, III. Scattering elastic waves from inhomogeneities embedded in a layered half-space. D, 1981, University of California, Berkeley. 169 p.

Hall, Robert G. Investigation of porosity of unconsolidated sands determined by various logging tools. M, 1959, University of Houston.

Hall, Robert Lynn. Spring stiffnesses for beam-column analysis of soil-structure interaction problems. D, 1984, Oklahoma State University. 121 p.

Hall, Rowland L. Radiological and environmental assessment of abondoned uranium mines in the Edgemont mining district; South Dakota. M, 1982, South Dakota School of Mines & Technology.

Hall, Roy Homes. The Devonian and associated rocks of parts of Cole and Moniteau counties, Missouri. M, 1921, University of Missouri, Columbia.

Hall, Russell Lindsay. Lower Bajocian (Jurassic) ammonoid faunas of the western Americas. D, 1976, McMaster University. 239 p.

Hall, S. T. Mineralogy, chemistry, and petrogenesis of some hypabyssal intrusions, Highland County, Virginia. M, 1975, Virginia Polytechnic Institute and State University.

Hall, Stephen Austin. Paleoecological interpretation of bison, mollusks, and pollen from the Hughes Peat Bed (Quaternary), Linn County, Iowa. M, 1971, University of Iowa. 75 p.

Hall, Stephen Austin. Stratigraphy and palynology of Quaternary alluvium at Chaco Canyon, New Mexico. D, 1975, University of Michigan. 87 p.

Hall, Stephen Harvey, Jr. The triclinic crystal structure of amesite. M, 1974, University of Wisconsin-Madison. 60 p.

Hall, Stephen J. Mineralogical and geochemical properties of the overburden and underclays associated with the Danville Coal Member (VII) in western Indiana. M, 1979, Indiana University, Bloomington. 42 p.

Hall, Steven D. Potential for selenium migration at a lignite power plant solid waste disposal facility. M, 1986, Texas A&M University. 198 p.

Hall, Susan Margaret. Metallogenesis and stable isotope systematics of stratabound lead/silver deposits in the Precambrian Noonday Dolomite, Death Valley, California. M, 1984, University of California, Davis. 133 p.

Hall, Sylvia Duncan. A microscopic study of the tin ores of the Huanuni Mine, Bolivia, South America. M, 1954, University of Michigan.

Hall, Ward Lee. Geology of Maverick area, Culberson and Reeves counties, Texas. M, 1952, University of Texas, Austin.

Hall, Wayne Everett. Structure and ore deposits of the Darwin Quadrangle, Inyo County, California. D, 1958, Harvard University.

Hall, William B. Geology of part of the upper Gallatin Valley of southwestern Montana. D, 1961, University of Wyoming. 239 p.

Hall, William B. Processes of slope retreat and pediment formation in the Rock Springs region of Wyoming. M, 1951, University of Cincinnati. 44 p.

Hall, William Douglas. Hydrogeologic significance of depositional systems and facies in lower Cretaceous sandstones, north-central Texas. M, 1974, University of Texas, Austin.

Hall, William Gordon. The kinetics of leaching of organic carbon from in-situ spent oil shale. D, 1982, University of California, Berkeley. 235 p.

Hall-Beyer, Barton MacNeill. Geochemistry of some ocean floor basalts of central B.C. M, 1976, University of Alberta. 103 p.

Hall-Beyer, Mryka Christine. Chemical petrology of some northern Saskatchewan granulites. M, 1976, University of Alberta. 120 p.

Hall-Burr, Marty Joanne. Geology of a portion of the Gun Creek area, northern Sierra Anchas, Gila County, Arizona. M, 1982, Northern Arizona University. 207 p.

Halla, Marilyn Margaret. Chemical changes during early diagenesis of benthic algal mats and their significance for petroleum formation. M, 1978, Colorado School of Mines. 45 p.

Halladay, Christopher R. Petrology and geochemistry of certain altered ultramafic deposits in southeastern Vermont. M, 1972, Lehigh University.

Hallager, William Sherman. Geology of Archean gold-bearing metasediments near Jardine, Montana. D, 1980, University of California, Berkeley. 140 p.

Hallam, Susan Lee. Deposition and diagenesis of the Abo Formation at the Empire Abo Field, Eddy County, New Mexico. M, 1982, University of Texas, Austin. 222 p.

Hallberg, G. R. Application of satellite remote sensing to Midwest Quaternary geology. D, 1975, University of Iowa. 192 p.

Hallberg, Jack Arthur. Small pegmatite bodies in the Hartland Formation of Connecticut. M, 1965, Stanford University.

Halle, Richard E. Stratigraphy and depositional environments of the Wilcox Group (Paleocene) at the Oxbow Mine site, northwestern Louisiana. M, 1981, University of North Dakota. 62 p.

Halleck, Phillip Michael. The compression and compressibility of grossular garnet, a comparison of X-ray and ultrasonic methods. D, 1973, University of Chicago. 82 p.

Hallee, Mark C. The geology of the central part of the Benton Range, east-central California. M, 1986, University of Nevada. 123 p.

Haller, Charles Regis. Devonian-Mississippian boundary relationships in central Missouri. M, 1957, University of Missouri, Columbia.

Haller, Charles Regis. Neogene foraminiferal faunas of Humboldt Basin, California. D, 1967, University of California, Berkeley. 204 p.

Haller, Kathleen Monger. Segmentation of the Lemhi and Beaverhead faults, east-central Idaho, and Red Rock Fault, Southwest Montana, during the late Quaternary. M, 1988, University of Colorado. 141 p.

Haller, M. C. The geology of the Amity Copper Property, Boston Creek, Ontario. M, 1928, University of Toronto.

Hallet, B. Nature and effects of chemical processes at the base of temperate glaciers. D, 1975, University of California, Los Angeles. 109 p.

Halley, Robert Bruce. Paleo-environmental interpretations of the upper Cambrian cryptalgal limestone of New York State. M, 1971, Brown University.

Halley, Robert Bruce. Repetitive carbonate bank development and subsequent terrigenous inundation; Cambrian Carrara Formation, southern Great Basin. D, 1974, SUNY at Stony Brook.

Hallford, C. M. Petrography and structure of the Saganaga granite (Precambrian), Saganaga-Northern Light Lakes area, Minnesota-Ontario (Canada). M, 1969, SUNY at Stony Brook.

Hallfrisch, Michael Paul. Unconfined sand aquifer characteristics of a forested and nonforested area, Maumee State Forest, Fulton County, Ohio. M, 1987, University of Toledo. 146 p.

Hallgarth, Walter Ervin. A foraminiferal study of the relationship of the Burditt Marl to the Austin Chalk and Taylor Marl in the vicinity of Austin, Texas. M, 1959, University of Tulsa. 33 p.

Halliday, Mark Everett. Gravity and ground magnetic surveys in the Monroe and Joseph Known Geothermal Resource areas and surrounding region, south-central Utah. M, 1978, University of Utah. 164 p.

Hallin, James S. Heat flow and radioactivity studies in Colorado and Utah, 1971-1972. M, 1973, University of Wyoming. 108 p.

Halliwell, Beverly Ann. Deep-water carbonate deposits of the southern margin of the Jurassic Central High Atlas Trough, Morocco. M, 1985, Colorado School of Mines. 235 p.

Hallman, Leon Charles. Landscape change through recreational use; Sam Rayburn and Toledo Bend reservoir lakes, East Texas. D, 1973, University of California, Los Angeles.

Hallock, Allan Richard. The geology of a portion of the Horseshoe Hills, Montana. M, 1955, Montana College of Mineral Science & Technology. 72 p.

Hallock, Donald H. V. Physiographic history of the Laramie Range. M, 1933, University of Wyoming. 50 p.

Hallock, Waite D. Tectonic history of a portion of Lincoln County, New Mexico. M, 1970, Virginia State University. 52 p.

Hallof, Phillip G. On the interpretation of resistivity and induced polarization results. D, 1957, Massachusetts Institute of Technology. 216 p.

Hallquist, John Berger. Micro-delay blasting and the resulting vibration phenomena in soil. M, 1961, Washington University. 23 p.

Halls, Henry Campbell. Geological interpretation of geophysical data from the Lake Superior region. D, 1970, University of Toronto.

Hallstein, William Weyrich. Insoluble residues of the Grand Tower Limestone. M, 1952, University of Illinois, Urbana. 14 p.

Halper, Fern Beth. The effect of storms on sediment resuspension and transport on the outer continental shelf, Northwest Gulf of Mexico. D, 1984, Texas A&M University. 145 p.

Halperin, Alan D. A marine/nonmarine transition in the Upper Devonian West Falls Group of south-central New York. M, 1986, SUNY at Binghamton. 72 p.

Halperin, Henry Ira. An investigation of mineral-kerogen interactions and their relation to petroleum genesis. D, 1981, University of California, Los Angeles. 278 p.

Halpern, Joseph B. Mineralization at Guanacevi, Durango, Mexico. M, 1940, Columbia University, Teachers College.

Halpern, Martin. Cretaceous sedimentation in Base O'Higgins area of Northwest Antarctic Peninsula. D, 1963, University of Wisconsin-Madison. 126 p.

Halpern, Martin. Sedimentation and stratigraphy of the Drayton Valley, North Pembina, Violet Grove, and Pembina blocks; Pembina oil field, west central Alberta. M, 1961, University of Wisconsin-Madison.

Halpern, Yvonne. The effect of lithology, diagenesis and low-grade metamorphism on the ultrastructure and surface sculpture of acritarchs from the late Proterozoic Chuar Group, Grand Canyon, Arizona. M, 1988, Tulane University. 133 p.

Halpin, David Lawrence. The geometry of folds and style of deformation in the Quabbin Hill area, Massachusetts. M, 1965, University of Massachusetts. 93 p.

Halsey, Jonathan Horace. Geology of parts of the Bridgeport, (California) and Wellington, (Nevada) quadrangles. D, 1953, University of California, Berkeley. 506 p.

Halsey, Ramond E. A study of the near-surface seismic velocities of the Melville area (Louisiana). M, 1947, Louisiana State University.

Halsey, Susan D. Late Quaternary geologic history and morphologic development of the barrier island system along the Delmarva Peninsula of the Mid-Atlantic Bight. D, 1978, University of Delaware. 608 p.

Halsey, Susan Dana. Distribution and significance of marine boring endolithic algae and fungi in sediments of the North and South Carolina continental margin. M, 1970, Duke University.

Halstead, Perry Neil. The geology of the northern third of the Lyon's Quadrangle, Oregon. M, 1955, University of Oregon. 87 p.

Haltenhoff, Frederick W. Geology of the Great Valley Sequence and related rocks in a portion of the Dublin 7 1/2-minute Quadrangle, California. M, 1978, San Jose State University.

Halter, Clarence R. Phosphate deposits of Florida. M, 1927, Columbia University, Teachers College.

Halter, Eric Francis. Glacial dispersal and surficial geological studies; Fish River Lake and Umsaskis Lake luadrangles, northern Maine. M, 1985, SUNY at Buffalo. 135 p.

Haltie, Ian Edward. The geology and geochemistry of a Robinson Creek gold occurrence, north of Amisk Lake, Saskatchewan. M, 1987, University of Saskatchewan. 176 p.

Halunen, A. J. Heat flow in the western equatorial Pacific Ocean. M, 1972, University of Hawaii. 42 p.

Halva, Carroll J. A geochemical investigation of "basalts" in southern Utah. M, 1961, University of Arizona.

Halvatzis, Gregory James. The use of pH to determine the percentage of portland cement required for stabilization of tropical and subtropical soils. M, 1975, University of Florida. 89 p.

Halverson, Mark O. Regional gravity survey of northwestern Utah and part of southern Idaho. M, 1961, University of Utah. 33 p.

Halverson, Ward D. A study of plasma columns in a longitudinal magnetic field. D, 1965, Massachusetts Institute of Technology.

Halvorson, Don Llewellyn. Geology and petrology of the Devils Tower, Missouri Buttes, and Barlow Canyon area, Crook County, Wyoming. D, 1980, University of North Dakota. 218 p.

Halvorson, James Walter. Depositional history and subsurface correlation of lithofacies of the Castle Reef Dolomite in the northern Montana disturbed belt. M, 1982, Montana College of Mineral Science & Technology. 52 p.

Halvorson, Phyllis Heather Fett. An exploration gravity survey in the San Pedro Valley, southeastern Arizona. M, 1984, University of Arizona. 69 p.

Halwas, David Bruce. The intrusion and deformation history of the Kinnaird Gneiss, southern British Columbia. M, 1986, University of Calgary. 155 p.

Ham, Cornelius Kimball. The geology of a portion of the Las Trampas Ridge Quadrangle, California. M, 1951, University of California, Berkeley. 97 p.

Ham, Herbert Hoover. The ground-water geology of the southwestern quarter of the Eugene Quadrangle, Oregon. M, 1961, University of Oregon. 90 p.

Ham, Linda J. The mineralogy, petrology, and geochemistry of the Halfway Cove-Queensport Pluton, Nova Scotia, Canada. M, 1988, Dalhousie University. 294 p.

Ham, William Eugene. Geology and petrology of the Arbuckle Limestone in the southern Arbuckle mountains, Oklahoma. D, 1950, Yale University. 229 p.

Ham, William Eugene. Origin and age of the Pawhuska Rock Plain of Oklahoma and Kansas. M, 1939, University of Oklahoma. 50 p.

Hamann, Richard John. Solubility of galena in alkaline sulfur-rich NaCl solutions at 25° C. M, 1973, University of Toronto.

Hamann, Walter Edward. Geology and geochemistry of the Big Horn gold mine, San Gabriel Mountains, Southern California. M, 1985, University of California, Los Angeles. 78 p.

Hamati, R. E. Effects of uniform and non-uniform seismic disturbances on a long multi-span highway bridge. D, 1976, University of California, Berkeley. 397 p.

Hamberg, Lawrence Roger. Paleontology and stratigraphy of two well cores in the Devonian rocks of Bay and Arenac counties in Michigan. M, 1953, University of Michigan.

Hamberger, Kimball Lee. The relation between porosity, permeability, and porous geometry in reservoir rock. M, 1957, University of Oklahoma. 106 p.

Hambleton, Arthur W. Interpretation of the palaeoenvironment of several Missourian carbonate sections in Socorro County, New Mexico by carbonate fabrics. M, 1959, New Mexico Institute of Mining and Technology. 87 p.

Hambleton, Harvey Jay. Petrographic studies of the Clinton carbonate rocks of western New York and the Niagara region of Ontario. M, 1958, Pennsylvania State University, University Park. 126 p.

Hambleton, Thomas. The geology of a portion of southeastern Callaway County, Missouri. M, 1951, University of Missouri, Columbia.

Hambleton, William W. A petrofabric study of layering in the Stillwater Complex, Montana. M, 1947, Northwestern University. 63 p.

Hambleton, William W. Petrographic study of Kansas coals. D, 1951, University of Kansas. 142 p.

Hamblin, Alden H. Paleogeography and paleoecology of the Myton Pocket, Uinta Basin, Utah (Uinta Formation-upper Eocene). M, 1987, Brigham Young University. 60 p.

Hamblin, Anthony P. Sedimentology of a prograding shallow marine slope and shelf sequence, Upper Jurassic Fernie-Kootenay transition, southern Front Ranges. M, 1978, McMaster University. 196 p.

Hamblin, Ralph Hugh. Stratigraphy and insoluble residues of the upper Paleozoic formations of Montana. M, 1939, Montana College of Mineral Science & Technology. 63 p.

Hamblin, Russell D. The stratigraphy and depositional environments of the Gebel El-Rus area, eastern Faiyum, Egypt. M, 1987, Brigham Young University. 83 p.

Hamblin, William Kenneth. Geology and groundwater of northern Davis County, Utah. M, 1954, Brigham Young University. 51 p.

Hamblin, William Kenneth. The Cambrian sandstone of northern Michigan. D, 1958, University of Michigan.

Hambrick, Dixie Ann. Geochemistry and structure of Tertiary volcanic rocks in the southwestern Monte Cristo Range, Nevada. M, 1984, University of Arizona. 140 p.

Hambrick, Gordon A., III. Effect of sediment pH and oxidation-reduction potential on microbial degradation of crude oil. M, 1979, Louisiana State University.

Hamburger, G. E. The crystal structure of tourmaline. D, 1948, Massachusetts Institute of Technology. 92 p.

Hamburger, Michael Wile. Seismicity of the Fiji Islands and tectonics of the Southwest Pacific. D, 1986, Cornell University. 332 p.

Hamburger, Michael Wile. Seismotectonics of the northern Philippine Island Arc. M, 1982, Cornell University. 67 p.

Hamdan, Abdul R. A. Ecology and distribution of Recent foraminifera on the Scotian Shelf. D, 1971, Queen's University. 190 p.

Hamdan, Abdul-Latif. Ground-water hydrology of Iroquois County (Illinois). M, 1970, University of Illinois, Urbana. 73 p.

Hamed, Jafar Gholi Arkani see Arkani Hamed, Jafar Gholi

Hamel, Cathy. Paleontology and biostratigraphy of the Mastapoka Group, (Aphebian) Richmond Gulf area, northern Quebec. M, 1985, Universite de Montreal.

Hamel, James Victor. Stability of slopes in soft, altered rocks. D, 1970, University of Pittsburgh. 342 p.

Hamell, Richard David. Stratigraphy, petrology, and paleoenvironmental interpretation of the Bertie Group (late Cayugan) in New York State. M, 1981, University of Rochester. 89 p.

Hames, Willis Emory. Timing and extent of Caledonian Orogeny within a portion of the Western Gneiss Terrane, Senja, Northern Norway. M, 1988, University of Georgia. 132 p.

Hamidzada, Nasir. Petrology, geochemistry and structure of the northeastern part of the Scituate Pluton, Rhode Island. M, 1988, University of Rhode Island.

Hamil, Brenton M.; Boydston, Donald and Santos, Elmer S. Geology of the east central portion of the Huerfano Quadrangle, Huerfano County, Colorado. M, 1954, University of Michigan.

Hamil, Brenton McCreary. Trace elements in accessory magnetite from Basin and Range quartz monzonites. D, 1967, University of Utah. 241 p.

Hamil, David F. A detailed chemical analysis for calcium and magnesium of the Sun Oil Company, Peterson-Howard Well core sample. M, 1961, Michigan State University. 34 p.

Hamil, Martha M. Breccias of the Manitou Springs area, Colorado. M, 1965, Louisiana State University.

Hamil, Martha M. Metamorphic and structural environment of Copper Mountain, Wyoming. D, 1971, University of Missouri, Columbia. 121 p.

Hamill, Gary Bruce. Quaternary interpretation of Southwest Tofino Basin, Pacific margin, Canada. M, 1982, University of Calgary. 148 p.

Hamilton, C. G. Structure in the Wabush Formation west of Wabush Lake with special reference to the Smallwood Mine. M, 1967, Carleton University. 82 p.

Hamilton, Carter John. Geology of the Maxon gas field, Pike County, Kentucky. M, 1958, University of Kentucky. 64 p.

Hamilton, Charles L. Geology of the Nancy Pegmatites, North Groton, New Hampshire. M, 1954, Dartmouth College. 42 p.

Hamilton, Charles L. Geology of the Peaks of Otter area, Bedford and Botetourt counties, Virginia. D, 1964, Virginia Polytechnic Institute and State University.

Hamilton, Charles T. A stratigraphic investigation of the Hot Springs Sandstone and Ten Mile Creek Formation in the southern Ouachita Mountains of Arkansas. M, 1973, Northern Illinois University. 93 p.

Hamilton, Daniel Kirk. Some solutional features of the limestone near Lexington, Kentucky. M, 1946, University of Kentucky. 29 p.

Hamilton, Daniel Kirk. The occurrence of ground water in the inner Bluegrass region, Kentucky. D, 1949, University of North Carolina, Chapel Hill. 80 p.

Hamilton, David Bennett. Plant succession and the influence of disturbance in the Okefenokee Swamp, Georgia. D, 1982, University of Georgia. 277 p.

Hamilton, David P. Hydrocarbon entrapment in Bradshaw gas field, Hamilton County, Kansas. M, 1976, West Texas State University. 29 p.

Hamilton, Douglas Holmes. Geology of the proposed Joel McCrea Reservoir site and vicinity, Ventura County, California. M, 1962, Stanford University. 130 p.

Hamilton, Douglas Holmes. The tectonic boundary of coastal Central California. D, 1984, Stanford University. 309 p.

Hamilton, Edward A. The sedimentary of the Upper Jurassic formations in the vicinity of Escalante, Utah. D, 1949, The Johns Hopkins University.

Hamilton, Edwin Lee. Sunken islands of the Mid-Pacific Mountains. D, 1951, Stanford University. 215 p.

Hamilton, H. E. Stratigraphy and depositional history of the Muddy Formation in the Freezeout Mountains area, northeastern Carbon County, Wyoming. M, 1964, University of Wyoming. 125 p.

Hamilton, Irving B. Ostracodes of the upper Atoka and Cherokee Group of northeastern Oklahoma. M, 1941, University of Tulsa. 34 p.

Hamilton, James A. Sedimentation of Hatches Harbor, Cape Cod, Massachusetts. M, 1978, University of Massachusetts. 96 p.

Hamilton, James Allen. Geology of the Caribou Mountain area, Bonneville and Caribou counties, Idaho. M, 1961, University of Idaho. 59 p.

Hamilton, James E. Use of pressure decline data. M, 1933, University of Pittsburgh.

Hamilton, Jean L. Shallow subsurface stratigraphy of Shackleford Banks, North Carolina. M, 1977, Duke University. 83 p.

Hamilton, John Bonar. Correlation of the Pennsylvanian rocks in the western part of the central Pennsylvanian Basin of New Brunswick by means of fossil spores. M, 1962, University of New Brunswick.

Hamilton, John Richard. Incipient metamorphism and the organic geochemistry of the Mancos Shale (Cretaceous) near Crested Butte, Colorado. M, 1972, Rice University. 64 p.

Hamilton, John Vernon. The structural and stratigraphic setting of gold mineralization in the vicinity of Larder Lake, southcentral Abitibi greenstone belt, northeastern Ontario. M, 1986, Queen's University. 156 p.

Hamilton, K. Hydrology and hydrogeochemistry of natural water from granitic terrane; central Sierra Nevada, Fresno County, California. M, 1978, California State University, Fresno.

Hamilton, Michael Miller. The paleoecology and stratigraphy of the Perth Limestone (Pennsylvanian) and underlying shales in Warren County, Indiana. M, 1975, Indiana University, Bloomington. 89 p.

Hamilton, Neil W. Geology of a portion of the Pinon Range, Pine Valley and Jiggs quadrangles, Nevada. M, 1956, University of California, Los Angeles.

Hamilton, Neil W. Geology of the Smith Creek area, Eureka and Elko counties, Nevada. M, 1956, University of California, Los Angeles.

Hamilton, Patricia A. Paleoecology; Devonian fauna. M, 1971, Colorado State University. 76 p.

Hamilton, Patrick. A geophysical study of the Manix Fault; its tectonic relationship to the western Mohave Desert, San Bernardino County, California. M, 1976, California State University, Los Angeles.

Hamilton, Paul Lawrence. Bottom sediments and the Cape Hatteras suspended sediment plume. M, 1973, Duke University. 200 p.

Hamilton, Peggy Kay. Mineral deposits of Mexico. M, 1947, Columbia University, Teachers College.

Hamilton, Richard C. Small non-marine gastropods from the Paleocene of Wyoming and Utah. M, 1955, University of Missouri, Columbia.

Hamilton, Robert Bruce. Geology of a portion of the Sonoma Volcanics near Mount Saint Helena. M, 1973, University of California, Berkeley. 39 p.

Hamilton, Robert D. The crystal structure of junitoite, $CaZn_2Si_2O_7\cdot H_2O$ and the crystallography and crystal chemistry of zinc silicates. D, 1977, Colorado School of Mines. 66 p.

Hamilton, Robert David. An investigation of two aspects of phosphorite genesis. M, 1973, Iowa State University of Science and Technology.

Hamilton, Robert Gilbert. Metamorphosed pre-Cambrian calcareous concretions in the Black Hills. M, 1933, University of Iowa. 24 p.

Hamilton, Robert Gilbert. Pre-Cambrian geology of the Keystone District, Black Hills, South Dakota. D, 1935, University of Iowa. 41 p.

Hamilton, Robert Morrison. The effects of frequency, temperature, and pressure on the electrical conductivity of periclase [MgO] and olivine [(Mg, Fe)$_2$SiO$_4$]. D, 1965, University of California, Berkeley. 123 p.

Hamilton, Samuel Clinton. Structure of southern Sierra Pilares, Municipio de Ojinaga, Chihuahua, Mexico. M, 1961, University of Texas, Austin.

Hamilton, Stanley Kerry. Copper mineralization in the upper part of the Copper Harbor Conglomerate (Precambrian) at White Pine, Michigan. M, 1965, University of Wisconsin-Madison.

Hamilton, Stanley Kerry. Factors influencing investment and production in the Peruvian mining industry 1940-1965. D, 1967, University of Wisconsin-Madison.

Hamilton, Thomas D. Quantitative study of mass movements, southwestern Wisconsin. M, 1963, University of Wisconsin-Madison.

Hamilton, Thomas Dudley. Geomorphology and glacial history of the Alatna Valley, northern Alaska. D, 1966, University of Washington. 264 p.

Hamilton, Thomas M. Groundwater flow in part of the Little Missouri Basin, North Dakota. D, 1970, University of North Dakota. 179 p.

Hamilton, Thomas M. Recent fluvial geology in western North Dakota. M, 1967, University of North Dakota. 99 p.

Hamilton, Timothy Scott. The petrology of the North Doherty mafic sill, Jefferson County, Montana. M, 1974, Indiana University, Bloomington. 117 p.

Hamilton, Walter M. Systematic variability of the petrography within a surface mine of the Waynesburg Coal. M, 1978, West Virginia University.

Hamilton, Warren Bell. Granitic rocks of the Huntington Lake area, Fresno County, California. D, 1951, University of California, Los Angeles.

Hamilton, Warren Bell. The geology of the Dessa Dawn Mountains, British Columbia. M, 1949, University of Southern California.

Hamilton, Wayne Lee. Microparticle deposition on polar ice sheets. D, 1969, Ohio State University.

Hamilton, William L. An investigation into modeling snow cover elements at Crater Lake National Park and surrounding environs as an improved "ground truth" method for satellite snow observations. D, 1981, Oregon State University. 112 p.

Hamilton, William, Jr. Areal geology of the Farview area, Major County, Oklahoma. M, 1961, University of Oklahoma. 73 p.

Hamilton, Wylie Norman. Mineralogy of Bearpaw sediments in the South Saskatchewan River valley. M, 1962, University of Saskatchewan. 50 p.

Hamilton-Smith, Terence. Sedimentation during the Taconic orogeny of late Ordovician and early Silurian rocks of the Siegas area, New Bruswick (Canada). M, 1969, Massachusetts Institute of Technology. 188 p.

Hamlen, Dale Alexander. Geology of the Middlesex Quadrangle, New York. M, 1959, University of Rochester. 102 p.

Hamlin, Herbert Scott. Depositional and groundwater flow systems of the Carrizo-upper Wilcox in South Texas. M, 1984, University of Texas, Austin. 142 p.

Hamlin, J. S. Structure of the upper Monterey submarine fan valley. M, 1974, Naval Postgraduate School.

Hamlin, Scott Norman. Platinum occurrence in serpentinite along the Melones fault zone near Foresthill, California. M, 1977, University of Nevada. 43 p.

Hamlin, Tracy L. Rhyolites in the western Mexican volcanic belt. M, 1986, University of New Orleans. 108 p.

Hamlin, William Henry. Geology and foraminifera of the Mount Walker-Quilcene-Leland Lake area, Jefferson County, Washington. M, 1962, University of Washington. 127 p.

Hammack, Joseph Leonard, Jr. Tsunamis; a model of their generation and propagation. D, 1972, California Institute of Technology. 261 p.

Hamman, Henry Royden. Geology of the Trinity and Fredericksburg groups, Erath County, Texas. M, 1963, University of Texas, Austin.

Hammatt, Hallett H. Late Quaternary stratigraphy and archaeological chronology in the Lower Granite Reservoir area, Lower Snake River, Washington. D, 1976, Washington State University. 272 p.

Hammel, David J. An application of the finite element method for rock slope stability analysis. D, 1971, University of Arizona.

Hammel, David J. The influence of mining and geologic variables on pit slope stability. M, 1967, University of Arizona.

Hammell, Laurence. Petrofabric studies in the Splitrock Pond area, Morris County, New Jersey. M, 1960, Rutgers, The State University, New Brunswick. 50 p.

Hammen, John Leo, III. Image processing of Landsat MSS data for buried valley detection in northern Indiana. D, 1988, Indiana State University. 167 p.

Hammer, Donald F. Geology and ore deposits of the Jackrabbit area, Pinal County, Arizona. M, 1961, University of Arizona.

Hammer, Jay A. Response of adenosine triphosphate pool and growth rate of marine microalgae to water solubles of crude and refined oils. M, 1978, Florida State University.

Hammer, Richard D. Soil morphology, soil water, and forest tree growth on three Cumberland Plateau landtypes. D, 1986, University of Tennessee, Knoxville. 319 p.

Hammer, W. R. A revision of the temnospondyl amphibian family Trematosauridae. D, 1979, Wayne State University. 128 p.

Hammerand, Veral Franklin. Geology and petrology of a part of Paradise Ridge in northwestern Idaho. M, 1936, University of Idaho. 25 p.

Hammermeister, Dale P. Water and anion movement in selected soils of western Oregon. D, 1978, Oregon State University. 270 p.

Hammerquist, Donald William. Relationship of seismic velocities and of certain mechanical properties of soils. M, 1952, Washington University. 25 p.

Hammerstrom, Lyle Thomas. Internally consistent thermodynamic data and phase relations in the CaO-Al$_2$O$_3$-SiO$_2$-H$_2$O system. M, 1981, University of British Columbia. 62 p.

Hammes, Richard Robert. Paleogeographic analysis of the Cretaceous in the Western Interior of the United States. M, 1958, University of Wisconsin-Madison. 71 p.

Hammes, Richard Robert. Stratigraphy of the Wesley Formation (Mississippian) western Ouachita Mountains, southeastern Oklahoma. D, 1965, University of Wisconsin-Madison.

Hammill, Gilmore Semmes, IV. Structure and stratigraphy of the Mount Shader Quadrangle, Nye County, Nevada-Inyo County, California. D, 1966, Rice University. 130 p.

Hammill, Gilmore Semmes, IV. The radioactivity, accessory minerals, and possibilities for absolute dating of bentonites. M, 1958, Rice University. 136 p.

Hammill, Robert W. Pleistocene mollusks of southeastern South Dakota. M, 1960, University of South Dakota. 78 p.

Hammitt, Jay W. Geology, petrology, and mineralization of the Paisley Mountains plutonic complex, Lake County, Oregon. M, 1977, Oregon State University. 162 p.

Hammitt, Ray Wesley. The geology and origin of copper deposits near Keating, Oregon. M, 1972, University of Oregon. 116 p.

Hammond, Barry M. The optimization of line spacings in prospecting. M, 1961, University of Western Ontario.

Hammond, Becky Jane. Analysis of the Grand Wash-Reef Reservoir-Gunlock fault zone, Washington County, Utah and Mohave County, Arizona. M, 1988, Brigham Young University. 57 p.

Hammond, Carol J. Effects of mica on unit weight and effective angle of internal friction of Ottawa silica sand. M, 1985, University of Idaho. 245 p.

Hammond, Charles M. Navajo Gap area between the Ladron Mountains and Mesa Sarca, Socorro County, NM; structural analysis. M, 1987, New Mexico Institute of Mining and Technology. 212 p.

Hammond, Charles R. The geology of the east flank of the Laramie Range in the vicinity of Iron Mountain, Wyoming. M, 1949, University of Wyoming. 63 p.

Hammond, D. A. Radiometric ages of selected Hawaiian corals. D, 1971, University of Hawaii. 166 p.

Hammond, David Richard. A petrofabric study of the Tintic Quartzite (lower and middle Cambrian) in the area of the Willard thrust, northern Utah. M, 1971, University of Utah. 511 p.

Hammond, Douglas E. Dissolved gases and kinetic processes in the Hudson River estuary. D, 1975, Columbia University. 161 p.

Hammond, Douglas E. The pressure-volume relationship of CsI up to 250 KB at room temperature. M, 1969, University of Rochester. 42 p.

Hammond, Janet G. Environmental significance of trace elements in the Eocene of the Ojai area, Ventura County, California. M, 1977, California State University, Long Beach. 106 p.

Hammond, Janet Louise Griswold. Late Precambrian diabase intrusions in the southern Death Valley region, California; their petrology, geochemistry, and tectonic significance. D, 1983, University of Southern California.

Hammond, L. L. The characterization and classification of the soils of Christmas Island. M, 1969, University of Hawaii.

Hammond, Patrick Allen. The kinetics of reequilibration of coexisting ilmenites and titano-magnetites. M, 1981, University of Tennessee, Knoxville. 42 p.

Hammond, Paul Ellsworth. Geology of the lower Santiago Creek area, San Emigdio Mountains, Kern

County, California. M, 1958, University of California, Los Angeles.

Hammond, Paul Ellsworth. Structure and stratigraphy of the Keechelus Volcanic Group and associated Tertiary rocks in the west-central Cascade Range, Washington. D, 1963, University of Washington. 264 p.

Hammond, Roger Darril. Geochronology and origin of Archean rocks in Marquette County, Upper Michigan. M, 1979, University of Kansas. 108 p.

Hammond, S. R. Paleomagnetic investigations of deep borings on the Ewa Plain, Oahu, Hawaii. M, 1970, University of Hawaii. 60 p.

Hammond, S. R. Paleomagnetism of western Equatorial Pacific sediment cores. D, 1974, University of Hawaii. 200 p.

Hammond, Scott Alan. Ground water iron concentration gradient and bedrock surface delineation in the Little Miami buried valley aquifer, Yellow Springs, Ohio. M, 1986, Wright State University. 177 p.

Hammond, W. P. Geology of the Monchalagan Lake area, Saguenay County, Quebec. M, 1946, University of Toronto.

Hammond, Weldon W. The Spiriferidae of the Madison Formation of the Logan Quadrangle, Utah. M, 1930, University of Missouri, Columbia.

Hammond, Weldon Woolf, Jr. Ground water resources of Matagorda County, Texas. M, 1969, University of Texas, Austin.

Hammond, Weldon Woolf, Jr. Hydrogeology of the lower Glen Rose Aquifer, south-central Texas. D, 1984, University of Texas, Austin. 344 p.

Hammuda, Khalifa Salem. A regional study of the Paleocene strata with emphasis on the oil producing horizons in western Sirte Basin, Libya. M, 1983, Ohio University, Athens. 91 p.

Hammuda, Omar Suleiman. Eocene biostratigraphy of the Derna area, Northeast Libya. D, 1973, University of Colorado.

Hammuda, Omar Suleiman. Jurassic and Lower Cretaceous geology of the central Jebel Nefusa, northwestern Libya. M, 1967, University of Colorado. 87 p.

Hamner, Edward J. The relation between closure and productivity of sands in the Mid-Continent and Gulf coast oil fields. M, 1932, University of Oklahoma. 54 p.

Hamon, J. Hill. Osteology and paleontology of the passerine birds of the Reddick Pleistocene. D, 1961, University of Florida. 293 p.

Hamou, Taric Aly Hadj see Hadj Hamou, Taric Aly

Hamp, Lonn P. The petrology of the late Proterozoic(?)-Early Cambrian Arumbera Sandstone, western MacDonnell Ranges, north-central Amadeus Basin, central Australia. M, 1985, Utah State University. 306 p.

Hampf, Andrew W. The geology of the Golden Rule Mine, Santa Cruz County, Arizona. M, 1972, University of Arizona.

Hampstead, Howard A. The geology of the southeastern portion of the Fulton Quadrangle, Missouri. M, 1953, University of Missouri, Columbia.

Hampton, Bret D. Carbonate sedimentology of the Manzanita Member of the Cherry Canyon Formation (middle Guadalupian, Permian), Guadalupe Mountains, West Texas. M, 1983, University of Wisconsin-Madison. 178 p.

Hampton, Donald Arthur. Geochemistry of the saline and carbonate minerals of Sevier Lake playa, Millard County, Utah. M, 1978, University of Utah. 75 p.

Hampton, Eugene R. Geology of the southeast one-quarter of the Tyee Quadrangle and part of the adjacent Sutherlin Quadrangle, Oregon. M, 1958, Oregon State University. 55 p.

Hampton, George Lee, III. Stratigraphy and archaeocyathans of Lower Cambrian strata of Old Douglas Mountain, Stevens County, Washington. M, 1978, Brigham Young University. 86 p.

Hampton, George R. Fresh water ice shelf fabrics of a nearshore area in Lake Michigan. M, 1974, University of Wisconsin-Milwaukee.

Hampton, Loyd Donald. Acoustic properties of sediments. D, 1967, Texas A&M University. 88 p.

Hampton, Mark W. Relationships between low flow stream conditions and geologic and physiographic characteristics of drainage basins in an alpine environment. M, 1986, University of Wisconsin-Milwaukee. 105 p.

Hampton, Monty Allen. Subaqueous debris flow and generation of turbidity currents. D, 1970, Stanford University. 180 p.

Hampton, N. F. A computerized methodology for multiobjective analyses as an aid to resource development planning. M, 1973, University of Arizona.

Hampton, O. Winston. Methods and cost of uranium exploration and mining on the Colorado Plateau. M, 1955, University of Colorado.

Hampton, O. Winston. Structural geology of the foothills region from Plainview to Golden (Jefferson County), Colorado. M, 1957, University of Colorado.

Hamric, Burt Ervin. Petrology and trace element geochemistry of the DeQueen Formation (Cretaceous), southwest Arkansas. D, 1965, University of Oklahoma. 166 p.

Hamric, Burt Ervin. Subsurface geology of the Hoffman gas area, Okmulgee and McIntosh counties, Oklahoma. M, 1961, University of Oklahoma. 52 p.

Hamrick, R. J. Dolomitizaion patterns in the Walker oil field, Kent and Ottawa counties, Michigan. M, 1978, Michigan State University. 86 p.

Hamroush, Hany Ahmed. Archaeological geochemistry of Hierakonpolis in the Nile Valley, Egypt. D, 1985, University of Virginia. 332 p.

Hamtak, Frank James. Correlation of strain and tilt episodes with local precipitation. M, 1976, University of Utah. 63 p.

Hamyouni, Ezzidin Ahmed. Subsurface study of the Upper Cretaceous Anf Formation, central Sirte Basin, Libya. M, 1981, South Dakota School of Mines & Technology. 84 p.

Hamza, Mokhtar S. A. Surface effect on isotopic fractionation between CO_2, water vapor and carbonate. D, 1972, Columbia University. 100 p.

Hamza, Valiya Mannathal. Vertical distribution of radioactive heat production in the Grenville geological province and the sedimentary sections overlying it. D, 1973, University of Western Ontario.

Hamzawi, Anwar T. Gravity survey of the Denver-Golden area, Colorado. M, 1966, Colorado School of Mines. 144 p.

Han, Daesuk. Mineralogy and paragenesis of the Cave-in-Rock fluorspar district, Hardin County, Illinois. M, 1967, University of Missouri, Rolla.

Han, De-hua. Effects of porosity and clay content on acoustic properties of sandstones and unconsolidated sediments. D, 1987, Stanford University. 219 p.

Han, Dongyup. Disintegration of Catskill-type graywacke. M, 1968, Rensselaer Polytechnic Institute. 39 p.

Han, G. C. Salt balance and exchange in the Rhode River, a tributary embayment to the Chesapeake Bay. D, 1974, The Johns Hopkins University. 200 p.

Han, Jong Hwan. Genetic stratigraphy and associated growth structures of the Vicksburg Formation, South Texas. D, 1981, University of Texas, Austin. 178 p.

Han, Myung Woo. Dynamics and chemistry of pore fluids in marine sediments of different tectonic settings; Oregon subduction zone and Bransfield Strait extensional basin. D, 1988, Oregon State University. 280 p.

Han, Soong Soo. Analysis of decoupling effect on particle velocity in underground blasting. M, 1987, Colorado School of Mines. 111 p.

Han, Tsu-ming. Ore deposits of Lo Aguirre, San Antonio and Africana mining prospects near Santiago, Chile. M, 1949, University of Cincinnati. 53 p.

Han, Uk. A preliminary evaluation of geothermal potential of the Republic of Korea with emphasis on the Haewundae hot spring. M, 1979, University of Utah. 90 p.

Han, Uk. Preliminary evaluation of geothermal potential of the Republic of Korea with emphasis on the Haewundae Hot Springs. M, 1979, University of Utah. 90 p.

Hanagan, E. J. Some Permian mollusks from Wyoming. M, 1956, University of Wyoming. 159 p.

Hanagan, Elizabeth Jean. Morphology and ontogeny of some ostracods from the Bromide Formation (Middle Ordovician) of the Simpson Group of Oklahoma. D, 1963, University of Illinois, Urbana. 174 p.

Hanai, Tetsuro. Studies on the Ostracoda from Japan; I, Subfamilies Leptocytherinae, n. subfam., "Toulminiinae," n. subfam., and Cytherurinae G. W. Muller. D, 1956, Louisiana State University. 142 p.

Hanan, B. B. Geochemistry and petrology of the Baltimore Complex. M, 1976, Virginia Polytechnic Institute and State University.

Hanan, Barry Benton. The petrology and geochemistry of the Baltimore mafic complex, Maryland. D, 1980, Virginia Polytechnic Institute and State University. 247 p.

Hanan, Mark Allen. Geochemistry and mobility in sediments of radium from oil-field brines, Grand Bay, Plaquemines Parish, Louisiana. M, 1981, University of New Orleans. 89 p.

Hanbury, Jonathan B. Hydrothermal alteration and mineralization of the Silver Creek porphyry copper deposit, Park County, Wyoming. M, 1982, Washington State University. 44 p.

Hanbury, Patricia Melling. A petrographic and petrochemical comparison of exotic gabbroic inclusions of the Sudbury sublayer and Nipissing Diabase-Sudbury Gabbro. M, 1982, Washington State University. 50 p.

Hance, James Harold. Geology and geography southwest of Abilene, Texas. D, 1918, University of Chicago. 261 p.

Hance, James Harold. Geology and mineral resources of the Wellsville Quadrangle, Ohio. D, 1923, University of Chicago.

Hancharik, Joan. Facies analysis and petroleum potential of the Upper Jurassic Smackover Formation, western and northern areas, East Texas basin. M, 1981, Baylor University. 88 p.

Hancock, James Martin. The Dale oil field, Caldwell County, Texas. M, 1930, University of Texas, Austin.

Hancock, James Martin, Jr. Areal geology of the Chewy-Watts area, Adair County, Oklahoma (including a part of western Arkansas). M, 1963, University of Oklahoma. 72 p.

Hancock, Kenneth J. An accuracy analysis of photogrammetric mapping in hilly wooded areas. M, 1961, Ohio State University.

Hancock, L. T. Diatomite. M, 1932, Acadia University.

Hancock, Michael Curtis. Analysis of water resources problems using electronic spreadsheets. M, 1986, University of Florida. 198 p.

Hancock, William Tarrant, Jr. Paleozoic geology of the Round Mountain area, Blanco Quadrangle, Texas. M, 1929, University of Texas, Austin.

Hancox, Gregory A. An investigation of temporal variability in the surface charge of suspended particles in the Oyster River, Strafford County, NH. M, 1986, University of New Hampshire. 102 p.

Hand, Bryce M. Marine geology of north end of San Diego Trough. M, 1962, University of Southern California.

Hand, Bryce Moyer. Hydrodynamics of beach and dune sedimentation. D, 1964, Pennsylvania State University, University Park. 172 p.

Hand, David. Geology of the Corral Canyon area, Churchill County, Nevada. M, 1955, University of Nevada. 32 p.

Hand, Forrest E., Jr. Ground water resources in the northern part of the Glen Canyon National recreation area and adjacent lands west of the Colorado and Green rivers, Utah. M, 1979, University of Wyoming. 44 p.

Hand, H. D. The stratigraphy of the Mississippian rocks of the southeastern part of Wyoming. M, 1938, University of Wyoming. 33 p.

Hand, John E. Radon flux at the earth-air interface. M, 1958, New Mexico Institute of Mining and Technology. 46 p.

Hand, Linda Mimura. Seismic model study of fault and fold geometries. M, 1988, Texas A&M University. 111 p.

Hand, Perry A. Geology of Inman-Webb-Sawmill Creek area, Bannock County, Idaho. M, 1978, Idaho State University. 49 p.

Handel (Tutt), Donna J. A petrographic study of the chert within the upper Ocala Limestone at a selected site in central Marion County, Florida. M, 1981, University of Florida. 88 p.

Handfield, Robert C. Archaeocyatha (early Cambrian, Sekivi Formation, Atan group) from the Mackenzie and Cassiar mountains, western Canada. D, 1969, Princeton University. 162 p.

Handford, Charles R. Sedimentology and diagenesis of the Monteagle Limestone (upper Mississippian); a high energy oolitic carbonate sequence in the southern Appalachian Mountains. D, 1976, Louisiana State University. 222 p.

Handford, Lincoln S. Rb-Sr whole rock age study of the Andover (northeastern Massachusetts) and Chelmsford granites. M, 1966, Boston University. 11 p.

Handin, John Walter. The source, transportation, and deposition of beach sediment in Southern California. D, 1949, University of California, Los Angeles.

Handley, Bruce. The structure and stratigraphy of the Hecla Hoek Sequence (Precambrian) of western Midterhuken Peninsula, Spitsbergen. M, 1983, University of Wisconsin-Madison. 97 p.

Handley, Howard W. Certain rocks of the Cascade Mountains (andesites, etc.). M, 1931, University of Oregon. 107 p.

Handman, Elinor H. An assessment of water quality in the Quinnipiac River basin, Connecticut. M, 1976, Wesleyan University.

Handschumacher, David W. Post-Eocene plate tectonics of the eastern Pacific. D, 1975, University of Hawaii. 81 p.

Handschy, James W. The geology and tectonic history of south-central Sierra del Cuervo, Chihuahua, Mexico. M, 1986, University of Texas at El Paso.

Handverger, Paul A. Geology of the Three R Mine, Palmetto mining district, Santa Cruz County, Arizona. M, 1963, University of Arizona.

Handwerk, Roger H. Basin analysis of upper Middle Ordovician strata in southwestern Virginia and northeastern Tennessee. M, 1981, Ohio University, Athens. 118 p.

Handy Barringer, Julia L. Geochemistry and petrology of mafic and intermediate meta-igneous rocks from the eastern Maryland Piedmont; structural and tectonic implications. D, 1983, University of Pennsylvania. 407 p.

Handy, Richard L. Petrography of selected southwestern Iowa loess samples. M, 1955, Iowa State University of Science and Technology.

Handy, Richard L. Stabilization of Iowa loess with Portland cement. D, 1956, Iowa State University of Science and Technology.

Handy, Walter A. Depositional history and diagenesis of lacustrine and fluvial sedimentary rocks of the Turners Falls and Mount Toby transition, north-central Massachusetts. M, 1976, University of Massachusetts. 115 p.

Haneberg, William C. Fractures in the Cambrian Rome Formation near Wytheville, Virginia. M, 1985, University of Cincinnati. 65 p.

Haner, Barbara E. Geomorphology and sedimentary characteristics of Redondo submarine fan, southern California. M, 1970, University of Southern California.

Haner, Barbara Elizabeth. Quaternary geomorphic surfaces on the northern Perris Block, Riverside County, California; interrelationship of soils, vegetation, climate and tectonics. D, 1982, University of Southern California.

Hanes, John A. An ^{40}Ar/^{39}Ar geochronological study of Precambrian diabase dykes. D, 1979, University of Toronto.

Hanes, John A. Petrographic and thermomagnetic study of Blake River Group mafic volcanics. M, 1973, University of Toronto.

Hanes, S. D. Structurally preserved Lepidodendracean cones from the Pennsylvanian of North America. D, 197?, [Ohio University, Athens]. 115 p.

Haney, Donald C. The soil profile-bedrock relationships of Garrard County, Kentucky. M, 1962, University of Kentucky. 71 p.

Haney, Donald Clay. Structural geology along a segment of the Saltville Fault, Hawkins County, Tennessee. D, 1966, University of Tennessee, Knoxville. 133 p.

Haney, Eileen M. Depositional history of the Upper Triassic Shinarump Member of the Chinle Formation, Circle Cliffs, Utah. M, 1987, Northern Arizona University. 102 p.

Haney, Joseph M. Geology of the McKinley Lake gold prospect area, Chugach National Forest, south-central Alaska. M, 1982, New Mexico Institute of Mining and Technology. 107 p.

Haney, Warren Dale. Petrology of the Chemung Formation above the Bradford third sand from core of the Summit Well (Pennsylvania). M, 1952, Pennsylvania State University, University Park. 197 p.

Hanford, Charles Robert. Geology of the Osage SW Quadrangle, Arkansas. M, 1969, University of Arkansas, Fayetteville.

Hang, Pham Thi. The hydrous magnesium nickel silicates; the garnerites. D, 1972, Pennsylvania State University, University Park. 111 p.

Hangari, Khaled. Mineralogy of the Wedi Shatti iron ore of southern Libya. M, 1974, University of Minnesota, Minneapolis. 98 p.

Hangari, Khaled M. Geology and stable isotope geochemistry of Upper Devonian ironstones from Wadi Shatti District, southern Libya. D, 1981, South Dakota School of Mines & Technology.

Hanger, Rex Alan. Biotic communities and brachiopod paleoecology of the Early Permian McCloud Formation, Northern California. M, 1986, Texas A&M University. 127 p.

Hanif, Muhammad. Statistical relations of hydrologic behavior in northern Pakistan. D, 1987, Colorado State University. 221 p.

Haniman, Kurt Christopher. Computer modeling of the Phillipston Reservoir magnetic anomaly, central Massachusetts. M, 1980, University of Massachusetts. 85 p.

Hanish, Mark Burton. Peridotite xenoliths from Bultfontein and Jagersfontein mines, South Africa; a case for pre-kimberlite serpentinization. M, 1984, Queen's University. 121 p.

Hank, Robert Allen. Strontium and oxygen isotope study of lower-crustal xenoliths, Royal Society Range - McMurdo Sound region, Antarctica. M, 1987, Northern Illinois University. 85 p.

Hanke, Harold Wayne. Subsurface stratigraphic analysis of the Cherokee Group (Pennsylvanian, Des Moines Series) in north central Creek County, Oklahoma. M, 1967, University of Oklahoma. 47 p.

Hankins, David D. Geologic setting of precious metals mineralization, King of Arizona District, Kofa Mountains, Yuma County, Arizona. M, 1984, San Diego State University. 135 p.

Hankins, Donald Wayne. Use of superposition and the extended pulse model to evaluate the contaminant transport parameters of variably source-loaded plumes. M, 1988, Texas A&M University. 102 p.

Hankins, John B. Geology and hydrogeology of an upland gravel-filled basin, Valentine Meadow, Mansfield, Connecticut. M, 1985, University of Connecticut. 168 p.

Hankinson, Peter Kent. Provenance of the upper Jackfork Sandstone of Arkansas and Oklahoma. M, 1986, University of New Orleans. 125 p.

Hanks, Catherine Leigh. The emplacement history of the Tom Martin ultramafic complex and associated metamorphic rocks, North-central Klamath Mountains, California. M, 1981, University of Washington. 112 p.

Hanks, Keith Lynn. Geology of the central House Range area, Millard County, Utah. M, 1962, Brigham Young University. 136 p.

Hanks, Teddy L. Geology and coal deposits, Ragged-Chair Mountain area, Pitkin and Gunnison counties, Colorado. M, 1962, Brigham Young University. 160 p.

Hanks, Thomas Colgrove. A contribution to the determination and interpretation of seismic source parameters. D, 1972, California Institute of Technology. 184 p.

Hanley, Christine Naomi. Modeling oceanic crustal magnetization in the Gulf of California. M, 1980, University of Washington. 74 p.

Hanley, J. B. Geology of the Poland Quadrangle. D, 1939, The Johns Hopkins University.

Hanley, J. H. Systematics, paleoecology, and biostratigraphy of nonmarine Mollusca from the Green River and Wasatch formations (Eocene), southwestern Wyoming and northwestern Colorado. D, 1974, University of Wyoming. 313 p.

Hanley, J. T. Structural geomorphology of the Catawba Mountain knolls, Roanoke County, Virginia. M, 1976, Syracuse University.

Hanley, Thomas Brainard. Structure and petrology of the northwestern Tobacco Root Mountains, Madison County, Montana. D, 1975, Indiana University, Bloomington. 289 p.

Hanlon, Edward A., Jr. Multiple element soil extractants and a data management system. D, 1983, [Oklahoma State University]. 222 p.

Hann, Megan. Petroleum potential of Niobrara Formation in Denver Basin. M, 1981, Colorado State University. 275 p.

Hanna, Abdulaziz Yalda. Effect of landscape position and aspect on available soil water and water recharge in Southeast Nebraska. D, 1982, University of Nebraska, Lincoln. 144 p.

Hanna, Augustine Booya. Mineralogical analyses of a Brown soil and a Chestnut soil of the Republic of Iraq. D, 1961, University of Wisconsin-Madison. 159 p.

Hanna, F. M. The geomorphology of clay surface playas. D, 1978, University of California, Los Angeles. 499 p.

Hanna, John Clark. Temporal framework of late Neogene pelagic and bioclastic sediments, Grand Cayman Island, B.W.I. M, 1978, Louisiana State University.

Hanna, Marcus Albert. An Eocene invertebrate fauna from the La Jolla Quadrangle, California. D, 1927, University of California, Berkeley. 398 p.

Hanna, Ronald E. Magnetic/nonmagnetic mineralogy and geochemistry of selected crystalline rocks from the San Andreas fault system. M, 1982, Kent State University, Kent. 117 p.

Hanna, Sara Ross. Hydrogeology and water resources of Bedford, Massachusetts. M, 1980, Boston University. 109 p.

Hanna, Thomas Murray. The geology and geochemistry of the Summit Creek molybdenum prospect, Custer County, Idaho. M, 1983, Western Michigan University.

Hanna, William Francis. Magnetic properties of selected volcanic rocks of southwestern Montana. D, 1965, Indiana University, Bloomington. 207 p.

Hannah, G. J. Raymond. Petrology of the Macquereau Series (Gaspe Peninsula, Quebec). D, 1954, Universite Laval.

Hannah, G. J. Raymond. The origin of the metasomatic iron formation at Old Chelsa, Quebec. M, 1952, Universite Laval.

Hannah, Judith Louise. Stratigraphy, petrology, paleomagnetism, and tectonics of Paleozoic arc complexes, northern Sierra Nevada, California. D, 1980, University of California, Davis. 323 p.

Hannah, William G. A geologic report of the Gypsum Canyon area in the northern Santa Ana Mountains of California. M, 1951, University of Southern California.

Hannan, Andrew E. A sedimentary and petrographic study of the Fort Payne Formation, Lower Mississippian (Osage), in south-central Kentucky. M, 1975, University of Cincinnati. 111 p.

Hannan, Selim Sarwar. An experimental study of fracture closure in elastically and non-elastically deformable rocks. D, 1988, University of Toronto.

Hanneman, Debbie L. Paleoecological study of the Banff Formation, Ram Mountain, Alberta. M, 1977, University of Calgary. 149 p.

Hanners, Albert James. The Pleistocene geology of Vilas County, Wisconsin. M, 1941, University of Wisconsin-Madison. 34 p.

Hanneson, Donna L. Deposition and diagenesis of a Middle to Upper Devonian carbonate buildup, Kaybob South area, Alberta. M, 1981, University of Manitoba. 234 p.

Hanneson, James Edward. The horizontal loop EM response of a thin vertical conductor in a conductive half-space. D, 1981, University of Manitoba.

Hannibal, Joseph Timothy. Systematics and functional morphology of the oniscomorph millipedes (Arthropoda, Diplopoda) from the Carboniferous of North America. M, 1980, Kent State University, Kent. 64 p.

Hannifan, Martin K. Caving process of Andes copper. M, 1948, University of Nevada - Mackay School of Mines. 36 p.

Hannington, Mark Donald. Geology, mineralogy, and geochemistry of a silica-sulfate-sulfide deposit, Axial Seamount, N.E. Pacific Ocean. M, 1986, University of Toronto.

Hannon, Willard James, Jr. Some effects of a layered system on dilational waves. D, 1964, St. Louis University. 114 p.

Hannoura, A. A. Numerical and experimental modelling of unsteady flow in rockfill embankments. D, 1978, [University of Windsor].

Hannum, Cheryl Ann. Clastic pipes and dikes of Kodachrome Basin area, Utah. M, 1979, Brigham Young University.

Hanor, Jeffrey S. The origin of barite. D, 1967, Harvard University.

Hanscom, Roger Herbert. The crystal chemistry and polymorphism of chloritoid. D, 1973, Harvard University.

Hansel, Ardith Kay. Sinkhole form as an indicator of process in karst landscape evolution. D, 1980, University of Illinois, Urbana. 175 p.

Hansell, James Myron. A cross-section in the Keweenawan of Wisconsin. M, 1926, University of Wisconsin-Madison.

Hansell, James Myron. Glacial geology of an area in the northwest corner of Wisconsin. D, 1930, University of Wisconsin-Madison.

Hanselman, David Henry. Depositional environments in the Upper Precambrian Ocoee Series of central eastern Tennessee. D, 1972, University of South Carolina. 130 p.

Hansen, Alan R. Regional stratigraphic analysis of the Uinta Basin, Utah and Colorado. M, 1952, Northwestern University.

Hansen, Alan Ray. Stratigraphic analysis of the Mississippian Mission Canyon Formation, Williston Basin. D, 1959, University of Utah. 95 p.

Hansen, Beauford Victor. The stratigraphy and petrogenesis of the Hogshooter Formation in northeastern Oklahoma. M, 1957, University of Tulsa. 198 p.

Hansen, Beverly J. Sedimentology and reservoir potential of the Lower Amaranth Member, southwestern Manitoba. M, 1988, University of Manitoba. 179 p.

Hansen, Brian G. Evaluating the hydrogeology of Meade County, Kansas, using vertical variability methods and numerical modeling. M, 1988, Colorado School of Mines. 173 p.

Hansen, Charles Allan. Subsurface Virgilian and lower Permian arkosic facies, Wichita Uplift-Anadarko Basin. M, 1978, Oklahoma State University. 63 p.

Hansen, Chris D. Geology of the Jump Creek 7 1/2′ Quadrangle, Carbon County, Utah. M, 1988, Brigham Young University. 53 p.

Hansen, Dan E. Subsurface correlations of the Cretaceous Greenhorn-Lakota interval in North Dakota; a study in facies. M, 1955, University of North Dakota. 86 p.

Hansen, Daniel Lloyd. The distribution of Cherokee Sandstones, Marion County, Iowa. M, 1978, Iowa State University of Science and Technology.

Hansen, David Ernest. A groundwater model with stochastic inflows for evaluating the feasibility of artificial groundwater recharge in the Weber Delta area, Ogden, Utah. D, 1983, Utah State University. 240 p.

Hansen, Diana Kay Thomas. Petrology, physical stratigraphy and biostratigraphy of the lower Sierra Madre Limestone, Cretaceous, west central Chiapas, Mexico. M, 1988, University of Texas, Arlington. 137 p.

Hansen, Don Allred. Electromagnetic prospecting with application to ground water problems. D, 1953, University of California, Los Angeles.

Hansen, Donald L. Distribution of Ostracoda in the Decorah Shale Formation at St. Paul, Minnesota. M, 1951, University of Minnesota, Minneapolis. 12 p.

Hansen, Donald S. Tritium movement in the unsaturated zone, Nevada Test Site. M, 1978, University of Nevada. 107 p.

Hansen, Donald V. Similarity solutions for salt balance and circulation in partially mixed estuaries. D, 1964, University of Washington. 76 p.

Hansen, Dorrell Reed, III. Sediment input parameters for ROSED, a road sediment erosion model. M, 1985, University of Idaho. 125 p.

Hansen, Edward Carlton. Strain facies of the metamorphic rocks in Trollheimen, Norway. D, 1963, Yale University.

Hansen, Edward Conrad, III. The granulite-grade metamorphism in southern Karnataka, South India; the role of the fluid base. D, 1983, University of Chicago. 275 p.

Hansen, Francis Dale. Semibrittle creep of selected crustal rocks at 1000 MPa. D, 1982, Texas A&M University. 243 p.

Hansen, George H. The Cretaceous geology of south central New Mexico and adjoining regions. D, 1927, George Washington University.

Hansen, Harry J. Pleistocene stratigraphy of the Salisbury area, Maryland, and its relationship to the Eastern Shore; a subsurface approach. D, 1967, Columbia University. 55 p.

Hansen, Harry J., III. Geology of the Big Creek area, Toiyabe Range, Lander County, Nevada. M, 1960, Columbia University, Teachers College.

Hansen, Henry Eugene. Study of the slope stability of a proposed road cut in a future route of U.S. 95. M, 1977, University of Idaho. 84 p.

Hansen, Henry Paul. Postglacial forest succession and climate in the Puget Sound region. D, 1937, University of Washington. 43 p.

Hansen, James C. A study of a portion of the post-Madison strata of the Moore area, central Montana. M, 1959, University of Oklahoma. 106 p.

Hansen, John Andrew, Jr. Geologic and engineering properties of till and loess, southeast Iowa. M, 1958, Iowa State University of Science and Technology.

Hansen, John Kenneth. The distribution of gamma radiation in the surficial deposits of the Florida Panhandle. M, 1988, University of Florida. 113 p.

Hansen, Kevin. A seismic study of aquifers in southern Delaware. M, 1988, University of Delaware. 150 p.

Hansen, Kirk S. Calculations of normal modes for the American Mediterranean Seas. M, 1974, University of Chicago. 51 p.

Hansen, Kirk S. Secular effects of oceanic tidal dissipation on the Moon's orbit and the Earth's rotation. D, 1981, University of Chicago. 110 p.

Hansen, Lawrence. The evolution of metamorphic textures in former clastic sedimentary rocks from the Black Hills, South Dakota. M, 1982, University of Akron. 59 p.

Hansen, Leon Alden. Geology and geochemical exploration at the Bristol silver mine, Lincoln County, Nevada. M, 1967, University of Utah. 105 p.

Hansen, Marcia Elaine. Optimization of sedimentation field-flow fractionation for the analyses of river water particulates. D, 1987, University of Utah. 209 p.

Hansen, Mayer G. A magnetite pegmatite deposit in St. Louis County, Minnesota. M, 1922, University of Minnesota, Minneapolis. 52 p.

Hansen, Michael Christian. Microscopic chondrichthyan remains from Pennsylvanian marine rocks of Ohio and adjacent areas; (Volumes I and II). D, 1986, Ohio State University. 559 p.

Hansen, Michael Christian. Paleoecology of the Portersville Shale in Athens County, Ohio. M, 1973, Ohio University, Athens. 163 p.

Hansen, Michael Wayne. Petrology of loesses in west-central Illinois. M, 1970, University of North Carolina, Chapel Hill. 64 p.

Hansen, Olav Louis. Thermal radiation from the Galilean satellites measured at 10 and 20 microns. D, 1972, California Institute of Technology. 93 p.

Hansen, Peter A. Geology of the Blackleaf Canyon area, Heart Butte Quadrangle, Montana. M, 1960, Washington State University.

Hansen, Peter Allen. Geology of the Blackleaf Canyon area, Heart Butte Quadrangle, Montana. M, 1962, Washington State University. 104 p.

Hansen, Peter Michael. Structure and stratigraphy of the Lemhi Pass area, Beaverhead Range, Southwest Montana and east-central Idaho. M, 1983, Pennsylvania State University, University Park. 112 p.

Hansen, Richard Otto. Isotopic distribution of uranium and thorium in soils weathered from granite and alluvium. D, 1965, University of California, Berkeley. 126 p.

Hansen, Robert F., Jr. Areal geology of the Southwest Mangum area, Oklahoma. M, 1958, University of Oklahoma. 101 p.

Hansen, Roger Alan. Observational study of terrestrial eigenvibrations. D, 1981, University of California, Berkeley. 135 p.

Hansen, Roger Dennis. A multivariate analysis of municipal water use in Utah. D, 1981, Utah State University. 90 p.

Hansen, Ronald Lee. Structural analysis of the Willard thrust fault near Ogden Canyon, Utah. M, 1980, University of Utah. 82 p.

Hansen, Spenst Mitchell. The geology of the Eldorado mining district, Clark County, Nevada. D, 1962, University of Missouri, Rolla. 328 p.

Hansen, Stephanie Thomas. A geochemical study of gem grossular garnets. M, 1986, University of New Orleans. 125 p.

Hansen, Steven C. The economic geology of the Wikieup prospect, Mohave County, Arizona. D, 1977, University of Idaho. 97 p.

Hansen, Steven Charles. Geology of the southwestern part of the Randolph Quadrangle, Utah-Wyoming. M, 1964, Utah State University. 50 p.

Hansen, Steven Michael. High-resolution study of geophysical methods; case study of the Spraberry Formation, Midland Basin, Texas. M, 1988, University of Texas of the Permian Basin. 96 p.

Hansen, Susan Sharp. Regional groundwater flow and its effect on tilted reservoir oil/water contacts in the Big Horn Basin of Wyoming and Montana. M, 1988, Stanford University. 95 p.

Hansen, T. A. Ecological control of evolutionary rates in Paleocene-Eocene marine molluscs. D, 1978, Yale University. 323 p.

Hansen, Terry Jay. Dissolution in the Hutchinson Salt Member of the Wellington Formation near Russell, Kansas. M, 1977, Kansas State University. 61 p.

Hansen, Thomas J. Re-analyses of ground water in Chase County, Kansas. M, 1969, Kansas State University. 28 p.

Hansen, Vicki L. Kinematic interpretation of mylonitic rocks in Okanogan Dome, north-central Washington, and implications for dome evolution. M, 1983, University of Montana. 53 p.

Hansen, Vicki Lynn. Structural, metamorphic, and geochronologic evolution of the Teslin suture zone, Yukon; evidence for Mesozoic oblique convergence outboard of the northern Canadian Cordillera. D, 1987, University of California, Los Angeles. 261 p.

Hansen, William Bradley. Spring water sampling in the vicinity of the North Coker Orebody, Southwest Wisconsin. M, 1979, University of Wisconsin-Madison.

Hansen, William Walter. The recharge and recovery constraints on optimal groundwater management. D, 1977, University of California, Berkeley. 191 p.

Hansen-Bristow, Katherine Jane. Environmental controls influencing the altitude and form of the forest-alpine tundra ecotone, Colorado Front Range. D, 1981, University of Colorado. 274 p.

Hanshaw, Bruce Busser. Membrane properties of compacted clays. D, 1962, Harvard University. 113 p.

Hanshaw, Bruce Busser. Structural geology of the west side of the Gore Range, Eagle County, Colorado. M, 1958, University of Colorado.

Hansink, James D. Geology of the Graniteville Granite (Precambrian), Missouri. M, 1965, St. Louis University.

Hansman, Robert Herbert. Coiled Pennsylvanian nautiloids of North America. D, 1958, University of Iowa. 494 p.

Hansman, Robert Herbert. Straight Pennsylvanian orthochoanitic nautiloids of North America. M, 1955, University of Iowa. 73 p.

Hansmire, W. H. Field measurements of ground displacements about a tunnel in soil. D, 1975, University of Illinois, Urbana. 357 p.

Hanson, Alfred Kenneth, Jr. The distribution and biogeochemistry of transition metal-organic complexes in marine waters. D, 1981, University of Rhode Island. 191 p.

Hanson, Alvin M. Geology of the southern Malad Range and vicinity in northern Utah. D, 1949, University of Wisconsin-Madison.

Hanson, Alvin M. Phosphate deposits in western Summit, Wasatch, Salt Lake, Morgan, and Weber counties, Utah. M, 1942, Utah State University. 44 p.

Hanson, B. V. The stratigraphy, provenance, age, and depositional environment of East central Iowa loesses. D, 1976, University of Iowa. 187 p.

Hanson, Bernold M. Geology of the Elkhorn Ranch area, Billings and Golden Valley counties, North Dakota. M, 1954, University of Wyoming. 152 p.

Hanson, Bradford C. A fracture pattern analysis employing small scale photography with emphasis on groundwater movement in Northwest Arkansas. M, 1973, University of Arkansas, Fayetteville.

Hanson, Bruce Vernor. Quaternary geology of the Willow Spring archaeological site, southwestern Wyoming. M, 1969, University of Wyoming. 60 p.

Hanson, C. Bruce. Variation in dentition of Hyracodon, (Badlands, South Dakota). M, 1968, South Dakota School of Mines & Technology.

Hanson, Carl R. Bedrock geology of the Shigar Valley area, Skardu, northern Pakistan. M, 1986, Dartmouth College. 124 p.

Hanson, Clifford Gail. Geochemical and mineralogical investigations and methods for detecting kimberlite in the area of the Stockdale Kimberlite, Riley County, Kansas. M, 1980, Pennsylvania State University, University Park. 118 p.

Hanson, David L. Rimforest landslides, San Bernardino Mountains, Southern California. M, 1988, Loma Linda University. 38 p.

Hanson, David W. Surface and subsurface geology of the Simi Valley area, Ventura County, California. M, 1982, Oregon State University. 112 p.

Hanson, Douglas Wade. Multi-parameter seismic inversion using the Born approximation. D, 1984, University of Wyoming. 169 p.

Hanson, E. Late Woodfordian drainage history in the lower Mohawk Valley. M, 1977, Rensselaer Polytechnic Institute.

Hanson, Gary M. Phosphorus distribution in unconsolidated sediments of the Great Bay estuary, New Hampshire. M, 1973, University of New Hampshire. 74 p.

Hanson, George. Some Canadian occurrences of pyritic deposits in metamorphic rocks. D, 1920, Massachusetts Institute of Technology. 113 p.

Hanson, George. The Star Lake gold district of Manitoba. M, 1915, University of Manitoba.

Hanson, George F. Some observations on the sediments of University Bay, Lake Mendota, Madison, Wisconsin. M, 1952, University of Wisconsin-Madison.

Hanson, Gilbert Nikolai. The contact metamorphic effect of the Duluth Gabbro upon the Rb-Sr age of the biotites of the Snowbank Stock. M, 1962, University of Minnesota, Minneapolis. 52 p.

Hanson, Gilbert Nikolai. The effect of contact metamorphism on mineral ages in the Snowbank Lake area, Minnesota, and in the Beartooth Mountains, Wyoming. D, 1964, University of Minnesota, Minneapolis. 130 p.

Hanson, Henry William Andrew, III. Petrogenesis of Graywacke in a mid-Paleozoic, northern Appalachian Epieugeosyncline. D, 1968, Pennsylvania State University, University Park. 177 p.

Hanson, Henry William Andrew, III. Petrography of the Dunn Brook Formation (Ordovician-Silurian) and geology of the Carlisle Pond area, Maine. M, 1965, Pennsylvania State University, University Park. 98 p.

Hanson, Hiram Stanley. A study of the relative frequency of feldspar twin types in crystalline rocks. M, 1958, Emory University. 46 p.

Hanson, Hiram Stanley. Petrography and structure of the Leatherwood Quartz Diorite, Santa Catalina Mountains, Pima County, Arizona. D, 1966, University of Arizona. 125 p.

Hanson, James Phillip. Modelling the hydraulics of flows and sediment transport of a Missoula flood in the Pasco Basin, Washington. M, 1986, Kent State University, Kent. 156 p.

Hanson, Jonathan M. Heat transfer effects in forced geoheat recovery systems. D, 1978, Oregon State University. 217 p.

Hanson, Kathryn Lee. The Quaternary and environmental geology of the Uncas-Port Ludlow area, Jefferson County, Washington. M, 1977, University of Oregon. 87 p.

Hanson, Kenneth E. A seismic refraction profile and crustal structure in central interior Alaska. M, 1968, University of Alaska, Fairbanks. 59 p.

Hanson, Larry. Size distribution of the White River Ash, Yukon Territory. M, 1965, University of Alberta. 59 p.

Hanson, Larry Gene. Bedrock geology of the Rainbow Mountain area, Alaska Range, Alaska. M, 1964, University of Alaska, Fairbanks. 82 p.

Hanson, Larry Gene. The origin and development of Moses Coulee and other scabland features of the Waterville plateau, Washington. D, 1970, University of Washington. 139 p.

Hanson, Lawrence G. A seismic refraction investigation of the hydrogeology of fractured rocks. M, 1988, University of Rhode Island.

Hanson, Lindley Stuart. Stratigraphy of the Jo-Mary Mountain area with emphasis on the sedimentary facies and tectonic interpretation of the Carrabassett Formation. D, 1988, Boston University. 330 p.

Hanson, M. W. Carbonate microfacies of the Monte Cristo Group (Mississippian), Arrow Canyon Range, Clark County, Nevada. D, 1975, University of Illinois, Urbana. 92 p.

Hanson, Peter James. Vertical distribution of radioactivity in the Columbia River estuary. M, 1967, Oregon State University. 77 p.

Hanson, Richard Eric. Petrology and geochemistry of the Carlton Rhyolite, southern Oklahoma. M, 1977, Oklahoma State University. 161 p.

Hanson, Richard Eric. Volcanism, plutonism, and sedimentation in a Late Devonian submarine island-arc setting, northern Sierra Nevada, California. D, 1983, Columbia University, Teachers College. 354 p.

Hanson, Robert F. A mineralogical study of selected raw and fired Missouri fireclays. M, 1953, University of Missouri, Columbia.

Hanson, Roy E. Empirical study of Rayleigh wave dispersion, South American earthquake of December 17, 1949. D, 1957, St. Louis University.

Hanson, Royce Brooks. Geology of Mesozoic metavolcanic and metasedimentary rocks, northern White Mountains, California. D, 1986, University of California, Los Angeles. 295 p.

Hanson, Stuart. Chemical controls on tetrahedral site occupancy in amphiboles. M, 1971, Michigan State University. 28 p.

Hanson, William B. Stratigraphy and sedimentology of the Cretaceous Nanaimo Group, Saltspring Island, British Columbia. D, 1976, Oregon State University. 339 p.

Hanss, Robert Edward. Domain structure of magnetite. D, 1965, Washington University. 128 p.

Hansuld, J. A. Oxidation potential of pyrite. D, 1962, McGill University.

Hansuld, John A. Oxidation potentials of pyrite. D, 1961, McGill University.

Hansuld, John Alexander. An experimental investigation of some factors influencing the rate of leaching of the Britannia ore. M, 1956, University of British Columbia.

Hanten, J. B. Determination of upper mantle structure by synthetic seismogram amplitude modeling. M, 1978, Purdue University. 45 p.

Hao, Wei Min. Geochemistry of alkaline earth elements in the Amazon River. M, 1979, Massachusetts Institute of Technology. 52 p.

Haper, H. G. The field operations of the Halross scintillometer. M, 1951, University of Toronto.

Happ, Stafford C. Geomorphic problems of the Minisink Valley. D, 1939, Columbia University, Teachers College.

Haq, K. E. The nature and origin of microseisms. D, 1954, Massachusetts Institute of Technology. 168 p.

Haq, Munir Ul. Thermoluminescence of shocked chondrites and regolith breccias. D, 1987, University of Arkansas, Fayetteville. 152 p.

Haq, Zubair Noor-ul. The geology of the NW quarter of the Roseburg Quadrangle, Oregon. M, 1975, University of Oregon. 75 p.

Hara, Elmer Hiroshi. A nuclear magnetic resonance proton magnetometer. M, 1960, University of British Columbia.

Harakal, J. E. Potassium argon ages of the Scituate granite gneiss, north-central Rhode Island. M, 1966, Brown University.

Haraldson, Harald C. Geomagnetic survey of parts of Pierce, Benson, Sheridan and Wells counties. M, 1953, University of North Dakota. 58 p.

Harbaugh, Dwight Warvelle. Depositional environments and provenance of Mississippian Chainman Shale and Diamond Peak Formation, Central Diamond Mountains, Nevada. M, 1980, Stanford University. 91 p.

Harbaugh, Jeffrey. Strain analysis of vein calcite in the Tonoloway Limestone. M, 1986, Bowling Green State University. 75 p.

Harbaugh, John W. Biogeochemical investigations in the Tri-State zinc and lead mining district. M, 1950, University of Kansas. 88 p.

Harbaugh, John W. Stratigraphy and paleontology of portions of the Klamath Mountains, California. D, 1955, University of Wisconsin-Madison.

Harbaugh, Marion D. Geology of a proposed tunnel in the Southern Appalachians. M, 1925, University of Wisconsin-Madison.

Harbaugh, Michael William. Lithofacies analysis of the Roubidoux Formation, south-central Missouri. M, 1983, University of Missouri, Rolla. 57 p.

Harber, D. L. Petrology and origin of three rock outcrops off the Texas continental shelf. M, 1974, Texas A&M University.

Harbert, William Perry. Tectonics of Alaska, plate tectonics of the Pacific Basin, and paleomagnetism of the Aleutian Arc. D, 1987, Stanford University. 327 p.

Harbin, Gary M. A gravity study over the Katahdin Batholith, central Maine. M, 1979, SUNY at Buffalo. 45 p.

Harbour, Jerry. Microstratigraphic and sedimentational studies of an early man site near Lucy, New Mexico. M, 1958, University of New Mexico. 111 p.

Harbour, Jerry. Stratigraphy and sedimentology of the upper Safford Basin sediments, (Pliocene? and Pleistocene), (Graham County, Arizona). D, 1966, University of Arizona. 285 p.

Harbour, Jerry L. The petrology and depositional setting of the middle dolomite unit, Metaline Formation, Metaline District, Washington. M, 1978, Eastern Washington University. 67 p.

Harbridge, William Frank. Hydrography and sedimentation in the Lake of Tunis, Tunisia. M, 1974, Duke University. 92 p.

Harcourt, G. A. Distribution of nickel in the Sudbury norite-micropegmatite. D, 1933, Queen's University. 53 p.

Harcourt, G. A. Minor constituents in the chemical composition of some igneous rocks. M, 1932, Queen's University. 49 p.

Harcourt, George Alan. The identification of opaque minerals by means of spectrographic analysis. D, 1936, Harvard University.

Hard, Edward Wihelm. Mississippian gas sands of the central Michigan area. D, 1937, University of Michigan.

Hard, Edward Wilhelm. The depositional environment of the Upper Devonian bituminous shales in New York State. M, 1929, Cornell University.

Hardardottir, Vigdis. The petrology of the Hengill volcanic system, southern Iceland. M, 1983, McGill University. 260 p.

Hardas, Avinash Vishnu. Stratigraphy of gypsum deposits south of Winkleman, Pinal County, Arizona. M, 1966, University of Arizona.

Hardaway, C. Scott. Shoreline erosion and its relationship to the geology of the Pamlico River estuary. M, 1980, East Carolina University. 116 p.

Hardeman, William D. Silurian residues of central Tennessee. M, 1941, Vanderbilt University.

Harden, Deborah R. Engineering geology and slope stability study of part of the town of Woodside, California. M, 1973, Stanford University.

Harden, Deborah Reid. The distribution and geometry of incised river meanders in the central Colorado Plateau. D, 1982, University of Colorado. 288 p.

Harden, James Thomas. Petrology of the Farmington gabbro-metagabbro complex, North Carolina. M, 1984, University of Tennessee, Knoxville. 64 p.

Harden, Jennifer Willa. A study of soil development using the geochronology of Merced River deposits, California. D, 1982, University of California, Berkeley. 247 p.

Harden, Rollin Wayne. A structural investigation in Sarpy County, Nebraska, and certain adjacent areas. M, 1959, University of Nebraska, Lincoln.

Harden, Stephen N. A seismic refraction study of west-central New Mexico. M, 1982, University of Texas at El Paso.

Harder, Edmund Cecil. Iron-depositing bacteria and their geologic relations. D, 1915, University of Wisconsin-Madison.

Harder, Edmund Cecil. The iron ores of the Iron Srings District, Utah. M, 1907, University of Wisconsin-Madison.

Harder, Leslie Frederick, Jr. Use of penetration tests to determine the cyclic loading resistance of gravelly soils during earthquake shaking. D, 1988, University of California, Berkeley. 465 p.

Harder, Paul Henry, II. Soil moisture estimation over the Great Plains with dual polarization 1.66 centimeter passive microwave data from Nimbus-7. D, 1984, Texas A&M University. 157 p.

Harder, Steven. Interpretation of wide angle reflections beneath the Green River basin, Wyoming. M, 1985, University of Wyoming. 58 p.

Harder, Steven Henry. Inversion of seismic velocities for the anisotropic elastic tensor. D, 1986, University of Texas at El Paso. 84 p.

Harder, Vicki M. Oil and gas potential of the Tularosa Basin-Otero Platform area, Otero County, New Mexico. M, 1982, University of Texas at El Paso.

Harder, Vicki Marie. Fission tracks in fluorite and apatite with geologic applications. D, 1987, University of Texas at El Paso. 94 p.

Hardesty, Alan F. The distribution of the Piatt Till Member of the Wedron Formation in Coles County, Illinois. M, 1982, Southern Illinois University, Carbondale. 66 p.

Hardey, Gordon Williams. Stratigraphy of the Topanga Formation in the eastern Santa Monica Mountains. M, 1958, University of California, Los Angeles.

Hardie, Charles Henning. The geology of a part of the Jeuco Mountains, Otero County, New Mexico. M, 1958, University of Illinois, Urbana.

Hardie, Lawrence Alexander. Phase equilibria involving minerals in the system $CaSO_4$- Na_2SO_4-H_2O. D, 1965, The Johns Hopkins University. 317 p.

Hardiman, Andrew L. The mineralogy and paragenesis of the tungsten-bearing quartz veins, Elkhorn mining district, Beaverhead County, Montana. M, 1975, Lehigh University. 56 p.

Hardin, Ernest L. Fracture characterization from attenuation and generation of tube waves. M, 1986, Massachusetts Institute of Technology. 98 p.

Hardin, Jack. A comparative study of porosity estimations derived from geophysical and petrographic analyses. M, 1983, San Jose State University. 91 p.

Hardin, Nancy S. Paleoecology of Ordovician conodonts of Southwest Wisconsin. M, 1972, University of Wisconsin-Madison.

Harding, Andrew George. Lower and Middle Jurassic stratigraphy and depositional environments in southern Tunisia. D, 1978, University of South Carolina. 193 p.

Harding, Andrew George. The stratigraphic analysis and significance of the Late Triassic to upper Lower Jurassic rocks of the western High Atlas Mountains in Southwest Morocco. M, 1975, University of South Carolina.

Harding, David John. Josephine Peridotite tectonites; a record of upper-mantle plastic flow. D, 1988, Cornell University. 352 p.

Harding, James L. The geology in the vicinity of Emory Gap, Roane County, Tennessee. M, 1957, University of Tennessee, Knoxville. 26 p.

Harding, James Lombard. Petrology and petrography of the Campeche lithic suite, Yucatan Shelf, Mexico. D, 1964, Texas A&M University. 247 p.

Harding, John D. Analysis of planktonic foraminifera and clay mineralogy of core V14-95, equatorial Indian Ocean. M, 1972, Duke University. 130 p.

Harding, John William, Jr. Geology of the southern part of the Pleasanton Quadrangle, California. M, 1942, University of California, Berkeley. 70 p.

Harding, Kenneth Stanley. The stratigraphy and lithology of the Indian Cave Sandstone in southeastern Nebraska and northeastern Kansas. M, 1950, University of Nebraska, Lincoln.

Harding, Lucy E. Petrology and tectonic setting relations of Cambrian quartzites in Arizona. M, 1978, University of Arizona.

Harding, Lucy Elizabeth. Tectonic significance of the McCoy Mountains Formation, southeastern California and southwestern Arizona. D, 1982, University of Arizona. 248 p.

Harding, Matthew B. The geology of the Wildrose Peak area, Panamint Mountains, California. M, 1987, University of Wyoming. 207 p.

Harding, Maynard W. The quantity and significance of boron in sea water. M, 1932, University of Southern California.

Harding, Michael J. A geophysical investigation of a portion of Santa Theresa County Park, Santa Clara County, California. M, 1984, San Jose State University. 78 p.

Harding, Norman C. A gravity investigation of Meteor Crater, Arizona. M, 1954, University of Wisconsin-Madison.

Harding, S. R. L. The geology of the lower Lorraine (Ordovician) in the vicinity of Montreal, (Quebec). M, 1943, McGill University.

Harding, Sherie Cerise. Distribution of selected trace elements in sediments of Pamlico River estuary, North Carolina. M, 1974, North Carolina State University. 40 p.

Harding, Steven Craig. The sedimentology of the Swift Formation, southeastern Alberta. M, 1985, University of Alberta. 197 p.

Harding, Tod P. Geology of the eastern Santa Monica Mountains between Dry Canyon and Franklin Canyon. M, 1952, University of California, Los Angeles.

Harding, W. D. Origin of the Greenville gneisses. M, 1930, Queen's University. 27 p.

Harding, W. R. An investigation of the effective stress-permeability relationship for a fractured clay till near Sarnia, Ontario. M, 1986, University of Waterloo. 245 p.

Harding, William Duffield. The structure and origin of the Porcupine porphyries. D, 1933, University of Wisconsin-Madison. 22 p.

Hardisty, Russell D. Geology of the igneous rocks of the Cienega Southwest Quadrangle, Presidio County, Texas. M, 1982, West Texas State University. 121 p.

Hardman, Charles F. Criteria for selecting an underground reservoir for natural gas storage. M, 1948, University of Pittsburgh.

Hardman, Elwood. Regional gravity survey of central Iron and Washington counties, Utah. M, 1964, University of Utah. 107 p.

Hardwick, James Fredrick, Jr. Epithermal vein and carbonate replacement mineralization related to caldera development, Cunningham Gulch, Silverton, Colorado. M, 1984, University of Texas, Austin. 130 p.

Hardy, Clyde Thomas. Stratigraphy and structure of a portion of the western margin of the Gunnison Plateau, Utah. M, 1948, Ohio State University.

Hardy, Clyde Thomas. Stratigraphy and structure of the Arapien Shale and the Twist Gulch Formation in Sevier Valley, Utah. D, 1949, Ohio State University.

Hardy, David Graham and Hentz, Max Ferdinand. Geology of the Pass Peak area, Sublette County, Wyoming. M, 1951, University of Michigan.

Hardy, Henry Reginald, Jr. The experimental investigation of the inelastic behavior of geologic materials. D, 1965, Virginia Polytechnic Institute and State University. 353 p.

Hardy, Hughey E. Igneous geology of the San Antonia area, Sierra Hechiceros, eastern Chihuahua, Mexico. M, 1987, West Texas State University. 128 p.

Hardy, James J., Jr. The structural geology, tectonics and metamorphic geology of the Arrastre Gulche Window, south-central Harquahala Mountains, Maricopa County, Arizona. M, 1984, Northern Arizona University. 99 p.

Hardy, Jenna-Lee. Stratigraphy, brecciation and mineralization, Gayna River, Northwest Territories. M, 1980, University of Toronto.

Hardy, L. Contribution à l'étude géomorphologique de la portion québecoise des basses terres de la Baie de James. D, 1976, McGill University. 264 p.

Hardy, L. R. The geology of an allochthonous Jurassic sequence in the Sierra de Santa Rosa, Northwest Sonora, Mexico. M, 1973, [University of California, San Diego].

Hardy, Lisa Steward. Mineralogy and petrology of the Haile gold mine, Lancaster County, South Carolina. M, 1987, University of Georgia. 165 p.

Hardy, Richard. Géologie de la région du lac des Chefs. M, 1968, Ecole Polytechnique. 244 p.

Hardy, Roy Paul. Conodonts from the Fernvale of eastern Missouri. M, 1947, University of Missouri, Columbia.

Hardy, Steven B. Provenance, tectonic setting, and depositional environment of Oligocene non-marine deposits along Watana Creek, Susitna Drainage, Alaska. M, 1987, University of Alaska, Fairbanks. 109 p.

Hardyman, R. F. Volcanic stratigraphy and structural geology of Gillis Canyon Quadrangle, northern Gillis Range, Mineral County, Nevada. D, 1978, University of Nevada. 377 p.

Hardyman, Richard F. The petrography of a section of the basal Duluth complex (Precambrian), Saint Louis County, northeastern Minnesota. M, 1969, University of Minnesota, Minneapolis.

Hare, Ben Dean. Experimental weathering of chlorite using soxhlet-extraction apparatus. D, 1973, University of Oklahoma. 77 p.

Hare, Ben Dean. Petrology of the Thurman sandstone (Desmoinesian) (Hughes and Coal counties, Oklahoma). M, 1969, University of Oklahoma. 133 p.

Hare, E. Matthew. Structure and petrography of the Tertiary volcanic rocks between Death Creek and Dairy Valley Creek (Box Elder Co.), Utah. M, 1982, Utah State University. 64 p.

Hare, Joseph E. Salt domes of the Gulf Coast and their relation to petroleum. M, 1922, Yale University.

Hare, Paul William. Geomorphic surfaces and vertical neotectonism of the Nicoya Peninsula, northwestern Costa Rica. M, 1984, Pennsylvania State University, University Park. 156 p.

Hare, Peter Edgar. The amino acid composition of the organic matrix of some Recent and fossil shells of some West Coast species of Mytilus. D, 1962, California Institute of Technology. 109 p.

Hares, E. M. The petrography and petrogenesis of the Blake's Ferry Pluton; Randolph County, Alabama. M, 1975, Memphis State University.

Hargadine, Gerald D. Clay mineralogy and other petrologic aspects of the Grenola Limestone Formation in the Manhattan, Kansas area. M, 1959, Kansas State University. 76 p.

Hargan, Bruce Alan. Regional gravity data analysis of the Papago Indian Reservation, Pima County, Arizona. M, 1978, University of Arizona.

Hargett, W. G. The hydrocarbon potential of the A. S. Johnson Mineral Trust Lands, Southwest Alabama. M, 1975, University of Alabama.

Hargis, D. Artificial recharge of storm runoff at Kahului, Maui. M, 1971, University of Hawaii. 71 p.

Hargis, D. R. Analysis of factors affecting water level recovery data. D, 1979, University of Arizona. 213 p.

Hargraves, Robert Bero. Petrology of the Allard Lake anorthosite suite, and paleomagnetism of the ilmenite deposits. D, 1959, Princeton University. 193 p.

Hargreaves, Roy. Morphology and chemistry of early Precambrian metabasalt flows, Utik Lake, Manitoba. M, 1978, University of Manitoba.

Hargrove, Howard Ralph. Geology of the southern portion of the Montana Mountains, McDermitt Caldera, Nevada. M, 1982, Arizona State University. 202 p.

Hariharan, Ganesan. Lithostratigraphy of sub-surface Cincinnatian strata in western and southern Ohio and adjacent areas. M, 1970, Ohio State University.

Haring, Alfred M., Jr. The Marlboro (green) Schists of Rhode Island. M, 1929, Brown University.

Hariri, Davoud. Relation between the bed pavement and the hydraulic characteristics of high gradient channels in noncohesive sediments. D, 1964, Utah State University. 109 p.

Harita, Yoichi. An investigation of the upper mantle P-velocity models using synthetic seismograms. M, 1968, Massachusetts Institute of Technology. 38 p.

Harita, Yoichi. Some applications of Cagniard's method to curved layer cases. M, 1969, Massachusetts Institute of Technology.

Harju, Hendric O. An application of statistics to the computation of ore reserves using sample data from the Roan Antelope copper mine, Zambia, Africa. M, 1966, Queen's University. 173 p.

Harker, David E. A crystallochemical study of garnets in Precambrian metamorphic rocks from the Ruby Range, southwestern Montana. M, 1984, Kent State University, Kent. 101 p.

Harker, George R. The delineation of flood plains using automatically processed multispectral data. D, 1974, Texas A&M University. 240 p.

Harker, Peter. Stratigraphy and paleontology of the Banff and associated Carboniferous formations of Western Canada. D, 1951, University of Michigan.

Harker, Stuart David. A survey of the stratigraphic distribution of organic-walled dinoflagellate cysts in the Cretaceous and Tertiary. M, 1975, University of Saskatchewan. 249 p.

Harker, Stuart David. Campanian organic-walled microplankton from the interior plains of Canada, Wyoming and Texas. D, 1978, University of Saskatchewan. 699 p.

Harkey, Donald A. Structural geology and sedimentologic analysis (Las Vigas Formation), Sierra San Ignacio, Chihuahua, Mexico. M, 1985, University of Texas at El Paso.

Harkrider, David Garrison. I, Propagation of acoustical gravity waves from an explosive source in the atmosphere; II, Rayleigh and Love waves from sources in multilayered elastic halfspace. D, 1963, California Institute of Technology. 146 p.

Harkrider, David Garrison. Seismic investigation of geological structure bordering the Caribbean island arc, Part III. M, 1957, Rice University. 44 p.

Harlan, Howard Marshall, II. Geology and ground magnetic survey of a portion of the Lampbright West area, Grant County, New Mexico. M, 1971, University of Arizona.

Harlan, J. Bruce. Geology and mineralization of Fireball Ridge, Churchill County, Nevada. M, 1984, Colorado State University. 201 p.

Harlan, Janis G. Glacial marine sedimentation in the Weddell sector of the East Antarctic Rise. M, 1981, Rice University. 53 p.

Harlan, John Lee. The geology of the east-central portion of the Jefferson City Quadrangle. M, 1951, University of Missouri, Columbia.

Harlan, Julie de Azevedo. The species duftite. M, 1971, University of Arizona.

Harlan, Ronald W. Pleistocene sediments along White Oak Bayou, northwestern Houston, Texas. M, 1962, University of Houston.

Harlan, Ronald Wade. A clay mineral study of Recent and Pleistocene sediments from the Sigsbee Deep, Gulf of Mexico. D, 1966, Texas A&M University. 140 p.

Harlan, Stephen Scott. Timing of deformation along the leading edge of the Montana disturbed belt, northern Crazy Mountains basin, Montana. M, 1986, Montana State University. 88 p.

Harlan, William Stephen. Signal/noise separation and seismic inversion. D, 1986, Stanford University. 165 p.

Harland, Gregg H. Phase relations in the system $CaCO_3$-H_2O. D, 1972, University of Minnesota, Minneapolis.

Harland, Rex. Dinoflagellates and acritarchs from the Bearpaw Formation (Cretaceous), southern Alberta. D, 1970, University of Alberta. 205 p.

Harle, David Sig. Geology of Babyshoe Ridge area, southern Cascades, Washington. M, 1974, Oregon State University. 71 p.

Harlem, Peter Wayne. Aerial photographic interpretation of the historical changes in northern Biscayne Bay, Florida; 1952-1976. M, 1979, University of Miami. 152 p.

Harlett, J. C. Daily changes in beach profile and sand texture on Del Monte beach, California. M, 1967, United States Naval Academy.

Harlett, John Charles. Sediment transport on the northern Oregon continental shelf. D, 1972, Oregon State University. 120 p.

Harley, D. N. Geology of the Half Mile Lake Zn-Pb-Cu deposit, New Brunswick. M, 1977, University of Western Ontario. 209 p.

Harley, George Townsend. The geology and ore deposits of northeastern New Mexico. M, 1935, New Mexico Institute of Mining and Technology.

Harley, William Frank and Salotti, Charles A. Geology of the upper Maes Creek area, Wet Mountains, Colorado. M, 1955, University of Michigan.

Harlin, J. M. Climatic variability and its influence on basin morphometry. D, 1975, University of Iowa. 141 p.

Harlow, David H. Volcanic earthquakes. M, 1971, Dartmouth College. 66 p.

Harlow, George. Stratigraphy and structure of Spruce Mt. area, Elko Co., Nevada. M, 1956, University of Washington. 72 p.

Harlow, George Edward, Jr. The structural geology of a part of the Pulaski thrust sheet near Boone Dam; Sullivan and Washington counties, Tennessee. M, 1987, University of Tennessee, Knoxville. 103 p.

Harlow, George Eugene. The anorthoclase structures; a room- and high temperature study. D, 1977, Princeton University. 220 p.

Harmala, John Clifford. Conodont biostratigraphy of some Mississippian rocks in northeastern Nevada and northwestern Utah. M, 1982, Arizona State University. 271 p.

Harman, John Warren. The geology and ore deposits of the Suyoc-Mankayan District, Suyoc, Mankayan, Mountain Province, Philippine Islands. M, 1944, University of California, Berkeley. 108 p.

Harman, Robert Allison. Distribution of foraminifera within the Santa Barbara Basin. M, 1962, University of Southern California.

Harmison, Lowell Thomas. Derivative neutron activation analysis of zinc. D, 1965, University of Maryland. 88 p.

Harmon, Carol Jean. X-ray radiographs of cores from the Hispaniola-Caicos Basin. M, 1973, Duke University. 158 p.

Harmon, Karen A. Abundance and distribution of tungsten in Archean iron-formations and in shales, metashales, and sandstones. M, 1975, McMaster University. 149 p.

Harmon, Kathryn Parker. Late Pleistocene forest succession in northern New Jersey. D, 1968, Rutgers, The State University, New Brunswick. 203 p.

Harmon, Russell S. Chemical and carbon isotopic evolution of carbonate waters in the Nittany Valley of central Pennsylvania. M, 1973, Pennsylvania State University, University Park.

Harmon, Russell Scott. Late Pleistocene paleoclimates in North America as inferred from isotopic variations in speleothems. D, 1975, McMaster University. 279 p.

Harmon, William Lloyd. Preliminary taxonomic study of the Ordovician Rafinesquinidae (Brachiopoda). M, 1974, University of Cincinnati. 334 p.

Harms, James E. Geology of the southeast one-quarter of the Camas Valley Quadrangle, Douglas County, Oregon. M, 1957, Oregon State University. 71 p.

Harms, John Conrad. Structural geology of the eastern flank of the southern Front Range, Colorado. D, 1959, University of Colorado. 165 p.

Harms, Richard William. Cirques of the Marble Mountains, northwestern California. D, 1983, University of California, Berkeley. 199 p.

Harms, Tekla A. The Newport Fault; low-angle normal faulting and Eocene extension, northeastern Washington and northwestern Idaho. M, 1982, Queen's University. 157 p.

Harms, Tekla Ann. Structural and tectonic analysis of the Sylvester Allochthon, northern British Columbia; implications for paleogeography and accretion. D, 1986, University of Arizona. 114 p.

Harned, C. Hal. The mineralogy and mechanical analysis of the mantle rock in the Manhattan area. M, 1940, Kansas State University. 119 p.

Harned, Wentworth V. Upper Cretaceous igneous activity in Mississippi. M, 1960, University of Mississippi. 47 p.

Harnett, Richard Allen. Sedimentary petrology of the Gallup Sandstone, (Cretaceous) San Juan County, New Mexico. M, 1962, University of Utah. various pagination p.

Harnish, David Emmanuel. Porphyry copper related mineralization in the Terre Neuve District, Haiti, West Indies. M, 1984, University of Wisconsin-Madison. 81 p.

Harnois, Luc. Geochemistry of the Ore Chimney Formation and associated metavolcanic rocks and gold deposits in the Flinton-Harlow area, Grenville Province, southeastern Ontario. D, 1988, Carleton University. 237 p.

Harnsberger, Wilbur Trout, Jr. Geology of the Bergton area, Northwest Rockingham County, Virginia. M, 1950, University of Virginia. 108 p.

Haroon, Mohammed A. Silurian conodonts and biostratigraphy of the Chimneyhill and Henryhouse formations, Arbuckle Mountains, Carter County, Oklahoma. M, 1981, West Texas State University. 62 p.

Harouaka, Abdallah. A new block sparse direct solution technique; application to hydrocarbon reservoir simulation. D, 1987, Pennsylvania State University, University Park. 250 p.

Harp, Brad D. Net shore-drift of Pierce County, Washington. M, 1983, Western Washington University. 170 p.

Harp, Edwin L. Thrust plate stratigraphy and structure of the Many Glacier area, Glacier National Park, Montana. M, 1971, Montana State University. 91 p.

Harp, Edwin Lynn. Fracture systems and tectonics of Mars. D, 1974, University of Utah. 116 p.

Harpel, Greg. Geology and tungsten mineralization of a portion of the Ragged Top mining district, Pershing County, Nevada. M, 1980, University of Nevada. 48 p.

Harper, Charles Thomas. The geology and uranium deposits of the central part of the Carswell Structure, northern Saskatchewan, Canada. D, 1983, Colorado School of Mines. 587 p.

Harper, Charles Thomas. The geology of the zinc-lead deposit at Sito Lake, northern Saskatchewan. M, 1975, University of Saskatchewan. 73 p.

Harper, Charles W., Jr. Rib branching in Atrypa reticularis. M, 1961, Massachusetts Institute of Technology. 211 p.

Harper, Charles Woods, Jr. The brachiopods of the Arisaig Series (Silurian-Lower Devonian) of Nova Scotia. D, 1964, California Institute of Technology. 467 p.

Harper, David Paul. Sedimentary dynamics of a disturbed estuary entrance sand shoal; the Shrewsbury entrance area of Sandy Hook Bay, New Jersey. M, 1975, Rutgers, The State University, New Brunswick. 49 p.

Harper, Delbert D. Structure of the Dumplin Valley fault system, Boyds Creek Quadrangle, Knox and Sevier cos., Tennessee. M, 1963, University of Tennessee, Knoxville. 53 p.

Harper, Denver. Selected trace elements in pyrite from coals of southern Illinois. M, 1977, Southern Illinois University, Carbondale. 66 p.

Harper, Edward S, III. Foraminifera of the Jackson Group (upper Eocene) in Caldwell and Catahoula parishes, Louisiana. M, 1965, Northeast Louisiana University.

Harper, Francis. A faunal reconnaissance in the Athabaska and Great Slave Lake region. D, 1925, Cornell University.

Harper, Gregory Don. Structure and petrology of the Josephine Ophiolite and overlying metasedimentary rocks, northwestern California. D, 1980, University of California, Berkeley. 281 p.

Harper, H. G. Radioactivity of rocks associated with Lake Athabaska pitchblende deposits. D, 1953, University of Toronto.

Harper, Herbert E. Preliminary report on the geology of the Molalla Quadrangle, Oregon. M, 1947, Oregon State University. 29 map p.

Harper, Howard E., Jr. The vertical distribution of phytoplankton remains in a deep-sea core from the Equatorial Pacific. M, 1974, Louisiana State University.

Harper, Howard Earl, Jr. Diatom biostratigraphy of the Miocene/Pliocene boundary in marine strata of the Circum-North Pacific. D, 1977, Harvard University.

Harper, J. N.; Davis, V. H. and Neish, J. F. A short-term study of beach sand migration adjacent to Monterey canyon (California). M, 1966, United States Naval Academy.

Harper, J. R. Cuspate spits, a shoreline response to longshore power gradients. M, 1976, Louisiana State University.

Harper, J. R. The physical processes affecting the stability of tundra cliff coasts. D, 1978, Louisiana State University. 228 p.

Harper, John A. Gastropods of the Gilmore City Limestone (Lower Mississippian) of Northcentral Iowa. D, 1977, University of Pittsburgh. 329 p.

Harper, John Andrew. Gastropods of the Suwannee Limestone, Oligocene, of peninsular Florida. M, 1972, University of Florida. 143 p.

Harper, John David. Stratigraphy, sedimentology and facies of the Rondout Formation (upper Silurian) of the Hudson valley region of New York State. D, 1969, Brown University.

Harper, John David. The sedimentary ecology of the Kirkfield Quarry, Ontario. M, 1964, University of Toronto.

Harper, John LeRoy. The geology of the southern Hartville Uplift area, Platte and Goshen counties, Wyoming. M, 1960, University of Nebraska, Lincoln.

Harper, Kennard R. Geology of the Hot Spring Quadrangle, Owyhee County, Idaho. M, 1963, University of Oregon. 107 p.

Harper, Margaret Francis. Problems in the origin of manganese. M, 1936, Smith College. 87 p.

Harper, Melvin Louis. Mechanics of basement deformation in Glenwood Canyon (Garfield and Eagle counties), Colorado. D, 1964, University of Colorado. 154 p.

Harper, Melvin Louis. The areal geology of Castle Creek valley, Utah. M, 1960, Texas Tech University. 121 p.

Harper, Stephen Brewer. The age and origin of granitic gneisses of the inner Piedmont, northwestern North Carolina. M, 1977, University of North Carolina, Chapel Hill. 92 p.

Harper, Wallace F. Structure of Precambrian metamorphic rocks west of the Ilse fault zone, Hardscrabble Quadrangle, Wet Mountains, Colorado. M, 1976, Louisiana State University.

Harper, William David. A study of the Bluffport Member of the Demopolis Chalk in Noxubee County, Mississippi. M, 1959, Mississippi State University. 47 p.

Harpin, Raymond Joseph. Hot-pressing of quartz powder to.5 GPA pressure and 1250°C. M, 1980, Massachusetts Institute of Technology. 50 p.

Harpine, Joseph E. Planning of open space using physical geographical features in the South Dry Sac drainage basin. M, 1984, Southwest Missouri State University.

Harpster, Robert Eugene. Geological application of soil mechanics to Del Rio Formation in Austin, Texas, area. M, 1957, University of Texas, Austin.

Harr, J. L. Paleobotany of a silicified peat. M, 1976, West Virginia University. 154 p.

Harrell, Glenn C. Paleostructural control of hydrocarbon accumulation, Murdock Pass field, Kennedy County, Texas. M, 1968, Texas A&M University.

Harrell, James Anthony. Grain size and shape distributions, grain packing, and pore geometry within sand laminae; characterization and methodologies. D, 1983, University of Cincinnati. 586 p.

Harrell, James Anthony. Relative mechanical durabilities of quartz and feldspar. M, 1976, University of Oklahoma. 128 p.

Harrell, Marshall Allen. Ground water in Indiana. D, 1933, Indiana University, Bloomington. 898 p.

Harrer, Joseph W., Jr. The depositional environments and paleoecology of benthic foraminifera in the Clayton Formation (Danian) of Alabama. M, 1986, Auburn University. 255 p.

Harrigan, Joseph A. The effect of channel morphology by variable sediment loads on three streams adjacent to highway construction sites in the Piedmont of South Carolina. M, 1977, University of South Carolina. 132 p.

Harriger, Ted. Impact on water quality by a coal ash landfill in northcentral Chautauqua County, New York. M, 1977, SUNY, College at Fredonia. 192 p.

Harrill, James R. Geology of the Davis Knolls and northern Big Davis Mountain area, Tooele County, Utah. M, 1962, University of Utah. 42 p.

Harrington, Anne. Hydrochemistry of the Columbia Formation, southwestern Kent County, Delaware. M, 1982, University of Delaware.

Harrington, Charles Dare. Differentiation of Alpine moraines and mass-wasted deposits utilizing fabric and textural properties, Colorado, Montana, Wyoming. D, 1970, Indiana University, Bloomington. 163 p.

Harrington, Charles Eston, Jr. Soil parameters and clay mineralogy of the Southeast Arlington area, Tarrant County, Texas. M, 1974, University of Texas, Arlington. 50 p.

Harrington, David Haymond. The geology of the southern Strawn Extension, Eastland County, Texas. M, 1953, University of Texas, Austin.

Harrington, Eldred R. Geologic report of the Shoshone region, Idaho. M, 1930, University of New Mexico. 63 p.

Harrington, Frederick Irving. Stratigraphy and structure of the northern two-thirds of the Hot Springs Quadrangle, South Dakota. M, 1954, University of Iowa. 106 p.

Harrington, James R. A finite element computer simulation of groundwater flow in the Pootatuck River valley, Newtown, Connecticut. M, 1985, University of Massachusetts. 61 p.

Harrington, John Ausman, Jr. Upper atmospheric controls, surface climate, and phytogeographical implications in the western Great Lakes region. D, 1980, Michigan State University. 247 p.

Harrington, John Mark. A geophysical investigation of Valle Trinidad, Baja California, Mexico. M, 1981, San Diego State University.

Harrington, John Wilbur. The geology and ore dressing problems of the Raleigh Graphite. M, 1946, University of North Carolina, Chapel Hill. 43 p.

Harrington, John Wilbur. The west border of the Durham Triassic basin. D, 1948, University of North Carolina, Chapel Hill. 106 p.

Harrington, Jonathan Waldo. Rhynchonellid brachiopods of New York Frasnian (Devonian). M, 1966, Cornell University.

Harrington, Jonathan Waldo. Taxonomy, evolution and paleoecology of the rhynchonellid brachiopods of the New York Senecan (Upper Devonian). D, 1968, Cornell University. 270 p.

Harrington, Mark Terrell. A study of the textures and of the deformation of the sphalerite at Friedensville, Pennsylvania. M, 1976, University of Michigan.

Harrington, R. Streamflow generation in a small Canadian Shield watershed. M, 1977, University of Waterloo.

Harrington, Robert B. The stratigraphy and structure of the southern part of the Manlius, New York 7-1/2 minute quadrangle. M, 1951, Syracuse University.

Harrington, Robert John. Geology and geochemistry of the Mt. Jefferson caldera complex, Nye County, Nevada. M, 1986, University of Colorado. 71 p.

Harrington, Robert Joseph. Depositional environments and ecologic gradients of the Upper Devonian Sultan Formation (Ironside Dolomite and Valentine Limestone members) and subjacent beds from the uppermost Mountain Springs Formation, near Mountain Springs, Clark County, Nevada. M, 1982, University of California, Riverside. 147 p.

Harrington, Robert Joseph. Growth patterns within the genus Protothaca (Bivalvia: Veneridae) from the Gulf of Alaska to Panama; paleotemperatures, paleobiogeography and paleolatitudes. D, 1986, University of California, Santa Barbara. 249 p.

Harrington, William Cornell. Geology of the Paskenta District, Tehama County, California. M, 1942, University of California, Berkeley. 38 p.

Harriott, Theresa A. Pyroxene relationships in the 4B mesosiderites Mincy and Budulan and implications for their origin. M, 1985, Rutgers, The State University, New Brunswick. 76 p.

Harris, A. B. Solution-mineral equilibria of ferrous iron, zinc, cadmium and lead carbonates and phosphates in interstitial water of the central basin of Lake Erie. M, 1977, Case Western Reserve University.

Harris, A. Wayne. Structural interpretation in the Rock Canyon area of southern Wasatch Mountains, Utah. M, 1936, Brigham Young University. 25 p.

Harris, Alan William. Dynamical studies of satellite origin. D, 1975, University of California, Los Angeles. 118 p.

Harris, Alfred Ray. Effect of deciduous, coniferous, and abandoned field cover on the hydrologic properties and frost morphology of frozen soil (Driftless area, Mississippi River valley). D, 1972, University of Minnesota, Minneapolis. 168 p.

Harris, Ann G. Structural geology of Austins Glen, Greene County, New York. M, 1958, Miami University (Ohio). 83 p.

Harris, Arthur Horne. The origin of the grassland amphibian, reptilian, and mammalian faunas of the San Juan-Chaco River drainage. D, 1965, University of New Mexico. 169 p.

Harris, Billy L. Genesis, mineralogy, and properties of Parkdale soils, Oregon. D, 1973, Oregon State University. 174 p.

Harris, Charles M. Agricultural gypsum for alkali land reclamation. M, 1956, University of Nevada - Mackay School of Mines. 32 p.

Harris, Charles Steven. Biostratigraphy and environmental distribution of conodonts from the lowermost Middle Ordovician in East Tennessee. M, 1982, University of Tennessee, Knoxville. 160 p.

Harris, Charles William. A sedimentological and structural analysis of the Proterozoic Uncompahgre Group, Needle Mountains, Colorado. D, 1988, Virginia Polytechnic Institute and State University. 261 p.

Harris, Charles William. An unconformity in the Carolina slate belt of central North Carolina; new evidence for the areal extent of the ca. 600 MA Virgilina Deformation. M, 1982, Virginia Polytechnic Institute and State University.

Harris, Claude Milner. Micropaleontology of lower Washita of southeastern Oklahoma. M, 1933, University of Oklahoma. 72 p.

Harris, Clayton D. Facies, depositional and diagenetic environments of the Golconda Group (Mississippian) in southwestern Illinois and southeastern Missouri. M, 1987, Southern Illinois University, Carbondale. 167 p.

Harris, Clyde E., Jr. Historical and structural geology of the Leeds Gorge area, Greene County, New York. M, 1958, Miami University (Ohio). 46 p.

Harris, Constance. The relationships of the Exshaw Formation in Alberta. M, 1964, University of Washington. 86 p.

Harris, Dahl Le Roy. Petrology of the Boundary Peak Adamellite Pluton (middle or late Mesozoic) in the Benton Quadrangle, Mono and Esmeralda counties, California and Nevada. M, 1967, University of California, Davis. 57 p.

Harris, Daniel R. Lower Cretaceous exhumed paleochannels; in the Cedar Mountain Formation near Green River, Utah. M, 1979, Brigham Young University.

Harris, David Milo. The concentrations of H_2O, CO_2, S, and Cl during pre-eruption crystallization of some mantle derived magmas; implications for magma genesis and eruption mechanisms. D, 1981, University of Chicago. 218 p.

Harris, David V. Late Quaternary alluviation and erosion in Box Elder Creek Valley, Larimer County, Colorado. D, 1959, University of Colorado.

Harris, David V. and Hoagland, Alan D. Tertiary igneous rocks of South Park, Colorado. M, 1935, Northwestern University.

Harris, David William. Crustal structure of northwestern Montana. M, 1985, University of Montana. 63 p.

Harris, DeVerle. The geology of the Dutch Peak area, Sheeprock Range, Tooele County, Utah. M, 1958, Brigham Young University. 82 p.

Harris, DeVerle Porter. An application of multivariate statistical analysis to mineral exploration. D, 1965, Pennsylvania State University, University Park. 278 p.

Harris, Donald Clayton. The application of X-ray spectroscopy to the study of three new ore minerals. D, 1966, University of Toronto.

Harris, Donald Clayton. The X-ray fluorescent analysis of rocks. M, 1961, University of Toronto.

Harris, Donald G. A study of Meramec and lower Chester strata in northeastern Oklahoma, southwestern Missouri, and northwestern Arkansas. M, 1956, University of Oklahoma. 112 p.

Harris, E. Donald, Jr. Depositional history and regional correlation of the Carrico Lake Formation, Lander County, Nevada. M, 1985, Brigham Young University. 104 p.

Harris, Elaine. The effect of ashfall from the May 18, 1980 eruption of Mount St. Helens on cryptograms. M, 1984, Washington State University. 50 p.

Harris, Ellwood Glendenning. Preliminary petrographic study of a portion of the coastal batholith of South America, Arequipa Quadrangle, Peru. M, 1952, University of Rochester. 106 p.

Harris, Eugene B., Jr. Trend surface analysis of some Upper Cretaceous rocks of the Clark County, Alabama, area. M, 1966, University of Alabama.

Harris, Forest Klaire. A dynamic electrometer for measuring the radioactivity of gases from oil and gas wells. M, 1923, University of Oklahoma. 16 p.

Harris, Frank Gaines, III. Geology of the Dry Creek area, Lewis and Clark County, Montana. M, 1963, University of Missouri, Columbia. 114 p.

Harris, Frank W. Textural and compositional variability in limestone beds from the Waynesville Formation (Upper Ordovician), Brookville, Indiana. M, 1977, Miami University (Ohio). 149 p.

Harris, Frederick Robson. Volcanic rocks of the Sunday Lake area (Charlotte County), New Brunswick. M, 1964, University of New Brunswick.

Harris, Gary D. Sedimentology and depositional history of a deltaic lower Atoka (Pennsylvanian) Sandstone, northwestern Arkansas. M, 1983, University of Arkansas, Fayetteville.

Harris, Harold Duane. Geology of the Birdseye area, Thistle Creek Canyon, Utah. M, 1953, Brigham Young University. 126 p.

Harris, Henry John Hayden. Hydrology and hydrogeochemistry of the south fork, Wright Valley, southern Victoria Land, Antarctica. D, 1981, University of Illinois, Urbana. 359 p.

Harris, Herbert. Geology of the Palomas Canyon-Castaic Creek area, Los Angeles County, California. M, 1950, University of California, Los Angeles.

Harris, Hobart Byron. The Greenville Fault area. M, 1948, Indiana University, Bloomington. 10 p.

Harris, I. McK. Geology of the Cobbs Arm area, New World Island, Newfoundland. M, 1966, Dalhousie University. 71 p.

Harris, J. J. The (Ordovician) Black River Group, in the vicinity of Montreal (Quebec). M, 1933, McGill University.

Harris, James Zack. The foraminifera of the Providence Sand. M, 1958, University of Alabama.

Harris, Jane. Statistical analysis of ground water quality; interaction of deterministic and stochastic components. D, 1988, Colorado State University. 159 p.

Harris, Janet M. The geology of ophiolitic and adjoining rocks of Chagnon Mountain, southern Quebec. M, 1984, SUNY at Albany. 113 p.

Harris, Jeanne Elizabeth. The development of regional gravity parameters to characterize tectonic provinces. M, 1975, University of Michigan.

Harris, John Elliott. Characterization of suspended matter in the Gulf of Mexico and northern Caribbean Sea. D, 1971, Texas A&M University. 212 p.

Harris, John F. Relationships of deformational fractures in sedimentary rocks to regional and local structure. M, 1959, University of Tulsa. 56 p.

Harris, John Michael. Oligocene vertebrates from western Jeff Davis County, Trans-Pecos Texas. M, 1967, University of Texas, Austin.

Harris, John O. Eocene-Oligocene foraminiferal biostratigraphy of holes 19 and 20, Leg 3, JOIDES Deep Sea Drilling Project. M, 1971, Indiana State University. 54 p.

Harris, John Richard. The petrology of the Sabinetown Formation, Wilcox Group, Bastrop County, Texas. M, 1957, University of Texas, Austin.

Harris, John Rodefer. The Devonian formations of Indiana (Part 2; structural conditions). M, 1940, Indiana University, Bloomington. 32 p.

Harris, Jonathan O. The emplacement and crystallization of the Cornelia Pluton, Ajo, Arizona; an analysis based on the compositional zoning of plagioclase and field relations. M, 1984, University of Arizona. 78 p.

Harris, Karen. Abundance and phenotypic variations of Globigerinoides sacculifer (Brady) in living populations. M, 1978, University of Miami. 169 p.

Harris, Kenneth L. Pleistocene geology of the Grand Forks-Bemidji area, northwestern Minnesota. D, 1975, University of North Dakota. 210 p.

Harris, Kenneth L. Pleistocene stratigraphy of the Red Lake Falls area, Minnesota. M, 1973, University of North Dakota. 117 p.

Harris, Lawrence A. A study of size and shape of quartz grains in a windblown silt. M, 1956, University of Tennessee, Knoxville. 19 p.

Harris, Leaman D. A statistical analysis of sediment dispersal in the alluvial valley of the lower Mississippi River. M, 1963, University of Kansas. 63 p.

Harris, Lloyd Addis. Some Lower Ordovician insoluble residues. M, 1940, Vanderbilt University.

Harris, M. K. Form-process relationships at Laguna Percebu, Baja California, Mexico. D, 1979, Louisiana State University. 104 p.

Harris, Mark T. Sedimentology of the Cutoff Formation (Permian), western Guadalupe Mountains, West Texas and New Mexico. M, 1982, University of Wisconsin-Madison. 186 p.

Harris, Mark Thomas. Margin and foreslope deposits of the Latemar carbonate buildup (Middle Triassic), the Dolomites, northern Italy. D, 1988, The Johns Hopkins University. 611 p.

Harris, Mary Katherine. Depositional environments, diagenesis and porosity development in the Mississippian Castle Reef Formation, Sawtooth Range, northwestern Montana. M, 1986, University of Idaho. 107 p.

Harris, Michael. Statistical treatment of selected trace elements in unoxidized gold ores of the Carlin gold deposit, Nevada. M, 1974, Stanford University.

Harris, N. B. Late Tertiary faults in South-central Alaska. M, 1977, Stanford University. 38 p.

Harris, N. B. Skarn formation near Ludwig, Yerington District, Nevada. D, 1980, Stanford University. 218 p.

Harris, Nancy Jensen. Diagenesis of upper Pleistocene strand-plain limestone, northeastern Yucatan Peninsula, Mexico. M, 1984, University of New Orleans. 130 p.

Harris, Paul B. Geology of the Tunis-Pastoria Creek area, Kern County, California. M, 1950, California Institute of Technology. 80 p.

Harris, Paul M. Holocene sediment and stratigraphy of marshes at Chincoteague Inlet, Virginia. M, 1973, West Virginia University.

Harris, Paul M. Sedimentology of the Joulters Cays ooids sand shoal, Great Bahama Bank. D, 1977, University of Miami. 452 p.

Harris, Quinton P., Jr. A summary of our knowledge of Oregon's igneous geology. M, 1935, Oregon State University. 179 p.

Harris, R. D. The physical hydrogeology at the Bruce nuclear power development radioactive waste operations, Site 2. M, 1981, University of Waterloo. 180 p.

Harris, R. E. Alteration and mineralization associated with sandstone uranium occurrences, Morton Ranch, Wyoming, and the Seboyeta area, New Mexico. M, 1982, University of Wyoming. 101 p.

Harris, R. J. Selection of surveying methods. M, 1975, Kansas State University.

Harris, Rae L., Jr. Geologic evolution of the Beartooth Mountains, Montana and Wyoming; Part 3, Gardner Lake area, Wyoming. D, 1959, Columbia University, Teachers College.

Harris, Reginald Wilson. Foraminifera from Jackson outcrops along Ouachita and Red rivers of Louisiana. M, 1926, Louisiana State University.

Harris, Reginald Wilson. Simpson (Ordovician) Ostracoda of the Arbuckle Mountains of Oklahoma. D, 1939, Harvard University.

Harris, Reginald Wilson, Jr. A faunal description and age determination of the Simpson Birdseye Limestone

(Ordovician) of the Criner Hills (Oklahoma). M, 1961, University of Oklahoma. 130 p.

Harris, Reginald Wilson, Jr. Palynology of the Sand Branch Member of the Caney Shale Formation (Mississippi) of southern Oklahoma. D, 1971, University of Oklahoma. 216 p.

Harris, Richard Huntington. A sulfur isotopic and major element study of the lower Chester Vein, Sunshine Mine, Idaho. M, 1974, University of Montana. 88 p.

Harris, Robert Alan. North American representatives of the Mesozoic nautiloid family Cymatoceratidae. M, 1943, University of Iowa. 40 p.

Harris, Robert D. Investigation of a possible mineralogical control of roof failure at the White Pine copper mine (Michigan). M, 1972, Michigan Technological University. 73 p.

Harris, Rodger S. Relation of titanium to basic dikes on the Marquette Range. M, 1955, University of Wisconsin-Madison.

Harris, Roger L. Depositional environments and diagenesis of the Ogdensburg Formation (Lower Ordovician) of New York State as determined from borehole samples. M, 1976, Rensselaer Polytechnic Institute. 80 p.

Harris, Ronald Albert. Paleomagnetism, geochronology and paleotemperature of the Yukon-Koyukuk Province, Alaska. M, 1985, University of Alaska, Fairbanks. 143 p.

Harris, Ruth Audrey. Oceanic geoid anomalies. M, 1984, Cornell University. 47 p.

Harris, Sandra. Oligocene evolution of the Pacific Plate off the California coast. M, 1981, Queen's University. 169 p.

Harris, Sherod A. Hydrocarbon accumulation in Meramec-Osage (Mississippian) rocks, Sooner Trend, northwest central Oklahoma. M, 1973, University of Oklahoma. 92 p.

Harris, Sidney L. Migration of the crestal plane in asymmetrical folds. M, 1931, University of Pittsburgh.

Harris, Stanley Edwards. Geology of the Mud Fork area, Tazewell County, Virginia. M, 1942, University of Iowa. 73 p.

Harris, Stanley Edwards. Subsurface stratigraphy of the Kinderhook and Osage series in southeastern Iowa. D, 1947, University of Iowa. 155 p.

Harris, Steven H. Weathering in the Illinoian glacial till near Batavia, Ohio. M, 1950, University of Cincinnati. 42 p.

Harris, Susan Frye. Kinetics of diffusion-controlled reactions in garnet amphibolite, Llano County, Texas. M, 1986, University of Texas, Austin. 121 p.

Harris, Thaddeus William. Origin of the stratified rocks of the New York series. D, 1890, Harvard University.

Harris, Therese. Key aspects of Indonesia's energy and mineral resource industry. M, 1987, University of Texas, Austin.

Harris, Thomas G. The role of bacteria in the fixation of Co, Fe, and Mn in Lake Oneida, N.Y. M, 1978, University of South Carolina.

Harris, Timothy Donovan. Geology of the Round Top porphyry copper-molybdenum deposit, west-central Alaska. M, 1985, University of Colorado.

Harris, W. Burleigh. Stratigraphy, petrology, and radiometric age (upper Cretaceous) of the Rocky Point Member, Peedee Formation, North Carolina. D, 1975, University of North Carolina, Chapel Hill. 189 p.

Harris, Walter Stephan. Geology of the southwestern portion of the Blanco Mountain Quadrangle, Inyo County, California. M, 1958, University of California, Los Angeles.

Harris, William. Geology of parts of the Ridge and Largent, West Virginia quadrangles. M, 1968, West Virginia University.

Harris, William Howard. Geology of the Post Corners area, Washington and Rensselaer counties, New York. M, 1961, University of Texas, Austin.

Harris, William Howard. Groundwater-carbonate rock chemical interactions, Barbados, West Indies. D, 1971, Brown University.

Harris, William Langseth. Stratigraphy and economic geology of the Great Falls-Lewistown coal field, (Montana). M, 1968, University of Montana. 126 p.

Harris, William Langseth. Upper Mississippian and Pennsylvanian sediments of central Montana. D, 1973, University of Montana. 252 p.

Harris, William Maurice. Biostratigraphy of the Upper Cretaceous Austin Group, Travis County, Texas. M, 1982, Texas A&M University. 100 p.

Harris, William Maurice, Jr. Organism interactions and their environmental significance, as exemplified by the Pliocene-Pleistocene fauna of the Kettleman Hills and Humboldt Basin, California. D, 1987, Texas A&M University. 278 p.

Harris, Willie Garner. Taxonomic, behavioral, and genetic significance of soil mica. D, 1984, Virginia Polytechnic Institute and State University. 161 p.

Harrison, Ben S. Dispersion of short period Rayleigh waves in the Atlantic Ocean. M, 1981, Texas Tech University. 43 p.

Harrison, C. J. The design and application of two dimensional digital filters. M, 1978, University of Western Ontario.

Harrison, E. M. Biodegradation of trace organics as a ground water reclamation scheme. M, 1985, University of Waterloo. 104 p.

Harrison, Earl P. Geology of the Hagan coal basin. M, 1949, University of New Mexico. 177 p.

Harrison, Earl Preston. Depositional history of Cisco-Wolfcamp Strata, Bend Arch, north-central Texas. D, 1973, Texas Tech University. 189 p.

Harrison, Edward Vernon. A faunal study of the Jolliff Limestone Member of the Dornick Hills Formation in the Ardmore Basin (Oklahoma). M, 1948, University of Oklahoma. 52 p.

Harrison, Ellen Zucker. Sedimentation rates, shoreline modification, and vegetation changes on tidal marshes along the coast of Connecticut. M, 1975, Cornell University.

Harrison, Frank W., III. The role of pressure, temperature, salinity, lithology, and structure in hydrocarbon accumulation in Constance Bayou, Deep Lake and Southeast Little Pecan Lake fields, Cameron Parish, Louisiana. M, 1979, Louisiana State University.

Harrison, Gary Clyde. Facies analysis of the Devonian in Black Mesa Basin. M, 1976, Northern Arizona University. 57 p.

Harrison, George B. Geology and near infra-red remote sensing of the Thunderbird Mine. M, 1972, Wright State University. 36 p.

Harrison, Hubert James. Subsurface geology, southeastern Travis County, Texas. M, 1957, University of Texas, Austin.

Harrison, J. A. The Carnivora and Camelidae of the Edson local fauna (Hemphillian), Sherman County, Kansas. D, 1979, University of Kansas. 272 p.

Harrison, J. M. Certain anorthosites in southeastern Ontario. D, 1943, Queen's University. 100 p.

Harrison, J. M. Metamorphism and origin of a part of the Wasekwan Series. M, 1941, Queen's University. 37 p.

Harrison, J. W. Pennsylvanian stratigraphy of the Laramie Range of southeastern Wyoming. M, 1938, University of Wyoming. 85 p.

Harrison, Jack Edward. Relationship between structure and mineralogy of the Sherman Granite, southern part of the Laramie Range, Wyoming-Colorado. D, 1951, University of Illinois, Urbana.

Harrison, Jack Lamar. Clay mineral stability and genesis during weathering. D, 1958, Indiana University, Bloomington. 56 p.

Harrison, Jack Lamar. Depositional environments of lower Chester rocks in Indiana. M, 1955, Indiana University, Bloomington. 24 p.

Harrison, James E. Mineralogy and petrology of the Vulcan iron formation and related rocks, Felch District, Dickinson County, Michigan. M, 1984, Bowling Green State University. 105 p.

Harrison, Jane. Geochemistry and petrology and pyroxenite xenoliths from Xalapasco de la Joya Honda, San Luis Potosi, Mexico. M, 1988, University of Houston.

Harrison, Jessica A. Mammals of the Wolf Ranch fauna, Saint David Formation (Pleistocene), Cochise County, Arizona. M, 1972, University of Arizona.

Harrison, John Albert. Subsurface Pennsylvanian studies in White County, Illinois. M, 1948, University of Illinois, Urbana.

Harrison, John Christopher. Petrology of the Ying Creek alkalic intrusion, Southeast Yukon. M, 1982, University of Toronto.

Harrison, John E. Proglacial drainage evolution and deglaciation of the Great Bend region, (northeastern Pennsylvania). M, 1966, SUNY at Binghamton. 71 p.

Harrison, John Edward. Quaternary geology of the North Bay–Mattawa region. D, 1971, Syracuse University. 94 p.

Harrison, Linda Kelley. Foraminifera of part of the Holocene Rio Grande Delta in Cameron County, Texas. M, 1972, University of Cincinnati. 100 p.

Harrison, Paul James. Continuous culture of the marine diatom Skeletonema costatum (Grev.) Cleve under silicate limitation. D, 1974, University of Washington. 140 p.

Harrison, Peter. The land water interface in an urban region; a spatial and temporal analysis of the nature and significances of conflicts between coastal uses. D, 1973, University of Washington. 182 p.

Harrison, Peter F. Paleoecology of the Oligocene Red Bluff Clay in Alabama and Mississippi. M, 1976, Northeast Louisiana University.

Harrison, Phillip W. A clay till fabric; its character and origin. D, 1956, University of Chicago. 33 p.

Harrison, Randolph S. Near-surface subaerial diagenesis of Pleistocene carbonates, Barbados, West Indies. D, 1974, Brown University. 350 p.

Harrison, Randolph Stephen. Petrology of some mottled and laminated middle Cambrian dolomites from Banff (Alberta) and Yoho (British Columbia) National Parks, (Canada). M, 1969, University of Calgary. 136 p.

Harrison, Richard W. Geology of the northwestern Datil Mountains, Socorro and Catron counties, NM. M, 1980, New Mexico Institute of Mining and Technology. 137 p.

Harrison, Samuel Sterrett. Relationship of the Turtle Forest, and Park rivers to the history of glacial Lake Agassiz. M, 1965, University of North Dakota. 50 p.

Harrison, Samuel Sterrett. The effects of groundwater seepage on stream regimen; a laboratory study. D, 1968, University of North Dakota. 69 p.

Harrison, Shane M. Albian foraminiferal biostratigraphy of a key borehole in northeastern British Columbia. M, 1988, University of Saskatchewan. 180 p.

Harrison, Stanley Cooper. Depositional mechanics of Permian Cherry Canyon sandstone tongue, Last Chance Canyon, New Mexico. M, 1966, Texas Tech University. 114 p.

Harrison, Stanley Cooper. The sediments and sedimentary processes of the Holocene tidal flat complex, Delmarva Peninsula (Delaware, Maryland, Virginia). D, 1971, The Johns Hopkins University.

Harrison, Steven Adam. The control of uranium concentrations in Pleistocene planktonic foraminifera. M, 1988, University of Miami. 88 p.

Harrison, Susan B. Geology of the Caulksville Quadrangle, Franklin and Logan counties, Arkansas. M, 1977, Northeast Louisiana University.

Harrison, Sylvia L. Sedimentology of Tertiary sedimentary rocks near Salmon, Idaho. D, 1985, University of Montana. 161 p.

Harrison, W. E. Experimental diagenetic study of a modern lipid-rich sediment. D, 1976, Louisiana State University. 178 p.

Harrison, W. G. Nitrogen budget of a North Carolina estuary. D, 1974, North Carolina State University. 183 p.

Harrison, William Baxter. Epigenetic growth of calcite-cemented nodules within a porous dolomite matrix; Avon Park Formation of central Florida. M, 1969, University of South Florida, Tampa. 91 p.

Harrison, William Baxter, III. Bivalvia (Pelecypoda) of the Brassfield Formation (Lower Silurian) of Kentucky, Indiana and Ohio. D, 1974, University of Cincinnati. 382 p.

Harrison, William Donald. Determination of the composition of metamorphic rocks by use of the point counter. M, 1954, McMaster University. 55 p.

Harrison, William Earl. Heavy mineral analysis of Horn Island, northern Gulf of Mexico. M, 1968, University of Oklahoma. 88 p.

Harrison, William J. Paleomagnetism of four Late Cretaceous plutons, North Cascades, Washington. M, 1984, Western Washington University. 106 p.

Harrison, William James. Clay mineral assemblages and chlorite polytypes in the basal Nonesuch Shale, White Pine, Michigan. M, 1971, Northern Illinois University. 84 p.

Harriss, Robert Curtis. Geochemical and (Visean-Namurian) studies on the weathering of granitic rocks. D, 1965, Rice University. 124 p.

Harriss, Robert Curtis. The transfer of strontium, iron, and magnesium from sea water to skeletal carbonate material. M, 1964, Rice University. 43 p.

Harriz, J. Kimberly. Fluvial depositional sequences in the Hanna Formation, Hanna Basin, Wyoming. M, 1985, Texas A&M University.

Harrold, Jerry. Geology of the North-central Pueblo Mountains, Harney County, Oregon. M, 1973, Oregon State University. 135 p.

Harron, Gerald Allan. Mercurometric investigations at Lake Dufault mines (Precambrian), northwestern Quebec (Canada). M, 1969, University of Western Ontario. 144 p.

Harrop-Williams, Kingsley Ormonde. Reliability of geotechnical systems. D, 1980, Rensselaer Polytechnic Institute. 281 p.

Harrover, Robin D. Stable oxygen isotope and crystallite size analysis of Alaskan cherts; a possible exploration tool for submarine exhalative deposits. M, 1981, New Mexico Institute of Mining and Technology. 55 p.

Harrower, Karen L. Geology and chronology of buried Precambrian basement rocks in central Kansas. M, 1976, University of Kansas. 43 p.

Harryok, Harry Jerrold Van *see* Van Harryok, Harry Jerrold

Harsh, John F. Correlation of Mississippian carbonate rocks by differential thermal analysis. M, 1964, South Dakota School of Mines & Technology.

Harsh, John Franklin. Relationships between streamflow and ground water in Humboldt River valley near Winnemucca, Nevada. D, 1969, University of Nevada - Mackay School of Mines. 114 p.

Harsha, Senusi. An interpretation of southern Georgia coastal plain velocity structure using refraction and wide-angle reflection methods. M, 1988, Georgia Institute of Technology. 66 p.

Harshbarger, John W. Petrography and stratigraphy of Upper Jurassic rocks of Central Navajo Reservation, Arizona. D, 1949, University of Arizona.

Harshbarger, John W. The Upper Jurassic stratigraphy of Black Mesa, Arizona. M, 1948, University of Arizona.

Harshman, Elbert N. Geology of the Belmont-Queen Creek area, Superior, Arizona. D, 1940, University of Arizona.

Harshman, Elbert N. Geology of the San Jose Hills, Los Angeles County, California. M, 1933, California Institute of Technology. 83 p.

Harston, Lee W. Geology of the Bear Creek-Beaver Creek area, Big Horn and Sheridan counties, Wyoming. M, 1959, University of Wyoming. 110 p.

Hart, Alan W. Geology of Sierra Hermosa Quadrangle (northern half) Zacatecas and San Luis Potosi, Mexico. M, 1979, University of Texas, Arlington. 110 p.

Hart, Brian R. Genesis of intra-till sorted sediment layers, Catfish Creek till, Bradtville, Ontario. M, 1988, University of Western Ontario. 219 p.

Hart, D. L. The relationship of the phytoplankton to petroleum-recovery activities in a Louisiana salt marsh. M, 1978, Louisiana State University.

Hart, Dabney Gardner. Ostracoda of zone 10 of the Calvert Formation (middle Miocene) of Maryland. M, 1970, Bryn Mawr College. 90 p.

Hart, Daniel Douglas. The geology and origin of the Green Mountain massive sulfide deposit, Mariposa County, California. M, 1978, University of Nevada. 107 p.

Hart, Dirk Van *see* Van Hart, Dirk

Hart, E. A. The geology of the Fontana gold mines property, Duvernay Township, Quebec. M, 1939, McGill University.

Hart, Earl W. Upper Mesozoic rocks near Atascadero, Santa Lucia Range, California with special reference to K-feldspar content of the sandstone. M, 1971, University of California, Berkeley. 70 p.

Hart, James J. P. The most economical method of treating the sulphide ores of Goldfield. M, 1911, University of Nevada - Mackay School of Mines. 15 p.

Hart, James Martin. The geology of a portion of the Santa Barbara Canyon area, northeastern Santa Barbara County, Southern California. M, 1959, University of California, Los Angeles.

Hart, Jeanne C. Remotely sensed distributions of suspended sediment in Lake Pontchartrain. M, 1978, University of New Orleans.

Hart, Lyman H. The velocity of conversion of organic matter of black Devonian oil shales to bitumen. M, 1925, University of Wisconsin-Madison.

Hart, Margaret Lynn. The geologic survey of Allegheny County, PA. M, 1973, Pennsylvania State University, University Park. 49 p.

Hart, Michael. Landslides (Pleistocene-Recent) of west-central San Diego County, California. M, 1972, [University of California, San Diego].

Hart, Orville Dorwin. A study of the Lower Pennsylvanian Wapanucka Formation of the frontal Ouachita Mountains, Southeast Oklahoma. M, 1961, University of Wisconsin-Madison.

Hart, Orville Dorwin. Geology of the eastern part of the Windingstair Range of the Ouachita Mountains in southeastern Oklahoma. D, 1963, University of Wisconsin-Madison. 155 p.

Hart, Pembroke J. Variation of velocity near the Mohorovicic discontinuity under Maryland and northeastern Virginia. D, 1955, Harvard University.

Hart, Philip. A hydrogeochemical study of Mississippian oil brines of western Kansas. M, 1977, Wichita State University. 111 p.

Hart, Richard M. The trace-element content of detrital quartz in part of the Mississippi River system. M, 1976, University of Missouri, Columbia.

Hart, Richard Royce. Biostratigraphic relations of the basal St. Peter Sandstone in Northeast Iowa and Southwest Wisconsin. D, 1963, University of Iowa. 220 p.

Hart, Robert Stuart. The distribution of seismic velocities and attenuation in the Earth. D, 1977, California Institute of Technology. 362 p.

Hart, Roger Dale. A fully coupled thermal-mechanical-fluid flow model for nonlinear geologic systems. D, 1981, University of Minnesota, Minneapolis. 360 p.

Hart, Stanley R. Mineral ages in metamorphism. D, 1960, Massachusetts Institute of Technology. 219 p.

Hart, Steven W. Hydrogeologic evaluation of a proposed sanitary landfill, Dougherty County, Georgia. M, 1982, Boston University. 133 p.

Hart, Suchit Suthirachartkul. The palynology of the Eocene Wilcox Group associated with Arkansas bauxite. M, 1976, University of Oklahoma. 146 p.

Hart, T. L. An ecological study of epipsammic diatoms from sediments associated with Juncus roemerianus in a Northwest Florida salt marsh. D, 1977, Florida State University. 202 p.

Hart, Thomas Allen. Areal geology and Cretaceous stratigraphy of northwestern Bryan County, Oklahoma. M, 1970, University of Oklahoma. 215 p.

Hart, Thomas C. Reduction of topographic shadow effects in Landsat data by division of mean brightness. M, 1978, Colorado State University. 87 p.

Hart, Thomas Robert. The geochemistry and petrogenesis of a metavolcanic and intrusive sequence in the Kamiskotia area, Timmins, Ontario. M, 1984, University of Toronto.

Hart, William D. Structural geology of the north portion of the Goosepond Mountain Quadrangle, Ouachita Mountains, Arkansas. M, 1985, University of Missouri, Columbia.

Hart, William E. Methods adaptable to field construction of hyperbolic and circular grids. M, 1960, Ohio State University.

Hart, William George. Microfacies analysis of the Permian reef complex (Guadalupian), Carlsbad Caverns, New Mexico. M, 1969, Texas Tech University. 88 p.

Hart, William J. E. The Panchimalco Tephra, El Salvador, Central America. M, 1981, Rutgers, The State University, New Brunswick. 101 p.

Hart, William Kenneth. Chemical, geochronologic and isotopic significance of low K, high-alumina olivine tholeiite in the northwestern Great Basin, U.S.A. D, 1982, Case Western Reserve University. 431 p.

Harter, Ty Andrew. Correlation of the lower and upper Millersburg coals to the Hymera (VI) Coal Member and the Danville (VII) Coal Member of Southwest Indiana using trace elements. M, 1985, Indiana University, Bloomington. 106 p.

Harth, Peter Marc. A gravity, seismic, and well log analysis of the Clarendon-Linden fault system in New York State. M, 1984, Syracuse University.

Harthill, Norman. Deep electromagnetic sounding; geological considerations. D, 1969, Colorado School of Mines. 132 p.

Hartig, Robert L. Minor petrographic constituents of some Permian rocks. M, 1954, Kansas State University. 48 p.

Hartke, Edwin Joseph. Source of acid mine drainage in the Monday Creek drainage basin, southeastern Ohio. M, 1974, Ohio University, Athens. 66 p.

Hartley, Alan H. The sulfides of the Bays-of-Maine Complex, Penobscot Township, Maine. M, 1972, University of Washington.

Hartley, Marvin Eugene, III. Ultramafic and related rocks in the Lake Chatuge area, Clay County, North Carolina, and Towns County, Georgia. M, 1971, University of Georgia. 99 p.

Hartley, P. D. The geology and mineralization of a Precambrian massive sulfide deposit at Vulcan, Gunnison County, Colorado. M, 1976, Stanford University. 86 p.

Hartley, Susan. Petrology of the Bays-of-Maine Complex near Penobscot Bay, Maine. M, 1972, University of Washington.

Hartline, B. K. Topographic forcing of thermal convection in a Hele-Shaw cell model of a porous medium. D, 1978, University of Washington. 177 p.

Hartline, Laurie Elizabeth. Illinoian stratigraphy of the Bond County region of west central Illinois. M, 1981, University of Illinois, Urbana. 103 p.

Hartman, B. A. The use of carbon and sulfur isotopic ratios and total sulfur content for identifying the origin of beach tars in Santa Monica Bay, California. M, 1978, University of Southern California.

Hartman, Blayne Alan. Laboratory and field investigations of the processes controlling gas exchange across the air-water interface. D, 1983, University of Southern California.

Hartman, C. M. Structure and preservation of tracheid walls in selected Paleozoic plant genera. D, 1979, Cornell University. 154 p.

Hartman, Donald A. Stratigraphy and structure of the Salt Wells Anticline area Sweetwater County, Wyoming. M, 1968, Oregon State University. 115 p.

Hartman, Donald Albert. Geology and low-grade metamorphism of the Greenwater River area, central Cascade Range, Washington. D, 1973, University of Washington. 99 p.

Hartman, Donald Carl. Geology of the Upper Wagon Road Canyon area, Southern California. M, 1957, University of California, Los Angeles.

Hartman, Frederick H. The geology of North Ute Pass, southern Douglas County, Colorado. M, 1951, Colorado School of Mines. 59 p.

Hartman, Harold Joseph, Jr. First report of a preliminary geothermal investigation of the Rio Grande Rift in New Mexico. M, 1972, New Mexico Institute of Mining and Technology.

Hartman, James A. Origin of the heavy minerals in Jamaican bauxite. M, 1955, University of Wisconsin-Madison.

Hartman, James Austin. Titanium mineralogy of some bauxites. D, 1956, University of Wisconsin-Madison. 95 p.

Hartman, Jane E. Z. Vertebrate paleontology of the lower part of the Polecat Bench Formation, southern Bighorn Basin, Wyoming. M, 1984, University of Wyoming. 202 p.

Hartman, Joseph H. Uppermost Cretaceous and Paleocene nonmarine Mollusca of eastern Montana and southwestern North Dakota. M, 1976, University of Minnesota, Minneapolis. 215 p.

Hartman, Joseph Herbert. Systematics, biostratigraphy, and biogeography of latest Cretaceous and early Tertiary Viviparidae (Mollusca, Gastropoda) of southern Saskatchewan, western North Dakota, eastern Montana, and northern Wyoming. D, 1984, University of Minnesota, Minneapolis. 998 p.

Hartman, Mary J. A study of groundwater age in the Bunker Hill Mine, Idaho. M, 1986, University of Idaho. 52 p.

Hartman, Russell T. Coal measures and coal mining in Iowa. D, 1898, Iowa State University of Science and Technology.

Hartman, Leo A. Petrology of Precambrian igneous and metamorphic rocks in a portion of the Rawah Batholith, Medicine Bow Mountains, Colorado. M, 1973, Colorado State University. 157 p.

Hartmann, William K. Radial structures surrounding lunar basins. M, 1964, University of Arizona.

Hartmetz, Christopher Pate. Thermoluminescence of annealed and shock-loaded feldspar. D, 1988, University of Arkansas, Fayetteville. 225 p.

Hartnagel, Florence A. Regional geography of Scotland. M, 1929, Washington University. 119 p.

Hartnell, Jill Ann. The vertebrate palaeontology, depositional environment and sandstone provenance of early Eocene rocks on Tornillo Flat, Big Bend National Park, Brewster County, Texas. M, 1980, Louisiana State University.

Hartner, John D. Depositional environments of the Salt Wash sandstones (Upper Jurassic) in portions of Rio Blanco and Moffat counties, Colorado. M, 1981, Colorado School of Mines. 110 p.

Hartness, Thomas Scott. Distribution and clay mineralogy of organic-rich mud sediments in the Pamlico River estuary, North Carolina. M, 1977, East Carolina University. 45 p.

Hartree, Ron. Polarographic determination of Pb and Zn in carbonate rocks. M, 1979, University of Ottawa. 62 p.

Hartsell, Mickey York. Niagara pinnacle reefs of western Michigan. M, 1982, Michigan State University. 109 p.

Hartshorn, David Robert. Soil geochemistry as a guide to mineralization in the Drum Mountains, Millard-Juab counties, Utah. M, 1980, California State University, Northridge. 76 p.

Hartshorn, Joseph Harold. Glacial geology of the Taunton Quadrangle, Massachusetts. D, 1955, Harvard University.

Hartsock, John Kaus. Submarine topography and bottom sediments off the southeast coast of Iceland. M, 1955, University of Cincinnati. 53 p.

Hartsog, William Smith. Soil erodibility prediction for excavated materials. D, 1988, Montana State University. 83 p.

Hartsook, Alan D. Petrology and diagenesis of the Pettet Interval, Sligo Formation (Lower Cretaceous), Bossier Parish, Louisiana. M, 1983, East Carolina University. 83 p.

Hartsough, Gregory Warren. The Mississippian-Pennsylvanian unconformity and post-Mississippian topography and areal stratigraphy of Knox County, Indiana. M, 1979, Indiana University, Bloomington. 64 p.

Hartung, Jack Burdair. Application of the potassium-argon method to the dating of shocked rocks. D, 1968, Rice University. 93 p.

Hartwell, Alan D. Hydrography and Holocene sedimentation of the Merrimack river estuary (Massachusetts). M, 1970, University of Massachusetts. 170 p.

Hartwell, J. M. Organometallic associations in sediment-sea water systems. M, 1978, Dalhousie University.

Hartwell, James N. A paleocurrent analysis of a portion of the Chuckanut depositional basin near Bellingham, Washington. M, 1979, Western Washington University. 85 p.

Hartwick, Wayde M. Taxonomy, morphology, and paleoecology of Strophomena planumbona (Hall), an articulate brachiopod from the Upper Ordovician of the Cincinnati, Ohio, area. M, 1987, Miami University (Ohio). 120 p.

Hartwig, Albert Ernest, Jr. Geology of the Mozo Quadrangle, Williamson County, Texas. M, 1952, University of Texas, Austin.

Hartwig, N. L. An anatomical study of Cycadeoidea dacotensis and Cycadeoidea mcbridei. D, 1976, University of Iowa. 139 p.

Hartz, Kenneth Eugene. Studies of methanogenesis in samples from landfills. D, 1980, University of Wisconsin-Madison. 236 p.

Hartzell, Stephan P. The Precambrian geology of the Kersey Lake area, South-central Beartooth Mountains, Montana. M, 1978, Northern Illinois University. 71 p.

Hartzell, Stephen H. Interpretation of earthquake strong ground motion and implications for earthquake mechanism. D, 1978, University of California, San Diego. 294 p.

Hartzog, Laurence David. The determination of rhenium by atomic absorption spectrophotometry. M, 1966, University of Nevada - Mackay School of Mines. 44 p.

Harun, Happy. Distribution and deposition of lower and middle Eocene strata in central San Joaquin Valley, California. M, 1984, Stanford University. 100 p.

Harvard, Charles Gentry. Geology of the southwestern part of the Van Horn Mountains, Trans-Pecos, Texas. M, 1949, University of Texas, Austin.

Harvard, Paul Odom. Time-rock correlations and biofacies of the Lower Cretaceous "Edwards" Limestone, south-central Texas. M, 1962, Texas Christian University.

Harvell, George R., Jr. A geologic study of the Chickamauga Formation of Raccoon Valley, Anderson County, Tennessee. M, 1954, University of Tennessee, Knoxville. 39 p.

Harvey, Andrew Frank, III. Surficial and environmental geology of the Sandpoint area, Bonner County, Idaho. M, 1984, University of Idaho. 136 p.

Harvey, Bruce A. Geology and petrology of the Sunlight Basin intrusions and surrounding area, Absaroka volcanic field, Park County, Wyoming. M, 1982, University of New Mexico. 130 p.

Harvey, Bruce Warren. The microscopic petrography and ore microscopy of the Boy Scout-Jones molybdenum prospect, Halifax County, North Carolina. M, 1974, North Carolina State University. 85 p.

Harvey, Colin Charles. A study of the alteration products of acid volcanic rocks from Northland, New Zealand. D, 1980, Indiana University, Bloomington. 322 p.

Harvey, Constance St. Clair. Petrography, structure, and trace element content of wall rock biotites from the Boyd and Calloway ore bodies, Ducktown, Tennessee. M, 1975, North Carolina State University. 75 p.

Harvey, Cyril Hingston, II. A paleoecological interpretation of the White River faunas of Sioux County, Nebraska. M, 1956, University of Nebraska, Lincoln.

Harvey, Cyril Hingston, II. Stratigraphy, sedimentation and environment of the White River Group of the Oligocene of northern Sioux County, Nebraska. D, 1960, University of Nebraska, Lincoln. 249 p.

Harvey, Danny James. A spectral method for computing complete synthetic seismograms. D, 1985, University of Colorado. 308 p.

Harvey, David B. Cassiterite mineralization in the Black Range tin district, Sierra and Catron counties, New Mexico. M, 1985, University of Texas at El Paso.

Harvey, E. T. A sand budget in the upper St. Mary's River. M, 1976, University of Waterloo.

Harvey, Edward J. The geology of the Avondale Quadrangle. M, 1948, University of Tennessee, Knoxville. 98 p.

Harvey, John Frank. Mississippian stratigraphy of Jasper Park, Alberta. D, 1953, University of Wisconsin-Madison.

Harvey, John Frank. Structural geology, northern Bridger Range, Montana. M, 1951, University of Wisconsin-Madison. 24 p.

Harvey, Joseph L. A geologic reconnaissance in the southwest Olympic Peninsula. M, 1959, University of Washington. 53 p.

Harvey, Michael David. Steepland channel response to episodic erosion. D, 1980, Colorado State University. 283 p.

Harvey, R. R. Detection of oceanic-induced variations in the telluric field at an island station. M, 1969, University of Hawaii. 88 p.

Harvey, R. R. Measurement of the vertical electric field in the deep ocean. D, 1972, University of Hawaii. 117 p.

Harvey, R. W. Electro static proton cyclotron harmonic waves observed with the Alouette II Satellite. M, 1969, University of British Columbia.

Harvey, Ralph Leon. Subsurface geology of a portion of southern Hughes County, Oklahoma. M, 1960, University of Oklahoma. 91 p.

Harvey, Richard David. Hydrothermal alteration in the Goldfield District, Nevada. D, 1960, Indiana University, Bloomington. 71 p.

Harvey, Roger Douglas. The electrical conductivity of the opaque minerals in polished section. D, 1927, Harvard University.

Harvey, Ronald William. Lead-bacterial interactions in an estuarine salt marsh microlayer. D, 1981, Stanford University. 184 p.

Harvey, Rose Mary. Geologic data for land use planning in Union Township, Madison County. M, 1974, Ball State University.

Harvey, Ruth S. Drainage and glaciation in the central Housatonic Basin (Connecticut). D, 1908, Yale University.

Harvey, T. J. The paleolimnology of Lake Mobutu Sese Seko, Uganda-Zaire; the last 28,000 years. D, 1976, Duke University. 113 p.

Harvey, Timothy William. Geology of the San Miguel fault zone, northern Baja California, Mexico. M, 1985, San Diego State University. 330 p.

Harvey, Y. Détermination de la structure cristalline de la néphéline hydrate I. M, 1974, Ecole Polytechnique.

Harvey, Yves. Détermination de la structure cristalline de la nepheline hydrate I. M, 1975, Ecole Polytechnique.

Harvey, Yves. Métallogénie de l'uranium pegmatitique dans le Grenville du Québec. D, 1983, Universite de Montreal.

Harvie, Charles Edmund. Theoretical investigations in geochemistry and atom surface scattering. D, 1981, [University of California, San Diego]. 293 p.

Harvie, Robert. Geology of a portion of Fabre Township, Pontiac County. D, 1911, University of Wisconsin-Madison.

Harvill, Lee L. A petrographic study of the Anacacho Limestone (Upper Cretaceous) of Texas. M, 1958, University of Houston.

Harvill, Lee Lon. Deformational history of the Pelona schist (formation age-Mesozoic(?); metamorphic age-Cretaceous (?)), northwestern Los Angeles County, California. D, 1969, University of California, Los Angeles. 167 p.

Harvill, Martin Lavell. Hydrothermal alteration in the Davis Mountains, Texas. M, 1961, University of Texas, Austin.

Harville, Donald G. The occurrence of clays and their bearing on evaporite mineralogy in the Salado Formation, Delaware Basin, New Mexico. M, 1985, Texas A&M University. 110 p.

Harwell, George Mathis, Jr. Stratigraphy of Sierra del Porvenir, Chihuahua, Mexico. M, 1959, University of Texas, Austin.

Harwell, Jeffrey W. Offset-amplitude seismic analysis; Bonny gas field, Yuma County, Colorado. M, 1987, Colorado School of Mines. 90 p.

Harwood, Christine Lee. Groundwater contaminant transport. D, 1988, Purdue University. 168 p.

Harwood, David. Geology of the Cupsuptic Quadrangle, Maine. D, 1967, Harvard University.

Harwood, David M. Oligocene-Miocene diatom biostratigraphy from the Equatorial to the Antarctic Pacific. M, 1982, Florida State University.

Harwood, David Michael. Diatom biostratigraphy and paleoecology with a Cenozoic history of Antarctic ice sheets; (Volumes I and II). D, 1986, Ohio State University. 653 p.

Harwood, James Robert. Compositional variations associated with carbonate aggregate-cement paste reactions. M, 1960, Iowa State University of Science and Technology.

Harwood, Peggy J. Stability and geomorphology of Pass Cavallo and its flood delta since 1856, central Texas coast. M, 1973, University of Texas, Austin.

Harwood, Robert James. Inorganic geochemistry of the lower Truckee River, Nevada. M, 1964, Stanford University.

Harwood, Roderick James. Community reconstruction in benthic paleoenvironments; trophic structure in living and dead macroinvertebrate associations, Corpus Christi and Aransas Bay systems, Texas. D, 1980, University of Texas, Austin. 270 p.

Harwood, T. A. Some aspects of Canadian mineral economics. M, 1953, University of Toronto.

Harwood, William E. The geology of the Salt Branch area, Mason County, Texas. M, 1959, Texas A&M University.

Harz, Mary Catherine. Paleoenvironmental interpretation of the Mississippian Chainman Formation in Nevada and Utah. M, 1982, Eastern Washington University. 82 p.

Hasan, Manzoor. Upper Bolarian and lower Trentonian conodonts from Herkimer County, New York. M, 1969, Boston University. 125 p.

Hasan, Mohammad Nurul. A new filtering technique for correcting time variations in magnetic data. M, 1988, Colorado School of Mines. 123 p.

Hasan, S. E. Standardization of cell size for environmental geology data base and generation of decision making criteria for land use planning. D, 1978, Purdue University. 277 p.

Hascall, Allan P. Stratigraphic palynology, vegetation dynamics and paleoecology of the Florissant Lake Beds (Oligocene), Colorado. M, 1988, Michigan State University. 149 p.

Hase, Donald H. The application of polarization figures and rotation properties to the identification of the telluride minerals. M, 1951, University of Wisconsin-Madison.

Hase, Donald H. Upper Huronian sedimentation in a portion of the Marquette Trough, Michigan. D, 1955, University of Wisconsin-Madison.

Hase, Harold W., Jr. Geological-geophysical site investigation of a portion of the Student Development Complex, Michigan Technological University, Houghton County, Michigan. M, 1973, Michigan Technological University. 35 p.

Hasegawa, Henery. Magneto-telluric studies in central Alberta. M, 1962, University of Alberta. 79 p.

Hasegawa, Henry S. A study of the attenuation of elastic waves in metals. D, 1965, University of British Columbia.

Haselau, Olivia V. A comparison of the Phytosauria and the Recent Crocodilia. M, 1948, Columbia University, Teachers College.

Haselow, John Stevens. Scaling dispersion during miscible displacement in heterogeneous porous media. D, 1988, Purdue University. 255 p.

Haselton, George Montgomery. Glacial geology of Muir Inlet, southeastern Alaska. D, 1967, Ohio State University. 242 p.

Haselton, Henry Trenholm. Calorimetry of synthetic pyrope-grossular garnets and calculated stability relations. D, 1979, University of Chicago. 98 p.

Haselton, Thomas M. An evaluation of earth resistivity techniques for study of earthquake related features in the Mississippi River alluvial plain. M, 1977, Vanderbilt University.

Haseman, Joseph Fish. An investigation of the use of heavy minerals in determining the origin and course of profile development in soils. D, 1944, University of Missouri, Columbia.

Haseman, Joseph Fish. The effect of different electrolytes on the A.F.A. clay determination. M, 1938, Cornell University.

Hasenaka, Toshiaki. The cinder cones of Michoacán-Guanajuato, central Mexico. D, 1986, University of California, Berkeley. 171 p.

Hasenmueller, Walter. Conodont biostratigraphy of the lower Triassic Thaynes Formation of the Confusion range, west central Utah. M, 1970, Ohio State University.

Hasenohr, E. J. Copper mineralization in the Lisbon Valley, San Juan County, Utah. M, 1976, Ohio State University. 79 p.

Hasenohr, Edward Joseph. Statistical analysis of trace element distributions in rocks and soils of the Breckenridge mining district, Summit County, Colorado. D, 1987, Ohio State University. 388 p.

Hasenpflug, Harry John, Jr. Anomalous energy distribution from alpha quarry blasts. M, 1955, St. Louis University.

Hash, Bender. A stratigraphic correlation of oil well sludge samples by spectrographic analysis. D, 1952, Stanford University. 137 p.

Hash, Troy M. Kirchhoff diffraction modeling; methods and seismic applications. M, 1983, Indiana University, Bloomington. 212 p.

Hashad, Ahmad Hassanain. Geochronological studies in the central Wasatch Mountains, Utah. D, 1964, University of Utah. 108 p.

Hashem, Fadel Musa. Groundwater flow theory for perforations in well casings and soil drain tubes. D, 1987, Iowa State University of Science and Technology. 229 p.

Hashim, Braik M. A seismic model of the upper Mississippi Embayment from P and S times of local earthquakes. M, 1977, St. Louis University.

Hashim, Hashim Mohammed. Development of curves of permeability vs. Reynolds number for flow in unconsolidated porous media. M, 1974, University of Idaho. 79 p.

Hashimoto, Isao. Differential dissolution analysis of clays and its application to Hawaiian soils. D, 1961, University of Wisconsin-Madison. 122 p.

Hashimoto, T. Iron silicate equilibria in the Cape Smith Belt, New Quebec. M, 1962, McGill University.

Hashimoto, Tsutomu. Mineral assemblages and phase equilibria in the metamorphosed silicate iron-formations of the Cape Smith Belt, New Quebec and the Labrador Trough (Quebec and Newfoundland). D, 1965, Universite Laval. 111 p.

Haskell, Barry S. The geology of a portion of the New York Mountains and Lanfair Valley. M, 1959, University of Southern California.

Haskell, Kenneth G. Rock discontinuity properties and ground water flow in an underground lead-zinc mine, Coeur d'Alene mining district, Idaho. M, 1987, University of Idaho. 130 p.

Haskell, N. L. Long Sand Shoal. D, 1977, University of Connecticut. 142 p.

Haskell, Norman Abraham. A study of the mechanics of deformation of the granitic rocks. D, 1936, Harvard University.

Haskell, Norman Leif. The petrography of the Tensleep Formation (Pennsylvanian and lower Permian) along the west flank of the Bighorn mountains, Bighorn County, Wyoming. M, 1969, Iowa State University of Science and Technology.

Haskin, Mark Allan. Archeological geology of Tulipe, Ecuador. M, 1982, University of Illinois, Urbana. 76 p.

Haskin, Richard Allen. Magnetic polarity stratigraphy and fossil Mammalia of the San Jose Formation, Eocene, New Mexico. M, 1980, University of Arizona. 74 p.

Haskins, Donald. Stratigraphy, petrology, and depositional environments of the Upper Devonian Devils Gate Limestone, Eureka County, Nevada. M, 1979, California State University, Fresno.

Haslam, Christopher R. S. Magnetotellurics in the eastern townships of Quebec. D, 1974, McGill University.

Haslett, James M. Depositional systems of the Tensleep Sandstone (Pennsylvanian) along the west flank of the Big Horn Mountains, Big Horn and Washakie counties, Wyoming. M, 1986, Northern Arizona University. 170 p.

Hass, W. H. Conodonts from the central mineral region, Texas. D, 1943, The Johns Hopkins University.

Hassan, Afifa Afifi. Geochemical and mineralogical studies on bone material and their implications for radiocarbon dating. D, 1975, Southern Methodist University. 123 p.

Hassan, Ahmad. Boron in the ground water in the Arvin-Edison area of California. M, 1959, Stanford University.

Hassan, Ahmad Amin Abdel Khalek. The distribution of exchangeable cations in some soils in the vicinity of Socorro, New Mexico. D, 1963, New Mexico Institute of Mining and Technology. 116 p.

Hassan, Farkhonda. Anisotropic color centers related to iron in amethyst quartz. D, 1970, University of Pittsburgh.

Hassan, Hassan-Hashim. Uranium and thorium distribution in the rocks of southwestern New Brunswick, Canada. D, 1984, University of New Brunswick.

Hassan, Ishmael. High resolution microscopy and x-ray study of cancrinite. M, 1980, McMaster University. 99 p.

Hassan, Ishmael. The crystal structure and crystal chemistry of sodalite and cancrinite groups of minerals. D, 1983, McMaster University. 236 p.

Hassan, Mamdouh Abdel-Ghafoor. On metamorphism of Jurrasic volcanic rocks in the foothills of the Sierra Nevada, California, with a review of stratigraphy and structure. D, 1968, University of California, Berkeley. 75 p.

Hassan, T. S. The properties and genesis of soils derived from Pahala Ash in Kau District, Hawaii. M, 1969, University of Hawaii. 72 p.

Hassanipak, Ali Asghar. Isotopic geochemical evidence concerning the origin of Georgia kaolin deposits. D, 1980, Georgia Institute of Technology. 217 p.

Hassanzadah, Siamak. Determination of depth to the magnetic basement using maximum entropy with application to the northern Chile Trench. M, 1976, Oregon State University. 64 p.

Hassanzadeh, Siamak. Seismic applications to reservoir characterization and shear-wave velocity determination. D, 1988, Columbia University, Teachers College. 78 p.

Hassell, Donald R. The geothermometric significance of Mg/Fe partitioning in coexistent garnet and biotite in metamorphosed pelitic sediments of New England. M, 1972, Miami University (Ohio). 126 p.

Hasseltine, George H. Geology of the San Miguel Syncline, Coahuila, Mexico. M, 1968, University of Missouri, Columbia.

Hassen-Bey, Tarak Mustafa. The use of microearthquakes in mapping the base of the low rigidity layer beneath Socorro, New Mexico. M, 1974, New Mexico Institute of Mining and Technology.

Hasser, Edward G. A study of Lake Iroquois and post Lake Iroquois sediments in a peatbog (New York). M, 1954, Syracuse University.

Hassinger, Jon Miller. Characteristics, dynamics, and paleoclimatic significance of rock glaciers in southern Victoria Land, Antarctica. M, 1981, University of New Hampshire. 134 p.

Hassinger, Russell Neal. Stratigraphy of the Morrison Formation of the Canon City Embayment, Colorado. M, 1959, University of Oklahoma. 116 p.

Hassler, H. Patricia. Burrow types and their distribution in the Helderberg Group (lower Devonian) of New York State. M, 1971, Brown University.

Hassler, Michael H. The correlation of core studies with geophysical field measurements for Green Bay sediments. M, 1984, University of Wisconsin-Milwaukee. 128 p.

Hasslock, Augusta T. The geology and geography of the area southwest of Abilene, Texas. M, 1910, University of Chicago. 67 p.

Hasson, Kenneth Owen. Lithostratigraphy and paleontology of the Devonian Harrell Shale along the Allegheny Front in West Virginia and adjacent states. M, 1966, University of Tennessee, Knoxville. 89 p.

Hasson, Kenneth Owen. Lithostratigraphy of the Grainger Formation (Mississippian) in Northeast Tennessee. D, 1972, University of Tennessee, Knoxville. 143 p.

Hasson, Mohey El-Din M. T. Effect of pressure and temperature on interfacial tensions for several water-hydrocarbon systems. D, 1953, Pennsylvania State University, University Park. 139 p.

Hasson, Phyllis Fairbanks. Studies in Cenozoic biostratigraphy based on microplankton. D, 1983, Princeton University. 289 p.

Hasson, Richard C. The sediments of the Madera Limestone, New Mexico. M, 1950, Texas Tech University. 42 p.

Hastie, L. M. The application of static elastic dislocation theory to thrust faulting. M, 1966, University of Toronto.

Hastings, David J. Analysis of geophysical data from the Point of Pines area, San Carlos Indian Reservation, Arizona. D, 1972, University of Arizona.

Hastings, Douglas. The sub-Eocene unconformity, Pine Mountain and Mount Pinos areas, central Transverse Ranges, California. M, 1977, University of California, Santa Barbara.

Hastings, James S. Geology of selected uranium-vanadium deposits of Long Park, Montrose County, Colorado. M, 1957, Colorado School of Mines. 93 p.

Hastings, John O. Structure, depositional environment, and pressure characteristics of the Vicksburg Formation; Javelina and East McCook Field, Hidalgo County, Texas. M, 1984, Texas A&M University.

Hastings, Susan Carol. Shallow-water sediment contribution to deep-sea deposits, Saint Croix, United States Virgin Islands. M, 1972, Duke University. 85 p.

Hastings, Thomas Worcester. A theoretical approach for assessing the role of rock and fluid properties in the development of abnormal fluid pressures. M, 1986, Texas A&M University.

Haston, Roger. Paleomagnetic results from Palau, West Caroline Islands; a constraint on Philippine Sea Plate motion. M, 1988, University of California, Santa Barbara.

Hatch, Floyd. Geology of the Elk Valley Quadrangle; Bear Lake and Caribou counties, Idaho and Lincoln County, Wyoming. M, 1980, Brigham Young University. 25 p.

Hatch, Gregory C. Geology of Caldwell Creek and adjacent areas of Fremont County, Wyoming. M, 1956, Miami University (Ohio). 90 p.

Hatch, Joseph R. Phase relationships in part of the system sodium carbonate-calcium carbonate-carbon dioxide-water at one atmosphere pressure. D, 1972, University of Illinois, Urbana. 85 p.

Hatch, Joseph Ray. Geochemical and petrographic trends in the limestones of the Oak Grove Member in west central Illinois (Pennsylvanian). M, 1968, University of Illinois, Urbana.

Hatch, Michael E. Neotectonics of the Agua Blanca Fault, Valle Agua Blanca, Baja California, Mexico. M, 1987, San Diego State University. 92 p.

Hatch, Norman Lowrie, Jr. The geology of the Dixville Quadrangle, New Hampshire. D, 1961, Harvard University.

Hatch, Robert Alchin. Phase equilibrium in the system $Li_2O \cdot Al_2O_3 - SiO_2$. D, 1942, University of Michigan.

Hatch, Robert Alchin. Pre-Cambrian crystalline rocks of the Wasatch Mountains, Utah. M, 1938, University of Michigan.

Hatchell, William O'Donnell. A stratigraphic study of the Navajo Sandstone (Upper (Triassic(?)–Jurassic), Navajo Mountain, Utah and Arizona. M, 1967, University of New Mexico. 121 p.

Hatcher, David A. Stratigraphy of the Pleasanton Group in Miami, Linn, and Bourbon counties, Kansas. M, 1961, University of Kansas. 65 p.

Hatcher, Robert. The petrology of the Hermitage Formation in central Tennessee. M, 1961, Vanderbilt University.

Hatcher, Robert Dean, Jr. Structure of the northern portion of the Dumplin Valley fault zone in East Tennessee. D, 1965, University of Tennessee, Knoxville. 168 p.

Hatcher, Roy Alvin, Jr. The gastropods of the loess of southern Boone County, Missouri. M, 1955, University of Missouri, Columbia.

Hatfield, Arlo Clark. A study of the Eagle Ford-Austin contact in Williamson, Travis, Hays, Comal, and Berax counties, Texas. M, 1932, University of Texas, Austin.

Hatfield, Craig Bond. Paleoecology of the Graneros Shale (Upper Cretaceous) in Kansas. M, 1961, Indiana University, Bloomington. 93 p.

Hatfield, Craig Bond. Stratigraphy and paleoecology of the Saluda Formation (Upper Ordovician) in Indiana, Ohio, and Kentucky. D, 1964, Indiana University, Bloomington. 112 p.

Hatfield, David M., Jr. Net shore-drift of Thurston County, Washington. M, 1983, Western Washington University. 120 p.

Hatfield, Harold Edmond. Gravity survey of southern Arbon and northern Curlew valleys, Oneida County, Idaho. M, 1983, Idaho State University. 45 p.

Hatfield, Kirk. Nonpoint source pollution management models for regional groundwater quality control. D, 1988, University of Massachusetts. 404 p.

Hatfield, Lloyd E. Stratigraphy of the Jacque Mountain and Whiskey Creek Pass formations; Pennsylvanian, Colorado. M, 1956, University of Colorado.

Hatfield, M. A. A land capability study of the proposed Stile Ranch site. M, 1978, Stanford University. 60 p.

Hatfield, Stanley Christopher. Mineralogy and chemistry of the tuff of Pritchards Station, east-central Nevada. M, 1983, University of Missouri, Rolla. 72 p.

Hatfield, Willis C. The geology of the Solwezi District, northern Rhodesia. D, 1937, Columbia University, Teachers College.

Hathaway, Allen Wayne. Lava tubes and collapse depressions. D, 1971, University of Arizona. 353 p.

Hathaway, Donald Joseph. The geology of a portion of the Red Knobs area, eastern Loudon County, Tennessee. M, 1957, University of Tennessee, Knoxville. 42 p.

Hathaway, John Cummins. Roundness and sphericity of the St. Peter Sandstone from Ottawa, Illinois. M, 1952, University of Illinois, Urbana.

Hathaway, Lawrence Robbins. Oxidation states of radiogenic leads in uranium minerals. D, 1963, University of Kansas. 84 p.

Hathaway, Wayne L. Geology of north part of the Corsicana shallow oil field, Ellis and Navaro counties, Texas. M, 1980, University of Texas, Arlington. 146 p.

Hatheway, Allen W. Engineering geology of subsidence at San Manuel Mine, Pinal County, Arizona. M, 1966, University of Arizona.

Hatheway, Richard B. Origin of the Saco Pluton, Saco, Maine. M, 1964, University of Missouri, Columbia.

Hatheway, Richard Brackett. Geology of the Wiscasset Quadrangle, Maine. D, 1969, Cornell University. 166 p.

Hathon, Eric Gene. Clay mineral distribution and elemental variability in near-surface, fine-grained sediments from the Aleutian Trench system. M, 1988, University of Missouri, Columbia.

Hatleberg, Eric Warner. Conodont biostratigraphy of the Lower Triassic at Van Keulenfjorden, Spitsbergen and the Thakkhola Valley, Nepal. M, 1982, University of Wisconsin-Madison.

Hatley, Allen Grady, Jr. Micropaleontology of a part of the Lower Cretaceous in Kent Quadrangle, Texas. M, 1955, Texas Tech University. 68 p.

Hatley, Michael D. Depositional environment of the Wayland Shale-Swastika (Avis) Sandstone (Upper Pennsylvanian) shelf-slope interval, west-central Texas. M, 1979, University of Texas, Arlington. 143 p.

Hatmaker, Paul C. Mining geology in the Rosiclare fluorspar district. M, 1927, University of Missouri, Rolla.

Hato, Masami. Optimal deconvolution using Wiener transform for compensation of time variance and non-minimum phase. M, 1986, Colorado School of Mines. 98 p.

Hatt, B. L. An interpretation of the carbonate geology exposed in the decline at Gays River, Nova Scotia. M, 1978, Dalhousie University. 134 p.

Hatt, Timothy A. Distinguishing depositional environments; an application of discriminant function analy-

sis to grain-size parameters. M, 1972, Rensselaer Polytechnic Institute. 43 p.

Hattin, Donald Edward. Depositional environment of the Wreford megacyclothem (Lower Permian) of Kansas. D, 1954, University of Kansas. 253 p.

Hattin, Donald Edward. The megascopic invertebrate fossils of the Carlile Shale of Kansas. M, 1952, University of Kansas. 122 p.

Hattner, John George. Upper Cretaceous calcareous nannofossil biostratigraphy of South Carolina. M, 1980, Florida State University.

Hatton, Josephine. A structural comparison of oceanic and continental shear zones. M, 1987, University of Rhode Island.

Hatton, Richard S. Aspects of marsh accretion and geochemistry; Barataria Basin, Louisiana. M, 1981, Louisiana State University.

Hatzikostantis, Nicholas G. Petrographic study of the Harriman Peridotite, Knox County, Maine, emphasizing variations in chlorite structure and composition. M, 1976, Boston University. 110 p.

Hau, Joseph A. Modeling the neutralization and groundwater responses to acid deposition in the non-saturated and saturated zones of phreatic aquifers in Lake and Wayne counties, Ohio. M, 1988, Kent State University, Kent. 132 p.

Haubold, Reiner G. A study of controlling salt water coning in aquifers. D, 1974, New Mexico Institute of Mining and Technology. 104 p.

Hauck, Anthony M. Deep structure resistivity in Massachusetts. M, 1960, Massachusetts Institute of Technology. 74 p.

Hauck, Rogers Austin. Stratigraphy and paleontology of Devonian rocks in part of west central Colorado; Pitkin, Eagle, and Garfield counties. M, 1954, University of Colorado.

Hauck, Samuel M. Geology of the southwest quarter of the Brownsville Quadrangle, Oregon. M, 1962, University of Oregon. 82 p.

Hauck, Steven A. Geology and petrology of the northwest quarter of the Bynum Quadrangle, Carolina slate belt, North Carolina. M, 1977, University of North Carolina, Chapel Hill. 146 p.

Hauck, Wayne Russell. Correlation and geochemical zonation of the mid-Tertiary volcanic and intrusive rocks in the Santa Teresa and northern Galiuro mountains, Arizona. M, 1985, University of Arizona. 140 p.

Hauf, C. B. Overthrusting in the upper Fontenelle-Labarge creeks area, Lincoln and Sublette counties, Wyoming. M, 1963, University of Wyoming. 75 p.

Haufler, J. B. Concentrations of selected trace elements in a wildlife food chain. D, 1979, Colorado State University. 154 p.

Haug, Frederick W. Post-glacial stratigraphy of the Great Bay estuary system. M, 1976, University of New Hampshire. 90 p.

Haug, Guido Alfredo. Early Silurian depositional environments of the lower unit of the Hidden Valley Dolomite, Great Basin. M, 1981, San Diego State University.

Haug, Jerry L. Geology of the Merry Widow and Kingfisher contact metasomatic skarnmagnetite deposits, northern Vancouver Island, British Columbia. M, 1977, University of Calgary. 117 p.

Haug, P. T. Ecological and economic system simulation for multiple-use decisions. D, 1975, Colorado State University. 208 p.

Haug, Patricia Ann. Applications of mass spectrometry to organic geochemistry. D, 1967, University of California, Berkeley. 306 p.

Hauge, Thomas Armitage. Geometry and kinematics of the Heart Mountain detachment fault, northwestern Wyoming and Montana. D, 1983, University of Southern California. 264 p.

Haugerud, Ralph A. The Shuksan Metamorphic Suite and Shuksan Thrust, Mt Watson area, North Cascades, Washington. M, 1980, Western Washington University. 125 p.

Haugerud, Ralph Albert. Geology of the Hozameen Group and Ross Lake shear zone, Maselpanik area, North Cascades, Southwest British Columbia. D, 1985, University of Washington. 263 p.

Haugh, Bruce Nilsson. Paleozoology of Mississippian camerate crinoids. D, 1973, University of California, Los Angeles.

Haugh, Bruce Nisson. Biostratigraphy of the Morrow Formation (Pennsylvanian), Tenkiller Ferry reservoir area, Oklahoma. M, 1968, University of Oklahoma. 236 p.

Haugh, Galen Rudolph. Late Cenozoic, cauldron-related silicic volcanism in the Twin Peaks area, Millard County, Utah. M, 1978, Brigham Young University.

Haugh, Ian. A petrofabric study of the Falcon Lake Stock, eastern Manitoba. M, 1962, University of Manitoba.

Haughey, William Henry. The relationship of quartz orientation to fluid flow structures found in acid extrusive rocks of the Royal Gorge Felsite, Missouri. M, 1952, Washington University. 84 p.

Haughton, David R. Solubility of sulphur in basaltic melts. D, 1971, Queen's University. 146 p.

Haughton, David Roderick. A mineralogical study of scapolite. M, 1968, McMaster University. 125 p.

Hauksson, Egill. Anomalous radon (222) emission and its association with earthquakes and tectonism. D, 1981, Columbia University, Teachers College. 161 p.

Haulenbeek, Roderick Beazley. Geology of the Sierra de Presidio, Chihuahua. M, 1970, University of Texas, Austin.

Haun, John D. Geology of a portion of the east flank of the Laramie Range. M, 1949, University of Wyoming. 40 p.

Haun, John D. Stratigraphy of Frontier Formation, Powder River basin, Wyoming. D, 1953, University of Wyoming. 149 p.

Hauntz, Charles E. Disseminated sulfide minerals (blue rock) in the Revett Formation, Lucky Friday Mine, Mullan, Idaho. M, 1982, University of Idaho. 75 p.

Haupt, Robert William. The effect of water diffusion on the magnetic field in a low-porosity crystalline rock. M, 1981, Pennsylvania State University, University Park. 68 p.

Hauptman, Charles McNerney. A thickness, structural and stratigraphic study of the Cherokee Group in the subsurface, Richardson and Nemaha counties, Nebraska. M, 1948, University of Nebraska, Lincoln.

Hauptman, Julie L. The sedimentology of the Wenatchee Formation; late Paleogene fluvial and lacustrine strata of the east-central Cascade Range, Washington State. M, 1983, University of Washington. 164 p.

Hausback, Brian Peter. Cenozoic volcanic and tectonic evolution of Baja California Sur, Mexico. D, 1984, University of California, Berkeley. 178 p.

Hausel, William Dan. Petrogenesis of lavas, Utah. M, 1974, University of Utah. 67 p.

Hausen, Donald M. Welded tuff along the Row River, western Oregon. M, 1951, University of Oregon. 98 p.

Hausen, Donald Martin. Fine gold occurrence at Carlin, Nevada. D, 1967, Columbia University. 166 p.

Hauser, E. E. Microcraters on lunar sample 12054,54. M, 1978, SUNY at Stony Brook.

Hauser, Ernest Clinton. Tectonic evolution of a segment of the West Spitsbergen Foldbelt in northern Wedel Jarlsberg Land. D, 1982, University of Wisconsin-Madison. 260 p.

Hauser, Ernest Clinton. The stratigraphy and structure of northern Chamberlindalen, Wedel Jarlsberg land, Svalbard. M, 1978, University of Wisconsin-Madison.

Hauser, K. A seismic reflection study in the Paradox Basin, southeastern Colorado. M, 1985, SUNY at Binghamton. 80 p.

Hauseux, M. A. Petrology and mineralogy of radioactive granitic rocks near Baie Johan Beetz, Quebec. M, 1976, McGill University. 85 p.

Hautau, Gordon H. The crystalline rocks of north central Rhode Island. M, 1939, Brown University.

Havard, Christina J. Sedimentary and early diagenetic cycles, Snipe Lake reef complex, Alberta. M, 1974, University of Calgary. 125 p.

Havard, Deborah Ann. Paleontological zonation of the Miocene of Hancock County, Mississippi. M, 1978, University of Southern Mississippi.

Havard, John F. R. Insoluble residues of Door County, Wisconsin, dolomites. M, 1936, University of Wisconsin-Madison.

Havard, Kenneth R. Mineralogy and geochemistry; Exshaw Formation (Mississippian), southern Alberta (Canada). M, 1967, University of Calgary. 126 p.

Havas, Magda. A study of the chemistry and biota of acid and alkaline ponds at the Smoking Hills, N.W.T. D, 1981, University of Toronto.

Have, Lewis Earl Ten *see* Ten Have, Lewis Earl

Haven, Elizabeth Lorraine. Gully erosion near Grapevine, California. M, 1980, Stanford University. 65 p.

Havenor, Kay Charles. Pennsylvanian framework of sedimentation in Arizona. M, 1958, University of Arizona.

Haveren, Bruce P. Van *see* Van Haveren, Bruce P.

Haverfield, John Joseph. The geology of the Melchior Islands. M, 1948, Miami University (Ohio). 77 p.

Haverland, Raymond Louis. Soil development on a granitic catena in southeastern Arizona. D, 1987, University of Arizona. 281 p.

Havers, Murray Hall. The Mississippian of the Canadian Rocky Mountains and adjacent plains. M, 1956, Northwestern University.

Haverslew, Roderick Edwin. Geology and genesis of the Ruttan Lake Deposit, Manitoba. M, 1976, University of Alberta. 149 p.

Havholm, Karen Gene. Dynamics of a modern draa, Algodones dune field, California. M, 1986, University of Texas, Austin. 99 p.

Haviland, Terrance. A structural and petrographic analysis of the Cedar Lake area, Stevens County, Washington and a review of the controversy surrounding the timing of deformation within the Kootenay Arc. M, 1983, Washington State University. 93 p.

Havlin, John LeRoy. Potassium and phosphorus chemistry and fertility in Calcareous soils. D, 1983, Colorado State University. 189 p.

Havryluk, Ihor. The geology and petrographic study of the Bean Station area, Grainger and Hawkins counties, Tennessee. M, 1963, University of Tennessee, Knoxville. 49 p.

Havskov, Jens. Plate tectonics and seismic evidence for mantle inhomogeneities. D, 1978, University of Alberta. 273 p.

Haw, Tong Chee. Petrology of Wayne Group (Silurian) carbonates along the Highland Rim, north central Tennessee. M, 1974, Memphis State University.

Haw, V. A. Further studies of nickel ores of the Sudbury range. M, 1948, Queen's University. 105 p.

Hawe, Robert Glen. A telluric current survey over two known geothermal areas. M, 1974, University of Montana. 45 p.

Hawes, Julian. A magnetic study of the Spavinaw Granite area, Oklahoma. D, 1950, Harvard University.

Hawes, R. A. A land evaluation methodology for natural resource planning in mountain regions in the American tropics; a case study of Rio Guanare, Venezuela. D, 1978, Cornell University. 197 p.

Hawes, R. A. A landscape approach to land classification and evaluation for regional land use planning, southern Okanagan Valley, British Columbia. M, 1974, University of British Columbia.

Hawes, Richard John. The glacial geomorphology of the Kananaskis Valley, Rocky Mountain Front Ranges, southern Alberta. M, 1977, University of Calgary. 204 p.

Hawisa, Ibrahim Sh. Depositional environment of the Bartlesville, the Red Fork, and the lower Skinner Sandstones (Pennsylvanian) in portions of Lincoln, Logan, and Payne counties, Oklahoma. M, 1965, University of Tulsa. 44 p.

Hawk, David Harold. Petrology of the Senora Formation (Pennsylvanian), Hughes and Pittsburg counties, Oklahoma. M, 1970, University of Oklahoma. 176 p.

Hawk, Jody M. Lithospheric flexure, overthrust timing, and stratigraphic modelling of the central Brooks Range and Colville Foredeep. M, 1985, Rice University. 179 p.

Hawk, Joseph H. Study of seismic methods in the detection of subsurface fracture zones. M, 1972, Pennsylvania State University, University Park. 72 p.

Hawke, Bernard R. Evolution of the early lunar crust; the role of impact melting, secondary cratering, and highland volcanism. D, 1979, Brown University. 108 p.

Hawke, Bernard Ray. Mixing model studies of the Apollo 17 regolith. M, 1978, University of Kentucky. 112 p.

Hawkes, H. E., Jr. Structural geology of the Plymouth-Rochester area, Vermont. D, 1940, Massachusetts Institute of Technology. 98 p.

Hawkins, Alfred C. The geology of western Rhode Island. D, 1916, Brown University.

Hawkins, Connie M. Microfacies of the Paleozoic rocks near Placer de Guadalupe, Chihuahua, Mexico. M, 1975, Texas Christian University. 64 p.

Hawkins, Daniel Ballou. Experimental hydrothermal studies bearing on rock weathering and clay mineral formation. D, 1961, Pennsylvania State University, University Park. 148 p.

Hawkins, David Wilfred. Emplacement, petrology and geochemistry of ultrabasic to basic intrusives at Aillik Bay, Labrador. M, 1977, Memorial University of Newfoundland. 236 p.

Hawkins, Fred Frost. Glacial geology and late Quaternary paleoenvironment in the Merchants Bay area, Baffin Island, N.W.T., Canada. M, 1980, University of Colorado. 146 p.

Hawkins, Gayne Patrick. The stratigraphy and mineralogy of the Lewisville member of the Cretaceous Woodbine Formation in the Arlington, Tarrant County, Texas area. M, 1970, University of Houston.

Hawkins, Glenn De Wayne. Tertiary foraminifera from Chiapas and Tabasco, Mexico. M, 1923, Columbia University, Teachers College.

Hawkins, H. G. Strike slip displacement along the Camp Rock Fault, central Mojave Desert, San Bernardino, California. M, 1976, University of Southern California. 63 p.

Hawkins, James E. Some bed-rock depth determinations, using seismic refraction and potential-gradient-ratio methods (T575). M, 1938, Colorado School of Mines. 38 p.

Hawkins, James E. The design of a dual coil ratiometer suitable for geophysical investigations. D, 1940, Colorado School of Mines. 42 p.

Hawkins, James Gregory, Jr. Processing and evaluation of explosive generated seismic reflection data for an energy source study in Greene County, Ohio. M, 1984, Wright State University. 133 p.

Hawkins, James Wilbur, Jr. Geology of the crystalline rocks of the northwestern part of the Okanogan Range, north central Washington. D, 1963, University of Washington. 173 p.

Hawkins, John O. The geology of the Lea Springs area, Grainger County, Tennessee. M, 1959, University of Tennessee, Knoxville. 41 p.

Hawkins, R. K. A neutron diffraction study of clay-water structure. D, 1979, University of Guelph.

Hawkins, Ralph D. The micropaleontology of the Abo Canyon section, New Mexico. M, 1950, Texas Tech University. 44 p.

Hawkins, Richard H. A study to predict storm runoff from storm characteristics and antecedent basin conditions. M, 1961, Colorado State University. 82 p.

Hawkins, Richard H. Water projects and watershed treatment. D, 1968, Colorado State University. 126 p.

Hawkins, Robert B. The geology and mineralization of the Jerritt Creek area, northern Independence Mountains, Nevada. M, 1973, Idaho State University. 104 p.

Hawkins, W. M. Geology of Goshen copper prospect, Goshen, New Brunswick. M, 1958, McGill University.

Hawkins, Wildon D. Stratigraphy of the Mississippian-Pennsylvanian boundary, north central Arkansas. M, 1983, University of Arkansas, Fayetteville.

Hawkins, William H. Paleozoic rocks, lower North Fork Canyon, northern Independence Range, Elko County, Nevada. M, 1960, Columbia University, Teachers College.

Hawks, Graham Parker. Increase A. Lapham [1811-1875]; Wisconsin's first scientist. D, 1960, University of Wisconsin-Madison. 314 p.

Hawks, Paul H. Selected properties of Little Tallahatchie Valley sediments by genetic types. M, 1973, University of Mississippi. 63 p.

Hawks, Ralph L. The stratigraphy and structure of the Cedar Hills, Sanpete County, Utah. M, 1979, Brigham Young University.

Hawksworth, Mark A. Alteration zoning and hydrothermal fluid characteristics associated with the Groundhog vein system, Grant County, New Mexico. M, 1984, Washington State University. 147 p.

Hawley, Bronson Waugh. Simultaneous inversion of local earthquake data for laterally-varying velocity structure and hypocenters. M, 1979, University of Utah. 142 p.

Hawley, Charles Caldwell. Geology of the Pikes Peak Granite and associated ore deposits, Lake George beryllium area, Park County, Colorado. D, 1963, University of Colorado. 346 p.

Hawley, Henry J. The stratigraphy and paleontology of a portion of the Santa Inez Mountains, Santa Barbara County, California. M, 1918, [Stanford University].

Hawley, James E. I, The geology and economic possibilities of the Sutton Lake area, District of Patricia, Ontario; II, Osmotic growth as a geologic phenomenon. D, 1926, University of Wisconsin-Madison.

Hawley, John Edward. Bicarbonate and carbonate ion association with sodium, magnesium and calcium at 25°C and 0.72 ionic strength. D, 1973, University of Oregon. 77 p.

Hawley, John William. The late Pleistocene and Recent geology of the Winnemucca segment of the Humboldt River valley, Nevada. D, 1962, University of Illinois, Urbana. 286 p.

Hawley, Katherine Taft. A study of the mafic rocks along the eastern flank of the Flint Creek Range, western Montana. M, 1974, University of Montana. 53 p.

Hawley, L. David. Ordovician shales and submarine slide breccias of the northern Champlain Valley, New York. D, 1953, Columbia University, Teachers College.

Hawley, Luther David. Structure of the Lower Cambrian in the vicinity of Burlington, Vermont. M, 1940, Cornell University.

Hawley, Nathan. An experimental investigation of flaser and wavy bedding. D, 1978, Massachusetts Institute of Technology. 205 p.

Hawley, Robert William. Phosphoria conodonts from southeastern Idaho. M, 1950, University of Idaho. 23 p.

Hawman, Robert Barrett. Crustal models for the Scranton and Kentucky gravity highs; regional bending of the crust in response to emplacement of failed rift structures. M, 1980, Pennsylvania State University, University Park. 107 p.

Hawman, Robert Barrett. Wide angle reflection studies of the crust and upper mantle beneath eastern Pennsylvania. D, 1988, Princeton University. 589 p.

Haworth, Charles C. The construction of an infra-red spectrometer and its application to the measurement of water vapors in fuel gases and in air. D, 1942, Pennsylvania State University, University Park. 96 p.

Haworth, Erasmus. A contribution to the Archean geology of Missouri. D, 1888, The Johns Hopkins University.

Haworth, Erasmus. A contribution to the geology of the lead and zinc mining district of Cherokee County, Kansas. M, 1884, University of Kansas. 47 p.

Haworth, Leah A. Holocene glacial chronologies of the Brooks Range, Alaska and their relationship to climate change. D, 1988, SUNY at Buffalo. 260 p.

Haworth, Randol A. Stratigraphy of the Cedar Mountain and Dakota formations (Cretaceous), Garfield County, Colorado. M, 1979, Colorado School of Mines. 93 p.

Haworth, Raymond Harrison. Biostratigraphic study of the Florena Shale. M, 1955, Columbia University, Teachers College.

Haworth, Roger Alan. Stratigraphy, paleontology, and depositional environments of Vaqueros Formation, lower Piru Creek, Los Angeles and Ventura counties, California. M, 1980, San Diego State University.

Haworth, William D. Geology of the northern part of the Diamond Range, Eureka and White Pine counties, Nevada. M, 1979, University of Nevada. 68 p.

Hawryszko, Julian W. The sedimentology of the Fernie shale formation. M, 1957, University of Kansas. 93 p.

Haws, W. J. The crystal chemistry of rare earth titanites and rare earth titanates. D, 1976, Purdue University. 244 p.

Hawthorne, Frank C. The crystal chemistry of the clino-amphiboles. D, 1973, McMaster University. 332 p.

Hawthorne, Hal W. Upper Morrow (Pennsylvanian) chert conglomerates and sandstones of the Reydon and Cheyenne fields, Roger Mills County, Oklahoma. M, 1984, Baylor University. 122 p.

Hawthorne, J. Michael. The stratigraphy and depositional environments of dinosaur tracks in the Edwards Plateau and Lampasas Cut Plain provinces of Texas. M, 1987, Baylor University. 303 p.

Haxby, Ronald L. In-bench relative rock hardness evaluation. M, 1966, University of Arizona.

Haxby, W. F. Studies of the state of stress, mechanical behavior, and density structure of the lithosphere. D, 1978, Cornell University. 91 p.

Haxel, Gordon. The Orocopia Schist and the Chocolate Mountain thrust, Picacho-Peter Kane Mountain area, southeasternmost California. D, 1977, University of California, Santa Barbara. 328 p.

Hay, Alexander Edward. Submarine channel formation and acoustic remote sensing of suspended sediments and turbidity currents in Rupert Inlet, B.C. D, 1981, University of British Columbia.

Hay, Bernward Josef. Depositional environment in the Late Middle Ordovician Taconic foreland basin (New York State); evidence from geochemical, sedimentological, and stratigraphic studies. M, 1982, Cornell University. 191 p.

Hay, Bernward Josef. Particle flux in the western Black Sea in the present and over the last 5,000 years; temporal variability, sources, transport mechanisms. D, 1987, Woods Hole Oceanographic Institution. 203 p.

Hay, Bradley W. B. The role of varying rates of local relative sea-level change in controlling the Holocene sedimentologic evolution of northern Casco Bay, Maine. M, 1988, University of Maine. 241 p.

Hay, David Evan. Structural, plutonic and mineralization history of Paramount Wash, eastern Mojave Desert. M, 1981, Vanderbilt University. 188 p.

Hay, Edward Alexander. Geology of the Cholame Ranch Quadrangle, California. M, 1961, University of California, Berkeley. 51 p.

Hay, Helen B. Lithofacies and formations of the Cincinnatian Series (Upper Ordovician), southeastern Indiana and southwestern Ohio. D, 1981, Miami University (Ohio). 236 p.

Hay, Helen B. Lithofacies classification for the Cincinnatian Series (upper Ordovician), southeastern Indiana. M, 1975, Miami University (Ohio). 150 p.

Hay, Howard William, Jr. Petrology of the Middle Cambrian Blacksmith Formation, north-central Utah. M, 1982, Utah State University. 157 p.

Hay, J. D. A comparative analysis of Cs-137 dynamics in two floodplain forests of a southeastern coastal plain stream. D, 1977, Emory University. 295 p.

Hay, K. L. La see La Hay, K. L.

Hay, Nicholas R. T. Geology of the Fincastle Syncline, Botetourt County, Virginia. M, 1950, University of Virginia. 134 p.

Hay, Peter W. The association of garnet and cordierite in regionally metamorphosed rocks from the Westport area, Ontario. M, 1962, Queen's University. 128 p.

Hay, Peter William. The stability and occurrence of cordierite in selected gneisses from the Canadian Shield. D, 1965, Stanford University. 118 p.

Hay, Randall Stuart. Chemically induced grain boundary migration and thermal grooving of calcite bicrystals. D, 1987, Princeton University. 223 p.

Hay, Richard Leroy. Petrography of the later Miocene volcanic rocks of the John Day River valley in Oregon. M, 1949, Northwestern University. 107 p.

Hay, Richard Leroy. Stratigraphy of the lower volcanic rocks in the southern part of the Absaroka Range, Wyoming. D, 1952, Princeton University. 213 p.

Hay, William Winn. A study of the Velasco Formation of northeastern Mexico. D, 1960, Stanford University. 420 p.

Hay, William Winn. Some Late Cretaceous and Tertiary Ostracoda from southwestern France. M, 1958, University of Illinois, Urbana.

Hay-Roe, Hugh. Geology of Wylie Mountains and vicinity, Trans-Pecos, Texas. D, 1958, University of Texas, Austin. 275 p.

Hay-Roe, Hugh. Structural geology of Wylie Mountains, Culberson County, Texas (H334). M, 1952, University of Texas, Austin.

Hayasaka, Ichiro. A new locality for Lyttonia richthofeni, Kayser, E. N., together with some observations concerning the world distribution of the genus Lyttonia. M, 1927, University of Pittsburgh.

Hayashi, Hiroshi. Static and seismic stability of cut slopes in terms of reliability. D, 1987, University of Illinois, Urbana. 139 p.

Hayatdavoudi, A. Theory of hydrocyclone operation and its modifications for application to the concentration of underwater heavy mineral sand deposits. D, 1974, University of Wisconsin-Madison. 159 p.

Hayatsu, A. Precise measurement of natural strontium isotope abundance ratios using double collection methods of mass spectrometry. D, 1965, University of Toronto.

Hayba, Daniel Owen. Characterization of scales formed from Salton Sea geothermal brines. M, 1979, Pennsylvania State University, University Park. 167 p.

Haycock, M. H. The geology of the Parrsboro area, Nova Scotia. M, 1926, Acadia University.

Haycock, Maurice Hall. The application of the quartz spectrograph to the study of opaque minerals. D, 1931, Princeton University. 33 p.

Haycock, Susan. Petroleum geology of the lower Eocene Metlaoui Group, east-central Tunisia. M, 1987, Baylor University. 165 p.

Haycocks, Christopher. Centrifugal model study and finite element analysis of stress near a fault. D, 1967, University of Missouri, Rolla. 158 p.

Hayden, Joseph M. A gravity and magnetic investigation of a microseismically active area in central Oklahoma. M, 1985, University of Oklahoma. 175 p.

Hayden, R. S. Alterations in the drainage density of small streams on the Georgia Piedmont due to road construction. D, 1979, University of Georgia. 251 p.

Hayden, Terry John. Petrology and structural geology of the southeast part of the White Cloud Stock. M, 1983, University of Idaho. 79 p.

Haydock, Samuel Rotch. Tectonic geology of the Waitsfield-Warren area, central Vermont. M, 1988, University of Vermont. 209 p.

Haydon, Osborne. Some possibilities of the deposition of galena from colloidal solutions. M, 1926, University of Michigan.

Haydon, Paul Richard. Ion filtration and stable oxygen isotopic fractionation resulting from the passage of sodium-calcium-chloride brine through compacted smectite layers at elevated temperatures and pressures. D, 1983, University of Illinois, Urbana. 195 p.

Haye, Jean La see La Haye, Jean

Hayes, Albert Orion. Wabana iron ore of Newfoundland. D, 1914, Princeton University.

Hayes, Arnell Saucier. Silurian foraminifera from Tennessee. M, 1956, Mississippi State University. 64 p.

Hayes, Arthur Wesley. Origin of the Tuscarora Formation (lower Silurian), southwestern Virginia. D, 1974, Virginia Polytechnic Institute and State University.

Hayes, Benjamin R. Application of a numerical model to brine contamination at the Beaver Creek oil field in Bond County, Illinois. M, 1988, Southern Illinois University, Carbondale. 150 p.

Hayes, Christopher George. The geology of the Shushan Quadrangle, New York-Vermont. M, 1978, Cornell University.

Hayes, David Thomas. Origin of fracture patterns and insoluble minerals in the Fort Dodge Gypsum, Webster County, Iowa. M, 1986, Iowa State University of Science and Technology. 76 p.

Hayes, David Wayne. A study of the distribution of the lanthanide elements in the Gulf of Mexico using neutron activation analysis. D, 1969, Texas A&M University. 179 p.

Hayes, Delvin Arnold. A case history approach to teaching earth science. M, 1973, California State University, Chico. 70 p.

Hayes, Dennis Edward. A geophysical investigation of the Peru-Chile Trench. D, 1966, Columbia University. 51 p.

Hayes, Frank A. Soils and soil-forest relationships in Shelterbelt Zone. D, 1936, University of Nebraska, Lincoln.

Hayes, Garry Fallis. Late Quaternary deformation and seismic risk in the northern Sierra Nevada-Great Basin boundary zone near the Sweetwater Mountains, California and Nevada. M, 1985, University of Nevada. 135 p.

Hayes, Graham Stephen. Geologic stability of Mystic Lake Dam, Gallatin County, Montana, and computer simulation of potential flood hazards from the failure of the dam. M, 1981, Montana State University. 120 p.

Hayes, J. B. Pre-Pennsylvanian unconformities of El Paso County, Colorado. M, 1957, Iowa State University of Science and Technology.

Hayes, James Frederick. Geology of northwest part of Cushing Quadrangle, northeastern Cherokee and southwestern Rusk counties, Texas. M, 1951, University of Texas, Austin.

Hayes, James Joseph. Quantitative analysis of the volatile components of selected granitic rocks. M, 1975, University of Tulsa. 143 p.

Hayes, John Bernard. Mississippian geodes of the Keokuk, Iowa, region. D, 1961, University of Wisconsin-Madison. 303 p.

Hayes, John F. A review of oil shale; the future of the industry. M, 1923, University of Oklahoma. 63 p.

Hayes, John H. The Anchor Mountain Mine, Black Hills, South Dakota. M, 1933, University of Minnesota, Minneapolis. 32 p.

Hayes, John Jesse. Geology of the Hodges Hills-Marks Lake area, northern Newfoundland. D, 1952, University of Michigan.

Hayes, John Jesse. Relative distance from source intrusive as a factor in pegmatite variation. M, 1948, University of Michigan.

Hayes, John R. Cretaceous stratigraphy in eastern Colorado. D, 1950, University of Colorado.

Hayes, Joseph J. Geochemical and hydrogeologic investigation of anomalous low-pH conditions in southcentral Rusk County, Texas. M, 1988, Stephen F. Austin State University. 119 p.

Hayes, Joseph Phillips. Pore pressure development and shallow groundwater in a colluvium-filled bedrock hollow. M, 1985, University of California, Santa Cruz.

Hayes, Julie Allison. Puerto Rico; reconnaissance study of the maturation and source rock potential of an ocean arc involved in a collision. M, 1985, Stanford University. 74 p.

Hayes, Larry Ross. The concentration and distribution of selected trace metals in the Maumee River Basin, Ohio, Michigan, and Indiana. M, 1973, Ohio State University.

Hayes, Lawrence Douglas. A petrographic study of the crystalline rocks of the Chapel Hill, North Carolina Quadrangle. M, 1962, University of North Carolina, Chapel Hill. 85 p.

Hayes, Lyman Neal. A study of the subsurface geology of the northeastern part of Comanche County, Oklahoma. M, 1952, University of Oklahoma. 67 p.

Hayes, Martha Anne. An aerial photographic investigation of barrier evolution; North Beach, Cape Cod, Massachusetts. M, 1981, University of Massachusetts. 113 p.

Hayes, Michael D. Conodonts of the Bakken Formation (Devonian and Mississippian), Williston Basin, North Dakota. M, 1984, University of North Dakota. 57 p.

Hayes, Michael John. Effects of surface-active fluids on the sliding behavior of Crab Orchard Sandstone. M, 1975, University of North Carolina, Chapel Hill. 58 p.

Hayes, Miles Oren. Petrology of the Krebs Subgroup (Desmoinesian) of western Missouri. M, 1960, Washington University. 61 p.

Hayes, Miles Oren. Sedimentation on a semiarid, wave-dominated coast (South Texas) with emphasis on hurricane effects. D, 1965, University of Texas, Austin.

Hayes, Pamela Dee. Minerals of Ohio (a complete literary survey). M, 1969, University of Virginia. 71 p.

Hayes, Philip T. Geology of the pre-Cambrian rocks of the northern end of the Sandia Mountains, Bernalillo and Sandoval counties, New Mexico. M, 1951, University of New Mexico. 54 p.

Hayes, Timothy S. Climate dependent geochemical mechanisms of copper, uranium, and vanadium transport and deposition in sandstone ores. M, 1982, Stanford University. 148 p.

Hayes, Timothy Scott. Geologic studies on the genesis of the Spar Lake strata-bound copper-silver deposits, Lincoln County, Montana. D, 1983, Stanford University. 340 p.

Hayes, William Clifton. Geology of the Ozark-Martin Mine area, Madison County, Missouri. M, 1947, University of Missouri, Rolla.

Hayes, William Clifton, Jr. Pre-Cambrian iron deposits of Missouri. D, 1951, University of Iowa. 141 p.

Hayes, William Errol, Jr. The smaller fossils from the Upper Cretaceous of the Starkville area, (Starkville, Mississippi). M, 1951, Mississippi State University. 147 p.

Hayes, William Harold. The geography of New Brunswick. M, 1929, Cornell University.

Hayfield, George H. Stratigraphy of the Mississippian formations between the Maury Shale and Pennington Formation, Bledsoe County, Tennessee. M, 1961, University of Tennessee, Knoxville. 56 p.

Hayford, Frank Sim. A petrographic analysis of the Great Sand Dunes of Colorado with conclusions as to source. M, 1947, University of Colorado.

Haygood, Christine Cricket. Melting relations in the system alkali basalt-H2O from 10 to 30 Kbars. M, 1973, Pennsylvania State University, University Park. 59 p.

Hayles, J. G. A study of the lead isotopic composition of galena at Manitcouwadge, Ontario. M, 1973, University of British Columbia.

Hayling, Kjell. Magnetization modelling in the North and Equatorial Atlantic Ocean using MAGSAT data. M, 1986, University of Miami. 21 p.

Hayling, Kjell Lennart. Heat flow and magnetization in the oceanic lithosphere. D, 1988, University of Miami. 129 p.

Hayman, Glenn A. Geology of a part of the Eagle Butte and Gateway quadrangles east of the Deschutes River, Jefferson County, Oregon. M, 1984, Oregon State University. 97 p.

Hayman, James W. The significance of some geologic factors in the Karst development of the Mount Tabor area, Montgomery County, Virginia. M, 1972, Virginia Polytechnic Institute and State University.

Haymes, David Edward. An isotopic study of East African and Canadian carbonatites. M, 1988, University of Illinois, Urbana. 86 p.

Haymon, Rachel Michal. Hydrothermal deposition on the East Pacific Rise at 21°N. D, 1982, University of California, San Diego. 291 p.

Haymond, Dan. Geology of the Longlick and White Mountains area, southern San Francisco Mountains, Beaver County, Utah. M, 1980, Brigham Young University.

Haynes, Caleb Vance, Jr. Quaternary geology of the Tule Springs area, Clark County, Nevada. D, 1965, University of Arizona.

Haynes, Charles W. and Browne, James L. Geology of the Mount Gratiot area, Keweenaw County, Michigan. M, 1956, Michigan Technological University. 44 p.

Haynes, Cynthia L. Sandstone diagenesis and development of secondary porosity, Shattuck Member, Queen Formation, Chaves County, New Mexico. M, 1978, University of Texas, Austin.

Haynes, Edward H. The geology of a portion of the east slope of the Sangre de Cristo Range, Huerfano County, Colorado (T. 29S., R. 69W.). M, 1952, University of Kansas. 102 p.

Haynes, Frederick Mitchell. Geologic and geochemical controls for sphalerite mineralization, Mascot-Jefferson City zinc district, East Tennessee. D, 1986, University of Michigan. 294 p.

Haynes, Frederick Mitchell. The evolution of fracture-related permeability within the Ruby Star Granodiorite, Sierrita porphyry copper deposit, Pima County, Arizona. M, 1980, University of Arizona. 48 p.

Haynes, John Bernard. Pre-Pennsylvanian unconformities of El Paso County, Colorado. M, 1957, Iowa State University of Science and Technology.

Haynes, John Tweedt. The stratigraphy and depositional setting of the Waynesboro Formation (Lower and Middle Cambrian) near Buchanan, Botetourt County, Virginia. M, 1985, University of Cincinnati. 291 p.

Haynes, Ronnie J. Some properties of coal spoilbank and refuse materials resulting from the surface-mining of coal in Illinois. D, 1976, Southern Illinois University, Carbondale. 126 p.

Haynes, S. J. Granitoid petrochemistry, metallogeny and lithospheric plate tectonics, Atacama Province, Chile. D, 1975, Queen's University. 330 p.

Haynes, Simon John. The conformable tin-bearing pyrrhotite-pyrite sills at Renison Bell, Tasmania. M, 1969, Carleton University. 143 p.

Haynes, Steven Anthony. Geomorphic development of Virgin River near Hurricane, Utah. M, 1983, University of Utah. 189 p.

Haynes, W. C. Geology of the Ozark-Mountain Mine area, Madison County, Missouri. M, 1947, University of Missouri, Rolla.

Haynes, Winthrop Perrin. A contribution of the geology of the region about Three Forks, Montana. D, 1914, Harvard University. 175 p.

Haynie, Anthon V., Jr. The Worm Creek Quartzite Member of the St. Charles Formation, Utah-Idaho. M, 1957, Utah State University. 39 p.

Hays, Frank Richard. A petrographic study of deeply buried sandstones from the Superior Pacific Creek unit No. 1 well, Sublette County, Wyoming. M, 1951, University of Cincinnati. 65 p.

Hays, James Douglas. A study of the South Flat and related formations of central Utah. M, 1960, Ohio State University.

Hays, James Douglas. Antarctic Radiolaria and the late Tertiary and Quaternary history of the Southern Ocean. D, 1964, Columbia University, Teachers College. 224 p.

Hays, James F. The system CaO-Al2O3-SiO at high pressure and high temperature. D, 1966, Harvard University.

Hays, James K. A study of the structures developed on the northeast flank of the Hugoton Embayment. M, 1961, University of Kansas. 40 p.

Hays, Norbert Alan. Geology of the Hurds Draw, Culberson and Jeff Davis counties, Texas. M, 1948, University of Texas, Austin.

Hays, Patricia E. Paleoceanography of the eastern Equatorial Pacific during the Pliocene; a high resolution radiolarian study. M, 1987, Oregon State University. 117 p.

Hays, Phillip Dean. Diagenesis of the Terry Sandstone Member of the Pierre Shale, Spindle Field, Weld County, Colorado. M, 1986, Texas A&M University. 121 p.

Hays, Ronny A. Monitoring technology; application to earth dams. M, 1981, University of Idaho. 184 p.

Hays, W. H. Geology of Cottonwood Springs and part of Coachella quadrangles (California). D, 1951, Yale University.

Hays, Walter Wesley. A paleomagnetic investigation of some of the Precambrian igneous rocks of Southeast Missouri. D, 1961, Washington University. 193 p.

Hays, Walter Wesley. A statistical analysis of magnetic susceptibilities and ore content of the Sanford Lake titaniferous magnetite deposit, New York. M, 1959, Washington University. 53 p.

Hays, William Henry. Geology of the central Mecca Hills, Riverside County, California. D, 1958, Yale University.

Hayslip, David L. Geochemistry of the bimodal Quaternary volcanism in the Medicine Lake Highland, northern California. M, 1973, New Mexico Institute of Mining and Technology. 129 p.

Hayter, Earl Joseph. Prediction of cohesive sediment movement in estuarial waters. D, 1983, University of Florida. 349 p.

Hayward, Chris. Alternative methods of displaying and interpreting common depth point seismic data; the stacking diagram time slice display. M, 1984, Baylor University. 197 p.

Hayward, Gary Lewis. Continental slope and rise sedimentary processes off eastern North America between Washington and Wilmington canyons. M, 1979, University of South Florida, St. Petersburg. 113 p.

Hayward, Jennifer A. Rb-Sr geochronology and the evolution of some "peraluminous" granites in New Hampshire. M, 1983, University of New Hampshire. 108 p.

Hayward, Oliver Thomas. The structural significance of the Bosque Escarpment, McLennan County, Texas. D, 1957, University of Wisconsin-Madison. 137 p.

Hayward, William. Depositional patterns of Upper Cambrian through Middle Devonian stratigraphy of the Greene-Potter Zone in western Pennsylvania. M, 1982, University of Pittsburgh.

Haywick, Douglas Wayne. Dolomite within the St. George Group (Lower Ordovician), western Newfoundland. M, 1984, Memorial University of Newfoundland. 281 p.

Haywood, Harry C. Depositional processes of the Casper Sandstone in the southernmost Laramie Basin as indicated by settling velocities of light and heavy minerals. M, 1973, University of Wyoming. 34 p.

Hayworth, Joel Stacey. Laboratory investigation of a method for determining in situ saturated hydraulic conductivity. M, 1988, University of Nevada, Las Vegas. 205 p.

Hazard, Allan W. A study of the heavy mineral and insoluble residue content of the limestones and sandstones of the Warm Springs area, Virginia. M, 1939, University of Virginia. 52 p.

Hazel, Harold F. The conodont fauna for the basal Clarksville Member of the Waynesville Formation. M, 1955, Miami University (Ohio). 51 p.

Hazel, James W. A study of biostratigraphy of the "Fairmont" in the southern Bluegrass area of Kentucky. M, 1973, University of Kentucky. 87 p.

Hazel, James W. A study of the bryozoan fauna of the Fairmount Formation (Upper Ordovician) in the southern Blue Grass of Kentucky. M, 1965, [University of Kentucky].

Hazel, Joe Ernest. The ostracode fauna of the Lomita Marl. M, 1960, University of Missouri, Columbia.

Hazel, Joseph Ernest. Part I, Systematics of some Gulfian trachyleberids from Texas and Arkansas; Part II, Ostracode biostratigraphy in some Austinian-Tayloran rocks. D, 1963, Louisiana State University. 151 p.

Hazell, S. Late Quaternary vegetation and climate of Dunbar Valley, British Columbia. M, 1979, University of Toronto. 101 p.

Hazelwood, Anna Marie. Fe-Ti oxide/silicate equilibria in the granulite facies, Stark and Diana complexes, Adirondack Mountains, New York. M, 1987, University of Wisconsin-Madison. 70 p.

Hazelworth, John Beemon. sP, SP, and PS waves of the deep-focus earthquake of May 25, 1944. M, 1949, University of Michigan.

Hazen, David Ralph. Sedimentology and paleogeography of the Late Triassic Higham Grit, southeastern Idaho and western Wyoming. M, 1985, Montana State University. 85 p.

Hazen, Richard Stewart. Discrimination of alluvial deposits near Vidal Valley, southeastern California, using coregistered Landsat MSS and Seasat SAR imagery. M, 1983, University of California, Los Angeles. 165 p.

Hazen, Robert Edward. Cadmium in an aquatic ecosystem. D, 1981, New York University. 145 p.

Hazen, Robert M. The effect of cation substitutions on the physical properties of trioctahedral micas. M, 1971, Massachusetts Institute of Technology. 88 p.

Hazen, Robert Miller. Effects of temperature and pressure on the crystal physics of olivine. D, 1975, Harvard University.

Hazenbush, George C. Geology of the eastern parts of the Dry Canyon and Las Flores quadrangles, Los Angeles County, California. M, 1950, University of California, Los Angeles.

Hazenbush, George Cordery. Stratigraphy and micropaleontology of the Mancos Shale (Cretaceous), Black Mesa Basin, Arizona. D, 1972, University of Arizona.

Hazimah, Ibrahim Anis. Phosphates. M, 1957, University of Michigan.

Hazlett, Jean. Petrology and provenance of the Triassic limestone conglomerate in the vicinity of Leesburg, Virginia. M, 1978, George Washington University.

Hazlett, Richard W. Geology and hazards of the San Cristobal volcanic complex, Nicaragua. M, 1977, Dartmouth College. 212 p.

Hazlett, Richard W. Geology of a Tertiary volcanic center, Mopah Range, San Bernardino County, California. D, 1986, University of Southern California.

Hazlett, William Henry, Jr. Structural geology of the Roanoke area, Virginia. D, 1967, Virginia Polytechnic Institute and State University. 317 p.

Hazneci, T. Hakan. Computer derived focal mechanisms for selected earthquakes off the south and west coast of Turkey. M, 1983, Virginia Polytechnic Institute and State University. 87 p.

Hazzaa, Abdullah F. Iron ore mining in the Saudi Arabian Shield. M, 1973, University of Arizona.

Hazzard, John C. The Paleozoic section in the Nopah and Resting Springs mountains, Inyo County, California. D, 1937, University of Southern California.

Hazzard, John Charles. Paleozoic and associated rocks of the Marble and Ship mountains, San Bernardino County, California. M, 1932, University of California, Berkeley. 97 p.

He, Guoqi. The effect of stress on the reflectance ellipsoid. M, 1987, West Virginia University. 84 p.

Hea, James Paul. Sedimentary geology of La Sierra Formation (Eocene) and Sierra de Perija, Venezuela. D, 1964, Pennsylvania State University, University Park. 484 p.

Heacock, Robert L. Stratigraphy and foraminifera of the upper part of the Nye Formation, Yaquina Bay, Oregon. M, 1952, Oregon State University. 47 p.

Head, C. M. Surface mining; a geomorphic process. D, 1975, University of Georgia. 201 p.

Head, James L. Petrology of some soils of Lubbock County, Texas. M, 1951, Texas Tech University. 46 p.

Head, James William, III. An integrated model of carbonate depositional basin evolution; late Cayugan (upper Silurian) and Helderbergian (lower Devonian) of the central Appalachians. D, 1969, Brown University. 430 p.

Head, Roger Wayne. Deposition and diagenesis of the middle Guadalupian Cherry Canyon Formation, Dimmitt Field, Loving County, Texas. M, 1987, Purdue University. 91 p.

Head, Thomas Franklin. Geology of the Madera County Quadrangle, Jeff Davis County, Texas. M, 1948, University of Texas, Austin.

Head, W. J. Heavy metal analysis of stream sediments in the James River basin, Missouri. M, 1973, University of Missouri, Rolla.

Head, William Burres. A study of the relationships between porosity and certain size parameters of uncemented natural sands. M, 1959, Rice University. 47 p.

Headington, Clare Wesley. Collette Creek Field, Victoria County, Texas (H342). M, 1941, University of Texas, Austin.

Headlee, Larry A. Geology of the coastal portion of the San Luis Range, San Luis Obispo County, California. M, 1965, University of Southern California.

Headley, Klyne. Stratigraphy and structure of the northwestern Guadalupe Mountains, New Mexico. M, 1968, University of New Mexico. 65 p.

Heal, George Edward Newton. The Wrigley-Lou and Polaris-Truro lead zinc deposits, Northwest Territories. M, 1976, University of Alberta. 172 p.

Heald, B. Patrick. Fabrication and evaluation of the demountable hollow cathode as a resonance cell for the spectral filtration of inductively coupled argon plasma spectra obtained from geochemical materials. M, 1985, Montana College of Mineral Science & Technology. 67 p.

Heald, Emerson Francis. A chemical study of Hawaiian magmatic gases. D, 1961, University of Hawaii. 115 p.

Heald, Milton Tidd. Stucture and petrology of the Lovewell Mountain Quadrangle, New Hampshire. D, 1949, Harvard University.

Healey, John M. A gravity interpretation of an aeromagnetic anomaly in Clay County, Iowa. M, 1975, University of Iowa. 110 p.

Healey, Neil D. Microstructure and ultrastructure of some cystoporate bryozoans. M, 1975, Southern Illinois University, Carbondale. 85 p.

Healing, David William. Geochemistry of the Gull Pond alteration zone, Gull Pond, central Newfoundland. M, 1980, University of Waterloo.

Healy, Henry G. Geologic structures of the Byram Gneiss along the New York State Thruway in western Rockland County. M, 1956, New York University.

Healy, James S. The Sherman Sandstone (Dunkard Series) of Meigs County, Ohio and Jackson County, West Virginia. M, 1959, Miami University (Ohio). 99 p.

Healy, John Helding. Geophysical studies of the basin sturctures along the eastern front of the Sierra Nevada. D, 1961, California Institute of Technology. 90 p.

Healy, Mary Jo. Dispersal of Mount St. Helens ash across the Washington continental shelf, 1980-1982. M, 1983, Lehigh University.

Healy, Michael P. Sedimentology and lithofacies of the Pilgrim Formation (Upper Cambrian), west-central Montana. M, 1986, University of Idaho. 146 p.

Healy, R. P. The sorption of copper on montmorillonite under various temperature conditions and in the presence of inorganic and organic ligands. D, 1979, University of Maryland. 327 p.

Healy-Williams, Nancy. Morphological and geochemical analysis of the planktonic foraminifera Neogloboquadrina pachyderma (Ehrenberg) from sediments of the southern Indian Ocean and the North Atlantic Ocean. D, 1988, University of South Carolina. 303 p.

Heaman, Larry M. Rb-Sr geochronology and Sr isotope systematics of some major lithologies in Chandos Township, Ontario. M, 1980, McMaster University. 141 p.

Heaman, Larry Michael. A geochemical and isotopic study of plutonic and high-grade metamorphic rocks from the Chandos Township area, Grenville Province, Ontario. D, 1986, McMaster University. 307 p.

Heaney, Michael J., III. Upper Mississippian bivalves from the Imo Formation of north-central Arkansas. M, 1985, Bowling Green State University. 99 p.

Heaney, Richard John. Frequency-wave number domain velocity filters and their application to seismic data. M, 1978, University of Utah. 103 p.

Heany, Franklin Maurice. An analysis of Amphizona aceta Kesling and Copeland. M, 1957, University of Michigan.

Heap, George. The structure of the Farmersburg region. M, 1939, Indiana University, Bloomington. 35 p.

Heard, Edward T. The Harrell Sand Reservoir of the Monroe Gas Field area, Louisiana. M, 1969, Northeast Louisiana University.

Heard, Garry John. The electromagnetic response of an Arctic bay. D, 1984, University of Victoria. 280 p.

Heard, Hugh Corey. The brittle to ductile transition in Solenhofen Limestone as a function of temperature, confining pressure, and interstitial fluid pressure. M, 1958, University of California, Los Angeles.

Heard, Hugh Corey. The effect of large changes in strain rate in the experimental deformation of rocks. D, 1962, University of California, Los Angeles.

Hearn, Bernard Carter, Jr. Geology of the northern half of the Rattlesnake Quadrangle, Bearpaw Mountains, Montana. D, 1959, The Johns Hopkins University. 195 p.

Hearn, Deborah J. Rays and waveforms in anelastic media. D, 1985, University of Calgary. 215 p.

Hearn, Douglas James. The application of ground penetrating radar to geologic investigations in karst terranes. M, 1987, University of Florida. 108 p.

Hearn, Frank. Depositional and diagenetic history of pre-Punta Gorda sediments from the Bass/Collier Company 12-2 Well, South Florida. M, 1984, University of Southwestern Louisiana. 131 p.

Hearn, P. P. Geochemistry and mineralogy of suspended and bottom sediments of the Rhode River estuary. M, 1977, George Washington University.

Hearn, Thomas Martin. Crustal structure in Southern California from array data. D, 1985, California Institute of Technology. 140 p.

Hearne, James H. Geology of the middle Tuscaloosa, lower Tuscaloosa and Dantzler formations as seen in the DuPont deNemours #1 Lester Earnest, Harrison County, Mississippi. M, 1987, University of Southwestern Louisiana. 175 p.

Hearty, David Joseph. Study of lateral variations in structure using slowness, azimuth, and travel time measurements of data recorded at the La Malbaie network and University of Western Ontario. M, 1977, University of Western Ontario.

Hearty, Paul Joseph. Age and aminostratigraphy of Quaternary coastal deposits in the Mediterranean Basin. D, 1987, University of Colorado. 227 p.

Heartz, William Thomas. Properties of a Piedmont residual soil. D, 1986, North Carolina State University. 279 p.

Heasler, Henry Peter. Heat flow in the Elk Basin oil field, northwestern Wyoming. M, 1978, University of Wyoming. 168 p.

Heasler, Henry Peter. Thermal evolution of coastal California with implications for hydrocarbon maturation. D, 1984, University of Wyoming. 85 p.

Heaslip, W. G. Fusulines of the Florena Shale (Permian) in Kansas. M, 1955, Columbia University, Teachers College.

Heaslip, William Graham. Cenozoic evolution of the alticostate venericards in Gulf and east coastal North America. D, 1963, Columbia University, Teachers College. 243 p.

Heath, C. G. Variabilities of in situ measured water transmissibility parameters. M, 1979, University of Guelph.

Heath, Christopher Peter Macclesfield. Microfacies of the Bird Spring Group (Pennsylvanian-Permian) Arrow Canyon, Clark County, Nevada. D, 1965, University of Illinois, Urbana. 163 p.

Heath, Christopher Peter Macclesfield. The mineralogy of tills in the Grand River glacial lobe in northeastern Ohio. M, 1963, University of Illinois, Urbana.

Heath, D. E. An environmental assessment of sewerage centralization with special reference to New York State. D, 1978, Syracuse University. 392 p.

Heath, Edward G. Geology along the Whittier Fault north of Horseshoe Bend, Santa Ana Canyon, California. M, 1954, Pomona College.

Heath, Edward W. Stratigraphy and structure of the Roach River Syncline, Piscataquis County, Maine. M, 1962, Massachusetts Institute of Technology. 125 p.

Heath, George Ross. Mineralogy of Cenozoic deep-sea sediments from the equatorial Pacific Ocean. D, 1968, University of California, San Diego.

Heath, Jenifer Sue. Toxicologic risk management at the community level; the case of chemically contaminated groundwater used as drinking water. D, 1987, Cornell University. 287 p.

Heath, Kathryn Carol. Distribution and sedimentology of periplatform sediment on a modern open-ocean carbonate slope; northern Little Bahama Bank. M, 1987, San Jose State University. 82 p.

Heath, Marla M. Geochronology of the Ayer Granite (late Paleozoic?) in the Wachusett-Marlborough Tunnel, Clinton, Massachusetts. M, 1965, Massachusetts Institute of Technology. 18 p.

Heath, Michael Thomas. Bedrock geology of the Monte Cristo area, northern Cascades, Washington. D, 1971, University of Washington. 164 p.

Heath, Michael Thomas. Mineralization of the Silver Star stock, Skamania County, Washington. M, 1966, University of Washington. 55 p.

Heath, Stanley A. Sr^{87}/Sr^{86} ratios in anorthosite and some associated rocks. D, 1967, Massachusetts Institute of Technology. 108 p.

Heathcote, I. W. Differential pollen deposition and water circulation in small Minnesota lakes. D, 1978, Yale University. 259 p.

Heathcote, Richard C. Fenitization of the Arkansas Novaculite and adjacent intrusive, Garland County, Arkansas. M, 1976, University of Arkansas, Fayetteville.

Heathcote, Richard Carl. Mica compositions and carbonatite petrogenesis in the Potash sulpur springs intrusive complex, Garland County, Arkansas. D, 1987, University of Iowa. 133 p.

Heathcote, Susan Kay Hudson. Geological interpretation of the Manchester geophysical anomaly, Delaware County, Iowa. M, 1979, University of Iowa. 144 p.

Heathcote, Susan Kay Hudson. Magnetic survey of the Wharton Arch area, central Madison County, Arkansas. M, 1976, University of Arkansas, Fayetteville.

Heather, Kevin B. The Aylwin Creek gold-copper-silver deposit, southeastern British Columbia. M, 1985, Queen's University. 273 p.

Heatherington, Ann Louise. Isotope systematics of volcanics from the south-central Rio Grande Rift and the western Mexican volcanic belt; implications for magmatic tectonic evolution of Cenozoic extensional regimes in western North America. D, 1988, Washington University. 207 p.

Heathman, J. H. Bentonite in Wyoming. M, 1939, University of Wyoming. 32 p.

Heatley, William Robert, Jr. Lead and cadmium fluxes through the Lake Issaqueena watershed. D, 1982, Clemson University. 213 p.

Heaton, Charles David. Gold ore deposits. M, 1909, University of Rochester. 75 p.

Heaton, Kevin Michael. K-Ar age patterns in the Cibbets Flat Pluton, Mount Laguna, California. M, 1981, San Diego State University.

Heaton, Kevin P. The hydrogeology of the City of Stow and Hudson Township, Summit County, Ohio. M, 1982, Kent State University, Kent. 251 p.

Heaton, Ross Leslie. Geology of the South Boulder District (Boulder County), Colorado. M, 1915, University of Colorado.

Heaton, Thomas Harrison. Generalized ray models of strong ground motion. D, 1979, California Institute of Technology. 304 p.

Heaton, Timothy H. The Quaternary paleontology and paleoecology of Crystal Ball Cave, Millard County, Utah; with emphasis on the mammals and the description of a new species of fossil skunk. M, 1984, Brigham Young University. illus.

Heaton, Timothy Howard. Patterns of evolution in Ischromys and Titanotheriomys (Rodentia; Ischyromyidae) from Oligocene deposits of western North America. D, 1988, Harvard University. 178 p.

Heatwole, David A. Geology of the Box Canyon area, Santa Rita Mountains, Pima County, Arizona. M, 1966, University of Arizona.

Heatwole, L. C. Variegated red beds in the Cathedral Bluffs Tongue of the Wasatch Formation, Wyoming. M, 1976, Ohio State University. 65 p.

Hebard, Edgar B. The Pleistocene of New Providence Island (Bahama Islands). M, 1966, New York University.

Hebb, David T. Mineral economic- and exploration architecture in developing countries; a preliminary study of South Viet-Nam's mineral potential with emphasis on heavy mineral sands. M, 1973, Colorado School of Mines. 195 p.

Hebberger, J. J., Jr. Recent laharic and glowing avalanche sediments, Guatemala. M, 1977, University of Missouri, Columbia.

Hebda, R. J. The paleoecology of a raised bog and associated deltaic sediments of the Fraser River delta. D, 1977, University of British Columbia.

Hébert, Claude. Contexte géologique régional du gisement aurifère de Chibex, Chibougamau, Québec. M, 1978, Universite du Quebec a Chicoutimi. 106 p.

Hebert, David Lawrence. A Mediterranean salt lens. D, 1988, Dalhousie University.

Hebert, Glenn P. Value of quasifunctional equations in determination of depositional environment relationships and classification of carbonate units. M, 1978, University of Illinois, Urbana. 134 p.

Hebert, Jean A. High-resolution seismic stratigraphy of the inner western Florida shelf west of Tampa Bay; evidence for a Miocene karst valley system. M, 1985, University of South Florida, St. Petersburg.

Hébert, Réjean. Etude pétrologique des roches ophiolitiques d'Asbestos et du Mont Ham (Ham sud), Québec. M, 1982, Universite Laval. 182 p.

Hebert, Roger L. Computer mapping in the Wilcox Group (Lower Eocene), east central Louisiana. M, 1972, Louisiana State University.

Hebert, Yves. Etude pétrologique du complexe ophiolitique de Thetford Mines, Québec. D, 1983, Universite Laval. 426 p.

Heberton, Richard P. Trace and minor element analyses of the Hymera Coal Member of Indiana (Coal VI). M, 1983, Indiana University, Bloomington. 107 p.

Hebertson, Keith M. Origin and composition of the Manning Canyon Formation in central Utah. M, 1950, Brigham Young University. 71 p.

Hebil, Keith Edmund. A petrographic study of the granite breccia, Levack Mine, Sudbury, Ontario. M, 1978, McGill University. 152 p.

Hebrew, Quey Chester. Ground-water geology of the Lehi, Utah area. M, 1950, Brigham Young University. 43 p.

Hebson, Charles S. Adaptive parameter estimation and filtering of partitioned hydrologic systems. M, 1983, Princeton University.

Hecht, Alan David. Morphological variation, diversity and stable isotope geochemistry of recent planktonic foraminifera from the North Atlantic. D, 1971, Case Western Reserve University. 242 p.

Hecht, Kurt. Magnetic susceptibility anisotropy of beach sand. M, 1962, Brown University.

Hecht, Paul David. Geology of Redwood Valley, California. M, 1970, University of Montana. 68 p.

Heck, Edward T. Correlation and stratigraphy of the Pottsville Series of eastern Fayette, southeastern Nicholas, and western Greenbrier counties (West Virginia). M, 1937, West Virginia University.

Heck, Edward T. Regional metamorphism of coal in southeastern West Virginia. D, 1942, West Virginia University.

Heck, Frederick Richard. Mesozoic foreland deformation and paleogeography of the western Great Basin, Humboldt Range, Nevada. D, 1987, Northwestern University. 226 p.

Heck, Jerome R. Engineering properties of sediments in the vicinity of Guide seamount (Pacific). M, 1970, United States Naval Academy.

Heck, Thomas J. Depositional environments and diagenesis of the Mississippian Bottineau interval (Lodgepole) in North Dakota. M, 1979, University of North Dakota. 227 p.

Heck, William J. Size distribution of residual quartz from weathered granites and gneisses. M, 1952, University of Wisconsin-Madison.

Heckard, John Martin. Geology of the Chester Series in the Grandview area, south-central Spencer County, Indiana. M, 1963, Indiana University, Bloomington. 29 p.

Hecke, Michael Clement Van see Van Hecke, Michael Clement

Heckel, Philip Henry. Stratigraphy, petrology and depositional environment of the Tully Limestone (Devonian) in New York State and adjacent region. D, 1966, Rice University. 448 p.

Hecker, Edward N. Subsurface correlation of Pleistocene deposit, East Baton Rouge Parish, Louisiana. M, 1949, Louisiana State University.

Hecker, Kenneth E., Jr. Effect of wind upon suspended sediment concentrations in waters east of the upper Florida Keys. M, 1973, University of Florida. 54 p.

Hecker, Stanley. A graphic analysis of current velocity, salinity, density and temperature during periods of ebb and flood in the entrance to Thimble Shoals channel, (Virginia). M, 1971, Old Dominion University. 115 p.

Hecox, Gary R. Engineering geology and geomorphology in Northeast Clear Creek County, Colorado. M, 1977, Colorado School of Mines. 131 p.

Hector, Scott T. Environmental geology of the Castle Rock Ridge area, Santa Cruz-Santa Clara counties, California. M, 1976, University of California, Davis. 98 p.

Hedberg, Hollis D. Stratigraphy of the Rio Querecual section of northeastern Venezuela. D, 1937, Stanford University. 68 p.

Hedberg, Hollis Dow. The effect of gravitational compaction on the structure of sedimentary rocks. M, 1926, Cornell University.

Hedberg, James D. A geological analysis of the Cameroon trend (Africa). D, 1969, Princeton University. 188 p.

Hedberg, James Dow. Differential movement between concentric beds during folding. M, 1961, Stanford University.

Hedberg, Kim E. Water chemistry of round and gravelly ponds, Hamilton, Massachusetts. M, 1988, University of New Hampshire. 115 p.

Hedberg, Leonard L. Clay mineralogy at the Brine-sediment interface in the south arm of Great Salt Lake, Utah. M, 1970, University of Utah. 55 p.

Hedberg, Ronald M. Dip determination in carbonate cores. M, 1960, University of Kansas. 35 p.

Hedberg, Ronald M. Stratigraphy of the Ovamboland Basin, South West Africa. D, 1975, Harvard University.

Hedberg, William H. Point bar sedimentation on Halfmoon Creek, Pennsylvannia. M, 1963, Pennsylvania State University, University Park. 32 p.

Hedberg, William Hollis. Pore-water chlorinities of subsurface shales. D, 1967, University of Wisconsin-Madison. 131 p.

Hedden, Albert H., Jr. (1) The geology of the Pinyon Peak area, East Tintic Mountains, Utah; (2) The geology of the upper Tick Canyon area, Los Angeles County, California. M, 1948, California Institute of Technology. 60 p.

Hedden, W. J. The stratigraphy and tectono-depositional history of the Smithville and Black Rock lithosomes (Canadian) of northeastern Arkansas with revisions of upper Canadian stratigraphy in the Ozarks. D, 1976, University of Missouri, Rolla. 501 p.

Hedden, William J. The geology of the Thayer area emphasizing the stratigraphy of the Cotter and the Jefferson City Formation. M, 1968, University of Missouri, Rolla.

Hedderly-Smith, David Arthur. Geology of the Sunrise breccia pipe, Sultan Basin, Snohomish County, Washington. M, 1975, University of Washington. 60 p.

Heddle, Duncan Walker. The relationship of dolomite and ore with special reference to the Jackpot Property, Ymir, BC. M, 1951, University of British Columbia.

Hedel, Charles. Late Quaternary faulting in western Surprise Valley, Modoc County, California. M, 1980, San Jose State University. 142 p.

Hedgcoxe, Reiffery H. Development of secondary faults between en echelon oblique-slip faults; small-fault systems in the Llano Uplift of central Texas. M, 1987, Texas A&M University.

Hedge, Carl E. Petrogenetic and geochronologic study of migmatites and pegmatites in the central Front Range (Colorado). D, 1969, Colorado School of Mines. 158 p.

Hedge, Carl E. Sodium-potassium ratios in muscovites as a geothermometer. M, 1960, University of Arizona.

Hedges, J. Lignin compounds as indicators of terrestrial organic matter in marine sediments. D, 1975, University of Texas, Austin. 148 p.

Hedges, Joseph W. Geology of Bear Creek Canyon area. M, 1949, University of Puget Sound. 51 p.

Hedges, Lynn S. The Pleistocene stratigraphy of Beadle County, South Dakota. M, 1966, University of South Dakota. 71 p.

Hedges, Robert Bruce. Hydrologic and geologic criteria in utilization of strip mines for solid waste disposal (Ohio). M, 1972, Ohio University, Athens. 69 p.

Hedin, Rae Ann. A subsurface study of the Bartlesville and Burgess sands, West Oolagah gas field, Rogers County, Oklahoma. M, 1982, University of Nebraska, Lincoln. 49 p.

Hedin, Robert Stewart. The consequences of strip mine reclamation; vegetation and economics of reclaimed and unreclaimed sites in west-central Pennsylvania. D, 1987, Rutgers, The State University, New Brunswick. 314 p.

Hedinger, Adam Stefan. Late Jurassic foraminifera from the Aklavik Range, Northwest Territories. M, 1979, University of Alberta. 427 p.

Hedley, Mathew S. Chemical changes in the contact metamorphism of argillaceous rocks. D, 1934, University of Wisconsin-Madison.

Hedley, Mathew S. Geology of non-ferrous metallic mineralization in the western mountain belt of British Columbia. M, 1932, University of Wisconsin-Madison.

Hedlund, David Carl. Graphic granites from selected zoned pegmatites of the Bryson City District, North Carolina. D, 1958, University of Wisconsin-Madison. 99 p.

Hedlund, Richard Warren. Microfossils of the Sylvan Shale (Ordovician) of Oklahoma. M, 1960, University of Oklahoma. 90 p.

Hedlund, Richard Warren. Palynology of the Red Branch Member of the Woodbine Formation (Upper Cretaceous) in Bryan County, Oklahoma. D, 1962, University of Oklahoma. 161 p.

Hedrick, O. F. An experimental and theoretical study of flaws or strike-slip faults of the Coeur D'Alene District, Idaho. M, 1924, University of Oklahoma. 26 p.

Hedstrom, Bradley L. Early Pleistocene vegetation of eastern Kansas and Nebraska. M, 1986, Emporia State University.

Heede, Burchard N. Gully development and control in the Rocky Mountains in Colorado. D, 1967, Colorado State University. 298 p.

Heelan, Patrick Aidan. Theory of elastic head wave propagation along a plane interface separating two solid media. D, 1952, St. Louis University.

Heele, Gordon L. A study of the lateral textural, mineralogical, and chemical variations in the Waldron Shale in southern Indiana, western Kentucky, and western Tennessee. M, 1963, Miami University (Ohio). 76 p.

Heeley, Richard William. Hydrogeology of wetlands in Massachusetts. M, 1973, University of Massachusetts. 129 p.

Heerden, Ivor Llewellyn van *see* van Heerden, Ivor Llewellyn

Heerden, Willem Maartens Van *see* Van Heerden, Willem Maartens

Hees, Edmond Harry Peter Van *see* Van Hees, Edmond Harry Peter

Heestand, Richard Lee. The effects of hotspots on the oceanic age-depth relation and the effect of a mantle plume on overlying lithosphere. D, 1982, Princeton University. 104 p.

Heeswijk, Marijke Van *see* Van Heeswijk, Marijke

Heet, Steve. Textural inversions in the Lamotte Sandstone of southeastern Missouri. M, 1981, University of Missouri, Columbia.

Heezen, Bruce C. Orleansville earthquake and turbidity currents. D, 1957, Columbia University, Teachers College.

Heezen, Bruce C. Turbidity currents and submarine slumps, and the 1929 Grand Banks, earthquake. M, 1952, Columbia University, Teachers College.

Hefferan, Kevin Patrick. The geology of the Bonneville Peak Quadrangle, southeastern Idaho. M, 1986, Bryn Mawr College. 69 p.

Heffern, Edward L. Geology of Mission Peak and vicinity, Alameda County, California. M, 1973, Stanford University.

Heffner, J. D. Precipitation of metastable aragonite from aqueous solution. D, 1976, University of Pennsylvania. 152 p.

Heffner, Larry B. Geology of the Grant Town 7 1/2′ Quadrangle, Marion County, West Virginia. M, 1966, West Virginia University.

Heffner, T. A. Quantitative analysis of the Coatesville pegmatite and pyrrhotite locality. M, 1977, Syracuse University.

Heflin, Larry Holden. Geology of the southern half of the Warrenton Quadrangle, Missouri. M, 1961, University of Missouri, Columbia.

Hefner, John Hardin. A rapid method for the correlation of fine-grained sediments with the aid of the spectrograph as applied to the Mississippian-Devonian sequence in Ogemaw County, Michigan. M, 1957, Michigan State University. 52 p.

Hegarty, Kerry Anne. Origin and evolution of selected plate boundaries; Part I, Tectonic and thermal history of Australia's southern margin; Part II, Development of the Caroline Plate region. D, 1985, Columbia University, Teachers College. 265 p.

Hegedus, Andreas Gerhard. The thermodynamics of narrow phase width solid solutions; dolomite, a case study. D, 1987, University of California, Berkeley. 99 p.

Hegemann, David Alan. Algal availability of soil phosphorus. D, 1981, University of Pennsylvania. 229 p.

Heger, Paul A. The direct detection of hydrocarbons; a quantitative approach. M, 1981, Indiana University, Bloomington. 140 p.

Heggen, R. J. Input/output energy analysis of regional water pollution control. D, 1978, Oregon State University. 230 p.

Heggeness, John O. The geology of Ragged Top Caldera. M, 1982, University of Nevada. 107 p.

Heggie, D. T. Copper in the sea; a physical-chemical study of reservoirs, fluxes and pathways in an Alaskan fjord. D, 1977, University of Alaska, Fairbanks. 228 p.

Hegre, Jo Ann B. The geology and petrology of Volcan Las Navajas, a Pleistocene pantellerite center in Nayarit, Mexico. M, 1985, Tulane University. 128 p.

Heichel, Kimberlee Sue. Microveinlet alteration and mineralization at the Sierrita porphyry copper deposit, Pima County, Arizona. M, 1981, University of Arizona. 104 p.

Heide, Kathleen Mae. Late Quaternary vegetational history of Northcentral Wisconsin, U.S.A.; estimating forest composition from pollen data. D, 1981, Brown University. 348 p.

Heidecker, Eric Joseph. Stratigraphic and structural relationships of the Judenan beds at Mount Isa, Australia. M, 1961, Queen's University. 103 p.

Heidecker, Eric Joseph. The tectonic significance of structures in some Grenville rocks (Precambrian, Ontario). D, 1963, Queen's University. 71 p.

Heider, Franz. Magnetic properties of hydrothermally grown Fe(3)O(4) crystals. D, 1988, University of Toronto.

Heidesch, Russell J. A wide angle seismic investigation of Pahute Mesa, Nevada. M, 1987, University of Texas at Dallas. 134 p.

Heidorn, Marjorie Arline. An assessment of structural control as an influence in paleo-valley development in central Auglaize County, Ohio. M, 1979, Wright State University. 69 p.

Heidrick, Tom Lee. Geology and ore deposits of the Ward mining district, White Pine County, Nevada. M, 1965, University of Colorado.

Heidt, J. Harmon. Geology of the Mount Theodore Roosevelt-Maitland area, Lawrence County, South Dakota. M, 1977, South Dakota School of Mines & Technology.

Heidtbrink, Werner Henry, Jr. A sedimentation and petrographic analysis of the Gering Formation of the Pine Ridge area of Northwest Nebraska. M, 1949, University of Nebraska, Lincoln.

Heigold, Paul Clay. The central structure of the St. Louis area determined from the phase velocity of Rayleigh waves. M, 1961, St. Louis University.

Heigold, Paul Clay. Theoretical analysis of regional groundwater flow in the Havana region, Illinois. D, 1969, University of Illinois, Urbana.

Heiken, Grant H. Tuff rings; examples from the Fort Rock, Christmas Lake valley basin, south central Oregon. D, 1972, University of California, Santa Barbara.

Heiken, Grant Harvey. Geology of Cerros Prietos, Municipio de Ojinaga, Chihuahua, Mexico. M, 1966, University of Texas, Austin.

Heikes, Brian Glenn. Atmospheric chemistry of H_2O_2, NO_3, HONO, and HNO_3; influence of aqueous phase and heterogeneous processes. D, 1984, University of Michigan. 226 p.

Heikes, Kenneth Eugene. An experimental study of convection in a rotating layer. D, 1979, University of California, Los Angeles. 89 p.

Heikkila, Henry H. Geology of the Bitterwater Creek area, Northwest Kern County, California. M, 1948, Stanford University. 55 p.

Heil, Darla Jo. Response of an accretionary prism to transform ridge collision south of Panama. M, 1988, University of California, Santa Cruz.

Heil, Richard John. Nitrate contamination of groundwater in northern Runnels County, Texas. M, 1972, University of Texas, Austin.

Heiland, James. Geology of Fall River County, southern Black Hills, South Dakota, as interpreted from Landsat satellite imagery. M, 1978, University of Iowa. 83 p.

Heilborn, George. Stratigraphy of the Woodbine Formation, McCurtain County, Oklahoma. M, 1949, University of Oklahoma. 48 p.

Heilbronner, Heinrich K. Three dimensional coordinates in geodesy. M, 1964, Ohio State University.

Heim, George Edward, Jr. The ground water geology of North Ann Arbor (Michigan). M, 1957, University of Michigan.

Heim, George Edward, Jr. The Pleistocene and engineering geology of the Hannibal-Canton area, Missouri. D, 1963, University of Illinois, Urbana. 137 p.

Heim, Herbert Carl. Petrography of the Dinwoody Formation of southwestern Montana. M, 1962, Indiana University, Bloomington. 53 p.

Heiman, Mary E. Casper Formation (Pennsylvanian and Permian) conodonts. M, 1971, University of Wyoming. 48 p.

Heiman, Mary E. Neogene deep-water benthonic foraminifera from the southwestern Pacific (DSDP legs 21 and 7). D, 1977, University of Wyoming. 286 p.

Heimann, William Henry. A morphometric and hydrogeologic survey of the Saline River paleovalley and valley fill, Ellis County, Kansas. M, 1987, Fort Hays State University. 100 p.

Heimburg, Klaus Frederick. Evapotranspiration; an automatic measurement system and a remote-sensing

method for regional estimates. D, 1982, University of Florida. 211 p.

Heimlich, Richard A. Structure and petrology of acid plutons in the Deer Lake area, northern Ontario, Canada. D, 1959, Yale University.

Heimmer, Donald. Dissolution of phosphate minerals in deionized water and diluted organic acids at 25°C and one atmosphere. M, 1972, University of South Florida, Tampa. 104 p.

Hein, Frances J. Deep-sea valley-fill sediments; Cap Enrage Formation, Quebec. D, 1979, McMaster University. 514 p.

Hein, Frances J. Gravel transport and stratification origins, Kicking Horse River, British Columbia. M, 1974, McMaster University. 135 p.

Hein, James Rodney. Part 1, Delgada submarine fan stratigraphy and provenance; a key to motion between California and the Pacific Plate; Part 2, Selected studies in marine diagenetic and authigenic processes. D, 1973, University of California, Santa Cruz.

Heinbokel, John F. Functional and numerical responses of coastal tintinnids; implications for the neritic food chain. D, 1977, University of California, San Diego. 189 p.

Heindl, Leopold A. Cenozoic alluvial deposits of the upper Gila River area, New Mexico and Arizona. D, 1958, University of Arizona.

Heine, Aida A. The structure and development of trilobites. M, 1905, Smith College. 46 p.

Heine, Christian J. Late Cretaceous dinoflagellate palynomorphs in northern Texas coastal plain strata. M, 1986, University of Tennessee, Knoxville. 145 p.

Heine, Richard Ralph. Geochemistry and mineralogy of the Permian red beds and related copper deposits, Payne, Pawnee, and Noble counties, Oklahoma. M, 1975, Oklahoma State University. 70 p.

Heine, Thomas Hermann. A palaeomagnetic study of the Catfish Creek Till. M, 1977, University of Windsor. 60 p.

Heinecke, Thomas A. Magnetostratigraphic correlation between terraces of the Green River and sediments within the passages of Mammoth Cave National Park, Kentucky. M, 1982, University of Pittsburgh.

Heineman, Robert E. S. The geology and ore deposits of the Johnson mining district, Arizona. M, 1927, University of Arizona.

Heinemeyer, G. R. Geology of the Razor Creek Dome area, northeastern San Juan Mountains, Saguache County, Colorado. M, 1973, Colorado State University. 125 p.

Heiner, Ted, Jr. Demonstration unit for demonstrating concepts of ground water. M, 1963, University of Utah. 30 p.

Heiner, Williams R. Geology and magnetic survey of Osborn Bank, southern California. M, 1969, University of Southern California.

Heiniger, Keith D. Measurement of crustal movement. M, 1964, Ohio State University.

Heinonen, Charles E. Stratigraphy, structural geology, and metamorphism of the Tacoma Lakes area, Maine. M, 1971, University of Maine. 80 p.

Heinrich, Eberhardt William. Geology of the Eight Mile Park pegmatite area, Colorado. D, 1948, Harvard University.

Heinrich, M. Allen. Geology of the Applegate Group (Triassic) in the Kinney Mountain area, southwest Jackson County, Oregon. M, 1966, University of Oregon. 107 p.

Heinrich, Nathan Daniel. Metamorphism and deformational history of a portion of the Campbell Mountain and Suches 7-1/2-minute quadrangles, northern Georgia. M, 1987, Auburn University. 126 p.

Heinrich, Paul Victor. Geomorphology and sedimentology of Pleistocene Lake Saline, southern Illinois. M, 1982, University of Illinois, Urbana. 145 p.

Heinrich, Ross Raymond. A contribution to the seismic history of Missouri. M, 1938, St. Louis University.

Heinrich, Silvia Maria. Geology and geochronology of the Zongo River valley, Cordillera Oriental, NW Bolivia. M, 1988, Queen's University. 188 p.

Heinrichs, Donald Frederick. Paleomagnetism of the Plio-Pleistocene Lousetown Formation, Virginia City, Nevada. D, 1966, Stanford University. 48 p.

Heins, Melburn E. A study of some lead-silver deposits, Galena, South Dakota. M, 1928, University of Minnesota, Minneapolis.

Heins, Vasco Ann M. Structural setting of Teton Pass with emphasis of fault breccia associated with the Jackson thrust fault, Wyoming. M, 1982, Montana State University. 64 p.

Heinsius, John Walter. The sediments and sedimentary environments of a barrier island, Cedar Island, Virginia. M, 1974, Western Michigan University.

Heintz, Greta Marie. Structural geology of Leverett Glacier area, Antarctica. M, 1980, Arizona State University. 158 p.

Heintz, Louis O. Seasonal magnetic variations in the beach sands at Malaga Cove, Los Angeles County, California. M, 1966, University of Southern California.

Heiny, Janet S. Sediment characteristics of rapidly retreating temperate valley glaciers. M, 1983, Northern Illinois University. 243 p.

Heinz, David Michael. Weathering of fossils in limestones. M, 1953, University of Illinois, Urbana.

Heinz, Dion Larsen. Experimental determination of the melting curve of magnesium silicate perovskite at lower mantle conditions, and its geophysical implications. D, 1986, University of California, Berkeley. 93 p.

Heinz, Heinrich Karl, Jr. Large cross section tunnels in soft ground. D, 1988, University of Alberta. 348 p.

Heinze, Daniel William. A geological application of Biot's theory. M, 1972, Texas A&M University.

Heinze, W. D. A geologic application of Biot's buckling theory. M, 1972, Texas A&M University.

Heinze, W. D. Distortion measurement between photographic images; a new rock-mechanic tool. D, 1977, Texas A&M University. 148 p.

Heinze, William D. Numerical simulation of stress concentration in rocks. M, 1973, Massachusetts Institute of Technology. 38 p.

Heinzelmann, Gerald Mathias, Jr. Mississippian rocks in the Stillwater-Chandler area. M, 1957, University of Oklahoma. 98 p.

Heinzler, C. Thomas. Suppression of almost periodic multiples using slant stacks. D, 1984, Stanford University.

Heiple, Linda J. A sedimentological study of the beach between Oceanside and San Clemente, Orange and San Diego counties, California. M, 1979, Colorado School of Mines. 99 p.

Heiple, Paul W. Primary trace element distribution of the Belden Shale (Pennsylvanian) from Trout Creek Pass to Eagle, Colorado; a possible source for ore deposits. M, 1980, Colorado School of Mines. 105 p.

Heirendt, Kenneth M. An analysis of ^{222}Rn soil gas concentrations in the Serpent Mound area, southwestern Ohio. M, 1988, University of Akron. 86 p.

Heironimus, Thurman L. Biostratigraphy, depositional environments, and paleogeography of Lower and Middle Devonian rocks, Death Valley area, California. M, 1982, Oregon State University. 167 p.

Heise, Bruce A. Structural geology of the Mt. Haggin area, Deer Lodge County, Montana. M, 1983, University of Montana. 77 p.

Heise, Roy H. Palynology of the Cardium Formation, west-central Alberta. M, 1987, University of Calgary. 101 p.

Heisey, Edmund L. Geology of the East Ferris Mountains, Carbon County, Wyoming. M, 1949, University of Wyoming. 87 p.

Heisterkamp, Warren Craig. A subsurface study of the Borden Group of rocks of Indiana. M, 1952, Indiana University, Bloomington. 29 p.

Heitkamp, George William. The origin and distribution of the loess-like material in the vicinity of Urbana, Illinois. M, 1914, University of Illinois, Urbana.

Heitman, Hal Louis. Distribution of benthic foraminifera in Tanner Basin, California continental borderland. M, 1979, University of Southern California.

Heitz, Leroy Fredrick. Hydrologic evaluation methods for hydropower studies. D, 1981, University of Idaho. 250 p.

Heitzman, Donald Paul. A gravitational interpretation of aeromagnetic anomalies in Blackhawk County, Iowa. M, 1972, University of Iowa. 112 p.

Heivilin, Fred G. The stratigraphy and lithology of the Ste. Genevieve formation (Mississippian) between White Hill, Illinois and Ste. Genevieve, Missouri. M, 1967, Southern Illinois University, Carbondale. 119 p.

Hejazi, Assadollah. Application of wave equations to pile driving analyses. D, 1983, Louisiana Tech University. 250 p.

Hejazi, Sayyed Hossein. Static and dynamic elastic properties of Denault Dolomite and Wishart Quartzite. M, 1980, University of Saskatchewan. 106 p.

Hekinian, Roger. Petrological and geochemical study of spilites and associated dike rocks from the Virgin Island core (Caribbean Island arc). D, 1969, SUNY at Binghamton. 216 p.

Hekinian, Roger. Rocks from the mid-oceanic ridge in the Indian Ocean. M, 1966, Columbia University. 48 p.

Helalia, Awad Mohamed Ahmed. Water quality and polymer type effects on soil infiltration. D, 1987, University of California, Riverside. 129 p.

Helander, Donald Peter. The effect of pore configuration, pressure, and temperature on rock resistivity. D, 1965, University of Oklahoma. 169 p.

Helbig, Steffan Reed. Investigation of a mineralization halo associated with the talc deposits near Winterboro, Alabama as expressed in the residual soils. M, 1983, Rutgers, The State University, Newark. 136 p.

Held, Harry L. Zircon distribution patterns in selected orbicular rocks. M, 1977, Texas Tech University. 53 p.

Helenek, Henry Leon. An investigation of the origin, structure and metamorphic evolution of major rock units in the Hudson highlands. D, 1971, Brown University. 363 p.

Helenek, Henry Leon. An investigation of the origin, structure and metamorphic evolution of major rock units in the Hudson Highlands (New York). M, 1965, Brown University.

Helenes-Escamilla, Javier. Stratigraphy, depositional environments and foraminifera of the Miocene Tortugas Formation, Baja California Sur, Mexico. M, 1980, Stanford University. 74 p.

Helenes-Escamilla, Javier. Studies on the morphology of fossil dinoflagellates, mainly from Baja California, Mexico. D, 1984, Stanford University. 392 p.

Helfenstein, Paul. Derivation and analysis of geological constraints on the emplacement and evolution of terrains on Ganymede from applied differential photometry. D, 1986, Brown University. 428 p.

Helfrich, Charles T. Silurian conodonts from the Wills Mountain Anticline, Virginia, West Virginia and Maryland. D, 1972, Virginia Polytechnic Institute and State University.

Helgason, Johann. Structural relationships and magnetostratigraphy of the volcanic succession and the Breiddalur dyke swarm in Reydarfjordur, eastern Iceland. D, 1983, Dalhousie University. 260 p.

Helgerson, Ronald N. Lateral distribution of clay minerals in the Tertiary Ludlow Formation, Perkins County, South Dakota. M, 1975, University of South Dakota. 56 p.

Helgesen, John O. Hydrogeology of outwash associated with the Antelope Moraine, southwestern Minnesota. D, 1967, Colorado State University. 59 p.

Helgeson, Harold Charles. Hydrothermal ore-forming solutions. D, 1962, Harvard University.

Helie, Robert G. Differentiation and genesis of diamictons on Somerset Island, N.W.T. M, 1981, McGill University. 109 p.

Heliker, C. Christina. Inclusions in the 1980-83 dacite of Mount St. Helens, Washington. M, 1984, Western Washington University. 185 p.

Heller, Frederick Klach. Geology and economics of the Crystal-Ferris oil and gas field (Michigan). M, 1936, University of Michigan.

Heller, Jeff C. Clay mineralogy of the southwest equatorial Atlantic sediments. M, 1970, University of California, Los Angeles.

Heller, John Lowell. The geology of the Blackoak Ridge area, Anderson County, Tennessee. M, 1959, University of Tennessee, Knoxville. 34 p.

Heller, Noah Russell. Geology of the Henson Springs Quadrangle, Alabama. M, 1984, Mississippi State University. 75 p.

Heller, Paul. Pleistocene geology and related landslides in the lower Skagit and Baker valleys, North Cascades, Washington. M, 1978, Western Washington University. 154 p.

Heller, Paul Lewis. Sedimentary response to Eocene tectonic rotation in western Oregon. D, 1983, University of Arizona. 343 p.

Heller, Robert Leo. Geology of the Marble Hill area, Bollinger County, Missouri. M, 1943, University of Missouri, Columbia.

Heller, Robert Leo. Stratigraphy and paleontology of the Roubidous Formation of Missouri. D, 1950, University of Missouri, Columbia.

Heller, Sara Anne. A hydrogeologic study of the Greenbrier Limestone karst of central Greenbrier County, West Virginia. D, 1980, West Virginia University. 204 p.

Hellerman, Joan. The fossil diatoms of the Mohonk Lake area, New York and their ecological significance. D, 1965, Rutgers, The State University, New Brunswick. 137 p.

Hellert, John R. Lithostratigraphy and soundness of the Tymochtee Formation (Silurian) in Adams, Pike, and Fayette counties, Ohio. M, 1972, Ohio State University.

Helley, Edward John. Sediment transport in the Chowchilla River Basin; Mariposa, Madera, and Merced counties, California. D, 1966, University of California, Berkeley. 189 p.

Hellier, Nancy W. Sanitary landfill in Massachusetts; a study of sixteen communities. M, 1972, Boston University. 67 p.

Hellinger, Steven J. The statistics of finite rotations in plate tectonics. D, 1979, Massachusetts Institute of Technology. 172 p.

Hellinger, Terry Scott. Natural microscopic deformation and annealing features of some massive sulfide ores, Sudbury, Ontario. M, 1977, University of Michigan.

Hellman, John Dale. Subsurface geology of the Joiner City Field, Carter County, Oklahoma. M, 1962, University of Oklahoma. 63 p.

Hellstern, Donald. Seismic stratigraphic processing and interpretation of the lower Miocene Robulus mayeri interval in the Matagorda Island, offshore Texas; with EM. M, 1987, University of Houston.

Helm, Kenneth Richard. The geology and geochemistry of the Davis Mine and surrounding Hawley Formation, Northwest Massachusetts. M, 1982, SUNY at Binghamton. 144 p.

Helm, Ronnie L. Depositional history and petrology of the Fort Union Formation (Paleocene) southern flank of the Rock Springs Uplift, Sweetwater County, Wyoming. M, 1985, San Jose State University. 145 p.

Helm, Sue. Prevention of water contamination through effective disposal of produced water generated by onshore oil and gas production. M, 1987, Southwest Missouri State University. 75 p.

Helman, Marc. Permian age evaporites of the Dolomite Mountains (Southern Alps), northern Italy. M, 1984, Queens College (CUNY). 106 p.

Helman, Ronald Paul. The petrology of the Cloverly Formation, Rawlins Uplift area, Carbon County, Wyoming. M, 1957, Miami University (Ohio). 93 p.

Helmberger, Donald Vincent. Head waves from the oceanic Mohorovicic discontinuity. D, 1967, University of California, San Diego. 190 p.

Helming, Bob Hager. Petrology of the Rogue Formation (Upper Jurassic), southwestern Oregon. M, 1966, University of Oregon. 82 p.

Helmke, Philip August. Rare-earth elements in the Steens Mountain basalts. D, 1971, University of Wisconsin-Madison. 171 p.

Helmold, K. Petrology of septarian concretions from the Purcell Shale (Devonian), Montgomery County, Washington and other selected concretions. M, 1974, George Washington University.

Helmold, Kenneth Paul. Diagenesis of Tertiary arkoses, Santa Ynez Mountains, California. D, 1980, Stanford University. 261 p.

Helms, Phyllis Borden. Lithogenesis of atoll sediments, Uliga Island, Majura Atoll, Marshall Islands. M, 1961, University of New Mexico. 105 p.

Helms, Thomas S. Petrology of the Laurel Creek Amphibolite, Georgia Blue Ridge. M, 1985, University of Tennessee, Knoxville. 83 p.

Helmstaedt, Herwart. Structural analysis of Beaver Harbour area (mid-Paleozoic), Charlotte County, New Brunswick (Canada). D, 1968, University of New Brunswick.

Helou, Amin Habib. Seismic analysis of submerged underwater oil storage tanks. D, 1981, North Carolina State University. 94 p.

Helper, Mark Alan. Structural, metamorphic and geochronologic constraints on the origin of the Condrey Mountain Schist, north central Klamath Mountains, Northern California. D, 1985, University of Texas, Austin. 264 p.

Helprin, Sydney. The geology of Minas de Matahambre, Province of Pinar del Rio, Cuba. M, 1930, Columbia University, Teachers College.

Helsel, D. R. Land use influences on heavy metals in an urban reservoir system. D, 1978, Virginia Polytechnic Institute and State University. 242 p.

Helsel, Dennis R. Lithologic control of streamwater chemistry in a portion of the Brandywine Creek, Pa. M, 1976, University of Delaware.

Helsel, Laura. Petrologic constraints on the Brennan Hill Thrust, southwestern New Hampshire. M, 1988, Dartmouth College. 125 p.

Helsen, Jan Nicolaas Walter. Geochemistry of tungsten. M, 1970, McMaster University. 50 p.

Helsen, Jan Nicolaas Walter. Geochemistry of tungsten in basalts and andesites. D, 1976, McMaster University. 203 p.

Helsey, Charles Everett. Geology of the British Virgin Islands. D, 1960, Princeton University. 282 p.

Helsinger, Marc H. Diagenetic modification and cementation at intertidal levels; beachrock and Oolitic limestones from the Red Sea and Jamaica. M, 1973, Rensselaer Polytechnic Institute. 126 p.

Helsinger, Marc H. Distribution and incorporation of trace elements in the bottom sediments of the Hudson River and tributaries. D, 1975, Rensselaer Polytechnic Institute. 127 p.

Helsley, Robert. Terry Sandstone Member of the Pierre Shale, Upper Cretaceous, Spindle Field, Denver Basin, Colorado. M, 1985, Texas A&M University. 63 p.

Helton, Walter L. The Silurian-Devonian stratigraphy of Pulaski County, Kentucky, and some features of the pre-mid-Devonian unconformity. M, 1963, University of Kentucky. 69 p.

Helton, Walter Lee. Lithostratigraphy of the Conasauga Group (Middle-Upper Cambrian) between Rogersville and Kingsport (Hawkins County), Tennessee. D, 1967, University of Tennessee, Knoxville. 96 p.

Helwa, Mohamed Fawzy. The channel network in hydrologic simulation; improved modeling and evaluation of significance. D, 1983, University of Maryland. 201 p.

Helweg, O. J. A salinity management strategy for stream-aquifer systems. D, 1975, Colorado State University. 190 p.

Helwick, S. J., Jr. Engineering properties of shallow sediments in West Delta and South Pass outer continental shelf lease areas, offshore Louisiana. M, 1977, Texas A&M University.

Helwick, S. J., Jr. Prediction of the geotechnical properties of late Quaternary Mississippi Delta deposits. D, 1979, Texas A&M University. 239 p.

Helwig, James Anthony. Stratigraphy and structural history of the New Bay area, north central Newfoundland. D, 1967, Columbia University. 214 p.

Helwig, Jo Wilson. Application of statistical analysis to grain size parameters in Recent sedimentary environments, N.C. coastal plain. M, 1969, University of North Carolina, Chapel Hill. 49 p.

Helz, George Rudolph. Hydrothermal solubility of magnetite. D, 1971, Pennsylvania State University, University Park. 110 p.

Helz, Rosalind Tuthill. The petrogenesis of the Ice Harbor Member, Columbia Plateau, Washington; a chemical and experimental study. D, 1978, Pennsylvania State University, University Park. 284 p.

Hemann, Mark Richard. Field evaluation of the relationships between transmissivity, permeability and particle size distribution in the Washita River alluvial aquifer, near Anadarko, Oklahoma. M, 1985, Oklahoma State University. 180 p.

Hemborg, Thomas Harold. The geology of the La Porte Quadrangle, Plumas County, California. M, 1966, University of California, Los Angeles.

Hembre, Donald R. A heavy mineral and sedimentary study of the Schoolhouse Tongue of the Weber Sandstone (Colorado). M, 1955, University of Wisconsin-Madison.

Hembree, Max Reed. Geology of the Fossil Hill area, Wind River Mountains, Fremont County, Wyoming. M, 1949, University of Missouri, Columbia.

Hemer, Darwin O. Correlation of the Damman and Rub'al Khali sections (Saudi Arabia). M, 1958, New York University.

Heming, Robert Frederick. The geology and petrology of Rabaul Caldera; an active volcano in New Britain, Papua New Guinea. D, 1973, University of California, Berkeley. 181 p.

Hemingway, Bruce S. A calorimetric determination of the stability, entropy, heat, and Gibbs energy of formation for the carbonate minerals huntite, nesquehonite, artinite and hydromagnesite. D, 1971, University of Minnesota, Minneapolis. 255 p.

Hemingway, Mark P. Mineralogy and geochemistry of the southern Amethyst vein system, Creede mining district, Colorado. M, 1986, New Mexico Institute of Mining and Technology. 91 p.

Heminway, Caroline Ella. Devono-Carboniferous differentia. M, 1928, Cornell University.

Heminway, Caroline Ella. The Tertiary foraminifera of Puerto Rico. D, 1941, Indiana University, Bloomington. 215 p.

Hemish, LeRoy A. Stratigraphy of the upper part of the Fort Union Group in southwestern McLean County, North Dakota. M, 1975, University of North Dakota. 160 p.

Hemlein, Kristin. The kinetics of formation of ordered dolomite from high magnesium calcite 250 to 350°C and 1000 bars. M, 1988, University of Cincinnati. 66 p.

Hemley, John Julian. A study of lead sulfide solubility and its relation to ore deposition. M, 1952, Northwestern University.

Hemley, John Julian. Some mineralogical equilibria in the system $K_2O-Al_2O_3-SiO_2-H_2O$. D, 1958, University of California, Berkeley. 115 p.

Hemming, N. Gary. Controls on the water chemistry of some springs in a volcanic terrain, Nayarit, Mexico. M, 1985, Tulane University. 121 p.

Hemmings, Charles David. Upper mantle structure in western Canada. M, 1969, University of Alberta. 116 p.

Hemphill, Gary Brian. Managing mine construction projects. M, 1982, University of Idaho. 112 p.

Hempkins, William B. Mathematical and statistical models for prediction of ore distribution in mines of the Witwatersrand Basin, South Africa. D, 1969, Northwestern University.

Hempkins, William Brent. Geology and petrography of the Sawtooth Mountain area, Jeff Davis County, Texas. M, 1962, University of Texas, Austin.

Hempton, Mark Robert. Structure of the northern margin of the Bitlis suture zone near Sivrice, southeastern Turkey. D, 1982, SUNY at Albany. 563 p.

Hempy, Daniel Willett. Fish Creek fluvial system and associated tectonically influenced morphology, Salton Trough, California. M, 1981, San Diego State University.

Hemstrom, Miles A. A Recent disturbance history of forest ecosystems at Mount Rainier National Park. D, 1979, Oregon State University. 67 p.

Hemud, Abdul Rahman. Geophysical and hydrogeological investigations of the Moscow landfill site, Latah County, Idaho. M, 1971, University of Idaho. 59 p.

Hemyari, Parichehr. A comparison of classical statistics and geostatistics for estimating soil surface temperature. D, 1984, Oklahoma State University. 69 p.

Hemzacek, Jean Marie. Replaced evaporites and the sulfur isotope age curve of the Precambrian. M, 1987, Northern Illinois University. 77 p.

Henage, Lyle Frederick. A definitive study of the origin of lamproites. M, 1972, University of Oregon. 140 p.

Henage, Lyle Frederick. The geology and compositional evolution of Mt. Olokisalie, Kenya. D, 1977, University of Oregon. 193 p.

Henao, Angela M. Modeling of multivariate streamflow series with gamma marginal distributions. D, 1987, Colorado State University. 186 p.

Henbest, Lloyd G. Fusulinellas from the Stonefort Limestone Member of the Tradewater Formation, Harrisburg Quadrangle, Illinois. M, 1927, University of Kansas.

Hench, Stephen Wayne. A reconnaissance gravity survey of the Ruby Pena Blanca area, Santa Cruz County, Arizona. M, 1968, University of Arizona.

Henckel, Elaine M. Paleomagnetic study of rhyolites and granites in the Fox River valley area of south-central Wisconsin. M, 1985, University of Wisconsin-Milwaukee. 175 p.

Hendershot, W. H. Surface charge properties of selected soils. D, 1978, University of British Columbia.

Henderson, Arnold R. Geology of the Powder Springs area, Union and Grainger counties, Tennessee. M, 1961, University of Tennessee, Knoxville. 30 p.

Henderson, Barry Keith. A mathematical model of some Pennsylvanian stratigraphy. D, 1969, Louisiana State University.

Henderson, Barry Keith. A model for Allegheny stratigraphy of northern West Virginia. M, 1965, Louisiana State University. 49 p.

Henderson, Barry Leon. Geologic map and structural analysis of the Sykes Spring area, northern Big Horn County, Wyoming, and southern Carbon County, Montana. M, 1985, University of Iowa. 100 p.

Henderson, Charles D. Minor structures of the High Plateau, northeastern Tucker County, West Virginia. M, 1973, West Virginia University.

Henderson, Charles F. Glaciation in the mountains bordering South Park, Colorado. M, 1936, Northwestern University.

Henderson, Charles Murray. Conodont paleontology of the Permian Sabine Bay, Assistance and Trold Fiord formations, northern Ellesmere Island, Canadian Arctic Archipelago. M, 1981, University of British Columbia. 135 p.

Henderson, David W. Structure of the Deer Trail Anticlinorium, Stevens County, Washington. M, 1983, Eastern Washington University. 43 p.

Henderson, Don K. Geology of Doty Mountain-Dad area, Carbon County, Wyoming, with emphasis on stratigraphy of the uppermost Cretaceous rocks. M, 1962, Colorado School of Mines. 162 p.

Henderson, Donald Munro. Geology and petrology of the eastern part of the Crawford Notch Quadrangle, New Hampshire. D, 1950, Harvard University.

Henderson, Donpaul. A preliminary hydrologic examination of an ephemeral stream basin in Union County, Illinois. M, 1976, Southern Illinois University, Carbondale. 102 p.

Henderson, Elizabeth Darrow. Stack unit mapping and Quaternary history Eddyville 7.5 minute quadrangle, southern Illinois. M, 1987, Southern Illinois University, Carbondale. 181 p.

Henderson, Eric P. Pleistocene geology of the Watino Quadrangle, Alberta. D, 1952, Indiana University, Bloomington. 92 p.

Henderson, Floyd Merl. Small scale land use mapping with radar imagery. D, 1973, University of Kansas.

Henderson, Frank A. Petrofabrics of the Nonesuch Formation (Precambrian) and its relationship to primitive stress directions in the White Pine (north Michigan) ore body. M, 1971, Michigan Technological University. 65 p.

Henderson, Frederick B., III. Deposition and oxidation of serpentinite-type mercury deposits. D, 1965, Harvard University.

Henderson, Frederick Bradley, III. Paleogeologic study of the central portion of the Permian basin of West Texas. M, 1960, Stanford University.

Henderson, Garry Couch. Computer analysis techniques applied to crustal studies of Campeche Bank, Mexico. D, 1965, Texas A&M University.

Henderson, Garry Couch. Crustal structure across the Campeche Escarpment, Gulf of Mexico, from gravity date. M, 1962, Texas A&M University.

Henderson, George G. Subsurface geology of the south part of the Blackwell Field, Oklahoma. M, 1923, University of Oklahoma. 18 p.

Henderson, Geral Vernon. The origin of pyrophyllite-rectorite in shales of north central Utah. D, 1969, University of Illinois, Urbana. 122 p.

Henderson, Gerald G. L. Structural studies in an area at the headwaters of McMurdo Creek, British Columbia. M, 1950, McGill University.

Henderson, Gerald Gordon Lewis. Geology of the Stanford Range, B.C. D, 1953, Princeton University. 197 p.

Henderson, Gerald J. Correlation and analysis of geologic time series. D, 1973, Indiana University, Bloomington. 289 p.

Henderson, Gerald V. Geology of the northeast quarter of the Soldier Summit Quadrangle, Wasatch and Utah counties, Utah. M, 1958, Brigham Young University. 40 p.

Henderson, Gordon William. Cenozoic geology of southern Ione Valley, Nevada. M, 1962, University of California, Berkeley. 85 p.

Henderson, Grant Stephen. An X-ray scattering and Raman spectroscopy study of Fe^{3+}, Ga^{3+} and Ge^{4+} substituted aluminosilicate glasses. D, 1983, University of Western Ontario. 143 p.

Henderson, J. F. The Dufault Lake Sill. M, 1933, Queen's University. 22 p.

Henderson, James F. The geology of the Granville Lake District, Manitoba. D, 1933, University of Wisconsin-Madison.

Henderson, James Henry. I, Quartz and cristobalite origin in selected soils and sediments and in relation to montmorillonite origin in bentonites determined by oxygen isotope abundance; II, Cation and silica relationships of vermiculite formation from mica in calcareous harps soil. D, 1971, University of Wisconsin-Madison. 208 p.

Henderson, Janet Elizabeth Rieder. Ostracod fauna of the Bell Shale Formation of Michigan. M, 1951, University of Michigan.

Henderson, Jeremy Robert. Numerical experiments on continental lithosphere extension. M, 1982, Massachusetts Institute of Technology. 52 p.

Henderson, Joe D. Geology of the Jasper Quadrangle (Paleozoic), Newton and Boone counties, Arkansas. M, 1972, University of Arkansas, Fayetteville.

Henderson, John B. Petrology and origin of the sediments of the Yellowknife group (Archean), Yellowknife, District of MacKenzie, Canada. D, 1970, The Johns Hopkins University.

Henderson, John R. Structural analysis of the Chickies Creek area, southeastern Pennsylvania. M, 1965, Northwestern University.

Henderson, John Russell. Structural and petrologic relations across the Grenville Province-Southern Province boundary, Sudbury District, Ontario (Canada). D, 1967, McMaster University. 119 p.

Henderson, Joseph C. Specific gravity determination of marine sediments. M, 1970, United States Naval Academy.

Henderson, L. Brooke. Geological study and mapping of Desmoinesian (Pennsylvanian) rocks in the Montrose Quadrangle, Henry County, Missouri. M, 1958, University of Iowa. 148 p.

Henderson, Laurel Jean. Motion of the Pacific Plate relative to the hotspots since the Jurassic and model of oceanic plateaus of the Farallon Plate. D, 1985, Northwestern University. 329 p.

Henderson, Lawrence. Recent sedimentation in the Southeast Indian Ocean. M, 1976, Brooklyn College (CUNY).

Henderson, Stephen William. Stratigraphy and petrography of the Rockford Limestone (Mississippian) of southeastern Indiana. M, 1974, Indiana University, Bloomington. 73 p.

Henderson, Stephen William. Taphonomy and distribution of shelled molluscs within the Duplin River and Doboy Sound, Sapelo Island, Georgia. D, 1984, University of Georgia. 306 p.

Henderson, Virginia W. Exterior controls on barrier island chain morphology and distribution; a global view. M, 1988, Duke University. 157 p.

Henderson, William Charles. A seismic analysis of reefs in the Traverse Limestone of Allegan County, Michigan. M, 1988, Western Michigan University.

Henderson, William Garth. Physico-chemical phenomena of soil materials. D, 1966, Oklahoma State University. 87 p.

Hendon, Bryon. The fauna of the type Umpqua Eocene of Oregon. M, 1926, University of Oregon. 87 p.

Hendrajaya, Lilik. Theoretical estimation of hydrocarbon reservoir pressure using seismic wave velocity. M, 1975, University of Utah. 40 p.

Hendren, Celia Faith. Paleoecology of the Brachiopoda (Sandia Mountains, New Mexico). M, 1949, University of New Mexico. 70 p.

Hendren, John Blair. Coral zonation of the Callaway Formation, Missouri. M, 1957, University of Missouri, Columbia.

Hendrick, John W. Submarine geography; three approaches to field survey techniques in shallow undersea environments. M, 1972, San Francisco State University.

Hendrick, Steven J. The geology of the Three Fingers Rock area, Owyhee Uplands, Malheur County, Oregon. M, 1978, South Dakota School of Mines & Technology.

Hendrick, Thomas K. The subsurface geology of the Anton-Irish Field, Lamb and Hale counties, Texas. M, 1948, University of Oklahoma. 36 p.

Hendricks, Charles E., Jr. Sedimentation and diagenesis of the Lower Silurian Brassfield Formation in southwestern Ohio. M, 1983, University of Cincinnati. 177 p.

Hendricks, Charles Leo. Subdivision and subsurface correlation of the Ellenburger Group in north-central Texas. D, 1942, University of Texas, Austin.

Hendricks, David M. Clay mineral genesis during andesite weathering and soil formation. D, 1966, University of California, Davis.

Hendricks, Herbert Edward. Geology of the Crooked Creek area, Missouri. D, 1949, University of Iowa. 218 p.

Hendricks, Herbert Edward. Geology of the Macks Creek Quadrangle, Missouri. M, 1942, University of Iowa. 122 p.

Hendricks, John D. Younger Precambrian basaltic rocks of the Grand Canyon, Arizona. M, 1972, Northern Arizona University. 122 p.

Hendricks, Leo. Systematic samples of well samples. M, 1930, Texas Christian University. 68 p.

Hendricks, Michael L. Stratigraphy and tectonic history of the Mesa Verde Group (Upper Cretaceous), east flank of the Rock Springs Uplift, Sweetwater County, Wyoming. D, 1983, Colorado School of Mines. 188 p.

Hendricks, Michael L. Stratigraphy of the Coalmont Formation near Coalmont, Jackson County, Colorado. M, 1977, Colorado School of Mines. 112 p.

Hendricks, Robert Craig. Trace-element distributions between coexisting minerals in metamorphic rocks from the Ruby Range, southwestern Montana. M, 1985, Kent State University, Kent. 120 p.

Hendricks, Thomas A. Some details of the sedimentation of Mesaverde Formation on the south side of the San Juan Basin, New Mexico. M, 1931, University of Colorado.

Hendricks, Walter J. Petrofabric investigation across the Daisy Lake Fault, Coniston, Ontario. M, 1958, Massachusetts Institute of Technology. 55 p.

Hendrickson, Brent R. Stratigraphic position, mineralogy, depositional environment, and gold distribution of the main reef at Morro do Cuscuz and Morro do Vento near Jacobina, Bahia, Brazil. M, 1984, South Dakota School of Mines & Technology.

Hendrickson, Charles R. Geometry and diagenesis of the J Sandstone, Comanche Creek Field area, Denver Basin, Colorado. M, 1983, Texas Tech University. 66 p.

Hendrickson, Denise M. Stratigraphy and sedimentology of the Pennsylvanian and Lower Permian Fountain Formation in Perry Park, Douglas County, Colorado. M, 1986, Colorado School of Mines. 267 p.

Hendrickson, Glen. Pre-Cambrian complex of upper Rist Canyon, Larimer County, Colorado. M, 1955, University of Colorado.

Hendrickson, Harald F. Formation of uranium minerals in petroleum and asphalt. M, 1959, University of Minnesota, Minneapolis. 81 p.

Hendrickson, M. A. The determination of seismic structure from teleseismic P waveforms on the Washington continental margin. M, 1986, University of Washington. 74 p.

Hendrickson, Ole Quist, Jr. Flux of nitrogen and carbon gases in bottomland soils of an agricultural watershed. D, 1981, University of Georgia. 225 p.

Hendrickson, Walter John. Depositional environment and petroleum geology of sandstones within the Bunger-Gunsight Formation (Pennsylvanian) of an area within north-central Archer County, Texas. M, 1976, University of Texas, Arlington. 97 p.

Hendrix, Bill. Structural study of an area surrounding the Tertiary Biedell volcanic center, Saguache County, Colorado. M, 1982, Wichita State University. 74 p.

Hendrix, Eric Douglas. Sedimentology and basin analysis of the upper Oligocene Vasquez Formation, Soledad Basin, Southern California. M, 1986, University of California, Los Angeles. 296 p.

Hendrix, Gary G. Structural and chemical changes across greenschists facies rocks in the Ocoee Gorge, Tennessee. M, 1977, University of Illinois, Chicago.

Hendrix, Marc S. Stratigraphy and sedimentology of phosphatic strata, Uinta Mountains, Utah. M, 1987, University of Wisconsin-Madison. 135 p.

Hendrix, Thomas Eugene. A petrologic study of two Lower Pennsylvanian sandstones from western Pennsylvania. M, 1957, University of Wisconsin-Madison. 76 p.

Hendrix, Thomas Eugene. Structural history of the East Gogebic iron range (Michigan-Wisconsin). D, 1960, University of Wisconsin-Madison. 109 p.

Hendrix, W. G. Regional ecosystem assessment; the ecological compatibility component of the metropolitan landscape planning model (Metland). D, 1977, University of Massachusetts. 225 p.

Hendrix, William Edwin. Ostracodes of the Olentangy Shale of central Ohio. M, 1939, Ohio State University.

Hendry, Charles W., Jr. Microfossils of the Moodys Branch Formation in Citrus and Levy counties, Florida. M, 1951, Florida State University.

Hendry, Lynne D. The geology and ore deposits of Central America. M, 1927, University of Minnesota, Minneapolis. 50 p.

Hendry, Michael James. Nitrate contributed by groundwater to a small stream in an agricultural watershed. M, 1977, University of Waterloo. 226 p.

Hendry, Michael James. Origin of groundwater sulfate in a fractured till in an area of southern Alberta, Canada. D, 1984, University of Waterloo. 226 p.

Hendry, N. W. and Wilson, H. D. Geology and quicksilver deposits of the Coso Hot Springs area, Inyo County, California. M, 1939, California Institute of Technology. 63 p.

Hendry, Robert D. Gravity study of subsurface geologic basement and crustal structures in the central Appalachian Rome Trough region, West Virginia. M, 1982, SUNY at Buffalo. 64 p.

Hendy, William James. A preliminary petrologic study of the Dakota Group in northeastern Nebraska. M, 1940, University of Nebraska, Lincoln.

Henes, Walter E. and Ogden, Maynard Blair. Geology of the Russell-Wagon Creek area, Costilla County, Colorado. M, 1960, University of Michigan.

Hengesh, James V. Metamorphic and depositional history of Archean migmatites and gneisses, central Horseshoe Lake Quadrangle, Wind River Mountains, Wyoming. M, 1987, Idaho State University. 72 p.

Henika, William Sinclair. Geology of the Buckingham injection complex (Buckingham County), Virginia. M, 1969, University of Virginia. 87 p.

Heninger, Steven G. Hydrothermal experiment on andalusite-sillimanite equilibrium. M, 1985, Pennsylvania State University, University Park.

Henk, Floyd (Bo) H., Jr. Biostratigraphic study of the Boggy Shale Formation; Franks Graben; Pontotoc County, Oklahoma. M, 1981, Texas Christian University. 120 p.

Henke, Jeffrey R. Hydrogeologic characterization of a surface mining-impacted watershed with implications for acid mine drainage abatement, Clarion County, Pennsylvania. M, 1985, Pennsylvania State University, University Park. 171 p.

Henke, Kevin R. Archean metamorphism in northwestern Ontario and southeastern Manitoba. M, 1984, University of North Dakota. 326 p.

Henke, Kim Ann. Origin of the late Paleocene-early Eocene Wilcox sandstones, Lobo Trend, Webb and Zapata counties, Texas. M, 1982, Texas A&M University. 198 p.

Henkel, Howard L. The solution of protores of iron in carbonated waters, and their effect on the formation of laterites and replacement deposits of iron. M, 1923, University of Minnesota, Minneapolis. 20 p.

Henkle, William R., Jr. Geology and engineering geology of eastern Flagstaff, Coconino County, Arizona. M, 1976, Northern Arizona University. 123 p.

Henne, Mark Siegfried. The dissolution of Rainier Mesa volcanic tuffs, and its application to the analysis of the groundwater environment. M, 1982, University of Nevada. 113 p.

Henneberger, Roger. Geology of the Bathtub-North Point area, San Mateo County, California. M, 1977, Stanford University.

Hennen, Gary James. Dunkard flora (Pennsylvanian and Permian) of Greene County, Pennsylvania. M, 1968, West Virginia University.

Hennessey, Russell B. Porosity development and diagenetic alteration in the Potsdam Sandstone, Keeseville Member (Late Cambrian), of New York State. M, 1982, Rensselaer Polytechnic Institute. 62 p.

Hennessy, Joe Allen. Geology and hydrothermal alteration of the Glen Oaks porphyry copper occurrence, Yavapai County, Arizona. M, 1981, University of Arizona. 103 p.

Hennessy, Joel. Isotopic variation in dolomite concretions from the Monterey Formation, California. M, 1983, Arizona State University. 50 p.

Hennet, Remy Jean-Claude. The effect of organic complexing and carbon dioxide partial pressure on metal transport in low-temperature hydrothermal systems. D, 1987, Princeton University. 318 p.

Hennier, Jeffrey H. Anticline, Bighorn Basin, Wyoming. M, 1984, Texas A&M University.

Hennigar, Terry W. Hydrogeology of the Salmon River and adjacent watersheds, Colchester County, Nova Scotia (Canada). M, 1968, Dalhousie University. 192 p.

Henniger, Bernard Robert. Petrology and stratigraphy of the Greenbrier Limestone and Upper Pocono Sandstones (Middle Mississippian), Wayne and Lincoln counties, West Virginia. D, 1972, West Virginia University.

Henniger, Bernard Robert. The type Waynesburg Sandstone (Upper Carboniferous) of southwestern Pennsylvania and adjacent areas. M, 1964, Miami University (Ohio). 101 p.

Henning, Leo G. Study of the Ireland Sandstone (Douglas Group, Upper Pennsylvanian) in east-central Kansas. M, 1985, Wichita State University. 103 p.

Henning, Roger John. The effect on the geologic environment by the Mentor municipal sanitary landfill. M, 1976, University of Akron. 178 p.

Henning, Roger John. The ground water-surface water interface in Ohio. D, 1978, Ohio State University. 543 p.

Henning, Russell J. A gravity survey of the Precambrian terrain in Southwest Marathon County, central Wisconsin. M, 1983, Eastern Kentucky University. 92 p.

Hennings, Peter Hill. Basement/cover rock relations of the Dry Fork Ridge Anticline termination, northeastern Bighorn Mountains, Wyoming and Montana. M, 1986, Texas A&M University. 118 p.

Hennings, Richard Armond. The Pleistocene geology of a portion of the Bucksport Quadrangle, Maine. M, 1961, University of Maine. 134 p.

Hennings, Ronald George. The hydrogeology of a sand plain seepage lake, Portage County, Wisconsin. M, 1978, University of Wisconsin-Madison.

Henningsen, Elmer Robert. The diagenesis of water in the Trinity aquifers of central Texas. M, 1960, Baylor University. 54 p.

Henningsen, Gary R. Deposition and diagenesis of the Flippen Limestone, Fisher and Jones counties, Texas. M, 1985, Texas Tech University. 92 p.

Henningsgaard, Jeffrey J. Petrology and diagenesis of the Codell Sandstone, Wattenberg gas field, Denver Basin, Colorado. M, 1986, Northern Arizona University. 103 p.

Henny, Robert Warren. The effect of the geologic setting on the distribution of ejecta from a buried nuclear detonation. D, 1977, Michigan State University. 325 p.

Henrich, Catherine. Depositional environment of the Twin Creek Limestone (Jurassic), Southeast Idaho. M, 1983, Idaho State University. 62 p.

Henrich, William J. Gravity survey of northern Marsh Valley. M, 1979, Idaho State University. 54 p.

Henricksen, Donald Anton. Eocene stratigraphy of the lower Cowlitz River, eastern Willapa Hills area, southwestern Washington. D, 1954, Stanford University. 223 p.

Henricksen, Raymond Milton. Placer mining methods in the Fairbanks District. M, 1937, [University of Nevada - Mackay School of Mines].

Henricksen, Thomas A. Geology and mineral deposits of the Mineral-Iron Mountain District, Washington County, Idaho, and of a metallized zone in western Idaho and eastern Oregon. D, 1974, Oregon State University. 205 p.

Henrickson, Eiler Leonard. A study of the metamorphism of the upper Huronian rocks of the western portion of the Marquette District, Northern Peninsula, Michigan. D, 1956, University of Minnesota, Minneapolis. 206 p.

Henriquez, F. J. Iron formation-massive sulfide relationships at Heath-Steele, Brunswick No. 6 (N.B.) and Matagami Lake, Bell Allard (Quebec). M, 1974, McGill University. 116 p.

Henrot, Jacqueline Francoise. Behavior of technetium in soil; sorption-desorption processes. D, 1988, University of Tennessee, Knoxville. 111 p.

Henry, Brian J. Filter theory for the Huntec facsimile seismograph. M, 1972, Queen's University. 99 p.

Henry, C. Wayne. Geologic field trip through the Ridge and Valley Province of central Pennsylvania. M, 1968, Virginia State University. 36 p.

Henry, Christopher Duval. Geology and geochronology of the granitic batholithic complex, Sinaloa, Mexico. D, 1975, University of Texas, Austin. 192 p.

Henry, Christopher Duval. K-Ar chronology of the granitic batholithic complex (Mesozoic and Tertiary) Sinaloa, Mexico. M, 1972, University of Texas, Austin.

Henry, D. M. Sedimentology and stratigraphy of the Baraboo Quartzite of South-central Wisconsin. M, 1975, University of Wisconsin-Madison.

Henry, Darrell James. Spinel peridotites of the Gold Beach area, southwestern Oregon. M, 1976, University of Wisconsin-Madison.

Henry, Darrell James. Sulfide-silicate relations of the staurolite grade pelitic schists, Rangeley Quadrangle, Maine. D, 1981, University of Wisconsin-Madison. 798 p.

Henry, Diana Louise. The crystal structure of cronstedtite-2H_2. M, 1974, University of Wisconsin-Madison. 46 p.

Henry, Donald Kenneth. Ore mineralogy and paragenesis of a portion of the Great Gossan lead district. M, 1977, Virginia Polytechnic Institute and State University.

Henry, E. Marie. Gravity survey over the Berea sandstone channels in Ashland County, Ohio. M, 1982, Bowling Green State University. 61 p.

Henry, Gary E. The petrology of the Bethel Formation (lower Chester) in Indiana. M, 1958, Indiana University, Bloomington. 22 p.

Henry, George, Jr. Geological investigations of a buried valley in the vicinity of Dike A, Caesar Creek Reservoir, Ohio. M, 1973, Wright State University. 52 p.

Henry, Hollis Earl. Two electrical problems with cylindrical symmetry; (I), the interpretation of IP data taken near a steel casing; (II), the interpretation of resistivity data taken in an underground tunnel. M, 1983, Pennsylvania State University, University Park. 95 p.

Henry, John Francis. The genus Hantkenina; its geographic and geologic distribution. M, 1949, University of Texas, Austin.

Henry, Joseph B. Filter theory for the Huntec facsimile seismograph. M, 1972, Queen's University.

Henry, M. Calcareous nannofossils of the Bonaire Trench, Caribbean Sea. M, 1974, Texas A&M University.

Henry, M. J. The unconsolidated sediment distribution on the San Diego County mainland shelf, California. M, 1976, San Diego State University.

Henry, Marilee. A new method for slant stacking refraction data and some applications. D, 1982, University of California, San Diego. 185 p.

Henry, Mitchell Earl. Marine petroleum prospecting and pollution monitoring with an airborne Fraunhofer line discriminator. D, 1982, Texas A&M University. 217 p.

Henry, Richard Lee. The Lac Sauvage volcanogenic iron formation near Chibogamou, Quebec, Canada; its petrology, geochemistry, structure and economic significance. M, 1978, University of Georgia.

Henry, Robert W. A study of clastics in the Negaunee Iron Formation near Palmer, Michigan. M, 1970, Michigan State University. 156 p.

Henry, Scott Duray. Distribution of density and selected trace elements in a flooded, bituminous coal mineshaft, Belmont County, Ohio. M, 1974, University of Akron. 66 p.

Henry, Stephen George. Paleomagnetism of the upper Keweenawan sediments, the Nonesuch Shale and Freda Sandstone. M, 1976, University of Michigan.

Henry, Steven George. Terrestrial heat flow overlying the Andean subduction zone. D, 1981, University of Michigan. 204 p.

Henry, Thomas W. Conodont biostratigraphy of the Morrow Formation (Lower Pennsylvanian) in portions of Cherokee, Sequoyah, Muskogee, and Adair counties, northeastern Oklahoma. M, 1970, University of Oklahoma. 170 p.

Henry, Thomas Wood. Brachiopod biostratigraphy and faunas of the Morrow Series (lower Pennsylvanian) of northwestern Arkansas and northeastern Oklahoma. D, 1973, University of Oklahoma. 515 p.

Henry, Vernon J., Jr. Correlation of late Quaternary shorelines, Galveston Bay area, Texas. M, 1955, Texas A&M University.

Henry, Vernon J., Jr. Recent sedimentation and related oceanographic factors in West Mississippi Delta area. D, 1961, Texas A&M University.

Henry, Waldo Henry. An investigation of subsurface reef conditions in the Traverse Group of Michigan. M, 1949, Michigan State University. 34 p.

Henry, William Edward. Environment of deposition of an organic laminated dolomite, anamosa facies of the Gower Formation, Silurian, Iowa. M, 1972, University of Wisconsin-Madison.

Henry, William Edward. Metamorphism of the pelitic schists in the Dixfield Quadrangle, NW Maine. D, 1974, University of Wisconsin-Madison. 147 p.

Hensarling, Larry Reid, Jr. Subsurface geology of upper Eocene, Beauregard Parish, Louisiana. M, 1981, University of Texas, Austin. 113 p.

Hensel, Bruce R. The spatial distribution of hydrogeologic conditions along the Lake Michigan shoreline in Wisconsin. M, 1984, University of Wisconsin-Milwaukee. 171 p.

Henshaw, Paul C. A Tertiary mammalian fauna from the Avawatz Mountains, California. M, 1938, California Institute of Technology. 76 p.

Henshaw, Paul C. A Tertiary mammalian fauna from the San Antonio Mountains near Tonopah, Nevada. D, 1940, California Institute of Technology. 173 p.

Henshaw, Paul C. Geology and mineral deposits of the Cargo Muchacho Mountains, Imperial County, California. D, 1940, California Institute of Technology. 72 p.

Henshaw, Paul Carrington, Jr. A study of the geochemistry of pelagic sediments; the interaction of trace metals and magnetic minerals with sedimentary environments. D, 1978, University of Washington. 236 p.

Henshaw, Paul Carrington, Jr. Paleomagnetic and geochemical interpretation of the deep-sea core. M, 1975, University of Washington.

Hensleigh, Diane E. Depositional setting of the Turonian Kamp Ranch Member, Eagle Ford Group, Northeast Texas. M, 1983, University of Texas, Arlington. 179 p.

Hensley, Carol. Seismic exploration of a buried valley. M, 1978, University of New Hampshire. 71 p.

Hensley, Frank S., Jr. Some macrofossils of the Pennsylvanian-Permian Casper Formation along the west flank of the Laramie Range, Albany County, Wyoming. M, 1956, University of Wyoming. 80 p.

Hensley, Perry John. A study of the Earth vibration velocities in the Arkansas area. M, 1957, University of Arkansas, Fayetteville.

Henson, Ivan Hendrix. Asymptotic theory and computational methods for the normal modes of a laterally heterogeneous Earth. D, 1986, Princeton University. 118 p.

Henson, Ivan Hendrix. Inversion for fault dislocation using teleseismic body waves. M, 1980, Pennsylvania State University, University Park. 52 p.

Henton, John Melvin, Jr. North American Mesozoic and Cenozoic Ostracoda of the family Cytheridae. M, 1949, University of Illinois, Urbana.

Hentz, Max Ferdinand and Hardy, David Graham. Geology of the Pass Peak area, Sublette County, Wyoming. M, 1951, University of Michigan.

Hentz, Tucker Fox. Sedimentology and structure of Culpeper Group lake beds (Lower Jurassic) at Thoroughfare Gap, Virginia. M, 1982, University of Kansas. 166 p.

Henyey, Thomas Louis. Heat flow near major strike-slip faults in central and southern California. D, 1968, California Institute of Technology. 421 p.

Hepburn, John Christopher. Geology of the metamorphosed Paleozoic rocks in the Brattleboro area, Vermont. D, 1972, Harvard University.

Hepler, Jeffrey Alan. Snow as an accumulator of acid pollutants in Colorado. M, 1983, [Colorado State University].

Hepp, Eric. A magnetotelluric profile of the Wind River Thrust. M, 1987, Michigan Technological University. 45 p.

Hepp, Mary Margaret. A Precambrian andesite dike swarm in the northeastern Front Range, Larimer County, Colorado. M, 1966, University of Colorado.

Hepp, Michael Arthur. Clay mineralogy of late Pleistocene sequences in northwestern Washington and southwestern British Columbia. M, 1972, Western Washington University. 48 p.

Heppard, Philip D. The Sundance-Morrison boundary in Natrona County, Wyoming. M, 1984, University of Akron. 291 p.

Hepworth, Richard Cundiff. Heaving in the subgrade of highways constructed on the Mancos Shale (Cretaceous). M, 1963, University of Utah. 98 p.

Herat, Samson T. Analysis error of the wild autograph A7-456. M, 1963, Ohio State University.

Herb, G. Diagenesis of deeply buried sandstones from the Scotian Shelf. D, 1975, Dalhousie University. 151 p.

Herbaly, Elmer L. Regional analysis of the Springer stratigraphic interval. M, 1950, Northwestern University.

Herber, Jon Philip. Holocene sediments under Laguna Madre, Cameron County, Texas. M, 1981, University of Texas, Austin. 664 p.

Herber, Lawrence J. Order-disorder in coexisting plagioclase and alkali feldspar from the Mineral Range of Southwest Utah. D, 1968, University of Nevada - Mackay School of Mines. 92 p.

Herber, Lawrence J. Structural petrology and economic features of the Precambrian rocks of La Joyita Hills. M, 1963, New Mexico Institute of Mining and Technology. 36 p.

Herberger, Don. Sedimentation and mineralization of a sandstone lead-zinc occurrence in the Helena Formation, Belt Supergroup, Lewis and Clark County, Montana. M, 1986, University of Montana. 82 p.

Herbert, Donald L. Studies of fractures in block caving by means of concrete models. M, 1933, Colorado School of Mines. 93 p.

Herbert, Elizabeth Lee. Clay mineralogy of granitic pegmatites in the Pala District, San Diego County, California. M, 1982, San Diego State University. 73 p.

Herbert, Frank J. The Nugget Sandstone (Lower Jurassic) of northwestern Fremont and southwestern Lincoln counties, Wyoming. M, 1958, Miami University (Ohio). 73 p.

Herbert, James C. Modeling of crustal structures in Southwest Georgia from magnetic data. M, 1980, Georgia Institute of Technology. 117 p.

Herbert, John R. Post-Miocene stratigraphy and evolution of northern Core Banks, North Carolina. M, 1978, Duke University. 133 p.

Herbert, Joseph J. A flat film Weissenberg template; an aid to rapid identification of single crystal specimens. M, 1975, Wright State University. 76 p.

Herbert, Louis. Crustal thickness estimate at AAE (Addis-Ababa, Ethiopia) and NAI (Nairobi, Kenya) using teleseismic P-wave conversions. M, 1983, Pennsylvania State University, University Park. 62 p.

Herbert, Paul, Jr. Stratigraphy of the Decorah Formation in western Illinois. D, 1949, University of Chicago. 177 p.

Herbert, Thomas Allan. An analysis of the physical and legal aspects of erosion on Lake Michigan, a case study at Saint Joseph, Michigan. D, 1974, Michigan State University. 251 p.

Herbert, Thomas Allan. Evaluation of three sampling methods used in gravel deposits. M, 1968, Michigan State University. 56 p.

Herbert, Timothy D. Eccentricity and precessional orbital periodicities in a Mid-Cretaceous deep-sea sequence; identification and application to quantitative paleoclimatology. D, 1987, Princeton University. 302 p.

Herbst, Emmett Lee. The geology of the northwestern portion of the Des Arc, Missouri Quadrangle. M, 1952, University of Missouri, Columbia.

Herczeg, Andrew Leslie. Carbon dioxide equilibria and $\delta^{13}C$ studies in some soft water lakes. D, 1985, Columbia University, Teachers College. 281 p.

Herd, Darrell G. Glacial and volcanic geology of the Ruiz-Tolima volcanic complex, Cordillera Central, Colombia. D, 1974, University of Washington. 78 p.

Herd, Darrell G. Sedimentary properties of Hawaiian pyroclastic breccias. M, 1972, University of Washington.

Herd, Howard Henry. Efficiency in data handling as applied to the petrology of the Sage Hen Pluton, White Mountains, Mono County, California. M, 1972, University of California, Davis. 73 p.

Herd, Leslie Lee. The silica budget of a modern tropical lagoon, Discovery Bay, Jamaica. M, 1985, University of Oklahoma. 79 p.

Herdendorf, Charles Edward, III. The geology of the Vermilion Quadrangle, Ohio. M, 1963, Ohio University, Athens. 182 p.

Herdman, David J. Chemical petrology of the Port Coldwell alkali intrusive, Marathon, Ontario. M, 1974, McMaster University. 88 p.

Herdrick, Melvin A. Geology of the Reindeer Hills, Alaska Range, Alaska. M, 1973, University of Idaho. 64 p.

Herece, Erdal Ibrahim. The Yenice-Gonen earthquake of 1953 and some examples of recent tectonic events in the Biga Peninsula of Northwest Turkey. M, 1985, Pennsylvania State University, University Park. 143 p.

Hereford, Richard. Cambrian-Devonian stratigraphy and Cambrian petrology in northern Yavapai County, Arizona. M, 1971, Northern Arizona University. 137 p.

Hergenroder, John D. Geology of the Radford area, Virginia. M, 1957, Virginia Polytechnic Institute and State University.

Hergenroder, John David. The Bays Formation (Middle Ordovician) and related rocks of the Southern Appalachians. D, 1966, Virginia Polytechnic Institute and State University. 432 p.

Herin, James Christopher. Numerical modeling of ground water flow and solute transport of an area near two landfills. M, 1986, Wright State University. 260 p.

Hering, Carl William. Geology and petrology of the Yamsay Mountain Complex, South-central Oregon; a study of bimodal volcanism. D, 1981, University of Oregon. 194 p.

Herkenham, Marjorie Watson. The petrography of the Guadalupe igneous complex, Mariposa County, California. M, 1946, University of California, Berkeley. 138 p.

Herkenhoff, Earl Frederic. Attenuation of seismic energy in the Earth's mantle. M, 1966, Stanford University.

Herlicska, E. Selected trace elements in Hawaiian lavas by atomic absorption spectrophotometry. D, 1967, University of Hawaii. 254 p.

Herlihy, Alan Tate. Sulfur dynamics in an impoundment receiving acid mine drainage. D, 1987, University of Virginia. 214 p.

Herlihy, D. R. Conflicts of interest over polymetallic sulfide minerals on the Gorda-Juan de Fuca Ridge system. M, 1985, University of Washington. 101 p.

Herlihy, Daniel M. Major element distribution in a folded gneiss, epidote amphibolite facies. M, 1975, University of New Hampshire. 51 p.

Herlinger, David L. Petrology of the Fall Creek Travertine, Bonneville County, Idaho. M, 1981, University of Houston.

Herlyn, Henry Traver. Lower Tertiary foraminifera from the western Santa Ynez Mountains, California. M, 1958, University of California, Berkeley. 260 p.

Herman, Bruce Meyer. The thermal evolution of oceanic lithosphere in the South Atlantic. D, 1988, Columbia University, Teachers College. 426 p.

Herman, Charles W. Subsurface geology of Phillips County, Kansas. M, 1957, Kansas State University. 67 p.

Herman, Gregory C. A structural analysis of a portion of the Valley and Ridge Province of Pennsylvania. M, 1984, University of Connecticut. 107 p.

Herman, Howard R. Cambrian and Ordovician structures and stratigraphy of the northwestern part of the Tomhannock and southeastern part of the Troy North quadrangles, New York. M, 1961, Rensselaer Polytechnic Institute. 45 p.

Herman, Janet Suzanne. The dissolution kinetics of calcite dolomite, and dolomitic rocks in the CO_2-water system. D, 1982, Pennsylvania State University, University Park. 218 p.

Herman, John D. Tests of sphalerite geobarometry at selected mining districts. M, 1985, University of Michigan. 33 p.

Herman, Julie D. Fossil preservation and the effects of groundwater leaching on fossils in the Yorktown Formation (upper Pliocene), Virginia. M, 1987, Virginia Polytechnic Institute and State University.

Herman, Marc Edward. Tidal fluctuations and ground-water dynamics in atoll island aquifers. M, 1984, University of Nevada. 150 p.

Herman, Theodore Coxon. Geology of the Placer Creek area, Huerfano and Costilla counties, Colorado. M, 1962, University of Michigan.

Hermance, J. F. Auroral zone geomagnetic variations in Iceland. D, 1967, University of Toronto.

Hermance, William. Relationship of faults, lineaments, and joints to regional structure; eastern Connecticut. M, 1983, Indiana State University. 63 p.

Hermann, Jon Michael. The features and significance of layering and shearing within the zone of mylonitization, northeastern Idaho Batholith. M, 1982, Western Michigan University. 75 p.

Hermann, Michael. Geology of the southwestern San Mateo Mountains, Socorro County, N.M. M, 1987, New Mexico Institute of Mining and Technology. 192 p.

Hermanrud, Christian. Determination of formation temperature from downhole measurements. D, 1988, University of South Carolina. 222 p.

Hermelin, Michel G. Petrology and structure of Precambrian crystalline and Tertiary volcanic rocks of the Chambers lake area, north-central Colorado. M, 1970, Colorado State University. 87 p.

Hermes, Oscar Don. A quantitative petrographic study of diabase in the Deep River [Triassic] basin, North Carolina. M, 1963, University of North Carolina, Chapel Hill. 37 p.

Hermes, Oscar Don. Geology and petrology study of the Mecklenburg Gabbro-Metagabbro Complex, North Carolina. D, 1967, University of North Carolina, Chapel Hill. 112 p.

Hernandez, Cristy R. A comparison of various methods of traverse adjustment. M, 1963, Ohio State University.

Hernandez, Federico. Tellurometer traverse. M, 1961, Ohio State University.

Hernandez, Gilbert Xavier. Geology of a part of the Black Butte Quadrangle, Elko County, Nevada. M, 1980, University of New Orleans. 86 p.

Hernandez, Heroel. Numerical modeling of the convection in a fault zone. M, 1980, Georgia Institute of Technology. 112 p.

Hernandez, Jose Jaime Gomez *see* Gomez Hernandez, Jose Jaime

Hernandez, Pedro Anmando. Investigation of seepage reduction by soil-water chemical reactions in irrigation canals. M, 1972, University of Idaho. 73 p.

Hernandez, Salvador. Study of the action of filters on a given signal. M, 1969, Rice University. 58 p.

Hernandez-Avila, Manuel L. Form-process relationships on island coasts. D, 1974, Louisiana State University. 134 p.

Herndon, Chesley Coleman, Jr. The inclined oil-water interface. M, 1952, University of Michigan.

Herndon, James M. The magnetization of carbonaceous meteorites. D, 1974, Texas A&M University. 124 p.

Herndon, Stephen D. Diagenesis and metamorphism in the Revett Quartzite (middle Proterozoic Belt), Idaho and Montana. M, 1983, University of Montana. 69 p.

Herndon, Thomas. Niagaran reefs. M, 1951, University of Michigan.

Herner, Robert R. An analysis of acid mine drainage problems in the Big Branch watershed, Sullivan, Greene, and Clay counties, Indiana. M, 1981, Indiana State University. 109 p.

Herness, Kermit. A manual of micro-chemical colorimetric mineral diagnosis. M, 1936, University of Minnesota, Minneapolis. 54 p.

Hernon, Robert Mann. Pegmatitic rocks of the Catalina-Rincon Mountains, Arizona. M, 1932, University of Arizona.

Hernon, Robert Mann. The Paradise Formation and its fauna. D, 1934, University of Arizona.

Herold, C. Lathrop. Preliminary report on the geology of the Salinas Quadrangle, California. M, 1935, University of California, Berkeley. 143 p.

Herold, Stanley C. The analytical effects of capillarity on the production of oil, gas, and water from wells. D, 1926, Stanford University. 425 p.

Heron, Stephen D., III. Depositional and diagenetic history of selected Mesozoic sediments of the South Florida Basin. M, 1982, University of Southwestern Louisiana. 90 p.

Heron, Stephen Duncan, Jr. Geology of the Irmo quadrangle. M, 1950, University of South Carolina. 24 p.

Heron, Stephen Duncan, Jr. The stratigraphy of the outcropping basal Cretaceous formations between the Neuse River, North Carolina, and Lynches River, South Carolina. D, 1958, University of North Carolina, Chapel Hill. 155 p.

Heroux, Yvon. Etude des lithofaciès de la Formation de Sayabec (Silurian), région de La Rédemption, Québec. D, 1971, Universite de Montreal.

Heroux, Yvon. Provenance des composants terrigènes des grès de la région de Beaumont-Saint-Michel, Comté de Bellechasse (Ordovicien), Québec, Canada). M, 1970, Universite de Montreal.

Heroy, William B., Jr. The geology of the Shell Canyon area, Bighorn Mts., Wyoming. D, 1941, Princeton University. 80 p.

Herpers, Henry F. The disappearance of the Wisconsin ice sheet from northern New Jersey. M, 1939, Massachusetts Institute of Technology. 93 p.

Herpfer, Marc Andreas. The silicate inclusions of group IAB iron meteorites; implications for metal-silicate segregation and core formation. M, 1988, Arizona State University. 70 p.

Herr, George Albert. Clay minerals in the Phosphoria Formation in Beaverhead County, Montana. M, 1955, Indiana University, Bloomington. 46 p.

Herr, Paul Edward. Possible deformation of a crystallizing pluton by stresses related to the McKinley strand of the Denali Fault, Central Alaska Range, Alaska. M, 1980, University of Wisconsin-Madison.

Herr, Randy Gerard. Sedimentary petrology and stratigraphy of the Lodore Formation (Upper Cambrian), Northeast Utah and Northwest Colorado. M, 1979, University of Utah. 129 p.

Herr, Stephen Richard. Biostratigraphy of the graptolite-bearing beds of the upper Ordovician Maquoketa Formation, Iowa. M, 1971, University of Iowa. 161 p.

Herrell, George Leslie. Geology of the Bald Mountain area, Fremont County, Wyoming. M, 1950, University of Missouri, Columbia.

Herrera, Amilcar Oscar. Geology of an area southwest of Florence, Fremont County, Colorado. M, 1951, Colorado School of Mines. 35 p.

Herrera, Leo John, Jr. Geology of the Tent Hills Quadrangle, California. M, 1951, University of California, Berkeley. 73 p.

Herrera, Luis Enrique. On the origin, propagation and mixing of Antarctic Intermediate Water in the Atlantic Ocean. D, 1973, New York University.

Herrera, Peter Ariel. Geology, alteration, and trace element distributions in the northern portion of the Bodie mining district, Mono County, California. M, 1988, Colorado School of Mines. 272 p.

Herrero-Bervera, Emilio. Some aspects of the geomagnetic field during polarity transitions. D, 1984, University of Hawaii. 219 p.

Herrick, David C. An isotopic study of the magnetite-chalcopyrite deposit at Cornwall, Pennsylvania. D, 1973, Pennsylvania State University, University Park. 95 p.

Herrick, Dean H. Gravity flow deposition on the Grenada Basin plain, southeastern Caribbean Sea. M, 1982, Duke University. 94 p.

Herrick, Robert. Analysis of gravity data over Aphrodite Terra, Venus. M, 1988, University of Houston.

Herrick, Rodney C. Authigenic minerals in the Pleistocene and Recent sediments of Lake Magadi, Kenya. M, 1972, University of Wyoming. 85 p.

Herrick, Stephen Marion. A lower Tertiary foraminiferal fauna from Costa Rica. D, 1933, Cornell University.

Herrick, Stephen Marion. Some fossil insects from the lithographic stone of Solnhofen, Germany (Upper Jurassic). M, 1929, University of Pittsburgh.

Herrin, Eugene Thornton, Jr. Correlation by spectrographic analysis of bentonite in the Gulf Series of Dallas area, Texas. M, 1953, Southern Methodist University. 27 p.

Herrin, Eugene Thornton, Jr. Geology of the Solitario area, Trans-Pecos, Texas. D, 1958, Harvard University.

Herrin, Moreland. Effects of aggregate shape on the stability of bituminous mixes. D, 1954, Purdue University.

Herrin, Toni Elizabeth. Large heteromyriads of the Miocene Texas Gulf Coastal Plain. M, 1977, Southern Methodist University. 126 p.

Herring, Bernard Geoffrey. Metamorphism and alteration of the basement rocks in the Carswell circular structure, Saskatchewan. M, 1976, University of British Columbia.

Herring, James R. Charcoal fluxes into Cenozoic sediments of the North Pacific. D, 1977, University of California, San Diego. 115 p.

Herring, John Charles. Changes in the geodetic coordinates due to a change in the reference ellipsoid. M, 1965, Ohio State University.

Herring, Maxwell, Jr. Geology of the Stairway Mountain area, Brewster County. M, 1964, Texas A&M University.

Herring, Thomas Abram. The precision and accuracy of intercontinental distance determinations using radio interferometry. D, 1983, Massachusetts Institute of Technology. 442 p.

Herringshaw, Dennis Charles. The germanium content of the lower Kittanning No. 5 coal of Ohio. M, 1981, University of Toledo. 131 p.

Herrington, Dawn. Cloud Chief Gypsum; petrography and depositional systems. M, 1981, Tulane University.

Herrington, Karen Laverne. Metamorphism, deformation and metasomatic alteration of an eclogite block in the Franciscan subduction complex, near Mt. Hamilton, central Diablo Range, California. M, 1985, University of Texas, Austin. 107 p.

Herrmann, Leo Anthony. The structural geology and petrology of the Stone Mountain-Lithonia District, Georgia. D, 1951, The Johns Hopkins University. 111 p.

Herrmann, Raymond. Shallow aquifers relative to surface waters, North Platte River valley, Goshen County, Wyoming. D, 1972, University of Wyoming. 194 p.

Herrmann, Robert B. Surface wave generation by central United States earthquakes. D, 1974, St. Louis University. 274 p.

Herrod, Wilson H. Relationship of Cottonwood Creek Field, Washakie County, Wyoming to carbonate facies of Permian Goose Egg Formation, eastern Bighorn Basin. D, 1980, Colorado School of Mines. 457 p.

Herron, Calvin Robert. Geology and structure of the Ace Hills, Inyo and Esmeralda counties, California-Nevada. M, 1981, California State University, Fresno.

Herron, Ellen Mary. A seismic study of the sediments in the Hudson River. M, 1967, Columbia University. 22 p.

Herron, Ellen Mary. Magnetic lineations and problems of plate tectonics in the eastern Pacific. D, 1974, Columbia University. 126 p.

Herron, Ellis Doyle. Establishing the presence of the Miocene on Crowley's Ridge, Arkansas. M, 1954, University of Arkansas, Fayetteville.

Herron, John E. Stratigraphy of the Miocene Agate Beach Formation in Lincoln County, Oregon. M, 1953, Oregon State University. 73 p.

Herron, Margaret J. Marine geology and geophysics of the western South Orkney Plateau, Antarctica; implications for Quaternary glacial history, tectonics, and paleoceanography. M, 1988, Rice University. 256 p.

Herron, Michael Myrl. The impact of volcanism on the chemical composition of Greenland Ice Sheet precipitation. D, 1980, SUNY at Buffalo. 143 p.

Herron, Susan Lynne. Physical properties of the deep ice core from Camp Century, Greenland. D, 1982, SUNY at Buffalo. 116 p.

Herron, Thomas Joseph. Detection and delineation of subsurface subsidence by seismic methods. M, 1956, Michigan Technological University. 56 p.

Herron, Thomas Joseph. Phase characteristics of geomagnetic micropulsations. D, 1966, Columbia University. 86 p.

Hersch, James Barry. Petrographic relationships among the host dolostone, gangue dolomite, and ore mineralization, in the Austinville-Ivanhoe lead-zinc district. M, 1978, University of Tennessee, Knoxville. 91 p.

Hersch, John Timothy. Origin of localized layering in the Twin Sisters Dunite, Washington. M, 1974, University of Washington. 65 p.

Hersey, David Ralph. Geophysical investigation of the eastern margin of the Espanola Basin, New Mexico. M, 1986, University of Texas at Dallas. 121 p.

Hersey, John Brackett. Gravity investigation of central-eastern Pennsylvania. D, 1942, Lehigh University.

Hershberg, Edward Leonard. Paleomagnetism of the Wolf River Batholith. M, 1976, University of Wisconsin-Madison.

Hershelman, William Lee. A permeability study of sand. D, 1940, University of Iowa. 190 p.

Hershelman, William Lee. A study of the physical properties of the Venango oil sands (New York, Pennsylvania). M, 1938, Syracuse University.

Hershey, Howard Garland. Structure and age of the Port Deposit granodiorite complex. D, 1936, The Johns Hopkins University.

Hershkowitz, Zoltan. Landfill leachates study; north shore, Jamaica Bay. M, 1984, Brooklyn College (CUNY).

Hershler, Robert. The systematics and evolution of the hydrobiid snails of the Cuator Cienegas Basin, Coahuila, Mexico. D, 1982, The Johns Hopkins University.

Hershner, Carlton H., Jr. Effects of petroleum hydrocarbons on salt marsh communities. D, 1977, College of William and Mary.

Hertel, Fritz. A comparison between Modern and Pleistocene vultures in the New World. M, 1986, University of Rochester. 97 p.

Hertel, John W. Geology of the Box Creek area, Sevier and Piute counties, Utah. M, 1980, Colorado State University. 103 p.

Hertig, Stephen Paul. Structure and stratigraphy of a part of the Coosa deformed belt, Alabama. M, 1983, University of Alabama. 66 p.

Herting, David Allen. The occurrence of ground water in a fractured crystalline rock aquifer of the South Carolina Piedmont. M, 1979, University of South Carolina.

Hertlein, Leo George. A revision of the Tertiary and Quaternary pectens of the Pacific Coast. M, 1923, Stanford University. 122 p.

Hertlein, Leo George. The geology and paleontology of the Pliocene of San Diego, California. D, 1929, Stanford University. 301 p.

Hertling, M. M. Origin of the Layer A and A' Eocene cherts and their correlatives from the Jicara Formation of Puerto Rico. M, 1976, Queens College (CUNY). 73 p.

Hertz, Terrance L. Organic geochemistry of the Wilkins Peak, Laney, and Fossil Butte members of the Green River Formation, Wyoming. D, 1984, The Johns Hopkins University. 271 p.

Herve, Miguel. Rb-Sr and K-Ar geochronology of the Patagonian Batholith, south of the Beagle and Ballenero channels and west of Long. 68°00'. M, 1984, University of Texas at Dallas. 53 p.

Hervet, Michel. Chronostratigraphie et pétrographie du complexe gneissique de Chicoutimi en bordure du complexe anorthositique du Lac St. Jean. M, 1987, Universite du Quebec a Chicoutimi. 403 p.

Hervig, Richard Lokke. Minor and trace element composition of mantle minerals; Ca-Mg exchange between olivine and orthopyroxene as geobarometer and the origin of harzburgites. D, 1979, University of Chicago. 130 p.

Herwig, Jonathan Charles. Stratigraphy, structure, and massive sulfide potential of late Mesozoic metamorphic rocks near El Cirian, Estado de Mexico, Mexico. M, 1982, University of Texas, Austin. 107 p.

Herz, Norman. The petrology of the Baltimore Gabbro and the petrography of the Baltimore-Patapsco Aqueduct. D, 1950, The Johns Hopkins University.

Herzberg, Peter Jansen. Geology of lode gold occurrences, Timberline Creek area, central Alaska. M, 1980, University of Alaska, Fairbanks. 120 p.

Herzen, Richard Pierre Von see Von Herzen, Richard Pierre

Herzer, Richard H. A geological reconnaissance of Bowie Seamount (Pacific Ocean, west of Queen Charlotte Islands). M, 1972, University of British Columbia.

Herzog, Dave. Geologic influence on sensitivity of watersheds in Rocky Mountain National Park to acidification. M, 1982, [Colorado State University].

Herzog, Leonard F., II. Natural variations in strontium isotope abundances in minerals; a possible geologic-age method. D, 1952, Massachusetts Institute of Technology. 135 p.

Herzog, R. H., Jr. Correlation of exploration variables to grout absorption in a karst terrain. M, 1975, University of Arizona.

Hesemann, Thomas. Hydrogeological/geophysical investigation of the Goodwater Creek watershed. M, 1979, University of Missouri, Columbia.

Hesler, Donald J. A hydrogeologic study of the Knox-Skull cave system, Albany County, New York. M, 1984, University of Connecticut. 91 p.

Hesler, James L. A quantitative analysis of discrimination of hummocky stagnation moraine from end moraine. M, 1968, Northern Illinois University.

Hesler, James Lewis. Within-order spatial and temporal variation in drainage network structure in developing rill systems. D, 1982, University of Iowa. 459 p.

Hesler, Roy Earl. Construction aggregate resources of Dane County, Wisconsin. M, 1978, University of Wisconsin-Madison.

Heslop, John B. Geology, mineralogy and textural relationships of the Coppercorp Deposit, Mamainse Point area, Ontario. M, 1970, Carleton University.

Hess, Alison Anne. Chertification of the Redwall Limestone (Mississippian), Grand Canyon National Park, Arizona. M, 1985, University of Arizona. 133 p.

Hess, Barry Samuel. Supercritical fluids; measurement and correlation studies of model coal compound solubility and the modeling of solid-liquid-fluid equilibria. D, 1987, University of Illinois, Urbana. 153 p.

Hess, David Filbert. Geology of pre-Beltian rocks in the central and southern Tobacco Root Mountains (Montana), with reference to superposed effects of the Laramide age Tobacco Root Batholith. D, 1967, Indiana University, Bloomington. 333 p.

Hess, Frank D. Rare earth elements in Red Sea geothermal deposits. M, 1970, University of Southern California.

Hess, Frank Devereaux. The geochemical cycle of mercury and the pollutional increment. D, 1974, University of Southern California.

Hess, Gordon R. Submarine fanfare; a comparison of modern and Miocene deep-sea fans. M, 1974, University of Minnesota, Minneapolis. 118 p.

Hess, Gordon Russell. Quaternary stratigraphy and sedimentation; northern Bering Sea, Alaska. D, 1985, Stanford University. 100 p.

Hess, Harry Hammond. Hydrothermal metamorphism of an ultra basic intrusive at Schuyler, Va. D, 1932, Princeton University.

Hess, James R. Geochemistry of the Elberton Granite and the geology of the Elberton West Quadrangle, Georgia. M, 1979, University of Georgia.

Hess, Jeffrey D. The geochemistry of silver in groundwater. M, 1985, Western Washington University. 47 p.

Hess, John Warren. Hydrochemical investigations of the central Kentucky karst aquifer system. D, 1974, Pennsylvania State University, University Park. 234 p.

Hess, K. W. A three dimensional numerical model of steady gravitational circulation and salinity distribution in Narragansett Bay. D, 1974, University of Rhode Island. 296 p.

Hess, Kathryn Marie. Reservoir sedimentation; Highland Creek Reservoir, Lake County, California. M, 1984, Stanford University. 137 p.

Hess, Lillian M. Volumetric changes related to barrier beach growth with reference to shoreline engineering, Rockaway, New York. M, 1980, Brooklyn College (CUNY).

Hess, Paul Charles. The metamorphic paragenesis of cordierite, garnet, and biotite in the Brimfield area, south central Massachusetts. D, 1969, Harvard University.

Hess, Paul Dennis. The geology of the northeast quarter of the Powers Quadrangle (Coos County), Oregon. M, 1967, University of Oregon. 91 p.

Hess, Susan Lynne. Geological education in the secondary schools and junior colleges of Florida. M, 1975, University of Florida. 43 p.

Hess, Thomas E. The observation of ultrasonic velocities and attenuation during pore pressure induced fracture. M, 1984, Massachusetts Institute of Technology. 94 p.

Hessa, Samuel Lyndon. The igneous geology of the Navajoe Mountains, Oklahoma. M, 1964, University of Oklahoma. 40 p.

Hesse, Curtis Julian. A vertebrate fauna from the type locality of the Ogallata Formation (Nebraska). M, 1933, University of California, Berkeley. 36 p.

Hessenbruch, John M. Salt tectonics of Winn Parish, Louisiana. M, 1975, Louisiana State University.

Hesser, Duane Harvey. Linear predictive processing and pattern recognition for automated classification of local seismic events. M, 1982, University of Washington. 241 p.

Hessert, Christian Van see Van Hessert, Christian

Hesslein, R. H. The fluxes of CH_4, ΣCO_2, and NH_3-N from sediments and their consequent distribution in a small lake. D, 1977, Columbia University, Teachers College. 199 p.

Hessler, Robert R. The Lower Mississippian proetid trilobites of North America. D, 1960, University of Chicago. 182 p.

Hester, B. W. Limestone resources of the Maritime Provinces and Newfoundland. M, 1954, University of Toronto.

Hester, Norman Curtis. A study of high-level valleys in southwest Hamilton County, Ohio. M, 1965, University of Cincinnati. 72 p.

Hester, Norman Curtis. The origin of the Cusseta Sand. D, 1968, University of Cincinnati. 219 p.

Hester, Patricia M. Depositional environments in an Upper Triassic lake, east-central New Mexico. M, 1988, University of New Mexico. 154 p.

Hester, William Christopher. Fracture patterns and flow orientation in the Carlton Rhyolite, Wichita Mountains area, Southwest Oklahoma. M, 1985, Oklahoma State University. 70 p.

Hesterberg, Dean L. R. Critical coagulation concentrations and rheological properties of illite. D, 1988, University of California, Riverside. 291 p.

Hestmark, Martin C. Development and evaluation of a Lagrangian one-dimensional steady flow trace metal surface water transport model. M, 1988, Colorado School of Mines. 137 p.

Heston, Deborah A. Geology of the Bradshaw vanadium prospect, Meadow Valley Mountains, Lincoln

County, Nevada. M, 1982, Colorado State University. 119 p.

Hestor, Brian W. Geology and economics of limestone. M, 1954, University of Toronto.

Hetherington, George Edward. Geology of the South Tapo Canyon area, Santa Susana Quadrangle, Ventura County, California. M, 1957, University of California, Los Angeles.

Hetherington, Martha J. The geology and mineralization at the McDermitt mercury mine, Nevada. M, 1983, University of Washington. 54 p.

Hetherington, Peter Alan. Stratigraphic relationships and depositional controls of the Mississippian (Meramecian and lower Chesterian) "Big Lime" in southeastern Kentucky; a subsurface petroliferous equivalent of the lower and middle Newman Group. M, 1981, University of Kentucky. 135 p.

Hetherly, David Christopher. Joint orientations and their relationships to structures and lithologies of rocks between Fine and Edwards, New York. M, 1986, SUNY at Buffalo. 67 p.

Hetrick, John. The clay mineralogy of the upper Porters Creek clay (Paleocene) in West Tennessee. M, 1968, University of Tennessee, Knoxville. 36 p.

Hetrick, John Harold, Jr. The crystal chemistry of mixed layer smectite/illites. D, 1973, Case Western Reserve University.

Hettenhausen, Roger L. Seismic migration by phase shift plus interpolation. M, 1988, Indiana University, Bloomington. 215 p.

Hetterman, John Leslie. A trace element study of fluorite from the central Kentucky and Kentucky-Illinois mineral districts. M, 1974, University of Kentucky. 147 p.

Heubeck, Christoph Egbert. Geology of the southeastern termination of the Cordillera Central, south-central Hispaniola, Dominican Republic. M, 1988, University of Texas, Austin. 333 p.

Heuberger, Mark Oscar. Late Cenozoic volcanism, precious-metal mineralization and evidence for a collapse caldera in the San Martin area, Altiplano of southern Peru. M, 1985, University of Nevada. 67 p.

Heuer, Edward. Structure of the Des Moines Series of south-central Iowa. M, 1948, University of Wisconsin-Madison. 116 p.

Heuer, Edward. The paleoautecology of the megafauna of the Pennsylvanian Wolf Mountain Shale in the Possum Kingdom area, Palo Pinto County, Texas. D, 1973, University of Wisconsin-Madison. 793 p.

Heuer, Ronald Eugene. Geology of the Soyalo-Ixtapa area, Chiapas, Mexico. M, 1965, University of Illinois, Urbana.

Heuer, Wolfgang C. Active faults in the northwestern Houston area. M, 1979, University of Houston.

Heuser, Robert Frederick. Upper Cretaceous sedimentation and tectonics in the Powder River basin, Wyoming. M, 1964, Michigan State University. 60 p.

Heusser, Calvin J. and Balter, Howard. Forest-soil relations on limestone and gneiss in southeastern New York and northern New Jersey. D, 1980, New York University. 177 p.

Heusser, Linda E. Florer. Late-Cenozoic palynology of sea cliffs of the western Olympic Peninsula, Washington. D, 1971, New York University. 94 p.

Heutmaker, D. L. Physical modelling of liquid injection into porous media containing a dynamic Ghyben-Herzberg lens system. M, 1976, University of Hawaii. 103 p.

Heuvel, Peter Van den *see* Van den Heuvel, Peter

Heuze, Francois. The design of room-and-pillar structures in competent jointed rock; example the Crestmore Mine, California. M, 1970, University of California, Berkeley.

Hevley, Richard Holmes. Pollen analysis of the Quaternary archaeological and lacustrine sediments from the Colorado Plateau. D, 1964, University of Arizona. 150 p.

Hewetson, J. P. An investigation of the groundwater zone in fractured shale at a landfill. M, 1985, University of Waterloo. 72 p.

Hewett, Donnel F. Geology and coal and oil resources of the Oregon Basin, Meeteetsee and Grass Creek Basin quadrangles (Wyoming). D, 1924, Yale University.

Hewett, Henry B. Geology as applied to the construction of airdromes. M, 1947, Washington State University.

Hewett, Robert Lewis. Geology of the Cerro la Zacatera area, Sonora, Mexico. M, 1978, Northern Arizona University. 99 p.

Hewins, Roger Herbert. The petrology of some marginal mafic rocks along the north range of the Sudbury Irruptive. D, 1971, University of Toronto.

Hewitt, Allan Edward. Decisions in the establishment of soil series. D, 1982, Cornell University. 188 p.

Hewitt, Charles Hayden. Geology and mineral deposits of northern Big Burro Mountains-Redrock area, Grant County, New Mexico. D, 1957, University of Michigan.

Hewitt, Clark S. A methodology for collecting land cover map boundary data. M, 1978, University of Virginia. 43 p.

Hewitt, David A. An experimental and field study of the progressive metamorphism of micaceous limestones. D, 1970, Yale University.

Hewitt, Donald F. A study of the Timiskaming series of the Kirkland Lake-Larder Lake Belt, Ontario. D, 1950, University of Wisconsin-Madison.

Hewitt, Edward Ringwood, II. The Cretaceous geology of the northeastern Chispa Quadrangle, Trans-Pecos, Texas. M, 1951, University of Texas, Austin.

Hewitt, Jeanne L. The textural evolution of a progressively dolomitized limestone. M, 1975, Michigan State University. 23 p.

Hewitt, L. W. Physical stratigraphy of the Cretaceous formations of Nebraska. M, 1931, University of Nebraska, Lincoln.

Hewitt, Marshall Cooper. Bryozoan reefs in the Middle Silurian of New York and Ontario; fistuliporoid bioherms on the Irondequoit-Rochester boundary in Niagara Gorge. M, 1983, Pennsylvania State University, University Park. 199 p.

Hewitt, Philip Cooper. Larger foraminifera of certain Eocene and Oligocene formations of Cuba. D, 1958, Cornell University. 84 p.

Hewitt, Phillip Cooper. A preliminary study of the microfossils of the Pennington Formation in eastern Tennessee. M, 1953, University of Tennessee, Knoxville. 34 p.

Hewitt, R. L. Geology and ore deposits of the Sturgeon River area. M, 1935, Queen's University. 82 p.

Hewitt, Robert Leigh. Experiments bearing on the relation of pyrrhotite to other sulphides. D, 1937, University of Minnesota, Minneapolis. 98 p.

Hewitt, Samuel L. Geology of the Fly creek quadrangle and north half of Round Butte dam quadrangle, Oregon. M, 1970, Oregon State University. 69 p.

Hewitt, William Paxton. Geology and mineralization of the San Antonio Mine, Santa Eulalia District, Chihuahua, Mexico. D, 1943, Columbia University, Teachers College.

Hewlett, C. G. Geology of Cameron Lake area, northwestern Quebec. M, 1951, Queen's University.

Hewlett, Cecil George. Significance of optical variations of some potash feldspars. D, 1954, University of Wisconsin-Madison.

Hewlett, James Scott. A seismic reflection and aeromagnetic study of central Fairfield County, Ohio. M, 1982, Wright State University. 77 p.

Hews, P. C. H. An Rb-Sr whole-rock study of some Proterozoic rocks, Richmond Gulf, Nouveau Quebec. M, 1976, Carleton University. 64 p.

Hey, Richard N. Tectonic evolution of the Cocos-Nazca Rise. D, 1975, Princeton University. 177 p.

Heyburn, Malcolm M. Geological and geophysical investigations of the Sanford Hill ore body extension, Tahawus, New York. M, 1960, Syracuse University.

Heydari Laibidi, Ezatoliah. 9 Valley region, eastern California. M, 1981, Pennsylvania State University, University Park. 139 p.

Heydenburg, Richard J. Stratigraphy and depositional environment of the Parkman Sandstone member, Mesaverde Formation (Cretaceous) near Midwest, Natrona County, Wyoming. M, 1966, University of Wyoming. 81 p.

Heydinger, Andrew Gerard. Analysis of axial single pile-soil interaction in clay. D, 1982, [University of Houston]. 257 p.

Heyl, Allen Van, Jr. The Upper Mississippi Valley zinc-lead district. D, 1950, Princeton University. 271 p.

Heyl, George Richard. Geology and mineral deposits of the Bay of Exploits area, Notre Dame Bay, Newfoundland. D, 1935, Princeton University. 219 p.

Heylmun, Edgar Baldwin. Systematic rock joints in parts of Utah, Colorado, and Wyoming, and a hypothesis on their origin. D, 1966, University of Utah. 245 p.

Heyman, Arthur M. Geology of the Peach-Elgin copper deposit, Helvetia District, Arizona. M, 1958, University of Arizona.

Heyman, Arthur Mark. Physical parameters in the development of peasant agriculture in the highland Guayana region, Venezuela. D, 1967, Columbia University, Teachers College. 373 p.

Heyman, Lou. Petrology of the Blackford Formation (middle Ordovician), Russell County, Virginia. D, 1970, Virginia Polytechnic Institute and State University.

Heyman, Louis. A loess-like silt in Ann Arbor, Michigan. M, 1949, University of Michigan.

Heyman, Oscar Glenn. Regional compression as the cause for Laramide deformation of the northwestern Uncompahgre Plateau, western Colorado and eastern Utah. D, 1983, University of Wyoming. 128 p.

Heyn, Teunis. Stratigraphic and structural relationships along the southwestern flank of the Sauratown Mountains Anticlinorium. M, 1984, University of South Carolina.

Heyse, J. V. The metamorphic history of the LL-group ordinary chondrites. D, 1979, SUNY at Stony Brook. 232 p.

Heywood, Charles Edward. Forearc deformation in southern Costa Rica; a consequence of the collision of the aseismic Cocos Ridge. M, 1984, University of California, Santa Cruz.

Heywood, William Walter. Geology of part of Ellef Ringnes Island in Canadian Arctic. M, 1954, University of Washington. 71 p.

Heywood, William Walter. Precambrian geology, Ledge Lake area, Manitoba and Saskatchewan. D, 1959, University of Washington. 140 p.

Hiatt, Cheryl Rae. A petrographic, geochemical, and well log analysis of the Utica Shale-Trenton Limestone transition in the northern Michigan Basin. M, 1985, Michigan Technological University. 146 p.

Hiatt, John Ludlow. The petrology of the Franconia Formation, Monroe to Adams counties, Wisconsin. M, 1970, Northern Illinois University. 96 p.

Hibbard, James Patrick. Evolution of anomalous structural fabrics in an accretionary prism; the Oligocene-Miocene portion of the Shimanto Belt at Cape Muroto, Southwest Japan. D, 1988, Cornell University. 241 p.

Hibbard, James Patrick. The southwest portion of the Dunnage Melange and its relationships to nearby groups. M, 1976, Memorial University of Newfoundland. 131 p.

Hibbard, Malcolm. Bedrock geology and petrology of Chopaka Mountain, Okanogan County, Washington. M, 1960, University of Washington. 64 p.

Hibbard, Malcolm. Geology and petrology of crystalline rocks of the Toats Coulee Creek region,

Okanogan County, Washington. D, 1962, University of Washington. 96 p.

Hibbert, Dennis Mark. Pollen analysis of late-Quaternary sediments from two lakes in the southern Puget Lowland, Washington. M, 1979, University of Washington. 37 p.

Hibbs, Roy Dean, Jr. Source effects in the three-dimensional electromagnetic induction problem with emphasis on transfer function analysis. D, 1977, University of Alberta. 288 p.

Hickcox, Alice Ellen. The relationship of cooling history and stability in two suites of igneous rocks. M, 1972, Rice University. 67 p.

Hickcox, Charles Atwood. Experimental studies on the settling velocities of some fine sediments. M, 1939, University of Oklahoma. 94 p.

Hickcox, Charles Woodbridge. The geology of a portion of the Pavant range allochthon, Millard County, Utah. D, 1971, Rice University. 105 p.

Hickcox, Charles Woodridge. Petrography of some ouachita-type graded bedding. M, 1968, University of Arkansas, Fayetteville.

Hickcox, David Hunter. Water and energy resource development in the Tongue River basin, southeastern Montana. D, 1979, University of Oregon. 336 p.

Hickenlooper, John W., Jr. Geology of the Slough Creek Tuff, northeastern Absaroka volcanic field, southwestern Montana. M, 1980, Wesleyan University. 89 p.

Hickey, Edward Paul. The geology of naturally-fractured petroleum reservoirs in the Monterey Formation, Cat Canyon area, Santa Barbara County, California. M, 1985, California State University, Northridge. 53 p.

Hickey, Edwin Weyman. Sedimentology and dolomitization in Eocene carbonate rocks, Gilchrist and Marion counties, Florida. M, 1976, University of Florida. 74 p.

Hickey, James C. Preliminary investigations of an integrative gas geochemical technique for petroleum exploration. M, 1986, Colorado School of Mines. 72 p.

Hickey, James Joseph. Stratigraphy, sedimentology, and petrology of the Jura-Cretaceous Eugenia Formation, Punta Eugenia area, Vizcaino Peninsula, Baja California Sur, Mexico. D, 1986, University of California, Santa Barbara. 661 p.

Hickey, John J. The ideal flow field in a small watershed and its relation to the drainage network. M, 1964, University of Arizona.

Hickey, Leo Joseph. The paleobotany and stratigraphy of the Golden Valley Formation (Eocene) in western North Dakota. D, 1967, Princeton University. 391 p.

Hickey, Michael Glenn. The partitioning of heavy metals in contaminated soils. D, 1982, Washington State University. 69 p.

Hickey, Rosemary Louise. Geochemistry of boninites and other low TiO_2 island arc volcanic rocks. D, 1983, Massachusetts Institute of Technology. 315 p.

Hickey, Thomas H. The use of factor analysis as an exploration tool in Puerto Rican porphyry copper deposits. M, 1986, Colorado School of Mines. 98 p.

Hickey, William K. Depositional environments of the lower Tullock Formation (Paleocene), Slope County, southwestern North Dakota. M, 1973, University of North Dakota. 103 p.

Hicklin, Richard Stuart. The stratigraphy of the McMillan Series on the eastern flank of the Cincinnati Arch in Kentucky. M, 1933, University of Kentucky. 63 p.

Hickling, Nelson Lawson. The mineralogy of allanite from the Boulder Creek Batholith, Colorado. M, 1965, American University. 48 p.

Hickman, A. Elizabeth Wenger. Spatial variations in tetrahedrite composition, Ontario Mine, Park City District, Utah. M, 1973, University of Wisconsin-Madison. 76 p.

Hickman, C. J. Bathyal gastropods of the family Turridae in the early Oligocene Keasey Formation in Oregon, with a review of some deep-water genera in the Paleogene of the eastern Pacific. D, 1975, Stanford University. 190 p.

Hickman, Carole Jean Stentz. The Oligocene marine molluscan fauna of the Eugene Formation in (Lane County), Oregon. M, 1968, University of Oregon. 211 p.

Hickman, Gary Thomas. Variations in the biodegradation potential of subsurface environments for organic contaminants. D, 1988, Virginia Polytechnic Institute and State University. 288 p.

Hickman, Paul R. The faunas of the Pickaway and Union limestones of southeastern West Virginia. M, 1951, West Virginia University.

Hickman, Robert G. Geology of the Paloma Creek area, Monterey County, California. M, 1968, Stanford University.

Hickman, Robert Gunn. Structural geology and stratigraphy along a segment of the Denali Fault System, central Alaska Range, Alaska. D, 1974, University of Wisconsin-Madison.

Hickman, Robert Gunn. The Denali Fault near Cantwell, Alaska. M, 1971, University of Wisconsin-Madison.

Hickmott, Donald Degarmo. Trace element zoning in garnets; implications for metamorphic petrogenesis. D, 1988, Massachusetts Institute of Technology. 449 p.

Hickok, William O. 1, Erosion surfaces in south-central Pennsylvania; 2, The iron ore deposits at Cornwall, Pennsylvania. D, 1932, Yale University.

Hickok, William O. The genesis of the ore deposits at Cornwall, Pennsylvania. M, 1929, Yale University.

Hickox, Charles Frederick, Jr. Geology of the central Annapolis valley, Nova Scotia. D, 1958, Yale University. 329 p.

Hickox, John E. Stratigraphy and petrology of a lagoonal tongue of the Capitan Reef complex. M, 1950, University of Kansas. 107 p.

Hicks, Benjamin Keith. Geology of a portion of the Rockford, Alabama 7 1/2' Quadrangle, Coosa County, Alabama. M, 1981, Auburn University. 130 p.

Hicks, Brian Douglas. Quartz dissolution features; an experimental and petrofabric study. M, 1985, University of Missouri, Columbia. 114 p.

Hicks, Bryan A. Geology, geomorphology, and dynamics of mass movement in parts of the Middle Santiam River drainage basin, Western Cascades, Oregon. M, 1982, Oregon State University. 169 p.

Hicks, Darryl Murray. Sand dispersion from an ephemeral river delta on the wave-dominated Central California Coast. D, 1984, University of California, Santa Cruz. 227 p.

Hicks, David Robert. Agricultural land use and related innovation and government assistance in Rio Grande do Sul, Brazil. D, 1980, Michigan State University. 375 p.

Hicks, Donald L. Geology of the southwest quarter of the Roseburg Quadrangle, Oregon. M, 1964, University of Oregon. 65 p.

Hicks, Forrest L. Photogrammetric measurements of Echo II satellite models. M, 1965, Ohio State University.

Hicks, H. S. A petrographic study of Precambrian rocks in northeastern Alberta. M, 1930, University of Alberta.

Hicks, Harold Smith. Geology of the Fitzgerald and northern portion of the Chipewan map areas, northern Alberta, Canada. D, 1932, University of Minnesota, Minneapolis. 82 p.

Hicks, Henry Thomas, Jr. Geology and ore genesis of the Buckhorn District iron deposits. M, 1982, North Carolina State University. 164 p.

Hicks, Jason F. A lacustrine delta lobe in the Paleocene Fort Union Formation, Belfry, Montana. M, 1988, University of Massachusetts. 221 p.

Hicks, John R. Hydrogeology of igneous and metamorphic rocks in the Shaffers Crossing area and vicinity near Conifer, Colorado. M, 1987, Colorado School of Mines. 174 p.

Hicks, Randall Thackery. Diagenesis of the Westwater Canyon Member, Morrison Formation, East Chaco Canyon Drilling Project, New Mexico. M, 1981, University of New Mexico. 130 p.

Hicks, Reginald V. Paleocurrent directions in the Vamoosa Formation (Pennsylvanian) of Oklahoma. M, 1962, University of Kansas.

Hickson, Catherine Jean. Quaternary volcanism in the Wells Gray-Clearwater area, east central British Columbia. D, 1987, University of British Columbia.

Hicock, Stephen Robert. Pre-Fraser Pleistocene stratigraphy, geochronology, and paleoecology of the Georgia Depression, British Columbia. D, 1980, University of Western Ontario. 230 p.

Hicock, Stephen Robert. Quaternary geology; Coquitlam-Port Moody area, British Columbia. M, 1976, University of British Columbia.

Hidalgo, Robert Valeriano. Inorganic geochemistry of coal, Pittsburgh Seam. D, 1974, West Virginia University.

Hidore, John Warren. Effects of ramp geometry on associated secondary deformation; a photoelastic modeling study. M, 1985, University of Oklahoma. 119 p.

Hiebert, Franz Kunkel. The role of bacteria in the deposition and early diagenesis of the Posidonienschiefer, a Jurassic oil shale in southern Germany. M, 1988, University of Texas, Austin. 125 p.

Hiemann, Mary Helen. Diagenesis of the Lower Pennsylvanian Morrow Sandstone, Empire South Field, Eddy County, New Mexico. M, 1987, Texas A&M University. 132 p.

Hiers, Miles Terry. Geology of Standing Stone State Park, Overton County, Tennessee. M, 1950, Vanderbilt University.

Hieshima, Glenn B. Sedimentology of Miocene Monterey Formation diatomites, California. M, 1987, University of Wisconsin-Madison. 113 p.

Higazy, Riad A. M. Petrogenesis of perthite pegmatites in the Black Hills, South Dakota. D, 1948, University of Chicago. 72 p.

Higdon, Charles E. Geology and ore deposits of the Sunshine area, Pima County, Arizona. M, 1933, University of Arizona.

Higginbotham, David R. Regional stratigraphy, environments of deposition, and tectonic framework of Mississippian clastic rocks between the Tuscumbia and Bangor limestones in the Black Warrior Basin of Alabama and Mississippi. M, 1985, University of Alabama. 177 p.

Higgins, Brenda Baer. Petrographic parameters of the beach sands of Northeast Florida. M, 1965, University of Florida. 75 p.

Higgins, Charles Graham, Jr. Geology of the lower Russian River, California. D, 1950, University of California, Berkeley. 187 p.

Higgins, Chris Thomas. Geology of the Twin Peaks area, Lake Tahoe region, California, with a survey of geologic hazards. M, 1977, University of California, Davis. 172 p.

Higgins, Daniel F., Jr. Granites of the Prince William Sound, Alaska. M, 1909, Northwestern University.

Higgins, David Thomas. Unsteady drawdown in an unconfined aquifer. D, 1968, University of Wisconsin-Madison. 145 p.

Higgins, Donald W. A review of Oligocene alligators from the Big Badlands of South Dakota. M, 1971, South Dakota School of Mines & Technology.

Higgins, Grove L., Jr. Saline ground water at Syracuse, New York. M, 1955, Syracuse University.

Higgins, James W. Structural petrology of the Pine Creek area, Dickinson County, Michigan. D, 1947, University of Chicago. 64 p.

Higgins, Janice M. Geology of the Champlin Peak Quadrangle, Juab and Millard counties, Utah. M, 1982, Brigham Young University. 58 p.

Higgins, Jerry Don. Seismic response mapping of Creve Coeur Quadrangle, St. Louis County, Missouri. D, 1980, University of Missouri, Rolla. 129 p.

Higgins, Jerry Don. The environmental geology of the Ebenezer Quadrangle, Greene County, Missouri. M, 1975, University of Missouri, Rolla.

Higgins, John Britt. Crystal chemistry, cation ordering and polymorphism in sapphirine. D, 1978, Virginia Polytechnic Institute and State University. 219 p.

Higgins, John Britt. The crystal chemistry of titanite. M, 1975, Virginia Polytechnic Institute and State University.

Higgins, Lee. Tectonic and kinematic analysis of the Rough Creek fault system, west central Kentucky. M, 1986, University of Kentucky. 45 p.

Higgins, Maurice J. Stratigraphic position of the coal seam near Porter, Wagoner County, Oklahoma. M, 1961, University of Oklahoma. 83 p.

Higgins, Michael Denis. Age and origin of the Sept Iles Anorthosite complex, Quebec. D, 1980, McGill University. 127 p.

Higgins, Michael W. The geology of Newberry Caldera, central Oregon. D, 1968, University of California, Santa Barbara. 320 p.

Higgins, Michael Wicker. The geology of part of Sandy Springs Quadrangle, Georgia. M, 1965, Emory University. 141 p.

Higgins, Neville Charles. The genesis of the Grey River tungsten prospect; a fluid inclusion, geochemical, and isotopic study. D, 1980, Memorial University of Newfoundland. 539 p.

Higgins, Ralph Edward. A chemical study of Cenozoic volcanism in the Los Angeles Basin and Santa Cruz Island, and the Mohave Desert. D, 1972, University of California, Santa Barbara.

Higgs, Donald V. Anorthosite and related rocks of the western San Gabriel Mountains, Southern California. D, 1950, University of California, Los Angeles.

Higgs, Nelson B. The geology of the southeastern part of the Jarbidge I Quadrangle, Elko County, Nevada. M, 1960, University of Oregon. 100 p.

Higgs, Nigel Gordon. Mechanical properties of ultrafine quartz, chlorite and bentonite in environments appropriate to upper-crustal earthquakes. D, 1981, Texas A&M University. 286 p.

Higgs, Roger Y. Provenance of Mesozoic and Cenozoic siliciclastic sediments of the Labrador and West Greenland continental margins. M, 1977, University of Calgary. 169 p.

Higgs, William Reginald. Foraminifera of the Mooreville Chalk in Montgomery County, Alabama. M, 1948, University of Alabama.

High, Lee Rawdon, Jr. Recent coastal sediments of British Honduras. D, 1967, Rice University. 206 p.

Highland, William Robert. The Reynold's number in problems of flow through porous media; a comparison of some theoretical and experimental techniques. M, 1975, University of Minnesota, Minneapolis. 98 p.

Highsmith, Patrick B. Chemical changes in chlorite and illite due to burial diagenesis in a Paleozoic shale. M, 1979, Georgia Institute of Technology. 146 p.

Hight, David H. Conodont fauna and sedimentary facies of the Middle Ordovician limestones, Lusters Gate area, Montgomery County, Virginia. M, 1979, Virginia Polytechnic Institute and State University.

Hight, Richard Parker. The differential thermal analysis of certain hydrated substances. D, 1959, Boston University. 112 p.

Hight, Robert, Jr. A small angle X-ray scattering study of some montmorillonite clay systems. D, 1962, University of Missouri, Columbia. 152 p.

Hightower, Charles Henry, Jr. Middle Cambrian stratigraphy of the Wah Wah Range, Beaver County, Utah. M, 1959, Southern Methodist University. 98 p.

Hightower, Jay Harold. The petroleum geology of the Rodessa and James limestones (Lower Cretaceous) in Henderson and Anderson counties, Texas. M, 1986, Baylor University. 148 p.

Hightower, Maxwell Lee. Structural geology of the Palestine salt dome, Anderson County, Texas. M, 1958, University of Texas, Austin.

Higinbotham, Larry R. Stratigraphy, depositional history, and petrology of the Upper Cretaceous(?) to middle Eocene Montgomery Creek Formation, Northern California. M, 1986, Oregon State University. 231 p.

Higuchi, Kazufumi. A global carbon cycle model. D, 1983, University of Toronto.

Hilali, Atika. Variation intraspecifique chez quelques Brachiopodes des calcaires superieurs de Gaspe (Devonien inferieur) de l'est de la Gaspesie. M, 1985, Universite de Montreal.

Hilbert, Eric George. Ultrasonic measurements of the elastic properties of single-crystal magnesium aluminate spinel, $MgAl_2O_4$. M, 1984, Pennsylvania State University, University Park. 59 p.

Hilchey, Gordon. Geology and economics of graphite. M, 1954, University of Toronto.

Hilchie, Douglas Walter. The effect of pressure and temperature on the resistivity of rocks. D, 1964, University of Oklahoma. 98 p.

Hild, Gregory Phillip. The relationship of Permian San Andres facies to the distribution of porosity and permeability in the Garza Field, Garza County, Texas. M, 1985, Baylor University. 122 p.

Hildebrand, Kanzira. Etude des complexes granito-gneissiques de Cyimbili, Mara et Mutara, République Rwandaise (Afrique Centrale). M, 1985, Universite du Quebec a Chicoutimi. 218 p.

Hildebrand, Robert S. A great continental volcanic arc of early Proterozoic age at Great Bear Lake, Northwest Territories. D, 1982, Memorial University of Newfoundland. 240 p.

Hildebrand, S. L. Alkali-carbonate reactivity in some Tennessee limestone aggregates. M, 1978, Memphis State University.

Hildebrand-Mittlefehldt, Nurit. The strain field near fault terminations. D, 1978, University of California, Los Angeles. 106 p.

Hildenbrand, T. G. Seismomagnetism. D, 1975, University of California, Berkeley. 166 p.

Hildick, Margaret E. The petrology, rank, and correlation of the Lower Pennsylvanian Blue Creek and Mary Lee coal seams in Jefferson and Walker counties, Alabama. M, 1982, Auburn University. 162 p.

Hildreth, Edward Wesley. The magma chamber of the Bishop Tuff; gradients in temperature, pressure, and composition. D, 1977, University of California, Berkeley. 328 p.

Hildreth, Gail Darice. The bedrock geology and stratigraphy of the Mississippian and Early Pennsylvanian rocks of the southeast flank, Armstead Anticline, Beaverhead County, Montana. M, 1981, Oregon State University. 144 p.

Hileman, James Alan. Part I, A contribution to the study of the seismicity of southern California; Part II, Inversion of phase times for hypocenters and shallow crustal velocities. D, 1978, California Institute of Technology. 237 p.

Hileman, Mary Esther. Stratigraphy and paleoenvironmental analysis of the Upper Jurassic Preuss and Stump formations, western Wyoming and southeastern Idaho. D, 1973, University of Michigan. 297 p.

Hileman, Mary Esther. Stratigraphy of the Preuss and Stump formations (Upper Jurassic) of western Wyoming and southeastern Idaho. M, 1969, University of Michigan.

Hilfiker, Kenneth G. Temperature gradients and heat flow in central and western New York. M, 1980, SUNY at Buffalo. 38 p.

Hill, Alan T. Systematics, biostratigraphy, and paleoenvironments of late Virgilian and early Wolfcampian corals, Bird Spring Group, Arrow Canyon Quadrangle, Clark County, Nevada. M, 1978, University of Illinois, Urbana. 105 p.

Hill, Benjamin Felix. The geology of Shoal Creek, with accompanying section. M, 1897, University of Texas, Austin.

Hill, Benjamin Harvey. A new method of classifying marine Gastropoda. M, 1925, Texas Christian University. 43 p.

Hill, Bernard Louis. Reclassification of winged Cythereis and winged Brachycythere. M, 1952, Washington University. 69 p.

Hill, Bradley M. Evaluation of hydrogeologic siting criteria for siting hazardous waste management facilities in Idaho. M, 1988, University of Idaho. 95 p.

Hill, Brittain Eames. Petrology of the Bend Pumice and Tumalo Tuff, a Pleistocene Cascade eruption involving magma mixing. M, 1985, Oregon State University. 109 p.

Hill, Carol A. Geology and mineralogy of cave nitrates. M, 1978, University of New Mexico. 125 p.

Hill, Claude P. T. The present shoreline sedimentation in the vicinity of Dingwall, Nova Scotia. M, 1952, Massachusetts Institute of Technology. 81 p.

Hill, Constance M. A diagenetic study of carbonate sediments from the Red Sea utilizing cathodoluminescence. M, 1976, Rensselaer Polytechnic Institute. 68 p.

Hill, D. O. A hydrodynamic and salinity model for Mobile Bay. D, 1974, University of Alabama. 339 p.

Hill, David B. The hydrology of an ice-contact deposit at Newmarket Plains, New Hampshire. M, 1979, University of New Hampshire. 93 p.

Hill, David Paul. Gravity survey in the western Snake River plain, Idaho. M, 1961, Colorado School of Mines. 52 p.

Hill, David Paul. High frequency wave propagation in the Earth; theory and observation. D, 1971, California Institute of Technology. 390 p.

Hill, David R. Geology of the Whetstone Anticline area, Teton County, Wyoming. M, 1964, Oregon State University. 118 p.

Hill, Donald Gardner. A laboratory investigation of conductivity and dielectric constant tensors of rocks. D, 1969, Michigan State University. 228 p.

Hill, Donald W. Gravity prediction by pseudo-density profiling. M, 1975, Pennsylvania State University, University Park.

Hill, E. Bratton, Jr. Geology of the Jamestown District (Boulder County), Colorado. M, 1933, University of Colorado.

Hill, F. C. Effects of the environment on animal exploitation by archaic inhabitants of the Koster Site, Illinois. D, 1975, University of Louisville. 210 p.

Hill, Frank Eugene. A petrographic study of the Overbrook Sandstone. M, 1955, University of Oklahoma. 72 p.

Hill, Frederick B. The soils of Jamaica, British West Indies; their origin and engineering significance. M, 1947, Purdue University.

Hill, G. F. Turquoise, its history and significance in the Southwest. M, 1938, University of Arizona.

Hill, Gary William. Infaunal and neoichnological characteristics of the South Texas outer continental shelf; Facies characteristics and patterns in modern size-graded shelf deposits, northwestern Gulf of Mexico; Facies characteristics and patterns in mid-estuary intertidal flat deposits, Willapa Bay, Washington. D, 1980, University of California, Santa Cruz. 437 p.

Hill, Gerhard William, Jr. An electrical resistivity logging study of the marine sediments at the offshore dredge disposal site, Galveston, Texas. M, 1976, Texas A&M University.

Hill, H. Stanton. Petrography of the Pelona schists of Southern California. M, 1939, Pomona College.

Hill, J. D. The petrology of the basic intrusions of the Bathurst-Newcastle area, New Brunswick. M, 1974, Acadia University.

Hill, J. M. Landsat assessment of estuarine water quality with specific reference to coastal land-use. D, 1978, Texas A&M University. 226 p.

Hill, J. M. Lower Ludlovian Chitinozoa from Arisaig, Nova Scotia and Baillie-Hamilton Island, Canadian Arctic Archipelago. M, 1977, University of Western Ontario. 123 p.

Hill, James Daniel. Mining districts of southwestern New Mexico. M, 1924, University of Colorado.

Hill, James David. Bluff erosion study; Ludington pumped storage project. M, 1970, University of Michigan.

Hill, James Gregory. A digital computer model study of the Breakneck Creek well field aquifer, Kent, Ohio. M, 1978, Kent State University, Kent. 73 p.

Hill, James M. Stratigraphy and paleoenvironment of the Miocene Monterey Group in the East San Francisco Bay region, California. M, 1978, San Jose State University. 113 p.

Hill, James Martin. Identification of sedimentary biogeochemical reservoirs in Chesapeake Bay. D, 1984, Northwestern University. 367 p.

Hill, Jeffrey. Sodium bisulfate fusion; application to the study of the black middle Marcellus Shale. M, 1983, University of Pittsburgh.

Hill, Jennie Lu Hill. Trace fossils as possible environmental indicators in the Dresbachian of the North American Rockies. M, 1975, University of Rochester.

Hill, John A. Sedimentology of delta-front sandstones, Cerro del Pueblo Formation (Upper Cretaceous), Parras Basin, Coahuila, Mexico. M, 1988, University of New Orleans.

Hill, John David. The structural development and crystallization of the Kenoran granitoid plutons in the Nose Lake-Back River area, Northwest Territories. D, 1980, University of Western Ontario. 239 p.

Hill, John Davis. Paleozoic stratigraphy of the Mud Springs Mountains, Sierra County, New Mexico. M, 1956, University of New Mexico. 72 p.

Hill, John Gilmore. Sandstone petrology and stratigraphy of the Stanley Group (Mississippian), southern Ouachita Mountains, Oklahoma. D, 1967, University of Wisconsin-Madison. 240 p.

Hill, John Gilmore. Some sedimentary structures in the (Pennsylvanian) Stanley Group, central Ouachita Mountains, Oklahoma. M, 1962, University of Wisconsin-Madison.

Hill, John Jerome. Petrology and structure of the Precambrian rocks of the Park range, of north-central Colorado. M, 1969, University of Oregon. 75 p.

Hill, John Jerome. The simulation of runoff from a logged watershed. M, 1976, Colorado State University. 85 p.

Hill, John R. Ice fabrics studies of the Plover River, Wisconsin. M, 1970, University of Wisconsin-Milwaukee.

Hill, Julie Anne. A finite difference simulation of seismic wave propagation and resonance in Salt Lake valley, Utah. M, 1988, University of Utah. 90 p.

Hill, Keith Charles. Sulfur in dolomite. M, 1985, Michigan State University. 84 p.

Hill, Kevin Charles. Structure and stratigraphy of the coal-bearing and adjacent strata near Cadomin, Alberta. M, 1980, University of Alberta. 191 p.

Hill, L. S. A petrographic study of a basic intrusive sheet in the Yellowknife area, Northwest Territories. M, 1940, McGill University.

Hill, Lawson Bruce. A tectonic and metamorphic history of the north-central Klamath Mountains, California. D, 1984, Stanford University. 282 p.

Hill, Lena Elizabeth. Slickrock Mountain; bimodal intrusive complex Big Bend National Park, Texas. M, 1985, Texas Tech University. 59 p.

Hill, Malcolm David. Volcanic and plutonic rocks of the Kodiak-Shumagin Shelf, Alaska; subduction deposits and near-trench magmatism. D, 1979, University of California, Santa Cruz. 279 p.

Hill, Malcom W., Jr. Geologic methods of subsurface mapping as an aid in solving the problems of petroleum overproduction. M, 1934, The Johns Hopkins University.

Hill, Mary Catherine. An investigation of hydraulic conductivity estimation in a ground-water flow study of northern Long Valley, New Jersey. D, 1985, Princeton University. 364 p.

Hill, Mary Louise. Geology of the Redcap Mountain area, Coast Plutonic Complex, British Columbia. D, 1985, Princeton University. 223 p.

Hill, Mason L. A contribution on the structure of the San Gabriel Mountains. M, 1929, Pomona College.

Hill, Mason L. Mechanics of faulting near Santa Barbara, California. D, 1932, University of Wisconsin-Madison.

Hill, Merton H., III. Albian-Cenomanian (Lower Cretaceous) calcareous nannofossils from Texas and Oklahoma. D, 1975, University of California, Los Angeles. 259 p.

Hill, Patrick Arthur. The geology, mineralization and leached outcrops of the Minas Carlota region, Las Villas, Cuba. D, 1958, Columbia University, Teachers College. 320 p.

Hill, Paul Lester, Jr. Wetland-stream ecosystems of the Western Kentucky Coalfield; environmental disturbance and the shaping of aquatic community structure. D, 1983, University of Louisville. 303 p.

Hill, Raymond Leslie. Pleistocene terraces in Georgia. M, 1966, University of Florida. 155 p.

Hill, Richard Bruce. Depositional environments of the Upper Cretaceous Ferron Sandstone south of Notom, Wayne County, Utah. M, 1982, Brigham Young University. 83 p.

Hill, Richard Lee. Strontium isotope composition of basaltic rocks of the Transantarctic mountains, Antarctica. M, 1969, Ohio State University.

Hill, Robert E. A scanning electron microscope study of quartz sand grains from some braided terraces and meander belts of the lower Mississippian alluvial valley. M, 1981, Memphis State University. 73 p.

Hill, Robert E. Stratigraphy and sedimentology of the middle Proterozoic Waterton and Altyn formations, Belt-Purcell Supergroup, Southwest Alberta. M, 1985, McGill University. 165 p.

Hill, Robert Ian. Petrology and pertogenesis of batholithic rocks, San Jacinto Mountains, Southern California. D, 1984, California Institute of Technology. 731 p.

Hill, Robert Lee. Geology and geochemistry of El Capitan mercury mine, Last Chance Range, Inyo County, California. M, 1972, University of California, Los Angeles.

Hill, Robert Lee. Tillage effects on selected soil physical properties. D, 1984, Iowa State University of Science and Technology. 143 p.

Hill, Robin E. T. The crystallization of basaltic melts as a function of oxygen fugacity. D, 1969, Queen's University. 150 p.

Hill, Roderic P. Structural and petrological studies in the Shuswap metamorphic complex near Revelstoke, British Columbia. M, 1975, University of Calgary. 147 p.

Hill, Roy Louis. Thermodynamic properties of Fe-Mg titanate spinels. M, 1985, Purdue University. 62 p.

Hill, Steve R. Geology of the Mining Mountain area, Cuba, NM. M, 1980, New Mexico Institute of Mining and Technology. 116 p.

Hill, Thomas G. Biostratigraphy of the upper Chickabally Member of the Budden Canyon Formation (Cretaceous), near Ono, Shasta County, California. M, 1975, University of California, Riverside. 115 p.

Hill, Vincent G. Phase transformation in the system zinc sulphide. D, 1956, University of Toronto.

Hill, Vincent G. The system gallia-alumina-water and its bearing on the stability and composition of diaspore. M, 1951, Pennsylvania State University, University Park. 66 p.

Hill, Virginia S. Mississippian Williams Canyon Limestone Member of the Leadville Limestone, south-central Colorado. M, 1983, Colorado School of Mines. 112 p.

Hill, Walter E., Jr. A geochemical study of the chert in the Plattsmouth Limestone Member of the Oread Limestone of northeastern Kansas. M, 1964, University of Kansas. 50 p.

Hill, Wayne Noe. Ostracoda from the Bloomsdale Member of the Plattin Formation (Middle Ordovician) of eastern Missouri. M, 1957, Washington University. 86 p.

Hill, William A. The lithostratigraphy and paleoenvironment of the Boggy Formation, Franks Graben, Pontotoc County, Oklahoma. M, 1981, Texas Christian University. 92 p.

Hill, William T. Petrography of some pebble horizons in the Southern Appalachians. M, 1951, University of Tennessee, Knoxville. 56 p.

Hill, William T. The geology, applied chemistry, and drilling results in the Puncheon Camp Creek area, Grainger County, Tennessee. D, 1971, University of Tennessee, Knoxville. 106 p.

Hill-Rowley, Richard. An evaluation of digital Landsat classification procedures for land use inventory in Michigan. D, 1982, Michigan State University. 262 p.

Hillard, David L. Regional study of the Inyan Kara Group of South Dakota. M, 1963, Michigan State University. 47 p.

Hillard, Patrick D. General geology and beryllium mineralization near Apache Warm Springs, Socorro County, New Mexico. M, 1967, New Mexico Institute of Mining and Technology. 58 p.

Hillary, Elizabeth M. Metamorphosed trondhjemitic basement beneath the Archean Favourable Lake volcanic complex, northwestern Ontario. M, 1980, University of Manitoba. 101 p.

Hille, Oscar Roy. The geology and mineral resources of the Ludwig area, Johnson County, Arkansas. M, 1951, University of Arkansas, Fayetteville.

Hillebrand, James R. Geology and ore deposits in the vicinity of Pitman Wash, Pinal County, Arizona. M, 1953, University of Arizona.

Hillemeyer, Frank Lloyd. Geometric and kinematic analysis of high- and low-angle normal faults within the Mojave-Sonoran detachment terrane. M, 1984, San Diego State University. 209 p.

Hillendbrand, Charles, III. Subsidence and fault activation related to fluid extraction, Sacet Field, Nueces County, Texas. M, 1985, University of Houston.

Hillerud, John Martin. Subfossil high plains bison. D, 1970, University of Nebraska, Lincoln.

Hillerud, John Martin. The Recent and fossil bison of Alberta, Canada. M, 1966, University of Nebraska, Lincoln.

Hilles, Robert and Choate, Raoul. The geology of the foothills west of Loveland, Larimer County, Colorado. M, 1954, University of Michigan.

Hillesland, Larry L. The geology, mineralization, and geochemistry of the Pine Creek area, Lemhi County, Idaho, Montana. M, 1982, Oregon State University. 97 p.

Hillhouse, Douglas Neil. Geology of the Piney River-Roseland titanium area, Nelson and Amherst counties, Virginia. D, 1960, Virginia Polytechnic Institute and State University. 169 p.

Hillhouse, Douglas Neil. Geology of the Veddar Mountain-Silver Lake area. M, 1956, University of British Columbia.

Hillhouse, J. W. Paleomagnetism of the Plio-Pleistocene sediments of Lake Tecopa, California and East Rudolf, Kenya; magnetic stratigraphy and polarity transitions. D, 1976, Stanford University. 235 p.

Hilliard, B. C. The weathering of the Methuen granite. M, 1973, University of Guelph.

Hilliard, Henry D. Topologic representations of channel networks. M, 1970, University of Missouri, Columbia.

Hillier, Mark R. A geochemical study of the latite of Government Well, Superstition Mountains, Arizona. M, 1978, Arizona State University. 69 p.

Hillis, Donuil M. The Randsburg mining district (California). M, 1924, Stanford University. 70 p.

Hillman, Barry Arthur. Hydrothermal activity as related to ore deposition at the Sierrita porphyry copper-molybdenite deposit, southwestern Arizona. M, 1972, University of Cincinnati. 69 p.

Hillman, C. Thomas. Geology of the Mount Margaret-Green River area, Washington. M, 1970, Miami University (Ohio). 52 p.

Hillman, Daniel Marc Jan. A study of small-scale deformation features associated with Embudo fault zone, north-central New Mexico. M, 1986, University of Oklahoma. 79 p.

Hillman, Douglas L. An analysis of the Cedar Bog hydrologic system through the use of a three-dimensional groundwater flow model. M, 1987, Wright State University. 156 p.

Hillman, H. F. Closure and adjustment of control survey nets. D, 1976, University of Arizona. 309 p.

Hillman, Bob. The geology and ore deposits of the Clayton silver mine, Custer County, Idaho. M, 1986, Eastern Washington University. 54 p.

Hills, Doris Volz. The petrography, mineral chemistry, and geochemistry of eclogites from the Koidu kimberlite complex, Sierra Leone. M, 1988, University of Massachusetts. 209 p.

Hills, Francis Allan. The Precambrian geology of the Glens Falls and Fort Ann quadrangles, southeastern Adirondack Mountains, New York. D, 1965, Yale University. 306 p.

Hills, John Moore. The method of insoluble residues as applied to the determination of the stratigraphy of Ohio. D, 1934, University of Chicago. 59 p.

Hills, Leonard Vincent. Glaciation, stratigraphy, structure and micropaleobotany of the Princeton Coalfield, British Columbia. M, 1962, University of British Columbia.

Hills, Leonard Vincent. Palynology and age of early Tertiary Basins, Interior British Columbia, Columbia. D, 1965, University of Alberta. 188 p.

Hills, R. G. Convection in the Earth's mantle due to viscous shear at the core-mantle interface and due to large-scale buoyancy. D, 1979, New Mexico State University, Las Cruces. 437 p.

Hills, Robert Chadwick. A comparison of the methods for the mechanical analysis of sediments. M, 1934, Cornell University.

Hills, Scott Jean. Community heterogeneity in a homogeneous habitat; mollusc populations from the lagoon shelf of Enewetak Atoll, Marshall Islands. M, 1982, University of Utah. 76 p.

Hills, Scott Jean. The analysis of microfossil shape; experiments using planktonic foraminifera. D, 1988, University of California, San Diego. 243 p.

Hilltrop, Carl Lee Roy. Silica behavior in aggregates and concrete. D, 1960, Iowa State University of Science and Technology. 104 p.

Hilmoe, Cynthia. Pleistocene geology of the Piedmont area near the Clarks Fork and Littlerock Creek canyons, Beartooth Mountains, Northwest Wyoming. M, 1980, Southern Illinois University, Carbondale. 149 p.

Hilmy, Mohamed Ezzeldin. Structural crystallographic relation between sodium sulfate and potassium sulfate and other sulfate minerals. D, 1952, University of Michigan.

Hilpert, Frederick Martin. An investigation of laccoliths by means of scale models. M, 1954, Montana College of Mineral Science & Technology. 108 p.

Hilpman, Paul Lorenz. Devonian rocks of Kansas and their epeirogenic significance. D, 1969, University of Kansas. 73 p.

Hilpman, Paul Lorenz. Geology of the Easy Ridge area, White Pine County, Nevada. M, 1956, University of Kansas.

Hiltabrand, Robert R. Experimental diagenesis of argillaceous sediment. D, 1970, Louisiana State University.

Hilterman, Fred J. Three-dimensional seismic modeling. D, 1970, Colorado School of Mines. 93 p.

Hilton, Allan Decou. Four Vaqueros sections in the Santa Cruz Mountains, Santa Cruz County, California. M, 1958, Stanford University.

Hilton, Allan Decou. Survey of the Cretaceous and pre-Cretaceous geology of New Mexico, the Texas Panhandle, and western Oklahoma. M, 1957, Stanford University.

Hilton, Don A. The petrology, sedimentology, and stratigraphy of the Late Cretaceous(?) and early Tertiary sedimentary rocks of southwest Markagunt Plateau, Utah. M, 1984, Kent State University, Kent. 137 p.

Hilton, E. R. Geology of the Horton sediments, Strait of Canso area (Nova Scotia). M, 1954, Massachusetts Institute of Technology. 66 p.

Hilton, George S. Geology of a portion of the northeastern Santa Ana Mountains (California). M, 1950, Pomona College.

Hilton, Glenn G. The depositional setting and diagenetic history of the Pennsylvanian (early Missourian) "Palo Pino Sandstone", Callahan County, Texas. M, 1986, University of Texas at Dallas. 145 p.

Hilton, Joanne. Heavy metal concentrations from abandoned mine drainage in Coal Creek, Colorado. M, 1984, Colorado State University. 132 p.

Hilton, R. P. The geology of the Ingot - Round Mountain area Shasta County, California. M, 1975, California State University, Chico. 83 p.

Hilton, S. W. The Elora-Fergus buried valley; further geophysical studies. M, 1978, University of Waterloo.

Hiltrop, Carl L. R. Relation of pore size distribution to the petrology of some carbonate rocks. M, 1958, Iowa State University of Science and Technology.

Hilty, Robert D. The hydrology and nutrient balance in a small pond located in the Oak Openings sand belt, Lucas County, Ohio. M, 1971, University of Toledo. 70 p.

Hilty, Robert Emil. Glaciological study of an ice cliff in Northwest Greenland. M, 1956, Ohio State University.

Himanga, James Carlo. Geology of the Sierra Chiltepins, Sonora, Mexico. M, 1977, Northern Arizona University. 99 p.

Himebaugh, John P. Petroleum potential of the Tilston Interval (Mississippian) of central North Dakota. M, 1979, University of North Dakota. 152 p.

Himes, David Madero. Analysis of gravity, magnetic and surface geologic data, northern Chihuahua, Mexico. M, 1968, Rice University. 52 p.

Himes, Gregory Tait. A preliminary study of the chemistry of icings and associated spring and surface waters of the northern Brooks Range, Alaska. M, 1980, Indiana University, Bloomington. 126 p.

Himes, Larry Douglas. A joint hypocenter velocity determination for the New Madrid seismic zone. M, 1987, St. Louis University.

Himes, Marshall D. Geology of the Pima Mine, Pima County, Arizona. M, 1972, University of Arizona.

Himmelberg, Glen Ray. Precambrian geology of the Granite Falls-Montevideo area, Minnesota. D, 1965, University of Minnesota, Minneapolis. 124 p.

Himmelberg, Glen Ray. Preliminary investigation of the geology and geochronology of the Precambrian core of the northern Wet Mountains, Colorado. M, 1960, Texas Tech University. 44 p.

Hims, A. G. A rock fracture analysis at three levels in the Kidd Creek Mine, Timmins, Ontario, Canada, with particular reference to an abnormality in the stress field. M, 1977, University of Waterloo.

Hinaman, Gary. Petrology of the Lower Water Island Formation (Early Cretaceous), Virgin Islands Deep Core. M, 1972, SUNY at Binghamton. 68 p.

Hinchey, Norman. The Plattin Formation between Herculaneum, Missouri and St. Albans, Missouri. M, 1928, Washington University. 109 p.

Hinchey, Norman S. The fauna and stratigraphy of the St. Louis Formation. D, 1934, Harvard University.

Hinchman, Judith Anne. Compositional gradients related to feldspar crystallization in natural liquids. M, 1984, University of Utah. 74 p.

Hinchman, Nancy. Clastic sedimentology, petrology, and paleohydrology of a sandstone facies in the Green River Formation, Southwest Wasatch Plateau, central Utah. M, 1986, Wayne State University. 102 p.

Hinchman, Steven B. Error analysis and design considerations for backpressure testing of gas wells. M, 1987, Colorado School of Mines. 91 p.

Hinckley, Bern Schmehl. Factors affecting off-road vehicle induced erosion acceleration in the Mojave Desert, California. M, 1980, Stanford University. 78 p.

Hinckley, David Narwyn. Mineralogical and chemical variations in the kaoline deposits of the coastal plain of Georgia and South Carolina. D, 1961, Pennsylvania State University, University Park. 215 p.

Hinckley, David Newyn. An investigation of the occurrence of uranium at Cameron (Coconino County), Arizona. M, 1957, University of Utah. 67 p.

Hinckley, T. K. Sequential visual imagery to assess change; West Midlands topographic sample, 1870-1961. D, 1979, University of Western Ontario.

Hindle, Robinson Joseph. A post-glacial pollen diagram from Kingston, Rhode Island. D, 1964, University of Rhode Island. 104 p.

Hindlet, Francois, J. F. Amplitude decomposition of thin layer offset versus amplitude data. D, 1986, University of Houston. 193 p.

Hindman, David Jerome. Mississippian structural movement of lineament blocks within the Williston Basin-central Montana area. M, 1984, University of Colorado. 69 p.

Hindman, James C. Properties of the system $CaCo_3$-Co_2-H_2O in sea water and sodium chloride solutions. D, 1943, University of California, Los Angeles. 148 p.

Hindman, James Richard. Hydrothermal equilibrium in the system CuO-SiO_2-H_2O. D, 1985, University of Utah. 177 p.

Hinds, George W. Geology of the Mike Spencer Canyon area, Bonneville County, Idaho. M, 1958, University of Wyoming. 41 p.

Hinds, Jim S. Controls on Paradox salt deposition in the area of Comb Monocline, San Juan County, Utah. M, 1960, University of New Mexico. 36 p.

Hinds, Norman Ethan Allen. The geology of Kauai and Niihau, Hawaiian Islands. D, 1924, Harvard University. 100 p.

Hinds, Robert Warren. Morphology and systematics of some tertiary and Recent Tubuliporinid Cyclostome Bryozoa. D, 1972, Columbia University. 166 p.

Hinds, Ronald C. A regional travel time and relative amplitude study using ray tracing on a subducted lithospheric slab model. M, 1983, University of Manitoba.

Hine, Albert C. Sand deposition in the Chatham Harbor estuary and on the neighboring beaches, Cape Cod, Massachusetts. M, 1972, University of Massachusetts. 187 p.

Hine, Albert C., III. Shallow carbonate bank margin structure and depositional processes; northwestern Little Bahama Bank; Bahamas. D, 1975, University of South Carolina. 225 p.

Hine, George T. Relation of fracture traces, joints, and ground water occurrence in the Bryantsville Quadrangle area, Kentucky. M, 1969, University of Kentucky. 38 p.

Hiner, John Edward. Geology of the Desert Peak geothermal anomaly, Churchill County, Nevada. M, 1979, University of Nevada. 84 p.

Hines, Gary Keith. Faunal succession and depositional environments within the Lodgepole Limestone (Early Mississippian), of Samaria Mt., Idaho. M, 1981, Utah State University. 118 p.

Hines, J. M. Concentration and distribution of selected trace elements in the ground water of the Maumee River basin, Ohio, Indiana, and Michigan. M, 1974, Ohio State University.

Hines, Mark Edward. Seasonal biogeochemistry in the sediments of the Great Bay estuarine complex, New Hampshire. D, 1981, University of New Hampshire. 151 p.

Hines, Robert A. Petrology of the Stony Gap Sandstone (Upper Mississippian), Mercer County, West Virginia. M, 1983, East Carolina University. 90 p.

Hines, Robert Arthur, Jr. Carboniferous evolution of the Black Warrior foreland basin, Alabama and Mississippi. D, 1988, University of Alabama. 403 p.

Hines, Stephen Anthony. Origins of ore-controlling folds in the Todilto Limestone, Grants mining district, New Mexico. M, 1976, New Mexico Institute of Mining and Technology. 141 p.

Hinkel, Kenneth Mark. Palsa formation in north-central Alaska. D, 1986, University of Michigan. 217 p.

Hinkley, Bruce F. The glacial geology of Upper Alder Creek and Iron Bog Creek, Butte and Custer counties, South-central Idaho. M, 1981, University of Wisconsin-Milwaukee. 96 p.

Hinkley, Carole M. An approach to the appraisal of groundwater law in the Ohio hydrologic environment. M, 1976, Wright State University. 82 p.

Hinkley, Everett A. The diagenetic and environmental significance of $^{87}Sr/^{86}Sr$ and geochemical variations in the Pennsylvanian limestones of Southeast Ohio. M, 1984, Miami University (Ohio). 93 p.

Hinkley, Todd King. Weathering mechanisms and mass balance in a high Sierra Nevada watershed; distribution of alkali and alkaline earth metals in components of parent rock and soil, snow, soil moisture and stream outflow. D, 1975, California Institute of Technology. 122 p.

Hinman, Eugene Edward. Jurassic Carmel-Twin Creek facies of northern Utah. M, 1954, Washington State University. 47 p.

Hinman, Eugene Edward. Silurian bioherms of eastern Iowa. D, 1963, University of Iowa. 199 p.

Hinman, Nancy Wheeler. Organic and inorganic chemical controls on the rates of silica diagenesis; a comparison of a natural system with experimental results. D, 1987, University of California, San Diego. 402 p.

Hinn, Helen B. The erosion history of the Southern Appalachians. M, 1924, Columbia University, Teachers College.

Hinners, Noel William. The solubility of sphalerite in aqueous solutions at 80° centigrade. D, 1963, Princeton University. 120 p.

Hinnov, Linda A. Effects of water storage on the Earth's wobble. M, 1985, University of Texas, Austin.

Hinojosa, Juan Homero. On the state of isostasy in the Central Pacific; static and dynamic compensation mechanisms. D, 1986, The Johns Hopkins University. 287 p.

Hinote, Russell E. Analysis of fossil communities in the Del Rio Formation, Upper Cretaceous, Texas. M, 1978, University of Texas, Austin.

Hinrichs, Edgar Neal. The pegmatites of the Errington-Thiel Mine, Elko County, Nevada. M, 1950, Cornell University.

Hinrichs, Frederick W. The occurrence of "A" mica in the Spruce Pine District, North Carolina. M, 1941, Northwestern University.

Hinshaw, Gaylord C. A petrographic study of the insoluble residues from four Fort Riley Limestone quarries in Kansas. M, 1960, Kansas State University. 75 p.

Hinson, C. A. Three-dimensional geologic structures from inversion of gravity anomalies. M, 1976, Texas A&M University.

Hinson, H. H. Reservoir characteristics of Rattlesnake oil and gas field, San Juan County, New Mexico. M, 1947, Texas Tech University. 41 p.

Hinterlong, Gregory Dale. Paleoecologic succession in the Corryville Member (McMillan Formation, Upper Ordovician), Stonelick Creek, Clermont County, Ohio. M, 1981, Miami University (Ohio). 119 p.

Hinthong, Chaiyan. Geological aspects and engineering analysis of some Louisiana soils. M, 1972, Louisiana Tech University.

Hinthorne, James Roscoe. Bedrock and engineering geology of the Mount Tom area (Hampshire County), Massachusetts. M, 1967, University of Massachusetts. 126 p.

Hinthorne, James Roscoe. The origin of sanbornite and related minerals. D, 1974, University of California, Santa Barbara.

Hintlian, Raymond Arthur. Three mid-Pleistocene vertebrate faunal assemblages from Sheridan County, Nebraska. M, 1975, University of Nebraska, Lincoln.

Hinton, Douglas. Geophysical studies of Southern Appalachian crustal structures. M, 1982, Georgia Institute of Technology. 111 p.

Hintze, Ferdinand F. A contribution to the geology of the Wasatch Mountains, Utah. D, 1913, Columbia University, Teachers College.

Hintze, Lehi F. An Ordovician section in Millard County, Utah. M, 1948, Columbia University, Teachers College.

Hintze, Lehi F. Ordovician stratigraphy from central Utah to central Nevada. D, 1950, Columbia University, Teachers College.

Hintzman, Davis Eugene. Geology and ore deposits of the Clinton mining district, Missoula County, Montana. M, 1961, University of Montana. 65 p.

Hintzmann, Kathleen Joyce. Stratigraphic correlation of the Mountain Springs Formation (Ordovician-Devonian), southern Great Basin. M, 1983, San Diego State University. 126 p.

Hinz, David W. Behavior of carbon isotopes during the hyperfiltration of calcium carbonate solutions through calcium bentonites. M, 1987, Texas A&M University. 74 p.

Hinzer, J. C. Geological and geochemical study of Lyon Lake and Creek ore zones, Sturgeon Lake area, northwestern Ontario. M, 1977, University of Western Ontario. 94 p.

Hipel, K. W. Contemporary Box-Jenkins modelling in hydrology. D, 1975, University of Waterloo.

Hipp, Thomas and McGee, Dean. Stream erosion. M, 1926, University of Kansas.

Hippe, Daniel J. Depositional systems and petrology of the Mesaverde Formation, southeastern Wind River basin, Wyoming. M, 1986, Colorado State University. 232 p.

Hippensteel, David Lee. Diagenesis, facies development, and depositional environment of the lower Tuscaloosa Formation, Thompson Field, Amite County, Mississippi. M, 1988, Northeast Louisiana University. 78 p.

Hipple, Dennis L. A study of the structural state and composition of the feldspars of the Bedford augen gneiss, New York. M, 1969, Miami University (Ohio). 49 p.

Hipple, Karl Walter. Genesis, classification and economics of deep loessial agricultural soils in Latah County, Idaho. D, 1981, University of Idaho. 231 p.

Hipple, Robert. A methodology for considering physiographic characteristics in making land use decisions in Southwest Missouri. M, 1988, Southwest Missouri State University.

Hipsch, G. Sedimentary aspects of uranium deposits. M, 1978, City College (CUNY).

Hipskind, Roderick Stephen. The three-dimensional wind structure around convective elements over a tropical island. M, 1976, University of Virginia. 55 p.

Hirekerur, Laxmikant Rangrao. Clay mineralogy and release and fixation of potassium in some soil series of Minnesota. D, 1963, [University of Minnesota, Minneapolis]. 156 p.

Hirsch, Alfred Martin. Biostratigraphy of the Mancos shale (lower and upper Cretaceous) in western New Mexico and western and central Colorado. D, 1971, Rutgers, The State University, New Brunswick. 175 p.

Hirsch, Jack. The origin of chert in the Tyrone Limestone of central Kentucky. M, 1935, University of Kentucky. 67 p.

Hirsch, Lee Mark. Electrical conductivity of olivine during high-temperature creep. D, 1987, University of California, Berkeley. 153 p.

Hirsch, R. M. A method for evaluating management options for an urbanizing watershed. D, 1977, The Johns Hopkins University. 262 p.

Hirsch, Robert M. Glacial geology and geomorphology of the upper Cedar River watershed, Cascade Range, Washington. M, 1975, University of Washington. 48 p.

Hirsch, Stuart. Depositional environment of Mississippian and Pennsylvanian strata along National Gypsum Company spur track, Martin County, Indiana. M, 1974, Indiana University, Bloomington. 67 p.

Hirschberg, David Jacob. Recent geochemical history of sedimentation in the northern Chesapeake Bay. M, 1979, SUNY at Stony Brook.

Hirschboeck, Katherine Kristin. Hydroclimatology of flow events in the Gila River basin, central and southern Arizona. D, 1985, University of Arizona. 354 p.

Hirschfeld, Sue Ellen. New megalonychid sloths from the Pliocene of Florida and Texas. M, 1965, University of Florida.

Hirschfield, Sue Ellen. Ground sloths and anteaters (Erentata, mammalia) from the Tertiary of Colombia, South America. D, 1971, University of California, Berkeley. 257 p.

Hirschmann, Marc M. Petrology of the transgressive granophyres from the Skaergaard Intrusion, East Greenland. M, 1988, University of Oregon. 144 p.

Hirschmann, Thomas Simon. Reconnaissance geology and stratigraphy of the Subinal Formation (Tertiary) of the El Progreso area, Guatemala. M, 1963, Indiana University, Bloomington. 67 p.

Hirst, Brian. Precambrian geology of the Cottonwood Creek area, Montana and Wyoming. M, 1976, Northern Illinois University. 76 p.

Hirst, Terence John. The influence of compositional factors on the stress-strain-time behavior of soils. D, 1968, University of California, Berkeley. 264 p.

Hirt, David S. Occurrence and biochronology of Middle Mississippian brachiopods of the Ramp Creek Formation and Harrodsburg Limestone, Indiana and Kentucky. M, 1988, Indiana University, Bloomington. 138 p.

Hirt, Warren. Transmission electron microscopy of antiphase boundaries in anorthite. M, 1974, [University of New Mexico].

Hirt, William Carl. Exploration characteristics of native copper mineralization in the Silver Bell Andesite at Las Guijas, Pima County, Arizona. M, 1978, University of Arizona.

Hiscott, Richard Nicholas. Sedimentology and regional implications of deep-water sandstones of the Tourelle Formation, Ordovician, Quebec. D, 1977, McMaster University. 542 p.

Hise, Charles R. Van *see* Van Hise, Charles R.

Hiseler, Robert Bruce. Variables controlling permeability of oolitic calcarenites of the Ste. Genevieve Limestone (Mississippian), southern Illinois. M, 1977, University of Illinois, Urbana.

Hisey, William Murphy. Preliminary faunal study of the Providence Sand. M, 1952, University of Alabama. 284 p.

Hiskey, Robert Marshall. The trophic-dynamics of an alkaline-saline Nebraska Sandhills lake. D, 1981, University of Nebraska, Lincoln. 115 p.

Hiss, W. L. Stratigraphy and ground-water hydrology of the Capitan aquifer, southeastern New Mexico and western Texas. D, 1975, University of Colorado. 501 p.

Hiss, William Louis. Ferromagnesian minerals in basic igneous rocks, Raggedy Mountains area, Wichita Mountains, Oklahoma. M, 1960, University of Oklahoma. 104 p.

Hissenhoven, René van *see* van Hissenhoven, René

Hissenhoven, Rene Van, S. J. Traveltime anomalies of P and S waves. M, 1967, Boston College.

Hitchcock, Charles B. Terraces of the Connecticut Valley. M, 1932, Columbia University, Teachers College.

Hitchcock, Kenneth Brent. Petrographic analysis of the Pennsylvanian Mineral Wells Formation, Strawn Group, north-central Texas. M, 1985, Stephen F. Austin State University. 175 p.

Hitchcock, Margaret R. A petrographic study of certain rocks found in Mongolia at mile 564 on the Urga Trail. M, 1925, Columbia University, Teachers College.

Hite, David Marcel. Carbonates of the Upper Triassic, Whitehorse area, Yukon Territory. M, 1964, University of Wisconsin-Madison.

Hite, David Marcel. Sedimentology of the upper Keweenawan Sequence of northern Wisconsin and adjacent Michigan (Precambrian). D, 1968, University of Wisconsin-Madison. 236 p.

Hite, John B. Investigation of molybdenum in southeastern Maine. M, 1965, University of Arizona.

Hite, Thomas H. The origin of certain clay deposits of Latah County, Idaho. M, 1930, University of Idaho. 25 p.

Hittinger, M. Numerical analysis of toppling failures in jointed rock. D, 1978, University of California, Berkeley. 297 p.

Hitzman, Murray W. Geology of the BT Claim Group SW Brooks Range Alaska. M, 1978, University of Washington. 80 p.

Hitzman, Murray Walter. Geology of the Cosmos Hills and its relationship to the Ruby Creek copper-cobalt deposit. D, 1983, Stanford University. 386 p.

Hively, Roger E. Stratigraphy and petrography of tuffaceous sedimentary rocks of the Duff Formation, Trans-Pecos, Texas. M, 1976, University of Texas, Arlington. 109 p.

Hixon, Robert Louis. Geopressured-geothermal energy investigations in the vicinity of Grand Isle, Lafourche and Jefferson parishes, Louisiana. M, 1979, Louisiana State University.

Hixon, Sumner Best. Facies and petrography of the Cretaceous Buda Limestone of Texas and northern Mexico. M, 1959, University of Texas, Austin.

Hixon, Sumner Best. Petrography of the Middle Devonian Bois Blanc Formation of Michigan and Ontario. D, 1964, University of Michigan. 118 p.

Hixson, Arthur Warren. Analysis of Iowa coals. M, 1915, University of Kansas.

Hixson, Harry C. Geology of the southwest quarter of the Dixonville Quadrangle, Oregon. M, 1965, University of Oregon. 97 p.

Hj-Elias, Mohd Rohani. Development of an innocuous coating to decrease absorption in aggregates used in bituminous pavement. M, 1987, University of Toledo. 196 p.

Hjortenberg, Erik. Microseisms in Alberta. D, 1964, University of Alberta. 124 p.

Hladky, Frank R. Geology of an area north of the narrows of Ross Fork Canyon, northernmost Portneuf Range, Fort Hall Indian Reservation, Bannock and Bingham counties, Idaho. M, 1986, Idaho State University. 110 p.

Hlava, Paul Frank. Unusual lavas from Molokai, Hawaii; alkalic olivine basalts transitional to hawaiites and strontium-rich mugearites. M, 1974, University of New Mexico. 170 p.

Hlavin, William J. A review of the vertebrate fauna of the upper Devonian (Famennian) Cleveland Shale; Arthrodira. M, 1973, Boston University. 155 p.

Hlavin, William J. Biostratigraphy of the late Devonian black shales on the cratonal margin of the Appalachian geosyncline. D, 1976, Boston University. 211 p.

Hluchanek, James Andrew. Radar Investigations of the Hockley salt dome (Harris County, Texas). M, 1973, Texas A&M University.

Hluchy, Michele Marie. Surficial geology and clay mineralogy of four Adirondack lake-watersheds. M, 1984, Dartmouth College. 87 p.

Hluchy, Michele Marie. The chemistry of clay minerals associated with evaporites in New York and Utah. D, 1988, Dartmouth College. 152 p.

Ho, Ching-Oh. Experimental study of plagioclase/liquid and clinopyroxene/liquid distribution coefficients for Sr and Eu in oceanic ridge basalt system. M, 1973, Columbia University. 147 p.

Ho, Diana Yunn. Mechanical analysis of heavy mineral suites on ancient and modern Erie beach sediments (western New York). M, 1967, SUNY at Buffalo.

Ho, Michael Man-Kai. Energy dissipation in a cohesive soil. D, 1964, Northwestern University. 267 p.

Ho, Phyllis Hang-Yin. A study of the convection pattern in a laterally homogeneous, compressible mantle with heat sources and the corresponding temperature profiles in the mantle. M, 1984, Pennsylvania State University, University Park. 141 p.

Ho, Sheng-Zong John. Three-dimensional constitutive modeling of a granular mine tailings material. D, 1987, West Virginia University. 258 p.

Ho, Tong-yun. Protein-nitrogen content in fossil shells as a new stratigraphic and paleoecologic indicator. D, 1964, University of Kansas. 49 p.

Ho, W. K. The distribution of chitin in the water and sediment columns in the Gulf of Mexico and its geochemical significance. M, 1977, Texas A&M University.

Ho-Liu, Phyllis Hang-Yin. I, Attenuation tomography; II, Modeling regional Love waves; Imperial Valley to Pasadena. D, 1988, California Institute of Technology. 166 p.

Ho-Tun, Edwin. Studies on eudialyte and eucolite. M, 1970, Carleton University. 114 p.

Hoadley, Carol R. Paleoecology of encrusting epifauna on echinoids and oysters of the Mid-Cretaceous. M, 1986, Baylor University. 86 p.

Hoadley, John W. Geology of the Zeballos map-area, Vancouver Island, British Columbia. D, 1950, University of Toronto.

Hoadley, John William. The metamorphism of the rocks of the Alderidge Formation, Kimberley, BC. M, 1947, University of British Columbia.

Hoag, Barbara Lillian. Channel erosion as a source of sediment in Newport Bay, California. M, 1983, University of California, Los Angeles. 76 p.

Hoag, R. B., Jr. Hydrogeochemistry of springs near the Eustis Mine, Quebec. D, 1975, McGill University.

Hoag, Robert Eugene. Characterization of soils on floodplains of tributaries flowing into the Amazon River in Peru. D, 1987, North Carolina State University. 185 p.

Hoag, Robert W., II. Estimation of the original shear strength of deep sediments from engineering index properties. M, 1970, United States Naval Academy.

Hoag, William Myrl. Porosity and permeability of various Paleozoic sediments in Missouri. M, 1957, University of Missouri, Columbia.

Hoagland, Alan D. and Harris, David V. Tertiary igneous rocks of South Park, Colorado. M, 1935, Northwestern University.

Hoagland, James A. An evaluation of the Marada salines deposit, Marada, Libya. M, 1968, University of Utah. 102 p.

Hoagland, James R. Petrology and geochemistry of hydrothermal alteration in borehole Mesa 6-2, East Mesa geothermal area, Imperial Valley, California. M, 1976, University of California, Riverside. 90 p.

Hoagland, Matthew. Hydrogeology and contamination investigation of the west branch of the Westport River watershed. M, 1988, Boston University. 129 p.

Hoaglund, John Robert, III. Origin and kinematic significance of linear structures in upper Proterozoic rocks of Chamberlindalen, Wedel Jarlsberg Land, Southwest Spitsbergen. M, 1987, University of Wisconsin-Madison. 169 p.

Hoak, Thomas E. Structural analysis across the northeast boundary of the Taconic Allochthon, west-central Vermont. M, 1987, SUNY at Albany. 236 p.

Hoang, Viet Thai. Estimation of in-situ thermal conductivities from temperature gradient measurements. D, 1980, University of California, Berkeley. 162 p.

Hoar, Richard James. Field measurement of seismic wave velocity and attenuation for dynamic analyses. D, 1982, University of Texas, Austin. 524 p.

Hoar, S. L. The structure and composition of basal ice at Camp Century, Greenland. M, 1977, SUNY at Buffalo. 56 p.

Hoare, Brian Stuart. Aeromagnetic interpretation and modeling of the Boulder Batholith, Montana. M, 1985, University of Wyoming. 90 p.

Hoare, Richard David. Desmoinesian Brachiopoda and Mollusca from Southwest Missouri. D, 1957, University of Missouri, Columbia. 409 p.

Hoare, Richard David. The ostracode genus Metacypris in North America. M, 1953, University of Missouri, Columbia.

Hoare, Thomas Bertram. The mirogastropods of the Putnam Hill Shale (Allegheny Group) in southeastern Ohio. M, 1973, Bowling Green State University. 61 p.

Hobart, Henry M., Jr. A subsurface study of the Simpson Group in Osage County, Oklahoma. M, 1958, University of Tulsa. 36 p.

Hobbet, Anna. The diabase of the western portion of the Mount Holyoke Range (Massachusetts). M, 1923, Smith College. 55 p.

Hobbet, Randall Douglas. Middle Frasnian age conodonts from the Wadleigh Limestone of southeastern Alaska. M, 1980, University of Oregon. 76 p.

Hobbs, Billy B. Structure and stratigraphy of the Argenta area, Beaverhead County, Montana. M, 1967, Oregon State University. 164 p.

Hobbs, Carl H., III. Sedimentary environments and coastal dynamics of a segment of the shoreline of Cape Cod Bay, Massachusetts. M, 1972, University of Massachusetts. 154 p.

Hobbs, Charles Roderick Bruce. Petrography and origin of dolomite-bearing carbonate rocks of Ordovician age in Virginia. D, 1957, Virginia Polytechnic Institute and State University.

Hobbs, Charles Roderick Bruce. Structural geology of the Sinking Creek area, Giles County, Virginia. M, 1953, Virginia Polytechnic Institute and State University.

Hobbs, Howard. Heavy minerals of glacial sediments in the area of Red Lake Falls, Minnesota. M, 1973, University of North Dakota. 45 p.

Hobbs, Howard C. Glacial stratigraphy of northeastern North Dakota. D, 1975, University of North Dakota. 42 p.

Hobbs, Robert Adelbert. The application of roundness and sphericity measurements to subsurface samples of the Marshall Formation of western Michigan. M, 1949, Michigan State University. 36 p.

Hobbs, Samuel W. Geology of the northern part of the Osgood Mountains, Humboldt County, Nevada. D, 1948, Yale University.

Hobbs, Susan Smith. Stratigraphy and micropaleontology of three wells in Bay and Franklin counties, Florida. M, 1939, Ohio State University.

Hobbs, Thomas M. C. Geology of the Square Peak volcanic series, northern Quitman Mountains, Hudspeth County, Texas. M, 1979, University of Texas at El Paso.

Hobbs, William H. The geology of the City Creek area, Gila County, Arizona. M, 1982, Northern Arizona University. 94 p.

Hobbs, William Henry. On the rocks occurring in the neighborhood of Ilchester, Howard County, MD. D, 1888, The Johns Hopkins University.

Hobday, David K. Upper Carboniferous lithofacies in north Alabama. D, 1969, Louisiana State University.

Hobert, Linda A. The validity of the relationship between plate tectonics and sand composition. M, 1984, Duke University. 104 p.

Hoblitt, Richard Patrick. Emplacement mechanisms of unsorted and unstratified deposits of volcanic rock debris as determined from paleomagnetically-derived emplacement-temperature information. D, 1978, University of Colorado. 206 p.

Hoblitzell, Timothy A. Joints along Interstate highway I-40, Smith and Wilson counties, Tennessee. M, 1970, Vanderbilt University.

Hobsen, Louis Arthur. The seasonal and vertical distribution of suspended particulate matter in an area of the Northeast Pacific Ocean. D, 1966, University of Washington. 107 p.

Hobson, John Peter. Stratigraphy of the northern belt of the Beekmantown Group in southeastern Pennsylvania. D, 1958, Pennsylvania State University, University Park. 510 p.

Hobson, Mary Michael. A preliminary geochemical study of Lake Maracaibo, Venezuela. D, 1979, University of Tulsa. 201 p.

Hobson, Richard D. A study of some phosphate-rich beach sands in Baja California. M, 1964, Northwestern University.

Hobson, Richard David. An environmental study of the Vaqueros Formation, Santa Ynez Mountains, California. D, 1966, Northwestern University. 158 p.

Hobson, Winston Edward. Inorganic phosphorus fractions and humic acid carbon to fulvic acid carbon ratios as differentiae for selected Alfisols and Mollisols. D, 1983, Iowa State University of Science and Technology. 385 p.

Hochella, M. F., Jr. High temperature crystal chemistry of hydrous Mg- and Fe-rich cordierites. M, 1977, Virginia Polytechnic Institute and State University.

Hochella, Michael Frederick, Jr. Aspects of structure and rheology of aluminosilicate melts. D, 1981, Stanford University. 215 p.

Hochstetler, Laurel Huggins. Stratigraphy and distribution of the Colma Formation, San Francisco, California. M, 1978, University of California, Berkeley. 110 p.

Hock, Philip F., Jr. Effect of the Pedernal axis (east-central New Mexico) on Permian and Triassic sedimentation. M, 1970, University of New Mexico. 51 p.

Hockens, Sidney N. Lithofacies study of Upper Pennsylvanian and Lower Permian rocks in northwestern Kansas and adjacent areas. M, 1959, University of Kansas.

Hockensmith, Brenda Louise. Hydrogeological application of the kriging technique to water quality data at the United Nuclear Corporation, N.E. Church Rock uranium mine tailings disposal site, New Mexico. M, 1985, Wright State University. 126 p.

Hockley, Glenn D. Stratigraphy and paleoenvironmental patterns of the Peyto-Mount Whyte sediments (lower-middle Cambrian) of the southwestern Canadian Rocky Mountains. M, 1973, University of Calgary. 117 p.

Hocq, Michel. Etudes minéralogiques et pétrologiques de la région Pipmuacan (Québec). D, 1971, Universite de Montreal.

Hocq, Michel. Le Précambrien de la Province de Grenville dans la région de Saint-Paulin (comtés de Maskinongé et de Saint-Maurice, P.Q., Canada). M, 1969, Universite de Montreal.

Hoda, Badrul. Feasibility of subsurface waste disposal in the Newcastle Formation, lower Dakota Group (Cretaceous) and Minnelusa Formation (Penn.), western North Dakota. M, 1977, Wayne State University.

Hoda, Syed Nurul. A study of the artificial alteration of trioctahedral mica to vermiculite. M, 1970, Southern Illinois University, Carbondale. 88 p.

Hoda, Syed Nurul. A thermal and compositional study of mineral assemblages in the system PbS-Ag$_2$S-Cu$_2$S- Sb$_2$S$_3$Bi$_2$S$_3$. D, 1973, Miami University (Ohio). 199 p.

Hodapp, Aloys Philip. The origin and accumulation of petroleum. M, 1914, University of Minnesota, Minneapolis. 64 p.

Hodder, Edwin J. Washability studies of north Thompson Creek coals. M, 1956, Colorado School of Mines. 83 p.

Hodder, Robert William. Alkaline rocks and niobium deposits near Nemagos, Ontario. D, 1959, University of California, Berkeley. 141 p.

Hodell, David Arnold. Major events in Neogene paleoceanography; faunal and isotopic evidence. D, 1986, University of Rhode Island. 319 p.

Hodgden, H. Jerry. The geology around the junction of the Roaring Fork and Frying Pan rivers, Eagle and Pitkin counties, Colorado. M, 1960, University of Kansas.

Hodge, Cara Jyl. Structure, petrology and geochemistry of the Aultman kimberlite diatremes, Albany Co., Wyo. M, 1983, Colorado State University. 155 p.

Hodge, Dennis S. Petrology and structural geometry of Precambrian rocks in the Bluegrass area, Albany County, Wyoming. D, 1967, University of Wyoming. 135 p.

Hodge, Dennis S. Polymetamorphism of Precambrian rocks in the southwestern Wind River Mountains, Fremont County, Wyoming. M, 1963, University of Wyoming. 49 p.

Hodge, Edwin T. The composition of waters in mines of sulphide ores. M, 1914, University of Minnesota, Minneapolis. 25 p.

Hodge, Edwin T. The geology of the Coamo-Gyayame District, Puerto Rico. D, 1917, Columbia University, Teachers College.

Hodge, John A. Erosion of the North Santee River delta and development of a flood-tidal complex. M, 1981, University of South Carolina. 151 p.

Hodge, Robert A. Regional geology, groundwater flow systems and slope stability. M, 1976, University of British Columbia.

Hodge, Steven McNiven. The movement and basal sliding of the Nisqually Glacier, Mount Rainier. D, 1972, University of Washington. 410 p.

Hodges, A. B. W. Report on Atlantic Mining Co., Houghton, Houghton Co., Mich. M, 1884, Washington University.

Hodges, Carroll Ann. Comparative study of S.P. and Sunset craters and associated lava flows, San Francisco volcanic field, Arizona. M, 1960, University of Wisconsin-Madison.

Hodges, Carroll Ann. Geomorphic history of Clear Lake, California. D, 1966, Stanford University. 222 p.

Hodges, D. J. Geology and geochemistry of the Croinor gold deposit, Pershing Township, Quebec. M, 1987, University of Waterloo. 218 p.

Hodges, David Alvin. Reconnaissance borings in Quaternary deposits of "Lake Willard", Huron County, Ohio. M, 1979, University of Akron. 47 p.

Hodges, Floyd N. Petrology, chemistry and phase relations of the Sierra Prieta nepheline-analcime syenite intrusion, Diablo Plateau, Trans-Pecos Texas. D, 1975, University of Texas, Austin. 209 p.

Hodges, Kip Vernon. Tectonic evolution of the Aefjord-Sitas area, Norway-Sweden. D, 1982, Massachusetts Institute of Technology. 192 p.

Hodges, L. T. Megafossil orientation of selected Silurian, Devonian, and Pleistocene reefs of North America. D, 1977, Loma Linda University. 142 p.

Hodges, P. A. Magnetometric prospecting and its application to the solution of faults. M, 1927, Massachusetts Institute of Technology. 65 p.

Hodges, Rex Alan. A petrologic study of the Lithium Corporation of America Mine in the tin-spodumene belt of North Carolina. M, 1983, University of North Carolina, Chapel Hill. 92 p.

Hodges, Steven Clarke. Interaction of aluminum species with natural exchangers and resins. D, 1980, Virginia Polytechnic Institute and State University. 96 p.

Hodges, Wade Allan. Igneous petrology, structural geology and mineralization of the central part of the Bayhorse mining district, Custer County, Idaho. M, 1978, Oregon State University. 310 p.

Hodges, William Kaufman. Experimental study of hydrogeomorphological processes in Dinosaur Badlands, Alberta, Canada. D, 1985, University of Toronto.

Hodgins, D. O. Salinity intrusion in the Fraser River, British Columbia. D, 1974, University of British Columbia.

Hodgins, Larry E. Morphology of the South Saskatchewan river valley, Outlook to Saskatoon (Canada). D, 1970, McGill University.

Hodgkins, Catherine E. Geochemistry and petrology of the Dry Hill Gneiss and related gneisses, Pelham Dome, central Massachusetts. M, 1985, University of Massachusetts. 174 p.

Hodgkinson, John. In situ determination of velocity anisotropy in sedimentary rocks by analysis of normal moveout data. M, 1970, University of Calgary. 81 p.

Hodgkinson, John Morris. The geology of the Kenora Airport area, Ontario (Canada). M, 1968, University of Manitoba.

Hodgkinson, Kenneth A. Permian stratigraphy of northeastern Nevada and northwestern Utah. M, 1961, Brigham Young University. 197 p.

Hodgkinson, Kenneth Allred. The late Paleozoic ammonoid families, Prolecanitidae and Daraelitidae. D, 1965, University of Iowa. 232 p.

Hodgkinson, R. J. On the craters of Mare Tyrrhenum (MC-22), Mars. M, 1974, University of Houston.

Hodgman, Robert F. The reduction of sulphates to sulphide ores. M, 1914, University of Minnesota, Minneapolis. 30 p.

Hodgson, Alexander Goldie. The geology of the Indin "Break" NWT. M, 1948, University of British Columbia.

Hodgson, Charles Jonathan. The mineralogy and structure of the New Broken Hill Consolidated Limited Mine, Broken Hill, New South Wales (Australia); 1 volume. D, 1968, University of California, Berkeley. 240 p.

Hodgson, Christopher J. Petrology of Monteregian dyke rocks (early Cretaceous), (southern Quebec, Canada). D, 1969, McGill University.

Hodgson, Christopher J. The Watten-Halkirk copper prospect, Rainy Lake area, western Ontario. M, 1962, McGill University.

Hodgson, Edward Askew. Petrogenesis of the lower Devonian Oriskany sandstone and its correlates in New York, with a note on their acritarchs. D, 1970, Cornell University.

Hodgson, Ernest A. A seismometric study of the Tango earthquake, Japan, March 7, 1927. D, 1932, St. Louis University.

Hodgson, Geffrey David. Structural studies of the gold-mineralized shear zone at Giant Mine, Yellowknife, Northwest Territories. M, 1976, University of Alberta. 128 p.

Hodgson, John H. A seismic survey in the Canadian Shield. D, 1951, University of Toronto.

Hodgson, Philip Richard. Application of micro-computer processing of shallow seismic reflection data. M, 1984, Purdue University. 161 p.

Hodgson, Robert A. Geology of the Wasatch Mountain front in the vicinity of Spanish Fork Canyon, Utah County, Utah. M, 1951, Brigham Young University. 60 p.

Hodgson, Robert Arnold. A regional study of jointing in the Comb Ridge-Navajo Mountain area, Arizona and Utah. M, 1959, Yale University.

Hodgson, Russell Beales. Reconnaissance of the Crescent Eagle structure and its connection with the Salt Valley Anticline, (Grand County, Utah). M, 1932, University of Utah. 44 p.

Hodgson, Walter Dale. An environmental study of some sands, silts, clays, and nodular limestone beds in Hood, Parker, and Jack counties, Texas. M, 1957, Texas Christian University. 163 p.

Hodison, Starlyn T. Ground water pollution in North Topeka, Kansas. M, 1975, Emporia State University. 53 p.

Hodler, Thomas W. Remote sensing applications in hydro-geothermal exploration of the northern Basin and Range Province. D, 1978, Oregon State University. 220 p.

Hodson, Floyd. The origin of bedded Pennsylvanian fire clays in the United States. M, 1922, Cornell University.

Hodson, Floyd. Venezuelan and Caribbean turritellas, with a list of Venezuelan type stratigraphic formations. D, 1926, Cornell University.

Hodson, Robert E. On the role of bacteria in the cycling of dissolved organic matter in the sea. D, 1977, University of California, San Diego. 171 p.

Hodson, Warren G. Geology and ground-water resources of Solomon River valley in Mitchell County, Kansas. M, 1956, University of Kansas. 172 p.

Hody, Harold Martin. Bouguer gravity anomalies of southeastern Michigan and the vertical gradient of gravity at Ann Arbor, Michigan. M, 1955, University of Michigan.

Hodych, Joseph Paul. The effect of axial compression on magnetization induced by weak fields in rock of low coercive force. D, 1971, University of Toronto.

Hoeffner, Steven Lewis. Laboratory simulation of radionuclide migration through basalt; applications to the proposed radioactive waste repository near Richland, Washington. D, 1983, University of Missouri, Columbia. 245 p.

Hoeft, David Ralph. The litho-stratigraphy of the Glenwood and Platteville formations of southeastern Minnesota. M, 1959, University of Minnesota, Minneapolis. 306 p.

Hoehl, Eberhard J. Geophysical investigation of the Limestone Mountain area of Baraga and Houghton counties, Michigan. M, 1981, Michigan Technological University. 141 p.

Hoeksema, Robert James. The geostatistical approach to the inverse problem in two-dimensional steady state groundwater modeling. D, 1984, University of Iowa. 152 p.

Hoekstra, Karl E. Cation distribution in some common (2-3) oxide spinels as a function of temperature. M, 1965, Miami University (Ohio). 70 p.

Hoekstra, Karl E. Crystal chemical considerations in the spinel structural group. D, 1969, Ohio State University.

Hoekzema, Robert B. The petrographic and structural characteristics of the Carville Basin Ayer pluton (Carboniferous?, Clinton, Massachusetts). M, 1971, Boston University. 73 p.

Hoel, Holly D. Goliad Formation of the South Texas Gulf Coastal Plain; regional genetic stratigraphy and uranium mineralization. M, 1982, University of Texas, Austin.

Hoelle, John Lowell. Structural and geochemical analysis of the Catalina Granite, Santa Catalina Mountains, Arizona. M, 1976, University of Arizona.

Hoen, E. W. The anhydrite diapirs and structure of central western Axel Heiberg Island, Canadian Arctic Archipelago. D, 1963, McGill University.

Hoen, Ernest L. W. B. The geology of Hornby Island. M, 1958, University of British Columbia.

Hoenig, Margaret A. Stratigraphy of the "Upper Silurian" and "Lower Devonian", Permian Basin, West Texas. M, 1976, University of Texas at El Paso.

Hoenstine, Ronald W., Jr. A sedimentologic study of Anclote anchorage with emphasis on littoral current

velocities, heavy mineral concentrations, and size analysis. M, 1974, University of Florida. 86 p.

Hoenstine, Ronald Woodrow. Biostratigraphy of the Hawthorn Formation in northeast and north central Florida. D, 1982, Florida State University. 268 p.

Hoerger, Steven Fred. Probabilistic and deterministic keyblock analyses for excavation design. D, 1988, Michigan Technological University. 276 p.

Hoerger, Steven Fred. Spatial analysis of rock joint orientations by geostatistical methods. M, 1985, Michigan Technological University. 89 p.

Hoernemann, M. J. Field observations of wave runup on a sand beach. M, 1967, United States Naval Academy.

Hoernle, Kaj. General geology and petrology of the Roque Nublo Volcanics on Gran Canaria, Canary Islands, Spain. M, 1987, University of California, Santa Barbara. 191 p.

Hoersch, A. L. Zoned calc-silicate nodules from the contact aureole of the Beinn an Dubhaich Granite, Isle of Skye, Scotland; a re-examination. D, 1978, The Johns Hopkins University. 209 p.

Hoerster, Mary Lou Cole see Cole Hoerster, Mary Lou

Hoeven, G. A. Van der see Van der Hoeven, G. A.

Hoeven, William Van see Van Hoeven, William, Jr.

Hoexter, D. F. Engineering geology and relative stability of part of Ladera, San Mateo County, California. M, 1975, Stanford University. 78 p.

Hoff, Jean Louise. The petrography of the Blowing Rock Gneiss, Grandfather Mountain Window, North Carolina. M, 1982, Duke University. 196 p.

Hoff, Jerald Herbert. Microfauna of the Oacoma facies of the Pierre Formation. M, 1960, University of South Dakota. 97 p.

Hoff, John Anderson. Strontium isotope and trace element variations through the upper marble, Balmat, New York. M, 1984, Miami University (Ohio). 130 p.

Hoffacker, Benjamin Franklin, Jr. The western extent of the Catskill red beds, facies in western Pennsylvania. M, 1948, University of Pittsburgh.

Hoffer, A. A comparative study of textures in orthoquartzites. M, 1957, University of Toronto.

Hoffer, Abraham. Low quartz and the geometry of its structure framework in terms of the directed bond. D, 1959, University of Chicago. 27 p.

Hoffer, Jerry Martin. Plagioclase variations in a porphyritic flow of the Columbia River Basalt. D, 1965, Washington State University. 127 p.

Hoffer, Jerry Martin. The geology and petrography of Deadman Mountain, South Dakota. M, 1958, University of Iowa. 80 p.

Hoffer, Roberta Lynne. Uranium geochemistry of selected rock units from the Marysvale volcanic field, Piute County, Utah. D, 1982, University of Texas at El Paso. 283 p.

Hoffer, Robin L. Contact metamorphism of the Precambrian Castner Marble, Franklin Mountains, El Paso County, Texas. M, 1976, University of Texas at El Paso.

Hoffman, Arnold Daniel. Miocene fossil plants from Metijoque, Venezuela. M, 1931, University of Chicago. 52 p.

Hoffman, Barry L. P. Geology of Bernard Mt. area, Tonsina, Alaska. M, 1974, University of Alaska, Fairbanks. 68 p.

Hoffman, Bradley C. Frequency domain synthetic seismograms; attenuation modeling. M, 1983, Indiana University, Bloomington. 194 p.

Hoffman, Bruce Frederick. A trace element distribution of the Mississippian Fort Payne Chert of Northeast Mississippi. M, 1978, University of Mississippi. 55 p.

Hoffman, Carlton Scott. A study of beach dynamics under varying wave environments at Cinnamon' Trunk, and Hawksnest bays, St. John, United States Virgin Islands. M, 1974, University of Virginia. 52 p.

Hoffman, Carlyle. A geological study of the Cobleskill Formation of western New York. M, 1949, SUNY at Buffalo.

Hoffman, Charles William. A stratigraphic and geochemical investigation of ferruginous bauxite deposits in the Salem Hills, Marion County, Oregon. M, 1981, Portland State University. 130 p.

Hoffman, D. Frederick. Paleogeographic implications of the south-west corner of Santa Cruz Island, California. M, 1979, University of California, Santa Barbara.

Hoffman, D. R. Petrogenesis of the Lawson Peak orbicular gabbro. M, 1975, San Diego State University.

Hoffman, Dale Sheridan. Tertiary vertebrate paleontology and paleoecology of a portion of the lower Beaverhead River basin, Madison and Beaverhead counties, Montana. D, 1972, University of Montana. 174 p.

Hoffman, David Gordon. Gravity study of northern Virgin Valley salt deposits, Clark County, Nevada. M, 1978, University of Washington. 63 p.

Hoffman, Douglas Weir. Crop yields of soil capability classes and their uses in planning for agriculture. D, 1973, University of Waterloo.

Hoffman, E. L. Phase relations of michenerite and merenskyite in the Pd-Bi-Te system; structural refinement of chalcanthite and a description of a new Pb-Sn-S mineral. M, 1975, McGill University. 92 p.

Hoffman, E. L. The platinum group element and gold content of some nickel sulphide ores. D, 1978, University of Toronto.

Hoffman, Edward Arthur, Jr. Pre-Chester Mississippian rocks of northwestern Oklahoma. M, 1964, University of Oklahoma. 26 p.

Hoffman, Emily J. Microbial endoliths from selected Cambrian substrates. M, 1981, Boston University. 69 p.

Hoffman, Floyd H. Geology of the Mosida Hills area, Utah County, Utah. M, 1951, Brigham Young University. 68 p.

Hoffman, Frank Owen, Jr. Environmental behavior of technetium in soil and vegetation; implications for radiological impact assessment. D, 1981, University of Tennessee, Knoxville. 103 p.

Hoffman, Frederic. The geology of the Black Mountain area, Mineral County, Nevada. M, 1969, University of Nevada. 54 p.

Hoffman, George Albert. The correlation by petrographic methods of dikes designated as numbers five and six in the Newport Mine, Ironwood, Michigan. M, 1950, Michigan State University. 52 p.

Hoffman, George L. Unconfined aquifer pumping tests. M, 1974, New Mexico Institute of Mining and Technology.

Hoffman, J. C. An evaluation of potassium uptake by Mississippi-River-borne clays following deposition in the Gulf of Mexico. D, 1979, Case Western Reserve University. 169 p.

Hoffman, James H. Stratigraphy and sedimentology of the Bliss Sandstone (Cambro-Ordovician), southern Franklin Mountains, Texas. M, 1976, Louisiana State University.

Hoffman, James Irvie, III. The electrochemical deposition of manganese from an aqueous solution. M, 1965, Michigan State University. 56 p.

Hoffman, James Irvie, III. The nature of calcium-organic complexes in natural solutions and its implications on mineral equilibria. D, 1969, Michigan State University. 103 p.

Hoffman, Janet L. Low-grade regional metamorphism of Cretaceous sedimentary units in the disturbed belt area of Montana. D, 1976, Case Western Reserve University. 266 p.

Hoffman, John Paul. Discontinuities in the Earth's upper mantle as indicated by reflected seismic energy. M, 1960, University of Utah. 56 p.

Hoffman, John W. Paleoecology of part of the upper Devonian of south-central New York. M, 1969, University of Rochester. 38 p.

Hoffman, Jonathan D. Water use, groundwater conditions, and slope failure on benchlands in western Montana; the Darby Slide example. M, 1980, University of Montana. 48 p.

Hoffman, Karen Sue. A reevaluation of the orthopyroxene isograd, Northwest Adirondack Mountains, New York. M, 1982, University of Michigan.

Hoffman, Kenneth Alan. Cation ordering, unmixing, and reverse thermoremanent magnetization in the ilmenite-hematite solid solution series. D, 1973, University of California, Berkeley. 175 p.

Hoffman, Lee Ellis. The effect of Thiobacillus ferrooxidans on argentite, stibnite, and a stibnite-silver ore. M, 1975, University of Nevada. 81 p.

Hoffman, Mark A. A study of some petrologic and structural aspects of the East Dover ultramafic bodies, South central Vermont. M, 1975, SUNY at Albany. 120 p.

Hoffman, Mark Allen. The Southern Complex; geology, geochemistry, mineralogy and mineral chemistry of selected uranium- and thorium-rich granites. D, 1987, Michigan Technological University. 541 p.

Hoffman, Mary Frances. A study of the feasibility of the punch-shear test for the determination of the shear strength of rock materials. M, 1987, Michigan Technological University. 149 p.

Hoffman, Melvin G. The geology and petrology of the Wichita Mountains (Oklahoma). D, 1930, University of Chicago. 184 p.

Hoffman, Monty E. Origin and mineralization of breccia pipes, Grand Canyon District, Arizona. M, 1977, University of Wyoming. 50 p.

Hoffman, Olive L. The Chouteau Limestone in central Missouri. M, 1927, University of Kansas.

Hoffman, Paul F. Stratigraphy and depositional history of a Proterozoic geosyncline, east arm of Great Slave lake, Northwest Territories, Canada. D, 1970, The Johns Hopkins University.

Hoffman, R. N. and Roberts, Clay. The occurrence and metallurgy of copper. M, 1913, University of Kansas.

Hoffman, Robert A. Foraminifera from the rocky intertidal zone of Barbados, West Indies. M, 1981, Bowling Green State University. 93 p.

Hoffman, S. A. Provenance, depositional environment and progressive diagenetic changes of sands from the West Mexican coastal plain. M, 1975, SUNY at Buffalo. 78 p.

Hoffman, Sarah Elizabeth. Alteration mineralogy and geochemistry of the Archaean Onverwacht Group, Barberton Mountain Land, South Africa. M, 1985, Oregon State University. 386 p.

Hoffman, Stanley Joel. Geochemical dispersion in bedrock and glacial overburden around a copper property in south central British Columbia. M, 1973, University of British Columbia.

Hoffman, Stanley Joel. Mineral exploration of the Nechako Plateau, central British Columbia, using lake sediment. D, 1976, University of British Columbia.

Hoffman, Thomas Frank. Origin and stratigraphic significance of pink layers in late Cenozoic sediments of the Arctic Ocean. M, 1972, University of Wisconsin-Madison. 74 p.

Hoffman, Victor J. Heavy mineral distribution in sands of the Tortolita Mountain pediment, southern Arizona. M, 1963, University of Arizona.

Hoffman, Victor Joseph. The mineralogy of the Mapimi mining district, Durango, Mexico. D, 1968, University of Arizona. 266 p.

Hoffmann, H. J. The (Ordovician) Chazy Group in the Saint Lawrence Lowlands (Quebec). D, 1962, McGill University.

Hoffmann, H. J. The occurrence and petrology of basic intrusion in the northern Mackenzie Mountains, Yukon and Northwest Territories. M, 1959, McGill University.

Hoffmann, John P. A statistical and geodetic approach to monitoring the formation and evolution of the shal-

low plumbing system at Puu Oo, Kilauea, Hawaii. M, 1988, University of Hawaii. 133 p.

Hoffmann, Joseph Walter. A thermal instability mechanism for glacier surges. M, 1972, University of British Columbia.

Hoffmeister, J. E. Some corals from the American Samoa and the Fiji Islands. D, 1923, The Johns Hopkins University.

Hoffmeister, W. S. The molluscan fauna of the Wilcox Group of Alabama. D, 1926, The Johns Hopkins University.

Hoffschwelle, John William. On the resistivity and moisture content of soil. M, 1952, St. Louis University.

Hofler, Vivian Estelle. A geological field trip in the Petersburg, Virginia and surrounding areas. M, 1974, Virginia State University. 35 p.

Hofmann, Albrecht Werner. Hydrothermal experiments on equilibrium partitioning and diffusion kinetics of Rb, Sr, and Na in biotite-alkali chloride solution systems. D, 1969, Brown University.

Hofmann, Albrecht Werner. Potassium-argon data on the detrital and postdepositional history of Pennsylvanian underclays. M, 1965, Brown University.

Hofmann, Douglas A. Synthesis and characterization of boehmite. M, 1978, Ohio State University.

Hofmann, Peter Michael. Dolomitization of the Hatch Hill arenites and the Burden iron ore. M, 1986, SUNY at Albany. 188 p.

Hofmann, Renner Bergene. Contribution of subsoil to microseismic interference at Florissant, Missouri. M, 1955, St. Louis University.

Hofmeister, Anne Marie. Spectroscopic and chemical study of the coloration of feldspars by irradiation and impurities, including water. D, 1984, California Institute of Technology. 425 p.

Hofstetter, Abraham. Effects of isostasy on large-scale geoid signal. D, 1986, University of Washington. 162 p.

Hofstetter, Abraham. Observations of volcanic tremor at Mount Saint Helens in April and May 1980. M, 1984, University of Washington. 87 p.

Hofstetter, Oscar Bernard, III. Graptolites and associated fauna of the Lebanon Limestone (Ordovician), central Tennessee. M, 1966, University of Tennessee, Knoxville. 70 p.

Hofstra, Albert H. Geology, alteration and genesis of the Ng alunite area, southern Wah Wah Range, Southwest Utah. M, 1984, Colorado School of Mines. 130 p.

Hofstra, Warren Elwin. The effects of washing oil well cuttings from rotary and cable tool wells as related to sample losses. M, 1949, Michigan State University. 84 p.

Hogan, Colleen A. Distribution of sorbed heavy metals in sediments from the rivers and estuaries of the Plum Island area, Massachusetts. M, 1977, University of Rhode Island.

Hogan, Gregory G. Dielectric properties of firn from Dye 3, Greenland. M, 1983, University of Wisconsin-Madison. 105 p.

Hogan, Howard R. The geology of the Nipissis River and Inpisso Lake map-areas, Quebec. D, 1953, McGill University.

Hogan, Howard R. The Mina Lake graywacke, Sawyer Lake, Labrador. M, 1950, McGill University.

Hogan, John P. Petrology of the Northport Pluton, Maine; a garnet bearing muscovite biotite granite. M, 1984, Virginia Polytechnic Institute and State University. 211 p.

Hogan, Patrick Joseph. A detailed geophysical and geological investigation at the intersection of the Black Mountain, Wingate Wash, and southern Death Valley fault zones, Death Valley, California. M, 1987, University of New Orleans. 101 p.

Hoganson, John W. Late Quaternary environmental and climatic history of the southern Chilean Lake region interpreted from coleopteran (beetle) assemblages. D, 1985, University of North Dakota. 377 p.

Hoganson, John William. The Spirulaea Vernoni zone of the Crystal River Formation (Upper Eocene) in Peninsular Florida. M, 1972, University of Florida. 70 p.

Hogarth, Craig G. Conodont color alteration, organic metamorphism, and thermal history of the "Trenton Formation," Michigan Basin. M, 1985, Michigan State University. 60 p.

Hogarth, D. D. A study of certain uranium bearing minerals. M, 1955, University of Toronto.

Hogarth, Donald David. A mineralogical study of pyrochlore and betafite. D, 1959, McGill University.

Hogarty, Barry Jane. The fate of molybdenum and other heavy metals in two Southeast Alaska fjords. M, 1985, University of Alaska, Fairbanks. 151 p.

Hogberg, Rudolph K. Mineralogy and petrography of iron formation at Lake Albanel, Quebec, Canada. M, 1957, Michigan State University. 46 p.

Hoge, Harry Porter. Neogene stratigraphy of the Santa Ana area, Sandoval County, New Mexico. D, 1970, University of New Mexico. 140 p.

Hoge, Harry Porter. The petrology of the Upper Grafton Sandstone in Athens County and vicinity. M, 1963, Ohio University, Athens. 79 p.

Hogenson, Glenmore Melvin. Geology of the Umatilla River basin area. M, 1956, Oregon State University. 56 p.

Hogg, Andrew Jenner Cowper. The petrology and geochemistry of the Prinsen Af Wales Bjerge, East Greenland. M, 1985, University of Toronto.

Hogg, G. M. The geology of the Decoeur-Garon Property, and its relationship to the Waite Akmulet area. M, 1952, Queen's University.

Hogg, J. E. An investigation of the possibility of natural variations in the abundance of the titanium isotopes. M, 1954, University of Toronto.

Hogg, J. E. Applications of counting techniques to the study of geologic ages. D, 1956, University of Toronto.

Hogg, Nelson. The geocolloid chemistry of mercury. M, 1940, Massachusetts Institute of Technology. 38 p.

Hogg, Norman Carroll. Shoshonitic lavas (Cenozoic) in west-central Utah. M, 1972, Brigham Young University. 184 p.

Hogg, W. A. Pleistocene deposits of Pictou County (Nova Scotia, Canada). M, 1953, Dalhousie University.

Hogg, William A. Building and industrial stones of eastern Canada. D, 1959, McGill University.

Hogg, William A. Geology of Red Island, Placentia Bay, Newfoundland. M, 1954, Dalhousie University.

Hogg, William Andrew. Mid-Devonian martiniid brachiopods from northern Canada. M, 1965, University of Saskatchewan. 149 p.

Hoggan, Roger D. Paleoecology of the Curtis Formation (upper Jurassic) in the Uinta mountains area, Daggett County, Utah. M, 1970, Brigham Young University. 65 p.

Hoggan, Roger D. Paleoecology of the Devonian Guilmette Formation in western Utah and east central Nevada. D, 1971, Brigham Young University.

Hoggatt, Leslie J. Residence time determination and classification of seven Indiana springs using tritium and chemical analyses. M, 1984, Indiana State University. 69 p.

Hogge, Curt Edward. Geology of the Paleozoic black shale sequence, Moberg Hill, Stevens County, Washington. M, 1982, Washington State University. 80 p.

Hogler, Jennifer Alice. Paleontology and stratigraphy of the middle Pleistocene, central Arctic Ocean. M, 1985, University of Wisconsin-Madison.

Hogue, John. Depositional and post-depositional study of Edwards Limestone (Lower Cretaceous) of cored interval, South Texas Syndicate Well II-5, Washburn Ranch Field, Lasalle County, Texas. M, 1985, University of Arkansas, Fayetteville.

Hogue, William C. Geology of the northern part of the Slate Mountains, Pinal County, Arizona. M, 1940, University of Arizona.

Hoheisel, Charles Richard. Geology of the Lower Pass Creek area (Wyoming). M, 1955, University of Michigan.

Hohenstein, C. G.; Jaeger, J. W. and Jones, D. L. A study of marked sand movement on Del Monte beach, Monterey bay, California. M, 1965, United States Naval Academy.

Hohl, Arthur Henry. Periglacial features and related surficial deposits (Quaternary) of Bull Creek basin, Henry Mountains, Utah. D, 1972, The Johns Hopkins University.

Hohl, Julia Catherine Beaman. Stability relations between calcite, rhodochrosite, and aqueous solutions from 300-400°C. M, 1967, Brown University.

Hohler, James Joseph. The geology of Perry Township, Hocking County, Ohio. M, 1950, Ohio State University.

Hohlt, Richard B. Aspects of the subsurface geology of South Louisiana. D, 1977, Rice University. 174 p.

Hohlt, Richard B. The nature and origin of carbonate porosity. M, 1948, Colorado School of Mines. 91 p.

Hohman, John Craig. Depositional model of coal-bearing, Upper Cretaceous Gallup Sandstone, Gallup Sag area, New Mexico. M, 1986, Colorado State University. 174 p.

Hohmann, Gerald Wayne. Electromagnetic scattering by two dimensional inhomogeneities in the Earth. D, 1971, University of California, Berkeley. 154 p.

Hohn, Michael Edward. Paleoecology and biostratigraphy of the Portland Point and Kashong members, Moscow Formation (Middle Devonian) of central New York state. M, 1975, Indiana University, Bloomington. 69 p.

Hohn, Michael Edward. Seed fatty acids; taxonomy and evolutionary significance in Recent and fossil seeds. D, 1976, Indiana University, Bloomington. 92 p.

Hoholick, D. John. Porosity, cement, grain fabric and water chemistry of the St. Peter Sandstone in the Illinois Basin. M, 1980, University of Cincinnati. 72 p.

Hohos, Edward F. Paleoenvironmental model of the upper Freeport coal seam in parts of Indiana and Armstrong counties, Pennsylvania. M, 1979, University of South Carolina.

Hoiles, Edwin K. Small index foraminifera in random thin section. M, 1960, New York University.

Hoiles, Harley Harold Kristjan. Nature and genesis of the Afton copper deposit, Kamloops, British Columbia. M, 1978, University of Alberta. 186 p.

Hoiles, R. G. Geology of McKenzie Red Lake gold mines. M, 1941, Queen's University. 37 p.

Hoiles, Randolph Gerald. Geology of the Bankfield vicinity, Little Long Lac area, Ontario. D, 1943, University of Arizona.

Hoin, Steven James. A high resolution ground magnetic study of part of the Albion-Scipio Trend. M, 1983, Michigan State University. 89 p.

Hoisch, Thomas David. Metamorphism in the Big Maria Mountains, southeastern California. D, 1985, University of Southern California.

Hoisington, W. David. Uranium and thorium distribution in the Conway Granite of the White Mountain Batholith. M, 1977, Dartmouth College. 107 p.

Hojnacki, Robert Stephen. Late Pennsylvanian cyclic sedimentary units of the Brownwood area, north-central Texas. M, 1986, Texas A&M University.

Hok, Charlotte I. Evaluation of linear feature mapping as a groundwater prospecting technique in the metamorphic terrane of Fairbanks, Alaska. M, 1986, University of Alaska, Fairbanks. 238 p.

Hokans, David Hamlin. Glacial water levels in the Housatonic and Naugatuck river valleys of western Connecticut. M, 1952, Clark University.

Hokanson, Claudia L. Sand layer characteristics in thirteen modern flat-floored basins; a comparison and basin plain classification. M, 1981, Duke University. 304 p.

Hokanson, Neil B. Separated earthquake location and its application to the Calaveras Fault. M, 1985, Indiana University, Bloomington. 101 p.

Hoke, John Humphrey. The origin of oil. M, 1951, University of Michigan.

Hokett, Samuel Lee. Depositional and post-depositional study of Edwards Limestone (Lower Cretaceous) of cored interval, South Texas Syndicate Well II-5, Washburn Ranch Field, Lasalle County, Texas. M, 1983, University of Arkansas, Fayetteville.

Hokkanen, Gary Elmer. Application of the alternating direction Galerkin technique to the simulation of contaminant transport at the Borden Landfill. M, 1984, University of Waterloo. 64 p.

Holail, Hanafy Mahmoud. Stable isotopic composition and its relation to origin and diagenesis of some Upper Cretaceous dolomites and dolomitic limestones from Egypt. D, 1987, Purdue University. 300 p.

Holasek, Raymond Joseph. Geology of Gozar area, Reeves County, Texas. M, 1952, University of Texas, Austin.

Holaway, Rose Mary White. A study of the Pleistocene gravel of western Ouachita Parish, Louisiana. M, 1967, Northeast Louisiana University.

Holbek, Peter Michael. Geology and mineralization of the Stikine Assemblage, Mess Creek area, northwestern British Columbia. M, 1988, University of British Columbia. 184 p.

Holbrook, Charles E. Stratigraphic relationships of the Silurian and Devonian in Clark, Powell, Montgomery and Bath counties. M, 1964, University of Kentucky. 79 p.

Holbrook, John M. Depositional history of Lower Cretaceous strata in northwestern New Mexico; implications for regional tectonics and sequence stratigraphy. M, 1988, University of New Mexico. 112 p.

Holbrook, Philip William. Geologic and mineralogic factors controlling the properties and occurrence of ladle brick clays. D, 1973, Pennsylvania State University, University Park. 318 p.

Holbrook, Philip William. The sedimentology and pedology of the Mauch Chunk Formation (Mississippian) at Pottsville (Schuylkill County), Pennsylvania and their climatic implications. M, 1970, Franklin and Marshall College. 111 p.

Holbrook, W. Steven. Wide-angle seismic studies of crustal structure and composition in Nevada, California, and southern Germany. D, 1988, Stanford University. 214 p.

Holcomb, D. J. A theoretical and experimental investigation of dilatancy, an aspect of nonlinear behavior in rock. D, 1978, University of Colorado. 294 p.

Holcomb, Derrold. Chemical characterization of swamp peat humic substances. M, 1978, Georgia Institute of Technology. 63 p.

Holcomb, Haden Ray. Modification of grain size distributions by flume and wave tank. M, 1972, University of Tulsa. 80 p.

Holcomb, R. T. Terraced depressions in lunar maria. M, 1975, University of Arizona.

Holcomb, Richard Alfred. Conodont biostratigraphy of Paleozoic carbonates near Bavispe, Sonora, Mexico. M, 1979, Texas Christian University. 99 p.

Holcomb, Robin Terry. Kilauea Volcano, Hawaii; chronology and morphology of the surficial lava flows. D, 1981, Stanford University. 383 p.

Holcomb, Samuel. Terranes of Northfield, Vermont. M, 1917, Syracuse University.

Holcombe, Horace Truman. Terrain effects in resistivity and magnetotelluric surveys. D, 1982, University of New Mexico. 175 p.

Holcombe, Lawrence J. Adsorption and desorption in mine drainages. M, 1977, Colorado School of Mines. 96 p.

Holcombe, Rodney John. Mesoscopic and microscopic analysis of deformation and metamorphism near Ducktown, Tennessee. D, 1973, Stanford University. 225 p.

Holcombe, Troy L. Roughness patterns and sea-floor geomorphology in the North Atlantic Ocean. D, 1972, Columbia University. 286 p.

Holcombe, Troy Leon. Geology of the Elk Creek area, Lewis and Clark County, Montana. M, 1963, University of Missouri, Columbia. 108 p.

Holcombe, Walter B. Petrology of the 1951 Kaw River flood deposits between Ogden and Manhattan, Kansas. M, 1957, Kansas State University. 75 p.

Holdaway, Michael Jon. Petrology and structure of metamorphic and igneous rocks of parts of Northern Coffee Creek and Cecilville quadrangles, Klamath Mountains, California. D, 1963, University of California, Berkeley. 202 p.

Holdeman, Timothy G. Evaluation of the Lower Freeport (No.6A) Coal as a waveguide for seismic energy. M, 1982, Wright State University. 83 p.

Holden, Edward Fuller. The cause of color in smoky quartz and amethyst. D, 1925, University of Michigan.

Holden, Frederick T. Lower and Middle Mississippian stratigraphy of Ohio. D, 1941, University of Chicago. 77 p.

Holden, G. S. The Slate Creek metamorphic terrain, Albany County, Wyoming; conditions of metamorphism and mineral equilibria in metapelites. D, 1978, University of Wyoming. 103 p.

Holden, Gregory Spry. Chemical and petrographic stratigraphy of the Columbia River Basalt in the lower Salmon River canyon, Idaho. M, 1974, Washington State University. 97 p.

Holden, John Clinton. Neogene ostracods from the drowned terraces in the Hawaiian Islands. M, 1964, San Diego State University. 101 p.

Holden, Kenneth D. Geology of the central Argus Range. M, 1976, San Jose State University. 65 p.

Holden, M. A. Study of the ores of Quirubilia and Milluachaqui, Peru. M, 1931, Massachusetts Institute of Technology. 25 p.

Holden, Mark Kellogg. Sand wave development and textural variation on a flood tidal delta. M, 1980, Boston College.

Holden, Peter Newhall. Porphyrin absorption bands isolated from whole rock reflectance spectra. M, 1988, Rensselaer Polytechnic Institute. 76 p.

Holden, Roy J. History of the iron ore industry in the United States. D, 1916, University of Wisconsin-Madison.

Holden, William F. Wave equation datuming and its application to land seismic data. M, 1986, Colorado School of Mines. 270 p.

Holden, William Robert. Permeability microstratification in natural sandstones. D, 1969, University of Texas, Austin. 115 p.

Holder, Grace Amelia McCarley. Geology and petrology of the intrusive rocks east of the Republic Graben in the Republic Quadrangle, Ferry County, Washington. M, 1986, Washington State University. 87 p.

Holder, Jon Thomas. Thermodynamic properties of crystals containing imperfections. D, 1968, University of Illinois, Urbana. 110 p.

Holder, Robert E. Surface geology of southern Union Parish, Louisiana. M, 1963, Louisiana Tech University.

Holder, Robert Wade. Emplacement and geochemical evolution of Eocene plutonic rocks in the Colville Batholith. D, 1986, Washington State University. 189 p.

Holdoway, Katrine A. Deposition of evaporites and red beds of the Nippewalla Group, Permian, western Kansas. D, 1976, University of Kansas. 225 p.

Holdoway, Katrine A. Petrofabric changes in heated and irradiated salt from Project Salt Vault, Lyons, Kansas. M, 1972, University of Kansas. 54 p.

Holdrege, Thomas J. Failure of limestone roof-beams in south-central Indiana caverns. M, 1986, Purdue University. 161 p.

Holdren, G. R. Distribution and behavior of manganese in the interstitial waters of Chesapeake Bay sediments during early diagenesis. D, 1977, The Johns Hopkins University.

Holdren, G. R., Jr. Factors affecting phosphorus release from lake sediments. D, 1977, University of Wisconsin-Madison. 185 p.

Holdship, S. A. The paleolimnology of Lake Manyara, Tanzania; a diatom analysis of a 56 meter sediment core. D, 1976, Duke University. 133 p.

Holdsworth, Gerald. An examination and analysis of the formation of transverse crevasses, Kaskawulsh Glacier, Yukon Territory, Canada. M, 1965, Ohio State University.

Holdsworth, Gerald. Mode of flow of Meserve glacier, Wright valley, Antarctica. D, 1969, Ohio State University.

Holdt, Laura Lynn Von *see* Von Holdt, Laura Lynn

Hole, Allen David. The Pleistocene geology of the Telluride (Colorado) Quadrangle. D, 1910, University of Chicago. 88 p.

Hole, Francis Doan. Correlation of the glacial border drift of north central Wisconsin. D, 1943, University of Wisconsin-Madison.

Hole, Gilbert L. A mineralogic analysis of some deep-sea sediments by differential thermal, X-ray, and petrographic methods. M, 1949, Columbia University, Teachers College.

Hole, Gilbert L. Clay-limonite cappings over sulphide mineralization in Southwest Virginia. D, 1951, Columbia University, Teachers College.

Holecek, Thomas J. The relation of morphology, structure and lithology to the emplacement of the Hot Creek rhyolite flow. M, 1978, George Washington University.

Holien, Christopher W. Origin and geomorphic significance of channel-bar gravel of the lower Kansas River. M, 1982, University of Kansas. 128 p.

Holifield, Billy Ray. Lower Tertiary Charophyta of North America. M, 1964, University of Missouri, Columbia.

Holke, Kenneth Arthur. Refractive index of chromite. M, 1950, Washington University. 52 p.

Hollabaugh, Curtis Lee. Experimental mineralogy and crystal chemistry of sphene in the system soda-lime-alumina-titania-silica-water. D, 1980, Washington State University. 107 p.

Holladay, Curtis O. Geology of northwestern Blue Holes Quadrangle, Fremont County, Wyoming. M, 1959, Miami University (Ohio). 27 p.

Holladay, John C. Geology of the northern Canyon Range, Millard and Juab counties, Utah. M, 1984, Brigham Young University. 28 p.

Holladay, John Scott, III. The generalized electrosounding method for sedimentary basin exploration. D, 1987, University of Toronto.

Holland, Ann Elizabeth. A petrographic and fluid inclusion study of the Julia Deposit, Mineral, Virginia. M, 1980, University of North Carolina, Chapel Hill. 84 p.

Holland, B. D. Geology of the Bull Fork area, White Pine and Nye counties, Nevada. M, 1956, University of Kansas.

Holland, Daniel Edward. The Masterson oil field, Pecos County, Texas. M, 1940, University of Texas, Austin.

Holland, Donna J. Little. A numerical study of Tertiary shales of the Los Angeles Basin and Gulf Coast region using geochemical data collected by Trask and Patnode. M, 1983, University of Texas, Arlington. 93 p.

Holland, Dwight Allen. Tidal gravity anomalies in southeastern North America. M, 1986, Virginia Polytechnic Institute and State University.

Holland, Frank Deleno, Jr. Stratigraphic details of Lower Mississippian rocks of northeastern Utah and southwestern Montana. M, 1950, University of Missouri, Columbia. 86 p.

Holland, Frank Delno, Jr. The Brachiopoda of the Oswayo and Knapp formations of the Penn-York Embayment. D, 1958, University of Cincinnati. 524 p.

Holland, Frank Richard. Some detailed sections of the New Albany Shale near North Vernon, Indiana. M, 1953, University of Cincinnati. 78 p.

Holland, Hans J. The extraction and purification of carbon from natural substances. M, 1951, Columbia University, Teachers College.

Holland, Heinrich D. Distribution of accessory elements in pegmatites. M, 1948, Columbia University, Teachers College.

Holland, Heinrich D. The geochemistry of uranium, ionium, and radium in the oceans. D, 1953, Columbia University, Teachers College.

Holland, Jasper L. Geological investigations of Honey Creek watershed. M, 1954, Texas Christian University. 34 p.

Holland, John Sylvester. Petrography and petrology of the igneous rocks of the Avery District, Shoshone County, Idaho. M, 1947, University of Idaho. 39 p.

Holland, Krista. Petrography and diagenetic history of dolomitized Lower Ordovician strata in Northwest Arkansas. M, 1987, University of Arkansas, Fayetteville.

Holland, Peter T. The petrology of Boyden Cave pendant (metasedimentary and metavolcanic rocks), (Sierra Nevada mts. Kings canyon area), California. M, 1970, University of Wisconsin-Madison.

Holland, Richard A. Depositional environment of the Horquilla Limestone, Sierra de Palomas, Chihuahua, Mexico. M, 1980, Texas Christian University.

Holland, Richard Rainey. The micro-fauna of the Clayton Formation in Oktibbeha County (Clayton-Paleocene). M, 1950, Mississippi State University. 60 p.

Holland, Stuart Sowden. Geology of the western half of the Vernon map area, B.C. D, 1933, Princeton University. 123 p.

Holland, Thelma H. Geology, vertebrate fossils and environmental controls, Lee Creek phosphate mine, North Carolina. M, 1974, Virginia State University. 68 p.

Holland, Walter Fox, Jr. Karst water evolution in Marion County, Tennessee; a synthesis of hydrochemical and isotopic evidence. M, 1973, University of Texas, Austin.

Holland, Wilbur Charles. The ostracods of the Nineveh Limestone of Pennsylvania and West Virginia. M, 1933, University of Pittsburgh.

Holland, Wilbur Charles. The physiography of Beauregard and Allen parishes, Louisiana. D, 1943, Louisiana State University.

Holland, William. The surficial geology of the Billerica Quadrangle, Middlesex County, Massachusetts. M, 1980, Boston University. 172 p.

Holland, William Thompson. A geophysical and geological survey of the northern part of the Deep River Triassic basin. M, 1930, University of North Carolina, Chapel Hill. 47 p.

Holland, Willis A., Jr. The geology of the Panola Shoals area, DeKalb County, Georgia. M, 1954, Emory University. 92 p.

Holland, Willis Algeon, Jr. The kinematic and metamorphic development of the Towaliga-Goat Rock mylonites, Georgia-Alabama. D, 1981, Florida State University. 116 p.

Hollander, David Jon. Origin of secondary carbonates, diagenesis of organic matter, and source bed evaluation, Tertiary basinal units, La Honda Basin, Santa Cruz Mountains, California. M, 1984, University of California, Santa Cruz.

Hollander, Eileen E. Geological assessment of the geopressured geothermal resource of South Louisiana. M, 1983, Tulane University.

Hollander, Margaret. Studies of phase transitions with particular reference to the polymorphs of KNO_3 and the system alumina-water. M, 1969, [University of California, Berkeley].

Hollands, Garrett G. Surficial geology of the Colden Quadrangle, western New York. M, 1975, University of Massachusetts. 206 p.

Holleman, M. A study on the distribution of sodium chloride and the predictability of its occurrence in the evaporites near Little Narrows, Victoria County, Nova Scotia. M, 1976, Acadia University.

Hollenbaugh, Donald William. Trace element geochemistry and mineralogy of septarian siderite concretions and enclosing shales in Columbiana County, Ohio. M, 1979, Kent State University, Kent. 60 p.

Hollenbaugh, Kenneth Malcolm. Geology of a portion of the north flank of the San Bernardino Mountains, California. D, 1968, University of Idaho. 109 p.

Hollenbaugh, Kenneth Malcolm. Geology of Lewiston and vicinity, Nez Perce County, Idaho. M, 1959, University of Idaho. 53 p.

Holler, D. P. Recognition and projection of the contact between the Mississippian Fort Payne Chert and Tuscumbia Limestone in Madison County, Alabama. M, 1976, University of Alabama.

Hollett, Douglas Whitlock. Petrological and physicochemical aspects of thrust faulting in the Precambrian Farmington Canyon complex, Utah. M, 1979, University of Utah. 99 p.

Hollett, K. J. Shoaling of Kaneohe Bay, Oahu. M, 1977, University of Hawaii. 145 p.

Holley, Carolayne Elizabeth. The lithology, environment of deposition, and diagenesis of the Queen Formation at McFarland North, and Magutex Queen fields, Andrews County, Texas. M, 1988, Texas A&M University. 157 p.

Holliday, Barry W. Observations on the hydraulic regime of the ridge and swale topography of inner Virginia shelf. M, 1971, Old Dominion University. 84 p.

Holliday, Joseph. The bedrock geology of the southeast part of Shasta Valley, Siskiyou County, California. M, 1983, Oregon State University. 165 p.

Holliday, Valerie E. Mechanisms of deposition of a carbonate mud spit; Ramshorn Spit, eastern Florida Bay. M, 1985, Lehigh University. 180 p.

Holliday, Vance Terrell. Morphological and chemical trends in Holocene soils at the Lubbock Lake archeological site, Texas. D, 1982, University of Colorado. 306 p.

Hollingshead, Stephen C. Hydrogeologic assessment of alternate landfill design configurations. M, 1985, Queen's University. 120 p.

Hollingsworth, J. A. C. and Berry, J. E. A seismic investigation in the Deer Creek area, Colorado (T 763). M, 1952, Colorado School of Mines. 48 p.

Hollingsworth, Richard Vincen. The Union Valley sandstone member of the Wapanucka Formation. M, 1933, University of Oklahoma. 118 p.

Hollingsworth, Terry J. Spheroidal weathering of sandstones in Nicholas County, West Virginia. M, 1978, West Virginia University.

Hollis, James Richard. Precision gravity reobservations at Yellowstone National Park, Wyoming, 1977-1987. M, 1988, University of Utah. 199 p.

Hollis, Stephen Hall. Geology of the Bunker Hills and vicinity, Lebanon County, Pennsylvania. M, 1974, Bryn Mawr College. 36 p.

Hollis, Thomas. Stratigraphy, lithology, depositional and diagenetic environments of the Middle Ordovician Antelope Valley Limestone at Lone Mountain and Ikes Canyon in central Nevada. M, 1986, California State University, Long Beach. 177 p.

Hollister, Charles D. Sediment distribution and deep circulation in the western North Atlantic. D, 1967, Columbia University. 472 p.

Hollister, Donald E. The geology of the cuestas of the Great Lakes region. M, 1926, University of Wisconsin-Madison.

Hollister, Lincoln Steffens. Electron microprobe investigations of metamorphic reactions and mineral growth histories, Kwoiek area, British Columbia. D, 1966, California Institute of Technology. 229 p.

Hollister, Victor F. Geology and ore deposits of the Shasta gold district. M, 1950, University of California, Berkeley.

Hollocher, Kurt T. Retrograde metamorphism of the Lower Devonian Littleton Formation in the New Salem area, West-central Massachusetts. M, 1981, University of Massachusetts. 268 p.

Hollocher, Kurt Thomas. Geochemistry of metamorphosed volcanic rocks in the Middle Ordovician Partridge Formation, and amphibole dehydration reactions in the high-grade metamorphic zones of central Massachusetts. D, 1985, University of Massachusetts. 348 p.

Hollod, Gregory J. Polychlorinated biphenyls (PCBs) in the Lake Superior ecosystem; atmospheric deposition and accumulation in the bottom sediments. D, 1979, University of Minnesota, Minneapolis. 262 p.

Holloway, Carleen D. Petrology of the Eocene volcanic sequence, Nez Perce and Blue Joint Creeks, southern Bitterroot Mountains, Montana. M, 1980, University of Montana. 129 p.

Holloway, D. C. The structural petrology of Precambrian volcanic-sedimentary rocks in north central Wisconsin. M, 1975, University of Wisconsin-Milwaukee.

Holloway, D. M. The mechanics of pile-soil interaction in cohesionless soils. D, 1975, Duke University. 301 p.

Holloway, Harold Deen. The geology of the Trinity aquifers of McLennan County. M, 1959, Baylor University.

Holloway, John N. Areal geology and contact relations of the basement complex and later sediments, west end of the San Gabriel Mountains (California). M, 1940, California Institute of Technology. 26 p.

Holloway, John Requa. Phase relations and compositions in the basalt-CO_2-H_2O system at high temperatures and pressures. D, 1970, Pennsylvania State University, University Park. 156 p.

Holloway, Perry Gregory. A subsurface study of the Waltersburg Sandstone in Gibson and Vanderburgh counties, Indiana. M, 1952, Indiana University, Bloomington. 38 p.

Holloway, Ralph Leslie, Jr. Some aspects of quantitative relations in the primate brain. D, 1964, University of California, Berkeley. 168 p.

Hollrah, Terry Lewis. Subsurface lithostratigraphy of the Hunton Group in parts of Payne, Lincoln, and Logan counties, Oklahoma. M, 1977, Oklahoma State University. 63 p.

Hollweg, William A. Permian evaporites overlying the Blaine Formation of western Kansas. M, 1964, University of Kansas. 76 p.

Holly, Dean E. Hydrogeology of northern Lancaster County, Nebraska. M, 1980, University of Nebraska, Lincoln.

Holly, James Benjamin. Stratigraphy and sedimentary history of Newnans Lake. M, 1976, University of Florida. 102 p.

Hollyday, Este F. A geohydrologic analysis of mine dewatering and water development, Tombstone, Cochise County, Arizona. M, 1963, University of Arizona.

Hollyfield, Charles E. Geology of Boston and Vinyard townships, Washington Couny, Arkansas. M, 1958, University of Arkansas, Fayetteville.

Holm, Bjarne. Bedrock geology of Mount Prindle, (Yukon-Tanana Highlands) Alaska. M, 1973, University of Alaska, Fairbanks. 55 p.

Holm, Daniel Keith. A structural investigation and tectonic interpretation of the Penokean Orogeny; east-central Minnesota. M, 1986, University of Minnesota, Duluth. 114 p.

Holm, E. Richard. The petrology and tectonic significance of the Green Pond Conglomerate at Green Pond, New Jersey. M, 1951, Columbia University, Teachers College.

Holm, Melody Ruth. Comparison of the Peterson and Draney limestones (Lower Cretaceous), Idaho and Wyoming, and the calcareous members of the Kootenai Formation, western Montana. M, 1977, Indiana University, Bloomington. 68 p.

Holm, Paul Eric. Strain rate and temperature controls on strain heterogeneity; its significance for deformational concepts. D, 1980, University of Illinois, Urbana. 145 p.

Holm, Richard F. Petrology and structural geology of the Dahomeya gneiss (Precambrian) in the western Accra plains, Ghana. D, 1969, University of Washington. 169 p.

Holm, Richard Frank. Geology of the Wild Horse Creek area, Custer County, Idaho. M, 1962, University of Idaho. 69 p.

Holman, Robert E., III. The ecology of four coastal lakes in North Carolina; tropic states measured from space imagery. D, 1978, North Carolina State University. 187 p.

Holman, W. R. The origin of sheeting joints; a hypothesis. D, 1976, University of California, Los Angeles. 86 p.

Holmberg, Russell L. A subsurface study of the Greenhorn and Graneros formations and the Dakota Group in selected deep wells from Northeast to Northwest Nebraska. M, 1954, University of Nebraska, Lincoln.

Holmer, Ralph C. A regular gravity survey of Colorado. D, 1954, Colorado School of Mines. 81 p.

Holmes, Allen Whitney. Upper Cretaceous stratigraphy of northeastern Arizona and south central Utah. M, 1954, University of Colorado.

Holmes, Andrew. Stability analysis of the Granite Lake pit based on displacement data generated by a surface extensometer system. M, 1976, University of British Columbia.

Holmes, Ann Elizabeth. Systematics and paleoecology of the stromatoporoids from the Lower Devonian Keyser Formation at Mustoe, Virginia. M, 1981, University of Alabama. 97 p.

Holmes, Charles R. Magnetic fields associated with igneous pipes in the central Ozarks. M, 1950, St. Louis University.

Holmes, Charles Robert. Dependence of resistivity of porous sandstones on fluid distribution. D, 1958, Pennsylvania State University, University Park. 98 p.

Holmes, Charles Ward. Rates of sedimentation in the Drake Passage (Antarctic). D, 1965, Florida State University.

Holmes, Charles Ward. Rates of sedimentation of the Ten Thousand Islands. M, 1962, Florida State University.

Holmes, Chauncey DeP. Pleistocene geology of the region south of Syracuse, New York. D, 1939, Yale University.

Holmes, Chauncey DeP. Special glacial features of the Cazenovia Quadrangle (New York). M, 1927, Syracuse University.

Holmes, Clifford Newton. Jurassic history and stratigraphy of Colorado. D, 1960, University of Utah. 396 p.

Holmes, David Allen. Cambrian-Ordovician stratigraphy of the northern Portneuf Range. M, 1958, University of Idaho. 58 p.

Holmes, David Brian. The surficial geology of parts of Benton and eastern Morrison counties, Minnesota. M, 1988, University of Wisconsin-Milwaukee. 180 p.

Holmes, Donald W. Moose Channel clastics (Late Cretaceous-Early Tertiary), Fish River, Northwest Territory. M, 1972, University of Calgary. 104 p.

Holmes, Douglas Allen and Lootens, Douglas Joseph. Geology of the Silver Mountain area, Huerfano County, Colorado. M, 1959, University of Michigan.

Holmes, Gary S. The environmental effects of causeway construction on the West River estuary, Prince Edward Island. M, 1974, Queen's University. 222 p.

Holmes, George William. Geology of the west central Wind River Mountains, Wyoming. D, 1949, Harvard University.

Holmes, Grace Bruce. A bibliography of the conodonts with descriptions of Early Mississippian species. D, 1925, George Washington University.

Holmes, Harvey N. Geological column and dip of the strata in the vicinity of Syracuse (New York). M, 1910, Syracuse University.

Holmes, James W. The depositional environment of the Mississippian Lewis Sandstone in the Black Warrior Basin of Alabama. M, 1981, University of Alabama. 172 p.

Holmes, John Ferrell. A seismographic study of mid-continental primary wave travel times. M, 1964, Kansas State University. 42 p.

Holmes, Kurt Quentin. Pebble shape and sorting development on the alluvial fans of the Deep Creek Mountains, Utah. M, 1982, University of Utah. 54 p.

Holmes, Leslie Arnold. The limestone oil fields of Wyoming. M, 1928, University of Illinois, Chicago.

Holmes, Leslie Arnold. Variations in coal tonnage production in Illinois, 1900-1940. D, 1942, University of Illinois, Urbana.

Holmes, Margaret A. Geology and mineralization of mines around the Cable Stock, Deer Lodge Co., Montana. M, 1982, Montana College of Mineral Science & Technology. 73 p.

Holmes, Mark Lawrence. Tectonic framework and geologic evolution of the southern Chukchi Sea continental shelf. D, 1975, University of Washington. 143 p.

Holmes, Mary Emilie. The morphology of the carinae under the septa of rugose corals. D, 1888, University of Michigan.

Holmes, Michael E. Correlation and variation studies of tills in southwestern Ohio. M, 1974, Miami University (Ohio). 42 p.

Holmes, Paul J. Infiltration uranium deposits in ash flow tuffs. M, 1972, University of Nevada - Mackay School of Mines. 65 p.

Holmes, Peter Winchester. McCabe-Gladstone and Rebel-Little Kicker base and precious metal occurrences; Big Bug District, Yavapai County, Arizona; significance to exploration. M, 1987, University of Western Ontario. 150 p.

Holmes, Ralph J. The higher mineral arsenides of cobalt, nickel, and iron. D, 1945, Columbia University, Teachers College.

Holmes, Rebecca Ann. The geology of a portion of the Hebgen Dam Quadrangle, Gallatin County, Southwest Montana. M, 1986, University of Idaho. 102 p.

Holmes, Richard D. Thermoluminescence dating of Quaternary basalts from the eastern margin of the Basin and Range Province, Utah and northern Arizona. M, 1977, Brigham Young University.

Holmes, Robert F., Jr. The geology of East Fork-Wind River junction area, Blue Holes Quadrangle, Fremont County, Wyoming. M, 1959, Miami University (Ohio). 39 p.

Holmes, Stanley W. A petrographic study of basic sills intruding the (Precambrian) Howse Series, Labrador. M, 1950, McGill University.

Holmes, Stanley Winchester. The geology and mineral deposits of the Fancamp-Hauy area, Abitibi County east, Quebec, Canada. D, 1953, Cornell University.

Holmes, Terence C. The geology of Hart Township, Ontario, and adjacent areas. D, 1936, University of Chicago. 46 p.

Holmes, Thomas Connor. Transformational superplasticity of the CsCl-RbCl solid solution. M, 1977, Pennsylvania State University, University Park. 64 p.

Holmgren, Dennis Arthur. Columbia river basalt patterns from central Washington to northern Oregon. D, 1969, University of Washington. 56 p.

Holmgren, Dennis Arthur. The Yakima-Ellensburg Unconformity, central Washington. M, 1967, University of Washington. 69 p.

Holmquest, Harold John, Jr. Statistical morphologic classification of the Paleozoic Ostracoda. M, 1952, Michigan State University. 32 p.

Holroyd, Michael Thomas. A system for automated compilation of earth science survey data with special provision for aerogeophysical data. D, 1984, University of Ottawa. 266 p.

Holser, William T. Geology of the Mint Canyon area, Los Angeles County, California. M, 1946, California Institute of Technology. 43 p.

Holser, William T. Metamorphism and associated mineralization in the Philipsburg region, Montana. D, 1950, Columbia University, Teachers College. 43 p.

Holsinger, S. L. The Sequatchie Anticline, an independently formed anticline. M, 1973, Wright State University. 49 p.

Holst, Norman Benjamin, Jr. The join diopside-ilmenite and its bearing on the incorporation of titanium into clinopyroxenes. M, 1978, University of Illinois, Chicago.

Holst, Timothy B. The determination of finite strain in rocks; the role of primary fabric. D, 1977, University of Minnesota, Minneapolis. 256 p.

Holstrom, Geoffrey Burwell. Elastic radiation from a propagating phase boundary. D, 1966, Stanford University. 99 p.

Holt, Charles Lee Roy, Jr. Geology of Kent Station area, Culberson and Jeff Davis counties, Texas. M, 1951, University of Texas, Austin.

Holt, Edward L. The Morrison and Summerville formations (Jurassic) of the Grand River valley (Mesa County, Colorado) and their vertebrate fauna. M, 1940, University of Colorado.

Holt, Henry E. Geology of the Lower Blue River area, Summit and Grand counties, Colorado. D, 1961, University of Colorado.

Holt, Jack Haston. A study of the physico-chemical, mineralogical and engineering properties of fine-grained soil in relation to their expansive characteristics. D, 1969, Texas A&M University. 128 p.

Holt, Larry E. Origin and trace element geochemistry of Arkansas wavellite and variscite. M, 1972, University of Arkansas, Fayetteville.

Holt, Olin R. A petrologic study of the Cypress, Tar Springs, and Mansfield sandstones. M, 1957, Indiana University, Bloomington. 42 p.

Holt, Richard D. The stratigraphy of the Arnheim Formation of the central basin of Tennessee. M, 1931, Vanderbilt University.

Holt, Richard J. A comparison of two methods of resistivity prospecting. M, 1956, Boston College.

Holt, Richard Wayne. A sedimentary study of the soils of the Double Mountain Fork of the Brazos River, Lubbock County, Texas. M, 1955, Texas Tech University. 95 p.

Holt, Robert Eugene. Structure and petrology of the diorite on the 800 level Mayflower Mine, Park City, Utah. M, 1953, University of Utah. 35 p.

Holt, Robert M. The depositional environments of the Late Permian Rustler Formation, in the vicinity of the Waste Isolation Pilot Plant (WIPP) site, southeastern New Mexico. M, 1988, University of Texas at El Paso.

Holt, William K. Streamflow synthesis and water allocation by water right priorities. M, 1980, Colorado State University. 139 p.

Holte, Karl Emrud. A floristic and ecological analysis of the Excelsior fen complex in Northwest Iowa. D, 1966, University of Iowa. 292 p.

Holter, Milton Edward. The middle Devonian prairie-evaporite of Saskatchewan (Canada). M, 1969, University of Saskatchewan. 161 p.

Holterhoff, F. K. Upper Ordovician lithostratigraphy and conodont biostratigraphy at Carrollton, Kentucky. M, 1977, Ohio State University. 68 p.

Holterhoff, Peter F. Paleobiology and paleoecology of crinoids from the lower Stanton Formation (Late Pennsylvanian, Missourian) of the Mid-continent, United

States. M, 1988, University of Nebraska, Lincoln. 137 p.

Holtje, R. Kenneth, Jr. Hydrologic trends of the Brandywine Creek, Pennsylvania. M, 1966, Colorado State University. 65 p.

Holtz, Kenneth Roger. Ground water of the Double Lakes area, Lynn County, Texas. M, 1980, Texas Tech University. 69 p.

Holtz, Robert Dean, II. Some effects of stress path and overconsolidation ratio on the shear strength properties of a Georgia kaolinite. D, 1970, Northwestern University. 221 p.

Holtzclaw, Mark John. Geology, alteration, and mineralization of the Red Mountain Stock, Grizzly Peak Cauldron Complex, Colorado. M, 1973, Oklahoma State University. 79 p.

Holtzclaw, Stuart R. Stratigraphy and sedimentology of the lower Frontier Formation (Cretaceous), southwestern Bighorn Basin, Wyoming. M, 1987, University of Texas at El Paso.

Holtzman, Alan McKim, Jr. The areal geology and Cretaceous stratigraphy of Northwest Marshall County, Oklahoma. M, 1978, University of Oklahoma. 97 p.

Holtzman, Allan F. Gravity study of the Manson "disturbed area" (Paleozoic, Mesozoic), Calhoun, Pocahontas, Humboldt and Webster counties, Iowa. M, 1970, University of Iowa. 63 p.

Holtzman, Richard Charles. Late Paleocene mammals of the Tongue River Formation, North Dakota. D, 1976, University of Minnesota, Minneapolis. 240 p.

Holway, Jeffrey V. Environmental geology of the Paradise Valley Quadrangle, Maricopa County, Arizona; Part I. M, 1977, Arizona State University. 64 p.

Holway, Orlando, III. The use of the torsion balance to measure the gradients of gravity in southern Ohio. M, 1963, Ohio State University.

Holway, William. The origin and occurrence of specular hematite (specularite) in the lower Huronian of the Marquette District, Michigan. M, 1952, Michigan State University. 46 p.

Holwerda, James Gerhardus. Geology of the Valyermo area, California. M, 1951, University of Southern California.

Holwerda, James Gerhardus. The ground water geology of Iraq. D, 1958, University of Southern California.

Holyk, Walter. Some geologic aspects of radioactivity; 1, The use of biotite in determination of geologic age by the strontium method; 2, The use of lepidolite in determination of geologic age by the calcium method; 3, The potassium content of ultramafic rocks and its heat contribution. D, 1952, Massachusetts Institute of Technology. 179 p.

Holzer, Robert Albert. Soil analysis by tristimulus color and statistical evaluations for forensic purposes. D, 1977, Rutgers, The State University, New Brunswick. 51 p.

Holzer, Thomas Lequear. Effect of a seismic loading on undrained creep of fine-grained sediment from San Francisco bay, California; an experimental study. D, 1970, Stanford University. 209 p.

Holzhausen, G. R. Sheet structure in rock and some related problems in rock mechanics. D, 1978, Stanford University. 532 p.

Holzheimer, Joanne M. Paleoenvironmental analysis of Upper Cretaceous strata deposited on the northern flank of the Precambrian Sioux Quartzite, Lake and Moody counties, eastern South Dakota. M, 1987, South Dakota School of Mines & Technology.

Holzhey, Charles Steven. Epipedon morphology of Xeralfs and some associated chaparral soils in Southern California. D, 1968, University of California, Riverside. 255 p.

Holzinger, Philip. Hydrologic budget for Hockessin area, Delaware. M, 1979, University of Delaware.

Holzman, Benjamin. Geophysical applications of radon measurements. M, 1933, California Institute of Technology. 81 p.

Holzman, Johnston Earl. Submarine geology of Cortes and Tanner banks. M, 1952, University of Southern California.

Holzwasser, Florrie. Clarification of the idea of geosyncline. M, 1918, Columbia University, Teachers College.

Holzwasser, Florrie. Geology of Newburgh and vicinity (New York). D, 1926, Columbia University, Teachers College.

Homan, Kimberly Sue. Structural analysis of the south flank of the Sweetwater Uplift, Carbon County, Wyoming. M, 1988, Baylor University. 160 p.

Homeister, Owen E. Geology of a section of the Darby Formation in Warm Spring Canyon, Fremont County, Wyoming. M, 1950, Miami University (Ohio). 55 p.

Homeniuk, Leonard Anthony. A near-vertical-incidence reflection survey conducted over the Precambrian shield of southeastern Manitoba. M, 1972, University of Manitoba.

Homme, Frank C. Contact metamorphism in the Tres Hermanas Mountains, Luna County, New Mexico. M, 1958, University of New Mexico. 88 p.

Homuth, Emil F. Earth strains; tidal and secular, observed near Bergen Park, Colorado. M, 1968, Colorado School of Mines. 53 p.

Hon, Kenneth Alan. Geologic and petrologic evolution of the Lake City Caldera, San Juan Mountains, Colorado. D, 1987, University of Colorado. 291 p.

Hon, René Aurel De see De Hon, René Aurel

Hon, Rudolph. Geology, petrology and geochemistry of Traveler Rhyolite and Katahdin Pluton; (northcentral Maine). D, 1976, Massachusetts Institute of Technology. 239 p.

Honapour, Mehdi. Chromatographic determination of hydrocarbons based on retention time data for squalene and tetracyanoethylated pentaerythritol columns. M, 1970, University of Missouri, Rolla.

Honda, Hiromi Rigakushi. Diagenesis and reservoir quality of the Norphlet Sandstone (Upper Jurassic), the Hatters Pond area, Mobile County, Alabama. M, 1981, University of Texas, Austin. 213 p.

Honderich, Jeff P. Origin and environment of deposition of Cenozoic chert gravel in eastern Kansas. M, 1970, University of Kansas. 117 p.

Honea, E. G. The effect of contact strength on the shape of buckle folds; including field examples in South Darwin Canyon, California. D, 1976, Stanford University. 101 p.

Honea, Elmont G. A model for deformation by gravity. M, 1969, University of Missouri, Columbia.

Honea, Robert Clair. Classification of Paleozoic Ostracoda. M, 1948, University of Illinois, Urbana.

Honea, Russel M. Volcanic geology of the Ruby Mountain area, Nathrop (Chaffee County), Colorado. M, 1955, University of Colorado.

Honea, Russell Morgan. Mineralogy of the uranium silicates. D, 1960, Harvard University.

Honess, A. P. The nature, origin and interpretation of etch figures on crystals. D, 1925, Princeton University.

Honess, Charles W. Geology of the southern Ouachita Mountains of Oklahoma. D, 1924, Columbia University, Teachers College.

Honess, Charles William. Studies in the Upper Devonian-Lower Carboniferous strata of the Appalachian sea. M, 1916, Cornell University.

Honey, James Gilbert. The paleontology of the Brown's Park Formation in the Maybell, Colorado area, and a taphonomic study of two fossil quarries Colorado and Arizona. M, 1977, University of Arizona.

Honeycutt, Floyd Mitchell. Petrology of the Balsam Gap ultramafic body, Jackson County, North Carolina. M, 1978, Kent State University, Kent. 46 p.

Honeycutt, Thomas K. Geophysical and geologic evidence regarding Cenozoic displacement along the Soda-Avawatz fault zone, San Bernardino County,

California. M, 1988, Southern Illinois University, Carbondale. 105 p.

Honeyman, Bruce Donald. Cation and anion adsorption at the oxide/solution interface in systems containing binary mixtures of adsorbents; an investigation of the concept of adsorptive additivity. D, 1984, Stanford University. 409 p.

Honeyman, Leslie R. Description and origin of carbonate minerals on the upper and lower cherty members of the Biwabik Formation, Minnesota. M, 1973, University of North Dakota. 66 p.

Hong, Chong-Huey. Development of a 2-D micellar/polymer simulator. D, 1982, University of Texas, Austin. 347 p.

Hong, Eason. Petrographic evidence on the paleoclimate and provenance of the Catskill, Pocono, and Pottsville formations, southeastern Pennsylvania. M, 1982, Pennsylvania State University, University Park. 99 p.

Hong, Gi Hoon. Fluxes, dynamics and chemistry of suspended particulate matter in a Southeast Alaskan fjord. D, 1986, University of Alaska, Fairbanks. 273 p.

Hong, Huasheng. Chemistry of iron in different marine environments and the binding of iron, copper, manganese and aluminum with particles in a microcosm system. D, 1984, University of Rhode Island. 262 p.

Hong, Kappyo. Assessment of seismic hazard using a probabilistic-fuzzy approach. D, 1988, University of Colorado. 225 p.

Hong, Ming-Ren. The inversion of magnetic and gravity anomalies and the depth to Curie isotherm. D, 1982, University of Texas at Dallas. 189 p.

Hong, Mingde. The elasticity of spodumene by Brillouin spectroscopy. M, 1988, University of Illinois, Chicago.

Hong, Sung Wan. Ground movements around model tunnels in sand. D, 1984, University of Illinois, Urbana. 449 p.

Hong, Tai Lin. Elastic wave propagation in irregular structures. D, 1978, California Institute of Technology. 105 p.

Hongnusonthi, A-ngoon. Geologic control of ground water quality and ground water treatment in the Khorat Plateau area, northeastern Thailand. M, 1971, Washington University. 77 p.

Honig, Cecily A. Estuarine sedimentation on a glaciated coast; Lawrencetown, Nova Scotia. M, 1987, Dalhousie University. 132 p.

Honjo, Norio. Petrology and geochemistry of the Magic Reservoir eruptive center, Snake River plain, Idaho. M, 1986, Rice University. 511 p.

Honkala, Adolf Uno. A study of Tertiary intrusives and associated mineralization in the Pillar Peak vicinity, Black Hills, South Dakota. M, 1949, University of Nebraska, Lincoln.

Honkala, Frederick Saul. Geology of the Auer Ranch area, Wyoming. M, 1942, University of Missouri, Columbia.

Honkala, Frederick Saul. Geology of the Centennial region, Beaverhead County, Montana. D, 1949, University of Michigan. 145 p.

Honma, Shigeo D. E. Finite element analysis of the injection and distribution of chemical grout in soils. D, 1984, University of Wisconsin-Milwaukee. 172 p.

Honnell, Pierre Marcel. An electromechanical transducer for the transient testing of seismographs. D, 1950, St. Louis University.

Hons, D. B. Thermophysical characterization of the surface tier of an organic soil at sub-zero temperatures. M, 1975, University of Guelph.

Hoobs, John H. Carboniferous island-arc and associated rocks from the Mision Calamajue area, Baja California, Mexico. M, 1985, San Diego State University. 122 p.

Hood, Edward John. Watershed organizations; impact on water quality management; an analysis of selected Michigan watershed councils. D, 1976, Michigan State University. 295 p.

Hood, H. C. Geology of Fortress Cliff Quadrangle, Randall County, Texas. M, 1977, West Texas State University. 123 p.

Hood, L. L. Lunar crustal magnetic anomalies detected by the Apollo subsatellite magnetometers. D, 1979, University of California, Los Angeles. 183 p.

Hood, Larry Quentin. Fuel-grade peat resources of the northern Everglades based on bulk density analysis. M, 1982, University of Florida. 160 p.

Hood, Lindsay Ann. The effect of depositional environment on framework mineralogy and diagenesis within a nonmarine-marine transition zone; the lower Fountain fan delta (Pennsylvanian), Manitou Springs, Colorado. M, 1987, Indiana University, Bloomington. 139 p.

Hood, Sandra Diane. Fenestrate bryozoan fauna from the lower part of the Cave Hill Member, Kinkaid Formation, Upper Chesterian (Mississippian) of southern Illinois. M, 1972, Southern Illinois University, Carbondale. 192 p.

Hood, William Calvin. Weathering of the Butte quartz monzonite near Butte, Montana. D, 1963, University of Montana. 87 p.

Hoodmaker, F. C. Regional stratigraphy of the Cloverly Formation, North Park Basin, Colorado. M, 1958, University of Wyoming. 139 p.

Hoof, Vertress Lawrence Vander see Vander Hoof, Vertress Lawrence

Hook, Donald L. Late Cenozoic stratigraphy and structure of a part of the Walnut Grove Basin, Yavapai County, Arizona. M, 1956, University of Arizona.

Hook, Harry Jerrold Van see Van Hook, Harry Jerrold

Hook, Joseph Frederick. On the theory of propagation of seismic waves in inhomogeneous isotropic elastic media. D, 1959, University of California, Los Angeles. 208 p.

Hook, Joseph S. Geology of the Pleasanton Quadrangle, California. M, 1913, [Stanford University].

Hook, Richard. The volcanic stratigraphy of the Mickey hot spring area, Harney County, Oregon. M, 1981, Oregon State University. 66 p.

Hook, Robert Warren. A paleoenvionmental model for the occurrence of vertebrate fossils in Carboniferous coal-bearing strata. D, 1985, University of Kentucky. 86 p.

Hook, Simon John. Integrated Landsat-MSS and TM and geophysical data applications in geological mapping and mineral exploration for W. Nfld. M, 1985, University of Alberta. 148 p.

Hook, Stephen Charles. Cephalopod faunas of latest Canadian age from southwestern United States. D, 1975, New Mexico Institute of Mining and Technology. 241 p.

Hook, Stephen Charles. The cephalopods of the Florida Mountains Formation. M, 1974, New Mexico Institute of Mining and Technology.

Hook, Thomas E. Niagaran pinnacle reefs of northern Michigan; a revised model. M, 1984, Wright State University. 53 p.

Hooke, Roger LeBaron. Alluvial fans. D, 1965, California Institute of Technology. 192 p.

Hooker, Andrew M. Structural geology along the Whiteoak Mountain Fault, Tranquility Quadrangle, Meigs County, Tennessee. M, 1964, University of Tennessee, Knoxville. 37 p.

Hooker, Andrew T. Interpretation of seismic record sections from the Cilician Basin, Northeast Mediterranean. M, 1981, New Mexico State University, Las Cruces. 83 p.

Hooker, Ellen Ostroff. The distribution and depositional environment of the Sprio Sandstone, Arkoma Basin, Haskell, Latimer and Pittsburg counties, Oklahoma. M, 1988, Oklahoma State University. 98 p.

Hooker, Marjorie M. Joints of the Ordovician sediments southwest of the Adirondacks (New York). M, 1933, Syracuse University.

Hooker, Richard A. Intraformational structural features of the Chase Group, Wolfcamp Series. M, 1956, Kansas State University. 73 p.

Hooks, J. David. Stratigraphy, depositional environments, and silica diagenesis of the Cambro-Ordovician carbonate sequence, Appalachian Valley and Ridge, Alabama. M, 1985, University of Alabama. 214 p.

Hooks, James E. A textural and mineralogical analysis of the Upper Cambrian Welge Sandstone, central Mineral region, Texas. M, 1961, Texas A&M University. 99 p.

Hooks, William Gary. The clay minerals and the iron oxide minerals of the Triassic "red beds" of the Durham Basin, North Carolina. M, 1953, University of North Carolina, Chapel Hill. 25 p.

Hooks, William Gary. The stratigraphy and structure of the Bucksville area, Alabama. D, 1961, University of North Carolina, Chapel Hill. 205 p.

Hookway, Lozell C. Geology of a portion of the Lompoc Quadrangle of Santa Barbara County, California. M, 1930, California Institute of Technology. 48 p.

Hooper, Jane. Field geology and petrology of Tertiary basalt intrusions, Routt County, Colorado. M, 1940, University of Rochester. 96 p.

Hooper, John Marten. Seismic refraction study of the Mississippian erosional surface and Pennsylvanian sediments in northern Marion County, Iowa. M, 1978, Iowa State University of Science and Technology.

Hooper, Medora L. Geology of the Thirteenth Lake Quadrangle of the State of New York. M, 1931, Columbia University, Teachers College.

Hooper, Richard H. Electrical resistivities of synthetic oil sands. M, 1936, California State University, Los Angeles.

Hooper, Robert James. Geologic studies at the East end of the Pine Mountain Window and adjacent Piedmont, central Georgia (Volumes I and II). D, 1986, University of South Carolina. 374 p.

Hooper, Robert Louis. Fission-track dating of the Mariano Lake uranium deposit, Grants mineral belt, New Mexico. D, 1983, Washington State University. 136 p.

Hooper, Robert Louis. Mineralogy of a coal burn near Kemmerer, Wyoming. M, 1982, Washington State University. 86 p.

Hooper, Warren G. Geology of the Smith and Morehouse-South Fork area, (Summit County), Utah. M, 1951, University of Utah. 55 p.

Hooper, William F. Petrology of the Lakota Conglomerate, Casper Arch area, Wyoming. M, 1959, University of Missouri, Columbia.

Hooper-Reid, N. M. Primary productivity, standing crop and seasonal dynamics of epiphytic algae in a southern Manitoba marsh pond. D, 1978, University of Manitoba.

Hoose, Lori A. A hydrogeochemical investigation of inorganic constituents in surface water and ground water in the vicinity of a disposal site in Clark County, Ohio. M, 1987, Wright State University. 130 p.

Hoose, Randolph Henry. Hydrochemical assessment of three streams draining the Dixie Bend Landfill, Pulaski County, Kentucky. M, 1986, Wright State University. 146 p.

Hooten, James E. Glacial geology of eastern Champaign County, Illinois. M, 1972, University of Illinois, Urbana. 58 p.

Hooten, John Albert. A preliminary investigation of the interrelationships of etch patterns (crystallography). M, 1952, Emory University. 68 p.

Hoots, Harold W. Geology of the Wheeler Ridge area, Kern County, California. D, 1925, Stanford University. 78 p.

Hoover, Amy L. Transect across the Salmon River suture, South Fork of the Clearwater River, western Idaho; rare earth element, geochemical, structural, and metamorphic study. M, 1987, Oregon State University. 138 p.

Hoover, Caroline. Determination of dispersivities from a natural-gradient dispersion test. M, 1985, Texas A&M University. 69 p.

Hoover, David Samuel. The development and evaluation of an automated reflectance microscope system for the petrographic characterization of bituminous coals. D, 1980, Pennsylvania State University, University Park. 282 p.

Hoover, Donald B. A two-dimensional seismic model study of a velocity transition zone. D, 1966, Colorado School of Mines. 92 p.

Hoover, Earl Gerald. Contributions to the stratigraphy and structure of the anticline east of Martinsburg, West Virginia. M, 1960, George Washington University.

Hoover, Elwin C. Stratigraphy of the lower Atoka Formation of southwestern Washington and northwestern Crawford counties, Arkansas. M, 1976, University of Arkansas, Fayetteville.

Hoover, James David. Basaltic volcanism during Lake Bonneville time, Black Rock Desert, Utah. M, 1974, Brigham Young University. 72 p.

Hoover, James David. Petrology of the Marginal Border Group of the Skaergaard Intrusion, East Greenland. D, 1982, University of Oregon. 709 p.

Hoover, John. Stratigraphy of the Monongahela Group (Pennsylvanian) in West Virginia, eastern Ohio and southwestern Pennsylvania. M, 1967, West Virginia University.

Hoover, John A. Ground-water resources of Sandusky County, Ohio. M, 1982, University of Toledo. 157 p.

Hoover, Jon R. A comparison of petrographic and geochemical aspects of some Pennsylvanian freshwater and marine carbonate rocks from Ohio and Pennsylvania. M, 1977, Bowling Green State University. 76 p.

Hoover, Julie Ann. Hydrodynamic and hydrologic investigation of a coastal landfill in southern Lake Michigan. M, 1988, Purdue University. 118 p.

Hoover, Karin A. Holocene paleohydrology and paleohydraulics of the Okanogan River, Washington. M, 1986, University of Washington. 128 p.

Hoover, Karl Victor. Joint patterns in the Chattanooga Shale of DeKalb County, Tennessee. M, 1954, University of Tennessee, Knoxville. 95 p.

Hoover, Linn. A summary of the stratigraphy and structural geology of Oregon and Washington. M, 1951, University of Michigan. 50 p.

Hoover, Linn. Geology of the Anlauf and Drain quadrangles, Douglas and Lane counties, Oregon. D, 1959, University of California, Berkeley. 110 p.

Hoover, Michael Thomas. Soil development in colluvium in footslope positions in the ridge and valley physiographic province of Pennsylvania. D, 1983, Pennsylvania State University, University Park. 286 p.

Hoover, P. R. The paleontology, taphonomy and paleoecology of the Palmarito Formation (Permian) of the Merida Andes, Venezuela. D, 1976, Case Western Reserve University. 648 p.

Hoover, Richard Alan. Areal geology and physical stratigraphy of a portion of the southern Santa Rosa Mountains, San Diego County, California. M, 1965, University of California, Riverside. 81 p.

Hoover, Richard Alan. Physiography and surface sediment facies of a recent tidal delta, Harbor Island, central Texas coast. D, 1968, University of Texas, Austin. 223 p.

Hoover, Steven Patrick. A detailed gravity survey of the Cedar Bog area, Ohio. M, 1976, Wright State University. 113 p.

Hoover, William Farrin. A highly weathered drift in the Kansas River valley between Manhattan and Kansas City, Kansas. M, 1932, University of Kansas. 44 p.

Hoover, William Farrin. A study of correlation value of insoluble residues of the Ste. Genevieve Limestone at selected localities in Illinois and adjacent areas. D, 1939, University of Illinois, Urbana.

Hope, Alvin C., Jr. Subsurface geology of the Claytonville area, Fisher County, Texas. M, 1956, Texas A&M University. 32 p.

Hope, Robert C. A paleobotanical analysis of the Sanford Triassic Basin, North Carolina. D, 1975, University of South Carolina. 81 p.

Hope, Robert C. Geomorphology of Sampson County, North Carolina. M, 1956, North Carolina State University. 54 p.

Hope, Roger Allen. Geology and structural setting of the eastern Transverse Ranges, Southern California. D, 1966, University of California, Los Angeles. 201 p.

Hopf, Robert W. Ostracodes and age of the Welden and Sycamore limestones, Mississippian, of Oklahoma. M, 1942, University of Missouri, Columbia.

Hopfinger, Carl. Textures and structures of Lake Erie sediment cores as revealed by X-ray radiography and continuous X-ray scanning. M, 1974, University of Toledo. 304 p.

Hopkins, David Moody. Quaternary geology of the Imuruk Lake area, Alaska. D, 1955, Harvard University.

Hopkins, Debbie L. A structural study of Durst Mountain and the north-central Wasatch Mountains, Utah. M, 1982, University of Utah. 50 p.

Hopkins, Don Eugene. Stratigraphic relations of the Nugget Sandstone, Wind River basin, Wyoming. M, 1962, University of Missouri, Columbia.

Hopkins, Edgar Member. Modification of relict coversands of the Florida Gulf Coast. D, 1974, University of Virginia. 152 p.

Hopkins, Edgar Member. Sedimentology of the Aguja Formation (Cretaceous) Big Bend National Park, Brewster County, Texas. M, 1965, University of Texas, Austin.

Hopkins, Henry Robert. Geology of western Louisa County, Virginia. D, 1960, Cornell University. 140 p.

Hopkins, Henry Robert. Magnetic intensities in the Lynchburg hematite and magnetite district, Virginia. M, 1957, University of Virginia. 104 p.

Hopkins, J. K. Water quality management planning for the Roaring Fork River basin in western Colorado. D, 1975, University of California, Los Angeles. 254 p.

Hopkins, James. Methods employed for exploration and development of a gold, silver, lead property in the Idaho Springs District, Colorado. M, 1927, University of Missouri, Rolla.

Hopkins, John Charles. Petrography, distribution and diagenesis of foreslope, nearslope and basin sediments, miette and ancient wall carbonate complexes (Devonian), Alberta. D, 1972, McGill University.

Hopkins, John Walter. Lithofacies and depositional environments of the Bigby-Cannon and adjacent formations (Ordovician) in North-central Tennessee. M, 1975, University of Kentucky. 79 p.

Hopkins, Kenneth Donald. Geology of the south and east slopes of Mount Adams volcano, Cascade Range, Washington. D, 1976, University of Washington. 143 p.

Hopkins, Kenneth Donald. Glaciation of Ingalls Creek Valley, east-central Casade Range, Washington. M, 1966, University of Washington. 79 p.

Hopkins, Kenneth W. Reconstruction of the paleoenvironments of Jameson (Strawn) reef field, Coke County, Texas. M, 1983, Texas A&M University.

Hopkins, L. A. Micropedology of a sequence of soils in the Turtle Mountain area. M, 1973, University of Manitoba.

Hopkins, M. E. The geology and coal resources of the Spadra District, Johnson County, Arkansas. M, 1951, University of Arkansas, Fayetteville.

Hopkins, M. E. The geology and petrology of the Anvil Rock Sandstone of southern Illinois. D, 1957, University of Illinois, Urbana. 98 p.

Hopkins, Oliver B. The Carboniferous Sphenophyllales, Equisetales and Lycopodiales of Maryland. D, 1912, The Johns Hopkins University.

Hopkins, Otho Neil. The geology of the Valley Mills and China Springs Quadrangle. M, 1961, Baylor University. 147 p.

Hopkins, Owen. Geological application of a programmable pocket calculator. M, 1977, Tulane University.

Hopkins, Ralph L. Depositional environments and diagenesis of the Fossil Mountain Member of the Kaibab Formation (Permian) Grand Canyon, Arizona. M, 1986, Northern Arizona University. 244 p.

Hopkins, Richard A. Regional seismic history of a power plant site in Northeast Mississippi. M, 1976, University of Tennessee, Knoxville. 37 p.

Hopkins, Robert T. Reservoir geology of the Captain Creek Limestone, Wilson Creek oil field, Ellsworth and Russell counties, Kansas. M, 1977, University of Kansas. 114 p.

Hopkins, Roy Marshall. A petrographic study of the Brassfield Limestone in southwestern Ohio. M, 1954, Ohio State University.

Hopkins, Theodor William. Microfracturing in Westerly Granite experimentally extended wet and dry at temperatures to 800°C and pressures to 200 mpa. M, 1986, Texas A&M University. 71 p.

Hopkins, Thomas. Currents off the continental shelf of Washington and Oregon. D, 1971, University of Washington. 207 p.

Hopkins, Thomas Cramer. Cambro-Silurian limonite ores of Pennsylvania. D, 1900, University of Chicago. 28 p.

Hopkins, Thomas Cramer. Marbles and other limestones. M, 1892, [Stanford University].

Hopkins, Willard N. Geology of the Newby Group and adjacent units in the southern Methow Trough, Northeast Cascades, Washington. M, 1987, San Jose State University. 95 p.

Hopkins, William H. An early Tertiary conglomerate in northwestern Wyoming. M, 1949, Syracuse University.

Hopkins, William Stephen, Jr. Palynology of Tertiary rocks of the Whatcom Basin, southwestern British Columbia and northwestern Washington. D, 1966, University of British Columbia. 184 p.

Hopkins, William Stephen, Jr. The geology of a portion of the Skagit Delta area, Skagit County, Washington. M, 1962, University of British Columbia. 135 p.

Hoppe, Wendel J. Origin and age of the Gabriel Peak Orthogneiss, North Cascades, Washington. M, 1984, University of Kansas. 79 2 plates e.

Hopper, George Steven. Magnetic properties of the olivine series Fe_2SiO_4–Mg_2SiO_4; Mn_2SiO_4–Mg_2SiO_4; Fe_2SiO_4–Mn_2SiO_4. D, 1968, Texas Christian University. 121 p.

Hopper, J. F. Long waves in and near the surf zone. M, 1967, United States Naval Academy.

Hopper, Jack T., Jr. Diagenesis and porosity development of lower Tuscaloosa sandstones in Wilkinson and Amite counties, Mississippi, and St. Helena Parish, Louisiana. M, 1988, Northeast Louisiana University. 95 p.

Hopper, John Wallace. The effects on rock properties of cycling heated compressed air in selected rocks, with emphasis on thermal properties. M, 1980, University of Wisconsin-Milwaukee.

Hopper, M. G. A study of liquefaction and other types of earthquake-induced ground failures in the Puget Sound, Washington, region. M, 1981, Virginia Polytechnic Institute and State University. 131 p.

Hopper, Margaret G. The determination of teleseismic earthquake magnitude at the Blacksburg, Virginia, seismograph observatory. M, 1972, Virginia Polytechnic Institute and State University.

Hopper, R. V. The manganese deposits of Tchiatouri, Georgia, Russia. M, 1929, McGill University.

Hopper, Richard H. A geologic section from the Sierra Nevada to Death Valley, California. D, 1939, California Institute of Technology. 122 p.

Hopper, Richard H. Electrical resistivities of synthetic oil sands. M, 1936, University of California, Los Angeles.

Hopper, Richard H. Magnetic studies in the Inglewood District (California). D, 1939, California Institute of Technology. 8 p.

Hopper, Sheridan Eileen. Paleoenvironment of the Middle Ordovician Guttenberg Formation in Southwest Wisconsin. M, 1978, University of Wisconsin-Madison.

Hopper, Walter Everett. The methods and cost of operating in the Caddo oil field, Louisiana. M, 1910, Cornell University.

Hopper, William M. Hydrogeology of southeastern Mercer County and northeastern Boyle County, Kentucky. M, 1985, University of Kentucky. 176 p.

Hoppie, James R. Morphologic responses of the Solomon River to geologic influences. M, 1980, University of Kansas. 183 p.

Hoppin, Richard A. Oscillations in the foraminifera of the Vicksburg Group from a well in George County, Mississippi. D, 1951, California Institute of Technology. 43 p.

Hoppin, Richard A. The geology of the Palen Mountains gypsum deposit, Riverside County, California. D, 1951, California Institute of Technology. 91 p.

Hoppler, Harl. Petrology and emplacement of the Long Potrero Pluton; a tail of one tonalite. M, 1983, San Diego State University. 44 p.

Hopson, Clifford Andrae. Petrology and structure of the Chelan Batholith, near Chelan, Washington; 2 volumes. D, 1955, The Johns Hopkins University.

Hoque, Mohammed Mozzammel. Numerical simulation of saltwater upconing in inland aquifers. D, 1983, Oklahoma State University. 163 p.

Hoque, Mominul. Stratigraphy, petrology, and paleogeography of the Mauch Chunk Formation in south-central and western Pennsylvania. D, 1965, University of Pittsburgh. 467 p.

Hoque, Monirul. Structure and petrology of the Gypsumville gypsum deposits. M, 1967, University of Saskatchewan. 94 p.

Hora, Marco Polo Pereira da Boa see da Boa Hora, Marco Polo Pereira

Horak, Ralph L. Structure of the Spring Mountain Paleozoic formations, Fremont County, Wyoming. M, 1952, Miami University (Ohio). 51 p.

Horall, Kenneth Bruce. The petrology of the Henderson Mountain Laccolith, Park County, Montana. M, 1966, Northern Illinois University. 85 p.

Horberg, Leland. The structural geology and physiography of the Teton Pass area, Wyoming. D, 1938, University of Chicago. 86 p.

Horcasitas, Gerardo Ruiz de la Pena see Ruiz de la Pena Horcasitas, Gerardo

Horck, Mark Patrick Vander see Vander Horck, Mark Patrick

Hore, R. C. A mechanical and heavy mineral analysis of some tills in the Trenton area, Ontario. M, 1962, University of Toronto.

Horen, Arthur. The manganese mineralization at the Merid Mine, Minas Gerais, Brazil. D, 1953, Harvard University.

Horick, Paul Joseph. Geology of the Warm Spring area, Wind River Mountains, Wyoming. M, 1948, University of Iowa. 120 p.

Horii, Hideyuki. Overall response and failure of brittle solids containing micro-cracks. D, 1983, Northwestern University. 163 p.

Horine, Robert Lee. Inversion and interpretation of gravity and magnetic data over a banded-iron formation near Atlantic City, Wyoming. M, 1986, University of Utah. 138 p.

Horino, Frank G. Geochemical-specrtographic prospecting at Malachite Mine, Jefferson County, Colorado. M, 1951, Colorado School of Mines. 72 p.

Horita, Masakuni. Modelling of cyclic behavior of saturated Monterey No. 0/30 sand. D, 1985, University of Colorado. 438 p.

Horkowitz, Kathleen O'Neill. Direct and indirect control of depositional fabric on porosity, permeability, and pore size and geometry; differential effect of sandstone subfacies on fluid flow, Cut Bank Sandstone, Montana. D, 1987, University of South Carolina. 136 p.

Horlacher, Craig F. Precambrian geology and gold mineralization in the vicinity of Ohio City, Gunnison County, Colorado. M, 1987, Colorado School of Mines. 223 p.

Horn, Ancel Dan. The sedimentary history of the Quinault Formation western Washington. M, 1969, University of Southern California. 179 p.

Horn, Clifford Layne Van see Van Horn, Clifford Layne

Horn, David Russell. Recent marine sediments and submarine topography, Sverdrup Islands, Canadian Arctic Archipelago. D, 1967, University of Texas, Austin.

Horn, James R. A snapshot of the Queen Charlotte fault zone obtained from P-wave refraction data. M, 1982, University of British Columbia. 87 p.

Horn, John E. Van see Van Horn, John E.

Horn, Marty. Physical model of pyroclastic clouds, Maricopa County, Arizona. M, 1986, Arizona State University. 125 p.

Horn, Michael D. Van see Van Horn, Michael D.

Horn, Myron K. A geological evaluation of porosity determination by radioactivity logging in the Permian Basin. M, 1958, University of Houston.

Horn, Myron Kay. A computer system for the geochemical balance of the elements. D, 1964, Rice University. 354 p.

Horn, Richard A. Neutron activation analysis of alkali metals in pegmatitic quartz and its fluid inclusions. M, 1972, Pennsylvania State University, University Park.

Horn, Robert Gary Van see Van Horn, Robert Gary

Horn, Robert Von see Von Horn, Robert

Horn, Stephen R. Van see Van Horn, Stephen R.

Horn, Toya D. Cordierite-anthophyllite-sulfide rocks at the Shepardson-Tapley Prospect, Brooksville, Maine. M, 1978, University of Washington.

Horn, William Lewis Van see Van Horn, William Lewis

Hornaday, Albert C. The geology of Nemaha County, Nebraska. M, 1932, University of Nebraska, Lincoln.

Hornaday, Gordon Raymer. Eocene biostratigraphy of the eastern Santa Ynez Mountains, California. D, 1970, University of California, Berkeley. 135 p.

Hornaday, Gordon Raymer. Upper Eocene foraminifera from south of Refugio Pass, California. M, 1954, University of California, Berkeley. 96 p.

Hornafius, John Scott. Paleomagnetism of the Monterey Formation in the western Transverse Ranges, California. D, 1984, University of California, Santa Barbara. 465 p.

Hornbacher, Dwight. Geology and structure of Kodachrome Basin State Reserve and vicinity, Kane and Garfield counties, Utah. M, 1985, Loma Linda University. 176 p.

Hornback, V. Quintin. The geology of the Salina-Ingram Gulch area, Boulder County, Colorado. M, 1956, University of Colorado.

Hornbaker, Allison Lynn. Structural and stratigraphic oil. M, 1947, University of Michigan.

Hornbeck, David Earl. Laboratory modelling of reinforced earth. D, 1982, Georgia Institute of Technology. 435 p.

Hornbeck, James M. Study and description of some upper Paleozoic corals from the Arco Hills, Southcentral Idaho. M, 1976, SUNY, College at Oneonta. 61 p.

Hornbeck, Ross Wright. Topographic and geological mapping of a part of the Morgan Creek basin. M,

1937, University of North Carolina, Chapel Hill. 28 p.

Hornberger, George Milton. Numerical studies of composite soil-moisture ground-water systems. D, 1969, Stanford University. 113 p.

Hornberger, Joseph, Jr. The geology of Throckmorton County, Texas (H783). M, 1932, University of Texas, Austin.

Hornberger, Roger J. Delineation of acid mine drainage potential of coal-bearing strata of the Pottsville and Allegheny groups in western Pennsylvania. M, 1985, Pennsylvania State University, University Park. 558 p.

Hornbostel, Scott. Iterative deconvolution using generalized "positivity". D, 1988, University of Houston.

Horne, Doyle Jackson. Currents and water characteristics around the West Flower Garden Bank. D, 1986, Texas A&M University. 184 p.

Horne, Gregory Stuart. Stratigraphy and structural geology of south-western New World Island area, Newfoundland. D, 1969, Columbia University. 280 p.

Horne, Jerry D. A field and petrographic study of the lower Whitsett Formation of the Jackson Group (Eocene; Texas). M, 1961, [University of Houston].

Horne, Jerry D. Geology of southeastern Atascosa County, Texas. M, 1961, University of Houston.

Horne, John Corbett. Detailed correlation and environmental study of some Late Pennsylvanian units of the Illinois Basin. D, 1968, University of Illinois, Urbana. 77 p.

Horne, John Corbett. Environmental study of the Bond Formation of the Illinois Basin and the Kansas City Group (Upper Pennsylvanian) of the northern and central Mid-continent. M, 1965, University of Illinois, Urbana.

Horne, Mary E. Environmental geology in York, Pa and the surrounding region. M, 1973, Pennsylvania State University, University Park.

Horne, Stewart Walsh. The stratigraphy of the Walnut Formation (Cretaceous) in Lampasas, Williamson, Travis, Hays, and Comal counties, Texas (H784). M, 1930, University of Texas, Austin.

Hornedo, Mercedes. Pleistocene carnivores of Dry Cave, Eddy County, New Mexico. M, 1972, University of Texas at El Paso.

Horner, Greg James. Stratigraphy and petrology of the Sylamore Sandstone (Devonian-Mississippian), north-central Arkansas. M, 1984, University of New Orleans. 202 p.

Horner, Rick Oliver. Petrography, diagenesis, and depositional environments, St. Joe Formation (Lower Mississippian), northern Arkansas. M, 1985, University of New Orleans. 144 p.

Horner, Timothy C. Depositional environments, diagenesis and porosity relationships in the Mission Canyon Formation, Elkhorn Ranch Field, Billings County, North Dakota. M, 1986, Texas Tech University. 114 p.

Horner, Wesley Pate. The Fox Hills-Laramie Contact (Upper Cretaceous) in Denver Basin; Colorado. M, 1954, University of Colorado.

Horner, William J. Paleocurrent studies of the middle and upper Keweenawan conglomerates of Michigan. M, 1960, University of Kansas.

Horney, William Rolland. The lithologic study of the Marsland Formation in Nebraska. M, 1941, University of Nebraska, Lincoln.

Horng, Fu-Wen. Anion retention mechanisms in four Forest soils; relevance to the leaching process. D, 1983, University of Washington. 168 p.

Hornig, Carl A. The barite deposits at Kings Creek, South Carolina. M, 1973, University of South Carolina. 51 p.

Horning, Bryan Lee. Asymmetric planetary electrical induction. D, 1975, University of California, Los Angeles. 206 p.

Horning, Thomas S. The geology, igneous petrology, and mineral deposit of the Ataspaca mining district,

Department of Tacna, Peru. M, 1988, Oregon State University. 402 p.

Hornsey, Edward Eugene. A theoretical investigation of elastic and Voigt transient spherical waves, and plane three-element viscoelastic waves. D, 1967, University of Missouri, Rolla. 143 p.

Hornstein, Owen Merle. A field and petrographic study of some extrusive and sedimentary rocks along the Carp and Little Carp rivers in Ontonagan and Gogebic counties, Michigan. M, 1950, Michigan State University. 55 p.

Hornung, Arthur G. Origin of sulphur deposits. M, 1933, University of Chicago. 53 p.

Horodyski, Robert Joseph. Bedrock geology of portions of Fish River Lake, Winterville, Greenlaw, and Mooseleuk Lake quadrangles, Aroostook County, Maine. M, 1968, Massachusetts Institute of Technology. 192 p.

Horodyski, Robert Joseph. Stromatolites and paleoecology of parts of the middle Proterozoic Belt Supergroup, Glacier National Park, Montana. D, 1973, University of California, Los Angeles. 264 p.

Horowicz, Leon. Two methods for separating higher modes of seismic surface waves. M, 1966, Massachusetts Institute of Technology. 54 p.

Horowitz, A. Fate of petroleum hydrocarbons in nearshore Arctic aquatic ecosystems. D, 1979, University of Louisville. 206 p.

Horowitz, Alan Stanley. Fauna of Glen Dean Limestone (Chester) in Indiana and northern Kentucky. D, 1956, Indiana University, Bloomington. 450 p.

Horowitz, Alan Stanley. Petrography of the Nunatarssuaq area, Northwest Greenland. M, 1954, Ohio State University.

Horowitz, Carol G. Benthic foraminifera of the Oligocene Suwannee Limestone, Georgia and northern Florida. M, 1979, University of Georgia.

Horowitz, Daniel Henry. Petrology of the Upper Ordovician and Lower Silurian rocks in the central Appalachians. D, 1965, Pennsylvania State University, University Park. 221 p.

Horowitz, M. R. An electrophoretic mobility study of suspended sediments in river and low salinity waters. M, 1976, Dalhousie University.

Horowitz, Marcie R. Diagenesis, metamorphism, and sulfide mineralization of the Balls Bluff Siltstone (Upper Triassic), Culpeper Basin, Virginia. M, 1988, Bryn Mawr College. 136 p.

Horowitz, Martin. The St. Peter-Glenwood problem in Michigan. M, 1961, Michigan State University. 59 p.

Horowitz, Seymour. The geology of the southwestern part of the Syracuse East, 7 1/2 minute quadrangle, New York. M, 1955, Syracuse University.

Horowitz, Warren Lee. A petrographic and structural comparison of five traverses across a transition zone of the Hackneyville Schist in Clay County, Alabama. M, 1980, Memphis State University.

Horrall, Kenneth Bruce. Mineralogical, textural, and paragenetic studies of selected ore deposits of the Southeast Missouri lead-zinc-copper district and their genetic implications. D, 1982, University of Missouri, Rolla. 714 p.

Horrell, Mark Alan. Stratigraphy and depositional environments of the Oregon Formation (Middle Ordovician) of central Kentucky. M, 1981, University of Kentucky. 121 p.

Horsburgh, Martha Sennett. The petrography, petrology and depositional environments of the Lucas Dolomite (Devonian) and adjacent rocks at its stratotype, Lucas County, Ohio. M, 1975, University of Nebraska, Lincoln.

Horscroft, F. D. M. The petrology of gabbroic sills in the volcanic series of Roy and McKenzie townships, Chibougamau region, Quebec. D, 1957, McGill University.

Horsecroft, F. D. The petrology of gabbroic sills in the volcanic series of Roy and McKenzie townships, Chibougamau region, Quebec. D, 1957, McGill University.

Horsfall, John Clayton. Studies in the beneficiation of Washington soapstones. M, 1952, University of Washington. 44 p.

Horsky, S. A study of pocket K-feldspar, Himalaya Pegmatite, Mesa Grande District, California. M, 1974, McGill University. 109 p.

Horstman, Arden William. Correlation of the Lower and Lower Upper Cretaceous rocks of a part of southwestern Wyoming, Utah, and Colorado. D, 1966, University of Colorado. 190 p.

Horstman, Arden William. On the origin of cone-in-cone structure. M, 1954, University of Cincinnati. 62 p.

Horstman, Elwood Louis. Distribution of lithium in volcanic rocks of West Texas. M, 1953, University of Minnesota, Minneapolis.

Horstman, Elwood Louis. The distribution of lithium, rubidium and cesium in igneous and sedimentary rocks. D, 1955, University of Minnesota, Minneapolis.

Horstman, Kevin C. Geology of the Club Ranch area, Mazatzal Mountains, Arizona. M, 1980, Northern Arizona University. 71 p.

Horstmann, Kent M. Estuarine-tidal flat depositional model for the Paleocene-Eocene Huber Formation, east-central Georgia. M, 1983, Duke University. 203 p.

Horton, Albert Bergen. Trace element content and distribution in the Wells Creek Formation of middle Tennessee. M, 1981, Vanderbilt University. 107 p.

Horton, Duane. Clay mineralogy and origin of the Huntingdon fire clays on Canadian Sumas Mountain, Southwest British Columbia. M, 1978, Western Washington University. 96 p.

Horton, Duane Gale. Argillic alteration associated with the amethyst vein system, Creede mining district, Colorado. D, 1983, University of Illinois, Urbana. 463 p.

Horton, Ernest Henderson. Some Lower Devonian Ostracoda from northern New Jersey. M, 1950, Rutgers, The State University, New Brunswick. 57 p.

Horton, Frank R. Pre-Cretaceous structural history of J-M/Brown-Bassett field area, Crockett, Terrell, Val Verde counties, Texas. M, 1977, Texas Christian University. 33 p.

Horton, Gary Walker. The morphology and evolution of the ostracod duplicature. M, 1961, University of Illinois, Urbana.

Horton, James Wright, Jr. Geology of the Kings Mountain and Grover quadrangles, North and South Carolina. D, 1977, University of North Carolina, Chapel Hill. 174 p.

Horton, James Wright, Jr. Geology of the Rosman area, Transylvania County, North Carolina. M, 1974, University of North Carolina, Chapel Hill. 63 p.

Horton, John W. The geology of the Mam-a-gah picnic area, Tucson Mountains, Pima County, Arizona. M, 1966, University of Arizona.

Horton, K. A. Geology of the upper Southwest Rift Zone of Haleakala. M, 1977, University of Hawaii. 115 p.

Horton, Leo V. The occurrence of petroleum in structures other than anticlines. M, 1922, Yale University.

Horton, Marc Allan. The role of the sediments in the phosphorus cycle of Lake Sammamish. M, 1972, University of Washington. 220 p.

Horton, Marvin Dean. Stratigraphy and structure of the Old Woman Anticline, Niobrara County, Wyoming. M, 1953, University of Nebraska, Lincoln.

Horton, Paul. A seismic reflection survey over the Wayne-25 oil field in Cass County, Michigan. M, 1987, Western Michigan University.

Horton, Robert A., Jr. Dolomitization and diagenesis of the Leadville Limestone (Mississippian), central Colorado. D, 1985, Colorado School of Mines. 178 p.

Horton, Robert A., Jr. Structural analysis of the Jacksboro Fault, East Tennessee. M, 1977, University of Tennessee, Knoxville. 79 p.

Horton, Robert B. Structural charge during diagenesis of illite-smectite mixed layer clays. M, 1981, University of Missouri, Columbia.

Horton, Robert Carlton. Statistical studies concerning the distribution of mining districts and mineral deposits in Nevada. M, 1965, University of Nevada - Mackay School of Mines. 22 p.

Horton, Roger Goldsmith. A study of the geologic history of the Potomac River and its environs (Maryland, Virginia, West Virginia). M, 1931, Catholic University of America. 81 p.

Horton, Wayne C. Foraminifera of the Cenozoic and Recent genus Sphaerogypsina Galloway. M, 1962, University of Missouri, Rolla.

Horvath, Allan Leo. Stratigraphy and paleontology of the Middle Devonian strata penetrated by core H-1A in Wayne County, Michigan. M, 1957, University of Michigan.

Horvath, Allan Leo. Stratigraphy of the Silurian rocks of southern Ohio and adjacent parts of West Virginia, Kentucky and Indiana. D, 1964, Ohio State University. 178 p.

Horvath, Edward Alexander. Preliminary fusulinid zonation of the Naco Formation in east-central Arizona. M, 1960, University of Utah. 104 p.

Horvath, Emilio Hubert. Spectral properties of Arizona soils and rangelands and their relationship to Landsat digital data. D, 1981, University of Arizona. 196 p.

Horvath, George. A sedimentologic study of the Pensacola Bay complex, northwestern Florida. M, 1968, Florida State University.

Horvath, George J. The geochemistry and transport of Mn, Fe, Co, Cu, Zn, Cd and Pb in the freshwater and estuarine environments of the Big Cypress-Everglades region of Florida. D, 1973, Florida State University.

Horvath, Peter. Analysis of lunar seismic signals; determination of instrumental parameters and seismic velocity distributions. D, 1979, University of Texas at Dallas. 229 p.

Horwood, C. H. Cross Lake map area, Manitoba. D, 1934, Massachusetts Institute of Technology. 173 p.

Horwood, H. C. Granite contact action in eastern Ontario. M, 1931, Queen's University. 54 p.

Horwood, H. C. The Cross Lake map area, Manitoba. D, 1934, University of Manitoba. 167 p.

Horzempa, L. M. The nucleation, growth, and flocculation of colloidal CuS sols. D, 1977, University of Maryland. 260 p.

Hose, David R. A conditional probability model of vegetation and climate relations in the southwestern United States. M, 1986, Kent State University, Kent. 219 p.

Hosein, I. Gravity studies over the Saint Barnabe fault. M, 1965, McGill University.

Hosek, Ronald Joseph. Quantitative hydrogeology and evaluation of water-management alternatives, Mira Valley area in Valley County, Nebraska. M, 1975, University of Nebraska, Lincoln.

Hosey, Lisa Elaine. Resident perception of shoreline recession in Nags Head, North Carolina. M, 1987, Southwest Missouri State University. 93 p.

Hosfeld, Richard K., Jr. Ground-water resources of Ottawa County, Ohio. M, 1984, University of Toledo. 112 p.

Hosford, Gregory F. Ground-water geology of Pleasant Grove, Utah, and vicinity. M, 1950, Brigham Young University. 45 p.

Hoshi, Kazuyoshi. Miocene ocean floor metamorphism during back-arc spreading in the Japan Sea. M, 1988, Stanford University. 172 p.

Hoskin, Charles Morris. Recent carbonate sedimentation on Alacran Reef, Yucatan, Mexico. D, 1962, University of Texas, Austin. 270 p.

Hosking, Peter Leighton. The relationship between lithology and surface form in a fluvially dissected landscape; an examination of the morphologic character of the upper portion of the Saint Francis River basin, Missouri,... to determine... lithologic influences on landscape evolution. D, 1967, Southern Illinois University, Edwardsville.

Hoskins, Benjamin Wayne. Computer-assisted joint study of Pennsylvanian carbonate banks in the Graford Formation, north central Texas. M, 1982, University of Texas, Arlington. 104 p.

Hoskins, Cortez William. Geology and paleontology of Coyote Hills, Orange County, California. M, 1955, Pomona College.

Hoskins, Cortez William. Paleoecology and correlation of the lowest emergent California marine terrace, from San Clemente to Halfmoon Bay. D, 1957, Stanford University. 226 p.

Hoskins, Donald Martin. Middle Ordovician silicified brachiopods from the Chaumont Formation of New York. M, 1954, University of Rochester. 91 p.

Hoskins, Donald Martin. The stratigraphy and paleontology of the Silurian Bloomsburg Formation (red beds) of Pennsylvania. D, 1960, Bryn Mawr College. 241 p.

Hoskins, Hartley. Seismic reflection observation on the Atlantic continental shelf, slope and rise southeast of New England. D, 1965, University of Chicago. 63 p.

Hoskins, John Richard. Design and construction of a basic geomechanics laboratory. D, 1962, University of Utah. 134 p.

Hosmer, Henry Liggett. A microscopic study of the tin ores of Morococala District, Bolivia. M, 1950, University of Michigan.

Hosmer, Henry Liggett. Geology and structural development of the Andean System of Peru. D, 1959, University of Michigan. 320 p.

Hossain, Aolad. The effect of urbanization on hydrology of watersheds with special reference to rainfall excess. D, 1974, Purdue University.

Hossain, Syed Abul. Application of numerical model for ground-water management of piedmont aquifer in central Dinajpur, Bangladesh. M, 1981, Oklahoma State University. 198 p.

Hosseini, Masood S. Bioherms of the Laborcita Formation, northern Sacramento Mountains, New Mexico. M, 1980, University of Southwestern Louisiana. 71 p.

Hossley, James Glenn. Use of salt marsh foraminifera in developing a model for the evolution of a barrier island; Plum Island, Massachusetts, U.S.A. M, 1986, Acadia University. 182 p.

Host, George Edward. Spatial patterns of forest composition, successional pathways and biomass production among landscape ecosystems of northwestern Lower Michigan. D, 1987, Michigan State University. 285 p.

Hosted, Joseph Orrin. Associations of platinum and chromite in basic magnesian rocks. M, 1919, University of Minnesota, Minneapolis.

Hostenske, Dale J. A subsurface study of the Berea Sandstone in a portion of Lincoln County, West Virginia. M, 1975, Bowling Green State University. 76 p.

Hostetler, Charles James. Equilibrium properties of some silicate materials; a theoretical study. D, 1982, University of Arizona. 89 p.

Hostetler, James M. The correlation of the lower bentonite bed in the Carlile Shale of Kansas. M, 1967, Wichita State University. 62 p.

Hostetler, Paul Blair. Low temperature relations in the system; MgO-SiO$_2$-CO$_2$-H$_2$O. D, 1961, Harvard University.

Hostetter, Heber P., III. Growth, reproduction, and survival of diatoms. D, 1969, University of Arizona.

Hotchkiss, Frances Luellen Stephenson. Observed circulation and inferred sediment transport in Hudson submarine canyon. D, 1982, Woods Hole Oceanographic Institution. 224 p.

Hotchkiss, Frederick. Studies on Paleozoic ophiuroids and ancestry of the Asterozoa. D, 1974, Yale University.

Hotchkiss, Samuel A. The geology of the Hell's Half Acre area, Southeast Idaho. M, 1976, Idaho State University. 51 p.

Hotchkiss, William Otis. Mineral land classification, showing indications of iron formation in parts of Ashland, Price, Oneida, Forest, Rusk, Barron, and Chippewa counties. D, 1916, University of Wisconsin-Madison.

Hotrabhavananda, Tachpong. Development of automated techniques for creating a natural resources information system. D, 1980, University of Missouri, Columbia. 252 p.

Hotson, Crispian John. The evolution and impact of government intervention in the South African gold mining industry. M, 1977, Stanford University. 87 p.

Hott, Albert C. Geology of the Sill Lake area, Wyoming. M, 1962, University of Colorado.

Hottman, W. E. Areal distribution of clay minerals and their relationship to physical properties, Gulf of Mexico. M, 1975, Texas A&M University.

Hottman, W. E. Physical properties of sediments from the continental margin of western Africa. D, 1978, Texas A&M University. 138 p.

Hotton, Carol Louise. Palynology of the Cretaceous-Tertiary boundary in central Montana, U.S.A., and its implications for extraterrestrial impact. D, 1988, University of California, Davis. 732 p.

Hotton, Nicholas, III. A survey of adaptive relationships of dentition to diet in the North American Iguanidae. D, 1950, University of Chicago. 68 p.

Hotz, Preston Enslow. Paleozoic volcanic rocks in the Medford Quadrangle, Oregon; a petrographic study. M, 1940, University of California, Berkeley. 40 p.

Hotz, Preston Enslow. Petrology and habit of some diabase sheets in southeastern Pennsylvania. D, 1949, Princeton University. 101 p.

Houck, James Edward. The potential utilization of scleractinian corals in the study of marine environments. D, 1978, University of Hawaii. 199 p.

Houck, Karen J. Petrography and petrology of lunar soils from the Apollo 16 Site. M, 1982, Indiana University, Bloomington. 66 p.

Houck, Richard Thomas. Application of the finite element method to potential field problems. M, 1974, Pennsylvania State University, University Park. 60 p.

Houck, Richard Thomas. Subsurface imaging with ground penetrating radar. D, 1984, Pennsylvania State University, University Park. 172 p.

Houde, Richard Francis. Sedimentology, diagenesis, and source bed geochemistry of the Spraberry Sandstone, subsurface Midland Basin, West Texas. M, 1979, University of Texas at Dallas. 198 p.

Houde, Robert. Etude des granulats à béton réactifs aux alcalis de la région de Trois-Rivières, Québec. M, 1986, Ecole Polytechnique. 221 p.

Hough, Alan N. A study of the thin-bedded volcaniclastic sediments of the Padre Miguel Group, southeastern Guatemala. M, 1980, University of Texas, Arlington. 117 p.

Hough, Jack L. The mechanical composition of the deposits of southern Lake Michigan. M, 1934, University of Chicago. 43 p.

Hough, Jack L. The sediments of Buzzards Bay and Cape Cod Bay, Massachusetts. D, 1940, University of Chicago. 83 p.

Hough, James Emerson. Regional subsurface structure, McLean County, Kentucky. M, 1958, University of Kentucky. 18 p.

Hough, Leo Willard. Petrographic comparison of Mississippi River and tributary bed material sands. M, 1937, Louisiana State University.

Hough, Margaret Jean. The auditory region of some fossil mustelines. M, 1942, University of Chicago. 26 p.

Hough, Ronald David. Little Sac Woods Metro Forest; a land management methodology and plan. M, 1981, Southwest Missouri State University. 60 p.

Hough, Susan Elizabeth. The attenuation of high frequency seismic waves. D, 1987, University of California, San Diego. 373 p.

Hough, Van Ness Dearborn. Joint orientations of the Appalachian Plateau in southwestern Pennsylvania. M, 1959, Pennsylvania State University, University Park. 82 p.

Houghton, Jonathan Parks. The intertidal ecology of Kiket Island, Washington, with emphasis on age and growth of Protothaca staminea and Saxidomus giganteus (Lamellibranchia; veneridae). D, 1973, University of Washington. 179 p.

Houghton, LeRoy Kingsbury, III. Selected trace elements and thermal stratification of Lake Lynn, West Virginia. M, 1973, University of Akron. 64 p.

Houghton, Marcia Lea. Geochemistry of the Proterozoic Hormuz Evaporites, southern Iran. M, 1980, University of Oregon. 85 p.

Houghton, Wendy Priestley. Structural analysis of the Salt Lake-Provo segment boundary of the Wasatch fault zone. M, 1986, University of Utah. 64 p.

Hougland, E. S. Air pollutant monitor network design using mathematical programming. D, 1977, Virginia Polytechnic Institute and State University. 365 p.

Hougland, Everett. The stratigraphy of the "Orient Gneiss". M, 1933, Washington State University. 37 p.

Houlday, Mark. Seismologic implications of post 1980 small magnitude earthquakes that occurred in regions of New York State that are characterized by low level rates of seismicity. M, 1983, Rutgers, The State University, Newark. 80 p.

Houle, Julie. Analysis of Moss Beach, San Mateo County, California. M, 1977, Stanford University.

Houle, Julie A. Depositional systems, sandstone diagenesis, and cobble study in the Lower Pennsylvanian Taos Trough, northern New Mexico. M, 1980, University of Texas, Austin.

Houlette, Kenneth N. The geology of McGill Anticline, Albany County, Wyoming. M, 1947, University of Wyoming. 39 p.

Houlik, Charles William. Carboniferous supratidal flat and desert sabkha sedimentation, western Wyoming. M, 1970, Rutgers, The State University, New Brunswick. 27 p.

Houlik, Charles William, Jr. Significance of carbonate-clastic transitions in the carboniferous of western Wyoming. D, 1972, Rutgers, The State University, New Brunswick. 58 p.

Houng, Kun- Huang. A study on the soils containing amorphous materials in the Island of Hawaii. D, 1964, University of Hawaii. 187 p.

Houng-Ming, Joung. The formulation and application of a generalized water quality index based on multivariate factor analysis of water quality from agricultural irrigation return flows. D, 1978, University of Nevada - Mackay School of Mines. 106 p.

Hounslow, Arthur William. Chemical petrology of some Grenville (Precambrian) schists near Fernleigh, Ontario. M, 1965, Carleton University. 87 p.

Hounslow, Arthur William. Crystal structures of two naturally occurring chlorapatites. D, 1968, Carleton University. 87 p.

Houpt, John Ronald. Field and laboratory investigations of the Elliott County, Kentucky peridotite dikes. M, 1968, Miami University (Ohio). 62 p.

House, Gordon D. Geology of the shale hosted zinc-lead deposits, Howard's Pass, Yukon Territory and District of MacKenzie, N.W.T. M, 1980, University of Alaska, Fairbanks. 139 p.

House, Larry A. A geophysical and subsurface investigation of the Illinois Basin. M, 1980, University of Texas at El Paso.

House, Leigh Scott. Studies of earthquakes and tectonics in the eastern Aleutians. D, 1982, Columbia University, Teachers College. 192 p.

House, Richard D. Radioactivity and geology of the Falcon Lake Stack, southern Manitoba. M, 1955, Northwestern University.

House, Valerie Hust. Pleistocene geology of Wyandot County, Ohio. M, 1985, Bowling Green State University. 91 p.

Householer, Earl R. Geology of Mohave County, Arizona. M, 1930, University of Missouri, Rolla.

Houseknecht, David W. Transportational and depositional history of the Lamotte Sandstone of southeastern Missouri. M, 1975, Southern Illinois University, Carbondale. 145 p.

Houseknecht, David Wayne. Petrology and stratigraphy of some Pottsville quartzites and graywackes of West Virginia. D, 1978, Pennsylvania State University, University Park. 255 p.

Houseman, Michel Dirk. Petrology and alteration of the Grouse Creek granodiorite porphyry. M, 1983, Washington State University. 166 p.

Houser, Brenda. Erosional history of the New River, Southern Appalachians, Virginia. D, 1980, Virginia Polytechnic Institute and State University. 309 p.

Houser, Frederick N. The geology of the Contention Mine area, Twin Buttes, Arizona. M, 1949, University of Arizona.

Houser, Gilbert L. Studies in the Cretaceous invertebrate fauna of the Atlantic, Gulf, and Rocky Mountains regions. M, 1891, Iowa State University of Science and Technology.

Houser, Gilbert Logan. Genera of Paleozoic corals of the Order Madreporaria; comprising a description of the more important North America genera and a conspectus of characteristic species. M, 1892, University of Iowa. 174 p.

Houser, John Foster. Structural geology of Threemile Hill area, Brewster County, Texas. M, 1967, University of Texas, Austin.

Houser, Kenneth L. Influence of depositional environments on texture and composition of the Muddy Sandstone, eastern Powder River basin, Wyoming. M, 1982, University of Missouri, Columbia.

Houseworth, James Evan. Longitudinal dispersion in nonuniform, isotropic porous media. D, 1984, California Institute of Technology. 252 p.

Housman, John J., Jr. The occurrence and distribution of selected heavy metals lead, cadmium and copper in the Ten Mile River system, Massachusetts and Rhode Island. M, 1975, Boston University. 110 p.

Houssiere, L. I. Studies in salt resistant drilling muds. M, 1941, Massachusetts Institute of Technology. 79 p.

Houston, Betty Green. Proterozoic Uncompahgre Formation; remnant of a Precambrian fold and thrust belt. M, 1983, University of Texas, Austin. 92 p.

Houston, Heidi Beth. Source characteristics of large-earthquakes at short periods. D, 1987, California Institute of Technology. 156 p.

Houston, Mark Harig, Jr. Numerical models of free convection in the Earth's upper mantle with radiogenic heating and variable viscosity. D, 1974, Rice University. 237 p.

Houston, Robert Stroud, Jr. Genetic study of some pyrrhotite deposits of Maine and New Brunswick. D, 1954, Columbia University, Teachers College.

Houston, Robert Stroud, Jr. Tungsten deposits in Cabarrus County, North Carolina. M, 1950, North Carolina State University. 56 p.

Houston, William Newton. Formation mechanisms and property interrelationships in sensitive clays. D, 1967, University of California, Berkeley. 183 p.

Houston, William Norman. The surface chemistry and geochemistry of feldspar weathering. M, 1972, McMaster University. 126 p.

Houten, Franklyn Bosworth Van *see* Van Houten, Franklyn Bosworth

Hovdebo, H. R. Structure of the Brule-Crossing Creek area, British Columbia. M, 1950, University of Saskatchewan. 46 p.

Hover, Frank Bryan. Geology of the east half of the Johnstown and Creighton quadrangles, Henry County, Missouri. M, 1958, University of Missouri, Columbia.

Hovey, Edmund O. Observations on some of the trap ridges of the East Haven-Branford region (Connecticut). D, 1889, Yale University.

Hoving, Sheryl J. The effects of different thicknesses of limestone and soil over pyritic material on leachate quality. M, 1982, Southern Illinois University, Carbondale. 90 p.

Hovis, Guy Leader. Thermodynamic properties of monoclinic potassium feldspars. D, 1971, Harvard University.

Hovland, David. Geology of the northwest part of the Lower Valley Quadrangle, Caribou County, Idaho. M, 1981, San Jose State University. 108 p.

Hovland, Nancy K. Geochemistry of dissolved and fine particulate matter in the Satilla River. M, 1980, Georgia Institute of Technology. 58 p.

Hovorka, Susan Davis. Stratigraphy and petrography of the Upper Chert and Shale Member, Caballos Formation, Brewster County, West Texas. M, 1981, University of Texas, Austin. 272 p.

Howard, Alan Dighton. A study of process and history in desert landforms near the Henry mountains, Utah. D, 1970, The Johns Hopkins University. 361 p.

Howard, Arthur D. History of the Grand Canyon of the Yellowstone (Wyoming). D, 1937, Columbia University, Teachers College.

Howard, C. Edward. Petrography of the Jackfork sandstone at De Gray Dam, Clark County, Arkansas. D, 1963, Louisiana State University. 213 p.

Howard, C. Scott. Geological and geophysical investigations in the Wilmington Complex/Wissahickon Formation boundary area, Delaware Piedmont. M, 1986, University of Delaware.

Howard, Calhoun L. H. X-ray irradiation of halite. M, 1958, Columbia University, Teachers College.

Howard, Calhoun Ludlow Harper. Irradiation and pressure effects in halite. D, 1963, Columbia University, Teachers College. 77 p.

Howard, Carolyn Kheboian. Selective extraction of aquatic sediments. D, 1988, University of New Hampshire. 129 p.

Howard, Charles Spaulding. Suspended matter in the Colorado River and its relation to the development of the river. D, 1928, American University. 104 p.

Howard, Clarence Edward. Petrography of the Sampson County area, North Carolina. M, 1955, North Carolina State University. 51 p.

Howard, Conrad B. Geology of the White Butte area and vicinity, Mitchell Quadrangle, Oregon. M, 1955, Oregon State University. 118 p.

Howard, David Ayers. Economic geology of Quartz Creek, King County, Washington. M, 1967, University of Washington. 48 p.

Howard, Edgar B. Evidence of early man in North America based on geological and archaeological work in New Mexico. D, 1935, University of Pennsylvania.

Howard, Edward Viet. A socioeconomic study of copper leaching at Santa Rita. M, 1963, New Mexico Institute of Mining and Technology. 120 p.

Howard, Ephraim Manasseh. Application of the Mossbauer effect in determination of phase diagrams. D, 1967, University of California, Davis. 82 p.

Howard, George Wilberforce. The meandering of streams. M, 1942, George Washington University. 48 p.

Howard, Hildegarde. A review of the fossil bird Parapavo californicus from the Pleistocene asphalt beds at Rancho La Brea. M, 1926, University of California, Berkeley. 57 p.

Howard, Hildegarde. The avifauna of Emeryville Shellmound. D, 1929, University of California, Berkeley. 144 p.

Howard, James Campbell. Simulation of salt dome forms. D, 1969, Stanford University. 258 p.

Howard, James D. Dispersal patterns in the Fountain Formation of Colorado. M, 1962, University of Kansas.

Howard, James Dolan. Upper Cretaceous Panther Sandstone Tongue of east-central Utah, its sedimentary facies and depositional environments. D, 1966, Brigham Young University. 215 p.

Howard, James F. The foraminifera and paleoecology of the type localities of the Waccamaw Formation (Pliocene?) of the Carolinas. M, 1963, University of Houston.

Howard, James Franklin. Biostratigraphy and paleoecology of the Duplin (late Miocene) and Waccamaw (Pliocene?) formations of North and South Carolina. D, 1966, Indiana University, Bloomington. 166 p.

Howard, James Hatten, III. Geochemical behavior of selenium in earth-surface environments. D, 1969, Stanford University. 364 p.

Howard, James Jennings. Diagenesis of mixed-layer illite-smectite in interlaminated shales and sandstones. D, 1979, SUNY at Binghamton. 156 p.

Howard, James Michael. Transition metal geochemistry and petrography of the potash Sulfur Springs intrusive complex, Garland County, Arkansas. M, 1974, University of Arkansas, Fayetteville.

Howard, James R. The relationship of the subsurface geology to the petroleum accumulation in Ellsworth County, Kansas. M, 1958, Kansas State University. 43 p.

Howard, Jeffrey Kellogg. The origin and significance of "grazing-step" terracettes. M, 1982, University of California, Davis. 137 p.

Howard, Jeffrey Lynn. Paleoenvironments, provenance and tectonic implications of the Sespe Formation, Southern California. D, 1987, University of California, Santa Barbara. 337 p.

Howard, Jesse James. A stratigraphic and paleontologic study of a well drilled in Stone County, Mississippi. M, 1944, University of Texas, Austin.

Howard, John Hall. Structural development of the Williams Range thrust, Colorado. D, 1961, Columbia University, Teachers College. 157 p.

Howard, John K. The stratigraphy and structure of the Cape Sebastian-Crook Point area, Southwest Oregon. M, 1961, University of Wisconsin-Madison. 52 p.

Howard, K. M. The successional pattern and development of benthic diatom assemblages in Fort Pond Bay, Montauk, New York. D, 1975, New York University. 263 p.

Howard, Keith Arthur. Structure of the metamorphic rocks of the northern Ruby Mountains, Nevada. D, 1966, Yale University. 302 p.

Howard, Kenneth Leon, Jr. Pyrophyllite-topaz alteration in the ore deposit at Butte [Central District], Montana. D, 1972, University of California, Berkeley. 162 p.

Howard, Kerry S. Surficial manifestations of deep-seated Permian salt bed dissolution near Loco Hills, New Mexico. M, 1987, Texas Tech University. 67 p.

Howard, Lauran L. An evaluation of inertial effects in a well-aquifer system. M, 1977, University of Nevada. 71 p.

Howard, Lawrence H. Internal stratigraphy of the Borden Formation (Lower Mississippian) in east-central Kentucky. M, 1987, Eastern Kentucky University. 47 p.

Howard, Leonard W. Englevale Sandstone (Pennsylvanian) of eastern Bourbon and northeastern Crawford counties, Kansas. M, 1959, University of Kansas. 96 p.

Howard, Peter Felix. Structure and rock alteration of the Elizabeth Mine, Vermont. D, 1957, Harvard University.

Howard, R. P. A study of inclusions in the Las Blancas Pluton, San Diego County, California. M, 1978, San Diego State University.

Howard, Richard Henry. Variations in cordierite composition, Laramie Range, Albany County, Wyoming. M, 1955, University of Illinois, Urbana.

Howard, Robert Bruce. Some aspects of braided stream geometry. D, 1974, University of California, Los Angeles.

Howard, Ronald Adrian. Upper Paleozoic stratigraphy of the area between Banff and Jasper, Alberta. M, 1954, University of Alberta. 127 p.

Howard, Terry R. Geology and engineering properties of the Melon Gravels. M, 1967, University of Idaho. 68 p.

Howard, W. Brant. The hydrogeology of the Raton Basin, southcentral Colorado. M, 1982, Indiana University, Bloomington. 95 p.

Howarth, Susan M. T. An investigation of dynamically measured elastic/mechanical properties for consolidated rocks and unconsolidated porous media including natural petroleum reservoir sands. D, 1987, Colorado School of Mines. 437 p.

Howatson, Charles Henry. Botany in relation to the subsurface geology. M, 1947, University of British Columbia.

Howd, Frank H. Geology of the northwest quarter of the Bolton Quadrangle, New York. M, 1953, University of Rochester. 167 p.

Howd, Frank Hawver. Geology and geochemistry of the wolframite deposits in southern Stevens County, Washington. D, 1956, Washington State University. 81 p.

Howd, Peter A. Beach foreshore response to long-period waves in the swash-zone. M, 1984, Oregon State University. 88 p.

Howdeshell, Jeffrey C. A geophysical investigation of the White Mountain region of New Hampshire and Maine. M, 1983, Tulane University.

Howe, Daniel Marshall. Correlation of the fauna from the Middle Permian section at Black Rock, northwestern Nevada. M, 1975, University of Nevada. 133 p.

Howe, Dennis M. The origin of the Hitchcock Lake banded gneiss, northern Waterbury Quadrangle, Connecticut. M, 1966, University of Wisconsin-Madison.

Howe, Dennis Milton. Post-Casper-Ingleside unconformity and related sediments of southeastern Wyoming and northcentral Colorado. D, 1970, University of California, Los Angeles. 353 p.

Howe, Ernest. The Pre-Cambrian intrusive rocks of the Animas Canyon, Colorado. D, 1901, Harvard University.

Howe, F. B. A study of certain loess soil in Iowa. M, 1916, Iowa State University of Science and Technology.

Howe, Henry V. The Miocene of Clatsop and Lincoln counties, Oregon. D, 1922, [Stanford University]. 186 p.

Howe, Herbert J. Surface geology of northwestern Williamson County and portions of eastern Burnet and southern Bell counties, Texas. M, 1953, Columbia University, Teachers College.

Howe, Herbert James. A contribution to the stratigraphy of the Montoya Group. D, 1960, Columbia University, Teachers College. 90 p.

Howe, James Robert. Tectonics, sedimentation, and hydrocarbon potential of the Reelfoot Aulacogen. M, 1985, University of Oklahoma. 109 p.

Howe, Jerry R. Geology of an area southwest of Rossie, St. Lawrence County, New York. M, 1957, Syracuse University.

Howe, John Alfred. The Oligocene rodent Ischyromys in relation to the Paleosols of the Brule Formation. M, 1956, University of Nebraska, Lincoln.

Howe, John Alfred. The Pleistocene horses of Nebraska. D, 1961, University of Nebraska, Lincoln. 285 p.

Howe, Martin R. The relationship of a potentiometric surface to conduit development in the carbonate lithologies of the Mitchell Plain, Harrison County, Indiana. M, 1981, Indiana State University. 154 p.

Howe, Milton W. Geologic studies of the Mesa group (Pliocene?), upper Magdalena valley, Colombia. D, 1970, Princeton University. 102 p.

Howe, Ralph H. Mineralization in the Silver Star area, Skamania County, Washington. M, 1938, University of Cincinnati. 32 p.

Howe, Robert. Geology of the Mokelumne Peak area. M, 1984, San Jose State University. 113 p.

Howe, Robert Crombie. Paleontology of the Mississippian Sunwapta Pass, Alberta. D, 1965, University of Wisconsin-Madison.

Howe, Robert Crombie. Type saline bayou (Eocene) Ostracoda of Louisiana. M, 1962, University of Wisconsin-Madison.

Howe, Robert Hsi Lin. Prediction of ground water conditions by airphoto interpretation. D, 1955, Purdue University. 257 p.

Howe, Roger. Major developments in the gold mining industry of the United States. M, 1988, University of Texas, Austin.

Howe, Stephen Sherwood. Mineralogy, fluid inclusions, and stable isotopes of lead-zinc occurrences in central Pennsylvania. M, 1981, Pennsylvania State University, University Park. 155 p.

Howe, Wallace Brady. Stratigraphy of pre-Marmaton Desmoinesian rocks in south-eastern Kansas. D, 1954, University of Kansas. 177 p.

Howe, Wallace Brady. The geology of the Oak Grove Quadrangle. M, 1948, University of Missouri, Columbia.

Howell, Barbara A. Organic carbon and nitrogen distributions and vertical gradations in the Seston and sediment of upper Delaware Bay. M, 1984, University of Delaware, College of Marine Studies.

Howell, Benjamin F., Jr. Ground vibrations near explosions. D, 1949, California Institute of Technology. 168 p.

Howell, Benjamin F., Jr. Some effects of geological structure on radio reception. M, 1942, California Institute of Technology. 40 p.

Howell, Benjamin F., Jr. Structural geology of the region between Pacioma and Little Tujunga canyons, San Gabriel Mountains, California. D, 1949, California Institute of Technology. 110 p.

Howell, Benjamin Franklin. The faunas of the Cambrian Paradoxides beds at Manuels, Newfoundland. D, 1920, Princeton University. 169 p.

Howell, Buford Fredrick. Sand movement along Carmel River State Beach, Carmel, California. M, 1972, United States Naval Academy.

Howell, David. Pennsylvanian "subdeltas" in Boone County, W VA. M, 1977, University of South Carolina.

Howell, David Adams. A thermodynamic study of the zeolite stilbite. M, 1987, Western Michigan University.

Howell, David Ernest. Geology of the southern two-thirds of the Celo 7.5′ Quadrangle, North Carolina. M, 1975, University of North Carolina, Chapel Hill. 60 p.

Howell, David G. Middle Eocene paleogeography of southern California and some speculations on the evolution of the San Andreas fault system. D, 1974, University of California, Santa Barbara. 249 p.

Howell, David J. A model for upper delta plain coals, south central West Virginia. D, 1980, University of South Carolina. 76 p.

Howell, E. C. Application of the magnetometric resistivity method in mapping a deeply buried geological structure (located near Superior, Arizona). M, 1975, University of Toronto.

Howell, Gary D. Environmental inventory of Hamilton County. M, 1974, Baylor University.

Howell, Jack. Analysis of temperature-time data from three-meter drillholes at Crystal Hot Springs, Utah. M, 1986, University of Utah. 63 p.

Howell, Jack W. A field guide to coastal Washington. D, 1977, University of Northern Colorado. 209 p.

Howell, James Douglas. The use of soils as a dating tool and climatic indicator in the Main Ranges, Alberta. M, 1977, University of Calgary.

Howell, James Robert, III. Stratigraphy and paleontology of the Fitchville Formation (Lower Mississippian) on Stansbury Island, Great Salt Lake, Utah. M, 1978, University of Utah. 134 p.

Howell, Jay Lee. Regional geology of the Southern Rocky Mountains. M, 1951, University of Michigan.

Howell, Jesse V. Occurrence and origin of the iron ores of Iron Hill near Waukon, Iowa. M, 1915, University of Iowa. 71 p.

Howell, Jesse V. Twin Lakes District of Colorado. D, 1922, University of Iowa.

Howell, John Edward. Silicification in the Fleming Formation of the Knob Lake Group of the Labrador iron belt. D, 1954, University of Wisconsin-Madison.

Howell, K. K. Geology and alteration of the Commonwealth Mine, Cochise County, Arizona. N, 1977, University of Arizona. 225 p.

Howell, L. W., Jr. A mathematical model for Lake Bonney, Antarctica. D, 1977, Virginia Polytechnic Institute and State University. 339 p.

Howell, Lamar Allan. Groundwater resources of Brimfield Township, Portage County, Ohio. M, 1976, Kent State University, Kent. 89 p.

Howell, Mark Joseph. Systematic fracture trends in the Allegheny Plateau and Valley and Ridge provinces, Pendleton County, West Virginia. M, 1988, Wright State University. 84 p.

Howell, Michael Wade. Formation of laminated and opal-rich sediments in the Mediterranean region during the late Neogene and Quaternary. D, 1988, University of South Carolina. 219 p.

Howell, Paul William. The Cenozoic geology of the Chetoh Country, Arizona and New Mexico. D, 1959, University of Arizona. 329 p.

Howell, Richard Shelby, Jr. Quaternary geology of Kent Quadrangle, Culberson, Reeves, and Jeff Davis counties, Texas. M, 1952, University of Texas, Austin.

Howell, Roger Lynn. Geology, alteration, and mineralization of Red Mountain, Custer County, Idaho. M, 1983, University of California, Santa Barbara. 106 p.

Howell, Roy Patton, III. Facies relations and depositional patterns of the upper West Spring Creek Formation, Arbuckle Group, Oklahoma. M, 1982, University of Texas at Dallas. 265 p.

Howells, K. D. M. Palaeoecological study of the Silurian of the Quinn Point section, northern New Brunswick. M, 1975, University of New Brunswick.

Howells, William C. The Windrum Lake area, Saskatchewan. D, 1940, McGill University.

Howells, William Crompton. A petrographical study of the heavy minerals in the Paskapoo Formation on the North Saskatchewan River. M, 1934, University of Alberta. 88 p.

Hower, James Clyde. Anisotropy of vitrinite reflectance in relation to coal metamorphism for selected United States coals. D, 1978, Pennsylvania State University, University Park. 356 p.

Hower, James Clyde. The paleomagnetism and structure of the Ordovician igneous rocks of Lebanon County, Pennsylvania. M, 1975, Ohio State University.

Hower, John, Jr. A telluric current measuring instrument. M, 1954, Washington University. 13 p.

Hower, John, Jr. The fixation of heavy metal cations by some clay minerals. D, 1955, Washington University. 65 p.

Howery, Sherrill D. Areal geology of northeastern Caddo County, Oklahoma. M, 1960, University of Oklahoma. 78 p.

Howes, Mary Rachel. The occurrence of sphalerite in Cherokee Group (Pennsylvanian) coals of southeastern and southcentral Iowa. M, 1983, University of Iowa. 150 p.

Howes, Ronald Clarence. Geology of the Wildcat Hills (Box Elder County), Utah. M, 1972, Utah State University. 43 p.

Howes, Susan Dawn. Changes in sediment characteristics and dilution of Stillwater Complex sediments in an expanding fluvial system, Beartooth mountain front, Montana. M, 1983, Southern Illinois University, Carbondale. 101 p.

Howes, Thomas B. A brief study of the geology and ground water conditions in the Pauma Valley area, San Diego County, California. M, 1955, California Institute of Technology. 64 p.

Howland, Arthur L. and Thayer, Thomas P. The geology of Gabamichigami Lake, Minnesota. M, 1931, Northwestern University.

Howland, Arthur Lloyd. Sulphide and metamorphic rocks at the base of the Stillwater Complex, Beartooth Plateau, Montana. D, 1933, Princeton University. 81 p.

Howland, Jonathan Dean. A simplified procedure for reliability analysis in geotechnical engineering. D, 1981, [Rensselaer Polytechnic Institute]. 226 p.

Howland, Mark Douglas. Hydrogeology of the Palo Alto Baylands, Palo Alto, California; with emphasis on the tidal marshes. M, 1977, Stanford University. 138 p.

Howle, Arlen G., Jr. Geology of the northern Del Carmen Mountains, Brewster County, Texas. M, 1964, Texas A&M University.

Howze, Bryn David. Stratigraphy and sedimentology of the Middle Ordovician Sevier Shale basin near Avens Bridge, Washington County, Virginia. M, 1987, University of Tennessee, Knoxville. 211 p.

Hoy, Robert B. Geology and ore deposits of the Capps gold mine, Mecklenburg County, North Carolina. M, 1939, California Institute of Technology. 51 p.

Hoy, Steven M. Stratigraphy and petrography of Tertiary volcanic rocks, El Muerto Peak Quadrangle, western Davis Mountains, Trans-Pecos, Texas. M, 1986, Baylor University. 117 p.

Hoy, T. Genesis of brucite and marbles near Wakefield, Quebec. M, 1970, Carleton University.

Hoy, Trygve. Structure and metamorphism of Kootenay Arc rocks, British Columbia. D, 1974, Queen's University. 202 p.

Hoya, H. Austin von der see von der Hoya, H. Austin, II

Hoyer, M. C. Quaternary valley fill of the abandoned Teays drainage system in southern Ohio. D, 1976, Ohio State University. 185 p.

Hoyer, Marcus Conrad. The Puget Peak avalanche and other effects of the March 27, 1964 Alaska earthquake, in Puget Bay, Alaska. M, 1968, Arizona State University. 62 p.

Hoyer, Peter Wlater. Taxonomy and biostratigraphy of dinoflagellates from the Barremian (Lower Cretaceous) stratotype at Angles, France. M, 1979, University of Toronto.

Hoyle, Blythe L. Lower Mississippian conodont biostratigraphy of the Chappel Limestone, central Texas, and Welden Limestone, southern Oklahoma. M, 1978, University of Texas, Austin.

Hoyle, Blythe Lynn. Suburban hydrogeology and ground-water geochemistry of the Ashport silt loam, Payne County, Oklahoma. M, 1987, Oklahoma State University. 278 p.

Hoyle, Lorraine E. Subsurface geology of the Shawnee Lake area, western Pottawatamie County, central Oklahoma. M, 1948, University of Oklahoma. 46 p.

Hoylman, Edward Wayne. The geology of the Poverty Hills area, Inyo County, California. M, 1974, University of California, Los Angeles. 84 p.

Hoyos-Patino, Fabian. Environmental geology of the Chandler Quadrangle, Maricopa County, Arizona, Part I. M, 1986, Arizona State University. 87 p.

Hoyt, Albert J. Laboratory study to simulate leachate migration through a clayey soil. M, 1987, University of Missouri, Kansas City. 160 p.

Hoyt, Brian R. The analysis of climatic effects of slopes on sedimentary rocks. M, 1981, Kent State University, Kent. 94 p.

Hoyt, D. H. Geology and Recent sediment distribution from Santa Barbara to Rincon Point, California. M, 1976, San Diego State University.

Hoyt, David Ellsworth. The paleogeographic interpretation of the Palmetto Complex of southwestern Nevada. M, 1974, University of Texas at Dallas. 134 p.

Hoyt, Gregory Dana. Nitrogen cycling in a southeastern coastal plain agricultural ecosystem. D, 1981, University of Georgia. 180 p.

Hoyt, John Harger. Historical geology laboratory manual. M, 1952, University of Michigan.

Hoyt, John Harger. Stratigraphy of the Pennsylvanian and Lower Permian of the northern Denver Basin, northeastern Colorado, southeastern Wyoming, and western Nebraska. D, 1960, University of Colorado. 245 p.

Hoyt, John W. Regional geology and tectonic setting for mineral deposits of southwestern Peru. M, 1965, University of Arizona.

Hoyt, Philip Munro. An investigation of pressure distribution in granular media by photoelastic means. D, 1966, Stanford University. 105 p.

Hoyt, Virginia. Heavy mineral studies of the New York Clinton (Silurian). D, 1943, University of Rochester. 193 p.

Hoyt, Virginia. Petrographic study of some Triassic sediments of the Connecticut Valley of Massachusetts. M, 1940, Smith College. 59 p.

Hoyt, William H. Long-distance turbidite correlations in the Horseshoe abyssal plain. M, 1976, SUNY at Albany. 136 p.

Hoyt, William Henry. Processes of sedimentation and geologic history of the Cape Henlopen/Breakwater Harbor area, Delaware. D, 1982, University of Delaware. 368 p.

Hozik, Michael Jacob. Brittle fracture history of the Narragansett Pier Granite, Rhode Island. D, 1981, University of Massachusetts. 321 p.

Hrabar, Stephanie Vladimira. Lower West Baden (Late Mississippian) sandstone body in Owen County, Indiana. M, 1967, Indiana University, Bloomington. 42 p.

Hrabar, Stephanie Vladimira. Stratigraphy and depositional environment of the St. Regis Formation of the Ravalli Group (Precambrian Belt Megagroup), northwestern Montana and Idaho. D, 1971, University of Cincinnati. 92 p.

Hradilek, P. J. Strong ground motion from point sources. D, 1978, University of California, Los Angeles. 294 p.

Hreggvidsdottir, Halldora. The greenschist to amphibolite facies transition in the Nesjavellir hydrothermal system, Southwest Iceland. M, 1987, Stanford University. 61 p.

Hriskevich, M. E. Geology of the Grand Falls area, Newfoundland. M, 1949, Queen's University. 43 p.

Hriskevich, Michael Edward. Petrology of the Nipissing diabase sheet of the Cobalt area of Ontario. D, 1952, Princeton University. 184 p.

Hromadka, T. V., II. Mathematical model of frost heave in freezing soils. D, 1980, University of California, Irvine. 178 p.

Hron, Marie Petra. Waves in inhomogeneous media. D, 1973, University of Alberta. 235 p.

Hrubec, John Anthony. Elastic moduli of MgO determined by Brillouin scattering. M, 1977, University of Rochester. 71 p.

Hruby, Alexander Joseph. Surface geology of northeastern Kay County, Oklahoma. M, 1955, University of Oklahoma. 72 p.

Hruska, Donald C. Geology of the Dry Range area, Meagher County, Montana. M, 1967, Montana College of Mineral Science & Technology. 89 p.

Hsi Chou T'an. Comprehensive study of mud cracks and other similar structures. M, 1927, University of Wisconsin-Madison.

Hsi, Ching-Kuo Daniel. Partition of oxygen isotopes and trace elements between carbonate and silicate melts at 1 kilobar, 800 degrees C and its bearing on the origin of carbonatite. M, 1978, Pennsylvania State University, University Park. 93 p.

Hsi, Ching-Kuo Daniel. Sorption of uranium (VI) by iron oxides. D, 1981, Colorado School of Mines. 154 p.

Hsi, Huey-rong. The effects of polyelectrolytes on flocculation of clay from clay deposits in Latah County, Idaho. M, 1960, University of Idaho. 51 p.

Hsia, Yu-ping. Molecular orbital studies on titanium-oxygen systems. D, 1967, Illinois Institute of Technology. 131 p.

Hsiao, Helmut Y. A. The "stress amplification" mechanism for intraplate earthquakes applied to the southeast United States. M, 1977, Georgia Institute of Technology. 84 p.

Hsieh, Chang-hsin. Empirical orthogonal function analysis of hydrologic data. D, 1988, Purdue University. 278 p.

Hsieh, Chia Yung. Origin of certain metamorphic foliated rocks. M, 1920, University of Wisconsin-Madison.

Hsieh, Hsii-Sheng. A non-associative Cam-Clay plasticity model for the stress-strain-time behavior of soft clays. D, 1987, Stanford University. 246 p.

Hsieh, Paul Anthony. A reservoir analysis of the Denver earthquakes; a case of induced seismicity. M, 1979, University of Arizona. 65 p.

Hsieh, Paul Anthony. Theoretical and field studies of fluid flow in fractured rocks. D, 1983, University of Arizona. 215 p.

Hsieh, Shuang-shi. Effects of bulk-components on the grindability of coals. D, 1976, Pennsylvania State University, University Park. 161 p.

Hsieh, Shuang-Shii. Analysis of ruthenium and osmium abundances in sulfide minerals from the Sudbury ores, Ontario. M, 1967, McMaster University. 68 p.

Hsiung, Ennchi David. Secondary flow and its effect on sediment transport. D, 1980, University of Pittsburgh. 171 p.

Hsu, Eugene Ying-Chih. The stratigraphy and sedimentology of the Late Precambrian Saint John's and Gibbetts Hill Formation and the upper part of the Conception Group (Late Precambrian) in the Torbay map area, Avalon Peninsula, Newfoundland. M, 1972, Memorial University of Newfoundland. 116 p.

Hsu, Fu-tzu. Geochemical exploration in the Nittany Valley area, Centre County, Pennsylvania. M, 1973, Pennsylvania State University, University Park. 108 p.

Hsu, I-Chi. Magnetic properties of igneous rocks in the northern Philippines. D, 1971, Washington University. 165 p.

Hsu, I-Chi. Paleomagnetic investigations of some of the Precambrian volcanic rocks in the St. Francois Mountains, Missouri. M, 1962, Washington University. 35 p.

Hsu, J. R. Analysis of soil deformation by elastic-plastic work-hardening model. D, 1977, Ohio State University. 201 p.

Hsu, Jeffrey Tsen-Jer. Emerged Quaternary marine terraces in southern Peru; sea level changes and continental margin tectonics over the subducting Nazca Ridge. D, 1988, Cornell University. 329 p.

Hsu, Ke-Chin. Geology of the tungsten deposits in southern Kiangsi, China. M, 1941, University of Minnesota, Minneapolis.

Hsu, Ke-Chin. Geology of tungsten deposits. D, 1944, University of Minnesota, Minneapolis. 311 p.

Hsu, Kenneth Jinghwa. A contribution to the problems of distribution of mountainbuilding movements in geologic time. M, 1950, Ohio State University.

Hsu, Kenneth Jinghwa. Petrology of the Cucamonga Canyon-San Antonio Canyon area, southeastern San Gabriel Mountains, California. D, 1954, University of California, Los Angeles.

Hsu, Kuan-Hsiung. Use of cluster analysis in interpolating faults detected in seismic exploration. M, 1979, Stanford University. 139 p.

Hsu, Liang-chi. Selected phase relationships in the system Al-Mn-Fe-Si-O-H, a model for garnet equilibria. D, 1966, University of California, Los Angeles. 178 p.

Hsu, Mao-Yang. Analysis of strain, shape, and orientation of the deformed pebbles in the Seine River area, Ontario. D, 1971, McMaster University. 179 p.

Hsu, Mao-Yang. Structural analysis along the Grenville Front, near Sudbury, Ontario. M, 1968, McMaster University. 104 p.

Hsu, Nien-Shieng. Optimum experimental design for parameter identification of a groundwater system. D, 1984, University of California, Los Angeles. 142 p.

Hsu, Rongshin. A weighting model for leveling networks based on constant correlation within a single line. D, 1988, University of Wisconsin-Madison. 172 p.

Hsu, S. I. Urbanization and its effects on the climate of Phoenix. D, 1979, Arizona State University. 216 p.

Hsu, Tsung Han. A comparison of the Maquoketa Formation and its fauna in eastern Wisconsin, Iowa, and Illinois. M, 1915, University of Illinois, Chicago.

Hsu, Tung. Reflection electron microscopy of crystal surfaces. D, 1983, Arizona State University. 126 p.

Hsu, Tung-Wen. Study of marine sediment drag forces on offshore pipelines. D, 1987, Texas A&M University. 186 p.

Hsu, Tzu-Li. Finite element analysis of the time-dependent earthwork design and construction problems in geotechnical engineering. D, 1982, SUNY at Buffalo. 107 p.

Hsu, Vindell. Seismic refraction survey of a power plant site, Northeast Mississippi. M, 1976, University of Tennessee, Knoxville. 48 p.

Hsue, Tien Shaing. Archeomagnetic intensity data for the Southwestern United States, 700-1900 A.D. M, 1978, University of Oklahoma. 124 p.

Hsueh, Chao-min. Organic and clay mineral diagenesis in Neogene sediments of western Taiwan. D, 1985, University of Missouri, Columbia. 153 p.

Hsui-Tsu-Fan, Paul. Subsurface Cambrian geology of Iowa. D, 1947, Iowa State University of Science and Technology.

Htoon, Myat. Geology of the Clinton Creek asbestos deposit, Yukon Territory. M, 1979, University of British Columbia.

Hu, Chung-Hung. The ontogeny and sexual dimorphism of Lower Paleozoic trilobites. D, 1968, University of Cincinnati. 446 p.

Hu, Hsien-Neng. The formation of lunar soil; chemistry of agglutinate particles. M, 1978, University of Tennessee, Knoxville. 52 p.

Hu, Liang-Zie. Imaging and processing borehole seismic data. D, 1987, University of Texas at Dallas. 304 p.

Hu, Nien-Tsu Alfred. Petrology and stratigraphy of a part of the central Death Valley volcanic field, eastern California. M, 1983, Pennsylvania State University, University Park. 159 p.

Hu, R. E. W. Influence of factors involved in the seismic stability analysis of an earth dam. D, 1977, University of California, Los Angeles. 357 p.

Hu, Tse-chuang. The crater ejecta component in the lunar regolith. M, 1971, University of North Carolina, Chapel Hill. 41 p.

Hu, Wenbao. A model study of the electromagnetic response of a channel, an island and a seamount in the South China Sea. D, 1987, University of Victoria. 148 p.

Hua, Zhang. A structural investigation of the "room temperature" phase transformation in maximum microcline. M, 1988, University of Toledo. 97 p.

Huaco, Daniel. Source parameters and the static field of earthquakes at near distances. D, 1977, St. Louis University. 140 p.

Huan-Zhang Lu *see* Lu Huan-Zhang

Huang, An-Bin. Laboratory pressuremeter experiments in clay soils. D, 1986, Purdue University. 251 p.

Huang, Chen Tair. Feasibility of longwall mining in the Spurgeon coal field, Pike and Warrick counties, Indiana. D, 1984, Purdue University. 245 p.

Huang, Chen-Feng. A study of biotites from the Boulder Batholith, Montana. M, 1973, Boston University. 70 p.

Huang, Cheng-yi. Late Neogene planktonic foraminiferal biostratigraphy and paleoclimatology of Southern California. M, 1980, California State University, Long Beach. 94 p.

Huang, Chi-I. An isotopic and petrologic study of the contact metamorphism and metasomatism related to copper deposits at Ely, Nevada. D, 1976, Pennsylvania State University, University Park. 188 p.

Huang, Chi-I. Cataclastic rocks in the Little Beaver creek area, Carbon County, Wyoming. M, 1970, University of Wyoming. 63 p.

Huang, Hann-Chen. Interactive and iterative seismic residual statics analyses. D, 1988, University of Houston. 114 p.

Huang, Hong-Hsi. Variations in the shape of the outer rise seaward of the central Aleutian Trench; implications for the rheology of oceanic lithosphere. M, 19??, Texas A&M University.

Huang, Hui-Lun. Stratigraphic investigations of several cores from the Tampa Bay area. M, 1977, University of South Florida, Tampa. 54 p.

Huang, I-Chen Eugene. Studies of the phase relationships, transformation mechanisms, and equations of state of Fe, Fe-Ni alloys, and Fe_3O_4 by synchrotron radiation in high-temperature diamond anvil cells; their geological implications. D, 1987, Cornell University. 211 p.

Huang, Jau-Inn. Exact solutions to one-dimensional inverse problems with arbitrary source functions. D, 1985, Columbia University, Teachers College. 147 p.

Huang, Kai-Yi. Application of Thematic Mapper data and Landsat data to eutrophication and water quality modeling of selected sites in Lake Michigan. M, 1984, Indiana State University. 125 p.

Huang, Kung. PDF of backscattered sound from live fish. M, 1977, University of Wisconsin-Madison.

Huang, Liang-Hsiung. Trapping and absorption of underwater sound. D, 1986, University of Iowa. 108 p.

Huang, Long-Cheng. Seismic water pressures on dams for arbitrarily shaped reservoirs. D, 1984, University of Iowa. 89 p.

Huang, Moh-Jiann. Investigation of local geology effects on strong earthquake ground motions. D, 1984, California Institute of Technology. 230 p.

Huang, Paul. Aftershocks of the 1968 Rampart, Alaska earthquake. M, 1979, University of Alaska, Fairbanks. 155 p.

Huang, Paul Yi-Fa. Focal depths and mechanisms of mid-ocean ridge earthquakes from body waveform inversion. D, 1986, Massachusetts Institute of Technology. 301 p.

Huang, Qilin. Movement on the core mantle boundary. M, 1987, University of Miami.

Huang, Scott Lin. Swelling behavior of Illinois coal shales. D, 1981, University of Missouri, Rolla. 209 p.

Huang, Scott Lin. The influence of the dispersion method on peak intensity of kaolinite, montmorillonite and illite clay standards. M, 1978, University of Kentucky. 98 p.

Huang, Shu-Li. Integrating scientific and institutional aspects of water resources management; a case study of the Brandywine Basin. D, 1983, University of Pennsylvania. 312 p.

Huang, Steve Kuo-Yi. Seismic data processing; a tool for energy source exploration. M, 1982, University of Nebraska, Lincoln.

Huang, Ter-Chien. A sedimentologic study of Charlotte Harbour, southwestern Florida. M, 1966, Florida State University.

Huang, Ter-Chien. The sediments and sedimentary processes of the eastern Mississippi cone, Gulf of Mexico. D, 1970, Florida State University. 135 p.

Huang, Wei Ta. Petrology and geology of the crystalline rocks of Chienchuan, Yunnan, China. D, 1949, Syracuse University.

Huang, Wei Ta. Volcanic rocks of Paricutin and vicinity, Mexico. M, 1946, University of California, Berkeley. 20 p.

Huang, Wen Hsing. Experimental studies of kinetics and mechanisms of simulated organo-chemical weathering of silicate minerals. D, 1970, University of Missouri, Columbia. 606 p.

Huang, Wen Yen. Sterols as the source indicator of sedimentary organic matters. D, 1975, Indiana University, Bloomington. 95 p.

Huang, Wu-Shung. The petrography of basement rocks in the northeastern Los Angeles Basin, California. M, 1971, Ohio University, Athens. 57 p.

Huang, Wuu-Liang. Water deficient melting relations of muscovite-granite, and related synthetic systems to 35 kilobars pressure with geological applications. D, 1973, University of Chicago. 151 p.

Huang, Y. F. Determination of uranium in minerals using fission tracks. M, 1975, University of Toronto.

Huang, Y. S. The ultrastructure, pigmentation, photosynthesis and genetic affinities of "Monodus sp.", an undescribed, unicellular yellow green alga from eutrophic brackish waters. D, 1977, University of North Carolina, Chapel Hill. 133 p.

Huang, Yih-Ping. Nonlinear, incremental, 2-D and 3-D finite element analysis of geotechnical structures using interactive computer graphics. D, 1983, Cornell University. 351 p.

Huang, Ying-Yan. Quantitative electrical study of the Laverty-Hoover Sand (Pennsylvanian), Harper County, Oklahoma. M, 1965, University of Oklahoma. 173 p.

Huang, Yue-Chain. Thermal history model of the Williston Basin. M, 1988, University of North Dakota. 117 p.

Huang, Zhixin. Microstructural analysis of the deformed rocks in the Tensleep-Beaver Creek fault zone, Bighorn Mountains, Wyoming. M, 1988, University of Iowa. 168 p.

Huang, Zhongxian. Time-space distribution of coda Q in fault zones and surface wave dispersion in eastern China. D, 1987, University of Colorado. 165 p.

Hubbard, Bela. The geology of the Lares District, Puerto Rico. D, 1923, Columbia University, Teachers College.

Hubbard, Bela. The Helderberg-Oriskany contact of the Broadhead Gap section. M, 1917, Columbia University, Teachers College.

Hubbard, D. K. Variations in tidal inlet processes and morphology in the Georgia Embayment. D, 1977, University of South Carolina. 90 p.

Hubbard, David Adam, Jr. The mineralogical composition and variation of heavy mineral bands of a washover fan at Assateague Island, Maryland. M, 1977, University of Virginia. 41 p.

Hubbard, Dennis K. Tidal inlet morphology and hydrodynamics of Merrimack Inlet, Massachusetts. M, 1974, University of South Carolina.

Hubbard, E. L. The effects of numerous old mill dams on lower basin stream development; 250 years of industrial use of the French River and its tributaries, southern Worcester County, Massachusetts. D, 1979, Clark University. 257 p.

Hubbard, Edwin L. General geology and petrology of the west ninth of the Spring Hill Quadrangle, Connecticut. M, 1966, University of Connecticut. 43 p.

Hubbard, Frank Steven. Calcareous nannofossil biostratigraphy of the Upper Cretaceous and lower Paleocene sediments of the New Jersey Coastal Plain. M, 1981, Ohio University, Athens. 140 p.

Hubbard, George David. Gold and silver mining as a geographic factor in the development of the United States. D, 1905, Cornell University.

Hubbard, John Edward. Cesium-137 in an alpine watershed. D, 1968, Colorado State University. 107 p.

Hubbard, Mary Syndonia. Thermobarometry, $^{40}Ar/^{39}Ar$ geochronology, and structure of the Main Central Thrust zone and Tibetan slab, eastern Nepal Himalaya. D, 1988, Massachusetts Institute of Technology. 169 p.

Hubbard, Norman Jay. Some trace elements in Hawaiian lavas. D, 1967, University of Hawaii. 123 p.

Hubbard, Perry, Jr. Hydrogeologic variability of the Floridan Aquifer in the Central Florida Phosphate District. M, 1981, University of Alabama. 90 p.

Hubbard, Philip Scott. Geology of the Saddle Butte Quadrangle, Washington. M, 1968, University of Hawaii. 75 p.

Hubbard, Richard Jon. Geological evolution of three rifted continental margins in the Americas. D, 1987, Stanford University. 386 p.

Hubbard, William E. The origin of petroleum. M, 1917, University of Minnesota, Minneapolis.

Hubbell, Joel M. Description of geothermal flow systems in the vicinity of the Caribou Range, southeastern Idaho. M, 1981, University of Idaho. 105 p.

Hubbell, Marion. Erosional terraces in Vermont. M, 1929, Smith College. 101 p.

Hubbell, Roger G. Stratigraphy of the Jelm, Nugget and Sundance formations of northern Carbon County, Wyoming. M, 1954, University of Wyoming. 131 p.

Hubbert, M. King. Theory of scale models as applied to the study of geologic structures. D, 1937, University of Chicago. 60 p.

Huber, Brian Thomas. Upper Campanian-Maastrichtian foraminifers of the high southern latitudes; ontogenetic morphometric systematics, biostratigraphy, and paleobiogeography. D, 1988, Ohio State University. 360 p.

Huber, Darrell Dean. Petrology of the Crouse Limestone (Permian) in the Manhattan, Kansas area. M, 1964, Kansas State University. 133 p.

Huber, Gary C. Stratigraphy and uranium deposits of the Lisbon Valley District, San Juan County, Utah. D, 1979, Colorado School of Mines. 210 p.

Huber, James Kenneth. A late Holocene vegetational sequence from the Southeast Missouri Ozarks. M, 1987, University of Minnesota, Duluth. 206 p.

Huber, James R. Sedimentary petrogenesis of the Yeso-Glorieta-San Andres Transition, Joyita Hills, Socorro County, New Mexico. M, 1961, University of New Mexico. 86 p.

Huber, Jeffrey A. The geology and mineralization of the Sukakpak Mountain area, Brooks Range, Alaska. M, 1988, University of Alaska, Fairbanks. 75 p.

Huber, M. E. A paleoenvironmental interpretation of the upper Cambrian Eau Claire Formation of West-central Wisconsin. M, 1975, University of Wisconsin-Madison.

Huber, Norman K. A study of environmental controls of iron mineral deposition and their relation to pre-Cambrian iron formations. M, 1952, Northwestern University.

Huber, Norman K. The environmental control of sedimentary iron minerals and its relation to the origin of the Ironwood iron formation. D, 1956, Northwestern University.

Huber, R. D. Application of the gravity method to the determination of high-level ground-water boundaries in the Pahala-Punaluu sections of the Kau District, Hawaii. M, 1970, University of Hawaii. 41 p.

Huber, Robert Evans. Frequency analysis of seismic pulses. M, 1958, Pennsylvania State University, University Park. 68 p.

Huber, Theodore A. The trend of the pound loss in gas well decline. M, 1928, University of Pittsburgh.

Huber, Thomas Patrick. A multi-step method for avalanche zone recognition and analysis. D, 1980, University of Colorado. 183 p.

Huber, Timothy P. Conodont biostratigraphy of the Bakken and lower Lodgepole formations (Devonian and Mississippian), Williston Basin, North Dakota. M, 1986, University of North Dakota. 274 p.

Huber, William Gregor. Stratigraphy of the Yorktown and St. Marys formations (Miocene) on the south bank of the James River, Virginia. M, 1972, University of Virginia. 83 p.

Hubert, C. M. Silurian Favositidae and stromatoporoids of Matapedia-Temiscouata regions, Quebec. M, 1963, McGill University.

Hubert, John F. Petrology of the Fountain and Lyons formations along the Colorado Front Range, Part 1-2. D, 1958, Pennsylvania State University, University Park. 467 p.

Hubert, John F. Structure and stratigraphy of an area east of Brush Creek, Eagle County, Colorado. M, 1954, University of Colorado.

Hubert, Kathleen Ann. Physical properties of selected basalt and greywacke and the inferred seismic structure of the Olympic Peninsula, Washington. M, 1979, University of Washington. 96 p.

Hubert, Loren Matthew. Structure and lithologic variation in the central core of the Broken Bow Uplift, Ouachita Mountains, Oklahoma. M, 1984, University of Texas at Dallas. 112 p.

Hubka, James Lewis, Jr. A structural and thickness study of Cretaceous and older sediments in Harlan, Furnas, Gosper and Phelps counties, Nebraska. M, 1949, University of Nebraska, Lincoln.

Huble, Christoph W. H. An investigation of the size and shape of quartz grains, Pedro Beach, California. M, 1957, University of California, Davis.

Hubred, G. L. Relationship of morphology and transition metal content of manganese nodules to an abyssal hill. M, 1970, University of Hawaii. 31 p.

Huck, Florence. Comparative lithological study of the materials in the drift of the Syracuse region (New York). M, 1920, Syracuse University.

Huckaba, William Arden. Subsurface geology of the Salt Creek and Sheep Springs oil fields. M, 1952, University of Southern California.

Huckabay, William. Correction of some back swamp clays and braided stream top stratum deposits by X-radiography; Little River diversion canal, east-central Louisiana. M, 1972, University of South Carolina.

Hudak, Curtis Martin. Paleoecology of an early Holocene faunule and an early Holocene florule from the Dows local biota of north-central Iowa. M, 1982, University of Iowa. 48 p.

Hudak, Curtis Martin. Quaternary landscape evolution of the Turkey River valley, northeastern Iowa. D, 1987, University of Iowa. 343 p.

Hudak, George. Carbonate diagenesis and depositional cycles of the Mission Canyon Limestone, Madison Group of southwestern Montana. M, 1985, University of Montana. 129 p.

Hudak, Larry J. Levels of groundwater contamination, Ipswitch River basin, Massachusetts. M, 1981, Boston University. 162 p.

Hudak, Paul F. Conjunctive use of ground water flow and mass transport models in simulation of contaminant migration, Butler County Landfill, Ohio. M, 1988, Wright State University. 160 p.

Hudder, Karen A. Diagenesis within the deep Tuscaloosa Formation, Profit Island Field, Louisiana. M, 1982, Texas A&M University. 146 p.

Huddle, John Warfield. Conodonts from the New Albany Shale of Indiana. D, 1934, Indiana University, Bloomington. 196 p.

Huddleston, James Herbert. Local soil-landscape relationships in eastern Pottawattamie County, Iowa. D, 1969, Iowa State University of Science and Technology. 261 p.

Huddlestun, Paul Francis. Correlation of upper Eocene and lower Oligocene strata between the Sepulga, Conecug, and Choctawahatchee rivers in southern Albama. M, 1965, Florida State University.

Huddlestun, Paul Francis. The Neogene stratigraphy of the central Florida Panhandle. D, 1984, Florida State University. 245 p.

Hudec, Peter P. The geology of Highstone Lake-Miniss Lake area, Ontario. M, 1960, Rensselaer Polytechnic Institute. 86 p.

Hudec, Peter Paul. The nature of water and ice in carbonate rock pores. D, 1965, Rensselaer Polytechnic Institute. 184 p.

Hudelson, Peter Marc. Sedimentology and paleogeography of the Kenilworth Member (Campanian), Blackhawk Formation, east-central Utah. M, 1984, University of Iowa. 152 p.

Hudgins, A. D. Geology of North Mountain in map-area, Baxters Harbour to Victoria Beach. M, 1957, Acadia University.

Hudnall, W. H. Genesis and morphology of secondary products in selected volcanic ash soils from the island of Hawaii. D, 1977, University of Hawaii. 362 p.

Hudson, Adonnis S. Depositional environment of the Red Fork and equivalent sandstones east of the Nemaha Ridge, Kansas and Oklahoma. M, 1969, University of Tulsa. 80 p.

Hudson, Andrew G. The chemistry of iron and manganese in submarine hydrothermal systems. M, 1980, Massachusetts Institute of Technology. 80 p.

Hudson, Ann Elizabeth. A geochemical analysis of the tungsten mineralization in the Boulder County tungsten district, Boulder County, Colorado. M, 1988, University of Colorado. 120 p.

Hudson, Belva D. Precambrian geology of the Front Range near the mouth of Big Thompson Canyon (Larimer County), Colorado. D, 1958, University of Colorado. 102 p.

Hudson, Carolyn Brauer. Determination of relative provenance contribution in sand samples using Q-mode factor analysis of Fourier grain shape data. M, 1979, University of South Carolina.

Hudson, D. D. A cost/benefit study of several land use mapping methodologies using remotely sensed data. M, 1976, University of Missouri, Rolla.

Hudson, Donald M. Geology and alteration of the Wedekind and part of the Peavine districts, Washoe County, Nevada. M, 1977, University of Nevada. 102 p.

Hudson, Donald McComb. Alteration and geochemical characteristics of the upper parts of selected porphyry systems, western Nevada. D, 1983, University of Nevada. 229 p.

Hudson, Edward Wallace. Geology of the Willow Creek area, Elko and Eureka counties, Nevada. M, 1958, University of California, Los Angeles.

Hudson, Frank Samuel. Geology of the Cuyamace region, California, with special reference to the origin of the nickeliferous pyrrhotite. D, 1920, University of California, Berkeley.

Hudson, Geofrey Robert. Talc-carbonate alteration and nickel mineralization of an extrusive ultramafic complex, Torquata, St. Ives, Western Australia. D, 1973, Queen's University. 237 p.

Hudson, Jeanna Sue. Geology and hydrothermal alteration of the San Simon mining district south of Steins Pass, Peloncillo Mountains, Hidalgo County, New Mexico. M, 1984, University of New Mexico. 125 p.

Hudson, John A. A critical examination of indirect tensile strength tests for brittle rocks. D, 1970, University of Minnesota, Minneapolis.

Hudson, John G. Minor structures in the Boston Mountain Monocline T.12N., R.31 and 32, W., Crawford County, Arkansas. M, 1955, University of Arkansas, Fayetteville.

Hudson, Jon P. Stratigraphy and paleoenvironments of the Creteceous rocks, North and South Pender islands, British Columbia. M, 1975, Oregon State University. 139 p.

Hudson, Karen A. Gold and base metal mineralization in the Nippers Harbour ophiolite, Newfoundland. M, 1988, Memorial University of Newfoundland. 305 p.

Hudson, Mark Ransom. Dispersed paleomagnetic data from the Kern Mountains, Northeast Nevada. M, 1983, Colorado School of Mines. 103 p.

Hudson, Mark Ransom. Paleomagnetic and structural evidence bearing on the tectonic history of the Dixie Valley region, west-central Nevada. D, 1988, Colorado School of Mines. 352 p.

Hudson, Michael R. Mineralogy, petrology, and structural geology of the Tatnic Hill Formation, Putnam, Connecticut. M, 1982, Indiana University, Bloomington. 301 p.

Hudson, Presley C. Interpretation of alinements visible on aerial photographs of central Texas. M, 1972, Baylor University. 97 p.

Hudson, Richard James. Genesis and depositional history of the Eaton Sandstone, Grand Ledge, Michigan. M, 1957, Michigan State University. 49 p.

Hudson, Richard M. Progradation of a secondary shelf margin, upper Salmon Peak Formation, Maverick Basin, Southwest Texas. M, 1986, University of Texas, Arlington. 120 p.

Hudson, Robert Frank. Structural geology of the Piney Creek thrust area, east flank Bighorn Mountains, Wyoming. D, 1965, University of Iowa. 173 p.

Hudson, Roy Browning. Mississippian and Lower Pennsylvanian stratigraphy of northwestern Oklahoma. M, 1949, University of Oklahoma. 65 p.

Hudson, T. L. Genesis of a zoned granite stock, Seward Peninsula, Alaska. D, 1977, Stanford University. 218 p.

Hudson, Thomas Allen. Geology of the Irish Creek tin district, Virginia Blue Ridge. M, 1982, University of Georgia.

Hudson, Thomas W. Deposition, environment, diagenesis of stromatoporoid paleoecology of Labechia huronensis beds in the Tanglewood Limestone and Stamping Ground Members of the Lexington Limestone (Ordovician), central Kentucky. M, 1984, University of Cincinnati. 299 p.

Hue, Nguyen Van. Effects of phosphorus levels and clays on the nitrification process. D, 1981, Auburn University. 116 p.

Hueber, Francis Maurice. Contribution to the fossil flora of the Onteora "red beds" (Upper Devonian) in New York State. D, 1960, Cornell University. 216 p.

Hueber, Francis Maurice. Fossil flora of the Onteora "red beds" (Upper Devonian) in New York State, a preliminary survey. M, 1959, Cornell University.

Huebner, J. Stephen. Stability relations of minerals in the system Mn-Si-C-O. D, 1967, The Johns Hopkins University. 298 p.

Huebner, Mark. Geochemistry of polymetamorphic rocks from the Beitridge region of the Limpopo Belt, Southern Africa. M, 1986, University of Saskatchewan. 86 p.

Huebschman, Richard Patrick. Correlation and environment of a key interval within the Precambrian Prichard and Aldridge formations, Idaho-Montana-British Columbia. M, 1972, University of Montana. 66 p.

Huedepohl, Anita. Measurements of lateral variations in seismic amplitudes. M, 1984, Stanford University.

Huedepohl, Earnest Brady. Subsurface structure of central Iowa. M, 1956, Iowa State University of Science and Technology.

Huehn, Bruce. Crustal structure of the Baja Peninsula between latitudes 22°N and 25°N. M, 1977, Oregon State University. 95 p.

Huene, Roland Ernest von *see* von Huene, Roland Ernest

Hueni, Camille D. Living benthonic foraminifera of the South Texas shelf; temporal variations and associated ecologic and paleoecologic implications. M, 1979, Rice University. 307 p.

Huerta, Raul. Seismic stratigraphic and structural analysis of Northest Campeche Escarpment, Gulf of Mexico. M, 1980, University of Texas, Austin.

Huertas, Fernando. Strain relief measurements in the Rocky and Star Range. M, 1980, University of Utah. 138 p.

Huested, Sarah S. Analyses of the sulfur system in waters from the Galapagos Ridge hydrothermal vents.

M, 1979, Massachusetts Institute of Technology. 71 p.

Huestis, Stephen P. Bounding the thickness of the oceanic magnetized layer. D, 1976, University of California, San Diego. 118 p.

Huey, Arthur Sidney. The geology of the Tesla Quadrangle of middle California. D, 1940, University of California, Berkeley. 115 p.

Huf, William Langley. Effects of dodecyl trimethyl ammonium bromide on the hydrofracturing of sandstone. M, 1976, University of North Carolina, Chapel Hill. 56 p.

Hufen, T. H. A geohydrologic investigation of Honolulu's basal waters based on isotopic and chemical analyses of water samples. D, 1974, University of Hawaii. 160 p.

Huff, Bryan Gregory. Brachiopod systematics, biostratigraphy and paleoecology of a proposed upper Atokan stratotype, Arrow Canyon Range, Clark County, Nevada. M, 1984, University of Illinois, Urbana. 160 p.

Huff, David W. Evidence for small-scale slumping on the continental slope in two topographically distinct areas off New Jersey. M, 1977, Lehigh University. 115 p.

Huff, Donald Ross. Transport of benzene, o-xylene, 2-methylnaphthalene, phenanthrene, and pyrene in vadose zone microcosms. D, 1988, University of Oklahoma. 139 p.

Huff, Glenn. The K-Ar geochemistry of carnallite; evaluation of the geochemical stability of salt cycle six of the Paradox Formation in Utah. M, 1984, Georgia Institute of Technology. 51 p.

Huff, Lyman Coleman. Sedimentology of an aggrading river; the glacial Chippewa. D, 1957, University of Chicago. 59 p.

Huff, Ray V. Theoretical and experimental investigation of the phenomenon of rock disaggregation. M, 1968, University of Tulsa. 53 p.

Huff, Warren David. A study of Middle Ordovician K-bentonites in Kentucky and southern Ohio. D, 1963, University of Cincinnati. 115 p.

Huff, William Jennings. New mid-Paleozoic ostracodes from western Tennessee. M, 1957, Rice University. 70 p.

Huff, William Jennings. The Jackson Eocene Ostracoda of Mississippi. D, 1960, Rice University. 324 p.

Huffington, Roy Michael. Geology of the northern Quitman Mountains, Trans-Pecos Texas. D, 1943, Harvard University.

Huffington, Terry L. Geology of Moss Beach, California. M, 1976, Stanford University.

Huffington, Terry Lynn. Faunal zonation and hydrothermal diagenesis of a Cenomanian (Middle Cretaceous) rudist reef, Paso del Rio, Colima, Mexico. M, 1981, University of Texas, Austin.

Huffman, Arlie C. Geology of the Big Salmon Complex (Mississippian) in the vicinity of Swift Lake (Cassiar Mts.), British Columbia. M, 1971, George Washington University.

Huffman, Arlie C. Jr. The geology of the crystalline rocks of northern Virginia in the vicinity of Washington, District of Columbia. D, 1974, George Washington University.

Huffman, David Patrick. Biostratigraphy of the upper Cretaceous White-Speckled shales in western Saskatchewan (Canada). M, 1969, University of Saskatchewan. 102 p.

Huffman, George Garrett. Middle Ordovician limestone from Lee County, Virginia, to central Kentucky. D, 1945, Columbia University, Teachers College.

Huffman, George Garrett. The geology of Winneshiek County, Iowa. M, 1941, University of Iowa. 128 p.

Huffman, Jerald Dwight. Oil and gas field producing from the Salina Formation (Upper Silurian) in Lockport Dolomite, east Kentucky. M, 1966, University of Kentucky. 106 p.

Huffman, Kenneth Jay. The Precambrian geology of the Coyote Creek area, Montana. M, 1977, Northern Illinois University. 104 p.

Huffman, Kenneth Paul. A study of the relationship between grain-size distribution and capillary pressure curves for some cores from (Pennsylvanian) Bartlesville Sand (northeastern Oklahoma). M, 1948, University of Pittsburgh.

Huffman, Marion Edward. Micropaleontology of lower portion of Boquillas Formation near Hot Springs, Big Bend National Park, Brewster County, Texas. M, 1960, Texas Tech University. 54 p.

Huffman, Mark E. Stratigraphy and paleoecology of the Silica Formation (Middle Devonian) of Lucas County, Ohio. M, 1978, University of Toledo. 51 p.

Huffman, Othis Frank. Upper Miocene geology, rocks, and the history of displacement of the late Miocene slip along the San Andreas Fault in central California. D, 1972, University of California, Berkeley. 137 p.

Huffman, Samuel Floyd. Rhabdomesid ectoprocts (Bryozoa) from the Edwardsville member, Muldraugh Formation (Mississippian) at Crawfordsville (Montgomery county), Indiana. D, 1970, Indiana University, Bloomington. 172 p.

Hufford, Walter R. Texture, composition, and diagenesis of the Burbank Sandstone, North Burbank Field, Tract 97, Osage County, Oklahoma. M, 1983, Texas A&M University. 121 p.

Huftile, Gary J. Geologic structure of the upper Ojai Valley and Chaffee Canyon areas, Ventura County, California. M, 1988, Oregon State University. 103 p.

Hugentobler, Michael Ned. Prediction of coal seam disturbances in the Upper Cretaceous Blackhawk Formation based on borehole geophysical interpretations, southern Wasatch Plateau, Utah. M, 1988, Southern Illinois University, Carbondale. 72 p.

Huggett, R. J. Copper and zinc in bottom sediments and oysters, Crassostrea virginica, from Virginia's estuaries. D, 1977, College of William and Mary. 125 p.

Huggett, Thomas K. Velocity-chemistry relationships in the Pageland Diabase Dike, Kershaw County, South Carolina. M, 1976, University of North Carolina, Chapel Hill. 72 p.

Huggins, Andrew. Modeling chemical balances of freshwater wetlands. D, 1982, State University of New York, College of Environmental Science and Forestry. 190 p.

Huggins, Camillus B. The migration of a chlorinated organic solvent through the groundwater of a shallow alluvial aquifer near Calvert City, Kentucky. M, 1987, University of Kentucky. 103 p.

Huggins, Francis Edward. Mossbauer studies of iron minerals under pressures of up to 200 kilobars. D, 1975, Massachusetts Institute of Technology. 358 p.

Huggins, Jonathan Wayne. Geology of a portion of the Painted Hills Quadrangle, Wheeler County, North-central Oregon. M, 1978, Oregon State University. 129 p.

Huggins, Michael James. Meramecian conodonts and biostratigraphy of the (Upper Mississippian) Greenbrier Limestone (Hurricane Ridge and Greendale synclines), south-western Virginia and southern West Virginia. M, 1983, Virginia Polytechnic Institute and State University. 305 p.

Hughart, Joseph L. Hydrologic investigations of hazardous waste landfills near South Charleston, West Virginia. M, 1982, Ohio University, Athens. 183 p.

Hughart, Richard D. Geology of the Osage, West Virginia-Virginia, 7 1/2 minute topographic quadrangle. M, 1967, West Virginia University.

Hughbanks, Julia A. Stock tank characteristics and performance in the Beaver Creek watershed, north-central Arizona. M, 1983, Northern Arizona University. 128 p.

Hughes, Conway Todd, III. Fracture analysis in the vicinity of the Beech Grove Lineament, east central Tennessee. M, 1983, Memphis State University. 47 p.

Hughes, David D. Nummulites (Foraminifera) of northern Libya. M, 1959, New York University.

Hughes, Dolores M. Petrography of the LaSalle Limestone (Pennsylvania) LaSalle County, Illinois. M, 1972, Northern Illinois University. 83 p.

Hughes, Donald Dudley. Microscopic methods in the correlation of oil field sediments. M, 1924, Stanford University. 16 p.

Hughes, Edward Stewart. Calcareous nannoplankton and biostratigraphy of late Miocene to late Pliocene sediments in the Conrad cores from the Blake Plateau, North Atlantic. D, 1973, Washington University. 87 p.

Hughes, Elizabeth. The hydrothermal alteration system at Daniels Mountain, northern Carolina Slate Belt, North Carolina. M, 1985, University of North Carolina, Chapel Hill. 86 p.

Hughes, Gary Claude. Cenozoic geology and geomorphology of the Dry Creek Valley, Gallatin County, Montana. M, 1980, Montana State University. 145 p.

Hughes, George Muggah. A study of Pleistocene Lake Edmonton and associated deposits. M, 1959, University of Alberta. 60 p.

Hughes, George Muggah. The glacial geology of the Redwater and Morinville areas, Alberta. D, 1962, University of Illinois, Urbana. 126 p.

Hughes, Gordon J., Jr. Petrology and tectonic setting of igneous rocks in the Henderson-Willow Creek igneous belt (Late Cretaceous-Early Paleocene), Granite County, Montana. D, 1971, Michigan Technological University. 236 p.

Hughes, Gordon J., Jr. Precambrian stratigraphy and structure in the Henderson-Willow creek igneous belt, Granite County, Montana. M, 1970, Michigan Technological University. 93 p.

Hughes, J. W. Stratigraphic analysis of the Pitkin, Hale, and Bloyd formations (Pennsylvanian, NW Arkansas) of the Grapevine Ridge-Liberty Hill area, Washington and Crawford counties, Arkansas. M, 1971, University of Arkansas, Fayetteville.

Hughes, Jerry L. Some aspects of hydrogeology of the Spring Mountains and Pahrump Valley, Nevada, and environs, as determined by spring evaluation. M, 1966, University of Nevada. 116 p.

Hughes, John David. Correlation and cyclicity analysis of the Jurassic-Cretaceous Kootenay Formation near Canmore, Alberta. M, 1975, University of Alberta. 191 p.

Hughes, John Derek. Physiography of a six quadrangle area in the Keweenaw Peninsula north of Portage Lake. D, 1964, Northwestern University. 261 p.

Hughes, John E. Stratigraphic relationships and depositional environments of the Hampton and Gilmore City formations, North-central Iowa. M, 1977, University of Iowa. 165 p.

Hughes, John Herbert. The geology of the Deer Creek Fault area, Perry County, Indiana. M, 1951, Indiana University, Bloomington. 31 p.

Hughes, John Michael. Geology and petrology of the Caldera Tzanjuyub, western Guatemala. M, 1978, Dartmouth College. 123 p.

Hughes, John Michael. The crystal chemistry of the copper vanadate and vanadium oxide bronze minerals from the fumaroles of Izalco Volcano, El Salvador. D, 1981, Dartmouth College. 147 p.

Hughes, John Patrick, Jr. Groundwater recharge calculations in San Diego County. M, 1982, San Diego State University. 221 p.

Hughes, Marie Jeanne. Iron staining of a carbonate dimension stone, Gasport Formation (middle Silurian), Queenston, Ontario. M, 1971, University of Western Ontario. 96 p.

Hughes, Mark A. A study of the Precambrian rocks of the Cow Creek area, Sierra Madre Mountains, Carbon County, Wyoming. M, 1973, University of Wyoming. 43 p.

Hughes, Owen L. Pleistocene geology of Nova Scotia. D, 1955, University of Kansas.

Hughes, Owen L. Surficial geology of Smooth Rock and Iroquois Falls map areas, Cochrane District, Ontario. D, 1959, University of Kansas.

Hughes, Paul W. The stratigraphy of the Supai Formation in the Chino Valley area, Yavapai County, Arizona. M, 1950, University of Arizona.

Hughes, Paul Warren. Stratigraphy of the Georgetown Formation. D, 1963, University of Illinois, Urbana. 109 p.

Hughes, Randall Edward. Mineral matter associated with Illinois coals. D, 1971, University of Illinois, Urbana. 145 p.

Hughes, Rhys Leckie. Sedimentology of the Sixtymile River placer gravels, Yukon Territory. M, 1986, University of Alberta. 226 p.

Hughes, Richard David. A revision of the stratigraphy of the Lea Park Formation in west central Saskatchewan. M, 1947, University of British Columbia.

Hughes, Richard David. Geology of portions of Sunwapta and Southesk map-areas, Jasper National Park, Alberta, Canada. D, 1953, University of Oklahoma. 164 p.

Hughes, Richard John, Jr. A study of the Lower Cretaceous section near Junction, Texas. M, 1948, University of Texas, Austin.

Hughes, Richard V. Geology of the Beartooth mountain front in Park County, Wyoming. D, 1933, The Johns Hopkins University.

Hughes, Robert Hilton. Use of Landsat multispectral scanner digital data for mapping suspended solids and salinity in the Atchafalaya Bay and adjacent waters, Louisiana. D, 1982, Louisiana State University. 115 p.

Hughes, Roger D. Petrochemistry of the Yellowstone-Absaroka region, Wyoming, Montana, and Idaho. M, 1979, Eastern Washington University. 205 p.

Hughes, Scott Stevens. Geology of the southwestern part of the Bill Williams Mountain Quadrangle, Coconino County, Arizona. M, 1978, Northern Arizona University. 79 p.

Hughes, Scott Stevens. Petrochemical evolution of High Cascade volcanic rocks in the Three Sisters region, Oregon. D, 1983, Oregon State University. 199 p.

Hughes, Steven. Facies anatomy of Lower Cambrian archaeocyathid biostrome complex, southern Labrador. M, 1979, Memorial University of Newfoundland. 241 p.

Hughes, Steven Allen. Beach and dune erosion during severe storms. D, 1981, University of Florida. 291 p.

Hughes, Susan Elaine McAlear. The paleogeography and subsurface stratigraphy of the Late Ordovician Queenston Coastal Complex in New York. M, 1976, Cornell University.

Hughes, Terence John. A comprehensive model of ionospheric-magnetospheric current systems during periods of moderate magnetospheric activity. D, 1978, University of Alberta. 364 p.

Hughes, Theresa M. The sedimentologic characteristics of the Union Lake - Maurice River system, New Jersey. M, 1982, Rutgers, The State University, New Brunswick. 141 p.

Hughes, Thomas Hastings. Early Pennsylvanian paleogeology of Oklahoma. M, 1955, Stanford University.

Hughes, Travis Hubert. Minor elements in sulfide minerals of the Hill Mine (Mississippian), Cave-In-Rock District, Illinois. D, 1967, University of Colorado. 79 p.

Hughes, Travis Hubert. The geology of the Gladeville Quadrangle, Tennessee. M, 1960, Vanderbilt University.

Hughes, William Brian. The Quaternary history of the lower Cache River valley, southern Illinois. M, 1987, Southern Illinois University, Carbondale. 181 p.

Hughes, William D. Peat resources and glacial geology of Chapman Swamp and adjacent area, Westerly, Rhode Island. M, 1982, University of Rhode Island.

Hughes, William Theodore. Geochemical evolution of basalts from Amboy and Pisgah lava fields, Mojave Desert, California. M, 1986, University of North Carolina, Chapel Hill. 113 p.

Hughson, Robert Carl. Upper Silurian carbonates of Lake Memphremagog and Lime Ridge areas, Quebec. M, 1987, McGill University. 166 p.

Hughston, Edward Wallace. Geology of northern half of Salt Draw Quadrangle, Culberson County, Texas. M, 1950, University of Texas, Austin.

Hughston, Mark D. Petrology and depositional environments of the Gunsight (Upper Pennsylvanian) shelf-margin carbonate bank, eastern shelf, west-central Texas. M, 1979, University of Texas, Arlington. 128 p.

Hughto, Richard John. Multi-constituent simulation and optimization models for water quality management. D, 1981, Cornell University. 242 p.

Hugman, Robert H. H., III. The effects of texture and composition on the mechanical behavior of experimentally deformed carbonate rocks. M, 1978, Texas A&M University.

Hugo, Ken J. Hydrodynamic flow associated with Leduc reefs. M, 1985, University of Calgary. 149 p.

Hugo, William D. Thermal controls on lithospheric strength and the evolution of the northern San Andreas fault system. M, 1986, Pennsylvania State University, University Park. 49 p.

Huguenin, Robert Louis. Photo-stimulated oxidation of magnetite, and an application to Mars. D, 1972, Massachusetts Institute of Technology. 218 p.

Huguley, Robert William. Environmental and geographic variations in populations in Modiolus demissus (Dillwyn). M, 1973, University of South Carolina.

Huh, Chih-An. Radiochemical and chemical studies of manganese nodules from three sedimentary regimes in the North Pacific. D, 1982, University of Southern California.

Huh, John Mun Suk. Geology and diagenesis of the Niagaran pinnacle reefs in the northern shelf of the Michigan Basin. D, 1973, University of Michigan.

Huh, John Mun Suk. Scanning and transmission microscopy of some ooids in oolitic limestones. M, 1970, Bowling Green State University. 58 p.

Huh, Oscar Karl, Jr. Geology of a portion of the Lemhi Range and Birch Creek basin, Lemhi, Clark and Butte counties, Idaho. M, 1963, Pennsylvania State University, University Park. 156 p.

Huh, Oscar Karl, Jr. Mississippian stratigraphy and sedimentology, across the Wasatch line, east-central Idaho and extreme south-western Montana. D, 1968, Pennsylvania State University, University Park. 230 p.

Huhn, Frank Jones. An atmospheric dispersion model for the Sudbury, Ontario, area. D, 1982, McMaster University. 353 p.

Huidobro N., Pablo. Utilization of heavy minerals in quantitative stratigraphy; the Pleistocene-Holocene transition in Massachusetts Bay sediments. M, 1980, Boston College.

Huie, James Powell. Subdivision of the Claiborne Group in eastern Atascosa County, Texas. M, 1948, University of Oklahoma. 97 p.

Huizenga, Douglas Lee. An examination of copper speciation in seawater. D, 1982, University of Rhode Island. 123 p.

Huizinga, Bradley James. The effect of the mineral matrix on laboratory-simulated catagenesis of kerogen. M, 1985, University of California, Los Angeles. 246 p.

Hula, Charles William. A structural interpretation of the Coke oil field, Wood County, Texas. M, 1950, Colorado College.

Hulbe, Christoph W. H. An investigation of the size and shape of quartz grains, Pedro Beach, California. M, 1957, Pennsylvania State University, University Park. 69 p.

Hulbert, Larry John. Geology of the Fraser Lake gabbro complex, Manitoba. M, 1978, University of Regina. 207 p.

Hulbert, Richard Charles, Jr. Linear discriminant analysis and variability of Pleistocene and Holocene Leporidae of Texas. M, 1979, University of Texas, Austin.

Hulbert, Richard Charles, Jr. Phylogenetic systematics, biochronology, and paleobiology of late Neogene horses (family Equidae) of the Gulf Coastal Plain and the Great Plains. D, 1987, University of Florida. 570 p.

Hulburt, Margery Ann. Hydrogeology of the Gillette area, Wyoming. M, 1979, University of Arizona. 145 p.

Hulen, Paul Leon. Determination of the source of barite in the Chamberlain Creek Deposit of Arkansas; Sr isotopic evidence. M, 1978, University of Kansas. 54 p.

Hulin, Carlton D. Geology and ore deposits of the Randsburg Quadrangle, California. D, 1925, University of California, Berkeley. 152 p.

Hulke, Steven D. Quantitative geomorphology of small drainage basins in central Texas. M, 1978, University of Texas, Austin.

Hull, Amy Berg. Kinetics of evaporite mineral-brine interactions; mathematical modeling and experimental determination of the effect of gamma radiation and threshold cyrstallization inhibition on Permian Basin brine composition. D, 1987, Northwestern University. 236 p.

Hull, Carter Dean. Low-to-moderate temperature geothermal resources potentially available to the City of Riverside, California; a preliminary assessment based on geochemistry. M, 1984, University of California, Riverside. 145 p.

Hull, Dennis N. A petrologic-geochronologic study of the host rocks and wall rocks at the Detour Prospect, a copper-zinc-silver discovery in Brouillan Township, Abitibi-West County, Quebec. M, 1976, Wright State University. 133 p.

Hull, Donald Albert. Geology of the Puzzle vein, Creede mining district (Mineral county), Colorado. D, 1970, University of Nevada. 200 p.

Hull, Douglas A. Stratigraphy and petrology of the Newton Creek and Alpena limestones (Middle Devonian) in the Huron Portland Quarry, Alpena, Michigan. M, 1980, Bowling Green State University. 176 p.

Hull, Harris Benjamin. Burial diagenesis and timing of reservoir development, North Haynesville Field, Louisiana. M, 1982, Texas A&M University.

Hull, Joseph Michael. Structural and tectonic evolution of Archean supracrustals, southern Wind River Mountains, Wyoming. D, 1988, University of Rochester. 280 p.

Hull, Joseph P. D. The development of soil classification. M, 1917, The Johns Hopkins University.

Hull, Joseph P. D., Jr. Deposition of the Permian Delaware Mountain Group, Texas. M, 1953, Columbia University, Teachers College.

Hull, Joseph P. D., Jr. Guadalupian sandstone facies of the Delaware Basin, Texas and New Mexico. D, 1955, Columbia University, Teachers College.

Hull, Laurence Charles. Mechanisms controlling the inorganic and isotopic geochemistry of springs in a carbonate terrane. D, 1980, Pennsylvania State University, University Park. 275 p.

Hull, Louis Vincent. Seismological studies. M, 1955, Southern Methodist University. 58 p.

Hull, M. J. Geochemical dispersal pattern around fluorspar deposits of Paisano Peak and Paisano Mine, Brewster County, Texas. M, 1977, West Texas State University. 39 p.

Hull, Marylee Witner. Microtechniques in Pb/U dating of Moroccan zircons. M, 1976, Massachusetts Institute of Technology. 38 p.

Hull, Paul W. Igneous rocks of the Cooperton area, Wichita Mountains, Oklahoma. M, 1951, University of Oklahoma. 58 p.

Hull, Robert. U. isotopic disequilibrium; its hydrological application to Floridian Aquifer waters of N.E. Fl. M, 1981, Florida State University.

Hull, Thomas Edward. Ground water production from alluvium and glacial gravels in the Pittsburgh District. M, 1948, University of Pittsburgh.

Hulme, James A. Stratigraphy of the Wells creek basin area, Stewart County, Tennessee. M, 1968, University of Tennessee, Knoxville. 145 p.

Hulmes, Leita Jean. Holocene stratigraphy and geomorphology of the Hills Beach/Fletcher Neck Tombolo System, Biddeford, Maine. M, 1980, University of Delaware.

Hulse, John Arthur. Microscopic investigation of the coral zone of the Jeffersonville Limestone (Middle Devonian), SE Indiana. M, 1963, University of Illinois, Urbana.

Hulse, Robert C. Stratigraphy and morphology of a modern transgression; Edisto Island, South Carolina. M, 1974, University of South Carolina. 39 p.

Hulse, Scott E. An investigation into the causes of steady state electrical potential differences occurring on the surface of the Earth. M, 1978, University of Arizona.

Hulse, William J. A geologic study of the Sallyards Field area, Greenwood County, Kansas. M, 1978, University of Kansas. 153 p.

Hulsey, Jess Dale. Beach sediments of eastern Lake Michigan. D, 1962, University of Illinois, Urbana. 165 p.

Hulsey, Jess Dale. Relations of settling velocity of sand-sized spheres and sample weight. M, 1960, University of Illinois, Urbana.

Hulsman, Robert B. The hystrichosphaerids of the Onondaga Formation, Devonian of eastern New York. M, 1956, University of Massachusetts.

Hulstrand, Richard F.; Tinker, Clarence N. and Wendt, Roy L. Geology of the Turkey Creek-Williams Creek area, Huerfano County, Colorado. M, 1955, University of Michigan.

Hultgren, Michael Charles. Tectonics and stratigraphy of part of the southern Galina-Archuleta Arch, French Mesa and Llaves quadrangles, Rio Arriba County, New Mexico. M, 1986, University of New Mexico. 123 p.

Hultman, John R., Jr. Paleoenvironmental reconstruction of the Bromide Formation (Middle Ordovician); southern Oklahoma and northern Texas. M, 1985, Stephen F. Austin State University. 120 p.

Hultman, John Richard. Cambrian sandstone, Marquette Quadrangle, Michigan. M, 1953, University of Michigan.

Hultman, William A. Geotechnical investigation of the Grouse Creek slope failure. M, 1988, University of Idaho. 131 p.

Humbertson, P. G. Valley morphologies related to process dominance in the Upper Allegheny. D, 1977, University of Pittsburgh. 177 p.

Humbertz, Jon M. A numerical solution to the non-linear problem of steady eastward barotropic flow past an island on the beta plane. D, 1974, Texas A&M University. 108 p.

Humble, Emmett Arl. Cenozoic history of northeastern Chispa Quadrangle, Trans-Pecos, Texas. M, 1951, University of Texas, Austin.

Hume, George S. Stratigraphy and geologic relations of the Paleozoic outlier of Lake Timiskaming, Ontario. D, 1920, Yale University.

Hume, Howard. The distribution of Recent foraminifera in Southeast Baffin Bay. M, 1972, Dalhousie University.

Hume, Howard Robertson. Geotechnical slope stability analysis, B & G sector, Bingham Canyon open pit, Utah, with a consequent investigation of methods for delineating rock mass fracture domains. D, 1983, Purdue University. 713 p.

Hume, Howard Robertson. Tunnel boring machines; some considerations and aspects. M, 1977, Purdue University.

Hume, James David and Leeder, Robert W. Geology of the Pike Mountain area, Gallatin County, Montana. M, 1950, University of Michigan. 45 p.

Hume, James David. Spectrochemical analyses of carbonate rocks [Michigan]. D, 1957, University of Michigan. 141 p.

Humes, Elmer C., Jr. Geology of portions of the Ithaca and Watkins Fifteen Minute quadrangles, New York. M, 1960, University of Rochester. 105 p.

Humiski, Robert Nicholas. The deformational history of the south central portion of the Munro Lake greenstone belt, northern Manitoba. M, 1974, University of Manitoba.

Hummel, Charles L. The structure and mineralization of a portion of the Bald Mountain mining district, Lawrence County, South Dakota. M, 1952, South Dakota School of Mines & Technology.

Hummel, Gary Alan. Interdune areas of the back-island dune field, North Padre Island, Texas. M, 1982, University of Texas, Austin. 94 p.

Hummel, J. M. Geology of the Pacific Creek area, Sublette and Fremont counties, Wyoming. M, 1958, University of Wyoming. 76 p.

Hummel, Judythe Ann. Stratigraphy of East River Mountain-Wolf Creek area, Virginia. M, 1972, University of Akron. 36 p.

Hummel, Richard. Texture and granulometry of planar cross-stratification from some unconsolidated, ancient and modern, quartz sand deposits. M, 1985, Florida State University.

Hummeldorf, Raymond George. The classification and evaluation of roof rocks for an underground coal mine by sonic logging. M, 1984, Purdue University. 123 p.

Hummert, B. A. Visitor's guide to the geology of Mount Diablo. M, 1977, San Jose State University. 39 p.

Humphreville, James A. Fluorspar mineralization on the ridge north of Daugherty Gulch near Challis, Custer County, Idaho. M, 1956, Cornell University.

Humphrey, Arthur G. The geology of Poorman Hill and the Poorman Mine, Boulder County, Colorado. M, 1955, University of Colorado.

Humphrey, Dana Norman. Design of reinforced embankments. D, 1986, Purdue University. 448 p.

Humphrey, Fred L. Geology of the Groom District, Lincoln County, Nevada. M, 1945, University of Nevada - Mackay School of Mines.

Humphrey, Fred L. Geology of the White Pine District, Nevada. D, 1956, University of California, Los Angeles.

Humphrey, Frederic Gavin. Geologic structure at Moss Beach, California. M, 1941, Stanford University.

Humphrey, John Dean. Processes, rates, and products of early near-surface carbonate diagenesis; Pleistocene mixing zone dolomitization and Jurassic meteoric diagenesis. D, 1987, Brown University. 271 p.

Humphrey, John Fitzgerald. Diagenesis and secondary porosity in the Vicksburg Formation, Lyda and North Rincon filelds, Starr County, Texas. M, 1985, Texas Tech University. 95 p.

Humphrey, Neil Frank. Basal hydrology of a surge-type glacier; observations and theory relating to Variegated Glacier. D, 1987, University of Washington. 206 p.

Humphrey, Neil Frank. Pore pressures in debris failure initiation. M, 1982, University of Washington. 169 p.

Humphrey, Ronald C. The geology of the crystalline rocks of Green and Hancock counties, Georgia. M, 1970, University of Georgia. 57 p.

Humphrey, Thomas M., Jr. The geology of the southeast third of the Susanville Quadrangle, Oregon. M, 1956, University of Oregon. 65 p.

Humphrey, William Elliot. Geology of the Sierra de Los Muertos area, Coahuila, Mexico, and Aptian cephalopods from the La Pena Formation. D, 1947, University of Michigan.

Humphrey, William Elliot. Revision of E. A. Strong's types from the Mississippian Point au Gres Limestone at Grand Rapids (Michigan). M, 1940, University of Michigan.

Humphrey, William R. The lithology of the Cambrio-Ordovician contact in the Mankato, Minnesota area. M, 1958, University of Minnesota, Minneapolis. 110 p.

Humphreys, Curtis H. Stratigraphy and petrography of the Lower Cretaceous (Albian) Salmon Peak Formation of the Maverick Basin, South Texas. M, 1984, University of Texas, Arlington. 138 p.

Humphreys, Cynthia. Uranium and radium isotopic distributions in ground waters and sediments of the land-pebble phosphate district and surrounding areas of west-central Florida. M, 1984, Florida State University.

Humphreys, Edwin W. An interpretation of the Comanche formations of the Black and Grand Prairie regions of Texas. M, 1906, Columbia University, Teachers College.

Humphreys, Eugene D. Telluric sounding and mapping in the vicinity of the Salton Sea geothermal area, Imperial Valley, California. M, 1978, University of California, Riverside. 149 p.

Humphreys, Eugene Drake. Studies of the crust-mantle system beneath Southern California. D, 1985, California Institute of Technology. 196 p.

Humphreys, Matthew. Sedimentary environments in Upper Devonian clastic rocks of north-central Pennsylvania. M, 1973, Rensselaer Polytechnic Institute. 71 p.

Humphreys, Richard. The characterization of the sulfide component in layered igneous and metasedimentary rocks in the Mouat Block of the Stillwater Complex. M, 1983, San Jose State University. 135 p.

Humphries, Stanley M. Seasonal variation in morphology at North Inlet, S.C. M, 1977, University of South Carolina. 97 p.

Humphris, Curtis Carlyle. Sedimentary processes along the shore of Lake Erie and Sandusky Bay, from Marblehead Lighthouse to Bay Bridge, Ottawa County, Ohio. M, 1952, Ohio State University.

Humphris, David D. Correlation of the central North American Rift system, Kansas, and the Hartville Uplift, Wyoming, using geophysical and subsurface data. M, 1987, University of Southwestern Louisiana. 73 p.

Humphris, Susan Elizabeth. The hydrothermal alteration of oceanic basalts by seawater. D, 1977, Massachusetts Institute of Technology. 247 p.

Humston, John. Stick-slip in Tennessee Sandstone. M, 1972, Texas A&M University. 68 p.

Hund, Erik A. U-Pb dating of granites from the Charlotte Belt of the Southern Appalachians. M, 1987, Virginia Polytechnic Institute and State University.

Hundley, Emily M. Microscopic characteristics of the sliding surface and fault gouge from sliding friction experiments in orthoquartzite. M, 1977, University of North Carolina, Chapel Hill. 162 p.

Huner, John, Jr. The geology of Caldwell and Winn parishes, Louisiana. D, 1939, Louisiana State University.

Hunerman, Aybars E. Miocene stratigraphy of the Central Plateau, Haiti. M, 1972, Louisiana State University.

Hung, C. Stochastic analysis of bedload particle movement. D, 1975, Colorado State University. 183 p.

Hung, C. C. Conodonts from the Devonian System in the Sacramento Mountains, Otero County, New Mexico. M, 1978, West Texas State University. 109 p.

Hunger, Arthur A. Distribution of foraminifera, Netarts Bay, Oregon. M, 1966, Oregon State University. 112 p.

Hunkins, Kenneth Leland. Elastic wave studies in the Arctic Ocean. D, 1960, Stanford University. 132 p.

Hunnewell, Dorothy S. The hydrogeology of two solid waste disposal sites in Berkley, Massachusetts. M, 1977, Boston University. 268 p.

Hunsaker, Carolyn Thomas. The effect of San Francisco's wastewater plan on water quality and aquatic biology. D, 1980, University of California, Los Angeles. 207 p.

Hunsaker, Ernest Leon, III. Geology and gold potential east of the Vulcan Mine, Gunnison and Sagauche counties, Colorado. M, 1988, Colorado State University. 124 p.

Hunsaker, Vaughn Edward. The formation of mixed magnesium-aluminum hydroxides in soils and solutions. D, 1969, University of California, Riverside. 70 p.

Hunsberger, Gloria Grace. Differentiation of the effects of pollution, sulfide-deposit weathering, and background weathering on stream chemistry in the Rocky River basin of North Carolina. M, 1984, North Carolina State University. 145 p.

Hunt, Adrian P. Stratigraphy, sedimentology, taphonomy and magnetostratigraphy on the Fossil Forest area, San Juan County, New Mexico. M, 1984, New Mexico Institute of Mining and Technology. 338 p.

Hunt, Allen Standish. Species of the tabulate coral Trachypora from the Middle Devonian Traverse Group of Michigan. M, 1957, University of Michigan.

Hunt, Allen Standish. Trilobite growth, variation, and instar development. D, 1964, Harvard University.

Hunt, Amanda M. Cenozoic *Fasciolaria* (Mollusca; Gastropoda) of the western Atlantic region. D, 1975, Tulane University. 202 p.

Hunt, C. D. The role of phytoplankton and particulate organic carbon in trace metal deposition in Long Island Sound. D, 1979, University of Connecticut. 282 p.

Hunt, Catherine Minna. A study of silver-lead-zinc ore mineralization from a Yukon suite of rocks. M, 1964, University of Toronto.

Hunt, Charles B. Pleistocene Lake Bonneville, ancestral Great Salt Lake as described in the notebooks of G. K. Gilbert, 1875-1880. M, 1982, Brigham Young University. 231 p.

Hunt, Christopher Paul. The acquisition of chemical remanent magnetization during the inversion of titanomaghemites in ocean basalt. M, 1986, University of Minnesota, Minneapolis. 111 p.

Hunt, David Gardiner. Clay mineralogy of the Vermejo Formation (Upper Cretaceous), Canon City, Colorado. M, 1963, University of Oklahoma. 126 p.

Hunt, David K. Geochemistry and geochronology of Archean metamorphic rocks of the eastern Beartooth Mountains, Montana. M, 1979, University of Florida.

Hunt, E. Lower and Middle Devonian formations near Everett, Bedford Co., Pennsylvania. M, 1941, Bryn Mawr College.

Hunt, Emily Lee. The modern sediments of Topsail Sound, North Carolina. M, 1980, University of North Carolina, Chapel Hill. 47 p.

Hunt, Gail S. Groundwater geology of Bristol and Cadiz valleys, San Bernardino County, California. M, 1966, University of Southern California.

Hunt, George Lewis, Jr. The foraminifera of the Byram Marl of Mississippi. M, 1957, Mississippi State University. 75 p.

Hunt, Graham Hugh. Petrology of the Purcell sills in the St. Mary Lake area, British Columbia. M, 1958, University of Alberta. 95 p.

Hunt, Graham Hugh. The Purcell eruptive rocks. D, 1961, University of Alberta. 139 p.

Hunt, Gregory L. Petrology of the Mt. Pennell central stock, Henry Mountains, Utah. M, 1983, Brigham Young University. 100 p.

Hunt, Hubert Bush. A regional magnetometer survey of the Pinckney area, Michigan. M, 1952, University of Michigan.

Hunt, Jesse L., Jr. The geology and origin of Gray's Reef, Georgia continental shelf. M, 1974, University of Georgia.

Hunt, Joel A. Analysis of recharge to an underground lead-zinc mine, Coeur d'Alene mining district, Idaho. M, 1984, University of Idaho. 89 p.

Hunt, John A. The stability of sphene; experimental redetermination and geologic implications. M, 1976, Pennsylvania State University, University Park.

Hunt, John Bancroft. Fore-reef petrography of a Silurian reef, Richvalley, Indiana. M, 1959, University of Illinois, Urbana.

Hunt, John Prior. Rock alteration, mica, and clay minerals in certain areas in the United States and Lark Mines, Bingham, Utah. D, 1957, University of California, Berkeley. 321 p.

Hunt, Lynn Bogue. Geology of the Mohonk Lake Quadrangle, New York. M, 1957, New York University.

Hunt, Mahlon Seymour. A technique of photogramography as utilized in geographic studies. M, 1952, Washington University. 132 p.

Hunt, Margo Elaine. Dynamics of soil algae in field and forest environments. D, 1976, Rutgers, The State University, New Brunswick. 117 p.

Hunt, Michael C. The environmental geology as related to the geohydrology of the Beaver Lake Reservoir area in Northwest Arkansas. M, 1973, University of Arkansas, Fayetteville.

Hunt, Patricia Kelly. Oxygen isotope geochemistry of Precambrian metamorphic iron-formations from southwestern Montana. M, 1980, Kent State University, Kent. 57 p.

Hunt, Paul Thomas. The metamorphic petrology and structural geology of the serpentinite-matrix melange in the Greenhorn Mountains, northeastern Oregon. M, 1985, University of Oregon. 127 p.

Hunt, Paula J. Determining aquifer transmissivity and storativity with recirculating pumping tests. M, 1988, Purdue University. 117 p.

Hunt, Randall James. A survey of the hydrogeochemistry of the Leopold Memorial Reserve, Baraboo, Wisconsin. M, 1987, University of Wisconsin-Madison. 170 p.

Hunt, Raymond Samuel. The geology and natural resources of an area in the vicinity of Bloomington, Indiana. M, 1925, Indiana University, Bloomington. 105 p.

Hunt, Robert Elton. The geology of the Dry Canyon region, Gunnison Plateau, Utah. M, 1948, Ohio State University.

Hunt, Robert Elton. The geology of the northern part of the Gunnison Plateau, Utah. D, 1950, Ohio State University.

Hunt, Robert Molyneaux, Jr. North American amphicyonids (Oligocene and Miocene) (Mammalia; carnivora). D, 1971, Columbia University. 664 p.

Hunt, Robert N. Oxygen isotope studies in sulphate. D, 1974, University of Alberta. 221 p.

Hunt, S. R. Bedrock stratigraphy as a tool in regional slope evaluation, upper Illinois River valley. M, 1975, University of Illinois, Urbana. 44 p.

Hunt, S. R. Surface subsidence due to coal mining in Illinois. D, 1980, University of Illinois, Urbana. 134 p.

Hunt, Sharon Barbara. Estimating groundwater travel times in heterogenous media. M, 1986, Washington State University. 71 p.

Hunt, Timothy J. Characterization of selected Ohio coals for industrial use; porosimetry, surface area and trace element content. M, 1982, University of Toledo. 301 p.

Hunt, Walter. Economic geology and geochemistry of part of the Montezuma District, Front Range, Colorado. M, 1980, Colorado School of Mines. 106 p.

Hunt, Walter Frederick. The origin of the sulfur deposits of Sicily. D, 1915, University of Michigan.

Hunter, A. Douglas. The geologic setting of the Aldermac copper deposit, Noranda, Quebec. M, 1979, Carleton University. 167 p.

Hunter, B. E. Fluvial sedimentation on an active volcanic continental margin; Rio Guacalate, Guatemala. M, 1976, University of Missouri, Columbia.

Hunter, Bruce Edward. Regional analysis of the Point Lookout Sandstone, Upper Cretaceous, San Juan Basin, New Mexico-Colorado. D, 1979, Texas Tech University. 118 p.

Hunter, Cindy Carothers. Geology and wall-rock alteration study of selected Bibb County barite deposits. M, 1978, University of Alabama.

Hunter, Craig Russell. Pedogenesis in Mazama tephra along a bioclimatic gradient in the Blue Mountains of southeastern Washington. D, 1988, Washington State University. 128 p.

Hunter, David Roy. Conodonts from the Cobbs Arm Formation (Middle Ordovician), North-central Newfoundland. M, 1978, Memorial University of Newfoundland. 180 p.

Hunter, Donald Reid. Petrographic and geochemical study of carbonate rocks in the Bangor-lower most Pennington interval, Monteagle Mountain, Tennessee. M, 1978, University of Georgia.

Hunter, Frank R. Geology of the Alabama tin belt. D, 1943, Cornell University.

Hunter, Gerhart Eugene. Late Pleistocene climatic cycles; evidence from planktonic foraminiferal assemblages. M, 1984, Miami University (Ohio). 120 p.

Hunter, Hugh E. Geological investigations of the Lynn Lake basic intrusive body, northern Manitoba. M, 1952, University of Manitoba.

Hunter, Hugh Edwards. Petrology of the Tow Lake Gabbro, Barrington Lake area, northern Manitoba. D, 1954, University of California, Los Angeles.

Hunter, Jack. Geology of the Casey Draw and Gozar quadrangles, Reeves and Jeff Davis counties, Texas. M, 1948, University of Texas, Austin.

Hunter, James A. M. Crust under Hudson Bay (Canada) from seismic refraction studies. M, 1967, University of Western Ontario.

Hunter, James A. M. Crustal seismic studies in northern Ontario and Manitoba from Project Early Rise data. D, 1971, University of Western Ontario.

Hunter, James C. Laramide synorogenic sedimentation in south-central New Mexico; petrologic evolution of the McRae Basin. M, 1986, Colorado School of Mines. 75 p.

Hunter, James M. Geology of the North Hans Peak area, Routt County, Colorado. M, 1955, University of Wyoming. 76 p.

Hunter, John Frederick. A study of the Pre-Cambrian rocks of the Gunnison River. D, 1912, The Johns Hopkins University.

Hunter, John Henry. Stability of simple cuts in normally consolidated clays. D, 1968, University of Colorado. 333 p.

Hunter, LaVerne D. Frontier Formation along the eastern margin of the Bighorn Basin, Wyoming. M, 1950, University of Wyoming. 123 p.

Hunter, Paul Kirk. Stratigraphy and facies analysis of the El Picacho Formation. M, 1986, University of Texas of the Permian Basin. 118 p.

Hunter, Philip M. The environmental geology of the Pine Grove Mills-Stormstown area, central Pennsylvania, with emphasis on the bedrock geology and ground water resources. M, 1977, Pennsylvania State University, University Park. 319 p.

Hunter, Ralph E. Iron sedimentation in the Clinton Group of the central Appalachian Basin. D, 1960, The Johns Hopkins University.

Hunter, Robert Bruce. Timing and structural relations between the Gros Ventre foreland uplift, the Prospect thrust system, and the Granite Creek Thrust, Hoback Basin, Wyoming. M, 1986, University of Wyoming. 115 p.

Hunter, Robert D. Volcanic stratigraphy and structural control of mineralization in the northeastern portion of the Patterson mining district, Mono County, California. M, 1976, University of California, Riverside. 135 p.

Hunter, William C. The Garnet Ridge and Red Mesa kimberlitic diatremes, Colorado Plateau; geology, mineral chemistry, and geothermobarometry. D, 1979, University of Texas, Austin. 157 p.

Hunter, William Clay. Petrography and petrology of a basanitoid flow from Hut Point Peninsula, Antarctica. M, 1974, Northern Illinois University. 91 p.

Hunter, William J. The heavy mineral assemblages of the Ogallala Group in southwestern Nebraska. M, 1955, University of Nebraska, Lincoln.

Hunter, William Patterson. Heavy mineral analysis of selected Monterey Bay cores. M, 1971, United States Naval Academy.

Hunter, Zena M. Geologic patterns in the foothills of the Front Range; Boulder-Lyons Area (Boulder County), Colorado. D, 1947, University of Colorado.

Huntington, Ellsworth. Changes in climate of Recent geological time. D, 1909, Yale University.

Huntington, George C. A sedimentary study of the Glorietta Sandstone of New Mexico. M, 1949, Texas Tech University. 34 p.

Huntington, Gordon Leland. Soil-land form relationships of portions of the San Joaquin River and Kings River alluvial depositional systems in the Great Valley of California. D, 1980, University of California, Davis. 246 p.

Huntington, Hope Davies. Anorthositic and related rocks from Nukasorsuktokh Island, Labrador. D, 1980, University of Massachusetts. 146 p.

Huntington, J. Craig. Mineralogy and petrology of metamorphosed iron-rich beds in the Lower Devonian Littleton Formation, Orange area, Massachusetts. M, 1975, University of Massachusetts. 106 p.

Huntley, David. Ground water recharge to the aquifers of northern San Luis Valley, Colorado; a remote sensing investigation. D, 1976, Colorado School of Mines. 298 p.

Huntoon, Jacqueline. Depositional environments and paleotopographic relief of the White Rim Sandstone (Permian) in the Elaterite Basin, Canyonlands National Park and Glen Canyon National Recreation Area. M, 1985, University of Utah. 143 p.

Huntoon, Peter W. Hydrogeology of the Tapeats amphitheater and Deer Basin, Grand Canyon, Arizona; a study in karst hydrology. M, 1968, University of Arizona.

Huntoon, Peter Wesley. The hydro-mechanics of the ground water system in the southern portion of the Kaibab plateau, Arizona. D, 1970, University of Arizona. 381 p.

Huntsberger, David V., II. Geology of the West Gold Creek Quadrangle, Pend Oreille Lake region, Idaho. M, 1973, University of Minnesota, Minneapolis. 144 p.

Huntsman, Brent Elliot. A geologic survey of possible sanitary landfill sites in Greene County, Ohio. M, 1975, Wright State University. 276 p.

Huntsman, Brent S. Structure of the eastern Red Rocks and Wind Ridge thrust faults, Wyoming; how a thrust fault gains displacement along strike. M, 1983, Texas A&M University. 70 p.

Huntsman, John Robert. Crystalline rocks of the Wagontown 7 1/2 minute quadrangle. M, 1975, Bryn Mawr College. 69 p.

Huntsman, John Robert. The geology and mineral resources of the Caribou Mountain area, southeastern Idaho. D, 1978, Bryn Mawr College. 146 p.

Huntsman, Scott Read. Determination of in-situ lateral pressure of cohesionless soils by static cone penetrometer. D, 1985, University of California, Berkeley. 460 p.

Huntting, Marshall Tower. Geology of the middle Tucannon area. M, 1942, Washington State University. 33 p.

Hunzicker, Ashley A. A laboratory study of antidune traction and the transportation and deposition of ellip-

tical disk-shaped pebbles. M, 1930, University of Wisconsin-Madison.

Hupp, William Ervin. The Bow Island sandstones in the Pakowki Lake area, southeastern Alberta, Canada. M, 1958, University of Colorado.

Huppert, G. N. Cave conservation in the United States; a historical perspective and analysis. D, 1979, University of Northern Colorado. 191 p.

Huppert, George N. Speleography of Papoose Cave, Idaho County, Idaho. M, 1972, University of Idaho. 122 p.

Huppert, Lawrence Norman. The P-velocity near the Tonga Benioff zone determined from traced rays and observations. M, 1980, SUNY at Binghamton.

Huppi, R. G. Geology of eastern Cape George, Nova Scotia. M, 1954, Massachusetts Institute of Technology. 58 p.

Huppunen, JoAnne Louise. Analysis and interpretation of magnetic anomalies observed in north-central California. D, 1984, Oregon State University. 248 p.

Huq, Syed Y. Potential sand-frac deposits in the basal Deadwood Formation, eastern Black Hills, South Dakota. M, 1983, South Dakota School of Mines & Technology.

Hurcomb, Douglas R. Chemical weathering in a small alpine drainage basin, North Cascades, Washington. M, 1984, University of Wyoming. 72 p.

Hurd, D. C. Determination of several species of soluble copper in seawater. M, 1969, University of Hawaii. 49 p.

Hurd, D. C. Interactions of biogenic opal, sediment, and seawater in the central Equatorial Pacific. D, 1972, University of Hawaii. 81 p.

Hurd, Donald W. Geology of Matchedash Lake area, Simcoe County, Ontario. M, 1950, McMaster University. 40 p.

Hurd, Michael L. The geochemistry of chromium in the Great Bay estuary and the Gulf of Maine. M, 1986, University of New Hampshire. 89 p.

Hurd, Robert James. Sandstone pockets in the upper St. Louis Limestone and associated rocks in the vicinity of Frenchburg, Kentucky. M, 1960, University of Kentucky. 45 p.

Hurd, Robert Lee. The deformational history and contact relationships in the central Hondo Syncline, Picuris Mountains, New Mexico. M, 1982, University of Texas at Dallas. 82 p.

Hurdle, David. The evaluation of contour current contribution to the lower rise near the Norfolk-Washington Canyon system. M, 1980, University of South Carolina.

Hurdle, E. J. Stratigraphy, structure and metamorphism of Archean rocks, Clan Lake, Northwest Territories. M, 1985, University of Ottawa. 146 p.

Hurich, Charles A. Gravity interpretation of the southern Wind River Mountains, Wyoming. M, 1981, University of Wyoming. 42 p.

Hurich, Charles A. Investigations of mylonite reflectivity and reflections in deep crustal reflection profiles. D, 1988, University of Wyoming. 144 p.

Hurlburt, Harley Ernest. The influence of coastline geometry and bottom topography on the eastern ocean circulation. D, 1974, Florida State University.

Hurlbut, Cornelius Searle, Jr. The structure and inclusions of the Bonsall Tonalite near Fallbrook, California. D, 1933, Harvard University.

Hurlbut, Elvin Millard, Jr. Geology of a portion of the Calistoga Quadrangle, California. M, 1948, University of California, Berkeley. 53 p.

Hurley, B. W. Geology of the Old Fort area, McDowell County, North Carolina. M, 1974, University of North Carolina, Chapel Hill. 64 p.

Hurley, Bruce William. The Metaline Formation-Ledbetter Slate contact in northeastern Washington. D, 1980, Washington State University. 141 p.

Hurley, James Patrick. Diagenesis of algal pigments in lake sediments. D, 1988, University of Wisconsin-Madison. 181 p.

Hurley, N. F. Facies mosaic of the lower Seven Rivers Formation (Permian), North McKittrick Canyon, Guadalupe Mountains, New Mexico. M, 1978, University of Wisconsin-Madison.

Hurley, Neal Lilburn. Groundwater motion and its relation to oil accumulation. D, 1953, Stanford University. 164 p.

Hurley, Neil Francis. Geology of the Oscar Range Devonian reef complex, Canning Basin, Western Australia. D, 1986, University of Michigan. 302 p.

Hurley, Patrick J. Subsurface geology of the North Dover area, Kingfisher County, Oklahoma. M, 1962, University of Oklahoma. 49 p.

Hurley, Patrick M. Investigation on the helium method of age determination. D, 1940, Massachusetts Institute of Technology. 134 p.

Hurley, Peter. The development and evaluation of a crosshole seismic system for crystalline rock environments. M, 1983, University of Toronto. 170 p.

Hurley, Robert Joseph. The geomorphology of the abyssal plains in the Northeast Pacific Ocean. D, 1959, University of California, Los Angeles. 173 p.

Hurley, Stephen C. Geohydrology of the Buffalo River basin and related land use problems. M, 1976, University of Arkansas, Fayetteville.

Hurley, Teresa Dawn. Petrology and geochemistry of the volcanic host rocks to the west and north pits of the Sherman Mine iron formation, Temagami, Ontario. M, 1985, McMaster University. 154 p.

Hurley, Timothy James. Depositional environments of Morrowan sandstones, Carlsbad Field, Eddy County, New Mexico. M, 1980, Texas Tech University. 81 p.

Hurley, William Daniel, Jr. Slope stability; Echo Canyon Conglomerate (Upper Cretaceous, northeastern Utah). D, 1972, University of Utah. 181 p.

Hurlow, Hugh A. Structural geometry, fabric, and chronology of a Tertiary extensional shear zone - detachment system, southwestern East Humboldt Range, Elko County, Nevada. M, 1987, University of Wyoming. 141 p.

Hurr, R. Theodore. Modeling steady-state and transient confined and unconfined ground-water flow by the finite-element-method. M, 1971, Colorado School of Mines. 79 p.

Hurry, Debra Jean. Genetic stratigraphy of the basal Cliff House Sandstone (Cretaceous), west-central San Juan Basin, New Mexico. M, 1985, University of Texas, Austin. 151 p.

Hursey, Michael J. An analysis of body-wave propagation across the Michigan Basin. M, 1974, University of Wisconsin-Milwaukee.

Hursky, M. J., Jr. Grain-size distribution and constituent particle analysis of nearshore carbonate sediments of Lower Matecumbe Key, Florida. M, 1977, University of Texas, Arlington. 93 p.

Hurst, C. H. The petrology and mineralization of the mafic complex of El Cajon Mountain. M, 1976, San Diego State University.

Hurst, Carolyn. Detailed gravity survey to delineate buried strike-slip faults in the Crawford Mountain portion of the Utah-Idaho-slip faults in the Crawford Mountain portion of the Utah-Idaho-Wyoming overthrust belt. M, 1983, Brigham Young University. 100 p.

Hurst, Donald J. Depositional environment and tectonic significance of the Tunp Member of the Wasatch Formation, Southwest Wyoming. M, 1984, University of Wyoming. 115 p.

Hurst, Donald Lindsay. The glacial and Pleistocene geology of the Dundas Valley, Hamilton. M, 1962, McMaster University. 185 p.

Hurst, Kenneth Joslin. The measurement of vertical crustal deformation. D, 1987, Columbia University, Teachers College. 208 p.

Hurst, M. E. Rock alteration and ore deposition. D, 1922, Massachusetts Institute of Technology. 78 p.

Hurst, Marc Vernon. Phytoliths as a source of silica in coals. M, 1984, University of Georgia. 44 p.

Hurst, R. W. Geochronologic studies in the Precambrian Shield of Canada; Part I, The Archaean of coastal Labrador; Part II, The Sudbury Basin, Sudbury, Ontario. D, 1975, University of California, Los Angeles. 144 p.

Hurst, Ray Eugene. Sediments from the Pre-Cambrian rocks of the Pedernal Hills and Los Pinos Mountains, New Mexico. M, 1949, Texas Tech University. 32 p.

Hurst, Stephen D. Geochemistry and geochronology of some Archean gneisses from around the Holenarasipur greenstone belt, Karnataka, India. M, 1981, University of North Carolina, Chapel Hill. 42 p.

Hurst, Thomas L. The potential for groundwater recharge via arid zone ephemeral wash channels. M, 1988, University of Nevada, Las Vegas. 61 p.

Hurst, Thomas Leonard. Refractory properties of chromite of the Pacific Northwest. M, 1937, University of Washington. 236 p.

Hurst, Vernon James. Geology of the Kennesaw Mountain-Sweat area, Cobb County, Georgia. M, 1952, Emory University.

Hurst, Vernon James. The stratigraphy and structure of the Mineral Bluff Quadrangle. D, 1954, The Johns Hopkins University. 179 p.

Hurt, Thomas Wayne. The geology of the Pryor area, Mayes County, Oklahoma. M, 1951, University of Oklahoma. 83 p.

Hurtubise, Donlon O. Geochemistry and petrology of the Talcot, Holyoke, and Hampden basalts, southern Hartford Basin, Connecticut. M, 1979, Rutgers, The State University, Newark. 76 p.

Hurwitz, Garvin L. The genesis and association of the gold and silver deposits of Colorado. M, 1930, Columbia University, Teachers College.

Husain, Athar. Differentiation between river and dune sand, Hunters Island and vicinity, Riley County, Kansas. M, 1964, Kansas State University. 114 p.

Husain, B. R. Semi-micro fossils of the Black River and Trenton rocks of Quebec. D, 1953, McGill University.

Husain, F. Stratigraphic type oil pools in the Minnelusa Formation of northeastern Wyoming. M, 1963, University of Wyoming. 33 p.

Husain, Mohammad Asghar. An alpha spectrometric method of analysis of radium in natural waters. M, 1977, Florida State University.

Husain, Sheikh Ansar. Sulphur exchange reactions. D, 1968, University of Alberta. 216 p.

Husband, Edna Maurine. The characteristics, distribution, and correlation of the Sylvan, Polk Creek, Carson, and Maquoketa shales. M, 1937, University of Oklahoma. 65 p.

Husband, Philip M. A sedimentary and chemical analysis of the Niagara Series in Michigan. M, 1958, Michigan State University. 37 p.

Husby-Coupland, Karen Joanne. Effect of dissolved oxygen depletion on the rate of sediment reworking by Stylodrilus heringianus. M, 1980, University of Michigan.

Husch, Jonathan Mark. Geology, petrology, structure, and geochemistry of anorthositic and related rocks associated with hypabyssal ring complexes, Air Massif, Republic of Niger. D, 1982, Princeton University. 231 p.

Huse, Scott M. Distribution of Recent benthonic foraminifera on the northeastern Gulf of Mexico continental shelf in relation to character of substrate. M, 1976, University of South Florida, Tampa.

Hushmand, Behnam. Experimental studies of dynamic response of foundations. D, 1984, California Institute of Technology. 317 p.

Husid, Raul. Gravity effects on the earthquake response of yielding structures. D, 1967, California Institute of Technology. 161 p.

Husk, Robert H. Stratigraphy and structure of a portion of the southeastern Inyo Mountains, California. M, 1979, San Jose State University. 60 p.

Huskinson, Edward July, Jr. Geology and fluorspar deposits of the Chise fluorspar district, Sierra County, New Mexico. M, 1975, University of Texas at El Paso.

Huspeni, Jeffrey Ralph. Petrology and geochemistry of rhyolitic volcanic rocks associated with tin mineralization in Mexico. M, 1982, University of Michigan.

Huss, Gary R. The matrix of unequilibrated ordinary chondrites; implications for the origin and subsequent history of chondrites. D, 1979, University of New Mexico. 139 p.

Huss, Gary Robert. The role of presolar dust in the formation of the solar system. D, 1987, University of Minnesota, Minneapolis. 263 p.

Hussain, Mahbub. Evaporite deposition and source rock evaluation of a Holocene-Pleistocene continental sabkha (salt flat playa) in West Texas. D, 1986, University of Texas at Dallas. 224 p.

Hussain, Mahbub. Textural and mineralogical analysis of Chignecto Bay sediments, Canada. M, 1980, Acadia University.

Hussain, Mehdi. The serpentine of Staten Island. M, 1948, Columbia University, Teachers College.

Hussain, Sultan M. A genetic study of the Gray Hydromorphic soils of the Hawaiian Islands. D, 1967, University of Hawaii. 252 p.

Hussein, Adel M. A hydrological study and evaluation of the relief well system at Bolivar Dam, northeastern Ohio. M, 1983, Ohio University, Athens. 231 p.

Hussein, Mohammad Hasan. Erosion and sediment yield prediction on farm fields. D, 1982, Purdue University. 293 p.

Husseini, Moujahed I. The fracture energy of earthquakes. D, 1975, Brown University. 103 p.

Husseini, Sadad Ibrahim. Abundance and distribution of high magnesium calcite on Andros platform, Great Bahama bank. M, 1970, Brown University.

Husseini, Sadad Ibrahim. Temporal and diagenetic modifications of the amino acid composition of Pleistocene coral skeletons. D, 1973, Brown University.

Hussey, Arthur Mekeel, II. Petrology and structure of three basic igneous complexes, southwestern Maine. D, 1961, University of Illinois, Urbana. 139 p.

Hussey, Eric Maurice. The stratigraphy, structure and petrochemistry of the Clode Sound map area, northwestern Avalon Zone, Newfoundland. M, 1979, Memorial University of Newfoundland. 312 p.

Hussey, Keith Morgan. Louisiana Cane River Eocene foraminifera. D, 1940, Louisiana State University.

Hussey, Keith Morgan. Some new genera and species of foraminifera from the Cane River Eocene formation of Louisiana. M, 1939, Louisiana State University.

Hussey, Russell Claudius. The Richmond Formation of Michigan. D, 1925, University of Michigan.

Hussin, Ismail Bin. A detailed paleomagnetic investigation of geomagnetic secular variation in Western Canada. M, 1978, University of Alberta. 134 p.

Hussin, James Joseph. A 1959 helicopter gravity survey in western Queensland, Australia. M, 1961, Michigan Technological University. 62 p.

Husson, Didier Emmanuel. Measurement of stress with acoustic waves. D, 1983, Stanford University. 207 p.

Hussong, D. A study of ground water configuration near Pahala, Hawaii, by the D.C. electrical resistivity method. M, 1967, University of Hawaii. 88 p.

Hussong, D. M. Detailed structural interpretations of the Pacific oceanic crust using ASPER and ocean-bottom seismometer methods. D, 1972, University of Hawaii. 165 p.

Husted, John E. A petrographic study of the rocks in a portion of western Nottoway County, Virginia. M, 1942, University of Virginia. 42 p.

Husted, John Edwin. Factors influencing occurrence of phosphorite in Georgia's plain sediments. D, 1970, Florida State University. 93 p.

Huston, Daniel Cliff. Interpretation of seismic signal and noise through line intersection mis-tie analysis. M, 1987, University of Texas, Austin. 223 p.

Huston, David Lowell. The significance of a widespread stream sediment copper anomaly in the Baltamote Mountains, Pima County, Arizona. M, 1984, University of Arizona. 163 p.

Huston, John. Implications of strath terrace levels along Grindstone and Stony creeks, Glenn County, California. M, 1973, Stanford University.

Huston, Ted Jay. Evolution and shock history of L and H chondrite meteorites; mobile trace elements and $^{40}Ar/^{39}Ar$ ages. D, 1982, Purdue University. 173 p.

Huston, W. J. The Steeprock manganiferous foot-wall paint. M, 1956, Queen's University. 76 p.

Hutasoit, Lambok M. Numerical simulation of thermohaline convection within a porous medium. M, 1986, University of Illinois, Urbana. 49 p.

Hutcheon, Ian E. The metamorphism of sulfide-bearing pelitic rocks from Snow Lake, Manitoba. D, 1977, Carleton University. 203 p.

Hutcheon, Ian E. The tremolite isograd near Marble Lake, Ontario. M, 1972, Carleton University. 74 p.

Hutcheson, Donald Wade. A textural study of the lower Kinkaid Limestone, southern Illinois and western Kentucky. M, 1957, University of Illinois, Urbana.

Hutcheson, Harvie Leon, Jr. Vegetation in relation to slope exposure and geology in the Arbuckle Mountains. D, 1965, University of Oklahoma. 43 p.

Hutcheson, Lewis Bryan. The stratigraphy and sedimentation of the northwest quarter of the Brundidge Quadrangle, Alabama. M, 1957, Emory University. 69 p.

Hutchings, Lawrence John. Modeling near-source earthquake ground motion with empirical Green's functions. D, 1987, SUNY at Binghamton. 180 p.

Hutchings, Roy Theodore, Jr. Textures and stratigraphy in the Lower Cambrian formations of Troy, New York. M, 1957, Rensselaer Polytechnic Institute. 78 p.

Hutchinson, Charles F. The digital use of Landsat data for integrated land resource survey; a study in the eastern Mojave Desert, California. D, 1978, University of California, Los Angeles. 277 p.

Hutchinson, Craig Brandt. Trend-surface analysis of lake distributions in a karst terrain, west-central Florida. M, 1971, University of South Florida, Tampa. 102 p.

Hutchinson, Deborah Ruth. An investigation of the structure and surficial geology of the central Lake Ontario basin. M, 1977, University of Toronto.

Hutchinson, Deborah Ruth. Structure and tectonics of the Long Island Platform. D, 1984, University of Rhode Island. 306 p.

Hutchinson, Murl W. Geology of the Butte Falls Quadrangle, Oregon. M, 1941, Oregon State University. 103 map p.

Hutchinson, Peter John. Stratigraphy and paleontology of the Bisti Badlands area, San Juan County, New Mexico. M, 1981, University of New Mexico. 219 p.

Hutchinson, R. Alan. Geology of the Burned mountain area, Rio Arriba County, New Mexico. M, 1968, Colorado School of Mines. 96 p.

Hutchinson, R. B. Report on the M.T. Key mine and mill, Pinos Altos mining camp, Grant County, New Mexico. M, 1889, Washington University.

Hutchinson, Richard William. Regional zonation of pegmatites near Ross Lake, Northwest Territories. D, 1954, University of Wisconsin-Madison.

Hutchinson, Richard William. The application of polarization figures to the identification of the sulphide, arsenide, and antimonide minerals of cobalt and nickel. M, 1957, University of Wisconsin-Madison.

Hutchinson, Robert D. The stratigraphy and correlation of the Cambrian sedimentary rocks of Cape Breton Island, Nova Scotia, Canada. D, 1950, University of Wisconsin-Madison.

Hutchinson, Robert Maskiell. Enchanted Rock Pluton, Llano and Gillespie counties, Texas. D, 1953, University of Texas, Austin.

Hutchinson, Robert Maskiell. Geology of the Browne Lake area, southwestern Montana. M, 1948, University of Michigan. 45 p.

Hutchinson, Robert O. Aeromagnetic profile of United States. M, 1956, Boston University.

Hutchinson, Roderick. Geologic setting of Sylvan Spring Geothermal Area, Yellowstone National Park. M, 1978, Iowa State University of Science and Technology.

Hutchinson, Thomas Weston. Upper Devonian and Lower Mississippian pectinoid pelecypods from Michigan, Ohio, Indiana, Iowa and Missouri. M, 1964, University of Michigan.

Hutchinson, Wayne Robert. A computer simulation of the glacial/carbonate aquifer in the Pequest Valley, Warren County, New Jersey. M, 1981, Rutgers, The State University, New Brunswick. 115 p.

Hutchison, David Malcolm. Provenance of sand in the Great Sand Dunes National Monument, Colorado. D, 1968, West Virginia University. 138 p.

Hutchison, David Malcolm. Volcanic ash in the northern part of the Bitterroot Valley, Ravalli County, Montana. M, 1959, University of Montana. 63 p.

Hutchison, Harold Christy. Geology and mineral resources of the Seelyville Quadrangle, Vigo County, Indiana. M, 1952, Indiana University, Bloomington. 46 p.

Hutchison, Hillary W. Porosity and permeability controls of the Muddy Sandstone. M, 1982, University of Missouri, Columbia.

Hutchison, J. L., Jr. Stratigraphy of Harral Quadrangle, Culberson County, Texas. M, 1952, University of Texas, Austin.

Hutchison, John Howard. The Talpidae (Insectivora, Mammalia); evolution, phylogeny, and classification. D, 1976, University of California, Berkeley. 234 p.

Hutchison, M. N. Refinement and application of the sphalerite geobarometer. D, 1978, University of Toronto.

Hutchison, W. W. A petrographic study of the quartz monzonite associated with the Holyrood Batholith, Newfoundland. M, 1959, University of Toronto.

Hutchison, William Watt. Conditions of metamorphism of certain rocks as indicated by solid inclusion decrepitation. D, 1962, University of Toronto. 306 p.

Hutnik, F. T. A. Petrography and diagenesis of the Mississippian Livingstone Formation (Rundle Group), southwestern Alberta. M, 1978, University of Manitoba. 102 p.

Hutsinpiller, Amy. Geochemistry of spring water from the Blackfoot Reservoir region, southeastern Idaho; application to geothermal potential. M, 1979, University of Utah. 88 p.

Hutson, Frederick John. Cement history of the Middle Devonian Cedar City Formation, central Missouri. M, 1985, University of Missouri, Columbia. 82 p.

Hutson, Osler C. Geology of the northern end of San Pedro Mountain, Rio Arriba and Sandoval counties, New Mexico. M, 1958, University of New Mexico. 55 p.

Hutson, Robert William. Preparation of duplicate rock joints and their changing dilatancy under cyclic shear. D, 1987, Northwestern University. 241 p.

Hutson, William H. Ecology and paleoecology of Indian Ocean planktonic foraminifera. D, 1976, Brown University. 442 p.

Hutt, G. M. The Whitemud sediments (Upper Cretaceous) of southern Saskatchewan. M, 1931, McGill University.

Hutt, Jeremy R. Relationships between thermal and electrical conductivities of ocean sediments and consolidated rocks. M, 1966, Oregon State University. 63 p.

Hutta, Joseph John. A study of the incorporation of uranium by synthesized crystals of lead sulfide. D, 1961, Pennsylvania State University, University Park. 108 p.

Hutta, Joseph John. Relationship of dimensional orientation of quartz grains to directional permeability in sandstones. M, 1956, Pennsylvania State University, University Park. 97 p.

Hutter, Adam Richard. Radon variability in soil gas over fracture traces. M, 1987, Pennsylvania State University, University Park. 156 p.

Hutter, Terry J. The biostratigraphy and taxonomy of chitinozoans, of the Leavenworth Limestone, Pennsylvanian (Virgilian) of eastern Kansas. M, 1976, Wichita State University. 162 p.

Hutto, Andrew Clifton, Jr. The diagnostic foraminifera of the Yazoo Clay of east-central Mississippi (Yazoo, lower Eocene). M, 1953, Mississippi State University. 93 p.

Hutton, Charles Wetherill. Geology of the Conneaut and Ashtabula quadrangles, Ohio. M, 1940, Ohio State University.

Hutton, Joan G. Bedrock control, sedimentation and Holocene evolution of the marsh archipelago coast, west-central Florida. M, 1986, University of South Florida, St. Petersburg.

Hutton, John R.; Guyton, J. Stephen and Sokolsky, George E. Geology of the Devil's Hole area, Custer and Huerfano counties, Colorado. M, 1960, University of Michigan.

Hutton, Joseph Gladden. The stratigraphy of the Devonian rocks of Calhoun and Jersey counties, Illinois, with a preliminary discussion of the physiography of the region. M, 1910, University of Illinois, Urbana.

Hutton, Robert A. Geology and uranium content of middle Tertiary ash-flow tuffs in the northern part of Dogskin Mountain, Nevada. M, 1978, University of Nevada. 103 p.

Huttrer, Gerald. Structure and stratigraphy of the central Grant Range, Nev. M, 1963, University of Washington. 59 p.

Huval, Isaac Martin. Petrography of the Atoka sandstones of Mayes, Wagoner, and Muskogee counties, Oklahoma. M, 1960, University of Tulsa. 95 p.

Huyck, Holly Louise. Lithologic controls of copper mineralization at the Noranda Lakeshore porphyry copper deposit, Pinal County, Arizona. D, 1986, University of California, Berkeley. 208 p.

Huycke, David T. Detailed study of magnetic reversals in the Lower Triassic Chugwater Formation, Wyoming. M, 1979, University of Wyoming. 68 p.

Huzarki, Richard George. Descriptive geometry in the geosciences. M, 1952, Texas Tech University. 127 p.

Huzarski, Jan Ralph. Petrology and structure of eastern Monte Largo Hills, (central) New Mexico. M, 1971, University of New Mexico. 45 p.

Huzzen, Carl Stewart. Occurrence, phase assemblages, and metamorphic facies of anthophyllite assemblages from the Haddam Quadrangle, Connecticut. M, 1962, University of Rochester. 57 p.

Hvolboll, Victor T. The effect of overburden pressure on relative permeability. M, 1955, University of Oklahoma. 98 p.

Hwang, Chung-Yung. Size and shape of airborne asbestos fibres in mining and mineral processing environments. D, 1981, McGill University. 468 p.

Hwang, Daekyoo. A probabilistic consolidation analysis for embankment foundations. D, 1980, University of Maryland. 379 p.

Hwang, Grace. Factors that affect the recovery of amplitude and waveform from sign-bit Vibroseis data. M, 1984, University of Houston.

Hwang, Horng-Jye. Lateral variation and frequency dependence of crustal Q_β. D, 1985, St. Louis University. 234 p.

Hwang, Jae-Young. Spectrochemical analysis of the moldavites (Ba, Li, Sr, & Rb). M, 1964, Massachusetts Institute of Technology. 26 p.

Hwang, Jiann-Yang. Mineralogy and petrology of the Chinkuashih area, Taiwan, and its associated gold-copper deposits. M, 1980, Purdue University. 175 p.

Hwang, L. F. Three-dimensional elastic and electromagnetic waves scattering and diffraction. D, 1978, Columbia University, Teachers College. 117 p.

Hwang, Li-San. Flow resistance of dunes in alluvial streams. D, 1965, California Institute of Technology. 157 p.

Hwang, Ralph Bang-Yen. The effects of soil stratification on underground seepage into tile drains. D, 1973, University of California, Davis. 214 p.

Hwong, Tzer Jong. 3D modeling of groundwater in the San Bernardino Valley, Southern California. M, 1987, University of California, Riverside. 124 p.

Hwu, Chen-Roon. Stress, seismicity, and initiation of subduction at passive continental margins. M, 1981, Carleton University. 91 p.

Hyatt, E. P. A study of Missouri flint clay. M, 1950, University of Missouri, Rolla.

Hyatt, Robert Allen. Hydrology and geochemistry of the Okefenokee Swamp basin. D, 1984, University of Georgia. 378 p.

Hyde, Bert Q. The economic geology of the Fifteen Mile Creek area, Stevens County, northeast Washington. M, 1985, Western Washington University. 87 4 plates p.

Hyde, David Edward. A structural and stratigraphic study of the Fairview-McMillan formational contact in the Cincinnati area. M, 1958, University of Cincinnati. 49 p.

Hyde, Jack Herbert. Late Quaternary volcanic stratigraphy, south flank of Mount St. Helens, Washington. D, 1973, University of Washington. 114 p.

Hyde, Jack Herbert. Structure and stratigraphy of the north central Grant Range, Nevada. M, 1963, University of Washington. 63 p.

Hyde, Jesse E. The Sciotoville (Ohio) fauna. M, 1907, Columbia University, Teachers College.

Hyde, Jimmie Collins. Mississippian rocks of the Drumright area, Oklahoma. M, 1957, University of Oklahoma. 57 p.

Hyde, Luter Willis. Geologic profile along Highway 69, Hale County, Alabama. M, 1960, University of Alabama.

Hyde, Matthew G. Stratigraphy, petrology, and depositional environment of the Upper Triassic Blomidon Formation, St. Mary's Bay, Nova Scotia. M, 1981, University of Massachusetts. 240 p.

Hyde, Michael Kevin. A study of the dolomite/calcite ratios relative to the structures and producing zones of the Kawkawlin oil field, Bay County, Michigan. M, 1979, Michigan State University. 92 p.

Hyde, Richard Franklin. Analysis of Landsat MSS land cover patterns using a geographic information system. D, 1986, Indiana State University. 163 p.

Hyde, Richard Stuart. Sedimentology, volcanology, stratigraphy, and tectonic setting of the Archean Timiskaming Group, Abitibi greenstone belt, northeastern Ontario, Canada. D, 1978, McMaster University. 422 p.

Hyde, Tinka G. Depositional environments, stratigraphy, and petrography of the Sarten Sandstone and related Lower Cretaceous formation, southwestern New Mexico. M, 1984, New Mexico State University, Las Cruces. 48 p.

Hyde, Victor Albert. A geographical survey of Knoxville, Tennessee. M, 1939, University of Tennessee, Knoxville. 113 p.

Hyde, William Thomas. An astronomical theory of the Pleistocene ice age. D, 1986, University of Toronto.

Hyden, Harold J. Uranium and other metals in crude oils. M, 1958, University of Tulsa. 123 p.

Hyer, Donald Eugene. The areal geology of the Salem Quadrangle and the subsurface geology of Washington County, Indiana. M, 1951, Indiana University, Bloomington. 33 p.

Hyers, Albert D. Mesoscale relationships of talus and insolation, San Juan Mountains, Colorado. D, 1980, Arizona State University. 255 p.

Hyers, Merlyn Eugene. A study of the fossil mollusk Mercenaria campechiensis. M, 1969, American University. 32 p.

Hyland, Mark R. The retreat of the late Wisconsinan ice sheet from New England and the Canadian Maritimes; a field and numerical reconstruction. M, 1986, University of Maine. 183 p.

Hylbert, David Kent. Development of geological structural criteria for predicting unstable mine roof rocks. D, 1976, University of Tennessee, Knoxville. 267 p.

Hylbert, David Kent. The geology of Lee Township, Carroll County, Ohio. M, 1963, Ohio University, Athens. 136 p.

Hylton, Alisa K. The petrology and environment of deposition of the Stone Corral Formation (Lower Permian, Kansas). M, 1986, Wichita State University. 76 p.

Hylton, Gary K. Geology of the Welch-Bornholdt pools area, Rice and McPherson counties, Kansas. M, 1960, Kansas State University. 109 p.

Hyman, David. Attenuation of harmonic disturbance in soil materials. M, 1975, Lehigh University.

Hyman, Marian. The origin of magnetite in carbonaceous chondrites. D, 1982, Texas A&M University. 82 p.

Hyndman, Donald William. Petrology and structure of Naksup map area, British Columbia. D, 1964, University of California, Berkeley. 150 p.

Hyndman, R. D. Electrical conductivity inhomogeneities in the Earth's upper mantle. M, 1963, University of British Columbia.

Hyne, Norman John, Jr. Sedimentary environments and submarine geomorphology of the continental shelf in the area of Choctawhatchee Bay, Florida. M, 1965, Florida State University.

Hyne, Norman John, Jr. Sedimentology and Pleistocene history of Lake Tahoe, California, Nevada. D, 1969, University of Southern California. 135 p.

Hynes, P. A shallow water carbonate platform classification using Landsat thematic mapper data; Great Bahama Bank, Bahamas. M, 1985, Murray State University. 60 p.

Hyodo, Hironobu. Paleomagnetic studies of rocks near the Grenville front. M, 1984, University of Toronto.

Hyrkas, Gerald Lee. The sedimentology and structural geology of the middle Precambrian Thomson Formation, central Carlton County, Minnesota. M, 1982, University of Minnesota, Duluth. 164 p.

Hyslop, Kevin D. Dinoflagellate biostratigraphy, Dawson Canyon Formation, Hibernia area, offshore Eastern Canada. M, 1986, University of Calgary. 110 p.

Hyslop, Ralph Craig. A field method of determining the susceptibility of rocks. M, 1939, Colorado School of Mines. 14 p.

Hyun, Byung Koo. Seismic energy absorption analysis by relaxation time. D, 1961, Colorado School of Mines. 101 p.

Hyun, I. Acoustic log-porosity corrections associated with compaction pattern, Texas Gulf Coast Basin. D, 1977, University of Texas, Austin. 151 p.

I'Anson, Lawrence W., Jr. Marine gravity discrepancies in the Solomon islands (southwestern Pacific). M, 1968, United States Naval Academy.

Iagmin, Paul Jean. Tertiary volcanic rocks south of Anaconda (Silver Bow County), Montana. M, 1972, University of Montana. 53 p.

Iams, William James. A study of the bioerosion of coastal limestones; a photogrammetric approach. D, 1977, Memorial University of Newfoundland. 278 p.

Iannacchione, Anthony. Geology of the lower Kittanning Coalbed and related mining problems in Cambria County, Pennsylvania. M, 1977, University of Pittsburgh.

Ianniello, Michael L. Pleistocene depositional environments of the Corinth and Gansevoort quadrangles, New York. M, 1985, Rensselaer Polytechnic Institute. 70 p.

Ibach, Darrell Henry. The structure and tectonics of the Blanco fracture zone. M, 1981, Oregon State University. 60 p.

Ibach, Lynne E. Johnson. The relationship between sedimentation rate and total organic carbon content in ancient marine sediments. M, 1980, Oregon State University. 46 p.

Ibanga, Iniobong Jimmy. The physical, chemical, and mineralogical properties of laterite samples formed in various environments. D, 1980, North Carolina State University. 129 p.

Ibe, Ralph Anthony. Quaternary palynology of five lacustrine deposits in the Catskill mountain region of New York. D, 1982, New York University. 210 p.

Ibenye, Ikechi S. Laboratory study on swell-consolidation properties of compacted Pierre Shale in the vicinity of Colorado Springs, Colorado. M, 1979, University of Idaho. 45 p.

Iberall, Eleanora R. Paleoecological studies from fecal pellets; Stanton Cave, Grand Canyon, Arizona. M, 1972, University of Arizona.

Ibiary, Nabil Yakout El *see* El Ibiary, Nabil Yakout

Ibrahim, Abd El Wahid. Areal P-n and P-6 velocity variation in central United States. M, 1963, University of Kansas.

Ibrahim, Abdelwahid. The application of the gravity method to mapping bedrock topography in Kalamazoo County, Michigan. D, 1970, Michigan State University. 109 p.

Ibrahim, Abou-Bakr K. Relation between compressional wave velocity and aquifer porosity. M, 1962, New Mexico Institute of Mining and Technology. 59 p.

Ibrahim, Abou-Bakr Khalil. Traveltime curves and upper mantle structure from long period S waves. D, 1967, St. Louis University. 157 p.

Ibrahim, Hassan Suliman. Phosphorus sorption in relation to time, temperature and plant availability. D, 1981, University of California, Riverside. 175 p.

Ibrahim, Ismail K. Use of thermodynamic data to evaluate iron availability from different iron materials in calcareous soils. D, 1981, Kansas State University. 173 p.

Ibrahim, Kamil E. Main physical principles of differential transient sounding. D, 1985, Colorado School of Mines. 174 p.

Ibrahim, Mohamed S. Subsurface geology and the chemical quality of ground water in Buckeye Valley, Arizona. M, 1962, University of Arizona.

Ibrahim, Noor Azim. Sedimentological and morphological evolution of a coarse-grained regressive barrier beach, Horseneck Beach, Massachusetts, USA. M, 1986, Boston University. 196 p.

Ibrahim, Yarub Khalid. Paleocurrents and sedimentary history of outcropping Cretaceous Middendorf sediments south of the Cape Fear River, North Carolina. D, 1973, University of North Carolina, Chapel Hill. 147 p.

Ice, Robert G. Geology of the northernmost Sierra de Catorce and stratigraphy and biostratigraphy of the Cuesta del Cura Formation in northeastern and central Mexico. M, 1979, University of Texas, Arlington. 162 p.

Ichara, Mark Josiah. Effects of a sand column in the wellbore during unsteady-state liquid flow. D, 1980, University of Texas, Austin. 295 p.

Ichimura, Vernon T. Solutions of the aquifer thermal energy storage concept utilizing an IBM operational version of the Intercomp deep well disposal numerical model. D, 1982, Washington State University. 396 p.

Ichimura, Vernon T. Uranium concentration in the ground waters of the Pullman-Moscow Basin, Washington, Idaho, by the nuclear track technique. M, 1978, Washington State University. 124 p.

Ide, Susan. Geology of mid-Tertiary volcanic rocks in the Laboricita-General Trias area, central Chihuahua, Mexico. M, 1986, University of Texas, Austin. 155 p.

Iden, Lee J. A study of the formation of surface ripple patterns during erosion by impingement. M, 1968, Rutgers, The State University, New Brunswick. 19 p.

Idike, Francis Igboji. Modeling the effects of conservation practices on soil moisture. D, 1981, University of Minnesota, Minneapolis. 192 p.

Idiz, Erdem Fahri. Studies on the interactions between organic matter, trace metals and sulfur in Recent and ancient sediments. D, 1987, University of California, Los Angeles. 318 p.

Idleman, Bruce D. Geology of the plutonic and hypabyssal rocks of the Betts Cove ophiolite complex, Newfoundland. M, 1981, SUNY at Albany. 139 p.

Idleman, Katrina A. J. The significance of shear zones within the plutonic section of North Arm Mountain, Bay of Islands ophiolite complex, Newfoundland. M, 1985, SUNY at Albany. 155 p.

Idowu, Adebayo Aderemi. Coordination of outer continental shelf petroleum and natural gas development policy; a multiphasic multilateral balanced interaction approach. D, 1982, University of Texas at Dallas. 313 p.

Idowu, Ayorinde O. Geology of the environment of Meji oil field, in the Niger Delta of southern Nigeria. M, 1977, Wright State University. 80 p.

Idris, Eltahir Osman. Geophysical investigation of diabase dikes in the Durham Triassic basin and their hydrological significance. M, 1980, North Carolina State University. 53 p.

Idris, Faisal. Reflection seismic measurements in the Old Bahama Channel north of Cuba. M, 1975, University of Miami. 41 p.

Idris, Faisal Mohamed. Cenozoic seismic stratigraphy and structure of the South Carolina lower coastal plain and continental shelf. D, 1983, University of Georgia. 176 p.

Iglehart, Charles F. Origin of the Crinkled Limestone of the Lykins Formation (Permian, Triassic), eastern Colorado. M, 1948, University of Colorado.

Ignatius, Heikki Gustaf. Late-Wisconsin stratigraphy in north-central Quebec and Ontario, Canada. D, 1956, Yale University. 140 p.

Iheme, Uzoma N. The stratigraphy and depositional environments of the upper Bloyd and lower Atoka formations (Pennsylvanian), in central Franklin, eastern Crawford and north eastern Sebastian counties, Arkansas. M, 1979, University of Arkansas, Fayetteville.

Ihle, Bethany A. Internal deformation in the Backbone thrust sheet, Sawtooth Range, Montana. M, 1988, Montana State University. 144 p.

Iivari, Thomas A. Cenozoic geologic evolution of the east-central Markagunt Plateau, Utah. M, 1979, Kent State University, Kent. 154 p.

Ijirigho, B. T. Pennsylvanian subsurface stratigraphy of the Black Mesa Basin and Four Corners area in northeastern Arizona. M, 1977, University of Arizona. 62 p.

Ijirigho, Bruce Tajinere. Secondary porosity and hydrocarbon production from the Ordovician Ellenburger Group of the Delaware and Val Verde basins, West Texas. D, 1981, University of Arizona. 489 p.

Ikawa, Haruyoshi. The weathering of Hawaiian volcanic glass and the transformation of amorphous and crystalline phases. D, 1968, Pennsylvania State University, University Park. 221 p.

Ikeagwuani, Frederick Duaka. Photogeology of the Picture Rock Pass area, Lake County, Oregon. M, 1965, University of Oregon. 75 p.

Ikeda, Keiichiro. Inverse problem for stress in the Earth based on geodetic data. D, 1981, Massachusetts Institute of Technology. 174 p.

Ikeda, Keiichiro. Three dimensional geodetic inversion method for stress modelling in the lithosphere. M, 1980, Massachusetts Institute of Technology. 92 p.

Ikeda, Margaret. Method development for anion analysis using ion chromatography and the application of these data to coal geochemistry. M, 1982, Montana College of Mineral Science & Technology. 100 p.

Ikingura, Justinian Rwezaula. Hydrothermal alteration and Cu-Zn sulfide mineralization in the D-68 Zone, Corbet Mine, Noranda District, Quebec, Canada. M, 1984, Carleton University. 326 p.

Ikins, William Clyde. Some echinoids from the Texas Cretaceous. M, 1939, University of Texas, Austin.

Ikins, William Clyde. Stratigraphy and paleontology of the Walnut and Comanche Peak formations. D, 1941, University of Texas, Austin.

Ikoku, Chinyere Ukeagumo. Transient flow of non-Newtonian power-law fluids in porous media. D, 1978, Stanford University. 257 p.

Ikola, Rodney Jacob. A geophysical investigation of the geologic structure of Carlton County, Minnesota. M, 1967, University of Minnesota, Minneapolis.

Ikpah, Azhinoto Ozodio. Oil and gas industry and environmental pollution; application of systems reliability analysis for the evaluation of the status of environmental pollution control in the Nigerian petroleum industry. D, 1981, University of Texas at Dallas. 357 p.

Ikpeama, Mmajuogu Onyelankea U. Strontium isotope composition of sediment and fossil shells from the Discovery Deep, Red Sea. M, 1971, Ohio State University.

Ikramuddin Ali, Syed. Study of Recent sediments of the beach and delta at the mouth of Alma River (Bay of Fundy), Albert County, New Brunswick. M, 1965, University of New Brunswick.

Ikramuddin, Mohammed. Geochemistry and geochronology of Precambrian dikes from Mysore State, India. D, 1974, Miami University (Ohio). 170 p.

Ikwuakor, Killian Chinwuba. Interrelations and integration or rock property measurements with problems. D, 1980, Colorado School of Mines. 283 p.

Ilavia, Piloo Eruchshaw. A comparative study of experimental experimental and computed compressibility factors of methane-nitrogen-carbon dioxide system. M, 1970, University of Missouri, Rolla.

Ilchik, Susan Emilie. Geology of part of the Black Range between Kingston and Hillsboro, Sierra County, New Mexico. M, 1982, New Mexico State University, Las Cruces. 49 p.

Ileri, Saldiray. Genesis and fabric study of stibnite ores at the Murchison Range, South Africa. D, 1973, Columbia University. 157 p.

Iles, Calvert D. Mineralization and geology of a portion of the Owl Head mining district, Pinal County, Arizona. M, 1966, University of Arizona.

Iles, D. L. Variations in the geometry and density of dolines, Alamakee County, Iowa. M, 1977, Iowa State University of Science and Technology.

Illangasekare, T. Influence coefficients generator suitable for stream-aquifer management. D, 1978, Colorado State University. 236 p.

Illes, Robert John. A geotechnical evaluation of abandoned strip mines for sanitary landfill purposes. M, 1987, Kent State University, Kent. 282 p.

Illfelder, H. M. J. Laboratory study of stick-slip friction. D, 1979, University of Wisconsin-Madison. 236 p.

Illich, Harold Aallen. Petrology and stratigraphy of the Flathead Formation (Middle Cambrian), Philipsburg-Drummond, Montana. M, 1966, University of Montana. 95 p.

Illsley, Charles Truman. An investigation of geochemical prospecting by testing stream waters. M, 1955, Pennsylvania State University, University Park. 71 p.

Ilsley, Ralph. Structural geology of eastern Massachusetts. D, 1934, Massachusetts Institute of Technology. 205 p.

Iltis, Steven T. Processing and interpretation of seismic reflection data from the Precambrian of the central Laramie Range, Albany County, Wyoming. M, 1983, University of Wyoming. 94 p.

Ilton, Eugene Saul. Base metal exchange between rock-forming silicates, oxides, and hydrothermal/metamorphic fluids. D, 1987, The Johns Hopkins University. 242 p.

Ilukewitsch, Alejandro G. Three-dimensional seismic modeling with source-receiver offset. M, 1977, University of Houston.

Imada, Jewelle Akie. Numerical modeling of the groundwater in the East rift zone of Kilauea Volcano, Hawaii. M, 1984, University of Hawaii at Manoa. 102 p.

Imam, Ali. Subsurface geology in the northern part of East Pakistan with special reference to the Garo-Rajmahal Gap. M, 1963, Wayne State University.

Imam, Hassan Fahmy El-Sayed. A viscoelastic analysis of mine subsidence in horizontally laminated strata. D, 1965, [University of Minnesota, Minneapolis]. 136 p.

Imasuen, Okpeseyi Isaac S. Kaolin-smectite transformations and soils of midwestern Nigeria. D, 1987, University of Western Ontario. 329 p.

Imbalzano, John Francis. Some chemical and biochemical studies of the accumulation of peat in southern Florida. D, 1970, Pennsylvania State University, University Park. 277 p.

Imbault, Paul E. The acidic plutonic rocks of the Iserhoff River area (Quebec). M, 1947, McGill University.

Imbault, Paul E. The Olga-Gealand Lake area, Abitibi-East County, Quebec. D, 1950, McGill University.

Imbrie, John. Protremate brachiopods of the Traverse Group "Devonian" of Michigan. D, 1951, Yale University.

Imbrigiotta, Thomas Edward. Trace metal distributions and retention factors in Lake Wingra sediments. M, 1982, University of Wisconsin-Madison. 173 p.

Imbt, William Clarence. A critical analysis of the structure of typical American oil fields. M, 1932, University of Chicago. 36 p.

Imlay, Ralph Willard. Geology of the Sierra de Cruillas, Tamaulipas, Mexico. M, 1931, University of Michigan.

Imlay, Ralph Willard. Stratigraphy and paleontology of the Upper Cretaceous beds along the eastern side of Laguna de Mayron, Coahuila, Mexico. D, 1933, University of Michigan.

Immega, Inda Proske. Mineralogy and petrology of some Precambrian iron-formations in southwestern Montana. D, 1976, Indiana University, Bloomington. 97 p.

Immega, Neal Terry. Environmental influences on trace element concentrations in some modern and fossil oysters. D, 1976, Indiana University, Bloomington. 183 p.

Immel, R. L. A reexamination of the U^{234}/U^{238} method of dating deep-sea sediments. M, 1974, Florida State University.

Immel, Robert L. Uranium isotope geochemistry of micromanganese nodules and related sedimentary components from Southern Ocean pelagic sediments. M, 1974, Florida State University.

Immitt, James Peter. Skarn and epithermal vein mineralization in the San Carlos Caldera region, northeastern Chihuahua, Mexico. M, 1981, University of Texas, Austin.

Imse, John P. Geology of the Smith Canyon area, Portneuf Range, Bannock County, Idaho. M, 1979, Idaho State University. 53 p.

Imsiler, James B. Structural geology of the Safford Peak area, Tucson Mountains, Pima County, Arizona. M, 1959, University of Arizona.

Inasi, James C. The geology, mineralization and associated features of the Eagle Mountain molybdenum prospect, Potato District, Guyana. M, 1975, University of New Brunswick.

Indares, Aphrodite. L'Evolution des conditions de température et de pression pendant le métamorphisme catazonal dans la région de Maniwaki, Province de Grenville, Bouclier Canadien. M, 1982, Universite de Montreal. 255 p.

Indeck, Jeff. Sediment analysis and mammalian fauna from Little Box Elder Cave, Wyoming. D, 1987, University of Colorado. 210 p.

Inden, Richard F. Paleogeography, diagenesis and paleohydrology of a Trinity Cretaceous carbonate beach sequence, central Texas. D, 1972, Louisiana State University.

Inden, Richard Francis. Petrographic analysis and environmental interpretation of the Breezy Hill Limestone in Illinois, Missouri, Kansas, and Oklahoma. M, 1968, University of Illinois, Urbana.

Inderbitzen, Anton L. Geology of part of the Santa Monica Mountains with special reference to the geological hazards. M, 1960, University of Southern California.

Inderbitzen, Anton Louis. Relationships between sedimentation rate and shear strength in recent marine sediments off southern California. D, 1970, Stanford University. 126 p.

Inderwiesen, Philip Leon. Direct inversion and Kirchhoff migration of unstacked seismic data using ray-theoretical methods (Volumes I and II). D, 1987, University of Houston. 769 p.

Indest, Daniel J. Depositional systems of the Upper Cretaceous Olmos Formation in Zavala County, Texas. M, 1982, University of Texas, Arlington. 88 p.

Indest, Stanley J. A petrographic and geochemical study of Wildhorse Mountain, Brewster County, Texas. M, 1978, University of Houston.

Indorf, Christopher P. Geology and mineralogy of the Silver Hill zinc deposit and associated deposits, central North Carolina. M, 1978, University of North Carolina, Chapel Hill. 142 p.

Indorf, Michael S. Uranium-phosphorus determinations for selected phosphate grains from the Miocene Pungo River Formation, North Carolina. M, 1982, East Carolina University. 90 p.

Indraratna, Buddhima Nalin. Application of fully grouted bolts in yielding rock. D, 1987, University of Alberta. 286 p.

Infanger, Michael F. Effects of the stacking sequence and water level on simulated coal mine effluent. M, 1980, Southern Illinois University, Carbondale. 85 p.

Ingall, Ellery D. Clay mineral characteristics in hydrothermally altered volcanic rocks, East Tintic, Utah. M, 1985, University of Utah. 53 p.

Ingebritsen, Steven Eric. Evolution of the geothermal system in the Lassen Volcanic National Park area. M, 1983, Stanford University. 90 p.

Ingebritsen, Steven Eric. Vapor-dominated zones within hydrothermal convection systems; evolution and natural state. D, 1986, Stanford University. 187 p.

Ingels, Jerome J. C. Study of a Silurian reef complex in northern Illinois. D, 1960, Northwestern University.

Ingels, Jerome J. C. The geology of the Lancaster Quadrangle of Dallas and Ellis counties, Texas. M, 1957, Southern Methodist University. 17 p.

Ingemansen, Dean Brian. Alteration and mineralization at the Chilco Mountain molybdenum-tungsten deposits, Bonner and Kootenai counties, Idaho. M, 1986, Washington State University. 141 p.

Ingen, L. B. Van *see* Van Ingen, L. B., III

Ingen, Robert Van *see* Van Ingen, Robert

Ingersoll, D. L. Assessing program impact; Water Resources Planning Act, Title III. M, 1975, University of Arizona.

Ingersoll, D. S. An investigation of possible mechanisms for New England seismicity. M, 1975, Boston College.

Ingersoll, Guy E. The origin of vein gypsum at Lookout Peak, Spearfish, South Dakota. M, 1918, University of Minnesota, Minneapolis. 7 p.

Ingersoll, Raymond Vail. Evolution of the Late Cretaceous fore-arc basin of northern and central California. D, 1977, Stanford University. 200 p.

Ingersoll, Robert George, Jr. A geochemical reconnaissance of the mining region from Corbin to Comet, Jefferson County, Montana. M, 1964, Montana College of Mineral Science & Technology. 46 p.

Ingerson, Fred Earl. Layered peridotitic laccoliths of the Trout River area, Newfoundland. D, 1934, Yale University.

Ingerson, M. J. and Bridge, J. The Middle Ordovician section in east central Missouri. M, 1922, University of Missouri, Rolla.

Ingham, Albert I. The zinc and lead deposits of Shawangunk Mountain, New York. M, 1939, Cornell University.

Ingham, Merton Charles. The salinity extrema of the world ocean. D, 1966, Oregon State University. 123 p.

Ingham, W. I. Wellington oil field (Larimer County), Colorado. M, 1934, Colorado School of Mines. 87 p.

Ingham, Walter Norman. Geology of the molybdenum deposits of North America. M, 1940, University of Toronto.

Ingham, Walter Norman. Structure and radioactivity of the Bourlamaque and Elzevir batholiths, Quebec. D, 1944, University of Toronto.

Inghram, Brent J. Stability of excavated slopes in horizontally bedded rock. M, 1981, University of Nevada. 155 p.

Ingle, James C., Jr. Paleoecologic, sedimentary and structural history of the late Tertiary Capistrano Embayment, California. M, 1962, University of Southern California.

Ingle, James Chesney, Jr. Facies variation and the Miocene-Pliocene boundary in southern California. D, 1967, University of Southern California. 351 p.

Ingle, Steven Carl. Basin analysis and paleodrainage, Dana Basin, Wyoming. M, 1977, South Dakota School of Mines & Technology.

Inglis, J. Mark. Nitrate contamination in a shallow unconfined aquifer in Perry, Ohio. M, 1982, Case Western Reserve University.

Ingraffea, A. R. Discrete fracture propagation in rock; laboratory tests and finite element analysis. D, 1977, University of Colorado. 374 p.

Ingraham, Jean. A textural and mineralogical study of the diabase in the vicinity of Mount Nonotuck, Massachusetts. M, 1964, Smith College. 84 p.

Ingraham, Mark G. Permafrost, hydrology, and management on the Arctic Coastal Plain, northern Alaska. M, 1978, [Colorado State University].

Ingraham, Neil L. Environmental isotope hydrology of the Dixie Valley geothermal system, Dixie Valley, Nevada. M, 1982, University of Nevada. 96 p.

Ingraham, Neil Layton. Light stable isotope systematics of large-scale hydrologic regimes in California and Nevada. D, 1988, University of California, Davis. 169 p.

Ingraham, Peter Curwood. A model study of optimum location and orientation of tensioned roof bolts in underground openings. M, 1985, University of Idaho. 85 p.

Ingram, Carey. Beach sands of the southern Delmarva Peninsula; patterns and causes. M, 1972, College of William and Mary.

Ingram, Frank Thompson. The stratigraphy and paleontology of the Ordovician System in Lookout Valley, Georgia. M, 1954, Emory University. 93 p.

Ingram, Gary R. Evaluation of Iowa phosphate deposits. M, 1975, University of Iowa. 111 p.

Ingram, Gregory D. Some approaches to the analysis and interpretation of wide-angle bottom loss data. M, 1981, University of Texas, Austin.

Ingram, John Jeffrey. Total sediment load measurement using point-source suspended-sediment data. D, 1988, Colorado State University. 140 p.

Ingram, Robert James. A sedimentary analysis of the Oriskany Sandstone, central Pennsylvania. M, 1954, University of Pittsburgh.

Ingram, Roy Lee. An experimental study of the influence of grain-size on the mark of oscillation ripplemark. M, 1943, University of Oklahoma. 16 p.

Ingram, Roy Lee. Fissility and non-fissility in shales and mudstones. D, 1948, University of Wisconsin-Madison.

Ingram, Ruth M. The application of deconvolution to field recorded CDP reflection data. M, 1979, University of Manitoba. 92 p.

Ingram, William Franklin. The inspection and orientation of piezoelectric quartz for use as oscillating plates. M, 1950, Emory University. 50 p.

Ingwell, Tim Harvey. Stratigraphy and structural geology of the Merriam Lake area, Lost River Range, Idaho. M, 1980, University of California, Los Angeles.

Ingwersen, James B. Downstream recovery of selected Colorado streams subject to acid drainage from abandoned metal mines. M, 1982, Colorado State University. 139 p.

Inman, Douglas L. Areal and seasonal variations in beach and nearshore sediments at La Jolla, California. D, 1952, University of California, Los Angeles. 170 p.

Inman, Joseph Robert, Jr. Direct interpretation of resistivity sounding. M, 1973, University of Utah. 107 p.

Inners, J. D. The stratigraphy and paleontology of the Onesquethaw Stage in Pennsylvania and adjacent states. D, 1975, University of Massachusetts. 737 p.

Innes, D. G. Proterozoic volcanism in the southern province of the Canadian Shield. M, 1977, Laurentian University, Sudbury. 161 p.

Innes, G. M. The economic significance of the distribution of metals through geologic time. M, 1958, University of Toronto.

Innes, Morris, J. S. Some structural features of the (Canadian) Precambrian Shield as revealed by gravity anomalies. D, 1952, University of Toronto.

Innis, John William. Sedimentology and stratigraphy of the Ordovician-Silurian Road River Formation, southern Richardson Mountains, Yukon Territory. M, 1980, University of Western Ontario. 196 p.

Insley, Herbert. The gabbros and associated intrusive rocks of Harford County. D, 1919, The Johns Hopkins University.

Insley, Robert Hiteshew. Studies in portions of the system $Na_2O-MgO-Al_2O_3-SiO_2$. M, 1952, Pennsylvania State University, University Park. 53 p.

Inthuputi, Boonmai. Geology and uranium occurrences of Elkins Mesa, McKinlay County, New Mexico. M, 1969, Wesleyan University.

Intraprasart, S. Experimental studies and analysis of compacted fills over a soft subsoil. D, 1978, Georgia Institute of Technology. 216 p.

Introne, Douglas Stuart. Amino acid racemization; an application for the determination of thermal activities along the Mid-Atlantic Ridge. M, 1983, University of Miami. 190 p.

Inverso, George Anthony. The scour, transport and deposition of Mt. St. Helens ash in a laboratory setting; a model. M, 1982, University of Idaho. 51 p.

Inyang, Aniefiok David. Environmental transport of heavy metals from soil amended with undigested sludge. D, 1982, University of Oklahoma. 153 p.

Ioannidou, Eleni I. Source parameters from inversion broadband accelerograms. M, 1984, SUNY at Binghamton. 105 p.

Ioannou, Christos. Distribution, transport and reclamation of abandoned mine tailings along the channel of the south fork of the Coeur d'Alene River and tributaries, Idaho. M, 1979, University of Idaho. 146 p.

Iovenitti, J. A reconnaissance study of jasperoid in the Kelly Limestone, Kelly mining district, New Mexico. M, 1977, New Mexico Institute of Mining and Technology.

Iqbal, Ghulam M. M. Cleanup of water based liquids from hydraulically fractured gas wells. D, 1988, University of Oklahoma. 174 p.

Iqbal, Mir Weseluddin Ahmed. Paleontology of the Ghazij Shale, Quetta-Kalat region, West Pakistan. M, 1963, University of California, Los Angeles.

Iqual, Javed. Sedimentology and distribution of Benthonic foraminifera in McClure Strait (Canadian Arctic Archipelago). M, 1973, Dalhousie University. 279 p.

Iradji, Amir Houshang. The geology of the northern Lenado area, Pitkin County, Colorado. M, 1955, Colorado School of Mines. 98 p.

Iranpanah, Assad. Petrology, origin and trace element geochemistry of the Ada Formation, Seminole and Pontotoc counties, Oklahoma. D, 1966, University of Oklahoma. 229 p.

Iranpanah, Assad. Structural geology of Burnet area, Burnet County, Texas. M, 1964, University of Texas, Austin.

Iranpanah, Touran Soltanzadeh. Pennsylvanian Chonetoidea of Oklahoma. M, 1966, University of Oklahoma. 144 p.

Ireland, H. A. The use of insoluble residues for correlation in Oklahoma. D, 1935, University of Chicago. 129 p.

Ireland, H. Andrew. Geology of Morgan Township, Township 2 South, Range 2 East, Murray and Carter counties, Oklahoma. M, 1927, University of Oklahoma. 43 p.

Ireland, Jarrette Lynn. Geology for land-use planning of western Rogers County and southern Washington County, Oklahoma. M, 1973, Oklahoma State University. 53 p.

Ireton, M. Frank. Passage formation, classification and modification of lava tubes. M, 1977, Boise State University.

Irion, Ronald L. Meramecian and Osagian rocks on the west flank of the central Kansas uplift. M, 1963, Wichita State University. 66 p.

Irisarri, A. M. de see de Irisarri, A. M.

Irish, Ernest J. W. The mineralogy of some of the gold mines of British Columbia. M, 1940, University of British Columbia.

Irish, Ernest James Wingett. The geology of the Moon Creek map area, west-central Alberta. D, 1949, University of Illinois, Urbana. 143 p.

Irish, Neil Frederick. Phase relations in the system Cu_5FeS_4-Cu_2S and hydrothermal ion-exchange data for the system Cu_2S-Ag_2S. M, 1987, Purdue University. 115 p.

Irish, Robert J. The geology of the Juniper Butte area, Spray Quadrangle, Oregon. M, 1954, Oregon State University. 73 p.

Iroe, Hindartono D. Evaluation of shaly sands, Sepinggan Field, Indonesia. M, 1981, Colorado School of Mines. 149 p.

Irons, L. A. A gravity survey of the Humboldt Fault and related structures in southeastern Nebraska. M, 1979, University of Nebraska, Lincoln.

Irrinki, R. R. The structural geology and regional metamorphism of the southeastern part of the Miramichi Zone (north). M, 1974, Acadia University.

Irtem, Oguz. Stratigraphy of the Manitou Formation between Aspen and Minturn, Colorado. M, 1973, Colorado School of Mines. 68 p.

Irtem, Oguz. Stratigraphy of the Minturn Formation (Pennsylvanian) between Glenwood Springs and Craig, Colorado. D, 1977, Colorado School of Mines. 385 p.

Irvin, Hollie F., Jr. The Yucca Formation of the Solitario Uplift. M, 1957, Southern Methodist University. 14 p.

Irvin, William Carl. The topographic expression of the sub-lacustrine geology of the Lake Superior basin. M, 1948, University of Illinois, Urbana. 18 p.

Irvine, Ben M. Geologic investigation of landslides in the Ralston Creek area, Jefferson County, Colorado. M, 1961, University of Colorado.

Irvine, Pamela Joe. The Posey Canyon Shale; a Pliocene lacustrine deposit of the Ridge Basin, Southern California. M, 1977, University of California, Berkeley. 97 p.

Irvine, T. N. and Fletcher, G. L. Areal geology of the Emo area, Rainy River District, 1 inch to 1 mile (Ontario). M, 1953, University of Manitoba.

Irvine, Thomas Neil. An investigation of the geology of a part of the Emo area, District of Rainy River, Ontario. M, 1955, University of Manitoba.

Irvine, Thomas Neil. The ultramafic complex and related rocks of Duke Island, southeastern Alaska. D, 1959, California Institute of Technology. 320 p.

Irvine, W. T. Nature, origin, and occurrence of the Garrison phosphate deposits. D, 1951, University of Toronto.

Irving, Earl Montgomery. The geology of a portion of the Corona and Riverside quadrangles, near Corona, California. M, 1935, University of California, Los Angeles.

Irving, John D. A contribution to the geology of the northern Black Hills. D, 1899, Columbia University, Teachers College.

Irving, Roland D. The geology of central Wisconsin. D, 1880, Columbia University, Teachers College.

Irving, Stephen Myles. An electron-optical study of the structure and morphology of sepiolite and palygorskite. M, 1966, Pennsylvania State University, University Park. 72 p.

Irwin, Arthur B. Geology of the Howson Creek area, Slocan mining division, British Columbia. D, 1950, McGill University.

Irwin, Arthur Bonshaw. Wallrock alteration in Pioneer and Bralorne mines, British Columbia. M, 1947, University of British Columbia.

Irwin, Barbara R. Interpretation of sedimentary structures in the upper Red Peak and Jelm formations (Triassic), southeastern Wyoming and northern Colorado. M, 1973, University of Wyoming. 117 p.

Irwin, Charles Dennis. Stratigraphic analysis of the upper Permian and lower Triassic strata in southern Utah. D, 1969, University of New Mexico. 158 p.

Irwin, David. Land use, Houston County, Tennessee. M, 1949, University of Tennessee, Knoxville.

Irwin, Don Dennis. A microlithological study of some Upper Pennsylvanian limestones. M, 1960, University of Nebraska, Lincoln.

Irwin, Dorothy W. Geology of the Fifty-Six Mine near Imlay, Nevada. M, 1934, Columbia University, Teachers College.

Irwin, Frank Albert. A mechanical model of early salt dome growth. M, 1988, Texas A&M University. 49 p.

Irwin, Gerald J. An investigation into the rock-magnetism of the Wilberforce (Ontario) Pyroxenite. M, 1963, University of Western Ontario.

Irwin, J. E. Petrographic analysis of a patch reef in the Edwards Limestone (Lower Cretaceous), Central Texas. M, 1977, Stephen F. Austin State University.

Irwin, James Joseph. Geology, K-Ar geochronology and paleomagnetism of parts of the coastal cordillera of central Chile, South America. D, 1986, University of California, Berkeley. 208 p.

Irwin, Joseph S. Oil possibilities in western Kansas and north central Wyoming. M, 1922, University of Missouri, Rolla.

Irwin, Melvin LeRoy. Straparollus and Porcellia of the Chester Series (Ir94). M, 1941, University of Illinois, Urbana.

Irwin, Randal L. Eocene lithofacies in the vicinity of Leucadia and Encinitas, San Diego County, California. M, 1986, San Diego State University. 124 p.

Irwin, Thomas D. The petrologic evolution of the North Mountain Stock, La Sal Mountains, Utah. M, 1973, University of Arizona.

Irwin, William H. Geology of the Fifty-Six Mine near Imlay, Nevada. M, 1934, Columbia University, Teachers College.

Irwin, William Harold. Geology of the rock foundation of Grand Coulee Dam, Washington. D, 1938, Columbia University, Teachers College. 26 p.

Irwin, William Kenneth Arthur. Depositional environments and diagenesis of the basal Belly River Sand, Strathmore area, Alberta, Canada. M, 1980, Texas Tech University. 76 p.

Irwin, William Porter. The Vasquez Series in the Upper Tick Canyon area, Los Angeles County, California. M, 1950, California Institute of Technology. 35 p.

Isaacs, Andrew Mansfield. An analytical electron microscopic study of a pyroxene-amphibole intergrowth. M, 1981, University of Michigan.

Isaacs, C. M. Diagenesis in the Monterey Formation examined laterally along the coast near Santa Barbara, California. D, 1980, Stanford University. 344 p.

Isaacs, Charles Manning, Jr. State College, Pennsylvania, crustal structure by modeling of long-period P-wave forms from teleseismic earthquakes. M, 1979, Pennsylvania State University, University Park. 82 p.

Isaacs, Kalman Nathan. Geology of northern portions of the Commatti Canyon and Grant Lake Quadrangle, San Luis Obispo County, California. M, 1951, University of California, Berkeley. 45 p.

Isaacs, Thelma. The geochemistry of nickel carbonates. D, 1962, University of Chicago. 67 p.

Isaacs, Thelma. The petrology of the Rensselaer Grit (Cambrian, New York). M, 1952, New York University.

Isaacson, Laurie Brown. A major negative gravity anomaly in central Colorado; its interpretation and significance. M, 1972, University of Wyoming. 43 p.

Isaacson, Laurie Brown. Paleomagnetics and secular variation of Easter Island basalts. D, 1974, Oregon State University. 69 p.

Isaacson, Peter Edwin. Devonian stratigraphy, paleobathymetry and brachiopod paleontology of Bolivia. D, 1975, Oregon State University. 543 p.

Isaak, Donald G. The elastic properties of almandine-spessartine. M, 1975, Pennsylvania State University, University Park. 40 p.

Isaaks, Edward Harold. Risk qualified mappings for hazardous waste sites; a case study in distribution free geostatistics. M, 1984, Stanford University. 85 p.

Isachsen, Clark. Geology, geochemistry, and geochronology of the Westcoast Crystalline Complex and related rocks, Vancouver Island, British Columbia. M, 1984, University of British Columbia. 144 p.

Isachsen, Yngvar W. Geology of the Rensselaer Falls Quadrangle, New York. D, 1953, Cornell University.

Isachsen, Yngvar William. Petrology of the Catoctin Formation in the Lovingston Quadrangle, central Virginia. M, 1949, Washington University. 95 p.

Isacks, Bryan L. Seismic waves with frequencies from 1 to 100 cycles per second recorded in a deep mine in northern New Jersey. D, 1964, Columbia University, Teachers College.

Isagholian, Varush. Geology of a portion of Horse Mesa and Fish Creek Canyon areas, central Arizona. M, 1983, Arizona State University. 73 p.

Isailovic, D. Optimal operation of coupled surface-underground storage. D, 1975, Colorado State University. 174 p.

Isarangkoon, Piphop. The distribution of zinc and lead in rock at the Big Bug Pluton, Big Bug mining district, Yavapai County, Arizona. M, 1978, South Dakota School of Mines & Technology.

Isarankura, Somsak. Seismic and resistivity methods applied to ground water studies at Hidden Water Creek coal strip-mine, Sheridan County, Wyoming. M, 1976, South Dakota School of Mines & Technology.

Isard, Scott Alan. Factors controlling soil moisture and evapotranspiration within alpine vegetation communities; Niwot Ridge, Colorado Front Range. D, 1984, Indiana University, Bloomington. 97 p.

Isberg, J. T. The geology of a portion of east-central Carbon County, north and west of Elk Mountain, Wyoming. M, 1937, University of Wyoming. 59 p.

Isby, John Scott. The petrology and tectonic significance of the Currant Creek Formation, north-central Utah. M, 1984, University of Utah. 134 p.

Isea, Andreina. Sedimentology and depositional model of the Holocene coastal sequence of Sinamaica, northwestern Venezuela. M, 1985, University of Cincinnati. 112 p.

Isenberger, Kenyon Jay. Properties of some selected Iowa carbonate aggregates. M, 1966, Iowa State University of Science and Technology.

Isenhower, Daniel Bruce. Siting of low-level radioactive waste disposal facilities in Texas. M, 1982, Texas A&M University. 136 p.

Ishag, Abudulla Bassan. A quantitative study of the potential recharge to principal aquifers in Kordofan Province of western Sudan. M, 1964, University of Arizona.

Isham, Julian C. Mercury uptake in Recent lake sediments. M, 1973, Michigan State University. 45 p.

Ishaq, A. M. Application of remote sensing to the location of hydrologically active (source) areas. D, 1974, University of Wisconsin-Madison. 241 p.

Isherwood, A. Quaternary geology and soil conditions; University of Waterloo campus. M, 1976, University of Waterloo.

Isherwood, D. Soil geochemistry and rock weathering in an arctic environment. D, 1975, University of Colorado. 188 p.

Isherwood, W. F. Gravity and magnetic studies of The Geysers-Clear Lake geothermal region, California. D, 1975, University of Colorado. 120 p.

Isherwood, William F. Regional gravity survey of parts of Millard, Juab, and Sevier counties, Utah. M, 1967, University of Utah. 31 p.

Ishibashi, Gary Duane. Stratigraphy of the Pennsylvanian-Permian sequence exposed in the Sublett Range of Southcentral Idaho. M, 1980, Idaho State University. 71 p.

Ishibashi, Isao. Torsional simple shear device, liquefaction and dynamic properties of sands. D, 1974, University of Washington. 143 p.

Ishihara, Shunso. Molybdenum mineralization at Questa Mine, New Mexico. M, 1963, Columbia University, Teachers College.

Ishii, H. Reflected wave propagation in a wedge. D, 1969, University of British Columbia.

Ishimoto, Toshio Tom. Serpentine soil and the foothill woodland community in Fresno County, California. M, 1952, California State University, Fresno.

Ishizaki, K. Effects of soluble organics on the flow through thin cracks in basaltic lava flow. M, 1967, University of Hawaii.

Isiorho, Solomon Akpoghenobor. Interactions between Lake Chad and the phreatic aquifer in the Southwest Chad Basin. D, 1988, Case Western Reserve University. 256 p.

Iskander, Atef Fanzi. A gravity survey of the Alkali Creek and the Bear Butte areas, Meade County, South Dakota. M, 1975, South Dakota School of Mines & Technology.

Iskander, Wilson. An appraisal of ground water resources of Zalengei area, Darfur Province, Sudan. M, 1967, University of Arizona.

Islam, Quazi Taufiqul. Structural analysis of a portion of the Reagan and Sulfur fault zones, Murray and Johnston counties, Oklahoma. M, 1985, University of Texas at Dallas. 68 p.

Islam, S. M. Nazrul. Well log analysis in Taber Mannville 'D' Pool, Taber Field, Alberta, Canada. M, 1972, University of Manitoba.

Islam, Shafiul. Thermal maturation patterns in Cambro-Ordovician flysch sediments of the Taconic Belt, Gaspe Peninsula. M, 1981, McGill University. 191 p.

Ismail, Azmi. Crustal structure in the vicinity of Yucca Mountain. M, 1986, University of Nevada. 106 p.

Ismail, Farouk Taha Ahmed. Artificial weathering of biotites and vermiculites and natural weathering of biotites to clay minerals in soils of arid and humid climates. D, 1966, University of California, Berkeley. 193 p.

Ismail, Mohamad I. B. A raypath explanation for R G waves from earthquakes. M, 1983, Texas Tech University. 89 p.

Ismail, Razimah. Stratigraphy, mineralogy and depositional environments of the mudrock facies of the Verde Formation, central Arizona. M, 1985, Northern Arizona University. 88 p.

Isokrari, O. F. Numerical simulation of United States Gulf Coast geothermal geopressured reservoirs. D, 1976, University of Texas, Austin. 312 p.

Isokrari, Ombo Ferguson. Relationship between geological engineering parameters in low permeability Red Fork shoestring oil reservoir in Oklahoma. M, 1974, University of Oklahoma. 77 p.

Isom, John William. Subsurface stratigraphic analysis, Late Ordovician to Early Mississippian, Southwest Oakdale-Campbell Trend, Woods, Major, and Woodward counties, Oklahoma. M, 1972, University of Oklahoma. 57 p.

Isotoff, Andrei. Russian contributions to the geographical knowledge of Alaska and the adjacent islands and seas. M, 1942, University of Oregon. 138 p.

Isphording, Wayne Carter. A study of the heavy minerals from the Hawthorne Formation and overlying sands exposed at the Devil's Mill hopper, Alachua County, Florida. M, 1963, University of Florida. 33 p.

Isphording, Wayne Carter. Petrology and stratigraphy of the Kirkwood Formation (Middle Miocene, eastern New Jersey). D, 1967, Rutgers, The State University, New Brunswick. 181 p.

Israel, Alan M. A sedimentologic study of a Holocene microtidal flood-tidal delta; San Luis Pass, Texas. M, 1983, Colorado State University. 200 p.

Israelsen, C. Earl. The effects of suspended-sediment, temperature, frequency, and dissolved salts on the dielectric properties of water. D, 1968, University of Arizona.

Isselhardt, Courtney Francis. Geology of parts of the Saint Helena rhyolite (Pliocene) near Saint Helena, California. M, 1969, University of California, Berkeley. 45 p.

Issler, Dale. The thermal and subsidence history of the Labrador Margin. D, 1987, Dalhousie University.

Ista, Jane Pohtilla. A correlation of coal mine roof fall rates with the daily solid Earth tide. M, 1980, University of Virginia. 70 p.

Istas, L. S. Chlorine distribution in the Idaho Batholith. M, 1976, Portland State University. 82 p.

Istas, Laurence Stewart. Trace elements in veins of the Bohemia mining district, Oregon. D, 1983, University of Washington. 127 p.

Isuk, Edet E. An experimental investigation on boudinage structures. M, 1972, Brooklyn College (CUNY).

Isuk, Edet Effing. Solubility of molybdenite in the system $Na_2O-K_2O-SiO_2-MoS_2-H_2O-CO_2$ with geologic application. D, 1976, University of Iowa. 101 p.

Italia, Santo. A new vibration meter. M, 1954, St. Louis University.

Itell, Karyn Marie. MicroComputer Assisted Model for Phosphorus Removal in Existing Wastewater Treatment Plants in the Chesapeake drainage basin (CAMPREP). D, 1987, Rensselaer Polytechnic Institute. 217 p.

Itkowsky, Francis A. Sensitivity of sample estimates to different Landsat classifications. M, 1978, Colorado State University. 187 p.

Ito, Emi. High-temperature metamorphism of plutonic rocks from the Mid-Cayman Rise; a petrographic and oxygen isotopic study. D, 1979, University of Chicago. 158 p.

Itson, Sonja P. The petrology of the mafic complex (Mesozoic) north of Jamul, San Diego County, California. M, 1970, University of San Diego.

Itter, Harry A. The geomorphology of the Wyoming-Lackawanna region (Pennsylvania). D, 1936, Columbia University, Teachers College.

Ivahnenko, Tamara I. An evaluation of point-source limestone introduction as an ameliorative procedure to reduce acidity in two low buffer capacity streams. M, 1988, West Virginia University. 260 p.

Iversen, Christine M. Computer-linked terrane analysis for landfill waste-disposal site selection. M, 1974, Michigan State University. 69 p.

Iversen, Gary M. Petrology of the Eagle Rock volcanic complex, Routt County, northwestern Colorado. M, 1976, Pennsylvania State University, University Park. 76 p.

Iverson, Louis Robert. The role of pioneering species on the reclamation of North Dakota surface mined lands. D, 1981, University of North Dakota. 218 p.

Iverson, Richard Matthew. Processes of accelerated pluvial erosion on desert hillslopes modified by vehicular traffic. M, 1979, Stanford University. 52 p.

Iverson, Richard Matthew. Unsteady, nonuniform landslide motion; theory and measurement. D, 1984, Stanford University. 303 p.

Iverson, William Paul. Processing and interpretation of Cocorp Southern Appalachian profiles. D, 1983, University of Wyoming. 253 p.

Ives, Ronald Lorenz. Glaciology of the Monarch Valley, Grand County, Colorado. M, 1937, University of Colorado.

Ives, William, Jr. Evaluation of acid etching of limestone. M, 1954, University of Kansas. 68 p.

Ivey, C. G., Jr. A gravity survey of Fort Ord, California. M, 1969, United States Naval Academy.

Ivey, John Barn. Geology of Cedars of Lebanon State Park and vicinity, Wilson and Rutherford counties, Tennessee. M, 1950, Vanderbilt University.

Ivey, Marvin Lee, Jr. The geologic history of the Swan Islands, Honduras. M, 1979, Texas Christian University. 49 p.

Ivosevic, Stanley Wayne. Geology and ore deposits of the Johnnie District, Nye County, Nevada. M, 1976, University of Nevada. 193 p.

Ivy, David. The biostratigraphy and paleoecology of the Permian strata exposed near Las Delicias, southwestern Coahuila, Mexico. M, 1975, Texas Christian University. 49 p.

Ivy, Logan Dudley. Systematics and biostratigraphy of the earliest North American Rodentia (Mammalia), latest Paleocene and early Eocene of the Clark's Fork Basin, Wyoming. M, 1982, University of Michigan.

Iwabuchi, Jotaro. The influence of cementation on liquefaction resistance of sands. D, 1986, Virginia Polytechnic Institute and State University. 215 p.

Iwai, Katsuhiko. Fundamental studies of fluid flow through a single fracture. D, 1976, University of California, Berkeley. 232 p.

Iwakuma, Tetsuo. Finite elastic-plastic deformation of polycrystalline metals and composites. D, 1983, Northwestern University. 87 p.

Iwamura, S. Plastic buckling of multi-layered beams. D, 1976, Texas A&M University. 158 p.

Iwasaki, Takeshi. Frictional properties between fine grained limestone, dolomite and sandstone along precut surfaces. M, 1970, Texas A&M University.

Iwashita, S. Chemical analysis of some volcanic rocks in Hawaii. M, 1940, University of Hawaii. 52 p.

Iwin, Francis R., Jr. A mechanical calculator for seismic refractions. M, 1955, Boston College.

Iwuagwu, Chukwumaeze Julian. Diagenesis of the basal Belly River Sandstone reservoir, Pembina Field, Alberta, Canada. M, 1979, University of Alberta. 175 p.

Iwuagwu, Chukwumaeze Julian. Lithofacies analysis, depositional environment and diagenesis of the basal Belly River Formation, southern and central foothills, Alberta, Canada. D, 1983, University of Alberta. 263 p.

Iyer, L. Srinivasa. The effect of freezing and thawing on Minnekahta Limestone and Sioux Quartzite used for concrete aggregates. D, 1974, South Dakota School of Mines & Technology. 102 p.

Iz, Huseyin Baki. An algorithmic approach to crustal deformation analysis. D, 1987, Ohio State University. 127 p.

Izaguirre, M. A. Zuniga *see* Zuniga Izaguirre, M. A.

Izard, John Emmette. X-ray study of the sedimentary pyrite and marcasite of western New York. M, 1967, SUNY at Buffalo. 61 p.

Izett, Glen Arthur. The Cretaceous-Tertiary boundary interval, Raton Basin, Colorado and New Mexico, and its content of shock-metamorphosed minerals; implications for the Cretaceous-Tertiary boundary impact-extinction theory. D, 1988, University of Alaska, Fairbanks. 201 p.

Izgi, Mehlika Fahri. Foraminifera from test wells in Adana, Turkey. M, 1940, University of Texas, Austin.

Izraeli, Ruth L. Water quality and hydrogeological investigations at the University of Connecticut waste disposal area. M, 1985, University of Connecticut. 108 p.

Izuka, Scot K. Biostratinomy of ostracode assemblages from a small reef flat in Maunalua Bay, Oahu, Hawaii. M, 1983, University of Kansas. 61 p.

Izuka, Scot Kiyoshi. The variation of magnesium concentrations in the tests of Recent and fossil benthic foraminifera. D, 1988, University of Hawaii. 115 p.

J., Nicolas G. Munoz *see* Munoz J., Nicolas G.

Jaacks, Jeffrey A. Meteorological influence upon mercury, radon and helium soil gas emissions. D, 1984, Colorado School of Mines. 170 p.

Jaafar, Idris Sir. Depositional and diagenetic history of the B-zone of the Red River Formation (Ordovician) of the Beaver Creek Field, Golden Valley County, North Dakota. M, 1980, West Texas State University. 67 p.

Jaayasinghe, Nimal Ranjith. Geology of the Wesleyville area, Newfoundland. M, 1976, Memorial University of Newfoundland. 228 p.

Jabali, Habib Hilmi. Dynamic stiffness of two layers in contact subjected to torsional oscillations. D, 1982, University of Miami. 57 p.

Jablonski, D. I. Paleoecology, paleobiogeography, and evolutionary patterns of Late Cretaceous Gulf and Atlantic Coastal Plain mollusks. D, 1979, Yale University. 616 p.

Jabro, Jalal David. Spatial variability of field-saturated hydraulic conductivity and simulation of water flow from a percolation test hole in layered soil. D, 1988, Pennsylvania State University, University Park. 151 p.

Jachens, Robert C. An experimental study of tidal gravity across the continental United States. D, 1971, Columbia University. 96 p.

Jack, Howard Corwin. Orientation of the valves of the pectinid Amusium ocalanum in the Crystal River Formation (Eocene), (northern and western Florida). M, 1970, University of Florida. 35 p.

Jack, Robert Norman. Quaternary sediments at Montara, San Mateo County, California. M, 1969, University of California, Berkeley. 133 p.

Jacka, Alonzo David. Bedding characteristics of the Platteville Formation of Wisconsin. M, 1957, University of Wisconsin-Madison. 41 p.

Jacka, Alonzo David. The environmental significance of stratification. D, 1960, Rice University. 185 p.

Jackimovicz, Joseph James. The petrology of two valley-fill sandstones in western Missouri. M, 1970, University of Missouri, Columbia.

Jackman, Albert Havens. Physiography of the Big Delta region, Alaska. D, 1953, Clark University.

Jackman, Anthony Edwin. Methods for assessing bioenergy potentials; their prospective management, energy efficiency requirements and ecological impacts in rural landscapes. D, 1983, University of Massachusetts. 356 p.

Jackman, Toni Kay. Geochemical correlation of ash-flow tuffs from the Platoro volcanic complex, Southeast Colorado. M, 1986, Wichita State University. 124 p.

Jackson, Alvin M. Petrography of the Homewood Sandstone, Fayette County, Pennsylvania. M, 1936, Northwestern University.

Jackson, Andrew Carlton. Optimization of well density in secondary recovery based on well pattern spacing and economics. M, 1977, University of Oklahoma. 46 p.

Jackson, Andrew Rupert Needham. Volcanism and genesis of Cu-Zn mineralization at Cook Lake, Snow Lake greenstone belt, Manitoba. M, 1983, University of Western Ontario. 155 p.

Jackson, Bethell H. Optical properties of crystals. M, 1906, University of Colorado.

Jackson, D. A. The urban and engineering geology of Montego Bay, Jamaica. M, 1977, University of Waterloo.

Jackson, Dale Robert. The fossil freshwater emydid turtles of Florida. D, 1977, University of Florida. 128 p.

Jackson, Dan Herman. Structure and petrography of the Precambrian rocks of the Yearling Head Mountain area, southern Llano County, Texas. M, 1972, Texas Christian University. 37 p.

Jackson, Dana Scott. The petrology, porosity and permeability of the Berea Sandstone (Mississippian), Perry Township, Ashland County, Ohio. M, 1985, University of Cincinnati. 112 p.

Jackson, David Diether. Grain boundary relaxations and the attenuation of seismic waves. D, 1969, Massachusetts Institute of Technology. 136 p.

Jackson, David Ernest. Petrogenesis of a shallow-level kimberlite from Taughannock Creek, New York. M, 1982, University of Tennessee, Knoxville. 71 p.

Jackson, Dicky Joe. Depositional environments of Pennsylvanian-Permian sequence exposed in Sublett Range of southeastern Idaho. M, 1979, Idaho State University. 65 p.

Jackson, E. L. Response to earthquake hazard; factors related to the adoption of adjustments by residents of three earthquake areas of the West Coast of North America. D, 1974, University of Toronto.

Jackson, Edward F. Free milling silver ores and their treatment. M, 1883, Washington University.

Jackson, Everett Dale. Primary textures and mineral associations in the ultramafic zone of the Stillwater Complex, Montana. D, 1960, University of California, Los Angeles.

Jackson, G. D. Petrographic study of part of the Potsdam Sandstone drill core from Mallet Well, Ste. Therese, Quebec. M, 1955, McGill University.

Jackson, G. D. The geology of the Neal (Virot) Lake area, west of Wabush Lake, Labrador, with special reference to iron deposits. D, 1963, McGill University.

Jackson, G. E. Global tectonics and the geoscientific revolution; the quest for an "interpenetrating ether" is critical to the meaning and understanding of the revolution in the Earth sciences. D, 1975, [Union College]. 68 p.

Jackson, Gerald Christopher Arden. The geology and structure of the West Kootenay composite batholith. M, 1926, University of British Columbia.

Jackson, James A. An electron microscope study of dolostone textures. M, 1968, Bowling Green State University. 50 p.

Jackson, James B. Petrography and stratigraphy of part of a Pitkin Reef Complex (Pennsylvanian, Arkansas). M, 1972, University of Arkansas, Fayetteville.

Jackson, James Milton. Geographic information system modeling of peak stream flows in small drainage basins. M, 1987, University of Colorado at Colorado Springs. 58 p.

Jackson, James R. A model study of the effects of small amplitude waves on the resuspension of fine-grained cohesive sediments. M, 1973, University of New Hampshire. 53 p.

Jackson, James Robert. Subsurface geology of the Sligo Formation (lower Cretaceous) in the Green-Fox field area, Marion and Harrison counties, Texas. M, 1969, University of Texas, Austin.

Jackson, James Roy, Jr. Subsurface Miocene foraminifera of Cameron County, Texas. M, 1940, University of Texas, Austin.

Jackson, James Streshley. Petrology of the Bull-of-the-Woods intrusive complex. M, 1978, Portland State University. 56 p.

Jackson, Jeffrey K. Geophysical study of permafrost drill core from Ross Island and Victoria Valley, Antarctica. M, 1975, Northern Illinois University. 60 p.

Jackson, Jeremy. A neontological and paleontological study of the autecology and synecology of the molluscan fauna of Fleets Bay, Virginia. M, 1968, George Washington University.

Jackson, John G. Physical property and dynamic compressibility analysis of a glacial lake deposit. D, 1971, University of Michigan.

Jackson, K. S. Geochemical dispersion of elements via organic complexing. D, 1975, Carleton University. 344 p.

Jackson, Kenneth E. Geology of the Circle Bar Lake Quadrangle, Fremont and Sweetwater counties, Wyoming. M, 1984, University of Wyoming. 73 p.

Jackson, Kenneth J. Paleohydraulic analysis of turbidity currents in an ancient canyon-channel system on the Indus deep-sea fan. M, 1979, University of Missouri, Columbia.

Jackson, Kenneth James. Chemical and thermodynamic constraints on the hydrothermal transport and deposition of tin. D, 1983, University of California, Berkeley. 104 p.

Jackson, Kern Chandler. A heavy mineral study of the Ajibik and Mesnard quartzites of Marquette County, Michigan. M, 1950, Michigan Technological University. 11 p.

Jackson, Kerne Chandler. The petrogenetic significance of quartz twins. D, 1951, University of Wisconsin-Madison.

Jackson, Lawson Erwin, Jr. A study of the Blackford Breccia in the Dalton Quadrangle (Georgia). M, 1951, Emory University. 44 p.

Jackson, Lionel E. Quaternary stratigraphy and terrain inventory of the Alberta portion of the Kananaskis Lakes 1:250,000 Sheet 82J. D, 1977, University of Calgary. 480 p.

Jackson, Lionel E. Use of the scanning electron microscope for sedimentary fabric studies of siltstones and very fine sandstones; a preliminary investigation. M, 1972, Stanford University.

Jackson, Mac Roy, Jr. The Timber Mountain magmato-thermal event; an intense widespread culmination of magmatic and hydrothermal activity at the southwestern Nevada volcanic field. M, 1988, University of Nevada. 46 p.

Jackson, Marie C. Physical controls on the formation of convolute laminations in Ouachita flysch sequences of Arkansas and Oklahoma. M, 1977, Louisiana State University.

Jackson, Marie Dolores. Deformation of host rocks during growth of igneous domes, southern Henry Mountains, Utah. D, 1987, The Johns Hopkins University. 184 p.

Jackson, Mary L. W. Geomorphology and sedimentology of experimental fan deltas. M, 1981, Colorado State University. 103 p.

Jackson, Michael Eldon. Thermoluminescence dating of Holocene paleoseismic events on the Nephi and Levan segments, Wasatch fault zone, Utah. M, 1988, University of Colorado. 149 p.

Jackson, Michael James. A paleomagnetic investigation of some Ordovician carbonates from cratonic North America. M, 1984, University of Michigan.

Jackson, Michael James. Secondary magnetizations in North American Ordovician rocks; observations and inferences. D, 1986, University of Michigan. 170 p.

Jackson, Michael Keith. Stratigraphic relationships of the Tillamook Volcanics and the Cowlitz Formation in the upper Nehalem River-Wolf Creek area, northwestern Oregon. M, 1983, Portland State University. 118 p.

Jackson, Michael Ralph. Stratigraphy and geochemistry of the Rusty Lake greenstone belt adjacent to the Ruttan Mine, Manitoba. M, 1979, University of Manitoba. 125 p.

Jackson, Neil A. A subsurface study of the Lower Pennsylvanian rocks of east central Oklahoma. M, 1948, University of Oklahoma. 38 p.

Jackson, P. A. The structure, stratigraphy and strain history of the Seine Group and related rocks near Mine Centre, northwestern Ontario. M, 1982, Lakehead University.

Jackson, Patrick A. A laboratory and field study of well screen performance and design. M, 1983, Ohio University, Athens. 189 p.

Jackson, Patrick Allan. Structural evolution of the Carpinteria Basin, western Transverse Ranges, California. M, 1981, Oregon State University. 107 p.

Jackson, Paul. The geology of the southern portion of Dutch Valley, Anderson County, Tennessee. M, 1956, University of Tennessee, Knoxville. 40 p.

Jackson, Philip Larkin. Digital simulation of seismic rays. D, 1970, University of Michigan. 92 p.

Jackson, Philip Richard. Geology and lithogeochemistry of the Flagstaff Mountain barite deposit and surrounding area, Stevens County, Washington. M, 1986, University of Nevada. 120 6 plates p.

Jackson, R. E. Hydrogeochemical processes affecting the migration of radionuclides in a shallow ground water flow system at the Chalk River Nuclear Laboratories. D, 1979, University of Waterloo.

Jackson, R. G., II. A depositional model of point bars in the lower Wabash River. D, 1975, University of Illinois, Urbana. 270 p.

Jackson, Richard A. Autochthon and allochthon of the Kent Quadrangle, western Connecticut. D, 1980, University of Massachusetts. 222 p.

Jackson, Richard A. Structural geology and stratigraphy of the Huntington area, Massachusetts. M, 1975, University of Massachusetts. 92 p.

Jackson, Richard Lance. The stratigraphy of the Gulfian Series (Upper Cretaceous), east-central Texas. M, 1983, Baylor University. 103 p.

Jackson, Robert Eugene, Jr. Sliding friction in foliated rocks. D, 1973, University of North Carolina, Chapel Hill. 77 p.

Jackson, Robert George. The application of lake sediment geochemistry to mineral exploration in the southern Slave Province of the Canadian Shield. M, 1975, Queen's University. 306 p.

Jackson, Robert L. The stratigraphy of the Supai Formation along the Mogollon Rim, central Arizona. M, 1951, University of Arizona.

Jackson, Robert Paul. Dolomitization and structural relations of the Deep River, North Adams and Pinconning oil fields, Michigan. M, 1958, Michigan State University. 57 p.

Jackson, Robert Reed. A petrographic study of the Middle Devonian limestones of central Ohio and the Bellefontaine Outlier. M, 1952, Ohio State University.

Jackson, Robert Tracy. Phylogeny of the Pelecypoda. D, 1889, Harvard University.

Jackson, Ronald Laverne. A mineralogical and geochemical study of the ferruginous bauxite deposits in Columbia County, Oregon and Wahkiakum County, Washington. M, 1974, Portland State University. 87 p.

Jackson, Stephen T. Late-glacial and Holocene vegetational changes in the Adirondack Mountains (New York); a macrofossil study. D, 1983, Indiana University, Bloomington. 182 p.

Jackson, Steven Leonard. Metamorphism and structure of the Laurie Lake area, northern Manitoba. D, 1988, Queen's University. 204 p.

Jackson, Stewart Albert. A study of Mississippi Valley type lead-zinc mineralization with special reference to sediment diagenesis. M, 1966, University of Toronto.

Jackson, Stewart Albert. The carbonate complex and lead-zinc ore bodies (Devonian), Pine Point, Northwest Territories, Canada. D, 1971, University of Alberta. 144 p.

Jackson, T. C. A physical inventory of Bosque County, Texas. M, 1975, Baylor University. 4 p.

Jackson, T. J. The value of Landsat data in urban water resources planning. D, 1976, University of Maryland. 251 p.

Jackson, Thomas Franklin. The description and stratigraphic relationships of fossil plants from the Lower Pennsylvanian rocks of Indiana. D, 1916, Indiana University, Bloomington. 35 p.

Jackson, Thomas Franklin. The fossil plants of the Mississippian and Pennsylvanian rocks of the Bloomington Quadrangle. M, 1914, Indiana University, Bloomington.

Jackson, Thomas Joseph. The biogeochemistry of thermal ecosystems. D, 1972, Indiana University, Bloomington. 203 p.

Jackson, Timothy J. Coastal processes and sedimentology of the Southeast Texas coast, Sea Rim State Park. M, 1979, Colorado State University. 230 p.

Jackson, Togwell A. The Gowganda Formation (Precambrian) of Canada and the theory of a Huronian ice age. M, 1963, University of Wisconsin-Madison.

Jackson, Togwell Alexander. The role of pioneer lichens in the chemical weathering of recent volcanic rocks on the island of Hawaii. D, 1969, University of Missouri, Columbia. 232 p.

Jackson, William. A pre-Pennsylvanian subcrop study as a guide to oil exploration along the Nemaha Ridge in Kansas. M, 1979, Wichita State University. 116 p.

Jackson, William Daniel. The stratigraphy and sedimentology of Muleshoe Mound, Sacramento Mountains, New Mexico. M, 1982, Texas Tech University. 87 p.

Jackson, William Ernest. Geology of the Brush Creek area and a petrographic study of the Morrison Formation, Gunnison County, Colorado. M, 1957, University of Kentucky. 58 p.

Jackson, William Henry. Areal geology of the Rocky Brook area, [York County] New Brunswick. M, 1954, University of Toronto.

Jackson, William Longstreth. Bed material routing and streambed composition in alluvial channels. D, 1981, Oregon State University. 163 p.

Jacob, Arthur Frank. Delta facies of the Green River Formation (Eocene), Carbon and Duchesne counties, Utah. D, 1969, University of Colorado. 288 p.

Jacob, Leonard, Jr. Geology of the Calera mining district, Chihuahua, Mexico. M, 1953, Columbia University, Teachers College.

Jacob, Thomas M. Reconnaissance geology of House Mountain, Elmore County, Idaho. M, 1985, University of Idaho. 134 p.

Jacobberger, Patricia Ann. Remote sensing in arid regions; three case studies (southwestern Kansas; Meatiq Dome, Eastern Desert, Egypt; and Kharga Depression, Western Desert, Egypt). D, 1982, Washington University. 122 p.

Jacobeen, Frank. Structure of the central Broad Top Synclinorium and contiguous areas. D, 1977, University of South Carolina.

Jacobeen, Frank H., Jr. Structure of the Broadtop Synclinorium and the contiguous Wills Mountain Anticlinorium and the Allegheny Frontal Zone. M, 1974, University of South Carolina. 55 p.

Jacobi, Louise Delano. Stratigraphy and structure of the Taconic Allochthon, northern Washington County, New York. M, 1978, SUNY at Albany. 191 p.

Jacobi, R. D. Geology of part of the terrane north of Lukes Arm Fault, North-central Newfoundland (Part I); Modern submarine sediment slides and their geological implications (Part II). D, 1980, Columbia University, Teachers College. 434 p.

Jacobs, Alan Korach. A mineralogic and textural study of ores from the Lick Mountain manganese district, Wythe County, Virginia. M, 1973, University of Tennessee, Knoxville. 112 p.

Jacobs, Alan Martin. Pleistocene proto-cirque hollows in the Cataract Creek Valley, Tobacco Root Mountains, Montana. D, 1967, Indiana University, Bloomington. 95 p.

Jacobs, Allan Samuel. The pre-stack migration of profiles. D, 1983, Stanford University. 113 p.

Jacobs, Bonnie Fine. Past vegetation and climate of the Mogollon Rim area, Arizona. D, 1983, University of Arizona. 180 p.

Jacobs, Brent B. Trench investigation of Quaternary movement along the east-bounding fault of the Shearer Graben of the Kentucky River fault system in southeastern Clark County, Kentucky. M, 1986, Eastern Kentucky University. 38 p.

Jacobs, Charles M. A subsurface study of the Tar Springs Sandstone in Hopkins County, Kentucky. M, 1961, University of Kentucky. 46 p.

Jacobs, Cyril. Some revisions in the interpretation of the geology of Becraft Mountain, Columbia County, New York. M, 1948, University of Wisconsin-Madison. 41 p.

Jacobs, David C. Geology and wallrock alteration of the Sunbird mercury mine, Santa Barbara County, California. M, 1970, University of California, Santa Barbara.

Jacobs, David Cal. Geochemistry of biotite in the Santa Rita and Hanover-Fierro stocks, Central mining district, Grant County, New Mexico. D, 1976, University of Utah. 230 p.

Jacobs, Elbridge C. The talc deposits of Vermont. M, 1914, Columbia University, Teachers College.

Jacobs, Elliott B. Shape parameters and distribution of macroborings; St. Croix, U.S. Virgin Islands. M, 1982, University of Georgia.

Jacobs, Gary Kermit. Experimental and thermodynamic analysis of metamorphic devolatilization equilibria in H_2O-CO_2-CH_4-NaCl fluids at elevated pressures and temperatures. D, 1981, Pennsylvania State University, University Park. 168 p.

Jacobs, Gary Wayne. Characterization and origin of the Garrard Siltstone (Upper Ordovician), central and east Kentucky. D, 1986, University of Kentucky. 178 p.

Jacobs, Gary Wayne. The subsurface structure, lithology, and stratigraphy of the Middle and Upper Ordovician rocks in the Cumberland Saddle area of south-central Kentucky. M, 1983, University of Kentucky. 143 p.

Jacobs, James Alan. Depositional and Quaternary history of the Red River in Northeast Texas. M, 1981, University of Texas, Austin. 107 p.

Jacobs, Karen Stine. Pond ecosystems and shoreline dynamics on Horn Island, Mississippi. M, 1980, University of Virginia. 34 p.

Jacobs, L. L., III. Small mammal fossils from Neogene Siwalik deposits, Pakistan. D, 1977, University of Arizona. 269 p.

Jacobs, Louis Leo, III. Small mammals of the Quiburis Formation, southeastern Arizona. M, 1973, University of Arizona.

Jacobs, Lucinda Ann. Metal geochemistry in anoxic marine basins. D, 1984, University of Washington. 217 p.

Jacobs, Marian Beckmann. Alteration studies and uranium emplacement near Moab, Utah. D, 1963, Columbia University, Teachers College. 244 p.

Jacobs, Michael A. Tertiary rhyolite intrusives of the northeastern Eagle Mountains, Hudspeth County, Texas. M, 1981, University of Texas at El Paso.

Jacobs, Richard C. Geology of the central front of the Fra Cristobal Mountains, Sierra County, New Mexico. M, 1956, University of New Mexico. 47 p.

Jacobs, Robert Sanger. Geochemistry and petrology of basalts from Kahoolawe Island, Hawaii. M, 1986, North Carolina State University. 101 p.

Jacobs, Stephen Emanual. Paleoecology of the Stull Shale (Upper Pennsylvanian) in southeastern Nebraska and southwestern Iowa. M, 1973, University of Nebraska, Lincoln.

Jacobsen, Alfred Thurl. Geology of the north fork and upper Duchesne River region (Duchesne County, Utah). M, 1941, University of Utah. 61 p.

Jacobsen, Edmund E. Net shore drift of Whatcom County, Washington. M, 1980, Western Washington University. 76 p.

Jacobsen, Eloise Tittle. Reconnaissance study of subsurface geology of northwestern Oklahoma. M, 1948, University of Oklahoma. 108 p.

Jacobsen, Lynn. Petrology of the Pennsylvania sandstones and conglomerates of the Ardmore Basin. D, 1953, Pennsylvania State University, University Park. 196 p.

Jacobsen, Lynn. Structural relations on the east flank of the Anadarko Basin, Cleveland and McClain counties, Oklahoma. M, 1948, University of Oklahoma. 34 p.

Jacobsen, S. I. The geochemistry of peat bogs over different bedrock types, Houghton County, Michigan. M, 1978, Michigan Technological University. 104 p.

Jacobsen, Stein Bjornar. Study of crust and mantle differentiation processes from variations in Nd, Sr, and Pb isotopes. D, 1980, California Institute of Technology. 300 p.

Jacobsen, W. L. Engineering properties of the Gila Conglomerate at Bagdad, Arizona. M, 1976, University of Arizona.

Jacobson, Carl Ernest. Deformation and metamorphism of the Pelona Schist beneath the Vincent Thrust, San Gabriel Mountains, California. D, 1980, University of California, Los Angeles. 276 p.

Jacobson, Carolyn. Major and rare-earth geochemistry of Mount Rougemont, Quebec. M, 1981, Boston University. 228 p.

Jacobson, David T. Paleomagnetism of Paleocene intrusives from central Montana. M, 1980, Western Washington University. 212 p.

Jacobson, Elizabeth Ann. A statistical parameter estimation method using singular value decomposition with application to Avra Valley aquifer in southern Arizona. D, 1985, University of Arizona. 321 p.

Jacobson, Gary Louis. Geology and geochemistry of the La Prosperidad banded ferromanganese deposit and other mineral deposits in the metavolcanic Fe-Cu province of Baja California, Mexico. M, 1982, San Diego State University. 171 p.

Jacobson, George Lloyd, Jr. A palynological study of the history and ecology of white pine in Minnesota. D, 1975, University of Minnesota, Minneapolis. 172 p.

Jacobson, Herbert S. Geology, geochemistry, and economic mineral possibilities of southern Cumberland and Colchester counties, Nova Scotia. M, 1955, Massachusetts Institute of Technology. 43 p.

Jacobson, Jimmy J. Deep electromagnetic sounding technique. D, 1969, Colorado School of Mines. 140 p.

Jacobson, John B. Ores of the San Francisco del Oro Mine, Chihuahua, Mexico. M, 1953, Columbia University, Teachers College.

Jacobson, John Martin. Vertical distribution of foraminifera in the lower chalk member of the Austin Formation, southern Dallas County, Texas. M, 1961, Southern Methodist University. 41 p.

Jacobson, Jule Marion. A description of some sandstone concretions from the Strawn Group, Lampasas County, Texas. M, 1941, University of Texas, Austin.

Jacobson, Mark Ivan. Petrologic variations in Franciscan sandstones from the Diablo Range, California. M, 1976, University of California, Berkeley. 186 p.

Jacobson, Marvin LeRoy. The northeast extension of the Knox Gneiss (Precambrian, Maine). M, 1963, University of Maine. 70 p.

Jacobson, Peter R. The surficial geoloy of the southwest half of the Mt. Holyoke Quadrangle, Massachusetts. M, 1981, University of Massachusetts. 170 p.

Jacobson, Randall Scott. Linear inversion of body wave data. D, 1980, University of California, San Diego. 104 p.

Jacobson, Robert B. Spatial and temporal distributions of slope processes in the upper Buffalo Creek drainage basin, Marion County, West Virginia (Volumes I and II). D, 1986, The Johns Hopkins University. 117 p.

Jacobson, Roger Leif. Controls on the quality of some carbonate ground waters; dissociation constants of calcite and $CaHCO_3$ from 0 to 50°C. D, 1973, Pennsylvania State University, University Park. 131 p.

Jacobson, Roger Leif. Movement of nitrate in subsurface water under feedlots in Missouri. M, 1969, University of Missouri, Columbia.

Jacobson, Rollyn Philip. Statistical analysis in the interpretation of electrical resistivity data. M, 1954, Washington University. 19 p.

Jacobson, Rudolph Harry. A computer model study of unsaturated flow in a Leach Dump, Socorro, New Mexico. D, 1972, New Mexico Institute of Mining and Technology. 72 p.

Jacobson, Russel James. Stratigraphic correlations of the Seelyville, Dekoven and Davis coal members-Spoon Formation (Illinois), Staunton Formation (Indiana), or beds-Carbondale Formation (western Kentucky). M, 1985, University of Illinois, Urbana. 110 p.

Jacobson, Russell Carl. Stratigraphy and fauna of two cores penetrating Middle Devonian formations of Missaukee County, Michigan. M, 1950, [University of Michigan].

Jacobson, Sara Sue. Petrology of the basic inclusions of the andesites and dacite domes of Crater Lake, Oregon. M, 1978, University of California, Los Angeles.

Jacobson, Stephen Richard. Acritarchs from Middle and Upper Ordovician rocks in New York State and the Cincinnati region in Ohio and Kentucky. D, 1978, Ohio State University. 314 p.

Jacobson, Susan Kay. The educational role of a developing country's national park system; an evaluation of Kinabalu Park, Malaysia. D, 1987, Duke University. 107 p.

Jacobvitz, Michael Alvin. A comparison of recharge estimates using a numerical flow model, Santa Margarita Aquifer, Scotts Valley, California. M, 1987, University of California, Santa Cruz.

Jacoby, Gordon Campbell, Jr. Fluvial flow as a function of drainage-basin morphology in the southeastern United States. D, 1971, Columbia University. 146 p.

Jacoby, Russel S. Petrogenesis and tectonic evolution of a perthitic gneiss complex in the central Grenville Province; the Baskatong reservoir area, Quebec (Canada). D, 1968, Queen's University. 193 p.

Jacome Villanueva, Enrique Osmar. Fate of fertilizer nitrogen as affected by soil pH, soil water content, and plant residue in a Florida Ultisol and three soils from the Sula Valley, Honduras. D, 1982, University of Florida. 188 p.

Jacques, Richard Dewey. Geology of the northwest corner of Yellowstone National Park, Montana. M, 1950, University of Michigan. 42 p.

Jacques, Theodore Emil. Microfossil distribution in the Hickory Creek Shale (Pennsylvanian), Wilson and Montgomery counties, Kansas. M, 1964, University of Kansas. 48 p.

Jacques-Ayala, César. Sierra el Chanate, northwestern Sonora, Mexico; stratigraphy, sedimentology and structure. M, 1983, University of Cincinnati. 143 p.

Jadkowski, Mark Andrew. Multispectral remote sensing of landslide susceptible areas. D, 1987, Utah State University. 223 p.

Jado, Abdul-Rasof. The stratigraphy and petrology of the Park City Formation in northwestern Colorado and northeastern Utah. D, 1976, Colorado School of Mines. 123 p.

Jaeger, David J. Paleomagnetic and geochemical correlation in the High Cascade, Indian Heaven volcanic field, south-central Washington. M, 1986, Michigan Technological University. 63 p.

Jaeger, J. W.; Hohenstein, C. G. and Jones, D. L. A study of marked sand movement on Del Monte beach, Monterey bay, California. M, 1965, United States Naval Academy.

Jaeger, Kenneth B. Structural geology and stratigraphy of the Elko Hills, Elko County, Nevada. M, 1987, University of Wyoming. 70 p.

Jaeger, Paul. The effects of depositional environment and diagenesis on the porosity and permeability in the Salem Limestone in Union and Monroe counties, southern Illinois. M, 1982, Southern Illinois University, Carbondale. 174 p.

Jaeger, Ralph Roger. Shock effects in and the origin of iron meteorites. D, 1967, Purdue University. 206 p.

Jaffe Duchmann, Peter Rudolf. The fate of toxic substances in rivers. D, 1981, Vanderbilt University. 89 p.

Jaffe, Daniel G. Petroleum potential and depositional environments of the Norphlet, Louann, Werner (Jurassic), and Eagle Mills (Triassic) formations along the north rim of the East Texas Basin. M, 1985, Baylor University. 168 p.

Jaffe, Gilbert. A field and laboratory investigaton of the oil-bearing Rhinestreet Shale, New York. M, 1950, SUNY at Buffalo.

Jaffe, Samuel. Wind velocity and eddy diffusivity of momentum profiles as applied to the diffusion equation. D, 1965, New York University.

Jaffer, Rebecca K. A study of the diagenesis of the overburden between the Badger and School coal seams, Dave Johnston coal field, Converse County, Wyoming. M, 1983, Portland State University.

Jafroudi, Siamak. Experimental verification of bounding surface plasticity theory for cohesive soils. D, 1983, University of California, Davis. 241 p.

Jaganathan, James. Speciation of arsenic in a Green River oil shale and in its retort oil and waters. D, 1986, Texas A&M University. 148 p.

Jagannadham, Gollakota. A comparative study of potash feldspar structural states from selected granites and augen gneisses. D, 1972, University of North Carolina, Chapel Hill. 49 p.

Jagannadham, Gollakota. A study of the feldspars of the Blowing Rock Gneiss, North Carolina (western). M, 1968, University of North Carolina, Chapel Hill. 116 p.

Jagannath, R. Hydrogeology and aquifer tests of Rodrigues Island. M, 1983, University of Waterloo.

Jager, Douglas John. Infiltration and sediment production on the Big Sage type near Eastgate, Nevada. D, 1972, University of Nevada. 182 p.

Jager, Eric Howard. The pre-Cretaceous topography of the western Edwards Plateau. M, 1941, University of Texas, Austin.

Jaggar, Thomas Augustus, Jr. I, A microsclerometer for determining the hardness of mineral thin sections; II, On the geologic evidence from fragmental inclusions contained in certain dikes of the Boston Basin (Massachusetts). D, 1897, Harvard University.

Jagiello, Keith James. Structural evolution of the Phoenix Basin, Arizona. M, 1987, Arizona State University. 157 p.

Jagnow, David Henry. Geologic factors influencing the speleogenesis in the Capitan reef complex, New Mexico and Texas. M, 1977, University of New Mexico. 201 p.

Jago, B. C. Mineralogy and petrology of the Ham Kimberlite, Somerset Island, Northwest Territories, Canada. M, 1982, Lakehead University.

Jagoda, John L. Hydrology and morphology of Hunter Creek. M, 1973, SUNY, College at Fredonia. 82 p.

Jahanbagloo, Iraj Cyrus. Part 1, The crystal structure of a hexagonal Al-serpentine; Part 2, Calculation X-ray powder patterns of minerals. D, 1967, University of Minnesota, Minneapolis. 217 p.

Jahedi, Jamshid. A fabric related constitutive model for granular materials. D, 1987, Illinois Institute of Technology. 246 p.

Jahn, Bor-ming. K-Ar mica ages and the margin of a regional metamorphism, Gravelly Range, Montana. M, 1967, Brown University. 37 p.

Jahn, Bor-ming. Strontium isotope and trace element studies of the lower Precambrian rocks from the Vermilion District, northeastern Minnesota. D, 1972, University of Minnesota, Minneapolis. 143 p.

Jahn, Jeanne E. An X-ray fluorescence study of biotites. M, 1951, Columbia University, Teachers College.

Jahn, Melvin Edward. A survey of endocranial anatomy of nine Rancho La Brea carnivores (families; Canidae, Felidae, and Ursidae). M, 1963, University of California, Berkeley. 96 p.

Jahnke, Richard Alan. Current phosphorite formation and the solubility of synthetic carbonate fluorapatite. D, 1981, University of Washington. 210 p.

Jahns, Richard H. Pre-Cambrian rocks of South Park, Colorado and Tertiary intrusions of the Chalmers area. M, 1937, Northwestern University.

Jahns, Richard H. Stratigraphy of the easternmost Ventura Basin, California, with a description of a new lower Miocene mammalian fauna from the Tick Canyon Formation. D, 1943, California Institute of Technology. 49 p.

Jahns, Richard Henry. Tactite rocks of the Iron Mountain District, Sierra and Socorro counties, New Mexico. D, 1943, California Institute of Technology. 75 p.

Jahrling, Chris E. Geophysical, structural and stratigraphic relations across the Monroe Line and related contacts in the Indian Stream area, northern New Hampshire. M, 1983, University of New Hampshire. 107 p.

Jain, Birendra Kumar. A low frequency electromagnetic prospecting system. D, 1978, University of California, Berkeley. 128 p.

Jain, Subhash Chandra. Evolution of sand wave spectra. D, 1971, University of Iowa. 102 p.

Jakes, Mary Clare. Surface and subsurface geology of the Camarillo and Las Posas Hills area, Ventura County, California. M, 1980, Oregon State University. 105 p.

Jakob, P. G. Relationship between water quality and regional ground-water flow in North Dakota. M, 1974, Ohio State University.

Jakobsson, Sigurdur. Melting experiments on basalts in equilibrium with a graphite-iron-wustite buffered C-O-H fluid. D, 1984, Arizona State University. 197 p.

Jakubiak, Annette Leisner. Mineralogy, ore microscopy, and petrography of a drill core from the Anorthosite I Subzone in the Frog Pond area of the Stillwater Complex, Montana. M, 1988, Montana College of Mineral Science & Technology. 162 p.

Jakway, G. E. The Pleistocene faunal assemblages of the middle Loup River terrace-fills of Nebraska. D, 1963, University of Nebraska, Lincoln.

Jalali-Yazdi, Younes. Pressure transient behavior of heterogeneous naturally fractured reservoirs. D, 1987, University of Southern California.

Jalichandra, Nithipatana. The volcanic rocks in the Wamsutta Formation of the Narragansett Basin (Massachusetts). M, 1946, Massachusetts Institute of Technology. 75 p.

Jam, L. Pedro. Subsurface temperatures in south Louisiana (Miocene-Oligocene). M, 1968, University of Tulsa. 121 p.

Jambor, John Leslie. Sulfosalts from Madoc, Ontario. D, 1966, Carleton University. 183 p.

Jambor, John Leslie. Vanadium bearing interlava sediment from the Campbell River area. M, 1960, University of British Columbia.

Jameossanaie, Abolfazl. Palynology and environments of deposition of the lower Menefee Formation (lower Campanian), South Hospah area, McKinley County, New Mexico. D, 1983, Michigan State University. 361 p.

James, A. H. Structure and stratigraphy of the southern Sierra de Pintas, Baja California, Mexico. M, 1973, [University of California, San Diego].

James, Alan Thomas. Clay mineral distribution and electron microscopy of the Excello shale (Pennsylvanian, Des Moines series) of the Illinois Basin and mid-continent. M, 1969, University of Illinois, Urbana.

James, Alan Thomas. Oxygen isotope exchange between illite and water at 22°C. D, 1974, Rice University. 57 p.

James, Allan H. Distribution of titanium, vanadium, chromium, cobalt and nickel in the magnetites of the Mount Hope Mine and the New Jersey Highlands. D, 1954, Massachusetts Institute of Technology. 96 p.

James, Arthur Darryl. The occurrence of water in the Precambrian crystalline rocks of the New Jersey Highlands. M, 1967, Rutgers, The State University, New Brunswick. 73 p.

James, Barry. Late Pliocene (Blancan) nonmarine and volcanic stratigraphy and microvertebrates of Lake Tecopa, California. M, 1985, University of California, Riverside. 89 p.

James, Bela Louis. A study of fusulinids of the Mid-continent area. M, 1937, University of Oklahoma. 96 p.

James, Bela Michael. Systematics and biology of the deep-water Palaeotaxodonta (Mollusca; Bivalvia) from the Gulf of Mexico. D, 1972, Texas A&M University.

James, Bruce Howard. Paleoenvironments of the Lower Triassic Thaynes Formation; near Diamond Fork in Spanish Fork Canyon, Utah County, Utah. M, 1979, Brigham Young University.

James, Bryan A. Finite-difference/polozhii decomposition resistivity modeling. M, 1986, Colorado School of Mines. 271 p.

James, Carolyn. A boundary between the Hudson Valley and New England provinces drawn from geomorphic differences. M, 1943, Columbia University, Teachers College.

James, Clarence Hubert Cavendish. Optimum preload rates for compressible normally consolidated soils. D, 1968, Northwestern University. 218 p.

James, Clifford M. A biofacies and lithofacies study of the Ramparts Formation (Middle and Upper Devonian), Mountain-Gayna River region, Northwest Territories. M, 1973, Carleton University. 224 p.

James, D. H. Geology of Adams Plateau. M, 1950, Queen's University. 67 p.

James, David Evan. Crustal structure of the Middle Atlantic States. D, 1966, Stanford University. 115 p.

James, David Paul. The sedimentology of the McMurray Formation, East Athabasca. M, 1977, University of Calgary. 198 p.

James, David Richard. The Ann Group copper deposit, Meridian Lake, Northwest Territories. M, 1972, University of Manitoba.

James, Dennis R. Geochemistry, petrology and paleontology of Indiana Coal V. M, 1977, Purdue University. 164 p.

James, Donald T. Origin and metamorphism of the Kisseynew Gneisses, Kisseynew Lake - Cacholotte Lake area, Manitoba. M, 1983, Carleton University. 188 p.

James, Ellen Louise. A new Miocene marine invertebrate fauna from Coos Bay, Oregon. M, 1950, University of Oregon. 75 p.

James, Eric William. Geochronology, isotopic characteristics, and paleogeography of parts of the Salinian Block of California. D, 1986, University of California, Santa Barbara. 193 p.

James, Eric William. Geology and petrology of the Lake Ann Stock and associated rocks. M, 1980, Western Washington University. 57 p.

James, Gerard W. Stratigraphic geochemistry of a Pennsylvanian black shale (Excello) in the mid-continent and Illinois Basin. D, 1970, Rice University. 102 p.

James, Gideon T. The paleontology and nonmarine stratigraphy of the Cuyame Valley badlands, California. D, 1961, University of California, Berkeley. 416 p.

James, Harold Edward, Jr. Sedimentology of the iron-oxide-bearing upper Miocene(?) Guayabo Group in the vicinity of Cucuta, Colombia. D, 1977, Princeton University. 214 p.

James, Harold Lloyd. Chromite deposits near Red Lodge, Montana. D, 1945, Princeton University. 139 p.

James, Harry Rudolph. Use of water table regression models, electrode potentials, and soil properties to explain soil forming processes in and between an artificially drained and undrained Clarion toposequence. D, 1981, Iowa State University of Science and Technology. 398 p.

James, Henry F. The geography of a portion of the Great Appalachian Valley and selected adjacent regions. M, 1920, University of Wisconsin-Madison.

James, Howard T. A study of the relations between the specific gravity and the metamorphism of coal. M, 1922, University of Wisconsin-Madison.

James, Howard Turnbull. Geology and ore deposits of the Britannia map-area, British Columbia. D, 1927, Harvard University.

James, Jack Alexander. Geology of the Berryman area, Washington County, Missouri (T 796). M, 1948, University of Missouri, Rolla.

James, Jack Alexander. The relationship of regional structural geology to the ore deposits in the southeastern Missouri mining district. D, 1951, University of Missouri, Rolla.

James, Jay R. Productive limits of the Speechley Sand in Pennsylvania. M, 1938, University of Pittsburgh.

James, Johnny. Paleoenvironments of the Upper Cambrian Lone Rock Formation of West-central and southwestern Wisconsin. M, 1977, University of Wisconsin-Madison.

James, Keith H. Numerical taxonomy of otoliths from five living species of fish. M, 1969, University of Houston.

James, L. Allan. Historical transport and storage of hydraulic mining sediment in the Bear River, California. D, 1987, University of Wisconsin-Madison. 304 p.

James, Laurence Pierson. Occurrence of silver and antimony in galena from Bingham, Utah. M, 1967, Stanford University.

James, Laurence Pierson. Zoned hydrothermal alteration and ore deposits in sedimentary rocks near mineralized intrusions, Ely area, Nevada. D, 1972, Pennsylvania State University, University Park. 280 p.

James, Matthew Joseph. Taxonomic revision of the gastropod family Turridae from the Dominican Republic Neogene. D, 1987, University of California, Berkeley. 479 p.

James, Michael N. G. A two-dimensional refinement of the crystal structure of sillimanite. M, 1963, University of Manitoba.

James, Noel P. Late Pleistocene reef limestones, northern Barbados, West Indies. D, 1972, McGill University. 155 p.

James, Noel P. Sediment distribution dispersal patterns on Sable Island and Sable Island Bank, (Nova Scotia). M, 1966, Dalhousie University.

James, O. L. Economic possibilities of Ramsey Quadrangle, Culberson and Reeves counties, Texas (J235). M, 1952, University of Texas, Austin.

James, Odette Francine Bricmont. Origin and emplacement of the ultramafic rocks of the Emigrant Gap area, California. D, 1967, Stanford University. 205 p.

James, Odette Francine Bricmont. Origin of lunar craters. M, 1964, Stanford University.

James, Odette Francine Bricmont. Petrographic study of Miocene basalts on either side of the San Andreas fault zone south of Palo Alto, California. M, 1964, Stanford University.

James, Richard Stephen. The properties of sodalite and its petrogenesis at the Princess Quarry, Bancroft, Ontario. M, 1965, McMaster University. 142 p.

James, W. Petrology of the Wishart Quartzite (Saskatchewan, Canada). M, 1953, University of Toronto.

James, Wesley P. Classification of Wisconsin ground moraine by airphoto interpretation. M, 1961, Purdue University.

James, William. Geology of Dungannon and Mayo townships in southeastern Ontario. D, 1957, McGill University.

James, William Calvin. Origin of nonmarine-marine transitional strata at the top of the Kootenai Formation (Lower Cretaceous), southwestern Montana. D, 1977, Indiana University, Bloomington. 433 p.

James, William Calvin. Petrology and regional relationships of the Ordovician Kinnikinic Formation and equivalents, central and southern Idaho. M, 1973, Utah State University. 268 p.

James, William Fleming. Geology of Duparquet maparea, Quebec. D, 1923, Princeton University.

James, William Fleming. The alteration of a quartz diabase dike at the Old Helen Mine (Michipicoten District, Ontario). M, 1921, McGill University.

James, William R. A study of grain-size distribution of sands on outer Cape Cod (Barnstable County, Massachusetts). M, 1967, Northwestern University.

James, William R. Development and application of nonlinear regression models in geology. D, 1968, Northwestern University. 131 p.

James, William Robert. Model choice; an operational comparison of stochastic streamflow models for droughts. D, 1981, Utah State University. 297 p.

Jameson, James Boyd. The stratigraphy of the Fredericksburg Division (Lower Cretaceous) of central Texas. M, 1959, Baylor University. 80 p.

Jameson, Maynard H. The effect of dipping strata on potential-drop ratio determination. M, 1936, Colorado School of Mines. 49 p.

Jamieson, Gordon Reginald. A preliminary study of the regional groundwater flow in the Meager Mountain geothermal area, British Columbia. M, 1981, University of British Columbia. 163 p.

Jamieson, Heather Edith. The distribution of magnesium and iron between olivine and spinel at 1300°C. D, 1982, Queen's University. 178 p.

Jamieson, Iain M. Heat flow in a geothermally active area; The Geysers, California. D, 1976, University of California, Riverside. 176 p.

Jamieson, John C. The free energy surfaces of calcite-aragonite as a function of pressure and temperature. D, 1952, University of Chicago. 22 p.

Jamieson, Rebecca Anne. The St. Anthony Complex, northwestern Newfoundland; a petrological study of the relationship between a peridotite sheet and its dynamothermal aureole. D, 1979, Memorial University of Newfoundland. 195 p.

Jamieson, William H., Jr. Depositional and diagenetic history of the Grayburg Formation, McElroy Field, Crane and Upton counties, Texas. M, 1985, Baylor University. 177 p.

Jamison, William Richard. Laramide deformation of the Wingate Sandstone, Colorado National Monument; a study of cataclastic flow. D, 1979, Texas A&M University. 181 p.

Jamison, William Richard. Numerical dynamic analysis of the McConnell thrust plate and associated structures. M, 1974, University of Calgary. 98 p.

Jamkhindikar, Suresh M. Petrography and clay mineralogy of the Lawrence shale formation (Pennsylvanian) in Kansas. D, 1969, University of Kansas. 151 p.

Jammallo, Joseph M. Delineation of bedrock fracture trace zones by remote sensing and magnetics and their hydrogeologic implications. M, 1983, University of Vermont. 274 p.

Jamnongpipatkul, Pichit. Remote sensing studies of some ironstone gravels and plinthite in Thailand. D, 1980, Cornell University. 336 p.

Jampton, Kathleen M. Delineation of abandonment procedures for the Bunker Hill and Crescent mines, Shoshone County, Idaho. M, 1985, University of Idaho. 64 p.

Jamsheed, Behsheed. The Fulton County tanning industry; a waste management proposal. D, 1984, Rensselaer Polytechnic Institute. 192 p.

Jan, M. Qasim. Geology of the McKenzie River Valley between the south Santiam highway and the McKenzie Pass highway, Oregon. M, 1967, University of Oregon. 70 p.

Jan, Michel Leopold Van Sint *see* Van Sint Jan, Michel Leopold

Jan, Ming-Ju. A gravity investigation of Brown's Gulch area of Butte North Quadrangle, Butte, Montana. M, 1974, Montana College of Mineral Science & Technology. 24 p.

Janak, Peter M. A comparison and analysis of seismic land source energy relationships and radiation patterns. M, 1982, Colorado School of Mines. 349 p.

Janakiramaiah, Bollapragada. Analysis of a reversed deep seismic refraction profile in the Atlantic Ocean, northeast of Puerto Rico. M, 1961, Texas A&M University. 46 p.

Janbaz, John Elisha, Jr. Petrology of the Upper Cretaceous strata of Sucia Island, San Juan County, Washington. M, 1972, Washington State University. 104 p.

Janbek, Tayseer T. The recharge-discharge aspects of Green valley, Pima County, Arizona. M, 1970, University of Arizona.

Janda, Richard John. Pleistocene history and hydrology of the upper San Joaquin River, California. D, 1966, University of California, Berkeley. 430 p.

Janders, David J. Comparative sedimentology, stratigraphy and economic potential of two Tertiary lacustrine deposits in Arizona. M, 1978, Arizona State University. 140 p.

Janecek, Thomas Raphael. The history of eolian sedimentation and atmospheric circulation during the late Cenozoic. D, 1983, University of Michigan. 186 p.

Janecky, David Richard. Serpentinization of peridotite within the oceanic crust; experimental and theoretical investigations of seawater-peridotite interaction at 200°C and 300°C, 500 bars. D, 1982, University of Minnesota, Minneapolis. 263 p.

Janes, Donald A. Geochemistry of altered volcanic and intrusive rocks of Bathurst District, New Brunswick. M, 1976, Universite de Montreal.

Janes, Donald A. Géochimie des roches volcaniques archéennes (Canada). M, 1971, Universite de Montreal.

Janes, Joseph Robert. Symbolic treatment of field and analytical data as applied to geological mapping. M, 1967, University of Toronto.

Janes, Stephen. Geology and paleontology of the Santa Susana Shale, Simi Valley, California. M, 1976, University of California, Santa Barbara.

Janes, Stephen Douglas. The evolution of a Permian shallow marine benthic community. D, 1982, University of California, Santa Cruz.

Janezic, Gary. Thermal and radiation-induced oxidation-reduction in pyroxene minerals. M, 1975, University of Pittsburgh.

Janik, Michael Garland. Cathodoluminescence of gangue dolomites near lead-zinc occurrences in Stevens County, northeastern Washington. M, 1982, University of Idaho. 88 p.

Janish, Jeanne R. Glaciation in the Kings River Canyon of the Sierra Nevada Mountains (California). M, 1925, Stanford University. 57 p.

Janish, Jeanne Russell. Glaciation in the King's River Canyon of the Sierra Nevada Mountains (California). M, 1926, Stanford University. 57 p.

Jank, Mary Ellen. Pressure solution and the development of cleavage in the Baraboo Quartzite. M, 1982, Michigan State University. 88 p.

Janke, Norman Charles. Effect of shape upon the settling velocity of regular geometric particles. D, 1963, University of California, Los Angeles.

Jannik, Nancy Olga. Recurved spit development and related beach processes on Horseshoe Spit (bayside), Sandy Hook, New Jersey. M, 1980, Rutgers, The State University, New Brunswick. 97 p.

Janoo, Vincent Clement. Drained and undrained behavior of sand under high pressures. D, 1986, University of Colorado. 614 p.

Janosky, R. A. Geology of the Hill City region, Pennington County, South Dakota. M, 1949, South Dakota School of Mines & Technology.

Janowiak, Matthew James. Variability of sedimentation rates, pollen influx and pollen percentages in varved lake sediments; Emrick Lake, Marquette County, Wisconsin. M, 1987, University of Wisconsin-Madison. 76 p.

Janowsky, Ronald E. Morphology and taxonomy of Athyris spiriferoides (Eaton) from the Middle Devonian of western New York State. M, 1965, SUNY at Buffalo.

Jansen, George James. The petrology and structure of the Gulf Mills-West Conshohocken area (Pennsylvania). M, 1952, Bryn Mawr College. 17 p.

Jansen, Gerhard Cyril Julius. Cambrian stratigraphy, Goodrich Ranch area, Burnet County, Texas. M, 1957, University of Texas, Austin.

Jansen, John Richard. A geophysical evaluation of a proposed landfill site in southeastern Wisconsin. M, 1983, University of Wisconsin-Milwaukee. 145 p.

Jansen, Lawrence T. Ground water conditions near the Elsinore fault zone in the Ocotillo-Coyote Wells Basin, Imperial County, California. M, 1983, San Diego State University. 143 p.

Jansen, Stephen T. Facies and depositional history of the Brigham Group, northern Bannock and Pocatello ranges, southeastern Idaho. M, 1986, Idaho State University. 142 p.

Jansen, Walrave. Quantitative compositional mapping with the electron microprobe. D, 1979, Colorado School of Mines. 273 p.

Jansma, Pamela Elizabeth. The tectonic interaction between an oceanic allochthon and its foreland basin during continental overthrusting; the Antler Orogeny, Nevada. D, 1988, Northwestern University. 331 p.

Jansons, Uldis. Petrography, petrology, and trace element relations of the Cooke City (Montana) Porphyry. M, 1963, University of Montana. 60 p.

Jansons, Uldis. Trace element abundances and variations in sulfide minerals from the Park City District, Utah. D, 1971, University of Utah. 215 p.

Janssen, Janelle L. Origin of vermicular and platy kaolinite crystals in Georgia kaolins as explained by $^{18}O/^{16}O$ isotopic ratios. M, 1985, Indiana University, Bloomington. 63 p.

Janssen, Raymond E. A key for the determination of Middle Pennsylvanian plant fossils of Illinois. M, 1937, University of Chicago. 54 p.

Janssen, Raymond E. A revision of fossil plant types of Illinois, augmented by description of new species. D, 1939, University of Chicago. 175 p.

Janssen, Robert J. Petrology of the lower Sawatch Formation (White River Plateau, Colorado). M, 1974, Colorado State University. 94 p.

Janssens, Adriaan. Analysis of the Paleozoic movements of the Cincinnati Arch. D, 1967, Ohio State University. 157 p.

Janssens, Adriaan. The subsurface geology of Champaign County, Ohio. M, 1964, Ohio State University.

Janssens, Arie. A contribution to the Pleistocene geology of Champaign County, Ohio. M, 1964, Ohio State University.

Janszen, Milton Hugo. Geology and vegetation of northern Hurd Draw Quadrangle, Culberson County, Texas. M, 1953, University of Texas, Austin.

Jantzen, R. E. Distribution of Recent coccolith-carbonate from the North Atlantic; carbonate minimas extrapolated to 225,000 yrs BP and to the Pliocene/Pleistocene boundary. M, 1974, Queens College (CUNY). 278 p.

Janzen, John H. Geology of the Heidegger Hill and Monumental Mountain area, northeastern Washington. M, 1981, Eastern Washington University. 35 p.

Janzon, Hans A. A joint analysis of oil shale with implications on mine design, Uinta Basin, Utah. M, 1980, Colorado School of Mines. 269 p.

Jaouni, Abdur-Rahim Khalil. The rhyolite tuff in the Berkeley Hills. M, 1975, University of California, Berkeley. 24 p.

Japakasetr, Thawat. Geology and uranium occurrences of Section 8 Mesa, McKinlay County, New Mexico. M, 1969, Wesleyan University.

Japy, Kate Elisabeth. The carbon cycling in an anoxic lake basin, Johnson Pond, north central Florida, using carbon isotopes as a tracer. M, 1988, University of Florida. 176 p.

Jaramillo Mejia, Jose Maria. Petrology and geochemistry of the Nevado del Ruiz Volcano, Northern Andes, Colombia. D, 1980, [University of Houston]. 182 p.

Jaramillo Torres, Wilson Fabian. Aggradation and degradation of alluvial-channel beds. D, 1983, University of Iowa. 191 p.

Jaramillo, Jose Maria. Volcanic rocks of the Rio Canca Valley, Colombia. M, 1976, Rice University. 38 p.

Jaramillo, L. Alteration and mineralization in the Jarilla Mountains, Otero County, New Mexico. M, 1973, New Mexico Institute of Mining and Technology.

Jard, Mustapha R. El *see* El Jard, Mustapha R.

Jardine, Donald Edwin. An investigation of brecciation associated with the Sullivan Mine orebody at Kimberley, British Columbia (Canada). M, 1967, University of Manitoba.

Jardine, W. G. A critical study of the statistical method in paleontology. M, 1950, McGill University.

Jardine, William George. Seismic study in the Ouachita System of southern Arkansas and the adjacent Gulf Coastal Plain. M, 1988, Purdue University. 112 p.

Jarjur, Salah Z. Spectrum analysis of seismic records in the study of the nature of noise. D, 1974, University of Kansas. 135 p.

Jarman, Clara Birchak. Clay mineralogy and sedimentary petrology of the Cretaceous Hudspeth Formation, Mitchell, Oregon. D, 1973, Oregon State University. 173 p.

Jarman, Gary D. Recent foraminifera and associated sediments of the continental shelf in the vicinity of Newport, Oregon. M, 1962, Oregon State University. 111 p.

Jarmell, Solomon. Gravity investigation in the Renovo area of central Pennsylvania. M, 1956, Pennsylvania State University, University Park. 60 p.

Jaroska, Robert Stanley. Population analysis of the Del Rio Formation in central and Southwest Texas. M, 1955, University of Texas, Austin.

Jarrard, Richard D. Pacific plate motions. D, 1975, University of California, San Diego. 324 p.

Jarre, G. A. Geology of the Sortehjorne area, Mesters Vig, Scoresby Land, East Greenland. M, 1962, University of Wyoming. 65 p.

Jarre, Guntram A. and Melrose, Thomas Graham. Reconnaissance geology of the Old Baldy Thrust, Alamosa, Costilla, and Huerfano counties, Colorado. M, 1959, University of Michigan.

Jarrell, Mary Kathryn. Conodonts from the Golconda Group (Chester) of the Illinois Basin. M, 1961, University of Houston.

Jarrett, J. T. A study of the hydrology and hydraulics of Pamlico Sound and their relation to the concentration of substances in the sound. M, 1966, North Carolina State University.

Jarrett, Marcus Lee. Heat flow of the Mississippi Embayment. M, 1982, University of Florida. 143 p.

Jarrett, Robert David. Flood hydrology of foothill and mountain streams in Colorado. D, 1987, Colorado State University. 246 p.

Jarrin, Keith Manuel. The adsorption of water vapor onto quartz surfaces at elevated temperatures and pressures. M, 1987, Pennsylvania State University, University Park. 61 p.

Jarroud, O. A. Ground water evaluation and cooling before utilization for Wadi Zam-Zam, Libya. M, 1977, University of Arizona.

Jarvi, Thomas Robert. Deposition and diagenesis of San Andres cores 16-69 and 20-69. M, 1982, Texas Tech University. 89 p.

Jarvis, Clarence Sylvester. Soils and erosional forms as affecting floods. D, 1927, American University. 63 p.

Jarvis, Daniel. A study of the possible time equivalence of the Tarrant Beds in northeastern Tarrant County. M, 1948, Texas Christian University. 19 p.

Jarvis, Daniel. Biostratigraphy of the type Wolfcamp. D, 1962, Stanford University. 69 p.

Jarvis, Gary Trevor. Thermal studies related to surging glaciers. M, 1973, University of British Columbia.

Jarvis, Harry Aydelotte, Jr. Geology of the Rio Pao-Rio Tiznados area, Cojedes and Guarico, Venezuela. D, 1964, Rice University. 93 p.

Jarvis, Linda Jane. Lower and Middle Devonian stratigraphy and depositional environments of the Sheep, Desert, Pintwater, and Spotted ranges, northern Clark County, Nevada. M, 1981, Oregon State University. 83 p.

Jarvis, W. Todd. Regional hydrogeology of the Paleozoic aquifer system, southeastern Bighorn Basin, Wyoming, with an impact analysis on Hot Springs State Park. M, 1986, University of Wyoming. 227 p.

Jarzabek, Dave. Microearthquake survey of the Dunes known geothermal resource area, Imperial Valley, southern California. M, 1977, University of Texas at Dallas. 514 p.

Jarzabek, Dianne Palmer. A geochemical reconnaissance of thermal waters along portions of the San Jacinto and San Andreas fault zones, southern California. M, 1980, San Diego State University.

Jarzen, David M. Evolutionary and paleoecological significance of Albian to Campanian angiosperm pollen from the Amoco B-1 Youngstown borehole, southern Alberta. D, 1974, University of Toronto.

Jasaitis, Richard A., Jr. Geology of pre-Mesozoic bedrock of the Amherst area, west-central Massachusetts. M, 1983, University of Massachusetts. 98 p.

Jasieniecki, Michael Simon. The flood hazard for Crab Orchard creek near Carbondale, Illinois. M, 1970, Southern Illinois University, Carbondale. 89 p.

Jasim, Rafid A. H. Origin and factors affecting distribution of solid oil in midwestern Missouri. M, 1966, University of Missouri, Rolla.

Jaske, Robert J. A study of the reservoir fluid and rock characteristics of the North Steamboat Butte Field, Wyoming, for the purpose of planning a future operating program. M, 1952, University of Tulsa. 60 p.

Jasmer, Rodney M. Hydrocompaction hazards due to collapsible loess in southeastern Idaho. M, 1987, Idaho State University. 129 p.

Jasper, Alan K. Stratigraphy and depositional environments of a late Pleistocene barrier island complex, southeastern Virginia. M, 1982, Old Dominion University. 142 p.

Jasper, John Paul. An organic geochemical approach to problems of glacial-interglacial climatic variability. D, 1988, Massachusetts Institute of Technology. 263 p.

Jaster, Marion Charlotte. The dwarf fauna of the oolitic Indiana Limestone. M, 1942, George Washington University. 53 p.

Jaumé, Steven C. The mechanics of the Salt Range-Potwar Plateau, Pakistan; qualitative and quantitative aspects of a fold-and-thrust belt underlain by evaporites. M, 1987, Oregon State University. 58 p.

Jaupart, Claude. On the mechanisms of heat loss beneath continents and oceans. D, 1981, Massachusetts Institute of Technology. 215 p.

Javaherian, Abdolrahim. Elimination of spurious reflections from finite-difference synthetic seismograms with applications to the San Andreas fault zone. D, 1982, University of Texas at Dallas. 445 p.

Javete, Donald Francis. A simple statistical approach to differential settlements on clay. D, 1983, University of California, Berkeley. 246 p.

Jaworowski, Cheryl C. Geomorphic mapping and trend analysis of Quaternary deposits with implications for late Quaternary faulting, central Wyoming. M, 1985, University of Wyoming. 109 p.

Jaworski, Bill L. The development and evaluation of a new marine seismic energy source; the HP water gun. M, 1976, University of Wisconsin-Milwaukee.

Jaworski, Gary William. An experimental study of hydraulic fracturing. D, 1979, University of California, Berkeley. 286 p.

Jaworski, John J. Use of coherence in seismic velocity determination. M, 1971, Colorado School of Mines. 47 p.

Jaworski, Michael John. Copper mineralization of the upper Moya Sandstone, Chupadero Mines area, Socorro County, New Mexico. M, 1973, New Mexico Institute of Mining and Technology. 102 p.

Jay, David Alan. Residual circulation in shallow, stratified estuaries. D, 1987, University of Washington. 173 p.

Jay, Jeremy Barth. The geology and stratigraphy of the Tertiary volcanic and volcaniclastic rocks, with special emphasis on the Deschutes Formation, from Lake Simtustus to Madras in central Oregon. M, 1983, Oregon State University. 119 p.

Jayakumar, Paramsothy. Modeling and identification in structural dynamics. D, 1987, California Institute of Technology. 204 p.

Jayaprakash, Gubbi P. Theoretical and experimental studies of the supergene alteration of lead minerals. M, 1970, University of California, Santa Barbara.

Jayaraman, K. N. Lower Cretaceous foraminiferal biostratigraphy and paleoecology of the Dalmiapuram and lower Uttatur formations, Tiruchirapalli District, Tamil Nadu, India. D, 1978, University of Washington. 272 p.

Jayaraman, Ramurthy. Comparative sedimentology across the base of the pterocephaliid biomere; Texas, Utah, and Nevada. M, 1979, SUNY at Stony Brook.

Jayasena, H. A. Hemachandra. Hydrogeology of fracture basement complexes; a case study from the Kurunegala District of Sri Lanka. M, 1988, Colorado State University. 134 p.

Jayasinghe, Nimal Ranjith. Granitoids of the Wesleyville area in northeastern Newfoundland; a study of their evolution and geological setting. D, 1979, Memorial University of Newfoundland. 350 p.

Jayatilaka, Chandrika Jayakanthi DeSilva. Development and testing of a deterministic-empirical model of the effect of the capillary fringe on near-stream area runoff. D, 1986, University of Waterloo. 156 p.

Jaycox, Robert Eugene. Sedimentary structures and flow regimes at Buckeye Quarry, south Amherst, Ohio. M, 1973, Bowling Green State University. 75 p.

Jayko, Angela Susan. Deformation and metamorphism of the Eastern Franciscan Belt, Northern California. D, 1984, University of California, Santa Cruz. 287 p.

Jayne, Douglas I. Depositional environment of the Precambrian Revett-St. Regis transition zone of northwestern Montana. M, 1978, Eastern Washington University. 61 p.

Jaynes, Dan Brian. Atmosphere and temperature within a reclaimed coal-stripmine and a numerical simulation of acid mine drainage from stripmined lands. D, 1983, Pennsylvania State University, University Park. 225 p.

Jaynes, William Frederick. Characterization and separation of soil clay minerals using ion exchange, lithium charge reduction, and density gradient techniques. D, 1988, Ohio State University. 244 p.

Jayyusi, G. S. Conodont paleoecology of Little Osage Formation (Desmoinesian) of Missouri. M, 1976, University of Missouri, Columbia.

Jean, Jiin-Shuh. Environmental geology of Northampton Township, Summit County, Ohio. M, 1982, University of Akron. 94 p.

Jean, Jiin-Shuh. Modeling ground-water flow by means of a hybrid trajectory image, boundary integral equation method. D, 1987, Purdue University. 229 p.

Jean, Joseph St. *see* St. Jean, Joseph

Jeanloz, Raymond. Physics of mantle and core minerals. D, 1980, California Institute of Technology. 412 p.

Jeanne, Richard A. Source area of the Brookline Member of the Roxbury Conglomerate as determined by textural analysis. M, 1976, Boston University. 63 p.

Jeary, Gene L. Areal geology of western Major County, Oklahoma. M, 1961, University of Oklahoma. 130 p.

Jecha, T. E. A gravimetric study of three plutons in Clay and Randolph counties, Alabama. M, 1978, Memphis State University.

Jeddeloh, George. The effect of glass composition on the coordination and chemical bonding of aluminum and the redox behavior of arsenic, cerium, europium and iron. D, 1984, University of Oregon. 140 p.

Jee, Jonathan Lucas. Stratigraphy and paleoenvironmental analysis of the Plattin limestones (Middle Ordovician), White River region, Independence, Izard, and Stone counties, Arkansas. M, 1981, University of New Orleans. 168 p.

Jefferies, Brenda K. Stratigraphy and depositional patterns of the Union Valley, Wapanucka and lower Atoka formations. M, 1982, University of Arkansas, Fayetteville.

Jefferies, Paula Therese. Eocene benthic foraminifera from the Sidney Flat Shale and Kellogg Shale, Contra Costa County, California. M, 1985, University of California, Davis. 104 p.

Jefferis, Lee H. An evaluation of radar imagery for structural analysis in gently deformed strata; a study in northeastern Kansas. M, 1970, University of Kansas. 132 p.

Jefferis, Robert Gilpin. Fracture analysis near the Mid-Atlantic Ridge boundary, Reykjavik-Hvalfjordur area, Iceland. M, 1980, Pennsylvania State University, University Park. 115 p.

Jeffers, John Douglas. Tectonic and sedimentary evolution of the Bransfield Basin, Antarctica. M, 1988, Rice University. 142 p.

Jeffers, Joseph William. The regional stratigraphy of the Upper Devonian Woodbend Group of Western Canada. M, 1955, Southern Methodist University. 54 p.

Jeffers, William Larry. The clay mineralogy of the Claiborne Formation in West Tennessee. M, 1982, Memphis State University.

Jefferson, Charles Wilson. Stromatolites, sedimentology and stratigraphy of parts of the Amundsen Basin, N.W.T. M, 1977, University of Western Ontario. 260 p.

Jefferson, Charles Wilson. The upper Proterozoic Redstone copper belt, Mackenzie Mountains, N.W.T. D, 1983, University of Western Ontario. 445 p.

Jefferson, Clinton Frank. An investigation of reactions involved in the preparation of ferrites. D, 1959, University of Michigan.

Jefferson, George Thomas. The Camp Cady local fauna from Pleistocene Lake Manix, Majave Desert, California. M, 1968, University of California, Riverside. 130 p.

Jeffery, D. A. Hydrogeological significance of the application of sewage effluent to agricultural land on the Frank Gray Farm, Lubbock County, Texas. M, 1978, West Texas State University. 45 p.

Jeffery, W. G. The geology of Campbell Chibougamau Mine, Quebec. D, 1959, McGill University.

Jeffords, Russell M. Lophophyllid corals of the Lower Pennsylvanian. M, 1941, University of Kansas. 114 p.

Jeffords, Russell M. Pennsylvanian lophophyllid corals. D, 1946, University of Kansas. 228 p.

Jeffrey, Alan William Adams. Thermal and clay catalysed cracking in the formation of natural gas. D, 1981, Texas A&M University. 135 p.

Jeffrey, Robert Graham, Jr. Rockbolt analysis for reinforcement and design in layered rock. D, 1985, University of Arizona. 309 p.

Jeffrey, Robert Graham, Jr. Shaft or borehole plug-rock mechanical interaction. M, 1981, University of Arizona. 145 p.

Jeffrey, W. G. The geology of the Campbell Chibougamau Mine, Ontario. D, 1959, McGill University.

Jeffreys, Stanley R. The foraminifera of the formation represented at Packard's Hill, Santa Barbara, Santa Barbara County, California. M, 1940, University of Southern California.

Jeffries, Charles D. 1, The mineralogical composition of some Pennsylvania soils; 2, Optical studies of feldspars. D, 1936, University of Wisconsin-Madison.

Jeffries, Dean Stuart. Ontario precipitation chemistry and heavy metal speciation. D, 1976, McMaster University. 125 p.

Jeffries, Edwin Lee. Areal geology of western Choctaw County, Oklahoma. M, 1965, University of Oklahoma. 51 p.

Jeffries, Norman W. The stratigraphy and structure of the Fordland Quadrangle, Missouri. M, 1955, University of Missouri, Rolla.

Jeffries, Norman William. Stratigraphy of the lower Marmaton rocks from Missouri. D, 1958, University of Missouri, Rolla. 355 p.

Jeffs, Donald N. A spectrographic study of elements in the ore and host rocks at Pine Point, Northwest Territories. M, 1955, Queen's University. 137 p.

Jefopoulos, Timothy. Evaluation of the relationship between wind velocity and erosion of the swash zone. M, 1987, Montclair State College. 44 p.

Jehn, James Lawrence. The delineation of a preglacial buried valley underlying a proposed reservoir complex in the vicinity of Harveysburg, Ohio. M, 1973, Wright State University. 55 p.

Jehn, Paul J. An environmental reconstruction of the Pitkin Formation of northern Arkansas. M, 1973, Northeast Louisiana University.

Jehn, Theresa C. The effectiveness of specific conductance as a groundwater monitoring tool by comparison of monitoring well designs in detecting contaminant plumes at Moraine, Ohio. M, 1985, Wright State University. 82 p.

Jekeli, Christopher. The downward continuation to the Earth's surface of truncated spherical and ellipsoidal harmonic series of the gravity and height anomalies. D, 1981, Ohio State University. 149 p.

Jelacic, Allan Joseph. Physical limnology of Green and Round lakes (Quaternary), Fayetteville (Onondaga county), New York. D, 1970, University of Rochester. 296 p.

Jellinger, M. Methods of detection and analysis of slope instability, Southeast Oahu, Hawaii. D, 1977, University of Hawaii. 279 p.

Jemmett, Joe Paul. Geology of the northern Plomosa Mountain Range, Yuma County, Arizona. D, 1966, University of Arizona. 169 p.

Jemmett, Joseph Paul. Geology of some of the phosphate deposits in the Centennial Mountains of Idaho and Montana. M, 1955, University of Idaho. 67 p.

Jemsek, John P. Heat flow and tectonics of the Ligurian Sea basin and margins. D, 1988, Woods Hole Oceanographic Institution. 488 p.

Jen, L.-S. Spatial distribution of crystals and phase equilibria in charnockitic granulites from the Adirondack Mountains, New York. D, 1975, University of Ottawa. 248 p.

Jen, Lo-Sun. Petrochemical study of amphibolitic gneisses (Precambriam, Grenville) from the Westport and Tichborne areas, southeast Ontario (Canada). M, 1967, Queen's University. 154 p.

Jenden, Peter Donald. Maturation of organic matter in the Paleocene-Eocene Wilcox Group, South Texas; relationship to clay diagenesis and sandstone cementation. D, 1983, University of California, Los Angeles. 294 p.

Jendrzejewski, John P. Archaemonadaceae of the Oamaru Diatomite (late Eocene) of New Zealand. M, 1972, University of Rhode Island.

Jendrzejewski, John P. Diatoms and other siliceous biogenic remains from surficial bottom sediments of the Gulf of Mexico. D, 1976, Louisiana State University. 406 p.

Jeness, John Lewis. Oceanography and physiography of the Canadian western Arctic. D, 1951, Clark University.

Jeng, Shian-Woei. Two dimensional random walk model for pollutant transport in natural rivers. D, 1986, University of Texas, Austin. 358 p.

Jeng, W.-L. Studies of the isotope chemistry of molecular oxygen in biological systems. D, 1976, University of Texas, Austin. 80 p.

Jeng, Yih. Analysis of an airgun source in marine seismic exploration. D, 1986, University of Connecticut. 101 p.

Jengo, John William. Paleoecology of molluscan assemblages in the Wenonah and Mt. Laurel formations (Upper Cretaceous) of New Jersey. M, 1983, University of Delaware. 173 p.

Jenik, Albert James. Facies and geometry of the Swan Hills Member (Devonian) in the Goose River field, Alberta. M, 1965, University of Alberta. 81 p.

Jenke, Arthur Louis. An investigation of the basic rocks in the Fredericktown Quadrangle, Missouri. M, 1948, Washington University. 114 p.

Jenke, Dennis R. Chemical characteristics of a typical mine waste water reclamation process involving the interaction of acid mine water and concentrator waste. M, 1978, Montana College of Mineral Science & Technology. 67 p.

Jenkins, Carl E. Geology of the Bates Creek-Corral Creek area, Natrona County, Wyoming. M, 1950, University of Wyoming. 80 p.

Jenkins, Cecilia L. Progressive hydrothermal alteration associated with gold mineralization of the zone 1 intrusion of the Callahan Property, Val d'Or region, Quebec. M, 1988, Ecole Polytechnique. 132 p.

Jenkins, Christine Maria. Depositional environments of the Middle Ordovician Greensport Formation and

Colvin Mountain Sandstone in Calhoun, Etowah, and St. Clair counties, Alabama. M, 1984, University of Alabama. 168 p.

Jenkins, Creties D., Jr. A preliminary tectonic analysis of the northeastern margin of the Bighorn Uplift; Buffalo to Dayton, Wyoming. M, 1986, South Dakota School of Mines & Technology.

Jenkins, David A. Geologic evaluation of the EPM mining claims, East Potrillo Mountains, Dona Ana Co., N.M. M, 1977, New Mexico Institute of Mining and Technology. 109 p.

Jenkins, David E. Geology of the Auburn 7 1/2′ Quadrangle, Caribou County, Idaho, Lincoln County, Wyoming. M, 1981, Brigham Young University. 15 p.

Jenkins, David Mark. Experimental phase relations pertinent to hydrated peridotites in the system H2O-CaO-MgO-Al2O3-SiO2. D, 1980, University of Chicago. 154 p.

Jenkins, David Maurice. A study of some Holocene foraminifera from the Cedar Keys area of Florida. M, 1966, University of Florida. 69 p.

Jenkins, Dennis Bruce. Petrology and structure of the Slate Creek ultramafic body, Yuba County, California. M, 1980, University of California, Davis. 164 p.

Jenkins, Dwight. Paleomagnetics of the eastern Ouachita Mountains, Arkansas, and their tectonic implications. M, 1983, University of Florida. 158 p.

Jenkins, Elmer Leroy. Reconnaissance mapping problem of the Page Mill Road-Fremont Hills area, Santa Clara County. M, 1957, Stanford University.

Jenkins, Evan Cramer. Hypabyssal rocks associated with the Christmas Mountains Gabbro, Brewster County, Texas. M, 1959, University of Texas, Austin.

Jenkins, Farish Alston, Jr. The postcranial skeleton of African cynodonts and problems in the evolution of mammalian postcranial anatomy. D, 1969, Yale University. 448 p.

Jenkins, Gale F. The structural geology of the Bear Creek area, Fremont County, Wyoming. M, 1977, Miami University (Ohio). 42 p.

Jenkins, Janice A. An isotopic study of Nile cone sediments and late Pleistocene sapropel formation in the eastern Mediterranean Sea. M, 1982, University of South Carolina. 84 p.

Jenkins, Jean. Seismic reprocessing and interpretation of profile WM-4, southwestern Whipple Mountains, San Bernardino County, California. M, 1988, University of Southern California.

Jenkins, Joe Earl. Albitization of Ca-plagioclase; an experimental study. M, 1980, University of North Carolina, Chapel Hill. 169 p.

Jenkins, John E. Geology and geochemistry of the Jones Camp magnetite deposits, Socorro County, NM. M, 1985, New Mexico Institute of Mining and Technology. 155 p.

Jenkins, John Stacy. Effects of the Western Boundary Undercurrent on sediment transport and deposition in the Pamlico Canyon off Cape Hatteras. M, 1980, Texas A&M University.

Jenkins, John T., Jr. The Pleistocene geology and paleoecology of the Fremont County Quarry, Fremont County, Iowa. M, 1972, University of Iowa. 55 p.

Jenkins, John Trevor. Anorthosite-ilmenite-pegmatite relations on the west bank of La Chaloupe River, Saguenay County, Quebec. M, 1956, McGill University.

Jenkins, Olaf P. A study of the Kreyenhagen Shale of California; stratigraphy, character, mode of formation, structure, extent, probable age, and economic aspects. D, 1930, Stanford University. 318 p.

Jenkins, Olaf P. The stratigraphy and geologic history of Tennessee to accompany the geologic map of Tennessee. M, 1915, [Stanford University].

Jenkins, Page T. The geology of a portion of the east side of the Laramie Range, Albany, Laramie and Platte counties, Wyoming. M, 1938, University of Wyoming. 40 p.

Jenkins, R. V. The origin of the rock reefs of the Everglades. M, 1984, Wichita State University. 137 p.

Jenkins, Robert Allen. Epigenetic sulfide mineralization in the Paleozoic rocks of eastern southern Wisconsin. M, 1968, University of Wisconsin-Madison.

Jenkins, Robert David. The geology of the Hillham Quadrangle, Martin, Orange, and Dubois counties, Indiana. M, 1956, Indiana University, Bloomington. 52 p.

Jenkins, Robert E., II. Geology, geochemistry and origin of mineral deposits in the Hill Gulch area, Jamestown, Colorado. D, 1979, Colorado School of Mines. 220 p.

Jenkins, Sidney Ford. A structural study of a portion of the Kernville Series. M, 1961, University of California, Berkeley. 84 p.

Jenkins, Steven Drexel. Anelastic transmission of seismic waves in the Austin Chalk, Pointe Coupee Parish, Louisiana. M, 1982, Indiana University, Bloomington. 172 p.

Jenkins, William Adrian. Geology of the Mercury Quadrangle, McCulloch County, Texas. D, 1952, University of Texas, Austin.

Jenkinson, Lewis Frank. Geology of the Drake Quadrangle, North Dakota. M, 1950, University of Iowa. 88 p.

Jenks, Margaret Dana. A morphometric analysis and classification of the Neogene basalt volcanoes in the south-central Snake River plain, Idaho. M, 1984, University of Idaho. 131 p.

Jenks, Maurice. Bedrock geology and garnet analysis in a portion of the Woronoco Quadrangle, Massachusetts. M, 1967, University of Vermont.

Jenks, Susan Elizabeth. Deposition and diagenesis of the Mississippian Lodgepole Formation, central Montana. M, 1972, Rice University. 50 p.

Jenks, William F. The geology of the alkaline stock at Pleasant Mountain, Maine. M, 1933, University of Wisconsin-Madison.

Jenks, William Furness. Geology of portions of the Libby and Trout Creek quadrangles, Montana and Idaho. D, 1936, Harvard University. 224 p.

Jenne, David Allen. Structural geology and metamorphic petrology of the Gold Mountain area, Snohomish County, Washington. M, 1978, Oregon State University. 177 p.

Jennemann, Vincent Francis. A meteorological investigation of microseisms at Corpus Christi, Texas. M, 1949, St. Louis University.

Jennemann, Vincent Francis. The enhancement of seismic velocity determination using interactive computer graphics. D, 1972, University of Tulsa. 120 p.

Jenner, G. A. Geochemistry of the upper Snooks Arm Group, Newfoundland. M, 1977, University of Western Ontario. 133 p.

Jenner, Gordon A., Jr. Tertiary alkalic igneous activity, potassic fenitization, carbonatitic magmatism, and hydrothermal activity in the central and southeastern Bear Lodge Mountains, Crook County, Wyoming. M, 1984, University of North Dakota. 232 p.

Jenner, John Slaten. Areal expansion of the City of St. Louis. M, 1939, Washington University. 139 p.

Jenness, M. I. Revegetation of a cold-desert grassland in northern Arizona. D, 1977, [Northern Arizona University]. 231 p.

Jenness, Stuart Edward. Geology of the Gander River ultrabasic belt, Newfoundland. D, 1955, Yale University. 212 p.

Jenness, Stuart Edward. Hydrothermal alterations at the Bordulac gold mine, Quebec. M, 1950, University of Minnesota, Minneapolis. 65 p.

Jennette, David C. Storm-dominated cyclic sedimentation of a intracratonic ramp; Kope-Fairview transition (Upper Ordovician), Cincinnati, Ohio region. M, 1986, University of Cincinnati. 203 p.

Jenney, Charles Phillip. Geology of the central Humboldt Range, Nevada. D, 1935, Columbia University, Teachers College.

Jenney, Charles Phillip. The Florida east coast. M, 1933, Columbia University, Teachers College.

Jenney, William Willis, Jr. The structure of a portion of the southern California Batholith, western Riverside County, California. D, 1968, University of Arizona. 167 p.

Jennings, A. A. Improved pathway descriptions for finite element models of subsurface solute transport. D, 1980, University of Massachusetts. 317 p.

Jennings, Albert Ray. Evaluation of selected radioisotopes as ground-water tracers. D, 1964, Texas A&M University. 57 p.

Jennings, Albert Ray. Geology of the Pontotoc northwest area, San Saba and Mason counties, Texas. M, 1960, Texas A&M University.

Jennings, Anne Elizabeth. Late Quaternary marine sediments from a transect of fiord and continental shelf environments; a study of piston cores from Clark Fiord and Scott Trough, Baffin Island, Canada. M, 1986, University of Colorado. 245 p.

Jennings, Arnold Harvey. The glacial geomorphology of the Sunwapta Pass area, Jasper National Park, Alberta, Canada. M, 1951, University of Iowa. 99 p.

Jennings, C. David. Iron-55 in Pacific Ocean organisms. D, 1968, Oregon State University. 81 p.

Jennings, Charles David. Radioactivity of sediments in the Columbia River estuary. M, 1966, Oregon State University. 62 p.

Jennings, Charles Williams. Geology of the southern part of the Quail Quadrangle, Los Angeles County, California. M, 1952, University of California, Los Angeles. 52 p.

Jennings, David S. Origin and metamorphism of part of the Hermon Group (late Precambrian) near Bancroft, Ontario. D, 1970, McMaster University. 225 p.

Jennings, E. A. A survey of the mainland and island belts, Thunder Bay, Silver District, Ontario, fluid inclusions, mineralogy and sulfur isotopes. M, 1987, Lakehead University.

Jennings, Edward Wallace. Radioactivity well logging, Edmonton District, Alberta. M, 1951, University of Alberta. 64 p.

Jennings, Joan K. Interaction of uranium with naturally occurring organic substances. M, 1976, Colorado School of Mines. 72 p.

Jennings, Kenneth. Reconstruction of the marine benthic paleo-communities of the marine Saugus Formation (Pleistocene) west of Ventura, California. M, 1980, University of California, Santa Barbara.

Jennings, Kenneth Van Baker. Technical and regulatory considerations involved in non-CERCLA response to uncontrolled hazardous waste sites, a case study; the Willco Landfill (Lynwood, California). D, 1985, University of California, Los Angeles. 142 p.

Jennings, Mark D. Geophysical investigations near subsidence fissures in northern Pinal and southern Maricopa counties, Arizona. M, 1977, Arizona State University. 102 p.

Jennings, Olin R. Economics of Nevada iron ores: a market study. M, 1967, University of Nevada. 116 p.

Jennings, Philip H. A microfauna from the Monmouth and basal Rancocas groups of New Jersey. D, 1936, Columbia University, Teachers College.

Jennings, Phillip H. Some ostracodes from the Morrison Formation, Fall River County, South Dakota. M, 1934, Columbia University, Teachers College.

Jennings, Robert Allen. Geology of the southeastern part of the Oat Mountain Quadrangle and adjacent parts of the San Fernando Quadrangle, Los Angeles County, California. M, 1957, University of California, Los Angeles.

Jennings, Robert H. Trace fossils and environments of deposition, Humboldt Basin, California. M, 1983, Texas A&M University. 207 p.

Jennings, Scott. Burial metamorphism in Plio-Pleistocene sediments of the Colorado River delta. M, 1983, University of Montana. 40 p.

Jennings, Ted Vernon. Faunal zonation of the Minnelusa Formation, Black Hills, South Dakota. M, 1958, University of Iowa. 65 p.

Jennings, Ted Vernon. Structural analysis of the northern Bighorn Mountains, Wyoming. D, 1967, University of Iowa. 224 p.

Jennison, Margo J. Zircon ages and feldspar mineralogy of the Navajo Sandstone. M, 1980, University of Utah. 20 p.

Jenq, Tzay-Rong T. Modeling of optimal posphorus pollution controls for use in regional water quality management with a case application to the Carnegie Lake watershed, New Jersey. D, 1982, Rutgers, The State University, New Brunswick. 273 p.

Jens, John Christian. Petrology of a mafic-layered intrusion near Lolo Pass, Montana-Idaho border area. M, 1972, University of Montana. 85 p.

Jensen, Arthur R. Computer simulation of surface water hydrology and salinity with an application to studies of Colorado River management. D, 1976, California Institute of Technology. 327 p.

Jensen, Christian. Studies of the early basic breccia of northeastern Yellowstone (Wyoming). M, 1931, University of California, Berkeley. 50 p.

Jensen, Clair Lynn. Matrix diffusion and its effect on the modelling of tracers returns from the fractured geothermal reservoir at Wairakei, New Zealand. M, 1983, Stanford University.

Jensen, David Edward. Some aspects of sedimentation of the Oriskany Formation of west-central New York. M, 1932, University of Rochester. 52 p.

Jensen, Fred S. The geology of the Nashua Quadrangle, Montana. D, 1951, The Johns Hopkins University. 86 p.

Jensen, J. R. The orthophoto and orthophotomap; characteristics, development and aspects of cartographic communication. D, 1976, University of California, Los Angeles. 176 p.

Jensen, Joseph. Some factors involved in the classification of coal lands. M, 1913, Columbia University, Teachers College.

Jensen, Joseph Matthew. The chronology and possible mechanics of failure of the upper Gros Ventre Landslide, Gros Ventre Mountains, Teton County, Wyoming. M, 1981, Iowa State University of Science and Technology. 64 p.

Jensen, Karen Grace. Fossil pollen and spores of the Jurassic-Cretaceous Great Valley Sequence, northwestern California. D, 1987, Loma Linda University. 333 p.

Jensen, Larry Sigfred. A petrogenic model for the Archaean Abitibi Belt in the Kirkland Lake area, Ontario. D, 1981, University of Saskatchewan. 520 p.

Jensen, Larry Sigfred. Geology and geochemistry of Melba and Bisley Townships, District of Timiskaming, Ontario. M, 1971, University of Saskatchewan. 80 p.

Jensen, Louis. Geology of the Sus Hills, Pima County, Arizona. M, 1973, University of Arizona.

Jensen, Lynn E. Effect of stress gradients on fracture of rock. M, 1972, University of Utah. 43 p.

Jensen, Mark E. Tertiary geologic history of the Slate Jack Canyon Quadrangle, Juab and Utah counties, Utah. M, 1986, Brigham Young University. 19 p.

Jensen, Martin. Major, minor, and trace elements (REE's) in apatite as recorders of pegmatite petrogenesis. M, 1984, South Dakota School of Mines & Technology.

Jensen, Mead Leroy. Diffusion in minerals. D, 1951, Massachusetts Institute of Technology. 113 p.

Jensen, Nolan R. Geology of the Fairview 7 1/2' Quadrangle, Sanpete County, Utah. M, 1985, Brigham Young University. 121 p.

Jensen, O. G. The construction of a feed-back seismograph station and an analysis of the long shot data from the Canadian seismograph stations. M, 1966, University of British Columbia.

Jensen, Oliver George. Linear systems theory applied to a horizontally layered crust. D, 1971, University of British Columbia.

Jensen, P. A. Development of a procedure to use relative spill dilution capacity to manage water pollution risk from the transportation of oil and hazardous materials. D, 1975, Texas A&M University. 274 p.

Jensen, Ronald Grant. Ordovician brachiopods from the Pogonip Group of Millard County, western Utah. M, 1967, Brigham Young University.

Jensen, Roy E. Palynostratigraphy of the Eocene Little River section, Grays Harbor County, Washington. M, 1983, Loma Linda University. 76 2 plates p.

Jensen, Stephen D. A Mossbauer effect study of synthetic titanomagnetites. D, 1973, University of Wyoming. 45 p.

Jensen, Thomas E. Electrical resistivity investigations in Wright and Taylor valleys, Antarctica. M, 1971, Northern Illinois University. 83 p.

Jensen, Tim R. Petrology of the Salt Wash Member of the Morrison Formation, Henry Basin, Utah. M, 1982, Northern Arizona University. 82 p.

Jensen, Wayne Gale. A geomorphic evaluation of natural and channelized stream segments. M, 1975, University of Nebraska, Lincoln.

Jensky, Wallace Arthur, II. Reconnaissance geology and geochronology of the Bahia de Banderas area, Nayarit and Jalisco, Mexico. M, 1974, University of California, Santa Barbara.

Jenson, John. Stratigraphy and facies analysis of the upper Kaibab and lower Moenkopi formations in southwestern Washington County, Utah. M, 1986, Brigham Young University. 43 p.

Jeong, Bongil. A study of the amplitude and energy of PS converted waves. M, 1963, University of Utah. 45 p.

Jeong, Sangman. Use of multivariate modeling to estimate impacts of groundwater withdrawals on streamflow for the Camas Creek basin. D, 1988, University of Idaho. 242 p.

Jephcoat, Andrew Philip. Hydrostatic compression studies on iron and pyrite to high pressures; the composition of the Earth's core and the equation of state of solid argon. D, 1986, The Johns Hopkins University. 227 p.

Jeppesen, Jon A. Petrology of part of the Wewoka Formation (Pennsylvanian) in Hughes County, Oklahoma. M, 1972, Kansas State University.

Jepsen, Anders Frede. Numerical modelling of resistivity induced polarization by the relaxation method. D, 1968, University of California, Berkeley. 187 p.

Jepsen, Glenn Lowell. The stratigraphy and paleontology of the Paleocene sediments of northeastern Park County, Wyoming. D, 1930, Princeton University. 146 p.

Jepsen, Karl Oscar. A petrographic study of the Star Point Sandstone and Blackhawk Formation (Upper Cretaceous), East Mountain, Emery County, Utah. M, 1987, Fort Hays State University. 77 p.

Jeran, Paul William. The structure and stratigraphy of the Poor Valley area near Rutledge, Grainger County, Tennessee. M, 1965, University of Tennessee, Knoxville. 24 p.

Jercinovic, Devon Eldridge. Geomorphic analysis of small watersheds affected by coal-surface mining in northwestern New Mexico. M, 1984, University of New Mexico. 244 p.

Jercinovic, Michael J. Alteration of basaltic glasses from British Columbia, Iceland, and the deep sea. D, 1988, University of New Mexico. 475 p.

Jerde, Eric A. A structural analysis of a local thrust fault system, Southeast Jefferson County, Montana. M, 1984, Washington State University. 98 p.

Jernigan, Bruce Lee. Geochemistry of chromium in surface environments. M, 1982, North Carolina State University. 93 p.

Jernigan, Dana Gregory. Comparative diagenetic histories of the Holston and Rockdell formations, Middle Ordovician of northeastern Tennessee. M, 1987, University of Tennessee, Knoxville. 234 p.

Jerome, Dominique Yves. Composition and origin of some achondritic meteorites. D, 1970, University of Oregon. 166 p.

Jerome, Norbert Hugh. Geology between Miller and Eightmile creeks, northern Sapphire Range, western Montana. M, 1968, University of Montana. 49 p.

Jerome, Stanley Everett. A reconnaissance geologic study of the Black Canyon schist belt, Bradshaw Mountains, Yavapai and Maricopa counties, Arizona. D, 1956, University of Utah. 138 p.

Jerpbak, Mark James. Seismic reflection profile across a domal structure and the Amana fault system in Linn County, Iowa. M, 1988, University of Iowa. 86 p.

Jerskey, Richard Garrard. A paleomagnetic study of the Bridger Formation, southern Green River basin, Wyoming. M, 1981, University of Wisconsin-Milwaukee. 113 p.

Jervey, Macomb T. Transportation and dispersal of biogenic material in the nearshore marine environment. D, 1974, Louisiana State University. 340 p.

Jesinkey, Christopher. Late Paleozoic paleomagnetizations from the Chilean Andes; constraints for Andean evolution. D, 1987, Rutgers, The State University, New Brunswick. 38 p.

Jespersen, Anna. Rocks and minerals of the District of Columbia. M, 1934, George Washington University. 35 p.

Jesse, Judith Mary. Activity centered laboratory investigations in sedimentation for introductory geology. M, 1970, University of Michigan.

Jesseau, Conrad Wayne. A structural-metamorphic and geochemical study of the Hunt River supracrustal belt, Nain Province, Labrador. M, 1976, Memorial University of Newfoundland. 211 p.

Jessell, Mark Walter. Dynamic grain boundary migration and fabric development; observations, experiments and simulations. D, 1986, SUNY at Albany. 287 p.

Jessen, Margaret Lynn. The geomorphic evolution of the Colorado River. M, 1977, Baylor University. 104 p.

Jessey, D. R. Mercury distribution in a massive sulfide ore deposit at Balmat, New York. M, 1974, University of Missouri, Rolla.

Jessey, David Ray. An investigation of the nickel-cobalt occurrence in the Southeast Missouri mining district. D, 1981, University of Missouri, Rolla. 230 p.

Jessup, David. Downward continuation of inversion of surface heat flow data from the interior Southwestern United States. M, 1985, Georgia Institute of Technology. 95 p.

Jessup, Donald David. Reconnaissance geology of the Chukchi Platform; west-central Chukchi Shelf, offshore Alaska. M, 1985, Michigan State University. 105 p.

Jessup, Donald Edward. The geology of a part of Jackson Township, Pike County, Ohio. M, 1951, Ohio State University.

Jester, Guy Earlscort. An experimental investigation of soil-structure interaction in a cohesive soil. D, 1969, University of Illinois, Urbana. 644 p.

Jestes, Edward Calvin. A stratigraphic study of some Eocene sandstones, northeastern Ventura Basin, California. D, 1963, University of California, Los Angeles. 376 p.

Jestes, Edward Calvin. Geology of the Wiley Canyon area, Ventura County, California. M, 1958, University of California, Los Angeles.

Jeter, Douglas DeL. The geo-economic potentialities of the North Carolina kaolin deposits. D, 1940, University of North Carolina, Chapel Hill. 69 p.

Jeter, Hewitt Webb. An investigation of high resolution dissolved oxygen profiles off the Oregon coast. D, 1973, Oregon State University. 169 p.

Jett, Guy A. Sedimentary petrology of the western melange belt, North Cascade Range, Washington. M, 1986, University of Wyoming. 85 p.

Jewell, John William. Geology of the northern two-thirds of the Dorton Quadrangle, Cumberland County, Tennessee. M, 1951, Vanderbilt University.

Jewell, Paul William. Chemical and thermal evolution of hydrothermal fluids, Mercur gold district, Tooele County, Utah. M, 1984, University of Utah. 77 p.

Jewell, Pliny. Morphodynamics of the Cape Romano shoals. M, 1987, University of South Florida, Tampa. 87 p.

Jewell, T. K. Urban stormwater pollutant loadings. D, 1980, University of Massachusetts. 280 p.

Jewell, Thomas Ross. Effects of geologic conductors upon audio-frequency magnetic fields. M, 1962, University of California, Berkeley. 82 p.

Jewell, Willard Brownell. Mineral deposits of the Hyder District, southeastern Alaska. D, 1926, Princeton University.

Jewell, William Franklin. A system of inventory for the shore-zones of the Great Lakes. M, 1947, Michigan State University. 30 p.

Jewett, David G. The use of computer-assisted instruction in introductory geology. M, 1983, Wichita State University. 124 p.

Jewett, John Mark. Notes on the fauna of the Iola Limestone. M, 1929, University of Kansas. 98 p.

Jewett, John Mark. Stratigraphy of the Marmaton Group, Pennsylvanian age, in Kansas. D, 1943, University of Kansas. 115 p.

Jewett, Peter D. The structure and petrology of the Slesse Peak area, Chilliwack Mountains, British Columbia, Canada. M, 1984, Western Washington University. 164 p.

Jewitt, Walter. Heavier accessory minerals of certain granite rocks of the Precambrian Shield. M, 1934, Queen's University. 48 p.

Jex, Garnet Wolseley. Water-repellency; high humidity origin in some sandy soils. D, 1984, University of Florida. 114 p.

Jeyapalan, Kanagasabai. Analyses of flow failures of mine tailings impoundments. D, 1981, University of California, Berkeley. 320 p.

Jezek, Kenneth Charles. Dielectric permittivity of glacier ice measured in situ by radar wide-angle reflection. M, 1977, University of Wisconsin-Madison.

Jezek, Kenneth Charles. Radar investigations of the Ross Ice Shelf, Antarctica. D, 1980, University of Wisconsin-Madison. 221 p.

Jezek, P. A. Geological studies in the area of the Fiji Plateau, Southwest Pacific. D, 1976, University of Massachusetts. 432 p.

Jha, K. Stabilization of Oklahoma shales. D, 1977, University of Oklahoma. 264 p.

Jha, Parmeshwari Prasad. Morphology, mineralogy and genesis of Williamson silt loam, a sol brun acid developed on silty lacustrine deposits. D, 1961, Cornell University. 199 p.

Jhaveri, Dilip Purshottamdas. Earthquake forces in tall buildings with setbacks. D, 1967, University of Michigan. 256 p.

Jiang, Ming-Jung. Calcareous nannofossils from the uppermost Cretaceous and the lowermost Tertiary of central Texas. M, 1981, Texas A&M University.

Jibson, Randall Wade. Landslides caused by the 1811-12 New Madrid earthquakes. D, 1985, Stanford University. 195 p.

Jicha, Henry Louis, Jr. Alpine lead-zinc ores of the Carinthian region, Europe. M, 1951, Columbia University, Teachers College.

Jicha, Henry Louis, Jr. Geology and mineral deposits of the Lake Valley Quadrangle, Sierra, Grant, and Luna counties, New Mexico. D, 1954, Columbia University, Teachers College.

Jillson, Willard Rouse. A preliminary report on the stratigraphy and paleontology of the Quimper Peninsula of the State of Washington. M, 1915, University of Washington. 65 p.

Jiménez-Mosquera, Carlos José Advances in time series analysis with hydrological applications. D, 1988, University of Western Ontario.

Jimenez-Salas, Oscar Hugo. Application of the Shuttle imaging radar-A (Sir-A) imagery to stratigraphic and tectonic studies in the Sierra Madre Oriental, northeastern Mexico. M, 1984, University of Texas at Dallas. 108 p.

Jin, Doo Jung. Groundwater exploration by geophysical methods. M, 1971, Stanford University.

Jin, Doo Jung. Surface wave studies of the Bering Sea and Alaska area. D, 1979, Southern Methodist University. 94 p.

Jin, J. S. Stabilization of dredged materials. D, 1976, Northwestern University. 305 p.

Jin, Jisuo. Early Silurian pentamerid brachiopods from Anticosti Island, Canada. M, 1984, Laurentian University, Sudbury. 151 p.

Jin, Jisuo. Late Ordovician and Early Silurian rhynchonellid brachiopods of Anticosti Island, Quebec. D, 1988, University of Saskatchewan. 366 p.

Jin, Zhenkui. Sedimentation and diagenesis of the Upper Devonian Cairn Formation, western Alberta, Canada. M, 1987, University of Calgary. 231 p.

Jindrich, Vladimir. Biogenic buildups and carbonate sedimentation, Dry Tortugas reef complex (Holocene), Florida. D, 1972, SUNY at Binghamton. 67 p.

Jindrich, Vladimir. Recent carbonate sedimentation by tidal channels in the lower Florida Keys. M, 1968, SUNY at Binghamton. 69 p.

Jinkins, Ronnie L. Textural parameters and heavy minerals as indicators of environment of some Late Cretaceous deposits in Alabama. M, 1967, University of Alabama.

Jinks, Douglas David. Genesis of some clays in the Altamaha River watershed of Georgia. D, 1966, University of Georgia. 103 p.

Jinks, Jimmie E. The Margaret Wash section of the Mogul Fault, Pinal County, Arizona. M, 1961, University of Arizona.

Jiracek, George. Radio sounding of an arctic ice. M, 1966, University of Wisconsin-Madison.

Jiracek, George R. Geophysical studies of electromagnetic scattering from rough surfaces and from irregularly layered structures. D, 1972, University of California, Berkeley. 167 p.

Jiricka, D. E. Geology of the Hart-McWatters nickel property. M, 1984, Laurentian University, Sudbury. 94 p.

Jirik, Richard Steven. Geology of the Takanis copper-nickel-cobalt prospect, Yakobi Island, southeastern Alaska. M, 1982, Washington State University. 190 p.

Jirsa, Mark Alan. The petrology and tectonic significance of interflow sediments in the Keweenawan North Shore Volcanic Group of northeastern Minnesota. M, 1980, University of Minnesota, Duluth.

Jiwani, Riyazali N. Contaminant hydrogeology of the Walpole Island Indian Reserve Lambton County, Ontario, Canada. M, 1983, University of Windsor. 250 p.

Jizba, Zdenek V. X-ray and optical studies of natural and heated plagioclase feldspars. D, 1953, University of Wisconsin-Madison.

Jizba, Zdenek Vaclav. Retardation method for determining the orientation of uniaxial crystals. M, 1950, Washington State University. 21 p.

Jo, Bong Gon. Dispersion and attenuation of mantle Rayleigh overtones. D, 1986, Yale University. 219 p.

Jobe, Billye Irene. Microfauna of the basal Midway outcrops near Hope, Arkansas. M, 1951, University of Oklahoma. 114 p.

Jobe, Thomas C. A preliminary study of the Ringwood oil pool, Major County, Oklahoma. M, 1951, University of Oklahoma. 57 p.

Jobe, Tracy Hutch. The depositional environment of the Morrison Formation in the West Poison Spider Field and surrounding areas, Southeast Wind River basin, Wyoming. M, 1986, Oklahoma State University.

Jobe, William T. The geological background of Marco Polo's travels. M, 1932, George Washington University. 24 p.

Jobidon, G. Crustal seismic studies across the Grenville front (Ontario). M, 1970, University of Western Ontario.

Jobin, Claude. Méthodes séismiques et électriques appliquées à l'étude des dépôts meubles de la Vallée du Saint-Laurent et à la détermination de l'épaisseur des glaciers contemporains (quaternaire), (Quebec, Canada). M, 1969, Universite de Montreal.

Jobin, Daniel Alfred. Provenance of detrital minerals from pre-Beltian areas, Montana. M, 1949, University of Michigan. 41 p.

Jobling, John Lloyd. Stratigraphy, petrography, and structure of the Laramide (Paleocene) sediments marginal to the Beartooth Mountains, Montana. D, 1974, Pennsylvania State University, University Park. 160 p.

Jobling, John Lloyd. The origin of shale chip deposits in southeastern Centre County, Pennsylvania. M, 1969, Pennsylvania State University, University Park. 51 p.

Jobson, Harvey Eugene. Vertical mass transfer in open channel flow. D, 1968, Colorado State University. 222 p.

Jochems, Theodore Paul. Geological, paleomagnetic and geophysical studies at Jones Camp Dike, Socorro County, NM. M, 1987, New Mexico Institute of Mining and Technology. 217 p.

Jochim, Candace L. Calcareous nannoplankton zonation of the type Yazoo Clay and Vicksburg Group sections, Mississippi. M, 1979, University of Colorado.

Jodry, Richard Louis. A rapid method for determining the magnesium/calcium ratios of well samples and its use as an aid in predicting structure and secondary porosity in calcareous formations. M, 1954, Michigan State University. 49 p.

Joeckel, Robert Matthew. Functional morphology and paleoecology of North American entelodonts (Artiodactyla, Entelodontidae). M, 1988, University of Nebraska, Lincoln. 189 p.

Joench-Clausen, T. Optimal allocation of water resources in an input-output framework. D, 1978, Colorado State University. 223 p.

Joerger, Arthur P. Recent Ostracoda from the South Pass Mudlump No. 5, Louisiana. M, 1959, Louisiana State University.

Joerger, Arthur Peter. Coccolithophorids and related nannoplankton from the upper Eocene to middle Oligocene of southwest Alabama. D, 1965, Washington University. 120 p.

Joesten, Raymond Leonard. Metasomatism and magmatic assimilation at a gabbro-limestone contact, Christmas Mountains, Big Bend region, Texas. D, 1974, California Institute of Technology. 397 p.

Joesting, Henry R. Magnetometer and direct-current resistivity studies in Alaska. D, 1941, The Johns Hopkins University.

Jogan, Brenda M. H. Subaerial laminated crusts of the Cambrian Allentown Dolomite of New Jersey. M, 1976, Rutgers, The State University, New Brunswick. 33 p.

Jogi, P. N. Estimation of the mechanical properties of fluid saturated rocks using measured wave motions. D, 1979, University of Texas, Austin. 144 p.

Johannesen, Dann. The geology and petrography of the Alum Creek area, Mineral County, Nevada. M, 1983, San Jose State University. 92 p.

Johannesen, Nils Poorbaugh. Geology of the northeast quarter of Bone Mountain Quadrangle (Oregon). M, 1972, University of Oregon. 98 p.

Johannesson, Helgi. Theory of river meanders. D, 1988, University of Minnesota, Minneapolis. 206 p.

Jóhannesson, Tómas. Dynamics of ice shelves covering subglacial lakes formed by geothermal heat flux. M, 1984, University of Washington. 79 p.

Johanns, Williams Mathias. Petrogenetic and trend surface analysis of Hell Canyon Pluton, Montana. M, 1973, Western Michigan University.

Johannsen, Albert. The serpentines of Harford County, MD. D, 1903, The Johns Hopkins University.

Johannson, Thora Marian. Petrology and metamorphism of calc-silicate rocks near Farmington and Madison, Maine. M, 1969, Bryn Mawr College. 62 p.

Johansen, Jeffrey R. Cryptogamic soil crusts; recovery from disturbance and seasonal variation in the West Desert, Utah, U.S.A. D, 1984, Brigham Young University. 99 p.

Johansen, K. A. Physical, chemical, and radiochemical characterization of Green Bay (Lake Michigan) ferromanganese nodules. D, 1979, University of Michigan. 540 p.

Johansen, Nils I. The influence of previous stress history upon strength and creep properties of limestone. D, 1971, Purdue University. 203 p.

Johansen, Steven John. Depositional environments of the Queantoweap Sandstone of northwestern Arizona and southern Nevada. M, 1981, University of Arizona. 213 p.

Johansen, Steven John. Provenance of the Mesaverde Group of West-central New Mexico. D, 1986, University of Texas, Austin. 323 p.

Johansing, Robert J. Physical-chemical controls of dolomite hosted sherman-type mineralization, Lake and Park counties, Colorado. M, 1982, Colorado State University. 173 p.

Johanson, David Bryan. Subsurface geology of Grand Lake Field, Cameron Parish, Louisiana. M, 1986, University of New Orleans. 141 p.

Johansson, Folke Carl, Jr. Micropaleontology of the Wesley, Johns Valley and Atoka formations of the Ouachita Mountains of Oklahoma. D, 1960, University of Wisconsin-Madison. 56 p.

Johansson, Folke Carl, Jr. Paleoecology of the Worland Limestone, (Pennsylvanian) upper Desmoinesian, of Iowa, Missouri, and Kansas. M, 1959, University of Wisconsin-Madison.

Johansson, Margaret V. A quantitative study of the conodonts of the Lake Neosho Shale, Desmoinesian, of Iowa, Missouri, and Kansas. M, 1960, University of Wisconsin-Madison.

Johari, Akbar. A predictive modeling study of long-term environmental impacts due to developments in the Scituate watershed. D, 1987, University of Rhode Island. 372 p.

John, Barbara Elizabeth. Structural and intrusive history of the Chemehuevi Mountains area, southeastern California and western Arizona. D, 1987, University of California, Santa Barbara. 344 p.

John, Billy Eugene St. *see* St. John, Billy Eugene

John, Chacko J. A petrological and general structural study of the Baltimore Gneiss at Glen Mills quarry, Chester County, Pennsylvania. M, 1973, University of Delaware.

John, Chacko J. Internal sedimentary structures, vertical stratigraphic sequences, and grain size parameter variations in a transgressive coastal barrier complex; the Atlantic Coast of Delaware. D, 1977, University of Delaware. 311 p.

John, Charles Bedford. The biostratigraphy and paleoecology of the Pliocene-Pleistocene strata of the High Island and Galveston offshore areas, Texas. M, 1974, Tulane University. 121 p.

John, David Allen. Evolution of hydrothermal fluids in intrusions of the central Wasatch Mountains, Utah. D, 1987, Stanford University. 258 p.

John, David Allen. Structure and petrology of pelitic schists in the Fremont Peak Pendant, northern Gabilan Range, California. M, 1979, Stanford University. 87 p.

John, Edward Charles. Petrology and petrography of the intrusive igneous rocks of the Levan area, Juab County, Utah. M, 1964, Brigham Young University. 96 p.

John, Jack W. St. *see* St. John, Jack W.

John, Ruth Nimmo Saint *see* Saint John, Ruth Nimmo

John, Viera. Coda-Q studies in the Indian Subcontinent. M, 1983, St. Louis University.

Johng, DuSik. Geophysical and geological investigations of ancient volcanic vents in Raymond, New Hampshire. M, 1975, University of New Hampshire. 120 p.

Johnpeer, Gary D. Reconnaissance geology and petrology of the Guaymas area, Sonora, Mexico. M, 1977, Arizona State University. 67 p.

Johns, David Ainslie. Geologic controls of massive sulfide mineralization in late Mesozoic metamorphic rocks, Tizapa, Estado de Mexico, Mexico. M, 1983, University of Texas, Austin. 117 p.

Johns, Hilary Desmond. Distribution and transportation of Halimeda sediments through the fringing reef system, Grand Cayman Island, British West Indies. M, 1980, Louisiana State University.

Johns, Kenneth Herbert. Geology of the Twelve Mile Pass area, Utah County, Utah. M, 1950, Brigham Young University. 100 p.

Johns, Mark William. Geotechnical properties of Mississippi River delta sediments utilizing in situ pressure sampling techniques. D, 1985, Texas A&M University. 115 p.

Johns, Michael E. Architectural element analysis and depositional history of the upper Petrified Forest Member of the Chinle Formation, Petrified Forest National Park, Arizona. M, 1988, Northern Arizona University. 163 p.

Johns, R. W. C. Economic geology and the airborne magnetometer. D, 1951, University of Toronto.

Johns, Robert Anthony. Injection through fractures. M, 1987, Stanford University.

Johns, Ronald Alan. Biostratigraphy, paleobiology, and taxonomy of articulate brachiopods from the Middle Ordovician Bromide Formation, southern Oklahoma. M, 1987, University of Texas, Austin. 392 p.

Johns, Warren Harvey. Vegetation history and paleoclimatology for the Quaternary of Isla de los Estados, Argentina. M, 1981, Michigan State University. 111 p.

Johns, Wendell S. Pre-lava geomorphology of the lower Coeur d'Alene and St. Joe River valleys, Kootenai and Benewah County, northern Idaho. M, 1961, University of Kansas.

Johns, William Davis. The mineralogy of flint clays and associated fireclays. D, 1952, University of Illinois, Urbana.

Johns, William Roy. The geology and quicksilver occurrences at Quartz Mountain, Oregon. M, 1949, University of Oregon. 76 p.

Johns, Willis Merle. Structure and mineralization of the southern Tidal Wave District, Madison County, Montana. M, 1958, Montana College of Mineral Science & Technology. 67 p.

Johnsen, John Herbert. The Schoharie Formation; a redefinition. D, 1957, Lehigh University. 178 p.

Johnsgard, Gordon Alexander. A simplified classification and map of the soils of Dutchess County, New York, and the relationship of some soil characteristics to land use. D, 1941, Cornell University.

Johnson Rozacky, Wendy. The petrology and sedimentation of the lower Proterozoic Barron Quartzite, NW Wisconsin. M, 1987, University of Minnesota, Duluth. 94 p.

Johnson, Alan Eugene. Mineralogy and textural relationships in the Lake Dufault ores, northwestern Quebec. M, 1966, University of Western Ontario. 187 p.

Johnson, Alan Eugene. Textural and geochemical investigation of Cyprus pyrite deposits (Jurassic-Cretaceous). D, 1970, University of Western Ontario. 190 p.

Johnson, Alan Roy. State space displacement analysis of the response of aquatic ecosystems to phenolic toxicants. D, 1988, University of Tennessee, Knoxville. 194 p.

Johnson, Allan Michael. Some chemical and physical-chemical properties of Michigan cherts. M, 1967, Michigan Technological University. 62 p.

Johnson, Allan Michael. The role of diagenesis in the redistribution of trace elements in the Permian reef complex (West Texas, New Mexico). D, 1971, Michigan Technological University. 90 p.

Johnson, Allen Harold. Stratigraphy and paleoenvironment of the Dinosaur Canyon Member of the Moenave Formation (Upper Triassic?) in the southern part of the Navajo and Hopi Indian reservations, Arizona. M, 1967, University of Arizona.

Johnson, Allen Harold. The paleomagnetism of the Jurassic rocks from southern Utah. D, 1972, Case Western Reserve University. 171 p.

Johnson, Alvin Charles, Jr. A magnetometer survey of the Webster-Addie ultra basic ring, Jackson County, North Carolina. M, 1958, University of Michigan.

Johnson, Alvin Charles, Jr. The geology of the Big Ben area, Cascade County, Montana. D, 1964, University of Michigan. 85 p.

Johnson, Andrea Marie. Plio-Pleistocene fluctuations of the Western Boundary Undercurrent; DSDP Site 533. M, 1987, Duke University. 90 p.

Johnson, Andrew Leigh. The effect of mineralizers on the fusibilities of quartz and silica gel. M, 1936, Rutgers, The State University, New Brunswick. 21 p.

Johnson, Ansel Grieg. Pore pressure changes associated with creep events on the San Andreas Fault. D, 1973, Stanford University. 177 p.

Johnson, Anthony. The Twiggs County earthquake swarm. M, 1984, Georgia Institute of Technology. 125 p.

Johnson, Anthony Gerard. Variation in mineralogy of Mississippi River suspended sediment. M, 1981, University of New Orleans. 89 p.

Johnson, Arthur H., Jr. Geochemistry of small streams in Vermont and New Hampshire. M, 1972, Dartmouth College. 47 p.

Johnson, Arvid M. Stratigraphy and lithology of the Deer Butte Formation, Malheur County, Oregon. M, 1961, University of Oregon. 144 p.

Johnson, Arvid Mauritz. A model for debris flow. D, 1965, Pennsylvania State University, University Park. 248 p.

Johnson, Barry Allen. Deep-sea fan-valley conglomerate; Cap Enrage Formation, Gaspe, Quebec. M, 1974, McMaster University. 108 p.

Johnson, Bernard T. Interpretation of total-intensity magnetic field data by Fourier analysis. M, 1960, University of Minnesota, Minneapolis.

Johnson, Brad T. Depositional environment of the Iron Springs Formation, Gunlock, Utah. M, 1984, Brigham Young University. 46 p.

Johnson, Bradford J. Geology of the Wilson Island Group, Great Slave Lake, Northwest Territories. M, 1987, Carleton University. 121 p.

Johnson, Bradford Knowlton. Geology of a part of the Manly Peak Quadrangle, southern Panamint Range, California. D, 1954, University of California, Los Angeles.

Johnson, Bradford Knowlton. Geology of the Castaic Creek-Elizabeth Lake canyon area. M, 1952, University of California, Los Angeles.

Johnson, Bradley Scott. Alleviation of compaction on fine-textured Michigan soils. D, 1987, Michigan State University. 227 p.

Johnson, Bruce Alan. Vertical sequence analysis of a deep-sea fan system, Santa Paula Creek, California. M, 1979, University of Southern California.

Johnson, Bruce D. Genetic stratigraphy and provenance of the Baca Formation, New Mexico and the Eagar Formation, Arizona. M, 1978, University of Texas, Austin.

Johnson, Bruce R. Migmatites along the northern border of the Idaho Batholith. D, 1975, University of Montana. 120 p.

Johnson, C. Branning. Characteristics and mechanics of formation of glacial arcuate abrasion cracks. D, 1975, Pennsylvania State University, University Park. 270 p.

Johnson, C. G. Geology of the Rocky Brook area. M, 1957, Acadia University.

Johnson, Cady Leonard. Correlation and origin of carnotite occurrences in the southern Nevada region. D, 1982, University of Nevada. 197 p.

Johnson, Carl Edward. I, CEDAR; an approach to the computer automation of short period local seismic networks; II, Seismotectonics of the Imperial Valley of Southern California. D, 1979, California Institute of Technology. 343 p.

Johnson, Carl Edward. Regionalized Earth models from linear programming methods. M, 1972, Massachusetts Institute of Technology. 225 p.

Johnson, Carla. Internal deformation features and finite strain in Little Grey's Anticline, Lincoln County, Wyoming. M, 1985, University of Colorado.

Johnson, Carlton Robert. Geology and ground-water resources of Logan County, Kansas. D, 1956, University of Iowa. 352 p.

Johnson, Carlton Robert. Hydrology of Scott Basin and vicinity, Scott County, Kansas. M, 1954, University of Iowa. 112 p.

Johnson, Charles. A petrographic comparison of the basal Arikaree volcanic ash exposures in western South Dakota. M, 1957, South Dakota School of Mines & Technology.

Johnson, Charles Craig. Stratigraphy of the Pennsylvanian Palo Pinto Limestone in the eastern part of the Midland Basin and adjacent areas. M, 1956, Southern Methodist University. 18 p.

Johnson, Charles Frederick. The structures and clastic dikes in the Sage Creek-Pinnacles area, Badlands National Monument, South Dakota. M, 1958, South Dakota School of Mines & Technology.

Johnson, Charles G. Depositional environment of Eocene Wilcox reservoir sandstones, North Milton Field, Harris County, Texas. M, 1983, Texas A&M University. 129 p.

Johnson, Charles Jerome. Petrography and paleoecology of the Sierra Blanca Limestone. M, 1968, University of California, Riverside. 52 p.

Johnson, Charles M. Occurrence and alteration of clay minerals in the Caribbean Sea. M, 1973, Texas A&M University.

Johnson, Charles T. L. Environment of deposition of the Pennsylvanian Bartlesville Sandstone, Labette County, Kansas. M, 1973, Texas A&M University. 72 p.

Johnson, Charlie Ernest. Permian Spiriferella from the Yukon Territory. M, 1963, University of Calgary. 69 p.

Johnson, Cheryl Elaine. Historic and geomorphic evidence of barrier dynamics and the origin of the sunken forest, south shore of Long Island, New York. M, 1982, University of Massachusetts. 190 p.

Johnson, Chris N. Geology and petro-chemistry of jasperoids, south-central Roberts Mountains, Eureka County, Nevada. M, 1986, Brigham Young University.

Johnson, Christopher A. A study of neotectonics in central Mexico from Landsat thematic mapper imagery. M, 1987, University of Miami. 112 p.

Johnson, Christopher Lee. Depositional environments, reservoir trends, and diagenesis of Red Fork sandstones in portions of Blaine, Caddo, and Custer counties, Oklahoma. M, 1984, Oklahoma State University. 122 p.

Johnson, Clarence Richard. Mineralogy and geochemistry of Wildcat Lake, Whitman County, Washington. M, 1980, Washington State University. 131 p.

Johnson, Clark Montgomery. The Questa magmatic system; petrologic, chemical and isotopic variations in cogenetic volcanic and plutonic rocks of the Latir volcanic field and associated intrusives, northern New Mexico. D, 1986, Stanford University. 336 p.

Johnson, Clark Montgomery. Triple junction magmatism; a geochemical study of Neogene volcanic rocks in western California. M, 1984, Stanford University. 67 p.

Johnson, Claudia C. Paleoecology, carbonate petrology and depositional environments of lagoonal facies, Cupido and El Abra formations, northeastern Mexico. M, 1984, University of Colorado. 147 p.

Johnson, Clayton Henry, Jr. Igneous metamorphism in the Orofino region, Idaho. D, 1943, Cornell University.

Johnson, Clayton Henry, Jr. Lower Pennsylvanian fusulinids of Boone County, Missouri. M, 1939, University of Missouri, Columbia.

Johnson, Clifford D. Microplankton zones, Savik Formation (Jurassic), Axel Heiberg and Ellesmere islands, District of Franklin. M, 1972, University of Calgary. 64 p.

Johnson, Cordell M. Conodonts of the "Passo Beds", Missouri. M, 1957, University of Missouri, Columbia.

Johnson, Craig Alden. The formation of garnet in olivine-bearing metagabbros from the Adirondacks. M, 1981, University of Michigan.

Johnson, Curtis Leonard. Microfauna of the Gregory Member of the Pierre Formation. M, 1961, University of South Dakota. 190 p.

Johnson, D. M. Atmospheric inputs of trace metals and nutrients to Saginaw Bay. D, 1976, University of Michigan. 184 p.

Johnson, Dane S. Foraminiferal distribution as a function of gas chromatographic analysis of the lower Mohnian Stage of the Monterey Formation, western Ventura Basin, California. M, 1974, University of Nevada. 74 p.

Johnson, Daniel L. Geotechnical characterization of Keweenaw copper sulfides for in-situ solution mining. M, 1988, Michigan Technological University. 136 p.

Johnson, Daniel Paul. Feasibility of extracting shear-wave information from P-wave reflection data. M, 1987, University of Utah. 91 p.

Johnson, David. A refraction and seismicity study of the northern Gorda Ridge. M, 1987, University of Washington. 120 p.

Johnson, David A. Unconsolidated sediments in the offshore zone near Plum Island, Massachusetts. M, 1966, Massachusetts Institute of Technology. 110 p.

Johnson, David Ashby. Studies of deep sea erosion using deep-towed instrumentation. D, 1971, University of California, San Diego. 188 p.

Johnson, David Bruce. Analysis of Lower Devonian conodont ecology, Eureka County, Nevada. D, 1978, University of Iowa. 186 p.

Johnson, David Bruce. Devonian stratigraphy of the southern Cortez Mountains, Nevada. M, 1972, University of Iowa. 55 p.

Johnson, David G. Ferromanganese concretions in Lake Champlain. M, 1969, University of Vermont.

Johnson, David Ian. Early Ordovician (Arenig) conodonts from St. Pauls Inlet and Martin Point, Cow Head Group, western Newfoundland. M, 1987, Memorial University of Newfoundland. 231 p.

Johnson, David Leslie. Arsenate and arsenite in sea water. D, 1972, University of Rhode Island. 172 p.

Johnson, David M. Deformational behavior of an anisotropic carbonate rock. M, 1977, Bowling Green State University. 52 p.

Johnson, David Perrin. The areal geology and stratigraphy of the Hillsboro Quadrangle, Wisconsin. M, 1947, University of Wisconsin-Madison. 64 p.

Johnson, Dean. A magnetometric survey of the Iron Horse magnetite deposit, Socorro County, New Mexico. M, 1953, New Mexico Institute of Mining and Technology. 43 p.

Johnson, Dennis V. Water wells in the Inyan Kara Group near Rapid City, South Dakota, M, 1973, South Dakota School of Mines & Technology.

Johnson, Diane Louise. Two-mica granites and metamorphic rocks of the east-central Ruby Mountains, Elko County, Nevada. M, 1981, Stanford University. 145 p.

Johnson, Diane Marilyn. Folding and faulting in the footwall of the Diversion Thrust, north-central Montana. M, 1988, Washington State University. 78 p.

Johnson, Donald H. Geology of Devil's Head Quadrangle, Douglas County, Colorado. D, 1961, Colorado School of Mines. 138 p.

Johnson, Donald Lee. Landscape evolution (Late Pleistocene-Holocene) on San Miguel Island, California. D, 1972, University of Kansas. 622 p.

Johnson, Donald Otto. A geophysical investigation of the bedrock surface in southwestern DuPage County, Illinois. M, 1967, Northern Illinois University. 59 p.

Johnson, Donald Otto. Stratigraphic analysis of the interval between the Herrin (No. 6) Coal and the Piasa Limestone in southwestern Illinois. D, 1972, University of Illinois, Urbana. 135 p.

Johnson, Donald Ray. The seasonal density structure and circulation on the continental shelf. D, 1974, University of Miami.

Johnson, Dorothy Bernice. Lower Pennsylvanian stratigraphy in northwestern Wyoming. M, 1945, University of Illinois, Urbana.

Johnson, Douglas Edward. Effect of rehabilitation practices on plant establishment in the Piceance Basin of Colorado. D, 1980, Colorado State University. 230 p.

Johnson, Douglas Martin. Near-surface physical properties of the Earth and Moon. D, 1978, University of Texas at Dallas. 184 p.

Johnson, Douglas Wade. Carbonate petrography, microfacies, and depositional history of the Marble Falls Formation (Pennsylvanian), Llano region, central Texas. M, 1983, University of Oklahoma. 258 p.

Johnson, Douglas Wilson. The geology of the Cerrillos Hills, New Mexico. D, 1903, Columbia University, Teachers College.

Johnson, Durwood Milton. Middle Jurassic of north central Montana and adjacent areas of Canada. M, 1961, University of Montana. 62 p.

Johnson, Eben Lennart. Geology of the pegmatites in the Hale Spring area, Wichita Mountains, Oklahoma. M, 1955, University of Oklahoma. 87 p.

Johnson, Eben Lennart. Precambrian geology of parts of Passaic County and Sussex County, New Jersey, and infrared absorption studies of biotite. D, 1968, Rutgers, The State University, New Brunswick. 265 p.

Johnson, Edward A. Geology and gold deposits of the confederate Gulch-White Gulch area, Broadwater County, Montana. M, 1973, Montana College of Mineral Science & Technology. 53 p.

Johnson, Edward A. Structural geology of the south Mazourka Canyon area, Inyo County, California. M, 1968, San Jose State University. 61 p.

Johnson, Edward Allison. Geology of a part of the southeastern side of the Cottonwood Mountains, Death Valley, California. D, 1971, Rice University. 107 p.

Johnson, Edward William. The geology of the region about Faribault, Minnesota. M, 1933, University of Minnesota, Minneapolis. 38 p.

Johnson, Edwin Lionel. Geochemistry of some Mississippian basin-slope facies, Utah - Nevada. M, 1982, University of Nevada. 118 p.

Johnson, Eldred. The structure and stratigraphy of the Frontier Formation in the Big Piney-La Barge area, Sublette and Lincoln counties, Wyoming. M, 1961, University of New Mexico. 85 p.

Johnson, Elizabeth A. Textural and compositional sediment characteristics of the southeastern Bristol Bay, continental shelf, Alaska. M, 1983, California State University, Northridge. 142 p.

Johnson, Eric Henry. Resistivity and induced polarization survey of a basalt flow in a geothermal environment, western Utah. M, 1975, University of Utah. 69 p.

Johnson, Eric Lee. The petrology and petrogeny of the diabase dikes of the Highland Range, southwestern Montana. M, 1985, SUNY at Binghamton. 42 p.

Johnson, Ernest. A one day geological field trip of the Greater Memphis area (Tennessee). M, 1967, Virginia State University. 37 p.

Johnson, F. O. Ground water geology of central Niobrara County, Wyoming. M, 1962, University of Wyoming. 62 p.

Johnson, Floyd R. Geology of the Quartzburg mining district, Grant County, Oregon. M, 1976, Oregon State University. 102 p.

Johnson, Francis Alfred. A petrographic study of the San Pablo Formation in the Nipomo Quadrangle of California. M, 1931, University of California, Berkeley. 85 p.

Johnson, Francis Alfred. The Merced (Pliocene) Formation north of San Francisco Bay, California. D, 1934, University of California, Berkeley.

Johnson, Frank Ernest. Depositional environment of the Arcola Member of the Mooreville Formation in Mississippi. M, 1976, Mississippi State University. 40 p.

Johnson, Frank Marion. Non-carbonate minerals in some Indiana limestones and dolomites. M, 1952, Indiana University, Bloomington. 43 p.

Johnson, Frank Melvin S. The application of the electrical resistivity method of geophysical exploration to indicate the structural value of concrete. D, 1937, Colorado School of Mines. 143 p.

Johnson, Frederick A. Fabric of limestones and dolomites. D, 1951, University of Chicago. 141 p.

Johnson, Fritz Kreisler. The sediments of the Newport River estuary, Morehead City, North Carolina. M, 1959, University of North Carolina, Chapel Hill. 36 p.

Johnson, Gary D. Small mammals of the middle Oligocene of the Big Badlands of South Dakota. M, 1966, South Dakota School of Mines & Technology.

Johnson, Gary Dean. Neogene molasse sedimentation in a portion of the Punjab-Himachal Pradesh Tertiary re-entrant, Himalayan foothill belt, India; a vertical profile of Siwalik deposition. D, 1971, Iowa State University of Science and Technology. 94 p.

Johnson, Gary Dean. Stratigraphic studies of the Siwalik Series (Middle Miocene-Pliocene), Punjab, India. M, 1967, Iowa State University of Science and Technology.

Johnson, Gary Dee. Early Permian vertebrates from Texas; Actinopterygii (Schaefferichthys), Chondrichthyes (including Pennsylvanian and Triassic Xenacanthodii), and Acanthodii. D, 1979, Southern Methodist University. 653 p.

Johnson, Gary Steven. Application of a numerical ground-water flow model to the Mud Lake area in southeastern Idaho. M, 1982, University of Idaho. 71 p.

Johnson, George. Mediterranean undercurrent and microphysiography west of Gibraltar. M, 1965, New York University.

Johnson, George D. Geology of the northwest quarter, Alvord Lake Three Quadrangle, Oregon. M, 1960, Oregon State University. 75 p.

Johnson, George Duncan. Geology of the mountain uplift transected by the Shoshone Canyon, Wyoming. D, 1934, The Johns Hopkins University.

Johnson, Gerald. Geodetic computations for lines halfway around the world. M, 1960, Ohio State University.

Johnson, Gerald Glenn, Jr. A mathematical analysis of terrestrial impact craters. D, 1965, Pennsylvania State University, University Park. 142 p.

Johnson, Gerald Homer. The stratigraphy, paleontology, and paleoecology of the Peoria Loess (upper Pleistocene) of southwestern Indiana. D, 1965, Indiana University, Bloomington. 200 p.

Johnson, Germaine P. Seismic stratigraphy of early Mesozoic deposits of the Southeast Neuquen Basin, Rio Negro Province, Argentina. M, 1985, North Carolina State University. 112 p.

Johnson, Glenn Wilbur. Pleistocene planktonic foraminiferal biostratigraphy and paleoecology; Northeast Gulf of Mexico. M, 1988, University of Delaware. 256 p.

Johnson, Gordon Harene. The influence of minor additions on the electrical conductivity of rutile. D, 1950, Pennsylvania State University, University Park. 77 p.

Johnson, Grace Phillips, II. Marine gastropods and pelecypods of the Panamic Province. M, 1956, Stanford University. 106 p.

Johnson, Gregory R. Frequency response of a geophone clamped to a borehole. M, 1986, Colorado School of Mines. 147 p.

Johnson, H. A. C. Rock alteration and metamorphism at Cochenour Willans gold mine, Red Lake, Ontario. M, 1948, University of Manitoba.

Johnson, H. Norton. An empirical examination of the outside direct differential method of electrical prospecting for subsurface geological structure. D, 1932, Harvard University.

Johnson, Hamilton McKee. An analysis of subsurface geophysical methods in determining lithology and fluid content. D, 1954, University of Oklahoma. 556 p.

Johnson, Harlan Paul. The low temperature oxidation of magnetite and titanomagnetite and the implications for paleomagnetism. D, 1972, University of Washington. 117 p.

Johnson, Harold F. The composition of the Earth's crust in Minnesota. M, 1932, University of Minnesota, Minneapolis. 21 p.

Johnson, Helgi. The stratigraphy and paleontology of (Silurian) cataract formations in Ontario. D, 1934, University of Toronto.

Johnson, Henry Derr, Jr. Mineralogical study of altered basic intrusive rocks, Wichita Mountains, Oklahoma. M, 1960, University of Oklahoma. 62 p.

Johnson, Henry Luther. A correlation of five oil wells in north central Texas. M, 1930, University of Iowa. 23 p.

Johnson, Herman F. The geological section at Cliffwood, New Jersey. M, 1906, Columbia University, Teachers College.

Johnson, Howard F. The upland erosional surfaces of the Catskills in relation to the present day drainage (New York). M, 1933, Cornell University.

Johnson, Hugh Nelson. Sequent occupance of the St. Francois mining region. D, 1950, Washington University. 244 p.

Johnson, Hugh Nelson. The stratigraphy of the Maquoketa Shale in Missouri and adjacent parts of Illinois. M, 1939, Washington University. 185 p.

Johnson, Ian Mayhew. Inversion of the geomagnetic secular variation; uniqueness and feasibility. D, 1972, University of British Columbia.

Johnson, J. A. Site and source effects on ground motion in Managua, Nicaragua. D, 1975, University of California, Los Angeles. 144 p.

Johnson, J. B. Mass balance and aspects of the glacier environment, Front Range, Colorado, 1969-1973. D, 1979, University of Colorado. 308 p.

Johnson, J. J. Effects of crude oils on representative species of marine phytoplankton. D, 1977, University of Southern Mississippi. 180 p.

Johnson, J. Kent. A study of the shell length of Mercenaria mercenaria in relation to bottom sediments of Little Bay, New Jersey. M, 1976, Montclair State College. 60 p.

Johnson, J. R. Geology of the Yukon Territory with special reference to Alaska. M, 1933, University of British Columbia.

Johnson, James Blake. Geochemistry of Belt Supergroup rocks, Coeur d'Alene district, Shoshone County, Idaho. D, 1971, University of Idaho. 673 p.

Johnson, James Elmer. The role of the oceans in the atmospheric cycle of carbonyl sulfide. D, 1985, University of Washington. 136 p.

Johnson, James F. Geology of the Marshall District, Boulder County, Colorado. M, 1935, University of Colorado.

Johnson, James F. H. Geology of the Beach Grove area, McLean County, Kentucky. M, 1959, University of Kentucky. 26 p.

Johnson, James Howe. A study of the heavy minerals in the Paleozoic rocks of the Blacksburg area. M, 1938, Virginia Polytechnic Institute and State University.

Johnson, James Kenneth. The development and geologic history of linear sand shoals in Green Bay, Wisconsin. M, 1981, University of Wisconsin-Milwaukee. 112 p.

Johnson, James Mark. Geology and mineralization of the Marietta area; Mineral County. M, 1978, University of Nevada - Mackay School of Mines. 56 p.

Johnson, James S. A contribution to the structure of northern Baffin Bay and Lancaster Sound (Canada). M, 1971, Dalhousie University.

Johnson, James T. A northern extension of the Magdalena mining district, Socorro County, New Mexico. M, 1955, New Mexico Institute of Mining and Technology. 46 p.

Johnson, James Weldon. A Bayesian analysis of excessive flood flows. D, 1981, George Washington University. 78 p.

Johnson, James Wesley. Critical phenomena in hydrothermal systems; state, thermodynamic, transport, and electrostatic properties of H_2O in the critical region. D, 1987, University of Arizona. 229 p.

Johnson, Jeffrey Alan. The sedimentology, paleohydrology, and paleogeography of the Brushy Basin Member, Morrison Formation in the Henry Basin, Wayne and Garfield counties, Utah. D, 1988, University of California, Berkeley. 117 p.

Johnson, Jeffrey Alan. The structural and sedimentary evolution of the North Pyrenean Basin, southern France. D, 1985, University of California, Los Angeles. 510 p.

Johnson, Jeffrey S. Structural and geophysical interpretation of Clark's Fork oil field, Montana. M, 1986, University of Nebraska, Lincoln. 94 p.

Johnson, Jesse Harlan. Tertiary deposits of South Park (Park County), Colorado, with a description of Oligocene algal limestones. D, 1937, University of Colorado.

Johnson, Jesse Harlan. The utilization of maps as an aid in the teaching of geological subjects. M, 1923, Colorado School of Mines. 96 p.

Johnson, John Burlin, Jr. Regional gravity survey of parts of Toole, Juab, and Millard counties, Utah. D, 1956, University of Utah. 34 p.

Johnson, John Burlin, Jr. Some induced polarization effects. M, 1954, New Mexico Institute of Mining and Technology. 45 p.

Johnson, John Clifton. Silicification of claystones from the Hawthorn Formation, Marion County, Florida. M, 1982, University of Florida. 106 p.

Johnson, John F. Paleoecological analysis of the Fremont Formation; Upper Ordovician, Canon City Embayment, Colorado. M, 1977, University of Wisconsin-Milwaukee.

Johnson, John Granville. Geology of the northern Simpson Park Range, Eureka County, Nevada. M, 1960, University of California, Los Angeles.

Johnson, John Granville. Great Basin Lower Devonian Brachiopoda. D, 1964, University of California, Los Angeles.

Johnson, John J. A subsurface study of the Ste. Genevieve Limestone (Mississippian) in the northeastern Warrick Co., Indiana. M, 1988, Indiana University, Bloomington. 275 p.

Johnson, John Thomas. Nonparametric analysis of variables influencing limestone ground water occurrence. M, 1970, University of Kentucky. 41 p.

Johnson, Joseph A. Determination of adsorption coefficients for trace organics in soil. M, 1985, University of California, Riverside. 92 p.

Johnson, Kathleen Elizabeth. Isotope geochemistry of Augustine Volcano, Alaska. M, 1986, Southern Methodist University. 144 p.

Johnson, Kathleen Esther. Origin of the Ag-Ni-Co-U mineralization, Black Hawk District, Grant County, New Mexico. M, 1981, University of Oregon. 121 p.

Johnson, Kathryn Olive. Geochemical model of the migration of trace metals from uranium mill tailings. D, 1986, South Dakota School of Mines & Technology.

Johnson, Keith Eric. Thermal springs of the western Transverse Ranges of southern California. M, 1980, San Diego State University.

Johnson, Kendall L. An analysis of state regulations of surface-groundwater development and use in Colorado. D, 1965, Colorado State University. 126 p.

Johnson, Kenneth Allen. An investigation of the mechanisms of debris flow initiation. D, 1987, University of California, Berkeley. 256 p.

Johnson, Kenneth Dee. Structure and stratigraphy of the Mount Nebo-Salt Creek area, southern Wasatch Mountains, Utah. M, 1959, Brigham Young University. 49 p.

Johnson, Kenneth F. The interrelationship of the lower Salina Group and Niagaran reefs in Saint Clair and Macomb counties, Michigan. M, 1971, Michigan State University. 34 p.

Johnson, Kenneth G. An aerial photographic study of the Glen Lake-Sleeping Bear Point area, Leelanau County, Michigan. M, 1957, Michigan State University. 32 p.

Johnson, Kenneth G. The Tully classic correlatives (Upper Devonian) of New York State; model for recognition of alluvial, dune (?), tidal, nearshore (bar and lagoon) and offshore sedimentary environments in a tectonic delta complex. D, 1968, Rensselaer Polytechnic Institute. 122 p.

Johnson, Kenneth S. Areal geology of the Sentinel-Gotebo area, Kiowa and Washita counties. M, 1962, University of Oklahoma. 99 p.

Johnson, Kenneth S. Ion association and activity coefficients in electrolyte solutions. D, 1979, Oregon State University. 320 p.

Johnson, Kenneth Sutherland. Stratigraphy of the Permian Blaine Formation and associated strata in southwestern Oklahoma. D, 1967, University of Illinois, Urbana. 265 p.

Johnson, Kenneth Walter. Occurrence and abundance of fishes in the intake and discharge areas of the Cedar Bayou Power Station before and during the first year of plant operation. D, 1973, Texas A&M University.

Johnson, Kent A. The petroleum geology of the Midal subinterval (Madison Formation-Mississippian) in north central North Dakota. M, 1971, University of North Dakota. 204 p.

Johnson, Kent Erwin. Paleocurrent study of the Tesnus Formation (Pennsylvanian) of the Marathon Basin of western Texas. M, 1961, University of Wisconsin-Madison.

Johnson, Kent Erwin. Sedimentary environment of Stanley Group of Ouachita Mountains of Oklahoma. D, 1963, University of Wisconsin-Madison. 125 p.

Johnson, Kent Raymond. Geology of the Gualán and southern Sierra de las Minas quadrangles, Guatemala. D, 1984, SUNY at Binghamton. 300 p.

Johnson, Kurt Warren. Petrology, diagenesis, and petroleum geology of Upper Minnelusa sandstones, Rainbow Ranch Field, Powder River basin, Wyoming. M, 1985, University of Colorado.

Johnson, Lane R. Measurements of mantle velocities of P waves with a large array. D, 1966, California Institute of Technology. 112 p.

Johnson, Lane R. Velocity anisotropy in drill cores. M, 1962, University of Minnesota, Minneapolis.

Johnson, Larry C. Structure and stratigraphy of an evolving salt ridge and basin complex, Louisiana continental shelf. M, 1980, University of Texas, Austin.

Johnson, Larry D. The geomagnetic secular variation pattern recorded in the stable remanence of Recent Wolf Creek, Minnesota sediments. M, 1974, University of Minnesota, Minneapolis. 63 p.

Johnson, Larry Douglas. Recent history of the Sebastian Inlet, Florida area. M, 1976, University of Florida. 49 p.

Johnson, Larry M. An overlap zone between a Laramide Rocky Mountain Foreland structure and Sevier-style thrust structures near Bannack, Montana. M, 1986, University of Montana. 47 p.

Johnson, Lawrence Clinton. The Ellison District; alteration-mineralization associated with a mid-Tertiary intrusive complex at Sawmill Canyon, White Pine County, Nevada. M, 1983, University of Arizona. 123 p.

Johnson, Leonard Evans. Inversion and inference for teleseismic ray data. D, 1971, University of California, San Diego.

Johnson, Linda Ann. General chemical composition of Precambrian crust in the Llano Uplift, central Texas. M, 1975, Rice University. 78 p.

Johnson, Lisa Kaye. Computer analysis of remote sensing and geologic datasets. M, 1988, Washington State University. 215 p.

Johnson, Lynn A. Geology of the Anthracite Range, West Elk Mountains, Gunnison County, Colorado. M, 1961, University of Kansas.

Johnson, M. T. A theoretical study of methods to improve the energy efficiency of the oil-in-methanol dispersion pipeline system. M, 1977, Stanford University. 34 p.

Johnson, Marcus W. Stratigraphy of the Sciponoceras gracile assemblage zone and adjacent strata within the Bridge Creek Limestone Member of the Colorado Formation (Upper Cretaceous), Cook's Range area, southwestern New Mexico. M, 1984, Indiana University, Bloomington. 179 p.

Johnson, Mark. Pennsylvanian-Permian deformation at 1,000-5,000 feet of overburden, Sacramento Mountains, New Mexico. M, 1985, Texas A&M University. 116 p.

Johnson, Mark Dale. Glacial geology of Barron County, Wisconsin. D, 1984, University of Wisconsin-Madison. 391 p.

Johnson, Mark Dale. Origin of the Lake Superior red clay and glacial history of Wisconsin's Lake Superior shoreline west of the Bayfield Peninsula. M, 1980, University of Wisconsin-Madison.

Johnson, Mark Galen. Clay mineralogy and chemistry of selected Adirondack Spodosols. D, 1986, Cornell University. 173 p.

Johnson, Mark James. Geology of the gold occurrences near Jackson's Gap, Tallapoosa County, Alabama. M, 1988, Auburn University. 172 p.

Johnson, Markes Eric. Community succession and replacement in Early Silurian platform seas; the Llandovery series of eastern Iowa. D, 1977, University of Chicago. 237 p.

Johnson, Martin. Computation of electrical relaxation spectra. M, 1971, University of Utah. 71 p.

Johnson, Martin. Induced polarization spectral analysis. D, 1971, University of Utah.

Johnson, Martin Chester. Mineral chemistry of four quartz-bearing monzonites in East-central Minnesota. M, 1978, University of Iowa.

Johnson, Martin S. The hydrogeology of an area near Marienthal, Wichita County, Kansas. M, 1978, Kansas State University. 65 p.

Johnson, Mary Lou. Natural resources of Story County, Iowa; a computer analysis. M, 1976, Iowa State University of Science and Technology.

Johnson, Mary Louise. The effect of substitutions on the physical properties of tetrahedrite. D, 1982, Harvard University. 352 p.

Johnson, Maureen G. Study of gypsum experimentally deformed at 5 kb. M, 1967, University of California, Berkeley. 68 p.

Johnson, Melvin C. Areal geology of the Wanship-Coalville area (Utah). M, 1952, University of Utah. 50 p.

Johnson, Michael. Stratigraphy and paleogeographic paleotectonic history of the middle Gebel el Rusas Formation; Red Sea Coast, Egypt. M, 1977, University of South Carolina.

Johnson, Michael David. The biostratigraphy and paleoecology of the Fossil Hill Formation on the Bruce Peninsula of Ontario. M, 1979, University of Windsor. 154 p.

Johnson, Michael G. Stratigraphy and sedimentology of the Bronson Sub-group (Missourian-Pennsylvanian) in southcentral Iowa. M, 1971, University of Wisconsin-Madison.

Johnson, Michael J. Geology, alteration and mineralization of a silicic volcanic center, Glass Buttes, Oregon. M, 1984, Portland State University. 129 p.

Johnson, Michael Lee. Carbon and oxygen isotope evolution in the Magnet Cove Complex, Arkansas. M, 1975, Rice University. 63 p.

Johnson, Mike Sam. Geology of the Twelvemile Canyon area, central Utah. M, 1949, Ohio State University.

Johnson, Milford Ronald. Foraminiferal biostratigraphy of portions of the Tuscumbia Limestone and Monteagle Limestone in northern Alabama. M, 1978, University of Georgia.

Johnson, Milo J. Tectonic transport of the Newport Allochthon, northeastern Washington and northern Idaho. M, 1981, University of Montana. 86 p.

Johnson, Nancy Perry. Experimental sulfide investigation; high pressure Cu-S phase equilibria. M, 1973, Purdue University.

Johnson, Neal Carter. Geometry, stratigraphy, and paleoecology of a Silurian reef at Bluffton, Indiana. M, 1981, Indiana University, Bloomington. 66 p.

Johnson, Neil Evan. A study of the vein copper mineralization of the Virgilina District, Virginia and North Carolina. M, 1983, Virginia Polytechnic Institute and State University. 195 p.

Johnson, Neil Evan. The crystal chemistry of tetrahedrite. D, 1986, Virginia Polytechnic Institute and State University. 216 p.

Johnson, Norma Grace. Early Silurian palynomorphs from the Tuscarora Formation in central Pennsylvania and their paleobotanical and geological significance. M, 1984, Pennsylvania State University, University Park. 97 p.

Johnson, Noye Monroe. Thermoluminescence in biogenic calcium carbonate. M, 1959, University of Wisconsin-Madison.

Johnson, Noye Monroe. Thermoluminescence in contact metamorphosed rock. D, 1962, University of Wisconsin-Madison. 252 p.

Johnson, Ollie Henry, Jr. The geology of the Baxterville oil field, Lamar and Marion counties, Mississippi (production from Tuscaloosa of Upper Cretaceous). M, 1951, Mississippi State University. 52 p.

Johnson, Paul Curtis. Mammalian remains associated with Nebraska Phase Earth lodges in Mills County, Iowa. M, 1972, University of Iowa. 71 p.

Johnson, Paul Howard. Fluidization, agitation, and euphide dissolution in an air-agitated autoclave. D, 1967, University of Utah. 222 p.

Johnson, Paul Richard. X-ray spectrographic analysis of loess deposits in Illinois. D, 1961, University of Illinois, Urbana. 140 p.

Johnson, Paula Ann. A petrographic and petrologic study of the Monument Peak Intrusion in the North Cascades, Washington State. M, 1974, University of Washington. 33 p.

Johnson, Peter Eric. The origin of the Chiwaukum Graben, Chelan County, Washington. M, 1983, Washington State University. 96 p.

Johnson, Peter P. Geology of the Red Rock Fault and adjacent Red Rock Valley, Beaverhead County, Montana. M, 1981, University of Montana. 88 p.

Johnson, Peter Roy. Petrology and environments of deposition of the Herrin (No. 6) Coal Member, Carbondale Formation at the Old Ben Coal Company Mine No. 24, Franklin County, Illinois. M, 1979, University of Illinois, Urbana. 169 p.

Johnson, Philip H. The surficial geology and Pleistocene history of the Milton Quadrangle (Vermont). M, 1970, University of Vermont.

Johnson, Philip W. The investigation of photogeologic fracture traces by electrical prospecting methods. M, 1966, Pennsylvania State University, University Park. 94 p.

Johnson, Pratt H. A study of large scale chemical variations in the Allende Meteorite. M, 1978, University of Houston.

Johnson, Priscilla L. The Cacachara epithermal silver deposit, Puno Department, southernmost Peru. M, 1986, Queen's University. 171 p.

Johnson, R. C. Precambrian geochronology and geology of the Boxelder Canyon area, northern Laramie Range, Wyoming. M, 1974, SUNY at Buffalo. 41 p.

Johnson, R. C. Sillimanite nodules in the Wissahickon Schist, Philadelphia. M, 1975, Temple University.

Johnson, R. D. Dispersal of Recent sediments and mine tailing in a shallow-silled fjord, Rupert Inlet, British Columbia. D, 1974, University of British Columbia.

Johnson, R. H. The synthesis of point data and path data in estimating sonar speed. D, 1968, University of Hawaii.

Johnson, Ragnar Edwin, Jr. Geology of the Richtex Quadrangle, South Carolina. M, 1951, University of South Carolina.

Johnson, Ray T. Geology of the Winslow-Schaberg area. M, 1963, University of Arkansas, Fayetteville.

Johnson, Raymond Larry. The geology of the northeastern quarter of Fillmore Quadrangle, Ventura County, California. M, 1960, University of California, Los Angeles.

Johnson, Rex J. Structure of the Schoonover sequence, Independence Mountains, Nevada; emplacement mechanisms for the Golconda Allochthon. M, 1986, Joint program, Idaho State Univ. and Boise State Univ. 95 p.

Johnson, Rex J. E. Late Precambrian and lower Paleozoic paleomagnetism of the Avalon Terrane in Nova Scotia. D, 1987, University of Michigan. 259 p.

Johnson, Rex J. E. Paleomagnetism and late diagenesis of Jurassic carbonates from the Jura Mountains, Switzerland and France. M, 1982, University of Michigan.

Johnson, Richard Alan. Conodont biostratigraphy of the Bangor Limestone (Mississippian), Monte Sano Mountain, Madison County, Alabama. M, 1974, University of Florida. 193 p.

Johnson, Richard C. Simulation of the behavior of the Kufra well fields. M, 1977, Ohio University, Athens. 245 p.

Johnson, Richard Foster. One-centimeter stratigraphy in foraminiferal ooze; theory and practice. D, 1980, University of California, San Diego.

Johnson, Richard Gustave. Copper-nickel mineralization in the basal Duluth Gabbro Complex, (Cook County), northeastern Minnesota; a case history. M, 1968, University of Iowa. 90 p.

Johnson, Richard Gustave. Economic geology of a portion of the basal Duluth complex (Precambrian), northeastern Minnesota. D, 1970, University of Iowa. 136 p.

Johnson, Richard K. Geomorphic and lithologic controls of diffuse-source salinity, Grand Valley, western

Colorado. M, 1982, Colorado State University. 109 p.

Johnson, Richard Lee, Jr. The groundwater transport of chlorophenolics through a highly fractured soil at Alkali Lake, Oregon. D, 1984, Oregon Graduate Institute of Science and Technology. 277 p.

Johnson, Robert. Advanced statistical analyses of cone penetration tests with other geotechnical parameters. D, 1986, Rensselaer Polytechnic Institute. 380 p.

Johnson, Robert. Potential effects of mineralogy of the Kope, Nancy, and Black Shale units on highway embankments. M, 1980, University of Kentucky.

Johnson, Robert Alfred. The design and characteristics of electro-kinetic transducers. M, 1956, Colorado School of Mines. 38 p.

Johnson, Robert B., Jr. Controls on Precambrian Cu-Pb-Zn mineralization at the Greenville Mine, Clark, CO. M, 1981, Colorado State University. 158 p.

Johnson, Robert Britten. Refraction seismic method for differentiating Pleistocene deposits in the Arcola and Tuscola quadrangles, Illinois. D, 1954, University of Illinois, Urbana. 168 p.

Johnson, Robert C. The geology of the southeastern portion of the Norris Quadrangle, Anderson-Knox counties, Tennessee. M, 1964, University of Tennessee, Knoxville. 51 p.

Johnson, Robert Crandall. Geological investigation and ore reserve estimation of the Copper Chief (Ruby Hill) Mine, Douglas County, Nevada. M, 1977, University of Nevada. 32 p.

Johnson, Robert Edwin. A regional lithofacies study of the Frio Formation in the upper Gulf Coast counties of Texas. M, 1957, University of Houston.

Johnson, Robert Eric. Petrogenesis of carbonate lithofacies along an evolving Paleozoic shelf, lower Chickamauga Group and equivalents, Middle Ordovician, northeastern Tennessee. D, 1988, University of Tennessee, Knoxville. 391 p.

Johnson, Robert Francis. Geology of the Masonic mining district, Mono County, California. M, 1951, University of California, Berkeley. 51 p.

Johnson, Robert Frederick. Processes of calcification in Strombus gigas. D, 1965, University of Miami. 106 p.

Johnson, Robert Kern. Subsurface geology of Northeast Cleveland County, Oklahoma. M, 1958, University of Oklahoma. 67 p.

Johnson, Robert L. The petrology, pebble morphology and origin of the Outer Conglomerate (Upper Member), Copper Harbor Conglomerate (Precambrian), Keweenaw Peninsula, Michigan. M, 1973, Michigan Technological University. 81 p.

Johnson, Robert Lane. Geology and environmental interpretation of the upper Cayugan Bass Islands Dolomite, southeastern Michigan. M, 1974, University of Michigan.

Johnson, Robert Post. Depositional environments of the upper part of the Sentinel Butte Formation, southeastern McKenzie County, North Dakota. M, 1973, University of North Dakota. 63 p.

Johnson, Robert William, Jr. Geology of the Commonwealth area, Florence County, Wisconsin. M, 1958, University of Wisconsin-Madison.

Johnson, Roderick H., Jr. Geology of the Medicine Springs area, Pushmataha County, Oklahoma. M, 1954, University of Oklahoma. 59 p.

Johnson, Rodney C. Geology and precious metal mineralization of the Silver Creek to Island Lake area, Marquette County, Michigan. M, 1987, Michigan Technological University. 138 p.

Johnson, Ronald Dwight. Pre-Jurassic sedimentation, tectonism and stratigraphy in southern Alberta and adjoining areas of British Columbia and Montana. M, 1954, University of British Columbia. 154 p.

Johnson, Ronald Frederick. Ice movement and structural characteristics of the Cathedral Glacier system, Atlin Provincial Park, British Columbia. M, 1983, Montana State University. 65 p.

Johnson, Ronald O. Lithofacies and depositional environments of the Rush Creek Member of the Woodbine Formation (Gulfian) of north-central Texas. M, 1974, University of Texas, Arlington. 160 p.

Johnson, Ross B. A petrographic study of the lower Madera Limestone in central New Mexico. M, 1948, University of New Mexico. 78 p.

Johnson, Roy A. Geophysical investigation of Precambrian crustal structure and Laramide effects across a Precambrian suture in southeastern Wyoming. D, 1984, University of Wyoming. 182 p.

Johnson, Russell Paul. Guidelines for and examples of an autotutorial program in physical geology. M, 1969, University of Toledo. 168 p.

Johnson, Ruth Helen. The petrology of the Exeter Granodiorite (New Hampshire). M, 1936, University of New Hampshire. 31 p.

Johnson, Samuel C. Ostracoda of the (Mississippian) Clore Limestone (Illinois, Kentucky). M, 1938, Columbia University, Teachers College.

Johnson, Samuel Yorks. Sedimentology, petrology, and structure of Mesozoic strata in the northwestern San Juan Islands, Washington. M, 1978, University of Washington. 105 p.

Johnson, Samuel Yorks. Stratigraphy, sedimentology, and tectonic setting of the Eocene Chuckanut Formation, northwest Washington. D, 1982, University of Washington. 221 p.

Johnson, Scot B. The karst of northern Door County, Wisconsin. M, 1987, University of Wisconsin-Green Bay. 122 p.

Johnson, Stephen A. The geology and geochemistry of the southeastern Bighorn Mountains, Johnson County, Wyoming. D, 1981, Colorado School of Mines. 180 p.

Johnson, Stephen A. The weathering of porphyritic leucogranite in Enchanted Rock Batholith, central Texas. M, 1974, University of Southern Mississippi.

Johnson, Stephen Edward. Tracer analysis of the Klamath Falls geothermal resource; a comparison of models. M, 1984, Stanford University.

Johnson, Stephen H. Crustal structures and tectonism in southeastern Alaska and western British Columbia from seismic refraction, seismic reflection, gravity, magnetic, and microearthquake measurements. D, 1972, Oregon State University. 129 p.

Johnson, Stephen Hans. Seismic investigation of the midcontinent gravity high in southeastern Minnesota. M, 1967, University of Minnesota, Minneapolis. 110 p.

Johnson, Stephen M. The geology and geochemistry of uranium and thorium in the Late Cretaceous intrusives and the enclosing Idaho Springs Formation, Black Hawk Quadrangle, Gilpin County, Colorado. M, 1977, Colorado School of Mines. 238 p.

Johnson, Steven A. Comparison of thermal conductivity with elastic properties for six igneous rock types. M, 1974, Purdue University.

Johnson, Steven B. Delayed yield in unconfined aquifers. M, 1977, Arizona State University. 42 p.

Johnson, Steven Carl. Astrapotheres from the Miocene of Colombia, South America. D, 1984, University of California, Berkeley. 191 p.

Johnson, Steven R. Impact of cattle grazing on the surface water quality of a Colorado Front Range stream. M, 1979, Colorado State University. 126 p.

Johnson, Steven T. The paragenesis of chert and dolomite in the formation of the Boyle Dolomite of east-central Kentucky. M, 1980, Eastern Kentucky University. 96 p.

Johnson, Stuart Donald. Trace element distribution and wall-rock alteration within the upper Mississippi Valley lead-zinc district. M, 1971, University of Wisconsin-Madison.

Johnson, Thomas C. The dissolution of siliceous microfossils in deep-sea sediments. D, 1975, University of California, San Diego. 181 p.

Johnson, Thomas F. Paleoenvironmental analysis and structural petrogenesis of the Carolina Slate Belt near

Columbia, South Carolina. M, 1972, University of South Carolina. 33 p.

Johnson, Thomas Lee. Water quality study of shallow aquifer at Keystone, South Dakota. M, 1975, South Dakota School of Mines & Technology.

Johnson, Thomas M. Features of the lake plain in Chautauqua County, New York. M, 1982, SUNY, College at Fredonia. 53 p.

Johnson, Thomas Mark. Surficial geology of a portion of South-central Walworth County, Wisconsin with planning implications. M, 1976, University of Wisconsin-Madison.

Johnson, Torrence Vaino. Albedo and spectral reflectivity of the Galilean satellites of Jupiter. D, 1970, California Institute of Technology. 92 p.

Johnson, Tracy L. A study of the dynamic and static properties of stick-slip friction of rock. D, 1974, Columbia University. 118 p.

Johnson, Vard H. The geology of the Helvetia mining district, Arizona. D, 1941, University of Arizona.

Johnson, Vard Hayes. A quantitative study of shortening during uplift of geosynclines. M, 1933, Brigham Young University. 17 p.

Johnson, Verner C. Fracture patterns along Pomona fault in Jackson and Union counties, Illinois. M, 1970, Southern Illinois University, Carbondale. 42 p.

Johnson, Verner Carl. Geophysical survey of the Yellow Creek area, Miss. D, 1975, University of Tennessee, Knoxville. 89 p.

Johnson, Vernon Gene. Decline of radioactivity in the Columbia River and estuary; rates and mechanisms. D, 1979, Oregon State University. 126 p.

Johnson, Vernon Gene. Retention of Zinc-65 by Columbia River sediment. M, 1966, Oregon State University. 56 p.

Johnson, W. C. The impact of environmental change on fluvial systems; Kickapoo River, Wisconsin. D, 1976, University of Wisconsin-Madison. 479 p.

Johnson, Wallace Ray. Structure and stratigraphy of the southeastern quarter of the Roseburg 15′ Quadrangle, Douglas County, Oregon. M, 1965, University of Oregon. 85 p.

Johnson, Wayne Lawrence. Copper-nickel sulphides in a layered ultramafic body (Precambrian), Renzy Mine, southwestern Quebec. M, 1972, University of Western Ontario. 109 p.

Johnson, Wayne S. The petrology of the Morgantown Sandstone of the Pittsburgh District. M, 1942, University of Pittsburgh.

Johnson, Wendell B. The mineralogy of some shales of the Lower Permian System of Riley County, Kansas. M, 1949, Kansas State University. 72 p.

Johnson, William Hilton. Stratigraphy and petrography of Illinoian and Kansan drift in central Illinois. D, 1962, University of Illinois, Urbana. 146 p.

Johnson, William Hilton. Weathering profile information on some Wisconsinan end moraines in east-central Illinois. M, 1961, University of Illinois, Urbana. 78 p.

Johnson, William James. The application of geophysical exploration techniques to a site investigation of the Houghton County Memorial Airport, Houghton, Michigan. M, 1973, Michigan Technological University. 28 p.

Johnson, William Jeffery. Conodont biostratigraphy, sedimentology, and depositional environments of the Etchart Limestone, north-central Nevada. M, 1987, University of Wisconsin-Madison. 100 p.

Johnson, William McDaniel. A remedial investigation work plan for Kem-Pest Laboratories in Cape Girardeau, Missouri. M, 1986, University of Missouri, Rolla. 68 p.

Johnson, William P. The physical and magnetic polarity stratigraphy of the Bunthang Sequence, Skardu Basin, northern Pakistan. M, 1986, Dartmouth College. 108 p.

Johnson, William R. Some Plattsmouth microfossils from the Snyderville Quarry, Cass County, Nebraska. M, 1931, University of Nebraska, Lincoln.

Johnson, William W. Interface waves in cubic symmetry media. D, 1965, University of Pittsburgh.

Johnson, William W. Regional gravity survey of part of Tooele County, Utah. M, 1958, University of Utah. 38 p.

Johnson, William Wallace. Instrumentation for a practical laboratory method to determine the dynamic tensil strength of rock. M, 1963, Michigan Technological University. 175 p.

Johnson, Wylie Bruce. Geology of the Lucas area, Collin County, Texas. M, 1948, Southern Methodist University. 41 p.

Johnsson, Mark. The thermal and burial history of south central New York; evidence from vitrinite reflectance, clay mineral diagenesis and fission track dating of apatite and zircon. M, 1984, Dartmouth College. 155 p.

Johnsson, Patricia. Magnetic polarity stratigraphy and age of the Rio Jachal and Mogna formations at the Sierra de Huaco, San Juan Province, Argentina. M, 1984, Dartmouth College. 49 p.

Johnston, A. C. An investigation of temporal P-velocity variation using travel time residuals at Hawaii and selected Circum-Pacific sites. D, 1979, University of Colorado. 258 p.

Johnston, A. W. Geology and ore deposits of the Bankfield gold mine. M, 1935, Queen's University. 29 p.

Johnston, Alan Dana. Mineralogy, petrology, and geochemistry of some unusually oxidized rocks from Kauai, Hawaii, and their entrained mantle xenoliths. D, 1983, University of Minnesota, Minneapolis. 210 p.

Johnston, Alan Dana. The mineralogy, petrology and origin of some spinel-lherzolite xenoliths from Papapapaholahola Hill, Kaui, Hawaii. M, 1978, University of Minnesota, Minneapolis. 82 p.

Johnston, Beatrice Bryant. Petrologic study of a Miocene gabbro emplaced during initial rifting in the Red Sea. M, 1978, Texas A&M University.

Johnston, Carl Hewitt. Areal geology of Red Bluff area, Eddy County, New Mexico. M, 1952, University of Texas, Austin.

Johnston, Carol Arlene. Effects of a seasonally flooded freshwater wetland on water quality from an agricultural watershed. D, 1982, University of Wisconsin-Madison. 146 p.

Johnston, Claud Stuart. The orientation of mineral particles during rock flowage. M, 1926, University of North Carolina, Chapel Hill. 46 p.

Johnston, Clifford. A Raman spectroscopic study of kaolinite. D, 1983, University of California, Riverside. 141 p.

Johnston, David A. Volatiles, magma mixing, and the mechanism of eruption of Augustine Volcano, Alaska. D, 1978, University of Washington. 177 p.

Johnston, David A. Volcanistic facies and implications for the eruptive history of the Cimarron Volcano, San Juan Mountains, SW Colorado. M, 1978, University of Washington. 117 p.

Johnston, David Dean. Petroleum geology of the upper Wilcox Group, (lower Eocene), Livingstone Field, Louisiana. M, 1986, University of California, Los Angeles. 235 p.

Johnston, David Earle. The effect of assumed source structure on inversion for earthquake source parameters; the eastern Hispaniola earthquake of 14 September, 1981. M, 1983, Pennsylvania State University, University Park. 34 p.

Johnston, David Hervey. The attenuation of seismic waves in dry and saturated rocks. D, 1979, Massachusetts Institute of Technology. 432 p.

Johnston, David Kent. Detailed subsurface geology and potential petroleum production of the Waltersburg Sandstone (Chester Series, Upper Mississippian) in Southwest Gibson County, Indiana. M, 1981, Ball State University. 135 p.

Johnston, Donald J. Capability mapping for sanitary landfill in southern Oconto County utilizing computer graphics. M, 1987, University of Wisconsin-Green Bay. 146 p.

Johnston, Francis N. Structure and stratigraphy of New Pass Range, Nevada. M, 1930, Stanford University. 53 p.

Johnston, Gregory Lamar. A seismic spectral discriminant for reservoir induced earthquakes in the Southeastern United States. M, 1980, Georgia Institute of Technology. 117 p.

Johnston, H. M. The effect of natural cesium and strontium concentrations on distribution coefficient determinations. M, 1979, University of Waterloo.

Johnston, Herbert C., Jr. Geology of the East Grants Ridge, Valencia County, New Mexico. M, 1953, University of New Mexico. 51 p.

Johnston, I. M. A detailed stratigraphic and environmental analysis of the San Rafael Group (Jurassic) between Black Mesa, Arizona, and the southern Kaiparowits Plateau, Utah. D, 1975, University of Arizona. 538 p.

Johnston, I. S. Functional ultrastructure of the skeleton and the skeletogenic tissues of the reef coral Pocillopora damicornis. D, 1978, University of California, Los Angeles. 225 p.

Johnston, Ian McKay. Eocene foraminifera from the lower Maniobra Formation, Orocopia Mountains, Riverside County, California. M, 1961, University of California, Berkeley. 94 p.

Johnston, J. C. Stratigraphy of the Dinwoody Formation of western Wyoming. M, 1939, University of Wyoming. 48 p.

Johnston, J. F. Geology of the Stratmat Group 61 ore zone (New Brunswick). M, 1959, University of New Brunswick.

Johnston, James S. Whole rock chemistry of granitic rocks as a guide to tectonic setting. M, 1979, Eastern Washington University. 131 p.

Johnston, James William Derek. The geology of the Valentine lake area, west central Newfoundland. M, 1950, Dalhousie University.

Johnston, Janet Catherine Pruszenski. An investigation of body wave magnitude using the new digital seismic research observatories plus conventional catalogue data. M, 1979, Massachusetts Institute of Technology. 84 p.

Johnston, Jay S. Depositional systems of the lower Atoka Formation in the Arkoma Basin, Arkansas. M, 1982, University of Arkansas, Fayetteville.

Johnston, John B. A magnetic survey of the Tertiary intrusives comprising the Circus Flats and Bear Butte structures, Meade County, South Dakota. M, 1987, Kent State University, Kent. 84 p.

Johnston, John E. Depositional systems in the Wilcox Group and the Carrizo Formation (Eocene) of central and South Texas and their relationship to the occurrence of lignite. M, 1977, University of Texas, Austin.

Johnston, Joseph Eggleston. Stratigraphy of the Stanley Group (Mississippian) in the central Ouachita Mountains of western Pushmataha County, Oklahoma. M, 1984, University of Arkansas, Fayetteville. 84 p.

Johnston, Laura M. A geochemical study of Deadman Bay near Kingston, eastern Lake Ontario. M, 1972, Queen's University. 101 p.

Johnston, Laura M. Geolimnological studies in the Kingston Basin-upper St. Lawrence River region. D, 1978, Queen's University. 243 p.

Johnston, Maureen Dawne. A geochemical study of boron, iron and manganese in the Deville Member of the lower Cretaceous McMurray Formation, Alberta. M, 1971, University of Western Ontario. 121 p.

Johnston, Paul Anthony Frederick. Late Cretaceous and Paleocene mammals from southwestern Saskatchewan. M, 1980, University of Alberta. 246 p.

Johnston, Paul Francis. Succession and distribution of Ostracoda in highway borrow-pit ponds of central Alberta, (Canada). M, 1966, University of Alberta. 115 p.

Johnston, Paul L. The geology of the Red Gulch area, Fremont County, Colorado. M, 1959, University of Kansas. 34 p.

Johnston, Paul Roche. Finite element consolidation analyses of tunnel behavior in clay. D, 1981, Stanford University. 236 p.

Johnston, R. H. Geology of the northern Leucite Hills, Sweetwater County, Wyoming. M, 1959, University of Wyoming. 83 p.

Johnston, Robert C. Areal geology around Parkdale, Fremont County, Colorado. M, 1953, University of Oklahoma. 101 p.

Johnston, Robert L. The geology of a portion of the western Verdugo Mountains. M, 1938, University of California, Los Angeles.

Johnston, S. C. Frequency domain analysis of AC dipole-dipole electromagnetic soundings. M, 1975, University of Wisconsin-Madison.

Johnston, S. W. Behavior of indigenous arsenic in flooded soils and sediments. M, 1978, Louisiana State University.

Johnston, Stedwell. Geology of a portion of the Calistoga Quadrangle, California. M, 1948, University of California, Berkeley. 46 p.

Johnston, Stephen Thomas. Structure of the Triangle Zone in the Rocky Mountain Foothills near Coalspur, Alberta. M, 1985, University of Alberta. 81 p.

Johnston, Steven E. A comparison of dating methods in laminated lake sediments in Maine. M, 1981, University of Maine. 79 p.

Johnston, W. G. The Maxam Lake-Cross Lake section of the Timiskaming-Grenville contact (Canada). D, 1950, Massachusetts Institute of Technology. 137 p.

Johnston, Wilbert E. A study of the gold-bearing telluride minerals. D, 1936, University of Toronto.

Johnston, William D., Jr. Ground water in the Paleozoic rocks of northern Alabama. M, 1933, George Washington University. 148 p.

Johnston, William George. A petrographic study of the relationship of the Timiskaming to the Grenville Subprovince. M, 1947, University of British Columbia.

Johnston, William Percy. The geology of ore deposits of the Copper Basin mining district, Yavapai County, Arizona. D, 1955, University of Utah. 132 p.

Johnstone, Robert M. Geology of the Stoughton-Roquemaure Group, Beatty and Munro townships, northwestern Ontario. M, 1987, Carleton University. 325 p.

Joity, John Frank. Paleoenvironmental interpretation of the upper Black Hand Conglomerate (Mississippian) in southeastern Ohio. M, 1973, University of Michigan.

Jokela, Arthur. Submarine geology of the Red Sea. M, 1965, Massachusetts Institute of Technology. 46 p.

Jokisaari, Allan O. The Latah County map series; a simplified method for local production of basic topographic mapping. M, 1982, University of Idaho. 99 p.

Joliat, Steven A. Petrology of selected coal seams of the Williams Fork Formation, Moffat County, Colorado. M, 1983, Southern Illinois University, Carbondale. 97 p.

Jolley, Richard Michael. The Clearville Siltstone Member of the Middle Devonian Mahantango Formation in parts of Pennsylvania, Maryland, West Virginia, and Virginia. M, 1983, University of North Carolina, Chapel Hill. 192 p.

Jolley, Ted R. The alteration of the feldspars. M, 1931, University of Minnesota, Minneapolis. 25 p.

Jolliff, Bradley L. Petrogenesis, geochemical relationships and internal evolution of granitic pegmatites in the Keystone area, Black Hills, South Dakota. D, 1987, South Dakota School of Mines & Technology.

Jolliff, Bradley L. Tourmaline as a recorder of pegmatite evolution; Bob Ingersoll Pegmatite, Black Hills,

South Dakota. M, 1985, South Dakota School of Mines & Technology.

Jolliffe, Alfred W. Pitchblende in a giant quartz vein, Beaverlodge Lake, Northwest Territories. D, 1935, Princeton University. 143 p.

Jolliffe, F. T. Laboratory investigation of gold-quartz vein formation. M, 1931, Queen's University. 46 p.

Jolly, Glenn Douglass. Correlation of the Woodford Formation in south-central Oklahoma using gamma-ray scintillation measurements of the natural background radiation. M, 1988, Stephen F. Austin State University. 153 p.

Jolly, James G. Geology, petrography, and sulfide mineralization of the Raddison Lake area, Itasca County, Minnesota. M, 1977, Bowling Green State University. 81 p.

Jolly, James H. The geology of the northwest third of the Monument Quadrangle, Oregon. M, 1957, University of Oregon. 55 p.

Jolly, Janice L. The geology of the southern third of the Monument Quadrangle, Oregon. M, 1957, University of Oregon. 76 p.

Jolly, Wayne Travis. Petrologic studies of the Robles Formation (upper Cretaceous), south central Puerto Rico. D, 1970, SUNY at Binghamton. 150 p.

Jolly, Wayne Travis. Zoned feldspars from the Town Mountain granite (Precambrian), Llano region, central Texas. M, 1970, University of Texas, Austin.

Jonah, Maxwell V. Investigations into the uses of variance-covariance analysis of error equation coefficients in aerial triangulation. M, 1961, Ohio State University.

Jonas, Anna Isabel and Bliss, Eleanora Frances. Relation of the Wissahickon micagneiss to the Shenandoah Limestone and to the Octoraro mica-schist of the Doe Run-Avondale District, Coatsville Quadrangle, Pa. D, 1912, Bryn Mawr College. 64 p.

Jonas, Edward Charles. The reversible dehydroxylization of three-layer clay minerals. D, 1954, University of Illinois, Urbana. 64 p.

Jonas, Edward Charles. Ultrasonic radiation techniques applied to clay mineralogy. M, 1952, University of Illinois, Urbana. 36 p.

Jonasson, Ralph George. Chemistry and geochemistry of the rare earth element phosphates. D, 1987, University of Western Ontario.

Joncas, Gilles. A study of pyrochlore from the Oka District. M, 1962, Ecole Polytechnique. 50 p.

Joneja, Danielle C. Sedimentary dynamics of Fire Island, New York. M, 1981, University of Massachusetts. 174 p.

Jones, A. E. Empirical studies of some of the seismic phenomena in Hawaii. M, 1938, University of Hawaii. 32 p.

Jones, A. E. Nyema. A chemical study of laterites and lateritic soils with emphasis on element distribution. D, 1962, University of Chicago. 110 p.

Jones, A. H. M. A preliminary study of southern Ontario tills. M, 1962, University of Toronto.

Jones, Alan M. A study of internal structure of fine-grained clastic rocks by x-radiography. M, 1969, Brigham Young University. 162 p.

Jones, Albert M., Jr. The application of prospecting with thermoluminescence in the Mayflower Mine, Heber, Utah. M, 1964, Tulane University. 48 p.

Jones, Albert Vincent. Foraminifera of the Vincentown Formation in New Jersey. M, 1952, Rutgers, The State University, New Brunswick. 124 p.

Jones, Alexander Gordon. Vernon map-area, British Columbia. D, 1960, Harvard University.

Jones, Alice J. Stratigraphy and fauna of the Late Ordovician strata of the Green Bay-Lake Winnebago region. M, 1930, Northwestern University.

Jones, Alice Jane. Spatial variability of hydraulic conductivity and related field soil water fluxes. D, 1982, Utah State University. 244 p.

Jones, Alison H. Paleomagnetism and rock magnetism of Powder River basin clinker, northeastern Wyoming

and southeastern Montana. M, 1983, Colorado School of Mines. 93 p.

Jones, Allen L. Spatial and temporal analysis for Idaho earthquake hazard assessments. M, 1988, University of Idaho. 73 p.

Jones, B. Facies and faunal aspects of the Silurian Read Bay Formation of northern Somerset Island, District of Franklin, Canada. D, 1974, University of Ottawa. 448 p.

Jones, B. E. The structural geology of the Owl's Head Peninsula, Halifax County, Nova Scotia. M, 1975, Acadia University.

Jones, Benjamin F. Geology of the San Luis Rey Quadrangle. M, 1959, University of Southern California.

Jones, Billy R. A sedimentary study of dune sands, Lamb and Bailey counties, Texas and White Sands National Monument, New Mexico. M, 1959, Texas Tech University. 95 p.

Jones, Billy Ray. Geology of southern Quitman Mountains and vicinity, Hudspeth County, Texas. D, 1968, Texas A&M University. 188 p.

Jones, Blair F. Hydrology and mineralogy of Deep Spring Lake, Inyo County, CA. M, 1963, The Johns Hopkins University.

Jones, Boone. The geology of the Wewoka area. M, 1922, University of Oklahoma. 29 p.

Jones, Bradley Blake. Depositional history of the basal Atoka Formation in northeastern Oklahoma as interpreted from primary sedimentary structures and stratification sequences. M, 1977, University of Oklahoma. 95 p.

Jones, Bradley William. Estimation of absorption from seismic reflection data. M, 1979, University of Utah. 114 p.

Jones, Brian A. A gravity survey and interpretation in northwestern Ontario. M, 1973, University of Toronto.

Jones, Brian K. Uranium and thorium in granitic and alkaline rocks in western Alaska. M, 1977, University of Alaska, Fairbanks. 123 p.

Jones, Bruce Charles. Data processing of a crooked seismic line in an area of thick glacial till; Clare County, Michigan. M, 1987, Wright State University. 87 p.

Jones, Byron K. A petrographic and structural comparison of the Hackneyville Schist and andalusite schist of the Wedowee Group in Clay and Tallapoosa counties, Alabama. M, 1980, Memphis State University.

Jones, C. Keith. Structure along the Whiteoak Mountain Fault near Kingston, Roane County, Tennessee. M, 1963, University of Tennessee, Knoxville. 27 p.

Jones, Carol C. The biology and evolution of Chione (Bivalvia; Veneridae) in the eastern coastal plains of North America. D, 1976, Harvard University.

Jones, Cecil L., Jr. An isopach, structural, and paleogeologic study of the pre-Desmoinesian units in north-central Oklahoma. M, 1959, University of Oklahoma. 81 p.

Jones, Charles Alan. A study of sediment transport and dispersal in the Sixes River Estuary, Oregon, utilizing fluorescent tracers. D, 1972, University of Oregon. 150 p.

Jones, Charles L. Crystallization process in solutions and the textures of igneous rocks. M, 1947, Northwestern University.

Jones, Charles L. The petrology of some Recent beachrock from the West Indies. M, 1961, Louisiana State University.

Jones, Clarke. The geology of the New Loyston area, Union and Anderson counties, Tennessee. M, 1962, University of Tennessee, Knoxville. 31 p.

Jones, Craig Howard. A geophysical and geological investigation of extensional structures, Great Basin, Western United States. D, 1988, Massachusetts Institute of Technology. 226 p.

Jones, Craig T. Development of a water quality model applicable to Great Salt Lake, Utah. D, 1976, Utah State University. 229 p.

Jones, D. L.; Hohenstein, C. G. and Jaeger, J. W. A study of marked sand movement on Del Monte beach, Monterey bay, California. M, 1965, United States Naval Academy.

Jones, D. L. and Sutton, J. S. Geology of Panoche Valley Quadrangle (California). M, 1953, [Stanford University].

Jones, Daniel Hubbard. Palynology of the Bevier Coal of Missouri. M, 1957, University of Missouri, Columbia.

Jones, Daniel J. Physiography of Greensboro, Hardwick, and Woodbury, Vermont. M, 1915, Syracuse University.

Jones, Daniel John. The conodont fauna of the (Pennsylvanian) Seminole Formation (Oklahoma). D, 1938, University of Chicago. 55 p.

Jones, Daniel John. The conodonts of the Nowata Shale. M, 1935, University of Oklahoma. 69 p.

Jones, Daphne L. Isotopic and petrographic evidence relevant to the origin of the Arkansas Novaculite. M, 1978, Louisiana State University.

Jones, Darrell King. Organic and inorganic carbon in Recent sediments of the open gulf, barrier island and bay environments, Mustang Island, Texas. M, 1960, University of Texas, Austin.

Jones, David E. Phase equilibria and conditions of metamorphism in the southwest portion of the Narragansett Basin. M, 1987, University of Rhode Island.

Jones, David Gordon. Geology of the iron formation and associated rocks of the Jackson County iron mine, Jackson County, Wisconsin. M, 1978, University of Wisconsin-Madison.

Jones, David Kerrell. A geochemical study of a mineralized breccia body in the central Tennessee zinc district. M, 1982, University of Kentucky. 90 p.

Jones, David L. A taxonomic review of the Cretaceous pelecypod subfamily Inoceraminae. D, 1956, Stanford University. 328 p.

Jones, David Laurence. Waulsortian facies of the Tin Mountain Limestone, Bat Mountain, California. M, 1988, University of California, Riverside. 151 p.

Jones, David Pierce. Volcanic geology of the Alegros Mountain area, Catron County, New Mexico. M, 1980, University of New Mexico. 76 p.

Jones, Donovan Deronda, Jr. Petrofabric and movement study of faults in Newton and Walton counties, Georgia. M, 1970, Emory University. 28 p.

Jones, Douglas. Geology of Boissier Parish, Louisiana. D, 1959, Louisiana State University.

Jones, Douglas Frank. The effect of vertical seawalls on longshore currents. D, 1975, University of Florida. 189 p.

Jones, Douglas Stephen. Annual cycle of shell growth and reproduction in the bivalves Spisula Solidissima and Arctica Islandica. D, 1980, Princeton University. 238 p.

Jones, Duke Forrest. Stratigraphy, environments of deposition, petrology, age, and provenance, basal red beds of the Argana Valley, western High Atlas Mountains, Morocco. M, 1975, New Mexico Institute of Mining and Technology. 148 p.

Jones, E. A. W. A comparison of some of the physical methods of age determination. M, 1954, University of Toronto.

Jones, Earl Verner. Niagaran series of southeastern Ohio and southwestern West Virginia. M, 1956, University of Pittsburgh.

Jones, Edward J. Paleogeography and depositional systems of the late Pleistocene Redbird Delta, Lake County, Ohio. M, 1985, University of Akron. 130 p.

Jones, Edward L. Geology of the district enclosing the forest of Dean iron mine in the highlands of New York. M, 1910, Columbia University, Teachers College.

Jones, Elbert Russ. Some minor structures in Arkansas River sand bars. M, 1962, University of Tulsa. 105 p.

Jones, Ernest Victor. A study of a Georgia clay from McDuffie County, Ga. M, 1909, Vanderbilt University.

Jones, Eugene L. Surface radioactivity of the Granite Mountain area, Pulaski County, Arkansas. M, 1952, University of Arkansas, Fayetteville.

Jones, Eugene Laverne. Plant microfossils of the laminated sediments of the lower Eocene Wilcox Group in south-central Arkansas. D, 1961, University of Oklahoma. 126 p.

Jones, Everett Bruce. Some aspects of ground-water development and management in the northern portion of the Colorado High Plains. D, 1964, Colorado State University. 157 p.

Jones, F. Ross. Structural geology of the northern Galice Formation, western Klamath Mountains, Oregon and California. M, 1988, SUNY at Albany. 211 p.

Jones, Frances Gwynn. The Hot Spring Limestone, Tertiary algal reefs from the Snake River plain, Idaho. M, 1978, University of Michigan.

Jones, Francis Hugh Melvill. Digital impulse radar for glaciology; instrumentation, modelling, and field studies. M, 1987, University of British Columbia. 110 p.

Jones, Frank Burdette. Study of the attenuation and azimuthal dependence of seismic wave propagation in the southeastern United States and evaluation of Q from the coda of southeastern United States earthquakes at ATL seismic observatory. D, 1983, Georgia Institute of Technology. 87 p.

Jones, G. M. The thermal interaction of the core and mantle. D, 1976, University of California, Berkeley. 81 p.

Jones, Gareth H. S. Strong motion seismic effects of the Suffield explosions. D, 1963, University of Alberta. 286 p.

Jones, Garnet W., Jr. Ground water resources of Lincoln Parish with emphasis on the southern part. M, 1962, Louisiana Tech University.

Jones, Garry. Seasonal ecology of Recent foraminifera in Samish Bay, Washington. M, 1977, Western Washington University. 96 p.

Jones, Garry Davis. Foraminiferal paleontology and geology of lower Claibornian rocks of the inner coastal plain of North Carolina. D, 1982, University of Delaware. 885 p.

Jones, George P., Jr. An electrical resistivity survey of a portion of the American Bottoms. M, 1952, St. Louis University.

Jones, Glenn A. Advective transport of sediments in the region of the Rio Grande Rise (SW Atlantic) during the Holocene/Pleistocene. D, 1983, Columbia University, Teachers College. 264 p.

Jones, Gregg W. Investigation of the saltwater-freshwater interface in a complex hydrogeologic environment, coastal Alafia River basin. M, 1986, University of South Florida, Tampa. 122 p.

Jones, Gregory L. Geology and ore deposits of the Reveille mining district, Nye County, Nevada. M, 1985, Brigham Young University. 84 p.

Jones, Hal Joseph. Experimental studies of the elasticity of rocks. D, 1950, University of Texas, Austin.

Jones, Hal Joseph. Some contemporary theories of mountain genesis. M, 1947, University of Cincinnati. 42 p.

Jones, Henry D. Stable isotope systematics in the Illinois-Kentucky fluorspar district. M, 1987, Iowa State University of Science and Technology. 129 p.

Jones, Hershel Leonard. Petrography, mineralogy, and geochemistry of the chamositic iron ores of north-central Louisiana. D, 1969, University of Oklahoma. 209 p.

Jones, Hershel Leonard. The geology of the iron ore deposits of the Sugar Creek area, Claiborne, Lincoln,

and Bienville parishes, Louisiana. M, 1962, Louisiana State University.

Jones, Ian Frederick. Analyses of global seismicity; earthquake migration; the frequency-magnitude relation. M, 1980, University of Western Ontario.

Jones, Ian Frederick. Applications of the Karhunen-Loève transform in reflection seismology. D, 1985, University of British Columbia.

Jones, Irene B. Carter. A study of the stability of Martian surface sample analog mineral phases under conditions of dry heat sterilization. M, 1976, University of Houston.

Jones, Islwyn Wyn. The microscopical and chemical nature of Alberta coals. D, 1928, University of Toronto.

Jones, J. A. Geology of the northern Kilbeck Hills and an adjacent portion of the Old Woman Mountains, eastern Mojave Desert, San Bernardino County, California. M, 1973, University of Southern California. 56 p.

Jones, J. Claiborne. Local environmental management, a case study; the Virginia Wetlands Act, 1972-74. M, 1976, College of William and Mary.

Jones, J. Claude. The geologic history of Lake Lahontan. D, 1923, University of Chicago. 79 p.

Jones, J. G. Inversion for crustal structure using the ratio of vertical to horizontal components of body wave spectra. M, 1982, SUNY at Binghamton. 46 p.

Jones, J. R. A multivariate analysis of dune and beach sediment parameters as possible indicators of barrier island migration. M, 1974, Boston University. 83 p.

Jones, J. R. An alternative hypothesis for barrier island migration. D, 1977, Boston University. 205 p.

Jones, J. Richard. A biostratigraphic model for the origin and development of salt marshes; Plum Island, Massachusetts. D, 1988, University of Pittsburgh. 131 p.

Jones, J. Robert. The etching of crystals by optically active solvents. M, 1934, Pennsylvania State University, University Park. 31 p.

Jones, Jackson G. Geology of the Ashland-Kiowa area, Pittsburg County, Oklahoma. M, 1957, University of Oklahoma. 126 p.

Jones, James Irvin. The ecology and distribution of living planktonic foraminifera of the West Indies and adjacent waters. D, 1964, University of Wisconsin-Madison. 199 p.

Jones, James Irwin. The significance of variability in Praeglobotruncana cretacea (d'Orbigny) 1840, from the Cretaceous Eagle Ford Group of Texas. M, 1960, University of Wisconsin-Madison.

Jones, James Ogden. Stratigraphy of the Walnut Formation (Lower Cretaceous), central Texas. M, 1966, Baylor University.

Jones, James Ogden. The Blaine Formation of north Texas. D, 1971, University of Iowa. 173 p.

Jones, James R. A study of some insoluble residues from the Elmwood beds at Kimber Spring Ravine (New York). M, 1948, Syracuse University.

Jones, James Winston. Depositional environment and morphology of Canyon sandstone reservoirs, central Midland Basin, Texas. M, 1980, Texas A&M University. 147 p.

Jones, Jay H. Leaf architectural and cuticular analyses of extant Fagaceae and "fagaceous" leaves from the Paleogene of southeastern North America. D, 1984, Indiana University, Bloomington. 328 p.

Jones, Jay W. Structural, hydraulic, and geophysical properties of granite near Oracle, Arizona. M, 1983, University of Arizona. 140 p.

Jones, Jayne Ann. Analysis of monoclinal folds associated with the Brittmore Fault in Northwest Houston, Texas. M, 1988, Texas A&M University. 160 p.

Jones, Jeffrey T. The geology and structure of the Canyon Creek Church Mountain area, North Cascades, Washington. M, 1984, Western Washington University. 125 p.

Jones, John Brett. Dispersion in trioctahedral micas. D, 1958, University of Wisconsin-Madison. 60 p.

Jones, John Frederick. Surficial geology and related problems, Beaverlodge District, northwestern Alberta. M, 1961, University of Western Ontario. 153 p.

Jones, John Hume. Studies of the geochemical similarity of plutonium and samarium and their implications for the abundance of [244]Pu in the early solar system. D, 1981, California Institute of Technology. 203 p.

Jones, John R. Eutrophication of some northwestern Iowa lakes. D, 1974, Iowa State University of Science and Technology. 92 p.

Jones, Jon Rex, Jr. Reservoir characterization for numerical simulation of Mesaverde meanderbelt sandstone, northwestern Colorado. M, 1988, University of Texas, Austin. 115 p.

Jones, Jonathan Wyn. A study of some low-grade regional metamorphic rocks from the Omineca crystalline belt, British Columbia. D, 1972, University of Calgary. 127 p.

Jones, Jonathan Wyn. Low-grade metamorphism of Proterozoic rocks from the Esplanade Range, British Columbia. M, 1969, University of Calgary. 112 p.

Jones, Joseph Maxfield. Triassic-Jurassic contact in central Wyoming. M, 1940, University of Missouri, Columbia.

Jones, Kathleen Ferris. Resonant interaction of internal waves in linear shear flow. M, 1974, University of Washington. 19 p.

Jones, Kathleen L. A field trip to analyze two geologic provinces of the Harrisburg, Pennsylvania area. M, 1976, Pennsylvania State University, University Park. 36 p.

Jones, Kenneth L. Martian obliterational history. D, 1974, Brown University. 125 p.

Jones, L. D. Homomorphic deconvolution of marine magnetic anomalies. M, 1976, Texas A&M University.

Jones, LaDon Carlos. Optimal control of nonlinear groundwater hydraulics using differential dynamic programming. D, 1986, University of California, Los Angeles. 82 p.

Jones, Larry LeRoy. Subsurface study of the Amacker-Tippett area, Upton County, Texas. M, 1960, University of Nebraska, Lincoln.

Jones, Leland Willard. The relation of sulphur to petroleum deposits. M, 1927, University of Michigan.

Jones, Leo David. Cepstral analysis of marine magnetic anomalies. M, 1976, Texas A&M University.

Jones, Lois Marilyn. The application of strontium isotopes as natural tracers; the origin of salts in the lakes and soils of southern Victoria Land, Antarctica. D, 1969, Ohio State University. 375 p.

Jones, Lucille Merrill. Field and laboratory studies of the mechanics of faulting. D, 1981, Massachusetts Institute of Technology. 106 p.

Jones, M. G. A study of the late-stage secondary fracture mineralization at sites across the Precambrian Canadian Shield. M, 1987, University of Waterloo. 283 p.

Jones, Margaret Grace McCorkle. Foraminifera in the lower Yazoo Member of the Jackson Formation (Mississippi). M, 1940, University of Mississippi.

Jones, Mark Lewis. An investigation of landslide hazards in northern Summit County, Ohio, with special emphasis on selected landslide locations. M, 1986, Kent State University, Kent. 192 p.

Jones, Mark M. The solubility of carbon dioxide in waters of low alkalinity. M, 1969, Oregon State University. 38 p.

Jones, Meridee. Paleozoic paleomagnetism of the Armorican Massif, France. M, 1978, University of Michigan.

Jones, Michael. Paleocurrents in lutaceous sediments (Ordovician, Silurian and Devonian, several localities in Tennessee, Virginia, West Virginia and Alabama). M, 1967, University of Tennessee, Knoxville. 46 p.

Jones, Michael B. Geology of Blue Hill mining district, Hancock County, Maine. M, 1969, University of Washington. 58 p.

Jones, Michael B. Hydrothermal alteration and mineralization of the Valley copper deposits, Highland Valley, British Columbia. D, 1975, Oregon State University. 262 p.

Jones, Murray L. G. Othniel Charles Marsh, palaeontologist. M, 1936, Columbia University, Teachers College.

Jones, N. O. Uranium, copper, and vanadium content of selected arenaceous sediments freom the lower Supai Formation, Mogollon Rim, Arizona. M, 1977, University of Arizona. 126 p.

Jones, Neil Owen. The development of piping erosion. D, 1968, University of Arizona. 219 p.

Jones, Norman Kenneth. Debris transport and deposition at Boundary Glacier, Banff National Park, Alberta. D, 1987, University of Waterloo. 314 p.

Jones, Norris William. The crystal chemistry of humite group. D, 1968, Virginia Polytechnic Institute and State University. 117 p.

Jones, Norris William. The relationships between the Duluth Gabbro and the dikes and sills in the vicinity of Hovland, Minnesota. M, 1963, University of Minnesota, Minneapolis. 90 p.

Jones, P. L. Petrology and petrography of beachrock (Pleistocene?) Sonoran coast, northern Gulf of California. M, 1975, University of Arizona.

Jones, Paul. Evidence from high resolution seismic reflection data for recurrent faulting in the New Madrid seismic zone. M, 1982, Purdue University. 142 p.

Jones, Paul H. Electric logging applied to groundwater exploration. M, 1950, Louisiana State University.

Jones, Paul Hastings. Hydrology of Neogene deposits in the northern Gulf of Mexico Basin. D, 1968, Louisiana State University. 163 p.

Jones, Paul M. The geology of Nashville and immediate vicinity. D, 1892, Vanderbilt University.

Jones, Paul R., III. Microearthquake studies of the Blanco fracture zone and Gorda Ridge using sonobuoy arrays. M, 1976, Oregon State University. 94 p.

Jones, Paul R., III. Seismic ray trace techniques applied to the determination of crustal structures across the Peru continental margin and Nazca Plate at 9° S. latitude. D, 1979, Oregon State University. 156 p.

Jones, Peter. Geology of the Flathead area, southeastern British Columbia, Canada. D, 1966, Colorado School of Mines. 209 p.

Jones, Peter John. The petroleum geochemistry of the Pauls Valley area, Anadarko Basin, Oklahoma. M, 1986, University of Oklahoma. 175 p.

Jones, Phyllis Case. Alteration of a nickel-bearing dunite by granitoid pegmatites, Democrat, Buncombe County, North Carolina. M, 1972, University of South Florida, Tampa. 63 p.

Jones, R. C. A critical study at the pegmatite stage. D, 1958, University of Toronto.

Jones, R. E. A critical study of the phase relations at the pegmatite stage. D, 1953, University of Toronto.

Jones, R. E. The Glendower iron mine, Frontenac Co., Ontario. M, 1953, Queen's University. 252 p.

Jones, R. T. Effect of clay mineral diagenesis on hydrocarbon-cracking ability of clay catalysts. M, 1977, University of Missouri, Columbia.

Jones, Ray S. The manner of occurrence of gold in the jarosite from Mary's Valley, Utah. M, 1924, Columbia University, Teachers College.

Jones, Rebecca Anne. Release of phosphate from dredged sediment. D, 1978, University of Texas at Dallas. 553 p.

Jones, Reece A. A geographical survey of Morgan County, Tennessee. M, 1940, University of Tennessee, Knoxville.

Jones, Richard Burton. Geochemistry as a guide in underground exploration for mercury ore at the New Idria Mine, San Benito County, California. M, 1972, University of Nevada. 100 p.

Jones, Richard D. Geology of the Derro Colorado mining district, Pima County, Arizona. M, 1957, University of Arizona.

Jones, Richard Edwin. A study of the Devonian brachiopod Spinocyrtia granulosa (Conrad) from the Hamilton Group of western New York. M, 1971, SUNY at Buffalo. 57 p.

Jones, Richard Edwin. Taxonomic treatment of dinoflagellates and acritarchs from the Mancos Shale (Upper Cretaceous) of the southwestern United States. D, 1976, University of Arizona.

Jones, Richard L. A foraminiferal fauna from the upper Senonian (Upper Cretaceous) Cafecongombe Shale of Angola, Portuguese West Africa. M, 1949, Syracuse University.

Jones, Richard Lewis. Determination of amounts of igneous, metamorphic, and sedimentary rocks on the Earth's surface. M, 1973, University of Oklahoma. 43 p.

Jones, Richard Lewis. Mineral dispersal patterns in the Pierre Shale. D, 1979, University of Oklahoma. 282 p.

Jones, Robert Alan. Geology and petrography of Ordovician volcanic rocks, Bathurst-Newcastle District, New Brunswick. D, 1964, University of Cincinnati. 213 p.

Jones, Robert Alan. The origin of the massive sulphide deposits in the Bathurst-Newcastle area, New Brunswick. M, 1960, University of New Brunswick.

Jones, Robert Douglas. Metamorphism across the English River gneissic belt (Precambrian) along the Red Lake Road, Ontario. M, 1973, University of Manitoba.

Jones, Robert Joseph. A vertical magnetometer survey of Gardner Butte area, Huerfano County, Colorado. M, 1957, University of Michigan.

Jones, Robert Lee. Geology of Township 1 South, Range 1 East, Murray County, Oklahoma. M, 1926, University of Oklahoma. 29 p.

Jones, Robert Lewis. Outwash terraces along Licking River, Ohio. M, 1959, Ohio State University.

Jones, Robert Sprague. Petrographic and other studies of Washington dunite refractories. M, 1942, University of Washington. 89 p.

Jones, Robert W. Lower Tertiary foraminifera from Waldport, Oregon. M, 1959, Oregon State University. 81 p.

Jones, Robert William. Geology, Finney Peak area, northern Cascades, Washington. D, 1959, University of Washington. 186 p.

Jones, Robert William. Petrology and structure, Higgins Mountain area, northern Cascades, Washington. M, 1957, University of Washington. 186 p.

Jones, Robert William. Structural evolution of parts of south-eastern Arizona. D, 1961, University of Chicago. 198 p.

Jones, Rollin C. The characteristics of X-amorphous phases of Arizona bentonite. D, 1971, University of Arizona.

Jones, Roy Meyrick Price. Micropaleontology of the Colorado Formation in Montana. M, 1941, Montana College of Mineral Science & Technology. 40 p.

Jones, Russell H. B. Geology and ore deposits of Hudson Bay Mountains, British Columbia. D, 1928, University of Wisconsin-Madison.

Jones, Sally. A lineament analysis of the Dale mining district, Pinto Mountains, California. M, 1986, California State University, Long Beach. 175 p.

Jones, Sandra Lynn. Contribution from deep sediments to the dissolved silica in the deep water of the Mediterranean Sea. M, 1977, University of South Florida, St. Petersburg. 93 p.

Jones, Shannon Elizabeth. Stratigraphy and petrology of the Wildcat Valley Sandstone (Lower Devonian), southwestern Virginia. M, 1982, University of North Carolina, Chapel Hill. 127 p.

Jones, Stephen Barr. Human occupance of the Bow-Kicking Horse region in the Canadian Rocky Mountains. D, 1934, Harvard University.

Jones, Steven Dennis. Provenance and paleoenvironment of the Cretaceous Cabrillo Formation, San Diego County, California. M, 1973, San Diego State University.

Jones, Steven K. Geology and mineralization in the zone of contact metamorphism associated with the Seligman Stock, White Pine mining district, White Pine County, Nevada. M, 1984, University of Nevada. 94 p.

Jones, Stewart M. Geology of Gatun Lake and vicinity, Panama. M, 1947, Oregon State University.

Jones, Terry Dean. Wave propagation in porous rock and models for crustal structure. D, 1983, Stanford University. 231 p.

Jones, Theodore Sidney. Geology of Sierra de la Pena and paleontology of the Indidura Formation, Coahuila, Mexico. D, 1935, University of Michigan.

Jones, Thomas. Origins of distribution and thickness of the Sewell Coal Bed, New River District, W. Virginia. M, 1975, University of South Carolina.

Jones, Thomas A. Jurassic quartz arenite-limestone conglomerate in northern Dixie Valley region, Churchill County, Nevada. M, 1968, Northwestern University.

Jones, Thomas Allen. A comparison of estimators of sediment grain-size distributions. D, 1969, Northwestern University. 107 p.

Jones, Thomas David. Pre-Pennsylvanian unconformity in parts of Knox, Sullivan, Greene, and Daviess counties, Indiana. M, 1953, Indiana University, Bloomington. 13 p.

Jones, Thomas W. Linear features study of Vermont using remote sensor data. M, 1979, Rensselaer Polytechnic Institute. 196 p.

Jones, Thomas Z. Petrography, structure, and metamorphic history of the Warrensville and Jefferson quadrangles, southern Blue Ridge, northwestern North Carolina. D, 1976, Miami University (Ohio). 130 p.

Jones, Tracy. Smectitic impurities in some commercial Georgia kaolins. M, 1988, University of Georgia. 60 p.

Jones, Verner Everett. Chromite deposits of Sheridan, Montana. M, 1930, Cornell University. 29 p.

Jones, Verner Everett. Origin of the Spring Hill gold ores, near Helena, Montana. D, 1933, Cornell University. 88 p.

Jones, Vernon K. Contributions to the geomorphology and neoglacial chronology of the Cathedral Glacier System, Atlin Wilderness Park, British Columbia. M, 1975, Michigan State University. 183 p.

Jones, Victor H. Certain phases of the Missouri series of southwestern Iowa. M, 1928, Iowa State University of Science and Technology.

Jones, Victor Harlan. Sedimentation in the Red River from Denison, Texas, to its mouth. D, 1933, University of Iowa. 148 p.

Jones, W. M., Jr. Iron-55 and zinc-65 in tissues of Columbia River carp. M, 1975, Western Oregon State College.

Jones, Walter B. The areal distribution of the Carboniferous of South America, with description of Carboniferous fossils from Peru and Bolivia. D, 1924, The Johns Hopkins University.

Jones, Walter Bryan. Geology of Northeast Alabama. M, 1920, University of Alabama.

Jones, Walter D. The effects of water content and density on the electrical resistivity of soil. M, 1970, Brooklyn College (CUNY).

Jones, Walter V. A study of groundwater movement in landslides. M, 1966, University of Idaho. 144 p.

Jones, William A. Killarney gneisses and xenoliths at Sudbury (Ontario). D, 1930, University of Toronto.

Jones, William Alfred. The petrography of the rocks of Hong Kong. M, 1927, University of British Columbia.

Jones, William C. General geology of the northern portion of the Ajo Range, Pima County, Arizona. M, 1974, University of Arizona.

Jones, William Charles. Geology of the Garnet Mountain-Aquila Ridge area, Ice River, BC. M, 1955, University of British Columbia.

Jones, William E. Trace element and stable isotopic trends associated with diagenesis of selected benthic foraminifers from Miocene sediments. M, 1988, East Carolina University. 147 p.

Jones, William R. The geology of the Sycamore Ridge area, Pima County, Arizona. M, 1941, University of Arizona.

Jones, Williams Maury. The political economy of natural resources; water scarcity in the High Plains region of the U. S. D, 1986, Columbia University, Teachers College. 343 p.

Jonescu, M. E. Natural vegetation and environmental aspects of strip mined land in the lignite coal fields of southeastern Saskatchewan. D, 1974, University of Saskatchewan. 248 p.

Jong, Bernardus Hermenigildus W. S. de *see* de Jong, Bernardus Hermenigildus W. S.

Jong, Hsing-Lian. A critical investigation of post-liquefaction strength and steady state flow behavior of saturated soils. D, 1988, Stanford University. 436 p.

Jong, Hsing-Lian. Jointing in rocks adjacent to the San Gregorio Fault, San Mateo and Santa Cruz counties, California. M, 1983, Stanford University. 48 p.

Jong, Ron S. Small push moraines in central coastal Maine. M, 1980, Ohio University, Athens. 75 p.

Jong, Sybren Hendrik de *see* de Jong, Sybren Hendrik

Jonge, E. J. Coen Kiewiet de *see* Kiewiet de Jonge, E. J. Coen

Jongedyk, Howard Albert. Some relations of electrokinetic phenomena to the hydraulic and electroosmotic permeability of uniform sands. D, 1968, University of Minnesota, Duluth. 382 p.

Jonson, David Carl. The geology of the Resurrection Mine area, Lake County, Colorado. M, 1955, Colorado School of Mines. 190 p.

Jonte, John Haworth. Studies of radioelement fractionation in hydrothermal transport processes and of the contribution of some nuclear reactions to hydrothermal activity [Arkansas]. D, 1956, University of Arkansas, Fayetteville. 72 p.

Jonte, John Haworth. The relationship of selenium and gold in ores from the Republic District. M, 1942, University of Washington. 33 p.

Joolazadeh, Mohammad. Finite element analysis of Hatwai Creek embankment. M, 1980, University of Idaho. 167 p.

Jooste, Rene F. Geology of the Bourget map area, Chicoutimi County, Quebec. D, 1949, McGill University.

Jopling, Alan Victor. An experimental study on the mechanics of bedding. D, 1961, Harvard University.

Joralemon, Peter. The occurrence of gold at the Getchell Mine, Nevada. D, 1949, Harvard University.

Jordan, Bradley C. Petrology of the East Greenwich Pluton, Rhode Island. M, 1983, University of Rhode Island.

Jordan, Carl Frederick. Quantity and composition of water in natural soil systems; a lysimeter study. D, 1966, Rutgers, The State University, New Brunswick. 70 p.

Jordan, Clifton F., Jr. Lower Permian stratigraphy of southern New Mexico and west Texas. D, 1971, Rice University. 136 p.

Jordan, Clifton F., Jr. Patch reefs off Bermuda. M, 1970, Rice University. 78 p.

Jordan, David Charles. The geology and geochemistry of the south-central portion of Ulysses Mountain Quadrangle, Lemhi County, Idaho. M, 1984, University of Idaho. 149 p.

Jordan, David Lohman. Late Givetian-early Frasnian (Devonian) stratigraphy in west central Alberta and east central British Columbia (Canada). M, 1967, University of Saskatchewan. 212 p.

Jordan, Douglas W. Trace fossils and stratigraphy of the Devonian black shale in east-central Kentucky. M, 1979, University of Cincinnati. 227 p.

Jordan, Eric K. A Pliocene fauna from Cedros Island and the Turtle Bay region, lower California. M, 1926, Stanford University. 44 p.

Jordan, Frank William. Genesis of carbonate concretions in the upper Ludlowville, Middle Devonian of Erie County, New York. M, 1968, McMaster University. 67 p.

Jordan, Jack Gerald. Correlation of reef calcarenites of the Pennsylvanian Paradox Formation, San Juan Canyon, Utah. M, 1957, University of New Mexico. 90 p.

Jordan, Jeffrey M. Facies and depositional environment of the Perdiz Conglomerate, Presidio County, Texas. M, 1979, University of Kansas. 68 p.

Jordan, Jimmie Lynn. Inert gas investigations of the Apollo 15 and 17 landing sites. D, 1975, Rice University. 151 p.

Jordan, John Edgar, Jr. A geological/geophysical study of a Pliocene fluvial delta in offshore Louisiana. M, 1981, Wright State University. 95 p.

Jordan, John M. Geothermal investigations in the San Luis Valley, south-central Colorado. M, 1974, Colorado School of Mines. 89 p.

Jordan, John T. Geology of the Cactus Mines, Rosamond, Kern County, Utah. M, 1941, California Institute of Technology. 30 p.

Jordan, Larry Eugene. The geology of the Kellytown Quadrangle, Georgia. M, 1974, Emory University. 69 p.

Jordan, Louise. A study of the Miocene foraminifera from Jamaica, the Dominican Republic, the republics of Panama, Costa Rica, and Haiti. D, 1939, Massachusetts Institute of Technology. 212 p.

Jordan, Louise. Foraminifera from the Pliocene of New Guinea. M, 1931, Massachusetts Institute of Technology. 111 p.

Jordan, M. R. Sedimentological examination of the Upper Devonian shales of northeastern Ohio. M, 1984, Case Western Reserve University.

Jordan, Mark Steven. Environmental and engineering geology of the Guadalupe Quadrangle, Maricopa County, Arizona; Part II. M, 1983, Arizona State University. 61 p.

Jordan, Martha Josephine Ellis. Stratigraphy, sedimentary petrology, and palaeontology of the Steinaker Draw locality in the Middle Jurassic Carmel Formation near Vernal, Utah. M, 1987, University of Minnesota, Minneapolis. 112 p.

Jordan, Michael Andrew. Garnetiferous metagabbro near Babyhead, Llano County, Texas. M, 1970, University of Texas, Austin.

Jordan, Michael Andrew. Geology of the Round Valley-Sanhedrin Mountain area, northern California Coast Ranges. D, 1978, University of Texas, Austin. 212 p.

Jordan, Patrick J. W. The environment of deposition of the Cane River Formation in a portion of Northwest Louisiana. M, 1976, Northwestern State University. 102 p.

Jordan, Paul J. Geology and hydrogeology of a proposed sanitary landfill site, Marion County, Indiana. M, 1986, Purdue University. 174 p.

Jordan, R. J. The deglaciation and consequent wetland occurrence on the Tug Hill Plateau, New York. D, 1978, Syracuse University. 282 p.

Jordan, Richard Hollister. The origin of barite deposits. M, 1939, Cornell University.

Jordan, Richard J. The nature and origin of drainage in Northwest Maine. M, 1976, Boston University. 64 p.

Jordan, Robert R. Columbia (Pleistocene) sediments of Delaware. D, 1964, Bryn Mawr College. 124 p.

Jordan, Robert R. Tertiary sediments of central Delaware. M, 1961, Bryn Mawr College. 47 p.

Jordan, Rudolph Henry. Some phases of the loess of Iowa with special reference to the relation of loess-like clay to the loess. M, 1921, University of Iowa. 64 p.

Jordan, Stephen James. Sedimentation and remineralization associated with biodeposition by the American oyster Crassostrea virginica (Gmelin). D, 1987, University of Maryland. 221 p.

Jordan, T. E. Evolution of the Late Pennsylvanian-Early Permian western Oquirrh Basin, Utah. D, 1979, Stanford University. 287 p.

Jordan, Thomas E. Gravity exploration of buried topography in the vicinity of Shumla, southwestern New York. M, 1986, SUNY, College at Fredonia. 97 p.

Jordan, Thomas Hillman. Estimation of the radial variation of seismic velocities and density in the Earth. D, 1973, California Institute of Technology. 199 p.

Jordan, William M. Sedimentary characteristics of selected Ordovician and Silurian stratigraphic sections, H and D Range, northeastern Nevada. M, 1961, Columbia University, Teachers College.

Jordan, William Malcolm. Regional environmental study of the Early Mesozoic Nugget and Navajo Sandstones. D, 1965, University of Wisconsin-Madison. 239 p.

Jordanovski, Ljupco R. Investigation of numerical methods in inversion of earthquake source mechanism. D, 1985, University of Southern California.

Jorden, Roger M. Relationship of gravel filtration to exhibited sediment charge. M, 1962, University of Arizona.

Jorden, Thomas E. Transformation to zero offset. M, 1987, Colorado School of Mines. 185 p.

Jordon, Larry Eugene. The geology of the Kelleytown Quadrangle, Georgia. M, 1971, Emory University. 69 p.

Jorgensen, David Wayne. Hydrology of surface-mined land; a determination of mine soil control on infiltration capacity, and runoff modeling of disturbed watersheds. M, 1985, Pennsylvania State University, University Park. 127 p.

Jorgensen, Donald Gene. Geology of the Elk Point Quadrangle, South Dakota-Nebraska-Iowa. M, 1958, University of South Dakota. 114 p.

Jorgensen, Gregory J. Geologic modeling of a gravity line from the central African Rift system, Sudan. M, 1987, Brigham Young University. 130 p.

Jorgensen, Per. Density, conductance and infrared studies of clay-water systems. D, 1967, Purdue University. 106 p.

Jorgensen, Ronald Wilbur. A reaction path model of the hydrochemistry of the Carbondale Aquifer, Southwest Indiana. D, 1983, Indiana University, Bloomington. 168 p.

Jorgenson, David Bruce. Petrology and origin of the Illinois River gabbro, a part of the Josephine peridotite-gabbro complex (upper Jurassic), Klamath mountains, southwestern Oregon. D, 1970, University of California, Santa Barbara. 266 p.

Jorgeson, Eric Charles. Surface and subsurface geology of the Dovesville and Mont Clare quadrangles in South Carolina. M, 1968, Duke University. 66 p.

Jorre, Louise de St. see de St. Jorre, Louise

Jorsch, James J. Stratigraphy of Upper Ordovician and Silurian rocks in south-central Idaho. M, 1982, University of Wisconsin-Milwaukee. 86 p.

Jorstad, Robert B. Geographic and stratigraphic distribution of Miocene palynomorphs in North Idaho. D, 1983, University of Idaho. 179 p.

Jorstad, Tom. Geoarchaeological investigations on Cinnamon Creek Ridge, McKenzie County, North Dakota. M, 1984, University of Pittsburgh.

Jory, Lisle Thomas. Mineralogical and isotopic relations in the Port Radium pitchblende deposit, Great Bear Lake, Canada. D, 1964, California Institute of Technology. 275 p.

Jose, Barrie Frederick. Feasibility of the piezoelectric exploration technique for quartz vein detection. M, 1979, University of British Columbia.

Joseph Shu Kay Lee see Shu Kay Joseph Lee

Joseph, Allan Jeffrey Anthony. Unsteady-state cylindrical, spherical and linear flow in porous media. D, 1984, University of Missouri, Rolla. 544 p.

Joseph, George John. Stochastic numerical optimization with applications to seismic exploration. D, 1988, Tulane University. 98 p.

Joseph, Nancy Lee. Epithermal veins in the Silver Bell District, Pima County, Arizona. M, 1982, University of Arizona. 49 p.

Joseph-Haakevitch, Phincas E. Primary or secondary origin of wulfenite. M, 1916, University of Arizona.

Josephson, John J. The mineral chemistry and phase relations in volcanogenic sediments metamorphosed to the pumpellyite-actinolite facies, Chatham Island, New Zealand Plateau, Southwest Pacific. M, 1985, University of Massachusetts. 156 p.

Josey, William L. Metamorphic petrology and structural geology of the Santa Barbara Quadrangle, Guatemala. M, 1970, Louisiana State University.

Joshi, Martand Shipadrado. The genesis of the granitic and associated rocks of the Box Springs Mountains, Riverside, California. D, 1967, University of California, Riverside. 169 p.

Joslin, Peter Schuyler. The stratigraphy and petrology of the Atoka Formation, West central Arkansas. M, 1980, Northern Illinois University. 123 p.

Josse, Genic Raymond. Rubidium-strontium age determinations from the File-Horton-Woosey lakes area of the Flin Flon volcanic belt, west central Manitoba. M, 1974, University of Manitoba.

Jost, Hardy. Geology and metallogeny of the Santana da Boa Vista region, southern Brazil. D, 1981, University of Georgia. 231 p.

Josten, Nicholas E. Precision gravity survey over a portion of the Bromley Oilfield, Denver Basin, Colorado. M, 1985, Colorado School of Mines. 90 p.

Joubin, Francis Renault. Geology of Babine Bonanza (Cronin) Mine, British Columbia. M, 1943, University of British Columbia.

Jouet, Cavalier H. An investigation of the rare earths of thorium, cerium, didymium, and lanthanum; Cretaceous rocks of Long Island. D, 1894, Columbia University, Teachers College.

Jourdain, Vincent. Analyse structurale et stratigraphie de la zone aurifère nord du Gisement de Montauban. M, 1987, Universite du Quebec a Chicoutimi. 77 p.

Journeay, John Murray. Stratigraphy, internal strain and thermo-tectonic evolution of northern Frenchman Cap Dome; an exhumed duplex structure, Omineca Hinterland, S. E. Canadian Cordillera. D, 1987, Queen's University. 404 p.

Jovanovich, Dushon Bogdan. The effect of Earth layering on elastic fields due to a dislocation source. M, 1972, Brown University. 137 p.

Jowett, A. A. Soil characteristics and water table relationships. M, 1974, University of Guelph.

Jowett, Edwin Craig. Acquisition and retention of magnetization in modern lime muds, ancient limestones and dolostones and carbonate-hosted sulphides. M, 1977, University of Toronto. 200 p.

Jowett, Edwin Craig. Timing and genesis of the Kupferschiefer Cu-Ag deposits in Poland. D, 1986, University of Toronto.

Jowett, Richard A. Seismic stratigraphy of the Long Island platform on the U.S. Atlantic continental margin. M, 1988, SUNY at Buffalo. 90 p.

Joy, James Anthony. The distribution and ecology of the benthic Ostracoda from the central Arctic Ocean. M, 1974, University of Wisconsin-Madison.

Joyce, David Brian. A phase equilibrium study in the system $NaAlSi_3O_8$-SiO_2-Al_2SiO_5-H_2O at 2 kilobars and petrogenetic implications. M, 1987, Pennsylvania State University, University Park. 60 p.

Joyce, Edwin A. A metallurgical investigation of ore from the North Star Mine, Blaine County, Idaho. M, 1926, University of Idaho. 37 p.

Joyce, James. High pressure-low temperature metamorphism and tectonic evolution of the Samana Pen-insula, Dominican Republic (Greater Antilles). D, 1985, Northwestern University. 270 p.

Joyce, John Edward. Preservation of aragonite in late Pleistocene sediments in the deep basin of the western Gulf of Mexico. D, 1984, Texas A&M University. 206 p.

Joyce, Michael J. Stratigraphy, clay mineralogy and pesticide analysis of Flathead Lake sediments, Flathead Lake, Montana. M, 1980, University of Montana. 86 p.

Joyce, Rosanne M. Geochemical implications on the provenance and fate of organic-rich deep-sea sediments. M, 1985, University of South Florida, St. Petersburg.

Joyner, William Blish. Gravity in New Hampshire and adjoining areas. D, 1958, Harvard University.

Ju, Chi-Rei. Temperature effects on the stability of fayalite under high pressure by YAG laser heating. M, 1976, University of Rochester. 25 p.

Ju, Fu-Shyong. Interpretation of aeromagnetic or gravity data by the method of difference mapping. D, 1973, University of Illinois, Urbana. 200 p.

Ju, Jiann-Wen. Constitutive modeling for inelastic materials including damage and finite strain effects. D, 1986, University of California, Berkeley. 203 p.

Juan, Francisco Claudio San see San Juan, Francisco Claudio, Jr.

Juan, Vei Chow. Mineral resources of China. D, 1945, University of Chicago. 94 p.

Juang, Charng-Hsein. Pore size distribution of sandy soils and the prediction of permeability. D, 1981, Purdue University. 251 p.

Juang, Franz H. T. The geochemical cycle of chlorine in the Hubbard Brook Watershed, New Hampshire. M, 1966, Dartmouth College. 71 p.

Juang, T. C. Genesis of secondary micas from basalts and related rocks in the Hawaiian Islands. M, 1965, University of Hawaii. 109 p.

Jubb, T. M. The validity of ecologic hypotheses in lower Eocene planktonic foraminifera. M, 1976, Washington University. 38 p.

Jucevic, Edward Paul. Technical and economic feasibility study for a copper mine and mill using segregation, leaching, solvent extraction and electrowinning. M, 1969, University of Nevada. 148 p.

Jud, Friedolina C. The determination of specimens of silicified wood from the Western United States. M, 1913, Columbia University, Teachers College.

Jud, William F. A study of crystal growth in gel. M, 1968, Washington University. 133 p.

Juda, Peter John. Shock induced melting; I, Experimental high pressure melting temperatures for tin, zinc, and pyrite and high pressure melting laws; II, Jetting for low velocity oblique impacts and the formation of silicate melt. D, 1979, University of California, Los Angeles. 172 p.

Judah, Othman Mohammed. Simulation of runoff hydrographs from natural watersheds by finite element method. D, 1973, Virginia Polytechnic Institute and State University.

Judd, Harl Elmer. A study of bed characteristics in relation to flow in rough, high-gradient, natural channels. D, 1963, Utah State University. 90 p.

Judd, J. B. The potential effects of selective dissolution on Oligocene planktonic foraminiferal biostratigraphy in the western and central equatorial Pacific Ocean. D, 1977, University of Colorado. 195 p.

Judd, James Brian. An analysis of the planktonic foraminiferal fauna from Core 6278, Tongue of the Ocean, Bahamas. M, 1969, Duke University. 106 p.

Judd, Robert William. Pennsylvanian coal ball flora of Indiana. D, 1968, Ball State University. 89 p.

Juddo, Edward Paul. An experimental study of the permeability of fracture intersections in Sioux Quartzite. M, 1986, Texas A&M University. 66 p.

Judge, Alan Stephen. Geothermal measurements in a sedimentary basin (southern Ontario and southern

Peninsula of Michigan). D, 1972, University of Western Ontario.

Judge, Anne Victoria. The relationship between plate curvature and elastic plate thickness; a study of the Peru-Chile Trench. M, 1988, Massachusetts Institute of Technology. 60 p.

Judge, Robert Michael. Cognitive strawman; public input to a water resources planning system. D, 1975, University of Arizona.

Judice, Philip C. An analysis of the environment of deposition of the Gray Sand of Terryville Field, Lincoln Parish, Louisiana. M, 1981, University of Southwestern Louisiana. 156 p.

Judkins, Thomas W. Selective sorting of hypersthene in the beach sands of southern Oregon. M, 1975, University of Oregon. 51 p.

Judson, Jack F. Geology of the LeBrun and Mint Canyon quadrangles, Los Angeles County, California. M, 1935, California Institute of Technology. 79 p.

Judson, S. Sheldon, Jr. Geology antiquity of the San Jose Site, eastern New Mexico. D, 1948, Harvard University.

Judy, J. R. Cenozoic stratigraphic and structural evolution of west-central Markagunt Plateau, Utah. M, 1974, Kent State University, Kent. 120 p.

Judziewicz, Emmet Joseph. Taxonomy and morphology of the tribe Phareae (Poaceae; Bambusoideae). D, 1987, University of Wisconsin-Madison. 557 p.

Juhan, Joe Paul. Stratigraphy of the Evacuation Creek Member and other related subjects in the Piceance Creek basin of northwestern Colorado. M, 1960, University of Tulsa. 143 p.

Juhas, Allan Paul. Geology and origin of copper-nickel sulphide deposits of the Bird River area of Manitoba. D, 1973, University of Manitoba.

Juhle, R. Werner. Iliamna Volcano and its basement. D, 1953, The Johns Hopkins University.

Jui, Pao Vung. The Mahopac iron ore deposits, New York. M, 1911, Columbia University, Teachers College.

Juilfs, John D. A stratigraphic correlation and lithofacies study of the Pennsylvanian and Permian sediments in the subsurface of south-central and southwestern Nebraska. M, 1953, University of Nebraska, Lincoln.

Juilland, Jean D. Mineralization of the Mount Bohemia intrusive (Precambrian), Keweenaw County, Michigan. M, 1965, Michigan Technological University. 82 p.

Julander, Dale Richard. Seismicity and correlation with fine structure in the Sevier Valley area of the Basin and Range-Colorado Plateau transition, south-central Utah. M, 1983, University of Utah. 142 p.

Julian, Bruce Rene. Regional variations in upper mantle structure beneath North America. D, 1970, California Institute of Technology. 216 p.

Julian, Edward L. Aplite-pegmatite-granite relations in the Castalia Quarry, Franklin County, North Carolina, and petrology of the surrounding granite. M, 1972, North Carolina State University. 61 p.

Julian, Louise Chandler. The Elberton orbicular adamellite, Elbert County, Georgia. M, 1972, North Carolina State University. 65 p.

Julien, Pierre St. *see* St. Julien, Pierre

Julius, Jonathan Fred. The uranium and thorium geochemistry of three selected pegmatites and their granites, South Platte pegmatite district, Jefferson County, Colorado. M, 1982, University of New Orleans. 100 p.

Jull, Robert Kingsley. Silurian Halysitidae from Western Canada. M, 1961, University of Alberta. 93 p.

Juma, Noorallah Gulamhusein. Dynamics of soil and fertilizer nitrogen. D, 1981, University of Saskatchewan.

Jumnongthai, Junya. Foraminifera of the Upper Cretaceous Belle Fourche Shale, Greenhorn, and Carlile formations, Weston County, Wyoming. M, 1979, South Dakota School of Mines & Technology.

Jumnongthai, Manit. Geology of the Sugarloaf Mountain area, Lead, South Dakota. M, 1979, South Dakota School of Mines & Technology.

Jumper, Robert S. Geology of the Claytonville carbonate buildup. M, 1978, University of Texas, Arlington. 155 p.

Juncal, Russell Wright. Exploration for geothermal resources using arsenic and mercury soil geochemistry, Dixie Valley, Nevada. M, 1980, University of Nevada. 88 p.

Juneidi, Mohmoud J. Proposed work plan Ziglab Watershed in Jordan. M, 1966, [Colorado State University].

Junemann, Paul Martin. Some sedimentary studies of the Lodi Shale in western Wisconsin. M, 1951, University of Wisconsin-Madison. 29 p.

Jung, Dorothy Anne. Mechanical analysis of sediments from Vermont. M, 1938, Columbia University, Teachers College.

Jung, George. The strontium distribution in Montastrea annularis by monochromatic X-ray radiography. M, 1966, University of Rochester. 90 p.

Jung, Heeok. Velocity and attenuation in young ocean crust. M, 1988, University of Washington. 124 p.

Jung, Jim Grant. A geologic and magnetic study of the Copper Bell property, Goldbar, Washington. M, 1967, University of Washington. 38 p.

Jung, Kwang Seop. Mathematical modeling of cavity growth during underground coal gasification. D, 1987, University of Wyoming. 150 p.

Jung, Woo-Yeol. Free-air gravity and geoid anomalies of the North Atlantic Ocean and their tectonic implications. D, 1985, Texas A&M University. 222 p.

Junge, W. R. Geology of the Islote area, North-central Puerto Rico. M, 1976, University of Southern California. 67 p.

Jungels, Pierre Henri. Modeling of tectonic processes associated with earthquakes. D, 1973, California Institute of Technology. 207 p.

Jungmann, William L. Stratigraphy of Labette County, Kansas. M, 1964, University of Kansas. 128 p.

Jungyusuk, Nikom. Petrology of the Christmas Mountains igneous rocks, Trans-Pecos Texas. M, 1977, University of Texas, Austin.

Junhavat, Suphachai. Origin of Paleozoic shale of Florida. M, 1976, Georgia Institute of Technology. 78 p.

Junnila, Randy Michael. Stratigraphy and sedimentology of the upper Gowganda Formation (early Proterozoic) in the Whitefish Falls area, Ontario. M, 1986, University of Western Ontario. 167 p.

Juo, Anthony Shiang-Ru. Chemical and physical factors affecting the relative availability of inorganic phosphorus in soils. D, 1967, Michigan State University. 113 p.

Juras, Dwight Stephen. Pre-Miocene geology of the northwest part of Olds Ferry Quadrangle, Washington County, Idaho. M, 1973, University of Idaho. 82 p.

Juras, Dwight Stephen. The petrofabric analysis and plagioclase petrography of the Boehl's Butte anorthosite. D, 1974, University of Idaho. 132 p.

Juras, Stephen Joseph. Alteration and sulphide mineralization in foot-wall felsic metapyroclastic and metasedimentary rocks, Brunswick No.12 Deposit, Bathurst, New Brunswick, Canada. M, 1981, University of New Brunswick.

Juras, Stephen Joseph. Geology of the polymetallic volcanogenic Buttle Lake Camp, with emphasis on the Price Hillside, central Vancouver Island, British Columbia, Canada. D, 1987, University of British Columbia.

Jurczyk, Gayle Katherine. Thermoelastic attenuation of Rayleigh waves. D, 1983, University of Connecticut. 172 p.

Jurdy, Donna M. A model for the temperature distribution around an intrusion. M, 1970, SUNY at Buffalo.

Jurdy, Donna Marie. A determination of the polar wander since the early Cretaceous. D, 1974, University of Michigan. 96 p.

Jure, Albert E. A study of some mineralogical relationships in and about the Sullivan ore body, Kimberley, British Columbia. M, 1928, University of Wisconsin-Madison.

Jure, Albert E. The petrography of the Purcell sills. D, 1930, University of Wisconsin-Madison. 41 p.

Jurewicz, Amy Jo Goldmintz. Effect of temperature, pressure, oxygen fugacity and composition on calcium partitioning, calcium-magnesium distribution and the kinetics of cation exchange between olivines and basaltic melts. D, 1986, Rensselaer Polytechnic Institute. 216 p.

Jurewicz, Stephen Richard. Melt infiltration and distribution in a granitic analog. M, 1982, Rensselaer Polytechnic Institute. 75 p.

Jurewicz, Stephen Richard. Partial melting related to the textural and chemical development of granites and migmatites. D, 1984, Rensselaer Polytechnic Institute. 144 p.

Jurich, David M. Analysis of soil slopes utilizing acoustic emissions. M, 1985, Colorado School of Mines. 168 p.

Jurik, Paul P. Insoluble residue study of the Comanche Peak and Edwards limestones of Kimble County, Texas. M, 1961, Texas A&M University. 98 p.

Jurkevics, Andrejs. A method for simulating and representing strong ground motion. M, 1978, University of British Columbia.

Jurko, Robert Clarence and Schultz, Arthur H. Stratigraphy and paleontology of a core from Kent County, Ontario, Canada. M, 1953, University of Michigan.

Jury, Harold Louie. An application of terrestrial photogrammetry to glaciology in Greenland. M, 1956, Ohio State University.

Jusbasche, Joachim Michael. Cross-plots and histograms for well log analysis in unfamiliar lithologies. D, 1977, Stanford University. 146 p.

Just, Evan. The origin of structural features in the northeast part of the Mid-continent oil field. M, 1925, University of Wisconsin-Madison.

Justen, John Joseph. An investigation of some factors affecting the radial filtration of bentonitic mud. M, 1951, University of Texas, Austin.

Juster, Thomas C. Very low-grade metamorphism of pelites associated with coal, northeastern Pennsylvania. M, 1984, University of Wisconsin-Madison.

Justice, Mahlon Gilbert. In situ determination of physical properties of residual soils and underlying dolomitic bedrock using seismic techniques. M, 1975, Pennsylvania State University, University Park. 96 p.

Justice, Mahlon Gilbert, Jr. The effect of water on high-temperature plastic deformation in olivine. D, 1982, Pennsylvania State University, University Park. 143 p.

Justice, Pamela Rose. Horizontal refraction of normal modes in shallow water. M, 1975, Pennsylvania State University, University Park. 101 p.

Justice, Pamela Rose Eckberg. A method for modeling the Earth's mantle using compositional and seismological constraints. D, 1986, Pennsylvania State University, University Park. 256 p.

Justus, Philip Stanley. Modal and textural zonation of diabase dikes, Deep River Basin, North Carolina. M, 1966, University of North Carolina, Chapel Hill. 76 p.

Justus, Philip Stanley. Structure and petrology along the Blue Ridge front and Brevard zone, Wilkes and Caldwell counties, North Carolina. D, 1971, University of North Carolina, Chapel Hill. 89 p.

Jutras, Marc. Establissement d'un modèle géostatistique à la Mine Doyon, Canton Bousquet, Abitibi, Québec. M, 1988, Ecole Polytechnique. 99 p.

Juul, Steve Thorvald Julius. A limnological assessment of Twin Lakes, Washington. M, 1986, Washington State University. 87 p.

Kaabar, Salah M. The geology of the Pinto Creek Quadrangle, Albany County, Wyoming. M, 1970, University of Wyoming. 61 p.

Kaal, Ayad Said. Analysis of hydrogeologic factors for the location of water wells in the granitic environment of Moscow Mountain, Latah County, Idaho. M, 1978, University of Idaho. 76 p.

Kaar, Robert Frederick. Lower Tertiary foraminifera from north central Tres Pinos, San Benito County, California. M, 1962, University of California, Berkeley. 148 p.

Kaarsberg, Ernest Andersen. Introductory studies of natural and artificial argillaceous aggregates by sound propagation and X-ray diffraction methods. D, 1957, University of Chicago. 25 p.

Kaarsberg, J. T.; Haas, James J. and Stephens, J. J. Geology of the Poison Canyon area, Huerfano County, Colorado. M, 1956, University of Michigan.

Kaasa, Robert A. Geophysical method for measuring glacial drift thickness. M, 1954, University of Minnesota, Minneapolis.

Kaback, D. S. The geochemistry of molybdenum in stream waters and sediments, central Colorado. D, 1977, University of Colorado. 294 p.

Kabala, Zbigniew Jan. Sensitivity analysis of flow in unsaturated heterogeneous porous media. D, 1988, Princeton University. 145 p.

Kabeiseman, William Joseph. An investigation of the control of hydraulic fracturing through the inclusions of prefractures. D, 1966, University of Missouri, Rolla. 167 p.

Kabir, Jobaid. Mathematical modeling of sediment discharge from irrigation furrows. D, 1983, Washington State University. 140 p.

Kabir, Muhammad Ismat. In-situ uranium mining; reservoir engineering aspects of leaching and restoration. D, 1982, University of Texas, Austin. 483 p.

Kablanow, Raynold I., II. Diagenesis and hydrocarbon generation in the Monterey Formation, Huasna Basin, California. M, 1983, University of Wyoming. 57 p.

Kablanow, Raynold Irvin, II. Influence of the thermal and depositional histories on diagenesis and hydrocarbon maturation in the Monterey Formation; Huasna, Pismo, and Salinas basins, California. D, 1986, University of Wyoming. 282 p.

Kacena, Jeffrey A. Facies, depositional environments, and dolomitization of a Middle Devonian carbonate shelf, eastern Nevada. M, 1984, Duke University. 127 p.

Kachanoski, Reginald Gary. Spatial and spectral relationships of soil and soil forming factors. D, 1984, University of California, Davis. 109 p.

Kachel, David G. Settling velocity and critical shear stress measurements of some Arctic foraminifera. M, 1971, University of Washington.

Kachel, Nancy Brandeberry. A time dependent model of sediment transport and strata formation on a continental shelf. D, 1980, University of Washington. 123 p.

Kachelmeyer, John Michael. Bedrock geology of the North Saanich-Cobble Hill area, British Columbia, Canada. M, 1979, Oregon State University. 153 p.

Kacira, Niyazi. Geology of chromitite occurrences and ultramafic rocks (Lower Paleozoic) of the Thetford Mines; Disraeli area, Quebec. D, 1972, University of Western Ontario. 248 p.

Kaczaral, Patrick Walter. Trace element analysis of L and LL chondrites; comparison of Antarctic and non-Antarctic meteorite populations. D, 1986, Purdue University. 149 p.

Kaczkowski, Peter. Damped least squares inversion applied to ducted propagation of acoustic energy in the ocean. M, 1986, Colorado School of Mines. 150 p.

Kaczmarek, Edward L., Jr. Depositional environments, sediment dispersal, and provenance of the Upper Cretaceous Colorado Formation, southwestern New Mexico. M, 1987, New Mexico State University, Las Cruces. 107 p.

Kaczmarek, Michael B. Geothermometry of selected Montana hot springs waters. M, 1974, Montana State University. 141 p.

Kaczmarowski, J. H., Jr. Geochemical analysis of soil samples from the Jewell Mine, Southwest Wisconsin; methods comparison and interpretation. M, 1975, University of Wisconsin-Madison.

Kaczor, Laurel. Petrology of the Mid-Cretaceous Aguas Buenas and Rio Maton limestones in southeast-central Puerto Rico. M, 1987, University of North Carolina, Chapel Hill. 67 p.

Kaczorowski, R. T. The Carolina Bays and their relationship to modern oriented lakes. D, 1977, University of South Carolina. 130 p.

Kaczorowski, Raymond T. Offset tidal inlets, Long Island, New York. M, 1972, University of Massachusetts. 150 p.

Kaddou, Nadheema Salih. Clay mineralogy of some alluvial soils of Iraq and Dubuque silt loam and underlying dolomitic limestone of Wisconsin. D, 1960, University of Wisconsin-Madison. 128 p.

Kadhi, Abdul. Structure of the Tom Mays park area, Franklin mountains, El Paso County, Texas. M, 1970, University of Texas at El Paso.

Kadib, Abdel-Latif Abdullah. A function for sand movement by wind. D, 1965, University of California, Berkeley. 102 p.

Kadinsky-Cade, Katharine A. Seismotectonics of the Chile margin and the 1977 Caucete earthquake of western Argentina. D, 1985, Cornell University. 268 p.

Kadko, David Charles. A detailed study of uranium series nuclides for several sedimentary regimes of the Pacific. D, 1981, Columbia University, Teachers College. 330 p.

Kadoch, Teresa Lynn. Field calibration and analysis of Rosed, a road sediment erosion model. M, 1983, University of Idaho. 115 p.

Kadri, Moinoddin Murtuzamiya. Structure and influence of the Tillamook Uplift on the stratigraphy of the Mist area, Oregon. M, 1982, Portland State University. 105 p.

Kaeding, Margeret E. Geochemistry of near-trench magmatic rocks from the Taitao Peninsula, southern Chile; implications for ridge subduction. M, 1986, Rutgers, The State University, New Brunswick. 86 p.

Kaegi, Dennis D. A coal rank profile of southeastern Illinois as determined by vitrinite reflectance. M, 1977, Southern Illinois University, Carbondale. 99 p.

Kaehler, Charles Alfred. Monitoring of responses to a local base-level change in an ephemeral stream. M, 1980, University of Arizona. 175 p.

Kaesler, Robert LeRoy. A quantitative re-evaluation of the ecology and distribution of Recent foraminifera and ostracoda of Todos Santos Bay, Baja California, Mexico. D, 1965, University of Kansas. 128 p.

Kaesler, Robert LeRoy. Recent marine and lagoonal ostracodes from the Estero de Tastiota region, Sonora, northeastern Gulf of California. M, 1962, University of Kansas.

Kafescioglu, Ismail A. A paleobathymetry study by means of specific diversity of planktonic foraminifera in the upper Gulf coastal region of east Texas, Arkansas and Louisiana. M, 1967, Case Western Reserve University.

Kafescioglu, Ismail Ali. Quantitative distribution of foraminifera on the continental shelf and uppermost continental slope off Massachusetts, eastern United States. D, 1969, Case Western Reserve University. 165 p.

Kafka, A. L. Caribbean tectonic processes; seismic surface wave source and path property analysis. D, 1980, SUNY at Stony Brook. 294 p.

Kafka, Frederick Thomas. Structural characteristics of normal fault areas significant in oil exploration. M, 1950, University of Pittsburgh.

Kafura, Craig John. The effect of fine-grained minerals on the Coulomb-parameter values of the debris-flow fluid-phase. M, 1988, Arizona State University. 70 p.

Kagel, Carla Turner. Application of modern instrumentation to the determination of gold in geological samples. D, 1984, University of Idaho. 241 p.

Kahane, S. W. The air quality and oil spill implications of Alaskan oil importation into southern California. D, 1978, University of California, Los Angeles. 253 p.

Kaharoeddin, Francis Amrisar. Regressive and transgressive phenomena in the Gallup Sandstone (upper Cretaceous), in the northwestern part of San Juan County, New Mexico. M, 1971, University of New Mexico. 81 p.

Kahil, Alain. A preliminary study of the White River terraces in central South Dakota. M, 1970, University of South Dakota. 73 p.

Kahl, Jeffrey S. Metal input and mobilization in two acid-stressed lake watersheds in Maine. M, 1982, University of Maine. 109 p.

Kahle, Charles F. A study of the tills of northwestern Ohio by size analysis. M, 1957, Miami University (Ohio). 77 p.

Kahle, Charles Franz. Diagenesis in oolitic and pellet-type limestones. D, 1962, University of Kansas. 149 p.

Kahle, James Edward. Megabreccias and sedimentary structures of the Plush Ranch Formation (Oligocene), northern Ventura County, California. M, 1966, University of California, Los Angeles.

Kahle, Michael Brinkman. The inflammability limits of Utah coal dust-methane mixtures and their detection. M, 1969, University of Utah. 98 p.

Kahn, Gail Anne Heffner. The ortho para hydrogen equilibrium in the Uranus atmosphere. M, 1979, Massachusetts Institute of Technology. 37 p.

Kahn, Jacob Henry. The role of hurricanes in the long-term degradation of a barrier island chain; Chandeleur Islands, Louisiana. M, 1980, Louisiana State University.

Kahn, James Steven. The measurement of packing in sandstones. M, 1954, Pennsylvania State University, University Park. 135 p.

Kahn, M. I. Non-equilibrium oxygen and carbon isotopic fractionation in tests of living planktic foraminifera from the eastern equatorial Atlantic Ocean. D, 1977, University of Southern California.

Kahn, Michael I. The significance of the intraspecific variability of living Globigerina dutertrei d'Orbigny, 1839, in a selected environmental range. M, 1969, Florida State University.

Kahn, N. M. Flotation mechanisms and ionic relations of the non-motile marine dinoflagellate Pyrocystis noctiluca Murray (1885). D, 1977, University of Rhode Island. 129 p.

Kahn, Peter A. Geology of Aden Crater, Dona Ana County, New Mexico. M, 1987, University of Texas at El Paso.

Kahn, Steven James. The analysis and distribution of packing in sediments. D, 1956, University of Chicago. 38 p.

Kahrs, Anna E. Petrographic province of New England and New York. M, 1924, Columbia University, Teachers College.

Kaighin, Hall Young. Cobalt in the waters of the Black Sea. D, 1969, [University of Miami].

Kailasam, Lakshmi Narayan. Geophysical investigations of near-surface geologic structure in the neighborhood of Saint Charles, Missouri. M, 1953, St. Louis University.

Kaiman, S. The crystal structure of rammelsbergite; NiAs$_2$. M, 1946, University of Toronto.

Kairo, Suzanne. Sedimentology, stratigraphy, and depositional history of the Cretaceous Split Rock Creek Formation, Minnehaha County, South Dakota. M, 1987, Purdue University. 122 p.

Kaiser, Charles John. Chemical and isotopic kinetics of sulfate reduction by organic matter under hydrothermal conditions. D, 1988, Pennsylvania State University, University Park. 132 p.

Kaiser, Charles John. Zoning and controls of mineralization in the Southeast Missouri barite district. M, 1983, University of Michigan.

Kaiser, Charles Philip. Stratigraphy of Lower Mississippian rocks in southwestern Missouri. D, 1946, University of Kansas.

Kaiser, Charles Philip. Stratigraphy of the Mississippian formations of the Osage River valley in western Missouri. M, 1945, University of Kansas.

Kaiser, Edward P. Metamorphic and tectonic study of a granite intrusive near Ganoyer, New Hampshire. M, 1936, Syracuse University.

Kaiser, Russell F. The composition and origin of the glacial till in the Mexico and Kasoag quadrangles, New York. M, 1939, Syracuse University.

Kaiser, Russell Florentine. The surficial geology of the southeastern segment of the Lake Ontario Plain, New York. D, 1957, Syracuse University. 319 p.

Kaiser, William R. The late Mesozoic geology of Pearse Peak Diorite, Southwest Oregon. M, 1962, University of Wisconsin-Madison. 75 p.

Kaiser, William Richard. Delta cycles in the Middle Devonian of central Pennsylvania. D, 1972, The Johns Hopkins University. 240 p.

Kaiteris, P. An evaluation of the apparent dissociation constants of carbonic acid in seawater. M, 1975, Queens College (CUNY). 116 p.

Kaktins, Teresa L. Fluvial terraces of the Juniata River valley in central Pennsylvania. M, 1986, Pennsylvania State University, University Park. 283 p.

Kaktins, Uldis. Hydrogeology and land-use planning in Townsend, Massachusetts. D, 1975, Boston University. 181 p.

Kaktins, Uldis. Stratigraphy and petrography of the ignimbrite flows (Devonian) of the Blue Hills area, Massachusetts. M, 1969, Syracuse University.

Kalafatis, Christodoulos. The stratigraphy and paleontology of core H-1A of Upper Silurian strata from Taylor Township, Wayne County, Michigan. M, 1958, University of Michigan.

Kalaitjis, Michael G. Effects of urbanization and natural processes in the sedimentation conditions of the Winton Woods drainage basin. M, 1975, University of Cincinnati.

Kalamarides, Ruth I. Evaluation of infrared determination of H2O+ in minerals. M, 1976, University of Massachusetts. 40 p.

Kalamarides, Ruth I. The oxygen isotope geochemistry of the Kiglapait layered intrusion. D, 1983, University of Massachusetts. 80 p.

Kalamarides-Berg, Ruth Irene. The oxygen isotope geochemistry of the Kiglapait layered intrusion. D, 1983, University of Massachusetts. 88 p.

Kalaswad, Sanjeev. Dynamic disequilibrium in the Utukok River-Outlook Ridge area, Alaska; a radar and map investigation. M, 1983, Indiana State University. 78 p.

Kalb, George William. Desorption isotherms of montmorillonite-organic complexes. D, 1969, Ohio State University. 113 p.

Kalbacher, Karl F. Diagenetic study of the upper member of the Smackover Formation (Upper Jurassic), Columbia County, Arkansas. M, 1985, Stephen F. Austin State University. 78 p.

Kaldenbach, Thomas. Geology of the Bannock Peak area, Deep Creek Mountains, Idaho. M, 1984, Colorado State University. 117 p.

Kaldon, Richard C. Investigation of tin spinels in the systems MgO-ZnO-SnO2, ZnO-NiO-SnO2, and MgO-NiO-SnO2. M, 1974, Miami University (Ohio). 31 p.

Kaldor, Michael. Sedimentology of Wando bar and related sediments (Pleistocene), Georgetown County, South Carolina. M, 1969, SUNY at Buffalo. 41 p.

Kaldy, Windsor John. Geology and geophysics of the Sherwill Dome of the southwestern Piedmont of Virginia. M, 1977, University of Kentucky. 46 p.

Kalesky, John F. Lithofacies, stratigraphy and cyclic sedimentation in a mixed carbonate and siliciclastic system, Red House Formation (Atokan), Sierra County, New Mexico. M, 1988, New Mexico State University, Las Cruces. 202 p.

Kalia, Hemendra Nath. Penetration in granite by shaped charge liners. D, 1970, University of Missouri, Rolla.

Kalil, E. K. The distribution and geochemistry of uranium in Recent and Pleistocene marine sediments. D, 1976, University of California, Los Angeles. 286 p.

Kalinec, James A. The structure and stratigraphy of the Seljehaugfjellet area, Wedel Jarlsberg Land, Spitsbergen. M, 1985, University of Wisconsin-Madison.

Kalinowski, Donald D. Meteorites of Ohio. M, 1972, Ohio State University.

Kalish, Phillip. Geology of the Water Canyon area, Magdalena Mountains, Socorro, New Mexico. M, 1953, New Mexico Institute of Mining and Technology. 48 p.

Kalish, Robert S. Structural geology and petrology of a portion of the Miners Delight Quadrangle, Wyoming. M, 1982, University of Missouri, Columbia.

Kalisky, Maurice. Coccolith Miocene-Pliocene stratigraphy of the Equatorial Pacific. M, 1970, Queens College (CUNY). 50 p.

Kalisz, Paul John. The longleaf pine islands of the Ocala National Forest, Florida; a soil study. D, 1982, University of Florida. 127 p.

Kalk, Thomas Raymond. The petrology of the Mitchell Dam amphibolite, Chilton and Coosa counties, Alabama. M, 1972, Memphis State University.

Kalkani, E. C. The rock slope stability problem and the application of two-dimensional finite element analysis to rock slope stability and comparison with actual failures. D, 1975, Purdue University. 684 p.

Kalkanis, George. Transportation of bed material due to wave action. D, 1965, University of California, Berkeley. 126 p.

Kallemeyn, Gregory William. Elemental fractionations among carbonaceous chondrites; implications for their classification and nebular formation. D, 1982, University of California, Los Angeles. 216 p.

Kallio, Thomas A. The stratigraphy of the Lower and Middle Silurian, Medina, Clinton, and Lockport groups along Ohio Brush Creek, Adams County, Ohio. M, 1976, Miami University (Ohio). 160 p.

Kalliokoski, Jorma Osmo Kalervo. Geology of Weldon Bay area, Manitoba. D, 1951, Princeton University. 177 p.

Kallsen, Clarence Edward. The stratigraphy and structure of the Mt. Darby area, Sublette County, Wyoming. M, 1952, University of Iowa. 154 p.

Kalman, Linda Susan. Absolute radiometric calibration of black and white film imagery with applications in remote sensing of suspended sediment in surface waters. D, 1982, University of Wisconsin-Madison. 448 p.

Kalogeropoulos, Stavros I. Chemical sediments in the hanging wall of volcanogenic massive sulfide deposits. D, 1982, University of Toronto.

Kalogeropoulos, Stavros Ilia. Geochemistry and mineralogy of the St. Lawrence pyrochlore deposit, Oka, P.Q. M, 1977, Queen's University. 125 p.

Kalra, A. K. Separation of strontium from barites and galenas for mass spectrometer analysis. M, 1967, University of British Columbia.

Kalra, Ashok. Wavefront reconstruction method for geophysical exploration. D, 1971, University of California, Berkeley. 153 p.

Kalstrom, Eric T. Late Cenozoic of the Glacier and Waterton parks area, northwestern Montana and southern Alberta and paleoclimatic implications. D, 1981, University of Calgary.

Kalthem, M. S. Evaluation of Riyadh City water supply and demand. M, 1978, University of Arizona.

Kam, Marlene Ngit Sim. Determination of Curie isotherm from aeromagnetic data in the Imperial Valley, California. M, 1980, University of California, Riverside. 110 p.

Kam, Moshe. Entropy and the basic percepts of system theory. D, 1987, Drexel University. 141 p.

Kamal, Rami A. Interpretive stratigraphy of the Lowville Limestone (medial Ordovician) of central New York. M, 1977, Boston University. 84 p.

Kamal-Aldin, Saad. Application of the paleomagnetic conglomerate test to the Huronian Gowganda Formation of northeastern Ontario. M, 1983, University of Windsor. 112 p.

Kamb, Hugo R. The measurement of the plasticity of clays. M, 1925, University of Minnesota, Minneapolis. 26 p.

Kambampati, Mohan V. Factor analysis in the interpretation of petrology of Rambler area, Newfoundland, Canada. M, 1986, Wichita State University. 97 p.

Kambampati, Mohan V. Geology, petrochemistry and tectonic setting of the Rambler area, Baie Verte Peninsula, Newfoundland. M, 1984, University of New Brunswick. 284 p.

Kamel, Asad. Response of base-isolated nuclear structures to earthquakes. D, 1981, Purdue University. 161 p.

Kamenka, Louis Anthony D. Natural and artificial weathering of bedrock overburden associated with open-pit coal mines, S.W. Alberta. M, 1983, University of Alberta. 194 p.

Kamens, John S. Sedimentation in the Lake of Bizerte, Tunisia. M, 1975, Duke University. 86 p.

Kamerling, Marc. Paleomagnetism of middle Miocene volcanics and tectonic rotation of the Santa Monica Mountains region, western Transverse Ranges, California. M, 1980, University of California, Santa Barbara.

Kamhi, Samuel. The crystal structure of vaterite, $CaCO_3$. D, 1963, Columbia University, Teachers College.

Kamilli, Robert J. Paragenesis, zoning, fluid inclusion and isotopic studies of the Finlandia Vein, Colqui District, central Peru. D, 1976, Harvard University.

Kamin, Thomas C. Stratigraphy of the Pony Express Limestone Member of the Wanakah Formation. M, 1968, Washington State University. 99 p.

Kamineni, D. C. Petrology and geochemistry of some Archean metamorphic rocks near Yellowknife, District of Mackenzie. D, 1973, University of Ottawa. 228 p.

Kaminski, Michael Anthony. Cenozoic deep-water agglutinated foraminifera in the North Atlantic. D, 1987, Woods Hole Oceanographic Institution. 262 p.

Kaminsky, Barney. Correlation of Landsat multispectral data to rock reflectance measurements. M, 1977, University of Wyoming. 55 p.

Kaminsky, P. D. The Canso strata at Knoydart, Nova Scotia. M, 1953, Massachusetts Institute of Technology. 85 p.

Kamis, James Edward. The petrology and reservoir capabilities of the Jurassic Nugget Sandstone and Cretaceous Tygee Member of the Bear River Formation in southeastern Idaho. M, 1976, Idaho State University. 59 p.

Kamm, John L. Trace-element concentration and distribution in the sediments of Antigua, British West Indies. M, 1981, Northern Illinois University. 78 p.

Kammer, Charles. A stratigraphic study of the (Devonian) Genesee Shale of western New York. M, 1952, SUNY at Buffalo.

Kammer, David. Petrology and geochemistry of the Patagonian Batholith between 69°W and 72°W longitude, south of the Strait of Magellan, southern Chile. M, 1987, University of Houston.

Kammer, Glenn D. Ionium and thorium in Colorado carnotite; recovery of ionium from carnotite; adsorption of thorium by barium sulfate. D, 1925, University of Pittsburgh.

Kammer, Heidi W. A hydrogeologic study of the Ravenna Arsenal, eastern Portage and western Trumbull counties, Ohio. M, 1982, Kent State University, Kent. 298 p.

Kammer, Thomas William. Fossil communities of the prodeltaic New Providence Shale Member of the Borden Formation (Mississippian), north-central Kentucky and southern Indiana. D, 1982, Indiana University, Bloomington. 301 p.

Kammer, Thomas William. Oxygen and carbon isotope variation in the Neogene foraminifera Globigerina and Uvigerina from DSDP Site 173 and the Centerville Beach section, California. M, 1977, Indiana University, Bloomington. 61 p.

Kamola, Diane L. Marginal marine and non-marine facies, Spring Canyon Member, Blackhawk Formation (Upper Cretaceous), Carbon County, Utah. M, 1987, University of Georgia. 186 p.

Kamp, Brad Douglas Vande *see* Vande Kamp, Brad Douglas

Kamp, Katherine Marland. The sinistral gastropod Conus (Contraconus) (Tertiary) and its relationship to dextral species. M, 1967, Tulane University. 113 p.

Kamp, Peter Cornelius Van de *see* Van de Kamp, Peter Cornelius

Kampf, Anthony R. The structure determination of a new barium silicate mineral, verplanckite, and the structure refinement of the mineral searlesite, a borosilicate. M, 1972, University of Illinois, Chicago.

Kampf, Anthony Robert. Structural studies on phosphate minerals containing octahedral cations; 1, Minyulite; II, Schoonerite. D, 1976, University of Chicago. 12 p.

Kampmueller, Elaine. A geochemical study of the role of magma mixing in the origin of the Marscoite Suite, Isle of Skye, Scotland. M, 1983, Michigan State University. 58 p.

Kan, David Lan-rong. Inorganic and organic influences on carbonate cementation. D, 1973, Texas A&M University.

Kan, David Lan-rong. Isotopic variations of dissolved inorganic carbon in the Gulf of Mexico. M, 1970, Texas A&M University.

Kan, Tze-Kong. Ray theory approximation in geoelectromagnetic probing. D, 1975, University of Wisconsin-Madison. 84 p.

Kan, Tze-Kong. Sonar mapping of underside of pack ice. M, 1973, University of Wisconsin-Madison.

Kana, Timothy W. Coastal processes and sediment transport at Price Inlet and North Inlet, South Carolina. M, 1976, University of South Carolina. 144 p.

Kana, Timothy W. Suspended sediment in breaking waves. D, 1979, University of South Carolina. 160 p.

Kanaan, Faisel Mohamed. The geology, petrology, and geochemistry of the granitic rocks of Jabal Alhawshah and vicinity, Jabal Alhawshah Quadrangle, Kingdom of Saudi Arabia. D, 1975, Colorado School of Mines. 269 p.

Kanasewich, Ernest Raymond. An interpretation of some gravity measurements in Canadian Cordillera. M, 1960, University of Alberta. 124 p.

Kanasewich, Ernest Raymond. Quantitative interpretations of anomalous lead isotope abundances. D, 1962, University of British Columbia.

Kanat, Leslie Howard. The stratigraphy, structure and metamorphic aspects of the pre-Carboniferous lithologies from southern St. Jonsfjorden, Oscar II Land, Svalbard. M, 1982, Wayne State University. 134 p.

Kanaya, Taro. Eocene diatom florules from California. M, 1955, Stanford University. 197 p.

Kanazawich, Michael F. The surficial geology, subsurface stratigraphy, and hydrogeology of Woods Corner, New York. M, 1985, University of Connecticut. 55 p.

Kanbergs, Karlis. Fracturing along the margins of a porphyry copper system, Silver Bell District, Pima County, Arizona. M, 1980, University of Arizona. 90 p.

Kandasamy, Kumarasamy. Three-dimensional dynamic analysis of gravity dam and reservoir systems. D, 1986, The Johns Hopkins University. 139 p.

Kandelin, John Jacob. Chemical and structural aspects of the elasticity of pyroxenes. D, 1988, SUNY at Stony Brook. 154 p.

Kandemir, Burhaneddin Hamit. A method for determination of magnetic permeability. M, 1970, New Mexico Institute of Mining and Technology.

Kandiah, Arumugam. Fundamental aspects of surface erosion of cohesive soils. D, 1974, University of California, Davis. 236 p.

Kane, Byron. Structural contour map, eastern South Dakota, using the Greenhorn Limestone as a datum plane. M, 1957, South Dakota School of Mines & Technology.

Kane, Byron L. Stratigraphy and faulting in the Sage Creek-Pinnacles area, Badlands National Monument, South Dakota. M, 1958, South Dakota School of Mines & Technology.

Kane, David George. The clay mineralogy of the Zilpha Formation in Panola County, Mississippi. M, 1982, Memphis State University. 50 p.

Kane, Douglas Lee. Hydraulic mechanism of aufeis growth. D, 1975, University of Minnesota, Minneapolis. 126 p.

Kane, Henry Edward. The Quaternary geology and geomorphology of the southeastern portion of the Canon City Embayment, Colorado. D, 1965, University of California, Los Angeles.

Kane, Julian. Some Recent pelagic micro-organisms. M, 1930, Columbia University, Teachers College.

Kane, Kevin J. Magnetic surveys conducted in eastern South Dakota to examine the Great Lakes tectonic zone and igneous intrusives. M, 1982, University of Wisconsin-Milwaukee. 101 p.

Kane, Martin Francis. Geophysical study of the tectonics and crustal structure of the Gulf of Maine. D, 1970, St. Louis University. 112 p.

Kane, Phillip S. The glacial geomorphology of the Lassen Volcanic National Park area. D, 1975, University of California, Berkeley.

Kane, Ward Thompson. Trace element geochemistry and 40Ar/39Ar geochronology of epithermal mineralization in Four Mile Canyon, Boulder County, Colorado. M, 1988, University of Colorado. 112 p.

Kane, William Theodore. The crystal structure of epistilbite. D, 1966, University of Missouri, Rolla. 159 p.

Kaneda, Ben Keith. Contact metamorphism of the Chiwaukum Schist near Lake Edna, Chiwaukum Mountains, Washington. M, 1980, University of Washington. 149 p.

Kanehiro, B. A hydrogeologic study of the Kiholo to Puako area on the Island of Hawaii. M, 1977, University of Hawaii.

Kanehiro, Y. Movement and availability of nitrogen in soil. M, 1948, University of Hawaii.

Kanehiro, Y. Status and availability of zinc in Hawaiian soils. D, 1964, University of Hawaii. 113 p.

Kaneps, Ansis Girts. Late Neogene (late Miocene to Recent) biostratigraphy (planktonic foraminifera), biogeography and depositional history, Atlantic Ocean, Caribbean sea, Gulf of Mexico. D, 1970, Columbia University. 179 p.

Kanes, William H. Facies and development of the delta of the Colorado River, Texas. D, 1965, West Virginia University.

Kanes, William Henry. A study of the upper Middle Mississippian of southeastern West Virginia. M, 1958, West Virginia University.

Kang, Hyo Jin. Cross-shore sediment transport in relation to waves and currents in a groin compartment. D, 1987, Old Dominion University. 183 p.

Kang, Joo-Myung. Finite element analysis of the heat and mass transfer in a magma body. D, 1981, University of Oklahoma. 113 p.

Kangas, Patrick Carl. Energy analysis of landforms, succession, and reclamation. D, 1983, University of Florida. 187 p.

Kania, Henry Joseph. The fate of mercury input into artificial stream systems. D, 1981, University of Georgia. 218 p.

Kania, Joseph Ernest Anthony. Origin of the pyritic copper deposits of the mesothermal type. D, 1930, Massachusetts Institute of Technology. 138 p.

Kania, Joseph Ernest Anthony. The geology and mineralogy of the western contact of the Coast Range Batholith. M, 1928, University of British Columbia.

Kanis, Oak K. A detailed magnetic survey of portions of northern Florida. M, 1981, University of Florida. 99 p.

Kanizay, Stephen P. Geology of Cross Mountain, Moffat County, Colorado. D, 1956, Colorado School of Mines. 129 p.

Kanizay, Stephen P. The Ordovician-Silurian problem in the Lanes Mill area, Oxford, Ohio. M, 1950, Miami University (Ohio). 31 p.

Kanji, Milton Assis. Shear strength of soil-rock interfaces. M, 1970, University of Illinois, Urbana. 69 p.

Kanna, Sanousi S. Subsurface study of the lower part of the Dakota Group in the northern Nebraska Panhandle. M, 1982, University of Nebraska, Lincoln. 74 p.

Kanschat, Katherine Ann. Diagenesis of the Bell Canyon and Cherry Canyon formations (Guadalupian), Coyanosa field area, Pecos County, Texas. M, 1981, University of Arizona. 196 p.

Kantaatmadja, Budi P. Subsurface study of the basal Chester sandstones in western Beaver County, Oklahoma Panhandle. M, 1987, West Texas State University. 71 p.

Kanter, Lisa Ruth. Paleomagnetic constraints on the movement history of Salinia. D, 1983, Stanford University. 166 p.

Kantner, David Arthur. Possible pre-Illinoian glacial outwash deposits in the Logan Quadrangle, Ohio. M, 1974, Ohio University, Athens. 97 p.

Kantner, Lynn M. Mined land reclamation, Wayne National Forest, Ohio. M, 1979, Ohio University, Athens.

Kantor, Joseph Alan. Assimilation and dike swarms in the Sugarloaf mountain area, Marquette County, Michigan. M, 1969, Michigan Technological University. 83 p.

Kantor, Tedral. Geology of the east-central portion of the Nelson Quadrangle, Clark County, Nevada. M, 1961, University of Missouri, Rolla.

Kantrowitz, Irwin Howard. An east-west structural traverse of western Ohio from Lancaster, Ohio, to Indiana. M, 1959, Ohio State University.

Kantrowitz, Ralph. An experimental study of natural convection in volcanic necks and stocks. M, 1976, Brooklyn College (CUNY).

Kantzas, Apostolos. A theoretical and experimental study of gravity assisted inert gas injection as a method of oil recovery. D, 1988, University of Waterloo. 395 p.

Kao, Ching-nan. A numerical investigation of explosive generated normal modes and leaking modes in an unsaturated surface layer overlying a saturated half-space. D, 1974, Michigan State University. 104 p.

Kao, Dominique Wen. The enhancement of signal to noise in magnetotelluric data. D, 1975, University of Alberta. 155 p.

Kao, Jason Chin-Sen. Automated interpolation of two-dimensional seismic profiles into three-dimensional data volumes. D, 1988, Colorado School of Mines.

Kao, Samuel E. Effect of urban street pattern on drainage. D, 1973, University of Arizona.

Kapaldo, David Wayne. A computer simulation model for use in petroleum resources management. D, 1983, Michigan State University. 259 p.

Kapchinske, John M. Petrology of the Coon Valley Member of the Jordan Formation near LaCrosse, Wisconsin. M, 1980, Northern Illinois University. 130 p.

Kaplafka, Nancy A. An alkalic rock suite from the Rocky Boy Stock (Tertiary), Bearpaw Mountains, Montana. M, 1971, University of Cincinnati. 130 p.

Kaplan, Allen Edward. The geomorphology of the McCullough alluvial fan, Clark County, Nevada. M, 1973, University of Cincinnati. 69 p.

Kaplan, David Mark. Sources of Holocene and late Pleistocene sediments on the continental shelf off Georgia. M, 1971, University of Georgia. 57 p.

Kaplan, Harvey I. Mineralogical changes in the Martinsburg formations (Upper Ordovician) of southeastern Pennsylvania. M, 1965, SUNY at Binghamton. 51 p.

Kaplan, Isaac Raymond. Sulfur isotope fractionations during microbiological transformations in the laboratory and in marine sediments. D, 1962, University of Southern California. 228 p.

Kaplan, Lazard Harold. The fauna of the Oak Grove Formation of the Alum Bluff Group of the Florida Miocene. M, 1941, Louisiana State University.

Kaplan, Louis Arnold. Dissolved organic matter in a rural Piedmont watershed. D, 1980, University of Pennsylvania. 243 p.

Kaplan, Paul Garry. A single-well tracer test for predicting advective flow velocity in a confined aquifer. M, 1985, Purdue University. 66 p.

Kaplan, Sanford. The distribution of sulfur in continental and nearshore coals from the Hanna and Green River basins of Wyoming. M, 1976, Lehigh University.

Kaplan, Sanford Sandy. The sedimentology, coal petrology, and trace element geochemistry of coal bearing sequences from Joggins, Nova Scotia, Canada, and southeastern Nebraska, USA. D, 1980, University of Pittsburgh. 323 p.

Kaplan, Seymour Fred. An X-ray diffraction study of some of the naturally occurring iodates. D, 1965, University of New Mexico. 186 p.

Kaplan, Terry. The Lassic Outlier; an outlier of Coast Range ophiolite. M, 1983, San Jose State University. 96 p.

Kaplan, W. A. Denitrification in a Massachusetts salt marsh. D, 1977, Boston University. 88 p.

Kaplin, John L. An engineering geology site investigation of the proposed Poudre Dam Site, Larimer County, Colorado. M, 1988, Colorado State University. 218 p.

Kaplin, Stephen, Jr. Chemical and oxidation reduction potential studies of Long Island Sound sediments. M, 1961, New York University.

Kaplowitz, Phyllis S. Clay mineral distribution and abundance within rocks of different carbonate lithologies of the Rondout and Manlius formations (Silurian and Devonian), Rosendale, New York. M, 1970, SUNY at Buffalo. 38 p.

Kapnicky, George. Igneous contact effects in Pendleton County, West Virginia. M, 1956, West Virginia University.

Kapnistos, Minas Michael. Refining the gravity method to locate pinnacle reefs in Lambton County, Ontario. M, 1979, University of Windsor. 117 p.

Kapp, Ronald Ormond. Pollen analytical investigations of Pleistocene deposits on the southern High Plains. D, 1963, University of Michigan. 256 p.

Kapp, Ulla. Paleoecology of stromatoporoid mounds, Middle Chazy (Middle Ordovician), Isle La Motte, Vermont. M, 1972, McGill University.

Kappel, Ellen Sue. Evidence for volcanic episodicity and a nonsteady state rift valley. D, 1985, Columbia University, Teachers College. 134 p.

Kappeler, Kristine Ann. Depositional environments of the Avenal Sandstone of Reef Ridge, Central California. M, 1984, California State University, Northridge. 86 p.

Kappelman, John Wesley, Jr. The paleoecology and chronology of the middle Miocene hominoids from the Chinji Formation of Pakistan. D, 1987, Harvard University. 318 p.

Kappler, Jonathan. Surface computer modelling of a mafic rock body near Clear Spring, Maryland. M, 1986, Bowling Green State University. 90 p.

Kappmeyer, Janet Carol. Quartz deformation in the Marquette and Republic troughs, Upper Peninsula of Michigan. M, 1982, University of Michigan.

Kapta, Mohammed Shafi. A gravity survey of Rio Grande Valley near San Acacia, New Mexico. M, 1971, New Mexico Institute of Mining and Technology.

Kapteyn, R. J. The molybdenum content of some Hawaiian soil families. M, 1963, University of Hawaii. 35 p.

Kar, Lakshmidhar. Earthquake behavior of arch dam-reservoir systems. D, 1972, South Dakota School of Mines & Technology.

Karabalis, Dimitris L. Dynamic response of three-dimensional foundations. D, 1984, University of Minnesota, Duluth. 337 p.

Karabatsos, George Tom. A provenance and durability study of heavy minerals in a small, modern river (Neuse River, North Carolina). M, 1959, University of Nebraska, Lincoln.

Karabinos, Paul M. S. Deformation and metamorphism of Cambrian and Precambrian rocks on the east limb of the Green Mountains Anticlinorium near Jamaica, Vermont. D, 1982, The Johns Hopkins University. 297 p.

Karachewski, John A. Paleomagnetism, sedimentry petrology, and paleogeography of the Oligocene Lincoln Creek Formation in Grays Harbor basin, southwestern Washington. M, 1983, Western Washington University. 181 p.

Karafiath, Leslie L. The effect of volume change characteristics of soils on the bearing capacity of shallow footings. D, 1969, Polytechnic University. 105 p.

Karakostanoglou, Iakovos. ESR isochron dating of speleothems. M, 1983, McMaster University. 209 p.

Karanja, Samuel W. A sedimentologic and stratigraphic study of Carboniferous through Jurassic strata of East Kenya. M, 1988, University of Windsor. 207 p.

Karanjac, Jasminko B. Elastic storage in aquifers. D, 1971, Princeton University. 200 p.

Karas, Paul A. Quaternary alluvial sequence of the upper Pecos River and a tributary Glorieta Creek, north-central New Mexico. M, 1988, University of New Mexico. 112 p.

Karasa, N. L. A gravity interpretation of the structure of the Granite Mountains area, central Wyoming. M, 1976, University of Wyoming. 63 p.

Karasaki, Kenzi. Well test analysis in fractured media. D, 1986, University of California, Berkeley. 255 p.

Karasek, Richard. A Pre-Aptian unconformity near Le Kef, Tunisia; its bearing on the evolution of Triassic-cored anticlines in the Tunisian Foreland. M, 1976, University of South Carolina.

Karasek, Richard Mark. Structural and stratigraphic analysis of the Paleozoic Murzuk and Ghadames basins, western Libya. D, 1981, University of South Carolina. 164 p.

Karasevich, Ellen Lee Richter. Radium and uranium contents of limonites from Pennsylvania and Wyoming. M, 1980, Pennsylvania State University, University Park. 106 p.

Karasevich, Lawrence Paul. Structure of the pre-Beltian metamorphic rocks of the northern Ruby Range, southwestern Montana. M, 1980, Pennsylvania State University, University Park. 172 p.

Karathanasis, Anastasios D. Characteristics of naturally acid soil smectites. D, 1982, Auburn University. 236 p.

Karboski, Frank Adam. Structure and stratigraphy of the Norcan Lake area, Grenville Province, southeastern Ontario. M, 1980, Carleton University. 100 p.

Kardos, William Gustave. A gamma ray log analysis of selected Chemung sands of west central Pennsylvania. M, 1958, University of Pittsburgh.

Kareth, Paul Edward. Color in shale; mineralogy versus bulk chemistry. M, 1984, University of Cincinnati. 92 p.

Karges, Burton E. A study of the insoluble residues from well samples of the Wisconsin Silurian. D, 1934, University of Wisconsin-Madison.

Karhi, Louis. Cone-in-cone in the Ohio Shale. M, 1948, Ohio State University.

Karig, Daniel E. Structural analysis of the Sangre de Cristo Range, Venable Peak to Creston Peak, Custer and Saguache counties, Colorado. M, 1964, Colorado School of Mines. 143 p.

Karig, Daniel Edmund. Marginal basins and their role in the development of island arc systems. D, 1970, University of California, San Diego. 153 p.

Karim, M. A. Optimization of water resources for irrigation in Dinajpur and Rangpur, East Pakistan. M, 1968, University of Arizona.

Karim, M. Fazle. Computer-based predictors for sediment discharge and friction factor of alluvial streams. D, 1981, University of Iowa. 74 p.

Karim, Mostafa Fahmy. Some geochemical methods of prospecting and exploration for oil and gas. D, 1964, University of Southern California. 227 p.

Karim, S. Abdul. Micropaleontology, biostratigraphy and paleoecology of the Serikagni Formation in the Jebel Gaulat area, NW Iraq. M, 1978, Queen's University. 99 p.

Karim, Usama Farhan. Large deformation analysis of penetration problems involving piles and sampling tubes in soils. D, 1985, SUNY at Buffalo. 177 p.

Karimpour, Mohammad Hassan. Petrology, geochemistry, and genesis of the A.O. porphyry copper complex in Jackson and Grand counties, northwestern Colorado. D, 1982, University of Colorado. 551 p.

Karish, Charles R. Mesozoic geology of the Ord Mountains, Mojave Desert; structure, igneous petrology, and radiometric dating of a failed incipient intra-arc rift. M, 1983, Stanford University. 112 p.

Kariyawasam, Hettigamage Cyril. Evaluation of hydropower potential in a river basin. D, 1980, University of Texas, Austin. 152 p.

Kark, Margaret Jeanne. A software approach to linear geostatistics. M, 1985, Stanford University. 144 p.

Karkanis, Basily George. Hydrochemical facies of ground water in the western provinces of Sudan. M, 1965, University of Arizona.

Karklins, Olgerts L. Ordovician Bryozoa of family Rhinidictydae from the Decorah Shale formation of Minnesota. M, 1960, University of Minnesota, Minneapolis. 127 p.

Karklins, Olgerts Longins. Cryptostome Bryozoa from Middle Ordovician Decorah Shale of Minnesota. D, 1966, University of Minnesota, Minneapolis. 275 p.

Karl, H. A. Processes influencing transportation and deposition of sediment on the continental shelf, southern California. D, 1977, University of Southern California.

Karl, Herman A. Depositional history of Dakota Formation (upper Cretaceous) sandstones, southeastern Nebraska. M, 1971, University of Nebraska, Lincoln.

Karl, Robert Otto. Insoluble residues of the lower Oneota Dolomite of the Madison, Wisconsin area. M, 1950, University of Wisconsin-Madison. 21 p.

Karl, Susan Margaret. Geochemical and depositional environments of upper Mesozoic radiolarian cherts from the northeastern Pacific rim and from Pacific DSDP cores. D, 1982, Stanford University. 245 p.

Karlen, Wibjorn. Late Holocene glacier and climatic fluctuations, Kebnekaise Mountains, Lappland, Sweden. M, 1972, University of Maine. 102 p.

Karleskind, Lorene Cora. The Ordovician of Sugar River, New York, and vicinity. M, 1931, University of Rochester. 63 p.

Karlin, Robert. Paleomagnetism, rock magnetism, and diagenesis in hemipelagic sediments from the Northeast Pacific Ocean and the Gulf of California. D, 1984, Oregon State University. 246 p.

Karlin, Robert. Sediment sources and clay mineral distributions off the Oregon coast; evidence for a poleward slope undercurrent. M, 1979, Oregon State University. 88 p.

Karlo, J. F. M. The geology and Bouguer gravity of the Hell's Half Acre area and their relation to volcano-tectonic processes within the Snake River plain rift zone, Idaho. D, 1977, SUNY at Buffalo. 152 p.

Karlo, John F. Structural geology of Paron-Fourche, Southwest Arkansas. M, 1973, University of Missouri, Columbia.

Karlsson, Haraldur R. Oxygen and hydrogen isotope geochemistry of zeolites. D, 1988, University of Chicago. 288 p.

Karlstrom, Adabell. A facies study of the Upper Ordovician Cincinnatian series. M, 1950, Northwestern University.

Karlstrom, Karl Edward. Geology of the Proterozoic Deep Lake Group, central Medicine Bow Mountains, Wyoming. M, 1977, University of Wyoming. 116 p.

Karlstrom, Karl Edward. Late Archean and early Proterozoic geologic history of metasedimentary rocks of the Medicine Bow Mountains, Wyoming. D, 1981, University of Wyoming. 240 p.

Karlstrom, Thor N. V. Geology and ore deposits of the Hecla mining district, Beaverhead County, Montana. D, 1953, University of Chicago. 87 p.

Karmen, Andrew A. Recognition of basin, slope, and shelf deposits in the Upper Devonian West Falls Group of New York State. M, 1968, University of Rochester. 58 p.

Karnauskas, Robert James. The hydrogeology of the Nepco Lake watershed in central Wisconsin with a discussion of management implications. M, 1977, University of Wisconsin-Madison.

Karner, Frank Richard. Petrology of the Tunk Lake granite pluton, southeastern Maine. D, 1963, University of Illinois, Urbana. 110 p.

Karner, Garry David. The thermo-mechanical properties of the continental lithosphere. D, 1983, Columbia University, Teachers College. 518 p.

Karnes, Kerri A. Siliciclastic influence on carbonate deposition in the Hermosa Formation, San Juan County, Colorado. M, 1985, Colorado School of Mines. 135 p.

Karnieli, Arnon. Storm runoff forecasting model incorporating spatial data. D, 1988, University of Arizona. 251 p.

Karns, Anthony Wesley Warren. Ophitic pyroxene from the Raggedy Mountains area, Wichita Mountains, Oklahoma. M, 1961, University of Oklahoma. 68 p.

Karp, Edwin. Aspects of strain history in experimental rock deformation. D, 1966, Columbia University. 77 p.

Karp, Edwin. Experimental deformation of (Lower Ordovician) Crown Point Limestone; a study of pre-straining and strain rate effects and the nature of recoverable deformation. M, 1963, Columbia University, Teachers College.

Karpa, John B. The Middle Ordovician Fincastle Conglomerate north of Roanoke, Virginia and its implications for Blue Ridge tectonism. M, 1974, Virginia Polytechnic Institute and State University.

Karpin, Timothy Lee. The relationship between earthquake swarms and magma transport; Kilauea Volcano, Hawaii. M, 1986, SUNY at Stony Brook. 86 p.

Karr, Leonard J. Physical and chemical controls of Zn-Pb-Cu-Ag mineralization at the Big Four Mine, Summit County, Colorado. M, 1986, Colorado State University. 122 p.

Karr, Michael Charles. Spatial and temporal variability of soil chemical parameters. D, 1988, Purdue University. 198 p.

Karren, J. R. An evaluation of aerial camera calibration by the multicollimator method. M, 1966, Ohio State University.

Karrenbach, Martin. Three dimensional migration of zero offset seismic data using Fourier transforms of time slices. M, 1988, University of Houston.

Karrow, Paul Frederick. Pleistocene geology of the Grondines map-area, Quebec. D, 1957, University of Illinois, Urbana. 123 p.

Karshenas, M. Modeling and finite element analysis of soil behavior. D, 1979, University of Illinois, Urbana. 294 p.

Karson, Jeffrey A. Geology of the northern Lewis Hills, western Newfoundland. D, 1977, SUNY at Albany. 474 p.

Karson, Jeffrey A. Structural studies in the mafic and ultramafic rocks of the Lewis Hills, western Newfoundland. M, 1975, SUNY at Albany. 125 p.

Karst, Gary B. Analysis of the northern Dixie Valley groundwater flow system using a discrete-state compartment model. M, 1987, University of Nevada. 100 p.

Karsten, Jill Leslie. An ion microplate determination of water diffusivity in rhyolite obsidian. M, 1980, University of Washington.

Karsten, Jill Leslie. Spatial and temporal variations in the petrology, morphology and tectonics of a migrating spreading center; the Endeavour Segment, Juan de Fuca Ridge. D, 1988, University of Washington. 329 p.

Karstrom, Adabell. A facies study of the Upper Ordovician Cincinnatian Series. M, 1950, Northwestern University.

Kartchner, Wayne E. The ores of the Bisbee District, Arizona. M, 1936, Stanford University. 61 p.

Kartchnor, Wayne E. The geology and ore deposits of the Harshaw District, Patagonia Mountains, Arizona. D, 1944, University of Arizona.

Karteris, Michael Apostolos. An evaluation of satellite data for estimating the area of small forestlands in the southern Lower Peninsula of Michigan. D, 1980, Michigan State University. 186 p.

Karubian, Ruhollah Y. Surface and subsurface geology of Montebello Hills (California). M, 1940, California Institute of Technology. 44 p.

Karvelot, Michael D. The Stigler Coal and collateral strata in parts of Haskell, LeFlore, McIntosh, and Muskogee counties, Oklahoma. M, 1972, Oklahoma State University. 93 p.

Karvinen, William O. Structure of Lardeau group rocks (Ordovician), Albert canyon, British Columbia. M, 1970, University of British Columbia.

Karvinen, William Oliver. Metamorphogenic molybdenite deposits in the Grenville Province. D, 1973, Queen's University. 280 p.

Karwoski, William James. A theoretical approach to the problem of determining ejecta volumes in a gravitational vacuumed half-space. D, 1966, University of Missouri, Rolla. 82 p.

Karya, Kim Aiko. Limitations of the magnetotelluric method as applied to the Pioche-Marysvale trend, Utah. M, 1986, University of Utah. 56 p.

Kasabach, Haig Frederick and Robinson, Paul T. A detailed study of the Old Baldy thrust fault, Huerfano and Costilla counties, Colorado. M, 1959, University of Michigan.

Kasadarli, Mustapha E. Stratigraphy and microfacies analysis of the Rancheria Formation (Meramec), Vinton Canyon, Franklin Mountains, El Paso County, Texas. M, 1977, University of Texas at El Paso.

Kasali, Gyimah. Three dimensional finite element analysis of shallow soil tunneling. D, 1982, Stanford University. 236 p.

Kasameyer, Paul William. Low-frequency magnetotelluric survey of New England. D, 1974, Massachusetts Institute of Technology. 204 p.

Kasapoglu, Kadri E. An aggregate quality investigation of the Meramec River gravels. M, 1968, University of Missouri, Rolla.

Kasapoglu, Kadri Ercin. Progressive failure in discontinuous rock masses subjected to shear deformation.

D, 1973, Pennsylvania State University, University Park. 165 p.

Kasey, Arthur R., III. A detailed field and laboratory investigation of a Northwest Alabama clay deposit. M, 1965, University of Tennessee, Knoxville. 72 p.

Kashatus, Gerard Paul. Depositional environment and reservoir morphology of Guadalupian Bell Canyon sandstones, Scott Field, Ward and Reeves counties, Texas. M, 1986, Texas A&M University.

Kashfi, Mansour S. Lithologic study of the Upper Cambrian of Foster Number 1 Well, Ogemaw County, Michigan. M, 1967, Michigan State University. 92 p.

Kashfi, Mansour S. Structure, stratigraphy and environmental sedimentology of the middle Ordovician Chickamauga Group of a segment of Monroe County, Tennessee. D, 1971, University of Tennessee, Knoxville. 100 p.

Kashirad, Ahmad. Characterization of tillage pans in selected coastal plain soils. D, 1966, University of Florida. 155 p.

Kashkuli, Heydar Ali. A numerical linked model for the prediction of the decline of groundwater mounds developed under recharge. D, 1981, Colorado State University. 155 p.

Kasim, A. G. A numerical solution for the stresses and deformations in a pseudo-elastic soil system. D, 1978, University of California, Berkeley. 313 p.

Kasim, S. A. A geologic study of the Jergins oil field. M, 1950, [University of Houston].

Kasino, Raymond Edward. Environment of deposition and diagenetic history of an upper Morrow Sandstone reservoir, Cimarron County, Oklahoma. M, 1979, Texas Tech University. 101 p.

Kasiraj, Iyadurai. Low-cycle fatigue damage in structures subjected to earthquake excitation. D, 1968, [University of New Mexico]. 100 p.

Kaska, Harold Victor and Galloway, Jesse J. The genus Pentremites and its species. M, 1952, Indiana University, Bloomington. 224 p.

Kasomekera, Zachary Mark. A seasonal conceptual model for watershed simulation. D, 1983, Colorado State University. 236 p.

Kasper, Andrew Edward, Jr. Pertica quadrifaria, a new genus of Devonian fossil plants from northern Maine. D, 1970, University of Connecticut. 73 p.

Kasper, David Conlin. Provenance of Paleogene terrigenous sandstones on Barbados; fine-grained Paleogene terrigenous turbidites on Barbados. M, 1985, Stanford University. 104 p.

Kasper, Robert B. Cation and oxygen diffusion in albite. D, 1975, Brown University. 156 p.

Kasper, Robert Basil. K-Ar ages of southeastern Massachusetts. M, 1970, Brown University.

Kasperski, Kim Louise. Studies of the physical chemistry of clays. D, 1988, University of Alberta. 184 p.

Kassander, Arno Richard, Jr. A regional structural study of the Nemaha Mountains of Oklahoma. M, 1943, University of Oklahoma. 45 p.

Kast, Joe Alex. Depositional environments of the Schuler Formation (Late Jurassic) in Upshur County, Texas. M, 1978, Southern Methodist University. 206 p.

Kastelic, Robert L., Jr. Precambrian geology and magnetite deposits of the New Jersey Highlands in Warren County, New Jersey. M, 1979, Lehigh University. 155 p.

Kasten, James A. Petrology and geochemistry of the calc-alkaline andesites within the Albuquerque Basin, Valencia County, New Mexico. M, 1977, University of New Mexico. 79 p.

Kasten, Terri Ann. Geology and metamorphism of Precambrian rocks in the Placitas area, northern Sandia Mountains, Sandoval County, New Mexico. M, 1980, University of New Mexico. 111 p.

Kastenhuber, Lynn Edward. Seismic processing and interpretation of reflection data collected in Canaan Township, Morrow County, Ohio. M, 1983, Wright State University. 82 p.

Kastens, Kim Anne. Structural causes and sedimentological effects of "cobblestone topography" in the eastern Mediterranean Sea. D, 1981, University of California, San Diego. 223 p.

Kastler, Everett J. Mineralogy of Fletcher Mine in Southeast Missouri. M, 1972, University of Missouri, Columbia.

Kastler, Neil Blair. Conodonts from Middle Pennsylvanian of southeastern Iowa. M, 1958, University of Iowa. 66 p.

Kastner, Miriam. Authigenic feldspars in carbonate rocks. D, 1970, Harvard University.

Kastner, Sidney Oscar. Central force models for crystals of calcite and aragonite type, derived from Raman and infrared frequencies. D, 1961, Syracuse University. 197 p.

Kastning, E. H. Cavern development in the Heldberberg Plateau, East-central New York. M, 1975, University of Connecticut. 102 p.

Kastning, Ernst H., Jr. Geomorphology and hydrogeology of the Edwards Plateau karst, central Texas. D, 1983, University of Texas, Austin. 714 p.

Kastrinsky, Alan Jay. Seismicity of the Wasatch Front, Utah; detailed epicentral patterns and anomalous activity. M, 1977, University of Utah. 138 p.

Kastritis, George John. A geophysical study of buried pre-Pleistocene drainage channels in the Yellow Springs Quadrangle, Ohio. M, 1977, Wright State University. 89 p.

Kasulis, Paul Francis. Carbonate apron deposition of the Bigfork Chert (Middle to Upper Ordovician), western Ouachita Mountains, Oklahoma. M, 1988, Southern Methodist University. 210 p.

Kasvinsky, J. Robert. Evaluation of erosion control measures on the lower Winooski River, Vermont. M, 1968, University of Vermont.

Kaszuba, John Paul. Polyphase deformation and metamorphism in the Penobscot Bay area, coastal Maine. M, 1986, Virginia Polytechnic Institute and State University.

Kat, Pieter W. Morphology, reproduction, genetics, and speciation among Atlantic slope unionids (Mollusca: Bivalvia). D, 1983, The Johns Hopkins University. 334 p.

Katahara, K. W. Pressure dependence of the elastic moduli of body-centered-cubic transition metals. D, 1977, University of Hawaii. 143 p.

Katahira, Yo. Comparison of frequency domain migration approaches with regard to vertical velocity variations. M, 1982, Colorado School of Mines. 167 p.

Kate, Frederick H. The geology of the Dayton-Evansville region, Rhea County, Tennessee. M, 1940, Ohio State University.

Kath, Randal Lee. Petrology of the contact metamorphic rocks beneath the Stillwater Complex, Montana; conditions and assemblages of metamorphism. M, 1986, University of Tennessee, Knoxville. 72 p.

Katham, Abd Al-Wahab N. Well-log analysis of the Upper Colorado (Turonian to Santonian) strata of southwestern Saskatchewan. M, 1982, University of Windsor. 175 p.

Katherman, C. E. Variations in porosity with depth for marine sediments. M, 1977, Texas A&M University.

Katherman, Vance Edward. The Flagstaff Limestone on the east front of the Gunnison Plateau of central Utah. M, 1948, Ohio State University.

Katich, Philip Joseph. The stratigraphy and paleontology of the pre-Niobrara Upper Cretaceous rocks of Castle Valley, Utah. D, 1951, Ohio State University.

Katigema, F. D. An interpretation of the anomalous magnetization in a Sudbury diabase dyke. M, 1977, University of Western Ontario.

Katili, Amanda Niode. Evaluation of environmental management of one nickel mining company in Indonesia with special emphasis on water quality. D, 1988, University of Michigan. 186 p.

Katiyar, Vidya. Rainfall and runoff relationships in a small Himalayan watershed. M, 1982, Colorado State University. 118 p.

Kato, T. T. The relationship between low-grade metamorphism and tectonics in the Coast Ranges of central Chile. D, 1976, University of California, Los Angeles. 278 p.

Kato, Terence Tetsuo. Tectonic contact in the Arroyo Valle region, California. M, 1972, University of California, Davis. 42 p.

Katock, Robert. Environmental report, Town of Vestal, New York; hydrology section. M, 1976, SUNY at Binghamton. 72 p.

Katopodes, Nikolas Demetrios. Two-dimensional unsteady flow through a breached dam by the method of characteristics. D, 1974, University of California, Davis. 157 p.

Katrinak, Karen Ann. Stable isotope studies of cherts from the Archean Swaziland Supergroup of South Africa. M, 1987, Arizona State University. 100 p.

Katrosh, Mark Ralph. Foraminiferal paleoecology and biostratigraphy of the Yorktown and Pungo River formations; Beaufort, Pamlico, Craven, and Carteret counties, North Carolina. M, 1981, East Carolina University. 161 p.

Katsambalos, Kostas Evangelos. Simulation studies on the computation of the gravity vector in space from surface data considering the topography of the Earth. D, 1981, Ohio State University. 146 p.

Katsaounis, A. Sediment interstitial dissolved mercury and carbon relationships in an artificial and natural marsh, James River, Virginia. M, 1977, Old Dominion University. 129 p.

Katsikas, C. A. Numerical models for soil liquefaction. D, 1979, University of Michigan. 310 p.

Katsura, Kurt Toshiro. The geology and epithermal vein mineralization at Champion Mine, Bohemia Mining District, Oregon. M, 1988, University of Oregon. 254 p.

Kattleman, Donald Franklin. Geology of the Desert Mountain Intrusives; Juab County, Utah. M, 1967, Brigham Young University.

Kattman, Robert J. Mineralogy of beach sands, western Puerto Rico. M, 1972, University of Wisconsin-Milwaukee.

Katuna, M. P. The sedimentology of Great Peconic Bay and Flanders Bay, Long Island, New York. M, 1974, Queens College (CUNY). 97 p.

Katuna, Michael P. Sedimentary structures of a modern lagoonal environment; Pamlico Sound, North Carolina. D, 1974, University of North Carolina, Chapel Hill. 117 p.

Katz, Arthur S. The mineralogy of the Sulphur Mine, Mineral, Virginia, including a history of the property. M, 1961, University of Virginia. 104 p.

Katz, B. G. The ability of selected soils to remove and retain molybdenum from industrial wastewaters. M, 1975, University of Colorado.

Katz, Barry. An application of amino acid racemization; the determination of paleoheat flow. D, 1979, University of Miami. 238 p.

Katz, Gary Joshua. A hydrogeological and thermal infrared investigation of the Blue Creek anomaly. M, 1987, University of Nebraska, Lincoln. 91 p.

Katz, M. Physio-chemical modelling of metamorphic reactions. D, 1963, McGill University.

Katz, Marvin. Geology and geochemistry of the southern part of the Cima volcanic field. M, 1981, University of California, Los Angeles. 126 p.

Katz, Michael Barry. The nature and origin of the granulites of the southern part of Mont Tremblant Park, Quebec (Canada). D, 1967, University of Toronto.

Katz, Samuel. Seismic refraction measurements in the Atlantic Ocean; Part VII, Atlantic Ocean bays, west of Bermuda. D, 1955, Columbia University, Teachers College.

Katz, Scott D. Fourier shape analysis and electron microscopy of detrital muscovite in two fluvial networks in North Carolina. M, 1983, Duke University. 71 p.

Katz, Solomon Stuart. An investigation of the variation of the temperature of inversion of quartz. M, 1975, Brooklyn College (CUNY).

Katz, Steven G. Hydrospires in Morrowan Pentremites (Blastoidea) from Oklahoma and Arkansas. D, 1975, University of Texas, Austin. 223 p.

Katz, Steven George. Ostracoda of the Lisbon Formation (middle Eocene) of (western) Alabama. M, 1972, Washington University. 54 p.

Katz, William Meyer. The correlation of the "Brookwood Coal Group" in the Black Warrior Basin of Alabama. M, 1982, University of Alabama. 123 p.

Katzmark, Robert Raymond. Elemental composition of tills in the Cuyahoga Valley National Recreation Area, northeastern Ohio. M, 1985, University of Akron. 97 p.

Kauahikaua, J. Electromagnetic transient soundings on the East Rift geothermal area of Kilauea Volcano, Hawaii; a study of interpretational techniques. M, 1976, University of Hawaii. 88 p.

Kauffman, Albert J. The identification of certain sepiolites (Yavapai County, Arizona). M, 1942, Columbia University, Teachers College.

Kauffman, Daniel F. The geology of the Jacks Creek area, Owyhee County, Idaho. M, 1987, University of Idaho. 93 p.

Kauffman, Erle Galen. Mesozoic paleontology and stratigraphy, Huerfano Park, Colorado; V. 1, Stratigraphy; V. 2, Macroinvertebrate paleontology, Appendix 2, and references cited. D, 1961, University of Michigan. 1467 p.

Kauffman, Erle Galen. The stratigraphy and paleontology of the Buss-Halb Unit No. 1 well core, Freedom Township, Washtenaw County, Michigan. M, 1956, University of Michigan.

Kauffman, John David. Geology of the Sklodowska region, lunar farside. M, 1974, University of Idaho. 147 p.

Kauffman, Marvin Earl. Geology of the Garnet-Bearmouth area, western Montana. D, 1960, Princeton University. 251 p.

Kauffman, Marvin Earl. Statistical analysis of certain characteristics of the Susquehanna River terrace deposits. M, 1957, Northwestern University.

Kaufman, Aaron. Th^{230}/U^{234} dating of carbonates from lakes Lahontan and Bonneville. D, 1964, Columbia University, Teachers College. 292 p.

Kaufman, Alan Jay. Covariance of $\delta^{13}C$ and $\delta^{18}O$ in microbanded carbonates from the Dales Gorge Member of the Brockman Iron Formation and its genetic implications. M, 1985, Indiana University, Bloomington. 182 p.

Kaufman, Darrel Scott. Morphometric analysis of Pleistocene glacial geology in the Kigluaik Mountains, northwestern Alaska. M, 1987, University of Washington. 50 p.

Kaufman, John W. Correlation of clay mineralogy and dolomitic rocks. M, 1963, Rensselaer Polytechnic Institute. 109 p.

Kaufman, John Warren. Phase equilibrium relations on the join forsterite-akermanite-gehlenite-spinel at one atmosphere pressure. D, 1971, Ohio State University. 73 p.

Kaufman, Matthew Ivan. Uranium isotope investigation of the Floridan aquifer and related natural waters of north Florida. M, 1968, Florida State University.

Kaufman, Ronald S. Factor analysis of ground water quality data from the Kingsford Mine, Polk and Hillsborough counties, Florida. M, 1979, University of South Florida, Tampa. 107 p.

Kaufman, William H. Structure, stratigraphy and ore deposits of the central Nacimiento Mountains, New Mexico. M, 1971, University of New Mexico. 87 p.

Kaufmann, Charles H., Jr. Velocity variation problems in seismic prospecting. M, 1952, University of Michigan.

Kaufmann, Harold E. Some monoclinic amphiboles and relation of their physical properties to chemical composition and crystal structure. M, 1963, Brigham Young University. 158 p.

Kaufmann, Karl W., Jr. The Pleistocene of Bermuda. M, 1969, Lehigh University.

Kaufmann, Philip Robert. Channel morphology and hydraulic characteristics of torrent-impacted forest streams in the Oregon Coast Range, U.S.A. D, 1988, Oregon State University. 235 p.

Kaufmann, Robert A. Selected physical properties of the low salt marsh, Skidaway Island, Georgia. M, 1981, University of Georgia.

Kaufmann, Robert F. The stratigraphy of northwestern Adams and northeastern Brown counties, Ohio. M, 1964, Ohio State University.

Kaufmann, Robert Frank. Hydrogeology of solid waste disposal sites in Madison, Wisconsin. D, 1970, University of Wisconsin-Madison. 436 p.

Kaufmann, Ronald Steven. Chlorine in ground water; stable isotope distribution. D, 1984, University of Arizona. 150 p.

Kaufmann, William Lawrence Martin. Petrology of the upper Madison, Southall area, Saskatchewan. M, 1961, University of Saskatchewan. 78 p.

Kaukonen, Everett Konstantine. Attenuation of seismic waves near an explosion. M, 1953, Pennsylvania State University, University Park. 47 p.

Kaul, Lisa Wells. Modeling estuarine nutrient geochemistry in a simple system. M, 1983, Florida State University.

Kaula, William M. Gravimetrically computed deflection of the vertical. M, 1953, Ohio State University.

Kaup, Carl B., III. Magnetic survey, Lexington game management area and vicinity, Cleveland County, Oklahoma. M, 1970, University of Oklahoma. 49 p.

Kauschinger, Joseph Lewis. Evaluation and implementation of Prevost's total stress model. D, 1983, University of Texas, Austin. 361 p.

Kaushal, S. K. Potassium-argon study of Deccan Traps. M, 1967, University of Hawaii.

Kautsky, Mark. Sorption of cesium and strontium by arid region desert soil. M, 1984, University of Nevada. 102 p.

Kautz, Steven Arthur. The importance of cryptic extension in scale models of normal faulting. M, 1987, University of Texas, Austin. 73 p.

Kautzman, Robert R. Sediment distribution and depositional processes involved in silting of an abandoned reservoir, Wake County, North Carolina. M, 1972, North Carolina State University. 67 p.

Kauwenbergh, James B. Van *see* Van Kauwenbergh, James B.

Kavanagh Yllarramendi, John A. A study of selected igneous bodies of the Norris-Red Bluff area, Madison County, Montana. M, 1965, Montana State University. 60 p.

Kavanagh, Paul E. Trace element patterns related to the origin of the Carr Fork Mine, Bingham District, Utah. M, 1981, Dartmouth College. 235 p.

Kavanagh, Paul Michael. Geology of the Hyland Lake area, New Quebec, Canada. D, 1954, Princeton University. 215 p.

Kavanaugh, Ernest G. Thermal history of fluorite deposition, Browns Canyon, Colorado. M, 1978, Northeast Louisiana University.

Kavanaugh, James. Petrography and structure of the Icarus diorite (Precambrian), Saganaga-Northern Light lakes area, Ontario (Canada). M, 1969, SUNY at Stony Brook.

Kavary, Emabeddin. Study of two new ostracods of the Middle Devonian genus Glyptopleura from Michigan and Ohio. M, 1959, University of Michigan.

Kavary, Emadeddin. Upper Cretaceous and lower Cenozoic foraminifera from west-central Iran. D, 1962, University of Missouri, Columbia.

Kavazanjian, E., Jr. A generalized approach to the prediction of the stress-strain-time behavior of soft

clay. D, 1978, University of California, Berkeley. 217 p.

Kaveh, F. Tile drainage on sloping land including unsaturated flow. D, 1979, Utah State University. 210 p.

Kawakatsu, Hitoshi. Double seismic zones; a first order feature of plate tectonics. D, 1985, Stanford University. 210 p.

Kawano, Y. The relationship of soil composition to rheological properties and the compactibility of some Hawaiian soils. M, 1957, University of Hawaii. 57 p.

Kawar, Kamel A. The relation between geologic structures and groundwater in Amman Zerqa area (Hashemite Kingdom of Jordan). M, 1967, University of Arizona.

Kawase, Yoshio. Lower Cambrian Archeocyatha from the Yukon Territory. M, 1956, University of British Columbia.

Kay Shu Joseph Lee *see* Shu Kay Joseph Lee

Kay, Anthony Edward. Effects of low temperature on the induced polarization response of mississippi valley type ore samples. M, 1981, University of Calgary. 121 p.

Kay, B. G. Recharge characteristics of the Oak Ridges aquifer complex; the role of Musselman Lake. M, 1986, University of Waterloo. 113 p.

Kay, Bruce D. Vein and breccia gold mineralization and associated igneous rocks at the Ortiz Mine, New Mexico, U.S.A. M, 1986, Colorado School of Mines. 179 p.

Kay, David William. Environmental investigation and remedial action at an industrial site; a case study. D, 1988, University of California, Los Angeles. 159 p.

Kay, Elizabeth Alexandra. A geochemical and fluid inclusion study of the arsenopyrite-stibnite-gold mineralization, Moreton's Harbour, Notre Dame Bay, Newfoundland. M, 1982, Memorial University of Newfoundland. 209 p.

Kay, George Frederick. The geology and ore deposits of Riddle's Quadrangle, Oregon. D, 1914, University of Chicago. 36 p.

Kay, George M. Stratigraphy of the Decorah Formation. D, 1929, Columbia University, Teachers College.

Kay, George Marshall. The paleontology of the Decorah Formation in Winneshiek County, Iowa. M, 1925, University of Iowa. 82 p.

Kay, Mary Ann. Biotic changes as a reflection of altered lake condition; a comparison of environmental indicators of Lake Michigan found prior to 1900 with those of the 1970's. M, 1982, Northeastern Illinois University. 185 p.

Kay, P. A. Post-glacial history of vegetation and climate in the forest-tundra transition zone, Dubawnt Lake region, Northwest Territories, Canada. D, 1976, University of Wisconsin-Madison. 154 p.

Kay, Robert Woodbury. The rare earth geochemistry of alkaline basaltic volcanics. D, 1970, Columbia University. 179 p.

Kay, W. T. Gastropods of the Kanopolis local fauna (Yarmouthian?; Pleistocene), Ellsworth County, Kansas. M, 1977, Kent State University, Kent. 101 p.

Kay, W. W. Mercury in the Terlingua District of Texas. M, 1938, University of Missouri, Rolla.

Kaya, M. H. The adsorption of phosphorus by Kaneohe Bay sediments. M, 1971, University of Hawaii. 139 p.

Kaye, John Morgan. Certain aspects of the geology of Lowndes County, Mississippi. M, 1955, Mississippi State University. 110 p.

Kaye, John Morgan. Pleistocene sediment and vertebrate fossil associations in the Mississippi Black Belt; a genetic approach. D, 1974, Louisiana State University.

Kayen, Robert E. The mobilization of Arctic Ocean landslides by sea level fall; induced gas hydrate decomposition. M, 1988, California State University, Hayward. 227 p.

Kayes, Douglas M. A gravity and seismic study of the buried Teays River, Benton, Tippecanoe, and Warren counties, Indiana. M, 1979, Indiana University, Bloomington. 170 p.

Kayler, Kyle L. Geology of the Humboldt Mountain area, Arizona. M, 1978, Arizona State University. 101 p.

Kaylor, Donald Charles. Facies and diagenesis of the Upper Devonian Palliser Formation, Front Ranges of the Southern Rocky Mountains, Alberta and British Columbia. M, 1988, McGill University. 176 p.

Kayode, Abiodum A. Thermal expansion and the effect of temperature on the angle of the rhombic section of plagioclase feldspars. D, 1964, University of Chicago. 86 p.

Kays, B. L. Relationship of soil morphology, soil disturbance and infiltration to stormwater runoff in the suburban NC Piedmont. D, 1979, North Carolina State University.

Kays, Marvin Allan. Gravity and resistivity measurements over circular mineralized deposits. M, 1958, Washington University. 31 p.

Kays, Marvin Allan. Petrography of the Sanford Hill titaniferous magnetite deposit, Essex County, New York. D, 1961, Washington University. 166 p.

Kayser, Robert Benham. Sedimentary petrology of the Nugget Sandstone (Jurassic), northern Utah, western Wyoming, and eastern Idaho. M, 1964, University of Utah. various pagination p.

Kazakos, George K. Geochemical study of trace metals in Jamaica Bay. M, 1984, Brooklyn College (CUNY).

Kazanowska, Maria. Suspended matter in Monterey Bay, California; some aspects of its distribution and mineralogy. M, 1971, United States Naval Academy.

Kazdal, Recep A. A study of the reservoir characteristics of the Niobrara Formation in South Dakota. M, 1985, South Dakota School of Mines & Technology.

Kazi, Wallid M. A gravity survey in the vicinity of the San Andreas Fault near Lake Hughes, California. M, 1985, University of California, Riverside. 83 p.

Kazmer, Carol. The Main Mantle thrust zone at Jawan Pass area; Swat, Pakistan. M, 1986, University of Cincinnati. 79 p.

Kazzaz, H. H. Crustal structure of the East Pacific Rise computed models for gravity data. M, 1972, University of Washington. 58 p.

Keables, Michael John. Spatial and temporal associations of midtropospheric circulation, precipitation, and maximum stream discharge in the upper Mississippi River basin. D, 1986, University of Wisconsin-Madison. 248 p.

Keach, R. William, II. Cenozoic active margin and shallow Cascades structures; COCORP results from western Oregon. M, 1986, Cornell University. 51 p.

Keady, Donald M. A regional study of the Cockfield Formation in the subsurface of west-central Mississippi. M, 1957, Mississippi State University. 40 p.

Keady, Donald Myron. Application of selected statistical methods to a study of the chemical quality of water in the Woodbine aquifer in Texas. D, 1970, Texas A&M University. 114 p.

Kealy, Charles D. Seepage patterns in mill tailings dams as determined by finite element models. M, 1970, University of Idaho. 113 p.

Kean, Alan E. A seismic refraction crustal study of the southeastern United States. M, 1978, Georgia Institute of Technology. 68 p.

Kean, Baxter Frederick. Stratigraphy, petrology and geochemistry of volcanic rocks of Long Island, Newfoundland. M, 1973, Memorial University of Newfoundland. 155 p.

Kean, William Francis, Jr. The effect of uniaxial compression of initial susceptibility of rocks as a function of grain size and composition of the titanomagnetite minerals. D, 1973, University of Pittsburgh.

Keany, J. Late Cenozoic Antarctic radiolarian distributions; their paleoclimatic and paleoceanographic implications. D, 1977, University of Rhode Island. 315 p.

Keany, John. Pliocene-early Pleistocene paleoclimatic history recorded in Antarctic-Subantarctic deep-sea cores. M, 1972, Florida State University.

Kearby, James Kimbro. The Cowbell Member of the Borden Formation (lower Mississippian) of the northeastern Kentucky; a delta front deposit. M, 1971, University of Kentucky. 89 p.

Kearfoot, Carl. Perthites of Virginia. M, 1935, University of Virginia. 47 p.

Kearl, Peter M. Water transport in desert alluvial soil. M, 1981, University of Nevada. 130 p.

Kearnes, James K. An evaluation of a method of microzonation by microseismic amplification spectra. M, 1976, University of Washington. 149 p.

Kearney, Michael Sean. Late Quaternary vegetational and environmental history of Jasper National Park, Alberta. D, 1981, University of Western Ontario.

Kearney, Terrence J. Analysis and formation mechanisms of N-halomethylamines; application to seawater chlorination. M, 1983, University of Hawaii at Manoa. 62 p.

Kearns, Lance E. A classification of sandy, coastal-plain coasts. M, 1973, University of Delaware.

Kearns, Lance E. The mineralogy of the Franklin Marble, Orange County, New York. D, 1977, University of Delaware. 235 p.

Kearsley, Fie. Concentration and δ C^{13} content of molecular size fractions of the dissolved organic matter in estuarine and marine waters. M, 1973, Florida State University.

Keasler, Walter Robin. Coal geology of the Chelsea Quadrangle in parts of Craig, Mayes, Nowata, and Rogers counties, Oklahoma. M, 1979, Oklahoma State University. 58 p.

Keathley, Frances Kathleen Stephens. Foraminifera from the Sundance (Jurassic) Formation of central Wyoming. M, 1944, University of Missouri, Columbia.

Keating, B. J. The pre-Carboniferous rocks of the Wentworth section of the Cobequid Hills, Nova Scotia. M, 1933, McGill University.

Keating, Barbara Helen. Biostratigraphy and magnetostratigraphy of Late Cretaceous sediments. M, 1975, University of Texas at Dallas. 83 p.

Keating, Barbara Helen. Contributions to paleomagnetism. D, 1976, University of Texas at Dallas. 211 p.

Keating, Pierre Benjamin. The inversion of time-domain airborne electromagnetic data using the plate model. D, 1988, McGill University. 173 p.

Keating, Robert William. A geochemical study of the Rickenbach Formation at the Friedensville zinc deposit, Lehigh County, Pennsylvania. M, 1983, Lehigh University.

Keatinge, Penelope Rosanna Gann. Late Quaternary till stratigraphy of southeastern Manitoba based on clast lithology. M, 1975, University of Manitoba.

Keaton, Jeffrey Ray. A probabilistic model for hazards related to sedimentation processes on alluvial fans in Davis County, Utah. D, 1988, Texas A&M University. 467 p.

Keats, Donna G. Geology and mineralization of the High Grade District, Modoc County, California. M, 1985, Oregon State University. 139 p.

Keats, Harvey Franklin. Geology and mineralogy of the pyrophyllite deposits, south of Manuels, Avalon peninsula, Newfoundland. M, 1970, Memorial University of Newfoundland. 77 p.

Keaveny, Joseph Michael. In-situ determination of drained and undrained soil strength using the cone penetration test. D, 1985, University of California, Berkeley. 395 p.

Keays, Reid Roderick. A neutron activation analysis technique for determination of the precious metals and its application to a study of their geochemistry. D, 1968, McMaster University. 249 p.

Kebert, Fay Dean. The geology of the Clifton City Quadrangle, Missouri. M, 1956, University of Missouri, Columbia.

Keck, Bradly Dwight. Thermoluminescence, cathodoluminescence and annealing studies of type 3 ordinary and carbonaceous chondrites. D, 1986, University of Arkansas, Fayetteville. 169 p.

Keck, David Alan. Trace element distribution in some Upper Mississippian and Lower Pennsylvanian shales of southern Illinois. M, 1973, Southern Illinois University, Carbondale. 61 p.

Keckler, Douglas James. Geology and ore deposits of the Seven Troughs mining district, Pershing County, Nevada. M, 1981, University of Utah. 71 p.

Kedzie, Laura L. High-precision ^{40}Ar/^{39}Ar dating of major ash-flow tuff sheets, Socorro, NM. M, 1984, New Mexico Institute of Mining and Technology. 197 p.

Keech, Dorothy Ann. The characterization of extracted soil and sludge fulvic acids as metal complexing agents in solution. D, 1979, University of California, Riverside. 219 p.

Keefer, C. M. An experimental study of synthetic titanomaghemite; synthesis, magnetic analysis, and crystallographic measurements. D, 1979, University of Wyoming. 89 p.

Keefer, D. K. Earthflow. D, 1977, Stanford University. 393 p.

Keefer, Keith Douglas. Phase separation and crystallization phenomena in silicate systems. D, 1981, Stanford University. 171 p.

Keefer, W. R. Geology of the DuNoir area, Fremont County, Wyoming. D, 1956, University of Wyoming. 252 p.

Keefer, W. R. Geology of the Red Hills area, Teton County, Wyoming. M, 1952, University of Wyoming. 99 p.

Keel, Raybon Thomas. The measurement of heavy metals in natural water; zinc dynamics in the southern basin of Lake Michigan. D, 1987, Clemson University. 177 p.

Keele, Joseph. The effect of feldspar on kaolin in burning. M, 1911, Cornell University.

Keeler, Charles Martyn. The relationships between the mechanical and other properties of a mountain snow cover, Alta, Utah, 1967. D, 1969, McGill University. 85 p.

Keeler, Gerald Joseph. A hybrid approach for source apportionment of atmospheric pollutants in the northeastern United States. D, 1987, University of Michigan. 294 p.

Keeler, Gordon T. Geology of the Algold Mine, Goudreau area, Ontario. M, 1940, University of Toronto.

Keeler, Jane V. A lithologic analysis of the carbonate rocks of north central Ohio. M, 1941, Oberlin College.

Keeler, John White. Study of the gold deposits of the Canadian Pre-Cambrian Shield of Ontario and Quebec. M, 1951, University of Michigan.

Keeler, Robert George. Stratigraphy and sedimentology, Lower Cretaceous Grand Rapids Formation, Wabasca A oil sand deposit area, Northeast Alberta, Canada. M, 1978, University of Calgary. 141 p.

Keeley, J. R. Observed and predicted longshore currents on Martinique Beach; a comparison of driving terms. M, 1975, Dalhousie University.

Keeley, James Chester. A geologic study of the Chattanooga Shale in the Elkmont Quadrangle, Alabama. M, 1968, University of Alabama.

Keeley, Joseph Francis. Trace metals in soils of the Coeur d'Alene River valley and their potential effects on water quality. M, 1979, University of Idaho. 116 p.

Keeling, D. L. Sulphur Banks fumaroles; gas sampling and analysis techniques and data correlations. M, 1973, University of Hawaii. 57 p.

Keeling, D. V. K-Ar dating; atmospheric argon contamination in volcanic rocks. D, 1974, University of Hawaii. 166 p.

Keeling, Theodore. The geology of the Pocahontas #11 coal seam in Southwest Virginia. M, 1978, University of South Carolina.

Keely, Joseph Francis, Jr. Ground-water contamination assessments. D, 1986, Oklahoma State University. 408 p.

Keen, Charlotte Elizabeth. A study of the physical properties of the oceanic crust and mantle. M, 1966, Dalhousie University. 50 p.

Keen, Douglass C. Remanent magnetization of some igneous rocks (Precambrian) from northwestern Iowa (Clay County). M, 1967, University of Iowa. 102 p.

Keen, Kerry Lee. Sand dunes on the Anoka sand plain. M, 1988, University of Minnesota, Minneapolis. 204 p.

Keen, Timothy R. The comparative sedimentology of two stranded bars and implications for their origin. M, 1987, Florida State University. 151 p.

Keenan, Everly Mary. Amino acid racemization dating; theoretical considerations and practical applications. D, 1983, University of Delaware. 369 p.

Keenan, Everly Mary. Hydrocarbon and fatty acid distribution in sediments of the Hudson Estuary. M, 1978, University of Delaware.

Keenan, James Edward. Ostracodes from the Maquoketa Shale of Missouri. M, 1940, University of Missouri, Columbia.

Keenan, Marvin F. The Eocene Sierra Blanca Limestone at the type locality in Santa Barbara County, California. M, 1932, Stanford University. 84 p.

Keenan, Steven J. Spectral analysis of signals received at the Bowling Green State University Seismic Observatory (BGO). M, 1979, Bowling Green State University. 160 p.

Keene, Arthur G. Heavy mineral correlations of Pleistocene aquifer material of the Los Angeles Basin (California), a feasibility investigation. M, 1965, University of Southern California.

Keene, D. G. Ground support for swelling ground; a case study. M, 1979, Stanford University. 148 p.

Keene, Donald F. A physical oceanographic study of the nearshore zone at Newport, Oregon. M, 1971, Oregon State University. 91 p.

Keene, Howard W. Salt marsh evolution and postglacial submergence in New Hampshire. M, 1970, University of New Hampshire. 87 p.

Keene, Jack R. Ground water resources of the western half of Fall River County, South Dakota. M, 1970, South Dakota School of Mines & Technology.

Keene, John B. The distribution, mineralogy, and petrography of biogenic and authigenic silica from the Pacific Basin. D, 1976, University of California, San Diego. 286 p.

Keene, Warren Elmer. Development and demonstration of a system for vitrifying high level radioactive waste in high silica glass. D, 1987, Catholic University of America. 285 p.

Keene, William Charles. Oxygen and nutrient dynamics in Lac de Tunis, a hypereutrophic subtropical lagoon. M, 1978, University of Virginia. 248 p.

Keeney, Joseph W. The microwave emission spectrum of colliding charged water drops. D, 1968, New Mexico Institute of Mining and Technology. 69 p.

Keeney-Kennicutt, Wendy Lisabeth. The geochemistry of trace metals in the Brazos River and the Brazos River estuary. D, 1982, Texas A&M University. 189 p.

Keenmon, Kendall Andrews. Geology of the Blacktail-Snowcrest region, Beaverhead County, Montana. D, 1950, University of Michigan. 246 p.

Keenmon, Kendall Andrews. Geology of the Red Creek area, Snake River Range, Wyoming. M, 1948, University of Michigan.

Keer, Frederick Rhoades. The sedimentary framework of a desert coastal lagoon, Bahiret el Bibane, Tunisia. M, 1976, Duke University.

Keesling, Stuart Allan. The submarine geology of Pago Bay, Guam, Marianas Islands. M, 1957, University of Southern California.

Keevil, A. R. Variations in radiogenic heat in igneous rocks. M, 1942, University of Toronto.

Keevil, Norman Bell, Jr. Exploration at the Craigmont mines, British Columbia. D, 1965, University of California, Berkeley. 163 p.

Keevil, Norman M. A laboratory investigation of induced polarization. M, 1961, University of California, Berkeley. 102 p.

Kegler, Vern L. The geology and related ore deposits of the Idaho Batholith. M, 1923, University of Minnesota, Minneapolis. 50 p.

Kehew, Alan E. Environmental geology of Lewiston, Idaho and vicinity. D, 1977, University of Idaho. 211 p.

Kehew, Alan E. Environmental geology of part of the West Fork Basin, Gallatin County, Montana. M, 1971, Montana State University. 58 p.

Kehle, Ralph Ottmar. Analysis of the deformation of the Ross Ice Shelf, Antarctica. D, 1961, University of Minnesota, Minneapolis. 193 p.

Kehlenbach, Richard W. Distribution and depositional environment of the Mississippian Cypress Formation in southern Illinois. M, 1969, Southern Illinois University, Carbondale. 54 p.

Kehlenbeck, Manfred M. The Lac Rouvray anorthosite mass (Mont Laurier area, Quebec); a study of deformation by cataclasis and recrystallization. D, 1971, Queen's University. 207 p.

Kehler, Philip Leroy. The lower Zuni sequence (Jurassic) in the southwestern United States. D, 1970, Southern Methodist University. 90 p.

Kehmeier, Richard J. The geology and geochemistry of the Broken Hill area, Hinsdale County, Colorado. M, 1973, Colorado School of Mines. 79 p.

Keho, Timothy H. The vertical seismic profile; imaging heterogenous media. D, 1986, Massachusetts Institute of Technology. 304 p.

Keho, Timothy Henson. Heat flow in the Uinta Basin determined from bottom hole temperature (BHT) data. M, 1987, University of Utah. 99 p.

Kehoe, J. R. de *see* de Kehoe, J. R., Jr.

Kehoe, James Mark. Environmental regulations governing geothermal exploration. M, 1982, California State University, Hayward. 96 p.

Kehrer, Harold Henry. Radiation patterns of seismic surface waves from nuclear explosions. M, 1969, Massachusetts Institute of Technology. 104 p.

Kehres, Cheryl A. The variation in hydraulic conductivity of heterogeneous drift deposits. M, 1984, Michigan State University. 53 p.

Kehrman, Robert F. The downhole suspension and testing of an electrodynamic shear wave exciter. M, 1968, University of Missouri, Rolla.

Keighin, Charles William. Phase relations in the system Ag-Sb-S. D, 1966, University of Colorado. 66 p.

Keighin, Charles William. Some aspects of feldspar geothermometry. M, 1960, University of Colorado.

Keigwin, L. D., Jr. Cenozoic stable isotope stratigraphy, biostratigraphy, and paleoceanography of deep-sea sedimentary sequences. D, 1979, University of Rhode Island. 169 p.

Keim, James Will. A study of photogeologic fracture traces over the Bisbee Quadrangle, Cochise County, Arizona. M, 1962, Pennsylvania State University, University Park. 42 p.

Keir, R. S. The dissolution kinetics of biogenic calcium carbonate; laboratory measurement and geochemical implications. D, 1979, Yale University. 303 p.

Keiser, Edward P. Geology of the north end of the Taconic Range in Vermont. D, 1941, Massachusetts Institute of Technology. 71 p.

Keisler, Ronnie S. Subsurface and ground magnetic survey of the Lone Elm gas field and adjacent areas of T. 10 N., R. 28 W., Franklin County, Arkansas. M, 1972, University of Arkansas, Fayetteville.

Keith, Brian D. Recognition of a Cambrian paleoslope and base-of-slope environment, Taconic Sequence, New York and Vermont. D, 1974, Rensselaer Polytechnic Institute. 196 p.

Keith, Brian D. Stratigraphic and sedimentologic study of Silurian and Devonian metasedimentary rocks, southern third of Bingham Quadrangle, westcentral Maine. M, 1971, Syracuse University.

Keith, Donald Alexander Walter. Sedimentology of the Cardium Formation (Upper Cretaceous), Willesden Green Field, Alberta. M, 1985, McMaster University. 241 p.

Keith, Douglas W. Observations on layering in the upper border series of the Skaergaard Intrusion, East Greenland. M, 1987, Dartmouth College. 115 p.

Keith, Jeffrey Davis. Magmatic evolution of the Pine Grove porphyry molybdenum system, southwestern Utah. D, 1982, University of Wisconsin-Madison. 229 p.

Keith, Jeffrey Davis. Miocene porphyry intrusions, volcanism, and mineralization, southwestern Utah and eastern Nevada. M, 1980, University of Wisconsin-Madison.

Keith, John L. A study of the benthic algae in the Kelp Bed off Del Monte Beach, Monterey, California. M, 1974, Naval Postgraduate School.

Keith, Laura A. A numerical compaction model of overpressuring in shales. M, 1982, Virginia Polytechnic Institute and State University. 80 p.

Keith, M. L. Gold deposit east of Lake Nipigon. M, 1936, Queen's University. 42 p.

Keith, M. L. Petrology of the alkaline intrusive at Blue Mountain, Ontario. D, 1939, Massachusetts Institute of Technology. 84 p.

Keith, Warren E. A gravity study of Big Hand and Columbus; Niagaran reef fields in St. Clair County, Michigan. M, 1967, Michigan State University. 106 p.

Keith, William J. Geology of the Red Mountain mining district, Esmeralda County, Nevada. M, 1975, San Jose State University. 75 p.

Keizer, Richard Paul. Volcanic stratigraphy, structural geology, and K-Ar geochronology of the Durango area, Durango, Mexico. M, 1974, University of Texas, Austin.

Kelafant, Jonathan Robert. A reinterpretation of the Burgess Shale paleoenvironment. M, 1987, George Washington University. 63 p.

Kelemen, Peter Boushall. Assimilation of ultramafic rock in fractionating basaltic magma. D, 1987, University of Washington. 414 p.

Kell, James Alexander. Magnetometer survey of possible fault, Northampton County, Pennsylvania. M, 1952, Lehigh University.

Kell, Scott Randolph. The distribution, ecology, and preservation of benthonic foraminifera in the salt marsh at Wachapreague, Virginia. M, 1979, Kent State University, Kent. 49 p.

Kellam, Jeffrey A. Neogene seismic stratigraphy and depositional history of the Tybee Trough area, Georgia-South Carolina. M, 1981, University of Georgia.

Kelleher, John A. Possible criteria for predicting earthquake locations and their application to major plate boundaries of the Pacific and the Caribbean. D, 1972, Columbia University. 109 p.

Kelleher, Patrick C. The Mono Craters-Mono Lake Islands volcanic complex, eastern California; evidence for several magma types, magma mixing, and a heterogeneous source region. M, 1986, University of California, Santa Cruz.

Kellenberger, Jack Eugene. The geology of Fairfield and southern Wesley townships, Washington County, Ohio. M, 1960, Ohio University, Athens. 130 p.

Keller, Allen Seely. Geology of the Mink Creek region. M, 1952, University of Utah. 37 p.

Keller, Allen Seely. Structure and stratigraphy behind the Bannock thrust in parts of the Preston and Montpelier quadrangles, Idaho. D, 1963, Columbia University, Teachers College. 239 p.

Keller, Barry. Seismic refraction, gravity anomalies, and the Peru Trench. M, 1978, University of Washington. 39 p.

Keller, Barry Ruland. Structural discontinuity within the Southern California continental margin; seismic and gravity models of the western Transverse Ranges. D, 1984, University of California, Santa Barbara. 108 p.

Keller, C. K. Hydrogeology of fractured clayey till confining a shallow sand aquifer near Saskatoon, Saskatchewan. M, 1985, University of Waterloo. 155 p.

Keller, Chester K. Report on the structural and sedimentary geology of Moss Beach, California. M, 1976, Stanford University.

Keller, Chester Kent. Controls on patterns of groundwater flow and major-ion occurrence in two deposits of clayey till near Saskatoon, Saskatchewan. D, 1987, University of Waterloo. 340 p.

Keller, Daniel James. A fracture study in the greater Cincinnati area. M, 1957, University of Cincinnati. 112 p.

Keller, David R. Structure and alteration as a guide to mineralization in the Secret area, Eureka County, Nevada. M, 1982, Brigham Young University. 116 p.

Keller, Dianne M. The chemistry and mineralogy of forest and rhizosphere soils in the Eastern United States. M, 1988, Colgate University.

Keller, Donald Frederick. An interpretation of the stream patterns in Pike and Sandy townships in southern Stark County, Ohio. M, 1975, University of Akron. 66 p.

Keller, Edward Anthony. Form and fluvial processes in alluvial stream channels. D, 1973, Purdue University.

Keller, Edward Anthony. Form and fluvial processes of Dry creek near Winters, (northern) California. M, 1969, University of California, Davis. 73 p.

Keller, Fred, Jr. A magnetic survey of the Canfield Estate, Mine Hill, Morris County, New Jersey. M, 1942, Rutgers, The State University, New Brunswick. 61 p.

Keller, Frederick Brian. Late Precambrian stratigraphy, depositional history, and structural chronology of part of the Tennessee Blue Ridge. D, 1980, Yale University. 422 p.

Keller, G. North Pacific biostratigraphy and paleoceanography of DSDP sites 173, 310, 296. D, 1978, Stanford University. 271 p.

Keller, George H. Sedimentary features of the "white quartzite" in the western Uinta Mountains, Utah. M, 1956, University of Utah. 55 p.

Keller, George Henrik. Sediments of the Malacca Strait, southeast Asia. D, 1966, University of Illinois, Urbana.

Keller, George Randy, Jr. Crustal structure of the Texas Gulf Coast. D, 1973, Texas Tech University. 125 p.

Keller, George Randy, Jr. Short period Rayleigh wave dispersion. M, 1969, Texas Tech University. 137 p.

Keller, George Vernon. A study of the applicability of the shielded mono-electrode to resistivity well logging. M, 1951, Pennsylvania State University, University Park. 108 p.

Keller, George Vernon. Dispersion of seismic waves near an explosion. D, 1954, Pennsylvania State University, University Park. 145 p.

Keller, Homer M. The magnetic polarity stratigraphy of an Upper Siwalik sequence in the Pabbi Hills of Pakistan. M, 1975, Dartmouth College. 110 p.

Keller, J. David. Acoustic wave propagation in composite, fluid saturated media. M, 1987, Colorado School of Mines. 91 p.

Keller, John David. The geologic setting of field sites; Summit County, Ohio. M, 1971, University of Akron. 57 p.

Keller, Kenneth Frank. Contact metamorphism near Alta (Wasatch, Utah, and Salt Lake counties, Utah). M, 1942, University of Utah. 38 p.

Keller, Marvin A. Stratigraphy of the pre-Cody Cretaceous rocks in the southeastern Wind River basin, Fremont County, Wyoming. M, 1957, University of Wyoming. 68 p.

Keller, Mary Ruth. Geologic framework of gravity anomaly sources in the central Piedmont of Virginia. M, 1983, Virginia Polytechnic Institute and State University. 83 p.

Keller, Paul Henry. Middle Devonian stromatoporoids of northwestern Ohio. M, 1963, Bowling Green State University. 50 p.

Keller, Peter Charles. Geology of the Sierra del Gallego area, Chihuahua, Mexico. D, 1977, University of Texas, Austin. 163 p.

Keller, Peter Charles. Mineralogy of the Tayoltita gold and silver mine, Durango, Mexico. M, 1974, University of Texas, Austin.

Keller, Roland Bradford. Stratigraphic distribution of middle and upper Eocene foraminifera from northwestern Marion County, Florida. M, 1962, University of Florida. 66 p.

Keller, Stephen M. Land application of secondary treated municipal wastewater in the town of Chautauqua, New York. M, 1974, SUNY, College at Fredonia. 110 p.

Keller, Walter David. A study of the green spots in red sediments. M, 1926, University of Missouri, Columbia.

Keller, Walter David. Earth resistivities at depths less than one hundred feet. D, 1933, University of Missouri, Columbia.

Kellett, Charles Richard. Subsurface geology of the Purcell area, Cleveland and McClain counties, Oklahoma. M, 1958, University of Oklahoma. 71 p.

Kelley, Alice A. Repsher. Provenance investigation of the glacial deposits of the Summit Creek and North Fork drainages and a portion of the Trail Creek drainage, Custer and Blaine counties, Idaho. M, 1981, Lehigh University.

Kelley, Arthur M. Measurement of deep crustal resistivities. M, 1963, Massachusetts Institute of Technology. 103 p.

Kelley, Barbara Ann. Quantitative mineralogy of Lake Michigan piston cores used for paleomagnetic determinations. M, 1978, University of Wisconsin-Milwaukee. 71 p.

Kelley, D. G. Areal geology of the Baddeck map-area, 1 inch to 1 mile (Canada). D, 1954, Massachusetts Institute of Technology.

Kelley, Dana Robineau. I, Urano-organic ore at Temple Mountain; II, Clay alteration and ore, Temple Mountain, Utah. D, 1959, Columbia University, Teachers College.

Kelley, Dana Robineau. Population studies on horses and hyopsodonts of the lower Eocene Lysite fauna. M, 1952, University of Massachusetts.

Kelley, Danford Greenfield. Mississippian stratigraphy and geologic history of central Cape Breton Island, Nova Scotia. D, 1959, Massachusetts Institute of Technology. 128 p.

Kelley, Deborah S. Two-phase separation and fracturing in mid-ocean ridge gabbros at temperatures greater then 700°C. M, 1987, University of Washington. 40 p.

Kelley, Frederic R. Stratigraphic allocation of Eocene foraminifera from the western Santa Ynez Mountains of California. M, 1941, Stanford University. 46 p.

Kelley, G. C. Late Pleistocene and Recent geology of the Housatonic River region in northwestern Connecticut. D, 1975, Syracuse University. 339 p.

Kelley, Gary M. An audio-frequency magnetotelluric survey in Marquette County and Baraga County, Michigan. M, 1974, Michigan Technological University. 170 p.

Kelley, Hiram. Depositional systems within the lower Atoka Formation of southwestern Washington and northeastern Crawford counties, Arkansas. M, 1977, University of Arkansas, Fayetteville.

Kelley, J. S. Structural geology of a portion of the southwest quarter of the New York Butte Quadrangle, Inyo County, California. M, 1973, San Jose State University. 93 p.

Kelley, James C. Least squares analysis of tectonite fabric data. D, 1966, University of Wyoming. 56 p.

Kelley, John Stewart. Environments of deposition and petrography of Lower Jurassic volcaniclastic rocks, southwestern Kenai Peninsula, Alaska. D, 1980, University of California, Davis. 304 p.

Kelley, Joseph A. The Pleistocene geology of the Octa and the Mt. Sterling quadrangles, Ohio. M, 1937, Ohio State University.

Kelley, Joseph T. Sediment and heavy metal distribution in a coastal lagoon complex, Stone Harbor, NJ. M, 1976, Lehigh University.

Kelley, Joseph Timothy. Sources of tidal inlet suspended sediment, Stone Harbor, New Jersey. D, 1980, Lehigh University. 188 p.

Kelley, L. M. Geology of Bear Butte, a Quaternary lava cone, Blaine County, Idaho. M, 1975, SUNY at Buffalo. 60 p.

Kelley, Lynn Irvin. Kaolinitic weathering zone on Precambrian basement rocks, Red River valley, eastern North Dakota and northwestern Minnesota. M, 1980, University of North Dakota. 85 p.

Kelley, Millard Lee. Physical stratigraphy of some upper Strawn and lower Canyon beds in Parker and Palo Pinto counties, Texas. M, 1958, Texas Christian University. 66 p.

Kelley, Patricia Hagelin. Mollusc lineages of the Chesapeake Group (Miocene). D, 1979, Harvard University.

Kelley, Peter Alexander. Pyrolytic characterization of the organic matter in selected coals and in the Devonian shales of southern West Virginia. D, 1980, West Virginia University. 275 p.

Kelley, Richard J., Jr. Geology of the Pickhandle Hills, San Bernardino Valley, Cochise County, Arizona. M, 1966, University of Arizona.

Kelley, Robert W. The geology of Cockburn Island, Ontario. M, 1949, Wayne State University.

Kelley, Shari Anne. Fission-track annealing systematics in apatite with geological applications. D, 1984, Southern Methodist University. 251 p.

Kelley, Stephen M. Body wave synthetic seismograms in anisotropic media with velocity gradients. M, 1983, University of Washington. 173 p.

Kelley, Van. Field determination of dispersivity of co-mingling plumes. M, 1985, Texas A&M University. 97 p.

Kelley, Vincent C. Geology and ore deposits of the Darwin silver-lead mining district, Inyo County, California. D, 1937, California Institute of Technology. 163 p.

Kelley, Vincent C. Geology of the Santa Monica Mountains west of the Malibu Ranch, Ventura County, California. M, 1932, California Institute of Technology. 55 p.

Kelley, Ward Wesley. An investigation of the movement of water and petroleum through rock under high pressure. M, 1918, University of Missouri, Columbia.

Kelley, William N., Jr. Geology and origin of the Woods Creek iron deposit (Precambrian), Ravalli County, Montana. M, 1967, Pennsylvania State University, University Park. 54 p.

Kellogg, Donald Walter. Areal geology of Butler County, Kansas. M, 1978, Wichita State University. 151 p.

Kellogg, Frederic H. The role of the clay minerals in soil mechanics. D, 1934, The Johns Hopkins University.

Kellogg, Frederic H. The titanium ore deposits; their geology, mineralogy and economic importance. M, 1929, The Johns Hopkins University.

Kellogg, Harold E. Stratigraphy and structure of the southern Egan Range, Nevada. D, 1959, Columbia University, Teachers College. 232 p.

Kellogg, James Nelson. The Cenozoic basement tectonics of the Sierra de Perija, Venezuela and Colombia. D, 1981, Princeton University. 236 p.

Kellogg, Karl Stuart. A paleomagnetic study of various Precambrian rocks in the northeastern Colorado Front Range and its bearing on Front Range rotation. D, 1973, University of Colorado.

Kellogg, Lee Olds. Notes on the geology of the Cochise mining district, Arizona. M, 1906, Columbia University, Teachers College.

Kellogg, Louise Helen. Chaotic mantle mixing and chemical geodynamics. D, 1988, Cornell University. 151 p.

Kellogg, Richard L. A geophysical investigation of the hydrogeological characteristics of the Udell Hills area, Manistee County, Michigan. M, 1964, Michigan State University. 74 p.

Kellogg, Richard L. An aeromagnetic investigation of the south peninsula of Michigan. D, 1971, Michigan State University. 161 p.

Kellogg, Thomas Bartlett. Late Pleistocene climatic record in Norwegian and Greenland Sea deep-sea cores. D, 1973, Columbia University. 545 p.

Kellough, Gene Ross. Biostratigraphic and paleoecologic study of Midway foraminifera along a section of Techuacana Creek, Limestone County, Texas. M, 1959, University of Houston.

Kells, Bruce Lynn. Paleogeography, depositional environments, and regional structure of a portion of the Pennsylvanian age Breathitt Formation, southeastern Kentucky. M, 1980, University of Kentucky. 90 p.

Kellum, L. B.; Daviess, S. N. and Swinney, C. M. Geology and oil possibilities of the southwestern part of the Wide Bay Anticline, Alaska. M, 1945, University of Michigan.

Kellum, Lewis B. Paleontology and stratigraphy of the (Eocene) Castle Hayne and (Miocene) Trent formations in North Carolina. D, 1924, The Johns Hopkins University.

Kelly, Anne O. Petroleum geology of the Casper Formation in the northern Laramie Basin. M, 1982, Colorado School of Mines. 112 p.

Kelly, Edward, Jr. Surface water hydrogeology of the Cold River, southwestern New Hampshire. M, 1987, Boston University. 166 p.

Kelly, F. Randolph. An environmental study of the subsurface Miocene of Jefferson County, Texas. M, 1965, Texas A&M University. 212 p.

Kelly, Gary G. Airborne gravimetry. M, 1964, Ohio State University.

Kelly, George A., Jr. Stratigraphic relationship of the Upper Cretaceous Taylor and Navarro groups of Northwest Louisiana. M, 1961, Louisiana State University.

Kelly, Glen Eric. Depositional environment and diagenesis of the Gloyd Member of the Rodessa Formation, Naconiche Creek Field, Nacogdoches County, Texas. M, 1988, Northeast Louisiana University. 65 p.

Kelly, Herbert A. and Connor, Mike. The geology of a portion of Rapid Canyon (South Dakota). M, 1957, South Dakota School of Mines & Technology.

Kelly, J. M. Geology, wall rock alteration and contact metamorphism associated with massive sulfide mineralization at the Amulet Mine, Noranda District, Quebec. D, 1975, University of Wisconsin-Madison. 290 p.

Kelly, James A. The petrography and petrology of the Stock Lake, Yellowknife, NWT. M, 1964, University of Montana. 60 p.

Kelly, James Michael. The geology of the Rattlesnake Park-Blue Mountain area, Larimer County, Colorado. M, 1967, University of Colorado.

Kelly, John C. Petrology and petrogenesis of alkali basalt and their associated inclusions from Elephant Butte area, Sierra County, New Mexico. M, 1988, University of New Mexico. 172 p.

Kelly, John Joseph and Malin, William John. Geology of West Pass Peak area, Sublette County, Wyoming. M, 1952, University of Michigan.

Kelly, John L. Variation of elastic wave velocity with water content in sedimentary rocks (K297). M, 1952, University of Texas, Austin.

Kelly, John M. Structural geology of the Indian Meadows area, northwestern Fremont County, Wyoming. M, 1955, University of Kansas. 146 p.

Kelly, John Martin. Relationship between sub-surface pressure studies and sub-surface correlations, Vacuum area, New Mexico. D, 1939, New Mexico Institute of Mining and Technology. 21 p.

Kelly, Joseph Allen. The Pleistocene geology of the Octa and the Mt. Sterling quadrangles, Ohio. M, 1937, Ohio State University.

Kelly, Joseph Michael. Mineralogy and petrography of the Basal Chuckanut Formation (Eocene) in the vicinity of Lake Samish, Washington. M, 1970, Western Washington University. 63 p.

Kelly, Kevin E. Stratigraphy and sedimentology of the Pennsylvanian Coffman Member of the Minturn Formation and Belden Formation, Mosquito Range, Colorado. M, 1984, Colorado School of Mines. 107 p.

Kelly, Martin Henry. Effects of mine drainage on the benthos and periphyton of the Bankston Fork of the Saline River. D, 1984, Southern Illinois University, Carbondale. 204 p.

Kelly, Michael A. Depositional environments of the Ordovician Steuben Limestone (Trenton Group) of northwestern New York. M, 1972, Boston University. 82 p.

Kelly, Patrick Vizard. The delineation of erosional highs on the Knox unconformity using the gravity method, Morrow County, Ohio. M, 1986, Wright State University. 88 p.

Kelly, Robert Bowen and Cooper, Jack Charles. Geology of a portion of the Santa Susana Quadrangle, Los Angeles and Ventura counties, California. M, 1941, University of California, Los Angeles.

Kelly, Stuart Mackenzie. Functional morphology and evolution of Iocrinus; an Ordovician disparid inadunate crinoid. M, 1978, Indiana University, Bloomington. 80 p.

Kelly, Stuart Mackenzie. Paleocology and paleontology of the Indian Springs Shale Member, Big Clifty Formation (middle Chesterian) in south-central Indiana. D, 1984, Indiana University, Bloomington. 343 p.

Kelly, Thomas E. Structural development of the Sedgwick Basin. M, 1961, University of Kansas. 31 p.

Kelly, Thomas Eugene. Geology of north-central Burleson County, Texas. M, 1955, Texas A&M University. 113 p.

Kelly, W. A. Geology of the Cadomin and Mountain Park areas, British Columbia. D, 1925, Princeton University.

Kelly, Walton Ross. The effects of methane perturbation on the chemistry and quality of groundwater systems. M, 1983, Case Western Reserve University.

Kelly, William Crowley. Preliminary report on zinc gossans in New Mexico. M, 1953, Columbia University, Teachers College.

Kelly, William Crowley. Selected aspects of the leached outcrop problem. D, 1954, Columbia University, Teachers College. 155 p.

Kelly, William J., Jr. An isotopic study of the Massabesic Gneiss, Southeast New Hampshire. M, 1980, University of New Hampshire. 121 p.

Kelly, William M. Occurrence of scapolite in the vicinity of Chapel Pond and Keene Valley, central Adirondacks, New York. M, 1974, University of Massachusetts. 65 p.

Kelly, William Morgan. Chemistry and genesis of titaniferous magnetite and related ferromagnesian silicates, Sanford Lake deposits, Tahawus, New York. D, 1979, University of Massachusetts. 236 p.

Kelm, Donald L. A gravity and magnetic study of the Laguna Salada area, Baja California, Mexico. M, 1972, [University of California, San Diego].

Kelm, James S. The propagation of elastic waves in heterogeneous media. D, 1980, University of Cincinnati. 142 p.

Kelsey, Graham Landers. Petrology of metamorphic rocks hosting volcanogenic massive sulphide deposits, Ambler District, Alaska. M, 1979, Arizona State University. 156 p.

Kelsey, Harvey Marion, III. Landsliding, channel changes, sediment yield and land use in the Van Duzen River basin, North coastal California, 1941-1975. D, 1977, University of California, Santa Cruz. 389 p.

Kelsey, Martin C., Jr. A study of the Devonian placoderm, Macropetalichthys, with particular reference to M. rapheidolabis. M, 1957, Miami University (Ohio). 46 p.

Kelsey, Richard Kelly. Modeling soil water extraction from rangeland vegetation in an experimental watershed. M, 1984, University of Idaho. 62 p.

Kelson, Keith Irvin. Long-term tributary adjustments to base-level lowering, northern Rio Grande Rift, New Mexico. M, 1986, University of New Mexico. 210 p.

Keltch, Brian. Depositional systems and reservoir quality of the Clinton Sandstone, Guernsey County, Ohio. M, 1985, University of Cincinnati. 73 p.

Kelty, Barbara M. The influence of exposed underlying, less resistant rock on the magnitude of caprock slopes; a quantitative analysis of caprock slopes developed in the presence and in the absence of exposed underlying, non-resistant strata. M, 1975, Indiana State University. 44 p.

Kelty, Kevin Blair. Stratigraphy, lithofacies, and environment of deposition of the Scappoose Formation in central Columbia County, Oregon. M, 1981, Portland State University. 81 p.

Kemeny, John McKenzie. Frictional stability of heterogeneous surfaces in contact; the mechanics of faulting and earthquake rupture. D, 1986, University of California, Berkeley. 165 p.

Kemerer, Thomas F. Barrier Island origin and migration (Holocene) near Wachapreague, Virginia. M, 1971, West Virginia University.

Kemme, Joseph W. Surfaces and their topology. M, 1942, Catholic University of America.

Kemmer, David Andrew. Characterization of kerogen from Gulf Coast Tertiary sediments. M, 1978, University of Missouri, Columbia.

Kemmerer, John L. Gilsonite. M, 1934, University of Utah. 61 p.

Kemmerer, Mahlon. Rock alteration at the Tintic Prince Mine, North Tintic District, Utah. M, 1935, University of Utah. 22 p.

Kemmerly, Phillip Randall. Environmental geology of the Mannford area, Oklahoma. D, 1973, Oklahoma State University. 96 p.

Kemmis, Timothy J. Properties and origin of the Yorkville Till Member at the National Accelerator Laboratory site, Northeast Illinois. M, 1979, University of Illinois, Urbana. 331 p.

Kemner, Mark. A Markovian computer simulation of conservative groundwater tracer behavior during transport. M, 1985, University of Houston.

Kemnitzer, Luis E. Geology of the San Nicolas and Santa Barbara islands, Southern California. M, 1933, California Institute of Technology. 45 p.

Kemnitzer, Luis E. Structural studies in the Whipple Mountains, south-eastern California. D, 1937, California Institute of Technology. 152 p.

Kemp, George Paul. Mud deposition at the shoreface; wave and sediment dynamics on the chenier plain of Louisiana. D, 1986, Louisiana State University. 163 p.

Kemp, Malcolm W. A preliminary study of Ordovician traverse in the south-western part of Big Valley, An-

derson County, Tennessee. M, 1950, University of Tennessee, Knoxville. 40 p.

Kemp, Peter Evans. The Ordovician rocks of the Wartrace Quadrangle, Bedford County, Tennessee. M, 1957, University of Tennessee, Knoxville. 34 p.

Kemp, Robert E. Structure and stratigraphy of the Harmony Group, northwestern Rhode Island. M, 1985, University of Kentucky.

Kemp, Thomas Earl. A study of the middle and lower-Upper Ordovician rocks of the Oak Ridge Valley between Elza Gate and Clinton; Anderson County, Tennessee. M, 1954, University of Tennessee, Knoxville. 47 p.

Kemp, Wayne Russell. Petrochemical affiliations of volcanogenic massive sulfide deposits of the Foothill Cu-Zn belt, Sierra Nevada, California. D, 1982, University of Nevada. 493 p.

Kemp, William Madison. Stable isotope and fluid inclusion study of the contact Al(Fe)-Ca-Mg-Si skarns in the Alta Stock aureole, Alta, Utah. M, 1985, University of Utah. 65 p.

Kempany, Ryan Glenn. Subsurface analysis of the Middle Devonian Sylvania Sandstone in the Michigan Basin. M, 1976, Michigan State University. 67 p.

Kemple, Harold F. Geology of Sherwin Point area, Latah County, Idaho. M, 1979, Eastern Washington University. 40 p.

Kempner, William. The magnetic properties of the Point Sal Ophiolite; a comparison with oceanic crust. M, 1977, University of California, Santa Barbara.

Kempter, Kirt Anton. Mid-Tertiary volcanic history of the Tomochic region, northern Sierra Madre Occidental, Chihuahua, Mexico. M, 1986, University of Texas, Austin. 134 p.

Kempton, John Paul. Outwash terraces of the Hocking River valley, Ohio. M, 1956, Ohio State University.

Kempton, John Paul. Stratigraphy of the glacial deposits in and adjacent to the Troy bedrock valley, northern Illinois. D, 1962, University of Illinois, Urbana. 141 p.

Kempton, Pamela Dara. Alkalic basalts from the Geronimo volcanic field; petrologic and geochemical data bearing on their petrogenesis; petrography, petrology and geochemistry of xenoliths and megacrysts from the Geronimo volcanic field, southeastern Arizona; and an interpretation of contrasting nucleation and growth histories from the petrographic analysis of pillow and dike chilled margins, Hole 504B, DSDP Leg 83. D, 1984, Southern Methodist University. 275 p.

Kempton, Pamela Dara. Quaternary terrace development along the Fall River Hot Springs area, South Dakota. M, 1980, Southern Methodist University. 195 p.

Kenaga, Steven Gerald. Refined interlobate stratigraphy of west-central Indiana and other tales of the Pleistocene. M, 1987, Purdue University. 414 p.

Kenah, Christopher. Mechanism and physical conditions of emplacement of the Quottoon Pluton, British Columbia. D, 1979, Princeton University. 196 p.

Kenaley, Douglas Scott. Petrology, geochemistry and economic geology of selected gold claims in rocks of the Wasekwan Lake area, Lynn Lake District, Manitoba, Canada. M, 1982, University of North Dakota. 309 p.

Kendall, Carol. Petrology and stable isotope geochemistry of three wells in the Buttes Area of the Salton Sea geothermal field, Imperial Valley, California, U.S.A. M, 1976, University of California, Riverside. 211 p.

Kendall, D. R. The role of macrobenthic organisms in mercury, cadmium, copper and zinc transfers in Georgia salt marsh ecosystems. D, 1978, Emory University. 254 p.

Kendall, Ernest W. Trend orebodies of the Section 27 mines, Ambrosia Lake uranium district, New Mexico. D, 1971, University of California, Berkeley. 169 p.

Kendall, George W. Some aspects of Lower and Middle Devonian stratigraphy in Eureka County, Nevada. M, 1975, Oregon State University. 199 p.

Kendall, Henry Madison. The central Pyrenean piedmont of France. D, 1933, University of Michigan.

Kendall, Hugh F. The Keweenawan diabase intrusives of northeastern Minnesota. M, 1928, University of Minnesota, Minneapolis. 48 p.

Kendall, John Manford. A study of the insoluble residues of the Paleozoic limestones of Minnesota. M, 1941, University of Minnesota, Minneapolis. 32 p.

Kendall, Richard Garsed. Rocks and minerals of the Soudan Mine 15th level. M, 1938, University of Minnesota, Minneapolis. 64 p.

Kendall, Robert Lee. Dolomitization and textural analysis of the Lower-Middle Devonian carbonates of north-central Ohio at Sandusky crushed stone Parkertown quarry, Erie County, Ohio. M, 1988, University of Akron. 177 p.

Kendorski, Francis S., III. Influence of jointing on engineering properties of San Manuel Mine rock (San Manuel, Pinal County, Arizona). M, 1971, University of Arizona.

Kendra, William R. Effects of volcanic ash on the composition, abundance, and vertical distribution of benthic macroinvertebrates in Chatcolet Lake, Idaho. M, 1983, University of Idaho. 40 p.

Kendrick, George C. Magma immiscibility in the Square Butte Laccolith of central Montana. M, 1980, University of Montana. 90 p.

Kendrick, Guy. A study of structural fabric in Grenville rocks of an area in southeastern Ontario (Canada). M, 1966, University of Toronto.

Kendrick, John W. Trace element studies of metalliferous sediments in cores from the East Pacific Rise and Bauer Deep, 10°S. M, 1974, Oregon State University. 117 p.

Kendrick, Michael Brian. A vertical seismic profiling study in Bath Township, Greene County, Ohio. M, 1986, Wright State University. 95 p.

Kendy, Eloise. Hydrogeology of the Wisconsin River valley in Marathon County, Wisconsin. M, 1986, University of Wisconsin-Madison. 218 p.

Kennard, Paul M. Volumes of glaciers on Cascade volcanoes. M, 1983, University of Washington. 151 p.

Kennedy, Allen Ken. The geochemitry of undersaturated arc lavas from the Tabar-Feni island groups, Papua New Guinea. D, 1988, Massachusetts Institute of Technology. 400 p.

Kennedy, Barbara Ann. An empirical formula for ray theory amplitudes in laterally inhomogeneous media. M, 1973, University of Toronto.

Kennedy, Burton Mack. Potassium-argon and iodine-xenon gas retention ages of enstatite chondrite meteorites. D, 1981, Washington University. 298 p.

Kennedy, Charles E., Jr. Geologic field trip along U.S. Route 322 from Bald Eagle Ridge to the Allegheny Front. M, 1969, Virginia State University. 40 p.

Kennedy, D. J. Conodonts from lower Ordovician rocks at Mount Arrowsmith, Northwest New South Wales, Australia. D, 1976, University of Missouri, Columbia. 117 p.

Kennedy, David Scott. A textural and chemical analysis of the Hanson Lake ore deposit (Precambrian), Saskatchewan. M, 1971, University of Saskatchewan. 100 p.

Kennedy, Denis Patrick Stephen. Geology of the Corner Brook Lake area, western Newfoundland. M, 1982, Memorial University of Newfoundland. 370 p.

Kennedy, Donald B. Geology and economic evaluation of part of the Utopia mining district, Beaverhead County, Montana. M, 1979, Eastern Washington University. 142 p.

Kennedy, Douglas S. Modern sedimentary dynamics and Recent glacial history of Marguerite Bay, Antarctic Peninsula. M, 1988, Rice University. 203 p.

Kennedy, Edward. Geologic studies in Attica Quadrangle, New York. M, 1954, University of Rochester. 149 p.

Kennedy, Edward Reynolds, Jr. Geology of Needle Peak area, Chispa Quadrangle, Trans-Pecos, Texas. M, 1949, University of Texas, Austin.

Kennedy, George Clayton. Geology, contact metamorphism, and mineral deposits of Jumbo Basin, Prince of Wales Island, southeastern Alaska. D, 1947, Harvard University.

Kennedy, George Lindsay. West American Cenozoic Pholadidae (Mollusca; Bivalvia). M, 1972, University of California, Davis. 288 p.

Kennedy, Henry David. Petrology of the Lance sandstones of northeastern Wyoming. M, 1961, University of Missouri, Columbia.

Kennedy, James A. Clay minerals of the Pennington Shales (Mississippian) near Rockwood, Tennessee; their influence on massive landslides along Interstate 40 and possible use as environmental indicators. M, 1971, University of Tennessee, Knoxville. 53 p.

Kennedy, James Lawrence, III. Hydrogeology of reclaimed Gulf Coast lignite mines. D, 1981, Texas A&M University. 302 p.

Kennedy, James Walton. Shape response of Pleistocene and Recent sediments, Chukchi Sea; a fourier analysis. M, 1972, Michigan State University. 35 p.

Kennedy, Jerry Wilson. Petrology and geochemistry of intermediate rocks in gabbro-granite contact zones, Wichita Province, Oklahoma. M, 1981, Rice University. 169 p.

Kennedy, John David. Conodonts from Lower Ordovician rocks at Mount Arrowsmith, Northwest New South Wales, Australia. D, 1976, University of Missouri, Columbia.

Kennedy, Joseph Max. The geology of the northwest quarter of the Huntington Quadrangle, Oregon. M, 1956, University of Oregon. 91 p.

Kennedy, K. G. The hydrology and hydrochemistry of a small Precambrian Shield watershed. M, 1974, University of Waterloo.

Kennedy, Kevin. Dikewater relationships to potential geothermal resources on leeward West Maui, State of Hawaii. M, 1985, University of Hawaii. 155 p.

Kennedy, Lawrence Patrick. The geology and geochemistry of the Archean Flavrian Pluton, Noranda, Quebec. D, 1985, University of Western Ontario. 469 p.

Kennedy, Luther Eugene. Geology and economic resources of the Illinois portion of the Vincennes Quadrangle. M, 1915, University of Illinois, Urbana.

Kennedy, Luther Eugene. The Cacaquabic Granite and porphyry and their contact effects. D, 1920, University of Illinois, Urbana.

Kennedy, M. C. The Quetico Fault in the Superior Province of the southern Canadian Shield. M, 1984, Lakehead University.

Kennedy, Michael Phipps. Lithologic and paleontologic facies, Eocene continental margin, San Diego, California. D, 1973, University of California, Riverside. 148 p.

Kennedy, Noel Lynne. Depth-gradient analysis of the Colony Creek cycle (Late Pennsylvanian) of North Texas. M, 1986, Texas A&M University.

Kennedy, Patrick J. The lithostratigraphy and paleontology of the Coils Creek Member of the McColley Canyon Formation North and West of Eureka, Nevada. M, 1978, University of California, Riverside. 131 p.

Kennedy, Richard R. Geology between Pine (Bullion) creek and Ten Mile Creek, eastern Tushar Range, Piute County, Utah. M, 1960, Brigham Young University. 58 p.

Kennedy, Richard Ray. Geology of Piute County, Utah. D, 1963, University of Arizona. 371 p.

Kennedy, S. K. Sedimentation in a glacier-fed lake. M, 1975, University of Illinois, Chicago.

Kennedy, Stephen Kenneth. Provenance and dispersal of sand and silt in a high gradient stream system on the west flank of the Bighorm Mountains, Wyoming; Fourier shape analysis. D, 1982, University of South Carolina. 90 p.

Kennedy, Vance Clifford. Geochemical studies of mineral deposits in the Lisbon Valley area, San Juan County, Utah. D, 1961, University of Colorado. 245 p.

Kennedy, Vance Clifford. Mineralization surrounding ore in the southwestern Wisconsin lead-zinc district. M, 1949, Pennsylvania State University, University Park. 59 p.

Kennedy, Virgil John. Stratigraphy of the upper Horse Creek area, Teton County, Wyoming. M, 1948, University of Illinois, Urbana. 88 p.

Kennedy, William Allen. Mid-Pleistocene horses from Sheridan County, Nebraska. M, 1983, Bowling Green State University. 225 p.

Kennedy, William David. Geophysical studies of the southern Albuquerque Basin of the Rio Grande Rift, New Mexico. M, 1982, University of Texas at Dallas. 231 p.

Kennerley, John Brian. An environmental and stratigraphic study of the Silurian System, Arrow Canyon Range, Nevada. M, 1960, University of Illinois, Urbana. 46 p.

Kenney, Leland Frederick. A regional environmental study of the Lansing Group (Pennsylvanian) in the northern midcontinent region. M, 1968, University of Illinois, Urbana.

Kenney, Robert John. Petrology and stratigraphy of the Upper Cambrian in northwestern Illinois. M, 1977, Northern Illinois University. 102 p.

Kennicutt, Mahlon Charles, II. Particulate and dissolved lipids in sea water. D, 1980, Texas A&M University. 234 p.

Kennimer, Mary Ann Y. Detection of sewage pollutants in the Indian River lagoon by photographic methods. M, 1972, Florida Institute of Technology.

Kennish, Michael Joseph. Effects of thermal discharges on mortality of Mercenaria mercenaria in Barnegat Bay, New Jersey. D, 1977, Rutgers, The State University, New Brunswick. 161 p.

Kennish, Michael Joseph. The effects of thermal addition on the microstructural growth of Mercenaria mercenaria. M, 1974, Rutgers, The State University, New Brunswick. 150 p.

Kenny, Ray. Reconnaissance environmental geology of the Tonto Foothills, Scottsdale, Arizona. M, 1986, Arizona State University. 158 p.

Kenoyer, Galen. Stratigraphy, texture, and microfossils of near-surface sediments from Taylor Valley and coastal McMurdo Sound, Antarctica. M, 1979, University of Maine. 148 p.

Kenoyer, Galen John. Groundwater/lake dynamics and chemical evolution in a sandy silicate aquifer in northern Wisconsin. D, 1986, University of Wisconsin-Madison. 192 p.

Kent, Deane F. The areal and structural geology of the north end of the Vermont Marble Valley. M, 1942, Northwestern University.

Kent, Dennis V. Magnetic mineralogy and magnetic properties of deep-sea sediments. D, 1974, Columbia University. 206 p.

Kent, Donald M. The Lloydminster oil and gas field, Alberta. M, 1959, University of Saskatchewan. 56 p.

Kent, Donald Martin Joseph. The geology of the Upper Devonian Saskatchewan Group and equivalent rocks in western Saskatchewan and adjacent area. D, 1968, University of Alberta. 383 p.

Kent, Douglas Bernard. On the surface chemical properties of synthetic and biogenic amorphous silica. D, 1983, University of California, San Diego. 449 p.

Kent, Douglas Charles. A late Pliocene faunal assemblage from Cheyenne County, Nebraska. M, 1963, University of Nebraska, Lincoln.

Kent, Douglas Charles. A preliminary hydrogeologic investigation of the upper Skunk River basin, central Iowa. D, 1969, Iowa State University of Science and Technology. 414 p.

Kent, George Robert. On four pegmatites in southwestern Nova Scotia. M, 1962, Dalhousie University.

Kent, Gretchen R. Temperature and age of precious metal vein mineralization and geochemistry of host rock alteration at the Eberle Mine, Mogollon mining district, southwestern New Mexico. M, 1983, Michigan Technological University. 84 p.

Kent, Harry Christison. Biostratigraphy of the Lower Mancos Shale (Cretaceous) in northwestern Colorado. D, 1965, University of Colorado. 173 p.

Kent, Kathleen. Two-dimensional gravity model of the Southeast Georgia Embayment-Blake Plateau. M, 1979, University of Delaware.

Kent, Leon Alfred. Structure and stratigraphy of Nevill Quadrangle, Culberson County, Texas. M, 1951, University of Texas, Austin.

Kent, Mavis Hensley. Stratigraphy and petrography of the Selah Member of the Ellensburg Formation in South-central Washington and North-central Oregon. M, 1978, Portland State University. 118 p.

Kent, P. Thermal study of the Ca-Mg-Fe carbonate minerals. M, 1955, Columbia University, Teachers College.

Kent, Ray Clarke. A strain gauge for use in drill holes. M, 1961, University of Utah. 51 p.

Kent, Richard C. The geology of the southeast quarter of the Bone Mountain Quadrangle, Oregon. M, 1972, Portland State University. 132 p.

Kent, W. Norman. Facies analysis of the Mississippian Redwall Limestone in the Black Mesa region. M, 1975, Northern Arizona University. 186 p.

Kenter, Richard J. Sea-level fluctuations recorded as rhythmic deposition in Northwest Providence Channel, Bahamas. M, 1985, Miami University (Ohio). 106 p.

Kenworthy, W. Judson. The interrelationship between seagrasses, Zostera marina and Halodule wrightii, and the physical and chemical properties of sediments in a mid-Atlantic Coastal Plain estuary near Beaufort, North Carolina (U.S.A.). M, 1981, University of Virginia. 114 p.

Kenyon, John Michael. Mo and U mineralization with special reference to a Mo (U) deposit at Carmi, B.C. M, 1978, University of Alberta. 188 p.

Kenyon, Kern. Microseisms and water waves. M, 1961, Massachusetts Institute of Technology. 90 p.

Kenyon, Patricia May. Studies of heat and mass transport in the Earth; mantle convection, magma mixing, and clastic sedimentation. D, 1986, Cornell University. 197 p.

Keogh, Richard J. Petrology of the limestones in the Park Shale Member, Gros Ventre Formation (Middle Cambrian), Dubois area, Wyoming. M, 1979, Miami University (Ohio). 99 p.

Keoughan, Kathleen M. Stratigraphy of the Pliocene and Pleistocene deposits of Cherry Point Marine Corps Air Station, North Carolina. M, 1988, University of North Carolina, Chapel Hill. 71 p.

Kepes, Gerald Joseph. Precambrian geology of the Kiowa Mountain area and Cerro Azul, north-central New Mexico. M, 1985, University of New Mexico. 94 p.

Kepferle, Roy. Stratigraphy, petrology and depositional environment, Kenwood Siltstone Member of the Borden Formation (Mississippian), Kentucky and Indiana. D, 1972, University of Cincinnati. 233 p.

Kepferle, Roy Clark. Geology of a portion of the White River badlands, Pennington County, South Dakota. M, 1954, South Dakota School of Mines & Technology.

Kephart, William W. Rare earth and radioactive mineralization at the Charlotte Prospect, Sussex County, New Jersey. M, 1962, American University. 39 p.

Kepkay, Paul E. Preliminary investigation of free gas as the control of a sub-bottom acoustic reflector in the fine-grained sediments of Halifax Harbour and St. Margaret's Bay, Nova Scotia. M, 1977, Dalhousie University.

Keplinger, Henry Ferdinand. Review of the geology and petroleum possibilities of Colombia, South America. M, 1965, University of Tulsa. 99 p.

Keppel, David. A study of the (Permian) bryozoan collection from the museum of the National Park Service at Grand Canyon, Arizona. M, 1934, Columbia University, Teachers College.

Keppel, David. Concentric patterns in the granites of the Llano-Burnet region, Texas. D, 1940, Columbia University, Teachers College.

Kepper, Jack C. Stratigraphy and structure, southern half of Fish Springs Range, Juab County, Utah. M, 1960, University of Washington. 92 p.

Kepper, John Charles. Stratigraphy and petrology of a middle and upper Cambrian interval in the Great Basin. D, 1969, University of Washington. 261 p.

Keppler, Belva Hudson. Structural variations in a single basalt flow of southeastern Washington. M, 1954, Washington State University. 41 p.

Kerans, Charles. Sedimentology and stratigraphy of the Dismal Lakes Group, Proterozoic, Northwest Territories. D, 1982, Carleton University. 304 p.

Kerba, Mona. Géochimie des roches archéennes granitiques (Canada). M, 1971, Universite de Montreal.

Kerby, Ernest Gordon. Geomorphic process and stress network development in miniature badland topography (Stafford County, Virginia). M, 1972, University of Virginia. 96 p.

Kerekgyarto, William L. Time-series study of beach morphology and sediment characteristics at Crane Creek State Park, Ohio, and Sterling State Park, Michigan, Lake Erie. M, 1977, University of Toledo. 146 p.

Kerfoot, Denis Edward. The geomorphology and permafrost conditions of Garry island, N.W.T. D, 1969, University of British Columbia. 308 p.

Kerhin, Randall Thomas. Time-series study of the foreshore zone in a non-tidal environment. M, 1970, Western Michigan University.

Kerin, L. John. The reconnaissance petrology of the Mt. Fairplay igneous complex. M, 1976, University of Alaska, Fairbanks. 95 p.

Kermabon, A. J. Study of some electro-kinetic properties of rocks. M, 1956, Massachusetts Institute of Technology. 41 p.

Kerman, Charles E. The application of the radio field intensity methods to mapping Precambrian structures in Lake Superior region. M, 1962, Michigan State University. 106 p.

Kermeen, James Seton. A study of some uranium mineralization in Athabasca Sandstone, near Stony Rapids, northern Saskatchewan, Canada. M, 1955, University of Saskatchewan. 49 p.

Kern, Billy Francis. Geology of the uranium deposits near Stanley, Custer County, Idaho. M, 1959, University of Idaho. 68 p.

Kern, Christian A. Petrology of the Narragansett Pier Granite, Rhode Island. M, 1979, University of Rhode Island.

Kern, Ernest Lee. Geology of Klump's Cave, Perry County, Missouri. S, 1974, Western Michigan University.

Kern, John Philip. Early Pliocene paleoecology of the eastern Ventura Basin, southern California. D, 1968, University of California, Los Angeles. 243 p.

Kern, John William. The effects of stress on the remanent magnetism of rocks. D, 1960, University of California, Berkeley. 86 p.

Kern, R. A. A systematic field test of growth and diffusion models of chemical zoning in garnet. D, 1977, University of Illinois, Urbana. 117 p.

Kern, Richard R. The geology and economic deposits of the Slate Creek area, Custer County, Idaho. M, 1972, Idaho State University. 135 p.

Kern, Ronald Arthur. Structure and petrology of the Vinalhaven Pluton, Maine. M, 1973, University of Illinois, Chicago.

Kernaghan, James S. Development of the metamorphic rocks in the Harney Peak region, Custer County, South Dakota. M, 1969, South Dakota School of Mines & Technology.

Kerns, Earl. Clay dikes in the Pittsburgh coal (Pennsylvanian) of southwestern Greene County, Pennsylvania. M, 1971, West Virginia University.

Kerns, John R. Geology of the Agua Verde Hills, Pima County, Arizona. M, 1958, University of Arizona.

Kerns, Raymond LeRoy, Jr. Structural charge site influence on the interlayer properties of expandable three-layer clay minerals. D, 1966, University of Oklahoma. 123 p.

Kerns, Raymond LeRoy, Jr. The accuracy of indirect determinations of epsilon in uniaxial minerals. M, 1961, Southern Illinois University, Carbondale. 24 p.

Kerr, Albert Ritz. Littoral erosion and deposition of the Santa Monica Bay. M, 1938, University of California, Los Angeles.

Kerr, Andrew. Late Archean igneous, metamorphic and structural evolution of the Nain Province at Saglek Bay, Labrador. M, 1980, Memorial University of Newfoundland. 267 p.

Kerr, Bobby G. Geology of the Pagoda area, Routt and Moffat counties, northwestern Colorado. M, 1958, Colorado School of Mines. 124 p.

Kerr, Daniel Ernest. Late Quaternary stratigraphy and depositional environments in the basin of the Richardson and Rae rivers, Northwest Territories. M, 1986, University of Ottawa. 249 p.

Kerr, Dennis R. Early Neogene continental sedimentation, western Salton Trough, California. M, 1982, San Diego State University. 138 p.

Kerr, Forrest Alexander. Geology of the Memphremagog (Quebec) map area. D, 1929, University of Chicago. 190 p.

Kerr, J. W. M. Siderite deposits in the Michipicoten District, Ontario. M, 1945, University of Toronto.

Kerr, James McKinnon, Jr. The volcanic and tectonic history of La Providencia Island, Colombia. M, 1978, Rutgers, The State University, New Brunswick. 52 p.

Kerr, James William. Paleozoic sequences in thrust slices of the Seetoya Mountains, northern Independence Range, Elko County, Nevada. D, 1960, Columbia University, Teachers College. 133 p.

Kerr, Joe Harriss. Stratigraphy and faunas of a core penetrating Middle Devonian and Upper Silurian formations of the Marblehead Peninsula, Ohio. M, 1950, University of Michigan.

Kerr, Paul F. The determination of opaque ore-minerals by X-ray diffraction patterns. D, 1923, [Stanford University].

Kerr, R. A. The isolation and partial characterization of the dissolved organic matter in seawater. D, 1977, University of Rhode Island. 195 p.

Kerr, Ralph S. Development and diagenesis of a Lower Cretaceous bank complex, Edwards Limestone, north-central Texas. M, 1976, University of Texas, Austin.

Kerr, Ronald. Physical and numerical modeling of an overthrust zone including radial trace profiles and dip moveout processing. M, 1986, University of Houston.

Kerr, Samuel Aubrey. The Tertiary sediments of Sumas Mountain. M, 1942, University of British Columbia.

Kerr, Steven Brent. Petrology of Pliocene(?) basalts of Curlew Valley (Box Elder Co.), Utah. M, 1987, Utah State University. 84 p.

Kerr, Stuart Duff. Early stream channels in Boone County, Kentucky. M, 1951, University of Cincinnati. 22 p.

Kerr-Lawson, D. E. Pleochroic haloes in biotite. D, 1928, University of Toronto.

Kerr-Lawson, L. Gastropods and plant microfossils from the Quaternary, Don Formation (Sangamonian interglacial, Toronto), Ontario. M, 1985, University of Waterloo. 202 p.

Kerrick, Derrill Maylon. Part I, Experiments on the stability of andalusite; Part II, Studies of contact metamorphism in the Sierra Nevada, California. D, 1968, University of California, Berkeley. 130 p.

Kerschner, David R. The groundwater resources of Green and Springfield townships, Summit County, Ohio. M, 1981, Kent State University, Kent. 148 p.

Kersey, James Doyle. An acid soluble percentage and textural study of subsurface Tertiary and Pleistocene formations in south-central Nebraska. M, 1949, University of Nebraska, Lincoln.

Kershaw, David M. Benthonic foraminifers as environmental indicators in the Shubuta Clay, Clark County, Mississippi. M, 1982, Memphis State University.

Kershner, James D. Unsteady flow in contiguous aquifers of different hydraulic properties. M, 1962, New Mexico Institute of Mining and Technology. 45 p.

Kerstetter, Frank Linwood, Jr. Subsurface geology of the eastern half of Texas County, Oklahoma. M, 1957, University of Oklahoma. 82 p.

Kersting, Cecil Carl. Stratigraphy of the Delray Core, Wayne County, Michigan. M, 1953, University of Michigan.

Kersting, Joseph Jeffrey. The petrology and petrography of the Mt. Simon and pre-Mt. Simon sandstones; evidence for possible Precambrian rifting in the central Midcontinent. M, 1982, University of Pittsburgh.

Kerswill, J. A. Geology and geochemistry of the Hotailuh Batholith and spatially associated volcanic rocks. M, 1975, University of Western Ontario. 191 p.

Kertis, Carla A. Recognition and prediction of coalbed discontinuities, Indiana and Armstrong counties, Pennsylvania. M, 1982, University of Missouri, Columbia.

Kerwin, James A. Classification and structure of the tidal marshes of the Poropotank River, Virginia, Fall 1964. M, 1966, College of William and Mary.

Kesebir, Mehmet. Relation of natural gas analyses to the geology and reservoir parameters in Chesterian Series (Mississippian) in Beaver County, Oklahoma. M, 1968, University of Oklahoma. 82 p.

Kesebir, Musa Mustafa. Hydrogeology of Hinckley Township, Medina County, Ohio. M, 1986, University of Akron. 104 p.

Keser, Judith. Wide-angle seismic refraction and reflection studies of the northern California and southern Oregon continental margins. M, 1979, Oregon State University. 103 p.

Keskinen, M. J. Experimental and field investigation of the stability relations of the manganese epidote, piemontite. D, 1979, Stanford University. 170 p.

Kesler, Stephen E.; Essere, Eric J. and Richardson, Stephen Vance. Origin and geochemistry of the Chapada Cu-Au deposit, Goias, Brazil; a metamorphosed wall rock porphyry copper deposit. M, 1984, University of Michigan.

Kesler, Stephen Edward. The geology and ore deposits of the Meme-Casseus District, Haiti. D, 1966, Stanford University. 192 p.

Kesler, Thomas Lingle. Geology of the Catawba Mountain (Virginia) area. M, 1930, University of North Carolina, Chapel Hill. 30 p.

Kesling, Robert Vernon. The morphology and ostracod molt stages. D, 1949, University of Illinois, Urbana. 40 p.

Kesling, Robert Vernon. Zonation of the larger foraminifera of the Upper Cretaceous Selma, Ripley, and Prairie Bluff formations in western Alabama (K481). M, 1941, University of Illinois, Urbana. 380 p.

Kesmarky, Susanna. Rates of migration of alkali and strontium ions in a heated quartz monzonite. M, 1977, University of Alberta. 107 p.

Kesse, Godfried Opang. Fauna of the Hidden Valley Dolomite (Silurian), Death Valley, California. M, 1963, University of Southern California.

Kessel, Richard H. The comparative morphology of inselbergs in different environments (Virginia, North and South Carolina, and Arizona) with emphasis on a humid temperature and an arid area. D, 1972, University of Maryland.

Kesselli, John E. Pleistocene glaciation in the valleys between Lundy Canyon and Rock Creek, eastern slope of the Sierra Nevada, California; 2 volumes. D, 1938, University of California, Berkeley.

Kessinger, Walter Paul, Jr. Cretaceous foraminifera of Lynn, Terry, Hockley, and Lamb counties, Texas. M, 1953, Texas Tech University. 83 p.

Kessinger, Walter Paul, Jr. Stratigraphic distribution of the Ostracoda of the Comanche (Cretaceous) Series of North Texas. D, 1974, Louisiana State University.

Kessler, Edgar M. Paleocurrent analysis and quantitative distinction of Magothy and Raritan formations, Cretaceous, in New York and New Jersey. M, 1974, Brooklyn College (CUNY).

Kessler, Edward Joseph. Rubidium-strontium geochronology and trace element geochemistry of Precambrian rocks in the northern Hualapai Mountains, Mohave County, Arizona. M, 1976, University of Arizona.

Kessler, Jane. Micro-structure of fire-clay brick after use. M, 1937, Virginia Polytechnic Institute and State University.

Kessler, Kirk J. Ground-water quality evaluation of Ottawa County, Ohio. M, 1986, University of Toledo. 130 p.

Kessler, L. Gifford, 2nd. Palynology and paleobotany of the Glen Rose Formation, north and central Texas. M, 1968, University of Texas, Austin.

Kessler, L. Gifford, II. Channel sequences and braided stream development in South Canadian River, Hutchinson, Roberts, and Hemphill counties, Texas. D, 1972, University of New Mexico. 143 p.

Kester, Dana Ray. Determination of the apparent dissociation constants of phosphoric acid in seawater. M, 1966, Oregon State University. 55 p.

Kester, Stephen Joseph. Abrasion, transport and distribution of sediment in selected streams of southern Ontario and western New York. M, 1986, Brock University. 112 p.

Kesterke, Donald G. Contact angle studies comparing xanthates and dithiocarbamates as collectors for sulphide minerals. M, 1959, University of Nevada - Mackay School of Mines. 60 p.

Ketani, Rapheal V. Petrology of the Magoffin beds of Morse (1931) in Morgan and Magoffin counties, Kentucky. M, 1980, Eastern Kentucky University. 130 p.

Ketchen, Harold G. A hydrographic survey in Pensacola Bay. M, 1979, Florida State University.

Ketcher, Austin. Concentration of selected metallic elements in the upper Illinois River. D, 1982, Oklahoma State University. 62 p.

Ketchum, Robert L. Reservoir capacity and flow data of the lower Sparta (Eocene) Sand underlying Lincoln Parish, Louisiana. M, 1971, Louisiana Tech University.

Ketelle, Martha J. Hydrogeologic considerations in liquid-waste disposal, with a case study in southeastern Wisconsin. M, 1970, University of Wisconsin-Madison.

Ketelle, Richard H. Characteristics of the mineral and metal content of suspended sediment, New River basin, Tennessee. M, 1977, University of Tennessee, Knoxville. 71 p.

Ketner, Keith Brindley. Mexico's mineral policy. M, 1952, University of Wisconsin-Madison.

Ketner, Keith Brindley. Ordovician siliceous sediments of the Cordilleran Geosyncline. D, 1968, University of Wisconsin-Madison. 140 p.

Keto, Lisette Scott. Nd and Sr isotopic evolution of the oceans of the past 800 million years. D, 1987, Harvard University. 289 p.

Ketrenos, Nancy Tompkins. The stratigraphy of the Scappoose Formation, the Astoria Formation, and the Columbia River Basalt Group in northwestern Columbia County, Oregon. M, 1986, Portland State University. 78 p.

Kettenacker, William Charles. Two-dimensional simulation of the Raft River geothermal reservoir and wells. M, 1977, University of Idaho. 96 p.

Kettenbrink, Edwin Carl, Jr. Depositional and post-depositional history of the Devonian Cedar Valley Formation, east-central Iowa. D, 1973, University of Iowa. 191 p.

Kettenbrink, Edwin Carl, Jr. Petrology and diagenesis of the cyclic Maquoketa Formation (upper Ordovician), Pike County, Missouri. M, 1970, University of Missouri, Rolla.

Kettenring, Kenneth Norman. The paleoenvironments and paleoecology of an Ordovician brachiopod community in southern Nevada and eastern California. M, 1976, University of California, Los Angeles.

Kettenring, Kenneth Norman, Jr. The trace metal stratigraphy and Recent sedimentary history of anthrogenous particulates on the San Pedro Shelf, California. D, 1981, University of California, Los Angeles. 173 p.

Ketterer, Walter P. A study of the strata immediately underlying the (Silurian) Olney Limestone in the vicinity of Syracuse, New York. M, 1940, Syracuse University.

Kettler, Richard M. Radioactive mineralization in the conglomerates and pyritic schists of the Kingston Peak Formation, Panamint Mountains, California. M, 1982, University of California, Los Angeles. 166 p.

Kettles, Inez M. Till stratigraphy of the Vandalia-Effingham-Marshall region, East-central Illinois. M, 1980, University of Illinois, Urbana. 124 p.

Kettles, Karen R. The Turgeon mafic volcanic associated Fe-Cu-Zn sulphide deposit in the ophiolitic Fournier Group, northern New Brunswick. M, 1987, University of New Brunswick. 213 p.

Kettren, Lee P. Relationship of igneous intrusions to geologic structures in Highland County, Virginia. M, 1970, Virginia Polytechnic Institute and State University.

Kety, Irvin. The impact of urban eaton on sediment character in Lake Maggiore, St. Petersburg, Florida. M, 1980, Florida State University.

Ketzlach, Norman. A study of olivine as a source of magnesium. M, 1944, University of Washington. 66 p.

Keuler, Ralph F. Coastal zone processes and geomorphology of Skagit County, Washington. M, 1979, Western Washington University. 127 p.

Keuren, Lewis Karl Van *see* Van Keuren, Lewis Karl, III

Kew, William Stephen Webster. Cretaceous and Cenozoic Echinoidea of the Pacific Coast of North America. D, 1917, University of California, Berkeley. 213 p.

Kew, William Stephen Webster. Tertiary echinoids of the Carrizo Creek region in the Colorado Desert. M, 1914, University of California, Berkeley. 21 p.

Kewen, T. J. Observations on the use of batch tests to determine cesium distribution coefficients in natural geologic materials. M, 1978, University of Waterloo.

Kewer, Robert Parker. The core length index; a method for the numerical classification of core for engineering purposes. M, 1974, Rutgers, The State University, New Brunswick. 94 p.

Key, Carlos Eduardo. Biostratigraphy of the Bitterwater-Packwood Creek area, Kern County, California. M, 1956, Stanford University. 139 p.

Key, Colin F. Stratigraphy and depositional history of the Amsden and lower Quadrant formations, Snowcrest Range, Beaverhead and Madison counties, Montana. M, 1987, Oregon State University. 187 p.

Key, John Ambrose. Physical and chemical properties of coke made from Washington and other coals. M, 1936, University of Washington. 58 p.

Key, M. D. Heavy metal concentrations and associated mobility parameters in the floodplain of the Dallas-Fort Worth region. D, 1976, University of Texas at Dallas. 217 p.

Key, Marcus M., Jr. Evolution of the halloporid clade (Bryozoa; Trepostomata) in the Ordovician Simpson Group of Oklahoma. D, 1988, Yale University. 312 p.

Key, Robert Marion. Examination of abyssal sea floor and near-bottom water mixing processes using Ra-226 and Rn-222. D, 1981, Texas A&M University. 239 p.

Key, Scott C. A high resolution coherency measure for event detection and velocity estimation in seismic reflection data. M, 1988, University of Wyoming. 54 p.

Keyes, C. R. The principal Mississippian section; a classification of the Lower Carboniferous rocks of the Mississippi Basin. D, 1892, The Johns Hopkins University.

Keyes, Scott Wellington. New Cyrtina (Brachiopoda) from the Traverse Group (Devonian) of Michigan. M, 1974, University of Massachusetts. 88 p.

Keyes, Steven Lynn. Sedimentology of the lower Strawn Formation (Atokan (?) - Desmoinesian), central Texas. M, 1982, Baylor University. 92 p.

Keys, David Gerald. Variation in the energy content of peat. M, 1984, University of New Brunswick. 126 p.

Keys, John N. An analysis of the Rend Lake fault system in southern Illinois. M, 1978, University of Illinois, Urbana. 59 p.

Keys, M. R. Comparative study of the mineralogy of the auriferous quartz veins encountered in the Holliger Mine, Porcupine District. M, 1938, Queen's University. 47 p.

Keys, Robert Gene. An application of the method of generalized linear inversion to the seismic reflection problem. D, 1983, University of Tulsa. 64 p.

Keys, Scott Walter. The geology of the May Ellen Mine, Hamilton District, White Pine County, Nevada. M, 1955, University of California, Los Angeles.

Keyser, Joseph Edward. Some subsurface limestone reefs in north-central Texas. M, 1948, University of Texas, Austin.

Keyser, Thomas Lee De *see* De Keyser, Thomas Lee

Keyte, I. Allen. The crinoid fauna of the Chouteau Formation. M, 1925, University of Missouri, Columbia.

Keyte, Wilbur Ross. The (Cretaceous) Frontier Formation of Wyoming. M, 1927, Colorado College.

Khadr, Hassan Ali Abdel-Aziz. A three-dimensional ground water flow model in a curvi-linear coordinate system. D, 1988, Colorado State University. 280 p.

Khafagi, Om Mohamed Ahmed. Studies on the belowground ecosystem in Missouri stripmines. D, 1980, University of Missouri, Columbia. 184 p.

Khaiwka, Moayrad Hamid. Geometry and depositional environments of Pennsylvanian reservoir sandstones, northwestern Oklahoma. D, 1968, University of Oklahoma. 279 p.

Khalaf, Mukhtar Hammali. Silica in three limestones of southeastern Ohio. M, 1973, Ohio University, Athens. 82 p.

Khalid, R. Dynamic properties of deep-sea sediments. D, 1976, University of Washington. 180 p.

Khalil, Kabiru. Groundwater model of the Chad Basin in Kano State (Nigeria). M, 1982, Ohio University, Athens. 186 p.

Khalvati, Mehdi. Finite element analysis of interacting soil-structure-fluid systems with local nonlinearities. D, 1981, University of California, Berkeley. 215 p.

Khamenehpour, Bahram. Seismic stability problems in earth dam design. D, 1983, University of California, Berkeley. 137 p.

Khamesra, Daulat Singh. Basal Belly River Sandstone (Upper Cretaceous) Pembina Field, Alberta, Canada. M, 1964, University of Alberta. 210 p.

Khan, A. K. M. Hamidur Rahman. Laboratory study of alluvial river morphology. D, 1971, Colorado State University. 208 p.

Khan, Abdul Qadir. Seismicity and structure in the vicinity of the Proposed Meramec Park reservoir. M, 1974, St. Louis University.

Khan, Abdur K. Watershed conditions, problems and research needs in West Pakistan. M, 1968, [Colorado State University].

Khan, M. H. Three-dimensional stress-strain relationships of unsaturated soil under different stress paths. D, 1979, University of Illinois, Urbana. 250 p.

Khan, Mohammad Javed. Magnetostratigraphy of Neogene and Quaternary Siwalik Group sediments of the Trans-Indus Salt Range, northwestern Pakistan. D, 1983, Columbia University, Teachers College. 231 p.

Khan, Mohammed Gulnawaz. Review of well log interpretation principles of shaly sands. M, 1980, University of Oklahoma. 159 p.

Khan, Mohammud Attaullah. Development of a new theory for determination of geopotential from the orbital motion of artificial satellites. D, 1967, University of Hawaii. 112 p.

Khan, Mohammud Attaullah. General geology and sulfide mineralization of Dry Canyon and vicinity, Gunnison Plateau, Sanpete County, Utah. M, 1967, University of Utah. 145 p.

Khan, Muhammad Aslam. Petrography of some Tertiary volcanic ash beds in western Nebraska. M, 1961, University of Nebraska, Lincoln.

Khan, Muhammad Yunus. Theory of some free surface groundwater seepage problems. D, 1973, Iowa State University of Science and Technology. 169 p.

Khan, Mumtaz Ahmed. The occurrence of shallow gas deposits in Harding County, South Dakota. M, 1984, South Dakota School of Mines & Technology.

Khan, Rashid Ali. Geochemical hydrology of the ground water in Baton Rouge, Louisiana. D, 1971, Louisiana State University. 122 p.

Khan, Shahid. Grain size distribution of some clayey sandstone; a comparison of grain mount, thin section and sieving techniques. M, 1969, University of Alberta. 122 p.

Khan, Shakeel Ahmed. Measuring the advantages of cooperation and integration in regional water supply management. D, 1988, Colorado State University. 183 p.

Khan, Sheraz M. Paleomagnetism of Archean rocks from the English River and Uchi subprovinces, northwestern Ontario, and polar wandering path for North America. M, 1982, University of Manitoba.

Khan, Subhotosh. Simulation of hydraulic fracturing of coal. D, 1981, West Virginia University. 344 p.

Khan, Suhail. Laboratory study of wettability characteristics of Bradford Sand. M, 1962, Stanford University.

Khan, T. R. Correlation of geochemical data on organic matter and metal ions in the Indian River. D, 1979, Carleton University.

Khandaker, Nazrul Islam. Turbidite and contourite deposits of the Nova Scotian continental margin, Canada. M, 1984, University of Rochester. 103 p.

Khandoker, Jalal Uddin. Three-dimensional response of a pile-supported multistory building to seismic disturbances. D, 1984, University of Missouri, Rolla. 179 p.

Khanna, Sat Dev. An experimental investigation of turbulence due to forms of bed-roughness. D, 1968, University of Connecticut. 115 p.

Khanna, Satish Kumar. Adsorption study of amphoteric surfactants on mineral surfaces using infrared spectroscopy. M, 1972, University of Nevada. 39 p.

Kharaka, Yousif Khoshu. Simultaneous flow of water and solutes through geological membranes; experimental and field investigations. D, 1971, University of California, Berkeley. 274 p.

Kharas-Khumbata, Nazneen. Geochemistry and petrology of the gabbroic rocks from Blow Me Down Mountain, Bay of Islands Complex; composition of the parent magma in liquids in equilibrium with gabbroic minerals. M, 1988, University of Houston.

Khatib, Abdulhamid Ahmad. Chemical interactions of wastewater in a soil environment. M, 1972, University of Idaho. 103 p.

Khatri-Chhetri, Tej Bahadur. Assessment of soil test procedures for available boron and zinc in soils of the Chitwan Valley, Nepal. D, 1982, University of Wisconsin-Madison. 318 p.

Khattab, Khattab Mansour M. Analysis of uranium and thorium and their daughter nuclides in uranium-thorium ores by high-resolution gamma-ray spectrometry. D, 1970, Pennsylvania State University, University Park. 120 p.

Khattab, Mohamed Mamdouh. Gravity and magnetic surveys of the Grouse Creek Mountains and the Raft River Mountains area and vicinity, Utah and Idaho. D, 1969, University of Utah. 368 p.

Khattak, Anwar S. Mechanical behavior of fibrous organic soils. D, 1978, Michigan State University. 279 p.

Khattri, Kailash Nath. Focal mechanism of six earthquakes in the Pacific coastal margin of South America. M, 1968, St. Louis University.

Khattri, Kailash Nath. Kinematic parameters of earthquakes. D, 1969, St. Louis University. 175 p.

Khawaja, Ikram Ullah. A mineralogic study of sphalerite from Cave-In-Rock fluorspar district, Illinois. M, 1967, Southern Illinois University, Carbondale. 36 p.

Khawaja, Ikram Ullah. Distribution, association and rate of oxidation of iron sulfide in Springfield Coal member (V) of Petersburg Formation (Pennsylvanian) in Sullivan County, Indiana. D, 1969, Indiana University, Bloomington. 63 p.

Khawlie, Mohamad R. Microfacies and geochemistry of the Brereton Limestone (middle Pennsylvanian) of southwestern Illinois. D, 1975, University of Illinois, Urbana. 110 p.

Khawlie, Mohamad R. Mineralogical studies of clays occurring in several Ordovician bentonites. M, 1972, University of Illinois, Urbana. 32 p.

Kheang, Lao. Evolution chimique des clinopyroxènes des laves mafiques de la ceinture métavolcanique de Rouyn-Noranda. M, 1978, Ecole Polytechnique.

Kheang, Lao. Thermo-géochimie appliquée aux inclusions fluides reliées au gisement volcanogène archéen de Millenbauch, Rouyn-Noranda, Québec. D, 1982, Ecole Polytechnique. 322 p.

Khedr, S. A. Residual characteristics of untreated granular base course and subgrade soils. D, 1979, Ohio State University. 298 p.

Khemici, Omar. Frequency domain corrections of earthquake accelerograms with experimental verifications. D, 1982, Stanford University. 118 p.

Kheoruenromne, Irb. A study of the flux of phosphorus, silica, iron and cations through two small streams on granitic terrain, Kershaw County, South Carolina. D, 1976, University of South Carolina. 116 p.

Kheradpir, Ahmad. Foraminiferal trends and paleo-oceanography in late Pleistocene-Recent cores, Tanner Basin, California. M, 1968, University of Southern California.

Kheradyar, Tara. Coccolith biostratigraphy of the Eocene Butano Sandstone, San Mateo County, California. M, 1987, University of California, Berkeley. 201 p.

Kherl, Dennis Donald. Hydrothermal investigations of some mixed-layer clays and kaolin. D, 1971, Case Western Reserve University.

Khilar, Kartic Chandra. The water sensitivity of Berea Sandstone. D, 1981, University of Michigan. 198 p.

Khin, Maung Aung. The geology of the district north of Indianola, Utah County, Utah. M, 1956, Ohio State University.

Khodair, Abdul-Wahab Abdul-Aziz. The accuracy of the Thelliers' technique for the determination of paleointensities of the Earth's magnetic field. D, 1978, University of California, Santa Cruz.

Khogia, Abdelhadi. The Villa Grove turquoise deposit, Saguache County, Colorado. M, 1967, Columbia University. 50 p.

Khoja, Elhadi Razzagh. Petrography and diagenesis of lower Paleocene carbonate reservoir rock, Dahra field, Libya. D, 1971, Rice University. 225 p.

Khondker, Sufian A. Bearing capacity and settlement characteristics of reinforced earth. D, 1982, Polytechnic University. 210 p.

Khourey, Christopher J. The source and transport of arsenic in northeastern Ohio ground water. M, 1981, Case Western Reserve University.

Khoury, H. N. Mineralogy and chemistry of some unusual clay deposits in the Amargosa Desert, southern Nevada. D, 1979, University of Illinois, Urbana. 185 p.

Khoury, M. A. Automatic identification of soil parent materials using quantitative terrain factors. D, 1977, University of Illinois, Urbana. 330 p.

Kiang, Wen Chao. Stabilities of plagioclase feldspars in dilute organic acids at room temperature and one atmosphere. M, 1971, University of South Florida, Tampa. 79 p.

Kiatta, Howard William. A provenance study of the Triassic deposits of northwestern Texas. M, 1960, Texas Tech University. 63 p.

Kick, J. F. An analysis of the bottom sediments of Lake Erie. M, 1962, University of Toronto.

Kick, John Frederick. A gravity study of the gneiss dome terrain of North-central Massachusetts. D, 1975, University of Massachusetts. 260 p.

Kick, Robert M. Carbonate sediments from Peterson Key Bank, Florida Bay. M, 1981, University of South Florida, Tampa. 101 p.

Kidd, Desmond F. Great Bear Lake-Coppermine River area, Mackenzie District, Northwest Territories. D, 1933, Princeton University.

Kidd, Donald J. The geochemistry of beryllium. D, 1951, University of Toronto.

Kidd, Gerald Daniel. Delineation of the bedrock surface underlying the Wright State University campus, Dayton, Ohio; using gravity, seismic refraction, and drilling investigations. M, 1980, Wright State University. 148 p.

Kidd, Jack James. Foraminifera of the Porters Creek Formation in Butler County, Alabama. M, 1971, University of Alabama.

Kidd, Stuart James. Lithology at Beaverhill Lake No. 2 Well, with special emphasis on the insoluble residues in the Paleozoic strata. M, 1948, University of Alberta. 200 p.

Kidda, Michael Lamond. Geology of part of the southern anthracite coal field in the vicinity of Tamaqua, Pennsylvania. M, 1953, University of Illinois, Urbana. 76 p.

Kidder, David Lee. Distribution and origin of mid-continent Pennsylvanian phosphorites. M, 1982, University of Iowa. 82 p.

Kidder, David Lee. Stratigraphy, micropaleontology, petrography, carbonate geochemistry, and depositional history of the Proterozoic Libby Formation, Belt Supergroup, NW Montana and NE Idaho. D, 1987, University of California, Santa Barbara. 202 p.

Kidder, Gerald. Swelling characteristics of hydroxy-aluminum interlayered clays. D, 1969, [University of Oklahoma]. 57 p.

Kidman, Mark. An integrated geophysical study of the southeast flank of the Espanola Basin, central Rio Grande Rift, northern New Mexico. M, 1985, Baylor University. 100 p.

Kidson, Evan Joseph. A biostratigraphic study of fossil silicoflagellates (Cretaceous-Recent) from California with observations on their evolution. M, 1965, Wichita State University. 178 p.

Kidson, Evan Joseph. Palynology and paleoecology of the Buck Tongue of the Mancos Shale (upper Cretaceous) from east central Utah and western Colorado. D, 1971, Michigan State University. 244 p.

Kidwai, Mohammed Ali. Development of graptolite reflectance as an indicator of paleotemperature and thermal maturation. M, 1986, Oklahoma State University. 72 p.

Kidwai, Zamir U. The relationship of groundwater to alluvium in the Tucson area. M, 1957, University of Arizona.

Kidwell, Albert Laws. Mesozoic igneous activity in the northern Gulf Coastal Plain. D, 1949, University of Chicago. 317 p.

Kidwell, Albert Laws. The igneous geology of Ste. Genevieve County, Missouri. M, 1942, Washington University. 83 p.

Kidwell, Susan Marie. Stratigraphy, invertebrate taphonomy and depositional history of the Miocene Calvert and Choptank formations, Atlantic Coastal Plain. D, 1982, Yale University. 531 p.

Kieckhefer, Robert Mariner. Geophysical studies of the oblique subduction zone in Sumatra. D, 1980, University of California, San Diego. 134 p.

Kiefer, Fred William, Jr. Influence of soil conditions on ground response during earthquakes. D, 1968, University of California, Berkeley. 171 p.

Kiefer, John David. Geology of a Devonian outcrop at Kentucky Lake, Marshall County, Kentucky. M, 1965, University of Illinois, Urbana.

Kiefer, John David. Pre-Chattanooga Devonian stratigraphy of Alabama and Northwest Georgia. D, 1970, University of Illinois, Urbana. 181 p.

Kiefer, Karen Bernice. Quaternary climatic cycles recorded in the isotopic record of peri-platform pelagic deposition; Northwest Providence Channel, Bahamas. M, 1983, Duke University. 107 p.

Kieffer, Hugh Hartman. Near infrared spectral reflectance of simulated Martian frosts. D, 1968, California Institute of Technology. 101 p.

Kieffer, Susan Werner. I, Shock metamorphism of the Coconino sandstone at Meteor Crater, Arizona; II, The specific heat of solids of geophysical interest. D, 1971, California Institute of Technology. 262 p.

Kiehl, Edwin L. A model field study in geology to augment the teaching of Earth science in the secondary schools of Lancaster County, Pennsylvania. M, 1972, Millersville University.

Kieller, Bernard John. Mineralogy of the No. 2 zone, Eldorado Mine, Port Radium, Northwest Territories. M, 1962, University of Alberta. 106 p.

Kiely, James M. Diagenesis and depositional setting of the fanglomerate of Little Florida Mountains (Miocene), southwestern New Mexico. M, 1987, University of Texas at El Paso.

Kienast, Val A. Structural geology of eastern part of Dairy Ridge Quadrangle and western part of Meachum Ridge Quadrangle, Utah. M, 1985, Utah State University. 41 p.

Kieniewicz, Paul Mary Michael. A gravity model of the Santa Maria Basin, California. M, 1985, University of California, Santa Barbara. 86 p.

Kienle, Clive Frederick, Jr. The Yakima Basalt in western Oregon and Washington. D, 1971, University of California, Santa Barbara. 172 p.

Kienle, Juergen. Gravity survey of Katmai National Monument. D, 1969, University of Alaska, Fairbanks. 163 p.

Kientop, Gregory Allen. Cenozoic evidence of displacements along the Meers Fault, southwestern Oklahoma. M, 1988, Texas A&M University. 112 p.

Kienzle, John Kenneth. Paleomagnetism of Plio-Pleistocene volcanic rocks from South Korea and studies of the Earth's ancient magnetic field. D, 1968, Washington University. 122 p.

Kier, Jerry Stephen. Silt and clay size carbonate mineralogy of piston cores from Tongue of the Ocean, Bahamas. M, 1968, Duke University. 103 p.

Kier, Porter Martin. Echinoderms of the Middle Devonian Silica Shale (Ohio). M, 1951, University of Michigan.

Kier, Robert Spencer. Carboniferous stratigraphy of eastern San Saba County and western Lampasas County, Texas. D, 1972, University of Texas, Austin. 509 p.

Kier, Robert Spencer. Stratigraphy of the Conococheague Group and Buffalo Springs Formation (Upper Cambrian) in the Bainbridge area, Pennsylvania. M, 1967, Franklin and Marshall College. 34 p.

Kieran, Mary. The history and development of some major problems of geology in Ohio, emphasizing economic aspects. M, 1934, Catholic University of America. 63 p.

Kierans, Martin Devalera. A sedimentation study of the Slocan Series, Sandon area, British Columbia. M, 1951, University of British Columbia.

Kiersch, George A. The geology and ore deposits of the Seventy Nine Mine area, Gila County, Arizona. D, 1947, University of Arizona.

Kies, Bouziane. Solute transport in unsaturated field soil and in groundwater. D, 1982, New Mexico State University, Las Cruces. 369 p.

Kies, Ronald Paul. Paleogene sedimentology, lithostratigraphic correlations and paleogeography, San Miguel Island, Santa Cruz Island, and San Diego, California. M, 1982, San Diego State University. 577 p.

Kiesewetter, Carl Herman. Significance of texture, constituent composition and insoluble residues in environmental relationships within the Pleistocene reef tracts on Barbados, West Indies. M, 1968, Brown University.

Kiesler, James Peter. Earth resistivity as a technique for monitoring drainfield pollution in Spokane outwash. M, 1973, Washington State University. 36 p.

Kiessling, Edmund. Geology of the southwest portion of the Lockwood Valley Quadrangle, Ventura County, California. M, 1958, University of California, Los Angeles.

Kiester, Jeffrey A. The Precambrian geology of the Goose Lake area of the Beartooth Mountains, Montana. M, 1984, Northern Illinois University. 89 p.

Kiester, Scott A. The mineralogy and sedimentology of the Cambrian strata of southeastern Minnesota. M, 1976, Northern Illinois University. 78 p.

Kietzman, Donald R. Paleomagnetic survey of the Touchet Beds in Burlingame Canyon of Southeast Washington. M, 1985, Eastern Washington University. 149 p.

Kiewiet de Jonge, E. J. Coen. Glacial water levels in the Saint John River valley. M, 1951, Clark University.

Kiff, I. T. Geology of the Lake Murray spillway. M, 1963, University of South Carolina. 35 p.

Kihm, Allen James. Early Eocene mammalian faunas of the Piceance Creek basin, northwestern Colorado. D, 1984, University of Colorado. 407 p.

Kihm, Allen James. Mammalian paleontology of the Yoder local fauna. M, 1975, South Dakota School of Mines & Technology.

Kihn, Gary Edward. Hydrogeology of the Bellevue-Castalia area, north-central Ohio, with an emphasis on Seneca Caverns. M, 1988, University of Toledo. 163 p.

Kiilsgaard, Thor H. The geology and ore deposits of Custer Mountain, Custer County, Idaho. M, 1949, University of California, Berkeley. 64 p.

Kilbane, N. A. Petrogenesis of the McClure Mountain mafic-ultramafic and alkalic complex, Fremont County, Colorado. M, 1978, Kansas State University.

Kilberg, James A. Petrology, structure, and correlation of the Upper Precambrian Ely's Peak basalts (Northeast Minnesota and Ontario, Canada). M, 1972, University of Minnesota, Duluth.

Kilbey, Thomas Ryan. Geology and structure of the Goldfield mining district, central Arizona. M, 1986, Arizona State University. 255 p.

Kilbourne, Deane Earl. Remanent magnetic properties of the Mesaverde Group (Upper Cretaceous), southwestern Wyoming and northeastern Utah. D, 1967, University of Arizona. 148 p.

Kilbourne, Deane Earle. The origin and development of the Howell Anticline in Michigan. M, 1947, Michigan State University. 120 p.

Kilbourne, John Lyle. Bauxite deposits of Babelthuap, Palau Islands. M, 1977, Northern Illinois University. 69 p.

Kilbourne, Richard T. Biostratigraphy and shell geochemistry of Quaternary planktonic foraminifera from the Cayman Trough. D, 1974, University of Georgia. 241 p.

Kilbourne, Richard T. Holocene foraminiferal zonation on the continental shelf off Georgia. M, 1970, University of Georgia. 141 p.

Kilbreath, Steven Perry. Geology and mineralization of the Silver Dyke Mine, Mineral County, Nevada. M, 1979, University of Nevada. 91 p.

Kilburg, James A. Geology of the Concepcion Tutapa Quadrangle, Guatemala, Central America. D, 1979, University of Pittsburgh. 284 p.

Kilburn, Chabot. Outcrop stratigraphy of the Lee Formation, Southeast Kentucky. M, 1956, Northwestern University.

Kilburn, Lionel Clarence. An investigation of the origin of pyrrhotite by synthesis with hydrogen sulfide and a microscopic study of some natural pyrrhotites from the N.W.T. M, 1954, University of Manitoba.

Kilburn, Lionel Clarence. The Ni, Co, Cu, Zn, Pb, and S content of some North American base metal sulphide ores. D, 1960, University of Manitoba.

Kilby, Ward Eldon. Structure and stratigraphy of the coal-bearing and adjacent strata near Mountain Park, Alberta. M, 1978, University of Alberta. 154 p.

Kilcommins, John Peter. Geology of the Grafton, West Virginia, Quadrangle. M, 1965, West Virginia University.

Kildal, Edwin and Drexler, James Michael. Geology of the Red Peak's area, Beaverhead County, Montana, and Clark County, Idaho. M, 1949, University of Michigan. 54 p.

Kildale, Malcolm B. Structure and ore deposits in the Tintic District, Utah. D, 1938, Stanford University. 200 p.

Kildale, Malcolm B. The arsenical type of cobalt-nickel ores. M, 1923, Stanford University. 91 p.

Kiley, Edward H. Geologic teaching resources of the Lebanon and Fredricksburg area, Lebanon County, Pennsylvania. M, 1971, Millersville University.

Kilfoil, Gerald Joseph. An integrated gravity, magnetic and seismic interpretation of the Carboniferous Bay St. George subbasin, western Newfoundland. M, 1988, Memorial University of Newfoundland. 172 p.

Kilgore, Brian Douglas. A study of the relationship between hydrocarbon migration and the formation of authigenic magnetite in the Triassic Chugwater Formation of southern Montana. M, 1987, University of Oklahoma. 59 p.

Kilgore, David L. An historical geomorphic study of the anomalous water gaps cut in Pine Ridge-Little Mountain by the north fork of the Holston River, southwestern Virginia. M, 1970, Indiana State University. 55 p.

Kilgore, John Elija. Ostracods of the families Leperditellidae, Drepanellidae, Glyptopleuridae, Kloedenellidae, Bairdiidae, Barychilinidae, and Thlipsuridae from the (Devonian) Genshaw Formation of Michigan. M, 1951, University of Michigan.

Kilgour, J. Mechanical and geological analysis of the (Ordovician) Beauharnois Clay (Ontario). M, 1951, University of Toronto.

Kilham, Susan Soltau. Deep sea bivalve molluscs; shell morphology, mineralogy, and geochemistry. D, 1971, Duke University. 216 p.

Kilian, Harry Stephen. Correlations within the Permian Big Lime of West Texas. M, 1932, University of Illinois, Urbana.

Kilian, Henry Martin. Geology of the Marble Mountains, San Bernardino County, California. M, 1964, University of Southern California.

Kilias, Stephanos. Genesis of gold deposits, Renabie area, Sudbury District, Ontario. M, 1984, University of Ottawa. 210 p.

Kilinc, Ishak Attila. An experimental study of redistribution of base metals in hydrothermally altered rocks. M, 1966, Pennsylvania State University, University Park. 117 p.

Kilinc, Ishak Attila. Experimental metamorphism and anatexis of shales and graywackes. D, 1969, Pennsylvania State University, University Park. 191 p.

Kilinski, Edward A. Petrographic notes on the siliceous gold ores of the Black Hills. M, 1923, Columbia University, Teachers College.

Killberg, Glen C. The crystal chemistry of pegmatitic perthites from the Black Hills, South Dakota. M, 1978, University of Toledo. 58 p.

Killeen, Katheryn Marie. Timing of folding and uplift of the Pismo Syncline, San Luis Obispo County, California. M, 1988, University of Nevada. 75 p.

Killeen, Patrick G. A gamma ray spectrometric study of the radioelement distribution on the Quirke Lake Syncline, Blind River area, Ontario. M, 1966, University of Western Ontario.

Killeen, Patrick G. The application of thick source alpha particle spectrometry to the detection of disequilibrium in the radioactive decay series of uranium 238 in rocks and minerals. D, 1971, University of Western Ontario.

Killeen, Pemberton L. The granites of western Rhode Island. M, 1931, Brown University.

Killeen, Terrence P. The geochemical behavior of uranium-thorium series nuclides in the South Atlantic Bight. M, 1988, University of South Carolina. 59 p.

Killen, David B. Stratigraphy and depositional environments of the Cherokee Group (Middle Pennsylvanian), Sedgwick Basin, south-central Kansas. M, 1986, University of Kansas. 165 p.

Killen, John Lippincott. Geology of a portion of the San Clemente Quadrangle, California. M, 1961, University of California, Los Angeles.

Killen, Rosemary Margaret. An analysis of longitude variations in the equatorial spectrum of Saturn. D, 1987, Rice University. 130 p.

Killey, Myrna Marie. Physical and mineralogical variations in the Yorkville Till Member, Grundy and adjacent counties, Illinois. M, 1980, Ball State University. 156 p.

Killey, R. W. D. Carbon geochemistry of an unconfined aquifer-lake system on the Canadian Shield. M, 1977, University of Waterloo.

Killian, Anna Mae. Ostracoda from the Haragan Formation (Devonian) of Murrav County, Oklahoma. M, 1961, University of Oklahoma. 93 p.

Killin, Alan Ferguson. A petrographic study of rocks from the Box Mine, Athabasca Lake. M, 1939, University of British Columbia.

Killin, Alan Ferguson. Geology of the Britannia Mine, British Columbia. D, 1951, University of Toronto.

Killingley, J. S. High temperature Knudsen cell/quadrupole mass spectrometric studies of magmatic volatiles. D, 1975, University of Hawaii. 157 p.

Killip, Colbeth. Geology of the Oxford and Tyner quadrangles, New York. M, 1961, University of Rochester. 57 p.

Killius, D. R. Petrology of a talc-serpentine complex near Blandford, Massachusetts. M, 1974, Boston University. 80 p.

Killman, Kathryn Susan. Geology and hydrothermal alteration of low-grade gold occurrences north of Weiser, Washington County, Idaho. M, 1981, University of Idaho. 83 p.

Killpack, Terry Joe. Apollo 17 specular power monitor data-passive mode. M, 1975, University of Utah. 237 p.

Kilmer, Frank Hale. Cretaceous and Cenozoic stratigraphy and paleontology, El Rosario area, Baja California, Mexico. D, 1963, University of California, Berkeley. 216 p.

Kilmer, Frank Hale. Stratigraphy of the "Diablo Formation". M, 1953, University of California, Berkeley. 68 p.

Kilmer, Kilmer W. The specific sorption of cobalt by the amphibole group minerals. M, 1969, Dartmouth College. 113 p.

Kilpatrick, Bruce E. Geology and geochemistry of the Wanamu-Blue mountains area, Waini SW, Guyana. M, 1968, Colorado School of Mines. 178 p.

Kilpatrick, John M. Data preparation and analysis techniques used in modeling the Sparta Aquifer in eastern Arkansas. M, 1987, University of Arkansas, Fayetteville.

Kilpatrick, John Thomas, III. A filtration analysis of the suspended particulate matter in the bottom waters over Nitinat Fan and the Washington continental slope. M, 1973, University of Washington. 39 p.

Kilps, James R. Actinocrinitidae of the Mississippian Lake Valley Formation, New Mexico. M, 1956, University of Wisconsin-Madison.

Kilsdonk, Bill. Deformation mechanisms in the Southeast Ramp region of the Pine Mountain Block, Tennessee. M, 1985, University of Michigan. 31 p.

Kilty, Kevin Thomas. An analysis of two hypotheses concerning the Cenozoic tectonism of the western United States. D, 1982, University of Utah. 215 p.

Kilty, Kevin Thomas. Aspects of forced convective heat transfer in geothermal systems. M, 1978, University of Utah. 61 p.

Kim, Adeline. The kinematics of brittle polyphase deformation within the Pioneer metamorphic core complex, Pioneer Mountains, Idaho. M, 1986, Lehigh University.

Kim, Ann Gallagher. Low temperature production of hydrocarbon gases from low rank coals. M, 1972, University of Pittsburgh.

Kim, Chang Shik. Interaction of long waves and nearshore barred topography; a mechanism of bar migration. D, 1987, College of William and Mary. 170 p.

Kim, Chang-Lak. Analyses of mass transfer from solid waste in a geologic environment. D, 1987, University of California, Berkeley. 176 p.

Kim, Chin Man. The effect of dynamic loading on confined Tennessee marble using the Hopkinson split-bar method. M, 1970, University of Minnesota, Minneapolis.

Kim, Chong Kwan. A gravity investigation of the Weed Sheet, northwestern California. D, 1974, University of Oregon. 145 p.

Kim, Chong Kwan. Gravity and magnetic surveys on Hole-in-the-Ground Crater, Lake County, central Oregon. M, 1968, University of Oregon. 66 p.

Kim, Choon-Sik. Geochemical aspects of Eocene-Oligocene volcanism and alteration in central Utah. M, 1988, University of Georgia. 106 p.

Kim, Chun-Soo. The influence of composition and microlithology on the weathering susceptibility of Ordovician mudrock in the Montreal, Quebec, area. D, 1984, McGill University. 314 p.

Kim, D. Decision analysis for the design of water resources systems under hydrologic uncertainty. D, 1976, University of Illinois, Urbana. 116 p.

Kim, Dae Choul. Diagenetic factors controlling physical, acoustic, and electrical properties of deep-sea carbonate sediments. D, 1985, University of Hawaii.

Kim, Daniel Yon Su. Theory of propagation of elastic waves through a porous granular medium saturated with fluid. D, 1955, University of Utah. 69 p.

Kim, Dong Jin. The attenuation of seismic waves in unconsolidated materials. M, 1983, University of Minnesota, Minneapolis.

Kim, Eul Soo. Reservoir simulation of in situ electromagnetic heating of heavy oils. D, 1987, Texas A&M University. 173 p.

Kim, Hae Soo. Polymetamorphism of metasedimentary rocks in the southern Sierras, California. D, 1972, Case Western Reserve University.

Kim, Haeyoun. Permian non-fusulinid foraminifera from the northern Yukon Territory. M, 1978, University of Saskatchewan. 199 p.

Kim, Hyung Keun. Late-Glacial and Holocene environment in central Iowa; a comparative study on pollen data from four sites. D, 1986, University of Iowa. 153 p.

Kim, Hyung Keun. Late-glacial and postglacial pollen studies from the Zuehl Farm site, north-central Iowa and the Cattail Channel Bog, northwestern Illinois. M, 1982, University of Iowa. 57 p.

Kim, Jin-Hoo. Forward modeling and inversion of responses for borehole normal and lateral electrode arrangements. D, 1986, Colorado School of Mines. 187 p.

Kim, Jin-Keun. Fractures and inelastic constitutive relations for concrete and geomaterials. D, 1985, Northwestern University. 233 p.

Kim, Jonathan J. Genesis of dioctahedral chlorites in the Miami Formation. M, 1984, University of South Florida, Tampa. 91 p.

Kim, Jonathan P. Distribution of dissolved copper in the Merrimack River estuary, Massachusetts. M, 1981, University of New Hampshire. 112 p.

Kim, Jonathan Philip. Volatilization and efflux of mercury from biologically productive ocean regions. D, 1987, University of Connecticut. 297 p.

Kim, Jong Dae. Mineralogy and trace elements of the uraniferous conglomerates, Nemo District, Black Hills, South Dakota. D, 1979, South Dakota School of Mines & Technology.

Kim, Joon Yol. Hydrostatic pressure effects on saturation remanent magnetization, susceptibility and magnetic hardness of magnetite. M, 1976, Michigan State University. 96 p.

Kim, Jung Joon. A crustal section of northern Central America as inferred from wide angle reflections from shallow earthquakes. D, 1981, University of Texas at Dallas. 98 p.

Kim, K. Deformation and failure of rock under monotonic and cyclic tension. D, 1976, University of Wisconsin-Madison. 117 p.

Kim, Kee Hyong. The phase equilibria in the system $Li_2O-B_2O_3-Al_2O_3-SiO_2$ and some of its subsidiary systems. D, 1961, Pennsylvania State University, University Park. 197 p.

Kim, Kee Hyun. Uranium-series nuclides in sediments and phosphate nodules from the Peruvian continental margin. D, 1984, Florida State University. 229 p.

Kim, Ki Young. Polarization studies of multicomponent seismic data. D, 1987, University of Oklahoma. 300 p.

Kim, Ki Young. Seismic studies near Crater Island in the Great Salt Lake Desert, Utah. M, 1985, University of Oklahoma. 73 p.

Kim, Ki-Tae. Experimental and theoretical studies of phase equilibria in the system $NaAlSiO_4-NaAlSi_3O_8-H_2O$ with special emphasis on the stability of analcite. D, 1970, McMaster University. 151 p.

Kim, Kun Deuk. Analytic continuation of electromagnetic fields. D, 1984, Colorado School of Mines. 122 p.

Kim, Kwang Jin. Finite element analysis of nonlinear consolidation. D, 1982, University of Illinois, Urbana. 220 p.

Kim, Kyung-Ryul. Methane and radioactive isotopes in submarine hydrothermal systems. D, 1983, University of California, San Diego. 229 p.

Kim, Moonkyum. A study of constitutive models for frictional materials. D, 1984, University of California, Los Angeles. 411 p.

Kim, Myoung Mo. Centrifugal model testing of soil slopes. D, 1980, University of Colorado. 189 p.

Kim, Ok Joon. Geology of the Wetmore-Beulah area, Custer and Pueblo counties, Colorado. M, 1951, Colorado School of Mines. 70 p.

Kim, Ran Young. An experimental investigation of creep and microseismic phenomena in geologic materials. D, 1971, Pennsylvania State University, University Park. 215 p.

Kim, Sang Wook. Geology of the middle Ordovician coticules of western New England. M, 1974, Wesleyan University.

Kim, So Gu. On the multipole expansion in the computation of gravity anomalies. M, 1971, Oregon State University. 61 p.

Kim, So Gu. Spectra scaling of earthquakes in some Eurasian aftershock sequences. D, 1976, St. Louis University. 219 p.

Kim, Soon Tae. A quantitative evaluation of Landsat for monitoring suspended sediments in a fluvial channel. D, 1980, Louisiana State University. 142 p.

Kim, Sung. Daily water supply and demand simulation for basinwide irrigation. D, 1987, University of Idaho. 615 p.

Kim, Tae In. Mass transport in laboratory water wave flumes. D, 1985, Oregon State University. 178 p.

Kim, Tae Moon. Three dimensional crack analysis by finite element method with boundary integral method at crack tip. D, 1985, University of Connecticut. 164 p.

Kim, Won Ho. Model study of seismic effects of explosions in prestressed media. D, 1966, St. Louis University.

Kim, Won Hyung. Biostratigraphy and depositional history of the San Gregorio and Isidro formations, Baja California Sur, Mexico. D, 1987, Stanford University. 224 p.

Kim, Won Sa. Phase relations in the system Pt-Pd-Sb-Te at mgagmatic and submagmatic temperatures (1000°C, 800°C and 600°C). D, 1984, Carleton University. 265 p.

Kim, Won Sa. Sulfide mineralization in the Coleman Member in the Cobalt area, Ontario. M, 1980, Carleton University. 112 p.

Kim, Woo Han. Gaussian beam synthetic body-wave seismograms using IPGT method with optimum beamwidths. M, 1986, University of Texas, Austin.

Kim, Yeadong. Ray theoretical traveltime inversion of seismic data in two dimensional plane dipping layers. D, 1987, Louisiana State University. 154 p.

Kim, Yong Sik. Centrifuge model study of an oil storage tank foundation. D, 1984, University of California, Davis. 161 p.

Kim, Yoo Bong. Conodont biostratigraphy of the Middle and Upper Ordovician of the Central Basin, Tennessee. D, 1988, Ohio State University. 244 p.

Kim, Young Il. Some geomorphic processes in the Perris Basin, Riverside County, California. D, 1967, University of California, Los Angeles. 286 p.

Kimball, B. A. Geochemistry of spring water, southeastern Uinta Basin, Utah and Colorado. M, 1978, University of Utah.

Kimball, Briant A. Trace elements in suspended sediment, southeastern Uinta Basin, Utah and Colorado. D, 1981, University of Wyoming. 83 p.

Kimball, Clark Gregory. The thermal, chemical and physical effects on the hydrogeologic system in the Sand Plain of Wisconsin from water source heat pump discharge via a return well. M, 1983, University of Wisconsin-Madison. 176 p.

Kimball, Colin Edward. A petrographic analysis of the basal carbonate beds of the Ferry Lake Anhydrite, Caddo-Pine Island Field, Caddo Parish, Louisiana. M, 1988, Northeast Louisiana University. 143 p.

Kimball, D. B. Public participation in the planning of coal-fired electric power development in the Southwest; a review and analysis. M, 1974, University of Arizona.

Kimball, Edgar W. The correlative value of foraminifera in the Pierre Shale (Upper Cretaceous) of Colorado. M, 1932, University of Colorado.

Kimball, Karen Lee. Phase relations in coexisting Fe-Mg and Ca amphiboles. D, 1981, University of Wisconsin-Madison. 246 p.

Kimball, Newton Scott, Jr. Geology of southwestern Palo Pinto County, Texas. M, 1967, Southern Methodist University. 50 p.

Kimball, Robert Vail. Structural interpretation of the Phosphoria Formation using geoelectrical methods. M, 1980, Idaho State University. 80 p.

Kimball, Royal Duane. Geology of the northcentral part of the Donner Pass Quadrangle, Sierra and Nevada counties, California. M, 1967, University of California, Davis. 70 p.

Kimberlein, Za Grant, Jr. The subsurface geology of Canadian County, Oklahoma. M, 1953, University of Oklahoma. 51 p.

Kimberley, Michael M. Origin of iron ore by diagenetic replacement of calcareous oolite. D, 1974, Princeton University. 784 p.

Kimberly, John Eli. Sedimentology of the Southwick Formation, Burnet County, Texas. M, 1961, University of Texas, Austin.

Kimbro, Charles. Geology of Winfrey-Low Gap area, Washington and Crawford counties, Arkansas. M, 1960, University of Arkansas, Fayetteville.

Kimbrough, C. W. Inorganic phosphorus species and transfer mechanisms in soils to sediments for two small Kansas watersheds. D, 1978, University of Kansas. 239 p.

Kimbrough, David Lee. Structure, petrology and geochronology of Mesozoic paleooceanic terranes on Cedros Island and the Vizcaino Peninsula, Baja California Sur, Mexico. D, 1982, University of California, Santa Barbara. 412 p.

Kimerling, A. J. Theoretical and practical relationships between remote sensing and cartography. D, 1976, University of Wisconsin-Madison. 309 p.

Kimler, Scott Thomas. A preliminary gravity study of the Guanacaste and Nicoya Peninsula regions of Costa Rica. M, 1983, Pennsylvania State University, University Park. 58 p.

Kimm, Diamond. The geology of the Dawn Mine District, San Gabriel Mountains, Los Angeles County, California. M, 1933, University of Southern California.

Kimmel, Bruce Lee, II. Nutrient transfers associated with seston sedimentation and sediment formation in Castle Lake, California. D, 1977, University of California, Davis. 166 p.

Kimmel, Garman O. Multi-stage separation of crude gas-oil mixtures. M, 1937, University of Oklahoma. 44 p.

Kimmel, Grant E. Ground-water resources and surficial geology of Colchester Township, Connecticut. M, 1964, Columbia University, Teachers College.

Kimmel, Margaret A. Uranium series ages of Pleistocene terraces on the island of Oahu, Hawaii. M, 1972, University of Southern California. 74 p.

Kimmel, Marion L. The development of a high pressure viscosimeter. M, 1952, Louisiana State University.

Kimmel, Peter Gerrit. Fishes of the Miocene-Pliocene Deer Butte Formation, Southeast Oregon. M, 1975, University of Michigan.

Kimmel, Peter Gerrit. Stratigraphy and paleoenvironments of the Miocene Chalk Hills Formation and Pliocene Glenns Ferry Formation in the western Snake River plain, Idaho. D, 1979, University of Michigan. 340 p.

Kimmel, Richard Elmer. Implications of photogeologic linears in the south Long Lake area, Alpena and Presque Isle counties, Michigan. M, 1973, Western Michigan University.

Kimmel, Tammy. Development of the spatial transport runoff environmental assessment model "STREAM". M, 1978, Rensselaer Polytechnic Institute. 85 p.

Kimpe, Nancy De *see* De Kimpe, Nancy

Kimura, H. S. A study of Lahaina silty Clay soils forming from different parent materials. M, 1966, University of Hawaii. 50 p.

Kimura, Hubert Satoshi. Alkaline dissolution of kaolin, montmorillonite, windy soil and saprolite clay minerals. D, 1969, University of California, Davis. 212 p.

Kimura, Shigeyuki. Phase equilibria at liquidus temperatures in the system Ca-Fe-Ti-O. D, 1968, Pennsylvania State University, University Park. 155 p.

Kimutis, Robert A. Geologic considerations in underground coal mining. M, 1978, West Virginia University.

Kimyai, Abbas. Palynology of the Raritan Formation (Cretaceous) in New Jersey and Long Island. D, 1964, New York University. 236 p.

Kimzey, Jo Ann. Petrography and geochemistry of the La Posta Granodiorite. M, 1982, San Diego State University. 81 p.

Kinard, John C. Jurassic of northern Claiborne Parishe, Louisiana, and vicinity. M, 1956, University of Oklahoma. 116 p.

Kinart, Kirk P. Geochemistry of part of the Loon Lake Batholith and its relationship to uranium mineralization at the Midnite Mine, northeastern Washington. M, 1980, Eastern Washington University. 200 p.

Kincaid, D. T. Theoretical and experimental investigations of Ilex pollen and leaves in relation to microhabitats in the southeastern United States. D, 1976, Wake Forest University. 350 p.

Kincaid, George Preston, Jr. Contemporary sources and geochemistry of tritium in the Gulf of Mexico and its distributive province. D, 1971, Texas A&M University. 257 p.

Kind, T. C. The Kentucky Bluegrass; cyclic or noncyclic topography. D, 1976, Indiana State University. 235 p.

Kinder, Anthony Stanley. A regional vertical magnetometer survey of Livingston County, Michigan. M, 1954, University of Michigan.

Kinder, T. H. The continental slope regime of the eastern Bering Sea. D, 1976, University of Washington. 272 p.

Kindle, Cecil Haldane. Geology of the south coast of Gaspe Co., Quebec. D, 1931, Princeton University.

Kindle, E. D. Origin of the Great Slave Lake lead-zinc deposits. M, 1931, Queen's University. 38 p.

Kindle, Edward Darwin. An analysis of the structural features of the Sudbury Basin and their bearing on ore deposition (Ontario). D, 1933, University of Wisconsin-Madison.

Kindle, Edward M. The Devonian and Lower Carboniferous faunas of southern Indiana and central Kentucky. D, 1899, Yale University.

Kindred, Fred R. Origin and diagenesis of carbonate mudstone, shallow to deeper shelf, Aurora Formation, Coahuila, Mexico. M, 1988, University of New Orleans.

Kindred, Valerie Prescott. The nature and origin of dolomite in the upper Fountain Formation (Pennsylvanian), east flank of Colorado Front Range, central Colorado. M, 1987, University of Colorado. 57 p.

Kindt, Eugene Anthony. Bivalvia of the Allegheny Group (Pennsylvanian) in Ohio. M, 1974, Bowling Green State University. 227 p.

Kinealy, James R. The metallurical treatment of New Mexico copper ores from the mines of the Avon Mining Company. M, 1883, Washington University.

Kinell, Carl B., III. Clay mineralogy of the Vilas Shale (Upper Pennsylvanian) in Wilson and Montgomery counties. M, 1964, University of Kansas. 38 p.

Kinerney, Eugene James. Geology of the Tarsney Quadrangle, Jackson County, Missouri. M, 1961, University of Missouri, Columbia.

King, Alan D. The use of computer analysis in interpreting gravity anomalies in glacial terrain. M, 1973, Wright State University. 48 p.

King, Alan G. Growth of galena crystals by sublimation. M, 1951, University of Utah. 31 p.

King, Arthur F. Geology of the Cape Makkovik Peninsula, Aillik, Labrador. M, 1963, Memorial University of Newfoundland. 130 p.

King, Brian Charles. Inferences drawn from clear and smoky quartz in granitic rocks. M, 1984, University of Kentucky. 178 p.

King, C. W. and Coryell, Lawrence Ritchie Brooke. A gravity study of the northern boundary of the Boston Basin. M, 1958, Massachusetts Institute of Technology. 44 p.

King, Charles Edward. Micropaleontology of the Gulfian of the Davis Mountain Front. M, 1958, Texas Tech University. 156 p.

King, Clifford L. A study of buried preglacial topography by the gravity method. M, 1974, Purdue University.

King, D. L. Phosphorous fixation in soils containing amorphous and crystalline aluminum and iron compounds. M, 1961, University of Hawaii.

King, D. R. The wave field on a shelf resulting from point source generation. D, 1978, University of British Columbia.

King, D. T., Jr. Stratigraphy and petrology of the upper part of the Riley Formation, Upper Cambrian of central Texas. M, 1976, University of Houston. 276 p.

King, D. Whitney. Spectrophotometric determination of pH and iron in seawater; equilibria and kinetics. D, 1988, University of Rhode Island. 261 p.

King, David Alexander. Controls of gold mineralization in the southern portion of the Hodson mining district, West Mother Lode gold belt, California. M, 1986, University of Montana. 60 p.

King, David Edward. A diagenetic history of the Waulsortian limestones, eastern Midlands and Dublin Basin, Republic of Ireland. M, 1984, SUNY at Stony Brook. 369 p.

King, David Thompson, Jr. Genetic stratigraphy of the Mississippian System in central Missouri. D, 1980, University of Missouri, Columbia. 129 p.

King, Edward Larnard. Post-Berea Mississippian stratigraphy in eastern Trumbull County, Ohio. M, 1952, Ohio State University.

King, Elbert Aubrey, Jr. Geology of northwestern Gonzales County. M, 1961, University of Texas, Austin.

King, Elbert Aubrey, Jr. Investigations of North American tektites. D, 1965, Harvard University.

King, Elbert Aubry, Jr. Geology of Dodge County, Georgia. M, 1962, Harvard University. 17 p.

King, Ellen Jean. Scapolite-plagioclase equilibria in metamorphosed carbonates of the Bluff Creek area, northern Idaho. M, 1984, University of Oregon. 91 p.

King, Evan Shelby, Jr. Sedimentation of Arroyo Blanco Bay. M, 1949, University of Iowa. 216 p.

King, F. P. Basic magnesian rocks associated with the corundum deposits of Georgia. D, 1897, The Johns Hopkins University.

King, G. M. The nature of methanogenesis in soils of a Georgia salt marsh. D, 1978, University of Georgia. 205 p.

King, Galen E., Jr. A trace element and clay mineral analysis of the Lower Cretaceous Kiowa Shale of southwestern Kansas and northwestern Oklahoma. M, 1972, Wichita State University. 33 p.

King, Gary L. Geometric analysis of minor folds in sedimentary rocks. M, 1973, West Virginia University.

King, George Anthony. Deglaciation and vegetation history of western Labrador and adjacent Quebec. D, 1986, University of Minnesota, Minneapolis. 325 p.

King, George Leslie, Jr. Stratigraphy of the Edwards Limestone, Coryell County, Texas. M, 1963, Baylor University.

King, Gordon Patrick. Chemical properties of an acid productive soil. D, 1981, Oklahoma State University. 142 p.

King, Guy Quintin. Morphometry of Great Basin playas. D, 1982, University of Utah. 137 p.

King, Guy Quintin. The late Quaternary history of Adrian Valley, Lyon County, Nevada. M, 1978, University of Utah. 88 p.

King, H. E., Jr. Phase transitions in FeS at high-temperatures and high-pressures. D, 1979, SUNY at Stony Brook. 192 p.

King, Harley Dee. Paleozoic stratigraphy of the James Peak Quadrangle, Utah. M, 1965, Utah State University. 62 p.

King, Harry J. The occurrence and origin of the vermiculite, Cerro de Pedregosa, Coahuila, Mexico. M, 1962, St. Louis University.

King, Harvey Dennis. Zircons; a petrogenic indicator in the San Isabel Batholith, Wet Mountains, Colorado. M, 1960, University of Texas, Austin.

King, Herman Leo. Subsurface Precambrian of Saskatchewan. M, 1966, University of Saskatchewan. 177 p.

King, Hobart Morse, II. The mode of occurrence and distribution of sulfur in West Virginia coals and Devonian shales. D, 1982, West Virginia University. 260 p.

King, J. D. The drying of marine sediments for water content determinations. M, 1969, United States Naval Academy.

King, J. R. Geology of the Boswell Creek area, Albany County, Wyo. M, 1961, University of Wyoming. 83 p.

King, James A., V. The petrography and structure of a portion of Soapstone Ridge, DeKalb and Clayton counties, Georgia. M, 1957, Emory University. 34 p.

King, James Edward. Late Pleistocene biogeography of the western Missouri Ozarks. D, 1972, University of Arizona.

King, James F. Reconnaissance geology of part of the Lemhi Range, Idaho. M, 1959, Northwestern University.

King, James Gagwane. Mise-a-la-masse study of prismatic bodies; model tank and field experiments. M, 1988, University of Western Ontario. 123 p.

King, James M. Ground-water resources of Williams County, Ohio. M, 1977, University of Toledo. 114 p.

King, James Michael. Hydrogeology and numerical modeling of the Flambeau Mine site, Rusk County, Wisconsin. D, 1983, Indiana University, Bloomington. 316 p.

King, Janet Elizabeth. Low-pressure regional metamorphism and progressive deformation in the eastern Point Lake area, Slave Province, N.W.T. M, 1982, Queen's University. 187 p.

King, Janet Elizabeth. Structure of the metamorphic-internal zone, northern Wopmay Orogen, Northwest Territories, Canada. D, 1985, Queen's University. 209 p.

King, Jerry Leon. Observations on the seismic response of sediment-filled valleys. D, 1981, University of California, San Diego. 310 p.

King, Jessie M. Glacial sculpture in the region of the north-flowing Finger Lakes of central New York. M, 1934, Columbia University, Teachers College.

King, John G. Sedimentation characteristics of the Horse Creek watersheds in North central Idaho. D, 1978, University of Idaho. 97 p.

King, John J. A field investigation of a select group of New Jersey Highland pegmatites. M, 1987, Rutgers, The State University, Newark. 275 p.

King, John Joseph. The micropaleontology of the lower formations of the Gulf Series of Texas. M, 1927, University of Texas, Austin.

King, John R. The geology of the southeastern Cuddy Mountain District western Idaho. M, 1971, Oregon State University. 78 p.

King, John S. Petrology and structure of the Precambrian and post Mississippian rocks of the northeastern Medicine Bow Mountains, Carbon County, Wyoming. D, 1963, University of Wyoming. 125 p.

King, John S. Study of some granitic pegmatites of the Edinburgh Quadrangle, New York. M, 1957, SUNY at Buffalo.

King, John S. The geology of the Buffalo Ridge area, Union County, Tennessee. M, 1960, University of Tennessee, Knoxville. 34 p.

King, John William. Geomagnetic secular variation curves for northeastern North America for the last 9,000 years B.P. D, 1983, University of Minnesota, Minneapolis. 206 p.

King, Jonathan K. Utility of Landsat imagery for recognition of Tertiary silicic and intermediate volcanic features in northern Nevada. M, 1984, University of Wyoming. 486 p.

King, K. S. Carbon isotope geochemistry of a landfill leachate. M, 1983, University of Waterloo. 120 p.

King, Keith. The Tampa Formation of peninsular Florida; a formal definition. M, 1979, Florida State University.

King, Kenneth R. The stratigraphy of the Interlake Group (Silurian) in Manitoba. M, 1964, University of Manitoba.

King, Kenneth Ross. Bayesian decision theory and computer simulation applied to multistage, sequential petroleum exploration. D, 1971, Pennsylvania State University, University Park. 214 p.

King, Kenneth, Jr. Amino acid composition of planktonic foraminifera; applications in evolution and geochronology. D, 1972, Columbia University. 173 p.

King, Lewis W. Studies on spontaneous combustion using Nova Scotia coals. D, 1955, Massachusetts Institute of Technology. 89 p.

King, Lowell Franklin. A petrographic study of the Garrard Siltstone in the southeastern Bluegrass region of Kentucky. M, 1959, University of Kentucky. 52 p.

King, M. J. The engineering land-use planning of the Harry S. Truman Dam site area, Benton County, Missouri. M, 1975, University of Missouri, Rolla.

King, Mark C. Occurrence and distribution of minerals in the Permian coals of the Parana Basin, Brazil. M, 1979, University of Toledo. 116 p.

King, Mary Anne. Application of a finite-difference ground-water flow model in the evaluation of the semi-confined carbonate aquifer of Hancock County, Ohio. M, 1988, University of Toledo. 186 p.

King, Norma Louise. Minerahgraphy of the dividend claim, Osoyoos Mines Limited and distribution of gold in Cariboo gold quartz tailings. M, 1942, University of British Columbia.

King, Norman J. Physical properties contributing to the relative erosional resistance of sandstone and shale-derived sediments. M, 1953, University of Utah.

King, Norman Ralph. Stratigraphy, biostratigraphic zonation, and depositional history of the Fort Hays Limestone and adjacent strata (Upper Cretaceous), southern Western Interior. D, 1972, Indiana University, Bloomington. 220 p.

King, Paul Hamilton. The movement of pesticides through soils. D, 1966, Stanford University. 223 p.

King, Philip Burke. The geology of the Glass Mountains, Texas. D, 1929, Yale University.

King, Philip Burke. The physiography of the Glass Mountains. M, 1927, University of Iowa. 67 p.

King, Ralph H. Stratigraphy and economic geology of the Phosphoria Formation (Permian), southern Wind River Mountains, Wyoming. D, 1956, University of Kansas. 179 p.

King, Robert B. Interpolation of deflections of the vertical using the gradients of gravity and comparison with the gravimetric method. M, 1963, Ohio State University.

King, Robert E. Invertebrate faunas of the Permian of Trans-Pecos Texas. D, 1929, Yale University.

King, Robert E. The subsurface structure and stratigraphy of the Bayou Pigeon, North Bayou Long, and Bayou Postillion areas, Iberia Parish, Louisiana. M, 1960, Louisiana State University.

King, Robert L. The Pierre-Fox Hills contact in west-central South Dakota. M, 1953, South Dakota School of Mines & Technology.

King, Robert Nephew. The Earth on which we live; a sourcebook for the Modern Elementary School Science program of the Science Manpower Project. D, 1965, Columbia University, Teachers College. 590 p.

King, Roger Hatton. Periglaciation on Devon island, Northwest Territories. D, 1969, University of Saskatchewan. 512 p.

King, Scott E. A seismic reflection study of Harlem Township, Delaware County, Ohio. M, 1983, Wright State University. 76 p.

King, Sharon Lynne. Biostratigraphic and paleoenvironmental interpretations from late Pleistocene Ostracoda, Charleston, South Carolina. M, 1981, University of North Carolina, Chapel Hill. 253 p.

King, Stagg Lipscomb. Microbiological studies of sulfate reduction and organic matter diagenesis in anoxic marine environments. D, 1984, University of Washington. 167 p.

King, Thomas J. The archaeological implications of the paleobotanical record from the Lucerne Valley region, Mojave Desert, California. M, 1976, University of California, Los Angeles.

King, Timothy Allen. Design, reclamation, and site selection for the Winding Gulf refuse pile, Raleigh County, West Virginia. M, 1986, Kent State University, Kent. 238 p.

King, Truxton W. Intergradation between the genera Merycodus and Capromeryx in the family Antilocapridae. M, 1952, University of California, Berkeley. 54 p.

King, Victor Hugo. A program for the conservation of the oil and gas resources of Ecuador, South America. M, 1949, Michigan State University. 65 p.

King, Victor LeRoy, Jr. Geology of the Mission Valley Quadrangle, Comal County, Texas. M, 1957, University of Texas, Austin.

King, W. Allan. The geomorphology and sedimentology of the lower reaches of the Attawapiskat River, James Bay, Ontario. M, 1980, University of Guelph.

King, Wendell Christopher. Development of a field method for subsurface remote imaging through implementation of seismic geophysical diffraction tomography. D, 1988, University of Tennessee, Knoxville. 187 p.

King, William Edward. Fusulinids of the type Marble Falls Limestone of Texas (Lower Pennsylvanian). D, 1959, University of Wisconsin-Madison. 134 p.

King, William Lyle. New Hope (Smackover) Field, Franklin County, Texas. M, 1957, University of Pittsburgh.

King, William Roy, Jr. The lower Chester limestones in Madison, Rockcastle, and Pulaski counties, Kentucky. M, 1950, University of Kentucky. 37 p.

Kingery, F. A. Textural analysis of shelf sands off the Georgia coast. M, 1973, [University of California, San Diego].

Kingery, Thomas LeRoy. The geology of the Triassic basin, Scottsville, Virginia. M, 1954, University of Cincinnati. 60 p.

Kingman, Celia Collins. Glacial history of Mount Monadnock, New Hampshire. M, 1932, Clark University.

Kingsbury, Joseph W. Paragenesis of the ores of the Minnie Moore and adjacent mines of Bellevue, Idaho. M, 1907, Columbia University, Teachers College.

Kingsbury, Richard Howard, Jr. Petrology and geochemistry of an ultramafic pluton, North Carolina. M, 1977, Kent State University, Kent. 39 p.

Kingsley, John. A geologic and gravity survey of the rocks underlying the San Gabriel Valley, Los Angeles County, California. M, 1963, University of California, Los Angeles.

Kingsley, Louise. Cauldron-subsidence of the Ossipee Mountains. D, 1931, Bryn Mawr College. 29 p.

Kingsley, Louise. The genera of the Oreodontidae, with special reference to Merychyus. M, 1924, Smith College. 50 p.

Kingston, Dave Russell. Paleozoic stratigraphy of the Tetsa-Halfway Rivers area, northeastern British Columbia, Canada. D, 1956, University of Wisconsin-Madison. 130 p.

Kingston, David. Geology and geochemistry of the Owl Creek gold deposit, Timmins, Ontario. M, 1987, Carleton University. 130 p.

Kingston, Jim. Long-term effects of in situ uranium leach mining restoration in the Oakville aquifer system near George West, Texas. M, 1987, Baylor University. 171 p.

Kingston, Paul W. E. The naturally occurring cobalt iron sulpharsenide minerals. D, 1969, Queen's University. 157 p.

Kingwell, Lorne. Multichannel filtering of small arrays. M, 1968, University of Toronto.

Kinley, Teresa May. The occurrence and timing of gold mineralization at the Red Pine Mine, western Tobacco Root Mountains, southwestern Montana. M, 1987, Montana State University. 129 p.

Kinnaman, Ross Lorrain. Geology of the foothills west of Sedalia, Douglas County, Colorado. M, 1954, University of Colorado.

Kinnan, Joseph E. The effect of density variation on the prediction of gravity anomalies. M, 1968, Ohio State University.

Kinne, Raymond Charles. The coking of Iowa coal. D, 1931, Iowa State University of Science and Technology.

Kinnebrew, Quin. Influence of a river valley constriction on upstream sedimentation. M, 1988, Texas A&M University.

Kinney, Douglas M. Geology of the Uinta River-Brush Creek area, Duchesne and Uintah counties, Utah. D, 1950, Yale University.

Kinney, Frederick Dawless. Geologic and trace element study of Mine One, Cave-in-Rock, Illinois. M, 1979, Kent State University, Kent. 97 p.

Kinney, Harry D. The basic dikes and associated rocks of Manhattanville Valley, Manhattan Island. M, 1910, Columbia University, Teachers College.

Kinney, Thomas E. Analysis of the groundwater quality management policies for Colorado and Wyoming. M, 1985, Colorado State University. 98 p.

Kinney, Vincent Lewis. Geology of the Mizpah Mine area, Latah County, Idaho. M, 1972, University of Massachusetts. 49 p.

Kinnison, John E. Geology and ore deposits of the southern section of the Amole mining district, Tucson Mountains, Pima County, Arizona. M, 1958, University of Arizona.

Kinnison, Phillip Taylor. A survey of the ground water of the State of Idaho. M, 1954, University of Idaho. 63 p.

Kinsel, Erick Paul. Petrology of silicic alkalic basalts, xenocrystic basaltic andesites, and crustal xenoliths from the Taos Plateau volcanic field, north-central New Mexico. M, 1986, Southern Methodist University. 138 p.

Kinser, James Hanford. A petrographic study of the Tory Hill Stock, Haliburton County, Ontario. M, 1935, University of Minnesota, Minneapolis. 37 p.

Kinsey, D. W. Carbon turnover and accumulation by coral reefs. D, 1979, University of Hawaii. 260 p.

Kinsland, Gary Lynn. Refractive index ratios of KCl, NaCl and AgCl at pressures up to 77 kb. M, 1971, University of Rochester. 65 p.

Kinsland, Gary Lynn. Yield strength under confining pressures to 300 kb in the diamond anvil cell. D, 1974, University of Rochester. 135 p.

Kinsley, Gerald W. Geology of the Leeton-Cornelia area, Johnson County, Missouri. M, 1960, University of Missouri, Columbia.

Kinsman, Daniel Francis. Some preliminary investigations on the organic soils of New York State. M, 1924, Cornell University.

Kintner, Henry B. A subsurface study of the "F" Zone (Pennsylvanian System) in Hitchcock County, Nebraska. M, 1984, University of Nebraska, Lincoln.

Kintner, Stephen S. Wastewater management plan for unincorporated Taney County, Missouri. M, 1983, Southwest Missouri State University. 82 p.

Kintzer, Frederick C. Geology and landslides at Calaveras Reservoir, Alameda and Santa Clara counties, California. M, 1980, California State University, Hayward. 138 p.

Kintzinger, Paul Raymond. A laboratory investigation of induced electrical polarization. D, 1956, New Mexico Institute of Mining and Technology. 71 p.

Kinyali, Samuel M. Long-term effects of variable water quality on some soil physical characteristics under field conditions. M, 1973, University of California, Riverside. 229 p.

Kinzelman, David J. The pre-upper Ordovician geology of the Birmingham-Erie field, Florence Township, Erie County, Ohio. M, 1968, Ohio State University.

Kinzler, Rosamond Joyce. A field, petrologic, and geochemical study of the Callahan lava flow, a basaltic andesite from Medicine Lake shield volcano, California. M, 1985, Massachusetts Institute of Technology. 100 p.

Kiouses, Stephan. A paleoclimatic interpretation from a petrologic comparison of Mid-Tertiary sandstones and Holocene stream sands for the southern Sierra Nevada. M, 1980, Southern Illinois University, Carbondale. 107 p.

Kiousis, Panagiotis Demetrios. Large strain theory as applied to penetration mechanism in soils. D, 1985, Louisiana State University. 108 p.

Kiphart, Kerry. The kinetics of sorption reactions at the goethite aqueous interface. D, 1983, University of Notre Dame. 254 p.

Kipp, James Alden. The response of hydrogeology following restoration at the Iowa Coal Project Demonstration Mine Number One. M, 1981, Iowa State University of Science and Technology. 227 p.

Kipphut, G. W. An investigation of sedimentary processes in lakes. D, 1978, Columbia University, Teachers College. 194 p.

Kirby, Edward George, III. Biology of diffusible pollen wall compounds. D, 1977, University of Florida. 74 p.

Kirby, Emery, Jr. A comparative study of two instructional methods; the activity-investigation method and the lecture-research method; content; measuring mineral properties at the eighth-grade level. M, 1968, Northern Illinois University. 129 p.

Kirby, James P. A stratigraphic study of the pre-Catheys Ordovician formations in the Pulaski Quadrangle, Giles County, Tennessee. M, 1957, Florida State University.

Kirby, Mark William. Sedimentology of the Middle Devonian Bellvale and Skunnemunk formations in the Green Pond Outlier in northern New Jersey and southeastern New York. M, 1981, Rutgers, The State University, Newark. 109 p.

Kirby, Stephen Homer. Creep of synthetic alpha quartz. D, 1976, University of California, Los Angeles.

Kirch, Annie B. The influence of geography upon primitive religions. M, 1914, University of Wisconsin-Madison.

Kircher, Dorcas Elizabeth. Volcanological investigation of the El Cajete Series, Jemez Mountains, New Mexico. M, 1988, University of Texas, Arlington. 192 p.

Kirchgasser, William Thomas. Paleontology and stratigraphy of the concretions and limestones of the Upper Devonian Cashaqua Shale Member, Sonyea Formation, New York. D, 1967, Cornell University. 223 p.

Kirchgasser, William Thomas. The Parrish Limestone (Upper Devonian) of west-central New York. M, 1965, Cornell University. 23 p.

Kirchgessner, David Arthur. A cross-bedding and heavy mineral analysis of the Buffalo Moraine in western New York. M, 1970, SUNY at Buffalo. 27 p.

Kirchgessner, David Arthur. Sedimentology and petrology of upper Devonian Greenland Gap Group along the Allegheny Front, Virginia, West Virginia and Maryland. D, 1974, University of North Carolina, Chapel Hill. 93 p.

Kirchner, Gail L. Field relations, petrology, and mineralization of the Linster Peak Dome, Fergus County, Montana. M, 1982, University of Montana. 115 p.

Kirchner, James G. The geology of the Windy Mountain area, Wyoming. M, 1962, Wayne State University.

Kirchner, James Gary. The petrography and petrology of the phonolite porphyry intrusions of the northern Black Hills, South Dakota. D, 1971, University of Iowa. 199 p.

Kirchner, Joseph F. Seismic refraction studies on the Ross Ice Shelf, Antarctica. M, 1978, University of Wisconsin-Madison.

Kirchoff, Scharine. Crustal structure beneath the Tularosa Basin in the White Sands Missile Range, New Mexico. M, 1986, Boston University. 79 p.

Kirchoff-Stein, Kimberly Susan. A gravity geologic study of Miami County, Ohio to locate buried valleys, two-dimensional modelling of the buried valleys. M, 1984, Wright State University. 95 p.

Kiremidjian, Anne Setian. Probabilistic hazard mapping; development of site dependent seismic load parameters. D, 1977, Stanford University. 233 p.

Kirk, Allan Robert. Petrology, structural geology and metamorphism of the Boothbay Harbor area, Maine. M, 1971, SUNY at Buffalo. 108 p.

Kirk, Bruce Glen. Gulf Series east of Austin, Texas. M, 1949, University of Texas, Austin.

Kirk, Charles Townsend. Condition of mineralization in the copper veins at Butte, Montana. D, 1911, University of Wisconsin-Madison. 46 p.

Kirk, Charles Townsend. The Pennsylvanian-Permian contact through Oklahoma. M, 1905, University of Oklahoma. 43 p.

Kirk, J. Norman. A regional study of radar lineament patterns in the Ouachita mountains, McAlester Basin-Arkansas valley, and Ozark regions of Oklahoma and Arkansas. M, 1970, University of Kansas. 46 p.

Kirk, John. Geology and hydrology of an area north of Donner Lake, California. M, 1974, California State University, Fresno.

Kirk, John S. Laboratory study of mineralogical changes during steam condensate flooding of Cold Lake oil sands. M, 1985, University of Alberta. 151 p.

Kirk, Mahlon V. The Cretaceous-Tertiary boundary in California. M, 1950, University of California, Berkeley. 46 p.

Kirk, Marcia W. Olson. Sensitivity analysis of groundwater contamination caused by immobilized gasoline hydrocarbons. M, 1987, University of Nevada. 101 p.

Kirk, Myrl Stuart, Jr. A subsurface section from Osage County to Okfuskee County, Oklahoma. M, 1956, University of Oklahoma. 54 p.

Kirk, Randolph Livingstone. I, Thermal evolution of Ganymede and implications for surface features; II, Magnetohydrodynamic constraints on deep zonal flow in the giant planets; III, A fast finite-element algorithm for two-dimensional photoclinometry. D, 1987, California Institute of Technology. 274 p.

Kirk, Stephen T. Analysis of the White River groundwater flow system using a deuterium-calibrated discrete-state compartment model. M, 1987, University of Nevada. 81 p.

Kirkby, Kent Charles. Deposition, erosion, and diagenesis of the upper Victorio Peak Formation (Leonardian), southern Guadalupe Mountains, West Texas. M, 1982, University of Wisconsin-Madison.

Kirker, Jill Kathleen. Geology, geochemistry and origin of Rusty Springs lead-zinc-silver deposit, Yukon Territory. M, 1982, University of Calgary. 159 p.

Kirkgard, Mark Mitchell. An experimental study of the three-dimensional behavior of natural normally

consolidated anisotropic clay. D, 1988, University of California, Los Angeles. 498 p.

Kirkham, Richard A. The geology of the northern portion of the Fox Range, Washoe County, Nevada. M, 1982, University of Nevada. 98 p.

Kirkham, Robert Marshall. Environmental geology of the western Carson City, Nevada area. M, 1976, University of Nevada. 97 p.

Kirkham, Rodney Victor. A mineralogical and geochemical study of the zonal distribution of ores in the Hudson bay range, British Columbia (Canada). D, 1969, University of Wisconsin-Madison. 162 p.

Kirkham, Rodney Victor. The geology and mineral deposits in the vicinity of the Mitchell and Sulphurets glaciers, Northwest British Columbia. M, 1963, University of British Columbia.

Kirkham, Virgil R. D. The geology and oil and gas possibilities of southwestern Idaho. D, 1930, University of Chicago. 253 p.

Kirkland, Brend L. Carbonate depositional environments and facies of the shelf margin and outer shelf, Lower Cretaceous Sligo Formation, South Texas. M, 1986, Texas A&M University. 150 p.

Kirkland, Douglas Wright. Paleoecology of the varved Rita Blanca Lake deposits, Hartley County, Texas. D, 1963, University of New Mexico. 119 p.

Kirkland, Douglas Wright. The environment of the Jurassic Todilto Basin, northwestern New Mexico. M, 1958, University of New Mexico. 69 p.

Kirkland, James I. Paleontology and paleoenvironments of the Greenhorn marine cycle, southwestern Black Mesa, Coconino County, Arizona. M, 1983, Northern Arizona University. 224 p.

Kirkland, James T. Location of the Valley Heads moraine in the Tully Valley region, New York. M, 1968, SUNY, College at Cortland. 37 p.

Kirkland, James Totten. Glacial geology of the western Catskills. D, 1973, SUNY at Binghamton. 126 p.

Kirkland, John K., Jr. Geology of the southeastern quarter of the Caledonia Quadrangle, Mississippi and Alabama. M, 1965, Mississippi State University. 38 p.

Kirkland, Kenneth John. Sulfide deposition (Cretaceous) at Noranda Creek, British Columbia. M, 1971, University of Alberta. 55 p.

Kirkland, Larry A. The gravel cover and catchment efficiency in the plastic-lined catchment. M, 1969, University of Arizona.

Kirkland, Michael John. Petrology and diagenesis of sandstones of the Bluefield Formation (Upper Mississippian), Southeast West Virginia. M, 1985, East Carolina University. 120 p.

Kirkland, Peggy L. Permian stratigraphy and stratigraphic paleontology of the Colorado Plateau. M, 1962, University of New Mexico. 245 p.

Kirkland, Robert W. A study of part of the Kaniapiskau system northwest of Attikamagen Lake, New Quebec. D, 1950, McGill University.

Kirkland, Robert W. The east zone of Giant Yellowstone Gold Mines, Limited, Northwest Territories. M, 1947, McGill University.

Kirkland, S. J. T. A study of the Tazin-Athabasca unconformity, northern Saskatchewan. D, 1953, Queen's University.

Kirkland, S. J. T. Structures of the Flin Flon-Weetago Bay area, northern Saskatchewan. D, 1956, Queen's University. 115 p.

Kirkland, Samuel John Thomas. Petrology of the granites and pegmatites, Charlebois Lake area, northern Saskatchewan. M, 1952, University of Saskatchewan. 79 p.

Kirkley, Melissa B. Peridotite xenoliths in Colorado-Wyoming kimberlites. M, 1980, Colorado State University. 298 p.

Kirkman, Roy C. Economic evaluation of fund resources, especially uranium, using probabilistic risk analysis. M, 1976, Colorado School of Mines. 51 p.

Kirkpatrick, Doug. Structure and stratigraphy of N. portion of Grant Range, E. central Nev. M, 1960, University of Washington. 76 p.

Kirkpatrick, Gerald. Statistical analysis of grain size data as a possible key to ancient depositional environments. M, 1982, Florida State University.

Kirkpatrick, Glen Edgar. Geology and ore deposits of the Big Creek area, Idaho and Valley counties, Idaho. M, 1974, University of Idaho. 92 p.

Kirkpatrick, John Curtis. A study of some marine middle Eocene formations in Southern California. M, 1958, University of California, Los Angeles.

Kirkpatrick, R. James. The kinetics of crystal growth in the system diopside-cats $(CaAl_2Si_2O_6)$ and the application of crystal growth theory to some geologic problems. D, 1972, University of Illinois, Urbana. 87 p.

Kirkpatrick, Samuel Roger. The geology of a portion of Stewart County, Georgia. M, 1959, Emory University. 79 p.

Kirkpatrick, Terrence D. A geologic study of Louisa County, Iowa. M, 1967, University of Iowa. 101 p.

Kirkwood, Donna. Géologie structurale de la région de Percé, Gaspésie. M, 1987, Universite de Montreal.

Kirkwood, Steven G. Stratigraphy and petroleum potential of the Cedar Mountain and Dakota formations, northwestern Colorado. M, 1977, Colorado School of Mines. 193 p.

Kirman, Z. M. Gamma-ray well-logging. M, 1942, Massachusetts Institute of Technology. 38 p.

Kirmani, Khalil-ullah. The Duhamel Reef, Alberta. M, 1962, University of Alberta. 173 p.

Kirn, Douglas J. Sandstone petrology of the Casper Formation, southern Laramie Basin, Wyoming and Colorado. M, 1972, University of Wyoming. 43 p.

Kirr, James N. Geology of the Blacksville, West Virginia-Pennsylvania 71/2 minute topographic Quadrangle. M, 1968, West Virginia University.

Kirr, James Neil. Laboratory study; the dispersion of sand-sized particles under conditions of axial and planar flow. D, 1974, West Virginia University.

Kirsch, Stephen Augustine. Structure of the metamorphic and sedimentary rocks of Mineral Ridge, Esmeralda County, Nevada. D, 1968, University of California, Berkeley. 79 p.

Kirschbaum, Charles Louis. Laser microprobe studies of rare gas isotopes in meteorites. D, 1986, University of California, Berkeley. 113 p.

Kirschke, William H. A petrographic core analysis of the Lower and Middle Ordovician rocks, Pulaski Field, Jackson County, Michigan. M, 1962, Michigan State University. 36 p.

Kirschner, Carolyn Elisabeth. Structural evolution of a Precambrian tidal flat, Taos County, New Mexico. M, 1979, University of Texas, Austin.

Kirschner, William A. Nonmarine molluscan paleontology and paleoecology of early Tertiary strata, Hanna Basin, Wyoming. M, 1984, University of Wyoming. 157 p.

Kirschvink, Joseph Lynn. I, A paleomagnetic approach to the Precambrian-Cambrian boundary problem; II, Biogenic magnetite; its role in the magnetization of sediments and as the basis of magnetic field detection in animals. D, 1979, Princeton University. 277 p.

Kirst, Paul. Petrology of metamorphic rocks from the equatorial Mid-Atlantic Ridge and fracture zone. D, 1976, University of Miami. 458 p.

Kirst, Paul William. Petrology and structure of Precambrian crystalline rocks of the southern half of the Rustic Quadrangle, Mummy Range, Colorado. M, 1967, SUNY at Buffalo.

Kirst, Timothy L. Authigenic cements related to the geopressured zone of the Manchester Field, Louisiana. M, 1977, Louisiana State University.

Kirstein, Dewey S., Jr. Some crystalline rocks north of Chapel Hill, North Carolina. M, 1957, University of North Carolina, Chapel Hill. 26 p.

Kirtley, David. Intertidal reefs of Sabellariidae (Annelida, Polychaeta) along the coasts of Florida. M, 1966, Florida State University.

Kirtley, David W. Geological significance of the polychaetous annelid family Sabellariidae. D, 1974, Florida State University. 322 p.

Kirumakki, Nagaraja Subraya. Stratigraphy, petrography and depositional environment of Greenhorn Formation (Upper Cretaceous), eastern Nebraska. M, 1976, University of Nebraska, Lincoln.

Kirwan, Laura. Inventory and assessment of salt water produced in conjunction with oil and gas for the Little Kanawha River basin of West Virginia. M, 1987, West Virginia University. 190 p.

Kirwin, Peter H. Subsurface stratigraphy of the upper Keweenawan red beds in southeastern Minnesota. M, 1963, University of Minnesota, Minneapolis.

Kiser, Nancy Louise. Stratigraphy, structure and metamorphism in the Hinkley Hills, Barstow, California. M, 1981, Stanford University. 70 p.

Kish, Stephen. A structural and metamorphic history of the northern terminus of the Murphy Belt. M, 1974, Florida State University.

Kish, Stephen Alexander. A geochronological study of deformation and metamorphism in the Blue Ridge and Piedmont of the Carolinas. D, 1983, University of North Carolina, Chapel Hill. 220 p.

Kishbaugh, James Wilbur. The geology and its relationship to the present landscape of Luzerne and surrounding counties. M, 1976, Pennsylvania State University, University Park. 38 p.

Kishida, Augusto. Hydrothermal alteration zoning and gold concentration at the Kerr-Addison Mine, Ontario, Canada. D, 1984, University of Western Ontario.

Kishk, Fawzy Mohamed. Chemical and physical properties of soil vermiculite clays as related to their origin. D, 1967, University of California, Berkeley. 185 p.

Kisling, D. C. Geology of the Antelope Ridge area, Fremont and Hot Springs counties, Wyoming. M, 1962, University of Wyoming. 69 p.

Kiss, Leslie. Geology of the Hart-Jaune River area (Quebec). D, 1965, Universite Laval. 78 p.

Kissick, Brian J. Structural geology of the Mosca Creek area, Alamosa County, Colorado. M, 1988, Wichita State University.

Kissin, S. A. Phase relations in a portion of the Fe-S system. D, 1974, University of Toronto.

Kissin, Stephen A. An investigation of the solubility of indium in hydrothermally synthesized galena, (PbS) and sphalerite (B-ZnS). M, 1968, Pennsylvania State University, University Park. 119 p.

Kissling, Don L. Lower Osagian (Mississippian) stratigraphy of east-central Missouri and adjacent Illinois. M, 1960, University of Wisconsin-Madison.

Kissling, Don Lester. Environmental history of lower Chesterian rocks in southwestern Indiana. D, 1967, Indiana University, Bloomington. 367 p.

Kissling, Randall Douglas. Oxide mineral petrogenesis in the evolution of the Fayette County kimberlite, Pennsylvania. M, 1981, University of Tennessee, Knoxville. 63 p.

Kisslinger, Carl. The effect of variations in chemical composition on the velocity of seismic waves in carbonate rocks. D, 1952, St. Louis University.

Kisslinger, Carl. The velocity of elastic waves in some Paleozoic formations in the vicinity of St. Louis, Missouri. M, 1949, St. Louis University.

Kister, Tom L. A geologic study of artesian thermal mineral water with emphasis on the Texas Gulf Coastal Plain. M, 1950, [University of Houston].

Kistler, Barbara R. The effect of sediment type and flooding on the channel characteristics of Hutchins Creek-Clear Creek, Union County, Illinois. M, 1983, Southern Illinois University, Carbondale. 80 p.

Kistler, James O. Regional stratigraphic analysis of the Simpson Group and equivalents, central United States. M, 1952, Northwestern University.

Kistler, R. B. Investigation of the ores of Morococha, Peru. M, 1958, Stanford University.

Kistler, Ronald Wayne. The geology of the Mono Craters Quadrangle, California. D, 1960, University of California, Berkeley. 133 p.

Kistner, David John. Fracture study of a volcanic lithocap, Red Mountain porphyry copper prospect, Santa Cruz County, Arizona. M, 1984, University of Arizona. 75 p.

Kistner, Frank B. Stratigraphy of the Bridger Formation in the Big Island-Blue Rim area, Sweetwater County, Wyoming. M, 1973, University of Wyoming. 174 p.

Kisucky, Michael J. Sedimentology, stratigraphy and paleogeography of the Mesa Rica Sandstone, Tucumcari Basin, east-central New Mexico. M, 1987, University of New Mexico. 173 p.

Kisvarsanyi, Eva B. Comparative petrographic, petrochemical and spectrographic analyses of the Precambrian granitic rocks of southeastern Missouri. M, 1960, University of Missouri, Rolla.

Kisvarsanyi, Geza. Geochemical and petrological study of the Precambrian iron metallogenic province of southeast Missouri. D, 1966, University of Missouri, Rolla. 224 p.

Kitani, Osamu. Stress-strain relationships for soil with variable lateral strain. D, 1966, Michigan State University. 97 p.

Kitchell, Jennifer A. Analysis and paleoecologic implications of Arctic and Antarctic deep-sea biogenic traces. D, 1978, University of Wisconsin-Madison. 271 p.

Kitchen, Earl William. Mississippian rocks in the subsurface of the Bartlesville area, Oklahoma. M, 1963, University of Oklahoma. 88 p.

Kitchen, J. C. Particle size distribution and the vertical distribution of suspended matter in the upwelling region off Oregon. M, 1978, Oregon State University. 118 p.

Kitchen, Lisa. Nitrate-N profiles of fine to medium textured sediments of the unsaturated zone of Southeast and south-central Nebraska. M, 1987, University of Nebraska, Lincoln. 134 p.

Kitchen, Mark R. A magnetotelluric transect of the Oregon Coast Range. M, 1988, Michigan Technological University. 112 p.

Kite, James Steven. Late Quaternary glacial, lacustrine, and alluvial geology of the upper St. John River basin, northern Maine and adjacent Canada. D, 1983, University of Wisconsin-Madison. 362 p.

Kite, James Steven. Postglacial geologic history of the middle St. John River valley. M, 1979, University of Maine. 136 p.

Kite, Lucille Eggborn. The Halifax County Complex; oceanic lithosphere in the northeastern Piedmont, North Carolina. M, 1982, North Carolina State University. 102 p.

Kite, William McDougall. Caldera-forming eruption sequences and facies variations in the Bandelier Tuff, central New Mexico. M, 1985, Arizona State University. 377 p.

Kiteley, Louise W. Facies analysis of the lower cycles of the Mesaverde Group (Upper Cretaceous) in northwestern Colorado. M, 1980, University of Colorado.

Kittelson, Roger. The amphibolite-to-granulite facies transition in the Franklin and Corbin Knob quadrangles, North Carolina Blue Ridge. M, 1988, University of Tennessee, Knoxville. 111 p.

Kittleman, Laurence Roy, Jr. Geology of the Owyhee Reservoir area, Oregon. D, 1962, University of Oregon. 174 p.

Kittleman, Lawrence. Post-Laramie sediments of the Denver-Colorado Springs region, Colorado. M, 1956, University of Colorado.

Kittleson, Kendell Lloyd. A gravity study of the Osborne magnetic anomaly, Clayton County, Iowa. M, 1975, University of Iowa. 105 p.

Kittredge, Tylor F. Formation of wave-ogives below the icefall on the Vaughan Lewis Glacier, Alaska. M, 1967, University of Colorado.

Kitz, Mary Beth. Petrology of a Lower Cretaceous (Neocomian) conglomerate sequence of the San Marcos Formation, Sierra Mojada, west-central Coahuila, Mexico. M, 1984, University of Texas, Arlington. 100 p.

Kitzmiller, John Michael, III. The geology of the Joes Valley Reservoir Quadrangle, Sanpete and Emery counties, Utah. M, 1981, Brigham Young University.

Kiureghian, Ahmen Der see Der Kiureghian, Ahmen

Kiven, Charles Wilkinson. Kinematics of deformation at the southwest corner of the Monument Uplift. M, 1976, University of Arizona.

Kiver, Eugene P. Geomorphology and glacial geology of the southern Medicine Bow Mountains, Colorado and Wyoming. D, 1968, University of Wyoming. 129 p.

Kivi, W. J. The stratigraphy of the Phosphoria Formation of western Wyoming, with notes on the occurrence of phosphate. M, 1940, University of Wyoming. 59 p.

Kivioja, Lassi Antti. The effect of topography and its isostatic compensation on free air gravity anomalies. D, 1963, Ohio State University. 146 p.

Kizis, Joseph Anthony, Jr. Mobilization of uranium, molybdenum, lithium, and fluorine in the Bates Mountain Tuff, central Nevada. M, 1979, University of Colorado.

Kjartansson, E. Attenuation of seismic waves in rocks and applications in energy exploration. D, 1980, Stanford University. 153 p.

Kjelleren, Gary Palmer. The source and deglacial implications of the Hollow Brook Delta; Champlain Valley, Vermont. M, 1984, University of Connecticut. 80 p.

Kjos, Einar Jarle. The feasibility of imaging near-vertical incidence lower crustal and upper mantle reflectors from seismic refraction data. M, 1988, University of Utah. 82 p.

Kladivko, Eileen Joyce. Soil N mineralization as affected by abiotic factors and tillage system. D, 1982, University of Wisconsin-Madison. 228 p.

Klaer, Fred H. The Peerless and related pegmatites, Keystone, South Dakota. M, 1937, Northwestern University.

Klanderman, David S. Stratigraphy, structure, and depositional environments of the Antelope Mountain Quartzite, Yreka, California. M, 1978, Oregon State University. 115 p.

Klanke, John Emil. An analysis of nitrate concentration in ground-water of Antwerp Township in Van Buren County, Michigan. M, 1981, Western Michigan University.

Klapheke, Jeffrey G. Pyrolysis and placticity of high-volatile bituminous coals of Kentucky. M, 1984, University of Kentucky. 84 p.

Klapper, Gilbert J. A late Devonian conodont fauna from the Darby Formation of the Wind River Mountains, Wyoming. M, 1958, University of Kansas. 30 p.

Klapper, Gilbert John. Upper Devonian and Lower Mississippian conodont zones in Montana, Wyoming, and South Dakota. D, 1962, University of Iowa. 203 p.

Klar, G. Geochronology of the El Manteco-Guri and Guasipati areas, Venezuelan Guiana Shield. D, 1979, Case Western Reserve University. 177 p.

Klara, Eugene W. Diagenetic textures and fabrics of the Crown Point Limestone (Chazy Group), lower middle Ordovician of northeastern New York State. M, 1977, Rensselaer Polytechnic Institute. 98 p.

Klare, Matthew William. Comparison of quantitative techniques for vegetational analysis in a Middle-Pennsylvanian coal. D, 1987, University of Iowa. 263 p.

Klas, Mieczyslawa. Diatoms and clay mineralogy in late Quaternary sediment from the Orca Basin. M, 1986, Rutgers, The State University, Newark. 237 p.

Klasik, John A. The silicification of Eocene deep-sea sediments in the Atlantic Ocean. D, 1976, Louisiana State University. 215 p.

Klasik, John Arthur. Sedimentation, under the influence of contour currents, on the middle Continental Rise, between the Hatteras Canyon system and the Blake Outer Ridge (Atlantic Ocean). M, 1972, Duke University. 95 p.

Klasner, John S. A study of buried bedrock valleys near South Haven, Michigan, by the gravity method. M, 1964, Michigan State University. 40 p.

Klass, Marcia Jean. Diagenesis and development of secondary porosity in the Vicksburg Sandstone, Mc-Allen Ranch Field, Hidalgo County, Texas. M, 1981, Texas A&M University.

Klassen, Rodney Alan. Geochemistry of lakes and lake sediments in the Kaminak Lake area, District of Keewatin, N.W.T. M, 1975, Queen's University. 209 p.

Klassen, Rodney Alan. Quaternary stratigraphy and glacial history of Bylot Island, N.W.T., Canada. D, 1982, University of Illinois, Urbana. 183 p.

Klassen, Rudolph Waldemar. A photogeologic study of selected group moraine areas; surface features and their significance. M, 1960, University of Alberta. 99 p.

Klassen, Rudolph Waldemar. The surficial geology of the Riding Mountain area, Manitoba-Saskatchewan. D, 1966, University of Saskatchewan. 240 p.

Klauk, Robert H. Stratigraphic and engineering study of the Lake Michigan shore-zone bluff in Milwaukee County, Wisconsin. M, 1978, University of Wisconsin-Milwaukee.

Klausing, Robert L. Basal Marmaton conodonts from Missouri. M, 1957, University of Missouri, Columbia.

Klauss, Thomas E. Petrographic variation along the Portneuf Valley lava flow. M, 1969, Idaho State University. 108 p.

Klavans, A. S. Petrology of the New Market Formation exposed along Interstate 70, Washington County, Maryland. M, 1974, George Washington University.

Klavaren, Richard William Van see Van Klavaren, Richard William

Klazynski, Ralph J. The geology and petrology of the Bluff Springs Pluton, Clay and Randolph counties, Alabama. M, 1977, Memphis State University.

Klebold, Thomas E. Remanent magnetism of sediments from the Cedarburg Bog. M, 1979, University of Wisconsin-Milwaukee. 58 p.

Kleck, Wallace D. The geology of some zeolite deposits in the southern Willamette Valley, Oregon. M, 1960, University of Oregon. 108 p.

Kleck, Wallace Dean. Chemistry, petrography, and stratigraphy of the Columbia River Group in the Imnaha River valley region, eastern Oregon and western Idaho. D, 1976, Washington State University. 203 p.

Klecker, Richard A. Stratigraphy and structure of the Dixon Mountain-Little Water Canyon area, Beaverhead County, Montana. M, 1981, Oregon State University. 223 p.

Kleesattel, David R. Petrology of the Beulah-Zap lignite bed, Sentinel Butte Formation (Paleocene), Mercer County, North Dakota. M, 1985, University of North Dakota. 188 p.

Kleffner, Mark Alan. Taxonomy and biostratigraphic significance of Wenlockian and Ludlovian (Silurian) conodonts in the Midcontinent outcrop area, North America. D, 1988, Ohio State University. 268 p.

Klefstad, Gilbert Eugene. Limitations of the electrical resistivity method for detecting landfill leachate in alluvial deposits. M, 1973, Iowa State University of Science and Technology.

Kleihege, Bernard W. Metamorphism of Paleozoic sediments in Val Verde and Terrell counties, Texas. M, 1949, University of Kansas. 59 p.

Kleijn, Willem Bastiaan. The physical chemistry of clays and oxides. D, 1981, University of California, Riverside. 174 p.

Kleiman, Laura Elena. An oxygen isotope and mineralogical study of host rocks, Midwest uranium deposit, northern Saskatchewan. M, 1982, University of Saskatchewan. 112 p.

Klein, Charles A. Attachment and epizoans of Cystiphylloides sp. in the "Spinatrypa Spinosa bed" of the Windom Shale of western New York. M, 1981, SUNY at Buffalo. 64 p.

Klein, Christopher William. Structure and petrology of a southeastern portion of the Happy Camp Quadrangle, Siskiyou County, Northwest California. D, 1975, Harvard University. 336 p.

Klein, Cornelis, Jr. Mineralogy and petrology of the Wabush Iron Formation, Labrador City area, Newfoundland. D, 1964, Harvard University.

Klein, Cornelius. Detailed study of the amphiboles and associated minerals in the Wabush Iron Formation, Labrador. M, 1960, McGill University.

Klein, D. P. Geomagnetic time-variations, the island effect, and electromagnetic depth sounding on oceanic islands; results from the analysis of data obtained in the frequency range of 0.5 to 10 cycles per hour on Oahu, Hawaii. M, 1972, University of Hawaii. 96 p.

Klein, D. P. Magnetic variations (2-30 cpd) on Hawaii Island and mantle electrical conductivity. D, 1976, University of Hawaii. 93 p.

Klein, Frederick W. Tidal triggering of earthquake swarms and the Reykjanes Peninsula, Iceland earthquake swarm of September 1972 and its tectonic and geothermal implications. D, 1976, Columbia University. 185 p.

Klein, G. E. The crystal structure of nepheline. M, 1947, Massachusetts Institute of Technology. 121 p.

Klein, George DeVries. Stratigraphy, sedimentary petrology and structures of Triassic sedimentary rocks, Maritime Provinces, Canada. D, 1960, Yale University.

Klein, George deVries. The geology of the Acadian Triassic in the type area, northeastern Annapolis-Cornwallis Valley, Kings County, Nova Scotia. M, 1957, University of Kansas. 143 p.

Klein, Helen M. The Beaver Creek Limestone sediment gravity flow fan deposits, Fort Payne Formation, Lower Mississippian (Osagean), south-central Kentucky, USA. M, 1974, University of Cincinnati. 121 p.

Klein, Ira. Lime-silicate rocks in a serpentine from the Santa Cruz Mountains. M, 1946, Stanford University.

Klein, Jack Jay, Jr. Paleoecology of the Aguja Formation, Brewster County, Texas. M, 1986, Stephen F. Austin State University. 119 p.

Klein, James David. A laboratory investigation of the nonlinear impedance of mineral-electrolyte interfaces. M, 1977, University of Utah. 93 p.

Klein, James David. Specific mineral identification in electrochemical logging of drill holes. D, 1980, University of Utah. 248 p.

Klein, John M. Geochemical behavior of silica in the artesian ground water of of the Closed Basin area, San Luis Valley, Colorado. M, 1971, Colorado School of Mines. 121 p.

Klein, John P. Stratigraphy, petrology, and depositional setting of Recent peat deposits, Yukon Delta, Alaska. M, 1980, University of Houston.

Klein, Kenneth Paul. The lithofacies and biofacies of the Formosa Reef Limestone (Eifelian) in Bruce and Huron counties of southwestern Ontario. M, 1980, University of Windsor. 233 p.

Klein, Philip Alan. Terracette morphology and soil characteristics, Santa Cruz Island, California. M, 1987, University of California, Los Angeles. 125 p.

Klein, Roger W. Subsurface structure of the Upper Devonian rocks in the Knapp Creek Quadrangle. M, 1982, Rensselaer Polytechnic Institute. 80 p.

Klein, Terence L. Geology and mineral deposits of the Silver Crown mining district, Laramie County, Wyoming. M, 1973, Colorado State University. 141 p.

Klein, Terry L. The geology and geochemistry of the sulfide deposits of the Seminoe District, Carbon

County, Wyoming. D, 1980, Colorado School of Mines. 232 p.

Kleinhampl, Frank Joseph. Sedimentary processes along the Lake Erie shore from 4 miles east of Lorain, Lorain County, to Huntington Beach Park, Cuyahoga County, Ohio. M, 1952, Ohio State University.

Kleinkopf, Merlin Dean. Trace element exploration of Maine lake water. D, 1955, Columbia University, Teachers College.

Kleinmann, Robert L. P. The biogeochemistry of acid mine drainage and a method to control acid formation. D, 1979, Princeton University. 104 p.

Kleinpell, Robert M. Miocene foraminifera from Reliz Canyon, Monterey County, California. D, 1934, Stanford University. 352 p.

Kleinpell, Robert Minssen. Miocene foraminifera from the Salinas Valley, California. M, 1928, Stanford University. 86 p.

Kleinrock, Martin Charles. Detailed structural studies of the propagator system near 95.5°W along the Galapagos spreading axis. D, 1988, University of California, San Diego. 334 p.

Kleinschmidt, John C. Bedrock petrology of at the Horse and Wall Creek drainages, Gravelly Range, Madison County, Montana. M, 1981, Montana College of Mineral Science & Technology. 73 p.

Kleinspehn, Karen Lee. Cretaceous sedimentation and tectonics, Tyaughton-Methow Basin, southwestern British Columbia. D, 1982, Princeton University. 184 p.

Kleist, John Raymond. Geology of the Coastal Belt, Franciscan Complex, near Ft. Bragg, California. D, 1974, University of Texas, Austin.

Kleist, John Raymond. The Denali Fault in the Canwell Glacier area, east-central Alaska Range. M, 1971, University of Wisconsin-Madison.

Kleist, Ronald J. Paleomagnetic study of the Lower Cretaceous Cupido Limestone, Coahuila, Mexico. M, 1980, [University of Houston].

Kleiter, Kathryn Jean. Paleochannel of the Tertiary White River Group near Fairburn, South Dakota. M, 1988, South Dakota School of Mines & Technology.

Klekamp, C. Thomas. Petrology and paleocurrents of the Ste. Genevieve Member of the Newman Limestone (Mississippian) in Carter County, northeastern Kentucky. M, 1971, University of Cincinnati. 38 p.

Klemic, Harry. A study of the uranium deposit near Mauch Chunk, Pennsylvania. M, 1952, Pennsylvania State University, University Park. 52 p.

Klemme, Daniel N. The Pre-Cambrian of Little Falls (New York); petrographic and structural studies of the gneiss in relation to a theory of origin. M, 1942, Syracuse University.

Klemme, Hugh Douglas. The geology of Sixteen Mile Creek area, Montana. D, 1949, Princeton University. 227 p.

Klemme, Michael. Economic considerations in underground mining equipment selection. M, 1982, University of Nevada. 183 p.

Klemperer, Simon Louis. The continental lower crust and Moho; studies using COCORP deep seismic reflection profiling. D, 1985, Cornell University. 159 p.

Klenk, Charlotte Dillon. The sedimentology and stratigraphy of the upper Wichita Group (Permian) on the eastern shelf in Baylor County, North-central Texas. M, 1981, University of Texas at Dallas. 185 p.

Klepacki, David Walter. Stratigraphy and structural geology of the Goat Range area. D, 1987, Massachusetts Institute of Technology. 268 p.

Klepper, Montis Ruhl. Geology of the southern Elkhorn Mountains, Jefferson and Broadwater counties, Montana. D, 1950, Yale University. 199 p.

Klepser, Harry John. A stratigraphic and paleontologic study of the (Devonian) Portage Group of Eighteen Mile Creek and vicinity (New York). M, 1933, Syracuse University.

Klepser, Harry John. The Lower Mississippian rocks of the eastern Highland Rim. D, 1937, Ohio State University.

Klett, William Young. The geology of the Talmo area, Jackson and Hall counties, Georgia. M, 1969, University of Georgia. 66 p.

Klewchuk, Peter. Mineralogy and petrology of some granitic rocks in the Canadian Shield north of Fort Chipewyan, Alberta. M, 1972, University of Calgary. 158 p.

Kleweno, Walter P. Permian stratigraphy of northern Utah, southeastern Idaho and southwestern Wyoming. M, 1958, Washington State University. 200 p.

Klewin, Kenneth Wade. Petrology of metamorphosed ultramafic rocks in the contact aureole of the Kiglapait Intrusion. M, 1982, Northern Illinois University. 140 p.

Klewin, Kenneth Wade. The petrology and geochemistry of the Keweenawan Potato River intrusion, northern Wisconsin. D, 1987, Northern Illinois University. 357 p.

Klich, Ingrid. Precambrian geology of the Elk Mountain-Spring Mountain area, San Miguel County, New Mexico. M, 1983, New Mexico Institute of Mining and Technology. 147 p.

Kligfield, R. Continental margin deformation in the northern Apennines (Italy); a structural study in the Alpi Apuane region. D, 1978, Columbia University, Teachers College. 214 p.

Klikoff, Waldemar A., Jr. Pressure transients in porous media with "dead-end" pore volume. M, 1960, University of California, Berkeley. 41 p.

Klim, D. G. Interactions between sea water and coral reefs in Kaneohe Bay, Oahu, Hawaii. M, 1969, University of Hawaii. 55 p.

Klima, Walter Francis, Jr. The kinetics of crystal growth of calcium sulfate dihydrate. D, 1983, SUNY at Buffalo. 330 p.

Klimberg, David M. The effect of tailings solutions on the stability and seepage characteristics of earth dams and ponds. M, 1981, University of Nevada. 130 p.

Klimentidis, R. Crystal chemistry and crystallography of anthophyllite; possible causes for fibrosity. M, 1980, Queens College (CUNY). 67 p.

Klimetz, Michael P. The pre-Tertiary geology and Mesozoic tectonic evolution of eastern China, Southeast Asia and adjacent regions. M, 1983, SUNY at Albany. 216 p.

Klimkiewicz, George C. Earthquake ground motion attenuation models for the northeastern United States. M, 1980, Boston College.

Klimowski, Richard Joseph. Development and application of a high-resolution mass spectrometer-computer system. D, 1969, Cornell University. 160 p.

Klimstra, Richard Kent. Petrographic and modal analysis of the Silvermine Granite, southeast Missouri. M, 1964, Southern Illinois University, Carbondale. 62 p.

Kline, Beryl Dale. The Hardinsburg Formation in Knox County, Indiana. M, 1952, Indiana University, Bloomington. 22 p.

Kline, Charles C. Gravity survey of Brevard zone in northwestern North Carolina. M, 1971, University of North Carolina, Chapel Hill. 31 p.

Kline, Gary L. Proterozoic budding bacteria from Australia and Canada. M, 1975, University of California, Santa Barbara.

Kline, Gheretein. A study of the distribution of the interstitial fauna of three beaches. M, 1969, [University of Miami].

Kline, James E. Pre-Cambrian rocks in the Chester-Califon area. M, 1957, Rutgers, The State University, New Brunswick. 56 p.

Kline, Jerry Robert. An investigation of copper and other selected elements in soils utilizing gamma radiations induced by thermal neutron activation. D, 1964, University of Minnesota, Minneapolis. 199 p.

Kline, John H. Natural and man-made factors that influence property damage due to swelling soils in

Southeast Jefferson County, Colorado. M, 1983, Colorado School of Mines. 182 p.

Kline, Mary-Cornelia. The thorium and uranium content of the Enchanted Rock Batholith. M, 1960, Rice University. 60 p.

Kline, Mortimer A., Jr. The structure and stratigraphy of Cretaceous rocks in northeastern Larimer County, Colorado. M, 1956, Colorado School of Mines. 123 p.

Kline, Robert J. Discrimination of geologic materials by thermal inertia mapping within the Raft River test site, Cassia County, Idaho. M, 1977, Colorado School of Mines. 165 p.

Kline, Stephen Warren. Metamorphic mineral chemistry, petrology, and sulfide mineralogy of the Dore Lake Complex, Chibougamau, Quebec. D, 1984, University of Georgia. 169 p.

Kline, Virginia Harriet. Revision of Alexander Winchell's types of brachiopods from the Middle Devonian Traverse Group of rocks of Michigan. M, 1933, University of Michigan.

Kline, Virginia Harriet. Stratigraphy and paleontology of the (Devonian) Silica Formation of southeastern Michigan. D, 1935, University of Michigan.

Kling, Donald Lee. Geology of the Wise Flat Quadrangle, Fremont County, Wyoming. M, 1962, University of Missouri, Columbia.

Kling, John F. Petrology of the Strong Creek Prospect, Albany Co., Wyoming, and in the system; Fe-Ti-Si-O-S. M, 1986, University of Wyoming. 109 p.

Kling, Stanley A. Permian fusulinids from Guatemala. M, 1959, Columbia University, Teachers College.

Kling, Stanley Arba. Castanellid and circoporid radiolarians; systematics and zoogeography in the eastern north Pacific. D, 1967, University of California, San Diego.

Klinge, David Michael. Age, chemistry, and petrography of the Wilton Quarry quartz monzonite. M, 1977, [University of North Carolina, Chapel Hill].

Klingel, Eric John. Mineral distributions and uranium rolls, North Walker and Sullivan mines, Shirley Basin, Wyoming. M, 1979, University of Akron. 57 p.

Klinger, Frederick Lindsley. Andalusite-corundum mineralization near Hawthorne, Nevada. M, 1952, University of Wisconsin-Madison.

Klinger, Frederick Lindsley. Geology and ore deposits of the Soudan Mine, Saint Louis County, Minnesota. D, 1960, University of Wisconsin-Madison. 117 p.

Klinger, Lee Francis. Successional change in vegetation and soils of Southeast Alaska. D, 1988, University of Colorado. 251 p.

Klingman, Darrell S. Depositional environments and paleogeographic setting of the Middle Mississippian section in eastern California. M, 1987, San Jose State University. 231 p.

Klingmueller, Lothar M. L. The recognition of inliers in the Wasatch Formation (Paleocene and Eocene) in parts of Rich County, Utah. M, 1967, University of Arizona.

Klingmueller, Lothar Max Ludwig. Geology of the eastern part of the Burr diapir, northern Flinders ranges, South Australia. D, 1971, University of Arizona. 302 p.

Klingsberg, Cyrus. A partial study of certain gravels near Phila., Pa. M, 1949, Bryn Mawr College.

Klingsberg, Cyrus. The system MnO-OH. D, 1958, Pennsylvania State University, University Park. 154 p.

Klinkenberg, Brian. Tests of a fractal model of topography. D, 1988, University of Western Ontario.

Klinkhammer, G. P. The distributions of manganese in the Pacific Ocean and some trace metals in pelagic pore waters. D, 1979, University of Rhode Island. 199 p.

Klins, Mark Albert. Numerical simulation of the immiscible carbon dioxide injection process. D, 1980, Pennsylvania State University, University Park. 239 p.

Klinzing, Susan L. The Labelle Clay of the Tamiami Formation. M, 1987, Florida State University. 80 p.

Klipfel, Paul Dexter. Geology of an area near Mt. Ogilvie northern Boundary Range, Juneau Icefield, Alaska. M, 1981, University of Idaho. 144 p.

Klise, David. Modern sedimentation on the California continental margin adjacent to the Russian River. M, 1984, San Jose State University. 120 p.

Klitgord, Kim D. Near-bottom geophysical surveys and their implications on the crustal generation process, sea floor spreading history on the Pacific and the geomagnetic time scales; 0 to 6 m.y.b.p. D, 1974, University of California, San Diego. 192 p.

Klobcar, Cheryl Louise. Petrogenesis of the Granite Harbour Intrusive at the Emlen Peaks, northern Victoria Land, Antarctica. M, 1984, Arizona State University. 139 p.

Kloc, Gerald J. Stratigraphic distribution of ammonoids from the Middle Devonian Ludlowville Formation in New York. M, 1983, SUNY at Buffalo. 78 p.

Klockenbrink, Thomas L. Depositional environments and petrology of the lower member of the Morrison Formation, Henry Mountains region, Utah. M, 1979, Northern Arizona University. 143 p.

Kloepfer, John Gerard. Viscous fingering in unconsolidated cores. M, 1975, University of Alberta. 91 p.

Klohn, Melvin Larry. Geology of the north-central part of the Coos Bay Quadrangle, Oregon. M, 1967, University of Oregon. 59 p.

Klonowski, John E. The water budget of Flanders Bay, N.Y. M, 1979, New York University.

Klonsky, Louis Farrell. A preliminary study of the origin of the physiographic boundary between the Nicaraguan Rise and Colombian Basin, Caribbean Sea. M, 1977, Rutgers, The State University, New Brunswick. 33 p.

Kloosterman, Bruce. Using a computer soil data file in the development of statistical techniques for the evaluation of soil suitability for land use. D, 1971, University of British Columbia.

Klopp, Helen C. Petrographic analysis of the Sunniland Formation, an oil-producing formation in South Florida. M, 1975, Brigham Young University.

Klosterman, Gregory. Structural relationship of certain gravels near Philadelphia, Pennsylvania. M, 1955, Bryn Mawr College.

Klosterman, Gregory Elmer. Structural relationship of mineral deposits of the Colorado Plateau. M, 1955, University of Michigan.

Klosterman, Keith Edward. An evaluation of the E.R.T.S. I system imagery in a structural study and a map application of E.R.T.S. I imagery in southeastern Washington, northeastern Oregon and parts of western Idaho. M, 1974, University of Washington. 84 p.

Klosterman, Michael Joseph. Structural analysis of olivine in pallasitic meteorites. M, 1971, Arizona State University. 61 p.

Klotz, Jack A. Nature and origin of the Maumee River terraces, northwestern Ohio. M, 1981, Bowling Green State University. 51 p.

Klouda, G. A. An investigation of the geochemical uniformity of an ice sheet. M, 1977, SUNY at Buffalo. 64 p.

Klovan, John E. Facies analysis of (Devonian) Redwater reef complex, Alberta, Canada. D, 1963, Columbia University, Teachers College.

Klovan, John E. Some Triassic terebratulacean brachiopods from northeastern British Columbia, Canada. M, 1958, Columbia University, Teachers College.

Klucking, Edward Paul. An Oligocene flora from the western Cascades. D, 1960, University of California, Berkeley.

Klucking, Edward Paul. The fossil Betulaceae of western North America. M, 1959, University of California, Berkeley. 170 p.

Klug, Curtis Robert. Conodonts and biostratigraphy of the Muscatatuck Group (Middle Devonian), south-

central Indiana and north-central Kentucky. M, 1983, University of Iowa. 75 p.

Kluger, Karen Lee. Paleomagnetic study of red beds from the Triassic Newark-Gettysburg basin; chemical and thermal demagnetization techniques and magnetic stratigraphy. M, 1977, Lehigh University. 263 p.

Klugman, Michael Anthony. An investigation of the origin and composition of sand and gravel deposits in the vicinity of certain Monteregian Hills of southern Quebec. M, 1953, McGill University.

Klugman, Michael Anthony. The geology of an area between Pigou and Sheldrake rivers, Saguenay County, Quebec, with a detailed study of the anorthosites. D, 1956, McGill University.

Kluit, Dirk Jacob Tempelman see Tempelman Kluit, Dirk Jacob

Klump, Jeffrey Val. Benthic nutrient regeneration and the mechanisms of chemical sediment-water exchange in an organic-rich coastal marine sediment. D, 1980, University of North Carolina, Chapel Hill. 160 p.

Klusman, Ronald William. Electron microprobe analysis of feldspars. D, 1969, Indiana University, Bloomington. 270 p.

Kluth, Charles F. Geology of the Elden Mountain area, Coconino County, Arizona. M, 1974, Northern Arizona University. 89 p.

Kluth, Charles Frederick. The geology and mid-Mesozoic tectonics of the northern Canelo Hills, Santa Cruz County, Arizona. D, 1982, University of Arizona. 364 p.

Kluth, Mary Jo Ann Morgan. Engineering geology of the South Carter Lake Anticline area. M, 1976, Colorado State University. 111 p.

Kluyver, Huybert M. Lower Paleozoic tectono-stratigraphy of the northern Appalachians and the Caledonides. D, 1971, Queen's University. 140 p.

Kmiecik, Jerome Gregory. Geology of the Loyal Valley, west area, Mason County, Texas. M, 1962, Texas A&M University.

Kmiecik, Jerome Gregory. Investigation of iron-silicate minerals in selected samples from the Weches Formation, Texas. D, 1964, Texas A&M University. 76 p.

Knaack, Edward Leslie. The origin of certain structures of the Minnekahta Formation in the Whitewood region, northern Black Hills, South Dakota. M, 1936, University of Iowa. 35 p.

Knabe, Robert George. Geology of Totuma Anticline, northeastern Venezuela. M, 1954, University of Texas, Austin.

Knadle, Marcia E. Petrology of the sandstones in the middle and upper members of the upper Precambrian(?) Deep Spring Formation, White-Inyo mountains, California. M, 1981, University of Montana. 99 p.

Knaebel, Carl Henry. Observation and deduction applicable to measurement of electrical resistivity of large volumes of earth in place. M, 1931, Michigan Technological University. 35 p.

Knaffle, Leonard Ludwig. Origin, nature, and character of salt dome cap rock. M, 1950, University of Michigan.

Knapik, James. Geology and minability of the upper Freeport Coal, southeastern Allegheny County, Pennsylvania. M, 1981, University of Pittsburgh.

Knapik, Leonard Joseph. Alpine soils of the Sunshine area in the Canadian Rocky Mountains. M, 1973, University of Alberta. 231 p.

Knapp, Caroline J. Late Ordovician solitary rugose corals of the Beaverfoot Formation, Southern Rocky Mountains, British Columbia and Alberta. M, 1985, University of Manitoba.

Knapp, Crawford. Carnotite in the southern Black Hills (South Dakota). M, 1957, South Dakota School of Mines & Technology.

Knapp, D. A. Structure and stratigraphy of the lower Paleozoic section along the east limb of the Berkshire

Massif, southwestern Massachusetts. M, 1977, University of Vermont.

Knapp, Douglas Alan. Ophiolite emplacement along the Baie Verte-Brompton Line at Glover Island, western Newfoundland. D, 1983, Memorial University of Newfoundland. 338 p.

Knapp, Esther Laura. The geography of the floodplain of the lower Missouri River. M, 1923, Washington University. 161 p.

Knapp, George Leroy. A diorite sill in the Lewis and Clark Range, Montana. M, 1963, University of Massachusetts. 63 p.

Knapp, Gregory Anthony. A magnetometer survey near House Springs, Missouri. M, 1955, St. Louis University.

Knapp, John Stafford, Jr. Seismicity, crustal structure, and tectonics near the northern termination of the San Andreas Fault. D, 1982, University of Washington. 316 p.

Knapp, John Stafford, Jr. Velocity changes associated with the Ferndale earthquake. M, 1976, University of Washington. 93 p.

Knapp, R. An analysis of the porosities of fractured crystalline rocks. M, 1975, University of Arizona.

Knapp, R. B. Consequences of heat dispersal from hot plutons. D, 1978, University of Arizona. 129 p.

Knapp, R. R. Depositional environments and diagenesis of the Nugget Sandstone, South-central Wyoming, Northeast Utah and Northwest Colorado. M, 1976, University of Wyoming. 67 p.

Knapp, Ralph William. Ellipticity of 0.4 to 2.4 Hertz Rayleigh waves with application to the study of near surface Earth structure. D, 1977, Indiana University, Bloomington. 180 p.

Knapp, Roy Marvin. The development and field testing of a basin hydrology simulator. D, 1973, University of Kansas.

Knapp, Steven. Gliding flow and recrystallization of halite gouge in experimental shear zones. M, 1983, Texas A&M University. 78 p.

Knapp, Steven A. Petrogenesis of Apollo 14 lunar breccia 14321. M, 1986, University of Tennessee, Knoxville. 118 p.

Knapp, Susan. The stratigraphic utility of Cretaceous small acritarchs. M, 1980, Queens College (CUNY). 102 p.

Knapp, Thomas Stevens. Geologic structure of a small area west of Mill Springs, Kentucky. M, 1932, University of Michigan.

Knapp, Vernon. Structural relations of Capitan and eastern border of Sierra Blanca Mountain groups in Lincoln County, New Mexico. M, 1933, University of Colorado.

Knapp, William Dale. Crinoid fauna from the Burgner Formation (Atokan) in Missouri. M, 1961, University of Missouri, Columbia.

Knapp, William Dale. Mississippian cephalopods of the Eastern Interior United States. D, 1965, University of Iowa. 208 p.

Knappe, Roy, Jr. The micropaleontology of a section of the Tepetate Formation, southern Baja California, and a paleobiogeographic comparison with equivalent foraminifera along the west coast of the United States. M, 1974, Ohio University, Athens. 114 p.

Knappen, Russel S. Geology and mineral resources of the Dixon Quadrangle, Illinois. D, 1926, Columbia University, Teachers College.

Knappen, Russell S. A summary of the present state of knowledge regarding subcrustal fusion. M, 1915, University of Wisconsin-Madison.

Knauer, Larry Craig. Geology of the Emerson Lake Quadrangle, San Bernardino County, California. M, 1982, University of California, Los Angeles. 190 p.

Knaup, William Wade. Sedimentology of the Pahrump Group and older strata, Old Dad Mountain Quadrangle, southeastern California. M, 1977, University of Southern California.

Knauss, K. G. Natural decay series isotopes in surface waters, bottom waters, and plankton from the East Pacific. D, 1976, University of Southern California.

Knauth, LeRoy Paul. Oxygen and hydrogen isotope ratios in cherts and related rocks. D, 1973, California Institute of Technology. 379 p.

Knebel, Harley John. Holocene sedimentary framework of the east-central Bering Sea continental shelf. D, 1972, University of Washington. 196 p.

Knecht, Carl Emil. The volcanics of the Silver Peak Range in Nevada. M, 1900, [Stanford University].

Knecht, Matthew D. Petrography, origin, and structural deformation of bent tubular chert of the Tosi Member, Phosphoria Formation (Upper Permian), in the Gros Ventre Mountains, Teton County, Wyoming. M, 1988, Miami University (Ohio). 147 p.

Knechtel, M. M. Cretaceous Ammonites of the Ellsworth Expedition to northern Peru. D, 1927, The Johns Hopkins University.

Kneiblher, Carolyn Ruth. Seismic refraction surveys of alluvium-filled washes, Yucca Mountain, Nevada. M, 1985, University of Nevada. 112 p.

Kneidel, Eliezer. Octahedral substitution in metamorphic allies. M, 1965, Dartmouth College. 52 p.

Kneisley, George W. The distribution in depth of the minerals of ore and gangue. M, 1908, Columbia University, Teachers College.

Knell, Gregory W. The sedimentology and petrology of the Cambria Coal, Newcastle, Weston County, Wyoming. M, 1985, South Dakota School of Mines & Technology.

Kneller, William Arthur. A geological and economic study of gravel deposits of Washtenaw County and vicinity, Michigan. D, 1964, University of Michigan. 226 p.

Kneller, William Arthur. Petrography and petrology of the southern group of the Rockefeller Mountains, King Edward VII Land, West Antarctica. M, 1955, Miami University (Ohio). 86 p.

Knepper, Daniel H. Tectonic analysis of the Rio Grande rift zone, central Colorado. D, 1974, Colorado School of Mines. 237 p.

Knepper, Daniel H., Jr. Stratigraphy and macroscopic structure of the metasedimentary rocks of the Blue Ridge area, Fremont County, Colorado. M, 1972, University of Kansas. 53 p.

Knetchtel, M. M. Cretaceous Ammonites of the Ellsworth expedition to northern Peru. D, 1927, The Johns Hopkins University.

Knewtson, Steve. Sedimentary of the Ste. Genevieve Limestone (Mississippian, Missouri). M, 1967, University of Missouri, Rolla.

Knibbe, Willem Gerard Johan. Potassium-calcium equilibria in clay fractions of some vertisols. D, 1968, Texas A&M University. 118 p.

Knight, Augustus S., Jr. Stratigraphy of the Carlin Canyon area, Nevada. M, 1954, Columbia University, Teachers College.

Knight, C. J. The geology of the Pater Mine (Blind River area, Ontario). M, 1963, University of Toronto.

Knight, Charles. Origin of pillow lava. D, 1959, University of Chicago. 88 p.

Knight, Cheryl L. Erickson. Critical properties of NaCl-H$_2$O solutions. M, 1988, Virginia Polytechnic Institute and State University.

Knight, Cole D. Pore-fluid chemistry and selected carbonate mudbanks and mangrove-fringed islands, Florida Bay. M, 1988, Wichita State University. 236 p.

Knight, Colin Joseph. A petrographic study of the Spragge Group and description of its correlation with the Sudbury Series (Precambrian), Ontario. M, 1965, University of Toronto.

Knight, Colin Joseph. Rubidium-strontium isochron ages of volcanic rocks on the north shore of Lake Huron, Ontario, Canada. D, 1967, University of Toronto.

Knight, Curtis Alan. Radio source positions through four antenna long baseline interferometry. D, 1979, Massachusetts Institute of Technology. 155 p.

Knight, David Cooper. Stratigraphy and mineralogy of a zinc-rich sillimanite gneiss near Maysville, Chaffee County, Colorado. M, 1981, University of Manitoba. 94 p.

Knight, F. J. Some aspects of diabase intrusion, upper Canyon Creek drainage, Gila and Navajo counties, Arizona. M, 1963, South Dakota School of Mines & Technology.

Knight, Garold L. A quantitative study of the mineral constituents of river waters. D, 1925, University of Wisconsin-Madison.

Knight, Ian. Geology of the Arkose Lake area (Proterozoic), Labrador. M, 1972, Memorial University of Newfoundland. 210 p.

Knight, Ian. Stratigraphy, sedimentology and paleogeography of Mississippian strata of the Bay St. George subbasin, western Newfoundland. D, 1983, Memorial University of Newfoundland. 430 p.

Knight, James A. Differential preservation of calcined bone at the Hirundo Site, Alton, Maine. M, 1985, University of Maine. 111 p.

Knight, James Brook. The St. Louis, Missouri, Pennsylvanian outlier with a detailed study of some of its gastropod fauna. D, 1931, Yale University.

Knight, James Brookes. Some Pennsylvanian ostracodes from the Henrietta Formation of eastern Missouri. M, 1928, University of Kansas. 92 p.

Knight, Jerry Eugene. A thermochemical study of alunite and copper-arsenic sulfosalt deposits. M, 1976, University of Arizona.

Knight, John Bruce. A microprobe study of placer gold and its origin in the lower Fraser River drainage basin, B.C. M, 1985, University of British Columbia. 197 p.

Knight, Jonathan Charles. An investigation of the general limnology of Georgetown Lake, Montana. D, 1981, Montana State University. 151 p.

Knight, Julia Baret. Eolian bedform reconstruction; a case study from the Page Sandstone (Jurassic), northern Arizona. M, 1986, University of Texas, Austin. 100 p.

Knight, Kenneth S. Lithostratigraphy, facies, and petrology of the Atoka Formation (Pennsylvanian), eastern Frontal Belt, Ouachita Mountains, Southeast Oklahoma. M, 1984, University of Arkansas, Fayetteville.

Knight, Kimbell Lee. Stratigraphy, depositional and diagenetic history of three Middle Pennsylvanian cyclothems (Breezy Hill and Fort Scott limestones), Midcontinent North America. D, 1985, University of Iowa. 340 p.

Knight, Lawrence W. Areal geology of the Bloomfield Quadrangle, southern Illinois. M, 1968, Southern Illinois University, Carbondale. 184 p.

Knight, Lester L. A preliminary heavy mineral study of the Ferron Sandstone. M, 1954, Brigham Young University. 31 p.

Knight, Louis Harold, Jr. Structural geology of the Cat Mountain Rhyolite (Tertiary) in the northern Tucson Mountains, Pima County, Arizona. M, 1970, University of Arizona. 251 p.

Knight, Louis Harold, Jr. Structure and mineralization of the Oro Blanco mining district, Santa Cruz County, Arizona. D, 1970, University of Arizona.

Knight, Michael Don. Stratigraphy and anisotropy of magnetic susceptibility of the Toba Ignimbrites, North Sumatra. M, 1985, University of Hawaii. 279 p.

Knight, Michael T. Deltaic sedimentation in the Nacatoch Formation (Late Cretaceous), Northeast Texas. M, 1984, University of Texas, Arlington. 202 p.

Knight, Pauline U. Former drainage of the Northern Appalachians. M, 1933, Columbia University, Teachers College.

Knight, R. John. Sediments, bedforms and hydraulics in a macrotidal environment, Cobequid Bay (Bay of Fundy), Nova Scotia. D, 1977, McMaster University. 693 p.

Knight, Raymond Louis. Permian fusulines from Nevada. M, 1952, University of Southern California.

Knight, Rosemary. Deposition on the pre-Cadomin unconformity surface in North-central Alberta. M, 1978, Queen's University. 93 p.

Knight, Rosemary Jane. The dielectric constant of sandstones, 5 Hz to 13 MHz. D, 1985, Stanford University. 127 p.

Knight, Ross. Sedimentology and stratigraphy of the Neohelikian Elwin Formation, uppermost Bylot Supergroup, Borden Rift basin, northern Baffin Island. M, 1988, Carleton University. 129 p.

Knight, Russell Vincent. Metamorphism of Belt Supergroup calc-silicate rocks in the northwest border zone of the Idaho Batholith. M, 1986, University of Utah. 126 p.

Knight, Samuel Howell. The Fountain and the Casper formations of the Laramie Basin; a study on genesis of sediments. D, 1929, Columbia University, Teachers College.

Knight, Thomas B. Mineralogy, petrology and stratigraphy of the Brewer-Robin Prospect, Pope County, Arkansas. M, 1985, University of Arkansas, Fayetteville.

Knight, Wilbur Clinton. Geology of Wyoming. D, 1901, University of Nebraska, Lincoln.

Knight, Wilbur H. Geology of the western part of Flat Top Anticline, Carbon County, Wyoming. M, 1944, University of Wyoming. 30 p.

Knight, William Victor. The historical and economic geology of Lower Silurian "Clinton" Sandstones of northeastern Ohio. M, 1968, University of Tulsa. 77 p.

Knightly, John Paul. The stratigraphy and sedimentology of the Precambrian Gowganda Formation near Matachewan, Ontario, Canada. M, 1987, Iowa State University of Science and Technology. 94 p.

Knighton, Philip M. The Devonian stratigraphy and structure of the Vincent gas storage field, Humboldt, Webster, and Wright counties, Iowa. M, 1967, Wichita State University. 88 p.

Knights, William Jay. A subsurface study of the Strawn and Atokan Series of the Pennsylvanian System, Southwest Jack County, Texas. M, 1984, Texas Christian University. 53 p.

Kniker, Hedwig Thusnelda. Comanchean and Cretaceous Pectinidae of Texas. M, 1917, University of Texas, Austin.

Knipe, Ralph Ernest. The limestone reservoir rocks of the Panhandle oil fields of Texas. M, 1929, University of Illinois, Urbana.

Knipling, Louis Henry. On the gravimetrical computation of the shape of the Earth. M, 1956, Ohio State University.

Knipling, Louis Henry, Jr. The metric cartographic potential of geostationary geosynchronous satellites. D, 1973, Ohio State University. 273 p.

Knirk, Ernest P. Environment and diagenesis of Salem limestone (Mississippian), Saint Louis County, Missouri. M, 1970, University of Missouri, Columbia.

Knirsch, Karen. Possible heat sources in geothermal waters at Edgemont, South Dakota. M, 1980, South Dakota School of Mines & Technology.

Knisel, Walter Gus, Jr. Response of karst aquifers to recharge. D, 1971, Colorado State University. 170 p.

Knitter, Clifford Charles. Metamorphism and structure of the Soards Creek area, British Columbia. M, 1979, University of Calgary. 152 p.

Knize, S. Marine deep seismic sounding off the coast of British Columbia. D, 1976, University of British Columbia.

Knochenmus, D. D. Alluvial history of the Nishnabotna Valley, southwestern Iowa. M, 1962, Iowa State University of Science and Technology.

Knock, Douglas G. Stratigraphy, petrography and depositional environment of the Kemik sandstones. M, 1987, University of Alaska, Fairbanks.

Knock, Kent K. Boron geochemistry and environmental chemistry. D, 1974, Arizona State University. 212 p.

Knodle, Robert Day. Textural features of water-bearing sands and gravels. M, 1948, University of Illinois, Urbana. 68 p.

Knoll, Andrew Herbert. Studies in Archean and early Proterozoic paleontology. D, 1977, Harvard University.

Knoll, Kenneth Mark. Chronology of alpine glacier stillstands, east-central Lemhi Range, Idaho. D, 1973, University of Kansas. 503 p.

Knoll, Kenneth Mark. Surficial geology of the Tolt River area, Washington. M, 1967, University of Washington. 91 p.

Knoll, Martin Albert. Tertiary basin evolution, eastern Mojave Desert. D, 1988, University of Texas at El Paso. 201 p.

Knoop, John William. The environment of deposition and sedimentation of the Chadron Formation in northwestern Nebraska. M, 1953, University of Nebraska, Lincoln.

Knopf, Adolph. Geology of the Seward Peninsula tin deposits, Alaska. D, 1909, University of California, Berkeley. 71 p.

Knopf, Adolph and Thelen, Paul. Sketch of the geology of Mineral King, California. M, 1905, University of California, Berkeley. 35 p.

Knopman, Debra S. Optimal design of sampling for parameter estimation and discrimination among one-dimensional models of transient solute transport in porous media. D, 1986, The Johns Hopkins University. 326 p.

Knopp, David A. A stratigraphic study of a portion of the lower Trinity Group in north-central Texas. M, 1957, Texas Christian University. 78 p.

Knorr, Jack H. Permian studies of Nevada (Clark, White Pine and Elko counties). M, 1967, University of Iowa. 59 p.

Knoth, Jeff. Analysis of longshore sediment transport on Bull Island, South Carolina, using a fluorescent tracer. M, 1978, University of South Carolina.

Knott, S. A. Geology of Moss Beach. M, 1977, Stanford University.

Knott, Stephanie Ann. Quaternary paleoceanography of the northeastern Pacific margin based on quantitative studies of planktonic foraminifera. M, 1986, Stanford University. 181 p.

Knowles, David M. Geology and petrology of the Wabush Lake iron formation, Labrador. M, 1955, Michigan Technological University. 96 p.

Knowles, David Martin. The structural development of Labrador Trough formations in the Grenville Province (Precambrian), Wabush Lake area, Labrador. D, 1967, Columbia University. 190 p.

Knowles, Leonard Ivison. The result of an artificial change in base level on Plum Creek, Brown County, Nebraska. M, 1959, University of Arkansas, Fayetteville.

Knowles, Raymond Robert. Geology of southern part of Leadore Quadrangle, east-central Idaho. M, 1961, Pennsylvania State University, University Park. 116 p.

Knowles, Raymond Robert. Geology of the Bedford-Everett-Saxton area, Bedford County, Pennsylvania. D, 1964, Pennsylvania State University, University Park. 254 p.

Knowles, Stephen C. Holocene geologic history of Sarasota Bay, Florida. M, 1983, University of South Florida, Tampa. 128 p.

Knowles, Steven Paul. Geology of the Schofield 7 1/2 minute quadrangle in Carbon, Emery and Sanpete counties, Utah. M, 1985, Brigham Young University. 100 p.

Knowling, Richard Dean. The geology and ore deposits of the Ocampo District, Municipio de Ocampo, Chihuahua, Mexico. M, 1977, University of Iowa. 167 p.

Knowlton, Sandra. Geomorphological history of tidal marshes, Eastern Shore, Virginia, from 1852 to 1966. M, 1971, University of Virginia. 192 p.

Knox, Alexander Walter. The geology and uranium mineralization of the Aphebian Amer Group, southwest of Amer Lake, District of Keewatin, N.W.T. M, 1980, University of Calgary. 207 p.

Knox, Burnal Ray. Bloyd-Atoka relationships in an area of Northwest Arkansas. M, 1957, University of Arkansas, Fayetteville.

Knox, Burnal Ray. Pleistocene and Recent geology of the southwest Ozark Plateaus. D, 1966, University of Iowa. 188 p.

Knox, Debra. Devonian paleoecology in central Arizona. M, 1979, Northern Arizona University. 138 p.

Knox, Ellis Gilbert. The bases of rigidity in fragipans. D, 1954, Cornell University.

Knox, James A. Mineralogy of the Ambrosia Lake uranium deposits, McKinley County, New Mexico. M, 1957, University of Minnesota, Minneapolis. 58 p.

Knox, James Clarence. Stream channel adjustment to physiographic factors in small drainage basins; iowa and southwestern Wisconsin. D, 1970, University of Iowa. 321 p.

Knox, John Harold. Selected trace elements and fixed carbon content of sediment from Lake Lynn, West Virginia. M, 1972, University of Akron. 69 p.

Knox, John Knox. Geology of the serpentine belt, Coleraine Sheet, Thetford-Black Lake mining district, Quebec. D, 1918, University of Chicago. 73 p.

Knox, Keith Sifton. The differentiation of the glacial tills along the north shore of Lake Erie (Ontario). M, 1952, University of Western Ontario.

Knox, Larry M. Microprobe whole rock analysis of igneous intrusives from the southwest corner of Cony Mountain Quadrangle, Wind River Mountains, Wyoming. M, 1976, University of Missouri, Columbia.

Knox, Larry William. Ostracods from the area of the type rocks of the Morrowan Series (lower Pennsylvanian), Arkansas and Oklahoma. D, 1974, Indiana University, Bloomington. 130 p.

Knox, Margaret S. A study of radioactivity with respect to alteration and ore location at Santa Rita, New Mexico and Gilman, Colorado. M, 1947, Columbia University. 27 p.

Knox, Newton Booth. The geology of the Mount Diablo Mine. M, 1938, University of California, Berkeley. 46 p.

Knox, Richard D. Geological and geophysical investigation of the Good Hope mining district, Elko County, Nevada. M, 1970, University of California, Riverside. 76 p.

Knox, Robert Charles. Effectiveness of impermeable barriers for retardation of pollutant migration. D, 1983, University of Oklahoma. 203 p.

Knox, Robert E. Cenozoic deposits of the Emigrant Canyon area, Panamint Range. M, 1963, University of Southern California.

Knox, William P. A comparison of the Schlumberger and Wenner geoelectrical sounding systems. M, 1964, University of Minnesota, Minneapolis. 188 p.

Knudsen, Harvey Peter. Development of a conditional simulation model of a coal deposit. D, 1981, University of Arizona. 121 p.

Knudson, Thomas D. Origin of the Jayville magnetite deposit, Northwest Adirondacks, New York. M, 1978, Kent State University, Kent. 84 p.

Knudtson, Lee Gardiner. Possible oil traps in the northern portion of the Denver Basin (Nebraka, South Dakota, Wyoming). M, 1966, Indiana University, Bloomington. 107 p.

Knuepfer, P. L. Geomorphic investigations of the Vaca and Antioch fault systems, Solano and Contra Costa counties, California. M, 1977, Stanford University. 53 p.

Knuepfer, Peter Louis Kruger. Tectonic geomorphology and present-day tectonics of the alpine shear system, South Island, New Zealand. D, 1984, University of Arizona. 509 p.

Knupke, James Albert. Lithology of the Sundance Formation, Dubois area, Fremont County, Wyoming. M, 1953, Miami University (Ohio). 41 p.

Knurr, Rick Allen. The crystal structures of $2M_1$-alurgite and 1M-manganophyllite. M, 1982, University of Wisconsin-Madison.

Knuth, Martin C. Determination of the source of salt contamination in a private water well in Hudson, Ohio. M, 1987, University of Akron. 63 p.

Knutsen, Gale Curtis. Friability of the Addy Quartzite, Stevens County, Washington. M, 1979, Washington State University. 87 p.

Knutson, Carroll Field. An investigation of some petrophysical aspects of the Third Grubb Zone, San Miguelito Field, Ventura County, California. D, 1959, University of California, Los Angeles.

Knutson, Clarence J. The derivation of copper deposits from different types of magmas. M, 1924, University of Minnesota, Minneapolis. 30 p.

Knutson, D. W. Use of strontium-90 as an environmental tracer in coral growth and structure. M, 1972, University of Hawaii. 63 p.

Knutson, Robert A. Activation analysis for silica in igneous rocks. M, 1954, University of Manitoba.

Knuttle, Stephen. Calcareous nannofossil biostratigraphy of the central East Pacific Rise, DSDP Leg 92; evidence for downshore transport of sediments. M, 1984, Florida State University.

Ko, Chong An. Geology and groundwater resources of the Hangman Creek drainage basin, Idaho-Washington. D, 1974, Washington State University. 150 p.

Ko, Hon-Yim. Static stress-deformation characteristics of sand. D, 1966, California Institute of Technology. 274 p.

Ko, Jachung. Controls on graywacke petrology in Middle Ordovician Cloridorme Formation; tectonic setting of source areas versus diagenesis. M, 1986, University of Toronto.

Ko, Jaidong. High-pressure phase transition in $MnTiO_3$ from the ilmenite to the $LiNbO_3$ structure. D, 1988, SUNY at Stony Brook. 263 p.

Ko, Kyung Chul. The failure criteria and deformational moduli of granular rock. D, 1970, University of Missouri, Rolla. 122 p.

Kobelski, Bruce Joseph. South African and Lesothan kimberlites, with emphasis on the variation of the stable carbon and oxygen isotopic composition of kimberlite carbonates. M, 1977, Pennsylvania State University, University Park. 164 p.

Koberle, A. Notes on mining and concentration in accordance with the scheme of work of the Summer School of Practical Mining of Washington University. M, 1988, Washington University.

Kobluk, David R. The paleoecology of stromatoporoids from the southeast margin of the Miette carbonate complex, Jasper National Park, Alberta. M, 1973, McGill University. 202 p.

Kobluk, David Ronald. Boring and cavity-dwelling algae; effects on cementation and diagenesis in marine carbonates. D, 1977, McMaster University. 224 p.

Kobre, N. A. Polyphase deformation in the Hartland Formation, Southeast Mount Vernon Quadrangle, New York. M, 1979, Queens College (CUNY). 114 p.

Koch, Allan James. Petrology of the "Hoh Formation" of Tertiary age in the vicinity of the Raft River, western Washington. M, 1968, University of Washington. 41 p.

Koch, Allan James. Stratigraphy, petrology, and distribution of Quaternary pumice deposits of the San Cristobal group, Guatemala City area, Guatemale. D, 1970, University of Washington. 80 p.

Koch, C. F. Evolutionary and ecological patterns of upper Cenomanian (Cretaceous) mollusk distribution in the Western Interior of North America. D, 1977, George Washington University. 226 p.

Koch, Carl Allinger. Debris slides and related flood effects in the 4-5 August 1938 Webb Mountain cloudburst; some past and present environmental geomorphic implications. M, 1974, University of Tennessee, Knoxville. 112 p.

Koch, Donald L. Shell Rock Formation (Devonian) of north central Iowa (Cerro Gordo County, parts of Butler, Floyd, Mitchell, Worth, Winnebago, and Hancock counties). M, 1967, University of Iowa. 169 p.

Koch, Ellis. Bedrock control of rainfall-runoff relations in the Peak Creek Watershed, Pulaski and Wythe counties, Virginia. M, 1967, Virginia Polytechnic Institute and State University.

Koch, Franklyn Gordon. The structure of the Mount Ellsworth Intrusion Henry Mountains, Utah. M, 1981, Stanford University. 66 p.

Koch, Fred. Evolutionary and ecological patterns of the upper Cenomanian (Cretaceous) mollusk distribution, Western Interior of North America. D, 1977, George Washington University.

Koch, George Schneider, Jr. Geologic structure of the Frisco Mine, Chihuahua, Mexico. D, 1955, Harvard University.

Koch, Gustave H. The hydrology of the Onondaga drainage basin (New York). M, 1932, Syracuse University.

Koch, Heinrich Louis. The igneous geology of the western half of the St. Francois Mountains. M, 1932, Washington University. 81 p.

Koch, Jo-Ann Major (Sherwin). A test of Biot's theory of folding of stratified viscous media. M, 1966, Brown University.

Koch, John Gerhard. Geology of the Humbug Mountain area, Southwest Oregon. M, 1960, University of Wisconsin-Madison. 33 p.

Koch, John Gerhard. Late Mesozoic orogenesis and sedimentation, Klamath Province, Southwest Oregon coast. D, 1963, University of Wisconsin-Madison. 304 p.

Koch, Michael Robert. Geologic evolution of the Mid-Cretaceous Copa Somerero Group - Lacones Basin, northwestern Peru. M, 1978, Northern Illinois University. 149 p.

Koch, Norris Gayle. Correlation of the Devonian Swan Hills Member, Alberta. M, 1959, University of Alberta. 119 p.

Koch, Philip Samuel. Rheology and microstructures of experimentally deformed quartz aggregates. D, 1983, University of California, Los Angeles. 495 p.

Koch, Richard J. Petrogeneis of the Precambrian Bevos and Musco groups, St. Francois Mountains igneous complex, Missouri. M, 1978, Kansas State University.

Koch, Robert Clement. Dinoflagellate biostratigraphy of Maestrichtian formations of the New Jersey coastal plain. D, 1975, Rutgers, The State University, New Brunswick. 116 p.

Koch, Robert Winfield. A structural study of the Venable Peak region of the Sangre de Cristo Range (Custer and Saguache counties), Colorado. M, 1963, Colorado School of Mines. 131 p.

Koch, Roy W. A physical-probabilistic approach to stochastic hydrology. D, 1982, Colorado State University. 204 p.

Koch, Thomas J. Geochemistry of ground water from two-mica granite, the effects of aquifer alteration on water chemistry and the concentration of dissolved radon. M, 1988, University of New Hampshire. 70 p.

Koch, William F., II. Brachiopod paleoecology, paleobiogeography, and biostratigraphy in the upper Middle Devonian of eastern North America; an ecofacies model for the Appalachian, Michigan, and Illinois basins. D, 1979, Oregon State University. 295 p.

Koch, William Frederick, II. Quantitative study of relationships between host brachiopods and epizoans in the middle Devonian Silica Formation. M, 1973, University of Michigan.

Koch, William Jerry. Lower Triassic lithofacies of the Cordilleran miogeosyncline in the western United States. D, 1969, Harvard University.

Kochan, Mark. The Burnt Gulch Formation; a Pliocene-Pleistocene tectonic arkose conglomerate, Wind River basin, Fremont Co., Wyoming. M, 1987, Miami University (Ohio). 100 p.

Kochanski, Mark Alan. Micro-NCRDS; the application of the microcomputer for calculating coal resources. M, 1984, Purdue University. 481 p.

Kochel, Robert Craig. Interpretation of flood paleohydrology using slackwater deposits, lower Pecos and Devils rivers, southwestern Texas. D, 1980, University of Texas, Austin. 387 p.

Kochel, Robert Craig. Interpretation of high-level gravel deposits and their significance to the erosional history of the Big Horn Basin, Big Horn Mountains, Wyoming. M, 1977, Southern Illinois University, Carbondale.

Kochem, Edward J. Diagenesis of the subsurface Miocene pinnacle reefs of Irian Jaya, Indonesia; a petrographic study. M, 1976, Rensselaer Polytechnic Institute. 106 p.

Kocher, Frederick. Character evolution and the Astarte thisphila; Astarte obruta (Bivalvia) transition in the Miocene Choptank Formation of Maryland. M, 1987, Queens College (CUNY).

Kochick, James P. Petroleum geology of the Misener Sandstone in parts of Payne and Lincoln counties, Oklahoma. M, 1978, Oklahoma State University. 63 p.

Kocis, Diane E. The contact relationships of the Narragansett Pier Granite in the Narragansett Basin area. M, 1981, University of Rhode Island.

Kocken, Roger James. Petrographic and mineralogic analysis of channel samples from the Kentucky No. 9 and Brookville coal seams. M, 1982, Iowa State University of Science and Technology. 224 p.

Kocurek, Gary. Petrology and environments of deposition of the Percha Formation, Upper Devonian, southwestern New Mexico. M, 1977, University of Houston.

Kocurek, Gary Alexander. Significance of bounding surfaces, interdune deposits, and dune stratification types in ancient erg reconstruction. D, 1980, University of Wisconsin-Madison. 373 p.

Kocurko, John M. The paleoecology of a late Pleistocene (Two Creekan) lake, southeastern Wisconsin. M, 1968, University of Wisconsin-Milwaukee.

Kocurko, Michael John. A paleoenvironmental investigation of San Andres Island, Colombia; a study of carbonate rocks. D, 1972, Texas Tech University. 169 p.

Kodama, K. P. Paleomagnetism of the Plio-Pleistocene sediments of Centerville Beach and Lake Waucoba, California, and the plastic deformation of an artificial sediment. D, 1977, Stanford University. 105 p.

Kodl, Edward. Surficial geology of the Amsterdam region, lower Mohawk Valley, New York. M, 1968, University of Vermont.

Kodosky, Lawrence Gerard. A detailed three-dimensional geochemical soil survey over the major anomalous gold region in the southern portion of the Aqueduct Prospect, Breckenridge mining district, Summit County, Colorado. M, 1985, University of Texas at Dallas. 216 p.

Kodybka, Richard Joseph. Erosion of Paleozoic bedrock in the terminal zone of Yoho Glacier, British Columbia. M, 1981, Memorial University of Newfoundland. 193 p.

Koechlein, Harold D. Carbonate sedimentation in a modern tidal delta complex, Windlay Harbor, Florida Keys. M, 1977, University of Toledo. 127 p.

Koederitz, Leonard Frederick. A three-dimensional mathematical simulator of multiphase systems in a petroleum reservoir. D, 1970, University of Missouri, Rolla.

Koehler, Adrienne. The geology of the Valentines area north of Lake Gaston, North Carolina and Virginia. M, 1982, East Carolina University. 91 p.

Koehler, Janet. Electrical resistivity as an approach to evaluating brine contamination of groundwater in the Walker oil field, Ottawa County, Michigan. M, 1988, Western Michigan University.

Koehler, R. P. The biostratigraphy of some Paleozoic and Mesozoic elasmobranch denticles. M, 1975, University of Wisconsin-Madison.

Koehler, Robert Paul. Sedimentary environment and petrology of the Ain Tobi Formation, Tripolitania, Libya. D, 1982, Rice University. 299 p.

Koehler, Steven William. Petrology and petrogeny of the diabase dikes of the Tobacco Root Mountains, southwestern Montana. M, 1973, Indiana University, Bloomington. 61 p.

Koehn, Henry Hans. The potential usefulness of quartz in geochemical exploration for copper deposits. M, 1977, New Mexico Institute of Mining and Technology. 73 p.

Koehn, Marsha A. Petrographical and paleoenvironmental study of the Glorieta Sandstone (Permian) near Rowe, New Mexico. M, 1972, New Mexico Institute of Mining and Technology. 132 p.

Koehnken, P. J. Petrology of anorthosites from two localities in northwestern Sonora, Mexico. M, 1976, University of Southern California. 99 p.

Koehr, J. E. and Rohrbough, R. D. Daily and quasi-weekly beach profile changes at Monterey, California. M, 1964, United States Naval Academy.

Koellner, Mark S. A study of the fluid inclusion, stable isotope and mineralogical characteristics of the Denton fluorspar deposit, Cave-in-Rock, Illinois. M, 1987, Iowa State University of Science and Technology. 110 p.

Koellner, Susan Elaine. The Stettin syenite complex, Marathon County, Wisconsin; petrography and mineral chemistry of olivine, pyroxene, amphibole, biotite, and nepheline. M, 1974, University of Wisconsin-Madison. 155 p.

Koelmel, Mark H. Interactions of hypersaline solutions with feldspars at 300°C to 350°C. M, 1978, Lehigh University. 175 p.

Koelsch, T. A. Relationship of acoustic emission and ultrasonic velocity to deformation mechanisms and dilatancy during the ductile deformation of marble. D, 1979, University of Illinois, Urbana. 128 p.

Koenen, Kenneth H. Geophysical studies in south central Wisconsin. M, 1956, University of Wisconsin-Madison.

Koenig, Afton A., Jr. Geology of the Troublesome Creek basin, Carbon County, Wyoming. M, 1952, University of Wyoming. 66 p.

Koenig, Brian A. Oxidation leaching and enrichment zones of a porphyry copper deposit; a quantitative mineralogic study. M, 1978, University of Arizona.

Koenig, James Bennett. The petrography of certain igneous dikes of Kentucky. M, 1956, Indiana University, Bloomington. 76 p.

Koenig, John Waldo. Fenestrate Bryozoa in the Chouteau Limestone of central Missouri. D, 1951, University of Kansas.

Koenig, Joseph Baldwin. A consideration of the Blanco River terraces north of San Marcos. M, 1940, University of Texas, Austin.

Koenig, Karl Joseph. Bridger Formation in the Bridger Basin, Wyoming. D, 1949, University of Illinois, Urbana. 119 p.

Koenig, Karl Joseph. Electric log study of the "Isabel" Sandstone, Louden Pool, Fayette County, Illinois. M, 1946, University of Illinois, Urbana. 30 p.

Koenig, Robert L. Stratigraphy and structural geology of the Snider Basin area in the overthrust belt, Sublette County, Wyoming. M, 1971, University of Wyoming. 79 p.

Koenigs, Robert Louis. Environmental gradients influencing vegetation on a serpentine soil; I, Principal components analysis of vegetation data; II, Chemical composition of foliage and soil. D, 1977, University of California, Davis. 67 p.

Koenigsberg, Andrew M. Particle-size analysis of fine-grained turbidites of the middle and lower Mississippi Fan, Gulf of Mexico. M, 1988, University of New Orleans.

Koening, Martin. Scale model studies of magnetic anomalies of geological structure. M, 1961, New York University.

Koeninger, C. Allan. Regional facies of the Caseyville Formation (Lower Pennsylvanian) south central Illinois. M, 1978, Southern Illinois University, Carbondale. 124 p.

Koenings, J. P. The metabolism of nonparticulate phosphorus in an acid bog lake. D, 1977, University of Michigan. 220 p.

Koepnick, Richard B. Paleoenvironmental analysis of the upper Cambrian (upper Dresbachian-lower Franconian) pterocephaliid biomere from West-central Utah. D, 1976, University of Kansas. 143 p.

Koepnick, Richard B. Statistical analysis of intraspecific variation of morphology of Triticites cullomemsis. M, 1970, University of Kansas.

Koerner, Harold E. An annotated catalogue of the fossil amphibians, birds, and mammals of Colorado. M, 1930, University of Colorado.

Koerner, Harold E. Geology and vertebrate paleontology of the (Miocene) Fort Logan and Deep River formations of Montana. D, 1939, Yale University. 141 p.

Koerner, Robert M. The behavior of cohesionless soils formed from various minerals. D, 1969, Duke University. 332 p.

Koerschner, William F., III. Cyclic peritidal facies of a Cambrian aggraded shelf; Elbrook and Conococheague formations, Virginia Appalachians. M, 1983, Virginia Polytechnic Institute and State University. 184 p.

Koesoemadinata, R. P. Stratigraphy and petroleum occurrence, Green River Formation, Red Wash Field, Utah; 2 volumes. D, 1967, Colorado School of Mines.

Koester, Edward Albert. Subsurface geology of Russell County, Kansas. M, 1929, University of Missouri, Columbia.

Koesterer, Mary Ellen. Archean history of the Medina Mountain area, central Wind River Range, Wyoming. M, 1986, University of Wyoming. 99 p.

Koesters, Baerbel. Geology of the Morrison Lake area, Montana, Idaho. M, 1963, Pennsylvania State University, University Park. 88 p.

Koesters, Donna Baird. A structural and hydrocarbon analysis of the Central Cascade Range, Washington. M, 1984, Texas Christian University. 120 4plates.

Koff, Leonid Roland. Tectonics of the Oklahoma City Uplift, central Oklahoma. M, 1978, University of Oklahoma. 64 p.

Kofoed, John W. Sedimentary environments in Apalachicola Bay and vicinity, Florida. M, 1961, Florida State University.

Kofron, Ronald J. Age and origin of gold mineralization in the southern portion of the Julian mining district, Southern California. M, 1984, San Diego State University. 75 p.

Kogan, Jerry. A seismic sub-bottom profiling study of Recent sedimentation in Flathead Lake, Montana. M, 1980, University of Montana. 98 p.

Koger, Curtis. Depositional and diagenetic history of the Austin Chalk, central Texas and its relationship to petroleum potential. M, 1981, Baylor University. 151 p.

Koh, In Seok. Geology of the Trepassey area, Avalon peninsula, Newfoundland. M, 1970, Memorial University of Newfoundland. 143 p.

Kohl, Barry. Paleobathymetry of the middle Miocene from a well in Assumption parish, Louisiana. M, 1969, University of Missouri, Rolla.

Kohl, Barry. The lower Pliocene benthic foraminifers from the Isthmus of Tehuantepec, Mexico. D, 1980, Tulane University. 477 p.

Kohl, C. P. Galactic cosmic ray produced radioactivity in lunar and chondritic materials. D, 1975, [University of California, San Diego]. 172 p.

Kohl, Karen Brummett. Mixed-volatile (H2O-CO2) equilibria in metamorphosed rocks, Blount Mountain, Llano County, Texas. M, 1976, Southern Methodist University. 57 p.

Kohl, Karl W. Regional study of the Muddy Sandstone of northeastern Wyoming. M, 1959, Michigan State University. 55 p.

Kohl, Martin Sanford. Tertiary volcanic rocks of the Jean-Sloan area, Clark County, Nevada, and their possible relationship to carnotite occurrences in caliches. M, 1978, University of California, Los Angeles.

Kohland, William Francis. Soils of Erie and Jefferson counties, Pennsylvania; a geographic study and comparison of the distribution and utilization of soils in a glaciated and a non-glaciated area. D, 1969, University of Tennessee, Knoxville. 300 p.

Kohler, Frederick William. The physiography of the Shenango Quadrangle. M, 1928, University of Pittsburgh.

Kohler, James F. Geology, characteristics, and resource potential of the low-temperature geothermal system near Midway, Wasatch County, Utah. M, 1980, Utah State University. 53 p.

Kohler, Martha Hansen. Sediment dispersal in Lake Michigan between Two Rivers and Two Creeks, Wisconsin. D, 1973, University of Wisconsin-Madison.

Kohles, Kevin Michael. Kinematics of polyphase deformation in the Valley and Ridge Province, Tennessee. M, 1985, University of Kentucky. 175 p.

Kohlmann, Nickolas Alfred John. The polymetamorphism of the Little Willow Formation, Wasatch Mountains, Utah. M, 1980, University of Minnesota, Duluth.

Kohls, Donald William. Lithostratigraphy of the Cedar Valley Formation of Minnesota and northernmost Iowa. D, 1961, University of Minnesota, Minneapolis. 256 p.

Kohls, Donald William. The geology of the Prescott Quadrangle. M, 1958, University of Minnesota, Minneapolis. 138 p.

Kohn, Jack Arnold. Directional variation of grinding hardness in silicon carbide (SiC). D, 1950, University of Michigan.

Kohn, Sara E. Mode of occurrence and distribution of inorganic elements in the Smith coal seam, Centralia Mine, Centralia, Washington. M, 1984, Washington State University. 151 p.

Kohout, Francis A. Relation of seaward and landward flow of ground water to the salinity of Biscayne Bay, (Florida). M, 1967, University of Miami. 98 p.

Kohsmann, James J. Qualitative correlation of seismic flux and free-air gravity with crustal structure of the Midcontinent of the United States. M, 1975, Northern Illinois University. 157 p.

Kohsmann, James Joseph. A computational study of two possible intraplate earthquake triggering mechanisms. D, 1983, St. Louis University. 348 p.

Kohut, Alan Peter. The geological and hydrological environment of the Whitewater Lake Basin, Manitoba. M, 1972, University of Manitoba.

Kohut, Joseph James. Quantitative analysis, taxonomy, and distribution of Middle and Upper Ordovician conodonts from the Cincinnati region of Ohio, Kentucky, and Indiana. D, 1967, Ohio State University. 162 p.

Koinm, David N. Growth faulting in the McAlester Basin of Oklahoma. M, 1966, University of Tulsa. 32 p.

Koivunen, Alan. Two magnetotelluric autopower bias reduction techniques. M, 1985, Michigan Technological University. 182 p.

Kojan, Eugene. Analysis of several gravitational mass movement processes; their incidence, morphology, mechanics and rates, with suggested procedure for regional hazard prediction. D, 1972, University of California, Berkeley. 211 p.

Kojic, Slobodan B. Earthquake response of arch dams to nonuniform canyon motion. D, 1988, University of Southern California.

Kokalis, Peter George. Terraces of the lower Salt River Valley, Arizona. M, 1971, Arizona State University. 104 p.

Kokcharoensup, Wichai. The distribution of metals in stream sediments at the Galena-Gilt Edge area, Lawrence County, South Dakota. M, 1979, South Dakota School of Mines & Technology.

Kokinos, John Peter. Late Cretaceous dinoflagellate cysts from the type Magothy Formation, Maryland. M, 1988, Stanford University. 79 p.

Kokli, Kewal Krishan. Prediction of surface subsidence profile due to underground mining in the Appalachian Coalfield. D, 1984, West Virginia University. 220 p.

Koksoy, Mumin. Geology of the northern part of Tincup mining district, Gunnison County, Colorado. M, 1961, Colorado School of Mines. 99 p.

Kolar, B. W. Physical and chemical characteristics of selected metabasites in West-central Vermont. M, 1975, University of Vermont.

Kolasa, William B. Far-infrared and VUV studies of vibrational spectra of rare earth doped fluorite crystals. D, 1982, [University of Windsor].

Kolata, Dennis Robert. Paleoecology and systematics of the echinoderm faunas of the middle Ordovician Platteville and lower Galena groups of north-central Illinois and south-central Wisconsin. D, 1973, University of Illinois, Urbana. 237 p.

Kolb, Charles R. Entrenched Valley of the lower Red River (Louisiana). M, 1949, Louisiana State University.

Kolb, Charles Rudolph. Distribution and engineering significance of soils bordering the Mississippi from Donaldsonville to the Gulf. D, 1962, Louisiana State University. 223 p.

Kolb, Grant. Geology of the New Fork Lakes area, Sublette County, Wyoming, with the application of remote sensing. M, 1983, University of Wyoming. 155 p.

Kolb, J. D. Pore size variation in fluid flow. M, 1954, Massachusetts Institute of Technology. 39 p.

Kolb, John Edward. Petrology of the Pre-cambrian complex in the northwestern Wind River Range, Fremont County, Wyoming. M, 1951, Miami University (Ohio). 49 p.

Kolb, K. K. Two new Pleistocene (Kansan) molluscan local faunas from Trego County, Kansas. M, 1975, Fort Hays State University. 32 p.

Kolb, Richard Alan. Geology of the Signal Hill Quadrangle, Hays and Travis counties, Texas. M, 1981, University of Texas, Austin. 82 p.

Kolbash, Ronald Lee. A study of Appalachia's coal mining communities and associated environmental problems. D, 1975, Michigan State University. 90 p.

Kolbe, E. R. The design and development of an ocean sediment probe. D, 1975, University of New Hampshire. 206 p.

Kolenbrander, Lawrence Gene. A method of evaluating landform classification systems for renewable resource assessment and planning. D, 1981, Colorado State University. 223 p.

Koler, Thomas Edward. Stratigraphy and sedimentary petrology of the northwest quarter of the Dutchman Butte Quadrangle, Southwest Oregon. M, 1979, Portland State University. 71 p.

Kolesar, John Charles. Geology of southwest quarter, Murphysboro Quadrangle, Illinois. M, 1964, Southern Illinois University, Carbondale. 83 p.

Kolesar, Peter Thomas, Jr. Factors affecting the magnesium content of calcite secreted by some articulated coralline algae. D, 1973, University of California, Riverside. 131 p.

Kolesar, Peter Thomas, Jr. Mineralogy, geochemistry, and petrography of fresh water carbonates. M, 1968, Rensselaer Polytechnic Institute. 50 p.

Kolich, Thomas M. Seismic reflection and refraction studies. M, 1974, Virginia Polytechnic Institute and State University.

Kolins, Warren B. Stratigraphy, depositional environments, and provenance of Lower Cretaceous sedimentary rocks, Pelocillo and Animas mountains, southwestern New Mexico. M, 1986, New Mexico State University, Las Cruces. 214 p.

Kolker, Allan. Petrology, geochemistry and occurrence of iron-titanium oxide and apatite (nelsonite) rocks. M, 1980, University of Massachusetts. 157 p.

Kolker, Oded. Caliche distribution and geomorphic relationships in the Southwestern United States. M, 1977, University of California, Los Angeles. 67 p.

Kollar, Frank. The precise intercomparison of lead isotope ratios. D, 1960, University of British Columbia.

Kolle, Jack John. Plastic deformation in two single crystal clinopyroxenes. D, 1980, University of Washington. 174 p.

Koller, G. R. Geophysical and petrologic study of the Lexington Batholith, West-central Maine. D, 1979, Syracuse University. 215 p.

Koller, G. R. Petrography, feldspar mineralogy, and petrology of the Lexington Batholith, western Maine. M, 1976, Syracuse University.

Kolm, K. E. Predicting the surface wind characteristics of southern Wyoming from remote sensing and eolian geomorphology. D, 1977, University of Wyoming. 174 p.

Kolm, K. E. Selenium in soils of the lower Wasatch Formation, Campbell County, Wyoming; geochemistry, distribution, and environmental hazards. M, 1975, University of Wyoming. 97 p.

Kolmer, Joseph R. Some physical properties of coal; porosity and pore size distribution. M, 1972, University of Tulsa. 64 p.

Kolodny, Carole Renee. Structural characteristics of the bryozoan Hallopora ramosa (d'Orbigny) in the Dillsboro Formation (southeastern Indiana) and paleoenvironmental implications. M, 1979, Boston College.

Kolodny, Yehoshua. Studies in geochemistry of uranium and phosphorites. D, 1969, University of California, Los Angeles. 235 p.

Kolpack, Ronald Lloyd. Oceanography and sedimentology of the Drake Passage, Antarctica. D, 1968, University of Southern California. 242 p.

Kolpack, Ronald Lloyd. Oligocene-Miocene sedimentology of the Tecolote Tunnel section of Southern California. M, 1962, University of Southern California.

Kolpin, Dana Ward. Indicators of pesticide contamination in shallow aquifers of Iowa. M, 1988, University of Iowa. 82 p.

Koltermann, Christine Rinzel. An LP embedded simulation model for conjunctive use in management optimization. M, 1983, University of Nevada. 170 p.

Koltermann, Howard H. Hydrogeochemical and environmental isotope investigation of groundwater recharge mechanisms in the Virginia City highlands, Nevada. M, 1984, University of Nevada. 149 p.

Kolva, David Allen. Exploratory palynology of a scabland Lake, Whitman County, Washington. M, 1975, Washington State University. 43 p.

Kolvoord, R. W. Spectrophotometric study of solutions at elevated temperature and pressure. D, 1975, University of Texas, Austin. 170 p.

Kolvoord, Roger Williams. The phase system BeO-Al$_2$O$_3$-H$_2$O. M, 1964, University of Utah. 17 p.

Komar, C. A. Statistical analysis of factors influencing fracture initiation and orientation in oil reservoir sandstone. M, 1972, West Virginia University.

Komar, Paul Douglas. Evaluation of methods of differentiating beach and dune sands by applicatin to Lake Michigan environments. M, 1965, University of Michigan.

Komar, Paul Douglas. The longshore transport of sand on beaches. D, 1969, University of California, San Diego. 158 p.

Komarkova, V. Alpine vegetation of the Indian Peaks area, Front Range, Colorado Rocky Mountains. D, 1976, University of Colorado. 682 p.

Komatar, Frank Donald. Geology of the Animikian metasedimentary rocks, Mellen Granite, and Mineral Lake Gabbro, west of Mellen, Wisconsin. M, 1972, University of Wisconsin-Madison. 70 p.

Komelasky, Michael Charles. Method for measuring the dynamic bulk modulus and loss factor in hydroacoustic materials. M, 1976, Florida Institute of Technology.

Komie, Earl Esar. Geology of Red Bluff area, Loving and Reeves counties, Texas. M, 1952, University of Texas, Austin.

Kominz, Michelle Anne. Geophysical modeling studies; Chapter 1, Oceanic ridge volumes and sea level change; Chapter II, Subsidence analyses of ancient miogeocline, Canadian Rocky Mountains; Chapter III, Part 1, Geophysical modeling of the thermal history of foreland basins; Part 11, Thermal modeling of foreland basins. D, 1986, Columbia University, Teachers College. 218 p.

Komjima, Russell Kei. Application of the Rayleigh-FFT technique to magnetotelluric modeling. M, 1985, San Diego State University. 175 p.

Kommeth, Bryan M. An examination of the fabric of modern submarine sediment slides using a scanning electron microscope. M, 1986, SUNY at Buffalo. 64 p.

Komor, Paul Stuart. Residential energy conservation; a descriptive model of individual choice. D, 1987, Stanford University. 183 p.

Komor, Stephen Charles. A detailed petrologic study of selected portions of the layered gabbro unit at the North Arm Mountain Massif, Bay of Islands ophiolite complex, Newfoundland; implications for crystallization processes in magma chambers underlying oceanic spreading centers. D, 1985, University of Houston. 412 p.

Komorowski, Jean-Christophe. Scanning electron microscope techniques for discrimination of magmatic and hydromagmatic pyroclasts; Vesuvius A.D. 79 deposits. M, 1988, Arizona State University. 396 p.

Kompanik, Gary Steven. A gravity and magnetic study of Clinton-Fayette county area; analysis of subsurface structure. M, 1978, Wright State University. 98 p.

Konan, Gilbert K. Etude des infiltrations d'eau dans le metro de Montreal. M, 1984, Ecole Polytechnique. 237 p.

Koncuk, Fatih. Mineralogy and petrology of scapolite and associated magnetite deposits, Buena Vista Hills, Nevada. M, 1980, Michigan Technological University. 88 p.

Kondelin, Robert J. Stratigraphy and microfacies analysis of the Ordovician System, North Franklin Mountains, Dona Ana County, New Mexico. M, 1984, University of Texas at El Paso.

Kondolf, George Mathias. Recent channel instability and historic channel changes of the Carmel River, Monterey County, California. M, 1982, University of California, Santa Cruz.

Konecny, Gottfried. Radial triangulation with convergent photography. M, 1955, Ohio State University.

Kong, Michael. Geophysical investigations of the southern continental margin of Australia and the conjugate sector of East Antarctica. D, 1980, Columbia University, Teachers College. 344 p.

Kong, Shun Tet. The development of the Hunan Mine in Leadville, Colorado. M, 1907, Columbia University, Teachers College.

Kong, V. William Tang see Tang Kong, V. William

Konicek, Daniela L. Geophysical survey in south-central Washington. M, 1974, University of Puget Sound. 35 p.

Konig, Michael. A magnetic profile across the Nemaha Anticline in Pottawatomie and western Jackson counties, Kansas. M, 1971, Kansas State University. 87 p.

Konig, Ronald Howard. Geology of the northwest flank of Mount Mansfield, Vermont. M, 1956, Cornell University.

Konig, Ronald Howard. Geology of the Plainfield Quadrangle, Vermont. D, 1959, Cornell University. 159 p.

Konigsberg, Richard Leonard. Geology along the San Francisquito Fault, Los Angeles County, California. M, 1967, University of California, Los Angeles.

Konigsmark, Theodore A. Uranium deposits in the Morrison Formation, northeast flank of the Zuni Uplift, New Mexico. M, 1956, University of California, Los Angeles.

Konigsmark, Theodore Albert. Geology of the northern Guarico-Lake Valencia area, Venezuela. D, 1958, Princeton University. 186 p.

Konikow, Leonard Franklin. Mountain runoff and its relation to precipitation ground water, and recharge to the carbonate aquifers of Nittany valley, Pennsylvania. M, 1969, Pennsylvania State University, University Park. 128 p.

Konikow, Leonard Franklin. Simulation of hydrologic and chemical quality variations in an irrigated stream-aquifer system, Arkansas River valley, Colorado. D, 1973, Pennsylvania State University, University Park. 87 p.

Konishi, Kenji. Geology of the Iles Dome area, Moffat and Rio Blanco counties, Colorado, and stratigraphic analysis of the Dakota Sandstone (Cretaceous) of northwestern Colorado. M, 1959, Colorado School of Mines. 194 p.

Konizeski, Richard L. Paleoecology of the middle Pliocene Deer Lodge local fauna, western Montana. D, 1953, University of Chicago. 19 p.

Konkel, David C. Heavy metal distributions of Lake St. Clair. M, 1979, Wayne State University.

Konkel, Phillip. The geology of the northeast portion of the Laramie Basin, Little Medicine District, Wyoming. M, 1935, University of Wyoming. 59 p.

Konkler, Jonathan L. Intensity, site discrimination, and seismic risk analysis in western Kentucky. M, 1981, University of Kentucky. 104 p.

Konkoff, Vladimir I. The genesis and geologic relations of gold ores in California. M, 1930, Columbia University, Teachers College.

Konopka, Edith Hoffman. Stratigraphy and sedimentology of the Butterfield Peaks Formation (Middle Pennsylvanian), Oquirrh Group, at Mt. Timpanogos, Utah. M, 1981, University of Wisconsin-Madison. 170 p.

Konstanty, Kevin Michael. Anelastic toroidal modes and seismic excitation. M, 1984, Arizona State University. 150 p.

Kontak, Daniel Joseph. Geology, geochronology, and uranium mineralization in the Central Mineral Belt of Labrador, Canada. M, 1980, University of Alberta. 380 p.

Kontak, Daniel Joseph. The magmatic and metallogenetic evolution of a craton-orogen interface; the Cordillera de Carabaya, central Andes, SE Peru. D, 1984, Queen's University. 714 p.

Kontis, Angelo L. A study of multi-level aeromagnetic survey data off Barking Sands, Kauai, Hawaii. M, 1970, Rensselaer Polytechnic Institute. 88 p.

Kontrovita, Hervin. A study of some Ostracoda of the Vaca Key, Florida Bay area. M, 1966, University of Florida. 77 p.

Kontrovitz, Mervin. Ostracoda (Holocene) of the central Louisiana continental shelf. D, 1971, Tulane University. 219 p.

Konwar, Lohit Narayan. Effects of bed roughness on the concentration of suspended clay in a salt water flow. M, 1976, Massachusetts Institute of Technology. 76 p.

Konya, Calvin Joseph. Spacing of explosive charges. M, 1968, University of Missouri, Rolla.

Konz, Leo Wilford. A plant-bearing horizon in the Permian of West Texas. M, 1932, University of Texas, Austin.

Koo, Ja Hak. Geology and mineralization in the Lorraine property area, Omineca mining division, British Columbia. M, 1968, University of British Columbia.

Koo, Jahak. Origin and metamorphism of the Flin Flon Cu-Zn sulfide deposit, northern Saskatchewan and Manitoba, Canada. D, 1973, University of Saskatchewan. 154 p.

Koo, Joseph Lok-shan. A gravity survey of the Flambeau Anomaly, Wisconsin. M, 1976, University of Wisconsin-Madison.

Kooi, Verna Vander *see* Vander Kooi, Verna

Kool, Jan Bart Jacobus. Parameter estimation for unsaturated flow models. D, 1987, Virginia Polytechnic Institute and State University. 174 p.

Koons, Donald L. Faulting as a possible origin for the formation of the Nemaha Anticline. M, 1956, Kansas State University. 33 p.

Koons, Edwin Donaldson. Geology of the Uinkaret Plateau, northern Arizona. D, 1945, Columbia University, Teachers College.

Koons, Edwin Donaldson. The origin of the Bay of Fundy and associated submarine scarps. M, 1941, Columbia University, Teachers College.

Koons, Frederic C. Origin of sand mounds of the pimpled plains of Louisiana and Texas. M, 1926, University of Chicago.

Koons, Gerald Jay. Some geological and engineering properties of the Pleistocene Ontonagon Clays at Victoria, Ontonagon County, Michigan. M, 1969, Michigan Technological University. 110 p.

Koons, R. D. Behavior of trace and major elements and minerals during early stages of weathering of diabase and granite in central Wisconsin. D, 1978, University of Wisconsin-Madison. 243 p.

Koop, W. J. Synthesis of pyrrhotite by hydrogen sulphide by iron bearing silicates. M, 1956, University of Manitoba.

Koopersmith, Craig Allen. Computer analysis of bank erosion on Lake Sharpe, South Dakota. M, 1980, South Dakota School of Mines & Technology.

Koopman, Donald Edward. The absorption of crystal violet lactone by kaolinite. D, 1969, Ohio State University. 113 p.

Kooser, Marilyn Ann. Stratigraphy and sedimentology of the San Francisquito Formation, Transverse Ranges, California. D, 1980, University of California, Riverside. 248 p.

Kooten, Gerald K. Van *see* Van Kooten, Gerald K.

Kopacz, M. A. A paleomagnetic study of coal and roof shale of Ohio, Kentucky, and West Virginia. M, 1976, Ohio State University. 103 p.

Kopania, Andrew A. Reconnaissance study of the deformation paths within the northern segment of the Western Thrust Belt. M, 1984, University of Michigan. 25 p.

Kopaska-Merkel, David Crispin. Paleontology and depositional environments of the Whirlwind Formation (Middle Cambrian), west-central Utah. D, 1983, University of Kansas. 215 p.

Kopel, Jerry H. Vein and amygdule minerals, Metchosin Formation (Eocene, lower (?) and middle), Vancouver island, British Columbia. M, 1970, University of San Diego.

Kopf, Rudolph. Paleontology of the (Devonian) Moscow Formation in the Leicester Quadrangle, New York. M, 1952, SUNY at Buffalo.

Kopicki, Robert J. Geology and ore deposits of the northern part of the Hansonburg District, Bingham, New Mexico. M, 1962, New Mexico Institute of Mining and Technology. 103 p.

Kopp, O. C. Directional hardness determinations on the minerals of Mohs' scale. M, 1955, Columbia University, Teachers College.

Kopp, Otto C. Differential thermal analysis of sulfides and arsenides. D, 1959, Columbia University, Teachers College.

Kopp, Richard A. Geothermal exploration of Presidio County, Texas. M, 1977, University of Texas at El Paso.

Kopp, Richard S. Petrology and structural analysis of the Orofino metamorphic unit. M, 1959, University of Idaho. 73 p.

Kopper, Randal W. Subsurface study of hydrocarbon accumulations in the Lansing and Kansas City groups (Pennsylvanian) in Gove County, Kansas. M, 1982, Wichita State University. 41 p.

Kopriva, Suzanne J. Shallow seismic reflection study over an underground coal gasification site, Hanna, Wyoming. M, 1981, University of Wyoming. 78 p.

Kopsick, Deborah A. Geochemistry of leachates from selected coal mining and combustion wastes. M, 1980, University of Kansas. 163 p.

Kor, Philip S. G. Heavy mineral analysis in late Wisconsinan tills of southeastern Manitoba. M, 1976, University of Manitoba.

Koral, Hayrettin. Folding of strata within shear zones; inferences from the azimuths of en echelon folds along the San Andreas Fault. M, 1983, Rensselaer Polytechnic Institute. 100 p.

Kordesch, Elizabeth Gierlowski. Sedimentology and trace fossil paleoecology of the Lower Jurassic East Berlin Formation, Hartford Basin, Connecticut and Massachusetts. D, 1985, Case Western Reserve University. 243 p.

Kordesh, Kathleen. Provenance of sandstones from the Belt Supergroup (middle Proterozoic), Montana. M, 1988, Tulane University. 177 p.

Korentajer, Leonid. Inorganic sulphur metabolism in soil. D, 1977, University of California, Berkeley. 119 p.

Korfiatis, George Panayiotis. Modeling the moisture transport through solid waste landfills. D, 1984, Rutgers, The State University, New Brunswick. 263 p.

Kormendy, Kenneth J. Geochemical and geologic controls on the distribution and release of heavy metals from coal mine overburden, above the No. 6 Coal in Carroll County, Ohio. M, 1982, Kent State University, Kent. 69 p.

Kornacki, Alan Stanley. The nature and origin of refractory inclusions in the Allende Meteorite. D, 1983, Harvard University. 241 p.

Kornas, Barbara Ellen. Chemical composition, mineralogy, and texture of tephra, El Chichón, Mexico, March 28-April 7, 1982, eruptions. M, 1983, University of Texas at El Paso. 84 p.

Kornbrath, Richard W. Late Pleistocene and Holocene geology of the Moe Site, New Town, North Dakota. M, 1975, University of North Dakota. 42 p.

Kornder, Steven Charles. Examination of the organic matter of Recent salt marsh sediments utilizing multivariate analysis of selected lipids. D, 1986, University of South Carolina. 253 p.

Kornegay, Francis Clyde. Kinetic energy budget analysis during interaction of Tropical Storm Candy (1968) with an extratropical frontal system. M, 1975, Purdue University. 67 p.

Kornegay, G. L. Lithologic, mineralogic and paleontologic variations in the Laney Member, Green River Formation, Sand Wash Basin and southernmost Washakie Basin, Colorado and Wyoming. M, 1976, University of Wyoming. 72 p.

Kornelsen, P. J. A numerical simulation of transient flow in viscoelastic polycrystalline materials. M, 1988, University of Waterloo. 224 p.

Korner, Lisa Ann. Radon in stream and ground waters of Pennsylvania as a reconnaissance exploration technique for uranium deposits. M, 1977, Pennsylvania State University, University Park. 151 p.

Kornfeld, Itzchak. Mineral chemical geothermometers. M, 1979, Brooklyn College (CUNY).

Kornfeld, Moses M. Recent littoral foraminifera of Texas and Louisiana. M, 1930, Stanford University. 36 p.

Kornhaus, James W. Petrography of the minor members of the nepheline syenite complex, Granite Mountain, Pulaski County, Arkansas. M, 1953, University of Arkansas, Fayetteville.

Kornicker, L. S. Distribution of ostracodes in the (Permian) Florena Shale (Kansas). M, 1954, Columbia University, Teachers College.

Kornicker, Louis S. Ecology and taxonomy of Recent marine ostracodes in the Bimini area, Great Bahama Bank. D, 1960, Columbia University, Teachers College.

Kornicker, William Alan. Interactions of divalent cations with pyrite and mackinawite in seawater and NaCl solutions. D, 1988, Texas A&M University. 212 p.

Kornik, Leslie J. A structural interpretation of the Russick Lake area, Manitoba. M, 1965, University of Manitoba.

Korompai, Americo E. Structure under the mid continent gravity high. M, 1969, University of Wisconsin-Madison.

Korosy, Marianne. Groundwater flow patterns as delineated by uranium isotope distributions in the Ochlocknee River basin area, SW Georgia and NW Florida. M, 1984, Florida State University.

Korotev, R. L. Geochemical modeling of the distribution of rare-earth and other elements in a basalt and grain-size fractions of soils from the Apollo 17 valley floor and a well-tested procedure for accurate instrumental neutron activation analysis of geologic materials. D, 1976, University of Wisconsin-Madison. 284 p.

Korpel, Joost Adrian. Depositional and diagenetic history of the Middle/Upper Ordovician Dunleith Formation in Northeast Iowa. M, 1983, University of Iowa. 91 p.

Korphage, M. L. Subsurface geology of southwestern Rush County, Kansas. M, 1973, Wichita State University. 104 p.

Korpi, Glen Kaye. Electrokinetic and ion exchange properties of aluminum oxide and hydroxide. D, 1965, Stanford University. 132 p.

Korpijaakko, Martti Jaakko. Studies on the hydraulic conductivity of peat. D, 1976, University of New Brunswick.

Korringa, Marjorie Kitchel Whallon. Vent area of the Soldier Meadow Tuff, an ash-flow sheet in northwestern Nevada. D, 1972, Stanford University. 105 p.

Kortemeier, Curtis Paul. Geology of the Tip Top District, Yavapai County, Arizona. M, 1984, Arizona State University. 138 p.

Kortemeier, Winifred Talbert. Ongonite and topazite dikes in the Flying W Ranch area, Tonto Basin, Arizona. M, 1986, Arizona State University. 94 p.

Kortenhof, Michael H. Natural hydraulic fracturing in a sedimentary basin. M, 1982, University of Missouri, Columbia.

Korth, W. W. Taphonomy of microvertebrate fossil assemblages. M, 1978, University of Nebraska, Lincoln. 95 p.

Korth, William Willard. A review of the geology of the northeastern part of the Wind River Formation, Wyoming, and the early evolution and radiation of rodents in North America. D, 1981, University of Pittsburgh. 279 p.

Kortis, Phillip C. Geochemical soil anomalies related to talc-mineralization in Lancaster County, Pennsylvania. M, 1988, Rutgers, The State University, Newark. 135 p.

Korzeb, Stanley L. Distribution and occurrence of minerals in Ohio coals from Licking, Perry, and Hocking counties. M, 1977, Miami University (Ohio). 74 p.

Korzendorfer, David Paul. The opaque minerals of two flows of the Imnaha Basalt formation, Columbia River Basalt group, in west-central Idaho. M, 1979, Washington State University. 104 p.

Kos, Charles George. The geology of the southern part of the Knobnoster, Missouri, Quadrangle. M, 1942, University of Iowa. 60 p.

Kosanke, Robert M. Contributions to Pennsylvanian paleobotany. D, 1953, University of Illinois, Urbana. 93 p.

Koschal, Gerald J. Petrology of the wall-rock alteration of the Uchi Orebody, Confederation Lake, Ontario, Canada. M, 1975, Wright State University. 94 p.

Koschmann, Albert H. The heavy residuals of the Tonopah rocks and their value in differentiating the various rock types. M, 1920, University of Wisconsin-Madison.

Koscielniak, Daniel E. Unusual eolian deposits on a volcanic terrain near Saint Anthony, Idaho. M, 1973, SUNY at Buffalo. 28 p.

Kosciusko, Kim Anne. Upper Devonian conodont biostratigraphy in Rocky Mountain Front Ranges of Alberta. M, 1987, University of Calgary. 161 p.

Kosco, Daniel Gregory. Part I, The Mount Edgecumbe volcanic field, Alaska, an example of tholeiitic and calc-alkaline volcanism; Part II, Characteristics of andesitic to dacitic volcanism at Katmai National Park, Alaska. D, 1981, University of California, Berkeley. 258 p.

Kose, Celal. Determination of gas content of a Waynesburg No. 11 coal bed in Belmont County, Ohio. M, 1984, University of Toledo. 92 p.

Kosich, Deborah Frances. Oriented cut sections of multilaminar cheilostome bryozoans from the modern Bermuda reefs. M, 1977, Pennsylvania State University, University Park. 196 p.

Kosiewicz, Stanley Timothy. Rare-earth elements in USGS rocks SCo-1 and STM-1, basalts from the Servilleta and Hinsdale formations, and rocks from the Stillwater and Muskox intrusions. D, 1973, University of Wisconsin-Madison. 135 p.

Kosiur, David Richard. Theoretical and experimental studies of mineral-seawater reactions. D, 1978, University of California, Los Angeles. 160 p.

Koski, John S. Internal correspondence as a method of combined gravity and magnetic analysis. M, 1977, Purdue University. 172 p.

Koski, R. A. Geology and porphyry copper-type alteration-mineralization of igneous rocks at the Christmas Mine, Gila County, Arizona. D, 1978, Stanford University. 268 p.

Koskinen, Victor K. Marker bed "F" in the Colorado Shale, Kevin-Sunburst Dome area, Toole County, Montana. M, 1951, Washington State University. 28 p.

Kosloff, Dan Douglas. Numerical models of crustal deformation. D, 1978, California Institute of Technology. 226 p.

Koslosky, Robert A. Analysis of folding in the Middle Ordovician foredeep facies of central New York and eastern Pennsylvania. M, 1986, SUNY at Buffalo. 142 p.

Kosobud, Ann Maxine. The influence of the burrowing mammal Tamias striatus on the soil creep process. D, 1985, University of Illinois, Urbana. 160 p.

Kososki, Bruce Alan. A gravity study of west Antarctica. M, 1972, University of Wisconsin-Madison. 78 p.

Kosro, P. M. The composition of crude oil and its relationship to reservoir oil viscosity. M, 1977, Stanford University. 68 p.

Koss, George Michael. Carbonate mass flow-turbidite sequences of the Permian Delaware Basin, West Texas. M, 1973, University of Wisconsin-Madison.

Kossoff, Martin Jay. The sedimentology of the medial Ordovician St. Paul Group in south-eastern Pennsylvania. M, 1986, SUNY at Binghamton. 60 p.

Kost, Linda Suzanne. Paleomagnetic and petrographic study of sandstone dikes and the Cambrian Sawatch Sandstone, eastern flank of the southern Front Range, Colorado. M, 1984, University of Colorado. 173 p.

Kostaschuk, Raymond A. Late Quaternary history of the Bow River valley near Banff, Alberta. M, 1980, University of Calgary.

Kostenko, James J. Fractionation of lead-210 in Lake Michigan sediments. M, 1982, University of Wisconsin-Milwaukee. 112 p.

Koster, E. H. Experimental studies of coarse-grained sedimentation. D, 1977, University of Ottawa. 221 p.

Koster, Samuel. Recent sediments and sedimentary history across the Pacific-Antarctic Ridge. M, 1966, Florida State University.

Kosters, Elisabeth Catharina. Parameters of peat formation in the Mississippi Delta. D, 1987, Louisiana State University. 281 p.

Kostov, Clement. Constrained interpolation in geostatistical applications. M, 1985, Stanford University. 101 p.

Kostura, John R. Stratigraphic and paleocurrent analysis of the Dakota Sandstone, Four Corners area of the San Juan Basin, New Mexico, Colorado, Utah, and Arizona. M, 1975, Bowling Green State University. 164 p.

Koszalka, E. J. The effects of high precipitation during 1970-74 on the ground-water reservoir of Long Island, New York. M, 1975, Queens College (CUNY)

Koterba, Michael Taylor. Differential influences of storm and watershed characteristics on runoff from ephemeral streams in southeastern Arizona. D, 1987, University of Arizona. 302 p.

Kothavala, Rustam Zal. A study of the remanent magnetism of Granite Mountain, Iron Springs District, Utah. M, 1960, University of Arizona.

Kothavala, Rustam Zal. Wall rock alteration at the United Verde Mine, Jerome, Arizona. D, 1964, Harvard University.

Kothe, Kenneth Ralph. Structural relations of the Paleozoic rocks in the Schunemunk Quadrangle, of southeastern New York. D, 1960, Cornell University. 105 p.

Kothman, Winnard Sidney. An investigation of primary features present in recent ephemeral braided stream deposits, Southern High Plains, Texas. M, 1963, Texas Tech University. 112 p.

Kotila, David Arthur. Algae and paleoecology of algal and related facies, Morrow Formation, northeastern Oklahoma. D, 1973, University of Oklahoma. 231 p.

Kotila, David Arthur. The bedrock geology of the Reily area, Butler County, Ohio. M, 1964, Miami University (Ohio). 119 p.

Kotila, Nancy Lee. The stratigraphy and depositional environment of the Lower Permian Moran Formation, north-central Texas. M, 1978, University of Texas, Arlington. 98 p.

Koto, Robert Y. Ductile and brittle deformation of Proterozoic rocks in the Allamuchy thrust sheet, Jenny Jump Horse and the Shades of Death thrust sheet, New Jersey Highlands. M, 1987, Rutgers, The State University, Newark. 84 p.

Kotoyantz, Alexander A. Geologic factors influencing oil production in Wabaunsee County. M, 1956, Kansas State University. 58 p.

Kotra, Ramakrishna. Organic analysis of the Antarctic carbonaceous chondrites. D, 1981, University of Maryland. 197 p.

Kotsch, Richard William. Effect of sedimentary environment on trace metal fixation; upper Lake Michigan. M, 1974, Michigan State University. 88 p.

Kotschar, Vincent F. Map projections and their applications. M, 1952, Columbia University, Teachers College.

Kotschevar, D. D. Geology and ore treatment of the Keep Cool Mine, Lakeview, Idaho. M, 1938, University of Washington.

Kott, Michael J. A comparison of fission track age determination techniques on mica. M, 1973, Rensselaer Polytechnic Institute. 30 p.

Kottlowski, Frank Edward. Geology of the Switz City and Coal City quadrangles, Indiana. D, 1951, Indiana University, Bloomington. 149 p.

Kottlowski, Frank Edward. Structure and stratigraphy of Towne Point fault block, Carbon County, Montana. M, 1949, Indiana University, Bloomington. 55 p.

Kottmeier, Steven Thornton. Biogeochemistry of silicon in a subtidal marsh pond at Mugu Lagoon, California. D, 1985, University of Southern California.

Kouassi, Frih. Etude stratigraphique et analyse de la dispersion des éléments traces dans le membre inférieur de la Formation d'Albanel, du Groupe de Mistassini, région du Lac Mistassini, Québec, Canada. M, 1979, Universite du Quebec a Chicoutimi. 112 p.

Koucky, Frank. The crystal chemistry of stannite. D, 1956, University of Chicago. 127 p.

Koulekey, Kodjo C. Modeling saltwater upconing in coastal aquifers. D, 1986, University of Florida. 247 p.

Kouns, Charles Wilmarth. Crystalline rocks of the Lake Barcroft area south of Falls Church, Virginia. M, 1966, George Washington University.

Kourse, Lauralee D. Silicoflagellate biostratigraphy of the upper Monterey Formation and lower Sisquoc Formation, Johnmansville Quarry, Lompoc, California. M, 1980, University of Texas at El Paso.

Kousparis, Dimitrios. Quantitative geophysical study of the Cleveland Sand reservoir (Pennsylvanian) in eastern part of Logan County, Oklahoma. M, 1975, University of Oklahoma. 205 p.

Kousparis, Dimitrios. Seismic stratigraphy and basin development; Nestos Delta area, northeastern Greece. D, 1979, University of Tulsa. 161 p.

Koutahi, Mohammed John. The movement of connate water in porous media under natural energy mechanism. M, 1969, University of Oklahoma. 94 p.

Kouther, Jameel H. Geology and mineralization of northwest part of Bonanza Quadrangle, Chaffee and Saguache counties, Colorado. M, 1969, Colorado School of Mines. 93 p.

Koutsibelas, Dimitrios A. Stabilization of slip in a fluid-infiltrated rock mass. D, 1988, Northwestern University. 133 p.

Koutsomitis, Dimitrios. Gravity investigation of the northern Triassic-Jurassic Newark Basin and Palisades Sill in Rockland County, New York. M, 1980, Rutgers, The State University, Newark. 103 p.

Koutz, Fleetwood R. Boron geochemistry of Central American volcanoes. M, 1971, Dartmouth College. 107 p.

Kovac, L. J. Upper Cretaceous oil shales of Manitoba and Saskatchewan; sedimentology, mineralogy, geochemistry and organic diagenesis. M, 1985, University of Manitoba.

Kovach, Edward Michael. A study of radon content of soil gas. D, 1946, Fordham University. 88 p.

Kovach, Jack. Stratigraphy and paleontology of the pentamerinid brachiopods of the Niagaran rocks of western Ohio and eastern Indiana. D, 1974, Ohio State University. 375 p.

Kovach, Jack. Whole-rock Rb-Sr age of the Gunflint Formation (Precambrian) of Ontario, Canada. M, 1967, Ohio State University.

Kovach, James Thomas. Geology of late Cenozoic sediments of the Borchers' Badlands, Meade County, Kansas. M, 1979, Fort Hays State University. 129 p.

Kovach, Linda Anne. Dynamics of magma ascent through the Sierra Nevada, California. D, 1984, The Johns Hopkins University. 306 p.

Kovach, Robert Louis. Geophysical investigations in the Colorado Delta region. D, 1962, California Institute of Technology. 84 p.

Kovach, Warren L. Dispersed plant remains from the Cenomanian of Kansas; systematic and paleoecologic approaches. D, 1987, Indiana University, Bloomington. 242 p.

Kovacik, Thomas Louis. Distribution of mercury in western Lake Erie water and bottom sediments. M, 1972, Bowling Green State University. 74 p.

Kovacs, Sandor. Design and application of a multipurpose management information system for drinking water quality assessment in distribution systems. D, 1974, Drexel University.

Koval, David B. Geophysical-structural analysis of the Lake Mary well field area with hydrologic interpretations, Coconino County, Arizona. M, 1976, Northern Arizona University. 123 p.

Kovaltchouk, A. G. Heat flow measurements in shallow boreholes. M, 1977, University of Southern California. 134 p.

Kovanick, Mark G. Geology of the Trout Creek area, Elko and Eureka counties, Nevada. M, 1957, University of California, Los Angeles.

Kovar, Erlece Paree. The petrography of some limestones from southern Iowa and southeastern Nebraska. M, 1967, Iowa State University of Science and Technology.

Kovarik, Mary Beth. Calculating constrained crustal geotherms. M, 1986, Pennsylvania State University, University Park.

Kovas, Edward J. The geology of the Sheep Basin Mountain area, northern Sierra Anchas, Gila County, Arizona. M, 1978, Northern Arizona University. 96 p.

Kovinick, Mark G. Geology of the Trout Creek area, Elko and Eureka counties, Nevada. M, 1957, University of California, Los Angeles.

Kovisars, Leons. Geology of the eastern flank of the La Culata massif, Venezuelan Andes. D, 1969, University of Pennsylvania. 262 p.

Kovschak, Anthony Andrew, Jr. Igneous and structural geology of the Grapevine Hills, Big Bend National Park, Brewster County, Texas. M, 1973, University of Texas, Arlington. 192 p.

Kowalczyk, Frank J. The geology of the Burroughs Creek-Horse Creek area, Fremont County, Wyoming. M, 1974, Miami University (Ohio). 85 p.

Kowalczyk, Gary R. Stratigraphy and paleoenvironment of the Cincinnatian Series (Upper Ordovician) of southeast Indiana. M, 1976, University of Toledo. 65 p.

Kowalik, Joseph, Jr. Geological, mineralogical, and stable isotope studies of a polymetallic massive sulfide deposit; Buchans, Newfoundland. D, 1979, University of Minnesota, Minneapolis. 168 p.

Kowalik, William Stephen. Atmospheric correction to Landsat data for limonite discrimination. D, 1981, Stanford University. 400 p.

Kowalik, William Stephen. Use of Landsat-1 imagery in the analysis of lineaments in Pa. M, 1975, Pennsylvania State University, University Park.

Kowall, S. J. Hydrogeology and geomorphology of two structurally dissimilar terranes in Pennsylvania. D, 1975, SUNY at Binghamton. 154 p.

Kowallis, Bart Joseph. Structure and stratigraphy of late Precambrian tillites near Kapp Lyell, Spitsbergen. M, 1979, University of Wisconsin-Madison.

Kowallis, Bart Joseph. Velocity behavior of rocks related to microcracks, micropores, and pore-fillings. D, 1981, University of Wisconsin-Madison. 160 p.

Kowalski, John Francis. The pre-Pleistocene geology of Thayer County, Nebraska. M, 1959, University of Nebraska, Lincoln.

Kowalski, Michael A. Analysis of deep electromagnetic sounding system. M, 1969, Colorado School of Mines. 78 p.

Kowalski, Richard G. Salt water intrusion into a coastal aquifer. M, 1985, University of Rhode Island.

Kowatch, John S. A gravimetric and magnetic survey of perlite domes at No Agua, New Mexico. M, 1974, Pennsylvania State University, University Park.

Kowsmann, Renato Oscar. Surface sediments of the Panama Basin; coarse components. M, 1973, Oregon State University. 73 p.

Kozak, Frank Daniel. The geology of the Pettit-Barber area, Cherokee County, Oklahoma. M, 1951, University of Oklahoma. 84 p.

Kozak, Samuel J., Jr. Foraminifera of the Mount Laurel-Navesink Formation (Upper Cretaceous) along the Chesapeake and Delaware Canal (Delaware, Maryland). M, 1958, Brown University.

Kozak, Samuel Joseph. Structural geology of the Cherry Creek basin area, Madison Mountains, Montana. D, 1961, University of Iowa. 136 p.

Kozar, Michael Glenn. The stratigraphy, petrology, and depositional environments of the Maryville Limestone (Middle Cambrian) in the vicinity of Powell and Oak Ridge, Tennessee. M, 1986, University of Tennessee, Knoxville. 242 p.

Kozarek, Robert James. Biostratigraphic analysis of ichthyoliths. M, 1978, University of Oregon. 227 p.

Kozary, Myron Theodore. Conglomerates associated with the Cubitas Plateau, Cuba. D, 1954, Columbia University, Teachers College.

Kozary, Myron Theodore. Investigations of streaming electropotentials in granular materials. M, 1947, Northwestern University.

Koziar, A. Applications of audio frequency magnetotellurics to permafrost, crustal sounding, and mineral exploration. D, 1976, University of Toronto.

Koziel, Andrea. Acitivity-composition relations of selected garnets determined by phase equilibrium experiments. D, 1988, University of Chicago. 131 p.

Kozimko, Leo M. Geology of the Greybull North Quadrangle, Wyoming. M, 1977, Iowa State University of Science and Technology.

Kozinski, Jane. Sedimentology and tectonic significance of the Nutzotin Mountains sequence, Alaska. M, 1985, SUNY at Albany. 132 p.

Koziol, Brenda L. Stratigraphy of the Viking Formation (Albian) and Newcastle Member (Albian) of the Ashville Formation in Saskatchewan. M, 1988, University of Saskatchewan. 238 p.

Kozo, Thomas. Internal wave study in the Tongue of the Ocean, Bahamas. M, 1969, Florida State University.

Kozusko, Raymond George. The petrology of the Cambridge Limestone in Athens, Morgan, and Perry counties, Ohio. M, 1968, Ohio University, Athens. 138 p.

Kraatz, Paul. Cleavage and deformation of strontium fluoride. D, 1972, University of Minnesota, Minneapolis. 423 p.

Kraatz, Paul. Rockwell hardness as an index property of rocks. M, 1964, University of Illinois, Urbana. 102 p.

Krabbenhoft, David Perry. Hydrologic and geochemical controls of freshwater ferromanganese deposit formation at Trout Lake, Vilas County, Wisconsin. M, 1984, University of Wisconsin-Madison.

Krabbenhoft, David Perry. Hydrologic and geochemical investigations of aquifer-lake interactions at Sparkling Lake, Wisconsin. D, 1988, University of Wisconsin-Madison. 226 p.

Kraemer, Bradley Robert. Conodont biostratigraphy of the northeastern limits of the Lower Triassic Dinwoody Formation and associated strata in southwestern Montana and adjacent areas in Wyoming and Idaho. M, 1988, University of Wisconsin-Milwaukee. 115 p.

Kraemer, Curtis Allen. A model study of compression and rebound of West Twin Lake sediments, Portage County, Ohio. M, 1976, Kent State University, Kent. 65 p.

Kraemer, John L. A traverse of the Big Blue Series from the Nebraska state line to Manhattan, Kansas. M, 1934, University of Nebraska, Lincoln.

Kraemer, Philip G. Tabulate corals of the Edgecliff Limestone (Middle Devonian) in New York State. M, 1968, SUNY at Binghamton. 98 p.

Kraemer, Susanne Margit Charlotte. Quaternary morphology, acoustic characteristics, and fan growth of the Conception Fan, Santa Barbara Basin, California continental borderland. M, 1986, California State University, Northridge. 205 p.

Kraemer, Thomas F. Geochemical investigation of Pleistocene sediments for the American Mediterranean. D, 1975, University of Miami. 154 p.

Kraemer, Thomas F. Rates of accumulation of iron, manganese, and certain trace elements on the East Pacific Rise. M, 1971, Florida State University.

Kraetsch, Ralph B., Jr. Stratigraphy of the Pennsylvanian-Lower Permian interval in southern Idaho and adjacent area. M, 1950, Northwestern University.

Kraeuter, John N. Descriptions of fecal pellets of some common invertebrates in the lower York River and lower Chesapeake Bay, Virginia. M, 1966, College of William and Mary.

Krafft, Alison D. Mineral and chemical composition of suspended and bottom sediments, Quinault submarine canyon. M, 1984, Lehigh University. 62 p.

Kraft, Gordon D. Two dimensional finite element direct current electrical resistivity modeling of axially symmetric structures. M, 1976, Pennsylvania State University, University Park. 91 p.

Kraft, Jennifer Lucille. The structural evolution of the Sunshine Springs Thrust area, Marathon Basin, Texas. M, 1984, University of Texas, Austin. 99 p.

Kraft, John Christian. A petrographic study of the Oneota-Jordan contact zone. M, 1952, University of Minnesota, Minneapolis. 115 p.

Kraft, John Christian. Morphologic and systematic relationships of some Middle Ordovician Ostracoda. D, 1955, University of Minnesota, Minneapolis. 248 p.

Kraft, Michael Thomas. Seismic stratigraphy in carbonate rocks; depositional history of the Natuna D-alpha reef complex, Natuna Sea, Indonesia. M, 1985, San Diego State University. 109 p.

Kraft, Richard P. The stratigraphy and structure of a portion of the Las Vegas Range, north-central Clark County, Nevada. M, 1988, California State University, Long Beach. 101 p.

Krage, Susan Marie. Metamorphic and fluid inclusion study of amphibolite-grade rocks, West Scotia, British Columbia. M, 1984, Bryn Mawr College. 98 p.

Kragness, Ned Low. A thesis on some aspects of the rise of a light fluid in a viscous body or fluid. M, 1939, University of Minnesota, Minneapolis. 25 p.

Krahulec, Kenneth A. Precambrian geology of the Roubaix District, Black Hills, South Dakota. M, 1981, South Dakota School of Mines & Technology. 46 p.

Kraig, David Harry. Emplacement of the Moxa Arch and interaction with the Western Overthrust Belt, Wyoming. M, 1986, Texas A&M University.

Kraig, Scott A. An electrical resistivity survey applied to the hydrogeology of unconsolidated glacial material in western Brimfield Township, Portage County, Ohio. M, 1982, Kent State University, Kent. 122 p.

Krail, Paul Michael. The reflected waveform of a spherical seismic wave. D, 1980, Virginia Polytechnic Institute and State University. 143 p.

Krajewski, Stephen A. The relationship between bedforms, occurrence, composition, texture, and physical properties of the Upper Devonian flagstones in northeastern Pennsylvania. M, 1972, Pennsylvania State University, University Park.

Krajewski, Stephen A. The source of compositional variation in late Pleistocene sands from Accomack County, Virginia. D, 1977, Pennsylvania State University, University Park. 407 p.

Krakker, Linda A. Crystallographic orientation in the stylophoran styloid element; quantum multimodality and styloid size. M, 1985, University of Michigan. 30 p.

Kral, Victor E. Mineral resources of Nye County, Nevada. M, 1951, University of Nevada - Mackay School of Mines. 223 p.

Krall, Donald Bowen. Fluvioglacial drainage between Skaneateles and Syracuse, New York. M, 1966, Syracuse University.

Krall, Donald Bowman. Till stratigraphy and Olean ice retreat (Pleistocene) in east central New York. D, 1972, Rutgers, The State University, New Brunswick. 96 p.

Kramberger, John Joseph. The structure and ore deposits of northern Manitoba (Canada). M, 1968, University of Michigan.

Kramer, Ann M. Paragenetic and fluid inclusion study of the Smith Vein, Smith Mine, Blackhawk, Colorado. M, 1984, Colorado School of Mines. 108 p.

Kramer, David J. Authigenic and detrital K-feldspars in the Franconia Formation. M, 1970, Northern Illinois University. 36 p.

Kramer, Frank Edward. The distribution of krypton in an anorthite-diopside-water system at five kilobars pressure. M, 1987, Washington University. 71 p.

Kramer, James Richard. The spectrochemical analysis of limestone and dolomite. M, 1954, University of Michigan.

Kramer, James Richard. The system; calcite-dolomite in sea water. D, 1958, University of Michigan. 86 p.

Kramer, Jerry Curtis. Geology and tectonic implications of the Coastal Belt Franciscan, Fort Bragg-Willits area, northern Coast Ranges, California. D, 1976, University of California, Davis. 128 p.

Kramer, John Howard. Petrology of the Mount Owen Stock, Gunnison County, Colorado; the genesis of K feldspar phenocrysts in porphyritic rocks. M, 1977, Pennsylvania State University, University Park. 83 p.

Kramer, Kenneth Francis. Oxygen compound acidity and oxygen polarization as a control in silicate weathering. D, 1967, Florida State University. 75 p.

Kramer, Mark Thomas. Prediction analysis applied to hypocenter determinations from the Southeast Missouri regional seismic network. M, 1976, St. Louis University.

Kramer, Matthew Joseph. Experimentally induced deformation mechanisms in single crystal sodic plagioclase. D, 1988, Iowa State University of Science and Technology. 144 p.

Kramer, Matthew Joseph. Origin of the Adirondack lineaments. M, 1982, University of Rochester. 118 p.

Kramer, Michael S. Contact metamorphism of the Mancos Shale associated with the intrusive at Cerrillos, New Mexico. M, 1981, Dartmouth College. 102 p.

Kramer, Steven Barker. Measurement of viscoelastic properties of some Recent marine sediments by a torsionally oscillating cylinder method. M, 1973, United States Naval Academy.

Kramer, Steven Lawrence. Liquefaction of sands due to non-seismic loading. D, 1985, University of California, Berkeley. 375 p.

Kramer, Terry M. The morphology and taxonomy of the Upper Ordovician, Cincinnatian Series, Lingulida (Brachiopoda) of Ohio, Indiana, and Kentucky. M, 1972, Miami University (Ohio). 116 p.

Kramer, Thomas L. The paleoecology of the postglacial Mud Creek Biota, Cedar and Scott counties, Iowa. M, 1972, University of Iowa. 69 p.

Kramer, Walter V. Geology of the Bishop Cap hills, Dona Ana County, New Mexico. M, 1970, University of Texas at El Paso.

Kramer, William B., III. Boulders from Bengalia (Caney Shale, southeastern Oklahoma). D, 1935, University of Chicago. 31 p.

Kramers, John William. Petrology and mineralogy of the Watt Mountain Formation (Devonian) in the Mitsue-Nipisi area, Alberta (Canada). M, 1967, University of Alberta. 92 p.

Kramers, John William. The Centerfield limestone (middle Devonian) of New York State and its clastic correlatives; a sedimentologic analysis. D, 1970, Rensselaer Polytechnic Institute. 135 p.

Kramm, Hugo E. Serpentines of the central Coast Ranges of California. M, 1909, [Stanford University].

Krammes, Jay Samuel. Hydrologic significance of the granitic parent material of the San Gabriel mountains, California. D, 1969, Oregon State University. 121 p.

Kramsky, Melvin Bernard. Petrography of certain Tertiary igneous intrusives of north-central Gunnison County, Colorado. M, 1958, University of Kentucky. 60 p.

Kran, Neil. Tidal controls on suspended sediment in a coastal lagoon, Stone Harbor, New Jersey. M, 1975, Lehigh University. 45 p.

Kranak, Peter Val. Petrography and geochemistry of the Butterly Dolomite, and associated sphalerite mineralization, Arbuckle Mountains, Oklahoma. M, 1978, Oklahoma State University. 116 p.

Krancer, Anthony Edward. Nannofossils of the Ozan Formation (Cretaceous), McCurtain County, Oklahoma. M, 1975, University of Oklahoma. 134 p.

Kranck, Svante Hakan. Chemical petrology of metamorphic iron formations and associated rocks in the Mount Reed area in northern Quebec. D, 1959, Massachusetts Institute of Technology. 150 p.

Kranck, Svante Hakan. Geology of the Stony Rapids norite area, northern Saskatchewan. M, 1955, McGill University.

Kranendonk, Martin Julian van see van Kranendonk, Martin Julian

Kranidiotis, Prokopis. Geology, geochemistry and hydrothermal alteration at the Phelps Dodge massive sulfide deposit, Matagami, Quebec. M, 1986, University of Toronto.

Krank, Kenneth D. The effects of weathering on the engineering properties of Sierra Nevada granodiorites. M, 1980, University of Nevada. 99 p.

Krans, Ainslie Earl Browen. The geology of the northwest quarter of the Bone Mountain Quadrangle, Coos County, Oregon. M, 1970, University of Oregon. 82 p.

Kransdorff, David. The geology of the Eureka Standard Mine, Tintic, Utah. D, 1935, Harvard University.

Krantz, David Eugene. Stable isotopes of oxygen and carbon in mollusk shell carbonate; interpretive models and applications. D, 1988, University of South Carolina. 364 p.

Krantz, Dennis V. Selected trace element distribution in galena and sphalerite, Magmont Mine, Iron County, Missouri. M, 1972, Southern Illinois University, Carbondale. 89 p.

Krantz, Gary Wayne. A comparison of sulfur dioxide adsorption percentage by selected calcium oxides. M, 1972, University of Florida. 91 p.

Krantz, Robert Warren. Detailed structural analysis of detachment faulting near Colossal Cave, southern Rincon Mountains, Pima County, Arizona. M, 1983, University of Arizona. 58 p.

Krantz, Robert Warren. The odd-axis model; orthorhombic fault patterns and three-dimensional strain fields. D, 1986, University of Arizona. 123 p.

Kranz, Dwight Stanley. Geological interpretation of the Pennsylvanian Bartlesville Sandstone in northeastern Oklahoma and southeastern Kansas, from continuous dipmeter and gamma-ray well logs. M, 1981, Texas A&M University. 105 p.

Kranz, Peter M. The anastrophic burial of bivalves and its paleoecological significance. D, 1972, University of Chicago. 177 p.

Kranz, R. L. The static fatigue and hydraulic properties of Barre Granite. D, 1979, Columbia University, Teachers College. 203 p.

Krarti, Moncef. Developments in ground-coupling heat transfer. D, 1987, University of Colorado. 238 p.

Krasowski, Dennis J. Geology and ore deposits of Burrows Park, Hinsdale County, Colorado. M, 1976, Colorado State University. 124 p.

Kraszewski, Stefan. Morphology and development of Palouse and related series. D, 1952, Washington State University. 105 p.

Kratchman, Jack. Sedimentation in a portion of New York Harbor. M, 1952, New York University.

Kratky, Mark Anthony. Carbonate tidal flat sedimentation in the Newman Limestone (Mississippian) of northeastern Kentucky. M, 1983, Wright State University. 111 p.

Kratochvil, Anthony L. Geology of the uraniferous Deep Gulch Conglomerate (late Archean), northwest-

ern Sierra Madre, Wyoming. M, 1981, University of Wyoming. 113 p.

Kratochvil, Gary L. Quaternary geology of the Kenosha Pass-Como area, Park County, Colorado. M, 1978, Colorado School of Mines. 313 p.

Kratz, Timothy Kellogg. The formation of a northern Wisconsin kettle-hole bog; a spatial, ecosystem modeling perspective. D, 1981, University of Wisconsin-Madison. 131 p.

Krauel, David Paul. A physical oceanographic study of the Margaree and Cheticamp river systems (Nova Scotia). M, 1969, Dalhousie University.

Kraus, Gregory Paul. A subsurface geologic study of petroleum possibilities in the Pennsylvanian rocks of Concho County, Texas. M, 1967, University of Missouri, Rolla.

Kraus, Mary J. The petrology and depositional environments of a continental sheet sandstone; the Willwood Formation, Bighorn Basin, Wyoming. M, 1979, University of Wyoming. 106 p.

Kraus, Mary Jean. Genesis of early Tertiary exotic metaquartzite conglomerates in the western Bighorn Basin, Northwest Wyoming. D, 1983, University of Colorado. 173 p.

Kraus, Paul S. The conodonts of the Grassy Creek Shale. M, 1931, University of Missouri, Columbia.

Krause, Alan Joel. Geologic site investigation of the Kaneohe-Kailua Flood Control Project, Kaneohe, Hawaii. M, 1980, University of Nevada. 109 p.

Krause, Dale Curtiss. Geology of the southern continental borderland west of Baja California, Mexico. D, 1961, University of California, San Diego. 205 p.

Krause, David James. The secret of the Keweenaw; native copper and the making of a mining district, 1500-1870. D, 1986, University of Michigan. 295 p.

Krause, David James. Theoretical gravity anomalies of some lunar surface features. M, 1970, Michigan State University. 143 p.

Krause, David Wilfred. Evolutionary history and paleobiology of early Cenozoic Multituberculata (Mammalia), with emphasis on the family Ptilodontidae. D, 1982, University of Michigan. 575 p.

Krause, Erwin Koerps. Pelecypods of the middle Eocene Stone City Beds of Texas. M, 1954, University of Texas, Austin.

Krause, Frederico F. Systematics and distributional patterns of inarticulate brachiopods on the Ordovician carbonate-mud-mound at Meiklejohn Peak, southwestern Nevada. M, 1974, University of Kansas. 282 p.

Krause, Frederico Fernando. Sedimentology and stratigraphy of a continental terrace wedge; the Lower Cambrian Sekwi and June Lake formations (Godlin River Group), Mackenzie Mountains, Northwest Territories, Canada. D, 1979, University of Calgary. 252 p.

Krause, Hans H. Geology of the Saddle Mountain-Carten Creek area, Powell County, Montana. M, 1964, University of Kansas. 57 p.

Krause, Jerome B. Chemical petrology of clinopyroxene gneisses from the Frontenac axis, Grenville Province, Ontario. D, 1971, Queen's University. 83 p.

Krause, Jerome Bernard. Thermal metamorphism of dolomitic limestones at Crestmore, California. M, 1966, University of Missouri, Rolla.

Krause, Karen Webber. Stratigraphy, geochemistry, and petrology of the Juchipila volcanic sequence and caldera complex, Juchipila, Zacatecas, Mexico. M, 1984, University of New Orleans. 167 p.

Krause, Kerwin J. Geology and mineralization of the Granite Creek area, eastern Alaska Range, Alaska. M, 1981, University of Nevada. 92 p.

Krause, Robert Georg Fritz. Provenance and sedimentology of Upper Cretaceous conglomerates in the Santa Ynez Mountains and their implications on the tectonic development of the western Transverse Ranges, Southern California. M, 1986, University of California, Los Angeles. 391 p.

Krauser, R. F. The sediment distribution and geomorphology of Brigantine Inlet, New Jersey. M, 1977, Queens College (CUNY). 78 p.

Krauskopf, Konrad Bates. The geology of the northwest quarter of the Osoyoos Quadrangle, Washington. D, 1938, Stanford University. 286 p.

Krausz, K. A study of diffusion in mica minerals. M, 1973, University of Ottawa. 94 p.

Kravik, Gerald Enestvedt. Properties of refractories from Cypress Island olivine. M, 1938, University of Washington. 103 p.

Kravits, Christopher M. The effects of overbank and splay deposition on the quality and maceral composition of the Herrin (No. 6) Coal (Pennsylvanian) of southern Illinois. M, 1980, Southern Illinois University, Carbondale. 202 p.

Kravitz, J. H. Textural and mineralogical characteristics of the surficial sediments of Kane Basin. M, 1975, George Washington University.

Kravitz, Joseph Henry. Sediments and sediment processes in a high Arctic glacial marine basin. D, 1983, George Washington University. 508 p.

Krawiec, Wesley. Recent sediments of the Louisiana inner continental shelf. D, 1966, Rice University. 139 p.

Krawiec, Wesley. Solution pits and solution breccias of the Florida Keys. M, 1963, University of Rochester. 30 p.

Kraybill, Richard Lancaster. Ground water quality, variation, and trends in Schantz Spring basin and adjacent areas of southwestern Lehigh County, Pennsylvania. M, 1977, Rutgers, The State University, New Brunswick. 69 p.

Kraye, Robert Frank. Sediments of the Bering and Chukchi seas. M, 1943, University of Illinois, Urbana. 30 p.

Kreamer, D. K. The effect of the Central Arizona Project on the thermal structure of Lake Havasu. M, 1976, University of Arizona.

Kreamer, David Kenneth. In situ measurement of gas diffusion characteristics in unsaturated porous media by means of tracer experiments. D, 1982, University of Arizona. 225 p.

Krebes, Elizabeth Ann. Acritarchs of the Waldron Shale (Middle Silurian) of southeastern Indiana. M, 1972, University of Cincinnati. 187 p.

Krebs, Robert Dixon. Seven soil profiles in northern New Jersey; a study of the factors in their genesis as shown by certain of their morphological, physical, chemical, and mineralogical characteristics. D, 1956, Rutgers, The State University, New Brunswick. 164 p.

Krebs, William Nelson. Ecology and preservation of neritic marine diatoms, Arthur Harbor, Antarctica. D, 1977, University of California, Davis. 216 p.

Krech, Warren Willard. The determination of the mechanical properties of rock and soil by means of a hollow cylinder loaded with external pressure. D, 1967, University of Minnesota, Minneapolis. 217 p.

Krecow, Frank C. Megascale variability of sediments on the U. S. Atlantic continental shelf and slope; a multivariate study. M, 1978, Rensselaer Polytechnic Institute. 99 p.

Kreczmer, Marek Jozef. The geology and geochemistry of the Fortuna mineralization, Fresnillo, Zacatecas, Mexico. M, 1977, University of Toronto.

Kreglo, James R. Clay mineralogy of the Pamunkey River basin. M, 1960, Virginia Polytechnic Institute and State University.

Kreider, John E. The Lund ash-flow tuff. M, 1970, University of Missouri, Rolla.

Kreidler, Eric Russell. Stoichiometry and crystal chemistry of apatite. D, 1967, Pennsylvania State University, University Park. 252 p.

Kreiger, E. William, Jr. A compositional comparison of recent eruptive materials from two tholeiitic volcanoes; Mount Hekla, Iceland and Kilauea, Hawaii. M, 1973, Millersville University.

Kreiger, Edgar William, Jr. Geology and petrology of the Two Buttes Intrusion. D, 1976, Pennsylvania State University, University Park. 116 p.

Kreighbaum, D. Utilization of Landsat thematic mapper data for structural geologic mapping in an arid region, Garfield County, Utah. M, 1987, Murray State University. 71 p.

Kreimendahl, Frank Alan. Precision selenodesy via radio interferometry. M, 1979, Massachusetts Institute of Technology. 26 p.

Kreis, Henry G. Basement porphyritic quartz monzonite and its relationship to ore near Ely, Nevada. M, 1973, University of Arizona.

Kreisa, Ronald D. The origin of the Price Formation (Mississippian) in the type area, Montgomery and Pulaski counties, Virginia. M, 1972, Virginia Polytechnic Institute and State University.

Kreisa, Ronald Dean. The Martinsburg Formation (Middle and Upper Ordovician) and related facies in southwestern Virginia. D, 1980, Virginia Polytechnic Institute and State University. 378 p.

Kreitler, Charles W. Determining the source of nitrate in groundwater by nitrogen isotope studies. D, 1974, University of Texas, Austin. 181 p.

Kreitler, Charles W. Nitrate contamination of groundwater in southern Runnels County, Texas. M, 1972, University of Texas, Austin.

Kreitner, Jerry D. The petrofabrics of aufeis in a turbulent Alaskan stream. M, 1969, University of Alaska, Fairbanks. 56 p.

Kremer, Benedict Peter. On the equation of motion of a damped horizontal pendulum excited by simple harmonic vibrations. M, 1939, St. Louis University.

Kremer, Dale Ernest. Geology of Preston-Mt. Si area, Washington. M, 1959, University of Washington. 103 p.

Kremer, Lois Antoinette. Petrographic study of the Thorold-Oneida area, New York. M, 1932, University of Rochester. 116 p.

Kremer, Marguerite C. Early Cretaceous foreland sedimentation of the Lower Clastic Unit, Kootenai Formation, southern Gallatin County, Montana. M, 1982, University of Montana. 57 p.

Kremer, Thomas. SEM and XRD investigation of mineralogical transformations during weathering. M, 1983, University of Georgia. 115 p.

Kremin-Smith, Denise J. Hydrogeology of a portion of the Ogallala Aquifer, south-central Todd County, South Dakota. M, 1984, South Dakota School of Mines & Technology.

Kremser, Daniel Timothy. High pressure phase transitions in europous oxide. D, 1982, Washington University. 127 p.

Krenik, Kevin J. A subsurface study of the Pennsylvanian detrital unit, Andrews County, Texas. M, 1985, Texas Tech University. 84 p.

Krenning, Erna Louise. The geography of Corsica. M, 1929, Washington University. 67 p.

Krentz, Daniel Hugh. Shallow transient electromagnetic sounding at the Val Gagne Test Site, Ontario. M, 1987, Queen's University. 152 p.

Kresan, P. L. Cadmium content in sphalerites, copper ores, soils and plants in southern Arizona. M, 1975, University of Arizona.

Kresl, Ronald J. Geology of eastern Wells County, North Dakota. M, 1964, University of North Dakota. 110 p.

Kress, Margaret R. Geomorphology of Tunica Bayou, Louisiana. M, 1979, Louisiana State University.

Kress, Victor Charles, II. Iron-manganese exchange in coexisting garnet and ilmenite. M, 1986, SUNY at Stony Brook. 42 p.

Kresse, Timothy M. A hydrogeological and geochemical site characterization of a proposed landfill site, Searcy County, Arkansas. M, 1987, University of Arkansas, Fayetteville.

Kresz, David. Evolution of an Archean greenstone belt in the Stormy Lake - Kawashegamuk Lake area (stratigraphy, structure and geochemistry); western Wabigoon Subprovince, Northwest Ontario. M, 1984, Brock University. 262 p.

Kretchmer, Andrea Gail. Petrology and provenance of modern sands from the Cascade Range forearc and Canadian Rocky Mountain retroarc. M, 1987, University of California, Los Angeles. 216 p.

Kretsch, Donald Lee. Conodonts from near the Devonian-Mississippian boundary of southern Indiana. M, 1955, University of Missouri, Columbia.

Kretschmar, U. Phase relations involving arsenopyrite in the system Fe-As-S and their application. D, 1973, University of Toronto.

Kretschmar, Ulrich Horst. A study of the opaque minerals in the Whitestone Anorthosite, Dunchurch, Ontario. M, 1968, McMaster University. 132 p.

Kretz, R. A. The petrology of dykes in the Terrace area, British Columbia. M, 1955, Queen's University. 161 p.

Kretz, Ralph Albert. Chemical equilibrium in garnet, biotite, and hornblende gneisses from an area of Quebec. D, 1958, University of Chicago. 31 p.

Kreutzberger, Melanie E. Behavior of illite and chlorite during pressure solution of shaly limestone of the Kalkberg Formation, Catskill, New York. M, 1986, University of Michigan. 26 p.

Kreutzfeld, James E. Pore geometry and permeability of the St. Peter Sandstone in the Illinois Basin. M, 1982, University of Toledo. 353 p.

Kreutzwiser, R. D. An evaluation of Lake Erie shoreline flood and erosion hazard policy. D, 1978, University of Western Ontario.

Krewedl, Dieter A. The geology and geochemistry of the Middle Inlet, Wisconsin, molybdenum prospect (Precambrian). M, 1967, Bowling Green State University. 52 p.

Krewedl, Dieter Anton. Geology of the central Magdalena Mountains, Socorro County, New Mexico. D, 1974, University of Arizona.

Krey, Frank. Geology of the Dongola Quadrangle. M, 1962, [Southern Illinois University, Carbondale].

Krey, Frank. Notes on the geology and manganiferous ore deposits of the Cuyuna Range. M, 1918, University of Minnesota, Minneapolis. 49 p.

Kreycick, Karen A. A medial Pleistocene faunal assemblage from Saunders County, Nebraska. M, 1969, University of Nebraska, Lincoln.

Krezoski, John Roman. The influence of zoobenthos on fine-grained particle reworking and benthic solute transport in Great Lakes sediments. D, 1981, University of Michigan. 116 p.

Krhoda, George Okoye. Flow behaviour in model open-channel bends and implications for lateral migration in rivers. D, 1985, Simon Fraser University. 384 p.

Kriausakul, Nivat. Investigation of isoleucine epimerization in model peptides and fossil protein. D, 1979, University of Texas at Dallas. 130 p.

Kribbs, Gary M. Joint analysis in the Valley and Ridge and Allegheny Plateau in portions of Mineral and Hampshire counties, West Virginia, and Allegany and Garrett Counties, Maryland. M, 1982, University of Toledo. 149 p.

Krick, Irving P. Topography vs. air masses; a discussion of the phenomena associated with the passage of air masses over the irregularities of the Earth's surface. M, 1933, California Institute of Technology. 54 p.

Krickenberger, K. R. The relationship between the exchangeable cation content and the flocculation of clay minerals. D, 1977, University of Maryland. 351 p.

Kridelbaugh, Stephen Joseph. Kinetics of the reaction calcite$_{(s)}$ + quartz$_{(s)}$ = wollastonite$_{(s)}$ + Carbon Dioxide$_{(g)}$ at elevated temperatures and pressures. D, 1971, University of Colorado.

Krieg, Lenny Albert. Mathematical modelling of the behavior of the LaCoste and Romberg "G" gravity meter for use in gravity network adjustments and data analyses. D, 1982, Ohio State University. 185 p.

Kriegel, Robert V. Study of the correlation between the zooxanthellae/host polyp symbiosis and the latitudinal extent of hermatypic coral formations in southern Florida coastal waters. M, 1972, Florida Institute of Technology.

Krieger, Medora H. Geology of the 13th Lake Quadrangle, New York. D, 1943, Columbia University, Teachers College.

Krieger, Phillip. Geology of the zinc-lead deposits at Pecos, New Mexico. D, 1932, Columbia University, Teachers College.

Krieger, Phillip. Notes on X-ray diffraction study of the series calcite-rhodochrosite. M, 1930, Columbia University, Teachers College.

Kriengsiri, Pirote. A methodology for estimating the regional flood frequencies for northeastern Thailand. D, 1976, Oklahoma State University. 141 p.

Kriens, Bryan Jon. Tectonic evolution of the Ross Lake area, Northwest Washington, Southwest British Columbia. D, 1988, Harvard University. 340 p.

Krier, Donathon James. Geology of the southern part of the Gila Primitive Area, Grant County, New Mexico. M, 1980, University of New Mexico. 112 p.

Krill, Allan George. Tectonics of N.E. Dovrefjell, central Norway. D, 1980, Yale University. 190 p.

Krill, Karl E. Geology of parts of the Maxwell and Hoosier Breccia Reefs, Boulder County, Colorado. M, 1948, University of Colorado.

Krimmel, Carl P. The serpentinite of Presque Isle, Marquette County, Michigan. M, 1941, Northwestern University.

Krimsky, Glenn A. Flow direction of volcanic rocks in the northern part of the Mogollon-Datil Province (Catron County), New Mexico. M, 1969, University of New Mexico. 41 p.

Kring, David Allen. The petrology of meteoritic chondrules; evidence for fluctuating conditions in the solar nebula. D, 1988, Harvard University.

Krinitzsky, Ellis Louis. A fault plane cavern. M, 1947, University of North Carolina, Chapel Hill. 34 p.

Krinitzsky, Ellis Louis. Some physical and chemical properties of loess deposits in the Lower Mississippi Valley. D, 1950, Louisiana State University.

Krinsley, David. Trace element and mineralogical composition of modern and fossil shell material. D, 1956, University of Chicago. 55 p.

Krish, Edward J. Relationship of trace element distribution to level of erosion in some porphyry copper deposits and prospects, southwestern U. S. and northwestern Mexico. M, 1974, Colorado School of Mines. 156 p.

Krisher, Daniel Lee. Coral zonation and morphology in a Silurian reef, eastern Wisconsin. M, 1978, Indiana University, Bloomington. 36 p.

Krishna, J. Hari. Runoff prediction and rainfall utilization in the semi-arid tropics. D, 1979, Utah State University. 145 p.

Krishna, Paul P., Jr. A sedimentologic and paleontologic investigation of the Tres Montes region, southern Chile. M, 1988, Purdue University. 233 p.

Krishnamurthi, N. Simulation of gravitational water movement in soil. D, 1975, Colorado State University. 138 p.

Krishnan, N. Gopala. Sulfation kinetics of sodium aluminite and reduced alunite. D, 1972, Stanford University.

Krishnan, T. K. Structural studies of the Schefferville mining district, Quebec-Labrador, Canada. D, 1976, University of California, Los Angeles. 241 p.

Krishnanath, Raghava. Textures and microstructures of the high-grade metamorphic rocks of Kilimanoor-Chadayamangalam area, Kerala State, India. M, 1979, Syracuse University.

Krishtalka, L. Systematics and relationships of early Tertiary Lipotyphla (Mammalia, Insectivora) of North America. D, 1975, Texas Tech University. 168 p.

Krissek, Lawrence Alan. Sources, dispersal, and contributions of fine-grained terrigenous sediments on the Oregon and Washington continental slope. D, 1982, Oregon State University. 226 p.

Krist, Hazel Fagley. Morphology and analysis of an unstable slope in the lower Cuyahoga River valley, Ohio. M, 1976, Kent State University, Kent. 32 p.

Kristjansson, Leo Geir. A magnetic study of Tertiary igneous rocks from Greenland, Baffin Island and Iceland. D, 1973, Memorial University of Newfoundland. 316 p.

Kristoffersen, Y. Labrador Sea; a geophysical study. D, 1977, Columbia University, Teachers College. 192 p.

Kristofferson, Ole Herman. Hydrothermal experiments with lead and zinc minerals. D, 1934, University of Minnesota, Minneapolis. 32 p.

Kritikos, William Paul. Engineering-management report on the application of a mini-computer system for data management and experimental control in an enhanced oil recovery laboratory. D, 1981, University of Kansas. 425 p.

Krivanek, Connie Mac. Areal geology of Lane NE Quadrangle, Pushmataha and Atoka counties, Oklahoma. M, 1961, University of Oklahoma. 113 p.

Krivanek, Kenneth R. Adsorption of heavy metals on clays at various pH; application to treatment of acid mine spoil material by sewage sludge. M, 1976, Southern Illinois University, Carbondale. 106 p.

Krivz, Andrea L. The structural petrology of the Park Dome, Black Hills, South Dakota. M, 1973, University of Illinois, Chicago.

Kriz, George James. Determination of confined and unconfined aquifer parameters by dimensional analysis. D, 1965, University of California, Davis. 152 p.

Kriz, Stanislav Jaroslav. Stratigraphy and structure of the Whitaker Peak-Reasoner Canyon area, Ventura and Los Angeles counties, California. D, 1947, Princeton University. 88 p.

Krizanich, Gary W. Landsat trophic state assessment of Fellows Lake. M, 1986, Southwest Missouri State University.

Kroeger, Glenn Charles. Synthesis and analysis of teleseismic body wave seismograms. D, 1987, Stanford University. 144 p.

Kroeger, Timothy J. Paleoenvironmental significance of Upper Cretaceous palynomorph assemblages in the Hell Creek Formation, Butte County, South Dakota. M, 1985, South Dakota School of Mines & Technology.

Kroenke, L. W. Geology of the Ontong Java Plateau. D, 1972, University of Hawaii. 119 p.

Kroenke, L. W. The crustal structure of the Scotia Sea from geomagnetic and other geological investigations. M, 1968, University of Hawaii.

Kroft, D. J. The strategic position of the United States with respect to uranium supply and demand in the foreseeable future. D, 1976, Stanford University. 285 p.

Kroft, David Jeffrey. Sand and gravel deposits in western King County, Washington. M, 1972, University of Washington. 62 p.

Krog, Marilyn K. Size analysis of fine material and related size-frequency distributions. M, 1963, Rice University. 31 p.

Kroger, Robert S. Moss Beach. M, 1941, Stanford University.

Krogh, Thomas Edward. The titaniferous magnetite deposit of the Newboro District (Ontario). M, 1960, Queen's University.

Krogh, Thomas Edward. The titaniferous magnetite deposit of the Newboro District, Ontario. M, 1961, Queen's University. 180 p.

Krogstad, Eirik Jens. Timing and sources of Late Archean magmatism, Kolar area, South India; implications for Archean tectonics. D, 1988, SUNY at Stony Brook. 224 p.

Krohn, Douglas H. Gravity survey of the Mogollon Plateau volcanic province, southwestern New Mexico. M, 1972, University of New Mexico. 32 p.

Krohn, I. M. Functional adaptation in Miocene Glyptodontidae (Edentata, Mammalia) from the Honda Group, Colombia, South America. D, 1978, University of California, Berkeley. 159 p.

Krohn, J. P. Engineering geology of Santa Ynez Canyon, Santa Monica Mountains, Los Angeles, California. M, 1976, University of Southern California. 116 p.

Krohn, Melvyn Dennis. Relation of lineaments to sulfide deposits and fractured zones along Bald Eagle Mountain; Centre, Blair, and Huntingdon counties, Pennsylvania. M, 1976, Pennsylvania State University, University Park. 104 p.

Krokosz, Michael. Evaluation of an aerial photographic film/filter technique for the geologic mapping of the Silver Bell Mountains and Eagle Tail Mountains, Arizona. M, 1981, Northern Arizona University. 76 p.

Kroll, Richard L. Bedrock geology (lower Paleozoic) of the Norwalk North Quadrangle, Connecticut-New York. D, 1971, Syracuse University.

Kroll, Richard Lawrence. Structure and petrology of the northern part of the Barndoor intrusion (Triassic), (north-central Connecticut). M, 1970, University of Massachusetts. 137 p.

Krolow, Mark R. Resistivity and seismic measurements in the detection of fracture direction. M, 1982, University of Wisconsin-Milwaukee. 209 p.

Krom, Thomas D. Induced polarization; a geophysical method for detecting metal contaminated ground water. M, 1988, University of Idaho. 115 p.

Kromah, Fodee. The geology and occurrences of iron deposits in Liberia and the impact of mining on the environment. D, 1974, Cornell University.

Kromah, Fodee. The mineralogy and petrography of the Nimba iron ore. M, 1971, Michigan State University. 57 p.

Kron, Donald Gordon. Miocene mammals from the central Colorado Rocky Mountains. D, 1988, University of Colorado. 377 p.

Kron, Donald Gordon. Oligocene vertebrate paleontology of the Dilts Ranch area, Converse County, Wyoming. M, 1978, University of Wyoming.

Kron, Terry Ray. Patterns in Early Pennsylvanian sedimentation, Sullivan County, Indiana. M, 1987, Indiana University, Bloomington. 41 p.

Kronberg, B. I. The geochemistry of some Brazilian soils and geochemical considerations for agriculture on highly leached soils. D, 1977, University of Western Ontario. 137 p.

Kronberg, Merrily. Socio-environmental problems related to Nevada's rural county subdivisions. M, 1975, University of Nevada. 80 p.

Krone, Steven. The mineralogy of two urano-organic ores from the La Ventana Mesa area, Sandoval County, New Mexico. M, 1980, Rutgers, The State University, Newark. 36 p.

Kronenberg, Andreas K. Flow strengths of quartz aggregates; carbon and oxygen diffusion in calcite. M, 1979, Brown University. 273 p.

Kronenfeld, Kathi R. Depositional history of a Holocene peat deposit near Wilmington, North Carolina. M, 1982, University of North Carolina, Chapel Hill. 67 p.

Kronfeld, Joel. A comparison of rates of sedimentation between the clay fraction and the concomitant foraminiferal fraction in deep-sea cores by the Th^{230}/Th^{232}. M, 1968, Florida State University.

Kronfeld, Joel. Hydrologic investigations and the significance of U^{234}/U^{238} disequilibrium in the ground waters of central Texas. D, 1972, Rice University. 74 p.

Kronig, Donald M. Cavity detection by geophysical methods. M, 1977, University of Wisconsin-Madison.

Kronman, George. Shoreline and river processes at and near the mouth of Salmon River, Lake Ontario. M, 1979, SUNY, College at Fredonia. 128 p.

Kroon, Harris M. Sedimentation of the Bow Island Sandstone (Upper Cretaceous), Alberta, Canada. M, 1951, Northwestern University.

Kroopnick, Peter M. Oxygen and carbon in the oceans and atmosphere; stable isotopes as tracers for consumption, production, and circulation models. D, 1971, University of California, San Diego.

Kropf, F. W. Inter-relation of key-factors for infiltration of liquid domestic waste into soil. D, 1975, University of Connecticut. 74 p.

Kropp, Walter Paul. An alteration and light stable isotope study of the Bingham Canyon, Utah, porphyry system. M, 1982, University of Utah. 148 p.

Kropschot, Robert E. A quantitative sedimentary analysis of the Mississippian deposits in the Michigan Basin. M, 1953, Michigan State University. 57 p.

Krosky, S. Tectonic analysis of the Forester Creek Anticline, Larimer County, Colorado. M, 1974, University of Wyoming. 46 p.

Kross, Burton Clare. Assessment of landfill leachate toxicity and contamination of groundwater in Iowa. D, 1987, University of Iowa. 292 p.

Krothe, Noel Calvin. Factors controlling the water chemistry beneath a floodplain in a carbonate terrane, central Pennsylvania. D, 1976, Pennsylvania State University, University Park. 292 p.

Krothe, Noel Calvin. Quality changes in water beneath a streambed in a carbonate terrane, central Pennsylvania. M, 1973, Pennsylvania State University, University Park. 223 p.

Krotser, Donald J. Improved high resolution seismic profiling. M, 1966, Massachusetts Institute of Technology. 51 p.

Krotzer, Chris J. Earth resistivity measurements as applied to the location and study of fault zones. M, 1964, University of Tennessee, Knoxville. 49 p.

Krowski, Stanley P. A petrographic study of the Leon Mountain analcite syenogabbro in the Terlingua area, Brewster County, Texas. M, 1963, University of Houston.

Krstanovic, Predrag Felix. Application of entropy theory to multivariate hydrologic analysis. D, 1988, Louisiana State University. 596 p.

Krstic, Dragan. Geochronology of the Charlebois Lake area, northeastern Saskatchewan. M, 1981, University of Alberta. 127 p.

Krstulovic L., G. Influence of boundary weakening techniques on the block caving mining method. D, 1974, Columbia University. 149 p.

Krueger, Allen Reed. Eocene carbonates of Mazarani number one, West Pakistan. M, 1960, University of Illinois, Urbana. 32 p.

Krueger, Harold W. The theory and techniques of Ar^{40}/K^{40} age determination. M, 1959, University of Minnesota, Minneapolis. 45 p.

Krueger, James P. Development of oriented lakes in the Eastern Rainbasin region of south central Nebraska. M, 1986, University of Nebraska, Lincoln. 75 p.

Krueger, Paul G. Seismic-stratigraphic study of the Red Fork Sandstone in Kay County, Oklahoma. M, 1986, Colorado School of Mines. 79 p.

Krueger, Robert Carl. A subsurface study of Mississippian rocks in the Tulsa area. M, 1957, University of Oklahoma. 96 p.

Krueger, Scott Raymond. Mixed-layer illite/smectite and vitrinite reflectance as paleotemperature indicators in Mesozoic clastic rocks from the Rocky Mountain foothills of Alberta, Canada. M, 1985, University of Illinois, Urbana. 74 p.

Krueger, William Charles. Sedimentary petrology of Schenectady and a portion of the Austin Glen in eastern New York. M, 1960, University of Rochester. 88 p.

Kruer, Stacie Ann. Hydrothermal alteration, mineral chemistry and the evolution of hydrothermal solutions. M, 1983, University of Utah. 97 p.

Krug, Donald J. Pennsylvanian Ostracoda from the Putman Hill limestone and shale (lower Allegheny) of eastern Ohio. M, 1986, Bowling Green State University. 95 p.

Krug, Edward Charles. Geochemistry of pedogenic bog iron and concretion formation. D, 1981, Rutgers, The State University, New Brunswick. 227 p.

Krug, Jack A. The effect of stress on the petrophysical properties of some sandstones. D, 1977, Colorado School of Mines. 174 p.

Kruge, Michael Anthony. Organic geochemistry and comparative diagenesis; Miocene Monterey Formation, Lost Hills oil field and vicinity, West San Joaquin Basin, California. D, 1985, University of California, Berkeley. 277 p.

Kruger, Anne Longsworth. Industrial waste disposition in New Jersey; an ecological perspective. D, 1981, Rutgers, The State University, New Brunswick. 482 p.

Kruger, Frederick C. A petrographic study of the red and gray drifts of Minnesota. M, 1936, University of Minnesota, Minneapolis. 30 p.

Kruger, Frederick C. The structure and metamorphism of the Bellows Falls Quadrangle of New Hampshire and Vermont. D, 1941, Harvard University.

Kruger, Gustav Otto, Jr. The development of dentition from a paleontological aspect. M, 1939, George Washington University. 72 p.

Kruger, John M. Late Proterozoic and Early Cambrian structural setting of western Montana and the stratigraphy and depositional environment of the Flathead Sandstone in west-central Montana. M, 1988, University of Montana. 109 p.

Kruger, Joseph Michael. Regional gravity anomalies in the Ouachita system and adjacent areas. M, 1983, University of Texas at El Paso. 197 p.

Kruger, Steven T. Studies of H-3, He-3, and He-4 production in light targets and meteorites by high energy protons. M, 1970, Rice University. 190 p.

Kruger, William Charles, Jr. Mineralogical composition and textural properties of river sediments from British Honduras. D, 1963, Rice University. 116 p.

Kruk, Taras. Paleocene-Eocene nannofossil biostratigraphy of southwestern Santa Cruz Island. M, 1987, California State University, Long Beach. 103 p.

Krukowski, Stanley T. Mineralogy and geochemistry of Upper Cretaceous clay-bearing strata, Torreon Wash/Johnson Trading Post areas, southeastern San Juan Basin, New Mexico. M, 1983, New Mexico Institute of Mining and Technology. 93 p.

Krulik, Joseph W. A hydrogeologic study of Franklin Township, Portage County, Ohio. M, 1982, Kent State University, Kent. 175 p.

Krum, William Mark. A petrologic study of the gray St. Cloud intrusive. M, 1935, University of Minnesota, Minneapolis. 78 p.

Krumbach, Keith Ronald. Paleoecology of Mid-Cretaceous calcareous nannofossils in the Atlantic and Indian oceans with a generic classification. M, 1982, University of Utah. 141 p.

Krumbein, William C. The mechanical analysis of related samples of glacial tills. D, 1932, University of Chicago. 26 p.

Krumenacher, Mark J. Detection of organic ground water contamination using the induced polarization method. M, 1987, University of Wisconsin-Milwaukee. 206 p.

Krumhansl, James Lee. Geochemistry of tungsten. D, 1977, Stanford University. 121 p.

Krumme, George W. Mid-Pennsylvanian source reversal on the Oklahoma Platform. D, 1975, University of Tulsa. 151 p.

Krummel, William J., Jr. The geology of a portion of the Weldon Spring Quadrangle. M, 1956, Washington University. 124 p.

Krumpolz, Bradley J. Paleoecology, biostratinomy and sedimentology of the Mt. Auburn Member (McMillan Formation, uppermost Maysvillian) in the area of Cincinnati, Ohio. M, 1980, University of Cincinnati. 196 p.

Krupa, John. Geophysical and structural investigation of adjacent sections of Hancock, Hardin, Allen and Putnam counties in northwestern Ohio. M, 1980, Wright State University. 129 p.

Krupka, Kenneth Michael. Neutron activation analysis of Na/K ratios in fluid inclusions in quartz from gold-quartz veins at the O'Brien Mine, Quebec, Canada. M, 1976, Pennsylvania State University, University Park. 42 p.

Krupka, Kenneth Michael. Thermodynamic analysis of some equilibria in the system $MgO \cdot SiO_2 \cdot H_2O$. D, 1984, Pennsylvania State University, University Park. 396 p.

Kruse, Curtis. A new method of nonlinear signal correlation using the instantaneous spectrum. M, 1988, Colorado School of Mines. 112 p.

Kruse, Fred A. Cause of color differences on Landsat color-ratio-composite images of limonitic areas in Southwest New Mexico. M, 1984, Colorado School of Mines. 205 p.

Kruse, Fred A. Use of high spectral resolution remote sensing to characterize weathered surfaces of hydrothermally altered rocks. D, 1987, Colorado School of Mines. 150 p.

Kruse, Henry Oscar. Some Eocene dicotyledonous woods from Eden Valley, Wyoming. D, 1952, University of Cincinnati. 64 p.

Krusekopf, Henry Herman, Jr. Geology of the Tendoy Range near Dell, Beaverhead County, Montana. M, 1948, University of Michigan. 53 p.

Krusekopf, Lily Marie Carter. Permian Phosphoria Formation in northwestern Wyoming and eastern Idaho. M, 1947, University of Michigan.

Krushensky, Richard D. Insoluble residues and stratigraphy of a Trenton-Black River section from the Northville Pool, Michigan. M, 1957, Wayne State University.

Krushensky, Richard Dean. Geology of the volcanic features of the Hurricane Mesa area, Park County, Wyoming. D, 1960, Ohio State University. 258 p.

Krutak, Paul Russell. Jackson Eocene Ostracoda from Keyser Hill, Alabama. M, 1960, Louisiana State University.

Krutak, Paul Russell. Structure, stratigraphy, and provincial relationships of Sierra de la Gavia, Coahuila, Mexico. D, 1963, Louisiana State University. 213 p.

Kruyter, Mark de *see* de Kruyter, Mark

Kryczka, Adam Alexander William. The Nikanassin Formation of the type near Cadomin, Alberta. M, 1959, University of Alberta. 136 p.

Kryger, Adolph H. Microclimate across the margin of the Sukkertoppen ice cap in Tasersiaq area, Greenland. M, 1965, Ohio State University.

Krygowski, Daniel. A general expression for body wave acoustic velocities in terms of bulk density, molecular weight, and melting temperatures. M, 1974, Colorado School of Mines. 55 p.

Krygowski, Daniel. The use of well logs to predict water saturation and recoverable hydrocarbon volumes in shaly sand reservoirs. D, 1977, Colorado School of Mines. 238 p.

Krynine, Paul D. Triassic sedimentary rocks of southern Connecticut. D, 1936, Yale University.

Krystinik, Jon G. Upper Minnelusa anhydrites; key to interpreting facies relationships. M, 1988, Baylor University. 170 p.

Krystinik, Lee Franklin. Pore-filling cements in turbidites, southern California; products of early diagenesis and dewatering of shale. D, 1981, Princeton University. 161 p.

Krywany, Joseph M. A biostratigraphic study of the Prout Limestone of north-central Ohio. M, 1982, University of Toledo. 150 p.

Kryza, Edward A., Jr. Petrology, depositional environment, and slake durability of some Middle Pennsylvanian coal measure shales. M, 1983, University of Cincinnati. 149 p.

Krzysztofowicz, Roman. Preferential short-range reservoir control. D, 1978, University of Arizona.

Ku, Chao-cheng. Paleomagnetic investigations of some of the Precambrian rocks in the Saint Francois Mountains, Missouri. M, 1965, Washington University. 86 p.

Ku, Chi Young. Data processing and geophysical interpretation of Magsat satellite magnetic anomaly data over mainland China. M, 1985, University of Iowa. 131 p.

Ku, Henry Fu Heng. Ground-water geology of the Augusta Township, Washtenaw County, and vicinity (Michigan). M, 1966, University of Michigan.

Ku, Kelly T. Sediment accretion and subduction structures in the Nankai Trough. M, 1985, Texas A&M University. 139 p.

Ku, Kong-Gyiu. Two new planimetric methods for torsion balance terrain corrections. M, 1936, Colorado School of Mines. 31 p.

Ku, Teh-Lung. Uranium series disequilibrium in deepsea sediments. D, 1966, Columbia University. 201 p.

Ku, Tsu-Wei. Fault detection and location in arraystructured VLSI. D, 1988, University of Washington. 148 p.

Ku, W.-C. Equilibrium adsorption of inorganic phosphate by lake sediments. D, 1975, University of Massachusetts. 205 p.

Kuan, Soong. The geology of Carboniferous volcanic rocks in the Harvey area, New Brunswick (Canada). M, 1970, University of New Brunswick.

Kuang, Jian. Intraplate stress and seismicity in the Southeastern United States. D, 1988, Georgia Institute of Technology. 135 p.

Kubera, Paula A. The relation of environment of deposition, clay mineralogy and hydrology to geopressure seals in the Hackberry Wedge, Calcasieu Parish, Louisiana. M, 1977, Louisiana State University.

Kubicek, Leonard. Facies distribution and oil accumulation in the Aux Vases Formation (upper Mississippian) Plumfield area, southwest Franklin County, Illinois. M, 1968, Southern Illinois University, Carbondale. 42 p.

Kubik, West T. Stratigraphy of the Wall Creek Member of the Frontier Formation, Laramie Basin, Wyoming. M, 1982, Colorado School of Mines. 127 p.

Kubilius, Walter Paris. Sulfur isotopic evidence for country rock contamination of granitoids in southwestern Nova Scotia. M, 1983, Pennsylvania State University, University Park. 103 p.

Kublick, Ernest E. Potassium-argon dating of slates from the (Lower Paleozoic) Meguna Group, Nova Scotia (Canada). M, 1972, Dalhousie University.

Kubota, H. The iodine content of the soils and rocks of Hawaii. M, 1936, University of Hawaii. 40 p.

Kubuck, E. E. Potassium-argon dating of slates from the Meguma Group, Nova Scotia. M, 1972, Dalhousie University.

Kucek, Leo. Geophysical investigation of the possible northeast extension of the New Madrid rift complex; magnetics. M, 1984, Purdue University. 108 p.

Kucera, Richard Edward. Geology of the Joes Valley and North Dragon area, central Utah. M, 1954, Ohio State University.

Kucera, Richard Edward. Geology of the Yampa District, Northwest Colorado. D, 1962, University of Colorado. 844 p.

Kuchenbuch, Pamela A. Petrology of some metagabbro bodies in the Mars Hill Quadrangle, western North Carolina. M, 1979, Eastern Kentucky University. 56 p.

Kuchs, Oscar M. Geology and ore deposits of southeastern Arizona. M, 1905, University of Kansas.

Kucinskas, D. P. The seismic stratigraphy of shoestring sands, southeastern Kansas. M, 1975, Boston College.

Kucinski, Russell. Geology and mineralization of the Ruby Tuesday Claim Block, Prince of Wales Island, Southeast Alaska. M, 1988, University of Alaska, Fairbanks. 105 p.

Kuck, P. H. The behavior of molybdenum, tungsten, and titanium in the porphyry copper environment. D, 1978, University of Arizona. 296 p.

Kuck, Peter H. The insonation of silicates and oxides selected for radioactive dating. M, 1969, University of Arizona.

Kucsma, Paul J. Geology of the northeast quarter of the Booneville Quadrangle, Logan County, Arkansas. M, 1978, Northeast Louisiana University.

Kucuk, Fikri. Transient flow in elliptical systems. D, 1978, Stanford University. 115 p.

Kucukcetin, Adnan Mehmet. The genera Idonearca and Trigonarca from the Comanchean of Texas. M, 1946, University of Texas, Austin.

Kuder, Harry Bruce. The petrology of the limestones of the Liberty and lower Whitewater formations (Cincinnatian Series), northeastern Warren County, Ohio. M, 1964, Miami University (Ohio). 82 p.

Kudlac, John J. Minor elements in quartz. M, 1965, Michigan State University. 51 p.

Kudlac, John Joseph. Tri- and diethanolamine clay complexes and their use in clay mineral identification. D, 1972, University of Pittsburgh. 115 p.

Kudo, Akira. Radioactivity transport in water-interaction between flowing water and bed sediment. D, 1969, University of Texas, Austin. 168 p.

Kudo, Albert Masakiyo. A study of the phase equilibrium of granitic melts. D, 1967, University of California, San Diego. 153 p.

Kudo, Albert Masakiyo. The discriminant function used in classifying some amphibolites. M, 1962, McMaster University. 121 p.

Kuecher, Gerald Joseph. Rhythmic sedimentation and stratigraphy of the Middle Pennsylvanian Francis Creek Shale near Braidwood, Illinois. M, 1983, Northeastern Illinois University. 143 p.

Kuechler, Adolph Harmon. A preliminary study of certain Montana clays. M, 1933, Montana College of Mineral Science & Technology. 38 p.

Kueffner, Mary H. E. Eskers. M, 1940, [Carleton University].

Kuehl, Steven Alan. Sediment accumulation and the formation of sedimentary structure on the Amazon continental shelf. D, 1985, North Carolina State University. 223 p.

Kuehn, Carl Anton. Geology and exploration geochemistry of the Big Creek and Kingston Canyon areas, Toiyave Range, Lander County, Nevada. M, 1984, Pennsylvania State University, University Park. 217 p.

Kuehn, Deborah Wilbur. Characterization of the organic structure of the Lower Kittanning Coal seam using Fourier transform infrared spectroscopy and optical properties. D, 1983, Pennsylvania State University, University Park. 404 p.

Kuehn, Deborah Wilbur. Offshore transgressive peat deposits of Southwest Florida; evidence for a late Holocene rise of sea level. M, 1980, Pennsylvania State University, University Park. 104 p.

Kuehn, Kenneth William. An automated microscopical method for the characterization of pyrite in coal. M, 1979, Pennsylvania State University, University Park. 105 p.

Kuehn, Kenneth William. The petrographic characterization of coals by automated reflectance microscopy and its application to the prediction of yields in coal liquefaction. D, 1982, Pennsylvania State University, University Park. 463 p.

Kuehn, William Jackson. Geology and petrographic studies of the marble canyon igneous complex (Tertiary), Culberson County, Texas. M, 1969, Texas Christian University. 60 p.

Kuehner, Irvin Verne. A geologic study of the soundness of limestone for use as concrete aggregate. M, 1956, Michigan State University. 70 p.

Kuehner, Scott Milton. Petrogenesis of ultrapotassic rocks, Leucite Hills, Wyoming. M, 1980, University of Western Ontario. 201 p.

Kuellmer, Frederick J. Geologic section of the Black Range at Kingston, New Mexico. D, 1952, University of Chicago. 188 p.

Kuentag, Chumpon. The plastic clay of the Wilcox Group, Henderson District, Rusk County, Texas. M, 1973, Northeast Louisiana University.

Kuentz, David C. The Otowi Member of the Bandelier Tuff; a study of the petrology, petrography, and geochemistry of an explosive silicic eruption, Jemez Mountains, New Mexico. M, 1986, University of Texas, Arlington. 168 p.

Kuenzi, Laurence M. Stratigraphy of the Pennsylvanian Amsden Formation, southwestern Montana. M, 1951, University of Wisconsin-Madison. 45 p.

Kuenzi, Wilbur David. Geology of the Kelly Hill area, Stevens County, Washington. M, 1961, University of Montana. 111 p.

Kuenzi, Wilbur David. Tertiary stratigraphy in the Jefferson River basin, Montana. D, 1966, University of Montana. 293 p.

Kuenzler, Howard W. An experimental in situ densitometer. M, 1968, Massachusetts Institute of Technology. 55 p.

Kues, Barry Stephen. The fauna and the paleoecology of the Oketo Member, Barneston Limestone (lower Permian), of Kansas and Nebraska. D, 1975, Indiana University, Bloomington. 351 p.

Kuespert, Jonathan Godard. The depositional environment and provenance of the Miocene Temblor Formation and associated Oligo-Miocene units in the vicinity of Kettleman North Dome, San Joaquin Valley, California. M, 1983, Stanford University. 105 p.

Kuest, Louis John, Jr. Genesis of upper Plattin carbonates. M, 1963, Washington University. 56 p.

Kufs, Charles T., Jr. A statistical analysis of the characteristics of slumps near Athens, Ohio. M, 1978, Ohio University, Athens. 128 p.

Kugler, Harry Wesley. The structure and stratigraphy of the Alfordsville area, Daviess County, Indiana. M, 1951, Indiana University, Bloomington. 19 p.

Kugler, Ralph Leonard. Regional petrologic variation, Jurassic and Cretaceous sandstone and shale, Neuquen Basin, west-central Argentina. D, 1987, University of Texas, Austin. 183 p.

Kugler, Ralph Leonard. Stratigraphy and petrology of the Bushnell Rock Member of the Lookingglass Formation, southwestern Oregon Coast Range. M, 1979, University of Oregon. 118 p.

Kuh, Hsien-Chien. Numerical model of a regional aquifer thermal energy storage system. D, 1982, Texas A&M University. 241 p.

Kuhaida, Andrew Jerome, Jr. Mass movements resulting from intense rainfall, Douglas County, Colorado; a morphometric analysis. D, 1980, University of Denver. 216 p.

Kuhl, Tim O. Precambrian geology of the Deerfield area, Pennington County, South Dakota. M, 1982, South Dakota School of Mines & Technology. 70 p.

Kuhleman, Milton Henry. Mississippian and Lower Pennsylvanian stratigraphy of portions of Stonewall and Atoka quadrangles, Oklahoma. M, 1948, University of Oklahoma. 52 p.

Kuhlman, Robert. Stratigraphy and metamorphism of the Wissahickon Formation in southeastern Chester County, Pennsylvania. M, 1975, Bryn Mawr College. 47 p.

Kuhlman, Steven Larry. Late Precambrian and Phanerozoic thermal and tectonic history of the western Canadian Shield, Manitoba and Ontario; evidence from fission-track analysis of apatite. M, 1987, University of Oklahoma. 82 p.

Kuhlthau, Richard Harold. Sensitivity analysis of a regional ground-water model of the Accomac-Onancock region of the Eastern Shore of Virginia. M, 1979, University of Virginia. 215 p.

Kuhn, Alan Karl. A geological study and engineering evaluation of the Strait of Gibraltar area. D, 1973, University of Illinois, Urbana. 107 p.

Kuhn, Alan Karl. Hydrogeology of the Fox Hills aquifer, north Kiowa-Bijou District, Colorado. M, 1968, Colorado State University. 61 p.

Kuhn, Bernard J. Stratigraphy and geologic evolution of Anclote Key, Pinellas County, Florida. M, 1983, University of South Florida, Tampa. 79 p.

Kuhn, Carlos Alfredo Clebsch. Environment of deposition and diagenesis of the Wolfcamp Formation, Nolley Field, Andrew County, Texas. M, 1986, Texas Tech University. 69 p.

Kuhn, Douglas Eugene. Stratigraphy and petrology of the lower half of the Everton Formation (Early-Middle Ordovician), Newton, Boone and Carroll counties, Arkansas. M, 1984, University of New Orleans. 188 p.

Kuhn, Jeffrey A. The stratigraphy and sedimentology of the middle Proterozoic Grinnell Formation, Glacier National Park and the Whitefish Range, NW Montana. M, 1987, University of Montana. 122 p.

Kuhn, Mark A. Sedimentology and sandstone petrogenesis of the Hartshorne Formation, southeastern Arkoma Basin, West-central Arkansas. M, 1981, University of Missouri, Columbia.

Kuhn, Matthew Randell. Micromechanical aspects of soil creep. D, 1987, University of California, Berkeley. 198 p.

Kuhn, Michael William. Ground water in the Santa Clara Valley, Santa Clara County, California. D, 1965, University of California, Los Angeles. 153 p.

Kuhn, Paul W. Magma immiscibility in the Box Elder Laccolith of north-central Montana. M, 1983, University of Montana. 86 p.

Kuhn, Thomas Alfred. The geology of a portion of the Berlin Quadrangle, Pennsylvania. M, 1952, University of Pittsburgh.

Kuhn, Truman H. Geology and ore deposits of the Copper Creek, Arizona, area. D, 1940, University of Arizona.

Kuhnert, Richard Franklin. Contact metamorphism of a mafic intrusion in Hartford County, Maryland. D, 1973, Millersville University.

Kuhnhenn, G. L. Carbonate microfacies of the Platteville Group (middle Ordovician), Lee and Lasalle counties, Illinois. D, 1976, University of Illinois, Urbana. 94 p.

Kuhnhenn, Gary L. The petrology of silica in the Boyle sequence of western Estill County, Kentucky. M, 1972, Eastern Kentucky University. 51 p.

Kuhnle, Roger Alan. Experimental studies of heavy mineral transportation, segregation, and deposition in gravel-bed streams. D, 1986, Massachusetts Institute of Technology. 209 p.

Kuhns, Mary Jo Pankratz. Late Cenozoic deposits of the Lower Clearwater Valley, Idaho and Washington. M, 1980, Washington State University. 71 p.

Kuhns, Roger James. Structural and chemical aspects of the Lochsa geothermal system near the northern margin of the Idaho Batholith. M, 1980, Washington State University. 103 p.

Kuhns, Roger James. The Golden Giant Deposit, Hemlo, Ontario; geologic and geochemical relationships between mineralization, alteration, metamorphism, magmatism and tectonism. D, 1988, University of Minnesota, Minneapolis. 457 p.

Kuich, Nicholas Franklin. Carboniferous stratigraphy of the Sloan area, San Saba County, Texas. M, 1964, University of Texas, Austin.

Kuiper, John L. Stratigraphy and sedimentary petrology of the Mascall Formation, eastern Oregon. M, 1988, Oregon State University. 153 p.

Kuivila, Kathryn Marie. Methane production and cycling in marine and freshwater sediments. D, 1986, University of Washington. 170 p.

Kujawa, Frank Benedict. A geometric analysis of stable and metastable phase equilibria in unary and binary systems. M, 1968, The Johns Hopkins University. 233 p.

Kulachol, Konthi. Development of offshore gas fields using barge-mounted methanol plants. M, 1980, Stanford University.

Kulander, Byron Rodney. A structural analysis of Browns Mountain Anticline in West Virginia. D, 1968, West Virginia University. 338 p.

Kulander, Byron Rodney. Geology of the Paw Paw and Oldtown, West Virginia, quadrangles. M, 1964, West Virginia University.

Kulansky, Gerald Harry. Sedimentology of the Fort Payne Formation (Osage), Scott County, Tennessee; case history of a waulsortian reservoir. D, 1981, Rensselaer Polytechnic Institute. 209 p.

Kulas-Adler, Helen A. Silicified Permian gastropods of the Park City Formation in Wyoming. M, 1984, University of Wyoming. 189 p.

Kulatilake, Pinnaduwa Howa S. W. Probabilistic approach to deformation and strength properties of shale mass. D, 1981, Ohio State University. 166 p.

Kulibert, Richard James. Delineation of the ancestral drainage paths of the Mad River, near Dayton, Ohio. M, 1979, Wright State University. 64 p.

Kulig, John Joseph. A sedimentation model for the deposition of glacigenic deposits in central Alberta. M, 1985, University of Alberta. 245 p.

Kulik, Joseph W. Stratigraphy of the Deadwood Formation, Black Hills, South Dakota and Wyoming. M, 1965, South Dakota School of Mines & Technology.

Kulkarni, Srikant N. Review of manganese ore deposits, production, metallurgy with special emphasis on India. M, 1973, Stanford University.

Kulla, J. B. Oxygen and hydrogen isotopic fractionation factors determined in experimental clay-water systems. D, 1979, University of Illinois, Urbana. 106 p.

Kulland, Roy E. Geology of the North Beulah, Center, and Glenharold lignite mines in Mercer and Oliver counties, southwestern North Dakota. M, 1975, University of North Dakota. 57 p.

Kullman, John D. Stratigraphy and structure of Floyd County, Iowa (Paleozoic). M, 1968, University of Iowa. 65 p.

Kulm, LaVerne Duane. Sediments of Yaquina Bay, Oregon. D, 1965, Oregon State University. 184 p.

Kulpanowski, Stephen E. Morphology and microstructure of nonmarine algal carbonates; North Horn and Flagstaff formations (Maestrichtian to Eocene), Gunnison Plateau, central Utah. M, 1987, Wayne State University. 94 p.

Kulpecz, Alexander A. Non-isothermal phosphide growth in Emery, a mesosiderite. M, 1978, Rutgers, The State University, New Brunswick. 49 p.

Kulvanich, Sermsakdi. Micas of the Panasqueira tin-tungsten deposits, Portugal. M, 1976, University of Michigan.

Kulyk, Valerie-ann. Holocene foraminifera of the eastern Nile Delta, Egypt. M, 1987, George Washington University. 57 p.

Kumamoto, Lawrence H. Microearthquake surveys of Snake River plain and Northwest Basin and Range geothermal areas. D, 1976, Colorado School of Mines. 171 p.

Kumamoto, Lawrence H. Redundant strain observations in southern Nevada; strain steps from the Handley nuclear test and secular strains. M, 1973, Colorado School of Mines. 73 p.

Kumanchan, Prasert. Mineralogy, petrology and field studies of the deposits of the Pleistocene Tule Basin, Briscoe and Twisher counties, Texas. M, 1972, Northeast Louisiana University.

Kumar, Ashok. Induced polarization in sedimentary rocks. M, 1962, University of California, Berkeley. 47 p.

Kumar, M. Monitoring of crustal movements in the San Andreas fault zone by a satellite-borne ranging system. D, 1976, Ohio State University. 152 p.

Kumar, Madhurendu Bhushan. Computer-aided subsurface structural analysis of the Miocene formations of the Bayou Carlin-Lake Sand area, South Louisiana. D, 1972, Louisiana State University. 231 p.

Kumar, Subodh. A study of the effects of simulated weathering and repeated loads on four lime stabilized Oklahoma shales. D, 1974, University of Oklahoma. 217 p.

Kumarapeli, P. S. The Saint Lawrence Valley system and its tectonic significance. D, 1974, McGill University. 394 p.

Kumbalek, Steven Charles. Digital processing and interpretation of a seismic reflection survey on the Mississippi River near New Madrid, Missouri. M, 1983, University of Wisconsin-Milwaukee. 114 p.

Kumbhojkar, Arvind Sadashiv. Theoretical and numerical analysis of geotechnical structures. D, 1987, SUNY at Buffalo. 222 p.

Kume, Jack. Investigation of the Bakken and Englewood formations Kinderhookian of North Dakota and northwestern South Dakota. M, 1960, University of North Dakota. 86 p.

Kuminecz, Cary Phillip. Lithologic and depositional variability within shoal-water deltaic and interdeltaic environments of the Perth Cyclothem and adjacent rocks (Atokan-Desmoinesian) of West-central Indiana. M, 1980, Indiana University, Bloomington. 344 p.

Kumkum, Ray. Fossiliferous calcareous concretions from the Rio Congo tidal flat, Gulf of San Miguel, western Gulf of Panama. M, 1973, University of Wisconsin-Milwaukee.

Kummel, Bernhard. Geological reconnaissance of the Contamana region, Peru. D, 1949, Columbia University, Teachers College.

Kummel, Henry Barnard. Lake Passaic, an extinct glacial lake. D, 1895, University of Chicago. 89 p.

Kump, Lee R. The global sedimentary redox cycle. D, 1986, University of South Florida, St. Petersburg. 261 p.

Kun, Peter. Geology of the Lakeview mining district, Idaho. M, 1970, University of Idaho. 135 p.

Kunar, Lloyd S. N. A marine seismic refraction system. D, 1970, University of Toronto.

Kunar, Lloyd S. N. A numerical ray-tracing method for seismology. M, 1968, University of Toronto.

Kunasz, Ihor Andrew. Geology and geochemistry of the lithium deposit in Clayton Valley, Esmeralda County, Nevada. D, 1970, Pennsylvania State University, University Park. 114 p.

Kunasz, Ihor Andrew. Significance of laminations in the upper Silurian evaporite deposit of the Michigan Basin. M, 1968, Pennsylvania State University, University Park. 62 p.

Kundert, Charles J. The geology of the Whittier-La Habra area, Southern California. M, 1951, Pomona College.

Kunen, S. M. Characterization of insoluble carbonaceous material in atmospheric particulates by pyrolysis/gas chromatography/mass spectrometry procedures. D, 1978, University of Arizona. 128 p.

Kung, Hsiang-Te. Geographic aspects of the urban hydrology of Knoxville, Tennessee. D, 1980, University of Tennessee, Knoxville. 241 p.

Kung, Samuel King-Jau. Soil erosion by raindrop impact. D, 1984, Cornell University. 99 p.

Kung, Shyh-Yuan. Effect of ground motion characteristics on the seismic response of torsionally coupled elastic systems. D, 1982, University of Illinois, Urbana. 230 p.

Kunishi, Harry Mikio. Factors affecting the release of soil potassium to exchangeable form on drying. D, 1963, University of Wisconsin-Madison. 96 p.

Kuniyoshi, Shingi. Geology of the Menzies Mountain area, British Columbia, Canada. M, 1966, University of California, Los Angeles.

Kuniyoshi, Shingi. Petrology of the Karmutsen (Volcanic) Group, northeastern Vancouver Island, British Columbia. D, 1972, University of California, Los Angeles.

Kunkel, James Robert. Analysis of a multipurpose water resource system in southeastern Mexico. D, 1974, University of Arizona.

Kunkel, Richard B. Linear regression models of sound velocity in the North Atlantic Ocean below a critical depth. D, 1970, University of Missouri, Rolla.

Kunkle, George Robert. Multiple glaciation in the Jago River area, northeastern Alaska. M, 1958, University of Michigan.

Kunkle, George Robert. The ground water geology and hydrology of Washtenaw County and the Upper Huron River basin. D, 1961, University of Michigan. 270 p.

Kunkle, Samuel H. Water quality of a mountain watershed in Colorado. M, 1967, Colorado State University. 150 p.

Kunst, Henery. Mesozoic stratigraphy of the Foothills region of Alberta. M, 1941, University of Alberta. 164 p.

Kunter, Richard Sain. Heavy media separation of western phosphorites. M, 1969, University of Idaho. 83 p.

Kuntz, Gregory Brent. Chert within the upper member of the Pennsylvanian Hermosa Formation; southeastern Utah. M, 1988, University of Nebraska, Lincoln. 62 p.

Kuntz, Mel A. Compositional variation in the Buck Creek Dunite, Clay County, North Carolina. M, 1964, Northwestern University.

Kuntz, Mel Anton. Petrogenesis of the Buckskin Gulch Intrusive Complex (latest Cretaceous) northern Mosquito Range, Colorado. D, 1968, Stanford University. 200 p.

Kuntz, Michael G. Effects of land use on ground water recharge and discharge rates. M, 1974, University of Toledo. 110 p.

Kuntz, Timothy. The geology of the Vanport Limestone (Pennsylvanian) in Elk County, Pennsylvania. M, 1986, University of Pittsburgh.

Kunz, Howard E. Subsurface geology of southwestern Lincoln County, Oklahoma. M, 1961, University of Oklahoma. 71 p.

Kunze, Adolf Wilhelm Gerhard. Creep response of the lunar crust in mare regions from an analysis of crater deformation. D, 1973, Pennsylvania State University, University Park. 139 p.

Kunze, Florence Mollie. Experimental modeling of the microscopic and mesoscopic structures of upper mantle diapirs. M, 1978, Rice University. 101 p.

Kunze, Florence Mollie. Silica preservation in the oceans. D, 1980, Rice University. 318 p.

Kunzer, Alexander Hourwich. Areal geology of eastern Jefferson County, Oklahoma. M, 1970, University of Oklahoma. 65 p.

Kunzinger, Frederick William, Jr. Climate and its relationship to the hydrology, morphometry, and soils of selected drainage basins in Virginia. M, 1981, Old Dominion University. 73 p.

Kunzler, Robert Henry. The aragonite-calcite transformation. D, 1969, Florida State University. 146 p.

Kuo, Ban-Yuan. Variations in physical properties of the oceanic crust and upper mantle. D, 1988, Brown University. 122 p.

Kuo, Ching-Liang. Modeling of dynamic deformation mechanisms for granular material. D, 1983, University of Massachusetts. 222 p.

Kuo, Frank Fu-Kwei. A land use map of central and eastern Oktibbeha County, Mississippi, 1958; its interpretation. M, 1971, Mississippi State University. 143 p.

Kuo, Hsiao-Yu. Palladium, iridium and gold in deep-sea cores. M, 1970, McMaster University. 94 p.

Kuo, Hsiao-Yu. Rare earth elements in the Sudbury nickel irruptive. D, 1975, McMaster University. 242 p.

Kuo, James Shaw-Han. On the nonlinear dynamic response of arch dams to earthquakes; I, Fluid-structure interaction; added-mass computations for incompressible fluid; II, Joint opening nonlinear mechanism; in-

terface smeared crack model. D, 1982, University of California, Berkeley. 202 p.

Kuo, John Tsung Fen. Theoretical and experimental study of seismic surface waves. D, 1958, Stanford University. 199 p.

Kuo, Kung Chia. Compression strength testing of the Springfield Coal, Coal V, Pike County, Indiana. M, 1986, Purdue University.

Kuo, Lung-Chuan Joseph. Kinetics of crystal dissolution in the system diopside-forsterite-silica. D, 1983, University of Illinois, Urbana. 124 p.

Kuo, Lung-Chuan Joseph. Morphology and zoning patterns of plagioclase in phyric basalts from DSDP legs 45 and 46, Mid-Atlantic Ridge. M, 1980, University of Illinois, Urbana. 74 p.

Kuo, Mao-Kuen. Crack kinking and crack forking under stress wave loading. D, 1984, Northwestern University. 143 p.

Kuo, Ming-Ching T. Shape factor correlations for transient heat conduction from irregular-shaped rock fragments to surrounding fluid. D, 1977, Stanford University. 195 p.

Kuo, Rong-Heng. Seepage from irrigation canals in Mesilla Valley, New Mexico. D, 1983, New Mexico State University, Las Cruces. 225 p.

Kuo, Say Lee. Geology and geochemistry of stratabound ore deposits in South-central Yukon Territory and southwestern District of Mackenzie, Northwest Territories. D, 1976, University of Alberta. 597 p.

Kuo, Say Lee. Uranium-lead geochronology of Kenoran rocks and minerals (Precambrian) of the Charles Lake area, Alberta. M, 1972, University of Alberta. 126 p.

Kuo, Shih-Yeng. Long wavelength static analysis using first breaks and generalized linear inversion. D, 1985, Colorado School of Mines.

Kuo, Tsai-Bao. Well log correlation using artificial intelligence. D, 1986, Texas A&M University. 150 p.

Kupecz, Julie A. Depositional environments, diagenetic history and petroleum entrapment in the Mississippian Mission Canyon Fm., Billings Anticline, North Dakota. M, 1984, Colorado School of Mines. 251 p.

Kupfer, Donald H. Structural geology of the Silurian hills, San Bernardino County, California. D, 1952, Yale University.

Kupfer, Donald Harry. Geology of the colemanite deposits near Stauffer, Ventura County, California. M, 1942, University of California, Los Angeles.

Kupsch, Walter Oscar. Geology of part of the Beaverhead Mountains, Nicholia Creek basin, Montana. M, 1948, University of Michigan. 86 p.

Kupsch, Walter Oscar. Geology of the Tendoy-Beaverhead area, Beaverhead County, Montana. D, 1950, University of Michigan. 251 p.

Kurash, George E., Jr. Subsurface geology of west-central Lincoln County, Oklahoma. M, 1961, University of Oklahoma. 59 p.

Kureth, Charles L., Jr. The bathymetry and magnetics of the Wilkes fracture zone of the East Pacific Rise at 9°S. M, 1980, University of Michigan.

Kurie, Andrew Edmunds. Fractures in Austin Group, southern Travis County, Texas. M, 1956, University of Texas, Austin.

Kurk, Edwin H. The problem of sampling heterogeneous sediments. M, 1941, University of Chicago. 37 p.

Kurkjy, Karen Anne. Experimental compaction studies of lithic sands. M, 1988, University of Miami. 131 p.

Kurrus, Andrew William, III. Geochemistry, geothermometry, and mineralogy of quartz and base metal vein deposits, Montgomery County, Arkansas. M, 1980, University of Arkansas, Fayetteville.

Kurshin, James. Geologic hazards of Indiana Township, Allegheny County, Pennsylvania. M, 1983, University of Pittsburgh.

Kurt, Vace H. Sedimentology of the Huronian Coleman and Firstbrook formations, Cobalt area, Ontario. M, 1973, Carleton University. 119 p.

Kurtak, Joseph M. Stratigraphy and structure in the Belmont mining district, Nye County, Nevada. M, 1975, University of Nevada. 68 p.

Kurth, Randall J. Subsurface evaluation of the geopressured-geothermal Chloe Prospect, Calcasieu Parish, Louisiana. M, 1981, University of Southwestern Louisiana.

Kurtis, Mehmet S. Determination and comparison of the C-factor of the autograph A-7 and the Kelsh plotter, by using different types of aerial cameras; the wild RC 5a and the Fairchild K-12. M, 1962, Ohio State University.

Kurtulus, Cengiz. Spectral analysis of blast vibrations from multi-hole surface commercial explosions. M, 1982, Michigan Technological University. 138 p.

Kurtz, Anna E. Formations adjacent to the Siluro-Devonic contact, Delaware Water Gap Quadrangle, New Jersey and Pennsylvania. M, 1911, Columbia University, Teachers College.

Kurtz, Dennis Darl. Sedimentology and stratigraphy of the Triassic Chinle Formation, eastern San Juan Basin, New Mexico. M, 1978, Rice University. 185 p.

Kurtz, Dennis Darl. Stratigraphy and genesis of early Proterozoic diamictites; North America. D, 1980, Rice University. 140 p.

Kurtz, Jeffrey Paul. A mechanism for the concentration of copper felsic magmas related to porphyry deposits. D, 1983, University of North Carolina, Chapel Hill. 196 p.

Kurtz, Jeffrey Paul. Geochemistry and tectonic setting of Tunisian metabasaltic rocks. M, 1979, University of North Carolina, Chapel Hill. 91 p.

Kurtz, John Cornell. The geology of the Indian Valley region of Plumas County, California. M, 1957, [California State University, Chico].

Kurtz, Peter. The bloating of Missouri's shales. M, 1953, University of Missouri, Rolla.

Kurtz, Ronald. Polarization of micropulsations. M, 1969, [University of Alberta].

Kurtz, Ronald D. A magnetotelluric investigation of eastern Canada. D, 1973, University of Toronto.

Kurtz, Timothy David. A high-temperature single-crystal/diffractometer study of bornite, Cu_5FeS_4. M, 1969, University of Michigan.

Kurtz, Vincent E. Ironton and lower Franconia of Southeast Minnesota. M, 1949, University of Minnesota, Minneapolis. 138 p.

Kurtz, Vincent Ellsworth. Stratigraphy and paleontology of the Elvins Formation, Southeast Missouri. D, 1960, University of Oklahoma. 218 p.

Kurtz, William L. Geology of a portion of the Coyote Mountains, Pima County, Arizona. M, 1955, University of Arizona.

Kuru, Durmus. Geology of the area surrounding the Reed Mine, Yolo County, California. M, 1962, University of California, Berkeley. 63 p.

Kurupakorn, Somchai. Preliminary investigation of upper Sabino Canyon Dam, Pima County, Arizona. M, 1973, University of Arizona.

Kury, Theodore William. Historical geography of the iron industry in the New York-New Jersey Highlands; 1700-1900. D, 1968, Louisiana State University.

Kuryliw, Chester J. The geology of the Trout Bay nickel deposits, Red Lake District, northwestern Ontario. M, 1966, University of Manitoba.

Kuryvial, Robert J. A study of x-ray methods for determining composition of alkali feldspars. M, 1969, Miami University (Ohio). 50 p.

Kuryvial, Robert J. Behavior of some minor elements in coexisting plagioclase, orthoclase, and biotite from three intrusive bodies (Tertiary) located in the central Wasatch Range (Utah). D, 1971, Ohio State University.

Kurz, Mark David. Helium isotope geochemistry of oceanic volcanic rocks; implications for mantle heterogeneity and degassing. D, 1982, Woods Hole Oceanographic Institution. 290 p.

Kurz, Steven L. Occurrence and geochemistry of pyrite from the Boulder Batholith, Montana. D, 1975, Boston University. 355 p.

Kurz, Steven L. Sulfide mineralogy in the Boulder Batholith (Cretaceous), Montana. M, 1972, Boston University. 149 p.

Kurzius, Elizabeth C. The significance of hornblende in the petrogenesis of the Cortlandt Complex, New York. M, 1983, Brooklyn College (CUNY).

Kushnir, Donald William. Sediments in the south basin of Lake Winnipeg, Manitoba. M, 1971, University of Manitoba.

Kusiak, John Robert. A detailed gravity and magnetic interpretation of the structure and deformational history of the Jacksonwald Syncline, Berks County, Pennsylvania. M, 1977, Pennsylvania State University, University Park. 87 p.

Kusky, Timothy M. Geology of the Frozen Ocean Lake-New Bay Pond area, north-central Newfoundland. M, 1985, SUNY at Albany. 214 p.

Kuslansky, Gerald H. Sedimentology of marine sediments; eastern North Atlantic. M, 1973, Rensselaer Polytechnic Institute. 156 p.

Kusmirski, Richard Taddeusz Michael. Metallogeny of the "East South C" ore zone in the Dickenson Mine, Red Lake, Ontario; evidence for syngenetic gold deposition. M, 1981, McMaster University. 187 p.

Kusnick, Judith Elaine. Biostratigraphy of Upper Campanian and Maestrichtian ammonites of the Great Valley Sequence, California; 1 volume. M, 1981, University of California, Davis.

Kussow, Roger G. Lower Permian Bryozoa from Carlin Canyon, Nevada. M, 1964, Bowling Green State University. 63 p.

Kussow, Roger Glenn. A study of the Decorah Formation (Middle Ordovician) in Missouri. D, 1972, University of Missouri, Columbia.

Kuster, Guy Thierry. Seismic wave propagation in two-phase media and its applications to the Earth's interior. D, 1972, Massachusetts Institute of Technology. 191 p.

Kustra, Clarence Ronald. Manganese-bearing minerals at the Cannon Mine, Iron River, Michigan. M, 1961, Michigan Technological University. 40 p.

Kuthy, Olga. Elevated shorelines as evidences of eustatic changes of sea level. M, 1911, Columbia University, Teachers College.

Kutlu, Nurettin. Application of modern aerial triangulation methods in Turkey for various purposes. M, 1959, Ohio State University.

Kutluk, Hatice. Megaspore palynology of the Bearpaw-Horseshoe Canyon Formation transition zone, Alberta. M, 1985, University of Calgary. 168 p.

Kutrubes, Doria Lee. Dielectric permittivity measurements of soil saturated with hazardous fluids. M, 1986, Colorado School of Mines. 300 p.

Kutschale, Henry Walter. Arctic Ocean geophysical studies; the southern half of the Siberia Basin. D, 1965, Columbia University. 75 p.

Kutz, Clarence A. A study of the opaque minerals in the Homestake ores. M, 1930, University of Minnesota, Minneapolis. 63 p.

Kutz, Keith Brian. A stable isotope and fluid inclusion study of minor upper mississippi valley-type sulfide mineralization in Iowa, Illinois, and Wisconsin. M, 1987, Iowa State University of Science and Technology. 86 p.

Kutz, William J. A quantitative study of ground water in Coldwater, Ohio. M, 1977, Ohio University, Athens. 206 p.

Kux, Otto. Depositional studies of two Chester cycles in Marion and Jefferson counties, Illinois. M, 1958, University of Wisconsin-Madison. 41 p.

Kuykendall, Michael Douglas. The petrography, diagenesis and depositional setting of the Glenn (Bar-

tlesville) Sandstone, William Berryhill Unit, Glenn Pool oil field, Creek County, Oklahoma. M, 1985, Oklahoma State University. 383 p.

Kuzela, Robert Christian. Miocene sedimentation in the Sigsbee abyssal plain, Gulf of Mexico. M, 1971, Texas A&M University.

Kuzila, Mark Steven. Genesis and morphology of soils in and around large depressions in Clay County, Nebraska. D, 1988, University of Nebraska, Lincoln. 199 p.

Kuzio, Michael Kay. A determination of the isothermal bulk modulus of stishovite and its first pressure derivative from static compression data. M, 1977, Pennsylvania State University, University Park. 69 p.

Kuzior, Jerry L. The geology of the basal sandstone-mudstone unit of the Blackhawk Landslide, Lucerne Valley, California. M, 1983, Texas A&M University. 121 p.

Kvaale, Sigurd O. Photogrammetric determination of single points from different flying heights. M, 1961, Ohio State University.

Kvale, Cindy Marie. Pleistocene pyroclastic deposits of the central Cagayan Valley, Luzon, Philippines. M, 1983, Iowa State University of Science and Technology. 113 p.

Kvale, Erik Peter. Mio-Pliocene deltaic facies and depositional environments of the Cagayan Basin, Luzon, Philippines. M, 1982, Iowa State University of Science and Technology.

Kvale, Erik Peter. Paleoenvironments and tectonic significance of the Upper Jurassic Morrison/Lower Cretaceous Cloverly formations, Bighorn Basin, Wyoming. D, 1986, Iowa State University of Science and Technology. 201 p.

Kvenvolden, Keith Arthur. Normal paraffin hydrocarbons in Recent sediments from San Francisco Bay, California. D, 1961, Stanford University. 111 p.

Kveton, Edward J. The value of teacher-prepared slides in a physical geology unit for the seventh grade. M, 1968, [Northern Illinois University].

Kvill, D. R. Glacial history of the Trout Creek basin, Summerland, British Columbia. M, 1976, University of Alberta. 80 p.

Kvill, D. R. The glacial geomorphology of the Brazeau River valley, foothills of Alberta. D, 1984, University of Alberta. 240 p.

Kwa, Boo Leong. A geostatistical study of the Alligator Ridge gold deposit. M, 1984, University of Nevada. 228 p.

Kwader, Thomas. Ground water flow in a limestone aquifer system of N.W. Florida using U^{234}/U^{238} disequilibrium analysis. M, 1979, Florida State University.

Kwader, Thomas. Interpretation of borehole geophysical logs in shallow carbonate environments and their application to ground water resources investigations. D, 1982, Florida State University. 337 p.

Kwak, Teunis Adrianus Pieter. A garnet bearing syenite near Kamloops, BC. M, 1964, University of British Columbia.

Kwak, Teunis Adrianus Pieter. Metamorphic petrology and geochemistry across the Grenville Province-southern Province boundary, Dill Township, Sudbury, Ontario (Canada). D, 1968, McMaster University. 200 p.

Kwan, Jonathan Tak Pui. Heavy organic components in fossil fuel conversions. D, 1981, University of Southern California.

Kwang, John Ako. Petrography and geochemistry of Lower Permian cornstones in southwestern Oklahoma. M, 1978, Oklahoma State University. 134 p.

Kwarteng, Andrews Mensah Yaw. Remote sensing applied to the exploration for uranium-mineralized breccia pipes in northwestern Arizona. D, 1988, University of Texas at El Paso. 212 p.

Kwasnica, Edward Anthony. Clay mineralogy and hydrocarbon production in a portion of the Forbes Formation, Grimes gas field, Sacramento Basin, California. M, 1986, San Diego State University. 127 p.

Kwiatowski, Peter. Numerical simulation of sea-water intrusion, Cutler Ridge, Florida. M, 1987, University of South Florida, Tampa. 85 p.

Kwok, Kai Ming. Gold mineralization at the Kremzar Property. M, 1988, Laurentian University, Sudbury. 125 p.

Kwolek, James Michael. Holocene stratigraphy and depositional history of Reckley Hill settlement pond, San Salvador Island, Bahamas. M, 1985, Indiana University, Bloomington. 238 p.

Kwon, Byung-Doo. Spectral analysis of geophysical logs for correlation. D, 1977, Indiana University, Bloomington. 255 p.

Kwon, Hyuck Jae. Source of Recent barrier island sediments along the northern Gulf coast of the United States. D, 1969, Louisiana State University. 89 p.

Kwon, Sung-Tack. Pb-Sr-Nd isotope study of the 100 to 2700 Ma old alkalic rock-carbonatite complexes in the Canadian Shield; inferences on the geochemical and structural evolution of the mantle. D, 1986, University of California, Santa Barbara. 253 p.

Kwon, T. The relationship between the gravity field and the geologic features in the Korean Peninsula. M, 1971, University of Hawaii. 50 p.

Kwong, Yan-Tat John. A new look at the Afton copper mine in the light of mineral distributions, host rock geochemistry and irreversible mineral-solution interactions. D, 1982, University of British Columbia. 121 p.

Kwong, Yan-Tat John. Distribution of gold in an Archaean greenstone belt as exemplified by the Kakagi Lake area, northwestern Ontario. M, 1975, McMaster University. 82 p.

Kwun, Soon-Kuk. A mathematical erosion model to simulate soil losses in agricultural watersheds. D, 1980, Iowa State University of Science and Technology. 261 p.

Kyle, Douglas Haig. Possible source rocks for oil in the Michigan Basin. M, 1958, University of Michigan.

Kyle, J. R. Development of sulfide-hosting structures and mineralization, Pine Point, Northwest Territories. D, 1977, University of Western Ontario. 226 p.

Kyle, James Richard. Preliminary investigation of brecciation, alteration, and mineralization in the upper Knox Group of Smith and Trousdale counties, Tennessee. M, 1973, University of Tennessee, Knoxville. 105 p.

Kyllonen, David P. Hydrogeology of the Inyan Kara, Minnelusa, and Madison aquifers of the northern Black Hills, South Dakota and Wyoming. M, 1984, South Dakota School of Mines & Technology.

Kypfer, Marvin Douglas. Environmental geology of ski area developments. M, 1979, University of Arizona. 102 p.

Kyselka, Will. The paragenesis of dumortierite. M, 1949, University of Michigan.

Kyser, Kathryn B. Interpretations of the varietal forms of Venericardia planicosta in America. M, 1908, Cornell University.

Kyser, T. Kurtis. Stable and rare gas isotopes and the genesis of basic lavas and mantle xenoliths. D, 1980, University of California, Berkeley. 198 p.

Kyte, Frank T. On the origin of extraterrestrial stratospheric particles; interplanetary dust or meteor ablation debris?. M, 1977, San Jose State University. 71 p.

Kyte, Frank Thomas. Analyses of extraterrestrial materials in terrestrial sediments. D, 1983, University of California, Los Angeles. 165 p.

Kyte, Harold F. A resurvey of Sullivan Flats, Morgan Bay, and Medomak Cove shore study areas, Maine; 2 volumes. M, 1955, University of Maine.

L'Egraye, Michael P. H. Study of some mining districts of Arizona and of the structure of some ore specimens. M, 1923, Stanford University.

L'Esperance, Robert. The geology of Duprat Township and some adjacent areas, Northwest Quebec. D, 1952, McGill University.

L'Esperance, Robert L. A study of the diabase dykes of the Canadian Shield. M, 1948, McGill University.

l'Etoile, Robert de *see* de l'Etoile, Robert

L'Heureux, David Maurice. A geophysical study of the Precambrian basement in central Montana. D, 1985, Purdue University. 325 p.

L'Orsa, Anthony Theophile. Geology and contact metasomatism at Tahsis, Vancouver Island, British Columbia. M, 1964, Tulane University. 37 p.

L., G. Krstulovic *see* Krstulovic L., G.

L., R. J. Estevez *see* Estevez L., R. J.

la Cruz, Luis A. De *see* De la Cruz, Luis A.

la Cruz, Nga de *see* de la Cruz, Nga

la Cruz, Servando De *see* De la Cruz, Servando

La Flure, Ernest. The hydrologic and morphologic response to urbanization of four small watersheds in the Pittsburgh, Pennsylvania region. M, 1978, SUNY at Binghamton. 110 p.

La Fon, Neal A. Sedimentology and depositional environments of the Semilla Sandstone Member (Mancos Shale, Upper Cretaceous), eastern side of the San Juan Basin, New Mexico. M, 1980, University of Kansas. 49 p.

La Forge, Laurence. The geology of Somerville, Massachusetts. D, 1903, Harvard University.

La Fortune, Irene A. An investigation of the Cabot Head Formation of southeastern Ontario along the Niagara Escarpment. M, 1982, Wayne State University. 132 p.

La Freniere, Jon Edmund. A gravity and geothermal study of the Presidio Bolson, Presidio County, Texas. M, 1983, University of Texas at El Paso. 68 p.

la Fuente Duch, M. F. F. de *see* de la Fuente Duch, M. F. F.

la Fuente Duch, Mauricio Fernando De *see* De la Fuente Duch, Mauricio Fernando

la Garza, Fernando Javier Rodriguez de *see* Rodriguez de la Garza, Fernando Javier

La Grange, John. Stratigraphy and structure of the (Silurian) Jamesville Formation in central New York State. M, 1954, Syracuse University.

La Hay, K. L. Early stages of the weathering of the Methuen granite. M, 1976, University of Guelph.

La Haye, Jean. Géologie du Cénozoique du Bassin de Santa Lucia, Uruguay; cartographie et sédimentologie. M, 1988, Universite du Quebec a Montreal. 150 p.

la Montagne, John de *see* de la Montagne, John

La More, Francis Ellsworth. A petrographic study of the Mamainse Diabase, Ontario, Canada. M, 1954, University of Cincinnati. 46 p.

La Mori, Phillip Noel. The determination of some solid transition pressures to 5,000 bars. M, 1963, University of California, Los Angeles.

La Motte, Robert S. Studies in Tertiary paleobotany. D, 1935, University of California, Berkeley. 257 p.

la Pena Horcasitas, Gerardo Ruiz de *see* Ruiz de la Pena Horcasitas, Gerardo

La Pena, Edward C. De *see* De La Pena, Edward C.

La Piana, Margo K. Geochemistry of the fine-grained sediments of the Falling Creek Member, Taylorsville Formation, Virginia. M, 1979, George Washington University.

La Point, Ronald. A mineralogical study of certain ores from four mines of the Iron River District, Michigan. M, 1964, Michigan Technological University. 106 p.

La Pointe, Paul Reggie. Structure and petrology of the Keweenaw volcanic rocks, Mellen-Grand View area, Wisconsin. M, 1976, University of Wisconsin-Madison.

La Salata, Frank Vincent Michael. Strain analysis of the Castle Creek area, South Fork of the Clearwater River, Idaho County, Idaho. M, 1982, Washington State University. 76 p.

la Torre Robles, Jorge de *see* de la Torre Robles, Jorge

La Tour, Timothy Earle. An examination of metamorphism and scapolite in the Skalkaho region, southern Sapphire Range, Montana. M, 1974, University of Montana. 95 p.

La Valle, Placido D. Areal variation of karst topography in southcentral Kentucky. D, 1965, University of Iowa. 204 p.

La Violette, John. Holocene and late Pleistocene displacement history of the western Garlock Fault. M, 1981, California State University, Long Beach. 72 p.

Laaksonen, Harry J. Basement lithology in Michigan as determined from well cuttings. M, 1971, Michigan State University. 34 p.

Laali, Hooman. Stratigraphy and petrology of the lower Cretaceous (Comanchean) Kiamichi and Duck Creek formations in north-central Texas. M, 1973, University of Texas, Arlington. 166 p.

Laan, Sieger Robbert van der *see* van der Laan, Sieger Robbert

Laasko, Raymond Kalervo. The geochemistry of certain Precambrian carbonate rocks. M, 1961, Queen's University. 127 p.

Laband, Beth Leah. Response of foreshore morphology to a changing wave climate. M, 1984, University of California, Santa Cruz.

Labandeira, Conrad Christopher. Paleobiology of the Dikelocephalidae (Trilobita, Upper Cambrian) and systematic revision of the genus Dikelocephalus (Owen), with special reference to changing species concepts in American paleontologic thought. M, 1985, University of Wisconsin-Milwaukee. 279 p.

Labat, Clevland. The foraminifera and sedimentation of the Moody's Branch Formation (Jackson, Eocene) of central Louisiana and Mississippi. M, 1965, University of Southwestern Louisiana.

Labbé, Jean-Yves. Mise en évidence de failles de chevauchement acadiennes dans la région de Weedon, Québec. M, 1988, Universite Laval. 27 p.

LaBerge, Gene L. Carbonate minerals in the iron-formation (Wisconsin) and their significance. D, 1963, University of Wisconsin-Madison.

LaBerge, Gene L. Optical characteristics of glasses of natural plagioclases. M, 1959, University of Wisconsin-Madison.

Labib, Tarik Mohamed. Pedogenic distribution of clay mineral species in soil profiles. D, 1968, University of California, Davis. 163 p.

Labno, Bruce A. The hydrogeological implications of chemical waste disposal in a glaciated terrain, Rosemount, Minnesota. M, 1973, University of Minnesota, Minneapolis. 133 p.

Labotka, Theodore Charles. Geology of the Telescope Peak Quadrangle, California and late Mesozoic regional metamorphism, Death Valley area, California. D, 1978, California Institute of Technology. 392 p.

Labovitz, Mark Larry. Unit regional value of the Dominion of Canada. D, 1978, Pennsylvania State University, University Park. 385 p.

Labovitz, Mark Larry. Unit regional value of the state of California. M, 1976, Pennsylvania State University, University Park. 348 p.

Labowski, James Lawrence. Conodont biostratigraphy and correlation of the middle Carboniferous of northeastern Alabama. M, 1975, University of Florida. 113 p.

LaBrake, Richard F. Microfacies analysis and correlation of the Thacher Limestone Member of the Manlius Formation of eastern New York. M, 1964, Rensselaer Polytechnic Institute. 63 p.

LaBrecque, J. L. A study of the marine magnetic anomaly pattern employing techniques based on the fast Fourier transform algorithm. D, 1977, Columbia University, Teachers College. 278 p.

Labrecque, Paul. Détection d'une augmentation de conductance par électromagnétisme. M, 1986, Ecole Polytechnique. 151 p.

LaBreque, Douglas John. Some studies of time and frequency-domain electromagnetic prospecting methods. M, 1984, University of Utah. 127 p.

Labson, Victor Franklin. Geophysical exploration with audio frequency magnetic fields. D, 1985, University of California, Berkeley. 140 p.

Labute, Gary James. Differential compaction over a Leduc (Devonian) reef, Wizard lake area, Alberta (Canada). M, 1968, University of Calgary. 61 p.

Labuz, Joseph F. A study of the fracture process zone in rock. D, 1985, Northwestern University. 199 p.

Labyak, P. S. An oceanographic survey of the coastal waters between San Francisco bay and Monterey bay, California. M, 1969, United States Naval Academy.

Laca, Ted Edwin De *see* De Laca, Ted Edwin

Lacabanne, W. David. Settling of heterogeneous particles. D, 1947, University of Minnesota, Minneapolis. 134 p.

Lacaze, John A., Jr. Structural analysis of the Petersburg Lineament in the eastern Appalachian Plateau Province, Tucker County, West Virginia. M, 1978, West Virginia University.

Lacazette, Alfred Julian. Structural geology of Chunky Gal Mountain, Clay County, North Carolina including a discussion of the mechanisms of accretion of the Piedmont and a computer program for generating fabric density diagrams on the reference hemisphere. M, 1986, University of Kentucky. 331 p.

Lace, Penny J. Viscosities and densities in basaltic melts from Kilauea, Hawaii. M, 1980, University of Cincinnati. 62 p.

Lacerda, W. A. Stress-relaxation and creep effects on soil deformation. D, 1976, University of California, Berkeley. 313 p.

Lacey, James Edward. Criteria for the distinction of autochthonous from allochthonous oolites; application to the Ste. Genevieve Limestone (Mississippian, Valmeyeran), southern Illinois. D, 1967, University of Illinois, Urbana. 104 p.

Lacey, James Edward. Cyclic sedimentation in the Silurian Wills Creek and Tonolomy formations at Mount Union, Pennsylvania. M, 1960, University of Pittsburgh.

Lachance, David J. Genesis of the White Mountain Magma Series. M, 1978, Eastern Washington University. 91 p.

Lachenbruch, Arthur Herold. Problems in the interpretation of thermal data in permafrost. D, 1958, Harvard University.

Lachmar, Thomas E. Glacial geology of Lundy, Virginia and Green canyons, Mono County, California. M, 1977, Purdue University. 88 p.

Lackey, Larry L. Petrography of metavolcanic and igneous rocks of Precambrian age in the Huston Park area, Sierra Madre, Wyoming. M, 1965, University of Wyoming. 78 p.

Lackey, Laurence Evan. Late Cenozoic history of the northern and eastern Hoback Basin, Wyoming. D, 1974, University of Michigan. 53 p.

Lacko, Peter J. Selected aspects of the geology of the Eoline and Cottondale formations (Upper Cretaceous) southeast of Fayette, AL. M, 1985, Mississippi State University. 86 p.

Lackoff, Martin Robert. Frequency domain seismic deconvolution filtering. D, 1974, University of Rhode Island.

Lacoste, Pierre. Aspects pétrographiques et géochimiques de zones minéralisées dans le Canton de Duberger (Chibougamau), dans le contexte métamorphique Grenville-supérieur. M, 1986, Universite du Quebec a Chicoutimi. 238 p.

Lacroix, A. V. Structure and contact relationships of the Marlboro Formation, Marlboro, Mass. M, 1968, Boston College. 83 p.

Lacroix, Pierre de *see* de Lacroix, Pierre

Lacroix, Robert. Géologie et géochimie de la propriété New Pascalis, Val d'Or, Québec. M, 1986, Ecole Polytechnique. 28 p.

Lacroix, Sylvain. Géologie et géochimie de la propriété New Pascalis, Val d'Or, Québec. M, 1986, Ecole Polytechnique. 128 p.

Lacroix, Sylvain. La Gîtologie et la genèse du Cu-Ni dans la région du Lac Aulneau, Fossé du Labrador. M, 1986, Ecole Polytechnique. 130 p.

Lacy, Robert J. The micro-chemical determination of sulphantiminite and sulpharsenite minerals. M, 1937, University of Minnesota, Minneapolis. 34 p.

Lacy, Willard C. Types of pyrite and their relations to mineralization at Cerro de Pasco, Peru. D, 1950, Harvard University.

Laczniak, Randell J. Analysis of the relationship between energy output and well spacing in a typical Atlantic Coastal Plain geothermal doublet system. M, 1980, Virginia Polytechnic Institute and State University.

Ladd, George Edgar. A preliminary study of the fire clays of the United States. D, 1894, Harvard University.

Ladd, Harry S. Stratigraphy and fauna of the (Ordovician) Maquoketa Shale of Jackson County, Iowa. M, 1924, Iowa State University of Science and Technology.

Ladd, Harry Stephen. Stratigraphy and fauna of the Maquoketa Shale of Iowa. D, 1925, University of Iowa. 221 p.

Ladd, J. W. South Atlantic sea-floor spreading and Caribbean tectonics. D, 1974, Columbia University. 251 p.

Ladd, John Howard. Mesozoic overthrusting of oceanic crust in south central British Columbia. M, 1979, Cornell University.

Ladd, John Walcott. Regional magnetic anomalies over the north Atlantic. M, 1969, Columbia University. 32 p.

Ladd, Robert Edward. The geology of Sheep Canyon Quadrangle, Wyoming. M, 1979, Iowa State University of Science and Technology.

Ladd, T. W. Stratigraphy and petrology of the Quiburis Formation near Mammoth, Pinal County, Arizona. M, 1975, University of Arizona.

Ladeira, Eduardo Antonio. Metallogenesis of gold at the Morro Velho Mine and in the Nova Lima District, Quadrilatero Ferrifero, Minas Gerais, Brazil. D, 1980, University of Western Ontario. 272 p.

Lader, Gary R. A sedimentological investigation of coastal cells from Cape San Blas to Indian Pass, Florida. M, 1974, Florida State University.

Laderoute, D. The petrography, geochemistry and petrogenesis of alkaline dyke rocks from the Coldwell alkaline complex, northwestern Ontario. M, 1988, Lakehead University.

Ladle, Garth H. The sedimentary petrography and sedimentation of the Deadwood Formation (Cambrian) in the Black Hills, South Dakota and Wyoming. M, 1972, [University of Houston].

Ladle, Garth Harrison. Scanning electron microscopy and petrography of glass particles produced by lava fountain eruptions. D, 1978, Brigham Young University. 203 p.

Ladwig, Kenneth J. Groundwater flow between the shallow glacial aquifer and the Cedarburg Bog wetland area, Ozaukee County, Wisconsin. M, 1981, University of Wisconsin-Milwaukee. 148 p.

Ladwig, Lewis R. A petrographic study of some coarse clastics from the Mitchell Butte Quadrangle, Oregon. M, 1957, University of Oregon. 83 p.

Ladzekpo, Doe Henry. Geochemistry and evaluation of petroleum potentials of the Tano Basin, Ghana, W. Africa. M, 1981, Syracuse University.

Laetz, Thomas J. Risk perceptions and tradeoffs; an assessment of federal hazard management. D, 1987, University of Washington. 268 p.

LaFave, John Irwin. Groundwater flow delineation in the Toyah Basin of Trans-Pecos, Texas. M, 1987, University of Texas, Austin. 159 p.

Lafe, Olurinde Ebenezer. Boundary integral solutions to nearly horizontal flows in multiply zoned aquifers. D, 1981, Cornell University. 158 p.

LaFehr, Thomas R. The eastern Snake River plain, Idaho; a gravity survey and interpretation. M, 1962, Colorado School of Mines. 80 p.

LaFehr, Thomas Robert. Gravity analysis of anomalous density distributions with applications in the southern Cascade Range. D, 1964, Stanford University. 108 p.

Laferriere, Alan Price. Investigation of the Carlile-Niobrara unconformity (Upper Cretaceous) in northeastern New Mexico and southeastern Colorado. M, 1981, Indiana University, Bloomington. 134 p.

Laferriere, Alan Price. Regional analysis of rhythmic bedding in the Fort Hays Limestone Member, Niobrara Formation (Upper Cretaceous), U.S. Western Interior. D, 1987, Indiana University, Bloomington. 278 p.

Lafferty, Mark Robert. A reconnaissance geochemical, geochronological, and petrological investigation of granitoids in the Big and Little Maria mountains and Palen Pass, Riverside County, California. M, 1981, San Diego State University.

Lafko, Eric M. The geology of the Indian Hot Springs area, Owyhee County, Idaho. M, 1984, University of Idaho. 109 p.

LaFleche, Marc. Petrochimie et volcanologie du complexe rhyolitique de Don, Rouyn-Noranda, Quebec. M, 1986, Universite de Montreal.

LaFleche, Paul Thomas. Underground UHM-EM transillumination; a feasibility study. D, 1985, McGill University. 226 p.

Lafleur, Pierre Jean. The Archean Round Lake Batholith, Abitibi greenstone belt; a synthesis. M, 1986, University of Ottawa. 245 p.

LaFleur, Robert George. Pleistocene geology of the Troy, New York, Quadrangle. D, 1961, Rensselaer Polytechnic Institute. 301 p.

LaFollette, Stephen G. Paleoenvironmental and paleo-ecological analysis of the Vestal Limestone in Knox County, Tennessee. M, 1974, University of Tennessee, Knoxville. 143 p.

Lafon, Guy Michel. A petrographic and geochemical study of the Nisku Formation (Devonian) in the Leduc-Wookbend field, Alberta. M, 1965, University of Alberta. 118 p.

Lafon, Guy Michel. Some quantitative aspects of the chemical evolution of the oceans. D, 1969, Northwestern University. 160 p.

Lafond, J. M. Influence de l'origine du dépôt et des structures sur le rapport de perméabilité de sédiments rythmiques. M, 1977, University of Ottawa. 225 p.

Lafontaine, Michel Albert Georges. Uranium-thorium deposits at the Yates Mine, Huddersfield Township, Quebec. M, 1979, University of Ottawa. 93 p.

Laforest, André La Formation de West Point á la marge septentrionale du Synclinorium de la Baie des Chaleurs, Gaspésie, Québec. M, 1987, Universite Laval. 33 p.

LaForge, R. Tectonic implications of seismicity in the Adak Canyon region, central Aleutians. M, 1977, University of Colorado.

Laforte, Marc-Antoine. Etude structurale d'argiles glaciaires de préconsolidation naturelle choisie. M, 1975, Ecole Polytechnique.

LaFountain, Lester J. A deformed differentiate (Precambrian) at Crystal Falls (Iron County), Michigan. M, 1966, University of Wisconsin-Madison.

LaFountain, Lester James, Jr. The metamorphic and structural geology of a portion of the Crystal Mountain Quadrangle, Colorado Front Range. D, 1973, University of Colorado. 160 p.

LaFreniere, Gilbert F. Contact metamorphism adjacent to the White Mountain magma series in New Hampshire. M, 1957, Dartmouth College. 126 p.

Lafrenz, W. B. Distribution and mobility of uranium and thorium in granitic rocks, central Beartooth Mountains, Montana. M, 1986, University of Florida. 127 p.

Lagace, Paul. The use of a digital model for simulation of a well field in an aquifer in stratified drift. M, 1982, Boston University. 57 p.

LaGarry, Hannan E. Taphonomic evidence of predation and scavenging of Teleoceras (Mammalia; Rhinocerotidae), with a description of the Camelidae from the Minium Quarry local biota of north-central Kansas. M, 1988, Fort Hays State University. 59 p.

Lagas, Philip Joseph. The glacial geology of the Gros Ventre Canyon, Teton County, Wyoming. M, 1984, Lehigh University.

Lagasse, P. F. Interaction of river hydraulics and morphology with riverine dredging operations. D, 1975, Colorado State University. 467 p.

Lager, G. A. Crystal chemistry of the olivines at elevated temperatures. D, 1976, University of British Columbia.

Lager, George A. The effect of O-P-O and P-O-P angle variations on bond overlap populations for some selected ortho- and pyrophosphates. M, 1972, Virginia Polytechnic Institute and State University.

Lager, James Lee. A petrographic study of rocks from the Manicougan-Mouchalagne lakes area, Saguenay County, Province of Quebec, Canada. M, 1969, Ohio State University.

Lageson, David Rodney. Depositional environments and diagenesis of the Madison Limestone, northern Medicine Bow Mountains, Wyoming. M, 1977, University of Wyoming. 130 p.

Lageson, David Rodney. Structural geology of the Stewart Peak Quadrangle, Lincoln County, Wyoming, and adjacent parts of the Idaho-Wyoming thrust belt. D, 1980, University of Wyoming. 378 p.

Lago, R. Semrau *see* Semrau Lago, R.

Lagoe, M. B. Recent benthic foraminifera from the central Arctic Ocean. M, 1975, University of Wisconsin-Madison.

Lagoe, Martin Brooks. Stratigraphy and paleoenvironments of the Monterey Formation and associated rocks, Cuyama Basin, California. D, 1982, Stanford University. 254 p.

Lagoni, Jack R. The geomorphology of the Truckee River delta, at Pyramid Lake, Nevada, and implications on the decline of Lake Lahontan. M, 1985, University of Nevada. 94 p.

LaGrange, Renee Leone. Organic geochemistry of the bitumen from the Ohio Shale by gas chromatography. M, 1984, University of Toledo. 156 p.

Laguna, Wallace de *see* de Laguna, Wallace

Laguros, George Andrew. Seismic-stratigraphic analysis of sedimentation processes in pelagic carbonate sequences of the Equatorial Pacific; Deep Sea Drilling Project Site 574. M, 1987, University of Texas, Austin. 99 p.

Lagus, Peter Leonard. The equations of state of hydrogen and argon; applications to the Jovian interior. D, 1974, California Institute of Technology. 135 p.

Lahabi, Ahmad-Ali. Development of field methods for engineering resistivity surveys. M, 1983, University of Idaho. 104 p.

Lahann, R. W. Molybdenum transport mechanisms in fresh water environments. D, 1975, University of Illinois, Urbana. 145 p.

Lahee, Frederick H. A study of metamorphism in the Carboniferous formation of the Narragansett Basin. D, 1911, Harvard University.

Lahey, Barry Armstrong Lloyd. Relationships between the Middle Devonian Prairie Evaporite Formation and the Salt-Free area, of south central Saskatchewan. M, 1965, University of Saskatchewan. 84 p.

Lahiere, Leon. Petrology of lacustrine deposits, Juchipila Quadrangle, Zacatecas, Mexico. M, 1982, University of New Orleans. 85 p.

Lahlou, Mourad. Highly accurate inversion methods for stratified media. D, 1982, University of Denver. 98 p.

Lahola, Irene. Mississippian conodonts of the Cuyahoga Formation. M, 1974, Kent State University, Kent. 74 p.

Lahoud, Joseph A. Determination of uranium concentrations of carbonate shell material by the fission track method. M, 1965, Rensselaer Polytechnic Institute. 30 p.

Lahr, John Clark. Detailed seismic investigation of Pacific-North American plate interaction in southern Alaska. D, 1975, Columbia University. 26 p.

Lahr, John Clark. The foreshock-aftershock sequence of the March 20, 1966 earthquake in the Republic of Congo. M, 1971, Columbia University. 141 p.

Lahr, Melvin M. Precambrian geology of a greenstone belt in Oconto County, Wisconsin and geochemistry of the Waupee metavolcanics (Precambrian). M, 1972, University of Wisconsin-Madison.

Lahti, Howard Reino. Factors contributing to secondary dispersion of trace elements in glacial soils, Saint Stephen area, New Brunswick. M, 1971, University of New Brunswick.

Lahti, Howard Reino. The reconnaissance and detailed rock geochemistry of Mykonos, Greece. D, 1978, University of New Brunswick.

Lahti, Victor R. Precambrian and structural geology of the Crevice Lake area, Southwest Beartooth Mountains, Montana and Wyoming. M, 1975, Northern Illinois University. 72 p.

Lai, Cheng-Hsien. Mathematical models of thermal and chemical transport in geologic media. D, 1986, University of California, Berkeley. 207 p.

Lai, David Yuekchung. Mean and time-dependent motions in the Blake Escarpment region. D, 1983, University of Rhode Island. 191 p.

Lai, Eong-Lip. Adsorption study of fatty acids on hematite surface. M, 1973, University of Nevada - Mackay School of Mines. 52 p.

Lai, Shang-Fei. Generalized linear inversion of 2.5 dimensional gravity and magnetic anomalies. D, 1984, University of Texas at Dallas. 190 p.

Lai, Shyh-Shiun. Analysis of the seismic response of prototype earth and rockfill dams. D, 1985, University of California, Berkeley. 286 p.

Lai, Tung-Ming. Pathways of cationic diffusion in clay minerals. D, 1967, Michigan State University. 75 p.

Laibidi, Ezatoliah Heydari *see* Heydari Laibidi, Ezatoliah

Laidig, Larry Wayne. Organic geochemical and clay mineral trends in sediments of the Catatumbo River delta area, Lake Maracaibo, Venezuela. M, 1977, University of Tulsa. 106 p.

Laidlaw, James Stuart. Tomographic techniques and their application to geotechnical and groundwater flow problems. M, 1987, University of British Columbia. 116 p.

Laidley, Richard Allan. An X-ray fluorescent analysis study of the distribution of selected elements within the Hopi Buttes volcanics, Navajo County, Arizona. D, 1966, University of Arizona. 136 p.

Laifa, Embarek. Les Gisements uranifères (Timgaouine et Abankor) du Hoggar (Algérie) et l'altération des épontes de leurs gîtes. M, 1977, Ecole Polytechnique.

Laifa, Embarek. Les gisements uranifères du Hoggar (Algérie) et l'altération des épontes de leurs gîtes. M, 1978, Ecole Polytechnique.

Laine, Edward Paul. Geological effects of the Gulf Stream system in the North American basin. D, 1977, Massachusetts Institute of Technology. 164 p.

Laine, R. P. Geological-geochemical relationships between porphyry copper and porphyry molybdenum ore deposits. D, 1974, University of Arizona. 342 p.

Laing, Eoghan MacRuaraidh. The relationship between surface geologic structure and aeromagnetic anomalies in east central Missouri. M, 1964, Washington University. 69 p.

Laird, Brien A. Geometry and kinematics of geologic structures, Pine Mountain-Ojai area, western Transverse Ranges, California. M, 1988, University of Nevada. 107 p.

Laird, Charles Elbert, Jr. One-dimensional magnetotelluric inversion techniques. D, 1970, University of Texas, Austin. 105 p.

Laird, H. C. The nature and origin of chert in the Lockport and Onondaga formations of Ontario. D, 1932, University of Toronto.

Laird, H. Scott. Geology of the Pelham Dome, near Montauge, west-central Massachusetts. M, 1974, University of Massachusetts. 84 p.

Laird, Jeffrey R. Complex-response of a drainage basin to geomorphically-effective fire, El Oso Creek, Arizona. M, 1986, Colorado State University. 217 p.

Laird, Jo. Phase equilibria in mafic schist and the polymetamorphic history of Vermont. D, 1977, California Institute of Technology. 464 p.

Laird, Joe Alex. A subsurface study of the Cockfield-Yegua Formation and the Wilcox (Eocene) Group in San Jacinto County, Texas. M, 1947, University of Oklahoma. 80 p.

Laird, John W. Diagenetic controls on reservoir characteristics and development in the Jurassic Norphlet Formation, Escambia County, Alabama. M, 1985, University of Alabama. 125 p.

Laird, Wilson Morrow. The stratigraphy and structure of the Martinsburg Formation near Harrisburg, Pennsylvania. M, 1938, University of North Carolina, Chapel Hill. 31 p.

Laird, Wilson Morrow. The stratigraphy of the Upper Devonian and Lower Mississippian of southwestern Pennsylvania. D, 1942, University of Cincinnati. 244 p.

Laity, Julie Ellen. A study of basin formation under ephemeral streamflow conditions. M, 1976, University of California, Los Angeles. 67 p.

Laity, Julie Ellen. Sapping processes and the development of theater-headed valleys on the Colorado Plateau, (with a discussion on Martian valleys). D, 1982, University of California, Los Angeles. 152 p.

Lajoie, J. R. The origin of the (Silurian) Val Brillant and Sayabec formations, Quebec. M, 1961, McGill University.

Lajoie, Jules J. The electromagnetic response of a conductive inhomogeneity in a layered Earth. D, 1973, University of Toronto.

Lajoie, Jules Joseph. The geomagnetic variation anomaly at Kootenay lake, British Columbia (Canada). M, 1970, University of British Columbia.

Lajoie, Kenneth Robert. Late Quaternary stratigraphy and geologic history of Mono Basin (Mono county), eastern California. D, 1968, University of California, Berkeley. 379 p.

Lajtai, Emery Zoltan. Pleistocene geology of the University Avenue subway route, Toronto, Ontario. M, 1961, University of Toronto.

Lajtai, Emery Zoltan. Pleistocene sediments of the Bloor-Danforth subway section, Toronto, Canada. D, 1966, University of Toronto.

Lajtai, N. V. The initiation of normal shear fractures. M, 1972, University of New Brunswick.

Lakatos, Stephen. Effects of water on the stability of fission tracks in mica and volcanic glass. D, 1971, Rensselaer Polytechnic Institute. 72 p.

Lakatos, Stephen. The nature of water in sepiolite and palygorskite; infrared and thermal studies. M, 1967, Universite Laval.

Lake, C. A. Adsorption of orthophoshate and tripolyphosphate by clay minerals. D, 1977, College of William and Mary. 105 p.

Lake, Carol A. Dissolution rates of silica sources in sea water. M, 1972, College of William and Mary.

Lake, Ellen A. Data processing techniques applied to a hidden layer problem in Jackson County, Indiana. M, 1978, Indiana University, Bloomington. 149 p.

Lake, J. L. The effects of petroleum on the salt marsh ecosystem. D, 1977, College of William and Mary. 145 p.

Lake, James L. A comparison of the chlorinated hydrocarbon content of surface and subsurface samples in the York River, Virginia. M, 1972, College of William and Mary.

LaKind, Judy Sue. Geochemical study of gold-quartz veins, Red Lake gold camp, Northwest Ontario. M, 1984, University of Wisconsin-Madison.

Lal, Ravindra Kumar. The petrology of the cordierite-gedrite rocks and associated gneisses on Fishtail Lake, Harcourt Township, Ontario, Canada. D, 1966, University of Toronto.

Laliberté, Jean-Yves. Etude des faciès de Bouleaux et d'Anse-à-la-Barbe de la Formation de West-Point à la Pointe-au-Bouleau, Baie des Chaleurs, Gaspésie, Québec. M, 1983, Universite Laval. 63 p.

Lalicker, Cecil Gordon. Micro-fauna from Cottonwood to Herington formations in central and southern Kansas. M, 1932, University of Oklahoma. 57 p.

Lalicker, Cecil Gordon. The foraminiferal family Textulariidae. D, 1935, Harvard University.

Lalko, Lynn-Edward. The chemistry, structure and mineralogy of light-dark ordinary chondrites. D, 1980, Arizona State University. 144 p.

Lall, Upmanu. Value of data in relation to uncertainty and risk. D, 1981, University of Texas, Austin. 299 p.

Lalla, Wilson. A stratigraphic study of the Osage-Layton Format in northeastern Oklahoma. M, 1975, University of Tulsa. 35 p.

LaLonde Schake, Celia May. Microfaunal study of the Upper Cretaceous Annona Chalk of southwestern Arkansas. M, 1959, Smith College. 185 p.

Lalonde, Andre E. The Baie-des-Moutons syenitic complex, La Tabatiere, Quebec. M, 1981, McGill University. 163 p.

Lalonde, Jean-Pierre. Fluorine and other vein elements as geochemical indicators of fluorite deposits in the Madoc area, Ontario. M, 1973, Carleton University. 154 p.

Lalonde, Kayron F. Sedimentology of a point bar sand associated with a middle Miocene meandering channel system in Southwest Louisiana. M, 1985, University of Southwestern Louisiana. 80 p.

Lalor, Jim. Geology and volcanism (Archean) in the Woman lake area (northwest Ontario). M, 1970, University of Manitoba.

Lam Do Sinh *see* Do Lam Sinh

Lam, Chi-Kin. Interpretation of state-wide gravity survey of Kansas. D, 1987, University of Kansas. 273 p.

Lam, Hing-Lan. Generation mechanisms for the transient ultra-low frequency oscillations (1-16 mHz) of the Earth's magnetic field in the morning sector. D, 1976, University of Alberta. 248 p.

Lam, Ronald Ka-Wei. Atoll permeability; calculated from ocean and ground water tides. D, 1971, University of California, San Diego. 108 p.

LaManna, John M. Mixed-layer illite/smectite in altered volcanic rocks at Bob Leroy Peaks, Iron County, Utah. M, 1983, Western Washington University. 139 p.

Lamar, Donald Lee. Geology of the Corona area, Orange, Riverside, and San Bernardino counties, California. M, 1959, University of California, Los Angeles. 95 p.

Lamar, Donald Lee. Structural evolution of the northern margin of the Los Angeles Basin. D, 1961, University of California, Los Angeles.

Lamar, F. S. The drift deposits of Clinton County, Ohio. M, 1900, Earlham College.

Lamarche, Robert Y. Etude des conglomérats de la Region d'Orford-Sherbrooke, Québec. M, 1962, Universite Laval. 71 p.

Lamarche, Robert Y. Geology of the Sherbrooke area (Quebec). D, 1965, Universite Laval.

Lamarche, Valmore C. Recent denudation of the Reed Dolomite (Precambrian), White Mountains, California. D, 1964, Harvard University.

Lamarre, Albert Leroy. Fluorite in jasperoid of the Salado Mountains, Sierra County, New Mexico; significance to metallogeny of the western United States. M, 1974, University of Western Ontario. 134 p.

Lamarre, Michele. Seismic hazard evaluation for sites in California; development of an expert system. D, 1988, Stanford University. 352 p.

Lamarre, R. A. Geology of the Alabama-Crystalline mines, Mother Lode gold belt, California. M, 1977, University of Western Ontario.

Lamb, Beth. Geomorphology of the upper Kurupa River valley system, Brooks Range, Alaska. M, 1984, SUNY at Buffalo. 151 p.

Lamb, Craig Forbes. Structural analysis of a similar-type fold; Booster Lake, Manitoba. M, 1974, University of Manitoba.

Lamb, Cynthia B. Analysis of the Ordovician Ellenburger Reservoir, Suggs Field, Nolan and Coke counties, Texas. M, 1987, University of Arkansas, Fayetteville.

Lamb, Garland Clayton. Bedrock geology of Fern Quadrangle, Crawford and Franklin counties, Arkansas. M, 1974, University of Arkansas, Fayetteville.

Lamb, George F. The Pennsylvanian limestones of northeastern Ohio. M, 1909, Ohio State University.

Lamb, George Marion. Biostratigraphy of the lower part of the Mancos Formation in the San Juan Basin. D, 1964, University of Colorado. 162 p.

Lamb, George Marion. Depositional features of the Silurian Red Mountain Formation in Northwest Georgia. M, 1954, Emory University. 69 p.

Lamb, Henry J. The use of multiple regression methods in predicting the gold values in Mina Panteon, El Limon, Nicaragua; an epithermal gold-silver vein type deposit. M, 1974, Dartmouth College. 77 p.

Lamb, John. The geology and mineralogy of the Brown McDade Mine. M, 1947, University of British Columbia.

Lamb, Kevin J. Evaluation and verification of the ROCKPACK/BACKPACK rock slope stability computerized analysis packages. M, 1988, University of Idaho. 265 p.

Lamb, Ralph C., Jr. Geology of the Dinwoody Creek area, Fremont County, Wyoming. M, 1956, University of Kansas. 109 p.

Lamb, Robert Alvis. A statistical investigation of sized zircon fractions from samples of Conway Granite (Devonian?) from several plutons located in New Hampshire. M, 1965, Brown University.

Lamb, Robert Odell. Liquid silts; their occurrence and distribution in loess. D, 1985, Iowa State University of Science and Technology. 99 p.

Lamb, Robert Reid. Studies of the lithologic variation of the Des Moines Series. M, 1948, University of Illinois, Urbana. 27 p.

Lamb, T. N. Geology of the Coronado Islands, Baja California, Mexico. M, 1974, San Diego State University.

Lamb, William Marion. Metamorphic fluids and granulite genesis. D, 1987, University of Wisconsin-Madison. 244 p.

Lamb, William Marion. The fluid phase in the granulite facies; evidence from the Adirondack Mountains, New York. M, 1983, Rice University. 67 p.

Lambe, Robert Noah. Crystallization and petrogenesis of the southern portion of the Boulder Batholith, Montana. D, 1981, University of California, Berkeley. 171 p.

Lamber, C. Kurt. Fossil and Recent beryciform otoliths; an adjunct to Ichthyological classifiction. M, 1963, University of Missouri, Rolla.

Lamberson, Michelle Noreen. Aspects of petrography, palynology and inorganic constituents of subbituminous coals from Canyon Creek, Alaska; a high latitude Tertiary coal field. M, 1987, Pennsylvania State University, University Park. 162 p.

Lambert, A. An anomaly in geomagnetic variations on the west coast of British Columbia. M, 1969, University of British Columbia.

Lambert, Anthony. A tilt meter study of the response of the Earth to ocean tide loading. D, 1969, Dalhousie University.

Lambert, David Dillon. Geochemical evolution of the Stillwater Complex, Montana; evidence for the formation of platinum-group element deposits in mafic layered intrusions. D, 1982, Colorado School of Mines. 274 p.

Lambert, David James. A detailed stratigraphic study of initial deposition of Tertiary lacustrine sediments near Mills, Utah. M, 1976, Brigham Young University. 35 p.

Lambert, Douglas Wade. A geophysical survey of a contaminated aquifer in Redlands, California. M, 1987, University of California, Riverside. 126 p.

Lambert, Earl Freeman. The geology of the Old Eureka Mine, Sutter Creek, Amador County, California. M, 1949, University of California, Berkeley. 71 p.

Lambert, Ellen E. Geology and petrochemistry of ultramafic and orbicular rocks, Zuni Mountains, Cibola County, New Mexico. M, 1983, University of New Mexico. 166 p.

Lambert, George. Deposition and diagenesis of the Edwards Limestone in north-central Texas. M, 1979, Baylor University.

Lambert, Gerald S. Geology of a portion of Mt. Hamilton Range, California. M, 1923, Stanford University.

Lambert, Henry D. The soils of the Chapel Hill region and their adaptation to crops. M, 1915, University of North Carolina, Chapel Hill. 13 p.

Lambert, Hubert C. Structure and stratigraphy in the southern Stansbury Mountains, Tooele County, Utah. M, 1941, University of Utah. 51 p.

Lambert, Joseph Michael, Jr. Transformation of pyrite to pyrrhotite and its implications in coal conversion processes. D, 1982, Pennsylvania State University, University Park. 191 p.

Lambert, Marc. Acquisition et traitement de signaux magnétotelluriques en temps réel. M, 1986, Ecole Polytechnique. 191 p.

Lambert, Marshall Brice. A study of natural illites by infrared absorption spectroscopy. M, 1974, University of Missouri, Columbia.

Lambert, Maurice Bernard. Geology of the Mount Brenner Stock near Dawson City, Yukon Territory. M, 1966, University of British Columbia.

Lambert, Maurice Bernard. The Bennett Lake cauldron subsidence complex, British Columbia and Yukon Territory. D, 1972, Carleton University. 317 p.

Lambert, Michael W. Copper sulfides in the Permian redbeds of Kansas. M, 1979, Indiana University, Bloomington. 99 p.

Lambert, Patricia Frost. Geology of the northeastern portion of the Fredonia Quadrangle, San Saba County, Texas. M, 1988, University of Texas of the Permian Basin. 89 p.

Lambert, Paul Wayne. Petrology of the Precambrian rocks of part of the Monte Largo area, New Mexico. M, 1961, University of New Mexico. 108 p.

Lambert, Paul Wayne. Quaternary stratigraphy of the Albuquerque area, New Mexico. D, 1968, University of New Mexico. 329 p.

Lambert, R. Etude des paramètres affectant l'impédance électrique de certains métaux et minéraux. D, 1975, McGill University.

Lambert, Ralph E. Shnabkaib Member of the Moenkopi Formation, depositional environment and stratigraphy near Virgin, Washington County, Utah. M, 1984, Brigham Young University. 65 p.

Lambert, Raymond S. Geology of the country east of the Santa Rita mining district, Grant County, New Mexico; the San Lorenzo area. M, 1973, University of New Mexico. 81 p.

Lambert, Rebecca Bailey. Environment of deposition and reservoir characteristics of Lower Pennsylvanian Morrowan sandstones, South Empire Field area, Eddy

County, New Mexico. M, 1986, Texas A&M University. 174 p.

Lambert, Robert A. Petrography of the upper Carboniferous-Permian sandstones of the northern half of the Dunkard Basin, Pennsylvania, Ohio, and West Virginia. M, 1969, Miami University (Ohio). 109 p.

Lambert, Roger. Etude spectrométrique de deux anomalies radiométriques. M, 1970, Ecole Polytechnique.

Lambert, Stephen P. Spectral amplitude studies of two-dimensional electrical measurements. M, 1985, University of Wisconsin-Milwaukee. 134 p.

Lambert, Stephen W. Subsurface study of sandstone geometry, distribution and depositional environments (Upper Mississippian and Lower Pennsylvanian), southern Illinois. M, 1974, Southern Illinois University, Carbondale. 79 p.

Lambert, Steven Judson. Stable isotope studies of some active hydrothermal systems. D, 1976, California Institute of Technology. 387 p.

Lambert, William Robert. Stress distribution in the Winters Pass thrust plate. M, 1978, Rice University. 97 p.

Lambeth, William Alexander. Notes on the geology of the Monticello area (Virginia). D, 1901, University of Virginia. 22 p.

Lambiase, Joseph J. Sediment dynamics in the macrotidal Avon River estuary, Nova Scotia. D, 1977, McMaster University. 415 p.

Lambiase, Joseph J., Jr. Distribution and movement of sediments in the narrows of the Pettaquamscutt River, Narragansett, Rhode Island. M, 1972, University of Rhode Island.

Lambie, John M. An experimental study of the stability of oscillatory-flow bed configurations. M, 1984, Massachusetts Institute of Technology. 103 p.

Lambo, W. A. Geology of the Silver Plains gypsum deposit (Jurassic), Manitoba. M, 1964, University of Manitoba.

Lamborg, Amy Davison. Subsurface geology and petrography of the Salem Limestone in portions of Hamilton, White, and Wayne counties, Illinois. M, 1986, University of Cincinnati. 290 p.

Lamborn, Helen Morningstar. The fauna of the Niagaran Series in Ohio. M, 1915, Ohio State University.

Lamerson, Paul R. The regional stratigraphy of lower Missourian rocks from eastern Kansas to central Iowa. M, 1956, University of Kansas. 132 p.

Lamey, Carl Arthur. Some metamorphic effects of the Duluth Gabbro. M, 1927, Northwestern University.

Lamey, Carl Arthur. The intrusive relations and metamorphic effects of the Republic Granite. D, 1933, Northwestern University.

Lamiotte, Louis. Subsurface lithostratigraphy of the Upper Jurassic Smackover-Bossier interval of the southwestern Arkansas. M, 1985, University of Arkansas, Fayetteville.

Lamme, Maurice A. On the specific gravities of niobium and tantalum pentoxides. D, 1909, Columbia University, Teachers College.

Lammers, Edward C. H. The structural geology of the Livingston Peak area, Montana. D, 1936, University of Chicago. 42 p.

Lammers, George. The taxonomic significance of the feudmastoid region with special reference to Smilodon. M, 1959, Florida State University.

Lammers, George Eber. The late Cenozoic Benson and Curtis Ranch faunas from the San Pedro valley, Cochise County, Arizona. D, 1970, University of Arizona. 214 p.

Lammers, Leo Joseph. Analysis of the core from drill hole No. 2, Wayne County Airport, Detroit, Michigan. M, 1956, University of Michigan.

Lammlein, David Raymond. Lunar seismicity, structure, and tectonics. D, 1973, Columbia University. 94 p.

Lammons, James Monroe. Disseminated phosphate content of the Bonneterre Formation (Upper Cambrian) of southeastern Missouri and its bearing on the paleoecology. M, 1961, Wichita State University. 105 p.

Lammons, James Monroe. The palynology and paleoecology of the Pierre shale (Campanian-Maestrichtian), of northwestern Kansas and environs. D, 1969, Michigan State University. 260 p.

Lamon, Kathryn Ann. Facies interpretation in the first marine terrace, Pajaro River, California. M, 1973, Stanford University.

Lamont, Norman. An analysis and model study of the stresses in drill columns. M, 1948, Louisiana State University.

Lamontagne, Yves. Applications of wideband, time domain, EM measurements in mineral exploration. D, 1975, University of Toronto.

Lamontagne, Yves. Model studies of the Turam electromagnetic method. M, 1970, University of Toronto.

LaMoreaux, Phillip E. The Hatchetigbee Formation in Choctaw County, Alabama. M, 1949, University of Alabama.

Lamoreaux, Scott B. Stratigraphy and development of the Ervine Creek marine algal bank (Shawnee Group, Upper Pennsylvanian) in Elk and Chautauqua counties. M, 1984, Wichita State University. 136 p.

Lamothe, Daniel. Analyse structurale du mélange ophiolitique du Lac Montjoie. M, 1978, Universite Laval. 63 p.

Lamothe, Michel. Lithostratigraphy and geochronology of the Quaternary deposits of the Pierreville and St. Pierre les Becquets areas, Quebec. D, 1985, University of Western Ontario. 227 p.

Lampe, Leslie Kent. A drought contingency manual for Kansas public water supplies. D, 1983, University of Kansas. 705 p.

Lampert, Jordan Keith. Measurement of trace cation activities by Donnan membrane equilibrium and atomic absorption analysis. D, 1982, University of Wisconsin-Madison. 189 p.

Lampert, Leon Max. Stratigraphy of Presidio area, Presidio County, Trans-Pecos, Texas. M, 1953, University of Texas, Austin.

Lamping, Neal Edward. The Mohorovicic discontinuity as a phase transition. M, 1970, Texas A&M University. 203 p.

Lampiris, N. Stratigraphy of the clastic Silurian rocks of central western Virginia and adjacent West Virginia. D, 1975, Virginia Polytechnic Institute and State University. 237 p.

Lampiris, Nicholas. Petrology of Bloomsburg interval rocks (Silurian), western Virginia and eastern West Virginia. M, 1969, George Washington University.

Lampshire, Wayne Gilbert. The Cretaceous stratigraphy and zonation in Cedar County, Nebraska. M, 1956, University of Nebraska, Lincoln.

Lan, Ching-Ying. Petrological study of dikes on Musconetcong Mountain, Bloomsbury Quadrangle, N.J. M, 1974, Rutgers, The State University, New Brunswick. 90 p.

Lanan, Holly Kay. Depositional framework, porosity and permeability in the San Andres Formation, Cato Field, Chaves County, New Mexico. M, 1981, University of Texas, Austin.

Lancaster, Mary Jane. Mineral resources and industries of the northwestern Canadian Shield. D, 1962, Columbia University, Teachers College. 292 p.

Lance, Donald M., Jr. Sulfide mineralization of the Midnite uranium mine, Stevens County, Washington. M, 1981, Washington State University. 167 p.

Lance, James Odell, Jr. Frequency domain analysis of least squares polynomial surfaces with application to gravity data in the Pedregosa Basin area. D, 1982, University of Texas at El Paso. 214 p.

Lance, John F. The origin of the Pioneer pyrophyllite deposit, San Diego County, California. D, 1949, California Institute of Technology. 107 p.

Lance, John Franklin. Evidence of termites in the Pleistocene asphalt of Carpinteria, California. M, 1946, California Institute of Technology. 14 p.

Lance, John Franklin. Phylogeny of the later Tertiary Equidae in the light of Pliocene horses from western Chihuahua, Mexico. D, 1949, California Institute of Technology. 109 p.

Lance, R. J. Heavy metals in the main streams of the James River basin, Missouri. M, 1973, University of Missouri, Rolla.

Lanchman, Gregory Joseph. Geochemistry of the Belford Dacite from St. Lucia, Lesser Antilles; trace element signatures in petrogenesis. M, 1986, Wright State University. 43 p.

Land, Cooper B. Stratigraphy of Fox Hills Sandstone (upper Cretaceous) and associated formations, Rock Springs uplift and Wamsutta Arch area, Sweetwater County, Wyoming; a shore-line-estuary sandstone model for the late Cretaceous. D, 1971, Colorado School of Mines. 111 p.

Land, Cooper B., Jr. A lithostratigraphic study of the Mississippian rocks in the southern Sangre de Cristo Mountains, New Mexico. M, 1959, University of Oklahoma. 86 p.

Land, David M. Geologic prospecting in the Val Verde Basin by the integration of Landsat, magnetics, gravity, and subsurface geology. M, 1983, Texas Christian University. 91 p.

Land, Gary F. Flourine-hydroxyl exchange in illites as applied to the Illinois-Kentucky fluorspar district. M, 1974, Southern Illinois University, Carbondale. 92 p.

Land, Lynton Stuart. Diagenesis of metastable skeletal carbonates. D, 1966, Lehigh University. 150 p.

Land, Ralph Joseph. Gravity and magnetic investigation of a segment of the Triassic Basin of northern Virginia. M, 1965, University of Virginia. 31 p.

Landa, Edward R. The behavior of technetium-99 in soils and plants. D, 1975, University of Minnesota, Minneapolis. 133 p.

Landau, David. Structural geology of the northern Shirley Mountains, Carbon County, Wyoming. M, 1966, University of Wyoming. 83 p.

Landau, H. G., Jr. Internal erosion of compacted cohesive soil. D, 1974, Purdue University. 256 p.

Lande, Andrew C. Stratigraphy and depositional environments of the Brigham Group, northeastern Bear River Range, Bear Lake County, Idaho. M, 1986, Idaho State University. 106 p.

Landenberger, Daniel Ross. Silification of Pleistocene plants and associated silica diagenesis. M, 1980, Texas Tech University. 93 p.

Lander, E. B. A review of the John Day oreodonts. M, 1972, University of California, Berkeley. 213 p.

Lander, E. B. A review of the Oreodonta (Mammalia, Artiodactyla); parts I, II and III. D, 1977, University of California, Berkeley. 476 p.

Lander, John French. Seismicity of Antarctica. M, 1962, American University. 88 p.

Landers, Ronald Alfred. Groundwater resources of the crystalline rocks, Llano area, Texas. M, 1972, University of Texas, Austin.

Landers, Thomas E. Elastic waves in laterally inhomogeneous media. D, 1971, Stanford University. 74 p.

Landes, Kenneth Knight. A study of the paragenesis of the pegmatites of central Maine. D, 1925, Harvard University.

Landes, Robert William. Stratigraphy and paleontology of the marine formations of the Montana Group, southeastern Alberta. D, 1937, Princeton University. 258 p.

Landin, William. Geology and mineral resources of the northern part of the Whitehall Quadrangle, Virginia. M, 1976, University of Akron. 59 p.

Landing, Edward William. Early Ordovician conodont-graptolite-chitinozoan biostratigraphy of the Taconic Allochthon, eastern New York. M, 1975, University of Michigan.

Landing, Edward William. Studies in Late Cambrian-Early Ordovician conodont biostratigraphy and paleoecology, northern Appalachian region. D, 1979, University of Michigan. 317 p.

Landis, Charles A., Jr. Geology of the Graphite Mountain-Tepee Mountain area, Montana-Idaho. M, 1963, Pennsylvania State University, University Park. 153 p.

Landis, Charles R. Changes in the fluorescence properties of selected Hartshorne Seam coals with rank. M, 1985, Southern Illinois University, Carbondale. 147 p.

Landis, Gary Perrin. Geologic, fluid inclusion, and stable isotope studies of a tungsten-base metal ore deposit; Pasto Bueno, northern Peru. D, 1972, University of Minnesota, Minneapolis. 195 p.

Landis, Sam Wallace. A petrographic study of the Kettleman Hills productive zones. M, 1941, Pennsylvania State University, University Park. 68 p.

Landisman, Mark G. The distortion of pulse-like earthquake signals by seismographs. D, 1959, Columbia University, Teachers College.

Landle, George Louis. Computer modeling of ground water and Bouguer gravity data in the northern Long Valley, west-central Idaho. M, 1986, Washington State University. 77 p.

Landman, Neil H. Ontogeny and evolution of Late Cretaceous (Turonian-Santonian) Scaphites. D, 1982, Yale University. 363 p.

Landmesser, C. W. Submarine geology of the eastern Coral Sea Basin, Southwest Pacific. M, 1974, University of Hawaii. 64 p.

Landon, Robert Emmanuel. Metamorphism and ore deposition in the Santa Rita-Hanover-Fierro area, New Mexico. D, 1929, University of Chicago. 32 p.

Landon, Ronald Arthur. The geology of the Gatesburg Formation in the Bellefonte Quadrangle, Pennsylvania, and its relationship to the general occurrence and movement of ground water. M, 1963, Pennsylvania State University, University Park. 88 p.

Landon, Susan. Environmental controls on growth rates in hermatypic corals from the lower Florida Keys. M, 1975, SUNY at Binghamton. 79 p.

Landress, Mark R. Geology, mineralogy and minor element distribution in a copper-uranium mineralized Shinarump channel, Deer Flat area, San Juan County, Utah. M, 1979, Michigan Technological University. 117 p.

Landress, R. A. Nature of the occurrence of mercury in soils of the Long Valley, California geothermal area. M, 1977, Colorado School of Mines. 68 p.

Landreth, John Orlin. Geology of the Rattlesnake Creek area, Lemhi County, Idaho. M, 1964, University of Idaho. 51 p.

Landreth, Robert Allen. Depositional environments and diagenesis of the lower Clear Fork Group, Mitchell County, Texas. M, 1977, Texas Tech University. 79 p.

Landro, Wanda-Lee de see de Landro, Wanda-Lee

Landrum, James Hanford. Fault detection in the northeastern North Carolina coastal plain utilizing remote sensing and geophysical methods. M, 1980, Old Dominion University. 147 p.

Landrum, Ralph Avery. The effects of some constant-k filter sections on a representative seismic wavelet. M, 1964, University of Tulsa. 142 p.

Landry, Richard G. Relationship between structure and sedimentation in the Marginulina Zone of the Anahuac in an area in Southwest Louisiana. M, 1980, University of Southwestern Louisiana. 78 p.

Landwehr, J. M. Water quality indices; construction and analysis. D, 1974, University of Michigan. 294 p.

Landwehr, Walter R. Factors of ore control. D, 1933, Stanford University. 178 p.

Landwer, William R. A computer assisted petroleum geophysical exploration game. M, 1977, Purdue University.

Landy, Richard Allen. The genetic Pottsville-Tuscarora relationship, eastern Pennsylvania. M, 1955, Pennsylvania State University, University Park. 142 p.

Landy, Richard Allen. Variation in chemical composition of rock bodies; metabasalts in the Iron Springs Quadrangle, South Mountain, Pennsylvania. D, 1961, Pennsylvania State University, University Park. 197 p.

Lane, Alfred Church. The geology of Nahant (Massachusetts). D, 1888, Harvard University.

Lane, Bernard O. The paleontology and stratigraphy of the "Meristella"-coral zone (Devonian) of eastern New York State. M, 1955, Brown University.

Lane, Bernard Owen. The paleontology and stratigraphy of the Ely Group in the Illipah area of Nevada. D, 1962, University of Southern California. 114 p.

Lane, Burke E. Sanitary landfill leachite interactions with a carbonate-rock derived soil in central Pennsylvania. M, 1969, Pennsylvania State University, University Park. 197 p.

Lane, Charles A. Geology, mineralogy, and fluid inclusion geothermometry of the El Paso gold mine, Cripple Creek, Colorado. M, 1976, University of Missouri, Rolla.

Lane, Charles F. Geology of the Fountain City Quadrangle, Knox County, Tennessee. M, 1945, University of Tennessee, Knoxville. 30 p.

Lane, Charles L. Pennsylvanian-Permian stratigraphy of west-central Arizona. M, 1977, Northern Arizona University. 120 p.

Lane, Charles L. Provenance and petrology of Tertiary arkoses of the Santa Monica Mountains, Southern California; implications for tectonic history. D, 1987, University of California, Los Angeles. 270 p.

Lane, Daniel Stephen. Tectonic implications of late Tertiary strata exposed along the Piankatank River, eastern Virginia. M, 1984, Old Dominion University. 73 p.

Lane, Diane Estelle. Sulfide mineralization in a sequence of metasedimentary and metavolcanic rocks, North Fork Lone Ranch Creek area, Ferry County, Washington. M, 1982, Washington State University. 71 p.

Lane, Donald Wilson. Studies of sedimentary environments in the Cretaceous Dakota Sandstone in northwestern Colorado. D, 1961, Rice University. 79 p.

Lane, Donald Wilson. Textural studies of the Oread Megacyclothem, Pennsylvanian, of the northern Midcontinent. M, 1958, University of Illinois, Urbana. 28 p.

Lane, Douglas L. Experimental studies on diffusion of volatiles and crystal growth in basaltic melts. M, 1978, University of Arizona.

Lane, Francis B. The Gold Hill mining district of North Carolina. D, 1908, Yale University.

Lane, Harold Richard. Conodonts from the Prairie Grove Member of the Hale Formation (Pennsylvanian) and the Brentwood Member of the Bloyd Formation (Morrowan) in Washington County, Arkansas. M, 1966, University of Iowa. 94 p.

Lane, Harold Richard. Morrowan (lower Pennsylvanian) conodonts from northwestern (Washington county) Arkansas and northeastern Oklahoma. D, 1969, University of Iowa. 130 p.

Lane, Harry Campbell. Radioactivity and geothermotics. M, 1934, University of Manitoba.

Lane, Jeffrey W. Relations between geology and mass movement features in a part of the East Fork Coquille River watershed, southern Coast Range, Oregon. M, 1987, Oregon State University. 107 p.

Lane, Jerry Leroy. An analysis of the tributary area of Billings, Montana. M, 1971, University of Utah. 102 p.

Lane, Joseph H., Jr. The stratigraphy and fauna of the Winterset Limestone of Jackson County, Missouri. M, 1939, University of Kansas.

Lane, L. J. Influence of simplifications of watershed geometry in simulation of surface runoff. D, 1975, Colorado State University. 214 p.

Lane, Larry Stephen. Deformation history of the Monashee Decollement and Columbia River fault zone, British Columbia. D, 1985, Carleton University. 240 p.

Lane, Larry Stephen. Structure and stratigraphy, Goldstream River-Downite Creek area, Selkirk Mountains, British Columbia. M, 1977, Carleton University. 150 p.

Lane, Margaret Lucille. Clay mineralogy and geochemical studies of the weathering of the Meade Peak Phosphatic Shale Member of the Phosphatic Shale Member of the Phosphoria Formation, Southeast Idaho. M, 1983, University of Idaho. 171 p.

Lane, Maurice Vincent. Geology of part of the upper Lewistown Valley of Mifflin County, Pennsylvania. M, 1956, Pennsylvania State University, University Park. 171 p.

Lane, Michael Arthur. Structural geology and structural analysis of part of the central Taconic region, eastern New York. D, 1970, Indiana University, Bloomington. 68 p.

Lane, Michael E., III. Microstructural ordering in synthetic opals; an X-ray diffraction and infrared and Raman spectral study. M, 1987, Wichita State University. 60 p.

Lane, Norman Gary. Environment of deposition of the Grenola Formation (Lower Permian) in southern Kansas. M, 1954, University of Kansas. 71 p.

Lane, Norman Gary. The monobathrid camerate crinoid family; Batocrinidae. D, 1959, University of Kansas.

Lane, Phillip Jene. A paleotectonic and paleogeologic study of the Mississippian System. M, 1950, University of Illinois, Urbana. 70 p.

Lane, Richard. Mineralogy and stable isotope geochemistry of serpentinized ultramafic rocks from Blow-Me-Down Mountain of the Bay of Islands ophiolite complex; implications from the oceanic Moho. M, 1988, University of Houston.

Lane, Robert Kenneth. Climate and heat exchange in the oceanic region adjacent to Oregon. D, 1965, Oregon State University. 115 p.

Lane, Robert W. Precambrian geology of the Rapid Creek-Bloody Gulch area near Rochford, South Dakota. M, 1951, South Dakota School of Mines & Technology.

Lane, Steven Dale. Relationship of the carbonate shelf and basinal clastic deposits of the Missourian and Virgilian series of the Pennsylvanian System in central Beaver County, Oklahoma. M, 1978, Oklahoma State University. 95 p.

Lane, Thomas James. Investigation in the fluorescence of minerals. M, 1935, Catholic University of America. 50 p.

Lane, W. L. Extraction of information on inorganic water quality. D, 1975, Colorado State University. 223 p.

Laney, Francis B. A report on the building and ornamental stones of North Carolina. M, 1905, University of Wisconsin-Madison.

Laney, Patrick T. Petrography of the Honey Creek Limestone (Upper Cambrian), north flank, Wichita Mountains, Oklahoma. M, 1982, Wichita State University. 113 p.

Laney, Randy T. A structural and petrographic study of the Kentland, Indiana impact structure. M, 1979, University of Kansas. 94 p.

Laney, Robert L. Knox Dolomite of Lawrence County, Indiana. M, 1960, Indiana University, Bloomington. 50 p.

Laney, Robert Lee. Weathering of the granodioritic rocks in the Rose Canyon lake area, Santa Catalina mountains, Arizona. D, 1971, University of Arizona. 267 p.

Laney, Stephen E. Study of a strata-bound barite deposit, Dempsey Cogburn Mine, Montgomery County, Arkansas. M, 1980, University of Arkansas, Fayetteville.

Lanford, Colleen Loretta. Deposition, diagenesis and porosity evolution of the Queen Formation, Winkler County, Texas. M, 1985, Texas Tech University. 101 p.

Lang, A. J. Geographical patterning of desert vegetation of the Deep Canyon Reserve, California. D, 1977, University of California, Los Angeles. 123 p.

Lang, Andrew J. Petrography and pegmatites, Four-Mile area, Custer Co., South Dakota. M, 1957, University of Washington. 129 p.

Lang, Arthur Hamilton. A brief study of the petrography of the Shuswap Beltian rocks of BC. M, 1928, University of British Columbia.

Lang, Arthur Hamilton. Geology and mineral deposits of the Owen Lake Mining Camp, British Columbia. D, 1930, Princeton University. 130 p.

Lang, Dorothy M. Origin of Mount Merino and shale (middle Ordovician), eastern New York State. M, 1969, SUNY at Albany. 64 p.

Lang, Edwin F. A quantification of parameters controlling beach face slope. M, 1985, Boston University. 122 p.

Lang, Harold R. Cretaceous (Albian-Turonian) foraminiferal biostratigraphy and paleogeography of northern Montana and southern Alberta. D, 1982, University of Calgary. 139 p.

Lang, Harold R. Late Cretaceous foraminiferal biostratigraphy of the Ladd Formation; southeastern Los Angeles Basin, California. M, 1976, California State University, Long Beach. 92 p.

Lang, Helen Marie. Metamorphism of pelitic rocks in the Snow Peak area, northern Idaho. D, 1983, University of Oregon. 224 p.

Lang, Robert Campbell, III. A preliminary report on the geology of the Elk City oil pool of western Oklahoma. M, 1951, University of Oklahoma. 73 p.

Lang, Roy. The utility of Landsat for petroleum exploration in South Arkansas. M, 1985, University of Arkansas, Fayetteville.

Lang, Thomas. Cenozoic calcareous nannofossils from DSDP leg 77; biostratigraphy and delineation of hiatuses. M, 1984, Florida State University.

Lang, Thomas Pursell. A study of the subsurface geology of Oceana County, Michigan. M, 1967, University of Michigan.

Lang, Walter B. Deformation by fracture. M, 1916, University of Minnesota, Minneapolis. 57 p.

Lang, William J. Oxidized uranium deposits of the southern Black Hills, South Dakota. M, 1963, University of Arizona.

Lang, William Joseph. The properties of clay-water systems under pressure. D, 1965, University of Illinois, Urbana.

Langan, Leon Verdin, Jr. Analysis of the subsurface geology near Stanford University from available water-well logs. M, 1957, Stanford University.

Langan, Robert Thomas. Kinematic thermal models of the Earth's mantle. D, 1981, Northwestern University. 84 p.

Langbein, J. O. Crustal deformation in central California; multi-wavelength geodetic observations and interpretations. D, 1979, University of Washington. 202 p.

Langdon, George Stanley. The petrology, geochemistry and petro genisis of the upper pillow lavas, Troodos ophiolite complex, Cyprus. M, 1982, Memorial University of Newfoundland. 168 p.

Lange, Alan Ulrich. Microfauna of the Verendrye Member of the Pierre Formation. M, 1962, University of South Dakota. 131 p.

Lange, Ian M. The K/Rb distribution of coexisting K feldspars and biotites. M, 1964, Dartmouth College. 26 p.

Lange, Ian Muirhead. Sulfur isotope geology of Butte, Montana. D, 1968, University of Washington. 41 p.

Lange, Marie L. Heavy mineral studies of the Fox Hills, Hell Creek and Cannonball sediments, Morton and Sioux counties, North Dakota. M, 1942, The Johns Hopkins University.

Lange, Nixon Richard. A paleomagnetic study of the Pliocene mudstones of the Verde Formation, northern Arizona. M, 1976, University of Arizona.

Lange, Peter C. Geology of the Telegraph Mine tectono-hydrothermal breccias, San Bernardino Co., California. M, 1988, Colorado State University. 206 p.

Lange, Robherd Edward. Decay of turbulence in stratified salt water. D, 1974, University of California, San Diego.

Lange, Rolf V. A petrographic study of the Backbone (Little Saline) Limestone (Lower Devonian) in southwestern Illinois and southeastern Missouri. M, 1983, Southern Illinois University, Carbondale. 110 p.

Lange, Stephen Stanley. The geology of the east side of the Lewis and Clark Pass area, Lewis and Clark County, Montana. M, 1963, University of Missouri, Columbia. 144 p.

Lange-Brard, Françoise. Etude des minéralisations, altérations et phases fluides associées au Gîte Devlin (Chibougamau, Québec, Canada). M, 1985, Universite du Quebec a Chicoutimi. 133 p.

Langenbahn, William Edward. Amherstburg facies (Lower Devonian) of central Wayne County, Michigan. M, 1960, University of Michigan.

Langendoerfer, Martha F. The geography of the Hermann (Missouri) region. M, 1930, University of Missouri, Columbia.

Langenheim, Ralph L., Jr. Geology of Black Tiger Gulch, Boulder County, Colorado. M, 1947, University of Colorado.

Langenheim, Ralph L., Jr. Pennsylvanian and Permian stratigraphy in the Crested Butte Quadrangle, Gunnison County, Colorado. D, 1951, University of Minnesota, Minneapolis. 179 p.

Langenheim, Virginia McCutcheon. Pennsylvanian and Permian paleontology and stratigraphy of Arrow Canyon, Arrow Canyon Range, Clark County, Nevada. M, 1960, University of California, Berkeley. 138 p.

Langenkamp, David F. A micro-earthquake study of the Elsinore fault zone. M, 1973, University of California, Riverside. 102 p.

Langer, Arthur M. Geology of the Manhattan Formation (Precambrian; New York and Connecticut). M, 1962, Columbia University, Teachers College.

Langer, Arthur M. Mineralogy and physical properties of Mojave Desert (California) playa crusts. D, 1965, Columbia University. 155 p.

Langer, C. J. The locations and characteristics of the aftershocks of the October 3, 1974, Peruvian earthquake. M, 1977, Virginia Polytechnic Institute and State University.

Langer, Laura L. Evaluation of water saturation in shaly sands. M, 1978, Stanford University.

Langer, Milton Friedrich. Subsurface study of the Beech Creek "Barlow" Limestone in south-central Illinois. M, 1955, University of Illinois, Urbana. 39 p.

Langer, William H. Clay deposits of the Connecticut River valley, Connecticut; a special problem in land management. M, 1972, Boston University. 39 p.

Langfelder, Leonard Jay. An investigation of initial negative pore water pressure in statically compacted cohesive soil. D, 1964, University of Illinois, Urbana. 132 p.

Langfield, Peter Michael. Geology of Ford Creek area, Sawtooth Range, Montana. M, 1967, University of Montana. 60 p.

Langford, Eldon Woodrow. Limestone dikes in sedimentary serpentine of southern Uvalde County, Texas. M, 1942, University of Texas, Austin.

Langford, Fred F. Geology of the Geco Mine, Manitouwadge area, district of Thunder Bay, Ontario. M, 1955, Queen's University. 77 p.

Langford, Fred Frazer. The geology of Levack Township, Ontario. D, 1960, Princeton University. 194 p.

Langford, George B. The Beardmore-Nezah gold area, Ontario. D, 1929, Cornell University.

Langford, Maxine. Study of a Pennsylvanian flora from Santo, Texas. M, 1939, Texas Tech University. 34 p.

Langford, Neal Gerald. Thermoluminescene of the apatite series. M, 1963, University of Florida. 49 p.

Langford, Othell Franklin. Collapsed caverns as modifiers of older structures in the vicinity of the Balcones zone of faults in Ulvade County, Texas. M, 1942, University of Texas, Austin.

Langford, Richard P. Depositional systems and geologic history of the lower part of the Fountain Formation, Manitou Embayment, Colorado. M, 1982, Indiana University, Bloomington. 300 p.

Langford, Richard Parker. Modern and ancient fluvial-eolian interactions. D, 1988, University of Utah. 336 p.

Langford, Stephen Arthur. Statistical method in optical crystallography; technique and application to rock forming minerals. D, 1972, University of Hawaii. 185 p.

Langford, Stephen Arthur. The surface morphology of the Tuscaloosa seamount (Hawaiian islands). M, 1969, University of Hawaii. 12 p.

Langhus, Bruce G. Lazoo foraminifera and depositional history, northeastern Gulf Coast (upper Eocene, Mississippi-Alabama). D, 1972, Dalhousie University. 108 p.

Langhus, Bruce Gunnar. Paleoecological and biostratigraphical zonation of Upper Cretaceous foraminifera, Vancouver Island. M, 1968, University of Calgary. 77 p.

Langill, Richard Francis. Properties of some Michigan cherts as used in concrete aggregate. M, 1964, Michigan Technological University. 134 p.

Langille, Andrew Benjamin. Sedimentology and palynology of Cretaceous and Tertiary strata, Southeast Baffin Island, Northwest Territories, Canada. M, 1987, Memorial University of Newfoundland. 163 p.

Langille, Gerald Burton. Earliest Cambrian; latest Proterozoic ichnofossils and problematic fossils from Inyo County, California. D, 1974, SUNY at Binghamton. 194 p.

Langley, Edward J. Unconsolidated sediments in the Black River area of New York State. D, 1952, Syracuse University.

Langlois, Joseph David. Hydrothermal alteration of intrusive igneous rocks in the Eureka mining district, Nevada. M, 1971, University of Arizona.

Langmaid, Kenneth K. A study of the organic fraction of three New Brunswick soils. M, 1950, University of New Brunswick.

Langman, James W., Jr. The Shinarump Member of the Chinle Formation in northern Arizona and southern Utah. M, 1976, Northern Arizona University. 106 p.

Langmuir, Charles Herbert, II. A major and trace element approach to basalts. D, 1980, SUNY at Stony Brook. 348 p.

Langmuir, Donald. Stability of carbonates in the system; $CaO-MgO-CO_2-H_2O$. D, 1965, Harvard University.

Langmyer, Kathryn. Foraminiferal biostratigraphy of portions of the Tuscumbia and Monteagle limestones in Northwest Georgia. M, 1984, University of Georgia. 73 p.

Langran, K. J. International water quality management; the Rhine River as a study in transfrontier pollution control. D, 1979, University of Wisconsin-Madison. 303 p.

Langrand, Edgar L. A petrologic study of the upper portion of Herrin (No. 6) Coal Member in relation to roof lithology. M, 1977, Southern Illinois University, Carbondale. 55 p.

Langseth, Marcus Gerhardt, Jr. Heat flow measurements in the Indian Ocean. D, 1964, Columbia University. 70 p.

Langstaff, George D. Investigation of Archean metavolcanic and metasedimentary rocks of Sellers Mountain, west-central Laramie Mountains, Wyoming. M, 1984, University of Wyoming. 386 p.

Langston, Charles Adams. Body wave synthesis for shallow earthquake sources; inversion for source and Earth structure parameters. D, 1976, California Institute of Technology. 222 p.

Langston, David J. The geology and geochemistry of the northeasterly gold veins, Sunnyside Mine, San Juan County, Colorado. M, 1978, Colorado School of Mines. 163 p.

Langston, Jackson Maurice. Areal geology of the Christie-Westville area, Adair County, Oklahoma. M, 1963, University of Oklahoma. 79 p.

Langston, Melana. Radiolarian biostratigraphy of cherts associated with ophiolites in the California Coast Ranges. M, 1979, University of Texas at Dallas. 78 p.

Langston, Robert Burlison. The nature of anauxite. D, 1967, University of California, Berkeley. 75 p.

Langston, Wann, Jr. A new genus and species of Cretaceous theropod dinosaur from the Trinity of Atoka County, Oklahoma. M, 1947, University of Oklahoma. 73 p.

Langston, Wann, Jr. The Permian vertebrates of New Mexico. D, 1952, University of California, Berkeley. 232 p.

Langton, Christine A. Titanate-silicate equilibria; parts of the system $Al_2O_3-TiO_2-SiO_2-CaO$. M, 1976, Pennsylvania State University, University Park. 75 p.

Langton, Claude M. Geology along the Lewis Thrust in the Schafer Meadows District, Montana. M, 1931, Cornell University. 41 p.

Langton, Claude M. Geology of the northwestern part of the Idaho Batholith and adjacent region in Montana. D, 1934, University of Chicago. 33 p.

Langtry, Tina M. Carbonate bodies within the basal Swift Formation (Jurassic) of northwestern North Dakota. M, 1982, University of North Dakota. 267 p.

Langway, Chester Charles, Jr. Stratigraphic analysis of a deep ice core from Greenland. D, 1965, University of Michigan. 242 p.

Lanham, James H. Mineral resources of Horse Creeks Valley (Tusculeosa Formation, Aiken County, South Carolina). M, 1972, Virginia State University. 31 p.

Lanham, Robert Evans. Petrography and diagenesis of low-permeability sandstones of the lower Almond Formation, southwestern Wyoming. M, 1980, University of Colorado. 113 p.

Lanier, Hershel Dale. Carbonate petrology of southwestern Mississippi loess. M, 1978, University of New Orleans.

Lanier, William Paul. Microstromatolites from the 2.3 G.A. Transvaal Sequence, South Africa. D, 1984, University of Arizona. 169 p.

Lanigan, Dennis Michael. A heavy mineral study of Mississippi Sound and Petit Bois Island. M, 1979, University of New Orleans.

Lanigan, John Carroll, Jr. Geology of the Beaver River Canyon area, Beaver County, Utah. M, 1980, Kent State University, Kent. 51 p.

Lankford, Robert Renninger. Micropaleontology of the Cretaceous-Paleocene, northeastern Utah. M, 1952, University of Utah. 75 p.

Lankford, Robert Renninger. Recent foraminifera from the nearshore turbulent zone, Western United States and Northwest Mexico. D, 1962, University of California, San Diego.

Lankston, Robert Wayne. A geophysical investigation in the Bitterroot Valley, western Montana. D, 1975, University of Montana. 112 p.

Lanmon, L. B. The emergence and development of dairying in Hopkins County, Texas, 1936-1970. M, 1971, East Texas State University.

Lanney, Nicholas Anthony. The erosional and depositional history of the Commerce Plain, southeastern Michigan. M, 1977, University of Michigan.

Lanning, David Roy. A study of the Oswego Sandstone and a consideration of its productive possibilities. M, 1947, University of Pittsburgh.

Lanning, Eric N. Estimating seismic attenuation using VSP measurements. D, 1985, University of Washington. 120 p.

Lanning, Eric N. Well log characteristics of the Columbia River basalts. M, 1981, University of Washington. 151 p.

Lanning, Robert Maye. An olivine tholeiite dike swarm in Lancaster County, Pennsylvania. M, 1972, Pennsylvania State University, University Park. 80 p.

Lannon, Mary Susan. The Quaternary history of surficial geology of the Stonefort 7.5 minute quadrangle, southern Illinois. M, 1988, Southern Illinois University, Carbondale. 178 p.

Lannon, Patrick Michael. The Quaternary stratigraphy and glacial history of the Duluth-Superior area. M, 1986, University of Minnesota, Duluth. 115 p.

Lanoix, Monique. Magnetic properties of the Allende meteorite. M, 1978, University of Toronto.

Lanphere, Marvin Alder. I, Geology of the Wildrose area, Panamint Range, California; II, Geochronologic studies in the Death Valley-Mohave Desert region. D, 1962, California Institute of Technology. 171 p.

Lant, Kevin J. Structure of the Aliso Canyon area, eastern Ventura Basin, California. M, 1977, Ohio University, Athens. 79 p.

Lanter, Robert B. Deposition and diagenesis of Abo carbonates, Lea County, New Mexico. M, 1985, Texas Tech University. 90 p.

Lanthier, L. R. Stratigraphy and structure of the lower part of the Precambrian Libby Creek Group, central Medicine Bow Mountains, Wyoming. M, 1978, University of Wyoming. 30 p.

Lantos, Etienne Alexandre. The uraniferous Matinenda Formation, Elliot Lake, Ontario; a braided river model. M, 1979, University of Windsor. 127 p.

Lantos, Julie Ann. Middle Devonian stratigraphy north of the Pine Point barrier complex, Pine Point, Northwest Territories. M, 1983, University of Alberta. 196 p.

Lantz, James R. Geology and kinematics of Clay-Rick land-slide with an undulatory slip surface. M, 1984, Texas A&M University. 132 p.

Lantz, Rik Earl. The influence of the geometry of the pluton-host rock interface on the orientations of thermally induced hydrofractures at the Cochise Stronghold Pluton, Cochise County, Arizona. M, 1984, University of Arizona. 112 p.

Lantzy, R. J. Global cycling behavior of trace metals. D, 1979, Northwestern University. 180 p.

Lanyk, James Louis. Geochemical exploration in the upper Mississippi Valley lead-zinc district (Iowa County, Wisconsin). M, 1968, University of Iowa. 47 p.

Lanz, Robert C. Development of Lockport (Niagaran) carbonate buildups and associated rocks at Genoa, Ohio. M, 1979, Bowling Green State University. 118 p.

Lanzirotti, Antonio. Geology and geochemistry of a Proterozoic supracrustal and intrusive sequence in the central Wet Mountains, Colorado. M, 1988, New Mexico Institute of Mining and Technology. 164 p.

Lao, Kheang. Evolution chimique des clinopyroxènes des laves mafiques de la ceinture métavolcanique de Rouyn-Noranda. M, 1978, Ecole Polytechnique.

Lao, Kheang. Thermo-géochimie appliquée aux inclusions fluides reliées au gisement volcanique archéen de Millenbach, Rouyn-Noranda, Québec. D, 1983, Ecole Polytechnique. 322 p.

Laosebikan, Samuel Cyebanji. Economic geology, geochemistry and geochronology of a granite in Jos-Bukuru Complex, central Nigeria. M, 1977, University of Pittsburgh.

Lapallo, Christopher M. Petrographic and geochemical relations between the rocks on the north and south limbs of the Chibougamau Anticline; assimilation of roof rocks, crystallization and residual liquid compositions in the Dore Lake Complex, Quebec, Canada. M, 1988, University of Georgia. 175 p.

LaPasha, Constantine Anthony. Paleoecology of the Lower Cretaceous Kootenai Formation flora in the Great Falls area, Montana. D, 1982, University of Montana. 222 p.

LaPensee, Earl Francis. Trace-element geochemistry of sandstones hosting uranium-vanadium deposits in the La Sal Mountains region of Utah and Colorado, U.S.A. M, 1986, University of California, Riverside. 170 p.

Lapham, D. M. A preliminary study of ore samples from Temple Mountain, Utah. M, 1955, Columbia University, Teachers College.

Lapham, Davis M. The nature of chromium chlorite. D, 1958, Columbia University, Teachers College.

Lapham, Kathryn Elizabeth. Pressure and isotopic effects on water diffusivity in a silicic melt. M, 1982, Arizona State University. 91 p.

Lapham, Wayne Wright. Conductive and convective heat transfer in sediments near streams. D, 1988, University of Arizona. 317 p.

Lapierre, D. P. Further developments of an inexpensive seismic detection-recording system. M, 1978, McGill University. 156 p.

Lapierre, Guy. Caractéristiques isotopiques (^{18}O) d'une tourbière a palses, et détermination des variations de composition isotopique des précipitations dans la région de Poste-de-la-Baleine, durant une année hydrologique. M, 1983, Universite du Quebec a Montreal. 97 p.

Lapinski, William James. The distribution of foraminifera off part of the Florida Panhandle coast. M, 1957, Florida State University.

Lapkowsky, Walter W. Minor elements in some carbonates. M, 1959, McMaster University. 114 p.

Laplante, Bernard Eric. Calc-silicate rocks in the Wollaston Lake area, Saskatchewan. M, 1974, University of Saskatchewan. 87 p.

Laplante, Richard. Etude de la mineralisation en Nb-Ta-U du complexe igné alcalin du Canton Crevier, Comte Roberval, Lac St-Jean, P.Q. M, 1980, Ecole Polytechnique. 51 p.

LaPoint, D. J. Geology, geochemistry, and petrology of sandstone copper in New Mexico. D, 1979, University of Colorado. 353 p.

LaPoint, Dennis John. Geology and geophysics of the southwestern Flathead Lake region (Precambrian) (Lake County), Montana. M, 1971, University of Montana. 110 p.

Lapointe, Bernard. Les relations structurales autour du lobe anorthositique de St. Fulgence, région de Saguenay. M, 1984, Universite du Quebec a Chicoutimi. 131 p.

Lapointe, Daniel. Le Granite des Monts Ste-Cécile et St-Sébastien. M, 1986, Universite Laval. 96 p.

LaPointe, Daphne D. Geology of a portion of the Eocene Sylvan Pass volcanic center, Absaroka Range, Wyoming. M, 1977, University of Montana. 60 p.

Lapointe, Guy. Study of a titaniferous iron deposit and surrounding country rocks in La Lievre area, Quebec. M, 1960, University of Manitoba.

Lapointe, Martine. Répartition stratigraphique des algues Ordoviciennes et Siluriennes de l'Ile d'Anticosti. M, 1987, Universite de Montreal.

Lapointe, P. Fenitization around hematite occurrences at the Haycock Mine, Hull and Templeton townships, Quebec. M, 1979, University of Ottawa. 102 p.

LaPointe, Paul Reggie. The suitability of sedimentary rock masses for annular superconductive magnetic energy storage units; feasibility studies, site evaluation techniques and site investigations. D, 1980, University of Wisconsin-Madison. 311 p.

Laporte, Jean. Etude du Précambrien de la région du Lac Beauport (Québec). M, 1952, Universite Laval.

Laporte, Leo F. Sedimentary facies of the Cottonwood Limestone (Permian), northern Midcontinent. D, 1960, Columbia University, Teachers College. 154 p.

Laporte, Pierre J. Geology of the Rankin Inlet area, Northwest Territories. M, 1976, Brock University. 147 p.

LaPorte, William D. The subsurface geology of the Pauls Valley area, townships 3 and 4 North, ranges 1

East and 1 West, Garvin County, Oklahoma. M, 1958, University of Oklahoma. 54 p.

Lapp, David B. The subduction geometry beneath western Washington from deconvolved teleseismic P-waveforms. M, 1987, University of Washington. 70 p.

Lapp, Eric Tod. Detailed bedrock geology of the Mt. Grant-South Lincoln area, central Vermont. M, 1986, University of Vermont. 114 p.

Lappin, Allen R. Metamorphism of basic rocks from the garnet and staurolite zones, Marquette Trough, Michigan. M, 1971, University of Illinois, Urbana. 68 p.

Lappin, Allen Ralph. Partial melting and the generation of quartz dioritic plutons at crustal temperatures and pressures within the Coast Range batholithic complex near the Khyex River, British Columbia. D, 1976, Princeton University. 143 p.

LaPrade, Kerby Eugene. Dust storm sediments of the Lubbock area, Texas. M, 1954, Texas Tech University. 73 p.

LaPrade, Kerby Eugene. Geology of Shackleton glacier area, Queen Maud Range, Transantarctic Mountains, Antarctica. D, 1969, Texas Tech University. 394 p.

Larabee, Peter A. Late-Quaternary vegetational and geomorphic history of the Allegheny Plateau at Big Run Bog, Tucker County, West Virginia. M, 1986, University of Tennessee, Knoxville. 115 p.

Laramore, Baylis Harriss. Geology of the Saginaw Quadrangle, Tarrant County, Texas. M, 1958, Southern Methodist University. 33 p.

Laravie, Joseph A. Geologic field studies along the eastern border of the Chiwaukum graben, central Washington. M, 1976, University of Washington. 56 p.

Laraway, William Harlan. Geology of the south fork area of the Ogden River areas, Weber County, Utah. M, 1958, University of Utah. 63 p.

Laraya, Rogelio Gotardo. Aeromagnetics in the search for porphyry copper deposits. D, 1973, Stanford University. 94 p.

Larbah, Mohamed Ali. Two-dimensional dynamic analysis of concrete gravity dams embedded in visco-elastic half space. D, 1987, University of Pittsburgh. 189 p.

Larberg, Gregory Martin. Hydrodynamic effect on oil accumulation in a stratigraphic trap, Kitty Field, Powder River basin, Wyoming. M, 1976, Texas A&M University. 180 p.

Larbi, Emmanuel Yaw. Deterioration of carbonate building stones due to acid solutions; measurement and prevention. M, 1988, University of Windsor. 172 p.

Lardner, James Edward. Petrology, depositional environment and diagenesis of Middle Pennsylvanian (Desmoinesian) "Lagonda Interval", Cherokee Group, in east-central Kansas. M, 1984, University of Iowa. 156 p.

Larese, R. E. Petrology and stratigraphy of the Berea Sandstone in the Cabin Creek and Gay-Fink trends, West Virginia. D, 1974, West Virginia University. 387 p.

Larese, Richard E. Experimental cementation of quartz sand and compaction of argillaceous sand. M, 1968, West Virginia University.

Larew, Hiram G. Use of field, laboratory, and theoretical procedures for analysing landslides. M, 1951, Purdue University.

Larimer, John William. The petrology of chondritic meteorites in the light of experimental studies. D, 1966, Lehigh University. 196 p.

Larimer, Ted Ray. Paleogeology of the pre-Mississippian surface, Bighorn Mountains, Wyoming. M, 1959, University of Iowa. 120 p.

Larkin, Brett James. Interactive generation of orebody section. M, 1988, Stanford University. 247 p.

Larkin, Randall George. Hydrogeologic controls on underflow in alluvial valleys; implications for Texas water law. M, 1988, University of Texas, Austin. 134 p.

Larner, Kenneth Lee. Near-receiver scattering of teleseismic body waves in layered crust-mantle models having irregular interfaces. D, 1970, Massachusetts Institute of Technology. 274 p.

Laroche, Bernard. Dynamique hydraulique et sédimentaire du delta actif de la Rivière Romaine. M, 1983, Universite du Quebec a Rimouski. 172 p.

Laroche, Paul. Etude de la densité, de la susceptibilité magnétique, et de la résistivité électrique des gisements sulfures de la Province de Québec. M, 1972, Universite Laval. 199 p.

Laroche, T. Matthew. Geology of the Gallinas Peak area, Socorro County, New Mexico. M, 1980, New Mexico Institute of Mining and Technology. 145 p.

Larochelle, Andre. A gravity survey in parts of St. Louis and Jefferson counties, Missouri. M, 1952, St. Louis University.

Larochelle, André A study of the palaeomagnetism of rocks from Yamaska and Brome Mountains, Quebec. D, 1959, McGill University.

Larocque, Cynthia. Geochronology and petrology of north-central Gaspé igneous rocks, Quebec. M, 1986, University of Toronto.

LaRocque, J. A. A study of the mineralogy of 13 and 14 veins and their influences on the gold-silver ratio at the Ross Mine, Ramore area, Ontario. M, 1952, University of Toronto.

LaRocque, Joseph Alfred Aurele. New genera and species of Mollusca from the Middle Devonian of Michigan and Manitoba. M, 1946, University of Michigan.

LaRocque, Joseph Alfred Aurele. Pre-Traverse Devonian pelecypods of Michigan. D, 1948, [University of Michigan].

LaRoge, Clifford Thomas. A study of the chemical composition and mineralogical relationships of sphalerite at Silver Mine, Madison County, Missouri. M, 1932, University of Missouri, Columbia.

Laronne, Jonathan B. Dissolution potential of surficial Mancos Shale and alluvium. D, 1977, Colorado State University. 141 p.

Larouche, Christiane. Recent deep water benthonic foraminifera of the Northwest Atlantic Ocean. M, 1979, Carleton University.

Larouche, Claude. Petrology of the granitic complex near Lac Remigny, Noranda area, Quebec. M, 1979, Carleton University. 176 p.

Larrain, Alberto P. The fossil and Recent shallow water irregular echinoids from Chile. D, 1984, University of Southern California.

Larrère, Marc H. A knowledge-based approach to full wave data processing. M, 1987, Massachusetts Institute of Technology. 118 p.

Larsen, Alfred L. A heavy mineral analysis of Pleistocene terrace sands in Liberty and Wakulla counties, Florida. M, 1958, Florida State University.

Larsen, Chris Robert. Stream sediment and soil geochemistry of an area of Paleozoic formations in southeastern Ontario. M, 1978, Queen's University. 350 p.

Larsen, David P. Growth control of a marine diatom by low and limiting levels of nitrate-nitrogen. D, 1975, Oregon State University. 121 p.

Larsen, Esper Signius. The areal geology of the Creede mining district, Colorado. D, 1918, University of California, Berkeley. 150 p.

Larsen, Esper Signius, III. The mineralogy and paragenesis of the variscite nodules from the vicinity of Fairfield, Utah. D, 1940, Harvard University.

Larsen, Everett Christian. Some influences of climate in Illinois (L329). M, 1931, [University of Illinois, Urbana].

Larsen, Frederick Duane. The surficial geology of the Mount Tom Quadrangle, Massachusetts. D, 1972, University of Massachusetts. 339 p.

Larsen, Jimmy Carl. Electric and magnetic fields induced by oceanic tidal motion. D, 1966, University of California, San Diego. 114 p.

Larsen, John H. Ground water conditions of a part of the Kendrick Project, Natrona County, Wyoming. M, 1951, University of Wyoming. 56 p.

Larsen, Kenneth G. Depositional environment of a part of the Fort Scott Formation in central Missouri. M, 1953, University of Missouri, Columbia.

Larsen, Leonard H. Lithology of the Cambro-Ordovician dolomite and its contact relations with the Hudson River series in Orange County, New York. M, 1953, Columbia University, Teachers College.

Larsen, Leonard Hills. Zircon studies in silicic igneous rocks. D, 1956, Columbia University, Teachers College. 67 p.

Larsen, Margaret Kreider. Geochemical relationships of Laramide intrusives of the Empire region, Colorado. D, 1968, University of Colorado. 86 p.

Larsen, Norbert William. Chronology of late Cenozoic basaltic volcanism; the tectonic implications along a segment of the Sierra Nevada and Basin and Range Province boundary. D, 1979, Brigham Young University. 95 p.

Larsen, Norbert William. Geology and ground-water resources of northern Cedar Valley, Utah County, Utah. M, 1959, Brigham Young University. 42 p.

Larsen, Norman R. Geology of the Lamb Canyon area near Beaumont, California. M, 1962, Pomona College.

Larsen, Richard A. Major and trace element characterization and correlation of the Sentinel Butte Ash/Bentonite (Paleocene), McKenzie County, North Dakota. M, 1988, University of North Dakota. 163 p.

Larsen, Richard K. Geology, alteration, and mineralization of the northern Cutting Stock, Lawrence County, South Dakota. M, 1977, South Dakota School of Mines & Technology.

Larsen, Roland M. A study of cathodoluminescent properties and the trace element distribution of the gangue calcite (and other phases) at the Elmwood Mine in Smith County, Tennessee. M, 1978, University of Tennessee, Knoxville. 87 p.

Larsen, Ronald Edward. Seismic designed backslopes and evaluation in a structurally disturbed basalt section. M, 1976, University of Idaho. 38 p.

Larsen, Veryl E. Clay mineralogy of the Dakota Group (Cretaceous) and adjacent sediments. D, 1953, University of Colorado.

Larsen, Willard N. Petrology and structure of Antelope Island, Davis County, Utah. D, 1957, University of Utah. 185 p.

Larsen, Willard N. Precambrian geology of the western Uinta Mountains, Utah. M, 1954, University of Utah. 53 p.

Larsen, William Robert. Hydrologic and geologic aspects of ground-water recharge along the Humboldt river near Winnemucca, Nevada. M, 1967, University of Nevada. 124 p.

Larsen, William Roger. Petrography of Pennsylvanian channel sand in Boone County, Iowa. M, 1956, Iowa State University of Science and Technology.

Larson, A. G. Origin of the chemical composition of undisturbed forested streams, western Olympic Peninsula, Washington State. D, 1979, University of Washington. 216 p.

Larson, Albert O. Mass-wasting in Tuckerman Ravine, Mt. Washington, New Hampshire. M, 1983, University of New Hampshire. 94 p.

Larson, Allan Richard. Stratigraphy and paleontology of Moenkopi Formation in southern Nevada. D, 1966, University of California, Los Angeles. 292 p.

Larson, Allan Richard. Tertiary geology of a portion of the northern Sierra Nevada, California. M, 1962, University of California, Los Angeles.

Larson, Carl Leonard, Jr. Subsurface correlation in southeastern Illinois. M, 1930, University of Illinois, Urbana.

Larson, Dana Christine. Depositional facies and diagenetic fabrics in the late Pleistocene Falmouth Formation of Jamaica. M, 1983, University of Oklahoma. 228 p.

Larson, Daniel M. Seismic detection of voids in coal seams. M, 1987, Colorado School of Mines. 112 p.

Larson, David Roy. The hydrogeology of the landfill at Lincoln, Nebraska. M, 1976, University of Nebraska, Lincoln.

Larson, David Warren. Paleoenvironment of the Mowry Shale (Lower Cretaceous), western and central Wyoming, as determined from biogenic structures. M, 1977, University of Wisconsin-Madison.

Larson, Douglas William. On reconciling lake classification with the evolution of four oligotrophic lakes in Oregon. D, 1970, Oregon State University. 159 p.

Larson, Edward Richard. Ordovician limestones of the Milheim area, Centre County, Pennsylvania. M, 1947, Columbia University, Teachers College.

Larson, Edward Richard. Stratigraphy of the (Ordovician) Plattin Group, southeastern Missouri. D, 1951, Columbia University, Teachers College.

Larson, Edwin Eric. The geology of the Potrero Seco area, Ventura County, California. M, 1958, University of California, Los Angeles.

Larson, Edwin Erie. The structure, stratigraphy, and paleomagnetics of the Plush area, southeastern Lake County, Oregon. D, 1965, University of Colorado. 193 p.

Larson, G. J. Meltwater storage in a temperate glacier. D, 1976, Ohio State University. 121 p.

Larson, George Delmore. Geology of the southwest portion of Sheridan Quadrangle, Wyoming. M, 1941, University of Iowa. 62 p.

Larson, Grahame J. Internal drainage of stagnant ice, Burroughs Glacier, Southeast Alaska. M, 1972, Ohio State University.

Larson, Jay L. The sulfosalt mineralogy of the Silver City mining district, central Black Hills, South Dakota. M, 1971, South Dakota School of Mines & Technology.

Larson, Jay Leo. A study of the diffusion, electrochemical mobility and removal of dissolved copper in a saturated porous medium. D, 1980, University of Colorado. 204 p.

Larson, Jeffery E. Stratigraphy and sedimentation of the Upper Cretaceous Montana Group of northwestern Montana, with detailed bedform analysis of Two Medicine Formation complexly cross-bedded sandstone. M, 1986, University of Montana. 105 p.

Larson, Jerome Valjean. A cross correlation study of the noise performance of electrostatically controlled La Coste and Romberg gravimeters. D, 1968, University of Maryland. 147 p.

Larson, Jill Marie. A three-dimensional simulation of the CECOS injection well near Lake Charles, Louisiana. M, 1987, Ohio University, Athens. 189 p.

Larson, John A. Pliocene and Pleistocene foraminifera populations in Arctic Ocean sediment cores. M, 1973, University of Wisconsin-Madison.

Larson, John A. Trace fossils, resedimented carbonates, and conodonts of the Wolfcampian portion of the eastern Oquirrh Formation, Utah. D, 1977, University of Wisconsin-Madison. 211 p.

Larson, John Edgar. Geology, geochemistry and volcanic history of Les Mines Selbaie, Quebec, Canada; an Archean epithermal system. D, 1987, Colorado School of Mines. 388 p.

Larson, John Edgar. Geology, geochemistry and wallrock alteration at the Magusi and New Insco massive sulfide deposits, Hebecourt Township, northwestern Quebec. M, 1983, University of Western Ontario. 173 p.

Larson, Kenneth A. A petrographic analysis of metasomatism in the Bedford Augen-Gneiss, Westchester County, New York. M, 1958, Miami University (Ohio). 82 p.

Larson, Kenneth Williams. The areal geology of the Rockport-Wanship (Summit County, Utah), area. M, 1951, University of Utah. 46 p.

Larson, Lawrence Tilford. Geology and mineralogy of certain manganese oxide deposits, Philipsburg, Montana. D, 1962, University of Wisconsin-Madison. 74 p.

Larson, Lawrence Tilford. The rotation properties of some anisotropic absorbing silver minerals. M, 1959, University of Wisconsin-Madison.

Larson, Mark J. An experiment using the HI-Cell for stress measurements in a Utah coal mine. M, 1987, University of Utah. 176 p.

Larson, Marvin E. Geology and ore deposits of the Silver Monument and New Era mines, Sierra County, New Mexico. M, 1975, University of Texas at El Paso.

Larson, Michael Kenneth. Origin of land subsidence and earth fissures, Northeast Phoenix, Arizona. M, 1982, Arizona State University. 151 p.

Larson, Murray Lloyd. Biostratigraphy of the Glenogle Formation (Ordovician) near Glenogle, British Columbia. M, 1965, University of Alberta. 45 p.

Larson, Paul Andrew. Petrology and depositional environments of a Lower Cretaceous (Albian) rudist reef, southeastern Oklahoma. M, 1982, University of Texas at Dallas. 156 p.

Larson, Peter Brennan. I, An $^{18}O/^{16}O$ investigation of the Lake City Caldera, San Juan Mountains, Colorado; II, $^{18}O/^{16}O$ relationships in Tertiary ash-flow tuffs from complex caldera structures in central Nevada and San Juan Mountains, Colorado. D, 1984, California Institute of Technology. 502 p.

Larson, Peter Brennan. The metamorphosed alteration zone associated with the Bruce Precambrian volcanogenic massive sulfide deposits, Yavapai County, Arizona. M, 1976, University of Arizona.

Larson, Richard Walter. The petrology of the Merom Sandstone at the type locality. M, 1955, Indiana University, Bloomington. 37 p.

Larson, Robert James. The Mott island conglomerate (Precambrian), Isle Royale National Park, Michigan. M, 1968, Michigan Technological University. 82 p.

Larson, Roger Lee. Near-bottom studies of the East Pacific Rise crest and tectonics of the mounth of the Gulf of California. D, 1970, University of California, San Diego. 180 p.

Larson, Ronald Gary. From molecules to reservoirs; problems in enhanced oil recovery. D, 1980, University of Minnesota, Minneapolis. 350 p.

Larson, Thomas A. Geology of T1N and T2N; R22E, R23E and R24E, Blaine and Butte Counties, south-central Idaho. M, 1974, University of Wisconsin-Milwaukee.

Larson, Thomas C. A stratigraphic study of the (Cretaceous) Inyan Kara Group of the Black Hills of South Dakota and Wyoming. M, 1955, University of Wisconsin-Madison.

Larson, Timothy Howe. Geologic structure of western McMurdo Sound; a seismic refraction study. M, 1980, Northern Illinois University. 53 p.

Larson, Wilbert Sanford. The post-Maple Mill subsurface geology and ground water possibilities of Lee County, Iowa. M, 1943, University of Iowa. 150 p.

Larson, William C. A study of particle parameters to determine the fragmentation history of test rock and mineral dust samples. M, 1972, University of Wisconsin-Milwaukee.

Larsson, Sven Y. Silurian paleontology and stratigraphy of the Hudson Bay Lowlands in western Quebec. M, 1984, McGill University. 188 p.

Larue, D. K. Sedimentary history prior to chemical iron sedimentation of the Precambrian X Chocolay and Menominee groups (Lake Superior region). D, 1979, Northwestern University. 180 p.

Larue, D. K. The sedimentary response to landsliding in the marine nearshore, Portuguese Bend landslide complex, California. M, 1976, Northwestern University.

LaRue, John W. The effect of rate on the recovery of oil by a water drive. M, 1950, University of Oklahoma. 71 p.

Lary, Brenda Brants. The ostracode fauna of the Pennsylvanian Wolf Mountain Shale of the Possum King-

dom area, Palo Pinto County, Texas. M, 1965, Texas Christian University.

Lary, H. N. Some phases of contact metamorphism in the southeastern Adirondacks. M, 1927, Massachusetts Institute of Technology. 62 p.

Laryea, K. B. Solute dispersion in soil. D, 1979, University of Guelph.

LaSalle, P. Field expression of some rock types with special reference to anorthosites. M, 1962, McGill University.

Lasca, Norman Paul and Volckmann, Richard Peter. Geology of the Music Pass area in the Sangre de Cristo Range, Colorado. M, 1961, University of Michigan.

Lasca, Norman Paul, Jr. The surficial geology of Skeldal, Mesters Vig, Northeast Greenland. D, 1965, University of Michigan. 86 p.

Lascelles, Peter A. Paleoenvironments of the Permian Elephant Canyon Formation, Canyonlands National Park, Utah. M, 1981, University of Wisconsin-Milwaukee. 97 p.

Lasemi, Y. Carbonate microfacies and depositional environments of the Kinkaid Formation (Upper Mississippian) of the Illinois Basin. D, 1980, University of Illinois, Urbana. 145 p.

Lasemi, Yaghoob. Subsurface geology and stratigraphic analysis of the Bayport Formation in the Michigan Basin. M, 1975, Michigan State University. 80 p.

Lasemi, Zakaria. Recognition of original mineralogy in microlites and its genetic and diagenetic implications. M, 1983, University of Illinois, Urbana. 131 p.

Laseski, Ruth Anne. Modern pollen data and Holocene climate changes in eastern Africa. M, 1977, Brown University. 284 p.

Lash, G. G. The structure and stratigraphy of the Pen Argyl Member of the Martinsburg Formation in Lehigh and Berks counties, Pennsylvania. M, 1978, Lehigh University. 218 p.

Lash, Gary George. The geology of the Kutztown and Hamburg 7 1/2′ minute quadrangles, eastern Pennsylvania. D, 1980, Lehigh University. 366 p.

Lasheen, M. R. M. W. Factors influencing metal uptake and release by sediments in aquatic environments. D, 1974, University of Michigan. 138 p.

Lashkari-Irvani, Bahman. Cumulative damage parameters for bilinear systems subjected to seismic excitations. D, 1983, Stanford University. 260 p.

Lashof, Daniel Abram. The role of the biosphere in the global carbon cycle; evaluation through biospheric modeling and atmospheric measurement. D, 1987, University of California, Berkeley. 314 p.

Lasker, Howard Robert. Intraspecific variability and its ecological consequences in the reef coral Montastrea cavernosa. D, 1978, University of Chicago. 141 p.

Lasker, Howard Robert. The measurement of paleoecological diversity. M, 1973, University of Rochester.

Laskin, John. An examination of selected ERDA-#9 drill core sections taken from the Permian Salado and Castile formations near Carlsbad, New Mexico. M, 1986, New Mexico Institute of Mining and Technology. 60 p.

Laskowski, Edward A. Sedimentary petrology and petrography of the Esopus, Carlisle Center, and Schoharie formations (Lower Devonian) in New York State, with a discussion of the Taonurus problem. M, 1956, Rensselaer Polytechnic Institute. 198 p.

Laskowski, Edward Albin. Anomalous distribution of toxic soils in the Castaic Valley, California; a study based on soil-chemical geography, geology, and geochemistry. D, 1968, University of California, Los Angeles. 314 p.

Laskowski, Erich Richard. Geology of the black shale belt of the Bruce Creek area, Stevens County, Washington. M, 1982, Washington State University. 113 p.

Laskowski, Keith A. The geology and petrochemistry of algoma-type iron formation in the western Vermilion District, Wawa Belt Subprovince, northeastern Minnesota. M, 1986, Colorado School of Mines. 256 p.

Laskowski, Thomas Edward. Rb-Sr, K-Ar, and $^{40}Ar/^{39}Ar$ systematics in Paleozoic glauconite from mississippi valley-type localities, and the dating of events during sediment diagenesis. D, 1982, Miami University (Ohio). 117 p.

Laskowski, Thomas Edward. The intrusive granitic bodies of the Crossnore plutonic-volcanic group; N.W. North Carolina. M, 1978, Miami University (Ohio). 108 p.

Lasky, Loren R. The thermal inertia of engineering geologic units. M, 1980, Colorado School of Mines. 248 p.

Lasky, Samuel G. Chalcocite-covellite solid solution and exsolution with an appendix on the copper ores of Kennecott, Alaska. M, 1929, Yale University.

Lasley, Bert A. Detailed study of thrust fault zones and the structural relationship of the Penitentiary and Carbon thrust faults, Krebs Quadrangle, Oklahoma. M, 1987, University of Texas at Dallas. 119 p.

LaSota, Kenneth Alan. The distribution of porosity in the sandstones of the Venango Group (Upper Devonian) of southwestern Pennsylvania. D, 1988, University of Pittsburgh. 162 p.

Lass, Garry L. Effects of hydrothermal contamination of strontium isotope systematics in rocks and minerals of the Point Sal Ophiolite, Santa Barbara Co., Cal. M, 1976, California State University, Los Angeles.

Lassin, Richard J. Clay mineral—low-temperature K-feldspar stability relationships in the St. Peter Formation of Wisconsin, eastern Iowa, and Minnesota. M, 1975, Northern Illinois University. 88 p.

Lassley, Richard Harold. An introduction to linear time-variant digital filtering of seismic data. D, 1965, University of Missouri, Rolla. 74 p.

Lassus, Roy E., Jr. Subsurface geology of Main Pass, Blocks 11 through 51, offshore Louisiana. M, 1974, University of New Orleans.

Last, William Michael. Clay mineralogy and stratigraphy of offshore Lake Agassiz sediments in southern Manitoba. M, 1974, University of Manitoba.

Last, William Michael. Sedimentology and post-glacial history of Lake Manitoba. D, 1980, University of Manitoba.

Lastrico, Roberto Mario. Effects of site and propagation path on recorded strong earthquake motions. D, 1970, University of California, Los Angeles. 205 p.

Lasuzzo, Anthony. Paleoecology of the Oligocene Bryam Marl in Mississippi. M, 1974, Northeast Louisiana University.

Laswell, Troy J. A textural analysis of the (Devonian or Mississippian) Bedford Shale of Lorain County, Ohio. M, 1948, Oberlin College.

Laswell, Troy James. Geology of the Bowling Green, Missouri Quadrangle. D, 1953, University of Missouri, Columbia.

LaTendresse, Henri L. Some observations on Old Faithful Geyser, and geyser action in general. M, 1923, University of Minnesota, Minneapolis. 28 p.

Latham, Alfred G. Paleomagnetism, rock magnetism and U-Th dating of speleothem deposits. D, 1981, McMaster University. 507 p.

Latham, Don Jay. The modulation of backscattered microwave radiation by oscillating water drops. D, 1968, New Mexico Institute of Mining and Technology. 114 p.

Latham, Everett B. Geological survey of the vicinity of Lehigh Gap in Pennsylvania. M, 1910, Columbia University, Teachers College.

Latham, Gary Vincent. Long-period seismic measurements on the ocean floor - Part I, Bermuda area. D, 1965, Columbia University. 134 p.

Latham, Margie Ann Patterson. Mixing of basaltic and rhyolitic magmas; the Borax Lake volcanic sequence, Clear Lake volcanic field, California. M, 1985, University of California, Davis. 405 p.

Lathan, Thomas Stanyer, Jr. Stratigraphy, structure, and geochemistry of Plio-Pleistocene volcanic rocks of the western Basin and Range Province, near Truckee, California. D, 1985, University of California, Davis. 341 p.

Lathram, Ernest H. A comparison of the physical properties and petrographic characteristics of some limestones of southeastern Minnesota. M, 1942, University of Minnesota, Minneapolis. 125 p.

Lathrop, Richard Gilbert, Jr. The integration of remote sensing and geographic information systems for Great Lakes water quality monitoring. D, 1988, University of Wisconsin-Madison. 243 p.

Latifi'naieni, Abdolhamid. Infiltration into layered soils and moisture redistribution in a closed column of layered soils. D, 1988, University of Mississippi. 159 p.

Latimer, Ira Sanders, Jr. An X-ray diffraction study of minerals in shale. M, 1958, West Virginia University.

LaTorraca, Gerald Alan. Differential tellurics with applications to mineral exploration and crustal resistivity monitoring. D, 1981, Massachusetts Institute of Technology. 182 p.

LaTour, T. E. The nature and origin of the Grenville Front near Coniston, Ontario; a reinterpretation. D, 1979, University of Western Ontario. 344 p.

LaTraille, S. L. Crustal structure of the Mariana island arc system and old Pacific Plate from seismic refraction data. M, 1978, University of Hawaii. 136 p.

Latshaw, Warren Leroy. An insoluble residue study of the Conococheague Dolomite near Leesport, Pennsylvania. M, 1950, University of Pittsburgh.

Latson, Rebecca Lynn. Ostracoda of the Oleneothyris biostrome of the Vincentown Formation (Paleocene) in central New Jersey. M, 1986, Northeast Louisiana University. 96 p.

Latta, James M. Radioactivity of subsurface igneous rocks of Southeast Arkansas. M, 1955, University of Arkansas, Fayetteville.

Latta, Lee Allen. The Des Moines Series of the Buckhorn area, Murray County, Oklahoma. M, 1942, University of Oklahoma. 52 p.

Latta, William Love, Jr. Law of ownership of oil in situ. M, 1947, University of Pittsburgh.

Lattanner, Alan V. Geological and geophysical prospecting at the Donner Mine, Calaveras County, California. M, 1975, University of California, Berkeley. 82 p.

Lattanzi, Robert David. Ecology and destruction of foraminifera in the Poquonock River and Baker Cove estuarine-marsh system, Groton, Connecticut. D, 1975, University of Michigan. 169 p.

Lattimore, Robert Kehoe. Two measured sections from the Mesozoic of northwestern Guatemala. M, 1962, University of Texas, Austin.

Lattman, Laurence Harold. Geomorphology of the Allegheny Mountains of east-central West Virginia. D, 1953, University of Cincinnati. 152 p.

Lattman, Laurence Harold. The sub-Eden beds of the Ohio Valley above Cincinnati. M, 1951, University of Cincinnati. 61 p.

Lattu, Andrew C. Missourian facies and lithofacies and depositional environments; Hockley County, Texas. M, 1976, University of Texas at El Paso.

Latuszynski, Felix Victor. Geology of the Southwest Pleasant Valley Quadrangle, Montana. M, 1962, University of Montana. 51 p.

Lau, Ka Ching. Horizontal drains in clay slopes. D, 1983, University of Toronto.

Lau, Meng Hoo Sebastian. Structural geology of the vein system in the San Antonio gold mine, Bissett, Manitoba, Canada. M, 1988, University of Manitoba. 154 p.

Laub, Donald. Copper-uranium mineralization in the Coyote mining district, Mora County, New Mexico. M, 1954, University of Utah. 27 p.

Laub, Mary G. The origin and movement of gravel bars in the intertidal zone of Parrsboro Harbour, Nova Scotia. M, 1968, University of Pennsylvania.

Laub, Richard Steven. The Auloporids of the New York upper Devonian. M, 1968, Cornell University.

Laub, Richard Steven. The corals of the Brassfield Formation (middle Llandovery) in the Cincinnati Arch region. D, 1976, University of Cincinnati. 677 p.

Laubach, James Taylor. A study of the Cow Run and lower Freeport sandstones in Athens County, Ohio. M, 1953, Ohio State University.

Laubach, Stephen Ernest. Polyphase deformation, thrust-induced strain and metamorphism, and Mesozoic stratigraphy of the Granite Wash Mountains, west-central Arizona. D, 1986, University of Illinois, Urbana. 181 p.

Laubach, Stephen Ernest. Structural processes in the Conway Rhyolite Dome; magmatic implications. M, 1983, University of Illinois, Urbana. 155 p.

Laubscher, Alan L. A basic investigation of perspective map projections. M, 1965, Ohio State University.

Laudati, Robert P. Computer simulation of secondary hydrocarbon migration and entrapment. M, 1988, Stanford University. 157 p.

Lauden, Edward J., Jr. Quantitative geomorphology of selected basins of the Louisiana coastal marsh. M, 1976, Virginia State University. 38 p.

Lauderback, Ralph Lewis. The geology of the Lyons area, Cherokee and Adair counties, Oklahoma. M, 1952, University of Oklahoma. 76 p.

Laudon, John Lowell. Origins of cleavage in the Wallace Formation, Superior, Montana. M, 1978, University of Montana. 63 p.

Laudon, Julie Ann. Fission-track dating of the Mount Taylor uranium deposit, Grants mineral belt, New Mexico. M, 1983, Washington State University. 79 p.

Laudon, Katherine Jean. Geophysical investigations of the Duck Lake ground water subarea near Omak, Washington. M, 1983, Washington State University. 77 p.

Laudon, Lowell Robert. The stratigraphy and paleontology of the northward extension of the Burlington Limestone. M, 1929, University of Iowa. 147 p.

Laudon, Lowell Robert. The stratigraphy of the Kinderhook Series of Iowa. D, 1930, University of Iowa. 185 p.

Laudon, Richard B. A study of Chesterian and Morrowan rocks in the McAlester Basin of Oklahoma. M, 1957, University of Wisconsin-Madison. 35 p.

Laudon, Richard Baker. Stratigraphy and zonation of the Stanley Shale of the Ouachita Mountains of Oklahoma. D, 1959, University of Wisconsin-Madison. 176 p.

Laudon, Robert C. A petrologic study of the Recent sands of the southern Oregon coast. M, 1968, University of Wisconsin-Madison. 60 p.

Laudon, Robert C. Stratigraphy and petrology of the Difunta Group, La Popa and eastern Parras basins, northeastern Mexico. D, 1975, University of Texas, Austin. 317 p.

Laudon, Thomas Stanzel. The geology of the Lochaber District, Queensland, Australia. D, 1963, University of Wisconsin-Madison. 193 p.

Lauer, Arnold W. Special problems in North American petroleum geology. D, 1917, Yale University.

Lauer, Timothy Campbell. Stratigraphy and structure of the Snowflake Ridge area, Gallatin County, Montana. M, 1967, Oregon State University. 182 p.

Laufer, Arthur R. Time lags in Geiger counters. D, 1950, New York University.

Lauffer, James Robert. A hydrochemical study of a shallow ground-water system peripheral to Rehoboth Bay. D, 1982, University of Delaware. 183 p.

Laughbaum, Lloyd Ronald. A paleoecologic study of the upper Denton Formation; Tarrant, Denton, and Cooke counties, Texas. M, 1959, Southern Methodist University.

Laughland, Matthew M. The use of density profiling as an indicator of maceral composition and coal rank. M, 1987, Southern Illinois University, Carbondale. 191 p.

Laughlin, A. William. Petrology of the Molino Basin area of the Santa Catalina Mountains, Arizona. M, 1960, University of Arizona.

Laughlin, Alexander William. Excess radiogenic argon in pegmatite minerals. D, 1969, University of Arizona.

Laughlin, C. H. The Carboniferous volcanic rocks of New Brunswick. M, 1960, University of New Brunswick.

Laughlin, Dwight J. Geology of northwestern Franklin County, Kansas. M, 1957, University of Kansas. 87 p.

Laughlin, George Ray. The effect of thermal treatment on properties of clays. M, 1958, University of Kentucky. 45 p.

Laughlin, Jefferson Edwin. A geochemical analysis of the carbon system in a modern tropical lagoon; Discovery Bay, Jamaica. M, 1985, University of Oklahoma. 128 p.

Laughlin, Kenneth J. Interpretation of refraction and reflection stack data over the Brevard fault zone in South Carolina. M, 1988, Virginia Polytechnic Institute and State University.

Laughon, Robert B. A study of weathering of terrace gravels along South Boulder Creek (Boulder County) Colorado. M, 1963, University of Colorado.

Laughon, Robert Bush. The crystal structure of a new hydrous calcium copper silicate mineral. D, 1970, University of Arizona. 51 p.

Laukel, Quinn C. Pennsylvanian foraminifera (exclusive of the Fusulinidae) of the Marmaton Group in southeastern Kansas and north-central Oklahoma. M, 1956, University of Kansas. 107 p.

Laule, Susan W. A reinterpretation of the Permian Wildcat Peak Formation, central Nevada. M, 1978, University of Nevada. 90 p.

Lauman, Gary W. Geology of the Iles Mountain area, Moffat County, northwestern Colorado. M, 1965, Colorado School of Mines. 129 p.

Laun, Philip Royal. Primary seismic waves (P) at 250-350 km compared to measured wave at 0.3 km from Gnome nuclear explosion. M, 1965, Oregon State University. 55 p.

Laurence, Robert Abraham. Lead and zinc deposits of the shallow vein zone. M, 1931, University of Cincinnati. 101 p.

Laurent, J. Scott. Structural geology of the foothills from Lefthand Canyon to Boulder (Boulder County), Colorado. M, 1958, University of Colorado.

Lauria, Jeffrey M. Mass flux measurement of sediment oxygen demand. D, 1981, Polytechnic University. 115 p.

Laurie, Archibald M. and Staub, Harrison L. Some studies of multiphase relative permeability in consolidated California oil sands as determined by the capillary pressure displacement method. M, 1950, University of Southern California.

Laurie, Robert John. Comparisons of metal demand during the twentieth century. D, 1988, Colorado School of Mines. 185 p.

Laurin, Andre. The problem of the boundary between Grenville and Keewatin-Timiskaming provinces (Ontario). D, 1955, Universite Laval.

Laurin, André Frédéric Joseph. Geology of Ducharme-Mignault map-areas, Quebec. D, 1957, Universite Laval. 114 p.

Laurin, André Frédéric Joseph. The sulphides and siderite of the Mathieu Property, Keewatin Lake area, District of Kenora, Ontario. M, 1954, McGill University.

Laurin, Priscilla Rehnquist. Interpretation of the depositional environment of the Hardinsburg Formation, Gibson County, Indiana. M, 1988, Indiana University, Bloomington. 74 p.

Lauritzen, Robert A. Foraminiferal isotopic and assemblage analysis across the Epoch 6 carbon shift, western Equatorial Pacific. M, 1987, University of Hawaii. 76 p.

Laury, Robert Lee. Geology of the type area, Canyon Group, north-central Texas. M, 1962, Southern Methodist University. 126 p.

Laury, Robert Lee. Sedimentology of the Pleasantview Sandstone (Pennsylvanian), southern Iowa and western Illinois. D, 1966, University of Wisconsin-Madison. 121 p.

Lausen, Carl. Geology of the Old Dominion Mine, Globe, Arizona. M, 1923, University of Arizona.

Lausen, Carl. Gold veins of the Catman and Katherine districts, Arizona. D, 1931, University of Arizona.

Lausten, Charles Dean. Gravity methods applied to the geology and hydrology of Paradise Valley, Maricopa County, Arizona. M, 1974, Arizona State University. 137 p.

Lautenschlager, Herman Kenneth. The geology of the central part of the Pavant Range, Utah. D, 1952, Ohio State University. 230 p.

Lautier, Jeffrey C. Geology of the subsurface Eocene Cockfield Formation, southern Allen Parish, Louisiana. M, 1980, University of Southwestern Louisiana. 60 p.

Laux, John Peter, III. Mineralization associated with the Quitman mountains intrusion (Oligocene?), Hudspeth County, Texas. M, 1969, University of Texas, Austin.

Lauzon, E. P. Analytical photogrammetry applied to cadastre. M, 1959, Ohio State University.

Lavado, Marcelo. Geological aspects of the occurrence of zinc-lead at the San Vicente Mine, San Ramon, Tarma, central Peru. M, 1980, University of Texas at Dallas. 115 p.

Laval, William Norris. An investigation of the Ellensburg Formation. M, 1948, University of Washington. 52 p.

Laval, William Norris. Stratigraphy and structural geology, portions of south-central Washington. D, 1956, University of Washington. 223 p.

Lavenue, Arthur Marsh. Preliminary assessment of regional dispersivity of the Hanford basalts. M, 1985, Texas A&M University. 81 p.

Laverdure, Louise. Gravimétrie de la ceinture volcanique de L'Abitibi. M, 1983, Ecole Polytechnique. 105 p.

LaVergne, Michel. Investigation of the cation exchange theory of induced electrical polarization by means of a radioactive tracer. M, 1956, New Mexico Institute of Mining and Technology. 53 p.

Lavery, Norman Garnsey. Zinc dispersion in the upper Mississippi valley zinc-lead district. D, 1968, Pennsylvania State University, University Park. 154 p.

Lavigne, Maurice Jean, Jr. Geological, geochemical and sulfur isotopic investigations of gold mineralization and sulfide facies banded iron formation at the Dickenson and Campbell Red Lake mines, Red Lake, Ontario. M, 1983, McMaster University. 324 p.

Lavin, Owen Patrick. Lithogeochemical discrimination between mineralized and unmineralized cycles of volcanism in the Sturgeon Lake and Ben Nevis areas of the Canadian Shield. M, 1976, Queen's University. 249 p.

Lavin, Peter M. Model studies of the effects of near-source velocity discontinuities on the first-motion patterns of P and S around different force systems. D, 1962, Pennsylvania State University, University Park. 125 p.

Lavin, Stephen J. Region perception variability on choropleth maps; pattern complexity effects. D, 1979, University of Kansas. 310 p.

Lavine, Irvin. Studies in the development of North Dakota lignite; 1, The aqueous tension of lignite; 2, The drying of lignite without disintegration. D, 1930, University of Minnesota, Minneapolis.

LaViolette, Paul Alex. Galactic explosions, cosmic dust invasions, and climatic change. D, 1983, Portland State University. 759 p.

Lavkulich, Leslie Michael. Soluble aluminum in soils and some factors affecting its magnitude. D, 1967, Cornell University. 143 p.

Lavoie, Clermont. Etude géophysique de trois gisements de sulfures sous de morts-terrains épais et conducteurs. M, 1968, Ecole Polytechnique. 135 p.

Lavoie, Dawn L. Sediments on the southeastern flank of the Bermuda Pedestal. M, 1982, University of New Orleans. 97 p.

Lavoie, Denis. Stratigraphie, géologie structurale, sédimentologie et paléomilieux de la bande silurienne supérieure des lacs Aylmer et Saint-François. M, 1985, Universite Laval. 119 p.

Lavoie, Denis. Stratigraphie, sédimentologie et diagenèse du Wenlockien (Silurien) du Bassin de Gaspésie-Matapédia. D, 1988, Universite Laval. 300 p.

Lavoie, Jacques S. Petrology of Tertiary lavas near Westover, B.C. M, 1971, University of Toronto.

Law, Benny E. Pennsylvanian-Permian conodont succession from the Bird Spring Formation (San Bernardino County), southeastern California. M, 1971, University of San Diego.

Law, Eric W. Petrologic, geochronologic, and isotopic investigation of the diagenesis and hydrocarbon emplacement in the Muddy Sandstone, Powder River basin. D, 1983, Case Western Reserve University. 360 p.

Law, K. T. Analysis of embankments on sensitive clays. D, 1975, University of Western Ontario.

Law, Kyin-kouk Hubert. Time-dependent bearing capacity of frozen ground. M, 1987, University of Alaska, Fairbanks. 129 p.

Law, L. K. The measurement and computation of anisotropic susceptibility of rock samples with a vibrational type magnetometer. M, 1966, University of Western Ontario.

Law, Lewis B. The iron oxide minerals. M, 1935, West Virginia University.

Law, Maureen Min Wu. Illite crystallinity and thermal history of Devonian-Mississippian shales of the Appalachian Basin. M, 1980, University of Cincinnati. 96 p.

Law, Michael S. Mapping of upland chert gravel deposits, east central Kansas. M, 1986, Emporia State University.

Law, S. L. Metals in the aqueous effluents from municipal incinerators and an incinerator-residue processing plant. D, 1976, University of Maryland. 212 p.

Lawal, Adetunji A. A statistical approach to the palynology of the Hazard Number 8 coal (Francis Coal) of eastern Kentucky. M, 1986, Eastern Kentucky University. 36 p.

Lawall, Charles E. A preliminary report on the limestones of Pennsylvania with special reference to their economic uses. M, 1921, Lehigh University.

Lawhorn, Thomas Warren. A seismic study of the shallow sub-bottom materials of the eastern Caribbean. M, 1957, Rice University. 74 p.

Lawler, Adrian R. Capsalids (Monogenea; Capsalidae) of some Australian fishes. M, 1964, College of William and Mary.

Lawler, Adrian R. Zoogeography and host-specificity of the superfamily Capsaloidea Price, 1936 (Monogenea; Monopisthocotyles). D, 1971, College of William and Mary.

Lawler, Jeanne Passante. Fluid inclusion evidence for ore-forming solutions; Phoenixville, Audubon and New Galena mine districts, Pa. M, 1981, Bryn Mawr College. 74 p.

Lawler, Kevin P. Iron partitioning and upper mantle composition. M, 1985, Rensselaer Polytechnic Institute. 147 p.

Lawler, Sydney Kent. Stratigraphy and petrology of the Mississippian (Kinderhookian) Chapin Limestone of Iowa. M, 1981, University of Iowa. 118 p.

Lawler, Thomas L. A field study of local magnetic disturbances from glacial drift in Michigan. M, 1962, Michigan State University. 167 p.

Lawless, Steven James. Subsurface geology of the upper member of the Minnelusa Formation in the southeastern corner of the Powder River basin, Wyoming. M, 1979, University of Colorado.

Lawlor, John Francis. The genesis of massive sulfide deposits in carbonate sediments at Silvermines, Ireland. D, 1970, Boston University. 273 p.

Lawrence, Christopher H. Dolomite and diagenesis of the Gasport Member of the Lockport Formation of southern Ontario and western New York. M, 1982, Wayne State University. 102 p.

Lawrence, D. E. Contribution to the petrology of the Great Dyke of Nova Scotia (Canada). M, 1966, Dalhousie University. 108 p.

Lawrence, D. P. Petrology and structural geology of the Sanarate-El Progreso area, Guatemala. D, 1975, SUNY at Binghamton. 322 p.

Lawrence, David C. Geology and revised stratigraphic interpretation of the Miocene Sucker Creek Formation, Malheur County, Oregon. M, 1988, Boise State University. 54 p.

Lawrence, David Parker. Structure and petrology of the Castle Peak stock, northeastern Cascade Mountains, Washington. M, 1967, University of Washington. 67 p.

Lawrence, David Reed. The oysters Crassostrea Gigantissima (Finch) (age of bearing beds; Oligocene) at Belgrade, North Carolina; a case study in paleoecology. D, 1966, Princeton University. 237 p.

Lawrence, David Trowbridge. Patterns and dynamics of Late Cretaceous marginal marine sedimentation; Overthrust Belt, southwestern Wyoming. D, 1984, Yale University. 602 p.

Lawrence, Edmond Francis. Antimony deposits of Nevada. M, 1967, University of California, Los Angeles.

Lawrence, Edmond Francis. Geological and geophysical investigations of the mineral deposits of the Calico area, Mineral County, Nevada. D, 1969, University of California, Riverside. 141 p.

Lawrence, Fred W. Identification of geochemical patterns in ground water by numerical analysis. M, 1976, University of South Florida, Tampa. 71 p.

Lawrence, Gregory Brad. Aluminum chemistry of headwater streams at the Hubbard Brook Experimental Forest, before and after whole trees harvesting. D, 1987, Syracuse University. 162 p.

Lawrence, J. C. Wasatch and Green River formations of the Cumberland Gap area, Lincoln and Uinta counties, Wyoming. M, 1962, University of Wyoming. 102 p.

Lawrence, James Robert. O^{18}/O^{16} and D/H ratios of soils, weathering zones and clay deposits. D, 1970, California Institute of Technology. 272 p.

Lawrence, John K. Geology of the southern third of the Sutherlin Quadrangle, Oregon. M, 1961, University of Oregon. 100 p.

Lawrence, John R. Geology of the Cerro del Grant area, Rio Arriba County, north-central New Mexico. M, 1979, University of New Mexico. 131 p.

Lawrence, Marc Arnold. The submerged forests of the Panama City, Florida, area; a paleoenvironment interpretation. M, 1974, University of Florida. 122 p.

Lawrence, Paul. The application of microgravity in the detection of subsurface cavities. M, 1978, Wright State University. 78 p.

Lawrence, Robert Dale. The Eightmile Creek fault, northeastern Cascade Range, Washington. D, 1968, Stanford University. 66 p.

Lawrence, Robert M. The growth and development of the Krotz Springs Dome in St. Landry Parish, Louisiana. M, 1959, Louisiana State University.

Lawrence, Thomas A. Hydrogeochemistry of the transition zone; Biscayne Aquifer, Dade County, Florida. M, 1988, University of South Florida, Tampa. 109 p.

Lawrence, Thomas Spencer. A comparison of metamorphic conditions for deformation; the Goat Rock and Bartletts Ferry faults. M, 1988, Georgia State University. 143 p.

Lawrence, Viki Ann. A study of the Indian Peaks tin-bearing rhyolite dome-flow complex, northern Black Range, New Mexico. M, 1985, University of Colorado.

Lawrence, William St. see St. Lawrence, William

Lawrence, William W., Jr. Sampling, geochemical, and statistical methods for soil geochemical surveys in the exploration for sediment-hosted, disseminated gold deposits; the Preble gold deposit as an example. M, 1986, University of Colorado. 127 p.

Lawrey, J. D. Litter decomposition and trace metal cycling studies in habitats variously influenced by coal strip-mining. D, 1977, Ohio State University. 113 p.

Laws, Bruce R. Magnetic properties of the sheeted dike sequence of the Samail Ophiolite near Ibra, Oman. M, 1980, University of California, Santa Barbara.

Laws, Richard Anthony. Paleoecology of Late Triassic faunas from Mineral County, Nevada and Shasta County, California. M, 1978, University of California, Berkeley. 149 p.

Laws, Richard Anthony. Quaternary diatom floras and Pleistocene paleogeography of San Francisco Bay. D, 1983, University of California, Berkeley. 363 p.

Lawson, Andrew C. The rocks of the Rainy Lake region. D, 1888, The Johns Hopkins University.

Lawson, Charles Alden. Magnetic and microstructural properties of minerals of the ilmenite-hematite solid solution series with special reference to the phenomenon of reverse thermoremanent magnetism. D, 1982, Princeton University. 369 p.

Lawson, D. E. Geology of the West Ferris Mountains, Carbon County, Wyoming. M, 1949, University of Wyoming. 103 p.

Lawson, D. E. Sedimentation in the terminus region of the Matanuska Glacier, Alaska. D, 1977, University of Illinois, Urbana. 294 p.

Lawson, David E. Sedimentology of the Boss Point Formation in southeastern New Brunswick. M, 1962, University of New Brunswick.

Lawson, Douglas A. Paleoecology of the Tornillo Formation (Upper Cretaceous), Big Bend National Park, Brewster County, Texas. M, 1972, University of Texas, Austin.

Lawson, Douglas Allan. Change in marine mollusc communities during the middle Eocene on the Pacific Coast. D, 1977, University of California, Berkeley. 119 p.

Lawson, Douglas R. Chemistry of the natural aerosol; a case study in South America. D, 1978, Florida State University.

Lawson, James Edward, Jr. Location of intermediate depth northern Colombian earthquakes with a small digital computer. M, 1967, University of Tulsa. 347 p.

Lawson, James Edward, Jr. Rate displacements of plate boundaries from moments of moderate and small earthquakes. D, 1972, University of Tulsa. 72 p.

Lawson, Jeffrey Thomas. Surficial and engineering geology of the Cayuga Inlet valley, Ithaca, New York. M, 1977, Cornell University.

Lawson, John Sheldon. Petrology, geochemistry, and tectonic setting of amphibolites of the Ducktown mining district, Tennessee. M, 1986, University of Tennessee, Knoxville. 108 p.

Lawson, Ralph W. Lithology of the Le Grand Beds. M, 1949, Iowa State University of Science and Technology.

Lawson, Ralph W. Physical attributes of the limestone members of some mid-continent Pennsylvanian cyclothems. D, 1953, University of Wisconsin-Madison.

Lawson, Robert A. Application of remote sensing techniques to determine the spectral radiance reflectance in the Fort Pierce Inlet area. M, 1977, Florida Institute of Technology.

Lawton, David Edward. Geology of the Hard Labor creek area in west-central Georgia. M, 1969, University of Georgia. 51 p.

Lawton, Dennis R. Ground water conditions in a shallow dolomite aquifer in northeastern Milwaukee and southeastern Ozaukee counties, Wisconsin. M, 1979, University of Wisconsin-Milwaukee. 152 p.

Lawton, Evert Carl. Wetting-induced collapse in compacted soil. D, 1986, Washington State University. 185 p.

Lawton, Jeffery L. Earthquake activity at the Kodiak continental shelf, Alaska, determined by land and ocean bottom seismograph networks. M, 1982, University of Texas, Austin.

Lawton, John Edward. Geology of the north half of the Morgan Valley Quadrangle and the south half of the Wilbur Springs Quadrangle [California]. D, 1956, Stanford University. 259 p.

Lawton, K. D. The Round Lake Batholith and its satellitic intrusions, Kirkland Lake, Ontario. D, 1954, University of Toronto.

Lawton, Kevin M. The use of Rayleigh wave inversion as a basis for determining sedimentary basin thickness. M, 1985, University of Wisconsin-Milwaukee. 155 p.

Lawton, Timothy Frost. Contact metamorphism associated with the Little Cottonwood Stock, Utah. M, 1980, Stanford University. 76 p.

Lawton, Timothy Frost. Tectonic and sedimentologic evolution of the Utah foreland basin. D, 1983, University of Arizona. 266 p.

Lawver, L. A. Heat flow in the Gulf of California. D, 1976, University of California, San Diego. 96 p.

Lawyer, Gary Frank. Sedimentary features and paleoenvironment of the Dakota Sandstone (Early Upper Cretaceous) near Hanksville, Utah. M, 1972, Brigham Young University. 120 p.

Lay, Thorne. Analysis of upper and lower mantle structures using shear waves. D, 1983, California Institute of Technology. 363 p.

Layas, Fathi Mohamed. Response and stability of oceanfloor soils under random waves. D, 1982, North Carolina State University. 169 p.

Layer, Paul William. Archean paleomagnetism of Southern Africa. D, 1986, Stanford University. 407 p.

Layfield, Moody E. Major causes and suggested solutions to the declining oil discovery rate. M, 1947, University of Pittsburgh.

Layla, Rasheed Ibrahim. Numerical analysis of transient salt/fresh-water interface in coastal aquifers. D, 1980, Colorado State University. 202 p.

Layman, E. B. Intrusive rocks in the near-trench environment; two localities in the northern California Franciscan Complex. M, 1977, Stanford University. 55 p.

Layman, Frederic G. Alteration of cordierite and the pinite problem. D, 1964, Harvard University.

Layman, Frederick G. A petrographic study of magnetite replacement in Pre-Cambrian rocks, Chester County, Pennsylvania. M, 1952, Lehigh University.

Layman, John W. Acid insoluble residues of the carbonate sediments of northwest Florida Bay, South Florida. M, 1977, University of Toledo. 78 p.

Layman, Thomas Bruce. Paleoenvironmental significance of benthic foraminiferal biofacies in the Yegua Formation (middle Eocene), Southeast Texas. M, 1987, University of Texas, Austin. 76 p.

Laymon, Charles Alan. Glacial geology of western Hudson Strait, Canada, with reference to Laurentide ice sheet dynamics. D, 1988, University of Colorado. 365 p.

Layne, Graham Donald. The JC tin skarn, southern Yukon Territory; a mineralogical, fluid inclusion and stable isotope study. D, 1988, University of Toronto.

Layton, Albert W., Jr. Geological study sites of the northern Cuyahoga Valley National Recreation Area. M, 1980, University of Akron. 94 p.

Layton, D. W. A computerized information system on the impacts of coal-fired energy development in the Southwest. D, 1975, University of Arizona. 118 p.

Layton, Donald W. Stratigraphy and structure of the south-western foothills of the Rincon Mountains, Pima County, Arizona. M, 1958, University of Arizona.

Layton, Michael C. Geophysical signature of Pliocene reef limestones using direct current and electromagnetic resistivity survey methods, Collier County, Florida. M, 1982, University of South Florida, Tampa. 83 p.

Lazaro, Rogelio Cruz. Adaptive real-time streamflow forecasting model for hydrosystem operational planning. D, 1981, Colorado State University. 230 p.

Lazarsky, Jennifer J. Petrographic analysis of synsedimentary and diagenetic features of the High Bridge Group (Middle Ordovician) in the subsurface, Fayette County, central Kentucky. M, 1983, Eastern Kentucky University. 83 p.

Lazarus, David B. Speciation and phyletic evolution in Pterocanium (Radiolaria). D, 1984, Columbia University, Teachers College. 292 p.

Lazarus, Norman H. Petrography and sedimentology of Wilcox sediments penetrated by a well in Allen Parish, Louisiana. M, 1985, University of Southwestern Louisiana. 175 p.

Lazier, Bruce Earl. A decision-making strategy for the appraisal of newly discovered oil fields. M, 1964, Stanford University. 91 p.

Laznicka, Peter. Quantitative aspects in the distribution of base and precious metal deposits of the world. D, 1970, University of Manitoba.

Laznicka, Peter. The distribution and mutual relations of copper, lead, zinc, gold and silver deposits of the world. M, 1968, University of Manitoba.

Lazo, Edward Nicholas. Determination of radionuclide concentrations of U and Th in unprocessed soil samples. D, 1988, University of Florida. 332 p.

Lazo, L. A. R. A subsurface study of rocks of the Morrow Series of the Pennsylvanian System in Cimarron County, Oklahoma. M, 1973, Texas Christian University.

Lazoff, Steven Barry. Deposition of diatoms and biogenic silica as indicators of Lake Sammamish productivity. M, 1980, University of Washington. 128 p.

Lazor, John David. Petrology and subsurface stratigraphy of the Traverse Formation (middle Devonian) in northern Indiana. D, 1971, Indiana University, Bloomington. 143 p.

Lazuk, Raymond. The relationship between electrical resistivity and hydraulic conductivity in two fractured bedrock aquifers in western Montana. M, 1988, University of Montana. 79 p.

Lazzeri, Joel Joseph. Stratigraphy and petrology of the Middle Jurassic La Joya Formation, Miquihuana, Aramberri-Mezquital, and Real de Catorce areas, Mexico. M, 1979, University of New Orleans.

Le Bras, Ronan J. Methods of multiparameter inversion of seismic data using the acoustic and elastic born approximations. D, 1985, California Institute of Technology. 152 p.

Le Clerk, R. V., II. A hydrogeologic analysis of a Class-I, liquid waste disposal facility using a groundwater model. M, 1974, Stanford University. 68 p.

Le Du, Raymonde. Etude géologique et géostatistique de la minéralisation en molybdène du gisement Mont Copper. M, 1976, Ecole Polytechnique.

Le Gallais, Christopher J. Stratigraphy, sedimentation and basin evolution of the Pictou Group (Pennsylvanian), Oromocto Sub-basin, New Brunswick, Canada. M, 1983, McGill University. 123 p.

Le Moine, Denis. Thorite deposits and general geology of the Hall Mountain area, Boundary County, Idaho. M, 1959, University of Idaho. 68 p.

Le Roux, Frederick Holmes. Alkali-ion sorption and exchange by soils and their size and mineral separates. D, 1963, University of California, Riverside. 103 p.

Le Seur, Linda Perkins. Hydrogeology and water quality of a fly ash landfill, Montpelier, Iowa. M, 1985, University of Iowa. 94 p.

Le, Duc. Size, grade and value characteristics of base metal deposits in Canada. M, 1983, Queen's University. 217 p.

Lea, Joseph W. The Ostracoda of the Red Bluff Group of eastern Mississippi. M, 1936, Louisiana State University.

Lea, Peter Donald. Pleistocene glaciation at the southern margin of the Puget Lobe, western Washington. M, 1984, University of Washington. 96 p.

Leach, Carl L. The petrology and structural geology of the Tolver Peak area, Gunnison County, Colorado. M, 1962, Wichita State University. 79 p.

Leach, David L. Investigation of fluid inclusions in Missouri barites. M, 1971, University of Missouri, Columbia.

Leach, David Lamar, Jr. A study of the barite-lead-zinc deposits of central Missouri and related mineral deposits in the Ozark region. D, 1973, University of Missouri, Columbia.

Leach, E. D. Maximum probable flood on the Brazos River in the City of Waco. M, 1978, Baylor University.

Leach, Elizabeth. The archeological geology of the Archaic Morrisroe Site, Tennessee River, western Kentucky. M, 1980, University of Minnesota, Minneapolis. 129 p.

Leach, Elizabeth Katrin. The archeological geology of the Archaic Morrisroe Site, Tennessee River, western Kentucky. M, 1981, University of Minnesota, Duluth.

Leach, Jerald Wayne. A study of lacustrine dolomite and associated sediments of Lake Mound, Lynn and Terry counties, Texas. M, 1969, Texas Tech University. 153 p.

Leachtenauer, Jon C. Geology of the northern half of the DeRuyter, New York, 7-1/2 minute, Quadrangle. M, 1959, Syracuse University.

Leaf, Howard Westley. A magnetometer survey of the magnetic highs to the east of Saint Clair, Missouri. M, 1955, St. Louis University.

Leagault, Jocelyne Andree. Chitinozoa and Acritarcha of the Hamilton Group (Middle Devonian) of southern Ontario. D, 1971, University of Oklahoma. 280 p.

Leahey, Alfred. A study of the inorganic phosphorus compounds of the soil. D, 1932, University of Wisconsin-Madison.

Leahy, P. Patrick. A comparison of three-dimensional, quasi-three-dimensional and two-dimensional finite-difference ground-water flow modeling techniques, with application to the principal aquifers of Kent County, Delaware. D, 1979, Rensselaer Polytechnic Institute. 190 p.

Leahy, Richard Gordon. Distribution of ionium and selected trace elements in deep-sea sediments. D, 1957, Harvard University.

Leak, Robert E. The geology of Coaling Grounds Ridge and vicinity, Roane County, Tennessee. M, 1957, University of Tennessee, Knoxville. 39 p.

Leake, Martha Alan. The intercrater plains of Mercury and the Moon; their nature, origin, and role in terrestrial planet evolution. D, 1981, University of Arizona. 768 p.

Leamer, Richard James. Petrology of some freshwater limestones from the Intermountain area. M, 1960, University of Utah. 123 p.

Leaming, S. F. Gold deposits on Eagle Lake. M, 1948, University of Toronto.

Lean, David Robert Samuel. Phosphorus compartments in lakewater. D, 1973, University of Toronto.

LeAnderson, P. James. The Pre-animikie Greenstone complex (Precambrian) of a small area in Marquette County, Michigan. D, 1969, Michigan State University. 100 p.

LeAnderson, Paul James. The metamorphism of "impure" marbles, calcareous schists and amphibolites in a portion of Limerick Township, Ontario. D, 1978, Queen's University. 400 p.

Leao, Zelinda. Morphology, geology and development history of the southernmost coral reefs of western Atlantic, Abrolhos Bank, Brazil. D, 1982, University of Miami. 216 p.

Leao, Zelinda Margarida de Andrade Nery *see* de Andrade Nery Leao, Zelinda Margarida

Leap, Darrell Ivan. The glacial geology and hydrology of Day County, South Dakota. D, 1974, Pennsylvania State University, University Park. 757 p.

Lear, Janet Marie. A geophysical investigation of the structure and hydrogeology of Twentynine Palms, California. M, 1987, University of California, Riverside. 310 p.

Lear, Paul Robert. The role of iron in nontronite and ferrihydrite. D, 1987, University of Illinois, Urbana. 128 p.

Learned, Robert Eugene. Paragenesis of mercury ore deposits. M, 1962, University of California, Los Angeles.

Learned, Robert Eugene. The solubilities of quartz, quartz-cinnabar and cinnabar-stibnite in sodium sulfide solutions and their implications for ore genesis. D, 1966, University of California, Riverside. 175 p.

Leary, David Austin. Diagenesis of the Permian (Guadalupian) San Andres and Grayburg formations, Central Basin Platform, West Texas. M, 1984, University of Texas, Austin. 129 p.

Leary, George Merilin. Petrology and structure of the Tuzo creek molybdenite prospect near Penticton, British Columbia. M, 1970, University of British Columbia.

Leary, Richard Lee. Experimental studies of strata deformation with special attention to two major theories of orogenesis. M, 1961, University of Michigan.

Leary, Richard Lee. Paleogeomorphology affecting Early Pennsylvanian floras in Rock Island County, Illinois. D, 1980, University of Illinois, Urbana. 153 p.

Lease, Robin Clair. Stratigraphy of the lower Morrison Formation along the Defiance Monocline, New Mexico and Arizona. M, 1967, University of New Mexico. 86 p.

Leask, Dennis M. The geochemistry of Precambrian argillites; Purgell system (Canada). M, 1967, University of Calgary. 129 p.

Leason, Jonathan Oren. Structural geology of the Horse Mountain area, Marathon Basin, Texas. M, 1983, University of Texas, Austin. 85 p.

Leatham, Stacey. Origin and distribution of soil nitrates on the Nevada Test Site. M, 1982, University of Nevada. 152 p.

Leatham, W. Britt. Conodont-based chronostratigraphy and conodont distribution across the Upper Ordovician western North American carbonate platform in the eastern Great Basin and a model for Ordovician-Silurian genesis of the platform margin based on interpretation of the Silurian Diana Limestone, central Nevada. D, 1987, Ohio State University. 275 p.

Leatherbarrow, Robert Wesley. Gneiss domes along the boundary between the Bear and Slave structural provinces (near Arseno Lake, N.W.T.). M, 1975, Carleton University. 77 p.

Leatherbarrow, Robert Wesley. Metamorphism of pelitic rocks from the northern Selkirk Mountains, southeastern British Columbia. D, 1981, Carleton University. 218 p.

Leatherman, Stephen Parker. Quantification of overwash processes. D, 1976, University of Virginia. 245 p.

Leathers, Michael R. Balanced structural cross section of the western Salt Range and Potwar Plateau, Pakistan; deformation near the strike-slip terminus of an overthrust. M, 1988, Oregon State University. 270 p.

Leatsler, Maynard E. The geology of the Viroqua Quadrangle, Wisconsin. M, 1930, University of Iowa. 124 p.

Leavell, Daniel Nelson. The geology of josephinite in the Josephine Peridotite, Southwest Oregon. D, 1983, University of Massachusetts. 222 p.

Leavens, Peter B. Rubidium and cesium in New England pegmatites. D, 1966, Harvard University.

Leavenworth, George. A contribution to the geology of Ste. Genevieve and Perry counties, Missouri. M, 1903, Columbia University, Teachers College.

Leaver, Donald S. A refraction study of the Oregon Cascades. M, 1982, University of Washington. 67 p.

Leaver, Donald S. Mixed stochastic and deterministic modeling of the crustal structure in the vicinity of Mount Hood, Oregon. D, 1984, University of Washington. 193 p.

Leaver, June. Sedimentology, mineralogy, and pore water chemistry of schizohaline pond sediments, Turks and Caicos islands, British West Indies. M, 1985, Duke University. 76 p.

Leavesley, George Haslam. A mountain watershed simulation model. D, 1973, Colorado State University. 184 p.

Leavitt, Eugene Millidge. The petrology, paleontology and geochemistry of the Carson Creek North Reef Complex (Middle Devonian), Alberta, (Canada). D, 1966, University of Alberta. 155 p.

Leavitt, Gene Millidge. Geology of the Precambrian Greenhead Group in the St. John area, New Brunswick. M, 1963, University of New Brunswick.

Leavitt, James Douglas. Geology of the Challis volcanic rocks near Basin Creek, Idaho. M, 1980, University of Oregon. 135 p.

Leavitt, Karen M. A comparison of techniques for the determination of sedimentation rates in Great Bay Estuary, New Hampshire. M, 1980, University of New Hampshire. 151 p.

Leavitt, Steven Warren. Inference of past atmospheric $\delta^{13}C$ and p_{CO2} from $^{13}C/^{12}C$ measurements in tree rings. D, 1982, University of Arizona. 235 p.

Leavitt, William Z. Solid state diffusion in the feldspar system. D, 1953, Massachusetts Institute of Technology. 102 p.

Leavy, Brian D. Petrology of lamprophyre dikes and mantle dried inclusions from Westerly, Rhode Island. M, 1979, University of Rhode Island.

Leavy, Brian David. Surface-exposure dating of young volcanic rocks using the in situ buildup of cosmogenic isotopes. D, 1987, New Mexico Institute of Mining and Technology. 167 p.

Leavy, Donald. Scattering of elastic waves around a compressional source. M, 1972, Massachusetts Institute of Technology. 188 p.

Leavy, Donald Lucien. The bulk lunar electrical conductivity. D, 1975, Massachusetts Institute of Technology. 200 p.

LeBaron, Philip Mallory. Geology of the White Belt Portis gold mine property of Franklin County, North Carolina. M, 1937, University of North Carolina, Chapel Hill. 27 p.

Lebauer, Lawrence Robert. Petrology of the Wolsey Shale and Meagher Formation (Middle Cambrian) of southwestern Montana. D, 1962, Indiana University, Bloomington. 70 p.

Lebauer, Lawrence Robert. Wall rock alteration in the Big Divide District, Esmeralda County, Nevada. M, 1958, Indiana University, Bloomington. 37 p.

Lebedin, J. The climatic and geologic controls on the recharge and discharge regime of an unconfined aquifer. M, 1975, University of Waterloo.

Lebel, Andre. An electronic analyzer of audio-magnetotelluric signals. M, 1971, Colorado School of Mines. 64 p.

Lebel, Jeanne. Gîtologie de la mine d'or Akasaba, Abitibi, Québec. M, 1987, Ecole Polytechnique. 209 p.

Lebel, John Laurence. An evaluation of the performance and applicability of battery-powered induced polarization systems. M, 1973, University of Manitoba.

Lebens, E. H. Report on the Chapin Mine at Iron Mountain. M, 1888, Washington University.

LeBlanc, Arthur E. A microfossil study of brackish and marine sediments near Rockport, Texas. M, 1954, University of Massachusetts.

LeBlanc, Gabriel. Andean and Caribbean surface waves. M, 1958, Boston College.

LeBlanc, Gabriel. Spectral analysis of short-period first arrivals of the April 13, 1963 Peruvian earthquake. D, 1966, Pennsylvania State University, University Park. 143 p.

LeBlanc, H. G. Radon gas activities in the ground waters of four major Nova Scotian aquifers. M, 1980, University of Waterloo.

LeBlanc, M. J. The effect of copper on phytoplankton. M, 1979, University of British Columbia.

LeBlanc, Robert C. Geologic evolution of carbonate shelf margin, Grand Cayman, B.W.I. M, 1979, Louisiana State University.

LeBlanc, Rufus J. Louisiana Eocene Midway Gastropoda. M, 1941, Louisiana State University.

Leblanc, Rufus Joseph, Jr. Environments of deposition of the Yegua Formation (Eocene), Brazos County, Texas. M, 1970, Texas A&M University.

Lebo, Marvel Hope. A compilation and evaluation of laboratory experiences in earth science for junior high school. M, 1966, Northern Illinois University. 177 p.

Leboeuf, Denis. Etude expérimentale et paramétrique de l'écoulement de l'eau souterraine dans les pentes argileuses du Québec. M, 1982, Universite Laval. 121 p.

Lebofsky, Larry Allen. Chemical composition of Saturn's rings and icy satellites. D, 1974, Massachusetts Institute of Technology. 108 p.

LeBret, George Curtis. Folded boudins within the Grand Forks Group, British Columbia. M, 1976, Washington State University. 76 p.

Lebreton, Claude Marie. The defect structure of wustite. D, 1983, Case Western Reserve University. 170 p.

Lebuis, Jacques. La Géologie des dunes de sable de la région du Lac Saint-Jean, P.Q. M, 1971, Universite de Montreal.

Lecheminant, Anthony Norman. Experimental control of H-O-S gas mixtures with applications to Fe-Ni sulfide-oxide-silicate reactions. D, 1973, University of British Columbia.

Lechler, Paul. The geochemistry of Cushetunk Mountain. M, 1978, Rutgers, The State University, Newark. 37 p.

Leck, Scott MacLeod. Rock bolting as coal mine roof reinforcement; a model study. M, 1981, University of Idaho. 68 p.

Leckie, Dale Allen. Sedimentology of the Moosebar and Gates formations (Lower Cretaceous). D, 1984, McMaster University. 515 p.

Leckie, Donald Gordon. Development of a nighttime cooling model for remote sensing thermal inertia mapping. D, 1980, University of British Columbia.

Leckie, George Gallie. Distribution of clay minerals in the Paleozoic rocks of southwestern Montana. M, 1962, Indiana University, Bloomington. 19 p.

Leckie, George Gallie. Petrology of the Big Snowy Group, Amsden Formation, and Quadrant Sandstone at Sappington Canyon, Jefferson County, southwest-

ern Montana. D, 1964, Indiana University, Bloomington. 118 p.

Leckie, Phyllis Gilmour. The mineralogy of the sands of the vicinity of Vancouver. M, 1936, University of British Columbia.

Leckie, R. Mark. Micropaleontology, biostratigraphy and paleoenvironmental studies of DSDP Site 270 (late Oligocene-Quaternary), Ross Sea, Antarctica. M, 1980, Northern Illinois University. 288 p.

Leckie, Robert Mark. Biostratigraphy and paleoecology of Mid-Cretaceous planktonic foraminifera off Northwest Africa. D, 1984, University of Colorado. 238 p.

Leclair, Alain Daniel. Low to medium grade metamorphism in the central part of the Hastings Basin southeasten Ontario; an evaluation of metamorphic conditions in chloritoid and staurolite bearing schists. M, 1982, Queen's University. 214 p.

Leclair, Alain Daniel. Polyphase structural and metamorphic histories of the Midge Creek area, Southeast British Columbia; implications for tectonic processes in the central Kootenay Arc. D, 1988, Queen's University. 264 p.

LeClair, Joseph Paul. Adsorption of copper and cadmium onto soils; influence of organic matter. D, 1985, University of California, Riverside. 267 p.

LeCompte, James R. Geology of the northeastern part of the Leach Range, Elko County, Nevada. M, 1978, San Jose State University. 71 p.

LeComte, Paul. Creep and internal friction of rock salt. D, 1960, Harvard University.

LeComte, Paul. Scale model study of some fracture systems. M, 1953, Queen's University. 72 p.

LeCount, Dorothy A. A study of the (Precambrian) Inwood Marble in the vicinity of Pleasantville, New York. M, 1947, Columbia University, Teachers College.

LeCouteur, Peter Clifford. A study of lead isotopes from mineral deposits in southeastern British Columbia and from the Anvil Range, Yukon Territory. D, 1973, University of British Columbia.

Ledbetter, M. T. Abyssal paleocirculation in the Vema Channel (SW Atlantic). D, 1977, University of Rhode Island. 130 p.

Ledbetter, Michael T. A Pennsylvanian-Permian shelf to craton transition, Azure ridge, Clark County, Nevada. M, 1970, Memphis State University.

Lederman, T. C. Calibration of a nutrient-limited, phytoplankton-growth model; parameter estimation and model discrimination. M, 1974, University of Virginia. 97 p.

Ledger, Ernest Broughton, Jr. Evaluation of the Catahoula Formation as a source rock for uranium mineralization, with emphasis on East Texas. D, 1981, Texas A&M University. 263 p.

Ledger, Ernest Broughton, Jr. Release of uranium and thorium from granitic rocks during in situ weathering and initial erosion. M, 1978, Texas A&M University.

Leding, Edward A., III. Regional distribution and reservoir potential of Jackfork Sandstones from facies analysis and petrography, central Ouachita Mountains, Oklahoma. M, 1986, University of Arkansas, Fayetteville.

Ledingham, G. W., Jr. Recent elevation changes, Los Angeles area, southern California. M, 1975, University of Southern California.

Ledje, Hakan Karl. Sedimentology, petrology and facies modelling of the Doshult Member (lower Sinemurian) in the Hoganas Basin, northwestern Scania, Sweden. M, 1985, University of California, Los Angeles. 125 p.

Ledoux, Robert Louis. Infrared studies of the hydroxyl groups in kaolinite; intercalated kaolinite complexes and deuteration. D, 1964, Purdue University.

Leduc, Maxime. Morphologie des faciès volcaniques et structures associées à des coulées basaltiques du Groupe de Kinojevis, canton d'Aiguebelle. M, 1981, Universite du Quebec a Chicoutimi. 169 p.

Ledwin, Jane M. Sedimentation and its role in the nutrient dynamics of a tidal freshwater marsh. M, 1988, College of William and Mary. 89 p.

Lee Moreno, Jose. Some mineralogical and chemical relations of granitic dikes to occurrence of ore deposits in West Cornwall (England). M, 1969, University of Arizona.

Lee Shu Kay Joseph see Shu Kay Joseph Lee

Lee, Alison Marian. Lithofacies and depositional environments of the Chickamauga Group in Jefferson County, north-central Alabama. M, 1983, University of Alabama. 209 p.

Lee, Barbara Jeanette. Recent faulting in California examined from remotely sensed data and seismicity. M, 1980, Pennsylvania State University, University Park. 86 p.

Lee, Bennon. Microwave, geodetic, survey system, AN-USQ-32. M, 1964, Ohio State University.

Lee, Bor-Jen. Seismic analysis of shells of revolution on interactive shallow or deep foundations. D, 1983, Washington University. 234 p.

Lee, Bryan Edward. Stratigraphy, sedimentology, and uranium potential of Virgillian-Leonardian strata of the Hollis-Hardeman Basin, Oklahoma and Texas. M, 1980, Oklahoma State University. 98 p.

Lee, Burdett W. Geology of the Red Wine Mountains, Labrador. D, 1954, McGill University.

Lee, Burdett W. The place of experimental work in the study of rock structures. M, 1946, McGill University.

Lee, C. A. The generation of unstable waves and the generation of transverse upwelling; two problems in geophysical fluid dynamics. D, 1975, University of British Columbia.

Lee, C. R. Hydrogeology of the Waipahu landfill area. M, 1973, University of Hawaii. 91 p.

Lee, Chang Kong. A study of the crustal structure of north central Georgia and South Carolina by analysis of synthetic seismograms. M, 1980, Georgia Institute of Technology. 121 p.

Lee, Chao-Shing. Origin and evolution of the West Philippine Basin. D, 1983, Texas A&M University. 135 p.

Lee, Charles A. and Borland, Gerald, C. The geology and ore deposits of the Cuprite mining district (Arizona). M, 1935, University of Arizona.

Lee, Charles Denard. The Silurian corals of the Upper Mississippi Valley, Lake Huron and Hudson Bay areas and their interpretation. M, 1930, University of Illinois, Urbana.

Lee, Charles Gordon. Skarn and hornfels petrogenesis at the San Pedro Mine, Santa Fe County, New Mexico. D, 1987, University of Colorado. 304 p.

Lee, Chen-Wah. Early Cretaceous Hedbergella (Foraminiferida) from Darwin, Australia; a biometrical approach. M, 1979, Carleton University. 136 p.

Lee, Cherylene Alice. Oxygen isotopes in the seasonal growth bands of the Pismo clam. M, 1979, University of California, Los Angeles. 102 p.

Lee, Chiekyo. A study on the recrystallization characteristics of high purity lead and its alloys. M, 1965, University of Nevada. 76 p.

Lee, Chong Do. Dynamic lateral earth pressures against retaining structures. D, 1981, University of Washington. 183 p.

Lee, Chong Y. Behavior of electric potential fields over randomly layered Earth models. D, 1977, Colorado School of Mines. 112 p.

Lee, Chong Y. The dipping layer problem in resistivity. M, 1973, Colorado School of Mines. 114 p.

Lee, Christopher W. Recent lithofacies and biofacies of North Santee Bay and the lower reaches of North Santee River, South Carolina, U.S.A. M, 1973, University of South Carolina. 105 p.

Lee, Chun Chi. The decomposition of organic matter in some shallow water, calcareous sediments of Little Black Water Sound, Florida Bay. D, 1969, University of Miami. 118 p.

Lee, Chun-sun. Pottsville gastropods (Pennsylvanian) of Ohio. M, 1971, Ohio University, Athens. 129 p.

Lee, Chung I. Stress condition in direct shear test. M, 1982, University of Illinois, Urbana. 141 p.

Lee, Chunsun. Lower Permian ammonoid faunal provinciality. D, 1975, University of Iowa. 253 p.

Lee, Cynthia L. Biological and geochemical implications of amino acids in sea water, wood, and charcoal. D, 1975, University of California, San Diego. 194 p.

Lee, Dar-Yuan. Dissolved organic matter-nonionic pesticide interaction and nonionic pesticide sorption by clay and soil as influenced by dissolved organic matter. D, 1988, University of California, Riverside. 81 p.

Lee, David Robert. Septic tank nutrients in groundwater entering Lake Sallie, Minnesota. M, 1972, University of North Dakota. 96 p.

Lee, David T. C. Structural analysis in Letang Peninsula and Frye Island, Charlotte County, New Brunswick (Canada). M, 1967, University of New Brunswick.

Lee, Der-Shing. Paleomagnetism of Middle Mississippian Greenbrier Group in West Virginia. M, 1982, University of Pittsburgh.

Lee, Derek Gordon. Stratigraphy of the Cornwallis Group (Ordovician) in the Abbott River District of Cornwallis Island, District of Franklin. M, 1976, University of Saskatchewan. 145 p.

Lee, Desmond Nyuk Hin. Diagenesis of clastic rocks from the Bird Fiord, Weatherall and basal Hecla Bay formations, Middle-Upper Devonian, Bathurst and Melville islands, Canadian Arctic Archipelago. M, 1986, University of Alberta. 119 p.

Lee, Don-Jin. Taxonomy and paleoecology of favositids from the Upper Silurian West Point Formation, Gaspe, Quebec. M, 1986, University of New Brunswick. 184 p.

Lee, Donald Edward. A mineralogical study of some manganese ores from Japan. D, 1954, Stanford University. 100 p.

Lee, Donald J. Bed configuration and sediment distribution; Matanzas Inlet, Florida. M, 1980, University of South Florida, Tampa.

Lee, Dong Soo. Bismuth, nickel, and palladium in Northeast Pacific waters; novel analytical methods in marine chemistry. D, 1983, University of California, San Diego. 59 p.

Lee, Doo-Sung. The influence of the surrounding medium in electromagnetic prospecting for an elongated conductor. D, 1979, Colorado School of Mines. 118 p.

Lee, Fitzhugh T. Geology of the Front Royal area, Warren County, Virginia. M, 1961, Virginia Polytechnic Institute and State University.

Lee, Florence Ling. The submicroscopic structure of wenkite; a transmission electron microscopy study. M, 1973, University of California, Berkeley. 93 p.

Lee, Gaylon Keith. Glaciation of the Red Mountain area, Klamath Mountains, California. M, 1972, Arizona State University. 91 p.

Lee, Gerald B. Lithostratigraphy of the Cincinnatian Series (Upper Ordovician) from Maysville, Kentucky to Dayton, Ohio. M, 1974, Miami University (Ohio). 127 p.

Lee, Glen A. A cross-section of the Borden Series between Columbus, Indiana, and Bloomington, Indiana. M, 1924, Indiana University, Bloomington. 140 p.

Lee, H. On the computational aspects of magnetometric resistivity and its application to the mapping of a sink. M, 1975, University of Toronto.

Lee, Han Yeang. Fe-Mg fractionation between garnet and orthopyroxene and application to geothermometry. M, 1985, University of Arizona. 63 p.

Lee, Han Yeang. Garnet-orthopyroxene equilibria in the FMAS system; experimental and theoretical studies, and geological applications. D, 1986, University of Arizona. 143 p.

Lee, Han-Lin. Stochastic dynamic programming for optimal reservoir control. D, 1987, University of Illinois, Urbana. 186 p.

Lee, Harry C. Geology of the Gaston area, Ouachita Mountains, Arkansas. M, 1959, University of Oklahoma. 70 p.

Lee, Harry William. Determination of infiltration characteristics of a frozen Palouse silt loam soil under simulated rainfall. D, 1983, University of Idaho. 115 p.

Lee, Hee Jin. Biostratigraphy and paleoecology of the Charlo-Upsalquitch Forks area, northern New Brunswick. M, 1976, University of New Brunswick.

Lee, Hei Yip. The change in soil stress state during and after the installation of a group of piles in clays. D, 1986, University of California, Berkeley. 251 p.

Lee, Hen-Chen. Analysis of satellite and aircraft remote sensor data as an aid for petroleum exploration in northeastern Oklahoma. M, 1981, University of Arkansas, Fayetteville.

Lee, Herbert Louis, Jr. Woodbine strata of Northwest Hill County, Texas. M, 1958, University of Texas, Austin.

Lee, Herbert V. Effect of shape of grain on the properties of molding sands. M, 1929, Cornell University.

Lee, Heungwon. Tungsten mineralization at the Wildhorse Mine, Custer County, Idaho. M, 1955, University of Idaho. 45 p.

Lee, Homa Jesse. Geotechnical properties of Northeast Pacific Ocean sediment and their relation to geologic processes. D, 1988, University of California, San Diego. 328 p.

Lee, Hulbert A. Two phases of till and other glacial problems in the Edmundston-Grand Falls region (New Brunswick, Quebec, and Maine). D, 1953, University of Chicago. 113 p.

Lee, Huyler Wells. Description of some new species together with notes on the stratigraphy of the upper Miocene and lower Pliocene of Lower Reliz Creek, Monterey County, California. M, 1923, University of California, Berkeley. 15 p.

Lee, Hyun-Ha. Model studies of a tunnel in stratified rock. D, 1974, McGill University.

Lee, In Mo. A probabilistic analysis of porewater predictions for unsteady groundwater flow on a sloping bed. D, 1986, Ohio State University. 172 p.

Lee, Irene. Shallow seismic stratigraphy of Tongue of the Ocean and Exuma Sound, Bahamas, based on single channel seismic reflection data. M, 1982, University of Delaware.

Lee, J. H. Investigation of correlation of the variations in hydrogen and helium compositions in the gases from fumaroles with eruptions in associated volcanoes. M, 1971, University of Hawaii. 80 p.

Lee, Jacqueline San Miguel. Morphology and origin of the Heck and Heckle seamount chains, Northeast Pacific. M, 1987, George Washington University. 73 p.

Lee, Jae K. Thematic mapper-based areal land-cover assessment; a comparative study of principal component analysis with factor analysis. M, 1987, Indiana State University. 106 p.

Lee, James Corbett. A three-component drillhole EM receiver probe. M, 1986, University of Toronto.

Lee, James William. A petrographic study of porphyry intrusives at Hedley, BC. M, 1949, University of British Columbia.

Lee, James William. The geology of Nickel Plate Mountain, British Columbia. D, 1951, Stanford University. 89 p.

Lee, Jaw-Fang. Finite element analysis of wave-structure interactions in the time domain. D, 1987, Oregon State University. 239 p.

Lee, Jeffrey Alan. Sand transport on a barchan dune. M, 1984, University of California, Los Angeles. 64 p.

Lee, Jia Ju. A three-dimensional ray method and its application to the study of wave propagation in crustal

structures with curved layers. D, 1983, Pennsylvania State University, University Park. 282 p.

Lee, Jiunn-Fwu. Retention of organic contaminants by soils and clays exchanged with organic cations. D, 1988, Michigan State University. 125 p.

Lee, John Clifford Hodges, III. Mississippian-Pennsylvanian boundary in the Soldier Quadrangle; implication in northeastern Kentucky. M, 1979, University of Cincinnati. 179 p.

Lee, John Scott. Mineralogy, paragenesis, and origin of three copper-bearing manganese deposits, Olympic Peninsula, Washington. M, 1976, Washington State University. 122 p.

Lee, Jose M. Geological and geochemical exploration characteristics of Mexican tin deposits in rhyolitic rocks. D, 1972, University of Arizona.

Lee, Jose M. Mineralogical and chemical investigations of granitic dikes as an aid to mineral exploration. M, 1968, University of Arizona.

Lee, Julie M. Sedimentology and diagenesis of the Avalon Member of the Mississauga Formation, Hibernia Field, Grand Banks of Newfoundland. M, 1987, University of Calgary. 135 p.

Lee, Jung Hoo. A study of the structure of one layer triclinic pyrophyllite using X-ray single crystal methods. M, 1980, University of Illinois, Chicago.

Lee, Jung Hoo. Mineralogical studies of phyllosilicates in a slaty cleavage development in the Martinsburg Formation near Lehigh Gap, Pennsylvania; Temaem study. D, 1984, University of Michigan. 236 p.

Lee, K. H. Electromagnetic scattering by two-dimensional inhomogeneity due to an oscillating magnetic dipole. D, 1978, University of California, Berkeley. 84 p.

Lee, K. W. Mechanical model for the analysis of liquefaction of horizontal soil deposits. D, 1975, University of British Columbia.

Lee, Keenan. Geology of the San Juan de Guadalupe Quadrangle (east half), Durango and Coahuila, Mexico. M, 1963, Louisiana State University.

Lee, Keenan. Infrared exploration for shoreline springs; a contribution to the hydrogeology of Mono Basin (Mono county), California. D, 1969, Stanford University. 216 p.

Lee, Kenneth. The effects of vanadium on phytoplankton; field and laboratory studies. D, 1982, University of Toronto.

Lee, Kenneth Lester. Triaxial compressive strength of saturated sand under seismic loading conditions. D, 1965, University of California, Berkeley. 532 p.

Lee, Kiehwa. The dispersion of Rayleigh waves in Eurasia. D, 1975, University of Pittsburgh. 22 p.

Lee, Kil Seong. A stochastic frequency analysis of multi-year hydrologic droughts. D, 1982, University of California, Los Angeles. 127 p.

Lee, Kuo-heng. Geopotential anomaly and geostrophic flow off Newport, Oregon. M, 1967, Oregon State University. 57 p.

Lee, Kwang-Yuan. A petrographic study of the Latest Cretaceous and earliest Tertiary formations of central Utah. D, 1953, Ohio State University. 235 p.

Lee, Kwang-Yuan. Petrography of the Price River Formation in the Sanpete Valley District, Utah. M, 1950, Ohio State University.

Lee, Kyoo-seock. Determination of soil characteristics from thematic mapper data and land use evaluation using a relational data base geographic information system. D, 1987, University of Wisconsin-Madison. 214 p.

Lee, L. Courtland. The economic geology of portions of the Tombstone-Charleston District, Cochise County, Arizona, in light of 1967 silver economics. M, 1967, University of Arizona.

Lee, Larry. Structural geology of Mazarn Synclinorium, Arkansas. M, 1965, University of Missouri, Rolla.

Lee, Larry Jack. Longitudinal profiles and related hydraulic properties of the Mississippi River system. D, 1972, Washington University. 147 p.

Lee, Li-Jien. High temperature reactions in oil shale. M, 1979, Texas Tech University. 69 p.

Lee, Li-Jien. Nature of the changes in clay minerals of the high temperature drilling fluids. D, 1984, Texas Tech University. 157 p.

Lee, Linda J. Origin of columnar jointing in Recent basaltic flows, Garibaldi area, Southwest British Columbia. M, 1988, University of Calgary. 135 p.

Lee, Luther. A study of anisotropic models in the resistivity method. M, 1976, University of California, Berkeley. 104 p.

Lee, Lyndon Charles. The floodplain and wetland vegetation of two Pacific Northwest river ecosystems. D, 1983, University of Washington. 268 p.

Lee, M. Y. Response of harbors with permeable breakwater to incident waves. D, 1978, University of Southern California.

Lee, M., Jose. Geological and geochemical exploration characteristics of Mexican tin deposits in rhyolitic rocks. D, 1972, University of Arizona.

Lee, Maxie Turner. Major and minor element geochemistry of the South Platte granite-pegmatite system, Jefferson County, Colorado. M, 1986, University of New Orleans. 130 p.

Lee, Michael Donald. Biodegradation of organic contaminants in the subsurface of hazardous waste sites. D, 1986, Rice University. 160 p.

Lee, Mingchou. Diagenesis of the Permian Rotliegendes Sandstone, North Sea; K/Ar, O^{18}/O^{16}, and petrologic evidence. D, 1984, Case Western Reserve University. 365 p.

Lee, Moon Joo. Clay mineralogy of Havensville Shale (Permian, Kansas). M, 1972, Kansas State University. 109 p.

Lee, Moon Joo. Geochemistry of the sedimentary uranium deposits of the Grants mineral belt, southern San Juan Basin, New Mexico. D, 1976, University of New Mexico. 241 p.

Lee, P. F. Y. Geophysical studies of buried bedrock channel in the Elora-Fergus region, southwestern Ontario. M, 1975, University of Waterloo. various pagination p.

Lee, Patricia D. Paleoenvironments at Waimanalo, Oahu, Hawaii. M, 1985, University of Hawaii. 117 p.

Lee, Patricia Tsean-Shu. Petrology of subsurface diabase sills encountered in Paleozoic rocks of Northwest Mississippi. M, 1978, University of Mississippi. 71 p.

Lee, Pei Jen. Applications of canonical correlation in geology. D, 1968, McMaster University. 124 p.

Lee, Pei Jen. Sedimentology of the Middle Ordovician Cobourg Limestone at Colborne, Ontario, Canada. M, 1965, University of Western Ontario. 72 p.

Lee, Randolph. Geology of the Strachan Creek area, British Columbia. M, 1958, University of British Columbia.

Lee, Raymond Frederick. Uptake of selected metals by fish at West Branch and Nimisila reservoirs (Ohio). M, 1973, University of Akron. 89 p.

Lee, Richard. Relationship between potential insolation and the orientation of watersheds with respect to evapotranspiration. D, 1962, [Colorado State University].

Lee, Richard A. A palynological analysis of the marshes at Plum Island, Massachusetts. M, 1981, University of Rhode Island.

Lee, Richard Cacy. Constraints on seismic velocity in the Earth's mantle. D, 1981, University of California, Berkeley. 77 p.

Lee, Richard Kenneth. Remanent magnetization of the Cambridge Limestone (Missourian, Pennsylvanian) within the Dunkard Basin. M, 1979, University of Pittsburgh.

Lee, Roger William. Nature of clay-kerogen associations in the oil shales from the Washakie Basin of the Green River Formation (Wyoming). M, 1975, Texas Tech University. 64 p.

Lee, S. Investigations of the functional morphology of angiosperm pollen. D, 1977, Duke University. 205 p.

Lee, S. Limiting mechanisms of dislocation motion in ice. D, 1979, University of Illinois, Urbana. 123 p.

Lee, S. J. Alkali solution treatment on sandstone cores. M, 1978, Texas A&M University.

Lee, S. S. Long water wave propagation under the influence of the depth variation; Part 1, Long water waves incident upon a transition zone of an exponential depth variation; Part 2, Long water waves incident upon a seamount model of a linearly cylindrical depth variation. D, 1979, Columbia University, Teachers College. 144 p.

Lee, Sa Ba. Use of index methodologies for predicting or evaluating pesticide pollution of ground water. D, 1988, University of Oklahoma. 660 p.

Lee, Samuel Bayard. Soils of the Jones-Ford Quadrangle. M, 1921, University of North Carolina, Chapel Hill. 20 p.

Lee, San Wei. Landslide control measures. M, 1975, [Colorado State University].

Lee, Sang Man. Geology of the South Hopewell Sound area, east of Hudson Bay. D, 1962, McGill University.

Lee, Sang Man. Mineral deposits of Korea. M, 1957, Michigan Technological University. 115 p.

Lee, Sang-Ho. Response of buried structures to ground shock loading. D, 1988, North Carolina State University. 168 p.

Lee, Sheng-Shyong. Secular variation of the intensity of the geomagnetic field during the past 3,000 years in North, Central, and South America. D, 1975, University of Oklahoma. 217 p.

Lee, Shu Kay Joseph. Multilayer gravity inversion using Fourier transforms. M, 1977, University of Alberta. 144 p.

Lee, Shu-Schung. Generalized ray analysis and its application to leaky modes and guided waves. D, 1982, Colorado School of Mines. 175 p.

Lee, Shuh-Chai. Adjustment of geodetic and engineering triangulation of fundamental figures. M, 1962, Ohio State University.

Lee, Steven Wendell. An interactive computer system for extraction of topography from digital stereo spacecraft imagery. M, 1979, Washington University. 127 p.

Lee, Steven Wendell. Eolian sediment transport on Mars; seasonal and topographic effects. D, 1984, Cornell University. 272 p.

Lee, Suk Jin. Sodium-hydroxide solution treatment on sandstone cores. D, 1984, Texas A&M University. 108 p.

Lee, Tai Sup. Theoretical investigation of smart seismic arrays. D, 1983, Colorado School of Mines. 121 p.

Lee, Tai Y. Fast method for particle size distribution determinations in sub-sieve sizes. M, 1975, University of Windsor. 77 p.

Lee, Teh-Quei. Paleomagnetic study of volcanic and volcaniclastic rocks from the Absaroka Mountains, Northwest Wyoming. M, 1983, University of Wyoming. 112 p.

Lee, Theodore David. Depositional environment of the Silurian Elder Sandstone in the North Shoshone Range, Nevada. M, 1978, Wayne State University.

Lee, Tien-Chang. Application of atomic absorption of spectroscopy to geochemical exploration in northern Idaho. M, 1969, University of Idaho. 89 p.

Lee, Tien-Chang. Heat flow and other geophysical studies in the southern California borderland. D, 1973, University of Southern California.

Lee, Tung-Yi. Seismic stratigraphy and tectonic evolution of Tungyintao Basin, offshore northern Taiwan. M, 1987, University of Texas, Austin. 107 p.

Lee, Victor J. B. Petrography, metamorphism and geochemistry of the Bermeja Complex and related rocks in southwestern Puerto Rico, and their significance in the evolution of the eastern Greater Antillian island arc. M, 1974, SUNY at Albany. 244 p.

Lee, Wah Seyle. Bibliography of the geology of China; with an outline of the geological features and statement of the problems connected with the geology of China. M, 1916, Stanford University. 425 p.

Lee, Wang Chih-ming. The petrology and origin of the Linkou Gravel in Taipei Shin, Taiwan, China. M, 1961, Florida State University.

Lee, Wilfred K. The petrography of the Camsell Lake area, Dist. of Mackenzie, NW. Territories, Canada. M, 1966, University of Washington. 49 p.

Lee, William Hung Kan. Thermal history of the Earth. D, 1967, University of California, Los Angeles. 364 p.

Lee, Willis Thomas. Stratigraphy of the coal fields of northern central New Mexico. D, 1913, The Johns Hopkins University.

Lee, Wook Bae. Simultaneous inversion of surface wave phase velocity and attenuation for continental and oceanic paths. D, 1978, Massachusetts Institute of Technology. 274 p.

Lee, Yong Il. Petrology and diagenesis of medium-grained clastic sediments in the back-arc basins of the western Pacific Ocean. D, 1984, University of Illinois, Urbana. 207 p.

Lee, Young-Hoon. Surface roughness characterization of rock masses using the fractal dimension and variogram. D, 1988, University of Missouri, Rolla. 203 p.

Lee, Young-Nam. Stress-strain-time relationship of Queenston Shale. D, 1988, University of Western Ontario.

Lee, Yuan-cheng. Analysis of sign-bit recording for synthetic vibrator data. D, 1984, Colorado School of Mines. 186 p.

Lee, Yuk Cheung. Conceptual models for geographic information systems. D, 1987, University of New Brunswick.

Lee-Hu, Chin-Nan. Rb-Sr age determinations on the younger Precambrian igneous intrusions, Needle Mountains, Colorado. M, 1967, University of California, Los Angeles.

Leech, Alice Payne. A reconnaissance: basic intrusive rocks of the Precambrian Shield, Canada. M, 1965, University of Alberta. 100 p.

Leech, G. B. Ultrabasic rocks and chromite deposits of the Shulaps Mountains, British Columbia. M, 1943, Queen's University. 60 p.

Leech, Geoffrey Bosdin. Petrology of the ultramafic and gabbroid intrusive rocks of the Shulaps Range, British Columbia. D, 1949, Princeton University. 220 p.

Leeden, Fritz Van der *see* Van der Leeden, Fritz

Leeden, John van der *see* van der Leeden, John

Leeder, Robert W. and Hume, James David. Geology of the Pike Mountain area, Gallatin County, Montana. M, 1950, University of Michigan. 45 p.

Leedom, Stephen H. Tertiary andesitic volcanic rocks of the northern Little Drum Mountains, Millard County, Utah. M, 1974, Brigham Young University. 108 p.

Leeds, Alan Robert. Rayleigh wave dispersion in the Pacific Basin. D, 1973, University of California, Los Angeles.

Leeds, Anna L. A geologic study of the Blanco gas field. M, 1954, University of Houston.

Leedy, Evert John. Geology of the Labarge area, Yukon. D, 1931, University of Toronto.

Leedy, Forrest Benton. The Clarksburg intrusions and their significance (northern Michigan). M, 1930, Michigan Technological University. 20 p.

Leedy, John B. Investigations into sources of sediment in streams with emphasis on agricultural versus channel-derived non-point sources, La Salle County, Illinois. M, 1979, Northern Illinois University. 87 p.

Leedy, Willard Page. Hydrothermal alteration of volcanic rocks in the Red Mountains District of the San Juan Mountains, Colorado. D, 1971, SUNY at Buffalo. 108 p.

Leefang, Willem Evert. Geophysical methods used in the design and inspection of dam embankments for earthquake resistance. M, 1978, University of Utah.

Leeflang, W. E. Geophysical methods used in the design and inspection of dam embankments for earthquake resistance. M, 1978, University of Utah.

Leelanitkul, S. Progressive settlements of foundations on cohesionless soils subjected to vertical vibratory loadings. D, 1980, University of Akron. 121 p.

Leeman, William P. Late Cenozoic basalts from the Basin-Range Province, western United States. M, 1969, Rice University. 83 p.

Leeman, William Prescott. Part I, Petrology of basaltic lavas from the Snake River Plain, Idaho; and Part II, Experimental determination of partitioning of divalent cations between olivine and basaltic liquid. D, 1974, University of Oregon. 337 p.

Leenheer, Mary Janeth. Use of lipids as indicators of diagenetic and source-related changes in Holocene sediments. D, 1981, University of Michigan. 259 p.

Leeper, Robert. Geochemical model for the hydrothermal formation of quartz diorite contact facies, Humboldt gabbroic complex (Jurassic (?) and Cretaceous), northwestern Nevada. M, 1971, Northwestern University.

Leeper, Robert H., Jr. Effect of organic decomposition on early diagenesis in anoxic marine sediments. D, 1975, Southern Methodist University. 230 p.

Leeper, Wayne S. Statistical measurement, analysis, and interpretation of primary bedding structures in southeastern Somerset County, Pennsylvania. M, 1960, University of Pittsburgh.

Lees, James Henry. The geological history of the Des Moines Valley (Iowa). D, 1915, University of Chicago. 190 p.

Lees, John Allen. Stratigraphy of the Lower Ordovician Axemann Limestone of the Beekmantown Group in central Pennsylvania. D, 1964, Pennsylvania State University, University Park. 283 p.

Lees, Jonathan. Tomographic inversion for 3-dimensional velocity variations in western Washington. D, 1988, University of Washington.

Lees, Laurence Fitch. Paleozoic stratigraphy and structure in Red Cedar Valley, Wisconsin. M, 1934, University of Iowa. 22 p.

Lees, William R. Geothermometric investigations on ores from the Magma Mine, Superior, Arizona. M, 1969, Texas Tech University. 50 p.

Leeson, Bruce Frank. Soils and associated natural resources as decision parameters in the regional planning precess. D, 1972, Montana State University.

Leeson, J. I. Petrofabric studies of Grenville series rocks from Shawbridge, Quebec. M, 1953, McGill University.

Leet, Lewis Don. Empirical investigation of surface waves generated by distant earthquakes. D, 1930, Harvard University.

Leetaru, H. Sedimentary processes in North Pond, eastern shore of Lake Ontario. M, 1978, Syracuse University.

Leete, Jeanette Helen. Sediment and phosphorus load to streamflow from natural and disturbed watersheds in northeastern Minnesota. D, 1986, University of Minnesota, Minneapolis. 158 p.

Leethem, John T. Facies distribution of Upper Cretaceous Woodbine Sandstone, southern Kurten Field, Brazos County, Texas. M, 1984, Texas A&M University. 128 p.

Leeuwen, Wim van see van Leeuwen, Wim

Leever, William H. Origin of the mineral deposits of North Santiam mining district, Oregon. M, 1941, Oregon State University. 98 p.

LeFebre, George B. Geology of the Chinks Peak area, Pocatello Range, Bannock County, Idaho. M, 1984, Idaho State University. 61 p.

LeFebre, George Bradburn. Tectonic evolution of Hanna Basin, Wyoming; Laramide block rotation in the Rocky Mountain foreland. D, 1988, University of Wyoming. 294 p.

LeFebre, Valerie S. Foraminiferal paleoecology of the lower Niobrara Formation at Springer, New Mexico. M, 1987, University of Wyoming. 162 p.

Lefebure, David V. Geology of the Nicola Group in the Fairweather Hills, British Columbia. M, 1976, Queen's University. 178 p.

LeFebvre, Ben Heywood. Petrology and biostratigraphy of the Lower Devonian (Lochkovian) McMonnigal Limestone and (Pragian) lower member of the Rabbit Hill Limestone, northern Toquima Range, Nye County, Nevada. M, 1988, University of California, Riverside. 211 p.

Lefebvre, David Victor. The Mina el Limon area and the Telica Complex; two examples of Cenozoic volcanism in northwestern Nicaragua, Central America. D, 1986, Carleton University. 269 p.

Lefebvre, Rene. A study of pre- and post-recovery cores from Aberfeldy thermal recovery pilots, Lloydminster area, Saskatchewan. M, 1984, University of Calgary. 212 p.

Lefebvre, Richard H. Joint patterns in the central part of the Hurricane fault zone, Washington County, Utah. M, 1961, University of Kansas. 35 p.

Lefebvre, Richard Harold. Variations of flood basalts of the Columbia River Plateau, central Washington. D, 1966, Northwestern University. 211 p.

Lefebvre, Yues. Le Carbone organique dans les séquences sédimentaires du Llandoverien de la Gaspésie et du nord-est du Nouveau-Brunswick. M, 1983, Universite Laval. 34 p.

LeFever, R. D. Settling velocities and sedimentary environments. D, 1979, University of California, Los Angeles. 155 p.

LeFever, Richard David. Sedimentology of the Cretaceous El Gallo Formation, Baja California. M, 1971, University of California, Los Angeles.

LeFevre, Elbert Walter, Jr. Soil plasticity dependency on surface area. D, 1966, Oklahoma State University. 56 p.

Leff, Craig Ernest. Aspects of the structure of Missouri based on digital image processing of gravity and aeromagnetic anomalies, remote sensing data, and drillcore logs. M, 1983, Washington University. 68 p.

Lefferts, Walter. Tidewater Maryland, an embayed coastal plain. D, 1918, University of Pennsylvania.

Leffingwell, Harry A. A study of the variation of some microfossil assemblages in the (Oligocene) Vicksburg Formation, Louisiana. M, 1958, New York University.

Lefkoff, Lawrence Jeffrey. The hydrologic and economic effects of water marketing on an irrigated stream-aquifer system subject to salinity degradation. D, 1988, Stanford University. 280 p.

Lefkowitz, Paul Allen. Paleoecology of a portion of a section of Blufftown Marl (upper Cretaceous) at Union Springs, Bullock County, Alabama. M, 1971, Emory University. 51 p.

LeFond, Joanne M. Estimating runoff and lake evaporation in the Western United States and Canada; an empirical method using climatic data. M, 1985, University of Massachusetts.

Leftwich, John Thomas. Structural geology of the West Camp area, Greene and Ulster counties, New York. M, 1973, University of Massachusetts. 88 p.

LeFurgey, Edoris Ann. Recent benthic foraminifera from Roanoke Sound, Croatan Sound, and northern Pamlico Sound, North Carolina. D, 1976, University of North Carolina, Chapel Hill. 383 p.

Legall, Franklyn David. Organic metamorphism and thermal maturation history of Paleozoic strata, southern Ontario. D, 1980, University of Waterloo.

LeGarde, Charles N. The petrographic structural geology and tectonic history of a portion of the Wachusett-Marlboro tunnel area (eastern Massachusetts). M, 1967, Boston College.

Legate, Carl Eugene. Gas chromatographic study of a clay mineral-organic system; determination of activity coefficients and heats of adsorption. D, 1958, Washington University. 68 p.

Legate, Carl Eugene. The application of the technique of microradiography to petrology; a preliminary investigation. M, 1953, Washington University. 37 p.

Legault, J. The conodont fauna of the Stonehouse Formation (Silurian), Arisaig, Nova Scotia. M, 1966, University of Ottawa. 71 p.

Legault, Jocelyne Andree. Chitinozoa and Acritarcha of the Hamilton Group (middle Devonian) of southern Ontario. D, 1971, University of Oklahoma. 295 p.

Legault, Marc H. The geology and alteration associated with the Genex volcanogenic Cu massive sulphide deposit, Godfrey Township, Timmins, Ontario. M, 1985, Carleton University. 222 p.

Legault, Raymond Z. Study of the Frank J. and Barbara Kammer No. 1 well in the Peters Reef, St. Clair County, Michigan. M, 1961, University of Michigan.

LeGendre, Gary Ralph. Removal of molybdenum by ferric oxyhydroxide in Clear Creek and Tenmile Creek, Colorado. M, 1973, University of Colorado.

Legendre, Kerry John. Stratigraphy and petrology of the Joachim Dolomite and Plattin Limestone (Middle Ordovician), Stone County, Arkansas. M, 1987, University of New Orleans. 325 p.

Leger, Albert Joseph. Transcurrent faulting history of southern New Brunswick. M, 1986, University of New Brunswick. 170 p.

Leger, Arthur R. Structure and tectonic history of the southwest part of Tanque Verde Ridge, Pima County, Arizona. M, 1967, University of Arizona.

Leger, William R. Salt-dome-related diagenesis of undifferentiated Miocene sediment, Black Bayou Field, Cameron Parish, Louisiana. M, 1988, University of New Orleans.

LeGeros, Racquel Zapanta. Crystallographic studies of the carbonate substitution in the apatite structure. D, 1967, New York University. 223 p.

Legg, Mark Randall. Geologic structure and tectonic of the inner continental borderland, offshore northern Baja California, Mexico. D, 1985, University of California, Santa Barbara. 618 p.

Legg, Thomas E. Palynology of middle Pinedale sediments in Devlins Park, Boulder County, Colorado. M, 1977, University of Iowa. 51 p.

Legge, John A. Paragenesis of the ore minerals of the Miami Mine, Arizona. M, 1939, University of Arizona.

Legge, Paul William. The bimodal basalt-rhyolite association west of and adjacent to the Pueblo Mountains, southeastern Oregon. M, 1988, Miami University (Ohio). 131 p.

Leggett, Bob D. Tertiary volcanics of the Northwest Eagle Mountains. M, 1979, University of Texas at El Paso.

Leggett, Sidney R. The South Heninga Lake copper-zinc deposit District of Keewatin, N.W.T. M, 1980, University of Manitoba. 73 p.

Leggitt, Shelley Maureen. Facies geometry and erosion surface in the Cardium Formation, Pembina Field, Alberta, Canada. M, 1987, McMaster University. 183 p.

Legler, June L. A submarine pyroclastic flow deposit in the Sierra Buttes Formation, northern Sierra Nevada, California. M, 1983, California State University, Hayward. 150 p.

Legowo, Eko. Estimation of water extractability and hydraulic conductivity in tropical Mollisols, Ultisols, and Andisols. D, 1987, University of Hawaii. 201 p.

LeGresley, Eric M. Holocene sedimentation on the western Grand Banks of Newfoundland. M, 1988, Queen's University. 307 p.

Legun, Andrew S. Sedimentology of the Clifton Formation (Upper Carboniferous) of northern New Brunswick; a semi-arid depositional setting for coal. M, 1980, University of Ottawa. 97 p.

Lehkemper, Leonard James. Petrology of Springerville area, Apache County, Arizona. M, 1956, University of Texas, Austin.

Lehle, Peter F. Deposition and development of Lockport and Salina (Silurian) rocks at west Millgrove,

Ohio. M, 1980, Bowling Green State University. 105 p.

Lehman, David H. Structure and petrology of the Hull Mountain area, northern California Coast Ranges. D, 1974, University of Texas, Austin. 156 p.

Lehman, Donald D. Geology of the Shell Canyon area, Bighorn Mountains, Wyoming. M, 1975, University of Iowa. 119 p.

Lehman, George Albert. The bedrock geology of a portion of the Cramer 15' Quadrangle, Lake County, Minnesota. M, 1980, University of Minnesota, Duluth.

Lehman, Jay A. Upper-crustal structure beneath Yellowstone National Park from seismic refraction and gravity observations. M, 1980, University of Utah.

Lehman, Linda L. Geochemical identification of an intermediate groundwater flow system in the Peace River basin, Florida. M, 1978, University of South Florida, Tampa. 74 p.

Lehman, N. E. The geology and pyrometasomatic ore deposits of the Washington Camp-Duquesne District, Santa Cruz County, Arizona. D, 1978, University of Arizona. 399 p.

Lehman, Norman E. Geology and mineralogy of the Fort Hall phosphate deposit (Lower Triassic), Idaho. M, 1966, University of Arizona.

Lehman, Russell John. The geology of Green Township, Harrison County, Ohio. M, 1954, Ohio State University.

Lehman, Thomas Mark. A ceratopsian bone bed from the Aguja Formation (Upper Cretaceous), Big Bend National Park, Texas. M, 1982, University of Texas, Austin. 210 p.

Lehman, Thomas Mark. Stratigraphy, sedimentology, and paleontology of Upper Cretaceous (Campanian-Maastrichtian) sedimentary rocks in Trans-Pecos Texas. D, 1985, University of Texas, Austin. 283 p.

Lehmann, David F. Paleontology and paleoecology of the Upper Ordovician Martinsburg Formation, Swatara Gap, Lebanon County, Pennsylvania. M, 1986, Miami University (Ohio). 150 p.

Lehmann, Elroy P. An analysis of the distribution of foraminifera from Recent sediments of Matagorda Bay, Texas, and adjacent environments. D, 1955, University of Wisconsin-Madison.

Lehmann, Elroy P. The microfauna of the (Pennsylvanian) Glen Eyrie Shale of Colorado. M, 1951, University of Wisconsin-Madison.

Lehmann, Patrick Jon. Deposition, porosity evolution and diagenesis of the Pipe Creek Jr. Reef (Silurian), Grant County, Indiana. M, 1978, University of Wisconsin-Madison.

Lehner, Florain K. A theory of transport processes in porous media. D, 1979, Princeton University. 136 p.

Lehner, Robert Eugene. A study of part of the gneissic complex north of the Felch Mountain District, Michigan. M, 1951, Ohio State University.

Lehocky, Alan John. Stratigraphy of a Carolina bay area, South Carolina. M, 1968, University of South Carolina.

Lehr, Jay H. Empirical studies of laminar flow in porous consolidated media. D, 1962, University of Arizona.

Lehre, Andre Kenneth. Sediment mobilization and production from a small mountain catchment; Lone Tree Creek, Marin County, California. D, 1982, University of California, Berkeley. 235 p.

Lehrer, Mark G. Glacial geology of the upper Bear Creek drainage, Park Range, Colorado. M, 1982, University of Wyoming. 123 p.

Lehrmann, Daniel J. Sedimentology and conodont biostratigraphy of the Road Canyon Formation (Permian), Glass Mountains, Southwest Texas. M, 1988, University of Wisconsin-Madison. 170 p.

Lehto, Douglas Andrew Warren. The application of a sequential partial extraction procedure to investigate uranium, copper, zinc, iron and manganese partitioning in Recent lake, stream and bog sediments, north-

ern Saskatchewan. M, 1981, Lakehead University. 164 p.

Lehtola, Kathleen Anne. Ordovician vertebrates from Ontario. M, 1971, Michigan State University. 29 p.

Lehtonen, Lee R. Late Paleozoic evolution of the Val Verde Basin, West Texas. M, 1987, University of Texas at El Paso.

LeHuray, Anne P. Lead and sulfur isotope systematics in sulfide deposits of the Piedmont and Blue Ridge provinces of the Southern Appalachians. D, 1983, Florida State University. 451 p.

Lei, Hsiang-Yuan. Porosity and hydrothermal alteration determined from wireline logs from the Salton Sea geothermal field, California, United States. M, 1987, University of California, Riverside. 168 p.

Lei, Wayne. Thorium mobilization in a terrestrial environment. D, 1984, New York University. 437 p.

Leibel, Robert John. A pore geometry study of the Mississippian Midale Carbonate of the Benson Field, southeastern Saskatchewan. M, 1981, University of Regina. 126 p.

Leible, K. A. The paleotectonic significance of two middle Ordovician polymictic conglomerates, near South Holston Dam, Tennessee, and Avens Bridge, Virginia. M, 1975, Virginia Polytechnic Institute and State University.

Leibold, Anne M. Mineral resource evaluation of the Browns Canyon area, Chaffee County, Colorado, using stream-sediment geochemistry. M, 1986, Colorado School of Mines. 165 p.

Leibold, Arthur W., III. Stratigraphy, petrography and depositional environment of the Bryantsville Breccia (Meramecian) of south-central Indiana. M, 1982, Indiana University, Bloomington. 171 p.

Leier-Engelhardt, Paula Jean. Middle Paleozoic strata of the Sierra las Pinta, northeastern Baja California Norte, Mexico. M, 1986, San Diego State University. 169 p.

Leifer, John C. Geology and precious metal mineralization near Washoke Canyon, Pershing County, Nevada. M, 1985, University of Colorado.

Leiggi, Peter A. Structure and petrology along a segment of the Shuksan thrust fault, Mount Shuksan area, Washington. M, 1986, Western Washington University. 207 6 plates p.

Leighton, Carl Winslow. Geochronology and metamorphic petrology of amphibolite blocks, Sierra Nevada foothills, California. M, 1986, SUNY at Stony Brook. 152 p.

Leighton, Cheryl D. Stratigraphy and sedimentology of the Pennsylvanian Gothic Formation in the Crested Butte area, Colorado. M, 1987, Colorado School of Mines. 195 p.

Leighton, Donald Lewis. Geology of a portion of the Dillon Quadrangle, Colorado. M, 1950, University of Iowa. 52 p.

Leighton, Freeman Beach. Contributions to the glaciology of the Seward ice field, Canada, and the Malaspina Glacier, Alaska. M, 1949, California Institute of Technology. 101 p.

Leighton, Freeman Beach. Geology of the vermiculite deposits, Gold Butte, southern Nevada. D, 1952, California Institute of Technology. 184 p.

Leighton, Freeman Beach. I, Ogives of the East Twin Glacier, Alaska; their nature and origin; II, Investigations in the Taku Glacier firn, Alaska. D, 1952, California Institute of Technology. 36 p.

Leighton, Helen Elizabeth. Distribution and character of the Oriskany Formation in the Northern Appalachians. M, 1934, University of Pittsburgh.

Leighton, Morris Morgan. Pleistocene history of the Iowa River valley north and west of Iowa City in Johnson County, Iowa. M, 1913, University of Iowa. 81 p.

Leighton, Morris Morgan. The Iowan drift; a review of the evidences of the Iowan Stage of glaciation. D, 1916, University of Chicago.

Leighton, Morris W. Petrogenesis of a gabbro-granophyre complex in northern Wisconsin. D, 1951, University of Chicago. 175 p.

Leighton, Van L. Depositional environments and petrography of the Trinidad Sandstone and related formations, Raton area, New Mexico. M, 1980, Colorado State University. 116 p.

Leighton-Puga, Tomas. Seismic survey of northeastern South Park, Park County, Colorado. M, 1969, Colorado School of Mines. 201 p.

Leighty, Robert D. Procedure for the evaluation of fluvio-glacial terraces from airphotos. M, 1956, Purdue University.

Leiker, Loren Michael. Deposition and diagenesis of the Smackover Formation, south-central Alabama. M, 1977, Texas Tech University. 110 p.

Leimer, Harold Wayne. The crystal structure of yugawaralite. D, 1969, University of Missouri, Columbia.

Lein, Carl A. Geology of Chestnut Ridge, Saylorsburg to Kunkletown, Monroe County, Pennsylvania. M, 1952, Lehigh University.

Leinbach, Alan Edward. A geochemical study of a layered portion of the Horoman Peridotite, southern Hokkaido, Japan. M, 1987, Massachusetts Institute of Technology. 308 p.

Leinen, Margaret. Biogenic silica sedimentation in the central equatorial Pacific during the Cenozoic. M, 1976, Oregon State University. 136 p.

Leinen, Margaret Sandra. Paleochemical signatures in Cenozoic Pacific sediments. D, 1979, University of Rhode Island. 331 p.

Leininger, Richard Keith. Chemical composition of a partially weathered Illinoian till. M, 1957, Indiana University, Bloomington. 55 p.

Leininger, Roland L. Cenozoic evolution of the southernmost Taos Plateau, New Mexico. M, 1982, University of Texas, Austin.

Leipertz, Steven Lee. Morphometrics and the evolutionary history of fishes of the teleost subfamily Pleuronectinae. D, 1987, University of Washington. 136 p.

Leipzig, Martin R. Stratigraphy, sedimentation and depositional environments of the Late Cretaceous Pictured Cliffs Sandstone, Fruitland Formation, Kirtland Shale and early Tertiary Ojo Alamo Sandstone, eastern San Juan Basin, New Mexico. M, 1982, University of Wisconsin-Milwaukee. 141 p.

Leis, Walter M. Geologic control of groundwater movement in a portion of the Delaware Piedmont. M, 1976, University of Delaware.

Leischer, Clayton Carter. The effect of coal fly-ash leachate from a landfill site on the shallow groundwater system near Saukville, Wisconsin. M, 1980, University of Wisconsin-Milwaukee.

Leischner, Lyle Myron. Border zone petrology of the Idaho Batholith in the vicinity of Lolo Hot Springs, Montana. M, 1959, University of Montana. 76 p.

Leister, G. L. Taxonomy and reproductive morphology of Iridaea cordata (Turner) Bory and Iridaea crispata Bory (Gigartinaceae, Rhodophyta) from southern South America. D, 1977, Duke University. 203 p.

Leitch, Catherine A. Mineralogy, petrology and origin of the unequilibrated enstatite chondrites. D, 1981, University of Chicago. 212 p.

Leitch, Henry Cedric Browning. Contributions to the geology of Bowen Island, BC. M, 1947, University of British Columbia.

Leite, Lourenildo Williame Barbosa. Application of optimization methods to the inversion of aeromagnetic data. D, 1983, St. Louis University. 305 p.

Leite, Michael B. Neogene stratigraphy and vertebrate paleontology of the north shore of Lake McConaughy, Keith County, Nebraska. M, 1986, University of Nebraska, Lincoln. 132 p.

Leith, Andrew. The application of mechanical structural principles in the Western Alps. D, 1931, University of Wisconsin-Madison.

Leith, Andrew. The chemical character of underground waters in the St. Peter Sandstone. M, 1927, University of Wisconsin-Madison.

Leith, Carlton J. Geology of the Quien Sabe Quadrangle, California. D, 1947, University of California, Berkeley. 78 p.

Leith, Charles K. Rock cleavage. D, 1901, University of Wisconsin-Madison.

Leith, Edward Issac. A stratigraphical study of the Coloradoan (Cretaceous) of the Manitoba escarpment with special reference to certain of the calcareous horizons. M, 1929, University of Manitoba.

Leith, Rory Marshall Montgomery. Contribution of GOES data to hydrologic regionalization in southern British Columbia. D, 1983, University of Waterloo. 190 p.

Leith, William Stanley. The Tadjik Depression, USSR; geology, seismicity and tectonics. D, 1984, Columbia University, Teachers College. 130 p.

Leithold, Elana Lynn. The relative roles of fluvial-sediment supply and marine processes in continental shelf sedimentation; a study of the modern Eel River and Pleistocene Rio Dell shelves. D, 1987, University of Washington. 253 p.

Leitinger, J. Hans. Investigation of displacement steps in a layered half-space by the finite-difference method. D, 1969, Colorado School of Mines. 67 p.

Leitner, Donald G. Geology of the Bald Hill area, Okmulgee County, Oklahoma. M, 1956, University of Oklahoma. 54 p.

Leitzke, Peter Andrew. Geology of the south half of the Hermon, New York, 7 1/2 minute Quadrangle. M, 1974, SUNY at Binghamton. 89 p.

Leja, Stanislaw, Jr. A magnetic and resistivity survey of Cork Street Landfill. M, 1983, Western Michigan University. 124 p.

Lekhakul, Somjintana. The effect of lime on chemical composition of surface-mined coal spoils, the growth of plants on spoils, and the leachate from spoils. D, 1981, University of Kentucky. 151 p.

Leland, George R. General geology and mineralization of the Mackay Stock area. M, 1957, University of Idaho. 71 p.

Leland, Rodney C. Paleozoic rocks in the vicinity of Ike's Canyon, Nye County, Nevada. M, 1952, Columbia University, Teachers College.

Lelek, Jeffrey John. The Skalkaho pyroxenite-syenite complex east of Hamilton, Montana, and the role of magma immiscibility in its formation. M, 1979, University of Montana. 130 p.

Lemaire, M. E. The quasi-piedmont characteristics of Marlborough, Southborough, and Westborough, Mass. D, 1935, Clark University.

LeMar, Harold K. The geology and structure of the Thornton Oil Company's Lot 133, Alma Township, Allegany County, New York. M, 1949, Syracuse University.

Lemastus, Steven Wayne. Stratigraphy of the Cason Shale (Ordovician-Silurian), northern Arkansas. M, 1979, University of New Orleans.

LeMasurier, W. E. Structural study of a Laramide fold involving shallow-seated basement rock, Larimer County, Colorado. M, 1961, University of Colorado.

LeMasurier, Wesley Ernest. Volcanic geology of Santa Rosa Range, Humboldt County, Nevada. D, 1965, Stanford University. 126 p.

LeMay, William Joseph. Mechanical analysis of glacial till from part of southeastern Minnesota. M, 1956, University of Michigan.

Lembach, Dixie Jane. Fauna of the Permian rocks near Quinn River Crossing, Nevada. M, 1964, Oregon State University. 77 p.

Lemerand, Martin. Land utilization in the Allegan State Forest; a historical geography. M, 1965, Western Michigan University.

Lemieux, Corinne Renee. Infiltration characteristics and hydrologic modeling of disturbed land, Moshannon, Pennsylvania. M, 1987, Pennsylvania State University, University Park. 174 p.

Lemine, James. Iron minerals in sedimentary phosphorites of the southeastern United States. M, 1986, Virginia Polytechnic Institute and State University.

Leming, Stephen L. Environmental interpretation of upper Caseyville (Lower Pennsylvanian) detrital sediments along Kinkaid Creek between Crisenberry Dam and the Big Muddy River, Jackson County (Illinois). M, 1973, Southern Illinois University, Carbondale. 62 p.

Lemire, Jerome A. Geology for land use planning of western Clark County, Ohio. M, 1972, Ohio State University.

Lemish, John. The geology of the Topia mining district, Topia, Durango, Mexico. D, 1955, University of Michigan.

Lemish, John. The geology of the west fork of the Madison River area, Montana. M, 1948, University of Michigan. 40 p.

Lemke, Karen A. Time series modeling of suspended sediment concentration in rivers. D, 1988, University of Iowa. 415 p.

Lemley, Irene Savanyo. The Lobo Formation and lithologically similar units in Luna and southwestern Dona Ana counties, New Mexico. M, 1982, New Mexico State University, Las Cruces. 95 p.

Lemley, Kenneth Ray. Paleocurrent analysis and paleoenvironmental interpretation of the Abo Formation, Abo Canyon area, Valencia, Torrance, and Socorro counties, New Mexico. M, 1984, New Mexico Institute of Mining and Technology. 91 p.

Lemmen, Donald Stanley. The glacial history of Marvin Peninsula, northern Ellesmere Island, and Ward Hunt Island, high Arctic Canada. D, 1988, University of Alberta. 176 p.

Lemmon, Dwight M. Geology of the andalusite deposits in the northern Inyo Range, California. D, 1937, Stanford University. 70 p.

Lemmon, Dwight M. Geology of the Hoge Mine, Nevada City, California. M, 1935, Stanford University. 33 p.

Lemmon, Harry W. Gravity, magnetics, and structure of the central Appalachian Plateau region, northern West Virginia. M, 1973, SUNY at Buffalo. 64 p.

Lemmon, Robert David. Petrographic and modal analysis of the Slabtown Granite, Southeast Missouri. M, 1964, Southern Illinois University, Carbondale. 42 p.

Lemmon, Robert Edgar. Geochemistry of the Salibury pluton, Rowan County, North Carolina. M, 1969, University of North Carolina, Chapel Hill. 79 p.

Lemmon, Robert Edgar. Geology of the Bat Cave and Fruitland quadrangles and the origin of the Henderson Gneiss, western North Carolina. D, 1973, University of North Carolina, Chapel Hill. 145 p.

Lemmons, Jacob E. Subsurface Marmaton of southeastern Kansas. M, 1946, University of Kansas. 68 p.

Lemoine, Stephen R. Correlation of the upper Wallace with the lower Missoula Group and resulting facies interpretations, Cabinet and Coeur d'Alene Mountains, Montana. M, 1979, University of Montana. 162 p.

Lemon, R. R. H. Plants from the (Devonian) Sextant Formation involving spore analysis (Ontario). M, 1953, University of Toronto.

Lemon, R. R. H. Proterozoic and Palaeozoic sediments of the Admiralty Inlet region, Baffin Island [Northwest Territories]. D, 1956, University of Toronto.

Lemon, R. R. H. The Upper Devonian limestones of southwestern Alberta, Canada. D, 1955, University of Toronto.

LeMone, David V. The Devonian stratigraphy of Cochise, Pima, Santa Cruz counties, Arizona, and Hidalgo County, New Mexico. M, 1959, University of Arizona.

LeMone, David VonDenburg. The Upper Devonian and Lower Mississippian sediments of the Michigan Basin and Bay County, Michigan. D, 1964, Michigan State University. 120 p.

Lemonnier, Thierry R. L. Estimate of the demand and supply of energy in France until 1985. M, 1978, Stanford University. 79 p.

Lemons, David Ray. Structural evolution of the Lower Cretaceous (Comanchean) Trinity Shelf. M, 1987, Baylor University. 301 p.

Lemos, Jose Antero Senra Vieira. A distinct element model for dynamic analysis of jointed rock with application to dam foundations and fault motion. D, 1987, University of Minnesota, Minneapolis. 307 p.

Lempke, Douglas A. Descriptive stratigraphy, nomenclature, and depositional environments of post Trenton lithologies in Wyandot County, Ohio. M, 1984, University of Toledo. 133 p.

Lenaers, W. Michael. Photochemical degradation of sediment organic matter; effect on Zn-65 release. M, 1972, Oregon State University. 56 p.

Lenaugh, Thomas C. Aerial distribution, and strontium contents of anhydrites peripheral to Kuroko massive sulphide deposits in Japan, and their implications for hydrothermal fluid circulation. M, 1984, Pennsylvania State University, University Park.

Lene, Gene Wilfred. Petrofabric and paleocurrent analysis of the Berea Sandstone (O(Devonian or Mississippian) at South Amherst, Ohio. M, 1966, Bowling Green State University. 73 p.

Leneman, M. Geomorphology and oceanography of Topanga Beach, California, in relation to a small-boat launching facility. M, 1976, University of Southern California. 149 p.

Leney, George Willard. Preliminary investigations of rock conductivity and terrestrial heat flow in southeastern Michigan. M, 1955, University of Michigan.

Lenhard, Robert James. Effects of clay-water interactions on water retention in porous media. D, 1984, Oregon State University. 145 p.

Lenhardt, Duane R. A statistical analysis of runoff and erosion on nonforested, noncultivated hillslopes in the Gowanda, New York area. M, 1973, SUNY at Buffalo. 117 p.

Lenhardt, Duane Rudolph. Characterization and strength analysis of representative Fragipans of New York State. D, 1983, Cornell University. 388 p.

Lenhart, Robert James. An evaluation of ERTS imagery for remote sensing of alluvial fans in Nevada. M, 1974, University of Cincinnati. 54 p.

Lenhart, Robert James. Nearshore marine bedforms formative processes, distribution, and internal structures. D, 1979, University of Cincinnati. 205 p.

Lenhart, Stephen W. Structural and paleographic control of Devonian carbonate lithostratigraphy on and adjacent to the Cincinnati Arch in south central Kentucky. M, 1985, University of Kentucky.

Lenhart, Stephen Wayne. Some upper Pennsylvanian and lower Permian fusulinids from the Bird Spring Group near Mountain Springs, Clark County, Nevada. M, 1975, University of Kentucky. 73 p.

Lenk, Cecilia. The post-glacial population dynamics of Fagus grandifolia Ehrh. in the region of its northern limit. D, 1982, Harvard University. 197 p.

Lenker, Earle Scott. A trace element study of selected sulfide minerals from the Eastern United States. D, 1962, Pennsylvania State University, University Park. 160 p.

Lennartz, Carl R. Engineering subsurface investigations by Earth resistivity method. M, 1953, Purdue University.

Lenney, Thomas William. The Cenozoic geology and alteration at Neals-Bully Creek, Oregon. M, 1980, SUNY at Binghamton. 67 p.

Lennon, G. P. The boundary integral equation method applied to free surface flow problems in porous media. D, 1980, Cornell University. 179 p.

Lennon, Russell Bert. A textural study of the Pennsylvanian limestones of southwestern Illinois. M, 1957, University of Illinois, Urbana. 79 p.

Lennox, D. H. Some terrestrial heat flow measurements in Alberta. M, 1960, University of Alberta. 57 p.

Lennox, R. J. An investigation of bottom changes in Monterey Harbor (California) (1932–1969). M, 1969, United States Naval Academy.

LeNoble, Michael J. Seismic precursors to icequakes; a feasibility study to determine the applicability of icequakes as a scale model for earthquake mechanism and prediction analysis. M, 1980, University of Wisconsin-Milwaukee. 154 p.

Lens, Larry F. Quaternary stratigraphy and sedimentary framework of the Sapelo Island area, coastal Georgia. M, 1981, University of Georgia.

Lent, Mary C. Diagenetic-aspects and development of stylolites in the Muddy Sandstone of Wyoming. M, 1983, University of Missouri, Columbia.

Lent, Robert Louis. Geology of the southern half of the Langlois Quadrangle, Oregon. D, 1969, University of Oregon. 189 p.

Lent, Robert M. The stable isotopic stratigraphy of the Pungo River Formation, Onslow Bay, North Carolina. M, 1985, North Carolina State University. 115 p.

Lent, Stephanie Jean. Classification of Piedmont and Coastal Plain Virginia soils by numerical methods. M, 1980, University of Virginia. 82 p.

Lentell, Randall Lynn. Depositional history of the Rio Tecolutla Esturary, Mexico. M, 1975, University of Florida. 44 p.

Lentell, Thomas L. Prediction of shoreline erosion at Clinton Lake on the basis of examples at Lake Perry. M, 1977, University of Kansas. 161 p.

Lentini, Michael Robert. Mineralogical and geochemical studies of arkose-brine interaction at 200°C and 500 bars total pressure; an experimental investigation. M, 1982, University of Wisconsin-Madison.

Lenton, Paul G. Mineralogy and petrology of the Buck Claim lithium pegmatite, Bernic Lake, southeastern Manitoba. M, 1979, University of Manitoba. 164 p.

Lentz, David Richard. Geology and depositional conditions of tin lodes at Trune Hill, New Brunswick. M, 1986, University of New Brunswick. 280 p.

Lentz, Leonard James. Lithological changes associated with a progradational deltaic sequence in southcentral West Virginia. M, 1983, North Carolina State University. 139 p.

Lentz, R. T. The petrology and stratigraphy of the Portland Hills Silt. M, 1977, Portland State University. 144 p.

Lentz, Robert C. Relation of Eocene depositional environments to sulfur content and quality of surface waters at lignite strip mines near Fairfield, Texas. M, 1975, University of Texas, Austin.

Lentz, Rodney Ward. Permanent deformation of cohesionless subgrade material under cyclic loading. D, 1979, Michigan State University. 217 p.

Lenz, Alfred Carl. Devonian stratigraphy and paleontology of lower Mackenzie Valley, Northwest Territories. D, 1959, Princeton University. 222 p.

Lenz, Alfred Carl. Ordovician and Silurian graptolitic fauna of the southern Richardson Mountains and adjacent areas, Yukon Territory. M, 1956, University of Alberta. 138 p.

Lenzer, Richard Charles. Geology and wallrock alteration at the Morey Mining District, Nye County, Nevada. D, 1972, University of Wisconsin-Madison. 155 p.

Lenzer, Richard Charles. Minor elements in the native copper from the Champion Mine, Painesdale, Michigan. M, 1968, University of Wisconsin-Madison.

Lenzi, Gary Wilson. Geochemical reconnaissance at Mercur, Utah. M, 1971, University of Utah. 51 p.

Leo, Gerhard William. The plutonic and metamorphic rocks of Ben Lomond Mountain, Santa Cruz County, California. D, 1961, [Stanford University]. 194 p.

Leo, Richard Francis. Silicification of wood. D, 1975, Harvard University.

Leo, Sandra Rose. A stratalogic analysis of the intra-Miocene Ochocoan Orogeny and the Walpapi Sequence in Washington and Oregon. M, 1979, University of Washington. 113 p.

Leon, Alfredo Aniano De see De Leon, Alfredo Aniano

Leon, Hernan Jose. Observations on the stratigraphy and structure of the Saratoga Gap area, California. M, 1958, Stanford University.

Leon, Jose G. Ponce de see Ponce de Leon, Jose G.

Leon, Luis Alfredo. Adsorption of aluminum and ferric ions by vermiculite. M, 1965, University of California, Riverside. 93 p.

Leon, Luis Alfredo. Chemistry of some tropical acid soils of Colombia, South America. D, 1967, University of California, Riverside. 191 p.

Leon, Ralph Richard. Provenance of sandstone in the Upper Ordovician Bald Eagle and Juniata formations, central Pennsylvania; implications for tectonic setting. M, 1985, University of Delaware. 137 p.

Leonard, A. G. The basic rocks of northeastern Maryland and their relation to the granite. D, 1898, The Johns Hopkins University.

Leonard, Arnold David. The petrology and stratigraphy of upper Mississippian Greenbrier limestones of eastern West Virginia. D, 1968, West Virginia University. 245 p.

Leonard, Arnold David. The Pocono Sandstone neighboring the northern anthracite basin, Pennsylvania. M, 1953, Pennsylvania State University, University Park. 153 p.

Leonard, Barbara June. Environment of deposition of the A-1 Carbonate, Salina Group, Michigan Basin. M, 1983, Western Michigan University. 83 p.

Leonard, Benjamin F., III. Magnetite deposits of the St. Lawrence County District, New York. D, 1951, Princeton University. 206 p.

Leonard, Benjamin Franklin. Oligocene stratigraphy of the Douglas area, Converse County, Wyoming. M, 1957, University of Nebraska, Lincoln.

Leonard, Clifford, Jr. The tectonic margins of the eastern Gulf Coast of North America. M, 1988, Northeast Louisiana University. 117 p.

Leonard, Eric Michael. Glaciolacustrine sedimentation and Holocene glacial history, northern Banff National Park, Alberta. D, 1981, University of Colorado. 287 p.

Leonard, Fred Andrew. Computerized system for open-pit drilling. M, 1969, University of Utah. 109 p.

Leonard, Jay E. Space-time sediment relationships in the nearshore zone; the case of storm conditions. D, 1978, Boston University. 557 p.

Leonard, John R. The Mississippian stratigraphy of the Gallatin Basin, Montana. M, 1946, University of Kansas. 65 p.

Leonard, Katherine Esther. Foreland fold and thrust belt deformation chronology, Ordovician limestone and shale, northwestern Vermont. M, 1985, University of Vermont. 138 p.

Leonard, Lynn Ann. An analysis of replenished beach design on the U. S. East Coast. M, 1988, Duke University. 130 p.

Leonard, Marc-Andre. Etudes stratigraphiques et sédimentologique du flysch de la région de Saint-Fabien (Comté de Rimouski). M, 1974, Universite de Montreal.

Leonard, Mark Steven. Estimation of residual statics corrections for three dimensional seismic data using the generalized matrix inverse technique. M, 1979, Indiana University, Bloomington. 343 p.

Leonard, Mary L. The geology of the Tres Hermanas Mountains, Luna County, New Mexico. M, 1982, University of Texas at El Paso.

Leonard, Patricia J. Calculated permeabilities of sands by the modified Kozeny equation. M, 1977, University of Wisconsin-Milwaukee.

Leonard, R. B. Structure and sedimentation in the Port Hood area, Nova Scotia. M, 1951, Massachusetts Institute of Technology. 47 p.

Leonard, Ralph Avery. Mica weathering in relation to structural and compositional chemistry. D, 1966, North Carolina State University. 176 p.

Leonard, Raymond C. An analysis of surface fracturing in Val Verde County, Texas. M, 1977, University of Texas, Austin.

Leonard, Raymond Jackson. The hydrothermal alteration of certain silicate minerals. D, 1926, University of Minnesota, Minneapolis. 39 p.

Leonard, Richard. Variable structural style, stratigraphy, total strain and metamorphism adjacent to the Purcell Thrust, near Blackman Creek, B.C. M, 1985, McGill University. 250 p.

Leonard, Robert Benjamin, Jr. Ground-water geology along the northwest foot of the Blue Ridge between Arnold Valley and Elkton, Virginia. D, 1963, Virginia Polytechnic Institute and State University. 336 p.

Leonard, Wendy C. Hydrogeologic constraints upon neutralization of acid precipitation; a study of Cadwell Creek watershed, west-central Massachusetts. M, 1984, University of Massachusetts. 150 p.

Leonardos, Othon Henry. Serpentinite contact relations in the Moccasin Quadrangle, California. M, 1966, University of California, Berkeley. 42 p.

Leonardson, Robert William. Petrology of the Bergland Rhyolite, the Firetower Rhyolite and associated rocks, Ontonagon County, Michigan. M, 1966, Michigan Technological University. 121 p.

Leone, John Michael, Jr. An investigation of finite element solutions to the advection equation in two dimensions. M, 1976, Iowa State University of Science and Technology.

Leone, Raymond John. Stratigraphic petrology of selected Pennsylvanian sedimentary rocks in Boone and Callaway counties, Missouri. M, 1956, University of Missouri, Columbia.

Leong, Eugene Yee. Air pollution control in California from 1970 to 1974; some comments on the implementation planning process. D, 1974, University of California, Los Angeles.

Leong, Wing K. Regional study of the (Cretaceous) Fall River Sandstone of the Powder River basin, Wyoming. M, 1962, Michigan State University. 73 p.

Leong, Wing Kwong. Sea floor topography and microtopography southwest of the Iberian Peninsula. D, 1973, University of Wisconsin-Madison. 91 p.

Leonhardt, Frederick H. Deformational history of Black Knob Ridge, southeastern Oklahoma. M, 1983, University of Texas at Dallas. 83 p.

Leonhardy, Frank Clinton. Artifact assemblages and archaeological units at Granite Point Locality 1 (45WT41), southeastern Washington. D, 1970, Washington State University. 247 p.

Leonhart, L. An analysis of combustion within surface mine spoils and of its consequent effects on the environment and reclamation practices. D, 1978, University of Arizona. 115 p.

Leonhart, Scott W. Correlation of the Tongue River Member of the Fort Union Formation throughtout the Canyon Creek area; Rosebud and Big Horn counties, Montana. M, 1975, University of Colorado. 61 p.

Leopold, Lawrence C. A study of seaward dipping internal structures within large scale ripple marks in the marine environment, (Monterey County) California. M, 1972, San Jose State University. 59 p.

Leopold, Luna B. The erosion problem of Southwestern United States. D, 1950, Harvard University.

Leopoldt, Winfried. Neogene geology of the central Mangas Graben, Cliff-Gila area, Grant County, New Mexico. M, 1981, University of New Mexico. 160 p.

Leosewski, John Fitzgerald. An investigation of a linear magnetic anomaly in east-central Indiana and its possible relationship to hydrocarbon accumulation. M, 1985, Purdue University. 86 p.

Lepage, Carolyn A. The composition and origin of the Pond Ridge Moraine, Washington County, Maine. M, 1982, University of Maine. 74 p.

LePain, David Lloyd. Olistoliths in the Yellow Breeches Member of the Wilhite Formation and their significance in determining the tectonostratigraphic position of the Ocoee Supergroup. M, 1987, Wright State University. 129 p.

Lepak, Robert James. Rb-Sr geochronology and rare-earth element geochemistry of Proterozoic leucogneisses from the northwestern Adirondacks, New York. M, 1983, Miami University (Ohio). 127 p.

Lepelletier, Thierry Georges. Tsunamis; harbor oscillations induced by nonlinear transient long waves. D, 1981, California Institute of Technology. 494 p.

Lepley, Larry K. Submarine geomorphology of eastern Ross Sea and Sulzberger Bay, Antarctica. M, 1964, Texas A&M University.

Lepp, Casey Louis. Depositional environments of Upper Cretaceous-lower Tertiary rocks, western Williston Basin, Montana. M, 1981, Texas Tech University. 83 p.

Lepp, Henry. An experimental study of interconversions among iron carbonates, oxides, and sulfates. D, 1954, University of Minnesota, Minneapolis. 108 p.

Leppaluoto, David Alan. Iron at pressures of the Earth's core; properties inferred from the significant structure theory of liquids. D, 1973, University of California, Berkeley. 163 p.

Leppert, Dave Eric. Differentiation of a shoshonitic magma at Snake Butte, Blaine County, Montana. M, 1985, University of Montana. 121 p.

Lepple, F. K. Eolian dust over the North Atlantic Ocean. D, 1975, University of Delaware, College of Marine Studies. 280 p.

Lepry, Louis Anthony, Jr. The structural geology of the Yauli Dome region, Cordillera Occidental, Peru. M, 1981, University of Arizona. 100 p.

Lepzelter, Carol. Calcareous algae and algal structures from the Lowville, Watertown, and Selby formations of the medial Ordovician Black River Group of northwestern New York and southeastern Ontario. M, 1981, Boston University. 145 p.

Lerbekmo, John Franklin. The character and origin of late Tertiary blue sandstones in Central California. D, 1956, University of California, Berkeley. 110 p.

Lerch, Christopher. Stratigraphy and paleomagnetism of upper Vaqueros-lower Topanqa formations (early to middle Miocene) at Point Mugu and La Jolla Canyon, western Santa Monica Mountains, Ventura County California. M, 1988, University of Southern California.

Lerch, Frederick George. Geology of the Red Cinder 7.5-minute quadrangle, Lassen and Plumas counties, California. M, 1987, Colorado School of Mines. 274 p.

Leree, Juan Antonio Cuevas see Cuevas Leree, Juan Antonio

Lerman, Abraham. Evolution and environment of Exogyra in the Late Cretaceous of the Southeastern United States. D, 1963, Harvard University. 152 p.

Lerner, David H. Microtektites from Gay's Cove, Barbados, West Indies. M, 1986, University of Delaware. 188 p.

Lerner-Lam, Arthur Lawrence. Linearized estimation of higher-mode surface wave dispersion. D, 1982, University of California, San Diego. 276 p.

LeRoux, Gay Breton, III. Sedimentological and environmental interpretation of the Chota-Sevier formations in easternmost Tennessee. M, 1974, University of Tennessee, Knoxville. 106 p.

Leroux, Jean Pierre. Calibration d'un système séismométrique ultrasensible. M, 1971, Universite Laval.

LeRoy, Frank L. Outlet history of peri-glacial Cayuga Lake (New York). M, 1938, Cornell University.

LeRoy, Leslie W. Stratigraphy of the Golden-Morrison area, Jefferson County, Colorado. D, 1944, Colorado School of Mines. 186 p.

Leroy, Paul G. Correlation of copper mineralization with hydrothermal alteration in the Santa Rita porphyry copper deposit (New Mexico). D, 1953, Columbia University, Teachers College.

Leroy, Paul G. Geology of the southern margin of the Santa Rita copper pit (New Mexico). M, 1951, Columbia University, Teachers College.

Leroy, Ronald. The mid-Tertiary to Recent lithostratigraphy of Putnam County, Florida. M, 1981, Florida State University.

LeRoy, S. D., Jr. A new method of sand grain shape measurement and its potential for application. M, 1975, University of Southern California.

LeRoy, Samuel David, Jr. Description of grain-size curve-form sequences; a new attempt at environmental differentiation. D, 1981, University of Southern California.

LeRoy, Tom E. Submarine topography of the Gulf Coast submerged continental platform; Southwest Pass, Mississippi Riber to Galveston Bay. M, 1941, Louisiana State University.

Lesack, K. A. Palynology of the Devonian sequence, northwestern Devon Island, Arctic Canada. M, 1988, University of Waterloo. 277 p.

Leschak, Pamela. Origin and distribution of sand types, northeastern U.S. Atlantic continental shelf. M, 1986, Texas A&M University.

Leschen, Melanie R. Stratigraphic control and geochemical zonation of the O'Carroll ore body, Christmas Mine, Gila County, Arizona. M, 1981, Dartmouth College. 200 p.

Leshchinsky, Dov. Theoretical analysis of the stability of three-dimensional slopes. D, 1982, University of Illinois, Chicago. 255 p.

Lesher, Carl Michael. Mineralogy and petrology of the Sokoman Iron Formation near Ardua Lake, Quebec. M, 1976, Indiana University, Bloomington. 62 p.

Lesher, Charles Edward. Thermal diffusion in silicate liquids. D, 1985, Harvard University. 203 p.

Leshner, Orrin. The calcareous nannofossil biostratigraphy of the Austin Group in the type area, central Texas. M, 1986, Tulane University. 385 p.

Lesht, Barry Mark. Field study of the bottom friction boundary layer on the inner continental shelf. D, 1977, University of Chicago. 251 p.

Leslie, Gordon Anthony. A comparison of the diagenetic and diagnostic features of the Sturgeon Lake, Normandville, and Clairmont reef complexes. M, 1955, University of Alberta. 77 p.

Leslie, John A. Contact alteration and mineralization in northern New Brunswick. M, 1963, University of Western Ontario. 88 p.

Leslie, Kenneth Campbell. Source area and abrasional history of the coarse fraction of Point Reyes beach sediment. M, 1975, University of California, Berkeley. 150 p.

Leslie, Louise E. Late glacial geology of the Finlay River valley, British Columbia. M, 1988, University of Alberta. 122 p.

Leslie, Robert James. Ecology and paleoecology of Hudson Bay (Canada) foraminifera. D, 1965, University of Southern California.

Leslie, Robert James. Sedimentology and foraminiferal trends of Hudson Bay, Canada. M, 1963, University of Southern California.

Leslie, Robin Bruce. Cenozoic tectonics of southern Chile; triple junction migration, ridge subduction, and forearc evolution. D, 1986, Columbia University, Teachers College. 284 p.

Leslie, Robin Bruce. Continuity and tectonic implications of the San Simeon-Hosgri fault zone, Central California. M, 1980, University of California, Santa Cruz.

Leslie, Teri Hall. 238U-230Th chronology of a basalt from Red Mt., California. M, 1980, University of California, Santa Cruz.

Leslie-Bole, Benjamin. Deltaic and lacustrine sediments from glacial Lake Sciota, Monroe County, Pennsylvania. M, 1986, University of Delaware.

Lesperance, Pierre J. Post-Taconic formations of the Temiscouata region, Quebec. D, 1961, McGill University.

Lesperance, Pierre Jacques. Drill hole number 8, Wayne County Major Airport, Wayne County, Michigan. M, 1957, University of Michigan.

Lessard, Denis. Minéralogie et microstructure d'échantillons d'argile de la mer de Champlain, Québec. M, 1984, Universite Laval. 162 p.

Lessard, Ghislain J. P. Biogeochemical phenomena in quick clays and their effects of engineering properties. D, 1981, University of California, Berkeley. 362 p.

Lessard, Robert Henry. Intertidal and shallow water foraminifera of the tropical Pacific Ocean. M, 1963, University of Southern California.

Lessard, Robert Henry. Micropaleontology and paleoecology of the Tunumk Member of the Mancos Shale. D, 1970, University of Utah. 78 p.

Lessenger, Margaret A. An expert system to evaluate geologic risk for onshore domestic petroleum exploration. M, 1988, Colorado School of Mines. 63 p.

Lessentine, Ross Henry. Areal and structural geology in Buckingham Valley, Pennsylvania. M, 1952, Lehigh University.

Lesser, Richard Peter. Major element and isotopic studies on the James Run/Port Deposit association, Maryland; tectonic analogues and Taconic deformation. M, 1982, Virginia Polytechnic Institute and State University. 198 p.

Lessig, Joseph Watson. The geology of the Norris Quadrangle, Anderson, Campbell, and Knox counties, Tennessee. M, 1949, University of Tennessee, Knoxville. 61 p.

Lessing, Peter. Petrology of the Poundridge Leptite (age uncertain), Westchester County, New York. D, 1967, Syracuse University. 86 p.

Lessing, Peter. Potassium and rubidium distribution in Hawaiian lavas. M, 1963, Dartmouth College.

Lessley, John C. Investigation of coal bumps in the Pocahontas No. 3 Seam, Buchanan County, Virginia. M, 1983, Virginia Polytechnic Institute and State University. 304 p.

Lessman, James Lamont. Geology and copper mineralization of the Coopers Hill District, Portland Parish, Jamaica, West Indies. M, 1979, University of Arizona. 73 p.

Lester, Barry Henry. The evaluation of elastic rebound in a three layered geologic system using potential theory. M, 1981, Pennsylvania State University, University Park. 88 p.

Lester, James George. Geology of the region around Stone Mountain, Georgia. D, 1938, University of Colorado. 147 p.

Lester, John Lawrence. Pennsylvanian stratigraphic studies of White and Hamilton counties, Illinois. M, 1939, University of Illinois, Urbana. 33 p.

Lester, Mark. Mineral exploration using geophysics; a model developed for use as a teaching learning tool. M, 1977, Purdue University.

Lester, Robert Worth. The geology of the Baywood Quadrangle, Putnam County, Florida. M, 1967, University of Florida. 61 p.

Lesure, Frank G. Geology of the Clifton Forge iron district, Virginia. D, 1955, Yale University.

Letargo, Maria Rosario R. Petrogenesis of alkaline pillow basalts of southwestern Panay, Philippines. M, 1988, University of Cincinnati. 83 p.

Letendre, Jacques. Révision des Phacopidae (Trilobita) du Silurien de l'est de l'Amérique du Nord. M, 1976, Universite de Montreal.

Letendre, Jacques. Systématique de Trilobites. M, 1971, Universite de Montreal.

LeTourneau, Nelson Joseph. A study of the earthquake phases Pa and Sa. M, 1968, Texas Tech University. 39 p.

LeTourneau, Peter Mark. The sedimentology and stratigraphy of the Lower Jurassic Portland Formation, central Connecticut. M, 1985, Wesleyan University. 247 p.

Letsch, Dieter K. Early Jurassic depositional history of the northern margin of the central High Atlas, Morocco. M, 1985, Colorado School of Mines. 171 p.

Lett, R. E. W. Secondary dispersion of transition metals through a copper-rich bog in the Cascade Mountains, British Columbia. D, 1979, University of British Columbia.

Letteney, Cole DeWitt. The anorthosite-charnockite series (Precambrian) of the Thirteenth Lake Massif, south central Adirondack Highlands, New York. D, 1967, Syracuse University. 188 p.

Lettis, William Robert. Late Cenozoic stratigraphy and structure of the western margin of the central San Joaquin Valley, California. D, 1982, University of California, Berkeley. 587 p.

Letts, Robert E. The relationship of compressive strength to petrology in the Belly River sandstones. M, 1973, University of Calgary. 102 p.

Letzsch, W. Stephen. Clay mineralogy and surface characteristics within a Holocene salt marsh, Sapelo Island, Georgia. D, 1986, Washington State University. 95 p.

Letzsch, W. Stephen. Erosion and deposition within a salt marsh, Sapelo Island, Georgia. M, 1978, University of Georgia.

Leu, David Jack. A procedure for merging remote sensing and field sampling methods to assess existing and historic environmental conditions of coastal wetlands. D, 1982, University of Delaware, College of Marine Studies. 223 p.

Leu, L.-K. Spectral analysis of gravity and magnetic anomalies. D, 1975, University of California, Berkeley. 176 p.

Leu, Ling-Ling Lillian. Three-dimensional velocity structure of the 1983 M7.3, Borah Peak, Idaho, earthquake areas using tomographic inversion of aftershock travel-times. M, 1986, University of Utah. 98 p.

Leu, Peih-Lin. Magnitude corrections for the Central Mississippi Valley Seismic Network. M, 1985, St. Louis University. 162 p.

Leu, Peih-Lin. Magnitude corrections for the central Mississippi Valley seismic network. M, 1986, St. Louis University.

Leuner, W. R. Geology of the west half of La Motte Township, Quebec. M, 1959, McGill University.

Leung, Irene Sheung-Ying. An optical and x-ray analysis of deformation structures in diopside. D, 1969, University of California, Berkeley. 126 p.

Leung, Jana C. Paleosecular variations in lake sediment cores from northern Ellesmere Island, Canada. M, 1988, University of Massachusetts. 181 p.

Leung, Kon Lim. Vibration isolation of structures from ground-transmitted waves in nonhomogeneous elastic soil. D, 1988, University of Minnesota, Minneapolis. 286 p.

Leung, Samuel Seh-Shue. A comparison of magnetites from ores and host rocks of different geologic occurrences. D, 1964, University of Illinois, Urbana. 107 p.

Leung, Samuel Seh-Shue. Fluorescent X-ray analysis of host rock pyroxene amphibolite from the Scott Mine, Sterling Lake, New York. M, 1960, University of Illinois, Urbana. 29 p.

Leung, Sydney Kwok-On. Coal in the Mannville Group (L. Cretaceous) of West central Saskatchewan. M, 1976, University of Alberta. 105 p.

Leung, W. H. The thermal properties of sea water and sea salts. D, 1974, University of Miami. 224 p.

Leuthart, C. A. Reclamation of orphan strip mined land in southern Illinois and western Kentucky; a field study of the Palzo Project of Williamson County,

Illinois and the Clear Creek swamp of Webster and Hopkins counties, Kentucky. D, 1975, University of Louisville. 236 p.

Leuty, Joseph L. Petroleum geology of the Upper Cretaceous Woodbine-Eagle Ford interval, southern East Texas Basin. M, 1987, Baylor University. 301 p.

Leutze, Willard. The stratigraphy and paleontology of the (Silurian) middle Salina in central New York. M, 1955, Syracuse University.

Leutze, Willard Parker. Stratigraphy and paleontology of the Salina Group in central New York. D, 1959, Ohio State University. 492 p.

Levan, Donald Clement and McLean, W. F. Structure and stratigraphy of the Red Canyon area, Gallatin County, Montana. M, 1951, University of Michigan. 66 p.

Levander, Alan R. The shear wave velocity structure of the lithosphere in central and Northern California. D, 1984, Stanford University. 237 p.

Levandowski, Donald William. Geology and mineral deposits of the Sheridan-Alder area, Madison County, Montana. D, 1956, University of Michigan. 318 p.

Levay, Joseph. The Precambrian geology of the Crevice Creek area, Southwest Beartooth Mountains, Montana and Wyoming. M, 1976, Northern Illinois University. 61 p.

Leve, Gilbert Warren. Geology of Red Bluff area, Eddy County, New Mexico. M, 1952, University of Texas, Austin.

Leveille, Gregory Paul. Geology of El Capitan, Sonora, Mexico. M, 1984, San Diego State University. 122 p.

Levendosky, W. T. The geology of the Barth Island layered intrusion, Labrador. M, 1975, Syracuse University.

Leventer, Amy. Relationships between anoxia, glacial meltwater, and microfossil preservation/productivity in the Orca Basin. M, 1982, University of South Carolina. 116 p.

Leventer, Amy Ruth. Recent biogenic sedimentation on the Antarctic continental margin. D, 1988, Rice University. 249 p.

Leventhal, B. A. Mapping change on urban rivers. M, 1978, SUNY at Binghamton. 80 p.

Leventhal, Joel Stephen. Chronology and correlation of young basalts by uranium-thorium-helium measurements. D, 1972, University of Arizona. 137 p.

LeVeque, Richard Alan. Stratigraphy and structure of the Palen Formation, Palen Mountains, southeastern California. M, 1981, University of Arizona.

Leverett, David Earl. Stratigraphy and depositional environments of the Mississippian-Pennsylvanian Parkwood Formation on a part of the northwest limb of the Cahaba Synclinorium, Jefferson and Bibb counties, Alabama. M, 1987, University of Alabama. 146 p.

Levert, Charles F., Jr. Lower Catahoula equivalents of Louisiana. M, 1959, Louisiana State University.

Leveson, David Jeffrey. Orbicular rocks of the Lonesome Mountain area, Beartooth Mountains, Montana and Wyoming. D, 1960, Columbia University, Teachers College. 219 p.

Levesque, Rene Joseph. Stratigraphy and sedimentology of Middle Cambrian to Lower Ordovician shallow water carbonate rocks, western Newfoundland. M, 1978, Memorial University of Newfoundland. 276 p.

Levet, Melvin N. Geology of the San Juan Canyon area, Orange County, California. M, 1940, California Institute of Technology. 42 p.

Levey, R. A. Characteristics of coarse-grained point bars, upper Congaree River, South Carolina. M, 1977, University of South Carolina. 61 p.

Levey, Raymond Allen. A depositional model for major coal seams in the Rock Springs Formation, Upper Cretaceous, Southwest Wyoming. D, 1981, University of South Carolina. 585 p.

Levi, Beatriz. Cretaceous volcanic rocks from a part of the Coast Range west from Santiago, Chile; a study in

lithologic variation and burial metamorphism in the Andean geosyncline. D, 1969, University of California, Berkeley. 124 p.

Levi, S. Some magnetic properties of magnetite as a function of grain size and their implications for paleomagnetism. D, 1974, University of Washington. 210 p.

Levich, Robert A. Geology and ore deposits of the Sierra de Santa Maria Dome, Velardena, Durango, Mexico. M, 1973, University of Texas, Austin.

Levien, L. Silicate minerals at pressure; crystal structures and elasticity. D, 1979, SUNY at Stony Brook. 228 p.

Levin, Douglas R. Sedimentation processes in Winthrop Harbor, Massachusetts. M, 1981, Boston University. 126 p.

Levin, Harold Leonard. The geology of north-central Ralls County, Missouri. M, 1952, University of Missouri, Columbia.

Levin, Harold Leonard. The micropaleontology of the Oldsmar Limestone of Florida. D, 1956, Washington University. 75 p.

Levin, Max. Geologic structure of part of Hurd Draw Quadrangle, Culberson County, Texas. M, 1951, University of Texas, Austin.

Levin, S. Benedict. Genesis of some Adirondack garnet deposits. D, 1948, Columbia University, Teachers College.

Levin, Samuel. Cenozoic geology of Kent Quadrangle, Culberson, Reeves, and Jeff Davis counties, Texas. M, 1952, University of Texas, Austin.

Levin, Stewart Arthur. Deconvolution with spatial constraints. D, 1987, Stanford University. 71 p.

Levine, Carol Alice. Internal structures of the northwestern portion of the Twin Sisters Dunite, North Cascades, Washington. M, 1981, University of Washington. 82 p.

Levine, Charles R. Character, geology, and origin of the Clear Creek tripoli deposits in southern Illinois. M, 1973, Southern Illinois University, Carbondale. 70 p.

Levine, Edward Neil. Alaskan geology and seismicity. M, 1962, Boston College.

Levine, Elissa Robin. Sensitivity of Pennsylvania soils to atmospheric deposition; an assessment model. D, 1984, Pennsylvania State University, University Park. 286 p.

Levine, Jeffrey Ross. Optical anisotropy of coals as an indicator of tectonic deformation, Broad Top coal field, Pennsylvania. M, 1981, Pennsylvania State University, University Park. 60 p.

Levine, Jeffrey Ross. Tectonic history of coal-bearing sediments in eastern Pennsylvania using coal reflectance anisotropy. D, 1983, Pennsylvania State University, University Park. 337 p.

Levine, Joseph Samuel. A study of the mechanism of water flooding. M, 1938, Pennsylvania State University, University Park. 93 p.

Levine, Joseph Samuel. An investigation of the displacement of oil by water in a porous medium. D, 1941, Pennsylvania State University, University Park. 140 p.

Levine, Norman Seth. The use of thermal infrared multispectral scanner data for geochronologic mapping of the Cima volcanic field, San Bernardino, California. M, 1988, Indiana State University. 171 p.

Levine, Paul Elliot. Sorption of lead, zinc, and cadmium on a glacial outwash soil. M, 1975, University of Washington. 136 p.

Levine, Saul R. Detection of faults in the Hollister Valley, California by thermal mapping techniques. M, 1972, San Jose State University. 68 p.

Levine, Stephen D. Provenance and diagenesis of the Cherokee sandstones, deep Anadarko Basin, western Oklahoma. M, 1984, Texas A&M University. 154 p.

Levine, Steven Joel. Genesis of soils derived from the Kaibab Formation of the Colorado Plateau, Arizona. D, 1987, University of Arizona. 162 p.

Levine, Steven L. Chemical remanent magnetization in iron ores and wall rocks of Cerro de Mercado, Durango, Mexico. M, 1975, University of Minnesota, Minneapolis. 124 p.

Levings, Gary Wayne. A groundwater reconnaissance study of the Upper Sugar Creek watershed, Caddo County, Oklahoma. M, 1971, Oklahoma State University. 107 p.

Levings, William S. Late Cenozoic erosional history of the Raton Mesa region; Las Animas County, Colorado. D, 1951, Colorado School of Mines. 198 p.

Levinson, Alfred Abraham. Mineralogy of the muscovite-lepidolite series. D, 1952, University of Michigan.

Levinson, Alfred Abraham. Petrography of pre-Beltian Cherry Creek marbles, southwestern Montana. M, 1949, University of Michigan. 46 p.

Levinson, Andrea R. Depositional environments of the shales and coals in the Dakota Sandstone and adjacent units of the San Juan Basin, Colorado and New Mexico. M, 1979, Bowling Green State University. 125 p.

Levinson, Richard A. Remote sensing applied to uranium exploration in Wyoming. M, 1979, University of Wyoming. 244 p.

Levinson, Stuart Alan. Studies of Paleozoic Ostracoda. D, 1951, Washington University. 67 p.

Levinson, Stuart Alan. The hingement of Paleozoic Ostracoda. M, 1949, Washington University. 31 p.

Levinton, Jeffrey S. The ecology of shallow water deposit feeding communities. D, 1971, Yale University.

Levish, Murray. Stratigraphic correlation of some Pennsylvanian limestones by thin section. M, 1955, University of Illinois, Urbana. 68 p.

Levitan, Arlette E. S. Orca Basin; a paleoclimatic study of the late Pleistocene in the Northwest Gulf of Mexico. M, 1981, Texas A&M University.

Levitan, Mark Leslie. Effects of hydrogeology on lignite recovery in the Manning Formation, Grimes County, Texas. M, 1976, Texas A&M University. 86 p.

Levitt, Stephen Robert. The vibrational spectroscopy and normal coordinate analysis of geological apatites. D, 1969, Alfred University. 234 p.

Levoie, Clermont. Une Combinaison des méthodes électromagnétiques a cadres horizontaux "Slingram" et Turam. D, 1972, McGill University.

Levorsen, A. I. A titaniferous magnetite deposit in Cook County, Minnesota. M, 1917, University of Minnesota, Minneapolis. 22 p.

Levorsen, Mark K. Stratigraphic analysis of the Gothic Formation (Desmoinesian), Pitkin and Gunnison counties, Colorado. D, 1987, Colorado School of Mines. 135 p.

Levorsen, R. D. Geology of the Las Llajas Canyon region (California). M, 1947, University of California, Los Angeles.

Levson, Victor Mathew. Quaternary sedimentation and stratigraphy of montane glacial deposits in parts of Jasper National Park, Canada. M, 1986, University of Alberta. 201 p.

Levy, A. S. Relations of Fordham and Manhattan formations near Katonah, New York. M, 1977, Queens College (CUNY). 107 p.

Levy, Alexandro Gustavo. Oil and gas maturation zones in the Jurassic and Cretaceous trends of Northwest Florida and adjacent parts of Georgia and Alabama. M, 1977, University of Florida. 79 p.

Levy, David J. Manganese mineralization hosted by the Rocky Gap Sandstone in Bland County, Virginia. M, 1985, Virginia Polytechnic Institute and State University.

Levy, J. S. Sblendorio see Sblendorio Levy, J. S.

Levy, Joel B. Comparison of texture, mineralogy, and organic content of suspended, accumulating, and bottom sediments within a coastal lagoon, Stone Harbor, New Jersey. M, 1978, Lehigh University. 69 p.

Levy, John Sanford. Suspended sediment distribution of Doboy Sound, Georgia. M, 1968, University of Georgia. 102 p.

Levy, Lawrence S. Radiolarian biostratigraphy of the Marca Shale Member, Moreno Formation, Fresno County, California. M, 1977, University of Texas at Dallas. 72 p.

Levy, Michael Arnold. Seismicity of the Earth, 1961-1967. M, 1970, University of Michigan.

Levy, Miguel Rudy Barbosa see Barbosa Levy, Miguel Rudy

Levy, Roy. Random processes for earthquake simulation. D, 1969, Polytechnic University. 194 p.

Levy, Shlomo. Inversion of reflection seismograms. D, 1985, University of British Columbia.

Levy, Shlomo. Wavelet estimation and debubbling using minimum entropy deconvolution and time domain linear inverse methods. M, 1979, University of British Columbia.

Levy, Stephen E. Removal of multiple reflections from seismic data using slant stack. M, 1988, University of New Orleans.

Levy, Susan S. Serpentinization textures in the Mashaba igneous complex, Rhodesia. M, 1975, University of Texas, Austin.

Levy, Thomas M. An age sequence in ancient rocks near Delhi, Minnesota. M, 1975, Northern Illinois University. 40 p.

Lew, Laurence Reed. The geology of the Osa Peninsula, Costa Rica; observations and speculations about the evolution of part of the Outer Arc of the southern Central American Orogen. M, 1983, Pennsylvania State University, University Park. 128 p.

Lew, Laurence Reed. The geology of the Santa Elena Peninsula, Costa Rica, and its implications for the tectonic evolution of the Central America-Caribbean region. D, 1985, Pennsylvania State University, University Park. 509 p.

Lewallen, Noble F., II. Structural geology of the northwestern portion of the Michigan Basin. M, 1983, Michigan State University. 77 p.

Lewan, Michael Donald. Geochemistry of vanadium and nickel in organic matter of sedimentary rocks. D, 1980, University of Cincinnati. 378 p.

Lewan, Michael Donald. Metasomatism and weathering of the Presque Isle serpentinized peridotite, Marquette, Michigan. M, 1972, Michigan Technological University.

Lewand, Raymond. The geomorphic evolution of the Leon River system (Coryell, Hamilton, Comanche, and Eastland counties, Texas). M, 1967, Baylor University. 74 p.

Lewark, James Edward. Bellerophontidae of the Chester Series. M, 1940, University of Illinois, Urbana. 19 p.

Lewchalermvong, Chettavat. Investigation and evaluation of the Royal Flush and Mex-Tex mines and adjacent area, Hansonburg mining district, Socorro County, New Mexico. M, 1973, New Mexico Institute of Mining and Technology. 102 p.

Lewellen, Dennis G. The structure and depositional environment of the Manastash Formation, Kittitas County, Washington. M, 1983, Eastern Washington University. 161 p.

Lewellen, Dennis Gilbert. A technique for analyzing geologic structures using Fourier analysis and Butterworth filters. D, 1985, University of Kentucky.

Lewen, Melvin C. Van see Van Lewen, Melvin C.

Lewis, Alvin. The distribution of lead in hypersolvus granites. M, 1978, Pennsylvania State University, University Park. 90 p.

Lewis, Amy Heywood. Depositional environment of the Harlem Coal Bed, Conemaugh Series, Pennsylvanian System, in Ohio. M, 1985, University of Toledo. 177 p.

Lewis, Anthony J. Geomorphic evaluation of radar imagery of southeastern Panama and northwestern Colombia. D, 1971, University of Kansas. 178 p.

Lewis, Arthur Edward. Geology and mineralization connected with intrusion of a quartz monzonite porphyry, Iron Mountain, Iron Springs District, Utah. D, 1958, California Institute of Technology. 75 p.

Lewis, Bernard A. A petrographic study of the relative durabilities of sand sized sediments from the upper Merced River, California. M, 1974, Southern Illinois University, Carbondale. 180 p.

Lewis, Brian Thomas Robert. An isostatic model for the U.S.A. derived from gravity and topographic data. D, 1970, University of Wisconsin-Madison. 52 p.

Lewis, C. L. The minor elements of the Sudbury ore minerals. M, 1950, Queen's University. 92 p.

Lewis, Catherine Louise. Computer contouring of orientation data and its application to paleocurrent analysis. M, 1987, University of Connecticut. 132 p.

Lewis, Charles Downing, Jr. A paleomagnetic investigation of three intrusions in Big Bend National Park; Brewster County, Texas. M, 1976, University of Oklahoma. 105 p.

Lewis, Charles Downing, Jr. Seismic models of geopressured natural gas reservoirs. D, 1983, University of Oklahoma. 92 p.

Lewis, Charles Frederick Michael. Reconnaissance geology of the Lake Erie basin. M, 1963, University of Toronto.

Lewis, Charles Frederick Michael. Sedimentation studies of unconsolidated deposits in the Lake Erie Basin. D, 1967, University of Toronto.

Lewis, D. M. The geochemistry of manganese, iron, uranium, lead-210 and major ions in the Susquehanna River. D, 1976, Yale University. 287 p.

Lewis, Dana Lyn. Pleistocene seismic stratigraphy of the Galveston South addition, offshore Texas. M, 1984, Rice University. 152 p.

Lewis, Daniel D. An investigation of the preferred orientation of phyllosilicates in the Martinsburg Slate, Lehigh Gap area, Pennsylvania. M, 1980, Indiana University, Bloomington. 96 p.

Lewis, David V. Relationships of ore bodies to dikes and sills. M, 1954, University of Minnesota, Minneapolis. 90 p.

Lewis, Dion A. The geochemistry of cadmium in selected NH ponds. M, 1984, University of New Hampshire. 70 p.

Lewis, Don W. Preliminary stratigraphy of the Pungo River Formation of the Atlantic continental shelf, Onslow Bay, North Carolina. M, 1981, East Carolina University. 75 p.

Lewis, Donald Austin. Subsidence and mine drainage consequences of underground coal mining. D, 1980, University of California, Los Angeles. 210 p.

Lewis, Donald W. Regional stratigraphic analysis of the "Morrow" unit of the western Anadarko Basin. M, 1959, Northwestern University.

Lewis, Douglas W. The paragenesis of a glauconite in the Bliss Formation, Silver City, New Mexico. M, 1961, University of Houston.

Lewis, Douglas Windsor. The Potsdam Sandstone (Cambrian), southern Quebec. D, 1965, McGill University.

Lewis, Eric S. Trace element content of surface and groundwaters in Northwest Ohio in relation to mississippi valley-type mineralization. M, 1980, Bowling Green State University. 115 p.

Lewis, Fletcher Sherwood. A reservoir study of the Blooming Grove Field, Black Warrior Basin. M, 1978, University of Oklahoma. 96 p.

Lewis, Fletcher Sherwood. Texture, clay mineralogy and biogenic composition of Taiwan shelf and slope sediments. M, 1975, University of Oregon. 166 p.

Lewis, George E. Siwalik fossil anthropoids (India and Pakistan). D, 1937, Yale University.

Lewis, Glenn Charles. Chemical and mineralogical study on slick spot soils. D, 1962, Purdue University. 119 p.

Lewis, J. F. Oceanic heat flow measurements over the continental margins of eastern Canada. M, 1975, [Dalhousie University].

Lewis, Jackson Ellis. Paleoecological study of a Pleistocene marine fauna, Flagler County, Florida. M, 1964, University of Florida. 221 p.

Lewis, James Albert. Flow of fluids through unconsolidated materials. M, 1932, Pennsylvania State University, University Park. 46 p.

Lewis, James Otis. A study of the geology of Preston (Boyle) ore in Bath County, Kentucky. M, 1949, University of Kentucky.

Lewis, Jean. Geology of Barillos Dome, Jeff Davis County, Trans-Pecos, Texas. M, 1949, University of Texas, Austin.

Lewis, Jean. Pleistocene hydrogeology of the dissected till plains, North central Missouri. M, 1982, University of Missouri, Columbia.

Lewis, Jerry D. K/Rb ratios of some Precambrian granulates; implication toward Rb depletion in the lower crust. M, 1971, Michigan State University. 42 p.

Lewis, John Hubbard. Petrology and diagenesis of Upper Cambrian rocks of central and western Colorado. D, 1965, University of Colorado. 201 p.

Lewis, John Richard. Structure and stratigraphy of the Rossie complex (Precambrian), northwest Adirondacks, New York. D, 1969, Syracuse University. 161 p.

Lewis, Jonathan C. Engineering characteristics of lacustrine deposits involving coal strip mining in Southwest Indiana. D, 1988, Purdue University. 238 p.

Lewis, Jonathan C. Structural geology and finite strain analysis of the Precambrian Thunderhead Sandstone along the Greenbrier Fault and the Roundtop Klippe; Great Smoky Mountains, Tennessee. M, 1988, University of Tennessee, Knoxville. 186 p.

Lewis, Joseph Thomas. Structural relationships of pegmatite to wall rock in the Southern Appalachians. M, 1958, University of Illinois, Urbana. 74 p.

Lewis, Laurel M. A gravity study of north-central New Mexico. M, 1980, University of Texas at El Paso.

Lewis, Laurence A. The relations of hydrology and geomorphology in a humid tropical stream basin - The Rio Grande de Manati, Puerto Rico. D, 1965, Northwestern University. 139 p.

Lewis, Linda L. Stratigraphic palynology of the uppermost Cretaceous and Paleocene formations near Golden, Colorado. M, 1978, Colorado School of Mines. 65 p.

Lewis, Lloyd A. Geology of the northern part of the Santa Ana Mountains, Orange County, California. M, 1941, California Institute of Technology. 66 p.

Lewis, Lloyd Allen. A microscopic study of vein carbonate. M, 1938, University of Minnesota, Minneapolis. 31 p.

Lewis, Lloyd F. Speed of sound in unconsolidated sediments of Boston Harbor. M, 1967, Massachusetts Institute of Technology. 62 p.

Lewis, M. A. Influence of an open-pit copper mine on the ecology of an upper Sonoran intermittent stream. D, 1977, Arizona State University. 122 p.

Lewis, Mordecai, II. Geology of the Catoctin Belt in the vicinity of Charlottesville, Albemarle County, Virginia. M, 1926, University of Virginia. 101 p.

Lewis, Norman M., Jr. The geology of the Little Rock Candy Mountain area, Yavapai County, Arizona. M, 1973, Northern Arizona University. 78 p.

Lewis, Patrick R. Structural geology of the northern Burnt Springs Range and Robber Roost Hills, Lincoln County, Nevada. M, 1987, Northern Arizona University. 90 p.

Lewis, Paul Heywood. Ultramafic inclusions in Limburgite, Hopi Buttes volcanic field, Arizona. M, 1973, Brigham Young University. 225 p.

Lewis, Paul J. Structure south of Wapiti River, Wapiti Lake area, British Columbia, Canada. M, 1948, University of Kansas. 62 p.

Lewis, Paul S. Igneous petrology and strontium isotope geochemistry of the Christmas Mountains, Brewster County, Texas. M, 1978, University of Texas, Austin.

Lewis, Peter D. Polyphase deformation and metamorphism in the western Cariboo Mountains near Ogden Peak, British Columbia. M, 1987, University of British Columbia. 132 p.

Lewis, Ralph S. Investigation of lead in the soils of Lancaster County, Pennsylvania. M, 1974, Franklin and Marshall College.

Lewis, Reed Stone. Geology of the Cape Horn Lakes Quadrangle, south-central Idaho. M, 1984, University of Washington. 84 p.

Lewis, Richard. Water quality in the three Cassadaga lakes, Chautauqua County. M, 1977, SUNY, College at Fredonia. 183 p.

Lewis, Richard Dale. Geochemical investigations of the Yellow Pine, Idaho and Republic, Washington, mining districts. D, 1984, Purdue University. 204 p.

Lewis, Richard Edwin. Geology of the Hackberry Mountain volcanic center, Yavapai County, Arizona. D, 1983, California Institute of Technology. 398 p.

Lewis, Richard Quintin. The geology of the southern Coburg Hills including the Springfield-Goshen area. M, 1950, University of Oregon. 58 p.

Lewis, Richard Timothy. Brachiopods of the Nancy Member of the Borden Formation (Mississippian), northeastern Kentucky. M, 1986, Wright State University. 276 p.

Lewis, Richard Wheatley, Jr. The geology, mineralogy, and paragenesis of the Castrovirreyna lead-zinc-silver deposits, Peru. D, 1964, Stanford University. 265 p.

Lewis, Robert E. Resistivity study of landslide near Alum Rock Park, Santa Clara County, California. M, 1967, San Jose State University. 81 p.

Lewis, Robert Harry. Electrical conductivity measurements of high temperature silicate melts. M, 1985, Washington University. 184 p.

Lewis, Roger James Gollan. Magnetotelluric studies in Montana and Wyoming. D, 1970, Princeton University. 91 p.

Lewis, Ronald Dale. Depositional environments and paleoecology of the Oil Creek Formation (Middle Ordovician), Arbuckle Mountains and Criner Hills, Oklahoma. D, 1982, University of Texas, Austin. 370 p.

Lewis, Ronald Dale. Studies in the inadunate crinoid family Pirasocrinidae. M, 1974, University of Iowa. 181 p.

Lewis, Sally Beth. The lithology, depositional environment and dolomitization of the Louisiana Formation. M, 1975, Washington University. 55 p.

Lewis, Sharon E. Geology of the southern part of the Riner Quadrangle, Montgomery and Floyd counties, Virginia. M, 1975, North Carolina State University. 106 p.

Lewis, Sharon Elizabeth. Geology of the Brevard Zone, Smith River Allochthon and Inner Piedmont in the Sauratown Mountains Anticlinorium, northwestern North Carolina. D, 1980, University of North Carolina, Chapel Hill. 131 p.

Lewis, Standley Eugene. Fossil insects of the Latah Formation (Miocene) of eastern Washington and northern Idaho. D, 1968, Washington State University. 97 p.

Lewis, Stanley Royce. Significance of the vertical and lateral changes in the clay mineralogy of the Dunbarton Triassic basin, South Carolina. M, 1974, University of North Carolina, Chapel Hill. 34 p.

Lewis, Stephen Dana. On the tectonics of small plate interactions; northern Philippines. D, 1982, Columbia University, Teachers College. 270 p.

Lewis, Sue Jane Lin see Lin Lewis, Sue Jane

Lewis, T. J. A geothermal survey at Lake Dufault, Quebec. D, 1975, University of Western Ontario.

Lewis, Thomas L. Geology of the Penn Yan Quadrangle, New York. M, 1958, University of Rochester. 105 p.

Lewis, Thomas Leonard. A paleocurrent study of the Potsdam Sandstone of New York, Quebec, and Ontario. D, 1963, Ohio State University. 174 p.

Lewis, W. L. A study of the angular veins from the Gold River gold district, Nova Scotia. M, 1960, Acadia University.

Lewis, Wardell L. Effect of aqueous surfactants on crack propagation rate in Crab Orchard Sandstone. D, 1976, University of North Carolina, Chapel Hill. 48 p.

Lewis, Wardell Lavon. Geology of the Knob Lick Field, Metcalfe County, Kentucky. M, 1960, University of Kentucky. 55 p.

Lewis, Warren S. Geology of uranium mineralization in the Browns Park Formation, Carbon County, Wyoming and Moffat County, Colorado. M, 1977, Colorado School of Mines. 85 p.

Lewis, William D. The geology of the upper Las Llajas Canyon area, Santa Susana Mountains, California. M, 1940, California Institute of Technology. 74 p.

Lewison, Maureen Ann. Fission track ages of two plutons in the central Klamath Mountains, California. M, 1984, University of Oregon. 76 p.

Lexa, David J. An analysis of the Bourbeuse River as a source of concrete aggregate. M, 1974, University of Missouri, Rolla.

Ley, Heber Cinco see Cinco Ley, Heber

Ley, J. W. Vander see Vander Ley, J. W.

Leybourne, Matthew Iain. Volcanism and geochemistry of parts of the Endeavour segment of the Juan de Fuca Ridge system and associated seamounts. M, 1988, Acadia University. 177 p.

Leyden, Barbara Wilhelmina. Late-Quaternary and Holocene history of the Lake Valencia Basin, Venezuela. D, 1982, Indiana University, Bloomington. 104 p.

Leyenberger, Terry. Precambrian geology of Cimarron Canyon, Colfax County, NM. M, 1984, University of New Mexico. 93 p.

Leyland, James G. Quaternary geology of the Campbellford, Trenton, Consecon, Tweed, Belleville, Wellington, Sydenham, Bath, and Yorkshire Island map-areas, Ontario. M, 1984, Brock University. 68 p.

Leytham, Keith Malcolm. Physical considerations in the analysis and synthesis of hydrologic sequences. D, 1982, University of Washington. 238 p.

Lezak, Jennifer Linn. Variation of tooth morphology in Sciurus niger and Citellus tridecemlineatus and Miocene sciurids from the Texas coastal plain. M, 1979, Southern Methodist University. 54 p.

Lhotka, Paul Gordon. Geology and geochemistry of gold-bearing iron formation in the Contwoyto Lake-Point Lake region, Northwest Territories, Canada. D, 1988, University of Alberta. 283 p.

Li, Ching-Chang E. Studies of ground water quality under a sanitary landfill. D, 1975, University of Oklahoma. 106 p.

Li, Ching-Yuan. Genesis of some ore deposits of southeastern Maine. D, 1941, Columbia University, Teachers College.

Li, Ching-Yuan. Preliminary study of the shoreline of China. M, 1938, Columbia University, Teachers College.

Li, Fu Shung. Analysis and application of new array techniques for seismic signal extraction. D, 1976, University of Minnesota, Minneapolis. 248 p.

Li, Fu-Shung. Interpretation of a gravity profile across the Midcontinent gravity high. M, 1971, University of Minnesota, Minneapolis. 81 p.

Li, Gordon Chi-Kwong. Free surface flow and stress analysis of earth dams. D, 1981, Virginia Polytechnic Institute and State University. 232 p.

Li, Huilin. Remagnetization of the Allouez Conglomerate in the Portage Lake Volcanics in Michigan. M, 1987, Michigan Technological University. 45 p.

Li, John Chien-Chung. Dynamic properties of frozen granular soils. D, 1979, Michigan State University. 316 p.

Li, Pun-yuk Daniel. Clay mineralogy at sites 33, 34, 40, and 42, Leg 5, Deep Sea Drilling Project. M, 1974, SUNY at Binghamton. 77 p.

Li, Shih Chang. The Miocene and Recent Mollusca of Panama Bay. M, 1926, Columbia University, Teachers College.

Li, T. M. Mantle conductivity models and long period magnetic variations. M, 1975, Boston College.

Li, Todd Ming Chun. Axisymmetric numerical simulation of hydrothermal systems including changes in porosity and permeability due to the quartz-water reaction. D, 1980, Pennsylvania State University, University Park. 270 p.

Li, Wan-Bing. Major and trace element geochemistry of Archean high grade metamorphic rocks from Inner Mongolia, China. M, 1987, University of Toronto.

Li, Yianping. Paleomagnetism of western China and the southern Sierra Nevada. D, 1988, Stanford University. 245 p.

Li, Yong-Gang. Seismic wave propagation in anisotropic media with applications to defining fractures in the Earth. D, 1988, University of Southern California.

Li, Yuesheng. Airgun signature estimation and wavelet processing of marine seismic data. M, 1987, Colorado School of Mines. 155 p.

Li, Zhenlin. Pivoting angles of gravel with applications in sediment threshold studies. M, 1986, Oregon State University. 101 p.

Li, Zhiming. Imaging steep-dip reflections by the linearly transformed wave equation method. D, 1986, Stanford University. 103 p.

Li, Zhongxue. Determining the size and life of underground coal mines. D, 1987, Virginia Polytechnic Institute and State University. 152 p.

Lian, Harold Maynard. The geology and paleontology of the Carpenteria District, Santa Barbara County, California. D, 1952, University of California, Los Angeles.

Liang, Dah-Ben. Ultrasonic wave velocity through some rocks under triaxial compression at various degrees of water saturation. M, 1971, University of Utah. 154 p.

Liang, Duohaw. A nonlinear frequency domain method for estimating acoustic attenuation in sediments. D, 1987, University of Rhode Island. 92 p.

Liang, George Ching-Chi. Evolutionary spectra for strong motion body and surface waves. D, 1980, University of California, Los Angeles. 454 p.

Liang, Huh-Yuan. Geologically-developed probabilistic seismic risk analysis. D, 1983, University of Missouri, Rolla. 242 p.

Liang, Long-Cheng. A geochemical study of some Texas lignites. M, 1982, University of Texas at Dallas. 154 p.

Liang, Luh-Cheng. Three-dimensional seismic modeling; velocity analysis and interpretation. D, 1981, University of Houston. 230 p.

Liang, Yueh. Strength of field compacted clayey embankments. D, 1981, Purdue University. 342 p.

Liao, Amy Hueymei. Anisotropy in the upper mantle of Eurasia. D, 1981, University of California, Los Angeles. 191 p.

Liao, Ching-Yi. 2-D surface to surface tomographic velocity inversion based on a polynomial parameterization using wave slowness data. M, 1985, University of Texas at Dallas. 150 p.

Liao, Jih-Sheng. Stability of near-surface excavations in weak rock and soil. D, 1988, University of Wisconsin-Madison. 208 p.

Liao, Kao Hsiung. Statistical models of porous media and other hydrological systems. D, 1969, University of Illinois, Urbana. 114 p.

Liao, Wen-Gen. The behavior of submerged, multiple bodies in earthquakes. D, 1982, University of California, Berkeley. 128 p.

Liao, Yu Jen. Metalliferous deposits of China. M, 1932, University of Minnesota, Minneapolis. 240 p.

Liard, A. C. A study of the energy-transport relationship and a computer simulation of Long Point, Lake Erie. M, 1975, University of Waterloo.

Liaskos, Dimitrios Anastasios. Resources estimation from historical data; upper Mississippi Valley, Tri-State, and mid-Tennessee base-metal districts; zinc, lead test cases. M, 1984, University of Wisconsin-Milwaukee. 167 p.

Liaw, Alfred L. Microseisms in geothermal exploration; studies in Grass Valley, Nevada. D, 1977, University of California, Berkeley. 188 p.

Liaw, Hong-Bing. Seismic velocity modeling from an ensemble of earthquakes. D, 1981, University of Texas at Dallas. 135 p.

Liaw, Liang-Chi. The effect of anaerobic algal decomposition on the interstitial water chemistry of Recent marine sediments. M, 1973, Southern Methodist University. 67 p.

Liaw, Zen-Sen. A modified cepstral method and its application to DWWSSN broadband data. M, 1984, Pennsylvania State University, University Park. 76 p.

Libby, Frederick Ernest. A study of caliche. M, 1951, Texas Tech University. 33 p.

Libby, Stephen Charles. The origin of potassic ultramafic rocks in the Enoree "vermiculite" district, South Carolina. D, 1975, Pennsylvania State University, University Park. 136 p.

Libby, Stephen Charles. The petrology of the igneous rocks of Putnam County, Georgia. M, 1971, University of Georgia. 99 p.

Libby, Willard G. Stratigraphic relationships in the Muddy Sandstone of Wyoming. M, 1959, Northwestern University.

Libby, Willard Gurnea. Petrography and structure of the crystalline rocks between Agnes Creek and the Methow Valley, Washington. D, 1964, University of Washington. 133 p.

Libert, John M. Shape characterization of loess; Fourier grain-shape analysis. M, 1982, University of Maryland.

Liberty, Bruce Arthur. A study of the family Unionidae from the Upper Cretaceous rocks of Western Canada (Saskatchewan and Alberta). M, 1949, University of Toronto.

Liberty, Bruce Arthur. Stratigraphy and palaeontology of the Lake Simcoe District, Ontario. D, 1954, University of Toronto.

Libes, Susan M. Stable isotope geochemistry of nitrogen in marine particulates. D, 1983, Massachusetts Institute of Technology. 288 p.

Libicki, Charles Melvin. Acoustic sensing of the vertical and temporal structure of sediment transport in the benthic boundary layer. D, 1986, Ohio State University. 236 p.

Libicki, Charles Melvin. Barium uptake by marine diatoms. M, 1978, Massachusetts Institute of Technology. 95 p.

Libra, Robert D. Hydrogeology and sulfur isotope variations of spring systems, South-central Indiana. M, 1981, Indiana University, Bloomington. 109 p.

Licari, Gerald Richard. Geology and amber deposits of the Simojovel area, Chiapas, Mexico. M, 1960, University of California, Berkeley. 76 p.

Licari, Gerald Richard. Paleontology and paleoecology of the Proterozoic Beck Spring dolomite of eastern California. D, 1971, University of California, Los Angeles. 193 p.

Licari, Joan Perusse. Environmental management of Pacific outer continental shelf oil and gas activities by the Minerals Management Service. D, 1983, University of California, Los Angeles. 249 p.

Licari, Joan Perusse. Foraminifera from the Simojovel region, Chiapas, Mexico. M, 1965, University of California, Berkeley. 106 p.

Licastro, Pasquale Hallison. Dielectric behavior of rocks and minerals. D, 1959, Pennsylvania State University, University Park. 167 p.

Licastro, Pasquale Hallison. Use of radio frequencies in the study of geologic structures. M, 1951, Pennsylvania State University, University Park. 47 p.

Lichaa, Pierre Michel. Rock properties using wave propagation techniques. D, 1970, University of Texas, Austin. 192 p.

Lichtler, William F. Groundwater resources of the Stuart area, Martin County, Florida. M, 1958, Syracuse University.

Lichtman, Grant S. Photogeologic mapping of the mid-ocean ridge; the East Pacific Rise, 21° N. M, 1980, Stanford University. 41 p.

Lickus, Robert John. Geology and geochemistry of the ore deposits at the Vauze Mine, Noranda District, Quebec. D, 1965, McGill University.

Lico, Michael. Lower Permian submarine sediment gravity flows in the Owens Valley Formation, southeastern California. M, 1983, San Jose State University. 80 p.

Lidback, M. M. Areal geology of the Attleboro, Massachusetts-Rhode Island, Quadrangle. D, 1977, Boston University. 242 p.

Liddell, Jessie Kelsey. A preliminary study of Frio Formation in the Rio Grande Embayment. M, 1943, University of Oklahoma. 35 p.

Liddell, William David. Biostratinomy and ecology of a middle Ordovician echinoderm assemblage from Kirkfield, Ontario. M, 1975, University of Michigan.

Liddicoat, Joseph C. Steady-state thermal gradients in New England lake sediments; their applicability for determining goethermal heat flow. M, 1970, Dartmouth College. 107 p.

Liddicoat, Joseph Carl. A paleomagnetic study of late Quaternary dry-lake deposits from the western United States and Basin of Mexico. D, 1976, University of California, Santa Cruz. 495 p.

Liddicoat, William Keith. Regional intensity zones and zoning by genetic affiliation in the Quebec-Ontario gold belt. M, 1953, University of Michigan.

Liddle, Susan Krongold. Trace-element analysis of the groundwater at a hazardous-waste landfill in the Piedmont of North Carolina. M, 1984, North Carolina State University. 83 p.

Lide, Chester Scott. Aftershocks of the May, 1980 Mammoth Lakes, California, earthquakes. M, 1984, University of Nevada. 78 p.

Lidgard, Scott. Taxonomic survivorship of late Cenozoic planktonic foraminifera. M, 1978, University of Rochester. 48 p.

Lidgard, Scott Harrison. Evolution of growth and form in encrusting cheilostome bryozoans. D, 1985, The Johns Hopkins University. 188 p.

Lidiak, Edward George. Petrology of the andesitic, spilitic and keratophyric lavas, north-central Puerto Rico. D, 1963, Rice University. 123 p.

Lidiak, Edward George. Precambrian geology of parts of the Little Llano River valley, Llano and San Saba counties, Texas. M, 1960, Rice University. 85 p.

Lidke, David J. Geology, structure, and geometrical analysis of structural relations along the south part of the Georgetown Thrust, Southwest Montana. M, 1985, University of Colorado. 104 p.

Lidstone, Christopher D. The development and distribution of alluvial placer deposits. M, 198?, Colorado State University. 208 p.

Lidstrom, John Walter, Jr. A new model for the formation of Crater Lake Caldera, Oregon. D, 1972, Oregon State University. 85 p.

Lidz, Louis. Investigation into the marine sedimentary environment. M, 1967, University of Miami.

Lidz, Louis. Sedimentary, environmental and foraminiferal parameters, Nantucket Bay, Massachusetts. M, 1963, University of Southern California.

Lie, G. B. Aspects of the ecology and physiology of freshwater macrophytes; phosphorus cycling by freshwater macrophytes; the case of Shagawa Lake. D, 1977, University of Minnesota, Minneapolis. 72 p.

Lieb, Carl Varney. Petroleum production of the world exclusive of Canada, United States and Mexico. M, 1942, University of Texas, Austin.

Liebe, Richard Milton. Conodonts from the Alexandrian and Niagaran series (Silurian) of the Illinois Basin. D, 1962, University of Iowa. 162 p.

Liebe, Richard Milton. Conodonts from the Renault Formation (Chester) of the Illinois Basin. M, 1959, University of Houston.

Liebe, William Mather. Petrographic and geochemical analysis of diagenesis, Pitkin Limestone, (Mississippian), Washington and Madison counties, Arkansas. M, 1983, University of Arkansas, Fayetteville. 81 p.

Liebelt, Michael F. Seasat SAR evaluation in New York, Pennsylvania and Arizona. M, 1981, University of Arkansas, Fayetteville.

Lieber, Paul. Relaxation phenomena and the origin of earthquakes. D, 1951, California Institute of Technology. 108 p.

Lieber, Paul. Temperature perturbations and their effect on the temperature maxima and minima in the interior of the Earth. M, 1941, California Institute of Technology.

Lieber, Robert Barry. Paleoenvironmental aspects of Lower Mississippian waulsortian type mounds of the Fort Payne Formation in northern Tennessee. M, 1978, University of Kentucky. 95 p.

Lieberman, Joshua Elliot. Metamorphic and structural studies of the Kigluaik Mountains, western Alaska. D, 1988, University of Washington. 191 p.

Lieberman, Joshua Elliot. Petrology and petrogenesis of marble and peridotite, Seiad ultramafic complex, California. M, 1983, University of Oregon. 120 p.

Lieberman, Kenneth Warren. The determination of bromine in terrestrial and extraterrestrial materials by neutron activation analysis. D, 1967, [University of Kentucky]. 180 p.

Lieberman, Marcus. The mineralogy and petrology of West Rock Ridge, Connecticut. M, 1974, Brooklyn College (CUNY).

Lieberman, S. H. Stability of copper complexes with seawater humic substances. D, 1979, University of Washington. 228 p.

Liebermann, Robert Cooper. Effect of iron content upon the elastic properties of oxides and some applications to geophysics. D, 1969, Columbia University, Teachers College. 173 p.

Liebes, Eric. Direct methods for determining the timing of magnetization in redbeds and isothermal remanent magnetization acquisition spectra for some naturally occurring hematites. D, 1981, University of Wyoming. 90 p.

Liebes, Eric. Structural implications of the paleomagnetism of the uppermost Catskill Formation, northeastern Pennsylvania. M, 1978, SUNY at Binghamton. 177 p.

Liebfreid, Doris Jean. An investigation of the fossils of the Allegheny Formation from selected sites in the Johnstown, Pennsylvania area. M, 1969, Indiana University of Pennsylvania. 57 p.

Lieblang, Sean. Sedimentology and petrology of conglomerate and sandstone in the Oligocene White River Formation, central Wyoming. M, 1983, Northern Arizona University. 177 p.

Liebling, Richard S. A thermal investigation of tetrahedrite. M, 1961, Columbia University, Teachers College.

Liebling, Richard Stephen. Glacial and postglacial quick clays. D, 1963, Columbia University, Teachers College. 73 p.

Lienert, B. R. Electrical conductivity in the crust of the western United States inferred from controlled source electromagnetic deep sounding data. D, 1976, University of Texas at Dallas. 201 p.

Lienhart, David A. The orientation-compensation method of mineral identification. M, 1965, University of Cincinnati. 26 p.

Lienkaemper, George William. Geomorphic and biological effects of accumulated debris in streams of the western Cascades. M, 1976, University of Oregon. 50 p.

Lierman, Robert Thomas. Environment of deposition and diagenetic history of the Warix Run and Paoli-Beaver Bend limestones (Upper Mississippian) of east-central Kentucky. M, 1984, Miami University (Ohio). 209 p.

Liesch, Aaron Robert. The depositional history of the lower Deese Group (Middle Pennsylvanian), Ardmore Basin, Oklahoma. M, 1988, University of Oklahoma. 220 p.

Liese, Homer C. Geology of the northern Mineral Range, Millard and Beaver counties, Utah. M, 1957, University of Utah. 88 p.

Liese, Homer C. Indirect geothermometric mineral studies of selected silicic igneous rocks. D, 1962, University of Utah. 93 p.

Lietzke, David A. The origin, distribution, and dynamic character of some chloritized vermiculite soil clays (Berrien County, Michigan). D, 1972, Michigan State University. 158 p.

Lieu, Junius A. Van see Van Lieu, Junius A.

Liew, M. J. C. Geochemical studies of the Goldenville Formation at Taylor Head, Nova Scotia. M, 1979, Dalhousie University.

Liew, M. Y.-C. Structure, geochemistry, and stratigraphy of Triassic rocks, north shore of Minas Basin, Nova Scotia. M, 1976, Acadia University.

Liew, Michael Wayne Van see Van Liew, Michael Wayne

Lifrieri, L. Sedimentology of the Upper Devonian upper Walton Formation, near Hancock, south-central New York State. M, 1983, SUNY at Binghamton. 102 p.

Lifschutz, Arthur Paul. Glacial geology of northeastern Tioga County, Pennsylvania. M, 1961, Pennsylvania State University, University Park. 59 p.

Lifshin, Arthur. A study of the relationship between the metamorphically recrystallized cherts of the Negaunee Iron Formation and metamorphic grade. M, 1963, Michigan State University. 75 p.

Lifshin, Arthur. Element migration across granitic dikes. D, 1969, Michigan State University. 125 p.

Ligasacchi, Attilio. A study of the genesis of the Krueger zinc deposit and the near-by barite deposits of the Potosi Quadrangle, Washington County, Missouri. M, 1959, University of Missouri, Rolla.

Ligasacchi, Giovanna R. A review of fossilization processes in different sedimentary environments. M, 1959, University of Missouri, Rolla.

Liggett, David Lee. Geology and geochemistry of a garnet-bearing granitoid in the southwestern Sierra Nevada, Tulare County, California. M, 1987, California State University, Northridge. 143 p.

Liggon, George Herbert. Petrology and depositional environments of some Triassic sediments in North Carolina. M, 1972, North Carolina State University. 99 p.

Light, Aaron Mitchell. A petrographic study of the sands of the Barnegat Bar area, New Jersey. M, 1947, University of Missouri, Columbia.

Light, Mitchell Arron. Glauconite of the New Jersey coastal plain. D, 1950, Rutgers, The State University, New Brunswick. 244 p.

Light, Thomas D. Geology of the Board Creek area, Yavapai County, Arizona. M, 1975, Northern Arizona University. 61 p.

Light, William George. Origin and development of the Ohio River. D, 1902, University of Chicago.

Lightcap, Dixon Samuel, II. Relationship between fold geometry, depth of cover and volatile matter content in the upper and lower Freeport coal seams in Cambria, Indiana, and Westmoreland counties of Pennsylvania. M, 1986, Indiana University of Pennsylvania.

Lightfoot, Peter Charles. The geology of the Tabankulu section of the Insizwa Complex, Transkei. M, 1982, University of Toronto.

Lightner, John Gwin, III. A mixed finite element procedure for soil-structure interaction including construction sequences. D, 1981, Virginia Polytechnic Institute and State University. 248 p.

Lightner, Jon T. Storm sediment transport as indicated by benthic foraminifera, north insular shelf, Puerto Rico. M, 1988, Duke University. 115 p.

Lighty, Robin Greg. Depositional and diagenetic history of an early Holocene relict shelf-edge coral reef; southeast coast of Florida. M, 1977, Duke University. 92 p.

Liias, Raimo Arnold. Geochemistry and petrogenesis of basalts erupted along the Juan de Fuca Ridge. D, 1986, University of Massachusetts. 293 p.

Likarish, Daniel Matthew. A magnetic profile of a Cascade volcano, Mount Baker, Washington. M, 1978, University of Washington. 59 p.

Lilburn, Ralph Anthony. Mineralogical, geochemical and isotopic evidence of diagenetic alteration attributable to hydrocarbon migration, Cement-Chickasha Field, Oklahoma. M, 1981, Oklahoma State University. 88 p.

Lile, Thomas Craig. The ostracods of the Bluffport Marl Member of the Demopolis Chalk at the type locality in Sumter County, Alabama. M, 1963, University of Alabama.

Lilga, Mary Colburn. Coastal process and change, Lake Erie, near Dunkirk, New York. M, 1984, SUNY, College at Fredonia. 140 p.

Lilienthal, Richard. A trace element analysis of Silurian carbonate rocks. M, 1972, Wayne State University.

Lilje, Anneliese. Quantitative estimates of compaction in the calcareous ooze, chalk, limestone sequence. M, 1986, University of California, Riverside. 67 p.

Lill, Gordon G. A glacio-fluvial terrace in Marshall and Washington counties, Kansas. M, 1946, Kansas State University. 84 p.

Lillard, Douglas Ray. A study of earthquake P phases reflected and diffracted from the Earth's outer core. M, 1966, Texas Tech University. 23 p.

Lillegraven, Jason A. The stratigraphy, structural geology, vertebrate paleontology and paleoecology of the Brule Formation, Slim Buttes, South Dakota. M, 1964, South Dakota School of Mines & Technology.

Lillegraven, Jason Arthur. The latest Cretaceous mammals of the upper part of the Edmonton Formation of Alberta, Canada, and a review of the marsupial-placental dichotomy in mammalian evolution. D, 1968, University of Kansas. 342 p.

Lilley, F. E. M. An analysis of the magnetic features of the Port Coldwell intrusive, Ontario. M, 1965, University of Western Ontario.

Lilley, Marvin Douglas. Studies on the marine chemistry of reduced trace gases. D, 1983, Oregon State University. 179 p.

Lilley, Wesley Wayne. Geology and depositional environments of the Cobble Conglomerate in the Scout Mountain Member of the Precambrian Pocatello Formation, Bannock County, Idaho. M, 1972, Idaho State University. 61 p.

Lilley, Wesley Wayne. Stratigraphy and paleoenvironments of the Pterocephaliid Biomere (upper Cambrian) of western Utah and eastern Nevada. D, 1976, University of Kansas. 187 p.

Lillie, John T. An analysis of selected gravity profiles on the Hanford Reservation, Richland, Washington. M, 1977, Wright State University. 87 p.

Lillie, R. J. Subsurface geologic structure of the Vale, Oregon, known geothermal resource area from the interpretation of seismic reflection and potential field data. M, 1977, Oregon State University. 52 p.

Lillie, Robert James. Early Paleozoic continental margin of North America; implications from seismic profiling of the Appalachian/Ouachita orogenic belt. D, 1984, Cornell University. 195 p.

Lillis, P. G. Petrography, geochemistry, and structure of the Corte Madera gabbro pluton, San Diego County, California. M, 1978, San Diego State University.

Lilly, Hugh Dalrymple. Geology of the Goose Arm-Hughes Brook area. M, 1961, Memorial University of Newfoundland. 123 p.

Lilly, Richard J. The geology of the Richmond-Sitting Bull Mine, South Dakota. M, 1923, University of Minnesota, Minneapolis. 31 p.

Lilo, Yehuda. Diagenetic processes affecting Pleistocene faviid and poritid corals from southern Sinai; a trace element and petrographic study. M, 1981, University of Kansas. 65 p.

Lim, Chin Huat. Amorphous materials, crystalline mixed layers, and impurities in clays. D, 1981, University of Wisconsin-Madison. 175 p.

Lim, Jose Bernardo R. Paleomagnetism of the Pomona-Weippe basalt flow in southeastern Washington and west-central Idaho. M, 1986, Washington State University. 181 p.

Lim, S. K. Effects of dehydration on physico-chemical properties of selected volcanic ash soils from Hawaii. M, 1976, University of Hawaii. 121 p.

Lim, Sheldon C. P. The effect of a porous layer on the kinetics of decomposition of calcite. M, 1975, University of California, Berkeley.

Lim, Sung J. An experimental investigation of pipeline stability in very soft clay. D, 1974, [University of Houston]. 149 p.

Lima, Edmilson Santos de *see* de Lima, Edmilson Santos

Limbach, Fred W. The geology of the Buena Vista area, Chaffee County, Colorado. M, 1975, Colorado School of Mines. 98 p.

Limentani, Giselle Beth. Chromatographic separation and identification of compounds in shale oil and oil shale with specific element detection. D, 1984, University of Massachusetts. 226 p.

Limerick, C. J., Jr. An intercontinental tie determined from observation of an artificial satellite. M, 1962, Ohio State University.

Limerick, Samuel Hazzard. Animal-sediment relationships on Tanner and Cortes banks, California continental borderland. M, 1978, University of Southern California.

Liming, Richard Brett. Geology and kinematic analysis of deformation in the Martinez Ranch area, Pima County, Arizona. M, 1974, University of Arizona.

Limoges, L. D. The ecological and economic impact of dredge and fill on Tampa Bay, Florida. D, 1975, University of Florida. 313 p.

Limper, Karl E. A study of the environment of the (Ordovician) Eden in Kentucky. D, 1953, University of Chicago. 155 p.

Lin Lewis, Sue Jane. A mathematical model and integrated simulation for urban flood and land use analysis. D, 1983, University of Oklahoma. 178 p.

Lin, Albert Niu. Experimental observations of the effect of foundation embedment on structural response. D, 1982, California Institute of Technology. 337 p.

Lin, Bea-Yeh. Paleoenvironments and depositional history of the Williamson Shale, Silurian, western New York. M, 1987, University of Rochester. 200 p.

Lin, Chang-Lu. Digital simulation of a stream-aquifer system. D, 1970, University of Illinois, Urbana. 78 p.

Lin, Chang-Lu. Factors affecting groundwater recharge in the Moscow Basin, Latah County, Idaho. M, 1967, Washington State University. 86 p.

Lin, Chen Hsin. Biodegradation of selected phenolic compounds in a simulated sandy surficial Florida aquifer. D, 1988, University of Florida. 182 p.

Lin, Chenfang. Modeling and simulation of phosphate reaction and transport in acid sandy soils. D, 1981, Rutgers, The State University, New Brunswick. 196 p.

Lin, Cheng-Leo George. Thermal evolution modelling of the Pismo Basin, California. M, 1987, University of Illinois, Urbana. 81 p.

Lin, Ching-Weei. Three dimensional photoelastic study of abrupt changes in strike associated with forced folding. M, 1988, University of Oklahoma. 91 p.

Lin, Chong-Pin. Clastic lenses in the Negaunee iron formation (Precambrian) at the Empire Mine, Palmer, Michigan. M, 1969, Bowling Green State University. 105 p.

Lin, Chung-Po. Turbidity currents and sedimentation in closed-end channels. D, 1987, University of Florida. 212 p.

Lin, Eugene Ching-Tsao. A study of micellar/polymer flooding using a compositional simulator. D, 1981, University of Texas, Austin. 501 p.

Lin, Feng-Chih. Dissolution kinetics of some phyllosilicate minerals. D, 1981, SUNY at Buffalo. 155 p.

Lin, Francis Chien-Ming. Statistical study of pressure gasification of coal. M, 1954, West Virginia University.

Lin, Gwo-Fong. Hydraulics of three dimensional flow in a curved channel. D, 1982, University of Pittsburgh. 100 p.

Lin, H. P. The life cycle of Bangia fuscopurpurea; an ultrastructural approach. D, 1977, Arizona State University. 204 p.

Lin, Hsi-Che. An investigation of the stability relations of $BaAl_2Si_2O_8$ (Celsian) in the system $BaO-Al_2O_3-SiO_2$. D, 1967, Ohio State University. 94 p.

Lin, Hsiuan. Permanent settlement prediction of layered systems under repeated surface load. D, 1987, University of Massachusetts. 329 p.

Lin, Hung-Liang. Analog modeling study of inverse scattering. D, 1981, Colorado School of Mines. 106 p.

Lin, Jia-Wen. A study of upper mantle structure in the Pacific Northwest using P waves from teleseisms. D, 1973, University of Washington. 98 p.

Lin, Jia-wen. Gravity and magnetic study, central Ozark uplift, Missouri. M, 1971, Washington University. 54 p.

Lin, Jian-Shengzhong. Determination of the partial pressure of sulfur over metal sulfide assemblages. M, 1969, Brown University.

Lin, Jian-Shengzhong. Potassium self-diffusion in microcline. D, 1971, Brown University.

Lin, Jin-Long. A study of the Watson-Merdler method for focal-depth determination of seismic disturbances from underground nuclear explosions. M, 1966, Pennsylvania State University, University Park. 95 p.

Lin, Jin-Lu. The apparent polar wander paths for the North and South China blocks. D, 1984, University of California, Santa Barbara. 264 p.

Lin, Joseph Tien-Chin. Hydrodynamic flow in Lower Cretaceous muddy sandstone, Gas Draw Field, Powder River basin, Wyoming. M, 1978, Texas A&M University. 102 p.

Lin, Li-Hua. Effect of biodegradation on tar sand bitumen of South Woodford area, Carter County, Oklahoma. M, 1987, University of Oklahoma. 91 p.

Lin, Liang Ching. The kinetics of the pyrolysis of tar sands and of the combustion of coked sands. D, 1988, University of Utah. 257 p.

Lin, Meei-Ling Teresa. Effect of soil compressibility on ocean wave-seafloor interaction. D, 1987, University of Texas, Austin. 219 p.

Lin, Ping-Sien. Compressibility of field compacted clay. D, 1981, Purdue University. 169 p.

Lin, Po-Ming. Investigation of subsidence damages above abandoned mine lands. D, 1988, West Virginia University. 208 p.

Lin, Roscow Ching-Hsing. Flow - shear interaction in open channel. D, 1982, University of Pittsburgh. 106 p.

Lin, Rui. The chemistry of coal maceral fluorescence; with special reference to the huminite/vitrinite group. D, 1988, Pennsylvania State University, University Park. 298 p.

Lin, Saulwood. An experimental study of the solubility and thermodynamic properties of nickel in the system $NiO-HCl-H_2O$. M, 1984, Texas A&M University.

Lin, Sze Chen. Secondary enrichment of tin deposits. M, 1923, University of Minnesota, Minneapolis. 15 p.

Lin, Szu-Bin. The crystal structure and crystal chemistry of scapolites. D, 1971, McMaster University. 232 p.

Lin, Szu-Bin. The system $CaF_2-CaMgSi_2O_6$. M, 1968, McMaster University. 59 p.

Lin, T. C. Minerals associated with the Number Eleven western Kentucky coal seam and their origin. M, 1975, University of Kentucky. 42 p.

Lin, Thomas T. H. Distribution of volcanic ash reflectors in the eastern Equatorial Pacific. M, 1981, University of Georgia.

Lin, Tso-Wang. Sedimentation and self weight consolidation of dredge spoil. D, 1983, Iowa State University of Science and Technology. 121 p.

Lin, Tung-Hung Thomas. Seismic stratigraphy and structure of the Sigsbee salt basin, south-central Gulf of Mexico. M, 1984, University of Texas, Austin. 102 p.

Lin, Tzeu-Lie. Simultaneous inversion of the three-dimensional velocity structure and microearthquake hypocenters in the Coso geothermal area and The Geysers-Clear Lake geothermal area, California. D, 1982, University of Texas at Dallas. 469 p.

Lin, Tzeu-Lie. The Lambertain assumption and Landsat data. M, 1978, [Colorado State University].

Lin, Wei-Hsiung. Development of a multi-factal analytic procedure for the quantification of shape and comparison to Fourier analysis. M, 1987, University of Pittsburgh.

Lin, Wu-Nan. A model study of the shooback electromagnetic Crome prospecting method; 1 volume. M, 1969, University of California, Berkeley.

Lin, Wu-Nan. Velocities of compressional wave in rocks of central California at high pressure and high temperature and applications to the study of the crustal structure of California Coast Ranges. D, 1977, University of California, Berkeley. 190 p.

Lin, Wuu-Jyh. Double beta-decay of selenium-82, tellurium-128 and tellurium-130. D, 1987, University of Missouri, Rolla. 91 p.

Lin, Zsay-Shing. Development of reservoir simulations which handle wellbore storage, infinite conductivity vertical fractures, skin effect, non-Darcy flow and Klinkenberg effect for various boundary conditions and production control policies. D, 1981, University of Kansas. 275 p.

Linclon, James Bruce. Freeze-dried clays. M, 1969, Ohio State University.

Lincoln, Beth Zigmont. A transmission electron microscope study of the development of cleavage in micaceous rocks. D, 1985, University of California, Los Angeles. 132 p.

Lincoln, Francis C. Certain natural associations of gold. D, 1911, Columbia University, Teachers College.

Lincoln, Francis C. Magmatic emanations. M, 1906, Columbia University, Teachers College.

Lincoln, Jonathan. Morphology, hydraulics and sediment transport patterns of southern Maine's small tidal inlets. M, 1985, Boston University. 151 p.

Lincoln, Rush B., III. Heavy mineral distribution in fluvial and marine environments; north coast, Puerto Rico. M, 1981, Duke University. 102 p.

Lincoln, T. N. The redistribution of copper during metamorphism of the Karmutsen Volcanics, Vancouver Island, British Columbia. D, 1978, University of California, Los Angeles. 251 p.

Lind, Aulis Olaf. Coastal landforms of Cat Island, Bahamas; a study of Holocene accretionary topography and sea-level change. D, 1968, University of Wisconsin-Madison. 284 p.

Lind, C. M. The Centralia, Illinois, oil field (L64). M, 1946, University of Texas, Austin.

Lind, Carl R. The geology of the Oxford-Morning Sun area, Butler and Preble counties, Ohio. M, 1957, Miami University (Ohio). 50 p.

Lindahl, David. The geology of the Brandy Camp Quadrangle, Elk County, Pennsylvania. M, 1977, University of Pittsburgh.

Lindau, Charles Wayne. Partitioning of selected micronutrients and trace elements and distribution of clay minerals in the substrate of one artificial and three natural marshes. D, 1980, Texas A&M University. 168 p.

Lindberg, Cheryl A. Stratigraphy and sedimentology of Upper Mississippian carbonate rocks in Cache, Weber and Rich counties, Utah. M, 1979, Colorado School of Mines. 182 p.

Lindberg, Craig Robert. Multiple taper spectral analysis of terrestrial free oscillations. D, 1986, University of California, San Diego. 206 p.

Lindberg, Jonathon W. A study of factors influencing phenetic variability in crinoid pluricolumnals. M, 1975, Washington State University. 91 p.

Lindberg, R. D. Effects of post-depositional compaction on the morphology of cross stratification. M, 1974, University of Colorado.

Lindberg, Ralph DeWitt. A geochemical appraisal of oxidation-reduction potential and interpretation of Eh measurements of ground water. D, 1983, University of Colorado. 491 p.

Lindberg, S. E. Mechanisms and rates of atmospheric deposition of selected trace elements and sulfate to a deciduous forest watershed. D, 1979, Florida State University. 548 p.

Lindbloom, Joseph T. Refinement of the crystal structure of hurlbutite. M, 1972, Virginia Polytechnic Institute and State University.

Lindecke, Joseph Werner. Geophysical exploration for caves near Zanesfield, Ohio. M, 1980, Wright State University. 72 p.

Lindemann, Richard H. Quantitative paleoecology of the Edgecliff Biostrome, Onondaga Formation in eastern New York. M, 1974, Rensselaer Polytechnic Institute. 71 p.

Lindemann, Richard Henry. Paleosynecology and paleoenvironments of the Onondaga Limestone in New York State. D, 1980, Rensselaer Polytechnic Institute. 131 p.

Lindemann, William Lee. Catahoula Formation, Duval County, Texas. M, 1962, University of Texas, Austin.

Linden, Karl Vonder see Vonder Linden, Karl

Linden, Ronald M. Strain energy and jointing in coal and adjacent sediments. M, 1987, New Mexico Institute of Mining and Technology. 123 p.

Lindenburg, George J. Factors contributing to the variance in the brines of the Great Salt Lake Desert and the Great Salt Lake. M, 1974, University of Utah. 70 p.

Linder, Gerhard Martin. Reconnaissance glacial geology of Avalanche Canyon, Juneau Icefield, Southeast Alaska. M, 1981, University of Idaho. 116 p.

Linder, Harold W. The origin of copper-nickel mineralization in gabbroic rocks of the northern Labrador Trough. D, 1966, University of Minnesota, Minneapolis. 176 p.

Linder, Henry D. An environmental study of the subsurface Miocene of Brazoria and Galveston counties, Texas. M, 1962, Texas A&M University.

Linder, Robert Andrew. Mid-Tertiary echinoids and Oligocene shallow marine environments in the Oregon central western Cascades. M, 1986, University of Oregon. 196 p.

Linderfelt, William R. Numerical analysis of infiltration and near-surface percolation in relation to Yucca Mountain, Nevada. M, 1987, University of Nevada. 64 p.

Lindgren, Donald W. An analysis of the mining industry's viewpoint on subsidies, tariffs, taxes, and stockpiling. M, 1951, University of Wisconsin-Madison.

Lindh, A. G. Detailed seismic studies of six moderate California earthquakes. D, 1980, Stanford University. 185 p.

Lindh, Thomas Bertil. Temporal variations in 13C, 34S and global sedimentation during the Phanerozoic. M, 1984, University of Miami. 98 p.

Lindholm, Gerald Franklin. Geologic and engineering properties of silts near Big Delta and Fairbanks, Alaska. M, 1957, Iowa State University of Science and Technology.

Lindholm, Rosanne M. Bivalve associations of the Cannonball Formation (Paleocene, Danian) of North Dakota. M, 1984, University of North Dakota. 184 p.

Lindholm, Roy Charles. Carbonate petrography of the Onondaga Limestone (Devonian), New York (Onondaga County). D, 1967, The Johns Hopkins University. 272 p.

Lindholm, Roy Charles. Structural petrology of the Ortega Quartzite, Rio Arriba County, New Mexico. M, 1963, University of Texas, Austin.

Lindley, Julia Ione. Chemical changes associated with the propylitic alteration of two ash-flow tuffs, Datil-Mogollon volcanic field, New Mexico. M, 1979, University of North Carolina, Chapel Hill. 197 p.

Lindley, Julia Ione. Potassium metasomatism of Cenozoic volcanic rocks near Socorro, New Mexico. D, 1985, University of North Carolina, Chapel Hill. 563 p.

Lindner, Ernest Norman. A constitutive and experimental investigation of load-history influences on the creep behavior of salt. D, 1983, University of Minnesota, Minneapolis. 313 p.

Lindner, Joseph Leicht. Artificial alteration of spodumene. M, 1930, University of Minnesota, Minneapolis. 62 p.

Lindner, Joseph Leicht. Hydrothermal experiments with alkali sulphide solutions. D, 1937, University of Minnesota, Minneapolis. 84 p.

Lindner, Robert Frederick. Detailed lithology of the carbonate members of the Elkhorn Formation in the crestal portion of the Cincinnati Arch. M, 1951, Miami University (Ohio). 24 p.

Lindorff, David E. Lithofacies and regional relationships in upper Ste. Genevieve and Girkin limestones (upper Mississippian), west-central Kentucky. M, 1969, University of Wisconsin-Madison.

Lindquist, Alec E. Structure and mineralization of the Whitehall mining district, Jefferson County, Montana. M, 1966, Montana College of Mineral Science & Technology. 104 p.

Lindquist, John Warren. Geology and paleontology of the Fork area, Dungeness and Graywolf rivers, Clallam County, Washington. M, 1961, University of Washington. 185 p.

Lindquist, P. Geology of the South Texas shelf banks. M, 1978, Texas A&M University.

Lindquist, Robert C. Slope processes and forms at Bryce Canyon National Park. D, 1980, University of Utah. 134 p.

Lindquist, Sandra J. Sandstone diagenesis and reservoir quality, Frio Formation (Oligocene), South Texas. M, 1976, University of Texas, Austin.

Lindquist, Tina Walburga. A provenance study of the Wall Creek Member of the Frontier Formation and the Turner Sandy Member of the Carlile Shale in the Powder River basin, Wyoming. M, 1986, Texas Christian University.

Lindsay, Charles S. The effects of urbanization on the water balance of the Fishtrap Creek basin, northwest Washington and south central British Columbia. M, 1988, Western Washington University. 65 p.

Lindsay, Curtis George. Purely ionic and molecular orbital modelings of the bonding in mineral crystal structures. D, 1988, Virginia Polytechnic Institute and State University. 105 p.

Lindsay, David Walter. The provenance of the Jacobsville Formation of the Upper Peninsula of Michigan through a petrographic study. M, 1986, Western Michigan University.

Lindsay, Donald R. Geology of the central part of the Solstice Canyon Quadrangle, Los Angeles County, California. M, 1952, University of California, Los Angeles.

Lindsay, Everett Harold, Jr. Biostratigraphy of the upper part of the Barstow Formation, Mojave Desert, California. D, 1967, University of California, Berkeley. 277 p.

Lindsay, John Francis. Stratigraphy and sedimentation of the lower Beacon rocks of the Queen Alexandra, Queen Elizabeth, and Holland Ranges, Antarctica, with emphasis on Paleozoic glaciation. D, 1968, Ohio State University. 316 p.

Lindsay, Louise Elizabeth. Capillary barriers; numerical and laboratory simulations of an engineered system for waste isolation. M, 1984, University of Waterloo. 55 p.

Lindsay, Robert F. Petrology and petrography of the Great Blue Formation at Wellsville Mountain, Utah. M, 1976, Brigham Young University. 136 p.

Lindsay, Robert R. Paleoecology of the Kiewitz Shale (upper Pennsylvanian) in southeastern Nebraska. M, 1971, University of Nebraska, Lincoln.

Lindsey, David Allen. The sedimentology of the Huronian Gowganda Formation (early Proterozoic), Ontario, Canada (with special reference to the Whitefish Falls area). D, 1967, The Johns Hopkins University. 464 p.

Lindsey, David S. Environmental significance of grain-size parameters, Bogue Inlet-White Oak Estuary area, North Carolina. M, 1969, Bowling Green State University. 99 p.

Lindsey, Douglas Dewitt. The structural geology of Barillas Quadrangle, Northwest Guatemala. M, 1975, University of Texas, Arlington. 91 p.

Lindsey, E. H. Geology and ore deposits of the Big Creek District, Idaho. M, 1958, University of California, Berkeley.

Lindsey, Kevin A. Character and origin of the Addy and Gypsy Quartzites, central Stevens and northern Pend Oreille Counties, northeastern Washington. D, 1987, Washington State University. 256 p.

Lindsey, Kevin A. The upper Proterozoic and Lower Cambrian Brigham Group, Oneida Narrows, southeastern Idaho. M, 1982, Idaho State University. 55 p.

Lindsey, Reavis Hall, Jr. A study of Recent foraminifera along a salinity gradient through the Mississippi Sound, Mississippi. M, 1962, University of Mississippi.

Lindskold, John Eric. Geology and petrography of the Gainesville, Virginia Quadrangle. M, 1961, George Washington University.

Lindsley, Donald H. Geology of the Spray Quadrangle, Oregon with special emphasis on the petrography and magnetic properties of the Picture Gorge Basalt. D, 1961, The Johns Hopkins University. 226 p.

Lindsley-Griffin, Nancy. Structure, stratigraphy, petrology, and regional relationships of the Trinity Ophiolite, eastern Klamath Mountains, California. D, 1982, University of California, Davis. 436 p.

Lindstedt, D. M. Effects of long term oil-recovery operations on macrobenthic communities near marsh-estuarine creek banks. M, 1978, Louisiana State University.

Lindstrom, David John. Experimental study of the partitioning of the transition metals between clinopyroxene and coexisting silicate liquids. D, 1976, University of Oregon. 201 p.

Lindstrom, Dean R. A study of the Pine Island and Thwaites glaciers. M, 1985, University of Maine. 136 p.

Lindstrom, Linda Jane. Stratigraphy of the South Platte Formation (Lower Cretaceous), Eldorado Springs to Golden, Colorado, and channel sandstone distribution of the J Member. M, 1978, Colorado School of Mines. 305 p.

Lindstrom, Marilyn Martin. Geochemical studies of volcanic rocks from Pinzon and Santiago islands, Galapagos Archipelago. D, 1976, University of Oregon. 186 p.

Lindstrom, Richard Mark. Radionuclides in meteorites and in the lunar surface. D, 1970, [University of California, San Diego]. 213 p.

Lindvall, Robert Marcus. Geology of the Kermit 3 Quadrangle, North Dakota. M, 1948, University of Iowa. 39 p.

Lindwall, Dennis. Deep crustal structure under and near the Hawaiian Islands. D, 1988, University of Hawaii. 122 p.

Lineback, Jerry A. The geology of northern Anderson County, Kansas. M, 1961, University of Kansas. 86 p.

Lineback, Jerry Alvin. Stratigraphy and depositional environment of the New Albany Shale (Upper Devonian and Lower Mississippian) in Indiana. D, 1964, Indiana University, Bloomington. 136 p.

Lineberger, David Howard, Jr. Geology of the Chatham fault zone, Pittsylvania County, Virginia. M, 1983, University of North Carolina, Chapel Hill. 79 p.

Lineberger, P. H. Sedimentary processes and pelagic turbidites in the eastern central Pacific Basin. M, 1975, University of Hawaii. 116 p.

Lineberger, Ralph D. Stratigraphy of the Chickamauga Group in Knoxville Belt, Loudon County, Tennessee. M, 1956, University of Tennessee, Knoxville. 38 p.

Lineberry, R. A. A manufacturing study of North Carolina brick clays. M, 1925, University of North Carolina, Chapel Hill.

Linehan, John M. Factors influencing production in the Toronto Limestone (Shawnee Group, Upper Pennsylvanian) of the Snake Creek field in Clark County, Kansas. M, 1986, Wichita State University. 146 p.

Lineman, David J. An expert system for well-to-well log correlation. M, 1987, Massachusetts Institute of Technology. 106 p.

Liner, Jeffery Lynn. Lithostratigraphy of the Boone Limestone (Lower Mississippian), Northwest Arkansas. M, 1979, University of Arkansas, Fayetteville.

Liner, Robert T. Lithostratigraphy and biostratigraphy of the Cane Hill Member, Hale Formation (Morrowan), northern Arkansas. M, 1979, University of Arkansas, Fayetteville.

Lines, L. R. Deconvolution and wavelet estimation in exploration seismology. D, 1976, University of British Columbia.

Lines, Laurence Richard. A numerical study of the perturbation of alternating geomagnetic fields near island and coastline structures. M, 1973, University of Alberta. 104 p.

Lines, Roland Arnold Granville. Radiocarbon dating using a proportional counter filled with carbon dioxide. M, 1962, Dalhousie University.

Lines, William B. Grain size analysis of the Hartshorne Sandstone in the Arkansas Valley. M, 1956, University of Arkansas, Fayetteville.

Lines, William B. Minor structures in the Boston Mountain Monocline in T. 9, 10, 11 N., R. 32 W., Crawford County, Arkansas. M, 1955, University of Arkansas, Fayetteville.

Ling, Hsin-Yi. A micropaleontologic and ecologic investigation of Recent sediments from two Gulf Coast cores. D, 1963, Washington University. 84 p.

Ling, S. C. Dynamic behavior of a NE Pacific pelagic clay. D, 1976, University of Washington. 225 p.

Lingamallu, Surya Narayana. Petroleum resources and national development; a case history of Saudi Arabia. M, 1976, Southern Illinois University, Carbondale. 230 p.

Linger, Robert E. Geology of the Berry's Ledge scheelite occurrence at Cornish, Maine. M, 1956, University of Maine. 81 p.

Lingle, C. S. Tidal flexure of Jakobshavns Glacier, West Greenland. M, 1978, University of Maine. 151 p.

Lingle, Craig Stanley. A numerical model of interactions between a polar ice stream, the ocean and the solid Earth; application to ice stream E, West Antarc-

tica. D, 1983, University of Wisconsin-Madison. 181 p.

Lingley, William S. Geology of the older Precambrian rocks in the vicinity of Clear Creek and Zoroaster Canyon, Grand Canyon, Arizona. M, 1973, Western Washington University. 78 p.

Lingner, David William. Trace element evidence for contrastive thermal histories of H4-6 and L4-6 chondrite parent bodies. D, 1985, Purdue University. 264 p.

Lingren, John E. Application of a groundwater model to the Boise Valley Aquifer in Idaho. M, 1982, University of Idaho. 88 p.

Lingrey, Steven Howard. Structural geology and metamorphism of the eastern Whipple Mountains, San Bernardino County, California. M, 1979, University of Southern California.

Lingrey, Steven Howard. Structural geology and tectonic evolution of the northeastern Rincon Mountains, Cochise and Pima counties, Arizona. D, 1982, University of Arizona. 248 p.

Linick, T. W. Uptake of bomb-produced carbon-14 by the Pacific Ocean. D, 1975, University of California, San Diego. 270 p.

Link, Arthur J. A physico-chemical study of a lagoonal environment of sedimentation, Baja California. M, 1964, Northwestern University.

Link, Arthur Jurgen. Inclusions of the half dome quartz monzonite (Cretaceous), Yosemite National Park, California. D, 1969, Northwestern University. 122 p.

Link, Christine Marie. A reconnaissance of organic maturation and petroleum source potential of Phanerozoic strata in northern Yukon and northwestern District of Mackenzie. M, 1988, University of British Columbia. 260 p.

Link, D. A. The factors controlling the suspended sediment in streams draining currently glaciated basins. D, 1977, Northwestern University. 120 p.

Link, J. E. Gold and copper mineralization in the McCoy Creek District, Skamania County, Washington. M, 1985, Washington State University. 176 p.

Link, John T. The toponomy of Nebraska. D, 1932, University of Nebraska, Lincoln.

Link, Martin H. Sedimentology, petrography and environmental analysis of the Matilija Sandstone (Eocene), north of the Santa Ynez fault (Santa Barbara County, California). M, 1971, University of California, Santa Barbara.

Link, Martin Hans. A sedimentologic history of Ridge Basin, Transverse Ranges, southern California. D, 1982, University of Southern California.

Link, Paul Karl. Geology of the upper Proterozoic Pocatello Formation, Bannock Range, southeastern Idaho. D, 1982, University of California, Santa Barbara. 273 p.

Link, Peter Karl. Stratigraphy of the Mount Whiteeastern Little Atlin Lake area, Yukon Territory, Canada. D, 1965, University of Wisconsin-Madison. 194 p.

Link, Peter Karl. The Cretaceous Pictured Cliffs Sandstone of the San Juan Basin, Colorado and New Mexico. M, 1955, University of Wisconsin-Madison.

Link, R. L. Computer analysis of normal pore distribution in selected cytheracean Ostracoda. D, 1975, University of Illinois, Urbana. 132 p.

Link, Theodore A. Three-dimensional experiments in earth deformation. D, 1927, University of Chicago. 227 p.

Linkletter, George O. Soluble particulates in an ice core from Greenland. M, 1967, Dartmouth College. 74 p.

Linkletter, George Onderdonk, II. Weathering and soil formation in Antarctic dry valleys. D, 1971, University of Washington. 121 p.

Linn, Anne. Depositional environment and hydrodynamic flow in Guadalupian Cherry Canyon Sandstone, West Fort and West Geraldine fields, Delaware Basin, Texas. M, 1985, Texas A&M University.

Linn, George Willison. Geology of Orange Hill, Alaska; 1 volume. M, 1973, University of California, Berkeley. various pagination p.

Linn, Kurt O. Geology of the Helena Mine area, Leadville, Colorado. D, 1964, Harvard University.

Linnen, Robert. Contact metamorphism, wallrock alteration, and mineralization at the Trout Lake stockwork molybdenum deposit, southeastern British Columbia. M, 1985, McGill University. 220 p.

Lins, Ibere Delmar Gondim. Improvement of soil test interpretations for phosphorus and zinc. D, 1987, North Carolina State University. 340 p.

Lins, Thomas Wesley. Origin and environment of the Tonganoxie Sandstone in northeastern Kansas. M, 1950, University of Kansas. 44 p.

Lins, Thomas Wesley. Phanerozoic batholiths; where and why. D, 1969, University of Kansas. 76 p.

Linsalata, Paul. Sources, distribution and mobility of plutonium and radiocesium in soils, sediments and water of the Hudson River estuary and watershed. D, 1984, New York University. 424 p.

Linscott, Jeffrey Parrish. A laboratory study of the effects of shear stress on fracture permeability. M, 1985, University of Oklahoma. 82 p.

Linscott, Robert Orrin. Petrography of the upper member of the Paradox Formation in the Four Corners region. M, 1962, Dartmouth College. 71 p.

Linsley, Robert Martin. Gastropods of the Middle Devonian Anderdon Limestone. D, 1960, University of Michigan. 256 p.

Linsley, Robert Martin. New gastropods from the Middle Devonian Anderdon Limestone of Michigan, Ohio, and Ontario. M, 1953, University of Michigan.

Linstedt, Kermit Daniel. Occurrence of vanadium in the Colorado River and its behavior in water treatment coagulation. D, 1968, Stanford University. 199 p.

Lintner, Stephen Francis. The historical physical behavior of the lower Susquehanna River, Pennsylvania, 1801-1976. D, 1983, The Johns Hopkins University. 222 p.

Linton, J. A. Gravity meter development and use in a search for core mode oscillations. D, 1977, York University.

Linton, John Alexander. Phase and amplitude variation of Chandler wobble. M, 1973, University of British Columbia.

Lintz, Joseph, Jr. A subsurface study of the Cherokee sands of southern Cleveland County, Oklahoma. M, 1947, University of Oklahoma. 31 p.

Lintz, Joseph, Jr. The fauna of the (Pennsylvanian) Ames and Brush Creek shales of the Conemaugh Formation of western Maryland. D, 1956, The Johns Hopkins University.

Lion, Leonard W. Cadmium, copper, and lead in estuarine salt marsh microlayers; accumulation, speciation, and transport. D, 1980, Stanford University. 386 p.

Lion, Thomas E. Engineering geology and the relative stability of ground for hillside development in part of Springfield Township, Hamilton County, Ohio. M, 1983, University of Cincinnati. 157 p.

Liongson, Leonardo Quesada. Operational hydrology and water quality investigations of the stream-reservoir system in the Upper Pampanga River Project. D, 1976, University of Arizona.

Liotta, Frederick P. Dissolved oxygen demand of reduced chemical species in the water column of Sebasticook Lake, Maine. M, 1979, University of Maine. 43 p.

Liou, C. P. A numerical model for liquefaction in sand deposits. D, 1976, University of Michigan. 218 p.

Liou, Jia-Shing. Magnetotelluric measurements in the western Trans-Pecos area of West Texas. M, 1982, University of Texas at Dallas. 75 p.

Liou, John Chwen-Haw. Atmospheric injection of radon daughters from the 1982 eruption of El Chichon

Volcano. D, 1983, University of Arkansas, Fayetteville. 97 p.

Liou, Juhn-Guang. Stability relations of zeolites and related minerals in the system CaO-Al$_2$O$_3$-SiO$_2$-H$_2$O. D, 1970, University of California, Los Angeles. 318 p.

Liow, Jeih-San. A two-dimensional finite-difference simulation of seismic wave propagation in elastic media. D, 1988, Georgia Institute of Technology. 220 p.

Lipchinsky, Zelek Lawrence. A study of the origin of limestone caverns in Florida. M, 1963, University of Florida. 62 p.

Lipin, Bruce Reed. Equilibrium relations among iron-titanium oxides in silicate melts; the system CaAl$_2$Si$_2$O$_8$-CaMgSi$_2$O$_6$-FeO-TiO$_2$ in contact with metallic iron. D, 1975, Pennsylvania State University, University Park. 177 p.

Lipinski, Paul William. The gamma member of the Kaibab Formation (Permian) in northern Arizona. M, 1976, University of Arizona.

Lipka, Joseph T., II. Stratigraphy and structure of the southern Sulphur Spring Range, Eureka County, Nevada. M, 1987, Oregon State University. 94 p.

Lipke, Audrey C. Geology and petrology of the northeastern quarter of the Anderson Ridge Quadrangle, Wyoming. M, 1978, University of Missouri, Columbia.

Lipman, Eric William. Systematics, biostratigraphy and paleoecology of Missourian and Virgilian brachiopods of the Bird Spring Group, Arrow Canyon, Arrow Canyon Range, Clark County, Nevada. M, 1982, University of Illinois, Urbana. 245 p.

Lipman, Leonard H., II. Formation and growth of a spit bar; a study using orientation and imbrication of clastic grains to show water flow directions. M, 1969, Rutgers, The State University, New Brunswick. 30 p.

Lipman, Peter W. Geology of the southeastern Trinity Alps, Northern California. D, 1962, Stanford University. 210 p.

Lipp, Edward George. Geology of an area east of Sheep Canyon, near Dell, Montana. M, 1948, University of Michigan. 59 p.

Lipp, Russell L. Paleoecological study of the Cambridge marine event in Harrison Township, Gallia County, Ohio. M, 1975, Ohio University, Athens. 69 p.

Lippert, James Brent. The stratigraphy and diagenesis of the Rierdon and Swift formations (Upper Jurassic), Missouri River headwaters basin. M, 1985, University of Montana. 161 p.

Lippert, Rudolph H. Investigation of structure and structural development by trend surface analysis. M, 1964, University of Kansas. 44 p.

Lippincott, Diane Kay. Recent water-level fluctuations in a well near the Garlock Fault, Fremont Valley, California. M, 1982, Stanford University. 67 p.

Lippitt, Clifford R. An X-ray fluorescence spectrometer for mineral exploration. M, 1980, Colorado School of Mines. 120 p.

Lippitt, Louis. Statistical analysis of regional facies change in Ordovician Cobourg Limestone in northwestern New York and southern Ontario. D, 1960, Columbia University, Teachers College.

Lippitt, Louis. Statistical treatment of the Cobourg Limestone, Trenton Group in northern Lewis County and Jefferson County, New York. M, 1953, Columbia University, Teachers College.

Lippmann, Marcelo J. Two dimensional stochastic model of a heterogeneous geologic system. D, 1974, University of California, Berkeley. 150 p.

Lippoth, Richard Edward. Geochemical factors controlling ore deposition, Trixie Mine, Tintic District, Utah. M, 1984, University of Utah. 111 p.

Lipps, Jere Henry. Cenozoic planktonic foraminifera; I, Wall structure, classification and phylogeny of genera; II, California mid-Cenozoic biostratigraphy and zoogeography. D, 1966, University of California, Los Angeles. 289 p.

Lippus, Craig Stephen. The seismic properties of mafic volcanic rocks of the Keweenawan Supergroup and their implications. M, 1988, Purdue University. 160 p.

Lipschultz, Fredric. Environmental factors affecting rates of nitrogen cycling. D, 1984, Harvard University. 186 p.

Lipsey, James Allen. An X-ray diffraction study of the Porters Creek Formation (lower Eocene) in Pontotoc County (Mississippi). M, 1964, University of Mississippi.

Lipshie, Steven Ross. Development of phyllosilicate preferred orientation in naturally and experimentally metamorphosed and deformed rocks. D, 1984, University of California, Los Angeles. 433 p.

Lipshie, Steven Ross. Surficial and engineering geology of the Mammoth Creek area, Mono County, California. M, 1974, University of California, Los Angeles.

Lipson, Joseph I. Potassium-argon dating of sediments. D, 1956, University of California, Berkeley.

Liptak, Alan Robert. Geology and differentiation of Round Butte Laccolith, central Montana. M, 1984, University of Montana. 49 p.

Lipten, Eric Jack Henry. The geology of Clover Hill and classification of the Wells tungsten prospect, Elko County, Nevada. M, 1984, Purdue University. 238 p.

Lis, Michael Gregory. The role of uniform flattening, deduced from the study of rotated garnets, in the formation of the Pudding Hill fold, Windsor County, Vermont. M, 1971, Brown University.

Lischer, Lowell K. Anatomy and sedimentary dynamics of an ancient subtidal sand sheet. M, 1974, University of Missouri, Columbia.

Lisco, Neil. Bedload transport in a small alluvial stream draining the Governors Square Mall construction site, Tall. Fl. M, 1979, Florida State University.

Liscum, F. Prediction of parameter values from physical basin characteristics for the U. S. Geological Survey rainfall-runoff model. D, 1978, Georgia Institute of Technology. 400 p.

Lisenbee, Alvis Lee. Geology of the Cerro Pelon-Arroyo de la Jara area, Santa Fe County, New Mexico. M, 1967, University of New Mexico. 112 p.

Lisenbee, Alvis Lee. Structural setting of the Orhaneli ultramafic massif near Bursa, northwestern Turkey. D, 1972, Pennsylvania State University, University Park. 271 p.

Liska, Robert D. The geology and biostratigraphy of Letterbox Canyon, San Diego County, California. M, 1964, San Diego State University.

Lisle, Barbara. Sandstone petrography and facies analysis of the Jackfork Group, frontal Ouachita, Arkansas. M, 1986, University of Arkansas, Fayetteville.

Lisle, Richard E. The geology and geochemistry of the Embargo intrusive center, Saguache County, Colorado. M, 1974, Colorado School of Mines. 85 p.

Lisle, Thomas E. Components of flow resistance in a natural channel. D, 1976, University of California, Berkeley. 86 p.

Lisle, Thomas Edwin. Sediment yield and hydrodynamic implications, West Fork of the Madison River, (Madison County) Montana. M, 1972, University of Montana. 81 p.

Lisowski, Michael. Geodetic strain measurements in central Vancouver Island. M, 1985, University of British Columbia. 100 p.

Lissey, Allan. Hydrology of the Regina Aquifer, Saskatchewan. M, 1964, University of Saskatchewan. 88 p.

Lissey, Allan. Surficial mapping of groundwater flow systems with application to the Oak River basin, Manitoba (Canada). D, 1968, University of Saskatchewan. 198 p.

Lissner, Frederick Gordon. Sources of littoral-zone sands in the vicinity of Gold Beach (Curry County), Oregon. M, 1971, University of Oregon. 86 p.

Lister, Gordon Frank. The composition and origin of selected iron-titanium deposits. D, 1965, University of Wisconsin-Madison.

Lister, James C. The sedimentology of Camas Prairie Basin and its significance to the Lake Missoula floods. M, 1981, University of Montana. 66 p.

Lister, Judith Smith. The crystal structure of two chlorites. D, 1966, University of Wisconsin-Madison. 91 p.

Lister, K. H. Paleoecology of Ostracoda from Quaternary sediments of the Great Salt Lake Basin, Utah. D, 1974, University of Kansas. 476 p.

Lister, Kenneth Henry. Paleoecology of insect bearing Miocene beds in Calico mountains, California. M, 1970, University of California, Los Angeles.

Listerud, W. H. Geology of a sulfide deposit in lower Precambrian metavolcanic-metasedimentary rocks near Birchdale, Koochiching County, Minnesota. M, 1974, University of Minnesota, Duluth.

Liston, Thomas C. Lithofacies and current direction studies of the Dunkard Group (Pennsylvanian-Permian) in Wirt County, West Virginia. M, 1962, Miami University (Ohio). 134 p.

Liszak, Jerry Lee. The Chilliwack Group on Black Mountain, Washington. M, 1982, Western Washington University. 104 p.

Liszewski, Michael Joseph. Distribution of inorganic elements in the Adaville #1 coal seam of Southwest Wyoming. M, 1982, Washington State University. 130 p.

Litaor, Michael Iggy. Geochemistry of alpine soils in the Colorado Front Range, with special reference to acid deposition. D, 1986, University of Colorado. 294 p.

Litchford R. F., Jr. Structural geology and stratigraphy of a part of the overthrust belt near Wyoming Peak, Lincoln and Sublette counties, Wyoming. M, 1966, University of Wyoming. 175 p.

Litehiser, J. J., Jr. Near-field seismograms from a two-dimensional propagating dislocation. D, 1976, University of California, Berkeley. 189 p.

Lithgow, Enrique W. An analysis of the factors controlling the occurrence of bloating shales in the Pennsylvanian System of western Pennsylvania. M, 1972, Pennsylvania State University, University Park. 209 p.

Litke, Gene Richard. The stratigraphy and sedimentation of Barillas Quadrangle, Department of Huehuetenango, Guat., C.A. M, 1975, University of Texas, Arlington. 196 p.

Litke, Richard Timothy. Transport of dissolved and particulate organic carbon in Rocky Mountain ephemeral streams. D, 1983, Idaho State University. 148 p.

Litsey, Linus R. Geology of the Hayden Pass-Orient area, Sangre de Cristo Mountains, Colorado. D, 1954, University of Colorado.

Litt, Donald D. Analysis and classification of airphotos as related to soil trafficability. M, 1952, Purdue University.

Little, Ann Carol. Hydrodynamic character of the Dundee Limestone in the central Michigan Basin. M, 1988, Western Michigan University.

Little, Gregory E. Subsurface analysis of the (Permian) Abo Formation in the Lucero region, west-central NM. M, 1987, New Mexico Institute of Mining and Technology. 182 p.

Little, Heward W. The ultrabasic and associated rocks of the Middle River Range, British Columbia. D, 1947, University of Toronto.

Little, Heward Wallace. Silver-lead relationships in British Columbia. M, 1940, University of British Columbia.

Little, Hower P. The physiographic features of Anne Arundel County. D, 1910, The Johns Hopkins University.

Little, John Marshall. Conodont faunas in the Hughes Creek Shale and Bennett Shale of Riley and Wabaunsee counties, Kansas. M, 1965, Kansas State University. 79 p.

Little, John Marshall. Palynology and micropaleontology of the Pitkin Limestone (Mississippian) in Washington County, Arkansas. M, 1964, Kansas State University.

Little, Lynne A. Geology and land-use investigation in the Pinnacle Peak area, Maricopa County, Arizona. M, 1975, Arizona State University. 102 p.

Little, Maynard N. The Saint Francis River; a study of sediment transport. M, 1975, Southern Illinois University, Carbondale. 104 p.

Little, Richard Douglas. Terraces of the Makran Coast of Iran and parts of West Pakistan. M, 1972, University of Southern California.

Little, Robert Lewis. A study of the foraminifera of the Bluff-Port Marl Member of the Demopolis Chalk (Upper Cretaceous) in Kemper Couty (Mississippi). M, 1959, University of Mississippi.

Little, Robert Lewis. Lithostratigraphy and structural geology of a portion of the Dunham ridge thrust block, Green and Washington counties, Tennessee. D, 1969, University of Tennessee, Knoxville. 130 p.

Little, Sarah Alden. Fluid flow and sound generation at hydrothermal vent fields. D, 1988, Massachusetts Institute of Technology. 152 p.

Little, Stephen W. Stratigraphy, petrology and provenance of the Cretaceous Gable Creek Formation, Wheeler County, Oregon. M, 1987, Oregon State University. 133 p.

Little, Thomas Marvin. Depositional environments, petrology, and diagenesis of the basal limestone facies, Green River Formation (Eocene), Uinta Basin, Utah. M, 1988, University of Utah. 154 p.

Little, Timothy Alden. Structure and metamorphism of upper Paleozoic rocks in the Mountain City Quadrangle, Elko County, Nevada. M, 1983, Stanford University. 392 p.

Little, Timothy Alden. Tertiary tectonics of the Border Ranges fault system, north-central Chugach Mountains, Alaska; sedimentation, deformation and uplift along the inboard edge of a subduction complex. D, 1988, Stanford University. 565 p.

Little, William. An engineering and economic feasibility study for diversion of central Arizona project waters from alternate sites. M, 1968, University of Arizona.

Little, William Meldrum. A study of inclusions in cassiterite and associated minerals. D, 1953, University of Toronto.

Little, William W. Geomorphology of Lake Bonneville deltas. M, 1987, Brigham Young University. 83 p.

Littlefield, James R. Evaluation of data for sinkhole development risk models; west central Florida. M, 1988, University of South Florida, Tampa. 75 p.

Littlefield, Max Sylvan. Bay muds at the mouth of the Mississippi River. M, 1923, Iowa State University of Science and Technology.

Littlefield, Max Sylvan. The origin and physical properties of natural bonded molding sands of Illinois. D, 1925, University of Iowa. 102 p.

Littlefield, Robert G. Structural analysis of the Long John Canyon area in the New York Butte Quadrangle, Inyo County, California. M, 1979, San Jose State University. 118 p.

Littlefield, Romaine Faye. A study of the inclusions in the quartz of some igneous and metamorphic rocks. M, 1949, University of Missouri, Columbia.

Littlejohn, Alastair Lewis. A comparative study of lherzolite nodules in basaltic rocks from British Columbia. M, 1972, University of British Columbia.

Littlejohn, John Joseph. Lower Cretaceous Comanchean stratigraphy of Texas. D, 1975, Harvard University.

Litwin, Ronald James. Fertile organs and in situ spores of ferns from the Late Triassic Chinle Formation of the Colorado Plateau (Arizona and New Mexico), with discussion of the associated dispersed spores. M, 1983, Pennsylvania State University, University Park. 84 p.

Litwin, Ronald James. The palynostratigraphy and age of the Chinle and Moenave formations, southwestern U.S.A. D, 1986, Pennsylvania State University, University Park. 282 p.

Liu, C. C.-K. Numerical evaluation of response functions of a nonlinear rainfall-runoff model; flood hydrograph analysis in the Chemung River basin. D, 1976, Cornell University. 117 p.

Liu, Char-Shine. Geophysical studies of the northeastern Indian Ocean. D, 1983, University of California, San Diego. 120 p.

Liu, Chi-Ching. Tectonic interpretation of leveling data in Southern California. D, 1985, University of California, Los Angeles. 220 p.

Liu, Chingju. A structural study of Hukou and Hsinchu areas, northwestern Taiwan. M, 1985, Rice University. 134 p.

Liu, Hok-Shing. The effect of impurity dopings and surface treatments on the electrokinetic behavior of silicon carbide. D, 1969, Pennsylvania State University, University Park. 229 p.

Liu, Hsi-Ping. Part I; Temperature dependence of single crystal spinel ($MgAl_2O_4$) elastic constants from 293K to 423K measured by light-sound scattering in the Raman-Nath region; Part II: Effect of anelasticity on periods of Earth's free oscillation (toroidal modes). D, 1974, California Institute of Technology. 172 p.

Liu, Hsin-Hsi. Diffraction events in seismic prospecting. M, 1967, University of Tulsa. 65 p.

Liu, Hsui-Lin. Interpretation of near-source ground motion and implications. D, 1983, California Institute of Technology. 198 p.

Liu, Jack Shan. Analyses of mechanically stabilized retaining walls. D, 1988, Utah State University. 210 p.

Liu, James Tsu-chien. Sediment patterns and shoreface bathymetry as a key to understanding shoreface dynamics; a case study of the south shore of Long Island, New York. D, 1987, SUNY at Stony Brook. 210 p.

Liu, Jeun-Shyang. Petrography of the Ballast Point, Brandon and Duette drill cores, Hillsborough and Manatee counties, Florida. M, 1978, University of Florida. 147 p.

Liu, John Lin-gun. Isothermal compression of magnesian garnets up to 300 kilobars and the implications to the Earth's mantle. M, 1968, University of Rochester.

Liu, Jue-Yu. Distributions of stress and strain near a stick-slip fault. D, 1982, Cornell University. 313 p.

Liu, Kam-Biu. Postglacial vegetational history of northern Ontario; a palynological study. D, 1982, University of Toronto.

Liu, Kannson T. H. Effect of sediment discharge on the performance of a V-type measuring flume. M, 1964, University of Idaho. 68 p.

Liu, Ko-Hui. Leaching of cations during displacement by acid solutions through columns of Cecil soil. D, 1987, University of Florida. 221 p.

Liu, Kon-Kee. Geochemistry of inorganic nitrogen compounds in two marine environments; the Santa Barbara Basin and the ocean off Peru. D, 1979, University of California, Los Angeles. 378 p.

Liu, Lin Gun. Isothermal compression of minerals pertinent to the Earth's mantle. D, 1970, University of Rochester. 156 p.

Liu, Mian. Migmatization and volcanic petrogenesis in the La Grande greenstone belt, Quebec. M, 1984, McGill University. 87 p.

Liu, Shih-Chi. Nondeterministic analysis of nonlinear structures subjected to earthquake excitations. D, 1967, University of California, Berkeley. 169 p.

Liu, Shih-Tseng. Pleistocene paleoenvironmental record of the planktonic foraminifera of the north-central Gulf of Mexico. M, 1988, Northern Illinois University. 86 p.

Liu, Shu-Wang. Piezomagnetic and thermomagnetic anomalies associated with fault slip. D, 1986, Case Western Reserve University. 175 p.

Liu, Teh-Ching. Liquidus phase relationships in the $CaO-MgO-Al_2O_3-SiO_2$ system at 1 atm and 20 kbar with implications to basalt petrogenesis and the occurrence of igneous sapphirine. D, 1987, University of Texas at Dallas. 148 p.

Liu, Tiebing. C-S-Fe correlation of shales hosting sedimentary manganese deposits. D, 1988, University of Cincinnati. 318 p.

Liu, Tze-Kung. A spectral density and response spectra analysis of some earthquake records. M, 1980, University of Washington. 95 p.

Liu, Wen David. Structure-foundation interactions under dynamic loads. D, 1984, University of California, Berkeley. 222 p.

Liu, Wen-Cheh. The sensitivity of selected soils from the Sierra Nevada to acidic deposition. D, 1988, University of California, Riverside. 116 p.

Liu, Win-Kay. Three-dimensional groundwater problems solved by direct and strongly implicit methods. D, 1985, Utah State University. 179 p.

Liu, Xing. The crystal structure and crystal chemistry of rare-earth cobaltites and a cobalt oxide. D, 1988, SUNY at Stony Brook. 166 p.

Liu, Yu-Lin. Effects of the intermediate principal stress on the strength and stiffness of a reinforced sand. D, 1988, University of Cincinnati. 176 p.

Livaccari, Richard F. Geology of the Lewisporte-Loon Bay area, Newfoundland, Canada. M, 1980, SUNY at Albany. 135 p.

Lively, John Robert. A subsurface study of the Sandusky area, Grayson County, Texas. M, 1956, University of Oklahoma. 46 p.

Lively, Richard S. Uranium-series disequilibrium investigations of three surficial uranium deposits. M, 1978, Michigan State University. 50 p.

Liverman, David Gordon Earl. Sedimentology and drainage history of a glacier dammed lake, St. Elias Mountains, Yukon Territory. M, 1981, University of Alberta. 219 p.

Liversage, Robert Richard. Development of a flow injection analysis/hydride generation system for the determination of arsenic by inductively coupled plasma optical emission spectrometry. M, 1984, University of Toronto.

Livesay, D. M. The hydrogeology of the upper Wanapum Basalt, upper Cold Creek valley, Washington. M, 1986, Washington State University. 159 p.

Livesay, Elizabeth Ann. Study on Tertiary ostracods. M, 1945, University of Illinois, Urbana.

Livieres, Ricardo. The geology and petrology of Volcan Sanganguey, Nayarit, Mexico. M, 1983, Tulane University. 98 p.

Livingston, Clifton Walter. Geology and vein mechanics of the Rambler gold prospect, Baie Verte, Newfoundland. M, 1942, Michigan Technological University. 54 p.

Livingston, Daniel A. The palaeolimnology of Arctic Alaska. D, 1953, Yale University.

Livingston, Donald E. Structural and economic geology of the Beaver Lake Mountains, Beaver County, Utah. M, 1961, University of Arizona.

Livingston, Donald Everett. Geochronology of older Precambrian rocks in Gila County, Arizona. D, 1969, University of Arizona. 277 p.

Livingston, John E., Jr. Structural analysis of the Great Smokey thrust sheet along the Little Tennessee River, Tennessee. M, 1978, University of Tennessee, Knoxville. 112 p.

Livingston, John Lee. Geology of the Brevard Zone and the Blue Ridge Province in southwestern Transylvania County, North Carolina. D, 1966, Rice University. 195 p.

Livingston, John Lee. Stratigraphic and structural relations in a portion of the Northwest Spring Mountains, Nevada. M, 1964, Rice University. 34 p.

Livingston, Neal D. Sediment distribution in thalwegs of Great Plains reservoirs. M, 1981, University of Kansas. 64 p.

Livingston, Vaughn Edward, Jr. Sedimentation and stratigraphy of the Humbug Formation in central Utah. M, 1955, Brigham Young University. 60 p.

Livingstone, C.E. Geomagnetic depth-sounding in the southwest U.S.A. and in southern British Columbia. M, 1967, University of British Columbia.

Livingstone, Jennie. Organic constituents of oil shales and related rocks; Garfield County, Colorado. M, 1927, University of Colorado.

Livingstone, Kent W. Geology of the Crawford bay map-area, southeastern British Columbia. M, 1970, University of British Columbia.

Livnat, Alexander. Metamorphism and copper mineralization of the Portage Lake Lava Series, northern Michigan. D, 1983, University of Michigan. 292 p.

Lix, Henry W. The composition and occurrence of linnaeite (siegenite). M, 1935, University of Missouri, Columbia.

Lizak, John B., Jr. The petrography and stratigraphy of sandstones within the Frontier and Mowry formations in the Crazy Mountains Basin, Montana. M, 1977, Purdue University. 86 p.

Lizanec, Theodore J. A DC resistivity investigation of the shallow aquifer east of Naples, Florida. M, 1985, University of South Florida, Tampa. 56 p.

Lizarraga, Arciniéga Jose R. Shoreline changes due to jetty construction on the Oregon coast. M, 1976, Oregon State University. 85 p.

Lizotte, Marcel Romuald. Structural and lithological controls on oil and gas producing zones in the Trenton Limestone, south-central Michigan, northern Indiana and Ohio. M, 1962, University of Michigan.

Llewellyn-Smith, Timothy M. The hydrogeology of Martha's Vineyard, Massachusetts. M, 1987, University of Massachusetts. 176 p.

Lloyd, George Perry, II. Geology of the north end of White River valley, White Pine County, Nevada. M, 1959, University of California, Los Angeles.

Lloyd, Orville Bruce. Bedrock geology of the western half of the Marlborough Quadrangle, Connecticut. M, 1963, University of Massachusetts. 177 p.

Lloyd, Robert A. A statistical analysis of the heavy minerals in the Atoka Formation outcrop area with emphasis on the tourmaline suite. M, 1962, University of Tulsa. 111 p.

Lloyd, Ronald LaVerne. The subsurface stratigraphy of the Gilmore City Formation (Mississippian) in (Pocahontas, Humboldt, and Hardin counties) Iowa. M, 1973, University of Iowa. 59 p.

Lloyd, Ronald Michael. A study of fine grained limestones. M, 1953, University of Illinois, Urbana.

Lloyd, Ronald Michael. The shell chemistry of some Recent and Pleistocene mollusks and its environmental significance. D, 1960, California Institute of Technology. 194 p.

Lloyd, Ruth E. Surface geology of Sulligent Quadrangle, Alabama-Mississippi; (concerned principally with the Upper Cretaceous Tuscaloosa, McShan, and Eutaw formations). M, 1980, Mississippi State University.

Lloyd, S. H. Distillation of oil shales under partial vacuum. M, 1921, University of Missouri, Rolla.

Lloyd-Morris, Anthony Edward. Geology of Hashemite, Kingdom of Jordan. M, 1960, University of California, Los Angeles.

Lo, Hoom-bin. A modified trend surface analysis model and its application. M, 1972, Northwestern University.

Lo, Hoom-bin. Reflectance indicatrix of vitrinite; model and its application. D, 1977, West Virginia University. 107 p.

Lo, Howard Hunghsin. Geochemistry of the Louis Lake pluton (Precambrian), (Wind River range, Wyoming). D, 1970, Washington University. 123 p.

Lo, Kwok-wai Kenneth. Comparisons between subjective analysis and the NCAR multivariate statistical objective analysis scheme. M, 1978, Purdue University. 81 p.

Lo, Kwong Fai Andrew. Estimation of rainfall erosivity in Hawaii. D, 1982, University of Hawaii at Manoa.

Lo, Su-Chu. Microbial fossils from the lower Yudoma Suite, earliest Phanerozoic, eastern Siberia. M, 1979, University of California, Santa Barbara.

Lo, Tien When. Seismic borehole tomography. D, 1988, Massachusetts Institute of Technology. 202 p.

Lo, Tien When. Ultrasonic laboratory tests of geophysical tomographic reconstruction. M, 1986, Massachusetts Institute of Technology. 40 p.

Lo, Yung-Kwong Terence. Geotechnical data bank for Indiana. D, 1980, Purdue University. 574 p.

Loan, Paul Rose Van see Van Loan, Paul Rose

Loar, Steven J. Joint and lineament study of the southern Appalachian Plateau near Elkhorn City, Kentucky. M, 1985, University of Cincinnati. 168 p.

Loayza, Oscar. Stability relations of anhydrite-alunite in hydrothermally altered rock of porphyry copper deposits. M, 1972, Stanford University.

Lobanoff-Rostovsky, Nikita. Geology of the Independence mining district, Park and Sweet Grass counties, Montana. M, 1960, Columbia University, Teachers College. 67 p.

Lobato, Fabiano Sayao. An interpretation method of seismic reflections from common-strike plane interfaces. D, 1965, Colorado School of Mines. 234 p.

Lobato, Lydia Maria. Metamorphism, metasomatism and mineralization at Lagoa Real, Bahia, Brazil. D, 1985, University of Western Ontario. 306 p.

Lobdell, Frederick K. Macropaleontology of the Gunn Member, Stony Mountain Formation (Upper Ordovician), Manitoba and North Dakota. D, 1988, University of North Dakota. 382 p.

Lobdell, Frederick K. The Ashern Formation (Middle Devonian) in the Williston Basin, North Dakota. M, 1984, University of North Dakota. 187 p.

Lobdell, John Little. A gravity study of Tarrant County, Texas. M, 1965, Southern Methodist University. 31 p.

Lobeck, Armin K. Physiography in Puerto Rico. D, 1917, Columbia University, Teachers College.

Lobo Guerrero U., Alberto. Notes on the expansivity of soils. M, 1968, Stanford University.

Lobo Guerrero U., Alberto. Slope stability of ground on Coal Mine Ridge, San Mateo County, California. M, 1964, Stanford University.

Lobo, C. F. Petrography and statistical analysis of the Tapeats Sandstone (late Precambrian-Cambrian), southeastern California. M, 1972, University of Southern California.

Lobo, Cyril Francis. Petrology and depositional history of late Precambrian-Cambrian quartzites in the eastern Mojave Desert, southeastern California. D, 1974, University of Southern California.

LoBue, Charles L. Depositional environments and diagenesis, Interlake Formation (Silurian), Williston Basin, North Dakota. M, 1983, University of North Dakota. 233 p.

LoCastro, Richard Peter. The influence of geology and agriculture on ground water quality in Clarke and Frederick counties, VA. M, 1988, University of Virginia. 172 p.

Locat, J. Quaternary geology of the Baie-des-Sables/Trois-Pistoles area, Quebec, with some emphasis on the Goldthwait Sea clays. M, 1976, University of Waterloo. 214 p.

Lochman, Christina. The fauna of the basal Bonneterre Formation of Missouri. D, 1933, The Johns Hopkins University.

Lochman, Christina. The faunal and stratigraphic relations of the (Cambrian) basal Bonneterre Formation (Missouri). M, 1931, Smith College. 285 p.

Lock, Frank Loren. The stratigraphy and structure of the southwest quarter of the Niangua Quadrangle, Missouri. M, 1954, University of Iowa. 83 p.

Lock, Susan. Salt tectonics; a structural and stratigraphic analysis of offshore Louisiana, Gulf of Mexico region. M, 1986, University of Houston.

Lockard, David W. Evaluation of the Golden Sunbeam gold deposit, Custer County, Idaho. M, 1970, University of Idaho. 78 p.

Locke, Augustus. The geology of El Oro and Tlalpujahua mining districts, Mexico. D, 1913, Harvard University.

Locke, Glenn Leslie. The increase in magnetite concentration of Recent sediments due to coal combustion. M, 1984, San Diego State University. 115 p.

Locke, James Leroy. Sedimentation and foraminiferal aspects of the Recent sediments of San Pablo Bay (California). M, 1971, San Jose State University. 100 p.

Locke, Kathleen A. Trace fossil assemblages in selected shelf sandstone. M, 1983, Texas A&M University. 136 p.

Locke, W. W., III. Etching of hornblende as a dating criterion for Arctic soils. M, 1976, University of Colorado.

Locke, William Willard, III. The Quaternary geology of the Cape Dyer area, southeasternmost Baffin Island, Canada. D, 1980, University of Colorado. 348 p.

Locker, Walter Augustine, Jr. The Hoffner Beds of Union and Johnson counties, Illinois. M, 1950, University of Illinois, Urbana.

Lockhart, Andrew. Gravity survey of the central Seward Peninsula. M, 1984, University of Alaska, Fairbanks. 83 p.

Lockhart, Elizabeth Blair. Nature and genesis of caymanite in the Oligocene-Miocene Bluff Formation of Grand Cayman Island, British West Indies. M, 1986, University of Alberta. 121 p.

Lockhead, William. Preglacial drainage of the upper Cayuga Basin (New York). M, 1895, Cornell University.

Locking, Tracy. Hydrology and sediment transport in an anastomosing reach of the upper Columbia River, B.C. M, 1983, University of Calgary.

Locklin, Jo Ann C. Recent foraminifera around petroleum production platforms on the Southwest Louisiana shelf. M, 1981, University of Houston.

Lockman, Dalton. Systematic fracture investigation of Valley and Ridge and Allegheny Plateau rocks in portions of Grant, Hampshire, Hardy, and Mineral counties, West Virginia. M, 1981, Wright State University. 78 p.

Lockner, David Avery. Sensitivity of a numerical circulation model for Lake Ontario to changes in lake symmetry and friction depth and to variable wind stress. M, 1973, University of Rochester.

Lockrem, Larry L. X-ray fluorescence analytical method development and geochemistry of the Rosebud coal seam in the Fort Union Formation. M, 1979, Montana College of Mineral Science & Technology. 210 p.

Lockrem, Timothy Mark. Geology and emplacement of the Slate Mountain volcano-laccolith, Coconino County, Arizona. M, 1983, Northern Arizona University. 103 p.

Lockwood, Charles W., Jr. Yorktown [Miocene] invertebrate fauna hitherto undescribed in the student series. M, 1949, University of Virginia. 138 p.

Lockwood, G. C. Pitzer ion interaction model; applied to nesquehonite solubility in NaCl solutions. M, 1988, University of Waterloo. 46 p.

Lockwood, John Paul. Geology of the Serrania de Jarara area, Guajira Peninsula, Colombia. D, 1965, Princeton University. 237 p.

Lockwood, Mary Glenn. An evaluation of the biofacies of the Francis Formation (Upper Pennsylvanian) in the vicinity of Ada, Oklahoma using cluster analysis. M, 1972, University of Oklahoma. 128 p.

Lockwood, Michael B. The petrogenetic and economic significance of chloritoid in the Wawa greenstone belt. M, 1987, Carleton University. 221 p.

Lockwood, Richard P. Petrologic study of syenites near Wausau, Wisconsin. M, 1967, University of Wisconsin-Madison.

Lockwood, Richard Patrick. Geochemistry and petrology of some Oklahoma and Texas redbed copper occurrences (Flowerpot Formation, Leonardian). D, 1972, University of Oklahoma. 125 p.

Lockyear, Eugene David. Urban geology of the Vanderbilt University campus and vicinity, Nashville, Tennessee. M, 1971, Vanderbilt University.

Loder, Theodore C. Distribution of dissolved and particulate organic carbon in Alaskan sub-polar and estuarine waters. D, 1971, University of Alaska, Fairbanks. 236 p.

Lodewick, Richard B. Geology and petrography of the Tijeras Gneiss, Bernalillo County, New Mexico. M, 1960, University of New Mexico. 63 p.

Lodewick, Richard Ballard. The petrology and stratigraphy of the Earp Formation (upper Pennsylvanian and Permian), Pima and Cochise counties, Arizona. D, 1970, University of Arizona. 228 p.

Lodge, Charles D. Simplified numerical models for the generation of synthetic reflection profiling seismograms and synthetic reflection/refraction seismograms. M, 1970, United States Naval Academy.

Lodge, Lynn. A groundwater reconnaissance of a portion of Nantucket Island, Massachusetts. M, 1975, University of Massachusetts. 87 p.

Lodha, Ganpat S. Quantitative interpretation of airborne electromagnetic response for a spherical model. M, 1973, University of Toronto.

Lodha, Ganpat S. Time domain and multifrequency electromagnetic responses in mineral prospecting. D, 1977, University of Toronto.

Lodise, Lisa. Petrology and geochemistry of Nisyros Volcano (Dodecanese, Greece). M, 1987, Wesleyan University. 245 p.

LoDuca, Steven T. Silurian conservation lagerstatten of New York. M, 1988, University of Rochester.

LoDuca, Steven T. Stratigraphy of the Middle Ordovician Sinnipee Group in northeastern Wisconsin. M, 1986, University of Wisconsin-Milwaukee. 53 p.

Loeb, Derek T. P-wave velocity structure of the crust-mantle boundary beneath Utah. M, 1986, University of Utah. 126 p.

Loeb, Robert Eli. An evaluation of the accuracy and reliability of the pollen record in representing regional forest change in the past century. D, 1984, New York University. 371 p.

Loeblich, Alfred R. Bryozoa from the Ordovician Bromide Formation, Oklahoma. D, 1941, University of Chicago. 23 p.

Loeblich, Alfred Richard, Jr. A study of McLish (Middle Ordovician) Bryozoa of the Arbuckle Mountains, Oklahoma. M, 1938, University of Oklahoma. 55 p.

Loeblich, Helen N. Tappan. Foraminifera from the (Cretaceous) Duck Creek Formation of Oklahoma and Texas. D, 1942, University of Chicago.

Loeffler, Bruce Marston. Major- and trace-element and strontium- and oxygen-isotope geochemistry of the Polvadera Group, Jemez volcanic field, New Mexico; implications for the petrogenesis of the Polvadera Group. D, 1984, University of Colorado. 271 p.

Loeffler, Peter T. Depositional environment and diagenesis, Birdbear Formation (Upper Devonian), Williston Basin, North Dakota. M, 1982, University of North Dakota. 268 p.

Loeffler, Richard J. The stratigraphy of a portion of the east side of the Laramie Range, Laramie County, Wyoming. M, 1939, University of Wyoming. 45 p.

Loeher, Larry Leonard. Fire as a natural hazard; Santa Monica Mountains, California. D, 1983, University of California, Los Angeles. 325 p.

Loel, Wayne F. The Vaqueros fauna of the lower Miocene (California). M, 1917, [Stanford University].

Loen, Jeffrey Scott. Origin of gold placers in the Pioneer District, Powell County, Montana. M, 1986, Colorado State University. 164 p.

Loep, Kenneth J. Upper Eocene foraminifera from the type section of the Yazoo Formation of Mississippi. M, 1970, University of Houston.

Loeser, Cornelius James. Some general principles useful in exploration for tungsten. M, 1947, University of Michigan.

Loeser, Cornelius James. The historical geography of Newport Beach, California. D, 1965, University of California, Los Angeles. 194 p.

Loetterle, Gerald John. Some foraminifera from the Niobrara Formation in Knox and Cedar counties, Nebraska. M, 1933, University of Nebraska, Lincoln.

Loetterle, Gerald John. The micropaleontology of the Niobrara Formation in Kansas, Nebraska, and South Dakota. D, 1937, Columbia University, Teachers College.

Loewenstamm, Heinz Adolph. Geology of the eastern Nazareth Mountains (Palestine). D, 1939, University of Chicago.

Loferski, Patricia J. Petrology of chromite-bearing metamorphosed ultramafic rocks from the Beartooth Mountains, Montana. M, 1980, Virginia Polytechnic Institute and State University.

Lofgren, Benjamin E. The Quaternary geology of southeastern Jordan Valley, Utah. M, 1947, University of Utah. 52 p.

Lofgren, David Carl. The bedrock geology of the southwestern part of the Kachess Lake Quadrangle, Washington. M, 1974, Portland State University. 73 p.

Lofgren, Donald L. Tertiary vertebrate paleontology, stratigraphy, and structure, North Boulder River basin, Jefferson County, Montana. M, 1985, University of Montana. 113 p.

Lofgren, Gary E. Effect of growth parameters of NaCl ice substructure. M, 1965, Dartmouth College. 63 p.

Lofgren, Gary Ernest. Experimental devitrification of rhyolite glass. D, 1969, Stanford University. 108 p.

Lofholm, Stephen T. Geology of the Old Dominion Mine prospect, Granite County, Montana. M, 1985, South Dakota School of Mines & Technology. 120 p.

Lofland, Darlane Kathryn. Petrology and depositional environment of the Middle Canyon Formation (Mississippian), south-central Idaho. M, 1986, Washington State University. 153 p.

Lofstrom, Dotty Mae. Fluid inclusion and stable isotope studies of the Silver Mine District Sn-W-Pb-Zn-Ag deposits, Southeast Missouri. M, 1987, University of Missouri, Columbia. 79 p.

Loft, Genivera E. Geographic influences in the economic development of Manitoba. D, 1925, University of Wisconsin-Madison.

Lofti, Hani Abdel-Latif. Development of a rational compaction specification for cohesive soils. D, 1984, University of Maryland. 489 p.

Loftin, L. K. Water chemistry of the pocosins of the Croatan National Forest, Carteret, Craven, and Jones counties, North Carolina. M, 1985, East Carolina University.

Loftsgaarden, Jan L. Paleoenvironment and economic geology of the Blackhawk Formation (Cretaceous) on North Horn Mountain, Wasatch Plateau, Utah. M, 1977, Brigham Young University.

Loftsson, M. Engineering geology of the Kettle Point Oil Shale. M, 1984, University of Waterloo. 149 p.

Logan, David Craig. Reconnaissance geology and structure of the Shuswap metamorphic complex north-west of Blue River, B.C., Canada. M, 1981, University of Idaho. 73 p.

Logan, Homer H. A ground-water recharge project associated with a flood protection plan in Hudspeth County, Texas; supportive geologic applications. M, 1984, Texas Christian University.

Logan, James Metcalfe. Geochemical constraints on the genesis of Ag-Pb and Zn deposits, Sandon, British Columbia. M, 1986, University of British Columbia. 166 p.

Logan, John Merle. A structural study of the northern margin of the Wet Mountains, Fremont County, Colorado. M, 1962, University of Oklahoma. 73 p.

Logan, John Merle. Structure and petrology of the eastern margin of the Wet Mountains, Colorado. D, 1966, University of Oklahoma. 283 p.

Logan, Katherine J. The geology and environment of deposition of the Kinneman Creek Interval, Sentinel Butte Formation (Paleocene), North Dakota. M, 1981, University of North Dakota. 137 p.

Logan, L. A. Stochastic estimation of states in models of hydrologic systems. D, 1979, University of Waterloo.

Logan, Lloyd Eugene. The paleoclimatic implications of the avian and mammalian faunas of the lower Sloth Cave, Guadalupe Mountains, Texas. M, 1977, Texas Tech University. 72 p.

Logan, Robert Ellis, III. The Boyer Gap; a record of Jurassic intra-arc fold and nappe tectonics in the northern Dome Rock Mountains of southwestern Arizona. M, 1986, San Diego State University. 176 p.

Logan, Robert L. Temporal trends in the geochemistry and petrology of the 1980 Mount St. Helens pyroclastic flow deposits. M, 1982, Western Washington University. 109 p.

Logan, Spencer R. Notes on geological survey methods. M, 1907, University of Kansas.

Logan, Stewart Michael. The feasibility of using a portable proton precession magnetometer in the location of subsurface drainage systems. M, 1977, University of Toledo. 117 p.

Logan, Thomas Francis, Jr. Pleistocene stratigraphy in Glynn and McIntosh counties, Georgia. M, 1968, University of Georgia. 103 p.

Logan, William Newton. A North American epicontinental sea of Jurassic age. D, 1900, University of Chicago. 32 p.

Logan, William Stevenson, IV. Geology of the Agalteca magnetite skarn deposit, central Honduras. M, 1983, University of Texas, Austin. 112 p.

Logel, John Duane. Use of an engineering seismograph for subsurface investigation along the Thurman-Redfield structural zone, Southwest Iowa. M, 1982, University of Iowa. 78 p.

Loggins, Susan Karleane Jones. A paleoenvironmental analysis of a new species of Permian rostroconch recovered from the upper Hueco Formation, Robledo Mountains, Dona Ana County, New Mexico. M, 1988, Northeast Louisiana University. 89 p.

Loghry, James D. Characteristics of favorable cappings from several southwestern porphyry copper deposits. M, 1972, University of Arizona.

Logie, Russell Moore. Some notes on the Sooke Formation, Vancouver Island. M, 1929, University of British Columbia.

Logothetis, John. The mineralogy and geochemistry of metasomatized granitoid rocks from occurrences in the South Mountain Batholith; New Ross area, southwestern Nova Scotia. M, 1985, Dalhousie University. 374 p.

Logsdon, Mark J. The aqueous geochemistry of the Lightning Dock Known Geothermal Resource Area, Animas Valley, Hidalgo County, New Mexico. M, 1981, University of New Mexico. 239 p.

Logsdon, Truman F. Some factors affecting the sedimentation of the Dakota Sandstone of the Southwest Powder River basin, Wyoming, and the relationship between structure, sedimentation and oil accumulation. M, 1954, Kansas State University.

Loh, Abraham Kwan-Yuen. Mechanism of friction and cohesion in clays. D, 1964, Michigan State University. 216 p.

Loh, Spencer E.-Y. Synthesis and fluorine-hydroxyl exchange in the amblygonite series. M, 1975, University of California, Santa Barbara.

Lohman, Clarence, Jr. Geology of the Whiteoak area, Craig and Rogers counties, Oklahoma. M, 1952, University of Oklahoma. 89 p.

Lohman, Henry William. Development of the Diamond Nimrod Mine, Park City, Utah. M, 1905, Columbia University, Teachers College.

Lohman, Kenneth E. Cenozoic nonmarine diatoms from the Great Basin. D, 1957, California Institute of Technology. 160 p.

Lohman, Kenneth E. Diatoms from the Modelo Formation (upper Miocene) near Girard, California. M, 1931, California Institute of Technology. 61 p.

Lohman, Stanley W. Ground water in northeastern Pennsylvania. M, 1938, California Institute of Technology. 312 p.

Lohmann, George P. Paleo-oceanography of the Oceanic Formation (Eocene-Oligocene), Barbados, West Indies. D, 1974, Brown University. 135 p.

Lohmann, K. C. Causative factors of the outer detrital belt House Embayment; a sedimentologic examination of a terrigenous-carbonate depositional system, early Upper Cambrian (Dresbachian), East-central Utah and West-central Nevada. D, 1977, SUNY at Stony Brook. 301 p.

Lohmar, John M. Shelf margin deposits of the Eocene San Diego Embayment. M, 1979, University of California, Santa Barbara.

Lohn, Cecil O. A study of the colloid obtained from weathered lignite. M, 1932, University of North Dakota. 29 p.

Lohnes, R. A. Petrography of Quaternary concretions from western Iowa. M, 1961, [University of Iowa].

Lohnes, Robert Alan. Quantitative geomorphology of selected drainage basins in Iowa. D, 1964, Iowa State University of Science and Technology. 113 p.

Lohr, Burgin Edison. Geology and soils of the Burhenglen Knob Quadrangle. M, 1922, University of North Carolina, Chapel Hill. 36 p.

Lohr, Jerrold R. Tidal channel sedimentation near Duck Key, Florida and Cat Cay, Bahamas. M, 1969, University of Wisconsin-Madison.

Lohr, Lewis Stillman. Geology of the Brock canyon area, Monitor Range, Eureka County, Nevada. M, 1965, University of Nevada - Mackay School of Mines. 44 p.

Lohrengel, Carl F. The geology of the Pleasant Hill Quadrangle of Jackson and Cass counties, Missouri. M, 1964, University of Missouri, Columbia.

Lohrengel, Carl Frederick. Palynology of the Kaiparowits Formation, Garfield County, Utah. D, 1978, Brigham Young University.

Lohse, Edgar Alan. Shallow-marine sediments of the Rio Grande Delta. D, 1952, University of Texas, Austin.

Lohse, Richard L. A heat flow study of Dona Ana County, southern Rio Grande Rift, New Mexico. M, 1980, New Mexico State University, Las Cruces. 101 p.

Loi, Kuong-Soon. Genesis and classification of selected soils intermediate between the red-yellow podzolic and lateritic groups in Sarawak, Malaysia. M, 1980, University of Guelph.

Loiacono, Nancy. Hydrologic and hydrochemical characterization of the shallow ground-water system of the Great Marsh, Indiana Dunes National Lakeshore. M, 1986, Purdue University. 118 p.

Loidolt, Lawrence H. Quartz-feldspar carbonate bodies of the Carrizo Mountains, Texas. M, 1970, University of Arizona.

Loiselle, James Richard. Development of detailed ground magnetic survey instrumentation and its role in mineral exploration. M, 1980, University of Manitoba. 108 p.

Loiselle, Marc Charles. Geochemistry and petrogenesis of the Belknap Mountains Complex and Pliny Range, White Mountain Series, New Hampshire. D, 1979, Massachusetts Institute of Technology. 302 p.

Lojek, Carole Ann. Petrology, diagenesis and depositional environment of the Skinner Sandstones, Desmoinesian, Northeast Oklahoma Platform. M, 1983, Oklahoma State University. 159 p.

Loken, Trygve. Hydrothermal alteration and oil show at the Summer Coon intrusive center, Saguache County, Colorado. M, 1983, Oregon State University. 91 p.

Lokke, Donald Henry. The paleoecology of the type Fresnal Group, Pennsylvanian of New Mexico. M, 1953, Texas Tech University. 100 p.

Lokken, John Carl. A metallurgical investigation of the ore of the Ramshorn Mine, Custer County, Idaho; and a study of classification and gravity concentration as applied to the Pecas ore. M, 1925, University of Idaho. 80 p.

Lolcama, J. L. An investigation of the geochemical evolution of acidic recharge through a non-calcareous soil. M, 1983, University of Waterloo. 155 p.

Lollar, Earl H. Geologic structure of the barrens of the Hollow Springs Quadrangle in Cannon and Coffee counties, Tennessee. M, 1924, Vanderbilt University.

Lollis, Joan Cullen. Chemical fingerprinting of five K-bentonite beds in the Nen Enterprise Member of the Salona Formation in central Pennsylvania. M, 1984, University of Cincinnati. 98 p.

Lomas, Margaret Kathleen. Synthetic seismograms related to detailed geology in the Celtic Field, Saskatchewan. M, 1984, University of Saskatchewan. 254 p.

Lomax, Francis Earl. Structure of the mixed layer and inversion layer associated with patterns of MCC during AMTEX 75. M, 1977, Purdue University. 123 p.

Lombana, Abdon Cortes see Cortes Lombana, Abdon

Lombardi, Leonard Volk. A quantitative study of Recent sediments in Halfmoon Bay, California. M, 1949, Stanford University. 64 p.

Lombardi, Leonard Volk. An analysis of the apparent absence of seismic reflectors in sedimentary geologic sections. D, 1953, Stanford University. 120 p.

Lombardi, Oreste W. Observations on the distribution of chemical elements in the terrestrial saline deposits of Saline Valley, California. M, 1957, New Mexico Institute of Mining and Technology. 51 p.

Lomenda, Melvin George. Cretaceous upper Bearpaw Formation in the Cypress Hills of Saskatchewan. M, 1973, University of Saskatchewan. 235 p.

Lomenick, Thomas F. The geology of the Chickamauga Group of a portion of Raccoon Valley, Knox County, Tennessee. M, 1958, University of Tennessee, Knoxville. 39 p.

Lomenick, Thomas Fletcher. Accelerated deformation of rock salt at elevated temperatures and pressure and its implications for high level radioactive waste disposal. D, 1968, University of Tennessee, Knoxville. 119 p.

Lomerson, William W. An occurrence of pinite rock. M, 1941, Columbia University, Teachers College.

Lomnitz, Cinna. Creep measurements in igneous rocks with some application to aftershock theory. D, 1955, California Institute of Technology. 111 p.

Londe, Michael David. The Colorado Plateau-Basin and Range transition zone in central Utah; thermomechanical modeling and spectral analysis of topographic and gravity data. D, 1986, University of Wyoming. 166 p.

Londergan, John Thomas. Transport parameter determination and modeling of sodium and strontium

plumes at the Idaho National Engineering Laboratory. M, 1987, Texas A&M University. 54 p.

London, David. Lithium mineral stabilities in pegmatites. D, 1981, Arizona State University. 259 p.

London, David. Occurrence and alteration of lithium minerals, White Picacho Pegmatites, Arizona. M, 1979, Arizona State University. 131 p.

London, William W. A geological and engineering study of the Mustang Pool, Canadian County, Oklahoma. M, 1973, University of Oklahoma. 78 p.

Londono, Oscar Ospino. Aluminum-organic matter complexes of the A horizon of soil acid soils. D, 1967, University of California, Riverside. 140 p.

Londry, John W. Paleomagnetism of Archean rock units and mineralization in the Noranda area, Quebec. M, 1975, University of Windsor. 81 p.

Lonergan, Stephen Colnon. A simulation/optimization model for resolving economic/ecological conflicts; philosophy, theory and application to the Chesapeake Bay. D, 1981, University of Pennsylvania. 207 p.

Loney, Robert Ahlberg. Geology of Crescent Bay area, Olympic Peninsula, Washington. M, 1951, University of Washington. 113 p.

Loney, Robert Ahlberg. Structure and stratigraphy of the Pybus-Gambier area, Alaska. D, 1961, University of California, Berkeley. 205 p.

Loney, Sabra Osborn. Stratigraphy and structure of the Castle Dome area, Cochise County, Arizona. M, 1958, University of California, Berkeley. 73 p.

Long, Aubrey Lamar. The relationship between the structure and geochemistry of the copper deposits of the Hillabee Greenstone in the Millerville region, Clay County, Alabama. M, 1981, University of Alabama. 233 p.

Long, Austin. Isotopic composition of lead from the Coeur d'Alene District, Idaho. M, 1959, Columbia University, Teachers College.

Long, Austin. Late Pleistocene and Recent chronologies of playa lakes in Arizona and New Mexico. D, 1966, University of Arizona. 161 p.

Long, Clarence Sumner, Jr. Basal Cretaceous strata, southeastern Colorado. D, 1966, University of Colorado. 495 p.

Long, Darrel Graham Francis. The stratigraphy and sedimentology of the Chibougamau Formation (Quebec, Canada). M, 1973, University of Western Ontario. 305 p.

Long, Darrel Graham Francis. The stratigraphy and sedimentology of the Huronian (lower Aphebian) Mississagi and Serpent formations. D, 1976, University of Western Ontario. 291 p.

Long, David Timothy. Hydrogeochemical study of carbonate groundwaters of an urban area in northeastern Illinois. M, 1973, University of Illinois, Chicago.

Long, David Timothy. Mobilization of selected trace elements from shales. D, 1977, University of Kansas. 194 p.

Long, E. G. Geology of the Enos Creek area, Bighorn Basin, Hot Springs County, Wyoming. M, 1957, University of Wyoming. 85 p.

Long, E. Tatum. Obvious conchological criteria for the differentiation of Devonian-Carboniferous border terranes. M, 1916, Cornell University.

Long, Eleanor T. Variation in the schist of northern Manhattan Island (New York). M, 1919, Columbia University, Teachers College.

Long, Emilie O. Phylogeny of the genus Turbo. M, 1910, Columbia University, Teachers College.

Long, George Henry. A Cocorp deep seismic reflection profile across the San Andreas Fault, Parkfield, California. M, 1981, Cornell University.

Long, J. S., Jr. Geology of the Phlox Mountain area, Hot Springs and Fremont Counties, Wyoming. M, 1959, University of Wyoming. 59 p.

Long, James Henry De see De Long, James Henry

Long, James Howard. The behavior of vertical piles in cohesive soil subjected to repetitive horizontal loading. D, 1984, University of Texas, Austin. 358 p.

Long, Jane C. S. Investigation of equivalent porous medium permeability in networks of discontinuous fractures. D, 1983, University of California, Berkeley. 293 p.

Long, Jerome Pillow, III. The Gay Head landslides, Martha's Vineyard, Massachusetts; causes and remedies. D, 1971, University of Connecticut. 133 p.

Long, John Douglas. Sedimentology of the Glenwood Member of the Middle Ordovician St. Peter Sandstone of southern Wisconsin. M, 1988, University of Wisconsin-Madison. 133 p.

Long, John F. Geology of a part of the Fish Creek Range, central Nevada. M, 1973, University of California, Riverside. 151 p.

Long, John M. Seismic stratigraphy of part of the Campeche Escarpment, southern Gulf of Mexico. M, 1978, University of Texas, Austin.

Long, Joseph Bacon. The paleogeology of central Michigan. M, 1952, Michigan State University. 17 p.

Long, Keith Richard. Estimating the number and sizes of undiscovered oil and gas pools. D, 1988, University of Arizona. 172 p.

Long, Keith Richard. Ground preparation and zinc mineralization in bedded and breccia ores of the Monte Cristo Mine, North Arkansas. M, 1983, University of Michigan.

Long, L. L. Mathematical modeling of river water temperatures. D, 1972, University of Missouri, Rolla.

Long, Leland Timothy. A study of short-period microseisms. M, 1964, New Mexico Institute of Mining and Technology. 54 p.

Long, Leland Timothy. Transmission and attenuation of the primary seismic wave, Δ=100 to 600 km. D, 1968, Oregon State University. 110 p.

Long, Leon E. Age of the metamorphism of the rocks of the Manhattan Prong (New York). M, 1958, Columbia University, Teachers College.

Long, Leon Eugene. Study of the metamorphic history of the New York City area (New York-New Jersey) using isotopic age methods. D, 1959, Columbia University, Teachers College. 146 p.

Long, Miner B. Origin of the Conococheague Limestone. D, 1953, The Johns Hopkins University.

Long, Miner Barton. Origin of the Conococheague Limestone (Cambrian, Maryland, Virginia, Pennsylvania). D, 1954, The Johns Hopkins University.

Long, Morris Andrew. Geology of the Millheim-Coburn-Aaronburg area, Pennsylvania. M, 1950, Lehigh University.

Long, P. E. Precambrian granitic rocks of the Dixon-Penasco area, northern New Mexico; a study in contrasts. D, 1977, Stanford University. 586 p.

Long, Persis Marian. Study of biological principles indicated by fossils and the planning of exhibits to illustrate these principles. M, 1936, University of Michigan.

Long, Richard Arthur. Origin and petrology of a portion of the southern complex near Palmer, Marquette County, Michigan. M, 1959, Michigan State University. 38 p.

Long, Richard F. de *see* de Long, Richard F.

Long, Robert Bryan. On generation of ocean waves by a turbulent wind. D, 1971, University of Miami. 202 p.

Long, Robert Edwin. The stratigraphy and paleontology of the type area of the Pancho Rico Formation, Salinas Valley, California. M, 1957, University of Southern California.

Long, Robert S. The stability of feldspar structures containing Ga, Fe and Ge under conditions of high pressure. D, 1966, University of Chicago. 113 p.

Long, Roney C. Geology and mineral deposits of the Red Ledge Mine area, Adams County, Idaho. M, 1976, Oregon State University. 115 p.

Long, William A. Geology of the north-central part of the Missoula Quadrangle, Montana. M, 1956, University of Montana. 33 p.

Long, William E. Stratigraphy of the Ohio Range, Horlick Mountains, Antarctica. D, 1964, Ohio State University.

Long, William Ellis. Geology of Mt. Glossopteris, central range of the Horlick Mountains, Antarctica. M, 1961, Ohio State University.

Longacre, Mark. Satellite magnetic investigation of South America. M, 1982, Purdue University. 57 p.

Longacre, Susan Ann Burton. Trilobites of the upper Cambrian ptychaspid biomere, Wilberns Formation, central Texas. D, 1968, University of Texas, Austin. 286 p.

Longacre, William A. A study of the problem of depth determination by means of earth resistivity measurements. M, 1940, Michigan Technological University. 35 p.

Longdale, John Tipton. Geology of the gold-pyrite belt of the northeastern Piedmont, Virginia. D, 1924, University of Virginia. 110 p.

Longden, Markham R. Structural geology and hydrocarbon potential of the east half of the Snedaker Basin 7.5 minute quadrangle, west-central Montana. M, 1986, University of New Mexico. 145 p.

Longe, Robert Vernon. A new approach to heavy mineral size distribution. M, 1965, McGill University.

Longenbaugh, R. A. Statistical techniques for predicting river accretions as applied to the South Platte River (Henderson-Fort Lupton). M, 1962, Colorado State University. 120 p.

Longgood, Theodore Edward, Jr. Geology of Chromo Peak area, Rio Arriba County, New Mexico. M, 1960, University of Texas, Austin.

Longhi, John A. Iron, magnesium and silica in plagioclase. D, 1976, Harvard University.

Longiaru, Samual Joseph. Structure and metamorphism of the Northeast Amisk Lake area, Saskatchewan. D, 1980, University of Saskatchewan. 119 p.

Longiaru, Samuel Joseph. Tectonic evolution of the Oak Creek volcanic roof pendant, eastern Sierra Nevada, California. D, 1987, University of California, Santa Cruz. 242 p.

Longley, William Warren. The geology of the Kamshigama map area, Abitibi District, Quebec. D, 1937, University of Minnesota, Minneapolis. 52 p.

Longman, Mark W. Depositional history, paleoecology, and diagenesis of the Bromide Formation (Ordovician), Arbuckle Mountains, Oklahoma. D, 1976, University of Texas, Austin. 327 p.

Longmire, Patrick. Geochemistry, diagenesis, and contaminant transport of uranium tailings, Grants mineral belt, New Mexico. M, 1984, University of New Mexico. 182 p.

Longoria-Trevino, Jose Francisco. Stratigraphic, morphologic and taxonomic studies of Aptian planktonic foraminifera. D, 1972, University of Texas at Dallas. 285 p.

Longshore, John David. Chemical and mineralogical variations in the Virgin Islands Batholith and its associated wall rocks. D, 1966, Rice University. 150 p.

Longshore, John David. Differentiation of a lamprophyre sill. M, 1959, Rice University. 52 p.

Longshore, Judith Clark. Comparative morphology of orthoclase crystals. M, 1965, Rice University. 94 p.

Longstaffe, Frederick John. The oxygen isotope and elemental geochemistry of Archean rocks from northern Ontario. D, 1977, McMaster University. 564 p.

Longwell, Chester Ray. Geology and mineralogy of the Wellington Mine, Breckenridge, Summit County, Colorado. M, 1916, University of Missouri, Columbia.

Longwell, Chester Ray. Geology of the Muddy Mountains, Nevada, with a section to the Grand Wash Cliffs in Arizona. D, 1920, Yale University.

Longyear, Robert D. The location of ore bodies by deflection of the plumb line. M, 1915, University of Wisconsin-Madison.

Lonker, S. W. Geology of the Mt. Zircon Quadrangle, NW Maine. M, 1975, University of Wisconsin-Madison.

Lonker, Steven Wayne. Conditions of metamorphism in high grade pelites from the Frontenac Axis, Ontario, Canada. D, 1979, Harvard University.

Lonn, Jeff. Structural geology of the Tarkio area, Mineral County, Montana. M, 1985, University of Montana. 51 p.

Lonsdale, John Tipton. Geology and ore deposits of Bedrock Gulch, La Plata County, Colorado. M, 1921, University of Iowa. 60 p.

Lonsdale, Peter Frank. Abyssal geomorphology of a depositional environment at the exit of the Samoan Passage. D, 1974, University of California, San Diego.

Lonsdale, Richard E. Water problems of the Simi Valley, California. M, 1953, University of California, Los Angeles. 117 p.

Lonsinger, Lu-Anne P. Lithostratigraphy and depositional systems of Pennsylvanian sandstones in the Arkoma Basin. M, 1980, University of Arkansas, Fayetteville.

Lontos, Jimmy T. The geology of the Coweta area, Wagoner, Muskogee, and Okmulgee counties, Oklahoma. M, 1952, University of Oklahoma. 55 p.

Loo, Walter Wei-To. The influence of vertical variations in lithology on a mathematical management model for the Ogallala Aquifer, Texas County, Oklahoma. M, 1972, Oklahoma State University. 69 p.

Loocke, Jack E. Growth history of the Hainesville salt dome, Wood County, Texas. M, 1978, University of Texas, Austin.

Loofbourow, John Stewart. Geology of a portion of the Santa Paula Quadrangle, Ventura County, California. M, 1941, University of California, Los Angeles.

Looff, Karl M. Petrology of the Higginsville limestone (Pennsylvanian), north-central Missouri. M, 1969, University of Missouri, Columbia.

Looff, Kurt M. Paleohydraulic interpretation of upper Abo paleochannel types in central and west-central New Mexico. M, 1987, University of New Mexico. 133 p.

Looker, Robert B. Rock terraces in the Connecticut River gorge, Connecticut. M, 1960, Clark University.

Lookingbill, John L. Stratigraphy and structure of the Gallina Uplift, Rio Arriba County, New Mexico. M, 1953, University of New Mexico. 118 p.

Loomis, Alden Albert. Petrology of the Fallen Leaf Lake area, California. D, 1961, Stanford University. 166 p.

Loomis, Dana Paul. Miocene stratigraphic and tectonic evolution of the El Paso Basin, California. M, 1985, University of North Carolina, Chapel Hill. 116 p.

Loomis, Edward Charles. The effects of anaerobically digested municipal sanitary sewage sludge on the oxidation of pyrite. M, 1981, Southern Illinois University, Carbondale. 129 p.

Loomis, Timothy Patrick. Metamorphic and structural effects of the emplacement of the Ronda ultramafic massif (Triassic-Miocene), southern Spain. D, 1971, Princeton University. 110 p.

Loon, J. C. Cartographic generalization of digital terrain models. D, 1978, Ohio State University. 214 p.

Looney, Hugh Marvin. Ostracodes from the Bear River Formation of southwestern Wyoming. M, 1948, University of Missouri, Columbia.

Looney, R. Michael. Late Quaternary geomorphic evolution of the Colorado River, Bastrop and Fayette counties, Texas. M, 1977, University of Texas, Austin.

Loop, Taylor H. Physical oceanography and sedimentary character of Little and Shark harbors of Catalina island, California. M, 1969, University of Southern California.

Loop-Avery, Mary Louise van der *see* van der Loop-Avery, Mary Louise

Loope, David Bittle. Deposition, deflation and diagenesis of upper Paleozoic eolian sediments, Canyonlands National Park, Utah. D, 1981, University of Wyoming. 180 p.

Loope, Walter Lee. Relationships of vegetation to environment in Canyonlands National Park. D, 1977, Utah State University. 142 p.

Loos, Kenneth Dingwell. A sedimentological study of the transition between the Pritchard Formation and Ravalli Group, Salish Mountains, northwestern Montana. M, 1985, University of Cincinnati. 153 p.

Lootens, Douglas Joseph and Holmes, Douglas Allen. Geology of the Silver Mountain area, Huerfano County, Colorado. M, 1959, University of Michigan.

Lootens, Douglas Joseph. Structure and petrography of the east side of the Sierrita Mountains, Pima County, Arizona. D, 1965, University of Arizona. 277 p.

Lopes, Thomas J. Hydrology and water budget of Owens Lake, California. M, 1987, University of Nevada. 128 p.

Lopes, Vincente Lucio. A numerical model of watershed erosion and sediment yield. D, 1987, University of Arizona. 162 p.

Lopez Correa, Victor Julio Lopez. Electromagnetic soundings in California, New Mexico and Wisconsin. D, 1981, University of Texas at Dallas. 216 p.

Lopez Escobar, Leopoldo. Plutonic and volcanic rocks from central Chile (33°-42°S); geochemical evidence regarding their petrogenesis. D, 1975, Massachusetts Institute of Technology. 270 p.

Lopez Eyzaguirre, Carlos J. Study of the weathering of basic, intermediate, and acidic rocks under tropical humid conditions. D, 1973, Colorado School of Mines. 113 p.

Lopez P., Wilfredo Armando. Cokriging gold-silver auriferous mineralization; a multivariate geostatistical approach. M, 1988, Colorado School of Mines. 234 p.

Lopez, Carlos. Elemental distributions in the components of metalliferous sediments from the Bauer and Roggeveen basins, Nazca Plate. M, 1978, Oregon State University. 154 p.

Lopez, Carlos Alberto. Thermal structure and hydrogeochemistry of the Momotombo geothermal field, Nicaragua, C.A. M, 1982, Kent State University. 174 p.

Lopez, Cynthia M. Early diagenesis of sands and sandstones from the Middle America Trench and trench slope, offshore Mexico and Guatemala. M, 1981, University of Texas, Austin.

Lopez, David A. Geology of the Datil area, Catron County, New Mexico. M, 1975, University of New Mexico. 72 p.

Lopez, David A. Stratigraphy of the Yellowjacket Formation of east-central Idaho. D, 1981, Colorado School of Mines. 252 p.

Lopez, Hector S. J. L. Environmental isotope and geochemical investigation of groundwater in Big Bend National Park, Texas. M, 1984, Texas A&M University. 114 p.

Lopez, Liisa Maki. Earthquake and nuclear explosion location using the global seismic network. D, 1983, University of Texas at Dallas. 196 p.

Lopez, Liisa Maki. Geomagnetic variation study in Hidalgo and Grant counties, southwestern New Mexico. M, 1978, University of Texas at Dallas. 83 p.

Lopez, Linares. Statistics of the elastic parameters and their possible correlation with sedimentary rocks. M, 1956, Massachusetts Institute of Technology. 28 p.

Lopez, Miguel Ramirez *see* Ramirez Lopez, Miguel

Lopez, Raymond. Chemical correlation between the upper bench, No. 8 coal and adjacent roof rock, Belmont Co., Ohio. M, 1978, [University of Toledo].

Lopez, Stephen G. The fauna and depositional environment of the Chesterian Helms Formation, southern New Mexico and West Texas. M, 1984, New Mexico State University, Las Cruces. 87 p.

Lopez, V. M. Primary mineralization at Chuquicamata, Chile, South America. D, 1937, Massachusetts Institute of Technology. 214 p.

Lopez, V. M. Primary sulphide mineralization at Chuquicamata, Chile, South America. M, 1936, Massachusetts Institute of Technology. 82 p.

Lopez-Escobar, Leopoldo. Appalachian rhyolites; geochemical data concerning their origin. M, 1972, Massachusetts Institute of Technology. 134 p.

Lopez-Martinez, Margarita. A $^{40}Ar/^{39}Ar$ geochronological study of komatiites and related rocks. D, 1985, University of Toronto.

Lopez-Rendon, Jorge E. Geology, mineralogy and geochemistry of the Cerro Matoso nickeliferous laterite, Cordoba, Colombia. M, 1986, Colorado State University. 378 p.

LoPiccolo, Robert D. Deposition of sediment and development of secondary structures in the Wildhorse Mountain Formation (Pennsylvanian) near Big Cedar, Oklahoma. D, 1977, Louisiana State University. 178 p.

LoPiccolo, Robert David. Sediment-topography relationships in the inner nearshore zone, southeastern Lake Michigan. M, 1972, Western Michigan University.

Lopik, Jack Richard Van *see* Van Lopik, Jack Richard

Lorandi, Francisco Vicente Vidal *see* Vidal Lorandi, Francisco Vicente

Lorandi, V. M. Vicente Vidal *see* Vicente Vidal Lorandi, V. M.

Lorber, Harvey Raymond. Local statistical variations in the composition of marine manganese nodules. M, 1965, University of California, Berkeley. 86 p.

Lorber, Peter Mark. The Kigoma Basin of Lake Tanganyika; acoustic stratigraphy and structure of an active continental rift. M, 1984, Duke University. 73 p.

Lorberg, Eberhard F. A landslide (Recent) near Edmonton (Alberta). M, 1971, University of Alberta. 119 p.

Lord, Arthur C. Glacial water levels in southeastern Massachusetts. M, 1959, Clark University.

Lord, C. J., III. The chemistry and cycling of iron, manganese, and sulfur in salt marsh sediments. D, 1980, University of Delaware, College of Marine Studies. 188 p.

Lord, C. S. Geology in the vicinity of Beresford Lake, Manitoba. D, 1937, Massachusetts Institute of Technology. 184 p.

Lord, Clifford Symington. A study of tetrahedrite in some British Columbia ores. M, 1933, University of British Columbia.

Lord, Gregory David. Stratigraphy, petrography and depositional environments of the Twin Creek Limestone-Arapien Shale, northern and central Utah. M, 1985, University of Utah. 87 p.

Lord, Jacques Passerat. Seismic stratigraphic investigation of the deep eastern Gulf of Mexico. M, 1986, Rice University. 290 p.

Lord, Mark Leavitt. Paleohydraulics of Pleistocene drainage development of the Souris, Des Lacs, and Moose Mountain spillways, Saskatchewan and North Dakota. M, 1984, University of North Dakota. 162 p.

Lord, Mark Leavitt. Sedimentology and stratigraphy of Glacial Lake Souris, North Dakota; effects of a glacial-lake outburst. D, 1988, University of North Dakota. 221 p.

Lord, Robert L. Heavy mineral suites of the northwestern shelf of the Gulf of Mexico. M, 1955, University of Oklahoma. 95 p.

Lorek, Edward G. Geochemistry and petrogenesis of the Proterozoic Pater metavolcanic suite, Spragge, Ontario. M, 1987, Brock University. 185 p.

Loren, J. D. Amplitude studies by model seismology of reflections from a free surface. M, 1961, University of Minnesota, Minneapolis.

Lorens, R. B. A study of biological and physical controls on the trace metal content of calcite and aragonite. D, 1978, University of Rhode Island. 411 p.

Lorenson, Thomas D. A study of acoustic sounder records at Gualala, California, 1982. M, 1984, San Diego State University. 103 p.

Lorenz, Brenna E. A petrological and geochemical study of the South Lake igneous complex, Newfoundland. M, 1980, SUNY at Buffalo. 54 p.

Lorenz, Brenna Ellen. A study of the igneous intrusive rocks of the Dunnage Melange, Newfoundland. D, 1985, Memorial University of Newfoundland. 220 p.

Lorenz, Donald Paul. The bedrock geology of Dakota and Dixon counties, Nebraska. M, 1956, University of Nebraska, Lincoln.

Lorenz, Douglas M. Integrated basin analysis of the Dunkard Group in Ohio, West Virginia and Pennsylvania. M, 1971, Miami University (Ohio). 215 p.

Lorenz, Douglas McNeil. Edenian (upper Ordovician) benthic community ecology in north-central Kentucky. D, 1973, Northwestern University.

Lorenz, Eleanor Mary. Physiographic provinces of northeastern United States. M, 1924, University of Cincinnati. 77 p.

Lorenz, Howard Wilhelm. Geology and ground water resources of the Helena Valley, Montana. M, 1949, University of Nebraska, Lincoln. 49 p.

Lorenz, John. Triassic sediments and basin structure of the Kerrouchen Basin, central Morocco. M, 1974, University of South Carolina.

Lorenz, John Clay. Sedimentary and tectonic history of the Two Medicine Formation, Late Cretaceous (Campanian), northwestern Montana. D, 1981, Princeton University. 215 p.

Lorenzen, D. J. Applications of trend-surface analysis for investigation of structure and prediction of gypsum occurrences in north-eastern Kansas. M, 1973, Kansas State University. 57 p.

Lorenzen, Robert M. The areal geology and stratigraphy of the southeastern quarter of the Richland Center Quadrangle (Wisconsin). M, 1952, University of Wisconsin-Madison.

Loretto, Thomas McLean. Spectral, pseudo-autocorrelation, and cepstral analysis for Pn-pPn and Pn-sPn delay time recovery. M, 1984, Indiana University, Bloomington. 92 p.

Lorig, Loren Jay. A hybrid computational model for excavation and support design in jointed media. D, 1984, University of Minnesota, Minneapolis. 246 p.

Loring, Anne K. Temporal and spatial distribution of basin-range faulting in Nevada and Utah. M, 1972, University of Southern California.

Loring, Arthur P. Distribution of planktonic foraminifera in the south Atlantic Ocean. D, 1966, New York University. 142 p.

Loring, D. H. Geology of the White Rock-Black River area, Nova Scotia. M, 1956, Acadia University.

Loring, Ralph C. Design and calibration of an astatic susceptibility meter. M, 1939, Colorado School of Mines. 14 p.

Loring, Richard Blake. The geology of the Martinsburg Formation (middle and upper Ordovician), northern Berks County, Pennsylvania. M, 1969, George Washington University.

Loring, William B. The geology and ore deposits of the Mountain Queen area, northern Swisshelm Mountains, Arizona. M, 1947, University of Arizona.

Loring, William Bacheller. Geology and ore deposits of the northern part of the Big Indian District, San Juan County, Utah. D, 1959, University of Arizona. 197 p.

Lorkowski, Robert Michael. Petrologic description and possible source areas for the Table Mountain Formation of southeastern San Diego County, California. M, 1981, San Diego State University.

Lorson, Richard C. Petrology of the Precambrian igneous and metamorphic rocks on western Casper Mountain, Wyoming. M, 1977, University of Akron. 68 p.

Lortie, Johanne. Attenuation of the effect of harmonic distortion on synthetic Vibroseis data using an "exact" wave-shaping filtering method. M, 1988, University of Calgary. 169 p.

Lortie, Ralph Burton. Stratabound copper deposits in rhyolitic ignimbrites and lavas, Copiapo District, Atacama, Chile. D, 1979, Queen's University. 429 p.

los Reyes, A. Guitron de *see* Guitron de los Reyes, A.

Losche, Craig Kendall. Soil genesis and forest growth on steeply sloping landscapes of the Southern Appalachians. D, 1967, North Carolina State University. 194 p.

Losee, Bruce Anthony. Rayleigh-wave dispersion applied to lithospheric structure in Canada. M, 1980, Purdue University. 85 p.

Losh, Steven Lawrence. Fluid migration and interaction in brittle-ductile shear zones, central Pyrenees, France. D, 1985, Yale University. 228 p.

Losher, Albert Justin. The geochemistry of sediments and mine tailings in the Alice Arm area. M, 1985, University of British Columbia. 189 p.

Losier, Lisanne. Interprétation aeromagnétique de la ceinture volcanique de l'Abitibi. M, 1985, Ecole Polytechnique. 105 p.

Loskot, Carole Lynn. Deposition of cave material in Wind Cave. M, 1973, South Dakota School of Mines & Technology.

Losonsky, George. The structural and lithologic setting of tectonic stylolites in the northern end of the Abbs Valley Anticline in Monroe County, West Virginia. M, 1983, University of Cincinnati. 105 p.

Lossman, Edward A. Study of the relationship between grain size and capillary pressure. M, 1954, Ohio State University.

Lothringer, Carl J. Geology and tectonics of an isolated Lower Ordovician allochthon, Rancho San Marcos, northwestern Baja California, Mexico. M, 1983, San Diego State University. 112 p.

Lotimer, A. R. Groundwater flow in a multiple aquifer system at Kitchener, Ontario. M, 1985, University of Waterloo. 118 p.

Lotspeich, Frederick B. A study of the Palouse catena [Whitman County, Washington]. M, 1952, Washington State University. 70 p.

Lott, Frederick S. Geology of the Monroe District, Orange County, New York. M, 1928, Columbia University, Teachers College.

Lott, Thomas L., Jr. Petrography, major element chemistry, and geology of uraniferous igneous rocks in the Turtle Lake Quadrangle, Washington. M, 1982, University of Georgia. 107 p.

Lou, Ken-An. On the numerical difficulties and their solutions for implementing plasticity models for soils. D, 1988, University of Akron. 282 p.

Lou, Wellington Coimbra. Mathematical modeling of earth dam breaches. D, 1981, Colorado State University. 235 p.

Lou, Y.-S. Stochastic simulation of earthquakes. D, 1975, University of Pennsylvania. 271 p.

Loubere, Paul Walter. Properties of the oceans reflected in the sea-bed distribution of Quaternary planktonic foraminifera; including a study of the limits of empirical paleo-oceanographic models and the recognition and interpretation of faunal assemblages lacking modern counterparts. D, 1981, Oregon State University. 110 p.

Loucks, Robert E. Pearsall Formation, Lower Cretaceous, South Texas; depositional facies and carbonate diagenesis and their relationship to porosity. D, 1976, University of Texas, Austin. 179 p.

Loucks, Robert R. Platinum-gold-copper mineralization, central Medicine Bow Mountains, Wyoming. M, 1977, Colorado State University. 315 p.

Loucks, Robert Ray. Zoning and ore genesis at Topia, Durango, Mexico (Volumes I and II). D, 1984, Harvard University. 628 p.

Loucks, Thomas A. Circular structures in the Tertiary volcanic rocks of eastern Guatemala. M, 1973, Dartmouth College. 98 p.

Loucks, Virginia L. Paleomagnetic and petrographic investigation of the Morgan Creek Limestone, central Texas. M, 1984, University of Oklahoma. 63 p.

Louden, Keith Edward. The origin and tectonic history of the Southwest Philippine Sea. D, 1976, Massachusetts Institute of Technology. 192 p.

Louden, Richard Owen. The Prout Limestone (Middle Devonian) and Plum Brook Shale (Middle Devonian); Description of lithologies and discussion of lithostratigraphic relationships with the Olentangy Shale (Upper Devonian), Ohio. M, 1965, Bowling Green State University. 112 p.

Louden, Robert James. The petrology and structure of the Wet Mountain metamorphics and the San Isabel Batholith margin along South Hardscrabble Creek, Custer County, Colorado. M, 1988, Fort Hays State University. 134 p.

Louderback, George Davis. On the origin of the glaucophane and associated schists of the Coast Ranges; a contribution to the theory of crystalline schists. D, 1899, University of California, Berkeley.

Loudin, Michael George. Structural-compositional models of the lunar interior and interpretation of observed estimates of the Moon's fundamental spheroidal free oscillations. M, 1979, Pennsylvania State University, University Park. 90 p.

Loudon, J. Russell. Petrographic criteria for the recognition of porphyritization. M, 1956, University of Toronto.

Loudon, J. Russell. The origin of the porphyry and porphyry-like rocks of Elbow, New Brunswick. D, 1960, University of Toronto.

Lougee, Richard Jewett. Geology of the Connecticut watershed. M, 1929, [University of Michigan].

Lougee, Richard Jewett. Physiography of the Quinnipac Farmington Lowland in Connecticut. D, 1939, Columbia University, Teachers College.

Lougheed, Milford Seymour. Radioactivity of the rocks at Port Radium, Great Bear Lake, Northwest Territories. D, 1953, Princeton University. 73 p.

Lougheed, Peter John. Geological and geochemical investigation of the Clinton-Colden Lake greenstone belt, Mackenzie mining district, N.W.T. M, 1986, University of Western Ontario. 147 p.

Loughlin, Gerald F. Contribution to the geology of eastern Connecticut. D, 1906, Yale University.

Loughlin, William Dornan. Hydrogeologic controls on water quality, ground water circulation, and collapse breccia pipe formation in the western part of the Black Mesa hydrologic basin, Coconino County, Arizona. M, 1983, University of Wyoming. 117 p.

Loughnane, Brian Keith. Further geologic investigation of a Superfund landfill site, Montgomery County, Pennsylvania. M, 1988, Purdue University. 221 p.

Loughridge, Michael Samuel. Fine-scale topography and magnetic anomalies of the deep sea floor off southern California. D, 1967, Harvard University.

Louie, John Nikolai. Seismic reflection experiments imaging the physical nature of crustal structures in Southern California. D, 1987, California Institute of Technology. 275 p.

Louis, Robert Michael St. *see* St. Louis, Robert Michael

Louisnathan, S. John. The nature of Mg-Al-Si ordering in melilites, $(Ca,Na)_2(Mg,Al)(Al,Si)_2O_7$ and the crystal structure of fresnoite $Ba_2TiOSi_2O_7$. D, 1969, University of Chicago. 122 p.

Loule, Jean-Pierre. Shallow marine sedimentary processes along the Late Devonian Catskill shoreline in Pennsylvania; storm versus tidal influence. M, 1987, Pennsylvania State University, University Park. 162 p.

Lounsbury, John Thomas. A study of the present bedrock surface of Kings Quadrangle, Illinois. M, 1959, Northern Illinois University. 51 p.

Lounsbury, Richard E. Geology of the Ashland City Quadrangle, Cheatham County, Tennessee. M, 1959, University of Tennessee, Knoxville. 47 p.

Lounsbury, Richard William. Petrology of the Nighthawk-Oroville area, Washington. D, 1951, Stanford University. 100 p.

Louque, Roland J. The sedimentology and petrophysics of the Deep Lake Field, Cameron Parish, Louisi-

ana. M, 1982, University of Southwestern Louisiana. 137 p.

Loureiro, Blanor Torres. Flow to a sink above a shallow water table. D, 1980, Colorado State University. 93 p.

Loureiro, Celso de Oliveira. Simulation of the steady-state radon transport from soil into houses with basements under constant negative pressure. D, 1987, University of Michigan. 403 p.

Loureiro, Daniel. Electrochemical determination of the oxygen fugacity of mantle-derived ilmenite megacrysts. D, 1986, SUNY at Albany. 393 p.

Lourenco, Jose S. Analysis of three component magnetic data. D, 1972, University of California, Berkeley. 153 p.

Loutit, Tom Stuart. Miocene stable isotope stratigrapy and paleoceanography. D, 1981, University of Rhode Island. 330 p.

Lovan, Norman Alan. Analysis of an interlobate boundary in the Wisconsinan drift of Kalamazoo County and adjacent areas in southwestern Michigan. M, 1977, Western Michigan University.

Lovato, Joseph. The geohydrologic characteristics of the glacial drift within the Water Quality Management Project Site, Michigan State University. M, 1979, Michigan State University. 136 p.

Love, Alonza H. Kerogen-isolation and thermal maturation study in sedimentary rock sections. M, 1977, Colorado School of Mines. 34 p.

Love, Charles M. The geology surrounding the headwaters of Nowlin, Flat, and Granite creeks, Gros Ventre Range, Teton County, Wyoming. M, 1968, Montana State University. 106 p.

Love, D. A. Petrology and alteration of the Rundle gold deposit, Newton Township, Porcupine Mining Division, Ontario. M, 1986, University of Waterloo. 192 p.

Love, D. L. Cavern development in the Muscatatuck regional slope of southeastern Indiana. M, 1975, Indiana State University. 98 p.

Love, David Waxham. Geology of the Rammel Mountain area, Teton County, Wyoming. M, 1971, University of New Mexico. 124 p.

Love, David Waxham. Quaternary geology of Chaco Canyon, northwestern New Mexico. D, 1980, University of New Mexico. 124 p.

Love, Edward G. Economic geology and chemical technology of clay and clay wares. D, 1878, Columbia University, Teachers College.

Love, Frank R. Foraminiferal biostratigraphy and spectral gamma ray profiling of the Tununk Shale in southeastern Utah. M, 1986, University of Wyoming. 201 p.

Love, George Edmond Wilson, Jr. Interpretation of the depositional environment of cyclic carbonate rocks of the Conococheague group (Cambrian) near Morgantown, southeastern Pennsylvania. M, 1969, Franklin and Marshall College. 75 p.

Love, John David. Geology of the southern margin of the Absaroka Range, Wyoming. D, 1938, Yale University.

Love, John David. The geology of the western end of the Owl Creek Mountains, Wyoming. M, 1934, University of Wyoming. 49 p.

Love, John Eric. A multivariate analysis of lithologic coloration within the Catskill Formation. M, 1977, Pennsylvania State University, University Park. 272 p.

Love, Karen. Petrology of Quaternary travertine deposits, Arbuckle Mountains, Oklahoma. M, 1985, University of Houston.

Love, Linda Lou A. The dacites of the Washakie Needles, Bunsen Peak, and the Birch Hills, Wyoming, and their relationship to the Absaroka-Gallatin volcanic province. M, 1972, University of New Mexico. 86 p.

Love, Michael A. Late Wisconsinan ice movements and deglaciation in the northeastern Porcupine Hills area, Alberta. M, 1977, University of Calgary.

Love, Timothy Christopher. Geochemical correlation of Salt Lake-equivalent pyroclastic deposits in Idaho and Wyoming. M, 1986, University of New Orleans. 114 p.

Love, William Wray. A contribution to the theory of oil accumulation. D, 1935, University of Illinois, Urbana.

Love, William Wray. Porosity and permeability of limestone reservoir rocks. M, 1929, University of Illinois, Chicago.

Lovegreen, J. R. Paleodrainage history of the Hudson Estuary. M, 1974, Columbia University. 152 p.

Lovejoy, Bill P. Paleontology and stratigraphy of the Jemez Springs area, Sandoval County, New Mexico. M, 1958, University of New Mexico. 101 p.

Lovejoy, Donald W. Overthrust Ordovician and the Nannie's Peak ring dike, Lone Mountain, Elko County, Nevada. D, 1958, Columbia University, Teachers College.

Lovejoy, Donald W. Preliminary report on the geology of Nannie's Peak (Lone Mountain), Nevada. M, 1955, Columbia University, Teachers College.

Lovejoy, Earl M. P. The geology of the Lower Clear Creek-Mount Zion area, Jefferson County, Colorado. M, 1951, Colorado School of Mines. 72 p.

Lovejoy, Earl Mark Paul. The Hurricane fault zone and the Cedar Pocket Canyon-Shebit-Gunlock Fault complex, southeastern Utan and northwestern Arizona. D, 1964, University of Arizona. 260 p.

Lovelace, Bobby G. An investigation into the structure and stratigraphy of Saint Martinville field, Saint Martin Parish, Louisiana. M, 1967, University of Southwestern Louisiana.

Lovelace, Kenneth A., Jr. Hydrogeology of crystalline rocks in the Colorado Front Range. M, 1980, Colorado State University. 157 p.

Loveless, A. J. Ion trajectories in mass spectrometer ion sources. M, 1967, University of Toronto.

Loveless, Arthur John. Isotopic composition of gadolinium, samarium and europium in the Abee meteorite, (originally discovered at Abee, Alberta). D, 1970, University of British Columbia.

Loveless, Janet Kay. CR^{3+} coordination in chlorites; refinement of the crystal structure of a chromian chlorite. M, 1978, University of Wisconsin-Madison.

Lovell, Howard Lawrence. Petrology, mineralogy, and trace-element chemistry of the Orvan Brook sulphide deposit, Restigouche County, New Brunswick. M, 1966, Carleton University. 92 p.

Lovell, Julian Patrick Bryan. Sandstones of the Eocene Tyee Formation, Oregon Coast Range. D, 1968, Harvard University. 422 p.

Lovell, Stephen Edd. Depositional and diagenetic characteristics of a phylloid algal mound, upper Palo Pinto Formation, Conley Field, Hardeman County, Texas. M, 1988, Texas A&M University. 111 p.

Lovely, C. J. Hydrologic modeling to determine the effect of small earthen reservoirs on ephemeral streamflow. M, 1976, University of Arizona.

Lovely, Daniel Arthur. The application of remote sensing to the detection of Niagaran pinnacle reefs in the Illinois Basin. M, 1983, Purdue University. 267 p.

Loveman, Michael H. Geology of the Phillips pyrites mine near Peekskill, New York. M, 1910, Columbia University, Teachers College.

Lovenburg, Mervin Frank. Kelp as a geological agent. D, 1971, University of California, Davis. 373 p.

Lovering, John Francis. Structural and compositional studies on selected phases of iron and stony-iron meteorites with new data concerning the origin of the meteorites. D, 1956, California Institute of Technology. 87 p.

Lovering, Thomas S. The solution and precipitation of silica in cold water. D, 1922, University of Minnesota, Minneapolis. 31 p.

Lovering, Thomas Seward. A report on the New World mining district, Park County, Montana. D, 1924, University of Minnesota, Minneapolis. 102 p.

Lovering, Thomas Seward. Petrology of the Henderson Mountain Stock. M, 1923, University of Minnesota, Minneapolis. 26 p.

Lovering, Tom G. The geology of a western portion of the Santa Rita Quadrangle, Grant County, New Mexico. M, 1953, University of Arizona.

Loversen, Robert I. Geology of the Las Llajas Canyon area, California. M, 1947, University of California, Los Angeles.

Lovett, Cole K. Paleomagnetism of the Touchet Beds of the Walla Walla River valley, southeastern Washington. M, 1984, Eastern Washington University. 175 p.

Lovett, Frank D. Areal geology of the Quartermaster area, Roger Mills and Ellis counties, Oklahoma. M, 1960, University of Oklahoma. 81 p.

Lovick, G. P. Petrography and sedimentation of the Upper Mississippian sandstones of the Goddard Formation and the Rod Club and Overbrook members of the Springer Formation in the Ardmore Basin, Oklahoma. M, 1977, University of Texas, Arlington. 66 p.

Lovingood, Daniel. The geology of the southern one-third Philomath and northern one-third of the Crawfordville, Georgia quadrangles. M, 1983, University of Georgia. 243 p.

Lovins, Eric E. An evaluation of the DRASTIC system in assessing groundwater pollution potential for parts of Marshall, Fayette and Madison counties, Kentucky. M, 1988, University of Kentucky. 135 p.

Lovison, Lucia Cecilia. Seismicity variations throughout the Middle America Trench subduction earthquake cycle. M, 1986, Harvard University.

Lovitt, Ronald L. The relation of a sediment size to a fossil plant distribution in strata exposed at Grannies Branch of Goose Creek, Clay County, Kentucky. M, 1980, Eastern Kentucky University. 97 p.

Lovrak, Peter William. A Paleoecological study of the Boggs Member (Pottsville Series) of northern Franklin Township, Tuscarawas County, Ohio. M, 1974, Ohio University, Athens. 98 p.

Lovrak, Steven R. The petrology of the Bisher Formation (Middle Silurian) of south-central Ohio. M, 1978, Miami University (Ohio). 68 p.

Lovseth, Timothy Peter. The Devils Mountain fault zone, northwestern Washington. M, 1975, University of Washington. 29 p.

Low, Bak Kong. Analysis of the behavior of reinforced embankments on weak foundations. D, 1985, University of California, Berkeley. 291 p.

Low, Barry M. The rationale of search for Silurian pinnacle reefs in southwestern Ontario. M, 1976, University of Windsor. 156 p.

Low, Dennis James. Geology of Whistler Mountain, Eureka County, Nevada. M, 1982, University of Nebraska, Lincoln. 127 p.

Low, John Hay. The geology of the Star gold mine and vicinity (southeastern Ontario). D, 1941, University of Toronto.

Low, Steven P. The response of a cylindrical target; implications for ground-penetrating radar. M, 1988, Pennsylvania State University, University Park. 106 p.

Lowden, James E. A combined geologic and gravity analysis of Walker Field, Michigan. M, 1964, Michigan State University. 45 p.

Lowder, Garry George. Studies of volcanic petrology; I, Talasea, New Guinea; II, Southwest Utah. D, 1970, University of California, Berkeley. 275 p.

Lowdon, Jack. Geology and hydrology of the Waynesboro area, Virginia. M, 1955, University of Virginia. 132 p.

Lowe, B. V. Mineralogy and geology of Inexco No. 1 Mine, Jamestown, Boulder County, Colorado. M, 1975, University of Southern California.

Lowe, Donald R. Geology of a part of the Stanford linear accelerator. M, 1964, Stanford University.

Lowe, Donald Ray. Logan Ridge Member of Venado Formation; origin and implications of an Upper Cretaceous slump deposit, Sacramento Valley, California. D, 1967, University of Illinois, Urbana. 75 p.

Lowe, Donald Wayne. An X-ray diffraction study of early frozen type I Portland cement. M, 1963, University of Mississippi.

Lowe, E. Charles, Jr. Engineering properties of the middle and upper Conemaugh Group, west-central, southwestern, and southern sections, Athens, Ohio. M, 1977, Ohio University, Athens. 59 p.

Lowe, Gary Duane. Hydrogeochemical investigation of low temperature geothermal resource potential near the Pine Creek Mine, Inyo County, California. M, 1981, San Diego State University.

Lowe, John Carl. Economic diversification and size of standard metropolitan statistical areas. D, 1969, Clark University. 104 p.

Lowe, Jurt Emil. Storm King Granite at Bear Mountain, New York. D, 1947, Columbia University, Teachers College.

Lowe, Kenneth Lance. Geology of the Ada area, Pontotoc County, Oklahoma. M, 1968, University of Oklahoma. 81 p.

Lowe, Kurt Emil. Silver mineralization at Namiquipa, Chihuahua, Mexico. M, 1938, Columbia University, Teachers College.

Lowe, Michael V. Surficial geology of the Smithfield Quadrangle, Cache County, Utah. M, 1987, Utah State University. 143 p.

Lowe, Nathan Ted. Distribution of heavy metals in the Tennessee River-Fort Loudon Lake system, East Tennessee. M, 1978, University of Tennessee, Knoxville. 30 p.

Lowe, Roger S. Depositional environments of the Salt Wash Member and Recapture Creek Member of the Morrison Formation in southern Montezuma Canyon, San Juan County, Utah. M, 1981, New Mexico Institute of Mining and Technology. 164 p.

Lowell, Gary Richard. Geologic relationships of the Salida area to the Thirtynine Mile Volcanic field of central Colorado. D, 1969, New Mexico Institute of Mining and Technology. 113 p.

Lowell, James Diller. Lower and Middle Ordovician stratigraphy in eastern and central Nevada. D, 1958, Columbia University, Teachers College. 123 p.

Lowell, James Diller. Lower Paleozoic stratigraphy of Hot Creek Canyon and adjacent area, Nye County, Nevada. M, 1957, Columbia University, Teachers College.

Lowell, Robert Paul. An approach to thermal convection problems in geophysics with application to the Earth's mantle and ground water systems. D, 1972, Oregon State University. 116 p.

Lowell, Thomas V. Late Wisconsin ice extent in Maine; evidence from Mount Desert Island and the Saint John River area. M, 1980, University of Maine. 180 p.

Lowell, Thomas Vinal. Late Wisconsin stratigraphy, glacial ice-flow, and deglaciation style of northwestern Maine. D, 1987, SUNY at Buffalo. 182 p.

Lowell, Wayne Russell. Glaciation in the Wallowa Mountains (Oregon). M, 1939, University of Chicago. 90 p.

Lowell, Wayne Russell. The paragenesis of some gold and copper ores of southwestern Oregon. D, 1942, University of Chicago. 97 p.

Lowenfels, Harold Stuart. Application of the finite element method to problems of crustal warping during deglaciation. M, 1974, University of Arizona.

Lowenhaupt, Douglas E. The high temperature crystal structure of Ag_2S-II. M, 1973, Pennsylvania State University, University Park.

Lowenstein, Glenn Robert. The environment of deposition of the Dolton Coal (Upper Pennsylvanian), Palo Pinto Co., Texas. M, 1986, Texas A&M University.

Lowenstein, Tim K. Deposition and alteration of an ancient potash evaporite; the Permian Salado Formation of New Mexico and West Texas. D, 1983, The Johns Hopkins University. 430 p.

Lower, S. R. Use of springs in analysis of the ground-water system at Mount Laguna, San Diego County, California. M, 1977, San Diego State University.

Lowery, Carol Janette. Sedimentation of Cenozoic deposits in western Salt River valley; Arizona. M, 1964, Arizona State University. 22 p.

Lowes, Brian Edward. Metamorphic petrology and structural geology of the area east of Harrison Lake, British Columbia, Canada. D, 1972, University of Washington. 162 p.

Lowes, Brian Edward. The geology of the old Casey Mine, Ontario. M, 1963, Queen's University. 89 p.

Lowey, Grant William. Depositional themes in a turbidite succession, Dezadeash Formation (Jura-Cretaceous), Yukon. M, 1980, University of Calgary. 149 p.

Lowey, Grant William. The stratigraphy and sedimentology of siliciclastic rocks, west-central Yukon, and their tectonic implications. D, 1984, University of Calgary. 175 p.

Lowey, James M. The Pleistocene geology of the Oriskany, New York, 7.5-minute quadrangle. M, 1983, Syracuse University. 66 p.

Lowey, Jennifer Fortune. A lithogeochemical study of the Akie Ba-Pb-Zn mineral district, northeastern British Columbia. M, 1984, University of Calgary. 620 p.

Lowey, Robert Francis. The detailed stratigraphy of northern Hansonburg mining district, Socorro County, New Mexico. M, 1984, New Mexico Institute of Mining and Technology. 108 p.

Lowman, Bambi M. Stratigraphy of the upper Benton and lower Niobrara formations (Upper Cretaceous) Boulder County, Colorado. M, 1977, Colorado School of Mines. 94 p.

Lowman, Paul Daniel, Jr. Geology of the Pine Creek area, Gilpin County, Colorado. D, 1963, University of Colorado. 156 p.

Lowman, Shepard Wetmore. Sedimentary facies in Gulf Coast. D, 1949, Columbia University, Teachers College.

Lowman, Shepard Wetmore. Some Calcaire grossier foraminifera from Grignon, France. M, 1928, Columbia University, Teachers College.

Lown, David J. Petrography, depositional environments and carbonate diagenesis of the Lower Ordovician Mascot Dolomite (K-2 to M-1 interval) near Woodbury, Cannon County, Tennessee. M, 1978, University of Tennessee, Knoxville. 99 p.

Lownes, Richard E. Geology of portions of the Santa Barbara and Goleta quadrangles, California. M, 1959, University of Southern California.

Lowney, Karen Anne. Certain Bear Gulch (Namurian A, Montana) Actinopterygii (Osteichthyes) and a re-evaluation of the evolution of the Paleozoic actinopterygians. D, 1980, New York University. 516 p.

Lowre, D. A. The possible economic significance of cross fold structures in Northwest Ontario. M, 1955, University of Toronto.

Lowrey, Ronald Ovel. Paleoenvironment of the Carmel Formation at Sheep Creek Gap, Daggett County, Utah. M, 1975, Brigham Young University.

Lowrey, William Stephen. Crustal structure of the Basin and Range Province using intermediate period surface wave dispersion. M, 1977, Purdue University.

Lowrie, William. The effects of internal stress on remanence and coercive force in nickel and magnetite. D, 1967, University of Pittsburgh. 180 p.

Lowright, Richard Henry. An analysis of factors controlling deviations in hydraulic equivalence in some modern sands. D, 1971, Pennsylvania State University, University Park. 224 p.

Lowry, Anthony R. A numerical modeling approach employing singularity removal applied to the evaluation of the Bristow interpretive technique for the detection of cavities. M, 1988, University of Wyoming. 58 p.

Lowry, Bruce E. A comparison of magnetic modeling techniques with application to southern New England. M, 1979, University of Rhode Island.

Lowry, Elizabeth J. The southwest end of the Mountain City Window, northeastern Tennessee. D, 1950, Yale University. 174 p.

Lowry, John C. A., Jr. Geology of the Concharty and Conjada Mountain areas of Tulsa, Wagoner, Muskogee, and Okmulgee counties, Oklahoma. M, 1955, University of Tulsa. 105 p.

Lowry, Patrick H. The stratigraphy and petrography of the Cambrian Leverett Formation, Antarctica. M, 1980, Arizona State University. 93 p.

Lowry, Philip. Stratigraphic framework and sedimentary facies of clastic shelf-margin; Wilcox Group (Paleocene-Eocene), central Louisiana. D, 1988, Louisiana State University. 415 p.

Lowry, Wallace D. Geology of the Bear Creek area, Crook and Deschutes counties, Oregon. M, 1940, Oregon State University. 79 map p.

Lowry, Wallace Dean. The geology of the northeast quarter of the Ironside Mountain Quadrangle, Baker and Malheur counties, Oregon. D, 1943, University of Rochester. 107 p.

Lowry-Chaplin, Barbara L. A proposed depositional model of a Lower Mississippian deltaic sequence (Cowbell Member, Borden Formation) in northeastern Kentucky. M, 1987, University of Texas, Arlington. 203 p.

Lowther, George K. Geology of an area near Shawinigan Falls, Province of Quebec. D, 1935, McGill University.

Lowther, George Kenneth. Geology and petrology of the Echo Bay region, Great Bear Lake, Northwest Territories. M, 1933, University of Alberta. 81 p.

Lowther, Harold C. The application of polarization figures and rotation properties to the identification of certain of the lead sulphantimonides. M, 1952, University of Wisconsin-Madison.

Lowther, Jack. The Solomon Sandstone in the foothills of central Alberta. M, 1957, University of Manitoba.

Lowther, John Stewart. A Cretaceous flora from northern Alaska. D, 1957, University of Michigan. 212 p.

Lowy, Robert Michael. Provenance and sediment-dispersal patterns, Westwater Canyon Sandstones, western San Juan Basin, New Mexico. M, 1982, University of New Mexico. 203 p.

Loy, William George. The coastal geomorphology of western Lake Superior. M, 1963, University of Minnesota, Minneapolis.

Loy, William George. The late Bronze Age landscape of the southwest Peloponnese. D, 1967, University of Minnesota, Minneapolis. 252 p.

Lozano, Efraim. Geology of the southwestern Garo area, South Park, Park County, Colorado. M, 1965, Colorado School of Mines. 115 p.

Lozano, Hernando. Trace elements in quartz as a geochemical prospecting tool in Cabarrus and Stanly Counties, North Carolina. M, 1974, University of North Carolina, Chapel Hill. 63 p.

Lozano, J. A. Antarctic sedimentary, faunal, and sea surface temperature responses during the last 230,000 years with emphasis on comparison between 18,000 years ago and today. D, 1974, Columbia University. 419 p.

Lozano, Jose A. Petrology of the Womble Shale and Bigfork chert formations, Montgomery County, Arkansas. M, 1963, University of Tulsa. 52 p.

Lozano-Chavez, Guillermo. Mineralogy and ceramic properties of refractory clays from Missouri and Mexico. M, 1986, University of Missouri, Columbia. 139 p.

Lozinsky, Richard Peter. Geology and late Cenozoic history of the Elephant Butte area, Sierra County, New Mexico. M, 1982, University of New Mexico. 142 p.

Lozinsky, Richard Peter. Stratigraphy and sedimentology of the Santa Fe Group in the Albuquerque Basin, north-central New Mexico. D, 1988, New Mexico Institute of Mining and Technology.

Lozo, Frank Edgar, Jr. Biostratigraphic studies of north Texas Trinity and Fredericksburg (Comanchean) foraminifera. D, 1941, Princeton University. 138 p.

Lozo, Frank Edgar, Jr. Stratigraphic and faunal studies of the Pawpaw Formation north of the Brazos River, with descriptions of new species of foraminifera. M, 1937, Texas Christian University. 96 p.

Lu Huan-Zhang. Genesis of tungsten ore deposits in South China. D, 1983, University of Pennsylvania. 271 p.

Lu, C. L. The effect of complexing agents on the environmental chemistry and bioavailability of aquatic cadmium. D, 1977, University of Michigan. 170 p.

Lu, C.-P. Tectonics, crustal and upper mantle structures of Taiwan. D, 1976, SUNY at Binghamton. 218 p.

Lu, Changsheng. Skarn formation at Tin Creek, Farewell District, Alaska. M, 1988, University of Oregon. 142 p.

Lu, Chih-Ping. A comparative study on the source mechanism of Borrego Mountain earthquake of 9 April 1968 and Parkfield earthquake of 28 June 1966. M, 1973, SUNY at Binghamton. 76 p.

Lu, J. C.-S. Studies on the long-term migration and transformation of trace metals in the polluted marine sediment-seawater system. D, 1976, University of Southern California.

Lu, Jau-Yau. Mathematical models for bed armoring, channel degradation and aggradation. D, 1984, Colorado State University. 228 p.

Lu, Lee. The relation of local pressure fluctuations to large-scale meteorology and the simulation of acoustic-gravity waves in inhomogeneous media. D, 1972, Stanford University. 82 p.

Lu, Ming. Genetic sequence analysis of carbonate and evaporite strata, Upper Mission Canyon (Madison Group), Mississippian, Williston Basin, North Dakota. M, 1986, Colorado School of Mines. 139 p.

Lu, Ming-Tar. Offset amplitude analysis of compressional seismic data for fractured reservoir exploration, Silo Field, Wyoming. D, 1988, Colorado School of Mines. 90 p.

Lu, Po-Yung. Model aquatic ecosystem studies of the environmental fate and biodegradability of industrial compounds. D, 1974, University of Illinois, Urbana.

Lu, Richard Shih-Ming. Magnetic and gravity interpretation of YALOC-69 data from the Cocos plate area. M, 1971, Oregon State University. 105 p.

Lu, Richard Shih-Ming. Perturbation methods in geophysics and oceanography. D, 1974, Oregon State University. 100 p.

Lu, Shan-tan. The role of minerals in the thermal alteration and hydrocarbon generating potential of black shale and humic coals. D, 1987, University of California, Los Angeles. 346 p.

Lubetkin, Lester Kenneth Cantelow. Late Quaternary activity along the Lone Pine Fault, Owens Valley fault zone, California. M, 1980, Stanford University. 85 p.

Lubke, E. Ronald. Hydrogeology of the Huntington-Smithtown area, Suffolk County, New York. M, 1962, New York University.

Lubner, Katherine E. A trace element study of ferric hydroxide produced by iron oxidizing bacteria. M, 1974, Wright State University. 59 p.

Lubowe, Joan K. Stream junction angles in the dentritic drainage patterns. M, 1961, Columbia University, Teachers College.

Luby, Thomas Patrick. An investigation of drift-covered environments using the shallow reflection seismic method, Kalamazoo County, Michigan. M, 1982, Western Michigan University. 110 p.

Luca, Anthony J. Upward continuation and modelling on a spherical Earth. M, 1978, Purdue University. 184 p.

Lucania, John A. The limestone cobble conglomerates and associated Ladentown Basalt, Rockland County, New York. M, 1974, Brooklyn College (CUNY).

Lucas, Elmer Lawrence. Petrographic character of sandstone members of the Springer Formation, with a supplementary study of sandstones of the overlying Pennsylvanian. D, 1934, University of Oklahoma. 132 p.

Lucas, Elmer Lawrence. The petroleum geology of the Princeton (Indiana) oil field. M, 1924, Indiana University, Bloomington. 54 p.

Lucas, Harold E. Geology of part of the Dunn Mountain and Arden 7 1/2-minute quadrangles, Stevens County, Washington. M, 1980, Eastern Washington University. 36 p.

Lucas, James R. Till lithology, Landsat data, and geomorphic surfaces in northwestern Iowa. D, 1977, University of Iowa. 405 p.

Lucas, Margaret Jennifer. The non-marine pelecypods of the Canso Group of the Carboniferous of Nova Scotia. M, 1955, University of Illinois, Urbana.

Lucas, Margaret Jennifer. Variation studies of non-marine pelecypods from the Upper Carboniferous of eastern North America. D, 1957, University of Illinois, Urbana. 200 p.

Lucas, Mark. Structural geology of the Mesozoic Jacksonwald Syncline, Berks County, Pennsylvania. M, 1985, Rutgers, The State University, Newark. 113 p.

Lucas, Peter Thomas. Environment of Salina salt deposition. M, 1954, University of Michigan.

Lucas, Philip E. Geology of the Osage Northeast Quadrangle, Arkansas. M, 1970, University of Arkansas, Fayetteville.

Lucas, Robert Charles. Fusulinids and Pennsylvanian stratigraphy of the Crested Butte area, Gunnison County, Colorado. M, 1959, University of Nebraska, Lincoln.

Lucas, Spencer George. Systematics, biostratigraphy and evolution of early Cenozoic Coryphodon (Mammalia, Pantodonta); (Volumes I and II). D, 1984, Yale University. 673 p.

Lucas, Stephen Ernest. Progressive cataclastic deformation in accretionary prisms; DSDP Leg 66, southern Mexico and onland examples from Barbados and Kodiak Islands. M, 1985, University of California, Santa Cruz.

Lucas-Clark, Joyce Emily. Studies of late Albian dinoflagellates from the Franciscan central belt, northern California. D, 1986, Stanford University. 225 p.

Lucchitta, Baerbel Koesters. Structure of the Hawley Creek area, Idaho-Montana. D, 1966, Pennsylvania State University, University Park. 235 p.

Lucchitta, Ivo. Cenozoic geology of the upper Lake Mead area adjacent to the Grand Wash Cliffs, Arizona. D, 1967, Pennsylvania State University, University Park. 274 p.

Luce, Philip G. The geology of the Mianus River gorge-Bargh Reservoir area (Connecticut). M, 1962, New York University.

Luce, Robert James. A theoretical study of fine-particle magnetization in rocks. D, 1980, University of Pittsburgh. 108 p.

Luce, Robert William. Dissolution of magnesium silicates. D, 1969, Stanford University. 101 p.

Luce, Robert William. Petrography of the igneous rocks, Isle au Haut, Maine. M, 1962, University of Illinois, Urbana.

Lucente, Michael Eugene. Geology of the Middle Pennsylvanian Virginia Formation in northeastern Minnesota. M, 1978, University of Kansas. 116 p.

Lucey, Keith J. Seasonal changes of subsurface water mass in the Mississippi Embayment. M, 1978, Northern Illinois University. 165 p.

Luchetti, Cynthia A. Geology, petrology and geochemistry of the Triassic mafic and chert terranes of the Seldovia Bay Complex, Kenai Peninsula, Alaska. M, 1985, University of Utah. 138 p.

Luchsinger, S. E. Epithermal lead-zinc deposits of Tunisia and Algeria. M, 1955, Columbia University, Teachers College.

Luchterhand, Dennis. The origin of the Housatonic Highlands Gneiss Complex (Precambrian) in the southern part of the South Canaan Quadrangle, Connecticut. M, 1967, University of Wisconsin-Madison.

Lucia, F. Jerry. Igneous geology of the Enger Tower area, Duluth, Minnesota. M, 1954, University of Minnesota, Minneapolis. 31 p.

Lucia, Patrick Chester. Review of experiences with flow failures of tailings dams and waste impoundments. D, 1981, University of California, Berkeley. 230 p.

Lucier, Wallace Anthony. The petrology of the Middle and Upper Devonian Kiskatom and Kaaterskill sandstones (southeastern New York); a vertical profile. D, 1966, University of Rochester. 92 p.

Luciuk, Gerald Michael. The hydrolytic reactions of aluminum as affected by the presence of monosilicic acid. M, 1973, University of Saskatchewan.

Lucius, Jeffrey E. Crustal geology of Ohio inferred from aeromagnetic and gravity anomaly analysis. M, 1985, Ohio State University. 131 p.

Lucke, John Becker. A study of Barnegat Inlet, New Jersey, and related shoreline phenomena. D, 1933, Princeton University. 157 p.

Lucken, John E. Tectonics of southern Phantom Canyon, Fremont County, Colorado. M, 1964, University of Kansas. 48 p.

Luckett, Michael A. Cretaceous and lower Tertiary stratigraphy along the Flint River. M, 1979, University of Georgia.

Luckey, Frederick J. Structural geology of the Paleozoic rocks of the Willsboro Quadrangle, New York. M, 1985, University of Massachusetts. 132 p.

Luckhardt, Paul G. L. New ostracodes from the (Mississippian) Glen Dean Formation (Illinois). M, 1940, University of Chicago. 52 p.

Lucking, John Chase. The role of regulatory delays in the development of geothermal energy in the United States. D, 1981, Stanford University. 308 p.

Luckman, Paul Gavin. Slope stability assessment under uncertainty; a first order stochastic approach. D, 1987, University of California, Berkeley. 486 p.

Lucotte, Marc Michel. The geochemistry of phosphorus in the Saint-Lawrence upper estuary. D, 1987, McGill University.

Lucy, Harold P. A study of some graywackes of northwestern Minnesota with a discussion of def. of graywacke and some related rocks. M, 1930, Northwestern University.

Ludden, Raymond W. Geology of the Campo Bonito area, Oracle, Arizona. M, 1950, University of Arizona.

Ludeman, Frank L. Precipitation chromatography; a possible new tool for the field geologist. M, 1962, Michigan Technological University. 78 p.

Ludena, S. E. A mineralogic study of the mineralization of the Maria Teresa, Venturosa, and Caridad ore deposits, Lima, Peru, South America. M, 1931, Ohio State University.

Ludington, S. D. Application of fluoride-hydroxyl exchange data to natural minerals. D, 1974, University of Colorado. 177 p.

Ludlam, Stuart Dietrich. The banded sediments of Cayuga Lake. D, 1964, Cornell University. 115 p.

Ludlum, Gloria King. Phase equilibrium in the join akermanite-anorthite-gehlenite. D, 1970, Ohio State University. 66 p.

Ludlum, John C. Continuity of the (Lower Cambrian) Hardyston Formation in the vicinity of Phillipsburg, New Jersey. M, 1939, Cornell University.

Ludlum, John C. Structural-stratigraphic interpretations of a part of the Bannock Range, Idaho. D, 1942, Cornell University.

Ludlum, Nathaniel Burroughs, Jr. The head shield of Macropetalichthys rapheidolabis and the shoulder girdle of an unknown euarthrodire. M, 1973, Miami University (Ohio). 47 p.

Ludman, Allan. Geology of the Mount Doherty igneous complex, Jefferson County, Montana. M, 1965, Indiana University, Bloomington. 48 p.

Ludman, Allan. Geology of the Skowhegan Quadrangle, Maine. D, 1969, University of Pennsylvania. 325 p.

Ludowise, Harry. The recognition, investigation, interpretation and treatment of landslides in the Pacific Northwest. M, 1974, Portland State University. 61 p.

Ludvigsen, Phillip John. Development of knowledge based expert systems to aid in hazardous waste management. D, 1987, Utah State University. 204 p.

Ludvigsen, Rolf. Brachiopods and dacryoconorid tentaculites of the Michelle Formation (Emsian), northern Yukon (lower Devonian). M, 1970, University of Western Ontario. 148 p.

Ludvigsen, Rolf. Middle Ordovician trilobites, South Nahanni River area, District of Mackenzie. D, 1975, University of Western Ontario. 524 p.

Ludvigson, Gregory Alan. Landsat-1 identified linears in Northeast Iowa and Southwest Wisconsin and their relation to the Ordovician stratigraphy, structure, and sulfide mineralization of the area. M, 1976, University of Iowa. 196 p.

Ludvigson, Gregory Alan. Petrology of fault-related diagenetic features in the Paleozoic carbonate rocks of the Plum River fault zone, eastern Iowa and Northwest Illinois. D, 1988, University of Iowa. 434 p.

Ludwick, Jimmy Donald. Tritium in meteorites; an investigation of the Norton County achondrite and the Ussuri (Sikhote Alin) siderite. D, 1958, Purdue University. 112 p.

Ludwick, John C. Deep water sand layers off San Diego, California. D, 1950, University of California, Los Angeles.

Ludwig, Claudia Petra. The micropaleontological boundary between the Holocene and Pleistocene sediments in the Gulf of Mexico. M, 1971, Texas A&M University.

Ludwig, Kenneth Raymond. Precambrian geology of the central Mazatzal Mountains, Arizona (Part I) and lead isotope heterogeneity in Precambrian igneous feldspars (Part II). D, 1974, California Institute of Technology. 363 p.

Ludwig, S. L. Sand within the silt; the source and deposition of loess in eastern Washington. M, 1987, Washington State University. 120 p.

Luebke, Laurence Orville. Oligocene stratigraphy of the Lance Creek area, Wyoming. M, 1964, University of Nebraska, Lincoln.

Lueck, Everett William. The geology of the Fulford mining district (Cenozoic), Eagle County, Colorado. M, 1970, University of Iowa. 79 p.

Lueck, Larry. Petrologic and geochemical characterization on the Red Dog and other base-metal sulfide and barite deposits in the Delong Mountains, western Brooks Range, Alaska. M, 1985, University of Alaska, Fairbanks. 156 p.

Lueck, S. L. Computer modelling of uranium species in natural waters. M, 1978, University of Colorado.

Luedemann, Lois Ann Weiser. A mineralogical study of several hydrous vanadates. D, 1956, Pennsylvania State University, University Park. 249 p.

Luedke, Robert George. Stratigraphy and structure of the Miners Mountain area, Wayne County, Utah. M, 1953, University of Colorado.

Lueninghoener, Gilbert Carl. A lithologic study of some typical exposures of the Ogallala Formation in western Nebraska. M, 1934, University of Nebraska, Lincoln.

Lueninghoener, Gilbert Carl. The post-Kansan geologic history of the lower Platte Valley area. D, 1947, University of Nebraska, Lincoln.

Luepke, Gretchen. Petrology and stratigraphy of the Scherrer Formation (Permian) in Cochise County, Arizona. M, 1967, University of Arizona.

Luessen, Michael J. Geology of the Massacre volcanic complex, eastern Snake River plain, Idaho. M, 1987, Idaho State University. 163 p.

Luetgert, James Howard. The Earth's crust beneath Lake Superior; an interpretation of cross-structure seismic refraction profiles. D, 1982, University of Wisconsin-Madison. 126 p.

Lueth, Virgil Walter. Comparison of copper skarn deposits in the Silver City mining region, southwestern New Mexico. M, 1984, University of Texas at El Paso.

Lueth, Virgil Walter. Studies of the geochemistry of the semimetal elements; arsenic, antimony, and bismuth. D, 1988, University of Texas at El Paso. 187 p.

Luethe, Ronald D. Petrology of the Ryan Canyon Stock, Mineral County, Nevada. M, 1974, University of Nevada. 88 p.

Luff, Glen Charles. Geology of the Beggs area, Okmulgee County, Oklahoma. M, 1957, University of Oklahoma. 83 p.

Luff, W. M. Structural geology, Brunswick No. 12 open pit. M, 1973, University of New Brunswick.

Lufholm, Peter Henry. The geophysical analysis of the Gray Mountain area, Coconino County, Arizona. M, 1975, Northern Arizona University. 54 p.

Lufkin, John L. Geology of the Stockton Stock and related intrusives, Tooele County, Utah. M, 1965, Brigham Young University.

Lufkin, John Laidley. Tin mineralization within rhyolite flow-domes (Tertiary), Black Range, New Mexico. D, 1972, Stanford University. 227 p.

Luft, Stanley Jeremie. Alteration and chromite mineralization in the Pennsylvania-Maryland state line serpentine belt. M, 1951, Pennsylvania State University, University Park. 151 p.

Lugar, Gary Lance. Elliptical evolution; a geometric model of shell shape and applications of the model to the evolution of form in three genera of patellacean limpets. D, 1988, University of California, Berkeley. 200 p.

Lugaski, Thomas P. Preliminary analysis of the physical stratigraphy, depositional environment, and paleoecology of the Miocene non-marine deposits, Stewart Valley, Nevada. M, 1986, University of Nevada. 234 p.

Lugay, Josefina. Transformation of asphaltenes through oxidation and thermal mechanisms. D, 1963, Fordham University. 250 p.

Lugenbeal, M. P. Evidences bearing on the time involved in the deposition of the fossil forests of the Specimen Creek area, Yellowstone National Park, Montana. M, 1968, Andrews University. 85 p.

Luginbill, Charles Philip. Paleoecology of the Dennison Formation, Tarrant and Denton counties, Texas. M, 1985, Baylor University. 145 p.

Lugn, Alvin Leonard. Geology of Lucas County. M, 1925, University of Iowa. 247 p.

Lugn, Alvin Leonard. Sedimentation in Mississippi River between Davenport, Iowa and Cairo, Illinois. D, 1927, University of Iowa. 132 p.

Lugn, Richard Victor. The heavy mineral assemblages of the Sheet Creek Beds in western Nebraska compared with similar assemblages in other formations in the same region. M, 1955, University of Nebraska, Lincoln.

Lugo, Hector Manuel. The effect of some mineralogical and physico-chemical properties of clay particles on soil strength. D, 1972, North Carolina State University.

Luh, Gary Gwo-Fea. An experimental study of the dynamic behavior of soils. D, 1980, University of Wisconsin-Madison. 148 p.

Luhn, Judith K. Geochemical investigation of the Buck Creek alpine-type intrusion, western North Carolina. M, 1968, Miami University (Ohio). 46 p.

Luhr, James Francis. The Colima volcanic complex, Mexico; I, Post-caldera andesites from Volcan Colima; II, Late-Quaternary cinder cones. D, 1980, University of California, Berkeley. 221 p.

Lui, Chung-Yao. Born inversion applied to reflection seismology. D, 1984, University of Tulsa. 105 p.

Lui, Chung-Yao. Microearthquakes in Red Willow County, Nebraska. M, 1981, University of Kansas. 53 p.

Luik, A. E. J. Van see Van Luik, A. E. J.

Lujan Sierraalta, Carlos Alberto. Three-phase flow analysis of oil spills in partially water-saturated soils. D, 1985, Colorado State University. 132 p.

Luk, Grace King Yan. Mathematical modelling of the two-dimensional mixing in natural streams. D, 1988, Queen's University.

Luk, King-sing. Stress relaxation function of a clayey soil. D, 1971, University of California, Los Angeles. 216 p.

Lukanuski, James N. Geology of part of the Mitchell Quadrangle, Jefferson and Crook counties, Oregon. M, 1963, Oregon State University. 90 p.

Lukas, Theodore Chris. Origin of bauxite, Eufaula District, Alabama. M, 1978, University of Florida. 130 p.

Lukasik, David M. Lithostratigraphy of Silurian rocks in southern Ohio and adjacent Kentucky and West Virginia. D, 1988, University of Cincinnati. 401 p.

Lukasik, David M. Stratigraphy, sedimentology, and paleoecology of the Kashong Shale (Middle Devonian) of New York. M, 1984, University of Cincinnati. 276 p.

Luke, Gene Edward. Structure of the area north of Horse Creek basin, Fremont County, Wyoming. M, 1955, Miami University (Ohio). 38 p.

Luke, Keith Joseph. Corals of the Devonian Guilmette Formation from the Leppy Range near Wendover, Utah-Nevada. M, 1974, Brigham Young University.

Luke, Robert Franklin. Structure of the eastern part of the Mill Creek Syncline. M, 1975, University of Oklahoma. 60 p.

Luken, Michael. Petrography and origin of limestones in the WPZ and TR belt, Klamath Mountains, California. M, 1985, San Jose State University. 130 p.

Luker, James A., Jr. Sedimentology of the Ellensburg Formation northwest of Yakima, Washington. M, 1985, Eastern Washington University. 184 p.

Luker, Richard Stephen. The lithostratigraphy, carbonate petrography, and depositional history of the Marble Falls Formation (Pennsylvanian) in the subsurface of Brown and Mills counties, central Texas. M, 1985, University of Oklahoma. 193 p.

Lukert, Louis Henry. Some foraminifera from the Niobrara Formation in the Republican River valley, Nebraska. M, 1934, University of Nebraska, Lincoln.

Lukert, Michael T. The Kaneville Esker (Illinois). M, 1962, Northern Illinois University. 37 p.

Lukert, Michael Thomas. The petrology and geochronology of the Madison area, Virginia. D, 1973, Case Western Reserve University. 218 p.

Lukesh, J. S. Application of Fourier's series to the crystal structure of nitrogen sulfide. M, 1938, Massachusetts Institute of Technology. 41 p.

Lukin, Craig G. Evaluation of sediment sources and sinks; a sediment budget for the Rappahannock River estuary. M, 1983, College of William and Mary. 204 p.

Lukk, Michael E. The geology and geochemistry of the Tertiary volcanic rocks of the northeastern half of the Clipper Mountains, eastern Mojave Desert, California. M, 1982, University of California, Riverside. 120 p.

Lukosius-Sanders, J. Petrology of syenites from Centre III of the Coldwell alkaline complex, northwest Ontario. M, 1988, Lakehead University.

Lukowicz, Leo Joseph. Geology of Mackay 3 SW and west part of Mackay 3 SE Quadrangle, Blaine County, Idaho. M, 1971, University of Wisconsin-Milwaukee.

Lull, John S. Petrology and structure of an Early Cretaceous dike swarm, northern Sierra Nevada, California. M, 1984, California State University, Hayward. 146 p.

Lull, Richard S. Fossil footprints of the Jura-Trias of North America with a preliminary revision of the eastern vertebrate fauna of the period. D, 1903, Columbia University, Teachers College.

Lulla, Kamlesh Parsram. Development of Landsat-based geobotanical stress index. D, 1982, Indiana State University. 184 p.

Lum, Daniel. Regional gravity survey of the north-central Wasatch Mountains and vicinity, Utah. M, 1957, University of Utah. 27 p.

Lum, L. W. K. The entrophication potential of Wahiawa Reservoir sediments. M, 1976, University of Hawaii. 363 p.

Luman, D. E. A multivariate analysis of the geometric form relationships among a set of geomorphometric variables describing valley-side asymmetry. D, 1978, University of Illinois, Urbana. 226 p.

Lumb, Wallace E. Geology of southeastern Shawnee County, Kansas. M, 1933, University of Kansas. 44 p.

Lumbers, Sydney Blake. Stratigraphy, plutonism, and metamorphism in the Ottawa River remnant in the Bancroft-Madoc area of the Grenville Province of southeastern Ontario, Canada. D, 1967, Princeton University. 374 p.

Lumbers, Sydney Blake. The geology of Steele, Bonis and Scapa townships, District of Cochrane, Ontario. M, 1960, University of British Columbia.

Lumino, Karen Marie. Deformation within the Diana Complex along the Carthage-Colton mylonite zone. M, 1987, University of Rochester. 104 p.

Lumpkin, G. R. Chemistry and physical properties of axinites. M, 1977, Virginia Polytechnic Institute and State University.

Lumsden, David N. The Grimsby Sandstone (Silurian) in New York. M, 1960, SUNY at Buffalo.

Lumsden, David Norman. Microfacies of the Middle Bird Spring Group, (Pennsylvanian-Permian), Arrow Canyon Range, Clark County, Nevada. D, 1965, University of Illinois, Urbana.

Lumsden, Jesse Beadles, III. Quadrupole splitting of Al-27 nuclear magnetic resonance in topaz. D, 1970, Ohio State University. 71 p.

Lumsden, William Watt, Jr. Geology of the southern White Pine Range and the northern Horse Range, Nye and White Pine counties, Nevada. D, 1964, University of California, Los Angeles. 355 p.

Lunardi, L. F. X-ray petrofabric analysis of mylonites from the northern Snake Range of East-central Nevada. M, 197?, University of Illinois, Chicago.

Lunceford, Robert A. Geology of a portion of the Pine Creek Quadrangle, Teton and Lincoln counties, Wyoming. M, 1976, Montana State University. 136 p.

Lund, Ernest H. Igneous and metamorphic rocks of the Minnesota River valley. D, 1950, University of Minnesota, Minneapolis.

Lund, John C. The geology and mineralogy of the 5500 level ore zones of Coast Copper Mine, Benson Lake, Vancouver Island, British Columbia. M, 1966, University of British Columbia.

Lund, John Casper. Structural geology of Empire Mine, Empire Development Company Limited, Port McNeil, British Columbia. M, 1966, University of British Columbia.

Lund, Karen. Geology of the Whistling Pig Pluton, Selway-Bitterroot Wilderness, Idaho. M, 1980, University of Colorado.

Lund, Karen Ivy. Tectonic history of a continent-island arc boundary; west-central Idaho. D, 1984, Pennsylvania State University, University Park. 242 p.

Lund, Kendall G. Erosional features of the northern Berkshire hills and Connecticut low lands, (Massachusetts). M, 1970, University of Massachusetts. 55 p.

Lund, Lamar. Further study of the bituminous sandstones of Santa Cruz County, California. M, 1957, Stanford University.

Lund, Lamar. Study on the bituminous sands of Santa Cruz, California. M, 1956, Stanford University.

Lund, Lanny Jack. Clay mineralogy of Recent lake sediments and their contributing soils (Indiana and Illinois). D, 1971, Purdue University. 153 p.

Lund, Richard. The cranial osteology of Elops saurus, with a comparison to some holostean fishes. M, 1963, University of Michigan.

Lund, Richard Jacob. A study of the waters in the Niagara Dolomite in Wisconsin, Illinois and Indiana. M, 1928, University of Wisconsin-Madison.

Lund, Richard Jacob. Differentiation in the Cape Spencer flow. D, 1930, University of Wisconsin-Madison.

Lund, Steven Phillip. Continental sediments as paleomagnetic records; evidence from the Upper Cretaceous Hill Creek Formation of the Williston Basin. M, 1976, University of Minnesota, Minneapolis. 137 p.

Lund, Steven Phillip. Late Quaternary secular variation of the Earth's magnetic field as recorded in the wet sediments of three North American lakes. D, 1981, University of Minnesota, Minneapolis. 314 p.

Lundahl, Arthur Charles. A shape-roundness study of the beach sands of Cedar Point, Ohio, Lake Erie. M, 1942, University of Chicago. 65 p.

Lundberg, Dennis LeRoy. Groin-associated rip currents measured using a new digital current meter. D, 1987, Old Dominion University. 186 p.

Lundberg, Neil Scott. Evolution of the forearc landward of the Middle America Trench, Nicoya Peninsula, Costa Rica and southern Mexico. D, 1982, University of California, Santa Cruz. 279 p.

Lundblad, Steven Paul. Stratigraphy and structure of the Skookoleel Creek area, southern Whitefish Range, Montana. M, 1988, University of Wisconsin-Madison. 124 p.

Lundby, William. The geology of the northern part of the Aurora District. M, 1957, University of Nevada. 36 p.

Lunde, Magnus. The Caulfield quartz diorite; modal determination and petrographic observations. M, 1942, University of British Columbia.

Lunde, Magnus. The Pre-Cambrian and Pleistocene geology of the Grondines map area, Quebec. D, 1953, McGill University.

Lundeen, Lloyd John. The use of digital simulation models to predict the effects of vegetation cover change on streamflow and downstream water use. D, 1977, Stanford University. 218 p.

Lundeen, Margaret Thompson. Structural evolution of the Ronda Peridotite and its tectonic position in the Betic Cordilleras, Spain. D, 1976, Harvard University.

Lundegard, Paul D. Sedimentology and petrology of a prodeltaic turbidite system; the Brallier Formation (Upper Devonian) western Virginia and adjacent areas. M, 1979, University of Cincinnati. 296 p.

Lundegard, Paul David. Carbon dioxide and organic acids; origin and role in burial diagenesis (Texas Gulf Coast Tertiary). D, 1985, University of Texas, Austin. 162 p.

Lundelius, Ernest L., Jr. Skeletal adaptation in two species of "Sceloporus" as determined by regression lines. D, 1954, University of Chicago. 50 p.

Lundell, L. L. Depositional environment of the Eocene Green River Formation, Piceance Creek basin, Colorado. D, 1977, University of Wyoming. 147 p.

Lundell, Leslie Lee. Secular variations of radiocarbon in the atmosphere. M, 1974, Rice University. 53 p.

Lundgren, Lawrence Williams. The geology of the Deep River area, Connecticut. D, 1958, Yale University.

Lundgren, Paul Randall. Rupture characteristics of complex earthquakes. D, 1988, Northwestern University. 172 p.

Lundin, Robert Folke. Morphology and ontogeny of Phanassymetria Roth from the Haragan Formation (Devonian) of Oklahoma. M, 1961, University of Illinois, Urbana.

Lundin, Robert Folke. Ostracodes from the Henryhouse Formation (Silurian) of Oklahoma. D, 1962, University of Illinois, Urbana. 99 p.

Lundquist, G. M. The frequency dependence of Q. D, 1979, University of Colorado. 281 p.

Lundquist, Gary M. Frequency-band and integrated magnitude using long period seismograms. M, 1969, Pennsylvania State University, University Park. 65 p.

Lundquist, John H. Sedimentology and stratigraphy of the Sweetwater Creek interbed, Lewiston Basin, Idaho-Washington. M, 1987, Washington State University. 133 p.

Lundquist, Susan Marya. Deformation history of the ultramafic and associated metamorphic rocks of the Seiad Complex, Seiad Valley, California. M, 1983, University of Washington. 167 p.

Lundstrom, Orville Glebe. Geology of the Pine Island-Mazeppa region. M, 1938, University of Minnesota, Minneapolis. 27 p.

Lundstrom, Scott C. Soil stratigraphy and scarp morphology studies applied to the Quaternary geology of the southern Madison Valley, Montana. M, 1986, Humboldt State University. 53 p.

Lundwall, Walter Raymond, Jr. Microfacies study of a Devonian bioherm, Columbus, Indiana. M, 1959, University of Illinois, Urbana.

Lundy, Curtis Lee. Geology of the Fairbanks Pool, Sullivan County, Indiana. M, 1958, University of Michigan.

Lundy, D. A. Hydrology and geochemistry of the Casper Aquifer in the vicinity of Laramie, Albany County, Wyoming. M, 1978, University of Wyoming. 76 p.

Lundy, Gerald W. Utility of Seasat SAR imagery for geologic analysis in Colorado, Wyoming, and Utah. M, 1982, University of Arkansas, Fayetteville.

Lundy, James Russell. Clues to structural history in the minor folds of the Soudan Iron Formation, NE Minnesota. M, 1985, University of Minnesota, Minneapolis. 144 p.

Lundy, William Leon. The stratigraphy and evolution of the Coconino Sandstone of northern Arizona. M, 1973, University of Tulsa. 106 p.

Luneau, Barbara Ann. Deformation and sedimentology of the Cambria slab; a Franciscan trench-slope basin. M, 1984, University of Texas, Austin. 86 p.

Lung, Han-Chuan. Migration of radionuclides through backfill in a nuclear waste repository. D, 1986, University of California, Berkeley. 189 p.

Lung, Richard. Geology of the South Mountain area, Ventura County, California. M, 1958, University of California, Los Angeles.

Lung, Richard Hai. Seismic analysis of structures embedded in saturated soils. D, 1981, City College (CUNY). 283 p.

Lung, W.-S. Modeling of phosphorus sediment-water interactions in White Lake, Michigan. D, 1975, University of Michigan. 246 p.

Lunsford, David L. Hydrogeological and hydrochemical investigation of water wells in Hays County, Texas. M, 19??, University of Arkansas, Fayetteville.

Luoma, Samuel N. Aspects of the dynamics of mercury cycling in a small Hawaiian estuary. D, 1974, University of Hawaii. 210 p.

Luongo, Ronald F. Analysis of seismic reflection data over the Sanford Triassic basin, Chatham and Lee counties, North Carolina. M, 1987, Virginia Polytechnic Institute and State University.

Lupe, Robert Douglas. Stratigraphy and petrology of the Swauk Formation in the Wenatchee Lake area, Washington. M, 1971, University of Washington. 27 p.

Lupher, Ralph L. Geology of the Silvies Canyon section, Ochoco Range, Oregon. D, 1930, California Institute of Technology.

Lupher, Ralph L. Stratigraphy and correlation of the marine Jurassic deposits of central Oregon. D, 1930, California Institute of Technology. 151 p.

Lupher, Ralph Leonard. Two new genera of pachyodonts from the Mesozoic of Oregon. M, 1927, University of Oregon. 38 p.

Lupindu, Kandidus P. Electrical resistivity studies near Bozeman Hot Springs of Gallatin County and White Sulphur Hot Springs of Meagher County, Montana. M, 1983, Montana College of Mineral Science & Technology. 59 p.

Lupo, Mark Joseph. Carbonyl transport of metal in meteorite parent bodies. M, 1981, Massachusetts Institute of Technology. 64 p.

Lupo, Mark Joseph. The flow of fluids from matrix to fractures in rock. D, 1987, Texas A&M University. 251 p.

Luppens, James Alan. Distribution of mercury in selected drill holes from the White Pine copper deposit, White Pine, Michigan. M, 1970, University of Toledo. 59 p.

Luppold, John Hugh. The extraction of manganese from low grade siliceous ores with sulfurous acid and sulphur dioxide. M, 1942, University of Washington. 52 p.

Lupton, D. Keith. The geology the Sheep Creek-Cold Springs area, Hell's Canyon, Idaho. M, 1951, Dartmouth College. 60 p.

Luquer, Lea M. The optical recognition and economic importance of the common minerals found in building stones; the relative effects of frost and the sulphate soda tests on building stones. D, 1894, Columbia University, Teachers College.

Lusch, David Paul. The origin and morphogenetic significance of patterned ground in the Saginaw Lowland of Michigan. D, 1983, Michigan State University. 154 p.

Lusk, Edwin Wallace. A study of the causes of local variations in rank of the lower Kittanning and lower Freeport coals in Clearfield County, Pennsylvania. M, 1958, University of Pittsburgh.

Lusk, John. Sulfur isotope abundances and base metal zoning in the Heath Steele B-1 orebody, Newcastle, New Brunswick. D, 1968, McMaster University. 261 p.

Lusk, Loren Douglas. Genesis of chert in the Middle Devonian Bois Blanc Formation of Michigan and southwestern Ontario. M, 1958, University of Michigan.

Lusk, Randolph Gordon. Geology and oil and gas resources of Gainesboro Quadrangle, Tennessee. D, 1927, Harvard University.

Lusk, Tracy W. Ground water investigations along Bogue Phalia between Symonds and Malvina, Bolivar County, Mississippi. M, 1951, University of Mississippi.

Luster, Gordon Ray. Lithologic variability of the Kings Mountain Pegmatite, North Carolina. M, 1977, Pennsylvania State University, University Park. 116 p.

Lustick, Charles Francis. Laser microscopy studies in the RbCl-KCl system. M, 1980, University of New Orleans. 79 p.

Lustig, Claire. Benthic foraminifera from the upper Luisian of Newport Bay, Orange County, California. M, 1984, University of California, Los Angeles. 113 p.

Lustig, Gary Norman. Geology of the Fox Orebody, northern Manitoba. M, 1979, University of Manitoba. 87 p.

Lustig, Lawrence Kenneth. The mineralogy and paragenesis of the Lone Star Deposit, Santa Fe County, New Mexico. M, 1957, University of New Mexico. 55 p.

Lustig, Lidia Diana. Middle Pennsylvanian chaetetes (Tabulata) from the Bird Spring Formation of southern Nevada. M, 1971, University of California, Los Angeles.

Lustwig, Lawrence Kenneth. Clastic sedimentation in the Deep Springs Valley, California. D, 1963, Harvard University.

Lusty, Quayle C. Geology and mineral deposits of the eastern part of the Elliston mining district, Lewis, Clark, and Powell counties, Montana. M, 1973, Colorado School of Mines. 105 p.

Lutenegger, A. J. Random-walk variable wind model for loess deposits. D, 1979, Iowa State University of Science and Technology. 392 p.

Luternauer, John Leland. Patterns of sedimentation in Queen Charlotte Sound, British Columbia. D, 1972, University of British Columbia.

Luternauer, John Leland. Phosphorite on the North Carolina continental shelf. M, 1966, Duke University. 46 p.

Lutes, Glenn Gordon. The geology of the northwestern part of the metamorphic aureole of the Poliok Batholith. M, 1981, University of New Brunswick.

Luth, Iris Annette. The petrology and petrography of the Trailside Complex, Bighorn Mountains, Wyoming. M, 1977, University of Iowa. 150 p.

Luth, Robert William. Hydrogen and the melting of silicates. D, 1985, University of California, Los Angeles. 141 p.

Luth, William Clair. Mafic and ultramafic rocks of the trailside area, Bighorn Mountains, Wyoming. M, 1960, University of Iowa. 138 p.

Luth, William Clair. The system KAlSiO₄-Mg₂SiO₄-SiO₂-H₂O from 500 to 3000 bars and 800 to 1200 C and its petrol- ical significance. D, 1963, Pennsylvania State University, University Park. 3363 p.

Luther, Edward Turner. Geology of the Spring Hill area, Spring Hill Quadrangle, Maury and Williamson County, Tennessee. M, 1951, Vanderbilt University.

Luther, Frank R. A petrologic investigation of the Precambrian crystalline rocks of the northwest portion of the Pingree Park Quadrangle, Mummy Range, north central Colorado. M, 1968, SUNY at Buffalo. 67 p.

Luther, Frank R. The petrological evolution of the garnet deposit at Gore Mountain, Warren County, New York. D, 1976, Lehigh University. 236 p.

Luther, Kathryn C. Proterozoic structures in north-central North Dakota; a gravity study. M, 1988, University of North Dakota. 104 p.

Luther, Lars Christian. Diffusion of helium in solids. D, 1964, Indiana University, Bloomington. 264 p.

Luther, Mark R. Deposition and diagenesis of a portion of the Frobisher-Alida Interval (Mississippian Madison Group), Wiley Field, North Dakota. M, 1988, University of North Dakota. 313 p.

Luthy, Stephen T. Petrology of Cretaceous and Tertiary intrusive rocks, Red Mountain - Bull Trout Point area, Boise, Valley, and Custer counties, Idaho. M, 1981, University of Montana. 109 p.

Lutter, William John. Refined crustal velocity models for the Mississippi Embayment area based on amplitude modeling of seismic refraction data. M, 1984, Purdue University. 109 p.

Lutton, Richard J. Some structural features of southern Arizona. M, 1958, University of Arizona.

Lutton, Richard Joseph. Geology of the Bohemia mining district, Lake County, Oregon. D, 1962, University of Arizona. 313 p.

Luttrell, Eric M. Sedimentary analysis of some Jackfork sandstones, (Pennsylvanian) Big Cedar, Oklahoma. M, 1965, University of Wisconsin-Madison.

Luttrell, Eric Martin. An analysis of the Silurian Keefer Sandstone of Pennsylvania. D, 1968, Princeton University. 111 p.

Luttrell, Pamela E. Carbonate diagenesis and facies distribution of the Anacacho Limestone associated with a Late Cretaceous volcano in Elaine Field, Dimmit County, Texas. M, 1977, University of Texas, Austin.

Luttrell, Patty Rubick. Basin analysis of the Kayenta Formation (Lower Jurassic), central portion Colorado

Plateau. M, 1987, Northern Arizona University. 217 p.

Luttrell, Stuart Paul. Ground-water flow characteristics in the Mud Lake area, southeastern Idaho. M, 1982, University of Idaho. 69 p.

Lutz, Charles Talbott. Structural geometries and carbonate deformation mechanisms in the Beaver Valley-Saltville transfer zone. D, 1987, University of Tennessee, Knoxville. 288 p.

Lutz, Delbert Henry. The geology of Lower Warm Spring Canyon, Fremont County, Wyoming. M, 1953, Miami University (Ohio). 45 p.

Lutz, George C. The Sobrante Sandstone and its fauna. M, 1950, University of California, Berkeley. 70 p.

Lutz, James F. The physico-chemical properties of soils affecting soil erosion. D, 1934, University of Missouri, Columbia.

Lutz, Katherine. Preliminary study of the application of the insoluble residue technique to the limestones of the Chester Valley, Norristown Quadrangle, Pennsylvania. M, 1948, Bryn Mawr College. 7 p.

Lutz, Katherine. The distribution of crustal shortening in the Appalachians of Pennsylvania, Maryland, and part of West Virginia. M, 1948, Bryn Mawr College. 11 p.

Lutz, T. M. Strontium and oxygen isotope relations and geochemistry of the Abu Khruq Complex, Egypt; implications for petrogenesis of the alkaline rocks of the Eastern Desert, Egypt. D, 1979, University of Pennsylvania. 321 p.

Lutzen, Edwin Earl. Geology of the Irene Quadrangle, South Dakota. M, 1957, University of South Dakota. 128 p.

Luvira, Somboon. A classification and analysis of natural watersheds. D, 1984, Colorado State University. 466 p.

Luwe, Randall Scott. The Mississippian erosional surface and its influence on Lower Pennsylvanian deltaic and sandstone sedimentation, Marion County, Iowa. M, 1983, Iowa State University of Science and Technology. 111 p.

Lux, Daniel R. A major element geochemical study of Laramide igneous rocks of the Colorado mineral belt. M, 1977, Rice University. 77 p.

Lux, Daniel R. Geochronology, geochemistry, and petrogenesis of basaltic rocks from the western Cascades, Oregon. D, 1981, Ohio State University. 182 p.

Lux, Gayle E. Bulk composition, mineralogy and petrology of chondrules in type H3 to H6 ordinary chondrites. M, 1978, University of New Mexico. 154 p.

Lux, Gayle Elizabeth. The solubility and diffusion of noble gases in silicate liquids and studies of noble gases in the Murchison carbonaceous chondrite. D, 1985, University of California, Berkeley. 74 p.

Lux, Jeffrey John. Selection of remedial engineering techniques for dioxin-contaminated soil sites, based on site geological characteristics. M, 1986, University of Missouri, Rolla. 99 p.

Lux, Richard Alan. The effect of a moving lithospheric plate on convection in the Earth's mantle. D, 1979, University of Rochester. 87 p.

Luyendyk, Bruce Peter. Geological and geophysical observations in an abyssal hill area using a deeply-towed instrument package. D, 1969, University of California, San Diego. 229 p.

Luz, Boaz. Late Pleistocene paleo-oceanography of the tropical Southeast Pacific. D, 1974, Brown University. 142 p.

Luza, Kenneth Vincent. A detailed structural analysis of the northwest corner of the Hill City Quadrangle, Black Hills, South Dakota. D, 1972, South Dakota School of Mines & Technology.

Luza, Kenneth Vincent. Origin, distribution, and development of bog iron in the Rochford District, north-central Black Hills, South Dakota. M, 1969, South Dakota School of Mines & Technology.

Luzardo, Manuel A. Ordovician subsurface geology of Cooke, Montague, and eastern Clay counties, Texas. M, 1971, University of Texas, Austin.

Luzier, James E. A petrographic evolution of calcareous sandstone for use as base course. M, 1961, West Virginia University.

Lyall, Anil K. A study of offshore sediment movement and differentiation of beach and dune sands in the Cape Sable island area, Nova Scotia. D, 1969, Dalhousie University.

Lyall, H. Bruce. Study of hornfels collar around Mount Bruno (Quebec). M, 1952, McGill University.

Lyall, H. Bruce. The geology of the Haevant-Champain area, Pontiac County, Quebec. D, 1958, Universite Laval. 180 p.

Lyatsky, Henry. Reflection seismic study of a shallow coal field in central Alberta. M, 1988, University of Calgary. 121 p.

Lybarger, James H. Subsurface study of the Rodessa Formation in southern Miller, Lafayette and Columbia Counties, Arkansas. M, 1960, University of Arkansas, Fayetteville.

Lyday, John Reed. A paleogeographic and environmental reconstruction of the Fort Union Formation (Paleocene) of northern Wyoming and adjacent areas. M, 1978, University of Tulsa. 98 p.

Lyday, Travis Quinton. The geology of the southern half of White Cross Quadrangle, North Carolina. M, 1974, University of North Carolina, Chapel Hill. 29 p.

Lydecker, William Frederick. Geology of the southern half of the Spy Rock Quadrangle, Mason and McCulloch counties, Texas. M, 1988, University of Texas of the Permian Basin. 58 p.

Lydon, John W. The significance of metal ratios of hydrothermal ore deposits. D, 1977, Queen's University. 598 p.

Lydon, Michael Thomas. Lg source parameter estimates in various Earth models. M, 1985, St. Louis University.

Lydon, Philip Andrew. Geology of the Butt mountain area, a source of the Tuscan Formation in northern California. D, 1968, University of Oregon. 198 p.

Lye, Joseph Alexander. The choice and application of selective extractants to elucidate the chemical forms of trace elements in some lake sediment samples. M, 1982, University of Toronto.

Lyford, Forest Parsons. Soil infiltration rates as affected by desert vegetation. M, 1968, University of Arizona.

Lyke, Frederick P. External morphology of some trepostomatous ectoprocts of the Bellevue Member, McMillan Formation, (middle Upper Ordovician, southwestern Ohio). M, 1982, Miami University (Ohio). 134 p.

Lyke, William LeRoy. The Minnehaha Member of the Upper Devonian Scherr and Brallier formations of the central Appalachians. M, 1981, University of North Carolina, Chapel Hill. 120 p.

Lyle, John. Facies geometry and carbonate petrology of the Bonneterre Formation in western Iron, northern Reynolds and eastern Dent Counties, Mo. M, 1973, University of Missouri, Columbia.

Lyle, John Hyer, III. Interrelationship of late Mesozoic thrust faulting and mid-Tertiary detachment faulting in the Riverside Mountains, southeastern California. M, 1982, San Diego State University. 81 p.

Lyle, Michael. Clay mineralogy of the Pungo River Formation, Onslow Bay, North Carolina continental shelf. M, 1984, East Carolina University.

Lyle, Mitchell. The formation and growth of ferromanganese oxides on the Nazca Plate. D, 1979, Oregon State University. 172 p.

Lyles, Bradley F. Time-variant hydrogeologic and geochemical study of selected thermal springs in western Nevada. M, 1985, University of Nevada. 203 p.

Lyman, John. Buffer mechanism of sea water. D, 1956, University of California, Los Angeles. 196 p.

Lyman, Robert M. The areal geology and structural analysis of the South Beaver Creek area of the northwest flank of the Bighorn Mountains, (Bighorn County), Wyoming. M, 1973, University of Iowa. 68 p.

Lynch, Bernard Walden. The Tertiary sediments of the Upper Horse Creek area, Teton County, Wyoming. M, 1948, University of Illinois, Urbana.

Lynch, Daniel James, II. Genesis and geochronology of alkaline volcanism in the Pinacate volcanic field, northwestern Sonora, Mexico. D, 1981, University of Arizona. 265 p.

Lynch, Daniel James, II. Reconnaissance geology of the Bernardino volcanic field, Cochise County, Arizona. M, 1972, University of Arizona.

Lynch, Dean W. The geology of the Esperanza Mine and vicinity, Pima County, Arizona. M, 1967, University of Arizona.

Lynch, Elizabeth Linfield. Late Quaternary fluctuations of the Western Boundary Undercurrent. M, 1986, Duke University. 140 p.

Lynch, F. Leo, III. The stoichiometry of the smectite to illite reaction in a contact metamorphic environment. M, 1985, Dartmouth College. 93 p.

Lynch, Hugh David, Jr. Numerical models of the formation of continental rifts by processes of lithospheric necking. D, 1983, New Mexico State University, Las Cruces. 309 p.

Lynch, John Douglas. Evolutionary relationships and osteology of the frog family Leptodactylidae (Volumes I and II). D, 1969, University of Kansas. 861 p.

Lynch, John Scott. Three-dimensional depth migration after stack with an implicit finite-differencing scheme. M, 1985, University of Utah. 93 p.

Lynch, Joseph Vincent Gregory. Mineralization and alteration zonation of the Kalzas wolframite vein deposit, Yukon Territory, Canada. M, 1985, Washington State University. 123 p.

Lynch, Matthew J. Geology of the Bull lake area, Fremont County, Wyoming. M, 1969, University of Missouri, Columbia.

Lynch, Maurice Butler. Remanent paleomagnetism in the Miocene basalts of North Idaho. M, 1976, University of Idaho. 150 p.

Lynch, S. The crustal structure of Winona Basin as determined by deep seismic sounding. M, 1977, University of British Columbia.

Lynch, Shirley Alfred. Studies in the distribution of Orbitolina walnutensis carsey. M, 1931, University of Missouri, Rolla.

Lynch, Thomas Wimp. Provenance of the Middle Devonian of the Permo-Triassic red beds on the northeast flank of the Bighorn Mountains, Wyoming. M, 1958, University of Illinois, Urbana.

Lynch, Vance M. A geophysical study of the Mustang Hill igneous mass, Uvalde County, Texas. M, 1959, University of Houston.

Lynch, William Charles. Chemical trends in the Ice Springs Basalt, Black Rock Desert, Utah. M, 1980, University of Utah. 408 p.

Lynch-Blosse, Michael A. Inlet sedimentation at Dunedin and Hurricane passes, Pinellas County, Florida. M, 1977, University of South Florida, Tampa. 170 p.

Lynd, Langtry E. A study of the mechanism of alteration of ilmenite. D, 1957, Rutgers, The State University, New Brunswick. 103 p.

Lynde, Harold William, Jr. Occurrence of tellurium in the United States with determinative mineral tables. M, 1964, University of Michigan.

Lyngberg, Erik. The Orpheus Graben, offshore Nova Scotia; palynology, organic geochemistry, maturation and time - temperature history. M, 1984, University of British Columbia.

Lynn, Alvin G. Slope and shoreline modification at Lake Meredith in the Texas Panhandle. M, 1975, West Texas State University. 51 p.

Lynn, C. G. Stratigraphic correlation of the El Paso and Montoya groups in the Victorio Mountains, the Snake Hills, and the Big Florida Mountains in southwestern New Mexico. M, 1975, University of Arizona.

Lynn, H. B. Migration and interpretation of deep crustal seismic reflection data. D, 1980, Stanford University. 170 p.

Lynn, Harold F. Microscopic study of the silver ores of the San Xavier District, Mexico. M, 1925, Stanford University. 50 p.

Lynn, Jeffrey S. The heavy mineral barite of the Dunkard Group (Upper Carboniferous) in Pennsylvania, Ohio, and West Virginia. M, 1975, Miami University (Ohio). 62 p.

Lynn, John R. The geology of the north margin of the Reshaw Hills, Platte County, Wyoming. M, 1947, University of Wyoming. 53 p.

Lynn, LaRee, Jr. Stratigraphic framework of some Tertiary coal-bearing alluvial strata, Powder River basin, Wyoming and Montana. M, 1980, North Carolina State University. 85 p.

Lynn, Leslie Michael. The late-Holocene successional complex of Little Cedar Pond, southeastern New York. D, 1983, New York University. 100 p.

Lynn, R. J., II. Petrology of the Upper Devonian Owen Member of the Lime Creek Formation, Iowa. M, 1978, University of Iowa. 83 p.

Lynn, W. S. Velocity estimation in laterally varying media. D, 1980, Stanford University. 83 p.

Lynn, Walter S. A geophysical analysis of the Orozco fracture zone and the tectonic evolution of the northern Cocos Plate. M, 1976, Oregon State University. 80 p.

Lynnes, Christopher Scott. I, Seismic velocity heterogeneity in the crust and upper mantle under the Nevada Test Site; II, Source processes of great intraplate earthquakes. D, 1988, University of Michigan. 198 p.

Lynnes, Christopher Scott. Paleomagnetism of the Cambro-Ordovician McClure Mountain alkalic complex, Colorado. M, 1984, University of Michigan.

Lynott, William John. Geology and mineral deposits of Warn Bay-Tofino Inlet map area, west coast of Vancouver Island, British Columbia. D, 1949, Princeton University. 165 p.

Lynts, George Willard. Distribution and model studies on foraminifera living in Buttonwood Sound, Florida. D, 1964, University of Wisconsin-Madison. 111 p.

Lynts, George Willard. Distribution of Recent foraminifera in upper Florida Bay and associated sounds. M, 1961, University of Wisconsin-Madison.

Lyon, Craig Alfred. Petrography of four northeastern Iowa loess samples. M, 1955, Iowa State University of Science and Technology.

Lyon, Denton Lloyd. Geology of the north flank, eastern Bridger Range, Hot Springs County, Wyoming. M, 1956, University of Wyoming. 95 p.

Lyon, Garth Monk. Subsurface stratigraphic analysis, lower "Cherokee" Group (Pennsylvanian), portions of Alfalfa, Major and Woods counties, Oklahoma. M, 1971, University of Oklahoma. 33 p.

Lyon, Henry Wortham. Application of analytical solute transport models in the initial evaluation of a hazardous waste disposal facility, Piedmont region, North Carolina. M, 1987, North Carolina State University. 129 p.

Lyon, John Grimson. The influence of Lake Michigan water levels on wetland soils and distribution of plants in the Straits of Mackinac, Michigan. D, 1981, University of Michigan. 139 p.

Lyon, Kenneth E. Retention of ^{137}Cs and ^{90}Sr by mineral sorbents in a shallow sand aquifer. M, 1981, Queen's University. 262 p.

Lyon, Robert B., Jr. Analysis of the planktonic foraminifera and mineralogy from Core E-14996, Hatteras Continental Rise. M, 1973, Duke University. 121 p.

Lyon, Robert P. Process response model for the formation of point bars on the Chippewa River, Wisconsin. M, 1981, University of Wisconsin-Milwaukee. 134 p.

Lyon, Ronald James Pearson. Studies in the geology of the western Sierra Nevada; I, Tectonic analysis of the Miles Creek area, Mariposa County, California; II, Mineralogic study of the ores of the southern Foothill copper belt in California. D, 1954, University of California, Berkeley. 186 p.

Lyon, Stephen R. Investigations in the systems SiO_2-H_2O and Co-SiO_2. M, 1964, Miami University (Ohio). 65 p.

Lyon, Stephen Reed. Phase reactions in the systems MgO-SiO_2-MgF_2, MgO-GeO_2-MgF_2, and MgO-GeO_2-$Mg(OH)_2$. D, 1968, Ohio State University. 151 p.

Lyon-Caen, Hélène. Deep structure of the Himalaya and Tibet from gravity and seismological data. D, 1986, Massachusetts Institute of Technology. 147 p.

Lyons, C. A. Petrography of four northeastern Iowa loess samples. M, 1955, Iowa State University of Science and Technology.

Lyons, Charles Gene. An analysis of waveform generation and transmission in longitudinal impact. D, 1968, Texas A&M University. 241 p.

Lyons, David James. Structural geology of the Boulder Creek metamorphic terrane, Ferry County, Washington. D, 1967, Washington State University. 115 p.

Lyons, Erwin John. Mafic and porphyritic rocks of the Niagara area (Wisconsin). D, 1947, University of Wisconsin-Madison.

Lyons, J. A. Application of pattern search optimization to geophysical model fitting. M, 1973, University of Western Ontario.

Lyons, James I., Jr. Volcanogenic iron ore of Cerro de Mercado and its setting within the Chupaderos Caldera, Durango, Mexico. M, 1975, University of Texas, Austin.

Lyons, John Bartholomew. Geology of the northern Big Belt Range, Montana. D, 1942, Harvard University. 144 p.

Lyons, Mark S. Interpretation of planar structure in drill-hole core. M, 1965, University of Nevada - Mackay School of Mines. 65 p.

Lyons, Nancy M. Structural history of the Edinburg Formation in a portion of the Shenandoah Valley, Virginia. M, 1985, Bowling Green State University. 65 p.

Lyons, Paul Christopher. Bedrock geology of the Mansfield Quadrangle, Massachusetts. D, 1969, Boston University. 326 p.

Lyons, Thomas R. Sedimentary petrology of the Dakota Sandstone in the San Juan Basin of New Mexico and Colorado. M, 1951, University of New Mexico. 94 p.

Lyons, Timothy Donald. A ground water management model for the Elk City Aquifer in Washita, Beckham, Custer, and Roger Mills counties, Oklahoma. M, 1981, Oklahoma State University. 88 p.

Lyons, W. B. Early diagenesis of trace metals in nearshore Long Island Sound sediments. D, 1979, University of Connecticut. 256 p.

Lyons, William S. Subsurface geology and geopressured/geothermal resource evaluation of the Lirette, Chauvin-Lake Boudreaux area, Terrebonne Parish, Louisiana. M, 1983, University of Southwestern Louisiana. 125 p.

Lyslo, Jeffery Allen. Seismic velocity analysis of shot point 16 from the PASSCAL Ouachita experiment. M, 1988, Purdue University. 182 p.

Lysonski, Joseph Charles. The IGSN 71 residual Bouguer gravity anomaly map of Arizona. M, 1980, University of Arizona. 74 p.

Lyth, Ambrose Lee. Reklaw Formation in western Bastrop County, Texas. M, 1949, University of Texas, Austin.

Lytle, Charles Russell. The copper complexation properties of dissolved organic matter from the Williamson River, Oregon. D, 1982, Portland State University. 208 p.

Lytle, Jamie Laverne. A microenvironmental study of an archaeological site, Arizona. M, 1971, University of Arizona.

Lytle-Webb, Jamie. The environment of Miami Wash, Gila County, Arizona. D, 1978, University of Arizona.

Lyttle, Norman A. Petrology and petrogenesis of basalts, from the Olympic Peninsula, Washington. M, 1972, Dalhousie University. 162 p.

Lyttle, Peter T. Petrology and structure of the Pierce Pond gabbroic intrusion and its metamorphic aureole, western Maine. D, 1976, Harvard University.

Lyttle, Thomas. Geochemistry of Bokan Mountain peralkaline system, Prince of Wales Island, Alaska. M, 1983, Colorado State University. 178 p.

Lytton, Gwyn B. The geologic structure of the igneous and sedimentary rocks of South Attleboro, Massachusetts. M, 1941, Brown University.

Lytton, Robert Leonard. Isothermal water movement in clay soils. D, 1967, University of Texas, Austin. 247 p.

Lytton, Rome Gaffney. A paleoecological analysis of the Plio-Pleistocene formations from Lake Waccamaw to Old Dock, North Carolina. M, 1981, University of Florida. 114 p.

Lytwyn, John N. Geochemistry and ages of Precambrian plutonic rocks from Southwest Beartooth Mountains, Yellowstone Park, Wyoming and Montana. M, 1982, Northern Illinois University. 111 p.

Lyzenga, Gregory Allen. Shock temperatures of materials; experiments and applications to the high pressure equation of state. D, 1980, California Institute of Technology. 208 p.

M'Gonigle, John William. A magnetic survey in the Rio Grande Depression. M, 1960, University of New Mexico. 56 p.

M'Gonigle, John William. Structure of the Maiden Peak area, Montana-Idaho. D, 1965, Pennsylvania State University, University Park. 146 p.

M'rah, Mustapha. New evidence on the formation of the Strait of Gibraltar. D, 1978, [University of Tulsa]. 172 p.

M., Armando Estrada see Estrada M., Armando

M., Henry O. Briceño see Briceño M., Henry O.

M., Jose Antonio Rial see Rial M., Jose Antonio

Ma Xueping. The Silurian brachiopod Eocoelia from Anticosti Island, Canada. M, 1984, Laurentian University, Sudbury. 79 p.

Ma, C. M. On the estimation of the friction velocity and the roughness parameter on the western Florida shelf. M, 1978, Florida State University.

Ma, Che-Bao. Phase equilibria and crystal chemistry in the system SiO_2-NiO-$NiAl_2O_4$. D, 1972, Harvard University.

Ma, Li. Regional tectonic stress in western Washington from focal mechanisms of crustal and subcrustal earthquakes. M, 1988, University of Washington. 84 p.

Ma, T. Models of carbon isotope equilibrium in aqueous carbon systems. M, 1975, University of Illinois, Urbana. 87 p.

Ma, Xiaochun. Shaley sand analysis using neutron and density logs. M, 1987, Colorado School of Mines. 86 p.

Maa, Peng-Yea. Erosion of soft muds by waves. D, 1986, University of Florida. 277 p.

Maala, Mohamed. Evaluation of reflection and transmission coefficients at a two-dimensional viscoelastic fluid-fluid interface as a function angle of incidence. M, 1987, University of Pittsburgh.

Maalel, Khlifa. Reliability analysis applied to modeling of hydrologic processes. D, 1983, University of Florida. 358 p.

Maalouf, George Y. Modelling and interpretation of a major gravity high in south-central Massachusetts. M, 1988, University of Massachusetts. 53 p.

Maarouf, Abdelrahman. Hydrogeology of glacial deposits in Tippecanoe County, Indiana. M, 1974, Purdue University.

Maarouf, Abdelrahman Mohammad Shafik. Morphostructural analyses of space imagery in the central Colorado Plateau. D, 1981, University of Utah. 130 p.

Maas, John P. A geophysical investigation of the lower Borrego Valley, California. M, 1973, University of California, Riverside. 78 p.

Maas, John P. Telluric mapping over the Mesa geothermal anomaly, Imperial Valley, California. D, 1976, University of California, Riverside. 152 p.

Maas, R. P. Characterization of water chemistry of Anakeesta pyrite-affected streams in Great Smoky Mountains National Park. M, 1979, Western Carolina University.

Maase, David Lawrence. An evaluation of polycyclic aromatic hydrocarbons from processed oil shales. D, 1980, Utah State University. 190 p.

Maasha, Ntungwa. Studies of the tectonics, seismicity and geothermics of the rift system of East Africa. D, 1975, Columbia University. 104 p.

Maass, Randall Steven. Structure and petrology of an early and middle Precambrian gneiss terrane between Stevens Point and Wisconsin Rapids, Wisconsin. M, 1977, University of Wisconsin-Madison.

Mabarak, Charles D. Heavy minerals in late Tertiary gravel and Recent alluvial-colluvial deposits in the Prairie Divide region of northern Larimer County, Colorado. M, 1975, Colorado State University. 101 p.

Mabbula, Sunand Shadrach. Metal sorption by carbonate rocks as influenced by their chemistry. M, 1978, University of Windsor. 106 p.

Mabee, S. B. The use of magnetite alteration as a relative age dating technique; preliminary results. M, 1978, University of Colorado.

Maberry, John O. Sedimentary features of the Blackhawk Formation (Cretaceous) at Sunnyside, Carbon County, Utah. M, 1968, Colorado School of Mines. 179 p.

Mabes, Deborah Lynn. The engineering properties of mill tailings. M, 1978, University of Idaho. 103 p.

Mabibi, Mohammad J. Subsurface study of the Tremont Sand, Wilcox Group, Caldwell Parish, Louisiana. M, 1979, Northeast Louisiana University.

Mabrey, P. R. Geological reconnaissance of the Coldwater River, Mississippi. M, 1944, University of Missouri, Rolla.

Mac Caskie, Dennis Raymond. Differentiation of the Nebo Granite (main Bushveld Granite), South Africa. D, 1983, University of Oregon. 155 p.

Mac Donald, Robert B. A petrological study of the Recent sands of the Delaware River. M, 1961, Rutgers, The State University, New Brunswick. 60 p.

Mac Nabb, Bert E. Stratigraphy and structure of Carboniferous sediments in southeastern New Brunswick. M, 1971, Northwestern University.

Macalpine, Steven. Jurassic-lower Cretaceous foraminifera from the Paskenta and Ono areas, northern California. M, 1969, University of Washington. 144 p.

Macar, Paul F. J. Effects of cut-off meanders on the longitudinal profiles of rivers. M, 1933, Columbia University, Teachers College.

Macarevich, Roger L. A shallow seismic reflection investigation of a portion of the buried Teays River valley, Champaign County, Ohio. M, 1988, Wright State University. 87 p.

Macauley, George. The Winnipeg Formation (middle or upper Ordovician) in Manitoba. M, 1952, University of Manitoba.

Macauley, George Raymond, Jr. Stratigraphy and paleontology of the Lower Ordovician Axemann Limestone of Kishacoquillas and Nittany valleys, central

Pennsylvania. M, 1952, Pennsylvania State University, University Park. 112 p.

Macauley, Terrence M. Geology of the Sherritt Gordon "B" Orebody, Lynn Lake, Manitoba. M, 1962, Michigan Technological University. 98 p.

Maccarone, Umberto. Heavy mineral behaviour in experimentally produced turbidite beds. M, 1972, McGill University. 98 p.

MacCarthy, Gerald R. The colors produced by iron in minerals and in the sediments. D, 1926, University of North Carolina, Chapel Hill. 74 p.

MacCarthy, Gerald Raleigh. The laccolith; a theoretical and experimental study. M, 1924, University of North Carolina, Chapel Hill. 117 p.

MacChesney, John Burnette. Phase equilibria in the system iron oxide-titania-silica. D, 1959, Pennsylvania State University, University Park. 136 p.

MacClintock, C. Upper part of Morrison Formation and Cloverly Formation, southeastern Big Horn Mountains, Wyoming. M, 1957, University of Wyoming. 142 p.

MacClintock, Copeland. Shell structure and muscle scars of cap-shaped archaeogastropods. D, 1964, University of California, Berkeley. 294 p.

MacClintock, Paul. The Pleistocene history of the lower Wisconsin River. D, 1920, University of Chicago. 115 p.

MacClure, Thomas William. The development of a functional magnetometer for measurement of remanent magnetization. M, 1970, Michigan State University. 106 p.

MacColl, Robert S. The structure and petrology of Rattlesnake Mountain Pluton, San Bernardino County, California. M, 1961, Pomona College.

MacCornack, Richard John. Geology and structure along a portion of the northern end of the Maxwell Reef, Boulder County, Colorado. M, 1944, University of Colorado.

MacCracken, Michael Calvin. Ice age theory analysis by computer model simulation. D, 1968, University of California, Davis. 193 p.

MacCracken, W. L. A new astatic gravimeter. M, 1955, University of Hawaii.

MacCutcheon, Murray. Some aspects of a digital analysis of seismic refraction data. M, 1971, Dalhousie University.

MacDaniel, R. P. Upper Ordovician sedimentary and benthic community patterns of the Cincinnati Arch area. D, 1976, University of Chicago. 181 p.

MacDiarmid, Roy A. Geology and ore deposits of the Bristol silver mine, Pioche, Nevada. D, 1960, Stanford University.

MacDiarmid, Roy Angus. Geology and ore deposits of the Bristol silver mine, Pioche, Nevada. D, 1959, Stanford University. 110 p.

MacDonald, Alan Stratton. The Salmo lead-zinc deposits; a study of their deformation and metamorphic features. D, 1973, University of British Columbia.

MacDonald, Alasdair James. Boss Mountain molybdenite deposit; fluid geochemistry and hydrodynamic considerations. D, 1984, University of Toronto.

MacDonald, Alasdair James. The geology and genesis of the Cueva Santa Branch silver, lead, zinc manto ore body, Fresnillo Mine, Mexico. M, 1979, University of Toronto.

MacDonald, Alistair P. T. Hydrogeologic parameters for zoning in Russell Township, Geauga County, Ohio. M, 1987, Kent State University, Kent. 170 p.

MacDonald, Andrew Harrington. Water diffusion rates through serpentinized peridotites; implications for reaction induced dynamic and chemical effects in ultramafic rocks. D, 1984, University of Western Ontario. 266 p.

MacDonald, Angus A. The Dumont dune system of the northern Mojave Desert, California. M, 1966, San Fernando Valley State University.

Macdonald, Calum. Inverting seismic data using reflection travel times and amplitudes. D, 1986, University of California, Los Angeles. 198 p.

MacDonald, Colin Campbell. Mineralogy and geochemistry of a Precambrian regolith in the Athabasca Basin. M, 1981, University of Saskatchewan. 151 p.

MacDonald, Donald E. Geology and resource potential of phosphates in Alberta and portions of southeastern British Columbia. M, 1985, University of Alberta. 254 p.

MacDonald, Donald J. Stratigraphy of the upper member of the Horton Bluff Formation (Lower Carboniferous) in the area of the type section near Hantsport, Nova Scotia. M, 1973, Acadia University.

MacDonald, Elizabeth Cora. Flow geometry in the Black Canyon sector of the Bitterroot Lobe of the Idaho Batholith, Idaho County, Idaho. M, 1986, University of Idaho. 89 p.

MacDonald, G. H. The Mississippian of Saskatchewan. D, 1953, University of Toronto.

MacDonald, Gilbert H. The Mississippian of Saskatchewan. D, 1954, University of Toronto.

MacDonald, Glen M. The post-glacial paleoecology of the Morley Flats and Kananaskis Valley region, southwestern Alberta. M, 1980, University of Calgary.

MacDonald, Glen Michael. Postglacial plant migration and vegetation development in the western Canadian boreal forest. D, 1984, University of Toronto.

MacDonald, Gordon Andrew. Geology of the western part of the Sierra Nevada between the Kings and San Joaquin rivers, California. D, 1938, University of California, Berkeley. 148 p.

MacDonald, Gordon Andrew. Sediments of Santa Monica Bay. M, 1934, University of California, Los Angeles.

Macdonald, Gordon James Fraser. A critical review of geologically important thermochemical data. D, 1954, Harvard University.

MacDonald, Harold Carleton. Geologic evaluation of radar imagery from Darien Province, Panama. D, 1969, University of Kansas. 141 p.

MacDonald, Harold Carleton. Post-glacial (Holocene) ostracodes from Lake Erie. M, 1962, University of Kansas.

MacDonald, I. M. Water-rock interaction in felsic rocks of the Canadian Shield. M, 1986, University of Waterloo. 190 p.

MacDonald, James Reid. The Pliocene carnivores of the Black Hawk Ranch fauna. M, 1947, University of California, Berkeley. 51 p.

Macdonald, James Reid. The Pliocene mammalian faunas of west-central Nevada. D, 1949, University of California, Berkeley. 116 p.

Macdonald, Kenneth Craig. Detailed studies of the structure, tectonics, near-bottom magnetic anomalies and microearthquake seismicity of the Mid-Atlantic Ridge near 37°N. D, 1975, Woods Hole Oceanographic Institution. 248 p.

MacDonald, M. V. The Aldermac syenite porphyry stock, Quebec. M, 1938, McGill University.

MacDonald, Neil William. Sulfate adsorption in Michigan forest soils. D, 1987, Michigan State University. 239 p.

Macdonald, R. D. Geology of the Pagwaqchuan Lake map area. M, 1938, Queen's University.

MacDonald, R. H. Occurrence and characteristics of polymineralic silicate inclusions in chromite grains, eastern Bushveld Complex. M, 1979, University of Wisconsin-Madison.

MacDonald, Ralph Crawford. The gossan of a lead deposit in limestone, Yukon Territory. M, 1947, University of British Columbia.

MacDonald, Robert D. Marine geology of upper Jervis Inlet. M, 1970, University of British Columbia.

Macdonald, Robert, III. Contour maps of expected bedrock acceleration from earthquakes; a critique. M, 1978, University of California, Los Angeles.

MacDonald, Roderick Dickson. The geology of the Wintering River map area. D, 1940, Princeton University. 89 p.

MacDonald, Roderick Dickson. The origin of gypsum in Alberta deposits. M, 1935, University of Alberta. 104 p.

MacDonald, Ruth Heather. Depositional environments of the Greenhorn Formation (Cretaceous), northwestern Black Hills. D, 1984, University of Wisconsin-Madison. 304 p.

MacDonald, Scott Edward. Sedimentologic and paleoceanographic implications of terrigenous deposits on the Maurice Ewing Bank, Southwest Atlantic Ocean. M, 1984, Rice University. 207 p.

MacDonald, William David. Geology of the Serrania de Macuira area, Guajira Peninsula, Colombia. D, 1965, Princeton University. 167 p.

MacDonald, William Delbert. A comparative study of the waterways and older formations of the McMurray area. M, 1947, University of Alberta. 107 p.

MacDougall, Craig S. A metallogenic study of polymetallic, granophile mineralization within the early Proterozoic upper Aillik Group, Round Pond area, Central Mineral Belt, Labrador. M, 1988, Memorial University of Newfoundland. 245 p.

MacDougall, J. Douglas. A study of the geochemistry of an arctic watershed. M, 1968, McMaster University. 84 p.

MacDougall, J. F. Experiments bearing on the genesis of sulphide deposits. D, 1957, McGill University.

Macdougall, J. F. The Birch Lake copper deposit, Saskatchewan. M, 1952, McGill University.

Macdougall, John Douglas. Particle track records in natural solids from oceans on Earth and Moon. D, 1972, University of California, San Diego. 256 p.

MacDougall, Robert Earl. A model study of an applied potential survey concerning the deep crust in Massachusetts. M, 1960, University of California, Los Angeles.

MacDowell, John Fraser. The Eh-pH fields of mercury mineral stability. M, 1959, University of Michigan.

Mace, Neal Wayne. A paleomagnetic study of the Miocene Alverson Volcanics of the Coyote Mountains, western Salton Trough, California. M, 1981, San Diego State University. 142 p.

Mace, Thomas Hooker. Microdensitometric analysis of aerial photographic imagery for detailed soils mapping. D, 1981, University of Wisconsin-Madison. 234 p.

MacEachern, James Anthony. Paleoenvironmental interpretation of the Lower Cretaceous Waseca Formation, upper Mannville Group, Lloydminster area, Saskatchewan. M, 1986, University of Regina. 237 p.

Macedo, Jamil. Preferential reduction of hematite over goethite in some Oxisols in Brazil. D, 1988, Cornell University. 56 p.

Macedo-Raa, Albino Reynaldo. Subsurface structural geology of the northeastern part of Crockett County, Texas. M, 1969, University of Texas, Austin.

Macek, Josef Jan. New determination curves for albite-twinned plagioclase feldspars based on an analysis of optical crystallographic scatter. D, 1979, University of Manitoba.

Macellari, Carlos Enrique. Late Cretaceous stratigraphy, sedimentology, and macropaleontology of Seymour Island, Antarctic Peninsula, (Volumes I and II). D, 1984, Ohio State University. 657 p.

Macer, Robert J. Fluid inclusion studies of fluorite around the Organ Cauldron, Dona Ana County, New Mexico. M, 1978, University of Texas at El Paso.

Macey, Gerald J. A sedimentological comparison of two Proterozoic red bed successions (the South Channel and Kazan formations of Baker Lake, Northwest Territories and the Martin Formation at Uranium City, Saskatchewan). M, 1973, Carleton University. 175 p.

Macfadden, B. J. Magnetic polarity stratigraphy and mammalian biochronology of the Chamita Formation stratotype (Mio-Pliocene) of Northcentral New Mexico. D, 1976, Columbia University, Teachers College. 116 p.

Macfadyen, John A., Jr. Structural geology of an area near Lehigh Gap, Pennsylvania. M, 1950, Lehigh University.

MacFadyen, John A., Jr. The geology of the Bennington area, Vermont. D, 1963, Columbia University, Teachers College.

MacFarlane, D. S. Hydrogeological studies of an abandoned landfill on a sandy aquifer; physical hydrogeology. M, 1980, University of Waterloo.

MacFarlane, Darryl B. An electron microprobe study of allanite from the Grenville Province, southeastern Ontario and southwestern Quebec. M, 1987, Queen's University. 84 p.

Macfarlane, Ian Charles. Sand variation of a beach and offshore area near Bodega Head, California. M, 1971, University of California, Davis. 129 p.

Macfarlane, N. D. The regional setting, primary mineralogy, and economic geology of the Nemeiben Lake ultramafic pluton. M, 1978, University of Saskatchewan. 312 p.

MacFarlane, Neil Daniel. The regional setting, primary mineralogy, and economic geology of the Nemeiben Lake ultramafic pluton. M, 1979, University of Saskatchewan. 182 p.

MacFarlane, P. Allen. Geologic constraints on land use in northeastern Morris County, Kansas. M, 1979, University of Kansas. 117 p.

MacFarlane, William T. Long wavelength gravity anomalies and implications concerning a descending lithospheric slab in the Pacific Northwest. D, 1975, Oregon State University. 81 p.

MacGaw, Bradford Kuhns. Ground water supplies in the Cedar Rapids, Iowa area. M, 1936, University of Iowa. 58 p.

MacGeachy, J. K. Boring by macro-organisms in the coral Montastrea annularis on Barbados reefs. M, 1975, McGill University. 83 p.

MacGeachy, J. K. Geological significance of boring sponges on Barbados reefs. D, 1978, McGill University. 207 p.

MacGeehan, P. J. The petrology and geochemistry of volcanic rocks at Matagami, Quebec, and their relationship to massive sulphide mineralization. D, 1979, McGill University. 414 p.

MacGill, Peter L. The stratigraphy and sedimentary petrology of the Renfro Member of the Borden Formation. M, 1973, Eastern Kentucky University. 51 p.

MacGill, Rotha A. Conodonts from the St. Louis Member of the Newman Formation of east-central Kentucky. M, 1973, Eastern Kentucky University. 51 p.

MacGinitie, Harry D. The flora of the Weaverville Beds, Trinity County, California. D, 1936, University of California, Berkeley. 131 p.

MacGowan, Donald B. Organic geochemistry and petroleum source potential of the Cretaceous Bear River Formation, northern Idaho-Wyoming thrust belt. M, 1982, Idaho State University. 65 p.

MacGregor, A. Roy. Chazy corals and reefs [Quebec]. D, 1954, McGill University.

Macgregor, A. Roy. Chazyan (Ordovician) reefs and reef builders. M, 1954, McGill University.

MacGregor, D. R. C. Weathering of oil spilled on cold, oceanic water with particular reference to sulfur containing compounds. M, 1975, Dalhousie University.

MacGregor, Ian D. A study of the contact metamorphic aureole surrounding the Mount Albert ultramafic intrusion. D, 1964, Princeton University. 195 p.

MacGregor, Ian Duncan. Geology, petrology and geochemistry of the Mount Albert and associated ultramafic bodies of central Gaspe, Quebec. M, 1962, Queen's University. 288 p.

Macgregor, J. E. An investigation of effects of X-radiation on the electrical properties of mica. M, 1949, Rice University.

MacGregor, James Donald. An isopachous map of the Pennsylvanian System of Oklahoma. M, 1941, University of Oklahoma. 33 p.

Mach, Leah E. Structural geology of the Howard Lake Quadrangle, Lincoln and Sanders counties, Montana. M, 1986, University of Idaho. 93 p.

Machacha, Tafilani P. Mineralogy and geochemistry of the iron dike in the Broken Boot Mine area west of Deadwood, Lawrence County, South Dakota. M, 1982, South Dakota School of Mines & Technology. 55 p.

Machado Brito, Ignacio Autrliano. Silurian and Devonian Acritarcha from Maranhao Basin, Brazil. M, 1966, Stanford University.

Machado Fernandes, Nuno. Géochimie de l'île de Sao Jorge (Açores). M, 1977, Universite de Montreal.

Machado, J. E. Geology and ore deposits of Pennsylvania Hill, Alma District, South Park, Park County, Colo. M, 1967, Colorado School of Mines. 153 p.

Machamer, J. F. The geology of the Forsyth and associated magnetite deposits, Hull Township, Province of Quebec. M, 1959, McGill University.

Machamer, Jerome Frank. Geology and origin of the iron ore deposits of the Zenith Mine, Ely, Minn. D, 1964, Pennsylvania State University, University Park. 172 p.

Machan-Castillo, Maria Luisa. Pliocene Ostracoda of the Saline Basin, Veracruz, Mexico. D, 1985, Louisiana State University. 271 p.

Machel, Hans-Gerhard. Facies and diagenesis of the Upper Devonian Nisku Formation in the subsurface of central Alberta. D, 1985, McGill University. 392 p.

Machemer, Steven Dean. Lithofacies and diagenesis of the Cardium Formation, Northeast Pembina area, Alberta. M, 1984, University of Calgary. 169 p.

Machenberg, Marcie Debra. Sand dune migration in Monahans Sandhills State Park, Texas. M, 1982, University of Texas, Austin. 127 p.

Machesky, Michael Lawrence. Calorimetric investigation of the goethite-water interface. D, 1986, University of Wisconsin-Madison. 244 p.

Machette, Michael N. The Quaternary geology of the Lafayette Quadrangle, Colorado. M, 1975, University of Colorado.

Machovec, Marvin Anthony. Regional stratigraphy of the lower Kansas City Group (Pennsylvanian) of Nebraska and adjacent region. M, 1963, University of Nebraska, Lincoln.

Macias-Chapa, Luis. Multiphase, multicomponent compressibility in petroleum reservoir engineering. D, 1985, Stanford University. 186 p.

MacIlvaine, Joseph Chad. Sedimentary processes on the continental slope off New England. D, 1973, Massachusetts Institute of Technology. 211 p.

MacInnes, Scott Charles. Lateral effects in controlled source audiomagnetotellurics. D, 1988, University of Arizona. 190 p.

Macintosh, Albert N. Stratigraphy of the Negaunee Iron Formation (Michigan). M, 1930, [Michigan Technological University].

Macintosh, James Alexander. The quartz deposits at Saint Donat, Quebec. M, 1956, McGill University.

MacIntyre, Donald George. Evolution of Upper Cretaceous volcanic and plutonic centers and associated porphyry copper occurrences, Tahtsa Lake area, British Columbia. D, 1976, University of Western Ontario. 149 p.

MacIntyre, Donald George. Zonation of alteration and metallic mineral assemblages, Coles Creek copper prospect, west central British Columbia. M, 1974, University of Western Ontario. 130 p.

MacIntyre, Giles Ternan. The Miacidae (Mammalia Carnivora); Part 1, The systematics of *Ictidopappus* and *Protictis*. D, 1964, Columbia University, Teachers College. 271 p.

Macintyre, Ian G. Recent sedimentation of the west coast of Barbados, West Indies. D, 1967, McGill University. 169 p.

Macintyre, Robert Mitchell. Studies in potassium-argon dating. D, 1966, University of Toronto.

MacIntyre, William G. The temperature variation of the solubility product of calcium carbonate in sea water. D, 1965, Dalhousie University.

MacIvor, Keith Alan. Geology of the Thousand Oaks area, Los Angeles and Ventura counties, California. M, 1956, University of California, Los Angeles.

Mack, Christopher Brown. Postglacial tephrochronology of the Indian Heaven area, south Cascade Range, Washington. M, 1980, Washington State University. 75 p.

Mack, Edward, Jr. The water content of coal, with some ideas of the genesis and nature of coal. D, 1916, [Princeton University].

Mack, Gregory Harold. Provenance and paleoclimatic interpretations from a petrographic comparison of Holocene sands and the Fountain Formation (Pennsylvanian) in the Colorado Range Front. M, 1975, Indiana University, Bloomington. 53 p.

Mack, Gregory Harold. The effects of depositional environment on detrital mineralogy; the Permian Cutler-Cedar Mesa facies transition, near Moab, Utah. D, 1977, Indiana University, Bloomington. 152 p.

Mack, Gregory Stebbins. Geology, ground water chemistry, and hydrogeology of the Murphy area, Josephine County, Oregon. M, 1983, Oregon State University. 130 p.

Mack, Harry. The nature of short period P wave signal variations at LASA. D, 1968, Southern Methodist University. 34 p.

Mack, John Erick, Jr. Reconnaissance geology of the Woodside Quadrangle, San Mateo County, California. M, 1959, Stanford University. 78 p.

Mack, John Wesley, Jr. A least-square method of gravity analysis and its applications in the study of sub-surface geology. D, 1963, University of Wisconsin-Madison. 135 p.

Mack, Lawrence Edward. Petrography and diagenesis of a submarine fan sandstone, Cisco Group (Pennsylvanian), Nolan County, Texas. M, 1984, University of Texas, Austin.

Mack, Leslie E. Evaluation of a conducting paper analog field plotter and a simple computer as aids in solving ground-water problems. M, 1957, University of Kansas.

Mack, Leslie E. Geology and water resources of Ottawa County, Kansas. D, 1959, University of Kansas. 164 p.

Mack, Pamela Diane Carpenter. Correlation and provenance of facies within the upper Santa Fe Group in the subsurface of the Mesilla Valley, southern New Mexico. M, 1985, New Mexico State University, Las Cruces. 137 p.

Mack, Seymour. Geology and ground water resources of Scott and Shasta Valleys, Siskiyou County, California. D, 1957, Syracuse University. 287 p.

Mack, Seymour. Petrography and structure of some non-pegmatite dikes in the Hudson Highlands (New York). M, 1950, Syracuse University.

Mack, Stanley Z. Stabilization of a universal airborne magnetometer. D, 1951, University of Toronto.

Mack, Tinsley. Geology of the Everona Formation (Lower Ordovician, Virginia). M, 1957, University of Virginia. 156 p.

MacKallor, Jules A. Geology of the western part of the Cobabi mining district, Pima County, Arizona. M, 1958, University of Arizona.

MacKallor, Jules A. The supply and demand of lead for the United States through 1983. D, 1976, Pennsylvania State University, University Park. 325 p.

Mackasey, William Oliver. Petrography and stratigraphy of a Lower Mississippian, pre-Horton volcanic succession in northwestern Cape Breton Island, Nova Scotia. M, 1963, Carleton University. 120 p.

Mackay, Angus James. Continuous seismic profiling investigation of the southern Oregon continental shelf between Cape Blanco and Coos bay. M, 1969, Oregon State University. 118 p.

MacKay, Bertram Reid. Beauceville map area, Quebec. D, 1920, University of Chicago. 105 p.

MacKay, D. G. Zinc deposits at Long Lake, Ontario. M, 1941, Queen's University. 36 p.

MacKay, Hugh James. Oil shales and the oil shale industry. M, 1918, University of Oklahoma. 86 p.

MacKay, Ian H. Geology of the Thomasville-Woods Lake area, Eagle and Pitkin counties, Colorado. D, 1953, Colorado School of Mines. 121 p.

Mackay, Thomas Stephen. Alteration and recovery of some detrital opaque heavy minerals; ilmenite and rutile resistates. D, 1972, Rice University. 210 p.

Macke, David L. Stratigraphy and sedimentology of experimental alluvial fans. M, 1977, Colorado State University. 116 p.

Macke, John Edward. Geochemistry of early Precambrian graywackes from the Fig Tree series, South Africa. M, 1969, Washington University. 38 p.

Macke, William Bernard. A faunal study of the Beechwood Limestone (Middle Devonian), Speed, Indiana. M, 1952, University of Cincinnati. 134 p.

MacKean, B. E. The geology of the Mount Reed area, Quebec. M, 1960, McGill University.

Mackensie, Duncan T., III. A comparative analysis of lower Tertiary calcareous nannofossils from southern oceans. M, 1977, Florida State University.

Mackenzie, Andrew N. The geology of the Republic of Nicaragua, Central America. M, 1922, [Stanford University].

Mackenzie, David Brindley. Geology of the north central Cojedes map-area, Venezuela. D, 1953, Princeton University. 268 p.

Mackenzie, Frederick Theodore. Paleocurrents in the Cloverly Group (Lower Cretaceous) of Wyoming. D, 1962, Lehigh University. 187 p.

Mackenzie, G. S. Some mineral deposits of post-Cambrian age in the St. Lawrence Basin. D, 1934, University of Toronto.

MacKenzie, George Donald. Minor elements in magnetite from central and northwestern Arizona. M, 1962, Pennsylvania State University, University Park. 72 p.

MacKenzie, Glenn Staghan. Stationary and non-stationary ground movements at frequencies from 1 to 200 millicycles per second. D, 1965, University of California, San Diego.

Mackenzie, John David. The geology of Graham Island, British Columbia. D, 1916, Massachusetts Institute of Technology. 10 p.

MacKenzie, Kevin Ralph. Crustal stratigraphy and realistic seismic data. D, 1984, University of California, San Diego. 140 p.

MacKenzie, Michael G. Sedimentation in the Moriches Inlet area, New York. M, 1962, New York University.

Mackenzie, Michael G. Stratigraphy and petrology of the Mesaverde Group, southern part of the Big Horn Basin, Wyoming. D, 1975, Tulane University. 241 p.

MacKenzie, Michael Vincent, Jr. A digital computer model of terrestrial heat flow refraction. M, 1967, University of Utah. 99 p.

Mackenzie, Robert John. Magnetic prospecting for uranium; southern Powder River basin, Wyoming. M, 1976, University of Wyoming. 71 p.

MacKenzie, W. Bruce. Investigation of minor elements in metamorphic biotites. M, 1958, Dartmouth College. 53 p.

MacKenzie, Wallace Bruce. Hydrothermal alteration associated with the Urad and Henderson molybdenite deposits (Tertiary), Clear Creek County, Colorado. D, 1970, University of Michigan. 280 p.

MacKenzie, Warren Stuart. The geology of the Mount Stornoway-Redan Mountain carbonate complex, Jasper Park, Alberta. M, 1962, University of Toronto.

MacKenzie, Warren Stuart. The geology of the Southesk Cairn Carbonate Complex (Devonian), Alberta, Canada. D, 1965, University of Toronto.

MacKenzie, Wayne Oliver. Geology and ore deposits of a section of the Beaverhead Range East of Salmon, Idaho. M, 1949, University of Idaho. 52 p.

Mackevett, Edward M. The geology of the Jurupa Mountains, San Bernardino and Riverside counties, California. M, 1950, California Institute of Technology. 41 p.

Mackey, Gary W. Radium-226 and strontium-90 in Iowa ground water. M, 1976, University of Iowa. 149 p.

Mackey, Ronald Taylor. Lithostratigraphy and depositional environment of the Perryville Member of the Lexington Limestone (Middle Ordovician, Kentucky). M, 1972, University of Kentucky. 94 p.

Mackey, Scudder Draper. Physical stratigraphy of Pleistocene sediments from the central Arctic Ocean. M, 1977, University of Wisconsin-Madison.

MacKidd, David G. Interpretation of gravity and magnetics north of Lake Superior. M, 1973, University of Toronto.

Mackie, B. W. Petrogenesis of the Lac Turgeon Granite and uranium occurrences near Baie Johan Beetz, Quebec. M, 1978, University of Manitoba. 24 p.

Mackie, Thomas L. Tectonic influences on the petrology, stratigraphy and structures of the Upper Cretaceous Golden Spike Formation, central-western Montana. M, 1986, Washington State University. 132 p.

Mackiewicz, Nancy E. The genesis of laminated ice-proximal glacimarine sediments in Muir Inlet, Alaska. M, 1983, Northern Illinois University. 168 p.

Mackin, J. Hoover. Erosional history of the Big Horn Basin, Wyoming. D, 1937, Columbia University, Teachers College.

Mackin, J. Hoover. River terraces in the Susquehanna Valley below Harrisburg (Pennsylvania). M, 1932, Columbia University, Teachers College.

Mackin, James E., II. The behavior of aluminum during early diagenesis in marine sediments. D, 1983, University of Chicago. 263 p.

Mackin, Richard L. A survey of earth science courses as a discipline in Montana secondary schools. M, 1970, Montana State University. 91 p.

MacKinney, John. The crystal structures of two clintonites. M, 1986, University of Wisconsin-Madison. 60 p.

MacKinnon, Cinda. A facies analysis of the late Miocene Puente Formation, Puente Hills, Southern California. M, 1984, California State University, Long Beach. 80 p.

MacKinnon, Paula. A study of echelon fracture sets in the Killarney igneous complex, Killarney, Ontario. M, 1988, McMaster University. 164 p.

MacKinnon, Thomas C. Cretaceous sedimentation, southern San Rafael Mountains, Santa Barbara County, California. M, 1976, University of California, Santa Barbara.

Mackintosh, Michael Edward. Auto-suspension of sediment; a test of the theory. M, 1975, Massachusetts Institute of Technology. 37 p.

Mackler, Anne. Chrysotile asbestos, an electron microscopic study (New York City). M, 1971, New York University.

Macknight, Franklin C. Flora of the (Pennsylvanian) Grape Creek coal at Danville, Illinois. D, 1938, University of Chicago. 118 p.

Macko, Stephen Alexander. Stable nitrogen isotope ratios as tracers of organic geochemical processes. D, 1981, University of Texas, Austin. 192 p.

Mackovjak, Dennis. Occurrence and origin of jasperoids on Lone Mountain, Independence Range, Elko County, Nevada. M, 1983, Bowling Green State University. 122 p.

Macksoud, Adrienne M. The nature and origin of mountain leather. M, 1939, Columbia University, Teachers College.

Macky, Tarek Ahmed Aly. Behavior of anisotropic clays subjected to cyclic loading. D, 1982, Case Western Reserve University. 298 p.

MacLachlan, Donald Claude. Geologic structure of a small area in Wayne, Pulaski, and Russell counties, Kentucky. M, 1927, [University of Michigan].

MacLachlan, Donald Claude. Warren shore line in Ontario and in the thumb of Michigan, and its formation. D, 1939, University of Michigan.

MacLachlan, James Crawford. Geology of the La Victoria area, Venezuela. D, 1952, Princeton University. 133 p.

Maclaren, Alexander S. Peridotites of northwestern Quebec. D, 1953, McGill University.

MacLaren, Alexander S. Some problems in correlation in north-western Quebec. M, 1950, McGill University.

Maclay, Robert Weaver. The ground water hydrology of the Elizabethton and Johnson City, Tennessee area. M, 1956, University of Tennessee, Knoxville. 92 p.

Maclean, Ann Louise. Evaluation of thematic mapper imagery for forest type mapping and its use in a computerized hazard-rating system. D, 1987, University of Wisconsin-Madison. 198 p.

MacLean, David Alexander. Depositional environments and stratigraphic relationships of the Glens Falls Limestone, Champlain Valley, Vermont and New York. M, 1986, University of Vermont. 170 p.

Maclean, Donald Wardrope. The (Devonian) Ghost River Formation between the Athabasca and Smoky rivers, Alberta. M, 1953, McGill University.

MacLean, Hugh James. Geology and mineral deposits of the Little Bay area, Newfoundland. D, 1940, Princeton University.

MacLean, J. Arthur. Some aspects of digital processing of seismic reflection data (from Baffin Bay). M, 1972, Dalhousie University.

MacLean, Wallace H. Liquidus phase relations in the FeS-FeO-Fe_3O_4-SiO_2 system, and their application in geology. D, 1968, McGill University. 153 p.

MacLean, William Finley. Postglacial uplift in the Great Lakes region. D, 1962, University of Michigan. 241 p.

MacLean, Willis John. Mineral analysis of concentrates from French Equatorial Africa. M, 1932, University of Minnesota, Minneapolis. 13 p.

MacLellan, Donald D. A San Diego fauna in the vicinity of Val Verde, California. D, 1936, California Institute of Technology. 10 p.

MacLellan, Donald D. Geology of the East Cochella Tunnel of the metropolitan water district of Southern California. D, 1936, California Institute of Technology. 90 p.

MacLellan, H. Elizabeth. Experimental study of dynamic crystallization in synthetic granitic melts at 1 kilobar. M, 1984, University of New Brunswick. 261 p.

MacLellan, Mary L. Geology of the Reany Lake area, Marquette County, Michigan. M, 1988, Michigan Technological University. 110 p.

MacLeod, Anne Jacquelyn. An investigation of two thermal models of the oceanic lithosphere based on topography, heat flow, and shear wave travel time residuals. M, 1983, Pennsylvania State University, University Park. 76 p.

Macleod, George M. The geology of the Packwood Creek area, Kern County, California. M, 1948, Stanford University. 72 p.

MacLeod, John. The metamorphism and chemical composition of the Littleton Series (Lower Devonian) of New Hampshire. M, 1952, McMaster University. 23 p.

MacLeod, N. The paleoecology of the Wolf Mountain Shale; community structure and trophic analysis. M, 1980, Southern Methodist University. 202 p.

MacLeod, Norman. Systematic, phylogenetic and morphometric analyses of the Jurassic radiolarian genus Perispyridium dumitrica. D, 1986, University of Texas at Dallas. 466 p.

Macleod, Norman Scott. Experimental investigation of three-dimensional kinematics at sharp terminations (corners) in forced folds. M, 1988, University of Oklahoma. 86 p.

MacLeod, Norman Stewart. Geology and igneous petrology of the Saddleback area, central Oregon Coast

Range. D, 1970, University of California, Santa Barbara. 205 p.

Macleod, R. B. The subsurface geology of the (Cretaceous) Blairmore Formation of east central Alberta. D, 1951, University of Toronto.

Macleod, Roderick J. The geology of the Gilt Edge area, northern Black Hills of South Dakota. M, 1986, South Dakota School of Mines & Technology.

MacLeod, William George. Rock fractures; an aerial photographic study. M, 1955, Cornell University.

MacMillan, John R. Late Paleozoic and Mesozoic tectonic events in west central Nevada. D, 1972, Northwestern University.

MacMillan, Logan. Stratigraphy of South Platte Formation (lower Cretaceous, Morrison-Weaver-Golden area, Jefferson County, Colorado). M, 1974, Colorado School of Mines. 131 p.

MacNab, Ronald Finlay. Gravity computations on a spherical Earth. M, 1968, Dalhousie University. 111 p.

MacNabb, Bert E. Stratigraphy and structure of Carboniferous sediments in southeastern New Brunswick. M, 1972, Northwestern University.

Macnae, James Charles. Geophysical prospecting with electric fields from an inductive EM source. D, 1981, University of Toronto.

Macnae, James Charles. The response of UTEM to a poorly conducting mineralized environment. M, 1977, University of Toronto.

MacNamara, Edlen E. Soils of the Howard Pass area, northern Alaska. D, 1965, Rutgers, The State University, New Brunswick. 239 p.

MacNaughton, David R. A structural and stratigraphical analysis at Bacon Ridge, Teton and Sublette counties, Wyoming. M, 1983, Miami University (Ohio). 41 p.

MacNeil, Donald Jonathan. The stratigraphy and structure of the Hillcrest coal field, Alberta. D, 1935, Princeton University. 43 p.

Macneil, Robert J. Geology of the Humphreys Station area, Los Angeles County, California. M, 1948, California Institute of Technology. 39 p.

MacNeill, R. H. Pleistocene geology of the Wolfville area, Nova Scotia. M, 1951, Acadia University.

MacNeille, P. R. A physical basis for the occurrence of radon anomalies before earthquakes. M, 1983, Case Western Reserve University.

MacNish, Robert Dick. Geomorphology of the western Sanpete Valley, Utah. M, 1962, University of Michigan.

MacNish, Robert Dick. The Cenozoic history of the Wet Mountain Valley, Colorado. D, 1966, University of Michigan. 159 p.

Macomber, Bruce E. Geology of the Soda Butte area, Yellowstone Park, Wyoming. M, 1956, Northwestern University.

Macomber, Bruce Edkins. Geology of the Cuale mining district, Jalisco, Mexico. D, 1962, Rutgers, The State University, New Brunswick. 387 p.

Macomber, Mark M. Geodetic computations on a projection plane. M, 1958, Ohio State University.

Macomber, Mark M. The influence of anomalous gravity on the performance of a mechanically perfect inertial navigation system. D, 1966, Ohio State University.

Macomber, Richard. Cambro-Ordovician stratigraphic relationships in the Upper Mississippi Valley. M, 1959, Northwestern University.

Macomber, Richard Wiltz. Articulate (Ordovician) brachiopods of the upper Bighorn Formation (Johnson and Sheridan counties), Wyoming. D, 1968, University of Iowa. 88 p.

Maconachie, James Roy Alexander. Petrological study of the dyke rocks of the Whitewater Creek and Lyle Creek area, Slocan District, British Columbia. M, 1940, University of British Columbia.

Macpherson, A. P. The coal deposits of North America. M, 1927, McMaster University.

Macpherson, Bruce A. The geology of the Halsey Quadrangle, Oregon. M, 1953, University of Oregon. 64 p.

MacPherson, Donald Stuart. Variations in the abundance of O^{18} in ice and snow from the Kaskawulsh Glacier. M, 1965, University of Alberta. 77 p.

MacPherson, Glenn Joseph. Geology and petrology of the Mesozoic submarine volcanic complex at Snow Mountain, northern Coast Ranges, California; an on-land seamount. D, 1981, Princeton University. 304 p.

MacPherson, Gwendolyn Lee. Low-temperature geothermal ground water in the Hosston/Cotton Valley hydrogeologic unit, Falls County Texas. M, 1982, University of Texas, Austin. 234 p.

MacPherson, H. G. A chemical and petrographic study of Precambrian sediments. D, 1955, University of Toronto.

MacPherson, H. G. A petrographic comparison of magmatic and granitized rocks. D, 1954, University of Toronto.

Macpherson, Louis Alan. A study of some Cambro-Ordovician oolites from central Pennsylvania. M, 1952, University of Pittsburgh.

MacPherson, Robert A. Pleistocene stratigraphy of the Winnipeg River in the Pine Falls-Seven Sisters Falls area, Manitoba (Canada). M, 1968, University of Manitoba.

MacQuarrie, K. T. B. Simulation of biodegradable organic contaminants in groundwater. M, 1988, University of Waterloo. 125 p.

MacQueen, J. Kenneth. Stratigraphy, structure and gold mineralization of the No. 5 vein/iron formation zone, Pickle Crow gold mines, Pickle Lake, Ontario. M, 1987, Carleton University. 166 p.

MacQueen, Jeffrey Donald. Linear inversion of gravity data; with geological and geophysical constraints. D, 1982, University of Washington. 137 p.

MacQueen, Peter A. A long core study of the effect of initial water saturation upon oil recovery by water flooding. M, 1952, University of Oklahoma. 125 p.

Macqueen, R. W. Sedimentation and diagenesis of the lower Fairholme Group in Jasper Park, Alberta. M, 1960, University of Toronto.

MacQueen, Roger Webb. Stratigraphy and sedimentology of the Mount Head Formation, Alberta, Canada. D, 1965, Princeton University. 182 p.

Macquown, William C. Structure and stratigraphy of the White River Plateau near Glenwood Springs, Colorado. D, 1943, Cornell University.

MacQuown, William Charles. Stratigraphy of the (Devonian) Oriskany Group in eastern West Virginia. M, 1940, University of Rochester. 93 p.

Macrae, John M. The morphological variation and autecology of the rhynchonellid brachiopod Hypothyridina venustula from the Devonian of New York. M, 1974, University of Calgary. 153 p.

Macrae, Leslie Blair. An investigation of Devonian rhynchonellids of the Great Western Basin. M, 1955, University of Alberta. 62 p.

MacRae, Neil Donald. Petrology and geochemistry of ultramafic-gabbroic intrusion in Abitibi area, Ontario. D, 1966, McMaster University. 163 p.

MacRae, Neil Donald. Petrology of the Centre Hill Complex, North Ontario. M, 1963, McMaster University. 102 p.

MacRae, William Edgar. Noble metal concentrations in selected komatiitic and tholeiitic Archean volcanic rocks from Munro Township, Ontario. M, 1982, McMaster University. 96 p.

Macrides, Costas. Interpretation of seismic refraction profiles in southern Saskatchewan. M, 1983, University of Alberta. 133 p.

MacRobbie, Paul. Stratigraphy, structure and volcanogenic sulphides of Sicker Group volcanic rocks on Mt. Sicker, Vancouver, B.C. M, 1988, Carleton University. 144 p.

MacTavish, John Nickolas. Faunal elements of part of the Lodgepole Formation, Madison Group (Mississippian), from Darby Canyon, Grand Teton Mountains, Wyoming. D, 1971, Case Western Reserve University. 252 p.

MacTavish, John Nickolas. The petrology of an iron ore body near Butternut, Wisconsin. M, 1963, Bowling Green State University. 32 p.

Macurda, Donald Bradford, Jr. Studies in the blastoid genus Orophocrinus. D, 1963, University of Wisconsin-Madison. 526 p.

MacVeigh, Edwin Lester. Mineralogic studies of some Pennsylvanian sandstones. M, 1932, University of Illinois, Chicago.

Macy, Jonathan S. Sediment yield of the upper San Lorenzo watershed using quantitative geomorphic and sediment sampling approaches. M, 1976, San Jose State University. 127 p.

Maczuga, David E. The petrology and geochemistry of the Fitchburg plutonic complex, central Massachusetts. M, 1981, University of Massachusetts. 128 p.

Madabhushi, Govindachari Venkata. Effects of temperature changes on chemical equilibria in groundwater due to groundwater heat pumps. D, 1984, Utah State University. 223 p.

Madalosso, Antonio. Stratigraphy and sedimentation of the Bambui Group, Paracatu region, MG, Brazil. M, 1979, University of Missouri, Columbia.

Madani, Mastaneh. Contribution to the petrogenesis of gabbro, Wichita Mountains, Oklahoma. M, 1967, Baylor University. 55 p.

Madani, Mostafa Seyed. Accuracy potential of non-metric cameras in close-range photogrammetry. D, 1987, Ohio State University. 211 p.

Madar, James Michael. Stratigraphic analysis of lower Conemaugh rocks (Pennsylvanian), Indiana and Armstrong counties, Pennsylvania. M, 1981, University of Pittsburgh. 42 p.

Madariaga Meza, Raul I. Free oscillations of the laterally heterogeneous Earth. D, 1971, Massachusetts Institute of Technology. 105 p.

Maddalena, Albert L. Armor coats, inverse grading, and streambed scour in selected streams of southern Ontario and western New York. M, 1985, Brock University. 154 p.

Madden, Cary T. Miocene mammalian fauna from Lothidok, Kenya. M, 1972, University of California, Berkeley. 52 p.

Madden, Cary Thomas. Mammoths of North America. D, 1981, University of Colorado. 289 p.

Madden, Dawn Hill. Geology of the Jurassic basement rocks, Santa Cruz Island, California. M, 1976, University of California, Santa Barbara.

Madden, Dawn J. Stratigraphy of Grand Hogback Coalfield, upper Campanian Mesaverde Group, Rifle Gap to New Castle, Garfield County, Colorado. D, 1983, Colorado School of Mines. 380 p.

Madden, H. Douglas. Stratigraphy and microfacies analysis of the Mississippian System, North Franklin Mountains, Dona Ana County, south-central New Mexico. M, 1984, University of Texas at El Paso.

Madden, Theodore R. Electrode polarization and its influence on the electrical properties of mineralized rocks. D, 1961, Massachusetts Institute of Technology. 176 p.

Madden, Victoria Hope. An approach to teaching environmental education through science and urban planning. D, 1981, Columbia University, Teachers College. 176 p.

Maddex, Robert M. Geology of the central portion of the Charlottesville Quadrangle. M, 1939, University of Virginia. 77 p.

Maddock, Marshall E. Geology of the Mount Boardman Quadrangle, California. D, 1955, University of California, Berkeley. 167 p.

Maddocks, Rosalie F. Living and subfossil distribution patterns of podocopid ostracodes in the Nosy Be area, northern Madagascar. D, 1966, University of Kansas.

Maddocks, Rosalie F. Recent ostracodes of Knysna Estuary, Cape Province, Union of South Africa. M, 1962, University of Kansas. 101 p.

Maddox, Eric. A spectrofluorometric investigation of calcite fluorescence. M, 1984, University of Kentucky. 44 p.

Maddox, George E. Subsurface geology along Northwest Rillito Creek. M, 1960, University of Arizona.

Maddox, George Edward. Geology and hydrology of the Roswell artesian basin, New Mexico. D, 1968, University of Arizona. 203 p.

Maddox, Gerald Caton. Late Cretaceous and Cenozoic changes of level in the Ozark region. M, 1925, University of Missouri, Columbia.

Maddox, Terrance. Geology of the southern third of the Marcola Quadrangle, Oregon. M, 1965, University of Oregon. 78 p.

Maddry, John W. Geologic history of coastal plain streams, eastern Pitt County, North Carolina. M, 1979, East Carolina University. 102 p.

Madej, Mary Ann. Response of a stream channel to an increase in sediment load. M, 1978, University of Washington. 111 p.

Madeley, Hulon M. Sedimentology of the Pre-Womble rocks of the Ouachita Mountains, Oklahoma and Arkansas. M, 1962, University of Oklahoma. 64 p.

Madeley, Hulon Matthews. Petrology of the Tuscaloosa Formation (Late Cretaceous) in west-central Georgia. D, 1972, Ohio State University. 89 p.

Madenwald, Kent A. Foraminifera from outcrops of the Niobrara Shale (Upper Cretaceous) of Emmons County, North Dakota. M, 1962, University of North Dakota. 154 p.

Mäder, Urs Karl. The Aley Carbonatite Complex. M, 1986, University of British Columbia. 104 p.

Madera, Ruford Francisco. The Slaughter Field, Hockley County, Texas. M, 1939, Texas Tech University. 155 p.

Maderak, Marion L. A facies study of the Des Moines Series in the Forest City Basin. M, 1960, Kansas State University. 50 p.

Maderazzo, Marc Matthew. The Viking inorganic analysis experiment; interpretation for petrologic information. M, 1977, Massachusetts Institute of Technology. 84 p.

Madi, Lutfi Ali. A quantitative model of the Al Khums-Zliten drainage basin (Libya). M, 1981, Ohio University, Athens. 323 p.

Madigan, Terence J. The igneous petrology and stratigraphy of the Marlow and Arnett mountains area, St. Francois Mountains, Madison County, Missouri. M, 1987, University of Toledo. 132 p.

Madigosky, Stephen Robert. Palynological and paleo-ecological assessment of a Pennsylvanian shale overlying the Danville Coal Member (VII) in Sullivan County, Indiana. D, 1987, Ball State University. 260 p.

Madin, Ian P. Structure and neotectonics of the northwestern Nanga Parbat-Haramosh Massif. M, 1987, Oregon State University. 160 p.

Madison, Frederick William. Genesis and mineralogy of selected soils and soil parent materials in the Valderan drift region of east-central Wisconsin. D, 1972, University of Wisconsin-Madison.

Madison, James Ambrose. Comparative geophysical surveys and lithology of a magnetic orebody. M, 1955, University of North Carolina, Chapel Hill. 17 p.

Madison, James Ambrose. Petrology and geochemistry of the Webster-Addie ultramafic body, Jackson County, North Carolina. D, 1968, Washington University. 139 p.

Madison, Patrick James. The feasibility of satellite remote sensing as a technique for evaluating coal mine surface features. D, 1984, Indiana State University. 133 p.

Madkour, Mohamed F. On the methods solving the upward continuation problem of gravity. M, 1963, Ohio State University.

Madkour, Mohamed Fathi. On the gravity anomaly above the Earth and its attenuation. D, 1966, Ohio State University. 178 p.

Madlem, Kathleen W. Particle size and matrix effects in X-ray spectrochemical analysis of light elements, zinc through oxygen. M, 1964, Pomona College.

Madley, Hulon Matthews. Petrology of the Tuscaloosa Formation in west-central Georgia. D, 1972, Ohio State University. 149 p.

Madole, Richard Frank. Glacial geology of upper South Saint Vrain Valley, Boulder County, Colorado. M, 1960, Ohio State University.

Madole, Richard Frank. Quaternary geology of Saint Vrain drainage basin, Boulder County, Colorado. D, 1963, Ohio State University. 309 p.

Madonna, James A. Zeolite occurrences of Alaska. M, 1973, University of Alaska, Fairbanks. 51 p.

Madrid, George A. Generation, testing, and filtering of p-tau transforms using the Fourier transform. M, 1988, Indiana University, Bloomington. 236 p.

Madrid, Raul John Jose. Stratigraphy of the Roberts Mountains Allochthon in north-central Nevada. D, 1987, Stanford University. 453 p.

Madrid, Victor Manuel. Magnetostratigraphy of the late Neogene Purisima Formation, Santa Cruz County, California. M, 1982, University of California, Davis. 104 p.

Madrid-Gonzalez, J. A. An empirical formula for Ray Theory amplitudes. M, 1972, University of Toronto.

Madrigal, Luis. Diffraction phenomena around salt masses. M, 1968, Rice University. 70 p.

Madsen, Jack William. Geology and ore deposits of the Spring Canyon area, Long Ridge, Utah. M, 1952, Brigham Young University. 92 p.

Madsen, James Henry. The fossil vertebrates of Utah, an annotated bibliography. M, 1979, Brigham Young University. 141 p.

Madsen, James Henry, Jr. Geology of the Lost Echo Canyon area, Morgan and Summit counties, Utah. M, 1959, University of Utah. 60 p.

Madsen, John Alfred. The isostasy and crustal structure of the East Pacific Rise and the morphotectonic fabric of the Orozco transform fault. D, 1988, University of Rhode Island. 229 p.

Madsen, Russel A. Geology of the Beverly Hills area, Utah. M, 1952, Brigham Young University. 39 p.

Madsen, Stanley Harold. The geology of a portion of the Salisbury Canyon area, northeastern Santa Barbara County, Southern California. M, 1959, University of California, Los Angeles.

Madson, M. A statistical analysis of heavy mineral variations in Upper Cretaceous black sandstone deposits of the Rocky Mountain states. M, 1978, University of Wyoming. 151 p.

Madzsar, Elizabeth Marie. Interpretation of the radioactive fallout record from the annual bands of Montastrea annularis, Broward County, Florida. M, 1987, University of North Carolina, Chapel Hill. 67 p.

Maebius, Jed Barnes. The Porter oil field (Michigan). M, 1935, University of Michigan.

Maehl, Richard H. Silurian of Pictou County, Nova Scotia. D, 1960, Massachusetts Institute of Technology. 164 p.

Maercklein, Douglas R. Analysis of deformation at Jim Falls, Wisconsin. M, 1974, University of Wisconsin-Milwaukee.

Maerz, N. H. A microprocessor controlled field data acquisition device. M, 1988, University of Waterloo. 143 p.

Maerz, R. H., Jr. Paleoautoecology of solitary rugose corals from the Beil Limestone of eastern Kansas. D, 1976, University of Kansas. 273 p.

Maerz, Richard Hugh. Paleoecology of the Poolville Member, Bromide Formation (Middle Ordovician), Criner Hills, Oklahoma. M, 1972, Texas Christian University. 89 p.

Maest, Ann Sharon. The geochemistry of metal transport in low and high temperature aqueous systems. D, 1984, Princeton University. 222 p.

Magalhaes, Antonio F. Charge characteristics of soils from the coastal region of Pernambuco (Brazil). D, 1979, University of California, Riverside. 91 p.

Magaw, Mary C. Ostracod faunas of the Elkhorn Formation. M, 1951, Miami University (Ohio). 38 p.

Magbee, Byron D. The geology of Tepee Trail Formation, Fremont County, Wyoming. M, 1950, Miami University (Ohio). 63 p.

Magdich, F. S. The Viking Formation in Saskatchewan. M, 1955, University of Saskatchewan. 38 p.

Mageau, Camille M. Foraminiferal test alterations resulting from ingestion by larger invertebrates. M, 1977, Dalhousie University.

Magee, Alfred W., III. Depositional environments of late Precambrian sediments from Liberia, Sierra Leone, and Senegal, West Africa. M, 1985, Old Dominion University. 276 p.

Magee, Andrew D. Investigation of shoaling problems at Westport River inlet and sedimentation processes at Horseneck and East Horseneck beaches. M, 1981, Boston University. 137 p.

Magee, John J. Big Snowy Group of the northern Great Plains, an isopach-lithofacies study. M, 1949, University of Colorado. 54 p.

Magee, Marian Eileen. Advanced magnetic modeling techniques applied to the Southern California continental borderland. M, 1985, University of California, Santa Barbara. 177 p.

Magee, Robert Wright. The lithology and depositional environments of the Denton Formation (Lower Cretaceous) of north-central Texas and south-central Oklahoma. M, 1974, University of Texas, Arlington. 120 p.

Magenheimer, Stewart J. Original quartz grain shape characteristics on the Recent sediments of the Colorado River drainage basin. M, 1985, Texas A&M University.

Magessis, Aberra. Petrologic study of Cu-Ni mineralization in the Duluth Complex in Northeast Minnesota. M, 1976, University of Minnesota, Duluth.

Magette, William Lawson. Wastewater treatment in soil; effect of residence time. D, 1982, Virginia Polytechnic Institute and State University. 459 p.

Magginetti, Robert T. Fusulinid biostratigraphy of the Darwin Canyon and Conglomerate Mesa areas, Inyo County, California. M, 1983, San Jose State University. 416 p.

Maggio, Carlos M. A foraminiferal and textural analysis of the Weches Formation in Robertson, Burleson, and Lee counties, Texas. M, 1961, Texas A&M University. 104 p.

Maggiore, Peter. Deformation and metamorphism on the floor of a major ash-flow tuff cauldron; the Emory Cauldron, Grant and Sierra Counties, New Mexico. M, 1981, University of New Mexico. 133 p.

Maghsoudi, Nosratollah. Mathematical modeling of unsteady open channel flow conservative models. D, 1982, Colorado State University. 116 p.

Magill, James Robert. Cenozoic tectonic rotations of Oregon and Washington. D, 1981, Stanford University. 311 p.

Magley, Wayne C. An analysis of heavy metals in the American oyster, Crassostrea virginica from four sites in the Tampa Bay region. M, 1978, Florida State University.

Magloughlin, J. F. Metamorphic petrology, structural history, geochronology, tectonics and geothermometry/geobarometry in the Wenatchee Ridge area, North Cascades, Washington. M, 1986, University of Washington. 343 p.

Magnell, Bruce A. Measurement of deep ocean currents with one-shot neutrally buoyant floats. M, 1968, Massachusetts Institute of Technology. 67 p.

Magness, Catherine Virginia. Some minerals in St. Louis and vicinity. M, 1934, Washington University.

Magnus, Keith R. Origin of microphysiography in southeastern Lake Michigan. M, 1977, Purdue University. 93 p.

Magnusson, Donald Harry. The Triassic sedimentary rocks of St. Martins, New Brunswick. M, 1955, University of New Brunswick.

Magnusson, Magnus Mar. The opening and closure of englacial conduits. M, 1985, University of Washington. 72 p.

Magnusson, Stefan G. The application of electrical and refraction seismic methods to the detection of abandoned lignite mine workings in East Texas. M, 1985, Colorado School of Mines. 88 p.

Magoon, Leslie B., III. Geology of T. 28 S., R. 11 W. of the Coquille and Sitkum quadrangles, Oregon. M, 1966, University of Oregon. 73 p.

Magorian, Thomas R. Stratigraphy and paleoecology of the (Ordovician) Mt. Hope Shale (Ohio, Kentucky, Indiana). D, 1952, University of Chicago. 73 p.

Magouirk, Deborah, A. Shoreline changes along the Southeast Texas coast, Galveston, Chambers and Jefferson counties. M, 1981, Stephen F. Austin State University. 74 p.

Magruder, George Lloyd. The chemical petrology of the corona growth; olivine + plagioclase = orthopyroxene + amphibole + spinel in an anorthositic gabbro of the Wilmington Complex. M, 1981, University of Delaware. 331 p.

Maguire, Thomas F. Combined gravity and magnetic interpretation of an extensive anomaly associated with Scituate Granite in western Rhode Island. M, 1983, University of Rhode Island.

Maguire, Timothy James. Gravity study of area adjacent to the Fall Zone of northwestern Delaware and northeastern Maryland. M, 1980, University of Delaware.

Magwood, James P. A. The ichnology of the Lower Cambrian Gog Group, Lake Louise, Alberta. M, 1988, University of Alberta. 325 p.

Mah, Anmarie Janice. The effect of Hurricane Allen on the Bellairs fringing reef, Barbados. M, 1984, McGill University. 213 p.

Mahadev, P. D. The process of land use evolution in Mysore City; a non-western example. D, 1973, University of Pittsburgh.

Mahaffey, Jack L. Study of hydraulic fracturing results in eastern Ohio. M, 1954, Ohio State University.

Mahaffy, M.-A. W. A three-dimensional numerical method for computing the load distribution of ice sheets with time. M, 1974, University of Colorado.

Mahai, Thomas. Infrared techniques in the study of drilling muds. M, 1962, Stanford University.

Mahajan, Amrish K. The mineral and sedimentary facies of the Pewabic West Conglomerate (Precambrian, Keweenaw Peninsula of Michigan). M, 1970, Michigan Technological University. 68 p.

Mahala, James. Net shore-drift along the Pacific Coast of Clallam and Jefferson Counties, Washington. M, 1985, Western Washington University. 73 2 plates p.

Mahamah, Dintie Shaibu. Numerical methods for assessing water quality in lakes and reservoirs. D, 1984, Washington State University. 122 p.

Mahan, Thomas Kent. Variations in the depositional environment of the lower Cincinnatian Kope Formation. M, 1981, Western Michigan University. 158 p.

Mahaney, William Cornelius. Soil genesis on deposits of neoglacial and late Pleistocene age in the Indian Peaks of the Colorado Front Range. D, 1971, University of Colorado. 322 p.

Mahar, Ddennis Lee. Geology and geochemistry of uranium deposits near Beaver Lake, Sullivan County, Pa. M, 1978, Pennsylvania State University, University Park. 142 p.

Mahar, J. W. The effect of geology and construction on behavior of a large, shallow, underground opening in rock. D, 1977, University of Illinois, Urbana. 333 p.

Mahar, James W. Permeability coefficients estimated from well logs. M, 1972, Colorado State University. 139 p.

Mahard, Richard H. Late Cenozoic chronology of the upper Verde Valley, Arizona. D, 1949, Columbia University, Teachers College.

Mahard, Richard H. The origin and significance of intrenched river meanders. M, 1941, Columbia University, Teachers College.

Maharidge, Allan D. The structural and tectonic history of a portion of the Felch Trough, central Dickinson County, northern Michigan. M, 1986, Bowling Green State University. 67 p.

Maharjan, Bhuwon D. Water quality problems in Nepal. M, 1983, Colorado State University.

Mahbubullah, A. K. M. Distribution of facies, depositional and diagenetic history of Knowles Limestone (Kimmeridgian), De Soto-Sabine Parish, Louisiana. M, 1988, Stephen F. Austin State University. 176 p.

Mahdavi, Azizeh. The thorium, uranium, and potassium contents of Atlantic and Gulf Coast beach sands. D, 1963, Rice University. 67 p.

Mahdy, Omar Rasheed El *see* El Mahdy, Omar Rasheed

Mahdyiar, Mehrdad. Attenuation properties of the Petatlan region, Mexico, and a local magnitude scale for microearthquakes in this area. D, 1984, University of Wisconsin-Madison. 270 p.

Maher, Alice Margaret. The Whittier, Chino, and Elsinore faults; a study of the relationship. M, 1982, California State University, Long Beach. 136 p.

Maher, Brian. Geology, geochemistry, and genesis of the Engineer Pass intrusive complex, San Juan Mountains, CO. M, 1983, Colorado State University. 240 p.

Maher, Harmon Droge. Structure and stratigraphy of Midterhuken Peninsula, Bellsund, West Spitsbergen. D, 1984, University of Wisconsin-Madison. 491 p.

Maher, Harmon Droge, Jr. Stratigraphy, metamorphism, and structure of the Kiokee and Belair belts near Augusta, Georgia. M, 1979, University of South Carolina.

Maher, John Bernard. Stratigraphy and petrology of the Pouch Cove-Cape Saint Francis area, Newfoundland. M, 1973, Memorial University of Newfoundland. 75 p.

Maher, John Charles. The calcium carbonate content of the Peorian Loess of Nebraska. M, 1937, University of Nebraska, Lincoln.

Maher, John Kelly. Application of a high pressure microbomb to hydrostatic pressurization of magnetite. M, 1977, Michigan State University. 46 p.

Maher, Kevin A. The structure and stratigraphy of lower Mesozoic strata, Midterhuken Peninsula, West Spitsbergen, Svalbard. M, 1987, University of Wisconsin-Madison. 287 p.

Maher, Linda M. Structural geology and stratigraphy near Panorama Mountain, central Alaska Range, Alaska. M, 1986, University of Wisconsin-Madison. 190 p.

Maher, Louis J. Jr. The geology of the east flank of the Bighorn Mountains near Story, Wyoming. M, 1959, University of Iowa. 130 p.

Maher, Louis J., Jr. Pollen analysis and postglacial vegetation history in the Animas Valley region, southern San Juan Mountains, Colorado. D, 1961, University of Minnesota, Minneapolis. 85 p.

Maher, R. V. An inquiry into the nature of biogeography. D, 1976, University of Western Ontario.

Maher, Stuart W. The geology of the John Sevier Quadrangle, Knox County, Tennessee. M, 1948, University of Tennessee, Knoxville. 71 p.

Maher, Thomas M. Rb-Sr systematics and rare-earth element geochemistry of Precambrian leucogneiss from the Adirondack Lowlands, New York. M, 1981, Miami University (Ohio). 125 p.

Mahfi, Achmad. A paleomagnetic study of Miocene and Eocene rocks from central Java, Indonesia. M, 1985, University of California, Santa Barbara. 99 p.

Mahfoud, Robert. Geology and petrology of the Montgomery Peak NE Quadrangle, Mineral and Esmeralda counties, Nevada. M, 1970, Tulane University. 65 p.

Mahfoud, Robert F. Paragenesis and mineralogy of the Burgin Mine, east Tintic District, Utah County, Utah. D, 1971, Brigham Young University. 104 p.

Mahgerefteh, Khosrow. Seepage and stability of earth dams. M, 1979, Virginia Polytechnic Institute and State University.

Mahin, Donald Alan. Analysis of groundwater flow in the Edwards Limestone aquifer, San Antonio area, Texas. M, 1978, University of Nevada. 49 p.

Mahjoory, Ramez. Clay mineralogy of some litho- and toposequences of soils in Michigan. D, 1971, Michigan State University. 138 p.

Mahlburg, Suzanne E. Precambrian geology of the Republic Trough, Marquette County, Michigan. M, 1972, University of Illinois, Urbana. 152 p.

Mahlburg, Suzanne E. The origin of antiperthites in anorthosites. D, 1975, Brown University. 261 p.

Mahler, Julianne Phyllis. Late-stage Alleghanian wrenching of the western Narragansett Basin, Rhode Island. M, 1988, University of Texas, Austin. 83 p.

Mahmood, Arshud. Fabric-mechanical property relationships in fine granular soils. D, 1973, University of California, Berkeley. 188 p.

Mahmood, Ramzi Jamil. Evaluation of enhanced mobility of polynuclear aromatic compounds in soil systems. D, 1988, Utah State University. 207 p.

Mahmoodian-Shooshtari, Mohamad. Investigation of a new method for measuring unsaturated soil hydraulic properties. D, 1980, Colorado State University. 153 p.

Mahmoud, Fida. Geochemistry and genesis of sedimentary marine carbonate fluorapatite. D, 1976, University of California, Santa Barbara.

Mahmoud, Idris Ahmed. Unsteady flow in an artesian aquifer in which elasticity and permeability depend on drawdown. D, 1966, Northwestern University. 140 p.

Mahon, Keith I. A numerical approach for determining the variable ascent velocity of a granitoid diapir. M, 1985, SUNY at Albany. 158 p.

Mahon, Sadie. Genera Lagena and Nodosaria of the Upper Cretaceous marls. M, 1926, Texas Christian University. 59 p.

Mahoney, Carroll F. A visual study of the flow of fluids through a porous medium. M, 1952, University of Oklahoma. 75 p.

Mahoney, J. Brian. The geology of the northern Smoky Mountains and stratigraphy of a portion of the Lower Permian Grand Prize Formation, Blaine and Camas counties, Idaho. M, 1987, Idaho State University. 143 p.

Mahoney, John. Stratigraphy and patterns of sedimentation of the Garden Valley Formation, in Eureka County, Nevada. M, 1979, Ohio University, Athens.

Mahoney, John J. Iron-titanium oxides in the Littleton Formation, New Hampshire. M, 1977, Boston University. 66 p.

Mahoney, John Joseph. Isotopic and chemical studies of the Deccan and Rajmahal traps, India; mantle sources and petrogenesis. D, 1984, University of California, San Diego. 205 p.

Mahoney, John Wells, Jr. Potassium-argon data on Pennsylvanian underclays and marine shales from the Ohio and Pennsylvania. M, 1969, Brown University.

Mahoney, Maureen. Use of electrical resistivity techniques in an evaluation of the geothermal potential of the Truth or Consequences, New Mexico, area. M, 1984, University of New Mexico. 98 p.

Mahoney, William Clement. Analysis of error factors in the use of Shoran statoscope-controlled photography for establishing horizontal ground control from a single stereo model. M, 1955, Ohio State University.

Mahoney, William Clement. Proposal, development, and testing of a system of analytical triangulation for medium-scale digital computers. D, 1961, Ohio State University.

Mahony, John Daniel. Reactions of He-3 with light elements; application to activation analysis. D, 1965, University of California, Berkeley. 66 p.

Mahood, Gail Ann. The geological and chemical evolution of a late Pleistocene rhyolitic center; the Sierra La Primavera, Jalisco, Mexico. D, 1980, University of California, Berkeley. 257 p.

Mahood, Richard O. Petrology, sedimentology, and diagenesis of volcaniclastics and sandstones from Leg 63 of the Deep Sea Drilling Project, Baja California and Southern California continental borderland. M, 1986, Oregon State University. 188 p.

Mahorney, James Robert. Geology of the Garrity Hill area, Deer Lodge County, Montana. M, 1956, Indiana University, Bloomington. 40 p.

Mahrer, K. D. Strike-slip faulting; models for deformation in a non-uniform crust. D, 1979, Stanford University. 200 p.

Mahrholz, Wolfgang Werner Ekkehardt. The microscopic study of ore minerals with transmitted near-infrared radiation. D, 1959, Stanford University. 212 p.

Mahtab, A. M. A study of field stress distribution around an elliptical hole under different loading conditions. M, 1965, McGill University.

Maiga, Bokary S. Halo effects around the metamorphosed Bleikvassli, Norway, polysulfide ore deposit. M, 1983, SUNY at Buffalo. 171 p.

Maiklem, William Robert. Clay minerals for some Upper Cretaceous bentonites, Southwest Alberta. M, 1962, University of Alberta. 72 p.

Maillet, James. Pétrographie et géochimie des dykes du camp minier de Chibougamau, Québec. M, 1978, Universite du Quebec a Chicoutimi. 150 p.

Maillet, Jean. Adsorption of Hg(II) from aqueous solutions onto activated carbon. M, 1972, Stanford University.

Maillis, A. Origin, emplacement, and tectonic significance of some selected ophiolites. M, 1976, Northeastern Illinois University.

Main, Frederic Hall. Geology and ore deposition, Inde-Cieneguillas District, Mexico. D, 1955, Columbia University, Teachers College.

Main, Frederic Hall. Preliminary report on the Cieneguillas mining district, Durango, Mexico. M, 1948, Columbia University, Teachers College.

Main, Linda Darlene. A structural interpretation of the Cove Fault and petrofabric study of the Tuscarora Sandstone, Fulton County, Pennsylvania. M, 1978, University of Oklahoma. 87 p.

Main, Talmage. The Canyon Reef Field of Scurry County, Texas. M, 1950, Texas Tech University. 121 p.

Mainville, Andre. The altimetry gravimetry problem using orthonormal base functions. D, 1987, Ohio State University. 214 p.

Mainwaring, P. R. The petrology of a sulfide-bearing layered intrusion at the base of the Duluth Complex, St. Louis County, Minnesota. D, 1976, University of Toronto.

Mainzer, George Frederick. Gravity and aeromagnetic patterns in the Basin and Range-Colorado Plateau transition in southwestern Utah and northwestern Arizona. M, 1978, University of Utah. 161 p.

Mainzinger, Brent David. An appraisal of the turbidites in the Ridge Basin, Southern California. M, 1986, University of California, Riverside. 156 p.

Maione, Steven J. Stratigraphy of the Frontier Sandstone Member of the Mancos Shale (upper Cretaceous) on the south flank of the eastern Uinta Mountains, Utah and Colorado. M, 1971, Colorado School of Mines. 126 p.

Mairaing, W. Penetration resistance of soils in relation to penetrometer shape. D, 1978, Iowa State University of Science and Technology. 170 p.

Mairs, Tom. A subsurface study of the Ferndale and Viola formations in the Oklahoma portion of the Arkoma Basin. M, 1962, University of Oklahoma. 80 p.

Maisano, Marilyn Dew. Petrology and depositional dynamics; fluvial Fort Union Group (Paleocene) silts, Williston Basin, southwestern North Dakota. M, 1975, Arizona State University. 184 p.

Maise, Charles Richard. Geology of Bear Creek Canyon, Arizona, New Mexico. M, 1955, University of Utah. 75 p.

Maitland, Michael R. The paleobathymetric and paleogeographic distribution of Middle Ordovician trilobites within a portion of the Valley and Ridge of East Tennessee. M, 1979, University of Tennessee, Knoxville. 140 p.

Maiurano, Karen. Stratigraphy and mineralogy of Woodfordian glacial deposits, Mohawk-Hudson Lowland. M, 1987, Rensselaer Polytechnic Institute. 57 p.

Majchszak, Frank L. Some economic and engineering properties of manufactured sand produced from selected quarries in northwest Ohio. M, 1974, University of Toledo. 134 p.

Majdi, Abbas. A study of tunnel stability and barrier pillar design in mines of the Cape Breton Coalfield. D, 1988, McGill University. 310 p.

Majedi, M. Subsurface study of the Detroit River group (middle Devonian) of southeast Michigan. M, 1969, Michigan State University. 74 p.

Majer, E. Seismological investigations in geothermal regions. D, 1978, University of California, Berkeley. 232 p.

Majewske, Otto P. Recognition of Invertebrate fossil fragments in rocks and thin sections. D, 1969, Louisiana State University. 390 p.

Majewski, David G. A paleontologic study of the Anna Shale using X radiographic techniques. M, 1973, Northern Illinois University. 139 p.

Majewski, Otto Paul. The Platteville Formation. M, 1953, University of Minnesota, Minneapolis. 111 p.

Majid, Abdul Hamid. Lithofacies, chemical diagenesis and dolomitization of the Tertiary carbonates of the Kirkuk oil field, Iraq. D, 1983, University of Ottawa. 270 p.

Majlis, Muhammad Ali Kahn. Some factors affecting the production and measurement of colors in montmorillonite and kaolinite clays and natural soils. D, 1967, Texas A&M University. 137 p.

Major, Charles Fredrick, Jr. Some Ostracoda from the Hackberry Formation of Iowa. M, 1954, University of Illinois, Urbana. 18 p.

Major, Jeffrey D. The use of Landsat thematic mapper (TM) and geophysical data for hydrocarbon exploration in the Illinois Basin, western Kentucky. M, 1988, Murray State University. 40 p.

Major, Jon Joseph. Geologic and rheologic characteristics of the May 18, 1980 southwest flank lahars at Mount St. Helens, Washington. M, 1984, Pennsylvania State University, University Park. 225 p.

Major, Maurice W. On elastic strain of the Earth in the period range 5 seconds to 100 hours. D, 1964, Columbia University, Teachers College.

Major, Millard Holland. Subsurface geology of McCulloch County, Texas. M, 1942, University of Texas, Austin.

Major, Richard Paul. Petrology and stratigraphy of the Allentown Dolomite (U. Cambrian), northwestern New Jersey. M, 1976, University of Connecticut. 148 p.

Major, Richard Paul. The Midway Atoll coral cap; meteoric diagenesis, amplitude of sea-level fluctuation and dolomitization. D, 1984, Brown University. 146 p.

Major, Rufus Orville. Geology of Toyah Field, Reeves County, Texas. M, 1950, University of Texas, Austin.

Major, Virginia L. Geological field trip in Fairfax County, Virginia. M, 1967, Virginia State University. 36 p.

Majtenyi, Steven Istvan. An application of stereo aerial photographs to the estimation of discharge proper-

ties of drainage basins. D, 1969, Cornell University. 327 p.

Majumdar, Amalendu J. Applicability of classical thermodynamics to solid-solid transitions. D, 1958, Pennsylvania State University, University Park. 153 p.

Majumdar, Arunaditya. Geochronology, geochemistry and petrology of the Precambrian Sandia Granite, Albuquerque, New Mexico. D, 1985, Louisiana State University. 212 p.

Majumdar, Dalim Kumar. Simplified approach to the problem of stability of soil slopes under horizontal earthquake and pore pressure. D, 1964, Utah State University. 90 p.

Mak, Eddy K. C. Ar^{40}/Ar^{39} dating of shock-metamorphosed rocks from the Mistastin Lake Meteorite impact crater. M, 1973, University of Toronto.

Makar, Laila. Radiation effects in titanium-silicate systems. D, 1978, University of Pittsburgh.

Makarim, Chaidir Anwar. Pressuremeter method for single piles subjected to cyclic lateral loads in overconsolidated clay. D, 1986, Texas A&M University. 332 p.

Makdisi, Faiz Isbir. Performance and analysis of earth dams during strong earthquakes. D, 1976, University of California, Berkeley. 241 p.

Makdisi, Richard. Application of aqueous geochemistry to deducing the origin and history of Mexicali Valley groundwaters, Baja California, Mexico. M, 1983, California State University, Hayward. 94 p.

Makeig, Kathryn. Hydrogeological evaluation of a drift-filled bedrock valley aquifer by means of a pumping test and finite element model. M, 1978, University of Minnesota, Duluth.

Makhrouf, Ali A. El see El Makhrouf, Ali A.

Maki, Mark Urho. Subsurface stratigraphy and structure of the Upper Cambrian Sauk Sequence of northern Ohio. M, 1986, Kent State University, Kent. 117 p.

Maknoon, Reza. Analysis of a conjunctively managed surface-ground water system. D, 1977, University of Washington. 184 p.

Mako, David Alan. The geology and genesis of the stratiform zinc-lead-barite mineralization of the Vulcan Property, Selwyn Basin, Northwest Territories, Canada. M, 1981, University of Wisconsin-Madison.

Makoju, C. A. Fractured gas well analysis; evaluation of in situ reservoir properties of low permeability gas wells stimulated by finite conductivity hydraulic fractures. M, 1978, Texas A&M University.

Makovicky, Emil. Cylindrite; crystallography, crystal structure and chemical composition. D, 1970, McGill University.

Makower, Jordan. Geology of the Prescott intrusive complex, Quabbin Reservoir Quadrangle, Massachusetts. M, 1964, University of Massachusetts. 91 p.

Makurath, J. H. The sedimentology and paleoecology of the Keyser Limestone (Silurian-Devonian) of the central Appalachians. D, 1975, The Johns Hopkins University. 110 p.

Malahoff, Alexander. Magnetic surveys over the Hawaiian Ridge and their geologic implications. D, 1965, University of Hawaii. 69 p.

Malanchak, John E. Geology of Atwater well field, Kalamazoo County, Michigan. M, 1973, Western Michigan University.

Malander, Mark William. A stable isotopic study of waters within and surrounding the Boundary Waters Canoe Area Wilderness, Minnesota. M, 1983, Northern Illinois University. 88 p.

Malarcher, Falvey L. Modeling of seismic multiples. M, 1974, Southern Methodist University. 41 p.

Malcolm, Daniel Connor. The ore deposits and structures of the Gratz District, Owen and Henry counties, Kentucky. M, 1952, University of Cincinnati. 36 p.

Malcolm, Frieda L. Petrology, mineral chemistry and microstructures of gabbros from the Mid-Cayman Rise spreading center. M, 1979, SUNY at Albany. 312 p.

Malcolm, Ronald Lee. Mobile soil organic matter and its interactions with clay minerals and sesquioxides. D, 1964, North Carolina State University. 157 p.

Malcolm, T. J. An examination of ex-solution ilmenite-hematite ores with special reference to Allard Lake (Ontario). M, 1962, University of Toronto.

Malcuit, Robert J. Petrogenesis of gneiss and associated rocks in the southern Bighorn Mountains, Wyoming (Precambrian). M, 1970, Kent State University, Kent. 167 p.

Malcuit, Robert Joseph. Implications of a lunar flyby encounter with Earth. D, 1973, Michigan State University. 50 p.

Maldonado, Florian. Geology of the northern part of the Sierra Cuchillo, Socorro and Sierra counties, New Mexico. M, 1974, University of New Mexico. 59 p.

Malecek, S. J. A marine deep seismic sounding survey in the region of Explorer Ridge. M, 1976, University of British Columbia.

Malek, Ali. Hydrologic aspects of water harvesting on processed oil shale, a saline medium. D, 1980, Utah State University. 208 p.

Malek, Debra Jean. Alteration of alkalic igneous rocks in the Payne's Waterhole area, Brewster County, Texas; evidence for selective trace element contamination. M, 1986, Texas Christian University.

Malek-Aslani, Morad K. Geology of the Beulah area, Pueblo County, Colorado. D, 1952, Colorado School of Mines. 106 p.

Malek-Aslani, Morad K. The geology of southern Perry Park, Douglas County, Colorado. M, 1950, Colorado School of Mines. 93 p.

Malekahmadi, Fatemah. Factors affecting the rheological properties of the bentonite fluids, and in-situ formation of cement materials in these fluids. M, 1984, Texas Tech University. 71 p.

Maletzke, Jeffrey D. Hydrogeology of the Hillsboro Landfill, Hillsboro, North Dakota. M, 1988, University of North Dakota. 186 p.

Maley, Elaine Gail. Diagenesis of Strawn Limestone, South Carlsbad Field, Eddy County, New Mexico. M, 1977, Texas Tech University. 104 p.

Maley, Kevin. A transgressive facies model for a shallow estuarine environment; the Delaware Bay nearshore zone, from Beach Plum Island to Fowler Beach, Delaware. M, 1982, University of Delaware.

Maley, Michael Paul. Depositional history of the upper Morrowan (Pennsylvanian) strata of the Ardmore Basin, Oklahoma. M, 1986, University of Oklahoma. 206 p.

Maley, Richard P. Shallow seismic refraction studies at the strong-motion seismograph stations in Chalome Creek valley, San Luis Obispo County, California. M, 1970, University of Southern California.

Maley, Terry Samuel. Structure and petrology of the lower Panther Creek area, Lemhi County, Idaho. D, 1974, University of Idaho. 130 p.

Malfait, Bruce. Distribution and transportation of marine sediments near Moss Landing, California. M, 1969, Stanford University.

Malfait, Bruce Terry. The Carnegie Ridge near 86° W.; structure, sedimentation and near bottom observations. D, 1975, Oregon State University. 131 p.

Malghan, Subhaschandra G. The dissolution kinetics of fibrous amphibole minerals in water. M, 1971, University of Nevada - Mackay School of Mines. 89 p.

Malhotra, Anil Kumar. Stochastic analysis of offshore tower structures. D, 1969, University of California, Berkeley. 200 p.

Malhotra, R. Veena. A field spectrometer and Landsat TM study of surface alteration due to hydrocarbon seepage; Sheep Mountain anticline, Bighorn Basin, Wyoming. M, 1988, Dartmouth College. 127 p.

Malhotra, Ramesh. K-Ar ages and petrography of mafic dikes (Precambrian), Vermillion District, Minnesota. M, 1970, SUNY at Stony Brook.

Malhotra, Renu. Some aspects of the dynamics of orbit-orbit resonances in the Uranian satellite system. D, 1988, Cornell University. 163 p.

Malick, Kenneth C. Petrology of the alaskites of the Boulder Batholith, Montana. M, 1977, University of North Dakota. 126 p.

Malicse, Jose Ariel Enriquez. The environment and diagenesis of the Shattuck Member of the Queen Formation (Guadalupian, Permian) at Caprock Queen Field, Chavez County, New Mexico. M, 1988, Texas A&M University. 159 p.

Malik, Om Parkash. Investigations in the Fe-Ni-S system; phase relations at liquidus temperatures (980 to 1030°C) and partitioning of Ni and Co between pyrrhotite and pentlandite (200 to 500°C). D, 1972, University of Saskatchewan. 256 p.

Malik, Om Parkash. Phase equilibria in the Fe-Ni-S system. M, 1968, University of Saskatchewan. 56 p.

Malik, Roberto F. Engineering geology of the Big Creek drainage basin; Cleveland, Ohio. M, 1980, Kent State University, Kent. 112 p.

Malila, William A. Information extraction and multiaspect techniques in remote sensing. D, 1974, University of Michigan. 194 p.

Malin, Eugene R. Castle Rock Conglomerate of the early Oligocene, Colorado. M, 1957, University of Nebraska, Lincoln.

Malin, Michael Charles. 1, Comparison of volcanic features of Elysium (Mars) and Tibesti (Earth); 2, Age of Martian channels; 3, Nature and origin of intercrater plains on Mars. D, 1976, California Institute of Technology. 184 p.

Malin, Peter Eric. A first order scattering solution for modeling lunar and terrestrial seismic codas. D, 1978, Princeton University. 101 p.

Malin, William John and Kelly, John Joseph. Geology of West Pass Peak area, Sublette County, Wyoming. M, 1952, University of Michigan.

Malinconico, Lawrence Lorenzo, Jr. SO₂ mass flow variations, Mount Etna Volcano, Sicily. M, 1978, Dartmouth College. 77 p.

Malinconico, Lawrence Lorenzo, Jr. Structure of the Himalayan suture zone of Pakistan interpreted from gravity and magnetic data. D, 1982, Dartmouth College. 128 p.

Malinconico, MaryAnn Love. The stratigraphy and structure of the southeastern Rumney 15 minute quadrangle, New Hampshire. M, 1982, Dartmouth College. 234 p.

Malinky, John M. Depositional environment of the Eudora Shale (Missourian, Upper Pennsylvanian) near Tyro, Kansas. M, 1980, Ohio University, Athens. 92 p.

Malinky, John Mark. Paleontology and paleoenvironment of "core" shales (Middle and Upper Pennsylvanian), Midcontinent North America. D, 1984, University of Iowa. 327 p.

Malinowski, M. J. Geology of the Kawaihae Quadrangle, Kohala Mountain, Island of Hawaii. M, 1977, University of Hawaii. 155 p.

Maliva, Robert G. Paleoecology and sedimentology of the Rockford Limestone and upper New Albany Shale (Lower Mississippian) in southern Indiana. M, 1984, Indiana University, Bloomington. 185 p.

Maliva, Robert George. Silicification of marine limestones. D, 1988, Harvard University. 173 p.

Malkames, Judith Ann. The petrology and petrochemistry of the Midlick Dellenite. M, 1974, Southern Illinois University, Carbondale. 96 p.

Malkin, Doris S. Miocene biostratigraphy and micropaleontology of New Jersey, Maryland and Virginia. D, 1953, Columbia University, Teachers College.

Malkin, Doris S. Some ostracods from the (Devonian) Widder Beds of the Hamilton, at Rock Glen, near Arkona, Ontario. M, 1935, Columbia University, Teachers College.

Malkoc, Selahaddin. The effect of well spacing on ultimate recovery. M, 1952, University of Tulsa. 89 p.

Malkoski, Mark. Geology of the Mount Gunstock iron deposit, Gilford, Belknap County, New Hampshire. M, 1976, University of New Hampshire. 103 p.

Malla, Prakash Babu. Characterization of layer charge, expansion behavior, and weathering of 2:1 layer silicate clays. D, 1987, Rutgers, The State University, New Brunswick. 205 p.

Mallard, Elliott A. The differentiation and classification of a sand underlying typical Trail Ridge sands near Screven, Wayne County, Georgia. M, 1988, University of Florida. 75 p.

Mallary, McKenzie. The usefulness of aeroradioactivity maps in locating heavy mineral deposits along portions of the Atlantic Coastal Plain of Georgia. M, 1988, Georgia State University. 94 p.

Mallery, Linda Leigh. Lead-zinc metallotects of the Ozark Dome. M, 1983, University of Missouri, Rolla. 135 p.

Mallette, Patrick M. Lithostratigraphic analysis of cyclical phosphorite sedimentation within the Miocene Pungo River Formation, North Carolina continental shelf. M, 1986, East Carolina University. 155 p.

Malley, Michael J. Characterization of airborne particulates from western Colorado oil shale lands by pyrolysis mass spectrometry. M, 1986, Colorado School of Mines. 70 p.

Mallick, Brian Charles. Radioisotope distribution, heat production and crustal temperatures in the eastern Adirondacks. M, 1979, SUNY at Buffalo. 73 p.

Mallick, K. A. Weathering of rocks and mobility of elements in soil profiles of Mount Saint Hilaire, Quebec (Canada). M, 1967, McGill University. 167 p.

Mallick, Subhasis. A vector reflectivity algorithm and synthesis of P and S. D, 1987, University of Hawaii.

Mallin, James Wilson. The Peebles Formation, a Niagaran reef deposit. M, 1951, University of Cincinnati. 34 p.

Mallinson, David J. Distribution and petrology of glauconitic sediments in the Miocene Pungo River Formation, Onslow Bay, North Carolina continental shelf. M, 1988, East Carolina University. 127 p.

Mallio, William Joseph. Geochemistry of pyrrhotite from metamorphosed pelitic rocks in western Maine. D, 1970, Boston University. 163 p.

Mallon, Alfred E. The Holyoke Range. M, 1915, University of Minnesota, Minneapolis. 45 p.

Mallon, Kenneth M. Precambrian geology of the northern part of the Los Pinos Mountains, New Mexico. M, 1966, New Mexico Institute of Mining and Technology. 88 p.

Mallory, Bob Franklin. Paleoecologic study of a brachiopod fauna from the Cerro Gordo Member of the Lime Creek Formation (Upper Devonian), north-central Iowa. D, 1968, University of Missouri, Columbia. 80 p.

Mallory, James A. Nodosaria sand environments in the eastern counties of the Texas Gulf Coast. M, 1959, University of Houston.

Mallory, M. J. Bouguer gravity map and a postulated basement structure configuration of the Nashville West and Oak Hill quadrangles, Tennessee. M, 1974, Vanderbilt University.

Mallory, Virgil Standish. Age and distribution of lower Tertiary foraminifera in the California Coast Ranges. D, 1952, University of California, Berkeley. 309 p.

Mallory, Virgil Standish. Eocene foraminifera from Media Agua Creek, Kern County, California. M, 1948, University of California, Berkeley. 63 p.

Mallory, William L. Stratigraphy of the Dolgeville facies (New York). M, 1946, Columbia University, Teachers College.

Mallory, William Mason. Lithostratigraphy of the Miocene Yorktown Formation in southeastern Virginia. M, 1967, University of North Carolina, Chapel Hill. 100 p.

Mallory, William Wyman. Pennsylvanian stratigraphy and structure, Velma Pool, Stephens County, Oklahoma. D, 1948, Columbia University, Teachers College.

Malloy, John. Distribution and ecology of Recent Ostracoda of the Bay of Naples, Italy. M, 1963, Florida State University.

Malloy, R. E. Carboniferous plant spores of the North 40 brine seam, Joggins, Nova Scotia. M, 1951, University of Massachusetts.

Malloy, Richard James. Marine geology of the Gulf of Maine. M, 1965, University of Southern California.

Malloy, Robert W. The geology west and north of the Yankee Fork-Salmon River confluence, Custer County, Idaho. M, 1979, University of Idaho. 124 p.

Malmberg, Glenn Thomas. Geology of the Sapello District, Las Vegas Quadrangle, New Mexico. M, 1950, University of Iowa. 143 p.

Malmquist, Kevin Lee. Paleoecology of late Quaternary molluscan ostracod assemblages from the Norwood site, southeastern Minnesota. M, 1979, University of North Dakota. 66 p.

Malo, Michel. Statigraphie et structure de l'anticlinorium d'Aroostook-Perce en Gaspesie, Quebec. D, 1986, Universite de Montreal.

Malone, Carl Hubert, Sr. Labeling of sand particles for sediment transport studies using stable isotopic tracers. D, 1969, Texas A&M University. 73 p.

Malone, Donald J. Stratigraphic relations and structural significance of the Blaine Formation in the subsurface of southwestern Kansas. M, 1962, Wichita State University. 72 p.

Malone, Gary B. Epithermal gold-silver veins of Central America and SO₂ emission from Hawaiian and Italian volcanoes. D, 1975, Dartmouth College. 150 p.

Malone, Gary Bruce. The geology of the volcanic sequence in the Horse Mesa area, Arizona. M, 1972, Arizona State University. 68 p.

Malone, Michael J. A pebble-cobble deposit in Monterey bay, California. M, 1970, United States Naval Academy.

Malone, Philip Garcin. Paleotemperature determinations from pelecypod skeletal carbonates from the upper Pleistocene of the Atlantic Coast. D, 1969, Case Western Reserve University. 208 p.

Malone, Rodney D. Stratigraphy of the Dunkard group (Pennsylvanian and Permian) in West Virginia, Pennsylvania and Ohio. M, 1969, West Virginia University.

Malone, Ronald F. Stochastic analysis of water quality. D, 1979, Utah State University. 124 p.

Malone, S. J. Stratigraphy of the Frontier Sandstone Member of the Mancos Shale (Upper Cretaceous) on the south flank of the eastern Uinta Mountains. M, 1971, Colorado School of Mines.

Malone, Stephen D. Earth strain measurements in Nevada and possible effects on seismicity due to the solid Earth tides. D, 1972, University of Nevada. 139 p.

Maloney, Moira N. Pennsylvanian palaeotextulariid foraminifera from the Appalachian Basin in east central and southeastern Ohio and western Pennsylvania. M, 1984, Bowling Green State University. 70 p.

Maloney, Neil J. Geology of the eastern part Beaty Butte Four Quadrangle, Oregon. M, 1961, Oregon State University. 87 p.

Maloney, Neil Joseph. Geology of the continental terrace off the central coast of Oregon. D, 1965, Oregon State University. 233 p.

Maloney, William Vincent. Grain-selecting pressures in fluvial environments and persistence of the provenance fingerprint. M, 1981, Syracuse University.

Malott, Burton Joseph. The entrenched meanders and associated terraces of the Muscatatuck near Vernon, Indiana. M, 1922, Indiana University, Bloomington.

Malott, Clyde Arnett. The "American Bottoms" region of eastern Greene County, Indiana; a type unit southern Indiana physiography. D, 1919, Indiana University, Bloomington. 61 p.

Malott, Clyde Arnett. The history of Glacial Lake Flatwood. M, 1915, Indiana University, Bloomington.

Malott, Mary Lou. Distribution of foraminifera in cores from Juan de Fuca Ridge, North East Pacific. M, 1981, University of British Columbia. 136 p.

Malouf, Stanley E. The geology of the Francoeur-Arntfiled District, Beauchastel Township, Quebec. D, 1942, McGill University.

Malouf, Stanley E. The petrology of a part of Westmount Mountain near Summit Circle, Montreal, Quebec. M, 1936, McGill University.

Malouta, D. N. Holocene sedimentation in Santa Monica Basin, California. M, 1978, University of Southern California.

Maloy, John A. Changes in grain-size distribution in a gravel-bed stream due to a point-source influx of fine sediment. M, 1988, Western Washington University. 50 p.

Malpas, John Graham. The petrochemistry of the Bull Arm Formation (Late Precambrian) near Rantem Station, Southeast Newfoundland. M, 1972, Memorial University of Newfoundland. 95 p.

Malpas, John Graham. The petrology and petrogenesis of the Bay of Islands ophiolite suite, western Newfoundland. D, 1976, Memorial University of Newfoundland. 432 p.

Malter, John A. Potential problems of oil pollution on Lake Champlain. M, 1973, University of Vermont.

Malterer, Thomas John. The genesis and characterization of some calcareous peatlands in west-central Minnesota. D, 1985, University of Minnesota, Minneapolis. 228 p.

Maltman, Alexander James. Structural geology of an area around Lake Bomoseen, west-central Vermont. M, 1971, University of Illinois, Urbana. 100 p.

Maltman, Alexander James. The serpentinites and related rocks of Angelsey, North Wales, United Kingdom. D, 1973, University of Illinois, Urbana. 259 p.

Maltz, Gary. Earthquake perception; the Borah Peak earthquake of October 28, 1983. M, 1985, University of Idaho. 53 p.

Maluf, Fred W. Stratigraphy of the Rockwall Member (new unit) of the Upper Cretaceous Pecan Gap Formation, Collin and Rockwall counties, Texas. M, 1975, University of Texas, Arlington. 65 p.

Maluski, Barbara Janine von *see* von Maluski, Barbara Janine

Mamah, Luke I. The field of the vertical magnetic dipole at the bottom of the ocean. D, 1981, Colorado School of Mines. 115 p.

Mamulas, Ned. Tectonics in northeastern Iceland; a remote sensing and analysis. M, 1981, Pennsylvania State University, University Park.

Man, K. T. A study of the chemical decomposition of rocks in Kilauea. M, 1940, University of Hawaii. 103 p.

Mana, A. I. Finite element analyses of deep excavation behavior in soft clay. D, 1978, Stanford University. 335 p.

Manahl, Kenneth A. Geology of the Shell Quadrangle, Wyoming. M, 1981, Iowa State University of Science and Technology. 100 p.

Manchester, Keith Stanton. Geophysical investigations between Canada and Greenland. M, 1964, Dalhousie University.

Manchester, Steven Russell. Fossil history of the Juglandaceae. D, 1981, Indiana University, Bloomington. 209 p.

Manchuk, Barry. Strain analysis of boudinage and ptygmatic folding in the Lily Pond Lake area, Manitoba. M, 1972, University of Manitoba.

Mancini, Ernest Anthony. Origin of micromorph faunas; Grayson Formation (upper Cretaceous, Texas). D, 1974, Texas A&M University. 345 p.

Mancuso, Christina M. Distribution of trace elements in the interstitial waters and sediments of a fresh water estuary. M, 1986, Bowling Green State University. 138 p.

Mancuso, James Dominic. Geology of Topliff Hill and the Thorpe Hills, Tooele and Utah counties, Utah. M, 1955, South Dakota School of Mines & Technology.

Mancuso, Joseph John. Geology and mineralization of the Mountain area, Wisconsin. M, 1957, University of Wisconsin-Madison. 32 p.

Mancuso, Joseph John. Stratigraphy and structure of the McCaslin District, Wisconsin. D, 1960, Michigan State University. 101 p.

Mancuso, Thomas Kaye. The origin of stratiform barite in Stevens County, Washington. M, 1983, University of Idaho. 115 p.

Mandal, Batakrishna. Observations and synthesis of seismic waves in anisotropic media. D, 1987, St. Louis University. 224 p.

Mandarino, Joseph Anthony. Pleochroism in synthetic ruby. M, 1951, Michigan Technological University. 25 p.

Mandarino, Joseph Anthony. Some optical and stress-optical properties of synthetic ruby. D, 1958, University of Michigan. 123 p.

Mandel, D. J., Jr. Neogene stratigraphy and micropaleontology of the southern San Diego area, California. M, 1974, San Diego State University.

Mandel, Peter, Jr. Resistivity studies of synthetic metalliferous cores. M, 1956, University of Utah. 37 p.

Mandell, Wayne A. Transition of environments, cementation, and diagenesis across the Ordovician-Silurian boundary on the west side of the Cincinnati Arch. M, 1975, Eastern Kentucky University. 65 p.

Mander, Marsha L. Morton. Petrography and environments of deposition of the Mission Canyon Formation Rough Rider Field, North Dakota. M, 1980, University of Colorado. 91 p.

Mandeville, Charles W. Tectonics of the Aalen area, central Norway; 1, Tectonic setting of the Tremadocian-Lower Ordovician Fundsjoe Group, central Norway. M, 1988, Virginia Polytechnic Institute and State University. 34 p.

Mandle, Richard J. A computer assisted recharge evaluation of a drift-bedrock aquifer system. M, 1975, Michigan State University. 60 p.

Mandra, York T. Eocene silicoflagellates from the Mount Diablo area, California. M, 1949, University of California, Berkeley.

Mandra, York Tooree. Fossil silicoflagellates from California. D, 1958, Stanford University. 158 p.

Manduca, Cathryn Clement Allen. Geology and geochemistry of the oceanic arc-continent boundary in the western Idaho Batholith near McCall. D, 1988, California Institute of Technology. 319 p.

Manera, Paul Allen. Impact of the ground water supply on the population of the Cave Creek area, Maricopa County, Arizona. D, 1982, Arizona State University. 214 p.

Manera, Thomas E. Sedimentology of southwest Pacific Ocean deep sea cores. M, 1969, University of Southern California.

Manes, Monna Lea. A zonation of calcareous nannofossils of the Fairport Chalk (Upper Cretaceous) in Kansas. M, 1973, Wichita State University. 113 p.

Manesh, Abdulkarim Nick. Stability of the working highwall in a strip mining operation and comparison of the failure in a physical model with that of a two-dimensional finite element analysis. D, 1983, University of Oklahoma. 113 p.

Maness, L. Van *see* Van Maness, L., Jr.

Maness, Lindsey V., Jr. A quantitative geomorphic study of stream valley symmetry in the Eden Shale-Outer Blue Grass area of northern Kentucky. M, 1977, Indiana State University. 122 p.

Maness, Timothy R. Ontogenetic changes in the carapace of Tyrrhenocythere amnicola (Sars), a hemicytherid ostracode. M, 1986, University of Kansas. 72 p.

Maneval, David R. Separation and analysis of bastnaesite rare earths. D, 1961, Pennsylvania State University, University Park. 103 p.

Manfredi-Mathews, Terri. Methods of soil and sediment oxide reduction. M, 1983, Old Dominion University. 51 p.

Manfredini, Antonio. Tank experiments on a model ore body of the vein type with the potential-drop-ratio method (T591). M, 1941, Colorado School of Mines. 25 p.

Manfrida, Jerry Lynn. Soil amplification studies using explosion impulse. M, 1977, University of Washington. 155 p.

Manfrino, Carrie. Stratigraphy and palynology of the upper Lewis Shale, Pictured Cliffs Sandstone and lower Fruitland Formation (Upper Cretaceous) near Durango, Colorado. M, 1984, Colorado School of Mines. 96 p.

Mangan, Margaret. Gold-silver mineralization at the London-Virginia Mine, Buckingham County, Virginia. M, 1983, Virginia Polytechnic Institute and State University. 65 p.

Mangen, Lawrence Raymond. Age relationships of igneous and metamorphic rocks of the Minnesota River valley. M, 1956, University of Minnesota, Minneapolis. 23 p.

Manger, George Edward. Some Pleistocene mollusks of San Quentin Bay and other localities from lower California. D, 1929, The Johns Hopkins University.

Manger, Katherine Chang. A structural analysis of the Roan Mountain mafic complex, North Carolina and Tennessee. M, 1981, University of Kentucky. 180 p.

Manger, Phillip Herrmann. A structural analysis along the Linville Falls thrust fault in western North Carolina. M, 1981, University of Kentucky. 118 p.

Manger, Walter L. The stratigraphy of the Hale Formation (Morrow; lower Pennsylvanian) in its type region, northwestern Arkansas. D, 1971, University of Iowa. 187 p.

Manger, Walter Leroy. The stratigraphy of the Logan and adjacent formations (Mississippian) of Ohio. M, 1969, University of Iowa. 188 p.

Mangham, J. R. The structure and petrology of the eastern Mellen intrusive complex, Iron County, Wisconsin. M, 1975, University of Wisconsin-Madison.

Manghnani, Murli Hukumal. Cation exchange capacities and infrared absorption properties of glauconites. D, 1962, University of Montana. 104 p.

Mangino, Stephen George. Teleseismic P-Su wave form modeling for crustal structure beneath Mina, Nevada and beneath Long Valley and Calden, California. M, 1987, SUNY at Binghamton. 126 p.

Mangion, Stephen M. Stratigraphy and sedimentary environments of the Medial Ordovician Kings Falls and Sugar River limestones (Trenton Group) of northwestern and central New York. M, 1972, Boston University. 98 p.

Mango, Helen N. A fluid inclusion and isotope study of the Las Rayas Ag-Au-Pb-Cu mine, Guanajuato, Mexico. M, 1988, Dartmouth College. 109 p.

Mangold, Kent M. A linear theory for the causal attenuation of seismic body waves. M, 1980, Indiana University, Bloomington. 144 p.

Mangum, Charles R. Geology of the Little Bluff Creek area, Mason County, Texas. M, 1960, Texas A&M University.

Mangun, Mark. A seismic refractory study of a buried valley near Peninsula, Ohio. M, 1980, University of Akron. 190 p.

Mangus, Marlyn. Slope angles in graded sands; a laboratory study. M, 1962, Columbia University, Teachers College.

Mangus, Marvin Dale. The type locality of celestite (Pennsylvania). M, 1946, Pennsylvania State University, University Park. 22 p.

Manhart, Thomas A. Model tank experiments and methods for interpretation of resistivity curves. M, 1932, Colorado School of Mines. 30 p.

Manheim, Frank. Variations in oxidation-reduction potential in the Jurassic of Morehouse Parish, Louisiana. M, 1953, University of Minnesota, Minneapolis. 46 p.

Manhoff, Charles N., Jr. Geology of the Hitchita area, Okmulgee and McIntosh counties, Oklahoma. M, 1957, University of Oklahoma. 50 p.

Mani, Philip C. Environment of deposition of the Upper Jurassic "gray" sandstones, Terryville Field, Lincoln Parish, Louisiana. M, 1983, Texas A&M University. 129 p.

Maniar, Papu Dayalal. Contributions to petrology of granites; (1) Modal analysis by quantitative X-ray diffraction; (2) Tectonic discrimination of granitoids; (3) Thermodynamic activity of oxides in granitoids; (4) Petrology of the Proterozoic granitoids of the Arbuckle Mountains, southern Oklahoma. D, 1987, University of Pittsburgh. 251 p.

Manifold, Albert Hedley. The mineralogy and geology of the Akaitcho area, Yellowknife, NWT. M, 1947, University of British Columbia.

Manings, Gordon C. Middle Silurian barrier island, Penfield Member, Lockport Formation, New York. M, 1982, SUNY, College at Fredonia. 67 p.

Maniocha, Michael L. Trace element distribution in sphalerite, galena, calcite and barite, Barnett Mine, Pope County, Illinois. M, 1974, Southern Illinois University, Carbondale. 63 p.

Manion, Ester Ann. Geography as treated in recent college textbooks. M, 1932, George Washington University.

Manion, L. J. Geology of the Irish Canyon-Vermillion Creek area, Moffat County, Colorado. M, 1961, University of Wyoming. 93 p.

Manion, R. E. Geology of an area east of Willow Creek Pass, Grand and Jackson counties. M, 1959, University of Wyoming. 42 p.

Maniw, John George. A study of chalcopyrite inclusions in sphalerite. M, 1967, University of Toronto.

Manjarres, Gilberto F. Cretaceous and Tertiary stratigraphy of the Cordillera Oriental, Colombia, South America. M, 1947, University of Chicago. 71 p.

Manka, Leroy Louis. A study of the sediments and depositional history of the delta of the Colorado river, Texas. M, 1970, Texas Christian University.

Manker, John P. Origin and distribution of silicate minerals in a carbonate environment, Florida Bay. M, 1969, University of South Florida, Tampa. 49 p.

Manker, John Phillip. Distribution and concentration of mercury, lead, cobalt, zinc, and chromium in suspended particulates and bottom sediments; upper Florida Keys, Florida Bay, and Biscayne Bay. D, 1975, Rice University. 125 p.

Mankiewicz, Carol. Biogeochemistry of Recent marine sediments; effects of the epibenthic holothurian Parastichopus parvimensis. M, 1980, University of California, Los Angeles.

Mankiewicz, Carol. Sedimentology and calcareous algal paleoecology of middle and upper Miocene reef complexes, near Fortuna (Murcia Province) and Nijar (Almeria Province), southeastern Spain. D, 1987, University of Wisconsin-Madison. 359 p.

Mankiewicz, David. Holocene sedimentation in Green Lake, Teton County, Wyoming; clastic sedimentation in a modern alpine lake. M, 1973, University of Wyoming. 38 p.

Mankiewicz, Paul Joseph. An organic geochemical investigation of a glacial sequence at Searles Lake, California. M, 1975, University of California, Los Angeles.

Mankiewicz, Paul Joseph. Hydrocarbon composition of sediments, water, and fauna in selected areas of the Gulf of Mexico and southern California marine environment. D, 1981, University of California, Los Angeles. 350 p.

Mankin, Charles John. Biostratigraphy of Tierra Vieja, Trans-Pecos, Texas. M, 1955, University of Texas, Austin.

Mankin, Charles John. Stratigraphy and sedimentary petrology of Jurassic and pre-Graneros Cretaceous rocks, northeastern New Mexico. D, 1958, University of Texas, Austin. 278 p.

Mankinen, Edward A. Paleomagnetism and potassium-argon ages of the Sonoma Volcanics, California. M, 1971, San Jose State University. 67 p.

Manley, Curtis Robert. Textural studies of young rhyolite flows. M, 1986, Arizona State University. 121 p.

Manley, Frederick. Heavy mineral study of the Upper Devonian Catskill facies of south-central New York. M, 1959, University of Rochester. 56 p.

Manley, Frederick Harrison, Jr. Clay mineralogy and clay-mineral facies of the Lower Cretaceous Trinity Group, southern Oklahoma. D, 1965, University of Oklahoma. 116 p.

Manley, K. The late Cenozoic history of the Espanola Basin, New Mexico. D, 1976, University of Colorado. 193 p.

Manley, Ronald D. Discussion and evaluation of ERTS-1 imagery in the central Bighorn Mountains, Wyoming. M, 1974, University of Iowa. 144 p.

Manley, Thomas R. A tectonic interpretation of the Keewatin Province, Ontario. M, 1958, Wayne State University.

Manley, Thomas Richard. The Niagaran rocks of Drummond Island, Michigan, and the related rocks of the Michigan Basin. D, 1964, Michigan State University. 114 p.

Manley, Walker D. Central peaks of lunar and Martian craters. M, 1977, University of Texas, Austin.

Manly, Robert L. The differential thermal analysis of certain phosphates. M, 1949, University of Minnesota, Minneapolis. 20 p.

Mann, Alan Eugene. The paleodemography of Australopithecus. D, 1968, University of California, Berkeley. 152 p.

Mann, C. John. Geology of Chandler Syncline, Fremont County, Colorado. M, 1957, University of Kansas. 86 p.

Mann, Charles F. The geology of Oliver Springs, Roane and Anderson counties, Tennessee. M, 1963, University of Tennessee, Knoxville. 54 p.

Mann, Christian John. Pennsylvanian stratigraphy of southwestern Wyoming, northwestern Colorado and northeastern Utah. D, 1961, University of Wisconsin-Madison. 132 p.

Mann, Daniel Hamilton. The Quaternary history of the Lituya glacial refugium, Alaska. D, 1983, University of Washington. 268 p.

Mann, Daven Craig. Clastic Laramide sediments of the Wasatch Hinterland, northeastern Utah. M, 1974, University of Utah. 112 p.

Mann, Donald M. The geology of an area near Eldorado Springs (Boulder County), Colorado. M, 1960, University of Colorado.

Mann, E. L. The geology of the Seal lake syncline, Central Labrador. D, 1959, McGill University.

Mann, Gary Dale. Multichannel signal enhancement techniques for reflection seismic records. M, 1979, University of Alberta. 164 p.

Mann, Henrietta. Algal uptake of U, Ba, Co, Ni and V; studies of natural and experimental systems. D, 1984, University of Western Ontario. 318 p.

Mann, Hugh Thomas. Geology of Barton Creek area, Earth County, Texas. M, 1951, University of Texas, Austin.

Mann, J. F. The sediments of Lake Elsinore. M, 1947, University of Southern California.

Mann, John Allen. The geology of part of the Gravelly Range area, Madison County, Montana. D, 1950, Princeton University. 157 p.

Mann, John Francis, Jr. Late Cenozoic geology of a portion of the Elsinore fault zone, Southern California. D, 1951, University of Southern California.

Mann, Keith Olin. Physiological, environmental, and mineralogical controls on Mg and Sr concentrations in the skeletal structures of Nautilus. D, 1987, University of Iowa. 297 p.

Mann, Keith Olin. The Sr, Mg and Ca chemistry of the mineralized structures of Nautilus. M, 1983, University of Texas, Arlington. 99 p.

Mann, Richard A. Cave development along selected areas of the western Cumberland Plateau escarpment. M, 1982, Memphis State University.

Mann, Robert Gordon. Seismic stratigraphy and salt tectonics of the northern Green Canyon area, Gulf of Mexico. D, 1987, Texas A&M University. 244 p.

Mann, Robin Carl. Mechanics of thrust fault formation; development of the Rocky Mountain foreland belt. M, 1979, Carleton University. 129 p.

Mann, Steven D. Petrography of the Upper Jurassic Olvido Limestone in Potrero Chico, Nuevo Leon, Mexico. M, 1982, University of Texas, Austin.

Mann, Virgil I. The relation of oxidation to the origin of soft iron ores of Michigan. D, 1950, University of Wisconsin-Madison.

Mann, Wallace. Subsurface geology of the Frank Graben, Pontotoc and Coal counties, Oklahoma. M, 1957, University of Oklahoma. 62 p.

Mann, William Paul. Cenozoic tectonics of the Caribbean; structural and stratigraphic studies in Jamaica and Hispaniola. D, 1983, SUNY at Albany. 777 p.

Mann, William Rhodes. The modification of suite of detrial rock-fragments within a portion of the Neuse River drainage basin, North Carolina. M, 1972, North Carolina State University. 38 p.

Mannai, Mohamed A. Digital-computer models of sedimentation in Angostura Reservoir, Fall River County, South Dakota. M, 1984, South Dakota School of Mines & Technology.

Mannard, George William. Geology of the Sungida kimberlite pipes, Tanganyika. D, 1963, McGill University.

Mannard, George William. The geology of the St. Pierre Prospect, Fort Chimo District, Quebec. M, 1956, McGill University.

Manner, Barbara Marras. Selected trace elements and stratification of West Branch (Ohio) Reservoir. M, 1971, University of Akron. 59 p.

Mannhard, Gregory W. Stratigraphy, sedimentology, and paleoenvironments of the La Ventana Tongue (Cliff House Sandstone) and adjacent formation of the Mesaverde Group (upper Cretaceous), southeastern San Juan Basin, New Mexico. D, 1976, University of New Mexico. 183 p.

Mannhard, Gregory William. Subsurface stratigraphic analysis of Morrow sandstones (Lower Pennsylvanian), portions of Harper and Woods counties, Oklahoma, and Clark and Comanche counties, Kansas. M, 1972, University of Oklahoma. 39 p.

Manni, Frederica M. Depositional environment, diagenesis, and unconformity identification of the Chimneyhill Subgroup in the western Anadarko Basin and northern shelf, Oklahoma. M, 1985, Oklahoma State University. 133 p.

Mannick, Matthew Lee. The geology of the northern flank of the upper Centennial Valley, Beaverhead and Madison counties, Montana. M, 1980, Montana State University. 86 p.

Manning, Christine. The Silurian of southeastern Indiana, southwestern Ohio, and northcentral and east-central Kentucky, with detailed discussion of the Brassfield in Wayne County, Indiana. M, 1932, Indiana University, Bloomington. 59 p.

Manning, Craig Edward. Hydrothermal clinopyroxenes of the Skaergaard Intrusion, East Greenland. M, 1985, Stanford University.

Manning, George A. The geology of the northern half of the Blue Lake Quadrangle, Humboldt County, California. M, 1947, University of California, Berkeley. 36 p.

Manning, Gerald E. Laboratory weathering of feldspar and investigations of cation content of streams draining essentially monolithic terrains. M, 1962, Colorado School of Mines. 59 p.

Manning, John Draige. A statistical-petrographic test of the aggregate-alkalic reactivity of concrete aggregate. D, 1951, Stanford University. 75 p.

Manning, John P., Jr. The relation of gold to other vein minerals. M, 1936, University of Minnesota, Minneapolis. 55 p.

Manning, Leslie D. Geology and structure of a portion of Southwest Mountain from Wolf Pitt Mountain to Turkey Sag Gap (Virginia). M, 1939, University of Virginia. 79 p.

Manning, Leslie Kay. Neodymium isotopes; a potential geochemical tracer for petroleum exploration. M, 1988, University of Wyoming. 141 p.

Manning, Paul De Vries. The genesis of oil shale and its relation to petroleum and other fuels. D, 1927, Columbia University, Teachers College.

Manning, Philip L. Surface mining potential of the coal semas of Lookout Mountain, Northeast Alabama. M, 1982, University of Georgia.

Manning, Retta A. A systematic analysis of several upper Ordovician brachiopods, (SE Indiana and NW Tennessee). M, 1970, Virginia State University. 53 p.

Mannion, Lawrence E. Geology of the La Grange Quadrangle, California. D, 1960, Stanford University. 188 p.

Mannion, Lawrence Edward. Geology of a part of the Blacktail Range, Beaverhead County, Montana. M, 1949, University of Michigan. 44 p.

Mannon, Leslie Susan. The real gas pseudo-pressure for geothermal steam. D, 1978, [Stanford University].

Manns, Francis Tucker. Stratigraphic aspects of the Silurian-Devonian sequence hosting zinc and lead mineralization near Robb Lake, northwestern B.C. D, 1981, University of Toronto. 345 p.

Manojlovic, Peter M. Retrogression of granulite facies gneisses in the Huntsville area, southwestern Grenville Province. M, 1987, Carleton University. 130 p.

Manoogian, Peter R. Late Neogene phosphorous accumulation in the Equatorial Pacific. M, 1979, Ohio University, Athens.

Manos, Constantine Thomas. Petrography and depositional environment of the Sparland Cyclothem (Pennsylvania). D, 1963, University of Illinois, Urbana. 106 p.

Manos, Constantine Thomas. Petrography of the Teays-Mahomet Valley deposits. M, 1960, University of Illinois, Urbana. 35 p.

Manry, John Phillips. The induration of quartz sandstones. M, 1949, University of Cincinnati. 36 p.

Manser, Richard J. An investigation into the movement of aldicarb residue in groundwater in the central Sand Plain of Wisconsin. M, 1983, University of Wisconsin-Madison. 193 p.

Mansfield, Charles Frederic. An x-ray study of kaolinite single crystals. M, 1967, University of Wisconsin-Madison.

Mansfield, Charles Frederic. Petrography and sedimentology of the Late Mesozoic (Cretaceous) Great Valley Sequence, near Coalinga, California. D, 1971, Stanford University. 71 p.

Mansfield, George R. The origin and structure of the Roxbury Conglomerate (Massachusetts). D, 1906, Harvard University.

Mansfield, S. P. Petrographic composition and sulfur content of selected Pennsylvania bituminous coal seams. M, 1965, Pennsylvania State University, University Park. 178 p.

Mansfield, Wendell Clay. The Miocene stratigraphy of Virginia, based on a study of the faunas. D, 1927, George Washington University.

Mansholt, Michael Scott. Micropaleontology, paleoecology and depositional environments of Wolfcampian carbonates in Arrow Canyon, Clark County, Nevada. M, 1984, Southern Illinois University, Carbondale. 106 p.

Mansilla, Enrique. Paragenesis in the Uchucchacua mining district, Peru. M, 1972, University of Minnesota, Minneapolis. 69 p.

Mansinha, Lalatendu. Radiation from tensile fractures. D, 1962, University of British Columbia.

Manske, Douglas Charles. Distribution of Recent foraminifera in relation to estuarine hydrography, Yaquina Bay, Oregon. D, 1968, Oregon State University. 176 p.

Manske, Douglas Charles. Geology of the Baldy Mountain area, Madison County, Montana. M, 1961, Oregon State University. 176 p.

Manske, Scott Lee. Fracturing events in the Ruby Star Granodiorite adjacent to the Esperanza porphyry copper deposit, Pima County, Arizona. M, 1980, University of Arizona. 102 p.

Mansker, William L. Petrogenesis of nephelinites and melilite nephelinites fron Oahu, Hawaii. D, 1979, University of New Mexico. 211 p.

Mansker, William L. Petrology of a southeastern Missouri ultramafic pipe. M, 1973, University of Missouri, Columbia.

Manson, Douglas Martin Vincent. Chemical variations in basaltic rocks. D, 1973, Columbia University. 122 p.

Manson, Marsden. Geological and solar climates; their causes and variations. D, 1893, University of California, Berkeley. 49 p.

Mansour, Wahid Omar. Development of a methodology for the design, construction and quality assurance of the core of rubble mound breakwaters. D, 1983, University of California, Berkeley. 413 p.

Mansouri, Tareg Ahmad. Dynamic response and liquefaction of earth structures. D, 1980, Colorado State University. 201 p.

Manspeizer, Warren. A study of the stratigraphy, paleontology, petrology and geologic history of the Canadaway and Conneaut groups in Allegany County, New York. D, 1963, Rutgers, The State University, New Brunswick. 354 p.

Mansur, Milud Abdulkrim. Geology and linears of Libya. D, 1981, University of Idaho. 213 p.

Mantei, Erwin J. K_2O/NaO content as a possible index to the chronological sequence of some Precambrian igneous rocks of Missouri. M, 1962, University of Missouri, Rolla.

Mantei, Erwin J. Variation in gold content of minerals of the Marysville quartz diorite stock, Montana. D, 1965, University of Missouri, Rolla. 83 p.

Mantuani, Mark Anthony. Sediment-water relations in lakes of the Lower Grand Coulee, Washington. D, 1969, Duke University. 136 p.

Manuel, Oliver Keith. The abundance pattern and isotopic composition of noble gases in the Fayetteville Meteorite. D, 1964, University of Arkansas, Fayetteville. 106 p.

Manuel, William Asbury. Colorado coals, Report "A"; some analyses and properties of coals mostly from the Western Slope. D, 1928, Colorado School of Mines. 133 p.

Manulik, Alexander John. Paleontology and stratigraphy of the Devonian section of a well core from Newaygo County, Michigan. M, 1951, University of Michigan.

Manus, Ronald W. The geology of the Condon area, Union County, Tennessee. M, 1963, University of Tennessee, Knoxville. 43 p.

Manus, Ronald Warren. Experimental chemical weathering of two alkali feldspars. D, 1968, University of Cincinnati. 109 p.

Manusmare, Purushottam. A study of the void-strengthening of aluminum and its nature. M, 1971, University of Missouri, Rolla.

Manwaring, Mark S. The use of side-looking airborne radar in the study of regional tectonics in West central Arkansas. M, 1981, Indiana State University. 117 p.

Manyak, David Michael. Calcified tube formation by the shipworm Bankia gouldi. D, 1982, Duke University. 265 p.

Manz, Richard P. Groundwater flow modeling of the Ojai Basin using the USGS 3 dimensional Modflow Model. M, 1988, California State University, Northridge. 185 p.

Manz, Ronald E. Experimental analysis of folding in simple shear. M, 1975, University of Oklahoma. 57 p.

Manzer, Geraald K., Jr. Petrology and geochemistry of Precambrian mafic dikes, Minnesota, and their bearing on the secular chemical variations in Precambrian basaltic magmas. D, 1978, Rice University. 230 p.

Manzer, Gerald K., Jr. Petrology of mafic and ultramafic rocks in the Burgess Junction-Tongue River area, Bighorn Mountains, Wyoming. M, 1972, Kent State University, Kent. 97 p.

Manzi, John Joseph. Temporal and spatial heterogeneity in diatom populations of the lower York River, Virginia. D, 1974, College of William and Mary.

Manzolillo, Claudio D. Stratigraphy and depositional environments of the Upper Cretaceous Trinidad Sandstone, Trinidad-Aguilar area, Las Animas County, Colorado. M, 1976, Colorado School of Mines. 147 p.

Mao Chen Ge. Optimization of transducer array geometry for acoustic emission/microseismic source location. D, 1988, Pennsylvania State University, University Park. 237 p.

Mao, Ho-Kwang. The pressure dependence of the lattice parameters and volume of ferromagnesian spinels, and its implications to the Earth's mantle. D, 1967, University of Rochester. 174 p.

Mao, Ho-Kwang. The pressure dependence of volume and lattice parameters of iron and iron-nickel alloy to 350 kilobars. M, 1966, University of Rochester. 45 p.

Mao, Liang-Tsi. Stochastic analysis and modeling of riverflow time series. D, 1982, Purdue University. 218 p.

Mao, Maurus Nai-Hsien. Geology and seismicity of Taiwan. M, 1962, Boston College.

Mao, Ming-Ling. Performance of vertically-fractured wells with finite-conductivity fractures. D, 1977, Stanford University. 177 p.

Mao, Nai-Hsien. Wave velocities in olivine rock and the equation of state for high compression. D, 1972, Harvard University.

Maos, Jacob O. The spatial organization of new land settlement in Latin America. D, 1974, The Johns Hopkins University.

Mapes, Royal H. Carboniferous and Permian Bactritoidae (Cephalopoda) in North America. D, 1977, University of Iowa. 386 p.

Mapes, Royal H. Geology of the Sulphur Springs area (Carboniferous), Benton County, Arkansas. M, 1968, University of Arkansas, Fayetteville.

Maples, Christopher Grant. Paleontology, paleoecology, and depositional setting of the lower part of the Dugger Formation (Pennsylvanian; Desmoinesian) in Indiana. D, 1985, Indiana University, Bloomington. 442 p.

Mapp, Marcus B. Foraminifera of the Ripley Formation, southeastern Oktibbeha County, Mississippi (Ripley-Upper Cretaceous). M, 1940, Mississippi State University. 87 p.

Maqtadir, Abdul. Three-dimensional nonlinear soil-structure interaction analysis of pile groups and anchors. D, 1984, University of California, Berkeley. 287 p.

Mar, Robert Del *see* Del Mar, Robert

Marafi, Hussein. The geology of the Newburg-south Westhope oil fields, Bottineau County, North Dakota. M, 1968, University of North Dakota. 154 p.

Marais, Jacobus Jan. An application of the varietal characteristics of zircon to the correlation of the Pre-Cambrian and early Paleozoic rocks of the lower Hudson region of New York State. D, 1941, University of Wisconsin-Madison.

Maranate, Srisopa. Petrogenesis of a layered amphibolite sill in the Nemo District, Black Hills, South Dakota. M, 1979, South Dakota School of Mines & Technology.

Marangakis, Andrew. Interpretation of reflection seismic data by analysis of cumulative energy spectra. M, 1983, Virginia Polytechnic Institute and State University. 78 p.

Marangunic, Cedomir Damianovic. Effects of a land-slide on Sherman Glacier, Alaska. D, 1968, Ohio State University. 222 p.

Marashi, Hamidedin. Depositional environment of the Strodes Creek Member (Middle Ordovician) of the Lexington Limestone. M, 1977, Eastern Kentucky University. 63 p.

Maratos, Alexander. Study of the near shore surface characteristics of windows and Langmuir circulation in Monterey Bay. M, 1971, Naval Postgraduate School.

Maravich, M. D. The geology of the Freezeout Mountain-Bald Mountain area, Carbon County, Wyoming. M, 1940, University of Wyoming. 20 p.

Marble, J. P. Petrology of a metagabbro body, Whitehouse Syncline, south-central Adirondacks, New York. M, 1974, SUNY at Binghamton. 74 p.

Marcantel, E. L. Conodont biostratigraphy and sedimentary petrology of the Gerster Formation (Guadalupian) in East central Nevada and West central Utah. D, 1975, Ohio State University. 215 p.

Marcantel, Emily L. Paleoecology and diagenetic fabrics of a Lower Cretaceous rudist reef complex in west-central Texas. M, 1968, Louisiana State University.

Marcantel, Jonathan. The origin and distribution of dolomites and associated limestones in the Fredericksburg Division (Cretaceous) of west-central Texas. M, 1968, Louisiana State University.

Marcantel, Jonathan Benning. Upper Pennsylvanian and lower Permian sedimentation in Northeast Nevada. D, 1973, Ohio State University.

Marcelletti, Nicholas. A statistical analysis of slope-profile development on contour surface coal mines, Clay County, Kentucky. M, 1987, Eastern Kentucky University. 54 p.

Marchand, Dennis Eugene. Chemical weathering, soil formation, geobotanical correlations in a portion of the White Mountains, Mono and Inyo counties, California. D, 1968, University of California, Berkeley. 409 p.

Marchand, Michael. A geochemical and geochronologic investigation of meteorite impact melts at Mistastin Lake, Labrador and Sudbury, Ontario. D, 1976, McMaster University. 142 p.

Marchand, Michael. Ultramafic nodules from the Ile Bizard kimberlite (Cretaceous), Quebec. M, 1970, McGill University. 73 p.

Marchel, Ronald Joseph. The depositional environment and paleogeography of the Lower Ordovician Goodwin Limestone in North central Nevada. M, 1978, Wayne State University.

Marchetti, John William, Jr. Magnetic study of basement configuration in south Cleveland and northeast McClain counties, Oklahoma. M, 1968, University of Oklahoma. 72 p.

Marchisio, Giovanni, B. Exact nonlinear inversion of electromagnetic induction soundings. D, 1985, University of California, San Diego. 187 p.

Marciano, Eugene. Seismic response and damage of retaining structures. D, 1986, Purdue University. 233 p.

Marcos, Zilmar Ziller. Morphologic and physical properties of fine-textured oxisols, State of São Paulo. D, 1971, Ohio State University. 216 p.

Marcott, Keith. The sedimentary petrography, depositional environment and tectonic setting of the Aldwell Formation, northern Olympic Peninsula, Washington. M, 1984, Western Washington University. 78 p.

Marcotte, Claude. Sedimentologie des conglomerats de la Formation de Cabano (Ashgillien-Llandoverien), a Cabano, comte de Temiscouata, Quebec. M, 1985, Universite de Montreal.

Marcotte, David L. An analysis of three component magnetic gradiometer data. M, 1986, University of Western Ontario. 198 p.

Marcotte, Denis. Le Krigeage bigaussien. D, 1987, Ecole Polytechnique. 280 p.

Marcotte, R. A. A paleostress analysis of a major fold in the southern part of the Hinesburg Synclinorium. M, 1975, University of Vermont.

Marcotty, Louise-Annette. The petrology of the magnetite paragneisses at Benson Mines, Adirondacks, New York. M, 1983, University of Michigan.

Marcou, John Andrew. Optimizing development strategy for liquid dominated geothermal reservoirs. M, 1985, Stanford University. 157 p.

Marcouiller, Barbara A. Overburden characterization and geotechnical studies of the Hanna IV UCG Site, Hanna, Wyoming. M, 1986, Colorado State University. 167 p.

Marcucci, Ettore. Distribution of elements in the Roraima Formation of the Amazonas Federal Territory (Venezuela). D, 1975, Rice University. 117 p.

Marcucci, Ettore. Origin and silica-carbonate relations of cherts of the Upper Cretaceous of Venezuela. M, 1973, Rice University. 81 p.

Marcus, A. L. The first-order drainage basin; a morphological analysis. D, 1976, Clark University. 153 p.

Marcus, Barry I. Geology, petrology and mineral chemistry of a garnet skarn, northern Sierra Nevada, California. M, 1988, California State University, Hayward. 204 p.

Marcus, David W. Formation, growth, and dissipation of ice in the nearshore region of Lake Erie in the area of Sturgeon Point, Angola, New York. M, 1986, SUNY at Buffalo. 134 p.

Marcus, Donald L. Igneous and metamorphic petrology of Barillas Quadrangle, northwestern Guatemala. M, 1974, University of Texas, Arlington. 116 p.

Marcus, Kim L. The rocks of Bulson Creek; Eocene-Oligocene sedimentation and tectonics in the Lake McMurray area, Washington. M, 1981, Western Washington University. 84 p.

Marcus, Leslie Floyd. The Bingara fauna; a Pleistocene vertebrate fauna from Murchison County, New South Wales, Australia. D, 1962, University of California, Berkeley. 380 p.

Marcus, Michael Dean. Ecology and management of headwater reservoirs. D, 1987, University of Wyoming. 235 p.

Marcus, Steven Roy. The rejuvenation of Hogtown Creek (Gainesville, Florida). M, 1971, University of Florida. 69 p.

Marcuson, William Frederick, III. The effects of time on the dynamic shear modulus and damping ratio of clay soils. D, 1970, North Carolina State University. 142 p.

Mardirosian, Charles Azad. Geochemical exploration of Crow Mine area, Lincoln County, New Mexico. M, 1964, University of Utah. 84 p.

Mardon, Duncan. Localization of pressure solution and the formation of discrete solution seams. D, 1988, Texas A&M University. 263 p.

Mare, Marius P. H. Alteration zones in the Flin Flon area, Manitoba. M, 1988, University of Manitoba. 113 p.

Marek, John M. Open pit slope design; Tenmile Cirque, Climax, Colorado. M, 1976, Colorado School of Mines. 150 p.

Marek, Norman J. Petrology and depositional environment of the Central Plain Group (Oligocene); Antigua, British West Indies. M, 1981, Northern Illinois University. 213 p.

Marentette, Kris Allen. Late Quaternary paleoceanography in Kane Basin, Canada and Greenland. M, 1988, University of Windsor. 67 p.

Maresca, Gerard P. Asbestos in water supplies of the northern New Jersey area; source, concentration, mineralogy, and size distribution. M, 1984, Rutgers, The State University, Newark. 192 p.

Maresca, J. W., Jr. Bluffline recession, beach change, and nearshore change related to storm passages along southeastern Lake Michigan. D, 1975, University of Michigan. 504 p.

Maresch, Walter Victor. The metamorphism and structure of northeastern Margarita Island, Venezuela. D, 1971, Princeton University. 274 p.

Mareschal, J.-C. Some geophysical implications of phase transitions inside the Earth. D, 1975, Texas A&M University. 259 p.

Mareschal, Marianne. Some problems associated with the inversion of polar magnetic sub-storm data recorded at the Earth's surface. D, 1975, Texas A&M University.

Maret, R. E. Sedimentary petrography of the Weber Formation at Rangely Field, Rio Blanco County, Colorado, and adjacent areas. M, 1956, University of Wyoming. 114 p.

Marfurt, K. J. Elastic wave equation migration-inversion. D, 1978, Columbia University, Teachers College. 185 p.

Margenau, Roy E., Jr. Geology of the Gold King deposits. M, 1953, Michigan Technological University. 16 p.

Margeson, G. Bradford. Iron-rich rocks of Gardner Mountain, New Hampshire, and their significance to base metal distribution. M, 1982, University of Western Ontario. 185 p.

Margheim, G. A. Water pollution from spent oil shale. D, 1975, Colorado State University. 148 p.

Margolis, Jacob. Structure and hydrothermal alteration associated with epithermal Au-Ag mineralization, Wenatchee Heights, Washington. M, 1987, University of Washington. 90 p.

Margolis, Stanley. Electron microscopy of surface features of minerals associated with the kaolin deposits of the southeastern U.S. M, 1966, Florida State University.

Margolis, Stanley. Surface microfeatures on quartz sand grains as paleoenvironmental indicators. D, 1971, University of California, Riverside. 182 p.

Margulies, Todd D. Soil and sediment chemistries in relation to water quality of Colorado mountain lakes. M, 1988, Colorado School of Mines. 90 p.

Marh, Bhupinder Singh. Geomorphology of the Ravi River near Chamba Town, Himachal Pradesh, India. D, 1985, Indiana State University. 137 p.

Marhadi. Physical modeling studies of thin beds. M, 1983, University of Houston.

Mari, David Lee. Correlation of the Absaroka sequence in the Rocky Mountain region. M, 1969, University of Washington. 71 p.

Mari, David Lee. Triassic and lower Jurassic stratigraphic patterns in the western United States. D, 1973, University of Washington. 126 p.

Marian, Melinda Lee. Sedimentology of the Beck Spring Dolomite, eastern Mojave Desert, California. M, 1979, University of Southern California.

Mariano, Anthony Nick. A study of chalcopyrite as a derivative structure of sphalerite. D, 1968, Boston University. 128 p.

Marianos, Andrew. Some Cretaceous ostracodes encountered in a well section from the Sacramento Valley in California. M, 1956, Louisiana State University.

Marie, James R. Stratigraphy and structure of the Millican Creek area, Sevier County, Tennessee. M, 1963, University of Tennessee, Knoxville. 37 p.

Marieau, Raymond Alban. A study of the relation of the Earth's field as presented on aeromagnetic maps to the geology in Beauce area, Quebec. M, 1956, McGill University.

Marin Rivera, Pedro A. Mineralogy and structure of the Stibnite Hill Mine, Thompson Falls, Montana. M, 1976, University of Montana. 68 p.

Marin, Carlos Mariano. Parameter uncertainty in water resource planning. D, 1983, Harvard University. 127 p.

Marin, Jon Randal. Structure and depositional environment of the Precambrian rocks in the McGee area, Black Hills, South Dakota. M, 1983, South Dakota School of Mines & Technology.

Marin, Luis Ernesto. Spatial and temporal patterns in the hydrogeochemistry of a bog-wetland system, northern Highlands Lake District, Wisconsin. M, 1986, University of Wisconsin-Madison. 85 p.

Marinai, Robert K. Petrography and diagenesis of the Ledge Sandstone Member of the Ivishak Formation (Permo-Triassic), Arctic National Wildlife Refuge, North Slope, Alaska. M, 1988, University of California, Santa Barbara.

Marincovich, Louie N. Pleistocene molluscan faunas from upper terrace deposits of the Palos Verdes hills (Los Angeles County), California. M, 1970, University of Southern California.

Marincovich, Louie Nick, Jr. Neogene to Recent Naticidae (Mollusca: Gastropoda) of the eastern Pacific. D, 1973, University of Southern California.

Marine, Ira Wendell. Geology and groundwater resources of Jordan Valley, Utah. D, 1960, University of Utah. 274 p.

Mariner, Robert Howard. Experimental evaluation of the authigenic minerals reaction in the Pliocene Moonstone Formation (central Wyoming). D, 1971, University of Wyoming. 170 p.

Mariner, Robert Howard. Volcanic geology of part of the western end of the Rabbit Ears Range, Jackson and Grand counties, Colorado. M, 1967, University of Wyoming. 67 p.

Maring, Hal Barton. The impact of atmospheric aerosols on trace metal chemistry in open ocean surface seawater. D, 1985, University of Rhode Island. 157 p.

Marino, Miguel A. Growth and decay of ground-water ridges in response to deep percolation. M, 1965, New Mexico Institute of Mining and Technology. 129 p.

Marino, Rrobert John. Paleomagnetism of two lake sediment cores from Seward Peninsula, Alaska. M, 1977, Ohio State University. 183 p.

Marintsch, Edward Joseph. Systematic paleontology, biostratigraphy, and paleoecology of trepostome Bryozoa from the Middle Ordovician Hermitage Formation of east-central Tennessee. D, 1986, SUNY at Stony Brook. 638 p.

Mario, Annette. The petrology of the Ellicott City Granodiorite. M, 1984, University of Pittsburgh.

Marion, Cecil Price, Jr. A study of the Recent marine sediments in the Biloxi-Ocean Springs area of the Mississippi Gulf Coast. M, 1951, Mississippi State University. 52 p.

Marion, Donat J. Sedimentology of the Middle Jurassic Rock Creek Member in the subsurface of west-central Alberta. M, 1982, University of Calgary. 128 p.

Marion, Roy Clarence. Neogene stratigraphy and hydrocarbon generation in the Salinas Basin, California. M, 1986, Stanford University. 104 p.

Mariotti, Philip A. K and Rb distributions in biotites from granitic and country rock members of the Poudre Canyon migmatite (Precambrian), Larimer County, Colorado. M, 1969, Wayne State University.

Mariotti, Philip Arno. A quantitative chemical test for the origin of the granitic portion of the Poudre Canyon Migmatite. D, 1975, Michigan State University. 50 p.

Maris, Cynthia Robin Parmalee. Chemical evidence for advection of hydrothermal solutions in ridge flank sediments. D, 1983, University of Rhode Island. 281 p.

Marius, Maré Alteration zones in the Flin Flon area, Manitoba. M, 1988, University of Manitoba.

Marjaniemi, Darwin Keith. Geologic history of an ash-flow sequence and its source area in the Basin and Range Province of southeastern Arizona. D, 1970, University of Arizona. 208 p.

Mark, Anson. Geology of the Illinois River-Buffalo Creek area, North Park (Jackson and Grand counties), Colorado. M, 1958, University of Colorado.

Mark, Christopher. Analysis of longwall pillar stability. D, 1987, Pennsylvania State University, University Park. 443 p.

Mark, Clara G. The Mercer Limestone and its associated rocks. M, 1910, Ohio State University.

Mark, D. M. Topological randomness of geomorphic surfaces. D, 1977, Simon Fraser University. 138 p.

Mark, David L. An evaluation of potential lateral saltwater intrusion in the Ocotillo-Coyote Wells groundwater basin, Imperial County, California. M, 1987, San Diego State University. 169 p.

Mark, Frazer John. An investigation of the soils within and beneath the middens of two Huron villages. D, 1984, University of Western Ontario.

Mark, Lawrence Edward. Petrology and metamorphism in the Marcus Hook Quadrangle, southeastern Pennsylvania. M, 1977, Bryn Mawr College. 58 p.

Mark, Norman. An analysis of the shallow seismic refraction method in its application to two minor geologic problems. M, 1968, Florida State University.

Mark, Robert Kent. Strontium isotopic study of basalts from Nunivak Island, (Pliocene and Pleistocene) Alaska. D, 1971, Stanford University. 50 p.

Mark, Roger Alan. Structural and sedimentary geology of the area north of Hot Springs Canyon, southern Galiuro Mountains, Cochise County, Arizona. M, 1985, University of Arizona. 96 p.

Markart, K. D. Opaque minerals and associated mafic silicates in differentiated and undifferentiated Keweenawan lavas. M, 1975, University of Wisconsin-Madison.

Markas, John Mitchell. Geology of northern McClain County, Oklahoma. M, 1965, University of Oklahoma. 128 p.

Markert, John Conrad. Mineralogical, geochemical, and isotopic evidence of diagenetic alteration, attributable to hydrocarbon migration, Raven Creek and Reel fields, Wyoming. M, 1982, Oklahoma State University. 126 p.

Markey, Daniel Gene. Depositional environment and diagenesis of the Viking Formation, Joffre Field, Alberta, Canada. M, 1970, Texas Tech University. 120 p.

Markgraf, Philip C. Environment of deposition and diagenesis of the San Andres Formation in the Slaughter Field, Cochran and Hockley counties, Texas. M, 1986, Texas Tech University. 98 p.

Markham, Deborah Kesselring. Quaternary loess deposits of Douglas County, central Washington. M, 1971, University of Washington. 23 p.

Markham, E. O. The geology of the Sturgeon Quadrangle. M, 1919, University of Missouri, Columbia.

Markham, John Joseph. The engineering geology of the Big Walker tunnel, Wytheville (Wythe County) Virginia. M, 1970, Cornell University.

Markham, Neville Lawrence. Phase relations in the system gold-silver-tellurium. D, 1957, Harvard University.

Markham, Thomas A. Depositional processes and environment of lower Womble Sandy Limestone beds in Montgomery, Garland, and Saline counties, Arkansas. M, 1976, Louisiana State University.

Markisohn, David B. Refraction residual analysis of a cryptoexplosion structure near Kentland, Indiana. M, 1988, Indiana University, Bloomington. 109 p.

Markl, R. G. Bathymetry, sediment distribution, and sea-floor spreading history of the southern Wharton Basin, eastern Indian Ocean. D, 1974, University of Connecticut. 94 p.

Markl, Rudi G. Structural comparison of a normal and an anomalous continental rise off the eastern U.S.A. M, 1970, Queens College (CUNY). 66 p.

Markland, Thomas Richard. Subsurface water geology of Spanish Fork Quadrangle, Utah County, Utah. M, 1964, Brigham Young University. 65 p.

Markle, Douglas F. Preliminary studies on the systematics of deep-sea Alepocephaloidea (Pisces: Salmoniformes). D, 1976, College of William and Mary.

Markle, Douglas Frank. Benthic fish associations on the continental slope of the Middle Atlantic Bight. M, 1972, College of William and Mary.

Markley, Joseph Hooker. Geology of parts of the Bonfils and Alton quadrangles in Missouri. M, 1926, Washington University. 99 p.

Markley, Lewis C. Mineralogy of cherts as a factor in the use of siliceous aggregates for concrete. M, 1953, Kansas State University. 49 p.

Marko, Joel. Seismic evidence for an equilibration mechanism between asthenosphere and surface temperatures. M, 1981, Northern Illinois University. 122 p.

Marko, Paul Joseph. Petrographic and modal analysis of the Knoblick area, Missouri. M, 1964, Southern Illinois University, Carbondale. 56 p.

Markos, Andrew George. Geology of the Cornucopia mining district, Elko County, Nevada. M, 1985, University of Nevada. 86 p.

Markos, G. Geochemical alteration of plagioclase and biotite in glacial and periglacial deposits. D, 1977, University of Colorado. 280 p.

Markovic, Francis Xaviare, Jr. Basal Cretaceous; a field study in Parker County, Texas. M, 1952, Texas Christian University. 29 p.

Markowitz, Gerald. A geochemical study of the Upper Devonian-Lower Mississippian black shales in eastern Kentucky. M, 1979, University of Kentucky. 226 p.

Markowitz, Philip E. Ostracode assemblages associated with South Florida coral reefs. M, 1979, Bowling Green State University. 121 p.

Marks, Danny Gregory. Climate, energy exchange, and snowmelt in Emerald Lake watershed, Sierra Nevada. D, 1988, [University of California, Santa Barbara]. 171 p.

Marks, Edward. Biostratigraphy of Jonah Quadrangle, Williamson County, Texas. M, 1950, University of Texas, Austin.

Marks, Gregory Thomas. Sedimentology of the Clinch Sandstone (Lower Silurian) at Powell Mountain, southwestern Virginia. M, 1987, University of Tennessee, Knoxville. 260 p.

Marks, Janet E. Multispectral remote sensing techniques applied to exploration for kimberlite diatremes, Laramie Range, Wyoming-Colorado. M, 1985, University of Wyoming. 164 p.

Marks, Jay G. Miocene stratigraphy and paleontology of southwestern Ecuador. D, 1951, Stanford University. 245 p.

Marks, Jay G. Stratigraphy of the (Eocene) Tejon Formation in its type area, Kern County, California. M, 1941, Stanford University.

Marks, Joel Harvey. Petrographic and geochemical study of a selected nepheline syenite in bauxite, Arkansas. M, 1977, University of Arkansas, Fayetteville.

Marks, John Wallace. Elemental composition and its stratigraphic significance in White Rockian to Givetian? rocks of the Arrow Canyon Range, Clark County, Nevada. M, 1966, University of Illinois, Urbana.

Marks, Larry Wayne. Computational topics in ray seismology. D, 1980, University of Alberta. 170 p.

Marks, Larry Wayne. Dynamic properties of headwaves near the critical point. M, 1976, University of Alberta. 147 p.

Marks, Milton R. Petrology of some sand and gravel deposits near Syracuse, New York. M, 1954, Syracuse University.

Marks, N. S. Sedimentation of new ocean crust; the Mid-Atlantic Ridge, 37° N. M, 1979, Stanford University. 64 p.

Marks, Thomas R. Correlation of geologic and geotechnical characteristics of the Paleocene Fort Union Formation. M, 1986, Colorado State University. 162 p.

Markun, Charles Daniel. Petrology and paleoenvironment of the Gunflint Iron Formation. M, 1978, University of Florida. 105 p.

Markward, Ellen L. Feldspathic materials in the sand of some Paleozoic sandstones in Missouri. M, 1952, University of Missouri, Columbia.

Markwell, Kenneth E. Petrology and structure of the northwest quarter of the Anderson Ridge Quadrangle, Wyoming. M, 1973, University of Missouri, Columbia.

Marland, Frederick Charles. An ecological study of the benthic macrofauna of Matagorda Bay, Texas. M, 1958, Texas A&M University.

Marland, Frederick Charles. The history of Mountain Lake, Giles County, Virginia; an interpretation based on paleolimnology. D, 1967, Virginia Polytechnic Institute and State University. 138 p.

Marland, Gregg Hinton. Phase relations in the system CaCO₃-H₂O. D, 1972, University of Minnesota, Minneapolis. 131 p.

Marlatte, Charles Raymond. The petrogenesis of the clastic materials of the Madras Formation. M, 1931, University of Oregon. 72 p.

Marlay, Lisa Emerick. Diagenesis of the lower Triassic Moenkopi Formation; Washington County, Utah, and Clark and Lincoln counties, Nevada. M, 1983, Duke University. 186 p.

Marleau, Raymond Alban. Geology of the Woburn, east Megantic, and Armstrong areas, Frontenac and Beauce counties, Quebec. D, 1958, Universite Laval. 184 p.

Marler, P. Petrographic study of the intermediate siltstone in the Sullivan Mine, Kimberley, British Columbia. M, 1953, McGill University.

Marlette, John William. The breakwater at Redondo Beach, California, and its effects on erosion and sedimentations. M, 1954, University of Southern California.

Marley, Walter Ellis. Stratigraphic controls of the coal-bearing portion of the Blackhawk Formation, Emery and Sevier counties, Utah. M, 1978, North Carolina State University. 91 p.

Marlian, Myron G. Stratigraphic position of the Lance Formation (upper Cretaceous) near Basin (Big Horn County), Wyoming. M, 1957, Wayne State University.

Marlow, Michael Stewart. Tectonic history of western Aleutian ridge-trench system. D, 1971, Stanford University. 102 p.

Marlowe, James I. Late Cenozoic geology of the lower Safford Basin on the San Carlos Indian Reservation, Arizona. D, 1961, University of Arizona.

Marmaduke, Richard C. Geology of Grand Marais township. M, 1941, University of Minnesota, Minneapolis. 30 p.

Maroney, D. A stratigraphic and paleoecologic study of some late Cenozoic sediments in the central Sand Hills Province of Nebraska. D, 1978, University of Nebraska, Lincoln. 326 p.

Maroney, David G. A population of the new species Ctenoconularia delphiensis from the New Albany Shale (Upper Devonian) at Delphi, Indiana. M, 1972, Ball State University.

Maroney, Michael H. Oxygen isotope study of the Ironwood Iron Formation, Gogebic Range, Wisconsin and Michigan. M, 1978, Northern Illinois University. 59 p.

Marozas, Dianne Catherine. The effects of mineral reactions on trace metal characteristics of groundwater in desert basins of southern Arizona. D, 1987, University of Arizona. 127 p.

Marozas, Dianne Catherine. The role of silicate mineral alteration in the supergene enrichment process. M, 1982, University of Arizona. 58 p.

Marple, M. L. Radium-226 in vegetation and substrates at inactive uranium mill sites. D, 1979, University of New Mexico. 74 p.

Marple, Mildred Fisher. Ostracodes from the Pottsville Formation in Ohio. D, 1950, Ohio State University.

Marquard, Randall Steven. Late Pleistocene paleoclimatology and paleoceanography of the northern Greenland Sea. M, 1984, University of Wisconsin-Madison.

Marquardson, Kent F. Preliminary diagram for the system K₂O-FeO-SiO₂. M, 1951, University of Utah. 26 p.

Marquardt, Tezz C. Age and paleoclimatic significance of a Lake Michigan beach-dune ridge complex located at Baileys Harbor, WI. M, 1986, University of Wisconsin-Green Bay. 129 p.

Marquez, Gustavo Enrique. The areal geology of the upper Bull Creek area. M, 1947, University of Texas, Austin.

Marquis, Robert. Etude tectono-stratigraphique à l'est de Val d'Or; essai de corrélation structurale entre les roches métasédimentaires des Groupes de Trivio et de Garden Island et application à l'exploration aurifère. M, 1984, Universite du Quebec a Chicoutimi. 185 p.

Marquis, Samuel Austin, Jr. Facies and diagenesis of the Goen Limestone cyclothem (early late Desmoinesian) Concho and Runnels county, central Texas. M, 1987, Southern Methodist University. 166 p.

Marquis, Urban Clyde. The relationship between the Fort Payne Formation and the Floyd Shale (Mississippian) in Northwest Georgia. M, 1958, Emory University. 111 p.

Marr, John. Petrology of the "northern granitic rocks" (Archean), Wanipigow river area, southeast Manitoba. M, 1970, University of Manitoba.

Marr, John D. Geology of the Pole Mountain, Buffalo Creek area, North Park, Jackson County, Colorado. M, 1931, Colorado School of Mines. 43 p.

Marr, John Donald. Geology of the western half of the Lake Toxaway Quadrangle, North Carolina. M, 1975, University of New Orleans.

Marr, Ronald Joseph. Geology of Lynch ranches, Catron and Valencia counties, New Mexico. M, 1956, University of Texas, Austin.

Marrall, Gerald E. A study of the (Cambrian) Jordan Sandstone in the area of the Wisconsin Arch. M, 1951, University of Wisconsin-Madison.

Marrett, David Joseph. Acid soil processes in the Okpilak Valley, Arctic Alaska. D, 1988, University of Washington. 177 p.

Marrin, Donn Louis. Remote detection and preliminary hazard evaluation of volatile organic contaminants in groundwater. D, 1984, University of Arizona. 140 p.

Marron, Donna Carol. Hillslope evolution and the genesis of colluvium in Redwood National Park, northwestern California; the use of soil development in their analysis. D, 1982, University of California, Berkeley. 212 p.

Marrone, Frank J. Optimization of offshore drilling platform location. M, 1987, Colorado School of Mines. 156 p.

Marrow, William Earl. Geology of the Leipers Fork Quadrangle, Williamson County, Tennessee. M, 1957, Vanderbilt University.

Marrs, Ronald W. Application of remote-sensing techniques to the geology of the Bonanza volcanic center (Southwest Colorado). D, 1973, Colorado School of Mines. 281 p.

Marrs, Thomas. Lithologic characteristics and depositional environments of the non-marine Benwood Limestone (Upper Pennsylvanian) in the Dunkard Basin, Ohio, Pennsylvania, and West Virginia. M, 1981, University of Pittsburgh.

Marryott, Robert Allen. Some petrologic and geochemical characteristics of the San Andreas Fault at the Stone Canyon Well, San Benito County, California. M, 1986, University of California, Riverside. 213 p.

Marsaglia, Kathleen Marie. Petrography, provenance, depositional environment, and diagenesis of the basal Salina Formation (lower Eocene) North-west Peru. M, 1982, University of Illinois, Urbana. 162 p.

Marsalis, W. E. The ostracode fauna of the Diploschiza cretacea zone of Alabama. M, 1966, University of Alabama.

Marschner, Arthur W. A method for the size analysis of sand on a frequency basis. M, 1952, Wayne State University.

Marsden, M. Origin and evolution of the Pleistocene Olorgesailie Lake Series; Kenya Rift valley. D, 1979, McGill University. 194 p.

Marsden, Ralph Walter. Accessory minerals of the Harney Peak Granite (South Dakota). M, 1933, University of Wisconsin-Madison.

Marsden, Ralph Walter. The application of accessory mineral methods to the Pre-Cambrian rocks of the Lake Superior region. D, 1939, University of Wisconsin-Madison.

Marsek, Frank A. Petrology of the metamorphosed ultramafic and related rocks on part of western Casper Mountain, Wyoming. M, 1978, University of Akron. 95 p.

Marsell, Ray Everett. Geology of the Jordan Narrows region, Traverse Mountains, Utah. M, 1932, University of Utah. 71 p.

Marsh, Bruce David. Aeromagnetic terrain effects. M, 1971, University of Arizona.

Marsh, Bruce David. Aleutian Island arc magmatism. D, 1974, University of California, Berkeley. 128 p.

Marsh, Charles W. Geology of water supplies and water analysis. D, 1882, Columbia University, Teachers College.

Marsh, Herman Earl. Geology of Triassic area southeast of Chapel Hill, North Carolina. M, 1920, University of North Carolina, Chapel Hill. 6 p.

Marsh, Jeffrey Robert. Facies development within algal-bryozoan carbonate mounds in the Brentwood Member, Bloyd Formation, Madison County, Arkansas. M, 1984, University of Arkansas, Fayetteville. 123 p.

Marsh, John W., Jr. Mobilization of selected trace metals by concentrated aqueous solutions; effects of solution composition, redox conditions, and metal partitioning. M, 1985, Michigan State University. 125 p.

Marsh, Leeda Elizabeth. Examination of a tephra deposit on the Great Rift of the Snake River plain in southeastern Idaho. M, 1984, SUNY at Buffalo. 104 p.

Marsh, Marion J. Some foraminifera and Ostracoda from a coquina marl, Rosefield, Louisiana. M, 1944, Columbia University, Teachers College.

Marsh, Owen Thayer. Geology of the Orchard Peak area, Kern, San Luis Obispo, Monterey, and Kings counties, California. D, 1955, Stanford University. 243 p.

Marsh, Philip Wienecke. Stratigraphy of the Dakota Group north-west flank of the Canon City embayment, Colorado. M, 1960, University of Oklahoma. 89 p.

Marsh, Phyllis Scudder. Sedimentation rates in small headwater reservoirs in Montana. D, 1974, University of Montana. 187 p.

Marsh, S. E. Quantitative relationships of surface geology and spectral habit to satellite radiometric data. D, 1979, Stanford University. 290 p.

Marshak, R. Stephen. A reconnaissance of Mesozoic strata in northern Yuma County, southeastern Arizona. M, 1979, University of Arizona. 110 p.

Marshak, Stephen. Aspects of deformation in carbonate rocks of fold-thrust belts of central Italy and eastern New York State. D, 1983, Columbia University, Teachers College. 280 p.

Marshall, Alan E. Geologic studies of the Proterozoic Bangemall Group, northwest Australia. D, 1968, Princeton University. 120 p.

Marshall, Alan Gould. An alluvial chronology of the lower Palouse River canyon and its relation to local archaeological sites. M, 1971, Washington State University. 73 p.

Marshall, Albert Ross. Kentucky petroleum and natural gas with map showing oil districts. M, 1902, [University of Kentucky].

Marshall, Brian David. Application of the potassium-calcium geochronometer to problems in geochronology and petrogenesis. D, 1984, University of California, Los Angeles. 242 p.

Marshall, Charles Harding. Geology of the Hazleton area, Bighorn Mountains, Wyoming. M, 1940, University of Iowa. 68 p.

Marshall, Clare Philomena. Cation arrangement in iron-zinc-chromium spinel oxides. M, 1983, University of California, Los Angeles. 54 p.

Marshall, Claude Monte, Jr. The magnetic properties and petrology of submarine pillow basalt. D, 1971, Stanford University. 97 p.

Marshall, David M. Structure and petrology of an Archean shear zone, Crescent Lake area, Wind River Mountains, Wyoming. M, 1987, University of Wyoming. 132 p.

Marshall, Donald J. Interfacial polarization in membranes and its significance in the induced polarization of geologic materials. D, 1959, Massachusetts Institute of Technology. 170 p.

Marshall, Donald R. Gravity gliding at Gray Mountain (Paleozoic), Coconino County, Arizona. M, 1972, Northern Arizona University. 83 p.

Marshall, Earl Elmore. The (Pennsylvanian) McCoy Formation of Eagle County, Colorado. M, 1939, University of Michigan.

Marshall, Ernest Willard. The geology of the Great Lakes ice cover. D, 1977, University of Michigan. 643 p.

Marshall, Frederick C. Geology of the Kent Window area, Wythe County, Virginia. M, 1959, Virginia Polytechnic Institute and State University.

Marshall, Frederick Charles. Lower and Middle Pennsylvanian fusulinids from the Bird Spring Formation near Mountain Springs Pass, Clark County, Nevada. D, 1967, Brigham Young University.

Marshall, H. I. The geology of the Desert Lake area. M, 1946, Queen's University. 69 p.

Marshall, Hollis D. Geology of the upper Schep Creek area, Mason County, Texas. M, 1959, Texas A&M University.

Marshall, John. The structure and age relations of the igneous-sedimentary complex of central Cecil County, Maryland. D, 1936, The Johns Hopkins University.

Marshall, John Harris. Geology in and adjacent to Cape Girardeau, Cape Girardeau County, Missouri. M, 1950, University of Missouri, Columbia.

Marshall, Larry G. Evolution of the Borhyaenidae, extinct South American predaceous marsupials. D, 1976, University of California, Berkeley. 205 p.

Marshall, Michael Cameron. Biostratigraphy and lithostratigraphy of the Middle Jurassic Tecocoyunca Group, Mexico. M, 1986, McMaster University. 281 p.

Marshall, Nissim Joseph. A study of the fundamental properties of Puget Sound glacial clays. M, 1955, University of Washington. 81 p.

Marshall, Paul Arthur. Selected partition ratios in feldspars and micas from the Bernic Lake (chemalloy) pegmatite (Precambrian), Manitoba. M, 1972, University of Western Ontario. 117 p.

Marshall, Paul Wellington. Resources and markets of sulphur and sulphur derivatives in the Pacific Northwest and Alaska. M, 1972, University of Washington. 137 p.

Marshall, Philip Schuyler. The linear coefficient of thermal expansion for frozen clays. M, 1977, University of Washington.

Marshall, Robert C. Structural evolution of the Starhope Canyon-Muldoon Canyon area, Copper Basin, Pioneer Mountains, south-central Idaho. M, 1983, Lehigh University.

Marshall, Robert Harden. Petrology of subsurface Mesozoic strata of the Yucatan Peninsula, Mexico. M, 1974, University of New Orleans.

Marshall, Royal Richard. I, The rate of recombination of iodine atoms; II, Some studies of the trace quantities of lead, uranium, and thorium in marine carbonate skeletons. D, 1955, California Institute of Technology. 117 p.

Marshall, S. The influence of adsorbed polyacrilimide on the cation-exchange behavior of kaolinite. M, 1979, Stanford University. 26 p.

Marshall, Timothy B. Compressed air storage in Paleozoic sandstones of southeastern Minnesota. M, 1979, University of Wisconsin-Milwaukee. 91 p.

Marshall, Timothy Robert. Biodegradation of petroleum wastes in soil; the microbial ecosystem and optimization of a treatment process. D, 1988, University of Southern California.

Marshall, W. S. Varve-like laminations in the Permian Bone Spring Limestone of western Texas. M, 1954, Columbia University, Teachers College.

Marshall, William Dustin. Depositional environment and reservoir characteristics of the lower Vicksburg sandstones, West Mcallen Ranch Field, Hidalgo County, Texas. M, 1978, Texas A&M University. 154 p.

Marshall, Willis Woodbury. Geography of the early Port of St. Louis. M, 1932, Washington University. 142 p.

Marsik, Dolores Dorothy. Laramide geology of the Jackson Hole area, Wyoming and adjacent areas. M, 1949, University of Michigan.

Marsters, Beverly Ann. Some unusual forms of pyrite from French Creek, Pensylvania. M, 1963, Bryn Mawr College. 22 p.

Marston, Richard Alan. The geomorphic significance of log steps in forest streams of the Oregon Coast Range. D, 1980, Oregon State University. 216 p.

Marszalek, Donald Stanley. Aspects of chamber formation by Archias angulatus, a foraminifer (Recent). D, 1969, University of Illinois, Urbana. 91 p.

Marszalek, Donald Stanley. Observations on the pseudopods of Iridia diaphana, a marine foraminifer (Recent). M, 1969, University of Illinois, Urbana.

Mart, Joseph. A tectonic model of an incipient spreading center and its margins; the northern Red Sea and the Dead Sea rift. D, 1984, Texas A&M University. 180 p.

Martel, Linda Marie Viglienzone. Radar observations of basaltic lava flows and fissure vents; Craters of the Moon, Idaho. M, 1984, Arizona State University. 61 p.

Martel, Richard. Modélisation de l'effet des sels déglaçants sur la qualité de l'eau souterraine à Trois-Rivières, Québec. M, 1986, Universite Laval. 91 p.

Martel, Stephen Joseph. Development of strike-slip fault zones in granitic rock, Mount Abbot Quadrangle, Sierra Nevada, California. D, 1987, Stanford University. 199 p.

Martel, Stephen Joseph. Late Quaternary activity on the Fish Springs Fault, Owens Valley fault zone, California. M, 1984, Stanford University. 112 p.

Martel, Yvon. The use of radiocarbon dating for investigating the dynamics of soil organic matter. D, 1972, University of Saskatchewan.

Martell, C. Michael. On dispersive continental shelf waves generated by alongshore variations in bottom topography. M, 1978, Oregon State University. 28 p.

Martell, Charles. Petrology and geochemistry of a progressively metamorphosed sedimentary formation in Big Thompson Canyon, Larimer County, Colorado. M, 1982, New Mexico Institute of Mining and Technology. 133 p.

Martell, Hildebrando Jose. Three Bar oil field, Andrews County, Texas. M, 1970, University of Texas, Austin.

Martell, J. Distribution of chlorinated hydrocarbons in Lake Anne, Virginia. M, 1973, George Washington University.

Martell, Paul W. Deposition and diagenesis of the Drinkard Formation, Lea County, New Mexico. M, 1985, Texas Tech University. 92 p.

Martello, Angela R. Petrography and geochemistry of Mesozoic diabase dikes of southern New England. M, 1986, University of Connecticut. 120 p.

Marten, Brian Ernest. The geology of the Western Arm Group, Green Bay, Newfoundland. M, 1971, Memorial University of Newfoundland. 72 p.

Marten, Brian Ernest. The relationship between the Aillik Group and the Hopedale Complex, Kaipokok Bay, Labrador. D, 1977, Memorial University of Newfoundland. 389 p.

Martens, Christopher. The inhibition of inorganic marine apatite precipitation by magnesium ions. M, 1969, Florida State University.

Martens, James H. A study of the basic dikes of the Ithaca region (New York). M, 1923, Cornell University.

Martens, James H. Some Pre-Cambrian rocks in northern Quebec. D, 1926, Cornell University.

Martens, Ronald Wayne. Classification of areas of suitability for sanitary landfill sites in the vicinity of Windsor, Colorado. M, 1973, University of Colorado.

Marthelot, Jean-Michel. Patterns of seismicity in the Vanuatu (New Hebrides) arc; regional variations and systematic evolution. D, 1983, Cornell University. 139 p.

Marti, J. Lateral loads exerted on offshore piles by subbottom movements. D, 1976, Texas A&M University. 125 p.

Martin, A. B. Growth conditions for single and mosaic crystals of zinc. M, 1939, University of Wyoming.

Martin, Anthony J. A paleoenvironmental interpretation of the "Arnheim" micromorph fossil assemblage from the Cincinnatian Series (Upper Ordovician), southeastern Indiana and southwestern Ohio. M, 1986, Miami University (Ohio). 221 p.

Martin, Archie H., III. Soil chronology and sedimentary petrology of the Canaveral Peninsula, Cape Canaveral, Florida. M, 1976, Eastern Kentucky University. 87 p.

Martin, Barton Sawyer. Paleomagnetism of basalts in northeastern Oregon and west-central Idaho. M, 1984, Washington State University. 151 p.

Martin, Ben Stephen. Petrography, depositional environments, and regional correlation of the lower member of the Cretaceous Quitman Formation, Hudspeth County, Texas. M, 1987, University of Texas, Arlington. 265 p.

Martin, Benjamin Frank, Jr. The petrology of the Corbin Gneiss. M, 1974, University of Georgia. 113 p.

Martin, Bruce. Pliocene of middle California. M, 1913, University of California, Berkeley. 59 p.

Martin, Bruce Delwyn. Microfauna of the Adaville Formation (Cretaceous), southwestern Wyoming. M, 1954, University of Utah. 62 p.

Martin, Bruce Delwyn. Monterey submarine canyon; genesis and relationship to continental geology (Monterey, California). D, 1963, University of Southern California.

Martin, Bruce J. Igneous intrusions of central Mississippi. M, 1979, Tulane University.

Martin, C. Structural evolution of the western margin of the Idaho Batholith in the Riggins, Idaho area. D, 1977, Pennsylvania State University, University Park.

Martin, Charles Arthur, Jr. Geology of the South Hardscrabble Creek area, Colorado. M, 1954, University of Kansas. 71 p.

Martin, Charles Wellington. Minor structures and their significance in the (Lower Cambrian) Hartland Formation, Connecticut. M, 1959, University of Wisconsin-Madison.

Martin, Charles Wellington. Petrology, metamorphism, and structure of the Hartland Formation in the central western Connecticut highlands. D, 1962, University of Wisconsin-Madison. 110 p.

Martin, Clyde D. The Cenozoic geology of Southwest Garza County, Texas. M, 1950, Texas Tech University. 40 p.

Martin, Conrad. The origin of crystalline magnesite deposits. M, 1958, University of Nevada - Mackay School of Mines. 77 p.

Martin, Daryl Lynn. Depositional systems and ichnology of the Bright Angel Shale (Cambrian), eastern Grand Canyon, Arizona. M, 1985, Northern Arizona University. 365 p.

Martin, David. Potential model for several mass anomalies associated with Nazca subduction. M, 1988, University of Washington. 57 p.

Martin, David B. Geology and petrology of the Caribou, New Brunswick area. M, 1957, Dartmouth College. 104 p.

Martin, David Carl. Effect of various communicative methods for controlling visitor removal of pumice at Mount St. Helens National Volcanic Monument. D, 1987, University of Washington. 142 p.

Martin, David L. Paleomagnetic results from Morocco, new results and survey, and a detailed study of the paleomagnetics of carbonates. D, 1976, University of South Carolina. 109 p.

Martin, David Lichty. Magnetic stratigraphy of the Columbus Limestone (middle Devonian), (Ohio). M, 1971, Ohio State University.

Martin, David Rolo. Geology of the western part of the Santa Susana Mountains, Ventura County, California. M, 1958, University of California, Los Angeles.

Martin, David W. Clastic to carbonate transitions; a modern example; the western Florida continental shelf. M, 1984, University of South Florida, St. Petersburg.

Martin, Dewayne C. H. The age relation of the Mount Tom hornblende gneiss of the dioritic gneisses of the Mount Prospect Complex, Connecticut. M, 1960, University of Wisconsin-Madison.

Martin, Edgar Keith. Biostratigraphy of the Upper Pennsylvanian Wayland Shale in McCullock-Coleman counties, central Texas. M, 1965, Rice University. 100 p.

Martin, Ellen Eckels. Diagenetic alteration of Pleistocene sediments beneath subaerial discontinuity surfaces, South Florida; petrography and geochemistry. M, 1987, Duke University. 116 p.

Martin, Etienne. Modèle de formation et de mise en place de la partie sud-ouest du complèxe anorthositique du Lac St. Jean. M, 1983, Universite du Quebec a Chicoutimi. 126 p.

Martin, Frank J. Relationship between diagenesis and depositional environments, lower Tuscaloosa Formation, southwestern Mississippi and southeastern Alabama. M, 1988, University of New Orleans.

Martin, G. C. The effect of physiography on the trade routes of East Tennessee. M, 1932, University of Tennessee, Knoxville.

Martin, G. C. The geology of northern Sequatchie Valley and vicinity. D, 1940, Ohio State University.

Martin, Gene B. Upper Ordovician Bryozoa of northwestern Alabama. M, 1954, Mississippi State University. 65 p.

Martin, George C. The Miocene gastropod fauna of Maryland. D, 1901, The Johns Hopkins University.

Martin, Glen Edward, II. Photosynthetic isotope fractionation; oxygen and carbon. D, 1982, Brigham Young University. 39 p.

Martin, H. L. Implementation of three-dimensional theories of consolidation. D, 1975, University of Colorado. 125 p.

Martin, Harold. Mosasaurus poultneyi; a South Dakota mosasaur. M, 1953, South Dakota School of Mines & Technology.

Martin, Harry. Reconnaissance sampling of the remanent magnetization of the New Oxford Formation near Thomasville, Pennsylvania. M, 1975, Lehigh University. 17 p.

Martin, Henry Jerome. A microearthquake survey in central New Hampshire. M, 1973, Brown University.

Martin, Irving Lee. Petrographic and micro-paleontological study of some marls of eastern North Carolina. M, 1928, University of North Carolina, Chapel Hill. 30 p.

Martin, J. A. Geology of the northwest quarter of the Washington Quadrangle (Missouri). M, 1956, University of Missouri, Rolla.

Martin, Jack Philip, Jr. Structural geology of Bayou Bouillon Field. M, 1973, University of Southwestern Louisiana.

Martin, James A., Jr. F-K filter for supression of surface waves in shallow seismic reflection. M, 1984, Ohio University, Athens. 100 p.

Martin, James Cook. The Pre-Cambrian rocks of the Canton Quadrangle, New York. D, 1913, Princeton University.

Martin, James Edward. Hemphillian rodents from northern Oregon and their relationships to other rodent faunas in North America. D, 1979, University of Washington. 265 p.

Martin, James Edward. Small mammals from the Miocene Batesland Formation of South Dakota. M, 1973, South Dakota School of Mines & Technology.

Martin, James Hosmer. Coal combustion residue; the effects upon shallow groundwater in an upland ravine landfill in southeastern Iowa. M, 1986, University of Iowa. 112 p.

Martin, James L., Jr. Claiborne Eocene species of the ostracode genus Cytheropteren. M, 1939, Louisiana State University.

Martin, James L., Jr. The geology of Webster Parish, Louisiana. D, 1943, Louisiana State University.

Martin, James Lee. Stratigraphic analysis of Pennsylvanian strata in the Lucero region of west central New Mexico. D, 1971, University of New Mexico. 196 p.

Martin, James O. A study of the drumlin area of New York State. M, 1903, Cornell University.

Martin, James Oliver. Geology of the central portion of the Mason Quadrangle, Mason County, Texas. M, 1986, University of Texas of the Permian Basin. 52 p.

Martin, James P. Determination of the deflection of the vertical by topographic and leveling methods. M, 1964, Ohio State University.

Martin, Jay John. The Webb Formation in the Roberts Mountains, central Nevada. M, 1985, University of California, Riverside. 75 p.

Martin, Jeffrey R. Pennsylvanian deltaic sedimentation in Grand Ledge, Michigan. M, 1982, Western Michigan University. 131 p.

Martin, John D. Geology and economic potential of the Gold Creek Quartzite, Bonner County, Idaho. M, 1982, Eastern Washington University. 85 p.

Martin, John Raymond. The determination of the porosities of subsurface geologic strata by analysis of the spectral distribution of gamma radiation. M, 1953, University of Pittsburgh.

Martin, Jonathan Bowman. Geochemistry of dolomites in Tertiary organic-rich sediments; Pisco Basin of Peru. M, 1987, Duke University. 214 p.

Martin, Joseph Paul. Embankment consolidation from unsaturated to saturated conditions. D, 1983, Colorado State University. 330 p.

Martin, Joseph Stewart. Geology of the Dry Canyon area in the eastern section of the Ventura area, California. M, 1947, California Institute of Technology. 44 p.

Martin, Kathleen M. Changes in fluid chemistry of the Auburn geothermal well. M, 1986, SUNY at Buffalo. 38 p.

Martin, Kenneth Glenn. Washita Group stratigraphy, south-central Texas. M, 1961, University of Texas, Austin.

Martin, Kyle. Oxygen isotope analysis, uranium-series dating, and paleomagnetism of speleothems in Gardner Cave, Pend Oreille County, Washington; pale-

oclimatic implications. M, 1988, Eastern Washington University. 101 p.

Martin, L. G. Structural geology of the Chelmsford Granite at Oak Hill. M, 1973, Boston College. 150 p.

Martin, Larry D. A medial Pleistocene fauna from near Angus, Nuckolls County, Nebraska. M, 1969, University of Nebraska, Lincoln.

Martin, Laura Hatch. Geology of the Stonington region, Connecticut. D, 1916, University of Chicago. 70 p.

Martin, Lawrence. Some features of glaciers and glaciation in College Fiord, Prince William Sound, Alaska. D, 1913, Cornell University.

Martin, Lawrence James. Potential for acid formation from tailings from the Coeur d'Alene mining area, Idaho. M, 1981, University of Idaho. 56 p.

Martin, Leonard John. Clearwater Shale foraminifera from the Athabasca River, Alberta. M, 1954, University of Alberta. 65 p.

Martin, Leonard John. Stratigraphy and depositional tectonics of the North Yukon-lower Mackenzie. D, 1958, Northwestern University.

Martin, Lewis. Biostratigraphy of the Moreno Gulf surface section, Panoche Hills, Fresno County, California. D, 1961, Stanford University. 380 p.

Martin, Lewis. Some Pliocene foraminifera from a portion of the Los Angeles Basin, California. M, 1951, University of Southern California.

Martin, Linda G. Preliminary foraminiferal biostratigraphy of the Smoky Hill Chalk Member of the Niobrara Chalk (Upper Cretaceous) in Trego, Gove, and Logan counties, Kansas. M, 1984, Colorado School of Mines. 120 p.

Martin, Lois Ticknor. Observations on living foraminifera from the intertidal zone of Monterey Bay, California. M, 1932, Stanford University. 66 p.

Martin, Mark W. The structural geology of the Worthington Mountains, Lincoln County, Nevada. M, 1987, University of North Carolina, Chapel Hill. 112 p.

Martin, Marshall A. Three-component seismic investigation of a fractured reservoir, Silo Field, Wyoming. D, 1987, Colorado School of Mines. 99 p.

Martin, Mary Margaret. Comparison of water quality from three surface coal mine spoils with different type and age of restoration, Clarion County, Pennsylvania. M, 1979, Pennsylvania State University, University Park. 198 p.

Martin, Michael David. La Salle Limestone (upper Pennsylvanian) conodonts of La Salle County, Illinois. M, 1974, University of Illinois, Urbana.

Martin, Michael W. Petrology of the lower and upper Sundance formations (Upper Jurassic), Wind River Basin, Wyoming. M, 1971, University of Missouri, Columbia.

Martin, Monty Gene. The structural conditions of the Crescent Pool. M, 1943, University of Oklahoma. 41 p.

Martin, Neill W. A gravimetric investigation of overburden depth and basement rock configuration at Crestmore, California. M, 1963, University of California, Riverside. 25 p.

Martin, Paul Felix. Glacial features of the township of Worcester, Massachusetts. M, 1941, Clark University.

Martin, Peter W. The geology of the Haystack Mountain area, northern Portneuf Range, Caribou County, Idaho. M, 1977, Idaho State University. 23 p.

Martin, Randolph J., III. Effect of pore pressure on the strength of low porosity crystalline rocks. M, 1968, Massachusetts Institute of Technology. 51 p.

Martin, Randolph J., III. Time-dependent crack growth in quartz and its applications to the creep of rocks. D, 1971, Massachusetts Institute of Technology. 158 p.

Martin, Ray Earl. Evaluation of the effectiveness of terbec soil stabilizer for use with West Virginia soils by the stabilometer method. D, 1971, West Virginia University.

Martin, Ray G. Geology of a portion of Franklin, Marion, and Winston counties, Alabama. M, 1965, University of Tennessee, Knoxville. 62 p.

Martin, Richard A. Geology of Devil's Racecourse boulderfield, Dauphin County, Pennsylvania. M, 1971, Millersville University.

Martin, Richard A. The effect of moisture on the compressive and tensile strength of a variety of rock materials. M, 1966, University of Missouri, Rolla.

Martin, Richard Harold. Geology of the Great Cacapon, West Virginia-Maryland, quadrangle. M, 1964, West Virginia University.

Martin, Richard Harold. The Recent delta of the Guadalupe River, Texas. D, 1967, West Virginia University. 147 p.

Martin, Richard Vernon. Petrology of the Carter Caves Sandstone, Mississippian, northeastern Kentucky. M, 1975, University of Cincinnati. 92 p.

Martin, Robert Allen. Fossil mammals of the Coleman IIA local fauna, Sumter County, Florida. D, 1969, University of Florida. 184 p.

Martin, Robert E. The ground water resources of Northeast Mississippi. M, 1955, Mississippi State University. 154 p.

Martin, Robert F. C. A chemical and petrographic study of the granite rocks of New Brunswick, Canada (Lenatuie). M, 1966, Pennsylvania State University, University Park. 120 p.

Martin, Robert Francois Churchill. Hydrothermal synthesis of low albite, orthoclase, and non-stoichiometric albite. D, 1969, Stanford University. 174 p.

Martin, Robert Joseph. The climate and physiographic aberrations of Washington, DC. M, 1936, George Washington University.

Martin, Robin. Structures in clayey sediments from a pre-Illinoian glacial lake, Claryville area, Campbell County, Kentucky. M, 1978, University of Cincinnati. 120 p.

Martin, Roger C. Vertical variations within some eastern Nevada ignimbrites. M, 1957, University of Idaho. 86 p.

Martin, Romeo Jarrett. The geology of the Thicketty Creek area, South Carolina. M, 1940, University of North Carolina, Chapel Hill. 22 p.

Martin, Ronald B. Relationship between quality of water in the Arbuckle Group (Upper Cambrian and Lower Ordovician) and major structural features in central and eastern Kansas. M, 1968, University of Kansas. 69 p.

Martin, Ronald Edward. Distribution and ecology of the foraminifera of John Pennekamp Coral Reef State Park, Key Largo, Florida, with emphasis upon the effects of turbid water produced by dredging. M, 1975, University of Florida. 205 p.

Martin, Sharon P. Trace elements in the Pennsylvanian-age Hushpuckney and Mecca Quarry shales, Mid-continent, U.S.A. M, 1982, University of Missouri, Kansas City. 76 p.

Martin, Sheila Shinn. Reexamination of the genus Ventilabrella Cushman (Upper Cretaceous). M, 1971, Southern Methodist University. 65 p.

Martin, Steven James. Chemical characteristics of dissolved organic matter in river water. M, 1973, Georgia Institute of Technology. 45 p.

Martin, Steven Lee. Hydrogeology of a dolomite aquifer and interaction with an overlying wetland in northeastern Waukesha County, Wisconsin. M, 1982, University of Wisconsin-Milwaukee. 188 p.

Martin, Viva Erma. Glaciation of the Two Medicine Valley, Glacier National Park, Montana. M, 1927, Indiana University, Bloomington. 81 p.

Martin, Walt M. The geochemistry and petrography of some veins and wallrocks hosting gold and silver in the Republic District, northeastern Washington. M, 1988, Eastern Washington University. 121 p.

Martin, Wayne Dudley. Petrology of the (Permian) upper Marietta and Hundred sandstones in southeastern Ohio. M, 1949, West Virginia University.

Martin, Wayne Dudley. The Hockingport Sandstone (Late Carboniferous) of southeastern Ohio. D, 1955, University of Cincinnati. 105 p.

Martin, William R. Transport of trace metals in nearshore sediments. D, 1985, Massachusetts Institute of Technology. 302 p.

Martin, William R., IV. A seismic refraction study of the northeastern Basin and Range and its transition with the eastern Snake River plain. M, 1978, University of Texas at El Paso.

Martindale, Kevin. Seismic wave generation and propagation from coal mine blasts at the Wright Mine, Warrick County, Indiana. M, 1982, Purdue University. 68 p.

Martindale, Robert David. The concentration and distribution of gold in the uraniferous conglomerates of Elliot Lake. M, 1968, McMaster University. 117 p.

Martindale, Steven. Stratigraphy and structural geology of the southern part of the Leach Range, Elko County, Nevada. M, 1981, San Jose State University. 117 p.

Martineau, Gismond. Géologie des terrasses fluviales du segment septentrional de la rivière Sainte Anne, Gaspésie. M, 1977, Universite de Montreal.

Martineau, Yvon Arthur. The relationships among rock groups between the Grand Lake Trust and Cabot Fault, West Newfoundland. M, 1980, Memorial University of Newfoundland. 150 p.

Martinek, Brian C. Geology and alteration along the margin of the Boulder Batholith, Hadley Park area, Jefferson County, Montana. M, 1985, Colorado School of Mines. 132 p.

Martinek, Charles Allen. Compressional wave speed and absorption measurements in a saturated kaolinite-water artificial sediment. M, 1972, United States Naval Academy.

Martinez D., Ignacio. Mexican lead-zinc mining industry. M, 1969, Stanford University.

Martinez Muller, Remigio. Economic geology of the Malpica Prospect, Sinaloa, Mexico. M, 1973, Colorado School of Mines. 115 p.

Martinez, Abraham. Foraminifera from the Pajuil Formation, lower Miocene, Colombia, South America. M, 1942, Columbia University, Teachers College.

Martinez, Fernando. On the transition from continental rifting to sea-floor spreading in the northern Red Sea. D, 1988, Columbia University, Teachers College. 251 p.

Martinez, Jaime Orlando. Neogene stratigraphy and sedimentary environments of Cumberland Island, Georgia. M, 1980, University of Georgia.

Martinez, Joseph Didier. The application of the photometer method in determining the crystallographic fabric of quartz in metamorphic quartzites. D, 1959, Louisiana State University. 126 p.

Martinez, Luis. Mineralization and hydrothermal alteration in porphyry copper deposit. M, 1968, Stanford University.

Martinez, Mario Luis Chavez see Chavez Martinez, Mario Luis

Martinez, Maximo. Geologic study of the Pirin area in the Departmento of Puno, Peru. M, 1968, Stanford University.

Martinez, Maximo. Geologic study of the Shasta Lake area, California. M, 1968, Stanford University.

Martinez, Norma. A regional subsurface study of the Austin Chalk, South Texas. M, 1982, Baylor University. 113 p.

Martinez, Paul Anthony. Simulation of sediment transport and deposition by waves; for simulation of wave-versus fluvial-dominated beach environments. M, 1987, Stanford University. 406 p.

Martinez, Paul Edwin. Conodont paleontology of the Lower Silurian (Llandovery) Brassfield Limestone, Independence, Izard, and Searcy counties, Arkansas. M, 1986, University of New Orleans. 118 p.

Martinez, Ricardo D. Tectonic synthesis of the Hat Six area, east end of Casper Mountain, Natrona County, Wyoming. M, 1988, University of Akron. 67 p.

Martinez, Rodolfo Ignacio. Simulation of floods in wide, flat overbank areas. D, 1982, Stanford University. 214 p.

Martinez, Ruben. Geology of the Pajarito Peak area, Sandoval County, New Mexico. M, 1974, University of New Mexico. 72 p.

Martinez, Ruben. Provenance study of the Westwater Canyon and Brushy Basin members of the Morrison Formation between Gallup and Laguna, New Mexico. D, 1979, University of New Mexico. 79 p.

Martinez-Garcia, Enrique. Geology of the Sharp Mountain area, Llano County, Texas. M, 1971, University of Texas, Austin.

Martini, Irendo Peter. The sedimentology of the Medina Formation outcropping along the Niagara escarpment (Ontario and New York State). D, 1966, McMaster University. 420 p.

Martini, Jose Alberto. Chemical, mineralogical and physical properties of seven surface soils from Panama with special reference to cation exchange capacity and potassium status. D, 1966, Cornell University. 199 p.

Martino, David S. A paleomagnetic study of some Texas intrusives. M, 1971, Texas Tech University. 53 p.

Martino, Ronald Layton. Sedimentology and paleoenvironments of the Maestrichtian Monmouth Group in the northern and central New Jersey coastal plain. M, 1976, Rutgers, The State University, New Brunswick. 94 p.

Martino, Ronald Layton. The sedimentology of the late Tertiary Bridgeton and Pensauken formations in southern New Jersey. D, 1981, Rutgers, The State University, New Brunswick. 285 p.

Martinoff, Alexander D. A systematic graphic investigation of the effect of convergence on the shape and position of anticlines. M, 1930, University of Pittsburgh.

Martins, Verónica E. de Sousa Carvalho see de Sousa Carvalho Martins, Verónica E.

Martinsen, Randi S. Geology of a part of the East Verde River canyon, near Payson, Arizona. M, 1975, Northern Arizona University. 117 p.

Martinson, Arthur David. Mountain in the sky; a history of Mount Rainier National Park. D, 1966, University of Washington. 173 p.

Martinson, Douglas George. An inverse approach to signal correlation with applications to deep-sea stratigraphy and chronostratigraphy. D, 1982, Columbia University, Teachers College. 351 p.

Martinson, Holly A. Channel changes on Powder River between Moorhead and Broadus, Montana. M, 1982, Colorado State University. 93 p.

Martinson, Timothy P. Geology of the Trail Creek area, Boulder Mountains, Blaine and Custer counties, Idaho. M, 1977, University of Wisconsin-Milwaukee.

Martison, N. W. Geology and ore deposits of Cochenour Williams gold mine. M, 1942, Queen's University. 59 p.

Martner, Samuel T. Geology of the Manila-Linwood area, Sweetwater County, Wyoming, and Daggett County, Utah. D, 1949, California Institute of Technology. 119 p.

Martner, Samuel T. Observations on seismic waves reflected at the Earth's core boundary. D, 1949, California Institute of Technology. 148 p.

Martorana, Anthony. Conodont biostratigraphy and paleoecology in the lower Mission Canyon Limestone, Williams County, North Dakota. M, 1988, Brigham Young University. 43 p.

Martt, Shirley Lee. Delineation of a sandstone channel within a coal seam using surface seismic reflection in Illinois. M, 1985, Wright State University. 114 p.

Marttila, Raymond K.   Fibrous minerals in the Wabush iron ore district, Labrador, Newfoundland, Canada. M, 1979, McMaster University. 128 p.

Martucci, Louis.   Determination of astro-geodetic vertical deflections for closely spaced points by transferring known astro-geodetic values through a gravimetric method. M, 1963, Ohio State University.

Marty, Richard Charles.   Formation and zonation of ferruginous bauxite deposits of the Chapman Quadrangle, Oregon. M, 1983, Portland State University. 127 p.

Martyn, Phillip F.   Oil and gas fields of the southern Gulf coastal plains of Texas. M, 1930, University of Missouri, Rolla.

Martz, Alan Matthew.   Petrology and chemistry of tuffs from the Shungura Formation, Southwest Ethiopia. M, 1979, University of Utah. 82 p.

Martz, Lawrence Wilfred.   Variability of net soil erosion and its association with topography in Canadian prairie agricultural landscapes. D, 1987, University of Saskatchewan.

Martz, Paul Warren.   The geology of a portion of the northern Warner mountains, Modoc County, California. M, 1970, University of California, Davis. 70 p.

Maruo, Jiro.   Mineral exploration in the Amazon region, Brazil. M, 1970, Stanford University.

Maruri, Raul D.   Geological investigations for dam foundations. M, 1977, University of Illinois, Urbana.

Marval, Francisco Rafael.   Simulated search strategies in oil exploration. M, 1977, Stanford University. 36 p.

Marvil, Joshua D.   Analysis and application of unsaturated zone instrumentation techniques and infiltration studies. M, 1987, University of Colorado. 122 p.

Marvin, Leslie Kenneth.   The geology of southwestern Howard County, Missouri. M, 1950, University of Missouri, Columbia.

Marvin, Mary Lewis.   Cone-in-cone. M, 1930, University of Kentucky.

Marvin, Peter R.   Regional heat flow based on the silica of groundwaters from north-central Mexico. M, 1984, New Mexico State University, Las Cruces.

Marvin, Raymond Glenn.   Jurassic stratigraphy of southwestern Colorado. M, 1953, University of Nebraska, Lincoln.

Marvin, Richard F.   Description and geologic history of selected areas in the vicinity of Gallatin Canyon, Gallatin County, Montana. M, 1952, Montana College of Mineral Science & Technology. 68 p.

Marvin, Theodore.   The small portable concentrating mill and its application to the Gothic, Colorado, mining district. M, 1923, Colorado School of Mines.

Marvin, Thomas Crockett.   The geology of the Hilton Ranch area, Pima County, Arizona. M, 1942, University of Arizona.

Marvin, Ursula Bailey.   Mineralogical studies of extraterrestrial materials. D, 1969, Harvard University.

Marvinney, Robert George.   Tectonic implications of stratigraphy, structure, and metamorphism in the Penobscot Lake region, northwestern Maine. D, 1986, Syracuse University. 306 p.

Marx, Archer H.   A study of malachite and azurite. M, 1929, University of Minnesota, Minneapolis. 43 p.

Marx, Pat (Washburn).   A dynamic model for an estuarine transgression based on facies variants in the nearshore of western Delaware Bay. M, 1982, University of Delaware.

Marz, Penny A.   Depositional history of the Mink River area, Door County, Wisconsin. M, 1986, University of Wisconsin-Milwaukee. 159 p.

Marzano, Michael S.   Rb-Sr whole rock geochronology of six rock suites from the Hopedale Block, Nain Province, Labrador, Canada. M, 1981, Miami University (Ohio). 103 p.

Marzke, Mary Ronald Walpole.   Evolution of the human hand. D, 1964, University of California, Berkeley. 289 p.

Marzolf, John E.   Evidence of changing environments during the depositional history of the Navajo Sand-

stone, Utah. D, 1970, University of California, Los Angeles.

Marzouki, F. M. H.   Petrogenesis of Al Hadah plutonic rocks, Kingdom of Saudi Arabia. D, 1977, University of Western Ontario. 267 p.

Mase, Charles W.   Geophysical study of the Monroe-Red Hill geothermal system. M, 1979, University of Utah. 89 p.

Mase, Charles William.   The effects of frictional heating on the thermal, hydrologic, and mechanical response of a fault. D, 1986, University of British Columbia.

Mase, Jack Edgar.   The study of a landslide in the Big Miami River valley near Miamitown, Ohio. M, 1952, University of Cincinnati. 29 p.

Mase, Russell Edwin.   The geology of the Indianola Embayment, Sanpete and Utah counties, Utah. M, 1957, Ohio State University.

Mashburn, Leslie Edwin.   Mineralized veins in the Franciscan Melange and Cambria Slab trench-slope basin, near San Simeon, California; a fluid inclusion analysis with implications for dewatering subducting and accreted sediments. M, 1986, University of Texas, Austin. 150 p.

Mashina, K. I.   An assessment of techniques for testing geothermal wells. M, 1975, University of Hawaii. 111 p.

Masias Echegaray, Juan A.   Morphology, shallow structure, and evolution of the Peruvian continental margin, 6° to 18°S. M, 1976, Oregon State University. 92 p.

Masiello, Remo Antonio.   A petrofabric study of the iron ore of Benson Mines, Star Lake, New York. M, 1967, University of Missouri, Rolla.

Masingill, John H., III.   Ground water of the Ripley Formation in Marengo and Wilcox counties, Alabama, with special reference to quality of water. M, 1978, University of Alabama.

Maske, N. D.   Geology of an area near Ingot, California. M, 1968, University of Hawaii.

Maslliwec, Anatolij.   Direct dating of ore minerals. M, 1981, University of Toronto.

Maslowski, Edith.   Pegmatites of the Batchellerville and Overlook area, southeastern Adirondacks, New York. M, 1944, Syracuse University.

Maslyn, Raymond M.   Late-Mississippian paleokarst in the Aspen, Colorado area. M, 1976, Colorado School of Mines. 96 p.

Mason, Arnold Caverly.   Geology of the Limestone Islands, Palau, western Caroline Islands. D, 1955, University of Illinois, Urbana. 180 p.

Mason, Arthur E.   Some foraminifera from the Terabratula bed of the (Eocene) Hornerstown Marl, New Egypt, New Jersey. M, 1933, Columbia University, Teachers College.

Mason, Charles Clifford.   The loess in St. Louis and St. Louis County, Missouri. M, 1928, Washington University. 59 p.

Mason, Chester Bowden.   The origin and history of Creve Coeur Lake (Saint Louis County, Missouri). M, 1971, Washington University. 57 p.

Mason, Clive S.   Geophysical investigations of universely magnetized rocks near Wilberforce, Ontario. M, 1959, University of Western Ontario.

Mason, Curtis.   Properties and stability of a Texas barrier beach inlet. M, 1971, Texas A&M University.

Mason, Curtis Calvin.   Sediments of Mustang Island, Texas. M, 1957, University of Texas, Austin.

Mason, David.   Pyritized microfossils from the Upper Devonian black shale of southwestern Ontario and southern Michigan. M, 1962, University of Western Ontario. 59 p.

Mason, Edward William.   Subsurface distribution and occurrence of Cherokee sandstones and their relationship to coal deposits at Madrid, Iowa. M, 1980, Iowa State University of Science and Technology.

Mason, Eric Paul.   The petrology, diagenesis and depositional environment of the Bartlesville Sandstone

in the Cushing oil field, Creek County, Oklahoma. M, 1982, Oklahoma State University. 150 p.

Mason, G. David.   Depositional environment of the Lorraine and Richmond Groups (Ordovician) in the Saint Lawrence Lowlands, P.Q. (Canada). M, 1967, McGill University. 69 p.

Mason, George David.   A stratigraphic and paleoenvironmental study of the upper Gaspe Limestone and lower Gaspe Sandstone groups, lower Devonian of eastern Gaspe Peninsula, Quebec. D, 1971, Carleton University. 194 p.

Mason, George William.   Interbasalt sediments of south central Washington. M, 1953, Washington State University. 116 p.

Mason, Glenn Michael.   Mineralogic aspects of stratigraphy and geochemistry of the Green River Formation, Wyoming. D, 1987, University of Wyoming. 390 p.

Mason, Herbert Louis.   A Pleistocene fauna from the Tomales Bay region and its bearing on the history of the coastal pine forests of California. D, 1932, University of California, Berkeley. 282 p.

Mason, Herbert Louis.   The genus Dodecatheon; a systematic and experimental study. M, 1923, University of California, Berkeley. 57 p.

Mason, Ian MacLean.   Petrology of the Whitestone anorthosite. D, 1969, McMaster University. 299 p.

Mason, James Leighton.   Condont biostratigraphy of the Lower Garden City Formation (Lower Ordovician) of northern Utah. M, 1975, University of Utah. 117 p.

Mason, James R., Jr.   Petrology of the Hampton Formation at Iowa Falls, Iowa. M, 1961, Iowa State University of Science and Technology.

Mason, James Trimble.   The geology of the Caballo Peak Quadrangle, Sierra County, New Mexico. M, 1976, University of New Mexico. 131 p.

Mason, John Frederick.   Contributions to the geology of the Cambrian in the southern Great Basin. D, 1941, Princeton University. 48 p.

Mason, John Frederick.   Paleontology and stratigraphy of the lower part of the Cambrian section of the Highland Range, Nevada. M, 1935, University of Southern California.

Mason, John M.   Hydrophilic organic compounds produced by underground coal gasification. M, 1984, University of Wyoming. 76 p.

Mason, Richard Harper.   The structure and stratigraphy of the Keefer Mountain area, Madison County, Arkansas. M, 1964, University of Arkansas, Fayetteville.

Mason, Robert Clifton.   Geology of the fire-clays in the lower Raritan River District, New Jersey. M, 1952, University of Missouri, Columbia.

Mason, Robert M.   The Afton soil developed in a fluvial sequence, in Doniphan Northeast Pit, Doniphan County, Kansas. M, 1971, University of Kansas. 66 p.

Mason, Robert Michael.   Landscape evolution of a portion of the Illinoian drift plain in central Illinois. D, 1973, University of Illinois, Urbana. 112 p.

Mason, Scott Edward.   Periodicities in color banding in sphalerite of the Upper Mississippi Valley district. M, 1987, Pennsylvania State University, University Park. 110 p.

Mason, Sharon A.   Seepage of groundwater into the St. Clair River near Sarnia, Ontario, Canada. M, 1987, University of Windsor. 187 p.

Mason, Shirley Lowell.   The geology of Asia Minor (Turkey in Asia) with special reference to Boyabad Ova and upper Tigris area. D, 1932, Harvard University.

Mason, Thomas Paxton.   Stratigraphic study of the Weber Formation (Permo-Pennsylvanian), northwest Colorado. M, 1958, University of Colorado.

Masood, Hamid.   Clay mineralogy and sedimentary environments of Eureka Sound Formation, Ellesmere Island, Canadian Arctic Archipelago. D, 1987, Purdue University. 116 p.

Masood, Hamid. Stratigraphy of Hothla Group (Jur. Cret.) of Hazara (Pakistan) with special emphasis on the Upper Cretaceous. M, 1984, Purdue University.

Masri, Fahad Isa. The electro-osmotic permeability of tailings from an Arizona porphyry copper mine. M, 1973, University of Arizona.

Masrous, Luis Felipe. Patterns of pressure in the Morrow sands (Lower Pennsylvanian) of central Oklahoma. M, 1973, University of Tulsa. 78 p.

Masry, Alaa Eldin Mohsen El see El Masry, Alaa Eldin Mohsen

Massa, Audrey Adams. The origin of sandy shore-parallel ridges and the formation of modern spits; examples from the western end of Fire Island, New York. D, 1988, Columbia University, Teachers College. 505 p.

Massa, Philip Joseph. Abundance and behavior of Tl, Rb, K, Ba, and Sr in gold-silver veins and associated volcanic rocks from the Como mining district, Lyon County, Nevada. M, 1984, Eastern Washington University. 96 p.

Massa, Vito, Jr. A geophysical and geological investigation of the Edison copper mine area, Edison, New Jersey. M, 1979, Rutgers, The State University, New Brunswick. 33 p.

Massad, Marilyn L. Anomalous remanent magnetization of deep-sea sediments in the Venezuelan Basin caused by self-reversal of precipitated magnetite. M, 1982, University of Georgia.

Massanat, Yousef M. Compressibility and rebound characteristics of compacted clays. D, 1973, University of Arizona.

Massanat, Yousif M. Evaluation of factors contributing to piping erosion. M, 1972, University of Arizona.

Massaquoi, Joseph George Momodu. Combustion in underground coal conversion. D, 1981, West Virginia University. 226 p.

Massar, Bruce Allen. Comparison of the relationships of the Hamilton to the Teays drainage systems using a gravity study. M, 1975, Wright State University. 116 p.

Massare, Judy Ann. Ecology and evolution of Mesozoic marine reptiles. D, 1984, The Johns Hopkins University. 200 p.

Massaro, David A. An interpretation of the glacial outwash terraces of Southeast Stark and Northeast Tuscarawas counties, Ohio. M, 1975, University of Akron. 85 p.

Masse, Ann Katherine. Estuary-shelf interaction; Delaware Bay and the inner shelf. D, 1988, University of Delaware, College of Marine Studies. 232 p.

Masse, Lucien. Investigation concerning the suitability of an induction varimeter as a portable magnetometer. M, 1938, Colorado School of Mines. 45 p.

Masse, Lucien. Theoretical studies in the magnetic field of some inductive systems. D, 1940, Colorado School of Mines. 119 p.

Masse, Robert Patrick. Seismic studies of the Earth's mantle; A: an investigation of the upper mantle compressional velocity distribution beneath the basin and range province; B: effects of observational errors on the resolution of surface waves at intermediate distances; C: contributions of theoretical seismograms to the study of modes, rays, and the Earth. D, 1972, Southern Methodist University. 73 p.

Massell, Wulf Friedrich. Inverse filtering of digitally recorded seismic refraction records. D, 1973, Indiana University, Bloomington. 163 p.

Massey, Kathleen Willis. Rubidium-strontium geochronology and petrography of the Hammamat Formation in the northeastern desert of Egypt. M, 1984, University of Texas at Dallas. 75 p.

Massey, Nicholas William David. The geochemistry of some Keweenawan metabasites from Mamainse Point, Ontario. D, 1980, McMaster University. 353 p.

Massey-Norton, John T. The alteration of the natural radiation of the Hocking River basin. M, 1980, Ohio University, Athens. 157 p.

Massie, L. R. An upland erosion model. D, 1975, University of Wisconsin-Madison. 294 p.

Massingill, G. L. Geology of Riley-Puertecito area, southeastern margin of Colorado Plateau, Socorro County, New Mexico. D, 1979, University of Texas at El Paso. 336 p.

Massingill, Gary L. Geology of the calcium carbonate materials used for manufacture of Portland cement, Bushland, Texas. M, 1975, West Texas State University. 39 p.

Massion, Peter J. Liquid immiscibility in silicate melts. M, 1974, University of Illinois, Chicago.

Massmann, Joel Warren. Groundwater contamination from waste-management sites; the interaction between risk-based engineering design and regulatory policy. D, 1987, University of British Columbia.

Massmann, Thomas A. The theory of Earth tides and the application of their measurement in Ohio. M, 1969, Ohio State University.

Masson, Arthur Guy. Deposition and diagenesis of the Salter Member, Lower Mount Head Formation, Southwest Alberta. M, 1978, University of Manitoba. 97 p.

Masson, Arthur Guy. The sedimentology of the upper Morien Group (Pennsylvanian) in the Sydney Basin east of Sydney Harbour, Cape Breton, Nova Scotia. D, 1986, University of Ottawa. 343 p.

Masson, Peter H. Geology of the Gunsight Peak District, Siskiyou County, California. M, 1949, University of California, Berkeley. 75 p.

Massoth, Terry Wayne. Depositional environments of some Upper Cretaceous coal-bearing strata at Trapper Mine, Craig, Colorado. M, 1982, University of Utah. 123 p.

Massoudi, Heidargholi. Hydraulics of river bed degradation, Willow Creek, Iowa. D, 1981, Iowa State University of Science and Technology. 237 p.

Massoudi, Nasser. Undrained behavior of fully and nearly saturated sand at high stresses under multiaxial loading conditions. D, 1988, University of Colorado. 826 p.

Mast, V. A. Distribution and engineering properties of landslides susceptible soils in Southeast Ohio. D, 1980, Ohio State University. 223 p.

Masten, Douglas Everett. A study of experimental faulting in granular materials. M, 1953, University of Michigan.

Master, Pilsum Phiroze. Relationship of pleochroism and specific refractivity of cupric compounds to cupric ion coordination. M, 1968, University of New Mexico. 68 p.

Master, Timothy D. Geological and geochemical evaluation of rock and mineral occurrences in the Precambrian of the Ferris Mountains, Carbon County, Wyoming. M, 1977, University of Wyoming. 85 p.

Masters, Bruce Allen. Eocene foraminifera from the Church Creek area, Santa Lucia Mountains, Monterey County, California. M, 1962, University of California, Berkeley. 94 p.

Masters, Bruce Allen. Stratigraphic and planktonic foraminifera of the upper Cretaceous Selma group, Alabama. D, 1970, University of Illinois, Urbana. 379 p.

Masters, Charles Day. Sedimentology of the Mesa Verde Group and of the upper part of the Mancos Formation (Upper Cretaceous), northwestern Colorado. D, 1966, Yale University. 329 p.

Masters, Charles Day. Structural geology of the Rabbit Mountain-Dowe Pass area; Boulder and Larimer counties, Colorado. M, 1957, University of Colorado.

Masters, John A. Frontier Formation (Wyoming). M, 1951, University of Colorado.

Masters, John Michael. Zircons in two Laramide porphyry dikes, Beartooth Mountains, Montana. M, 1966, University of Cincinnati. 52 p.

Masters, Kenneth E. Geology of the Prague area, Lincoln and Pottawatomie counties, Oklahoma. M, 1955, University of Oklahoma. 37 p.

Masterson, James Arthur. Some variations of large Cyrtina alpensis Hall and Clarke. M, 1951, University of Michigan.

Masterson, Robert P., Jr. Sediment movement in Tubbs Inlet, North Carolina (Brunswick County, Recent). M, 1973, North Carolina State University. 108 p.

Masterson, Tina K. Anhydrite precipitation in the presence of organic additives. M, 1983, Iowa State University of Science and Technology. 53 p.

Masterson, Wilmer Dallam, IV. Epithermal gold mineralization in the Velvet District, Pershing County, Nevada. M, 1981, University of Texas, Austin. 70 p.

Mastin, Gary Arthur. Computer analysis of coastal ocean features in synthetic aperture radar imagery. D, 1983, Louisiana State University. 255 p.

Mastin, Larry Garver. Stress, surface deformation, and phreatic eruptions above a shallow dike, Inyo Craters, Long Valley Caldera, California. D, 1988, Stanford University. 197 p.

Mastin, Larry Garver. The development of borehole breakouts in sandstone. M, 1984, Stanford University. 118 p.

Masuoka, Edward Jay. Modelling an Ordovician marine community succession. M, 1978, University of Tennessee, Knoxville. 149 p.

Masuoka, Penny McFarlan. Analysis of photogeologic fracture traces and lineaments in the northern portion of the Wartburg Basin, Tennessee. M, 1981, University of Tennessee, Knoxville. 91 p.

Masursky, Harold. Geology of the western Owl Creek Mountains, Wyoming. D, 1956, Yale University. 214 p.

Mat, Bruce T. Magnetic properties of rocks associated with the New Cornelia porphyry copper deposit (early Tertiary-Laramide), Pima County, Arizona. D, 1968, University of Arizona.

Matalucci, Rudolph Vincent. The microstructure of loess and its relationship to engineering properties. D, 1969, [University of Oklahoma]. 155 p.

Matarese, Joseph Richard. Topographic reconstruction from radar imagery. M, 1988, Massachusetts Institute of Technology. 121 p.

Matava, Timothy. Settlement of thawing subsea permafrost at Prudhoe Bay, Alaska. M, 1986, University of Alaska, Fairbanks. 115 p.

Mateker, Emil Joseph, Jr. Generation of Rayleigh waves by contained explosions. D, 1964, St. Louis University. 199 p.

Mateker, Emil Joseph, Jr. Some gravity and magnetic interpretation problems in eastern Missouri and vicinity. M, 1959, St. Louis University.

Matel, Joel E. Sedimentology of the Bliss Sandstone, Franklin Mountains, Texas. M, 1982, University of Texas at El Paso.

Matesich, Charles O. Saratoga foraminifera from the type locality in southwestern Arkansas. M, 1957, University of Oklahoma. 263 p.

Matheney, Ronald K. Rb-Sr geochronologic studies of plutonic rocks, Florida Mountains, New Mexico. M, 1984, University of New Mexico. 141 p.

Matheny, James Paul. Sedimentology and depositional environments of the Emery Sandstone, Emery and Sevier counties, Utah. M, 1982, University of Utah. 100 p.

Mather, Gordon Scott. Cobalt; key aspects of a critical and strategic metal. M, 1986, University of Texas, Austin.

Mather, Kirtley Fletcher. The fauna of the (Pennsylvanian) Morrow Group of Arkansas and Oklahoma. D, 1915, University of Chicago. 225 p.

Mather, Terry James. Stratigraphy and paleontology of the Permian Kaibab Formation, Mogollon Rim region (east central), Arizona. D, 1970, University of Colorado. 187 p.

Mather, Thomas T. The deep-sea sediments of the Drake Passage and Scotia Sea. M, 1966, Florida State University.

Mather, William B. The geology and paragenesis of the gold ores of the Howey Mine, Red Lake, Ontario, Canada. D, 1936, University of Chicago. 48 p.

Matherne, Anne Marie. Paleoceanography of the Gulf of California; a 350-year diatom record. M, 1982, Oregon State University. 111 p.

Mathers, Lewis John. Flume study of boulder-strewn streams under flood discharge conditions using geometric roughness elements in transverse rows. D, 1966, Pennsylvania State University, University Park. 172 p.

Matheson, Archie Farquhar. Michipicoten River area, District of Algoma, Ontario, Canada. D, 1932, University of Minnesota, Minneapolis. 96 p.

Matheson, Gordon M. Geologic and mining factors that control subsidence above abandoned underground coal mines along the Colorado Front Range. D, 1987, Colorado School of Mines. 312 p.

Matheson, Marion Henderson. Some effects of coal mining upon the development of the Nanaimo area. M, 1950, University of British Columbia.

Mathew, David. Geology of the Beckley coal seam in the Eccles no. 5 Mine near Eccles, West Virginia. D, 1977, University of South Carolina. 87 p.

Mathews, Asa A. Lee. Marine Lower Triassic beds of Utah. M, 1924, Stanford University. 44 p.

Mathews, Asa A. Lee. The Lower Triassic cephalopod fauna of the Fort Douglas area, Utah. D, 1929, University of Chicago. 46 p.

Mathews, David Lane. The clay mineralogy of Mojave Desert playas. M, 1960, Indiana University, Bloomington. 23 p.

Mathews, Edward B. The granites and derived gneisses of the Pikes Peak folio of the geologic atlas of the United States. D, 1894, The Johns Hopkins University.

Mathews, Frank Samuel. The electrical conductivity of Atlantic type pyromagmas from Mount Etna, Sicily. D, 1969, Oregon State University. 131 p.

Mathews, Geoffrey William. The petrology and geochemical variability of the Audubon-Albion stock, Boulder County, Colorado. D, 1970, Case Western Reserve University. 272 p.

Mathews, Haywood. Primary production measurements on an artificial reef. M, 1966, Florida State University.

Mathews, James Clay. The depositional environment and diagenesis of the lower Tuscaloosa Formation, Newtonia Field, Wilkinson County, Mississippi. M, 1987, Northeast Louisiana University. 91 p.

Mathews, Jane E. Mineral content and petrography of the Springfield (V) Coal in southwestern Indiana. M, 1981, Indiana University, Bloomington. 81 p.

Mathews, Mark A. Electromagnetic sounding on the ocean floor. M, 1970, Colorado School of Mines. 135 p.

Mathews, Melissa J. The formation of sulfides in coastal marine sediments. M, 1982, Pennsylvania State University, University Park. 70 p.

Mathews, Thomas Delbert. The contemporary geochemistry of radiocarbon in the Gulf of Mexico. D, 1972, Texas A&M University. 125 p.

Mathews, Wilbert L. Depositional environment of the Plattsmouth limestone member of the Oread Formation (Univerity of Pennsylvania) in northeastern Kansas. M, 1978, University of Kansas. 47 p.

Mathews, William Henry. Geology of the Ironmask Batholith. M, 1941, University of British Columbia.

Mathews, William Henry. Geology of the Mount Garibaldi map area, southwestern British Columbia. D, 1948, University of California, Berkeley. 229 p.

Mathewson, Christopher C. Engineering analysis of subaerial and submarine geomorphology along the north coast of Molokai Island, Hawaii. D, 1971, University of Arizona. 123 p.

Mathewson, Christopher C. The engineering significance of some sediments from the Hudson submarine canyon region southeast of Long Island, New York. M, 1966, University of Arizona.

Mathewson, David C. Structures related to the Osburn Fault, Coeur d'Alene mining district, Idaho. M, 1972, University of Idaho. 57 p.

Mathey, Bernard. Etude sédimentologique du flysch de la région de Saint-Simeon de Rimouski. M, 1970, Universite de Montreal.

Mathez, Edmond Albigese. Geology and petrology of the Logan intrusives (Precambrian) of the Hungry Jack Lake Quadrangle, Cook County, Minnesota. M, 1971, University of Arizona.

Mathez, Edmond Albigese. The geochemistry of sulfur and carbon in basaltic melts. D, 1981, University of Washington. 133 p.

Mathias, David L., Jr. Petrography and structural significance of the mafic igneous rocks in the Ishpeming-Negaunee area, Marquette Iron Range, Michigan. D, 1959, Columbia University, Teachers College. 124 p.

Mathias, Donald Ernest. The geology of the northern part of the Elk Mountains, Elko County, Nevada. M, 1959, University of Oregon. 76 p.

Mathias, Henry Edwin. Pyrite concretions in the Pennsylvanian shales. M, 1924, University of Missouri, Columbia.

Mathias, Kenneth E. Shallow seismic survey of the Kansas State University campus area. M, 1972, Kansas State University. 33 p.

Mathias, William F. Detrital mineral studies of some Cenozoic sediments, Safford Valley, Arizona. M, 1960, University of Arizona.

Mathieson, Elizabeth Lincoln. Late Quaternary activity of the Madison Range Fault along its 1959 rupture trace, Madison County, Montana. M, 1983, Stanford University. 169 p.

Mathieson, Gillian Ann. The Cantung E-zone scheelite skarn orebody, Tungsten, N.W.T., Canada; a revised genetic model. M, 1982, Queen's University. 156 p.

Mathieson, Neil Alexander. Geology and mineralisation in the area of the east south "C" ore zone, Dickenson Mine, Red Lake, northwestern Ontario. M, 1982, Queen's University. 155 p.

Mathieson, Scott Alan. Pre- and post-Sangamon glacial history of a portion of Sierra and Plumas counties, California. M, 1981, California State University, Hayward. 258 p.

Mathieu, R. J. The morphometrics of glacially derived lakes and swamps. D, 1977, University of Georgia. 189 p.

Mathiot, R. K. A preliminary investigation of the land use limitations of the major landforms along a portion of the Lincoln County coast, Oregon. M, 1973, Portland State University. 83 p.

Mathis, Harry Leon, Jr. Structural and sedimentological study of the accumulation of the No. 9 Coal to No. 13 Coal zone interval in the western Kentucky coal field. M, 1983, University of Kentucky. 103 p.

Mathis, Robert L. Carbonate sedimentation and diagenesis of reef and associated shoal-water facies, Sligo Formation (Aptian), Black Lake field, Natchitoches Parish, Louisiana. M, 1978, Rensselaer Polytechnic Institute. 214 p.

Mathis, Robert Warren. Heavy minerals of Colorado River terraces. M, 1942, University of Texas, Austin.

Mathisen, M. E. A provenance of environment analysis of the Plio-Pleistocene sediments in the East Turkana Basin, Lake Turkana, Kenya. M, 1977, Iowa State University of Science and Technology.

Mathisen, Mark Evan. Plio-Pleistocene geology of the central Cagayan Valley, northern Luzon, Philippines. D, 1981, Iowa State University of Science and Technology. 218 p.

Mathison, Joseph Edward. Stratigraphy and ostracodes of the Manitoba and Saskatchewan groups of the northern Williston Basin. M, 1983, University of Saskatchewan. 236 p.

Mathisrud, Gordon C. The magnetic separation of minerals. M, 1939, University of Minnesota, Minneapolis. 30 p.

Mathur, Jagdish Narain. Analysis of the response of earth dams to earthquakes. D, 1969, University of California, Berkeley. 172 p.

Mathur, Shashi. One-dimensional consolidation of saturated-unsaturated compressible porous media with variable total stress. D, 1984, University of Delaware, College of Marine Studies. 353 p.

Mathur, Surendra Pratap. Recognition of seismic pulses by studies of their frequency spectra. M, 1956, Pennsylvania State University, University Park. 44 p.

Mathur, Surendra Pratap. Standardization of gravity and bouguer anomalies in India. D, 1969, University of Hawaii. 62 p.

Mathur, Uday Prakash. Study of the continental structure of Southeastern United States by dispersion of Rayleigh waves. M, 1971, Georgia Institute of Technology. 167 p.

Matis, John P. Hydrogeology of the Sells area, Papago Indian Reservation (Pima county) Arizona. M, 1970, University of Arizona.

Matisoff, G. Early diagenesis of Chesapeake Bay sediments; a time series study of temperature, chloride, and silica. D, 1978, The Johns Hopkins University. 235 p.

Matlack, Keith S. The emplacement of clays in sand by infiltration. M, 1986, University of Missouri, Columbia.

Matlack, William Fuller. Geology and sulfide mineralization of Duluth Complex-Virginia Formation contact, Minnamax Deposit, St. Louis County, Minnesota. M, 1980, University of Minnesota, Duluth.

Matlick, Joseph S., III. A new geothermal exploration method using mercury. M, 1975, Arizona State University. 83 p.

Matlock, Joseph Franklin. Internal sediments within ore bodies of the Mascot-Jefferson City zinc district, East Tennessee. M, 1987, University of Tennessee, Knoxville. 125 p.

Matlock, William Gerald. The effect of silt-laden water on infiltration in alluvial channels. D, 1965, University of Arizona. 112 p.

Matondo, Jonathan Ihoyelo. A methodology for risk analysis of on-farm water management under stochastic irrigation water supply. D, 1983, Colorado State University. 107 p.

Matos, Jose Francisco. Deep structural geology (Paleozoic) of Denton, Wise, and eastern Jack counties, Texas. M, 1971, University of Texas, Austin.

Matos, Milton Martins de *see* de Matos, Milton Martins

Matsch, Charles L. Pleistocene geology of the St. Paul Park and Prescott quadrangles (Minn.). M, 1962, University of Minnesota, Minneapolis. 49 p.

Matsch, Charles Leo. Pleistocene stratigraphy of the New Ulm region, southwestern Minnesota. D, 1971, University of Wisconsin-Madison. 91 p.

Matson, Adrian L. Dissolved silicate in waters offshore Oregon and in four adjacent rivers. M, 1964, Oregon State University. 98 p.

Matson, Dean W. A mass spectrometric investigation of volatiles in mantle-derived amphiboles and micas and a Raman spectroscopic study of silicate glass structures. D, 1984, University of Hawaii. 317 p.

Matson, Ernest Augustus. The use of tetrazolium salts in studies of carbon and energy dynamics in anoxic sediments. D, 1982, University of Connecticut. 95 p.

Matson, George C. A contribution to the study of the interglacial gorge problem. M, 1903, Cornell University.

Matson, George Charlton. Phosphate deposits of Florida. D, 1920, University of Chicago. 98 p.

Matson, Neal A., Jr. Geology of the Royal Mountain and Chief Mountain area (Precambrian), Summit County, Colorado. M, 1967, University of Colorado.

Matson, Robert Ernest. Petrography and petrology of Smoky Butte intrusives, Garfield County, Montana. M, 1960, University of Montana. 74 p.

Matsui, Kunio. Pressure and stress distribution around a pipeline buried in a poro-elastic seabed. D, 1982, [University of Houston]. 124 p.

Matsui, T. Elasticity of single-crystal Fe-Si alloys as function of temperature and pressure. M, 1976, University of Hawaii. 102 p.

Matsusaka, Y. Dehydration curves and differential thermal curves of clays from Hawaiian soils. M, 1952, University of Hawaii. 72 p.

Matsutsuyu, Bruce Akira. Geology of the southern halves of the Crocker Mountain and Dixie Mountain 7.5' quadrangles, Plumas County, California. M, 1979, University of California, Davis. 169 p.

Matt, C. Diane. Quaternary history of the Ghost River area, Alberta. M, 1975, University of Calgary. 87 p.

Matten, Lawrence Charles. Contributions to the Upper Devonian flora of New York. D, 1965, Cornell University. 155 p.

Matteo, Brett. Paleoenvironments and depositional history of the Hartshorne Formation in East-central Oklahoma. M, 1981, University of Missouri, Columbia.

Matter, Conrad F. Separation, analysis and test maceral constituents from selected samples of Hornshaw bed coal, Wharton No. 2 Mine, Boone County, West Virginia. M, 1957, University of Pittsburgh.

Matter, Philip, III. Petrochemical variations across some Arizona pegmatites and their enclosing rocks (Santa Catalina and Tortolita mountains in southern Arizona and Wickenburg mountains in central Arizona). D, 1969, University of Arizona. 205 p.

Mattern, Joel K. Comparative study of coal reserve estimation methods. M, 1980, Southern Illinois University, Carbondale. 75 p.

Matters, Seth Eugene. Drift aquifer geometry and interconnection defined by numerical simulation. M, 1983, Pennsylvania State University, University Park. 159 p.

Mattes, B. W. Diagenetic history of the Upper Devonian Miette carbonate buildup, Jasper National Park, Alberta; with an emphasis on dolomitization. M, 1980, McGill University. 322 p.

Matteson, Charles. Geology of the Slate Creek area, Mount Hayes A-2 Quadrangle, Alaska. M, 1973, University of Alaska, Fairbanks. 66 p.

Matteson, Jane S. The stratigraphy of the (Pennsylvanian) Minnelusa Formation (South Dakota, Wyoming). M, 1939, Smith College. 117 p.

Matteucci, Thomas D. High resolution seismic stratigraphy of the North Carolina continental margin; the Cape Fear region; sea level cyclicity, paleobathymetry, and Gulf Stream dynamics. M, 1984, University of South Florida, St. Petersburg.

Matthew, William D. Effusive and dyke rocks near St. John, New Brunswick. D, 1895, Columbia University, Teachers College.

Matthews, Burton Clare. The fixation and release of soil potassium. D, 1952, Cornell University.

Matthews, Claude W. The mineralogy of some soils and shales from Saline County, Kansas. M, 1949, Kansas State University. 79 p.

Matthews, Curtis B., III. Geology of the central Vanocker Laccolith area, Meade County, South Dakota. M, 1979, South Dakota School of Mines & Technology.

Matthews, James Coert. Nature and origin of potassic alteration of the Ordovician Spechts Ferry Formation in the Upper Mississippi Valley. M, 1988, University of Illinois, Urbana. 152 p.

Matthews, James Emory. Geology of the northeastern Caribbean region. M, 1970, University of Utah. 84 p.

Matthews, Jerry L. Fusulinid species associated with the Pennsylvanian-Permian contact of the Manhattan, Kansas, area. M, 1959, Kansas State University. 66 p.

Matthews, Jerry Lee. Sedimentation of the coastal dunes at Oceano, California. D, 1966, University of California, Los Angeles. 138 p.

Matthews, John Joseph. Paleozoic stratigraphy and structural geology of the Wheeler Ridge area, northwestern Mohave County, Arizona. M, 1976, Northern Arizona University. 145 p.

Matthews, John V., Jr. A paleoenvironmental analysis of the three late Pleistocene coleopterous assemblages from Fairbanks, Alaska. M, 1968, University of Alaska, Fairbanks. 59 p.

Matthews, John V., Jr. Quaternary environments at Cape Deceit (Seward Peninsula, Alaska); a study of tundra ecosystem evolution. D, 1973, University of Alberta. 141 p.

Matthews, L. A gamma-ray spectrometer survey of two British Columbia porphyry copper-molybdenum deposits. M, 1977, University of Western Ontario.

Matthews, Larry Edwin. A study of the structure of the Ridley Limestones (Middle Ordovician) in the Gladeville Quadrangle in central Tennessee. M, 1971, Vanderbilt University.

Matthews, Leo Gerard. Evidence for an offset crustal block in the Southern Appalachians. M, 1982, Pennsylvania State University, University Park. 86 p.

Matthews, Martin David. Flocculation as exemplified in the turbidity maximum of Acharon Channel, Yukon River Delta, Alaska. D, 1973, Northwestern University.

Matthews, Martin David. Petrology, stratigraphy and structure of the (Mississippian) Greenbrier Limestone in a part of Wayne County, West Virginia, and its relationship to production of gas. M, 1963, West Virginia University.

Matthews, Neffra Alice. A petrographic and chemical study of a late Precambrian dike swarm in southeastern New Brunswick, Canada. M, 1985, University of Kentucky. 116 p.

Matthews, Robert Norman. Copper, zinc and lead distribution in the rocks near Revelstoke, B.C. M, 1978, University of Manitoba. 66 p.

Matthews, Robley Knight. Continuous seismic profiles of a shelf-edge bathymetric prominence in northern Gulf of Mexico. M, 1963, Rice University. 21 p.

Matthews, Robley Knight. Genesis of Recent lime mud in southern British Honduras. D, 1965, Rice University. 139 p.

Matthews, Ross Butler. Aspect of stratigraphy, diagenesis, and potential reservoir characteristics of the Monterey Formation in the Point Arguello area, California. M, 1984, Texas Christian University. 110 p.

Matthews, Roy Edgar. Subsurface geology of a part of the north flank of the Wichita Mountains near Hobart, Oklahoma, and southern Union and Columbia counties, Oklahoma. M, 1956, [University of Oklahoma].

Matthews, Steven N. Thermal maturity of Carboniferous strata, Ouachita thrust fault belt. M, 1982, University of Missouri, Columbia.

Matthews, Truitt F. The petroleum potential of "serpentine plugs" and associated rocks, central and South Texas. M, 1985, Baylor University. 150 p.

Matthews, Vincent, III. Geology and petrology of the pegmatite district in southwestern Jasper County, Georgia. M, 1967, University of Georgia. 68 p.

Matthews, Vincent, III. Geology of the Pinnacles Volcanic Formation and the Neenach Volcanic Formation and their bearing on the San Andreas Fault problem. D, 1973, University of California, Santa Cruz.

Matthews, Wilfred J. Geology of the Cameron Lake property, Quebec. M, 1952, Michigan Technological University. 79 p.

Matthews, William Henry, III. Some aspects of reef paleontology and lithology in the Edwards Formation of Texas. M, 1949, Texas Christian University. 94 p.

Matthews, William K., III. Potassium-argon age and petrography of the Sierra Blanca Peak igneous intrusives, Hudspeth County, Texas. M, 1983, Rice University. 32 p.

Matthias, Cheryl Louise. A gas chromatographic determination of tributyltin species in estuarine water and sediment using hydride derivatization and flame photometric detection. D, 1988, University of Maryland. 190 p.

Matti, J. C. Depositional history of middle Paleozoic carbonate rocks deposited at an ancient continental margin, central Nevada. D, 1979, Stanford University. 526 p.

Matti, Jonathan C. Stratigraphy and conodont biostratigraphy of Lower Devonian limestones, Copenhagen Canyon, Nevada. M, 1971, University of California, Riverside. 141 p.

Mattigod, S. V. Muscovite-gibbsite equilibrium solubility and the $K_2O-Al_2O_3-SiO_2-H_2O$ system at $25°C$ and 1 atmosphere. D, 1976, Washington State University. 107 p.

Mattinson, Cyril R. A study of certain Canadian building and monumental stones of igneous origin. M, 1952, McGill University.

Mattinson, Cyril R. The geology of the Mount Logan area, Gaspé, Quebec. M, 1958, McGill University.

Mattinson, James Meikle. Uranium-lead geochronology of the northern Cascade mountains, Washington. D, 1970, University of California, Santa Barbara. 80 p.

Mattioli, Glen Steven. Activities and volumes in $MgAl_2O_4-Fe_3O_4-\gamma Fe_{8/3}O_4$ spinels and their implications for upper mantle oxygen fugacity. D, 1987, Northwestern University. 139 p.

Mattis, Allan. The petrology and sedimentation of the basal Keweenawan sandstones (Precambrian) of the north and south shores of Lake Superior. M, 1971, University of Minnesota, Duluth.

Mattis, Allen Francis. Nonmarine Triassic sedimentation, central High Atlas Mountains, Morocco. D, 1975, Rutgers, The State University, New Brunswick. 75 p.

Mattison, George David. The chemistry, mineralogy and petrography of the Pine Valley Mountains, southwestern Utah. D, 1972, Pennsylvania State University, University Park. 149 p.

Mattison, Willie W. Geology of the western half of the Church Road Quadrangle, southeastern Virginia. M, 1971, Virginia State University. 46 p.

Mattox, Richard B. A discussion and method of study of the porosity of carbonate reservoir rock. M, 1954, Miami University (Ohio). 26 p.

Mattox, Richard Benjamin. A study of sand dunes. D, 1954, University of Iowa. 149 p.

Mattox, Robert M. A study of consolidation and swell characteristics of undisturbed Mississippi clays. M, 1968, Mississippi State University. 120 p.

Mattox, W. A. Relative heat flow mapping by means of horizontal gradiometry. M, 1974, San Diego State University.

Mattraw, Harold Claude, Jr. Paleoenvironment of a Pennsylvanian age bioherm, San Juan Mountains, Colorado. M, 1969, Washington State University. 78 p.

Mattson, John Lyle. A contribution to Skagit prehistory. M, 1971, Washington State University. 205 p.

Mattson, Louis Arthur. Structure and stratigraphy of the Thomson Formation, Carlton-Thomson area, Carlton County, Minnesota. M, 1959, University of Minnesota, Minneapolis. 96 p.

Mattson, Peter Humphrey. Geology of the Mayaguez area, Puerto Rico. D, 1957, Princeton University. 170 p.

Mattson, Stephanie Margaret. Optical expressions of ion-pair interactions in minerals. D, 1985, California Institute of Technology. 176 p.

Mattson, Steven R. Magma mixing in Iceland. D, 1984, Michigan State University. 144 p.

Mattson, Steven R. The Austerhorn and Vesturhorn acidic and basic complexes in southeastern Iceland; examples of magma mixing. M, 1981, Michigan State University. 88 p.

Matty, David Joseph. Petrology of deep crustal xenoliths from the eastern Snake River plain, Idaho. D, 1984, Rice University. 284 p.

Matty, David Joseph. The geology and geochemistry of the North Fork Stock, northeastern Oregon. M, 1979, Portland State University. 145 p.

Matty, Jane Miller. Sediments and sedimentary processes in Lake Houston, Texas. M, 1984, Rice University. 124 p.

Matulevich, Myrna Rae Monk. External inversion of seismic reflection data from the 1986 PASSCAL Basin and Range seismic experiment. M, 1988, University of Utah. 180 p.

Matus, I. The geology of the lower French Creek area, Carbon County, Wyoming. M, 1958, University of Wyoming. 38 p.

Matuszak, David R. Reduction of electric log data for stratigraphic and trend analysis of Pennsylvanian rocks, Carter County, Oklahoma. D, 1961, Northwestern University.

Matuszak, David Robert. Trilobites from the Fort Sill Formation (Upper Cambrian). M, 1957, University of Oklahoma. 55 p.

Matuszczak, Roger A. Studies of insoluble residues of Cambro-Ordovician contact in western Wisconsin. M, 1951, University of Wisconsin-Madison.

Matz, David B. Stratigraphy and petrography of the Beacon Sandstone, southern Victoria Land, Antarctica. M, 1968, University of Massachusetts. 110 p.

Matzen, Thomas A. Pelletal lithofacies of the Simpson Group of south-central Kansas. M, 1985, Wichita State University. 116 p.

Matzke, Richard H. Fracture trace and joint patterns in western Centre County, Pennsylvania. M, 1961, Pennsylvania State University, University Park. 39 p.

Matzko, John Rodney. Geologic interpretation of Landsat and RB-57 imagery of the Hot Springs area, Black Hills, South Dakota. M, 1979, University of Iowa. 88 p.

Matzner, David Marc. Metamorphism of early Paleozoic island arc and Mesozoic plutonic rocks intruding the Trinity Peridotite, eastern Klamath Mountains, Northern California. M, 1986, University of California, Los Angeles. 172 p.

Matzner, Ingrid Adelheid Maria. Differential and isothermal dehydration states of retgersite-nickel sulfate hexahydrate. M, 1969, University of Utah. 33 p.

Matzner, Robert A. Hydrogeologic and geophysical investigations of the Springfield-Blackfoot area, Idaho. M, 1982, University of Idaho. 165 p.

Mauch, Elizabeth Ann. A seismic stratigraphic and structural interpretation of the middle Paleozoic Ikpikpuk-Umiat Basin, National Petroleum Reserve in Alaska. M, 1985, Rice University. 220 p.

Mauch, Joseph James. The late Paleozoic tectono-sedimentary history of the Marfa Basin, West Texas. M, 1982, Texas Christian University. 94 p.

Maucini, Joseph J. The geology of the vehicular tunnel across the Hudson River. M, 1924, Columbia University, Teachers College.

Mauck, Abram Vardiman. A quantitative study of variation in the fossil brachiopod Platystrophia biforata Schl. M, 1901, Indiana University, Bloomington. 8 p.

Maud, Randall Lee. Stratigraphy and depositional environments of the carbonate-terrigenous member of the Crystal Spring Formation, Death Valley, California. M, 1979, Pennsylvania State University, University Park. 177 p.

Maud, Randall Lee. Stratigraphy, petrography and depositional environments of the carbonate-terrigenous member of the Crystal Spring Formation, Death Valley, California. D, 1983, Pennsylvania State University, University Park. 240 p.

Mauer, Kenneth. Soil moisture of the Oak Openings Sand, Lucas County, Ohio. M, 1978, [University of Toledo].

Mauffette, P. Geology of the Calumet Mines Limited (Quebec). M, 1941, McGill University.

Mauger, Lucy L. Benthonic foraminiferal paleoecology of the Yorktown Formation at Lee Creek Mine, Beaufort County, North Carolina. M, 1979, East Carolina University. 198 p.

Mauger, Richard Leroy. A petrographic and geochemical study of Silver Bell and Pima mining districts, Pima County, Arizona. D, 1966, University of Arizona. 156 p.

Maughan, John Bohan. A microscopic examination of the aqueous phase in heterogeneous fluid flow through porous media. M, 1951, University of Oklahoma. 72 p.

Mauk, Frederick John. A tectonic-based Rayleigh wave group velocity model for prediction of dispersion character through ocean basins. D, 1977, University of Michigan. 260 p.

Mauk, Frederick John. On triggering of volcanic eruptions by solid Earth tides. M, 1972, University of Michigan.

Mauk, Jeffrey L. Stratigraphy and sedimentation of the Proterozoic Burke and Revett formations, Flathead Reservation, western Montana. M, 1983, University of Montana. 91 p.

Mauldin, Randall A. Foraminiferal biostratigraphy, paleoecology, and correlation of the Del Rio Clay (Cenomanian) from Big Bend National Park, Brewster County, and Dona Ana County, New Mexico. M, 1985, University of Texas at El Paso.

Maulsby, Joe. Geology of the Rancho de Lopez area east of Socorro, New Mexico. M, 1981, New Mexico Institute of Mining and Technology. 85 p.

Maung, Tun U. Polarization with variation of water saturation in core samples. M, 1960, Colorado School of Mines. 51 p.

Maurath, Garry. Heat generation and terrestrial heat flow in northwestern Pennsylvania. M, 1980, Kent State University, Kent. 156 p.

Maureau, Gerrit T. F. R. Crustal structure in Western Canada. M, 1965, University of Alberta. 79 p.

Maurel, Laure Elisabeth. The nature and significance of the Barrier Peak thrust stack, Rocky Mountain Front Ranges, Alberta, Canada. D, 1987, University of Calgary. 278 p.

Maurer, Donald Leo. Biostratigraphy of Buck Mountain member and adjacent units in Winthrop area, Washington. M, 1958, University of Washington. 111 p.

Maurer, Robert Eugene. Geology of the Cedar mountains (Tooele County), Utah. D, 1970, University of Utah. 186 p.

Maurer, Robert J. The classification of bedding contact in the Richmondian stage of the Cincinnatian Series (Upper Ordovician). M, 1974, Miami University (Ohio). 149 p.

Mauret, Kevin de *see* de Mauret, Kevin

Maurice, Charles S. Leuco-tonalite and granite pegmatites of the Spruce Pine District, North Carolina. D, 1936, The Johns Hopkins University.

Maurice, Charles Stewart. An investigation of a residual clay for use as a filler in rubber or other compounds. M, 1931, University of North Carolina, Chapel Hill. 33 p.

Maurice, Ovide Dollard. Effet de la chaleur sur les minéraux métalliques. M, 1944, Universite Laval.

Maurice, Ovide Dollard. Geochemistry of zirconium. D, 1948, University of Toronto.

Mauritsen, Mark Vernon. Studies of diagenesis of Bermuda limestone; 1, The calcretes; 2, Modern marine cement in a Pleistocene eolianite. M, 1983, Washington State University.

Mauritsen, Mark Vernon. Studies of diagenesis of Bermuda limestone; 1. The calcretes; 2. Modern marine cement in a Pleistocene eolianite. M, 1984, Washington State University. 104 p.

Maurmeyer, Evelyn M. Analysis of short-and long-term elements of coastal change in a simple spit system; Cape Henlopen, Delaware. M, 1974, University of Delaware.

Maurmeyer, Evelyn M. Geomorphology and evolution of transgressive estuarine washover barriers along the western shore of Delaware Bay. D, 1978, University of Delaware. 293 p.

Mauro, Gene Louis Del *see* Del Mauro, Gene Louis

Maurrasse, Florentin J.-M. R. Biostratigraphy, paleoecology, biofacies variations of middle Paleogene sediments in the Caribbean deep sea. D, 1973, Columbia University. 424 p.

Maury, Carlotta J. A comparison of the Oligocene of Western Europe and the Southern United States. D, 1902, Cornell University.

Maus, Daniel Albert. Ore controls at the Golden Rule Mine, Cochise County, Arizona. M, 1988, University of Arizona. 114 p.

Mauser, John Kemmer. Origin of solutes in groundwaters of clastic silicate aquifers. M, 1981, Texas Tech University. 114 p.

Mausser, Herbert F. Stratigraphy and sedimentology of the Cleveland Shale (Devonian) in Northeast Ohio. M, 1982, Case Western Reserve University.

Mavko, B. B. Crustal and upper mantle structure of the Sierra Nevada. D, 1980, Stanford University. 56 p.

Mavko, G. M. Time dependent fault mechanics and wave propagation in rocks. D, 1977, Stanford University. 181 p.

Mavor, Matthew John. Transient pressure behavior of naturally fractured reservoirs. D, 1978, Stanford University. 56 p.

Mavris, George. Petrology of the lithofacies of a deltaic plain (Allegheny Group, Middle Pennsylvanian) near Ironton, Ohio and Ashland, Kentucky. M, 1981, Miami University (Ohio). 146 p.

Maw, George Glayde. Lake Bonneville history in Cutler Dam Quadrangle, Cache and Box Elder counties, Utah. M, 1968, Utah State University. 58 p.

Mawby, John Evans. Cranial anatomy of the late Cenozoic machairodonts. M, 1960, University of California, Berkeley. 41 p.

Mawby, John Evans. Pliocene vertebrates and stratigraphy in Stewart and Ione valleys, Nevada. D, 1965, University of California, Berkeley. 211 p.

Mawdsley, James Buckland. The anorthosites and associated illemite in the vicinity of St. Urbain, Quebec. D, 1924, Princeton University.

Mawdsley, James Cleugh. Paleontology and distribution of the Beavertail (Devonian) Limestone of the Mackenzie River valley, Northwest Territories. M, 1954, University of Alberta. 102 p.

Mawer, Malcolm Frank. A fluid inclusion study of the Highland Valley porphyry copper deposits. M, 1977, University of Alberta. 155 p.

Mawla, Haulam. The relationship of ground water to alluvium in Dinajpur and Rangpur, East Pakistan. M, 1968, University of Arizona.

Max, Michael D. Gypsum deposits in the northeastern part of the Bighorn Basin, Big Horn County, Wyoming. M, 1965, University of Wyoming. 63 p.

Maxell, Arthur Eugene. The outflow of heat under the Pacific Ocean. D, 1958, University of California, Los Angeles. 128 p.

Maxey, George Burke. Cambrian stratigraphy in the northern Wasatch region. M, 1941, Utah State University. 64 p.

Maxey, George Burke. Lower and Middle Cambrian stratigraphy in western Utah and southeastern Idaho. D, 1951, Princeton University. 148 p.

Maxey, James Roy. Permian Ostracoda from the Foraker to the Herington formations of southern Kansas and northern Oklahoma. M, 1932, University of Oklahoma. 79 p.

Maxey, Julian S. Geology of a portion of Lawrence County, Ohio. M, 1940, Ohio State University.

Maxey, Larry R. A study of the apatite mineralogy and apatite in phosphorite. M, 1963, University of Florida. 65 p.

Maxey, Lawrence R. Metamorphism and origin of Precambrian amphibolites of the New Jersey Highlands. D, 1971, Rutgers, The State University, New Brunswick. 156 p.

Maxey, Marilyn Helen. Stratigraphy and petroleum geology of the Middle Ordovician Ottawa Limestone Supergroup in northwestern Ohio. M, 1979, Kent State University, Kent. 112 p.

Maxfield, E. Blair. Foraminifera from the Mancos Shale of east-central Utah. D, 1976, Brigham Young University. 170 p.

Maxfield, E. Blair. Sedimentation and stratigraphy of the (Pennsylvanian) Morrowan Series in central Utah. M, 1956, Brigham Young University. 46 p.

Maxfield, Ray A. Geology of zinc-lead deposits, Ruby Creek District, Idaho; with suggestions of methods of concentrating the ores. M, 1933, Washington State University.

Maxfield, William Kinsey. Evaluation of the ground-water resources and present water system, Yellow Springs, Ohio. M, 1975, Wright State University. 84 p.

Maxis, Ike. Delineation of the bedrock topography of Warren County, Ohio, using spatial-frequency filtering for regional removal. M, 1986, Wright State University. 76 p.

Maxon, Jonathan Rolfe. The age of the Tres Piedras Granite, New Mexico; A case of large scale isotopic homogenization. M, 1976, Florida State University.

Maxson, Anne E. Variations in shell form in the gastropod genus Diodora. M, 1982, Virginia Polytechnic Institute and State University. 180 p.

Maxson, John H. A Tertiary mammalian fauna from the Mint Canyon Formation, Southern California. M, 1928, California Institute of Technology. 26 p.

Maxson, John H. Geology of the western Siskiyou Mountains, northwestern California. D, 1931, California Institute of Technology. 226 p.

Maxwell, Brian. Crustal seismic reflection survey in the Vermilion Bay-Red Lake region, with emphasis on defining the proposed "mid-crustal" layer. M, 1986, University of Manitoba.

Maxwell, Charles H. Pleonaste crystals from an olivine basalt, Caballo Mountains, New Mexico. M, 1952, University of New Mexico. 43 p.

Maxwell, Dwight Thomas. Diagenetic-metamorphic trends of layer silicates in the Precambrian Belt Series (western Montana and northern Idaho). D, 1965, University of Montana. 94 p.

Maxwell, Frank Kristian. Rayleigh and Stoneley waves on cylindrical boundaries. D, 1982, University of Western Ontario.

Maxwell, Frank Kristian. The scattering of elastic waves by a cylindrical cavity. M, 1974, University of Western Ontario.

Maxwell, Garry S. Heavy mineral analysis of the Cassel Hill Member, Catahoula Formation, Louisiana. M, 1979, Northeast Louisiana University.

Maxwell, Gary P. Size, shape, and distribution of microscopic pyrite in selected Ohio coals. M, 1982, University of Toledo. 158 p.

Maxwell, Geraldine. Influence of Poisson's ratio on head waves from thin layers. M, 1988, University of Houston.

Maxwell, Glenn B. Depositional environment of the Chemard Lake Lignite (Paleocene) of DeSoto Parish, Louisiana. M, 1979, Northeast Louisiana University.

Maxwell, J. A. A mercury cathode-polarographic method for the determination of titanium in rocks. M, 1950, McMaster University.

Maxwell, James Christie. Quantitative geomorphology of the San Dimas Experimental Forest, California. D, 1962, Columbia University, Teachers College. 201 p.

Maxwell, James M. A petrographic study of the Lyons Formation (Permian), Colorado. M, 1934, University of Colorado.

Maxwell, John A. Geochemical study of chert and related deposits. D, 1953, University of Minnesota, Minneapolis. 124 p.

Maxwell, John Crawford. Accessory minerals of the Mesabi iron ore and protore. M, 1937, University of Minnesota, Minneapolis. 31 p.

Maxwell, John Crawford. The geology of Tobago, British West Indies. D, 1946, Princeton University. 185 p.

Maxwell, John R. The relationship of the distribution coefficient K_d, to surface area in microcline and albite feldspars. M, 1968, Texas A&M University. 87 p.

Maxwell, Michael George. Isotopic identification of subglacial processes. D, 1986, University of British Columbia.

Maxwell, R. J. A study of rubidium, strontium and strontium isotopes in some mafic and sulphide minerals. M, 1976, University of British Columbia.

Maxwell, Riley G. The Warsaw Formation in the vicinity of St. Louis, Missouri. M, 1929, Washington University. 58 p.

Maxwell, Robert. Origin and geomorphic history of high fluvial terraces in Wilcox City, Alabama. M, 1973, Florida State University.

Maxwell, Ross A. The (Silurian-Devonian) Hunton Formation of the Arbuckle Mountains (Oklahoma). M, 1931, University of Oklahoma. 36 p.

Maxwell, Ross A. The stratigraphy and areal distribution of the "Hunton Formation" in Oklahoma. D, 1936, Northwestern University.

Maxwell, Theodore Allen. Paleohydrology and depositional environment of the Duchesne River formation (Eocene-Oligocene?) near Roosevelt, Utah. M, 1973, University of Utah. 60 p.

Maxwell, Theodore Allen. Stratigraphy and tectonics of southeastern Serenitatis. D, 1977, University of Utah. 144 p.

May, A. Brent. Depositional environments, sedimentology and stratigraphy of the Dockum Group (Triassic) in the Texas Panhandle. M, 1988, Texas Tech University. 180 p.

May, Bruce Tipton. Magnetic properties of rocks associated with the new Cornelia porphyry copper deposit, Pima County, Arizona. D, 1968, University of Arizona. 185 p.

May, Charles A. Measurement of thermal conductivity in rocks. M, 1971, Case Western Reserve University.

May, Clifford John. Microgravity exploration for a sediment-filled channel intersecting a near-surface lignite seam. M, 1984, Texas A&M University. 79 p.

May, Daniel J. General and structural geology of a portion of Moss Beach, San Mateo County, California. M, 1976, Stanford University.

May, Daniel Joseph. Amalgamation of metamorphic terranes in the southeastern San Gabriel Mountains, California. D, 1986, University of California, Santa Barbara. 377 p.

May, Dann Joseph. The paleoecology and depositional environment of the late Eocene-early Oligocene Toutle Formation, southwestern Washington. M, 1980, University of Washington. 110 p.

May, David William. Holocene alluviation, soil genesis, and erosion in the South Loup Valley, Nebraska. D, 1986, University of Wisconsin-Madison. 239 p.

May, F. E. Dinoflagellate cysts of the Gymnodiniaceae, Peridiniaceae, and Gonyaulacaceae from the upper Cretaceous Monmouth Group, Atlantic Highlands, New Jersey. D, 1976, Virginia Polytechnic Institute and State University. 376 p.

May, Fred E. Palynology of the Dakota Sandstone (Middle Cretaceous) near Bryce Canyon National Park, southern Utah. M, 1972, Pennsylvania State University, University Park.

May, Geoffrey R. The contact between the Upper Jurassic Stump Formation and the basal mudstones of the Ephraim Conglomerate at five localities in the Idaho-Wyoming thrust belt. M, 1985, Idaho State University. 51 p.

May, Glenn M. The flexural wavelength due to isostatic rebound of Lake Bonneville. M, 1988, SUNY at Buffalo. 49 p.

May, H. M. Aluminum determination, aluminum hydrolysis and the solubilities of some common aluminous minerals in near-neutral aqueous solutions. D, 1978, University of Wisconsin-Madison. 174 p.

May, J. E. Mineral composition in relation to particle size for a Missouri plastic fire clay. M, 1951, University of Missouri, Rolla.

May, James H. Interpretation of post Oligocene depositional cycles in the Mendenhall West Quadrangle, Mississippi. M, 1980, University of Southern Mississippi.

May, James Herbert. Geologic and hydrodynamic controls on the mechanics of knickpoint migration. D, 1988, Texas A&M University. 214 p.

May, James P. Sedimentary and geomorphic response to systematic variation of wave energy in the nearshore zone. D, 1973, Florida State University. 222 p.

May, James P. Stream valley geometry as an index to the evolution of the piedmont surface. M, 1966, University of North Carolina, Chapel Hill. 61 p.

May, Jeffrey A. Endolithic infestation of carbonate substrates above and below the sediment/water interface. M, 1977, Duke University. 162 p.

May, Jeffrey Allyn. Basin-margin sedimentation; Eocene La Jolla Group, San Diego County, California. D, 1982, Rice University. 434 p.

May, Karen Anne. Archean geology of a part of the northern Gallatin Range, Southwest Montana. M, 1985, Montana State University. 91 p.

May, Michael T. Sedimentology and diagenesis of Pingelap Atoll, eastern Caroline Islands, Micronesia. M, 1986, University of Kansas. 116 p.

May, Milton E. The De Queen Limestone formation of the (Cretaceous) Trinity Group in McCurtain County, Oklahoma. M, 1950, University of Oklahoma. 48 p.

May, Paul Russell. Geology of the Monument Mountain-Gallatin River area, Yellowstone National Park, and Gallatin County, Montana. M, 1950, University of Michigan. 43 p.

May, Paul Russell. Stratigraphic section of the Plum Brook, Huron, and Chagrin shales of Middle and Upper Devonian age in Lorain County, Ohio. M, 1950, University of Michigan.

May, R. J. Thermoluminescence dating of Hawaiian basalts. D, 1975, Stanford University. 169 p.

May, Ronald William. Cu, Ni, Zn, Cr, CaO, MgO content of Wisconsin tills in southern Ontario. D, 1971, University of Western Ontario. 191 p.

May, Ronald William. Geology of the Peace River, Viking, Joli Fou and Notikewin formations, Daybob area, northern Alberta. M, 1968, University of Calgary. 99 p.

May, S. Judson. Geology of the Almont Graben, Gunnison County, Colorado. M, 1974, Eastern Kentucky University. 82 p.

May, S. Judson. Neogene geology of the Ojo Caliente-Rio Chama area, Espanola Basin, New Mexico. D, 1980, University of New Mexico. 204 p.

May, Steven R. Geology and mammalian paleontology of the Horned Toad Hills, Mojave Desert, California. M, 1981, University of California, Riverside. 290 p.

May, Steven Robert. Paleomagnetism of Jurassic volcanic rocks in southeastern Arizona and North American Jurassic apparent polar wander. D, 1985, University of Arizona. 243 p.

May, Suzette Kimball. Electrical resistivity methods in the unconsolidated glacial sediments of Delaware County, Indiana. M, 1981, Ball State University. 194 p.

May, Suzette Kimball. Regional wave climate and shorezone response. D, 1983, University of Virginia. 102 p.

May, Thomas Patrick. The geology of a portion of the Marshall Lake District, Idaho County, Idaho. M, 1984, University of Idaho. 101 p.

May, Thomas Wayne. Holographic analysis and interpretation of the 5 megahertz radar imagery of the Apollo Lunar Sounder Experiment in Maria

Serenitatis and Crisium. M, 1976, University of Utah. 91 p.

May, Timothy Crawford. A study of some problems involved in the oil shale industry (Colorado). M, 1930, Catholic University of America. 45 p.

May, Timothy Crawford. The constitution of Pinnacle Bed coal from Hayden Mine, Haybro, Routt County, Colorado. D, 1938, Catholic University of America. 29 p.

May, William P., Jr. Maximum entropy spectral analysis and linear prediction of the Susquehanna drainage basin runoff. M, 1981, Rensselaer Polytechnic Institute. 71 p.

Mayberry, J. W. Oklahoma building stones. M, 1906, University of Kansas.

Maybin, Arthur H., III. The geology of the Opawica River Complex, Quebec, Canada. M, 1976, University of Georgia.

Maycock, Ian D. The Ordovician limestones of the Kingston (Ontario) District. M, 1959, Queen's University. 240 p.

Maycotte, Jorge I. Engineering geology and slope stability assessment of the Caracol Hydroelectric Project, Mexico. M, 1980, University of Illinois, Urbana. 128 p.

Maye, Peter Robert, III. Some important inorganic nitrogen and phosphorus species in Georgia salt marsh. M, 1972, Georgia Institute of Technology. 60 p.

Mayer, Anton Bernard. Structural and volcanic geology of the Salado Mountains-Garcia Peaks area, Sierra County, New Mexico. M, 1987, New Mexico State University, Las Cruces. 61 p.

Mayer, Edward A. The structural trend lines in the western end of the Baraboo Syncline (Wisconsin). M, 1934, University of Wisconsin-Madison.

Mayer, Edward William. Atmospheric methane; concentration, swamp flux and latitudinal source distribution. D, 1982, University of California, Irvine. 204 p.

Mayer, Gale G. Characterization and utilization of high- and low-sodium lignite of the Beulah-Zap Bed (Paleocene), Indian Head Mine, southwestern North Dakota. M, 1988, University of North Dakota. 144 p.

Mayer, Harold Melvin. The geography of the Port of St. Louis. M, 1937, Washington University. 202 p.

Mayer, James Roger. Amplitude studies of COCORP deep seismic reflection profiling data from the Basin and Range and Colorado Plateau. M, 1986, Cornell University. 62 p.

Mayer, Larry. Quantitative tectonic geomorphology with applications to neotectonics of northwestern Arizona. D, 1982, University of Arizona. 596 p.

Mayer, Larry. The geology and geomorphology of the Buckhead Mesa area, Gila County, Arizona. M, 1979, University of Arizona. 112 p.

Mayer, Larry A. The origin and geologic setting of high-frequency acoustic reflectors in deep-sea carbonates. D, 1979, University of California, San Diego. 178 p.

Mayer, Lawrence M. Aluminosilicate sedimentation in Lake Powell. M, 1973, Dartmouth College. 30 p.

Mayer, Lawrence M. Geochemistry of silica in Lake Powell. D, 1976, Dartmouth College. 93 p.

Mayer, Maurice J., Jr. A contribution to the geology of Bienville Parish, Louisiana. M, 1932, Louisiana State University.

Mayer, Peter W. Uniaxial compressive strengths of a quartzite, granite, and graywacke and their dependence on sample length to diameter ratio. M, 1979, SUNY at Binghamton. 14 p.

Mayer, Richard R. The computation of a first-order triangulation net. M, 1960, Ohio State University.

Mayer, Tatiana. Phosphorus mobility in lacustrine sediments upon lake acidification. M, 1984, McMaster University. 136 p.

Mayer, Terry Ann. Engineering geology of a mudslide at Bracebridge Inlet, Bathurst Island, Northwest Territories, Canada. M, 1980, Texas A&M University. 71 p.

Mayer, Victor J. Stratigraphy and paleontology of the Mississippian formations of Moffat County, Colorado. M, 1960, University of Colorado.

Mayers, Ian Richard. An analysis of continuous seismic profiles from the Strait of Juan de Fuca. M, 1971, University of Washington. 123 p.

Mayes, Catherine Lynn. Tectonic history and new isochron chart of the South Pacific. M, 1988, University of Texas, Austin. 117 p.

Mayes, John Wilmot. A further study of the Tepee Creek Formation, Wichita Mountains, Oklahoma. M, 1947, University of Oklahoma. 45 p.

Mayewski, Paul Andrew. Glacial geology and late Cenozoic history of the Transantarctic Mountains, Antarctica. D, 1973, Ohio State University.

Mayfield, Charles F. Geology of the Thompson Creek area, Custer County, Idaho. M, 1973, University of Idaho. 101 p.

Mayfield, Darrell G. Magmatic variation in the El Salvador segment of the Middle America Arc. M, 1978, Rutgers, The State University, New Brunswick. 57 p.

Mayfield, Jack Hastings, Jr. Subsurface Pennsylvanian geology, eastern Coke County, Texas. M, 1965, University of Texas, Austin.

Mayfield, Michael Wells. Variations in streamflow among watersheds of the Cumberland Plateau, Tennessee. D, 1984, University of Tennessee, Knoxville. 171 p.

Mayfield, Ricky L. Deposition and diagenesis of the Gilmer Limestone (Jurassic), Leon and Freestone counties, Texas. M, 1983, Texas Tech University. 56 p.

Mayfield, Samuel M. Geology of Fordsville and Connelton quadrangles (Kentucky). D, 1932, University of Chicago. 181 p.

Mayher, Andrea M. Middle Devonian Cryptostomata (Bryozoa) from the Silica Formation, Lucas County, Ohio. M, 1965, Bowling Green State University. 107 p.

Mayhew, George Herbert. An application of electrical resistivity depth profiling to lake shore subsurface problems. M, 1960, Ohio State University.

Mayhew, George Herbert. Seismic reflection study of the subsurface structure in western and central Ohio. D, 1969, Ohio State University. 89 p.

Mayhew, John D. Geology of the eastern part of the Bonanza volcanic field, Saguache County, Colorado. M, 1969, Colorado School of Mines. 94 p.

Mayhew, Michael Allen. Marine geophysical measurements in the Labrador sea; relation to Precambrian geology and sea-floor spreading. D, 1969, Columbia University. 100 p.

Maynard, B. R. Form and sediment characteristics in alluvial fans of southwestern United States. D, 1976, University of Pittsburgh. 244 p.

Maynard, D. E. Geomorphic constraints to urban residential development in the Seymour area, district of North Vancouver, B.C. M, 1978, University of British Columbia.

Maynard, Donald. Surficial geology and glacial history of the Little Falls Quadrangle, central Mohawk Valley, New York. M, 1988, Rensselaer Polytechnic Institute. 41 p.

Maynard, Douglas George. Transformations and dynamics of soil available sulfur. D, 1983, University of Saskatchewan.

Maynard, G. L. Development of seismic instrumentation for remote islands with adverse environmental conditions. M, 1967, University of Hawaii.

Maynard, G. L. Seismic wide angle reflection and refraction investigation of the sediments on the Ontong Java Plateau. D, 1973, University of Hawaii. 156 p.

Maynard, J. E. An experimental contribution to the origin of the Pre-Cambrian banded iron formations. D, 1979, University of Manitoba.

Maynard, J. E. The origin of the Precambrian banded iron formations. D, 1928, University of Toronto.

Maynard, James Barry. Kinetics of silica sorption by kaolinite. D, 1972, [Harvard University].

Maynard, James E. The clays of the Lake Agassiz Basin. M, 1925, University of Manitoba.

Maynard, LeRoy Carson. Geology of Mount McLoughlin (Oregon). M, 1974, University of Oregon. 139 p.

Maynard, Robert G. Geology of the Tropico Mine, Rosamond, California. M, 1947, University of California, Los Angeles.

Maynard, Stephen R. Base-metal and silver vein in a volcanic environment. M, 1985, University of New Mexico. 155 p.

Maynard, T. P. The Corrigan Formation of Maryland. D, 1909, The Johns Hopkins University.

Maynes, A. O. The determination of trace amounts of lead in igneous rocks. D, 1956, University of Toronto.

Maynor, Gregory Keith. Middle Paleozoic subsurface geology of the Rockcastle River uplift area, southeastern Kentucky. M, 1984, University of Kentucky. 126 p.

Mayo, A. L. Geology and ore deposits of the South-central Calico Mountains. M, 1972, [University of California, San Diego].

Mayo, Alan Lee. Ground water flow patterns in the Meade thrust allochthon, Idaho-Wyoming thrust belt, southeastern Idaho. D, 1982, University of Idaho. 192 p.

Mayo, Curtis Ray. Petrography of a Middle Devonian bioherm, southwestern Ontario, Canada. M, 1964, University of Nebraska, Lincoln.

Mayo, Evans B. Petrography of a portion of the Sierra Nevada Batholith, California. D, 1932, Cornell University.

Mayo, Evans B. The geology of the Mammoth Lakes area, Mono County, California. M, 1929, Stanford University. 81 p.

Mayo, G. D., Jr. Correlation of the properties and freeze thaw testing of some Tennessee aggregates. M, 1978, Memphis State University.

Mayo, Lawrence Rulph. 1961 meteorology and mass balance, Gulkana Glacier, central Alaska Range, Alaska. M, 1963, University of Alaska, Fairbanks. 52 p.

Mayo, Robert Truitt. Stratigraphy of eastern half Gentry Quadrangle, Culberson County, Texas. M, 1950, University of Texas, Austin.

Mayor, Jerrold N. Structural geology of the northern part of Oxford Quadrangle, Idaho. M, 1979, Utah State University. 76 p.

Mayotte, Timothy. An investigation of a soil gas sampling technique and its applicability for detecting gaseous PCE and TCA over an unconfined granular aquifer. M, 1988, Western Michigan University.

Mayou, Taylor Vinton. Facies distribution and animal-sediment relationships in Doboy Sound, a Georgia estuary. D, 1972, University of Iowa. 217 p.

Mayou, Taylor Vinton. Paleontology of the Permian Loray Formation in White Pine County, Nevada. M, 1967, Brigham Young University.

Mayr, U. Subdivisions and correlation of the Tully Formation (Upper Devonian) in New York state and Pennsylvania on the basis of a statistical and stratigraphical analysis of selected conodonts. D, 1966, University of Ottawa. 103 p.

Mays, Linda Lowry. Carbon, sulfur, and phosphorus relations in oceanic Neogene sediments. M, 1987, Texas A&M University.

Mays, Major Dewayne. A comparison of Mollisols from three different climatic regions in the United States. D, 1982, University of Nebraska, Lincoln. 203 p.

Maytum, James R. Areal geology of southeast portion (Triassic and early Cretaceous) Chanchelulla Peak Quadrangle, California (Shasta-Tehama counties). M, 1967, University of San Diego.

Mayu, Philippe Henri. Determining parameters for stiff clays and residual soils using the self-boring pressuremeter. D, 1987, Virginia Polytechnic Institute and State University. 337 p.

Mayuga, Manuel Nieva. The geology and ore deposits of the Helmet Peak area, Pima County, Arizona. D, 1942, University of Arizona.

Mayuga, Manuel Nieva. The geology of the Empire Peak area, Pima County, Arizona. M, 1940, University of Arizona.

Maywood, Paul S. Stratigraphic model of the southern portion of the Jim Bridger coal field, Sweetwater County, Wyoming. M, 1987, Portland State University. 128 p.

Mazaheri, Seyed Ahmad. The petrology and metamorphism of ultramafic and mafic rocks near Pulga, Butte County, California. M, 1982, University of California, Davis. 139 p.

Mazalan, Paul Alan. Magnetic and paleomagnetic properties of the metamorphic rocks of the Carmichael Canyon area, Madison County, Montana. M, 1975, Indiana University, Bloomington. 108 p.

Mazariegos Alfaro, Ruben Alberto. Seismic stratigraphic analysis of the Cretaceous carbonates of Paso Caballos Basin, Guatemala. M, 1988, University of Oklahoma. 173 p.

Mazaris, George Michael. The development and application of an LHD underground face simulator. M, 1981, University of Arizona. 165 p.

Maze, William Bronson. Jurassic La Quinta Formation in the Sierra de Perija, northwestern Venezuela; geology, tectonic environment, paleomagnetic data, and copper mineralization on red beds and volcanics. D, 1983, Princeton University. 331 p.

Mazid, Mohamad. Flood forecasting using remote sensed information. D, 1987, Pennsylvania State University, University Park. 197 p.

Mazierski, Paul F. A geologic and petrologic study of Pine and Crater buttes; two basaltic constructs on the eastern Snake River plain, Idaho. M, 1988, SUNY at Buffalo. 131 p.

Mazurak, Robert E. Economic geology of the Warm Spring talc deposit, Inyo County, Calif. M, 1976, Pennsylvania State University, University Park. 65 p.

Mazurek, Monica Ann. Geochemical investigations of organic matter contained in ambient aerosols and rainwater particulates. D, 1985, University of California, Los Angeles. 372 p.

Mazurski, Marcia Ann. Geology and estimates of P-T-X_{CO2} conditions of metamorphism of part of the Foxe fold belt (NTS 47A/3), Melville Peninsula, Northwest Territories. M, 1980, Queen's University. 126 p.

Mazza, Antonio Gennaro. Modelling of the liquid-phase thermal cracking kinetics of Athabasca Bitumen and its major chemical fractions. D, 1987, University of Toronto.

Mazza, Thomas A. Factors controlling the frequency of seismic waveforms used in the "petite sismique" method of rock mass assessment. M, 1982, Pennsylvania State University, University Park. 157 p.

Mazzella, A. T. Deep resistivity study across the San Andreas fault zone. D, 1976, University of California, Berkeley. 145 p.

Mazzella, Frederick E. A thermal and gravity model of a geothermal anomaly near Marysville, Montana. M, 1973, Southern Methodist University. 35 p.

Mazzella, Frederick E. The generation of synthetic seismograms for laterally heterogeneous models using the finite difference technique. D, 1979, Purdue University. 225 p.

Mazzo, Carl R. The petrology and stratigraphy of the Port Ewen Formation in the Kingston, N.Y., vicinity. M, 1981, Rensselaer Polytechnic Institute. 73 p.

Mazzone, Peter. Amphibole dehydration and correlated isotopic and geochemical variations in the Garner Mountain Andesite, Siskiyou County, California. M, 1983, Miami University (Ohio). 78 p.

Mazzone, Peter. Petrography, mineral chemistry and geochemistry of peraluminous xenoliths from the Jagersfontein Kimberlite. D, 1988, University of Massachusetts. 232 p.

Mazzucchelli, Vincent George. Qualitative terrain representation in cartography; an historical evaluation of methods of landform rendering in relation to changing technology. D, 1974, University of California, Los Angeles. 367 p.

Mazzullo, Elsa K. Depositional environment of Lower Cretaceous Mitchell Sandstone, St. Mary and Duty fields, Lafayette County, Arkansas. M, 1983, Texas A&M University.

Mazzullo, James. Preliminary stratigraphic analysis of the St. Peter Sandstone; Fourier grain shape analysis. M, 1979, University of South Carolina.

Mazzullo, James Michael. Grain shape variation in the St. Peter Sandstone; a record of eolian and fluvial sedimentation in a early Paleozoic cratonic sheet sand. D, 1981, University of South Carolina. 71 p.

Mazzullo, L. J. Abyssal tholeiites from the Nazca Plate; DSDP Leg 34. M, 1975, SUNY at Stony Brook.

Mazzullo, Salvatore J. Deltaic facies in the Hamilton Group (middle Devonian), southeastern New York State. M, 1971, Brooklyn College (CUNY).

Mazzullo, Salvatore J. Sedimentology and depositional environments of the Cutting and Fort Ann formations (lower Ordovician) in New York and adjacent southwestern Vermont. D, 1974, Rensselaer Polytechnic Institute. 203 p.

Mbagwu, Joe Sonne Chinyere. Studies on soil loss-productivity relationships of Alfisols and Ultisols in southern Nigeria. D, 1981, Cornell University. 334 p.

Mburu, S. G. Vertical temperature and chemical gradients in groundwater in the Tucson Basin, Arizona. M, 1975, University of Arizona.

McAdams, Frederick W. The accessory minerals of the Wolf Mountain Granite, Llano County, Texas. M, 1935, Texas A&M University.

McAdams, Mark Patrick. The mapping of flood inundation using Landsat multispectral scanner imagery. M, 1980, University of Iowa. 111 p.

McAdoo, D. C. Inelastic deformation of lithospheres with terrestrial and Martian applications. D, 1976, Cornell University. 186 p.

McAdoo, Richard Lee. Structural relationships of the Precambrian Carrizo Mountain Group, Carrizo Mountains, Trans-Pecos Texas. M, 1979, Texas Tech University. 80 p.

McAlary, J. D. Marginal phase of the Pennsylvanian sediments in northern New Brunswick. M, 1952, University of Toronto.

McAlaster, Penelope. The geology and genesis of jasperoid in the northern Swisshelm Mountains, Cochise County, Arizona. M, 1980, University of Arizona. 78 p.

McAleer, Joseph Francis. Ore minerals and wall rock alteration on deep levels at Butte, Montana. M, 1966, University of California, Berkeley. 75 p.

McAlester, Arcie Lee. Pelecypod faunas of the Late Devonian Chemung Stage, central New York. D, 1960, Yale University.

McAlister, Randall Lee. Late Quaternary landforms and associated deposits, Palo Duro Canyon State Park, Texas. M, 1983, West Texas State University. 206 p.

McAllen, William R. Petrology and diagenesis of the Esplanade Sandstone, central Arizona. M, 1984, Northern Arizona University. 90 p.

McAllister, Arnold L. A cobalt-tungsten deposit in the Sudbury District (Ontario). M, 1948, McGill University.

McAllister, Arnold L. The geology of the Ymir map-area, British Columbia. D, 1950, McGill University.

McAllister, James F. Rocks and structure of the Quartz Spring area, northern Panamint Range, California. D, 1951, Stanford University. 110 p.

McAllister, James R. Glacial history of an area near Lake Tahoe (California). M, 1936, Stanford University. 78 p.

McAllister, John J. Detection of induced polarization. M, 1958, Massachusetts Institute of Technology. 54 p.

McAllister, Raymond Francis, Jr. Clay minerals of Recent marine sediments to the west of the Mississippi Delta. D, 1958, Texas A&M University.

McAllister, Raymond Francis, Jr. Varved clays of the Goulais River valley of Ontario. M, 1951, University of Illinois, Urbana.

McAllister, Ronald Eric. Seismic crustal studies in eastern Canada; Atlantic coast of Nova Scotia. M, 1963, Dalhousie University. 57 p.

McAlpin, Archie Justus. Paleopsephurus wilsoni, a new polyodontid fish from the Upper Cretaceous of Montana, with a discussion of allied fish, living and fossil. D, 1941, University of Michigan.

McAndrew, John. A study of some rare and uncertain ore minerals. D, 1952, University of Toronto.

McAndrews, Harry. Origin of chert in (Devonian) Helderberg formations. M, 1956, West Virginia University.

McAndrews, John Henry. Postglacial vegetation history of the Prairie-Forest transition of northwestern Minnesota. D, 1964, University of Minnesota, Minneapolis. 161 p.

McAndrews, Kevin P. Copper-molybdenum potential of parts of the Colville Indian Reservation, north central Washington. M, 1986, Colorado State University. 100 4 plates p.

McAndrews, Martin George. Petrography of upper Miocene sandstones, southern Santa Cruz County, California. M, 1948, University of California, Berkeley. 32 p.

McAneny, John Maurice. Analysis and fusibility of ash from certain Washington coals. M, 1928, University of Washington. 31 p.

McAnulty, William Noel. Geology of Cathedral Mountain Quadrangle, Brewster County, Texas. D, 1953, University of Texas, Austin.

McAnulty, William Noel. The vertebrate fauna and geologic age of some Pleistocene terraces in Henderson County, Texas. M, 1948, University of Oklahoma. 102 p.

McAnulty, William Noel, Jr. Geology of the Fusselman Canyon area, Franklin Mountains, El Paso County, Texas. M, 1967, University of Texas, Austin.

McAnulty, William Noel, Jr. Geology of the northern Nacozori District, Sonora, Mexico. D, 1970, University of New Mexico. 103 p.

McArdle, John E. A numerical (computerized) method for quantifying Aooarcheological comparisons. M, 1974, University of Illinois, Chicago.

McArthur, David Samuel. Sand movement in relation to beach topography. D, 1969, Louisiana State University. 47 p.

McArthur, John Gilbert. The geology of the Stirling copper property, Springdale, Newfoundland. M, 1973, Memorial University of Newfoundland. 92 p.

McArthur, Richard Earl. Geologic study of a submarine channel of Illinoisan-Sangamon age in the East Cameron area, Northwest Gulf of Mexico. M, 1979, Texas Tech University. 58 p.

McAtee, Christopher L. Palynology of late-glacial and postglacial sediments in Georgian Bay, Ontario, Canada, as related to the Great Lakes history. M, 1977, Brock University. 153 p.

McAuliffe, James Michael. Marine depositional environments in the Lower Silurian of East Tennessee and adjacent areas. M, 1986, Pennsylvania State University, University Park. 201 p.

McAuly, Roger J. Uppermost Ordovician and lower-most Silurian stratigraphy and solitary rugose corals of the east-central United States. M, 1985, University of Manitoba.

McAvey, Michael B. Computer analysis of the water budget of the proposed Oak Openings New Town, Lucas County, Ohio. M, 1978, University of Toledo. 113 p.

McBean, Alan Johnston, II. The Proterozoic Gunsight Formation, Idaho-Montana; stratigraphy, sedimentology and paleotectonic setting. M, 1983, Pennsylvania State University, University Park. 235 p.

McBee, William, Jr. Devonian stratigraphy and paleontology of the Wapiti Lake area, British Columbia, Canada. M, 1948, University of Kansas.

McBeth, Paul Edward, Jr. Hydrogeologic significance of Landsat thematic mapper lineament analyses in the Great Basin. M, 1986, University of Nevada. 154 p.

McBirney, Alexander Robert. Geology of a part of the central Guatemalan Cordillera. D, 1961, University of California, Berkeley. 118 p.

McBrayer, Michael A. Seismic refraction and Earth resistivity studies of hydrogeologic problems of the Brazos River alluvium, Texas A & M Plantation, Burleson County, Texas. M, 1966, Texas A&M University. 73 p.

McBride, Barry Christopher. Geometry and kinematics of the central Snowcrest Range; a Rocky Mountain foreland uplift in southwestern Montana. M, 1988, Western Michigan University.

McBride, D. E. The structure and stratigraphy of the B-zone, Heath Steele mines, Newcastle, New Brunswick. D, 1976, University of New Brunswick.

McBride, David James. Paleoenvironment in the outer detrital belt facies of the Dunderbergia Zone in the central Great Basin. D, 1977, University of Kansas. 181 p.

McBride, David James. Phenetic variation and inferred cladistic pathways in the Apsotreta-Linnarssonella Complex (Brachiopoda, upper Cambrian). M, 1974, University of Kansas. 88 p.

McBride, Derek E. The Macex deposit, British Columbia; a geological and geochemical study of a porphyry copper-molybdenum deposit. M, 1972, Queen's University. 159 p.

McBride, Donald David. The depositional and geologic history of two Late Pennsylvanian carbonate banks in eastern Oldham County, Texas. M, 1988, West Texas State University. 86 p.

McBride, Earle F. Certain energies of cation exchange of four Missouri fire clays. M, 1956, University of Missouri, Columbia.

McBride, Earle F. Martinsburg flysch of the Central Appalachians. D, 1960, The Johns Hopkins University.

McBride, John Henry. Geophysical studies of regional crustal structure in the southern Appalachian orogen and crustal studies from seismic reflection, gravimetric, and satellite remote sensing observations. D, 1987, Cornell University. 262 p.

McBride, John Henry. Ground magnetic and gravity study of the central portion of the Arkoma Basin, Arkansas. M, 1981, University of Arkansas, Fayetteville.

McBride, John M. Measurement of ground water flow to the Detroit River, Michigan and Ontario. M, 1987, University of Wisconsin-Milwaukee. 107 p.

McBride, Karen. Properties of Cotton Valley sandstone reservoirs (Upper Jurassic), Terryville Field, Lincoln Parish, Louisiana. M, 1982, Texas A&M University. 92 p.

McBride, Katherine Kretow. Hydrogeochemistry and occurrence of selenium in the Garber-Wellington Aquifer, central Oklahoma. M, 1985, Oklahoma State University. 129 p.

McBride, Mark Stuart. Hydrology of Lake Sallie, northwestern Minnesota, with special attention to ground water-surface water interactions. M, 1972, University of Minnesota, Minneapolis. 62 p.

McBride, Raymond Allan. Agronomic and engineering soil interpretations from water retention data. D, 1983, University of Guelph.

McBride, Robert T. J. Distribution of Recent sediments in Maumee Bay, western Lake Erie. M, 1975, University of Toledo. 155 p.

McBride, Sandra L. A K-Ar study of the Cordillera Real, Bolivia, and its regional setting. D, 1977, Queen's University. 231 p.

McBride, Sandra L. A potassium-argon age investigation of igneous and metamorphic rocks from Catamarca and La Rioja provinces, Argentina. M, 1972, Queen's University. 101 p.

McBride, William Joseph. The surface geology of Hamilton County, Texas. M, 1953, University of Houston.

McBrinn, Geraldine E. Structure beneath Trinidad using teleseismic P-wave conversions. M, 1982, Pennsylvania State University, University Park. 63 p.

McBroom, Mark N. Mineral sources of dissolved constituents in Illinois streams. M, 1976, Northern Illinois University. 118 p.

McBryde, J. C. Pennsylvanian (middle Desmoinesian) coarse-grained fluvial and fluvial-deltaic deposits, Taos Trough, North-central New Mexico. M, 1979, University of Texas, Austin.

McBryde, Thomas J. The areal geology of the Locust Grove area, Mayes County, Oklahoma. M, 1952, University of Oklahoma. 57 p.

McCabe, Chad Law. Paleomagnetic results from the upper Keweenawan Chequamegon Sandstone and implications for red bed diagenesis and late Precambrian apparent polar wander of North America. M, 1982, University of Michigan.

McCabe, Charles Law. Paleomagnetism of North American Paleozoic sedimentary carbonates. D, 1985, University of Michigan. 179 p.

McCabe, Hugh R. Mississippian stratigraphy of the Williston Basin area. D, 1956, Northwestern University.

McCabe, Hugh R. Regional stratigraphic analysis of the Mississippian Madison Group, Williston Basin area. D, 1961, Northwestern University. 214 p.

McCabe, Hugh Ross. Lyleton and Amaranth red beds in southwestern Manitoba. M, 1956, University of Manitoba.

McCabe, Kirk. Geology and botany of Stoneman Lake, Coconino County, Arizona. M, 1971, Northern Arizona University. 104 p.

McCabe, Louis Cordell. Some physical evidence of rank development exhibited by vitrain. D, 1937, University of Illinois, Urbana. 65 p.

McCabe, Louis Cordell. The lithological and botannical constituents of Coal Number Six at Nashville, Illinois. M, 1933, University of Illinois, Urbana.

McCabe, Marcella R. The coal deposits of the Lee Ranch coal mine, McKinley County, New Mexico. M, 1987, University of Texas at El Paso.

McCabe, Peter Joseph. The role of sediments in hypereutrophic lakes; factors effecting phosphorus exchange. D, 1977, University of Notre Dame. 280 p.

McCabe, Robert Joseph. Magmatic affinity of the late Paleozoic Taylor Formation, northern Sierra Nevada; an island arc derived sequence of marine pyroclastic flows. M, 1983, University of California, Santa Barbara. 100 p.

McCabe, Steven Lee. Evaluation of a structural response and damage resulting from earthquake ground motion. D, 1987, University of Illinois, Urbana. 295 p.

McCabe, W. M. Acoustic emission in coal; a laboratory study. D, 1979, Drexel University. 223 p.

McCabe, William Stokes. Petrology of Herrin (6) Coal near Edgemont Station, St. Clair County, Illinois. M, 1935, University of Illinois, Urbana. 60 p.

McCaffrey, R. J. A record of the accumulation of sediment and trace metals in a Connecticut, U.S.A., salt marsh. D, 1977, Yale University. 167 p.

McCaffrey, Robert. Crustal structure and tectonics of the Molucca Sea collision zone, Indonesia. D, 1981, University of California, Santa Cruz. 174 p.

McCague, John Joseph. Porosity determination in well consolidated sandstone. M, 1980, Boston College.

McCaig, Andrew Malcolm. Dynamothermal aureoles of ophiolites and ultramafic bodies in the Canadian Appalachians. M, 1980, University of Western Ontario.

McCain, John C. The Caprellidae (Crustacea: Amphipoda) of Virginia and a partial revision of Mayer's varieties of Caprella acutifrons Latreille. M, 1964, College of William and Mary.

McCain, Ronald Gordon. Relationship between water loss from stream channels and gravity and seismic measurements; Beaver Creek watershed 7, Coconino County, Arizona. M, 1976, Northern Arizona University. 101 p.

McCaleb, James A. The late Morrowan-Atokan stratigraphy and paleontology of Crawford County, Arkansas. M, 1961, University of Arkansas, Fayetteville.

McCaleb, James Abernathy. Lower Pennsylvanian ammonoids from the Bloyd Formation of Arkansas and Oklahoma. D, 1964, University of Iowa. 223 p.

McCaleb, Stanley Bert. Morphological, physical, chemical, and mineralogical studies of a gray-brown podzolic-brown podzolic soil sequence of New York State. D, 1950, Cornell University.

McCall, Mary A. Lower Tertiary subsurface foraminiferal sequence of McDonald Island, California. M, 1961, University of California, Berkeley. 175 p.

McCall, Robert R. A study of the effect of viscosity ratio on the displacement of miscible hydrocarbon liquids in porous media. M, 1952, University of Oklahoma. 67 p.

McCallister, Dennis Lee. Alteration of exchangeable cation distribution and associated chemical changes in acidifying surface mined soils. D, 1981, Texas A&M University. 140 p.

McCallister, Phyllis Grace. An application of cokriging for the estimation of tripartite response spectra. M, 1984, University of Missouri, Rolla.

McCallister, Robert Hood. An experimental study of diopside-enstatite exsolution. D, 1972, Brown University. 79 p.

McCallister, Robert Hood. Kinetics of precipitation in the nepheline-kalsilite system. M, 1968, Brown University.

McCallum, Henry DeRosset. The Darst Creek oil field, Guadalupe County, Texas. M, 1932, University of Texas, Austin.

McCallum, Ian S. Equilibrium relationships among the co-existing minerals in the Stillwater complex (Precambrian), Montana. D, 1968, University of Chicago. 175 p.

McCallum, John S. Heavy minerals of the pre-Matawan Cretaceous sediments of the New Jersey Coastal Plain. M, 1958, Lehigh University.

McCallum, M. E. Petrology and structure of the Precambrian and post-Mississippian rocks of the east-central portion of the Medicine Bow Mountains, Albany and Carbon counties, Wyoming. D, 1964, University of Wyoming. 164 p.

McCallum, Malcolm E. The geology of the western half of the Jacksboro Quadrangle, Campbell County, Tennessee. M, 1958, University of Tennessee, Knoxville. 72 p.

McCallum, Marjorie Louise. A petrographic investigation of vertical deposition within the Mason Esker relative to its origin (Michigan). M, 1949, Michigan State University. 39 p.

McCalpin, James. Quaternary geology and neotectonics of the west flank of the northern Sangre de Cristo Mountains, south-central Colorado. D, 1981, Colorado School of Mines. 287 p.

McCamis, John Graham. Anhydritization in the Mississippian Souris Valley Beds of the Broadview area, Saskatchewan. M, 1958, University of Saskatchewan. 34 p.

McCammon, Helen Mary. Fauna of the Manitoba Group, from Manitoba, Canada. D, 1959, Indiana University, Bloomington. 171 p.

McCammon, James William. A mineralogical study of some granites from the east half of the Smithers map sheet. M, 1939, University of British Columbia.

McCammon, John Henry, Jr. The problem of petroleum reserves in the United States. M, 1942, University of Texas, Austin.

McCammon, Richard Baldwin. Sedimentology and origin of the alluvial terraces along the Wabash Valley. D, 1959, Indiana University, Bloomington. 54 p.

McCammon, Richard Baldwin. Stratigraphy and paleontology of a core penetrating Upper Silurian and Middle Devonian strata in Wayne County, Michigan. M, 1956, University of Michigan.

McCampbell, John. Age of "St. Mary's" Formation of North Carolina. M, 1936, Vanderbilt University.

McCampbell, John Caldwell. Further geomagnetic evidences as to the origin of the Carolina Bays. D, 1944, University of North Carolina, Chapel Hill. 67 p.

McCampbell, William Gibson, Jr. Geology of the Willow City-Eckert area, Gillespie County, Texas. M, 1940, University of Texas, Austin.

McCamy, Keith. An investigation and application of the crustal transfer ratio as a diagnostic for explosion seismology. D, 1967, University of Wisconsin-Madison. 68 p.

McCandless, David Oliver. A reevaluation of Cambrian through Middle Ordovician stratigraphy of the southern Lemhi Range. M, 1982, Pennsylvania State University, University Park. 157 p.

McCandless, Garrett Clair, Jr. Geology of the Indian Fault area, Llano County, Texas. M, 1957, University of Texas, Austin.

McCandless, Richard Melvin. Measurement of bulking in landslides on the basis of topographic form and density changes in landslide debris. M, 1976, University of Cincinnati. 63 p.

McCandless, Susan L. Subsurface interpretation of the Bradford Sand zones in the Punxsutawney 15-minute quadrangle, Pennsylvania. M, 1981, Indiana University of Pennsylvania. 53 p.

McCandless, Tom Elden. The mineralogy, morphology, and chemistry of detrital minerals of a kimberlitic and eclogitic nature, Green River basin, Wyoming. M, 1982, University of Utah. 107 p.

McCann, A. James. Les Minéralisations en niobium et en terres rares des migmatites du Lac Walker, Comté Duplessis, Qué M, 1986, Ecole Polytechnique. 224 p.

McCann, Allan Mervyn. Structural and stratigraphic relationships in Silurian rocks of the Port Albert-Horwood area, Twillingate-Fogo districts, Newfoundland. M, 1973, Memorial University of Newfoundland. 102 p.

McCann, James. Les minéralisations en niobium et en terres rares des migmatites du Lac Walker, Comte Duplessis, Québec. M, 1986, Ecole Polytechnique. 224 p.

McCann, Martin William, Jr. A Bayesian geophysical model for seismic hazard. D, 1980, Stanford University. 339 p.

McCann, Michael R. Hydrogeology of Northeast Woodford County, Kentucky. M, 1978, University of Kentucky. 104 p.

McCann, Thomas P. Evaporites and sedimentary features of the Upper Satanka Shale (Permian) of the Box Elder Creek-Sand Creek region, Larimer County, Colorado. M, 1949, University of Colorado.

McCann, Tommy. Sedimentology and ichnology of the Cretaceous Kodiak Formation, Alaska. M, 1985, University of New Brunswick.

McCann, William Richard. Large- and moderate-size earthquakes; their relationship to the tectonics of subduction. D, 1980, Columbia University, Teachers College. 194 p.

McCann, William Sidney. Geology and mineral deposits of the Bridge River map-area, British Columbia. D, 1920, Yale University.

McCardle, Michael F. Survey of electrical geophysical techniques. M, 1972, Stanford University.

McCarl, Henry Newton. A study of the Lower Kittanning Under-Clay near Curwensville, Pennsylvania. M, 1964, Pennsylvania State University, University Park. 120 p.

McCarl, Henry Newton. The mineral aggregate industry in the vicinity of Baltimore, Maryland. D, 1969, Pennsylvania State University, University Park. 266 p.

McCarley, Lon Allen. An autoregressive process model for constant Q attenuation. M, 1979, University of Texas, Austin.

McCarn, Steve T. Petrology of modern stream sands and Upper Triassic sandstones in the eastern Piedmont of North Carolina; a clue to paleoclimate interpretation. M, 1980, Southern Illinois University, Carbondale. 81 p.

McCarron, Kathryn R. PS converted reflections; a feasibility study using vertical receivers. M, 1984, Virginia Polytechnic Institute and State University. 67 p.

McCarter, Michael Kim. A correlation of strength and dynamic properties of some clastic sedimentary rocks. D, 1972, University of Utah. 206 p.

McCarter, Paul. Geology and mineralization of the Lateral Lake Stock, District of Kenora, northwestern Ontario. M, 1981, Oregon State University. 141 p.

McCarter, William Blair. The Woodbine Formation. M, 1928, University of Texas, Austin.

McCarthy, Brian P. Sedimentation of the Austin Glen Graywacke Member of the Normanskill Formation, southeastern New York State. M, 1985, SUNY, College at Fredonia. 73 p.

McCarthy, Conrad Joseph. Continuity equation process-response models of landform evolution. M, 1977, University of Washington.

McCarthy, Conrad Joseph. Sediment transport by rainsplash. D, 1980, University of Washington. 215 p.

McCarthy, Eugene Desmond. Treatise on organic geochemistry. D, 1967, University of California, Berkeley. 290 p.

McCarthy, Francine Marie Gisele. Late Holocene water levels in Lake Ontario; evidence from Grenadier Pond. M, 1986, University of Toronto.

McCarthy, James M. Optimal pump test design for parameter estimation and prediction in groundwater hydrology. D, 1988, University of California, Los Angeles. 152 p.

McCarthy, Jeremiah Francis. Cretaceous ammonites of Shafter area, Presidio County, Trans-Pecos, Texas. M, 1953, University of Texas, Austin.

McCarthy, Jill. Seismic reflection imaging of plate boundaries; examples from the central Aleutians, the North Atlantic, and the Basin and Range Province of the Western U.S. D, 1987, Stanford University. 188 p.

McCarthy, John Patrick. Alkalinity, sulfate, and chloride production in the streams of Illinois. M, 1972, Northern Illinois University. 89 p.

McCarthy, Michael Martin. Towards understanding environmental impact; monitoring and analyzing levels of existing land alteration. D, 1973, University of Wisconsin-Madison.

McCarthy, R. J. Geology of the southern Maverick Springs Range, White Pine County, Nevada. M, 1974, San Diego State University.

McCarthy, Susan Mary. A crustal and upper mantle study using the Gauribidanur array in southern India. M, 1982, University of North Carolina, Chapel Hill. 150 p.

McCarthy, Thomas Richard. The metamorphic petrology of the sideroplesite and cummingtonite schist facies of the Homestake Formation, Homestake Mine, Lead, South Dakota. M, 1976, University of Wisconsin-Madison.

McCarthy, William R. Stratigraphy and structure, Funlock-Motoqua area, Washington County, Utah. M, 1959, University of Washington. 41 p.

McCartney, Garnet Chester. A petrographic study of the (Mississippian) Chester sandstones of Indiana. D, 1931, University of Wisconsin-Madison.

McCartney, Garnet Chester. A petrographic study of various sand horizons of Manitoba and eastern Saskatchewan. M, 1928, University of Manitoba.

McCartney, M. Carol L. Statistical reliability of surficial materials maps in a portion of Dane County, Wisconsin. M, 1976, University of Wisconsin-Madison.

McCartney, M. Carol L. Stratigraphy and compositional variability of till sheets in part of northeastern Wisconsin. D, 1979, University of Wisconsin-Madison. 159 p.

McCartney, Merle G. Depositional facies, petrology, and diagenesis of the Frontier Formation, Whiskey Buttes gas unit, Wyoming. M, 1985, University of Texas, Austin. 164 p.

McCartney, Richard F. Origin and diagenesis of Paleosols in the Bahamas. M, 1987, Miami University (Ohio).

McCartney, William Douglas. Areal geology of the Holyrood map-area, Newfoundland, 1 inch to 1 mile. D, 1952, Harvard University.

McCartney, William Douglas. Geology of the north-central Avalon Peninsula, Newfoundland. D, 1959, Harvard University.

McCartney, William H. Method of sampling diamond-drill core. M, 1922, University of Missouri, Rolla.

McCarty, Dana G. Application of the heat-flow equation to the problem of unsteady flow of fluids in petroleum reservoirs. M, 1951, [University of Houston].

McCarty, Harry Brinton, Jr. Stable carbon isotope distributions of thermocatalytically generated low molecular weight hydrocarbon gases. D, 1984, University of Rhode Island. 132 p.

McCarty, J. E. The distribution and relationship between copper, lead, and zinc in an oyster reef and its peripheral sediments in St. Louis Bay, Mississippi. M, 1973, University of Southern Mississippi.

McCarty, Kevin L. Sedimentology of the Carboniferous West Bay Formation, Nova Scotia, Canada. M, 1980, University of Nebraska, Lincoln.

McCarty, Rose Mary. Structural geology and petrography of part of the Vadito Group, Picuris Mountains, New Mexico. M, 1983, University of New Mexico. 159 p.

McCarty, Tedford A. The geology of a portion of southwestern Callaway County, Missouri. M, 1951, University of Missouri, Columbia.

McCarty, Thomas Richard. A field study of water flow over and through a shallow, sloping, heterogeneous soil. D, 1980, Cornell University. 173 p.

McCary, Charles Edgar Little. Pre-Catheys geology of the Aspen Hill Quadrangle, Giles County, Tennesse. M, 1957, Vanderbilt University.

McCaskey, Hiram Dryer. Geology of the iron ore deposits of Durham, Pennsylvania. M, 1907, Lehigh University.

McCaskey, Michael D. Rock deformation associated with thrust transverse ramps. M, 1982, Texas A&M University.

McCasland, Ross Duncan. Subsurface geology of the Dalhart Basin, Texas Panhandle. M, 1980, Texas Tech University. 147 p.

McCaslin, John. A problem to check and develop theoretical resistivity curves in the laboratory with a model tank. M, 1952, New Mexico Institute of Mining and Technology. 52 p.

McCauley, Charles Anthony. Management of subsiding lands; an economic evaluation. D, 1973, University of Arizona.

McCauley, James R. Reservoir water quality monitoring with orbital remote sensors. D, 1977, University of Kansas. 111 p.

McCauley, James R. Surface configuration as an explanation for lithology-related cross-polarized radar image anomalies. M, 1973, University of Kansas.

McCauley, James Weymann. Control of nucleation, crystal growth, and doping by various calcium carbonate phases by the gel technique. M, 1965, Pennsylvania State University, University Park. 151 p.

McCauley, John F. Preliminary report on the sedimentary uranium occurrences in the State of Pennsylvania. M, 1957, Columbia University, Teachers College.

McCauley, Marlene Louise. Phase relations of the sodic amphibole crossite. D, 1986, University of California, Los Angeles. 139 p.

McCauley, Marvin Leon. Environmental geologic analysis of the Emerald Bay slide. M, 1973, University of California, Davis. 46 p.

McCauley, Ronald Arthur. Crystal chemistry and luminescence of pyrochlores. D, 1969, Pennsylvania State University, University Park. 152 p.

McCauley, Victor T. Zinc mines near Shullsburg, Wisconsin. M, 1954, University of Wisconsin-Madison.

McCauslin, S. E. A computer simulation model of caldera-forming silicic magmatic activity in the southwestern United States. M, 1985, Kent State University, Kent. 150 p.

McCave, Ian Nicholas. A stratigraphical and sedimentological analysis of a portion of the Hamilton Group (Middle Devonian) of New York State. D, 1967, Brown University. 315 p.

McChesney, Robert Douglass Ross. Stress investigation conducted on the shaft pillar Ahmeek Mines, number three shaft, Calumet and Hecla Incorporated, Ahmeek, Michigan. M, 1956, Michigan Technological University. 73 p.

McChesney, Stephen Michael. Geology of the Sultana Vein, Bohemia mining district, Oregon. M, 1987, University of Oregon. 160 p.

McClain, Anthony. Depositional environment of lower Green River Formation sandstones (Eocene), Red Wash Field (Uinta Basin), Uintah County, Utah. M, 1985, Texas A&M University. 116 p.

McClain, Donald Schofield, Jr. Geophysical exploration on the coastal plain of Georgia in Baker County, Georgia. M, 1953, Emory University. 41 p.

McClain, J. S. The implication of Pn amplitudes for velocity gradients in the uppermost oceanic mantle. D, 1979, University of Washington. 96 p.

McClain, Kevin John. A geophysical study of accretionary processes on the Washington continental margin. D, 1981, University of Washington. 141 p.

McClain, Linda K. A petrogenetic model for a Precambrian metamorphic garnetite unit, southwestern Montana. M, 1977, Southern Illinois University, Carbondale. 76 p.

McClain, Shannon. Oceanic lithospheric cooling of the South Atlantic; depth anomalies and geoid height signature. M, 1984, Purdue University. 110 p.

McClain, T. J. Digital simulation of the Ogallala Aquifer in Sherman County, northwestern Kansas. M, 1970, Kansas State University. 29 p.

McClain, William R. Petrography and stable oxygen isotope compositions of companion grainstones and rudstones, Northeast Providence Channel, Bahamas; ODP Leg 101 Hole 634A. M, 1987, University of Georgia. 57 p.

McClannahan, Kevin M. Paleomagnetic investigation of mid-Tertiary rocks from the Mogollon-Datil volcanic field, southwestern New Mexico. M, 1984, Michigan Technological University. 65 p.

McCleary, Jefferson Rand. Geology of the Carbon Ridge area, Eureka County, Nevada, with emphasis on the Diamond Peak Formation. M, 1974, University of Nevada. 123 p.

McCleary, John T. Geology of the northern part of the Fra Cristobal Range, Sierra and Socorro counties, New Mexico. M, 1960, University of New Mexico. 59 p.

McCleery, Raymond Scott. Morphology and stratigraphy of paleochannels on the Louisiana continental shelf. M, 1987, University of New Orleans. 98 p.

McClellan, Bruce S. Geochemical analysis of the Bridgeport Dike, Connecticut. M, 1986, Rutgers, The State University, Newark. 110 p.

McClellan, Elizabeth A. Geologic history of a portion of the eastern Blue Ridge, Southern Appalachians; Tray Mountain and Macedonia 7 1/2' quadrangles, Georgia. M, 1988, University of Tennessee, Knoxville. 179 p.

McClellan, Guerry Hamrick. Geology of Attapulgus Clay in North Florida and Southwest Georgia. D, 1964, University of Illinois, Urbana. 127 p.

McClellan, Guerry Hamrick. Identification of clay minerals from the Hawthorne Formation, Devil's Mill Hopper, Alachua County, Florida. M, 1962, University of Florida. 38 p.

McClellan, R. D. Geology of the San Juan Islands (Washington). D, 1927, University of Washington.

McClellan, Thomas Stewart. Permian carbonate facies of the Franson Member, Phosphoria Formation, in southwestern Montana. M, 1973, University of Montana. 85 p.

McClellan, Thurman Ralph. Subsurface geology of the Ballevue Field, Bossier Parish, Louisiana. M, 1959, University of Arkansas, Fayetteville.

McClellan, William Alan. Arenaceous foraminifera from the Waldron Shale (Niagaran) of Southeast Indiana. M, 1965, University of Cincinnati. 120 p.

McClellan, William Alan. Siluro-Devonian microfaunal bistratigraphy in Nevada. D, 1969, University of Washington. 199 p.

McClelland, H. W. Evolution in some Pliocene foraminifera of Southern California. M, 1926, Massachusetts Institute of Technology. 115 p.

McClelland, John E. The effect of time, temperature, and particle size on the release of bases from some common soil-forming minerals of different crystal structure. D, 1949, Iowa State University of Science and Technology.

McClelland, John Edward A. A chemical and mineralogical study of Saskatchewan boulder clay. M, 1945, University of Saskatchewan.

McClelland, Lindsay R. Some glowing avalanche deposits from the 1974 eruption of Volcan Fuego, Guatemala. M, 1976, Dartmouth College. 113 p.

McClelland, Steven W. The crystallography and petrology of kammererite from the Day Book body, Yancey County, North Carolina. M, 1973, University of Iowa. 32 p.

McClennen, Charles Eliot. Nature and origin of the New Jersey continental shelf topographic ridges and depressions. D, 1973, University of Rhode Island. 103 p.

McClernan, Henry G. Geology of the Sheep Creek area, Meagher County, Montana. M, 1969, Montana College of Mineral Science & Technology. 51 p.

McClernan, Henry G. Metallogenesis of mineral deposits in Cambrian carbonate rocks, southwestern Montana. D, 1977, University of Idaho. 122 p.

McClincy, Matthew John. Tephrostratigraphy of the middle Eocene Chumstick Formation, Cascade Range, Douglas County, Washington. M, 1986, Portland State University. 125 2 plates p.

McClish, Richard F. Lithostratigraphy and conodont biostratigraphy of the Richmond Group of southwestern Ohio and southeastern Indiana. M, 1965, Ohio State University.

McClory, Joseph Patrick. Carbon and oxygen isotopic study of Archaean stromatolites from Zimbabwe. D, 1988, University of Rhode Island. 139 p.

McCloy, Cecelia. Studies in micropaleontology and stratigraphy; planktonic foraminifera and euthecosomatus pteropods in the surface waters of the North Atlantic; stratigraphy and depositional history of the San Jose del Cabo Trough, Baja California Sur, Mexico. M, 1984, Stanford University. 42 p.

McCloy, James Murl. Morphologic characteristics of the Blow River delta, Yukon Territory, Canada. D, 1969, Louisiana State University. 176 p.

McClung, David M. Avalanche defense mechanics. D, 1974, University of Washington. 103 p.

McClung, Esther Carroll. The geologic section in Fayette County, Texas. M, 1930, University of Texas, Austin.

McClung, Wilson S. Marine offshore to alluvial plain transitions within the "Chemung"-Hampshire interval (Upper Devonian) of the southern Central Appalachi-

ans. M, 1983, Virginia Polytechnic Institute and State University. 257 p.

McClure, D. Interpretation of long aeromagnetic profiles. M, 1963, University of Alberta. 81 p.

McClure, Daniel Victor. Late Cretaceous sedimentation, southern Santa Lucia Range, California. M, 1969, University of California, Santa Barbara.

McClure, Dennis. Stream patterns of the Calamus River, Nebraska. M, 1987, University of Illinois, Chicago.

McClure, James Edward. A seismic investigation in the Yorkton area of Saskatchewan. M, 1973, University of Saskatchewan. 409 p.

McClure, James Graham. Physicochemical investigation of shale slaking. D, 1980, University of California, Berkeley. 314 p.

McClure, John W. Depositional environments of the Upper Cambrian Corset spring shale and Sneakover Limestone in the House Range, western Utah. M, 1978, University of Kansas. 65 p.

McClure, M. E. Application of geophysical methods to the determination of certain horizons in the Pleistocene deposits of the London area, Ontario. M, 1954, University of Western Ontario.

McClure, Paul Frederick. Earth dynamic filtering for earthquake prediction and love number determination. D, 1972, Colorado State University. 161 p.

McClure, R. K. Recognition of shallow water and tidal flat aspects of the Abrigo Formation (Cambrian). M, 1977, University of Arizona. 132 p.

McClure, Robert I. Geology and utilization of the Sharon Conglomerate of Jackson County, Ohio. M, 1939, University of Cincinnati. 75 p.

McClure, William C. Geology of the Enoree area, South Carolina. M, 1963, University of South Carolina. 39 p.

McClurg, Dai C. Coda-Q at Mt. St. Helens; implications for volcanic seismology. M, 1987, University of Washington. 75 p.

McClurg, James Edson. Teaching geology; a sourcebook for the elementary and secondary teacher. M, 1959, University of Michigan.

McClurg, Larry William. Source rocks and sediments in drainage area of north Eden creek, Bear lake plateau, Utah (Rich County) and (Bear Lake County) Idaho. M, 1970, Utah State University. 84 p.

McCluskey, James M. The role and magnitude of eolian processes in the barrier island environment. D, 1987, Rutgers, The State University, New Brunswick. 339 p.

McClymonds, Neal E. Stratigraphy and structure, Waterman Mountains, Pima County, Arizona. M, 1957, University of Arizona.

McClymont, Gordon Lee. Expert systems applications in hydrogeology. M, 1988, University of Alberta. 136 p.

McCoard, David. Structure of the Last Chance thrust in the Last Chance range, California. M, 1970, University of California, Los Angeles.

McCobb, Harry W. A geologic map of South America. M, 1927, University of Pittsburgh.

McColl, Kathryn Margaret. Geology of Britannia Ridge, East Section, Southwest British Columbia. M, 1987, University of British Columbia. 162 p.

McCollister, Linda Suzanne. Aeration of anaerobic marsh sediments by the marsh crabs Uca and Sesarma spp. M, 1974, University of Virginia. 44 p.

McColloch, Samuel. Mauch Chunk series (Mississippian) in southwestern West Virginia. M, 1957, West Virginia University.

McCollom, Robert Lloyd. An array study of upper mantle seismic velocity in Washington State. M, 1972, University of Washington. 62 p.

McCollom, Robert Lucien, Jr. Lithofacies study of the (Miocene) Vaqueros Formation, Santa Cruz Mountains, California. M, 1959, Stanford University. 48 p.

McCollor, Douglas Clayton. Power transmission harmonic current and its use in geophysical exploration. M, 1982, University of British Columbia. 136 p.

McCollough, Edward Heron. A preliminary report of the geology of a part of Grant County, North Dakota. M, 1925, University of Oklahoma. 48 p.

McCollough, Edward L. The geology of the Finley area, Pushmataha County, Oklahoma. M, 1954, University of Oklahoma. 48 p.

McCollough, William F. Stratigraphy, structure, and metamorphism of Permo-Triassic rocks along the western margin of the Idaho Batholith, John Day Creek, Idaho. M, 1984, Pennsylvania State University, University Park. 141 p.

McCollough, William Matthew. A chemical analysis of the core from a deep test well, Wood County, West Virginia. M, 1957, University of Pittsburgh.

McCollum, Audrey Britton. Geology of the McNeil, Watters Park area, Travis County, Texas. M, 1932, University of Texas, Austin.

McCollum, Jack H. A subsurface study of the Northeast Elmore Field, Garvin County, Oklahoma. M, 1949, University of Oklahoma. 35 p.

McCollum, L. B. Distribution of marine faunal assemblages in a Middle Devonian stratified basin, lower Ludlowville Formation, New York. D, 1980, SUNY at Binghamton. 176 p.

McCollum, Morris J. Petrography of the Midale Subinterval in the Bottineau and Renville counties area, North Dakota. M, 1962, University of North Dakota. 107 p.

McCollum, Robert Andrew. Paleomagnetic and rock magnetic investigation of the relationship between hydrocarbons and authigenic magnetic minerals in the Permian Lyons Sandstone, northern Front Range and Denver Basin, Colorado. M, 1988, University of Oklahoma. 81 p.

McColly, Robert A. Geology of the Saguaro National Monument area, Pima County, Arizona. M, 1961, University of Arizona.

McComas, C. H., III. Nonlinear interaction of internal gravity waves. D, 1975, The Johns Hopkins University.

McComas, Murray Ratcliffe. Environmental control of inorganic water quality near Severence (Weld County), Colorado. M, 1966, Colorado State University. 61 p.

McComas, Murray Ratcliffe. Pleistocene geology and hydrogeology of the middle Illinois Valley. D, 1969, University of Illinois, Urbana. 130 p.

McComb, Ronald. Petrology, paleodepositional environments, biostratigraphy and paleontology of the Decatur and Rockhouse limestones (Upper Silurian-Lower Devonian), west-central Tennessee. M, 1987, University of Tennessee, Knoxville. 309 p.

McComb, Thomas D. Structural geology of southeastern Sierra El Batamote (northwestern Sonora, Mexico). M, 1987, University of Cincinnati. 111 p.

McConachy, Timothy Francis. Hydrothermal plumes over spreading ridges and related deposits in the Northeast Pacific Ocean; the East Pacific Rise near 11 degrees north and 21 degrees north, Explorer Ridge and J. Tuzo Wilson Seamounts. D, 1988, University of Toronto.

McConn, Virginia Barr. Disequilibrium study of Recent volcanic rocks. M, 1967, Columbia University. 31 p.

McConnaughey, Paul Kevin. Transient microsite models of denitrification; theory and experiment. D, 1983, Cornell University. 311 p.

McConnaughey, Ted Alan. Oxygen and carbon isotope disequilibria in Galapagos corals; isotopic thermometry and calcification physiology. D, 1986, University of Washington. 340 p.

McConnel, Denis B. Gamma-ray scattering and neutron-capture gamma-ray detection. M, 1958, University of Western Ontario.

McConnell, C. L. Mineral variations in phosphatic slimes. M, 1973, University of South Florida, St. Petersburg.

McConnell, Cary Lewis. Groundwater salinity and resource evaluation by spontaneous potential mea-surements; eastern Jefferson and Carter counties, Oklahoma. D, 1981, University of Oklahoma. 278 p.

McConnell, D. R. Bedrock topography and paleogeomorphology northeast of Joliet, Illinois. M, 1975, University of Illinois, Chicago.

McConnell, David Alan. Paleozoic structural evolution of the Wichita Uplift, Southwest Oklahoma. D, 1987, Texas A&M University. 366 p.

McConnell, David Alan. The mapping and interpretation of the structure of the northern Slick Hills, Southwest Oklahoma. M, 1983, Oklahoma State University. 131 p.

McConnell, Duncan. A structural investigation of the isomorphism of the apatite group. D, 1937, University of Minnesota, Minneapolis. 46 p.

McConnell, Duncan. Garnets from Sierra Tlayacac, Morelos, Mexico. M, 1932, Cornell University.

McConnell, Elliott Bonnell, Jr. Laboratory studies of the self-potential of reservoir sands. M, 1953, Pennsylvania State University, University Park. 49 p.

McConnell, G. W. An analysis of the economic potential of the Bathurst District base metal deposits New Brunswick. M, 1959, University of Toronto.

McConnell, Harold Lee. Some quantitative aspects of slope inclination in portions of the glaciated Upper Mississippi Valley. D, 1964, Iowa State University of Science and Technology. 201 p.

McConnell, John Wilson. Geochemical dispersion in wallrocks of Archean massive sulphide deposits. M, 1976, Queen's University. 230 p.

McConnell, Keith I. Geology of the late Precambrian Flat River Complex and associated volcanic rocks near Durham, North Carolina. M, 1974, Virginia Polytechnic Institute and State University.

McConnell, M. D. The geology of the Clark-Hinman Park area, Routt County, Colorado. M, 1960, University of Wyoming. 71 p.

McConnell, R. K., Jr. Theoretical studies in prospecting for massive geologic bodies using elastic waves. M, 1960, University of Toronto.

McConnell, Robert. Lithostratigraphy and petrography of the Upper Cambrian Maynardville Formation within the Copper Creek fault belt of East Tennessee. M, 1967, University of Tennessee, Knoxville. 92 p.

McConnell, Robert. The Apache Group (Proterozoic) of central Arizona. D, 1972, University of California, Santa Barbara.

McConnell, Roger Harmon. Petrology of a quartzite-quartz diorite contact zone near Harpster, Idaho. M, 1936, University of Idaho. 24 p.

McConnell, Wallace R. Geography of southwestern Wisconsin. M, 1918, University of Wisconsin-Madison.

McCool, Kevin E. Taphonomy of the Valentine Railway Quarry "B" bone bed (late Barstovian), north-central Nebraska. M, 1988, University of Nebraska, Lincoln. 115 p.

McCord, Daniel Lee. Hecla Hoek stratigraphy and structure, South Chamberlindalen, Spitsbergen. M, 1978, University of Wisconsin-Madison.

McCord, Thomas Bard. Color differences on the lunar surface. D, 1968, California Institute of Technology. 181 p.

McCord, Virgil Alexander Stuart. A new computerized X-ray densitometric system for tree-ring analysis. M, 1984, University of Arizona. 150 p.

McCord, Wallace R. Geology of the Upper Creek area of Gunnison County, Colorado, including a petrographic study of the Paleozoic sedimentary rocks. M, 1957, University of Kentucky. 49 p.

McCord, William K. Effects of the power plant discharge into Monterey Bay at Moss Landing. M, 1971, Naval Postgraduate School.

McCorkell, Robert H. Fundamentals of fission track dating. M, 1973, Carleton University. 143 p.

McCorkle, Daniel Charles. Stable carbon isotopes in deep sea pore waters; modern geochemistry and paleoceanographic applications. D, 1987, University of Washington. 209 p.

McCormack, John Kevin. Paragenesis and origin of sediment-hosted mercury ore at the McDermitt Mine, McDermitt, Nevada. M, 1986, University of Nevada. 97 p.

McCormack, Martin R. H. The Cardium Formation (Upper Cretaceous); Morley and Jumpingpound map areas, Alberta. M, 1972, University of Calgary. 164 p.

McCormack, Michael David. Two-dimensional modeling and migration using a hybrid Kirchhoff-Trorey approach. D, 1980, University of Houston. 91 p.

McCormack, Robert Keith. Quaternary and Tertiary geology of the Millville area in southern New Jersey. M, 1955, Rutgers, The State University, New Brunswick. 121 p.

McCormick, C. L. Salt marshes of Merrimack and Parker River estuaries. D, 1967, University of Massachusetts. 115 p.

McCormick, Charles D. The stratigraphy and petrology of Gartra Member (Triassic), Uinta Mountain area, Utah and Colorado. M, 1968, University of Nebraska, Lincoln.

McCormick, Charles Larry. Petrology of the Waynesville Formation (Cincinnatian Series) in the Ft. Ancient-Oregonia region, Warren County, Ohio. M, 1964, Miami University (Ohio). 89 p.

McCormick, Dennis Joseph. Intrabasinal and eustatic controls on the deposition of the Watton Canyon Member, Twin Creek Limestone, Lincoln County, Wyoming. M, 1983, University of Oklahoma. 71 p.

McCormick, George Robert. An investigation of the compatibility relations in the system MgO-GeO_2-MgF_2-LiF principally at 1000°C. D, 1964, Ohio State University. 122 p.

McCormick, George Robert. Petrology of pre-Cambrian rocks in some wells in Ohio. M, 1960, Ohio State University.

McCormick, J. W. Transmission electron microscopy of experimentally deformed synthetic quartz. D, 1977, University of California, Los Angeles. 184 p.

McCormick, John Murray. Seasonal distribution of foraminifera from Tomales Bay, Marin County, California. M, 1987, University of California, Davis. 132 p.

McCormick, Kelli A. Sources of clasts in terrestrial and lunar impact melts. M, 1988, University of New Mexico. 90 p.

McCormick, Louis M. Bedrock geology of the Norhtwest quarter of the Dongola Quadrangle, Illinois. M, 1967, Southern Illinois University, Carbondale. 114 p.

McCormick, Michael. Facies analyses and paleoenvironmental interpretation of the Upper Triassic Hosselkus Limestone, Shasta County, California. M, 1986, University of California, Berkeley. 147 p.

McCormick, Robert B. Composition and properties of melilite. M, 1932, University of Wisconsin-Madison.

McCormick, Robert B. The ternary system hardystonite-akermanite-gehlenite. D, 1936, University of Wisconsin-Madison.

McCormick, Tamsin Cordner. Crystal chemistry and breakdown reactions of aluminous mantle-derived omphacites. D, 1984, Arizona State University. 122 p.

McCormick, Tamsin Cordner. Exsolution in alkali feldspars and implications for the cooling history of the Battleship Rock Tuff, northern New Mexico. M, 1980, University of New Mexico. 99 p.

McCormick, Wade Lowery. A study of the McDermott Formation, Upper Cretaceous age in the San Juan Basin, southwestern Colorado and northern New Mexico. M, 1953, University of Illinois, Urbana.

McCormick, William Vincent, III. The geology, mineralogy, and geochronology of the Sierra San Pedro Martir Pluton, Baja California, Mexico. M, 1986, San Diego State University. 123 p.

McCorquodale, Ross J. Structures obtained by etching polished ore mineral specimens. M, 1939, University of Minnesota, Minneapolis. 40 p.

McCourt, George H. Quaternary palynology of the Bluefish Basin, northern Yukon Territory. M, 1982, University of Alberta. 178 p.

McCowan, Douglass William. Dynamic finite element analysis with applications to seismological problems. D, 1975, Pennsylvania State University, University Park. 197 p.

McCoy, A. W. Artesian water of Boone County, Missouri. M, 1914, University of Missouri, Columbia.

McCoy, David L. A description and interpretation of the Permian Minnekahta Limestone of the Black Hills, South Dakota. M, 1985, Kent State University, Kent. 94 p.

McCoy, Floyd. Geology of Ofu and Olosega islands, Manu's Group, American Samoa. M, 1966, University of Hawaii.

McCoy, Floyd W., Jr. Late Quaternary sediments in the eastern Mediterranean Sea. D, 1974, Harvard University.

McCoy, Gail. Analysis of fractures and speleogenesis at Lilburn Cave, Tulare County, California. M, 1983, San Jose State University. 83 p.

McCoy, Henry J. An electric log and sample study of the Citronella oil field, Mobile County, Alabama. M, 1958, Florida State University.

McCoy, John J. The fossil birds of the Itchtuchknee River, Florida. D, 1960, University of Florida. 96 p.

McCoy, M. R. Pre-Whitewood Ordovician stratigraphy of the Black Hills. M, 1952, University of Wyoming. 83 p.

McCoy, Melinda Delle. Paleoclimatic implications of Mississippi loess. M, 1987, University of New Orleans. 110 p.

McCoy, Robert L. The vertical gradients of gravity and the geoid in Maine and Southeast Canada. M, 1964, Ohio State University.

McCoy, Roger Michael. An evaluation of radar imagery as a tool for drainage basin analysis. D, 1967, University of Kansas. 112 p.

McCoy, Scott, Jr. A description of the limestone blocks of the Tucson Mountain chaos, Pima County, Arizona. M, 1964, University of Arizona.

McCoy, William Dennis. Quaternary aminostratigraphy of the Bonneville and Lahontan basins, Western U.S., with paleoclimatic implications. D, 1981, University of Colorado. 616 p.

McCracken, A. D. Late Ordovician and Early Silurian conodonts from Anticosti Island, Quebec. M, 1978, University of Waterloo.

McCracken, Alexander Duncan. Middle Ordovician to Silurian (Wenlock) conodont taxonomy and biostratigrphy from basinal strata of the Road River Formation in the Richardson Mountains, northern Yukon Territory. D, 1985, University of Western Ontario. 576 p.

McCracken, Ralph Joseph. Soil geography of Palau and Saipan. M, 1951, Cornell University.

McCracken, Weaver H., Jr. Geology of Hurds Draw area near Kent, Culberson, and Jeff Davis counties, Texas. M, 1948, University of Texas, Austin.

McCracken, Willard A. Petrology of the Catahoula Formation (Miocene) in northeastern Gonzales County, northwestern Lavaca County, and southwestern Fayette County, Texas. M, 1967, University of Houston.

McCracken, Willard Alton. Paleocurrents and petrology of Sespe Sandstones and conglomerates (Oligocene), Ventura Basin, California. D, 1972, Stanford University. 192 p.

McCrae, Robert O. Geomorphic effects of the 1951 Kansas River flood. M, 1954, University of Kansas. 68 p.

McCrary, Megan Marie. Depositional history and petrography of the Todilto Formation (Jurassic), New Mexico and Colorado. M, 1985, University of Texas, Austin. 184 p.

McCraw, David Jackson. A phytogeographic history of Larrea in southwestern New Mexico illustrating the

historical expansion of the Chihuahuan Desert. M, 1985, [University of New Mexico]. 137 p.

McCrea, Maureen. Evaluation of Washington State's coastal management program through changes in port development. D, 1980, University of Washington. 316 p.

McCreary, Gary B. A subsurface study of the Tar Springs Sandstone bodies in a portion of Webster County, Kentucky. M, 1957, University of Kentucky. 56 p.

McCreery, Robert Atkeson. Mineralogy of the Palouse and related series. D, 1954, Washington State University. 120 p.

McCreesh, Catherine A. Determination of pore type-throat size relationships in sandstones and their implications in terms of diagenesis and petrophysics. D, 1987, University of South Carolina. 120 p.

McCrehan, Richard E. Colorimetric analytic methods applied to silicate rocks. M, 1956, Dartmouth College. 62 p.

McCrevey, John Alfred. Fossil traces of the Whitestone Limestone and associated strata of the Walnut Formation, Lower Cretaceous, south-central Texas. M, 1974, Rice University. 105 p.

McCrink, Marie Taaffe. Diagenesis in the Creede Formation, San Juan Mountains, Creede, Colorado. M, 1982, New Mexico Institute of Mining and Technology. 113 p.

McCroden, Thomas J. Geology of a portion of the Gabilan Range, California. M, 1949, Stanford University. 81 p.

McCrone, Alistair William. The Red Eagle cyclothem (Lower Permian). D, 1960, University of Kansas. 268 p.

McCrone, Alistair William. The Red Eagle Formation in Nebraska. M, 1955, University of Nebraska, Lincoln.

McCrory, Patricia Alison. Late Cenozoic history of the Humboldt Basin, Cape Mendocino area, California. D, 1987, Stanford University. 266 p.

McCrory, Roy. Pressure waves generated by nuclear explosions. M, 1965, New Mexico Institute of Mining and Technology. 56 p.

McCrory, Thomas Alan. Joint and lineament origins; evaluation of several possible causes in a region of northwestern Wyoming. M, 1982, University of Massachusetts. 52 p.

McCrossan, Robert G. A study of the Ireton inter-reef member of the Upper Devonian Woodbend Formation of central Alberta. D, 1957, University of Chicago. 121 p.

McCrum, Michael Arthur. A chemical mass balance of the Ester Creek and Happy Creek watersheds on Ester Dome, Alaska. M, 1985, University of Alaska, Fairbanks. 119 p.

McCrumb, Dennis R. Geologic factors in mountain water pollution. M, 1973, Colorado State University. 155 p.

McCuaig, James A. A copper-nickel occurrence in Pardee Township, Thunder Bay District, Ontario. M, 1950, McGill University.

McCuaig, James A. Experimental studies in rheomorphism. D, 1953, McGill University.

McCubbin, Donald Gene. Basal Cretaceous of southwestern Colorado and southeastern Utah. D, 1961, Harvard University.

McCue, J. J. Facies changes within the Phosphoria Formation in the southeast portion of the Big Horn Basin, Wyoming. M, 1953, University of Wyoming. 88 p.

McCulla, Michael. The genesis of the Los Ochos uranium deposits, Saguache County, Colorado. M, 1980, University of Nevada. 128 p.

McCulla, Michael S. Geology and metallization of the White River area, King and Pierce counties, Washington. D, 1987, Oregon State University. 213 5 plates p.

McCullar, Dan Brett. Seismic refraction analysis of the crust and upper mantle beneath the Rio Grande

Rift in southern New Mexico. M, 1977, University of Wyoming. 127 p.

McCulley, Bryan L. Source of nitrate in Arroyo Grande Basin, California. M, 1978, University of Texas, Austin.

McCulloch, Charles Malcolm. A trace element study of selected conodonts of middle Ordovician age (from Lexington Limestone, Clearmont County, Ohio, and Clays Ferry Formation, Madison County, Kentucky). M, 1971, University of Kentucky. 71 p.

McCulloch, David Sears. Late Cenozoic erosional history of Huerfano Park, Colorado. D, 1963, University of Michigan. 182 p.

McCulloch, David Sears. Vacuole disappearance temperatures of laboratory-grown hopper halite crystals. M, 1958, University of Michigan.

McCulloch, John Snyder. The geology of the Oceanic (Pennsylvanian) Field, Howard and Borden counties, Texas. M, 1959, University of Oklahoma. 63 p.

McCulloch, Malcolm Thomas. Part I, Sm-Nd and Rb-Sr chronology of crustal formation; Part II, Ba, Nd and Sm isotopic anomalies in the Allende meteorite. D, 1980, California Institute of Technology. 400 p.

McCulloch, Paul D. Biostratigraphy of the Upper Mississippian Kennetcook Limestone in the Fundy Epieugeosyncline, Maritime provinces. M, 1973, Acadia University.

McCulloch, Richard. Report of the Atlantic Mine, Atlantic Michigan. M, 1888, Washington University.

McCulloch, Richard. Report of the Col. Sellers concentration mill, Leadville, Colorado. M, 1890, Washington University.

McCullogh, Debia Hershelle Fine. Petrologic analysis of surficial sediments, Gray's Reef National Marine Sanctuary, Georgia continental shelf, U.S.A. M, 1985, Emory University. 269 p.

McCulloh, Richard P. Geology of the Yellow Hill Quadrangle, Brewster County, Texas. M, 1977, University of Texas, Austin.

McCulloh, Thane Hubert. Geology of the southern half of the Lane Mountain Quadrangle, California. D, 1952, University of California, Los Angeles.

McCullough, Charles R. Airphoto interpretation of soils and drainage of Rush County, Indiana. M, 1948, Purdue University.

McCullough, Douglas L. Geology of the Parkdale area, Fremont County, Colorado. M, 1959, University of Kansas. 44 p.

McCullough, Edgar J. Resistivity methods in depth determinations. M, 1955, West Virginia University.

McCullough, Edgar Joseph, Jr. A structural study of the Pusch Ridge-Romero Canyon area, Santa Catalina Mountains, Arizona. D, 1963, University of Arizona. 107 p.

McCullough, Edward Allen. The petrology of certain igneous rocks of eastern New Mexico. M, 1932, Texas Tech University. 71 p.

McCullough, J. D., Jr. Mineralogy and petrology of the Tallahatta Formation (middle Eocene) in Southwest Alabama. M, 1977, Memphis State University.

McCullough, John. Pressure-temperature microscopy of petroleum-derived asphaltenes from Venezuela. M, 1984, University of Pittsburgh.

McCullough, L. A. Playa-lake and clastic deposition in Paleocene-Eocene Flagstaff Limestone, Wasatch Plateau, Utah. M, 1977, Ohio State University. 116 p.

McCullough, Louis Marshall. The Amusium ocalanum biozone, (Crystal River Formation, which is late Eocene (Jacksonian) age), (Peninsular Florida). M, 1969, University of Florida. 53 p.

McCullough, Patrick Terrence Peter. Geology of the Britannia mineralized district, British Columbia, west section. M, 1968, University of Illinois, Urbana. 115 p.

McCullough, Patrick Terrence Peter. Origin of the Ragged Top gneisses, Laramie Range, Wyoming. D, 1974, University of Illinois, Urbana. 138 p.

McCullough, Sahar A. Significance of textures in granites, Somesville Pluton, Mount Desert Island, Maine. M, 1971, University of Illinois, Urbana. 67 p.

McCullough, Thomas Richard. The geology of the Timber Canyon area, Santa Paula Peak Quadrangle, Ventura County, California. M, 1957, University of California, Los Angeles.

McCullough, Warren D. Gold mineralization at the Giltedge Property, Judith Mountains, Fergus County, Montana. M, 1978, Bowling Green State University. 133 p.

McCurdy, Harland R. Pennsylvanian and Permian stratigraphy along the eastern flank of the Laramie Range, southeastern Wyoming. M, 1941, University of Wyoming. 42 p.

McCurdy, Maureen. The effect of a petroleum waste on the Beaumont Clay. M, 1986, University of Southwestern Louisiana.

McCurdy, Robert. Water motion and sediments of northeast San Pedro Bay, California. M, 1965, University of Southern California.

McCurley, Earl B. Cu-Mo-W skarns of the Seven Devils mining district, Adams County, Idaho. M, 1986, University of Idaho. 111 p.

McCurry, Michael Owen. The petrology of the Woods Mountains volcanic center, San Bernardino County, California. D, 1985, University of California, Los Angeles. 467 p.

McCurry, Wilson G. Mineralogy and paragenesis of the ores, Christmas Mine, Gila County, Arizona. M, 1971, Arizona State University. 47 p.

McCusker, Robert. Geology of the Soledad Mountain volcanic complex, Mojave Desert, California. M, 1983, San Jose State University. 113 p.

McCuskey, Sue Ann. Demography and behavior of one-male groups of yellow baboons (Papio cynocephalus). M, 1975, University of Virginia. 64 p.

McCutchen, William T. Petrography and structural geology of the Piedmont in Egdefield County, South Carolina. M, 1961, Florida State University.

McCutchen, Wilmont R. The behavior of rocks and rock masses in relation to military geology (T642). M, 1948, Colorado School of Mines. 74 p.

McCutcheon, Fletcher Snead. Mineralogy of dolomites quarried by the Radford Limestone Company, Radford, Virginia. M, 1955, Virginia Polytechnic Institute and State University.

McCutcheon, Janice A. A geophysical investigation of the Precambrian rocks in the Llano Uplift region, Texas. M, 1982, University of Texas at El Paso.

McCutcheon, Kirk. Environmental stratigraphy of post-Dunderberg carbonate strata, Nopah Formation, (Upper Cambrian), southern Great Basin. M, 1988, California State University, Long Beach. 220 p.

McCutcheon, Tim. The petrology and geochemistry of the Precambrian granites in the northern Franklin Mountains. M, 1982, University of Texas at El Paso.

McDade, Joel A. Application of surface geophysical methods to the evaluation of waste disposal sites in North Carolina. M, 1984, North Carolina State University. 78 p.

McDade, Laddie Burl. The sedimentation and petrography of the lower Calvin Sandstone of Hughes County, Oklahoma. M, 1953, University of Oklahoma. 99 p.

McDaniel, Alice. Reinvestigation of the structure of nacrite. M, 1966, University of Wisconsin-Madison.

McDaniel, Charles Russell, Jr. Geology of the Pine Mountain window and adjacent Uchee Belt in eastern Lee County, Alabama. M, 1984, Auburn University. 105 p.

McDaniel, Gary A. Isopachous and paleogeologic studies of Southwest Oklahoma. M, 1959, University of Oklahoma. 85 p.

McDaniel, George O., Jr. Parameters of subsurface structural reconnaissance in the Simpson Group (Ordovician), South Norman area, Oklahoma. M, 1962, University of Oklahoma. 73 p.

McDaniel, P. A. Group B Streptococci; isolation, physiology, and pathogenicity. M, 1974, East Texas State University.

McDaniel, Robert C. Sedimentologic and stratigraphic significance of boulder layers in the outer coastal plain of southeastern Virginia. M, 1985, Old Dominion University. 98 p.

McDaniel, Ronald Dean. Application of a hot spring-fumarole alteration model to the genesis of the pyrophyllite deposits of the Carolina Slate Belt. M, 1976, North Carolina State University. 75 p.

McDaniel, Scott Byron. Permian-Triassic source bed analysis at Quinn River Crossing, Humboldt County, Nevada. M, 1982, University of Nevada. 120 p.

McDaniel, Willard R. Environmental study of the Wind River Formation, Fremont County, Wyoming. M, 1957, Miami University (Ohio). 55 p.

McDavid, J. D. Geology of the Muddy Creek, Carbon County, Wyoming. M, 1953, University of Wyoming. 91 p.

McDermott, A. A. Conditions of deposition of the crinoidal limestones of the Carboniferous rocks of southwestern Alberta. M, 1955, University of Toronto.

McDermott, Robert W. Depositional processes and environments of the Permian sandstone tongue of the Cherry Canyon Formation and the upper San Andres Formation, Last Chance Canyon, New Mexico. M, 1984, University of Texas, Austin.

McDermott, Robert Wayne. Depositional processes and environments of the Permian Sandstone Tongue of the Cherry Canyon Formation and the upper San Andres Formation, Last Chance Canyon, southeastern New Mexico. M, 1983, University of Texas, Austin. 179 p.

McDermott, Vincent J. Determinative mineralogy of the Randolph Intrusions, Riley County, Kansas. M, 1967, Kansas State University.

McDevitt, Marybeth C. Petrology and diagenesis of the Terryville sandstones (Cotton Valley Group), northwestern Louisiana. M, 1983, University of New Orleans. 80 p.

McDivitt, James Frederick. Economic aspects of mineral resources development in southern Idaho. D, 1954, University of Illinois, Urbana.

McDole, Robert Elroy. Loess deposits adjacent to the Snake River flood plain in the vicinity of Pocatello, Idaho. D, 1969, University of Idaho. 216 p.

McDole, Robert Elroy. Some properties of loess in the Bannock Peak transect in southeastern Idaho. M, 1968, University of Idaho. 162 p.

McDonald, Barrie Clifton. Pleistocene events and chronology in the Appalachian region of southeastern Quebec, Canada. D, 1967, Yale University. 215 p.

McDonald, Bruce Walter Robert. Geology and genesis of the Mount Skukum Tertiary epithermal gold-silver vein deposit, southwestern Yukon Territory; (NTS 105D SW). M, 1987, University of British Columbia. 170 p.

McDonald, Cecilia Louise. A reconnaissance study of the bioavailability of copper, iron, lead, magnesium, manganese, silver and zinc on the polymetallic Aqueduct Prospect, Breckenridge, Colorado. M, 1984, University of Texas at Dallas. 205 p.

McDonald, D. G. Gravity studies in the Saint Lawrence Lowlands. M, 1965, McGill University.

McDonald, David C. Gold and silver telluride mineralization at the Reid Mine, Shasta County, California. M, 1986, University of Tennessee, Knoxville. 136 p.

McDonald, David Wilson. Sedimentary petrology and paleontology of part of the Hermosa Group (Pennsylvanian) between Durango and Silverton, Colorado. M, 1984, Northeast Louisiana University. 107 p.

McDonald, Dean William Arthur. Fragmental pyrrhotite ore at Heath Steele Mines, New Brunswick. M, 1983, University of New Brunswick. 251 p.

McDonald, Eric V. Correlation and interpretation of the stratigraphy of the Palouse Loess of eastern Washington. M, 1987, Washington State University. 218 p.

McDonald, Hugh Gregory. A systematic review of the Plio-Pleistocene scelidotherline ground sloths (Mammalia; Xenarthra; Mylodontidae). D, 1988, University of Toronto.

McDonald, James M. Sediments and structure of the Nicobar Fan, Northeast Indian Ocean. D, 1977, University of California, San Diego. 162 p.

McDonald, James Vernon. Glaciation of the Seven Devils Mountain as an example of Pleistocene glaciation in central Idaho. M, 1954, University of Idaho. 42 p.

McDonald, John A. A petrological study of the Cuthbert Lake ultrabasic and basic dyke swarm; a comparison of the Cuthbert Lake ultrabasic rocks to the moak lake-type serpentinite. M, 1960, University of Manitoba.

McDonald, John Angus. Evolution of part of the lower critical zone, Farm Ruighoek, western Bushveld. D, 1963, University of Wisconsin-Madison. 190 p.

McDonald, John Harlan. Coping with environmental complexity; a computer simulation methodology. M, 1973, University of Arizona.

McDonald, Kathleen. Three-dimensional analysis of Pleistocene and Holocene coastal sedimentary units at Bethany Beach, Delaware. M, 1982, University of Delaware.

McDonald, Kent Charles. Nubia Formation, Quesir-Safaga area, Eastern Desert, Egypt. M, 1979, University of New Orleans.

McDonald, Kirk W. Structural style, thermal maturity, and sedimentary facies of the frontal Ouachita thrust belt, Le Flore County, eastern Oklahoma. M, 1986, University of Missouri, Columbia. 211 p.

McDonald, Nicholas Grant. Fossil fishes from the Newark Group of the Connecticut Valley. M, 1975, Wesleyan University. 250 p.

McDonald, Ralph L. Stratigraphy of Little Horse Creek area, Fremont County, Wyoming. M, 1953, Miami University (Ohio). 36 p.

McDonald, Robert B. A petrological study of the Recent sands of the Delaware River. M, 1961, Rutgers, The State University, New Brunswick. 60 p.

McDonald, Robert Joseph. Chemical properties of and crop production on mine spoils in western Washington. M, 1979, Washington State University. 37 p.

McDonald, Robert Lacy, Jr. Stratigraphic control of coal deposition in some Upper Carboniferous strata of south central West Virginia. M, 1978, North Carolina State University. 81 p.

McDonald, Sandra D. Use of geophysical measurements to assess cinder-aggregate potential of volcanic cinder cones. M, 1975, Northern Arizona University. 104 p.

McDonald, Sister Mary Aquin. A possible mathematical approach to the problem of pyroelectricity in tourmaline crystals. M, 1966, American University. 36 p.

McDonald, Stanley M. A study of the regional isopachous pattern of the subsurface Eocene Wilcox sediments in central Mississippi and the significance of local thickness variations. M, 1941, Louisiana State University.

McDonald, Susan. Tantalum and columbium. M, 1987, University of Texas, Austin.

McDonald, Thomas Joseph. Volatile organic compounds in Gulf of Mexico sediments. D, 1988, Texas A&M University. 250 p.

McDonald, William L. The relation of structure and value contours in some ore deposits. M, 1939, University of Toronto.

McDonald, William P. Influence of organic matter on the geotechnical properties and consolidation characteristics of northern Oregon continental slope sediments. M, 1983, Oregon State University. 69 p.

McDonnell, Daniel E. Limits of the Devonian Hampshire Formation in parts of Virginia and West Virginia. M, 1981, University of North Carolina, Chapel Hill. 128 p.

McDonnell, Sheila Louise. A gravity and magnetic survey of the Parras Basin, Mexico. M, 1987, University of New Orleans. 169 p.

McDonough, Daniel T. Structural evolution of the southeastern Scapegoat Wilderness, west-central Montana. M, 1985, University of Montana. 125 p.

McDonough, Michael Robert. Structural evolution and metamorphism of basement gneisses and Hadrynian cover, Bulldog Creek area, British Columbia. M, 1984, University of Calgary. 162 p.

McDonough, Scott David. Sedimentology of the Eocene Las Palmas Gravels, Baja California, Mexico. M, 1981, San Diego State University.

McDonough, William F. The geochemistry and petrology of a trachyte-comendite suite from the Oligocene Paisano Volcano, West Texas. M, 1983, Sul Ross State University. 86 p.

McDougald, William D. Geology of Beaver Creek and adjacent areas, Utah. M, 1953, University of Utah. 541 p.

McDougall, David J. Radioactive minerals at Otter Rapids, Ontario. M, 1949, McGill University.

McDougall, David J. The pegmatites of Otter Rapids area, Ontario. D, 1953, McGill University.

McDougall, Donald S. Carbonate microfacies of the upper Monte Cristo limestone (Mississippian) and the lower Birdsprings (Pennsylvanian) Group of Mountain Springs, Clark County, Nevada. M, 1970, University of Southern California.

McDougall, Dristin Ann. Paleoecological evaluation of late Eocene biostratigraphic zonations on the West Coast. D, 1979, University of Southern California.

McDougall, Gillian Frances Ellen. Trace element bedrock geochemistry around the Cantung skarn-type scheelite deposits at Tungsten, Northwest Territories. M, 1977, Queen's University. 458 p.

McDougall, J. F. Experiments bearing on the genesis of sulphide deposits. D, 1957, McGill University.

McDougall, J. F. The Birch Lake copper deposits, Saskatchewan. M, 1952, McGill University.

McDougall, James John. The telescoped silver lead zinc deposits of the Contact Group mineral claims, Cassiar District, BC. M, 1953, University of British Columbia.

McDougall, James William. Geology and geophysics of the foreland fold-thrust belt of northwestern Pakistan. D, 1988, Oregon State University. 139 2 plates p.

McDougall, James William. Geology and structural evolution of the Foss River-Deception Creek area, Cascade Mountains, Washington. M, 1980, Oregon State University. 86 p.

McDougall, John E. Petrographic correlation of sands in the Cameron Meadows oil field, Cameron Parish, Louisiana. M, 1939, Louisiana State University.

McDougall, K. A. Paleoecological evaluation of late Eocene biostratigraphic zonations on the West Coast. D, 1979, University of Southern California.

McDougall, Kristin Ann. The Narizian-Refugian boundary in the Twin River Formation (Eocene, Oligocene) of the northern Olympic Peninsula, Washington. M, 1972, University of Washington. 223 p.

McDowell, Alfred N. The origin of the structural depression above Gulf Coast salt domes with particular reference to Clay Creek Dome, Washington County, Texas. M, 1951, Texas A&M University.

McDowell, Fred Wallace. Potassium-argon dating of Cordilleran intrusives. D, 1966, Columbia University. 280 p.

McDowell, John P. A paleocurrent study of the Mississagi Quartzite along the north shore of Lake Huron. M, 1963, The Johns Hopkins University.

McDowell, John Parmelee. Early Tertiary geology of northeastern Utah, northwestern Colorado, and southwestern Wyoming. M, 1955, Dartmouth College. 44 p.

McDowell, Kenneth Otto. Lower Ordovician El Paso Group depositional system and diagenesis, southern Franklin Mountains, El Paso County, Texas. M, 1983, University of Texas, Austin. 169 p.

McDowell, Patricia Frances. Holocene fluvial activity in Brush Creek watershed, Wisconsin. D, 1980, University of Wisconsin-Madison. 273 p.

McDowell, Robert C. Geology of the Argonia slate quarries, Buckingham County, Virginia. M, 1964, Virginia Polytechnic Institute and State University.

McDowell, Robert Carter. Structural geology of Macks Mountain area, (Pulaski, Wythe, Floyd and Carroll counties, Virginia). D, 1967, Virginia Polytechnic Institute and State University.

McDowell, Ronald R. Paleogeography, depositional environments, and petroleum potential of the Middle Ordovician Kanosh Formation. D, 1987, Colorado School of Mines. 410 p.

McDowell, Stephen. Analysis and interpretation of a magnetic and gravity anomaly near Sumerville, Tennessee. M, 1978, Memphis State University.

McDowell, Stewart Douglas. The intrusive history of the Little Chief Granite Porphyry Stock (Mesozoic), central Panamint Range, California; I, Structural relationships; II, Petrogenesis, based on electron microprobe analyses of the feldspars. D, 1967, California Institute of Technology. 319 p.

McDowell, Theodore R. Geographic variations in water quality and recreational use along the upper Wallowa River and selected tributaries. D, 1980, Oregon State University. 199 p.

McDowell, William Hunter, II. Mechanisms controlling the organic chemistry of Bear Brook, New Hampshire. D, 1982, Cornell University. 166 p.

McDuff, Russell E. Conservative behavior of calcium and magnesium in the interstitial waters of marine sediments; identification and interpretation. D, 1978, University of California, San Diego. 201 p.

McDuffie, Roger H. Mississippian rocks in the subsurface of Garfield and western Noble counties, Oklahoma. M, 1958, University of Oklahoma. 59 p.

McEachern, R. G. The Cranbrook Formation near Marysville, British Columbia. M, 1953, University of British Columbia.

McEachern, Ronald Graham. Cranbrook Formation near Marysville, British Columbia. M, 1942, University of British Columbia.

McEachern, Slater E., Jr. Ostracoda of the Upper Cretaceous Selma Group near Tupelo, Mississippi. M, 1962, Pennsylvania State University, University Park. 155 p.

McEachran, David Ballard. Structural geometry and evolution of the basal detachment in the Hudson Valley fold-thrust belt north of Kingston, New York. M, 1985, University of Illinois, Urbana. 97 p.

McEdwards, Donald George. Multiwell variable rate well test analysis. D, 1979, University of California, Berkeley. 153 p.

McEldowney, Roland C. Geology of the northern Sierra Pinta (Paleozoic and Tertiary), Baja California, Mexico. M, 1970, University of San Diego.

McElhaney, David A. Depositional environments and provenance of the lower Tertiary Ferris and basal Hanna formations, Hanna Basin, southeastern Wyoming. M, 1988, University of Wyoming. 200 p.

McElhaney, Matthew Stuart. Petrogenesis of the Chunky Gal Mountain mafic-ultramafic complex. M, 1981, University of Tennessee, Knoxville. 87 p.

McElroy, James Ralph. Geology of the Derby Creek area, Eagle, Routt, and Garfield counties, Colorado. M, 1953, University of Colorado.

McElroy, John. Nickel; key aspects of a strategic and critical ferrous metal. M, 1987, University of Texas, Austin.

McElroy, John Joseph, Jr. In situ stress determination in rock using the acoustic emission technique. D, 1985, Drexel University. 369 p.

McElroy, Marcus Nelson. Isopach and lithofacies study of the Desmoinesian Series of north-central Oklahoma. M, 1961, University of Oklahoma. 78 p.

McElroy, Marcus Nelson. Lithologic and stratigraphic relationships between the Reagan Sandstone (Upper Cambrian) and sub-Reagan and supra-Reagan rocks in western Kansas. D, 1965, University of Kansas. 260 p.

McElroy, Thomas Alan. The hydrogeology of selected aquifers supplying fish hatcheries in western Massachusetts. M, 1987, University of Massachusetts. 133 p.

McEntee, Robert A. The accurate modeling and removal of regional sources of the gravitational field. M, 1987, Colorado School of Mines. 184 p.

McEntire, John A., III. Geology along a portion of Highway 23, Madison County, Arkansas. M, 1964, University of Arkansas, Fayetteville.

McEuen, Robert Blair. Voltage- and frequency-dependent impedance variations for synthetic metalliferous ore. D, 1959, University of Utah. 82 p.

McEvers, Lloyd K. Stratigraphic and petrographic analysis of the Fusselman Dolomite (Lower to Middle Silurian), North Franklin Mountains, Dona Ana County, New Mexico. M, 1984, University of Texas at El Paso.

McEvilly, Thomas Vincent. Crustal and upper mantle structure of the central United States from surface wave and body wave studies. D, 1964, St. Louis University.

McEvoy, Thomas D. A subsurface investigation of the St. Peter Formation (Ordovician) in southwestern Wisconsin. M, 1963, University of Wisconsin-Madison.

McEwan, Eula (Davis). A study of the brachiopod genus, Platystrophia. D, 1918, Indiana University, Bloomington.

McEwan, Eula (Davis). The Ordovician-Silurian boundary. M, 1914, Indiana University, Bloomington.

McEwen, Alfred S. Volcanic hot spots on Io; correlation with low-albedo calderas. M, 1985, Northern Arizona University. 96 p.

McEwen, Alfred Sherman. Topics in planetary science. D, 1988, Arizona State University. 239 p.

McEwen, John H. Geological setting of the Gallen volcanogenic massive sulphide deposit, Rouyn-Noranda area, Quebec. M, 1987, Carleton University. 94 p.

McEwen, Michael C. Sedimentary facies of the Trinity River delta, Texas. D, 1963, Rice University. 100 p.

McEwen, Michael C. Textural properties of some source materials of clastic sediments. M, 1959, Rice University. 58 p.

McEwen, Robert Barlow. Orbital photogrammetry; a theoretical study of auxiliary data application. D, 1968, Cornell University. 178 p.

McFadden, C. P. A critical evaluation of hypotheses on the nature of the Earth's core. M, 1965, University of British Columbia.

McFadden, C. P. The effect of a region of low viscosity on thermal convection in the Earth's mantle. D, 1969, University of Western Ontario.

McFadden, Leslie D. Soils of the northern Canada Del Oro Valley, southern Arizona. M, 1978, University of Arizona.

McFadden, Leslie David. The impacts of temporal and spatial climatic changes on Alluvial soils genesis in Southern California. D, 1982, University of Arizona. 449 p.

McFadden, Maureen. Petrology of porcellanites in the Hawthorn Formation, Hamilton County, Florida. M, 1982, University of South Florida, Tampa. 113 p.

McFadden, Stephen S. Flow energy in rivers; middle Oconee and Mulberry rivers, Georgia. M, 1981, University of Georgia.

McFall, Clinton Carew. Permian geology of northeastern Chispa Quadrangle, Trans-Pecos, Texas. M, 1952, University of Texas, Austin.

McFall, Clinton Carew. The geology of the Escalante-Boulder area, Garfield County, Utah. D, 1956, Yale University.

McFarlain, Tommy. Lower Miocene subsurface geology; Lake Verret area, Iberia, St. Martin, and Assumption parishes, Louisiana. M, 1972, University of Southwestern Louisiana.

McFarlan, Arthur Crane. The bryozoan faunas of the (Mississippian) Chester Series of Illinois and Kentucky. D, 1924, University of Chicago. 188 p.

McFarlan, Edward, Jr. Geology of the Casey Draw and Gozar quadrangles, Reeves and Jeff Davis counties, Texas. M, 1948, University of Texas, Austin.

McFarland, Carl R. Geology of the West Canyon area, northwestern Utah County, Utah. M, 1955, Brigham Young University. 21 p.

McFarland, Gregory J. Stratigraphy of the Taylor and Navarro groups of the East Texas Basin and their petroleum potential. M, 1986, Baylor University. 193 p.

McFarland, John Barnett. A geophysical investigation of the Cornwall copper mines. M, 1953, Washington University. 39 p.

McFarland, John D., III. Lithostratigraphy and conodont biostratigraphy of Lower Mississippian strata, Northwest Arkansas. M, 1975, University of Arkansas, Fayetteville.

McFarland, LaRue Buzan. The Ulvade Formation and associated deposits. M, 1939, University of Texas, Austin.

McFarland, Mark Lee. Selective placement disposal of drilling fluids in West Texas. D, 1988, Texas A&M University. 159 p.

McFarland, Veronica T. The hydrodynamic mechanisms of some Pennsylvanian orthocerid nautiloids; physiologic, ontogenetic and taxonomic implications. M, 1986, University of Texas, Arlington. 207 p.

McFarland, William Douglas. Development of a reliable method for resource evaluation of deep-sea manganese nodule deposits using bottom photographs. M, 1980, Washington State University. 146 p.

McFarlane, Deborah Nyal. Oreana tungsten-bearing pegmatite and related Rocky Canyon Stock, Pershing County, Nevada. M, 1981, University of Nevada. 98 p.

McFarlane, James J. Silurian strata of the eastern Great Basin. M, 1955, Brigham Young University. 53 p.

McFarlane, Michael James. Geology of the Moonlight Valley porphyry copper deposit, Lights Creek, Plumas County, California. M, 1981, University of Nevada. 81 p.

McFarlane, Robert Craig. Unsteady-state distributions of fluid compositions in two-phase petroleum reservoirs undergoing gas injection. D, 1966, Stanford University. 122 p.

McFaul, Michael. A geomorphic and pedological interpretation of the mima-mounded prairies, South Puget Lowland, Washington State. M, 1979, University of Wyoming. 70 p.

McGammon, Norman R. Experimental investigation of the rate of frost penetration in clay. M, 1961, Purdue University.

McGannon, Donald E., Jr. A lithofacies study of the St. Lawrence Formation of Minnesota. M, 1957, University of Minnesota, Minneapolis. 88 p.

McGannon, Donald E., Jr. A study of the St. Lawrence Formation in the Upper Mississippi Valley. D, 1960, University of Minnesota, Minneapolis. 353 p.

McGarr, Arthur Francis. Transmission and reflection of Rayleigh waves at vertical boundaries and amplitude variations of Rayleigh waves-propagation across a continental margin-horizontal refraction. D, 1968, Columbia University. 169 p.

McGarvie, Scott Douglas. Utilization of selected soil properties to estimate saturated hydraulic conductivity

in the Truckee Meadows. M, 1979, University of Nevada. 69 p.

McGary, Etta Gaynell. White Knob Limestone (Mississippian) of Pecks Canyon, northern White Knob Mountains, south-central Idaho. M, 1982, Washington State University. 94 p.

McGaughey, W. John. Microcomputer processing of shallow seismic reflection data; a case study. M, 1986, Queen's University. 149 p.

McGavock, Cecil B., Jr. Descriptions of the majority of the pelecypods of the (Miocene) Yorktown Formation of Virginia. M, 1935, University of Virginia. 125 p.

McGavock, Edwin Harris. An evaluation of biogeochemical prospecting for zinc in the Shenandoah Valley, Virginia. M, 1962, University of Virginia. 51 p.

McGeary, David Fitz Randolph. Sediments of the Vema fracture zone (Atlantic off Amazon river). D, 1969, University of California, San Diego. 74 p.

McGeary, David Fitz Randolph. Size analysis of nearshore sediments, southeastern Lake Michigan. M, 1964, University of Illinois, Urbana. 41 p.

McGeary, Susan Emily. Oceanic plateaus, anomalous subduction, and the accretion of buoyant features at convergent margins. D, 1984, Stanford University. 133 p.

McGee, David C. Late Cretaceous Foraminiferida and paleoecology, northwest Baja California, Mexico. M, 1967, University of San Diego.

McGee, David Thomas. Lithostratigraphy and depositional history of the upper Dornick Hills Group (early Desmoinesian, Pennsylvanian) of the Ardmore Basin, Oklahoma. M, 1985, University of Oklahoma. 348 p.

McGee, Dean and Hipp, Thomas. Stream erosion. M, 1926, University of Kansas.

McGee, Edward Franklin. Stratigraphy of Wylie Mountains, Culberson County, Texas. M, 1952, University of Texas, Austin.

McGee, John B. Geology of the northwest quarter of the Chunky, Mississippi, Quadrangle. M, 1966, Mississippi State University. 129 p.

McGee, Joseph William. Depositional environments and inarticulate brachiopods of the lower Wheeler Formation, east-central Great Basin, Western United States. M, 1978, University of Kansas. 140 p.

McGee, Kenneth A. Stability constants for MgOH and brucite below 100°C. M, 1973, University of Missouri, Columbia.

McGee, Kenneth Ray. Lithostratigraphy and depositional systems of upper Bloyd and lower Atoka strata in western Arkansas and eastern Oklahoma. M, 1979, University of Arkansas, Fayetteville.

McGee, Linda C. Laramide sedimentation, folding and faulting, southern Wind River Range, Wyoming. M, 1983, University of Wyoming. 92 p.

McGee, Patricia. Depositional environment of Upper Devonian sandstones in Westmoreland County, southwestern Pennsylvania. M, 1985, Texas A&M University.

McGehee, Richard Vernon. Precambrian geology of the southeastern Llano Uplift, Texas. D, 1963, University of Texas, Austin.

McGehee, Thomas Lee. Igneous-hydrothermal mineral deposits in the Balcones Fault trend in Uvalde County, Texas. D, 1987, University of Texas at Dallas. 361 p.

McGerrigle, Harold W. Geology of the Lacolle area, Quebec. D, 1930, Princeton University.

McGetchin, Thomas Richard. Bottom sediments and fauna of western Narragansett Bay, Rhode Island. M, 1961, Brown University.

McGetchin, Thomas Richard. The Moses Rock Dike; geology, petrology and mode of emplacement of a kimber-like beam breccia dike, San Juan County, Utah. D, 1968, California Institute of Technology. 440 p.

McGhee, George Rufus, Jr. Geometric and functional analysis of shell morphology in the Pentamerida, Rhynchonellida, Spiriferida, and Terebratulida

(Articulata; Brachiopoda). D, 1978, University of Rochester. 161 p.

McGhee, George Rufus, Jr. Late Devonian benthic marine communities along the Allegheny Front in Virginia, West Virginia and Maryland. M, 1975, University of North Carolina, Chapel Hill. 126 p.

McGibbon, Douglas H. Origin and paragenesis of ore and gangue minerals, La Paz mining district, San Luis Potosi, Mexico. M, 1979, University of Texas, Arlington. 86 p.

McGill, Daniel W. A correlation of certain Washita formations with their central Texas equivalent, the Georgetown Formation. M, 1956, Texas Christian University. 249 p.

McGill, George Emmert. Geology of the northwest flank of the Flint Creek Range, western Montana. D, 1958, Princeton University. 193 p.

McGill, George Emmert. The geology of the Elbern and Judson open pit mines, with special reference to the Cretaceous, (Mesabi Range, Minn.). M, 1955, University of Minnesota, Minneapolis. 73 p.

McGill, Glen C. A study of ambiguities in magnetic interpretation. M, 1985, University of Western Ontario. 119 p.

McGill, John Thomas. Geology of a portion of the Las Flores and Dry Canyon quadrangles. M, 1948, University of California, Los Angeles.

McGill, John Thomas. Quaternary geology of the north-central San Emigdio Mountains, California. D, 1951, University of California, Los Angeles.

McGill, Kathryn A. Tectonic history of the Blue Ridge and Brevard Zone near Old Fort, North Carolina. M, 1980, University of North Carolina, Chapel Hill. 68 p.

McGill, Mary Margaret. Study of the diagenesis of a portion of the Holston Limestone; paragenesis of cements and evaluation of geochemistry. M, 1982, University of Tennessee, Knoxville. 103 p.

McGill, W. J. Geology and ore deposits of the Gold Eagle Mine, Red Lake, Ontario. M, 1939, Queen's University. 75 p.

McGill, William Peter. A study of the contact print method of determination and localization of metallic minerals. M, 1944, University of Western Ontario.

McGillis, John L. The silver content of caliche on alluvial fans as a regional guide to areas of silver and gold mineralization in the Basin and Range Province. M, 1967, University of Nevada. 30 p.

McGillivray, James G. Lithofacies control of porosity trends, Leduc Formation (upper Devonian), Golden Spike reef complex, Alberta. M, 1971, McGill University.

McGilvery, Thomas A. Lithostratigraphy of the Brentwood and Woolsey members, Bloyd Formation (type Morrowan) in Washington and western Madison counties, Arkansas. M, 1982, University of Arkansas, Fayetteville.

McGimsey, Debra Hanson. Structural geology of the Wolf Gun Mountain area, Glacier National Park, Montana. M, 1982, University of Colorado. 74 p.

McGimsey, Robert Gamewell. The Purcell Lava, Glacier National Park, Montana. M, 1985, University of Colorado. 283 p.

McGinley, John Robert. A study of gravimeter recorded Earth tide data using a normal curve filter and a determination of the gravimetric factor. M, 1963, University of Tulsa. 102 p.

McGinley, John Robert, Jr. A comparison of observed permanent tilts and strains due to earthquakes with those calculated from displacement dislocations in elastic earth models. D, 1969, California Institute of Technology. 300 p.

McGinn, Carl Wilson. Possible petrogenetic relationships between the Pee Dee Gabbro and Lilesville Granite, Anson and Richmond counties, North Carolina. M, 1988, University of Tennessee, Knoxville. 87 p.

McGinn, R. A. Alluvial fan geomorphic systems; the Riding Mountain Escarpment model. D, 1979, University of Manitoba.

McGinn, Stephen R. Facies distribution in a microtidal barrier lagoon system Ninigret Pond, Rhode Island. M, 1982, University of Rhode Island.

McGinnis, John Patrick. Gravity investigations of the St. Louis Arm of the New Madrid rift complex. M, 1984, Purdue University. 80 p.

McGinnis, Lyle David. Crustal movements in northeastern Illinois. D, 1965, University of Illinois, Urbana.

McGirk, Lon S., Jr. The examination of ferromagnetic minerals in polished section. D, 1953, Stanford University. 119 p.

McGirr, Robert R. Acid soluble lead concentrations in the sediments of Lake Powell, Arizona-Utah. M, 1977, Dartmouth College. 56 p.

McGlade, William George. The geology of the northcentral rectangle of the Wind Gap, Pennsylvania, Quadrangle. M, 1952, Pennsylvania State University, University Park. 73 p.

McGlasson, James A. Geology of central Knight Island, Prince William Sound, Alaska. M, 1976, Colorado School of Mines. 136 p.

McGlasson, Robert H. Foraminiferal biofacies around Santa Catalina Island. M, 1957, University of Southern California.

McGlew, Peter J. The hydrogeologic investigation of a hazardous waste site in southern New Hampshire. M, 1984, Boston University. 85 p.

McGlynn, J. C. Petrology and correlation of the acidic intrusives of Darling Township, Lanark County, Ontario. M, 1949, Queen's University. 43 p.

McGlynn, John C. A study of the rocks of Elbow-Heming lakes, Manitoba. D, 1953, University of Chicago. 165 p.

McGonigal, Michael Henry. The Gander and Davidsville groups; major techtonostratigraphic units in the Gander Lake area, Newfoundland. M, 1973, Memorial University of Newfoundland. 121 p.

McGookey, D. P. Pennsylvanian and Early Permian stratigraphy of the southeast portion of the east flank of the Laramie Range, Wyoming, and Colorado. M, 1952, University of Wyoming. 45 p.

McGookey, Donald Paul. Geology of the northern portion of the Fish Lake Plateau, Utah. D, 1959, Ohio State University.

McGookey, Douglas A. The Queen Formation of the Ozona Arch area of Crockett County, Texas; structure and depositional systems. M, 1986, University of Texas, Austin.

McGovney, J. E. E. Deposition, porosity evolution, and diagenesis of the Thornton Reef (Silurian), northeastern Illinois. D, 1978, University of Wisconsin-Madison. 531 p.

McGovney, James E. The diagenesis and sedimentological history of a Silurian-to-Devonian bank-to-basin transition facies in the Hot Creek Range, Nevada. M, 1977, University of California, Riverside. 139 p.

McGowan, Clyde Ronald. A study of Recent point bar sediments of the lower Ouachita River (Louisiana). M, 1967, Louisiana State University.

McGowan, E. B. A field examination of radioactive occurrences in the Caddy Lake area of Manitoba. M, 1981, University of Manitoba. 117 p.

McGowan, Francis Herbert. The Pettus Field, Bee County, Texas. M, 1932, University of Texas, Austin.

McGowan, K. I. Geochemistry of alteration and mineralization of the Wind River gold prospect, Skamania County, Washington. M, 1985, Portland State University. 136 p.

McGowan, Michael. The Feltville Formation of the Watchung Syncline, Newark Basin, New Jersey. M, 1981, Rutgers, The State University, Newark. 135 p.

McGowan, Michael Francis. A structural and stratigraphic study of a portion of the Batesville manganese district, Arkansas. M, 1981, University of Arkansas, Fayetteville.

McGowen, Clyde B. A study of Recent point bar sediments of the lower Ouachita River. M, 1967, Northeast Louisiana University.

McGowen, Joseph H. The stratigraphy of the Harpersville and Pueblo formations, southwestern Stephens County, Texas. M, 1964, Baylor University. 440 p.

McGowen, Joseph Hobbs. Gum Hollow fan delta, Nueces bay, Texas; mode of development and sediments. D, 1969, University of Texas, Austin. 286 p.

McGrail, David W. Mechanisms of sedimentation on the continental shelf of Liberia and Sierra Leone. D, 1976, University of Rhode Island. 129 p.

McGrail, David W. The mineralogy of sands from the shelf and upper slope off Guinea, Portuguese Guinea, and Sierra Leone. M, 1972, University of Rhode Island.

McGrain, Preston. The St. Louis and Ste. Genevieve limestone of Harrison County, Indiana. M, 1942, Indiana University, Bloomington. 28 p.

McGrane, Daniel J. Geology of the Idol City area; a volcanic-hosted, disseminated, precious-metal occurrence in east-central Oregon. M, 1985, University of Montana. 88 p.

McGrath, Bernard D. Geology of the Fredonia area, Texas. M, 1952, Texas A&M University.

McGrath, Dennis G. Evaluation of two sand-height gage systems for monitoring coastal sediment migration. M, 1973, Old Dominion University. 68 p.

McGrath, Dennis J. n-alkane distributions in natural gases; evidence for migration in ground water. M, 1984, Indiana University, Bloomington. 63 p.

McGrath, Peter H. Effect of bedrock relief on gravity surveys. M, 1962, University of Western Ontario.

McGraw, Maryann Margaret McDonough. Carbonate facies and diagenesis of the upper Smackover Formation (Jurassic); Paup Spur-Mandeville fields, Miller County, Arkansas. M, 1983, University of Texas, Austin. 163 p.

McGraw, Patricia A. A petrological/geochemical study of rocks from the Sao Miguel drillhole, Sao Miguel, Azores. M, 1976, Dalhousie University. 87 p.

McGraw, R. B., Jr. Geology in the vicinity of the Copper King Mine, Laramie County, Wyoming. M, 1954, University of Wyoming. 68 p.

McGregor, Bonnie A. Topographic and structural features at the southern terminus of the Sohm abyssal plain. M, 1967, University of Rhode Island.

McGregor, Bonnie Ann. Crest of Mid-Atlantic Ridge at 25°N; topographic and magnetic patterns. D, 1975, University of Miami. 93 p.

McGregor, Catherine R. Characterization of granitic and pegmatitic K-feldspars from Lac du Bonnet, S.E. Manitoba and Dryden, N.W. Ontario by rapid X-ray diffraction and chemical methods. M, 1984, University of Manitoba.

McGregor, Dan R. Micropaleontology of the Comanche Peak Limestone of the Llano Estacado of Texas. M, 1962, Texas Tech University. 134 p.

McGregor, Don L. Stratigraphy and micropaleontology of Kiamichi Formation equivalents Trans-Pecos Texas. M, 1962, Texas Tech University. 183 p.

McGregor, Duncan D. Engineering interpretation of agricultural soils maps and correlation with airphoto patterns. M, 1956, Purdue University.

McGregor, Duncan J. Stratigraphic analysis of Upper Devonian and Mississippian rocks in the Michigan Basin. D, 1953, University of Michigan.

McGregor, Duncan J. The geology of the gypsum deposits near Sun City, Barber County, Kansas. M, 1948, University of Kansas. 60 p.

McGregor, J. P. Copper mineralization in the diabase near Elk Lake, Ontario. M, 1941, University of Toronto.

McGregor, Jackie Delaine. Geology of the Ellsworth Quadrangle and vicinity, Maine. D, 1964, University of Illinois, Urbana. 116 p.

McGregor, Jackie Delaine. Petrography of some diabase-felsite dikes, Mount Desert, Maine. M, 1959, University of Illinois, Urbana. 31 p.

McGregor, James D. An isopachous map of the Pennsylvanian System of Oklahoma. M, 1941, University of Oklahoma. 31 p.

McGregor, Michael Andrew. Volatile organic compounds in Narragansett Bay; composition, distribution and seasonal variation. D, 1983, University of Rhode Island. 187 p.

McGrew, Alan R. Stratigraphy and mineralogy of the Blue Point Member of the Wiggins Formation (Oligocene), southeast Absaroka Range, Park County, Wyoming. M, 1965, University of Wyoming. 74 p.

McGrew, Allen J. Deformational history of the southern Snake Range, Nevada, and the origin of the southern Snake Range decollement. M, 1986, Stanford University. 46 p.

McGrew, Bill Judson. Petrography of Candelaria area, Presidio County, Trans-Pecos, Texas. M, 1955, University of Texas, Austin.

McGrew, Gloria. Lineament analysis related to fracture and mineralization trends in southeastern Utah and northwestern New Mexico. M, 1979, Purdue University.

McGrew, L. W. The geology of the Grayrocks area, Platte and Goshen counties, Wyoming. M, 1953, University of Wyoming. 94 p.

McGrew, Paul O. The Aplodontoidea. D, 1942, University of Chicago. 30 p.

McGrew, Paul Orman. The Burge fauna, a lower Pliocene mammalian assemblage from Nebraska. M, 1935, University of California, Berkeley. 27 p.

McGroder, Michael F. Stratigraphy and sedimentation of the Proterozoic Garnet Range Formation, Belt Supergroup, western Montana. M, 1984, University of Montana. 107 p.

McGroder, Michael F. Structural evolution of the eastern Cascades foldbelt; implications for late Mesozoic accretionary tectonics in the Pacific Northwest. D, 1988, University of Washington. 140 p.

McGrossan, R. A facies study of the Woodbend Formation of Alberta. D, 1953, University of Chicago.

McGuchen, John G. Honesdale paleocurrents in northeastern Pennsylvania. M, 1959, Lehigh University.

McGuiggan, Patricia Marie. Forces between mica surfaces in aqueous solutions. D, 1987, University of Minnesota, Minneapolis. 205 p.

McGuinness, Charles L. The minerals of New Mexico. M, 1936, University of New Mexico. 346 p.

McGuire, Anne Vaughan. Petrology of mantle and crustal inclusions in alkali basalts from western Saudi Arabia; implications for formation of the Red Sea. D, 1987, Stanford University. 312 p.

McGuire, D. M. Terrigenous conglomerates and sands cored on Midway Islands. M, 1968, University of Hawaii.

McGuire, Donn. Geophysical survey of the Anna, Ohio area. M, 1975, Bowling Green State University. 79 p.

McGuire, Douglas Joseph. Stratigraphy, depositional history, and hydrocarbon source-rock potential of the Upper Cretaceous-lower Tertiary Moreno Formation, central San Joaquin Basin, California. D, 1988, Stanford University. 331 p.

McGuire, Emily. Geology of the Rye Quadrangle, Pueblo and Huerfano counties, Colorado. M, 1978, Wichita State University. 46 p.

McGuire, Michael Dale. Thermal springs of the Elsinore fault zone; relation to groundwater recharge and structural geology. M, 1980, San Diego State University.

McGuire, Michael James. Geology for land-use planning of southeastern Osage, eastern Pawnee, northern Creek, and western Tulsa counties, Oklahoma. M, 1974, Oklahoma State University. 66 p.

McGuire, Odell S. Altered ilmenite from New Jersey. M, 1958, Columbia University, Teachers College.

McGuire, Odell S. Population studies in the ostracode genus Polytlyites from the Chester Series. D, 1962, University of Illinois, Urbana. 113 p.

McGuire, Robert Hillary, Jr. The Lower Cretaceous Kootenai Formation in Granite and Powell counties, Montana. M, 1957, University of Montana. 107 p.

McGuirt, James H. Louisiana Tertiary Bryozoa. D, 1938, Louisiana State University.

McGuirt, James H. Tertiary Bryozoa of Louisiana. M, 1934, Louisiana State University.

McGurk, Bruce James. A comparison of four rainfall-runoff methods. D, 1982, Utah State University. 257 p.

McHam, Robert M. Geologic criteria for the selection of unconfined dredged material disposal sites in estuaries and lagoons. M, 1977, Texas A&M University. 135 p.

McHargue, Lanny Ray. Late Quaternary deposition and pedogenesis of the Aguila Mountains piedmont, southwestern Arizona. M, 1981, University of Arizona. 132 p.

McHargue, Timothy Reed. Stratigraphy, petrology, depositional and diagenetic environments of Lower-Middle Ordovician boundary beds, Arbuckle Mountains, Oklahoma. D, 1981, University of Iowa. 174 p.

McHaro, B. A. An evaluation of certain geophysical methods used in exploring for massive sulphides at key Anacon mines. M, 1975, University of New Brunswick.

McHone, J. G. Petrochemistry and genesis of Champlain Valley dike rocks. M, 1975, University of Vermont.

McHone, James Gregory. Lamprophyre dikes of New England. D, 1978, University of North Carolina, Chapel Hill. 187 p.

McHone, John F., Jr. Morphologic time series from a submarine sand ridge on the South Virginia coast. M, 1972, Old Dominion University. 59 p.

McHone, John Frank, Jr. Terrestrial impact structures; their detection and verification with two new examples from Brazil. D, 1986, University of Illinois, Urbana. 218 p.

McHorgue, Timothy Reed. Conodonts of the Joins Formation (Ordovician) Arbuckle Mountains, Oklahoma. M, 1974, University of Missouri, Columbia.

McHugh, Alice Ellen. Styles of diagenesis in Norphlet Sandstone (Upper Jurassic) onshore and offshore Alabama. M, 1987, University of New Orleans. 164 p.

McHugh, Brian. X-ray exploration for talc deposits. M, 1985, Eastern Washington University. 64 p.

McHugh, Margaret H. Landslide occurrence in the Elk and Sixes river basins, Southwest Oregon. M, 1987, Oregon State University. 106 p.

McHugh, Mary Lopina. Reevaluation of the southwestern end of the Salem Synclinorium. M, 1986, Virginia Polytechnic Institute and State University.

McHugh, S. L. Short period tilt events and episodic slip on the San Andreas Fault. D, 1977, Stanford University. 256 p.

McHuron, Eric Jay. Biology and paleobiology of modern invertebrate borers. D, 1976, Rice University. 290 p.

McIlreath, Ian A. Initial dip and compaction of limestones adjacent to Precambrian topographic highs in the Kingston area, Ontario. M, 1971, Queen's University. 192 p.

McIlreath, Ian Alexander. Stratigraphic and sedimentary relationships at the western edge of the Middle Cambrian carbonate facies belt, Field, British Columbia. D, 1977, University of Calgary. 259 p.

McIlvride, William Allen. Surficial geology of the northeastern part of the Mt. Holyoke Quadrangle, Massachusetts. M, 1982, University of Massachusetts. 105 p.

McIlwaine, William Hardy. Age and origin of the Perry Formation (Devonian), Charlotte County, New Brunswick, (Canada). M, 1968, University of New Brunswick.

McInnes, Brent Ian Alexander. Geological and precious metal evolution at Freegold Mountain, Dawson Range, Yukon. M, 1987, McMaster University. 230 p.

McIntire, Pamela Ellen. Groundwater seepage and sulfur diagenesis in acidified lake sediments. D, 1988, University of Virginia. 206 p.

McIntire, William L. The thermodynamic treatment of trace element distribution in geologic systems with applications to geologic thermometry. D, 1958, Massachusetts Institute of Technology.

McIntire, William Leigh. Geology of Vieja Pass area, Tierra Vieja Mountains, Trans-Pecos, Texas. M, 1950, University of Texas, Austin.

McIntosh, Allen. The depositional environments and diagenesis of sediments from the Arabian Gulf. M, 1983, University of Houston.

McIntosh, D. S. On an occurrence of tin ores and associated minerals in Nova Scotia, with a comparative study of tin deposits in other parts of the world. M, 1908, McGill University.

McIntosh, George Clay. A pseudoplanktonic crinoid colony from the Upper Devonian of western New York. M, 1971, University of Michigan.

McIntosh, George Clay. Review of the Devonian cladid inadunate crinoids; suborder Dendrocrinina (volumes 1 and 2). D, 1983, University of Michigan. 538 p.

McIntosh, Joseph Arthur. Geology of the Blowout Canyon area, Snake River Range, Idaho. M, 1947, University of Michigan.

McIntosh, Ronald Alexander. Geology and geochemistry of the mining zone at Esterhazy, Saskatchewan (Canada). M, 1967, University of Saskatchewan. 87 p.

McIntosh, William. An individualized approach to the development of a major river system; the Colorado River as an example. M, 1977, Pennsylvania State University, University Park. 79 p.

McIntyre, Andrew. The Coccolithophoridae of the Atlantic Ocean. D, 1967, Columbia University. 178 p.

McIntyre, Colleen. Diffusion of tritiated water in basalt-bentonite mixtures. M, 1985, University of Idaho. 217 p.

McIntyre, David Harry. Cenozoic geology of the Reynolds Creek Watershed, Owyhee County, Idaho. D, 1966, Washington State University. 248 p.

McIntyre, Donald David. The size and proportions of some mineral grains from a beach, Lorraine, Ontario. M, 1957, Pennsylvania State University, University Park. 102 p.

McIntyre, James R. Lower Paleozoic faunas of Plowman's Valley, California. M, 1949, University of California, Berkeley. 47 p.

McIntyre, John Francis, III. Structural and metamorphic interrelationships across the Copper Hill Anticline, Picuris Mountains, New Mexico. M, 1981, University of Texas, Austin. 76 p.

McIntyre, Kenneth E., Jr. Geology of the Sulphur Rock Quadrangle, Independence County, Arkansas with special reference to the history of manganese mining. M, 1972, Northeast Louisiana University.

McIntyre, Loren B. Geology of the Marshall Butte area and vicinity, Mitchell Quadrangle, Oregon. M, 1953, Oregon State University. 96 p.

McIver, Norman L. Sedimentation of the Upper Devonian marine sediments of the Central Appalachians. D, 1961, The Johns Hopkins University.

McJennet, George Stanley. Geology of the Pyramid Lake-Red Rock Canyon area, Washoe County, Nevada. M, 1957, University of California, Los Angeles.

McJunkin, Herbert Henry, Jr. The stratigraphy of the Grayson Formation in Tarrant County, Texas. M, 1955, Southern Methodist University. 40 p.

McJunkin, Richard Dean. Geology of the central San Bernardino Mountains, San Bernardino County, California. M, 1976, California State University, Los Angeles.

McKague, H. Lawrence. The petrology of the Hatchery Creek serpentinite, Chelan County, Washington. M, 1960, Washington State University. 42 p.

McKague, Herbert Lawrence. The geology, mineralogy, petrology, and geochemistry of the State Line Serpentinite and associated chromite deposits. D, 1964, Pennsylvania State University, University Park. 167 p.

McKallip, Curtis Jr. Newkirk Field; the geology of a shallow streamflood project in Guadalupe County, NM. M, 1984, New Mexico Institute of Mining and Technology. 85 p.

McKallip, Thomas E. Burial diagenesis and specific catalytic activity of illite-smectite clays. M, 1985, University of Missouri, Columbia. 61 p.

McKay, D. B. Aspects of the gallium geochemistry in upper mantle-derived lherzolite xenoliths and continental alkaline volcanic rocks. M, 1987, Lakehead University.

McKay, David Stewart. A chemical study of coexisting metamorphic muscovite and biotite from eastern New York and western Connecticut. D, 1964, Rice University. 87 p.

McKay, David Stuart. Mechanism of crystallization of amorphous silica. M, 1960, University of California, Berkeley. 82 p.

McKay, Dennis A. A determination of surface currents in the vicinity of the Monterey submarine canyon (California) by the electromagnetic method. M, 1970, United States Naval Academy.

McKay, E. D., III. Stratigraphy and zonation of Wisconsinan loesses in southwestern Illinois. D, 1977, University of Illinois, Urbana. 251 p.

McKay, E. D., III. Stratigraphy of glacial tills in the Gibson City Reentrant, central Illinois. M, 1975, University of Illinois, Urbana. 59 p.

McKay, E. J. Geology of Red Canyon Creek area, Fremont County, Wyo. M, 1948, University of Wyoming. 68 p.

McKay, Gordon Alan. Petrology of the Seven Springs Formation, Davis Mountains, Texas. M, 1969, Rice University. 25 p.

McKay, Gordon Alan. The petrogenesis of titanium-rich basalts from the lunar maria and KREEP-rich rocks from the lunar highlands. D, 1977, University of Oregon. 243 p.

McKay, James Hughes, Jr. Measuring the change of storage of ground water. D, 1967, The Johns Hopkins University. 184 p.

McKay, Mary Winifred. Mississippian foraminifera. M, 1961, University of Alberta. 114 p.

McKay, Richard H. A depositional model for the Aux Vases Formation and the Joppa Member of the Ste. Genevieve Formation (Mississippian) in southwestern Illlinois and southeastern Missouri. M, 1980, Southern Illinois University, Carbondale. 184 p.

McKay, Rodney H. Texture and mineralogy of Oregon beach sand. M, 1962, University of Montana. 70 p.

McKay, Sheila Mahan. A geochemical investigation of the Precambrian Knife Lake series, northeastern Minnesota. M, 1969, Rice University. 20 p.

McKay, Wayne Irving. Effects of temperature on the absolute permeability of consolidated sandstone. M, 1983, Stanford University.

McKeague, Gordon Clark. An incremental method for oil and gas exploration program evaluation. M, 1968, University of Tulsa. 62 p.

McKean, James A. Density slicing of aerial photography applied to slope stability study. M, 1977, Colorado State University. 101 p.

McKechnie, Deborah Jean. Transport of soluble pollutants in a run-of-the-river impoundment. D, 1988, University of Iowa. 287 p.

McKecknie, Neil Douglas. The geology of the Beltian rocks of the Cordillera in Canada. M, 1933, University of British Columbia.

McKee, Brent Andrew. The fate of particle-reactive radionuclides on the Amazon and Yangtze continental shelves. D, 1986, North Carolina State University. 250 p.

McKee, Bryce J. The relationship of Permian San Andres facies to the distribution of porosity and permeability in the ODC Field, Gaines County, Texas. M, 1986, Baylor University. 216 p.

McKee, Edwin H. The stratigraphy and structure of a portion of the Magruder Mountain-Soldier Pass quadrangles, California-Nevada. D, 1962, University of California, Berkeley. 111 p.

McKee, Elliott B., Jr. The geology of the Pacheco Pass area, California. D, 1959, Stanford University. 88 p.

McKee, Howard Harper. Geologic studies on the northwestern slope of the San Juan Mountains of Colorado. M, 1912, University of Chicago. 41 p.

McKee, J. A. Laboratory investigation of fluid flow through intact and fractured core specimens. M, 1986, University of Waterloo. 198 p.

McKee, James M. Petrology, alteration and mineralization of the Poorman Creek-Silver Bell stock prophyry Cu-Mo deposit. M, 1978, University of Montana. 97 p.

McKee, James W. Petrology of some Pottsville (Pennsylvanian) rocks of the Cahaba coal basin of Alabama. M, 1964, Louisiana State University.

McKee, James Walker. Pennsylvanian sediment-fossil relationships in part of the Black Warrior Basin of Alabama. D, 1967, Louisiana State University. 76 p.

McKee, John H. A geophysical study of microearthquake activity near Bowman, South Carolina. M, 1973, Georgia Institute of Technology. 65 p.

McKee, Kathryn Merkle. A fabric study of the suspended sediment of the lower Brazos River, Texas. M, 1979, Texas A&M University. 71 p.

McKee, L. H. The geology, stratigraphy, and hydrothermal alteration in the Nesbit gold mine area, Union County, North Carolina. M, 1985, University of North Carolina, Chapel Hill.

McKee, Larry Douglas. An investigation of the mid-Michigan geophysical anomaly. M, 1985, Purdue University. 94 p.

McKee, Mary Eileen. Microearthquake studies across the Basin and Range—Colorado Plateau transition zone in central Utah. M, 1982, University of Utah. 117 p.

McKee, T. R. Ferromanganese remobilization in Recent sediments of the Gulf of Mexico. D, 1977, Texas A&M University. 218 p.

McKeegan, Kevin Daniel. Ion microprobe measurements of H, C, O, Mg, and Si isotopic abundances in individual interplanetary dust particles. D, 1987, Washington University. 201 p.

McKeel, Daniel Royce. Systematics, biostratigraphy and paleoceanography of Tertiary planktonic foraminiferal assemblages, Oregon coast range. M, 1972, University of California, Davis. 118 p.

McKeever, Douglas. Volcanology and geochemistry of the south flank of Mount Baker, Cascade Range, Washington. M, 1977, Western Washington University. 126 p.

McKeever, Lauren Joann. Geographical variation in the genus Astarte (Phylum Mollusca; Class Bivalvia) from the Yorktown and Jackson Bluff formations (early Pliocene) of the Atlantic Coastal Plain. M, 1985, Virginia Polytechnic Institute and State University.

McKellar, Barbara J. The role of giant fields in future U.S. oil production. D, 1988, University of Wisconsin-Madison. 292 p.

McKellar, Ronald Lawrence. The type Lea Park Formation of the Upper Cretaceous series in the Western Interior plains. M, 1977, University of Saskatchewan. 117 p.

McKellar, Tommy R. The Codell Sandstone (Upper Cretaceous) of Kansas. M, 1962, University of Kansas. 92 p.

McKelvey, Gregory Ellis. The lithofacies of the Wallace and related formations of the Belt Series (Precambrian), (western Montana; eastern Idaho). M, 1967, Franklin and Marshall College. 48 p.

McKelvey, Vincent E. Stratigraphy of the phosphatic shale member of the (Permian) Phosphoria Formation in western Wyoming, southeastern Idaho, and northern Utah. D, 1947, University of Wisconsin-Madison.

McKelvey, Vincent Ellis. Stream and valley sedimentation in the Coon Creek drainage basin, Wisconsin. M, 1939, University of Wisconsin-Madison. 122 p.

McKenna, James W. Geology of the Conger Mine, Boulder County, Colorado. M, 1940, University of Colorado.

McKenna, John J. Buehman Canyon Paleozoic section, Pima County, Arizona. M, 1965, University of Arizona.

McKenna, Malcolm Carnegie. Fossil Mammalia from the early Wasatchian Four Mile fauna, Eocene of Northwest Colorado. D, 1958, University of California, Berkeley. 338 p.

McKenna, Robert Daniel. Petroleum geology of the Mississippi Lime in parts of Payne and Pawnee counties, Oklahoma. M, 1982, Oklahoma State University. 71 p.

McKennitt, D. B. The Middle Devonian Elk Point Group in Manitoba. M, 1961, University of Manitoba.

McKenny, Jere Wesley. Subsurface geology of northeastern Logan County, Oklahoma. M, 1952, University of Oklahoma. 46 p.

McKenzie, C. B. Petrology of the South Mountain Batholith, western Nova Scotia. M, 1974, Dalhousie University.

McKenzie, Garry Donald. Glacial history of Adams Inlet, southeastern Alaska. D, 1968, Ohio State University. 227 p.

McKenzie, Harvey Kenneth. A tectonic interpretation of geothermal parameters in northern Mexico. M, 1981, University of Florida. 95 p.

McKenzie, John C. The geology of Little Mountain, South Carolina. M, 1962, University of South Carolina. 49 p.

McKenzie, Kathleen Jane. Sedimentology and stratigraphy of the southern Sustut Basin, north-central British Columbia. M, 1985, University of British Columbia. 103 p.

McKenzie, William Frank. Theoretical calculation of equilibrium constraints in geochemical processes to 900°C. D, 1980, University of California, Berkeley. 105 p.

McKeon, J. B. Delineation of central Ohio glacial deposits by computer processing of multispectral satellite data. D, 1975, Ohio State University. 166 p.

McKeon, John B. Late-glacial wind direction in west-central Maine. M, 1972, University of Maine. 151 p.

McKeown, David Alexander. Spectroscopic study of silica-rich glasses and selected minerals within the sodium aluminosilicate system. D, 1985, Stanford University. 142 p.

McKeown, Rosalyn Rae. Regional variation of streamflow distributions in Tennessee. D, 1986, University of Oregon. 233 p.

McKereghan, Peter Fleming. Digital sumulation of chloride transport in the Silurian dolomite aquifer in Door County, Wisconsin. M, 1988, University of Wisconsin-Milwaukee. 132 p.

McKibben, Mark Eugene. Ferroan carbonate cements in limestones and sandstones of southeastern Kansas; late diagenetic processes and conditions. M, 1986, University of Kansas. 82 p.

McKibben, Michael Andersen. Kinetics of aqueous oxidation of pyrite by ferric iron, oxygen, and hydrogen peroxide from pH 1-4 and 20-40°C. D, 1984, Pennsylvania State University, University Park. 171 p.

McKibben, Michael Andersen. Ore minerals in the Salton Sea geothermal system, Imperial Valley, California, U.S.A. M, 1979, University of California, Riverside. 90 p.

McKillop, John H. Geology of the Corner Brook area, Newfoundland, with emphasis on the carbonate rocks. M, 1961, Memorial University of Newfoundland. 102 p.

McKiness, John Paul. The Quaternary System of the eastern Spokane Valley and the lower, northern slopes of the Mica Peak uplands, eastern Washington and northern Idaho. M, 1988, University of Idaho. 100 1 plate p.

McKinlay, Phillip France. Welded tuff and related rocks from near Lone Hill, Santa Clara County, California. M, 1949, Stanford University. 37 p.

McKinlay, Ralph Harold. A study of the Pilot Knob, Travis County, Texas. M, 1940, University of Texas, Austin.

McKinley, George Alvin. Estimation of dispersion curves for Rayleigh waves complicated by multipath effects. M, 1975, Southern Methodist University. 52 p.

McKinley, Glenn Ernest. Igneous rocks of the Lake Altus-Little Bow Mountain area, Wichita Mountains, Oklahoma. M, 1950, University of Oklahoma. 62 p.

McKinley, James P. Chemistry and petrology of Apollo 16 rock samples; impact metal sheets, nature of the Cayley Plains, and Descartes Mountains and geologic history. M, 1983, University of New Mexico. 277 p.

McKinley, Myron Earnest. The geology and petrology of the Arbuckle and Timbered Hills groups in the southeastern Arbuckle Mountains, Oklahoma. M, 1954, University of Oklahoma. 200 p.

McKinley, Susan G. Petrology and classification of 145 small meteorites from the Allan Hills, Antarctica. M, 1984, University of New Mexico. 100 p.

McKinney, Barbara A. The spring 1976 erosion of Siletz Spit, Oregon, with an analysis of the causative wave and tide conditions. M, 1977, Oregon State University. 66 p.

McKinney, Curtis Ross. An evaluation of uranium series disequilibrium dating of fossil teeth. M, 1977, University of Florida. 65 p.

McKinney, D. Brooks. Origin of the comb layered and orbicular rocks near Fisher Lake, Sierra Nevada Batholith, California. D, 1986, The Johns Hopkins University. 350 p.

McKinney, Frank Kenneth. Nonfenestrate Bryozoa of the Bangor Limestone, (Chester) of Alabama. M, 1967, University of North Carolina, Chapel Hill. 182 p.

McKinney, Frank Kenneth. Trepostomatous ectoproct from the lower Chickamauga group (middle Ordovician), Wills valley, Alabama. D, 1970, University of North Carolina, Chapel Hill. 262 p.

McKinney, James S. Petrographic analysis of the Croweburg Coal and its associated sediments. M, 1959, University of Oklahoma. 124 p.

McKinney, Marjorie Jackson. Faunal analysis of the Fort Payne Chert (Lower Mississippian) near Trussville, Alabama. M, 1968, University of North Carolina, Chapel Hill. 139 p.

McKinney, Michael L. Ontogeny, phylogeny, and post-depositional alteration of the oligopygoid echinoids of the Ocala Limestone. M, 1982, University of Florida. 128 p.

McKinney, Michael Lyle. Heterochrony and its environmental correlates in Cenozoic echinoids of the coastal plain and Caribbean areas. D, 1985, Yale University. 296 p.

McKinney, R. D. The effect of a political boundary on an urbanized system; a case study of the Texarkana, Arkansas-Texas urbanized area. M, 1974, East Texas State University.

McKinney, R. E. A study of ground-water quality of Brighton, Colorado. M, 1963, University of Kansas.

McKinney, Richard Bowen. Subsurface stratigraphy of Late Jurassic (?) through middle Eocene strata in a portion of the North Carolina coastal plain. M, 1985, North Carolina State University. 88 p.

McKinney, Robert Geers. Petrology of eruptive rocks of Porvenir area, Presidio County, Trans-Pecos, Texas. M, 1957, University of Texas, Austin.

McKinney, Roy Franklin. Environmental geology of Southeast Carson City, Nevada. M, 1976, University of Nevada. 135 p.

McKinney, Thomas F. Geology along the Great Smoky Fault, Ground Hog Mountain area, Tellico Plains, Tennessee. M, 1964, University of Tennessee, Knoxville. 49 p.

McKinney, Thomas Francis. Continental shelf sediments off Long Island, New York. D, 1969, Rensselaer Polytechnic Institute. 102 p.

McKinney, W. N., Jr. Carboniferous stratigraphy of the San Saba area, San Saba County, Texas. M, 1963, University of Texas, Austin.

McKinnie, Diana B. Some stratigraphic and taxonomic revision of the middle Cambrian in Montana. M, 1975, University of Washington. 66 p.

McKinnie, Nancy Jayne. A bedding and cyclicity study of the upper member of the Swan Peak Formation, southeastern Idaho. M, 1976, Idaho State University. 45 p.

McKinnon, Frederick Allan. A study of the Cambrian-Devonian contact at the front of the mountains in Bow Valley, Alberta. M, 1942, University of Alberta. 67 p.

McKinnon, William Beall. Large impact craters and basins; mechanics of syngenetic and postgenetic modification. D, 1981, California Institute of Technology. 240 p.

McKinstry, Brian William. An experimental examination of the reaction; tremolite + forsterite = 5 enstatite + 2 diopside + fluid. M, 1980, Carleton University. 121 p.

McKinstry, Herbert Alden. The thermal expansion of certain alkali halides and their solid solutions. D, 1960, Pennsylvania State University, University Park. 213 p.

McKinstry, Hugh Exton. I, Supergene and hypogene mineralization in certain Cordilleran silver deposits; II, Qualitative microchemical and magnetic tests in the identification of opaque minerals. D, 1926, Harvard University.

McKinstry, Hugh Exton. Petrology of the granites and associated pegmatites of Rockport, Massachusetts. M, 1921, Massachusetts Institute of Technology. 93 p.

McKirgan, Bruce Stephen. Geology of the Mississippian System of the Oakvale area of West Virginia. M, 1971, University of Akron. 34 p.

McKitrick, William Ernest. The geology of the Suplee Paleozoic Series of central Oregon. M, 1934, Oregon State University. 89 map x-section p.

McKniff, Joseph Michael. Geology of the Highlands-Cashiers area, North Carolina, South Carolina, and Georgia. D, 1967, Rice University. 167 p.

McKnight, Brian Keith. Petrology and sedimentation of Cretaceous and Eocene rocks in the Medford-Ashland region, southwestern Oregon. D, 1971, Oregon State University. 177 p.

McKnight, Cleavy L. Descriptive geomorphology of the Guadalupe Mountains, south-central New Mexico and West Texas. M, 1983, Baylor University. 172 p.

McKnight, Edwin Thor and Ward, Alfred H. Geology of Snohomish Quadrangle. M, 1925, University of Washington. 95 p.

McKnight, John Forrest. Geology of Bofecillos Mountains area, Trans-Pecos Texas. D, 1968, University of Texas, Austin. 216 p.

McKnight, John Forrest. Igneous rocks of Sombreretillo area, northern Sierra de Picachos, Nuevo Leon, Mexico. M, 1963, University of Texas, Austin.

McKnight, Randy Henry. Geologic interpretation of Apollo orbital photographs of Texas. M, 1971, Northwestern State University. 41 p.

McKnight, William M., Jr. The distribution of foraminifera off parts of the Antarctic Coast. M, 1952, Florida State University.

McKnight, William Ross. A paleomagnetic study of Recent Cascade Lavas. M, 1968, Oregon State University. 46 p.

McKone, Thomas Edward. Chemical cycles and health risks of some crustal nuclides. D, 1981, University of California, Los Angeles. 310 p.

McKosky, John A. Quartz in provenance examination. M, 1975, Michigan State University. 52 p.

McKown, David Melvin. Application of gamma-gamma coincidence counting techniques to the non-destructive activation analysis of meteoritic materials. D, 1969, [University of Kentucky]. 165 p.

McKoy, Mark L. Bioturbation of sediments in upper Chesapeake Bay. M, 1984, Dartmouth College. 206 p.

McKyes, Shirley Edward. Yielding of a remoulded clay under stress states. D, 1969, McGill University.

McLachlin, D. Development of a prototype for point load testing of aggregates. M, 1985, University of Waterloo. 127 p.

McLafferty, Susan W. Depositional environments in the transition from Mancos Shale to Mesa Verde Group near Carthage, Socorro County, NM. M, 1979, New Mexico Institute of Mining and Technology. 119 p.

McLain, Jay Forman. Tertiary reefs of southern Louisiana. D, 1957, Harvard University.

McLain, John P. Petrographic analysis of lineaments in the San Francisco volcanic field, Coconino and Yavapai counties, Arizona. M, 1965, University of Arizona.

McLain, William Henry. Geothermal and structural implications of magnetic anomalies observed over the southern Oregon Cascade Mountains and adjoining Basin and Range Province. M, 1982, Oregon State University. 151 p.

McLamore, Roy Travis. Strength-deformation characteristics of anisotropic sedimentary rocks. D, 1966, University of Texas, Austin. 277 p.

McLamore, Vernon Reid. Paleozoic geology of the Spring Gulch-Box Canyon area, Fremont County, Colorado. M, 1958, Texas Tech University. 93 p.

McLane, Charles F., III. Channel network growth; an experimental study. M, 1978, Colorado State University. 109 p.

McLane, Michael John. Phanerozoic detrital rocks at the north end of the Tobacco Root Mountains, southwestern Montana; a vertical profile. D, 1971, Indiana University, Bloomington. 253 p.

McLaren, Digby Johns. Devonian stratigraphy and correlation of the Alberta Rocky Mountains, with descriptions of the brachiopod family Rhynchonellidae. D, 1951, University of Michigan.

McLaren, Donald B. Stratigraphy and areal geology of northeastern Leavenworth County, Kansas. M, 1958, University of Kansas. 99 p.

McLaren, G. P. Minor elements in sphalerite and their implications for metallogenesis of carbonate-hosted zinc-lead deposits of the Yukon Territory and adjacent District of Mackenzie. M, 1978, University of British Columbia.

McLaren, James Peter. Model calculations of regional network locations of subduction zone earthquakes. M, 1984, University of Texas, Austin. 157 p.

McLaren, Patrick. Cirque analysis as a method of predicting the extent of a Pleistocene ice advance. M, 1972, University of Calgary. 91 p.

McLaren, Patrick. The coasts of eastern Melville and western Byam Martin islands; coastal processes and related geology of a High Arctic environment. D, 1977, University of South Carolina. 317 p.

McLaren, R. G. A coupled numerical-analytical model for mass transport in multiple aquifer systems. M, 1981, University of Waterloo.

McLaren, Shirley Anne. Quaternary seismic stratigraphy and sedimentation of the Sable Island sand body,

Sable Island Bank, Outer Scotian Shelf. M, 1988, Dalhousie University. 95 p.

McLarnan, Timothy J. Some counting problems in crystallography. D, 1980, University of Chicago. 74 p.

McLaughlin, Dennis B. A distributed parameter state space approach for evaluating the accuracy of groundwater predictions. D, 1985, Princeton University. 217 p.

McLaughlin, Donald H. The occurrence and significance of bornite. D, 1917, Harvard University.

McLaughlin, Donald Hamilton, Jr. Geology of the Warthan Canyon-Upper Jacalitos Creek District, Fresno County, California. M, 1954, University of California, Berkeley. 103 p.

McLaughlin, Franklin B., III. Influence of discharge, season, and urbanization on the concentration, speciation, and bioavailability of trace metals in the Raritan River basin, New Jersey. M, 1988, Rutgers, The State University, New Brunswick. 227 p.

McLaughlin, Jeffrey Donald. Organic geochemistry of the kerogen from the Ohio Shale by pyrolysis-gas chromatography. M, 1984, University of Toledo. 156 p.

McLaughlin, Keith Lynn. Spatial coherency of seismic waveforms. D, 1983, University of California, Berkeley. 275 p.

McLaughlin, Kenneth P. Interpretation of the Dutchtown conodont fauna. M, 1941, University of Missouri, Columbia.

McLaughlin, Kenneth Phelps. Pennsylvanian stratigraphy of the Colorado Springs Quadrangle, Colorado. D, 1947, Louisiana State University.

McLaughlin, Robert Everett. Plant microfossils from the (Eocene) Wilcox in Northeast Louisiana. M, 1952, [Tulane University].

McLaughlin, Robert Everett. Plant microfossils from the Bruhn Lignite [Tennessee]. D, 1957, University of Tennessee, Knoxville. 183 p.

McLaughlin, Robert J. Geology of the Sargent fault zone in the vicinity of Mount Madonna, Santa Clara and Santa Cruz counties, California. M, 1973, San Jose State University. 131 p.

McLaughlin, Thad G. The pegmatite dikes of the Bridger Mountains, Wyoming. D, 1937, University of Kansas.

McLean, D. C. The mineral association in manganese ores. M, 1941, University of Missouri, Rolla.

McLean, D. J. Geometry of facies packages and E5 erosion surface in the Cardium Formation, Ferrier Field, Alberta. M, 1987, McMaster University. 144 p.

McLean, Dewey Max. Organic-walled phytoplankton from the lower Tertiary Pamunkey Group of Virginia and Maryland. D, 1971, Stanford University. 239 p.

McLean, F. H. Paleontological problems of the Silurian Arisaig Series. D, 1917, Yale University.

McLean, Gordon William. Retention and release of nickel by clays and soils. D, 1966, University of California, Riverside. 93 p.

McLean, Hugh. Petrography and sedimentology of the Blakeley Formation, Kitsap County, Washington. M, 1968, University of Washington. 56 p.

McLean, Hugh. Stratigraphy, mineralogy, and distribution of the Sumpango group (Quaternary) pumice deposits in the volcanic highlands of Guatemala. D, 1970, University of Washington. 90 p.

McLean, James D. Later Tertiary foraminiferal zones of the Gulf Coast. M, 1947, Louisiana State University.

McLean, James Ross. The upper Cretaceous Judith River Formation in the Canadian Great Plains; its history and lithostratigraphy. D, 1970, University of Saskatchewan. 446 p.

McLean, Stephen Russell. Mechanics of the turbulent boundary layer over sand waves in the Columbia River. D, 1976, University of Washington. 214 p.

McLean, Steven Arthur. The distribution and genesis of sepiolite and attapulgite on the Llano Estacado. M, 1969, Texas Tech University. 71 p.

McLean, Thomas Richard. Structural style of the Wichita Mountains of southern Oklahoma. M, 1983, University of Oklahoma. 94 p.

McLean, W. F. and Levan, Donald Clement. Structure and stratigraphy of the Red Canyon area, Gallatin County, Montana. M, 1951, University of Michigan. 66 p.

McLean, Willis John. The crystal structures of the tetrahydrates of tetramethylammonium fluoride and tetramethylammonium sulfate, a refinement of the crystal structure of enargite and the crystal structure of hemihedrite. D, 1968, University of Pittsburgh. 88 p.

McLearn, Frank H. The stratigraphy and correlation of the Arisaig Silurian series (Nova Scotia). D, 1917, Yale University.

McLellan, P. J. A. Investigation of some rock avalanches in the Mackenzie Mountains. M, 1983, University of Alberta. 281 p.

McLellan, Robert Bryant. A petrographic study of granitization in the norite at Dinty Lake, northern Saskatchewan. M, 1940, University of British Columbia.

McLellan, Robert Charles. Stratigraphy of Chugwater Group (Triassic), east side of Bighorn Mountains, Wyoming and Montana. M, 1968, University of Nebraska, Lincoln. 212 p.

McLellan, Roy D. A study of properties of crystal detectors. M, 1922, University of Washington. 25 p.

McLellan, Roy Davison. The geology of the San Juan Islands. D, 1927, University of Washington. 185 p.

McLellan, Russell R. Geology of the Gold Hill area (petrology and mining), Boulder County, Colorado. M, 1948, University of Colorado.

McLelland, Douglas. Geology of the basement complex, Thorvald Nilsen Mountains, Antarctica. M, 1967, University of Nevada. 54 p.

McLelland, James M. Mechanics of dike formation and the tectonic environment of igneous rocks. D, 1961, University of Chicago. 131 p.

McLemore, Virginia T. Geology of the Precambrian rocks of the Lemitar Mountains, Socorro County, NM. M, 1980, New Mexico Institute of Mining and Technology. 169 p.

McLemore, William Hickman. The geology and geochemistry of the Mississippian system in Northwest Georgia and Southeast Tennessee. D, 1971, University of Georgia. 251 p.

McLemore, William Hickman. The geology of the Pollard's Corner area, Columbia County, Georgia. M, 1965, University of Georgia. 149 p.

McLen, P. C. The petrography of the diabase rocks in the Thunder Bay District, Ontario. M, 1951, University of Toronto.

McLennan, John David. Hydraulic fracturing; a fracture mechanics approach. D, 1980, University of Toronto.

McLennan, S. M. Geochemistry of some Huronian sedimentary rocks (north of Lake Huron) with emphasis on rare earth elements. M, 1977, University of Western Ontario. 138 p.

McLeod, Arthur A. The geology of the Silver Hill District, Spokane County, Washington. D, 1923, University of Washington. 33 p.

McLeod, J. A. The Giant Mascot Ultramafite and its related ores. M, 1975, University of British Columbia.

McLeod, John David. The systematics, paleoecology and correlation of Lower Ordovician (Canadian) bryozoans from the Ozark Uplift area. M, 1979, Northern Illinois University. 98 p.

McLeod, Malcolm John. The geology of Campobello Island, southwestern New Brunswick. M, 1979, University of New Brunswick.

McLeod, Norman Stuart. Monitoring of contaminant behaviour in discharging groundwater in lakebeds or streambeds. M, 1982, University of Waterloo. 79 p.

McLeod, Paul J. The depositional history of the Deer Lodge Basin, western Montana. M, 1987, University of Montana. 61 p.

McLeod, Robert Andy. Development and application of a new method for collecting soil moisture samples for chemical analysis. M, 1980, Queen's University. 217 p.

McLeod, William George. Rock fractures; an aerial photographic study. M, 1955, Cornell University.

McLeroy, Carol Ann Chmura. Upper Cretaceous (Campanian-Maastrichtian) angiosperm pollen from the western San Joaquin valley, California. D, 1970, Stanford University. 394 p.

McLeroy, Donald F. and Clemons, Russell E. Geology of Torreon and Pedricenas quadrangles, Coahuila and Durango, Mexico. M, 1962, University of New Mexico. 182 p.

McLeroy, Donald Frazier. Geology and origin of the Precambrian banded iron deposits at Cleveland Gulch, Iron Mountain, and Canon Plaza, Rio Arriba County, New Mexico. D, 1966, Stanford University. 302 p.

McLerran, Richard D. Conodont biostratigraphy of Hueco Mountains, El Paso County, West Texas. M, 1983, Texas A&M University. 82 p.

McLimans, Roger. Archean conglomerates (Precambrian) of the Vermillion District (Minnesota); their provenance and tectonic significance. M, 1971, University of Minnesota, Duluth.

McLimans, Roger Kenneth. Geological, fluid inclusion, and stable isotope studies of the Upper Mississippi Valley zinc-lead district, Southwest Wisconsin. D, 1977, Pennsylvania State University, University Park. 187 p.

McLin, Stephen G. Lime stabilization effects on some undisturbed and remolded clay-rich Pleistocene glacial lake and associated hydraulic fill deposits of Lucas County, Ohio. M, 1973, University of Toledo. 86 p.

McLinn, Gene. An improved borehole dilution device for the in situ determination of the magnitude and direction of groundwater flux. M, 1987, University of Wisconsin-Milwaukee.

McLoughlin, Thomas F. The geological significance of Landsat imagery lineament analysis in selected areas of eastern Kentucky. M, 1979, Eastern Kentucky University. 81 p.

McLure, John William. A history of the Illinois State Geological Survey, 1851-1875. M, 1962, University of Illinois, Urbana. 175 p.

McMackin, Matthew Robert. Extension tectonics in the southeastern Kingston Range and northern Mesquite Mountains; reinterpretation of the Winters Pass "thrust" fault. M, 1987, University of California, Davis. 96 p.

McMackin, Samuel Carl. The geology of a portion of Calhoun County, Illinois. M, 1928, University of Illinois, Urbana.

McMahan, Gregory Lee. A geochemical and thermodynamic investigation of the Great Salt Plains groundwater, Alfalfa County, Oklahoma. M, 1977, University of Tulsa. 107 p.

McMahan, Troy. Geology of representative manganese deposits in the Batesville District, Arkansas. M, 1958, University of Arkansas, Fayetteville.

McMahon, B. E. Paleomagnetic investigation of Late Paleozoic and Early Triassic redbeds of central Colorado. D, 1966, University of Colorado.

McMahon, Barry K. A statistical study of rock slopes in jointed gneiss, central Colorado. D, 1968, Colorado School of Mines. 171 p.

McMahon, Beverly Edith. Paleomagnetic investigation of late Paleozoic and early Triassic redbeds of central Colorado. D, 1967, University of Colorado. 176 p.

McMahon, Brendan Michael. Petrologic redox equilibria in the Benfontein kimberlite sills and in the Allende Meteorite, and the T-fO₂ stability of kimberlitic ilmenite from the Monastery Diatreme. D, 1984, University of Massachusetts. 222 p.

McMahon, David A., Jr. Deep subsurface structural geology of Reeves County, Texas. M, 1977, University of Texas, Austin.

McMahon, Peter B. Computer simulation of mass transport in groundwater; affect of macroscopic heterogeneities in hydraulic conductivity. M, 1984, University of Texas, Austin. 111 p.

McManis, K. L. The effects of conventional soil sampling methods on the engineering properties of cohesive soils in Louisiana. D, 1975, Louisiana State University. 378 p.

McMannis, William Junior. Geology of the Bridger Range area, Montana. D, 1952, Princeton University. 338 p.

McManus, Dean Alvis. Stratigraphy and depositional history of the Kearny Formation (Lower Pennsylvanian) in western Kansas. D, 1959, University of Kansas. 274 p.

McManus, Dean Alvis. Stratigraphy of the Upper Pennsylvanian Merriam Limestone in eastern Kansas. M, 1956, University of Kansas. 196 p.

McManus, Jeffrey. Petrography and structure along the Vermilion fault, Lake Vermilion, Minnesota. M, 1970, University of Missouri, Columbia.

McManus, Kathleen M. The aqueous aragonite to calcite transformation; rate, mechanisms, and its role in the development of neomorphic fabrics. M, 1982, Virginia Polytechnic Institute and State University. 64 p.

McMaster, Glenn Edward. Archean volcanism and geochemistry, Washeibamaga-Thundercloud lakes area, Wabigoon Subprovince; Superior Province, Northwest Ontario. M, 1979, McMaster University. 222 p.

McMaster, Larry. The effects of hydrothermal alteration and weathering in the Morning Mist area (Tertiary), Idaho City, Idaho. M, 1971, Bowling Green State University. 49 p.

McMaster, Natalie Dawn. A preliminary ⁴⁰Ar³⁹Ar study of the thermal history and age of gold in the Red Lake greenstone belt. M, 1987, University of Toronto.

McMaster, Robert Luscher. Petrography and genesis of the New Jersey beach sands. D, 1953, Rutgers, The State University, New Brunswick. 156 p.

McMaster, William M. The geology of East Fork Ridge and Pilot Knob, Oak Ridge, Anderson County, Tennessee. M, 1957, University of Tennessee, Knoxville. 30 p.

McMasters, Catherine R. Geology of Moss Beach. M, 1976, Stanford University.

McMasters, John H. The Eocene Llajas Formation, Ventura County, California. M, 1932, Stanford University. 77 p.

McMath, Vernon Everett. The geology of the Taylorsville area, Plumas County, California. D, 1958, University of California, Los Angeles.

McMechan, G. Depth limits in body wave inversions. M, 1971, University of Toronto.

McMechan, Margaret Evaline. Stratigraphy, structure and tectonic implications of the middle Proterozoic Purcell Supergroup in the Mount Fisher area, southeastern British Columbia. D, 1980, Queen's University. 280 p.

McMechan, Robert Douglas. Stratigraphy, sedimentology, structure and tectonic implications of the Oligocene Kishenehn Formation, Flathead Valley Graben, southeastern British Columbia. D, 1981, Queen's University. 327 p.

McMenamin, Mark Allan Schulte. Paleontology and stratigraphy of Lower Cambrian and upper Proterozoic sediments, Caborca region, northwestern Sonora, Mexico. D, 1984, University of California, Santa Barbara. 229 p.

McMichael, Lawrence Bradley. Geology of the northeastern Olympic Peninsula. M, 1946, University of Washington. 33 p.

McMillan, D. K. Crystallization and metasomatism of the Cuchillo Mountain Laccolith, Sierra County, New Mexico. D, 1979, Stanford University. 298 p.

McMillan, Donald Theodore. Geology and ore deposits of the contact area at Silver Star, Montana. M, 1939, Montana College of Mineral Science & Technology. 54 p.

McMillan, Gordon Warner. Ostracods of the family Hollinidae from the Bell Shale of Michigan. M, 1950, University of Michigan.

McMillan, N. J. A study of platy structure and concretions in Saskatchewan soils by use of thin sections. M, 1951, University of Saskatchewan.

McMillan, Nancy Jeanne Stoll. Petrology and geochemistry of andesites and dacites of the Taos Plateau volcanic field, northern New Mexico. D, 1986, Southern Methodist University. 175 p.

McMillan, Neil John. Petrology of the Nodaway Underclay (Pennsylvanian), Kansas. D, 1955, University of Kansas. 134 p.

McMillan, Paul Francis. A structural study of aluminosilicate glasses by Raman spectroscopy. D, 1981, Arizona State University. 399 p.

McMillan, Richard C. Geology of the Lookout Mill area, Castle Creek, Pennington County, South Dakota. M, 1977, South Dakota School of Mines & Technology.

McMillan, Ronald Hugh. A comparison of the (Ordovician) geological environments of the base metal sulphide deposits of the "B" zone and "North Boundary" zone at the Heath Steele Mine, New Brunswick, (Canada). M, 1969, University of Western Ontario. 198 p.

McMillan, Ronald Hugh. Petrology, geochemistry and wallrock alteration at Opemiska; a vein copper deposit crosscutting a layered Archean ultramafic-mafic sill. D, 1973, University of Western Ontario. 169 p.

McMillan, T. Britt. Leachate monitoring in naturally saline groundwater, Chesapeake Landfill, Chesapeake, Virginia. M, 1985, Old Dominion University. 130 p.

McMillan, William John. Geology of Vedder Mountain, near Chilliwack, British Columbia. M, 1966, University of British Columbia.

McMillan, William John. Petrology and structure of the west flank, Frenchman's Cap Dome, near Revelstoke, B.C. D, 1969, Carleton University. 150 p.

McMillen, Dan E., Jr. Geology of the Shawnee Bend Quadrangle, Benton County, Missouri. M, 1958, University of Missouri, Columbia.

McMillen, Daniel David. The structure and economic geology of Buckhorn Mountain, Okanogan County, Washington. M, 1979, University of Washington. 68 p.

McMillen, Hugh O. Study of the relationship between the Cottonwood Limestone and Neva Limestone structures. M, 1947, Kansas State University. 30 p.

McMillen, Kenneth James. Ecology, distribution, and preservation of polycystine Radiolaria in the Gulf of Mexico and Caribbean Sea. D, 1977, Rice University. 135 p.

McMillen, Kenneth James. Quaternary deep-sea lebensspuren and their relationship to depositional environments in the Caribbean Sea, the Gulf of Mexico, and the eastern and central North Pacific Ocean. M, 1975, Rice University. 147 p.

McMillen, Ralph E. The Minnekahta Limestone of the Black Hills. M, 1939, University of Minnesota, Minneapolis. 24 p.

McMillin, Steven. Geology, alteration and sulfide mineralization of Sheep Mountain in the Matanuska Valley, southcentral Alaska. M, 1984, University of Alaska, Fairbanks. 112 p.

McMinn, Paul Meloy. The Pennsylvanian stratigraphy of Daggett County, Utah. M, 1948, University of Wisconsin-Madison.

McMonagle, Anne Linette. Stable isotope and chemical compositions of surface and subsurface waters in Saskatchewan. M, 1987, University of Saskatchewan. 108 p.

McMoran, William Dalton. Geology of the Gaither Quadrangle (Carboniferous), Boone County, Arkansas. M, 1968, University of Arkansas, Fayetteville.

McMullen, Cindy Anne. Seismicity and tectonics of the northeastern Sea of Okhotsk. M, 1985, Michigan State University. 107 p.

McMullen, L. D. Systems analysis applied to the protection and enhancement of the water quality of Lake MacBride. D, 1975, University of Iowa. 193 p.

McMullen, Richard J., Jr. The effect of geothermal gradients on nonlinear creep deformation in the lithosphere. M, 1978, SUNY at Buffalo. 150 p.

McMullen, Robert Michael. Sedimentary petrology of the Cardium Formation, west-central Alberta. M, 1959, University of Alberta. 91 p.

McMullen, Terrence Leigh. The mammals of the Duck Creek local fauna, late Pleistocene of Kansas. M, 1974, Kansas State University. 29 p.

McMullin, W. Dennis. Subsurface geology of the Grimsby (Clinton) Sandstone of Ashtabula County, Ohio. M, 1976, University of Texas, Arlington. 160 p.

McMurchy, Robert Connell. The crystal structure of the chlorite minerals. D, 1934, University of Minnesota, Minneapolis. 14 p.

McMurdie, Dennis Stoddard. An investigation of a reported occurrence of fossil crabs from the Upper Cretaceous Tropic Shale Formation in southern Utah. M, 1967, University of Utah. 30 p.

McMurray, Jay Maurice. Geology of the Freezeout Mountain area, Malheur County, Oregon. M, 1962, University of Oregon. 87 p.

McMurray, Keith S. Lithofacies study of Lower and Middle Pennsylvanian rocks in northwestern Kansas and adjacent states. M, 1962, University of Kansas. 56 p.

McMurray, M. Geology and organic geochemistry of saline A-1 carbonate oil source-rock lithofacies (Upper Silurian) southwestern Ontario, Canada. M, 1985, University of Waterloo. 236 p.

McMurrough, Hugh Kenneth. Stratigraphy of lower Atoka and upper Bloyd strata in western Arkansas and eastern Oklahoma. M, 1980, University of Arkansas, Fayetteville.

McMurry, H. V. Periodicity of deep focus earthquakes. D, 1938, Massachusetts Institute of Technology. 105 p.

McMurry, Jude B. Petrology and Rb-Sr geochemistry of the Monte das Gameleiras and Dona Ines plutons, northeastern Brazil. M, 1982, University of Texas, Austin. 180 p.

McMurtry, G. M. Geochemical investigations of sediments across the Nazca Plate at 12°S. M, 1975, University of Hawaii. 59 p.

McMurtry, Robert Paul. The subsurface geology of the Penick area, Jones County, Texas. M, 1953, University of Oklahoma. 47 p.

McNaboe, Gerald Joseph. Paleomagnetic study of selected Holocene, Miocene, Mesozoic, and Ordovician rocks from Baja California, Mexico. M, 1987, San Diego State University. 88 p.

McNair, Andrew Hamilton, Jr. Cryptostomatous Bryozoa from the Middle Devonian Traverse rocks of Michigan. D, 1935, University of Michigan.

McNair, Andrew Hamilton, Jr. Some cryptostomatous Bryozoa from the Traverse Group of Michigan. M, 1933, University of Montana. 60 p.

McNairn, William H. Growth of etch figures. D, 1916, University of Toronto.

McNally, Joseph T. Application of groundwater geochemistry as an exploration tool for carbonate-hosted zinc-lead deposits in Pennsylvania. M, 1984, Pennsylvania State University, University Park. 228 p.

McNally, Karen Cook. Spatial, temporal, and mechanistic character in earthquake occurrence; a segment of the San Andreas Fault in central California. D, 1976, University of California, Berkeley. 140 p.

McNamara, Susan Joan. Quaternary geology of the Phelps Lake area (64-M) Saskatchewan. M, 1987, University of Saskatchewan. 114 p.

McNary, Samuel W. Petrography and field studies of late Cenozoic basalt flows and intrusions in the Orofino-Elk River area, Idaho. M, 1976, University of Idaho. 135 p.

McNaughton, D. C. M. A hydrogeological, geochemical, and isotopic study of an uncontaminated flow system, Perch Lake basin, Ontario. M, 1975, University of Waterloo.

McNaughton, Duncan A. Geology of the Lytle Canyon area, eastern San Gabriel Mountains (California). M, 1934, California Institute of Technology. 71 p.

McNaughton, Duncan Anderson. Geologic guides to oil accumulations in metamorphic rocks. D, 1950, University of Southern California.

McNaughton, Kenneth C. A fluid inclusion study of the Nanisivik lead-zinc deposit, Baffin Island, N.W.T. M, 1983, University of Windsor. 95 p.

McNay, Lewis Morris. Soil clay minerals from selected Middle Atlantic States river valleys. M, 1967, American University. 66 p.

McNeal, Brian Lester. Effects of solution composition on the hydraulic conductivity of fragmented soils. D, 1965, University of California, Riverside. 120 p.

McNeal, James E. Analysis of textural variations in nearshore bay and gulfside sands, Dauphin Island (Mobile County) and Fort Morgan (Baldwin County), Alabama. M, 1965, University of Alabama.

McNeal, James Marr. Abundance and occurrence of mercury and other trace metals in rocks, soil, and stream sediments in Pennsylvania. D, 1975, Pennsylvania State University, University Park. 184 p.

McNeely, John B. Stratigraphy of subsurface Mississippian carbonates of Warsaw age in northern Sumner County, Kansas. M, 1974, University of Kansas.

McNeely, Warren L. Gravity survey of the Shelton-Grays Harbor area, Washington. M, 1969, University of Puget Sound. 38 p.

McNeese, Linda Roberts. The stromatolites of Storr's Lake, San Salvador, Bahamas. M, 1988, University of North Carolina, Chapel Hill. 95 p.

McNeil, Andrew Malcolm. Experimental studies of Na metasomatism of a model mantle pyrolite; implications for the natural system. M, 1986, University of Western Ontario. 116 p.

McNeil, D. H. The Cretaceous System in the Manitoba Escarpment. D, 1977, University of Saskatchewan. 733 p.

McNeil, Mary Deligant. The geology of the eastern half of the Victorville Quadrangle, San Bernardino County, California. M, 1963, University of California, Los Angeles.

McNeill, Albert Russell. Geology of the Hoodoo mining district, Latah County, Idaho. M, 1972, University of Idaho. 98 p.

McNeill, R. J. A biogeochemical investigation of the Pb and Zn contents in Eastern hemlock, Tsuga canadensis (L.) Carr, in northeastern United States. M, 1978, Queens College (CUNY). 93 p.

McNeillie, Jennifer I. A sedimentological and paleontological analysis of a large scale gravity flow, DSDP site, Leg 96. M, 1987, University of South Florida, St. Petersburg.

McNeish, Jerry A. Indirect evaluation of hydrogeologic spatial correlation length; an approach to quantification of hydraulic conductivity variance reduction using synthetic hydraulic tests. M, 1987, University of Texas, Austin. 120 p.

McNelis, David N. Aerosol formation from gas phase reactions of ozone and olefin in the presence of sulfur dioxide. D, 1974, University of North Carolina, Chapel Hill. 232 p.

McNellis, Jesse M. Control of procedural errors in thermoluminescence studies. M, 1959, University of Kansas.

McNevin, Thomas F. Use of archived soils and historical data as indicators of acid precipitation effects on New Jersey soils. D, 1985, Rutgers, The State University, New Brunswick. 162 p.

McNew, Gregory E. Tactite alteration and its late stage replacement in the southern half of the Rosemont mining district, Arizona. M, 1981, University of Arizona. 80 p.

McNew, Mark. Evaluating the effects of grain size distribution on the porosity of unconsolidated grain assemblages by computer modeling. M, 1988, University of Houston.

McNichol, Ann P. A study of the remineralization of organic carbon in nearshore sediments using carbon isotopes. D, 1986, Massachusetts Institute of Technology. 225 p.

McNiel, Norman. A geographical study of the Baton Rouge Fault, Baton Rouge, Louisiana. M, 1961, Louisiana State University.

McNitt, James Raymond. Geology and mineral deposits of the Kelseyville S.E. Quadrangle, Sonoma County, California. D, 1961, University of California, Berkeley. 128 p.

McNitt, James Raymond. Inclusions in quartz in igneous and metamorphic rocks of central New Hampshire. M, 1954, University of Illinois, Urbana. 28 p.

McNully, Claude V. Grain size studies of the Wedington Sandstone (Mississippian) of the Fayetteville Formation (Arkansas). M, 1966, University of Arkansas, Fayetteville.

McNulty, Charles L., Jr. Insoluble residues of Lower Ordovician limestone along the Black River, New York. M, 1948, Syracuse University.

McNulty, Charles Lee, Jr. Foraminifera of the Austin Group in Northeast Texas. D, 1955, University of Oklahoma. 213 p.

McNulty, Edmund Gregory. Consolidation with axisymmetric flow in soils having radial and vertical variation of properties. D, 1982, University of Texas, Austin. 168 p.

McNutt, Gordon Russell. A correlation of the Cretaceous of the United States. M, 1938, University of Texas, Austin.

McNutt, Gordon Russell. Geology of southern Delaware Mountains and northwestern Apache Mountains, Texas. D, 1948, University of Oklahoma. 83 p.

McNutt, James R. A. The petrology of the Goodwin Lake Gabbro (New Brunswick). M, 1962, University of New Brunswick.

McNutt, Marcia K. Continental and oceanic isostasy. D, 1978, University of California, San Diego. 206 p.

Mcnutt, Robert H. A Study of the strontium redistribution under controlled conditions of temperature and pressure. D, 1965, Massachusetts Institute of Technology.

McNutt, Stephen Russell. The eruptive activity, seismicity, and velocity structure of Pavlof Volcano, eastern Aleutians. D, 1985, Columbia University, Teachers College. 214 p.

McNutt, V. H. A study of the precipitation of iron disulphide and its relation to certain types of deposits. M, 1912, University of Missouri, Rolla.

McNutt, William Paul, Jr. Depositional environments of the Tombigbee Sand Member of Cretaceous Age as indicated by particle size analysis. M, 1963, University of Alabama.

McOnie, A. W. The stability of intermediate chlorites of the clinochlore-daphnite solid solution series at 2 kilobars total pressure. M, 1972, University of Toronto.

McParland, Brian J. The interpretation of a gravity anomaly in the Canadian Arctic. M, 1975, Colorado School of Mines. 63 p.

McPartland, John T. Snowpack accumulation in relation to terrain, and meteorological factors in southwestern Montana. M, 1971, Montana State University. 106 p.

McPeek, Lawrence A. Geology of the El Vado-Nutria Canyon area, Rio Arriba County, New Mexico. M, 1962, Colorado School of Mines. 82 p.

McPhail, Derry Campbell. The stability of Mg-chlorite. M, 1985, University of British Columbia. 45 p.

McPhail, Robert Louis. A survey of the heavy mineral distribution in East Goose Creek watershed (Lafayette County, Mississippi). M, 1963, University of Mississippi.

McPhearson, Ronald Dean. Depositional environment and diagenesis of the Benbrook Member of the Goodland Formation (Lower Cretaceous), Sabine Parish, Louisiana. M, 1988, Northeast Louisiana University. 111 p.

McPhee, D. S. Geology of the Eric Lake area, Saguenay County, Quebec. M, 1958, McGill University.

McPhee, James P. Regional gravity analysis of the Anna, Ohio seismogenic region. M, 1983, Purdue University.

McPherron, Donald S. Clay mineralogy of some of the Permian shales. M, 1956, Kansas State University. 57 p.

McPherson, Constance Barbara. A fossil fungal spore assemblage from the Punta Carballo Formation (Miocene), Barranca, Costa Rica. M, 1980, Kent State University, Kent. 76 p.

McPherson, Harold J. Glacial geomorphology of the upper Red Deer Valley, Alberta. M, 1963, University of Calgary.

McPherson, Robert A. Pleistocene geology of the Beausejour area, Manitoba. D, 1970, University of Manitoba.

McPherson, Robert A. Pleistocene stratigraphy of the Winnipeg River in the Pine Falls-Seven Sisters Falls area, Manitoba. M, 1968, University of Manitoba.

McPherson, Roger Ian. Geology of the metavolcanic belt to the east of Lake Athabaska, Saskatchewan. M, 1961, Montana College of Mineral Science & Technology. 133 p.

McPherson, Ronald Bruce. Static response of flexible domes buried in sand. D, 1968, West Virginia University. 235 p.

McPherson, William J. An engineering study of Round Mountain oil field, Kern County, California. M, 1951, University of Saskatchewan.

McPherson, William John. An engineering study of Round Mountain oil field, Kern County, California. D, 1949, Stanford University. 123 p.

McPherson, William John. Critical review of criteria used in the nomenclature and classification of sandstones. M, 1947, McGill University.

McQuade, B. N. Petro-chemistry of the Bachelor Lake volcanic complex and associated ore deposits, Abitibi East, N.W. Quebec. D, 1981, University of Ottawa. 338 p.

McQueen, Donald James. A paleomagnetic investigation of the charnockitic massif in the vicinity of Mirnyy, Antarctica. M, 1971, Washington University. 64 p.

McQueen, Henry S. Determining the rate of movement of oil through various sands. M, 1922, University of Missouri, Columbia.

McQueen, Jereld Edward. Geology and fracture patterns of southern Burnett County, Texas. M, 1963, University of Texas, Austin.

McQueen, Kay C. Subsurface stratigraphy and depositional systems of the Hartshorne Formation, Arkoma Basin, Oklahoma. M, 1982, University of Arkansas, Fayetteville.

McQuillan, Kirk A. Petrology of nodular chert in the Burlington Limestone (Mississippian), central Missouri. M, 1979, University of Missouri, Columbia.

McQuillan, Michael W. Geology of Davis-Bronson Pool, Allen and Bourbon counties, Kansas. M, 1968, Kansas State University. 117 p.

McQuillan, Michael William. Contemporaneous faults; a mechanism for the control of sedimentation in the southwestern Arkoma Basin, Oklahoma. D, 1977, University of Oklahoma. 228 p.

McQuillen, Daniel. Sand body geometry and facies distribution of a Holocene washover fan complex, Brown Cedar Cut, East Matagorda Bay, Texas. M, 1984, University of Houston.

McQuillian, Tom Alan. The areal extent and correlation of the Deer Valley Limestone of Somerset County, Pennsylvania. M, 1960, University of Pittsburgh.

McQuiston, Ian Brice. Ostracoda of Holocene deposits from the Belfast area, Northern Ireland. M, 1965, University of Illinois, Urbana.

McQuown, M. Scott. An engineering geophysical and geological investigation of the sand and gravel resources of part of the Kent kame complex, Portage County, Ohio. M, 1988, Kent State University, Kent. 154 p.

McRae, Edward Walton. The geology and mineral resources of the eastern part of the Philpott coal field. M, 1950, University of Arkansas, Fayetteville.

McRae, Lee E. Sedimentology and paleomagnetics of the basal Potsdam Sandstone in the Adirondack border region, New York State, southwestern Quebec, and southeastern Ontario. M, 1985, Dartmouth College. 178 p.

McRae, Otis M. Geology of the northern part of the Ortiz Mountains, Santa Fe County, New Mexico. M, 1958, University of New Mexico. 112 p.

McRee, David E. Paleozoic stratigraphy and structural geology of the Nebaj Quadrangle, Guatemala (Central America). M, 1969, Louisiana State University.

McReynolds, J. Carroll. Structural geology of Southwest Austin area, Travis County, Texas. M, 1958, University of Texas, Austin.

McReynolds, J., Jr. The geology of the Maynardville area, Union County, Tennessee. M, 1962, University of Tennessee, Knoxville. 38 p.

McReynolds, Joseph A. Paleoenvironments and facies relations of the Lower Cambrian Rome Formation along Haw Ridge on the U.S. Department of Energy Reservation, Oak Ridge area in Roane and Anderson counties, Tennessee. M, 1988, University of Tennessee, Knoxville. 121 p.

McRoberts, Gordon D. An integrated geological, geochemical and petrogenetic study of a part of the Archean Larder Lake Group at the Adams Mine, northeastern Ontario. M, 1986, McMaster University. 386 p.

McSaveney, E. C. Terraces of the Hocking River basin, Ohio; particle morphology and paleohydrology. D, 1976, Ohio State University. 193 p.

McSaveney, Eileen Craven. The surficial texture of rockfall talus. M, 1971, Ohio State University.

McSaveney, M. J. The Sherman Glacier rock avalanche of 1964; its emplacement and subsequent effects on the glacier beneath it. D, 1975, Ohio State University. 426 p.

McSwain, James L. Stratigraphic analysis of the Queen and Grayburg formations, southeastern Chaves County, New Mexico. M, 1963, University of New Mexico. 53 p.

McSween, Harry Y., Jr. Petrological and geochemical studies in the Coronaca area, Greenwood County, South Carolina. M, 1969, University of Georgia.

McSween, Harry Younger, Jr. Petrologic and chemical studies of the (C3) carbonaceous chondritic meteorites. D, 1977, Harvard University.

McSwiggin, P. L. Stratigraphy, structural geology and metamorphism of the northeast extension of the Liberty-Orrington Antiform, South-central Maine. M, 1978, University of Maine. 128 p.

McTaggart, K. C. The Cretaceous rocks of Lytton, British Columbia. M, 1946, Queen's University. 55 p.

McTaggart, Kenneth C. The belt of the Lower Cretaceous rocks along Fraser River, southwestern British Columbia. D, 1948, Yale University.

McTaggart-Cowan • Medford

McTaggart-Cowan, G. H. A digital model of fluctuations in wells produced by fluctuations in nearby surface waters. M, 1967, University of British Columbia.

McTague, Stephen B., Jr. Petrology and petrography of the Mount Rogers Volcanic Series (Precambrian) of Grayson, Smyth, and Washington counties, Virginia. M, 1967, Virginia Polytechnic Institute and State University.

McThenia, Andrew W., Jr. Geology of the Madison River canyon area, north of Ennis, Montana. M, 1960, Columbia University, Teachers College. 40 p.

McTigue, D. F. A nonlinear continuum model for flowing granular materials. D, 1979, Stanford University. 176 p.

McUsic, James Michael. Geology of the red conglomerate peaks area, Beaverhead County, Montana, and Clark County, Idaho. M, 1949, University of Michigan. 47 p.

McWhorter, James G. A preliminary water budget and reconnaissance of the hydrogeology of the Paulinskill drainage basin, Warren and Sussex counties, New Jersey. M, 1974, Rutgers, The State University, New Brunswick. 94 p.

McWilliams, David Bruce. Depositional facies, diagenesis and porosity relationships of the Lower Devonian Thirtyone Formation of the Permian Basin. M, 1985, Texas Tech University. 90 p.

McWilliams, Edward B. Paleoenvironmental analysis of the Chapman Ridge Sandstone (middle Ordovician) in the area of Knoxville, Tennessee. M, 1975, University of Tennessee, Knoxville. 109 p.

McWilliams, George Robert. Sedimentary analyses of Conemaugh sandstones in the Pittsburgh area. M, 1955, University of Pittsburgh.

McWilliams, Michael O. Paleomagnetism of Precambrian metamorphic rocks from Magnetawan, Ontario and apparent polar wander of the Grenville Province. M, 1974, University of Toronto.

McWilliams, Richebourg Gaillard, Jr. A gravity investigation of the Klepac Dome, McIntosh salt dome, and Jackson Fault in Southwest Alabama. M, 1970, University of Alabama.

McWilliams, Robert G. Paleocene stratigraphy and biostratigraphy of central-western Oregon. D, 1968, University of Washington. 138 p.

McWilliams, Robert Gene. Geology and biostratigraphy of the Lake Crescent area, Clallam County, Washington. M, 1965, University of Washington. 165 p.

Mdala, Chisengu L. Magnetotelluric investigation in the San Luis Valley, Colorado. D, 1980, Colorado School of Mines. 174 p.

Meacham, James F. Ostracoda from shales in the Saint Laurent limestone (middle Devonian) of Perry County, Missouri. M, 1968, Southern Illinois University, Carbondale. 84 p.

Meacham, Reid Phillip. The Leipers fauna around Nashville. M, 1925, Vanderbilt University.

Mead, Edwin Ruthven. Problems of separation and identification of uranium minerals. M, 1956, University of Minnesota, Minneapolis. 21 p.

Mead, James Glen. Paleontology of the Sid McAdams Locality (Permian), southern Taylor County, Texas. M, 1971, University of Texas, Austin.

Mead, Jim I. Harrington's extinct mountain goat (Oreamnos harringtoni) and its environment in the Grand Canyon, Arizona. D, 1983, University of Arizona. 232 p.

Mead, Judson. Investigation of deep crustal structure by seismic reflection. D, 1949, Massachusetts Institute of Technology. 99 p.

Mead, Richard George. Geology of a portion of Point Dume Quadrangle, California. M, 1952, University of Southern California.

Mead, Richard H. The use of helium in uranium exploration. D, 1980, Colorado School of Mines. 240 p.

Mead, Warren J. Occurrence and origin of the bauxite deposits of Arkansas. D, 1926, University of Wisconsin-Madison.

Meade, Carroll Wade. Lithology and distribution of the Cook Mountain Formation (Eocene) in Lincoln Parish, Louisiana. M, 1961, Louisiana Tech University.

Meade, Grayson Eichelberger. The (Pliocene) Blanco fauna (Texas). D, 1946, University of Chicago. 47 p.

Meade, Grayson Eichelberger. The Camelidae of the Central Plains (Nebraska, Colorado and Wyoming). M, 1937, University of Nebraska, Lincoln.

Meade, H. D. Petrology and metal occurrences of the Takla Group and Hogem and Germansen batholiths, North central British Columbia. D, 1977, University of Western Ontario. 355 p.

Meade, James Sherwood. A joint intensity and isostatic gravity anomaly study across the Jemez Lineament near Grants, New Mexico. M, 1985, North Carolina State University. 57 p.

Meade, Robert Francis. Molluscan paleoecology of the Fernando Group (Pliocene and Pleistocene) of the southern Ventura Basin, California. D, 1967, University of California, Los Angeles. 196 p.

Meade, Robert Herber, Jr. Compaction and development of preferred orientation in clayey sediments. D, 1961, Stanford University. 73 p.

Meade, Ronald Bartholomew. Evidence of reservoir induced macroseismicity. D, 1981, Purdue University. 247 p.

Meader, N. M. Paleoecology and paleoenvironments of the Upper Devonian Martin Formation in the Roosevelt Dam-Globe area, Gila County, Arizona. M, 1977, University of Arizona. 124 p.

Meader, Robert Wooten. Stratigraphy and limnology of a hard water lake in western Minnesota. M, 1956, University of Minnesota, Minneapolis. 57 p.

Meader, Sally Jo. Paleoecology of the Upper Devonian Percha Formation of south-central Arizona. M, 1976, University of Arizona.

Meador, John P. The geology and petrography of eastern Lunenburg County, Virginia. M, 1949, University of Virginia. 56 p.

Meador, Karen Jean. Geologic evolution of the northern Newfoundland Basin. M, 1988, University of Texas, Austin. 111 p.

Meadows, George Richard. Petrology of Mesozoic age diabase dikes in the Georgia Piedmont. M, 1978, Emory University.

Meadows, Guy Allen. A field investigation of the spatial and temporal structures of longshore currents. D, 1977, Purdue University. 168 p.

Meadows, James Lawson. A study of the geological section across the southern portion of Travis County. M, 1930, University of Texas, Austin.

Meadows, Mark Allen. The inverse problem of a transversely isotropic elastic medium. D, 1985, University of California, Berkeley. 397 p.

Meagher, David Pope. Louisiana lignite outcrops. M, 1941, Louisiana State University.

Meagher, Edward Patrick. The crystal structure and polymorphism of cordierite. D, 1967, Pennsylvania State University, University Park. 123 p.

Meaker, Harold N. The stratigraphy and structure of the eastern half of the 7-1/2 Minute Jamesville Quadrangle (New York). M, 1958, Syracuse University.

Meaney, William R. Sediment transport and sediment budget in the fore-reef zone of a fringing coral reef, Discovery Bay, Jamaica. M, 1973, Louisiana State University.

Means, Alan Hay. Geology and ore deposits of Red Cliff, Colorado. M, 1914, Massachusetts Institute of Technology. 73 p.

Means, Jeffrey Lynn. Geochemical controls on trace metal transport in aqueous environmental systems. D, 1981, Princeton University. 260 p.

Means, John Albert. A population study of the Cretaceous foraminiferal genus Orbitolina. M, 1948, University of Texas, Austin.

Means, John Brittian, Jr. The subsurface geology and oil possibilities of the Love Ranch area of Kerr and Real counties, Texas. M, 1941, University of Texas, Austin.

Means, John H. Geology of the (Pennsylvanian) lower coal measures of Arkansas. M, 1892, [Stanford University].

Means, Kendall D. Sediments and foraminifera of Richardson Bay, California (San Francisco Bay area). M, 1965, University of Southern California.

Means, Winthrop Dickinson. Structure and stratigraphy in the central Toiyabe Range, Nevada. D, 1960, University of California, Berkeley. 169 p.

Mear, Charles Eugene. Quaternary geology of upper Sabinal River valley, Uvalde and Bandera counties, Texas. M, 1953, University of Texas, Austin.

Meara, Joseph Edwin. A study of the geology of the Kankakee River valley, Kankakee County, Illinois. M, 1938, Catholic University of America. 60 p.

Mears, Arthur Irvin. Glacial geology and crustal properties in the Nedlukseak Fiord region, East Baffin Island, Canada. M, 1972, University of Colorado.

Mears, Brainerd. Cenozoic faults, gravels, and volcanics of Oak Creek Canyon, Arizona. D, 1950, Columbia University, Teachers College.

Meaux, R. P. and Brennan, J. F. Observation of the nearshore water circulation off a sand beach. M, 1964, United States Naval Academy.

Mebane, R. Alan. The geology of the South Harriman Area, Roane County, Tennessee. M, 1957, University of Tennessee, Knoxville. 30 p.

Mecham, Brent H. Petrography and geochemistry of the Fish Haven Formation (Ordovician) and lower part of the Laketown Formation (Orodovician-Silurian), (Cache and Rich counties, Utah), Bear River Range, Utah. M, 1973, Utah State University. 64 p.

Mecham, Derral F. The structure of Little Rock Canyon, central Wasatch Mountains, Utah. M, 1948, Brigham Young University. 59 p.

Mechergui, Mohamed. Stochastic modeling of the water table in the vicinity of drainage tiles. D, 1984, University of California, Davis. 207 p.

Mechler, Lina S. Recent foraminifera of St. Andrew Bay, Florida. M, 1984, University of Texas, Arlington. 200 p.

Mechling, George William. Geology and water resources of Lancaster County. M, 1931, University of Nebraska, Lincoln.

Mecionis, Robert. Description and interpretation of calcrete profiles and related features in the Warix Run and Paoli-Beaver Bend members of the Newman Limestone (Mississippian), northeastern Kentucky. M, 1984, Bowling Green State University. 220 p.

Medall, Sheldon E. Geology of the Castle Mountains, California. M, 1963, University of Southern California.

Medaris, Levi Gordon, Jr. Geology of the Seiad Valley area, Siskiyou County, California, and petrology of the Seiad Ultramafic Complex. D, 1966, University of California, Los Angeles. 395 p.

Medary, Tom A. Replacement and recrystallization textures in Bonneterre dolostones. M, 1986, University of Missouri, Columbia.

Meddaugh, William Scott. Age and origin of uraninite in the Elliot Lake, Ontario uranium ores. D, 1983, Harvard University. 225 p.

Meddaugh, William Scott. The distribution of uranium and thorium in the Wolf River Batholith. M, 1978, University of Wisconsin-Milwaukee.

Mederos H., Alfredo. Lower Tertiary coccolithophorids and discoasterids of the Mount Diablo area, Central California. M, 1961, Stanford University.

Medford, G. A. Geology and thermal history of an area near Okanagan Lake, southern British Columbia. D, 1976, University of British Columbia.

Medford, Gary A. Calcium diffusion in a mugearite melt. M, 1970, McGill University. 38 p.

Medford, Richard M. Mesoscopic analysis of folds in the Briggs Formation, Malone Mountains, Texas. M, 1983, Texas Tech University. 60 p.

Medhani, Rezene Gurmu. Stabilization of Ponca City shale. D, 1982, University of Oklahoma. 226 p.

Medina, Diana Magdalena Diez de see Diez de Medina, Diana Magdalena

Medina, Julian Michael. The geology and geochemistry of the San Jose iron deposit and selected geochemistries from the iron copper region of north-central Baja California. M, 1983, San Diego State University. 117 p.

Medina-Melo, Francisco Jorge. Modelling of soil-structure interaction by finite and infinite elements. D, 1981, University of California, Berkeley. 48 p.

Meditz, Richard Donald. Stratigraphy and micropaleontology of Barnegat City well. M, 1955, Rutgers, The State University, New Brunswick. 91 p.

Medlin, Jack Harold. Comparative petrology of two igneous complexes in the South Carolina Piedmont. D, 1968, Pennsylvania State University, University Park. 328 p.

Medlin, Jack Harold. Geology and petrography of the Bethesda Church area, Green County, Georgia. M, 1964, University of Georgia. 100 p.

Medlin, Linda Karen. Community analysis of epiphytic diatoms from selected species of macroalgae collected along the Texas coast of the Gulf of Mexico. D, 1983, Texas A&M University. 165 p.

Medlin, W. Eric. Modeling local thermal anomalies; constraints from conductivity, gravity and heat flow. M, 1983, University of Wyoming. 114 p.

Medlock, Patrick Lee. Depositional environment and diagenetic history of the Frisco and Henryhouse formations in central Oklahoma. M, 1984, Oklahoma State University. 146 p.

Medlyn, Gary Wayne. The influence of soil erosion on crop productivity and surface soil characteristics for five selected Oklahoma soils sown to winter wheat. D, 1983, Oklahoma State University. 47 p.

Medwedeff, Donald Arthur. Kinematic and dynamic relationships between microscopic and mesoscopic fabric elements and macroscopic folds and faults, Grotto Creek valley, southwestern Alberta. M, 1983, Queen's University. 157 p.

Medwedeff, Donald Arthur. Structural analysis and tectonic significance of late-Tertiary and Quaternary, compressive-growth folding, San Joaquin Valley, California. D, 1988, Princeton University. 208 p.

Mee, Cathleen E. A pedological study of the Trout Creek basin, Porcupine Hills. M, 1972, University of Calgary.

Meeder, C. A. The structure of the oceanic crust off southern Peru determined from an OBS experiment. M, 1977, University of Washington. 31 p.

Meeder, John Frank. The Mollusca from the Inglis Formation (upper Eocene, Florida) and their zoogeographic implications. M, 1976, University of Florida. 104 p.

Meeder, John Frank. The paleoecology, petrology and depositional model of the Pliocene Tamiami Formation, Southwest Florida (with special reference to corals and reef development). D, 1987, University of Miami. 774 p.

Meehan, Kenneth Tillotson. Analyses and spatial-classification of multivariate geologic and remotely sensed data using empirical discriminant analysis. D, 1978, University of Idaho. 605 p.

Meehan, Michael. Geologic factors affecting the mining of the upper Freeport Coal seam in west-central Preston County, West Virginia. M, 1980, University of Pittsburgh.

Meek, Burk D. The effects of organic matter, flooding time, and temperature on the dissolution of iron and manganese from soil in situ. M, 1967, University of California, Riverside. 40 p.

Meek, Frederick Barber, III. The lithostratigraphy and depositional environments of the Springer and lower Golf Course formations (Mississippian-Pennsylvan-

ian) in the Ardmore Basin, Oklahoma. M, 1983, University of Oklahoma. 212 p.

Meek, Reed Harold. Conodont biostratigraphy and biofacies of the Norian (Upper Triassic) strata of western Nevada. M, 1983, University of Wisconsin-Madison. 101 p.

Meek, Richard M. Rayleigh and Love wave dispersion in eastern Kansas and western Missouri. M, 1962, University of Kansas. 41 p.

Meek, Robert A. The geology of the Onapa-Council Hill area, Muskogee and McIntosh counties, Oklahoma. M, 1957, University of Oklahoma. 67 p.

Meek, Ward B. Factors governing the localization of gold-sulfide ores in the Paracale-Jose Panganiban mining district of the Philippine Islands. D, 1947, University of Wisconsin-Madison.

Meek, Ward B. The heavy accessory minerals of the Palms quartz slate of the Penokee-Gogebic Range (Minnesota). M, 1936, University of Wisconsin-Madison.

Meeker, Kimberly A. The emission of gases and aerosols from Mount Erebus Volcano, Antarctica. M, 1988, New Mexico Institute of Mining and Technology. 172 p.

Meekins, Keith Leroy. A study of granite pegmatite at Captain Cook Quarry in Fairfield County, Connecticut. M, 1974, Rutgers, The State University, Newark. 44 p.

Meeks, R. L. A laboratory based earth science course for McLean Middle School. M, 1972, East Texas State University.

Meeks, Yvonne Joyce. Hydrogeology of the Twin Cities Basin with numerical emphasis on the Mount Simon-Hinckley Aquifer. D, 1986, Stanford University. 118 p.

Meen, James Kenneth. The origin and evolution of a continental volcano; Independence, Montana. D, 1985, Pennsylvania State University, University Park. 877 p.

Meen, Victor B. The determination of the specific gravity of minerals by use of index liquids. M, 1933, University of Toronto.

Meen, Victor B. The temperature of formation of vein quartz and some associated minerals. D, 1936, University of Toronto.

Meenaghan, Susan Lee. Brines; mechanisms of concentration and possible origins. M, 1980, Texas Tech University. 74 p.

Meents, Richard O. Tectonic examinations of the western Arbuckle Mountains. D, 1930, University of Oklahoma. 82 p.

Meers, Ronald B. Spatial distribution of zircons in porphyroblastic gneisses. M, 1976, Texas Tech University. 46 p.

Meertens, Charles Mangelaar. Tidal and secular tilt at Adak Island, Alaska. M, 1980, University of Colorado. 95 p.

Meertens, Charles Mangelaar. Tilt tides and tectonics at Yellowstone National Park. D, 1987, University of Colorado. 253 p.

Mees, Ronald L. Geophysical investigation of the northern Palo Verde Valley, Blythe, California. M, 1978, University of California, Riverside. 41 p.

Meese, Debra A. The chemical and structural properties of sea ice in the southern Beaufort Sea. D, 1988, University of New Hampshire. 294 p.

Mefford, James. Geology of the Weirton, West Virginia, Ohio, Pennsylvania 7 1/2 minute quadrangle. M, 1969, West Virginia University.

Megard, Robert Ordell. The biostratigraphic history of Dead Man Lake, Chuska Mountains, New Mexico. D, 1962, Indiana University, Bloomington. 47 p.

Megathlin, Gerrard R. Faulting in the Mohawk Valley. D, 1933, Cornell University.

Megathlin, Gerrard R. The pegmatite dikes of the Gilsum area, New Hampshire. M, 1928, Cornell University.

Megaw, Peter Kenneth McNeill. Volcanic rocks of the Sierra Pastorias Caldera area, Chihuahua, Mexico. M, 1979, University of Texas, Austin.

Meger, Steven Anthony. Mercury pollution in sediment of wilderness lakes of northern Minnesota. M, 1984, University of Delaware. 119 p.

Megerisi, Mohamed Fadlalla. Geological and petrographical study of phosphate in the Gharyan and Al Khurmat al Hamra areas, Libyan Arab Republic. M, 1976, Ohio University, Athens. 62 p.

Meghji, M. H. Water quality model for small agricultural watershed. D, 1975, West Virginia University. 142 p.

Megivern, Katherine Jean. Paleoclimatic significance of the Pleistocene Insectivora and Rodentia of Trench 24, Peccary Cave, Newton County, Arkansas. M, 1982, University of Iowa. 56 p.

Megivern, Stephen James. A structural and finite element analysis of the Badwater and Deep Creek faults, southern Bighorn Mountains, Wyoming. M, 1983, University of Iowa. 83 p.

Meglen, Joseph Francis. Geology of the Perma area, Sanders County, Montana. M, 1975, University of Oklahoma. 80 p.

Meglio, Joanne Teresa. The oxidation and titanium-enrichment mechanism of "altered ilmenite" grains in the Tertiary Kirkwood and Cohansey formations of New Jersey. M, 1979, Lehigh University. 101 p.

Meglis, Andrew J. Bioaggregates and their role in inorganic sediment transport in suspended sediments of a coastal lagoon complex near Stone Harbor, New Jersey. M, 1987, Lehigh University. 90 p.

Meglis, Irene Llewellyn. Ultrasonic velocity and porosity in the Kent Cliffs, N.Y., test well cores and the application to in situ stress determination. M, 1987, Pennsylvania State University, University Park. 102 p.

Megrue, George H. An infrared study of dioctahedral clay minerals. M, 1960, Columbia University, Teachers College.

Megrue, George Henry. Summerville alteration associated with uranium mineralization, Laguna and Grants, New Mexico; Part 1, Breccia pipes of the southern Laguna area; Part 2, Uranium mineralization and Summerville alteration, Grants, New Mexico. D, 1962, Columbia University, Teachers College. 136 p.

Meguid, Fayek. Gravity and magnetic anomalies over Lake Ontario, N.Y., and their relation to the surrounding areas. M, 1973, University of Rochester.

Mehdi, Purnendu K. A geologic study of the Bontatoc Mine area, Pima County, Arizona. M, 1964, University of Arizona.

Mehegan, James M. Secondary mineralization and hydrothermal alteration in the Reydarfjordur drill core, eastern Iceland. M, 1982, University of California, Riverside. 145 p.

Mehelich, Miro. High alumina clay; the Olson Deposit. M, 1950, University of Idaho. 54 p.

Mehenni, Mourad. Subsurface geology of Cecilia Field area, Southwest Louisiana. M, 1975, University of Southwestern Louisiana. 81 p.

Mehl, Maurice Goldsmith. The Phytosauria of the Trias. D, 1915, University of Chicago.

Mehl, Robert L. Subsurface Paleozoic geology of northwestern Kansas, southwestern Nebraska and northeastern Colorado. M, 1959, University of Kansas. 56 p.

Mehner, David T. The geology of the Whiterocks Mountain alkalic complex in south-central British Columbia. M, 1982, University of Manitoba.

Mehringer, Peter J., Jr. Pollen analysis (late Quaternary) of the Tule Springs site, (ten miles north of Las Vegas, Nevada) Nevada. D, 1968, University of Arizona. 176 p.

Mehrotra, P. N. Middle Devonian stromatoporoids from Yukon Territory. M, 1967, McGill University. 202 p.

Mehrotra, Vikram Pratap. Pelletization of coal fines; kinetic and strength aspects. D, 1980, University of California, Berkeley. 192 p.

Mehrtens, Charlotte Jean. A paleoenvironmental reconstruction of a shelf margin; the Caradoc (Middle Ordovician) of southern Quebec. D, 1979, University of Chicago. 268 p.

Mehta, Sudhir. Sulfate and monofluorophosphate bearing apatite. D, 1973, Lehigh University. 99 p.

Mehuys, Guy Robert. Influence of stones on isothermal and thermally induced movement of water through relatively dry desert soils. D, 1973, University of California, Riverside. 86 p.

Meibos, Lynn Clark. Structure and stratigraphy of the Nephi NW 7 1/2 minute Quadrangle, Juab County, Utah. M, 1981, Brigham Young University.

Meidav, Tsvi. Analysis of the viscoelastic properties of some rheological models as applied to earth materials. D, 1960, Washington University. 96 p.

Meidav, Tsvi. Location of buried channels in Grundy County, Missouri, by means of electrical resistivity. M, 1956, Washington University. 43 p.

Meier, Adolph Ernest. An association of harmotome and barium feldspar at Glen Riddle, Pa. M, 1939, Bryn Mawr College.

Meier, Dudley R. Geophysical investigations in the Trenton-Old Bridge area. M, 1949, Princeton University. 48 p.

Meier, Laurence F. Geology of the Crow Peak area, Lawrence County, South Dakota. M, 1981, South Dakota School of Mines & Technology. 69 p.

Meier, Mark Frederick. Glaciers of the Gannett Peak-Fremont Peak area, Wyoming. M, 1951, University of Iowa. 159 p.

Meier, Mark Frederick. Mode of flow of Saskatchewan Glacier, Alberta, Canada. D, 1957, California Institute of Technology. 159 p.

Meier, P. An estimation of the accuracy of quantitative drainage basin analysis from Landsat MSS data, Kentucky; Mississippi Embayment, Eastern coal field. M, 1984, Murray State University. 66 p.

Meier, Robert William. Geology of the Britton Quadrangle, Dallas, Ellis, Johnson, and Tarrant counties, Texas. M, 1964, Southern Methodist University. 24 p.

Meier, Thomas Allan. Multispectral and geomorphic investigations of the surface of Europa. M, 1985, University of Houston. 360 p.

Meierding, T. C. Age differentiation of till and gravel deposits in the Upper Colorado River basin. D, 1977, University of Colorado. 416 p.

Meijer, Arend F. A strontium tracer study of Tertiary epizonal plutons in the Cascade Mountains of Washington and Oregon. M, 1971, University of California, Santa Barbara. 48 p.

Meijer, Arend F. A study of the geochemistry of the Mariana island arc system and its bearing on the genesis and evolution of volcanic arc magmas. D, 1974, University of California, Santa Barbara. 229 p.

Meijer, Willem Otto Jan Groeneveld *see* Groeneveld Meijer, Willem Otto Jan

Meike, Annemarie. A study of deformation-enhanced dissolution in theory, experiment and nature based on microstructural evidence. D, 1986, University of California, Berkeley. 166 p.

Meikle, Brian Keith Michael. Experiments with copper sulphides at elevated temperatures. D, 1959, McGill University.

Meikle, Brian Keith Michael. The geology of the Little River area, Baie d'Espoir, Newfoundland. M, 1955, McGill University.

Meilliez, Francis. Structure of the southern Solitude Range, Rocky Mountains, British Columbia. M, 1972, University of Calgary. 112 p.

Meiman, James R. Influence of coarse fragments on soil moisture and soil temperature. D, 1962, Colorado State University. 165 p.

Meinecke, L., III. A geochemical reconnaissance of Puerto Rican beach sands. M, 1972, University of Missouri, Rolla.

Meinert, Joseph G. Areal geology of the Starvation Creek area, Roger Mills and Beckham counties, Oklahoma. M, 1961, University of Oklahoma. 66 p.

Meinert, Lawrence David. Skarn, manto, and breccia pipe formation in sedimentary rocks in the Cananea District, Sonora, Mexico. D, 1980, Stanford University. 264 p.

Meinert, Richard Joseph. Island Lake Series (Precambrian); East Island Lake, Manitoba, Canada. M, 1957, University of Cincinnati. 46 p.

Meinke, Deborah Kay. Structure and development of the dermal skeleton of Polypterus and fossil osteichthyans and acanthodians. D, 1980, Yale University. 326 p.

Meints, Joyce P. Statistical characterization of fractures in the Museum and Rocky Coulee flows of the Grande Ronde Formation, Columbia River Basalts. M, 1986, Washington State University. 238 p.

Meintzer, Robert Ells. The mineralogy and geochemistry of the granitoid rocks and related pegmatites of the Yellowknife pegmatite field, Northwest Territories. D, 1987, University of Manitoba.

Meintzer, Robert Ells. Thorium and rare earth element migration incipient to the weathering of an allanite pegmatite, Amherst County, Virginia. M, 1981, University of Virginia. 107 p.

Meinwald, Javan N. Extensional faulting in the Mina region; study of an Oligocene basin, west-central Nevada. M, 1982, Rice University. 79 p.

Meinzer, Oscar Edward. Occurrence of ground water in the United States, with a discussion of principles. D, 1923, University of Chicago. 321 p.

Meis, Philip J. Aeromagnetic compensation. M, 1986, Colorado School of Mines. 278 p.

Meisel, Kent E. The use of electrical resistivity to delineate a brine contamination plume in the Walker oil field, Kent County, Michigan. M, 1985, Western Michigan University.

Meisen, Daniel S. A statistical evaluation of the clay minerals in a portion of the Upper Devonian Cashaqua Shale of western New York. M, 1973, SUNY at Buffalo. 46 p.

Meisenheimer, James Kenneth. An investigation of the application of photoelastic coatings to soil studies. M, 1970, University of Missouri, Rolla.

Meiser, E. William, Jr. Ground-water geology and a digital simulation of sustained yield potential of the Altoona area, Pennsylvania. D, 1975, Pennsylvania State University, University Park. 291 p.

Meiser, Edward William, Jr. The geology and water resources of the Bellefonte-Mingoville area, Pennsylvania. M, 1971, Pennsylvania State University, University Park. 113 p.

Meisler, Harold. Resume of the geology and tectonics of southeastern Newfoundland. M, 1953, University of Michigan.

Meisling, Kristian E. Neotectonics of the north frontal fault system of the San Bernardino Mountains, Southern California; Cajon Pass to Lucerne Valley. D, 1984, [University of California, San Diego]. 471 p.

Meissner, Fred F. The geology of Spring Creek Park, Gunnison County, Colorado. M, 1954, Colorado School of Mines. 150 p.

Meister, Laurent Justin. Seismic refraction study of Dixie Valley, Nevada. D, 1967, Stanford University. 81 p.

Meistrell, Frank Joseph. The spit-platform concept; laboratory observation of spit development. M, 1966, University of Alberta. 46 p.

Meitzke, Jane Ellen. Pleistocene terraces in Florida. M, 1962, University of Florida. 62 p.

Meixner, Richard Eugene. Investigation of the redox parameter, pe+pH in soil suspensions. D, 1988, Colorado State University. 161 p.

Mejia Angel, Jose M. On the generation of multivariate sequences exhibiting the Hurst Phenomenon and

some flood frequency analyses. D, 1971, Colorado State University. 149 p.

Mejia, Daisy. Facies and diagenesis of Permian lower San Andres Formation, Yoakum County, West Texas. M, 1977, University of Texas, Austin.

Mejia, Jorge I. Restrepo *see* Restrepo Mejia, Jorge I.

Mejia, Jose Maria Jaramillo *see* Jaramillo Mejia, Jose Maria

Mejia, Lelio Hernan. Three-dimensional dynamic response analysis of earth dams. D, 1981, University of California, Berkeley. 237 p.

Mejstrick, Peter Francis. Petrogenesis of the Purcell Sill, Glacier National Park, Montana. M, 1975, University of Montana. 74 p.

Mekaru, T. Anion adsorption in Hawaiian soils. M, 1969, University of Hawaii. 81 p.

Meko, David Michael. Applications of Box-Jenkins methods of time series analysis to the reconstruction of drought from tree rings. D, 1981, University of Arizona. 162 p.

Melack, J. M. Limnology and dynamics of phytoplankton in equatorial African lakes. D, 1976, Duke University. 480 p.

Meland, Norman. A contribution ot the study of the red beds of Oklahoma. M, 1922, University of Oklahoma. 41 p.

Melas, John P. De *see* De Melas, John P.

Melby, John Harold. The stratigraphy of the Sunset Point Sandstone (Cambrian) in western Wisconsin. M, 1967, University of Wisconsin-Madison.

Melcer, Allen. Sedimentology of four late Pleistocene glacilacustrine deltas, St. Lawrence Lowland, New York. M, 1988, University of Illinois, Chicago.

Melcher, Charles Francis. Texture of oil sands with relation to the production of oil. D, 1922, George Washington University.

Melchin, Michael Jerome. Late Ordovician and Early Silurian graptolites, Cape Phillips Formation, Canadian Arctic Archipleago. D, 1987, University of Western Ontario. 762 p.

Melchin, Michael Jerome. Ordovician chitinozoans from the Simcoe Group, southern Ontario. M, 1982, University of Western Ontario.

Melchior, Daniel C. The application of ion-interaction theory of electrolytes to the solubilities of copper and of some alkaline-earth sulfates in brines. D, 1984, Colorado School of Mines. 91 p.

Melchior, Daniel Carl, III. A Moessbauer study of the transformations of the iron minerals in oil shale during retorting. M, 1980, Colorado School of Mines. 86 p.

Melchior, Robert Charles. Pollen and spores of the Shawmut Anticline, Montana. D, 1963, University of Minnesota, Minneapolis. 153 p.

Melchor, James R. Surface water turbidity in the entrance to Chesapeake Bay, Virginia. M, 1972, Old Dominion University. 67 p.

Melcon, P. Z. Age and origin of iron indurations in Dane County, Wisconsin. D, 1979, University of Wisconsin-Madison. 190 p.

Melcon, Zenas K. A preliminary study of the geology and ground-water of the Kings River area, California. M, 1932, [Stanford University].

Meldahl, Elmer G. Geology of the Grassy Butte area, McKenzie County, North Dakota. M, 1956, University of North Dakota. 42 p.

Meldgin, Neil J. Simulated acid-rain weathering. M, 1978, Kent State University, Kent. 25 p.

Meldrum, R. D. An application of feedback to electromagnetic seismometers. M, 1965, University of British Columbia.

Mele, Thomas Anthony. The occurrence of hydrocarbons in the Berea Sandstone in southeastern Ohio. M, 1981, Ohio University, Athens. 80 p.

Melear, John David. The petrology and ore deposits of the Seafoam mining district, Custer County, Idaho. M, 1953, University of Idaho. 36 p.

Meleen, Elmer E. The Pennsylvanian formations in the vicinity of St. Louis, Missouri. M, 1929, Washington University. 96 p.

Meleen, N. H. Strip mines and fluvial systems; geomorphic effects and environmental impact in northeastern Oklahoma. D, 1977, Clark University. 264 p.

Melenberg, R. R. Vibroseis refraction profiling of the Troy Valley of southeastern Wisconsin. M, 1979, University of Wisconsin-Madison.

Melendres, Mariano M., Jr. Pre-Cretaceous paleogeology isopach and lithofacies of the Cretaceous of South Dakota and parts of Minnesota and Iowa. M, 1957, Stanford University.

Melhorn, Wilton Newton. A quantitative analysis of Silurian sediments in the Michigan Basin. M, 1951, Michigan State University. 58 p.

Melhorn, Wilton Newton. Valders glaciation of the southern peninsula of Michigan. D, 1954, University of Michigan.

Melia, Michael B. Late Wisconsin deglaciation and postglacial vegetation change in the upper Susquehanna River drainage of east-central New York. M, 1975, SUNY, College at Oneonta. 139 p.

Melia, Michael Brendan. Distribution and provenance of palynomorphs in Northeast Atlantic aerosols and bottom sediments. D, 1980, Michigan State University. 215 p.

Melihercsik, Stephen J. Geology and petrology of the Precambrian in the Portneuf area (Quebec). M, 1949, Universite Laval.

Melihercsik, Stephen J. Petrology of the Charney formations (Quebec). D, 1952, Universite Laval.

Melik, James Charles. The hingement and contact margin structure of palaeocopid ostracodes from some Middle Devonian formations of Michigan, southwestern Ontario, and western New York. D, 1963, University of Michigan. 144 p.

Melillo, Allan Joseph. Late Miocene (Tortonian) sea-level events of Maryland-New Jersey coastal plain. M, 1982, Rutgers, The State University, New Brunswick. 98 p.

Melillo, Allan Joseph. Late Oligocene to Pliocene sea-level cycle events in the Baltimore Canyon trough and western North Atlantic Basin. D, 1985, Rutgers, The State University, New Brunswick. 245 p.

Melim, Leslie A. The sedimentary petrology and sedimentology of the unnamed middle Eocene sandstones of Scow Bay, Indian and Marrowstone islands, Northwest Washington. M, 1984, Western Washington University. 123 p.

Melius, Douglas James. Weathering of feldspars in modern soils developed on granitic terrain. M, 1982, University of Texas, Austin. 103 p.

Melka, Timothy E. Physical and chemical hydrogeology of the Iron Bog alpine wetland, Custer County, Idaho. M, 1986, University of Wisconsin-Milwaukee. 132 p.

Mellars, Gillian. Deglaciation of the Pouch Cove area, Avalon Peninsula, Newfoundland; a palynological approach. M, 1981, Memorial University of Newfoundland. 202 p.

Mellegard, Andy. Structure and geology of Djebel Zaghouan. M, 1979, University of South Carolina.

Mellen, James Vedrey. Pre-Cambrian sedimentation in the Northeast part of Cohutta Mountain Quadrangle, Georgia. M, 1956, Cornell University. 42 p.

Mellen, Michael H. Offset vertical seismic profiling; two-dimensional forward modeling with asymptotic ray theory. M, 1984, Massachusetts Institute of Technology. 85 p.

Mellett, James Silvan. Fossil mammals from the Oligocene Hsanda Gol Formation, Mongolia; Part I, Insectivora, Rodentia and Deltatheridia, with notes on the paleobiology of Cricetops dormitor. D, 1966, Columbia University. 275 p.

Melling, David R. Geological setting, structure, and alteration associated with gold-pyrite mineralization in mafic volcanic rocks at Cameron Lake, Wabigoon Subprovince, northwestern Ontario. M, 1986, Carleton University. 112 p.

Melling, Patricia Hanbury. A petrographic and petrogeochemical comparison of exotic gabbroic inclusion of the Sudbury sublayer and Nipissing Diabase-Sudbury Gabbro. M, 1982, Washington State University.

Mellinger, Michel. Etude de l'altération de laves mafiques archéennes en pillows, dans la région de Rouyn-Noranda (Abitibi, Québec). M, 1976, Ecole Polytechnique.

Mellman, George Robert. A method for waveform inversion of body-wave seismograms. D, 1979, California Institute of Technology. 121 p.

Mello, James F. Stratigraphy and micropaleontology of the upper Pierre Shale (Cretaceous) in north-central South Dakota. D, 1962, Yale University.

Mellon, George Barry. Age and origin of the McMurray Formation. M, 1955, University of Alberta. 84 p.

Mellon, George Barry. The petrology of the Blairmore Group, Alberta, Canada. D, 1959, Pennsylvania State University, University Park. 298 p.

Mellon, Steven Allen. Stratigraphic controls on scheelite-bearing tactite within the Amsden Group at the Lost Creek mining district, Beaverhead County, Montana. M, 1978, South Dakota School of Mines & Technology.

Mellor, Edgar I., Jr. A structural and petrographic study of Permian rocks near Villa Aldama, Chihuahua, Mexico. M, 1978, Texas Christian University. 44 p.

Mellor, Jack Conrad. Bathymetry of Alaskan Arctic lakes; a key to resource inventory with remote-sensing methods. D, 1982, University of Alaska, Fairbanks. 360 p.

Mellott, James Charles. Preliminary ground-water flow system analysis in the Columbia River basalts. M, 1973, Washington State University. 63 p.

Mellott, Mark G. Geochemical, petrologic, and isotopic investigation of andesites and related volcanic rocks in the Santa Rosa Range and Bloody Run Hills, Nevada; tectonic implications. M, 1987, Miami University (Ohio). 164 p.

Melnychenko, Paul. Geology of the Iron Bog Creek area, Butte and Custer counties, south-central Idaho. M, 1978, University of Wisconsin-Milwaukee.

Melnyk, David H. Biofacies and ostracode biostratigraphy of the Permo-Carboniferous of central and north-central Texas; the application of numerical techniques to the recognition of biozones and biofacies in a depositionally complex sequence. D, 1985, University of Houston. 456 p.

Meloche, John Dennis. Evolution of biogeochemical element cycles with emphasis on the role of metal-organic interactions in the accumulation of heavy metals in organic-rich sediments. D, 1981, University of Western Ontario.

Meloche, Marvin J. The origin of some copper and copper-nickel sulphide orebodies in granite. M, 1965, Queen's University. 214 p.

Meloy, David Urey. Depositional history of the (Silurian) northern carbonate bank of the Michigan Basin. M, 1974, University of Michigan.

Melrose, John Walter. The metamorphism of certain igneous dikes near Northport, Stevens County, Washington. M, 1934, Washington State University. 15 p.

Melrose, Thomas Graham and Jarre, Guntram A. Reconnaissance geology of the Old Baldy Thrust, Alamosa, Costilla, and Huerfano counties, Colorado. M, 1959, University of Colorado.

Melson, William Gerald. Geology of the Lincoln area, Montana and contact metamorphism of impure carbonate rocks. D, 1964, Princeton University. 153 p.

Melton, Douglas C. Base metal mineralization in south central Missouri. M, 1978, University of Missouri, Rolla.

Melton, Frank Armon. Structural and stratigraphic studies in the Sangre de Cristo Range of Colorado. D, 1924, University of Chicago. 72 p.

Melton, Mark A. An analysis of the relations among elements of climate, surface properties and geomorphology. D, 1958, Columbia University, Teachers College.

Melton, Richard W. The regional geohydrology of the Roubidoux and Gasconade Formation, Arkansas and Missouri. M, 1976, University of Arkansas, Fayetteville.

Melton, Robert A. Paleoecology and paleoenvironment of the upper Honaker Trail Formation (Pennsylvanian) near Moab, Utah. M, 1972, Brigham Young University. 88 p.

Melton, Robert A. Study of geologic factors effecting roof conditions in the Pocahontas No. 3 coal seam, southern West Virginia. D, 1978, University of South Carolina. 583 p.

Meltz, Robert E. Geothermometry and geobarometry in the contact aureoles of the Nain Complex, Labrador. M, 1982, Northern Illinois University. 115 p.

Meltzer, Anne S. Scattering of P-waves beneath the large seismic array SCARLET, Southern California. M, 1982, University of North Carolina, Chapel Hill. 51 p.

Meltzer, Bernard David. The effect of connate water in the water flooding of porous media. M, 1950, University of Oklahoma. 100 p.

Melvin, James W. Cretaceous stratigraphy in the Jornada del Muerto region, including the geology of the Mescal Creek area, Sierra County, New Mexico. M, 1963, University of New Mexico. 121 p.

Melvin, John Harper. The geology of a portion of the Piketon, Ohio, Quadrangle. M, 1933, Ohio State University.

Melvin, Judith L. Subsurface stratigraphy and depositional systems of the Richfield Member of the Lucas Formation, lower Middle Devonian, Clare and Gladwin counties, Michigan. M, 1984, University of Arkansas, Fayetteville. 76 p.

Melvin, Norman Wayne. Detrital minerals of Tijeras Canyon, Bernalillo County, New Mexico. M, 1962, University of New Mexico. 81 p.

Melvin, Robert L. The surficial geology of Ben Lomond area, Saint John and Kings counties, New Brunswick (Canada). M, 1966, University of New Brunswick.

Memarian, H. Past and present stability and evolution of the Niagara Gorge in the vicinity of the Niagara Glen, Niagara Falls, Ontario. M, 1975, University of Waterloo.

Menard, Henry W. The geology of the Agua Dulce Canyon area (Los Angeles County, California). M, 1947, California Institute of Technology. 33 p.

Menard, Henry W., Jr. Transportation of bed-load by running water. D, 1949, Harvard University.

Mencenberg, Frederick E. Groundwater geology of the (Pennsylvanian) Saginaw Group in the Lansing area, Michigan. M, 1963, Michigan State University. 38 p.

Mench, Patricia Anne. Diagenesis of the Reynales Formation (Middle Silurian), Monroe and Wayne counties, New York. M, 1973, Duke University. 183 p.

Mencher, Ely. Sedimentary study of the (Devonian) Catskill facies in New York State. D, 1938, Massachusetts Institute of Technology. 164 p.

Mendeck, M. F. A history of metal transport to reservoir sediments in New Haven, Connecticut. M, 1976, San Diego State University.

Mendelson, Carl Victor. Studies in micropaleontology; Proterozoic microfossils, Ordovician microphytoplankton, and Recent agglutinated foraminifera. D, 1981, University of California, Los Angeles. 255 p.

Mendelson, J. D. Petroleum source rock logging. M, 1985, Massachusetts Institute of Technology. 96 p.

Mendenhall, Arthur J. Structural geology of eastern part of Richmond and western part of Naomi Peak quadrangles, Utah-Idaho. M, 1975, Utah State University. 45 p.

Mendenhall, Gerald Vernon. The bedrock geology of Boyd and northern Holt counties, Nebraska. M, 1953, University of Nebraska, Lincoln.

Mendenhall, Maurice Elvin. Conodonts and fish remains of the Douglas and Shawnee groups of the Virgil Series (Pennsylvanian) of Nebraska. M, 1951, University of Nebraska, Lincoln.

Mendenhall, Richard A. Surface geology of Bala, Riley County, Kansas. M, 1958, Kansas State University. 44 p.

Mendez, Andres J. Forward modeling and inversion of near-source earthquake ground motion. D, 1988, University of California, San Diego.

Mendiguren, Jorge Andres. Source mechanism of a deep earthquake from analysis of world wide observations of free oscillation. D, 1972, Massachusetts Institute of Technology. 140 p.

Mendoza Sanchez, V. Geology of the Suapure River area, NW Guiana Shield, Venezuela. D, 1975, SUNY at Binghamton. 275 p.

Mendoza, Carlos. An investigation into the possibility of slow-rate triggering of low magnitude earthquakes by nuclear explosions. M, 1978, University of Wisconsin-Milwaukee.

Mendoza, Carlos. Study of aftershock source properties using digital surface-wave data; the December 1979 Colombia earthquake. D, 1985, Colorado School of Mines. 166 p.

Mendoza, Herbert A. Stratigraphy of the Howard Limestone (Virgilian) between the Kansas River and Neosho River valleys, Kansas. M, 1959, University of Kansas. 161 p.

Mendoza, Jorge Segundo. Modelling deformation, porosity and elastic constants in porous rocks. D, 1987, Stanford University. 219 p.

Mendoza, Vicente. Petrology and structural geology of the Hailesboro area (NW Adirondacks, New York State). M, 1970, SUNY at Binghamton. 131 p.

Meneley, Robert Allison. Nisku Formation in Saskatchewan. M, 1958, University of Saskatchewan. 42 p.

Meneley, W. A. Theory of microfabric analysis. M, 1960, University of Saskatchewan. 34 p.

Meneley, William Allison. Geology of the Melfort Area (73-A), Saskatchewan. D, 1964, University of Illinois, Urbana. 193 p.

Menell, Richard. An epithermal disseminated precious metals prospect at White Horse Mountain, Elko County, Nevada. M, 1982, Stanford University. 128 p.

Menendez, Alfredo. Geology of the Tinaco area, north central Cojedes, Venezuela. D, 1962, Princeton University. 275 p.

Menendez, Fernando Osacar Ricart y see Ricart y Menendez, Fernando Osacar

Meneses-Rocha, Javier de Jesus. Tectonic evolution of the strike-slip fault province of Chiapas, Mexico. M, 1985, University of Texas, Austin. 315 p.

Meng, H. M. Mineralogical study of some specimens from Gedang Ihi Mine, Sumatra. M, 1927, Massachusetts Institute of Technology. 34 p.

Meng, Haiyan. Some considerations on the numerical reservoir simulation of a pilot waterflood in China. M, 1987, Colorado School of Mines. 212 p.

Meng, Yanxi. Tourmaline from the Isua Supracrustal Belt, southern West Greenland. M, 1988, Washington University. 196 p.

Menge, J. L. Effect of herbivores on community structure of the New England rocky intertidal region; distribution, abundance and diversity of algae. D, 1975, Harvard University. 165 p.

Mengel, Flemming Cai. Thermotectonic evolution of the Proterozoic-Archaean boundary in the Saglek area, northern Labrador. D, 1988, Memorial University of Newfoundland. 349 p.

Mengel, Joseph Torbitt, Jr. The cherts of the Lake Superior iron-bearing formations. D, 1963, University of Wisconsin-Madison. 151 p.

Mengel, Joseph Torbitt, Jr. The relationship of clastic sediments to iron formation in the vicinity of Palmer, Michigan. M, 1956, University of Wisconsin-Madison.

Mengel, Martin. Computerized two-dimensional shape sieving of carbonate sands using the multivariate rotation method. M, 1985, Lehigh University. 90 p.

Menges, Christopher M. The tectonic geomorphology of mountain-front landforms in the northern Rio Grande Rift near Taos, New Mexico. D, 1988, University of New Mexico. 339 p.

Menges, Christopher Martin. The Sonoita Creek Basin; implications for late Cenozoic tectonic evolution of basins and ranges in southeastern Arizona. M, 1981, University of Arizona.

Menheim, Frank. Variations in oxidation-reduction potential in the Jurassic of Morehouse Parish, Louisiana. M, 1953, University of Minnesota, Minneapolis.

Menke, Kathleen Patricia. Subsurface study of the Hunton Group in the Cheyenne Valley field, Major County, Oklahoma. M, 1986, Oklahoma State University. 91 p.

Menke, William H. Studies of the long range P_n phase. D, 1982, Columbia University, Teachers College. 114 p.

Menke, William Henry. Lateral variation of P velocity in tha Himalayan crust and upper mantle; a study based on observations of teleseisms at the Tarbela seismic array. M, 1976, Massachusetts Institute of Technology. 72 p.

Mennen, Gary. The properties and interrelationships of nine central Appalachian coals. M, 1968, West Virginia University.

Mennicke, Christine M. Stratigraphy and paleontology of Pliocene to Recent sediment from the eastern Alpha Ridge, central Arctic Ocean. M, 1985, University of Wisconsin-Madison. 87 p.

Mensah, M. K. Ecology of the Kirkfield Quarry (Ordovician, Ontario). M, 1962, University of Toronto.

Mensah, Winterford W. Rare earth element distribution in low and high temperature fluorite. M, 1985, University of Kentucky. 57 p.

Mensah-Dwumah, Francis Kwabena. Finite element modeling of brick masonry building behavior in response to deformations induced by shallow soft-ground tunneling. D, 1984, Stanford University. 223 p.

Mensing, Teresa Marie. Geology and petrogenesis of the Kirkpatrick Basalt, Pain Mesa and Solo Nunatak, northern Victoria Land, Antarctica. D, 1987, Ohio State University. 391 p.

Mentzer, Thomas Cartwright. Composition trends in a folded gneissic layer, Sussex County, New Jersey. D, 1963, Lehigh University. 107 p.

Menut, D. Charles. Correlation of the Pennsylvanian Kickapoo Creek inlier with the Brazos River section, Hood and Parker counties, Texas. M, 1957, Texas Christian University. 86 p.

Menzer, Fred J. Sulfide mineralization in the Battle Lake area, Grand Encampment mining district, Carbon County, Wyoming. M, 1981, Colorado State University. 142 p.

Menzer, Frederick John, Jr. Geology of the crystalline rocks west of Okanogan, Washington. D, 1964, University of Washington. 64 p.

Menzie, Thomas Eugene. The geology of the Box Springs Mountains, Riverside County, California. M, 1962, Stanford University. 50 p.

Menzie, W. David, II. Sedimentology of upper Chemung and Catskill sediments, along the Appalachian Front in central Pennsylvania. M, 1974, Pennsylvania State University, University Park. 206 p.

Menzie, W. David, II. The unit regional value of the Republic of South Africa. D, 1977, Pennsylvania State University, University Park. 251 p.

Menzies, Anthony J. Flow characteristics and relative permeability functions for two phase geothermal reservoirs from a one dimensional thermodynamic model. M, 1982, Stanford University.

Menzies, Douglas G. Deformation of the metavolcanic rocks around the eastern nose of the Aulneau Batholith district of Kenora, Ontario Canada. M, 1978, University of Manitoba. 75 p.

Menzies, Morris McCallum. Geology and mineralogy of the Strangward copper property, South Tetsa River. M, 1951, University of British Columbia.

Mercado, Abraham. The kinetics of mineral dissolution in aquifers and their use for hydrologic investigations. D, 1972, New Mexico Institute of Mining and Technology. 221 p.

Mercado, Edward John. An application of Poisson's relation. M, 1958, Washington University. 20 p.

Mercado, Edward John. Stress propagation in nonlinear viscoelastic materials. D, 1963, Rensselaer Polytechnic Institute. 109 p.

Mercer, Barry P. Areal geology, brecciation, and mineralization west of Trout Creek Pass, Chaffee County, Colorado. M, 1976, University of Missouri, Rolla.

Mercer, David A. Paleomagnetism of the Baraboo Quartzite. M, 1984, University of Wisconsin-Milwaukee. 294 p.

Mercer, David Morris. Variations in point bar deposits from Georgia fluvial and estuarine environments. M, 1984, University of Georgia. 99 p.

Mercer, James Wayne, Jr. Finite element approach to the modeling of hydrothermal systems. D, 1973, University of Illinois, Urbana. 106 p.

Mercer, Jerry Wayne. Inorganic water quality of the Little South Poudre Watershed, with a section on the Precambrian petrology of the upper Fall Creek area, Larimer County, Colorado. M, 1966, Colorado State University. 80 p.

Mercer, Mark F. The origins of three volcanic constructs on the South-central Snake River plain, Idaho. M, 1979, SUNY at Buffalo. 94 p.

Mercer, Richard B. Geochemical reconnaissance for copper on the upper Tellico River and its tributaries. M, 1978, Wright State University. 54 p.

Mercer, William. A solubility model for deposition of stratiform massive sulphide deposits, New Brunswick, and its relation to the distribution of gold and palladium. D, 1975, McMaster University. 225 p.

Merchant, Abdul Rashid. Convergence of material balance in mathematical simulation of petroleum reservoirs. D, 1973, University of Missouri, Rolla.

Merchant, Dean Charles. Beam deflection measurements by application of photogrammetry. M, 1955, Ohio State University.

Merchant, Dorothy. Structure and development of the Cephalopoda. M, 1921, Smith College. 67 p.

Merchant, James William, Jr. Employing spatial logic in classification of Landsat thematic mapper data. D, 1984, University of Kansas. 484 p.

Merchant, John Stines. Mineralogy and ore deposition at Colquiri, Bolivia. M, 1952, University of Michigan.

Mercier, J. C. C. Natural peridotites; chemical and rheological heterogeneity of the upper mantle. D, 1977, SUNY at Stony Brook. 700 p.

Mercier, John Michael. Petrology of the Upper Cretaceous strata of Stuart Island, San Juan County, Washington. M, 1977, Washington State University. 157 p.

Mercier, Michael J. A chemical weathering study of overburden materials from three surface coal mines in southern Illinois and western Kentucky. M, 1978, Southern Illinois University, Carbondale. 111 p.

Mercure, S. Use of the homomorphic deconvolution to obtain crustal absorption. M, 1975, McGill University. 186 p.

Mercurio, Richard Nicholas. Subsurface studies of the Tar Springs Formation, Union-Henderson counties, Kentucky. M, 1954, University of Illinois, Urbana.

Merdler, Stephen C. Estimation procedure for focal-depth determination of seismic disturbances. M, 1964, Pennsylvania State University, University Park. 94 p.

Meredith, John C. Diagenesis of Holocene-Pleistocene (?) travertine deposits; Fritz Creek, Clark County, and Fall Creek, Bonneville County, Idaho. M, 1980, University of Houston.

Merenbach, Simon Eugene. Some leading structural features of eastern Santa Monica Mountains, Los Angeles, California. M, 1931, University of Southern California.

Mereu, Robert F. Methods of converting the kinetic energy of falling weight into seismic energy. D, 1962, University of Western Ontario.

Merewether, Edward Allen. The geology of the lower Sprague River area, Klamath County, Oregon. M, 1953, University of Oregon. 62 p.

Mergner, Marcia. Geology of Long Valley, Lassen County, California, and Washoe County, Nevada. M, 1978, Colorado School of Mines. 59 p.

Merguerian, Charles Michael. Stratigraphy, structural geology, and tectonic implications of the Shoo Fly Complex and the Calaveras-Shoo Fly Thrust, central Sierra Nevada, California. D, 1985, Columbia University, Teachers College. 307 p.

Mericle, James E. Gravity study of crustal structures in the South Florida Embayment. M, 1979, University of South Florida, Tampa. 57 p.

Merifield, Paul Milton. Geologic information from hyperaltitude photography. D, 1963, University of Colorado. 208 p.

Merifield, Paul Milton. Geology of a portion of the southwestern San Gabriel Mountains, San Fernando and Oat Mountain quadrangles, Los Angeles County, California. M, 1958, University of California, Los Angeles.

Meriney, Paul E. Sedimentology and diagenesis of Jurassic lacustrine sandstones in the Hartford and Deerfield basins, Massachusetts and Connecticut. M, 1988, University of Massachusetts. 401 p.

Merino, Enrique. Diagenetic mineralogy and water chemistry in Tertiary sandstones from Kettleman North Dome, West San Joaquin Valley, California; 1 volume. D, 1973, University of California, Berkeley.

Merite, John B. The igneous rocks of the Raton Mesa region of New Mexico and Colorado. D, 1911, The Johns Hopkins University.

Merk, George P. Dune form and structure at Great Sand Dunes National Monument; Saguache County, Colorado. M, 1962, University of Colorado.

Merk, George Philip. Provenance and tectonic inferences concerning the Keweenawan interflow sediments of the Lake Superior region. D, 1972, Michigan State University. 126 p.

Merkel, David C. The development and ground testing of the U.S. Geological Survey airborne sixty-hertz reconnaissance system. M, 1986, Colorado School of Mines. 125 p.

Merkel, Gregory Albert. Thermodynamic mixing properties of binary analbite-sanidine feldspars. D, 1984, Pennsylvania State University, University Park. 163 p.

Merkel, Richard H. Petential field development from steady-state current flow with buried sources in an inhomogeneous half-space. D, 1970, Pennsylvania State University, University Park. 171 p.

Merkel, Richard H. Time series analysis as applied to continental margin refraction data. M, 1967, Pennsylvania State University, University Park. 89 p.

Merker, Robert Randall. Pegmatite emplacement in the Cribbenville area, Petaca District, Rio Arriba County, New Mexico. M, 1981, University of New Mexico. 73 p.

Merkl, Roland S. Petrographic and depositional characteristics of the Hymera and Danville coal members in southwestern Indiana. M, 1985, Indiana University, Bloomington. 67 p.

Merkle, Arthur B. Anistropism in garnets. M, 1961, University of New Mexico. 51 p.

Merkle, Arthur Beiser. The crystal structure of heulandite. D, 1967, University of Missouri, Columbia. 144 p.

Mero, John L. An economic analysis of mining deep-sea phosphorite. D, 1959, University of California, Berkeley.

Mero, William Edward. The geology of Black Lion Mountain and a portion of Canyon Creek, Beaverhead County, Montana. M, 1962, University of California, Santa Barbara. 56 p.

Merrell, Harvey Webb. Petrology and sedimentation of significant (Pennsylvania) Paradox shales. M, 1957, University of Utah. 40 p.

Merriam, Charles W. Fossil turritellas from the Pacific Coast region, North America. D, 1932, University of California, Berkeley. 250 p.

Merriam, Daniel Francis. Tertiary geology of the Piceance Basin, northwestern Colorado. M, 1953, University of Kansas.

Merriam, Daniel Francis. The geologic history of Kansas; Vol. 1-2. D, 1961, University of Kansas. 475 p.

Merriam, J. B. The dissipation of tidal energy in the solid Earth. D, 1976, York University.

Merriam, Martha. A microearthquake study of a region east of The Geysers, Northern California; operation of a network of seismic recorders, data processing, and interpretations in view of regional and local tectonic regimes. M, 1986, University of California, Davis. 300 p.

Merriam, Patricia D. Geology of the El Segundo sand hills. M, 1950, University of Southern California.

Merriam, Richard Holmes. Geology of the southwestern part of the Ramona Quadrangle, San Diego County, California. D, 1941, University of California, Berkeley. 139 p.

Merriam, Robert Willis. A Madison bioherm, Big Snowy Mountains, Montana. M, 1958, Washington State University. 87 p.

Merrick, Margaret Anne. Geology of the eastern part of the Regina Quadrangle, Sandoval and Rio Arriba counties, New Mexico. M, 1980, University of New Mexico. 91 p.

Merrill, Bertha. Some problems of off-shore bars and tidal inlets. M, 1916, Columbia University, Teachers College.

Merrill, Douglas E. Glacial geology of the Chiwaukum Creek drainage basin and vicinity, Washington. M, 1966, University of Washington. 36 p.

Merrill, Frederick J. The origin of the serpentines in the vicinity of New York. D, 1890, Columbia University, Teachers College.

Merrill, Glen K. Allegheny (Pennsylvanian) conodonts. D, 1968, Louisiana State University.

Merrill, Glen Kenton. Zonation of platform conodont genera in Conemaugh strata of Ohio and vicinity. M, 1964, University of Texas, Austin.

Merrill, John D. Geology of the lower part of Buck Mountain Quadrangle, Nevada. M, 1960, University of Southern California.

Merrill, John R. Beryllium geochemistry related to age determination with beryllium-10. D, 1958, Princeton University. 97 p.

Merrill, Laura. Accumulation of heavy metals in Sunset Lake, Delaware. M, 1978, University of Delaware.

Merrill, Leo. Crystallographic studies of the metastable high pressure phases of calcium carbonate, $CaCO_3(II)$ and $CaCO_3(III)$. D, 1973, University of Rochester. 78 p.

Merrill, Milford S., Jr. Mathematical modeling of water quality in a vertically stratified estuary. D, 1974, University of Washington. 186 p.

Merrill, R. J. The geology and ore deposits of Akaitcho Yellowknife gold mines. M, 1947, Queen's University. 36 p.

Merrill, Richard C. Geology of the Mill Fork area, Utah. M, 1972, Brigham Young University. 88 p.

Merrill, Robert D. Geology at the southern terminus of the Sawtooth Range, Montana. M, 1965, University of Massachusetts. 72 p.

Merrill, Robert D. Geomorphology of terrace remnants of the Greybull River, Big Horn Basin, northwestern Wyoming. M, 1973, University of Texas, Austin.

Merrill, Robert David. Geomorphology of terrace remnants of the Greybull River, Big Horn Basin, northwestern Wyoming. D, 1974, University of Texas, Austin.

Merrill, Robert J. Reconnaissance geology of the basic rocks of northern Maine. D, 1949, The Johns Hopkins University.

Merrill, Robert Kimball. The glacial geology of the Mount Baldy area, Apache County, Arizona. M, 1970, Arizona State University. 118 p.

Merrill, Robert Kimball. The late Cenozoic geology of the White Mountains, Apache County, Arizona. D, 1974, Arizona State University.

Merrill, Ronald Thomas. Origin of the remanent magnetization and magnetic reversals in the Bucks Pluton, California. D, 1967, University of California, Berkeley. 170 p.

Merrill, Russell B. Hydrous minerals in the Upper Mantle; experimental studies of titanoclinohumite, kaersutite, and kaersutite eclogite under water-excess and water-deficient conditions with geologic applications. D, 1973, University of Chicago. 228 p.

Merrill, William G. A geological and legal analysis of groundwater resources in the West Divide Basin, Garfield County, Colorado. M, 1983, Colorado State University. 106 p.

Merrill, William M. The geology of northern Hocking County, Ohio. D, 1950, Ohio State University.

Merrill, William Meredith. The geology of Green and Ward townships, Hocking County, Ohio. M, 1948, Ohio State University.

Merrill, William Meredith. The geology of northern Hocking County, Ohio. D, 1978, Ohio State University.

Merriman, Ray Warren. Microscopic study of capillary interchange. M, 1927, University of Pittsburgh.

Merrin, Seymour. Experimental investigations of epidote paragenesis. D, 1962, Pennsylvania State University, University Park. 116 p.

Merrin, Seymour. The Cretaceous stratigraphy and mineral deposits of the east face of Black Mesa, Arizona. M, 1954, University of Arizona.

Merrit, P. C. Certain molybdenite occurrences of Maine. M, 1954, Columbia University, Teachers College.

Merritt, Andrew Hutcheson. Engineering classification for in situ rock. D, 1968, University of Illinois, Urbana.

Merritt, Andrew Hutcheson. Stratigraphy and mineralogy of the glacial tills in the Quincy and Mendon quadrangles, Illinois. M, 1965, University of Illinois, Urbana.

Merritt, Clifford A. A microscopic study of the ores of Austin, Nevada. D, 1928, University of Chicago. 53 p.

Merritt, Clifford A. The function of gels in the formation of pegmatites and of quartz and carbonate veins. M, 1924, University of Manitoba.

Merritt, David H. Advective circulation in Lake Powell, Utah-Arizona. M, 1976, Dartmouth College. 88 p.

Merritt, Donald W. Gravity investigation of the Scipio oil field, Hillsdale County, Michigan, with a related study for obtaining a variable elevation factor. D, 1968, Michigan State University. 120 p.

Merritt, Donald W. Velocity anisotropy studies of Precambrian lamellar formations. M, 1961, Michigan State University. 41 p.

Merritt, Eleanor Walton. Petrography of the Somesville Pluton (Devonian), Mount Desert Island, Maine. M, 1968, University of Illinois, Urbana.

Merritt, Gary L. The hydrogeology of the Toms Run drainage basin, Clarion County, Pennsylvania. M, 1969, West Virginia University.

Merritt, James C. Surficial geology of the southern half of Griggs County, North Dakota. M, 1966, University of North Dakota. 78 p.

Merritt, John Wesley. Structural and metamorphic geology of the Hanover District of New Hampshire. D, 1917, University of Wisconsin-Madison.

Merritt, John Wesley. The geology of a portion of the Desert Mountains, Nevada. M, 1912, Northwestern University.

Merritt, Linda Carol. Sandstone diagenesis of Olmos, San Miguel, and Upson formations (Upper Cretaceous), northern Rio Escondido Basin, Coahuila, Mexico. M, 1980, University of Texas, Austin.

Merritt, Michael Louis. Subsurface geology of the Madill-Cumberland-Aylesworth area, Marshall County, Oklahoma. M, 1978, University of Oklahoma. 109 p.

Merritt, Philip L. Problem of the Seine, Coutchiching, and Keewatin (Ontario). D, 1934, Columbia University, Teachers College.

Merritt, Philip L. The origin and occurrence of the metallic ore deposits of Idaho. M, 1930, Columbia University, Teachers College.

Merritt, Roy Dale. Geochemical analysis of Middle Pennsylvanian coal-bearing strata in eastern Kentucky; its relation to paleoenvironments and application to reclamation. M, 1978, Eastern Kentucky University. 212 p.

Merritt, Zen S. Tertiary stratigraphy and general geology of the Alpine, Idaho-Wyoming area. M, 1958, University of Wyoming. 94 p.

Merritts, Dorothy J. Segmentation on the Surprise Canyon alluvial fan, Panamint Valley, southeastern California. M, 1983, Stanford University. 57 p.

Merritts, Dorothy Jane. Geomorphic response to late Quaternary tectonism; coastal Northern California, Mendocino triple junction region. D, 1987, University of Arizona. 211 p.

Merriweather, Annie Pearl. The kaolin clays of Langley and Graniteville, South Carolina. M, 1972, Virginia State University. 33 p.

Merry, Carolyn J. The correlation and quantification of airborne spectroradiometer data to ground turbidity measurement at Lake Powell, Utah. M, 1976, Dartmouth College. 54 p.

Merry, Ray D. Precambrian geology, shear zones and associated mineral deposits of the Hog Park area, Carbon County, Wyoming. M, 1964, University of Wyoming. 73 p.

Merryman, Raleigh J. Geology of the Winkler Area, Riley County, Kansas. M, 1957, Kansas State University. 34 p.

Merschat, Carl. Mineralogy of certain west Tennessee ceramic clays (Tertiary). M, 1967, University of Tennessee, Knoxville. 48 p.

Merschat, Walter R. Lower Tertiary paleocurrent trends, Santa Cruz Island, California. M, 1971, Ohio University, Athens. 77 p.

Merselis, William B. Late Pleistocene shorelines of Southern California. M, 1962, University of Southern California.

Mersereau, Terence Gerard. Secondary dispersion of the metals nickel, cobalt and copper near the Saint Stephen gabbro (Ordovician), Charlotte County, New Brunswick, (Canada). M, 1969, University of New Brunswick.

Mershon, Robert E. A sedimentary analysis of the core from a deep test well, Walker District, Wood County, West Virginia. M, 1956, University of Pittsburgh.

Mersky, Ronald Lee. Resoumetric analysis of the domestic primary copper industry. D, 1985, University of Pennsylvania. 468 p.

Mersmann, Mark A. Metamorphism of a gabbronorite dike; corona development and amphibolitization, Hellroaring Lakes area, Beartooth Mountains, Montana. M, 1981, University of Cincinnati. 136 p.

Mertes, Leal Anne Kerry. Floodplain development and sediment transport in the Solimoes-Amazon River, Brazil. M, 1985, University of Washington. 108 p.

Mertie, John B., Jr. The igneous rocks of the Raton Mesa region of New Mexico and Colorado. D, 1911, The Johns Hopkins University.

Mertikas, Stilianos P. A statistical investigation into reliable and efficient accuracy measures in positioning. D, 1988, University of New Brunswick.

Mertz, Karl Anton. Sedimentology of the Upper Triassic Blomidon and Wolfville formations, Gerrish Mountain, north shore of the Minas Basin, Nova Scotia. M, 1980, University of Massachusetts. 198 p.

Mertz, Karl Anton, Jr. Origin and depositional history of the Sandholt Member, Miocene Monterey Formation, Santa Lucia Range, California. D, 1984, University of California, Santa Cruz. 40 p.

Mertzman, Stanley Arthur, Jr. The geology and petrology of the Summer Coon Volcano, Colorado. D, 1972, Case Western Reserve University.

Merwin, Herbert E. Mineralogical and petrographical researches, with special reference to the stability ranges of the alkali feldspars. D, 1911, Harvard University.

Merz, B. Potassium-argon dating, paleomagnetism and geochemistry of the Sudbury diabase dykes. M, 1976, University of Western Ontario.

Merz, Joy J. The geology of the Union Hill area, Silver Bell District, Pima County, Arizona. M, 1967, University of Arizona.

Merzbacher, Celia Irene. The structure of alkaline earth aluminosilicate glasses and melts; a spectroscopic study. D, 1987, Pennsylvania State University, University Park. 158 p.

Merzbacher, Celia Irene. Water-saturated and -undersaturated phase relations of the Mount St. Helens dacite magma erupted on May 18, 1983, and an estimate of the pre-emptive water content. M, 1983, Pennsylvania State University, University Park. 56 p.

Mesard, Peter Morris. The alteration and mineralization of the Poplar copper-molybdenum porphyry deposit, West-central British Columbia. M, 1979, University of British Columbia.

Meschede, Louis Henry. Possible mobility of residuals of ammonium nitrate fuel oil blasting agents in desert soil and ground water. M, 1979, University of Arizona. 101 p.

Mescher, Paul A. Insoluble residues of the Madison Limestone in south-central Wyoming. M, 1950, University of Wyoming. 83 p.

Mescher, Paul K. Structural evolution of southeastern Michigan; Middle Ordovician to Middle Silurian. M, 1980, Michigan State University. 120 p.

Mesecar, Roderick Smit. Oceanic vertical temperature measurements across the water-sediment interface at selected stations west of Oregon. D, 1968, Oregon State University. 99 p.

Mesenbrink, Joseph H. Slope processes on Mount Tamalpais, Marin County, California. M, 1968, [University of California, Berkeley].

Meserve, Clement Dann. A study of the faunal and stratigraphic relations of the middle and lower Miocene of the Santa Ana Mountains, Southern California. M, 1923, University of California, Berkeley. 42 p.

Meshkov, Alexandra. Magnetic properties of some young basalts from the East Pacific Rise at 21°N, 109°W. M, 1983, University of California, Santa Barbara. 77 p.

Meshref, Wafik M. Aeromagnetic study of the regional geology of the western half of the northern Peninsula of Michigan. D, 1968, Michigan State University. 132 p.

Meshri, Indurani Dayal. Deposition and diagenesis of glauconite sandstone, Berrymore-Lobstick-Bigoray area, south-central Alberta; a study of physical chemistry of cementation. D, 1981, University of Tulsa. 130 p.

Mesolella, Kenneth Joseph. The uplifted reefs of Barbados (West Indies); physical stratigraphy, facies relationships and absolute chronology. D, 1968, Brown University.

Mesoloras, Nancy. Paleomagnetism of the Black Dyke and Mina formations, southwestern Nevada. M, 1986, Western Washington University. 113 p.

Mesri, Gholamreza. Engineering properties of montmorillonite. D, 1969, University of Illinois, Urbana. 96 p.

Messa, John Francis. A geochemical characterization of Bone Spring and San Andres crude oils from the northern Delaware Basin and northwestern shelf of southeastern New Mexico. M, 1988, University of Tulsa. 154 p.

Messenger, Harold M., III. Geology and fluorspar deposits of the southern Zuni Mountains fluorspar district, Valencia County, New Mexico. M, 1979, University of Texas at El Paso.

Messenger, Jane A. Wisconsinan paleogeography of Alaska. M, 1977, Arizona State University. 98 p.

Messer, Jay James. Nitrogen mass balances in Florida ecosystems. D, 1978, University of Florida. 416 p.

Messfin, Derbew. Seismic approaches for mineral exploration in the Sudbury Basin of Ontario, Canada. M, 1984, University of Manitoba.

Messin, Gerard M. L. Geology of the White Horse Pluton, Elko County, Nevada. M, 1973, University of Nebraska, Lincoln.

Messina, Angela R. The geologic occurrence of the emerald. M, 1935, Columbia University, Teachers College.

Messina, Mario Leo. Expansion of fractionated montmorillonites under various relative humidities. M, 1962, University of Texas, Austin.

Messineo, Anthony Vincent. An insoluble residue study of the Annville and upper "Stones River" limestones, south central Pennsylvania. M, 1955, University of Pittsburgh.

Messinger, Curtis. A geomorphic study of the dunes of the Provincetown Peninsula, Cape Cod, Massachusetts. M, 1958, University of Massachusetts. 150 p.

Messinger, Donald J. Form and change of a recurved sandspit, Presque Isle, Erie, PA. M, 1977, SUNY, College at Fredonia. 122 p.

Messmer, William J. Magnetic recording in frequency analysis of strong earth motion. M, 1961, St. Louis University.

Mesticky, L. J. The geology of the Nassau Brick Company clay deposit, Old Bethpage, Long Island. M, 1977, Queens College (CUNY). 56 p.

Meszoely, Charles Aladar Maria. North American fossil anguid lizards. D, 1967, Boston University. 135 p.

Metarko, Thomas A. Porosity, cement, grain fabric and water chemistry of the Mt. Simon Sandstone in the Illinois Basin. M, 1980, University of Cincinnati. 89 p.

Metcalf, Artie Lou. Fishes of the Kansas River System in relation to zoogeography of the Great Plains. D, 1964, University of Kansas. 400 p.

Metcalf, Charles T. A study of the use of sandstone in bituminous surface courses. M, 1949, Purdue University.

Metcalf, Linda Anne. Tephrostratigraphy and potassium - argon age determinations of seven volcanic ash layers in the Muddy Creek Formation of southern Nevada. M, 1982, University of Nevada. 187 p.

Metcalf, Richard Carl. Physical and chemical processes associated with the erosional energy of the Nisqually Glacier. M, 1977, University of Washington. 109 p.

Metcalf, Rodney Virgil. A study of uranium and thorium in the Precambrian Old Rag Granite, Madison County, Virginia. M, 1984, University of Kentucky. 80 p.

Metcalf, Thomas P. Intraplate tectonics of the Appalachians in post-Triassic times. M, 1982, Wesleyan University. 223 p.

Metcalf, William James, III. Investigation of paleotemperatures in the vicinity of the Washita Valley

Fault, southern Oklahoma. M, 1985, University of Oklahoma. 94 p.

Metcalfe, A. P. The igneous rocks of the Ruppert Coast, West Antarctica. M, 1978, University of Wisconsin-Madison.

Metcalfe, Cynthia Watson. Gravity study of the Big Bend region, Brewster County, Texas. M, 1980, Rice University. 85 p.

Metcalfe, Paul. Petrogenesis of Quaternary alkaline lavas in Wells Gray Provincial Park, B.C. and constraints on the petrology of the Subcordilleran mantle. D, 1987, University of Alberta. 416 p.

Metcalfe, Paul. Petrogenesis of the Klondike Formation, Yukon Territory. M, 1981, University of Manitoba. 103 p.

Metcalfe, Susan Judd. Acid-insoluble residues in Chickamauga equivalents (middle Ordovician) in East Tennessee; an aid to paleogeographic reconstruction. M, 1974, University of Tennessee, Knoxville. 64 p.

Methot, Robert Leo. Internal geochronologic study of two large granitic pegmatites, Connecticut. D, 1973, Kansas State University. 135 p.

Méthot, Yves. Pétrographie et géochimie du minerai et de l'altération reliée au gîte aurifère Eldrich, Abitibi, Québec. M, 1987, Ecole Polytechnique. 232 p.

Metrin, Deborah B. Geochemical significance of selected ions in Eocene carbonate rocks of Peninsular Florida. M, 1979, University of Florida. 64 p.

Metry, Amir Alfi. Mathematical modeling of pollutants migration in an unconfined aquifer. D, 1973, Drexel University. 204 p.

Metter, Raymond Earl. Sedimentary processes along the Lake Erie shore, Cedar-Point to Huron, Erie County, Ohio. M, 1952, Ohio State University.

Metter, Raymond Earl. The geology of a part of the southern Wasatch Mountains, Utah. D, 1955, Ohio State University. 262 p.

Metternich, Viola B. Brown County, Indiana; a physiographic study. M, 1940, University of Cincinnati. 28 p.

Mettes, Kim J. The contact between the late Paleozoic Tensleep Sandstone and Phosphoria Formation, and origin of the boundary beds, in northwestern Wyoming. M, 1980, University of Wyoming. 122 p.

Mettler, Don E. Dune sands of the Syracuse area in Kansas. M, 1956, University of Kansas. 95 p.

Mettner, Francis E. The Pre-cambrian basement complex of Kansas. M, 1936, University of Kansas. 59 p.

Metts, David. Structures and stratigraphy of the Sonora Quadrangle (Washington and Benton counties, Arkansas). M, 1961, University of Arkansas, Fayetteville.

Metz, Cheryl Lynn. Stratigraphy and facies analysis of the San Carlos Formation type section, Presidio County, Texas. M, 1987, University of Texas of the Permian Basin. 110 p.

Metz, Clyde Thomas. Nature and distribution of faunas in well cores from the Bradford oil field. M, 1952, Pennsylvania State University, University Park. 141 p.

Metz, Harold L. A petrographic comparison of the Loveland and Peoria loesses of northern Kansas. M, 1954, Kansas State University. 45 p.

Metz, Harold L. Geology of the El Pilar fault zone of Sucre, Venezuela. D, 1964, Princeton University. 105 p.

Metz, Jenny. Petrology of the Santa Rita Flat Pluton, Inyo Mountains, California. M, 1978, San Jose State University. 95 p.

Metz, Jenny. Physical and chemical evolution of Glass Mountain; pre-caldera high-silica rhyolites from the Long Valley magma system. D, 1987, Stanford University. 224 p.

Metz, Jerry P. A petrologic study of some Jurassic (?) sediments located at North Creek, Custer and Pueblo counties, Colorado. M, 1959, Kansas State University.

Metz, Michael C. The geology of the Snowbird deposit, Mineral County, Montana. M, 1971, Washington State University. 93 p.

Metz, Paul. Geology of the central portion of the Valdez C-2 Quadrangle, Alaska. M, 1975, University of Alaska, Fairbanks. 65 p.

Metz, Rebecca W. The effects of variation in discharge on the stream chemistry of the Christina River, Delaware. M, 1975, University of Delaware.

Metz, Robert. Stratigraphy and structure of the Cambridge Quadrangle, New York. D, 1967, Rensselaer Polytechnic Institute. 118 p.

Metz, Robert. The petrography of the Pantano beds in the Cienega Gap area, Pima County, Arizona. M, 1963, University of Arizona.

Metz, Robert Louis. Heavy minerals in some glacial tills of northwestern Illinois. M, 1971, Northern Illinois University. 93 p.

Metz, Simone. Metal enrichment processes in the marine environment. D, 1986, Florida Institute of Technology. 157 p.

Metzgar, Craig R. The petrology and structure of the Edgefield 7 1/2′ Quadrangle, South Carolina Piedmont. M, 1977, University of South Carolina. 51 p.

Metzger, Bernhard Hugo. Management of health risk from groundwater contamination. D, 1987, Harvard University. 311 p.

Metzger, Charles Frederick. A study of sorption of cesium by montmorillonite. D, 1965, University of Illinois, Urbana.

Metzger, Charles Frederick. The petrography of the North Sullivan Pluton, Maine. M, 1964, University of Illinois, Urbana.

Metzger, Chris W. Geology, mineralization, and geochemistry of the upper Illinois River drainage basin, Grand and Jackson counties, Colorado. M, 1974, Colorado School of Mines. 92 p.

Metzger, Ellen Pletcher. Investigation of partial melting at Ledge Mountain, central Adirondacks, New York. M, 1980, Syracuse University.

Metzger, Ellen Pletcher. Structure, lithologic succession, and petrology of the Stony Creek area, Warren Co., southeastern Adirondacks, New York. D, 1984, Syracuse University. 300 p.

Metzger, Fredrick William. Scanning electron microscopy of daughter minerals in fluid inclusions. M, 1976, University of Michigan.

Metzger, Robert. The Precambrian geology of the Little Buffalo Creek area, Southwest Beartooth Mountains, Montana and Wyoming. M, 1978, Northern Illinois University. 94 p.

Metzger, Robert J. The environmental geology of Hancock County, Ohio. M, 1984, Bowling Green State University. 276 p.

Metzger, Ronald Allen. Upper Devonian conodont biostratigraphy in the subsurface of north-central Iowa and Southeast Nebraska. M, 1988, University of Iowa. 116 p.

Metzger, Stacy Lynn. A subsurface study and paleoenvironmental analysis of the Medina Group, Chautauqua County, New York. M, 1981, SUNY, College at Fredonia. 102 p.

Metzger, William John. Pennsylvanian stratigraphy of the Warrior Basin, Alabama. D, 1961, University of Illinois, Urbana. 160 p.

Metzger, William John. Petrography of the Bar Harbor Series, Mount Desert Island, Maine. M, 1959, University of Illinois, Urbana.

Metzler, Christopher Virgil. Constraints on deposition and diagenesis of Jurassic carbonates from Italy and Switzerland. D, 1987, University of California, San Diego. 200 p.

Metzler, Jean M. Bedrock and glacial geology of Bath Township, Summit County, Ohio. M, 1967, Kent State University, Kent. 71 p.

Metzner, David Craig. Kinematic analysis of deformation in the Valley and Ridge Province north of Knoxville, Tennessee. M, 1983, University of Kentucky. 86 p.

Metzner, Ron. A magnetotelluric investigation of Augustine Island Volcano. M, 1975, University of Alaska, Fairbanks. 116 p.

Metzsch, Ernst Hans von *see* von Metzsch, Ernst Hans

Meyden, Hendrik Jan van der *see* van der Meyden, Hendrik Jan

Meyer, Alfred Herman Ludwig. A geological survey of the Vermont Quadrangle, Illinois. M, 1923, University of Illinois, Chicago.

Meyer, Beatrice I. Depositional environment and source rock potential, (Jurassic) Smackover Formation, Van Zandt County, Texas. M, 1984, University of Texas at Dallas. 151 p.

Meyer, Brenda S. The Great Salt Lake; variations in salt water intrusion with lake and ground water fluctuation. M, 1984, University of Utah.

Meyer, Charles. Hydrothermal wall rock alteration at Butte, Montana. D, 1950, Harvard University. 270 p.

Meyer, Charles. The geology of the Pilot Knob, Missouri iron mineralization. M, 1939, Washington University. 206 p.

Meyer, Charles Richard. Tungsten mineralization in the southern Grouse Creek Mountains, Utah. M, 1981, Texas Tech University. 43 p.

Meyer, Charles, Jr. Sputter-condensation of silicates. D, 1969, University of California, San Diego. 159 p.

Meyer, Dann. The hydrothermal origin of spilites; an experimental study. M, 1979, University of North Carolina, Chapel Hill. 182 p.

Meyer, David F. Factors influencing the yields of dissolved solids in Illinois streams. M, 1976, Northern Illinois University. 113 p.

Meyer, David Frederick. The significance of sediment transport in arroyo development. D, 1986, Colorado State University. 152 p.

Meyer, Franz Oswald. Depositional environment, taphonomy, and paleoecology of a Maclurites magnus assemblage in the Crown Point Formation, New York. M, 1975, University of Michigan.

Meyer, Franz Oswald. Middle Devonian lagoon patch reef complex of Michigan; stratigraphy and physical and biological determinants of reef structure. D, 1979, University of Michigan. 223 p.

Meyer, Gary Dean. The surficial geology of the Guthrie North Quadrangle, Logan County, Oklahoma. M, 1975, Oklahoma State University. 60 p.

Meyer, Gary N. Applied geology in the Beulah-Hazen area, Mercer County, West-central North Dakota. M, 1979, University of North Dakota. 216 p.

Meyer, Gary Peter. Late Pleistocene water mass boundaries in the southwest Equatorial Pacific based on planktonic foraminiferal distributions. M, 1976, Rutgers, The State University, Newark. 91 p.

Meyer, George L. Pliocene-Quaternary geology of eastern Santa Cruz Island, California. M, 1967, University of California, Santa Barbara.

Meyer, Grant Arnold. Genesis and deformation of Holocene shoreline terraces, Yellowstone Lake, Wyoming. M, 1986, Montana State University. 94 p.

Meyer, Harvey John. Petrography of the Catskill Sandstone facies in central Pennsylvania. D, 1964, Pennsylvania State University, University Park. 146 p.

Meyer, Herbert William. An evaluation of the methods of estimating paleoaltitudes using Tertiary floras from the Rio Grande Rift vicinity, New Mexico and Colorado. D, 1986, University of California, Berkeley. 217 p.

Meyer, Howard J. A combined magnetic and gravity analysis of the Sauble Anomaly, Lake County, Michigan. M, 1963, Michigan State University. 55 p.

Meyer, J. Lithospheric deflection under the Necker Ridge. M, 1977, University of Hawaii. 112 p.

Meyer, James Arthur. Bryozoa from the Cambridge Limestone (Pennsylvanian). M, 1972, Bowling Green State University. 56 p.

Meyer, Jane Doris E. Geology of the Hudson Dome area, Fremont County, Wyoming. M, 1946, University of Missouri, Columbia.

Meyer, Jeffrey Wayne. Alteration and mineralization of the Grasshopper Prospect, Beaverhead County, Montana. M, 1980, University of Arizona. 91 p.

Meyer, Joachim Dietrich. Geology of the Ahuachapan area, western El Salvador, Central America. M, 1961, University of Texas, Austin.

Meyer, John E. The stratigraphy of the Gypsum Spring Formation (Middle Jurassic), northwestern Bighorn Basin, Wyoming. M, 1984, University of Wyoming. 94 p.

Meyer, Jurg Walter. Clay mineralogy of Recent marine sediments in the Gulf of Mexico. D, 1958, University of Illinois, Urbana. 106 p.

Meyer, Kevin S. Structural and stratigraphic framework of lower Mesozoic and upper Paleozoic strata, Northeast Texas. M, 1981, University of Texas, Arlington. 107 p.

Meyer, Martha Grose. Shallow-outer-ramp limestones of the Zuloaga Formation (Oxfordian), Astillero Canyon, Zacatecas, Mexico. M, 1982, University of New Orleans. 107 p.

Meyer, Marvin Phillip. Pre-St. Peter areal geology of northern Illinois. M, 1946, University of Illinois, Urbana.

Meyer, Michael. Geohydrology of the glaciofluvial aquifer of the lower Floyd River basin, Plymouth County, Iowa. M, 1978, University of Iowa. 93 p.

Meyer, Michael R. Projectile penetration in dry sand. M, 1988, Washington State University. 60 p.

Meyer, Michael T. Geochronology and geochemistry of the Wathaman Batholith, the remnant of an early Proterozoic continental-arc in the Trans-Hudson Orogen, Saskatchewan, Canada. M, 1987, University of Kansas. 107 p.

Meyer, Paul Eaton. Hydrothermal alteration of greenschist facies schist surrounding gold-bearing quartz veins, Cleary Hill area, Fairbanks mining district, Alaska. M, 1984, University of Idaho. 114 p.

Meyer, Peter Sheafe. Petrology and geochemistry of Iceland basalts; spatial and temporal variations. D, 1984, University of Rhode Island. 203 p.

Meyer, Richard Burt. A mechanical analysis of beach sediments of the south shore of Long Island, New York. M, 1955, University of Oklahoma. 58 p.

Meyer, Richard Fastabend. Geology of Pennsylvanian and Wolfcampian rocks in southeastern New Mexico. D, 1968, University of Kansas. 208 p.

Meyer, Robert Lee. Late Cretaceous elasmobranchs from the Mississippi and East Texas embayments of the Gulf Coastal Plain. D, 1974, Southern Methodist University. 419 p.

Meyer, Robert Paul. The geologic structure of the Cape Fear axis (Atlantic Coastal Plain) as revealed by refraction seismic measurements. D, 1957, University of Wisconsin-Madison. 176 p.

Meyer, Rudolf. A strain study in the Kona Formation, Marquette County, Michigan. M, 1983, Michigan State University. 65 p.

Meyer, Scott C. Origin of the Keefer Sandstone (Middle Silurian) of northeastern West Virginia and western Maryland. M, 1987, University of North Carolina, Chapel Hill. 173 p.

Meyer, Thomas. Sedimentary petrology of the Monroe gas rock and Arkadelphia formations (Cretaceous), Monroe gas field area, Louisiana. M, 1972, Northeast Louisiana University.

Meyer, Thomas Scott. A subsurface study of the Goodhope Sandstone (Mississippian), north central West Virginia. M, 1984, University of Wisconsin-Madison. 171 p.

Meyer, W. C. Late Pleistocene and Holocene paleoceanography of the Red Sea. M, 1973, University of Southern California. 118 p.

Meyer, Wallace Harold, Jr. Computer modelling of electromagnetic prospecting methods. D, 1976, University of California, Berkeley. 155 p.

Meyer, Walter F. Paleodepositional environments of the Stafford Limestone (Middle Devonian) across

New York State. M, 1985, SUNY, College at Fredonia. 67 p.

Meyer, William. Radiative properties of rock types in the Harquahala Plains area, Arizona and possible meteorological implications. M, 1965, University of Arizona.

Meyer, William Vincent. The Wichita magnetic low, southeastern Kansas. M, 1987, Purdue University. 95 p.

Meyer, Willis George. Recurrent coral faunas of the Ordovician. M, 1933, University of Cincinnati. 34 p.

Meyer, Willis George. Stratigraphy and historical geology of Gulf Coastal Plain in the vicinity of Harris County, Texas. D, 1941, University of Cincinnati. 122 p.

Meyerhoff, Arthur A. A study of leaf venation in the Betulaceae, with its application to paleobotany. D, 1952, Stanford University. 248 p.

Meyerhoff, Howard A. Geology of Puerto Rico. D, 1935, Columbia University, Teachers College.

Meyerhoff, Howard A. The (Silurian) Niagaran series of Ohio; a problem of correlation. M, 1922, Columbia University, Teachers College.

Meyerhoff, James C. The nature of the Cretaceous-Tertiary contact, central Texas. M, 1983, Baylor University. 257 p.

Meyerhoff, Lisa H. The petroleum potential of the Cotton Valley Troy Lime, north-central Louisiana. M, 1983, Baylor University. 74 p.

Meyerholtz, Keith A. Enhanced geotomography and its application to the Kentland, Indiana impact site. M, 1988, Indiana University, Bloomington. 133 p.

Meyers, Alan E. Mountain water pollution from road reconstruction and wildfire. M, 1968, Colorado State University. 79 p.

Meyers, Bernard M. Structure and stratigraphy of the Triassic System in the Bernardsville Quadrangle, New Jersey. M, 1960, New York University.

Meyers, Clay Kenton. The red beds (Pennsylvanian) of the Pittsburgh (Pennsylvania) quadrangle. M, 1935, University of Pittsburgh.

Meyers, Darwin. The depositional environment of the Bedford Shale (Devonian–Mississippian) in northeastern Ohio. M, 1967, [Case Western Reserve University].

Meyers, David C. The mineralogy of the red clay and its relation to slope stability in Douglas County, Wisconsin. M, 1977, Miami University (Ohio). 96 p.

Meyers, Harold G. The (Upper Devonian) Genundewa Limestone (New York). M, 1937, Cornell University.

Meyers, James B. Soil geochemistry in the vicinity of the Howell zinc prospect, Jefferson County, West Virginia. M, 1974, University of Toledo. 95 p.

Meyers, James Harlan. Stratigraphy and petrology of the Maywood Formation (upper Devonian), southwestern Montana. D, 1971, Indiana University, Bloomington. 274 p.

Meyers, Joseph Duncan. The geochemistry of water with special reference to mineral composition. M, 1952, University of Oregon. 123 p.

Meyers, R. E. The geology and origin of the New Insco copper deposit, Noranda District, Quebec. M, 1980, McGill University. 134 p.

Meyers, Stephen Douglas. Triradiality in the Echinodermata; occurrence and implications. M, 1977, University of Cincinnati. 281 p.

Meyers, Theodore Ralph. Geology of Jefferson and Bedford townships, Coshocton County, Ohio. M, 1929, Ohio State University.

Meyers, W. C. Geology of the Tom Creek-Keg Springs area, Centennial Range, Idaho and Montana. M, 1968, Idaho State University.

Meyers, William C. Environmental analysis of Almond Formation (Upper Cretaceous) from the Rock Springs Uplift, Wyoming. D, 1977, University of Tulsa. 324 p.

Meyers, William Cady. Microfossils of the Henryetta Coal. M, 1963, University of Tulsa. 59 p.

Meyers, William D. Insoluble residues from Humble Well #15 Gulf Coast Realities Corporation, Collier County, Florida. M, 1952, Florida State University.

Meyers, William John. Chertification and carbonate cementation in the Mississippian Lake Valley Formation, Sacramento Mountains, New Mexico. D, 1974, Rice University. 352 p.

Meyers, William John. Geology of a portion of the Carrizo Mountains, northeastern Arizona. M, 1963, University of California, Berkeley. 155 p.

Meyerson, Arthur Lee. Pollen and paleosalinity analyses from a Holocene tidal marsh sequence, Cape May County, New Jersey. D, 1971, Lehigh University. 160 p.

Meyertons, Carl Theile. Mineralogical investigation of coal mine roof shales in part of the Southern Appalachian coal field. M, 1956, Virginia Polytechnic Institute and State University.

Meyertons, Carl Theile. The geology of the Danville Triassic Basin of Virginia. D, 1959, Virginia Polytechnic Institute and State University. 240 p.

Meylan, M. A. Marine sedimentation and manganese nodule formation in the southeastern Pacific Ocean. D, 1978, University of Hawaii. 325 p.

Meyland, Maurice A. The mineralogy and geochemistry of manganese nodules from the Southern Ocean. M, 1968, Florida State University.

Meza, Raul I. Madariaga *see* Madariaga Meza, Raul I.

Mezcua, Julio. Attenuation of rayleigh waves in Europe and Eastern North America. M, 1972, St. Louis University.

Mezga, Lance Joseph. Hydrology and water balance of a partially strip-mined watershed, WSD-19 of the Little Mill Creek Basin, Coshocton, Ohio with emphasis on flow through spoil material. M, 1973, Kent State University, Kent. 82 p.

Miall, Andrew. The sedimentary history of the Peel Sound Formation, (lower to middle Devonian) Prince of Wales island, Northwest Territories (Canada). D, 1969, University of Ottawa. 279 p.

Miao, Desui. Skull morphology of Lambdopsalis bulla (Mammalia, Multituberculata) and its implications to mammalian evolution. D, 1987, University of Wyoming. 363 p.

Miatech, Gerald James. An investigation of a magnetic discontinuity in the Iron River District of Michigan. M, 1956, St. Louis University.

Michael, Andrew Jay. Regional stress and large earthquakes; an observational study using focal mechanisms. D, 1986, Stanford University. 167 p.

Michael, Eugene Donald. The geology of the Cache Creek area, Kern County, California. M, 1960, University of California, Los Angeles.

Michael, Fouad Yousry. Studies of foraminifera from the Comanchean Series (Cretaceous) of (north central) Texas. D, 1971, Southern Methodist University. 86 p.

Michael, Gerald Eric. Effect of biodegradation upon porphyrin biomarkers in Upper Mississippian tar sands and related oils, southern Oklahoma. M, 1987, University of Oklahoma. 118 p.

Michael, M. O. Circum-Pacific periodicity in volcanic activity. M, 1970, University of Hawaii. 33 p.

Michael, M. O. Fluctuations in Circum-Pacific volcanic activity and in the seismicity of South America. D, 1973, University of Hawaii. 130 p.

Michael, Peter John. Emplacement and differentiation of Miocene plutons in the foothills of the southernmost Andes. D, 1983, Columbia University, Teachers College. 344 p.

Michael, Peter Robert. Subsidence over abandoned bituminous coal mines in the Appalachian coal basin; an analysis of subsidence parameters and three case studies. M, 1984, SUNY at Binghamton. 138 p.

Michaelis, Sara E. Long-term energy availability from artificial geothermal wells. M, 1979, University of Oklahoma. 51 p.

Michaels, Anthony Francis. Acantharia in the carbon and nitrogen cycles of the Pacific Ocean. D, 1988, University of California, Santa Cruz. 238 p.

Michaels, Paul. An application of the generalized linear inverse method to the location of microearthquakes and simultaneous velocity model determination. M, 1973, University of Utah. 187 p.

Michaelson, Caryl Ann. Three-dimensional velocity structure of the crust and upper mantle in Washington and northern Oregon. M, 1983, University of Washington. 100 p.

Michalek, D. D. Pre-Cambrian geology of Jelm Mountain, Albany County, Wyoming. M, 1952, University of Wyoming. 53 p.

Michalek, Daniel D. Fanlike features and related periglacial phenomena of the southern Blue Ridge. D, 1969, University of North Carolina, Chapel Hill. 197 p.

Michalkow, Albert. Structural geology and petrology of the Lily Pond Area, Manitoba. M, 1954, University of Manitoba.

Michalski, Paul J. The distribution of minor elements between coexisting phases in the Gasport Member of the Lockport Formation (Silurian), Lockport, New York. M, 1969, SUNY at Buffalo. 53 p.

Michaud, Denis Paul. Depositional and diagenetic history of the Upper Jurassic Rierdon Formation, southwestern Montana. M, 1986, Indiana University, Bloomington. 124 p.

Michaud, Marion Catharine. Numerical simulation of reservoirs. D, 1980, Brown University. 123 p.

Michel, F. A. Hydrogeologic studies of springs in the central Mackenzie Valley, Northwest Territories, Canada. M, 1977, University of Waterloo. 424 p.

Michel, Fred A., Jr. Geology of the King Mine, Helvetia, Arizona. M, 1959, University of Arizona.

Michel, Frederick A. Isotope investigations of permafrost waters in northern Canada. D, 1982, University of Waterloo. 424 p.

Michel, Jacqueline M. Mobilization of uranium and thorium decay series isotopes in the hydrologic cycle. D, 1980, University of South Carolina. 113 p.

Michel, Jaqueline M. Ground water pollution and geochemical variations in leachate from solid waste disposal. M, 1976, University of South Carolina. 69 p.

Michel, Sandra Jo Hagni. The petrologic relationships between magnetite and hematite in the Pilot Knob iron deposit, southeastern Missouri. M, 1981, Washington University. 133 p.

Michel, Sylvestre Georges. Caractérisation de l'indice aurifère de Mont-Organisé (Haiti) (secteurs Maman Noël et Grenier). M, 1986, Ecole Polytechnique. 229 p.

Michels, Donald E. Processes in mineral replacement. D, 1966, Colorado School of Mines. 68 p.

Michels, Donald E. Replacement of carbonate minerals and rocks by cerussite in dilute aqueous solutions at low temperature and pressure. M, 1962, Colorado School of Mines. 89 p.

Michels, Joseph William. Lithic serial chronology through obsidian hydration dating. D, 1965, University of California, Los Angeles. 296 p.

Michelson, J. E. Drumlins of the Crandon area, northeastern Wisconsin. M, 1960, University of Kansas. 54 p.

Michelson, Peter C. Orientation analysis of Trough Cross stratification; case study of Cambrian sandstones in western Wisconsin. M, 1971, University of Wisconsin-Madison.

Michelson, Ronald Wayne. Tundra relief features near Barrow, Alaska. M, 1965, Iowa State University of Science and Technology.

Michener, Charles Edward. Minerals associated with large sulphide bodies of the sudbury type. D, 1940, University of Toronto.

Michener, Charles Edward. The Sweetgrass Arch, Alberta, Canada. M, 1932, Cornell University.

Michener, Stuart Reid. Bedrock geology of the Pelham-Shutesbury Syncline, Pelham Dome, west-

central Massachusetts. M, 1983, University of Massachusetts. 101 p.

Michie, Joanna. Geology of a portion of the northern Kings Mountain Belt, North Carolina. M, 1985, Texas A&M University. 115 p.

Michkofsky, R. N. Magnetics of Bowie seamount, (Pacific Ocean, west of Queen Charlotte islands, 53° 18N, 134° 41W). M, 1969, University of British Columbia.

Michlik, David Michael. Petrographic and mapping study of the subsurface "Oswego" Limestone in part of the Putnam Trend, T 15-16 N, R 15-17 W, Dewey and Custer counties, Oklahoma. M, 1980, Oklahoma State University. 71 p.

Michlik, Rudolph R. Symmetry improvement of the magnetic anomalies and spreading-rate ratio determination for mid-oceanic ridges. M, 1970, Pennsylvania State University, University Park.

Michrina, Barry Paul. Studies of extracted soil organic matter fractions used to determine soil nitrogen availability. D, 1981, Pennsylvania State University, University Park. 105 p.

Mickel, Charles Joseph. Gastropods of the Fox Hills Formation in its type area. M, 1962, University of South Dakota. 85 p.

Mickel, Edward G. Wasta fauna of the Pierre Formation. M, 1961, University of South Dakota. 134 p.

Mickelson, David M. A chronological investigation of a kettle-hole peat bog, Cherryfield, Maine. M, 1968, University of Maine. 62 p.

Mickelson, David Melvin. Glacial geology of the Burroughs Glacier area, southeastern Alaska. D, 1971, Ohio State University. 238 p.

Mickelson, John Chester. Reclassification of the Pleistocene Loveland Formation of Iowa. D, 1949, University of Iowa. 97 p.

Mickelson, John Chester. The geology of the northern portion of the Steelville Quadrangle, Missouri. M, 1948, University of Iowa. 66 p.

Mickey, Michael B. Upper Cretaceous biostratigraphy of a portion of northwestern Baja California, Mexico. M, 1971, University of San Diego.

Mickle, David Grant. The design of exploration programs for economic mineral deposits. D, 1964, University of Arizona. 97 p.

Mickle, James Earl. Stems of Paleozoic Marattialean ferns; growth, development, and taxonomic delimitation. D, 1983, [Ohio University, Athens]. 178 p.

Micklin, Richard F. Fusulinids and conodonts of a Pennsylvanian-Permian section in the northern Dragoon mountains, Cochise County, Arizona. M, 1969, University of Arizona.

Middelaar, Wilhelmus T. van see van Middelaar, Wilhelmus T.

Middendorf, Robert P. High level stream channels in northwestern Boone County, Kentucky. M, 1952, University of Cincinnati. 42 p.

Middlebrook, John. Geology at Tungsten, Nevada, emphasizing structural aspects. M, 1957, University of Nevada. 41 p.

Middlebrooks, Peter Kendrick. A shallow seismic reflection study of the Mad River buried valley at Huffman Dam, Ohio. M, 1987, Wright State University. 98 p.

Middlekauff, Bryon Douglas. Relict periglacial morphosequences in the northern Blue Ridge. D, 1987, Michigan State University. 172 p.

Middleman, Allen. Trace fossils of the Logan Formation (Lower Mississippian) in northern Hocking County, Ohio. M, 1976, Ohio University, Athens. 132 p.

Middleman, Bruce H. Sedimentology and tectonic significance of the Lower Cretaceous Cloverly Formation, southwestern Wind River basin, Wyoming. M, 1987, Indiana University, Bloomington. 91 p.

Middleton, Bruce Donald. A preliminary investigation into the application of coal petrography in the blending of anthracite and bituminous coals for the produc-

tion of metallurgical coke. M, 1961, Pennsylvania State University, University Park. 85 p.

Middleton, Dennis L. Regional examination of sites of tufa deposition along highways in northeast Ohio. M, 1981, Kent State University, Kent. 102 p.

Middleton, Jack A. Diagenesis of reef materials as seen through cathodoluminescence. M, 1979, Rensselaer Polytechnic Institute. 42 p.

Middleton, John L. Geology of Spruce Gulch area, and the mineral deposits of the Deadwood lead and zinc mines. M, 1923, University of Minnesota, Minneapolis.

Middleton, Kenneth Douglas. The geology of the Chisos Mine NE quadrangle, Big Bend Country, Brewster County, Texas. M, 1974, University of Texas, Arlington. 122 p.

Middleton, Larry T. Depositional environments of the Glenns Ferry Formation near Jackass Butte, Idaho. M, 1976, Idaho State University. 59 p.

Middleton, Larry Thomas. Sedimentology of Middle Cambrian Flathead Sandstone, Wyoming. D, 1980, University of Wyoming. 198 p.

Middleton, Michael D. Early Paleocene vertebrates of the Denver Basin, Colorado. D, 1983, University of Colorado. 421 p.

Middleton, Robert Stuart. Remanent magnetism and magnetic susceptibility in the interpretation of ground magnetics, Jamieson township, district of Cochrane, Ontario. M, 1969, Michigan Technological University. 96 p.

Middour, E. S. Radiolaria from the Porters Creek Formation (Paleocene, Midway Group) of Stoddard County, Mo. M, 1951, University of Missouri, Rolla.

Miedema, Oene. Mid-Devonian solitary rugose corals from the Northwest Territories. M, 1961, University of Saskatchewan. 131 p.

Miele, Martin J. A magnetotelluric profiling and geophysical investigation of the Laguna Salada Basin, Baja California. M, 1986, San Diego State University. 154 p.

Mielenz, Richard Childs. The geology of the southwestern part of San Benito County, California. D, 1940, University of California, Berkeley. 295 p.

Mielke, James Edward. Geochemical study of the sediments of the Potomac River estuary. D, 1974, George Washington University.

Mielke, James Edward. Trace element investigation of the "Turkey Track" porphyry, southeastern Arizona. M, 1964, University of Arizona.

Mielke, R. Geology of the Porcupine conglomerates and associated rocks at the Dome Mine. M, 1985, University of Waterloo. 134 p.

Miers, John Harlow. Ultramafic dikes on Jumbo mountain, Snohomish County, Washington. M, 1970, University of Washington. 54 p.

Mies, Jonathan Wheaton. An investigation of the Proterozoic basement-Ashe Formation boundary west of the Grandfather Mountain Window, northwestern North Carolina. M, 1987, University of North Carolina, Chapel Hill. 74 p.

Miesch, Alfred Thomas. Geochemistry of the Frenchy Incline uranium deposits, San Miguel County, Colorado. D, 1961, Northwestern University. 110 p.

Miesch, Alfred Thomas. Geology of the Socorro manganese area, Socorro County, New Mexico. M, 1954, Indiana University, Bloomington. 43 p.

Miesen, David Lee. Crustal movement and topography. M, 1986, Cornell University. 35 p.

Miesfeldt, Mark Alan. Facies relationships between the Parkwood and Bangor formations in the Black Warrior Basin. M, 1985, University of Alabama. 158 p.

Miesner, James F. The pre-Missourian sedimentary rocks of Decatur County, Kansas. M, 1967, Wichita State University. 54 p.

Miesse, J. V. Quaternary geology of a basin near Linn, (Washington County), Kansas. M, 1974, Kansas State University. 122 p.

Mifflin, Martin D. Geology of a part of the southern margin of the Gallatin Valley, Southwest Montana. M, 1963, Montana State University. 107 p.

Mifflin, Martin David. Delineation of ground-water flow systems in Nevada. D, 1968, University of Nevada. 213 p.

Migliore, John J., Jr. Petrography of some Champlain Valley dikes, (Vermont). M, 1959, University of Vermont.

Mignardot, Eddie Roy. Stratigraphy and depositional environments of the upper half of the Frontier Formation in south central Fremont County, Wyoming. M, 1984, West Texas State University. 163 p.

Mignery, Florence Perkins. Terrace history of the Kennebec River in the Waterville-Fairfield region, Maine. M, 1943, University of Michigan.

Mignery, Thomas J. An electrical resistivity survey applied to the hydrogeology of glacial and shallow bedrock aquifers in northeastern Ohio. M, 1987, Kent State University, Kent. 126 p.

Mihalasky, Mark John. A database and mineral deposit models for the giant lode gold camps of North America. M, 1988, Eastern Washington University. 289 p.

Mihalich, John P. Granitic pegmatites in the southeastern Adirondacks; their use as indicators of temperature, pressure, and fluid conditions during a late stage of the Grenville Orogeny. M, 1987, SUNY at Albany. 123 p.

Mihalyi, Dale L. Petrology and dolomitization of the Sunniland Limestone (Lower Cretaceous), a subsurface unit of Peninsular Florida. M, 1976, University of Rochester. 57 p.

Mihalyi, Dale Lynn. Petrology and dolomitization of the Sunniland Limestone (Lower Cretaceous), a subsurface unit of peninsular Florida. M, 1977, University of Rochester.

Mihalynuk, Mitchell George. Metamorphic, structural and stratigraphic evolution of the Telkwa Formation, Zymoetz River area (NTS 103 I/8 and 93 L/5), near Terrace, British Columbia. M, 1987, University of Calgary. 128 p.

Mihelcic, James Robert. Microbial degradation of polycyclic aromatic hydrocarbons under denitrification conditions in soil-water suspensions. D, 1988, Carnegie-Mellon University. 211 p.

Mihm, Richard J. Quartzite, chert, and Ordovician shale in Chicken creek cirque, northern Independence Range, Elko County, Nevada. M, 1960, Columbia University, Teachers College.

Mihoren, Jerry John. Late Quaternary invertebrate macrofossils and microfossils from the central St. Lawrence Lowland, Canada. M, 1987, University of Windsor. 170 p.

Mihychuk, Maryann. The surficial geology, sedimentology and geochemistry of the late glacial sediments and Paleozoic bedrock in the Campbellford area, Ontario, with special reference to the Dummer Complex. M, 1984, Brock University. 101 p.

Miille, Michael James. The photochemical transformations of organic compounds in seawater. D, 1980, University of California, Davis. 77 p.

Mika, James E. The structure of pre-Pennsylvanian strata in a portion of Payne County, Oklahoma. M, 1982, Wright State University. 48 p.

Mikami, Harry M. Granitic intrusives and metamorphic gneisses of the East Haven-Guilford area, Connecticut. D, 1945, Yale University.

Mikan, Frank M. A paleoenvironmental interpretation of the Lower Silurian "Clinton Sands", western Guernsey County, Ohio. M, 1973, Ohio State University.

Mikesh, David L. Correlation of Devonian strata in northwestern Wyoming. M, 1965, University of Iowa. 98 p.

Mikesh, David Leonard. Permian goniatitid ammonoids; family Paragastrioceratidae. D, 1968, University of Iowa. 274 p.

Miklausen, Anthony J. Contributions to the fossil flora of the roof shales of the Pittsburgh coal vein of south-western Pennsylvania. D, 1949, University of Pittsburgh.

Miklius, Asta. Petrogenesis and source character of Taal Volcano, Philippines. M, 1988, University of Illinois, Chicago.

Mikroudis, George Konstantinos. GEOTOX; a knowledge-based surrogate consultant for evaluating waste disposal sites. D, 1987, [Lehigh University]. 329 p.

Mikucki, Edward J. Petrographic, fluid inclusion, and stable isotope studies of the Red Hill Cu-Mo-Au prospects, Ord Mountain, California. M, 1986, Pennsylvania State University, University Park.

Mikulic, Donald George. The paleoecology of Silurian trilobites with a section on the Silurian stratigraphy of southeastern Wisconsin. D, 1980, Oregon State University. 864 p.

Mikulich, Matthew Jonathan. Seismic reflection and aeromagnetic surveys of the Great Salt Lake, Utah. D, 1971, University of Utah. 233 p.

Milam, Robert Wilson. Biostratigraphy and sedimentation of the Eocene and Oligocene Kreyenhagen Formation, Central California. D, 1985, Stanford University. 253 p.

Milam, Robert Wilson. Distribution and ecology of Recent benthonic foraminifera of the Dumont d'Urville Sea, Antarctica. M, 1981, Rice University. 109 p.

Milan, C. S. Accumulation of petroleum hydrocarbons in a salt marsh ecosystem exposed to steady state oil input. M, 1978, Louisiana State University.

Milavec, Gary John. The Nonesuch Formation; Precambrian sedimentation in an intracratonic rift. M, 1986, University of Oklahoma. 142 p.

Milbauer, John A. The historical geography of the Silver City mining region of New Mexico. D, 1983, University of California, Los Angeles.

Milbert, Dennis Gerard. Treatment of geodetic leveling in the integrated geodesy approach. D, 1988, Ohio State University. 247 p.

Mileff, Robert John. Geology for land-use planning of the Muskogee area, Muskogee County, Oklahoma. M, 1976, Oklahoma State University. 57 p.

Milender, Kenneth Westcott. An engineering geology study of lake bluff stability, Shorewood, Wisconsin. M, 1987, University of Wisconsin-Milwaukee. 224 p.

Miles, Alfred. Geology of the Hendersonville Quadrangle, Sumner and Wilson counties, Tennessee. M, 1955, Vanderbilt University.

Miles, C. M. Paleoenvironmental interpretations in the Salina Formation using isotopes (southwestern Ontario and Michigan). M, 1985, University of Waterloo. 190 p.

Miles, Charles H. The geology of the Pass Creek-Wolford Mountain area, Grand County, Colorado. M, 1961, Colorado School of Mines. 89 p.

Miles, Charles Hammond. Metamorphism and hydrothermal alteration in the Lecheguilla Peak area of the Rincon Mountains, Cochise County, Arizona. D, 1965, University of Arizona. 104 p.

Miles, Christine E. Surficial geology (Quaternary) of the northern part of Red Lodge Creek drainage basin, southern Montana. M, 1971, Franklin and Marshall College. 78 p.

Miles, Daniel John. North-south trends in the Northern Appalachian foreland. M, 1948, University of Pittsburgh.

Miles, Deborah R. A gravity survey in the haulage tunnel of the Henderson Mine near Berthoud Pass, Colorado. M, 1977, Colorado School of Mines. 37 p.

Miles, Gregory Allen. Living planktonic foraminifera in the Northeast Pacific Ocean. M, 1973, University of Oregon. 131 p.

Miles, Gregory Allen. Planktonic foraminifera of the lower Tertiary Roseburg, Lookingglass, and Flournoy formations, Southwest Oregon. D, 1977, University of Oregon. 360 p.

Miles, Paul R. The geomagnetic Sq field at Weston Observatory (Boston, Massachusetts). M, 1965, [Boston University].

Miles, Phil Middleton. Geology of the Burning Springs gas field. M, 1939, University of Kentucky. 7 p.

Miles, R. C. Modifications of sediment composition and texture in two fluvial systems, Guatemala. M, 1977, University of Missouri, Columbia.

Miles, Randall Jay. Development of soils on terraces associated with the Brazos River in Young and Throckmorton counties, Texas. D, 1981, Texas A&M University. 263 p.

Miles, Roy G. Carbonate petrology of the Bankston Fork Limestone of southern Illinois. M, 1958, Northwestern University.

Miles, Thomas O. Physical model studies of the Turam time domain electromagnetic method in minerals exploration. M, 1985, Colorado School of Mines. 282 p.

Milewski, R. G. A study of pollen deposition in Crystal Lake, Hartstown, Pennsylvania with special reference to the decomposition of pollen. D, 1978, University of Pittsburgh. 89 p.

Milford, John Calverley. Geology of the Apex Mountain Group, north and east of the Similkameen River, south-central British Columbia. M, 1984, University of British Columbia. 108 p.

Milfred, Clarence James. Pedography of three soil profiles of Wisconsin representing the Fayette, Tama and Underhill series. D, 1966, University of Wisconsin-Madison. 233 p.

Milholland, E. Cheney Snow. Compositional variations of coexisting phases during upper mantle melting; experimental partial melting of garnet lherzolite at 12 and 20 kilobars. M, 1985, University of Hawaii. 244 p.

Milholland, Phillip Delbert. Geoacoustic model for deep-sea carbonate sediment. M, 1978, University of Hawaii. 109 p.

Milholland, Phillip Delbert. Seismicity at the Galapagos 95.5°W propagating rift. D, 1984, University of Hawaii.

Milici, Alfred W. Paleoenvironmental interpretation of the Middle Ordovician High Bridge Group and Wells Creek Dolomite in the subsurface, Cumberland County, Kentucky. M, 1986, Eastern Kentucky University. 83 p.

Milici, Robert Calvin. The geology of the Sequatchie Valley overthrust block, Sequatchie Valley, Tennessee. D, 1960, University of Tennessee, Knoxville. 83 p.

Milici, Robert Calvin. The structural geology of the Harriman Corner, Roane County, Tennessee. M, 1955, University of Tennessee, Knoxville. 34 p.

Milionis, Peter Nicholas. Rare earth element geochemistry of lavas from Central America; constraints for basalt petrogenesis. M, 1987, Rutgers, The State University, New Brunswick. 105 p.

Millage, Andra H. Mineralogy of the Victor Claim, New Idria District, San Benito County, California. M, 1981, Stanford University. 172 p.

Millage, Clayton Dodge. Contact metamorphism at Dutch Flat, Hot Springs Range, Humboldt County, Nevada. M, 1981, Stanford University. 68 p.

Millar, William Winston. Foraminiferal biostratigraphy of a portion of the Chesterian Series (Upper Mississippian) in cores from Blount and Chandler mountains, Alabama. M, 1988, University of Georgia. 83 p.

Millard, Alban Willis, Jr. Geology of the southwestern quarter of the Scipio North (15 minute) Quadrangle, Millard and Juab counties, Utah. M, 1982, Brigham Young University.

Millard, MacDonald John. Quaternary geology and drift prospecting, Dawn Lake area, northern Saskatchewan. M, 1988, University of Saskatchewan. 171 p.

Millard, Richard C. Structure and petrology of the Empire Stock, Clear Creek County, Colorado. M, 1960, University of Colorado.

Millard, William J. Geologic notes on the Belgian Congo. M, 1914, Columbia University, Teachers College.

Millberry, Kimberlee Whitney. Fan-delta variations and associated shelf-bars, lower member of the Honaker Trail Formation (Desmoinesian), southwestern Colorado. M, 1984, University of Texas, Austin. 62 p.

Millen, Timothy M. Stratigraphy and petrology of the Green River Formation (Eocene), Gunnison Plateau, central Utah. M, 1982, Northern Illinois University. 221 p.

Millendorf, S. A. A comparison of biostratigraphic methods for the quantification of assemblage zones. M, 1977, Syracuse University.

Miller, A. R. Petrology and geochemistry of the 2-3 ultramafic sill and related rocks, Cape Smith-Wakeham Bay fold belt, Quebec. D, 1977, University of Western Ontario.

Miller, Andrew J. Shore erosion processes, rates, and sediment contributions to the Potomac tidal river and estuary. D, 1983, The Johns Hopkins University. 363 p.

Miller, Andrew Michael. Comparison of manual baseflow separation techniques to a computer baseflow separation program and application to six drainage basins. M, 1984, Oklahoma State University. 161 p.

Miller, Ann-Alberta Louise. Elphidium excavatum (Terquem); paleobiological and statistical investigations of infraspecific variation. M, 1983, Dalhousie University. 372 p.

Miller, Arnold I. Spatio-temporal development of the class Bivalvia during the Paleozoic era. D, 1986, University of Chicago. 399 p.

Miller, Arthur K. The (Ordovician) Bighorn Formation of the Wind River Mountains of Wyoming and its fauna. D, 1930, Yale University.

Miller, Arthur K. The development of armor in some of the reptile groups. M, 1925, University of Missouri, Columbia.

Miller, Barry Bennett. A late Pleistocene molluscan faunule from Meade County, Kansas. M, 1960, University of Michigan.

Miller, Barry Bennett. Five Illinoian molluscan faunas from the southern Great Plains. D, 1963, University of Michigan. 191 p.

Miller, Ben L. The geology of Prince George's County, Maryland and the District of Columbia. D, 1903, The Johns Hopkins University.

Miller, Brent L. Design and application of a multi-frequency data collection system for the study of rock properties. M, 1988, Montana College of Mineral Science & Technology. 112 p.

Miller, Bruce Calvin. Physical stratigraphy and facies analysis, Lower Cretaceous, Maverick Basin and Devils River Trend, Uvalde and Real counties, Texas. M, 1983, University of Texas, Arlington. 217 p.

Miller, Bruce E. A study of the authigenic minerals in the Blairmore Group, southern Alberta foothills. M, 1972, University of Calgary. 96 p.

Miller, Buford Maxwell. Cambrian stratigraphy of northwestern Wyoming. D, 1936, Columbia University, Teachers College.

Miller, Burford Maxwell. The fauna and age of the (Devonian) Haragan Shale of Oklahoma. M, 1931, Columbia University, Teachers College.

Miller, Buster W. The geology of the western Potato Hills, Pushmataha and Latimer counties, Oklahoma. M, 1955, University of Oklahoma. 55 p.

Miller, C. Petrologic study of the intrusive granitic rocks of the Sweetwater Wash area, Old Woman Mts., California. M, 1973, George Washington University.

Miller, C. F. Alkali-rich monzonites, California; origin of near silica-saturated alkaline rocks and their significance in a calc-alkaline batholithic belt. D, 1977, University of California, Los Angeles. 298 p.

Miller, C. K. Geology of the Mindamar (Sterling) deposits, Nova Scotia. M, 1978, Dalhousie University. 223 p.

Miller, Carl John. A petrographic study of the (Mississippian) Warsaw Formation of the St. Louis area (Missouri). M, 1939, St. Louis University.

Miller, Carol Pomering. Leakage through clay liners (groundwater, landfills, soils). D, 1984, University of Michigan. 292 p.

Miller, Carolyn Ann. Sediment and nutrient inputs to the marshes surrounding Fourleague Bay, Louisiana. M, 1983, Louisiana State University. 68 p.

Miller, Cass Timothy. Modeling of sorption and desorption phenomena for hydrophobic organic contaminants in saturated soil environments. D, 1984, University of Michigan. 435 p.

Miller, Charles David. Alteration and geochemistry of the Monroe known geothermal resource area. M, 1976, University of Utah. 120 p.

Miller, Charles Elden, Jr. A reconnaissance survey of bryozoan distribution within the Keyser Limestone (Silurian-Devonian) of central Pennsylvania. D, 1979, Pennsylvania State University, University Park. 226 p.

Miller, Charles N. A study of the volcanic intrusives of the White Mountain Complex (Tertiary), Park County, Wyoming. M, 1971, Wayne State University.

Miller, Charles Nash, Jr. The evolution of the fern family Osmundaceae. D, 1965, University of Michigan. 229 p.

Miller, Charles Parker. Geochemical and biogeochemical prospecting for nickel. D, 1957, Stanford University. 56 p.

Miller, Clarence Edmund. Certain problems connected with the mineralogical analysis of kaolin clays and the introduction of a new method of mechanical analysis. M, 1926, University of North Carolina, Chapel Hill. 83 p.

Miller, Clarence J. Geology of portions of the Red Mountain and San Francisquito quadrangles, California. M, 1952, University of California, Los Angeles.

Miller, Cliff Q., Jr. The subsurface geology of Vermilion Parish, Louisiana. M, 1954, University of Houston.

Miller, Clifford Daniel. Chronology of neoglacial moraines in the Dome Peak area, north Cascade Range, Washington. M, 1967, University of Washington. 37 p.

Miller, Clifford Daniel. Quaternary glacial events in the northern Sawatch range, Colorado. D, 1971, University of Colorado. 139 p.

Miller, Cydney Michele. Mesozoic stratigraphy of the Big Maria, Little Maria, Arica and Riverside mountains, and the Palen Pass area, Riverside County, California. M, 1981, San Diego State University.

Miller, Cynthia Kay. Thermal-mechanical controls on seismicity in the San Andreas fault zone of Northern and Central California. M, 1988, Pennsylvania State University, University Park. 50 p.

Miller, D. J. An acetylene inhibition method for the in situ measurement of denitrification rates in aquifers. M, 1986, University of Waterloo. 79 p.

Miller, D. N. An ecological study of foraminifera of Mason Inlet, North Carolina (T951). M, 1950, University of Missouri, Rolla.

Miller, Dale Everett. Seismic investigation of geological structure bordering the Caribbean island arc. M, 1957, Rice University. 57 p.

Miller, Dale Everett. Two dimensional seismic model study of near-vertical reflections from thin layers. D, 1962, University of Utah. 156 p.

Miller, Daniel D. Holocene sediments and microfauna of Cat Island, Mississippi. M, 1972, University of Nebraska, Lincoln.

Miller, Daniel J. Transverse isotropy; some consequences for travel time inversion and models of the oceanic crust. M, 1987, University of Hawaii. 54 p.

Miller, Daniel Newton, Jr. Petrology of Pierce Canyon redbeds; Delaware Basin, Texas and New Mexico. D, 1955, University of Texas, Austin.

Miller, David. Inversion of surface waves for upper mantle structure using phase velocities and waveforms. M, 1987, Purdue University.

Miller, David I. Foraminiferal and lithologic characteristics through the zone of Midway-Wilcox contact in Bastrop, Williamson, and Milam counties, Texas. M, 1962, Texas A&M University. 64 p.

Miller, David MacArthur. Deformation associated with Big Bertha Dome, Albion Mountains, Idaho. D, 1978, University of California, Los Angeles. 323 p.

Miller, David Martin. An evaluation of miscible displacement as a method for determining sorption equilibria and kinetics in the phosphate/goethite system. D, 1988, University of Georgia. 193 p.

Miller, David W. Potential ground-water storage capacity of the upper Pleistocene deposits in Kings County, New York. M, 1953, Columbia University, Teachers College.

Miller, David Wayne. Petrology and origin of migmatites in the Bryant Pond area, northwestern Maine. M, 1979, University of Wisconsin-Madison.

Miller, Debra Janel. Ridge and swale distribution of foraminifera on the continental shelf. M, 1979, University of Virginia. 82 p.

Miller, Delmon W. Distribution of silver, lead, and copper within the Curdsville Member of the Lexington Limestone (Middle Ordovician) of central Kentucky. M, 1969, University of Kentucky. 56 p.

Miller, Don Adair, Jr. Geology of the Leonia Knob area, Boundary County, Idaho. M, 1973, University of Idaho. 103 p.

Miller, Don E. Stratigraphy of the outcropping Pennsylvanian rocks in Miami County, Kansas. M, 1963, University of Kansas. 177 p.

Miller, Don John. Petrology of the Waynesburg Sandstone in the Dunkard Basin. M, 1942, University of Illinois, Urbana.

Miller, Don R. Stratigraphy of the lower Conemaugh Group in part of the North Potomac Basin. M, 1964, West Virginia University. 66 p.

Miller, Donald David. A study of transverse anisotropy in a Devonian shale. M, 1979, University of Illinois, Urbana. 99 p.

Miller, Donald George, Jr. Investigation of a carbonate sediment core from Exuma Sound, Bahamas. M, 1969, University of Illinois, Urbana.

Miller, Donald Spencer. The isotopic geochemistry of uranium, lead and sulfur in the Colorado Plateau uranium ores. D, 1960, Columbia University, Teachers College. 143 p.

Miller, Donn William. Hydrogeologic analysis of shallow hole temperatures at Allen Springs and Lee Hot Springs, Churchill County, Nevada. M, 1978, University of Nevada. 73 p.

Miller, Donna L. Porosity, diagenesis, and the source of silica for cement in the Clinton Sandstone (Silurian) in a core taken from the subsurface of Stark County, Ohio. M, 1982, Bowling Green State University. 126 p.

Miller, Douglas Robert. The effect of diversion water on the channel morphology of Fountain Creek, Colorado. D, 1987, University of Colorado. 128 p.

Miller, Duane Jay. Carbonatite genesis; a geochemical link between carbonatite and silicate magmas from Brava, Cape Verde Islands. M, 1988, University of Texas, Arlington. 163 p.

Miller, Edward Buford. The transportation of sand and finer clastics by running water. M, 1934, University of Oklahoma. 57 p.

Miller, Edward James. A stress strain time model for soil. D, 1977, University of California, Davis. 168 p.

Miller, Edward T. Geomagnetic measurements in the Gulf of Mexico and in the vicinity of Caryn Peak. D, 1956, Columbia University, Teachers College.

Miller, Edward T. In-shore marine magnetic investigations; the area from New Jersey to Cape Cod. M, 1952, Columbia University, Teachers College.

Miller, Edwin L. Report of the Matahambre Mine, Pinar del Rio, Cuba. M, 1925, University of Missouri, Rolla.

Miller, Edwin Lawrence, Jr. The origin, occurrence and mining of talc in Macon, Cherokee and Swain counties, North Carolina. M, 1946, North Carolina State University. 29 p.

Miller, Eldon S. Optimum sizes for urban complexes and components of an urban hierarchy. D, 1968, George Washington University.

Miller, Elizabeth Jean. The Buckeye Pluton; a peraluminous two-mica granite. M, 1987, Arizona State University. 106 p.

Miller, Elizabeth Louise. Geology of the Victorville region, California. D, 1978, Rice University. 292 p.

Miller, Elizabeth Louise. Structural and stratigraphic relationships between the Numidian Formation and underlying formations in the westernmost Mogod Mountains, northern Tunisia. M, 1976, Rice University. 95 p.

Miller, Elizabeth V. Graphical and digital slope stability analyses for Giles County, Virginia. M, 1985, Virginia Polytechnic Institute and State University.

Miller, Elliot White. Geology and petrology of an area intersected by latitude 28, 40'N., and longitude 102, 30'W., Coahuila, Mexico. M, 1963, Massachusetts Institute of Technology. 72 p.

Miller, Eric G. The geology of a core complex in the southern Snake Range. M, 1984, Stanford University.

Miller, Erick W. B. Laramide basement deformation in the northern Gallatin Range and southern Bridger Range, Southwest Montana. M, 1987, Montana State University. 78 p.

Miller, Ernest George. Subsurface stratigraphic and statistical analysis, Cherokee Marmaton, and Kansas City-lansing groups (Pennsylvanian), Major and Woods counties, Oklahoma. M, 1970, University of Oklahoma. 58 p.

Miller, Floyd E., Jr. The atmosphere of Jupiter; experimental formation of Jovian colors. D, 1976, Pennsylvania State University, University Park. 85 p.

Miller, Forrest J. The relation of geology to coal stripping in southeastern Kansas. M, 1918, University of Kansas. 28 p.

Miller, Francis Xavier. Spore analysis of the Dawson Coal. M, 1961, University of Tulsa. 98 p.

Miller, Franklin S. The petrology of the San Marcos Mountain gabbro, San Luis Rey Quadrangle in California. D, 1934, Harvard University.

Miller, Fred Key. Structure and petrology of the southern half of the Plomosa Mountains, Yuma County, Arizona. D, 1966, Stanford University. 173 p.

Miller, Frederick Powell. Physical, chemical, and mineralogical properties related to the micromorphology of the Canfield silt loam; a Fragiudalf. D, 1965, Ohio State University. 228 p.

Miller, G. H. Glacial and climatic history of northern Cumberland Peninsula, Baffin Island, Canada, during the last 10,000 years. D, 1975, University of Colorado. 253 p.

Miller, Gardner Burnham. Geology of area east of Page Mill Road. M, 1942, Stanford University.

Miller, George H. Geology of the Bee Branch-Mill Creek area, Mason County, Texas. M, 1957, Texas A&M University.

Miller, Gerald Matthew. Post-Paleozoic structure and stratigraphy, Blue Mt. Area, S. W. Utah. M, 1958, University of Washington. 58 p.

Miller, Gerald Matthew. The pre-Tertiary structure and stratigraphy of southern Wah Wah Mountains, southwestern Utah. D, 1960, University of Washington. 170 p.

Miller, Gerard Roland, Jr. Fluoride in sea water; distribution in the North Atlantic Ocean and formation of ion pairs with sodium. D, 1974, University of Rhode Island.

Miller, Glen A. The geology of the Middle Ordovician of the Knoxville Belt, Concord Quadrangle, Loudon County, Tennessee. M, 1953, University of Tennessee, Knoxville. 61 p.

Miller, Gregory Radford. Petrology and diagenesis of the lower Wilcox sandstones, North Milton Field area, Harris County, Texas. M, 1982, Texas A&M University. 152 p.

Miller, Greta E. Sedimentology, depositional environment and reservoir characteristics of the Kern River Formation, southeastern San Joaquin Basin. M, 1986, Stanford University. 80 p.

Miller, H. F., II. Debris flows in the vicinity of Boulder, Colorado. M, 1979, University of Colorado.

Miller, H. N. An ecological study of foraminifera of Mason Inlet, North Carolina. M, 1950, University of Missouri, Rolla.

Miller, H. T. Sedimentology and petrology of the Javelina Formation, Big Bend National Park, Brewster County, Texas. M, 1978, West Texas State University. 71 p.

Miller, Halsey W., Jr. Stratigraphic and paleontologic studies of the Niobrara Formation (Cretaceous) in Kansas. D, 1958, University of Kansas. 180 p.

Miller, Harold Ellis, Jr. A gravity survey in the Triassic lowlands, Berks and North Chester counties, Pa. M, 1963, Pennsylvania State University, University Park. 83 p.

Miller, Hayden Daniel. A magnetic survey of the Sardinia, Ohio, Quadrangle. M, 1951, University of Cincinnati. 44 p.

Miller, Henry J. Calibration of long period seismographs at thirteen stations throughout the world. M, 1963, Columbia University. 112 p.

Miller, Henry Joseph. Oklahoma earthquake of April 9, 1952. M, 1953, St. Louis University.

Miller, Howard W. The geology of the Dakota Hogback, South Park, Colorado. M, 1937, Northwestern University.

Miller, Hugh G. A gravity survey of eastern Notre Dame bay, Newfoundland. M, 1970, Memorial University of Newfoundland. 84 p.

Miller, Hugh Gordon. An analysis of geomagnetic variations in western British Columbia. D, 1973, University of British Columbia.

Miller, Iori P. An electron microprobe survey and Hume-Rothery application to naturally occurring tetrahedrite minerals. M, 1986, Queen's University. 146 p.

Miller, J. A. The suspended sediment system in the Bay of Fundy. M, 1966, Dalhousie University.

Miller, J. K. Geochemical dispersion over massive sulphides within the zone of continuous permafrost, Bathurst, District of Mackenzie, N.W.T. M, 1978, University of British Columbia.

Miller, J. W., Jr. The ore mineralogy of the Cofer property, Louisa County, Virginia; a volcanogenic massive sulfide deposit. M, 1978, Virginia Polytechnic Institute and State University.

Miller, Jack Edward. Mapping the glacial drainage patterns of the Saginaw Lobe from Landsat imagery. M, 1981, Old Dominion University. 116 p.

Miller, Jack H. L. Geology of the central part of the Callaghan Creek Pendant, southwestern British Columbia. M, 1979, University of British Columbia.

Miller, Jacquelin N. The present and the past molluscan faunas and environments of four Southern California coastal lagoons. M, 1966, University of California, San Diego.

Miller, James A. Quaternary history of the Sangamon River drainage system, central Illinois. M, 1972, University of Illinois, Urbana. 68 p.

Miller, James A. Stratigraphic and structural setting of the middle Miocene Pungo River Formation of North Carolina. D, 1971, University of North Carolina, Chapel Hill. 82 p.

Miller, James Anderson. Geology of the southern half of the Farrington, North Carolina, Quadrangle. M, 1963, North Carolina State University. 71 p.

Miller, James Andrew. Nacatoch (Upper Cretaceous) foraminifera of High Bluff at Arkadelphia, Arkansas. M, 1965, Louisiana Tech University.

Miller, James B. Sedimentation studies in the Sabino Canyon area near Tucson, Arizona. M, 1961, University of Arizona.

Miller, James Bennett. The flint-like slates of the Jones-Ford Quadrangle. M, 1921, University of North Carolina, Chapel Hill. 17 p.

Miller, James Duane, Jr. The geology and petrology of anorthositic rocks in the Duluth Complex, Snowbank Lake Quadrangle, northeastern Minnesota; (Volumes I and II). D, 1986, University of Minnesota, Minneapolis. 583 p.

Miller, James E. An evaluation of gamma ray well logging in glacial materials and bedrock of southwestern Ohio and southeastern Indiana. M, 1976, Miami University (Ohio). 86 p.

Miller, James Frederick. Conodont evolution and biostratigraphy of the upper Cambrian and lowest Ordovician (Texas, Oklahoma, Nevada and Utah). D, 1971, University of Wisconsin-Madison.

Miller, James Fredrick. Conodont fauna of the Notch Peak Limestone (Cambro-Ordovician) from the House Range, Utah. M, 1968, University of Wisconsin-Madison.

Miller, James H. Subsurface geology of the East Columbia area, Kingfisher and Logan counties, Oklahoma. M, 1959, University of Oklahoma. 64 p.

Miller, James K. Multistage dolomitization of the Portoro Limestone, Liguria, Italy. M, 1988, University of Texas, Austin. 215 p.

Miller, James L. The geology and geochemistry of late Eocene lignites and associated sediments in Tipton and Lauderdale counties, Tennessee. M, 1981, Vanderbilt University. 106 p.

Miller, James R. Geotechnical evaluation procedures for nuclear power plant siting, southern Basin and Range Province. M, 1977, Colorado School of Mines. 135 p.

Miller, Jeffery Allen. Uranium potential of Lower Permian arkosic facies, northern Kiowa County, Oklahoma. M, 1981, Oklahoma State University. 65 p.

Miller, Jeffrey Peter. Very low frequency electromagnetics and geochemistry in exploration for chromite mineralization, Soldiers Delight chromite district, Maryland. M, 1987, George Washington University. 129 p.

Miller, Jerry Russell. Sediment storage, transport, and geochemistry in semiarid fluvial and eolian systems; applications to long-term stability and potential dispersal patterns of uranium tailings in the Grants mineral belt, New Mexico. M, 1985, University of New Mexico. 160 p.

Miller, Jesse W. Drumlins in the Oswego, Weedsport, and Auburn, New York, quadrangles. D, 1970, Syracuse University.

Miller, Jim Patrick. A preliminary investigation of the distribution of miospore assemblages in the Summit Coal of north-central Missouri. M, 1964, University of Missouri, Columbia.

Miller, John. Geology and groundwater resources of Saline County, Missouri (Paleozoic). M, 1967, University of Missouri, Rolla.

Miller, John Agnew, III. Occurrences of vivianite in Minnesota lake sediments. M, 1978, University of Minnesota, Minneapolis. 50 p.

Miller, John Albert. A hydrodynamic approach to hydrothermal ore deposits in sedimentary environments. M, 1984, University of New Brunswick. 117 p.

Miller, John Charles. Geologic conditions affecting the accumulation of oil and gas in the Beggs District, Oklahoma. M, 1921, University of Missouri, Rolla.

Miller, John Charles. The North and South McCallum Anticline, Jackson County, Colorado. D, 1933, George Washington University.

Miller, John Collins. An investigation of the use of the spectrograph for correlation in limestone rock. M, 1953, University of Texas, Austin.

Miller, John David. Determination of temperature of fluorite formation by fluid inclusion thermometry, east Tennessee Zinc District. M, 1968, University of Tennessee, Knoxville. 41 p.

Miller, John David. The formation of endellite and halloysite clay in argillic hydrothermal and supergene alteration processes. D, 1979, University of Utah. 164 p.

Miller, John Donley. Paleomagnetism applied to tectonic problems in the Appalachians of North America. D, 1988, Columbia University, Teachers College. 153 p.

Miller, John J. The distribution, abundance, and nature of radioactive materials in the subsurface of the McCaslin Formation, McCaslin District, northeastern Wisconsin. M, 1980, Bowling Green State University. 67 p.

Miller, John P. A portion of the system calcium carbonate-carbon dioxide-water, with geological implications. D, 1951, Harvard University.

Miller, John William, Jr. Statistical modeling of the Austinville Mine; a guide to exploration. D, 1985, University of Georgia. 155 p.

Miller, Joseph H. Geology of the area southwest of Canon City, Fremont County, Colorado. M, 1951, Colorado School of Mines. 48 p.

Miller, Judith M. Growth-form analysis and paleoecology of the corals of the Lower Mississippian Lodgepole Formation, Bear River Range, north-central Utah. M, 1977, Utah State University. 166 p.

Miller, Julia Mary Gertrude. Stratigraphy and sedimentology of the upper Proterozoic Kingston Peak Formation, Panamint Range, eastern California. D, 1983, University of California, Santa Barbara. 417 p.

Miller, Keith Brady. A temporal hierarchy of paleoecologic and depositional processes across a Middle Devonian epeiric sea. D, 1988, University of Rochester. 253 p.

Miller, Keith Brady. Paleomicroecologic analysis of depositional environments and sequences, "Pleurodictyum Zone", Ludlowville Formation, Middle Devonian of western New York. M, 1982, SUNY at Binghamton. 131 p.

Miller, Keith Ray. Petrology, hydrothermal mineralogy, stable isotope geochemistry and fluid inclusion geothermometry of Borehole Mesa 31-1, East Mesa geothermal field, Imperial Valley, California. M, 1980, University of California, Riverside. 113 p.

Miller, Kenneth George. Late Paleogene (Eocene to Oligocene) paleoceanography of the northern North Atlantic. D, 1983, Massachusetts Institute of Technology. 92 p.

Miller, Kenneth Joseph. Paleontological study of an Appalachian deep test well, Wood County, West Virginia. M, 1957, University of Pittsburgh.

Miller, Kenneth W. Subsolidus phase relations in the system $BaCO_3$ - $CaCO_3$. M, 1979, University of Houston.

Miller, Laurence S. Subsurface investigation of Early Pennsylvanian strata, SW Nebraska. M, 1966, University of Nebraska, Lincoln.

Miller, Lawrence F. Geology of part of Okmulgee County, Oklahoma. M, 1957, Florida State University.

Miller, Lee Durward. Steaming and warm ground in Yellowstone National Park; their location, geophysics, vegetation and mapping with aerial multispectral imagery. D, 1968, University of Michigan. 198 p.

Miller, Leo J. The chemical environment of pitchblende. D, 1955, Columbia University, Teachers College.

Miller, Leroy C. A study of peneplain slopes. M, 1924, Columbia University, Teachers College.

Miller, Lewis Ruthardt. Englacial structures of the Vaughan Lewis icefall, and related observations on the Juneau icefield, Alaska. M, 1970, Michigan State University. 82 p.

Miller, Loye Holmes. Contributions to the avian paleontology from the Pacific Coast of North America. D, 1912, University of California, Berkeley. 54 p.

Miller, Lucile. Geology of the filled valleys of western Kentucky. M, 1937, University of Cincinnati. 39 p.

Miller, M. Stratigraphy and depositional systems of carboniferous rocks in Southwest Virginia. M, 1973, West Virginia University.

Miller, M. A. Maysvillian (Upper Ordovician) chitinozoans from the Cincinnati region of Ohio, Indiana, and Kentucky. M, 1976, Ohio State University. 251 p.

Miller, M. A. Stratigraphic relations of the Austin Chalk (Upper Cretaceous) in central Texas. M, 1978, Baylor University. 81 p.

Miller, M. B. F. Utility of trace fossils in paleoenvironmental interpretation; the distribution of biogenic structures in three modern and ancient nearshore environments. D, 1977, University of California, Los Angeles. 389 p.

Miller, M. F. Paleoenvironmental analysis of the Tonoloway and lower Keyser Formation. M, 1971, George Washington University.

Miller, Marilyn Sue. Ammonoid biostratigraphy of the Gene Autry Shale (Morrowan-Pennsylvanian), south-central Oklahoma. M, 1985, University of Arkansas, Fayetteville. 130 p.

Miller, Mark A. Effect of temperature on oil-water relative permeabilities of unconsolidated and consolidated sands. D, 1983, Stanford University. 93 p.

Miller, Mark E. Hydrogeologic characteristics of central Oahu subsoil and saprolite; implications for solute transport. M, 1987, University of Hawaii. 231 p.

Miller, Mark E. The use of quartz grain cathodoluminescent colors for interpreting the provenance of the Jackfork Sandstone, Arkansas. M, 1985, University of Cincinnati. 145 p.

Miller, Mark L. Cell parameter systematics of the binary silicate olivines; methods for the determination of composition and intracrystalline cation ordering. M, 1985, Virginia Polytechnic Institute and State University.

Miller, Martin C. Laboratory and field investigations on the movement of sand tracer under the influence of water waves; ripple development and longitudinal spreading of tracer material. D, 1978, Oregon State University. 123 p.

Miller, Mary Helen Alexander. Mechanical and statistical analysis of grain-size distribution and heavy mineral study of some Tertiary sediments from west-central Wyoming. M, 1954, University of Missouri, Columbia.

Miller, Mary Meghan. Tectonic evolution of late Paleozoic island arc sequences in the Western U.S. Cordillera; with detailed studies from the eastern Klamath Mountains, Northern California. D, 1987, Stanford University. 261 p.

Miller, Maynard M. Observations on the regimen of the glaciers of Icy Bay and Yakutat Bay, Alaska. M, 1948, Columbia University, Teachers College.

Miller, Michael Allan. Dolomitization and porosity evolution. D, 1988, Michigan State University. 168 p.

Miller, Michael Byron. A petrographic and fluid inclusion study of some scheelite-bearing quartz veins, Cabarrus County, North Carolina. M, 1982, University of North Carolina, Chapel Hill. 84 p.

Miller, Michael Carl. Uranium occurrence in syenitic rocks of central Arkansas. M, 1987, Stephen F. Austin State University. 81 p.

Miller, Michael E. Chemical and mechanical manifestations by surface active agents and brine pore fluid environments in stable sliding of orthoquartzite. M, 1986, Indiana University, Bloomington. 96 p.

Miller, Michael E. The chemical mechanism of subcritical crack growth in synthetic quartz. D, 1988, Indiana University, Bloomington. 113 p.

Miller, Michael Eugene. Uranium roll front study in Atascosa County, Texas. M, 1979, Texas A&M University. 71 p.

Miller, Michael J. Devonian stratigraphy of the northern Providence Mountains, San Bernardino County, California. M, 1983, University of California, Riverside. 121 p.

Miller, Michael L. A petrographic study of the Devonian-Lower MissISippian black shales in eastern Kentucky. M, 1978, University of Kentucky. 108 p.

Miller, Michael Schas. The bedrock geology of the southeast quarter of Mount Steel quadrangle, Washington. M, 1967, University of Washington. 78 p.

Miller, Michael Vernon. Soil-geomorphology and soil-stratigraphy in the lower Pomme de Terre River valley, southwestern Missouri. D, 1984, University of Illinois, Urbana. 222 p.

Miller, Michele Gean. Stratigraphy of the Paleogene sequence, eastern Santa Rosa Hills, western Santa Ynez Mountains, Santa Barbara County, California. M, 1983, University of California, Santa Barbara. 339 p.

Miller, Murray L. Geology of the St. Simeon-Tadoussac map area, Quebec. D, 1953, University of Minnesota, Minneapolis. 182 p.

Miller, Murray L. Stratigraphy of the Rouge River-Oshawa area, Ontario. M, 1942, University of Toronto.

Miller, Nancy Jill. Stratigraphy, structure and metamorphism of the Calaveras Complex in part of Amador and Calaveras counties, California. M, 1982, University of California, Santa Cruz.

Miller, Nathaniel R. The effect of humic materials on the trace element chemistry of ooids. M, 1985, Bowling Green State University. 144 p.

Miller, Nevin Lane. The iron absorption index; a baseline-formulated technique for the assessment of limonite in thematic mapper imagery. M, 1985, Stanford University. 144 p.

Miller, Norman F. Geochemistry of the Academy Pluton, Fresno County, California. M, 1976, University of California, Santa Barbara.

Miller, Norton George. Late- and postglacial vegetation change in southwestern New York state. D, 1969, Michigan State University. 325 p.

Miller, P. Preferred orientation in chloritoid. M, 1976, Ohio State University. 56 p.

Miller, Paul M. The evolution of tin bearing fluids in the Devonian granitoid rocks of the Halifax Pluton, Nova Scotia. M, 1979, University of Windsor. 131 p.

Miller, Paul Melby. Stratigraphy and petrography of the Greenfield and Tymochtee formations of southern Ohio. M, 1956, Ohio State University.

Miller, Paul R. Mid-Tertiary stratigraphy, petrology, and paleogeography of Oregon's central western Cascade Range. M, 1984, University of Oregon. 187 p.

Miller, Paul Theodore. Pleistocene gravels of northeastern Iowa. M, 1930, University of Iowa. 172 p.

Miller, Paul Theodore. The Pleistocene gravels of Iowa. D, 1932, University of Iowa. 373 p.

Miller, Peter L. Some coccolithophorids and related nannofossils from the Upper Cretaceous of southwestern Arkansas. M, 1962, University of Minnesota, Minneapolis. 77 p.

Miller, R. A. A system for the quantification of environmental geology in land-use planning as applied to the Kingston Springs Quadrangle, Tennessee. M, 1974, Vanderbilt University.

Miller, R. H. Silurian conodont biostratigraphy of the southwestern Great Basin. D, 1975, University of California, Los Angeles. 123 p.

Miller, R. W. Mercury in freshwater sediments. D, 1975, University of Georgia. 85 p.

Miller, Ralph L. Stratigraphy of the (Ordovician) Jacksonburg Limestone, with an appended chapter on Martinsburg limestones in eastern Pennsylvania. D, 1937, Columbia University, Teachers College.

Miller, Ralph Rillman, III. Determination of natural exchange constants of oxygen in oceanic sulphates. M, 1971, United States Naval Academy.

Miller, Ralph Wayne. Geology of the Freedom Quadrangle, Owen and Greene counties, Indiana. M, 1960, Indiana University, Bloomington. 60 p.

Miller, Randall Francis. Palaeoentomological analysis of a postglacial site in northwestern New York State. M, 1980, University of Waterloo.

Miller, Randall Francis. Stable isotopes of carbon and hydrogen in the exoskeleton of insects; developing a tool for paleoclimatic research. D, 1984, University of Waterloo. 194 p.

Miller, Randy Robert. Age and petrological relationships of some igneous-textured and gneissic alkaline rocks in the Haliburton-Bancroft area. D, 1985, University of Toronto.

Miller, Randy Robert. Petrology of nepheline-bearing rocks in Glamorgan and Monmouth townships, Ontario. M, 1979, University of Toronto.

Miller, Randy Vernon. The environmental significance of shale properties. M, 1973, University of Tulsa. 103 p.

Miller, Raymond F. Variations of Seminula argentea (Pennsylvanian, eastern Kansas). M, 1912, University of Kansas.

Miller, Richard C. A thermally convecting fluid heated non-uniformly from below. D, 1968, Massachusetts Institute of Technology. 182 p.

Miller, Richard D. Geology of a part of the Afton fluorite mining district, San Bernardino County, California. M, 1950, Stanford University. 26 p.

Miller, Richard E. Regression analysis of carbonate diagenesis; an application to the the Miocene of Saipan. M, 1988, University of Maryland.

Miller, Richard George. The metallogeny of uranium in the Great Bear Batholith complex, Northwest Territories. D, 1982, University of Alberta. 272 p.

Miller, Richard H. Possible recharge to the Dakota Sandstone aquifer from the Juro-Cretaceous Sandstone, Black Hills area (South Dakota). M, 1972, South Dakota School of Mines & Technology.

Miller, Richard Harry. Silurian conodonts from the Starcke Limestone of central Texas. M, 1967, University of California, Los Angeles.

Miller, Richard Lincoln. The physiological ecology of an estuarine clone of the marine diatom Nitzschia americana. D, 1984, North Carolina State University. 192 p.

Miller, Richard Nelson. Geology of the South Moccasin Mountains, Fergus County, Montana. M, 1954, Montana College of Mineral Science & Technology. 108 p.

Miller, Robert Bruce. Structure and petrology of the Ingalls Peridotite and associated pre-Tertiary rocks southwest of Mount Stuart, Washington. M, 1976, University of Washington. 90 p.

Miller, Robert Bruce. Structure, petrology and emplacement of the ophiolitic Ingalls Complex, North-central Cascades, Washington. D, 1980, University of Washington. 422 p.

Miller, Robert C. Lithofacies analysis of sandstones in the Dowelltown member of the Chattanooga shale (Devonian), northern Highland Rim, Macon County, Tennessee. M, 1968, University of Tennessee, Knoxville. 52 p.

Miller, Robert E. Geochemical study of bio-organomontmorillonite and synthetic sea water and fresh water system. D, 1968, Texas A&M University.

Miller, Robert E. Thermoluminescence and its relationship to the carbonate reservoir of the Hunton Group, Vaege Pool, Riley County, Kansas. M, 1963, University of Tulsa. 85 p.

Miller, Robert J. M. Geology and ore deposits of the Cedar Bay Mine area, Chibougamau District, Quebec. D, 1957, Universite Laval.

Miller, Robert Norman. A geochemical study of the inorganic constituents in some low-rank coals. D, 1977, Pennsylvania State University, University Park. 319 p.

Miller, Robert R. Cenozoic geology of the Dixon area, north-central New Mexico. M, 1956, University of Kansas. 106 p.

Miller, Robert S. Pre-Pennsylvanian rocks of New Mexico. M, 1955, Stanford University.

Miller, Roger Arthur. Sedimentary microstructures of surficial sediments in lower Green Bay, Lake Michigan. M, 1986, University of Wisconsin-Milwaukee. 254 p.

Miller, Roger Glenn. The geology of southeastern Callaway County, Missouri. M, 1951, University of Missouri, Columbia.

Miller, Roswell, III. The structure and petrology of the Webster-Addie ultrabasic ring, Jackson County, N.C. D, 1951, Princeton University. 39 p.

Miller, Roy Malcolm. Iso-positional substitution in crystalline substances. D, 1960, Michigan State University. 107 p.

Miller, S. D. Chemistry of a pyritic strip-mine spoil. D, 1979, Yale University. 201 p.

Miller, Sarah B. History of the glacial landforms in the Deblois region, Maine. M, 1986, University of Maine. 87 p.

Miller, Stanley Mark. Determination of spatial dependence in fracture set characteristics by geostatistical methods. M, 1979, University of Arizona. 111 p.

Miller, Stanley Mark. Statistical and Fourier methods for probabilistic design of rock slopes. D, 1982, University of Wyoming. 221 p.

Miller, Stephen P. Observations and interpretation of the pole tide. M, 1973, Massachusetts Institute of Technology. 97 p.

Miller, Steven B. Application of complex trace attributes to reflection seismic data near Charleston, South Carolina. M, 1985, Virginia Polytechnic Institute and State University.

Miller, Susan T. Geology and mammalian biostratigraphy of a portion of the northern Cady Mountains, Mojave Desert, California. M, 1978, University of California, Riverside. 189 p.

Miller, Thomas Edward. Hydrochemistry, hydrology and morphology of the Caves Branch karst, Belize. D, 1982, McMaster University. 280 p.

Miller, Thomas N. A comparison of waterflood performance characteristics for Pennsylvanian sands. M, 1971, University of Oklahoma. 92 p.

Miller, Thomas P. A study of the Sioux Formation in the New Ulm area, Minnesota. M, 1961, University of Minnesota, Minneapolis. 75 p.

Miller, Thomas Patrick. Petrology of the plutonic rocks of west-central Alaska. D, 1971, Stanford University. 176 p.

Miller, Timothy Robert. The chemical variation in the upper Hance coal seam, southeastern Kentucky. M, 1984, University of Kentucky. 130 p.

Miller, Victor C. A quantitative geomorphic study of drainage basin characteristics in the Clinch Mountain area, Virginia and Tennessee. D, 1953, Columbia University, Teachers College.

Miller, Victor C. Pediments and pediment-forming processes near House Rock, Arizona. M, 1948, Columbia University, Teachers College.

Miller, Victor Van. The sedimentology and provenance of the Upper Cretaceous Rosario Formation and the upper Paleocene-lower Eocene Sepultura Formation, Baja California, Mexico. M, 1987, San Diego State University. 240 p.

Miller, Wade E. Mammalian fauna of the Pleistocene Palos Verdes Formation, California. M, 1962, University of Arizona.

Miller, Wade E. Pleistocene vertebrates of the Los Angeles Basin and vicinity (exclusive of Rancho La Brea). D, 1971, University of California, Berkeley. 124 p.

Miller, Wayne Davis. Pre-Cenozoic stratigraphy of Porvenir area, Presidio County, Trans-Pecos, Texas. M, 1957, University of Texas, Austin.

Miller, Wesley L. Hydrologic resources of the Everglades City area, Collier County, Florida. M, 1973, Wright State University. 60 p.

Miller, William C., III. Ecological units, paleocommunity structure, and ecological history of late Pleistocene deposits in Dare County, North Carolina. M, 1978, Duke University.

Miller, William Charles. Paleosynecologic history of the middle Pleistocene Flanner Beach Formation, eastern North Carolina; a study in community replacement. D, 1984, Tulane University. 382 p.

Miller, William Donald. Geology of the Cox Formation, Trans-Pecos, Texas. D, 1963, University of Missouri, Columbia. 144 p.

Miller, William Donald. The general geology of Moab Valley, Utah. M, 1959, Texas Tech University. 121 p.

Miller, William Frank. Pennsylvanian stratigraphy of western Kansas. M, 1955, University of Illinois, Urbana.

Miller, William Gossett. Relationships between minerals and selected trace elements in some Pennsylvanian age coals of northwestern Illinois. M, 1974, University of Illinois, Urbana.

Miller, William J. The crystalline limestones of Baltimore County, Maryland. D, 1905, The Johns Hopkins University.

Miller, William Lawrence. Petrology of the Nomlaki Tuff (upper Pliocene) in Yolo and Tehama counties, California. M, 1967, University of California, Davis. 89 p.

Miller, William Paul. Sequential extraction of Cu from soil components and Cu-amended soils. D, 1981, Virginia Polytechnic Institute and State University. 171 p.

Miller, William R. Chemical weathering and related controls on surface water chemistry in the Absaroka Mountains, Wyoming. D, 1974, University of Wyoming. 128 p.

Miller, William R. Geology of the Indian Rocks area, Carbon County, Wyoming, with particular reference to the evolution of mafic dikes. M, 1972, University of Wyoming. 46 p.

Miller, William Roger. A gravity investigation of the Porcupine Mountains and adjacent area, Ontonagon and Gogebic counties, Michigan. M, 1966, Michigan State University. 67 p.

Miller-Hoare, Martha Lynn. Gabbroic with associated culmate mafic and ultramafic rocks, a probable ophiolitic slice, near Stevens Creek Canyon, California. M, 1980, Stanford University. 108 p.

Millet, Marion T. Glaciation in the headwaters of Middle Boulder Creek (Boulder County), Colorado. M, 1956, University of Colorado.

Millett, John. Clay minerals of Lake Champlain (Vermont). M, 1967, University of Vermont.

Millette, Jacques Armand. Evolution of a newly reclaimed organic soil in southwestern Quebec. D, 1984, McGill University.

Millgate, Marvin Leroy. Geology of the Haystack Range, Goshen and Platte counties, Wyoming. M, 1964, University of Utah. 96 p.

Millholland, Madelyn Ann. Mineralogy and petrology of Precambrian metamorphic rocks of the Gravelly Range, southwestern Montana. M, 1976, Indiana University, Bloomington. 134 p.

Millhollen, Gary Lloyd. Melting and phase relations in nepheline syenites with H_2O and $H_2O + CO_2$. D, 1970, Pennsylvania State University, University Park. 93 p.

Millhollen, Gary Lloyd. The petrography of the basalts of the Cow Creek Lakes area, Malheur County, Oregon. M, 1965, University of Oregon. 72 p.

Millians, Robert Wilson. Drainage basin shape as a measurement of physiographic differences. M, 1963, University of Georgia. 103 p.

Millican, Richard S. The geology and petrology of the Mount Riley-Cox Pluton, (post-Cretaceous, probably Tertiary, Dona Ana County, New Mexico). M, 1971, University of Texas at El Paso.

Millice, Roy. Stratigraphy of the Green River Formation in the southeastern Bridger Basin, Wyoming. M, 1959, University of Wyoming. 89 p.

Milligan, Donald B. Marine geology of the Florida Straits. M, 1962, Florida State University.

Milligan, Donald Bristowe. The Kara Sea; geologic structure and water characteristics. D, 1981, College of William and Mary. 114 p.

Milligan, George Clinton. The geology of the Lynn Lake District, northern Manitoba. D, 1961, Harvard University.

Milligan, James Homer. Optimizing conjunctive use of groundwater and surface water. D, 1969, Utah State University. 165 p.

Milligan, S. D. Grouping of marine sediments using a multivariate analysis of seismic profiles. D, 1977, University of Rhode Island. 71 p.

Millikan, Carl E. The fissure system of El Potosi Mine... Santa Eulalia, Chihuahua, Mexico. M, 1925, University of Missouri, Rolla.

Millikan, Gregory Robert. Geology of a portion of the Sherman Mountain Quadrangle, Nevada. M, 1979, San Diego State University.

Milliken, Jeffrey V. Late Paleozoic and early Mesozoic geologic history of the northwestern Gulf Coast. M, 1988, Rice University.

Milliken, Kitty Lou. Petrology and burial diagenesis of Plio-Pleistocene sediments, northern Gulf of Mexico. D, 1985, University of Texas, Austin. 130 p.

Milliken, Kitty Lou. Silicified evaporate modules from the Mississippian rocks of southern Kentucky and northern Tennessee. M, 1977, University of Texas, Austin.

Milliken, Mark D. Geology of the Peter Dan Creek area, Okanogan County, Washington, with a gravity survey across the Koontzville Lineament. M, 1981, Eastern Washington University. 76 p.

Milliman, John Douglas. Recent marine sediments in Gray's Harbor, Washington. M, 1963, University of Washington. 172 p.

Milliman, John Douglas. The marine geology of Hogsty Reef, a Bahamian atoll. D, 1966, University of Miami. 307 p.

Milling, Marcus Eugene. Morphometric analysis of Clear Creek and Old Man Creek, Iowa and Johnson counties, Iowa. M, 1964, University of Iowa. 109 p.

Milling, Marcus Eugene. Petrology and petrography of glacial tills and their weathering profiles in southeastern Iowa. D, 1968, University of Iowa. 331 p.

Millington, Berton R. The physiography of Rhode Island as affected by the retreat of the late Wisconsin ice-sheet. M, 1930, Brown University.

Millison, Clark Drury. Geology of the Wellington Quadrangle of south-central Kansas. M, 1927, University of Oklahoma. 34 p.

Millitante, Priscilla Juan. A systematic study of the gastropod family Rissoinidae. D, 1961, Stanford University. 89 p.

Millman, Dean Beardsley. The stratigraphy and structure of the Caplinger Mills Quadrangle, Missouri. M, 1954, University of Iowa. 122 p.

Millon, Eric R. Water pollution, Red Feather lakes area (Larimer county), Colorado. M, 1970, Colorado State University. 77 p.

Mills, Alison M. The effects of aquifer dewatering on two ephemeral streams in southwestern San Juan Basin, New Mexico. M, 1985, Pennsylvania State University, University Park. 143 p.

Mills, B. A. Selective kinetic extractions of some northeastern equatorial Pacific pelagic sediments. M, 1978, University of Wisconsin-Madison.

Mills, Bradford Alan. Crystal chemistry and paragenesis of pegmatitic beryl. M, 1979, Stanford University. 149 p.

Mills, Donald E. Petrogenesis of some perlite in Peralta Canyon, Sandoval County, New Mexico. M, 1952, University of New Mexico. 50 p.

Mills, Earl Lee. The geology of the Parkhill area, Cherokee County, Oklahoma. M, 1951, University of Oklahoma. 83 p.

Mills, Earl R. Effects of water dispersions and water-soluble fractions of two crude and two processed oils on three marine algal species. D, 1974, Texas A&M University. 102 p.

Mills, Gary Lawrence. The chemical nature and geochemistry of dissolved copper-organic complexes in the Narragansett Bay estuary. D, 1982, University of Rhode Island. 164 p.

Mills, Herbert Cornell. Relative abundances of selected trace elements in sediments of streams draining specific rock types near Wilton in Granville County, North Carolina. M, 1966, North Carolina State University. 60 p.

Mills, Hugh H., Jr. The variation of drumlin form. M, 1972, University of North Carolina, Chapel Hill. 95 p.

Mills, Hugh Harrison. Sediment characteristics of some small temperate glaciers. D, 1975, University of Washington. 144 p.

Mills, J. W. A spectrographic study of wall rock alteration, Pachuca, Mexico. D, 1942, Massachusetts Institute of Technology. 157 p.

Mills, James Gordon, Jr. The geology and geochemistry of volcanic and plutonic rocks in the Hoover Dam 7 1/2 minute quadrangle, Clark County, Nevada, and Mohave County, Arizona. M, 1985, University of Nevada, Las Vegas. 119 p.

Mills, John Peter. Petrography of selected speleothems of carbonate caverns. M, 1965, University of Kansas. 47 p.

Mills, John Peter. Petrological studies in the Sakami-Lake greenstone belt of northwestern Quebec. D, 1974, University of Kansas. 220 p.

Mills, John Robert. Secondary clay minerals from a monzonite from northeast Louisiana. M, 1972, Northeast Louisiana University.

Mills, John Ross. A study of lakes in northeastern Vermont. M, 1949, Lehigh University.

Mills, Joseph Henry. A determination of the possible direction of the source of sediments in the Brereton Cycle of the Pennsylvanian System in Streator Quadrangle, Illinois. M, 1934, University of Kentucky. 66 p.

Mills, Lloyd Clarence. Some upper Eocene Ostracoda from the Conroe oil field, Montgomery County, Texas. M, 1936, University of Nebraska, Lincoln.

Mills, Patrick Clarence. Geology of the Lower Permian rocks of the Bird Spring Group. M, 1984, University of Illinois, Urbana. 162 p.

Mills, Rodger K. Petrography of uranium ore rolls in the Powder River basin, Wyoming. M, 1965, University of Wyoming. 65 p.

Mills, Thomas. Inner continental shelf sediments off New Hampshire. M, 1977, University of New Hampshire. 64 p.

Mills, W. C. Coupling stochastic and deterministic hydrologic models for decision-making. D, 1979, University of Arizona. 219 p.

Mills, William Glenn. Paleocene and Eocene climate of Southern California. M, 1985, San Diego State University. 94 p.

Millsop, Mark D. A quantitative analysis of shoreline erosion processes, Lake Sakakawea, North Dakota. M, 1985, University of North Dakota. 290 p.

Millward, Lewis G. Contributions to Canadian paleontology. M, 1928, University of British Columbia.

Millward, William. A new locality for Lyttonia richtofeni, Kayser, E. M., together with some observations concerning the world distribution of the genus Lyttonia. M, 1927, University of Pittsburgh.

Millwood, Lynn. Primary and secondary sources of quartz sand in the black shell turbidite, Hatteras abyssal plain; Fourier grain shape analysis. M, 1978, University of South Carolina.

Milne, Beulah L. The relation between earthquakes and weather. M, 1930, Columbia University, Teachers College.

Milne, Bonnie L. Petrology and depositional environment of the Spergen Formation, southeastern Iowa. M, 1979, University of Iowa. 156 p.

Milne, Ivan Herbert. A study of certain uranium-bearing minerals and compounds. D, 1951, University of Toronto.

Milne, Ivan Herbert. A study of chloritoid and ottrelite. M, 1948, University of Toronto.

Milne, Peter C. Prograde and retrograde metamorphism in the Carboniferous rocks of the Narragansett Bay area, Rhode Island. M, 1972, University of Rhode Island.

Milne, V. G. The non-radioactive heavy minerals of the Mississagi Conglomerate, Blind River, Ontario. M, 1959, University of Toronto.

Milne, Victor Gordon. The petrography and alteration of some spodumene pegmatites near Beardmore, Ontario. D, 1962, University of Toronto. 399 p.

Milne, W. G. Earthquake risk in Canada. D, 1965, University of Western Ontario.

Milne, Wendy. A comparison of reconstructed lake-level records since the mid-1800's of some Great Basin lakes. M, 1987, Colorado School of Mines. 207 p.

Milne-Home, William Alexander. Modelling mass transport in clastic sands and sandstones. D, 1985, University of Alberta. 347 p.

Milner, Carlos E., Jr. Geology and ore deposits of the Princess Blue Ribbon Mine, Camas County, Idaho. M, 1950, University of Idaho. 45 p.

Milner, Carlos E., Jr. The geology and ground water in Northumberland County, Virginia. M, 1948, University of Virginia.

Milner, Charles Porter. The application of surface active chemicals to water floods of water wet porous media. M, 1953, University of Texas, Austin.

Milner, Darwin Quigley. A restudy of existing graphic methods of interpreting magnetic data and their applications to interpreting the results of magnetic surveys across the Los Angeles Basin. M, 1949, California Institute of Technology. 26 p.

Milner, M. W. Geomorphology of the Klondike placer goldfields, Yukon Territory. D, 1977, McGill University.

Milner, Michael. Structure and metamorphism of the Merry Mountain area, Quebec, (Canada). M, 1969, Dalhousie University. 123 p.

Milner, Robert L. Geology and ore deposits of Barry Lake map-area, northern Quebec. D, 1940, McGill University.

Milner, Sam. Sedimentology of a sandstone-carbonate transition, lower San Andreas Formation (Middle Permian), Lincoln County, New Mexico. M, 1974, University of Wisconsin-Madison. 156 p.

Milner, William Collier. Petrology, stratigraphy and structure of the basal section of the Greenbrier Limestone (Upper Mississippian) in the Vadis field in Lewis and Gilmer County, West Virginia. M, 1968, West Virginia University.

Milnes, Peter Treadwell. Structural geology and metamorphic petrology of the Illabot Peaks area, Skagit County, Washington. M, 1977, Oregon State University. 118 p.

Milske, Jodi A. Stratigraphy and petrology of clastic sediments in Mystery Cave, Fillmore County, Minnesota. M, 1982, University of Minnesota, Minneapolis. 111 p.

Milton, Arnold Powell. Geology of Cajoncito area in Municipio de Guadalupe, Chihuahua, and Hudspeth County, Texas. M, 1964, University of Texas, Austin.

Milton, Arthur Alvern. Geology of the Circle Ridge-Spring Mountain area, Owl Creek Mountains, Wyoming. M, 1942, University of Iowa. 70 p.

Milton, Charles. Nepheline syenites and related rocks of the Franklin Furnace Quadrangle, New Jersey. D, 1929, The Johns Hopkins University.

Milton, Daniel Jeremy. Geology of the Old Speck Mountain Quadrangle, Maine. D, 1961, Harvard University.

Milton, Gwendolyn Margaret. Uranium series disequilibrium in rock-water systems of the Canadian Shield. D, 1985, University of Waterloo. 233 p.

Milton, Jesse W. The distribution of Recent foraminifera south of St. George Island, Florida. M, 1958, Florida State University.

Milton, William Billingslea, Jr. The Walnut Formation of central Texas. M, 1928, University of Texas, Austin.

Mims, Charles Van Horn. Anisotropy of magnetic susceptibility as a petrofabric and strain indicator in the Falls lineated gneiss and adjacent Raleigh Belt gneisses, Wake County, North Carolina. M, 1988, University of North Carolina, Chapel Hill. 69 p.

Mims, Jimmie Floyd. Location and evaluation of borrow material for beach nourishment, Melbourne Beach, Florida. M, 1975, University of Florida. 84 p.

Mims, Robert Lewis, Jr. Microfacies analysis of the La Tuna Formation (Morrowan) (lower Pennsylvanian), Vinton Canyon, El Paso County, Texas. M, 1971, University of Texas at El Paso.

Min, K. D. Analytical and petrofabric studies of experimental faulted drape-folds in layered rock specimens. D, 1974, Texas A&M University. 100 p.

Min, Kyoung Won. A geochemical study of granitic rocks from the Turtle Lake Quadrangle and vicinity, northeastern Washington. D, 1986, Colorado School of Mines. 120 p.

Min, Maung Myo. Petrography and alteration of the Kitt Peak area, Pima County, Arizona. M, 1965, University of Arizona.

Minard, Claude Russell. Late Quaternary beaches and coasts between the Russian River and Drakes Bay, California. D, 1971, University of California, Berkeley. 206 p.

Minard, James Pierson. The geology of Peapack-Ralston Valley in north central New Jersey. M, 1959, Rutgers, The State University, New Brunswick. 104 p.

Minatidis, Demitris George. A comparative study of trace element geochemistry and mineralogy of some uranium deposits of Labrador, and evaluation of some uranium exploration techniques in a glacial terrain. M, 1976, Memorial University of Newfoundland. 216 p.

Minch, John A. Stratigraphy and structure of the Tijuana-Rosarito Beach area, northwestern Baja California, Mexico. M, 1966, [University of California, San Diego].

Minch, John Albert. The late Mesozoic-early Tertiary framework of continental sedimentation, northern Peninsular Ranges, Baja California, Mexico. D, 1972, University of California, Riverside. 192 p.

Mincher, Albert Russel. Fauna of the Pascagoula Formation (Mississippi, Louisiana, Alabama). M, 1939, Louisiana State University.

Minck, Kathleen Marie. Facies analysis and depositional setting of the middle member of the Miocene Blanca Formation, Santa Cruz Island, California. M, 1982, University of California, Santa Barbara. 209 p.

Minck, Robert J. A seismic study of the Atlantic outer continental shelf, slope, and upper rise east of Delaware. M, 1978, University of Delaware.

Mindevalli, Oznur. Crust and upper mantle structure of Turkey and the Indian Sub-continent from surface wave studies. D, 1988, St. Louis University.

Mindheim, Bruce K. Trilobites and lithofacies of the Saukia Zone, southern Oklahoma. M, 1987, University of Missouri, Columbia. 114 p.

Mindling, Anthony Leo. Investigation of the relationship of the physical properties of fine grained sediments to land subsidence, Las Vegas Valley, Nevada. M, 1965, University of Nevada. 90 p.

Minear, John Wesley. Spectra and bi-spectra of seismic signals. D, 1964, Rice University. 122 p.

Miner, Ernest Lavon. Paleobotanical examinations of Cretaceous and Tertiary coals; 1, Cretaceous coals from Greenland; 2, Cretaceous and Tertiary coals from Montana. D, 1934, University of Michigan. 90 p.

Miner, Neil Alden. Pleistocene glaciation of the Gardiner, Mammoth Hot Springs, and Lava Creek regions, Yellowstone National Park, Wyoming. D, 1937, Iowa State University of Science and Technology.

Miner, Neil Alden. The origin and history of Green and Round lakes in Green Lake State Park at Fayetteville, New York. M, 1933, Syracuse University.

Minero, Charles John. Sedimentary environments and diagenesis of the El Abra Formation (Cretaceous), Mexico. D, 1983, SUNY at Binghamton. 367 p.

Minervini, George B. Quartz porphyry intrusives and associated propylitic breccias in the Modoc area, Butte District, Montana. M, 1975, Montana College of Mineral Science & Technology. 96 p.

Ming, Douglas Wayne. Chemical and crystalline properties of clinoptilolite in South Texas soils. D, 1985, Texas A&M University. 272 p.

Ming, James. Theoretical studies of refraction effects in terrestrial heat flow measurements. M, 1968, Virginia Polytechnic Institute and State University.

Ming, James D. Digital computer study of the heat flow due to buried intrusives. M, 1969, Virginia Polytechnic Institute and State University.

Ming, Li-chung. High pressure phases in the system of $FeO\text{-}MgO\text{-}SiO_2$ and their geophysical implications. D, 1974, University of Rochester. 134 p.

Ming, Li-chung. The green phlogopite in Talcville, New York. M, 1971, University of Rochester. 22 p.

Ming, Pam P. Chou. High pressure Debye-Scherrer X-ray diffraction method with full 360° dispersion. M, 1973, University of Rochester. 24 p.

Minick, J. N. Tertiary stratigraphy of southeastern Wyoming and northeastern Colorado. M, 1951, University of Wyoming. 53 p.

Minicucci, David Andrew. Physical stratigraphy of lower Pliocene to Pleistocene sediment from the eastern Alpha Cordillera, Amerasian Basin, central Arctic Ocean. M, 1979, University of Wisconsin-Madison.

Minier, Jeffrie D. A geothermal study in west-central New Mexico. D, 1987, New Mexico Institute of Mining and Technology. 229 p.

Mink, Leland Leroy. Evaluation of settling ponds as a mining wastewater treatment facility. D, 1972, University of Idaho. 250 p.

Mink, Leland Leroy. Water quality of the Coeur d'Alene River basin (northern Idaho). M, 1971, University of Idaho. 95 p.

Mink, Robert M. Geology of the Helena and Cahaba Valley faults from Lake Purdy to New Hope, Alabama. M, 1981, University of Alabama. 68 p.

Minke, Joseph Garrett. Geology of the Hancock and Stotler's Crossroads quadrangles in West Virginia. M, 1964, West Virginia University.

Minke, Joseph Garrett. Petrology of dolomites in the Everton Formation, Marion County, Arkansas. D, 1969, University of Missouri, Columbia. 107 p.

Minkel, Donald H. Geometric analysis of an analog fold model. M, 1982, Washington State University.

Minkin, Steven C. Pennsylvanian deltaic sediments of the northern Cumberland Plateau, Tennessee. M, 1977, University of Tennessee, Knoxville. 141 p.

Minkofsky, Anna. The metamorphism of the (Precambrian-Ordovician) Hudson River slates from Poughkeepsie to Dover Furnace, New York. M, 1937, Columbia University, Teachers College.

Minnery, Gregory A. Bioerosion of branching hermatypic corals; a comparison of the coral-bearing fauna in varying reef subenvironments. M, 1980, University of Cincinnati. 126 p.

Minnery, Gregory Andrew. Distribution, growth rates, and diagenesis of coralline algal structures on the Flower Garden Banks, northwestern Gulf of Mexico. D, 1984, Texas A&M University. 193 p.

Minnich, Gene W. Processing and interpretation of Vibroseis and downhole air gun data to evaluate seismic energy propagation through clinker, Campbell County, Wyoming. M, 1988, University of Wyoming. 73 p.

Minnick, Edward. Stratigraphy and structure of the Vinini Formation in the Tyrone Gap area, Eureka County, Nevada. M, 1975, Ohio University, Athens. 55 p.

Minning, Gretchen V. The significance of till-fabric analysis in the Puget Lowland, Washington. M, 1967, University of Washington. 30 p.

Minning, Robert C. A ground water pollution study of the Silurian aquifer at Millbury Ohio. M, 1970, University of Toledo. 106 p.

Minnis, Steffi Ann. Stable isotope profiles of hermatypic corals; indicators of changing environmental conditions in upwelling and non-upwelling regions of the eastern tropical Pacific. M, 1986, Rice University. 122 p.

Minor, John A. Petrology of the Tonganoxie sandstone (Pennsylvanian), Kansas-Missouri. M, 1969, University of Missouri, Columbia.

Minor, Scott Alan. Stratigraphy and structure of the western Trout Creek and northern Bilk Creek Mountains, Harney County, Oregon, and Humboldt County, Nevada. M, 1986, University of Colorado. 177 p.

Minshew, Velon H., Jr. Geology of the Scott Glacier and Wisconsin Range areas, central Transantarctic Mountains, Antarctica. D, 1967, Ohio State University.

Minshew, Velon Haywook, Jr. Stratigraphy and sedimentation of the Meridian Sand in east-central Mississippi. M, 1962, University of Nebraska, Lincoln.

Minster, Jean-Bernard Honore. Elastodynamics of failure in a continuum. D, 1974, California Institute of Technology. 520 p.

Minter, Larry Lane. Impact of water resource development on the hydrology and sedimentology of the Brazos River system. M, 1976, Texas A&M University.

Minton, Morris Cresswell. The geology of the Raynor Creek Quadrangle, California. M, 1941, University of California, Berkeley. 79 p.

Mintz, Leigh Wayne. The origins, phylogeny, and descendants of the echinoid family Disasteridae A. Gras, 1848. D, 1966, University of California, Berkeley. 314 p.

Mintz, Leigh Wayne. The species of the crinoid Dolatocrinus from the Middle Devonian Dock Street Clay of Michigan. M, 1962, University of Michigan.

Mintz, Milton. The geology of Westchester County, New York. M, 1974, Virginia State University. 33 p.

Mintz, Yale. Slate Mountain volcano-laccolith, Arizona. M, 1943, Columbia University, Teachers College.

Mir Mohamad Sadeghi, Ali. Remote sensing of the water status of a bare soil using microwave and hydrologic techniques. D, 1984, University of Arkansas, Fayetteville. 173 p.

Miranda Barbosa, Aluzio Liciniode. Geology of Warrior Ridge, Huntingdon County, Pennsylvania. M, 1947, Pennsylvania State University, University Park. 29 p.

Miranda G., Miguel A. Geology of the Sierra Los Arados Cenozoic volcanic field, Chihuahua, Mexico. M, 1986, University of Texas at El Paso.

Miranda, Antonio Nunes de *see* de Miranda, Antonio Nunes

Miranda, Leandro J. Insoluble residues of the (Jurassic) Calera Limestone from its type locality, Calera Valley, San Mateo County, California. M, 1947, Stanford University. 51 p.

Miranda, Roger M. Geochemical variations in sedimentary organic matter within a one hundred meter shale core of uniform lithology and thermal history (middle Tuscaloosa, Upper Cretaceous). M, 1988, University of Texas at Dallas. 220 p.

Mirbaba, Mehdi H. Lateral variations of carbonate rock adjacent to ore deposits in the upper Mississippi Valley zinc-lead district, Wisconsin, Illinois, Iowa. M, 1968, University of Missouri, Rolla.

Mirbagheri-Firoozabad, Seyed Ahmad. Characterization and computer simulation modelling of suspended sediment transport in Colusa Basin drain, California. D, 1981, University of California, Davis. 317 p.

Mirecki, June Elizabeth. Amino acid racemization dating of some Coastal Plain sites, southeastern Virginia and northeastern North Carolina. M, 1985, University of Delaware. 118 p.

Mirkin, Adam Nicholas. Three-dimensional ray tracing in the determination of local source structure effects or teleseismic P-waves from buried explosions; application to Yucca Flat, Nevada. M, 1985, Pennsylvania State University, University Park. 59 p.

Mirkin, Andrew S. Structural analysis of the East Eagle Creek area, southern Wallowa Mountains; northeastern Oregon. M, 1986, Rice University. 116 p.

Miron, Mark J. The paleoecological significance of the Receptaculida of the Galena Formation. M, 1978, University of Wisconsin-Milwaukee.

Miron, Sam. A geologic study of the Livingston Field, Texas. M, 1949, [University of Houston].

Mirowka, Jack P. Some relationships between surface hydrology and geomorphology in the Walker River basin, California. D, 1974, California State University, Los Angeles.

Mirsky, Arthur. Jurassic rocks of the Lucero Uplift, northwestern New Mexico. M, 1955, University of Arizona.

Mirsky, Arthur. Stratigraphy of the nonmarine Upper Jurassic and Lower Cretaceous rocks, southern Big Horn Mountains, Wyoming. D, 1960, Ohio State University. 192 p.

Mirynech, E. A mineralogical study of some drift soil profiles of southern Ontario. M, 1959, University of Toronto.

Mirynech, E. The Pleistocene geology of the Trenton-Campbellford map area, Ontario. D, 1962, University of Toronto.

Mirza, K. Late Ordovician to Late Silurian stratigraphy and conodont biostratigraphy of the eastern Canadian Arctic Islands. M, 1976, University of Waterloo.

Misener, Donald James. Cation diffusion in olivine at 1400°C and 35 kb. D, 1973, University of British Columbia.

Misener, Donald James. Diffusion in Fe-Mg olivine at elevated temperatures. M, 1971, University of British Columbia.

Miser, Donald Evans. Microstructures in natural and synthetic dolomite. D, 1987, University of Texas, Austin. 339 p.

Mishkin, L. A. The need for nuclear energy. M, 1973, Northeastern Illinois University.

Mishler, Harry Michael. Collinear dipole-dipole resistivity technique applied to the siting of crystalline rock water wells. M, 1982, San Diego State University. 247 p.

Mishra, Srikanta. On the use of pressure and tracer test data for reservoir description. D, 1987, Stanford University. 157 p.

Mishu, Louis Petrous. A study of stresses and strains in soil specimens in the triaxial test. D, 1966, Purdue University. 160 p.

Mishu, Louis Petrous. Collapse in one-dimensional compression of compacted clay on wetting. M, 1963, Purdue University.

Misiaszek, Edward Thomas. Engineering properties of Champaign-Urbana (Illinois) subsoils. D, 1960, University of Illinois, Urbana. 301 p.

Miska, William S. The Gratz-Lockport vein, Owen and Henry counties, Kentucky. M, 1960, Miami University (Ohio). 48 p.

Miskell, Kimberlee Jeanne. Accumulation of opal in deep sea sediments from the Mid-Cretaceous to the Miocene; a paleocirculation indicator. M, 1983, University of Miami. 169 p.

Misko, Ronald Michael. Petrology and provenance of the Upper Cretaceous-lower Tertiary sands in southern Saskatchewan. M, 1977, University of Saskatchewan. 159 p.

Misko, T. An analysis of Soviet iron ore, its benefication and role in the iron and steel industry of the Comecon nations. D, 1978, Kent State University, Kent. 227 p.

Misra, Brij Raj. A comparative study of experimental and computed compressibility factors of ethane-carbon dioxide-nitrogen system. M, 1970, University of Missouri, Rolla.

Misra, Kiran Shanker. Integrated remote sensing of the Amisk Lake-Wekusko Lake area in Manitoba. D, 1983, University of Manitoba.

Misra, Krishna Kant. Stratigraphy, sedimentation and petroleum possibilities of the Middle Ordovician (Kimmswick-Galena) rocks of Missouri, Illinois, and Iowa. D, 1964, University of Missouri, Rolla. 303 p.

Misra, Kula Chandra. Phase relations in the Fe-Ni-S system. D, 1972, University of Western Ontario. 213 p.

Misra, Manoranjan. Physical separation of bitumen from Utah tar sands. D, 1981, University of Utah. 217 p.

Misra, Shiva Balak. Geology of the Biscay bay, Cape Race area, Avalon peninsula, Newfoundland, (Canada). M, 1969, Memorial University of Newfoundland. 139 p.

Missallati, Amin A. Geology and ore deposits of Mt. Hope mining district, Eureka County, Nevada. D, 1973, Stanford University. 235 p.

Missallati, Amin Abdulla. The King turquoise deposit, Manassa, Colorado. M, 1967, Columbia University. 47 p.

Missimer, Thomas M. The depositional history of Sanibel Island, Florida. M, 1973, Florida State University.

Mistler, Alvin Jess. The geology of the Radnor Lake area, Nashville Quadrangle, Tennessee. M, 1937, Vanderbilt University.

Mistretta, Suzanne Barrerre. Carbonate petrology and lithostratigraphy of the Lecompton Member (Pawhuska Formation), Jennings-Shamrock area, Oklahoma. M, 1975, University of Oklahoma. 110 p.

Mitcham, Thomas W. A study of possible laboratory techniques of approach to an accessory mineral project on the (Coeur d'Alene) silver belt (Idaho). M, 1949, Columbia University, Teachers College.

Mitcham, Thomas W. Indicator minerals, Coeur d'Alene silver belt (Idaho). D, 1953, Columbia University, Teachers College.

Mitchel, R. The structure of the upper mantle of western North America from multimode Rayleigh wave dispersion. D, 1977, University of California, Los Angeles. 132 p.

Mitchell, A. Wallace. Geology of bedded barite deposits, north-central Nevada. M, 1977, University of Nevada - Mackay School of Mines. 58 p.

Mitchell, Brian James. Electrical and seismic properties of the Earth's crust in the southwestern Great Plain. D, 1970, Southern Methodist University. 53 p.

Mitchell, Brian James. Surface wave dispersion and crustal structure across the central United States. M, 1965, University of Minnesota, Minneapolis. 81 p.

Mitchell, Bruce T. A determination of upper mantle structure under Southern California using a forward raytracing technique. M, 1985, University of North Carolina, Chapel Hill. 78 p.

Mitchell, Burke M. An engineering analysis of the May 1983 rock slope failure on Slide Mountain, Nevada. M, 1986, University of Nevada. 176 p.

Mitchell, C. E. Middle and Upper Ordovician strophomenids (Brachiopoda) from the central Mackenzie Mountains, Northwest Territories. M, 1978, University of Western Ontario. 305 p.

Mitchell, Charles Clifford, Jr. The sulfur fertility status of Florida soils. D, 1980, University of Florida. 178 p.

Mitchell, Charles Dale, Jr. Detection of vegetation stress related to hydrocarbon microseepage at the Velma Oilfield, Oklahoma, using field spectroradiometer, geochemical and Landsat thematic mapper. M, 1988, University of Texas at Dallas. 225 p.

Mitchell, Charles Emerson. Astogeny and phylogeny of the Diplograptina (Graptoloidea). D, 1983, Harvard University. 364 p.

Mitchell, Charles Leonard. Petrology of selected carbonate rocks from the Hawthorn Formation (Miocene) Devil's Millhopper, Alachua County, Florida. M, 1965, University of Florida. 53 p.

Mitchell, Christopher B. Geochemistry of the granite-gabbro complex on Vinalhaven Island, Maine; an investigation of the commingling of mafic and felsic magmas. M, 1988, University of Massachusetts. 111 p.

Mitchell, David L. Influence of regional syndepositional faulting on sedimentation patterns and stratigraphic hydrocarbon traps in Cretaceous strata, San Juan Basin, New Mexico. M, 1987, Purdue University. 182 p.

Mitchell, David Laurie. An on-line computer assisted mass spectrometer. M, 1971, University of British Columbia.

Mitchell, Edward D. Eastern-North Pacific fossil pinnipeds. D, 1967, University of California, Berkeley. 352 p.

Mitchell, Francis J. The Chestnut River gorge of Cheat River (West Virginia). M, 1960, West Virginia University.

Mitchell, Gareth D. A petrographic classification of solid residues for the evaluation of coal performance during the hydrogenation of bituminous coals. M, 1977, Pennsylvania State University, University Park. 98 p.

Mitchell, Gary C. Geology of the Argyle Quadrangle, Denton County, Texas. M, 1973, University of Texas, Arlington. 123 p.

Mitchell, Gary Clark. Gravity survey and crustal structure of east-central Missouri. M, 1971, University of Missouri, Columbia.

Mitchell, George Dampier. The Santa Cruz earthquakes of October, 1926. M, 1927, University of California, Berkeley. 167 p.

Mitchell, George Scott. Abandoned coal and clay mines; subsidence potential in Summit and Portage counties, Ohio. M, 1978, Kent State University, Kent. 93 p.

Mitchell, Gerald George. The search for intermodulation coupling in the ground. M, 1978, University of British Columbia.

Mitchell, Graham J. A geological cross-section of New York City. M, 1913, Columbia University, Teachers College.

Mitchell, Graham J. The geology of the Ponce District, Puerto Rico. D, 1918, Columbia University, Teachers College.

Mitchell, Harold Delong. Cardiidae of the Pacific Coast. M, 1934, University of Washington. 48 p.

Mitchell, Harold G. A geological report on the Dry Canyon Division of the Ophir mining district, Utah. M, 1925, Stanford University.

Mitchell, J. G. Variations in the composition of granitic rocks as a guide to ore. M, 1960, University of Toronto.

Mitchell, James. Photogrammetric measurements of spherical balloons used in connection with the Echo Satellite Program. M, 1964, Ohio State University.

Mitchell, James F. Stress model for the Danville area during the earthquake sequence of summer, 1970. M, 1970, Stanford University.

Mitchell, James K. Delineation of buried river valleys in Butler County, Ohio; using gravity. M, 1984, Wright State University. 82 p.

Mitchell, James Michael. The geochemical significance of the alkanes in bat guano. D, 1972, Indiana University, Bloomington. 32 p.

Mitchell, James Porter. Paleodynamics of the Green Pond Outlier, New Jersey Highlands; evidence for noncoaxial deformation during late Paleozoic orogenesis. M, 1985, Rutgers, The State University, New Brunswick. 129 p.

Mitchell, Jay Preston. Shallow seismic reflection data acquisition and processing techniques applied to the delineation of buried bedrock topography. M, 1984, Purdue University. 72 p.

Mitchell, Jeffrey Leonard. Sediment differentiation in the Altamaha River-estuary-marine system. M, 1972, Emory University. 74 p.

Mitchell, Jeffrey Todd. Submarine lithification of a Holocene reef hardground; Discovery Bay, Jamaica. M, 1987, University of Texas, Austin. 198 p.

Mitchell, John C. Geology and petrology of the Wateree Lake area, Kershaw County, South Carolina. M, 1970, University of South Carolina. 53 p.

Mitchell, John Charles. Biostromes and bioherms of the Solon Member of the Cedar Valley Limestone, Middle Devonian, eastern Iowa. M, 1977, University of Iowa. 179 p.

Mitchell, John Charles. Stratigraphy and depositional history of the Iola Limestone, Upper Pennsylvanian (Missourian), northern Midcontinent U.S. D, 1981, University of Iowa. 364 p.

Mitchell, Judson T. The crystal structure refinement of actinolite. M, 1971, Virginia Polytechnic Institute and State University.

Mitchell, K. Daniel. A comparative study of trace element analyses by argon plasma emission and atomic absorption spectrophotometry. M, 1978, University of Windsor. 208 p.

Mitchell, Lane. Mineral and colloidal constitution of some Georgia kaolins. D, 1941, Pennsylvania State University, University Park. 114 p.

Mitchell, Mark W. The foraminifera of the Cook's Mountain Eocene of Texas. M, 1932, University of Southern California.

Mitchell, Martha Jeanne. Equilibrium vapor pressure study of Bi_2S_3 (bismuthinite). D, 1975, Stanford University. 134 p.

Mitchell, Martha Jeanne. Vapor pressure of mercury sulfide and mercury selenide. M, 1970, San Jose State University. 92 p.

Mitchell, Martin Lane. Geologic responses to late Cenozoic marine transgressions in the Poropotank River estuary, Virginia. M, 1984, College of William and Mary. 154 p.

Mitchell, Michael Harold. Depositional environment and facies relationships of the Canyon Sandstone, Val Verde Basin, Texas. M, 1975, Texas A&M University. 211 p.

Mitchell, Michael M. Fundamentals of spatial coordinates in terrestrial geodesy. M, 1963, Ohio State University.

Mitchell, Michael M. Geometric and mesofabric analysis of the Hossfeldt Anticline-Eustis Syncline and the Lombard thrust zone near Three Forks, southwestern Montana. M, 1987, University of Tennessee, Knoxville. 159 p.

Mitchell, Nolan William Ralph. Direct shear tests on thin samples of remolded shales from the Bighorn Mountains, Wyoming. M, 1965, University of Illinois, Urbana.

Mitchell, Peter Ashley. Geology of the Hope-Sunrise (gold) mining district, North-central Kenai Peninsula, Alaska. M, 1979, Stanford University. 123 p.

Mitchell, Peter S. Deep magnetotelluric profiling of the southern Albuquerque-Belen Basin, New Mexico. M, 1983, San Diego State University. 315 p.

Mitchell, Phillip Dwight. An interpretation of a "Birdseye" limestone sequence of the McLish Formation (Ordovician) of southern Oklahoma. M, 1973, University of Wisconsin-Madison. 126 p.

Mitchell, Raymond Weatheral, III. The comparative sedimentology of the bahama-type shelf carbonates of the Middle Ordovician St. Paul Group of the Central Appalachians. D, 1982, The Johns Hopkins University. 510 p.

Mitchell, Richard Scott. Polytypism of cadmium iodide and its relationship to screw dislocations. D, 1956, University of Michigan.

Mitchell, Robert C. Methods of studying physical characteristics of crude oil within the reservoir. M, 1936, University of Kansas.

Mitchell, Robert H. The geology of Richland Township, Belmont County, Ohio. M, 1929, Ohio State University.

Mitchell, Robert J. Paleomagnetism of the Indian Heaven volcanic field, southern Washington. M, 1986, Michigan Technological University. 57 p.

Mitchell, Roger Howard. The isotopic composition of strontium in South African kimberlites and in alkaline rocks of the Fen area, S. Norway. D, 1969, McMaster University. 293 p.

Mitchell, S. D. Geology of the eastern Alcova-Bear Creek area, Natrona County, Wyoming. M, 1957, University of Wyoming. 71 p.

Mitchell, S. L. Structure and petrology of Bentley-Siddon Lakes area, Bancroft, Ontario, with special reference to rocks of Alkaline affinity. M, 1976, University of Waterloo.

Mitchell, Stephen M. Geology of Sierra Gomez, Chihuahua, Mexico. M, 1980, University of Texas at El Paso.

Mitchell, Steven W. Biostratigraphy of the lower Traverse Group (middle Devonian), southern Michigan Basin. M, 1971, Wayne State University.

Mitchell, Terry Edward. Uranium mineralization of the metamorphic aureole of the Spirit Pluton, Stevens County, Washington. M, 1981, Oregon State University. 109 p.

Mitchell, Thomas M. A parametric model for a dense plume near a stream bed. D, 1974, Texas A&M University. 161 p.

Mitchell, Thomas M. The dispersion of dense effluent from an inclined jet discharging into still fluid. M, 1970, Texas A&M University.

Mitchell, Victoria E. Geology of the lower Basin Creek-lower East Basin Creek area, Custer County, Idaho. M, 1980, University of Idaho. 157 p.

Mitchell, Wendy J. Structure and stratigraphy of the Warsaw Mountain area, British Columbia. M, 1976, University of Calgary. 164 p.

Mitchell, William Louis. A detailed study of the Silurian stratigraphy in Walker County, Georgia. M, 1950, Emory University. 71 p.

Mitchell, William Sutherland. The Sooke gabbro. M, 1973, University of British Columbia.

Mitchell, William, Jr. Origin and occurrence of black manganese in Montana. M, 1942, Montana College of Mineral Science & Technology. 64 p.

Mitchell, Wilson Doe. Some tetrahedrite silver ores. D, 1941, Harvard University.

Mitchell, Wilson Doe. The geology of a portion of Lanark and Carleton counties, Ontario. M, 1936, University of Wisconsin-Madison.

Mitchell-Tapping, Hugh J. The mechanical breakdown of Recent carbonate sediment in the coral reef environment. D, 1978, Florida State University. 463 p.

Mitchell-Tapping, Hugh J. The origin and distribution of marine sediment around Water Island, U.S. Virgin Islands. M, 1976, Florida State University.

Mitchum, Robert Mitchell, Jr. Pottsville strata (Pennsylvanian) of part of the Central Appalachian coal field. D, 1954, Northwestern University. 405 p.

Mitchum, Robert Mitchell, Jr. The Dycus disturbance, Jackson County, Tennessee. M, 1951, Vanderbilt University.

Mitera, Zygmut. A theoretical and experimental examination of the potential-drop ratio method. D, 1933, Colorado School of Mines. 108 p.

Mithal, Rakesh. Evidence for a basal low velocity zone in oceanic crust, and new methods of phase analysis and extremal inversion. D, 1986, Columbia University, Teachers College. 204 p.

Mitlin, Lucille List. The historical development of land use in Starkville, Mississippi, a small university city. M, 1975, Mississippi State University. 394 p.

Mitock, Joanne R. Structural geology of a portion of the northwest flank of the Gros Ventre Uplift, upper Gros Ventre River area, Teton County, Wyoming. M, 1985, Miami University (Ohio). 100 p.

Mitra, Devi. A bibliography of geomechanics. M, 1962, University of Utah. 210 p.

Mitra, G. The mechanical processes of deformation of granitic basement, and the role of ductile deformation zones in the deformation of Blue Ridge basement rocks in northern Virginia. D, 1976, The Johns Hopkins University. 374 p.

Mitra, Rabindranath. A study of metamorphic facies at the New Calumet Mine, Quebec. D, 1954, University of Toronto.

Mitra, S. Studies on deformation mechanisms and finite strain in quartzites and their relation to structures of various scales within the South Mountain Anticline. D, 1977, The Johns Hopkins University. 292 p.

Mitronovas, Walter. Seismic velocity anomalies in the upper mantle beneath the Tonga-Kermadec island arc. D, 1969, Columbia University. 117 p.

Mitten, Hugh T. Some aspects of differential thermal analysis experiments applied to the Antrim and Bedford-Berea shales (Devonian-Mississippian) from Oakland County, Michigan. M, 1957, Michigan State University. 86 p.

Mitterer, Richard Max. Amino acid and protein geochemistry in mollusk shells. D, 1966, Florida State University. 161 p.

Mittlefehldt, David Wayne. The differentiation history of small bodies in the solar system; the howardite and mesosiderite meteorite parent bodies. D, 1978, University of California, Los Angeles. 223 p.

Mittler, Peter Robert. Storm related sediment flux and equilibrium in a barred nearshore; Kouchibouguac Bay, New Brunswick, Canada. D, 1981, University of Toronto.

Mittler, Richard Wayne. Geology and geochemistry of the Turkey Hill gold mine area, Lumpkin County, Georgia. M, 1987, Northeast Louisiana University. 108 p.

Mittsdarffer, Alan. Hydrodynamics of the Mission Canyon Formation in the Billings Nose area, North Dakota. M, 1985, Texas A&M University. 162 p.

Miura, Ryosuke. A colorimetric method for the analysis of coprostanol, an indicator of water pollution. D, 1973, University of Florida. 112 p.

Mix, Alan Campbell. Late Quaternary paleoceanography of the Atlantic Ocean; foraminiferal faunal and stable-isotopic evidence. D, 1986, Columbia University, Teachers College. 754 p.

Mix, Sidney E. The geology of a portion of Guernsey and Muskingum counties. M, 1918, Ohio State University.

Mixon, Robert B. Jurassic formations of the Ciudad Victoria area, Tamaulipas, Mexico. M, 1958, Louisiana State University.

Mixon, Robert Burnley. Geology of the Huizachal Redbeds, Sierra Madre Oriental, Mexico. D, 1963, Louisiana State University. 134 p.

Miyajima, Melvin H. Subantarctic region, Southeast Indian Ocean; absolute chronology of upper Pleistocene calcareous nannofossil zones and paleoclimatic history determined from silicoflagellate, coccolith, and carbonate analyses. M, 1975, Florida State University.

Miyasaki, Brent. TSRA5B, a Fortran IV computer program for trend surface analysis, with analysis of gravity data in the Sutter Buttes area, California. M, 1979, University of California, Davis.

Miyazaki, J. M. The evolution of a Tertiary scallop and Haeckel's law. M, 1978, SUNY at Stony Brook.

Miyazaki, Yoshinori. Analysis of aeromagnetic anomalies; mapping of Curie isothermal surface at Long Valley, California. D, 1985, Stanford University. 207 p.

Mize, Thomas R. Magnetic survey across the Niagara Fault in northern Wisconsin. M, 1986, University of Wisconsin-Milwaukee. 84 p.

Mizell, Nancy Brent Hunt. Quaternary geology of the central Truckee Meadows, Nevada. M, 1975, University of Nevada. 68 p.

Mizell, Stephen A. A preliminary study of flow-sediment transport in mountainous watersheds. M, 1975, University of Nevada. 79 p.

Mizula, Joseph William. A cross-stratification study of the Omadi Sandstone in eastern Nebraska and northcentral Kansas. M, 1960, University of Nebraska, Lincoln.

Mizumura, K. Study of sediment transport in natural rivers. D, 1976, University of Pittsburgh. 162 p.

Mizuno, Eiji. Plasticity modeling of soils and finite element applications. D, 1981, Purdue University. 237 p.

Mkumba, J. T. K. δ^{34}S and δ^{18}O variations of aqueous sulfates in groundwater systems of Winnipeg and Kitchener-Waterloo. M, 1983, University of Waterloo. 174 p.

Mloszewski, M. J. Some specularite and associated rocks, Blough Lake area, northern Quebec. M, 1956, University of Toronto.

Moammar, Mustafa Omar. Marine diagenesis of hydrothermal sulfide. D, 1985, University of California, San Diego. 241 p.

Mobasseri, Shahpur. Fault plane solutions of recent Iranian earthquakes. M, 1985, University of Akron. 139 p.

Moberly, Ralph Moon, Jr. Mesozoic Morrison, Cloverly, and Crooked Creek formations, Bighorn Basin, Wyoming and Montana. D, 1956, Princeton University. 150 p.

Mobley, Bruce Justin. Geology of the southwestern quarter of the Bates Quadrangle, Oregon. M, 1956, University of Oregon. 66 p.

Mochizuki, S. A theory of the elastic wave propagation in partially saturated porous rocks. D, 1979, University of Texas, Austin. 172 p.

Mochtar, Indrasurya Budisatria. An experimental study of skin friction and creep of piles in clay. D, 1985, University of Wisconsin-Madison. 348 p.

Mochtar, Noor Endah. Compression of peat soils. D, 1985, University of Wisconsin-Madison. 225 p.

Mock, Ralph G. Correlation of land surfaces in the Truckee River valley between Reno and Verdi, Nevada. M, 1972, University of Nevada. 91 p.

Mock, S. J. Topological properties of some trellis pattern channel networks. D, 1975, Northwestern University. 100 p.

Mock, Steven J. Accumulation and snow temperatures of the Thule lobe of the Greenland ice sheet. M, 1965, Dartmouth College. 54 p.

Mockbee, Gael A. Geochemistry of subsurface saline waters in Seminole County, Oklahoma. M, 1979, University of Texas, Arlington. 218 p.

Modarresi, Hassan Ghavami. Statistical analysis to be applied to the petrographic and reservoir properties of the "First Venango Oil Sand", Warren County, Pennsylvania. M, 1963, Pennsylvania State University, University Park. 244 p.

Modarrsei, Hassan Ghavami. Relationships of sedimentary bed forms, petrology, and hydraulic equivalence properties of an ancient point-bar sandstone deposit. D, 1968, Pennsylvania State University, University Park. 227 p.

Moddle, D. A. On a crystal of augite from the commercial deposit of andalusite at White Mountain, California. M, 1941, University of Toronto.

Moddle, P. M. Evaluation of the efficiency and chemistry of reactions between cementitious grout and carbon dioxide. M, 1987, University of Waterloo. 204 p.

Mode, William Niles. Quaternary stratigraphy and palynology of the Clyde foreland, Baffin Island, N.W.T., Canada. D, 1980, University of Colorado. 233 p.

Mode, William Niles. The glacial geology of a portion of North-central Wisconsin. M, 1976, University of Wisconsin-Madison.

Modell, David Isaiah. The geology and petrology of the Belknap Mountains area, New Hampshire. D, 1933, Harvard University.

Modene, Janet S. Origin and sulfur isotope geochemistry of the Grum Deposit, Anvil Range, Yukon Territory, Canada. M, 1982, University of Wisconsin-Madison. 158 p.

Modisi, Motsoptse P. The geology and geochemistry of the Iron Mountain Ironstone, Pennington and Custer counties, South Dakota. M, 1982, South Dakota School of Mines & Technology. 70 p.

Modreski, Peter John. The phase relationships of phlogopite in the system $K_2O\text{-}MgO\text{-}CaO\text{-}Al_2O_3\text{-}SiO_2\text{-}H_2O$ to 35 kilobars pressure; a model for the stability of mica in the upper mantle of the Earth. D, 1972, Pennsylvania State University, University Park. 109 p.

Modreski, Peter John. The stability of phlogopite + enstatite to 35 Kb; a model for phlogopite in the mantle of the Earth. M, 1971, Pennsylvania State University, University Park.

Modzeleski, Vincent E. Solvent extraction of organic compounds from the Navesink Formation (Cretaceous) of New Jersey. M, 1963, New York University.

Moebs, Noel N. The petrology of the Chain Bridge Gneiss (near Washington, D.C.). M, 1951, West Virginia University.

Moecher, David Paul. Determination of late Archean granulite facies metamorphic conditions, Granite Falls, Minnesota. M, 1984, University of Wisconsin-Madison.

Moecher, David Paul. Scapolite phase equilibria and carbon isotope variations in high grade rocks; tests of the CO_2-flooding hypothesis of granulite genesis. D, 1988, University of Michigan. 298 p.

Moed, Barbara A. Preliminary determination of the radon-222 baseline in Houston, Texas. M, 1979, Rice University. 134 p.

Moen-Vaziri, Nasser. Investigation of scattering and diffraction of plane seismic waves through two dimensional inhomogeneities. D, 1984, University of Southern California.

Moeglin, T. D. Geomorphic development on the Log Cabin Batholith, Larimer County, Colorado. D, 1978, University of Nebraska, Lincoln. 109 p.

Moeglin, Thomas Dean. Lateral variation in composition of Nebraskan and Kansas tills in eastern Nebraska and southwestern Iowa. M, 1975, University of Nebraska, Lincoln.

Moehl, William R. A study of the Lenoir Limestone (Middle Ordovician) in the Knoxville-Friendsville Belt, Knox and Blount counties, Tennessee. M, 1965, University of Tennessee, Knoxville. 107 p.

Moehlman, Robert S. Late Tertiary ore deposits near Ouray, Colorado. D, 1935, Harvard University.

Moench, Allen F. A transient heat-flow method for determination of thermal constants. M, 1962, Pennsylvania State University, University Park. 79 p.

Moench, Allen Forbes. An evaluation of heat transfer coefficients in moist porous media. D, 1969, University of Arizona. 157 p.

Moench, Robert H. The geology of the Phillips Quadrangle, Maine. D, 1954, Boston University. 289 p.

Moeng, Bruce C. Wall rock alteration as an indicator of bedded replacement ore deposits in the Park City District, Utah. M, 1974, University of Iowa. 96 p.

Moezzi, Bahman. Studies in X-ray crystallography; a new method of absorption correction. D, 1987, University of California, Davis. 232 p.

Moffat, Ian William. Geometry and mechanisms of transverse faulting, Rocky Mountain Front Ranges, Canmore, Alberta. M, 1980, University of Calgary. 193 p.

Moffat, Ian William. The nature and timing of deformational events and organic and inorganic metamorphism in the northern Groundhog Coalfield; implications for the tectonic history of the Bowser Basin. D, 1986, University of British Columbia.

Moffett, James Robert. Geology of the East Bastian Bay Field, Plaquemines Parish, Louisiana. M, 1963, Tulane University. 27 p.

Moffett, James William. A limnological investigation of the dynamics of a barren, sandy, wave-swept shoal in Douglas Lake, Michigan. D, 1939, University of Michigan.

Moffett, Joanne Lynn. Distribution of certain heavy metals in cores from Lake St. Clair and western Lake Erie. M, 1980, Wayne State University.

Moffett, Tola B. Hydrogeology of the Taburiente Caldera, La Palma, Canary Islands, Spain. M, 1972, University of Missouri, Columbia.

Moffett, Tola Burton. Water-resources management in Alabama; a proposed system based upon a study of agency activities. D, 1983, University of Alabama. 570 p.

Moffitt, John. Depositional history of late Paleozoic rocks in the southeastern Darwin Hills. M, 1978, San Jose State University. 66 p.

Mogekwu, Emmanuel. Hydrology and water resources of Nachoba Brook watershed, Massachusetts. M, 1981, Boston University. 111 p.

Mogensen, P. Fort Union Formation, east flank of the Rock Springs Uplift, Sweetwater County, Wyoming. M, 1959, University of Wyoming. 86 p.

Moger, Seth R. The geology of the west central portion of the Patagonia mountains, Santa Cruz County, Arizona. M, 1970, [University of California, Davis].

Mogharabi, Ataolah. Carbonate petrology of the Foraker Formation (Lower Permian), north-central Oklahoma. D, 1966, University of Oklahoma. 204 p.

Mogharabi, Ataolah. Petroleum geology of T. 19N., R. 6W., Hennesey area, Kingfisher County, Oklahoma. M, 1962, University of Oklahoma. 51 p.

Mogk, David William. Contact metamorphism of carbonate rocks at Cave Ridge, Snoqualmie Pass, Washington. M, 1978, University of Washington. 110 p.

Mogk, David William. The petrology, structure and geochemistry of an Archean terrane in the North Snowy Block, Beartooth Mountains, Montana. D, 1984, University of Washington. 440 p.

Mogolesko, Fred J. Development of an analytical technique for the design of a submerged thermal discharging system in a tidal estuary. D, 1978, New York University. 188 p.

Mohajer-Ashjai, Arsalan. Method for the determination of the magnetic polarity of unoriented submarine basalts. M, 1971, Florida State University.

Mohamad Sadeghi, Ali Mir *see* Mir Mohamad Sadeghi, Ali

Mohamad, Daud Bin. A study of uranium in ground water around Greyhawk Mine, Bancroft, Ontario. M, 1980, McMaster University. 119 p.

Mohamad, Kamel B. Multivariate analysis of interstitial water chemistry of the Chesapeake Bay. M, 1987, Bowling Green State University. 150 p.

Mohamad, Ramli Bin *see* Bin Mohamad, Ramli

Mohamed, Abdel-Mohsen Onsy. Performance of anisotropic clays under variable stresses. D, 1987, McGill University.

Mohamed, Magzoub Ahmed. Ground water conditions in Elgeteina area, Sudan. M, 1978, North Carolina State University. 76 p.

Mohamed, Fawzi Said. Effect of drain tube openings on water table drawdown. D, 1982, North Carolina State University. 131 p.

Mohammad, Omar Mohammad Joudeh. Evaluation of the present and potential impacts of open pit phosphate mining on groundwater resource system in south-eastern Idaho phosphate field. D, 1976, University of Idaho. 165 p.

Mohammad, Omar Mohammad Joudeh. Hydrogeology of the Boise Ridge area. M, 1970, University of Idaho. 66 p.

Mohammadi, Hossein K. Geology of the area east of Bagdad, Yavapai County, Arizona. M, 1984, Arizona State University. 64 p.

Mohammadi, Mohammad. Olfactometric in situ soil exploration; development of the electro-odo-cone. D, 1986, Louisiana State University. 276 p.

Mohammed, K. The original phase of Castile calcium sulphate; a study of the classic outcrops, southeastern New Mexico and West Texas. M, 1988, Sul Ross State University.

Mohammed, Mahdi. Physical-chemical characteristics of some carbonate aggregates from central and western New York. M, 1963, Rensselaer Polytechnic Institute. 74 p.

Mohan, Madhukar. Crustal interpretations of Magsat satellite data over the Indian Subcontinent region. M, 1987, University of Iowa. 169 p.

Mohanty, Priya Ranjan. Migration by the diffraction stack method. M, 1984, Memorial University of Newfoundland. 112 p.

Mohar, John, Jr. Geology of the Rinconada Canyon area, Valencia County, New Mexico. M, 1956, University of New Mexico. 55 p.

Mohayej, Zeinalabedin. Subsurface study of Tuscaloosa Group in Tensas Parish, Louisiana. M, 1979, Northeast Louisiana University.

Mohd-Nurin, Abdul Razak. Ground-water resources of Defiance County, Ohio. M, 1986, University of Toledo. 152 p.

Mohl, Gregory Blaine. Bouguer gravity investigation of the cratonic margin; southeastern Washington. M, 1985, Washington State University. 67 p.

Mohler, Amy Szumigala. The Susanville earthquake sequence of 1976. M, 1979, University of Nevada - Mackay School of Mines. 250 p.

Mohler, Charles Edwin. A field study of the Inola Formation of Osage and Tulsa counties. M, 1942, University of Oklahoma. 45 p.

Mohn, Jean Doris. La Luna (Upper Cretaceous) foraminifera from the State of Trujillo, Venezuela. M, 1958, University of Oklahoma. 100 p.

Mohn, Kenneth William. Stratigraphy and the tectonic development of the Parowan Gap area, Iron County, Utah. M, 1986, Stephen F. Austin State University. 129 p.

Mohon, J. P., Jr. Comparative geothermometry for the Monte Cristo Pegmatite, Yavapai County, Arizona. M, 1975, University of Arizona.

Mohorich, Leroy Martin. The urban geology of a portion of Tarrant County, Texas. M, 1971, Texas Christian University.

Mohorjy, Abdullah Mustafa. Methodology for comprehensive water reuse planning. D, 1987, Colorado State University. 262 p.

Mohr, David W. Stratigraphy, structure, and metamorphism of the eastern part of the Fontana Lake Reservoir, Great Smoky Mountains, North Carolina. D, 1972, University of Chicago. 284 p.

Mohr, David Wildred. Regional setting and intrusion mechanics of the Stone Mountain Pluton (Permian in Georgia). M, 1965, Emory University. 68 p.

Mohr, Eileen T. A geochemical reconnaissance study of groundwater from an eighteen county area of northwestern Ohio. M, 1983, Kent State University, Kent. 103 p.

Mohr, Elizabeth B. The geology of the Bass Islands. M, 1931, Ohio State University.

Mohr, Richard Earl. Some aspects of the theory of petrogenetic grids and an application to cherty ironformation. D, 1978, University of Minnesota, Minneapolis. 128 p.

Mohrenschildt, George Sergius de *see* de Mohrenschildt, George Sergius

Mohring, Eric H. A study of subsurface water flow in a southeastern Minnesota karst drainage basin. M, 1983, University of Minnesota, Minneapolis. 99 p.

Mohsen, Lotfi A. F. M. Distribution of manganese in the Philipsburg Batholith (Tertiary) Montana, and its relationship to associated manganese ore deposits. D, 1969, Boston University. 109 p.

Mohsen, M. F. N. Gas migration from sanitary landfills and associated problems. D, 1975, University of Waterloo.

Mohsenisaravi, Mohsen. Forecasting subsurface water flow and storage on forested slopes using a finite element model. D, 1981, University of Idaho. 162 p.

Moila, Richard James. Late Cenozoic geology of the northern Silver Peak region, Esmeralda County, Nevada. D, 1969, University of California, Berkeley. 162 p.

Moine, Denis Le *see* Le Moine, Denis

Moir, Gordon James. Depositional environments and stratigraphy of the Cretaceous rocks, southwestern Utah. D, 1974, University of California, Los Angeles.

Moir, Rory Douglas. Biogeochemical exploration for Mo in central-south B.C. M, 1978, University of Alberta. 101 p.

Moise, Theodore, Jr. The event stratigraphy of a portion of the Upper Cretaceous of central Alabama. M, 1986, University of New Orleans. 89 p.

Moisseeff, Alexis Nicolas. The geology and the geochemistry of the Wilbur Springs quicksilver district, Colusa and Lake counties, California. D, 1966, Stanford University. 242 p.

Mojab, Fathollah. Rudistid fossils (Cretaceous) of Darab area, south Iran. M, 1967, University of Florida. 90 p.

Mojica, Iran H. Effects of changes in land use on the streamflow of the Reventazon River, Costa Rica. D, 1972, University of Washington. 185 p.

Mojtahedi, Soheil. Earthquake analysis of arch damfoundation systems. D, 1976, University of California, Berkeley. 136 p.

Mok, Chee Sun. Calcitic limestones, their properties, origin, and classification. M, 1948, University of Michigan.

Mok, Wai Man. Chemical speciation of arsenic and antimony in natural water systems and its applications to environmental problems. D, 1988, University of Idaho. 177 p.

Mok, Young Jin. Analytical and experimental studies of borehole seismic methods. D, 1987, University of Texas, Austin. 296 p.

Moke, Charles B. Petrology, structure, and metamorphism of the Plymouth Quadrangle, New Hampshire. D, 1948, Harvard University.

Mokhtar, Talal Ali. Seismic velocity and Q model for the shallow structure of the Arabian Shield from short period Rayleigh waves. D, 1987, St. Louis University. 185 p.

Mokhtar, Talal Ali. The relationship between the seismicity and late Cenozoic tectonics in Arizona. M, 1979, University of Arizona. 53 p.

Moklestad, T. Charles. Dielectric properties of sulfate minerals. M, 1973, University of Oregon. 47 p.

Mokma, Delbert L. Cation exchange properties of weathered micas and mineralogy of eolian dusts, sediments, and two soil chronosequences (New Zealand and Australia). D, 1971, University of Wisconsin-Madison.

Molander, Gene Emery. Lower Tertiary stratigraphy and foraminifera of the eastern Santa Rosa Hills, Santa Barbara County, California. M, 1956, University of California, Berkeley. 114 p.

Moldovanyi, Eva Paulette. Isotopic recognition of successive cementation events within the phreatic environment, Lower Cretaceous Sligo and Cupido formations. M, 1982, University of Michigan.

Molina, Carlos Rodriguez *see* Rodriguez Molina, Carlos

Molina-Cruz, Adolfo. Late Quaternary oceanic circulation along the Pacific coast of South America. D, 1978, Oregon State University. 246 p.

Molina-Cruz, Adolfo. Paleo-oceanography of the subtropical southeastern Pacific during Late Quaternary; a study of Radiolaria, opal and quartz contents of deep-sea sediments. M, 1976, Oregon State University. 179 p.

Molinari, Mark Philip. Late Cenozoic geology and tectonics of Stewart and Monte Cristo valleys, west-central Nevada. M, 1984, University of Nevada. 124 p.

Molinari, Robert L. The effect of topography on the Yucatan current. M, 1968, Texas A&M University.

Molinda, Gregory. Investigations of methane occurrence and emission in the Cote Blanche domal salt mine, LA. M, 1988, University of Pittsburgh.

Moline, Myrna Marie. The effects of shallow burial on the clay mineralogy of the Oligocene Bucatunna Formation (Vicksburg Group) in Mississippi. M, 1987, University of New Orleans. 148 p.

Molinelli, Jose Antonio. Geomorphic processes along the Autopista las Americas in north central Puerto Rico; implications for highway construction, design, and maintenance. D, 1983, Clark University. 184 p.

Molinsky, Linda. The biostratigraphy and paleoecology of the Oligocene and Miocene of the Canadian Atlantic continental margin off Nova Scotia. M, 1973, Queen's University. 169 p.

Molitor, William C. A radioactive survey of sediments of Long Island Sound (New York), analysis of variation with grain, size, depth and harbor erosion. M, 1960, New York University.

Moll, Elizabeth Jean. Geology and geochemistry of the Sierra el Virulento area, southeastern Chihuahua, Mexico. M, 1979, University of California, Santa Cruz.

Moll, J. Gregory. Magnetic investigation of the Waukesha Fault, Wisconsin. M, 1987, University of Wisconsin-Milwaukee. 94 p.

Moll, Nancy Eileen. The structure and petrology of the gabbro unit and the mafic-ultramafic contact, Table Mountain, Bay of Islands Complex, Newfoundland. D, 1981, University of Washington. 158 p.

Moll, Robert F. Clay mineralogy of the north Bering sea shallows (Alaska and Siberia). M, 1969, University of Southern California.

Moll, Stanton H. Discrimination of rock units in north-central Colorado by cluster analysis of airborne gamma spectrometry. M, 1988, Colorado State University. 181 p.

Moll, William Francis, Jr. The orientation of some cyclic amine cations on vermiculite. D, 1963, Washington University. 152 p.

Moll-Stalcup, Elizabeth Jean. The petrology and Sr and Nd isotopic characteristics of five Late Cretaceous-early Tertiary volcanic fields in western Alaska. D, 1987, Stanford University. 326 p.

Mollard, John D. Airphoto interpretation of soils and drainage of Montgomery County, Indiana. M, 1947, Purdue University.

Mollard, John Douglas Ashton. Aerial photographic studies on the central Saskatchewan irrigation project. D, 1952, Cornell University.

Mollazal, Yazdan. Petrology and petrography of Ely Limestone in part of eastern Great Basin. M, 1961, Brigham Young University. 35 p.

Moller, Stuart A. Structural analysis of the Grandfather Mountain window and vicinity, North Carolina. M, 1976, University of North Carolina, Chapel Hill. 66 p.

Molling, Philip Andrew. Petrology and mineralogy of a drill core (DDH 295) from the Partridge River Troctolite of the Duluth Complex, Minnesota. M, 1979, Miami University (Ohio). 150 p.

Mollison, Richard Allen. Petrology and depositional environments of the Pitkin Formation (Chesterian), Washington and Madison counties, Northwest Arkansas. M, 1983, University of Arkansas, Fayetteville.

Molloy, Martin W. A comparative study of ten monazites. M, 1958, Columbia University, Teachers College.

Molloy, Martin William. Tertiary volcanism in the Tushar Range, Utah. D, 1960, Columbia University, Teachers College. 212 p.

Molnar, James Stephen. Structure and Petrology of the northern Shadow Mountains, San Bernardino County, California. M, 1973, University of Southern California.

Molnar, Paul S. Correlation of thermal conductivity with physical properties obtained from geophysical well logs. M, 1982, SUNY at Buffalo. 52 p.

Molnar, Peter H. Three studies of the structure and dynamics of the lithosphere; I, Lateral variation of attenuation in the upper mantle and discontinuities in the lithosphere; II, Tectonics of the Caribbean and middle America regions from focal mechanisms and seismicity; III, Mantle earthquake mechanisms and the sinking of the lithosphere. D, 1970, Columbia University. 199 p.

Molnar, Ralph E. Jaw musculature and jaw mechanics of the Eocene Crocodilian Sebecus icaeorhinus. M, 1969, University of Texas, Austin.

Molnia, Bruce Franklin. Pleistocene ice rafting in the North Atlantic Ocean. D, 1972, University of South Carolina. 110 p.

Molnia, Bruce Franklin. The origin and distribution of calcareous fines on the North and South Carolina continental margin. M, 1969, Duke University. 127 p.

Molson, J. W. H. Three-dimensional numerical simulation of groundwater flow and contaminant transport at the Borden Landfill. M, 1988, University of Waterloo. 169 p.

Momayezzadeh, Mohammed. An inexpensive system of geophysical data acquisition. M, 1987, McGill University. 184 p.

Momen, Hassan Mostafa. Modeling and analysis of cyclic behavior of sands. D, 1980, University of Illinois, Urbana. 443 p.

Monaghan, Marc Courtney. Be-10 in the atmosphere and soils. D, 1984, Yale University. 421 p.

Monahan, Edward James. Pore pressure development in a bearing capacity test on an overconsolidated clay model. D, 1968, Oklahoma State University. 120 p.

Monahan, Robert H. Sedimentology of an unnamed sandstone within the Mowry Shale of central Montana. M, 1985, University of Wisconsin-Madison. 86 p.

Monastero, Francis C. Tasman Basin (Pacific Ocean) sedimentation; patterns and processes. M, 1972, Florida State University.

Moncure, George Kinser. Depositional environment of the Green River Formation in the vicinity of the Douglas Creek Arch, Colorado and Utah. M, 1979, University of Wyoming. 55 p.

Moncure, George Kinser. Zeolite diagenesis below Pahute Mesa, Nevada Test Site. D, 1980, University of Wyoming. 97 p.

Mondschein, Herman F. A procedure for predicting flow at Keokuk Dam (Iowa). M, 1958, St. Louis University.

Mondy, Holland H. The areal geology of the Greenleaf area, Cherokee and Muskogee counties, Oklahoma. M, 1950, University of Oklahoma. 72 p.

Monell, Joseph. The metallurgy of nickel and cobalt with special reference to the Missouri ores. M, 1881, Washington University.

Monet, William Francis. A geoacoustic model of seafloor reflectors between the Murray and Mendocino fracture zones in the Northeast Pacific Ocean. M, 1987, George Washington University. 152 p.

Money, Nancy R. A seismic investigation of the North Golden area, Jefferson County, Colorado. M, 1977, Colorado School of Mines. 56 p.

Money, Peter L. The Precambrian geology of the Needle Falls area, Saskatchewan (Canada). D, 1967, University of Alberta. 251 p.

Money, Peter Lawrence. The geology of Hawkesbury Island, Skeena mining division, British Columbia. M, 1959, University of British Columbia.

Moneymaker, Beren C. The caves of East Tennessee. M, 1929, University of Tennessee, Knoxville. 63 p.

Monger, Hugh Curtis. Geochemical and mineralogical properties of Copper Ridge and Chepultepec regolith at the Oak Ridge National Laboratory Reservation-West Chestnut Ridge Site. M, 1986, University of Tennessee, Knoxville. 112 p.

Monger, James W. H. Stratigraphy of the Kereford Limestone in eastern Kansas. M, 1961, University of Kansas. 107 p.

Monger, James William Heron. The stratigraphy and structure of the type area of the Chilliwack Group (Late Pennsylvanian–Late Permian), southwestern British Columbia. D, 1967, University of British Columbia.

Monger, James William Heron. The stratigraphy and structure of the type-area of the Chilliwack Group, southwestern British Columbia. D, 1966, University of British Columbia.

Mongillo, Joseph C. Recent sediments at Hammonasset Beach, (Madison, Connecticut). M, 1968, Southern Connecticut State College.

Moniz, Antonio Carlos. Formation of an Oxisol-Ultisol transition in Sao Paulo, Brazil. D, 1980, North Carolina State University. 208 p.

Monk, George D. Analytical study of some factors affecting the permeability and porosity of unconsolidated sands. M, 1941, University of Chicago. 53 p.

Monk, Wilfred Jerale. A faunal study of the Fremont Formation in the Canon City Embayment, Colorado. M, 1954, University of Oklahoma. 87 p.

Monks, Edwin Tod. Hydrogeochemistry of two small streams, Bobcat Creek and Tributary Creek, in monolithologic basins at low flow, Gravelly Range, southwestern Montana. M, 1988, Wright State University. 158 p.

Monks, Katherine Schauder. Geochemical analysis of Cold Creek and Blayne Spring, Ennis, Montana. M, 1987, Wright State University. 173 p.

Monmonier, Mark Stephen. Upland accordance in the Ridge and Valley section of Pennsylvania. M, 1967, Pennsylvania State University, University Park. 58 p.

Monnens, Lee Edwin. The stratigraphy of the Cenozoic deposits in the Cave Hills, northwestern South Dakota. M, 1980, Iowa State University of Science and Technology.

Monnett, Victor Brown. The (Mississippian) Marshall Formation of Michigan. D, 1947, University of Michigan.

Monnett, Victor E. The geology of the northern Black Hills. M, 1912, University of Oklahoma.

Monnett, Victor E. The glacial physiography of the Skaneateles Basin (New York). D, 1922, Cornell University.

Monney, Neil Thomas. Engineering aspects of the ocean floor. D, 1967, University of Washington. 133 p.

Monrad, John R. Peedee River floodplain development, Marlboro and Darlington counties, South Carolina. M, 1972, Duke University.

Monrad, John Raymond. Geochemistry and geochronology of sialic terranes in the Hassan District, southern India. D, 1983, University of North Carolina, Chapel Hill. 131 p.

Monreal, Rogelio. Lithofacies, depositional environments, and diagenesis of the Mural Limestone (Lower Cretaceous), Lee siding area, Cochise County, Arizona. M, 1985, University of Arizona. 101 p.

Monroe, Charles Jr. A model for distribution of mineable Pocahontas No. 6 seam in southern West Virginia. M, 1980, Eastern Kentucky University. 42 p.

Monroe, Eugene Allen. Genesis of bentonite. M, 1959, University of Illinois, Urbana.

Monroe, Eugene Allen. Mineralogical reactions between fireclay ladle bricks and basic slags at high temperature. D, 1961, University of Illinois, Urbana. 68 p.

Monroe, Frederick Fales. Geography, geology and potential resources of the ocean floor. M, 1970, American University. 103 p.

Monroe, James Stewart. Vertebrate paleontology, stratigraphy and sedimentation of the upper Ruby River basin, Madison County, Montana. D, 1976, University of Montana. 301 p.

Monroe, John Napier. Origin of the clastic dikes of Northeast Texas. M, 1949, Southern Methodist University. 43 p.

Monroe, Scott, C. Gold solubilities and transport mechanisms near Sheep Mountain, Beaverhead County, Montana. M, 1980, Western Washington University. 43 p.

Monroe, Sheila A. Petrographic and geochemical characteristics of bedded cherts within the Jurassic Otter Point Formation, southwestern Oregon. M, 1987, University of Oregon. 120 p.

Monroe, William Allen. Petrology of the Domengine Formation (middle Eocene), west side of the Sacramento Valley, California. M, 1965, University of California, Davis. 110 p.

Monsalve, Obdulio Alfonso. Geology of the San Arroyo gas field, Grand County, Utah. M, 1972, University of Utah. 59 p.

Monsees, James Eugene. Design of support systems for tunnels in rock. D, 1970, University of Illinois, Urbana. 265 p.

Monsen, Susan Ann. Structural evolution and metamorphic petrology of the Precambrian-Cambrian strata, Northwest Bare Mountain, Nevada. M, 1983, University of California, Davis. 66 p.

Monson, K. David. Intramolecular patterns of isotopic order in biosynthesized fatty acids. D, 1981, Indiana University, Bloomington. 186 p.

Monson, Lawrence Milton. The mineralogy and chemistry of natural pseudoleucite. M, 1979, University of Utah. 110 p.

Monsour, Edward. Fossil corals of the genus Turbinolia from the Gulf Coast. M, 1942, Louisiana State University.

Monsour, Eli Thomas. Range and distribution of the family Verneuilinidae in the Tertiary of the Gulf Coast. M, 1932, Louisiana State University.

Monsour, Emil. Faunal zonation of the Jackson Eocene of Mississippi. M, 1936, Louisiana State University.

Montag, Rafael L. Rb-Sr geochronology of selected glauconite morphologies of the Upper Cretaceous (Navesink Fornmation) of New Jersey. M, 1978, Brooklyn College (CUNY).

Montagne, Clifford. Quaternary and environmental geology of part of the West Fork Basin, Gallatin County, Montana. M, 1971, Montana State University. 89 p.

Montagne, Clifford. Slope stability evaluation for land capability reconnaissance in the northern Rocky Mountains. D, 1976, Montana State University. 229 p.

Montagne, H. W. Crust development in a titaniferous ferruginous Latosol on Kauai, Hawaii. M, 1970, University of Hawaii. 104 p.

Montagne, John de la *see* de la Montagne, John

Montalbetti, James. Analysis of data from a broad-band-three-component digital seismic system. M, 1969, University of Alberta. 117 p.

Montanari, Alessandro. Event stratigraphy of Cretaceous and Tertiary pelagic limestones from the Northern Apennines, Italy. D, 1986, University of California, Berkeley. 287 p.

Montazer, P. M. Engineering geology of upper Bear Creek area, Clear Creek County, Colorado. M, 1978, Colorado School of Mines. 229 p.

Montazer, Parviz. Permeability of unsaturated, fractured metamorphic rocks near an underground opening. D, 1982, Colorado School of Mines. 617 p.

Monte, J. A. The impact of petroleum dredging on Louisiana's coastal landscape; a plant biogeographical analysis and resource assessment of spoil bank habitats in the Bayou Lafourche Delta. D, 1978, Louisiana State University. 352 p.

Monte, J. L. One-dimensional mathematical model for large strain consolidation. D, 1975, Northwestern University. 73 p.

Monte, Lois Del *see* Del Monte, Lois

Monteath, G. M. Environmental analysis of the sediments of southern Monterey bay, California. M, 1965, United States Naval Academy.

Montellano, Marisol. Mammalian fauna of the Judith River Formation (Late Cretaceous, Judithian), northcentral Montana. D, 1986, University of California, Berkeley. 196 p.

Montenyohl, Victor I. Tensile strength anisotropy of the Winnsboro Granite. M, 1974, University of South Carolina. 33 p.

Montenyohl, Victor I. Tensional anisotropy of the Ridd Adamellite. D, 1976, University of South Carolina. 77 p.

Montero, Felipe J. Aerial triangulation using auxiliary data. M, 1959, Ohio State University.

Montero, Felipe J. The application of some adjustments procedures to a block triangulation. D, 1963, Ohio State University.

Montes, Hernan A. Ionospheric background motions; relevance to upper atmospheric structure and geomagnetic micropulsations. D, 1970, Columbia University. 113 p.

Monteverde, Donald H. Tonalite and trondhjemite plutonism in the western Chugach Mountains, So. Alaska; an example of near trench magmatism. M, 1984, Lehigh University. 82 p.

Montgomery, A. N. Water movement in selected fine textured soils in the region of Haldimand-Norfolk, southern Ontario. M, 1983, University of Guelph.

Montgomery, Arthur. Pre-Cambrian geology of the Picuris Range, north-central New Mexico. D, 1951, Harvard University.

Montgomery, Carla Paige Westlund. Uranium-lead isotopic investigation of the Archean Imataca Complex, Guayana Shield, Venezuela. D, 1977, Massachusetts Institute of Technology. 261 p.

Montgomery, Carla Westlund. Contact metamorphism at Mont Rougemont, Quebec. M, 1974, Dartmouth College. 66 p.

Montgomery, Edward Sharar. Geology of part of the southern Butte Mountains, White Pine County, Nevada. M, 1963, Stanford University. 61 p.

Montgomery, Eric Lee. Facies development and porosity relationships in the Dundee Limestone of Gladwin County, Michigan. M, 1986, Western Michigan University.

Montgomery, Errol Lee. Determination of coefficient of storage by use of gravity measurements. D, 1971, University of Arizona. 173 p.

Montgomery, Errol Lee. The geology and ground water investigation of the Tres Alamos Dam site area of the San Pedro River, Cochise County, Arizona. M, 1963, University of Arizona.

Montgomery, Gerald Edward. Aeromagnetic study of part of the Ross Island and Taylor Glacier quadrangles, Antarctica. M, 1972, Northern Illinois University. 49 p.

Montgomery, Homer Albert, Jr. Paleozoic paleogeography of northeastern Mexico. D, 1988, University of Texas at Dallas. 223 p.

Montgomery, Hugh Brinton. Geology of the Mayburn Mines property, Kenora District, Ontario. M, 1957, Pennsylvania State University, University Park. 92 p.

Montgomery, James Alan. The geomorphic evolution of the Taylor Black Prairie between the Trinity and Colorado rivers, central Texas. M, 1986, Baylor University. 148 p.

Montgomery, James H. The pre-Chattanooga strata of the Illinois River valley, northeastern Oklahoma. M, 1950, University of Tulsa. 61 p.

Montgomery, Jerry R. Regional gravity study of western Utah. D, 1973, University of Utah. 142 p.

Montgomery, Joel Kenneth. Geology of the Nimrod area, Granite County, Montana. M, 1958, University of Montana. 61 p.

Montgomery, John R. Structural relations of the southern Quesnel Lake Gneiss, Isosceles Mountain area, southwest Cariboo Mountains, British Columbia. M, 1985, University of British Columbia.

Montgomery, Joseph Hilton. A study of barium minerals from the Yukon Territory. M, 1960, University of British Columbia.

Montgomery, Joseph Hilton. Petrology, structure, and origin of the Copper Mountain intrusions (about Triassic-Jurassic boundary) near Princeton, British Columbia. D, 1968, University of British Columbia.

Montgomery, Keith. History, form and function in fluvial systems; the Quaternary evolution of the lower Grand River, Ontario. D, 1986, University of Waterloo. 374 p.

Montgomery, Keith. The synthesis and dissolution of sodalite; implications for nuclear waste disposal. M, 1986, University of Alberta. 145 p.

Montgomery, Michael. Geology and landsliding along the coastline between Lechuza and Seqult Point, Los Angeles County, California. M, 1981, California State University, Long Beach. 184 p.

Montgomery, R. T. Environmental and ecological studies of the diatom communities associated with the coral reefs of the Florida Keys. D, 1978, Florida State University. 549 p.

Montgomery, Scott Lyons. Structural and metamorphic history of the Lake Dunford map area, Cariboo Mountains, British Columbia; ophiolite obduction in the southeastern Canadian Cordillera. M, 1978, Cornell University.

Montgomery, Sonya Paris. Actinolite (byssolite) and stilpnomelane from French Creek, Pennsylvania. M, 1963, Bryn Mawr College. 41 p.

Montgomery, William Willson. Deformation of the Tyler Slate (middle Precambrian) in northern Wisconsin and western upper Michigan. M, 1977, University of Wisconsin-Madison.

Monti, Joseph. Seismic mapping of alluvial fans and sub-fan bedrock in Big Bend National Park, Texas. M, 1984, Texas A&M University. 52 p.

Montoya, Rodrigo Araya *see* Araya Montoya, Rodrigo

Monty, Claude Leopold Victor. Geological and environmental significance of Cyanophyta. D, 1965, Princeton University. 598 p.

Montz, Melissa Jean. Environmental geology of the Rio Salado Development District, central part. M, 1982, Arizona State University. 66 p.

Monz, David J. Cross-structural development at the southwestern termination of Walker Mountain, Virginia. M, 1985, Virginia Polytechnic Institute and State University.

Monzon, Felipe G. A uranium anomaly in the Silver Bell prospect, Sweetwater County, Wyoming. M, 1969, University of Arizona.

Moo, Charles Anthony. Nonlinear interactions of acoustic-gravity waves. D, 1976, Massachusetts Institute of Technology. 123 p.

Moodie, Roy Lee. A contribution to a monograph of the extinct Amphibia of North America; new forms from the Carboniferous. D, 1908, University of Chicago. 44 p.

Moody, David Wright. Coastal morphology and processes in relation to the development of submarine sand ridges off Bethany Beach, Delaware. D, 1964, The Johns Hopkins University. 168 p.

Moody, Donald S. Trace element distribution and geochemistry of the Kern mines. M, 1973, California State University, Fresno.

Moody, Dwight Millington. The geology of the Rock Island (No. 1) coal in northwestern Illinois; Rock Island, Henry, Mercer and Warren counties. M, 1959, University of Illinois, Urbana.

Moody, J. B. Serpentinization of iron-bearing olivines; an experimental study. D, 1974, McGill University.

Moody, Jack R. Dolomite bodies in the St. Louis Limestone Member of the Newman Limestone (Mississippian) of east-central Kentucky. M, 1982, Eastern Kentucky University. 45 p.

Moody, John Drummond. The subsurface geology of southeastern Colorado. M, 1947, Colorado School of Mines. 70 p.

Moody, Marjorie. Characteristics and genesis of a Devono-Carboniferous debris flow and bounding facies; West Morocco. M, 1973, University of South Carolina.

Moody, Richard H., Jr. The geology of the Oceanographer transform fault. M, 1982, SUNY at Albany. 163 p.

Moody, Ula Laura. Distribution and characterization of tephra from the May 18, 1980, Mount Saint Helens eruption in northern Idaho and western Montana. D, 1982, University of Idaho. 386 p.

Moody, Ula Laura. Late Quaternary stratigraphy of the Channeled Scabland and adjacent areas. D, 1987, University of Idaho. 419 p.

Moody, Ula Laura. Microstratigraphy, paleoecology, and tephrochronology of the Lind Coulee Site, central Washington. D, 1978, Washington State University. 273 p.

Moody, William Clyde, Jr. Origin of the Plumbago Mountain mafic-ultramafic pluton in the Rumford Quadrangle, Maine, U.S.A. M, 1974, University of Wisconsin-Madison. 90 p.

Mooers, Howard DuWayne. Quaternary history and ice dynamics of the late Wisconsin and Superior lobes, central Minnesota. D, 1988, University of Minnesota, Minneapolis. 262 p.

Moog, Polly Lu. The hydrogeology and freshwater influx of Buttermilk Bay, Massachusetts with regard to the circulation of coliform and pollutants; a model study and development of methods for general application. M, 1987, Boston University. 166 p.

Mook, Anita Louise. Petrography of the Athol Quadrangle, Massachusetts. M, 1967, University of Michigan.

Mook, Charles C. A statistical study of variation in Spirifer mucronatus. M, 1914, Columbia University, Teachers College.

Moomaw, Benjamin Franklin. The manganese deposits on the Buckthorn survey of Giles and Bland counties, Virginia. M, 1932, Virginia Polytechnic Institute and State University.

Moon, Charles Gardley. A study of the igneous rocks of Travis County, Texas. M, 1942, University of Texas, Austin.

Moon, Charles Gardley. Geology of Agua Fria Quadrangle, Brewster County, Texas. D, 1950, University of Texas, Austin.

Moon, David Earl. The genesis of three Podzol-like soils occurring over a climatic gradient on Vancouver Island. D, 1982, University of British Columbia.

Moon, Evangeline A. The development of Sphenotrochus crispus. M, 1910, Columbia University, Teachers College.

Moon, Hyunkoo. Elastic moduli of well-jointed rock masses. D, 1987, University of Utah. 283 p.

Moon, Hyunkoo. Study of modeling materials for physical modeling of Utah coal mines. M, 1983, University of Utah. 117 p.

Moon, Morgan Ray. Development cost analysis for a small mine. M, 1969, University of Utah. 751 p.

Moon, Thomas Scott. Trans-domain TRM. D, 1985, University of Washington. 195 p.

Moon, W. The variational formulation and finite element type solution of free oscillation of rotating laterally heterogeneous Earth. D, 1976, University of British Columbia.

Moon, Warren D. Acoustical transmission of an ocean bottom sedimentary layer. M, 1966, Massachusetts Institute of Technology. 46 p.

Moon, William A. Geology of the Poplar Hill area, Giles County, Virginia. M, 1961, Virginia Polytechnic Institute and State University.

Mooney, Albert Russell. Geology of the Torrey Creek area, Fremont County, Wyoming. M, 1952, University of Iowa. 172 p.

Mooney, Harold Morton. A study of the energy contained in the seismic waves P and pP. D, 1950, California Institute of Technology. 129 p.

Mooney, Thomas Rodney. The stratigraphy of the Floridan Aquifer east and northeast of Lake Okeechobee, Florida. M, 1979, Florida State University.

Mooney, Tom D. Procedure for the operation of the Fayetteville seismograph station and interpretation of its records (Arkansas). M, 1961, University of Arkansas, Fayetteville.

Mooney, W. D. Seismic refraction studies of the Western Cordillera, Colombia and an East Pacific-Caribbean Ridge during the Jurassic and Cretaceous and the evolution of western Colombia. D, 1979, University of Wisconsin-Madison. 93 p.

Moor, Amanda. Stratigraphy and structure of Potosi Anticline, Nuevo Leon, Mexico. M, 1980, University of Texas, Austin.

Moor, Ann Lynnette. The geology of the Denton East Quadrangle, Denton County, Texas. M, 1973, University of Texas, Arlington. 68 p.

Mooradian, Michael Minas. The Ina Road Landfill as a source of ground-water pollution. M, 1983, University of Arizona. 86 p.

Moore, Arthur Howard. Lower Ordovician conodonts from the upper Jefferson City Formation of central and eastern Missouri. M, 1970, University of Missouri, Columbia.

Moore, Bernard N. Cretaceous paleontology of the Santa Ana Mountains. D, 1930, California Institute of Technology.

Moore, Bernard N. Geology of the southern Santa Ana Mountains, Orange County, California. D, 1930, California Institute of Technology. 139 p.

Moore, Beth Anne. Geophysical investigations of the Gable Mountain Pond-West Lake area, Hanford Site, south-central Washington. M, 1982, University of Idaho.

Moore, Billy R. Investigation of a felsic flow in northern Moore County, North Carolina. M, 1980, North Carolina State University. 70 p.

Moore, Bonnie K. Controls on ground-water availability and quality, the Bridger Canyon area, Bozeman, Montana. M, 1984, Montana State University. 187 p.

Moore, Brian Keith. Environments of deposition and diagenesis of the upper Clear Fork Group, Yoakum County, Texas. M, 1987, Texas Tech University. 133 p.

Moore, Bruce H. The application of airphotos to foundation problems. M, 1956, Purdue University.

Moore, Calvin Turner. Drainage districts of southeastern Nebraska. M, 1915, University of Nebraska, Lincoln.

Moore, Carl Allphin. Cephalopods from the Carboniferous Morrow Group of northern Arkansas and Oklahoma. M, 1938, University of Iowa. 60 p.

Moore, Carl Allphin. The Morrow Group of Adair County, Oklahoma. D, 1940, University of Iowa. 216 p.

Moore, Charles Aurelius, Jr. Mineralogical and pore fluid influences on deformation mechanisms in clay soils. D, 1968, University of California, Berkeley. 199 p.

Moore, Charles H., Jr. The geology and mineral resources of the Amherst Quadrangle, Virginia. D, 1940, Cornell University.

Moore, Charles H., Jr. The staurolite belts of Patrick and Henry counties, Virginia. M, 1937, University of Virginia. 69 p.

Moore, Clarence Victor. A petrographic study of the Columbus and Delaware limestones in Franklin and Delaware counties, Ohio. M, 1951, Ohio State University.

Moore, Clyde Herbert, Jr. Stratigraphy of the Fredericksburg division, south-central Texas. D, 1961, University of Texas, Austin.

Moore, Clyde Herbert, Jr. Stratigraphy of the Walnut Formation, south-central Texas. M, 1959, University of Texas, Austin.

Moore, Craig H. Geology and petrology of the Mosca Pass area, Sangre de Cristo Mountains, Colorado. M, 1981, University of Akron. 72 p.

Moore, Craig Hayden. Simulation of natural geochemical systems using a coupled transport/reaction model; an evaluation using the fluorine zone of quartz-cassiterite greisens as an example. D, 1987, Indiana University, Bloomington. 199 p.

Moore, D. E. Sedimentary, deformational, and metamorphic history of Franciscan conglomerates of the Diablo Range, California. D, 1978, Stanford University. 277 p.

Moore, Dan. Subsurface geology and potential for geopressured-geothermal energy in the Turtle Bayou Field area, Terrebonne Parish, Louisiana. M, 1982, University of Southwestern Louisiana.

Moore, Darrell C. Devonian and Mississippian depositional patterns in southwestern New Mexico. M, 1984, Texas Tech University. 109 p.

Moore, David. An evaluation of the use of primary sedimentary fabrics for the determination of strain in metaconglomerates, southeastern California. M, 1980, San Jose State University. 115 p.

Moore, David G. The marine geology of San Pedro Shelf. M, 1952, University of Southern California.

Moore, David John. A finite element model of a glacier. M, 1977, Rensselaer Polytechnic Institute. 84 p.

Moore, David L. Geophysical signatures to photolinears at the Cross-Bar Wellfield, Pasco County, Florida. M, 1981, University of South Florida, Tampa. 73 p.

Moore, David Lafayette. Geology of the Santa Rosalia mine area, District of Arizpe, Sonora, Mexico. M, 1948, University of Arizona.

Moore, David Ross. The Silurian fossils of Franklin County, Indiana. M, 1886, Indiana University, Bloomington.

Moore, David W. Burial depth and stratigraphic controls on shale diagenesis. M, 1983, Texas A&M University. 94 p.

Moore, David Warren. Geomorphology of the Deep River drainage basin and Carthage area, North Carolina. M, 1972, University of North Carolina, Chapel Hill. 81 p.

Moore, David Warren. Stratigraphy of till and lake beds of late Wisconsinan age in Iroquois and neighboring counties, Illinois. D, 1981, University of Illinois, Urbana. 211 p.

Moore, David William. Geology and geochemistry of a gold-bearing iron formation and associated rocks, Back River, N.W.T. M, 1977, University of Toronto.

Moore, Dennis Patrick. Rubble Creek landslide, Garibaldi, British Columbia. M, 1976, University of British Columbia.

Moore, Donald B. A subsurface investigation of the Gilbertown oil field (Choctaw County, Alabama). M, 1965, University of Alabama.

Moore, Donald Bruce. Recent coastal sediments, Double Point to Point San Pedro, California. M, 1965, University of California, Berkeley. 86 p.

Moore, Donald James. Ordovician and Cambrian fossils from a well core, Delta County, Michigan. M, 1962, Michigan State University. 52 p.

Moore, Donald P. Distribution of trace elements in selected samples of wallrock from the Boyd orebody, Ducktown, Tennessee. M, 1964, Virginia Polytechnic Institute and State University.

Moore, Duane Milton. Mineralogy of the White Pine Shale, eastern central Nevada. M, 1961, University of Illinois, Urbana.

Moore, Duane Milton. Selected trace elements of the White Pine Group, eastern Nevada. D, 1963, University of Illinois, Urbana. 88 p.

Moore, Dwight G., Jr. The niobium and tantalum content of some alkali igneous rocks. M, 1965, [University of New Mexico].

Moore, Dwight Garrison. Geology, mineralogy, and origin of feldspar rocks associated with alkalic carbonatitic complexes, northern Wet Mountains, Colorado. D, 1969, University of Michigan.

Moore, E. James. Determination of crustal structure by the dispersion of Rayleigh waves. D, 1951, University of Michigan.

Moore, Elmer Glendon. Niagaran residues of central Tennessee. M, 1942, Vanderbilt University.

Moore, Elwood S. Geology of the Onaman iron range area, District of Thunder Bay, Ontario. D, 1909, University of Chicago. 60 p.

Moore, Fred Edward. Petrology of the Green River oil shales. M, 1950, St. Louis University.

Moore, Fred Edward. The geomorphic evolution of the east flank of the Laramie Range, Colorado and Wyoming. D, 1959, University of Wyoming. 123 p.

Moore, Fred H. Geology of a portion of the Piedmont in the vicinity of Carter's Bridge, Virginia. M, 1931, University of Virginia. 96 p.

Moore, G. F. Structural geology and sedimentology of Nias Island, Indonesia; a study of subduction zone tectonics and sedimentation. D, 1978, Cornell University. 152 p.

Moore, Gary Lance. Diagenesis of deep marine carbonates; Deep Sea Drilling Project Sites. M, 1977, Memphis State University.

Moore, George E. Geology of the Williams Creek area, Hinsdale and Archuleta counties. M, 1964, Texas A&M University.

Moore, George Emerson, Jr. Stratigraphy of the northern half of the Columbia Quadrangle, Boone County, Missouri. M, 1938, University of Missouri, Columbia.

Moore, George Emerson, Jr. Structure and metamorphism of the Keene-Brattleboro area, New Hampshire-Vermont. D, 1948, Harvard University.

Moore, George Thomas. Intrusions in the Middle Cambrian formations, Cottonwood Canyon, Jefferson County, Montana. M, 1954, Indiana University, Bloomington. 45 p.

Moore, George Thomas. The geology of the Mount Fleecer area, Montana. D, 1956, Indiana University, Bloomington. 88 p.

Moore, George W. Origin and chemical composition of evaporite deposits. D, 1960, Yale University. 217 p.

Moore, Gilbert Parvin. The Devonian of Ralls County, Missouri. M, 1920, University of Missouri, Columbia.

Moore, Harry Leander, III. A systematic and paleoecologic review of the Coon Creek Fauna. M, 1974, University of Tennessee, Knoxville. 187 p.

Moore, Helen Louise. Pleistocene fauna of Centinela Park, Inglewood, California. M, 1937, University of Southern California.

Moore, Henry John, II. Transportation of heavy-metal sulfides as particulate matter. D, 1965, Stanford University. 120 p.

Moore, Howard Earl. The geochemistry of the inert gases in natural gas. D, 1960, University of Arkansas, Fayetteville. 89 p.

Moore, Ian D. Infiltration into tillage affected soils. D, 1979, University of Minnesota, Minneapolis. 233 p.

Moore, J. Casey. Geologic studies of the Cretaceous? flysch, southwestern Alaska. D, 1971, Princeton University. 130 p.

Moore, James Allan. Mineralogy, geochemistry and petrogenesis of the lavas of Mount Erebus, Antarctica. M, 1986, New Mexico Institute of Mining and Technology. 277 p.

Moore, James F., Jr. Geology of the northeast quarter of Camas Valley Quadrangle, Douglas County, Oregon. M, 1957, Oregon State University. 95 p.

Moore, James G. Geology of Mt. Rose area, Nevada. M, 1952, University of Washington. 87 p.

Moore, James G. Geology of the Sierra Nevada front near Mt. Baxter. D, 1955, The Johns Hopkins University.

Moore, James H. The geology of the Shooks Gap Quadrangle, Knox, Sevier, and Blount counties, Tenn. M, 1948, University of Tennessee, Knoxville. 47 p.

Moore, James L. A chemical and petrographic study of limestone-dolomite sequence in the Knox group (lower, middle, and upper Cambrian and lower Ordovician) of east Tennessee. M, 1968, University of Tennessee, Knoxville. 83 p.

Moore, James Leslie. Surficial geology of the southwestern Olympic Peninsula. M, 1965, University of Washington. 63 p.

Moore, Jamison S. Classification of unconsolidated formations and glacial features from regression modified ratio combinations. M, 1985, Indiana State University. 89 p.

Moore, John Byron. The structure and stratigraphy of the Ordovician limestones in Mill Creek Valley (Georgia). M, 1954, Emory University. 55 p.

Moore, John C. G. Classification of gold deposits of the Canadian Shield based on structural control. M, 1940, Cornell University.

Moore, John C. G. Geology of the Courageous-Matthews lakes gold belt, Northwest Territories, Canada. D, 1955, Harvard University.

Moore, John Ezra. Petrography of late Wisconsin tills in the northern part of the southern peninsula of Michigan. M, 1958, University of Illinois, Urbana. 50 p.

Moore, John Ezra. Petrography of northeastern Lake Michigan bottom sediments. D, 1960, University of Illinois, Urbana.

Moore, John M. Phase relations in the contact aureole of the Onawa Pluton, Maine. D, 1960, Massachusetts Institute of Technology. 229 p.

Moore, Johnnie Nathan. Stratigraphic comparison of the Precambrian Wyman and Johnnie formations in the western Great Basin, California. M, 1973, University of California, Los Angeles.

Moore, Johnnie Nathan. The Poleta Formation; a tidally dominated, open coastal and carbonate bank depositional complex, western Great Basin. D, 1976, University of California, Los Angeles. 312 p.

Moore, Joseph Neal. Mixed-volatile equilibria in calcareous rocks of three contact metamorphic aureoles in the western U.S. D, 1976, Pennsylvania State University, University Park. 89 p.

Moore, Joseph Neal. The northeast breccia dike; a petrologic study of a diatreme fissure from the Oka area, Quebec. M, 1972, Pennsylvania State University, University Park.

Moore, Kim. The petrology of the Lower Gallatin and Upper Gros Ventre formations (Cambrian), Dubois area, Wyoming. M, 1977, Miami University (Ohio). 111 p.

Moore, Larry Joe. Isotopic fractionation in Hawaiian volcanic gases. D, 1968, University of Hawaii. 122 p.

Moore, Leonard Vanard. Palynology of the H.T. Butte Lignite (Tongue River Formation) and the superadjacent Sentinel Butte Formation of western North Dakota (Paleocene). M, 1974, Southern Methodist University. 110 p.

Moore, Leslie Ray. Some Middle Ordovician Ostracoda from the Marathon Mountains. M, 1957, University of Oklahoma. 165 p.

Moore, Linda J. Petrology of the Paoli-Beaver Bend Member of the Newman Formation (Upper Mississippian) of southeast-central Kentucky. M, 1978, Eastern Kentucky University. 48 p.

Moore, M. B. An investigation of the permeability of a sand aquifer to determine the contaminant mass flux of selected contaminants from a landfill plume. M, 1986, University of Waterloo. 305 p.

Moore, Marcus Harvey. A restudy of the type localities of the upper Washita. M, 1928, Texas Christian University. 97 p.

Moore, Mark Power. The sediments of Lake Izabal, Guatemala, C.A. M, 1977, University of Florida. 75 p.

Moore, Michael. Stratigraphy and depositional environments of the Carboniferous Calico Bluff Formation, east central Alaska. M, 1980, University of Alaska, Fairbanks. 93 p.

Moore, Michael C. Age and depositional environment of the earliest Mississippian (Kinderhookian) rocks in western Illinois. M, 1970, University of Illinois, Urbana. 108 p.

Moore, Nelson Kinzly. Carbonate petrography and stratigraphy of the Nashville Group (Ordovician) in eastern Williamson County, Tennessee. M, 1972, Vanderbilt University.

Moore, Nelson Kinzly. Taxonomy and paleoenvironmental distribution of the benthic algal flora from Middle Ordovician carbonate rocks of East Tennessee. D, 1978, University of Tennessee, Knoxville. 148 p.

Moore, P. N. Scanning electron microscope examination of some archaeologically significant cherts of the western Tennessee River valley. M, 1978, Memphis State University.

Moore, Patricia D. A study of the interlayer stacking sequence of the Unst-type serpentine. M, 1972, University of Wisconsin-Madison.

Moore, Patrick Albert. The administration of pollution control in British Columbia; a focus on the mining industry. D, 1974, University of British Columbia.

Moore, Paul Brian. The crystal structure of laueite $MnFe_2(PO_4)_2$ $(H_2O)_6$ $\cdot 2H_2O$ and metastrengite, Fe^{+3} (PO_4) $(H_2O)_2$ and a crystallochemical classification of Fe-Mn orthophosphate. D, 1965, University of Chicago. 68 p.

Moore, Philip Alderson, Jr. Metal behavior in acid sulfate soils of Thailand. D, 1987, Louisiana State University. 161 p.

Moore, Prentiss D. The bryozoan fauna of the Arnheim Formation of Kentucky and Ohio. M, 1921, Indiana University, Bloomington.

Moore, R. L. Metamorphic petrology of the area between Mattawa, North Bay and Temiscaming, Ontario. D, 1976, Carleton University. 158 p.

Moore, Raymond A. The sands of the lower south Canadian River basin. M, 1925, University of Oklahoma. 62 p.

Moore, Raymond C. Early Mississippian formations in Missouri. D, 1916, University of Chicago. 283 p.

Moore, Raymond Kenworthy. Effects of crystallization, pressure, irradiation and trace elements on the thermoluminescence of synthetic fluorite. M, 1965, University of Florida. 98 p.

Moore, Raymond Kenworthy. Spectroscopic properties of the natural silicate garnets. D, 1969, Pennsylvania State University, University Park. 165 p.

Moore, Reginald George. A Paleocene fauna from the Hobak Formation, Wyoming. D, 1960, University of Michigan. 176 p.

Moore, Reginald George. Mollusks from the (Paleocene) Hoback Formation, western Wyoming. M, 1954, University of Michigan.

Moore, Return F. Geology of the pre-Cretaceous rocks in a portion of the Santa Ana Mountains (California). M, 1948, California Institute of Technology. 50 p.

Moore, Richard. Evidence indicative of former grounding-lines in the Great Bay region of New Hampshire. M, 1978, University of New Hampshire. 149 p.

Moore, Richard B. Geology, petrology and geochemistry of the eastern San Francisco volcanic field, Arizona. D, 1974, University of New Mexico. 350 p.

Moore, Richard B. Petrography of xenolith zones in the Black Face-Ames plutons, western San Juan mountains, Colorado. M, 1970, University of North Dakota. 216 p.

Moore, Richard C. Modeling hydraulic roughness in steep mountain streams. M, 1980, Colorado State University. 109 p.

Moore, Richard F. Magmatic and hydrothermal history of the Emigrant Gulch igneous complex, Park County, Montana. M, 1982, University of Montana. 155 p.

Moore, Richard Lee. Ordovician stratigraphy of the northern Black Hills, South Dakota. M, 1960, University of Iowa. 113 p.

Moore, Richard T. Geology of northwestern Mohave County, Arizona. M, 1958, University of Arizona.

Moore, Richard Thomas. A structural study of the Virgin and Beaverdam mountains, Arizona. D, 1966, Stanford University. 137 p.

Moore, Richard Thomas. A structural study of the Virgin and Beaverdam Mountains, Arizona. D, 1967, Stanford University. 137 p.

Moore, Richard W. Heavy mineral dispersal patterns of the abyssal plain and Louisiana inner shelf of the Gulf of Mexico. M, 1969, Texas A&M University.

Moore, Robert A. Cretaceous (?) stratigraphy of the southeast flank of the Empire Mountains, Pima County, Arizona. M, 1960, University of Arizona.

Moore, Roger Kent. Pre-Pleistocene topography, lithology and glacial drift thickness of Livingston and Shiawassee counties, Michigan. M, 1959, Michigan State University. 39 p.

Moore, Rosalie Carol. The geology of a portion of the west-central Markagunt Plateau, southwestern Utah. M, 1982, Kent State University, Kent. 72 p.

Moore, Rossie E., Jr. Geology of the Big Level Mountain-Vesuvius District, Augusta and Rockbridge counties, Virginia. M, 1952, University of Virginia. 114 p.

Moore, Ruth Albertine. Analytical methodology for the study of trace organic pollutants in Trinidad waterways. D, 1984, University of Waterloo. 212 p.

Moore, S. B. A petrographic study of the Big Injun Sandstone in Roane County, West Virginia. M, 1976, West Virginia University. 109 p.

Moore, Sally Pennington. The petrology and geochemistry of the Bisher Limestone and its associated zinc occurrences, Lewis County, Kentucky. M, 1978, University of Georgia.

Moore, Samuel L. Geology of a portion of the Apex Stock, Gilpin County, Colorado. M, 1952, University of Colorado.

Moore, Stanley Ralston. Report on Polaris Mine. M, 1906, University of Missouri, Rolla.

Moore, Stephen C. Geology and thrust fault tectonics of parts of the Argus and Slate ranges, Inyo County, California. D, 1976, University of Washington. 127 p.

Moore, Stephen Carlisle. Layered igneous rocks and intrusive units of the Tapto Lakes area, Whatcom County, Washington. M, 1972, University of Washington. 104 p.

Moore, Steven. Geology of a part of the southern Monte Cristo Range, Esmeralda County, Nevada. M, 1981, San Jose State University. 157 p.

Moore, T. Geology of the Baie d'Espoir-Kaegudeck region, Newfoundland. D, 1953, McGill University.

Moore, Teresa L. Foraminiferal biostratigraphy and paleoecology of the Miocene Pungo River Formation, central Onslow Embayment, North Carolina continental margin. M, 1986, East Carolina University. 180 p.

Moore, Theodore Carlton, Jr. Deep-sea sedimentation and Cenozoic stratigraphy in the central equatorial Pacific. D, 1968, University of California, San Diego.

Moore, Thomas E. Structure and petrology of the Sierra de San Andres Ophiolite, Vizcaino Peninsula, Baja California Sur, Mexico. M, 1976, San Diego State University.

Moore, Thomas Edward. Geology, petrology, and tectonic significance of the Mesozoic paleooceanic terranes of the Vizcaino Peninsula, Baja California Sur, Mexico. D, 1984, Stanford University. 418 p.

Moore, Thomas G. The geology and ore deposits of Morococha, Peru. D, 1936, Harvard University.

Moore, Thomas H. Geochemistry of waters of the Hog Creek Basin; Hamilton, Coryell, and Bosque counties, Texas. M, 1968, Baylor University. 106 p.

Moore, Thomas Howard. A new calorimetric method for determining heats of solution of minerals, and its application. D, 1955, McGill University.

Moore, Thomas Howard. Igneous dyke rocks of the Aillik-Makkovik area, Labrador. M, 1951, McGill University.

Moore, Thomas R. Genetic lithostratigraphy of the Dunkard Group, southwestern Pennsylvania and northern West Virginia. M, 1981, University of Missouri, Columbia.

Moore, Timothy Allen. Characteristics of coal bed splitting in the Anderson-Dietz coal seam (Paleocene), Powder River basin, Montana. M, 1986, University of Kentucky. 109 p.

Moore, Timothy Joseph. Grain sequences of samples from the West Farrington Pluton, North Carolina. M, 1974, Northwestern University.

Moore, V. Scott. The geomorphology of Frost Lake basin, Lynn-Terry counties, Texas. M, 1985, Texas Tech University. 194 p.

Moore, Vinton Aubrey. The Cretaceous stratigraphy and structure of northwestern Nebraska with special attention to the Chadron Dome. M, 1954, University of Nebraska, Lincoln.

Moore, Vinton Aubry. Relation of stratigraphy to basement rock patterns along the Chadron arch-Cambridge arch trend (Nebraska). D, 1970, University of Nebraska, Lincoln. 306 p.

Moore, W. S. Oceanic concentrations of radium-228 and a model for its supply. D, 1969, SUNY at Stony Brook.

Moore, Walter Leroy. Pennsylvanian foraminifera from the Big Saline Formation of the Llano Uplift of Texas. D, 1959, University of Wisconsin-Madison. 184 p.

Moore, Walter Leroy. The (Pennsylvanian) Virgilian fusulinids of Mockingbird Gap, New Mexico. M, 1954, University of Wisconsin-Madison.

Moore, Walter Richard. Heavy mineral dispersal patterns of the abyssal plain and Louisiana inner shelf of the Gulf of Mexico. M, 1969, Texas A&M University.

Moore, Walter Richard. Sedimentary history of the Casper Formation (Wolfcampian), Powder river basin, Wyoming. D, 1970, Texas A&M University. 280 p.

Moore, Wayne E. Some foraminifera from the (Upper Devonian) Tully Limestone of central New York. M, 1948, Cornell University.

Moore, Wayne E. The geology of Jackson County, Florida. D, 1950, Cornell University.

Moore, Wayne Ewing. Turonian megafauna of northeastern British Columbia. M, 1959, University of Alberta. 97 p.

Moore, Willard S. Uranium and thorium series inequilibria in sea water. M, 1964, Columbia University, Teachers College.

Moore, William A., Jr. Dimensions, flow rates, and rupture strength of Nautilus siphuncular tube. M, 1981, Brooklyn College (CUNY).

Moore, William Halsell, Jr. The detailed stratigraphy and paleontology of the Mississippian System of the area between Cooper Heights and Trenton, Georgia. M, 1954, Emory University. 53 p.

Moore, William Joseph. Igneous rocks in the Bingham mining district, Utah; a petrologic framework. D, 1970, Stanford University. 197 p.

Moore, William Joseph. Studies of carbonate aggregate reactions; expansion behavior, environmental effects, concrete matrix investigations. M, 1963, Iowa State University of Science and Technology.

Moore, William S. DeGaspe. Trace metal concentrations in water and sediments of the Trinity River system, Tarrant County, Texas. M, 1971, Texas Christian University.

Moorefield, Thomas P. Geologic processes and history of the Fort Fisher coastal area, North Carolina. M, 1978, East Carolina University. 100 p.

Moores, Eldridge Morton, III. Geology of the Currant area, Nye County, Nevada. D, 1963, Princeton University. 122 p.

Moores, Eugene A. Configuration and surface velocity profile of the Gulkana Glacier, central Alaska Range. M, 1962, University of Alaska, Fairbanks.

Moores, Eugene Albert. Regional drainage basin morphometry. D, 1966, Iowa State University of Science and Technology.

Moores, Richard C., II. The geology and ore deposits of a portion of the Harshaw District, Santa Cruz County, Arizona. M, 1972, University of Arizona.

Moorhead, Johnny Bob. The Crockett oil field, Crockett County, Texas. M, 1939, University of Texas, Austin.

Moorhouse, Michael David. The geology of a part of the California Lake area, Manitoba. M, 1957, University of Manitoba.

Moorhouse, Walter W. Geology of the zinc-lead deposit on Calumet Island, Quebec. D, 1941, Columbia University, Teachers College.

Moorhouse, Walter W. The stratigraphy and fauna of the (Ordovician) Trenton Limestone between Montreal and Quebec. M, 1936, University of Toronto.

Mooring, Carol Elizabeth. Petrogenesis of the Russian Peak ultramafic complex, northern California. M, 1978, University of Wisconsin-Madison.

Moorman, Mary. Paleontology of a nannoplankton assemblage from the Late Proterozoic Hector Formation, Alberta, Canada. M, 1972, University of California, Santa Barbara.

Moorshead, Frank Arthur. An investigation of stream infiltration in the carbonate Nittany Valley of South-central Pennsylvania. M, 1975, Pennsylvania State University, University Park. 95 p.

Moos, Daniel. Velocity, attenuation, and natural fractures in shallow boreholes. D, 1983, Stanford University. 110 p.

Moos, Milton. The age and significance of a paleosol in Fayette County, Ohio. M, 1970, Ohio State University.

Moose, Louis. Gravity study of Buckingham Mountain, Bucks County, Pennsylvania. M, 1967, Lehigh University.

Moose, Roger David. High resolution seismic reflection profiles in middle Delaware Bay. M, 1973, University of Delaware.

Mora, A. Roland. Geometry of Mesozoic folding and faulting in the Cerro Gordo area, Inyo Mountains, southeastern California. M, 1983, University of California, Los Angeles. 112 p.

Mora, Claudia Ines. Fluid-rock interaction in scapolite bearing Belt Group metasediments, northwest of the Idaho Batholith. D, 1988, University of Wisconsin-Madison. 312 p.

Mora, Claudia Ines. The temperature/pressure conditions of Grenville-age granulite-facies metamorphism of the Oaxacan complex, southern Mexico. M, 1983, Rice University. 81 p.

Mora, Jorge. Chatham fault zone, Old Chatham-East Chatham, New York; mesostructures and microstructures; their spatial and age relationships. M, 1982, SUNY at Albany. 145 p.

Mora, Peter Ronald. Elastic wavefield inversion. D, 1987, Stanford University. 160 p.

Mora, Stephen John de *see* de Mora, Stephen John

Morabbi, Mohammad. The geology and ore deposits of Cherry Creek Pluton, White Pine County, Nevada. M, 1980, Idaho State University. 92 p.

Morabit, Almoundir. The Lias reefs of the Middle Atlas Platform, Morocco. M, 1974, University of South Carolina.

Moraes, Joao A. De *see* De Moraes, Joao A.

Moraes, Joao A. P. Hydrogeology of the McBaine area, central Missouri. M, 1969, University of Missouri, Columbia.

Moraes, Joao A. P. de *see* de Moraes, Joao A. P.

Moraes, Jose Francisco Valente. Effect of phosphate on zinc adsorption on aluminum and iron hydrous oxides and in soils. D, 1982, University of California, Riverside. 184 p.

Morahan, George Thomas. Structural relationships of Paleozoic and Mesozoic rocks, northeastern Placer County, California. M, 1977, Texas Tech University. 76 p.

Morain, Stanley Alan. Ecological segregation of phytogeographic elements, upper Burdekin valley, north Queensland. D, 1970, University of Kansas. 214 p.

Morais, Francisco Ilton de Oliveira *see* de Oliveira Morais, Francisco Ilton

Morales C., Enrique. First step to obtain a single correlation characterizing the different patterns in waterflooding. M, 1975, Stanford University.

Morales, Frias, Gustavo. Ecology, distribution and taxonomy of Recent Ostracoda of Laguna de Terminos, Campeche, Mexico. D, 1965, Louisiana State University. 229 p.

Morales, Gustavo Adolfo. The Devonian and Lower Mississippian Charophyta of North America. M, 1962, University of Missouri, Columbia.

Morales, Pedro Augusto. A contribution to the knowledge of the Colombian Devonian strata and their faunas. M, 1965, Cornell University.

Morales-Frias, Gustavo Adolfo. Ecology, distribution and taxonomy of Recent Ostracoda of Laguna de Terminos, Campeche, Mexico. D, 1965, Louisiana State University. 229 p.

Moran, Andrew I. Allochthonous carbonate debris in Mesozoic flysch deposits in the Santa Ana Mountains, California, and their implications to the regional geology of Southern California. M, 1973, University of California, Riverside. 99 p.

Moran, David Rick. The effect of local geologic conditions on observed seismic intensities. M, 1985, Texas A&M University. 133 p.

Moran, Jean Elizabeth. Heat flow and the thermal evolution of Cascadia Basin. M, 1986, University of Washington. 96 p.

Moran, John L. Structure and stratigraphy of the Sheep Mountain area Centennial Range, Montana-Idaho. M, 1971, Oregon State University. 175 p.

Moran, Martin V. A hydrogeologic study of the Blackhawk Reclamation Site, Vigo County, Indiana. M, 1987, Indiana University, Bloomington. 150 p.

Moran, Mary E. Structural geology of the Commissary Ridge area, Bonneville County, Idaho. M, 1981, University of Wyoming. 127 p.

Moran, Mary Shanks. Differential aquifer development within the Ft. Payne Formation along the eastern Highland Rim, central Tennessee. M, 1977, Vanderbilt University.

Moran, Michael James. Oligocene calcareous nannofossil biostratigraphy, sedimentation rates and unconformities from Site 540, DSDP Leg 77. M, 1988, University of Nebraska, Lincoln. 62 p.

Moran, Robert E. Trace element content of a stream affected by metal-mine drainage, Bonanza, Colorado. D, 1974, University of Texas, Austin. 178 p.

Moran, Sidney Stuart. Structure of Candelaria area, Presidio County, Trans-Pecos, Texas. M, 1955, University of Texas, Austin.

Moran, Stephen Royse. Geology of the Hudson bay area, Saskatchewan, (Canada). D, 1969, University of Illinois, Urbana. 283 p.

Moran, Stephen Royse. Stratigraphy of Titusville Till in the Youngstown region, eastern Ohio. M, 1967, University of Illinois, Urbana.

Moran, William Edward. Fossil amphibian and reptilian occurrences in the Pennsylvanian and Permian strata of the Appalachian Geosyncline. M, 1942, George Washington University.

Morasse, Suzanne. Geological setting and evolution of the Lac Shortt gold deposit, Waswanipi, Quebec, Canada. M, 1988, Queen's University. 221 p.

Moravec, David. Study of the Concordia fault system near Jerico, Chiapas, Mexico. M, 1983, University of Texas, Arlington. 155 p.

Moravec, George Frank. The development of karren karst forms on the Newala Limestone in Dry Valley, Shelby County, Alabama. M, 1975, University of Alabama.

Morden, Audley D. Stratigraphy and insoluble residues of a Paleozoic section exposed near Cody, Wyoming. M, 1950, Wayne State University.

Mordoff, Richard Alan. The influence of large bodies of water on the temperatures of the surrounding country. M, 1918, Cornell University. 161 p.

More, Francis Ellsworth La *see* La More, Francis Ellsworth

More, Syver Wakeman. The geology and mineralization of the Antler Mine and vicinity, Mohave County, Arizona. M, 1980, University of Arizona. 149 p.

Morea, Michael Frank. On the species of Hoplophoneus and Eusmilus (Carnivora, Felidae). M, 1975, South Dakota School of Mines & Technology.

Morea, Michael Frank. The Massacre Lake local fauna (Mammalia, Hemingfordian) from northwestern Washoe County, Nevada. D, 1981, University of California, Riverside. 262 p.

Moreau, Alain. Contribution des méthodes géostatistiques récentes (1983) à l'évaluation des gisements d'or du Bouclier Canadien. M, 1987, Ecole Polytechnique. 223 p.

Moreau, Peter Allan. Dynamic analysis of the quartz microfabric in the Dry Creek quartzites, "S" structure, Jefferson County, Montana. M, 1978, Indiana University, Bloomington. 139 p.

Moredock, Duane E. Regional variations of hydrocarbons in the Edwards Limestone (Cretaceous) of South Texas. M, 1963, University of Houston.

Morehouse, David Frank. The Iowa bituminous coal industry. M, 1970, Iowa State University of Science and Technology.

Morehouse, George E. Geology of the Malachite Mine, Jefferson County, Colorado. M, 1950, Colorado School of Mines. 55 p.

Morehouse, Jeffrey Allen. A synopsis of the geologic and structural history of the Randsburg mining district. M, 1988, University of Arizona. 57 p.

Morel, J. D. A paleomagnetic investigation of the Lake Owens Complex, Medicine Bow Mountains, Wyoming. D, 1978, University of Wyoming. 73 p.

Morelan, Alexander Edward. The Avenal Formation (Eocene) of Reef Ridge, central California; a transgressive shelf facies succession. M, 1985, University of California, Los Angeles. 218 p.

Morell, Douglas J. Crystal chemistry of the Group II-B tungstates. M, 1978, Miami University (Ohio). 87 p.

Morell, Douglas Jeffrey. Analysis of groundwater impacts from selected waste management practices at base metal and precious metal waste disposal facilities. D, 1985, University of Idaho. 344 p.

Morelock, Jack. Sedimentation and mass physical properties of marine sediments, western Gulf of Mexico. D, 1967, University of Texas, Austin. 156 p.

Moreman, Walter L. Stratigraphy and paleontology of the Eagle Ford Formation of north and central Texas. D, 1931, University of Kansas. 75 p.

Moreman, Walter L. The Eagleford-Austin Chalk transition zone. M, 1926, Texas Christian University. 46 p.

Morency, Maurice L. Determination of strain effects of crystalline rocks by thermoluminescence. D, 1973, University of Kansas. 68 p.

Morency, Robert E., Jr. The geophysical expression of structures between the Clinton-Newbury and Bloody Bluff fault zones near Newbury, Massachusetts. M, 1986, University of New Hampshire. 96 p.

Moreno Agreda, Francisco. Contribution to the geology of the area between Pescadero Beach and Ano Nuevo Point, San Mateo County, California. M, 1958, Stanford University.

Moreno Agreda, Francisco. Pre-Cretaceous paleogeology of Mississippi and Alabama and parts of Tennessee and Kentucky. M, 1957, Stanford University.

Moreno, Jose Lee *see* Lee Moreno, Jose

Moreno-Hentz, Pedro E. Sedimentology of modern alluvial fans, Baja California, Mexico. M, 1985, McMaster University. 107 p.

Moresi, Cyril Killian. Geology of Lafayette and St. Martin parishes, Louisiana. M, 1932, Louisiana State University.

Moreton, Christopher. A geological, geochemical and structural analysis of the Lower Ordovician Turks Hill Cu-Zn-(Pb) volcanogenic massive sulphide deposit, central Newfoundland, Canada. M, 1984, Memorial University of Newfoundland. 323 p.

Moreton, E. Peter. The Boya tungsten-molybdenum stockwork-skarn deposit, northeastern British Columbia. M, 1984, Queen's University. 273 p.

Moretti, Frank Joseph. Petrographic study of the sandstones of the Jackfork Group, Ouachita Mountains, southeastern Oklahoma. D, 1958, University of Wisconsin-Madison. 135 p.

Moretti, Frank Joseph. Physical characteristics of some (Pennsylvanian) Desmoinesian limestones from Iowa. M, 1956, University of Wisconsin-Madison.

Moretti, George, Jr. Reference section for Paleozoic rocks in eastern Wisconsin; Van Driest No. 1, Sheboygan County, Wisconsin. M, 1971, University of Wisconsin-Milwaukee.

Moretz, Leonard C. Diagenesis of benthic foraminifera in the Miocene Pungo River Formation of Onslow Bay, North Carolina continental shelf. M, 1988, East Carolina University. 94 p.

Morey, Booker Williams. Dry conditioning for semi-taconite flotation by chemisorption of alcohols onto quartz. D, 1969, Stanford University. 123 p.

Morey, Caroll A. Glaciation in the Arapahoe and Albion valleys (Boulder County), Colorado. M, 1927, University of Colorado.

Morey, Erol D. A paleoenvironmental analysis of the Mississippian Caballero and lower Lake Valley formations, Sacramento Mountains, Otero County, New Mexico. M, 1985, Texas A&M University. 100 p.

Morey, Glenn B. Geology of the Keweenawan sediments near Duluth, Minn. M, 1960, University of Minnesota, Minneapolis. 111 p.

Morey, Glenn Bernhardt. The sedimentology of the Precambrian Rove Formation in northeastern Minnesota. D, 1965, University of Minnesota, Minneapolis. 336 p.

Morey, Phillip Stockton. Ostracoda of the Sylamore Sandstone of central Missouri. M, 1933, University of Missouri, Columbia.

Morgan, Allan V. Lithological and geomorphological studies near the erratics train, Calgary region, Alberta. M, 1966, University of Calgary.

Morgan, Arthur M. Correlation of the Cambrian of the Cordilleran Geosyncline. M, 1932, Columbia University, Teachers College.

Morgan, Arthur M. Geology and shallow water resources of the Roswell artesian basin, New Mexico. D, 1938, Columbia University, Teachers College.

Morgan, Baylus K. An analysis of controls on hydrocarbon occurrences in the Berea Sandstone, Lawrence County, Kentucky. M, 1983, Eastern Kentucky University. 44 p.

Morgan, Benjamin A., III. Geology of the Valencia area, Carabobo, Venezuela. D, 1967, Princeton University. 220 p.

Morgan, Benjamin Arthur, III. A gravity study of a ring-dike near Concord, North Carolina. M, 1963, University of North Carolina, Chapel Hill. 51 p.

Morgan, Bill Eugene. Foraminifera of the Pecan Gap Formation (Cretaceous) in northeastern Texas. M, 1962, University of Oklahoma. 169 p.

Morgan, Bill Eugene. Palynology of a portion of the El Reno Group (Permian) of southwestern Oklahoma. D, 1967, University of Oklahoma. 139 p.

Morgan, Bill R. Geology of southwestern Lincoln County, Oklahoma. M, 1958, University of Oklahoma. 129 p.

Morgan, Brenda E. The role of the atmosphere as the major source of sulfate in Illinois surface water. M, 1978, Northern Illinois University. 58 p.

Morgan, C. J. Exposure age dating of lunar features; lunar heavy rare gases. D, 1975, Washington University. 165 p.

Morgan, C. L. Field and laboratory examination of soil erosion as a function of erosivity and erodibility for selected hillslope soils from southern Ontario. D, 1979, University of Toronto.

Morgan, C. L. Nucleation and accumulation of marine ferromanganese deposits. D, 1975, University of Wisconsin-Madison. 161 p.

Morgan, Cecil G. Genesis and morphology of three Southwest Virginia soils which were developed from material weathered from limestone. M, 1941, Virginia Polytechnic Institute and State University.

Morgan, Cedwyn. Geochemistry of basalts from the Cobb Hotspot astride the Juan de Fuca Ridge. M, 1985, University of Massachusetts. 138 p.

Morgan, Charles Orville. The geology and groundwater resources of Washington County, Iowa. M, 1956, University of Iowa. 210 p.

Morgan, David. Paleomagnetism of tillites on the Avalon Peninsula, Newfoundland. M, 1980, Memorial University of Newfoundland. 134 p.

Morgan, David J. The Mississippi River delta; legal geomorphology evaluation of historic shoreline. M, 1973, Louisiana State University.

Morgan, David S. Hydrogeology of the Stillwater geothermal area, Churchill County, Nevada. M, 1979, [Stanford University].

Morgan, Douglass H. Lower Paleozoic chitinozoans and scolecodonts from North Dakota. M, 1964, University of North Dakota. 155 p.

Morgan, Frank Dale Oliver. Electronics of sulfide minerals; implications for induced polarization. D, 1981, Massachusetts Institute of Technology. 137 p.

Morgan, Gary Scott. Late Pleistocene fossil vertebrates from the Cayman Islands, British West Indies. M, 1977, University of Florida. 273 p.

Morgan, George B., Jr. Geology of Williams Canyon area, north of Manitou Springs, El Paso County, Colorado. M, 1950, Colorado School of Mines. 80 p.

Morgan, George Beers, VI. Alteration of amphibolitic wallrocks around the Tanco rare-element pegmatite, southeastern Manitoba. M, 1986, University of Oklahoma. 181 p.

Morgan, George Beers, VI. The igneous and metamorphic geochemistry of boron. D, 1988, University of Oklahoma. 408 p.

Morgan, George D. A contribution to the geology of Starr and Zapata counties, Texas. M, 1920, Columbia University, Teachers College.

Morgan, George D. Geology of the Stonewall Quadrangle, Oklahoma. D, 1924, Columbia University, Teachers College.

Morgan, J. R. Structure and stratigraphy of the northern part of the south Virgin Mountains, Clark County, Nevada. M, 1968, University of New Mexico. 103 p.

Morgan, James Cyrus, III. Depositional environment and diagenetic history of the Pettet Limestone (Lower Cretaceous) in the Rischers Store Field, Freestone County, Texas. M, 1987, University of Texas, Arlington. 98 p.

Morgan, James Leland. The correlation of certain Desmoinesian coal beds of Oklahoma by spores. M, 1955, University of Oklahoma. 118 p.

Morgan, James P. Mudlumps at the mouths of the Mississippi River. D, 1951, Louisiana State University.

Morgan, James Thornton. Petrography of the Verden Sandstone. M, 1955, University of Oklahoma. 72 p.

Morgan, John. Structure of the Finlayson Lake greenstone belt. M, 1978, University of Toronto.

Morgan, John. Three-dimensional strain in centrifuge models and an Archean greenstone belt. D, 1986, University of Toronto.

Morgan, John Harold. The application of accessory mineral methods to the Pre-Cambrian rocks of the Oxford House area. D, 1940, University of Manitoba.

Morgan, John Harrison. Bedrock geology of the southwest one-quarter of the Marion Quadrangle (Illinois). M, 1960, Southern Illinois University, Carbondale. 42 p.

Morgan, John Henry, II. Design of an instrument to measure the shear modulus of soft sediments. M, 1972, United States Naval Academy.

Morgan, John Milton, III. Computer-assisted application of the universal soil loss equation as a first approximation of the physical carrying capacity of backcountry areas for dispersed recreational use. D, 1980, University of Maryland. 205 p.

Morgan, Joseph K. Geology of the Otter Creek area, Washakie County, Wyoming. M, 1951, University of Wyoming. 55 p.

Morgan, K. M. Application of remote sensing airphoto interpretation to cropland erosion studies. D, 1979, University of Wisconsin-Madison. 124 p.

Morgan, Kirk A. Glacial marine sediment of the Central Arctic Basin and other high-latitude seas; a comparative study based on silt-clay grain size distributions. M, 1978, University of Wisconsin-Madison.

Morgan, Lisa Ann. Explosive rhyolitic volcanism on the eastern Snake River plain. D, 1988, University of Hawaii. 191 p.

Morgan, Nabil Assad. Geophysical studies in Lake Erie by shallow marine seismic methods. D, 1964, University of Toronto. 222 p.

Morgan, R. E. The effect of several environmental parameters on the phytotoxicity of triallate. M, 1973, University of Saskatchewan.

Morgan, Ralph Archie. Palynology of the Ozan Formation (Cretaceous), McCurtain County, Oklahoma. M, 1967, University of Oklahoma. 121 p.

Morgan, Richard. Geological aspects of Ohio archaeology. M, 1929, Ohio State University.

Morgan, Robert R. Methods of determining electric and magnetic fields of a dipole source across a fault. M, 1968, Colorado School of Mines. 64 p.

Morgan, Scott R. Geology and petrography of the Paleocene strata at Devils Slide, San Mateo County, California. M, 1981, Western Washington University. 224 p.

Morgan, Stanley Sherwood. Stratigraphy and petrography of the Pearlette Volcanic Ash in south-central Nebraska. M, 1962, University of Nebraska, Lincoln.

Morgan, Susan K. Petrology of passive-margin epeiric sea sediments; the Garden City Formation, north-central Utah. M, 1988, Utah State University. 168 p.

Morgan, T. H. A single correlation characterizing areal sweep for the different patterns in waterflooding after breakthrough. M, 1977, Stanford University. 47 p.

Morgan, Thomas G. Lithostratigraphy and paleontology of the Red Hill area, Eureka County, Nevada. M, 1974, University of California, Riverside. 112 p.

Morgan, Thomas Glen. The Middle Devonian fish faunas of central Nevada. D, 1980, University of California, Berkeley. 250 p.

Morgan, Thomas Richard. Three-dimensional migration velocity analysis in the space-frequency domain. D, 1981, University of Houston. 133 p.

Morgan, Warren P., Jr. Clay mineral descriptions and distributions, Altamaga, Doboy, and Sapelo sounds, Georgia. M, 1977, University of Georgia.

Morgan, William Andrew. Stratigraphy and paleoenvironments of three Permian units in the Cassia Mountains, central southern Idaho. M, 1977, University of Wisconsin-Madison.

Morgan, William Jason Phipps *see* Phipps Morgan, William Jason

Morgan, William Tony. Sedimentological, hydrological, and engineering controls on slope form, Indiana Dunes National Lakeshore. M, 1983, Indiana University, Bloomington. 105 p.

Morganelli, Daniel. Depositional environment and trend of the uppermost part of the Vamoosa Formation and Lecompton Limestone in the eastern part of north-central Oklahoma. M, 1976, Oklahoma State University. 68 p.

Morganti, John Michael. Geology and ore deposits of the Seven Devils Volcanics, Seven Devils Mining district, Hell's Canyon, Idaho, Idaho. M, 1972, Washington State University. 153 p.

Morganti, John Michael. The geology and ore deposits of the Howards Pass area, Yukon and Northwest territories; the origin of basinal sedimentary stratiform sulphide deposits. D, 1979, University of British Columbia.

Morganti, John Michael. The geology and ore deposits of the Howards Pass area, Yukon and Northwest Territories; the origin of basinal sedimentary stratiform sulphides deposits. D, 1980, University of British Columbia.

Morganwalp, David William. The geochemistry of the Julian Mine coal waste site in Knox County, Indiana. M, 1986, Indiana University, Bloomington. 323 p.

Morgenstein, M. E. Sedimentary diagenesis and manganese accretion on submarine platforms, Kauai Channel, Hawaii. D, 1974, University of Hawaii. 172 p.

Morgenstein, Maury. The composition and development of palagonite in deep-sea sediments (Recent) from the Atlantic and Pacific Oceans. M, 1969, Syracuse University.

Morgenstern, Lillian. The evolution of some rocks west and south of Boston, Massachusetts. D, 1970, Boston University. 132 p.

Morgridge, Dean L. The (Mississippian) Sappington Formation of southwestern Montana. M, 1955, University of Wisconsin-Madison. 67 p.

Mori, James Jiro. Localized stress drops within the rupture areas of great thrust earthquakes; importance to strong ground-motion. D, 1984, Columbia University, Teachers College. 148 p.

Mori, Kenji. Factors affecting the liquefaction characteristics of sands. D, 1977, University of California, Berkeley. 216 p.

Mori, Phillip Noel La *see* La Mori, Phillip Noel

Moriarty, Thomas D. Estimation of the direction of remanent magnetization; an inverse method using the phase spectrum of a magnetic anomaly. M, 1988, Texas A&M University. 96 p.

Morilla, Alberto Garcia. A preliminary assessment of the feasibility of using a shallow ground-water system for the cooling cycle of a geothermal power plant. M, 1976, University of Idaho. 117 p.

Morin, Claude. Anatomie d'un complexe péritidal en climat aride, Silurien supérieur, Gaspésie. M, 1986, Universite Laval. 80 p.

Morin, George C. Transverse dispersion through non-uniform porous media. M, 1968, University of Arizona.

Morin, James A. A study of the petrology of granitic rocks in the Tustin-Bridges area, northwestern Ontario. D, 1979, University of Saskatchewan. 191 p.

Morin, James Arthur. Petrology of the Clotty granite, Perrault Falls, Ontario (Archean). M, 1970, University of Manitoba.

Morin, Karen Marie. Analysis and definition of Campanian (Late Cretaceous) radiolarian populations characteristic of tropical to subtropical latitudes. D, 1982, University of Texas at Dallas. 375 p.

Morin, Kevin. The geohydrology and hydrogeochemistry of the proposed Garrison lignite mine. M, 1979, University of North Dakota. 104 p.

Morin, Kevin Andrew. Prediction of subsurface contaminant transport in acidic seepage from uranium tailings impoundments. D, 1983, University of Waterloo. 714 p.

Morin, Lorraine. High chloride ground water of the central Sierra Nevada, California. M, 19??, California State University, Fresno.

Morin, Marcel. Geology of the Lavrieville map-area, Saguenay County, Quebec. D, 1956, Universite Laval.

Morin, Marcel. Study of some conglomerates from southeastern Quebec. M, 1954, Universite Laval. 174 p.

Morin, Roger Henri. Thermophysical properties of deep sea sediments and their influence upon oceanic heat flow. D, 1982, University of Rhode Island. 176 p.

Morin, Ronald W. Late Quaternary biostratigraphy of cores from beneath the California Current. M, 1971, University of Southern California.

Morin, Wilbur Joseph. Foundation conditions; Cart Creek Bridge, Daggett County, Utah. M, 1963, University of Utah. 55 p.

Morin, Wilbur Joseph. Geotechnical properties of Ethiopian volcanic soils. D, 1969, University of Utah. 190 p.

Morin-Jansen, Ann. A study of the principal lineament and associated lineaments, Idaho National Engineering Laboratory. M, 1987, Idaho State University. 79 p.

Morisawa, Marie. A subsurface study of the basal Ordovician of the Powder River basin, Wyoming. M, 1952, University of Wyoming. 32 p.

Morisawa, Marie E. Relation of quantitative geomorphology to stream flow in representative watersheds of the Appalachian Plateau province. D, 1960, Columbia University, Teachers College.

Morison, Stephen Robert. Sedimentology of White Channel placer deposits, Klondike area, west-central Yukon. M, 1985, University of Alberta. 161 p.

Moriton, William Thomas. Stratigraphic investigations of selected sediments, Southwest Gulf of Mexico. M, 1969, Texas A&M University.

Moritz, C. A. A descriptive survey of the head of Carmel submarine canyon (California). M, 1968, United States Naval Academy.

Moritz, Carl A. Mesozoic stratigraphy of a portion of southwestern Montana. D, 1950, Harvard University. 136 p.

Moritz, Gail. Combined geochemical and temperature study in a near-shore environment as a petroleum exploration technique. M, 1982, University of Rochester. 50 p.

Moritz, Harold W. Muscovite-paragonite relationships in altered and mineralized meta-anorthosite at Chibougamau, Quebec, Canada. M, 1975, University of Georgia.

Moritz, Robert Peter. Geological and geochemical studies of the gold-bearing quartz fuchsite vein at the Dome Mine, Timmins area. D, 1988, McMaster University. 280 p.

Mork, Andrew R. Geology of Boulder Mountain Quadrangle, Cache County, Utah; with special reference to the petrography of the Middle Cambrian Ute Formation. M, 1987, Eastern Washington University. 69 p.

Morley, David Patterson. Lake-margin deposition in an ensialic rift basin; the Miocene Chalk Hills Formation of the southwestern Snake River plain, Idaho. M, 1983, University of Michigan.

Morley, Earl R., Jr. Geology of the Borrego Mountain Quadrangle and the western portion of the Shell Reef Quadrangle, San Diego County, California. M, 1963, University of Southern California.

Morley, J. J. Upper Pleistocene climatic variations in the South Atlantic derived from a quantitative radiolarian analysis; accent on the last 18,000 years. D, 1977, Columbia University, Teachers College. 359 p.

Morley, L. C. Magnetotellurics with a vertical magnetic dipole artificial source. M, 1975, University of Toronto.

Morley, Laurence Charles. Predictive techniques for marine multiple suppression. D, 1982, Stanford University. 73 p.

Morley, Lawrence W. Correlation of the susceptibility and remanent magnetism with the petrology of rocks from some Precambrian areas in Ontario. D, 1952, University of Toronto.

Morley, Lloyd Albert. Electroosmotic stabilization of mine materials. D, 1972, University of Utah. 134 p.

Morley, Raymond L. Joint pattern and trace element distribution in the igneous rocks of the Jarilla Mountains, NM. M, 1977, New Mexico Institute of Mining and Technology. 89 p.

Morneau, Richard A. A comparative study of seismic modeling techniques. M, 1983, Purdue University. 52 p.

Morningstar, Helen. The fauna of the Pottsville Formation of Ohio below the lower Mercer Limestone. D, 1921, Bryn Mawr College. 92 p.

Morogan, Viorica. Fenitization and ultimate rheomorphism of xenoliths from the Oldoinyo Lengai carbonatitic volcano, Tanzania. M, 1982, McGill University. 193 p.

Morra, Franco Piero. Geology and U deposits of the Charlebois-Higginson Lake area, northern Saskatchewan. M, 1977, University of Alberta. 202 p.

Morra, Matthew John. Gaps in the sulfur cycle; biogenic hydrogen-sulfide production and atmospheric deposition. D, 1986, Ohio State University. 261 p.

Morreale, Steve. Depositional environments of the lower Middle Ordovician Antelope Valley Limestone, central Nevada. M, 1981, University of Missouri, Columbia.

Morrell, R. P. Geology and mineral paragenesis of Franciscan metagraywacke near Paradise Flat, northwest of Pacheco Pass, California. M, 1978, Stanford University. 73 p.

Morrey, Margaret. The Lagenidae from the lower Tertiary of Mantua, Ecuador. M, 1928, Columbia University, Teachers College.

Morrice, Martin Gray. Mineralogy, petrology, and geochemistry of the Sangihe Arc; volcanism accompanying arc-arc collision in the Molucca Sea, Indonesia. D, 1982, University of California, Santa Cruz. 378 p.

Morrice, Martin Gray. The occurrence of tin at the Dickstone No. 2 orebody, northern Manitoba. M, 1974, University of Manitoba.

Morrill, David Currier. Recent sediments in the Merrimack River estuary and adjacent beaches, Massachusetts. M, 1958, University of Illinois, Urbana.

Morris, Arthur. Genesis of the sulphide mineralization at the Big Ledge Property, BC. M, 1948, University of British Columbia.

Morris, Arthur Edward. Phase equilibria in the system MnO-Mn2O3-SiO2. D, 1965, Pennsylvania State University, University Park. 81 p.

Morris, Bradley Allen. Analysis of controls of hydrocarbon production in limestone reservoirs of the Sligo Formation (Pettet Limestone), Rodessa Field, Cass County, Northeast Texas. M, 1985, Northeast Louisiana University. 73 p.

Morris, Charles Brady. Foraminifera of the Del Rio Shale (Cretaceous) of South Texas. M, 1955, University of Texas, Austin.

Morris, David Albert. A study of Lower Pennsylvanian cyclothems of Ohio. M, 1961, University of Kansas. 137 p.

Morris, David Albert. Lower Conemaugh (Pennsylvanian) depositional environments and paleogeography in the Appalachian Coal Basin. D, 1967, University of Kansas. 800 p.

Morris, David G. Depositional facies and diagenetic features of the Flippen Limestone, central Jones County, Texas. M, 1983, Baylor University. 90 p.

Morris, David Gordon. Streamflow synthesis employing a multi-zone hydrologic model with distributed rainfall and distributed parameters. D, 1977, Oklahoma State University. 241 p.

Morris, Donald Arthur. Foraminifera from the Oligocene Donnay Formation of Saipan. M, 1948, University of Nebraska, Lincoln.

Morris, Drew. Quadripole mapping near the Fly Ranch geothermal prospect, Northwest Nevada. M, 1975, Colorado School of Mines. 100 p.

Morris, Elliot C. Geology of the Big Piney area, Summit County, Utah. M, 1953, University of Utah. 66 p.

Morris, Elliot C. Mineral correlations of some Eocene sandstones of Central California. D, 1962, Stanford University. 85 p.

Morris, Frank R., IV. Karsy hydrogeology of Cedar Creek and adjacent basins in east-central Pulaski County, Kentucky. M, 1983, Eastern Kentucky University. 92 p.

Morris, Frederick K. Central Asia in Cretaceous time. D, 1936, Columbia University, Teachers College.

Morris, Frederick K. The faunules of the (Devonian) Hamilton beds of Morse's Creek (New York). M, 1910, Columbia University, Teachers College.

Morris, Frederick W., IV. Hydraulic measurements, data analysis, and rational design procedures for residential tidal canal networks. D, 1978, University of Florida. 603 p.

Morris, Gary R. Geology of the Dicks Peak area, Park County, Colorado. M, 1969, Colorado School of Mines. 69 p.

Morris, Gerald B. Comparison of changes in acoustic impedance in the ground with seismic reflections. M, 1957, Texas A&M University.

Morris, Gerald Brooks. The effect of rock density in synthesizing seismic reflection records. M, 1962, Texas A&M University. 43 p.

Morris, Gerald Brooks. Velocity anistrophy and crustal structure of the Hawaiian Arch. D, 1969, University of California, San Diego. 152 p.

Morris, H. R. Surface geology of the Farady uranium mine, Bancroft, Ontario. M, 1956, University of Toronto.

Morris, Hal T. Igneous rocks of East Tintic mining district. M, 1947, University of Utah. 30 p.

Morris, J. H. The geology of the Upper Beaver Mine, Gauthier Township, Ontario. M, 1974, University of Waterloo.

Morris, James Robert. A preliminary study of the importance of hydrothermal reactions on the temperature history of a hot, dry rock geothermal reservoir. M, 1975, Pennsylvania State University, University Park. 79 p.

Morris, Joe Lockhardt, Jr. Attenuation estimates from seismograms recorded in a deep well. M, 1979, Texas A&M University.

Morris, Julie Dianne. Enriched geochemical signatures in Aleutian and Indonesian arc lavas; an isotopic and trace element investigation. D, 1984, Massachusetts Institute of Technology. 320 p.

Morris, Kent D. Location of septic system failures in Southwest Missouri using false color infrared photography. M, 1981, Southwest Missouri State University. 38 p.

Morris, Kimberly Peck. The geology and petrology of the Circle volcanic-plutonic complex, near Circle City, east-central Alaska. M, 1985, University of California, Santa Barbara. 144 p.

Morris, Mark Steven. Biodegradation of organic contaminants in subsurface systems; kinetic and metabolic considerations. D, 1988, Virginia Polytechnic Institute and State University. 243 p.

Morris, Mark W. Study of rubidium, barium, and strontium fractionation in synthetic feldspars. M, 1973, Stanford University.

Morris, Marvin. Quantitative factors in mineral exploration. M, 1971, University of Arizona.

Morris, Mary Walker. A review of soil mapping and a descriptive index of soil maps of countries outside the United States. M, 1947, University of Michigan.

Morris, Michael M. Physical and biogenic sedimentary structures as depositional indicators in the Card-

ium Formation (Upper Cretaceous), southwestern Alberta. M, 1982, University of Georgia.

Morris, P. A. The bryozoan family Hippothoidae (Cheilostomata-Ascophora), with emphasis on the genus Hippothoa. D, 1975, University of California, Berkeley. 303 p.

Morris, P. G. A chemical, optical and X-ray study of certain zeolites. D, 1957, McGill University.

Morris, Peter Gerald. A petrological study of intrusive rocks along the Fraser Canyon near Hells Gate, BC. M, 1955, University of British Columbia.

Morris, Richard W. Geology and mineral deposits of the northern Cooks Range, Grant County, New Mexico. M, 1974, University of Texas at El Paso.

Morris, Robert Clarence. A geologic investigation of the Jackfork Group of Arkansas. D, 1965, University of Wisconsin-Madison. 242 p.

Morris, Robert Clarence. Geology of the Acorn and Y City quadrangles, Arkansas. M, 1962, University of Wisconsin-Madison.

Morris, Robert Hall. Heavy mineral analysis of sedimentary rocks of northern Alaska. M, 1952, Rutgers, The State University, New Brunswick. 74 p.

Morris, Robert Jones. Stratigraphy and sedimentary petrology of the Dakota Group, central Colorado Foothills. M, 1961, University of Oklahoma. 110 p.

Morris, Robert William. An Ordovician graptolite faunule from shales above the Table Head Limestone near Port au Port, Newfoundland. M, 1965, Columbia University. 39 p.

Morris, Robert William. Microfaunal analysis of the upper part of the Mancos Formation, Mesaverde group (upper Cretaceous), and basal Lewis Formation (upper Cretaceous), in northwestern Colorado. D, 1969, Columbia University. 242 p.

Morris, Robert Wynn. A restudy of the Waldron fauna at Hartsville, Indiana, with descriptions of new species. M, 1949, Miami University (Ohio). 33 p.

Morris, Sandra Lee. Diagenesis of the Permo-Triassic Ivishak Formation of the Sadlerochit Group, Prudhoe Bay Field, Alaska. M, 1987, University of Oregon. 211 p.

Morris, Sandra Plumlee. Variational normal mode computations for laterally heterogeneous Earth models. D, 1986, Stanford University. 192 p.

Morris, Scott Edward. The surficial debris cascade and hillslope evolutionary tendencies in the Colorado Front Range foothills. D, 1983, University of Colorado. 156 p.

Morris, Stephen John Samuel. An asymptotic method for determining the transport of heat and matter by creeping flows with strongly variable viscosity; fluid-dynamic problems motivated by island arc volcanism. D, 1981, The Johns Hopkins University. 165 p.

Morris, Stuart K. The geology and ore deposits of Mineral Mountain, Washington County, Utah. M, 1980, Brigham Young University.

Morris, Thomas Henry. Stable isotope stratigraphy and paleoceanography of the Arctic Ocean. D, 1986, University of Wisconsin-Madison. 206 p.

Morris, Thomas Henry. The stratigraphy and late Pleistocene sedimentological history of the Lomonosov Ridge-Makarov Basin, central Arctic Ocean. M, 1983, University of Wisconsin-Madison.

Morris, Thomas J. The nature of the contact of the Navarro and Midway groups at four localities in Falls, Milam, Travis, and Bastrop counties, Texas. M, 1959, Texas A&M University. 84 p.

Morris, W. J. Geochemistry and origin of the Yellow Dog Plains Peridotite, Marquette County, northern Michigan. M, 1977, Michigan State University. 75 p.

Morris, William Joseph. Part I; Eocene stratigraphy of the Washakie Basin, Wyoming and Colorado; Part III; a new species of anaptomorphid. D, 1951, Princeton University. 79 p.

Morris, William Lee. Geology of the Gordon Spring area, Whitfield, Catoosa, and Walker counties, Georgia. M, 1986, Emory University. 80 p.

Morris, William R. Sedimentological evolution of a Late Cretaceous submarine canyon; stratigraphy, petrology, and depositional processes in the Rosario Formation at San Carlos, Mexico. M, 1987, University of California, Santa Barbara. 129 p.

Morrisey, Norman S. The insoluble residues of the Madison Limestone, Wind River basin, Wyoming. M, 1950, University of Tulsa. 120 p.

Morrison, Andrew D. Growth of pure and doped single crystals of yttrium aluminum garnet. M, 1967, Boston University. 25 p.

Morrison, Anne Kranek. Hystrichospheres of the Onesquethaw Stage (Devonian) in parts of West Virginia, Virginia, and Maryland. M, 1964, University of Illinois, Urbana. 46 p.

Morrison, Bradford Crary. Stratigraphy of the Eau Claire Formation (Upper Cambrian) of west-central Wisconsin. M, 1968, University of Wisconsin-Madison.

Morrison, Charles E. Highway engineering. D, 1909, Columbia University, Teachers College.

Morrison, Charles Michael. Permian uranium-bearing sandstones on the Muenster-Waurika Arch and in the Red River area. M, 1977, Oklahoma State University. 60 p.

Morrison, Donald A. Reconnaissance geology of the Lochsa area, Idaho County, Idaho. D, 1968, University of Idaho. 126 p.

Morrison, Donald Allen. Geology and ore deposits of Kantishna District, Alaska. M, 1965, University of Alaska, Fairbanks. 109 p.

Morrison, Ernest Robert. Subsurface study of the Lansing Group in Gove and Trego counties, Kansas. M, 1979, West Texas State University. 53 p.

Morrison, Euen Ritchie. A study of Porphyry Mountain, Holland Township, Quebec. M, 1955, McGill University.

Morrison, Garrett L. The chemical composition and structure of staurolite. M, 1968, University of South Carolina.

Morrison, Garrett Louis. The structure of the interlayer water in montmorillonite. D, 1974, University of Oklahoma. 85 p.

Morrison, Greg. Geology of the La Negra Mine vicinity, State of Querétaro, Mexico. M, 1982, Colorado School of Mines. 118 p.

Morrison, Gregg William. Setting and origin of skarn deposits in the Whitehorse copper belt, Yukon. D, 1981, University of Western Ontario. 308 p.

Morrison, Gregory D. Depositional and diagenetic history of the San Andres Formation, Cornell Unit Wasson Field, Yoakum County, Texas. M, 1986, Texas A&M University. 150 p.

Morrison, H. F. Seismic investigations in the Sverdrup Basin, Queen Elizabeth Islands (Arctic Archipelago, Northwest Territories, Canada). M, 1961, McGill University.

Morrison, Huntley Frank. A magnetotelluric profile across the state of California. D, 1967, University of California, Berkeley. 359 p.

Morrison, James Douglas. Bedrock geology of the Ponca Quadrangle, Newton County, Arkansas. M, 1971, University of Arkansas, Fayetteville.

Morrison, Jean. Paleomagnetism of the Silurian and Ordovician Lower Red Mountain, Catheys and Sequatchie formations from the Valley and Ridge Province. M, 1983, University of Georgia.

Morrison, Jean. Petrology and stable isotope geochemistry of the Marcy anorthosite massif, Adirondack Mountains, New York. D, 1988, University of Wisconsin-Madison. 211 p.

Morrison, Joan O. Molluscan carbonate geochemistry and paleoceanography of the Late Cretaceous Western Interior Seaway of North America. M, 1986, Brock University. 214 p.

Morrison, Lawrence S. Gravity survey of the Ralston Dike, Jefferson County, Colorado (T735). M, 1952, Colorado School of Mines. 54 p.

Morrison, Lowell Russell. Surface sediment and seismic reflection profiling, outer Newport submarine canyon, southern California continental borderland. M, 1982, California State University, Northridge. 150 p.

Morrison, Melvin E. The petrology of Phelps Ridge-Red Mountain area, Chelan County, Washington. M, 1954, University of Washington. 95 p.

Morrison, Michael L. Structure and petrology of the southern portion of the Malton Gneiss, British Columbia. D, 1982, University of Calgary. 314 p.

Morrison, Michael L. Structure and stratigraphy of the Shuksan metamorphic suite in the Gee-Point-Finney Peak area, North Cascades. M, 1977, Western Washington University. 69 p.

Morrison, Robert Fairchild. The pre-Tertiary geology of the Snake River canyon between Cache Creek and Dug Bar, Oregon-Idaho boundary. D, 1963, University of Oregon. 195 p.

Morrison, Robert Rex. Geology of the Sand Canyon-Placerta Canyon area, parts of the Humphreys and Sylmar quadrangles, Los Angeles County, California. M, 1958, University of California, Los Angeles.

Morrison, Roger B. Celestite and fluorite deposits in the (Silurian) upper Niagara Dolomite at Clay Center, Ohio. M, 1934, Cornell University.

Morrison, Roger Barron. Soil stratigraphy; principles, applications to differentiation and correlation of Quaternary deposits and landforms, and applications to soil science. D, 1964, University of Nevada. 178 p.

Morrison, Ronald C. Variances in P and S wave velocities before earthquakes. M, 1974, New Mexico Institute of Mining and Technology.

Morrison, Scott. Structure and stratigraphy of the Shuksan metamorphic suite in the Gee Point-Finney Peak area, North Cascades. M, 1977, Western Washington University.

Morrison, Stan Jay. Chemical and thermal evolution of diagenetic fluids and the genesis of uranium and copper ore in and adjacent to the Paradox Basin with emphasis on the Lisbon Valley and Temple Mountain areas, Utah and Colorado. D, 1986, University of Utah. 198 p.

Morrison, Stanley J. Electron optical study of brown bodies in some Ordovician trepostome bryozoans. M, 1975, Michigan State University. 23 p.

Morrissey, Arthur Michael. Element partitioning in feldspars and apatite fission track ages from seven intrusive bodies of the Park City mining district, Utah. M, 1980, University of Iowa. 188 p.

Morritt, Robin Frederick Charles. Landsat and radar; a critical evaluation of a geologic application in Brazil. M, 1984, Stanford University. 124 p.

Morritt, Robin Frederick Charles. The radar structure of Brazil south of the Amazon Basin with special emphasis on the structural setting of the Quadrilatero Ferrifero, Minas Gerais State. D, 1988, Queen's University. 273 p.

Morro, Richard D. Geochemistry and petrography of Pugh Quarry, Ohio. M, 1975, Bowling Green State University. 262 p.

Morrow, A. L. A study of the invertebrate fauna of the Colorado Group. M, 1931, University of Kansas. 72 p.

Morrow, Aubrey Lyndon. The stratigraphy and invertebrate paleontology of the Colorado Group in Kansas. D, 1941, Yale University. 357 p.

Morrow, Carolyn Alexandria. Electrical resistivity changes in tuffs. M, 1979, Massachusetts Institute of Technology. 93 p.

Morrow, David L. Cretaceous stratigraphy and microtextures on Little Horse Creek, Fremont County, Wyoming. M, 1949, Miami University (Ohio). 51 p.

Morrow, David Watts. Stratigraphy and petrography of the Elk Point group (Devonian), northeast British Columbia (Canada). M, 1970, University of Texas, Austin.

Morrow, David Watts. Stratigraphy and sedimentology of lower Paleozoic formations near and on

Grinnell Peninsula, Devon Island, Northwest Territories. D, 1973, University of Texas, Austin. 383 p.

Morrow, H. F. Geology and ore deposits of the Macleod Cockshutt gold mine. M, 1940, Queen's University.

Morrow, Harold Francis. The geology of the Macleod-Cockshutt gold mine, Geraldton, Ontario. D, 1950, McGill University.

Morrow, Harold Julin. Structural analysis of an outcrop of folded Cuesta del Cura Limestone (Upper Cretaceous), Nuevo Leon, Mexico. M, 1965, Tulane University. 78 p.

Morrow, Hyland B. Structural geology of the southern half of Cruso Quadrangle, North Carolina. M, 1977, University of North Carolina, Chapel Hill. 67 p.

Morrow, John George. The beneficiation and utilization of Deer Harbor soda feldspar. M, 1948, University of Washington. 83 p.

Morrow, William Bruce. Simple mathematical model of a vapor dominated geothermal reservoir. M, 1973, Stanford University.

Morrow, William Earl. Geology of the Leipers Fork Quadrangle. M, 1957, Vanderbilt University.

Morse, D. G. Paleogeography and tectonic implications of the Late Cretaceous to middle Tertiary rocks of the southern Denver Basin, Colorado. D, 1979, The Johns Hopkins University. 365 p.

Morse, Earl L. Structural geology of the Coal Bank Pass-Molas Lake area, San Juan County, Colorado. M, 1984, New Mexico State University, Las Cruces. 94 p.

Morse, Edwin W. Study of subsurface Pleistocene drift at Saginaw, Michigan. M, 1970, Wayne State University.

Morse, J. G. Rainfall infiltration characteristics for a semi-arid watershed soil. M, 1976, University of Arizona.

Morse, James Donald. Deformation in ramp regions of thrust faults; experiments with rock models. M, 1978, Texas A&M University. 138 p.

Morse, John Wilbur. The dissolution kinetics of calcite; a kinetic origin for the lysocline. D, 1973, Yale University.

Morse, Margaret L. The micro-fauna from the Pennsylvanian strata near La Salle, Illinois. M, 1935, Northwestern University.

Morse, Margaret Louise. Conodonts from the (Devonian) Norwood and Antrim shales of Michigan. D, 1938, University of Michigan.

Morse, R. H. The surficial geochemistry of radium, radon and uranium near Bancroft, Ontario with applications to prospecting for uranium. D, 1970, Queen's University. 186 p.

Morse, Robert Harold. Potassium, rubidium, and cesium contents of some ocean floor rocks. M, 1967, Columbia University. 40 p.

Morse, Robert K. The geologic controls on porosity and permeability in the upper Minnelusa Formation, East-central Powder River basin, Wyoming. M, 1983, University of Missouri, Columbia.

Morse, Roy Robert. The geological structure and stratigraphy of Tsin-Ling-Shan, China. D, 1922, University of California, Berkeley. 6 p.

Morse, Stearns A. Chemistry, mineralogy and metamorphism of the Standing Pond Amphibolite, Hanover Quadrangle, New Hampshire, Vermont. M, 1958, McGill University.

Morse, Stearns Anthony. Geology of the Kiglapait layered intrusion, coast of Labrador, Canada. D, 1962, McGill University.

Morse, W. C. Paleozoic rocks of Mississippi. D, 1927, Massachusetts Institute of Technology. 244 p.

Morse, William C. The Maxville Limestone. M, 1908, Ohio State University.

Mort, F. P. The geology of an area between La Bonte and La Prele creeks, Converse County, Wyoming. M, 1939, University of Wyoming. 16 p.

Mortensen, James Kenneth. Age and evolution of the Yukon-Tanana Terrane, southeastern Yukon Territory. D, 1983, University of California, Santa Barbara. 172 p.

Mortensen, James Kenneth. Stratigraphic, structural, and tectonic setting of an Upper Devonian-Mississippian volcanic-sedimentary sequence and associated base metal deposits in the Pelly Mountains, southeastern Yukon Territory. M, 1979, University of British Columbia.

Mortensen, Kay Sherman. A fundamental study of cratering in granular materials with an application to lunar craters. D, 1967, University of Utah. 132 p.

Mortensen, Martin Eckert. Granulometry and morphology of the Defiance and Fort Wayne moraines in Lenawee County and vicinity, Michigan. M, 1975, University of Michigan.

Mortensen, Paul Stuart. Stratigraphy and sedimentology of the Upper Silurian strata on eastern Prince of Wales Island, Arctic Canada. D, 1985, University of Alberta. 375 p.

Mortensen, Thomas William. The flooding potential and geomorphology of five selected Arctic rivers, Arctic coastal plain, Alaska. M, 1982, University of Alaska, Fairbanks. 103 p.

Mortenson, John J. Stratigraphy and sedimentology of the Mazomanie Formation (Upper Cambrian) in southwestern Wisconsin. M, 1981, University of Wisconsin-Madison.

Mortera Gutierrez, Carlos Angel Q. Time term analysis of seismic data from the East Pacific Rise (Project Rose). M, 1984, University of Hawaii at Manoa. 327 p.

Mortgat, Christian Pierre. A Bayesian approach to seismic hazard mapping; development of stable design parameters. D, 1977, Stanford University. 335 p.

Mortimer, Nicholas. Petrology and structure of Permian to Jurassic rocks near Yreka, Klamath Mountains, California. D, 1984, Stanford University. 113 p.

Mortimer, Robert E. Design and implementation of an automatic settling tube for sand grain size analysis with semi-automated data acquisition system. M, 1978, University of Arizona.

Mortimore, Morris Edmon. Correlation of Paleozoic sandstones by sedimentary analysis. D, 1927, University of Iowa. 37 p.

Mortimore, Morris Edmon. Geology of the Cushing region, Oklahoma. M, 1924, University of Iowa. 40 p.

Morton, Douglas Maxwell. Petrology of the Lakeview Mountains Pluton and adjacent area, Riverside County, California. D, 1966, University of California, Los Angeles. 218 p.

Morton, Jack Andrew. The Yellowhead zinc-lead deposits; origin, post-depositional history, and comparisons with similar deposits in the Metaline mining district, Washington. M, 1974, Washington State University. 159 p.

Morton, Janet Lee. Oceanic spreading centers; axial magma chambers, thermal structure, and small scale ridge jumps. D, 1984, Stanford University. 110 p.

Morton, John Phillip. Isotopic dating of Paleozoic glauconites from the Llano region of central Texas. M, 1977, University of Texas, Austin.

Morton, John Phillip. Rb-Sr dating of clay diagenesis. D, 1983, University of Texas, Austin. 246 p.

Morton, Loren B. Geology of the Mount Ellen Quadrangle, Henry Mountains, Garfield County, Utah. M, 1984, Brigham Young University. 95 p.

Morton, Marc K. Evaluation of seismic refraction profiling of alluvial aquifers, Johnson County, Iowa. M, 1987, University of Iowa. 147 p.

Morton, Maurice Warner. A study of the Simpson Group of southeastern New Mexico and a part of West Texas. M, 1955, Texas Tech University. 53 p.

Morton, Penelope. Archean volcanic stratigraphy, and petrology and chemistry of mafic and ultramafic rocks, chromite, and the Shebandowan Ni-Cu Mine, Shebandowan, northwestern Ontario. D, 1982, Carleton University. 346 p.

Morton, Penelope Cane. Geochemistry of bedrock and soils in the vicinity of the Anvil Mine, Yukon Territory. M, 1973, University of British Columbia.

Morton, Peter S. Tectonic breccia of metamorphosed intrusive igneous rocks in an Acadian shear zone, Brooks Village, north-central Massachusetts. M, 1985, University of Massachusetts. 61 p.

Morton, R. L. Alkalic volcanism and copper deposits of the Horsefly area, central British Columbia. D, 1976, Carleton University. 196 p.

Morton, Robert. A study of some ancient, modern, and experimental deltaic sediments (Texas, West Virginia). D, 1972, West Virginia University.

Morton, Robert A. Clay mineral distribution of some Recent sediments. M, 1966, West Virginia University.

Morton, Robert Brading. Geology of the Round Mountain District, Nelson County, Virginia. M, 1943, University of Illinois, Urbana. 39 p.

Morton, Robert Whelden. Spatial and temporal observations of suspended sediment; Narragansett Bay and Rhode Island Sound. M, 1967, Duke University. 70 p.

Morton, Robert Wheldon. The effect of carbonate sediments on the acoustic reflectivity of the Whiting Basin (Puerto Rico). D, 1972, George Washington University.

Morton, Ronald. The geology and hydrothermal alteration of the Independence porphyry deposit (Mesozoic), British Columbia. M, 1970, McGill University. 140 p.

Morton, William R. An insoluble residue study of the Cretaceous Cow Creek Limestone of central Texas (Travis, Blanco and Comal counties, Texas). M, 1967, Texas A&M University. 79 p.

Mortonson-Liedle, Judith D. Evaluating cost-effectiveness of remedial actions at an uncontrolled hazardous waste site in New York. D, 1987, University of California, Los Angeles. 241 p.

Morzenti, Stephen P. A new high-pressure hydrous phase in the system MgO-SiO_2-H_2O. M, 1971, Lehigh University.

Mosburg, Shirley Krauthausen. A discussion of exsolved material in galena from the Idarado Mine (San Juan Mountains, southwest Colorado). M, 1972, University of Colorado.

Moscati, A. F., Jr. Environmental issues in the siting, construction and operation of new coal-fired and nuclear power plants. D, 1975, University of California, Los Angeles. 197 p.

Moschella, Victor Charles. Experimental albitization of bytownite. M, 1986, University of New Orleans. 82 p.

Mosconi, Carlos Eduardo. River-bed variations and evolution of armor layers. D, 1988, University of Iowa. 257 p.

Mosconi, Deborah Anne. Numerical modelling of chloride migration in a fractured bedrock aquifer, Bear Creek, Pennsylvania. M, 1987, Pennsylvania State University, University Park. 215 p.

Mosconi, Louis S. Tectonic history of the Black Mesa-Lowes Valley area, Brewster County, Texas. M, 1984, Stephen F. Austin State University. 72 p.

Moscoso, Belisario A. A thick section of Jurassic "flysch" in the Santa Ana Mountains (Orange County), southern California. M, 1967, University of San Diego.

Mose, Douglas George. Chronology of Precambrian volcanic rock units in the center of the Saint Francois Mountains, (southeast) Missouri. D, 1971, University of Kansas.

Mose, Douglas George. Precambrian geochronology in the Unaweep Canyon, west-central Colorado. M, 1968, University of Kansas.

Moseley, Cecil Robert. A petrographic study of Pennsylvanian sandstones at Mountainburg and Dyer, Arkansas. M, 1962, University of Arkansas, Fayetteville.

Moseley, Craig G. The geology of a portion of the northern Cady Mountains, Mojave Desert, California. M, 1978, University of California, Riverside. 131 p.

Moseley, John R. Ordovician-Silurian contact in eastern Pennsylvania. D, 1949, Harvard University.

Moseley, Marianne G. Geochemistry and metamorphic history of the Whitt Metagabbro, Llano County, Texas. M, 1977, University of Texas, Austin.

Moser, Erwin Leroy, Jr. A porosity study of the Allentown Dolomite near Bethlehem, Pennsylvania. M, 1952, University of Pittsburgh.

Moser, Frank. The Michigan Formation; a study in the use of a computer oriented system in stratigraphic analysis. D, 1963, University of Michigan. 103 p.

Moser, Fredrika C. The storage and transport of sediments, pesticides, and PCB's in two impounded fluvial systems in southern New Jersey. M, 1985, Rutgers, The State University, New Brunswick. 180 p.

Moser, John Archer. The hydrogeology of Agawam, Longmeadow, East Longmeadow and Hampden, Massachusetts. M, 1975, University of Massachusetts. 100 p.

Moser, John Christian. Sedimentation and accumulation rates of Nazca Plate metalliferous sediments by high resolution Ge(Li) gamma-ray spectrometry of uranium series isotopes. M, 1980, Oregon State University. 65 p.

Moser, Joseph Arthur. Stratigraphy of the Jefferson Formation at King Mountain in the southern Lost River Range, south-central Idaho. M, 1981, Washington State University. 86 p.

Moser, Kenneth Robert. Structural analysis of Cambrian rocks in the Bowen Lake area Stevens County, Washington. M, 1978, Washington State University. 64 p.

Moser, Michael Anthony. The response of stick-slip systems to random seismic excitation. D, 1987, California Institute of Technology. 149 p.

Moser, Paul H. Geology of a portion of southeastern Madison County, Kentucky. M, 1960, University of Kentucky. 50 p.

Moses, Alfred J. Methods of blowpipe analysis. D, 1890, Columbia University, Teachers College.

Moses, Carl Owen. Kinetic investigations of ferrous iron and pyrite oxidation. D, 1988, University of Virginia. 184 p.

Moses, Carl Owen. Kinetic investigations of sulfide mineral oxidation in sterile aqueous media. M, 1982, University of Virginia. 100 p.

Moses, Clarence F. The geology of the Bellefontaine Outlier. M, 1922, Ohio State University.

Moses, Clarice Gayle. Pool morphology of Redwood Creek, California. M, 1985, University of California, Santa Barbara. 117 p.

Moses, John H. The identification of opaque minerals by their reflecting power as measured photo-electrically. D, 1936, Harvard University.

Moses, Lynn Jane. A study of the Neogene Mehrten Formation; stratigraphy, petrology, paleomagnetism, southwestern Placer County, California. M, 1985, University of California, Davis. 163 p.

Moses, Michael J. The gravity field over the Bane Dome in Giles County, Virginia. M, 1988, Virginia Polytechnic Institute and State University. 84 p.

Moses, Selma. Laboratory techniques and statistical methods in pollen analysis. M, 1949, Stanford University. 82 p.

Mosesso, Michael Angelo. Two reactions involving phlogopite formation below two kilobars water pressure. M, 1974, University of Arizona.

Mosher, Byard William. The atmospheric biogeochemistry of selenium. D, 1986, University of Rhode Island. 238 p.

Mosher, Charles Clinton. Magnitude calibration for a small earthquake recording array. M, 1979, University of Minnesota, Minneapolis.

Mosher, Charles Clinton. Signal processing techniques applied to a small circular seismic array. D, 1980, University of Minnesota, Minneapolis. 291 p.

Mosher, David Cole. Late Quaternary sedimentology and sediment instability of a small area of the Scotian Slope. M, 1987, Memorial University of Newfoundland. 249 p.

Mosher, Loren Cameron. Conodonts from the Middle Triassic Prida Formation of northwestern Nevada. M, 1964, University of Wisconsin-Madison.

Mosher, Loren Cameron. Conodonts from the Triassic of western North America and Europe and their correlation. D, 1967, University of Wisconsin-Madison.

Mosher, S. Pressure solution as a deformation mechanism in the Purgatory Conglomerate. D, 1978, University of Illinois, Urbana. 186 p.

Moshier, S. Depositional regimes in the Upper Cambrian Richland Formation, Lebanon Valley, Pennsylvania. M, 1979, SUNY at Binghamton. 195 p.

Moshier, Stephen Oakley. On the nature and origin of microporosity in micritic limestones. D, 1987, Louisiana State University. 369 p.

Moshiri-Yazdi, Reza. Geology of the oil producing strata of the Gulf Coast oil province. M, 1959, University of Wisconsin-Madison.

Moshiri-Yazdi, Reza. Paleozoic and younger folded rocks near Demavend Town, east of Teheran, Iran. D, 1964, Columbia University, Teachers College. 126 p.

Moskal, Thomas Eugene. Examination of scaling criteria for nuclear reactor thermal-hydraulic test facilities. D, 1987, Carnegie-Mellon University. 243 p.

Moskow, Michael Gideon. Flint kaolins in the Georgia Coastal Plain. M, 1988, University of Georgia. 79 p.

Moskowitz, B. M. Numerical analysis of electrical fluid and rock resistivity in hydrothermal systems. M, 1977, University of Arizona. 95 p.

Moskowitz, Bruce Matthew. A laboratory study of the magnetic properties of synthetic titanomaghemites; implications for the magnetic properties on ocean basalts. D, 1980, University of Minnesota, Minneapolis. 191 p.

Moslem, Kaazem. Effects of soil structure interaction on the response of buildings during the strong earthquake ground motions. D, 1983, University of Southern California.

Mosley, M. Paul. An experimental study of channel confluences. D, 1975, Colorado State University. 231 p.

Mosley, M. Paul. An experimental study of rill erosion. M, 1972, [Colorado State University].

Moslow, Thomas F. Quaternary evolution of Core Banks, North Carolina from Cape Lookout to New Drum Inlet. M, 1977, Duke University. 171 p.

Moslow, Thomas Francis. Stratigraphy of mesotidal barrier islands of South Carolina. D, 1980, University of South Carolina. 265 p.

Moss, Albert E. Geology of the Siscoe gold mine, Siscoe, Quebec. D, 1940, McGill University.

Moss, Albert E. Microscopical investigation of certain Quebec ores; Part A, Technique of investigation. M, 1937, McGill University.

Moss, C. M. The surficial and environmental geology of the French River Quadrangle, St. Louis County, Minnesota. M, 1977, University of Minnesota, Duluth.

Moss, Frank Ambrose. Geology of the Mother Lode in the vicinity of Carson Hill, California. M, 1924, University of California, Berkeley. 175 p.

Moss, Frank Ambrose. The geology of the Mother Lode in the vicinity of Carson Hill, Calaveras County, California. D, 1927, University of California, Berkeley. 175 p.

Moss, John H. Evidence of multiple glaciation in New England from the tills of the Concord Quadrangle, Massachusetts. M, 1943, Massachusetts Institute of Technology. 34 p.

Moss, John H. Glaciation in the southern Wind River Mountains and its relation to early man in the Eden Valley, Wyoming. D, 1949, Harvard University.

Moss, John Lawrence. The morphology and phylogenetic relationships of the lower Permian tetrapod Tseajaia campi Vaughn. D, 1971, University of California, Los Angeles. 191 p.

Moss, Kenneth Lee. Geological and geostatistical elements of ore-grade control at the Candelaria Mine, Mineral County, Nevada. M, 1987, University of Nevada. 49 p.

Moss, Michael James. Some pegmatites near Gwinn, Michigan. M, 1975, Western Michigan University.

Moss, Muriel Ellen. Foraminifera from the Campagrande (Lower Cretaceous) Formation, Finlay Mountains, Texas. M, 1948, Southern Methodist University. 78 p.

Moss, Neil E. Depositional and diagenetic characteristics of the Smackover Formation in the Mississippi Interior salt basin of Southwest Alabama. M, 1987, University of Alabama. 149 p.

Moss, Rycroft Gleason. Configuration of the present surface of the buried Pre-Cambrian rocks east of the Rocky Mountains. D, 1933, Cornell University.

Moss, Rycroft Gleason. The geology of Ness and Hodgeman counties, Kansas. M, 1931, University of Kansas. 67 p.

Moss, Steven A. Correlation between the resistivity and the seismic velocity of rocks. M, 1959, University of Utah. 50 p.

Moss, Thomas Allen. A comparative study of the trace element distribution in limestones, dolostones, and breccias in the upper Knox Group of the Copper Ridge zinc district, East Tennessee. M, 1982, University of Tennessee, Knoxville. 174 p.

Mossaad, Mostafa El-Sayed. A stochastic model for soil erosion. D, 1981, Ohio State University. 189 p.

Mossel, Leroy Gene. Zeolite distribution and stratigraphic relations in the Columbia River Basalt (Miocene) near Ritter Hot Springs (Grant County), Oregon. M, 1967, University of Oregon. 64 p.

Mossesso, Michael Angelo. Two reactions involving phlogopite formation below two kilobars water pressure. M, 1974, University of Arizona.

Mossler, John H. Facies and diagenesis of Swope limetones (upper Pennsylvanian), southeast Kansas. D, 1970, University of Iowa. 228 p.

Mossler, John Hamilton. Ordovician potassium bentonites of Iowa. M, 1964, University of Iowa. 115 p.

Mossman, Brian John. A comparative study of Holocene sand and Cretaceous sandstone derived from the Peninsular Ranges, California, and Baja California Norte, Mexico; evidence for deep dissection of the Peninsular Ranges magmatic arc. M, 1986, San Diego State University. 78 p.

Mossman, David John. The orientation and distribution of dike rocks in the eastern part of Northern Rhodesia. M, 1964, Dalhousie University. 60 p.

Mossman, Malcolm H. A quantitative X-ray technique for the determination of clay minerals. M, 1955, University of Wisconsin-Madison.

Mossman, Reuel Wallace. Foraminifera from the Charco Azul Formation, Pliocene, of Panama. M, 1939, Columbia University, Teachers College.

Mossop, Grant D. Origin of the peripheral rim, Redwater Reef (upper Devonian), Alberta. M, 1971, University of Calgary. 89 p.

Mostafa, Abd-elmonem Sayed-ahmad. Movement of phosphorus in soils as influenced by chelates and soil types. D, 1966, Utah State University. 114 p.

Mostafa, Ayman A. Propagation in complex anisotropic layered media. D, 1988, University of Maryland. 211 p.

Mostaghel, Mohammad Ali. Genesis and distribution of mississippi valley-type ore assemblages in Middle Silurian strata, Niagara Peninsula, Ontario. M, 1978, Brock University. 164 p.

Mostaghimi, Saied. Trickle irrigation; a study of moisture distribution in the soil profile. D, 1982, University of Illinois, Urbana. 139 p.

Mosteller, M. A. The subsurface stratigraphy of the Comanchean series (lower and upper Cretaceous) in east central Texas. M, 1970, Baylor University. 64 p.

Mosteller, Stanley A. The geology of the North Fredonia area, McCulloch and San Saba counties, Texas. M, 1957, Texas A&M University.

Moster, Neal H. Porosity and surface area measurements of some low rank coals of the Western United States. M, 1976, University of Toledo. 82 p.

Motamedi, Saadi. Celestite mineralization in the Lockport Dolomite (Niagaran) at Genoa, Ohio. M, 1982, Bowling Green State University. 132 p.

Motamedi, Shoaullah. The Keweenawan lavas in the City of Duluth. M, 1984, University of Minnesota, Duluth. 140 p.

Motan, Eyup Sabri. Influence of soil suction on the behavior of unsaturated soils under repetitive loads. D, 1981, University of Wisconsin-Madison. 231 p.

Motazed, Behnam. Inference of ferrous cylinders from magnetic fields. D, 1984, Carnegie-Mellon University. 91 p.

Mote, Richard H. The geology of the Maury Mountain region, Crook County, Oregon. M, 1940, Oregon State University. 79 map x-section p.

Motes, Arthur Glenn, III. A sedimentological study of the middle and upper Minnelusa Formation, Crook County, Wyoming. M, 1984, South Dakota School of Mines & Technology.

Mothersill, John S. The Halfway Formation (Triassic) of the Milligan Creek area, British Columbia. D, 1967, Queen's University. 325 p.

Motooka, P. S. The use of 0.1N HCl-extractable zinc in assessing available and fixed zinc in Hawaiian soils. M, 1962, University of Hawaii. 37 p.

Motsch, Aaron Sherrill. Castiles, Culberson County, Texas. M, 1951, University of Texas, Austin.

Mott, Charles James. The geology of the Putnam Hall Quadrangle, Putnam County, Florida. M, 1967, University of Florida. 54 p.

Mott, L. V. Origin of the uraniferous phosphatic zones of the Wilkins Peak Member, Green River Formation, Wyoming. M, 1978, University of Wyoming.

Mott, Richard P., Jr. Relationship of microearthquake activity to structural geology for the region surrounding Socorro, New Mexico. M, 1976, New Mexico Institute of Mining and Technology.

Motta, Christopher J. The sedimentology and hydrology of the lower and middle reaches of the Raritan River estuary, New Jersey. M, 1984, Rutgers, The State University, New Brunswick. 179 p.

Motte, Robert S. La *see* La Motte, Robert S.

Motten, Roger H., III. The bedrock geology (Precambrian) of the Thunder Mountain area, Wisconsin. M, 1972, Bowling Green State University. 59 p.

Mottern, Hugh Henry, Jr. Pre-Tertiary geology of a portion of Cedar Mountain, Nevada. M, 1962, University of California, Berkeley. 64 p.

Mottl, Michael James. Chemical exchange between sea water and basalt during hydrothermal alteration of the oceanic crust. D, 1976, Harvard University.

Motto, Harry Lee. Total analysis of soils by emission spectroscopy. D, 1964, University of Illinois, Urbana. 116 p.

Motts, Ward Sundt. Geology and ground water resources of the Carlsbad area, New Mexico. D, 1957, University of Illinois, Urbana. 120 p.

Motumah, Linus Kiambati. Estimation of earthquake-induced permanent displacements in embankment dams. D, 1987, University of California, Davis. 230 p.

Motyka, Roman John. Increases and fluctuations in thermal activity at Mount Wrangell, Alaska. D, 1983, University of Alaska, Fairbanks. 368 p.

Motzer, Mary Ellen Benson. Paleoenvironment of the lower Lakeview Limestone (Middle Cambrian), Bonner County, Idaho. M, 1980, University of Idaho. 174 p.

Motzer, William Erhardt. Tertiary epizonal plutonic rocks of the Selway-Bitterroot Wilderness, Idaho County, Idaho. D, 1985, University of Idaho. 467 p.

Motzer, William Erhardt. Volcanic stratigraphy of an area east of the White Cloud peaks, Sawtooth National Recreation Area, Custer County, Idaho. M, 1978, University of Idaho. 115 p.

Mou, Ching-Hua. Soil suction approach for the evaluation of swelling potentials of expansive clays. D, 1981, University of South Carolina. 120 p.

Mou, Duenchien C. Diagenesis of Tensleep Sandstone in Lost Soldier Field, Wyoming; effects upon reservoir characteristics and well-log signatures. M, 1980, University of Iowa. 144 p.

Mouat, Malcolm M. The mineralogy of certain manganese deposits in the Artillery Mountains region, Arizona. M, 1962, University of Wisconsin-Madison.

Moughamian, J. M. The U. S. oil import program and its effect on the American petroleum industry. E, 1974, Stanford University. 72 p.

Mould, John Calvin, Jr. Stress induced anisotropy in sand and the evaluation of a multi-surface elasto-plastic material model. D, 1983, University of Colorado. 368 p.

Moulthrop, James S. Pleistocene geology and ground water of Kansas River valley between Manhattan and Junction City, Kansas. M, 1963, Kansas State University. 42 p.

Moulton, David Richard, Jr. Geology of the upper Bull Creek area, northern Hoback Range, Teton County, Wyoming. M, 1981, Idaho State University. 53 p.

Moulton, Floyd C. Ground-water geology of Cedar Valley and western Utah Valley. M, 1951, Brigham Young University. 50 p.

Moulton, Gail Francis, Jr. An investigation into the nature of the basal quartz Cretaceous sands of the Bellshill Lake trend of central Alberta, Canada, to develop criteria for the recognition of oil prospects. M, 1960, Michigan Technological University. 40 p.

Moultrie, William A. Computer simulation of erosional precesses in a natural watershed; a study of drainage basin development. D, 1971, University of Iowa. 173 p.

Moultrop, Kendall. Airphoto boundary delineation of loess or loess-like soils in southwestern Indiana. M, 1953, Purdue University.

Mounce, Donald D. Geology of the Camp San Saba, west area, Mason and McCulloch counties, Texas. M, 1957, Texas A&M University.

Mound, Michael Charles. Arenaceous foraminifera from cores from the Silurian rocks of northern Indiana. D, 1963, Indiana University, Bloomington. 214 p.

Mound, Michael Charles. Arenaceous foraminifera from the Brassfield Limestone (Albion) of southeastern Indiana. M, 1961, Indiana University, Bloomington. 67 p.

Mount, Donald Lee. Geology of the Wanship-Park City region, (Summit County, Utah). M, 1952, University of Utah. 35 p.

Mount, Jack Douglas. Late Paleozoic biostratigraphy of the Pancake Range, Nye County, Nevada. M, 1972, University of California, Los Angeles.

Mount, James Russell. A petrofabric analysis of the Cox Sandstone, Hudspeth County, Texas. M, 1960, Texas Tech University. 67 p.

Mount, Jeffrey Frazer. The environmental stratigraphy and depositional systems of the Precambrian(?)-Cambrian Campito Formation, eastern California and western Nevada. D, 1980, University of California, Santa Cruz. 246 p.

Mountain, David. A study of the Bering Sea flow into the Arctic Ocean. D, 1974, University of Washington. 153 p.

Mountain, Gregory Stuart. Stratigraphy of the western North Atlantic based on the study of reflection profiles and DSDP results. D, 1981, Columbia University, Teachers College. 331 p.

Mountjoy, E. W. Structure and stratigraphy of the Miette and adjacent areas, eastern Jasper National Park, Alberta. D, 1960, University of Toronto.

Mounzer, Maroun C. Slope failure on the seaward flank of the Santa Monica Mountains, Los Angeles County, California. D, 1987, University of California, Los Angeles.

Mourad, Asa George. An interpretation of two gravity anomalies in Ohio. M, 1959, Ohio State University.

Mourdock, R. E. Paleontology and stratigraphy of a Silurian (Niagaran) reef-flank bed at Francesville, Indiana. M, 1975, Ball State University. 92 p.

Moushegian, Richard. The oil import policy of the United States. M, 1947, University of Pittsburgh.

Moussa, Mounir Tawfik. Geology of the Soldier Summit Quadrangle, Utah. D, 1965, University of Utah. 129 p.

Moussa, Osama Moursy. Satellite data based sediment-yield models for the Blue Nile and the Atbara River watersheds. D, 1987, Ohio State University. 397 p.

Moussavi-Harami, Reza. Areal geology of the Cordell area, central Washita County. M, 1977, University of Oklahoma. 76 p.

Moussavi-Harami, Reza. Stratigraphy, petrology and depositional environments of sandstone associated with the Stanton Formation (Upper Pennsylvanian) in southeastern Kansas. D, 1980, University of Iowa. 286 p.

Moustafa, Adel Ramadan. Analysis of Laramide and younger deformation of a segment of the Big Bend region, Texas. D, 1983, University of Texas, Austin. 278 p.

Moustafa, Mary Sue Jablonsky. Swash induced zonation of a foreshore; sediment size distribution. M, 1988, College of William and Mary. 139 p.

Moustafa, Youssef S. Structure and affinities of the Permian amphibian Parioxys ferricolus, Cope. D, 1950, Harvard University.

Mousuf, Abdul K. Determination of the branching ratio for the K-capture process of potassium 40, and its application to geological age determination. D, 1952, University of Toronto.

Mowatt, Thomas Charles. An investigation of some geochemical relationships in the Stillwater complex, (Precambrian) (Sweet Grass County,) Montana. D, 1965, University of Montana. 199 p.

Mowery, Dale Harris. Pleistocene molluscan faunas of the Jewell Hill Deposit, Logan County, Ohio. M, 1959, Ohio State University.

Mowrey, Gary Lee. Development and application of a computer analysis system to process microseismic data associated with an underground natural-gas storage reservoir. D, 1980, Pennsylvania State University, University Park. 317 p.

Moxham, Robert Lynn. Minor element distribution in some pyroxenes of metamorphic origin. M, 1958, McMaster University. 98 p.

Moxham, Robert Lynn. Minor elements in hornblendes and biotites. D, 1963, University of Chicago. 96 p.

Moxley, Frances M. An analysis of heavy minerals in sediment of Delaware bay. M, 1970, Millersville University.

Moya, Eduardo A. Structural and stratigraphic analysis of Precambrian rocks, North Clear Creek area, Jefferson and Gilpin counties, Colorado. M, 1969, Colorado School of Mines. 86 p.

Moye, Falma Jean. Geology and petrochemistry of Tertiary igneous rocks in the western half of the Seventeenmile Mountain 15' Quadrangle, Ferry County, Washington. D, 1984, University of Idaho. 242 p.

Moye, Robert Josephus, Jr. The Bayleaf mafic-ultramafic belt, Wake and Granville counties, North Carolina. M, 1981, North Carolina State University. 122 p.

Moyer, Carol B. The depositional environment of the Waynesburg Coal, upper Monongahela-lower Dunkard groups of West Virginia. M, 1978, West Virginia University.

Moyer, Chris. Subsurface stratigraphy of the Prairie Grove Member, Hale Formation (Morrowan), in Franklin and Logan counties, Arkansas. M, 1984, University of Arkansas, Fayetteville.

Moyer, Dorothy A. The relation of a fossil foraminifera fauna from Lomita Quarry, San Pedro, California, to a recent foraminifera fauna from off the coast of San Pedro, California. M, 1929, University of Southern California.

Moyer, Grant Luke. Cretaceous geology of the San Martine Quadrangle, Reeves and Culberson counties, Texas. M, 1952, University of Texas, Austin.

Moyer, Paul T., Jr. Nature and origin of chert in the (Devonian) Onondaga Limestone at LeRoy and Oaks Corners, New York. M, 1956, University of Rochester. 120 p.

Moyer, Paul Tyson, Jr. The geology and amphibolites of the Brock-Saint Urcisse area; Albitibi and Mistassini territories, Quebec (Canada). D, 1968, University of Michigan.

Moyer, Raymond B. Thermal maturity and organic content of selected Paleozoic formations; Michigan Basin. M, 1982, Michigan State University. 62 p.

Moyer, Ronald D. Paleomagnetism of the Tertiary rocks of the northern Olympic Peninsula, Washington, and its tectonic implications. M, 1985, Western Washington University. 154 p.

Moyer, Thomas Carl. The Pliocene Kaiser Spring (AZ) bimodal volcanic field; geology, geochemistry, and petrogenesis. D, 1986, Arizona State University. 320 p.

Moyer, Thomas Carl. The volcanic geology of the Kaiser Spring area, SE Mohave County, Arizona. M, 1982, Arizona State University. 220 p.

Moyle, Richard W. Ammonoids of Wolfcampian age from the Glass Mountains and contiguous areas in West Texas. D, 1963, University of Iowa. 327 p.

Moyle, Richard W. Paleoecology of the Manning Canyon Shale in central Utah. M, 1958, Brigham Young University. 86 p.

Moyse, David Wayne. Trace metal levels in the forest floor at Saddleback Mountain, Maine. M, 1986, University of Maine. 153 p.

Moysey, D. G. Ostracoda and associated fauna of the lower Walnut Formation (lower Cretaceous) of Travis and Williamson counties, Texas. M, 1975, University of Houston.

Mozer, Robert J. The hydrogeology and geochemistry of Woodhaven Park, West Oneonta, New York. M, 1983, SUNY, College at Oneonta. 86 p.

Mozeson, Charles Edward. Inverse magnetotelluric analysis by the method of sequential layering. M, 1971, University of Alberta. 53 p.

Mozeto, Antonio A. Carbon isotope exchange in aqueous systems; a field and laboratory investigtion. D, 1981, University of Waterloo.

Mozley, Edward Clarence. An investigation of the conductivity distribution in the vicinity of a Cascade volcano. D, 1982, University of California, Berkeley. 386 p.

Mozley, Peter Snow. Origin of kaolinite in the Dakota Group (Cretaceous age), northern Front Range foothills, Colorado. M, 1985, University of Colorado.

Mozley, Peter Snow. Petrography and diagenesis of the Sag River and Shublik formations in the National Petroleum Reserve, Alaska and topics in siderite geochemistry. D, 1988, University of California, Santa Barbara. 272 p.

Mozola, Andrew J. A survey of ground water resources in Oakland County, Michigan. D, 1954, Syracuse University.

Mozola, Andrew J. Contributions on the origin of the (Silurian) Vernon Shale (New York). M, 1938, Syracuse University.

Mozumdar, Bijoy Kumar. A mathematical model of ground movement due to underground mining. D, 1974, Pennsylvania State University, University Park. 145 p.

Mrakovich, John Vincent. New technique for stratigraphic analysis and correlation; fourier grain shape analysis, Louisiana offshore Pliocene. D, 1974, Michigan State University. 136 p.

Mrakovich, John Vincent. Sedimentary structures and depositional environment of the Sharon conglomerate (Pennsylvanian) in western Summit, eastern Medina, and northeastern Wayne counties, Ohio. M, 1969, Kent State University, Kent. 92 p.

Mraz, Joseph R. A gravity and subsurface investigation of the Presidio Bolson area, Texas. M, 1977, University of Texas at El Paso.

Mrkvicka, Steven Robert. The mechanical and thermal subsidence history of the Williston Basin. M, 1982, University of Oklahoma. 131 p.

Mrotek, Kathryn Anastasia. The mineralogical and textural aspects of ferromanganese nodules from the South Atlantic, South Pacific and North Pacific oceans. M, 1981, Rutgers, The State University, Newark. 159 p.

Mrowka, Jack Peter. An analysis of longitudinal stream profiles of low order coastal drainage basins in a heterogeneous lithologic and tectonically unstable environment. M, 1969, University of California, Los Angeles.

Mrowka, Jack Peter. Some relationships between surface hydrology and geomorphology in the Walker River basin, California-Nevada. D, 1974, University of California, Los Angeles.

Mrozowski, Cary Louis. On the tectonic evolution of the Philippine Plate; marine geophysical and geological studies of the Parece Vela Basin, the West Philippine Basin and the Mariana fore arc. D, 1981, Columbia University, Teachers College. 214 p.

Mruk, Denise Helen. Cementation and dolomitization of the Capitan Limestone (Permian), McKittrick Canyon, West Texas. M, 1985, University of Colorado.

Msek, Salahaddin Akif. Petrographic study of the Clinton Sandstone, southeastern Ohio. M, 1973, Ohio University, Athens. 68 p.

Mtundu, Nangantani Davies Godfrey. The stochastic behavior of soil moisture and its role in catchment response models. D, 1987, Portland State University. 222 p.

Mualchin, Lalliana. The descending slab beneath the Kurile-Kamchatka Arc and its influence on ray paths of body waves. D, 1974, St. Louis University. 222 p.

Muan, Arnulf Ingau. Phase equilibria at liquidus temperatures in the system MgO-FeO-Fe$_2$O$_3$-SiO$_2$. D, 1955, Pennsylvania State University, University Park. 115 p.

Muangnoicharoen, Nopadon. The geology and structure of a portion of the northern Piedmont, east-central Alabama. M, 1975, University of Alabama.

Muangnoicharoen, Nopadon. The Saltville Thrust decollement; deformation of Maccrady Formation evaporites. D, 1978, University of North Carolina, Chapel Hill. 122 p.

Mucci, Alfonso. The solubility of calcite and aragonite and the composition of calcite overgrowths in seawater and related solutions. D, 1981, University of Miami. 251 p.

Muchow, Charlotte I. Aspects of precipitation measurement relevant to landslide prediction. M, 1980, University of California, Los Angeles.

Muck, Karl L. A petrographic study of the Binnewater Formation in New York State. M, 1978, SUNY at Buffalo. 161 p.

Mudambi, Anand Rajagopal. Photochemistry of mirex in Lake Ontario. D, 1987, SUNY at Albany. 170 p.

Mudd, S. W. An ore containing copper, gold and silver, and its treatment. M, 1883, Washington University.

Mudd, Wayne Adrian. Geochemical properties of problematic clay-shales in the greater Waco area. M, 1977, Baylor University. 95 p.

Mudge, L. Taylor. Iron-titanium oxides in igneous rocks. M, 1975, Boston University. 50 p.

Mudge, Melville R. The pre-Quaternary stratigraphy of Riley County, Kansas. M, 1949, Kansas State University. 247 p.

Mudgett, Philip Michael. Regional survey of parts of Beaver and Millard counties, Utah. M, 1964, University of Utah. 19 p.

Mudie, Peta J. Palynology of later Quaternary marine sediments, eastern Canada. D, 1980, Dalhousie University. 638 p.

Mudrey, Michael George, Jr. Petrology of the Northern Light gneiss (Precambrian), Northern Light Lake, Thunder Bay District, Ontario, Canada. M, 1969, Northern Illinois University. 66 p.

Mudrey, Michael George, Jr. Structure and petrology of the sill on Pigeon Point, Cook County, Minnesota. D, 1973, University of Minnesota, Minneapolis. 310 p.

Mudroch, Alena. The feasibility of using dredged bottom sediments as an agricultural soil. M, 1974, McMaster University. 79 p.

Muehlberg, Gary E. Structure and stratigraphy of Tertiary and Quaternary strata, Heceta Bank, central Oregon shelf. M, 1971, Oregon State University. 78 p.

Muehlberger, Eric William. The structure and general stratigrahy of the western half of the Payson Basin, Gila County, Arizona. M, 1988, Northern Arizona University. 86 p.

Muehlberger, Eugene Bruce. Pismo Beach-Point Sal dune field, California. M, 1955, University of Kansas. 106 p.

Muehlberger, William Rudolf. Deposition and deformation in the Soledad Basin, Los Angeles County, California. D, 1954, California Institute of Technology. 87 p.

Muehlberger, William Rudolf. Mode of emplacement of the Barre Granite, Vermont. M, 1949, California Institute of Technology. 34 p.

Muehlhauser, Helmut C. Subsurface geology of Pratt County, Kansas. M, 1958, Kansas State University. 66 p.

Mueke, Gunter Kurt. Fracture analysis in the Canadian Rocky Mountains. M, 1965, University of Alberta. 117 p.

Muela, Pedro, Jr. The geology of the northern portion of Sierra Santa Rita, Chihuahua, Mexico. M, 1985, University of Texas at El Paso.

Muellenhoff, William P. Effects of pressure and deposit thickness on the stabilization rate of benthic marine sludge deposits. D, 1977, Oregon State University. 131 p.

Mueller, Amy F. Geology and physiography of the Wisconsin state parks. M, 1927, University of Wisconsin-Madison.

Mueller, Bruce Elmo. The morphology and growth characteristics of Metablastus bipyramidalis. M, 1960, University of Illinois, Urbana. 63 p.

Mueller, Carl Vincent. Seismic velocities versus soil properties. M, 1954, Washington University. 20 p.

Mueller, Charles Scott. The influence of site conditions on near-source high-frequency ground motion; case studies from earthquakes in Imperial Valley, CA., Coalinga, CA., and Miramichi, Canada. D, 1987, Stanford University. 239 p.

Mueller, David K. The effect of subdivision development on soil moisture and groundwater recharge in the Colorado Front Range. M, 1979, Colorado State University. 98 p.

Mueller, Donald. An analysis of the rural Lake Michigan shoreline and its adjacent land use development in Sheboygan County, Michigan. M, 1973, Indiana State University. 192 p.

Mueller, Frederick W. The sedimentary petrography and correlation of well samples from Frio County, Texas. M, 1934, Texas A&M University.

Mueller, G. V. Experimental work bearing on the origin of hydrous nickel-magnesium silicate minerals. D, 1954, McGill University.

Mueller, H. E. Geology of the north half of the Meramec Spring Quadrangle, Missouri. M, 1951, University of Missouri, Rolla.

Mueller, Harry W., III. Centrifugal progradation of carbonate banks; a model for deposition and early diagenesis, Ft. Terrett Formation, Edwards Group, lower Cretaceous, central Texas. D, 1975, University of Texas, Austin. 316 p.

Mueller, Ivan Istvan. The gradients of gravity and their applications in geodesy. D, 1960, Ohio State University. 227 p.

Mueller, James Manning. A seismic study of an axial valley-type plate boundary at 97°W on the Galapagos spreading center. M, 1979, Duke University. 121 p.

Mueller, Joseph Charles. Differential compaction study of Pennsylvanian channel sandstones in Jefferson County, Illinois. M, 1955, University of Illinois, Urbana.

Mueller, Joseph Fred, Jr. The Upper Carboniferous to Upper Permian strata of the Inner Mongolia Autonomous Region, China. M, 1988, University of North Carolina, Chapel Hill. 142 p.

Mueller, Joseph W. Geology and ore controls of uranium deposits in Oligocene and Miocene sediments, northern Hartville Uplift, Wyoming. M, 1976, Colorado School of Mines. 119 p.

Mueller, Karl Jules. Neotectonics, alluvial history and soil chronology of the southwestern margin of the Sierra los Cucapas, Baja California Norte. M, 1984, San Diego State University. 363 p.

Mueller, Paul Allen. Geochemistry and geochronology of the mafic rocks of the southern Beartooth mountains, Montana and Wyoming. D, 1971, Rice University. 58 p.

Mueller, Paul M. Geology of Cross Mountain, Colorado, north of the Yampa River. M, 1957, Washington State University. 136 p.

Mueller, Raymond George. Soil genesis as influenced by three-dimensional landsurface form in two low-order drainage basins in the Flint Hills of Kansas. D, 1982, University of Kansas. 211 p.

Mueller, Robert Emerson. The Cheyenne Belt, southeastern Wyoming; Part I, Descriptive geology and petrography. M, 1982, University of Wyoming. 98 p.

Mueller, Robert F. Origin and relative age of certain specular hematites of northeastern New York. M, 1955, University of Wisconsin-Madison.

Mueller, Robert Frances. Compositional characteristics of and equilibrium relations in mineral assemblages of a metamorphosed iron formation. D, 1959, University of Chicago. 83 p.

Mueller, Steven Wayne. Three-layered models of Ganymede and Callisto; structure and evolution. M, 1985, Washington University. 101 p.

Mueller, Tanya L. A seismic-stratigraphic analysis of Starkey and Winters gas reservoirs, using amplitude-offset studies from the Crossroads Field, Sacramento Valley, California. M, 1987, Colorado School of Mines. 131 p.

Mueller, Wayne Paul. The distribution of cladoceran remains in surficial lacustrine sediments from three northern Indiana lakes. M, 1962, Indiana University, Bloomington. 122 p.

Mueller, Wolfgang H. T. The mineralogy, geology and paragenesis of the 1208 Oxide Stope, Defiance Workings, Darwin Mine, Darwin, California. M, 1974, University of California, Riverside. 88 p.

Muench, Robin. The physical oceanography of the northern Baffin Bay region. D, 1970, University of Washington. 150 p.

Muench, Robin D. Exchange of water between Arctic Ocean and the Baffin Bay via Nares Strait. M, 1966, Dartmouth College. 89 p.

Muenzinger, John W. Williams Canyon Formation; Devonian-Mississippian, Colorado. M, 1956, University of Colorado.

Muerdter, David Robert. Late Quaternary biostratigraphy and paleoceanography of the Strait of Sicily and the eastern Mediterranean Sea. D, 1982, University of Rhode Island. 336 p.

Muessig, Karl Walter. The central Falcon igneous rocks, northwestern Venezuela; their origin, petrology, and tectonic significance. D, 1979, Princeton University. 281 p.

Muessig, Siegfried Joseph. Geology of a part of Long Ridge, Utah. D, 1951, Ohio State University.

Muffler, Leroy John Patrick. Geology of the Frenchie Creek Quadrangle, north central Nevada. D, 1962, Princeton University. 205 p.

Muffler, Steven A. Sedimentation in the Early Cretaceous foreland basin; the Cloverly Formation, Maverick Springs Dome, Wind River basin, Wyoming. M, 1986, Indiana University, Bloomington. 75 p.

Mugel, Douglas N. Geology of the Blanket lead-zinc deposit, Buick Mine, Viburnum Trend, Southeast Missouri. M, 1983, University of Missouri, Rolla. 155 p.

Mugler, Dorothy S. The geology of Hastings-on-the-Hudson and vicinity (New York). M, 1932, Columbia University, Teachers College.

Mugridge, Samantha-Jane. Carbonate cementation and dolomitization in the Silurian (Llandovery C3-Wenlock), northern New Brunswick and eastern Gaspe Peninsula, Canada. M, 1984, University of New Brunswick. 280 p.

Muhaimeed, Ahmad Saleh. Soil property relationships on selected landscape segments under cultivated vs. rangeland conditions. D, 1981, Colorado State University. 190 p.

Muhammad, Mir Jan. Geology of the till in a part of southeastern Manitoba. M, 1973, University of Manitoba.

Muhammed, Shah. Efficiency in the use of saline water for the reclamation of sodic soils. D, 1968, University of California, Riverside. 75 p.

Muhs, Daniel R. Relationship of diagenetic patterns to depositional facies in the Winterset Limestone (U. Penn.) of southeastern Kansas. M, 1979, University of Kansas. 88 p.

Muhs, Daniel Robert. Quaternary stratigraphy and soil development, San Clemente Island, California. D, 1980, University of Colorado. 255 p.

Mui, Kwoon Chuen. Stability of displacement fronts in porous media; large finger analyses. D, 1981, Carnegie-Mellon University. 137 p.

Muilenberg, Garrett Anthony. The petrology of the Wisconsin drift of the Lake Okoboji region, Dickinson County, Iowa. M, 1913, University of Iowa. 127 p.

Muilenburg, Garret A. Geology of the Tarryall District, Park County, Colorado. D, 1925, Columbia University, Teachers College.

Muir, Clifford Donald. Stability of slopes with seepage. D, 1968, Colorado State University. 290 p.

Muir, Iain D. Devonian Hare Indian and Ramparts formations, Mackenzie Mountains, Northwest Territories; basin-fill, platform and reef development. D, 1988, University of Ottawa. 593 p.

Muir, Iain D. Sedimentology of the Snowblind Bay Formation, Cornwallis Island, Northwest Territories. M, 1982, University of Ottawa. 109 p.

Muir, James Lawrence. Sedimentary studies of the Blaine Formation. M, 1933, University of Oklahoma. 132 p.

Muir, John E. A preliminary study of the petrology and ore genesis of the giant Mascot 4600 ore body, Hope, British Columbia. M, 1971, University of Toronto.

Muir, Mark P. The role of pre-existing, corrugated topography in the development of stone stripes. M, 1988, University of Washington. 189 p.

Muir, Nancy Jean. Depositional environments and diagenesis of the lower San Andres Formation, Roosevelt and Quay counties, New Mexico. M, 1978, Texas Tech University. 76 p.

Muir, Stephen G. Holocene deformed sediments of the southern San Joaquin Valley, Kern County, California. M, 1981, California State University, Northridge. 209 p.

Muir, Thomas L. A petrological study of the ultramafic and related rocks of the Shaw Dome, southeast of Timmins, Ontario. M, 1975, Queen's University. 271 p.

Muir, W. Electromagnetic modelling of air-borne (electromagnetic) methods. M, 1954, University of Western Ontario.

Muir, William. The structure and stratigraphy of a portion of the Treasure Hill Quadrangle, White Pine County, Nevada. M, 1984, California State University, Long Beach. 94 p.

Mukasa, Samuel Benjamin. Comparative Pb isotope systematics and zircon U-Pb geochronology for the Coastal, San Nicolas and Cordillera Blanca batholiths, Peru. D, 1984, University of California, Santa Barbara. 383 p.

Mukherjee, Amar Chandra. Structural analysis of the Missi metasedimentary rocks (Archean), near Flin Flon, Manitoba (Canada). M, 1968, University of Saskatchewan. 64 p.

Mukherjee, Amar Chandra. The Precambrian geology of the Flin Flon area, northern Saskatchewan and Manitoba, Canada. D, 1971, University of Saskatchewan. 161 p.

Mukherjee, Nilendu S. Geology and mineral deposits of the Galena-Gilt Edga area, northern Black Hills, South Dakota. D, 1968, Colorado School of Mines.

Mukherji, Kalyan Kumar. Petrology of the Black River Limestones (Middle Ordovician) in southwestern Ontario (Canada). D, 1968, University of Western Ontario. 412 p.

Mukhopadhyay, Bimal. Rubidium-strontium whole-rock geochronology and strontium isotope geology of the Madera Formation near Albuquerque, New Mexico. D, 1974, University of New Mexico. 182 p.

Mulcahey, Michael Thomas. The geology of Fidalgo Island and vicinity, Skagit County, Washington. M, 1975, University of Washington. 49 p.

Mulcahy, Marjorie. Normal mode excitation by moment tensors and other phenomenological descriptions of seismic sources. D, 1976, Princeton University. 227 p.

Muldoon, Maureen A. Hydrogeologic and geotechnical properties of pre-late Wisconsin till units in western Marathon County, Wisconsin. M, 1987, University of Wisconsin-Madison. 251 p.

Muldoon, William James. Stratigraphic facies of Cretaceous coal-bearing coastal zone strata of the Wasatch Plateau, Ferron, Utah. M, 1980, North Carolina State University. 68 p.

Mulhern, K. Petrography and structure of the northern margin of the Barth layered structure, Labrador. M, 1974, Syracuse University.

Mulhern, M. E. Physicochemical characterization of sediment facies and paleoclimatic inferences, California continental borderland. M, 1976, University of Southern California. 131 p.

Mulholland, James Willard. Surficial geology of the Ware Quadrangle, Worchester and Hampshire counties, Massachusetts. D, 1975, University of Massachusetts. 291 p.

Mulholland, Malcolm M. A study of the optical and chemical properties of a plagioclase feldspar in some quartz-bearing plutonites from Vermont. M, 1938, Syracuse University.

Mulholland, P. L. Organic carbon cycling in a swamp-stream ecosystem and export by streams in eastern North Carolina. D, 1979, University of North Carolina, Chapel Hill. 152 p.

Mulica, Walter S. The hydrology of the Cedarburg Bog area. M, 1973, University of Wisconsin-Milwaukee.

Mulky, Francis P. The subsurface geology of the Bristow Township. M, 1922, University of Oklahoma. 14 p.

Mull, Bert Hathaway. Miocene volcanics of the San Jose Hills. M, 1934, University of California, Berkeley. 95 p.

Mull, Charles G. Geology of the Grand Hogback Monocline near Rifle (Garfield County), Colorado. M, 1960, University of Colorado.

Mullarkey, Peter William. CONE; an expert system for interpretation of geotechnical characterization data from cone penetrometers. D, 1985, Carnegie-Mellon University. 161 p.

Mullen, Christopher Edward. Structure and stratigraphy of Triassic rocks in the Immigrant Canyon area, Northeast Elko County, Nevada. M, 1986, University of Wisconsin-Milwaukee. 61 p.

Mullen, Donna Marie. Structure and stratigraphy of Triassic rocks in the Long Canyon area, northeastern Elko County, Nevada. M, 1986, University of Wisconsin-Milwaukee. 75 p.

Mullen, E. D. Geology of the Greenhorn Mountains, northeastern Oregon. M, 1979, Oregon State University. 372 p.

Mullen, Ellen D. Petrology and regional setting of peridotite and gabbro of the Canyon Mountain Complex, Northeast Oregon. D, 1983, Oregon State University. 277 p.

Mullen, John C. Environmental geology of Milton, Westford, and Underhill, (Chittenden County), Vermont. M, 1973, University of Vermont.

Mullenax, Arthur Craig. Deformation features within the Martinsburg Formation in the St. Clair and Narrows thrust sheets, Giles County, Virginia. M, 1981, Virginia Polytechnic Institute and State University.

Mullens, Rockne Lyle. Stratigraphy and environment of the Toroweap Formation (Permian) north of Ash Fork, Arizona. M, 1967, University of Arizona.

Muller, A. B. Desalination by salt replacement and ultrafiltration. M, 1974, University of Arizona.

Muller, Charles Julian. Geology and origin of the iron ore deposits of the Richards Mine, Wharton, New Jersey. M, 1920, Massachusetts Institute of Technology. 45 p.

Muller, Charles Julian. Origin of the New Jersey magnetite ores. D, 1923, Massachusetts Institute of Technology. 195 p.

Muller, David S. Glacial geology and Quaternary history of Southeast Meta Incognita Peninsula, Baffin Island, Canada. M, 1980, University of Colorado.

Muller, Eric Charles. What hath man wrought? Effect of man upon the botanical landscape, Belvezet, France. D, 1973, George Washington University.

Muller, Ernest Hathaway. The glacial geology of the Naknek District, the Bristol Bay region, Alaska. D, 1952, University of Illinois, Urbana.

Muller, François. A kinetic approach to zinc speciation in marine and estuarine waters. D, 1988, University of Rhode Island. 206 p.

Muller, Frederick Lorenz, Jr. Growth analysis of kummerform and normalform phenotypes of the species Globigerinoides ruber from the North Atlantic. M, 1976, Rutgers, The State University, New Brunswick. 65 p.

Muller, James Louis. Earthquake source parameters, seismicity, and tectonics of the Oceanographer transform fault. M, 1982, Massachusetts Institute of Technology. 82 p.

Muller, Leigh Neville. Sedimentological aspects of bioturbation in abyssal oozes in the South Central Atlantic. M, 1983, University of Utah.

Muller, Otto Helmuth. Offset dike contacts and the stress field in the Spanish Peaks region, Colorado. D, 1974, University of Rochester. 129 p.

Muller, Otto Helmuth. The Scranton (Pennsylvania) gravity high. M, 1972, University of Rochester.

Muller, P. M. H. Some aspects of the ecology of several large, symbiont-bearing foraminifera and their contribution to warm, shallow-water biofacies. D, 1977, University of Hawaii. 191 p.

Muller, Peter Dale. Geology of the Los Amates Quadrangle and vicinity, Guatemala, Central America. D, 1979, SUNY, College at New Paltz. 326 p.

Muller, Remigio Martinez *see* Martinez Muller, Remigio

Muller, Sean Conroy. The geology and distribution of base metal deposits in the Fort Hall mining district, Bannock County, Idaho. M, 1978, Idaho State University. 194 p.

Muller, Siemon W. Recognition of new faunal horizons in the Triassic and Jurassic systems of the Pilot Mountains, Mineral County, Nevada. M, 1929, Stanford University. 26 p.

Muller, Siemon W. Triassic of Gabb's Valley Range, Nevada. D, 1930, Stanford University. 79 p.

Müller-Henneberg, Matthias. Diffusion in zinc selenide. D, 1970, Stanford University. 126 p.

Mullett, Douglas J. The geology of the Albion Group of Tuscarawas County, Ohio. M, 1982, Wright State University. 104 p.

Mullican, Jerry W. Investigation of relationship between select mortality rates and ground water chemistry. M, 1975, Texas Tech University. 123 p.

Mullican, William Franklin, III. The stratigraphy and diagenesis of the Lake Valley Formation (Mississippian), San Andres Mountains, New Mexico. M, 1981, Texas Tech University. 163 p.

Mulligan, Kevin Reilley. The movement of transverse coastal dunes, Pismo Beach, California. M, 1985, University of California, Los Angeles. 94 p.

Mulligan, Patrick John. Quantitative mineralogical analysis and scanning electron microscopy techniques for the study of argillaceous formations which are potential candidates for the geologic disposal of high-level radioactive waste. M, 1987, University of Tennessee, Knoxville. 109 p.

Mulligan, Robert. Geology of the Nelson and adjoining parts of Salmo map areas, British Columbia. D, 1951, McGill University.

Mulligan, Robert. Geology of the northern part of the east shore of Great Bear Lake, Northwest Territories. M, 1948, McGill University.

Mullin, Peter R. Facies of North Santee Inlet and contiguous areas. M, 1973, University of South Carolina. 133 p.

Mullin, Rosemary Patricia. The deformation of solid carbon dioxide. M, 1974, University of Michigan.

Mullineaux, Donald R. and Stark, William James. The glacial geology of the City of Seattle. M, 1950, Washington State University. 89 p.

Mullineaux, Donald Ray. Geology of Renton, Auburn and Black Diamond quadrangles, Washington. D, 1961, University of Washington. 202 p.

Mulliner, Beulah A. Development of physiography in American textbooks. M, 1911, Cornell University.

Mullins, Allen T. A study of marine terrigenous sediments from the Gulf of Mexico. M, 1959, Florida State University.

Mullins, B. M. Geochemical aspects of atmospherically transported trace metals over the Georgia Bight. M, 1978, Georgia Institute of Technology. 66 p.

Mullins, H. T. Deep carbonate bank margin structure and sedimentation in the northern Bahamas. D, 1978, University of North Carolina, Chapel Hill. 166 p.

Mullins, Henry T. Stratigraphy and structure of Northeast Providence Channel, Bahamas and origin of the northwestern Bahama Platform. M, 1975, Duke University. 203 p.

Mullins, John. Geology of the Noel Paul's Brook area, central Newfoundland. M, 1961, Memorial University of Newfoundland. 128 p.

Mullins, L. E. A terrain study of the Bloodland Quadrangle, Pulaski County, Missouri. M, 1965, University of Missouri, Rolla.

Mullins, Robert L. Ostracod occurrence in the Perdue Hill Section, Alabama. M, 1962, Louisiana State University.

Mullock, J. E. Biogeochemical and geochemical research in the Amisk Lake and Flin Flon region. M, 1952, University of Saskatchewan. 63 p.

Mulry, Christopher J. Geology of the Buckfield 7.5′ Quadrangle, Maine. M, 1986, University of Maine. 89 p.

Mulryan, Henry. Geology and ores of the Leona Heights pyrite deposit, Oakland, California. M, 1925, Stanford University. 59 p.

Mulsow, Miriam H. Petrography of the Lower Cretaceous Cupido Formation; San Lorenzo Canyon, Saltillo, Coahuila, Mexico. M, 1983, University of Southwestern Louisiana. 69 p.

Mulsow, Randall R. Effects of clay mineralogy on the SP log response. M, 1984, University of Southwestern Louisiana. 137 p.

Multer, Harold Gray. Some structural and physical properties of the (Pennsylvanian) Richburg Sand, Allegany County, New York. M, 1951, Syracuse University.

Multer, Harold Gray. Stratigraphy, structure, and economic geology of Pennsylvanian rocks in Wayne County, Ohio. D, 1955, Ohio State University. 222 p.

Mumby, Joyce I. Oligocene stratigraphy and foraminifera of the Porter Bluffs area, Washington. M, 1959, University of Washington. 137 p.

Mumby, Joyce Ione. Upper Cretaceous foraminifera from the marine formations along the Chesapeake and Delaware Canal. D, 1962, Bryn Mawr College. 202 p.

Mumcu, Hasan H. Study of the various coal mines in Utah, Colorado, and Wyoming. M, 1950, University of Utah. 79 p.

Mumma, Martin Dale. Palynology of selected coals of north-central Missouri. M, 1960, University of Missouri, Columbia.

Mumma, Martin Oale. Vicksburgian ostracode biostratigraphy of Mississippi and Alabama. D, 1965, Louisiana State University.

Mumme, Stephen T. Geopressure in the Houma and Hollywood fields, Louisiana. M, 1979, Louisiana State University.

Mummert, Mark Christopher. Quantifying uncertainty in nitrate pollution from land application of sewage sludge. D, 1987, Cornell University. 229 p.

Mummery, Robert Craig. Coronite amphibolites from the Whitestone area, Parry Sound, Ontario. D, 1973, McMaster University. 202 p.

Mumpower, Douglas S. Regional fracture investigation of the Brallier Formation and its transient fracture morphology in Mineral and Hampshire counties, West Virginia, Allegany County, Maryland, and Bedford County, Pennsylvania. M, 1982, Wright State University. 92 p.

Mumpton, Frederick Albert. Stability studies of the zirconthorite group and the effect of related oxides. D, 1958, Pennsylvania State University, University Park. 117 p.

Mumpton, Frederick Albert. The influence of controlled compositional variation on the hydrothermal stability of synthetic montmorillonoids. M, 1956, Pennsylvania State University, University Park. 60 p.

Mumtazaddin, M. The geology of the area between Carol Lake and Wabush Lake, Labrador. M, 1958, McGill University.

Mumtazuddin, M. Geology of the area between Carol Lake and Wabush Lake, Labrador. M, 1958, McGill University.

Munasifi, Wasim G. A. Maximum entropy spectral analysis of surface wave dispersion. M, 1979, Georgia Institute of Technology. 121 p.

Muncaster, Neill K. The paragenesis of galena and sphalerite. M, 1960, University of Minnesota, Minneapolis. 84 p.

Munchrath, Marvin A. A sedimentary study of the sandstone member of the Ordovician Oil Creek Formation in Cooke and Grayson counties, Texas. M, 1956, Southern Methodist University. 38 p.

Muncill, Gregory Ernest. Igneous petrologic evolution near a plutonic-metasedimentary contact in the Peninsular Ranges Batholith, Southern California. M, 1984, San Diego State University. 112 p.

Muncill, Gregory Ernest. Igneous plagioclase feldspar; kinetics of crystal growth and the origin of complex compositional zoning. D, 1985, Pennsylvania State University, University Park. 259 p.

Mundi, Emmanuel Kengnjisu. Elastic and viscoelastic behavior of two sedimentary rocks. M, 1970, Pennsylvania State University, University Park. 94 p.

Mundi, Emmanuel Kengnjisu. The physical characteristics of some fractured aquifers in central Pennsylvania and a digital simulation of their sustained yields. D, 1972, Pennsylvania State University, University Park. 342 p.

Mundorff, Maurice J. The geology of the Salem Quadrangle, Oregon. M, 1939, Oregon State University. 79 map p.

Mundorff, Norman L. The geology of Alkali Lake basin, Oregon. M, 1947, Oregon State University. 77 map x-section p.

Mundt, Philip Amos. A regional study of the Amsden Formation. D, 1956, Stanford University. 374 p.

Mundt, Philip Amos. Geologic considerations involved in a secondary-recovery program for the Palestine Sand, Maunie South Field, White County, Illinois. M, 1953, Washington University. 130 p.

Mundy, Brian Roy. An environmental model for the deposition of the Upper Jurassic limestones of the Djebel Zaghouan region of northern Tunisia. M, 1976, University of South Carolina.

Munger, Robert D. Geology of the Spread Eagle Peak area, Sangre de Cristo Mountains (Saguache and Custer counties), Colorado. M, 1959, University of Colorado.

Mungi, Julio C. Castañada. Statistical treatment of geochemical data with the help of the computer. M, 1977, Montana College of Mineral Science & Technology. 56 p.

Munguia-Orozco, Luis. Strong ground motion and source mechanism studies for earthquakes in the northern Baja California-Southern California region. D, 1983, University of California, San Diego. 167 p.

Munha, Jose Manuel Urbano. Igneous and metamorphic petrology of the Iberian pyrite belt volcanic rocks. D, 1982, University of Western Ontario. 712 p.

Muniz, Paul Francisco. Precious metal vein deposits of the Buffalo Hump District, west-central Idaho. M, 1985, University of Wisconsin-Madison. 84 p.

Munk, Walter H. Increase in the period of waves traveling over large distances; with application to tsunamis, swell, and seismic surface waves. D, 1946, University of California, Los Angeles.

Munly, Walter C. Ultrasonic determination of the elastic properties of aluminum oxynitride spinel (ALON) as a function of pressure and temperature. M, 1984, Pennsylvania State University, University Park. 96 p.

Munn, James Knox. The Wellman Field of south-central Terry County, Texas. M, 1954, Texas Tech University. 58 p.

Munne, Aarne Iivari. Petrography and facies of Candeias–Ilhas Sandstones, lower Cretaceous of Bahia, Brazil. M, 1971, Ohio University, Athens. 82 p.

Munoz Bravo, Jorge Oswaldo. Evolution of Pliocene and Quaternary volcanism in the segment of the Southern Andes between 38° and 39°S. D, 1988, University of Colorado. 172 p.

Munoz J., Nicolas G. Magdalena turbidites in deep-sea sediments. M, 1965, Columbia University, Teachers College. 211 p.

Munoz, J. Nicholas G. Magdalena turbidites in deep-sea sediments. M, 1964, Columbia University, Teachers College.

Munoz, J. W. DeVilbiss *see* DeVilbiss Munoz, J. W.

Munoz, James Loomis. Synthesis and stability relations of lepidolites. D, 1966, The Johns Hopkins University. 245 p.

Munoz, Jose Ramirez *see* Ramirez Munoz, Jose

Munro, D. S. Energy exchange on a melting glacier. D, 1975, McMaster University.

Munro, I. Ordovician conodonts of the Lake Timiskaming Paleozoic outlier. M, 1975, University of Waterloo.

Munroe, Henry. Yesso coals. D, 1877, Columbia University, Teachers College.

Munroe, Robert John. A heat flow determination near Yerington, Nevada. M, 1965, Stanford University. 34 p.

Munsart, C. A. An analysis of the lower Silurian "Clinton" Sandstone Reservoir in Lenox Field, Ashtabula, Ohio. M, 1975, Queens College (CUNY). 91 p.

Munsell, Charles E. Parallelism between the chemical compositions and the geological positions of the carbon minerals. D, 1884, Columbia University, Teachers College.

Munsey, Gorson Cloyd, Jr. A Paleocene ostracod fauna from the Coral Bluff Marl Member of the Naheola Formation of Alabama. M, 1951, Washington University. 102 p.

Munsey, Jeffrey W. Focal mechanism analysis for Recent (1978-1984) Virginia earthquakes. M, 1984, Virginia Polytechnic Institute and State University. 214 p.

Munsey, John Sal. Late Quaternary geotechnical stratigraphy of North Texas continental shelf. M, 1986, Texas A&M University. 108 p.

Munsil, John Michael. Depositional environments and diagenesis in the Bromide Sandstones, Caddo and Comanche counties, Southwest Oklahoma. M, 1983, Oklahoma State University. 183 p.

Munson, Michael G. Tidal delta facies relationships, Harbor Island, Texas. M, 1975, University of Texas, Austin.

Munson, Robert C. An investigation of the seismicity in the vicinity of Rangely, Colorado. M, 1968, Colorado School of Mines. 86 p.

Munson, Timothy Wayne. Depositional, diagenetic, and production history of the upper Morrowan Buckhaults Sandstone, Farnsworth Field, Ochiltree County, Texas. M, 1988, West Texas State University. 117 p.

Munter, James Arnold. Groundwater modeling of three lake/aquifer systems in Wisconsin. M, 1979, University of Wisconsin-Madison.

Munthe, Lynn Kathleen. The osteology of the Miocene rodent Schizodontomys. M, 1975, University of California, Berkeley. 110 p.

Munthe, Lynn Kathleen. The skeleton of the Borophaginae (Carnivora; Canidae); morphology and function. D, 1979, University of California, Berkeley. 278 p.

Munts, Steven Rowe. Geology and mineral deposits of the Quartzville mining district, Linn County, Oregon. M, 1978, University of Oregon. 213 p.

Muntzert, James K. Geology and mineral deposits of the Brattain District, Lake County, Oregon. M, 1969, Oregon State University. 70 p.

Munyan, Arthur Claude. A report on the geology of Clark County, Kentucky. M, 1931, University of Cincinnati. 134 p.

Munyan, Arthur Claude. Geology of the Dalton Quadrangle, Georgia-Tennessee. D, 1931, University of Cincinnati. 179 p.

Muramoto, Frank Shigeki. The use of well logs to determine the effects of hydrothermal alteration and other reservoir properties of the Salton Sea and Westmoreland geothermal systems in the Imperial Valley, California, U.S.A. M, 1982, University of California, Riverside. 219 p.

Murany, Ernest Elmer. Subsurface stratigraphy of the Wasatch Formation (Eocene) Uinta Basin, Utah. D, 1963, University of Utah. 364 p.

Muraro, Theodore W. Stratigraphy, structural geology and mineralization at Duncan Mine, Lardeau District, British Columbia. M, 1962, Queen's University. 174 p.

Murathan, Mustafa. The effect of depth on seismic migration. M, 1983, Texas A&M University.

Murchie, Bonnie. An analysis of jointing in folded and faulted sedimentary rocks of the northern Tobacco Root Mountains, Montana; a computer implemented study. M, 1979, Indiana University, Bloomington. 183 p.

Murchie, Scott Lawrence. 210Pb dating and Recent geologic history of two bays of Lake Minnetonka, Minnesota. M, 1984, University of Minnesota, Duluth. 160 p.

Murchie, Scott Lawrence. The tectonic and volcanic evolution of Ganymede and its implications for the satellite's internal structure and evolution. D, 1988, Brown University. 346 p.

Murchison, David K. A test of two-feldspar geothermometry. M, 1980, Miami University (Ohio). 95 p.

Murchison, Roderick Goldston, Jr. The geology of Las Monas oil-field, Colombia, South America. M, 1950, University of North Carolina, Chapel Hill. 56 p.

Murck, Barbara Winifred. Factors influencing the formation of chromite seams; Part I, The effects of temperature and oxygen fugacity on the behaviour of chromium in basic and ultrabasic melts; Part II, The petrology and geochemistry of the G & H chromite seams in the Mountain View area of the Stillwater Complex, Montana. D, 1986, University of Toronto. 157 p.

Murcy, Richard James. Thickness and character of the sub-Trenton interval in the Appalachian Basin. M, 1948, University of Pittsburgh.

Murdaugh, Daniel J. Stratigraphy of the Witts Springs Formation (Morrowan) in its type area, north central Arkansas. M, 1983, University of Arkansas, Fayetteville.

Murday, Mayloganaden. Beach erosion in West Africa. D, 1986, University of South Carolina. 115 p.

Murdoch, Joseph. The microscopic determination of the opaque minerals; a contribution to the study of ores. D, 1915, Harvard University.

Murdoch, Lawrence Corlies, III. The characteristics of some breccia structures associated with mineralized igneous rocks of felsic to intermediate composition. M, 1983, University of Cincinnati. 152 p.

Murdoch, P. S. Water budget comparison of two headwater lake basins subjected to low pH precipitation in the western Adirondack Mountains, New York. M, 1983, SUNY at Binghamton. 98 p.

Murdock, Clair N. Geology of the Weston Canyon area, Bannock Range, Idaho. M, 1961, Utah State University. 57 p.

Murdock, Don M. Geology of the central Sierra del Carmen Mountains of Brewster County, Texas. M, 1964, Texas A&M University.

Murdock, James Neil. The physical composition of the Sylamore Sandstone. M, 1931, University of Missouri, Columbia.

Murdy, Richard James. Thickness and character of the sub-Trenton interval in the Appalachian Basin. M, 1948, University of Pittsburgh.

Murdy, William. Environment of deposition and basinal analysis of a part of the Anakeesta Formation, western North Carolina. M, 1985, Texas A&M University.

Murgatroyd, Carolyn Drake. Significance of silicified carbonate rocks near the Devonian-Mississippian boundary, Ouachita Mountains, Oklahoma. M, 1980, University of Oklahoma. 65 p.

Murin, Timothy. Sedimentology and structure of the first Bradford Sandstone in the Pennsylvania Plateau Province. M, 1988, University of Pittsburgh.

Murk, Ronald Clarence. The geology of the Bannock Range west of Hawkins Basin, Power County, Idaho. M, 1971, Idaho State University. 94 p.

Murley, William H. The lithology, structure, paleoenvironment and paleontology of the White Hills, Jeff Davis County, Texas. M, 1987, Sul Ross State University. 75 p.

Murlin, Jack Ronald. Stratigraphy and depositional environments of the Arlington Member, Woodbine Formation (Upper Cretaceous), Northwest Texas. M, 1975, University of Texas, Arlington. 213 p.

Murnane, Richard James. Opal Ge/Si and the germanium geochemical cycle. D, 1988, Princeton University. 219 p.

Murosko, John E. Tabular diabase intrusives on Midterhuken Peninsula, Spitsbergen. M, 1981, University of Wisconsin-Madison.

Murowchick, James Bernard. Preliminary investigation of mineral precipitation from the Salton Sea geothermal brines. M, 1979, Pennsylvania State University, University Park. 112 p.

Murowchick, James Bernard. The formation and growth of pyrite, marcasite, and cubic FeS. D, 1984, Pennsylvania State University, University Park. 182 p.

Murphey, Clifford W. Areal geology of the Erick area, Beckman and Greer counties, Oklahoma. M, 1958, University of Oklahoma. 102 p.

Murphey, Joseph Bledsoe. A quantitative geomorphologic analysis of the drainage net in a small North Mississippi watershed (Pigeon Roost Watershed, Marshall County, Mississippi). M, 1965, University of Mississippi.

Murphey, Leslie V. A bibliographic index of fossils from New Mexico which have been described and/or illustrated in the geologic literature. M, 1940, University of New Mexico. 250 p.

Murphy Rohrer, William Lyman. Migration of meanders in alluvial rivers. D, 1982, University of Minnesota, Minneapolis. 159 p.

Murphy, Allen Emerson. Lithofacies investigations of the Middle and Upper Devonian rocks of the Allegheny Synclinorium [Appalachian Basin]. D, 1955, Syracuse University. 95 p.

Murphy, Allen Emerson. Sedimentary petrography of the (Devonian) Oriskany-Helderberg contact in northwestern West Virginia. M, 1948, West Virginia University.

Murphy, Andrew J. A comprehensive study of long-period (20-200 sec) earth noise at the high-gain world-wide seismograph stations. D, 1975, Columbia University. 141 p.

Murphy, Barbara Ann. Quantitative analysis of the distribution of Holocene benthic foraminifera on the continental shelf off South Carolina. M, 1987, University of Georgia. 129 p.

Murphy, Brendan E. Geology of the Limekiln Knoll Quadrangle, north-central Utah. M, 1983, Bryn Mawr College. 52 p.

Murphy, C. J., III. Modeling the northeastern North Carolina gravity low. M, 1972, University of North Carolina, Chapel Hill.

Murphy, Catherine Marie. Stratigraphy and metamorphic petrography of a late Pre-cambrian diamictite, Deep Creek Mountains, Utah. M, 1975, University of Nebraska, Lincoln.

Murphy, D. J. Structure geology of the Glenvar area, Virginia. M, 1968, Virginia Polytechnic Institute and State University.

Murphy, D. K. Littoral drift studies at Scarborough, Ontario. M, 1963, University of Toronto.

Murphy, Daniel Lawson. An investigation of the Graydon Formation in portions of Callaway County, Missouri. M, 1955, University of Missouri, Columbia.

Murphy, Daniel Lawson. Precambrian geology of the Lake Carheil area, Saguenay electoral district, Quebec. D, 1961, University of Michigan. 198 p.

Murphy, David Andrew. Sedimentation of the upper Hell Creek Formation (Upper Cretaceous), Carter County, Southeast Montana. M, 1986, San Diego State University. 104 p.

Murphy, David Hazlett. Carbonate deposition and facies distribution in a central Michigan marl lake. M, 1978, University of Michigan.

Murphy, Don R. Fauna of the Morrowan rocks of central Utah. M, 1954, Brigham Young University. 64 p.

Murphy, Donald Currie. Fabric transitions in the formation of mylonites in Columbia River fault zone, B. C. M, 1980, Stanford University. 111 p.

Murphy, Donald Currie. Stratigraphy and structure of the east-central Cariboo Mountains, British Columbia, and implications for the geological evolution of the southeastern Canadian Cordillera. D, 1986, Carleton University. 337 p.

Murphy, Donald J., Jr. Soils and rocks; composition, confining level and strength. D, 1971, Duke University.

Murphy, Donald James. The petrology and deformational history of the basement complex (Precambrian), Wright Valley, Antarctica, with special reference to the origin of the augen gneisses. D, 1971, University of Wyoming. 114 p.

Murphy, Edward C. The effect of oil-and-gas well drilling fluids on shallow groundwater in western North Dakota. M, 1983, University of North Dakota. 242 p.

Murphy, Edward Joseph. Always convergent iterative deconvolution for acoustic non-destructive evaluation. M, 1986, University of New Orleans. 133 p.

Murphy, Eleanor F. Laboratory experiments on the development of shoreline features and processes. M, 1938, Cornell University.

Murphy, Ellyn Margaret. Carbon-14 measurements and characterization of dissolved organic carbon in ground water. D, 1987, University of Arizona. 199 p.

Murphy, Franklin Mac. Geology and ore deposits of a part of the Panamint Range (California). M, 1929, California Institute of Technology. 122 p.

Murphy, J. B. The stratigraphy and geological history of the Forchu Group, South eastern Cape Breton, Nova Scotia. M, 1977, Acadia University.

Murphy, J. D. Surface and groundwater chemistry of the Sage Creek drainage basin, Southwest Wyoming; significance for trans-basin water diversion. D, 1979, University of Wyoming. 167 p.

Murphy, James Brendan. Tectonics and magmatism in the northern Antigonish Highlands, Nova Scotia. D, 1982, McGill University. 243 p.

Murphy, James D. The geology of Eagle Cove Basin at Bruneau, Idaho. M, 1973, SUNY at Buffalo. 77 p.

Murphy, James Howard. Geology of the Upper Cement Creek area of Gunnison County, Colorado and a petrological study of the Maroon Formation. M, 1951, University of Kentucky. 49 p.

Murphy, James W. Joint orientations and their relationships to structures and lithologies of rocks between Harrisville and Natural Bridge, New York. M, 1979, SUNY at Buffalo. 84 p.

Murphy, John F. Petrography and geochemistry of the Cape Neddick gabbro (Devonian?), Maine. M, 1972, University of Rhode Island.

Murphy, John F. The origin of pegmatites in the Grafton District, New Hampshire. M, 1949, Dartmouth College. 53 p.

Murphy, John R. A focal mechanism study of the Aleutian teleseism of February 4, 1965. M, 1966, [Boston University].

Murphy, Kathleen. Eolian origin of upper Paleozoic red siltstones at Mexican Hat and Dark Canyon, southeastern Utah. M, 1987, University of Nebraska, Lincoln. 128 p.

Murphy, Kenneth Robert. An investigation of rubidium concentrations in fresh water reaches of the Hudson River. D, 1969, Rensselaer Polytechnic Institute. 129 p.

Murphy, Kim Marie. Transition metal and rare earth element fluxes at two sites in the eastern tropical Pacific; relationship to ferromanganese nodule genesis. M, 1985, Oregon State University. 120 p.

Murphy, Mark Thomas. Strontium-isotope and trace-element geochemistry of the Platoro caldera complex, Colorado. M, 1985, University of New Mexico. 118 p.

Murphy, Michael Arthur. The paleontology and stratigraphy of the Horsetown stage of Northern California. D, 1954, University of California, Los Angeles.

Murphy, Michael Joseph. An X-ray study of some mineral sulfosalts. M, 1953, University of California, Berkeley. 64 p.

Murphy, Peter J. Ground water-surface water interactions in the Nashotah-Nemahbin-Nagawicka lake district, Waukesha County, Wisconsin. M, 1979, University of Wisconsin-Milwaukee. 119 p.

Murphy, Philip Joseph. Holocene diagenesis of the Anastasia Formation, northeastern Florida. M, 1973, Duke University. 104 p.

Murphy, R. J. An engineering and economic analysis of the Soldotna Creek Unit, Swanson River oil field, Alaska. M, 1977, Stanford University. 148 p.

Murphy, Raymond E. The glacial geology of an area in northwestern Wisconsin. M, 1926, University of Wisconsin-Madison.

Murphy, Richard W. Geology of the Slavonia-Diamond Park area, Routt County, Colorado. M, 1958, University of Wyoming. 54 p.

Murphy, Robert E. Petrographic study of the Lower Hygiene Sandstone members of the Upper Cretaceous Pierre shales of northeastern Colorado. M, 1932, University of Colorado.

Murphy, Robert Edward. Paleontology and stratigraphy of the Middle and Upper Ordovician limestone in Rabbit Valley, Georgia. M, 1953, Emory University. 96 p.

Murphy, Robert Parsons. Investigation of potential source rocks for the Granite Wash hydrocarbons, Washita County, Oklahoma. M, 1985, University of Texas at Dallas. 99 p.

Murphy, Russell King. Geochemistry of garnets and pyroxenes from skarn orebodies, Mines Gaspe, Quebec. M, 1986, University of Missouri, Columbia. 191 p.

Murphy, Sean C. Geology of the northern half of the Metasville 7 1/2 minute quadrangle, Georgia. M, 1984, University of Georgia. 130 p.

Murphy, Sean E. Upper Silurian ostracodes from the Roberts Mountains, central Nevada. M, 1975, University of California, Berkeley.

Murphy, Susan Hope. The geochemistry of molybdenum in the environment at Raytown, Georgia. M, 1982, University of Georgia.

Murphy, T. Dennis. Strandplain and deltaic deposits and related coals of the middle Cherokee Subgroup of southwestern Kansas and western Missouri. M, 1978, University of Texas, Austin.

Murphy, Thomas C. A study of the basal Cherokee in the Rolla area. M, 1929, University of Missouri, Rolla.

Murphy, Thomas M. The geology of the Nicolai Mountain-Gnat Creek area, Clatsop County, northwestern Oregon. M, 1981, Oregon State University. 355 p.

Murphy, Timothy B. Intergranular pressure solution in the Tuscarora orthoquartzite of central Pennsylvania. M, 1983, University of Missouri, Columbia.

Murphy, Vincent J. A magnetic reconnaissance of the Connecticut River valley in Massachusetts. M, 1957, Boston College.

Murphy, W. Dale. Stratigraphy and depositional environments of the Muddy Sandstone in North and Middle Parks basin, Jackson and Grand counties, Colorado. M, 1982, University of Colorado. 164 p.

Murphy, William Francis, III. Effects of microstructure and pore fluids on the acoustic properties of granular sedimentary materials. D, 1982, Stanford University. 269 p.

Murphy, William Marshall. An experimental study of solid-liquid equilibria in the albite-anorthite-diopside system. M, 1977, University of Oregon. 88 p.

Murphy, William Marshall. Thermodynamic and kinetic constraints on reaction rates among minerals and aqueous solutions. D, 1985, University of California, Berkeley. 165 p.

Murphy, William Owen, Jr. Seismic velocity studies of synthetic sandstone cores. M, 1957, University of Utah. 57 p.

Murrah, William Eugene. Geology of Cherry Canyon area, Tierra Vieja Mountains, Trans-Pecos, Texas. M, 1950, University of Texas, Austin.

Murray, Albert Nelson. Some whiteware and refractory clays of Colorado. M, 1924, University of Colorado.

Murray, Albert Nelson. The limestones and dolomites of northeastern United States as reservoir rocks for petroleum. D, 1928, University of Illinois, Chicago.

Murray, Bruce. Data on the Southwest Mabou River, (Mississippian) Horton strata (Nova Scotia). M, 1954, Massachusetts Institute of Technology. 36 p.

Murray, Bruce. Stratigraphy of the (Mississippian) Horton Group in parts of Nova Scotia. D, 1956, Massachusetts Institute of Technology. 238 p.

Murray, Cecil George. Petrologic studies of zoned ultramafic complexes in northern Venezuela and southeastern Alaska. D, 1972, Princeton University. 194 p.

Murray, Charles R. Geology and ore deposits of the Adit Tunnel and Columbia Mine, Ward (Boulder County), Colorado. M, 1934, University of Colorado.

Murray, Christopher J. Heavy metal distribution in the sediments of Flathead Lake, Montana. M, 1982, University of Montana. 53 p.

Murray, Clyde L. Structure and stratigraphy of the Jefferson Mountain area, Centennial Range, Idaho-Montana. M, 1973, Oregon State University. 108 p.

Murray, D. W. Part I, On the oxidation of some Nova Scotian coals; Part II, Viscosities of ethylenediomine extract of Joggins coal. M, 1955, Mount Allison University.

Murray, Daniel A. Petrology and paleoecology of the Williams Point Limestone (Carboniferous), Antigonish County, Nova Scotia. M, 1971, Dalhousie University. 109 p.

Murray, Daniel Patrick. An investigation of origin, structure, and metamorphic evolution of the Precambrian rocks of the West Point Quadrangle, New York. M, 1968, Brown University.

Murray, Daniel Patrick. Chemical equilibrium in epidote-bearing calc-silicates and basic gneisses, Reading Prong, New York. D, 1976, Brown University. 256 p.

Murray, David H., Jr. The effect of thermal metamorphism on quartz shape; Fourier-series analysis. M, 1982, Wichita State University. 44 p.

Murray, David L. Calibration of Little Beaver Creek watershed. M, 1968, Colorado State University. 94 p.

Murray, David Lloyd. The structural relationship between rocks of the George River and Fourchu groups in the Ingonish River-Clyburn Brook area, Cape Breton Island, Nova Scotia. M, 1977, Queen's University. 65 p.

Murray, David W. Paleo-oceanography of the Gulf of California based on silicoflagellates from marine varved sediments. M, 1982, Oregon State University. 129 p.

Murray, David William. Spatial and temporal variations in sediment accumulation in the central tropical Pacific. D, 1987, Oregon State University. 343 p.

Murray, Donald James. A synergetic study of seismic reflections and critical refractions in the Great Salt Lake Desert, Utah. M, 1984, University of Oklahoma. 95 p.

Murray, E. H., Jr. Recognition of cyclic patterns of species dominance in the Delphi Station Member of the Skaneateles Formation (Middle Devonian), New York State. M, 1977, Queens College (CUNY). 54 p.

Murray, Eugene J. Trace fossils of the Brassfield Formation, lower Silurian, in South-central Ohio and

North-central Kentucky. M, 1975, Western Michigan University.

Murray, F. E. Geology of the Nowood area, Washakie County, Wyoming. M, 1957, University of Wyoming. 115 p.

Murray, Faye Helen. Geochemical trends in lignite bearing Cretaceous sediments from the Mattagami Formation (James Bay Lowlands). M, 1984, University of Western Ontario. 104 p.

Murray, Frederick N. Stratigraphy and structural geology of the Grand Hogback Monocline (Rio Blanco, Garfield, Pitkin, and Gunnison counties), Colorado. D, 1965, University of Colorado.

Murray, Frederick N. The geology of the Grand Hogback Monocline near Meeker (Rio Blanco County), Colorado. M, 1962, University of Colorado.

Murray, Gene L. Digital computer simulation of a pumping well profile. M, 1984, University of Nebraska, Lincoln.

Murray, Grover E., Jr. Claiborne Eocene species of the ostracode genus Loxoconcha. M, 1939, Louisiana State University.

Murray, Grover E., Jr. Geology of DeSoto and Red River parishes, Louisiana. D, 1942, Louisiana State University.

Murray, Harrison F. Stratigraphic study of Pennsylvanian of McCoy area (Routt, Grand, and Eagle counties), Colorado. M, 1950, University of Colorado.

Murray, Hayden Herbert. The structure of kaolinite and its relation to acid treatment. D, 1951, University of Illinois, Urbana.

Murray, Haydn Herbert. Contact phenomena of the Bethlehem Gneiss and Littleton Formation in New Hamsphire. M, 1950, University of Illinois, Urbana.

Murray, J. C. Geothermal system at Kilauea Volcano, Hawaii. D, 1974, Colorado School of Mines. 86 p.

Murray, James P. Kohl Spring (Missouri); biogeochemistry of a blackwater (river) system. M, 1972, University of Missouri, Columbia.

Murray, James Wolfe. Some stratigraphic and paleoenvironmental aspects of the Swan Hills and Waterways formations, Judy Creek, Alberta, Canada. D, 1964, Princeton University. 199 p.

Murray, James Wray. The interaction of metal ions at the hydrous manganese dioxide-solution interface. D, 1973, Massachusetts Institute of Technology. 274 p.

Murray, Jane. Stratigraphy, structure and metamorphism of Precambrian rocks in Belmont and southern Methuen townships, southeastern Ontario. M, 1982, Carleton University. 128 p.

Murray, Jay Dennis. The structure and petrology of the San Jose Pluton, northern Baja California, Mexico. D, 1978, California Institute of Technology. 795 p.

Murray, John C. An investigation of the grain size dependence of the thermal remanent magnetization of magnetite. M, 1970, Michigan Technological University. 75 p.

Murray, Joseph Buford. Silurian and lower Devonian deposition in the southern Appalachian Mountains; a stratigraphic and environmental analysis. D, 1971, Case Western Reserve University. 422 p.

Murray, Joseph Buford. The geology of portions of the Powell and Clinton quadrangles, Anderson County, Tennessee. M, 1960, University of Tennessee, Knoxville. 66 p.

Murray, Kent Stephen. Geology of the Woody Mountain volcanic field, Coconino County, Arizona. M, 1973, Northern Arizona University. 89 p.

Murray, Kent Stephen. Tectonic implications of space-time patterns of Cenozoic volcanism in the Palo Verde Mountain volcanic field, southeastern California. D, 1981, University of California, Davis. 132 p.

Murray, L. G. Wall-rock alteration around sulphide deposits, southern Quebec. D, 1954, McGill University.

Murray, Louis Charles, Jr. Modeling solute transport in porous media with transient porewater velocities. D, 1988, Duke University. 320 p.

Murray, Marc Michael. Petrology, structure, and origin of the San Isabel Batholith (Precambrian), Wet mountains, Colorado. D, 1970, Rice University. 54 p.

Murray, Margaretha E. Geology of the east-central Quinn Canyon Range, Nye County, Nevada. M, 1985, University of North Carolina, Chapel Hill. 107 p.

Murray, Raymond C. The petrology of the Cary and Valders tills of northeastern Wisconsin. M, 1952, University of Wisconsin-Madison.

Murray, Raymond C. The Recent sediments of three Wisconsin lakes. D, 1955, University of Wisconsin-Madison.

Murray, Robert Cozzens. Checkerboard chalcedony in a paleosilcrete, North Texas. M, 1985, University of Texas, Austin. 82 p.

Murray, S. M. Accumulation rates of sediments and metals off southern California as determined by Pb^{210} method. D, 1977, University of Southern California.

Murray, Sharon Ann. The Marshall Lake mines; their history and development. M, 1979, University of Idaho. 97 p.

Murray, Stephen P. Effects of particle size and wave state on grain dispersion. D, 1966, University of Chicago. 76 p.

Murray, Steven M. Diagenesis of sediment in the Santa Barbara Basin of the California borderland. M, 1972, University of Southern California.

Murray, Steven M. Geochemistry and diagenesis of Santa Barbara Basin sediments off southern California. M, 1973, University of Southern California.

Murray, Thomas H., Jr. The anorthosite, ilmenite-magnetite, and associated rocks of a portion of the Laramie anorthosite, Albany County, Wyoming. M, 1961, Colorado School of Mines. 79 p.

Murray, William B. Landslides and other mass-wasting phenomena in the Little Mountain area, Southwest Wyoming. M, 1984, University of Wyoming. 126 p.

Murray, William Gerard. Interpretation of washover laminations. M, 1978, University of Virginia. 43 p.

Murray, William L. The atmospheric effect in remote sensing of earth surface reflectivity. M, 1973, Purdue University. 63 p.

Murray, William Wallace. The origin of clay materials in Keystone Lake, Oklahoma. M, 1972, University of Tulsa. 54 p.

Murray-Rust, Douglas Hammond. Irrigation water management in Sri Lanka; an evaluation of technical and policy factors affecting operation of the main channel system. D, 1983, Cornell University. 374 p.

Murrell, Michael Tildon. Cosmic-ray produced radio-nuclides in extraterrestrial material. D, 1980, [University of California, San Diego]. 166 p.

Murrie, Gary Wayne. The distribution of selective trace elements of magnetite in the Pilot Knob ore body, Iron County, Missouri. M, 1973, Southern Illinois University, Carbondale. 54 p.

Murrin, Theresa Eileen. Landsat interpretation of desert landforms, northwestern Australia. M, 1983, University of California, Los Angeles. 133 p.

Murrish, Charles H. An integrated geologic-geophysics study of the Auburndale area, Wood County, Wisconsin. M, 1966, Michigan State University. 63 p.

Murrow, Patricia J. Cyclical variation in historic eruptions from Mount Etna, Sicily. M, 1979, Dartmouth College. 96 p.

Murry, Danny H. Economic mineral potential of the northern Quitman Mountains, Hudspeth County, Texas. M, 1978, University of Texas at El Paso.

Murry, Phillip Anthony. Biostratigraphy and paleoecology of the Dockum Group (Triassic) of Texas. D, 1982, Southern Methodist University. 459 p.

Mursky, Gregory. Mineralogy, petrology and geochemistry of Hunter Bay area, Great Bear Lake, N.W.T., Canada. D, 1963, Stanford University. 195 p.

Murtaugh, J. G. Geology of the Manicouagan cryptoexplosion structure. D, 1975, Ohio State University. 332 p.

Murtaugh, John Graham. Geology of Craters of the Moon National Monument, Idaho. M, 1961, University of Idaho. 99 p.

Murtha, Patricia Ellen. Seismic velocities in the upper part of the Earth's core. D, 1985, University of California, Berkeley. 121 p.

Murthy, Gummuluru S. Design and calibration of an astatic magnetometer and the study of remanent magnetization of some Newfoundland rocks. M, 1967, Memorial University of Newfoundland. 177 p.

Murthy, Gummuluru Satyanarayana. Paleomagnetic studies in the Canadian shield. D, 1969, University of Alberta. 127 p.

Murthy, Nallur Prahlada. Mohr's envelope for rocks under uniaxial compression. M, 1966, University of Utah. 60 p.

Murthy, Varanasi Rama. Bedrock geology of the East Barre area, Vermont. D, 1957, Yale University.

Murty, Rama C. Gamma ray investigation of heterogeneous solids. D, 1962, University of Western Ontario.

Murty, Vadali Venkata Narasimha. A finite element model for miscible displacement in ground water aquifers. D, 1975, University of California, Davis. 200 p.

Musa, Nagieb S. Hydrogeology of the alluvial aquifer in eastern Rapid City, Pennington County, South Dakota. M, 1984, South Dakota School of Mines & Technology.

Muscallo, David. A petrological and structural study of the Precambrian crystalline rocks, northeast Pingree Park Quadrangle, Mummy range, Larimer County, Colorado. M, 1969, SUNY at Buffalo. 58 p.

Muse, Paul Stephen. Groundwater resource evaluation of Southwest Washington County, Arkansas. M, 1982, University of Arkansas, Fayetteville.

Musgrave, Albert W. Wavefront charts and construction of raypath plotter. D, 1952, Colorado School of Mines. 79 p.

Musgrave, David L. Penetrative convection in sediments. D, 1983, University of Alaska, Fairbanks. 125 p.

Musgrave, John A. Geochemistry of ore deposition, Sultan Mountain Mine, western San Juan Mountains, Colorado. M, 1986, Colorado State University. 113 p.

Musgrove, Carl D. Petroleum geology of the Bryant area, Okmulgee, Okfuskee, and McIntosh counties, Oklahoma. M, 1963, University of Oklahoma. 37 p.

Musgrove, Frank William. A geophysical investigation of the mid-Michigan gravity high. M, 1983, University of Michigan.

Musgrove, Lee Ann. Geophysical investigation of the southern Angola Basin; SE Atlantic Ocean. M, 1982, University of Texas, Austin.

Mushake, William I., Jr. The Antiquity Sandstone (Permo-Carboniferous) of southeastern Meigs County, Ohio. M, 1956, Miami University (Ohio). 65 p.

Mushinsky, Edward Stephen. A paleomagnetic study of the cumberlandite intrusive body near Diamond Hill, Rhode Island. M, 1974, Virginia State University. 31 p.

Musiker, Laurie B. Origin of an ice-channel filling, Spring Hill and Willimantic quadrangles, Connecticut. M, 1984, University of Connecticut. 111 p.

Muskat, Judd. Geologic interpretations of Seasat-A radar images and Landsat MSS images of a portion of the southern Appalachian Plateau; Virginia, Kentucky, West Virginia. M, 1983, California State University, Northridge. 104 p.

Muskatt, Herman S. Petrology of the origin of the Clinton group (Silurian) of east-central New York and its relationship to the Shawangunk Formation (Silurian) of southeastern New York. D, 1969, Syracuse University.

Muskatt, Herman Solomon. Structure and stratigraphy of the Lower Knob area, Fulton County, Pennsylvania. M, 1959, Rutgers, The State University, New Brunswick. 57 p.

Musolf, Gene Emil. The geography of the lower Wisconsin River valley with emphasis on soil resources of the fluvial terraces. D, 1970, University of Wisconsin-Madison. 276 p.

Mussard, Donald E. Petrology and geochemistry of selected Precambrian felsic plutons, southern Medicine Bow Mountains, Wyoming. M, 1982, Colorado State University. 259 p.

Musselman, Elmer Thomas. Selective solubility effects in limestone. M, 1949, Colorado School of Mines. 67 p.

Musselman, George Abraham. Chester production in the Louden Field, Fayette County, Illinois. M, 1940, University of Texas, Austin.

Musselman, George Hayes. Ground water resources of the Ann Arbor area (Michigan). M, 1953, University of Michigan.

Musselman, Thomas E. A modified crustal source for the Colorado mineral belt; implications for REE buffering in CO2-rich fluids. M, 1987, Colorado School of Mines. 127 p.

Mussels, J. H. Reflection profiling investigations of the East Caroline Basin. M, 1972, University of Hawaii. 74 p.

Musselwhite, Donald Stanley. An isotopic and trace element study of the Woods Mountains caldera complex; a bimodal basalt/rhyolite volcanic center in the eastern Mojave Desert. M, 1984, University of California, Los Angeles. 127 p.

Musser, Kathryn Jeanne. Conical structures in the middle Precambrian Michigamme Formation. M, 1981, Michigan State University. 84 p.

Mussett, Jack D. Geology and mineral resources of central Logan County, Arkansas. M, 1952, University of Arkansas, Fayetteville.

Mussman, William J. The Middle Ordovician Knox Unconformity, Virginia Appalachians; transition from passive to convergent margin. M, 1982, Virginia Polytechnic Institute and State University. 158 p.

Mustafa, Mukhtar Ahmed. The influence of nonionic surfactants on unsteady water flow in unsaturated porous media. D, 1969, University of California, Riverside. 58 p.

Mustard, Peter S. Sedimentology of the Lower Gowganda Formation Coleman Member (Early Proterozoic) at Cobalt, Ontario. M, 1986, Carleton University. 143 p.

Mustart, David Alexander. Phase relations in the peralkaline portion of the system Na2O-Al2O3-SiO2-H2O. D, 1972, Stanford University. 202 p.

Mustoe, George Edward. Biochemical origin of coastal weathering features in the Chuckanut Formation of northwest Washington. M, 1971, Western Washington University. 67 p.

Mustonen, E. D. A micro-geothermal survey, Lake Dufault, Quebec (Canada). M, 1967, University of Western Ontario.

Muszynski, Isabelle. The dynamics of coupled marine ice stream-ice shelf systems and implications for the Quaternary ice ages. D, 1987, Northwestern University. 85 p.

Mutch, A. D. The effect of pressure in the process of mineral transport and deposition. M, 1950, University of Toronto.

Mutch, Thomas A. Geology of the northeast flank of the Flint Creek Range, Montana. D, 1960, Princeton University. 159 p.

Mutch, Thomas Andrew. The petrology of a portion of the Ice River, British Columbia, igneous complex. M, 1957, Rutgers, The State University, New Brunswick. 59 p.

Mutchler, W. H. The ores, methods of production, and metallurgical applications of beryllium. M, 1932, George Washington University.

Muth, Lorant Andreas. A study of frozen flux fluid motions at the surface of Earth's liquid core as deduced from magnetic observations at Earth's surface. D, 1980, University of Colorado. 131 p.

Muthig, Michael Gregory. Control on oil accumulation in the Aux Vases Formation, south-central Illinois. M, 1984, University of Kentucky. 112 p.

Muthig, Paul Joseph. Geological and statistical predictors for coal seam thickness of the Rocky Mountain basins. M, 1982, University of Kentucky. 107 p.

Mutis-Duplat, Emilio. Precambrian geology of a portion of the Purdy Hill Quadrangle, Mason County, Texas. M, 1969, Texas A&M University. 89 p.

Mutis-Duplat, Emilio. Stratigraphic sequence and structure of Precambrian metamorphic rocks in Purdy Hill Quadrangle, Mason County, Texas. D, 1972, University of Texas, Austin. 177 p.

Mutlu, Hamlin. The use of the Pitzer equations to determine saturation of trona and other associated saline minerals; a case study of the Green River trona deposits, Wyoming. M, 1988, South Dakota School of Mines & Technology.

Muto, Paul. Geology and mineralization of the Willard mining district, Pershing County, Nevada. M, 1980, University of Nevada. 62 p.

Mutschler, Felix Ernest. Geology of the Canjilon cauldron sink near Bernalillo, Sandoval County, New Mexico. M, 1956, University of New Mexico. 41 p.

Mutschler, Felix Ernest. Geology of the Treasure Mountain Dome, Gunnison County, Colorado. D, 1968, University of Colorado. 353 p.

Mutter, John Colin. Rifting of the Norwegian margin and young ocean basin accretion dynamics in the Norwegian-Greenland Sea. D, 1982, Columbia University, Teachers College. 289 p.

Mutti, Laurence Joseph. Structure and metamorphism of the Cranberry region, Thor-Odin gneiss dome, Shuswap metamorphic complex, British Columbia. D, 1978, Harvard University.

Muza, Jay P. Paleogene oxygen isotope record for DSDP Sites 511 and 512, sub-Antarctic South Atlantic Ocean; paleotemperature paleoceanographic changes. M, 1983, Florida State University.

Muza, Richard E. A numerical simulation of groundwater flow and contaminant migration at the LaBounty Landfill, Charles City, Iowa. M, 1986, Ohio University, Athens. 220 p.

Muzyka, L. J. Pb-210 chronology in a core from the Flax Pond marsh, Lond Island. M, 1976, SUNY at Stony Brook.

Mwanang'onze, E. H. B. Bare metal distribution in Gossans of the Kakagi Lake Archean volcanic rocks, northwestern Ontario. M, 1974, University of Manitoba.

Mwanang'onze, E. H. B. Stratigraphy and petrochemistry of the host rocks of copper-zinc deposits in the Flin Flon-Snow Lake greenstone belt. D, 1978, University of Manitoba.

Mwangi, Martin Peter. Isotopic and geochemical study of groundwater in the Eastern Province of Kenya. M, 1988, University of Windsor. 161 p.

Mwenifumbo, Campbell Jonathan. Interpretation of mise-a-la-masse data for vein type bodies. D, 1980, University of Western Ontario.

Mychkovsky, George. The origin of fluorite, sulfate and sulfide minerals in northwestern Ohio. M, 1978, Kent State University, Kent. 74 p.

Myer, George Henry. The mineralogy of epidote and zoisite. D, 1965, Yale University. 127 p.

Myer, Larry Richard. A physical model study of stand-up time of tunnels in squeezing ground. D, 1977, University of California, Berkeley. 187 p.

Myers, Addison Reid. The geology of the Ford South oil pool, Posey County, Indiana. M, 1953, University of Cincinnati. 64 p.

Myers, Allen Cowles. Sediment reworking, tube building, and burrowing in a shallow subtidal marine bottom community; rates and effects. D, 1973, University of Rhode Island.

Myers, Arthur John. Geology of Harper County, Oklahoma. D, 1957, University of Michigan.

Myers, Arthur John. The Terrenceville Conglomerate and the general geology of the Fortune Bay area, Newfoundland. M, 1949, Michigan Technological University. 42 p.

Myers, Bruce Eric. The formation of zoned metasomatic veins and massive skarn in dolomite, southern Sierra Nevada, California. M, 1988, University of Arizona. 125 p.

Myers, C. P. Aeromagnetic reconnaissance survey of Lake Erie. M, 1977, Ohio State University. 172 p.

Myers, Carl Weston, II. Geology of the Presley's Mill area, Northwest Putnam County, Georgia. M, 1968, University of Georgia. 67 p.

Myers, Carl Weston, II. Yakima basalt flows near Vantage, and from core holes in the Pasco Basin, Washington. D, 1973, University of California, Santa Cruz. 147 p.

Myers, Clay Kenton. The red beds of the Pittsburgh Quadrangle (Pennsylvania). M, 1935, University of Pittsburgh.

Myers, David Arthur. The geology and hydrogeology of the Albion Basin, Cassia County, Idaho. M, 1967, University of Idaho. 31 p.

Myers, Delbert Edward, Jr. Evaluation of zinc chloride complexes at 28 and 80° C from smithsonite solubilities. M, 1972, University of Toronto.

Myers, Dennis E. A stratigraphic study of the Pennsylvanian Virgilian Series in southwestern Chautauqua County, Kansas. M, 1968, Wichita State University. 241 p.

Myers, Gary A. Structural analysis of foliated Proterozoic metadiabase dikes in the Marquette-Republic region of northern Michigan. M, 1984, Michigan State University. 78 p.

Myers, Genne Marie. Tungsten occurrences in Arizona and their possible relationship to metallogenesis. M, 1983, University of Arizona. 96 p.

Myers, Georgianna. Fluid expulsion during the underplating of the Kodiak Formation; a fluid inclusion study. M, 1987, University of California, Santa Cruz.

Myers, Gregory L. Geology and geochemistry of the iron-copper-gold skarns of Kasaan Peninsula, Alaska. M, 1985, University of Alaska, Fairbanks. 165 p.

Myers, Ingrid A. Geology and mineralization at the Cyclopic Mine, Mohave County, Arizona. M, 1984, University of Nevada, Las Vegas. 64 p.

Myers, J. D. Geology and petrology of the Edgecumbe volcanic field, southeastern Alaska; transform fault volcanism and magma mixing. D, 1980, The Johns Hopkins University. 306 p.

Myers, James Edward. Geologic interpretations of the engineering properties of unconsolidated near-surface sediments in Kansas. M, 1968, Wichita State University. 105 p.

Myers, James M. Applications of photoacoustic microscopy in the study of inertinite macerals and to the detection of oxidation in vitrinite macerals. M, 1985, Southern Illinois University, Carbondale. 176 p.

Myers, Jed Anthony. Reduction in exchangeable magnesium upon liming acid soils of Ohio. D, 1985, Ohio State University. 145 p.

Myers, John B. Permian patch reefs in the Finlay mountains, west Texas. M, 1969, University of Texas at El Paso.

Myers, John D. Environment of deposition of the Morrison Formation of Quay County, New Mexico. M, 1956, Mississippi State University. 49 p.

Myers, Jonathan. Temperature-oxygen fugacity equilibria for the cobalt-cobalt oxide buffer, and in the magnetite-ilmenite deposit at Iron Mountain, Wyoming. M, 1978, University of Wyoming. 69 p.

Myers, Jonathan. Thermodynamic properties of minerals in the system Fe-Si-O. D, 1982, University of Wyoming. 119 p.

Myers, Keith. Petrology and depositional environments of the Lake Valley Formation, Lower Mississippian, southwestern New Mexico. M, 1983, University of Houston.

Myers, Lesley Louise. Geochemistry of the Crandon massive sulfide deposit, Wisconsin; sulfur isotope and fluid inclusion data. M, 1983, University of Wisconsin-Madison. 88 p.

Myers, Mark D. The stratigraphy and sedimentology of the upper Wonewoc Formation; a critique of the Ironton Member. M, 1981, University of Wisconsin-Madison.

Myers, Nathan Cebren. Marine geology of the western Ross Sea; implications for Antarctic glacial history. M, 1982, Rice University. 234 p.

Myers, Otto Jay. Petrographic study of Miocene basalts near Granby (Grand County), Colorado. M, 1942, Colorado School of Mines. 40 p.

Myers, Paul Benton, Jr. Geology of the Vermont portion of the Averill Quadrangle. D, 1960, Lehigh University. 123 p.

Myers, Paul E. A petrographic and statistical study of the "crystalline" constituents of central New York glacial deposits. M, 1958, Syracuse University.

Myers, Paul Edward. The geology of the Harpster Quadrangle and vicinity, Idaho. D, 1968, University of Michigan.

Myers, Philip B. The (Lower Cambrian) Hardyston Formation in Lehigh and Northampton counties, Pennsylvania. M, 1934, Lehigh University.

Myers, Ralph Lawrence, II. Biostratigraphy of the Cardenas Formation (Cretaceous) San Luis Potosi, Mexico. D, 1965, University of Texas, Austin. 76 p.

Myers, Richard Lee. Magnetic profiles in the central Wasatch region, Utah. M, 1955, University of Utah. 40 p.

Myers, Robert. Late Cenozoic sedimentation in the northern Labrador Sea; a seismic-stratigraphy analysis. M, 1986, Dalhousie University. 268 p.

Myers, Robert C. Stratigraphy of the Frontier Formation (Upper Cretaceous), Kemmerer area, Lincoln County, Wyoming. M, 1976, Colorado School of Mines. 152 p.

Myers, Robert Errol. Fusulinid fauna from Rhodes Canyon, San Andres Range, Socorro County, New Mexico. M, 1955, University of Illinois, Urbana.

Myers, Robert G. The petrology of the hypabyssal igneous intrusions of the Mariscal Mountain area, Big Bend National Park, Brewster County, Texas. M, 1978, Kent State University, Kent. 83 p.

Myers, Robert L. Dynamic phenomena of sediment compaction in Matagorda County, Texas. M, 1963, University of Houston.

Myers, Ronald E. Geologic factors of the Garfield Field area, Pawnee County, Kansas, in relation to petroleum accumulation. M, 1959, Kansas State University. 73 p.

Myers, Ronald Lewis. The ecology of low diversity palm swamps near Tortuguero, Costa Rica. D, 1981, University of Florida. 300 p.

Myers, Sharon A. Significance of the chloritic breccia zone, Bitterroot Dome, western Montana. M, 1986, University of Montana. 72 p.

Myers, W. G. Geology of the Sixmile Gap area, Albany and Carbon counties, Wyoming. M, 1958, University of Wyoming. 74 p.

Myers, Wallace Darwin, II. The sedimentology and tectonic significance of the Bayfield Group (upper Keweenawan?), Wisconsin and Minnesota. D, 1971, University of Wisconsin-Madison. 314 p.

Myers, William Marsh. Geology and economics of the exploitation of the copper ores. D, 1933, University of Michigan.

Myers-Bohlke, Brenda. A characterization of deep weathering profiles in foliated, metamorphic rocks for

tunnelling and shaft sinking. D, 1983, University of California, Berkeley. 179 p.

Myerson, Bertram L. A study of mineralization in the vicinity of Jim Thorpe, Carbon County, Pennsylvania. M, 1955, New York University.

Myhrman, Matts A. Hydro-physical aspects of soil treated with hexadecanol. M, 1967, University of Arizona.

Myint, Khin Maung. Tertiary intrusive activity and mineralization in the Empire mining district, Grand, Gilpin and Clear Creek counties, Colorado. D, 1988, Colorado School of Mines. 191 p.

Myles, James Robert. Petrology of a Granitic Terrain in Northeast Wisconsin. M, 1972, University of Wisconsin-Madison.

Mylroie, J. E. Speleogenesis and karst geomorphology of the Helderberg Plateau, Schoharie County, New York. D, 1977, Rensselaer Polytechnic Institute. 336 p.

Myrand, D. Diffusion of volatile organic compounds in natural clay deposits. M, 1987, University of Waterloo. 82 p.

Myrick, A. C., Jr. Variation, taphonomy, and adaptation of the Rhabdosteidae (=Eurhinodelphidae) (Odontoceti, Mammalia) from the Calvert Formation of Maryland and Virginia. D, 1979, University of California, Los Angeles. 437 p.

Myrow, Paul Michael. A paleoenvironmental analysis of the Cheshire Formation, west-central Vermont. M, 1983, University of Vermont. 177 p.

Myrow, Paul Michael. Sedimentology and depositional history of the Chapel Island Formation (late Precambrian to Early Cambrian) Southeast Newfoundland. D, 1987, Memorial University of Newfoundland. 510 p.

Mysen, Bjoorn Olav. Melting of a hydrous mantle; phase relations of natural peridotite to 30 kilobars and 1250°C with controlled activities of water, carbon dioxide and hydrogen. D, 1974, Pennsylvania State University, University Park. 219 p.

Mysore, R. K. Finite element analysis of sand as a hypoelastic material. D, 1978, SUNY at Buffalo. 149 p.

Mysyk, Walter K. The chemistry and mineralogy of the Homewood (Manitoba) meteorite. M, 1978, University of Manitoba. 16 p.

Mytton, James W. An insoluble residue study of the Madison Formation of the northern part of the Laramie Range, Wyoming. M, 1951, University of Wyoming. 94 p.

N'Guessan, Yao. High-resolution seismic reflection survey in Athens County; data acquisition and processing. M, 1987, Ohio University, Athens. 166 p.

N., Pablo Huidobro see Huidobro N., Pablo

Na, J. Thermal convection in a Hele-Shaw cell. D, 1976, Florida State University. 214 p.

Naas, David Hugh. Petrography of some Mesozoic basalts from northwestern Sicily. M, 1980, Wright State University. 67 p.

Naas, S. L. Flow behavior in alluvial channel bends. D, 1977, Colorado State University. 189 p.

Nabb, Bert E. Mac see Mac Nabb, Bert E.

Nabelek, John L. Determination of earthquake source parameters from inversion of body waves. D, 1984, Massachusetts Institute of Technology. 24 p.

Nabelek, Peter Igor. Nucleation and growth of plagioclase and the development of textures in high-alumina basaltic melt. M, 1978, University of Tennessee, Knoxville. 52 p.

Nabelek, Peter Igor. The geochemical evolution of the inversely zoned Notch Peak granitic stock, Utah. D, 1983, SUNY at Stony Brook. 286 p.

Nabighian, Misac N. The wedge problem in geophysics. D, 1967, Columbia University. 189 p.

Nace, Raymond L. Feasibility of groundwater features of the alternate plan for the mountain home project, Idaho. D, 1960, Columbia University, Teachers College.

Nace, Raymond L. Summary of the Upper Cretaceous and early Tertiary stratigraphy of Wyoming. M, 1936, University of Wyoming. 277 p.

Nachlas, Jesse. Dissolution kinetics of spinel; $NiFe_2O_4$ and $NiAl_2O_4$. M, 1983, Pennsylvania State University, University Park.

Nachman, Daniel A. Geology of the Duzel Rock area, Yreka Quadrangle, California. M, 1977, Oregon State University. 153 p.

Nacht, Steve Jerry. Urban geology for planning in Madison Township, Lake County, Ohio. M, 1973, Kent State University, Kent. 67 p.

Nack, Nissa Louise. Brachiopod biostratigraphy of the Lodgepole Formation, Swimming Woman Canyon, Montana. M, 1984, Washington State University. 80 p.

Nackowski, Matthew Peter. Geochemical prospecting applied to the Illinois-Kentucky fluorspar area. D, 1952, University of Missouri, Rolla.

Nackowski, Matthew Peter. Geology and mineralization of the Minerva Mine No. 1, Hardin County, Illinois. M, 1949, University of Missouri, Rolla.

Nadeau, Andre. Interprétation quantitative d'anomalies magnétiques. M, 1971, Universite de Montreal.

Nadeau, Joseph. Temperature of fluorite mineralization by fluid inclusion thermometry, Sweetwater barite district (Ordovician), (McMinn and Monroe counties), east Tennessee. M, 1967, University of Tennessee, Knoxville. 36 p.

Nadeau, Joseph Edward. The stratigraphy of the Leadville Limestone, central Colorado. D, 1971, Washington State University. 144 p.

Nadeau, Léopold. Deformation of leucogabbroic rocks at Parry Sound, Ontario. M, 1984, Carleton University. 191 p.

Nadeau, Paul H. UV radiational effects on Martian regolith water. M, 1977, Dartmouth College. 89 p.

Nadeau, Paul Henry. Burial and contact metamorphism in the Mancos Shale. D, 1980, Dartmouth College. 200 p.

Nadjmabadi, Siavash. Paleo-environment of the Guilmette Limestone (Devonian) near Wendover, Nevada. M, 1967, Brigham Young University.

Nadler, Carl Theodore, Jr. River metamorphosis of the South Platte and Arkansas rivers, Colorado. M, 1978, Colorado State University. 164 p.

Nadolski, John A. A study of bedload and total sediment from the East-central Sierra Nevada. M, 1979, University of Nevada. 94 p.

Nadon, Gregory Crispian. The stratigraphy and sedimentology of the Triassic at St. Martins and Lepreau, New Brunswick. M, 1981, McMaster University. 279 p.

Nadon, R. Impact of groundwater conditions on underground mining and space development in the Niagara Escarpment area. M, 1981, University of Waterloo.

Naegele, Orville Dale. Insoluble residues of the Cynthiana Formation in Kentucky. M, 1960, University of Cincinnati. 68 p.

Naegeli, Faith I. Petrofabrics at the falls of the Sturgeon, Dickson County, Michigan. M, 1959, Michigan State University. 46 p.

Naert, Karl Achiel. Geology and extrusion history of the No Agua perlite domes. M, 1973, Pennsylvania State University, University Park.

Naert, Karl Achiel. Geology, extrusion history and analysis of characteristics of perlites from No Agua, New Mexico. D, 1974, Pennsylvania State University, University Park. 236 p.

Naeser, Charles W. Neutron activation analysis of sodium and magnesium in silicate rocks and minerals. M, 1964, Dartmouth College. 28 p.

Naeser, Charles Wilbur. Fission track age relationships in a contact zone, Eldora (Boulder County), Colorado. D, 1967, Southern Methodist University. 85 p.

Naeve, Valarie Adrienne. Postglacial environmental history of a Marl Lake site in Kalamazoo County, southwestern Michigan. M, 1979, Western Michigan University.

Naff, John Davis. Geology and paleontology of the Upper Boggy drainage area, Coal and Pontotoc counties, Oklahoma. D, 1962, University of Kansas. 470 p.

Naff, John Davis. The Tuscaloosa-Pottsville contact in Tuscaloosa County, Alabama. M, 1940, University of Alabama.

Naff, Richard Louis. Hydrogeology of the southern part of Amargosa Desert in Nevada. M, 1973, University of Nevada. 207 p.

Nafziger, Ralph Hamilton. Equilibrium phase compositions and thermodynamic properties of solid solutions in the system $MgO\text{-}FeO^{2++'\text{-}}SiO_2$. D, 1966, Pennsylvania State University, University Park. 122 p.

Nagai, Richard Brian. A magnetic survey of parts of Hardin and Pope counties, Illinois. M, 1973, Southern Illinois University, Carbondale. 72 p.

Nagaraj, Benamanahalli Kempegowda. Modelling of normal and shear behavior of interface in dynamic soil-structure interaction. D, 1986, University of Arizona. 305 p.

Nagaraja, Hebbur Narasimhamurthy. Application studies of scanning electron microscope photographs for micro-measurements and three dimensional mapping. D, 1974, Ohio State University.

Nagata, Masato. Bifurcations in nonlinear problems of hydrodynamic instability of plane parallel shear flows. D, 1983, University of California, Los Angeles. 107 p.

Nagel, David. Evolution of the Ascension submarine canyon; central California continental margin. M, 1983, San Jose State University. 88 p.

Nagel, Fritz G. Regional stratigraphic analysis of Upper Cretaceous rocks of the Rocky Mountain region and adjacent areas. M, 1952, Northwestern University.

Nagel, Joe Jochen. The geology of part of the Shulaps ultramafite; near Jim Creek, southwestern British Columbia. M, 1979, University of British Columbia.

Nagel, Steven P. Geochemistry and petrology of xenoliths in the Elberton Batholith. M, 1981, University of Georgia.

Nagelberg, J. Leo. The superficial geology at Franklin Furnace, New Jersey. M, 1916, Columbia University, Teachers College.

Nagell, Raymond Harris. Geology of the Philadelphia Quadrangle, Jefferson County, New York. M, 1952, University of Rochester. 125 p.

Nagell, Raymond Harris. Structural geology of the Morococha District, Peru. D, 1957, Stanford University. 152 p.

Nagengast, Timothy John. An explanation for the lineation in production on the southern portion of the Central Oklahoma Platform. M, 1981, University of Oklahoma. 90 p.

Nageotte, Alton Lee. Lithostratigraphy and depositional environments of the Pitkin Limestone (Chesterian, Mississippian) in portions of Cherokee, Muskogee, and Sequoyah counties, Oklahoma. M, 1981, University of Oklahoma. 197 p.

Nagi, David Michael. Serine and cysteine thermal decomposition with respect to fossil dating and carbonaceous meteorites. M, 1984, University of Arizona. 24 p.

Nagle, Frederick, Jr. Geology of the Puerto Plata area, Dominican Republic. D, 1967, Princeton University. 192 p.

Nagy, Bartholomew S. The formation of the lead-silver ores of the Coeur d'Alene District in Idaho. M, 1950, Columbia University, Teachers College.

Nagy, Bartholomew Stephen. Mineralogy of the serpentine group. D, 1953, Pennsylvania State University, University Park. 194 p.

Nagy, Kathryn Louise. The solubility of calcite in NaCl and Na-Ca-Cl brines. D, 1988, Texas A&M University. 232 p.

Nagy, Richard Michael. Geochemistry of a portion of the Servilleta basalts, Taos County, New Mexico. M, 1973, University of North Carolina, Chapel Hill. 44 p.

Nagy, Richard Michael. Geochemistry of the Raba el Garrah Pluton, Egypt. D, 1978, Rice University. 78 p.

Nahama, Rodney. Geology of the northeast quarter of the Treasure Hill Quadrangle, Nevada. M, 1961, University of Southern California.

Nahhas, Tariq Mohammed. Dynamics of earth dams. D, 1987, University of Southern California.

Nahm, Jay W. A study of sonic transit time-to-porosity transform in shaly formations. M, 1986, Baylor University. 261 p.

Nahring, Eldon L. Geology and alteration of Tutiai Mountain and Fawnie Nose, Central Interior, British Columbia. M, 1971, University of Idaho. 88 p.

Naidoo, Devamonie D. A petrographic and geochemical study of the metavolcanic rocks from the Proterozoic Bou Azzer ophiolite complex, Morocco. M, 1988, Duke University. 101 p.

Naidu, Janakiram Ramaswamy. Radioactive zinc (^{65}Zn), zinc, cadmium, and mercury in the Pacific Hake, Merluccius productus (Ayres), off the west coast of the United States. D, 1974, Oregon State University. 159 p.

Naiknimbalkar, Narendra M. Sedimentary petrography and environments of deposition of the Plattin Group, Cliffidale Hollow, Jefferson County, Missouri. M, 1963, University of Missouri, Rolla.

Naiknimbalker, N. M. Textural properties, provenance, and environments of formation of the Rimrock Sandstone of the filled sinkholes northeast of Rolla, Missouri. M, 1971, University of Missouri, Rolla.

Naim, Shamim. Identification of irrigation practices using photographic and optical-mechanical scanning remote sensing techniques. D, 1981, Oregon State University. 266 p.

Naiman, Ellen Rose. Sedimentation and diagenesis of a shallow marine carbonate and siliciclastic shelf sequence; the Permian (Guadalupian) Grayburg Formation, southeastern New Mexico. M, 1982, University of Texas, Austin. 197 p.

Naimi, Ali Ibrahim. Fluoride in ground water. M, 1963, Stanford University.

Naing, W. Photogeology of the Caledonian area of southern New Brunswick. D, 1977, University of New Brunswick.

Naini, B. R. A geological and geophysical study of the continental margin of western India, and the adjoining Arabian Sea including the Indus Cone. D, 1980, Columbia University, Teachers College. 182 p.

Nair, G. P. Response of soil-pile systems to seismic waves. D, 1975, McMaster University.

Nairn, James P. Depositional environment of the DeCew Member of the Lockport Formation in New York and Ontario. M, 1973, SUNY, College at Fredonia. 61 p.

Najjar, Ismail Muhammad. Distribution of trace elements in the ground water of the Moscow-Pullman Basin, Idaho-Washington. M, 1972, University of Idaho. 189 p.

Najjar-Bawab, M. Mummtaz. Application of computer lithostratigraphic correlation and three-dimensional configuration. D, 1980, University of Oklahoma. 221 p.

Najjar-Bawab, Moumtaz Mohammad. The Barker Dome oil field, Custer County, South Dakota. M, 1977, South Dakota School of Mines & Technology.

Najjor, Abdullatif. Dolomitized Pahasapa Limestone (Mississippian), northeastern sector of the Black Hills, South Dakota, petrography and geological setting. M, 1971, University of Missouri, Rolla.

Nakai, Theresa Sigl. Stratigraphy of the Payette Formation, Washington County, Idaho. M, 1979, University of Idaho. 187 p.

Nakaki, David Kiyoshi. Uplifting response of structures subjected to earthquake motions. D, 1987, University of California, Los Angeles. 212 p.

Nakamura, Eizo. The geochronology and geochemistry of Cenozoic alkaline basalts from Japan, Korea and China. D, 1986, University of Toronto.

Nakamura, M. T. The distribution of chromium in the Latosols of the Hawaiian Islands. M, 1957, University of Hawaii. 27 p.

Nakamura, Yosio. Frequency spectra of refraction arrivals and the nature of the Mohorovicic discontinuity. D, 1963, Pennsylvania State University, University Park. 148 p.

Nakanart, Araya. The Buffalo oil field, Harding County, South Dakota. M, 1977, South Dakota School of Mines & Technology.

Nakanishi, K. K. The deep structure of the upper mantle from the dispersion of long-period Rayleigh waves. D, 1978, University of California, Los Angeles. 269 p.

Nakashima, Lindsay D. Spatial and temporal variations in barred and non-barred topographies, Sandy Hook, New Jersey. D, 1984, Rutgers, The State University, New Brunswick. 205 p.

Nakashiro, Masaykui. An X-ray investigation of quartz grains exhibiting undulose extinction. M, 1964, McGill University.

Nakata, John K. Distribution and petrology of the Anderson-Coyote Reservoir volcanic rocks. M, 1977, San Jose State University. 105 p.

Nakato, Tatsuaki. Wave-induced sediment entrainment from rippled beds. D, 1974, University of Iowa. 209 p.

Nakayama, Eugene. Geology of southeastern Payne County, Oklahoma. M, 1955, University of Oklahoma. 69 p.

Nakayama, Kazuo. A simulation model of petroleum migration in clastic sediments. M, 1979, University of Houston.

Nakayama, Kazuo. Two-dimensional basin analysis for petroleum exploration. D, 1987, University of South Carolina. 241 p.

Nakhinbodee, Veerasak. Foraminifera of the Moodys Branch Marl (Eocene) at Creole Bluff, Montgomery, Louisiana. M, 1971, Louisiana Tech University.

Nakiboglu, Sadik Mete. Earth's surface and outer gravity field by buried point-mass sets. D, 1974, Cornell University.

Nakornthap, Kurujit. Numerical simulation of multiphase fluid flow in naturally fractured reservoirs. D, 1982, University of Oklahoma. 238 p.

Naldrett, Anthony J. Ultrabasic rocks of the Porcupine and related nickel deposits. D, 1964, Queen's University. 264 p.

Naldrett, Anthony James. The geochemistry of cobalt in the ores of the Sudbury district (Ontario). M, 1961, Queen's University. 152 p.

Naldrett, Dana L. Aspects of the surficial geology and permafrost conditions, Klondike Goldfields and Dawson City, Yukon Territory. M, 1981, University of Ottawa. 169 p.

Naldrett, Dana L. Glacigenic clays of the Ottawa Valley. D, 1987, University of Ottawa. 218 p.

Nalepa, Randolph. The use of computer enhanced Landsat imagery for geologic mapping of the Tascotal region, Trans-Pecos, Texas. M, 1981, Texas Christian University.

Nalewaik, Gerald Guy. Petrology of the Upper Permian Cloud Chief Formation of western Oklahoma. D, 1968, University of Oklahoma. 134 p.

Nalewaik, Gerald Guy. The geology of the Pitts Springs area, Knox County, Tennessee. M, 1961, University of Tennessee, Knoxville. 36 p.

Nalwalk, Andrew Jerome. Geology of a portion of the southern Portneuf Range, Idaho. M, 1959, University of Idaho. 54 p.

Nalwalk, Andrew Jerome. Geology of the north wall of the Puerto Rico Trench. D, 1967, University of Pittsburgh. 231 p.

Nam-Koong, Wan. Removal of phenolic compounds in soil. D, 1988, University of Texas, Austin. 260 p.

Namazie, Mizra Hussain Ali. Hingement structure in Upper Cretaceous ostracodal genera. M, 1954, University of Oklahoma. 123 p.

Namdarian, Faridoon Ardeshire. Analysis of trace elements in river and spring waters. M, 1967, University of Missouri, Rolla.

Namiq, Laith Ismail. Stress-deformation study of a simulated lunar soil. D, 1970, University of California, Berkeley. 218 p.

Nammah, Hassan Audah. Effect of Mount St. Helens volcanic ash on soil erosion and water quality. M, 1983, University of Idaho. 72 p.

Namminga, Harold Eugene. Heavy metals in water, sediments, and chironomids in a stream receiving domestic and oil refinery effluents. D, 1975, Oklahoma State University. 120 p.

Namoglu, A. Coskun. Geology and petroleum potential of the Minnelusa Formation along the Cottonwood Creek Anticline and the surrounding area, Fall River County, South Dakota. M, 1985, South Dakota School of Mines & Technology.

Namowitz, Samuel N. The great inland scarp of Long Island. M, 1937, Columbia University, Teachers College.

Namson, Jay Steven. Studies of the structure, stratigraphic record of plate interaction and role of pore-fluid pressure in the active fold and thrust belt of Taiwan and a study of manganese deposits from northern California. D, 1982, Princeton University. 379 p.

Namy, Dominique. An investigation of certain aspects of stress-strain relationships for clay soils. D, 1970, Cornell University. 265 p.

Namy, Jerome Nicholas. Stratigraphy of the Marble Falls group (lower and middle Pennsylvanian), southeast Burnet County, Texas. D, 1969, University of Texas, Austin.

Nance, Hardie Seay, III. Facies relations and controls on Artesia Group deposition in the Matador Arch area, Texas. M, 1988, University of Texas, Austin. 142 p.

Nance, Richard Leon. Caddo oil field, Carter County, Oklahoma. M, 1958, University of Oklahoma. 52 p.

Nance, Roger B. Petrology of the Saint Laurent Limestone (Middle Devonian) of southeastern Missouri. M, 1968, Southern Illinois University, Carbondale. 132 p.

Nance, Steven. The role of suspended matter on the trace metal transport in an estuarine environment. M, 1974, Georgia Institute of Technology. 36 p.

Nanda, Atul. Finite element analysis of elastic-plastic anisotropic soils. D, 1987, Virginia Polytechnic Institute and State University. 191 p.

Naney, James Wesley. The determination of the impact of an earthen-fill dam on the ground-water flow using a mathematical model. M, 1974, Oklahoma State University. 119 p.

Naney, M. T. Phase equilibria and crystallization in iron- and magnesium-bearing granitic systems. D, 1978, Stanford University. 244 p.

Naney, Michael Terrance. A study of the chemical and mineralogical variation of the Academy Pyroxene-Quartz Diorite, Fresno County, California. M, 1971, University of California, Davis. 148 p.

Nania, Jay C. The structure and stratigraphy along a major Proterozoic unconformity Dunderdalen to Brevassdalen, Wedel Jarlsberg Land, Spitsbergen. M, 1987, University of Wisconsin-Madison. 112 p.

Nankervis, Jeffrey Chambers. A thermal study of the mineral phases present in lignites from the Beulah-Zap coal bed, Mercer County, North Dakota. M, 1979, Wayne State University.

Nanna, Richard F. Precambrian geology of central Casper Mountain, Natrona County, Wyoming. M, 1981, University of Akron. 109 p.

Nanson, G. C. Channel migration, floodplain formation, and vegetation succession on a meandering-river floodplain in N.E. British Columbia, Canada. D, 1977, Simon Fraser University. 349 p.

Nantais, Philip Thomas. Lithostratigraphy and sedimentology of the Late Jurassic-Early Cretaceous Mic Mac and lower Missisauga formations, offshore Nova Scotia, Canada. M, 1986, University of Windsor. 290 p.

Nantel, Suzanne. Le problème des équilibres cordiérite-grenat et orthopyroxène-clinopyroxène en catazone; application à la géothermo-barométrie dans le sud de la Province de Grenville. M, 1977, Universite de Montreal.

Nanz, Robert Hamilton. Composition and abundance of fine-grained Pre-Cambrian sediments of the southern Canadian Shield. D, 1952, University of Chicago. 188 p.

Napoleoni, Jean-Gerard Pascal. The dynamics of iceberg drift. M, 1979, University of British Columbia.

Napp, Donald E. The geology of the Gus area, Burleson County, Texas. M, 1956, Texas A&M University. 54 p.

Napper, Jack E. The geology of the northern two-thirds of the Courtrock Quadrangle, Oregon. M, 1958, University of Oregon. 44 p.

Naqvi, Ikram Husain. The Belloy Formation (Permian), Peace river area, northeast British Columbia, and northwest Alberta, Canada. M, 1969, University of Calgary. 124 p.

Narahara, Gene Masao. Rock property measurements and measurement techniques for low permeability cores. D, 1987, Texas A&M University. 150 p.

Narain, Kedar. A study of clay minerals from certain producing and nonproducing oil sands of Hamilton County, Illinois. M, 1959, University of Illinois, Urbana.

Narans, Harry Donald, Jr. Sub-basement seismic reflections in northern Utah. M, 1959, University of Utah. 71 p.

Narasimha Chary, K. An isotopic and geochemical study of gold-quartz veins (Precambrian) in the Con-Rycon Mine, Yellowknife, (Northwest Territories). M, 1971, University of Alberta. 90 p.

Narasimhan, Tyagarajan. Eocene discoasters and coccolithophores from Central California. D, 1961, Stanford University. 254 p.

Narayana, V. V. Dhruva. Analog computer simulation of the runoff characteristics of an urban watershed. D, 1969, Utah State University. 174 p.

Narbonne, G. M. Stratigraphy, reef development and trace fossils of the Upper Silurian Douro Formation in the southeastern Canadian Arctic Islands. D, 1981, University of Ottawa. 259 p.

Narbut, Susan Margaret. Determination of fault-related stress changes using the piezomagnetic effect. D, 1983, Pennsylvania State University, University Park. 183 p.

Nardi, Guiseppi. Tectonic and magnetostratigraphic significance of Middle and Upper Jurassic rocks from western Sicily. D, 1982, University of South Carolina. 121 p.

Nardin, Barbara A. Suspended sediments in continental shelf water of Cape Hatteras, North Carolina. M, 1971, University of Illinois, Chicago.

Nardin, T. R. Late Cenozoic history of the Santa Monica Bay area, California. M, 1976, University of Southern California. 189 p.

Nardin, Thomas Richard. Seismic stratigraphy of Santa Monica and San Pedro basins, California continental borderland; late Neogene history of sedimentation and tectonics. D, 1981, University of Southern California.

Nardone, Craig D. Study of sediment provenance at DSDP Site 379A, Black Sea, based on the isotopic composition of strontium. M, 1978, Ohio State University.

Narendra, Choudary Y. B. An investigation of the influence of grain size on stress wave attenuation and dispersion in rock like materials. D, 1973, University of Missouri, Rolla.

Narod, B. B. Ultra high frequency radio echo sounding of glaciers. M, 1975, University of British Columbia.

Narod, B. B. Ultrahigh frequency radio echo sounding of Yukon glaciers. D, 1979, University of British Columbia.

Narr, Wayne Mark. The origin of fractures in Tertiary strata of the Altamont Field, Uinta Basin, Utah. M, 1978, University of Toronto.

Narramore, Rebecca Lynn. Stream channel enlargement due to urbanization in the White Rock Prairie, north-central Texas. M, 1981, Baylor University. 198 p.

Narten, Parry Foote, Jr. The Pottsville-Pocono contact in western Pennsylvania. M, 1948, Washington University. 65 p.

Naruk, Stephen John. Kinematic significance of mylonitic foliation. D, 1987, University of Arizona. 85 p.

Narupon, Anant S. The effect of vegetation, soil, and rock type on runoff from some small watersheds in Missouri. M, 1965, University of Missouri, Rolla.

Nascimbene, Giovanni Giuseppe. Bentonites and the geochronology of the Bearpaw Sea (Upper Cretaceous; Alberta, Montana). M, 1963, University of Alberta. 78 p.

Naseer, S. A. Sewage effluent; a partial solution to Riyadh's water program. M, 1978, University of Arizona.

Nash, Barry Stuart. Adsorption of dissolved organic carbon by particulate detritus. M, 1980, University of Virginia. 70 p.

Nash, David Byer. The evolution of abandoned, wave-cut bluffs in Emmet County, Michigan. D, 1977, University of Michigan. 269 p.

Nash, David Byer. The relative age of the escarpments in the Martian polar laminated terrain based on morphology. M, 1974, University of Michigan.

Nash, Douglas B. Contact metamorphism at Birch Creek, Blanco Mountain Quadrangle, Inyo County, California. M, 1962, University of California, Berkeley.

Nash, John Thomas. Geology and uranium deposits of the Jackpile Mine area, Laguna, New Mexico. D, 1967, Columbia University. 216 p.

Nash, John Thomas. Introductory study of the Jackpile Sandstone and uranium mineralization, Valencia County, New Mexico. M, 1965, Columbia University. 45 p.

Nash, K. G. Geochemistry of selected closed basin lakes in Sheridan County, Nebraska. M, 1978, University of Nebraska, Lincoln.

Nash, Robert E. Historical changes in the mean high water shoreline and nearshore bathymetry of South Georgia and North Florida. M, 1977, University of Georgia.

Nash, Sandra Lee. Taxonomic and phylogenetic studies of Heterohelix, Pseudoguembelina, Kastalina, Planoglogulima, Pseudotextularia and Racemiguembelina. M, 1982, University of Texas at Dallas. 104 p.

Nash, Tamie Rene. The biotic and sedimentologic development of the Red Buoy patch reef, Discovery Bay, Jamaica. M, 1982, Texas Christian University. 90 p.

Nash, Victor. An experimental investigation on the surface reactions of feldspars. D, 1955, University of Missouri, Columbia. 156 p.

Nash, Walter A. Geochemistry of the Saint Stephen (New Brunswick) Intrusive Complex. M, 1967, University of New Brunswick.

Nash, William Purcell. Studies on the petrogenesis of alkaline igneous rocks. D, 1971, University of California, Berkeley. 188 p.

Nasim, Mushtaq Ahmad. Lateral impedance of contact pile foundations. D, 1987, Rensselaer Polytechnic Institute. 210 p.

Naslund, Howard Richard. Part I, Petrology of the Upper Border Group of the Skaergaard Intrusion, East Greenland; Part II, An experimental study of liquid immiscibility in iron-bearing silicate melts. D, 1980, University of Oregon. 346 p.

Naslund, Howard Richard. The geology of the Hyatt Reservoir and Surveyor Mountain quadrangles, Oregon. M, 1977, University of Oregon. 127 p.

Nasmith, Hugh Wallis. The mineralogy and physical properties of soil in the foundation of the chemistry building at Washington University. M, 1951, Washington University. 53 p.

Nason, Geoffrey W. Geology of a portion of the northern Newbery Mts., San Bernardino Co., Cal. M, 1978, California State University, Los Angeles.

Nason, Robert Dohrmann. Investigation of fault creep slippage in northern and central California. D, 1971, University of California, San Diego.

Nasr, Athanacios Nabeh. A study on interpretation of pressuremeter test. D, 1986, Wayne State University. 338 p.

Nassar, Ibrahim Nassar. Soil thermal diffusivity and water transport in unsaturated, nonisothermal, salty soil. D, 1988, Iowa State University of Science and Technology. 120 p.

Nassar, M. M. A. Gravity field and levelled heights in Canada. D, 1977, University of New Brunswick.

Nassereddin, M. T. Hydrogeological analysis of groundwater flow in Sonoita Creek Basin, Santa Cruz County, Arizona. M, 1967, University of Arizona.

Nassichuk, Walter William. Pennsylvanian ammonoids from Ellesmere Island, Canadian Arctic Archipelago. D, 1965, University of Iowa. 257 p.

Nassichuk, Walter William. Permian ammonoids from the Canadian Arctic Archipelago and eastern Greenland. M, 1963, University of Iowa. 80 p.

Nassr, Mohammad Nashaat Gad. Contrasts between quartz grain shapes of the Cambrian and the Carboniferous "Nubia" facies in Gabal Abu-Durba, Sinai, Egypt; Fourier grain shape analysis. D, 1985, University of South Carolina. 394 p.

Nast, Heidi J. The geology and petrochemistry of the Sisson Brook W-Cu-Mo deposit, New Brunswick. M, 1985, McGill University. 208 p.

Nasu, Mitsuru. Geometric processing for digital mapping with multiseries remote sensing data. D, 1976, University of California, Berkeley. 198 p.

Nasu, Noriyuki. Significance of the separation of sediments into sand and mud fractions during common processes of transportation as illustrated by modern and Tertiary sediments. D, 1955, University of California, Los Angeles. 161 p.

Natali, Patricia March. Paleoecologic interpretation of the Castle Hayne Limestone in North Carolina utilizing bryozoan zoarial forms. M, 1985, Kent State University, Kent. 151 p.

Nataraj, Mysore Subbarao. Settlement and bearing capacity of footings on reinforced sand. D, 1984, [Vanderbilt University]. 163 p.

Natarajan, Palamadai S. Effect of walls on structural response to earthquakes. D, 1970, Michigan State University. 104 p.

Natenstedt, Christopher J. Sedimentology and paleogeographic implications of the Eocene South Point and Cozy Dell formations, Santa Rosa Island, southern California. M, 1983, San Diego State University. 216 p.

Nater, Edward Arthur. Pedogenic weathering; the formation of kaolinite on feldspar surfaces. D, 1987, University of California, Davis. 112 p.

Nathan, Harold Decantillon. The geology of a portion of the Duluth complex (Precambrian), Cook County, (Minnesota). D, 1969, University of Minnesota, Minneapolis. 236 p.

Nathman, Douglas Robert. Rayleigh wave scattering across step discontinuities. M, 1980, Massachusetts Institute of Technology. 149 p.

Nathman, Neal J. Epicenter determination of California earthquakes using single-station data. M, 1988, Pennsylvania State University, University Park. 51 p.

Nations, Brenda K. Palynology of a Miocene deposit of the Xian Feng Basin, Yunnan Province, China. M, 1987, University of Iowa. 84 p.

Nations, Darrell L. A study of the depositional environment of the Athens Shale. M, 1978, University of Toledo. 52 p.

Nations, Jack Dale. The Black Prince Limestone of southeastern Arizona. M, 1961, University of Arizona.

Nations, Jack Dale. The family Cancridae and its fossil record on the west coast of North America. D, 1969, University of California, Berkeley. 252 p.

Natishan, Joseph John. Environmental pollutants; a field investigation and measurements of solubility and soil migration. D, 1988, SUNY at Binghamton. 145 p.

Natland, James H. Petrologic studies of linear island chains; Part 1, The Samoan islands; Part 2, The Line Islands. D, 1975, University of California, San Diego. 405 p.

Natland, M. L. Pleistocene and Pliocene stratigraphy of Southern California. D, 1952, University of California, Los Angeles.

Natur, F. S. Finite difference solution for drainage of heterogeneous sloping lands. D, 1974, Utah State University. 180 p.

Nau, James Michael. An evaluation of scaling methods for earthquake response spectra. D, 1982, University of Illinois, Urbana. 350 p.

Naughton, Margaret M. A hydrogeological study of the upper Kilmanagh River basin, Republic of Ireland. M, 1978, University of Alabama.

Naugler, Frederick Paist. Recent sediments of the East Siberian Sea. M, 1967, University of Washington. 71 p.

Nault, Kenneth James De *see* De Nault, Kenneth James

Naumann, Terry Richard. Geology of the central Boulder Canyon Quadrangle, Clark County, Nevada. M, 1987, University of Nevada, Las Vegas. 68 p.

Nauss, Anne L. Lithiotis bioherms in the Pliensbachian (Lower Jurassic) of North America. M, 1986, University of British Columbia. 102 p.

Nauss, Arthur W. Stratigraphy of the Vermilion area, Alberta. D, 1943, Stanford University. 102 p.

Nauta, Robert. A three-dimensional groundwater flow model of the Silurian dolomite aquifer of Door County, Wisconsin. M, 1987, University of Wisconsin-Madison. 105 p.

Nautiyal, Avinash Chandra. Frasnian (Upper Devonian) acritarcha and biostratigraphy of the interior Plains Region, Canada. D, 1972, University of Saskatchewan. 421 p.

Nautiyal, Avinash Chandra. The Cambro-Ordovician sequence in the southeastern part of the Conception Bay area, eastern Newfoundland. M, 1967, Memorial University of Newfoundland. 334 p.

Nautiyal, Chandra Mohan. Seismic wave propagation in a layer over a half-space. M, 1972, Massachusetts Institute of Technology. 108 p.

Nava Pichardo, Fidencio Alejandro. Study of seismic wave excitation for two earthquakes in northern Baja California. D, 1980, University of California, San Diego. 293 p.

Nava, David Francis. Atomic absorption analysis for copper and zinc in iron and stony meteorites. D, 1968, Arizona State University. 252 p.

Nava, Susan Jane. The New Madrid seismic zone; a test case for naturally induced seismicity. M, 1983, Memphis State University.

Navarro, Enrique Farran. Petrogenesis of the eclogitic rocks of Isla de Margarita, Venezuela. D, 1974, University of Kentucky. 237 p.

Navarro, Enrique Farran. Petrogenesis of the Pre-Cambrian metasedimentary rocks of the Almont area, Gunnison County, Colorado. M, 1971, University of Kentucky. 107 p.

Navas, Jaime. Geology of the Como area, South Park (Park County), Colorado. M, 1966, Colorado School of Mines. 145 p.

Nave, Floyd Roger. Geology of a portion of the Bridger Range, Montana. M, 1952, University of Iowa. 108 p.

Nave, Floyd Roger. Pleistocene Mollusca of southwestern Ohio and southeastern Indiana. D, 1968, Ohio State University.

Navias, Robert Alexander. The mineralogy of Campbell mines, Bisbee, Arizona. M, 1952, Pennsylvania State University, University Park. 61 p.

Navolio, Michael Edward. Digital convolution and fast Fourier transform methods for calculation of potential field anomalies. M, 1975, University of California, Berkeley. 110 p.

Navoy, Anthony S. Assessment of management strategies for drought mitigation through a conjunctive use irrigation system in Smith Valley, Lyon County, Nevada. M, 1978, University of Nevada. 149 p.

Nawab, Z. A. H. Evolution of the Al Amar-Idsas region of the Arabian Shield; Kingdom of Saudi Arabia. D, 1978, University of Western Ontario. 174 p.

Nawrocki, Michael Andrew. Hydrogeologic characteristics of shallow bedrock aquifers in the vicinity of Norman Creek, central Phelps County, Missouri. M, 1967, University of Missouri, Rolla.

Nayak, U. B. On the functional design and effectiveness of groins in coastal protection. D, 1976, University of Hawaii. 205 p.

Naylor, Bruce Gordon. The systematics of fossil and Recent salamanders (Amphibia; Caudata), with special reference to the vertebral column and trunk musculature. D, 1978, University of Alberta. 857 p.

Naylor, L. M. A statistical study of the variations in Des Moines River water quality. D, 1975, Iowa State University of Science and Technology. 253 p.

Naylor, Richard Stevens. A field and geochronologic study of mantled gneiss domes in central New England. D, 1967, California Institute of Technology. 131 p.

Naylor, Rrichard G. Geologic inputs to a liquid waste management plan, Essex Quadrangle, Connecticut. M, 1974, Wesleyan University.

Naymik, T. G. A digital computer model for estimating bedrock water resources, Maumee River basin, NW Ohio. D, 1977, Ohio State University. 300 p.

Naymik, Thomas G. Scanning electron microscope study of copper and its soil reaction products. M, 1974, Louisiana State University.

Nayudu, Y. Rammohanroy. Recent sediments of the Northeast Pacific. D, 1959, University of Washington. 217 p.

Nayyeri, Cyrus. Surficial geology of Cimarron River valley from one mile east of Perkins eastward to Oklahoma Highway 18, north-central Oklahoma. M, 1979, Oklahoma State University. 72 p.

Nazar, Edward. Effects of population expansion and physical forces on the western Connecticut shoreline. M, 1980, Western Connecticut State University. 33 p.

Nazarian, Mohammad H. Bryozoans of the upper part of the Bird Spring Group, Clark County, Nevada. M, 1980, Eastern Washington University. 52 p.

Nazarian, Soheil. In situ determination of elastic moduli of soil deposits and pavement systems by spectral-analysis-of-surface-waves method. D, 1984, University of Texas, Austin. 485 p.

Nazer, Naji M. Rhynchonellide brachiopods of the middle and upper Jurassic of the Tuwaiq mountains, Saudi Arabia. M, 1970, University of Kansas.

Nazikoglu, Zekai. Electrical resistivity survey in Keciborlu, Turkey. M, 1968, Stanford University.

Nazli, Kazim. Geostatistical modelling of microfossil abundance data in Upper Jurassic shale, Tojeira sections, central Portugal. M, 1988, University of Ottawa. 369 p.

Nazy, David John. A seismic refraction study of a portion of the northeastern margin of the Tualatin Valley, Oregon. M, 1987, Portland State University. 81 p.

Nchako, Felix N. Chemical weathering rate and a geochemical mass balance model for the Flyin-W Catchment, Greenwood County, Kansas. M, 1987, Wichita State University. 89 p.

Ndombi, J. M. Geology, gravity and resistivity studies of Olkaria geothermal field, Kenya. D, 1978, Stanford University. 157 p.

Neace, Thomas Foster. Eruptive style, emplacement, and lateral variations of the Mesa Falls Tuff, Island Park, Idaho, as shown by detailed volcanic stratigraphy and pyroclastic studies. M, 1986, Idaho State University. 98 p.

Neal, Charles William. Effects of primary condensation, secondary metamorphism and tertiary shock reheating on the chemistry of L and LL chondrites and a mineralogical survey of the Cumberland Falls chondritic inclusions. D, 1980, Purdue University. 284 p.

Neal, Donald Arthur. A gravity and magnetic survey in southeastern Missouri. M, 1956, St. Louis University.

Neal, Donald W. Palynology of the coals of the lower tongue of the Breathitt Formation (lower Pennsylvanian) of eastern Kentucky. M, 1975, Eastern Kentucky University. 53 p.

Neal, Donald Wade. Subsurface stratigraphy of the Middle and Upper Devonian clastic sequence in southern West Virginia and its relation to gas production. D, 1979, West Virginia University. 144 p.

Neal, H. E. The geology of the Hook Lake area, New Quebec, with special reference to the iron formation. M, 1949, University of Toronto.

Neal, Henry Percy. The use of native stone in the construction of small houses. M, 1932, Mississippi State University.

Neal, James T. Stratigraphy and structure of a portion of the iron formation at Lake Albanel, Quebec. M, 1959, Michigan State University. 74 p.

Neal, K. G. A tectonic study of a part of the northern Eagle Cap Wilderness area, northeastern Oregon. M, 1973, Portland State University. 85 p.

Neal, Ronald E. Paraecology of Holocene Ostracoda in the northwestern Gulf of Mexico. D, 1977, Louisiana State University.

Neal, William Joseph. Heavy mineral petrology of Wisconsin and post-glacial deep-sea sands and silts, western North Atlantic. M, 1964, University of Missouri, Columbia.

Neal, William Joseph. Petrology and paleogeography of the Blackjack Creek Formation (Pennsylvanian), western Missouri. D, 1968, University of Missouri, Columbia. 214 p.

Neal, William L. Geochemical relationships between gold, silver, antimony, arsenic, and bismuth in gold mineralization of the Goldville District, Tallapoosa County, Alabama. M, 1986, Auburn University. 151 p.

Neal, William Scott. Structure and geochemistry of the southeastern portion of the Chloride mining district, Sierra County, New Mexico. M, 1987, Washington State University. 136 p.

Neale, Ernest R. W. Geology of the Bethoulat Lake area, Mistassini Territory, Quebec. D, 1952, Yale University. 388 p.

Neale, John. A sedimentological study of the Gulf Coast of Cayo-Costa and N. Captiva Island, Florida. M, 1980, Florida State University.

Neale, Kathryn L. Stratigraphy and basin evolution of the upper Hayes River and Cochrane Bay groups, Island Lake greenstone belt, northeastern Manitoba. M, 1984, Carleton University. 177 p.

Neale, Patrick S. The Pleistocene stratigraphy of the Laura and Brookville quadrangles, southwestern Ohio. M, 1979, Miami University (Ohio). 138 p.

Nealey, Lorenza David. Geology of Mount Floyd and vicinity, Coconino County, Arizona. M, 1980, Northern Arizona University. 144 p.

Nealon, Dennis J. A hydrological simulation of hazardous waste injection in the Mt. Simon, Ohio. M, 1982, Ohio University, Athens. 285 p.

Neame, Peter Austin. Benthic oxygen and phosphorus dynamics in Castle Lake, California. D, 1975, University of California, Davis. 234 p.

Nearhoof, Elmer G. Geologic field teaching resources of the Palmyra-Indiantown gap area, Lebanon County, Pennsylvania. M, 1969, Millersville University.

Neary, Thomas Hubert. The analysis of linears in the location of mineral deposits. M, 1969, University of Idaho. 65 p.

Neasham, John West. Lithology and stratigraphy of the Willwood Formation (lower Eocene), southwest Big Horn County, Wyoming. M, 1967, Iowa State University of Science and Technology.

Neasham, John West. Sedimentology of the Willwood Formation (lower Eocene); an alluvial molasse facies in northwestern Wyoming, U.S.A. D, 1970, Iowa State University of Science and Technology.

Neathery, Orphie, III. Characteristics of strength properties of sedimentary rocks. M, 1965, Texas A&M University.

Neathery, Thornton Lee. Paragenesis of the Turkey Heavy Mountain kyanite deposits, Cleburne County, Alabama. M, 1964, University of Alabama.

Neauhauser, Kenneth. A structural and petrofabric analysis of several ultramafics of western North Carolina. D, 1974, University of South Carolina.

Neauhauser, Kenneth. A structural and petrographic analysis of the Bank's Creek serpentinite. M, 1971, University of South Carolina.

Neave, Kendal Gerard. Glacier seismology. M, 1968, University of Toronto.

Neave, Kendal Gerard. Icequake seismology. D, 1971, University of Toronto.

Neavel, Kenneth Edward. Structural features and kinematic history of the Del Rio District, central-eastern Tennessee. M, 1985, University of Kentucky. 124 p.

Neavel, Richard Charles. Some aspects of the petrography of western lignites. M, 1957, Pennsylvania State University, University Park. 81 p.

Neavel, Richard Charles. Sulfur in coal; its distribution in the seam and in mine products. D, 1966, Pennsylvania State University, University Park. 351 p.

Nebel, Mark Louis. Stratigraphy, depositional environment, and alteration of Archean felsic volcanic rocks, Wawa, Ontario. M, 1982, University of Minnesota, Duluth. 117 p.

Nebel, Merle Louis. The Duluth Gabbro and its contact metamorphism in the vicinity of Gabamichigami Lake, Vermilion iron-bearing district, Minnesota. D, 1917, University of Illinois, Chicago.

Neblett, Sidney S. Engineering geology of the Dana Point Quadrangle, California. M, 1966, University of Southern California.

Nebrija, E. L. Offshore mineral exploration around the Keweenaw Peninsula in Lake Superior. D, 1979, University of Wisconsin-Madison. 354 p.

Necioglu, Altan. Study of S wave spectral properties. M, 1969, St. Louis University.

Necioglu, Altan. Surface wave attenuation in North America east of the Rocky Mountains. D, 1974, St. Louis University. 218 p.

Neder, Irving R. Geology of a central part of the Bird Spring Range, Clark County, Nevada. M, 1967, University of California, Los Angeles.

Neder, Irving Robert. Conodont biostratigraphy and depositional history of the Mississippian Battleship Wash Formation, southern Nevada. D, 1973, University of California, Los Angeles.

Neder, Reinhard Bernhard Wilhelm. Sensitivity of convergent-beam electron diffraction to F-OH order-

disorder in topaz. M, 1986, Arizona State University. 141 p.

Nedland, Daniel E. Stratigraphy of the Chickasaw Creek Formation, upper Stanley Group (Mississippian), central Ouachita Mountains, Oklahoma. M, 1971, University of Wisconsin-Madison.

Neece, Neal, Jr. Implications of mathematical treatment of the forces involved in the buckling of the Earth's crust. M, 1950, Southern Methodist University. 31 p.

Need, Edward Adams. Till stratigraphy and glacial history of Wisconsin's Lake Superior shoreline; Wisconsin Point to Bark River. M, 1980, University of Wisconsin-Madison.

Needham, Bruce Harry. The self-scouring aspects of Rudee Inlet, Virginia Beach, Virginia. M, 1973, Old Dominion University. 100 p.

Needham, Claude Ervin. Contributions to the subsurface geology of northern Illinois, between the outcrops of the St. Peter and Dresbach formations, with special reference to the New Richmond Formation. D, 1932, Northwestern University.

Needham, Claude Ervin. The Pacific-Rocky Mountain field trip in geology; summer 1924. M, 1924, Mississippi State University. 76 p.

Needham, Daniel L. Failure of asperities by hydraulically induced fatigue; a model for the generation of intraplate seismicity. M, 1987, Virginia Polytechnic Institute and State University.

Needham, Edward Keith. Effects of urbanization of peak runoff, Mission Valley, San Diego River, California. M, 1985, University of California, Los Angeles.

Needham, Paul E. Tellurometer trilateration applied to a small control survey. M, 1960, Ohio State University.

Needham, Robert Edmund. The geology of the Murray County, Georgia talc district. M, 1972, University of Georgia. 107 p.

Neef, George Herman. Sedimentation of the Todilto Limestone in San Miguel County, New Mexico. M, 1950, Texas Tech University. 21 p.

Neel, Robert. Stratigraphy and glacial geology of the Portage Quadrangle, New York. M, 1951, University of Rochester. 102 p.

Neel, Robert H. Geology of the Tillamook Head-Necanicum Junction area, Clatsop County, Northwest Oregon. M, 1976, Oregon State University. 204 p.

Neel, Thomas Howard. Geology of the lower Tularcitos Creek-Cachagua grade area, Jamesburg Quadrangle, California. M, 1963, Stanford University.

Neeland, W. D. The petrology of the Grenville Limestone (Precambrian) contacts with certain intrusives, St. Jerome, Province of Quebec. M, 1935, McGill University.

Neeley, Don Hitt. Subsurface stratigraphy of the Tar Springs Sandstone and adjacent units (upper Chesterian) in part of western Kentucky. M, 1982, University of Kentucky. 104 p.

Neely, Dorothy G. The stratigraphy and facies of the Stanley Group of the western central Ouachitas of Oklahoma. M, 1985, University of Arkansas, Fayetteville.

Neely, Florence. Small petrified seeds from the Pennsylvanian of Illinois. D, 1951, University of Illinois, Urbana.

Neely, Joseph. The geology of the north end of the Medicine Bow Mountains, Carbon County, Wyoming. M, 1934, University of Wyoming. 18 p.

Neely, Kenneth Wray. Stratigraphy and paleoenvironments of the Late Mississippian White Knob Formation and equivalent rocks in the White Knob Mountains, South-central Idaho. M, 1981, University of Idaho. 142 p.

Neely, Laura Lea. Biostratigraphic study of the Horquilla Limestone, Big Hatchet Mountains, Hidalgo County, New Mexico. M, 1982, Texas Christian University. 117 p.

Neese, D. G. Facies mosaic of the upper Yates and lower Tansill formations (Upper Permian), Walnut Canyon, Guadalupe Mountains, New Mexico. M, 1979, University of Wisconsin-Madison.

Neese, Douglas G. Ooid diagenesis; Holocene geochemistry of the rock/water/atmospheric system. D, 1986, University of Oklahoma. 359 p.

Neese, Michael Charles. Delineation of the bedrock topography of Champaign County, Ohio, using the gravity-geologic method. M, 1978, Wright State University. 84 p.

Neet, Kerrie Elise. A stable isotopic investigation of the Mazatzal Peak Quartzite; implications for source terranes. M, 1988, Arizona State University. 109 p.

Nefe, Erick C. Comparison of the Q_{LSi} ellipsoid and ellipsoidal reduction spots used to determine finite strain in the Precambrian Kona Slate Member; Marquette, Michigan. M, 1980, Michigan State University. 95 p.

Neff, Arthur W. A study of the fracture patterns of Riley County, Kansas. M, 1949, Kansas State University. 47 p.

Neff, Everett Richard. Subsurface geology of McIntosh County, Oklahoma. M, 1961, University of Oklahoma. 54 p.

Neff, Jerry W. Geologic studies of the Gaysport and Skelley limestones in eastern Ohio. M, 1965, Ohio University, Athens. 94 p.

Neff, Linda M. Economic geology of part of the Gold Brick District, Gunnison County, Colorado. M, 1988, Colorado State University. 117 p.

Neff, Nancy Ann. The basicranial anatomy of the Nimravidae (Mammalia: Carnivora); character analyses and phylogenetic inferences. D, 1983, City College (CUNY). 660 p.

Neff, Nancy E. Stratigraphy, depositional environments, and paleoecology of the Upper Silurian (Pridoli) Sneedville Formation in Hancock County, Tennessee. M, 1983, University of Tennessee, Knoxville. 201 p.

Neff, Thomas Rodney. Petrology and structure of the Buffalo mountain pluton (probably Late Permian), Humboldt County, Nevada. D, 1969, Stanford University. 210 p.

Neff, Thomas Rodney. Petrology and structure of the Little Willow Series, Wasatch Mountains, Utah. M, 1962, University of Utah. 83 p.

Negas, Taki. Phase equilibria studies in the systems $PbO\text{-}Cr_2O_3\text{-}O_2$ and $PbO\text{-}SrO\text{-}O_2$. D, 1966, Miami University (Ohio). 147 p.

Negas, Taki. Solid state reactions, synthesis, and polymorphism in the system $PbO\text{-}SiO_2$. M, 1963, Miami University (Ohio). 69 p.

Negev, Moshe. A sediment model on a digital computer. D, 1967, Stanford University. 122 p.

Negi, Balwant Singh. A saturation core magnetometer. M, 1948, Colorado School of Mines. 64 p.

Negri, Daniel R. An error analysis of depth determination for finely and coarsely sampled aeromagnetic data using the technique of Werner deconvolution. M, 1975, Pennsylvania State University, University Park. 94 p.

Negrini, Robert Mark. The middle to late Pleistocene geomagnetic field as recorded in fine-grained sediments from the northwestern Basin and Range, Nevada and Oregon. D, 1986, University of California, Davis. 280 p.

Negro, Arsenio, Jr. Design of shallow tunnels in soft ground. D, 1988, University of Alberta. 1480 p.

Negron, Jenaro R. Map projections in Puerto Rico. M, 1962, Ohio State University.

Negus, Kenneth D. A study of the accelerated methods for the observation and computation of geodetic position and direction of gravity. M, 1961, Ohio State University.

Negus-de Wys, J. The Eastern Kentucky gas field. D, 1979, West Virginia University. 224 p.

Negussey, Dawit. An experimental study of the small strain response of sand. D, 1985, University of British Columbia.

Neibling, William-Howard. Transport and deposition of soil particles by shallow flow on concave slopes. D, 1984, Purdue University. 189 p.

Neidy, Carrie L. The geography and economic development of Black Hawk County, Iowa. M, 1924, University of Wisconsin-Madison.

Neighbor, Frank. Limestone caverns of the Black Hills region of South Dakota. M, 1940, University of Colorado.

Neiheisel, James. Origin and heavy minerals of the Isle of Palms, South Carolina. M, 1958, University of South Carolina. 32 p.

Neiheisel, James. Source of detrital heavy minerals in estuaries of the Atlantic Coastal Plain. D, 1973, Georgia Institute of Technology. 126 p.

Neill, Jerry R. The geology of Cedar County, Missouri; a guide for planners. M, 1987, Southwest Missouri State University. 81 p.

Neill, William Marshall. Geology of the southeastern Owyhee Mountains and environs, Owyhee County, Idaho. M, 1975, Stanford University.

Neilson, Frank Murray. The geometry of stable bed forms under oscillatory flow. D, 1969, Georgia Institute of Technology. 121 p.

Neilson, James M. Geology of the Lake Mistassini region, northern Quebec. D, 1950, University of Minnesota, Minneapolis. 155 p.

Neilson, James M. Stratigraphy and structure of the Mistassini Series (Cambrian) in the Lake Albanel area (Quebec). M, 1947, McGill University.

Neish, J. F.; Davis, V. H. and Harper, J. N. A short-term study of beach sand migration adjacent to Monterey canyon (California). M, 1966, United States Naval Academy.

Neitzel, Thomas William. Geology of the Van Stone mine, Stevens County, Washington. M, 1972, Washington State University. 47 p.

Nekritz, Richard. A thickness and lithofacies study of the Dakota Group (Cret.) in southwestern Nebraska. M, 1953, University of Nebraska, Lincoln.

Nekut, A. G., Jr. Crustal electrical conductivity structure in the Adirondacks. D, 1979, Cornell University. 125 p.

Nekvasil, Hanna. A theoretical thermodynamic investigation of the system Ab-Or-An-Qz(-H_2O) and implications for melt speciation. D, 1986, Pennsylvania State University, University Park. 285 p.

Nelimark, John H. Geology of ore deposits and associated rocks in Boulder-Leadville-Gunnison Belt, Colorado. M, 1924, University of Minnesota, Minneapolis. 40 p.

Nelis, Mary Karen. Deformation and metamorphism of the Rough Ridge Formation, Llano County, Texas. M, 1984, University of Texas, Austin. 101 p.

Nelligan, Frederick M. Geology of the Newhall area of the eastern Ventura and western Soledad basins, Los Angeles County, California. M, 1978, Ohio University, Athens. 117 p.

Nelligan, John D. Petroleum resources analysis within geologically homogeneous classes. D, 1980, Clarkson University. 104 p.

Nellis, David A. Tantalum in the Volney Pegmatite, Tinton, South Dakota, a geochemical study. D, 1973, Boston University. 284 p.

Nellis, Jose Carlos. Influence of paleostructure on paleochannels and uranium deposits in the Salt Wash Member of the Morrison Formation, San Rafael Swell, Utah. M, 1979, University of Utah. 117 p.

Nellis, Marvin Duane. Water management in the North Deschutes Unit Irrigation District; geographic perspectives and remote sensing applications. D, 1980, Oregon State University. 217 p.

Nellist, William Edward. Stratigraphy, petrography, and depositional environment of the Kokomo Dolomite Member, Wabash Formation (Silurian, Ludlovian-Pridolian) of Cass, Howard, and Miami counties,

Indiana. M, 1986, Indiana University, Bloomington. 137 p.

Nelms, Jerry L. The Fort Sill Formation of the Wichita Mountains, Oklahoma. M, 1958, University of Oklahoma. 91 p.

Nelms, Katherine Currier. Sedimentary and faunal analysis of a marginal marine section, the Stone City Member (middle Eocene), Crockett Formation, Burleston County, Texas. M, 1979, Texas A&M University.

Nelridge, Richard Alan. X-ray diffraction data for talc content in carbonate containing soils and their use in prospecting for talc ore deposits. M, 1987, Rutgers, The State University, Newark. 94 p.

Nelsen, Craig J. The geology and blueschist petrology of the western Ambler schist belt, southwestern Brooks Range, Alaska. M, 1979, University of New Mexico. 123 p.

Nelsen, J. J. A model for the movement of selenium in a closed aquatic system. D, 1974, University of Oklahoma. 253 p.

Nelsen, John Edward, Jr. Hypocentral trend surface analysis; an improved algorithm for investigating the morphology of Wadati-Benioff zones. M, 1988, North Carolina State University. 168 p.

Nelsen, K. P. A steady-state finite difference model of the groundwater of the Salinas Valley, California. M, 1975, Stanford University. 98 p.

Nelsen, Terry Allen. Density separation of clay minerals. M, 1971, Oregon State University. 59 p.

Nelsen, Terry Allen. The nature of general and mass-movement sedimentary processes on the outer-shelf, slope, and upper-rise northeast of Wilmington Canyon. D, 1981, University of Miami. 319 p.

Nelson, A. Graham. The geology of the northwest portion of Soledad Mountain, Kern County, California. M, 1940, University of Southern California.

Nelson, A. R. Quaternary glacial and marine stratigraphy of the Qivitu Peninsula, northern Cumberland Peninsula, Baffin Island, Canada. D, 1978, University of Colorado. 331 p.

Nelson, Adrian Marian. The geology of the southeastern quarter of the Bates Quadrangle, Grant County, Oregon. M, 1956, University of Oregon. 50 p.

Nelson, Afred M. Olivine for foundry sand. M, 1952, University of Washington. 96 p.

Nelson, Alan Robert. Age relationships of the deposits of the Wisconsin Valley and Langlade glacial lobes of north-central Wisconsin. M, 1973, University of Wisconsin-Madison.

Nelson, Ann B. The effectiveness of the National Flood Insurance Program as a land use management tool. M, 1978, University of Rochester. 72 p.

Nelson, Anthony. Cementation and dolomitization of Mississippian limestones, Kentucky and Virginia. M, 1985, Virginia Polytechnic Institute and State University.

Nelson, Anthony S. Upper Cretaceous depositional environments and provenance indicators in the central San Rafael Mountains, Santa Barbara County, California. M, 1979, University of California, Santa Barbara.

Nelson, Arthur E. Heavy minerals of the Kingsport Formation. M, 1955, University of Tennessee, Knoxville. 44 p.

Nelson, Bradford E. Lithofacies of the subsurface "Packer Shell" (Brassfield Limestone-Rochester Shale) interval in eastern Ohio. M, 1983, Kent State University, Kent. 83 p.

Nelson, Bruce Howard. Geophysical investigation and analysis of sedimentary units in the Indianola Field, Erath County, Texas. M, 1987, University of Texas at Dallas. 82 p.

Nelson, Bruce K. Uranium-lead isotopic systematics in the northern border zone of the Idaho Batholith, Bitterroot Range, Montana. M, 1982, University of Kansas. 82 p.

Nelson, Bruce Kert. Samarium-neodymium and rubidium-strontium isotopic studies of the origin and evolution of continental crust. D, 1985, University of California, Los Angeles. 204 p.

Nelson, Bruce R. The morphologic and sedimentologic response of four small watersheds to urbanization, Pittsburgh, PA. M, 1979, SUNY at Binghamton. 88 p.

Nelson, Bruce Warren. Mineralogy and stratigraphy of the pre-Berea sedimentary rocks exposed in northern Ohio. D, 1955, University of Illinois, Urbana. 113 p.

Nelson, Bruce Warren. The serpentine-amesite join in the system MgO-Al$_2$O$_3$-SiO$_2$-H$_2$O and classification of the chlorite minerals. M, 1954, Pennsylvania State University, University Park. 63 p.

Nelson, Carl E. Structure and petrology of the Thorndike Pond area, southwestern New Hampshire. M, 1975, Dartmouth College. 76 p.

Nelson, Carl Owen. Radiolarian biostratigraphic and paleoceanographic studies of Monterey-like rocks of the Humboldt Basin, Northern California. D, 1987, Rice University. 175 p.

Nelson, Carlton Hans. Geological limnology of Crater Lake, Oregon. M, 1961, University of Minnesota, Minneapolis. 175 p.

Nelson, Carlton Hans. Marine geology of Astoria and deep-sea fan. D, 1968, Oregon State University. 287 p.

Nelson, Carol Jeanne. Compression of nickel silicate spinel in a diamond anvil cell. M, 1987, University of Washington. 85 p.

Nelson, Carolynn. Transition elements in alkali-aluminosilicate melts; spectroscopy and thermodynamics of glass analogues. D, 1981, Pennsylvania State University, University Park. 196 p.

Nelson, Cheryl Ann. Petrology of the Mesaverde Formation (upper Cretaceous), southwest Casper arch, Wyoming. M, 1970, University of Missouri, Columbia.

Nelson, Clemens A. A petrographic study of the lower Keweenawan sandstones of northern Minnesota. M, 1942, University of Minnesota, Minneapolis. 46 p.

Nelson, Clemens A. Cambrian stratigraphy of the St. Croix Valley. D, 1949, University of Minnesota, Minneapolis.

Nelson, Clifford M. Gypsum Spring and the "lower Sundance" formations, eastern Big Horn Mountains, Wyoming and Montana. M, 1963, Michigan State University. 127 p.

Nelson, Clifford Melvin. Evolution of the late Cenozoic gastropod Neptunea (Gastropoda, Buccinacea). D, 1974, University of California, Berkeley. 802 p.

Nelson, Craig V. Sedimentological and foraminiferal characterization of a Holocene island slope (130-240m), North Jamaica. M, 1986, Utah State University. 145 p.

Nelson, David Ezra. Geology of the Fishhawk Falls, Jewell area, Clatsop County, Northwest Oregon. M, 1985, Oregon State University. 360 p.

Nelson, David L. Petrology and paragenesis of the Spring Hill algal mound; Plattsburg Formation, Upper Pennsylvanian at Neodesha, Kansas. M, 1978, University of Iowa. 181 p.

Nelson, Deborah J. Geochemistry of the Phi Kappa Formation, Blaine and Custer counties, Idaho. M, 1986, University of Idaho. 107 p.

Nelson, Dennis O. Strontium isotopic and trace element geochemistry of the Saddle Mountains and Grande Ronde basalts of the Columbia River Basalt Group. D, 1980, Oregon State University. 224 p.

Nelson, DeVon O. Evaluation of a land classification in Nepal. D, 1981, University of Wisconsin-Madison. 268 p.

Nelson, Diane Marie. The Drummond Mine Limestone; Mississippian Basin plain carbonate turbidites and hemipelagites in the Pioneer Mountains, Blaine County, South-central Idaho. M, 1979, University of Southern California.

Nelson, Douglas DeWayne. Late Pleistocene and Holocene clay mineralogy and sedimentation in the Pamlico Sound region, North Carolina. D, 1973, University of South Carolina.

Nelson, Edward T. The molluscan fauna of the later Tertiary of Peru. D, 1869, Yale University.

Nelson, Eleanor. The geology of Picketpost Mountain, Pinal County, Arizona. M, 1966, University of Arizona.

Nelson, Elmer R. Notes on the osteology of the skull of Ictops dakotensis (Oligocene, South Dakota). M, 1941, University of Chicago. 39 p.

Nelson, Eric Bruce. The geology of the Fairview-McKinley area, central Coos County, Oregon. M, 1966, University of Oregon. 59 p.

Nelson, Eric G. The distribution and structure of nearshore Holocene sediments from Provincetown to Cuttyhunk, Massachusetts. M, 1974, University of New Hampshire. 123 p.

Nelson, Eric Paul. Analysis of foreland basement deformation associated with the Clark Mountain thrust complex, southeastern California. M, 1977, Rice University. 73 p.

Nelson, Eric Paul. Geologic evolution of the Cordillera Darwin orogenic core complex, Southern Andes. D, 1981, Columbia University, Teachers College. 256 p.

Nelson, Frank J. The geology of the Pena Blanca and Walker Canyon areas, Santa Cruz County, Arizona. M, 1963, University of Arizona.

Nelson, Frederick Edward. Spatial properties of cryoplanation terraces and associated deposits in northwestern North America. D, 1982, University of Michigan. 313 p.

Nelson, Garry Richard. Ground water modeling in the Fountain Creek drainage. M, 1984, University of Colorado at Colorado Springs. 35 p.

Nelson, Gerald E. Origin and paleoenvironmental significance of pediments in the Bighorn Canyon area of southcentral Montana. D, 1983, University of Kansas. 175 p.

Nelson, Gordon C. Petrology and structure of Precambrian rocks in the Hesse Mountain area, Bighorn Mountains, Wyoming. M, 1970, Kent State University, Kent. 100 p.

Nelson, Gordon L. A reconnaissance; the petrology and diagenesis of the Step Conglomerate, east central Alaska. M, 1972, University of Alaska, Fairbanks. 53 p.

Nelson, Grant E. Alteration of footwall rocks at Brunswick Number 6 and Austin Brook deposits, Bathurst, New Brunswick, Canada. M, 1983, University of New Brunswick. 276 p.

Nelson, Harry Eugene. Lightweight aggregate for concrete from Montana shales. M, 1947, Montana College of Mineral Science & Technology. 37 p.

Nelson, Henry F. Heavy minerals of the Permo-Triassic red beds in Wyoming and South Dakota. D, 1952, University of Wisconsin-Madison.

Nelson, James. Geotechnical and geologic properties of the Santa Clara Formation claystone, Saratoga foothills, Santa Clara County, California. M, 1985, San Jose State University. 120 p.

Nelson, James Douglas. Ground water monitoring strategies to support community management of on-site home sewage disposal systems. D, 1980, Colorado State University. 143 p.

Nelson, James Warren. Soil properties related to vegetation and time on the Kautz Creek flood area, Mount Rainier, Washington. M, 1958, University of Washington. 56 p.

Nelson, Jeffrey Carter. Late Quaternary sedimentation on the continental rise between the Hudson and Albemarle-Transverse Canyon systems. M, 1979, Duke University. 124 p.

Nelson, Joanne Lee. Origin of Georgia Depression; the coast plutonic complex insular belt province boundary on Hardwicke and West Thurlow islands, B.C. M, 1976, University of British Columbia.

Nelson, John R. Sedimentology and stratigraphy of the late Paleozoic rocks of the Mountain Pine Ridge, Belize. M, 1984, SUNY at Binghamton. 52 p.

Nelson, John Wayne. Structural and geomorphic controls of the karst hydrogeology of Franklin County, Alabama. M, 1988, Mississippi State University. 165 p.

Nelson, Jon Sherwood. Geochemical study of micas from a small metaperidotite body and related rocks in the Skagit Gneiss, northern Cascades, Washington. M, 1972, University of Washington. 26 p.

Nelson, Jonathan Mark. Mechanics of flow and sediment transport over nonuniform erodible beds. D, 1988, University of Washington. 227 p.

Nelson, K. D. Geology of the Badger Bay-Seal Bay area, North-central Newfoundland. D, 1979, SUNY at Albany. 256 p.

Nelson, Karin E. Variations of late Neogene diatom assemblages from Deep Sea Drilling Project Leg 75, Site 532, cores 12, 13, 24, and 25. M, 1984, Boston University. 143 p.

Nelson, Karl R. The geology and geochemistry of the Poison Ridge intrusive center (Tertiary), Grand and Jackson counties, Colorado. M, 1971, Colorado School of Mines. 79 p.

Nelson, Karl Russell. Effects of channel roughness and permeability on hydraulic jump characteristics. D, 1981, University of Colorado. 118 p.

Nelson, Katherine Helen. Estuarine and anastomosing fluvial systems of the lower Mesaverde Group, northwestern Colorado. M, 1984, University of Texas, Austin. 151 p.

Nelson, Lloyd A. A partial index and bibliography of the Devonian faunas of North America. M, 1929, University of Colorado.

Nelson, Lloyd A. Gastropoda from the Pennsylvanian (Magdelena) of the Franklin Mountains of West Texas. D, 1937, University of Colorado.

Nelson, Mark. Geology and ore genesis of the Sixteen-to-One Mine, Esmeralda County, Nevada. M, 1984, University of Idaho. 93 p.

Nelson, Mark T. Geology of a part of the Laramie Anorthosite, Albany County, Wyoming. M, 1965, Colorado School of Mines. 67 p.

Nelson, Martin Andrew. Geology and fluorspar deposits of the southern Caballo Mountains, Sierra and Dona Ana counties, New Mexico. M, 1974, University of Texas at El Paso.

Nelson, Michael E. Studies of the Medicine Root Gravel (Quaternary), of southwestern South Dakota. M, 1967, University of South Dakota. 87 p.

Nelson, Michael Earl. Stratigraphy and paleontology of Norwood Tuff and Fowkes Formation, northwestern Utah and southwestern Wyoming. D, 1971, University of Utah. 181 p.

Nelson, Michael Peter. Tertiary stratigraphy and sedimentation in the Young's River-Lewis and Clark River area, Clatsop County, Oregon. M, 1978, Oregon State University. 242 p.

Nelson, Michael Ray. Areal geology of Cement-Cyril area, southeastern Caddo County, Oklahoma. M, 1983, University of Oklahoma. 64 p.

Nelson, Michael Roy. I, Paleomagnetism and crustal rotations along a shear zone, Las Vegas Range, southern Nevada; II, Seismotectonics of the Tien Shan, central Asia. D, 1988, Massachusetts Institute of Technology. 225 p.

Nelson, Murray Robert. Petrology, diagenesis and depositional environment of the Lagonda Interval, Cabaniss Subgroup, Cherokee Group, Middle Pennsylvanian, in north-eastern Kansas. M, 1985, University of Iowa. 176 p.

Nelson, Paul D. The reflection of the basement complex in the surface structures of the Marshall-Riley County area of Kansas. M, 1952, Kansas State University. 73 p.

Nelson, Paul Hugh. Geology of Floyd County, Iowa. M, 1939, University of Iowa. 158 p.

Nelson, Pennelope Conover. Community structure and evaluation of trophic analysis in paleoecology, Stone City Formation (middle Eocene, Texas). M, 1975, Texas A&M University.

Nelson, Priscilla. Tunnel boring machine performance in sedimentary rock. D, 1983, Cornell University. 468 p.

Nelson, R. Fluid inclusion study of the Fernvale Limestone, northern Arkansas. M, 1974, University of Missouri, Columbia.

Nelson, R. C. Geology of the Magnet Consolidated Mine. M, 1951, University of Toronto.

Nelson, Ralph L. Facies and depositional environments of the Hosston Formation (Lower Cretaceous) south-central Texas. M, 1978, University of Texas, Arlington. 148 p.

Nelson, Rex W. Stratigraphy and paleontology of the Kearsarge area, Inyo County, California. M, 1959, University of California, Berkeley. 107 p.

Nelson, Richard C. Structure and mineralogy of the C-JD-5 Mine, Montrose County, Colorado. M, 1979, Michigan Technological University. 70 p.

Nelson, Richard Newman. The geology of the hydrographic basin of the upper Santa Ynez Basin, California. D, 1923, University of California, Berkeley. 107 p.

Nelson, Richard V., Jr. Pirate cove (Sitka sound, Alaska); a study of cold water littoral carbonate deposition (Recent). M, 1970, University of Alaska, Fairbanks. 41 p.

Nelson, Robert. Spectrophotometric study of the satellites of Jupiter. D, 1977, University of Pittsburgh.

Nelson, Robert B. Stratigraphy and structure of region surrounding Currie, Elko County, Nevada. M, 1956, University of Washington. 66 p.

Nelson, Robert B. Stratigraphy and structure of Snake Range and Kern Mts., Nevada and W. Utah. D, 1959, University of Washington. 165 p.

Nelson, Robert Edward. Late Quaternary environments of the western Arctic Slope, Alaska. D, 1982, University of Washington. 146 p.

Nelson, Robert Edward. Quaternary environments of the Arctic slope of Alaska. M, 1979, University of Washington. 141 p.

Nelson, Robert H., Jr. Evolution of Laborcita Formation fan-deltas, Sacramento Mountains, New Mexico. M, 1985, University of Texas, Austin. 108 p.

Nelson, Robert Harry. A structural investigation in eastern Cass County and northeastern Otoe County, Nebraska. M, 1958, University of Nebraska, Lincoln.

Nelson, Robert Leslie. A study of the seismic waves SKS and SKKS. D, 1952, California Institute of Technology. 123 p.

Nelson, Robert Leslie. Glacial geology of the Frying Pan River drainage, Colorado. D, 1952, California Institute of Technology. 65 p.

Nelson, Robert Stanley, Jr. The glaciation of the headwaters area of Clear creek, Bighorn mountains, Wyoming. D, 1970, University of Iowa. 111 p.

Nelson, Robert Stanley, Jr. The gravels and surfaces west of Buffalo (Johnson County), Wyoming. M, 1968, University of Iowa. 71 p.

Nelson, Ronald Alan. Fracture permeability in porous reservoirs; an experimental and field approach. D, 1975, Texas A&M University. 185 p.

Nelson, Ronald Alan. The geochemistry and petrogenesis of the Tertiary igneous rocks of the Eagle Mountains, Van Horn, Texas. M, 1972, Texas A&M University. 93 p.

Nelson, Ronald G. Investigation of copper ion sorption by soil mineral particles. M, 1959, University of Minnesota, Minneapolis. 72 p.

Nelson, Ronald Harry. Mineralogy of the solution cavities in selected Devonian limestone areas of Ohio. M, 1967, Ohio University, Athens. 91 p.

Nelson, Rosie. The origin of the variations of the granite and associated rocks of Northampton and vicinity (Massachusetts). M, 1925, Smith College. 62 p.

Nelson, Samuel J. Ordovician palaeontology and stratigraphy of the Churchill and Nelson rivers, Manitoba. D, 1952, McGill University.

Nelson, Samuel James. Ecology of American Paleozoic sponges. M, 1950, University of British Columbia.

Nelson, Stephen Allen. The geology and petrology of Volcan Ceboruco, Nayarit, Mexico and partial molar volumes of oxide components of silicate liquids. D, 1979, University of California, Berkeley. 202 p.

Nelson, Steven Wayne. The petrology of a zoned granitic stock, Stillwater Range, Churchill County, Nevada. M, 1975, University of Nevada. 102 p.

Nelson, Terry A. The nature of general and mass-movement sedimentary processes on the outer-shelf, slope, and upper-rise northeast of Wilmington Canyon. D, 1981, University of Miami. 302 p.

Nelson, Thomas Arthur. A study of parameters important to soil-structure interaction in seismic analyses of nuclear power plants. D, 1983, University of California, Davis. 213 p.

Nelson, Thomas M. Biogenic silica and organic carbon accumulation in the Bransfield Strait, Antarctica. M, 1988, North Carolina State University. 65 p.

Nelson, Thomas Mason. A study of the Chugwater Formation of the Lander Wyoming region. M, 1925, University of Missouri, Columbia.

Nelson, Vincent E. The structural geology of the Cache Creek area, Gros Ventre Mountains, Wyoming. D, 1942, University of Chicago. 71 p.

Nelson, Walter John, Jr. Paleoenvironmental analysis of Chaetetes biostromes (Pennsylvanian) of the Arrow Canyon Quadrangle, Clark County, Nevada. M, 1973, University of Illinois, Urbana.

Nelson, Warren Lee. A seismic study of North Boulder Valley and other selected areas, Jefferson and Madison counties, Montana. M, 1962, Indiana University, Bloomington. 33 p.

Nelson, Wilbur A. Report of a reconnaissance of the Tennessee coal field south of the Tennessee central railroad. M, 1915, [Stanford University].

Nelson, Willis H. Petrology of early Tertiary volcanic rocks, Iron Springs Disctrict, Utah. M, 1950, University of Washington. 42 p.

Nelstead, Kevin Torval. Correlation of tephra layers in the Palouse Loess. M, 1988, Washington State University. 80 p.

Nemcek, David Francis. A paleo-environmental study of two localities in the Wanakah Shale (middle Devonian) (western New York). M, 1971, SUNY at Buffalo. 65 p.

Nemchak, Frank M. Sedimentology of some late Pleistocene outwash deposits, Sheboygan County, Wisconsin. M, 1978, University of Wisconsin-Milwaukee.

Nemeth, Donald F. Morphology, grain size characteristics, and fluvial processes of two bars, Colville River delta, Alaska. D, 1977, Louisiana State University. 146 p.

Nemeth, Donald F. Sedimentary and topographic characteristics of scree at three California localities. M, 1968, University of Southern California.

Nemeth, Kenneth E. Petrography of the Lower Volcanic Group, Tayoltita-San Dimas District, Durango, Mexico. M, 1976, University of Texas, Austin.

Nemetz, Arthur C. The finite element method as applied to structural geology. M, 1970, Rensselaer Polytechnic Institute.

Nemickas, Bronius. Geohydrologic digital computer simulation model of the Wenonah-Mount Laurel aquifer system in the coastal plain of New Jersey. D, 1975, Rutgers, The State University, New Brunswick. 142 p.

Nemser, Katherine B. Western coal mine production and distribution patterns; 1970 and 1980. M, 1983, University of Idaho. 169 p.

Nentwich, Franz W. Sedimentology and stratigraphy of the Reindeer Formation (early Tertiary) in the subsurface, Mackenzie Delta - Beaufort Sea area, Northwest Territories. M, 1980, Carleton University. 185 p.

Nentwich, Franz Werner. Stratigraphy and sedimentology of the Ordovician and Silurian Brodeur Group, northern Brodeur Peninsula, Baffin Island. D, 1987, University of Alberta. 457 p.

Nepomuceno, Francisco Filho. Effects of offset on seismic reflection data in Amazonas Basin, Brazil. M, 1982, University of Texas, Austin. 177 p.

Nerhot, Antonio Valentin Segovia see Segovia Nerhot, Antonio Valentin

Nery Leao, Zelinda Margarida de Andrade see de Andrade Nery Leao, Zelinda Margarida

Nesbit, Lee C. Hydrothermal uranium mineralization genesis of the Gillis Lease, Spokane Indian Reservation, Stevens County, Washington. M, 1979, Eastern Washington University. 34 p.

Nesbit, Robert A. The triassic rocks of the Dayville Quadrangle, Central Oregon. M, 1951, Oregon State University. 39 p.

Nesbitt, B. I. Some ultramafic rocks of northern British Columbia. M, 1941, Queen's University. 67 p.

Nesbitt, Bruce Edward. Fluid and magmatic inclusions in the carbonatite at Magnet Cove, Arkansas. M, 1976, University of Michigan.

Nesbitt, Bruce Edward. Regional metamorphism of the Ducktown, Tennessee, massive sulfides and adjoining portions of the Blue Ridge Province. D, 1979, University of Michigan. 259 p.

Nesbitt, Elizabeth Anne. Paleoecology and biostratigraphy of Eocene marine assemblages from western North America. D, 1982, University of California, Berkeley. 222 p.

Nesbitt, Robert H. The geology of the Cumberland Ohio district. M, 1930, Ohio State University.

Nesemeier, Bradley D. Stratigraphy and sedimentology of the Sentinel Butte Formation (Paleocene) near Lost Bridge, Dunn County, West-central North Dakota. M, 1981, University of North Dakota. 67 p.

Ness, Deborah Lee. Depositional systems in the Dockum Group (Upper Triassic), Scurry and Mitchell counties, Texas. M, 1988, Texas Christian University.

Ness, Gordon Everett. Late Neogene tectonics of the mouth of the Gulf of California. D, 1982, Oregon State University. 143 p.

Ness, Gordon Everett. The structure and sediments of Surveyor deep-sea channel. M, 1972, Oregon State University. 77 p.

Ness, Mark William. A land cover change study utilizing Landsat multispectral scanner data in Lancaster County, Pennsylvania. D, 1988, Pennsylvania State University, University Park. 206 p.

Ness, Norman F. Resistivity interpretation in geophysical prospecting. D, 1959, Massachusetts Institute of Technology. 236 p.

Nesse, W. D. Geology and metamorphic petrology of the Pingree Park area, Northeast Front Range, Colorado. D, 1977, University of Colorado. 265 p.

Nest, Julieann Van *see* Van Nest, Julieann

Neterwala, M. P. Areal and genetic relations of the metalliferous deposits in the India-Burma region. D, 1944, Massachusetts Institute of Technology.

Nethercott, Mark A. Geology of the Deadman Canyon 7 1/2 minute quadrangle, Carbon County, Utah. M, 1986, Brigham Young University. 85 p.

Netkowski, Thomas Frank. A historical review of Triassic nomenclature and stratigraphy and stratigraphic and sedimentological interpretations of the Nugget Sandstone in western Wyoming and southeastern Idaho. M, 1970, University of Michigan.

Netoff, D. I. Soil clay mineralogy of Quaternary deposits in two Front Range-Piedmont transects, Colorado. D, 1977, University of Colorado. 178 p.

Netolitzky, Ronald Kort. Geology of the Kipahigan Lake copper-nickel sulphide deposit, Saskatchewan (Canada). M, 1967, University of Calgary. 138 p.

Netschert, Bruce Carlton. The mineral foreign trade of the United States in the twentieth century; a study in mineral economics. D, 1949, Cornell University.

Nett, Allan R. A quantitative analysis of production in the Oak Park Field, Moorpark, California, using sedimentological and structural criteria. M, 1972, University of Southern California.

Netterville, John A. Quaternary stratigraphy of the lower God's River region, Hudson Bay Lowlands, Manitoba. M, 1974, University of Calgary. 79 p.

Nettles, James Edward. Study of foraminifera in the Clayton Formation (Paleocene) near Ft. Gaines, Georgia. M, 1959, Florida State University. 52 p.

Nettles, Norton Sandy. Intertidal calcitic muds along the west coast of Florida. M, 1976, University of Florida. 110 p.

Nettleton, Wiley Dennis. Pedogenesis of certain Aquulitic and Aquic Normudultic soils of the North Carolina coastal plain. D, 1966, North Carolina State University. 296 p.

Netto, A. Sergio. Petroleum and reservoir potentialities of the Agua Grande Member (Cretaceous), Reconcavo Basin, Brazil. M, 1975, University of Texas, Austin.

Netzband, Michael K. Focal depth and other sources characteristics of several eastern North American earthquakes through the analysis of Rayleigh surface waves. M, 1984, University of Kentucky.

Netzband, William F. The role of geology in prospecting for lead and zinc in the Tri-State District. M, 1927, University of Missouri, Rolla.

Netzband, William Ferdinand. The arch theory of diastrophism. M, 1923, University of Illinois, Chicago.

Netzler, Bruce William. The description and origin of beds of secondary quartz in the Mississippian St. Louis Limestone of eastern Missouri. M, 1981, University of Missouri, Rolla. 119 p.

Neubeck, William Sidney. Baseline study of the hydrology and morphology of Little Sewickley Creek, Sewickley, Pa. M, 1979, SUNY at Binghamton. 88 p.

Neuberger, Daniel John. Swastika (Upper Pennsylvanian) shelf-margin deltas and delta-fed turbidites, Flowers "Canyon Sand Field" area, Stonewall County, Texas. M, 1987, University of Texas, Austin. 344 p.

Neuder, Gary Leslie. Controls on peat accumulation in the No. 9 to No. 13 interval in western Kentucky. M, 1984, University of Kentucky. 80 p.

Neuerburg, George Joseph. Optical figures obtained with the reflecting microscope. M, 1947, University of California, Los Angeles.

Neuerburg, George Joseph. Petrology of the pre-Cretaceous rocks of the Santa Monica Mountains, California. D, 1951, University of California, Los Angeles.

Neugebauer, Henry Edwin Otto. Lithology and structure of the late Paleozoic rocks of the Apex mountain area, British Columbia, (Canada). M, 1965, University of Oregon. 46 p.

Neuhart, Donna Marie. Alteration of saponite to chlorite within the vesicles of volcanic rocks. M, 1984, University of California, Santa Barbara. 117 p.

Neuhauser, K. R. A structural and petrographic analysis of the Bank's Creek (NC) serpentinite. M, 1971, University of South Carolina.

Neuhauser, Kenneth Reed. A structural and petrofabric analysis of several ultramafics of western North Carolina. D, 1974, University of South Carolina.

Neuman, Cheryl Lynn. Aeolian processes and landforms in south Pangnirtung Pass, Southeast Baffin Island, N.W.T., Canada. D, 1988, Queen's University.

Neuman, Lawrence Donald. The deep structure of the Atlantic Ocean. D, 1973, Columbia University. 184 p.

Neuman, Lawrence Donald. Thermal history of the Moon. M, 1966, Columbia University. 44 p.

Neuman, Robert B. The St. Paul Group; a revision of the "Stones River" Group of Maryland and adjacent states. D, 1949, The Johns Hopkins University.

Neuman, Shlomo Peter. Transient flow of ground water to wells in multiple-aquifer systems. D, 1968, University of California, Berkeley. 104 p.

Neumann, Andrew Conrad. Processes of recent carbonate sedimentation in Harrington Sound, Bermuda. D, 1963, Lehigh University. 157 p.

Neumann, Conrad Andrew. The configuration and sediments of Stetson Bank, northwestern Gulf of Mexico. M, 1958, Texas A&M University.

Neumann, Else Rahnhild. Comparison of preferred orientations of calcite and dolomite in experimentally and naturally deformed rocks. M, 1967, University of California, Los Angeles.

Neumann, Fred Robert. The origin of the Cretaceous white clays of South Carolina. D, 1926, Cornell University.

Neumann, Lynda L. The hydrogeology of Menomonee Falls, Wisconsin. M, 1982, University of Wisconsin-Milwaukee. 164 p.

Neumann, Peter C. Fracture toughness of Westerly Granite using a round compact tension specimen. M, 1982, Michigan Technological University. 152 p.

Neumann, Scott Nelson. Lower and middle Precambrian geology of the Denham area, northern Pine County east-central Minnesota. M, 1985, University of Minnesota, Duluth. 180 p.

Neumann, William Henry. The fluvial geomorphology of a bedrock channel, Cedar Ceek, Union County, Illinois. M, 1976, Southern Illinois University, Carbondale. 98 p.

Neumeier, Donald P. Geology of the Woolsey area, Washington County, Arkansas. M, 1959, University of Arizona.

Neurauter, Thomas William. Bed forms on the western Florida shelf as detected with side scan sonar. M, 1979, University of South Florida, St. Petersburg. 144 p.

Neurauter, Thomas William. Mud mounds on the continental slope, northwestern Gulf of Mexico and their relation to hydrates and seafloor instability. D, 1988, Texas A&M University. 254 p.

Neuschafer, Gregory F. Fluvial metal contribution and the ferromanganese nodules of Green Bay, Lake Michigan. M, 1975, University of Wisconsin-Milwaukee.

Neustadt, Walter, Jr. Regional investigations of the Pennsylvanian System in northwestern Oklahoma and south-central Kansas. M, 1941, University of Oklahoma. 32 p.

Neuweld, Mark Adam. The petrology and geochemistry of the Great Sitkin Suite; implications for the genesis of calc-alkaline magmas. M, 1987, Cornell University. 174 p.

Neuzil, Christopher Eugene. Fracture leakage in the Cretaceous Pierre Shale and its significance for underground waste disposal. D, 1980, The Johns Hopkins University. 167 p.

Nevero, Ann Bernadette. A gravity investigation of the Teays Stage, Hamilton drainage system, in Preble County, Ohio. M, 1985, Wright State University. 69 p.

Nevers, George Morrison. Time variations of self potential and fluid resistivities in drill holes. M, 1957, Indiana University, Bloomington. 49 p.

Neves, A. S. The generalized magnetotelluric method. D, 1957, Massachusetts Institute of Technology. 123 p.

Neville, Allen Sneed. Order-disorder features of plagioclase from the Adirondack anorthosite. M, 1974, Wright State University.

Neville, Colleen Ann. Magnetostratigraphy and magnetic properties of the Pliocene Glenns Ferry Formation of Southwest Idaho. D, 1981, Columbia University, Teachers College. 222 p.

Neville, Scott L. Late Miocene alkaline volcanism, northeastern San Bernardino Mountains and adjacent Mojave Desert. M, 1983, University of California, Riverside. 156 p.

Neville, William D. Subsurface geology of the Nelson oil field, Andrew County, Texas. M, 1959, University of Wisconsin-Madison.

Nevin, Andrew Emmet. Geology of the paragenesis of the east flank of the Kaniksu Batholith, Boundary County, Idaho. D, 1966, University of Idaho. 59 p.

Nevin, Andrew Emmet. Late Cenozoic stratigraphy and structure of the Benton area, Mono County, California. M, 1963, University of California, Berkeley. 65 p.

Nevin, Charles M. A study of certain physical properties of foundry sands. M, 1923, Cornell University.

Nevin, Charles M. Some physical properties of molding sands. D, 1925, Cornell University.

Nevins, Barbara B. Structural evolution of the Russell Range oil field and vicinity, southern Coast Ranges, California. M, 1983, Oregon State University. 69 p.

Nevins, Judith B. Determination of manganese in estuarine pore waters by electron paramagnetic resonance spectroscopy. M, 1978, University of New Hampshire. 68 p.

New, Robert A. A geochemical study of ash-flow tuffs from the Mogollon-Datil volcanic field of southwestern New Mexico. M, 1981, University of Pittsburgh.

New, William Randal. The origin of Goose Egg Mountain, Yakima County, Washington. M, 1940, Washington State University. 40 p.

Newbauer, Thomas Raymond. Geomagnetic depth-sounding in the State of Washington. M, 1974, University of Washington. 49 p.

Newberg, Donald W. Geology and mineralogy of chrysocolla-bearing gravels. D, 1965, Harvard University.

Newberry, James Tyler. Seismicity and tectonics of the far western Aleutian Islands. M, 1983, Michigan State University. 93 p.

Newberry, R. J. J. Polytypism in molybdenite; mineralogy, geology, and geochemistry. M, 1978, Stanford University. 59 p.

Newberry, Rainer Jerome Joachim. The geology and chemistry of skarn formation and tungsten deposition in the central Sierra Nevada, California. D, 1980, Stanford University. 342 p.

Newberry, Ralph Jeffrey. The structural and metamorphic history of the Wissahickon Formation within Washington, D.C. M, 1974, University of Illinois, Urbana.

Newberry, Spencer B. Formation of coal. D, 1881, Columbia University, Teachers College.

Newberry, William Bohning. Oil, gas, and water possibilities of Kent Station Quadrangle, Culberson County, Texas. M, 1952, University of Texas, Austin.

Newbill, Thomas J., Jr. Geology of the White Raven Mine, Colorado. M, 1935, University of Colorado.

Newbry, Brooks Walter. Water quality management using the hazard assessment approach. D, 1981, Colorado State University. 582 p.

Newbury, Robert William. The Nelson River; a study of subarctic river processes. D, 1968, The Johns Hopkins University. 372 p.

Newby, McInnis S. Frio (Hackberry) deposition and deformation near the Newton-Orange county line, southeastern Texas. M, 1963, University of Houston.

Newby, Ray A. Quantitative variations of clay minerals in Pearl River estuarine sediment. M, 198?, University of New Orleans. 90 p.

Newcomb, Edward L. Structure and petrography of the Greenwood Lake Quadrangle (New York). D, 1949, Cornell University.

Newcomb, Ester Hollis; Wyman, Anne F. and Wyman, R. Geology of the northern Snake River Range, Idaho and Wyoming. M, 1949, University of Michigan.

Newcomb, Frederick A. Geology of a portion of southwestern Brewster County, Trans-Pecos, Texas. M, 1978, University of Houston.

Newcomb, John Hartnell. Geology of Bat Cave Quadrangle (south-central Texas). M, 1971, University of Texas, Austin.

Newcomb, Joseph Judge. Petroleum geology of Cretaceous rocks in South Dakota. M, 1974, University of Tulsa. 85 p.

Newcomb, Reuben Clair. Geology of a portion of the Okanogan Highlands, Washington. M, 1937, Washington State University. 62 p.

Newcomb, W. E. Geology, structure, and metamorphism of the Chuacus Group, Rio Hondo Quadrangle and vicinity, Guatemala. D, 1975, SUNY at Binghamton. 214 p.

Newcombe, Robert John Burgoyne. Deposition and structural features of the Michigan synclinal basin. D, 1931, University of Michigan.

Newcome, Roy, Jr. Kyanite at Henry Knob, South Carolina. M, 1949, University of South Carolina. 23 p.

Newcomer, Darrell R. Detailed water-level data and an evaluation of mathematical approaches for near-river monitoring. M, 1988, Montana College of Mineral Science & Technology. 69 p.

Newcomer, Earle Seifried. Subsurface geology of the Rockport oil and gas field, Spencer County, Indiana. M, 1951, Indiana University, Bloomington. 27 p.

Newcomer, Robert W. Geology, hydrothermal alteration and mineralization of the northern part of the Sugarloaf Peak quartz monzonite, Dona Ana County, New Mexico. M, 1984, New Mexico State University, Las Cruces. 108 p.

Newdale, Karen Marie. The Pleistocene stratigraphy of the Trenton and southern portions of the Middletown quadrangles of southwestern Ohio. M, 1980, Miami University (Ohio). 155 p.

Newell, Hildreth Adele. Size analysis of tills from some east-central Illinois moraines. M, 1954, University of Illinois, Urbana.

Newell, James H. A preliminary study of the Coccolithophoridae and related calcareous nannofossils from the Mooreville Formation (Upper Cretaceous), Lowades County, Mississippi. M, 1968, Mississippi State University. 145 p.

Newell, John G. The geomorphic development of the Tug Hill Plateau (New York). M, 1940, Syracuse University.

Newell, K. D. Stratigraphy and structural geology of the Moose Creek area, central Alaska Range, Alaska. M, 1975, University of Wisconsin-Madison.

Newell, Norman D. The stratigraphy and paleontology of the upper part of the (Pennsylvanian) Missouri series in eastern Kansas. D, 1933, Yale University.

Newell, Norman Dennis. Stratigraphy of Johnson County, Kansas. M, 1931, University of Kansas. 125 p.

Newell, R. A. Exploration geology and geochemistry of the Tombstone-Charleston area, Cochise County, Arizona. D, 1975, Stanford University. 280 p.

Newell, Roger A. Geology and geochemistry of the northern Drum Mountains, Juab County, Utah. M, 1971, Colorado School of Mines. 115 p.

Newell, W. R. Stratigraphy, structure, and petrology of the Bingham Quadrangle, West-central Maine. M, 1978, Syracuse University.

Newell, Wayne L. Saprolite development in the Sleepers River Watershed, Danville, Vermont. M, 1966, Dartmouth College. 71 p.

Newell, Wayne Linwood. Surficial geology of the Passumpsic valley, northeastern Vermont. D, 1970, The Johns Hopkins University. 224 p.

Newhall, Christopher George. Geology and petrology of Mayon Volcano, southeastern Luzon, Philippines. M, 1977, University of California, Davis.

Newhall, Christopher George. Geology of the Lake Atitlan area, Guatemala; a study of subduction zone volcanism and caldera formation. D, 1980, Dartmouth College. 364 p.

Newham, W. D. N. An experimental investigation of intergrowth phenomena in bornite and chalcopyrite. M, 1963, McGill University.

Newhart, Joseph A. Gravity and magnetic geophysical investigations of Sandusky, Seneca, and portions of Hancock and Wood counties, Ohio. M, 1975, Bowling Green State University. 75 p.

Newhart, Richard Eugene. Carbonate facies of the Middle Ordovician Michigan Basin. M, 1976, Michigan State University. 50 p.

Newhouse, W. H. Ore deposits of Kokomo, Colorado. M, 1923, Massachusetts Institute of Technology. 65 p.

Newhouse, W. H. Paragenesis of certain occurrences of marcasite. D, 1926, Massachusetts Institute of Technology. 159 p.

Newkirk, Deborah J. Downhole electrode resistivity interpretation with three-dimensional models. M, 1983, University of Utah. 107 p.

Newkirk, Steven Ross. Petrology of the Ono Metaperidotite, south central Klamath Mountains, California. M, 1976, Southern Methodist University. 60 p.

Newland, Bernard Terence A. On the diffusion of radiogenic argon from potassium feldspar. M, 1963, University of Alberta. 119 p.

Newman, Bradford Scott. Geology and water chemistry, North Fork drainage, Bishop Creek, California. M, 1976, University of California, Los Angeles.

Newman, Brent D. Experimental determination of reactive tracer suitability for groundwater field tests. M, 1988, University of Texas at El Paso.

Newman, Daniel Benjamin. Near-surface in-situ stress measurements in the Anna, Ohio earthquake zone. M, 1977, University of Michigan.

Newman, David Alan. The design of coal mine roof support and yielding pillars for longwall mining in the Appalachian Coalfield. D, 1985, Pennsylvania State University, University Park. 419 p.

Newman, Donald Hughes. Paleoenvironment of the Lower Triassic Thaynes Formation near Cascade Springs, Wasatch County, Utah. M, 1974, Brigham Young University. 96 p.

Newman, Ewa N. Paleoenvironments and stratigraphy of the Middle Ordovician carbonates in the lower Mohawk Valley. M, 1980, Boston University. 90 p.

Newman, Frederick George. Stratigraphy and paleontology of a core from Silurian and Devonian strata of Wayne County, Michigan. M, 1955, University of Michigan.

Newman, Gary James. Conodonts and biostratigraphy of the Lower Mississippian in western Utah and eastern Nevada. M, 1972, Brigham Young University.

Newman, Gregory Alex. Three-dimensional transient electromagnetic modeling for exploration geophysics. D, 1987, University of Utah. 242 p.

Newman, Harry E., III. The biostratigraphy of the Bluffport marl member of the Demopolis chalk, Cretaceous of (western) Alabama. M, 1969, University of Alabama.

Newman, James W. The geology of the Tracy Canyon area, Saguache County, Colorado. M, 1976, Colorado State University. 117 p.

Newman, Jerry Savrda. Site surveys of the central and southern Ninetyeast Ridge for the Ocean Drilling Program, Leg 121. M, 1987, University of Texas, Austin. 105 p.

Newman, Karl Robert. Micropaleontology and stratigraphy of Late Cretaceous and Paleocene formations, northwestern Colorado. D, 1961, University of Colorado. 116 p.

Newman, Karl Robert and Woodhams, Richard L. Stratigraphy and paleontology of a core from Loraine County, Ohio. M, 1954, University of Michigan.

Newman, Peter Vincent. Geology of the Round Spring Canyon area, northwestern Ventura County, California. M, 1959, University of California, Los Angeles.

Newman, Richard. The stone sculpture of India; a petrographic study of the materials used by Indian sculptors from ca. 2nd century to the 16th century. M, 1983, Boston University. 228 p.

Newman, Sally. ^{230}Th-^{238}U disequilibrium systematics in young volcanic rocks. D, 1983, University of California, San Diego. 302 p.

Newman, Stephen Lars. Sedimentology and depositional history of shallow shelf deposits within the Cretaceous Blackhawk Formation and Mancos Shale, east-central Utah. M, 1985, University of Utah. 166 p.

Newman, Stephen Miller. A paleoenvironmental analysis of the upper Demopolis and lower Ripley formations in the Rock Hill area, Oktibbeha County, Mississippi. M, 1975, Mississippi State University. 88 p.

Newman, Thomas Stell. Toleak Point; an archaeological site on the north-central Washington coast. M, 1959, University of Washington. 138 p.

Newman, Walter S. Geological significance of recent borings in the vicinity of Castle Harbor, Bermuda. M, 1959, Syracuse University.

Newman, Walter S. Late Pleistocene paleoenvironments of western Long Island Sound. D, 1966, New York University. 189 p.

Newman, William Alexander. Wisconsin glaciation of northern Cape Breton Island, Nova Scotia, Canada. D, 1971, Syracuse University. 158 p.

Newman, William J. Tungsten mining at the Getchell Mine. M, 1955, University of Nevada - Mackay School of Mines. 43 p.

Newman, William Roy. Geology and microscopic features of the Phalen Seam, Sydney coal field, Sydney, Nova Scotia. D, 1934, University of Toronto.

Newmann, Fred Robert. Origin of the colors of sedimentary rocks. M, 1923, University of Chicago.

Newmarch, Charles Bell. Geology of the Crowsnest coal basin with special reference to the Fernie coal area (British Columbia). D, 1951, Princeton University. 133 p.

Newmark, Robin L. In-situ geophysical investigations into the nature of the oceanic crust. D, 1984, Columbia University, Teachers College. 236 p.

Newmark, Robin Lee. Tectonic interaction in the Philippines; calculated apparent slip along the Philippine Fault. M, 1980, University of California, Santa Cruz.

Newmeyer, Amel James, Jr. A study of cyclical sedimentation adjacent to the Pittsburgh coal seam, Southeast Pittsburgh. M, 1948, University of Pittsburgh.

Newport, Roy Leo. Acritarchs of the Lime Creek Formation (Upper Devonian) of Iowa. M, 1973, University of Texas at Dallas. 136 p.

Newport, Roy Leo. Palynology, depositional environment, and organic metamorphism of the hydrocarbon producing interval (Cretaceous) of Polk County, Texas. D, 1978, University of Texas at Dallas. 199 p.

Newsom, Horton E. Metal fractionation patterns in the Benculban Meteorite. M, 1978, University of Arizona.

Newsom, Horton Elwood. The experimental partitioning behavior of tungsten and phosphorus; implications for the composition and formation of the Earth, Moon and eucrite parent body. D, 1982, University of Arizona. 84 p.

Newsom, John Flesher. A geologic and topographic section across southern Indiana, from the Ohio River at Hanover to the Wabash River at Vincennes, with a discussion of the general distribution and character of the Knobstone Group in the State of Indiana. D, 1901, [Stanford University].

Newsom, John Flesher. The thickness of the Carboniferous sediments in North America. M, 1892, Stanford University.

Newsom, Steven Wayne. Middle Ordovician paleogeography of the Appalachian Valley and Ridge Province, central Pennsylvania. M, 1983, University of Delaware. 235 p.

Newson, Ralph. The structural geology of the Perth Road Syncline (Grenville Province, Precambrian Shield, Frontenac County, Ontario). M, 1970, Queen's University. 90 p.

Newton, A. C. Distribution of radioactivity and zirconium in the Athona Stock, Lake Athabaska, Alberta-Saskatchewan. M, 1951, University of Toronto.

Newton, Albert N. Surface and subsurface geology of Union Parish, Louisiana. M, 1967, Northeast Louisiana University.

Newton, Carl Adams. An investigation of Rayleigh wave ellipticity with applications to Earth structure. D, 1973, Pennsylvania State University, University Park. 201 p.

Newton, Cathryn Ruth. Biofacies analysis and paleoecology of the Onesquethaw Stage (Lower to Middle Devonian) in the Virginias. M, 1979, University of North Carolina, Chapel Hill. 121 p.

Newton, Cathryn Ruth. Norian (Late Triassic) molluscs of Cordilleran allochthonous terranes; paleoecology and paleozoogeography. D, 1983, University of California, Santa Cruz. 184 p.

Newton, Garth David. Application of a simulated model to the Snake Plain Aquifer. M, 1978, University of Idaho. 82 p.

Newton, Geoffrey B. Fauna of the Jefferson Formation (Upper Devonian), south central Montana and northwest Wyoming. M, 1967, Wayne State University. 41 p.

Newton, Geoffry Bruce. The Rhabdomesidae of the Wreford megacyclothem (Wolfcampian Permian) of Nebraska, Kansas, and Oklahoma. D, 1970, Pennsylvania State University, University Park. 232 p.

Newton, George Denson. Ostracodes and the Silurian stratigraphy of northwestern Alabama. M, 1967, Arizona State University. 58 p.

Newton, John Gordon. Geology of formations cropping out along State Highway 25, Marengo County, Alabama. M, 1960, University of Alabama.

Newton, John LeBaron. The Canada Basin; mean circulation and intermediate scale flow features. D, 1973, University of Washington. 157 p.

Newton, Joseph. Effect of reagents on sedimentation rate of various finely pulverized minerals. M, 1931, University of Idaho. 20 p.

Newton, Mark Shepard. An experimental study of the P-V-T-S relations of sea water. D, 1964, University of California, Los Angeles. 48 p.

Newton, Maury Claiborne, III. A late Precambrian resurgent cauldron in the Carolina slate belt of North Carolina, U.S.A. M, 1983, Virginia Polytechnic Institute and State University. 111 p.

Newton, Morgan Roe. Porosity evolution in the Salem Limestone (Mississippian), Monroe and northern Lawrence counties, Indiana. M, 1985, University of Missouri, Columbia. 119 p.

Newton, R. M. Stratigraphy and structure of some New England tills. D, 1978, University of Massachusetts. 256 p.

Newton, Ralph James. The geology northwest of Dublin, California, in the vicinity of Divide Ridge. M, 1948, University of California, Berkeley. 38 p.

Newton, Robert. Deglaciation (Wisconsinan) of the Ossipee Lake area, New Hampshire. M, 1972, SUNY at Binghamton. 87 p.

Newton, Robert Chaffer. Some equilibrium reactions in the system $CaAl_2Si_2O_8$-H_2O. D, 1963, University of California, Los Angeles. 58 p.

Newton, Robert Chaffer. The Malibu Bowl fault area, Santa Monica Mountains, California. M, 1958, University of California, Los Angeles.

Newton, Robert Stirling. Geology of southwestern Eastland County, Texas. M, 1963, University of Texas, Austin.

Newton, William Albert. Pennsylvanian stratigraphy of southeastern Illinois. M, 1937, University of Illinois, Urbana.

Newville, Harold Lee. Stratigraphy of the Lecompton megacyclothem (Upper Pennsylvanian) in southeastern Nebraska and adjacent regions. M, 1958, University of Nebraska, Lincoln.

Ney, Charles S. The Cretaceous of Southwest BC with special reference to the Pasayten Series in the east half of the Hope area. M, 1942, University of British Columbia.

Neybert, Daniel Steven. Petrography and depositional environments of the Upper Cretaceous (Cenomanian-Turonian) Ojinaga Formation, southern Quitman Mountains, Hudspeth County, Texas. M, 1985, University of Texas, Arlington.

Neyestani, Mohammad. The effect of water application rate on infiltration and wetting front characteristics of unsaturated silt loam soil. D, 1968, Utah State University. 105 p.

Neyman, Percy. The determination of conditions for the transformation of hydrocarbons of the fatty series into those of the aromatic; origin and condition of carbonaceous matter in bituminous shales. D, 1884, Columbia University, Teachers College.

Nezafati, Hooshang. Salt release from suspended sediments as a source of Colorado River salinity. D, 1982, Utah State University. 242 p.

Ng'ang'a, Patrick. Sedimentation of deltas and border faults in northern Lake Malawi; evidence from high-resolution acoustic remote sensing and gravity cores. M, 1988, Duke University. 93 p.

Ng, Albert Tung-Yiu. Multiple channel maximum entropy spectral estimator and its application. M, 1977, Massachusetts Institute of Technology. 58 p.

Ng, Carolyn Yee-Han. Combined use of wavenumber analysis of Landsat digital imagery and seismic data to infer the orientation of tectonic stress in the Haggar region in Africa. M, 1983, Pennsylvania State University, University Park. 120 p.

Ng, Chihang Amy. Ancient Arctic ice does not contain large excesses of natural lead; II, Chronological variations in lead and barium concentrations and lead isotopic compositions in sediments of four southern California off-shore basins. D, 1982, California Institute of Technology. 113 p.

Ng, David Tai Wai. A subsurface study of the lower Atoka (Lower Pennsylvanian) clastics in portions of Jack, Palo Pinto, Parker and Wise counties, north central Texas. M, 1974, Texas Christian University. 103 p.

Ng, Elliot Kin. Efficiency/equity analysis of water resources problems; a game theoretic approach. D, 1985, University of Florida. 160 p.

Ng, Kwok-Choi Samuel. Stratigraphy, sedimentology and diagenesis of the Upper Mississippian to Permian strata of the Talbot Lake area, Jasper National Park, Alberta. M, 1985, University of Alberta. 257 p.

Ng, Robert Man Chiu. Ground reaction and behaviour of tunnels in soft clays. D, 1984, University of Western Ontario.

Ng, Tai-Ping. Genetic classification of Pc3 and Pc4 geomagnetic pulsations in mid-latitudes. D, 1969, University of British Columbia.

Ngah, Khalid Bin. Stratigraphic and structural analyses of the Penyu Basin, Malaysia. M, 1975, Oklahoma State University. 52 p.

Ngoddy, Adaeze. Crustal thickness across the Southern Appalachians. M, 1984, Georgia Institute of Technology. 55 p.

Nguene, Francois Roger. Geology and geochemistry of the Mayo-Darle tin deposit, west-central Cameroon, central Africa. D, 1982, New Mexico Institute of Mining and Technology. 179 p.

Ngunjiri, Philip Gichonge. Environmental impact assessment practices in the sub-Saharan Africa; cases from Kenya. D, 1987, University of California, Berkeley. 230 p.

Nguyen, B. T. Analysis and interpretation of seismic reflection data from Lambton County, south Western Ontario. M, 1970, University of Western Ontario.

Nguyen, Bao Van. Determination of surface-wave source parameters from search procedure and from linear moment tensor inversion. D, 1988, St. Louis University. 220 p.

Nguyen, Bao Van. Surface wave focal mechanisms, magnitudes and energies for some eastern North American earthquakes with tectonic implications. M, 1985, St. Louis University. 275 p.

Nguyen, Chau Trung. Depositional facies stratigraphy of the Lower Ordovician (Tremadocian) platform carbonates of the Central Appalachians. D, 1986, The Johns Hopkins University. 324 p.

Nguyen, K. K. K^{40}-Ar^{40} isotopic age determination of the Nelson Batholith (Lower or Middle Cretaceous), British Columbia. M, 1968, University of British Columbia.

Nguyen, Son Ngoc. The nature and distribution of Ni(II) complexes in oil-sand asphaltenes. D, 1986, Washington State University. 145 p.

Nguyen, Sy. Suppression of long-period multiple events from marine seismic reflection data. M, 1985, California State University, Long Beach. 98 p.

Ni Ta Pe. Upper Cretaceous arenaceous foraminifera from Lowndes and Oktibbeha counties, Mississippi. M, 1956, University of Kansas. 74 p.

Ni, James Fu. Seismicity and active tectonics of the Himalayas and Tibetan Plateau. D, 1984, Cornell University. 280 p.

Ni, Sheng-Huoo. Dynamic properties of sand under true triaxial stress states from resonant-column torsional shear tests. D, 1987, University of Texas, Austin. 447 p.

Niay, Robert A. The principles of modern mineral exploration; a review. M, 1972, [Stanford University].

Niazi, Mansour. Partition of seismic energy. D, 1964, University of California, Berkeley. 86 p.

Niazy, Adnan Mohammed. An exact solution for a finite moving dislocation in an elastic half-space, with application to the San Fernando earthquake of 1971. D, 1974, Massachusetts Institute of Technology. 163 p.

Nibbe, Rod K. The application of the direct inversion of electrical data to hydrogeologic investigations. M, 1985, University of Wisconsin-Milwaukee. 123 p.

Nibbelink, Kenneth A. Depositional environments of the Fox Hills Sandstone (Upper Cretaceous), Cheyenne Basin, Colorado. M, 1983, Colorado State University. 245 p.

Nibbelink, Mark P. The paleomagnetic stratigraphy of the Pliocene age Quiburis Formation near Mammoth, Arizona. M, 1972, Dartmouth College. 84 p.

Nibler, Gerald John. The use of upper air data in the estimation of snow melt. M, 1973, University of Arizona.

Niccum, Marvin Richard. Geology and permeable structures in basalts of the east central Snake River Plain near Atomic City, Idaho. M, 1969, Idaho State University. 137 p.

Nice, David E. The role of skeletal diagenesis in carbonate fabric development, Thaynes Formation (Lower Triassic), Northeast Utah and Southeast Idaho. M, 1985, University of Wyoming. 122 p.

Nichol, Ian. Trace element study of contemporaneous sulphides, pyrite, pyrrhotite and chalcopyrite. M, 1958, Queen's University. 191 p.

Nichol, Michael R. Rock fall protection methods. M, 1983, University of Nevada. 178 p.

Nichol, Robert F. Geology and mineral resources of the Tinker Mountain-Fincastle, Virginia area. M, 1959, Virginia Polytechnic Institute and State University.

Nicholaichuk, W. A soil moisture and temperature prediction model under evaporative conditions. D, 1974, University of Guelph.

Nicholas, D. E. Underground mine pillar design utilizing rock mass properties, Marble Peak, Pima County, Arizona. M, 1976, University of Arizona.

Nicholas, John. Late Pleistocene palynology of southeastern New York and northern New Jersey. D, 1968, New York University. 115 p.

Nicholas, Raymond H. The subsurface structure and stratigraphy related to petroleum accumulation in Rice

County, Kansas. M, 1954, Kansas State University. 46 p.

Nicholas, Richard Ludlam. Stratigraphy and sedimentation of the Conococheague Formation (Cambrian) in the Shenandoah Valley, Virginia. M, 1954, University of Kansas. 132 p.

Nicholeris, N. A plant microfossil investigation of three Nova Scotian coal seams. M, 1951, University of Massachusetts.

Nicholl, Michael John. Computer assisted analysis of discontinuous rock masses. M, 1987, University of Nevada. 178 p.

Nicholls, Elizabeth L. Marine vertebrates of the Pembina Member of the Pierre Shale (Campanian, Upper Cretaceous) of Manitoba and their significance to the biogeography of the Western Interior Seaway. D, 1988, University of Calgary. 337 p.

Nicholls, James Watson. Studies of volcanic petrology of the Navajo-Hopi area, Arizona. D, 1969, University of California, Berkeley. 107 p.

Nicholos, Billy F. Knox Dolomite as an underground natural gas storage reservoir at Unionport, Indiana. M, 1966, University of Mississippi.

Nichols, Bruce MacKenzie. Hydrothermal sulfide and arsenide deposits associated with ultramafic and mafic rocks, Snohomish County, Washington. M, 1970, University of Washington. 41 p.

Nichols, Charles W. Trace element correlation using neutron activation analysis. M, 1971, Wichita State University. 122 p.

Nichols, Chester E. Geology of the southern half of the Stoutland, Missouri, quadrangle. D, 1977, University of Missouri, Rolla. 183 p.

Nichols, Chester Encell. Geology of a segment of Deep Creek fault zone southern Bighorn Mountains (Washakie County), Wyoming. M, 1965, University of Iowa. 216 p.

Nichols, Christine C. Analysis of the New Brunswick 1982 earthquake sequence with inferences on source parameters from multi-mode surface wave dispersion and spectral excitation. M, 1984, Pennsylvania State University, University Park.

Nichols, Clayton Ralph. Alteration of igneous rocks in the Lugert area, Kiowa County, Oklahoma. M, 1968, University of Oklahoma. 140 p.

Nichols, Clayton Ralph. The geology and geochemistry of the Pathé geothermal zone, Hidalgo, Mexico. D, 1970, University of Oklahoma. 178 p.

Nichols, David A. A genetic classification of land forms and illustrations of several of the more important forms. M, 1923, Columbia University, Teachers College.

Nichols, David Ryden. Bedrock geology of the Narragansett Pier Quadrangle, Rhode Island. M, 1954, Cornell University.

Nichols, Douglas J. Paleoecological analysis of the Merchantville Formation (Upper Cretaceous) in the New Jersey coastal plain. M, 1966, New York University.

Nichols, Douglas James. Palynology in relation to depositional environments of lignite in the Wilcox group (early Tertiary) in Texas. D, 1970, Pennsylvania State University, University Park. 467 p.

Nichols, Elizabeth Ann. Geology of the Monte Cristo Vein area, Black Rock mining district, Yavapai County, Arizona. M, 1983, University of Arizona. 63 p.

Nichols, George H. Thermal gradients in the metamorphosed Arkansas Novaculite of the Chamberlin Creek Syncline, Magnet Cove, Arkansas. M, 1973, Northeast Louisiana University.

Nichols, Gerald. The seismic structure of the Pt. Sal Ophiolite and its relationship to oceanic crustal structure. M, 1978, University of California, Santa Barbara.

Nichols, John Conner. Stratigraphy of Sierra de los Fresnos, Chihuahua, Mexico. M, 1958, University of Texas, Austin.

Nichols, Kathryn Marion. Triassic depositional history of China Mountain and vicinity, north-central Nevada. D, 1972, Stanford University. 196 p.

Nichols, Lee C. Hydrologic balance of a complex drainage area (near Shefferville) in New Quebec, Canada. M, 1967, Syracuse University.

Nichols, Maynard Meldrim. Composition and environment of Recent transitional sediments on the Sonoran Coast, Mexico. D, 1965, University of California, Los Angeles. 419 p.

Nichols, Paul Harry. Periglacial ventifacts in New Jersey. M, 1953, Rutgers, The State University, New Brunswick. 46 p.

Nichols, Paul Harry. The stratigraphy of the Trinity Group of southeastern Oklahoma, southwestern Arkansas and northeastern Texas. D, 1956, Rutgers, The State University, New Brunswick. 175 p.

Nichols, Ralph. Early Miocene mammals from the Lemhi Valley of Idaho. M, 1976, University of Montana. 74 p.

Nichols, Renny Roger. A study of the piezometric surface of the Grand Falls Formation (Mississippian) in the Dunwet-Oronogo mining belt east of Joplin, Missouri. M, 1965, University of Missouri, Rolla.

Nichols, Robert Leslie. The mechanism of flow in basaltic lavas. D, 1940, Harvard University.

Nichols, Roland Franklin. Archeomagnetic study of Anasazi-related sediments of Chaco Canyon, New Mexico. M, 1975, University of Oklahoma. 111 p.

Nichols, Thomas Chester, Jr. Deformations associated with relaxation of residual stresses in the Barre Granite (Devonian) of Vermont. M, 1972, Texas A&M University.

Nichols, Thomas Chester, Jr. The state of stress in the Barre Granite of Vermont and other near-surface crystalline rocks of North America; paleo and present-day stress fields and the significance to engineering practice. D, 1980, University of Colorado. 208 p.

Nichols, William David. Petrology and structural geology of the Bennett Hill area, Clayville Quadrangle, Rhode Island. M, 1965, Syracuse University.

Nichols, William Marvin. A geologic section across the middle Salinas Valley of California, from the San Antonio River to San Lorenzo Creek. M, 1924, University of California, Berkeley. 37 p.

Nicholson, Craig. Seismicity of the New Madrid fault system; a microearthquake study in western Tennessee. M, 1981, St. Louis University.

Nicholson, Craig Claverie. Crustal structure and fault kinematics from the analysis of microearthquake data. D, 1986, Columbia University, Teachers College. 163 p.

Nicholson, Frank. The ore of the Cornwall copper mines and its treatment. M, 1880, Washington University.

Nicholson, Frank Herbert. Sedimentology of the Clinch Sandstone, northeastern Tennessee and southwestern Virginia. M, 1978, University of Kentucky. 65 p.

Nicholson, G. E. Geology of the upper portion of the Millet Range Quadrangle, Nevada. M, 1978, San Diego State University.

Nicholson, Jane E. Lithology and paleoecology of the Cottonwood Limestone (Early Permian) in southeastern Nebraska. M, 1977, University of Nebraska, Lincoln.

Nicholson, John Hirston. The areal geology of the lower Bull Creek area. M, 1947, University of Texas, Austin.

Nicholson, Philip David. I, Tidal synchronization of the rotation of early main sequence stars in close binaries; II, The rings of Uranus; results of the 1978 April 10 occultation; III, On the resonance theory of the rings of Uranus. D, 1979, California Institute of Technology. 205 p.

Nicholson, Ronald Vincent. Hydrochemical patterns and processes in the groundwater near an abandoned landfill; with emphasis on redox-sensitive species. M, 1980, University of Waterloo.

Nicholson, Ronald Vincent. Pyrite oxidation in carbonate-buffered systems; experimental kinetics and control by oxygen diffusion in a porous medium. D, 1984, University of Waterloo. 176 p.

Nicholson, Roy J. Long line computation in relation electronic surveying. M, 1961, Ohio State University.

Nicholson, Suzanne Warner. Petrography and geochemistry of the Illinois River gabbros, Kalmiopsis wilderness area, southwestern Oregon. M, 1981, University of Massachusetts. 137 p.

Nicholson, T. J. Geomorphology and hydrogeology of Wild Horse Valley, Foothills Park, Palo Alto, California. M, 1976, Stanford University. 106 p.

Nicholson, Toni Jost. Geology and the accumulation of hydrocarbons in the "Big Lime" and Borden Group (Mississippian) and pre-Chattanooga (Silurian-Devonian) of Knox, Laurel, and Whitley counties, Kentucky. M, 1983, University of Kentucky. 188 p.

Nicholson, Walter Allen. The analysis and fusibility of the ash from certain Washington coals. M, 1929, University of Washington. 16 p.

Nick, Kevin E. Depositional environments of the Miocene Esmeralda Formation, Stewart Basin, Stewart Valley, Nevada. M, 1983, Loma Linda University. 112 p.

Nickel, Brian K. The hydration and alteration of the perlite, pitchstone, and upper pyroclastic unit at Ruby Mountain, Nathrop, Colorado. M, 1987, Bowling Green State University. 84 p.

Nickel, Ernest H. The distribution of major and minor elements among co-existing pyrite and ferromagnesian silicates. D, 1953, University of Chicago. 63 p.

Nickel, Ernest Henry. The geology of the Bessemer magnetite deposit. M, 1951, McMaster University. 46 p.

Nickell, Clarence Oliver. The Coleman Junction horizon in Archer and Clay counties, Texas. M, 1932, University of Texas, Austin.

Nickell, Frank A. Geology of the Soledad Quadrangle, Central California. D, 1931, California Institute of Technology. 131 p.

Nickell, Frank A. Geology of the southwestern part of the Elizabeth Lake Quadrangle between San Francisquito and Bouquet canyons (California). M, 1928, California Institute of Technology. 29 p.

Nickelsen, B. H. Depositional environment of the Twilight Park Conglomerate. M, 1983, SUNY at Binghamton. 72 p.

Nickelsen, Richard P. The geology of northwestern Loudoun County, Virginia. D, 1953, The Johns Hopkins University.

Nickens, Dan Alan. Surface textures of quartz sand grains from Trail Ridge and adjacent areas. M, 1977, University of Florida. 100 p.

Nickerson, Theodore Dean. A late Eocene vertebrate fauna from Carlsbad, California. M, 1969, University of California, Riverside. 91 p.

Nickey, David Allen. Ostracodes from the Middle Ordovician Nealmont Limestone of central Pennsylvania. D, 1966, Pennsylvania State University, University Park. 174 p.

Nickias, Peter N. NMR studies and synthesis of Mo and W organometallic complexes. D, 1987, University of Kentucky. 197 p.

Nickle, Neil LeRoy. Geology of the southern part of the Buena Vista Hills, Churchill County, Nevada. M, 1968, University of California, Los Angeles.

Nicklin, Michael Earl. The hydrogeology of the regolith aquifer supplying the Iowa State University well field. M, 1974, Iowa State University of Science and Technology.

Nickmann, Rudy J. The palynology of Williams Lake fen, Spokane County, Washington. M, 1979, Eastern Washington University. 71 p.

Nicknish, John. Investigation of the basal ash of the Arikaree Formation in northern Shannon County, South Dakota. M, 1957, South Dakota School of Mines & Technology.

Nicks, A. D. Stochastic generation of hydrologic model inputs. D, 1975, University of Oklahoma. 142 p.

Nicks, Linda P. The study of the glacial stratigraphy and sedimentation of the Sheldon Point Moraine, Saint John, New Brunswick. M, 1988, Dalhousie University. 171 p.

Nicol, Alan Boswell. Development of an instrument to measure minute rotational movements in unstable slopes. M, 1969, University of Utah. 59 p.

Nicol, David. A brief study of some pelecypod genera of the Goodland Formation. M, 1939, Texas Christian University.

Nicol, David. Classification and evolution of the pelecypod family Glycymeridae. D, 1947, Stanford University. 180 p.

Nicol, David. Statistical analysis of new foraminiferal species of Elphidium from the west coast of North America. M, 1942, Stanford University. 43 p.

Nicol, Dorian L. The origin and evolution of sulfur in an Archean volcano-sedimentary basin, Deer Lake area, Minnesota. M, 1980, Indiana University, Bloomington. 53 p.

Nicolais, Stephen M. Geology of the South Ophir District, San Juan and San Miguel counties, Colorado. M, 1975, Colorado School of Mines. 174 p.

Nicolaou, Anthony. Errors and approximations in the seismic refraction method. M, 1958, University of Minnesota, Minneapolis.

Nicolaysen, Gerald. Geology of the Coaldale area, Fremont County, Colorado. M, 1972, Colorado School of Mines. 58 p.

Nicolaysen, Louis Otto. Age-determinations of African Precambrian minerals. D, 1954, Princeton University. 101 p.

Nicoletis, Serge. A comparative study of induced-polarization and electromagnetic effects for a grounded electric dipole on a half-space. M, 1981, Colorado School of Mines. 287 p.

Nicoll, G. A. Geology of the Hutton Lake Anticline area, Albany County, Wyoming. M, 1963, University of Wyoming. 80 p.

Nicoll, Larry D. Petrogenesis of the Highwood and Alnoite petrographic subprovinces of the Highwood Mountain area, Montana. M, 1974, University of Calgary. 123 p.

Nicoll, Robert Sherburne. Stratigraphy and conodont paleontology of the Brassfield Limestone (Lower Silurian) and Salamonie Formation (Middle Silurian) in southeastern Indiana and adjacent Kentucky. M, 1967, Indiana University, Bloomington. 86 p.

Nicoll, Robert Sherburne. Stratigraphy and conodont paleontology of the Sanders Group (Mississippian) in Indiana and adjacent Kentucky. D, 1971, University of Iowa. 88 p.

Niebauer, Timothy Michael. New absolute gravity instruments for physics and geophysics. D, 1987, University of Colorado. 166 p.

Nieber, J. L. Hillslope runoff characteristics. D, 1979, Cornell University. 275 p.

Niebuhr, Walter W. Paleoecology of the Eurekasprifer pinyoensis Zone, Eureka County, Nevada. M, 1974, Oregon State University. 151 p.

Niebuhr, Walter Ward, II. Biostratigraphy and paleoecology of the Guilmette Formation (Devonian) of eastern Nevada. D, 1980, University of California, Berkeley. 246 p.

Niederreither, Michael Scott. Sedimentation and hydrology of the Boca Chica Pool and other areas on Margarita Island, Venezuela. M, 1988, North Carolina State University. 104 p.

Niedoroda, Allen W. The generation of currents associated with transverse bars. D, 1972, Florida State University. 96 p.

Niehaus, James R. Investigation of amino acids in Precambrian rocks. M, 1969, University of Minnesota, Minneapolis. 75 p.

Nieland, Connie Lynn. Migration of chloride and sulfate within the peat of the Cedarburg Bog. M, 1988, University of Wisconsin-Milwaukee. 113 p.

Nielsen, Arne Rudolph. A microfaunal study of the Shaftesbury Formation. M, 1950, University of Alberta. 111 p.

Nielsen, David M. Environmental geology for land-use planning in Wood County, Ohio. M, 1977, Bowling Green State University. 96 p.

Nielsen, Dennis Niels. Quaternary geology of Sargent County, North Dakota. D, 1973, University of North Dakota. 85 p.

Nielsen, Dennis Niels. Washboard moraines in northeastern North Dakota. M, 1969, University of North Dakota. 51 p.

Nielsen, Eric Richard. A sedimentological study of the bottom sediments of southern Portage Lake (Michigan). M, 1973, Michigan Technological University. 78 p.

Nielsen, Erik. The composition and origin of Wisconsinan till in mainland Nova Scotia. D, 1976, Dalhousie University. 256 p.

Nielsen, Hans Peter. Geology, rainfall, and groundwater associated with several debris flows in Santa Cruz County, California. M, 1984, University of California, Santa Cruz.

Nielsen, K. C. Tectonic setting of the northern Okanagan Valley at Mara Lake, British Columbia. D, 1978, University of British Columbia.

Nielsen, Kent Christopher. Structural evolution of the Picuris Mountains, New Mexico. M, 1972, University of North Carolina, Chapel Hill. 47 p.

Nielsen, Mark Andrew. Depositional and diagenetic study of the Exline Limestone, Pleasanton Group, Upper Pennsylvanian (Missourian) of the northern Midcontinent. M, 1987, University of Iowa. 87 p.

Nielsen, Merrill Longhurst. Mississippian cephalopods from western Utah. M, 1950, University of Idaho. 39 p.

Nielsen, Mitchell Frederic. Some Late Mississippian pleurotomarian gastropods from Nevada and Utah. M, 1957, University of Nebraska, Lincoln.

Nielsen, Norman C. A cumulative hydrologic impact assessment of the Wyodak-Anderson coal seam in the southern Powder River basin, Wyoming. M, 1987, University of Colorado. 352 p.

Nielsen, Peter. Regional metamorphism and metamorphic isograds in the northwest lowlands of the Adirondacks. M, 1971, SUNY at Binghamton. 59 p.

Nielsen, Richard Leroy. Geology of the Pilot Mountains and vicinity, Mineral County, Nevada. D, 1964, University of California, Berkeley. 161 p.

Nielsen, Soren Bom. The continuous temperature log; method and applications. D, 1986, University of Western Ontario.

Nielsen, Thomas W. Stratigraphy and microfacies analysis of the Colina Limestone (Lower Permian), Hidalgo County, New Mexico. M, 1978, University of Texas at El Paso.

Nielson, Dennis Lon. Contact metamorphism and molecular diffusion at Ascutney Mountain, Vermont. M, 1972, Dartmouth College. 101 p.

Nielson, Dennis Lon. The structure and petrology of the Hillsboro Quadrangle, New Hampshire. D, 1974, Dartmouth College. 254 p.

Nielson, Dianne Ruth Gerber. A critical analysis of the potassium content-versus-depth relationship of andesitic lavas in subduction zones. M, 1972, Dartmouth College. 44 p.

Nielson, Dianne Ruth Gerber. Metamorphic diffusion in New Hampshire; soapstone bodies and flecky gneisses. D, 1974, Dartmouth College. 262 p.

Nielson, Dru R. Depositional environments and petrology of the Middle to Upper Jurassic Carmel Formation in the Gunlock area, Washington County, Utah. M, 1988, Brigham Young University. 215 p.

Nielson, Eric S. Sedimentary petrology, depositional environment, and diagenesis of the Middle and Upper Cambrian sequence, Two Mile Canyon, northern Malad Range, Southeast Idaho. M, 1983, University of Idaho. 96 p.

Nielson, Grant Leroy. Groundwater resources of the Blindman River valley. M, 1963, University of Alberta. 109 p.

Nielson, Grant Leroy. Hydrogeology of the irrigation study basin, Oldman river drainage, Alberta, Canada. D, 1970, University of Alberta.

Nielson, Jamie Adler. The surface processes, internal structure, and net deposits of eolian dunes and sand sheets. D, 1986, University of Texas, Austin. 224 p.

Nielson, Peter Alfred. Metamorphic petrology and mineralogy of the Arseno Lake area, Northwest Territories. D, 1977, University of Alberta. 233 p.

Nielson, R. LaRell. The geomorphic evolution of the Crater Hill volcanic field of Zion National Park. M, 1976, Brigham Young University. 70 p.

Nielson, Roger L. Pyroxene-melt equilibria. M, 1978, University of Arizona.

Nielson, Russell LaRell. Stratigraphy and depositional environments of the Toroweap and Kaibab formations, southwestern Utah. D, 1981, University of Utah. 1015 p.

Niem, Alan Randolph. Stratigraphy and origin of tuffs in the Stanley Group (Mississippian), Ouachita Mountains, Oklahoma and Arkansas. D, 1971, University of Wisconsin-Madison. 201 p.

Nieman, Timothy Lynn. Teleseismic mislocations of earthquakes in island arcs; theoretical results. M, 1985, Michigan State University. 54 p.

Nieman, William George. Petrochemistry and structural geology of the Fish Creek Mountains volcanic center, Lander County, Nevada. M, 1980, Washington State University. 83 p.

Niemann, James Cottier. Regional cementation associated with unconformity-sourced aquifers and burial fluids, Mississippian Newman Limestone, Kentucky. M, 1984, Virginia Polytechnic Institute and State University.

Niemann, Knut Olaf. DEM drainage as ancillary data to enhance digital Landsat classification accuracies. D, 1988, University of Alberta. 191 p.

Niemann, Nancy L. Upper mantle P-velocity structure between eastern North America and Hispaniola. M, 1986, Pennsylvania State University, University Park. 45 p.

Niemann, Robert Leslie. Occurrence of traces of strontium in some Wyoming bentonites. M, 1959, University of Illinois, Urbana.

Niemann, Robert Leslie. X-ray spectrochemical investigations of a Pennyslvanian underclay. D, 1961, University of Illinois, Urbana. 165 p.

Niemann, William L. Stratigraphy, depositional history, and diagenesis of the Lost City Limestone Member of the Hogshooter Formation (Missourian, Upper Pennsylvanian) in northeastern Oklahoma. M, 1986, University of Iowa. 192 p.

Niemi, Leslie Owen. Economics and geology of the world, United States of America and California manganese. M, 1978, [Stanford University].

Niemi, Tina Marie. Late Holocene paleoenvironmental history of the submerged ruins of Thronion, northern Euboean Gulf coastal plain, central Greece. M, 1988, Stanford University. 90 p.

Niemi, Warren Lee. The identification and stratigraphic correlation of basalt aquifers in the southern half of the Quincy Basin, Grant County, Washington, using borehole geophysics. M, 1981, University of Idaho. 105 p.

Niemitz, J. W. Tectonics and geochemical exploration for heavy metal deposits in the southern Gulf of California. D, 1978, University of Southern California.

Nienaber, James H. Shallow marine sediments offshore from the Brazos River, Texas. D, 1958, University of Texas, Austin. 192 p.

Nienaber, Wilfred. A laboratory analogue model and field station study of electromagnetic induction for an island situated near a continent. D, 1978, University of Victoria. 139 p.

Nienkerk, Monte M. Regional distribution of the major dissolved solids in the streams of Illinois. M, 1975, Northern Illinois University. 139 p.

Niermeier, V. D. Geology of Dinosaur Quarry 7 1/2' quadrangle, Uintah County, Utah. M, 1976, Fort Hays State University.

Niesen, Preston L. Stratigraphic relationships of the Florissant Lake Beds to the Thirtynine Mile Volcanic Field of central Colorado. M, 1970, New Mexico Institute of Mining and Technology. 65 p.

Nieset, Rev. C. F. Geological investigation of soil conditions in Jasper County, Indiana. M, 1928, Catholic University of America. 45 p.

Nieswand, George Heinz. The conjunctive use of surface and ground waters in the Mullica River basin, New Jersey; a chance constrained linear programming approach. D, 1970, Rutgers, The State University, New Brunswick. 221 p.

Nietert, Thomas Christian. A study of the lateral textural, mineralogical, and chemical variations in the Clarksville Shale in Indiana, Ohio, and Kentucky. M, 1963, Miami University (Ohio). 77 p.

Nieto-Antunez, Antonio. Tentative approach for design of pillars in veins. M, 1970, Stanford University.

Nieto-Pescetto, A. S. Experimental study of the shear stress-strain behavior of clay seams in rock masses. D, 1974, University of Illinois, Urbana. 207 p.

Nieto-Pescetto, Alberto Santiago. Some physical properties of a massive silt (loess). M, 1964, Washington University. 36 p.

Nieuwenhuis, Carol Ann. Alluvial history of Gypsum Canyon, southeastern Utah. M, 1978, University of California, Berkeley. 38 p.

Nieuwenhuise, Donald S. Van see Van Nieuwenhuise, Donald S.

Nieuwenhuise, Robert Van see Van Nieuwenhuise, Robert

Nieuwenhuyse, Ulrich Eric van see van Nieuwenhuyse, Ulrich Eric

Niever, Emanuel J. A study of type sediments and foraminifera of Long Island Sound (New York, Connecticut). M, 1957, New York University.

Niewendorp, Clark Alan. Possible role of petroliferous materials in sulfide precipitation at the Frank R. Millikan Mine, Southeast Missouri. M, 1987, Western Michigan University.

Niewoehner, Walter B. Devonian-Mississippian boundary formations, Missouri. M, 1955, University of Missouri, Columbia.

Niewold, Cary L. The rubidium-strontium age of the Rockville Granite and associated rocks of central Minnesota. M, 1973, Northern Illinois University.

Nightingale, William Thomas. Oil shales of Washington. M, 1924, University of Washington. 36 p.

Nigra, John O. A statistical study of the metapodial of the Dire Wolf from the Pleistocene of Rancho La Brea (California). M, 1946, California Institute of Technology. 39 p.

Nigrini, Andrew. Prediction of ionic fluxes in rock alteration processes at elevated temperatures. D, 1969, Northwestern University.

Nigrini, Andrew. Stratigraphy of the Stump Sandstone (Jurassic) of southeastern Idaho and southwestern Wyoming. M, 1967, Northwestern University.

Nigro, Danny Michael. Dissolved arsenic, tin, and antimony in thermal waters of southern California, and northern Baja California, Mexico. M, 1981, San Diego State University.

Niimony, Kunitaro. The magnetite deposits of Manchuria. D, 1924, Cornell University.

Niinomy, Kunitaro. Preliminary notes on magnetic deposits in Manchuria. M, 1923, Cornell University.

Niizeki, N. Structural studies of the mineral sulphosalts. D, 1957, Massachusetts Institute of Technology. 299 p.

Nijak, Walter F. Reconnaissance petrology of the Loon Lake Batholith, northeastern Washington. M, 1979, Eastern Washington University. 132 p.

Nik Wan, N. M. B. Classification criteria of Ultisols and Oxisols as applied to six Malayan soils. M, 1974, University of Guelph.

Nikhanj, Yashvir A. Petrology of granitic Echo-Pond complex, N.E. Vermont. M, 1970, Massachusetts Institute of Technology. 74 p.

Nikias, Peter A. Investigation of the wetting properties of the Bradford Sand. M, 1961, Stanford University.

Nikiforuk, Zan Frank. Lower Cretaceous microfauna from Bear Villa No. 1, Alberta. M, 1956, University of Alberta. 112 p.

Nikolaidis, Nikolaos P. Modeling the direct versus delayed response of surface waters to acid deposition in northeastern United States. D, 1987, University of Iowa. 288 p.

Nikolic, Slobodan. Metal distribution in the Strathcona nickel-copper deposit. M, 1979, Laurentian University, Sudbury. 52 p.

Nikols, Carol A. Geology of the Clyde Forks mercury antimony-copper deposit and surrounding area, Lanark County, Ontario. M, 1972, Queen's University. 116 p.

Nikravesh, Rashel. Microfauna of the type Allen Valley Shale, Upper Cretaceous, Sanpete County, Utah. M, 1963, Ohio State University.

Nikravesh, Rashel. The Foraminifera and paleoecology of the Blufftown Formation (Upper Cretaceous) of Georgia and eastern Alabama. D, 1967, Louisiana State University. 159 p.

Nili-Esfahani, Alireza. Investigation of Paleocene strata, Point Lobos, Monterey County, California. M, 1965, University of California, Los Angeles.

Nilsen, Tor Helge. Geology of the Animikean Pine River (Breakwater) Quartzite Conglomerate and the Keyes Lake Quartzite (Precambrian), Florence County, Wisconsin. M, 1964, University of Wisconsin-Madison.

Nilsen, Tor Helge. The relationship of sedimentation to tectonics in the Solund Devonian District of southwestern Norway. D, 1967, University of Wisconsin-Madison.

Nilson, Ariplinio Antonio. The nature of the Americano do Brasil mafic-ultramafic complex and associated sulfide mineralization, Goias, Brazil. D, 1981, University of Western Ontario. 460 p.

Nilsson, Harold Daniel. Coastal and submarine morphology of eastern Cape Cod Bay (Mass.). M, 1973, University of Massachusetts. 178 p.

Nilsson, Harold Daniel. Multiple longshore sand bars; environments of deposition and a model for their generation and maintenance. D, 1979, University of Massachusetts. 148 p.

Nilsson, Kristen. Plutonic and volcanic inclusions in rocks from the northern Mariana Island arc; crustal assimilation in an intraoceanic arc. M, 1987, Duke University. 110 p.

Nimick, David Acheson. Glacial geology of Lake Wenatchee and vicinity, Washington. M, 1977, University of Washington. 52 p.

Nimick, Karol Gillespie. Geology and structural evolution of the east flank of the Ladron Mountains, Socorro County, New Mexico. M, 1986, University of New Mexico. 98 p.

Nimickas, Bronius. Abrasion of granules in a spheregrinder. M, 1966, West Virginia University.

Nimpagaritse, Gérard. Pétrographie, minéralogie et géochimie de l'indice vanadifère de Mukanda, massif gabbroïque de Buhoro (Burundi). M, 1987, Ecole Polytechnique. 86 p.

Nimri, Faris Tawfiq. Hydrogeological application of the kriging technique to transmissivity data from the Gallup Sandstone aquifer, New Mexico. M, 1984, Wright State University. 98 p.

Nimsic, Thomas L. Control of montmorillonite swelling with electrolyte solutions. M, 1977, University of Nevada. 51 p.

Nine, Ogden Wells, Jr. A microfauna from the Upper Cretaceous Navesink Formation in New Jersey. D, 1954, Rutgers, The State University, New Brunswick. 259 p.

Nisbet, Bruce W. Structural studies in the northern Chester Dome of East-central Vermont. D, 1976, SUNY at Albany. 265 p.

Nishenko, Stuart Paul. Seismic hazards evaluation in interplate and intraplate environments. D, 1983, Columbia University, Teachers College. 233 p.

Nishida, Atsuhiro. World wide changes in geomagnetic field. D, 1962, University of British Columbia.

Nishihara, George Shikataro. The rate of reduction of acidity of descending waters by certain ore and gangue minerals, and its bearing upon secondary sulphide enrichment. M, 1914, University of Minnesota, Minneapolis. 19 p.

Nishimori, Richard K. The petrology and geochemistry of gabbros from the Peninsular Ranges Batholith, California, and a model for their origin. D, 1976, University of California, San Diego. 272 p.

Nishimura, Clyde Edwin. Velocity structure of the upper mantle in the Pacific determined by Love and Rayleigh wave dispersion data. D, 1986, Brown University. 174 p.

Nishimura, Katsuyoshi. Airphoto pattern study of the Erie lobe recessional moraines in Indiana. M, 1952, Purdue University.

Nishioka, Gail Keiko. Copper occurrences in stromatolites of the Copper Harbor Conglomerate, Keweenaw Peninsula, northern Michigan (USA). M, 1983, [University of Michigan].

Niskanen, Keith A. Petrology of the Maxville Limestone (upper Mississippian) of East-central Ohio and correlatives in Northeast Kentucky. M, 1975, Eastern Kentucky University. 60 p.

Niski, James T. A sedimentary core analysis of late Pleistocene to Recent sediments in a portion of Bellingham Bay, Washington. M, 1972, Western Washington University. 59 p.

Nissen, Arvid E. An investigation of argentiferous galena ores. M, 1914, University of Minnesota, Minneapolis. 9 p.

Nissen, Thomas C. Field and laboratory studies of selected periglacial wedge-polygons in southern Wyoming. M, 1985, University of Wyoming. 165 p.

Nissenbaum, Arie. Studies in the geochemistry of the Jordan River-Dead Sea system. D, 1969, University of California, Los Angeles. 305 p.

Nitchman, Steve P. Tectonic geomorphology and neotectonics of the San Luis Range, San Luis Obispo County, California. M, 1988, University of Nevada. 120 p.

Nitecki, Matthew. Systematic division of North American cyclocrinitids. D, 1968, University of Chicago. 182 p.

Nitsan, Uzi. Electronic structure and transport properties of dense silicates. D, 1973, Harvard University.

Nittrouer, Charles Albert. The fate of a fine-grained dredge spoils deposit in a tidal channel of Puget Sound, Washington. M, 1974, University of Washington.

Nittrouer, Charles Albert. The process of detrital sediment accumulation in a continental shelf environment; an examination of the Washington shelf. D, 1978, University of Washington. 243 p.

Nivargikar, Vasantrao R. The influence of soil structure on the shear strength characteristics of compacted kaolinite. D, 1970, North Carolina State University. 111 p.

Niven, David W. Determination of porosity and permeability of selected sandstone aquifers of South Dakota. M, 1967, South Dakota School of Mines & Technology.

Nivens, William. The distribution of several metals and cation exchange capacity in sediment fractions from an artificial and natural marsh, James River, Virginia. M, 1978, Old Dominion University. 150 p.

Nix, Joe Franklin. A neutron activation analysis of uranium in stone meteorites. D, 1966, University of Arkansas, Fayetteville. 79 p.

Nixon, Achilles Harry. Benedum Field of Upton and Reagan counties, Texas. M, 1951, Texas Tech University. 32 p.

Nixon, C. M. Petrographic and chemical study of a diabase dike in Gatineau Park, Quebec. M, 1976, University of Ottawa. 92 p.

Nixon, Edward Calvert. Geology of the Harper's Crossroads area, southwestern Chatham County, North Carolina. M, 1955, North Carolina State University. 37 p.

Nixon, Gail Alice. An analysis of geophysical anomalies in north-central Oklahoma and their relationship to the Midcontinent geophysical anomaly. M, 1988, University of Oklahoma. 118 p.

Nixon, Graham Tom. Contributions to the geology and petrology of the Trans-Mexican volcanic belt. D, 1986, University of British Columbia. 301 p.

Nixon, Graham Tom. Late Precambrian (Hadrynian) ash-flow tuffs and associated rocks of the Harbour Main Group near Colliers, Avalon Peninsula, S. E. Newfoundland. M, 1974, Memorial University of Newfoundland. 301 p.

Nixon, Kenneth Ray. Metamorphic reactions in the rocks of Craggy Gardens in North Carolina. M, 1978, Ohio State University.

Nixon, Robert Paul. Geology of the Hilliard Flat area, Uinta County, Wyoming. M, 1955, University of Utah. 50 p.

Nixon, Robert Paul. Petroleum source beds in the Cretaceous Mowry Shale of the northwestern Interior, United States. D, 1972, Brigham Young University. 24 p.

Nixon, Roy Arthur, III. Geomorphological effects of the June 9, 1972 flood on Victoria Creek, Black Hills, South Dakota. M, 1973, South Dakota School of Mines & Technology.

Nixon, Roy Arthur, III. Rates and mechanisms of chemical weathering in an organic environment at Panola Mountain, Georgia. D, 1981, Emory University. 182 p.

Niyogi, Dipankar. Slope stability in geomorphology. M, 1956, University of Illinois, Urbana.

Njus, I. J. An investigation of environmental factors affecting the near-bottom currents in the Monterey (California) submarine canyon. M, 1968, United States Naval Academy.

Nnaji, Soronadi. Simulation of stream pollution under stochastic loading. D, 1981, University of Arizona. 132 p.

Nnolim, Chude Austine. Geology, geochemistry and mineralization of the Faymar gold mine, Timmins, Ontario, Canada. M, 1983, Carleton University. 151 p.

Noack, Richard Eric. Sources of ground water recharging the principal alluvial aquifers in Las Vegas Valley, Nevada. M, 1988, University of Nevada, Las Vegas. 160 p.

Noah, Calvin G. Interpretation of surface gamma radiation with respect to subsurface petroleum deposits. M, 1956, Wichita State University. 51 p.

Noakes, John E. An electromagnetic method of geophysical prospecting for application to drill holes. D, 1951, University of Toronto.

Noakes, John Edward. Natural radiocarbon measurements by liquid scintillation counting. D, 1963, Texas A&M University. 144 p.

Nobes, David Charles. The magnetometric off-shore electrical sounding (Moses) method and its application in a survey of upper Jervis Inlet, British Columbia. D, 1984, University of Toronto.

Noble, Calvin Athelward. Effect of temperature on strength of soils. D, 1968, Iowa State University of Science and Technology. 139 p.

Noble, Clyde S. Investigation of the inert gas content of Hawaiian inclusions that exhibit anomalous ages. D, 1969, University of Hawaii. 115 p.

Noble, David Frederick. Origin of the expandable clay minerals in the twigs of Eocene age. M, 1962, Florida State University. 85 p.

Noble, Donald Charles. Mesozoic geology of the southern Pine Nut Range, Douglas County, Nevada. D, 1962, Stanford University. 251 p.

Noble, E. A. Water of compaction as an ore-forming fluid. D, 1961, University of Wyoming. 99 p.

Noble, Edwin A. Geology of the southern Ladron Mountains, Socorro County, New Mexico. M, 1950, University of New Mexico. 81 p.

Noble, Ian A. Magsat anomalies and crustal structure of the Churchill-Superior boundary zone. M, 1983, University of Manitoba.

Noble, James A. Geology of the Homestake gold mine. D, 1939, Harvard University.

Noble, James Eugene. A limited geologic reconnaissance of Clark and Lewis counties, Missouri. M, 1957, University of Missouri, Columbia.

Noble, James Peter Allison. A paleoecologic and paleontologic study of an Upper Devonian reef in the Miette area, Jasper National Park, Alberta, Canada. D, 1966, Case Western Reserve University. 312 p.

Noble, John H. The Petrology of the conglomerate zone of the Freda Formation of northern Michigan. M, 1965, Miami University (Ohio). 125 p.

Noble, Levi F. The geology of the Shinumo area, Grand Canyon, Arizona. D, 1909, Yale University.

Noble, Marlene Ann. Shelf circulation studies on the northeastern United States continental shelf. D, 1984, University of Rhode Island. 156 p.

Noble, Raymond Lee. Depositional and directional features of a braided-meandering stream. M, 1973, Oklahoma State University. 78 p.

Noble, Robert A. Geology of the Chriesman-Milano area, Burleson and Milam counties, Texas. M, 1956, Texas A&M University. 103 p.

Noble, S. R. Petrology and fluid inclusion study of W-Mo mineralization at the Logtung Deposit, south-central Yukon Territory. M, 1983, University of Toronto. 286 p.

Nobles, Laurence H. Glacial geology of the Mission Valley, western Montana. D, 1952, Harvard University. 125 p.

Nobles, Melvin A. Mineralogy and properties of typical Texas clays. D, 1946, University of Texas, Austin.

Noblett, J. B. Volcanic petrology of the Eocene Clarno Formation on the John Day River near Cherry Creek, Oregon. D, 1980, Stanford University. 209 p.

Nocita, B. W. Clay-size $CaCO_3$ and clay mineralogy of Recent marine sediments; southern California continental borderland. M, 1977, San Diego State University.

Nocita, Bruce William. Sedimentology and stratigraphy of the Fig Tree Group, west limb of the Onverwacht Anticline, Barberton greenstone belt, South Africa. D, 1986, Louisiana State University. 196 p.

Nock, Harvey. Geology of the Capon Bridge, West Virginia, Virginia Quadrangle. M, 1968, West Virginia University.

Nodeland, Steven K. Cenozoic tectonics of Cretaceous rocks in Northeast Sierra de Juarez, Chihuahua, Mexico. M, 1977, University of Texas at El Paso.

Noe, David Charles. Variations in shoreline sandstones from a Late Cretaceous interdeltaic embayment, Sego Sandstone (Campanian), northwestern Colorado. M, 1984, University of Texas, Austin. 127 p.

Noel, G. A. Copper bearing syenite of Omineca Batholith and its relation to the United States porphyry copper deposits. M, 1951, University of Toronto.

Noel, J. R. Paleomagnetism of late Precambrian metavolcanic rocks from the Carolina slate belt. M, 1986, University of Georgia.

Noel, James Arthur. Geology of the Beaver Mine, Thetford Mines, Quebec (Canada). M, 1951, Dartmouth College. 47 p.

Noel, James Arthur. The geology of the east end of the Anaconda Range and adjacent areas, Montana. D, 1956, Indiana University, Bloomington. 74 p.

Noel, Stephen D. Subsurface stratigraphy and water resources of Cass County, Indiana. M, 1978, Purdue University. 95 p.

Noffsinger, Kent Eugene. Statistical stacking of common-depth-point seismic data using the principle of maximum entropy. D, 1988, University of Wyoming. 316 p.

Nogami, H. H. An investigation of the absorption of alpha-rays emitted from thick mineral sources. M, 1947, Massachusetts Institute of Technology. 57 p.

Nogami, T. Soil-pile interaction under vibratory loading. D, 1977, University of Western Ontario.

Nogan, Donald Stanley. Micropaleontology, stratigraphy and paleoecology of the Aquia Formation of Maryland and Virginia. D, 1962, Rutgers, The State University, New Brunswick. 350 p.

Noggle, Karen Sue. Stratigraphy and structure of the Leavitt Reservoir Quadrangle, Bighorn County, Wyoming. M, 1986, Iowa State University of Science and Technology. 102 p.

Noguchi, Naohiko. Quantitative geomorphology and relative rate of erosion, Pescadero Creek basin (Tertiary), San Mateo County, California. M, 1972, University of California, Santa Cruz.

Nogueira, Alexandrino Cosme. Mineralogy and geochemistry of contact metasomatic iron deposits at Jones Camp, Socorro County, New Mexico. M, 1972, New Mexico Institute of Mining and Technology. 101 p.

Nogueira, Vicente de Paulo Queiroz. A mathematical model of progressive earth dam failure. D, 1984, Colorado State University. 149 p.

Noguera Urrea, Victor Hugo. Geology and diagenetic history of overpressured siliciclastic reservoirs in the lower Mississauga-Mic Mac formations of the Venture gas field, Scotian Shelf, Nova Scotia. M, 1987, Dalhousie University. 228 p.

Nogues, DeWitt Collier. Areal geology of the Nevill Quadrangle, Culberson County, Texas. M, 1950, University of Texas, Austin.

Nohara, Tomohide. Microfauna of the Upper Mississippian Great Blue Limestone near Morgan, Utah. M, 1966, University of Utah. 84 p.

Nokes, Charles Mormon, Jr. The igneous rocks of Utah. M, 1912, University of Utah. 43 p.

Nokleberg, Warren. Geology of the Strawberry Mine roof pendant, central Sierra Nevada (Madera county), California. D, 1970, University of California, Santa Barbara.

Nolan, Donny Ray. A study of sediments from Recent bar deposits of the Ouachita River. M, 1967, Northeast Louisiana University.

Nolan, Erich. An engineering geologic impact analysis of hydraulic dredging for lignite in Texas alluvial valleys. M, 1985, Texas A&M University. 141 p.

Nolan, Francis J. Heavy mineral analysis of the beach sands of Nova Scotia. M, 1963, Dalhousie University. 131 p.

Nolan, George W. A study of a seismic phase from 10 degrees to 30 degrees. M, 1954, Boston College.

Nolan, Grace Margaret. Sedimentation of the (Pennsylvanian) Oquirrh Formation, West Mountain. M, 1950, Brigham Young University. 39 p.

Nolan, K. Michael. Flood hazard mapping in the Bitterroot Valley, Montana. M, 1973, University of Montana. 56 p.

Nolan, Thomas B. Geology of the northwest portion of the Spring Mountains, Nevada. D, 1924, Yale University.

Noland, Anne Vinson. Revisions of selected Silurian arenaceous foraminifera from north-central Kentucky and southeastern Indiana. D, 1969, University of Louisville. 213 p.

Nold, John L. Geology of the northeastern border zone of the Idaho Batholith, Idaho and Montana. D, 1968, University of Montana. 189 p.

Nold, John Lloyd. Geology of basic dikes and sills in the southern Wind River Mountains, Wyoming. M, 1964, University of Missouri, Columbia.

Nolen-Hoeksema, Richard Clarence. A heat-flow investigation of the Lepontine Alps, in the Valle Maggia region, Ticino Canton, Switzerland. D, 1983, Yale University. 395 p.

Nolet, Daniel Arthur. Mossbauer study of temperature-dependent intervalence charge transfer in ilvaite. M, 1978, Massachusetts Institute of Technology. 84 p.

Nolet, Gilbert J. Benthic foraminiferal response to sapropel deposition in the late Quaternary eastern Mediterranean. M, 1987, Duke University. 157 p.

Nolf, Bruce Owen. Structure and stratigraphy of part of the northern Wallowa Mountains, Oregon. D, 1966, Princeton University. 193 p.

Noll, Charles Richard, Jr. Geology of southeastern Kay County, Oklahoma. M, 1955, University of Oklahoma. 88 p.

Noll, John H. Geology of the Picacho Colorado area, northern Sierra de Cobachi, central Sonora, Mexico. M, 1981, Northern Arizona University. 169 p.

Noll, Mark R. Geochemistry and petrogenesis of the alkaline lavas and their associated xenoliths, Mount Overload, northern Victoria Land, Antarctica. M, 1984, New Mexico Institute of Mining and Technology. 129 p.

Noll, Philip D., Jr. Geochemistry and tectonic setting of the 1700 Ma Alder and Red Rock groups from Tonto Basin, Arizona. M, 1988, New Mexico Institute of Mining and Technology. 206 p.

Nolley, Janis Mergele. Geology and hydrocarbon production, Sweetwater Field, Fisher and Nolan counties, Texas. M, 1987, Texas Christian University. 52 p.

Nollsch, David Allen. Diagenesis of Middle Creek and Bethany Falls limestones, Swope Formation, Upper Pennsylvanian (Missourian), Midcontinent North America. M, 1983, University of Iowa. 168 p.

Nolte, Clifton Jerry. Geology of subsurface formations of northern Logan County, Oklahoma. M, 1951, University of Oklahoma. 47 p.

Nolting, Richard M., III. Pennsylvanian-Permian stratigraphy and structural geology of the Orient-Cotton creek area, Sangre de Cristo mountains, Colorado. M, 1970, Colorado School of Mines. 102 p.

Nolting, Richard Massie, III. Absolute stress measurement in rock by overcoring cast-in-place epoxy inclusions. D, 1980, University of California, Berkeley. 153 p.

Nomland, Jorgen O. Relation of the invertebrate to the vertebrate faunal zones of the Etchegoin and Jacalitos in the Coalinga oil district, California. M, 1914, University of California, Berkeley. 15 p.

Nomland, Jorgen O. The Etchegoin Pliocene of middle California. D, 1916, University of California, Berkeley. 86 p.

Noon, Patrick L. Surface to subsurface stratigraphy of the Dakota Sandstone (Cretaceous) and adjacent units along the eastern flank of the San Juan Basin, New Mexico and Colorado. M, 1980, Bowling Green State University. 133 p.

Noonan, Albert F. A description of the Forest Vale meteorite, (about twelve miles northeast of Tullibigeal, New South Wales, Australia). M, 1968, University of Tennessee, Knoxville. 55 p.

Nooncaster, John R. Lithostratigraphy, petrology, and facies, Atoka Formation (Pennsylvanian), western frontal Ouachita Mountains, southeastern Oklahoma. M, 1985, University of Arkansas, Fayetteville.

Nooner, Daryl Wilburn. Alkanes in meteorite and terrestrial samples. D, 1966, [University of Houston]. 357 p.

Noor, Iqbal. An application of differential strain analysis to study of fractures in reservoir sandstones. M, 1983, University of Toronto.

Noort, Peter John van *see* van Noort, Peter John

Norbeck, Peter M. Water table configuration and aquifer and tailings distribution, Coeur d'Alene Valley, Idaho. M, 1974, University of Idaho. 97 p.

Norbisrath, Nans. The geology of the Mount Vernon area. M, 1939, University of Washington. 28 p.

Norburn, Martha Elizabeth. The influence of the physiographic features of western North Carolina on the settlement and development of the region. D, 1932, University of North Carolina, Chapel Hill. 110 p.

Norby, John W. Geology and geochemistry of Precambrian amphibolites and associated gold mineralization, Tinton District, Lawrence County, South Dakota, and Crook County, Wyoming. M, 1984, South Dakota School of Mines & Technology.

Norby, Philip Arthur. A subsurface interpretation using three-dimensional seismic methods of a portion of the Erawan gas/condensate field, Gulf of Thailand. M, 1983, California State University, Northridge. 90 p.

Norby, Rodney D. Conodont apparatuses from Chesterian (Mississippian) strata of Montana and Illinois. D, 1976, University of Illinois, Urbana. 305 p.

Norby, Rodney Dale. Conodont biostratigraphy of the Mississippian rocks of southeastern Arizona. M, 1971, Arizona State University. 195 p.

Nord, Gordon L., Jr. Imbricate thrusting in the Illinois Peak area, Shoshone County, Idaho. M, 1967, University of Idaho. 95 p.

Nord, Gordon Ludwig. The origin of the Boehl's Butte Anorthosite and related rocks, Shoshone County, Idaho. D, 1973, University of California, Berkeley. 159 p.

Nordby, George Roy. Time rate considerations in consolidation of Palouse Loess. M, 1966, Washington State University. 85 p.

Nordeck, Robert E. Geology of the Rivesville, West Virginia, 7 1/2 minute Quadrangle. 1967, West Virginia University.

Nordeng, Stephan C. The internal structure of some Pennsylvanian and Permian crinoid stems. D, 1954, University of Wisconsin-Madison.

Nordeng, Stephan H. A preliminary study of the relationship between pore geometry and mass transport in porous media using fractal geometry and digital image analysis. M, 1988, Michigan Technological University. 80 p.

Nordeng, Stephen Carl. Occurrence and paleoecology of stromatolites in the Oneota Dolomite of Wisconsin. M, 1951, University of Wisconsin-Madison. 41 p.

Nordin, Carl F., Jr. Statistical properties of dune profiles. D, 1968, Colorado State University. 152 p.

Nordlie, Bert E. The composition of the basaltic gas phrase. D, 1967, University of Chicago. 146 p.

Nordlie, Bert Edward. Contact metamorphism of limestone and dolomite at three Colorado areas. M, 1965, University of Colorado.

Nordquist, Gregg Anson. A study of the low frequency magnetotelluric response of a conductive outcropping ellipsoidal body. M, 1984, University of Utah. 73 p.

Nordquist, Ronald W. Origin, development, and facies of a young hurricane washover fan on southern Saint Joseph Island, central Texas Coast. M, 1972, University of Texas, Austin.

Nordstog, Kim Thomas. Petrology and petroleum reservoir potential of the upper Tensleep Sandstone (Pennsylvanian age), Rawlins Uplift/Ferris Mountains area, Carbon County, Wyoming. M, 1982, University of Colorado. 93 p.

Nordstrom, Charles E. A stratigraphic and petrographic study of an unnamed conglomerate unit (late Cretaceous) near Rancho Santa Fe, California. M, 1967, University of San Diego.

Nordstrom, D. K. Hydrogeochemical and microbiological factors affecting the heavy metal chemistry of an acid mine drainage system. D, 1977, Stanford University. 230 p.

Nordstrom, Harold Edward. Mesoscopic and macroscopic structural analysis of tectonites in the Northport district, Washington. M, 1972, Washington State University. 77 p.

Nordstrom, John Eric. An algorithm for contouring geologic spatial data with known discontinuities. M, 1984, Wichita State University. 140 p.

Nordstrom, Karl Fredrik. Beach response rates to cyclic wave regimes at Sandy Hook, New Jersey. D, 1975, Rutgers, The State University, New Brunswick. 174 p.

Nordstrom, Paul M. Trace element distribution in gold deposits of the North Moccasin mining district, Fergus County, Montana. M, 1985, Eastern Washington University. 131 p.

Norford, Brian Seeley. Paleozoic stratigraphy and paleontology of the Turnagain River map area, northern British Columbia. D, 1959, Yale University.

Nork, Diane M. The analysis of water level fluctuations in a shallow, unconfined aquifer in Owens Valley, California. M, 1987, University of Nevada. 61 p.

Nork, William Edward. The occurrence and use of groundwater in a sampling of wells in Niagara County, New York. M, 1962, SUNY at Buffalo.

Norland, William D. Thermal maturation of the Mesilla Valley Shale (late Albian) on the north and east flanks of the Cerro de Cristo Rey Pluton, Dona Ana County, New Mexico. M, 1986, University of Texas at El Paso.

Norling, Donald Leonard. Geology of Morgan County, Ohio. D, 1957, Ohio State University.

Norman, Carl Edgar. Microfractures in brittle rocks; their relationship to larger scale structural features and existing ground stresses. D, 1967, Ohio State University.

Norman, Carl Edgar. Stratigraphy and petrology of the upper Eden strata in the Ohio Valley. M, 1959, Ohio State University.

Norman, Charles Darrel. Petrology of the Monteagle Limestone (Upper Mississippian), Northeast central Tennessee. M, 1981, Memphis State University. 106 p.

Norman, David Irwin. Geology and geochemistry of Tribag Mine, Batchawana Bay, Ontario. D, 1977, University of Minnesota, Minneapolis. 269 p.

Norman, Elizabeth A. S. The structure and petrology of the Summit Valley area, Klamath Mountains, California. M, 1984, University of Utah. 148 p.

Norman, Emmerson Kirkpatrick. Petrography of some Pennsylvanian underclay carbonate beds in Illinois. M, 1959, University of Illinois, Urbana.

Norman, Franklin John. The iron content of sphalerites from the Manitouwadge area, Ontario (Canada). M, 1968, University of Toronto.

Norman, George W. H. The geology of the Lake Ainslie Quadrangle, Inverness County, Cape Breton, Nova Scotia. D, 1929, Princeton University. 113 p.

Norman, Linda S. The provenance of mud in Old Woman Creek Estuary, Erie County, Ohio. M, 1987, Bowling Green State University. 95 p.

Norman, Lonnie Dale. Subsurface Pleistocene stratigraphy of the Barrington area and its relationship to planning. M, 1974, Northern Illinois University. 37 p.

Norman, Lonnie Dale. The grain size distribution of tailings and other solids in the Bunker Hill central improvement area and its relationship to the occurrence and control of leakage and seepage. M, 1977, University of Idaho. 74 p.

Norman, Marc D. Glass particles in the Luna 24 core and magma evolution at Mare Crisium. M, 1979, University of Tennessee, Knoxville. 58 p.

Norman, Marc Douglas. Geology, geochemistry, and tectonic implications of the Salmon Creek volcanic sequence, Owyhee Mountains, Idaho. D, 1987, Rice University. 247 p.

Norman, Mark. An analysis of the shallow seismic refraction method in its application to two minor geologic problems. M, 1968, Florida State University.

Norman, Mark Daniel. Structural analysis of the Precambrian rocks of the Long Canyon-Gold Hill area, Taos Range, northern New Mexico. M, 1984, University of Texas at Dallas. 81 p.

Norman, Ryburn E. Geology and petrochemistry of ophiolitic rocks of the Baie Verte Group exposed at Ming's Bight, Newfoundland. M, 1973, Memorial University of Newfoundland. 123 p.

Norman, William Robert. Nitrate levels in the groundwater of Kalamazoo County, Michigan. M, 1982, Western Michigan University. 99 p.

Norman-Gregory, Gillian Margaret. Volume change behavior of granular materials subjected to vibration. D, 1986, University of Massachusetts. 315 p.

Normand, David Ernest. Precambrian geology of the northern Sangre de Cristo Range, Chaffee, Fremont, and Saguache counties, Colorado. D, 1972, Texas Tech University. 88 p.

Normand, David Ernest. The geology of the Whitehorn Stock area, Chaffee, Fremont and Park counties, Colorado. M, 1968, Texas Tech University. 61 p.

Normand, Diane Lynn. Petrography and diagenetic history of the Alma Sandstone, upper Atoka Formation, in the Arkoma Basin of Arkansas. M, , University of Arkansas, Fayetteville.

Normand, Dianna Lynn. Petrography and diagenetic history of the Alma Sandstone, upper Atoka Formation, in the Arkoma Basin of Arkansas. M, 19??, University of Arkansas, Fayetteville.

Normark, William R. Growth patterns of deep-sea fans. D, 1969, University of California, San Diego.

Normark, William Raymond. Development of a portable short-base mercury-level tiltmeter. M, 1965, Stanford University.

Norrany, Iraj. Study of the shear strength characteristics of undisturbed saturated clays. D, 1964, University of California, Berkeley. 222 p.

Norris, Arnold Willy. A study of the genus Atrypa of Western Canada. D, 1953, University of Toronto.

Norris, Arnold Willy. Some cutinized microfossils from Western Canada. M, 1951, University of Alberta. 182 p.

Norris, Cynthia R. Buried karst and geology in north-central Ohio. M, 1982, Kent State University, Kent. 41 p.

Norris, Donald Kring. Structural conditions and violent stress relief in coal mines in the southern Canadian Cordillera. D, 1953, California Institute of Technology. 135 p.

Norris, Earl G. Conodonts and biostratigraphy of the Brazer Dolomite. M, 1979, Brigham Young University.

Norris, Gary Martin. The drained shear strength of uniform quartz sand as related to particle size and natural variation in particle shape and surface roughness. D, 1977, University of California, Berkeley. 523 p.

Norris, James Richard. Fracturing, alteration, and mineralization in oxide pit, Silver Bell Mine, Pima County, Arizona. M, 1981, University of Arizona. 72 p.

Norris, Janice G. Layer charge magnitude and homogeneity and their relationship to the thixotropic properties of bentonites. M, 1987, SUNY at Buffalo. 111 p.

Norris, John W. Seismic characteristics of coal-mine rockbursts in Utah. M, 1967, University of Utah. 79 p.

Norris, Malcolm Stewart. Geology of the Nifty Carbonate Member, Broadhurst Formation, Paterson Province, Western Australia. M, 1987, University of Western Ontario. 295 p.

Norris, Marc J. Depositional environments of the upper Ashlock Formation, Upper Ordovician of east-central Kentucky. M, 1979, Eastern Kentucky University. 44 p.

Norris, Mary Lillian. The geology of Southern California. M, 1939, George Washington University.

Norris, Paula Jean. Metamorphism at the base of the Trinity Peridotite, Coffee Creek and southern Trinity Alps areas, Klamath Mountains, Northern California. M, 1983, University of California, Los Angeles. 109 p.

Norris, Robert M. Marine geology of the San Nicholas Island region, California. D, 1951, University of California, Los Angeles. 124 p.

Norris, Robert N. The geology of a portion of the Santa Ynez Range, Santa Barbara County, California. M, 1949, University of California, Los Angeles.

Norris, Robert Peter. The origin and sedimentation of Wilson Canyon, Caddo County, Oklahoma. M, 1951, University of Kansas.

Norris, T. L. Total nitrogen content of deep sea basalts with implications for the origin of the Earth's atmosphere. D, 1979, SUNY at Stony Brook. 141 p.

Norris, Will Victor. The oil shale industry. M, 1920, Texas Christian University. 164 p.

Norrish, Winston A. The effects of organisms on sediment deposition in St. Joseph Bay, Florida. M, 1985, University of Cincinnati. 137 p.

Nortey, Peter Alphonsus. Control of non-point sources of water pollution within an ecological framework; the case of the Tri-County region, Michigan. D, 1976, Michigan State University. 187 p.

North, Beatrice Ruth. Foraminifera from cretaceous Bearpaw formation in southern Saskatchewan. M, 1961, University of Saskatchewan. 72 p.

North, Jon W. The stratigraphy, structure, geochemistry, and metallogeny of the Moran Lake Group Central Mineral Belt, Labrador. M, 1988, Memorial University of Newfoundland. 202 p.

North, Robert. Determination of textural signature and their relation to paleoenvironment for nine fluvial channel sequences from the Upper Cretaceous Mesaverde Group of Piceance Creek Basin, northwestern, Colorado. M, 1986, University of New Mexico. 165 p.

North, William Benjamin. Coastal landslides in northern Oregon. M, 1964, Oregon State University. 85 p.

North, William Gordon. Lower Devonian stratigraphy of Illinois interpreted from well data. M, 1965, University of Illinois, Urbana.

North, William Gordon. The stratigraphy of the formation at and beneath the middle-upper Devonian boundary in southern Illinois. D, 1969, University of Illinois, Urbana.

Northam, Mark Alexander. The organic geochemistry of lipids extracted from Orca Basin sediment. D, 1981, University of Texas, Austin. 129 p.

Northcote, Kenneth Eugene. Distribution of sulphur, iron, copper and zinc in modern marine sediments of Mud Bay, Crescent Beach, BC. M, 1961, University of British Columbia.

Northcote, Kenneth Eugene. Geology and geochronology of the Guichon Creek Batholith (Early Jurassic), British Columbia. D, 1968, University of British Columbia.

Northrop, Harold R. Oxygen isotopic analysis of silicates and oxides using both bromine pentafluoride and fluorine gas as reagents. M, 1979, University of New Mexico. 63 p.

Northrop, Harold Roy. Origin of the tabular-type vanadium-uranium deposits in the Henry structural basin, Utah. D, 1982, Colorado School of Mines. 340 p.

Northrop, John. An investigation of the relation between source characteristics and T phases in the North Pacific area. D, 1968, University of Hawaii. 134 p.

Northrop, Stuart. Geology of the Port Daniel-Gascons area, Quebec. D, 1929, Yale University.

Northrup, John. Ocean bottom photographs and mechanical analysis of Recent and Eocene sediments exposed off the eastern coast of North America. M, 1948, Columbia University, Teachers College.

Northrup, John I. A study of the histology of the stem of the wax, Hoya carnosa L. B. R. fossil leaves from Budgeton, New Jersey. D, 1888, Columbia University, Teachers College.

Northwood, Thomas D. Model seismology; propagation of an elastic pulse over the free surface of a solid. D, 1951, University of Toronto.

Norton, Annette H. The relationship between refractive index and chemical composition in certain selected rocks. M, 1969, American University. 40 p.

Norton, Arthur Randolph. Quaternary geology of the Itasca-St. Croix moraine interlobate area, north-central Minnesota. M, 1982, University of Minnesota, Duluth. 119 p.

Norton, Charles Warren. Foraminiferal distribution and paleogeography of the Brush Creek marine event (Missourian; Pennsylvanian) in the Appalachian Basin. D, 1975, University of Pittsburgh. 153 p.

Norton, David Lee. The economic geology of the petroleum and iron ore deposits south of Jacksonville, Texas. M, 1954, University of Oklahoma. 53 p.

Norton, Denis Locklin. Geological and geochemical investigations of stibnite deposits. D, 1964, University of California, Riverside. 116 p.

Norton, Dorita A. X-ray spectrographic analysis of cryolite. M, 1956, Columbia University, Teachers College.

Norton, Dorita Anne. A mineralogical and geochemical study of clinopyroxenes from southeastern Pennsylvania and Delaware. D, 1958, Bryn Mawr College. 97 p.

Norton, Hiram A., Jr. Trace element geochemistry of the Cape Ann Granite, eastern Massachusetts. M, 1974, University of Kentucky. 105 p.

Norton, James Austin. Petrology of the Lingle Limestone (Middle Devonian), Union County, Illinois. M, 1966, Southern Illinois University, Carbondale. 101 p.

Norton, James J. and Staatz, Mortimer H. The Precambrian geology of the Los Pinos Range, New Mexico. M, 1942, Northwestern University.

Norton, James Jennings. Geology of the Precambrian rocks of the Keystone pegmatite district, southern Black Hills, South Dakota. D, 1958, Columbia University, Teachers College. 235 p.

Norton, Lloyd Darrell. Loess distribution and pedogenesis of loess-derived soils in East-central Ohio. D, 1981, Ohio State University. 245 p.

Norton, Marc A. Hydrogeology and potential reclamation procedures for an uncontrolled mine waste deposition site, Kellogg, Idaho. M, 1980, University of Idaho. 138 p.

Norton, Margaret Marilyn. Guide fossils of the Middle Jurassic Tuxedni Sandstone, Alaska. M, 1946, University of Michigan.

Norton, Matthew Frank. Mineralogy of the Puerto Rico Trench, an environmental study. D, 1958, Columbia University, Teachers College. 128 p.

Norton, Norman James. Palynology of the Upper Cretaceous and lower Tertiary in the type locality of the Hell Creek Formation. D, 1963, University of Minnesota, Minneapolis. 176 p.

Norton, Richard Drake. Ecological relations of some smaller foraminifera; the deposition of the Byram calcareous marl at Vicksburg and Brandon, Mississippi. M, 1927, University of California, Berkeley. 64 p.

Norton, Stephen Allen. Geology of the Windsor Quadrangle, Massachusetts. D, 1967, Harvard University.

Norton, Willard Eugene. Middle Ordovician graptolites of Greene County, Tennessee. M, 1983, University of Tennessee, Knoxville. 136 p.

Norville, Charles R. Petrographic analysis and provenance determination of sand layers from DSDP forearc and backarc sites in the Aleutian area, northern Pacific. M, 1984, University of Missouri, Columbia. 152 p.

Norwick, Stephen A. The potash feldspars of the Kinsman quartz monzonite New Hampshire. M, 1967, Dartmouth College. 77 p.

Norwick, Stephen Allan. The regional Precambrian metamorphic facies of the Prichard Formation of western Montana and northern Idaho. D, 1972, University of Montana. 129 p.

Norwood, Daniel Lee. Aqueous halogenation of aquatic humic material; a structural study. D, 1985, University of North Carolina, Chapel Hill. 241 p.

Norwood, Edward M., Jr. Geology of the Evitts Creek 7 1/2 minute topographic quadrangle (Maryland, Pennsylvania, West Virginia). M, 1957, West Virginia University.

Nosker, Richard Ernest. Stratigraphy, structure, geophysics, and water chemistry of the Jersey Valley area Pershing and Lander counties, Nevada. M, 1981, University of Nevada. 88 p.

Nosker, Sue Anderson. Stratigraphy and structure of the Sou Hills, Pershing County, Nevada. M, 1981, University of Nevada. 60 p.

Noson, Linda Jeanne. A paleomagnetic study of three granitic plutons exposed in the Cascade Mountains, Washington. M, 1973, Western Washington University. 53 p.

Nosow, Edmund. A faunal study of the McMillan Formation in the vicinity of Shelbyville, Shelby County, Kentucky. M, 1951, University of Kentucky. 134 p.

Nossaman, Leslie Norene. Deposition and diagenesis of the Hunton Group, Mathers Ranch Field, Hemphill County, Texas. M, 1981, Texas Tech University. 91 p.

Nosseir, Mostafa Kamel. Monitoring erosion features affected by land use from remotely sensed data (1938-1976). D, 1980, Ohio State University. 178 p.

Nostrand, Amy K. Van see Van Nostrand, Amy K.

Nostrand, Timothy Stuart van see van Nostrand, Timothy Stuart

Notestein, Frank B. Some chemical experiments bearing on the origin of certain uranium-vanadium ores. M, 1918, University of Minnesota, Minneapolis.

Notley, Donald Frances. Geology of the west end of the Tensleep Fault, Wyoming. M, 1947, University of Michigan.

Notley, Keith Roger. Analysis of the Springhill mine disaster (October 23, 1958). D, 1980, Queen's University.

Nott, Jerry Alan. Pleistocene stratigraphy and landscape evolution in the Long and Crooked Creek drainage basins of Southeast Iowa. M, 1981, University of Iowa. 227 p.

Nottingham, Larry Curtis. Use of quasi-static friction cone penetrometer data to predict load capacity of displacement piles. D, 1975, University of Florida. 553 p.

Nottingham, Marsh Whitney. Pennsylvanian stratigraphy of the Canon City Embayment, Fremont County, Colorado. M, 1957, University of Colorado.

Nour-el-Din, Mohamed Mohamed. A finite element model for salinity management in irrigated soils. D, 1986, University of California, Davis. 241 p.

Nourbehecht, Bijan. Irreversible thermodynamic effects in inhomogeneous media and their application in certain geoelectric problems. D, 1963, Massachusetts Institute of Technology. 142 p.

Nourse, Susan Marie. A laboratory investigation of mass transfer from hydrocarbons to water in porous media. M, 1986, University of Minnesota, Minneapolis. 166 p.

Novacek, Michael John. Evolution and relationships of the Leptictidae (Eutheria; Mammalia). D, 1978, University of California, Berkeley. 302 p.

Novak, Gary A. Petrography and mineralogy of a "Rapakivi" quartz monzonite pluton, Eagle Mountain Quadrangle, California. M, 1967, Pennsylvania State University, University Park. 69 p.

Novak, Gary Alan. The crystal chemistry of the silicate garnets. D, 1971, Virginia Polytechnic Institute and State University. 78 p.

Novak, Irwin Daniel. The origin of the beach ridges in Fort Clinch State Park, Florida. M, 1968, University of Florida. 51 p.

Novak, Irwin Daniel. The origin, distribution, and transport of gravel on Broad Cove beach, Appledore island, Maine. D, 1971, Cornell University. 170 p.

Novak, James Michael. Metamorphic petrology, mineral equilibria, and polymetamorphism in the Augusta Quadrangle, South-central Maine. M, 1978, Southern Methodist University. 76 p.

Novak, Mark Thomas. Sedimentological effects of bioturbation in deep-sea calcareous ooze. M, 1980, University of Utah. 97 p.

Novak, Michael David. The moisture and thermal regimes of a bare soil in the lower Fraser Valley during spring. D, 1981, University of British Columbia.

Novak, Robert James. The mineralogy and elemental composition of different aged soils formed on the Martinsburg Formation. D, 1970, Rutgers, The State University, New Brunswick. 369 p.

Novak, Robert M. Origin and distribution of selenium in the upper Cretaceous Niobrara and Pierre formations, northeastern North Dakota. M, 1971, University of North Dakota. 49 p.

Novak, S. M. d'O. A study of nitrogen and phosphorus in a region of the upper Ohio River Basin. D, 1974, University of Pittsburgh. 143 p.

Novak, Stephanie Anne. A methodology for identification of aquifer contamination by gas/oil brines. M, 1986, Kent State University, Kent. 91 p.

Novak, Stephen W. Contact metamorphism of the Lucerne Pluton, Hancock County, Maine. M, 1979, Virginia Polytechnic Institute and State University.

Novak, Steven William. Geology and geochemical evolution of the Kane Springs Wash volcanic center, Lincoln County, Nevada. D, 1985, Stanford University. 220 p.

Novakovic, B. The scale of groundwater flow systems in Big Otter and Big Creek drainage basins, southern Ontario. M, 1973, University of Waterloo.

Novakovich, Bruce D. Analysis of photolinear elements and structure of Elk Mountain, Carbon County, Wyoming. M, 1987, University of Wyoming. 85 p.

Novakowski, Kentner Stephen. Field investigations of the capillary-fringe effect on water-table response. M, 1982, University of Waterloo. 52 p.

Novelli, Paul C. The biogeochemistry of molecular hydrogen in sulfate-reducing sediments. D, 1987, SUNY at Stony Brook. 253 p.

Novich, Bruce Eric. The composition and rheology of the Florida phosphatic waste clay slurries; geotechnical implications. D, 1983, Massachusetts Institute of Technology. 270 p.

Novick, Jonathan S. Metamorphism of the Weeks Limestone, Notch Peak, Utah; the low grade reactions and fluid inclusions. M, 1988, University of Tennessee, Knoxville. 102 p.

Novillo, Mary Muscarella. Petrological evolution of the Ruby Mountain garnetiferous gneiss, southeastern Adirondack Mountains, New York. M, 1981, Lehigh University.

Novitsky-Evans, Joyce Marie. Geology of the Cowhole Mountains, southeastern California; structural, stratigraphic, and geochemical studies. D, 1978, Rice University. 156 p.

Novitsky-Evans, Joyce Marie. Petrochemical study of the Clarno Group; Eocene-Oligocene continental margin volcanism of North-central Oregon. M, 1974, Rice University. 96 p.

Novotny, James R. The geology of the northeastern Shadow Mountains, western San Bernardino County, California. M, 1955, University of California, Los Angeles.

Novotny, Robert F. Bedrock geology of the Dover-Exeter region, New Hampshire. D, 1963, Ohio State University. 220 p.

Novotny, Robert F. Structure and sedimentation of the (Pennsylvanian) Richburg Sands in part of Wirt Township, Allegany County, New York. M, 1952, Syracuse University.

Novotny, Robert T. A gravity and magnetic study of Antelope Island, Great Salt Lake, Davis County, Utah. M, 1958, University of Utah. 24 p.

Nowack, Robert L. Wave propagation in laterally varying media and iterative inversion for velocity. D, 1985, Massachusetts Institute of Technology. 255 p.

Nowak, Frank John. Geology of the Baldy Mount Norite, Albany County, Wyoming. M, 1970, University of Illinois, Urbana. 50 p.

Nowak, Frank John. Microfacies of the upper Bird Spring Group (Pennsylvanian-Permian), Arrow Canyon Range, Clark County, Nevada. D, 1972, University of Illinois, Urbana. 73 p.

Nowak, Gregory. The geology and uranium occurrences of the Washington mining district, Lyon County, Nevada. M, 1979, University of Nevada. 92 p.

Nowak, Michael. Archeological dating by means of volcanic ash strata. D, 1968, University of Oregon. 210 p.

Nowak, Robert Lars. Petrology of some lower Cretaceous sandstones of the northern Richardson Mountains near Stony Creek, Northwest Territories. M, 1971, University of Alberta. 75 p.

Nowak, Ronald P. Clay mineralogy of pre-Coldwater (Mississippian) argillaceous sediments in the State-Foster number 1 well, Ogemaw County, Michigan. M, 1978, Michigan State University. 66 p.

Nowatzki, Edward Alexander. Fabric changes accompanying shear strains in a cohesive soil. D, 1966, University of Arizona. 101 p.

Nowell, A. R. M. Turbulence in open channels; an experimental study of turbulence structure over boundaries of differing hydrodynamic roughness. D, 1975, University of British Columbia.

Nowell, William Benjamin. The petrology and alteration of the Kirwin mineralized area, Park County, Wyoming. M, 1971, University of Montana. 72 p.

Nowicki, V. An investigation of the Kitchener aquifer system using the stable isotopes ^{34}S and ^{18}O. M, 1976, University of Waterloo.

Nowina-Zlotnicki, Stephan F. Petrofabric and dielectric anisotropy in rock. M, 1979, University of Toronto.

Nowlan, G. A. Genesis of manganese-iron oxides in stream sediments of Maine. D, 1976, University of Colorado. 365 p.

Nowlan, G. S. Late Cambrian to Late Ordovician conodont evolution and biostratigraphy of the Franklinian Miogeosyncline, eastern Canadian Arctic Islands. D, 1976, University of Waterloo.

Nowlan, Godfrey Shackleton. Conodonts from the Cow Head Group, western Newfoundland. M, 1973, Memorial University of Newfoundland. 183 p.

Nowlan, James P. The Silurian stratigraphy of the Niagaran Escarpment in Ontario. D, 1935, University of Toronto.

Nowroozi, Ali Asghar. Terrestrial eigenvibrations following the Great Alaskan Earthquake, March, 1964. D, 1964, University of California, Berkeley. 119 p.

Noyer, John Milford De see De Noyer, John Milford

Noyes, Alvin Peter, Jr. Geology of Purgatory Creek area, Hays and Comal counties, Texas. M, 1957, University of Texas, Austin.

Noyes, Harold James. Petrogenesis of Sierran plutons; a petrologic and geochemical investigation into the origin and differentiation of granodioritic plutons of the central Sierra Nevada Batholith, California. D, 1978, Massachusetts Institute of Technology. 325 p.

Noyes, J. E. An evaluation of hydraulic conductivity tests and data. M, 1975, University of Vermont.

Nozdryn-Plotinicki, Michael John. On the short-term forecasting of spring floods in real time. D, 1980, Queen's University.

Nozette, Stewart David. The physical and chemical properties of the surface of Venus. D, 1983, Massachusetts Institute of Technology. 188 p.

Nriagu, Jerome O. Distribution of sulfur and iron in Lake Mendota sediments (Upper Cambrian) (near

Madison, Wisconsin). M, 1967, University of Wisconsin-Madison.

Nriagu, Jerome Okonkwo. Solubility of galena under hydrothermal conditions. D, 1970, University of Toronto.

Ntiamoah-Adjaquah, R. J. Obuasi gold deposits of Ghana; their genesis in light of comparison to selected deposits. M, 1974, University of Western Ontario. 180 p.

Ntiamoah-Agyakwa, Yaw. Geology, hydrothermal mineralization, and geochemical exploration; New York Mountains and northern mid hills areas, San Bernardino County, California. D, 1987, University of California, Los Angeles. 377 p.

Ntokotha, Enock Mangwiyo. Properties and classification of "Mopanosols" occurring in the upper Shire Valley, Malawi. D, 1984, University of Florida. 157 p.

Nualchawee, K. Spatial land cover inventory, modeling, and projection, northern Thailand. D, 1979, Colorado State University. 282 p.

Nuccio, R. M. Sedimentology of the Beechers Bay Formation, Santa Rosa Island, California. M, 1977, San Diego State University.

Nuckels, Clarence Edward. Geothermal studies in northwestern Mexico. M, 1976, University of Florida. 117 p.

Nuckels, Mark Gordon. Diagenesis of West-central Florida chert. M, 1981, University of South Florida, Tampa. 123 p.

Nuckolls, Helen Marie. Geology of the Bootstrap Mine, Nevada; a sediment-hosted disseminated gold deposit. M, 1985, Stanford University. 62 p.

Nuckols, John Robert. The influence of atmospheric nitrogen influx upon the stream nitrogen profile of two relatively undisturbed forrested watersheds in the Cumberland Plateau of the Eastern United States. D, 1982, University of Kentucky. 283 p.

Nuelle, Laurence M. Geology and mineralization, Ohio and Mt. Baldy districts, Marysvale, Piute County, Utah. M, 1979, University of Missouri, Rolla.

Nufer, Janet Ann. Criteria for the recognition of depositional environments ranging from littoral to abyssal in the Cretaceous of Texas. M, 1979, Texas A&M University. 167 p.

Nuffield, Edward Wilfrid. A study of some rare ore minerals. D, 1944, University of Toronto.

Nugent, Lawrence E., Jr. Structure and stratigraphy of the Kern River salient (California). D, 1941, Cornell University.

Nugent, R. Michael. A Galerkin finite element model for transient, saturated-unsaturated flow in a hillslope. D, 1987, University of Missouri, Columbia. 332 p.

Nugent, Robert Charles. Geology of the Elmira, Waverly and Owego quadrangles, New York. M, 1960, University of Rochester. 96 p.

Nugent, Robert Charles. Jointing in a Quaternary basalt, Buckboard Mesa (Nye County), Nevada, and its effect on cratering experiments. D, 1967, Northwestern University. 265 p.

Nuhfer, Edward. Efflorescent minerals associated with coal. M, 1967, West Virginia University.

Nuhfer, Edward B. Temporal and lateral variations in the geochemistry, mineralogy, and microscopy of seston collected in automated samplers (from selected lakes in Ohio, Pennsylvania, and New Mexico). D, 1979, University of New Mexico. 390 p.

Numbere, D. T. Correlations for the physical properties of petroleum reservoir brines. M, 1977, Stanford University. 63 p.

Numbere, Daopu Thompson. A general streamline modelling technique for homogeneous and heterogeneous porous media, with application to streamflood prediction. D, 1982, University of Oklahoma. 344 p.

Nunan, Adrienne Nichola. The geology of the magnetite deposits of the Kings Mountain Belt, North Carolina. M, 1983, University of North Carolina, Chapel Hill. 86 p.

Nunan, Walter Edward. Stratigraphy of the lower Devonian rocks of northwestern Georgia. M, 1971, Emory University. 89 p.

Nunan, Walter Edward. Stratigraphy of the Weverton Formation, northern Blue Ridge Anticlinorium. D, 1980, University of North Carolina, Chapel Hill. 215 p.

Nunes Correia, Francisco Carlos da Graca. OMEGA; a watershed model for simulation, parameter calibration and real-time forecast of river flows. D, 1984, Colorado State University. 166 p.

Nunes, Arturo de F. Geology of the Island of Orleans, Montmorency County, Quebec. D, 1958, Universite Laval. 216 p.

Nunes, Paul Donald. U-Pb mineral ages of the Stillwater igneous complex and associated rocks, Montana. D, 1970, University of California, Santa Barbara. 83 p.

Nunez del Arco, Eugenio. Regional and local geochemical variation in the Devonian black shales of eastern Kentucky. M, 1980, University of Kentucky. 235 p.

Nunez, Luis. Low-temperature geochemistry of Eocene sedimentary rocks along Sespe Creek, Ventura County, California. M, 1978, California State University, Long Beach. 117 p.

Nunn, Jeffrey Allen. Thermal contraction and flexure of intracratonic basins; a three-dimensional study of the Michigan Basin. D, 1981, Northwestern University. 363 p.

Nunn, Jerald Ralph. The petrology and paleogeography of the Starved Rock Sandstone in southeastern Iowa, western Illinois, and northeastern Missouri. M, 1986, University of Iowa. 144 p.

Nunn, Susan Christopher. The political economy of institutional innovation; coalitions and strategy in the development of groundwater law. D, 1986, University of Wisconsin-Madison. 334 p.

Nunnally, Nelson Rudolph. Flood plain morphology along the lower Ohio. D, 1965, University of Illinois, Urbana. 127 p.

Nur, Amos Michael. Effects of stress and fluid inclusions on wave propagation in rock. D, 1969, Massachusetts Institute of Technology.

Nurkowski, John Ronald. Coal quality, coal rank variation and its relation to reconstructed overburden; Upper Cretaceous and Tertiary plains coals, Alberta, Canada. M, 1984, University of Alberta. 129 p.

Nurmi, Roy D. Stratigraphy and sedimentology of the lower Salina Group (upper Silurian) in the Michigan Basin. D, 1975, Rensselaer Polytechnic Institute. 261 p.

Nurmi, Roy David. Upper Ordovician stratigraphy of the southern peninsula of Michigan. M, 1972, Michigan State University. 48 p.

Nurse, Leonard Alfred. Development and change on the Barbados leeward coast; a study of human impact on the littoral environment. D, 1986, McGill University.

Nusbaum, Robert L. A collapse-caldera boundary in the Precambrian St. Francois Mountains, southeastern Missouri. M, 1981, University of Kansas.

Nusbaum, Robert L. The Wah Wah Springs Tuff; compositional variations in a dominant volume tuff. D, 1984, University of Missouri, Rolla. 173 p.

Nussmann, David George. Ecology and pyritization of the silica formation, Middle Devonian, of Lucas County, Ohio. M, 1961, University of Michigan.

Nussmann, David George. Trace elements in the sediments of Lake Superior. D, 1965, University of Michigan. 252 p.

Nutalaya, Prinya. Geology of Cottonwood Ridge area, Larimer County, Colorado. M, 1964, University of Colorado.

Nutalaya, Prinya. Metamorphic petrology of a part of the northeastern Front Range, Larimer County, Colorado. D, 1966, University of Colorado.

Nute, Alton John. A numerical simulation of pressure distribution and radius of drainage in infinite radial aquifers-constant rate case. D, 1969, University of Missouri, Rolla.

Nute, Alton John. Some aspects of transient flow behavior in artesian aquifers and in hydrocarbon reservoirs surrounded by artesian aquifers. M, 1967, University of Missouri, Rolla.

Nutent, Lawrence E., Jr. Oil and gas of West Virginia. M, 1936, West Virginia University.

Nutini, John, Jr. A short-term trend analysis of recent coastal sedimentation subsequent to Hurricane Alicia on Galveston and Foletts islands, Texas. M, 1985, University of Texas at Dallas. 249 p.

Nutt, C. J. The Escondido mafic-ultramafic complex; a concentrically zoned body in the Santa Lucia Range, California. M, 1977, Stanford University. 90 p.

Nutt, William H. Post Pleistocene depositional history of Pigeon Creek, San Salvador Island, Bahamas, using Ostracoda in selected cores. M, 1985, University of Akron. 130 p.

Nuttall, Brandon Duncan. The Nantahala-Ocoee Contact in North Georgia. M, 1951, University of Cincinnati. 32 p.

Nuttall, Jeffrey Clarke. A trace element and replacement analysis of some trepostome bryozoans. M, 1968, Michigan Technological University. 77 p.

Nutter, Brian L. Small-pebble and heavy-mineral composition of glacial deposits in northeastern Kansas. M, 1988, Emporia State University. 77 p.

Nutter, Larry J. Stratigraphy of the Newton Hamilton Formation (Onondagan) of south central Pennsylvania. M, 1962, University of Minnesota, Minneapolis. 88 p.

Nutter, Neill Hodges. The compilation and evaluation of instructional objectives for introductory geology courses taught by the audio-tutorial approach at institutions of higher learning in the United States. D, 1971, Michigan State University. 162 p.

Nutter, Neill Hodges. The use of infrared spectrophotometric analysis in the correlation of different units within Dundee Limestone, Devonian age, Rogers City, Michigan. M, 1958, Michigan State University. 47 p.

Nuttli, Otto William. A study of the seismic P wave in the shadow zone of the Earth's core. D, 1953, St. Louis University.

Nuttli, Otto William. The western Washington earthquake of April 13, 1949. M, 1950, St. Louis University. 56 p.

Nuzzo, M. L. The adsorption chemistry of thorium and protactinium in the marine environment. M, 1977, Texas A&M University.

Nwabuokei, Samuel Onyeabor. Compressibility and shear strength characteristics of impact compacted lacustrine clay. D, 1984, Purdue University. 579 p.

Nwachukwu, Joseph Iheanacho. Organic geochemistry of the Orinoco Delta, Venezuela; a study of Recent sediments and their size fractions. D, 1981, University of Tulsa. 275 p.

Nwachukwu, Silas Ogo Okonkwo. The geologic significance of geomagnetic measurements in the Lake Huron Basin and adjacent areas. D, 1964, University of Toronto. 384 p.

Nwadialo, Bernard-Shaw Emeje. Morphological, chemical, and mineralogical properties of soils and the effects of acid sulfate weathering in the Copper Basin of Tennessee. D, 1982, University of Tennessee, Knoxville. 168 p.

Nwangwu, Uka. Deltaic and interdeltaic stratigraphy and sedimentology of the Fox Hills and associated formations in the central Front Range and Colorado Springs area, Colorado. D, 1976, Colorado School of Mines. 172 p.

Nwangwu, Uka. Stratigraphy and sedimentology of upper Cretaceous Pierre, Fox Hills, Laramie and lower Arapahoe formations south of Golden, Colorado. M, 1974, Colorado School of Mines. 123 p.

Nwankor, G. I. A comparative study of specific yield in a shallow unconfined aquifer. M, 1982, University of Waterloo.

Nwankwo, Linus N. Origin of the sand deposit in western Lake Erie between Monroe, Michigan and West Sister Island, Ohio. M, 1979, Bowling Green State University. 85 p.

Nwankwor, G. I. A comparative study of specific yield in shallow unconfined aquifer. M, 1982, University of Western Ontario.

Nwankwor, Godwin Ifedilichukwu. Delayed yield processes and specific yield in a shallow sand aquifer. D, 1985, University of Waterloo. various pagination p.

Nwaochei, Ben Nnaemeka. Geophysical investigations of the Nicaraguan Rise. D, 1981, Rutgers, The State University, New Brunswick. 80 p.

Nwaogazie, Ifeanyi Lawrence. Finite element modeling of streamflow routing. D, 1982, Oklahoma State University. 161 p.

Nyagah, Kivuti. A stratigraphic and sedimentologic study of the Cretaceous and Tertiary strata of East Kenya. M, 1988, University of Windsor. 207 p.

Nybakken, Bette Helene Halvorsen. The paleoecology of Southwest Umnak Island and Southwest Kodiak Island, Alaska. D, 1966, University of Wisconsin-Madison. 112 p.

Nyberg, Albert Victor, Jr. Contributions to micropaleontology; Proterozoic stromatolitic chert and shale-facies microfossil assemblages from the western United States and the Soviet Union; morphology and relationships of the Cretaceous foraminifer Colomia Cushman & Bermudez. D, 1982, University of California, Los Angeles. 265 p.

Nyblade, Andrew A. Timing volcanic events by secular variation and thermal modeling. M, 1985, University of Wyoming. 76 p.

Nychas, Anastaios Emmanuel. Nitrogen in sedimentary materials. D, 1978, University of California, Davis. 76 p.

Nye, Christopher. The Teklanika Formation in the Calico Creek area, Mt. McKinley National Park, Alaska. M, 1978, University of Alaska, Fairbanks. 68 p.

Nye, Christopher John. Petrology and geochemistry of Okmok and Wrangell volcanoes, Alaska. D, 1983, University of California, Santa Cruz. 215 p.

Nye, Osborne Barr, Jr. Generic revision and skeletal morphology of some cerioporid cyclostomes (Bryozoa). D, 1972, University of Cincinnati. 383 p.

Nye, Roger K. Causes of observed variations in strain and tilt at the Granite Mountain Records Vault, Salt Lake County, Utah. M, 1977, University of Utah. 200 p.

Nye, Thomas Spencer. Geology of the Apex uranium mine, Lander County, Nevada. M, 1958, University of California, Berkeley. 47 p.

Nye, Thomas Spencer. The relationship of structure and alteration to some ore bodies in the Bisbee (Warren) district, Cochise County, Arizona. D, 1968, University of Arizona. 244 p.

Nyerges, Timothy Lee. Modeling the structure of cartographic information for query processing. D, 1980, Ohio State University. 211 p.

Nygreen, Paul Wallace. Stratigraphy of the lower Oquirrh Formation in the type area and near Logan, Utah. M, 1955, University of Nebraska, Lincoln.

Nygren, Walter E. An outline of the general geology and physiography of the Grand Valley district (Mesa County), Colorado. M, 1935, University of Colorado.

Nyland, Edo. A determination of absolute stress in the crust of the Earth. M, 1964, Dalhousie University. 58 p.

Nyland, Edo. Low-rate aspects of focal mechanisms from permanent deformation in the near field. D, 1970, University of California, Los Angeles. 189 p.

Nyman, Dale James. Ute Pass Fault and related structures of El Paso County, Colorado. M, 1958, Iowa State University of Science and Technology.

Nyman, Douglas Christian. Stacking and stripping for normal mode eigenfrequencies. D, 1973, [University of California, San Diego].

Nyman, Matthew William. Petrology and geologic relationships of metagabbro shear zones, southeastern Adirondack Mountains, Whithall/Ft. Anne, New York. M, 1987, SUNY at Binghamton. 159 p.

Nyobe, Jean Blaise. A geological and geochemical study of the Fongo-Tongo and areally related bauxite deposits, western highlands, Republic of Cameroon. D, 1987, Lehigh University. 380 p.

Nyobe, Jean Blaise. Lower Cambrian clastic rocks of the Reading Prong and its structural extensions in Pennsylvania, New Jersey, New York, and Maryland. M, 1974, Lehigh University. 353 p.

Nyong, Eyo Etim. A paleoslope model for Campanian to lower Maestrichtian foraminifera of the North American Basin and continental margin. D, 1983, Rutgers, The State University, New Brunswick. 179 p.

Nyong, Eyo Etim. Campanian-early Maestrichtian benthic foraminiferal paleoecology and paleobathymetry of the New Jersey and northern Delaware Atlantic margin. M, 1981, Rutgers, The State University, New Brunswick. 86 p.

Nyquist, David. Eutrophication trends of Bear Lake, Idaho-Utah and their effect on the distribution and biological productivity of zooplankton. D, 1968, Utah State University. 225 p.

Nyquist, Jonathan Eugene. Thermal and mechanical models of the Mid-Continent Rift. D, 1986, University of Wisconsin-Madison. 204 p.

Nyquist, Laurence Elwood. The cosmic ray record in the metallic phase of chondrites. D, 1969, University of Minnesota, Minneapolis. 165 p.

Nystrom, Paul G. Geology of the Catarrh NW quadrangle. M, 1972, University of South Carolina. 49 p.

Nyumbu, Inyambo Liyambila. System decomposition approach for design of conjunctive ground-surface water storages. D, 1981, Colorado State University. 138 p.

Nyunt, U. The Canutillo (late Middle Devonian), Percha (Late Devonian), Las Cruces (Middle Mississippian) formations of the Spike "S" Ranch, southern Hueco Mountains, Hudspeth County, Texas. M, 1985, University of Texas at El Paso.

O'Bannon, Charles Edward. Stabilization of montmorillonite clay by electro-osmosis and base exchange of ions. D, 1971, Oklahoma State University. 131 p.

O'Bara, Jeffrey Brian. Magnetic analysis of metamorphic rocks; a comparison of two gneiss domes (Old Lyme and Bristol) in Connecticut. M, 1979, Wesleyan University. 113 p.

O'Beirne, A. M. Geology of the Gillis Mountain Pluton, Cape Breton Island, Nova Scotia. M, 1979, Acadia University. 168 p.

O'Bert, Lawrence Kay. Geology of a portion of the Dry Canyon and Las Flores quadrangles, Santa Monica Mountains, California. M, 1948, University of Southern California.

O'Brien, Ann Marie. Stable isotope study of Pennsylvanian age material. M, 1981, Ohio University, Athens. 94 p.

O'Brien, Arnold Leo. Hydrologic investigations of two wetlands in Lincoln, Massachusetts. D, 1973, Boston University. 264 p.

O'Brien, B. H. The geology of parts of the Coldbrook Group, southern New Brunswick. M, 1976, University of New Brunswick.

O'Brien, Bob Randolph. Geology of Cienega Amarilla area, Catron County, New Mexico and Apache County, Arizona. M, 1956, University of Texas, Austin.

O'Brien, Brian E. Geology of east-central Caddo County, Oklahoma. M, 1963, University of Oklahoma. 72 p.

O'Brien, C. Nearshore processes and sedimentation at Queens and Richmond bays, Tobago, W.I. M, 1986, University of Waterloo. 260 p.

O'Brien, D. E. Stratigraphy, petrology, and depositional sequence of the Marmaton, Pleasanton, and lowermost Kansas City groups (late Middle - early

Upper Pennsylvanian) in a core from South-central Iowa. M, 1977, University of Iowa. 120 p.

O'Brien, Dennis C. Lithofacies and current direction study of Dunkard Group sediments (Pennsylvanian-Permian) in the northern Dunkard area. M, 1964, Miami University (Ohio). 64 p.

O'Brien, Dennis Craig. The geology of the southeast Silvertip Quadrangle, Montana (Precambrian-Belt Series). D, 1971, University of Massachusetts. 121 p.

O'Brien, Dennis J. Kinetics of the oxidation of reduced sulfur species in aqueous solution. D, 1974, University of Maryland. 135 p.

O'Brien, Donald Edward. Osteology of Kindleia fragosa Jordan (Holostei: Amiidae) from the Edmonton Formation (upper Cretaceous), (Maestrichtian) of Alberta, Canada. M, 1969, University of Alberta. 125 p.

O'Brien, Douglas Patrick. A theoretical investigation into certain micropulsations in the auroral zone. D, 1966, University of California, Berkeley. 96 p.

O'Brien, Edward J., III. On the validity of the geostrophic approximation for the Florida current. M, 1967, [University of Miami].

O'Brien, Felicity Heather Claire. The stratigraphy and paleontology of the Clam Bank Formation, and the upper part of the Long Point Formation of the Port au Port Peninsula, on the west coast of Newfoundland. M, 1973, Memorial University of Newfoundland. 156 p.

O'Brien, Harry Deforest, Jr. Lithotope analysis of Holocene sediments in the shallow substrate of Mississippi Sound. M, 1984, University of Mississippi. 101 p.

O'Brien, James J. Minerals in bauxite. M, 1946, Columbia University, Teachers College.

O'Brien, Jimmy Steven. Physical processes, rheology and modeling of mud flows. D, 1986, Colorado State University. 172 p.

O'Brien, John M. The stratigraphy of the upper part of the Richmond group (Ordovician) of Indiana, Kentucky, and Ohio. M, 1970, Miami University (Ohio). 155 p.

O'Brien, John Malcolm. Narizian-Refugian (Eocene-Oligocene) sedimentation, western Santa Ynez Mountains, Santa Barbara County, California. D, 1973, University of California, Santa Barbara.

O'Brien, Kathryn Gronberg. Stratigraphy and paleontology of the Silverhorn Dolomite (Middle Devonian) of Dutch John Mountain, Lincoln County, Nevada. M, 1963, University of Illinois, Urbana. 76 p.

O'Brien, Keith M. The applicability of artificial groundwater recharge in the Ipswich River basin. M, 1982, University of Massachusetts. 141 p.

O'Brien, Lawrence Edward. A detailed plan for teaching a junior high school earth science course in the St. Louis area. M, 1972, University of Michigan.

O'Brien, Leslie J. Theoretical calibration of the Kelsh plotter. M, 1954, Ohio State University.

O'Brien, Michael. An experimental study of drainage network modifications. M, 1984, Colorado State University. 79 p.

O'Brien, Michael. Multiple bar systems in Chesapeake Bay, (east coast of Maryland and Virginia). M, 1968, College of William and Mary.

O'Brien, Neal Ray. A study of fissility in argillaceous rocks. D, 1963, University of Illinois, Urbana. 88 p.

O'Brien, Neal Ray. The use of underclay mineralogy in stratigraphic correlation. M, 1961, University of Illinois, Urbana.

O'Brien, P. J. The geology of Schaghticoke and environs, Rensselaer County, New York. M, 1960, Rensselaer Polytechnic Institute.

O'Brien, P. N. S. Model seismology; the critical reflection of waves. D, 1954, University of Toronto.

O'Brien, Patrick W. Problems associated with determination of well efficiencies in four Ogallala Aquifer wells, northeastern Colorado. M, 1983, Colorado School of Mines. 133 p.

O'Brien, Peter N. S. Model seismology; the critical reflection of elastic waves. D, 1955, University of Toronto.

O'Brien, Philip Joseph. Aquifer transmissivity distribution as reflected by overlying soil temperature patterns. D, 1970, Pennsylvania State University, University Park. 247 p.

O'Brien, Sean James. Volcanic stratigraphy, petrology and geochemistry of the Marystown Group, Burin Peninsula, Newfoundland. M, 1979, Memorial University of Newfoundland. 253 p.

O'Brien, Thomas Francis. Evidence for the nature of the lower crust beneath the central Colorado Plateau as derived from xenoliths in the Buell Park-Green Knobs diatremes. D, 1983, Cornell University. 223 p.

O'Brien, William. The role of source, depositional environment, and age on the composition of Quaternary sands, Monterey Bay, California. M, 1986, University of Houston.

O'Brien, William Jay. Analysis of the carbon-oxygen log. M, 1982, Stanford University.

O'Brien, William P. Petrochemistry and K, U, and Th distribution in the Graniteville-Granite, Iron County, Missouri. M, 1978, University of Missouri, Rolla.

O'Brien, William R. The partial subsurface distribution of the Oolagah Formation in northeastern Oklahoma. M, 1950, University of Tulsa. 44 p.

O'Brient, James David. Hypabyssal crystallization and emplacement of the rhyolitic Rabb Park Complex, Grant County, New Mexico. D, 1980, Stanford University. 250 p.

O'Bright, David Edward. Significance of the Lower-Upper Cretaceous unconformity to oil and gas accumulations in east-central Louisiana and adjacent Mississippi. M, 1986, University of Wisconsin-Milwaukee. 53 p.

O'Bryan, James William. The geochemistry and mineralogy of manganese nodules from the South Atlantic (southern Argentine Basin). M, 1975, Rutgers, The State University, Newark. 46 p.

O'Byrne, Thomas James. The geology of the Anchor area, Hot Springs and Fremont counties, Wyoming. M, 1941, University of Missouri, Columbia.

O'Clair, C. E., Jr. The effects of uplift on intertidal communities at Amchitka Island, Alaska. D, 1977, University of Washington. 198 p.

O'Connell, Anne F. Origin of the Glen Eyrie Member of the Fountain Formation. M, 1981, Indiana University, Bloomington. 174 p.

O'Connell, Daniel Robert. Seismic velocity structure and microearthquake source properties at The Geysers, California, geothermal area. D, 1986, University of California, Berkeley. 213 p.

O'Connell, Daniel Triggott. A revision of the geology of the eastern part of the County of the Bronx, State of New York, and related areas. D, 1932, New York University.

O'Connell, Dennis B. Paleoecology and environments of deposition of the Brereton Limestone Member (Pennsylvanian, Desmoinesian) in southwestern Illinois. M, 1983, Southern Illinois University, Carbondale. 121 p.

O'Connell, James F. Study of Ordovician rocks from deep wells in the Hillsdale, Northville, and adjacent areas in Southeast Michigan. M, 1958, Michigan State University. 58 p.

O'Connell, Marjorie. Studies of Paleozoic fossils; 1, Revision of the genus Zaphrentis among the Madrepora Rugosa; 2, The habitat of the eurypterids. M, 1912, Columbia University, Teachers College.

O'Connell, Marjorie. The habitat of eurypterids. D, 1916, Columbia University, Teachers College.

O'Connell, Michael R. A theoretical analysis of Rayleigh wave generation by impact sources. M, 1976, University of Wisconsin-Milwaukee.

O'Connell, Richard John. Part 1, Dynamic response of phase boundaries in the Earth to surface loading; Part 2, Pleistocene glaciation and the viscosity of the lower mantle. D, 1969, California Institute of Technology. 279 p.

O'Connell, Rita Marie. Environmental considerations as part of the effluent limitation variance decision-making process for point source direct discharges under wastewater discharge permit programs in Minnesota. D, 1988, University of California, Los Angeles. 407 p.

O'Connell, Shaun C. The Viking Formation (Lower Cretaceous) of southeastern Saskatchewan. M, 1982, University of Windsor. 191 p.

O'Connell, Suzanne. Geology of the mafic/ultramafic transition, Table Mountain, western Newfoundland. M, 1979, SUNY at Albany. 78 p.

O'Connell, Suzanne Bridget. Anatomy of modern submarine depositional and distributary systems. D, 1986, Columbia University, Teachers College. 313 p.

O'Conner, Franklin Austin. A subsurface analysis of the lower Chesterian Lewis Sandstone of the Black Warrior Basin in Mississippi and Alabama. M, 1984, Oklahoma State University. 111 p.

O'Conner, P. Short-term sea-level anomalies at Monterey, California. M, 1964, United States Naval Academy.

O'Connor, Bruce James. A petrologic and electron microprobe study of pelitic mica schists in the vicinity of the staurolite-disappearance isograd in Philadelphia, Pennsylvanian and Waterbury, Connecticut. D, 1973, The Johns Hopkins University.

O'Connor, Howard G. Geology and ground-water resources of Douglas County, Kansas. M, 1959, University of Kansas. 143 p.

O'Connor, John Edward. Geology of part of the Blacktail Range, Beaverhead County, Montana. M, 1949, University of Michigan. 33 p.

O'Connor, Joseph Tappan. The structural geology and Precambrian petrology of the Horsetooth Mountain area, Larimer County, Colorado. D, 1961, University of Colorado. 167 p.

O'Connor, Joyce. A study of the influence of shoreline and diagenesis on the mineralogy of the Ames Limestone and shale (Pennsylvania). M, 1986, University of Pittsburgh.

O'Connor, Kevin Myles. Distinct element modeling and analysis of mining-induced subsidence. D, 1988, Northwestern University. 191 p.

O'Connor, Lynn D. Gravity study of crustal structures in northeast peninsular Florida. M, 1984, University of South Florida, Tampa. 62 p.

O'Connor, M. J. Structural geology of Dugout Mountain, Marathon, Texas. M, 1988, Sul Ross State University.

O'Connor, Michael J. Differential compaction in the Woodbend Group (Upper Devonian) of central Alberta. M, 1972, University of Calgary. 160 p.

O'Connor, Michael Peter. Stratigraphy and petrology across the Precambrian Piegan Group-Missoula Group boundary, southern Mission and Swan ranges, Montana. D, 1968, University of Montana. 269 p.

O'Connor, Thomas E. The structure and stratigraphy of an area west of Taylor Park, northeast Gunnison County, Colorado. M, 1961, University of Colorado.

O'Connor, Thomas Mark. A preliminary study of the hydrogeology of the Dakota Formation in Douglas, Sarpy and Washington counties, eastern Nebraska. M, 1987, University of Nebraska, Lincoln. 147 p.

O'Connor, Thomas Patrick. The adsorption of copper and cobalt from aqueous solution onto illite and other substrates. D, 1974, University of Rhode Island.

O'Day, Michael Stephen. The structure and petrology of the Mesozoic and Cenozoic rocks of the Franciscan Complex, Leggett-Piercy area, northern California Coast Ranges. D, 1974, University of California, Davis. 152 p.

O'Day, Peggy Anne. Sedimentary rocks of the Barama-Mazaruni Supergroup, northern Guyana; sedimentology, petrography, and geochemistry. M, 1984, Cornell University. 154 p.

O'Dell, Catherine Hobbs. A laboratory investigation of the relationship between groundwater velocity and mass transfer from a simulated aromatic hydrocarbon spill at the water table interface. M, 1988, University of Minnesota, Minneapolis. 233 p.

O'Donnell, Edward. A regional study of the Brassfield Limestone as exposed in southeastern Indiana. M, 1963, University of Cincinnati. 66 p.

O'Donnell, Edward. The lithostratigraphy of the Brassfield Formation (Lower Silurian) in the Cincinnati Arch area. D, 1967, University of Cincinnati. 143 p.

O'Donnell, George Anthony. A new vector wave motion theory. D, 1935, St. Louis University.

O'Donnell, James. Buoyant plumes; a numerical model. M, 1981, University of Delaware, College of Marine Studies. 80 p.

O'Donnell, Lynn Louise. Characterization of the nature of deformation and metamorphic gradient across the Grenville Front tectonic zone in Carlyle Township, Ontario. M, 1986, McMaster University. 199 p.

O'Donnell, Michael Raymond. Regressive shelf deposits in the Pennsylvanian Arkoma Basin, Oklahoma and Arkansas. M, 1983, University of Oklahoma. 79 p.

O'Donnell, Neil Dennis. Glacial indicator trains near Gullbridge, Newfoundland. M, 1973, University of Western Ontario. 259 p.

O'Donnell, Paul J., Jr. Fusulinids of the Virgil Series in southeastern Nebraska. M, 1956, University of Nebraska, Lincoln.

O'Donnell, Terrence T. Depositional systems in the lower Claiborne (Eocene) of central Mississippi. M, 1974, University of Mississippi.

O'Donnell, Thomas Henry. Chemical and petrographic features of basalts from 25°N on the Mid-Atlantic Ridge. M, 1974, University of Texas at Dallas. 62 p.

O'Driscoll, Cyril Francis. Geology, petrology and geochemistry of the Hermitage Peninsula, southern Newfoundland. M, 1977, Memorial University of Newfoundland. 144 p.

O'Flaherty, K. F. Analysis of structures of Canadian ore deposits. M, 1951, University of Toronto.

O'Gara, William Thomas. Some Washita spatangid echinoids of North Texas. M, 1940, Texas Christian University. 130 p.

O'Gorman, John R. A computer analysis to determine the feasibility of using a gravity survey to study buried river valleys in Blackford County and surrounding areas. M, 1974, Ball State University.

O'Grady, Martin Dempsey. Paleobathymetry of the Bass River Formation and its implications. M, 1976, Rutgers, The State University, New Brunswick. 103 p.

O'Haire, Daniel P. Geology and hydrology of hot springs in the upper Jefferson Valley, Montana. M, 1978, Montana State University. 100 p.

O'Halloran, Daniel John. Relationship between earthquakes and tectonics. M, 1949, University of Michigan.

O'Hanley, David Sean. The origin and the mechanical properties of asbestos. D, 1986, University of Minnesota, Minneapolis. 94 p.

O'Hanley, Hilda Nevius. Chemical changes during diagenesis of calcareous ooze; an experimental study. M, 1985, University of Minnesota, Minneapolis. 73 p.

O'Hara, Kieran D. The Alleghanian Orogeny and its role in the late Paleozoic tectonic evolution of the Avalon Zone in Southeast New England. D, 1985, Brown University. 99 p.

O'Hara, Norbert Wilhelm. A statistical and mechanical analysis of the Marshall Sandstone in western Michigan to determine the environmental pattern of the deposit. M, 1954, Michigan State University. 142 p.

O'Hara, Norbert Wilhelm. An aeromagnetic survey and geophysical interpretation of the Precambrian framework and tectonic structure of the eastern Lake Superior region. D, 1967, Michigan State University. 260 p.

O'Hara, P. The economic geology, petrology and origin of the Willsboro wollastonite deposit, New York. M, 1976, Queens College (CUNY). 126 p.

O'Hara, Patrick Francis. Metamorphic and structural geology of the northern Bradshaw Mountains, Yavapai County, Arizona. D, 1980, Arizona State University. 145 p.

O'Hare, Andrew T. A geochemical study of the partitioning of elements during retorting of oil shales from Lewis County, Kentucky. M, 1982, University of Kentucky. 130 p.

O'Harra, Cleophas C. The geology of Allegany Co., MD. D, 1898, The Johns Hopkins University.

O'Hayre, Arthur P. A hydronomic analysis of resource management alternatives; a case study of Itasca County. D, 1976, University of Minnesota, Minneapolis. 175 p.

O'Hern, D. W. The Trilobita, Mollusca, and Echinodermata of the paleo-Devonian of Maryland. D, 1907, The Johns Hopkins University.

O'Hirok, Linda Susan. Barrier beach formation and breaching, Santa Clara River mouth, California. M, 1985, University of California, Los Angeles. 165 p.

O'Kane, John Anthony, Jr. Silicoflagellates of Monterey Bay, California. M, 1970, San Jose State University. 92 p.

O'Keefe, Arthur Francis Xavier. Characteristics of a thrust fault displacement transfer zone. M, 1980, Texas A&M University. 77 p.

O'Keefe, J. D. Impact phenomena on the terrestrial planets. D, 1977, University of California, Los Angeles. 190 p.

O'Keefe, Jane Frances. A porosity study on Cretaceous rocks of the Great Valley Sequence, California, and its implications for their burial history. M, 1981, University of Washington. 36 p.

O'Keefe, Monica Elizabeth. Structural geology and mineral deposits of Red Top Mountain, Stevens County, Washington. M, 1980, Washington State University. 152 p.

O'Kelley, Robert V. Late Quaternary vertebrate fossils and pollen of Edisto Island, South Carolina. M, 1976, University of Georgia.

O'Laskey, Robert H. The petrology and mineral chemistry of the Ste. Dorothée Sill, Quebec. M, 1982, University of Massachusetts. 174 p.

O'Leary, Dennis. A study of the genesis of the pyritic gneiss in the Grenville Series (Precambrian), Saint Lawrence and Jefferson counties, New York. M, 1967, University of Missouri, Rolla.

O'Leary, Dennis William. The form, structure, and evolution of the Allegheny Front in Centre County, Pennsylvania. D, 1972, Pennsylvania State University. 164 p.

O'Leary, Ellen Frances. Petrology and chemistry of the Malene ultramafic rocks, Godthab region, Southwest Greenland. M, 1988, Washington University. 175 p.

O'Leary, Michael J. Thermodynamic properties of tetrahedrite-tennantites in natural environments. M, 1987, Purdue University. 56 p.

O'Leary, William J. Magmatic paragenesis of the Fish Canyon ash-flow tuff, central Jose Mountains, Colorado. M, 1981, University of Georgia.

O'Loughlin, Sharon Beth. Bedrock geology of the Mt. Abraham-Lincoln Gap area, central Vermont. M, 1986, University of Vermont. 164 p.

O'Mahoney, Laurence. Geologic factors affecting roof stability at the Dilworth coal mine, Greene County, Pennsylvania. M, 1978, Lehigh University. 68 p.

O'Malley, David P. Trilobite biostratigraphy of the Gordon Shale in the southern part of the Libby Trough, Sanders County, Montana. M, 1985, Washington State University. 144 p.

O'Malley, Frank Ward. Determination of the elastic behavior of rocks by means of the electrical strain gage. M, 1948, Stanford University. 67 p.

O'Malley, Peg Ann. Quaternary geology and tectonics of the Waucoba Wash 15-minute quadrangle, Saline Valley, Inyo County, California. M, 1980, University of Nevada. 142 p.

O'Malley, Robert Thomas. Temporal variations of gravity in Northwest Montana near Lake Koocanus. M, 1981, University of Washington. 83 p.

O'Neal, Jill Evans. A facies analysis of the Queen City Formation (Eocene), northeastern Texas. M, 1982, Stephen F. Austin State University. 82 p.

O'Neal, Marc A. Sources, transport, and dispersal history of southern Oregon continental shelf sediments; Fourier grain-shape analysis. M, 1986, Wichita State University. 107 p.

O'Neal, Marianne Victoria. A petrographic investigation of the Windsor Formation, St. Ann's Basin; implications concerning the Cretaceous tectonic history of northern Jamaica. M, 1984, University of Oklahoma. 159 p.

O'Neil, Caron. Characterization of northwesterly-trending lineaments, French Creek, northwestern Pennsylvania. M, 1986, University of Pittsburgh.

O'Neil, J. R. pH, conductivity, alkalinity, zinc, cadmium, lead and copper in the precipitation of Morgantown, West Virginia, November 1975 through April 1976. M, 1976, West Virginia University. 72 p.

O'Neil, Robert Lester. A study of trace element distribution in the Chattanooga Shale. M, 1956, Pennsylvania State University, University Park. 62 p.

O'Neil, Thomas J. Chemical interactions due to subsurface mixing of fresh and marine waters in a Pleistocene reef complex, Rio Bueno, Jamaica. M, 1974, Louisiana State University.

O'Neil, Thomas J. Estimating minimum copper price levels through production cost projections. D, 1972, University of Arizona.

O'Neill, Aloysius Joseph. The form of occurrence of uranium in urano-organic deposits. M, 1976, Rutgers, The State University, Newark. 87 p.

O'Neill, B. J. Pliocene and Pleistocene benthic foraminifera from the central Arctic Ocean. M, 1979, University of Wisconsin-Madison.

O'Neill, Bernard J., Jr. Geology of the anorthosite massif in Chester County, Pennsylvania. M, 1952, California Institute of Technology. 35 p.

O'Neill, Charles W. Sedimentology of East Key, Dry Tortugas, Florida. M, 1976, University of South Florida, Tampa. 74 p.

O'Neill, Dennis Charles. Implementation of a frequency-domain phase-difference polarization filter and its application to a determination of crustal structure in parts of Eastern North America. M, 1983, Pennsylvania State University, University Park. 145 p.

O'Neill, J. M. The geology of the Mt. Richthofen Quadrangle and adjacent Kawuneechee Valley, Northcentral Colorado. D, 1976, University of Colorado. 219 p.

O'Neill, John Michael. Geology of the southern Pilot Range, Elko County, Nevada, and Box Elder and Tooele counties, Utah. M, 1968, University of New Mexico. 113 p.

O'Neill, Margaret M. Crystal growth velocities, habit, and dehydration in copper sulphate pentahydrate. M, 1973, Boston University. 95 p.

O'Neill, Patrick P. An economic, metamorphic, structural and geochemical study of the Isle aux Morts Prospect, Southwest Newfoundland. M, 1985, Memorial University of Newfoundland. 263 p.

O'Neill, Robert L. Superposition of Cenozoic extension on Mesozoic compressional structures of the Pioneer Mountains, central Idaho. M, 1985, Lehigh University.

O'Neill, Thomas Francis. The artificial alteration of feldspars under conditions of elevated temperature

and pressure. D, 1942, University of Minnesota, Minneapolis. 46 p.

O'Neill, Thomas Francis. The geology of the Stayton Quadrangle, Oregon. M, 1939, Oregon State University. 79 map p.

O'Niell, Charles A., III. Isopach map and expansion index study of growth faults and the structural evolution of Belle Isle Salt Dome, Saint Mary Parish, Louisiana. M, 1966, Louisiana State University.

O'Niell, John J. Geology and petrography of the Beloeil and Rougemont mountains, Quebec. D, 1912, Yale University.

O'Nions, Robert Keith. Geochronology of the Bamble sector of the Baltic shield (Precambrian), South Norway. D, 1969, University of Alberta. 232 p.

O'Nour, Ibrahim. Gravity anomalies of central Georgia. M, 1982, Georgia Institute of Technology. 69 p.

O'Quinn, Edgar Byron. The Mooreville Formation in the Tibbee Creek area, Clay and Lowndes counties, Mississippi. M, 1961, Mississippi State University. 37 p.

O'Reilly, George A. Geology and geochemistry of the Sangster Lake and Larrys River plutons, Guysborough County, Nova Scotia. M, 1988, Dalhousie University. 290 p.

O'Reilly, Thomas Clark. Behavior of trace elements in layered igneous intrusions. M, 1982, Washington University. 137 p.

O'Rourke, Edward Joseph. The physiography of the "Clinton Sand" (Silurian; Ohio). M, 1919, Ohio State University.

O'Rourke, Elizabeth Frances. A chemographic analysis of glaucophane- and chloritoid-bearing metamorphic rocks. M, 1983, University of Minnesota, Duluth. 97 p.

O'Rourke, John T. Geotechnical investigation, Carmel Valley Ranch, Monterey County, California. M, 1980, Stanford University. 66 p.

O'Rourke, Joseph Edward. The stratigraphy of the metamorphic rocks of the Rio de Pedras and Gandarela quadrangles, Minas Gerais, Brazil. D, 1958, University of Wisconsin-Madison. 166 p.

O'Rourke, Mary Kay. An absolute pollen chronology of Seneca Lake, New York. M, 1976, University of Arizona.

O'Rourke, Mary Kay. The implications of atmospheric pollen rain for fossil pollen profiles in the arid Southwest. D, 1986, University of Arizona. 189 p.

O'Rourke, T. D. A study of two braced excavations in sands and interbedded stiff clay. D, 1975, University of Illinois, Urbana. 273 p.

O'Rourke, T. J. Shear and compressional analysis of VSP in Silo Field, Denver-Julesburg Basin. M, 1986, Colorado School of Mines. 93 p.

O'Rourke, Terence Lee. An analysis of the performance of Texas state agencies in protecting the environment. M, 1972, Rice University. 70 p.

O'Shea, K. J. Diagenesis of the Lower Silurian Cataract Group sandstones of southern Ontario. M, 1988, University of Waterloo. 251 p.

O'Shields, Richard L. Sand exclusion in producing wells. M, 1951, Louisiana State University.

O'Sullivan, John Blandford. Geology and bituminous stabilization of soil materials at Point Barrow, Alaska. M, 1958, Iowa State University of Science and Technology.

O'Sullivan, John Blandford. Quaternary geology of the Arctic Coastal Plain, northern Alaska. D, 1961, Iowa State University of Science and Technology. 204 p.

O'Sullivan, Robert Brett. Geology and mineralogy of the Fierra-Hanover District, Grant County, New Mexico. M, 1953, University of New Mexico. 76 p.

O'Sullivan, Terence Patrick. Geophysical well log response to fractured and altered zones in the Raft River, Idaho, geothermal reservoir. M, 1979, Wright State University. 65 p.

O'Sullivan, William H., Jr. Petrographic and stratigraphic analysis of the Fort Worth Limestone and its equivalents (Lower Cretaceous) of central Texas. M, 1983, Stephen F. Austin State University. 119 p.

O'Sullivan, William Joseph. Trinity division of Smithwick area, Burnet County, Texas. M, 1967, University of Texas, Austin.

O'Toole, Frederick S. Petrology of the Cenozoic phonolites and related rocks of the Houston area, near Bear Lodge Mountains, Wyoming. M, 1981, University of North Dakota. 112 p.

O'Toole, Patrick Brian. Provenance evidence for inwash and subglacial fluvial origins for the debris in Connecticut Valley heads of outwash. M, 1988, Lehigh University. 173 p.

O'Toole, Walter Leonard. Geology of the Keetley-Kamas volcanic area (Wasatch Mountains, Utah). M, 1951, University of Utah. 38 p.

Oakes, Chandler A. Environmental significance of variations in first-order bounding surfaces in eolian deposits; Entrada Sandstone, southeastern Utah. M, 1984, University of Texas, Austin.

Oakes, Charles Steger. Evidence for replacement of dilute hydrothermal solutions by hot, hypersaline brine in the northeastern part of the Salton Sea geothermal system, California; a fluid inclusion and oxygen isotope study. M, 1988, University of California, Riverside. 115 p.

Oakes, David Thomas. Electrokinetic phenomena in colloidal clays. M, 1951, University of Oklahoma. 112 p.

Oakes, Edward H. Geology of the northern Grapevine Mountains, northern Death Valley, California. M, 1977, University of Wyoming. 113 p.

Oakes, Edward L. The Woodfordian moraines of Rock County, Wisconsin. M, 1960, University of Wisconsin-Madison.

Oakes, Malcolm C. The helium problem, theories of the origin of helium in natural gas. M, 1922, University of Oklahoma. 122 p.

Oakes, Millis Henry. Upper Cretaceous of Cabo Cabras, Baja California. M, 1954, University of California, Berkeley. 86 p.

Oakes, Ramsey L. Recent sediments, New Orleans, Louisiana. M, 1947, Louisiana State University.

Oakeshott, Gordon Blaisdell. A detailed geologic section across the western San Gabriel Mountains of California. D, 1936, University of Southern California.

Oakeshott, Gordon Blaisdell. The petrography of the Stanley Mountain Franciscan of the Nipomo Quadrangle, California. M, 1929, University of California, Berkeley. 70 p.

Oakley, Stewart M. The geochemical partitioning and bioavailability of trace metals on marine sediments. D, 1981, Oregon State University. 92 p.

Oakman, Marial. Seafloor alteration of hydrothermal sulphide deposits. M, 1988, University of Rhode Island.

Oaks, Robert Quincy, Jr. Post-Miocene stratigraphy and morphology, outer coastal plain, southeastern Virginia. D, 1965, Yale University. 423 p.

Oaksford, Edward T. Determination of cation exchange constants for copper (II)-calcium and copper (II)-magnesium exchange reactions on kaolinitas. M, 1973, SUNY at Buffalo. 72 p.

Oates, Kenneth Michael. Measurement of exchangeable aluminum, its use as a predictor for lime requirement, and its neutralization as affected by limestone particle size. D, 1982, North Carolina State University. 145 p.

Oates, N. D. Geology of the western portion of the Cañada Gobernadora Quadrangle, Orange County, California. M, 1960, University of Southern California.

Obaoye, Michael Olajide. Interpretation of detailed gravity traverses across northeastern Georgia. M, 1979, Georgia Institute of Technology. 88 p.

Obata, Masaaki. Petrology and petrogenesis of the Ronda high-temperature peridotite intrusion, southern Spain. D, 1977, Massachusetts Institute of Technology. 247 p.

Obeda, Barbara A. Watershed disturbance and its contribution to accelerated eutrophication; a case study of Shadow Lake, Ridgefield, Conn. M, 1981, Western Connecticut State University. 124 p.

Obelenus, Thomas J. Depositional environments and diagenesis of carbonates and associated evaporites, Frobisher-Alida Interval, Madison Group (Mississippian), Williston Basin, northwestern North Dakota. M, 1985, University of North Dakota. 313 p.

Oben-Nyarko, K. A strategy for decision making in water resources planning for developing countries. D, 1979, University of Arizona. 218 p.

Oben-Nyarko, K. Urban stormwater runoff management; a model study. M, 1976, University of Arizona.

Obenshain, Karen R. Amphibolite anatexis within a contact aureole. M, 1981, University of Georgia.

Obenson, Gabriel Francis. Prediction of mean gravity anomalies of large blocks from sub-block means. M, 1968, Ohio State University.

Obenson, Gabriel Francis També Direct evaluation of the Earth's gravity anomaly field from orbital analysis of artificial earth satellites. D, 1970, Ohio State University.

Oberdorfer, June Ann. Wastewater injection; near-well processes and their relationship to clogging. D, 1983, University of Hawaii. 204 p.

Oberg, Clayton John. Quaternary paleomagnetic/geomagnetic studies in Western Canada. M, 1978, University of Alberta. 132 p.

Oberg, Rolland. The conodont fauna of the Viola Formation (Ordovician), (Oklahoma). D, 1966, University of Iowa. 186 p.

Oberg, Rolland. The conodont fauna of the Winnipeg Formation in the type area. M, 1964, University of Iowa. 129 p.

Oberhansley, Gary. Geology of the Fairview Lakes Quadrangle, Sanpete County, Utah. M, 1980, Brigham Young University.

Oberlander, Phil Louis. Development of a quasi three-dimensional groundwater model for a portion of the Nevada Test Site. M, 1978, University of Nevada. 89 p.

Oberlander, Theodore Marvin. The origin of the structurally discordant drainage of the central Zagros Mountains, Iran. D, 1963, Syracuse University. 376 p.

Oberlindacher, H. P. Geology of the southern part of the Wooley Range, Caribou County, Idaho. M, 1983, San Jose State University. 110 p.

Oberling, John James. Shell structure of western American Pelecypoda. D, 1955, University of California, Berkeley. 408 p.

Oberling, John James. Tertiary assemblages from southern Colombia. M, 1951, University of California, Berkeley. 77 p.

Obermiller, Walter A. Geologic, structural and geochemical features of basaltic and rhyolitic volcanic rocks of the Smith Rock/Gray Butte area, central Oregon. M, 1987, University of Oregon. 189 p.

Obernyer, Stanley L. Stratigraphy of the Fort Union and Wasatch formations (Tertiary) in the Buffalo-Lake de Smet area, Johnson County, Wyoming. M, 1978, Colorado School of Mines. 106 p.

Oberste-Lehn, D. Slope stability of the Lomerias Muertas area, San Benito County, California. D, 1976, Stanford University. 247 p.

Oberste-Lehn, Deane. Passive microwave study of geologic materials in a volcanic province. M, 1970, Stanford University. 107 p.

Oberste-Lehn, Deane. Slope stability of the Lomerias Muertas area, San Benito County, California. D, 1977, Stanford University. 231 p.

Obert, G. E. Distribution of organic matter in sediments about a tropical mangrove island. M, 1985, SUNY at Binghamton. 30 p.

Obert, Karl Richard. Geology of the Sheep Mountain-Gray Jockey Peak area, Beaverhead County, Mon-

tana. M, 1962, University of California, Berkeley. 80 p.

Oberts, Gary Leonard. The chemistry and hydrogeology of dry valley lakes, Antarctica. M, 1973, Northern Illinois University. 55 p.

Obi, Adeniyi Olubunmi. Isotope studies on crop utilization and soil fixation of nitrogen from urea, calcium nitrate and ammonium sulphate in several Manitoba soils. D, 1981, University of Manitoba.

Obi, Curtis Mitsuru. Mechanisms of basement deformation at the Five Springs Range front, Bighorn Mountains, Wyoming. M, 1986, University of Massachusetts. 121 p.

Obinna, F. C. The geology and some genetic aspects of Fox Mine mineralization, northern Manitoba. M, 1974, University of Manitoba.

Obolewicz, David. Geochemical study of the Limekiln Hill area, a part of the Highland Mountains, Southwest Montana. M, 1978, Montana College of Mineral Science & Technology. 145 p.

Oborne, Juli G. Stratigraphy and depositional environments of the Vaqueros Formation, central Santa Monica Mountains, California. M, 1987, California State University, Northridge. 95 p.

Oborne, Mark Stephen. Stratigraphy of Early to Middle (?) Triassic marine to continental rocks, southern Inyo Mountains, California. M, 1983, California State University, Northridge. 90 p.

Obradovic, Milan Mitch. An isotopic and geochemical study of runoff in the Apex River watershed, Baffin Island, N.W.T. M, 1986, University of Windsor. 189 p.

Obradovich, John Dinko. Problems in the use of glauconite and related minerals for radioactivity dating. D, 1964, University of California, Berkeley. 93 p.

Oca, G. R. Engineering geology problems at Ambuclao damsite, Bokod, Benguet, Philippine Islands. M, 1953, Massachusetts Institute of Technology. 63 p.

Ochoa, Rafael Eugenio. Depositional environments of the Newcastle Formation (Cretaceous), interpreted from outcrops in the Black Hills area, northeastern Wyoming. M, 1970, University of Wyoming. 116 p.

Ochs, Allan Michael. Comparative petrology of lower Tertiary sandstones, southern Piceance Creek basin, Colorado. M, 1978, Southern Illinois University, Carbondale. 147 p.

Ochs, Steffen. Stratigraphy, depositional environments, and petrology of the lowermost Moenkopi Formation, southeastern Utah. M, 1988, University of Utah. 202 p.

Ochsenbein, C. Douglas. An investigation of the miospores in coals of the lower tongue of the Breathitt Formation and the Stearns coal zone in south-central Kentucky. M, 1987, Eastern Kentucky University. 32 p.

Ochsenbein, Gary Dean. Origin of caves in Carter Caves State Park, Carter County, Kentucky. M, 1974, Bowling Green State University. 64 p.

Ockerman, John William. A petrographic study of the Madison and Jordan sandstones in southern Wisconsin. D, 1929, University of Wisconsin-Madison.

Ockert, Donn Lee. A photogrammetric calibration of a radio telescope. M, 1958, Ohio State University.

Ocola, Leonidas. A non-linear least squares method for seismic refraction mapping. D, 1971, University of Wisconsin-Madison. 358 p.

Ocola, Leonidas. Seismic activity of Peru; spatial distribution. M, 1965, University of Wisconsin-Madison.

Oddo, John Edward. Models to optimize the evaluation and extraction of uranium from sedimentary ore deposits with applications to in-situ leaching. D, 1980, Rice University. 128 p.

Oddo, John Edward. Some partitioning relations of selected trace elemets for sediments of the Maumee River system near Toledo, Ohio. M, 1971, University of Toledo. 93 p.

Oddy, Richard William. Trace elements in Cretaceous rocks of Manitoba. M, 1966, University of Manitoba.

Odegard, M. E. Upper mantle structure of the North Pacific. D, 1975, University of Hawaii. 271 p.

Odegard, Mark E. Gravity interpretation using the Fourier integral. M, 1965, Oregon State University. 47 p.

Odem, Wilbert I. The delta of the diverted Brazos River of Texas. M, 1953, University of Kansas. 112 p.

Oden, Arlo Leigh. The occurrence of Mississippian brachiopods in Michigan. M, 1952, Michigan State University. 69 p.

Oden, James Russell. Microfacies and correlation of the San Andres Formation, in El Paso Products Company's mine No. 1 well, Ector County, Texas. M, 1976, University of Texas at El Paso.

Oden, Josh Winters. Carboniferous stratigraphy of the Leonard Ranch area, San Saba County, Texas. M, 1958, University of Texas, Austin.

Oden, Thomas Ellsworth. Geology of the Sunrise Quadrangle, Hickman County, Tennessee. M, 1958, Vanderbilt University.

Oder, A. Louis. A preliminary survey of the fauna and flora of a series of morainal lakes in the Montane Zone near Silver Lake (Boulder County), Colorado. M, 1932, University of Colorado.

Oder, Charles Rollin Lorain. The geology of the north-central and northeastern portions of the Harrisonburg Quadrangle, Virginia. M, 1929, University of Virginia. 178 p.

Oder, Charles Rollin Lorain. The stratigraphy, structure, and paleontology of the zinc-bearing Knox Dolomite, between Jefferson City, Missouri and Bristol, Tennessee. D, 1933, University of Illinois, Urbana.

Oderkick, Jerry Ray. Lead alpha age determinations of five Utah rocks. M, 1963, University of Utah. 31 p.

Odien, Robert J. The geology and geomorphology of the south-eastern flank of the Wet Mountains, Huerfano and Pueblo counties, Colorado. M, 1972, Colorado School of Mines. 110 p.

Odom, Arthur Leroy. A Rb-Sr isotopic study; implications regarding the age, origin and evolution of a portion of the southern Appalachians, western North Carolina, southwestern Virginia, and northeastern Tennessee. D, 1971, University of North Carolina, Chapel Hill. 92 p.

Odom, Arthur LeRoy. Rb-Sr geochronology of Precambrian ignimbrites in the Saint Francois mountains, southeast Missouri. M, 1970, University of Kansas.

Odom, Ira Edgar. Clay mineralogy and clay mineral orientation of shales and claystones overlying coal seams in Illinois. D, 1963, University of Illinois, Urbana. 137 p.

Odom, Ira Edgar. Geology of the southeast protion of the Cahokia Quadrangle, in Illinois. M, 1958, University of Illinois, Urbana.

Odom, Robert Irving, Jr. Experimental and theoretical investigations of ocean-earth acoustic coupling. D, 1980, University of Washington. 175 p.

Odouli, Khalil. Geology of the northeastern Larimer and northwestern Weld counties, Colorado. M, 1966, Colorado School of Mines. 135 p.

Odt, David Albert. Geology and geochemistry of the Sterling gold deposit, Nye County, Nevada. M, 1983, University of Nevada. 100 p.

Odum, Howard Thomas. The biogeochemistry of strontium; with discussion on the ecological integration of elements. D, 1950, Yale University. 383 p.

Oduolowu, Olusegun Akinyemi. Recoverable hydrocarbon estimates from well logs and early production histories. M, 1971, Colorado School of Mines. 100 p.

Oefelein, Rosalie Teresa. A study of the mineral composition and origin of loess with special reference to the loess of the St. Louis area. M, 1933, St. Louis University.

Oehler, Dorothy Zeller. Carbon isotopic and electron microscopic studies of organic remains in Precambrian rocks. D, 1973, University of California, Los Angeles.

Oehler, E. T. Geology of the Black Creek area, Dickinson County, Michigan. M, 1947, University of Chicago. 48 p.

Oehler, John Harlan. Morphological and biochemical changes in blue-green algae during simulated fossilization in synthetic chert; a guide to the interpretation of Precambrian microfossils. D, 1973, University of California, Los Angeles.

Oertel, Allen O. Effects of surface coal mining on ground water quality and quantity, Southwest Perry County, Illinois. M, 1980, Southern Illinois University, Carbondale. 141 p.

Oertel, George Frederick, II. Processes and structures in a marine-estuarine environment, (north shore Long Island, New York). M, 1969, University of Iowa. 149 p.

Oertel, George Frederick, Jr. Sediment-hydrodynamic interrelationships at the entrance of the Doboy Sound estuary, Sapelo Island, Georgia. D, 1971, University of Iowa. 172 p.

Oesleby, T. W. Uplift and deformation of the Uncompahgre Plateau; evidence from fill thickness in Unaweep Canyon, West-central Colorado. M, 1978, University of Colorado.

Oesterling, William Arthur. Geology of a portion of the Snake Range, Wyoming. M, 1947, University of Illinois, Urbana.

Oestreich, Ernest Sebastian. Geology of Tassajara Quadrangle, California. M, 1958, University of California, Berkeley. 83 p.

Oestrike, Richard Wilson, Jr. Inferences on the structure of aluminosilicate glasses; a high resolution solid-state silicon-29 and aluminum-27 nuclear magnetic resonance study. D, 1985, University of Illinois, Urbana. 149 p.

Oestrike, Richard Wilson, Jr. The Endion Sill, Duluth, Minnesota; mineralogy and petrology of a composite intrusion. M, 1983, University of Illinois, Urbana. 142 p.

Oetking, Philip F. The relation of the lower Paleozoic to the older rocks in the Northern Peninsula of Michigan. D, 1952, University of Wisconsin-Madison.

Ofeegbu, Goodluck Iroanya. The analysis of thermal stresses in a heated viscous hydrocarbon reservoir. D, 1985, University of Toronto.

Offer, Stuart Adam. Determination of recharge rates to a drift aquifer using bomb tritium within the saturated zone. M, 1982, Michigan State University. 53 p.

Officer, Charles B., Jr. Seismic refraction measurements in the Atlantic Ocean. D, 1952, Columbia University, Teachers College.

Offield, Terry Watson. Bedrock geology of the Goshen and Greenwood Lake quadrangles, New York. D, 1962, Yale University. 292 p.

Offield, Terry Watson. Mineralogic variation in quartz monzonite gneiss, Sterling Lake, New York. M, 1955, University of Illinois, Urbana.

Offutt, Patrick A. Sericitic clays of the Carolina slate belt of North and South Carolina and their association with Coastal Plain sediments. M, 1969, Duke University. 80 p.

Ofoh, Ebere Paulinus. Development of suitable approximation algorithms to be used in the description of heterogeneous reservoirs for secondary recovery studies. D, 1981, University of Oklahoma. 122 p.

Ofrey, Ofiafate. Use of radial and vertical magnetic field components in time-domain electromagnetic sounding. M, 1975, Colorado School of Mines. 113 p.

Ogawa, Hisashi. Evaluation methodologies for the flood mitigation potential of inland wetlands. D, 1982, University of Massachusetts. 232 p.

Ogawa, Toyokazu. Elasto-plastic, thermo-mechanical and three-dimensional problems in tunnelling. D, 1986, University of Western Ontario.

Ogbonlowo, David Babajide. Analysis of surface coal mining using computer simulation. D, 1983, West Virginia University. 296 p.

Ogbukagu, Ikechukwu Nwafo. The petrography and ceramic properties of some Pennsylvanian shales of southeastern Nebraska. M, 1956, University of Nebraska, Lincoln.

Ogburn, Reuben Walter, III. Phosphorus dynamics in an acidic, soft-water Florida lake. D, 1984, University of Florida. 152 p.

Ogden, A. E. The hydrogeology of the central Monroe County karst, West Virginia. D, 1976, West Virginia University. 324 p.

Ogden, Duncan G. Geology and origin of the kaolin at East Monkton, Vermont. M, 1960, University of Vermont.

Ogden, George Malcolm. Depositional environment of the fuller's earth clays of Northwest Florida and southwest Georgia. M, 1978, Florida State University.

Ogden, James Gordon, III. Wisconsin vegetation and climate of Martha's Vineyard, Massachusetts. D, 1958, Yale University. 177 p.

Ogden, Joseph Cornelius. Stratigraphic interpretation of some Quaternary sediments, Barrow area, Alaska. M, 1964, Iowa State University of Science and Technology.

Ogden, Julius Sterly. The growth of stannic oxide crystals. M, 1970, Ohio State University.

Ogden, Lawrence. Permian-Jurassic facies, Colorado Front Range and adjacent area; Fremont, Pueblo, and El Paso counties. D, 1958, Colorado School of Mines. 191 p.

Ogden, Maynard Blair and Henes, Walter E. Geology of the Russell-Wagon Creek area, Costilla County, Colorado. M, 1960, University of Michigan.

Ogden, P. R., Jr. The geology, major element geochemistry, and petrogenesis of the Leucite Hills volcanic rocks, Wyoming. D, 1979, University of Wyoming. 213 p.

Ogden, Palmer Raphael, Jr. The mineralogy, petrology and petrogenesis of selected units of the Two Medicine Formation, Lewis and Clark County, Montana. M, 1974, University of Florida. 59 p.

Ogg, James George. Sedimentology and paleomagnetism of Jurassic pelagic limestones ("Ammonitico Rosso" facies). D, 1981, University of California, San Diego. 219 p.

Ogidan, Richardson D. Petrography of caliche. M, 1976, Rensselaer Polytechnic Institute. 52 p.

Ogier, Stephen Hahn. Stratigraphy of the Upper Cretaceous Tokio Formation, Caddo Parish, Louisiana. M, 1962, Louisiana State University.

Ogilvie, Ida H. Geology of the Paradox Lake Quadrangle, New York. D, 1903, Columbia University, Teachers College.

Ogilvie, Jeffrey. Modeling of seismic coda, with application to attenuation and scattering in southeastern Tennessee. M, 1988, Georgia Institute of Technology. 139 p.

Ogilvie, Thomas F. Propagation of waves over an obstacle in water of finite depth. D, 1959, University of California, Berkeley.

Ogle, Burdette Adrian. Geology of the Eel River valley area, Humboldt County, California. D, 1951, University of California, Berkeley. 398 p.

Ogle, Burdette Adrian. The geology of the southern half of the Blue Lake Quadrangle, Humboldt County, California. M, 1947, University of California, Berkeley. 113 p.

Ogle, Robert Allen. The short time Fourier transform and its uses in seismic signal processing. D, 1983, University of Tulsa. 126 p.

Ogle, Ronald Kent. A study of the effect of temperature and pressure on drill hole fluid resistivity. M, 1959, Indiana University, Bloomington. 38 p.

Oglesby, Chris A. Depositional and dissolution of the Middle Devonian Prairie Formation, Williston Basin,

North Dakota and Montana. M, 1988, Colorado School of Mines. 79 p.

Oglesby, Gayle Arden. Geology of the Brentwood Limestone in Crawford and Washington counties, Arkansas. M, 1952, University of Arkansas, Fayetteville.

Oglesby, T. W. A model for the distribution of manganese, iron, and magnesium in authigenic calcite and dolomite cements in the upper Smackover Formation in eastern Mississippi. M, 1976, University of Missouri, Columbia.

Oglesby, Woodson R. Differential growth of a species of Cypridae. M, 1951, Columbia University, Teachers College.

Ogley, David S. Eruptive history of the Pine Canyon Caldera, Big Bend National Park, Texas. M, 1978, University of Texas, Austin.

Ogren, David Ernest. Stratigraphy of the Paleozoic rocks of central Oregon. M, 1958, Oregon State University. 48 p.

Ogren, David Ernest. Stratigraphy of the Upper Mississippian rocks of northern Arkansas. D, 1961, Northwestern University. 159 p.

Ogren, John Addison. Elemental carbon in the atmosphere. D, 1983, University of Washington. 134 p.

Ogryzlo, Stephen Peter. Hydrothermal experiments with gold. D, 1934, University of Minnesota, Minneapolis. 34 p.

Ogujiofor, Ikechukwu Jonathan. Active convergent margins of Northwest South America. M, 1985, University of Hawaii. 100 p.

Ogunade, Samuel Olumuyiwa. Electromagnetic response of a sphere embedded in a layered conducting Earth. D, 1973, University of Victoria. 189 p.

Ogunyomi, Olugbenga. Depositional environments, Foremost Formation (Late Cretaceous), Milk River area, southeastern Alberta. M, 1976, University of Calgary. 119 p.

Ogunyomi, Olugbenga. Diagenesis and deep-water depositional environments of lower Paleozoic continental margin sediments in the Quebec-City area, Canada. D, 1980, McGill University. 246 p.

Ogushwitz, P. R. Determination of earthquake source characteristics by computer simulation of teleseismic P-wave pulses. D, 1978, University of Connecticut. 125 p.

Ogwada, Richard Ayoro. Kinetics and mechanisms of ion exchange on soil constituents. D, 1986, University of Delaware, College of Marine Studies. 108 p.

Oh, Jin S. A gravity survey and its interpretation in the vicinity of the intersection of the San Andreas and Garlock faults in southern California. M, 1970, University of Southern California.

Ohaeri, Uche Charles. Unsteady state flow and pressure build-up behavior of a well produced at a constant bottom hole pressure in a naturally fractured reservoir. D, 1982, University of Tulsa. 222 p.

Ohashi, Yoshikazu. High temperature structural crystallography of synthetic clinopyroxenes, $(Ca,Fe)SiO_3$. D, 1973, Harvard University.

Ohkuma, Hiroshi. Numerical simulation of geopressured-geothermal aquifer phenomena. D, 1986, University of Texas, Austin. 949 p.

Ohland, Grant Lawrence. A water resource appraisal of the Skunk Creek Aquifer in Moody and Minnehaha counties, South Dakota. M, 1986, Oklahoma State University. 107 p.

Ohlander, Coryell A. Effects of rehabilitation treatments on the sediment production of granitic road materials. M, 1964, Colorado State University. 78 p.

Ohle, Ernest L. The geology of the Ozark lead mine, Fredericktown, Missouri. M, 1940, Washington University. 148 p.

Ohle, Ernest L. The influence of permeability on ore deposition in limestone and dolomite. D, 1950, Harvard University.

Ohlen, Henry R. The Steinplatte reef complex of the Alpine Triassic (Rhaetian) of Austria. D, 1959, Princeton University.

Ohlen, Robert Henry. A subsurface study of a portion of the Strawn Series in Jack, Palo Pinto, Parker and Wise counties, Texas. M, 1956, Texas Christian University. 35 p.

Ohlhber, Robert F. Robinson Ranch Field, Crook County, Wyoming. M, 1960, Michigan State University. 46 p.

Ohlhorst, Sharon Lee. Jamaican coral reefs; important biological and physical parameters. D, 1980, Yale University. 163 p.

Ohlman, James R. A structural analysis of the Butte-Eminence fault system, eastern Grand Canyon, Coconino County, Arizona. M, 1982, Northern Arizona University. 427 p.

Ohlsen, Violet E. On the Tertiary corals of Puerto Rico. M, 1928, Columbia University, Teachers College.

Ohmoto, Hiroshi. The Bluebell Mine (east shore of Kootenay lake in southeastern British Columbia), Canada. D, 1969, Princeton University. 100 p.

Ohotnicky, Raymond E. Heavy mineral residue and feldspar content of the Paleozoic rocks occurring along the Pennsylvania Turnpike. M, 1957, Miami University (Ohio). 68 p.

Ohr, Matthias. Geology, geochemistry, and geochronology of the Lems Ridge olistostrome, Klamath Mountains, California. M, 1987, SUNY at Albany. 278 p.

Ohrbom, Richard R. Structural geology of the Franciscan rocks of the San Francisco area, a re-interpretation. M, 1967, University of California, Riverside. 39 p.

Ohrenschall, Robert D. The phosphate deposits of middle Tennessee. D, 1929, The Johns Hopkins University.

Oinonen, Russell Lee. A study of selected Salina salt beds in northeastern Ohio. M, 1965, Ohio University, Athens. 60 p.

Oivanki, Stephen M. Depositional environments in the Zuloaga Formation (Upper Jurassic) in northern Mexico. M, 1973, Louisiana State University.

Oja, Reino Verner. Experiments in anatexis. D, 1959, McGill University.

Oja, Reino Verner. The porphyry-greenstone area of the Preston East Dome mines, Limited (Ontario). M, 1960, Queen's University. 164 p.

Ojakangas, Dennis R. Depauperate fauna from the Maquaketa Formation of Iowa and Illinois. M, 1959, University of Missouri, Columbia.

Ojakangas, Dennis Roger. Mathematical simulation of oil trap development. D, 1967, Stanford University. 161 p.

Ojakangas, Gregory Wayne. I, Episodic volcanism of tidally heated satellites with application to Io; II, Thermal state of an ice shell on Europa; III, Polar wander of a synchronously rotating satellite with application to Europa. D, 1988, California Institute of Technology. 186 p.

Ojakangas, Richard W. The stratigraphy and petrology of the Lamotte Formation in Missouri. M, 1960, University of Missouri, Columbia.

Ojakangas, Richard Wayne. Petrology and sedimentation of the Cretaceous Sacramento Valley sequence, Cache Creek, California. D, 1964, Stanford University. 190 p.

Ojala, Gary L. Geology of the Oregon King Mine and vicinity, Jefferson County, Oregon. M, 1964, University of Kansas. 123 p.

Ojanuza, Abayomi G. A study of soils and soil genesis in the southwestern upland of Nigeria. D, 1971, University of Wisconsin-Madison.

Ojo, Samuel Bakare. An investigation of P-wave scattering in the crust and upper mantle using travel-time fluctuations and array signal coherence. D, 1981, University of Western Ontario.

Okada, Airton Hiroshi. Facies, petrographic, and engineering analysis of Lower Cretaceous Upanema reservoirs, Ubarana Field, Potiguar Basin, Brazil. M, 1982, University of Texas, Austin. 192 p.

Okagbue, Celestine Obialo. Geologic and engineering aspects concerning slope stability of surface coal mine spoils. D, 1981, Purdue University. 368 p.

Okagbue, Celestine Obialo. Grain size analysis and petrographic evaluation of gravel deposit to discuss origin and determine engineering quality. M, 1979, Purdue University. 137 p.

Okal, Emile Andre. I, Application of normal mode theory to seismic source and structure problems; II, Seismic investigations of upper mantle lateral heterogeneity. D, 1978, California Institute of Technology. 258 p.

Okandan, Ender. Effect of temperature on oil-water capillary pressure curve of Boise Sandstone. M, 1970, Stanford University.

Okandan, Ender. The effect of temperature and fluid composition on oil-water capillary pressure curves of limestone and sandstones and measurement of contact angle at elevated temperatures. D, 1973, Stanford University. 86 p.

Okaya, David Akiharu. Seismic reflection studies in the Basin and Range Province and Panama. D, 1985, Stanford University. 147 p.

Oke, Christopher. Structure and metamorphism of Precambrian basement and its cover in the Mount Blackman area, British Columbia. M, 1982, University of Calgary. 123 p.

Oke, William Crompton. The Pleistocene section on Don and Little Don rivers, Ontario. M, 1964, University of Toronto.

Okereke, Victor Onuzurike Irokanulo. Time series and transect analysis of water quality in Saylorville and Red Rock reservoirs. D, 1986, Iowa State University of Science and Technology. 229 p.

Okerlund, Maeser D. A study of the calcite-aragonite deposits of Lake Mountain, Utah County, Utah. M, 1951, Brigham Young University. 44 p.

Okhravi, Rasool. Carbonate microfacies and depositional environments of the Joachim Dolomite (Middle Ordovician), Southeast Missouri and southern Illinois, U.S.A. D, 1983, University of Illinois, Urbana. 151 p.

Oki, Malemi. The biostratigraphy and paleomagnetic stratigraphy of the San Miguel and Olmos formations, Piedras Negras, Coahuila, Mexico. M, 1976, University of Southern California.

Okita, Patrick Masao. Geochemistry and mineralogy of the Molango manganese orebody, Hidalgo State, Mexico. D, 1987, University of Cincinnati. 379 p.

Okla, Saleh Mohamed. Subsurface stratigraphy and sedimentation of middle and upper Silurian rocks of northern Indiana. D, 1976, Indiana University, Bloomington. 143 p.

Okland, Howard E. A geomagnetic survey of Pembina, Grand Forks, and eastern Walsh counties, North Dakota. M, 1978, University of North Dakota. 51 p.

Okland, Linda E. Paleoecology of a late Quaternary biota at the McClusky Canal site, central North Dakota. M, 1978, University of North Dakota. 97 p.

Okolo, Stephen Anago. Subsurface stratigraphy of the Maxon sands and adjacent Upper Mississippian units, Leslie County, eastern Kentucky. M, 1977, University of Kentucky. 70 p.

Okon, Edem Effiong. Overburden profile studies in glaciated terrain as an aid to geochemical exploration for base metals in the Babine Lake area, British Columbia. M, 1974, University of Calgary. 74 p.

Okonkwo, Ignatius Okechukwu. Durability of particulate grouts; for isolation of low-level radioactive wastes. D, 1988, Purdue University. 490 p.

Okonkwo, Ignatius Okechukwu. Geologic and engineering evaluation of Reservoir 29 Dam, Greene-Sullivan State Forest, Indiana. M, 1984, Purdue University. 300 p.

Okonny, Isaac Peri. Geologic analysis of remote sensing imagery of the eastern Niger Delta, Nigeria. D, 1981, Purdue University. 125 p.

Okoye, Christian Udokwu. An experimental investigation of tertiary recovery of oil by alkaline steam flooding. D, 1982, University of Oklahoma. 159 p.

Okoye, David Mobike. Use of finite element method to simulate the performance of geopressured-geothermal aquifers undergoing elastic and linear viscoelastic deformations. D, 1980, University of Texas, Austin. 244 p.

Okubo, Paul G. Experimental and numerical model studies of frictional instability seismic sources. D, 1986, Massachusetts Institute of Technology. 162 p.

Okubo, S. A thermomagnetic study of lava. M, 1939, University of Hawaii.

Okulewicz, Steven C. The petrology of the Staten Island alpine ultramafic body. M, 1979, Brooklyn College (CUNY).

Okulitch, Andrew V. Geology of Mount Kobau near Oliver, British Columbia. D, 1970, University of British Columbia.

Okulitch, Vladimir. The geology of the (Ordovician) Black River Group in the vicinity of Montreal. D, 1934, McGill University.

Okulitch, Vladimir J. Ore deposits of the eastern side of the Coast Range Batholith with special reference to Atlin District. M, 1932, University of British Columbia.

Okuma, Angelo Frederick. Geology and mineral resources of Path valley, Franklin County, Pennsylvania. M, 1968, Pennsylvania State University, University Park. 75 p.

Okuma, Angelo Frederick. Structure of the southwestern Ruby Range, near Dillon, Montana. D, 1971, Pennsylvania State University, University Park. 122 p.

Okumura, Terrence A. Geology of the Lost-Lake Duling Pass area, Sangre de Cristo Mountains, Colorado. M, 1979, University of Colorado. 104 p.

Okusami, Temitope Abayomi. Land and pedogenetic characterization of selected wetlands in West Africa with emphasis for rice production. D, 1981, University of Minnesota, Minneapolis. 373 p.

Okwueze, Emeka Emmanuel. Geophysical investigations of the bedrock and the groundwater-lake flow system in the Trout Lake region of Vilas County, northern Wisconsin. D, 1983, University of Wisconsin-Madison. 144 p.

Olade, M. A. D. Bedrock geochemistry of porphyry copper deposits, Highland Valley, British Columbia. D, 1975, University of British Columbia.

Olade, Moses. Studies on the lower Aphebian (Precambrian), East Arm of Great Slave Lake, Norhtwest Territories, Canada. M, 1972, University of Alberta. 131 p.

Oladipo, Emmanuel Olukayode. On the spatial and temporal characteristics of drought in the interior plains of North America; a statistical analysis. D, 1983, University of Toronto.

Olaechea, Julio M. Cretaceous systems in northeastern Peru. M, 1960, University of California, Los Angeles.

Olafsson, Magnus. Partial melting of peridotite in the presence of small amounts of volatiles, with special reference to the low-velocity zone. M, 1980, Pennsylvania State University, University Park. 59 p.

Olander, Harvey Chester, Jr. Paleogeology of southeastern New Mexico. M, 1960, Stanford University.

Olander, Jon David. Stratigraphy and sedimentology of the Eocene Tatman Formation, Bighorn Basin, Wyoming. M, 1987, Iowa State University of Science and Technology. 133 p.

Olander, Paul A., Jr. Authigenic mineral reactions in tuffaceous sedimentary rocks, Buckhorn, New Mexico. M, 1979, University of Wyoming. 87 p.

Olaniyan, Olufemi. Multivariate analyses of sedimentological data at two early Bronze Age sites, Bab edh-Dhra and Numeira, Southeast Dead Sea Graben, Jordan. D, 1984, University of Pittsburgh. 281 p.

Olaniyan, Olufemi. Sedimentation of the archaeological site 22TS784; a sandstone rockshelter in the Bay

Springs segment of the Tennessee-Tombigbee Waterway, Tishomingo County, Mississippi. M, 1982, University of Pittsburgh.

Olarewaju, Joseph Shola. Analytical modeling of composite and layered reservoir systems. D, 1988, Texas A&M University. 267 p.

Olbinski, James S. Geology of the Buster Creek/Nehalem Valley area, Clatsop County, Oregon. M, 1983, Oregon State University. 204 p.

Olcay, Kaya Yilmay. Subsurface analysis of southern Pottawatamie County, Oklahoma. M, 1968, University of Oklahoma. 35 p.

Olcott, Perry Gail. Geology and water resources of Winnebago County, Wisconsin. M, 1965, University of Wisconsin-Madison.

Oldach, Frederick Maier. Sulfide mineralization and its development at Falls of French Creek Mines, Chester County, Pennsylvania. D, 1929, University of Pennsylvania.

Oldale, H. R. Geology of the Albert Formation in the Sussex and Elgin areas of New Brunswick. M, 1959, Acadia University.

Oldani, Martin J. Regional stratigraphy of the Paleocene Midway Group, East Texas Basin. M, 1988, Baylor University. 212 p.

Oldenburg, Douglas. Separation of magnetic substorm fields for mantle conductivity studies in the western United States. M, 1969, University of Alberta. 153 p.

Oldenburg, Douglas William. Geophysical modelling of oceanic ridges. D, 1974, University of California, San Diego.

Older, Kathy. Some mineralogic aspects of spoil material, North central Missouri. M, 1982, University of Missouri, Columbia.

Oldfield, James H. A study of the Sidney interstadial weathering profile and middle Wisconsinan till in portions of Preble and Butler counties, southwestern Ohio. M, 1977, Miami University (Ohio). 115 p.

Oldham, Albert. Geologic history of Oklahoma; Paleozoic era. M, 1927, University of Chicago. 46 p.

Oldham, Charles H. G. Gravity and magnetic investigations along the Alaska Highway and in southeastern Ontario. D, 1955, University of Toronto.

Oldham, David Martin. F-K migration of multichannel seismic data from the Yucatan Basin, Caribbean Sea. M, 1987, Texas A&M University. 132 p.

Oldham, David Wayne. Analysis of the relationships between composition and the occurrence of bloating shales and clays in the Pennsylvanian System of western Pennsylvania. M, 1979, Pennsylvania State University, University Park. 167 p.

Oldham, Richard Lewis. Structural geomorphic analysis of the Virginia Mountains, Washoe County, Nevada. M, 1971, University of Nevada. 120 p.

Oldnall, R. J. Possible sources of metals in pelagic sediments; with special reference to the Bauer Basin. M, 1975, University of Hawaii. 72 p.

Oldow, J. S. Structure and kinematics of the Luning Allochthon, Pilot Mountains, western Great Basin, USA. D, 1978, Northwestern University. 382 p.

Oldroyd, John David. Biostratigraphy of the Cambrian Glossopleura Zone, west-central Utah. M, 1973, University of Utah. 104 p.

Oldroyd, John David. Paleocommunities and environments of the Dillsboro and Saluda formations (Upper Ordovician), Madison, Indiana. D, 1978, University of Michigan. 340 p.

Olds, Edward B. Methods and costs of sinking Bunker Hill and Sulivan Mining and Concentrating Company's crescent shaft (Nevada). M, 1955, University of Nevada - Mackay School of Mines. 21 p.

Olds, Michael Warren. Ground water recharge to coal mines in eastern Kentucky. M, 1978, Ohio State University.

Olds, T. S. Occasional rapid decline and draining of Lake Bradford, Tallahassee, Florida. M, 1961, Florida State University.

Olea, Ricardo Antonio. Systematic approach to sampling of spatial functions. D, 1982, University of Kansas. 395 p.

Oleck, Robert Francis, Jr. The development of a nonlinear soil element using a modified endochronic material model. D, 1982, Syracuse University. 114 p.

Oles, Keith Floyd. The geology and petrology of the crystalline rocks of the Becker River-Mason Ridge area, Washington. D, 1956, University of Washington. 192 p.

Oles, Keith Floyd. The petrology of the Stevens Pass-Mason Ridge area, Washington. M, 1951, Washington State University. 92 p.

Olesen, James. Foraminiferal biostratigraphy and paleoecology of the Mancos Shale (Late Cretaceous) at Blue Point, Arizona. M, 1987, Northern Arizona University. 133 p.

Olesen, Marc H. Structure and petrology of Tertiary volcanic rocks in parts of Toms Cabin Spring and Dairy Valley quadrangles (Box Elder Co.), Utah. M, 1984, Utah State University. 84 p.

Oleson, Nan E. Petroleum geology of the Eocene lower Green River Formation, Duchesne and Uintah counties, Utah. M, 1986, Baylor University. 173 p.

Oleynek, Fred J. Cornulitids in the Hamilton Group of western New York. M, 1983, SUNY at Buffalo. 50 p.

Olgaard, David LeClair. Grain growth and mechanical processes in two-phased synthetic marbles and natural fault gouge. D, 1985, Massachusetts Institute of Technology. 204 p.

Olhoeft, G. R. The electrical properties of permafrost. D, 1975, University of Toronto.

Olimpio, J. C. A chemical study of garnet growth in the Moine rocks of western Scotland; evidence of the spatial extent of equilibrium. D, 1979, University of Illinois, Urbana. 167 p.

Olimpio, J. C. The compositional variation in zoned garnets from South Morar, Inverness-shire, Scotland. M, 1976, University of Illinois, Urbana. 129 p.

Olinger, Barton. The effects of high pressure on bonding and on structure stability in transition metals. D, 1970, University of Chicago. 88 p.

Oliphant, Charles W. Comparison of field and laboratory measurements of seismic velocities in sedimentary rock. D, 1948, Harvard University.

Oliphant, Jerrelyn. Geology of the igneous rocks in southeastern Penobscot township, Hancock County, Maine. M, 1969, University of Washington.

Oliphant, Joseph Lawrence. The thermodynamic properties of mixtures of water and Na-montmorillonite. D, 1980, Purdue University. 85 p.

Olive, Robert Southerland. Surface geology of the Pleasant Grove Quadrangle, Colbert and Franklin counties, Alabama. M, 1987, [Mississippi State University]. 102 p.

Olive, Wilds W. The Spotted Horse coal field, Sheridan and Campbell counties, Wyoming. D, 1953, Louisiana State University.

Oliveira Morais, Francisco Ilton de see de Oliveira Morais, Francisco Ilton

Oliveira, Jose A. The petrology of the Chibougamau greenstone belt volcanics. M, 1973, University of Georgia.

Oliveira, Jose Auto Lancaster. The stratigraphy, petrology, and paleomagnetics of the Moss Flat area, Lake County, Oregon. D, 1981, University of Nevada. 230 p.

Oliveira, Michael E. Geology of the Fish Springs mining district, Fish Springs Range, Utah. M, 1975, Brigham Young University.

Oliveira, Ruy Bruno Bacelar de see Bacelar de Oliveira, Ruy Bruno

Oliver, Adolph A., III. Gravity study of the Sonora junction area, Mono County, California. M, 1974, Stanford University.

Oliver, Dean Stuart. Benard convection with strongly temperature-dependent viscosity. D, 1980, University of Washington. 149 p.

Oliver, Deborah M. A paleomagnetic investigation of Miocene volcanic rocks in Organ Pipe Cactus National Monument, south-central Arizona. M, 1984, San Diego State University. 130 p.

Oliver, Donald McCreery and Cotey, Bradford James. The geology of the southern part of Sheep Mountain, Freemont County, Wyoming. M, 1935, University of Missouri, Columbia.

Oliver, Earl Davis. Chlorination of olivine. M, 1947, University of Washington. 33 p.

Oliver, Elisabeth Margaret. Late Cretaceous (Santonian-Campanian) foraminiferal studies, eastern Vancouver Island. M, 1979, University of Calgary. 191 p.

Oliver, Elizabeth Sumner. A Miocene flora from the Blue Mountains of Oregon. M, 1934, University of California, Berkeley. 68 p.

Oliver, Garnet W. The geology of part of the Santa Paula Quadrangle, California. M, 1940, University of California, Los Angeles.

Oliver, Gary Earl. Biostratigraphy and paleoenvironments of the Tuscahoma marls (Paleocene) in Southwest Alabama. M, 1979, University of Alabama.

Oliver, George Rick. Effects of tillage and soil properties on infiltration in a Histosol. D, 1982, Ohio State University. 127 p.

Oliver, Harold L. Geology of the Chickamauga Group in the vicinity of Andersonville, Anderson County, Tennessee. M, 1960, University of Tennessee, Knoxville. 30 p.

Oliver, J.; Dorman, Henry J. and Ewing, M. Study of shear-velocity distribution in the upper mantle by mantle Rayleigh waves. D, 1962, Columbia University, Teachers College.

Oliver, Jack. Crustal structure and surface wave dispersion; Part IV, The Atlantic and Pacific Ocean basins. D, 1953, Columbia University, Teachers College.

Oliver, Joseph W. Hydrogeochemical variations in ground water of the Ogallala Aquifer in an area of the High Plains in Oklahoma and Kansas. M, 1981, Indiana University, Bloomington. 178 p.

Oliver, Kenneth L. Gravity study of the Hanna Basin (southeast Wyoming). M, 1970, University of Wyoming. 51 p.

Oliver, L. A. Geology and mineralization in the Republic District, Ferry County, Washington. M, 1986, University of Washington. 97 p.

Oliver, Lonnie G. The Cowley shale facies of Elk County, Kansas. M, 1975, Wichita State University. 111 p.

Oliver, Lynne. The stack unit mapping, Quaternary stratigraphy, and engineering properties of the surficial geology of the Waltersburg 7.5 minute quadrangle, Pope County, Illinois. M, 1988, Southern Illinois University, Carbondale. 149 p.

Oliver, Ricky Dean. Depositional environments of the upper Wilcox Group (Eocene) in Shelby County and portions of Panola and Harrison counties, Texas. M, 1980, Stephen F. Austin State University.

Oliver, Robert L. Origin and geologic significance of the Snake River Lineament, northern Idaho-Wyoming thrust belt. M, 1982, University of Wyoming. 56 p.

Oliver, Steffenie Anne. Sediment color and lithofacies of the Southeastern United States Atlantic continental shelf and related land use plan recommendations. M, 1977, Duke University. 106 p.

Oliver, Thomas Albert. A study of the effect of uralization upon the chemical composition of the Sudbury (Ontario) Norite. M, 1949, University of Manitoba.

Oliver, Thomas Albert. Geology of the McGavock Lake area, northern Manitoba. D, 1952, University of California, Los Angeles.

Oliver, William Albert, Jr. Coral beds of the Middle Devonian (Hamilton) of central New York. M, 1950, Cornell University.

Oliver, William Albert, Jr. Stratigraphy of the (Devonian) Onondaga Limestone in central New York. D, 1952, Cornell University.

Oliver, William Benjamin, IV. Depositional systems in the Woodbine Formation (upper Cretaceous), northeast Texas. M, 1970, University of Texas, Austin.

Olivier, Jacques M. Sedimentation survey of Lake Lyndon B. Johnson, Texas. M, 1977, University of Texas, Austin.

Olivier, Wendell Gregory. An analysis of sediment transport and deposition of the Grays Lake basin, Bonneville and Caribou counties, Idaho. M, 1980, Idaho State University. 116 p.

Olivieri, Cesar A. Borges see Borges Olivieri, Cesar A.

Olle, John Michael. A geological and economic study of the sand deposits of south central Lenawee County, Michigan, and of north central Fulton County, Ohio. M, 1971, University of Toledo. 102 p.

Ollenburger, Ronald D. Source and stratigraphy of the Livermore Gravels, Alameda County, California. M, 1987, California State University, Hayward. 218 p.

Ollerenshaw, N. C. Stratigraphic problems of the western Shickshock Mountains in the Gaspe Peninsula (Quebec). D, 1963, University of Toronto.

Ollerenshaw, N. C. The Simcoe diamond drill core section of the Paleozoic rocks of southwestern Ontario. M, 1959, University of Toronto.

Ollerenshaw, Neil C. The Simcoe diamond drill core section of the Palaeozoic rocks of southwestern Ontario. M, 1958, University of Toronto.

Ollila, Paul William. Bedrock geology of the Santanoni Quadrangle, New York. D, 1984, University of Massachusetts. 280 p.

Olling, C. R. The seasonal variation of the zonal velocity of the Atlantic equatorial undercurrent. M, 1977, Texas A&M University.

Olm, Mark C. Electromagnetic scale model study of the dual frequency differencing technique. M, 1981, Colorado School of Mines. 291 p.

Olmore, Stephen Duane. Style and evolution of thrusts in the region of the Mormon Mountains, Nevada. D, 1971, University of Utah. 213 p.

Olmore, Stephen Duane. The structure and stratigraphy of East Mormon range and vicinity, southern Nevada. M, 1969, University of Utah. 98 p.

Olmstead, Dennis L. Clay mineralogy of the Washington continental slope. M, 1972, University of Washington. 40 p.

Olmstead, Elizabeth Warren. The physiographic development of the Reading Prong (Pennsylvania). M, 1933, Smith College. 87 p.

Olmstead, Franklin H. Geology of the Little Puente Hills and western San Jose Hills, California. M, 1948, Pomona College.

Olmstead, Lewis Bertie. Moisture relations of soils in centrifugal force fields. D, 1933, American University. 26 p.

Olmsted, Franklin Howard. Geology of the pre-Cretaceous rocks of the Pilot Hill and Rocklin quadrangles, California. D, 1961, Bryn Mawr College. 193 p.

Olmsted, James Frederick. A study of graphitic slates from northern Michigan. M, 1962, Michigan State University. 66 p.

Olmsted, James Frederick. Petrology of a differentiated anorthositic intrusion in northwestern Wisconsin. D, 1966, Michigan State University. 155 p.

Olmsted, Richard Warren. Geochemical studies of uranium in south-central Oklahoma. M, 1975, Oklahoma State University. 116 p.

Olney, Sylvie L. An investigation of the relationship between the coefficient of permeability and effective grain size of unconsolidated sands. M, 1983, Boston University. 61 p.

Olorunfemi, Biodun Elijah Nathaniel. The geochemistry and mineralogy of Recent sediments of the Niger Delta. D, 1983, University of Western Ontario. 325 p.

Olowomeye, Richard Boluwaji. The management of solid waste in Nigerian cities. D, 1987, Howard University. 279 p.

Olsen, Barbara A. Petrogenesis of the Concord gabbro-syenite complex, Cabarrus County, North Carolina. M, 1982, University of Tennessee, Knoxville. 72 p.

Olsen, Ben L. Geology of the Baldy area, west slope of Mount Timpanogos, Utah County, Utah. M, 1955, Brigham Young University. 33 p.

Olsen, Bruce Michael. Stratigraphic occurrence of argillaceous beds in the St. Peter Sandstone, Twin City Basin. M, 1976, University of Minnesota, Minneapolis. 89 p.

Olsen, C. R. Radionuclides, sedimentation and the accumulation of pollutants in the Hudson Estuary. D, 1979, Columbia University, Teachers College. 263 p.

Olsen, Clayton E. Geostatistical and geochemical investigations of the uranium distributions in stream sediments and in surface and ground waters from the Estancia Valley, the Black Hawk mining district, and an area north of the Grants mineral belt, New Mexico; application to methods of geochemical exploration. M, 1982, University of New Mexico. 417 p.

Olsen, Donald R. A differential thermal X-ray and optical analysis of some chlorites. M, 1951, University of Utah. 26 p.

Olsen, Donald R. Geology and mineralogy of the Delno mining district and vicinity, Elko County, Nevada. D, 1960, University of Utah. 120 p.

Olsen, Edward John, Jr. Inorganic formation of graphite-quartz deposits. D, 1959, University of Chicago. 135 p.

Olsen, John Roger. A gravity study of the Nampa-Caldwell area, Canyon County, Idaho. M, 1979, Brigham Young University.

Olsen, Mikael Per Jexen. A mathematical model for predicting frost penetration in saturated porous materials. D, 1982, University of Illinois, Urbana. 154 p.

Olsen, Paul Eric. Comparative paleolimnology of the Newark Supergroup; a study of ecosystem evolution (Volumes I and II). D, 1984, Yale University. 756 p.

Olsen, Rebecca Sarah. Interpretation of depositional environment and morphology of sandstones in the Jurassic Smackover, Thomasville Field, Rankin County, Mississippi. M, 1980, Texas A&M University. 145 p.

Olsen, Robert C. Geology of the northwestern Inyo Mountains, Inyo County, California. M, 1970, San Jose State University. 73 p.

Olsen, Robert Roger. Petrology and fission-track ages of the North Fork Pluton and associated rocks, northeastern Oregon. M, 1972, University of Oregon. 116 p.

Olsen, Roger L. The crystal and molecular structures of acetamidinium, tetrachlorocobaltate and acetamidinium tetrachlorocuprate. D, 1979, Colorado School of Mines. 188 p.

Olsen, Royce W. Seismic time-distance relationships from P-wave arrivals at Socorro, (New Mexico). M, 1965, New Mexico Institute of Mining and Technology. 31 p.

Olsen, Sakiko Nakaya. Petrology of the Baltimore Gneiss. D, 1972, The Johns Hopkins University. 357 p.

Olsen-Heise, Katrina Edith Desa. Depositional systems of the Mississippian MC-3 (Alida Beds), Pierson Field, Manitoba. M, 1988, University of Calgary. 147 p.

Olson, Allen Hiram. Forward simulation and linear inversion of earthquake ground motions. D, 1982, University of California, San Diego. 123 p.

Olson, Boyd E. On the abyssal temperatures of the world oceans. D, 1968, Oregon State University. 151 p.

Olson, Carolyn G. A mechanism for the origin of Terra rossa in southern Indiana. D, 1979, Indiana University, Bloomington. 150 p.

Olson, Carolyn G. The stratigraphy and morphology of loess and related Paleosols in southwestern Indiana. M, 1977, Indiana University, Bloomington. 95 p.

Olson, Christopher J. Late Paleogene and Neogene radiolarian biostratigraphy of the Southern Ocean. M, 1980, Northern Illinois University. 139 p.

Olson, Clifford M. A study of the diagenetic history of the Point Pleasant Formation. M, 1983, University of Cincinnati. 184 p.

Olson, Dale R. Subsurface geology of McPherson County, Kansas. M, 1956, Kansas State University. 53 p.

Olson, Dan E. A study of the local and regional controls on the composition and mineralogy of manganese nodules from the northeastern Equatorial Pacific. D, 1986, University of Wisconsin-Madison. 530 p.

Olson, Dan E. Microchemical and textural studies of selected manganese nodules from a siliceous ooze site in the Northeast equatorial Pacific. M, 1979, University of Wisconsin-Madison.

Olson, Daniel J. Surface and subsurface geology of the Santa Barbara-Goleta metropolitan area, Santa Barbara County, California. M, 1983, Oregon State University. 71 p.

Olson, David N. Hydrogeology and hydraulic characteristics of a fractured dolomite aquifer at an EPA Superfund Site, Ogle County, Illinois. M, 1988, Northern Illinois University. 148 p.

Olson, David Peck. The petrology of the Park City Formation (Permian), northwestern Wind River Basin, Wyoming. M, 1982, Miami University (Ohio). 138 p.

Olson, Donald L. Biometric analysis of two large benthonic foraminifera, Kingshill Marl, St. Croix, U.S. Virgin Islands. M, 1976, Northern Illinois University. 139 p.

Olson, Edwin Andrew. The problem of sample contamination in radiocarbon dating. D, 1963, Columbia University, Teachers College. 332 p.

Olson, Eric Robert. Oxygen and sulphur isotope geochemistry of marine evaporites. D, 1975, McMaster University. 254 p.

Olson, Everett C. The dorsal axial skeleton and inferred musculature of certain primitive Permian tetrapods. D, 1935, University of Chicago. 87 p.

Olson, Everett C. The Upper Mississippian formations of North America. M, 1933, University of Chicago. 132 p.

Olson, George Harrison. Differentiation of Lake Lahontan sediments (Pleistocene) in western Nevada (Churchill, Pershing, Storey, and Washoe counties) by grain size parameters. M, 1970, University of Nevada. 40 p.

Olson, Gregory A. Geology of the Grouse Creek area, China Mountain Quadrangle, California. M, 1978, Oregon State University. 129 p.

Olson, Harry J. The geology and tectonics of the Idaho Porphyry Belt (Tertiary) from the Boise Basin to the (Casto) Quadrangle. D, 1968, University of Arizona. 214 p.

Olson, Harry J. The geology of the Glove Mine, Santa Cruz County, Arizona. M, 1961, University of Arizona.

Olson, Henry David. Mechanism of transverse petroleum migration. M, 1965, Rice University. 39 p.

Olson, Hilary Clement. Middle Tertiary stratigraphy, depositional environments, paleoecology and tectonic history of the southeastern San Joaquin Basin, California. D, 1988, Stanford University. 382 p.

Olson, James P. Geology and mineralization of the North Santiam mining district, Marion County, Oregon. M, 1979, Oregon State University. 135 p.

Olson, Jean Marie. The sedimentation and petrology of the lower Proterozoic McCaslin Formation, northeastern Wisconsin. M, 1981, University of Minnesota, Duluth.

Olson, Jerry. Vegetation substrate relations in Lake Michigan sand dune development. D, 1951, University of Chicago.

Olson, Jerry Chipman. Geologic setting of the Mountain Pass rare earth deposits San Bernardino County, California. D, 1953, University of California, Los Angeles.

Olson, Karin Elizabeth. Greenschist-amphibolite metabasites at the northern margin of the Cape Smith foldbelt, Ungava, Quebec. M, 1983, McGill University. 114 p.

Olson, Kurt Nathaniel. Assessment of Upper Mississippi River floodplain changes with sequential aerial photography. D, 1981, University of Minnesota, Minneapolis. 280 p.

Olson, Lawrence John, Sr. Geology of eastern Bryan County, Oklahoma. M, 1965, University of Oklahoma. 64 p.

Olson, Norman Keith. Depositional factors of the Upper Cretaceous Eagle Formation, south-central Montana. M, 1961, University of Iowa. 128 p.

Olson, P. E. The stratigraphy, structural geology and geochemistry of the Fox Lake massive sulfide deposit. M, 1987, University of Manitoba.

Olson, Peter Lee. Internal waves and hydromagnetic induction in the Earth's core. D, 1977, University of California, Berkeley. 136 p.

Olson, R. A. Geology and genesis of zinc-lead deposits within a late Proterozoic dolomite, northern Baffin Island, N.W.T. D, 1977, University of British Columbia.

Olson, Reginald Arthur. Statistical analysis of geochemical data from Cape Rosier, Maine (Paleozoic). M, 1971, University of Western Ontario. 126 p.

Olson, Richard Hubbell. Geology of the Promontory Range, Box Elder County, Utah. D, 1960, University of Utah. 378 p.

Olson, Robert K. The stratigraphy and petrology of the Eau Claire Formation in northern Illinois. M, 1973, Northern Illinois University. 84 p.

Olson, Robert Kenneth. Factors controlling uranium distribution in Upper Devonian-Lower Mississippian black shales of Oklahoma and Arkansas. D, 1982, University of Tulsa. 209 p.

Olson, Robert Laurence. Engineering geology and relative stability of ground adjacent to the Sawyer Place area, Cincinnati, Ohio. M, 1988, University of Cincinnati. 102 p.

Olson, Robert Wendell. Stratigraphy and structure of the western Munzur Mountains, eastern central Turkey. D, 1977, Rice University. 212 p.

Olson, Roger K. Bedrock geology of the southwest sixth of the Saponac Quadrangle, Penobscot and Hancock counties, Maine. M, 1972, University of Maine. 61 p.

Olson, Roger W. Valley morphology and landslides, Roan Creek and Parachute Creek basins, western Colorado. M, 1974, Colorado State University. 111 p.

Olson, Russel D. A study of the time-dependent deformation of sedimentary rock. M, 1977, University of Houston.

Olson, Steven Frederick. Geology of the Potrerillos District, Atacama, Chile. D, 1984, Stanford University. 296 p.

Olson, Terrilyn M. Sedimentary tectonics of the Jalipur Sequence, Northwest Himalaya, Pakistan. M, 1982, Dartmouth College. 152 p.

Olson, Tes Lewis. Earthquake surveys of the Roosevelt Hot Springs and the Cove Fort areas, Utah. M, 1976, University of Utah. 80 p.

Olson, Theodore M. The geology and groundwater resources of part of the Hangman and Marshall Creek drainage basins, Spokane County, Washington. M, 1975, Eastern Washington University. 70 p.

Olson, Thomas L. Petrology and stratigraphy of the carbonates in the Benwood-Arnoldsburg Limestone interval of the Monongahela Group (Pennsylvanian) in western Noble County, Ohio. M, 1978, University of Akron. 80 p.

Olson, Timothy John. Sedimentary evolution of the Miocene-Pliocene Camp Davis Basin, northwestern Wyoming. M, 1987, Montana State University. 62 p.

Olson, Todd Rowland. Evolutionary and adaptive significance of cranial specializations in the Tarsiiformes Gregory, 1915. M, 1974, University of California, Berkeley. 72 p.

Olson, Victor Emanuel. The geology of the Smedley Quadrangle, Washington County, Indiana. M, 1952, Indiana University, Bloomington. 24 p.

Olson, Walter Sigfrid. The origin and associations of magnesite. M, 1925, University of Minnesota, Minneapolis. 30 p.

Olson, Walter Sigfrid. The range of brachiopods in the Cherokee Formation near Columbia, Missouri. M, 1931, University of Missouri, Columbia.

Olson, William R. Seismic investigation of the Round Butte area, Northcentral Colorado. M, 1977, Colorado School of Mines. 42 p.

Olsson, Richard K. The Cretaceous Tertiary boundary in New Jersey. M, 1955, Rutgers, The State University, New Brunswick. 58 p.

Olsson, Richard Keith. Late Cretaceous-early Tertiary stratigraphy of New Jersey. D, 1958, Princeton University. 354 p.

Olsson, William A. Fracture pattern analysis along the Rattlesnake Ferry Fault, Jackson and Union counties, Illinois. M, 1968, Southern Illinois University, Carbondale. 52 p.

Olsson, William Arthur. Deformational behavior of intact and prefaulted Crown Point Limestone at elevated temperatures and pressures. D, 1973, University of Illinois, Urbana. 101 p.

Olszewski, Gregory P. An interpretation of the sedimentary environments of the Michigan Formation (Mississippian) in Michigan. M, 1978, Wayne State University.

Olszewski, William John. A geochronologic study of metamorphic rocks in northeastern Massachusetts. D, 1978, Massachusetts Institute of Technology. 295 p.

Oltz, Donald F., Jr. Numerical analysis of palynological data from Cretaceous and early Tertiary sediments in east-central Montana. D, 1968, University of Montana. 291 p.

Olup, Bernard J., Jr. Petrology of glacial drift in northern Marathon County, Wisconsin. M, 1969, University of Wisconsin-Milwaukee.

Olver, Richard. Resistivity investigation of a buried valley in Summit County, Ohio. M, 1981, University of Akron. 128 p.

Olynyk, Max. Some physical rock parameters of New York City water tunnel No. 3 and their relation to ground water inflows. M, 1976, Brooklyn College (CUNY).

Olyphant, G. A. Models of cirque development and postglacial modification. D, 1979, University of Iowa. 100 p.

Oman, Carl Henry. A bio-stratigraphic analysis of the type Moody's Branch Formation. M, 1960, Rutgers, The State University, New Brunswick. 52 p.

Oman, Carl Henry. A biostratigraphic study of the Lisbon Formation (Eocene) of Alabama. D, 1965, Florida State University.

Oman, Charles Lee. Geology of the Fieldhouse Cave, (outside of Franklin, West Virginia). M, 1966, American University. 79 p.

Omana, Miguel Angel Alvarado. Gravity and crustal structure of the south-central Gulf of Mexico, the Yucatan Peninsula, and adjacent areas from 17°30′N to 26°N and from 84°W to 93°W. M, 1987, Oregon State University.

Omar, Gomaa Ibrahim. Phanerozoic tectono-thermal history of the Nubian Massif, Eastern Desert, Egypt, and its relationship to the opening of the Red Sea as revealed by fission-track studies. D, 1985, University of Pennsylvania. 164 p.

Omar, Mazmumah Mamudah. Shallow seismic reflection and refraction in the Athens City area. M, 1982, Ohio University, Athens. 145 p.

Omara-Ojungu, Peter Hastings. Resource management in mountainous environments, the case of the East Slopes region, Bow River basin, Alberta, Canada. D, 1980, University of Waterloo.

Omari, Sid'Ali N. Integrated geophysical study of the central Algerian Sahara. M, 1976, Colorado School of Mines. 74 p.

Omer, Mekki A. Measurement and simulation of mulch and tillage effects on soil water conservation. D, 1987, Washington State University. 159 p.

Omernik, John Beebe. The stratigraphy of the Ordovician Galena Dolomite in southwestern Wisconsin. M, 1958, University of Wisconsin-Madison.

Omole, Olusegun. Determination of the optimum solvent bank size requirement during CO_2 flooding. D, 1983, Texas A&M University. 204 p.

Omolo, Fenner O. Impacts of abandoned surface coal-mined lands on the local population in southwestern Indiana. M, 1984, Indiana State University. 66 p.

Omoregie, Osazuwa Sunday. Steam drive; definition and enhancement. D, 1981, University of Southern California.

Omran, Abdelmoneim Ben see *Ben Omran, Abdelmoneim*

Omre, Karl Henning. Alternative variogram estimators in geostatistics. D, 1985, Stanford University. 281 p.

Onasch, Charles M. Analysis of the minor structural features in the north-central portion of the Pelham Dome. M, 1973, University of Massachusetts. 87 p.

Onasch, Charles Martin. Structural evolution of the western margin of the Idaho Batholith in the Riggins, Idaho, area. D, 1977, Pennsylvania State University, University Park. 296 p.

Onasch, Christine Condon. Analysis of the relationships between the fundamental properties and the derived qualities for 44 samples of lower Kittanning underclay. M, 1977, Pennsylvania State University, University Park. 86 p.

Ondrick, Charles William. Petrography and geochemistry of the Rensselaer Graywacke, Troy, New York. D, 1968, Pennsylvania State University, University Park. 237 p.

Ondrick, Charles William. Statistical comparison of the Keener and Big Injun sands (Mississippian), Pleasants County, West Virginia. M, 1965, Pennsylvania State University, University Park. 185 p.

Oneacre, John William. Geology and engineering characteristics of Pennsylvanian redbed residual soils and soil slips in southeastern Ohio. M, 1978, Kent State University, Kent. 56 p.

Onega, Lawrence Kerokadho. Geology of the Roberts Mine in the Edison magnetite region, Sussex County, New Jersey. M, 1970, Rutgers, The State University, New Brunswick. 62 p.

Onesti, Lawrence Joseph. Stream sinuosity as it relates to the network hierarchy of the Pecatonica River in southwestern Wisconsin. D, 1973, University of Wisconsin-Madison.

Ong, Han-Ling. The effectiveness of natural organic matter in concentrating metals; Part I, Adsorption of copper by peat, lignite, and bituminous coal; Part II, Translocation and immobilation of metals in humus solution. D, 1967, Colorado School of Mines. 216 p.

Ongley, Jennifer S. A petrologic and oxygen isotopic study of eclogite xenoliths from the Roberts Victor kimberlite pipe, South Africa. M, 1986, University of Rochester. 148 p.

Ongley, Lois K. Crystallization temperatures and oxygen fugacities of magmas from the Southeast Indian Ocean Ridge system. M, 1977, Texas A&M University. 70 p.

Onions, Diane. Dimensional grain orientation of Ordovician turbidite greywackes. M, 1965, McMaster University. 102 p.

Onken, Beth Renee. Lidaconus palmettoensis, n. gen., n. sp.; an enigmatic new fossil from the Lower Cambrian of western Nevada. M, 1987, University of California, Davis. 55 p.

Onoda, Kiyoka. The correlation of the Newcastle and associate sandstones. M, 1948, South Dakota School of Mines & Technology.

Onofryton, Jerry K. The relationship between size and shape of sand and silt using Fourier shape analysis. M, 1973, Michigan State University. 39 p.

Onstott, Gregory Erle. Processing and display of offset dependent reflectivity in reflection seismograms. M, 1984, University of Texas, Austin. 113 p.

Onstott, Tullis Cullen. Paleomagnetism of the Guyana Shield, Venezuela and its implications concerning Proterozoic tectonics of South America and Africa. D, 1980, Princeton University. 430 p.

Ontuna, Kazim Ates. Experimental and analytical description of stress-induced anisotropy in sand. D, 1984, University of Colorado. 330 p.

Onuonga, Isaac Oriechi. The geology and geochemistry of Macalder Mine, Kenya. M, 1983, Carleton University. 118 p.

Onuschak, Emil. Carbonate compounds in some alluvial fans of northern Grass Valley, Nevada. M, 1960, University of Nevada. 91 p.

Onyeagocha, Anthony Chukwuma. Optical determination of composition, twinning, and structural state of the Adirondack anorthosite plagioclase, (eastern New York). M, 1970, Iowa State University of Science and Technology.

Onyeagocha, Anthony Chukwuma. Petrology and mineralogy of the Twin Sisters Dunite, Washington. D, 1973, University of Washington. 135 p.

Onyegam, Emmanuel I. The stratigraphy of the Buda Limestone on the western margin of the East Texas Basin. M, 1983, Baylor University. 133 p.

Onyejekwe, Okey Oseloka. A Galerkin finite element model for solution of nonconservative water quality transients in an estuarine system. D, 1983, University of California, Davis. 291 p.

Onyia, Ernest Chijioke. An investigation to determine the causes of low groundwater yield from the Brassfield Limestone in a portion of Greene County, Ohio. M, 1981, Wright State University. 116 p.

Oostdam, Bernard Lodewijk. Suspended sediment transport in Delaware Bay. D, 1971, University of Delaware.

Opalka, Richard B. Lithofacies of the Smackover Formation (Upper Jurassic) of the Teas, Carter Bloxom, and Box Church gas fields in East Texas. M, 1980, Miami University (Ohio). 89 p.

Opell, Douglas A. Hydraulic fracture of sandstone and limestone in the presence of surfactants. M, 1986, Indiana University, Bloomington. 99 p.

Openshaw, Ronald E. The low temperature heat capacities of analbite, low albite, microcline, and sanidine. D, 1974, Princeton University. 325 p.

Opera, John L. A sedimentological study of the Thorold, Grimsby, and Kodak sandstones, Silurian, western New York. M, 1965, SUNY at Buffalo.

Ophori, Duke Urhobo. A numerical simulation analysis of regional groundwater flow for basin management; plains regions, Alberta. D, 1986, University of Alberta. 342 p.

Ophori, Duke Urhobo. Evaluation of the Forwell induced infiltration site, Kitchener-Waterloo area. M, 1982, University of Waterloo. 77 p.

Opitz, Dale A. Gravity anomalies in the Sandy Lake area of Mercer County, Pennsylvania. M, 1972, Slippery Rock University. 49 p.

Opland, Homer N. The design of a long period horizontal station seismometer employing a resistance bridge transducer. M, 1949, Colorado School of Mines. 45 p.

Opp, Albert G. A magnetometer survey of the Keene Dome, McKenzie County, North Dakota. M, 1955, St. Louis University.

Oppel, Richard E. Geology of the Sheep Mountain area, Jackson County, Colorado. M, 1953, Colorado School of Mines. 85 p.

Oppel, Theodore Wells. The unconformity between the Fresnal Group and the Bursum Formation, Sacramento Mountains, New Mexico. M, 1957, University of Wisconsin-Madison. 36 p.

Oppenheimer, Joan Mary. Gravity modeling of the alluvial basins, southern Arizona. M, 1980, University of Arizona. 81 p.

Opper, Steven Carl. The hydrogeology of Lake Wauberg and vicinity, Alachua County, Florida. M, 1982, University of Florida. 119 p.

Oppliger, Gary Lee. Three-dimensional terrain effects in electrical and magnetometric resistivity surveys. D, 1982, University of California, Berkeley. 253 p.

Opstad, Erik Alan. The application of mole component systems to chemical variation studies in volcanic rocks. M, 1978, University of Iowa. 90 p.

Or, Arthur Chunchiu. Buoyancy-driven instabilities in a rapidly rotating cylindrical annulus. D, 1985, University of California, Los Angeles. 158 p.

Orajaka, Ifeanacho Paul. Mineralogy and uranium geochemistry of selected volcaniclastic sediments in the western United States; an exploration model. D, 1981, University of Texas at El Paso. 393 p.

Orajaka, Stephen. Geology of the Obudu area, eastern Nigeria. D, 1963, Universite Laval. 218 p.

Orajaka, Stephen Onyebueke. A geological, geochemical, and radiometric investigation of the Huron River and part of the Huron Mountain area, Houghton, Michigan. M, 1951, Michigan Technological University. 82 p.

Oralratmanee, Komol. A gravity survey in the northern end of Socorro Basin, Rio Grande rift zone, New Mexico. M, 1972, New Mexico Institute of Mining and Technology.

Orange, Daniel Lewis. Metamorphic petrology, pressure-temperature paths, and tectonic evolution of the Mount Cube Quadrange, New Hampshire and Vermont. M, 1985, Massachusetts Institute of Technology. 275 p.

Oray, Erdogan. Regional gravity investigation of the eastern portion of the northern peninsula of Michigan. D, 1971, Michigan State University. 86 p.

Orazulike, Donatus Maduka. Hydrothermal alteration associated with porphyry copper mineralization in Fawn Peak Stock, Okanogan County, Washington. M, 1979, Washington State University. 71 p.

Orazulike, Donatus Maduka. Igneous petrology and petrochemical variation in the Coyote Creek Pluton, Colville Batholith, and its relation to mineralization at Squaw Mountain, Colville Indian Reservation, Washington. D, 1982, Washington State University. 275 p.

Orback, C. J. Geology of the Fox Creek area, Albany County, Wyoming. M, 1960, University of Wyoming. 62 p.

Orchard, David Merle. Geology of the Robinson Creek-Ukiah area, northern Coast Ranges, California. M, 1979, University of Texas, Austin.

Orcutt, John A. Structure of the oceanic crust and upper mantle. D, 1976, University of California, San Diego. 203 p.

Ord, Alison. Determination of flow stress from microstructures of mylonitic rocks. D, 1981, University of California, Los Angeles. 239 p.

Ordan, Laura G. Diagenesis of Pleistocene carbonate sediments in a borehole; Mediterranean coastal plain of Isreal. M, 1974, Rensselaer Polytechnic Institute. 87 p.

Ordonez, José Luis. Preparation and properties of barium clinoptilolite. M, 1970, University of Alaska, Fairbanks. 86 p.

Ordonez, Steve. Permian (Guadalupian) shelf deposition and diagenesis; the Tansill Formation of the Cheyenne (Capitan) Field, Winkler County, Texas. M, 1981, Northern Arizona University. 83 p.

Ordway, Richard J. Geology and structure of the Buffalo Mountain-Cherokee Mountain area in northeastern Tennessee. D, 1948, Yale University.

Ore, H. T. The braided stream depositional environment. D, 1963, University of Wyoming. 205 p.

Ore, Henry Thomas. Geology of a portion of the Heart Butte Quadrangle, Sawtooth Mountains, Montana. M, 1959, Washington State University. 105 p.

Orem, William Henry, V. Organic matter in anoxic pore water from Great Bay, New Hampshire. D, 1982, University of New Hampshire. 386 p.

Organ, David William. Pleistocene gravels of the Red River valley (Manitoba). M, 1953, University of Manitoba.

Orgill, Jeffrey R. The Permian-Triassic unconformity and its relationship to the Moenkopi, Kaibab, and White Rim formations in and near the San Rafael Swell, Utah. M, 1971, Brigham Young University.

Orgren, April Hoefner. Lithostratigraphy and depositional environments of the Pitkin Limestone and Fayetteville Shale (Chesterian) in portions of Wagoner, Cherokee and Muskogee counties, Oklahoma. M, 1979, University of Oklahoma. 144 p.

Orgren, Mark David. Lithostratigraphy and depositional environments of the Morrowan Series (Lower Pennsylvanian) in parts of Cherokee, Wagoner and Mayes counties, Oklahoma. M, 1979, University of Oklahoma. 163 p.

Orheim, Olav. A 200-year record of Glacier Mass Balance at Deception Island, Southwest Atlantic Ocean, and its bearing on models of global climatic change. D, 1972, Ohio State University.

Orians, Kristin Jean. The marine geochemistry of the hydrolysis elements, aluminum and gallium. D, 1988, University of California, Santa Cruz. 117 p.

Oriard, Lewis L. A seismic investigation of subsurface geology in the Missouri River flood-plain near St. Charles, Missouri. M, 1951, St. Louis University.

Oriel, Steven S. Geology of the Basin Quadrangle, Idaho; Albion Mountains and Middle Mountain metamorphic core complex and surrounding Cenozoic. M, 19??, University of British Columbia. 99 p.

Oriel, Steven S. Geology of the Hot Springs window, North Carolina. D, 1949, Yale University.

Oriel, William Michael. Detailed bedrock geology of the Brenda copper-molybdenum mine, Peachland, British Columbia. M, 1972, University of British Columbia.

Orkan, Nebil I. Regional joint evolution and paleothermometry-barometry from fluid inclusions in the Valley and Ridge Province of Pennsylvania in relation to the Alleghany Orogeny. M, 1986, Pennsylvania State University, University Park.

Orlander, Peter R. Petrology and mineralization of the Snow Creek area, Humboldt County, Nevada. M, 1981, Eastern Washington University. 60 p.

Orlando, M. E. Late Pleistocene mammalian extinctions in North Africa; extent, theorized causes, and implications. M, 1977, University of Minnesota, Duluth.

Orlando, Robert C. Sedimentology and bedform morphology in lower Cook Inlet, Alaska. M, 1983, San Jose State University. 75 p.

Orlansky, Ralph. A stratigraphic study of fusulinid foraminifera in the Cherokee Group (Penn.) of western Mo. and Kansas. M, 1951, University of Missouri, Rolla.

Orlansky, Ralph. Palynology of the Upper Cretaceous Straight Cliffs Sandstone, Garfield County, Utah. D, 1967, University of Utah. 199 p.

Orlean, Howard M. Stratigraphy and structural analysis of the Gold Hill area, Inchelium Quadrangle, Stevens County, Washington. M, 1981, Eastern Washington University. 57 p.

Orlich, Michael S. Physical and chemical properties of overburden materials overlying coals V and VI in western Greene County, Indiana. M, 1977, Indiana University, Bloomington. 45 p.

Orlin, Hyman. Gravity meter observations aboard a surface vessel and their geodetic applications. D, 1962, Ohio State University. 151 p.

Orlopp, Donald Easton. Environmental study of stages within the Gimlet and Exline cyclothems of the Eastern Interior Basin and part of the Western Interior Basin. M, 1962, University of Illinois, Urbana.

Orlopp, Donald Easton. Regional paleoenvironmental study of some Middle Pennsylvanian strata of the Midcontinent region and textural analysis of included limestones. D, 1964, University of Illinois, Urbana. 166 p.

Orlowski, Louis Allen. A stochastic model for nearshore coastal processes. M, 1974, Michigan State University. 36 p.

Orlowski, Wayne C. Paleogeomorphology and sedimentary analysis of the Brunswick Formation, N.Y. Thruway, Rockland County, N.Y. M, 1979, Rutgers, The State University, Newark. 87 p.

Ormiston, Allen R. Lower and Middle Devonian trilobites of the Canadian Arctic Islands. D, 1964, Harvard University.

Ormiston, Allen R. The regional stratigraphy and paleontology of the Devonian rocks of Bathurst Island, Northwest Territories. M, 1960, Dartmouth College. 158 p.

Ormond, John. Empirical magnetic modeling of Cretaceous submarine volcanic mounds in South Texas. M, 1984, Texas Tech University. 136 p.

Ormsby, Marka R. Probability that another intensity X event could occur in the S.E. during a 200 year period. M, 1980, Georgia Institute of Technology. 100 p.

Ormsby, Walter B. The crystal structure of manganese dioxide. M, 1930, University of Arizona.

Orndorff, Harold Anton. A study of mineral diagenesis occurring in the Mather Sandstone lentil (Pennsylvanian-Permian), northern West Virginia and southwestern Pennsylvania. M, 1980, Miami University (Ohio). 133 p.

Orndorff, Randall C. Conodont biostratigraphy of the Cambrian-Ordovician boundary interval in the northern Shenandoah Valley of Virginia, U.S.A. M, 1985, Old Dominion University. 106 p.

Ornelas, Richard Henry. Clay deposits of Utah County, Utah. M, 1953, Brigham Young University. 80 p.

Ornsby, Walter B. The crystal structure of manganese dioxide. M, 1930, University of Arizona.

Ornstein, Peter. Numerical simulation of nondispersive contaminant transport. M, 1982, University of Nevada. 197 p.

Oro, Fe Haresco. The carbon and oxygen isotopic composition of Late Cretaceous fossils from the Bearpaw Formation, Saskatchewan. M, 1977, University of Saskatchewan. 106 p.

Oropeza, Romolo Marquez. The hydrogeology and development of the groundwater resources in the El Asentamiento Campesia, El Cortez, estado Argua, Venezuela. M, 1970, University of Arizona.

Oros, Robert. Sediment-water interactions in a eutrophic lake. M, 1976, Kent State University, Kent. 98 p.

Orpwood, Timothy Gordon. An evaluation of methods used to measure groundwater flow and transport characteristics in clayey deposits at two sites in Essex County, Ontario. M, 1984, University of Windsor. 204 p.

Orr, Cynthia Dolores. A seismotectonic study and stress analysis of the Kermit seismic zone, Texas. D, 1984, University of Texas, Austin. 289 p.

Orr, Donald G. Geology of the Mt. Pittsburg Quadrangle, El Paso, Fremont and Pueblo counties, Colorado. M, 1976, Colorado School of Mines. 171 p.

Orr, Elizabeth Decker. Beach and glacial particle morphogeny, Montauk Peninsula, Long Island, New York. M, 1981, University of Texas, Austin. 191 p.

Orr, G. Daniel. Niagaran reefs, northwestern Michigan. M, 1984, Michigan State University. 54 p.

Orr, Harold D. Structural geology of the San Saba area, San Saba County, Texas. M, 1962, University of Texas, Austin.

Orr, James Conrad. ^{222}Rn, ^{226}Ra, and ^{228}Ra as tracers for the evolution of warm core rings. D, 1988, Texas A&M University. 339 p.

Orr, James M. An investigation of the geological occurrence and use of titanium with special reference to the San Gabriel titanium deposits, California. M, 1938, California Institute of Technology. 86 p.

Orr, John Barrie Bain. Ordovician bentonites from Ontario. M, 1959, University of Alberta. 89 p.

Orr, John F. W. Mineralogy and computer-oriented study of mineral deposits in Slocan City camp, Nelson mining division, British Columbia. M, 1971, University of British Columbia.

Orr, Kristin Elizabeth. Structural features along the margins of Okanogan and Kettle domes, northeastern Washington and southern British Columbia. D, 1985, University of Washington. 109 p.

Orr, Patricia Lynne. Experimental dolomitization and aragonitization at low temperatures and pressures. D, 1971, Michigan State University. 33 p.

Orr, Robert William. Biostratigraphic zonation and correlation based on conodonts of Middle Devonian strata of southern Illinois and adjacent states. M, 1964, University of Texas, Austin.

Orr, Robert William. Conodonts from Middle Devonian strata of the Michigan Basin. D, 1967, Indiana University, Bloomington. 169 p.

Orr, William N. Campanian foraminifera of the Santa Ana Mountains, California. M, 1964, University of California, Riverside. 175 p.

Orr, William Norton. The distribution of planktonic Foraminifera in the northwest Gulf of Mexico. D, 1968, Michigan State University. 173 p.

Orrell, S. Andrew. Evidence of transpressional tectonics, southeastern terminus of East Pryor Mountain, Pryor Mountains, Montana and Wyoming. M, 1988, University of Massachusetts. 54 p.

Orrell, Suzanne. Petrologic studies in the basement of the upper plate of the Whipple detachment fault, Whipple Mountains, southeastern California. M, 1988, University of Southern California.

Orris, Greta J. A quantitative investigation of landslides in a portion of the Santa Clara Formation, Santa Clara County, California (1979). M, 1984, San Jose State University. 80 p.

Orsen, David A. Correlation of selected Nevada lignite deposits by pollen analysis. M, 1977, University of Nevada. 173 p.

Ortalda, Robert A. The geology of the northern part of the Morgan Hill Quadrangle, California. M, 1950, University of California, Berkeley. 55 p.

Ortega Guerrero, A. Analysis of regional groundwater flow and boundary conditions in the Basin of Mexico. M, 1988, University of Waterloo. 45 p.

Ortega, Jose F. A regional geologic review in the subsurface of the Clinton Sandstone in northeastern central Ohio. M, 1978, Ohio University, Athens. 93 p.

Ortigoza Cruz, Felipe. The volcano-sedimentary deposits of La Minita, Michoacan, Mexico. M, 1988, University of Alberta. 150 p.

Ortiz Vertiz, Salvador. Mexican fluorspar mining industry. M, 1973, Stanford University.

Ortiz, Gustavus A. Mineralogy of the suspended material in the Missouri River. M, 1955, University of Missouri, Columbia.

Ortiz, Keith. On the stochastic modeling of a fatigue crack growth. D, 1985, Stanford University. 120 p.

Ortiz, N. V. Artificial ground water recharge with capillarity. D, 1977, Colorado State University. 102 p.

Ortiz, Terri S. Megacrysts and mafic and ultramafic inclusions in the southern West Portillo basalt field, Dona Ana County, New Mexico. M, 1979, University of Texas at El Paso.

Ortiz-Ramirez, Jaime. Two-phase flow in the geothermal wells; development and uses of a computer code. M, 1983, Stanford University.

Ortman, Dale. Controlling seepage from the Bunker Hill central impoundment area, Kellogg, Shoshone County, Idaho. M, 1978, University of Idaho. 101 p.

Orville, Philip Moore. The composition of some un-zoned pegmatites in the Keystone District, South Dakota. D, 1958, Yale University.

Orwig, Eugene R. The Vaqueros Formation, west of Santa Barbara, California. D, 1957, University of California, Los Angeles.

Orwig, Robert E. The (Miocene) Vaqueros Formation west of Santa Barbara, California. D, 1957, University of California, Los Angeles.

Orwig, T. L. Geology of the eastern central Pacific Basin. M, 1975, University of Hawaii. 64 p.

Orzech, Mary Ann Terese. The long-term model of salinity intrusion into the estuarine rivers. M, 1972, College of William and Mary.

Orzeck, John Joseph. Provenance partitioning of beach, river, and cliff sands in San Diego County, California using fourier shape analysis. M, 1972, Michigan State University. 22 p.

Orzol, Leonard Lee. Explosion structures in Grande Ronde Basalt of the Columbia River Basalt Group, near Troy, Oregon. M, 1987, Portland State University. 220 p.

Osa'-Idahosa, A. Theoretical prediction of trace element distribution during sequential melting with relevance to basaltic magma genesis. M, 1977, Laurentian University, Sudbury. 198 p.

Osadetz, Kirk Gordon. Stratigraphy and Eurekan (Tertiary) structure of the Tanquary Fiord-Ekblaw Lake area, northern Ellesmere Island, Canadian Arctic Archipelago. M, 1983, University of Toronto. 132 p.

Osaimi, Aayed Eid. Finite element analysis of time dependent deformations and pore pressures in excavations and embankments. D, 1977, Stanford University. 259 p.

Osakada, Did. Study of slope stability by limit-equilibrium methods. M, 1985, University of Utah.

Osamor, Chukwuka Azubuike. Oil slick dispersal mechanics; spreading, dissolution, and chemical dispersion. D, 1981, Rutgers, The State University, New Brunswick. 426 p.

Osanik, Alec N. Lower and middle Jackson stratigraphy of central East Texas. M, 1942, Louisiana State University.

Osanloo, Gholi Morteza. Chemical comminution for deep and thin bituminous coal by using CO_2+H_2O as a solvent. D, 1982, University of Oklahoma. 280 p.

Osatenko, Myron John. Mineralogy, geochemistry and petrology of a pyrochlore-bearing carbonatite at Seabrook Lake, Ontario (Canada). M, 1967, University of British Columbia.

Osberg, Philip Henry. The Green Mountain anticlinorium in the vicinity of Rochester and East Middlebury, Vermont. D, 1952, Harvard University.

Osborn, Donald R., Jr. Flood control and conservation in the North River watershed (Rockingham and Augusta counties, Va.). M, 1951, Catholic University of America.

Osborn, Elburt F. and Rainwater, Edward H. Geology of the Calumet, Colorado mining district. M, 1934, Northwestern University.

Osborn, Elburt F. Micrometric and petrofabric studies of the Val Verde Tonalite, Southern California. D, 1938, California Institute of Technology. 76 p.

Osborn, Gerald Davis. Quaternary geology and geomorphology of the Uinta Basin and the south flank of the Uinta Mountains, Utah. D, 1973, University of California, Berkeley. 266 p.

Osborn, James B. Subsurface study of the Marmaton Group in southwestern Kansas. M, 1962, University of Kansas. 60 p.

Osborn, Noel Irene. Disconformities and paleo-oceanography in the Southeast Indian Ocean during the last 5.4 million years. M, 1981, University of Georgia.

Osborn, Roger T. Methods adaptable to determining the accuracy of electronic surveying systems. M, 1961, Ohio State University.

Osborne, David H. Arsenic as an indirect geochemical guide to epithermal precious metal deposits. M, 1963, University of Nevada. 69 p.

Osborne, Freleigh Fitz. Certain magmatic titaniferous iron ores and the problem of their origin. D, 1928, Yale University.

Osborne, Freleigh Fritz. The magnetite occurrences of the west coast of Vancouver Island, BC; their contact metamorphism and ore genesis. M, 1925, University of British Columbia.

Osborne, Lesslie W., Jr. Fluid inclusions and geochemistry of selected veins and mantos in the Leadville District, Colorado. M, 1982, Colorado State University. 104 p.

Osborne, Margaret Eleanor. Some ore deposits in western Massachusetts. M, 1930, Cornell University.

Osborne, Margery D. Crystal chemistry of chromium-aluminium spinels. D, 1983, University of Western Ontario. 150 p.

Osborne, Marla A. Alteration and mineralization of the northern half of the Aurora mining district, Mineral County, Nevada. M, 1985, University of Nevada. 93 p.

Osborne, Paul F. Microscopic characteristics of the Pierre and Foxhills formations (Upper Cretaceous) of Fort Morgan (Morgan County), Colorado, with special reference to the foraminifera. M, 1932, University of Colorado.

Osborne, Paul S. Analysis of well losses pertaining to artificial recharge. M, 1968, University of Arizona.

Osborne, Robert Edward. Geology of the Sentinal Mountain area, North Park, Jackson County, Colorado. M, 1957, Colorado School of Mines. 67 p.

Osborne, Robert Howard. Bedrock geology and limestone petrology of eastern Hamilton County, Ohio. D, 1966, Ohio State University. 165 p.

Osborne, Robert Howard. Geology of the northwest quarter of the Cave Mountain Quadrangle, Montana. M, 1963, Washington State University. 88 p.

Osborne, Steven D. Late Cenozoic movement on the central Wasatch Fault, Utah. M, 1978, Brigham Young University.

Osborne, Thomas Cramer. Petrography and petrogenesis of the Bird river chromite-bearing sill (southeastern Manitoba). M, 1950, University of Manitoba.

Osborne, Walter Edward. Depositional environments and dispersal system of the Parkwood Formation (Carboniferous) on the northwest limb of the Cahaba Syncline, Jefferson County, Alabama. M, 1985, University of Alabama. 160 p.

Osborne, Wiley Wilson. The isostatic reduction of a gravity profile off the west coast of South America. M, 1955, Ohio State University.

Osborne, Willis Williams. Geology of the Coquihalla Serpentine Belt between Spuzzum and Boston Bar, British Columbia. M, 1966, University of British Columbia.

Osbun, Erik. Geology of the Sveadal area, southern Santa Cruz Mountains, California. M, 1975, San Jose State University. 156 p.

Osburn, William L. The sediments and sedimentary transport processes of the Chilean continental margin between 37°27' and 41°00'S. M, 1972, Florida State University.

Osby, Donald R. Paleohydrology of Taylor Valley, Antarctica; inferred from pore ice in permafrost core. M, 1977, Northern Illinois University. 89 p.

Oscar, Aguilar M. Study of the copper uranium deposits at Vilcabamba, Department of Cuzco, Peru. M, 1962, University of Missouri, Rolla.

Oschman, Kurt Patrick. Slope stability for a strip mine box cut, Williamson County, Illinois. M, 1984, Purdue University. 270 p.

Oser, Robert K. Textural analysis of fine-grained sediments; pelagic sediments of the northwest Pacific. M, 1972, Oregon State University. 54 p.

Osgood, John O. The bedrock geology of Stony Brook ravine, Suffield, Connecticut. M, 1966, University of Connecticut. 98 p.

Osgood, Richard Grosvenor, Jr. Trace fossils of the Cincinnati area. D, 1965, University of Cincinnati. 537 p.

Osgood, Wayland. Geology of the Gogebic Range in the vicinity of Potato River Gap. M, 1923, University of Wisconsin-Madison.

Oshchudlak, Martha E. Petrology of Mesozoic sandstones in the Newark Basin. M, 1985, University of Massachusetts. 134 p.

Oshin, Igbekele Oyeyemi. The abundances and geochemistry of some noble metals in Thetford Mines ophiolites, PQ. D, 1981, McMaster University. 391 p.

Oshinowo, Babatunde Oluwasegun. A combined adjustment of photogrammetric, geodetic and Global Positioning System (GPS) data. D, 1988, University of Washington. 303 p.

Oshiro, K. The properties and genesis of a sequence of soils on Kohala Mountain, Hawaii. M, 1969, University of Hawaii. 67 p.

Osiecki, Richard Alan. Textural development of pegmatite, aplite, and associated rock types in the Mason-Milford Granite. D, 1982, Stanford University. 175 p.

Osiensky, James Leo. A system for the hydrogeologic analysis of uranium mill waste disposal sites. D, 1983, University of Idaho. 445 p.

Osiensky, James Leo. Reconnaissance hydrogeology of the potential mining sites in the Woodrat Mountain area, Idaho County, Idaho. M, 1978, University of Idaho. 169 p.

Osking, Erick B. Sedimentation controls and resulting geomorphology of microtidal, low energy, freshwater influenced shelf embayments, west-central Florida. M, 1985, University of South Florida, St. Petersburg.

Oskoorouchi, Ali Mohammad. Drum centrifuge modeling of overconsolidated clay slopes. D, 1981, University of California, Davis. 168 p.

Oskvarek, Jerome David. Oxygen isotope variations in selected samples from the Isua area, West Greenland. M, 1975, Northern Illinois University. 52 p.

Osleger, David A. Stratigraphy and microfacies analysis of the La Tuna Formation (Morrowan-Atokan) Franklin Mountains, Texas and New Mexico and Bishop Cap Hills, New Mexico. M, 1981, University of Texas at El Paso.

Oslund, Jeffrey S. Beneficiation of the Danville Coal Member (VII) of Indiana using high intensity magnetic separation and froth flotation. M, 1979, Indiana University, Bloomington. 54 p.

Osman, Abdelaziz A. Mineralization and rock alteration at the Reed gold mine, Cabarrus County, North Carolina. M, 1978, North Carolina State University. 76 p.

Osman, Mohamed Akode. Channel width response to changes in flow hydraulics and sediment load. D, 1985, Colorado State University. 184 p.

Osman, R. W. Origin, development and maintenance of the marine epifaunal community at Woods Hole, Massachusetts. D, 1975, University of Chicago. 132 p.

Osmani, Ikramuddin Ahmad. Component magnetization of the iron formation and deposits at the Griffith Mine near Red Lake, Ontario. M, 1982, University of Windsor. 137 p.

Osment, Frank Carter. Sedimentation studies of the Selma, Ripley, and Prairie Bluff formations of western Alabama (Os53). M, 1941, University of Illinois, Urbana. 38 p.

Osmond, John C. Dolomites in the Silurian and Devonian of east-central Nevada. D, 1954, Columbia University, Teachers College.

Osmond, John C., Jr. Stratigraphy of the Cloverly Formation, Thermopolis Shale, and Muddy Sandstone in the Laramie Basin, Wyo. M, 1950, University of Wyoming. 89 p.

Osmond, John K. Bentonites as time horizons in the Specht's Ferry Shale, Decorah Formation (Ordovician), southwestern Wisconsin. M, 1952, University of Wisconsin-Madison.

Osmond, John K. Radioactivity of bentonites. D, 1954, University of Wisconsin-Madison.

Osolin, Robert Lyle. Geology of Ledges State Park, Boone, Iowa. M, 1983, Iowa State University of Science and Technology. 192 p.

Osorio, Gustave A. Pennsylvanian ostracodes from the Nowata Shale (Oklahoma). M, 1931, Columbia University, Teachers College.

Ospanik, Laddy Franklin. Quaternary stratigraphy of the west side of the Cuyahoga Valley between Peninsula, Ohio, and State Route 82. M, 1983, University of Akron. 87 p.

Ospovat, Alexander Meier. Abraham Gottlob Werner and his influence on mineralogy and geology. D, 1960, University of Oklahoma. 259 p.

Oss, Hendrik G. van see van Oss, Hendrik G.

Ossege, John. Quantitative determination for the shape of sand grains. M, 1980, Wright State University. 44 p.

Ossi, Edward John, III. Mesoscopic structures and fabric within the thrust sheets between the Cumberland Escarpment and Saltville Fault. M, 1979, University of Tennessee, Knoxville. 98 p.

Ossi, John C. A new petrographic method for interpreting coal-forming environments of deposition. M, 1984, University of Maryland.

Ossian, Clair Russell. Fishes of a Pleistocene lake in South Dakota. M, 1970, Michigan State University. 58 p.

Ossian, Clair Russell. Paleontology, paleobotany and facies characteristics of a Pennsylvanian delta in southeastern Nebraska. D, 1974, University of Texas, Austin.

Ossing, Henry A. Atlantic Rayleigh-wave dispersion. M, 1961, Boston College.

Ossinger, Richard A. Relationship of systematic outcrop fractures and lineaments to the geology and geophysics of western Pulaski County, Kentucky. M, 1983, Wright State University. 130 p.

Ossman, Robert M. Seismotectonics of southeastern New York State. M, 1984, Pennsylvania State University, University Park. 105 p.

Ostby, M. The methanol alternative for producing potential gas fields north of the 62nd parallel on the Norwegian continental shelf. M, 1976, Stanford University. 172 p.

Osten, Erimar Alfred von der. Age and correlation of the Barranquín Formation of northeastern Venezuela. D, 1956, Stanford University. 228 p.

Osten, Erimar Alfred von der see von der Osten, Erimar Alfred

Osten, Lawrence William. The use of pectoral fin spines of channel catfish (Ictalurus Punctatus) in order to estimate paleoclimates. M, 1977, Southern Methodist University. 27 p.

Osten, Mark Allen. The subsurface stratigraphy, paleoenvironmental interpretation and petroleum geology of the Albion Group (Lower Silurian), southeastern Ohio. M, 1983, Kent State University, Kent. 166 p.

Ostendorf, Paul. Geology of the Highland Lakes area, Alpine County, California. M, 1981, San Jose State University. 89 p.

Ostenso, Ned Allen. Geophysical investigations of the Arctic Ocean basin. D, 1962, University of Wisconsin-Madison. 202 p.

Ostenso, Nile. Soil moisture detection using 35mm aerial photography. M, 1979, University of Wisconsin-Milwaukee. 63 p.

Oster, L. D. Stratigraphy of the Cloverly Formation, the Thermopolis Shale, and the Muddy Sandstone in northeasten Carbon and southeastern Natrona counties, Wyoming. M, 1952, University of Wyoming. 86 p.

Osterberg, Mark Warren. Subaqueous pyroclastic volcanism in the vicinity of the Helen Mine, Wawa, Ontario. M, 1982, University of Minnesota, Duluth. 77 p.

Osterberg, Steven Arvid. Stratigraphy and hydrothermal alteration of Archean volcanic rocks at the Headway-Coulee massive sulfide prospect, northern

Onaman Lake area, northwestern Ontario. M, 1985, University of Minnesota, Duluth. 114 p.

Ostergard, Deborah. Subsurface stratigraphy of the Morrow Formation in southeastern Texas County, Oklahoma. M, 1979, University of Colorado.

Osterholt, William Russell B. The origin of the main physiographical features of Borrego Valley, California. M, 1934, University of Southern California.

Osterkamp, W. R. The role of sediment in determining the geometry of alluvial stream channels. D, 1976, University of Arizona. 201 p.

Osterkamp, Waite R. Variation and geologic significance of water quality in the Judith River Formation (upper Cretaceous), north-central Montana. M, 1970, University of Arizona. 71 p.

Osterlund, David Paul. A subsurface study of the Mississippian Chappel Limestone, Young County, Texas. M, 1984, Texas Christian University. 75 p.

Osterman, Lisa Ellen. Late Quaternary history of southern Baffin Island, Canada; a study of foraminifera and sediments from Frobisher Bay. D, 1982, University of Colorado. 402 p.

Osterwald, Doris B. Metamorphism as related to structure in western Jefferson County, Colorado. M, 1949, University of Wyoming. 52 p.

Osterwald, Edward J. The Beidell and Sanderson volcanic centers, Saguache County, Colorado. M, 1977, Colorado State University. 158 p.

Osterwald, Frank W. Relation of granitization and mineral facies to structure in northeastern Albany County, Wyoming. M, 1947, University of Wyoming. 58 p.

Osterwald, Frank W. Structural and petrologic studies in the northern Bighorn Mountains, Wyoming. D, 1951, University of Chicago. 155 p.

Ostic, Ronald George. Isotopic investigation of conformable lead deposits. D, 1963, University of British Columbia.

Ostler, J. Sedimentation on the margin of an Archean felsic volcanic complex (Slave structural province). M, 1977, Carleton University. 130 p.

Ostos-Rosales, Marino. Structural interpretation of the Tinaquillo Peridotite and its country rock, Cojedes State, Venezuela. M, 1984, Rice University. 135 p.

Ostrander, Alan. Geography of the southern Illinois Coalfield. D, 1938, Washington University.

Ostrander, Allen R. General geology and structure of Hamilton Mounds, Adams County, Wisconsin. M, 1931, University of Wisconsin-Madison.

Ostrander, Charles C. Geology of the Ooltewah, Tennessee, Quadrangle. M, 1966, Emory University. 91 p.

Ostrander, Gregg. Mammalia of the early Oligocene (Chadronian) Raben Ranch local fauna of northwestern Nebraska. M, 1980, South Dakota School of Mines & Technology.

Ostrander, Robert Earl. The geology of the west-central portion of the Promontory Butte Quadrangle, Arizona. M, 1950, University of Iowa. 66 p.

Ostrander, William J. Special estimation of signal and noise power and power ratios for reflection seismograms. M, 1966, Pennsylvania State University, University Park. 84 p.

Ostro, Steven Jeffrey. The structure of Saturn's rings and the surfaces of the Galilean satellites as inferred from radar observations. D, 1978, Massachusetts Institute of Technology. 242 p.

Ostrom, John H. Cranial morphology of the North American Hadrosauridae. D, 1960, Columbia University, Teachers College. 396 p.

Ostrom, Meredith Eggers. Clay mineralogy of some carbonate rocks of Illinois. D, 1959, University of Illinois, Urbana. 167 p.

Ostrom, Meredith Eggers. Electric log studies of Pennsylvanian limestones in the Illinois structural basin. M, 1954, University of Illinois, Urbana.

Ostry, R. C. An analysis of some tills in Scarborough Township and vicinity, Ontario. M, 1962, University of Toronto.

Ostrye, T. F. An approach to defining geometrical beach shape changes at Lotus Bay, N. Y. on Lake Erie. M, 1975, SUNY at Buffalo. 138 p.

Osuch, Lawrence Theodore. Geology of the Three Sisters Quadrangle, California. M, 1970, University of California, Berkeley. 60 p.

Oswald, Mary L. Faunal zones of the Pierre Formation (Upper Cretaceous) in the foothills of northern Colorado. D, 1944, University of Colorado.

Osweiler, Donna Jean. A geological interpretation of the Stratford geophysical anomaly on the Midcontinent paleorift zone, central Iowa. M, 1982, University of Iowa. 166 p.

Otalora, Guillermo. Geology of the Barranquitas Quadrangle, Puerto Rico. D, 1961, Princeton University. 187 p.

Othberg, Kurt L. Paleomagnetism of late Pleistocene sediments, Puget Lowland, Washington. M, 1973, Western Washington University. 54 p.

Otis, Marlène. Etude des contrôles de la distribution latérale et verticale de la concentration de l'uranium dans les sédiments de lacs. M, 1988, Universite du Quebec a Chicoutimi. 62 p.

Otis, Robert Michael. Interpretation and digital processing of seismic reflection and refraction data from Yellowstone Lake, Wyoming. D, 1975, University of Utah. 223 p.

Otooni, M. A. Upper Ordovician dolomite sequence of the southern Egan Range, Nevada. M, 1961, Columbia University, Teachers College.

Ott, D. W. Comparative analysis of adjacent vegetated and bare strip mine spoils. D, 1978, University of Tennessee, Knoxville. 70 p.

Ott, Henry Louis. Stratigraphic distribution of Charophyta in the (Jurassic) Morrison Formation of Colorado and Utah. M, 1958, University of Missouri, Columbia.

Ott, Henry Louis. Stratigraphic distribution of Charophyta in the Morrison Formation of Colorado and Utah. M, 1957, University of Missouri, Columbia.

Ott, Kyle R. Landslide susceptibility; an investigation of the Binghamton area. M, 1979, SUNY at Binghamton. 20 p.

Ott, Lawrence E. Economic geology of the Wenatchee mining district, Chelan County, Washington. D, 1988, University of Idaho. 270 p.

Ott, Lawrence E. Geology and ore localization at the Northumberland gold mine, Nye Co., Nevada. M, 1983, Montana College of Mineral Science & Technology. 52 p.

Ott, Valen D. Geology of the Woodruff Narrows Quadrangle, Utah-Wyoming. M, 1979, Brigham Young University. 17 p.

Otte, Carel, Jr. Geology of the upper Tick Canyon area, Los Angeles County, California. M, 1950, California Institute of Technology. 53 p.

Otte, Carel, Jr. Late Pennsylvanian and Early Permian stratigraphy of the northern Sacramento Mountains, Otero County, New Mexico. D, 1954, California Institute of Technology. 240 p.

Otte, Lee J. Genotypic, ecologic and diagenetic variation in Labechia huronensis (Billings) 1865, from the Millersburg Member, Lexington Limestone (Middle Ordovician) of Kentucky. M, 1977, University of North Carolina, Chapel Hill. 114 p.

Otte, Lee James. Petrology of the exposed Eocene Castle Hayne Limestone of North Carolina. D, 1981, University of North Carolina, Chapel Hill. 183 p.

Otten, William John. The geology of the southeast portion of the Millersburg Quadrangle, Callaway County, Missouri. M, 1955, University of Missouri, Columbia.

Ottensman, Donald Clay. A facies and paleoenvironmental analysis of part of the Bliss Formation (Cambrian-Ordovician) in southern New Mexico and western Texas. M, 1982, University of Texas at Dallas. 208 p.

Ottensman, Vicki Vieroski. Minor element analysis of nautiloid skeletons from asphaltic portions of the Boggy Formation, Oklahoma. M, 1981, University of Texas, Arlington. 70 p.

Ottenstein, Robert Paul. The subsurface geology of eastern Baylor County, Texas. M, 1955, University of Oklahoma. 70 p.

Ottmann, Jeffry D. Hydrodynamic entrapment of petroleum in the "J" Sandstone of the Denver Basin, northeastern Colorado. M, 1981, University of Houston.

Otto, Bruce Richard. Structure and petrology of the Sheepeater Peak area, Idaho primitive area, Idaho. M, 1978, University of Montana. 68 p.

Otto, David Arthur. A study of the Middle Pennsylvanian flora found in Kansas coal balls. D, 1967, University of Missouri, Columbia. 245 p.

Otto, Ellen E. Engineering and environmental geology of Clinton County, Indiana. M, 1977, Purdue University. 118 p.

Otto, Ernest Paul. Sedimentology of the Entrada Sandstone (Jurassic), northeastern Utah and northwestern Colorado. M, 1973, University of Utah. 87 p.

Otto, George H. An interpretation of the glacial stratigraphy of the City of Chicago. D, 1942, University of Chicago. 116 p.

Otto, Jens. Melting relations in some carbonate-silicate systems; sources and products of CO_2-rich liquids. D, 1984, University of Chicago. 212 p.

Otto, Judith E. Sedimentology of the Sixteen Mile Creek Lagoon, Niagara Peninsula, Ontario, Canada. M, 1984, Brock University. 167 p.

Ottolini, Richard Albert. Migration of reflection seismic data in angle-midpoint coordinates. D, 1983, Stanford University. 106 p.

Otton, Edmond George. The subsurface geology of Sullivan County, Indiana. M, 1947, University of Illinois, Urbana.

Otton, James Keith. Geology of the central Black Mountains, Death Valley area, California. D, 1977, Pennsylvania State University, University Park. 155 p.

Ottoni, Theophilo Benedicto Filho. Soil moisture and the water balance in a border-irrigated field. D, 1984, University of Arizona. 264 p.

Otts, Charlotte. Paleontology and paleoecology of some Pleistocene nearshore deposits, Little River, South Carolina, with considerations of taphonomic problems. M, 1983, North Carolina State University. 173 p.

Ottum, Margaret G. Paleoecology of the Brush Creek (Pennsylvanian), of Columbiana County, Ohio. M, 1969, Bowling Green State University. 86 p.

Ou, Joyce Ling-Mei. Petrography, porosity and permeability of Chesterian sandstones (Upper Mississippian) southern Illinois, U.S.A. M, 1983, University of Illinois, Urbana. 88 p.

Ouchark, William F. Conodonts of the Upper Devonian and Lower Mississippian strata in Powell Valley, Southwest Virginia. M, 1964, Virginia Polytechnic Institute and State University.

Ouchi, Shunji. Response of alluvial rivers to active tectonics. D, 1983, Colorado State University. 222 p.

Oudomugsorn, Prakal. Geology of the Buckhorn Field, Tensas Parish, Louisiana. M, 1971, Louisiana Tech University.

Ouellet, Eric. Evolution tectono-métamorphique de la continuité lithologique des roches vertes du supérieur dans la zone orogénique de la Province du Grenville. M, 1988, Universite du Quebec a Chicoutimi. 363 p.

Ouellet, Guy. Etude de l'interaction des animaux benthiques avec les sédiments du chenal Laurentien. M, 1982, Universite du Quebec a Rimouski. 200 p.

Ouellet, Rodrigue. Détermination des contrôles de la mise en place d'indices minéralisés dans la partie ouest du Pluton de Chibougamau. M, 1986, Universite du Quebec a Chicoutimi. 136 p.

Ouellette, Dorice J. Sediment and water characteristics, South Pass, Mississippi River. M, 1969, Louisiana State University.

Ouellette, Robert. Upper crustal structure in the Newport, (Jackson county) Arkansas area. M, 1970, Washington University. 64 p.

Ougland, Ronald M. Quantitative X-ray determination of phases in fired whiteware bodies and their shrinkage and porosity characteristics. M, 1961, Pennsylvania State University, University Park. 59 p.

Oulgout, Bassou. Preliminary groundwater study and use of resistivity sounding for determining some hydrologic parameters of glacial outwash in Clear Lake, Sherburne County, MN. M, 1984, University of Minnesota, Duluth. 223 p.

Oung, Jung Nan. Tricyclic terpanes and related compounds in crude oils and source rock extracts. M, 1987, University of Oklahoma. 138 p.

Ousey, John Russell, Jr. Safe-yield estimates of a water supply reservoir on an ungaged stream. D, 1980, Pennsylvania State University, University Park. 92 p.

Outlaw, Donald Elmer. A geologic and hydrologic study of Shackham watershed, New York State. D, 1953, Pennsylvania State University, University Park. 175 p.

Outlaw, Donald Elmer. The geology of the McNeil area in Travis County, Texas. M, 1947, University of Texas, Austin.

Outler, Brenda. The stratigraphy and environment of deposition of the Tuscaloosa Formation in part of Panhandle, Florida. M, 1979, Florida State University.

Ouyang, Shoung. Land subsidence due to gypsum solution in the western part of Rapid City, South Dakota. M, 1983, South Dakota School of Mines & Technology.

Ovalles, Francisco Antonio. Selection of important properties to evaluate the use of geostatistical analysis in selected Northwest Florida soils. D, 1986, University of Florida. 208 p.

Ovejero, Julio Alamos *see* Alamos Ovejero, Julio

Ovenden, Lynn Elise. Hydroseral histories of the Old Crow Peatlands, northern Yukon. D, 1985, University of Toronto.

Ovenshine, Alexander Thomas. Sedimentary structures in portions of the Gowganda Formation (Precambrian), north shore of Lake Huron, Canada. D, 1965, University of California, Los Angeles.

Over, D. Jeffrey. Some Silurian conodonts from the southern Mackenzie Mountains, Northwest Territories. M, 1985, University of Alberta. 250 p.

Overbeck, Robert M. The copper ores of Maryland. D, 1915, The Johns Hopkins University.

Overland, James Edward. A model of salt intrusion in a partially mixed estuary. D, 1973, New York University.

Overman, William C. Aeromagnetic study of the Kiernan Sills of Iron County, Michigan. M, 1978, Bowling Green State University. 110 p.

Overpeck, Jonathan Taylor. Time series analysis of Holocene pollen data; paleoclimatological and paleoecological applications. D, 1986, Brown University. 240 p.

Oversby, Brian Sedgwick. An early Antlerian (Mississippian) orogenic pulse and post-Antlerian emplacement of allochthonous rocks in northeastern Nevada. D, 1969, Columbia University. 152 p.

Oversby, Brian Sedgwick. The geology of upper Black Island, Bay of Exploits, Newfoundland. M, 1967, Columbia University. 68 p.

Oversby, Virginia M. Lead isotope composition in recent volcanic rocks from islands in the Atlantic Ocean and from the troilite phase of iron meteorites. D, 1969, Columbia University. 193 p.

Overshine, Alexander T. Geology of the Spruce Run Mountain area, Giles County, Virginia. M, 1961, Virginia Polytechnic Institute and State University.

Overstreet, William C. Methods in the determination of graphite. M, 1942, University of Virginia. 38 p.

Overton, Deborah J. A mineralogical study of the red clay in Douglas County, Wisconsin. M, 1978, University of Wisconsin-Milwaukee.

Overton, Theodore D. Mineral resources of Douglas, Ormsby, and Washoe counties. M, 1947, University of Nevada - Mackay School of Mines. 91 p.

Overymyer, Dale Owen. Geology of the Pleasant Grove area, Dallas County, Texas. M, 1953, Southern Methodist University. 10 p.

Oviatt, Charles G. Glacial geology of the Lake Marie area, Medicine Bow Mountains, Wyoming. M, 1977, University of Wyoming. 82 p.

Oviatt, Charles Gifford. Lake Bonneville stratigraphy at the Old River Bed and Leamington, Utah. D, 1984, University of Utah. 122 p.

Owe, Manfred. The distribution of heavy metals and petroleum residues in the soil from urban surface runoff. D, 1981, State University of New York, College of Environmental Science and Forestry. 193 p.

Oweis, Issa Sebeitan. A solution of nonlinear plane strain problems in dynamic soil mechanics. D, 1969, University of Texas, Austin. 173 p.

Owen, Claudia. The petrogenesis of blueschist facies ironstones in the Shuksan and Easton schists, North Cascades, Washington. D, 1988, University of Washington. 290 p.

Owen, David J. Abnormal electrical resistivity effects over buried bedrock fractures. M, 1987, University of Rhode Island. 117 p.

Owen, Dennis Ray. Occurrence, petrography, and genesis of sandstone dikes, Capitan Limestone, New Mexico. M, 1983, Sul Ross State University. 148 p.

Owen, Donald Edward. Stratigraphy of bioherms and other deposits of the Upper Pennsylvanian Bern Limestone in east-central Kansas. M, 1959, University of Kansas. 185 p.

Owen, Donald Edward. The Dakota Formation of the San Juan Basin, New Mexico and Colorado. D, 1963, University of Kansas. 388 p.

Owen, Donald Eugene. Geologic structure of Seven Heart Gap, Culberson County, Texas. M, 1951, University of Texas, Austin.

Owen, Donald Eugene. Lower Cretaceous stratigraphy, eastern Great Plains area. D, 1964, University of Wisconsin-Madison. 146 p.

Owen, Edgar Wesley. The geology of the Breckenridge, Colorado, District. M, 1916, University of Missouri, Columbia.

Owen, Jerry A. A lithogeochemical survey of the Pinos Altos Pluton, Silver City, New Mexico. M, 1983, University of Georgia. 172 p.

Owen, John Victor. Tectono-metamorphic evolution of the Grenville Front zone, Smokey Archipelago, Labrador. D, 1985, Memorial University of Newfoundland. 382 p.

Owen, John Wallace. The distribution, relationship and character of the Swan Peak Quartzite within the Logan Quadrangle, Utah. M, 1931, University of Missouri, Columbia.

Owen, Lawrence B. A gravity and magnetic study of an anorthosite complex; southern part of the Laramie range, Wyoming. M, 1968, University of Tennessee, Knoxville.

Owen, Lawrence Barry. Age determinations by the lutetium-176/hafnium-176 method. D, 1974, Ohio State University. 300 p.

Owen, M. T. The Paluxy sand in North-central Texas. M, 1977, Baylor University. 214 p.

Owen, Michael Rainey. Sedimentary petrology and provenance of the upper Jackfork Sandstone (Morrowan), Quachita Mountains, Arkansas, U.S.A. D, 1984, University of Illinois, Urbana. 167 p.

Owen, Michael Rainey. Sedimentology of thinly laminated rhythmites of the Port Askaig Tillite (late Precambrian), Southwest Scotland. M, 1980, University of Illinois, Urbana. 65 p.

Owen, Philip C. An examination of the Clarno Formation in the vicinity of the Mitchell Fault, Lawson Mountain and Stephenson Mountain area, Jefferson and Wheeler counties, Oregon. M, 1978, Oregon State University. 166 p.

Owen, R. M. Sources and deposition of sediments in Chagvan Bay, Alaska. D, 1975, University of Wisconsin-Madison. 214 p.

Owen, Roy W. Red Sea algal sediments and the Hoyt limestones of New York; a comparison of Recent and Cambrian algal deposition. M, 1973, Rensselaer Polytechnic Institute. 121 p.

Owen, Sandra Joan. Petrology of the Deadwood Formation from the Ragged Top Mountain cores, Black Hills, South Dakota. M, 1975, South Dakota School of Mines & Technology.

Owen, Vaux, Jr. The stratigraphy and lithology of Webster County, Georgia. M, 1957, Emory University. 82 p.

Owen, Victor J. Petrography of leucocratic segregations in the migmatitic Old Gneiss Complex east of Chicoutimi, Quebec. M, 1981, Universite du Quebec a Chicoutimi. 172 p.

Owen, William Patrick. A geophysical study of potential ground water contamination at Norton Air Force Base, California. M, 1988, University of California, Riverside. 112 p.

Owens, Anthony D. Mineralogy and petrology of iron-formation and associated lithologies in the Owl Creek Mountains, central Wyoming, and the Beartooth Mountains, southern Montana. M, 1983, Indiana University, Bloomington. 185 p.

Owens, Brent Edward. Xenoliths and autoliths in the Kiglapait Intrusion, Labrador. M, 1986, University of Massachusetts. 149 p.

Owens, Don Ray. Bedrock geology of the V intrusion, Garland County, Arkansas. M, 1967, University of Arkansas, Fayetteville.

Owens, Edward H. The geodynamics of two beach units in the Magdalen Islands, Quebec, within the framework of coastal environments of the southern Gulf of St. Lawrence. D, 1975, University of South Carolina. 581 p.

Owens, Eric O. Precambrian geology and precious metal mineralization of the Fire Center area, Marquette County, Michigan. M, 1986, Michigan Technological University. 152 p.

Owens, George V. Sedimentary rocks of lower Mill Creek, San Bernardino Mountains, California. M, 1959, Pomona College.

Owens, Gordon L. The petrology and structure of the Wills Mountain Anticline area, West Virginia. M, 1961, Ohio University, Athens. 89 p.

Owens, James P. The Cowlesville marcasite. M, 1949, SUNY at Buffalo.

Owens, John Snowden. The geology of parts of Colusa and Lake counties, California. M, 1940, University of California, Berkeley. 56 p.

Owens, L. D. Oil and gas in Kent and southern Lambton counties, Ontario, Canada. M, 1947, University of Michigan.

Owens, Michael. Petrologic study of talc mineralization in Murphy marble (lower Cambrian) in southwestern North Carolina. M, 1968, University of Tennessee, Knoxville. 78 p.

Owens, Owen E. Areal geology of the Labrador Trough south of Leaf Bay, northern Quebec. D, 1953, McGill University.

Owens, Owen E. The quartz deposits of the Watshishou Knoll area on the north shore of the St. Lawrence River (Quebec). M, 1951, McGill University.

Owens, Robert N. Petrologic analysis of the Mississinewa Member of the Wabash Formation and the effect of reef proximity on interreef sedimentation. M, 1981, Ball State University. 83 p.

Owens, Stephen M. Stratigraphy and sedimentology of flat-pebble conglomerates of the upper Lone Rock Formation (Upper Cambrian), western Wisconsin. M, 1985, University of Wisconsin-Madison. 125 p.

Owens, Thomas Joseph. Determination of crustal and upper mantle structure from analysis of broadband teleseismic P-waveforms. D, 1984, University of Utah. 159 p.

Owens, Thomas Joseph. Flexure and normal faulting in lithospheric plates with application to the Wasatch Front, Utah. M, 1980, University of Utah. 38 p.

Owens, W. B. A numerical study of mid-ocean meso-scale eddies. D, 1975, The Johns Hopkins University.

Owens, Willard G. Occurrence of mineralized ground water in southern St. Louis, Jefferson County, Missouri. M, 1960, University of Missouri, Rolla.

Oworu, Oyewola Oyeniyi. Reconnaissance geochemical study of surficial sediments of the Nova Scotian shelf. M, 1978, University of Windsor. 137 p.

Owsiacki, Leonede. The geology of the C Zone, Heath Steele Mine, New Brunswick. M, 1979, University of New Brunswick.

Owusu, John. On the geophysical well log studies of subsurface fracture zones in a granitic batholith. M, 1984, University of Manitoba.

Owusu, Joseph Kwame. Velocities and drillability profiles derived from three dimensional seismic data. D, 1981, University of Houston. 182 p.

Owusu, Lawrence A. Numerical simulation of oil recovery performance in complex naturally fractured reservoirs. D, 1988, University of Southern California.

Oxburgh, Ernest Ronald. Geology of the eastern Carabobo area, Venezuela. D, 1960, Princeton University. 225 p.

Oxley, David R., Jr. Magnetic and seismic refraction surveys on Jackson Lake, Wyoming. M, 1975, University of Wisconsin-Milwaukee.

Oxley, Marvin L. A subsurface study of the Hunton in northwestern Oklahoma. M, 1958, University of Oklahoma. 67 p.

Oxley, Philip. Chazyan stratigraphy of southern Isle La Motte, Vermont. M, 1948, Columbia University, Teachers College.

Oxley, Philip. Chazyan stratigraphy west of the Champlain Thrust, New York and Vermont. D, 1953, Columbia University, Teachers College.

Oxley, Philip. Ordovician Chazyan Series of Champlain Valley, New York and Vermont, and its reefs. D, 1960, Columbia University, Teachers College.

Oyer, W. Brian. Geometric analysis of shell coiling in brachiopods. M, 1969, University of Rochester.

Oyibo, Chamberlain Oruwari. Sedimentology of the lowermost Mannville Sandstone units (Lower Cretaceous) in the Chin Coulee oil field, southern Alberta. M, 1972, University of Calgary. 126 p.

Oyler, Daniel Leland. Stratigraphy and textural analyses of a beach ridge complex, Cape Henry, Virginia. M, 1984, Old Dominion University. 107 p.

Ozard, John Malcolm. Solid source lead isotope studies with application to rock samples from the Superior Geological Province, (Rice lake-Beresford lake area, Ontario; and Vogt-Hobbs area in Ontario). D, 1970, University of British Columbia. 106 p.

Ozcandarli, Tevfik Demir. The use of entire apparent resistivity curves for interpretation of normal resistivity logs. M, 1969, Colorado School of Mines. 104 p.

Ozier, Ronald L. Geology of the Teriary-pre-Tertiary angular unconformity, McGraw Creek to Grande Rhonde River, Snake River Canyon, Oregon. M, 1972, Indiana State University. 86 p.

Ozkaya, Ismail. The mio-eugeosynclinal thrust interface and related petroleum implications in the Sason-Baykan area, Southeast Turkey. D, 1972, University of Missouri, Rolla.

Ozkol, Sedat. Swelling characteristics of Permian clay. D, 1965, Oklahoma State University. 175 p.

Ozol, Michael A. A petrographic study of lower Allegheny and upper Pottsville sandstones in southern Somerset County, Pennsylvania. M, 1958, University of Pittsburgh.

Ozol, Michael Arvid. Alkali reactivity of cherts and stratigraphy and petrology of cherts and associated limestones of the Onondaga Formation of central and western New York. D, 1963, Rensselaer Polytechnic Institute. 228 p.

Ozoray, Judit. Serpentinization and metamorphism in the Proterozoic Cape Smith Foldbelt, New Quebec. M, 198?, McGill University. 124 p.

Ozsvath, David Lynn. Glacial geomorphology and late Wisconsinan deglaciation of the western Catskill Mountains, New York. D, 1985, SUNY at Binghamton. 219 p.

Ozsvath, David Lynn. Modelling heavy metal sorption from subsurface waters with the n-power exchange function. M, 1979, Pennsylvania State University, University Park. 60 p.

P., Raul Ramirez *see* Ramirez P., Raul

P., Wilfredo Armando Lopez *see* Lopez P., Wilfredo Armando

Paape, Donald W. Description and evolutionary study of the Mississippian Lake Valley platycrinitids. M, 1955, University of Wisconsin-Madison.

Paarlberg, Norman. A study in the erosion and weathering of Badlands. M, 1970, Northern Illinois University. 112 p.

Paba-Silva, Fernando. Microscopic studies of Eocene coal from Cundinamarca, Colombia. M, 1945, University of Chicago. 64 p.

Pabalan, Roberto Tuason. Solubility of cassiterite (SnO_2) in NaCl solutions from 200°C - 350°C, with geologic applications. D, 1986, Pennsylvania State University, University Park. 151 p.

Pabian, Roger Karr. Pennsylvanian and Permian trilobites of southeastern Nebraska. M, 1970, University of Nebraska, Lincoln.

Pabst, Adolf R. Observations on inclusions in the granitic rocks of the Sierra Nevada. D, 1928, University of California, Berkeley. 70 p.

Pabst, Larry Dean. The Mississippian stratigraphy and structure of the Vincent Anticline, Webster County, Iowa. M, 1965, University of Iowa. 49 p.

Pabst, Marie Bertha. Ferns of the Clarno (Eocene and Oligocene) Formation (Oregon). M, 1948, University of California, Berkeley. 36 p.

Pabst, Marie Bertha. Flora of the Chuckanut Formation (Eocene) of northwestern Washington; the Equisetales, Filicales and Conferales. D, 1962, University of California, Berkeley. 158 p.

Pac, Floyd. A provenance study of the Lower Cretaceous Yucca Bed. M, 1963, Texas Tech University. 26 p.

Pac, Timothy J. Equivalence and curve interpretation. M, 1985, University of Rhode Island.

Pace, Forrest Wilson, Jr. Sedimentology of a Holocene platform margin carbonate lagoon, San Salvador Island, Bahamas. M, 1987, Mississippi State University. 116 p.

Pace, Karen Klusmeyer. Interpretation of $^{87}Sr/^{86}Sr$ ratios and chemical compositions of the Kirkpatrick basalts, Mt. Falla, Queen Alexandra Range, Antarctica. M, 1977, Ohio State University. 113 p.

Paces, James Bryant. Magmatic processes, evolution and mantle source characteristics contributing to the petrogenesis of Midcontinent Rift basalts; Portage Lake Volcanics, Keweenaw Peninsula, Michigan. D, 1988, Michigan Technological University. 432 p.

Pacesova, Magdalena G. K-Rb-Tl relationship in some gneissic rocks. M, 1973, McMaster University. 100 p.

Pachernegg, Sheila M. Ground water flow and solute transport modeling of the lakes area, Thurston County, Washington. M, 1987, Washington State University. 85 p.

Pachman, Jerrold Marvin. An evaluation of quartz orientation in folded strata. M, 1958, University of Pittsburgh.

Pachman, Jerrold Marvin. Interrelations among petrographic, textural, and oil reservoir properties in the Chipmunk Sandstone. D, 1961, Pennsylvania State University, University Park. 104 p.

Pacht, J. A. Depositional environments and diagenesis of the Nugget Sandstone; western Wyoming and North central Utah. M, 1976, University of Wyoming. 107 p.

Pacht, Jory Allen. Sedimentology and petrology of the Late Cretaceous Nanaimo Group in the Nanaimo Basin, Washington and British Columbia; implications for Late Cretaceous tectonics. D, 1980, Ohio State University. 380 p.

Pachut, Joseph F., Jr. Environmental stability and morphogenetic relaxation in bryozoan colonies from the Eden Shale (Ordovician, Ohio Valley); a developmental explanation of stability-diversity-variation hypotheses. D, 1977, Michigan State University. 61 p.

Pack, Donald David. A mineralogic and petrologic study of the Cranberry magnetite mine, Cranberry, North Carolina. M, 1976, University of Tennessee, Knoxville. 126 p.

Pack, Fred James. Farmington gneiss. M, 1905, Columbia University, Teachers College.

Pack, Fred James. Geology of Pioche, Nevada, and vicinity; fossils of the Pioche Mountains. D, 1906, Columbia University, Teachers College.

Pack, Robert Taylor. Multivariate analysis of relative landslide susceptibility in Davis County, Utah. D, 1985, Utah State University. 233 p.

Packard, Earl Leroy. Faunal studies in the Cretaceous of the Santa Ana Mountains of Southern California. D, 1915, University of California, Berkeley. 52 p.

Packard, Earl Leroy. The Tertiary and Quaternary Mactrinae of the Pacific Coast. M, 1912, University of Washington. 59 p.

Packard, F. A. The hydraulic geometry of a discontinuous ephemeral stream on a bajada near Tucson, Arizona. D, 1974, University of Arizona. 142 p.

Packard, Frank A. The stratigraphy of the Upper Mississippian Paradise Formation of southeastern Arizona and southwestern New Mexico. M, 1955, University of Wisconsin-Madison.

Packard, Jeffrey J. The Upper Silurian Barlow Inlet Formation, Cornwallis Island, Arctic Canada. D, 1985, University of Ottawa. 144 p.

Packard, John A., Jr. Paleoenvironments of the Cretaceous rocks, Gabriola Island, British Columbia. M, 1972, Oregon State University. 101 p.

Packariyangkun, Adisorn. The spreading of impulse response wave propagating in an elastic contrast-Q models. M, 1988, University of Pittsburgh.

Packer, Bonnie Marcia. Provenance and petrology of Deep Sea Drilling Project sands and sandstones from the Japan and Mariana forearc and backarc regions. M, 1985, University of California, Los Angeles. 181 p.

Packer, Duane Russell. Paleomagnetism of the Mesozoic in Alaska. D, 1972, University of Alaska, Fairbanks. 172 p.

Packer, Ethel. Arizona; some adjustments to an arid environment. M, 1932, Columbia University, Teachers College.

Pacquett, Arthur Leon, Jr. Mineralogy and petrochemistry of the Mackay dike swarm, Mackay, Idaho. M, 1971, University of Wisconsin-Milwaukee.

Padan, Ady. Clay mineralogy of the bedded salt deposits in the Paradox Basin, Gibson Dome Well No. 1, Utah. D, 1984, Georgia Institute of Technology. 289 p.

Paddock, David Ray. A gravity investigation of eastern Iron County, Michigan. M, 1982, Michigan State University. 110 p.

Paddock, Robert Edwards. Geology of the Newfoundland Mountains, Box Elder County, Utah. M, 1956, University of Utah. 101 p.

Paden, Edward A. Conodont biostratigraphy and depositional environments of the Hidden Valley Dolomite (Early Silurian through Early Devonian), Death Valley, California. M, 1980, San Diego State University.

Paderes, Fidel Calimoso, Jr. Geometric modeling and rectification of satellite scanner imagery and investigation of related critical issues. D, 1986, Purdue University. 265 p.

Padgett, Guy. Carboniferous sedimentation, McDowell and Wyoming counties. M, 1972, University of South Carolina.

Padgett, Guy V., Jr. Sedimentary analysis of the Rabat Basin, Morocco. D, 1975, University of South Carolina. 96 p.

Padgett, Jeffrey Thomas. The nature and occurrence of pseudovitrinite in the Herrin (No. 6) Coal seam of southern Illinois. M, 1980, Southern Illinois University, Carbondale. 133 p.

Padgett, Joel P. Exploration for kimberlite in the Green Mountain-Magnolia area, Boulder County, Colorado. M, 1985, Colorado State University. 287 p.

Padgett, Michael F. Statistical analysis of residential damage in an area of underground coal mining, Boulder County, Colorado. M, 1987, Colorado School of Mines. 97 p.

Padgett, Philip C. O. Petrology, diagenesis and depositional environment of the Tonkawa Sandstone in southwestern Dewey County, Oklahoma. M, 1988, Oklahoma State University. 105 p.

Padgham, John B. The structural geology of the Mound City-Pleasanton area, Linn County, Kansas. M, 1957, University of Kansas. 152 p.

Padgham, William Albert. The geological structure of the Lac La Ronge region, Saskatchewan, (Canada). D, 1969, University of Wisconsin-Madison. 448 p.

Padgham, William Albert. The geology of the Ecstall-Quaal rivers area, British Columbia. M, 1958, University of British Columbia.

Padian, Kevin. Studies of the structure, evolution, and flight of pterosaurs (Reptilia; Pterosauria). D, 1980, Yale University. 326 p.

Padick, Clement. Control and conservation of natural runoff water in the San Fernando Valley, California. M, 1956, University of California, Los Angeles. 178 p.

Padilla y Sanchez, Ricardo Jose. Geologic evolution of the Sierra Madre Oriental between Linares, Concepcion del Oro, Saltillo, and Monterrey, Mexico. D, 1982, University of Texas, Austin. 274 p.

Padilla, Francisco. Analyse théorique de l'infiltration de l'eau à trois phases; modèle d'application aux sols gelés. D, 1986, Universite Laval. 240 p.

Padilla, Francisco. Ecoulement en milieu saturé et non saturé pour une nappe souterraine à faible profondeur; modèle ponctuel d'application à la fonte de la neige. M, 1982, Universite Laval. 153 p.

Padilla, Washington Augusto. Relationship between erosion and soil physical properties of temperate and tropical soils. D, 1984, University of Minnesota, Minneapolis. 181 p.

Padmanabhan G. Stochastic analysis and modeling of hydrologic and climatologic time series. D, 1980, Purdue University. 386 p.

Padovani, E. L. R. Granulite facies xenoliths from Kilbourne Hole maar, New Mexico and their bearing on deep crustal evolution. D, 1977, University of Texas at Dallas. 158 p.

Paduana, Joseph Anthony. The effect of type and amount of clay on the strength and creep characteristics of clay-sand mixtures. D, 1966, University of California, Berkeley. 203 p.

Paerl, Hans William. The regulation of heterotrophic activity by environmental factors at Lake Tahoe, California-Nevada. D, 1973, University of California, Davis. 139 p.

Paeth, Robert Carl. Depositional origin of mima mounds. M, 1967, Oregon State University. 61 p.

Paeth, Robert Carl. Genetic and stability relationships of four western Cascade soils. D, 1970, Oregon State University. 126 p.

Pagano, Timothy Samuel. Hydrogeologic and geomorphic properties of several tributary drainage basins of the Susquehanna River basin, New York. M, 1987, SUNY at Binghamton. 153 p.

Page, Ben M. The Pleasant Valley fault zone, Nevada. M, 1934, Stanford University. 45 p.

Page, Ben Markham. Geology of a part of the Chiwaukum Quadrangle, Washington. D, 1939, Stanford University. 203 p.

Page, C. E. The geology of the Cullaton Lake B-zone gold deposit, Northwest Territories. M, 1983, University of Waterloo. 140 p.

Page, Eric J. A hydrogeologic investigation of the groundwater resources in Richardson County, Nebraska. M, 1987, University of Nebraska, Lincoln. 136 p.

Page, Frederick W. Geochemistry of subsea permafrost at Prudhoe Bay, Alaska. M, 1978, Dartmouth College. 110 p.

Page, G. W., III. Toxic substances in water; patterns of contamination and policy implications. D, 1980, [University of Michigan].

Page, Gary Walter. Occurrence, mineralogy, and geochemistry of barite in the Keweenaw Peninsula, Michigan. M, 1971, Michigan Technological University. 41 p.

Page, George Ava. Studies of Pacific Northwest kaolins. M, 1931, University of Washington. 159 p.

Page, Gordon B. Beach erosion and composition of sand dunes, Playa del Ray-El Segundo area, California. M, 1950, University of California, Los Angeles.

Page, Kenneth G. The subsurface geology of southern Noble County, Oklahoma. M, 1955, University of Oklahoma. 69 p.

Page, Lincoln R. Some experiments on the solution and precipitation of metallic copper. M, 1932, University of Minnesota, Minneapolis. 14 p.

Page, Lincoln R. The geology of the Rumney Quadrangle, New Hampshire. D, 1937, University of Minnesota, Minneapolis. 150 p.

Page, Norman J. Mineralogy and chemistry of the serpentine group minerals and the serpentinization process. D, 1966, University of California, Berkeley. 353 p.

Page, Norman John. Carbonate replacement of detrital quartz in Upper Cambrian dolomites of Warren County, New Jersey. D, 1961, University of Illinois, Urbana. 69 p.

Page, Oliver S. Feasibility of subsurface correlation at the Atomic Energy Commission's National Reactor Testing Station, Idaho. M, 1966, Colorado State University. 132 p.

Page, Richard Adams. Micropaleontology and stratigraphy of the Brightseat Formation. D, 1959, Rutgers, The State University, New Brunswick. 166 p.

Page, Richard James. A preliminary petrographic examination of the Gabriola, Geoffrey, Decourcy and Comox formations, Nanaimo Group, Vancouver Island and Gulf Islands, British Columbia. M, 1972, University of Washington.

Page, Richard James. Sedimentology and tectonic history of the Esowista and Ucluth peninsulas, west coast, Vancouver Island, British Columbia. D, 1974, University of Washington. 97 p.

Page, Richard Owens. Stratigraphy and structure of the Quaternary Malpais Maar Volcano, Dona Ana County, New Mexico. M, 1973, University of Texas at El Paso.

Page, Robert Alan, Jr. Aftershocks and microaftershocks of the great Alaska Earthquake of 1964. D, 1967, Columbia University. 137 p.

Page, Robert Hull. Alteration-mineralization history of the Butte, Montana ore deposit, and transmission electron microscoy of phyllosilicate alteration phases. D, 1979, University of California, Berkeley. 234 p.

Page, Roger Henry. Solving differential and numerical models of systems with uncertain material parameters; applications to convection-dispersion equations. D, 1983, Princeton University. 264 p.

Page, Roland C. Geology of the Martinsburg Quadrangle, West Virginia. M, 1963, West Virginia University.

Page, Tench C. Geology, structural setting, and gold deposits of the Argus District, Inyo County, California. M, 1988, University of Nevada. 142 p.

Page, William Delano. Reconnaissance geology of the north quarter of the Horse Plains Quadrangle, Montana. M, 1963, University of Colorado. 66 p.

Page, William Delano. The geological setting of the archaeological site at Oued el Akarit and the paleoclimatic significance of gypsum soils, southern Tunisia. D, 1972, University of Colorado. 201 p.

Pagenhart, Thomas Harsha. Water use in the Yuba and Bear river basins, California. D, 1969, University of California, Berkeley. 219 p.

Pagiatakis, Spiros Demitris. Ocean tide loading on a self-gravitating, compressible, layered, anisotropic, viscoelastic and rotating Earth with solid inner core and fluid outer core. D, 1988, University of New Brunswick.

Pagoaga, Mary Katherine. The crystal chemistry of the uranyl oxide hydrate minerals. D, 1983, University of Maryland. 179 p.

Paguirigan, Francisco. Geologic investigation of the San Pedro Point area, California. M, 1941, Stanford University. 51 p.

Pai, Miao-Li M. A knowledge based system approach for hydrocarbon prospect and play analysis. D, 1988, University of South Carolina. 207 p.

Paick, John N. Stratigraphy, structure, and composition of cement materials in north central California. D, 1959, University of Arizona.

Paige, David Stanley. A rapid method for the correlation of fine grained sediments with the aid of the spectrograph as applied to the Mississippian-Devonian sequence in Clare County, Michigan. M, 1952, Michigan State University. 35 p.

Paige, Lennon Troy. A geomorphic interpretation of the Thousand Oaks corridor, Los Angeles and Ventura counties. M, 1956, University of California, Los Angeles.

Paige, Richard E. The morphology, depositional setting, and evolution of the Paleocene Lavaca submarine canyon system, Northwest Gulf of Mexico. M, 1988, University of Texas, Austin. 124 p.

Paige, Russell. Tertiary geology, Cheyenne Creek area, Alaska. M, 1959, University of Washington. 66 p.

Paijitprapapon, Vivat. Paleomagnetism of Mesozoic plutonic, and volcanic rocks in the White Mountains (New Hampshire), with emphasis on the influence of tectonic deformations. M, 1980, Wesleyan University.

Paik, Sun Mok. A molecular dynamics study of surface kinetics and epitaxial crystal growth. D, 1988, University of Maryland. 188 p.

Paik, Y. S. Field and laboratory study of bedload transport using the Bogardi bedload sampler. D, 1977, Rensselaer Polytechnic Institute. 128 p.

Pailoor, Govind. Variations in cation exchange capacities of some representative Michigan soils with analytical procedures and their relationships to acidity, clay mineralogy and organic matter. D, 1969, Michigan State University. 102 p.

Paine, Alasdair D. M. Canyon and terrace formation near Mount St. Helens, Washington. M, 1984, Colorado State University. 166 p.

Paine, Francis W. Notes on the Keweenaw copper deposits. M, 1911, University of Wisconsin-Madison.

Paine, Jack W. The subsurface geology of T.4N., R.4 and 5E., Pontotoc County, Oklahoma. M, 1958, University of Oklahoma. 46 p.

Paine, Jeffrey G. Crustal structure of volcanic arcs based on physical properties of andesites, volcaniclastic rocks, and inclusions in the Mt. St. Helens lava dome. M, 1982, University of Washington. 138 p.

Paine, Michael Henry. A reconnaissance study of the Gastropod Limestone (lower Cretaceous) in southwest Montana. M, 1970, Indiana University, Bloomington. 72 p.

Paine, William Rhodes. Stratigraphy of the Phosphoria Formation in Montana. M, 1952, Montana College of Mineral Science & Technology. 124 p.

Painter, Alice. Recent Cytheracean ostracodes from the Chukchi Sea and the northeastern Pacific including Puget Sound, Haro Strait and adjacent areas, Strait of Juan de Fuca, Silver Bay, and Gulf of Alaska. M, 1965, University of Kansas. 165 p.

Painter, Brian D. A three-dimensional hydrologic model of Lee County, Florida. M, 1984, Ohio University, Athens. 642 p.

Pair, D. Reconstruction and history of ice retreat, proglacial lakes, and the Champlain Sea, central St. Lawrence Lowlands, New York. M, 1985, University of Waterloo. 201 p.

Pait, Eugene D. The stratigraphy and facies relationships of some coal-bearing alluvial plain strata, Powder River basin, Montana and Wyoming. M, 1981, North Carolina State University. 40 p.

Pak, Dorothy Kim. The late Pleistocene stratigraphy and sedimentology of the Fram Basin, central Arctic Ocean. M, 1987, University of Wisconsin-Madison. 113 p.

Pakalnis, Rimas C. Thomas. Empirical stope design at the Ruttan Mine, Sherritt Gordon Mines, Ltd. D, 1986, University of British Columbia.

Pakkong, Mongkol. Ground water of the Boulder Park area, Lawrence County, South Dakota. M, 1979, South Dakota School of Mines & Technology.

Paktunc, A. Dogan. Petrology and geochemistry of some ultramafic rocks of the Thompson nickel belt and the Cuthbert Lake dikes of the Pikwitonei region, northern Manitoba. D, 1983, University of Ottawa. 182 p.

Pal, Badal Kanti. Geomagnetic induction studies in eastern Newfoundland. M, 1983, Memorial University of Newfoundland. 93 p.

Pal, Sakti K. Stochastic analysis for response, and stability of horizontal soil sites and earth dams under seismic loading. D, 1987, North Carolina State University. 215 p.

Palacas, James G. Lithocharacters and stratigraphy of the upper New Market, Row Park and lower Beekmantown beds in the vicinity of Greencastle, Pennsylvania. M, 1957, Pennsylvania State University, University Park. 102 p.

Palacas, James George. Geochemistry of carbohydrates. D, 1959, University of Minnesota, Minneapolis. 131 p.

Palache, Charles. The geology of the (Pliocene) Grizzly Peak; volcanic series, California. D, 1894, University of California, Berkeley.

Palacios, Alejandro. The theory and measurement of energy transfer during Standard Penetration Test sampling. D, 1977, University of Florida. 391 p.

Palacios-Velez, Enrique. Strategies to improve water management in Mexican irrigation districts; a case study in Sonora. D, 1976, University of Arizona.

Palacky, George Joseph. Computer assisted interpretation of multichannel airborne electromagnetic measurements. D, 1972, University of Toronto.

Palafox, Hector. Study of multiple reflection with the use of synthetic seismogram. M, 1968, Rice University. 82 p.

Palais, Julie Michelle. Tephra layers and ice chemistry in the Byrd-Station ice core, Antarctica. D, 1985, Ohio State University. 545 p.

Palal, Vistasp R. Characterization of the Arbuckle Reservoir (Upper Cambrian to Lower Ordovician) of the Lyons underground gas storage field, Rice County, Kansas. M, 1987, Wichita State University. 129 p.

Palaniappan, E. A. C. Shear modulus and damping characteristics of soils. D, 1976, Georgia Institute of Technology. 175 p.

Palczuk, Nicholas C., Jr. A stable isotope evaluation of the Blake Outer Ridge (BOR) deep sea deposit. M, 1988, North Carolina State University. 104 p.

Palen, Frank S. Stratigraphy of the Hurwal (Upper Triassic) Formation of the Wallowa Mountains, Oregon. M, 1952, University of Wisconsin-Madison. 58 p.

Palen, Walter Albert. The roles of pore pressure and fluid flow in the hydraulic fracturing process. D, 1980, University of California, Berkeley. 211 p.

Palencia, Cesar M. Economic analysis of the Surigao nickel deposit, the Republic of the Philippines. M, 1976, Colorado School of Mines. 172 p.

Palensky, John Joseph. Study of a subsurface Ordovician sandstone in southeastern Iowa. M, 1955, University of Iowa. 37 p.

Palestino, Robert. Gibbs free energy of water at high temperatures and pressures and its application to phase equilibria. M, 1982, Brooklyn College (CUNY). 183 p.

Palispis, Jaime Rafael. Direct shear strengths of dolomite-dolomite and dolomite-concrete interfaces. D, 1985, University of Wisconsin-Milwaukee. 618 p.

Pality, George D., Jr. The petrology and petrography of the Lower Precambrian metavolcanic rocks of the Upson area, Wisconsin. M, 1973, Bowling Green State University. 120 p.

Palke, Dale Robert. An investigation of the chelating abilities of LIX-26, LIX-54, LIX-70, LIX-622 and P-50. M, 1984, Montana College of Mineral Science & Technology. 48 p.

Palko, Gregory Jonathan. Environments of deposition and diagenesis of Jurassic upper Smackover Formation in the Lincoln Parish area, Louisiana. M, 1980, Texas A&M University.

Palladino, Deanna L. Structural analysis of a portion of the Washita Valley fault zone, Arbuckle Mountains, southern Oklahoma. M, 1986, Baylor University. 161 p.

Palladino, Donald Joseph. Slope failures in an overconsolidated clay, Seattle, Washington. D, 1971, University of Illinois, Urbana. 188 p.

Pallesen, Thomas J. Petrogenesis of the Silvermine Granite; the role of liquid fractionation and K-metasomatism in the formation of a W-Sn deposit. M, 1988, University of Missouri, Columbia.

Pallister, John Stith. Magmatism at an oceanic spreading-ridge; examples from the Samail Ophiolite, Oman. D, 1980, University of California, Santa Barbara. 375 p.

Palm, Richard Steven. Glacial history; an indicator of the spatial variation of topographic slope residual to functional relationship. M, 1968, Northern Illinois University.

Palma, Vicente Vladimir. The San Rafael tin-copper deposit, Puno, S. E. Peru. M, 1981, Queen's University. 226 p.

Palmer, A. The Quaternary history of Basin Head Harbour, Kings County, P. E. I. M, 1975, Dalhousie University. 143 p.

Palmer, Allison R. The fauna of the Upper Cambrian Riley Formation in central Texas. D, 1950, University of Minnesota, Minneapolis.

Palmer, Amanda Ann. Neogene radiolarians of the U.S. Mid-Atlantic Coastal Plain; biostratigraphic and paleoenvironmental analysis, and implications to shelf paleoceanography and depositional history. D, 1984, Princeton University. 295 p.

Palmer, Andrew James Malcolm. Implications of diatom biostratigraphy and biogeography in the Eocene North Atlantic ocean. D, 1984, University of South Carolina. 286 p.

Palmer, Arthur H., Jr. The Dubuque Formation. M, 1939, Northwestern University.

Palmer, Arthur Nicholas. A hydrologic study of the Indiana karst. D, 1969, Indiana University, Bloomington. 181 p.

Palmer, Beth Ann. Deposition of the Oligocene-Miocene Mount Dutton Formation; accumulation of lahar-dominated aprons. D, 1987, University of Kansas. 274 p.

Palmer, C. Stratigraphy, petrology, and depositional environments of the Ione Formation in Madera County, California. M, 1978, California State University, Fresno.

Palmer, Carl David. Hydrogeological implications of various wastewater management porposals for the Falmouth area of Cape Cod, Massachusetts. M, 1977, Pennsylvania State University, University Park. 142 p.

Palmer, Carl David. Modelling hydrogeochemical processes with the mass transfer model WATEGM-SE. D, 1983, University of Waterloo. 375 p.

Palmer, Carl Riley. Petrology of the Triassic Thaynes Formation, southeastern Idaho and southwestern Wyoming. M, 1981, Idaho State University. 88 p.

Palmer, Catherine Grace. Pennsylvanian bellerophontacean gastropods from the Appalachian Basin. M, 1986, Bowling Green State University. 94 p.

Palmer, Curtis Allyn. The distribution of trace elements in minerals found in coal. D, 1983, Washington State University. 214 p.

Palmer, David Paul. Clay mineralogy of Permian sabkha sequences, Palo Duro Basin, Texas. M, 1981, University of Texas, Austin. 94 p.

Palmer, Dennis Erwin. Geology of Stansbury Island, Tooele County, Utah. M, 1970, Brigham Young University. 27 p.

Palmer, Donald Frank. Geology and conditions of metamorphism at Benson mines, Saint Lawrence County, New York. D, 1968, Princeton University. 123 p.

Palmer, Dorothy Bryant (Kemper). A fauna from the middle Eocene shales near Vacaville, California. M, 1923, University of California, Berkeley. 29 p.

Palmer, Elizabeth Ann. The structure and petrology of Precambrian metamorphic rock units, northwestern Marathon County, Wisconsin. M, 1980, University of Minnesota, Duluth.

Palmer, Eugene C. Geology of the Laramie Anorthosite-Syenite Complex in the Poe Mountain area, Wyoming. M, 1973, SUNY at Buffalo.

Palmer, Francis Henry, Jr. Development of a marine monitoring program in California. D, 1981, University of California, Los Angeles. 135 p.

Palmer, Harold D. Wave-induced scour around natural and artificial objects. D, 1969, University of Southern California.

Palmer, Harold Dean. Marine geology of Rodríguez Seamount. M, 1963, University of Southern California.

Palmer, Harold S. The South Moccasin Mountains, Fergus County, Montana. D, 1923, Yale University. 252 p.

Palmer, Henry Currie. Geology of the Moncion-Jarabacoa area, Dominican Republic. D, 1963, Princeton University. 282 p.

Palmer, Irven France. Geology of the Council Mountain area, Adams County, Idaho. M, 1963, University of Idaho. 54 p.

Palmer, J. H. L. Supported excavations in weak clay. D, 1974, University of Toronto.

Palmer, James Edward. Geology of an area near Rosiclare, Hardin, and Pope counties, Illinois. M, 1956, University of Illinois, Urbana.

Palmer, James Edward. Some geological and magnetic characteristics of buried and resurrected Precambrian hills of southeastern Mo. D, 1966, University of Missouri, Rolla. 296 p.

Palmer, Jeffrey John. Stacked shoreline and shelf sandstones of the La Ventana Tongue (Campanian), northwestern New Mexico. M, 1982, University of Texas, Austin. 58 p.

Palmer, Joel Oleen. Petrology of the Clayton Peak Stock, a zoned pluton near Brighton, Utah. M, 1974, Brigham Young University. 214 p.

Palmer, Katherine V. The Veneridae of eastern America; Cenozoic and Recent. D, 1925, Cornell University.

Palmer, L. M. Ridge junctions and hexagonal hierarchies. M, 1977, SUNY at Binghamton. 96 p.

Palmer, Leonard Arthur. Marine terraces of California, Oregon and Washington. D, 1967, University of California, Los Angeles. 379 p.

Palmer, Leonard Arthur. Pleistocene and Recent geology western foothills of Mount Rainier. M, 1960, University of Washington. 65 p.

Palmer, Margaret V. Ground-water flow patterns in limestone solution conduits. M, 1976, SUNY, College at Oneonta. 150 p.

Palmer, Mark A. Gravity and magnetic survey of the Weiser Warm Spring area, Weiser, Idaho. M, 1983, Idaho State University. 69 p.

Palmer, Paul W. Geology of the Lake Winona area, frontal Ouachita Mountains, Arkansas. M, 1971, Northern Illinois University. 37 p.

Palmer, Raleigh A. The geology of the Blaine Valley area, Grainger, Knox and Union counties, Tennessee. M, 1961, University of Tennessee, Knoxville. 51 p.

Palmer, Richard B. Triassic tectonics in Maryland. D, 1949, The Johns Hopkins University.

Palmer, Robert H. The rudistids of southern Mexico. D, 1925, Stanford University. 137 p.

Palmer, S. E. Organic geochemistry of modern marginal marine sediments of the Mississippi Gulf Coast. D, 1975, University of Nebraska, Lincoln. 114 p.

Palmer, Stephen Philip. Fracture detection in crystalline rock using ultrasonic reflection techniques. D, 1982, University of California, Berkeley. 343 p.

Palmer, Steven Dale. Linear analysis of the West-central Black Hills, South Dakota. M, 1977, University of Iowa. 103 p.

Palmer, William James. Quartz carbonate veins of the Wabush lake region (Precambrian, Labrador). M, 1961, Dalhousie University. 67 p.

Palmer-Rosenberg, Paul S. Himalayan deformation and metamorphism of rocks south of the Main Mantle Thrust, Karakar Pass area, southern Swat, Pakistan. M, 1986, Oregon State University. 68 p.

Palmieri, Francesco. A study of a climosequence of soils derived from volcanic rock parent material in Santa Catarina and Rio Grande do Sul states, Brazil. D, 1986, Purdue University. 276 p.

Palmiter, D. B. Geology of the Koloa Volcanic Series of the south coast of Kauai, Hawaii. M, 1975, University of Hawaii at Manoa. 88 p.

Palmore, Robert Donald. Geology of the Forkland Quadrangle, Greene County, Alabama. M, 1959, University of Alabama.

Palmquist, John Charles. Petrology and structure of the Horn area, Bighorn Mountains, Wyoming. D, 1961, University of Iowa. 188 p.

Palmquist, Robert Clarence. The geomorphic development of a part of the Driftless area, Southwest Wisconsin. D, 1965, University of Wisconsin-Madison. 201 p.

Palombo, D. A. The hydrogeology of a proposed nuclear power plant site near Sandusky Bay, Ohio. M, 1974, Ohio State University.

Palombo, Kevin M. Ground-water resources of Henry County, Ohio. M, 1983, University of Toledo. 164 p.

Palomino Cardenas, Jack Roger. Environmental mapping of the Macoupin and Shoal Creek cyclothems. M, 1963, University of Illinois, Urbana.

Palomino Cardenas, Jack Roger. Sedimentological and environmental study of the fluvio deltaic Cabo Blanco Sandstone Member, Echinocyamus Formation, lower Eocene, Talara Basin, N.W. Peru. D, 1976, University of Illinois, Urbana. 159 p.

Palonen, Pentti Arnold. Sedimentology and stratigraphy of the Gog Group sandstones in southern Canadian Rockies. D, 1976, University of Calgary. 210 p.

Palonen, Pentti Arnold. Stratigraphy and depositional environment of the Mississagi Formation (Precambrian), Ontario. M, 1971, University of Calgary. 103 p.

Palubniak, Daniel S. Paleoecology of Upper Cretaceous oysters in the Fox Hills Formation, North Dakota and comparisons with Recent oyster ecology. M, 1972, Kent State University, Kent. 119 p.

Paludan, C. T. N. Geographic research from Earth orbit; with special emphasis on land-use. D, 1975, University of Denver. 166 p.

Paluzzi, Peter Ronald. Evolution of Sheep Creek Fan. M, 1979, University of Southern California.

Pamenter, Charles Beverly. Cephalopods and other fauna from the Exshaw Formation Jasper Park, Alberta. M, 1957, McMaster University. 64 p.

Pampe, William Riley. Biostratigraphy of the Devonian rocks of western Colorado. D, 1967, University of Nebraska, Lincoln. 313 p.

Pampe, William Riley. Phylogeny of the Pentremites. M, 1948, University of Illinois, Urbana.

Pampeyan, Earl H. Geology of Cajalco area, Riverside County, California. M, 1952, Pomona College.

Pamukcu, Sibel. Low strain shear measurements of soft sediments using triaxial vane device. D, 1986, Louisiana State University. 219 p.

Pan, Cheh. A preliminary study of the relation between polar wandering and seismicity. M, 1963, Massachusetts Institute of Technology. 59 p.

Pan, Chih-Wei. Subsurface geology of Saline County, Kansas. M, 1959, Kansas State University. 57 p.

Pan, Chung-Hsiang. The geology and metallogenetic provinces of eastern Asia. D, 1946, University of Minnesota, Minneapolis. 388 p.

Pan, Gee-Shang. Full waveform inversion of plane wave seismograms. D, 1987, Princeton University. 136 p.

Pan, Jeng-Jong. Nonlinear inverse methods applied to interpreting gravity anomalies produced by multi-interfaced geologic bodies. D, 1984, University of Connecticut. 102 p.

Pan, Kuan-Chou. Gravity survey of the north part of the Reardan Quadrangle, Washington. M, 1981, Eastern Washington University. 58 p.

Pan, Kuo-Liang. Regional geologic analysis of the Black Hills of South Dakota and Wyoming from remote sensing data. D, 1978, University of Iowa. 186 p.

Pan, Naide. The double slant stack with application to seismic data analysis. D, 1987, University of Houston. 94 p.

Pan, Poh-Hsi. Direct location of oil and gas by the seismic reflection method. D, 1969, Rice University. 76 p.

Pan, W. S. Thermally driven magma migration in the crust. M, 1977, SUNY at Buffalo. 79 p.

Pan, Yii-Wen. An inelastic constitutive model for soils. D, 1986, University of Washington. 199 p.

Pan, Yuh-Shyi. The genesis of the Mexican type tin deposits in acidic volcanics. D, 1974, Columbia University. 286 p.

Panayiotou, A. Geology and geochemistry of the Limassol Forest plutonic complex and the associated Cu-Ni-Co-Fe sulphide and chromite deposits, Cyprus. D, 1977, University of New Brunswick.

Pancoast, Laurence Edwin. Geology of the east flank of Alligator Ridge, White Pine County, Nevada. M, 1986, University of Idaho. 162 p.

Pandey, Sheo Ji. Prediction and comparison of properties of Hawaiian and Indian red earths using automatic data processing techniques. D, 1969, University of Hawaii. 497 p.

Pandit, Ashok. Numerical simulation of contaminant transport problems in groundwater using the finite element method. D, 1982, Clemson University. 274 p.

Pandit, Bhaskar Iqbal. An A.C. bridge for the measurement of magnetic susceptibility of rocks in low fields. M, 1967, Memorial University of Newfoundland. 170 p.

Pandit, Bhaskar Iqbal. Experimental studies on the mechanism of internal friction (Q^{-1}) of rocks. D, 1971, University of Toronto.

Pandolfi, John M. Late Ordovician colonial corals and depositional environments of the eastern Great Basin, Utah and Nevada. M, 1982, University of Wisconsin-Milwaukee. 163 p.

Pandolfi, John Michael. Paleobiological studies of colonial marine animals. D, 1987, University of California, Davis. 266 p.

Pandya, Dinesh N. A study of the bauxitic deposits along the Midway-Wilcox contact in Oktibbeha County, Mississippi. M, 1973, Mississippi State University. 62 p.

Pane, Vincenzo. Sedimentation and consolidation of clays. D, 1985, University of Colorado. 317 p.

Panek, Louis A. Stresses about mine openings in a homogeneous rock body. D, 1951, Columbia University, Teachers College.

Panfil, Daniel John. Comparative high temperature rheology of drilling fluids containing saponite, montmorillonite, and a saponite/sepiolite mixture. M, 1987, Texas Tech University. 142 p.

Panian, John. Topographically modified stresses and foreland thrust propagation. M, 1987, University of Pittsburgh.

Panian, Thomas F. Spatial variability of hydraulic properties in an undisturbed alluvial soil. M, 1986, University of Nevada. 156 p.

Panigrahi, Bijay Kumar. A stochastic approach to estimate the locations and strengths of ground water pollution sources. D, 1985, Drexel University. 241 p.

Pankiwyskyj, Kost A. Geology of the Dixfield Quadrangle, Maine. D, 1964, Harvard University.

Panko, A. W. Some aspects of the geochemistry of a post-glacial sedimentary organic sequence. M, 1977, University of Waterloo.

Panko, Andrew William. Mercury mass balance and uptake by fish in acid lake ecosystems. D, 1985, McMaster University. 170 p.

Pannella, Giorgio. Palynology of the Dakota Group and Graneros Shale (Cretaceous) of the Denver Basin; Colorado. D, 1965, University of Colorado.

Panno, Samuel V. Structural and volcanic stratigraphic controls on ore emplacement at the Pilot Knob iron ore mine, Iron County, Southeast Missouri. M, 1978, Southern Illinois University, Carbondale. 199 p.

Panozzo, Renée Heilbronner. Deformation mechanisms associated with mesoscopic kinkbands, Jura Mountains (Switzerland) and Chaines Subalpines (France). D, 1984, Texas A&M University. 348 p.

Pansza, Arthur J., Jr. Geology and ore deposits of the Silver City-De Lamar-Flint region, Owyhee County, Idaho. D, 1971, Colorado School of Mines. 151 p.

Pantall, Jack Travis, Jr. An electrical resistivity survey of some typical Cheltenham clay pits. M, 1954, St. Louis University.

Pantano, John James. Sensitivity analysis and application of multiple thermal indicator inversion for basin analysis. D, 1988, University of South Carolina. 259 p.

Panteleyev, A. Geologic setting, mineralization, and aspects of zoning at the Berg porphyry copper-molybdenum deposit, central British Columbia. D, 1976, University of British Columbia.

Panteleyev, Andrejs. Mineralization of the Driftwood property, McConnell creek district, British Columbia, (Canada). M, 1969, University of British Columbia.

Pantin, Jose H. Insoluble residues of the (Jurassic) Calera Limestone in Santa Clara County, California. M, 1946, Stanford University. 87 p.

Pantin, Ronald C. The isothermal compressibility of natural gases. M, 1977, [Stanford University].

Pantoja Alor, Jerjes. A geological reconnaissance of the San Pedro del Gallo area, Durango, Mexico. M, 1963, University of Arizona.

Panttaja, Susan Kay. Provenance and tectonic significance of the lower Paleozoic Douglas Conglomerate, northern Churchill Mountains, Antarctica. M, 1988, University of Nevada, Las Vegas. 82 p.

Panus, Elizabeth Alice. A geochemical, mineralogic and petrographic study of the Middle Ordovician Trenton Group in New York State. M, 1988, SUNY at Buffalo. 79 p.

Panuska, Bruce C. Paleomagnetism of the Wrangellia and Alexander terranes and the tectonic history of southern Alaska. D, 1984, University of Alaska, Fairbanks. 199 p.

Panuska, Bruce C. Stratigraphy and sedimentary petrology of the Kiska Harbor Formation and its relationship to the Near Island-Amchitka Lineament, Aleutian Islands. M, 1980, University of Alaska, Fairbanks. 90 p.

Pao, Gloria Ai-yi. Sedimentary history of California continental borderland basins as indicated by organic carbon content. M, 1977, University of Southern California.

Paola, Christopher. Flow and skin friction over natural rough beds. D, 1983, Woods Hole Oceanographic Institution. 347 p.

Paolini, Michael Joseph. Experimental hydrothermal alteration of feldspar; the influence of solid composition of alteration. M, 1985, Texas A&M University. 52 p.

Paolo, William Dominic Di *see* Di Paolo, William Dominic

Papacharalampos, Demetrios. Determining the applicability of ridge regression to the definition of mineral-deposit anomalies. M, 1984, University of Georgia. 102 p.

Papadakis, James and Raza, Saiyid. Geology of the Bare Hill and Montreal River areas, Keweenaw County, Michigan. M, 1954, Michigan Technological University. 83 p.

Papadopoulos, Haralambos I. Pore water pressures in saturated clays. D, 1974, Wayne State University. 211 p.

Papadopoulos, Panayiotis Charilaou. Long-term soil-structure interaction and design of large-span flexible culverts. D, 1987, University of Florida. 224 p.

Papadopulos, Istavros S. Hydromechanics of collector wells. M, 1962, New Mexico Institute of Mining and Technology. 81 p.

Papageorge, George Elefterios. Design of an electrical analog for spherical wave propagation in solid elastic media. M, 1964, Oregon State University. 43 p.

Papantonis, Dimosthenis. Isothermal compression of Fe-Co wt. 40% and pressure determination of alpha-to-epsilon phase transition for pressure calibration. M, 1975, University of Rochester. 53 p.

Paparizos, Leonidas G. Some observations on the random response of hysteretic systems. D, 1987, California Institute of Technology. 181 p.

Papaspyros, A. G. Geology of the Martinsburg Formation (Middle and Upper Ordovician), Harrisburg West Quadrangle, Pennsylvania. M, 1967, George Washington University.

Papazachos, Basil C. Angle of incidence and amplitude ratio of P and PPP waves. M, 1963, St. Louis University.

Papcke, David E. Strawberry Hill iron deposits (Lawrence County, South Dakota). M, 1958, South Dakota School of Mines & Technology.

Pape, Donald A. Terraced alluvial fills in Contra Costa County, California. M, 1978, University of California, Berkeley. 60 p.

Pape, Edwin Henry, III. A drifter study of the Lagrangian mean circulation of Delaware Bay and adjacent shelf waters. M, 1981, University of Delaware, College of Marine Studies. 150 p.

Pape, Lance W. Geology and mineralogy of the Trump fissure-fault ore body, Trixie Mine, East Tintic District, Utah. M, 1970, Brigham Young University.

Papenfus, E. B. Copper deposits of Cumberland County, Nova Scotia, and allied types. M, 1929, Massachusetts Institute of Technology. 79 p.

Papenguth, Hans William. Diagenesis of orthoconic nautiloids from an asphaltic limestone; an integrated physical and chemical analysis. M, 1983, University of Texas, Arlington. 172 p.

Papesh, Henry. A regional geophysical study of the Southern Oklahoma Aulacogen. M, 1983, University of Texas at El Paso.

Papezik, V. S. Geology of the Deer Horn Prospect. M, 1954, University of British Columbia.

Papezik, V. S. Trace elements in anorthosites. D, 1961, McGill University.

Papezik, Vladimir Stephen. Geology of the Deer Horn prospect, Omineca mining district, British Columbia. M, 1957, University of British Columbia.

Papike, James Joseph. The crystal structure and crystal chemistry of scapolite. D, 1964, University of Minnesota, Minneapolis. 64 p.

Papirchuk, W. An investigation of Lake Winnipeg beach sands. M, 1956, University of Manitoba.

Papish, Jacob. Germanium. D, 1921, Cornell University.

Papke, Keith G. Geology and ore deposits of the eastern portion of the Hilltop Mine area, Cochise County, Arizona. M, 1952, University of Arizona.

Papp, Alexander R. Extractable cations in claystones and shales of the Pennsylvanian Breathitt Formation in eastern Kentucky as related to depositional environments. M, 1982, Eastern Kentucky University. 85 p.

Pappajohn, Steven. Description of Neogene marine section at Split Mountain, easternmost San Diego County, California. M, 1980, San Diego State University.

Pappas, Thomas. Ostracods from the Silurian drift of North Germany. M, 1954, Washington University. 51 p.

Papson, Ronald P. Mineralogy and geochemistry of carbonatites from the Gem Park Complex, Fremont and Custer counties, Colorado. M, 1981, Colorado State University. 82 p.

Papusch, Richard G. Depositional history of the lower McAlester Formation, Arkoma Basin, East-central Oklahoma. M, 1983, University of Missouri, Columbia.

Paquette, Douglas Edward. Depositional and diagenetic history of a Devonian coral and stromatoporoid biostrome, falls of the Ohio River, Louisville, Kentucky. M, 1987, Wright State University. 110 p.

Paquette, Jeanne. Sedimentology and diagenesis of the Levis slope conglomerates near Quebec City; remnants of a Cambro-Ordovician carbonate platform margin. M, 1986, McGill University. 241 p.

Paradeses, William D. Geology of the Blythewood, South Carolina, quadrangle. M, 1962, University of South Carolina. 51 p.

Pararas-Carayannis, G. The barium content in the calcareous skeletal materials of some Recent and fossil corals of the Hawaiian Islands. M, 1967, University of Hawaii. 63 p.

Parashivamurthy, Agasanapura Subbanna. Potassium-calcium exchange equilibria in sandy soils containing interstratified micaceous clays. D, 1971, Texas A&M University. 119 p.

Paravincini, P. Guido. Geophysical exploration in the Colquiri mining district, Bolivia. M, 1965, Washington University. 80 p.

Parbery, David. Petrogenesis of the Seabrook Lake carbonatite alkaline complex, N.W. Ontario. M, 1984, University of Western Ontario. 175 p.

Parcel, Rodney F. Geology of part of the Santa Rosa shear zone, Riverside County, California. M, 1972, University of California, Riverside. 98 p.

Parcher, James Vernon. Critical evaluation of procedures used to analyze the stability of slopes. D, 1968, University of Arkansas, Fayetteville. 218 p.

Parchman, Mark Alan. Precambrian geology of the Hell Canyon area, Manzano Mountains, New Mexico. M, 1981, University of New Mexico. 108 p.

Parchman, William. Conodonts of the (Pennsylvanian) Marmaton Group in south-central Iowa. M, 1955, University of Wisconsin-Madison.

Parchure, Trimbak Mukund. Erosional behavior of deposited cohesive sediments. D, 1984, University of Florida. 321 p.

Pardi, Richard Raymond. The origin of carbonate concretions within glaciolacustrine and glaciomarine sediments of the Northeast United States. D, 1983, University of Pennsylvania. 376 p.

Pardieck, Daniel Lee. Biodegradation of phenols in aquatic culture by soil-derived microorganisms, with reference to their fate in the subsurface. D, 1988, University of Arizona. 217 p.

Pardini, Charles Holliger. Petrological and structural analysis of the Sierra City Melange, northern Sierra Nevada, California. M, 1986, San Diego State University. 89 p.

Parduhn, Nancy Louise. The ecology and distribution of Bacillus cereus and other microorganisms in soils associated with gold deposits. D, 1987, Colorado School of Mines. 193 p.

Paredes, Carmela Hernandez. Determination of radioactivity in and radon emanation coefficient of selected building materials and estimation of radiation exposure from their use. D, 1984, Purdue University. 183 p.

Paredes, Manuel. Paleozoic paleogeology of North Texas. M, 1959, Stanford University.

Paredes, Manuel. Significance of the sedimentary structures of the Pigeon Point Formation between Pescadero and Bean Hollow Beach, California. M, 1960, Stanford University.

Pareja, German J. A long period Rayleigh wave experiment in the Vancouver Island region. M, 1975, University of British Columbia.

Parent, Allan A. The depositional environment and diagenetic history of the Pennsylvanian Pottsville Formation in the Black Warrior Basin of Mississippi. M, 1987, University of Texas at Dallas. 103 p.

Parent, Michel. Late Pleistocene stratigraphy and events in the Asbestos-Valcourt region, southeastern Quebec. D, 1987, University of Western Ontario. 320 p.

Parham, Walter Edward. Clay mineral facies of certain Pennsylvanian underclays. D, 1962, University of Illinois, Urbana. 132 p.

Parham, Walter Edward. The petrology of the underclay of the Illinois No. 2 coal, Pennsylvanian, in the Eastern Interior Basin. M, 1958, University of Illinois, Urbana.

Parhial, Leimo I. The Great Valley of East Tennessee; relation of structure and lithology to topography. M, 1941, University of Tennessee, Knoxville. 80 p.

Pariente, Vita. Distribution of dinoflagellate cysts in sediments of Santa Monica Basin, California. M, 1988, University of Southern California.

Parikh, Upendra J. Textural properties, provenance, and environments of formation of the "rimrock" sandstone of the filled sinkholes northeast of Rolla, Missouri. M, 1970, University of Missouri, Rolla.

Paris, Chester E. Petrography, lithofacies, and depositional setting of the Kuparuk River Formation, North Slope, Alaska. M, 1981, University of Alaska, Fairbanks. 95 p.

Paris, Gabriel. Geology of the east-central part of Mineral Quadrangle, Washington County, Idaho. M, 1969, University of Idaho. 106 p.

Paris, Oliver L. A study of Precambrian rocks between Big Cottonwood and Little Willow canyons in the central Wasatch Mountains, (Utah). M, 1935, University of Utah. 44 p.

Paris, Randy Max. Developmental history of the Howell Anticline. M, 1977, Michigan State University. 76 p.

Paris, T. A. The geology of the Lincolnton 7 1/2′ Quadrangle, Georgia-South Carolina. M, 1976, University of Georgia. 184 p.

Parish, Kenneth L. A stratigraphic study of the insoluble residues of the Council Grove Group limestones of the Manhattan, Kansas, area. M, 1952, Kansas State University. 125 p.

Parizek, Eldon Joseph. A Lower Permian ammonoid fauna from Southeast New Mexico. M, 1947, University of Iowa. 35 p.

Parizek, Eldon Joseph. The geology of the Tift and Vineland quadrangles of Southeast Missouri. D, 1949, University of Iowa. 225 p.

Parizek, Richard Rudolph. Glacial geology of the Willow Bunch Lake area, Saskatchewan. D, 1961, University of Illinois, Urbana. 285 p.

Park, Allan Morey. Geology of the Trancos Woods area and San Andreas Fault, California. M, 1963, Stanford University.

Park, Byong Kwon. Mineralogy of recent sediments of North Carolina sounds and estuaries. D, 1971, University of North Carolina, Chapel Hill. 157 p.

Park, Charles Frederick, Jr. Geology of the San Xavier District. M, 1929, University of Arizona.

Park, Charles Frederick, Jr. Hydrothermal experiments with copper compounds. D, 1931, University of Minnesota, Minneapolis. 37 p.

Park, Choon-Byong. Migration of shallow reflection data. M, 1988, Ohio University, Athens. 190 p.

Park, Darrell G. The origin and nature of brecciation of Devonian strata in Wood Buffalo National Park. M, 1984, University of Alberta. 365 p.

Park, David Eugene. Petrology of the Tertiary Anchor Canyon Stock, Magdalena Mountains, New Mexico. M, 1971, New Mexico Institute of Mining and Technology. 92 p.

Park, David Eugene. The origin of bedded silicates with particular reference to the Caballos and Arkansas novaculite formations. D, 1961, Rice University. 82 p.

Park, Debra Ann. Further investigations of spectra and spectral interference due to group A elements in ICP spectroscopy; groups IVA and VA. D, 1987, Arizona State University. 274 p.

Park, Duk-Won. Application of laser holographic interferometry to the analysis of ground movement above underground openings. D, 1975, University of Missouri, Rolla.

Park, Edward C., Jr. Mineralogical and chemical study of some calcareous sediments from southeastern Gulf of Mexico. M, 1969, Texas A&M University. 98 p.

Park, Edwena Kay Eger. Well water and aquifers west of Twolick Creek Reservoir, Indiana County, Pennsylvania. M, 1976, Indiana University of Pennsylvania. 38 p.

Park, Frederick. A petrographic correlation study of the crystalline rocks of the Reading hills (Precambrian) of Berks County, Pennsylvania. M, 1961, University of Pittsburgh.

Park, Frederick B. Genesis of the Marmoration pyrometasomatic iron deposit, Marmora, Ontario. D, 1966, Queen's University. 131 p.

Park, Frederick Blair. Gravity investigation of the Soquel Creek area, California. M, 1961, Stanford University.

Park, Gerald. Some geochemical studies of beryllium deposits in western Utah. M, 1968, University of Utah. 105 p.

Park, Hong Bong. Regional metamorphism of marble and contact of the marble with quartz diorite at Conklin Quarry, Lincoln, Rhode Island. M, 1961, Brown University.

Park, Inbo. Numerical simulation of aggradation and degradation of alluvial-channel beds. D, 1987, University of Iowa. 159 p.

Park, J. K. Cluster analysis based on density estimates and its application to Landsat imagery. D, 1979, Colorado State University. 151 p.

Park, Jack Melvin. Biostratigraphy of the Upper Cretaceous White Speckled shales in central Saskatchewan. M, 1965, University of Saskatchewan. 78 p.

Park, Jeffrey John. Applications of the Galerkin formalism in the coupling of the Earth's free oscillations. D, 1985, University of California, San Diego. 227 p.

Park, John Keith. A vibrating sample magnetometer and the rock magnetic properties of olivines. M, 1978, University of Alberta. 233 p.

Park, Jong-Sim. Conditions of regional and contact metamorphism, Llano Uplift, Texas. M, 1986, University of Wisconsin-Madison. 52 p.

Park, Kapsong. Degradation and transformation of polycyclic aromatic hydrocarbons in soil systems. D, 1987, Utah State University. 154 p.

Park, Lawrence A. Crustal structure of part of the East-Continental Gravity High from the analysis of magnetic and gravity data. M, 1986, Kent State University, Kent. 103 p.

Park, Lee Brown. New analytical tests for use in determinative mineralogy. M, 1932, University of Oklahoma. 24 p.

Park, Nam. Geophysical study north of Big Bear Lake, San Bernardino Mountains, California. M, 1983, California State University, Long Beach. 120 p.

Park, Richard Avery, IV. Paleoecology of the genus Venericardia (Pelecypoda) in the Atlantic and Gulf Coastal Province. D, 1967, University of Wisconsin-Madison. 161 p.

Park, Robert Gene. The electrophoretic separation of mixtures of pure clays and the electrophoretic separation and characterization of a soil allophane. D, 1969, University of Idaho. 65 p.

Park, Sam, III. A petrographic study of the Edwards and associated limestones (Lower Cretaceous) of Real County, Texas. M, 1959, [University of Houston].

Park, Scott Gregory. Deposition, diagenesis, and porosity development of the Middle Devonian, Lucas Formation in the West Branch oil field, Ogemaw County, Michigan. M, 1987, Western Michigan University.

Park, Seung O. Simulation of lava flow with small scale models. D, 1981, Iowa State University of Science and Technology. 178 p.

Park, Seung Woo. Modeling soil erosion and sedimentation on small agricultural watersheds. D, 1981, University of Illinois, Urbana. 270 p.

Park, Stephen Keith. Three-dimensional magnetotelluric modelling and inversion. D, 1983, Massachusetts Institute of Technology. 185 p.

Park, Steven Lynn. Paleomagnetic stratigraphy, geochemistry and source areas of Miocene ash-flow tuffs and lavas of the Badger Mountain area, northwestern Nevada. M, 1983, University of Nevada. 112 p.

Park, T. Dynamic soil behavior under cyclic loading conditions. D, 1975, University of Chicago. 166 p.

Park, Won Choon. Stylolites and sedimentary structures in the Cave-in-Rock fluorspar district, southern Illinois. M, 1962, University of Missouri, Rolla.

Park, Yong Ahn. Petrography and depositional environments of the Triassic border conglomerates (eastern Virginia). M, 1966, Brown University.

Park, Young, Jr. Seismic damage analysis and damaging-limiting design for R/C structures. D, 1985, University of Illinois, Urbana. 179 p.

Park-Jones, Rosann. Sedimentology, structure, and geochemistry of the Galice Formation; sediment fill of a back-arc basin and island arc in the western Klamath Mountains. M, 1988, SUNY at Albany. 166 p.

Parkash, Barham. Depositional mechanism of greywackes, Cloridorme Formation (middle Ordovician), Gaspe, Quebec. D, 1969, McMaster University. 238 p.

Parke, Craig D. Geoelectric estimation of specific yield. M, 1984, University of Rhode Island.

Parke, David L. Paleoecology and stratigraphy of the lower Wildcat Group, Humboldt County, California. M, 1976, University of California, Davis. 118 p.

Parke, Michael E. Global numerical models of the open ocean tides M2, S2, K1 on an elastic Earth. D, 1978, University of California, San Diego. 164 p.

Parke, Robert Preston. Geology of the Southwest Mason-Llano River, Texas. M, 1953, Texas A&M University.

Parker Plaut, Lynda Marjorie. Experimental studies of selected plutonic and volcanic undersaturated alkaline rocks (Tertiary to Recent). M, 1971, University of Western Ontario. 153 p.

Parker, Albert John. Environmental and temporal patterns along compositional and structural gradients in two montane conifer forests of western North America. D, 1980, University of Wisconsin-Madison. 295 p.

Parker, Ben Hutchinson. A review of the geology of Colorado with a contribution to orogeny. D, 1934, Colorado School of Mines. 132 p.

Parker, Ben Hutchinson. A study of Rocky Mountain Cretaceous geosyncline and its relation to diastrophism. M, 1932, Colorado School of Mines. 26 p.

Parker, Ben Hutchinson, Jr. The Geology of the gold placers of Colorado. D, 1961, Colorado School of Mines.

Parker, Boyd Kent. The effects of solution composition and substrate on the hydrothermal formation of dolomite. M, 1976, Wright State University. 94 p.

Parker, Calvin A. A thin section study of some Upper Pennsylvanian limestones from southwestern Iowa. M, 1956, University of Wisconsin-Madison.

Parker, Calvin Alfred. Paleoecology of the Tiawah limestone, Middle Pennsylvanian of northeastern Oklahoma. D, 1958, University of Wisconsin-Madison. 138 p.

Parker, Dale Edward. I, use of remote sensing techniques for soil mapping in northern Wisconsin landscapes; II, Delineation of a flood plain by interpretation of panchromatic and color aerial photographs. D, 1970, University of Wisconsin-Madison. 142 p.

Parker, Dana C. A petrological investigation of the greenstone flow at Tamarack location, Houghton County, Michigan. M, 1959, Michigan Technological University. 95 p.

Parker, Donald Andrew. The effect of vibration on the shear strength of sand. D, 1967, University of Connecticut. 116 p.

Parker, Donald James. Petrology of selected volcanic rocks of the Harney Basin, Oregon. D, 1974, Oregon State University. 119 p.

Parker, Donnie F., Jr. Petrology and eruptive history of an Oligocene trachytic shield volcano, near Alpine, Texas. D, 1976, University of Texas, Austin. 214 p.

Parker, Donnie F., Jr. Stratigraphy, petrography, and K-Ar geochronology of volcanic rocks, northeastern Davis Mountains, Trans-Pecos Texas. M, 1972, University of Texas, Austin.

Parker, Douglar M. A comparison of deep reef foraminiferal assemblages in the eastern Gulf of Mexico. M, 1982, University of South Florida, St. Petersburg.

Parker, Erich Charles. Teleseismic deep sounding in the Rio Grande Rift of central New Mexico. M, 1984, University of California, Los Angeles. 44 p.

Parker, F. L. Foraminifera of the east coast of South America. M, 1930, Massachusetts Institute of Technology. 83 p.

Parker, Frank Z. The geology and mineral deposits of the Silver District, Trigo Mountains, Yuma County, Arizona. M, 1966, [University of California, San Diego].

Parker, G. G. Geology and groundwater of the Kissimmee River, Lake Okeechobee area, Florida. M, 1946, University of Washington. 76 p.

Parker, Gary. Theoretical aspects of fluvial meandering and braiding. D, 1974, University of Minnesota, Minneapolis. 58 p.

Parker, H. M. The geostatistical evaluation of ore reserves using conditional probability distributions; a case study for the Area 5 prospect, Warren, Maine. D, 1975, Stanford University. 362 p.

Parker, Harold Barnes. The Bighorn Dolomite. M, 1950, Miami University (Ohio). 71 p.

Parker, Herbert F. R. A sedimentary and petrographic study of the phosphatic limetones of central Kentucky. M, 1932, University of Kentucky. 59 p.

Parker, J. B. Stratigraphic analysis of uranium deposit variability, New Mexico and Utah. M, 1973, [University of California, San Diego].

Parker, James M. Holocene geology of Punta Caiman, northwestern Venezuela. M, 1981, University of Texas, Arlington. 83 p.

Parker, John Arthur. Depositional systems of the lower Tuscaloosa Formation (Cretaceous), north-central Gulf Coast Basin. M, 1983, University of New Orleans. 92 p.

Parker, John Charles. Water adsorption, microstructure, and volume change behavior of clay minerals and soils. D, 1980, Virginia Polytechnic Institute and State University. 113 p.

Parker, John Mason, III. Faulting in the Nittany Arch in central Pennsylvania. M, 1933, Cornell University.

Parker, John Mason, III. Regional systematic jointing in gently dipping sedimentary rocks. D, 1935, Cornell University.

Parker, John Stephen. A preliminary seismic investigation of Tertiary basin fill in the Jefferson Island Quadrangle, Montana. M, 1961, Indiana University, Bloomington. 35 p.

Parker, John Stephen Dawson. Geological relationships of the basement to cover, Musquash Harbour to Black River, Saint John area, southern New Brunswick. M, 1984, University of New Brunswick. 354 p.

Parker, John William. Ion ratios, water analyses, and subsurface fluid relationships, East Texas Basin. D, 1967, University of Texas, Austin. 350 p.

Parker, John Williams. Paleontology and stratigraphy of a well core from Lorain, Ohio. M, 1950, University of Michigan.

Parker, Jonathan David. Lower Paleogene geology of the Simi Valley area, Ventura County, California. M, 1985, California State University, Northridge. 96 p.

Parker, Kenneth L. A study of Hillje and South Hillje oil fields, Wharton County, Texas. M, 1950, [University of Houston].

Parker, Lee Ross. The paleoecology and flora of the Blackhawk Formation (upper Cretaceous) from central Utah. D, 1976, Michigan State University. 241 p.

Parker, Margaret Ann. Punched card analysis of lithologic data in regional facies studies of the Pennsylvanian System. M, 1953, University of Illinois, Urbana.

Parker, Mary. Calcareous nannofossil biostratigraphy of DSDP Leg 82, Sites 558 and 563, Azores Triple Junction. M, 1984, Florida State University.

Parker, Michael. Studies on the distribution of cobalt in lakes. D, 1966, University of Wisconsin-Madison. 94 p.

Parker, Neal M. A sedimentologic study of Perdido Bay (west Florida and east Alabama) and adjacent offshore environments. M, 1968, Florida State University.

Parker, Patrick LeGrand. The geochemistry of the stable isotopes of chlorine. D, 1960, University of Arkansas, Fayetteville. 97 p.

Parker, Pierce D. The relation of intensity gradient to persistence in ore deposits. M, 1956, University of Wisconsin-Madison.

Parker, Pierce Dow. Some physical chemical aspects of ore deposition. D, 1960, University of Wisconsin-Madison. 97 p.

Parker, Pierre Edward. Fossil and Recent pelecypods of the genus Chione on the west coast of North America. M, 1941, University of California, Berkeley. 51 p.

Parker, R. A. Radiocarbon dating of marine sediments. D, 1977, Texas A&M University. 127 p.

Parker, R. Jay. Magnetic and subsurface investigation of structures influencing the accumulation of natural gas in a portion of Northwest Arkansas. M, 1983, University of Arkansas, Fayetteville.

Parker, Randolph S. Experimental study of basin evolution and its hydrologic implications. D, 1977, Colorado State University. 351 p.

Parker, Raymond Lawrence. Alunitic alteration at Marysvale, Utah. D, 1954, Columbia University, Teachers College. 139 p.

Parker, Raymond Lawrence. Mineralogy of the alunitized volcanic rocks near Marysville, Utah. M, 1949, Indiana University, Bloomington. 46 p.

Parker, Robert Alan. Archeomagnetic secular variation. M, 1976, University of Utah. 26 p.

Parker, Robert W. Gravity analysis of the subsurface structure of the Santa Cruz Valley, Santa Cruz County, Arizona. M, 1978, University of Arizona.

Parker, Ronald Alvin. Geology of the transition between the Central and Southern Appalachians in West Virginia and adjacent Virginia. M, 1984, University of North Carolina, Chapel Hill. 101 p.

Parker, Ronald Bruce. Petrology and structure of pre-Tertiary rocks in western Alpine County, California. D, 1959, University of California, Berkeley. 172 p.

Parker, Ronald Lewis Michael. Lithofacies, paleoenvironments and tectonic history of the Deschambault Limestone, southeastern Quebec. M, 1986, University of Vermont. 174 p.

Parker, Steven E. Tectonostratigraphic relationships of the Upper Cretaceous and lower Tertiary strata of the Bighorn Basin, Wyoming. M, 1986, University of Wyoming. 125 p.

Parker, T. S. The sedimentology and petrography of the Keeler Canyon Formation at Ubehebe Mine Canyon, California. M, 1976, Stanford University. 124 p.

Parker, Thomas Reilly. The Ordovician rocks of north-central Illinois. M, 1933, University of Illinois, Urbana.

Parker, Timothy Scott. The sedimentology and petrography of the Keeler Canyon Formation at Ubehebe Mine Canyon, California. M, 1977, Stanford University. 128 p.

Parker, Travis Jay. Dirks oil field, Bee County, Texas. M, 1939, University of Texas, Austin.

Parker, Travis Jay. Model studies of salt-dome tectonics. D, 1952, University of Texas, Austin.

Parker, W. W. Simulation of the strength characteristics of pneumatically compacted fine grained soils using the gyratory testing machine. D, 1976, West Virginia University. 346 p.

Parker, William C. Fossil ecological successions in Paleozoic level bottom brachiopod-bryozoan communities. D, 1983, University of Chicago. 217 p.

Parker, William Charles. Paleoecological significance of population structures and dynamics of selected middle Ordovician brachiopods. M, 1975, University of Tennessee, Knoxville. 93 p.

Parkhill, Thomas A. Geology of the Tertiary igneous rocks in the Richmond Hill intrusive complex, northern Black Hills, South Dakota. M, 1976, South Dakota School of Mines & Technology.

Parkhurst, J. I. A multivariate approach to the classification and correlation of till; a case study of tills of the Wedron Formation in Illinois. D, 1975, University of Illinois, Urbana. 211 p.

Parkhurst, Robert W. Surface to subsurface correlations and oil entrapment in the Lansing and Kansas City groups (Pennsylvanian) in northwestern Kansas. M, 1959, University of Kansas. 71 p.

Parkin, Gary William. Sedimentology of a Pleistocene glacigenic diamicton sequence near Campbell River, Vancouver Island, British Columbia. M, 1986, University of Western Ontario. 194 p.

Parkin, Kathleen Marie. Electronic absorption spectra of minerals at elevated temperatures. D, 1979, Massachusetts Institute of Technology. 155 p.

Parkins, William George. Onondaga chert; geological and palynological studies as applied to archaeology. M, 1977, Brock University. 104 p.

Parkinson, Charles R. The Flathead Formation (Middle Cambrian) of northwestern Wind River Mountains, Wyoming. M, 1958, Miami University (Ohio). 106 p.

Parkinson, Craig Leonard. Geology of the Gem, Dago Peak, and Murray stocks, Coeur d'Alene mining district, Idaho. M, 1984, University of Idaho. 126 p.

Parkinson, David Lamon. U-PB geochronometry and regional geology of the southern Okanagan Valley, British Columbia; the western boundary of a metamorphic core complex. M, 1985, University of British Columbia.

Parkinson, Lucius J., Jr. Geology of an area east of Boulder (Boulder County), Colorado. M, 1955, University of Colorado.

Parkinson, R. N. A study of rock alteration associated with silver mineralization at Cobalt, Ontario. M, 1951, University of Toronto.

Parkinson, Randall W. Holocene sedimentation and coastal response to rising sea level along a subtropical low energy coast, Ten Thousand Islands, Southwest Florida. D, 1987, University of Miami. 244 p.

Parkinson, Randall William. Petrology, conodont distribution, and depositional interpretation of the Lenapah Formation, Middle Pennsylvanian, southeastern Kansas and northeastern Oklahoma. M, 1982, University of Iowa. 89 p.

Parkinson, Robert J. Genesis and classification of Arctic Coastal Plain soils, Prudhoe Bay, Alaska. M, 1977, Ohio State University.

Parkinson, Robert W. The stratigraphy and structure of the northern half of the Tully 7-1/2 Minute Quadrangle, New York. M, 1960, Syracuse University.

Parkison, Gary Alden. Tectonics and sedimentation along a Late Jurassic (?) active continental margin, western Sierra Nevada foothills, California. M, 1976, University of California, Berkeley. 170 p.

Parks, Donald A. Heavy mineral analysis of the Cretaceous Red Bank Formation of New Jersey. M, 1959, University of Massachusetts. 65 p.

Parks, James Marshall. The use of thermoluminescence of limestones in subsurface stratigraphy. D, 1951, University of Wisconsin-Madison.

Parks, John T. Seismic investigation of Carteret County, North Carolina. M, 1983, University of North Carolina, Chapel Hill. 85 p.

Parks, Louis Steven. Distribution and a possible mechanism of uranium accumulation in the Catahoula Tuff, Live Oak County, Texas. M, 1979, Texas A&M University. 107 p.

Parks, Oattis Elwyn. Geomorphic description and history of the terraces and meanders of the Brazos River valley in Parker County, Texas. M, 1956, Texas Christian University. 35 p.

Parks, Pamela Hennis. Gravity interpretation of the Wind River Mountains, Wyoming. M, 1979, Texas A&M University. 107 p.

Parks, Peter. Cranial anatomy and mastication of the Triassic reptile Trilophosaurus. M, 1969, University of Texas, Austin.

Parks, Sandra Moffett. One-dimensional finite difference model of the effects of high water stages of the Miami River on groundwater levels in the Dayton Aquifer. M, 1981, Wright State University. 86 p.

Parks, Thomas. Dolomitization in the (Ordovician) Trenton of southwestern Ontario. M, 1948, University of Toronto.

Parks, Thomas. Reservoir conditions in the (Devonian) Norfolk (Big Lime) Formation of southwestern Ontario. D, 1950, University of Toronto.

Parks, W. A. The Huronian of the basin of the Moose River (Ontario). D, 1900, University of Toronto.

Parks, William Scott. Petrology of the Morrison Formation in the Las Vegas Basin, San Miguel County, New Mexico. M, 1957, Mississippi State University. 56 p.

Parks, William Scott. The clay minerals of the Ocmulgee River (Georgia). M, 1971, Emory University. 82 p.

Parks, William Scott, Jr. Multivariate analysis of Georgia coastal plain ground waters. D, 1982, Georgia Institute of Technology. 143 p.

Parman, Lynn. Thermal maturity of Carboniferous strata around the Ozark Dome. M, 1988, University of Missouri, Columbia. 92 p.

Parmelee, E. Bruce. A physical and chemical study of stratigraphic section of the Ellenburger Group, Llano County, Texas. M, 1946, Texas A&M University. 133 p.

Parmenter, Guy Norris. Glacial water levels in Narragansett Basin and the Blackstone River Valley [Rhode Island-Massachusetts]. D, 1956, Clark University. 124 p.

Parmentier, Edgar M. Studies of thermal convection with application to convection in the Earth's mantle. D, 1975, Cornell University. 157 p.

Parmley, Walter C. On the structural geology of a portion of the Wasatch Mountains, near Ogden, Utah. M, 1893, University of Wisconsin-Madison.

Parmley, William P. Geochemical characterization and shard morphology of volcanic glass from the Oligocene Catahoula Formation, Jasper County, Texas. M, 1987, Stephen F. Austin State University. 114 p.

Parnell, Robert H. A study of old Tombigbee River terrace deposits in eastern Lowndes County, Mississippi and western Pickens and Lamar counties, Alabama. M, 1962, Mississippi State University. 42 p.

Parnell, Roderic Alan, Jr. Aluminum migration and chemical weathering in subalpine and alpine soils and tills, Mt. Moosilauke, N.H.; the effects of acid rain. D, 1981, Dartmouth College. 286 p.

Parodi, Margaret. Petrology, structure, and geochemistry of the Dana Hill metagabbro, Russell, New York. M, 1978, SUNY at Binghamton. 134 p.

Parolini, Joseph R. Debris flows within a Miocene alluvial fan, Lake Mead region, Clark County, Nevada. M, 1986, University of Nevada, Las Vegas. 120 p.

Parr, Andrew Joseph. Geology, alteration, and mineralization at the Western World Lakes Cu-Zn prospect, Yuba County, California. M, 1987, University of Nevada. 112 p.

Parr, Clayton Joseph. A study of primary sedimentary structures around the Moab Anticline, Grand County, Utah. M, 1965, University of Utah. 102 p.

Parr, David F. Mineralogy of Pugh Quarry, Wood County, Ohio. M, 1975, Miami University (Ohio). 209 p.

Parr, Deborah L. Jones. The effects of season and rain events on the chemistry of selected springs waters in limestone terrain, Washington County, Arkansas. M, 1987, University of Arkansas, Fayetteville.

Parr, John Thomas. A study of black magnetic spherules from certain Precambrian meta-sediments. M, 1966, Brown University.

Parra, Jorge. A least-squares fitting method for the interpretation of magnetic anomalies caused by two-dimensional structures. M, 1972, Colorado School of Mines. 52 p.

Parra-Mata, Juan Jose. Process-response systems; a practical approach to watershed classification considering water and sediment yields. D, 1981, Colorado State University. 565 p.

Parraga, Felipe. Geological study of the B5×3, B5×4 and B3×14 oil reservoirs west Eocene area, Lake Maracaibo, Venezuela. M, 1983, Indiana State University. 104 p.

Parraras-Carayannis, G. An investigation of anthropogenic sediments in the New York Bight. D, 1975, University of Delaware, College of Marine Studies. 288 p.

Parrillo, Daniel G. Precambrian geology of the Wanaque-Butler area. M, 1960, Rutgers, The State University, New Brunswick. 60 p.

Parris, David C. Morphology and relationships of the Prosciurinae (late Oligocene, late Miocene) (Slim Buttes, South Dakota). M, 1968, South Dakota School of Mines & Technology.

Parris, Frank G. Relative reliability in structure contour maps made from comparative elevations and from dip readings. M, 1930, University of Pittsburgh.

Parris, Thomas Martin. Petrologic and diagenetic patterns in the Kaibab Formation (Periman), Toroweap Valley, Arizona. M, 1984, Texas Christian University. 160 p.

Parrish, Christopher C. Dissolved and particulate lipid classes in the aquatic environment. D, 1986, Dalhousie University. 259 p.

Parrish, David. Quantitative study of two folds. M, 1969, University of Missouri, Columbia.

Parrish, David Keith. A nonlinear finite element fold model. D, 1972, Rice University. 138 p.

Parrish, Irwin S. The Roy, Monze and Wold Creek meteorites. M, 1960, Indiana University, Bloomington. 54 p.

Parrish, Jay Bennett. The relationship of geophysical and remote sensing lineaments to regional structure and kimberlite intrusions in the Appalachian Plateau of Pennsylvania. M, 1978, Pennsylvania State University, University Park. 65 p.

Parrish, Jay Bennett. The use of geophysical, geobotanical, and remotely sensed data in a low-cost hydrocarbon exploration strategy for the Appalachians. D, 1985, Pennsylvania State University, University Park. 148 p.

Parrish, John George. Quaternary terraces of the lower Colorado river, Texas. M, 1970, University of Houston.

Parrish, Judith Totman. Problems in the biogeography of Recent and fossil benthic marine invertebrates. D, 1979, University of California, Santa Cruz. 596 p.

Parrish, Randall Richardson. Cenozoic thermal and tectonic history of the Coast Mountains of British Columbia as revealed by fission track and geological data and quantitative thermal models. D, 1982, University of British Columbia.

Parrish, Randall Richardson. Structure, metamorphism and geochronology of the Northern Wolverine Complex near Chase Mountain, Aiken Lake map area, British Columbia. M, 1977, University of British Columbia.

Parrish, Walter C. Paleoenvironmental analysis of a Lower Permian bonebed and adjacent sediments, Wichita County, Texas. M, 1975, University of Texas, Austin.

Parrish, William. The reflectance of opaque minerals. D, 1940, Massachusetts Institute of Technology. 229 p.

Parrott, Emory W. Geology of the northern portion of the Hightown Anticline, Monterey Quadrangle, Highland County, Virginia. M, 1948, University of Virginia. 102 p.

Parrott, Jack D. Shallow ground water denitrification associated with an oxidation-reduction zone in Hall County, Nebraska. M, 1988, University of Nebraska, Lincoln. 79 p.

Parrott, Mark H. Interpretation of MAGSAT anomalies over South America. M, 1985, Purdue University. 95 p.

Parrott, Richard J. E. $^{40}Ar/^{39}Ar$ dating of the Labrador Sea volcanics and their relation to sea-floor spreading. M, 1976, Dalhousie University. 166 p.

Parry, John Powell. Groundwater flow patterns of some prairie sloughs. M, 1968, University of Saskatchewan. 106 p.

Parry, Marshall Eugene. A sandstone channel in the Mesaverde Group near Cuba, New Mexico. M, 1957, University of New Mexico. 90 p.

Parry, Robert J. Integral equation formulations of scattering from two-dimensional inhomogeneities in a conductive earth. D, 1969, [University of California, Berkeley].

Parry, Steven Elliott. Volcanism, hydrothermal alteration and copper-zinc mineralization at the Four Corners property, Noranda, Quebec. M, 1979, University of Western Ontario. 157 p.

Parry, William Thomas. Cation substitutions in biotites from Basin and Range quartz monzonites. D, 1961, University of Utah. 129 p.

Parry, William Thomas. Trace elements in pyrite from the U.S. Lark Mine, Lark, Utah. M, 1959, University of Utah. 45 p.

Parsley, G. Peter. Stratigraphy and carbonate petrology of the Lower Ordovician Garden City Formation, Southeast Idaho and northern Utah. M, 1988, University of Idaho. 134 p.

Parsley, Matthew Jay. Deposition and diagenesis of a late Guadalupian barrier-island complex from the middle and upper Tansill Formation (Permian), East Dark Canyon, Guadalupe Mountains, New Mexico. M, 1988, University of Texas, Austin. 247 p.

Parsley, Robert M. Late Cretaceous through Eocene paleocurrent directions, paleoenvironment and paleogeography of San Miguel Island, California. M, 1972, Ohio University, Athens. 135 p.

Parsley, Ronald L. North American Soluta. M, 1963, University of Cincinnati. 104 p.

Parsley, Ronald Lee. Studies in Middle Ordovician primitive Echinodermata. D, 1969, University of Cincinnati. 353 p.

Parson, Charles Grady. Rock glaciers and site characteristics on the Blanca Massif in South-central Colorado. D, 1980, University of Iowa. 165 p.

Parsons, Barbara Mae. A palynological investigation of glacial Lake Monongahela sediments. D, 1969, University of Tennessee, Knoxville. 77 p.

Parsons, Brian E. Geologic factors influencing recharge of the Baton Rouge ground water system, with emphasis on the Citronelle Formation (PlioPleistocene) (Louisiana). M, 1967, Louisiana State University.

Parsons, Clifford C. A federal coal leasing model. M, 1976, Colorado School of Mines. 111 p.

Parsons, Ian Dennis. The application of the multigrid method to the finite element solution of solid mechanics problems. D, 1988, California Institute of Technology. 267 p.

Parsons, John Edward. Development and application of a three-dimensional water management model for drainage districts. D, 1987, North Carolina State University. 492 p.

Parsons, Karla Moreau. Taphonomy, stratigraphy, and depositional processes in the Middle Devonian Windom Shale (Moscow Formation, Hamilton Group) of New York. M, 1987, University of Rochester. 88 p.

Parsons, Kenneth E. Regional variations in chemistry of the oil brines of the "Clinton" Sandstone in eastern Ohio. M, 1982, Kent State University, Kent. 105 p.

Parsons, Marion Grace. The Middle Cambrian, Upper Cambrian, and lower Tremadoc acritarchs of Random Island, Trinity Bay, southeastern Newfoundland. M, 1986, Memorial University of Newfoundland. 546 p.

Parsons, Marshall Clay. Geology of Core Canyon area, Colorado. M, 1954, Colorado School of Mines. 102 p.

Parsons, Michael Raymond. Spatial and temporal changes in stream network topology; post-eruption drainage, Mount St. Helens. D, 1985, Oregon State University. 179 p.

Parsons, Michael W. Distribution and origin of elongate sandstone concretions, Bullion Creek and Slope formations (Paleocene), Adams County, North Dakota. M, 1980, University of North Dakota. 133 p.

Parsons, Myles Lyle. Geochemistry of ground water in the Upper Notukeu Creek area, southwestern Saskatchewan. M, 1964, University of Saskatchewan. 62 p.

Parsons, Myles Lyle. Groundwater movement and subsurface temperatures in a glacial complex, Cochr-ane District, Ontario, (Canada). D, 1969, University of Michigan.

Parsons, Robeert L. Geology and ore deposits of the Neenach mining district of California. M, 1937, University of Southern California.

Parsons, Roger Clark. Structure and stratigraphy of the George River series (Proterozoic), Craignish hills, Cape Breton island (Nova Scotia). M, 1964, Dalhousie University. 63 p.

Parsons, W. H. The geology of the Camsell River map area and the general correlation of Precambrian rocks of Mackenzie District, Northwest Territories. M, 1948, Queen's University. 43 p.

Parsons, Willard Hall. Geology and ore deposits of the Sunlight area, Park County, Wyoming. D, 1936, Princeton University. 310 p.

Parsons, William H. Fluorescence of minerals in monochromatic ultra-violet radiation. M, 1938, Union College. 21 p.

Parsons, Willie Frank. The stratigraphy of the Arcola Limestone in Lowndes County, Mississippi (Arcola-Upper Cretaceous). M, 1951, Mississippi State University. 59 p.

Partain, Bruce Robert. Morphology and development of the Cape Tribulation fringing reefs, Great Barrier Reef, Australia. M, 1988, University of Texas of the Permian Basin. 218 p.

Partin, Elizabeth. Ferric/ferrous determinations in synthetic biotite. M, 1984, Virginia Polytechnic Institute and State University. 38 p.

Partlow, Deborah Paruso. Optical studies of biaxial aluminum-related color centers in smoky quartz. D, 1985, University of Pittsburgh. 63 p.

Partowidagdo, Widjajono. An oil and gas supply and economics model for Indonesia. D, 1987, University of Southern California.

Partridge, Lloyd R. The Permian Phosphoria and Triassic Dinwoody formations, northern Bighorn Basin, Wyoming and Montana. M, 1949, University of Wyoming. 75 p.

Paruso, Debbie. An investigation of elemental fractionation during sputter deposition. M, 1977, University of Pittsburgh.

Parviainen, Esko Atso Uolevi. The sedimentation of the Huronian Ramsay Lake and Bruce formations, north shore of Lake Huron, Ontario. D, 1973, University of Western Ontario. 426 p.

Parvis, Merle. Airphoto interpretation of soils and drainage of Parke County, Indiana. M, 1946, Purdue University.

Paschal, Lawrence W., Jr. and Cetrone, Ronald. Correlation between a well in Fallon County, Montana, and a well in Harding County, South Dakota. M, 1957, South Dakota School of Mines & Technology. 35 p.

Paschis, James A. Geology of the Eureka Mine, Boulder County, tungsten district, Boulder County, Colorado. M, 1973, Colorado School of Mines. 82 p.

Pascoe, H. L. The geology of the Combined Metals Mine, Pioche, Nevada. M, 1932, Columbia University, Teachers College.

Pascual, Ruben. A comparison of multi-fold seismic data with CDP gathers recorded along the strike versus the dip of a two-dimensional overthrust model. M, 1984, University of Houston.

Pasecznyk, Michael J. Comparison of platinum group element and graphite mineralization in two zones of the Stillwater ultramafic complex. M, 1987, Montana College of Mineral Science & Technology. 82 p.

Pashin, Jacob Charles. Paleoenvironmental analysis of the Bedford-Berea sequence (Devonian-Mississippian), northeastern Kentucky and south-central Ohio. M, 1985, University of Kentucky. 104 p.

Pashley, Emil Frederick, Jr. Structure and stratigraphy of the central, northern, and eastern parts of the Tucson Basin, Arizona. D, 1966, University of Arizona. 366 p.

Pashley, Emil Frederick, Jr. The geology of the western slope of the Wasatch Plateau between Spring City and Fairview, Utah. M, 1956, Ohio State University.

Pasho, David W. Character and origin of marine phosphorites. M, 1972, University of Southern California.

Pashuck, R. J. Palynology of Magnolia Petroleum Company's No. 78 Honaker Well, Electra Field, Wichita County, Texas. M, 1975, East Texas State University.

Pasitschniak, Anna. The sulfur content and sulfur isotopic composition of Archean basaltic rocks at Matagami, Quebec, and their relationship to massive sulfides. M, 1981, McGill University. 131 p.

Pask, Joseph Adam. Properties and use of talc and soapstone of Pacific Northwest. M, 1935, University of Washington. 113 p.

Paska, Michael A. Petroleum geology of the Early Silurian Albion Group in Morgan, Athens, Hocking and Vinton counties, Ohio. M, 1981, Ohio University, Athens. 86 p.

Paslay, Jack Duane. The stratigraphy and structure of the Tiffin Quadrangle, Missouri. M, 1953, University of Iowa. 62 p.

Pasley, Douglas C. Tabulate corals of the Keyser Formation, West Harrisburg Quadrangle, Pennsylvania. M, 1967, George Washington University.

Pasley, Mark A. Fluorescent spectral types of the liptinite macerals from selected Colorado bituminous coals. M, 1987, Southern Illinois University, Carbondale. 139 p.

Pasqualetti, Martin J. Energy in an oasis; geothermal resource development in the Imperial Valley of California. D, 1977, University of California, Los Angeles.

Pasquali-Zanin, Jean. Fluorite formation from dilute aqueous solutions of F- on carbonate minerals and rocks at room temperature and pressure. M, 1962, Colorado School of Mines. 59 p.

Pasquali-Zanin, Jean. Interpretation of the soil geochemical expression of the mineralization within the El Callao gold mining district, Venezuela. D, 1972, Colorado School of Mines. 59 p.

Pasquini, Thomas. A provenance investigation of glacial deposits in the Copper Basin, Idaho. M, 1976, Lehigh University. 79 p.

Passalacqua, Herminio. Electromagnetic fields due to a thin resistive layer. D, 1980, Colorado School of Mines. 189 p.

Passel, Charles Fay. Sedimentary rocks of the Edsel Ford mountain range, Marie Byrd Land, Antarctica. M, 1942, Indiana University, Bloomington. 147 p.

Passero, Richard N. A study of lateral mineralogical variations in the Fulton Mudstone of Kentucky. M, 1961, Miami University (Ohio). 116 p.

Passey, Quinn R. Viscosity structure of the lithospheres of Ganymede, Callisto, and Enceladus, and of the Earth's upper mantle. D, 1982, California Institute of Technology. 373 p.

Passmore, Gary William. Subsidence-induced fissures in Las Vegas, Nevada. M, 1975, University of Nevada. 105 p.

Passmore, Virginia L. Subsurface stratigraphy of the Supai Formation (Pennsylvanian and Permian) in east central Arizona. M, 1969, University of Arizona.

Pasta, Dave. Geology of the Las Posas-Camarillo Hills area, Ventura County, California. M, 1958, University of California, Los Angeles.

Paster, Theodore Phillip. Petrologic variations within submarine basalt pillows of the south Pacific-Antarctic Ocean. D, 1968, Florida State University.

Pasteris, Jill Dill. Opaque oxide phases of the De Beers Pipe kimberlite (Kimberley, South Africa) and their petrologic significance. D, 1980, Yale University. 483 p.

Pasternack, E. S. Thermoluminescence of ordered and thermally disordered albites. D, 1978, University of Pennsylvania. 344 p.

Pasternack, Stephen C. Nonlinear finite element analysis using an independent, two surface material model with strain softening. D, 1984, University of Akron. 149 p.

Pastirik, George Paul. The sedimentology of unconsolidated deltaic and aeolian sediments east of Dunnville, Ontario. M, 1985, Brock University. 213 p.

Pastrana, Jose Manuel. A study of magnetic properties of iron-containing minerals. D, 1971, University of Minnesota, Minneapolis. 198 p.

Pastula, Edward J. The ecology and distribution of Recent foraminifera of Choctawhatchee bay, Florida. M, 1967, Florida State University.

Pastuszak, Robert A. Geomorphology of part of the La Plata and San Juan rivers, San Juan County, New Mexico. M, 1969, University of New Mexico. 84 p.

Patch, Susan. Petrology and stratigraphy of the Epitaph dolomite (Permian) in the Tombstone hills, Cochise County, Arizona. M, 1969, University of Arizona.

Patch, W. R. The Waterloo Quaternary Entomology Database user's guide. M, 1985, University of Waterloo. 131 p.

Patchen, Allan D. Stratigraphy and petrology of the Lake Owen layered mafic complex, SE Wyoming. M, 1987, University of Wyoming. 163 p.

Patchen, Douglas G. Petrology of the Oswego (Upper Ordovician), Queenston (Upper Ordovician), and Grimsby (Lower Silurian) formations, Oswego County, New York. M, 1965, SUNY at Binghamton. 191 p.

Patchen, Douglas Gene. Stratigraphy and petrology of the Williamsport Sandstone (Upper Silurian), West Virginia. D, 1972, Syracuse University.

Patchett, Joseph Edmund. A study of the radioactive minerals of the uraniferous conglomerates, Blind River area (Ontario). D, 1960, University of Toronto.

Patchett, Joseph Edmund. The distribution of accessory minerals as a criterion of the origin of gneisses. M, 1954, University of Toronto.

Patchick, Paul F. Economic geology of the Bullion mining district, San Bernardino County, California. M, 1959, University of Southern California.

Pate, Christopher Scott. A subsurface study of the Queen City Formation in Nacogdoches and Angelina counties. M, 1987, Stephen F. Austin State University. 160 p.

Pate, Colin Roger. Cretaceous compaction and tectonic subsidence of the Alberta Basin. M, 1986, University of Alberta. 264 p.

Pate, David L. Evaluation of amino acid stereochemical measurements as chronological tools in mollusks from the South Pacific, Morocco, and Barbados. M, 1978, University of Delaware.

Pate, James D., Jr. A geological engineering study of the Sooner Trend, middle Layton Sand unit (Pennsylvanian), Kingfisher County, Oklahoma. M, 1972, University of Oklahoma. 86 p.

Pate, James Durwood. The geology of Cotton County, Oklahoma. M, 1947, University of Oklahoma. 59 p.

Pate, Joe Henry. Subsurface geology of Carter Knox oil field, Stephens and Grady counties, Oklahoma. M, 1953, University of Oklahoma. 59 p.

Patek, John Mark. The wettability and floatability of silicate minerals. M, 1933, Michigan Technological University. 29 p.

Patel, Arunkumar Kalidas. Determination of compositions and structural states of alkali feldspars from shallow instrusions in the Terlingua District, Texas. M, 1973, [University of Houston].

Patel, Bharat A. Geophysical study of the talc deposit in the State Line serpentinite, Lancaster, Pennsylvania. M, 1983, Rutgers, The State University, Newark. 130 p.

Patel, C. B. Mineragraphy of the copper ores of the Encampment District, Wyoming. M, 1950, University of Wyoming.

Patel, Dinesh. Distribution of trace elements in host rock limestone in the Tri-State zinc-lead district. M, 1973, University of Missouri, Rolla.

Patel, Jayantilal P. Rock magnetic and paleomagnetic properties of the Nipissing Diabase (Precambrian, southern Ontario). M, 1971, University of Western Ontario.

Patel, Kishore N. A mineralogical and petrological study of a slightly radioactive Pyritized Conglomerate at Shut-In Creek, near Hot Springs, North Carolina. M, 1973, University of Tennessee, Knoxville. 50 p.

Patenaude, Robert W. Results of a regional aeromagnetic survey of a part of upper and lower Michigan. M, 1962, University of Wisconsin-Madison.

Patenaude, Robert William. A regional aeromagnetic survey of Wisconsin. D, 1967, University of Wisconsin-Madison. 56 p.

Patera, Edward Smyth, Jr. Phase equilibria of the upper Martian mantle; theoretical calculations and experiments. D, 1982, Arizona State University. 147 p.

Paterek, James Robert. Ecology of methanogenesis in two hypersaline biocoenoses; Great Salt Lake and a San Francisco Bay saltern. D, 1983, University of Florida. 136 p.

Paternoster, Benoit J. Effects of layer boundaries on full waveform acoustic logs. M, 1985, Massachusetts Institute of Technology. 137 p.

Paterson, Ian Arthur. The geology of the Pinchi Lake area, central British Columbia. D, 1973, University of British Columbia.

Paterson, Lorraine Ann. Petrology of the Tsu Lake Gneiss in the Fort Smith area. M, 1985, University of Alberta. 331 p.

Paterson, Norman R. Model seismology; elastic wave propagation in granular media. D, 1955, University of Toronto.

Paterson, Philip G. Ore deposits associated with the Coast Range Batholith. M, 1927, University of British Columbia.

Paterson, Raymond G. On the determination of magnetic susceptibility of rocks in situ. M, 1937, Colorado School of Mines. 30 p.

Paterson, Scott Robert. Strains, metamorphism, and structural evolution of tectonostratigraphic terranes in the Western Metamorphic Belt, Sierra Nevada, California. D, 1986, University of California, Santa Cruz. 222 p.

Paterson, W. Stanley B. Observations on Athabaska Glacier (British Columbia) and their relation to the theory of glacier flow. D, 1962, University of British Columbia.

Patet, Alix. A subsurface study of the foraminiferal fauna of the Vaqueros, Rincon and lower Monterey formations (Miocene) from the Elwood oil field, Santa Barbara County, California. M, 1972, University of California, Berkeley. 67 p.

Patias, Petros Georgios. Application of random field theory in mapping problems. D, 1987, Ohio State University. 178 p.

Patku, B. Comparison of theoretical results and experimental results with models in response obtained from a cylindrical conductor situated in an electromagnetic field. M, 1953, University of Toronto.

Patmore, William Henry. Pseudo-pyrometasomatic gold at Hedley, B.C. D, 1941, Princeton University. 196 p.

Patnode, Homer Whitman. Petrographic study of an intraseptum intrusion in the Sierra Nevada. M, 1935, Cornell University.

Patraw, James M. A study of the distribution of selenium in the Niobrara Formation of western South Dakota. M, 1963, South Dakota School of Mines & Technology.

Patrick, Brian E. Phase relations between jadeitic pyroxene and coexisting minerals in Franciscan metagraywackes, northern Diablo Range, California. M, 1984, University of California, Davis. 105 p.

Patrick, Brian Ellsworth. Petrological and structural studies in the Seward Peninsula blueschist terrane, Alaska. D, 1987, University of Washington. 122 p.

Patrick, David Maxwell. Mineralogy and geochemistry of volcanic ash and bentonite in the Ogallala Formation (Tertiary) of western Oklahoma. D, 1972, University of Oklahoma. 153 p.

Patrick, David Maxwell. The Cheltenham Clay of Boone County, Missouri. M, 1964, University of Missouri, Columbia.

Patrick, G. C. A natural gradient tracer experiment of dissolved benzene, toluene and xylenes in a shallow sand aquifer. M, 1986, University of Waterloo. 176 p.

Patrick, Joseph L. The effects of Lake Superior shore currents on Recent sediments. M, 1955, Michigan State University. 40 p.

Patrick, R. R. Morphologic variation in the ostracod Haplocytheridea montgomeryensis. M, 1975, Indiana State University. 53 p.

Patrick, T. O. H. Sericite in granitic feldspar. D, 1954, University of Wisconsin-Madison.

Patrick, W. C. Creep and stiffness of full column resin grouted roof bolts. M, 1975, University of Missouri, Rolla.

Patrick, W. F. Memoir of a visit to the Ducktown copper mines, August 1873. M, 1873, Washington University.

Pattarozzi, Michael. Economic and environmental aspects of lignite strip mining, Bastrop County, Texas. M, 1975, University of Texas, Austin.

Patten, Harvey L. A postglacial pollen diagram from Lake Carlson, Dakota County, southern Minnesota. M, 1959, University of Minnesota, Minneapolis. 35 p.

Patten, Philip R. The (Miocene) San Pablo Formation north of Mount Diablo, California. M, 1948, University of California, Berkeley. 67 p.

Patten, Richard C. Metalliferous deposits of Alaska. M, 1922, University of Minnesota, Minneapolis. 60 p.

Pattengill, Maurice Glenn. Variations of some fenestrate bryozoans of the Gearyan Series in eastern Kansas. M, 1964, Kansas State University. 81 p.

Patterson, Archibald Balfour, III. Check list of the most characteristic foraminifera of the Taylor Group in Travis County, Texas. M, 1941, University of Texas, Austin.

Patterson, Ben Arnold. A study of the basal member of the Chinle Formation in the inter-river area, Grand and San Juan counties, Utah. M, 1957, University of Illinois, Urbana.

Patterson, Brooks Alan. Characterization and correlation of biodegraded oils from Kern County, California. M, 1984, University of California, Riverside. 109 p.

Patterson, Charles G. Geochemistry of Boulder Creek, Boulder, Jefferson, and Gilpin counties, Colorado. M, 1980, University of Colorado.

Patterson, Charles M. Alteration in the Santa Rita copper deposits, Santa Rita, New Mexico. D, 1947, Columbia University, Teachers College.

Patterson, Charles M. The geology of Lake Cobbosseecontee, Augusta, Maine. M, 1942, Columbia University, Teachers College.

Patterson, Dale Duane. Correlation of the Montana Group of the Upper Cretaceous between Salt Creek, Wyoming and Billings, Montana. M, 1956, University of Illinois, Urbana. 59 p.

Patterson, Daniel J. A structural analysis of a portion of the Seneca Fault, Mayes County, Oklahoma. M, 1986, University of Missouri, Columbia. 92 p.

Patterson, Deborah. The foraminiferal biostratigraphy and paleoecology of the type Rosario Formation, El Rosario, Baja California del Norte, Mexico. M, 1978, University of California, Santa Barbara.

Patterson, Deborah Lynn. The Valle Formation; an integrated stratigraphy of turbidite deposits and the reconstruction of the Late Cretaceous paleogeography of the Vizcaino Basin, Baja California, Mexico. D, 1984, University of California, Santa Barbara. 603 p.

Patterson, Elmer Davisson. Geologic interpretation of the occurrence of glacial drift in Boone County, Missouri. M, 1950, University of Missouri, Columbia.

Patterson, G. T. The influence of soil capability on economic returns to grain corn production. M, 1975, University of Guelph.

Patterson, George Cameron. The geology of the Kapkichi Lake ultramafic-mafic bodies and related Cu-Ni mineralization, Pickle Lake, Ontario. D, 1980, Carleton University. 331 p.

Patterson, J. L. A geophysical study of the Waldoboro Pluton, South-central Maine. M, 1976, SUNY at Buffalo. 61 p.

Patterson, J. Wilfred. The manto type limestone replacement deposits of northern Mexico. D, 1932, California Institute of Technology. 260 p.

Patterson, Jacqueline Woodman. Recognition and application of wave base as a critical factor in Pennsylvanian sedimentation. M, 1951, University of Illinois, Urbana. 66 p.

Patterson, James William, Jr. A ground-water management model of the Washita River alluvial aquifer in Grady, McClain, Garvin, Murray, Carter, and Johnston counties in south-central Oklahoma. M, 1984, Oklahoma State University. 285 p.

Patterson, Joel M. Reservoir properties and petrographic properties of the Cattleman Sandstone (Cabaniss Formation, Cherokee Group, Pennsylvanian) Montgomery County, Kansas. M, 1985, Wichita State University. 114 p.

Patterson, John Murray. The geology of the Canterbury west half (New Brunswick) map area. M, 1957, University of New Brunswick.

Patterson, Joseph E. Exploration potential and variations in shelf plume sandstones, Navarro Group (Maestrichtian), east central Texas. M, 1983, University of Texas, Austin. 92 p.

Patterson, Joseph Gilbert. Geology of the Parnassus area east of the North Mountain Fault (Virginia). M, 1958, University of Virginia. 46 p.

Patterson, Joseph M. The Douglas Group of the Pennsylvanian system in Douglas and Leavenworth counties, Kansas. M, 1933, University of Kansas.

Patterson, Judith Gay. Amer Lake; an Aphebian fold and thrust complex. M, 1981, University of Calgary. 106 p.

Patterson, Judith Gay. Tectonic evolution of distal portions of the late Proterozoic-early Paleozoic continental margin to ancestral North America. D, 1987, Virginia Polytechnic Institute and State University. 218 p.

Patterson, Kenneth D. Origin of graphite in western Platte County, Wyoming. M, 1950, University of Wyoming. 86 p.

Patterson, Logan Reid. Structure associated with a portion of Pine Mountain. M, 1959, University of Kentucky. 40 p.

Patterson, Luther Edwin, Jr. The Pennsylvanian overlap in the Seminole area. M, 1932, University of Oklahoma. 29 p.

Patterson, Marcus Brent. Structure and acoustic stratigraphy of the Lake Tanganyika rift valley; a single-channel seismic survey of the lake north of Kalemie, Zaire. M, 1983, Duke University. 90 p.

Patterson, Neil Bruce. Changes in channel morphology on the Nolichucky River caused by fluvial aggradation resulting from a rise in base. M, 1987, University of Cincinnati. 101 p.

Patterson, Orus Fuquay, III. The depositional environment and paleoecology of the Pekin Formation (upper Triassic) of the Sanford Triassic basin, North Carolina. M, 1969, North Carolina State University. 104 p.

Patterson, Patricia L. Numerical modeling of hydrothermal circulation at ocean ridges. M, 1976, Georgia Institute of Technology. 84 p.

Patterson, Patricia Lynn. Resolving the heat-flow anomaly at the Galapagos spreading center. D, 1980, Georgia Institute of Technology. 174 p.

Patterson, Penny Ellen. Petrographic, geochemical, and paleomagnetic study of color banding in the Cathedral Bluffs Tongue of the Wasatch Formation (Eocene), Washakie Basin, Wyoming. M, 1981, University of Colorado. 151 p.

Patterson, Peter Vosper. Geology of the northern third of the Glide Quadrangle, Oregon. M, 1961, University of Oregon. 83 p.

Patterson, Robert L. Geology of part of the northeast quarter of the Mitchell Quadrangle, Oregon. M, 1966, Oregon State University. 97 p.

Patterson, Ronald James. Hydrology and carbonate diagenesis of a coastal sabkha in the Persian Gulf. D, 1972, Princeton University. 473 p.

Patterson, Roy. Tectonic geomorphology and neotectonics of the Santa Cruz Island faults, Santa Barbara County, California. M, 1979, University of California, Santa Barbara.

Patterson, Roy Timothy. Late Oligocene to Pleistocene benthic foraminifera from DSDP Site 357 (Leg 39) on the Rio Grande Rise in the South-west Atlantic Ocean. D, 1986, University of California, Los Angeles. 418 p.

Patterson, Sam Hunting. Conodonts from the Prospect Hill Member of the Hannibal Formation of Iowa. M, 1947, University of Iowa. 67 p.

Patterson, Sam Hunting. Geology of the northern Black Hills bentonite mining district. D, 1955, University of Illinois, Urbana. 152 p.

Patterson, Terence Edward. Paleomicroflora of the lower part of the Ashville Formation (Cretaceous) of Saskatchewan. M, 1970, University of Alberta. 236 p.

Patterson, Thomas Lee. The cycle of atmospheric cadmium in the remote North Pacific. D, 1988, University of Rhode Island. 249 p.

Patterson, W. Ray. Leaching of dolines. M, 1932, University of Minnesota, Minneapolis. 16 p.

Patterson, W. Samuel, Jr. Facies geometry and constituent composition of a Lower Pennsylvanian ooid grainstone shoal deposits; effects of source and transport. M, 1987, Texas A&M University.

Patterson, William A. A formula for the computation of the gravity effect of radially symmetrical bodies. M, 1951, Washington University. 22 p.

Patterson, William Dean, II. Geology of Permian rocks near Ascencion, northern Chihuahua, Mexico. M, 1978, Texas Christian University. 70 p.

Pattey, Phillip D. An analytical technique for estimating the deflection profile, mechanical thickness, and failure of the oceanic lithosphere. M, 1988, Southern Methodist University.

Pattin, Virginia L. A standard X-ray diffraction pattern for opal. M, 1938, Columbia University, Teachers College.

Pattison, Edward Foyer. Coexisting micas in igneous and metamorphic rocks. M, 1965, McGill University.

Pattison, Halka M. The stratigraphy and paleontology of the (Permian) Kaibab Formation at Walnut Canyon, Arizona. M, 1947, Stanford University. 77 p.

Pattison, Linda. Petrofabric analysis of experimentally folded multi-lithologic, layered rocks. M, 1972, Texas A&M University.

Pattison, Simon Alan James. Relative sea level control of incised shoreface sediments in the Burnstick Member, Cardium Formation, Upper Cretaceous, Alberta. M, 1987, McMaster University. 206 p.

Patton, Delmar Keith. Investigation of stream alluvium on the middle fork of the White River near Elkins, Arkansas (sub-Recent). M, 1968, University of Arkansas, Fayetteville.

Patton, E. C. National balance method of evaluating pressure-maintenance by internal gas injection in tight volumetrically controlled reservoirs (P278). M, 1946, University of Texas, Austin.

Patton, Franklin Davis. Multiple modes of shear failure in rock and related materials. D, 1966, University of Illinois, Urbana. 293 p.

Patton, G. D. Studies in the system Al_2O_3-H_2O; boehmite-gibbsite relationships. M, 1974, University of Guelph.

Patton, Howard John. Source and propagation effects of Rayleigh waves from central Asian earthquakes. D, 1978, Massachusetts Institute of Technology. 342 p.

Patton, Howard Lewis. The general geology of the upper Horse Creek region, Teton County, Wyoming. M, 1948, University of Illinois, Urbana.

Patton, Jacob Luther. The paleontology of the Austin Chalk in Travis and Williamson counties, Texas. M, 1932, University of Texas, Austin.

Patton, Jean J. Depositional history of the Terry and Hygiene sandstone members, Cheyenne Basin. M, 1980, Colorado State University. 101 p.

Patton, Jeffrey Connor. Map design for children; an evaluation of planimetric and plan-oblique symbols to represent the environment. D, 1980, University of Kansas. 151 p.

Patton, John Barratt. Geology of Griffin oil field, Indiana. M, 1940, Indiana University, Bloomington. 37 p.

Patton, John Barratt. Limestone resources of southern Indiana. D, 1954, Indiana University, Bloomington. 385 p.

Patton, LeRoy Thompson. Geology of the Beaver River valley between Beaver Falls and Wampum, Pennsylvania. M, 1916, University of Iowa. 90 p.

Patton, LeRoy Thompson. The geology of a portion of western Oklahoma between the Canadian River and the Wichita Mountains. D, 1923, University of Iowa. 167 p.

Patton, Peter C. Geomorphic criteria for estimating the magnitude and frequency of flooding in central Texas. D, 1977, University of Texas, Austin. 238 p.

Patton, Peter C. Gully erosion in the semiarid west. M, 1973, Colorado State University. 143 p.

Patton, Thomas Charles. Economic geology of the L-D Mine, Wenatchee, Washington. M, 1967, University of Washington. 29 p.

Patton, Thomas Charles. Geology and hydrothermal alteration of the Middle Fork copper prospect, King County, Washington. D, 1971, University of Washington. 83 p.

Patton, Thomas Hudson, Jr. Fossil vertebrates from Miller's Cave, Llano County, Texas. M, 1962, University of Texas, Austin.

Patton, Thomas Hudson, Jr. Miocene and Pliocene artiodactyls, Texas Gulf Coastal Plain. D, 1966, University of Texas, Austin. 177 p.

Patton, Thomas L. and Robinson, Mark. Bedrock geology, geochemistry and geophysics of Brooks Mountain, Seward Peninsula, Alaska. M, 1975, University of Alaska, Fairbanks. 106 p.

Patton, Thomas Lewis, III. Normal-fault and fold development in sedimentary rocks above a pre-existing basement normal fault. D, 1984, Texas A&M University. 224 p.

Patton, Thomas W. Quaternary and Tertiary geology of the Turner-Hogeland Plateau, north-central Montana. M, 1987, Montana College of Mineral Science & Technology. 173 p.

Patton, William John Hudson. Carboniferous fauna of the South Nahanni River area. M, 1954, University of Alberta. 103 p.

Patton, William Wallace, Jr. Geology of the Clayton area, Custer County, Idaho. M, 1948, Cornell University.

Patton, William Wallace, Jr. Geology of the upper Killik-Itkillik region, Alaska. D, 1959, Stanford University. 237 p.

Pattridge, Katherine Amanda. The Gannett Peak Lineament; a passive element during Laramide uncoupling of the Wyoming Cordilleran Foreland. M, 1976, University of Michigan.

Patzewitsch, Wwndy W. Tectonic framework of Georges Bank, offshore Massachusetts. M, 1976, Southern Methodist University. 50 p.

Patzke, Jeffrey A. Evaluation of natural groundwater recharge rates within the upper and middle Mahoning

River basins, northeastern Ohio. M, 1986, Kent State University, Kent. 357 p.

Patzkowsky, Mark E. Analysis of inferred water flow patterns in Fistulipora M'Coy (Cystoporata, Bryozoa) (Chesterian, Mississippian). M, 1986, Indiana University, Bloomington. 132 p.

Patzold, Raymund Rainer. High temperature magnetic susceptibility bridge. M, 1972, Memorial University of Newfoundland. 58 p.

Pau, Joseph H. K. Stratigraphy, microfacies analysis, and thermal maturation of the Mississippian Rancheria and Helms formations, Spike "S" Ranch, southern Hueco Mountains, Hudspeth County, Texas. M, 1985, University of Texas at El Paso.

Pauken, Robert J. Paleoecological study of the Ostracoda of the Silica Formation (Middle Devonian, Ohio). M, 1964, Bowling Green State University. 95 p.

Pauken, Robert J., Jr. A population study of the Pleistocene Molluscan faunas in loess of the Missouri river basin in Missouri. D, 1969, University of Missouri, Columbia.

Paukert, Gary William. Geophysical study of Precambrian basement fault structure and related Cretaceous stratigraphic variation in southern Alberta. M, 1982, University of Calgary. 114 p.

Paukstaitis, Eric John. Computer simulation of the alluvial aquifer along the north fork of the Red River in southwestern Oklahoma. M, 1981, Oklahoma State University. 111 p.

Paul, Alexander H. Geology and ore deposits of the Camp Bird Mine, Ouray County, Colorado. M, 1974, Colorado School of Mines. 89 p.

Paul, Bradley Compton. Geostatistical evaluation of the Mahogany Oil Shale zone in the eastern Uinta Basin of Utah using a rotating cartesian coordinate system. M, 1985, University of Utah. 112 p.

Paul, Dennis A. Investigation of late Tertiary to Recent movement along northwest trending faults within the Kentucky River fault system in Northeast Madison and southern Clark counties, Kentucky. M, 1982, Eastern Kentucky University. 63 p.

Paul, Donald Lee. The onset of convection in highly viscous fluid spheres with application to planetary interiors. M, 1969, Massachusetts Institute of Technology. 79 p.

Paul, Duane G. The effect of construction, installation, and development techniques on the performance of monitoring wells in fine-grained glacial tills. M, 1987, University of Wisconsin-Milwaukee. 230 p.

Paul, E. Kenneth, Jr. Tin deposits of the Galena District, Blaine County, Idaho. M, 1981, University of Idaho. 112 p.

Paul, Harriet E. Geology of the Richfield oil producing zones. M, 1956, Wayne State University.

Paul, J. Bryan. The Saltville Fault at Sharp Gap; geometries, mesofabrics, and microfabrics associated with a major thrust. M, 1986, University of Tennessee, Knoxville. 143 p.

Paul, R. W. The age and origin of the Decaturville structure, Camden and Laclede counties, Missouri. M, 1970, University of Kansas.

Paul, Rick Lee. Chemical analyses of howardites, eucrites, and diogenites (HED); Antarctic/non-Antarctic comparisons and evolution of the eucrite parent body. D, 1988, Purdue University. 301 p.

Paul, Santosh Kumar. Gravity survey of the South Boulder area, Boulder County, Colorado. M, 1956, Colorado School of Mines. 53 p.

Paul-Douglas, Gabrielle. The flow of water in a particulate medium. M, 1971, McGill University. 46 p.

Paulding, Bartlett W. Crack growth during brittle fracture in compression. D, 1965, Massachusetts Institute of Technology. 184 p.

Paulet, Manuel R. An interpretation of reservoir sedimentation as a function of watershed characteristics. D, 1971, Purdue University. 178 p.

Pauli, David Allen. Evidence of trophic group interactions from the Middle Ordovician Spechts Ferry Formation. M, 1985, University of Wisconsin-Madison. 114 p.

Paull, Charles. The structure, stratigraphy and development of the Florida-Hatteras slope and inner Blake Plateau. M, 1978, University of Miami. 88 p.

Paull, Charles Kerr. I, Florida Escarpment; chemosynthetic communities, geochemical processes and geological consequences; II, Stable isotopic signal carriers in fine pelagic carbonates. D, 1986, University of California, San Diego. 218 p.

Paull, Rachel Kay. Geological resource guide to the Riverridge Nature Center, Ozaukee County, Wisconsin. M, 1970, University of Wisconsin-Milwaukee.

Paull, Rachel Krebs. Conodont biostratigraphy of the Lower Triassic Dinwoody Formation in northwestern Utah, northeastern Nevada, and southeastern Idaho. D, 1980, University of Wisconsin-Madison. 202 p.

Paull, Richard Allen. Depositional history of the Muddy Sandstone, Bighorn Basin, Wyoming. D, 1957, University of Wisconsin-Madison. 270 p.

Paull, Richard Allen. Stratigraphy of the (Mississippian) Dessa Dawn, Rundle, and (Pennsylvanian) Rocky Mountain formations in Jasper Park, Alberta. M, 1953, University of Wisconsin-Madison.

Paulsen, Gerald W. Preservation and stratigraphical distribution of pigments in Minnesota lake sediments. M, 1962, University of Minnesota, Minneapolis. 66 p.

Paulsen, Steven George. Contributions of sediment denitrification to the nitrogen cycle in Castle Lake, California. D, 1987, University of California, Davis. 113 p.

Paulsen, Thomas Arne. Nondestructive determination of the compressive strength of some rocks. D, 1971, University of Utah. 126 p.

Paulson, Edwin G. Origin of Pluma Hildago titanium (Oaxaca, Mexico). M, 1962, Massachusetts Institute of Technology. 54 p.

Paulson, Gary David. Major and trace element geochemistry of the Carolina slate belt volcanics, Southern Appalachians, and its tectonic implications. M, 1980, University of Georgia.

Paulson, Gerald Raymond. The mammals of the Cudahy fauna (Pleistocene; Meade County, Kansas). M, 1960, University of Michigan.

Paulson, James D. Ground-water resources of Wood County, Ohio. M, 1981, University of Toledo. 178 p.

Paulson, Oscar L., Jr. The fauna of the Ross Limestone. M, 1955, Mississippi State University. 90 p.

Paulson, Oscar Lawrence, Jr. Ostracoda and stratigraphy of Austin and Taylor equivalents of Northeast Texas. D, 1960, Louisiana State University. 127 p.

Paulsson, Bjorn Nils Patrick. Seismic velocities and attenuation in a heated underground granitic repository. D, 1983, University of California, Berkeley. 663 p.

Paulus, Fred. Geology of Grindstone Quadrangle, Jefferson County, New York. M, 1950, University of Rochester. 143 p.

Paulus, George Edmund. The petrography of the Rice Lake Batholith (southern Manitoba, Canada). M, 1968, University of Manitoba.

Pauly, George. Lipids of methanogenic bacteria in swamp and deep sea sediments. M, 1985, University of South Florida, St. Petersburg. 60 p.

Pauly, Harold Porter. The structure of the area between the junction of Burroughs Creek and Horse Creek, Younts Peak Quadrangle, Wyoming. M, 1950, Miami University (Ohio). 31 p.

Pauschke, Joy Marie. Dynamic response of the Imperial County services building during the 1979 Imperial Valley earthquake. D, 1982, Stanford University. 222 p.

Pautsch, Richard Joseph. A statistical analysis of hydrologic drought in the Humboldt Basin, Nevada. M, 1978, University of Nevada. 104 p.

Pavelka, Anne. Trends in the variations of metal ratios with ore type on stratigraphic position in the Kuroko deposits, Japan. M, 1984, Pennsylvania State University, University Park. 160 p.

Pavey, Richard R. Late Wisconsinan till stratigraphy in west-central Indiana. M, 1983, Purdue University. 149 p.

Pavich, M. A study of saprolite buried beneath the Atlantic Coastal Plain in South Carolina. D, 1974, The Johns Hopkins University. 214 p.

Pavish, Marie. Stratigraphy and chronology of Holocene alluvium between the Cascade Crest and the Columbia River in central Washington. M, 1973, University of Washington. 38 p.

Pavlak, Stephen John. Stratigraphy and sedimentary petrology of the Moss Back Member of the Late Triassic Chinle Formation, North Temple Wash-San Rafael Desert area, Emery County, Utah. M, 1979, University of Southern California.

Pavlakis, Parissis P. Biochronology, paleoecology and biogeography of the Plio-Pleistocene fossil mammal faunas of the Western Rift (east-central Africa) and their implication for hominid evolution. D, 1987, New York University. 547 p.

Pavletich, Joseph P. Amino-acid racemization in siliceous sediments from the Southern Ocean. M, 1987, California State University, Hayward. 126 p.

Pavlicek, John A. Conodont distribution and correlation of the Tacket Shale (Missourian, Upper Pennsylvanian) in southeastern Kansas and eastern Oklahoma. M, 1986, University of Iowa. 132 p.

Pavlicek, Meeyoun In. Upper Devonian conodont biostratigraphy in the subsurface of south-central and southeastern Iowa. D, 1986, University of Iowa. 142 p.

Pavlides, Louis. The structure and stratigraphy of the limestone quarry at Verplanck, New York, and vicinity. M, 1947, Columbia University, Teachers College.

Pavlik, Hannah Flora. Geochemical modeling and evaluation of lithium and fluoride distribution coefficients using batch and column methods; oil shale leachate in contact with Uinta Formation Sandstone, Piceance Creek basin, Colorado. D, 1987, University of Colorado. 226 p.

Pavlik, John D. Determination of trace metals in water by X-ray fluorescence. M, 1986, Wright State University. 126 p.

Pavlik, Marcy Lynn. The origin and distribution of pyrite in coal of parts of the Des Moines and Mystic districts of Iowa. M, 1976, University of Iowa. 121 p.

Pavlin, Gregory B. Applicability of a two-dimensional, digitally integrating, silicon vidicon system in the detection of natural resources. M, 1973, Massachusetts Institute of Technology. 146 p.

Pavlin, Gregory Byron. Source parameter inversion of a reservoir-induced seismic sequence, Lake Kariba, Africa; September 1963 - August 1974; a reassessment of triggering mechanisms. D, 1981, Pennsylvania State University, University Park. 293 p.

Pavlis, Erricos C. On the geodetic applications of simultaneous range-differencing to Lageos. D, 1983, Ohio State University. 230 p.

Pavlis, Gary Lee. Progressive inversion. D, 1982, University of Washington. 295 p.

Pavlis, Terry Lynn. Deformation along a late Mesozoic convergent margin; the Border Ranges fault system, southern Alaska. D, 1982, University of Utah. 177 p.

Pavlis, Terry Lynn. Stress history during growth of a noncylindrical fold. M, 1979, University of Utah. 93 p.

Pavona, Kennon Vincent. Relation of groundwater chemistry to lithology in a Kentucky carbonate aquifer. M, 1971, University of Kentucky. 38 p.

Pawel, David T. The petrology, fossil communities and environment of deposition of the Church Limestone Member of the Howard Formation, southeastern Kansas. M, 1975, Wichita State University. 84 p.

Pawlewicz, Mark J. Stratigraphy, environments of deposition and petrography of selected coals of the

Upper Cretaceous Menefee Fm. near Durango, Colorado. M, 1983, Colorado School of Mines. 116 p.

Pawling, John W. Morphometric analysis of the southern peninsula of Michigan. D, 1970, Michigan State University. 143 p.

Pawliw, Paul Andrew. A proposed design for a spinner-type magnetometer. M, 1969, University of Saskatchewan. 81 p.

Pawlowicz, Edmund Frank. An isostatic study of northern and central Greenland based on gravity values and airborne radar-ice thickness measurements. D, 1969, Ohio State University.

Pawlowicz, Richard M. Stratigraphy of the Marshall Formation (Mississippian, Osagian) in Michigan. M, 1969, University of Toledo. 43 p.

Pawlowicz, Richard Melvin. Discrimination among depositional environments based on element abundance in upper Cretaceous rocks of southern Alberta. D, 1974, New Mexico Institute of Mining and Technology. 360 p.

Pawlowski, Michael Raymond. Geology and exploration geochemistry of the Magruder Corridor, Idaho County, Idaho. M, 1982, University of Idaho. 129 p.

Pawlowski, Robert S., Jr. Gravity anomaly separation by wiener filtering. M, 1987, Colorado School of Mines. 170 p.

Paxson, Kevin B. Petrology of the Reynales Formation (Silurian; Llandoverian) of western New York and the Niagaran Peninsula of Ontario. M, 1985, Miami University (Ohio). 175 p.

Paxton, Stanley Turner. Petrography and diagenetic evolution of cements in some Pleistocene glaciofluvial deposits from southwestern Ohio, southeastern Indiana, and northern Kentucky. M, 1980, Miami University (Ohio). 185 p.

Paxton, Stanley Turner. Relationships between Pennsylvanian-age lithic sandstone and mudrock diagenesis and coal rank in the Central Appalachians. D, 1983, Pennsylvania State University, University Park. 526 p.

Payawal, Pacifico Cruz. Vegetation and modern pollen rain in a tropical rain forest, Mt. Makiling, Philippines. D, 1981, University of Arizona. 103 p.

Paydar, Zahra. Hysteretical model of one-dimensional transient unsaturated flow in a soil-water-plant system. D, 1981, Utah State University. 139 p.

Payette, Christine. The melt inclusions in quartz phenocrysts of the quartz-feldspar porphyry, Harvey Station, New Brunswick. M, 1985, McGill University. 99 p.

Payette, Francine. Géochimie des claystones du flysch des régions de Bic, Saint Fabien et Trois Pistoles, P.Q. M, 1977, Universite de Montreal.

Payken, C. L. Residual error analysis of a direct method solution in a petroleum reservoir mathematical simulation model. M, 1976, University of Missouri, Rolla.

Paylor, Earnest D., II. Investigation of surface features interpreted from remote sensing data of the Table Rock gas field, Sweetwater County, Wyoming. M, 1983, University of Wyoming. 81 p.

Payne, Anthony. Hydrothermal dolomitization of Cambrian sediments, Tintic District, Utah. M, 1950, University of Utah. 48 p.

Payne, Anthony L. Geology and uranium deposits of the Colorado Plateau. D, 1959, Stanford University. 245 p.

Payne, Billie Rex. A check list of the invertebrate macrofauna of the Lower Cretaceous of Texas. M, 1941, University of Texas, Austin.

Payne, Charles Marshall. Engineering aspects of the St. Peter Sandstone in the Minneapolis-St. Paul area of Minnesota. M, 1967, University of Arizona.

Payne, Craig William Charles. Petrography, geochemistry and structure of the Timmins area, Ontario. M, 1979, Brock University. 155 p.

Payne, Curtis M. Geology of a part of the Nemo District in the Black Hills of South Dakota. M, 1979, South Dakota School of Mines & Technology.

Payne, Franklin Russell, Jr. A gamma ray log study of the Mississippian sands in the Elders Ridge Quadrangle, Armstrong and Indiana counties, Pennsylvania. M, 1958, University of Pittsburgh.

Payne, Gary. Diagenesis of silica, clays and zeolites in the Tallahatta Formation, east-central Mississippi. M, 1985, Tulane University.

Payne, James Norman. Subsurface geology of the Marseilles, Ottawa, and Streator quadrangles, and vicinity, Illinois. D, 1938, University of Chicago. 71 p.

Payne, James Norman. The geology of the southeastern quarter of the Fayetteville Quadrangle of Northwest Arkansas. M, 1933, University of Arkansas, Fayetteville.

Payne, Janie Hopkins. Sedimentation and pedogenesis of the Lower Cretaceous Hensel Formation, central Texas. M, 1982, University of Texas, Austin. 136 p.

Payne, John Beckwith. Anastomosing and meandering fluvial systems, Mesaverde Group, (Campanian), northwestern Colorado. M, 1982, University of Texas, Austin. 144 p.

Payne, John Garfield. Geology and geochemistry of the Blue Mountain nepheline syenite body. D, 1966, McMaster University. 183 p.

Payne, John Garnett. The Twillingate Granite and its relationships to surrounding country rocks. M, 1974, Memorial University of Newfoundland. 159 p.

Payne, Kathryn Lynn. Late Neogene planktonic foraminiferal biostratigraphy of the North Philippine Sea (DSDP LEG 58). M, 1979, Washington University. 39 p.

Payne, Kenneth Armstrong. Pennsylvanian Ostracoda from Sullivan County, Indiana. M, 1936, Indiana University, Bloomington. 92 p.

Payne, Lawrence H. Sediments and morphology of continental shelf off southeast Virginia. M, 1970, Columbia University. 71 p.

Payne, Michael A. The origin of the magnetization of various lithologies from the lower Conemaugh Group in southwestern Pennsylvania. D, 1979, University of Pittsburgh. 276 p.

Payne, Myron William. Basinal sandstone facies in the Delaware Mountain Group, West Texas and Southeast New Mexico. D, 1973, Texas A&M University.

Payne, Myron William. Geomorphology and sediments of the Santee Delta. M, 1970, University of South Carolina.

Payne, R. R. Radiolaria in Brunhes sediment of the Southeast Indian Ocean south of Australia. D, 1977, University of South Carolina. 309 p.

Payne, Richard Allan. A gravity investigation of the Boulder Baldy Quadrangle, Big Belt Mountains, central, Montana. M, 1986, Montana College of Mineral Science & Technology. 102 p.

Payne, Robert A. Groundwater resource potential of diabase dikes in the Durham Triassic Basin. M, 1984, North Carolina State University. 87 p.

Payne, Robert Ridley. Turbidite sedimentation off the Antarctic continent. M, 1972, University of South Carolina.

Payne, Thomas G. Stratigraphic analysis and environmental reconstruction. D, 1942, University of Chicago. 102 p.

Payne, Thomas Gibson. Coral zones of the (Devonian) Hamilton Group of western New York. M, 1939, University of Rochester. 134 p.

Payne, William Downes. The role of sulphides and other heavy minerals in copper anomalous stream sediments. D, 1971, Stanford University. 119 p.

Payne, William Ross. The geology of the Edwards Limestone of Bosque County, Texas. M, 1960, Baylor University. 77 p.

Payne, William W., III. Paleoenvironmental study and geologic map of Lower and Middle Ordovician strata near Knoxville, Tennessee. M, 1975, University of Tennessee, Knoxville. 69 p.

Payson, Harold, Jr. Investigation of bottom sediment probing by 12 kilocycle sound pulses reflected from shallow water bottom sediment layers. M, 1963, Massachusetts Institute of Technology. 65 p.

Payton, Charles Ellis. Petrology and paleogeography of the Middle Creek, Bethany Falls and Winterset (Pennsylvanian) limestones. D, 1964, University of Missouri, Columbia. 266 p.

Payton, Charles Ellis. The petrology of some Pennsylvanian black shales. M, 1958, Iowa State University of Science and Technology.

Payton, Clifford Charles. The geology of the middle third of the Sutherlin Quadrangle, Oregon. M, 1961, University of Oregon. 81 p.

Payton, J. Wayne. Paleoenvironmental determination of the lower Tuscaloosa at South Carlton Field, Baldwin and Clarke counties, Alabama. M, 1984, University of Alabama. 180 p.

Payton, Patrick Herbert. A search for the solar wind in the lunar soil. D, 1974, University of California, Los Angeles.

Paz, Jose Gabriel. Altitude of magnetic measurement; an important factor in aeromagnetic survey. M, 1976, University of Oklahoma. 150 p.

Pazdersky, G. J. Factor analysis of the sedimentary textures and fauna of the Becraft Formation (lower Devonian) of New York State. M, 1976, Temple University.

Pe Ni Ta *see* Ni Ta Pe

Peabody, Carey Evans. Geology and petrology of a tungsten skarn; El Jaralito, Baviacora, Sonora, Mexico. M, 1979, Stanford University. 90 p.

Peabody, David M. Morphologic variation of Osmundaceae spores. M, 1964, University of Arizona.

Peabody, Frank Elmer. Reptile and amphibian trackways from the Lower Triassic Moenkopi Formation of Arizona. D, 1946, University of California, Berkeley.

Peabody, Frank Elmer. Trackways of Pliocene and Recent salamandroids of the Pacific Coast of North America. M, 1940, University of California, Berkeley. 109 p.

Peabody, William Wirt. Geology of the Waxahachie Quadrangle, Ellis County, Texas. M, 1958, Southern Methodist University. 17 p.

Peace, H. W., II. The Springer Group (Mississippian and Pennsylvanian) of the southeastern Anadarko Basin in Oklahoma. M, 1964, University of Oklahoma. 37 p.

Peace, Jerry. Geology and mineral occurrences of the Porcupine Lake area, northeastern Brooks Range, Alaska. M, 1979, University of Alaska, Fairbanks. 72 p.

Peach, Peter A. Some pegmatites from eastern Ontario and their geological environment. D, 1951, University of Toronto.

Peach, Peter A. The genetical association of accessory minerals; apatite. M, 1947, University of Toronto.

Peacock, C. Herschel. Geology of the Government Hill area, a part of Long Ridge, Utah. M, 1953, Brigham Young University. 96 p.

Peacock, David N. A mathematical study of Voigt viscoelastic Love wave propagation. M, 1966, University of Missouri, Rolla.

Peacock, David Nuse. Discrimination of signal and noise events on seismic recordings by linear threshold estimation theory. D, 1970, University of Missouri, Rolla.

Peacock, Edward W. Paleomagnetic and geomagnetic studies in the Lake Mary area, Coconino County, Arizona. M, 1978, Northern Arizona University. 83 p.

Peacock, Kenneth L. The design of discrete nonadaptive operators for seismic data processing. D, 1977, University of Tulsa. 121 p.

Peacock, Simon M. The systematics of sulfide mineralogy in the regionally metamorphosed Ammonoosuc Volcanics. M, 1981, University of Waterloo.

Peacock, Simon Muir. Thermal and fluid evolution of the Trinity thrust system, Klamath Province, Northern California; implications for the effect of fluids in subduction zones. D, 1985, University of California, Los Angeles. 354 p.

Peacock, Steve. The post Eocene stratigraphy of southern Collier County, Fl. M, 1981, Florida State University.

Peacor, Donald R. Crystal structure of marsarsukite. M, 1960, Massachusetts Institute of Technology. 64 p.

Peacor, Donald R. The structures and crystal chemistry of bustamite and rhodonite. D, 1962, Massachusetts Institute of Technology. 254 p.

Peake, J. S. Seasonal change in the microclimates of the Colville Delta, Alaska. D, 1977, Louisiana State University. 152 p.

Peale, Robert Newton. Geology of the area southeast of Yellowjacket, Lemhi County, Idaho. M, 1982, University of Idaho. 129 p.

Peale, Rodgers. Some ore deposits of the Rouyn District, northwestern Quebec, Canada. D, 1930, Harvard University.

Pealey, Annette. Garnet genesis in the Sugarloaf Gabbro. M, 1973, Boston University. 62 p.

Pear, James Lewis. Factors controlling stratigraphic variation within the Mississippian "Big Lime" of eastern Kentucky; a petroliferous subsurface equivalent of the lower Newman Group (Meramecian, lower Chesterian). M, 1980, University of Kentucky. 98 p.

Pearce, Andrew John. Mass and energy flux in physical denudation, defoliated areas, Sudbury. D, 1973, McGill University. 126 p.

Pearce, Edward Wayne. An evaluation of the use of amorphous silica as a ground water tracer in carbonate aquifer systems. M, 1984, University of South Florida, Tampa. 61 p.

Pearce, G. W. Magnetism and lunar surface samples. D, 1973, University of Toronto.

Pearce, George William. Design and calibration of apparatus for the alternating field demagnetization of rocks. M, 1967, Memorial University of Newfoundland. 134 p.

Pearce, R. W. Paleoecology of some Upper Pennsylvanian benthic invertebrates. M, 1973, Kansas State University. 90 p.

Pearce, Suzanne M. Glacial and surficial geology of the upper Big Wood River drainage system, Blaine County, Idaho. M, 1988, Lehigh University. 126 p.

Pearce, Thomas H. Petrology and chemistry of the Crows Nest Volcanics (Upper Cretaceous) (Crowsnest Pass, Alberta). D, 1967, Queen's University. 181 p.

Pearce, Thomas Hulme. Quartz-feldspar porphyry, Bathurst, New Brunswick. M, 1963, University of Western Ontario. 61 p.

Peare, Robert Kunkel. A study of iron oxide pseudomorphs after pyrite porphyroblasts in the Lynchburg Gneisses (Precambrian) of Virginia. M, 1959, University of Virginia. 76 p.

Peargin, Thomas R. Ordovician to Silurian stratigraphy of part of the North-central Monitor Range, Nye County, Nevada. M, 1979, Oregon State University. 107 p.

Pearl, James E. Petrology of Tertiary sedimentary rocks in the northwesternmost part of the Olympic Peninsula, Washington. M, 1977, San Jose State University. 91 p.

Pearl, Richard Howard. Quaternary physiographic development and surficial deposits of the Morrison Quadrangle, Colorado. M, 1963, University of Missouri, Columbia.

Pearl, Richard M. The turquoise of Colorado. M, 1940, University of Colorado.

Pearn, William Charles. Thermoluminescence ages of the igneous rocks of Marble Point, Antarctica. D, 1963, University of Kansas. 156 p.

Pearn, William Charles. Thermoluminescence applied to the dating of certain tectonic events. M, 1959, University of Kansas. 74 p.

Pearring, Jerome Richard. A study of basic mineralogical, physical-chemical, and engineering index

properties of laterite soils. D, 1968, Texas A&M University. 133 p.

Pearsall, William G. The geological structure of the Sunday Lake area, Gogebic County, Michigan. M, 1912, University of Wisconsin-Madison.

Pearse, Gary H. Origin of the band of quartzo-feldspathic rocks (Archean) along the north shore of Falcon Lake, Manitoba. M, 1969, University of Manitoba.

Pearson, Carl A. A geophysical investigation of the East Greenwich, RI, group of rocks. M, 1985, University of Rhode Island.

Pearson, Christopher F. Seismic refraction study of Augustine Volcano. M, 1977, University of Alaska, Fairbanks. 131 p.

Pearson, Corrinne Dorset. Selective silicification of skeletal carbonates in some Mississippian and Devonian limestones. M, 1981, University of Illinois, Urbana. 81 p.

Pearson, D. R. Surface and shallow subsurface sediment regime of the nearshore inner continental shelf, Nags Head and Wilmington areas, NC. M, 1980, East Carolina University.

Pearson, Daniel Bester, III. Carbon and stable carbon isotopes in mantle derived material. M, 1973, Rice University. 49 p.

Pearson, Daniel Bester, III. Experimental simulation of thermal maturation in sedimentary organic matter. D, 1981, Rice University. 569 p.

Pearson, Daniel L. Palynology of the Middle and Upper Seminole coals (Pennsylvanian) of Tulsa County, Oklahoma. M, 1975, University of Oklahoma. 75 p.

Pearson, Daniel R. Surface and shallow subsurface sediment regime of the nearshore inner continental shelf, Nags Head and Wilmington areas, North Carolina. M, 1979, East Carolina University. 120 p.

Pearson, David Victor. A geological engineering report of a gravity slide in the Fall River Formation (Lower Cretaceous), south of Rapid City, South Dakota. M, 1967, South Dakota School of Mines & Technology.

Pearson, Elizabeth M. Sedimentology and diagenesis of the Warwick Reef (Silurian), Lambton County, southwestern Ontario. M, 1980, University of Waterloo.

Pearson, Eugene Favre. Origin and diagenesis of the Owl Canyon and lower Goose Egg formations (Permian) of southeastern Wyoming and adjacent Colorado. D, 1971, University of Wyoming. 182 p.

Pearson, Frederick Joseph, Jr. Ground-water ages and flow rates by the carbon-14 method. D, 1966, University of Texas, Austin. 105 p.

Pearson, Frederick Joseph, Jr. Origin and interlayer bond strength of Texas mines vermiculite. M, 1962, University of Texas, Austin.

Pearson, G. Raymond. A geochemical study of sillimanite and andalusit and kyanite. M, 1955, McMaster University. 53 p.

Pearson, George Raymond. Granitic gneisses around the Clare River Syncline, Ontario. D, 1958, Queen's University. 291 p.

Pearson, Jerome. Laboratory techniques for measurement of seismic velocities in rock cores compared to field measurements. M, 1977, Wright State University. 92 p.

Pearson, John G. The relationship of granitic and metasedimentary rocks near lower Foster Lake, Saskatchewan. M, 1979, University of Saskatchewan. 85 p.

Pearson, John W. Genesis of the vein deposits along the Great Master Lode, Phillipsburg area, northern Black Range, Sierra and Catron counties, New Mexico. M, 1988, University of Texas at El Paso.

Pearson, Mark F. Geology, mineralogy, and sulfur isotope studies of the Real de Angeles silver deposit, Zacatecas, Mexico. M, 1985, University of Texas at El Paso.

Pearson, Monte L. Geomorphological analysis of North Fork Toutle River, Washington; 1980-1984. D, 1985, Oregon State University. 177 p.

Pearson, Monte Laurence. Carbonate petrology of the "upper Hasmark", Cambrian, of southwestern Montana. M, 1974, University of Montana. 87 p.

Pearson, R. An analysis of shuttle imaging radar-B data, Landsat multispectral scanner data, and coregistered SIR-B/MSS data for the detection of glacial features in western Lake County, Illinois. M, 1986, Murray State University. 62 p.

Pearson, Robert L. Design of field spectrophotometer lab. M, 1971, Colorado State University. 111 p.

Pearson, Robert Stanley. An examination of some metal-nonmetal relations in geochemical prospecting. D, 1964, Kansas State University. 79 p.

Pearson, Steven Gerald. Depositional environments, diagenesis and barite mineralization of the Middle Devonian Wapsipinicon Formation in Fayette County, Iowa. M, 1982, University of Iowa. 203 p.

Pearson, Vital. Pétrographie, géochimie et interprétation d'un assemblage à cordiérite-anthophyllite dans les roches mafiques archéennes de Macanda, Canton Beauchastel, Noranda, Québec. M, 1988, Universite du Quebec a Chicoutimi. 168 p.

Pearson, W. C. A gravity study of the Juan de Fuca Ridge and Sovanco fracture zone in the northeast Pacific Ocean. D, 1975, University of Washington. 135 p.

Pearson, Walter John. Origin of the kyanite occurrences in the Wanipitei and Crocan Lake areas of Ontario. D, 1959, Queen's University. 336 p.

Pearson, Walter John. The origin and history of the Neagle Lake Pluton in the Amisk Lake area. M, 1951, University of Saskatchewan. 74 p.

Pearson, William Norman. Copper metallogeny, north shore region of Lake Huron, Ontario. D, 1980, Queen's University. 397 p.

Pearson, William Norman. The Minto copper deposit, Yukon Territory; a metamorphosed orebody in the Yukon crystalline terrane. M, 1977, Queen's University. 195 p.

Peart, J. E. Deposits of sulfur hot springs along the northeast coast of Baja California. M, 1978, San Diego State University.

Pease, Maurice Henry, Jr. Geology of the Sobrante Anticline and vicinity (Contra Costa County, California). M, 1954, University of California, Berkeley. 95 p.

Pease, Robert Charles. Scarp degradation and fault history south of Carson City, Nevada. M, 1979, University of Nevada. 90 p.

Pease, Rodney Wayne. Electron-optical investigations on the morphological variations in dioctahedral aluminum smectites. M, 1975, Texas Tech University. 93 p.

Peatfield, Giles Russum. Geologic history and metallogeny of the "Boundary District", southern British Columbia and northern Washington. D, 1978, Queen's University. 250 p.

Peattie, Roderick. Geographic conditions of the lower St. Lawrence Valley (Quebec). D, 1920, Harvard University.

Peavy, Samuel Thomas. A gravity and magnetic interpretation of the Bay St. George Carboniferous subbasin in western Newfoundland. M, 1985, Memorial University of Newfoundland. 207 p.

Pecci, Anthony Salvatore. Calculations on the state of the crust, lithosphere and upper mantle of Venus. M, 1984, University of Connecticut. 68 p.

Pechmann, James Christopher. The relationship of small earthquakes to strain accumulation along major faults in Southern California. D, 1983, California Institute of Technology. 184 p.

Peck, Albert Becker. Changes in the constitution and microstructure of andalusite, cyanite, and sillimanite at high temperatures in industrial practice and their significance. D, 1925, University of Michigan.

Peck, Benejhar J. Hydrothermal alteration of ferromagnesian minerals. M, 1931, University of Minnesota, Minneapolis. 25 p.

Peck, Charles W. A geochemical and fluid inclusion study of the mineral deposits of the Platoro fault zone, Platoro Caldera, San Juan Mountains, Colorado. M, 1982, Colorado School of Mines. 143 p.

Peck, Craig Jonathan. Paleomagnetism and diagenesis of the Upper Cambrian Peerless Formation, central Colorado. M, 1985, University of Oklahoma. 58 p.

Peck, Curtis Allen. Reconnaissance study of the sediments of Krause Lake, Sand Hills, Nebraska. M, 1980, Iowa State University of Science and Technology.

Peck, Dallas Lynn. Geological reconnaissance of the western Cascades of Oregon north of latitude 43 degrees. D, 1960, Harvard University. 232 p.

Peck, Dallas Lynn. Geology of Paradox No. 3 Mine area (Randsburg, California). M, 1953, California Institute of Technology. 35 p.

Peck, David C. The geology and geochemistry of the Cartwright Lake area; Lynn Lake greenstone belt, northwestern Manitoba. M, 1986, University of Windsor. 270 p.

Peck, Donald. The adsorption-desorption of diuron by freshwater sediments. M, 1977, University of California, Riverside. 48 p.

Peck, Douglas M. Biostratigraphy and paleoecology of the Tamiami Formation in Lee County, Florida. M, 1976, Florida State University.

Peck, Gregory Erman. The ecology and recolonization of benthic foraminifera from the continental shelf of New Jersey. M, 1979, University of Virginia. 96 p.

Peck, John H. The magnetic intensity of pyrrhotite. M, 1962, Dartmouth College. 53 p.

Peck, Joseph Howard, Jr. Geology of the Merrit Pass area, Fremont and Hot Springs counties, Wyoming. M, 1941, University of Missouri, Columbia.

Peck, Lindamae. Stress corrosion and crack propagation in Sioux Quartzite. D, 1982, Yale University. 224 p.

Peck, Nathan Russell. Geology, petrology, geochemistry, and genesis of the late Proterozoic iron-formations, central Eastern Desert, Egypt. M, 1986, Washington University. 168 p.

Peck, Raymond E. The Blastoidea of the Chouteau Limestone. M, 1928, University of Missouri, Columbia.

Peck, Raymond E. The North American trochilisids Paleozoic Charophyta. D, 1932, University of Missouri, Columbia.

Peck, Timothy Joseph. Threshold conditions for channel alteration in small urbanized watersheds, southeastern Pennsylvania. M, 1986, University of Delaware. 115 p.

Peckenpaugh, J. M. Alluvial ground water quality alteration as related to solid waste disposal sites in Iowa. M, 1973, Iowa State University of Science and Technology.

Peckham, Alan Embree. Petrographic analysis and stratigraphic relationships of the heavy minerals of the Harrison Formation of northwestern Nebraska. M, 1955, University of Nebraska, Lincoln.

Peckham, David Arthur. Seismic and tectonic study of a region in the Cascades west of Mount Rainier. M, 1982, University of Washington. 82 p.

Peckins, Eric L. The computer generation of synthetic seismograms from an Oakland County, Michigan, continuous velocity log. M, 1981, Bowling Green State University. 117 p.

Peckman, D. A. Seismic and tectonic study of a region in the Cascades west of Mt. Rainier. M, 1982, University of Washington. 83 p.

Pecora, William C. Bedrock geology of the Blacktail Mountains, southwestern Montana. M, 1981, Wesleyan University. 203 p.

Pecora, William Thomas, II. Petrology and mineralogy of the western Bearpaw Mountains, Montana. D, 1940, Harvard University.

Peczkis, Jan. Anastrophic deposition in the Silurian dolomites of the Chicago area. M, 1981, Northeastern Illinois University. 117 p.

Peddada, Anantaramam. Petrology of the Nemeiben Lake ultramafic and associated nickel-sulphide deposits (Saskatchewan, Canada). M, 1972, SUNY at Albany. 166 p.

Pedersen, D. E. Alluvial morphology of the Little Sioux River valley in western Iowa. M, 1963, Iowa State University of Science and Technology.

Pedersen, Jens R. A magnetotelluric investigation of the deep electrical structure of the Rio Grande Rift and adjacent tectonic provinces. D, 1980, Brown University. 325 p.

Pedersen, Kenneth. Sulfide mineralization of the northwestern part of the Sunlight mining region, Park County, Wyoming. M, 1968, University of Wyoming. 50 p.

Pedersen, Steven. Geology and geothermal investigations of the Crazy Hills area, northern Skamania County, Washington. M, 1978, Portland State University.

Pederson, Bernhardt L. Geochemical studies in the Bighorn Basin, Wyoming. M, 1985, University of Utah. 62 p.

Pederson, Dale Russell. Aeromagnetic study of the Ross Island and Taylor Glacier quadrangles, Antarctica. M, 1975, Northern Illinois University. 103 p.

Pederson, Darryll T. Erosion and sedimentation in Lake Ashtabula, southeastern North Dakota. D, 1971, University of North Dakota. 154 p.

Pederson, David E. Petrology of some Allegheny Pennsylvanian rocks near Ashland (Boyd county) Kentucky. D, 1970, Louisiana State University. 126 p.

Pederson, Edward Peter. Sedimentology and stratigraphy of basin-fill sediments of the Payson Basin, Gila County, Arizona. M, 1969, Arizona State University. 136 p.

Pederson, Robert J. Geology of the upper Rock Creek drainage, Granite County, Montana. M, 1976, Montana College of Mineral Science & Technology. 238 p.

Pederson, Roger Lynn. Abundance, distribution, and diversity of buried seed populations in the Delta Marsh, Manitoba, Canada. D, 1983, Iowa State University of Science and Technology. 104 p.

Pederson, Selmer Lane. Stratigraphy of the Fountain and Casper formations of southeastern Wyoming and north central Colorado. M, 1953, University of Wyoming. 87 p.

Pedlow, George W., III. A peat island hypothesis for the formation of thick coal. D, 1977, University of South Carolina. 181 p.

Pedone, Vicki A. Petrography, chemistry, and crystallography of Baroque Dolomite, Kingsport, Tennessee. M, 1978, University of Texas, Austin.

Pedora, John Michael. Mineralization of the High Lake Pluton and adjacent country rocks. M, 1976, University of Manitoba.

Pedrick, Jane Nuli. A microprobe study of mafic inclusions from the Delegate Pipes, New South Wales, Australia. M, 1986, Iowa State University of Science and Technology. 165 p.

Pedrosa, Oswaldo Antunes, Jr. Use of hybrid grid in reservoir simulation. D, 1985, Stanford University. 274 p.

Pedrotti, Daniel A. The geology of the Roan's Prairie area, Grimes County, Texas. M, 1958, Texas A&M University. 105 p.

Pedry, John Joseph. The geology of Chagrin Falls Township, Cuyahoga County, and Bainbridge Township, Geauga County, Ohio. M, 1951, Ohio State University.

Peebles, John Kevan. The geological formations in the vicinity of Petersburg, Virginia, with special reference to the (Pleistocene and Pliocene) Appomattox Formation. D, 1890, University of Virginia. 19 p.

Peebles, Mark Whitney. Taphonomy of common shallow-water benthic foraminifera from San Salvador, the Bahamas. M, 1988, Auburn University. 134 p.

Peebles, Pamela C. Late Cenozoic landforms, stratigraphy and history of sea level oscillations of southeastern Virginia and northeastern North Carolina. D, 1984, College of William and Mary. 227 p.

Peebles, R. W. Flow recession in the ephemeral stream. D, 1975, University of Arizona. 97 p.

Peebles, Roger Waite. Gravity studies of buried bedrock topography in Northwest Oakland County, Michigan. M, 1969, Michigan State University. 61 p.

Peek, Bradley C. Geology and mineral deposits of the Niblack Anchorage area, Prince of Wales Island, Alaska. M, 1975, University of Alaska, Fairbanks. 50 p.

Peel, Frederick A. New interpretations of Pennsylvanian and Permian stratigraphy and structural history, northern Sangre de Cristo Range, Colorado. M, 1971, Colorado School of Mines. 75 p.

Peeler, James A. Reservoir characterization of the Mississippian "Chat," Hardtner Field, southern Barber County, Kansas. M, 1985, Wichita State University. 132 p.

Peeling, Gordon Roderick. Petrography and geochemistry of sedimentary rocks of the Yellowknife Supergroup (Archean), Slave Province, Northwest Territories. M, 1974, Carleton University. 89 p.

Peeples, Joseph D. Structural studies of non-producing sediments in selected Mississippi oil fields. M, 1951, University of Mississippi.

Peeples, Vernon. Triassic depositional environments of Jebel Rehoch in southern Tunisia. M, 1975, University of South Carolina.

Peeples, Wayne. Magneto-telluric profiling over a deep structure. D, 1969, University of Alberta. 161 p.

Peery, Trusten Edwin. The conodonts of the (Upper Devonian) Snyder Creek Shale (Missouri). M, 1932, University of Missouri, Columbia.

Peery, Trusten Edwin. The stratigraphy of the western half of the Fulton Quadrangle. D, 1940, University of Missouri, Columbia.

Peery, William M. Migration and degradation of dissolved gasoline in a highly transmissive, unconfined, gravel and cobble aquifer; a study of the Champion Missoula sawmill spill, Missoula, Montana. M, 1988, University of Montana. 222 p.

Peerzada, Michael. Tectonophysics of the Himalayas. M, 1963, Boston College.

Pees, Samuel T. The Pucara (Triassic, Jurassic) and Chapiza (Jurassic) formations, Peru. M, 1959, Syracuse University.

Pefley, David R. Geology of the Stanley and Rushville quadrangles, New York. M, 1956, University of Rochester. 131 p.

Pegau, Arthur August. The pegmatite dikes of America, Goochland, and Ridgeway areas, Virginia. D, 1924, Cornell University.

Pegram, W. J. Strontium isotope stratigraphy of a core from the Tipton Shale Member of the Green River Formation, Wyoming. M, 1977, Ohio State University. 102 p.

Pegram, William Joseph. The isotope, trace element, and major element of the Mesozoic Appalachian tholeiite province. D, 1986, Massachusetts Institute of Technology. 625 p.

Pegrum, Reginald Herbert. The alkaline syenites and associated rocks of the French River region, Ontario. D, 1927, Princeton University. 127 p.

Pehme, Peeter Enn. Identification of Quaternary deposits with borehole geophysics in the Waterloo region. M, 1984, University of Waterloo. 124 p.

Peikert, Ernest William. Petrological study of a group of porphyroblastic rocks in the Precambrian of northeastern Alberta. D, 1961, University of Illinois, Urbana. 181 p.

Peikert, Ernest William. The Mount Tallac roof pendant, Sierra Nevada, southwest of Lake Tahoe. M, 1958, University of California, Berkeley. 63 p.

Peindo, Jorge Fernando. Moment-magnitude relations in the intermountain seismic belt from regional seismic data; applications of an indirect network calibration scheme. M, 1986, University of Utah. 95 p.

Peiper, John Christopher. Thermodynamics of geochemically important aqueous electrolytes. D, 1983, University of California, Berkeley. 215 p.

Peirce, Frederick Lowell. Structure and petrography of part of the Santa Catalina Mountains. D, 1958, University of Arizona. 192 p.

Peirce, Howard Wesley. Geologic studies of the Phosphoria Formation in restricted areas, Melrose phosphate field, Montana. M, 1952, Indiana University, Bloomington. 36 p.

Peirce, Howard Wesley. Stratigraphy of the De Chelly Sandstone of Arizona and Utah. D, 1958, University of Arizona.

Peirce, John Wentworth. The origin of the Ninetyeast Ridge and the northward motion of India, based on DSDP paleolatitudes. D, 1977, Massachusetts Institute of Technology. 103 p.

Peirce, Robert W. Ultrastructure and biostratigraphy of the conodonts from the Monte Cristo group (Lower and Upper Mississippian), Arrow Canyon range, Clark County, Nevada. D, 1969, University of Illinois, Chicago.

Pekarek, Alfred H. The structural geology and igneous petrology of the Rattlesnake Hills, Wyoming. D, 1974, University of Wyoming. 150 p.

Pekas, Bradley S. Brine contamination of the Hell Creek Aquifer from an unlined drilling-fluid pit in northern Harding County, South Dakota. M, 1987, South Dakota School of Mines & Technology.

Pekkan, Ahmet. The genesis of chromite. M, 1945, Stanford University.

Pelizza, Mark S. Environmental and surficial geology in East-central Clear Creek County, Colorado. M, 1978, Colorado School of Mines. 102 p.

Pelka, Gary Jerome. Geology of the McCoy and Palen mountains, southeastern California. D, 1973, University of California, Santa Barbara.

Pelke, Paul A. Petrology and geochemistry of granitic rocks, Cape Ann, Massachusetts. M, 1972, Massachusetts Institute of Technology. 72 p.

Pell, Jennifer. Stratigraphy, structure and metamorphism of Hadrynian strata in the southeastern Cariboo Mountains, B.C. D, 1984, University of Calgary. 185 p.

Pellegrin, Freddie John. Late Pleistocene-Holocene depositional systems, southern Hancock County, Mississippi. M, 1978, University of Mississippi. 98 p.

Pellerin, Louise Donna. Use of transient electromagnetic soundings to correct static shifts in magnetotelluric data. M, 1988, University of Utah. 100 p.

Pelletier, Bernard R. Pocono paleocurrents. D, 1957, The Johns Hopkins University.

Pelletier, Bernard R. The (Silurian) Grimsby Sandstone of the Niagara Peninsula (Ontario). M, 1953, McMaster University.

Pelletier, Karen. Geology of the Ellington volcanic-sedimentary complex, Great Bear magmatic zone, Northwest Territories. M, 1988, Carleton University.

Pelletier, Lise. Pétrologie des roches volcaniques siluro-dévoniennes de Dalhousie, Nouveau-Brunswick. M, 1986, Universite Laval. 98 p.

Pelletier, Marc. Evolution sédimentologique de l'estuaire fluvial du Saint-Laurent. M, 1982, Universite du Quebec a Rimouski. 259 p.

Pelletier, Michael. A laboratory study of the response of dark gray shales to varying acid leachants with implications for overburden management in coal strip mines in southern Pennsylvania. M, 1979, University of South Carolina.

Pelletier, Willis Joseph. Paleontology and stratigraphy of the Clarendonian continental beds west of Tracy, California. M, 1951, University of California, Berkeley. 91 p.

Pelline, Joseph Emmett. The geology of adjacent parts of the Las Flores and Topanga quadrangles, Santa Monica Mountains, California. M, 1951, University of California, Los Angeles.

Pelowski, Sandra M. Behavior of selected trace metals in sediments from the continental shelf of the Amazon River. M, 1983, Michigan State University. 134 p.

Pels, Robert John. Sediments of Albemarle Sound, North Carolina. M, 1967, University of North Carolina, Chapel Hill. 73 p.

Peltier, Edward J. Pennsylvanian fusulinids from the Sandia Mountains, New Mexico. M, 1958, University of Kansas. 152 p.

Peltier, Louis Cook. Pleistocene terraces of the Susquehanna River, Pennsylvania. D, 1948, Harvard University.

Peltier, Louis Cook. The glacial geology of lower Clove Creek Valley, New York. M, 1939, Columbia University, Teachers College.

Pelto, Chester Robert. Petrology of the Gatesburg Formation of central Pennsylvania. M, 1942, Pennsylvania State University, University Park. 60 p.

Pelto, Mauri S. The mass balance and climatic sensitivity of North Cascade, Washington and Coast Range, Southeast Alaska glaciers. M, 1988, University of Maine. 115 p.

Pelton, Harold A. Geology of the Loomis-Blue Lake area, Okanogan County, Washington. M, 1957, University of Washington. 92 p.

Pelton, John R. The analysis of deformation-induced variations in orthometric height and gravity with an application to recent crustal movements in Yellowstone National Park. D, 1979, University of Utah. 212 p.

Pelton, Peter John. Mississippian rocks of the southwestern Great Basin, Nevada and California. D, 1966, Rice University. 146 p.

Pelton, William Harvey. Interpretation of induced polarization and resistivity data. D, 1977, University of Utah. 272 p.

Peltonen, Dean R. A geochemical and biogeochemical survey of a part of Iron County, Wisconsin. M, 1978, University of Wisconsin-Milwaukee.

Peltzer, Edward T., III. Geochemistry of hydroxy and dicarboxylic acids. D, 1979, University of California, San Diego. 227 p.

Pelzer, Ernest Edward. Mineralogy, geochemistry and stratigraphy of the Besa River Shale (NE British Columbia). D, 1965, University of Alberta. 238 p.

Pemberton, Roger H. Gravity study of Des Plaines disturbance, Cook County, Illinois. M, 1954, University of Wisconsin-Madison.

Pemberton, Stuart George. Deep bioturbation by Axius serratus in the Strait of Canso, Nova Scotia. M, 1976, McMaster University. 225 p.

Pemberton, Stuart George. Selected studies in lower Paleozoic ichnology. D, 1979, McMaster University. 516 p.

Pemsler, Paul. Diffusion of heavy water into hydrated crystalline zeolites; the mobility of water in zeolites. D, 1954, New York University. 138 p.

Pena Horcasitas, Gerardo Ruiz de la *see* Ruiz de la Pena Horcasitas, Gerardo

Pena, Edward C. De La *see* De La Pena, Edward C.

Penas, Carlos. AI determination of seismic field parameters. M, 1985, Colorado School of Mines. 86 p.

Pencak, Michael Stanley. The petrology of the Snowflake troctolite zone in the Hettasch Intrusion, Labrador. M, 1982, Northern Illinois University. 76 p.

Pence, Jennifer Joan. Sedimentation and tectonics of the upper Oligocene to middle Miocene Temblor Formation of the northern Temblor Range, Kern and San Luis Obispo counties, California. M, 1985, Stanford University. 92 p.

Pendala, Krishnamurthy. Experimental deformation of ores. M, 1967, McGill University.

Pendell, Ray. A petrographic study of the Brassfield Limestone (Silurian) in western Ohio. M, 1952, Ohio State University.

Pender, Jeffrey Thomas. An inventory and evaluation of the geologic and hydrologic environment of Yellow Springs, Ohio, for land use planning purposes. M, 1976, Wright State University. 66 p.

Pendergast, Margaret A. Devonian and Mississippian stratigraphy of the Swales Mountain area, Elko County, Nevada. M, 1981, Oregon State University. 113 p.

Pendergrass, James M. Recent sediments and sedimentation of Laguna Salada, Baffin Bay, Texas. M, 1975, Texas Christian University.

Pendergrass, T. Michael. Analysis of the Cactus Ridge Syncline and related thrust faults, Gila and Yavapai counties, Arizona. M, 1984, Northern Arizona University. 148 p.

Pendery, Eugene Christian, III. Stratigraphy and carbonate petrography of the Baline Formation, north central Texas. M, 1962, Texas Tech University. 164 p.

Pendexter, Charles. A study of metamorphic formations in southwestern Maine. M, 1949, Washington University. 82 p.

Pendexter, Charles. Genesis of the Westbrook Granite. D, 1951, Washington University. 134 p.

Pendexter, William Sands. Structural and stratigraphic relationships of the low grade metamorphic rocks in the Marble Valley area, Coosa and Chilton counties, Alabama. M, 1985, University of Alabama. 83 p.

Pendleton, J. A. Geology and the conduct of local government; an urban geologist's viewpoint. D, 1978, University of Colorado. 727 p.

Pendleton, Margaret Meda. An investigation of the mineral content of Pennsylvanian sandstones. M, 1948, University of Illinois, Urbana. 42 p.

Pendleton, Martha Warren. Cemented Pleistocene gravels of northern New Jersey. M, 1973, Rutgers, The State University, New Brunswick. 21 p.

Pendrel, J. V. The maximum entropy principle in two-dimensional spectral analysis. D, 1978, York University.

Peng, C. Y. A diagnostic model of continental shelf circulation. D, 1976, Florida State University. 136 p.

Peng, Chi-jui. The Mountain Maid ore body, Bisbee, Arizona. M, 1949, University of Arizona.

Peng, Chi-Jui. Thermal analysis study of the natrolite group. D, 1953, Columbia University, Teachers College.

Peng, Syh-Deng. Fracture and failure of Chelmsford granite (Devonian, Westford, Massachusetts). D, 1970, [Stanford University]. 344 p.

Penha, Lala. The nature of the inclusions in gem corundum and their relations to asterism. M, 1942, Columbia University, Teachers College.

Penley, Gary N. Lower and Middle Cambrian inarticulate brachiopods of southeastern Newfoundland. M, 1975, University of Kansas.

Penley, H. Michael. The geology along a portion of the Saltville Fault, Fountain City Quadrangle, Knox County, Tennessee area. M, 1973, University of Tennessee, Knoxville. 41 p.

Penn, Sheldon H. A preliminary investigation of the sediments of the Great South bay, Long Island, New York. M, 1969, Long Island University, C. W. Post Campus.

Penn, William Y. Upper Pennsylvanian fossils of the Sacramento Mountains, New Mexico. M, 1932, Stanford University. 203 p.

Pennell, Ray, Jr. A petrographic study of the Brassfield Limestone in western Ohio. M, 1952, Ohio State University.

Pennequin, Didier Franz Edgar. Groundwater circulation and groundwater budget for Lake Wingra, Madison, Wisconsin. M, 1982, University of Wisconsin-Madison. 252 p.

Penner, Alvin Paul. Rubidium-strontium age determinations in the Bird River area, southeastern Manitoba. M, 1970, University of Manitoba.

Penner, Lynden A. Fracture patterns and their influence on the strength and hydraulic conductivity of floral till, Saskatchewan. M, 1986, University of Saskatchewan. 140 p.

Penney, Dolores Jeanne. Index of Cenozoic foraminifera in California; with notes on the compositional changes of the faunas. M, 1949, Stanford University. 186 p.

Penneypacker, N. R. and Evans, Morris De B. The geology, mining, and development of the Tonopah gold and silver district. M, 1906, Lehigh University.

Pennifill, Roger Alan. Availability of aggregate in Bonner County and Boundary County, Idaho. M, 1978, University of Idaho. 84 p.

Pennington, Dennis Ira. Chromium and nickel in soil as geochemical indicators for chromite deposits in the State Line District, Pennsylvania. M, 1973, Pennsylvania State University, University Park. 62 p.

Pennington, Erwin K. A stratigraphic study of the Upper Cambrian of the Perry-Wooden No. 1 deep test-well, Cass County, Michigan. M, 1967, Michigan State University. 64 p.

Pennington, Jack, Jr. Stratigraphy and structure of the Horse Draw area, Albany and Platte counties, Wyoming. M, 1947, University of Missouri, Columbia.

Pennington, Jerry B. A survey of modern peritidal stromatolitic mats of the Yucatan Peninsula and a depositional model of a carbonate tidal flat in Rio Legartos, Yucatan, Mexico. M, 1987, Tulane University. 198 p.

Pennington, Richard L. Magnetic survey of a portion of the Felch District, central Dickinson County, Michigan. M, 1986, Bowling Green State University. 129 p.

Pennington, W. D. The subduction of the eastern Panama Basin and the seismotectonics of northwestern South America. D, 1979, University of Wisconsin-Madison. 138 p.

Pennington, Wayne David. Analysis of short-period waveforms from deep-focus earthquakes near the Fiji Islands. M, 1976, Cornell University.

Pennino, James D. Some physical, chemical, and engineering properties of a dolostone from east-central Cass County, Indiana. M, 1974, University of Toledo. 62 p.

Pennock, Daniel John. Soil landscape evolution in the Highwood River basin, southern Alberta. D, 1984, Queen's University.

Pennock, Edward S. Structural interpretation of seismic reflection data from the eastern Salt Range and Potwar Plateau, Pakistan. M, 1988, Oregon State University. 78 p.

Penoyer, Peter E. Geology of the Saddle and Humbug Mountain area, Clatsop County, northwestern Oregon. M, 1977, Oregon State University. 232 p.

Penrose, Richard A. F., Jr. Nature and origin of deposits of phosphate of lime. D, 1886, Harvard University.

Pense, Glenn Martin. The volcanic stratigraphy of the Banco Bonito Vitrophyre; a petrochemical study of a rhyolite lava flow. M, 1977, Texas Christian University.

Penso, Sharon Marie Hirt. Geophysical surveys of several selected ultramafic bodies in western North Carolina. M, 1981, Kent State University, Kent. 73 p.

Pentecost, David C. Fracture study of the Paleozoic bedrock in a portion of east-central Indiana. M, 1978, Ball State University.

Pentland, Arthur G. The heavy minerals of the Franconia and Mazomanie formations, Wisconsin. D, 1930, University of Wisconsin-Milwaukee.

Pentony, Kevin John. Neogene flysch provenance variations, Sea of Japan. M, 1977, Michigan State University. 86 p.

Pentsill, Benjamin Kobina. Geology and mineral deposits of West Africa. M, 1960, University of Michigan.

Penttila, William C. Stratigraphy and structure of the Horseshoe Gallup, and Berde Gallup pools, San Juan County, New Mexico. M, 1962, Colorado School of Mines. 140 p.

Penzo, Michael Anthony. The frequency and distribution of debris avalanches in selected areas of the White Mountains, New Hampshire, with a predictive model. M, 1981, SUNY at Binghamton.

Peoples, Darrell D. Sedimentology and petrography of the Swauk Formation, Blewett Pass area, Washington. M, 1984, Washington State University.

Peoples, Joe Webb. Geology of the Stillwater igneous complex. D, 1932, Princeton University. 180 p.

Peoples, Joe Webb. The stratigraphy of the Middle Devonian of the Tennessee Central Basin. M, 1929, Vanderbilt University.

Peoples, M. W. Determination of interval velocity, dip, strike, and thickness of subsurface layers from conventional CGP seismic data. M, 1977, Texas A&M University.

Pepe, Philip John. Bioerosion by a polychaete annelid, Eunice afra Peters, at Puerto Peñasco, Gulf of California, Mexico. D, 1983, University of Southern California.

Peper, John Dunkak. Geology of the Tinnie Fold belt, Lincoln County, New Mexico. M, 1964, University of Massachusetts. 87 p.

Peper, John Dunkak. Stratigraphy and structure of the Monson area, Massachusetts, Connecticut. D, 1966, University of Rochester. 127 p.

Pepin, Robert Osborne. Isotopic anomaly patterns in meteoritic xenon. D, 1964, University of California, Berkeley. 110 p.

Pepper, Gail Louise. Hydrogeology of reclaimed central Texas lignite mine. M, 1980, Texas A&M University. 77 p.

Pepper, James Franklin. Some features in the occurrence of marl noted during the field season of 1925. M, 1926, University of Michigan.

Pepper, James Franklin. The effect of the Taconic Orogeny upon lateral deformations in the Hudson River region. D, 1934, Cornell University.

Pepper, Miles Warren. The geology of the Sumatra Quadrangle, Montana. M, 1955, Montana College of Mineral Science & Technology. 87 p.

Pepperberg, Roy V. Preliminary report of the Carboniferous flora of Nebraska, coal in Nebraska. M, 1906, University of Nebraska, Lincoln.

Peppers, Russel Allen. Palynology of the McLeansboro Group of Illinois and equivalent strata of western Kentucky. D, 1961, University of Illinois, Urbana. 273 p.

Peppers, Russel Allen. Stratigraphy of the Muddy (New Castle) Formation of the Powder River basin, Wyoming and Montana. M, 1959, University of Illinois, Urbana. 183 p.

Peppin, William Alan. The cause of the body-wave surface-wave discriminant between earthquakes and underground nuclear explosions at near-regional distances. D, 1974, University of California, Berkeley. 315 p.

Pequegnat, J. E. Trace metals in phytoplankton from an area of coastal upwelling. D, 1975, Oregon State University. 100 p.

Peralta, Tobias Requejo. Hydrologic problems, Philippine Iron Mines, Inc., Larap, Jose Panganiban, Camerines Norte, Philippines. M, 1963, Washington University. 72 p.

Peralta-Cardenas, Gale. Structural analysis of northwestern Peru, South America. M, 1968, Cornell University.

Percious, Donald Joseph. Aquifer dispersivity by recharge-discharge of a dye tracer through a single well. M, 1968, University of Arizona.

Percious, Judith K. Geochemical investigation of the Del Bac Hills Volcanics, (middle Tertiary), Pima County, Arizona. M, 1968, University of Arizona.

Percival, John Allan. Geological evolution of part of the central Superior Province based on relationships among the Abitibi and Wawa subprovinces and the Kapuskasing structural zone. D, 1981, Queen's University. 300 p.

Percival, John Allan. Stratigraphy, structure and metamorphism of the Hackett River gneiss dome, District of Mackenzie, N.W.T. M, 1978, Queen's University. 129 p.

Percival, Stephen F. Brachiopods, mollusks, and tentaculitids from the Lower Devonian Shriver Chert of central Pennsylvania. M, 1959, Pennsylvania State University, University Park. 120 p.

Percival, Stephen F., Jr. Changes in calcareous nannoplankton in the Cretaceous-Tertiary biotic crisis at Zumaya, Spain. D, 1972, Princeton University. 99 p.

Percival, Tim. The Tomcat Mine roof pendant, Drum Valley, Tulare County, California. M, 1978, [California State University, Chico].

Percival, Timothy Jerold. Geology and geochemistry of the Tom Cat Mine pendant, Drum Valley, Tulare County, California. M, 1977, California State University, Fresno.

Percy, Cynthia W. A petrographic and field study of the Belchertown Tonalite (Massachusetts). M, 1955, University of Massachusetts. 58 p.

Perdue, Elizabeth Ann. The petrology and geochemistry of lavas from the west flank of Mauna Kea Volcano, Hawaii. M, 1982, University of California, Santa Barbara. 171 p.

Perdue, H. S. The geology of the area extending from Corbin to Weary Creek along the British Columbia-Alberta boundary line. M, 1930, McMaster University.

Perdue, Henry Stewart. Couchiching, Kashabowie Lake, Ontario. D, 1938, University of Chicago. 25 p.

Peredery, Walter V. A study of cordierite and anthophyllite rocks northwest of Lake Mistassini, Quebec. M, 1966, McGill University.

Peredery, Walter Volodymyr. The origin of rocks at the base of the Onaping Formation, Sudbury, Ontario. D, 1972, University of Toronto.

Peregrine, Keith. Potential field enhancement and two-dimensional gravity and magnetic modeling of the northeast extension of the New Madrid tectonic feature. M, 1982, Purdue University. 217 p.

Pereira da Cunha, Roberto. Geologic reactivation in Northeast Brazil and West Africa; an example of basement controlled tectonics. D, 1987, University of Kansas. 197 p.

Pereira V., Jesus Orangel. A one-dimensional seismic model. M, 1974, University of Tulsa. 64 p.

Pereira, Enio Bueno. Some problems concerning the migration and distribution of helium-4 and radon-222 in the upper sediments of the crust, a theoretical model; and the development of a quadrupole ion filter for measuring helium at the soil-air interface. D, 1980, Rice University. 120 p.

Pereira, Helton. Preliminary evaluation and exploration planning for mineral development in the Amazon region, Brazil. M, 1973, Stanford University.

Pereira-Soarez, Orlando. Influence of depth on physical properties of sandstones. M, 1968, University of Oklahoma. 77 p.

Perelman, David S. Seismic response of structure-foundation systems. D, 1968, Northwestern University. 117 p.

Peretsman, Gail Sue. Geochemical and petrographic analysis of early Mesozoic evaporites from Morocco; implications for the evolution of the North Atlantic Rift. M, 1985, University of Oregon. 87 p.

Pereus, Steven Charles. Rock mass characterization and slope failure mechanisms in the phosphate sequence of southeastern Idaho. M, 1983, University of Idaho. 20 p.

Perez A., Omar J. Spatial-temporal-energy characteristics of seismicity occurring during the seismic cycle; a reappraisal. D, 1983, Columbia University, Teachers College. 136 p.

Perez Guzman, Ana Maria. Biostratigraphic and paleoceanographic reconstruction of the late Miocene in Baja California and Tres Marias Islands, Mexico. D, 1983, Rice University. 216 p.

Perez, A. V. Economic geology of the Alamos mining district, Sonora, Mexico. M, 1975, University of Arizona. unpaginated p.

Perez, Adalberto Vasquez *see* Vasquez Perez, Adalberto

Perez, C. E. Rodriguez *see* Rodriguez Perez, C. E.

Perez, Emmanuel DeJesus. Magnetic surveys of selected ultramafic deposits in western North Carolina. M, 1979, Kent State University, Kent. 43 p.

Perez, Francisco Luis. Geomorphic slope processes, soils, and their relationship to the distribution of Espeletiinae cuatr., (Compositae) in the high Andean Paramo de Piedras Blancas, Venezuela. D, 1985, University of California, Berkeley. 619 p.

Perez, Humberto Ramon. Geology and geochemical exploration of the gold-silver deposits at Soledad Mountain (Mojave, Kern County, California). M, 1978, University of California, Los Angeles.

Perez, Jorge M. Geological history and sedimentary environment of the Muskogee Oilfield, Oklahoma. M, 1988, University of Oklahoma. 112 p.

Perez, K. R. Diagenesis of the Shannon Sandstone, Southwest Powder River basin, Wyoming. M, 1978, University of Wyoming. 124 p.

Pérez, Libardo Aquiles. The kinetics of crystallization, flocculation and phase transformation of some alkaline-earth salts. D, 1987, SUNY at Buffalo. 297 p.

Perez, Olivia Ramoz. Relationship between microfauna and lithology of the lower Taylor Marl, central Texas. M, 1975, Baylor University.

Perez, Omar. Spectral analysis of accelerograms recorded during Nicaraguan earthquakes. M, 1977, Colorado School of Mines. 85 p.

Perez, Stephanie. Geophysical and hydrogeological investigation of a buried river valley in Munson Township, Geauga County, Ohio. M, 1979, Kent State University, Kent. 80 p.

Perez, Vinicio Suro *see* Suro Perez, Vinicio

Perez-Ramirez, Gerardo Antonio. Probabilistic analysis of uncertainties in the disposal of nuclear waste. D, 1987, Purdue University. 188 p.

Perfetti, Jose N. Differential thermal studies of the manganese oxides. M, 1949, Columbia University, Teachers College.

Perfit, M. R. The petrochemistry of igneous rocks from the Cayman Trench and the Captains Bay Pluton, Unalaska Island; their relation to tectonic processes at plate margins. D, 1977, Columbia University, Teachers College. 189 p.

Perhac, Ralph Matthew. Geology and mineral deposits of the Gallinas Mountains, New Mexico. D, 1961, University of Michigan. 259 p.

Perhac, Ralph Matthew. Petrogenesis of the Voluntown and Oneco quadrangles. M, 1952, Cornell University.

Perigo, Russell Edward. Palaeomagnetism of late Precambrian-Cambrian volcanics and intrusives from the American Massif, France. M, 1982, University of Michigan.

Perillo, Gerardo Miguel Eduardo. Geomorphology and dynamics of a sand wave in lower Chesapeake Bay, Virginia. D, 1981, Old Dominion University. 229 p.

Pering, Katherine Lundstrom. A geochemical evaluation of hydrocarbon characteristics as criteria for the abiogenic origin of naturally occurring organic matter. D, 1971, Stanford University. 128 p.

Perini, Vincent C., Jr. Oil reconnaissance in northwest Colorado; Moffat and Routt counties. M, 1920, University of Colorado.

Perissoratis, C. Jutland Klippe; a Taconic type allochthon in western New Jersey. M, 1974, Queens College (CUNY). 117 p.

Perkins, Bobby Frank. Biostratigraphic studies in the Comanche (Cretaceous) series of northern Mexico and Texas. D, 1956, University of Michigan. 298 p.

Perkins, Bobby Frank. Studies of Upper Cretaceous corals. M, 1950, Southern Methodist University. 49 p.

Perkins, Clarence Michael. A bibliography of kimberlites and related rocks of central and Northeast United States with some reference on possible related structures; 1785-1974. M, 1975, University of Toledo. 146 p.

Perkins, Dexter, III. Application of new thermodynamic data to grossular phase relations. M, 1976, University of Michigan.

Perkins, Dexter, III. Application of new thermodynamic data to mineral equilibria. D, 1979, University of Michigan. 223 p.

Perkins, Edward P. The origin of the Dighton Conglomerate of the Narragansett Basin of Massachusetts and Rhode Island. D, 1919, Yale University.

Perkins, Elizabeth Gregory. 1, Geology of Mt. Warner (North Hadley, Massachusetts); 2, Structure and development of the Brachiopoda. M, 1915, Smith College. 62 p.

Perkins, Ernest Henry. A reinvestigation of the theoretical basis for the calculation of isothermal-isobaric mass transfer in geochemical systems involving an aqueous phase. M, 1980, University of British Columbia.

Perkins, Ernest Henry. The theoretical basis for the modelling of chemical reactions in rock-water systems with specific reference to the heat flow, fluid flow and solute transport laws. D, 1986, University of British Columbia.

Perkins, George David. Cretaceous and Devonian strata near Saskatoon. M, 1962, University of Saskatchewan. 37 p.

Perkins, Hamilton C. The use of insoluble residues in the study of limestone beds in the Shawnee Group of the Kansas River valley. M, 1952, University of Kansas. 215 p.

Perkins, James A. Provenance of the upper Miocene and Pliocene Etchegoin Formation; implications for paleogeography of the late Miocene of Central California. M, 1987, San Jose State University. 121 p.

Perkins, James Morgan. Geology of the Greenhorn Quadrangle and the northwest portion of the Whitney Quadrangle, Baker and Grant counties, Oregon. M, 1976, University of Oregon. 98 p.

Perkins, Jerome Hunt. Geology of the Decide Pool, Clinton County, Kentucky. M, 1954, University of Kentucky. 47 p.

Perkins, Max Allen. The geology of the Jacksboro and Bartons Chapel quadrangles, Jack County, Texas. M, 1964, Texas Christian University.

Perkins, Michael. Geology and mineral deposits of the Unaweep mining district, Mesa County, Colorado. M, 1975, University of Colorado.

Perkins, Michael. Geology and petrology of the East Bay Outlier of the late Mesozoic Great Valley Sequence, Alameda County, California. M, 1974, [University of New Mexico].

Perkins, Michael John. Structural geology and stratigraphy of the northern big bend of the Columbia River, Selkirk Mountains, B.C. D, 1983, Carleton University. 238 p.

Perkins, Philip Laurence. Mandibular mechanics and feeding groups in the Dipnoi. D, 1972, Yale University. 406 p.

Perkins, R. M. Geology and mineral deposits of the Unaweep mining district, Mesa County, Colorado. M, 1975, University of Colorado.

Perkins, Richard F. Structure and stratigraphy of the lower American Fork-Mahogany Mountain area, Utah County, Utah. M, 1955, Brigham Young University. 38 p.

Perkins, Robert Allen. Trace metal geochemistry and hydrothermal alteration of three molybdenum-bearing stocks, Gunnison and Pitkin counties, Colorado. M, 1973, Oklahoma State University. 78 p.

Perkins, Robert Lee. Geology of the Charles Town Quadrangle, West Virginia. M, 1963, West Virginia University.

Perkins, Robert Lee. Tectonites from the Blue Ridge and Great Valley of West Virginia and northwestern Virginia. D, 1967, West Virginia University. 139 p.

Perkins, Roderick L. The late Cenozoic geology of West-central Minnesota from Moorhead to Park Rapids. M, 1977, University of North Dakota. 99 p.

Perkins, Rodney K. Depositional environments of a portion of the Bullion Creek Formation (Paleocene), western Billings County, North Dakota. M, 1987, University of North Dakota. 244 p.

Perkins, Ronald Dee. Lithogenesis of the Pennsylvanian Madera Formation of Palomas Peak, Sandia Mountains, New Mexico. M, 1959, University of New Mexico. 76 p.

Perkins, Ronald Dee. Petrology of the Jeffersonville Limestone (Middle Devonian) of southeastern Indiana. D, 1962, Indiana University, Bloomington. 138 p.

Perkins, Russell Edward. Sedimentology of the Upper Triassic redbeds of King's County, Nova Scotia. M, 1981, University of Massachusetts. 196 p.

Perkins, Russell W. Geology of the Pebble Creek area, Caribou County, Idaho. M, 1977, Idaho State University. 50 p.

Perkins, T. W. Textures, fossil content, and conditions of formation of some middle Pennsylvanian coal balls, southeastern Kansas and northeastern Oklahoma. M, 1974, University of Kansas. 79 p.

Perkins, Warren W. A hydrogeologic investigation of South Deerfield, MA. M, 1985, University of Massachusetts. 75 p.

Perkins, William A. Factors controlling mineral deposition in contact metamorphic zones. M, 1933, Yale University.

Perkins, William D. Petrology and mineralogy of Quaternary basalts, Gem Valley and adjacent Bear River Range, southeastern Idaho. M, 1979, Utah State University. 91 p.

Perkins, William Enfield. Deep crustal reflections on land and at sea. D, 1970, Princeton University. 203 p.

Perkinson, Floyd. The microfauna of the Francis Formation in the vicinity of Ada, Oklahoma. M, 1934, University of Oklahoma. 95 p.

Perkinson, Mary C. Interpretations of the depositional environments for the Province Limestone Member of the Sturgis Formation (Upper Pennsylvanian) in Muhlenburg and Ohio counties, Kentucky. M, 1981, Eastern Kentucky University. 73 p.

Perley, Philip Charles. Geology of upper Basin Creek,-Upper West Fork Yankee Fork area, Custer County, Idaho. M, 1982, University of Idaho. 82 p.

Perlikos, Panayotis. A study of gold deposits in Rouyn-Noranda area, Quebec. M, 1977, University of Western Ontario. 179 p.

Perlman, Stephen H. The origin and mode of deposition of upper Kanawha and lower Allegheny sandstones on Bolt Mountain, Boone, Raleigh, and Wyoming counties, West Virginia. M, 1976, University of South Carolina.

Perlman, Vicky A. Relationship of composition and original thickness of lithologic layers to geometries developed during folding. M, 1973, Northwestern University. 25 p.

Perlmutter, Barry. Conodonts from the uppermost Wabaunsee Group (Pennsylvanian) and the Admire and Council Grove groups (Permian) in Kansas. D, 1971, University of Iowa. 121 p.

Perlmutter, Martin A. The recognition and reconstruction of storm sedimentation in the nearshore, Southwest Florida. D, 1982, University of Miami. 230 p.

Perlmutter, Nathaniel M. Geologic correlation of logs of wells in Long Island, New York. M, 1953, Columbia University, Teachers College.

Perman, Roseanne Chambers. Stratigraphy and sedimentology of the Lewis Shale and the Fox Hills Formation in south-central Wyoming. D, 1988, University of California, Berkeley. 266 p.

Pernichele, Albert D. Microfacies analysis of the Duperow Formation (upper Devonian) in the Beaver Lodge field, Williams County, North Dakota. M, 1964, University of North Dakota. 108 p.

Pernsteiner, Robert K. Distribution and behavior of fluorine in uranium-bearing granitic rocks, northeastern Washington. M, 1979, Eastern Washington University. 72 p.

Perotta, Anthony J. Some alumino-silicate framework structures. D, 1965, University of Chicago. 95 p.

Perras, Danielle. Modification d'un modèle hydrologique de type déterministe dans le but d'en améliorer la fiabilité M, 1988, Universite du Quebec a Montreal. 199 p.

Perrault, Guy S. Areal geology of the western margin of the Labrador Trough, Quebec. D, 1955, University of Toronto.

Perrault, Guy S. Stratigraphy and sedimentation of the western section of the Scarborough Bluffs (Ontario). M, 1951, University of Toronto.

Perreault, Serge. Géothermométrie, géobarométrie et nature des fluides métamorphiques dans le gneiss à cordiérite de la région de St. Augustin, Saguenay, province de tectonique de Grenville, Québec. M, 1987, Universite de Montreal.

Perrin, Nancy A. Depositional environments and diagenesis, Winnipegosis Formation (Middle Devonian), Williston Basin, North Dakota. D, 1987, University of North Dakota. 634 p.

Perrin, Nancy Ann. The distribution of the inarticulate brachiopods in the Big Horse Limestone Member, Orr Formation (Upper Cambrian), House Range, Utah. M, 1978, University of Kansas. 154 p.

Perrin, Shannon E. Depositional environment and subsurface geometry of oolitic grainstones in the Ste. Genevieve Limestone (Valmeyeran), Owensville North Field, Gibson County, Indiana. M, 1986, Indiana University, Bloomington. 76 p.

Perrine, Irving. The (Eocene) Claiborne Pelecypoda of the southern states. M, 1911, Cornell University.

Perrine, Irving. The Claiborne pelecypod fauna of the Gulf province. D, 1912, Cornell University.

Perrone, Emily F. The reworking of deep-sea sediments as indicated by the vertical dispersion of microtektites. M, 1980, University of Delaware.

Perrot, Jeannine A. Characteristics of polycrystalline quartz/chert in the Stanley Shale (Mississippian) during diagenesis/low-grade metamorphism, Ouachita Mountains, Arkansas. M, 1986, University of Oklahoma. 115 p.

Perruzza, Albert. Carbonate sedimentation and the climatic history of the equatorial Atlantic Ocean. M, 1971, Columbia University. 67 p.

Perry, A. O. Engineering geology of the northern portion of the Illinois shore of Lake Michigan. D, 1977, Purdue University. 152 p.

Perry, Albert J. The geology of the Emancipation Hill area, Boulder County, Colorado. M, 1956, University of Colorado.

Perry, Bernard James. Slope and topographic textures analysis and an application in Oklahoma. M, 1943, University of Oklahoma. 58 p.

Perry, Bobbie L. Permeability study of algal reef beds within the Bonneterre Formation, National Mines, St. Francois, Mo. M, 1958, University of Missouri, Rolla.

Perry, Christopher. Coal exploration and overburden evaluation, northeastern Pike County, Indiana; a stratigraphic and engineering geologic analysis. M, 1982, Purdue University. 224 p.

Perry, Christopher L. The Wagwater Belt, Jamaica; the tectonic-geologic development of an aulacogen. M, 1984, University of Oklahoma. 152 p.

Perry, Clinton W. Pleistocene deposits of a section of the larger Cicero Swamp (Syracuse, New York). M, 1912, Syracuse University.

Perry, David G. Age and faunas of the Ogilvie Formation (Devonian), northern Yukon. M, 1971, University of Western Ontario. 161 p.

Perry, David G. Paleontology and biostratigraphy of Delorme Formation (Siluro-Devonian), Northwest Territories. D, 1975, University of Western Ontario. 682 p.

Perry, David V. Genesis of the contact rocks at the Abril Mine, Cochise County, Arizona. M, 1964, University of Arizona.

Perry, David Vinson. Genesis of the contact rocks at the Christmas Mine, Gila County, Arizona. D, 1968, University of Arizona. 253 p.

Perry, Edward Adams, Jr. Burial diagenesis in Gulf Coast pelitic sediments. D, 1969, Case Western Reserve University. 131 p.

Perry, Elwyn Lionel. The geology of Bridgewater and Plymouth townships, Vermont. D, 1927, Princeton University. 105 p.

Perry, Esther p. Profile studies of the more extensive primary soils derived from granitic rocks in California. D, 1939, University of Southern California.

Perry, Eugene Carleton, Jr. Aluminum substitution in quartz, a study in geothermometry. D, 1963, Massachusetts Institute of Technology. 95 p.

Perry, Eugene Sheridan. Some extraordinary drainage features of southeastern Ohio. D, 1927, University of Chicago.

Perry, Eugene Sheridan. Structural geology of eastern Kentucky. M, 1923, University of Kentucky.

Perry, Frank Vinton. The evolution of magmatic systems during lithospheric extension; geologic and geochemical studies of volcanic rocks from the Rio Grande Rift region. D, 1988, University of California, Los Angeles. 174 p.

Perry, Frederick Welford. Variability of mullite as a function of conditions of crystallization and subsequent heat treatment. D, 1960, Pennsylvania State University, University Park. 161 p.

Perry, Harry A. Geology of the northern part of the Bonanza Volcano Field (Tertiary), Saguache County, Colorado. M, 1971, Colorado School of Mines. 72 p.

Perry, Harry Mcaughton. Geology of the Moose River Sandstone and associated formations of upper Enchanted Township, Maine. M, 1950, University of Missouri, Columbia.

Perry, John Kent. Neutron activation analysis applied to a geochemical study in the Big Five mines, Idaho Springs, Clear Creek County, Colorado. M, 1963, Colorado School of Mines. 216 p.

Perry, Kenneth, Jr. High-grade regional metamorphism of Precambrian gneisses and associated rocks, Paradise Basin Quadrangle, Wind River Mountains, Wyoming. D, 1965, Yale University.

Perry, L. Geology of Deep Springs Valley. M, 1954, University of California, Los Angeles.

Perry, Lawrence Dean. The petrography and depositional history of the Nellie Bly Formation in northeastern Oklahoma. M, 1959, University of Tulsa. 112 p.

Perry, LeRoy J. Geology of the east-central portion of the Blanco Mountain Quadrangle, Inyo County, California. M, 1955, University of California, Los Angeles.

Perry, Louis M. The Cretaceous subsurface geology in the vicinity of Cape Hatteras, North Carolina. M, 1949, University of Minnesota, Minneapolis. 116 p.

Perry, Mary J. Dynamics of phosphate utilization by marine phytoplankton in chemostat cultures and in oligotrophic waters of the central North Pacific Ocean. D, 1974, University of California, San Diego. 135 p.

Perry, Michael. Correlation analyses of Seasat SAR in the Southern Appalachians. M, 1982, University of Kansas.

Perry, Norton. Ontogenetic study of some Upper Devonian polygnathids. M, 1957, University of Missouri, Columbia.

Perry, Norton R. Ontogenetic study of some Upper Devonian polygnathids. M, 1958, University of Missouri, Columbia.

Perry, Patricia Lynn. An interpretation of the depositional setting for the Sugarite coal zone of the Raton Formation, located near the city of Raton, N.M. M, 1987, New Mexico Institute of Mining and Technology. 261 p.

Perry, R. V. The mineralogy and geology of the Buffalo Boy Mine and surrounding area, San Juan County, Colorado. M, 1977, University of Colorado.

Perry, Raymond Clair. An investigation of the effects of sphericity and roundness on the permeability of unconsolidated sediments. M, 1951, Michigan State University. 40 p.

Perry, Richard B. A study of the marine sediments of the Canadian eastern Arctic Archipelago. M, 1959, Texas A&M University.

Perry, Richard Baker. Submarine geology of the Aleutian arc. D, 1971, George Washington University. 343 p.

Perry, Richard Michael. Geology and mineral deposits of the northern half of the Mt. Tobin mining district, Pershing County, Nevada. M, 1985, University of Nevada. 130 p.

Perry, Robert Gayle. Seismic hazard analysis for the central United States. M, 1981, St. Louis University. 175 p.

Perry, S. C. The petrography and mineralogy of Adams Island and vicinity, Charlotte County, New Brunswick. M, 1932, University of Toronto.

Perry, Sandra Linthicum. Lineaments of the northern Denver Basin and their paleotectonic and hydrocarbon significance. M, 1985, Colorado School of Mines. 111 p.

Perry, Stanley James. Fluid inclusions and stable isotope study of a mesothermal gold deposit, Boryeon Mine, Republic of Korea. M, 1988, University of Missouri, Columbia. 86 p.

Perry, Stephen Kenneth. Structural geometry and tectonic evolution of the southwestern Gulf of Suez, Egypt. D, 1986, University of South Carolina. 605 p.

Perry, Thomas C. Economic geology and petrology of the Mississippian gypsum/anhydrite of Iowa. M, 1971, University of Iowa. 152 p.

Perry, Thomas G. J. An interpretation of the bryozoan genus Fistulipora. D, 1951, University of Toronto.

Perry, Thomas G. J. Preliminary study of the cyclostomatous bryozoan genus Fistulipora as represented at Arkona, Ontario. M, 1948, University of Toronto.

Perry, Vincent D. Some Tertiary Mollusca of Chiapas, Mexico. M, 1924, Columbia University, Teachers College.

Perry, William J., IV. The structural development of the Nittany Anticlinorium in Pendleton County, West Virginia. D, 1971, Yale University. 367 p.

Perry, William James, Jr. Regional aspects of the stratigraphy and facies relationships in the Lower Silurian of central and southern West Virginia and neighboring counties of Virginia, Kentucky, and Ohio. M, 1960, University of Michigan.

Persaud, Naraine. Influence of dispersion, exclusion, and metathetical sorption on the transport of inorganic solutes in a calcium-saturated porous medium. D, 1978, University of Florida. 97 p.

Pershouse, Jonathan Ralph. The use of Landsat imagery for resource mapping in the Rim Rock region of West Texas. M, 1981, Texas Christian University. 85 p.

Persico, John L. Behavior, speciation, and environmental impact of selenium at Bosque del Apache National Wildlife Refuge and Poison Canyon, New Mexico. M, 1988, University of New Mexico. 114 p.

Person, C. P. The middle Jurassic flora of Oaxaca, Mexico. D, 1976, University of Texas, Austin. 210 p.

Person, Donald W. Structural geology of the Peak District, Idaho. M, 1951, Washington State University. 32 p.

Person, Jennifer Ann. Petrology and depositional environment of the Kinderhookian Series in Southeastern Iowa. M, 1976, University of Iowa. 89 p.

Personius, Stephen Francis. Geologic setting and geomorphic analysis of Quaternary fault scarps along the Deep Creek Fault, upper Yellowstone Valley, south-central Montana. M, 1982, Montana State University. 77 p.

Persons, Jeffrey L. The delineation, lithology, and susceptibility to vertical saline communication of a fresh water lens, Cape Eleuthera, Eleuthera, Bahamas. M, 1974, Wright State University.

Persons, Philip. Pre-"Chattanooga" paleogeology of Kentucky and Tennessee. M, 1956, Stanford University.

Persson, Lars Evar. The geology of the Vine Quadrangle, Tennessee. M, 1960, Vanderbilt University.

Pertl, David Joseph. Geology of the Carrizo Mountains, Lincoln County, New Mexico. M, 1984, West Texas State University. 129 p.

Perttu, Janice C. An analysis of gravity surveys in the Portland Basin, Oregon. M, 1980, Portland State University. 106 p.

Perttu, R. K. Structural geology of the northeast quarter of the Dutchman Butte Quadrangle, Southwest Oregon. M, 1976, Portland State University. 62 p.

Pertusio, Serge M. A microfauna from the Vaqueros Formation, lower Miocene, Simi Valley, Ventura County, California. M, 1941, Columbia University, Teachers College.

Perucca, Melissa A. Stratigraphy and environmental analysis of the Siberia Limestone Member (Tobinsport Formation) and the Leopold Limestone Member (Branchville Formation) of Perry and Dubois counties, Indiana. M, 1988, Indiana University, Bloomington. 101 p.

Perucchio, Renato S. An integrated boundary element analysis system with interactive computer graphics for three-dimensional linear-elastic fracture mechanics. D, 1984, Cornell University. 244 p.

Perumalswami, P. R. Interaction between dam and foundation during earthquakes. D, 1968, University of Minnesota, Minneapolis. 98 p.

Perusek, Cyril J. Origin of the sediments in the Sangre de Cristo Formation in the Upper Pecos Valley of New Mexico. M, 1947, Texas Tech University. 27 p.

Pérusse, Jacques. Optical properties of opaque minerals in reflected light. M, 1953, Universite Laval.

Peryam, Richard Calvin. Geology of the Annette Quadrangle, San Luis Obispo and Kern counties, California. M, 1949, University of California, Berkeley. 47 p.

Peryea, Francis Joseph. Reclamation and regeneration phenomena in high boron soils. D, 1984, University of California, Riverside. 143 p.

Pescador, P. Yield response to zinc and the assessment of three extracting solutions for their estimation of "available" zinc in Hawaiian soils. M, 1963, University of Hawaii. 33 p.

Pesonen, Lauri Juhani. On the magnetic properties and paleomagnetism of aome Archean volcanic rocks from the Kirkland Lake area. M, 1973, University of Toronto.

Pesonen, Lauri Juhani. Paleomagnetic, paleointensity and paleosecular variation studies on Keweenawan igneous and baked contact rocks. D, 1978, University of Toronto.

Pesret, F. Kinetics of carbonate-seawater interactions. M, 1972, University of Hawaii.

Pessagno, Emile Anthony, Jr. Geology of the Ponce-Coama area, Puerto Rico. D, 1960, Princeton University. 170 p.

Pessagno, Emile Anthony, Jr. Preliminary analysis of stratigraphy, paleoecology, and micropaleontology of the older sequence, Mayaguez-Yauco District, Puerto Rico. M, 1957, Cornell University.

Pessl, Fred, Jr. Recession of the Illecillewaet Glacier (British Columbia) and the relationship between recession and annual snowfall. M, 1958, University of Michigan.

Pestana, Edith M. Geochemistry and tectonic significance of the volcanic rocks associated with the Rensselaer graywacke of east-central New York. M, 1985, Rutgers, The State University, Newark. 124 p.

Pestana, Harold R. Stratigraphy and paleontology of the Johnson Spring Formation, Middle Ordovician, Independence Quadrangle, California. M, 1959, University of California, Berkeley. 71 p.

Pestana, Harold Richard. Stromatoporoids of the Coralville Member of the Middle Devonian Cedar Valley Limestone (Johnson County, Iowa). D, 1965, University of Iowa. 65 p.

Pestrong, Raymond. Bedrock geology of the southern half of East Lee Quadrangle, Massachusetts. M, 1961, University of Massachusetts. 100 p.

Pestrong, Raymond. The development of drainage patterns on tidal marshes. D, 1965, Stanford University. 135 p.

Peteet, Dorothy Marie. Holocene vegetational history of the Malaspina Glacier District, Alaska. D, 1983, New York University. 181 p.

Petefish, David Michael. Metamorphism in the Marysville geothermal area, Marysville, Montana. M, 1975, Southern Methodist University. 46 p.

Peteghem, James Karl Van *see* Van Peteghem, James Karl

Petelka, Martin Frank. Leaching of radioactive waste forms under saturated and unsaturated flow conditions. D, 1987, Georgia Institute of Technology. 188 p.

Peter, George and Peters, James F. Geology and geophysics of the Venezuelan continental margin between Blanquilla and Orchilla islands. D, 1971, George Washington University. 222 p.

Peter, Jan Matthias. Genesis of hydrothermal vent deposits in the southern trough of Guaymas Basin, Gulf of California; a mineralogical and geochemical study. M, 1987, University of Toronto.

Peter, Kathy Dyer. Hydrochemistry of the Lower Cretaceous aquifers of the northern Great Plains. M, 1982, South Dakota School of Mines & Technology. 113 p.

Peterfreund, Alan Richard. Contemporary aeolian processes on Mars; local dust storms. D, 1985, Arizona State University. 247 p.

Petering, George Wilfred. New Pass; facies and characteristics of a tidal inlet on the southwest Florida Coast. M, 1974, University of Texas, Austin.

Peterman, Bruce D. Hydrogeology of the proposed Northern Great Plains superconducting Super Collider Site, Miner, Hanson, and Sanborn counties, South Dakota. M, 1987, South Dakota School of Mines & Technology.

Peterman, Zell E. Petrology of the metasediments of the Rainy Lake region. M, 1959, University of Minnesota, Minneapolis. 62 p.

Peterman, Zell Edwin. Precambrian basement of Saskatchewan and Manitoba. D, 1962, University of Alberta. 317 p.

Peters, Christopher Scott. Long term geomorphic evolution and recession models for the Lake Michigan bluffs in Wisconsin. M, 1982, University of Wisconsin-Madison. 375 p.

Peters, Colen R. Peat resources of selected wetlands on Block Island, Rhode Island. M, 1981, University of Rhode Island.

Peters, David Cornelius. Hypocenter location and crustal structure inversion of seismic array travel-times. D, 1973, University of Washington. 127 p.

Peters, Douglas C. Discrimination of alteration related to uranium mineralization, based upon mineralogy, using airborne multispectral scanner data in the southern Powder River basin, Converse County, Wyoming. M, 1981, Colorado School of Mines.

Peters, Dusty. The sedimentologic history of the sandstones of Tempe Butte, Arizona. M, 1979, Arizona State University. 197 p.

Peters, H. R. Areal geology of the Stony Lake area, central Newfoundland. M, 1953, Dalhousie University.

Peters, Herbert N., Jr. Chemical and physical properties of native Oklahoma asphalts. M, 1930, University of Oklahoma. 65 p.

Peters, James F. and Peter, George. Geology and geophysics of the Venezuelan continental margin between Blanquilla and Orchilla islands. D, 1971, George Washington University. 222 p.

Peters, James F. Stratigraphy and structure of the Rock Creek area, Beaverhead County, Montana. M, 1971, Oregon State University. 112 p.

Peters, Janet. Ontogeny and taxonomy of some Pennsylvanian crinoids from the Millersville Limestone Member (Bond Formation), Missourian, in Coles County, Illinois. M, 1985, Indiana University, Bloomington. 103 p.

Peters, John F. Multispectral analysis of ERTS imagery by color enhancement. M, 1973, University of Missouri, Rolla.

Peters, John Fredrick. Constitutive theory for stress-strain behavior of frictional materials. D, 1983, University of Illinois, Chicago. 337 p.

Peters, Joseph A. The geology of a greenstone belt in southern Iron County, Wisconsin. M, 1980, University of Wisconsin-Milwaukee. 134 p.

Peters, Joseph John. Petrology and geochemistry of a diabase contact suite, Cornwall, Pennsylvania. M, 1976, Rutgers, The State University, Newark. 73 p.

Peters, Kenneth E. Conversion of humic acid to kerogen under conditions simulating geothermal maturation. M, 1975, University of California, Santa Barbara.

Peters, Kenneth Eric. Effects on sapropelic and humic proto-kerogen during laboratory-simulated geothermal maturation experiments. D, 1978, University of California, Los Angeles. 186 p.

Peters, Lisa. Origin of vertical rhythmic layering in the marginal border group of the Skaergaard Intrusion, East Greenland. M, 1987, University of Texas at El Paso.

Peters, Norman Edward. An evaluation of environmental factors affecting the chemical composition of streams in the United States. D, 1982, University of Massachusetts. 117 p.

Peters, Robert Henry. Phosphorus regeneration by zooplankton. D, 1972, University of Toronto.

Peters, Thomas J. Geology of the Huronian sequence in parts of Fraleck, Grigg, Stobie, and Telfer townships, Ontario. M, 1969, Bowling Green State University. 59 p.

Peters, Walter G. A Paleoecological study of Pennsylvanian black shales using radiographic and photographic techniques. M, 1970, Northern Illinois University. 130 p.

Peters, William C. Geology and ore deposits of the Sunset District, Boulder County, Colorado. M, 1948, University of Colorado.

Peters, William C. The geologic environment of fluorspar deposits in the Western United States. D, 1956, University of Colorado.

Petersen, Carl Frank. Shock wave studies of selected rocks. D, 1969, Stanford University. 102 p.

Petersen, Carol Ann. Geology and geothermal potential of the Roosevelt hot springs area, Beaver County, Utah. M, 1975, University of Utah. 50 p.

Petersen, Cheryl L. Alternatives to septic system home wastewater disposal in Northwest Arkansas. M, 1977, University of Arkansas, Fayetteville.

Petersen, David Ward. Geochemistry and geology of the Notch Peak tungsten deposits, Millard County, Utah. M, 1976, University of Utah. 73 p.

Petersen, Edward Arnt. Shawangunk talus topography and clast distribution, Delaware Water Gap area, New Jersey and Pennsylvania. M, 1975, Rutgers, The State University, Newark. 56 p.

Petersen, Erich Ulrech. The Oxec copper deposit, Guatemala; an ophiolite copper occurrence. M, 1979, Dartmouth College. 116 p.

Petersen, Erich Ulrich. Metamorphism and geochemistry of the Geco massive sulfide deposit and its enclosing wallrocks. D, 1984, University of Michigan. 205 p.

Petersen, Gary Gene. Geologic interpretation of a ground magnetic survey of Johnson County and part of Iowa County, Iowa. M, 1966, University of Iowa. 41 p.

Petersen, Harry W. Structural and petrological relationships of the Bear Mountain Intrusive, Silver City Range, Grant County, New Mexico. M, 1979, University of Houston.

Petersen, Herbert Neil. Structure and Paleozoic stratigraphy of the Currant Creek area near Goshen, Utah. M, 1953, Brigham Young University. 60 p.

Petersen, James Frederick. Topographic profile analysis of piedmont scarps, northern Wasatch Front, Utah. D, 1981, University of Utah. 125 p.

Petersen, John W. Geology of the Tienditas creek-La Junta canyon area, Taos and Colfax counties, New Mexico. M, 1969, University of New Mexico. 82 p.

Petersen, Kenneth Lee. 10,000 years of climatic change reconstructed from fossil pollen, La Plata Mountains, southwestern Colorado. D, 1981, Washington State University. 197 p.

Petersen, Lee Edward. Metacopa and Platycopa (Ostracoda) from the Wenlock Series (Silurian) of the Welsh Borderland and central England. D, 1975, Arizona State University. 199 p.

Petersen, Lee Edward. Ostracodes and stratigraphy of the Birdsong Formation (Devonian) of western Tennessee. M, 1972, Arizona State University. 180 p.

Petersen, M. E. Chemical composition of dolomite from the Green River Formation, Wyoming. M, 1977, University of Missouri, Columbia.

Petersen, Mark A. Geology of the Quartz Creek tungsten deposit, Yellow Pine, Idaho. M, 1984, Kent State University, Kent. 94 p.

Petersen, Morris Smith. Devonian strata of central Utah. M, 1956, Brigham Young University. 37 p.

Petersen, Morris Smith. Upper Devonian (Famennian) ammonoids from the Fitzroy Basin, Western Australia. D, 1962, University of Iowa. 147 p.

Petersen, N. F. The bearing of trace element covariance and mineralogy of associated sediments on the origin of manganese nodules. M, 1967, University of Pennsylvania.

Petersen, Richard G. A section of the (Upper Cambrian) Allentown Formation near Freemansburg, Pennsylvania, with descriptions of the oolite beds. M, 1948, Lehigh University.

Petersen, Richard Randolph. A paleolimnological study of the eutrophication of Lake Erie. D, 1971, Duke University. 122 p.

Petersen, Robert M. Patterns of Quaternary denudation and deposition at Pipes Wash (Mojave Desert), California. D, 1976, University of California, Los Angeles. 279 p.

Petersen, Scott Walter. Geology and petrology around Titus Ridge, north-central Klamath Mountains, California. M, 1982, University of Oregon. 73 p.

Petersen, Ulrich B. Genesis of ore deposits in the Andes of central Peru. D, 1963, Harvard University.

Petershagen, John Haynes. Experimental nucleation and cavitation in a synthetic granitic melt from 1 kb to 0.5 kb. M, 1981, Iowa State University of Science and Technology. 174 p.

Petersn, Daniel Wayne. Taphonomy and community analysis of a restricted subtropic lagoon; Long Key Lake, Long Key, USA. M, 1988, University of Cincinnati. 155 p.

Peterson, Allen R. Paleoenvironments of the Colton Formation, Colton, Utah. M, 1975, Brigham Young University.

Peterson, Arthur F. Geology of the Shirley Basin and Bates Hole regions, Carbon and Natrona counties, Wyoming. M, 1935, University of Wyoming. 70 p.

Peterson, Benjamin Leland. Stratigraphy and structure of the Antelope Peak area, Snake Mountains, Elko County, northeastern Nevada. M, 1968, University of Oregon. 78 p.

Peterson, Carol Audrey. An empirical evaluation of the predictive capabilities of geophysical logging techniques in the Reydarfjordur, Iceland, Drillhole. M, 1980, Dalhousie University. 408 p.

Peterson, Caroline. Petrographic study of the Oligocene Arida sandstone formation, Sirte Basin, Libya. M, 1983, University of Houston.

Peterson, Carolyn Pugh. Geology of the Green Mountain-Young's River area, Clatsop County, Northwest Oregon. M, 1984, Oregon State University. 215 p.

Peterson, Christine Mary. Late Cenozoic stratigraphy and structure of the Taos Plateau, northern New Mexico. M, 1981, University of Texas, Austin. 58 p.

Peterson, Christoph R. Topical petrographic study of four porphyry southwestern copper deposits of the United States. M, 1964, Columbia University, Teachers College.

Peterson, Curt Daniel. Sedimentation in small active-margin estuaries of the northwestern United States. D, 1984, Oregon State University. 158 p.

Peterson, D. A. Effects of fine volcanic ash on surface irrigation in central Washington. M, 1982, Washington State University. 111 p.

Peterson, Dallas O. Structure and stratigraphy of the Little Valley area, Long Ridge, Utah. M, 1953, Brigham Young University. 96 p.

Peterson, Dallas Odell. Regional stratigraphy of the Pennsylvanian System in northeastern Utah, western Wyoming, northwestern Colorado, and southeastern Idaho. D, 1959, Washington State University. 252 p.

Peterson, Daniel Eric. Application of geothermometry, geobarometry and oxygen barometry to LIL/REE depleted granulite facies terrane in the Bamble sector, Southeast Norway. M, 1988, University of Wisconsin-Madison. 308 p.

Peterson, David. Hillslope erosion processes related to bedrock soils, and topography of the Three Peaks area, Marin County, California. M, 1979, San Jose State University. 94 p.

Peterson, David Harland. A study of modern sedimentation at Malakoff Diggins State Historic Park, Nevada County, California. M, 1980, University of California, Davis. 87 p.

Peterson, David Holmen. Fatty acid composition of certain shallow-water marine sediments. D, 1967, University of Washington. 71 p.

Peterson, David Michael. The influence of selected precipitation and land use characteristics upon the water quality of urban runoff. M, 1976, University of Nevada. 100 p.

Peterson, Dennis E. Earth fissuring in the Picacho area, Pinal County, Arizona. M, 1962, University of Arizona.

Peterson, Deverl J. Stratigraphy and structure of the West Loafer Mountain-Upper Payson Canyon area, Utah County, Utah. M, 1956, Brigham Young University. 40 p.

Peterson, Don H. The geology of the middle Beaver Creek area, Mason County, Texas. M, 1959, Texas A&M University.

Peterson, Donald Neil. Glaciological investigations on the Casement glacier, southeast Alaska. D, 1969, Ohio State University. 196 p.

Peterson, Donald W. Structural geology of the Peck District, Idaho. M, 1951, Washington State University.

Peterson, Donald William. Dacitic ash-flow sheet near Superior and Globe, Arizona. D, 1961, Stanford University. 178 p.

Peterson, Earl Thomas. The fauna of the Oolagah Limestone, Pennsylvanian, of northeastern Oklahoma. M, 1951, University of Tulsa. 113 p.

Peterson, Eric Thomas. Studies of the subduction process; seismic moment release rates and absolute plate motions. D, 1986, Stanford University. 146 p.

Peterson, Eunice. The (Cambrian) Dresbach Formation of Minnesota. D, 1927, University of Minnesota, Minneapolis.

Peterson, Eunice. The Cambrian geology of the lower St. Croix valley; Osceola to Stillwater. M, 1924, University of Minnesota, Minneapolis. 56 p.

Peterson, Frank Lynn. Measurement of short-term subsidence around a pumped and injected well. M, 1965, Stanford University.

Peterson, Fred. Cretaceous sedimentation and tectonism in the southeastern Kaiparowits region, Utah. D, 1969, Stanford University. 439 p.

Peterson, Frederick F. Solodized Solonetz soils occurring on the uplands of the Palouse loess. D, 1961, University of Washington. 280 p.

Peterson, Gary Lee. Physical stratigraphy and structure of type Horsetown Formation, Nor. Calif. M, 1961, University of Washington. 60 p.

Peterson, Gary Lee. Regional Cretaceous sequences in Northern California and Oregon. D, 1963, University of Washington. 103 p.

Peterson, Gerald Edwin. Petrographic analysis of the crystalline rocks of the Maryland Piedmont. M, 1951, Miami University (Ohio). 47 p.

Peterson, Gilbert M. Pollen analysis of cave and surface sediments. M, 1974, University of Wisconsin-Madison. 65 p.

Peterson, Gilbert Moseley. Holocene vegetation and climate in the western USSR. D, 1983, University of Wisconsin-Madison. 387 p.

Peterson, Harold. A comparison of the lithologic units in Utah, southeastern Idaho, and western Wyoming. M, 1929, Utah State University.

Peterson, Hazel Agnes. Interval maps of Cretaceous sediments of the United States. M, 1942, University of Texas, Austin.

Peterson, Hjalmer V. Occurrence and distribution of selenium in North Dakota ground waters. M, 1938, University of North Dakota. 26 p.

Peterson, J. E. Geology of the Noachis Quadrangle, Mars. M, 1974, University of Colorado.

Peterson, J. J. Compressional wave velocity characteristics of rocks from the Mings Bight-Betts Cove ophiolite complex, Newfoundland; a model for the oceanic crust. M, 1975, Queens College (CUNY). 45 p.

Peterson, Jahn Jean. Stratigraphy and historical geology of the Conant Creek area, Fremont County, Wyoming. M, 1940, University of Missouri, Columbia.

Peterson, James A. Stratigraphy and micropaleontology of the Sundance Group, eastern Wyoming. D, 1952, University of Minnesota, Minneapolis. 148 p.

Peterson, James B. A petrologic study of the Fumarole Butte volcanic complex, Utah. M, 1979, University of Utah. 63 p.

Peterson, James Carl. Geology of the Sweetwater Canyon area and origin of interbasinal canyons, southwestern Montana. M, 1974, Western Michigan University. 89 p.

Peterson, James Eugene. Cenozoic stratigraphy of Candelaria area, Presidio County, Trans-Pecos, Texas. M, 1955, University of Texas, Austin.

Peterson, James Leonard. Heavy minerals of the Judith River Formation in Musselshell and Golden Valley counties of central Montana. M, 1961, University of Kansas. 45 p.

Peterson, Janet L. Interpretation of electrical soundings and self potential measurements in the Norris Hot Springs area, Madison County, Montana. M, 1984, Montana College of Mineral Science & Technology. 31 p.

Peterson, John Christian. Water content determinations for dried marine sediments. M, 1971, United States Naval Academy.

Peterson, John E. Chemical evolution; an alternate hypothesis. M, 1971, San Diego State University.

Peterson, John Edward, Jr. The application of algebraic reconstruction techniques to geophysical problems. D, 1986, University of California, Berkeley. 198 p.

Peterson, John Ellis. Water resources and hydrogeology of the San Onofre Basin, San Diego County, California. M, 1978, San Diego State University. 109 p.

Peterson, John K. Quantitative nuclear well logging in permafrost for geotechnical purposes. M, 1985, University of Alaska, Fairbanks. 173 p.

Peterson, John L. Radiometric characterization of an arid range land site in New Mexico. M, 1985, [University of New Mexico]. 88 p.

Peterson, John Robert; Brant, Russell Alan; Elmer, Nixon and Gillespie, W. A. Geology of the Armstead area, Beaverhead County, Montana. M, 1949, University of Michigan. 118 p.

Peterson, John W. Geology of the southern part of the Ortiz Mountains, Santa Fe County, New Mexico. M, 1958, University of New Mexico. 115 p.

Peterson, Joseph D. Depositional systems associated with the Pittsburgh Number 8 Coal Seam in the Upper Pennsylvanian of southeastern Ohio. M, 1981, Ohio University, Athens. 74 p.

Peterson, Kent A. Geologic interpretations of ground water conditions in the Silurian and Cambrian-Ordovician aquifers of western Lake and western Cook counties. M, 1978, Northern Illinois University. 404 p.

Peterson, Lance Eric. A ground water contamination susceptibility map for the Dayton, Ohio, vicinity. M, 1988, Wright State University. 91 p.

Peterson, Larry Curtis. Late Quaternary deep-water paleoceanography of the eastern equatorial Indian Ocean; evidence from benthic foraminifera, carbonate dissolution, and stable isotopes. D, 1984, Brown University. 438 p.

Peterson, Larry Lynn. Impact of seawater intrusion control on the liquefaction susceptibility of the coastal plain, Ventura County, California. M, 1982, California State University, Northridge. 120 p.

Peterson, Lee Louis. The propagation of sunlight and the size distribution of suspended particles in a municipally polluted ocean water. D, 1974, California Institute of Technology. 174 p.

Peterson, Lorenz August. Geology of a portion of Fremont County, Colorado. M, 1941, University of Iowa. 59 p.

Peterson, M. L. Hydroacoustic fish stock assessment. M, 1975, University of Wisconsin-Madison.

Peterson, Mark Andrew. The distribution of and sources for quartz silt deposited on the northern Gulf of Mexico continental shelf. M, 1988, Texas A&M University. 54 p.

Peterson, Mark E. Chemical composition of dolomite from the Green River Formation, Wyoming. M, 1977, University of Missouri, Columbia.

Peterson, Mark P. The geology of the southwest quarter of the Avon 15-minute quadrangle, Powell County, Montana. M, 1985, Montana College of Mineral Science & Technology. 39 p.

Peterson, Martin Spencer. Geology of the Coachella Fanglomerate, San Gorgonio Pass, California. M, 1973, University of California, Santa Barbara.

Peterson, Marvin L. The study of a bentonite-like layer in the Spechts Ferry Member of the Decorah Formation. M, 1951, Northwestern University.

Peterson, Melvin N. A. Stratigraphy and petrography of the Big Clifty Formation of Tennessee. M, 1956, Northwestern University.

Peterson, Melvin Norman Adolph. The mineralogy and petrology of Upper Mississippian carbonate rocks of the Cumberland Plateau in Tennessee. D, 1960, Harvard University.

Peterson, Michael John. Fractionation characterization and comparison of organic acid and neutral components from various water systems and soil extracts. M, 1979, Colorado School of Mines. 187 p.

Peterson, Michael L. Vertical distributions and rates of deposition of barium in sediments of the Pacific. M, 1974, University of Washington. 41 p.

Peterson, Michael Paul. Map, image and mind; a pattern quality test of some graduated symbol maps. D, 1982, SUNY at Buffalo. 201 p.

Peterson, Morris Smith. Upper Devonian (Famennian) ammonoids from the Fitzroy Basin, Western Australia. D, 1962, University of Iowa. 147 p.

Peterson, Nathan N. Carbonate petrology, structure, and stratigraphy of the Middle Ordovician carbonates in the vicinity of Kingston, Ontario. D, 1969, Queen's University. 129 p.

Peterson, Nels P. Geology and ore deposits of the Mammoth Mining Camp, Pinal County, Arizona. D, 1938, University of Arizona.

Peterson, Nels P. The use of concrete in underground mine structures. M, 1932, University of Arizona.

Peterson, Norman Vernon. The geology of the southeast third of the Camas Valley Quadrangle, Oregon. M, 1957, University of Oregon. 89 p.

Peterson, Parley Royal. Geology of the Thistle area, Utah. M, 1952, Brigham Young University. 72 p.

Peterson, Paula S. Tephra of the Laguna de Ayarza calderas of southeastern Guatemala and its correlation to units of the Guatemalan highlands. M, 1980, Michigan Technological University. 108 p.

Peterson, R. M. Biofacies analysis of the uppermost Hamlin Shale and the Americus Limestone (Permian, Wolfcampian) in northeastern Kansas. D, 1978, University of Kansas. 253 p.

Peterson, Raymond A. Results of gravity measurements in Southern California. D, 1935, California Institute of Technology. 36 p.

Peterson, Raymond Judd. Bedrock Geology of the Benton Complex (Cretaceous), Saline County, Arkansas. M, 1972, University of Arkansas, Fayetteville.

Peterson, Reed H. Microfossils and correlation of part of the Frontier Formation, Coalville, Utah. M, 1950, University of Utah. 54 p.

Peterson, Rex Marion. Ostracods of the family Hollinidae from the Middle Devonian Jeffersonville Limestone at the falls of the Ohio (Kentucky-Indiana). M, 1957, University of Michigan.

Peterson, Rex Marion. Ostracods of the family Quasillitidae from the Middle Devonian strata of Michigan, Ohio, New York and Ontario. D, 1961, University of Michigan. 193 p.

Peterson, Richard C. Structural geology of the Sabino Canyon Fold, Santa Catalina Mountains, Pima County, Arizona. M, 1963, University of Arizona.

Peterson, Richard Charles. A structural study of the east end of the Catalina Forerange, Pima County, Arizona. D, 1968, University of Arizona. 193 p.

Peterson, Richard Frank. Conodonts from the Maple Mill Formation of southeastern Iowa. M, 1947, University of Iowa. 71 p.

Peterson, Richard Robert. The geology of the Buckhorn Creek area, Yavapai County, Arizona. M, 1985, Northern Arizona University. 109 p.

Peterson, Robert E. A study of suspended particulate matter; Arctic Ocean and northern Oregon continental shelf. D, 1977, Oregon State University. 122 p.

Peterson, Robert E. Calcium carbonate, organic carbon, and quartz in hemipelagic sediments off Oregon; a preliminary investigation. M, 1970, Oregon State University. 44 p.

Peterson, Robert Howard. The tectonic geomorphology of the northern Sangre de Cristo Mountains, near Villa Grove, Colorado. M, 1979, University of Arizona. 99 p.

Peterson, Robert Michael. The paleoecology of the Ames Limestone (Conemaugh Group) in east-central Ohio. M, 1973, Bowling Green State University. 81 p.

Peterson, Robert W. A study of bed load sediments from the Lynches river of South Carolina. M, 1969, SUNY at Buffalo. 33 p.

Peterson, Ronald Charles. Bonding in minerals; I, Charge density of the aluminosilicate polymorphs; and II, Molecular orbital studies of distortions in layer silicates. D, 1980, Virginia Polytechnic Institute and State University. 233 p.

Peterson, Ronald Milton. Lithofacies of the St. Louis Limestone (late Valmeyeran), Southeastern Iowa. M, 1970, Iowa State University of Science and Technology.

Peterson, Shirley J. Diagenesis and porosity distribution in deltaic sandstone, Strawn Series (Pennsylvanian) north-central Texas. M, 1977, University of Texas, Austin.

Peterson, Stanley Ross. A chemical kinetic-equilibrium simulation model of salt release from Mancos Shale and Mancos Shale-derived soils. D, 1982, Utah State University. 189 p.

Peterson, Stephen L. Geology of the Apache No. 2 mining district, Hidalgo County, New Mexico. M, 1976, University of New Mexico. 86 p.

Peterson, Steven D. Modal analysis of seismic guided waves in coal seams. D, 1979, Colorado School of Mines. 111 p.

Peterson, Thomas Charles. Transport of copiotrophic bacteria in oligotrophic coarse soils; a Monte Carlo analysis. D, 1987, Colorado State University. 196 p.

Peterson, Thomas L. The time of growth of the Opelika Dome, Henderson and Van Zandt counties. M, 1958, Texas A&M University.

Peterson, Tony Douglas. The petrogenesis and evolution of nephelinite-carbonatite magmas. D, 1987, The Johns Hopkins University. 406 p.

Peterson, Victor E. A study of the geology and ore deposits of the Asbrook silver mining district, Utah. D, 1941, University of Chicago. 70 p.

Peterson, Victor E. The geology of a part of the Bear River Range and some relationships that it bears with the rest of the range. M, 1936, Utah State University. 71 p.

Peterson, Virginia L. The structure and stratigraphy of the bedrock in the Ashburnham-Ashby area, north-central Massachusetts. M, 1984, University of Massachusetts. 240 p.

Peterson, Warren S. The recovery of gallium from a Virginia feldspar ore. M, 1940, University of Virginia.

Peterson, William R. Proposed plan for the mechanized mining of a coal property in McDowell County, West Virginia. M, 1950, Ohio State University.

Petkewich, Richard Mathew. Tertiary geology and paleontology of the northeastern Beaverhead and lower Ruby River basins; (Madison and Beaverhead County) southwestern Montana. D, 1972, University of Montana. 365 p.

Peto, Peter S. The petrology of the Similkameen Batholith (Cretaceous/Jurassic?), (S.W. British Columbia). M, 1970, University of Alberta. 98 p.

Petocz, Ronald George. Biostratigraphy and Lower Permian Fusulinidae of the upper Delta River area, east central Alaska Range. D, 1968, University of Alaska, Fairbanks. 292 p.

Petrachenko, William Terry. Evaluation of a satellite phase link for use in long baseline radio interferometry. D, 1983, York University.

Petraitis, Michael John. Carbon dioxide in the unsaturated zone of the Southern High Plains of Texas. M, 1981, Texas Tech University. 134 p.

Petrak, J. A. Some theoretical implications of strike-slip faulting. M, 1965, University of British Columbia.

Petrakis, Emmanuel. Micromechanical modeling of granular soil at small strain by arrays of elastic spheres. D, 1987, Rensselaer Polytechnic Institute. 1987 p.

Petraske, A. K. The mechanics of emplacement of layered basic intrusions. M, 1976, SUNY at Buffalo. 96 p.

Petree, David Hoke, Jr. The influence of grain size on the clay mineral composition of sediments in the Neuse River estuary, North Carolina. M, 1974, University of North Carolina, Chapel Hill. 28 p.

Petricca, Ann M. Sedimentology and diagenesis of the interbedded carbonate and siliciclastic rocks of the Big Snowy, Amsden, and Quadrant formations, Tobacco Root Mountains region, southwestern Montana. M, 1985, Indiana University, Bloomington. 160 p.

Petrick, Audrey B. A microfauna of the Pliocene of Florida. M, 1941, Columbia University, Teachers College.

Petrick, Glen. The andalusite and sillimanite deposits of the Harney Peak region, Black Hills, South Dakota. M, 1935, University of Iowa. 37 p.

Petrick, William Robert. A fully internal hybrid technique for calculating electromagnetic scattering from three dimensional bodies in the Earth. D, 1984, University of Utah. 163 p.

Petrie, Gregg M. A gravity survey and analysis of the Mount Stuart block of Washington State. M, 1978, Western Washington University. 105 p.

Petrie, J. M. Field response of a clay till in a layered aquifer system at Waterloo, Ontario. M, 1985, University of Waterloo. 76 p.

Petrie, John David. Fluorine and chlorine in metamorphic minerals from two amphibolite facies terrains and an evaluation of hydrous metamorphic minerals as a possible source of hydrothermal mineralizing solutions. M, 1985, Bryn Mawr College. 93 p.

Petrie, Mark Alan. Morphologic and lithologic influences on recharge in a glaciated basin. M, 1984, Michigan State University. 38 p.

Petrie, William Leo. Geology of the southeast portion of the Sullivan Quadrangle, Missouri. M, 1951, University of Iowa. 74 p.

Petrini, Rudolf Harald Wilhelm. A postmortem assessment of environmental compliance of high-level radioactive waste repository, Hanford Site, Washington. M, 1988, Texas A&M University. 156 p.

Petro, William L. Major element chemistry and tectonic setting of plutonic rock suites. M, 1977, Michigan State University. 45 p.

Petrone, Anthony. The Moses Lake sand dunes. M, 1970, Washington State University. 89 p.

Petrovic, Radomir. Alkali ion diffusion in alkali feldspars. D, 1972, Yale University. 164 p.

Petrovski, David M. Chemomechanical weakening of natural and synthetic quartz. M, 1983, Indiana University, Bloomington. 110 p.

Petrowski, Nila Chari. Light element geochemistry at the Frau Mauro region of the Moon. M, 1974, University of California, Los Angeles. 151 p.

Petroy, David Edward. Historical seismicity and plate kinematics of the Northeast Indian Ocean. M, 1988, Washington University. 142 p.

Petruk, W. Petrofabric analysis of the Amisk and Missi sediments in the Amisk and Hanson Lakes area. M, 1956, University of Saskatchewan. 46 p.

Petruk, William. The Clearwater copper-zinc deposit (New Brunswick) and its setting, with a special study of mineral zoning around such deposits. D, 1959, McGill University.

Petrus, Carolyn Ann. Investigation of fracture traces and underground roof fall fatalities in the Southern Anthracite Field, Pennsylvania. M, 1979, Pennsylvania State University, University Park. 126 p.

Petrus, Richard T. Geology and petrography of the southwestern portion of Cony Mountain Quadrangle

and adjacent areas, Wyoming. M, 1976, University of Missouri, Columbia.

Petry, Thomas Merton. Identification of dispersive clay soils by a physical test. D, 1974, Oklahoma State University. 179 p.

Petryk, Allen Alexander. Lower Carboniferous foraminifera and biostratigraphy of southwestern Alberta. D, 1969, University of Saskatchewan. 422 p.

Petryk, Allen Alexander. Some Silurian stromatoporoids from northwestern Baffin Island (Canada). M, 1965, McGill University.

Petsrillo, Ira. Special and statistical analysis of the subsurface structure of the Oriskany Sandstone in the gas fields of western and northern Pennsylvania. M, 1980, SUNY at Buffalo. 194 p.

Pett, John Woodfull. The morphology of flutes. M, 1970, McMaster University. 121 p.

Petta, Timothy J. Diagenesis and paleohydrology of a rudist reef complex (Cretaceous), Bandera County, Texas. D, 1976, Louisiana State University. 225 p.

Petta, Timothy Joseph. Application of a Holocene model to the depositional environment of the Tepee Zone of the Pierre Shale, Pubelo County, Colorado. M, 1973, Wichita State University. 82 p.

Pettengill, James G. Structural analysis of Coconino Point, Coconino County, Arizona. M, 1970, Northern Arizona University. 86 p.

Petter, Charles K., Jr. Late Tertiary history of the upper Little Missouri River, North Dakota. M, 1956, University of North Dakota. 49 p.

Petters, Sunday W. Subsurface upper Cretaceous stratigraphy and foraminiferal biostratigraphy of the Atlantic Coastal Plain of New Jersey. D, 1975, Rutgers, The State University, New Brunswick. 258 p.

Petticord, David V. Implementation and evaluation of the NESA program at Glacier National Park, Montana. M, 1975, Montana State University. 59 p.

Pettigrew, Robert J., Jr. Geology and flow systems of the Hickory Aquifer in the San Saba County area, Texas. M, 1988, Baylor University. 132 p.

Pettigrew, Robert William. Geology of Kickapoo Creek area, Erath, Hood, Palo Pinto, and Parker counties, Texas. M, 1954, University of Texas, Austin.

Pettijohn, Francis J. A study of the conglomerate of Abram Lake, Ontario, and its extensions; a study in pre-Cambrian sedimentation and structure. D, 1930, University of Minnesota, Minneapolis. 183 p.

Pettijohn, Francis J. Phosphate pebbles of the Twin City Ordovician and their geologic significance. M, 1925, University of Minnesota, Minneapolis. 52 p.

Pettingill, Henry S. Age and origin of anorthosites, charnockites, and granulites in the central Virginia Blue Ridge; Nd and Sr isotopic evidence. M, 1983, Virginia Polytechnic Institute and State University. 49 p.

Pettis, Rani Hathaway. Pleistocene benthic foraminifera of the Benham Rise, western Philippine Basin, western Pacific Ocean, DSDP Site 292 (Leg 31) and a taxonomic revision of the unilocular foraminifera. M, 1985, University of California, Los Angeles. 181 p.

Pettus, David S. Ultramafic xenoliths from Llera de Canales, Tamaulipas, Mexico. M, 1979, University of Houston.

Petty, Andrew J., Jr. Biostratigraphy of the Graford Formation, Missourian, Wise County, Texas. M, 1975, University of Texas at El Paso.

Petty, John Kirkpatrick. The Aylor Bluff Member of the Big Saline Formation of Marble Falls Group in San Saba County, Texas. M, 1947, University of Texas, Austin.

Petty, Rebecca J. Applications of Landsat imagery to geologic mapping, Ross County, Ohio. M, 1985, Ohio University, Athens. 97 p.

Petty, Steven Matthew. The geology of the Laurel Creek mafic-ultramafic complex in Northeast Georgia; intrusive complex or ophiolite?. M, 1982, Florida State University.

Petty, Van Alvin, Jr. Datum planes used in reflection seismograph exploration. M, 1941, University of Texas, Austin.

Pettyjohn, Wayne A. The stratigraphy of the Dakota Sandstone in South Dakota. M, 1959, University of South Dakota. 89 p.

Pettyjohn, Wayne Arvin. Geology of a part of west-central South Dakota. D, 1965, Boston University. 450 p.

Petuch, Edward James. A re-analysis of Neogene Caribbean provinciality with reference to the discovery of a relict caenogastropod fauna off northern South America. D, 1980, University of Miami. 174 p.

Petzel, Gerald J. Evaluation of data from the first Earth Resources Technology Satellite for the purpose of structural analysis in the Anadarko Basin, Oklahoma and Texas. M, 1974, University of Oklahoma. 107 p.

Petzold, Daniel D. Paleontology and paleoecology of an unnamed nonmarine interval of the Dugger Formation in Warrick County, Indiana. M, 1987, Indiana University, Bloomington. 103 p.

Petzold, Donald Emil. Synoptic investigations of the summer climate and lake evaporation in Quebec-Labrador. D, 1980, McGill University.

Peurifoy, Raymond E. Petrology of the Upper Jurassic Smackover Limestone in the B.F. Hare No. 1, Columbia County, Arkansas. M, 1985, University of Arkansas, Fayetteville. 101 p.

Pevear, David R. Clay mineral relationships in recent river, nearshore marine, continental shelf, and slope sediments of the southeastern United States. D, 1968, University of Montana. 164 p.

Pevear, David R. Phosphatic and oolitic sediments of the Georgia continental shelf. M, 1967, University of Montana. various pagination p.

Pew, Elliott. Seismic structural analysis of deformation in the southern Mexican ridges. M, 1982, University of Texas, Austin.

Péwé, Troy Lewis. Geology of the Red Peak area, Owl Creek Mountains, Wyoming. M, 1942, University of Iowa. 118 p.

Péwé, Troy Lewis. Geomorphology of the Fairbanks area, Alaska. D, 1953, Stanford University. 220 p.

Peyton, Robert Lee, Jr. Solute transport in overland flow during rainfall. D, 1985, Colorado State University. 262 p.

Peyton, T. O. The input and distribution of aerially deposited heavy metals in an urban aquatic ecosystem. D, 1975, Purdue University. 272 p.

Pezzetta, John Mario. Recent sediments in the Scotian Shelf (Nova Scotia). M, 1962, Dalhousie University. 46 p.

Pezzetta, John Mario. The Saint Clair River delta (Michigan). D, 1968, University of Michigan. 193 p.

Pfaff, Bruce Justin. Sedimentologic and tectonic evolution of the fluvial facies of the Upper Cretaceous Castlegate Sandstone, Book Cliffs, Utah. M, 1985, University of Utah. 124 p.

Pfaff, Dieter. Facies analysis of the Keg River Formation (Middle Devonian) in Rainbow B Pool, Alberta (Canada). M, 1967, University of Alberta. 115 p.

Pfaff, Virginia J. On forms of folds. D, 1986, University of Cincinnati. 583 p.

Pfaff, Virginia Josette. Geophysical and geochemical analyses of selected Miocene coastal basalt features, Clatsop County, Oregon. M, 1981, Portland State University. 150 p.

Pfaffman, George A. The geology of the Martinez Formation of the Tejon and Elizabeth Lake quadrangles, California. M, 1941, University of Southern California.

Pfann, H. D. A paleomagnetic study of the Moxie Pluton, west-central Maine. M, 1978, SUNY at Binghamton. 184 p.

Pfau, Gerchard Edmund. Separation and deconvolution of vertical seismic profiles. M, 1984, Indiana University, Bloomington. 121 p.

Pfau, Mark A. Geology of the Emigrant Gulch porphyry copper-molybdenum complex, Park County, Montana. M, 1981, University of Idaho. 89 p.

Pfeffer, Beverley James. Relationships between textures and depositional environment in Mississippian limestones of the Alida area, Saskatchewan. M, 1962, University of Saskatchewan. 79 p.

Pfeffer, Helmut W. Petrogenesis of the dioritic rocks (metadiabases) of the O'Sullivan Lake area, Ontario. D, 1951, University of Toronto.

Pfeffer, Tad. Enhancement of radiative absorptance of a glacier surface due to 10-meter scale surface roughness. M, 1981, University of Maine. 114 p.

Pfeffer, William Tad. Structure and deformation in a propagating surge front. D, 1988, University of Washington. 134 p.

Pfeifer, Mary Catherine. Multicomponent underground DC resistivity study at the Waste Isolation Pilot Plant, Southeast New Mexico. M, 1987, Colorado School of Mines. 96 p.

Pfeiffer, Dan E. The stratigraphic chemistry and mineralogy of the Negaunee iron formation (Precambrian), sections 7, 8, and 18, T. 47 N., R. 26 W., Marquette County, Michigan. M, 1972, Bowling Green State University. 75 p.

Pfeiffer, Deborah Susan. Temperature variations and their relation to groundwater flow, South Texas, Gulf Coast Basin. M, 1988, University of Texas, Austin. 199 p.

Pfeil, R. W., Jr. Stratigraphy and sedimentology, Cambrian Shady Dolomite, Virginia. M, 1977, Virginia Polytechnic Institute and State University.

Pferd, Jeffery William. Engineering and related physical properties of the coastal salt marshes in McIntosh County, Georgia. M, 1970, University of Georgia. 91 p.

Pferd, Jeffrey William. The bedrock geology of the Colrain Quadrangle, Massachusetts-Vermont 1980. D, 1981, University of Massachusetts. 367 p.

Pfiefer, James E., Jr. Geology of T. 15 N., R. 27, 28 W., Washington and Madison counties, Arkansas. M, 1967, University of Arkansas, Fayetteville.

Pfirman, Richard S. Stratigraphy and history of the Toroweap Formation (Permian) between Grindstone Canyon and Sycamore Canyon, Coconino County, Arizona. M, 1968, University of Arizona.

Pfirman, Stephanie Louise. Modern sedimentation in the northern Barents Sea; input, dispersal and deposition of suspended sediments from glacial meltwater. D, 1984, Woods Hole Oceanographic Institution. 376 p.

Pflucker, Eduardo Cabieses. A review of the geology of the oil fields of South America with emphasis on prospective areas. M, 1945, University of Texas, Austin.

Pfluke, John Henry. A comparison of the properties of elastic waves generated by explosion and impact seismic sources. M, 1961, St. Louis University.

Pfluke, John Henry. Seismic model studies of first motions produced by an actual fault. D, 1963, Pennsylvania State University, University Park. 108 p.

Pflum, Charles E. The distribution of foraminifera in the eastern Ross Sea, Amundsen Sea, and Bellingshausen Sea, Antarctica. M, 1963, Florida State University.

Phadke, Suhas. Imaging crustal diffraction zones and seismic tomography. D, 1988, University of Alberta. 183 p.

Phair, George. Petrology of the southwestern part of the Long Range, Newfoundland. D, 1949, Princeton University. 165 p.

Phair, George. The geology of the Shell Mountain area, Park County, Montana. M, 1942, Rutgers, The State University, New Brunswick. 54 p.

Phair, Ronald Leslie. Seismic stratigraphy of the Lower Cretaceous rocks in the southwestern Straits of Florida, southeastern Gulf of Mexico. M, 1984, University of Texas, Austin. 319 p.

Phalakarakula, Charas. Deflections of the vertical from gravity anomalies. M, 1956, Ohio State University.

Phalen, W. C. Study of shrinkage in clay. M, 1902, Massachusetts Institute of Technology. 66 p.

Phamwon, Sanguan. Network model for optimal management of stream-aquifer systems. D, 1982, Colorado State University. 284 p.

Phanartzis, Christos Apostolou. Spatial variability of precipitation in the San Dimas Experimental Forest and its effect on simulated streamflow. M, 1972, University of Arizona.

Phares, Rod S. Depositional framework of the Bartlesville Sandstone (Pennsylvanian). M, 1969, University of Tulsa. 59 p.

Phariss, Edward Irvin. Geology and ore deposits of the Alpine mining district, Esmeralda County, Nevada. M, 1974, University of Nevada. 114 p.

Pharo, Christopher Howard. Sediments of the central and southern Strait of Georgia, British Columbia. D, 1972, University of British Columbia. 287 p.

Pharo, Christopher Howard. Sediments of the central and southern strait of Georgia, British Columbia. D, 1973, University of British Columbia.

Phasukyud, Prapon. Trench safety consultation system. D, 1987, University of Missouri, Columbia. 406 p.

Pheasant, David R. The glacial geomorphology of the Ya-Ha Tinda Ranch area, Alberta. M, 1968, University of Calgary.

Pheasant, David Richard. The glacial chronology and glacio-isostasy of the Narpaing/Quajon Fiord area, Cumberland Peninsula, Baffin Island. D, 1971, University of Colorado.

Pheifer, R. N. The paleobotany and paleoecology of the unnamed shale overlying the Danville Coal Member (VII) in Sullivan County, Indiana. D, 1979, Indiana University, Bloomington. 295 p.

Phelan, Janet Meredith. Volcanoes as a source of volatile trace elements in the atmosphere. D, 1983, University of Maryland. 154 p.

Phelan, Kevin. Glacial geology in the eastern Vestal area, New York. M, 1981, SUNY at Binghamton. 65 p.

Phelan, Michael Joseph. Crustal structure in the central Mississippi Valley earthquake zone (Missouri). D, 1969, Washington University. 166 p.

Phelan, Michael Joseph. Gravity and magnetic survey of the Wichita Mountains, Oklahoma. M, 1965, Washington University. 47 p.

Phelan, Patrick John. Potassium exchange equilibria in some potassium deficient soils. D, 1987, University of California, Riverside. 262 p.

Phelps, Daniel Craig. Heavy mineral dispersal on the northern North Carolina continental shelf. M, 1979, Duke University. 71 p.

Phelps, David William. Petrology, geochemistry, and structural geology of Mesozoic rocks in the Sparta Quadrangle and Oxbow and Brownlee Reservoir areas, eastern Oregon and western Idaho. D, 1978, Rice University. 241 p.

Phelps, David William. Phase chemistry of the layered series, Raggedy Mountain gabbro group, Oklahoma. M, 1976, Rice University. 121 p.

Phelps, Dorothy A. A singular land use in the California Desert. M, 1981, University of California, Los Angeles.

Phelps, George B. Geology of the Newlan creek area, Meagher County, Montana. M, 1969, Montana College of Mineral Science & Technology. 56 p.

Phelps, Gertrude Gunia. Estimates of recharge to the principal artesian aquifer in northeast Florida. M, 1975, University of North Carolina, Chapel Hill. 41 p.

Phelps, James Carl. Stratigraphy and structure of the northeastern Doonerak Window area, central Brooks Range, northern Alaska. D, 1987, Rice University. 293 p.

Phelps, Lee Barry. Minimization of groundwater contamination in surface mine backfills. D, 1981, Pennsylvania State University, University Park. 146 p.

Phelps, Richard T. Geologic hazards in the Gould area, Jackson County, Colorado. M, 1978, Colorado State University. 154 p.

Phelps, Robert Karl. A fault-controlled Pb-Zn occurrence in Essex and Kent counties, Ontario. M, 1978, University of Windsor. 137 p.

Phelps, Willard B. The petrology of the eastern part of Worchester County, Massachusetts. M, 1936, Ohio State University.

Phelps, William Eugene. The geology of the Sandy Hook Field, Marion County, Mississippi. M, 1957, Mississippi State University. 43 p.

Phetteplace, Thurston Mason. Geology of the Chesaw area, Okanogan County, Washington. M, 1954, University of Washington. 82 p.

Phiancharoen, Charoen. Interpretation of the chemical analysis of the groundwater of the Khorat Plateau, Thailand. M, 1962, University of Arizona.

Phibbs, John. Gravimetric study of Lone Pine Canyon, San Bernardino County, California. M, 1988, California State University, Long Beach. 126 p.

Phienweja, Noppadol. Ground response and support performance in a sheared shale, Stillwater Tunnel, Utah. D, 1987, University of Illinois, Urbana. 530 p.

Philbrick, Charles Russell. Studies of centers produced in sapphire and ruby by gamma radiation. D, 1967, North Carolina State University. 107 p.

Philbrick, Shailer Shaw. The contact metamorphism of the Onawa Pluton, Piscataquis County, Maine. D, 1933, The Johns Hopkins University. 380 p.

Philen, O. D., Jr. Relationships between surface charge density of layered silicates and competitive adsorption of two divalent organic cations. D, 1972, North Carolina State University.

Philip, Aldwyn Thomas. Independent model aerial triangulation; refinements and error studies. D, 1973, Ohio State University.

Philips, K. A. Geology and mineral deposits of the Waverly Island District, Amisk Lake, Saskatchewan. M, 1950, Queen's University. 54 p.

Philips, Richard P. Measurement of the radioactivity of the soil across the Hayward fault zone, near Irvington, California. M, 1956, Stanford University.

Phillabaum, Stephen D. A geomorphic inventory of Whatcom County marine shoreline with considerations for its management. M, 1973, Western Washington University. 176 p.

Philley, John C. A stratigraphic study of the Richland Valley area, Grainger County, Tennessee. M, 1961, University of Tennessee, Knoxville. 27 p.

Philley, John Calvin. The environmental stratigraphy of some Mississippian (Renfro-Saint Louis-Ste. Genevieve) carbonates in northeastern Kentucky. D, 1971, University of Tennessee, Knoxville. 152 p.

Phillion, George William. Fracture pattern analysis of the Keweenaw Peninsula (Michigan). M, 1962, Michigan Technological University. 87 p.

Phillips, A. M., III. Packrats, plants, and the Pleistocene in the lower Grand Canyon. D, 1977, University of Arizona. 137 p.

Phillips, Andrew. Prodeltaic sedimentation of an ice-contact delta, Riggs Glacier, Glacier Bay, Alaska. M, 1988, University of Illinois, Chicago. 80 p.

Phillips, Bert. Phase equilibria in the system CaO-FeO-Fe$_2$O$_3$-SiO$_2$. D, 1959, Pennsylvania State University, University Park. 106 p.

Phillips, Bruce Edwin. Sedimentology of a rapidly aggrading fluvial system; Cloverly Formation (Cretaceous), Sheridan County, Wyoming. M, 1986, University of Illinois, Urbana. 89 p.

Phillips, Charles Heulan. Geology of the La Madera area, Sandia Mountains, New Mexico. M, 1964, University of New Mexico. 75 p.

Phillips, Chase A. Geology of the bauxite deposits of Paranam, Surinam, South America. M, 1955, University of Arkansas, Fayetteville.

Phillips, D. P. Geology of the Sheep Ridge area, Hot Springs and Fremont counties, Wyoming. M, 1958, University of Wyoming. 135 p.

Phillips, D. S. Precipitation in star sapphire. D, 1978, Case Western Reserve University. 155 p.

Phillips, David Lee. A study of the Cowrun Sandstone and related sedimentary rocks in an area in southeastern Ohio. M, 1951, University of Cincinnati. 34 p.

Phillips, Donald T., II. Deformational history of the Kentucky River Fault system. M, 1976, Eastern Kentucky University. 65 p.

Phillips, Edward Hayden. Coaxial refolding in Southeast San Diego County, California. M, 1964, San Diego State University.

Phillips, Edward Lindsey, Jr. The Salisbury Adamellite Pluton, North Carolina. M, 1965, University of North Carolina, Chapel Hill. 47 p.

Phillips, Franklin Jay. Miocene foraminifera from Caliente Mountain, San Luis Obispo County, California. M, 1976, University of California, Berkeley. 112 p.

Phillips, Fred Melville. Hydrology of the Bird's Nest Aquifer, Uintah County, Utah. M, 1979, University of Arizona. 165 p.

Phillips, Fred Melville. Noble gases in ground water as paleoclimatic indicators. D, 1981, University of Arizona. 204 p.

Phillips, Gerald R. A shallow resistivity survey of the north-central portion of Michigan's Upper Peninsula. M, 1974, Michigan Technological University.

Phillips, Harris Edwards. The geology of the Starr Mountain area, Southeast Tennessee. M, 1952, University of Tennessee, Knoxville. 62 p.

Phillips, Howard Cottrell. The petrology of cement-aggregate reaction in concrete. M, 1953, Virginia Polytechnic Institute and State University.

Phillips, Irvine Lewis. A study of the geology and soils of the Oceanside Quadrangle (California). M, 1940, University of California, Berkeley. 58 p.

Phillips, J. D. Statistical analysis of magnetic profiles and geomagnetic reversal sequences. D, 1975, Stanford University. 143 p.

Phillips, James Richard. "Middle" Cretaceous metasedimentary rocks of La Olvidada, northeastern Baja California, Mexico. M, 1984, San Diego State University. 107 p.

Phillips, John Asa. A grain size frequency distribution analysis of selected beach and dune sands of Dauphin Island, Alabama. D, 1973, University of Southern Mississippi.

Phillips, John Edward. Same aspects of phosphorus metabolism in lakes as studied by the use of radioactive phosphorus. M, 1957, Dalhousie University.

Phillips, John Stephen. Origin and significance of subsidence structures on carbonate rocks overlying Silurian evaporites in Onondaga County, New York. M, 1956, Syracuse University.

Phillips, John Stephen. Sandstone-type copper deposits of the Western United States. D, 1960, Harvard University.

Phillips, Johnnie O., Jr. Petrology of the late Proterozoic(?) - Early Cambrian Arumbera Sandstone and the late Proterozoic Quandong Conglomerate, east-central Amadeus Basin, central Australia. M, 1986, Utah State University. 403 p.

Phillips, Jonathan David. A spatial analysis of shoreline erosion, Delaware Bay, New Jersey. D, 1985, Rutgers, The State University, New Brunswick. 195 p.

Phillips, Jonathan Wilton. A gravity survey of the Glenn Pool area (Oklahoma). M, 1952, University of Tulsa.

Phillips, Joseph D. Geology of the southern portion of Sierra de Palomas, northwestern Chihuahua, Mexico. M, 1986, University of Texas at El Paso.

Phillips, Joseph Daniel. A paleomagnetic study of the Devonian Catskill Formation of central Pennsylvania. D, 1966, Princeton University. 340 p.

Phillips, Kenneth A. The mining geology of the Mt. Nebo District, Utah. M, 1940, Iowa State University of Science and Technology.

Phillips, Kent D. Stratigraphy, depositional environments and petroleum-reservoir potential of the Lyons Sandstone (Permian), east-central Colorado. M, 1983, Colorado School of Mines. 107 p.

Phillips, Leonard. Photographic studies of the interactions of bubbles streaming through silicone fluids. M, 1972, Brooklyn College (CUNY).

Phillips, Loren F. The effect of slope on experimental drainage patterns; possible application to Mars. M, 1986, Colorado State University. 128 p.

Phillips, Mark Paul. Geology of Tumamoc Hill, Sentinel Peak and vicinity, Pima County, Arizona. M, 1976, University of Arizona.

Phillips, Michael W. Structural and crystal chemical studies of the feldspars. D, 1972, Virginia Polytechnic Institute and State University.

Phillips, Ned H. The age of the Forest Hill-Red Bluff Formation in Mississippi (lower Oligocene is conclusion). M, 1951, Mississippi State University. 54 p.

Phillips, R. L. A dendrochronological interpretation of the pre-settlement and historic climate of St. Joseph, Missouri. D, 1976, University of Nebraska, Lincoln. 275 p.

Phillips, Richard Porter. Geophysical investigations in the Gulf of California. D, 1964, University of California, San Diego. 324 p.

Phillips, Robert Arthur. Effects of atmospheric acid deposition on water chemistry and benthic invertebrate distributions in some Northern Appalachian streams. D, 1988, SUNY at Buffalo.

Phillips, Robert Laurence. Compressional wave velocities to 2000 bars of basalts from the Columbia Plateau, Washington. M, 1980, Southern Methodist University. 71 p.

Phillips, Robert Lawrence. Structure and stratigraphy of the northern quarter of the Langlois Quadrangle, Oregon. M, 1968, University of Oregon. 91 p.

Phillips, Roberts Lawrence. Depositional environments of the Santa Margarita Formation in the Santa Cruz Mountains, California. D, 1981, University of California, Santa Cruz. 544 p.

Phillips, Roger Jay. The investigation of dipole radiation in the lunar environment. D, 1968, University of California, Berkeley. 291 p.

Phillips, Ross M. The general geology of part of the Priest Valley Quadrangle, California. M, 1938, University of California, Berkeley. 88 p.

Phillips, Sandra. Interpretation of reservoir morphology in the Guadalupian Cherry Canyon Formation, Indian Draw Field, Eddy County, New Mexico. M, 1981, Texas A&M University. 173 p.

Phillips, Sandra. Shelf sedimentation and depositional sequence stratigraphy of the Upper Cretaceous Woodbine-Eagle Ford groups, East Texas. D, 1987, Cornell University. 507 p.

Phillips, Sanford Ingels. Residues from the hydrofluoric acid treatment of Paleozoic shales. M, 1956, University of Illinois, Urbana. 40 p.

Phillips, Scott Harlan. Subsurface studies of the Cypress Formation, Richland County area, Illinois. M, 1954, University of Illinois, Urbana.

Phillips, Stanley Michael. A comparative analysis of Recent deep-sea sediments of the upper continental rise off the Eastern United States. M, 1973, Florida State University.

Phillips, Stephen E. Geology and economic mineral potential of the Klondike Mine area, Saguache County, Colorado. M, 1982, University of Texas at El Paso.

Phillips, Stephen T. Fan-delta sedimentation, Waltman Member of the Fort Union Formation, Wind River basin, Wyoming. M, 1981, University of Missouri, Columbia.

Phillips, Therese C. The structure and stratigraphy of the Grisryggen area, Wedel Jarlsberg Land, West Spitsbergen. M, 1986, University of Wisconsin-Madison. 154 p.

Phillips, Thomas L. The ordering of CR^{+3} in chlorite; one-dimensional projections and a three-dimensional refinement of the crystal structure of Cr-chlorite. M, 1978, University of Wisconsin-Madison.

Phillips, Tommy Lee. American species of *Botryopteris* from the Pennsylvanian. D, 1961, Washington University. 133 p.

Phillips, Victor Duzerah, III. Responses by alpine plants and soils to microtopography within sorted polygons. D, 1982, University of Colorado. 219 p.

Phillips, Walter T., Jr. Petrology of the Ahtell Creek Pluton, eastern Alaska Range. M, 1968, University of Alaska, Fairbanks. 77 p.

Phillips, Wayne Jude. Electrical conductivity structure of the San Andreas Fault in Central California. D, 1981, Cornell University. 119 p.

Phillips, Wesley M. The Michigamme Intrusion, a case study of the importance of remanent magnetization in magnetic interpretations, Marquette County, Michigan. M, 1979, Michigan State University. 82 p.

Phillips, William Morton. Structural geology and Cambrian stratigraphy of the Crown Creek-Bowen Lake area, Stevens County, Washington. M, 1979, Washington State University. 82 p.

Phillips, William Revell. A crystal chemical classification of the chlorite minerals. D, 1954, University of Utah. 132 p.

Phillips, William Revell. A projector to aid in the determination of theoretical structures. M, 1951, University of Utah. 56 p.

Phillips, William Scott. The separation of source, path and site effects on high frequency seismic waves; an analysis using coda wave techniques. D, 1985, Massachusetts Institute of Technology. 195 p.

Philpotts, A. R. Experimental study of reactions in a quartzite mica schist under high temperature and pressure. M, 1960, McGill University.

Philpotts, J. A. Experiments on thermochemical methods of producing high temperatures and pressures. M, 1961, McGill University.

Philpotts, John Aldwyn. The chemical compositions and origin of moldavites. D, 1965, Massachusetts Institute of Technology. 113 p.

Phinney, Robert A. Later refraction arrivals in layered liquids. M, 1959, Massachusetts Institute of Technology. 83 p.

Phinney, Robert Alden. I, Propagation of leaking modes in a plane seismic waveguide; II, Propagation of leaking interface waves. D, 1961, California Institute of Technology. 85 p.

Phinney, William C. Phase equilibrium in the metamorphic rocks of St. Paul Island and Cape Breton, Nova Scotia. D, 1959, Massachusetts Institute of Technology. 168 p.

Phinney, William C. Structural relationships around the southern extension of the Mabou Highlands, Inverness County, Cape Breton, Nova Scotia. M, 1956, Massachusetts Institute of Technology. 75 p.

Phippen, Peter D. Hydrogeologic assessment of the Lowell Quadrangle, Massachusetts and New Hampshire. M, 1981, Boston University. 185 p.

Phipps Morgan, William Jason. Dynamics of mid-ocean ridges. D, 1986, Brown University. 116 p.

Phipps, C. V. G. Petrology and environment of the alkaline rocks of the Blue Mountain area of Ontario. D, 1954, University of Toronto.

Phipps, C. V. G. The petrology and structure of the alkaline rocks of the Blue Mountain area of Ontario. D, 1955, University of Toronto.

Phipps, Donald. The geology of the unconsolidated sediments of Boston Harbor. M, 1964, Massachusetts Institute of Technology. 53 p.

Phipps, James B. Geology of the area north of New Castle (Garfield County), Colorado. M, 1961, University of Colorado.

Phipps, James Benjamin. Sediments and tectonics of the Gorda-Juan de Fuca Plate. D, 1974, Oregon State University. 118 p.

Phipps, Rodney T. Geology of the Monte Cristo mining area, California. M, 1951, California Institute of Technology. 36 p.

Phipps, Stephen Paul. Mesozoic ophiolitic olistostromes and Cenozoic imbricate thrust faulting in the northern California Coast Ranges; geology of the Mysterious Valley area, Napa County. D, 1984, Princeton University. 359 p.

Phipps, William Mason. The nature of organic matter in calcareous and non-calcareous soils. M, 1925, Cornell University.

Phlainen, John A. Outline of research for field studies of piles in permafrost. M, 1952, Purdue University.

Phleger, Fred B., Jr. Lichadian trilobites. D, 1936, Harvard University.

Phleger, Fred B., Jr. Notes on certain Ordovician faunas of the Inyo Mountains (California). M, 1932, California Institute of Technology. 68 p.

Phongbetchara, R. An application of infrared spectrophotometry for studies of the "structure" amorphous components of some Hawaiian soils. M, 1966, University of Hawaii. 46 p.

Phongprayoon, Pongsak. Subsurface geology (Paleozoic) of Wilbarger and Baylor counties, Texas. M, 1972, University of Texas, Austin.

Phuphatana, Amorn. The fluoride ion-specific electrode as applied to geochemical prospecting for fluorspar deposits. M, 1974, Kent State University, Kent. 57 p.

Phyfer, Daniel Wade. The geochemistry and petrology of the Day Book (Ordovician) (North Carolina) Dunite. M, 1968, University of South Carolina.

Pi-Sunyer, James. Stratigraphy of the Harbor well, Ashtabula City, Ohio. M, 1955, University of Michigan.

Piaggio, Arthur D. The Los Nogales Member of the Taraises Formation (Lower Cretaceous) near Monterrey, Nuevo Leon, Mexico. M, 1961, Tulane University. 82 p.

Piaggio, Arthur Donald. Sedimentary petrology of the Paleozoic carbonate sequence in the Nopah mountains, Inyo County, California. D, 1970, [Stanford University].

Piana, Margo K. La *see* La Piana, Margo K.

Piatt, David Allan. Pattern recognition of porphyry copper deposits in New Mexico and Texas. M, 1984, Texas A&M University. 66 p.

Piatt, Larry L. Geology of the Signal Mountain area, southeastern Pushmataha County, Oklahoma. M, 1962, University of Oklahoma. 79 p.

Piazza, Idelso Antonio. A modified Wiener filter approach to inverse filtering reflection seismograms. M, 1979, University of Texas, Austin.

Piburn, Michael D. Metamorphism and structure of the Villa de Cura Group (early Cretaceous), northern Venezuela. D, 1967, Princeton University. 206 p.

Picard, Meredith Dana. Paleomagnetic correlation of the unit within the Chugwater Formation, west-central Wyoming. D, 1963, Princeton University. 117 p.

Piccola, Larry J. Feasibility of developing sanitary landfills at abandoned surface mines in Georgia. M, 1972, Georgia Institute of Technology. 36 p.

Piccoli, Philip. Petrology and geochemistry of the Old Rag Granite on Old Rag Mountain. M, 1987, University of Pittsburgh.

Picha, Mark Gregory. Structure and stratigraphy of the Montezuma Salient-Hagan Basin area, Sandoval County, New Mexico. M, 1982, University of New Mexico. 248 p.

Pichardo, Fidencio Alejandro Nava *see* Nava Pichardo, Fidencio Alejandro

Piché, Mathieu. La Formation de Haüy à l'ouest de Chapais; volcanisme sub-aérien en milieu fluviatile. M, 1985, Universite du Quebec a Chicoutimi. 188 p.

Pichtel, John Robert. Influence of complex organic amendments on the oxidation of pyritic mine spoil. D, 1987, Ohio State University. 240 p.

Pichulo, R. O. Polymorphism and phase relations in FeS at high pressure. D, 1979, Columbia University, Teachers College. 178 p.

Pickard, Frank Robert. Bedrock geology of the Gorham area. M, 1963, Southern Illinois University, Carbondale. 70 p.

Pickart, David. Physical stratigraphy and geologic evolution of the Providence River, Narragansett Bay, Rhode Island. M, 1987, University of Rhode Island. 183 p.

Picken, Cyrus Seeley, Jr. Trace-metal distribution in the sediments of Rochester Harbor, New York and vicinity. M, 1975, Michigan State University. 147 p.

Pickens, Caroline M. The Dakota Formation, Black Mesa Basin, Arizona; a deltaic complex. M, 1974, Arizona State University. 130 p.

Pickens, Craig A. Facies analysis of an Early Cretaceous rudist biostrome, East Potrillo Mountains, Dona Ana County, New Mexico. M, 1986, University of Texas at El Paso.

Pickens, John Franklin. Field, theoretical and modeling studies of scale-dependent dispersivity in saturated granular geologic materials. D, 1982, University of Waterloo.

Pickens, Kathleen L. A study of the Hebgen Lake, Montana, area fault scarps and knickpoints twenty years after the 1959 earthquake. M, 1980, SUNY at Binghamton. 61 p.

Pickens, William Robert III. Carboniferous stratigraphy of the Jackson Ranch area, Lampasas County, Texas. M, 1959, University of Texas, Austin.

Pickering, Ann. Stratigraphy and depositional environment of bone beds at the North Vernon Limestone-New Albany Shale (Devonian) contact in southeastern Indiana. M, 1979, Indiana University, Bloomington. 117 p.

Pickering, Dennison John. A simple shear machine for soil. D, 1969, University of British Columbia.

Pickering, Ranard Jackson. An analysis of selected Indiana coals by the particle count method. M, 1952, Indiana University, Bloomington. 28 p.

Pickering, Ranard Jackson. Some solubility studies of alumina, ferric oxide, and silica and their relation to laterization. D, 1961, Stanford University. 123 p.

Pickering, Samuel Marion, Jr. Stratigraphy and paleontology of portions of Perry and Cochran quadrangles, Georgia. M, 1966, University of Tennessee, Knoxville. 89 p.

Pickering, Warren. A study of the insoluble residues of the Paleozoic limestones of the central Black Hills, South Dakota. M, 1941, University of Minnesota, Minneapolis. 59 p.

Pickett, George R. Theories of seismic wave propagation and pressure measurements in the vicinity of an explosion. D, 1955, Colorado School of Mines. 166 p.

Pickett, Jacob Wayne. A geological and geochemical study of the Skidder basalt and Skidder trondhjemites; and the geology, ore petrology and geochemistry of the Skidder Prospect and its accompanying alteration zone, Buchans area, central Newfoundland. M, 1988, Memorial University of Newfoundland. 543 p.

Pickett, Kendell. In-situ leaching studies of South Texas uranium ores. M, 1985, Texas A&M University. 243 p.

Pickett, Robert Lee. Numerical modeling investigation of cyclonic Gulf Stream eddies. D, 1971, College of William and Mary. 95 p.

Pickett, T. E. Stratigraphy of the Dan River Triassic basin in North Carolina. M, 1962, University of North Carolina, Chapel Hill.

Pickett, Thomas Ernest. The modern sediments of Pamlico Sound, North Carolina. D, 1965, University of North Carolina, Chapel Hill. 135 p.

Pickford, Peter John. Geology of the southeastern portion of the Richwoods Quadrangle, Missouri. M, 1952, University of Iowa. 79 p.

Picking, Larry Webb. A study of shales associated with landslides in southeastern Ohio. M, 1965, Ohio University, Athens. 57 p.

Pickle, John D. Dynamics of clastic sedimentation and watershed evolution within a low-relief karst drainage basin, Mammoth Cave region, Kentucky. M, 1985, University of New Mexico. 147 p.

Picklyk, Donald D. Overthrust faulting; finite element models of the initial stages in the development of the southern Canadian Rockies, a foreland thrust and fold belt. D, 1973, Queen's University. 161 p.

Pickthorn, William Joseph. Stable isotope and fluid inclusion study of the Port Valdez gold district, southern Alaska. M, 1982, University of California, Los Angeles. 66 p.

Pidcoe, William W., Jr. Investigation of the hydrologic parameters of a topographic basin in relation to the modeling of the associated groundwater system. M, 1977, University of Nevada. 109 p.

Piegat, James Jan. Glacial geology of central Nevada. M, 1980, Purdue University. 105 p.

Piegat, James Jan. Post-Laramide history of north central Wyoming. D, 1984, Purdue University. 121 p.

Piekarski, Leonard L. Relative age determination of Quaternary fault scarps along the southern Wasatch, Fish Springs, and House Ranges, Utah. M, 1979, Brigham Young University.

Piekenbrock, Joseph Robert. The structural and chemical evolution of phyllic alteration at North Silver Bell, Pima County, Arizona. M, 1983, University of Arizona. 95 p.

Piel, Kenneth Martin. Palynology of middle and late Tertiary sediments from the central interior of British Columbia, Canada. D, 1969, University of British Columbia.

Pien, Natalie Chen-Hsi. The effect of acid mine drainage on anaerobic respiration in fresh water lake sediments. M, 1981, University of Virginia. 81 p.

Pienaar, Petrus J. Stratigraphy and petrography of the Ventersdorp System in the Orange Free State goldfield, South Africa. M, 1956, Queen's University. 190 p.

Pienaar, Petrus J. Stratigraphy, petrography, and genesis of the Elliot Group, Blind River area, Ontario. D, 1959, Queen's University. 323 p.

Piepgras, Donald J. The isotopic composition of neodymium in different ocean masses. M, 1980, Oregon State University. 56 p.

Piepgras, Donald John. The isotopic composition of neodymium in the marine environment; investigations of the sources and transport of rare earth elements in the oceans. D, 1984, California Institute of Technology. 294 p.

Piepul, R. G. Analysis of jointing and faulting at the southern end of the Eastern border fault, Connecticut. M, 1975, University of Massachusetts. 109 p.

Pier, Katherine D. Cephalopods of Saint Joseph Island. M, 1937, Columbia University, Teachers College.

Pieracacos, Nicholas J. Conodont biostratigraphy of the Boyle Dolomite (Middle Devonian) of east-central Kentucky. M, 1983, Eastern Kentucky University. 112 p.

Pierce, Andrew. Post-glacial denudation of Mont Saint Hilaire, Quebec. M, 1971, McGill University.

Pierce, Arthur P. The geology of eastern La Veta Pass, Huerfano County, Colorado. M, 1953, Colorado School of Mines. 54 p.

Pierce, Carlos R. Geology of the southern Little Drum Mountains, Utah. M, 1974, Brigham Young University. 129 p.

Pierce, Douglas Stanley. Petrology of the Copper Boy Cu-Zn skarn deposit, Fremont County, Colorado. M, 1970, University of Michigan.

Pierce, Guy Russell. Foraminifera of the (Upper Cretaceous) Prairie Bluff Formation of Mississippi and Alabama. M, 1940, Northwestern University.

Pierce, Guy Russell. Foraminiferal zones of the Upper Cretaceous of Mississippi. D, 1942, University of Iowa. 128 p.

Pierce, Harold George. Diversity of late Cenozoic gastropods on the Southern High Plains. D, 1975, Texas Tech University. 267 p.

Pierce, Harold George. The Blanco Beds; mineralogy and paleoecology of an ancient playa. M, 1973, Texas Tech University. 93 p.

Pierce, Jack Warren. Recent stratigraphy and geologic history of the Core Banks region, North Carolina. D, 1964, University of Kansas. 134 p.

Pierce, Jack Warren. Structural history of the Uinta Mountains, Utah (P611). M, 1950, University of Illinois, Urbana. 71 p.

Pierce, Jack William. Salt water infiltration into the Alameda County water district. M, 1948, Stanford University. 48 p.

Pierce, Marcus Lacy. Kinetics of recovery and grain growth in hydrostatically annealed quartz aggregates. M, 1987, University of California, Los Angeles. 116 p.

Pierce, Richard H., Jr. A study of the mechanism of the adsorption of chlorinated hydrocarbons in marine sediments. D, 1973, University of Rhode Island. 109 p.

Pierce, Richard L. Mohnian foraminifera and fish from Benedict Canyon, Sherman Oaks, California. M, 1956, University of Southern California.

Pierce, Richard LeRoy. Early Upper Cretaceous plant microfossils of Minnesota. D, 1957, University of Minnesota, Minneapolis. 193 p.

Pierce, Robert Remsen. Application of a resource analysis technique to a subtropical coastal area, South Brevard County, Florida. D, 1982, Florida Institute of Technology. 241 p.

Pierce, Robert W. Coccoliths and related calcareous nannofossils from surficial bottom sediments of the Gulf of Mexico. D, 1975, Louisiana State University. 546 p.

Pierce, Robert William. Stratigraphy and paleontology of a portion of the Pogonip Group (Lower and Middle Ordovician), Arrow Canyon Range, Nevada. M, 1966, University of Illinois, Urbana.

Pierce, Robert William. Ultrastructure and biostratigraphy of the conodonts of the Monte Cristo group, Arrow Canyon range, Clark County, Nevada. D, 1969, University of Illinois, Urbana. 118 p.

Pierce, S. E. Provenance and paleoclimatology of the Mission Valley Formation, San Diego County, California. M, 1974, San Diego State University.

Pierce, Stephen Davis. Gulf Stream velocity structure through combined inversion of hydrographic and acoustic Doppler data. M, 1987, Massachusetts Institute of Technology. 65 p.

Pierce, Thomas A. Petrology of dolerite-metadolerite dikes of southeastern New England. M, 1976, University of Rhode Island.

Pierce, Walter. Geology and Pennsylvanian-Permian stratigraphy of Howard area, Fremont County, Colorado. M, 1969, Colorado School of Mines. 129 p.

Pierce, Walter H. Pennsylvanian and Lower Permian stratigraphy and history of part of the Virgin Mountains area, Northwest Arizona and Southeast Nevada. D, 1980, Colorado School of Mines. 394 p.

Pierce, William G. Rosebud Creek Coal Field, Rosebud and Custer counties, Montana. D, 1931, Princeton University. 107 p.

Pieri, D. C. Geomorphology of Martian valleys. D, 1979, Cornell University. 293 p.

Pieri, Robert Victor. Size effects in linear elastic fracture mechanics. D, 1987, Carnegie-Mellon University. 101 p.

Piermattei, Rodolfo. Dispersion of Rayleigh waves for purely oceanic paths in the Pacific. M, 1969, Columbia University. 46 p.

Pierson, Bernard J. Cyclic sedimentation, limestone diagenesis and dolomitization in upper Cenozoic carbonates of the southeastern Bahamas. D, 1982, University of Miami. 343 p.

Pierson, Bernard J. The control of cathodo-luminescence in dolomite by iron and manganese. M, 1977, University of Kentucky. 85 p.

Pierson, Beverly A. Morphology and grain size patterns of a small overbank deposit of the Brandywine Creek, Pennsylvania. M, 1988, University of Delaware. 148 p.

Pierson, Frederick Barker, Jr. Spatial variability of aggregate stability in the Palouse region of Washington. D, 1988, Washington State University. 119 p.

Pierson, John R. Petrology and petrogenesis of the Bokan granite complex, Alaska. M, 1980, Colorado State University. 97 p.

Pierson, Lowell Craig. Depositional environments of the Shinarump and Monitor Butte members of the Chinle Formation in the White Canyon area of southern Utah. M, 1984, Northern Arizona University. 106 p.

Pierson, Milton Lee. The limits of solubility between jadeite and enstatite at 25 kbar pressure. M, 1987, Iowa State University of Science and Technology. 123 p.

Pierson, Richard Edwin. Possible stratigraphic relationships of the (Eocene) Sandersville Limestone to the Ocala Limestone of West Georgia. M, 1951, Emory University. 98 p.

Pierson, Thomas Charles. Factors controlling debris-flow initiation on forested hillslopes in the Oregon Coast Range. D, 1977, University of Washington. 166 p.

Pierson, Thomas Charles. Petrologic and tectonic relationships of Cretaceous sandstones in the Harts Pass area, North Cascade Mountains, Washington. M, 1972, University of Washington. 37 p.

Pierson, William R. A geophysical study of the contact between the greenstone-granite terrain and the gneiss terrain in central Minnesota. M, 1984, University of Wyoming. 84 p.

Pietenpol, David John. Structure and ore deposits of the Silver District, La Paz County, Arizona. M, 1983, University of Arizona. 67 p.

Pieters, Carle Ellen. Characterization and distribution of lunar mare basalts types using remote sensing techniques. D, 1977, Massachusetts Institute of Technology. 348 p.

Pieters, Carle Ellen. Wavelength dependence of the polarization of light reflected from a particulate surface in the spectral region of a transition metal absorption band. M, 1972, Massachusetts Institute of Technology. 85 p.

Pietrobon, Vincent J., Jr. A geologic and magnetic investigation of the Challis Volcanics, MacKay Quadrangle 2NW and 2SW, Standhope Peak Quadrangle, and Harry Canyon Quadrangle, Custer County, Idaho. M, 1978, Lehigh University. 92 p.

Pietschker, Harold L. Saratoga Ostracoda from the type locality in southwestern Arkansas. M, 1952, University of Oklahoma. 90 p.

Piette, Carl R. Geology of Duck Creek ridges, east-central Wisconsin. M, 1963, University of Wisconsin-Madison.

Piette, Marjorie A. Heavy metals investigation of ground water near several landfills and sewage lagoons in New Hampshire. M, 1982, University of New Hampshire. 91 p.

Piette, Robyn A. Geology and petrology of the Lake Juliette region, East Juliette Quadrangle, Georgia Piedmont. M, 1988, Georgia State University. 71 p.

Piety, William Duncan. Surface sediment facies and physiography of a recent tidal delta, Brown Cedar Cut, central Texas coast. M, 1972, University of Houston.

Pigage, L. C. Metamorphism and deformation on the north-east margin of the Shuswap metamorphic complex, Azure Lake, British Columbia. D, 1979, University of British Columbia.

Pigage, Lee Case. Metamorphism southwest of Yale, British Columbia. M, 1976, University of British Columbia.

Pigg, John H., Jr. The lower Tertiary sedimentary rocks in the Pilot Rock and Heppner areas, Oregon. M, 1961, University of Oregon. 67 p.

Pigg, Kathleen Belle. Anatomically preserved Glossopteris and Dicroidium from the Transantarctic Mountains. D, 1988, Ohio State University. 258 p.

Piggott, Guido M., Jr. Some experiments on the replacement theory for the migration of oil and the interpretation of their results. M, 1941, University of Tulsa. 55 p.

Pignolet, Susanne. Rb-Sr geochronology of the Honey Hill Fault area, eastern Connecticut. M, 1981, Miami University (Ohio). 71 p.

Pignolet-Brandom, Susanne. Mineralogy, paragenesis and electron microprobe analysis of the cobalt-nickel and nickel minerals in the mississippi valley-type deposits of Southeast Missouri. D, 1988, University of Missouri, Rolla. 245 p.

Pigott, John D. Interstitial water chemistry of Jamaican reef sediments; I, Early diagenesis of dissolved sulfur and nitrogen species (determined by in situ sampling); II, Redox model for submarine cementation. M, 1977, University of Texas, Austin.

Pigott, John Dowling. Global tectonic control of secular variations in Phanerozoic sedimentary rock/ocean/atmosphere chemistry. D, 1981, Northwestern University. 211 p.

Pike, Dale P. The geology of the Montauban area, P.Q. (Canada). D, 1967, McGill University.

Pike, Jane E. N. Intrusions and intrusive complexes in the San Luis Obispo Ophiolite; a chemical and petrologic study. D, 1974, Stanford University. 228 p.

Pike, Jane Ellen Nielson. Petrology and genesis of the Webster-Addie Ultramafic body (late Paleozoic; either Pennsylvanian or Permian), Jackson County, North Carolina. M, 1968, University of Michigan.

Pike, John David. Feldspar diagenesis in the Yowlumne Sandstone, Kern County, California. M, 1981, Texas A&M University. 132 p.

Pike, Richard Joseph, Jr. Meteoritic origin and consequent endogenic modification of large lunar craters; a study in analytical geomorphology. D, 1968, University of Michigan. 440 p.

Pike, Ruthven W. Geological studies in the Bryson and Cape San Martin quadrangles, Monterey County, California. M, 1925, Stanford University. 31 p.

Pike, Ruthven W. The geology of a portion of the Crystal City Quadrangle, Missouri. D, 1928, University of Chicago. 277 p.

Pike, Stanley F. Computer simulation of configuration and sealevel changes in Florida Bay. M, 1987, Wichita State University. 160 p.

Pike, Stephen Joseph. Piezomagnetic stress sensitivity measurements of rocks from the vicinity of the San Andreas Fault. D, 1981, University of Southern California.

Pike, Stewart J. Contribution on the origin and occurrence of the Little Falls or Herkimer quartz "diamonds" (New York). M, 1949, Syracuse University.

Pike, Thomas Mace. Structural geometry of the Tully body of Monson gneiss (age uncertain - Precambrian or Cambrian-Ordovician, Massachusetts). M, 1969, University of Massachusetts. 48 p.

Pike, W. S., Jr. Correlation of the Upper Cretaceous between McCarty and Alamosa Creek, New Mexico. D, 1933, The Johns Hopkins University.

Pikul, Joseph Lawrence, Jr. Simulation and field validation of heat and water flow during soil freezing and thawing. D, 1988, Oregon State University. 170 p.

Pikul, Mary Frances. Numerical studies of linked soil-moisture and groundwater systems. D, 1973, Stanford University. 92 p.

Pilatzke, Richard H. Geology of the western portion of the Dogtooth Pluton, Lake of the Woods region, southwestern Ontario. M, 1976, University of North Dakota. 81 p.

Pilch, Peter George Henry. Dispersion of some trace elements in soils of glacial origin near the Armstrong "A" sulphide deposit (Bathurst area, New Brunswick, Canada). M, 1970, University of New Brunswick.

Pilcher, Benjamin Luther, Jr. The Paleozoic formations north of the San Saba River in Mason County, Texas. M, 1931, University of Texas, Austin.

Pilcher, Stephen H. Geology and geochemistry of the Orphan Boy Mine, Park County, Colorado. D, 1968, Colorado School of Mines.

Pilcher, Stephen H. Rock alteration and vein mineralization at the Buffalo Mine, Grant County, Oregon. M, 1959, Oregon State University. 83 p.

Piles, Charles Foster. A heavy mineral study of the Hartshorne Sandstone in the Arkansas Valley; Section two. M, 1955, University of Arkansas, Fayetteville.

Pilger, Rex H., Jr. Structural geology of part of the northern Toano Range, Elko County, Nevada. M, 1972, University of Nebraska, Lincoln.

Pilger, Rex Herbert, Jr. Structure of Santa Cruz-Catalina Ridge and adjacent areas, Southern California continental borderland from reflection and magnetic profiling; implications for late Cenozoic tectonics of Southern California. D, 1976, University of Southern California.

Pilgrim, Alan Thomas. Spatial variability of hydrologic response on naturally vegetated hillslopes in a semi-arid environment. D, 1981, University of Oklahoma. 190 p.

Pilkey, Orrin H. The effect of environment on the concentration of magnesium and strontium in certain Recent echinoid tests. M, 1959, University of Montana. 48 p.

Pilkey, Orrin H. Trace elements in recent and fossil mollusk shells. D, 1963, Florida State University. 131 p.

Pilkington, Edgar M. The gravels of the (Pennsylvanian) Pottsville of western Kentucky. M, 1926, Vanderbilt University.

Pilkington, Harold Dean and Pillmore, Charles L. Petrography and petrology of a part of Mount Sopris Stock, Pitkin County, Colorado. M, 1954, University of Colorado.

Pilkington, Harold Dean. Structure and petrology of a part of the east flank of the Santa Catalina Mountains, Pima County, Arizona. D, 1962, University of Arizona. 155 p.

Pilkington, Mark. Determination of crustal interface topography from potential fields. D, 1985, McGill University. 216 p.

Pilkinton, Edgar. The gravels of the Pottsville of western Kentucky. M, 1926, Vanderbilt University.

Pillars, William Wynn. The crystal structure of a beta eucryptite as a function of temperature. M, 1973, University of Michigan.

Pillay, K. K. Sivasankara. Characteristic X-rays from (n,γ) products and their utilization in activation analysis. D, 1965, Pennsylvania State University, University Park. 117 p.

Pillmore, Charles L. and Pilkington, Harold Dean. Petrography and petrology of a part of Mount Sopris Stock, Pitkin County, Colorado. M, 1954, University of Colorado.

Pillmore, Kathryn A. Geology and mineralization of the Paradise Pass area, Pitkin County, Colorado. M, 1984, Colorado School of Mines. 120 p.

Pillsbury, Norman H. A system for landslide evaluation on igneous terrane. D, 1976, Colorado State University. 122 p.

Pillsbury, Stephen W. Ultra-structure, strength, geometry and flow rate of Nautilus siphuncular tube and its paleobiological significance. M, 1983, Brooklyn College (CUNY).

Pilny, J. Paleoecology of two late Pleistocene sites in southern Ontario. M, 1985, University of Waterloo. 227 p.

Pilote, Pierre. Stratigraphie et significations des minéralisations dans le secteur du Mont Bourbeau, Canton de McKenzie, Chibougamau. M, 1987, Universite du Quebec a Chicoutimi. 167 p.

Pilskalin, Cynthia Hughes. The fecal pellet fraction of oceanic particle flux. D, 1985, Harvard University. 262 p.

Pilson, Michael L. California coastal Monterey relationships. M, 1975, University of Washington. 24 p.

Pimentel, Nelly R. Potassium-argon age determination of a mica-peridotite dike in western Pennsylvania. M, 1971, University of Pittsburgh.

Pimentel-Klose, Mario Rafael. Neodymium isotopic studies of Precambrian banded iron formations. D, 1986, Harvard University. 293 p.

Pina, Jon J. The effect of limestone as a coal slagging flux for use in high pressure coal gasification. M, 1978, Indiana University of Pennsylvania. 68 p.

Pinault, C. Thomas. Structure, tectonic geomorphology and neotectonics of the Elsinore fault zone between Banner Canyon and the Coyote Mountains, Southern California. M, 1984, San Diego State University. 231 p.

Pinch, James Jeffrey. Sedimentology and stratigraphy of Wisconsinan deposits in the McKittrick site and Beaver River gorge, Clarksburg, Ontario. M, 1979, University of Western Ontario. 103 p.

Pincha, Pamela M. Limestones in the Nicola Group; Iron Mountain, Merritt, British Columbia. M, 1982, Western Washington University. 154 p.

Pinchin, E. A study of the deconvolution of reflection seismograms. M, 1967, University of Toronto.

Pinchock, John. The late-stage accessory minerals of the Pilot Knob iron orebody, Iron County, Southeast Missouri. M, 1974, Southern Illinois University, Carbondale. 61 p.

Pinckney, Darrell Mayne. Structure in the Precambrian Coal Creek Series, Coal Creek Canyon (Jefferson County), Colorado. M, 1953, University of Colorado.

Pinckney, Darrell Mayne. Veins in the northern part of the Boulder Batholith, Montana. D, 1965, Princeton University. 208 p.

Pinckney, Linda Ruth. Five- and seven-membered single chain pyroxenoids. D, 1982, Harvard University. 314 p.

Pincomb, Arthur Chesney. A correlation of remote sensing, biogeochemical, and geochemical techniques in prospecting for ore deposits west of Butte, Montana. M, 1975, University of Montana. 91 p.

Pincus, Howard J. Statistical methods applied to the study of rock fractures; quantitative comparative analysis of fractures in gneisses and overlying sedimentary rocks of northern New Jersey. D, 1951, Columbia University, Teachers College.

Pincus, Howard J. Statistical techniques applied to the study of rock fractures. M, 1948, Columbia University, Teachers College.

Pincus, Scott D. Composition of Holocene sand derived from the Peninsular Ranges, Southern California; implications for provenance-discrimination diagrams and Cretaceous tectonism. M, 1986, San Diego State University. 85 p.

Pincus, William J. An evaluation of drainage sediments and spring waters as geochemical prospecting media in arid environments. M, 1982, Colorado School of Mines. 131 p.

Pindell, James Lawrence. Permo-Triassic reconstruction of western Pangea and the evolution of the Gulf of Mexico-Caribbean region. M, 1981, SUNY at Albany. 121 p.

Pinder, George Francis. A numerical technique for aquifer evaluation. D, 1968, University of Illinois, Urbana. 83 p.

Pine, Clyde A. Subsurface structure and stratigraphy of the Lake Arthur Field, Jefferson Davis Parish, Louisiana. M, 1963, Tulane University. 21 p.

Pine, F. W. The relationship between land use characteristics and the trace metal concentration in sub-

merged vascular plants. D, 1980, The Johns Hopkins University. 116 p.

Pine, Gordon L. Sedimentation studies in the vicinity of Willcox Playa, Cochise County, Arizona. M, 1963, University of Arizona.

Pine, Gordon Leroy. Devonian stratigraphy and paleogeography in Gila, Graham, Greenlee, and Pinal counties, Arizona. D, 1968, University of Arizona. 361 p.

Pine, Keith A. Glacial geology of the Tonasket - Spectacle Lake area, Okanogan County, Washington. M, 1985, Western Washington University. 130 p.

Pinel, Mark J. Stratigraphy of the upper Carlile and lower Niobrara formations (Upper Cretaceous), Fremont and Pueblo counties, Colorado. M, 1977, Colorado School of Mines. 111 p.

Pineo, Charles C. Conceptual study for maximizing the information from analysis of drill hole samples. M, 1973, Stanford University.

Pinero, Edwin. The depositional history of the Midway-Wilcox section, New Olm Field, Austin County, Texas. M, 1982, Texas A&M University. 84 p.

Piñero, Joanne Louise. Characteristics and development of the Rojo Caballos detachment slide, Pecos County, Texas, and its comparison to Heart Mountain, Park County, Wyoming. M, 1985, University of Texas, Arlington. 125 p.

Pines, Philip Jacques. Model seismic studies of intensely scattering media. M, 1974, Massachusetts Institute of Technology. 85 p.

Pinet, Paul Raymond. Petrology of the upper division of the Beacon sandstone (Devonian-Triassic), Dry Valley region, south Victoria Land (Antarctica). M, 1970, University of Massachusetts. 113 p.

Pinet, Paul Raymond. Structural development of the northern continental margin of Honduras and the adjacent sea floor, northwestern Caribbean Sea. D, 1972, University of Rhode Island. 123 p.

Pinette, Steven R. Stratigraphy, structure, metamorphism and volcanic petrology of islands in East Penobscot Bay, Maine. M, 1983, University of Maine. 141 p.

Ping, Kuo. Preparation of active carbon from Pennsylvania coals. D, 1940, Pennsylvania State University, University Park. 84 p.

Ping, Russell Gordon. Stratigraphic and structural relationships of the Stoney Fork Member of the Breathitt Formation in southeastern Kentucky. M, 1978, University of Kentucky. 125 p.

Pingarron, Luis Alberto Barba *see* Barba Pingarron, Luis Alberto

Pingitore, Nicholas Elias, Jr. Diagenesis and porosity modification in Acropora palmata, Pleistocene of Barbados, West Indies. M, 1968, Brown University.

Pingitore, Nicholas Elias, Jr. Vadose and phreatic diagenesis; processes, products and their recognition, Pleistocene of Barbados, West Indies. D, 1973, Brown University.

Piniazkiewicz, Robert J. Geology of and exploration techniques for pre-Beltian talc deposits on the Malesich Range, Ruby Range, Madison County, Montana. M, 1984, University of Arizona. 101 p.

Pinker, Robert. The sedimentary petrology of the Logan Formation (Mississippian) in Licking County, Ohio. M, 1970, Ohio State University.

Pinkerton, James B. Structural geology of the south-central part of the Vallecitos, San Benito County, California. M, 1967, San Jose State University. 133 p.

Pinkerton, Roger Parrish. Rayleigh wave model of crustal structure of northeastern Mexico. M, 1978, Texas Tech University. 50 p.

Pinkley, Victoria Elizabeth. A paleoecological investigation of Pitkin lime mud mounds, Durham, Arkansas. M, 1982, University of Arkansas, Fayetteville.

Pinkus, Joel R. Hydrogeologic assessment of three solid waste disposal sites in the Brazos River alluvial deposits. M, 1987, Baylor University. 157 p.

Pinnell, Michael Lu. Geology of the Thistle Quadrangle (Utah). M, 1972, Brigham Young University. 130 p.

Pinney, Reese Bruner. Emplacement of the Palisades detachment, Idaho-Wyoming thrust belt. M, 1984, University of New Orleans. 62 p.

Pinney, Robert I. The stratigraphy and paleontology of the Meppen Formation in western Illinois and eastern Missouri. M, 1962, Ohio University, Athens. 106 p.

Pinney, Robert Ivan. A preliminary survey of Mississippian biostratigraphy (Conodonts) in the Oquirrh Basin of central Utah. D, 1965, University of Wisconsin-Madison. 199 p.

Pino, Henry. Reservoir characterization; a strategy for history matching studies of petroleum reservoirs. M, 1988, University of Oklahoma. 181 p.

Pinsak, Arthur Peter. A regional chemical mineralogical study of surficial sediments in the Gulf of Mexico. D, 1958, Indiana University, Bloomington. 63 p.

Pinsak, Arthur Peter. High-silica sand potentialities of the Ohio River Formation. M, 1953, Indiana University, Bloomington. 68 p.

Pinsent, Robert Hugh. Precambrian geology along the Fraser River between Mount Robson and Tete Jaune Cache, British Columbia. M, 1971, University of Alberta. 146 p.

Pinsof, John David. The Pleistocene vertebrate fauna of South Dakota. M, 1986, South Dakota School of Mines & Technology.

Pinson, William Hamet, Jr. Geology of Polk County, Georgia. M, 1949, Emory University. 178 p.

Pinson, William Hamet, Jr. Trace elements in meteorites and rocks and the origin of meteorites. D, 1952, Massachusetts Institute of Technology. 174 p.

Pinsonnault, François. Granulats á hautes performances; application aux cas de dix-sept exploitations québécoises de matériaux granulaires. M, 1988, Universite Laval. 129 p.

Pinta, James, Jr. A model for the origin of calc-alkaline andesites at Crater Lake, Oregon. D, 1981, Pennsylvania State University, University Park. 225 p.

Pinta, James, Jr. A reconnaissance geochemical survey in parts of Oneida, Price, and Vilas counties, Wisconsin. M, 1975, University of Wisconsin-Milwaukee.

Pintado, Galo Yanez see Yanez Pintado, Galo

Pinter, Thomas J. The geology of an area from Whiskey Mountain to Moon Lake, Fremont County, Wyoming. M, 1959, Miami University (Ohio). 47 p.

Pinto-Auso, Montserrat. Geochemistry, petrology, and mineralogy of the pelagic/hemipelagic sequence overlying the Josephine Ophiolite, Klamath Mountains, California. M, 1984, University of Utah. 109 p.

Piombino, Joseph J. Depositional environments and petrology of the Fort Union Formation near Livingston, Montana; an evaluation as a host for sandstone type uranium mineralization. M, 1979, University of Montana. 84 p.

Piotrowicz, S. R. Studies of the sea to air transport of trace metals in Narragansett Bay. D, 1977, University of Rhode Island. 182 p.

Piotrowski, J. The Woodstock drumlin field, southern Ontario; Quaternary geology, paleogeomorphology and mechanisms of formation. M, 1985, University of Waterloo. 165 p.

Piotrowski, Joseph Martin. Part I - Melting relations in selected undersaturated alkaline rocks; part II - Phase relations in the system NaAlSi$_3$O$_8$ (albite) LiAlSi$_2$O$_6$ (B-spodumene) - SiO$_2$ (silica) - H$_2$O at 2,070 bars P(H$_2$O). D, 1967, University of Western Ontario. 199 p.

Piotrowski, Robert G. Carbonate sedimentation on U. S. Atlantic and Eastern Gulf of Mexico beaches. M, 1974, Duke University. 102 p.

Piotruszczewicz, Michael. An analysis of a drill core from western Clark County, Wisconsin. M, 1978, University of Wisconsin-Milwaukee.

Pipentacos, John. Analysis of gravity anomalies, Terre Haute, Indiana. M, 1984, Indiana State University. 61 p.

Piper, Arthur Maine. Paragenesis of some primary ores from the Rex and Success mines, Coeur d'Alene District, Idaho. M, 1925, University of Idaho. 58 p.

Piper, David Zink. Distributions of several trace elements in the water and sediments of a Norwegian anoxic fjord. D, 1969, University of California, San Diego. 125 p.

Piper, David Zink. Lithostratigraphy of the Brazeau Formation (Upper Cretaceous), Cripple Creek, Alberta, Canada. M, 1963, Syracuse University.

Piper, Kenneth Allen. Earthquake locations and seismic velocity structure in the Los Angeles area, Southern California. D, 1985, University of Southern California.

Piper, Larry Dean. The effect of gas column thickness on primary oil recovery from a horizontal reservoir. D, 1984, Texas A&M University. 113 p.

Pipes, Jeffrey W. The hydrogeologic framework of the Paleozoic aquifers of the Papio Natural Resources District, Nebraska. M, 1987, University of Nebraska, Lincoln. 106 p.

Pipes, William Vaughn. Recent carbonate sedimentary environments and sedimentation, Utila Island, Bay Islands, Honduras. M, 1978, Dartmouth College. 165 p.

Pipiringos, George Nicholas. Stratigraphy of the Sundance Formation, the Nugget (?) Sandstone and the Jelm Formation (restricted), in Laramie Basin, Wyoming. M, 1948, University of Wyoming. 143 p.

Pipiringos, George Nicholas. Uranium-bearing coal in the central part of the Great Divide Basin, Sweetwater County, Wyoming. D, 1956, The Johns Hopkins University.

Pipkin, Bernard W. Geology of the south third of the Green Springs Quadrangle, Nevada. M, 1956, University of Southern California.

Pipkin, Bernard Wallace. Clay mineralogy of the Willcox Playa and its drainage basin, Cochise County, Arizona. D, 1965, University of Arizona. 178 p.

Pippen, Fred M., Jr. Proposed mechanism for beach cusp formations. M, 1974, Northwestern State University. 36 p.

Pirc, Simon. Uranium and other elements in the Catskill Formation of East-central Pennsylvania. D, 1979, Pennsylvania State University, University Park. 321 p.

Pires, Fernando Roberto Mendes. Structural geology and stratigraphy at the junction of the Curral Anticline and the Modea Syncline, Quadrilatero Ferrifero, Minas Gerais, Brazil. D, 1979, Michigan Technological University. 260 p.

Pires, Jose Antunes. Stochastic analysis of seismic safety against liquefaction. D, 1983, University of Illinois, Urbana. 149 p.

Pirie, Ian David. Lithogeochemical dispersion in the area of the Norbec Deposit, Noranda, Quebec. M, 1980, Queen's University. 189 p.

Pirie, James. Mobilization in a migmatite from the Grenville Province, Quebec. D, 1971, Queen's University. 189 p.

Pirie, Robert Gordon. Petrology and physical-chemical environment of bottom sediments of the Riviere Bonaventure-Chaleur Bay area, Quebec, Canada. D, 1963, Indiana University, Bloomington. 182 p.

Pirie, Robert Gordon. The general geology of the upper Cobalt Group, Bruce Mines District, Ontario, Canada. M, 1961, Indiana University, Bloomington. 117 p.

Pirkel, Fredric Lee. Evaluation of possible source regions of Trail Ridge sands. M, 1974, University of Florida. 58 p.

Pirkle, Earl Conley, Jr. Pebble phosphate of Alachua County, Florida. D, 1956, University of Cincinnati. 203 p.

Pirkle, Earl Conley, Jr. The penetration of gamma rays through various soil materials. M, 1947, Emory University. 32 p.

Pirkle, Frederic Lee. Characterization of variation in grain size and heavy mineral content of Trail Ridge sediments. D, 1977, Pennsylvania State University, University Park. 245 p.

Pirkle, William Arthur. The offset course of the Saint Johns River,. M, 1970, University of North Carolina, Chapel Hill. 46 p.

Pirkle, William Arthur. Trail Ridge, a relict shoreline feature of Florida and Georgia. D, 1972, University of North Carolina, Chapel Hill. 90 p.

Piroshco, Darwin W. Relationship of hydrothermal alteration to structure and stratigraphy at the Coniaurum and Davidson Tisdale gold deposits, Tisdale Township, northeastern Ontario. M, 1985, Queen's University. 102 p.

Pirson, Sylvain J. Pt. I, Effect of pressure during consolidation on the elasticity of sedimentary rocks; Pt. II, Forms of decline curves extrapolated by the loss ratio method. M, 1930, University of Pittsburgh.

Pirson, Sylvain J. Study of an adjustable wave filter suitable for the reception of reflected seismic waves. D, 1931, Colorado School of Mines. 36 p.

Pirtle, George William. The geology of Jessamine County accompanied by a geologic map of the southern portion. M, 1925, University of Kentucky. 61 p.

Pisasale, Eugene T. Surface and subsurface depositional systems in the Escondido Formation, Rio Grande Embayment, South Texas. M, 1980, University of Texas, Austin.

Pischke, Gary Michael. Paleomagnetic study of the Neogene tectonic history of Baja California, Mexico. M, 1979, San Diego State University.

Pisciotto, Kenneth A. Petrographic study of the Lyre Formation and equivalent strata along the northern Olympic Peninsula, Washington. M, 1972, Stanford University.

Pisciotto, Kenneth Anthony. Basinal sedimentary facies and diagenetic aspects of the Monterey Shale, California. D, 1978, University of California, Santa Cruz. 468 p.

Pish, Timothy A. Palynology and paleoecology of the Cambria Coal, Weston County, Wyoming. M, 1988, South Dakota School of Mines & Technology.

Pisias, N. G. Model of late Pleistocene-Holocene variations in rate of sediment accumulation; Panama Basin, eastern equatorial Pacific. M, 1974, Oregon State University. 77 p.

Pisias, N. G. Paleoceanography of the Santa Barbara Basin and the California Current during the last 8000 years. D, 1978, University of Rhode Island. 236 p.

Piskin, Kemal. Ground water investigation in the southern half of Franklin County, Missouri. M, 1962, University of Missouri, Rolla.

Piskin, Rauf. Hydrogeology of the University of Nebraska field laboratory at Mead, Nebraska. D, 1972, University of Nebraska, Lincoln.

Piskin, Rauf. Non-destructive measurement of bulk density in marine sediment cores by gamma radiation and calculation of water content. M, 1968, University of Illinois, Urbana.

Pissarides, Andreas Savva. The interaction of phosphate with hydroxy-aluminum interlayers of montmorillonite. D, 1969, University of Saskatchewan. 161 p.

Pistek, Pavel. Conductivity of the ocean crust. D, 1977, University of California, San Diego. 116 p.

Pisutha-Arnond, Visut. Sulfur isotope study of sulfate and sulfide minerals in some hydrothermal ore deposits. M, 1978, Pennsylvania State University, University Park. 122 p.

Pisutha-Arnond, Visut. Thermal history, chemical and isotopic compositions of the ore-forming fluids responsible for the Kuroko massive sulfide deposits in the Hokuroku District of Japan. D, 1982, Pennsylvania State University, University Park. 168 p.

Pita, Frank. Zinc, lead and cadmium distribution and mode of occurrence in Oklahoma Reservoir sediments. M, 1972, University of Tulsa. 72 p.

Pitakpaivan, Kasana. Ecology and distribution of Recent foraminifera and ostracodes from Baffin Bay, Texas. M, 1988, Texas A&I University. 174 p.

Pitard, Alden McLellan. Geology of the Mazama area, Methow Valley, Washington. M, 1958, University of Washington. 61 p.

Pitcher, Grant Grow. Geology of the Jordan Narrows Quadrangle, Utah. M, 1957, Brigham Young University. 47 p.

Pitcher, Jacob J. A study of the mineralogy and petrology of White Mountain, Park County, Wyoming. M, 1972, Wayne State University.

Pitcher, Max Grow. Evolution of Chazyan (Ordovician) reefs of Eastern United States and Canada. D, 1964, Columbia University, Teachers College. 112 p.

Pitcher, Max Grow. Fusulinids of the Cache Creek Group (Permian), Stikine River area, Cassiar District, British Columbia, Canada. M, 1960, Brigham Young University. 64 p.

Pitkin, James A. Geology of the Palmer Quadrangle, Ellis County, Texas. M, 1959, Southern Methodist University. 23 p.

Pitlick, John. The effect of a major sediment influx on Fall River, Colorado. M, 1985, Colorado State University. 138 p.

Pitlick, John Charles. The response of coarse-beds rivers to large floods in California and Colorado. D, 1988, Colorado State University. 147 p.

Pitman, Walter Clarkson, III. Magnetic anomalies in the Pacific and ocean floor spreading. D, 1967, Columbia University. 82 p.

Pitrat, Charles W. The coral fauna of a biostrome in the Cedar Valley Limestone (Devonian) of Iowa. M, 1951, University of Wisconsin-Madison.

Pitrat, Charles William. The correlation of the (Mississippian) Madison Limestone by the thermoluminescence method. D, 1953, University of Wisconsin-Madison.

Pitt, John Michael. Deformation restraint and the mechanics of soil behavior. D, 1981, Iowa State University of Science and Technology. 204 p.

Pitt, Robert Ervin. Small storm urban flow and particulate washoff contributions to outfall discharges. D, 1987, University of Wisconsin-Madison. 539 p.

Pitt, William D. The Lonsdale Limestone. M, 1950, Northwestern University.

Pitt, William D. The stratigraphy and structure of portions of the Ouachita Mountains of Oklahoma and Arkansas. D, 1955, University of Wisconsin-Madison.

Pittenger, Gary C. Geochemistry, geothermometry and mineralogy of Cu, Pb, Zn and Sb deposits, Sevier County, Arkansas. M, 1974, University of Arkansas, Fayetteville.

Pittinger, Lyndon Frank. Potential to increase U.S. oil and gas production in an emergency. M, 1981, Stanford University. 116 p.

Pittman, Edward Dale. Geology of the northwest portion of the Blanco Mountain Quadrangle, California. M, 1958, University of California, Los Angeles.

Pittman, Edward Dale. Plagioclase feldspar as an indicator of provenance in sedimentary rocks. D, 1962, University of California, Los Angeles.

Pittman, Franklin T. A sedimentary study of the limestone of Jasper County, Mississippi. M, 1958, Mississippi State University. 23 p.

Pittman, Gardner M. Pennsylvanian stratigraphy of Tecolate area, New Mexico. M, 1951, Texas Tech University. 97 p.

Pittman, Kate L. Rock quality as a guide to fragmentation. M, 1984, University of Nevada. 89 p.

Pittman, William Gene. The reinterpretation of the structure and stratigraphy of the sedimentary rocks. M, 1951, University of Arkansas, Fayetteville.

Pitts, Gerald Stephen. Interpretation of gravity measurements made in the Cascade Mountains and adjoining Basin and Range Province in central Oregon. M, 1979, Oregon State University. 186 p.

Pitts, P. D. Temperature relations at Oswaldo Mine, Santa Rita, New Mexico. M, 1949, University of Toronto.

Pitz, Charles Forrest. A remote sensing and structural analysis of the Trans-Idaho discontinuity near Lowell, Idaho. M, 1985, Washington State University. 130 p.

Pitzak, A. N. A study of the relationship of sulfur content to color reversal in natural hackmanite. M, 1976, Wayne State University.

Pitzer, Carroll D. The occurrence of the Moody's Ranch Formation at four surface localities between the Red River and the Brazos River. M, 1960, Texas A&M University. 56 p.

Pitzrick, Raymond August. The climatic factor as it affects the occurrence, discovery, and development of mineral deposits. M, 1943, Cornell University.

Pivetz, Bruce E. Suspended sediment and bacterial transport in the east branch, Westport River, Massachusetts. M, 1986, Boston University. 229 p.

Pivnik, David. Compositional and sedimentological trends; Late Cretaceous Little Muddy Creek and Sphinx conglomerates as signatures of timing and style of thrust-related deformation. M, 1988, University of Rochester. 122 p.

Pivonka, Lee J. Geochemical detection of sulfide-bearing vein structures in a disturbed surficial environment, Central City District, Colorado. M, 1985, Colorado School of Mines. 139 p.

Pivorunas, August. Allometry in the limbs and sail of Dimetrodon. M, 1970, University of Illinois, Chicago.

Piwinski, Alf J. Experimental study of rocks from a zoned pluton. M, 1965, Pennsylvania State University, University Park. 118 p.

Piwinskii, Alfred J. Studies of batholithic feldspars (Sierra Nevada, California) and experimental studies of igneous rock series (central Sierra Nevada, California). D, 1968, University of Chicago. 40 p.

Pixler, Everett T. The origin of Gulf Coast salt dome sulphur. M, 1927, University of Minnesota, Minneapolis. 53 p.

Pizinger, D. D. and Gatje, P. H. Bottom current measurements in the head of Monterey submarine canyon (California). M, 1965, United States Naval Academy.

Pizzaferri, L. Some mathematical aspects of the inverse problem for Love waves. M, 1978, University of Chicago. 38 p.

Pizzuto, James Eugene. Channel-forming processes of straight sandbed streams. D, 1982, University of Minnesota, Minneapolis. 190 p.

Place, Jeannie Theresa. Geology and geochemistry of the Mariano Lake uranium mine, McKinley County, New Mexico. M, 1981, University of New Mexico. 100 p.

Place, John Louis. Man's role in geomorphic change on the shoreline of Los Angeles County, California. D, 1970, University of California, Los Angeles. 252 p.

Placher, George A. Petrologic and paleobiologic analysis of basin-slope cores, Islas de Aves, Venezuela. M, 1974, Northern Illinois University. 109 p.

Plafcan, Maria. Evaluating the significance of stream-aquifer interaction in the Mississippi alluvial plain of eastern Arkansas using hydrograph analysis and a numerical ground-water flow model. M, 1987, University of Arkansas, Fayetteville.

Plafker, George. Geology of the southwest part of the Kaweah Quadrangle, California. M, 1956, University of California, Berkeley. 46 p.

Plafker, George. The Alaskan earthquake of 1964 and Chilean earthquake of 1960; implications for arc tectonics. D, 1971, Stanford University. 180 p.

Plafker, Lloyd. Geologic reconnaissance of the Cenozoic Walnut Grove Basin, Yavapai County, Arizona. M, 1956, University of Arizona.

Plain, Donald Robert. Impact of the Alaska gas conditioning facilities project on the Prudhoe Bay environment. D, 1983, University of California, Los Angeles. 154 p.

Plamondon, Michael P. Land use capabilities in the Georgetown area, Williamson County, Texas. M, 1975, University of Texas, Austin.

Planalp, Roger Newton. Stratigraphy of the Oregon Quadrangle, Missouri. M, 1954, University of Missouri, Columbia.

Planck, Robert F. and Davis, Robert Irving. Geology of the McKenzie Canyon area, Beaverhead County, Montana. M, 1949, University of Michigan. 46 p.

Plane, Michael Dudley. The stratigraphy and fauna of the Otibanda Formation (Pliocene), New Guinea. M, 1965, University of California, Berkeley. 188 p.

Planinsek, Frances. The application of Lang X-ray topography to the investigation of magnetic domain structure in naturally occurring crystals of magnetite. M, 1972, University of Pittsburgh.

Plank, Adrian Van der *see* Van der Plank, Adrian

Plank, Robert Forrest. Geology of the McKenzie Canyon area, Beaverhead County, Montana. M, 1949, University of Michigan. 31 p.

Planke, Sverre. Cenozoic structures and evolution of the Sevier Desert basin, west-central Utah, from seismic reflection data. M, 1987, University of Utah. 163 p.

Planner, Harry N. An experimental investigation of highly undercooled magnesium silicate chondrule-like spherules. M, 1974, University of New Mexico. 89 p.

Planner, Harry N. Chondrule thermal history implied from olivine compositional data. D, 1979, University of New Mexico. 150 p.

Plante, Langis. Modélisation géophysique des cratères météoritiques du Lac à l'Eau-Claire, Nouveau-Québec. M, 1986, Universite Laval. 172 p.

Plantevin, Jean Paul. Some wettability characteristics of an unconsolidated sand. M, 1958, Stanford University.

Plants, H. F. Paleoecology of the Martinsburg Formation at Catawba Mountain, Virginia. M, 1977, Virginia Polytechnic Institute and State University.

Plants, Kenneth D. Coal gasification correlation. M, 1956, West Virginia University.

Plappert, John Wesley. Processing and interpretation of near vertical and wide angle seismic reflection data from the PASSCAL Ouachita lithospheric seismic study. M, 1987, Purdue University. 134 p.

Plasil, Georg. X-ray crystallographic studies on ilmenite. M, 1970, University of California, Berkeley. 93 p.

Plaster, Rodger W. Geologic, mineralogic, and engineering studies on three Virginia soil profiles. M, 1968, University of Virginia. 109 p.

Platco, Nicholas L., Jr. Earth science field trip guidebook for Chester County. M, 1972, Pennsylvania State University, University Park. 89 p.

Plato, Phillip Alexander. Predicting the movement of a radionuclide through the soil. D, 1968, Iowa State University of Science and Technology. 154 p.

Platt, J. B. Franciscan blueschist facies metaconglomerates, Diablo Range, California. M, 1974, Stanford University. 77 p.

Platt, John Paul. The petrology, structure, and geologic history of the Catalina terrain, Southern California. D, 1973, University of California, Santa Barbara.

Platt, Lucian. Structure and stratigraphy of the Cassayuna area, New York. D, 1960, Yale University.

Platt, Richard Garth. Residual liquid trends in the system nepheline-diopside-sanidine and their significance to alkaline rock genesis. D, 1969, University of Western Ontario. 124 p.

Platt, Robert M. Lead and zinc occurrence in the (Silurian) Lockport Dolomite of New York State. M, 1949, University of Rochester. 53 p.

Platt, Ronald Lorne. The western margin of the lower Mannville Group, Lower Cretaceous, central Alberta. M, 1960, University of Alberta. 54 p.

Platt, Wallace S. Land-surface subsidence in the Tucson area. M, 1963, University of Arizona.

Plaus, Peggy. Tourmaline-rich rocks in the Gile Mountain Formation near the Elizabeth Mine, Vermont copper belt; field relations, petrography, petrochemistry, metamorphism and inferred origin. M, 1983, University of Cincinnati. 209 p.

Plaut, Lynda Marjorie Parker *see* Parker Plaut, Lynda Marjorie

Plaut, M. G. A reflection seismic survey, Lambton County, southwestern Ontario. M, 1970, University of Western Ontario.

Plavidal, Kay Rosalie. The geology and petrology of the eastern Keg Mountains, Juab County, Utah. M, 1987, University of Utah. 137 p.

Plawman, Thomas Leon. Crustal structure of the continental borderland and the adjacent portion of Baja California between latitudes 30°N and 33°N. M, 1978, Oregon State University. 72 p.

Player, Gary Farnsworth. Petrography and origin of the phosphorite member of the Munson Creek Formation (upper Miocene), Ventura County, California. M, 1966, University of California, Los Angeles.

Playford, Phillip Elliott. Geology of the Egan Range, near Lund, Nevada. D, 1962, Stanford University. 249 p.

Plebuch, Raymond Otto. Tectonics of the Mississippian System of Western United States. M, 1957, University of Illinois, Urbana.

Pledge, Neville Stewart. Paleoenvironments and paleoecology of part of the lower Bridger Formation (Eocene), south of Opal, Wyoming. M, 1969, University of Wyoming. 81 p.

Plenge, Gustavo Carlos. Evaluation of demagnetizing field effects in geomagnetic data interpretation. M, 1978, Purdue University. 69 p.

Plescia, Jeffrey B. A gravity and magnetic study of the Tehachapi Mountains, California. D, 1985, University of Southern California.

Pleskot, Larry Kenneth. The opposition effect of particulate mineral surfaces and condensates; applications to Saturn's rings. D, 1981, University of California, Los Angeles. 210 p.

Plevniak, John E. Gravity survey of buried valleys in Hiram and Nelson townships, Portage County, Ohio. M, 1980, Kent State University, Kent. 82 p.

Plichta, C. A comparison of surficial exploration methods for uranium in the Marshall Pass District, Colorado. M, 1976, Queens College (CUNY). 144 p.

Plikk, Martin. Resolution of geophysical surveys for shallow work. M, 1981, Colorado School of Mines. 124 p.

Pliler, Richard. The distribution of thorium and uranium in sedimentary rocks and oxygen content of the pre-Cambrian atmosphere. M, 1957, Rice University. 39 p.

Pliler, Richard. The distribution of thorium, uranium, and potassium in a Pennsylvanian weathering profile and the Mancos Shale. D, 1959, Rice University. 59 p.

Plint, Heather Elizabeth. Metamorphism of the Jumping Brook metamorphic suite, western Cape Breton Highlands, Nova Scotia; microstructure, P-T-t paths, and tectonic implications. M, 1987, Dalhousie University. 368 p.

Pliva, Gustav L. Crustal structure of northeastern Ontario. M, 1973, University of Toronto.

Ploch, Richard Allen. A new species of ostracod from the New Albany Shale (Upper Devonian, Floyd County, Indiana). M, 1960, University of Michigan.

Plocharczyk, Elise J. The seismicity of southern New England; an investigative survey of historical data, current research, and related environmental considerations. M, 1980, Central Connecticut State University.

Ploessel, Michael R. Geology and geologic hazards of the sea cliffs of Southern California; Malaga Cove to Dana Point. M, 1972, University of Southern California.

Ploger, Louis W. Crustal disturbances in the Cattaraugus Creek region (New York). M, 1922, Syracuse University.

Ploger, Sheila Lynn Wagner. Effect of Laramide folding on previously folded Precambrian metamorphic rocks, Madison County, Montana. M, 1967, Indiana University, Bloomington. 27 p.

Plomer, James R. A comparative study of seismic explosive sources in a thick glacial till area. M, 1986, Wright State University. 91 p.

Plongeron, A. The United States production demand relationships to the world supply of phosphates, sulfur and potash. M, 1978, Stanford University. 133 p.

Plopper, Christopher Steven. Hydraulic sorting and longshore transport of beach sand, Pacific Coast of Washington. D, 1978, Syracuse University. 194 p.

Plopper, Christopher Stevens. Intergranular lattice coherency controls on limestone texture. M, 1972, Michigan State University. 23 p.

Plotnick, Roy E. Geomagnetic reversals and the history of life. M, 1978, University of Rochester. 48 p.

Plotnick, Roy Elliott. Patterns in the evolution of the eurypterids. D, 1983, University of Chicago. 411 p.

Plouff, Michael Thomas. Trace element and stable isotope geochemistry and diagenesis in Cenozoic Mineta Formation limestones, southeastern Arizona. M, 1983, University of Arizona. 64 p.

Pluenneke, Judith Louise. Comparative analysis of debrites, turbidites, and contourites. M, 1976, Texas A&M University.

Pluijm, Bernardus Adrianus Van Der *see* Van Der Pluijm, Bernardus Adrianus

Pluim, Scott B. Sedimentology of the lower Parrsboro Formation (Carboniferous), Parrsboro, Nova Scotia, Canada. M, 1980, University of Nebraska, Lincoln.

Plumb, Richard A. Surface deformation of the active rift zones in Iceland. M, 1974, Dartmouth College. 140 p.

Plumb, Richard Allen. The relationship between fractures and in situ stress. D, 1982, Columbia University, Teachers College. 157 p.

Plumley, Patrick S. Volcanic stratigraphy and geochemistry of the Hole in the Ground area, Owyhee Plateau, southeastern Oregon. M, 1986, University of Idaho. 161 p.

Plumley, Peter W. Paleomagnetism of Tertiary intrusive rocks in the Oregon Coast Range; timing and mechanism of tectonic rotation. M, 1980, Western Washington University. 239 p.

Plumley, Peter William. A paleomagnetic study of the Prince William terrane and Nixon Fork terrane, Alaska. D, 1984, University of California, Santa Cruz. 334 p.

Plumley, William J. Black Hills terrace gravels; a study of sediment transportation. D, 1948, University of Chicago. 111 p.

Plummer, Charles Carlton. Geology of the crystalline rocks, Chiwaukum mountains and vicinity, Washington Cascades. D, 1969, University of Washington. 137 p.

Plummer, Charles Carlton. The geology of the Mount Index area of Washington State. M, 1964, University of Washington. 62 p.

Plummer, David A. The petrology and structure of Proterozoic rocks northeast of Salida, Colorado. M, 1986, University of New Mexico. 88 p.

Plummer, Helen S. Some foraminifera of the Midway Formation in northeast Texas. M, 1925, Northwestern University.

Plummer, Leonard N. Rates of mineral-aqueous solution reactions. D, 1972, Northwestern University.

Plummer, Leonard Niel, Jr. Distribution of barium in host rock adjacent to two barite veins of the Central (Kentucky) mineral district. M, 1969, University of Kentucky. 69 p.

Plummer, Roger Sherman, Jr. The geology of west central Bastrop County, Texas. M, 1949, University of Texas, Austin.

Plunkett, J. D. Testing of Missouri shales for light weight aggregate production. M, 1954, University of Missouri, Rolla.

Plusquellec, Paul Lloyd. Coastal morphology and changes of an area between Brigantine and Beach Haven Heights, New Jersey. M, 1966, University of Illinois, Urbana.

Plusquellec, Paul Lloyd. Some ostracod genera of the subfamily Campylocytherinae. D, 1968, University of Illinois, Urbana. 157 p.

Plut, Frederick W. Geology of the Eagle Peak-Hells Gate area, Happy Valley Quadrangle, Cochise County, Arizona. M, 1968, University of Arizona.

Plymate, Thomas George. Equations of state for the polymorphs of Sn and Fe_2SiO_4 determined by in situ energy-dispersive X-ray diffraction in a heated diamond-anvil pressure cell; (Volumes I and II). D, 1986, University of Minnesota, Minneapolis. 419 p.

Plytus, Michael. The adsorption of lithium by clay minerals and the geochemical cycle of lithium. M, 1982, University of Minnesota, Minneapolis. 68 p.

Poag, Claude Wylie. Age, biostratigraphy, and paleoecology of the Chickasawhay Formation (late Oligocene), Alabama and Mississippi, with emphasis on ostracodes and planktonic foraminifers. D, 1971, Tulane University. 371 p.

Pober, Patricia Taylor. Modal analysis of airphoto linears. M, 1967, University of Idaho. 67 p.

Poborski, Stanislaw Jozef. The (Triassic) Virgin Formation of the St. George area, southwestern Utah. M, 1952, The Johns Hopkins University.

Pobran, Vernon Stephen. Laboratory and field testing of an acoustic borehole logging system. M, 1974, University of Saskatchewan. 66 p.

Poche, David John. Petrology of the Palo Pinto Limestone (Pennsylvanian) in the type area of the Canyon Group (Upper Pennsylvanian), north-central Texas. M, 1966, Southern Methodist University. 47 p.

Poche, David John. Selective sorting of sediment by waves; the influence of grain shape. D, 1973, University of Virginia. 99 p.

Pocock, Yvonne Patricia. Devonian schizophoriid brachiopods from northern Canada. M, 1962, University of Saskatchewan. 131 p.

Podell, Mark Edward. The interrelationship of early colony development, monticules, and branches in Paleozoic bryozoans. M, 1978, Michigan State University. 37 p.

Podio-Lucioni, Augusto. Experimental determination of the dynamic elastic properties of anisotropic rocks, ultrasonic pulse method. D, 1968, University of Texas, Austin. 200 p.

Podkhlebnik, Yvette. Continuation vers le bas et inversion des données de flux de chaleur. M, 1988, Universite du Quebec a Montreal. 120 p.

Podoff, Nedda. Microstructure of modern deep-water corals (Flabellidae and Parasmiliinae). M, 1976, SUNY at Binghamton. 92 p.

Podolak, Wilfred E. Drift dispersal in central southern Ontario. M, 1984, Brock University. 158 p.

Podolsky, T. X-ray investigation of rock fabric. D, 1954, Massachusetts Institute of Technology.

Podolsky, T. M. Timiskaming-Pontiac relationships in Rouyn Township, Quebec. M, 1950, Queen's University. 97 p.

Podorsky, Robert A. The stratigraphy and mammalian paleontology of the Wasatch Formation (Eocene), Fossil Butte National Monument, Lincoln County, Wyoming. M, 1981, Fort Hays State University. 68 p.

Podosek, Frank Anthony. Early solar system abundances of iodine-129 and plutonium-244 from analysis of xenon in neutron-irradiated meteorites. D, 1969, University of California, Berkeley. 72 p.

Podrebarac, Thomas Joseph. Trace fossils of the Brigham Quartzite (Lower and Middle Cambrian) in northern Utah and southeastern Idaho. M, 1976, University of Utah. 157 p.

Podruski, James Allan. Petrology of the upper plate crystalline complex in the Whipple Mountains, San Bernardino County, California. M, 1979, University of Southern California.

Podsen, Donald Wayne. Diagenesis of the Nugget Sandstone in the Overthrust Belt of southwestern Wyoming. M, 1981, University of Colorado. 72 p.

Pody, Robert Dale. A survey of the depositional environments and paleoecology of the upper Pottsville Formation in the Black Warrior Basin, along Alabama State Highway 69. M, 1987, Mississippi State University. 153 p.

Poe, Thomas I., III. The intrusive sequence of igneous rocks in the Gallinas Mountains, New Mexico. M, 1965, New Mexico Institute of Mining and Technology. 28 p.

Poehls, Kenneth Allen. Geomagnetic variations in the North Atlantic; implications for the electrical resistivity of the oceanic lithosphere. D, 1976, Massachusetts Institute of Technology. 134 p.

Poehls, Kenneth Allen. Geomagnetic variations in the Northwest Atlantic; implications for the electrical resistivity of the oceanic lithosphere. D, 1975, Woods Hole Oceanographic Institution. 134 p.

Poel, Washinton I. Van der see Van der Poel, Washinton I., III

Poelchau, Harald S. Holocene silicoflagellates of the north Pacific; their distribution and use for paleotemperature determination. D, 1974, University of California, San Diego. 178 p.

Poelchau, Harold S. Geology of the Gore Creek area, Eagle County, Colorado. M, 1963, University of Colorado.

Poeter, Eileen. Mathematical models of the gamma-gamma and neutron-epithermal neutron geophysical logging processes. D, 1980, Washington State University. 184 p.

Poetzl, Kenneth G. Seismic activity before and after great earthquakes and large nuclear detonations. M, 1973, University of Wisconsin-Milwaukee.

Poey, Jean-Luc. Stratigraphy and depositional environments of an Upper Ordovician to Lower Devonian shelf-to-basin transition, Svendsen Peninsula, Ellesmere Island, N. W. T. M, 1988, University of Ottawa. 245 p.

Poffenberger, Michael Robert. Petrology of the middle Tokio sandstones (Grafton and Carterville), North Webster Parish, Louisiana. M, 1986, Louisiana Tech University. 112 p.

Pogoncheff, Nicholas C. The effects of precipitation on the quality of ground water and leachate seeps at a stabilizing landfill. M, 1982, Western Michigan University. 92 p.

Pogorzelski, Brett Katherine. Petrochemistry and petrogenesis of the Highlandcroft plutonic series, NH, VT, and ME. M, 1983, Dartmouth College. 109 p.

Pogue, Jesse B. A survey of the Ordovician Edrioasteroidea of the Cincinnati area. M, 1954, University of Cincinnati. 112 p.

Pogue, Joseph E. The Cid mining district of Davidson County, North Carolina; a region of ancient volcanic rocks. D, 1909, Yale University.

Pogue, Kevin R. The geology of the Mt. Putnam area, northern Portneuf Range, Bannock and Caribou counties, Idaho. M, 1984, Idaho State University. 106 p.

Pogue, Peggy Todd. Misener Sandstone in portions of Grant and Garfield counties, north-central Oklahoma. M, 1987, West Virginia University. 71 p.

Pohana, Richard Edward. Engineering geologic and relative stability analysis of a portion of Anderson Township. M, 1983, University of Cincinnati. 132 p.

Pohanka, Susan J. Controls of mineral weathering upon groundwater chemistry in Cadwell Creek watershed, central Massachusetts. M, 1985, University of Massachusetts. 124 p.

Pohl, Demetrius Christmus. Experimental hydrothermal geochemistry; basalt glass-seawater reactions. D, 1985, Stanford University. 240 p.

Pohl, Erwin Robert. Faunal study of the Wisconsin Devonian. D, 1928, George Washington University.

Pohler, Suzanne Margarete Luise. Conodont biofacies and carbonate lithofacies of Lower Ordovician megaconglomerates, Cow Head Group, western Newfoundland. D, 1987, Memorial University of Newfoundland. 545 p.

Pohlman, John Carl. A study of a shallow-focus earthquake cluster possibly related to volcanism. M, 1982, University of Colorado. 206 p.

Pohlmann, Karl. Investigation of the ground water resources in the Wabash Valley glacial deposits near West Lafayette, Indiana. M, 1987, Purdue University. 148 p.

Pohlo, Ross H. Geology of T. 16 N., R. 30 W., Washington County, Arkansas. M, 1958, University of Arkansas, Fayetteville.

Pohowsky, Robert Alexander. An introduction to the paleontology of shell penetrating Ctenostomata (Ectoprocta). M, 1969, University of Cincinnati. 109 p.

Pohowsky, Robert Alexander. The boring ctenostomate Bryozoa; taxonomy and paleobiology based on cavities in calcareous substrata. D, 1974, University of Cincinnati. 399 p.

Poincloux, P. A. The aluminum, copper, lead, and zinc industries in France. M, 1977, Stanford University. 63 p.

Poindexter, Edward Haviland. Piezobirefringence in diamond. D, 1956, University of Michigan.

Poindexter, Marian Elizabeth. Behavior of subaqueous sediment mounds; effect on dredged material disposal site capacity. D, 1988, Texas A&M University. 211 p.

Point, Ronald La see La Point, Ronald

Pointe, Paul Reggie La see La Pointe, Paul Reggie

Pointer, Gary Neal. Taxonomic study of fossil and Recent otoliths of certain cuskeels. M, 1965, University of Missouri, Rolla.

Poirier, C. A. The carrying down of free oil in sea water by settling sediments. M, 1940, University of Minnesota, Minneapolis. 62 p.

Poirier, Ghislain. Etude métallogénique de gîtes de nickel, cuivre et platinoides de l'ouest de la Province de Grenville, Québec. M, 1988, Universite du Quebec a Montreal. 299 p.

Poitras, Alain. Caractérisation géochimique du Complexe de Cummings, région de Chibougamau-Chapais, Québec. M, 1985, Universite du Quebec a Chicoutimi. 172 p.

Pojeta, John. North American Ambonychiidae. D, 1963, University of Cincinnati. 436 p.

Pojeta, John. The pelecypod genus Byssonychia as it occurs in the Cincinnatian at Cincinnati, Ohio. M, 1961, University of Cincinnati. 158 p.

Pokorny, Harvey Dreifuss. Petrogenesis of the Sheep Canyon Basalt, Buck Hill Volcanic Series, West Texas. M, 1972, Northern Illinois University. 43 p.

Pokras, Edward M. Biostratigraphy and paleobathymetry of benthic foraminifera, D.S.D.P. Site 232, eastern Gulf of Aden. M, 1977, University of Wyoming. 111 p.

Pokras, Edward Mathew. Paleoclimatic investigation of late Quaternary diatoms from the Equatorial Atlantic. D, 1984, Columbia University, Teachers College. 252 p.

Pol, James Campalans. Sedimentation and diagenesis of an Upper Pennsylvanian (Virgilian) mixed carbonate-clastic sequence, Hueco Mountains, El Paso County, Texas. M, 1982, University of Texas, Austin. 212 p.

Polan, Kevin Patrick. The allochthonous origin of "bioherms" in the Early Devonian Stuart Bay Formation of Bathurst Island, Arctic Canada. M, 1982, McGill University. 99 p.

Poland, F. B. The geology of the rocks along the James River between Sabot and Cedar Point, Virginia. M, 1976, Virginia Polytechnic Institute and State University.

Poland, Joseph F. Ground-water conditions in Ygnacio Valley, California. M, 1935, Stanford University. 83 p.

Poland, Joseph Fairfield. The occurrence and control of land subsidence due to ground-water withdrawal with special reference to the San Joaquin and Santa Clara valleys, California. D, 1981, Stanford University. 172 p.

Polanshek, D. H. A mathematical analysis of the time-displacement characteristics of fault-creep events recorded in central California. M, 1975, University of Arizona.

Polasek, John. Onondaga Limestone facies and reef trends in eastern New York State. M, 1978, SUNY at Binghamton. 79 p.

Polasky, Mark Edward. Characterization of extracted shale oil crudes. D, 1988, University of Akron. 245 p.

Poleschook, Daniel, Jr. Stratigraphy and channel discrimination of the J Sandstone, Lower Cretaceous Dakota Group, south and west of Denver, Colorado. M, 1978, Colorado School of Mines. 226 p.

Poley, Denise F. Acquisition and processing of high resolution reflection seismic data from permafrost affected areas of the Beaufort Sea continental shelf. D, 1987, University of Calgary. 261 p.

Polikar, Marcel. Relative permeability measurements in oil sands. D, 1987, University of Alberta. 249 p.

Polivka, David R. Quaternary volcanology of the West Crater-Soda Peaks area, southern Washington Cascade Range. M, 1984, Portland State University. 78 p.

Poljak, Marijan. Lineament tectonics of Central Montana. M, 1980, Purdue University. 78 p.

Polk, Ted P. Geology of the West Mason area, Texas. M, 1952, Texas A&M University.

Polk, Thomas Robb. A study of the igneous rocks of the Devils Canyon Mountain Group, Wichita Mountains, Oklahoma. M, 1948, University of Oklahoma. 87 p.

Poll, Henk Wouter van de see van de Poll, Henk Wouter

Pollack, Henry N. Stratigraphy of the Dakota Group in the southern Front Range Foothills, Colorado. M, 1960, University of Nebraska, Lincoln.

Pollack, Henry Nathan. Elastic wave studies. D, 1963, University of Michigan. 42 p.

Pollack, Jerome Marvin. A faunal study of the Harding Sandstone in the Canon City Embayment, Colorado. M, 1951, University of Oklahoma. 79 p.

Pollack, Jerome Marvin. Significance of compositional and textural properties of South Canadian River channel deposits, New Mexico, Texas, and Oklahoma. D, 1959, University of Oklahoma. 114 p.

Pollack, Sidney Solomon. The mineralogy of a gray-brown podzolic soil and a humic-gley soil of southeastern Wisconsin. D, 1956, University of Wisconsin-Madison. 107 p.

Pollak, Henry. Some results of Earth tide data analysis in the Pittsburgh, Pennsylvania area. D, 1971, University of Pittsburgh.

Pollak, Miriam B. A study of the gold-silver mineralization at Maguarichic, Chihuahua, Mexico. M, 1937, Columbia University, Teachers College.

Pollak, Robert. Lead isotope composition of sulfide ores from the Central Andes. M, 1977, University of California, Santa Barbara.

Pollard, Charles Oscar, Jr. Attempted correlation of bond nature with anisotropic anion polarizability in rutile-type minerals. D, 1967, Florida State University. 75 p.

Pollard, Craig D. Regional depositional framework of the Lower Cretaceous James Limestone and its relationship to hydrocarbon accumulations in East Texas. M, 1985, Baylor University. 170 p.

Pollard, Dalton Leon. Geology of the Hasley Canyon area, Los Angeles County, California. M, 1958, University of California, Los Angeles.

Pollard, David Dierker. Deformation of host rocks during sill and laccolith formation. D, 1968, Stanford University. 153 p.

Pollard, Dwight D. The source and distribution of beach sediments, Santa Barbara County, California. D, 1979, University of California, Santa Barbara. 293 p.

Pollard, Lin Davis. Sedimentation rate determination on ocean bottom cores by gamma ray spectrometry. D, 1967, Florida State University. 99 p.

Pollard, Richard Mark. Distribution and preservation of pteropods in Pleistocene sediments of the Sigsbee Plain, Gulf of Mexico. M, 1979, Texas A&M University. 151 p.

Pollard, W. S. Rate and time effects in cyclic triaxial testing. D, 1979, Cornell University. 397 p.

Pollard, William D. Stratigraphy and origin of Winchell limestone (upper Pennsylvanian) in Possum Kingdom area, north-central Texas, and role of phylloid algae in carbonate sedimentation. M, 1970, University of Kansas.

Pollard, William S., Jr. Airphoto interpretation of soils and drainage of Henry County, Indiana. M, 1948, Purdue University.

Pollastro, R. M. A reconnaissance analysis of the clay mineralogy and major element geochemistry in the Silurian and Devonian carbonates of western New York; a vertical profile. M, 1977, SUNY at Buffalo. 120 p.

Polley, Mark R. Petrology of the Ohio Shale; effect of clay fabric, mineralogy and carbon content on fissility. M, 1982, Bowling Green State University. 97 p.

Pollman, Keith S. Brittle-ductile deformation of feldspar in the Garlock fault zone, Quail Mountains, California. M, 1983, University of Texas, Austin.

Pollock, Clifford Ralph. Ground-water hydrogeology and geochemistry of a reclaimed lignite surface mine. M, 1982, Texas A&M University. 152 p.

Pollock, David W. Numerical simulation of energy transport in shallow aquifers subjected to a thermal stress from high temperature energy storage in the unsaturated zone. M, 1977, University of Minnesota, Minneapolis. 149 p.

Pollock, David Warren. Fluid flow and energy transport in a high-level radioactive waste repository in unsaturated alluvium. D, 1982, University of Illinois, Urbana. 96 p.

Pollock, Donald William Thomas. The geology of the Addington-Preston area [Quebec]. D, 1957, McGill University.

Pollock, Donald William Thomas. The mineralogy of the Eastern Metals nickel-copper deposit, Quebec. M, 1955, McGill University.

Pollock, Gerald D. Age determination of granitic rocks from Manitoba and northwestern Ontario by the lead-alpha method. M, 1960, University of Manitoba.

Pollock, Gerald D. Petrology, mineralogy and structural geology of the Duval Lake area, Manitoba. D, 1965, University of Manitoba.

Pollock, J. Michael. Geology and geochemistry of hydrothermal alteration, eastern portion of the North Santiam mining area. M, 1985, Portland State University. 100 p.

Pollock, J. P. The geocolloid chemistry of mercury. M, 1940, Massachusetts Institute of Technology. 88 p.

Pollock, J. R. The structure of the Rocky Mountain trench (British Columbia). M, 1926, University of British Columbia.

Pollock, S. J. Bedrock geology of the Tiverton Quadrangle, Rhode Island-Massachusetts. M, 1956, Brown University.

Pollock, S. P. The isotopic geochemistry of the Prairie Lake carbonatite complex, Ontario. M, 1987, Carleton University.

Pollock, Stephen G. Paleoenvironments and dispersal patterns of the Matagamon Sandstone, Lower Devonian, northern Maine. M, 1972, University of Maine. 76 p.

Pollock, Stephen Garrett. Stratigraphy, sedimentation and basin development of the Jacksonburg Limestone and Martinsburg Formation, Ordovician, northern New Jersey. D, 1975, Rutgers, The State University, New Brunswick. 45 p.

Pollock, Stephen Matthew. Structure, petrology and metamorphic history of the Nome Group blueschist terrane, Salmon Lake area, Seward Peninsula, Alaska. M, 1982, University of Washington. 222 p.

Polo, Jesus Miguel. Analysis of settlements of pile groups. D, 1982, Duke University. 186 p.

Polovina, Joseph Stanley. Geology and mineral deposits of the Bagdad Chase Mine and vicinity, Stedman District, San Bernardino County, California. M, 1980, University of California, Los Angeles.

Polski, William. Foraminiferal biofacies off the North Asiatic Coast. M, 1959, University of Southern California.

Polugar, Morton. A phylogenetic study of Sakesaria and Lockartia. M, 1954, New York University.

Pomerene, J. B. Geology of the Einstein-Apex tungsten mine area. M, 1947, University of Missouri, Rolla.

Pomerening, James Albert. An analysis of soil maps prepared by conventional and aerial photograph interpretative methods. M, 1956, Cornell University.

Pomeroy, Paul W. Long period seismic waves from large near-surface nuclear explosions. D, 1963, Columbia University, Teachers College.

Pomes, Michael L. Stratigraphy, paleontology, and paleobiogeography of lower vertebrates from the Cedar Mountain Formation (Lower Cretaceous), Emery County, Utah. M, 1988, Fort Hays State University. 87 p.

Ponader, Carl Wilson. Trace metals in silicate glasses and melts; coordination environments, halogen complexes, and Soret diffusion. D, 1988, Stanford University. 135 p.

Ponader, Heather Boek. An EXAFS' and X-ray diffraction study of the local Ca environment in anorthite ($CaAl_2Si_2O_8$) crystal and glass. M, 1984, Stanford University. 110 p.

Ponce de Leon, Jose G. Investigation of the CDP method by means of synthetic seismograms. M, 1969, Rice University. 67 p.

Ponce, Benjamin F. Caldera development and economic mineralization in the Zacatecas mining district, Zacatecas, Mexico. M, 1985, University of Texas at El Paso.

Ponce, David. Gravity investigations at the proposed Wahmonie Nuclear Waste Storage Site, Nevada Test Site, Nye County, Nevada. M, 1981, San Jose State University. 76 p.

Pond, Adela. The geology of the Clendening Creek area, Giles County, Virginia. M, 1926, Smith College. 36 p.

Ponder, Herman. The geology, mineralogy and genesis of selected fireclays from Latah County, Idaho. D, 1959, University of Missouri, Columbia. 181 p.

Pongsapich, Wasant. A petrographic reconnaissance of the Swauk, Chuckanut and Roslyn formations (Eocene), Washington. M, 1970, University of Washington. 63 p.

Pongsapich, Wasant. The geology of the eastern part of the Mount Stuart Batholith, central Cascades, Washington. D, 1974, University of Washington. 170 p.

Ponsetto, Louis R. Geology of the area of Smith Mills South consolidated and Midway oil pools, Henderson County, Kentucky. M, 1958, University of Kentucky. 30 p.

Ponsler, Harley E. The geology and mineral deposits of the Garfield District, Mineral County, Nevada. M, 1977, University of Nevada. 73 p.

Ponti, Daniel J. Geology of the Moss Beach area, San Mateo County, California. M, 1976, Stanford University.

Ponti, Daniel John. Stratigraphy and engineering characteristics of upper Quaternary sediments in the eastern Antelope Valley and vicinity, California. M, 1980, Stanford University. 157 p.

Pontier, Nancy Kilbridge. Magmatic evolution of Izalco Volcano and relation to the Santa Ana Complex, El Salvador. M, 1979, Rutgers, The State University, New Brunswick. 72 p.

Pontigo, Felipe A., Jr. Petrology and stable isotope geochemistry of the Smackover Formation in the Apalachicola Embayment. M, 1982, Florida State University.

Pontius, David C. and Cowan, A. Gordon. The geology of a portion of Crescent Valley and Hilltop quadrangles, Nevada. M, 1950, University of California, Los Angeles.

Pontius, Jeffrey Allan. The geology of the south-central Bayhorse mining district, Custer County, Idaho. M, 1982, University of Idaho. 87 p.

Ponton, James D. Structural analysis of the Little Water Syncline, Beaverhead County, Montana. M, 1983, Texas A&M University. 165 p.

Pontoriero, Pasquale. Magnetic and physical characteristics of clastic sediment from the caves of Mammoth Cave National Park, Kentucky, U.S.A. M, 1981, University of Pittsburgh. 63 p.

Poobrasert, Suparb. Lithologic and micropaleontologic characteristics of the Midway-Wilcox contacts of northeastern Milam and Falls counties, Texas. M, 1961, Texas A&M University. 49 p.

Pool, Alexander S. Geology of the Homer Martin Ranch area, Mason County, Texas. M, 1960, Texas A&M University.

Pool, J. R. Morphology and recharge potential of certain playa lakes on the Edwards Plateau of Texas. M, 1976, Baylor University. 43 p.

Pool, R. Harold. The Devonian hiatus in the Arbuckle Mountains of Oklahoma. M, 1922, University of Oklahoma. 22 p.

Pool, Raymond John. A study of the vegetation of the Sandhills of Nebraska. D, 1913, University of Nebraska, Lincoln.

Poole, David M. and Daley, A. Cowles. A geologic section in east-central California eastward from Donner Pass. M, 1949, University of California, Los Angeles.

Poole, Donald Hudson. Slope failure forms; their identification, characteristics, and distribution as depicted by selected remote sensor returns. D, 1969, University of Georgia. 203 p.

Poole, Forrest Graham. Geology of the southern Grand Hogback area, Garfield and Pitkin counties; Colorado. M, 1954, University of Colorado.

Poole, James Leroy. The trace element content of the Knox Group (Upper Cambrian and Lower Ordovician) from three wells in middle Tennessee. M, 1972, University of Tennessee, Knoxville.

Poole, Joe Lester. The geology of the Harriman Quadrangle, Roane and Morgan counties, Tennessee. M, 1949, University of Tennessee, Knoxville. 66 p.

Poole, Leslie Ann. Sedimentology, structural style, and thermal maturity of the Lynn Mountain Formation, Frontal Ouachitas, Latimer and Le Flore counties, Oklahoma. M, 1985, University of Missouri, Columbia. 189 p.

Poole, Russel Wayne. Gastropoda and Pelecypoda of the Moody's Branch marl (Eocene) in Louisiana. M, 1969, Louisiana Tech University.

Poole, Thomas Craig, Jr. Carbonate petrology and paleoenvironment of the Riepe Spring Limestone, east central Nevada. M, 1987, California State University, Fresno. 147 p.

Poole, Vickie Lynn. Water resources of basal Pennsylvanian sandstones in Perry, Jackson, and Randolph counties, Illinois. M, 1985, Purdue University. 149 p.

Poole, William Hope. The geology of the Cassiar Mountains in the vicinity of the Yukon-British Columbia boundary. D, 1956, Princeton University. 279 p.

Pooler, Michael Lee. Location-allocation model effectiveness in public facilities location. M, 1983, Oklahoma State University. 101 p.

Pooley, Robert Neville. Basement configuration and subsurface geology of eastern Georgia and southern South Carolina as determined by seismic-refraction measurements. M, 1960, University of Wisconsin-Madison. 47 p.

Poopath, Visharn. Elasto-plastic incremental approach in slope stability analysis. D, 1971, Michigan State University. 255 p.

Poor, Russell Spurgeon. Stratigraphic and sedimentation studies in the Galesburg Quadrangle, Illinois. D, 1927, University of Illinois, Chicago.

Poor, Russell Spurgeon. The stratigraphy of the Mississippian System in the Alto Pass Quadrangle, Illinois. M, 1925, University of Illinois, Chicago.

Poore, Clark Alan. Ground-water flow in a heavily exploited buried channel aquifer, Souris River basin, North Dakota. M, 1987, Oklahoma State University. 70 p.

Poore, Richard Zell. Late Cenozoic planktonic foraminiferal biostratigraphy and paleoclimatology of the North Atlantic Ocean; DSDP Leg 12. D, 1975, Brown University. 207 p.

Poore, Richard Zell. The Leroy bioherm; onondaga limestone (middle Devonian), western New York. M, 1969, Brown University.

Poorman, Stephen E. Environmental framework, structural evolution and petroleum potential of the Cambrian Wilberns Formation, west-central Texas. M, 1984, Baylor University. 79 p.

Poort, Jon Michael. Stratigraphic patterns in the Sauk sequence (upper Proterozoic to upper Cambrian) in the western craton. D, 1969, Southern Methodist University. 45 p.

Pooser, William K. Geology of the Fort Jackson North Quadrangle. M, 1957, University of South Carolina.

Pooser, William K. Geology of the Fort Jackson North quadrangle. M, 1958, University of South Carolina. 38 p.

Pooser, William Kenneth. Cenozoic biostratigraphy and Ostracoda of South Carolina. D, 1962, University of Kansas. 317 p.

Pope, Allen J. Strain analysis of repeated triangulation for the investigation of crustal movement. M, 1966, Ohio State University.

Pope, Christiana Sheldon. The taxonomy and paleoecology of the stromatoporoid fauna of the Silurian West Point Formation, Gaspe Peninsula, Quebec. M, 1986, University of New Brunswick. 222 p.

Pope, Clifton Washington, Jr. Geology of the lower Verde River valley, Maricopa County, Arizona. M, 1974, Arizona State University. 104 p.

Pope, David E. The Harang fauna of the southeastern Louisiana coastal parishes. M, 1948, Louisiana State University.

Pope, David M. Geology of the Foster Gabbro, West central Rhode Island. M, 1975, University of Rhode Island.

Pope, Frederick J. Investigation of magnetic iron-ores from eastern Ontario. D, 1899, Columbia University, Teachers College.

Pope, Joan. Relationship between sediment cation content and chemical environment; Point Judith Pond, Rhode Island. M, 1976, University of Rhode Island.

Pope, John Keyler. Comparative morphology and shell histology of the Ordovician Strophomenacea (Brachiopoda). D, 1966, University of Cincinnati. 270 p.

Pope, John Keyler. Some gastropods from the Upper Cretaceous Pugnellus Sandstone of Huerfano Park, Colorado. M, 1956, University of Michigan.

Pope, John L., Jr. The concentration of a complex silver ore. M, 1890, Washington University.

Pope, Kevin Odell. Late Quaternary alluviation and soil formation in the southern Argolid, Greece. M, 1984, Stanford University. 43 p.

Pope, Kevin Odell. Palaeoecology of the Ulua Valley, Honduras; archaeological perspective. D, 1986, Stanford University. 224 p.

Pope, Leslie Anne. An experimental investigation into the effects of fluid composition on certain geothermometry methods. M, 1985, Texas A&M University. 52 p.

Pope, N. M. Petrology and structure of the late Precambrian mafic sills east of Silver Creek, Lake County, Minnesota. M, 1976, University of Minnesota, Duluth.

Pope, Robert William. An analysis of the carbonate facies of the Hermosa Formation (Pennsylvanian) of northeastern Arizona. M, 1976, Northern Arizona University. 144 p.

Pope, Stephen Van Wyck. Antimony-124 in the lower Columbia River. M, 1970, Oregon State University. 57 p.

Popek, John P. An analysis of the Precambrian Mount Rogers Formation, Grayson Highlands State Park, Grayson County, Virginia. M, 1974, University of South Florida, Tampa. 98 p.

Popelar, Stanley James. Miocene geology of the southwestern portion of the San Rafael Wilderness, Santa Barbara County, California. M, 1988, California State University, Northridge. 109 p.

Popenoe, Frank Wallace. Geology of the southeastern portion of the Indio Hills, Riverside County, California. M, 1960, University of California, Los Angeles.

Popenoe, W. P. and Findlay, Willard A. Transposed hinge structure in lamellibranchs. D, 1940, California Institute of Technology.

Popenoe, Willis P. An analysis and comparison of the (Cretaceous) Trabuco and Baker conglomerates of the Santa Ana Mountains (California). D, 1936, California Institute of Technology. 26 p.

Popkin, Barney Paul. Effect of mixed-grass cover and native-soil filter on urban runoff quality. M, 1973, University of Arizona.

Poplin, Jack Kenneth. A model study of dynamically loaded square footings on dry sand. D, 1968, North Carolina State University. 349 p.

Popoff, Constantine C. The geology of the Rosemont mining camp, Pima County, Arizona. M, 1941, University of Arizona.

Popovich, Daniel Eugene. Distribution of certain elements in the major rock units at the Cornwall and Morgantown mines, Pennsylvania. M, 1965, Pennsylvania State University, University Park. 73 p.

Popovich, Michael Joseph. A study of the relationship between grain size and capillary pressure curves for consolidated sands. M, 1947, University of Pittsburgh.

Popowich, James Leslie. Bin flow principles and mine draw control. M, 1969, University of Saskatchewan. various pagination p.

Popp, Brian Nicholas. Coordinated textural, isotopic, and elemental analyses of constituents in some Middle Devonian limestones. M, 1981, University of Illinois, Urbana. 136 p.

Popp, Brian Nicholas. The record of carbon, oxygen, sulfur, and strontium isotopes and trace elements in late Paleozoic brachiopods. D, 1986, University of Illinois, Urbana. 199 p.

Popp, John T. The geological factors controlling the migration, retention, and emission of methane in the Beckley coalbed. M, 1974, Southern Illinois University, Carbondale. 65 p.

Popp, R. K. Iron-magnesium amphiboles; synthesis and stability with respect to temperature, pressure, oxygen fugacity, and sulfur fugacity. D, 1975, Virginia Polytechnic Institute and State University. 132 p.

Popp, Robert K. The stability relations of sodic pyroxenes at low pressure. M, 1971, Virginia Polytechnic Institute and State University.

Poppe, James. Coccolith biostratigraphy of the Cretaceous of southwestern Minnesota. M, 1979, University of Minnesota, Minneapolis. 115 p.

Poppelreiter, Barbara Savage. A structural analysis of metasedimentary rocks of the Ocoee Supergroup above the Great Smoky overthrust in Southeast Tennessee. M, 1980, University of Tennessee, Knoxville. 115 p.

Poppendeck, Mark C. Subsurface petroleum study of the Medina sandstones (Silurian), southwestern Chautauqua County, New York. M, 1979, SUNY at Buffalo. 37 p.

Popper, George H. P. Paleobasin analysis and structure of the Anguille group (Mississippian), west-central Newfoundland. D, 1970, Lehigh University. 225 p.

Popper, George H. P. Stratigraphic and tectonic history of the Memramcook terrestrial red beds (Mississippian), of New Brunswick, Canada. M, 1965, University of Massachusetts. 129 p.

Popper, Robert J. Microforaminifera from the early Tertiary of the Gulf Coast. M, 1957, New York University.

Poprik, Lee Albert. Sedimentation in the Saint Clair river delta, Muscamoot bay area, Michigan. M, 1968, Wayne State University.

Porcher, Eric N. Lithofacies and geochemistry of interreef carbonates, Middle Silurian, Michigan Basin. M, 1985, Western Michigan University.

Poreda, Robert Joseph. Helium, neon, water, and carbon in volcanic rocks and gases. D, 1983, University of California, San Diego. 235 p.

Porsch, Herman W., Jr. Biometry of Enallaster texanus. M, 1965, University of Texas, Austin.

Porter. The sedimentology, paleontology and paleoecology of the Jones Creek Long Reach Formation (Silurian). M, 1973, University of New Brunswick.

Porter, Charles Earnest. Geology of east central part of Cushing Quadrangle, southern Rusk County, Texas. M, 1951, University of Texas, Austin.

Porter, Charles O. Electrical resistivity investigations over limestone caverns. M, 1966, Texas A&M University. 120 p.

Porter, Christopher. Statistical method for the determination of water saturations from well logs. M, 1968, Colorado School of Mines. 140 p.

Porter, Darrell D. A role of the borehole pressure in blasting; the formation of cracks. D, 1970, University of Minnesota, Minneapolis.

Porter, Edward J. Geology and hazards of the Bohlman Road region, Santa Clara County, California. M, 1978, San Jose State University. 136 p.

Porter, Elise White. Petrographic, geochemical and isotopic investigation of the Golden Sunlight Deposit, Jefferson County, Montana. D, 1983, Indiana University, Bloomington. 166 p.

Porter, Elise White. Structural analysis of the Sandy Hollow Decollement. M, 1977, Indiana University, Bloomington. 61 p.

Porter, Gerald. Sand provenance of southern Mentevey Bay California. M, 1978, University of South Carolina.

Porter, J. Stratigraphy and depositional environments of the Norwalk Member of the Jordan Formation (Upper Cambrian), southwestern Wisconsin. M, 1978, University of Wisconsin-Madison.

Porter, James W. Biological, physical, and historical forces structuring coral reef communities on opposite sides of the Isthmus of Panama. D, 1973, Yale University.

Porter, John B. The iron ores of the region of southern Tennessee and the surrounding states. D, 1884, Columbia University, Teachers College.

Porter, John Robert, Jr. Clay mineral geochemistry of the Kramer borate district, Kern County, California. M, 1964, University of Oklahoma. 85 p.

Porter, John Seaman. Studies of the Aux Vases Formation in a portion of Hamilton County, Illinois. M, 1954, University of Illinois, Urbana.

Porter, Joseph A. Bedrock geology of the part of the Gorham and Wolf Lake quadrangles, Illinois. M, 1963, Southern Illinois University, Carbondale. 85 p.

Porter, Karen W. Stratigraphic model for the Upper Cretaceous (Campanian) Hygiene Member, Pierre Shale, West Denver Basin, Colorado. D, 1976, Colorado School of Mines. 142 p.

Porter, Lee. Ecology of a late Pleistocene (Wisconsin) ungulate community near Jack Wade, east-central Alaska. M, 1979, University of Washington. 85 p.

Porter, Lee. Late Pleistocene fauna of Lost Chicken Creek, Alaska. D, 1984, Washington State University. 200 p.

Porter, Leonard A. Biofacies analysis of middle Trenton limestones of New York and Vermont. M, 1970, Rensselaer Polytechnic Institute.

Porter, Michael L. Sedimentology and petrology of the Chalk Hills Formation, southwestern Idaho. M, 1982, Northern Arizona University. 141 p.

Porter, Michael Lowry. Sedimentology and petrology of ancient erg margin; Aztec Sandstone, southern Nevada and Southern California. D, 1985, University of Wisconsin-Madison. 237 p.

Porter, Philip Weldon. Geology of the lower Sucker Creek area, Mitchell Butte Quadrangle, Oregon. M, 1953, University of Oregon. 81 p.

Porter, Richard W. Geology of the Facey Rock area, Etna Quadrangle, California. M, 1974, Oregon State University. 87 p.

Porter, Robert Bowden. Geology of west half of Foster Quadrangle, Culberson County, Texas. M, 1951, University of Texas, Austin.

Porter, Samuel G. The geology along a portion of the Wills valley fault, DeKalb and Etowah counties, Alabama. M, 1968, University of Tennessee, Knoxville. 54 p.

Porter, Stephen Cummings. Geology of Anaktuvuk Pass, central Brooks Range, Alaska. D, 1962, Yale University. 402 p.

Porter, Suzanne. Factors influencing stream/aquifer interactions in the Neponset River basin, Westwood, Norwood, and Canton, Massachusetts. M, 1988, Boston University. 329 p.

Porter, William M. Structural and geophysical studies in the Delaware-Pennsylvania Piedmont. M, 1976, University of Delaware.

Portig, Elisabeth R. An investigation of the chemical stratigraphy of the Ordovician bentonites in central Kentucky. M, 1983, University of Kentucky. 74 p.

Portman, Mark E. Computer enhancement techniques for a seismic reflection line in an area of thick glacial till. M, 1986, Wright State University. 89 p.

Portnoy, Michael B. A study of two isolated Cambrian stromatolitic outcrops, Mason and McCulloch counties, Texas. M, 1987, Texas Tech University. 126 p.

Porto, Everaldo Rocha. An economic evaluation of selected soil and water management technologies for rainfed agriculture; a study case in the arid zones of Brazil. D, 1988, University of Arizona. 345 p.

Portugal Tejada, Jorge A. General geology and structure of the Hump area between Ray and Superior, Pinal County, Arizona. M, 1961, University of Cincinnati. 45 p.

Portugal Tejada, Jorge A. Geology of the Puno-Santa Lucia area, Department of Puno, Peru. D, 1964, University of Cincinnati. 141 p.

Porturas-Plaza, Antonio. Granitization in the Lake Pelesier area, formerly known as Lake Mary, Marquette County, Michigan. M, 1945, Michigan Technological University. 65 p.

Posada, Jorge H. Karst-related features and controls on ore mineralization in the Leadville Formation, central Colorado. M, 1973, Colorado School of Mines. 107 p.

Posadas, Veronica Gomez de. Rb-Sr whole rock age in the Imataca Complex, Venezuela. M, 1966, Massachusetts Institute of Technology. 32 p.

Posamentier, H. W. Glaciation of the Schobergruppe, East Tyrol, Austria. D, 1976, Syracuse University. 213 p.

Posehn, Gary Arnold. The environment of deposition of the Booster Lake conglomerate and greywacke. M, 1976, University of Manitoba.

Posen, Harold. The application of linear operators to geophysical problems. M, 1956, Massachusetts Institute of Technology. 90 p.

Posey, Donald Rue. Petrology of Mississippian cherts, Story County, Iowa. M, 1955, Iowa State University of Science and Technology.

Posey, Ellen. The Hunton of Kansas. M, 1932, University of Oklahoma. 25 p.

Posey, Harry H. Regional characteristics of strontium, carbon and oxygen isotopes in salt dome cap rocks of the western Gulf Coast. D, 1986, University of North Carolina, Chapel Hill. 248 p.

Posey, Harry Howard. Brecciation, mineralization and facies relationships of Cambro-Ordovician carbonates in the Arbuckle Mountains and Southern Oklahoma Aulacogen, Oklahoma. M, 1979, University of Missouri, Rolla.

Posner, Alex. The effect of progressive urbanization of the chemical variability and hydrogeologic regime of several streams on Long Island, NY. M, 1980, Indiana State University. 74 p.

Posnick, Allan Edward. A floristic study of the Upper Devonian Perry Formation in Maine. M, 1982, Rutgers, The State University, Newark. 128 p.

Possin, Boyd Nelson. The hydrogeology of Mirror and Shadow lakes in Waupaca, Wisconsin. M, 1973, University of Wisconsin-Madison. 84 p.

Post, Edwin Vaulton, Jr. Paragenesis of the lead-zinc ores of Pend Oreille and Stevens counties, Washington. M, 1953, Washington State University. 38 p.

Post, Eugene. Conodont biostratigraphy of the St. Joe Formation (Lower Mississippian), north-central Arkansas. M, 1982, University of Arkansas, Fayetteville.

Post, Jeffrey Edward. Characterization of particles in the Phoenix aerosol, and structure refinements of hollandite minerals. D, 1981, Arizona State University. 340 p.

Post, Paul T. A phase of the metamorphism of the Kekequabic Granite, Lake Co., Minnesota. M, 1924, Northwestern University.

Post, Richard Edward. Ground-water contamination near a road salt storage site in Coventry Township, Summit County, Ohio. M, 1980, University of Akron. 163 p.

Post, Robert Louis, Jr. Flow laws of Mount Burnett Dunite (Southeast Alaska) at 700°C - 1400°C. D, 1972, University of California, Los Angeles.

Post, Robert Louis, Jr. The flow laws of Mount Burnett Dunite. D, 1973, University of California, Los Angeles.

Post, Tim E. Geology and mineralogy of the Calliham uranium-vanadium mine, Sage Plains, southeastern Utah. M, 1981, New Mexico Institute of Mining and Technology. 103 p.

Postawko, Susan Elaine. Martian paleoclimate. D, 1983, University of Michigan. 147 p.

Postel, Albert Williams. Hydrothermal emplacement of granodiorite near Philadelphia. D, 1939, University of Pennsylvania.

Postel, Albert Williams. The petrology and correlation of some Recent deposits of archaeological interest near Clovis, New Mexico. M, 1935, University of Pennsylvania.

Postell, William Dosite. Contributions to the geology of the Front Range of Colorado. M, 1932, Louisiana State University.

Postlethwaite, Clay Edward. The structural geology of the Red Cloud thrust system, southern eastern Transverse Ranges, California. D, 1988, Iowa State University of Science and Technology. 149 p.

Postlethwaite, Clay Edward. The structural geology of the western Rand Mountains, northwestern Mojave Desert, California. M, 1983, Iowa State University of Science and Technology. 91 p.

Potamianos, Socrates N. A laboratory study of the effect of wettability on residual oil saturation in artificial cores. M, 1964, University of Missouri, Rolla.

Poth, Charles Warner. A study of the Pleistocene sediments and glacial geology in the valley of Ischua Creek, New York. M, 1953, SUNY at Buffalo.

Poth, Charles Warner. Geohydrology of the Mercer Quadrangle in Northwest Pennsylvania. D, 1964, Pennsylvania State University, University Park. 241 p.

Poth, Stephen. Structural transition between the Santiago and Del Carmen mountains in northern Big Bend National Park, Texas. M, 1979, University of Texas, Austin.

Pothacamury, Innaiah. Magnetic properties of the Boulder Batholith near Helena, Montana, and their use in magnetic interpretation. M, 1970, Michigan State University. 64 p.

Potisat, Somsak. Copper and uranium deposits in the red beds of the Connecticut Valley. M, 1978, Wesleyan University. 123 p.

Potluri, Ramamohan Rao. Petrology of Atlantic Coastal Plain phosphate deposits. M, 1971, University of Georgia. 76 p.

Potochnik, Mark. Petrology and depositional environment of the Lower Jurassic Moenave/Kayenta-equivalent strata, southeastern Spring Mountains, southern Nevada. M, 1985, Southern Illinois University, Carbondale. 101 p.

Potocki, Sigmund R. Geology of a portion of Black Mountain (Park County, Wyoming). M, 1962, Wayne State University.

Potosky, Robert. The application of cathodoluminescence in the field of petrology. M, 1967, University of Tennessee, Knoxville. 51 p.

Potratz, Victoria Yeko. Ground-water geochemistry of Ogalla Aquifer in the southern High Plains of Texas and New Mexico. M, 1980, Texas Tech University. 107 p.

Potter, Alfred Warren. Stratigraphy and selected Ordovician brachiopods from the Horseshoe Gulch and Gregg Ranch areas, eastern Klamath Mountains, Northern California. D, 1988, Oregon State University. 367 17 plates 8 fig 2 tables p.

Potter, Benjamin A. Secular trends in the petrology of some Mesozoic sandstones, Wind River basin, Wyoming. M, 1977, University of Missouri, Columbia.

Potter, C. W. Lower Ordovician conodonts of the upper West Spring Creek Formation, Arbuckle Mountains, Oklahoma. M, 1975, University of Missouri, Columbia.

Potter, Christopher D. Petrology and chemistry of modified quartz metadiabase intrusives and their associated wall rocks, Quirke Syncline, Blind River-Elliot Lake District, Ontario. M, 1987, Bowling Green State University. 144 p.

Potter, Christopher John. Geology of the Bridge River Complex, southern Shulaps Range, British Columbia; a record of Mesozoic convergent tectonics. D, 1983, University of Washington. 192 p.

Potter, Darrell L. The lower and lower middle Silurian of the Michigan Basin. M, 1975, Michigan State University. 36 p.

Potter, Dean Edward George. Zinc–lead mineralization in the Wollaston Group stratigraphy, Sito—Fable lakes area, Saskatchewan. M, 1980, University of Regina. 118 p.

Potter, Delbert E. Spores and pollen in a Cretaceous coal in the Omadi Formation (Cretaceous), Dakota Group, of Cimarron County, Oklahoma. M, 1963, New York University.

Potter, Donald Brandreth. Geology of the Fortune-Grand Bank area, Burin Peninsula, Newfoundland, Canada. M, 1949, Brown University.

Potter, Donald Brandreth. High-alumina metamorphic rocks of the Kings Mountain District, North Carolina and South Carolina. D, 1954, California Institute of Technology. 204 p.

Potter, Donald Brandreth, Jr. The Chesler Canyon Lineament, Canyonlands National Park, Utah; a structural analysis. M, 1979, University of Massachusetts. 82 p.

Potter, Donald Brandreth, Jr. The stratigraphy and structure of the Loon Pond Syncline, Adirondack Mountains, New York State. D, 1985, University of Massachusetts. 226 p.

Potter, Elizabeth Anne. Silicate liquid inclusions in olivine crystals from Kilauea, Hawaii. M, 1975, Queen's University. 92 p.

Potter, Eric C. Paleozoic stratigraphy of the northern Hot Creek Range, Nye County, Nevada. M, 1976, Oregon State University. 129 p.

Potter, Franklin C. Scolecodonts from the upper Richmond (Ordovician) of Illinois. D, 1933, University of Chicago. 104 p.

Potter, Franklin C. The physiography of the Superior Highland. M, 1928, Northwestern University.

Potter, Jared Michael. Experimental permeability studies at elevated temperature and pressure of granitic rocks. M, 1978, University of New Mexico. 101 p.

Potter, Jared Michael. Experimental rock-water interactions at temperatures to 300°C; implications for fluid flow, solute transport, and silicate mineral zoning in crustal geothermal systems. D, 1982, Stanford University. 188 p.

Potter, John Claude. Modeling and updating site characterization for risk analysis of offshore structures. D, 1982, Ohio State University. 143 p.

Potter, K. W. A stochastic model of the Hurst phenomenon; non-stationarity in hydrologic processes. D, 1976, The Johns Hopkins University. 104 p.

Potter, Kenneth L. Petrology and correlation of the Delleker Formation in the Haskell Peak area, Sierra County, California. M, 1986, California State University, Hayward. 98 p.

Potter, Kenneth Neil. Surface conditions effect on energy exchange at the soil surface. D, 1985, Iowa State University of Science and Technology. 117 p.

Potter, Lee Shefte. Petrology and petrogenesis of Tertiary igneous rocks, Chico Hills, New Mexico. M, 1988, Iowa State University of Science and Technology. 179 p.

Potter, Lloyd D. The sedimentation of the Scenic Member of the Brule Formation in the Big Badlands of South Dakota. M, 1958, South Dakota School of Mines & Technology.

Potter, Lloyd Dean and Smith, Joseph Blake. Some uranium mines with production in the Black Hills of Wyoming and South Dakota. M, 1958, South Dakota School of Mines & Technology.

Potter, Noel, Jr. Dolomite and heavy mineral distribution in Devonian sediments, Bathurst Island, Northwest Territories. M, 1963, Dartmouth College. 92 p.

Potter, Noel, Jr. Rock glaciers and mass-wastage in the Galena creek area, northern Absaroka mountains, Wyoming. D, 1969, University of Minnesota, Minneapolis. 167 p.

Potter, Paul Edwin. Petrology and origin of the (Cretaceous and Tertiary) Lafayette Gravel (Southeastern United States). D, 1952, University of Chicago. 137 p.

Potter, Phillip M. The geology of the Carlton Quadrangle, Inner Piedmont, Georgia. M, 1981, University of Georgia.

Potter, Ralph Richard. Correlation of geology and aeromagnetic results in the Province of New Brunswick. M, 1959, University of New Brunswick.

Potter, Ralph Richard. The geology of the Burnt hill area (New Brunswick) and ore controls of the Burnt hill tungsten deposit (Devonian). D, 1969, Carleton University. 124 p.

Potter, Robert K. The effect of the reference ellipsoid on the accuracy of long line measurement. M, 1961, Ohio State University.

Potter, Robert W., II. Geochemical, geothermetric, and petrographic investigation of the Rush Creek Mining Distict, Arkansas. M, 1970, University of Arkansas, Fayetteville.

Potter, Robert William, II. The systematics of polymorphism in binary sulfides; I, Phase equilibria in the system mercury-sulfur; II, Polymorphism in binary sulfides. D, 1973, Pennsylvania State University, University Park. 72 p.

Potter, Russell Marsh. The tetravalent manganese oxides; clarification of their structural variations and relationships and characterization of their occurrence in the terrestrial weathering environment as desert varnish and other manganese oxide concentrations. D, 1979, California Institute of Technology. 254 p.

Potter, Steven C. Geology of Baca Canyon, Socorro County, New Mexico. M, 1970, University of Arizona.

Pottinger, Michael H. The source, fate and movement of herbicides in an unconfined, sand and gravel aquifer in Missoula, Montana. M, 1988, University of Montana. 172 p.

Pottmeyer, J. A. The effects of Mt. St. Helens tephra on the water relations and growth of Verbascum thapsus L. D, 1984, Washington State University. 187 p.

Pottorf, Robert J. Interpretation of discordant U-Pb zircon ages. M, 1973, Northern Illinois University. 66 p.

Pottorf, Robert John. Hydrothermal sediments of the Red Sea, Atlantis II Deep; a model for massive sulfide-type ore deposits. D, 1980, Pennsylvania State University, University Park. 205 p.

Pottorff, Edward J. A new approach for simulating heat transfer and groundwater flow in the Leach Hot Springs hydrothermal system, Pershing County, Nevada. M, 1988, University of Nevada. 153 p.

Pottratz, S. W. Lithified carbonate dunes of Oahu, Hawaii. M, 1968, University of Hawaii. 49 p.

Potts, Mark John. Petrochemistry of the Preacher Creek ultramafic body (Precambrian) (eastern border of Laramie Range, southeastern Wyoming). D, 1971, Washington University. 152 p.

Potts, Ray Horton. Cationic and structural changes in Missouri River clays when treated with ocean water. M, 1959, University of Missouri, Columbia.

Potucek, Tony Lee. Preliminary investigation and exploration in the Ranch El Rodeo area, Sonora, Mexico. M, 1978, University of Arizona.

Pough, Frederick Harvey. A study of the morphology and paragenesis of phenacite. D, 1935, Harvard University.

Pough, Frederick Harvey. The Rueppele iron mine. M, 1932, Washington University. 57 p.

Poujol, Michel. High resolution seismic refraction study of the uppermost oceanic crust near the Juan de Fuca Ridge. M, 1988, Oregon State University. 93 p.

Poulin, M. Groundwater contamination near a liquid waste lagoon, Ville Mercier, Quebec. M, 1977, University of Waterloo.

Poulin, R. La Pétrologie et la minéralogie du complexe meta-igne du Lac Masten, Québec. M, 1974, Ecole Polytechnique.

Poulin, Richard. La pétrologie et la minéralogie du complexe meta-igné du Lac Masten, Québec. M, 1975, Ecole Polytechnique.

Pouliot, G. The thermal history of the Monteregian intrusives based on a study of the feldspars. D, 1962, McGill University.

Poulsen, Knud Howard. Archean tectonics and mineralization at Rainy Lake, northwestern Ontario. D, 1985, Queen's University. 341 p.

Poulsen, Knud Howard. The stratigraphy, structure and metamorphism of Archean rocks at Rainy Lake, Ontario. M, 1980, Lakehead University.

Poulter, Glenn Joseph. The geology of the Georgetown thrust area southwest of Philipsburg, Montana. D, 1957, Princeton University. 279 p.

Poulton, T. P. Jurassic Trigoniidae of Western Canada and United States, and a review of the family. D, 1974, Queen's University. 412 p.

Poulton, Terrence Patrick. Stratigraphy and sedimentology, Horsethief Creek Formation (late Proterozoic), northern Dogtooth Mountains (Rogers Pass area, southeastern British Columbia). M, 1970, University of Calgary. 118 p.

Pound, James Hannon, Jr. Recent stream sedimentation in the vicinity of Stone Mountain, DeKalb County, Georgia. M, 1957, Emory University. 75 p.

Pound, Wayne R. The geology and hydrocarbon potential of the Dawson Bay Formation carbonate unit (Middle Devonian), Williston Basin, North Dakota. M, 1985, University of North Dakota. 320 p.

Pourzadeh-Bousheri, Jalil. Geochemical aspects of paleoecology of Plio-Pleistocene of Florida. M, 1973, University of Florida. 39 p.

Powdysocki, Melvin Henri. A petrographic and chemical study of coal dikes intruding lamprophyre sills in the Purgatoire river valley of Colorado. M, 1968, Pennsylvania State University, University Park. 87 p.

Powdysocki, Melvin Henri. The relationships of fracture traces to geologic parameters in flat-lying sedimentary rocks; a statistical analysis. D, 1974, Pennsylvania State University, University Park. 27 p.

Powe, George Robert. A study of the brachiopod fauna of certain Mississippian formations of central Montana. M, 1937, Montana College of Mineral Science & Technology. 77 p.

Powe, Walker H., III. Mineralogical studies on coexisting saponite and sepiolite. M, 1977, Texas Tech University. 84 p.

Powell, Benjamin Neff. Diatreme near Red Mesa, Utah. M, 1966, Columbia University. 47 p.

Powell, Benjamin Neff. Petrology and chemistry of mesosiderites. D, 1969, Columbia University. 188 p.

Powell, Boyd DeWitt Hartley, Jr. The subsurface geology of Woodward County, Oklahoma. M, 1953, University of Oklahoma. 64 p.

Powell, Christine A. Mantle heterogeneity; evidence from large seismic arrays. D, 1976, Princeton University. 340 p.

Powell, Clarence Cave. The geology of the Bunch area, Adair and Sequoyah counties, Oklahoma. M, 1951, University of Oklahoma. 80 p.

Powell, Darron Lee. The structure and stratigraphy of the Early Cretaceous of the southernmost East Potrillo Mountains, Dona Ana County, New Mexico. M, 1983, University of Texas at El Paso. 126 p.

Powell, Dean Keith. The geology of southern House Range, Millard County, Utah. M, 1959, Brigham Young University. 49 p.

Powell, Edward Reed. Olympic manganese ores. M, 1917, University of Washington. 23 p.

Powell, Harriet. Archeomonadaceae from lower beds of the Calvert Formation (Miocene) from Calvert County, Maryland. M, 1983, University of Rhode Island.

Powell, Heidi Sara. Decomposition of organic matter in estuarine sediments by sulfate reduction; a field study from Yaquina Bay and sediment incubation experiments. M, 1980, Oregon State University. 173 p.

Powell, James Adrian. A statistical study of earthquake occurrence. D, 1972, St. Louis University.

Powell, James Daniel. Microfauna of the Lower portion of the Gaptank Formation. M, 1958, Texas Tech University. 27 p.

Powell, James Daniel. Stratigraphy of Cenomanian-Turonian (Cretaceous) strata, northeastern Chihuahua and adjacent Texas. D, 1961, University of Texas, Austin.

Powell, James L. Strontium isotopic composition and origin of carbonatites. D, 1962, Massachusetts Institute of Technology. 276 p.

Powell, Joe Douglas. Geology of the Blackrock Ridge area, Hot Springs and Fremont counties, Wyoming. M, 1957, University of Wyoming. 118 p.

Powell, John Edwin. Geology of the Columbia Hills, Klickitat County, Washington. M, 1982, University of Idaho. 56 p.

Powell, John R. A hydrogeologic investigation of the Fort Hunter-Liggett Military Reservation, Monterey County, California. M, 1987, San Jose State University. 75 p.

Powell, John S. Stratigraphy and sedimentation of the Gallegos (formerly Ojo Alamo) Sandstone (late Cenozoic), San Juan Basin, New Mexico. M, 1972, University of Arizona.

Powell, L. R. Criteria for coal reserve evaluation of some Breathitt Formation coals in eastern Kentucky using depositional systems analysis. D, 1978, West Virginia University. 235 p.

Powell, L. R. Major causes of variations in suspended sediment concentrations in Chincoteague Bay, Delmarva Peninsula. M, 1974, Millersville University.

Powell, Louis Harvey. A study of the Ozarkian faunas of southeastern Minnesota. D, 1933, University of Minnesota, Minneapolis. 150 p.

Powell, Louise M. Calcium carbonate-magnesium carbonate ratios in the (Devonian) Rogers City and Dundee formations of the Pinconning Field (Michigan). M, 1950, University of Michigan.

Powell, Lynn Gladieux. Petrology and diagenesis of the upper Smackover Formation, Jay Fields, Santa Rosa County, Florida. M, 1984, University of New Orleans. 153 p.

Powell, Michael A. Geochemistry of the brine and related sediment from within the ephemeral saline lake subenvironment of a Holocene playa; Valle El Sobaco, State of Coahuila, Mexico. M, 1985, Wayne State University.

Powell, Michael A. The inorganic geochemistry of two Western U.S. coals; Emery coal field, Utah and Powder River coal field, Wyoming. D, 1987, University of Western Ontario. 254 p.

Powell, Mildred A. Experimental crystallization of the Stannern Meteorite. D, 1981, Harvard University.

Powell, R. L. Some geomorphic and hydrologic implications of jointing in carbonate strata of Mississippian age in South-central Indiana. D, 1976, Purdue University. 169 p.

Powell, R. M. A test of the half-life of Rb^{87}. M, 1957, Massachusetts Institute of Technology. 34 p.

Powell, Raina Rae. Depositional environment and reservoir morphology of the Frio sandstones, Nine Mile Point Field, Aransas County, Texas. M, 1976, Texas A&M University. 193 p.

Powell, Richard Conger. The interaction of chromium ions in ruby crystals. D, 1967, Arizona State University. 245 p.

Powell, Richard Justin. Stratigraphic and petrologic analysis of the middle Eocene Santee Limestone, South Carolina. M, 1981, University of North Carolina, Chapel Hill. 182 p.

Powell, Robert Edward. Geology of the crystalline basement complex, eastern Transverse Ranges, southern California; constraints on regional tectonic interpretation. D, 1981, California Institute of Technology. 504 p.

Powell, Robert Leslie. Deterministic models and uncertainty in the management of regional ground water flow and contaminant transport systems. D, 1983, University of Maryland. 380 p.

Powell, Ross David. Holocene glacimarine sediment deposition by tidewater glaciers in Glacier Bay, Alaska. D, 1980, Ohio State University. 420 p.

Powell, Thomas. The sea-floor spreading history of the East Indian Ocean. M, 1978, University of California, Santa Barbara.

Powell, Thomas Edward, Jr. The geology of the St. Mark's Quadrangle. M, 1924, University of North Carolina, Chapel Hill. 13 p.

Powell, William E. Physiography of a portion of the Cherokee Plains, Oklahoma. M, 1958, University of Arkansas, Fayetteville.

Powell, William Frank. A comparison of some Eocene and modern sediments by coarse fraction analysis. D, 1959, Rice University. 112 p.

Powell, William Frank. A petrologic study of a portion of the lower Beaumont Clay. M, 1957, Rice University. 37 p.

Powell, William I., Jr. and Ryan, Wallace. Establishing high order control points for the Ohio State University campus area. M, 1960, Ohio State University.

Powell, Wyveta. The foraminifera of the (Cretaceous) Goodland Formation of Oklahoma. M, 1941, Smith College. 71 p.

Power, Bruce Andrew. Depositional environments of the Lower Cretaceous (Albian) Viking Formation Joarcam Field, Alberta, Canada. M, 1987, McMaster University. 165 p.

Power, Harry H. The mechanism of polytropic flow in condensate gas wells. D, 1946, University of Pittsburgh.

Power, Jeanne Denise. The Devil's Gate Limestone of the northern Roberts Mountains, central Nevada. M, 1984, University of California, Riverside. 98 p.

Power, Kathleen M. Time variations in energy released through earthquakes in the Mississippi Embayment region. M, 1980, Northern Illinois University. 77 p.

Power, M. Mass movement, seismicity and neotectonics in the northern St. Elias Mountains, Yukon. M, 1988, University of Alberta.

Power, Paul C., Jr. Optical diffraction analysis of petrographic thin sections. M, 1973, University of Wisconsin-Milwaukee.

Power, Peter Edward. Climatic significance of some (Pennsylvanian) Paleosols in western Colorado; Eagle County and vicinity. D, 1963, University of Colorado. 131 p.

Power, Sara Glen. The "tops" of porphyry copper deposits; mineralization and plutonism in the western Cascades, Oregon. D, 1985, Oregon State University. 234 p.

Power, W. Robert, Jr. Geology and petrology of Haiwee Ridge, Inyo County, California. D, 1959, The Johns Hopkins University.

Power, William Laurence. Mechanics of low-angle extensional faulting in the Riverside Mountains, southeastern California. M, 1986, University of California, Santa Barbara. 111 p.

Powers, Brian Kenneth. Depositional environment of Oligocene Hackberry sandstones; southern Jefferson County, Texas. M, 1980, Texas A&M University. 105 p.

Powers, Delmer Lance. The pre-Tertiary cherts of the Pacific Ocean border. M, 1924, Stanford University. 109 p.

Powers, Dennis Wayne. Geology of Mio-Pliocene sediments of the lower Kerio River valley, Kenya. D, 1980, Princeton University. 191 p.

Powers, E. Lloyd. An investigation of the relation between interfacial tension and minimum water saturation in a section of the (Pennsylvanian) Bartlesville Sand (Oklahoma). M, 1948, University of Pittsburgh.

Powers, Elliot Holcomb. Paleozoic stratigraphy and structure in the valleys of Willow and Apple Rivers, Wisconsin. M, 1932, University of Iowa. 121 p.

Powers, Elliot Holcomb. The Prairie du Chien problem. D, 1935, University of Iowa. 123 p.

Powers, Eric R. Diatom biostratigraphy and paleoecology of the Miocene Pungo River Formation, North Carolina continental margin. M, 1987, East Carolina University. 240 p.

Powers, Harold Auburn. Diatomite deposits of southwestern Idaho. M, 1947, University of Idaho. 47 p.

Powers, Howard Adorno. The geology and petrology of the Modoc Lava-bed Quadrangle, California. D, 1929, Harvard University.

Powers, Howard Adorno. The history and petrography of the Siskiyou Batholith, Oregon. M, 1926, University of Oregon. 50 p.

Powers, James. Paleomagnetic analysis of Eocene rocks from the Peninsular Ranges terrane, San Diego, California. M, 1988, University of Southern California.

Powers, Jonathan Andrew. Mineralization and alteration associated with breccia pipe structures, Haw Branch copper deposit, Moore County, North Carolina. M, 1985, North Carolina State University. 105 p.

Powers, Laura J. Ore mineralogy and paragenesis of the Midwest Deposit, northern Saskatchewan. M, 1985, University of Saskatchewan. 156 p.

Powers, Mark William. Sand and gravel deposits in parts of the Spokane (SE and SW) quadrangles, Washington. M, 1976, Eastern Washington University. 52 p.

Powers, Maurice Cary. Black Creek Cretaceous deposits along the Cape Fear River, North Carolina. M, 1951, University of North Carolina, Chapel Hill. 100 p.

Powers, Maurice Cary. The adjustment of land derived clays to the marine environment. D, 1955, University of North Carolina, Chapel Hill. 59 p.

Powers, R. W. Arabian Upper Jurassic carbonate reservoir rocks. D, 1961, Yale University.

Powers, Richard Blake. Geology of the Woods Chapel Quadrangle, Missouri. M, 1952, University of Missouri, Columbia.

Powers, Richard James. The paleoecology of the St. David Shale (Penn.) in western and northern Illinois. M, 1957, University of Illinois, Urbana.

Powers, Russell S. Geology of the Summit Mountains and vicinity, Grant County, New Mexico, and Greenlee County, Arizona. M, 1976, University of Houston.

Powers, Sandra L. Jasperoid and disseminated gold at the Ogee-Pinson Mine, Humboldt County, Nevada. M, 1978, University of Nevada. 112 p.

Powers, Sidney. Geology of the Diamond Hill, Cumberland District, Massachusetts-Rhode Island. M, 1913, Massachusetts Institute of Technology. 105 p.

Powers, Sidney. The Acadian Triassic. D, 1915, Harvard University.

Powers, Stephen John. Stratigraphy and structure in the area of the Cooper Creek Anticline, Virginia. M, 1977, University of Kentucky. 48 p.

Powers, William Edwards. Metamorphism induced by the Kekequabic Granite, at Kekequabic Lake, Lake County, Minnesota. M, 1928, Northwestern University.

Powley, D. Devonian stratigraphy of central Saskatchewan. M, 1951, University of Saskatchewan. 96 p.

Pownell, Leland D. Surface geology of northwestern Lincoln County, Oklahoma. M, 1957, University of Oklahoma. 65 p.

Poyner, William Donald. Geology of the San Guillermo area and its regional correlation, Ventura County, California. M, 1960, University of California, Los Angeles.

Pozzobon, Joseph G. Sedimentology and stratigraphy of the Viking Formation, Eureka Field, southwestern Saskatchewan. M, 1987, McMaster University. 161 p.

Prabhu, Mohan Keshav. Geology, geochemistry and genesis of Montauban lead-zinc deposits. D, 1982, McGill University. 261 p.

Pradham, Bi-swa M. and Singh, Yogendra L. Geology of the area between Virden and Red Rock, Hidalgo and Grant counties, New Mexico. M, 1960, University of New Mexico. 75 p.

Praditan, Surawit. Subsurface geology and fluid migration of Sunrise Field area in Terrebonne Parish, Louisiana. M, 1982, University of Southwestern Louisiana.

Prado, Connie A. de see de Prado, Connie A.

Praetorius, H. W. Stratigraphy of the Permian Phosphoria Formation of northwestern Colorado (Moffat County) and northeastern Utah. M, 1956, University of Colorado.

Prager, Ellen Joyce. The growth and structure of calcareous nodules (for-algaliths) on Florida's outer shelf. M, 1987, University of Miami. 87 p.

Prager, Gerald David. The structure and stratigraphy of the Gardner Mountain area, New Hampshire. D, 1971, University of Cincinnati. 132 p.

Prahl, Fredrick George. The geochemistry of polycyclic aromatic hydrocarbons in Columbia River and Washington coastal sediments. D, 1982, University of Washington. 209 p.

Prajmovsky, Atalia. The geology of an area along Route 20 near Judd's Falls, New York. M, 1960, Smith College. 143 p.

Prall, John Russell. Geology of Rancho Jasay-Las Margaritas, Baja California, Mexico. M, 1981, San Diego State University.

Prammani, Prapath. Geology of the east-central part of the Malad Range, Idaho. M, 1957, Utah State University. 60 p.

Pranger, Harold S. The planimetric patterns of experimental tension fractures and their geomorphic significance to Mars. M, 1988, Colorado State University. 193 p.

Pranschke, Frank A. Stratigraphy of Lake Michigan Formation in Lake Michigan's Southern Basin. M, 1980, Northeastern Illinois University.

Prapaharan, Sinnadurai. Effects of disturbance, strain rate, and partial drainage on pressuremeter test results in clay. D, 1987, Purdue University. 286 p.

Prasad, Jagat Nandan. A paleomagnetic investigation of early Paleozoic rocks in western Newfoundland. D, 1986, Memorial University of Newfoundland. 336 p.

Prasad, Jagat Nandan. Paleomagnetism of the Mesozoic lamprophyre dikes in north-central Newfoundland. M, 1981, Memorial University of Newfoundland. 119 p.

Prasad, Mithilesh Nandan. The geochemistry of inter flow sedimentary rocks and felsic to mafic-ultramafic volcanics from Munro Township, Ontario; the possible role of liquid immiscibility in petrogenesis. D, 1984, McMaster University. 328 p.

Prasad, Nirankar. The study of carbon and oxygen isotope ratios in some Silurian carbonate rocks. M, 1967, McMaster University. 132 p.

Prasad, Schindra. Microsucrosic dolomite from the Hawthorn Formation (Miocene) of Florida; distribution and development. M, 1985, University of Miami. 163 p.

Prasetyo, Hardi. Marine geology and tectonic development of the Banda Sea region, eastern Indonesia; a model of an "Indo-borderland" marginal basin. D, 1988, University of California, Santa Cruz. 450 p.

Prasse, Eric Martin. Uranium and its relationship to host rock mineralogy in an unoxidized roll front in the Jackson Group, south Texas. M, 1978, Texas A&M University. 51 p.

Prat, Giovanni C. Da *see* Da Prat, Giovanni C.

Prather, Barry W. Seismic anisotropy in the Vaughan Lewis Glacier, Juneau Icefield, Alaska, 1969. M, 1972, Michigan State University. 72 p.

Prather, Bradford E. Petrology and diagenesis of the D-zone cyclothem of the Lansing-Kansas City groups, Hitchcock County, Nebraska. M, 1981, University of New Orleans. 97 p.

Prather, Jesse Preston. The geology of eastern Monroe County, Georgia. M, 1971, University of Georgia. 82 p.

Prather, John K. The Cretaceous clays at Atlantic Highlands, New Jersey. M, 1905, Columbia University, Teachers College.

Prather, Thomas Leigh. Geology of the North Italian Mountain area, Gunnison County, Colorado. M, 1961, University of Colorado.

Prather, Thomas Leigh. Stratigraphy and structural geology of the Elk Mountains (Pitkin and Gunnison counties), Colorado. D, 1964, University of Colorado. 153 p.

Pratson, Lincoln. Recent sedimentation on the continental rise off the Eastern United States. M, 1987, University of Rhode Island.

Pratt, Alan Rogers. The geomorphology and geomorphic history of Bean Bossom Valley, Monroe and Brown counties, Indiana. M, 1960, Indiana University, Bloomington. 77 p.

Pratt, Ann E. The taphonomy and paleoecology of the Thomas Farm local fauna (Miocene, Hemingfordian), Gilchrist County, Florida. D, 1986, University of Florida. 487 p.

Pratt, Brian Richard. The St. George Group (Lower Ordovician), western Newfoundland; sedimentology, diagenesis, and cryptalgal structures. M, 1979, Memorial University of Newfoundland. 254 p.

Pratt, Daniel Allen. The hydrogeology of a mixed pine and cypress commercial forest in Bradford County, Florida. M, 1978, University of Florida. 103 p.

Pratt, David E. Sedimentation in the North San Clemente Basin, California continental borderland. M, 1979, Rice University. 110 p.

Pratt, Ernest George. Petrologic and economic evaluation of Blanchet Island, Northwest Territories. M, 1974, University of Alberta. 80 p.

Pratt, Ernest S. The Tonkawa Field of north-central Oklahoma. M, 1923, University of Oklahoma. 91 p.

Pratt, Franklin Pierce. Marine geology of Long Bay (North and South Carolina continental shelf). M, 1970, Duke University. 88 p.

Pratt, Howard Riley. Progressive metamorphism of calc-silicate rocks, Connecticut. D, 1965, University of Rochester. 99 p.

Pratt, Howard Riley. The petrology of the Lyme Granite Gneiss (Connecticut). M, 1962, University of Rochester. 78 p.

Pratt, J. Lynn. Crystal structures of some pyrite-type minerals. M, 1977, University of Calgary. 76 p.

Pratt, Jennifer Adams. Paleoenvironment of the Eocene/Oligocene Hancock Mammal Quarry of the upper Clarno Formation, Oregon. M, 1988, University of Oregon. 104 p.

Pratt, Joseph H. Northupite; pirssonite, a new mineral; gaylussite; and hanksite, from Borax Lake, San Bernardino County, California. D, 1896, Yale University.

Pratt, Lisa M. The stratigraphic framework and depositional environment of the Lower to Middle Silurian Massanutten Sandstone. M, 1979, University of North Carolina, Chapel Hill. 76 p.

Pratt, Lisa Mary. A paleo-oceanographic interpretation of the sedimentary structures, clay minerals, and organic matter in a core of the Middle Cretaceous Greenhorn Formation drilled near Pueblo, Colorado. D, 1982, Princeton University. 176 p.

Pratt, Richard Murray. Geology of the Deception Pass area, Chelan, King, and Kittitas counties, Washington. M, 1954, University of Washington. 58 p.

Pratt, Richard Murry. The geology of the Mount Stuart area, Washington. D, 1958, University of Washington. 229 p.

Pratt, Sandra. Clay mineral authigenesis as related to pore geometry of sandstones in the Medina Group (Lower Silurian), New York State. M, 1980, SUNY at Binghamton. 72 p.

Pratt, Thomas Lee. A geophysical investigation of a concealed granitoid beneath Lumberton, North Carolina. M, 1982, Virginia Polytechnic Institute and State University. 86 p.

Pratt, Thomas Lee. A geophysical study of the Earth's crust in central Virginia with implications for lower crustal reflections and Appalachian crustal structure. D, 1986, Virginia Polytechnic Institute and State University. 81 p.

Pratt, Walden Penfield. Geology of the Marble Mountains area, Siskiyou County, California. D, 1964, Stanford University. 116 p.

Pratt, Willis Layton, Jr. Geology of the northwest corner of the Joaquin Rocks Quadrangle. M, 1951, University of California, Berkeley. 63 p.

Pratt, Willis Layton, Jr. The origin and distribution of glauconite from the sea floor off California and Baja California. D, 1963, University of Southern California. 296 p.

Pratte, J. Frances. The paleomagnetism of Recent sediments; an indicator of past geomagnetic intensity variations. M, 1971, Rensselaer Polytechnic Institute. 69 p.

Prave, Anthony Robert. An interpretation of Upper Cretaceous sedimentation and tectonics and the nature of Pyrenean deformation in the northwestern Basque Pyrenees. D, 1986, Pennsylvania State University, University Park. 298 p.

Prave, Anthony Robert. Stratigraphy, sedimentology, and petrography of the Lower Cambrian Zabriskie Quartzite in the Death Valley region, southeastern California and southwestern Nevada. M, 1984, Pennsylvania State University, University Park. 193 p.

Pray, Lloyd C. Stratigraphy of the escarpment of the Sacramento Mountains, Otero County, New Mexico. D, 1952, California Institute of Technology. 370 p.

Pray, Lloyd C. Studies of certain Sierran concrete aggregates. M, 1943, California Institute of Technology. 74 p.

Pray, Lloyd C. The Mocam bastnaesite deposit, San Bernardino County, California. D, 1952, California Institute of Technology. 50 p.

Preble, Harold Douglas. Paleontology and stratigraphy of a well core from Garfield Township, Newaygo County, Michigan. M, 1951, University of Michigan.

Preece, Richard Kellar, III. Paragenesis, geochemistry, and temperatures of formation of alteration assemblages at the Sierrita Deposit, Pima County, Arizona. M, 1979, University of Arizona. 106 p.

Pregger, Brian H. Geology and origin of the Cobachi bedded barite deposit, northern Sierra de Cobachi, Sonora, Mexico. M, 1984, Northern Arizona University. 125 p.

Pregill, Gregory K. Late Pleistocene herpetofaunas from Puerto Rico. D, 1979, University of Kansas. 198 p.

Prehmus, Cynthia Anne. Classification and distribution of carbonate sediments on the U. S. Virgin Islands platform using Fourier shape analysis. M, 1981, Duke University. 98 p.

Prehoda, W. P. The investigation of two lineaments in the Mount Marcy, New York region as possible fault zones. M, 1988, SUNY at Binghamton. 71 p.

Preisig, Joseph Richard Mark. Relationships of earthquakes (and earthquake-associated mass movements) and polar motion as determined by Kalman filtered, very-long-baseline-interferometry group delays. D, 1988, SUNY at Binghamton. 241 p.

Prelat, Alfredo Eduardo. Statistical estimation of wildcat well outcome probabilities by visual analysis of structure contour maps of Stafford County, Kansas. D, 1974, Stanford University. 103 p.

Prell, Warren L. Late Pleistocene faunal, sedimentary, and temperature history of the Colombia Basin, Caribbean Sea. D, 1974, Columbia University. 517 p.

Premo, Bette Jayne. A model for concentration of total phosphorus and chlorophyll A in a small, eutrophic lake. D, 1982, Michigan State University. 91 p.

Premo, Wayne R. U-Pb zircon geochronology of some Precambrian rocks of the Sierra Madre Range, Wyoming. M, 1984, University of Kansas. 106 p.

Prenosil, Wolfgang Peter. A geochemical study of thermal and non-thermal groundwater in the vicinity of Cahuilla Spring, Southern California. M, 1988, University of California, Riverside. 78 p.

Prensky, S. E. Carbonate stratigraphy and related events, California continental borderland. M, 1973, University of Southern California. 926 p.

Prentice, Michael L. Surficial geology and stratigraphy in central Wright Valley, Antarctica; implications for Antarctic Tertiary glacial history. M, 1982, University of Maine. 248 p.

Prentice, Michael Lanman. The deep sea oxygen isotopic record; significance for Tertiary global ice volume history, with emphasis on the latest Miocene/early Pliocene. D, 1988, Brown University. 576 p.

Prentki, R. T. Phosphorus cycling in tundra ponds. D, 1976, University of Alaska, Fairbanks. 293 p.

Presant, Edward W. A trace element study of some selected soil profiles from Bathurst District of New Brunswick. M, 1963, Carleton University.

Presch, William Frederick, Jr. The evolution of macroteid lizards; an osteological interpretation. D, 1970, University of Southern California. 266 p.

Prescott, Basil. The main mineral zone of the Santa Eulalia District, Chihuahua (Mexico). M, 1915, Stanford University. 43 p.

Prescott, Glenn C. Geology and ground-water resources of Lane County, Kansas. M, 1950, Brown University.

Prescott, John Whitman. Petrogenesis of ultramafic xenoliths from the Canadian Cordillera and Alaska. M, 1983, McGill University. 186 p.

Prescott, Max W. Geology of the northwest quarter of the Soldier Summit Quadrangle, Utah. M, 1958, Brigham Young University. 44 p.

Prescott, William Herbert. The accommodation of relative motion along the San Andreas fault system in California. D, 1981, Stanford University. 199 p.

Presley, Bobby Joe. Chemistry of interstitial water from marine sediments. D, 1969, University of California, Los Angeles. 241 p.

Presley, M. W. A depositional systems analysis of the upper Mauch Chunk and Pottsville groups in northern West Virginia. D, 1977, West Virginia University. 186 p.

Presley, Mark Whitehead. Igneous and metamorphic geology of the Willow Creek drainage basin, southern Sapphire Mountains (Ravalli County), Montana. M, 1971, University of Montana. 64 p.

Presley, Marsha Ann. The origin and history of surficial deposits in the central equatorial region of Mars. M, 1986, Washington University. 77 p.

Presley, Olan Dee. Ostracoda of the Pecan Gap (Cretaceous) Formation of northeastern Texas. M, 1965, University of Oklahoma. 81 p.

Presley, Susan. Morphometric analysis of Glabrocingulum (Gastropoda, Pleurotomariacea) from Pennsylvanian strata of the Appalachian Basin. M, 1983, University of Pittsburgh.

Presnall, Dean Carl. The join Mg_2SiO_4-$CaMgSi_2O_6$-iron oxide at oxygen pressures from 0.21 to 10^{-8} atmospheres. D, 1963, Pennsylvania State University, University Park. 142 p.

Presnell, Ricardo Davis. Structural model for the Sevier Desert and environs, Utah. M, 1983, University of Michigan.

Press, Frank. Theoretical magnetic anomalies of three oceanic structures. M, 1946, Columbia University, Teachers College.

Press, Frank. Two applications of normal mode sound propagation in the ocean. D, 1949, Columbia University, Teachers College.

Pressburger, Alexander. Silicate (chlorite)-sulphide (Po-Py) relationships. M, 1970, McGill University. 67 p.

Pressler, Edward D. Contribution to the paleontology and stratigraphy of the upper part of the Fernando Group of the Las Posas-South Mountain District, Ventura County, California. M, 1928, University of California, Berkeley. 23 p.

Prest, Victor Kent. An investigation of the pre-Cambrian volcanic centres of the Flin Flon area (Manitoba). M, 1936, University of Manitoba.

Prest, Victor Kent. The Precambrian of the Miminiska-Fort Hope area (Canada). D, 1941, University of Toronto.

Prestegaard, Karen. Relative importance of roughness components and their influence on channel slope. M, 1979, University of California, Berkeley.

Prestegaard, Karen Leah. Variables influencing water surface slope in gravel and coarse sand streams. D, 1982, University of California, Berkeley. 158 p.

Presto, Vittorio Annibale Guiseppe. Structural relations between the Shuswap Terrane and the Cache Creek Group in southern British Columbia. M, 1964, University of British Columbia.

Preston, Charles Dean. The paleocurrents of the Red Mountain Formation (Silurian) of Georgia. M, 1965, Emory University. 45 p.

Preston, John Kante. The origin of the Tichka plutonic massif, Morocco. M, 1975, Michigan State University. 46 p.

Preston, Michael B. Quantitative geomorphology and hydrology of Chippewa Creek basin, Cuyahoga County, Ohio. M, 1987, University of Akron. 76 p.

Preston, Michael M. Geology of a portion of the Byron Hot Springs Quadrangle, Contra Costa County, California. M, 1965, University of Southern California.

Preston, Ralph J. A gravity study of a portion of the Rio Grande Embayment, Texas. M, 1970, Texas A&M University. 89 p.

Prestridge, Jefferson D. A subsurface stratigraphic study of the Sycamore Formation in the Ardmore Basin. M, 1957, University of Oklahoma. 62 p.

Prete, Anthony Del *see* Del Prete, Anthony

Prete, Anthony del *see* del Prete, Anthony

Preto, Victor A. G. Structure and petrography of the Grand Forks Group (Precambrian or Paleozoic, Shuswap terrane), British Columbia. D, 1967, McGill University.

Pretorius, Eugene B. Equilibrium measurements with a bearing on the distribution of nickel between crystalline and liquid phases in silicate systems. M, 1986, Pennsylvania State University, University Park. 114 p.

Pretzer, Elizabeth J. Geology of the vicinity of St. Albans Bay, Vermont. M, 1945, Columbia University, Teachers College.

Preuss, Charlotte. The American Piedmont and the Norwegian strandflat; a comparison. M, 1934, Columbia University, Teachers College.

Prevec, Stephen Anthony. The petrology, geochronology and geochemistry of the White Bear Arm Complex and associated units, Grenville Province, eastern Labrador. M, 1987, McMaster University. 155 p.

Prevey, John Leo. The petrology of some loess from Missouri and Massachusetts. M, 1954, University of Missouri, Columbia.

Prevost, Dana Victor. Geology of a transect of the northern Argus Range, California. M, 1984, California State University, Northridge. 105 p.

Prewit, Billie Neil. Subsurface geology of the Cretaceous coastal plain, southern Oklahoma. M, 1961, University of Oklahoma. 82 p.

Prewitt, Charles T. Crystal structure of gaynite, $Ca_2B(OH)_4AsO_4$. M, 1960, Massachusetts Institute of Technology. 83 p.

Prewitt, Charles T. Structures and crystal chemistry of wollastonite and pectolite. D, 1962, Massachusetts Institute of Technology. 303 p.

Prewitt, Ronald H. Crustal thickness in central Texas as determined by Rayleigh wave dispersion. M, 1969, Texas Tech University. 33 p.

Prezbindowski, Dennis. Comparative value of Fourier shape, mosaic-structural, and binary skeletal characters in a Late Devonian bryozoan fauna (Three Forks Fm, Montana). M, 1974, Michigan State University. 113 p.

Prezbindowski, Dennis Robert. Carbonate rock-water diagenesis, Lower Cretaceous, Stuart City Trend, South Texas. D, 1981, University of Texas, Austin. 252 p.

Preziosi, Gregory Joseph. Velocity structure of the Precambrian basement in southeastern and south-central Kentucky determined from seismic refraction profiling. M, 1985, University of Kentucky. 135 p.

Price, Alan Paul. Mescalero Sandhills of Cochran and Yoakum counties, Texas. M, 1987, Texas Tech University. 246 p.

Price, Annette. An investigation of the effects of modified streamflow of the Sangamon River, Illinois, on the adjacent flood plain. M, 1974, University of Illinois, Urbana.

Price, Arthur S. The fauna of the middle portion of the (Permian) Phosphoria Formation on the east side of the Wind River Mountains, Wyoming. M, 1927, University of Missouri, Columbia.

Price, Barbara Ann. Equatorial Pacific sediments; a chemical approach to ocean history. D, 1988, University of California, San Diego. 394 p.

Price, Barry James. Minor elements in pyrites from the Smithers map area, British Columbia and exploration application of minor element studies. M, 1972, University of British Columbia.

Price, C. Allen. Geochemical behavior of cadmium under simulated marsh conditions. M, 1977, Louisiana State University.

Price, Carol A. Fluid pressure analysis of sandstones in the overpressured Forbes Formation; southern Grimes gas field, Sacramento Valley, California. M, 1986, San Diego State University. 257 p.

Price, Chadderdong. Glacial and drainage history of the upper Cow Creek drainage, Sierra Madre Range, Wyoming. M, 1973, University of Wyoming. 90 p.

Price, Charles Edgar, Jr. Surficial geology of the Las Vegas Quadrangle, Nevada. M, 1966, University of Utah. 60 p.

Price, Charles Errol. Gravity and seismic survey of a karst region located in Harrill Heights Subdivision, Knoxville, Tennessee. M, 1976, University of Tennessee, Knoxville. 54 p.

Price, Charles Raines, Jr. Transportational and depositional history of the Wedington Sandstone (Mississippian), Northwest Arkansas. M, 1981, University of Arkansas, Fayetteville.

Price, Curtis Yarnay. Discrimination of lithologic units on the basis of botanical data and Landsat TM spectral data from the Ridge and Valley Province, Pennsylvania. M, 1985, Dartmouth College. 135 p.

Price, David Tennyson. Graphical slope stability design method for highway cuts in rock. D, 1972, University of Utah. 169 p.

Price, Donald R. Hydrogeologic study of groundwater-surface water interactions at Topaz Lake, Nevada. M, 1981, University of Nevada. 187 p.

Price, Edwin Henry. Stratigraphy, structure, and metamorphic history of the southern half of the Nottely Dam Quadrangle, Georgia-North Carolina. M, 1977, University of Georgia.

Price, Edwin Henry. Structural geometry, strain distribution, and mechanical evolution of eastern Umtanum Ridge and a comparison with other selected localities within Yakima fold structures, South-central Washington. D, 1982, Washington State University. 196 p.

Price, F. T. A sulfur isotope study of Illinois Basin coals. D, 1977, Purdue University. 215 p.

Price, Floyd Ray. The experimental annealing of artificially formed galena aggregates at temperatures from 200°C to 700°C. M, 1973, University of Michigan.

Price, Garry L. The clay mineral and element distribution of the limestone beds in the Magoffin Member of the the Breathitt Formation (Middle Pennsylvanian) in Morgan and Magoffin counties, Kentucky. M, 1980, Eastern Kentucky University. 60 p.

Price, Georgina. Geology and mineralisation, Taylor-Windfall gold prospect, British Columbia, Canada. M, 1986, Oregon State University. 144 p.

Price, Jack Rex. Structure and stratigraphy of the Slate Jack Canyon area, Long Ridge, Utah. M, 1951, Brigham Young University. 75 p.

Price, James Arra. The geology of the coal fields of Indiana. M, 1899, Indiana University, Bloomington.

Price, James Edward. Consideration of environmental quality in Bureau of Reclamation planning. D, 1979, Stanford University. 306 p.

Price, James Kennedy. A new approach to the Rensselaer Graywacke problem (New York). M, 1956, Cornell University.

Price, James N. Geology for planning in eastern Middle Park, Grand County, Colorado. M, 1976, Colorado State University. 164 p.

Price, John E. A comparison of traditional and automatic methods for retrieval of geological well data, Washington County, Iowa. M, 1967, University of Iowa. 79 p.

Price, Jonathan Greenway. Geological history of alteration and mineralization at the Yerington porphyry copper deposit, Nevada. D, 1977, University of California, Berkeley. 171 p.

Price, Katherine H. A geologic investigation of the Turkey Run-Shades area, Indiana. M, 1975, DePauw University.

Price, Kevin Paul. Detection of soil erosion with thematic mapper (TM) satellite data within pinyon-juniper woodlands. D, 1987, University of Utah. 197 p.

Price, L. G. Ostracode communities from Lake Chichancanab, Yucatan, Mexico. M, 1974, Washington University. 49 p.

Price, Larry Wayne. Morphology and ecology of solifluction lobe development, Ruby range, Yukon Territory. D, 1970, University of Illinois, Urbana. 337 p.

Price, Leigh Charles. The solubility of hydrocarbons and petroleum in water as applied to the primary migration of petroleum. D, 1973, University of California, Riverside. 312 p.

Price, Maurice Carlton. Geology of the southeastern Puente Hills (California). M, 1953, Pomona College.

Price, Michael Glyn. A study of sediments from the Juan de Fuca Ridge, Northeast Pacific Ocean; with special reference to hydrothermal and diagenetic components. M, 1981, University of British Columbia. 141 1 plate p.

Price, Michael Lee. Trace element distribution in the native copper from the Centennial Number 3 Mine, Houghton County, Michigan. M, 1977, Michigan State University. 106 p.

Price, Michael Louis. The stratigraphy and environments of deposition of the Birmingham Shale (upper Pennsylvanian) of the Pittsburgh area. M, 1970, University of Pittsburgh.

Price, Myron W. Underthrust faulting at Clayton, Oklahoma. M, 1963, University of Wisconsin-Madison.

Price, P. The geology and ore deposits of the Horne Mine, Noranda, Quebec. D, 1933, McGill University.

Price, Pamela J. Mineral and chemical analysis of the subaerially exposed Paoli-Beaver Bend Member of the Newman Limestone, northeastern Kentucky, Olive Hill. M, 1986, Bowling Green State University. 68 p.

Price, Patricia E. Preliminary survey of sources of geoscientific information in and about the Pacific Island Nations. M, 1985, University of Hawaii at Manoa. 61 p.

Price, Paul Holland. The Appalachian structural front. D, 1930, Cornell University.

Price, Paul Holland. The coprolite limestone horizon of the (Pennsylvanian) Conemaugh Series in and around Morgantown, West Virginia. M, 1926, West Virginia University.

Price, Peter. Ore deposits of southeastern British Columbia. M, 1926, University of British Columbia.

Price, Peter Elliot. The calculation and interpretation of geothermal gradients in the Commonwealth of Kentucky. M, 1978, University of Kentucky. 113 p.

Price, R. C. Lithologic differentiation by use of remote sensing in Tuscaloosa County, Alabama. M, 1977, University of Alabama.

Price, Raymond Alex. Structure and stratigraphy of the Flathead North map-area (east half), British Columbia and Alberta. D, 1958, Princeton University. 363 p.

Price, Rene Marie. Geochemical investigation of salt water intrusion along the coast of Mallorca, Spain. M, 1988, University of Virginia. 186 p.

Price, Rex Clayton. Stratigraphy, petrography, and depositional environments of the Pawnee Limestone,

Middle Pennsylvanian (Desmoinesian), Midcontinent North America. D, 1981, University of Iowa. 279 p.

Price, Robert C. Shallow coal bed delineation with resistivity mapping in the Illinois Basin. M, 1985, Indiana University, Bloomington. 42 p.

Price, Ronald Harlow. The effect of anhydrite on the mechanical behavior of rock salt. M, 1980, Texas A&M University. 47 p.

Price, Susan Alys Medlicott. An evaluation of dike-flow correlations indicated by geochemistry, Chief Joseph Swarm, Columbia River Basalt. D, 1978, University of Idaho. 343 p.

Price, Susan Gay. Utilization of $^{87/86}$Sr ratios to trace the groundwater flowpath into Cedar Bog Memorial Swamp, Champaign County, Ohio. M, 1987, Wright State University. 43 p.

Price, Vaneaton, Jr. Distribution of trace elements in plutonic rocks of the southeastern Piedmont. D, 1969, University of North Carolina, Chapel Hill. 89 p.

Price, Vaneaton, Jr. Ground water chemistry as a tool for geologic investigations in the southeastern Piedmont. M, 1967, University of North Carolina, Chapel Hill. 29 p.

Price, Vicky Irene. Deposition and diagenesis of the Mississippian Leadville Formation at Molas Lake, San Juan County, Colorado. M, 1981, University of Texas, Austin.

Price, W. A., Jr. The invertebrate fauna of the Pennsylvanian of Maryland. D, 1913, The Johns Hopkins University.

Price, William E., Jr. Rim rocks of Sycamore Canyon, Arizona. M, 1948, University of Arizona.

Price, William Evans. A random-walk simulation model of alluvial-fan disposition. D, 1972, University of Arizona.

Price, William H. The feasibility of exploration by pedogeochemistry in the Kellogg area, Coeur d'Alene mining region, Idaho. M, 1971, Miami University (Ohio). 70 p.

Prichard, Gordon H. Authigenic kaolinite in the Bear Den Member (Paleocene) of the Golden Valley Formation, in southwestern North Dakota. M, 1980, University of North Dakota. 174 p.

Prichett, Wilson, III. Solar power plant impacts; a study of the environmental and economic implications of large scale solar thermal-electric power generation in the southwestern United States. M, 1975, University of Virginia. 174 p.

Prichinello, Katherine A. Earliest Eocene mammalian fossils from the Laramie Basin of southeast Wyoming. M, 1971, University of Wyoming. 40 p.

Priddy, Charles Parrish. A sedimentary analysis of the Cox Formation of Trans-Pecos Texas. M, 1956, Texas Tech University. 42 p.

Priddy, Richard Randall. A petrographic study of the Niagaran rocks of southwestern Ohio and southeastern Indiana. D, 1938, Ohio State University.

Priddy, Richard Randall. Geology of Cranmore Cove and vicinity, Rhea County, Tennessee. M, 1936, Ohio State University.

Pride, C. R. Rare earth element studies of a granulite facies terrain; the Lewisian of N.W. Scotland. D, 1978, Dalhousie University. 240 p.

Pride, Douglas E. Size and heavy mineral studies of the New Richmond Sandstone of Lower Ordovician age (Wisconsin, Illinois, Iowa, and Minnesota). M, 1966, University of Wisconsin-Madison.

Pride, Douglas Elbridge. Geochemical and petrologic investigation of the Seminoe, Atlantic City and Copper Mountain iron formation deposits, Wyoming. D, 1969, University of Illinois, Urbana. 109 p.

Pridmore, Cynthia Lee. The genetic association of mid-Tertiary sedimentation, detachment-fault deformation, and antiformal uplift in the Baker Peaks-Copper Mountains area of southwestern Arizona. M, 1983, San Diego State University. 127 p.

Pridmore, Donald Francie. Three-dimensional modelling of electric and electromagnetic data using the fi-

nite element method. D, 1978, University of Utah. 268 p.

Priesmeyer, Steven T. Geology and gold mineralization on the New Year Mine and North Giltedge properties, Fergus County, Montana. M, 1986, University of Idaho. 84 p.

Priest, George R. Eruptive history and geochemistry of the Little Walker volcanic center, East central California. M, 1974, University of Nevada. 99 p.

Priest, George R. Geology and geochemistry of the Little Walker volcanic center, Mono County, California. D, 1980, Oregon State University. 253 p.

Priest, Tim. The Marmaton Group (Desmoinesian, Pennsylvanian) in south-central Kansas. M, 1982, Wichita State University. 50 p.

Priestaf, Iris Gail. Sacramento River seepage; alternative mitigating measures. D, 1983, University of California, Berkeley. 298 p.

Priestley, Keith F. Earth strain observations in the western Great Basin. D, 1974, University of Nevada. 192 p.

Priestley, Keith F. Earth strains observed at the Cascade (Washington) geophysical site using a long base laser interferometer strain meter. M, 1971, University of Washington. 65 p.

Prieto Cedraro, Rodulfo. Seismic stratigraphy and depositional systems of the Orinoco Platform area, northeastern Venezuela. D, 1987, University of Texas, Austin. 170 p.

Prieto, Corine. An interactive iterative scheme for gravity and magnetic interpretation. M, 1973, University of Toronto.

Prieto-Portar, Luis A. Earth anchors; load transfer analysis using photoelastic, analytic and finite element methods. D, 1978, Princeton University. 371 p.

Prigmore, Susan Marcheta. Mineralogical, petrological and geochemical aspects of Pleistocene low temperature rhyolites of the Tewa Group, Jemez Mountains, New Mexico. D, 1978, Southern Methodist University. 83 p.

Prigmore, Susan Marcheta. Tephrochronology of late Quaternary volcanics in the main Ethiopian Rift valley. M, 1975, Southern Methodist University. 70 p.

Prihar, Douglas W. Geology and mineralization of the Seaman Gulch area, East Shasta mining district, Shasta County, California. M, 1988, Oregon State University. 129 p.

Prijambodo, Raden. Well test analysis for wells producing layered reservoirs. D, 1981, University of Tulsa. 140 p.

Prime, Garth A. The Antigonish Basin of maritime Canada; a sedimentary tectonic history of a late Paleozoic fault-wedge basin. M, 1987, Dalhousie University. 223 p.

Primmer, Stanley Russell. The type Kirker (Oligocene, (Oligocene western California). M, 1966, University of California, Berkeley. 247 p.

Prince, Alan T. A study of Canadian sphene. M, 1938, University of Toronto.

Prince, Alan T. The system albite-anorthite-sphene. D, 1941, University of Chicago.

Prince, Arthur M. Solid solutions of niobium and tantalum in rutile. M, 1971, Queens College (CUNY).

Prince, Donald. Mississippian coal cyclothems in the Manning Canyon Shale of central Utah. M, 1963, Brigham Young University. 103 p.

Prince, R. A. Deformation in the Peru Trench, 6°-10°S. M, 1974, Oregon State University. 88 p.

Prince, Roger Allan. Gravity studies of the Vema fracture zone in the Equatorial Atlantic Ocean. D, 1988, Brown University. 246 p.

Principal, P. A. A geophysical study of the Sala y Gomez Ridge. M, 1974, University of Hawaii. 66 p.

Principe, Paul A. A late Pliocene (Ringold Formation) fish fauna from south-central Washington and its distributional significance to the Columbia-Snake drainage system. M, 1977, University of Massachusetts. 97 p.

Pringle, G. J. A technique for the determination of the feldspar crystallization history in a tholeiite sheet. M, 1972, University of New Brunswick.

Pringle, Mary Katherine Williams. Determination of the rate of DBCP volatilization from soils of central Oahu, Hawaii. M, 1984, University of Hawaii at Manoa. 123 p.

Pringle, Patrick. Geologic study sites of the Cuyahoga Valley National Recreation Area, southern sector. M, 1982, University of Akron. 181 p.

Prinz, Martin. Geologic evolution of the Beartooth Mountains, Montana and Wyoming; Part 4, Mafic dike swarms of the southern Beartooth Mountains. D, 1963, Columbia University, Teachers College. 145 p.

Prinz, Martin. The origin of the muscovite in the Alleghenian sandstones of Indiana. M, 1957, Indiana University, Bloomington. 65 p.

Prinz, William Charles. Geology of the igneous complex, north-eastern Wisconsin and adjacent Michigan. D, 1959, Yale University.

Prinz, William Charles. The geology of a portion of the Quinnesec igneous complex south of Iron Mountain, Michigan. M, 1952, Ohio State University.

Prior, Scott William. Geology and biostratigraphy of a portion of western San Luis Obispo County, California. M, 1974, University of California, Los Angeles.

Priore, William J. Microtextural investigation of the Black Butte diamicton, Gravelly Range, Madison County, Montana. M, 1984, Wright State University. 120 p.

Priovolos, George Jim. Gravity field approximation using the predictors of Bjerhammar and Hardy. D, 1988, Ohio State University. 146 p.

Prisbrey, Keith A. Sonic particle size analysis. M, 1970, Stanford University.

Pritchard, Charles L. Petrology of Pleistocene loess from northwestern Missouri. M, 1971, University of Missouri, Columbia.

Pritchard, D. W. The physical structure, circulation, and mixing in a coastal plain estuary. D, 1951, University of California, Los Angeles. 60 p.

Pritchard, George Flory. Trace element studies of glacial deposits in southwestern Ohio. M, 1980, Miami University (Ohio). 90 p.

Pritchard, James I. Theory of electromagnetic soundings in the frequency domain. D, 1971, Colorado School of Mines. 256 p.

Pritchett, Frank Ide. The geology of the western portion of the Mitchell Butte Quadrangle, Malheur County, Oregon. M, 1953, University of Oregon. 81 p.

Privett, Donald R. Petrography and structure of diabase dikes in central South Carolina. M, 1963, University of South Carolina. 33 p.

Privett, Donald Ray. Alkali reactivity and poor construction practices; causes of deterioration of concrete bridges in central Tennessee. D, 1968, University of Tennessee, Knoxville. 160 p.

Privette, Robert W. Petrology of the Weir Sand (Lower Mississippian) in the Ashland-Clark Gap gas field of southern West Virginia. M, 1983, East Carolina University. 89 p.

Privrasky, Norman Calvin. Geology of the northeast portion of the Mitchell Butte Quadrangle, Oregon. M, 1953, University of Oregon. 87 p.

Priz, Paula Marie. Chemical analysis and X-ray diffraction study of vanadium bearing mixed-layered clay from potash, Sulphur Springs, Arkansas. M, 1977, University of Iowa. 44 p.

Probandt, William Taylor. Regional geological aspects of the Moab Valley area, Grand County, Utah. M, 1959, Texas Tech University. 86 p.

Probst, David Arthur. Sedimentary-stratigraphic analysis of the Oficina, Venezuela. D, 1953, Northwestern University. 190 p.

Probst, David Arthur. The Ellenburger dolomite of West Texas and New Mexico. M, 1945, University of Pittsburgh.

Prochaska, E. J. Foraminifera from two sections of the Cody Shale in Fremont and Teton counties, Wyoming. M, 1960, University of Wyoming. 79 p.

Prochaska, Kevin M. A geochemical study of the Sudbury Breccia. M, 1981, Western Michigan University. 70 p.

Prochnau, John F. Distribution and mode of occurrence of gold in the Chibougamau District, Quebec. M, 1971, McGill University. 134 p.

Prochnow, Suzanne. Differentiation of facies in the Captain Creek Limestone Member (Upper Pennsylvanian), Douglas and Johnson counties, Kansas. M, 1982, University of Kansas. 118 p.

Procter, Richard Malcolm. Quantitative clay mineralogy of the Vanguard and Blairmore formations, southwestern Saskatchewan. D, 1959, University of Kansas. 141 p.

Proctor, B. L. Chemical investigation of sediment cores from four minor Finger Lakes of New York. D, 1978, SUNY at Buffalo. 240 p.

Proctor, Christian Jennings. A subtle stratigraphy in the Cretaceous kaolins of Georgia; evidence for weathering of a previously deposited sediment. M, 1987, Duke University. 139 p.

Proctor, Cleo V. The north Bosque County inventory of a drainage basin, Texas. M, 1967, Baylor University. 100 p.

Proctor, David D. Paleontology and paleoenvironment of the Janes Gravel Quarry, Crosby County, Texas. M, 1980, Texas Tech University. 82 p.

Proctor, Kenneth E. The petrography and gas-reservoir potential of Stanley-Jackfork sandstones, Ouachita Mountains, Arkansas and Oklahoma. M, 1974, Northern Illinois University. 151 p.

Proctor, N. S. Effects of shorebirds on dispersal and growth of algae in Connecticut. D, 1976, University of Connecticut. 78 p.

Proctor, Paul Dean. Geology of the Harrisburg (Silver Reef) mining district, Washington County, Utah. D, 1949, Indiana University, Bloomington. 167 p.

Proctor, Paul Dean. The geology of the Bulley Boy Mine, Piute County, Utah. M, 1943, Cornell University.

Proctor, R. M. Palaeontology of the (Ordovician) Stony Mountain Formation (Manitoba). M, 1954, University of Manitoba.

Proctor, Richard James. Geology of the Desert Hot Springs area, Little San Bernardino Mountains, California. M, 1958, University of California, Los Angeles.

Proctor, Richard M. Trilobites, cephalopods, and brachiopods of the Stony Mountain Formation, Manitoba. M, 1957, University of Manitoba.

Proett, B. A. Hydrogeology of Marcellus disappearing lake. M, 1978, Syracuse University.

Proffett, John Maddon, Jr. Nature, age and origin of Cenozoic faulting and volcanism in the Basin and Range Province (with specific reference to the Yerrington District, Nevada). D, 1972, University of California, Berkeley. 77 p.

Proffitt, J. R., III. The petrography and petrogenesis of the Milltown Pluton; Chambers County, Alabama. M, 1975, Memphis State University.

Prokop, Christopher Jon. The engineering properties of alluvial/colluvial fault scarp soils in the western Basin and Range, and their influence on fault scarp morphology/dating. M, 1983, University of Nevada. 154 p.

Prombo, Carol Ann. Nitrogen contents and isotopic compositions of metal-rich meteorites. D, 1984, University of Chicago. 120 p.

Prommool, Suthep. Hydrologic effects of shifting cultivation in northern Thailand; a review and study plan. M, 1974, [Colorado State University].

Pronko, Peter Paul. Simultaneous argon diffusion and electrical conduction measurements in microcline feldspar. D, 1966, University of Alberta. 159 p.

Pronold, Thomas George. The one-atmosphere phase relations of three volcanic rocks from the Taos Pla-

teau volcanic field, North-central New Mexico. M, 1979, Southern Methodist University. 55 p.

Propes, Russell. Crustal velocity variation in the Southern Appalachians. M, 1985, Georgia Institute of Technology. 113 p.

Prosen, Barbara J. Natural brine contamination of groundwater in the western Upper Peninsula, Michigan. M, 1988, Michigan Technological University. 113 p.

Prosh, Eric C. A Lower Devonian reef sequence and fauna, Disappointment Bay Formation, Canadian Arctic Islands; Volume 1. D, 1988, University of Western Ontario. 446 p.

Prossard, R. L. The geology of the Nixa area (Missouri). M, 1942, University of Missouri, Columbia.

Prosser, Charles Smith. A study of the fossil faunas of the Middle and Upper Devonian along the Unadilla River, New York. M, 1886, Cornell University.

Prosser, Charles Smith. The classification and distribution of the (Devonian) Hamilton and Chemung series of central and eastern New York. D, 1907, Cornell University.

Prosser, Jerome T. The geology of Poas Volcano, Costa Rica. M, 1983, Dartmouth College. 165 p.

Prosser, Rex Michael. Static lattice models for higher-order elastic properties in face-centered cubic crystals, implications for the lower mantle. M, 1980, Pennsylvania State University, University Park. 81 p.

Prost, Gary Leo. Evaluation of Skylab photographs over central Colorado for locating indicators of mineralization. M, 1975, Colorado School of Mines. 111 p.

Prost, Gary Leo. Jointing in relatively undeformed strata; relation to basement and exploration implications. D, 1986, Colorado School of Mines. 291 p.

Prostka, Harold J. Structure and petrology of the Sparta Quadrangle, Oregon. M, 1963, The Johns Hopkins University. 236 p.

Prothero, Donald Ross. Medial Oligocene magnetostratigraphy and mammalian biostratigraphy; testing the isochroneity of mammalian biostratigraphic events. D, 1982, Columbia University, Teachers College. 297 p.

Protzman, Donald LeRoy. The facies relationships of the Sespe and Alegria formations, Santa Barbara County, California. M, 1960, University of California, Los Angeles.

Protzman, Gretchen Marie. The emplacement and deformation history of the Meade thrust sheet, southeastern Idaho. M, 1985, University of Rochester. 117 p.

Proudfoot, David Nelson. A lithostratigraphic and genetic study of Quaternary sediments in the vicinity of Medicine Hat, Alberta. D, 1985, University of Alberta. 265 p.

Proudy, William F. Manlius Limestone (Silurian) of Onondaga County (New York). M, 1904, Syracuse University.

Proulx, I. Analysis of the contaminant plume in the Oak Ridges Aquifer at the Stouffville landfill. M, 1988, University of Waterloo. 134 p.

Proust, Rodrigo Diez. Characterization of a sandstone reservoir using seismic methods; Yowlumne Field, Kern County, California. M, 1988, Texas A&M University. 178 p.

Prouty, Chilton E. Lower Middle Ordovician of southwestern Virginia and northeastern Tennessee. D, 1948, Columbia University, Teachers College.

Prouty, Chilton E. Petrographic analyses of the Chepultepec sandstones of the Norris Reservoir region, Tennessee. M, 1938, University of Missouri, Rolla.

Prouty, W. F. The Niagara and Clinton formations of Maryland. D, 1906, The Johns Hopkins University.

Province, Harold Edward. A study of possible contemporaneous deformation in sedimentary rock. M, 1952, University of Cincinnati. 74 p.

Provins, Dean Allen. Interpretation scheme formulation for DPM series of EM prospecting tools using a simple model. M, 1971, University of Manitoba.

Provo, Linda Jeanne. Interpretation of a turbidite; sedimentology of the West River Shale and the Ithaca Formation (Upper Devonian), N.Y. M, 1973, University of Illinois, Urbana.

Provo, Linda Jeanne. Stratigraphy and sedimentology of radioactive Devonian-Mississippian shales of the central Appalachian Basin. D, 1977, University of Cincinnati. 177 p.

Provost, Gilles. Les rhyolites du complexe "Don" région de Rouyn-Noranda, Abitibi-ouest. M, 1979, Ecole Polytechnique.

Provost, Gilles. Les Rhyolites du Complexe "Don", région de Rouyn-Noranda, Abitibi-Ouest. M, 1978, Ecole Polytechnique.

Prowant, S. O. A finite element analysis comparison with scale models of elastic deformation of geologic structure. M, 1975, Ball State University. 56 p.

Prowell, David Cureton. Geology of selected Tertiary volcanics in the central Coast Range mountains of California and their bearing on the Calaveras and Hayward fault problems. D, 1974, University of California, Santa Cruz. 283 p.

Prowell, David Cureton. Ultramafic plutons in the central Piedmont of Georgia. M, 1972, Emory University. 83 p.

Prowell, Sarah Eddings. Stratigraphic and structural relations of the north-central Superstition volcanic field. M, 1984, Arizona State University. 123 p.

Pruatt, Martin Aaron. The Southern Oklahoma Aulacogen; a geophysical and geological investigation. M, 1975, University of Oklahoma. 62 p.

Prucha, Christopher P. Lineament study from satellite imagery of the Promontory Mountains in Utah. M, 1988, University of Oklahoma. 76 p.

Prucha, John James. A petrogenetic study of the Hermon Granite in a part of the northwest Adirondacks. D, 1950, Princeton University. 113 p.

Prud'homme, Michel. Caractérisation des granulats de carrières dans la région de Montréal. M, 1981, Ecole Polytechnique. 99 p.

Pruden, Jimmy Lee. The foraminiferal zones of the Mooreville Formation in Lowndes and Oktibbeha counties, Mississippi. M, 1955, Mississippi State University. 59 p.

Prueher, Elizabeth M. The geology and geochemistry of thirteen cinder cones at Crater Lake National Park, Oregon. M, 1985, University of Oregon. 158 p.

Pruett, Frank Donald. A study of the magnetic properties of some residual soils. M, 1959, Indiana University, Bloomington. 58 p.

Pruett, Robert J. The origin of kaolin contained in the Whitemud Formation, southern Saskatchewan, Canada. M, 1988, Indiana University, Bloomington. 144 p.

Prufert, Leslie E. Seasonal variations of iron and manganese diagenesis in an anoxic marine basin. M, 1985, University of North Carolina, Chapel Hill. 104 p.

Prugger, Arnfinn F. Microseismicity related to potash mining. M, 1985, University of Saskatchewan. 156 p.

Pruit, John Dave. Structure and stratigraphy of Harry Canyon (Hailey 1 NE) Quadrangle, Custer County, Idaho. M, 1971, University of Wisconsin-Milwaukee.

Pruitt, Earl Joseph, Jr. A study of the Trinity Group, Santa Anna Mountains, Coleman County, Texas. M, 1948, University of Texas, Austin.

Pruitt, Glenn N. Geology of the Rhea Springs Window, Rhea County, Tennesee. M, 1962, University of Tennessee, Knoxville. 35 p.

Pruitt, Jacqueline Davis. Lead adsorption characteristics in soils along U.S. Interstate 35E, Ellis County, Texas. M, 1977, Southern Methodist University. 45 p.

Pruitt, Maria Pankos. A provenance investigation of glacial terraces along the Big Lost River, Custer County, Idaho. M, 1984, Lehigh University. 175 p.

Pruitt, Robert Grady, Jr. The Brevard Zone of northeastern Georgia. M, 1952, Emory University. 71 p.

Prunier, A. R., Jr. Calculation of temperature-oxygen fugacity tables for H_2-CO_2 gas mixtures at one atmosphere total pressure and an investigation of the zoisite-clinozoisite transition. M, 1978, Virginia Polytechnic Institute and State University.

Prunty, Merle C., Jr. An evaluation of airport locations in the Kansas City metropolitan area. M, 1940, University of Missouri, Columbia.

Prusak, Deanne. Provenance, areal distribution, and contemporary sedimentation of quartz sand and silt types on the mid-Atlantic continental shelf. M, 1985, Texas A&M University.

Prusok, Ridi Albin. The stratigraphy and structure of the Fayette Quadrangle, Howard County, Missouri. M, 1961, University of Iowa. 71 p.

Pruss, E. F. A paleomagnetic study of basalt flows from the Absaroka Mountains, Wyoming. D, 1977, University of Wyoming. 68 p.

Prusti, Bansi D. Geology of O'Connor Lake area, Northwest Territories, with special reference to the mineral deposits. D, 1954, McGill University.

Prutzman, John M., Jr. Compensation for geometrical spreading effects in seismic exploration. M, 1978, University of Minnesota, Duluth.

Prutzman, William James. The geology of the Craigsville area, Augusta and Rockbridge counties, Virginia. M, 1953, University of Virginia. 58 p.

Pryce-Harvey, Jacqueline Simone. Estimating recharge to the Liguanea Aquifer of Kingston, Jamaica. D, 1985, University of Tennessee, Knoxville. 157 p.

Pryor, M. The glacial history of the Langford, New York 71/2' quadrangle and the southern half of the Hamburg, New York 71/2' quadrangle. M, 1975, SUNY at Buffalo. 58 p.

Pryor, Stanley J. Geology of the limestone, Walnut area, Newton County, Arkansas. M, 1967, University of Arkansas, Fayetteville.

Pryor, Wayne Arthur. Cretaceous geology and petrology of the Upper Mississippi embankment. D, 1959, Rutgers, The State University, New Brunswick. 194 p.

Pryor, Wayne Arthur. Sandstone aquifers in the upper part of the Pennsylvanian System, White County, Illinois. M, 1954, University of Illinois, Urbana.

Pryslak, Anthony Paul. Structural geology of Tustin and Bridges townships, District of Kenora, northwestern Ontario. M, 1971, University of Manitoba.

Przybyl, Bruce J. The regimen of Grand Union Glacier and the glacial geology of the northeastern Kigluaik Mountains, Seward Peninsula, Alaska. M, 1988, SUNY at Buffalo. 106 p.

Przygocki, Robert. Bedrock identification by Fourier grain shape analysis of fluvial quartz in the North Carolina Blue Ridge Province. M, 1977, University of South Carolina.

Przywara, Mark S. Determination of the sediment plumes of the Detroit and Maumee rivers in western Lake Erie using grain-size and heavy-mineral analyses. M, 1978, Bowling Green State University. 238 p.

Psutka, J. F. Structural setting of the Downie Slide, northeast flank of Frenchman Cap gneiss dome, Shuswap Complex, southeastern British Columbia. M, 1978, Carleton University. 70 p.

Psuty, Norbert Phillip. The geomorphology of beach ridges in Tabasco, Mexico. D, 1966, Louisiana State University.

Psycharis, Ioannis N. Dynamic behavior of rocking structures allowed to uplift. D, 1982, California Institute of Technology. 234 p.

Pszenny, Alexander A. P. Atmospheric deposition of nitrate to the ocean surface. D, 1987, University of Rhode Island. 286 p.

Ptacek, Anton Donald. Cenozoic geology of the Seven Devils Mountains, Idaho. D, 1965, University of Washington. 96 p.

Ptacek, Anton Donald. Structure and stratigraphy of the Horse Range, Nevada. M, 1962, University of Washington. 70 p.

Ptacek, C. J. A comparison of methods for measuring retardation coefficients of trace organic solutes in hydrogeologic regimes. M, 1985, University of Waterloo. 92 p.

Puccetti, A. Contemporary volcanic CO_2 emission; a study employing dilution of atmospheric C-14 as a quantitative tracer. D, 1973, University of Hawaii. 128 p.

Pucci, Amleto Arthur, Jr. A model of aldicarb transport in the Town of Southold, L.I., N.Y. D, 1983, SUNY at Stony Brook. 168 p.

Puccini, Piero Miguel. Mechanical behavior of cohesionless soils; mathematical modelling and experimental verification. D, 1988, Case Western Reserve University. 249 p.

Puchlik, Kenneth Phillip. Nickel mineralization associated with early ultramafic intrusions of the Sierra Nevada (California). M, 1972, University of Nevada. 80 p.

Puchstein, Richard L. A study of some species of the orthid brachiopod genus Platystrophia from the Upper Ordovician of Kentucky and Ohio. M, 1972, Memphis State University.

Puchy, Barbara J. Mineralogy and petrology of lava flows (Tertiary-Quaternary) in southeastern Idaho and at Black Mountain, Rich County, Utah. M, 1981, Utah State University. 73 p.

Puckett, James Carl, Jr. Petrographic study of a quartz diorite stock near Superior, Pinal County, Arizona. M, 1970, University of Arizona.

Puckett, James L. Geophysical study of shear zones in the east-central Medicine Bow Mountains, Wyoming, and kimberlitic diatremes in northern Colorado and southern Wyoming. M, 1971, Colorado State University. 101 p.

Puckett, Terry Markham. Biofacies analysis of ostracodes from the Pride Mountain interval in Colbert, Lawrence and Jefferson counties, northern Alabama. M, 1987, Mississippi State University. 202 p.

Puckette, William L. Stratigraphy and petrology of the Kessler Limestone Member of the Bloyd Formation in western Madison County, Arkansas. M, 1976, University of Arkansas, Fayetteville.

Pucovsky, G. M. The influence of agricultural practices on the nitrate concentration in the unsaturated and shallow saturated groundwater zone. M, 1977, University of Waterloo.

Puech, J. F. C. G. Mississippian overlap near Cape George, Nova Scotia. M, 1950, Massachusetts Institute of Technology. 60 p.

Pueschel, C. M. Tetrasporogenesis and other cytological features of the red alga Palmaria palmata (Rhodymenia palmata). D, 1978, Cornell University. 187 p.

Puffer, E. L. Potential megafossils of the Rockport, Texas, area, and their distribution. M, 1953, University of California, Berkeley. 144 p.

Puffer, John H. The petrology of a portion of the Mellon Gabbro Complex and its relationship to the Keweenawan Basalts and the Tyler Formation (Precambrian) (Michigan-Wisconsin). M, 1964, Michigan State University. 69 p.

Puffer, John Harold. Magnetite and ilmenite in the granitic rocks of the Pegmatite points area (Park county), Colorado. D, 1969, Stanford University. 182 p.

Pugh, C. E. Influence of surface area and morphology on the oxidation of pyrite from Texas lignite. D, 1978, Texas A&M University. 177 p.

Pugh, Griffith Thompson. Pleistocene deposits of South Carolina, with an especial attempt at ascertaining what must have been the environmental conditions under which the Pleistocene Mollusca of the state lived. D, 1905, Vanderbilt University.

Pugh, Lewis E. Geology along portion of the Cross Mountain Fault near Blountville, Sullivan County, Tennessee. M, 1966, University of Tennessee, Knoxville. 66 p.

Pugh, William Emerson. Temperature compensation on a new magnetic system for Schmidt vertical balances. M, 1932, Colorado School of Mines. 38 p.

Puglio, Donald. Geology of the lower Kittanning coalbed in portions of Indiana; Westmoreland and Armstrong. M, 1977, University of Pittsburgh.

Pugmire, Ralph U. The geology of Bill Williams Mountain, Coconino County, Arizona. M, 1977, Northern Arizona University. 97 p.

Pugsley, Robert L. Chemical relationships among the unoxidized, oxidized, and ore zones of a Wyoming "roll-type" uranium deposit, Shirley Basin, Wyoming. M, 1983, Bowling Green State University. 114 p.

Puig, Joseph Albert. Thermal analyses of brannerite. M, 1954, New York University.

Pujol, Jose M. Rayleigh waves spectral studies of some Alaskan intraplate earthquakes. M, 1982, University of Alaska, Fairbanks. 78 p.

Pujol, Jose M. Vertical seismic profiling; processing and interpretation, velocity determination and attenuation measurements. D, 1985, University of Wyoming. 167 p.

Pulanco, Demetrio Hidalgo. Differential thermal analyses of some aluminous laterites in Surigao, Mindanao, Philippines. M, 1962, University of Illinois, Urbana.

Pulchan, Kalidas. Organic geochemical comparisons of Fortune Bay and Bay d'Espoir. M, 1987, Memorial University of Newfoundland. 219 p.

Pulfrey, Robert John. Geology and geochemistry of the Mt. Antero Granite and contiguous units, Chaffee County, Colorado. M, 1971, Oklahoma State University. 84 p.

Pulido, Oscar H. Geology and alteration of the Quilmaná porphyry copper prospect, Lunahuana, Department of Lima, Perú M, 1983, University of Nevada. 56 p.

Pulju, Hugo James. Geology of the Black Butte area, Cascade County, Montana. M, 1964, Montana College of Mineral Science & Technology. 50 p.

Pullen, Michael John Leslie Thomas. The Pleistocene geology of Toronto as seen from the Bloor--Danforth subway cut. M, 1966, University of Toronto.

Pullen, Milton William, Jr. Geologic aspects of radio wave transmission. D, 1950, University of Illinois, Urbana.

Pulley, Thomas E. Marine mollusks of the Texas coast. M, 1950, [University of Houston].

Pulli, Jay J. Seismicity, earthquake mechanisms, and seismic wave attenuation in the Northeastern United States. D, 1983, Massachusetts Institute of Technology. 390 p.

Pulliam, James Millard. The stratigraphy and structure of the southeast quarter of the Stafford Quadrangle, Missouri. M, 1954, University of Missouri, Columbia.

Pulling, David Michael. Subsurface stratigraphic and structural analysis, Cherokee Group, Pottawatomie County, Oklahoma. M, 1976, University of Oklahoma. 73 p.

Pullman, Steven A. The petrography and petrology of a portion of the northern Cedar Mountains, Mineral County, Nevada. M, 1983, University of Nevada. 130 p.

Pulpan, Hans. Stress distribution around certain underground inhomogeneities. D, 1968, University of Illinois, Urbana. 96 p.

Puls, David Donald. Geometric and kinematic analysis of a Proterozoic foreland thrust belt, northern Mazatzal Mountains, central Arizona. M, 1986, Northern Arizona University. 102 p.

Puls, Robert William. Adsorption of heavy metals on soil clays. D, 1986, University of Arizona. 152 p.

Pulse, Richard Reid. Conodonts from the Upper Ordovician Fairview Formation, in Cincinnati region,

Ohio, Kentucky, and Indiana. M, 1959, Ohio State University.

Pumphrey, Phillip L. Landuse conflicts in Tulsa County Oklahoma; urban expansion vs. petroleum development. M, 1988, Southwest Missouri State University.

Punatar, Gajendra Kesshavlal. An investigation of the flow-duration characteristics of selected streams of South Dakota. M, 1975, South Dakota School of Mines & Technology.

Punches, Richard K. Ostracoda of the (Upper Mississippian) Golconda Limestone (Illinois, Kentucky). M, 1939, Columbia University, Teachers College.

Pungrassami, Thongchai. Geology of the western Detroit reservoir area Quartzville and Detroit quadrangles, Linn and Marion counties, Oregon. M, 1970, Oregon State University. 76 p.

Punongbayan, Raymundo Santiago. Geology of the Rattlesnake Reservoir area (Precambrian-Paleozoic), Larimer County, Colorado and a redefinition of the second period of regional metamorphism in northeastern Front Range, Colorado. D, 1972, University of Colorado. 202 p.

Punwasee, J. D. N. Geochemical investigations in the Groete-Black creeks areas of the tropical rainforests of Guyana. M, 1973, University of New Brunswick.

Pupa, Diane Marie. Crystalline rocks of the Honey Brook 7 1/2-minute Quadrangle, southeastern Pennsylvania. M, 1988, Bryn Mawr College. 76 p.

Puppolo, David G. A study of fold behavior by means of microstructural features. M, 1979, Rensselaer Polytechnic Institute. 156 p.

Purcell, Charles Wilson. Eolian transport; a process supplying carbonate material for caliche formation. M, 1973, University of Cincinnati. 66 p.

Purcell, David Richard James. Structure of the Upper Cretaceous Ojinaga Formation, foothills area, southern Quitman Mountains, Hudspeth County, Texas. M, 1987, University of Texas, Arlington. 193 p.

Purcell, Francis A., Jr. The geology of the western half of the Hurricane Quadrangle, Utah. M, 1961, University of Southern California.

Purcell, Thomas E. The Mesaverde Formation of the northern and central Powder River basin, Wyoming. M, 1960, University of Oklahoma. 89 p.

Purdie, J. J. Certain pre-ore dykes, Copper Mountain Mine, British Columbia. M, 1953, University of British Columbia.

Purdom, William Berlin. Geology of La Minera Occidental, Bosch, S.A., and the Coto Francisco, Pinar del Rio, Cuba. D, 1960, University of Arizona. 157 p.

Purdue, Gary Lynn. Geology and ore deposits of the Blackbird District, Lemhi County, Idaho. M, 1975, University of New Mexico. 49 p.

Purdy, C. A. Pleistocene geology of the Kentville area, Nova Scotia. M, 1951, Acadia University.

Purdy, Edward George. Recent calcium carbonate facies of the Great Bahama Bank. D, 1960, Columbia University, Teachers College. 181 p.

Purdy, Joel W. Paleomagnetism and tectonic interpretation of the Crescent and Blakeley Formations of Kitsap Peninsula, Washington. M, 1987, Western Washington University. 138 p.

Purdy, John Winston. A rubidium-strontium and potassium-argon isotopic age investigation within the Superior Province of the Precambrian Canadian Shield. D, 1967, University of Toronto.

Purdy, John Winston. Rubidium-strontium isotopic studies (granite-gneisses, Red Lake, Ontario). M, 1964, University of Toronto.

Puri, Harbans S. Correlation of the Miocene of the Florida Panhandle with the west and central Gulf states. D, 1953, Louisiana State University.

Puri, Vijay Kumar. Liquefaction behavior and dynamic properties of loessial (silty) soils. D, 1984, University of Missouri, Rolla. 320 p.

Purington, Paul Richard. Geology of the Indian Spring area, northern Washoe County, Nevada. M, 1985, University of Nevada. 94 p.

Purisinsit, Pitsamai. Evaluation of two hydrologic models for the North Carolina Blacklands. D, 1982, North Carolina State University. 195 p.

Purnell, Guy. Measurement and analysis of wave attenuation and dispersion using a physical modeling system. M, 1983, University of Houston.

Purnell, James A. A study of the relation of the burning characteristics of several coals to their geological occurence and their general physical structure. M, 1935, [University of Kentucky].

Purrington, Wealthy. Stratigraphic distribution of microfossils (exclusive of the Fusulinidae) in the lower part of the Shawnee Group in the vicinity of Lawrence and Lecompton, Kansas. M, 1948, University of Kansas. 81 p.

Pursell, Victoria Jane. The petrology and diagenesis of Pleistocene and Recent travertines from Gardiner, Montana, and Yellowstone National Park, Wyoming. M, 1985, University of Texas, Austin. 153 p.

Purson, John D. Uranium migration; a geochemical examination of sediments and their source rocks in McKinley County, New Mexico. M, 1985, New Mexico Institute of Mining and Technology. 124 p.

Purucker, Michael E. Oolitic ironstones and banded iron-formation; controls on chemical sedimentation. D, 1984, Princeton University. 169 p.

Purushothaman, Krishnier. Transport of ^{85}Sr and ^{137}Cs under induced clay suspensions. D, 1968, University of Texas, Austin. 135 p.

Purvance, David Thomas. A comprehensive approach to the geophysical inverse problem. M, 1975, University of Utah. 69 p.

Purves, W. J. Paleoenvironmental evaluation of Mississippian age carbonate rocks in central and southeastern Arizona. D, 1978, University of Arizona. 802 p.

Purves, William John. Stratigraphic control of the ground water through Spokane Valley (Washington). M, 1969, Washington State University. 213 p.

Puryear, Sam M. A study of jointing in the area of the Wells creek structure, Houston, Montgomery, Stewart, Dixon counties, Tennessee. M, 1968, Vanderbilt University.

Purzer, James J. Lithologic variations in the Lamar fore-reef limestone tongue of the Capitan reef complex. M, 1954, University of Kansas. 55 p.

Pusc, Steve W. Hydrogeology of the lower Bruno Creek area, Custer County, Idaho. M, 1982, University of Idaho. 206 p.

Puscas, George. The study of the Aux Vases Formation in the Monroe City oil field, Knox County, Indiana. M, 1953, Indiana University, Bloomington. 31 p.

Pusey, Richard Downing. A study of the sediments of the (Eocene) lower Huerfano Formation (Colorado). M, 1957, University of Michigan.

Pusey, Walter Carroll, III. Recent calcium carbonate sedimentation in northern British Honduras. D, 1964, Rice University. 255 p.

Pushkar, Paul Demitru. The isotopic composition of strontium in volcanic rocks from island arcs. D, 1966, University of California, San Diego. 211 p.

Putcha, Sastry Purnanjaneya. Crust strength as influenced by simulated rainfall and drying time. D, 1983, North Carolina State University.

Putimanitpong, Supalak. Pyrite morphotypes as paleoenvironment indicators from ten lignite cores from the Eocene of Northwest Louisiana. M, 1983, Northeast Louisiana University. 112 p.

Putlitz, Fritz H. Significant Eocene sections of California. M, 1939, Stanford University. 70 p.

Putman, George Wendell. A study in the distribution of trace elements in some igneous rocks of northwestern and central Arizona. D, 1961, Pennsylvania State University, University Park. 142 p.

Putman, George Wendell. The geology of some wollastonite deposits in the eastern Adirondacks, New York. M, 1958, Pennsylvania State University, University Park. 107 p.

Putnam, Bruce McCormick. Subsurface study of the Las Animas Arch area, Colorado and Kansas, with special reference to its future oil and gas potential. M, 1960, Stanford University. 73 p.

Putnam, Burleigh John, III. Geology and geochemistry of mercury occurrences in the Horse Creed area, Siskiyou County, California. M, 1974, University of California, Los Angeles. 124 p.

Putnam, P. C. Reconnaissance among some volcanoes of Central America (Vol. 1); Geological literature on Central America 1529-1924 (Vol. 2). M, 1924, Massachusetts Institute of Technology. 24 p.

Putnam, Peter Edward. Reservoir origin and controls of hydrocarbon distributions interpreted with a computerized data base, Lower Cretaceous Mannville Group, west-central Saskatchewan. D, 1985, University of Calgary. 209 p.

Putnam, Peter Edward. The sedimentology of the Colony Formation, East-central Alberta. M, 1979, University of Calgary. 150 p.

Putnam, William C. Geology of the Mono Craters, California. D, 1937, California Institute of Technology. 206 p.

Putnam, William C. Physiography of the Ventura region, California. D, 1937, California Institute of Technology.

Putnam, William C. Terrace levels in the Ventura District, California. M, 1930, Stanford University. 116 p.

Putney, Kevin Lee. Structural analysis of the western extension of the Florence Pass Lineament, Bighorn Mountains, Wyoming. M, 1984, University of Iowa. 160 p.

Putney, Thomas R. Geology, geochemistry, and alteration of the Seligman and Monte Cristo stocks, White Pine mining district, White Pine County, Nevada. M, 1985, University of Nevada. 152 p.

Putt, David J. Phase relations on the join $Mg_{0.6}Ca_{0.4}SIO - Fe_{0.6}Ca_{0.4}SIO_3$. M, 1968, University of Manitoba.

Putzier, Paul. Surface resistivity delineation of the shallow aquifer, St. Lucie County, Florida. M, 1987, University of South Florida, Tampa. 144 p.

Putzig, Nathaniel E. Modeling wide-angle seismic data from the central California margin. M, 1988, Rice University. 141 p.

Puumala, Paava Pellervo. A geological reconnaissance of the igneous activity in the Mount Fleecer area, Silver Bow County, Montana. M, 1948, Montana College of Mineral Science & Technology. 33 p.

Puwakool, Suchit. Geology of the northern part of the Hosta Butte (7.5 Min) Quadrangle, McKinley County, New Mexico. M, 1971, University of Wyoming. 43 p.

Pwa, Aung. Regional rock geochemical exploration, Bathurst District, N.B. M, 1978, University of New Brunswick.

Pybas, Gerald Wayne. Petroleum geology of southwestern Pottawatomie County, Oklahoma. M, 1962, University of Oklahoma. 42 p.

Pybas, Kevin M. Geology of the Turner Falls area, with emphasis on the Collings Ranch Conglomerate, Arbuckle Mountains, Oklahoma. M, 1987, Oklahoma State University. 62 p.

Pye, Edgar G. A petrographic study of the textures of basic and ultrabasic igneous rocks. D, 1954, University of Toronto.

Pye, Edgar George. A petrographic study of the textures of basic and ultrabasic igneous rocks. D, 1953, University of Toronto.

Pye, Graham D. Heat flow measurements in Baffin Bay and the Labrador Sea (Canada). M, 1971, Dalhousie University.

Pye, Willard Dickison. A comparison of records from the linear strain and pendulum seismographs. M, 1937, California Institute of Technology. 14 p.

Pye, Willard Dickison. The (Upper Mississippian) Bethel Sandstone of south central Illinois. D, 1942, University of Chicago. 151 p.

Pyeatt, Lloyd M. Smooth and reticulate members of the genus Cythere in the Gulf Coast Tertiary. M, 1931, Louisiana State University.

Pyke, Anne Rutherford. The geology of the Hampshire Hills, northeastern New Hampshire and western Maine. M, 1985, University of Vermont. 102 p.

Pyke, Dale Randolph. Plagioclase compositions in the Amisk Lake area, Saskatchewan. M, 1961, University of Saskatchewan. 36 p.

Pyke, Murray W. Microclines from a Precambrian granodiorite. M, 1958, University of Saskatchewan. 50 p.

Pyle, C. A. Late Quaternary geologic history of the South Texas continental shelf. M, 1977, Texas A&M University.

Pyle, Phillip F. Geology and hydrothermal alteration of the Matterhorn Peak Stock, Hinsdale County, Colorado. M, 1980, University of Texas, Austin.

Pyle, Thomas Edward. Micropaleontology and mineralogy of a Tertiary sediment core from the Sigsbee Knolls, Gulf of Mexico. M, 1966, Texas A&M University.

Pyle, Thomas Edward. Structure of the West Florida platform, Gulf of Mexico. D, 1972, Texas A&M University. 199 p.

Pyle, William D., Jr. Aquifer geometry based on surface data. M, 1969, Colorado State University. 71 p.

Pyles, Marvin Russell. The undrained shearing resistance of cohesive soils at large deformations. D, 1981, University of California, Berkeley. 172 p.

Pyrak, Laura J. Refraction of isotherms; applications to define rift basin geometry. M, 1983, Virginia Polytechnic Institute and State University. 77 p.

Pyrak, Laura Jeanne. Seismic visibility of fractures. D, 1988, University of California, Berkeley. 166 p.

Pyrih, Roman Z. Determination of trace mercury in soil, rock and vegetation. M, 1970, Colorado School of Mines. 33 p.

Pyrih, Roman Z. Ferric iron oxidation of pyrites from Gilman, Colorado. D, 1974, Colorado School of Mines. 123 p.

Pyron, Arthur J. Geology and geochemistry of the El Paso tin deposit, Franklin Mountains, El Paso County, Texas. M, 1980, University of Texas at El Paso.

Pytlak, Shirley Ruth. Geology of the Blanca Lake area, Snohomish County, Washington. M, 1970, University of Washington. 45 p.

Pytte, Anthony Mark. The kinetics of the smectite to illite reaction in contact metamorphic shales. M, 1982, Dartmouth College. 78 p.

Pywell, H. R., III. The effect of changes in land use and vegetative cover on streamflow from a major municipal watershed. D, 1977, University of Massachusetts. 244 p.

Pyzik, A. J. The kinetics and mechanism of sedimentary iron sulfide formation. D, 1976, University of Maryland. 158 p.

Qabazard, F. A. Management of brackish water supply and demand in Kuwait. M, 1977, University of Arizona.

Qahwash, Abdellatif Ahmad. An application of mathematical models for determining the optimum pattern of a geophysical exploration program. D, 1974, University of Arizona.

Qahwash, Abdullatif A. An electrical resistivity survey in the Avra Valley, Pima County, Arizona. M, 1972, University of Arizona.

Qamar, Anthony. Seismic wave velocity in the Earth's core; a study of PKP and PKKP. D, 1971, University of California, Berkeley. 204 p.

Qandil, Yacoub A. Arenaceous foraminifera of the Bronson Subgroup (Missourian Series) in Kansas. M, 1961, University of Kansas. 56 p.

Qayyun, M. Abdul. Mechanical properties of the middle Waseca Sand and their influence on heavy oil recovery. M, 1986, University of Saskatchewan. 127 p.

Qing, Hairuo. Diagenesis and sedimentology of Rainbow F and E buildups (Miette Devonian), northwestern Alberta. M, 1986, University of Toronto.

Quaah, Amos Ofori. Gravity and magnetic studies of the Everett and Mason-Dixon lineaments in Southcentral Pennsylvania. M, 1977, Pennsylvania State University, University Park. 79 p.

Quade, Jack Gehring. An evaluation of multifrequency systems at the Crow Springs mining district, Esmeralda County, Nevada. M, 1973, University of Nevada. 94 p.

Quade, Jay. Quaternary geology of the Corn Creek Springs area, Clark County, Nevada. M, 1983, University of Arizona. 135 p.

Quadri, Shah M. G. J. Geology and mining operations on Calumet Island (Ontario). M, 1951, University of Toronto.

Quaide, William Lee. Geology of the central Peloncillo Mountains, Hidalgo County, New Mexico. M, 1953, University of California, Berkeley. 89 p.

Quaide, William Lee. Petrography and clay mineralogy of Pliocene sedimentary rocks from the Ventura Basin, California. D, 1956, University of California, Berkeley. 81 p.

Qualls, Robert Ralph. Crustal study of Oklahoma. M, 1965, University of Tulsa. 82 p.

Quam, Louis O. Morphology of landscape of the Estes Park area, Colorado. D, 1938, Clark University.

Quam, Louis Otto. Geology of the Rabbit Mountain area (Boulder and Larimer counties), Colorado. M, 1932, University of Colorado.

Quan, Choon Kooi. The characteristics of radial strain propagation induced by explosive impact in Jefferson City Dolomite. M, 1964, University of Missouri, Rolla.

Quan, Richard A. Chemical analyses of halite trend inclusions from the Granisle porphyry copper deposit, British Columbia. M, 1985, University of Michigan. 29 p.

Quaraishi, A. A. Petrography of a quartz-diorite stock, Elbow Lake, Manitoba, Canada. M, 1967, University of Manitoba.

Quarles, Miller, Jr. Geology of the Repetto and Montebello hills (California). M, 1941, California Institute of Technology. 63 p.

Quas, Marilyn. Morphological and geochemical study of glass shards from the Grenada Basin, Lesser Antilles. M, 1988, University of Illinois, Chicago.

Quaschnick, Ralph Kohler. The geology of the Marine Quadrangle and the Falls Creek area, Minnesota. M, 1959, University of Minnesota, Minneapolis. 89 p.

Quattlebaum, David M. Stratigraphy, regional correlation and depositional environment of the Bonner Formation (Precambrian Missoula Group, Southwest Montana). M, 1980, University of Montana. 104 p.

Quay, P. D. An experimental study of turbulent diffusion in lakes. D, 1977, Columbia University, Teachers College. 208 p.

Quay, Paul C. The use of resistivity techniques and subsurface sampling to determine the lateral movement of leachate from Mentor sanitary landfill, Mentor, Ohio. M, 1975, University of Akron. 38 p.

Quearry, M. W. Continental volcanic sediments in the region of Volcan de Fuego, Guatemala. M, 1975, University of Missouri, Columbia.

Queen, Elizabeth Bolton. A depositional environmental analysis of the Moenave Formation, Zion National Park, Springdale, Utah. M, 1988, Stephen F. Austin State University. 80 p.

Queen, J. M. Carbonate sedimentology and ecology of some pelleted muds west of Andros Island, Great Bahama Bank. D, 1978, SUNY at Stony Brook. 401 p.

Queneau, Augustin L. J. Size of grain in igneous rocks. M, 1901, Columbia University, Teachers College.

Quensen, John F., III. The effect of low O₂ levels on the filtration efficiency and pumping rate of Crassostrea virginica. M, 1975, College of William and Mary.

Querau, Edmund C. A study of the Alps. M, 1893, Northwestern University.

Querol, Sune Francisco. K, Na distribution and unit cell parameters of alkali feldspars. M, 1968, Massachusetts Institute of Technology. 62 p.

Querol-Sune, Francisco. The genesis of the antimony deposits at Wadley, San Luis Potosi, Mexico; field investigations and stibnite solubility studies. D, 1974, Stanford University. 180 p.

Querry, Jamie L. Subsurface geology of south central Kay County, Oklahoma. M, 1957, University of Oklahoma. 70 p.

Quesada, Antonio E. Content and solubility of trace elements in veins of Ouachita Mountains. M, 1965, Massachusetts Institute of Technology. 23 p.

Quevedo, Ermel B. Engineering properties of clay interbeds in the vicinity of Lewiston, Idaho. M, 1972, University of Idaho. 109 p.

Quick, Allen N. Structural geology of Hackettstown Quadrangle (New Jersey). M, 1960, New York University.

Quick, Arthur William. Component magnetization of the iron formation and deposits at the Adams Mine, Kirkland Lake, Ontario. M, 1981, University of Windsor. 118 p.

Quick, James Edward. Part I, Petrology and petrogenesis of the Trinity Peridotite, northern California; Part II, Petrogenesis of lunar breccia 12013. D, 1981, California Institute of Technology. 513 p.

Quick, Jay Dudley. Contact alteration and mineralization of stratified rocks in southeastern Arizona. M, 1976, University of Arizona.

Quick, Jeffrey Charles. The use of dyes as an aid to coal petrography. M, 1984, University of Toledo. 124 p.

Quick, Kurt. The geology of the southern portion of the Maynardville Quadrangle, Union County, Tennessee. M, 1960, University of Tennessee, Knoxville. 58 p.

Quick, Ray A. A hydrogeological and hydrochemical investigation of the Edwards Aquifer in the San Marcos area, Hays County, Texas. M, 1985, University of Arkansas, Fayetteville.

Quick, Robert Carl, III. Gravity-magnetic survey of portions of Wood and Lucas counties, Ohio. M, 1976, Bowling Green State University. 81 p.

Quick, Thomas J. Interrelationships and low-temperature oxidation of the properties of coal from the Pittsburgh Seam. M, 1983, University of Akron. 75 p.

Quiett, Frederick T. The geology of the Plainview area, Jefferson County, Colorado. M, 1951, Colorado School of Mines. 80 p.

Quigley, Claud Merle, Jr. Conodonts from the Sylamore Sandstone of southwestern Missouri and northwestern Arkansas. M, 1942, University of Missouri, Columbia.

Quigley, Darwin M. A study of the optical properties of isometric opaque minerals. M, 1947, Northwestern University.

Quigley, Donald Walker. Earth pressures on conduits and retaining walls. D, 1978, University of California, Berkeley. 419 p.

Quigley, Milner D. A restudy of existing graphical methods of interpreting magnetic data and their application to interpreting the results of magnetic surveys across Los Angeles Basin. M, 1950, California Institute of Technology.

Quigley, R. M. A study of the occurrences, properties and origins of varved clays [Ontario and Connecticut]. M, 1956, University of Toronto.

Quigley, Robert M. Composition and engineering properties of some vermiculitic products of weathering. D, 1961, Massachusetts Institute of Technology. 239 p.

Quillam, William E. An investigation of the mechanics of oil migration from the source beds to the reservoir. M, 1928, University of Missouri, Rolla.

Quillin, Michael Edward. A statistical study of fracture orientation and spacing on the East Kaibab Monocline, Arizona. M, 1983, University of Oklahoma. 86 p.

Quillin, Robert Lynn. Microearthquake survey of the Radium Spring KGRA, south central New Mexico. M, 1977, University of Texas at Dallas. 155 p.

Quin, Andrew. Crevasse splay deposits; a study of the geomorphology and sedimentology in an anastomosing fluvial system. M, 1982, University of Calgary.

Quine, Richard Lyle. Geology of Lower Chrome Ridge, Josephine County, Oregon. M, 1977, University of Oregon. 105 p.

Quinlan, Alician V. Seasonal and spatial variations in the water mass characteristics of Muir inlet, Glacier bay, Alaska. M, 1969, University of Alaska, Fairbanks. 145 p.

Quinlan, James F. Types of karst, with emphasis on cover beds in their classification and development. D, 1978, University of Texas, Austin. 342 p.

Quinlivan, William D. Geology of the Herring Park area northeast of Salida (Park and Fremont counties), Colorado. M, 1959, University of Colorado.

Quinn, Alonzo Wallace. Normal faults of the Lake Champlain region (New York). D, 1931, Harvard University.

Quinn, Alonzo Wallace. Some features of correlative value in a deep well section in western Kansas. M, 1927, University of Iowa. 25 p.

Quinn, Christopher F. Depositional history and diagenesis of the Sherwood and Bluell beds (Mississippian) southwestern Renville County, North Dakota. M, 1986, University of North Dakota. 254 p.

Quinn, H. A. Some phases of wallrock alteration at certain Ontario gold mines. M, 1942, Queen's University.

Quinn, Harold Arthur. Geology and gold deposits of the Gianque Lake section, Yellowknife area, Canada. D, 1950, Cornell University.

Quinn, Harold Edward, III. Sedimentary facies, structure, palynology of the lower and middle members of the Atoka Formation, central Arkansas. M, 1986, University of Missouri, Columbia. 157 p.

Quinn, Harry M. Precambrian, Eocambrian rocks of the Basin and Range Province of eastern California. M, 1968, University of Southern California.

Quinn, Howard E. The solubility of galena. M, 1918, University of Minnesota, Minneapolis. 13 p.

Quinn, Howard Edmond. Sabinal District. M, 1926, University of Minnesota, Duluth.

Quinn, Howard Edmond. Some structural and genetic details of gold occurrence at the Hollinger Mine, Porcupine, Ontario. D, 1932, Harvard University.

Quinn, James Harrison. Miocene Equidae of the Texas Gulf Coastal Plain. D, 1954, University of Texas, Austin.

Quinn, James P. Geology and biostratigraphy of the Bopesta Formation, southern Sierra Nevada mountains, Kern County, California. M, 1984, University of California, Riverside. 237 p.

Quinn, Kenneth Elmus. Selection of potential mineral deposits in southwestern Sinai using Landsat imagery. M, 1983, Oklahoma State University. 97 p.

Quinn, Kenneth J. A computer evaluation of the Piketon water well field, South-central Ohio. M, 1980, Ohio University, Athens. 73 p.

Quinn, Laughlin C. Paleofacies of the Late Mississippian carbonate complex, Southeast Idaho and western Wyoming. M, 1985, Idaho State University. 186 p.

Quinn, Louise A. The Humber Arm allochthon at South Arm, Bonne Bay, with extensions in the Lomond area, western Newfoundland. M, 1985, Memorial University of Newfoundland. 188 p.

Quinn, M. J. The glacial geology of Ross County, Ohio. D, 1974, Ohio State University. 293 p.

Quinn, Michael J. A scanning electron microscope study of the microstructure of dispersed and flocculated kaolinite clay taken out of suspension. M, 1980, University of Florida. 85 p.

Quinn, Michael J. The glacial geology of Champaign County, Ohio. M, 1972, Ohio State University.

Quinn, O. P. Bomb tritium in a sandy recharge zone. M, 1981, University of Waterloo.

Quinn, Reay Pullar. An investigation of the physical and chemical properties of Manitoba and Saskatchewan sands. M, 1928, University of Manitoba.

Quinn, Terrence Michael. Evolution of selected Holocene mangrove islands of west-central Florida Bay. M, 1984, Wichita State University. 114 p.

Quinn, William H. A petrographical analysis of the sandstones in the Dakota Group in the Wind River basin of Wyoming. M, 1942, University of Missouri, Columbia.

Quintana, C. W. Ice structure in a vertical core at the margin of South Cascade Glacier, Washington. M, 1983, University of Washington. 97 p.

Quintana, Miguel Alfredo. Facies analysis and environment of deposition for the Jurassic "A" zone of the "Mulussa" (Dolaa) Group in the Homs Block, Syria. M, 1988, Texas A&M University. 94 p.

Quinterno, Paula J. Distribution of Recent Foraminifera in central and south San Francisco Bay (California). M, 1968, San Jose State University. 83 p.

Quintus-Bosz, Robert L. Petrology and distribution of phosphate in the lower Salada Formation, Santa Rita, Baja California Sur, Mexico. M, 1980, Colorado School of Mines. 126 p.

Quirk, Bruce Kenneth. Comparison of land cover maps prepared by common remote sensing methods. D, 1981, University of Wisconsin-Madison. 351 p.

Quirke, Terence Thomas. Geology of Espanola District, Ontario, Canada. D, 1915, University of Chicago. 92 p.

Quirke, Terence Thomas. Geology of the Killdeer Mountains, Dunn County, North Dakota. M, 1913, University of North Dakota. 41 p.

Quirke, Terence Thomas, Jr. Mineralogy and stratigraphy of the Temiscamie iron formation, Lake Albanel Iron Range, Mistassini Territory, Quebec, Canada. D, 1958, University of Minnesota, Minneapolis. 147 p.

Quirke, Terence Thomas, Jr. Preliminary investigation of the geology of the North Michigamme area, Marquette Iron Range, Michigan. M, 1953, University of Minnesota, Minneapolis. 76 p.

Quirt, David Hulse. A comparative study of major element rock analyses by argon plasma emission and atomic absorption/emission spectrophotometry. M, 1978, University of Windsor. 257 p.

Quirt, G. Stewart. A potassium-argon geochronological investigation of the Andean mobile belt of north-central Chile. D, 1972, Queen's University. 240 p.

Quist, Earl Francis. The geography of the Chain of Rocks by-pass canal. M, 1950, Washington University. 81 p.

Quist, Lawrence G. Groundwater resources in the Red Deer area, Alberta. M, 1969, Queen's University. 158 p.

Quittmeyer, Richard Charles. Seismicity and tectonics of Pakistan and surrounding regions. D, 1982, Columbia University, Teachers College. 305 p.

Quitzau, Robert P. Regional gravity survey of the back valleys of the Wasatch Mountains and adjacent areas of Utah, Idaho, and Wyoming. M, 1961, University of Utah. 48 p.

Quon, Charles. Electromagnetic fields of elevated dipoles on a two-layer earth. M, 1963, University of Alberta. 250 p.

Quon, Shi Haung. Geochemistry and paragenesis of carbonatitic calcites and dolomites. D, 1965, University of Michigan. 260 p.

Quraishi, Raziuddin. Laboratory determination of the electrical properties of mineralized porphyry by the induced polarization method. M, 1966, University of Arizona.

Qureshi, Riffat Mahmood. Counting of natural carbon-14 for hydrological studies using a commercially available CO_2 absorber. M, 1982, University of Waterloo. 232 p.

Qureshi, Riffat Mahmood. The isotopic composition of aqueous sulfate (a laboratory investigation). D, 1986, University of Waterloo. 353 p.

Qureshy, Mohammed N. Gravity anomalies and computed variation in thickness of the Earth's crust in Colorado. D, 1958, Colorado School of Mines.

Qutub, Musa Y. The hydrogeology of the northeast quarter of Marion County, Iowa. M, 1966, Colorado State University. 73 p.

Qutub, Musa Yacub. The objectives of the Earth Science curriculum project; an evaluation of their achievement. D, 1969, Iowa State University of Science and Technology. 102 p.

R. F., Jr. Litchford *see* Litchford R. F., Jr.

R., J. Alfredo Cervantes *see* Cervantes R., J. Alfredo

R., L. M. Echeverria *see* Echeverria R., L. M.

Raab, Enrique Pedro Gentzsch. Properties of hydroxy-aluminum interlayers in clays and of peats, major components of very weak acid sources in soils. D, 1970, [University of Kentucky]. 140 p.

Raab, James Michael. A finite strain study of the Baraboo Quartzite; observation from quartz grain shape. M, 1985, Michigan State University. 65 p.

Raab, Werner Joseph. Solubilities of stibnite, orpiment and realgar in borate, carbonate and hydroxide solutions as function of temperature and pressure and their implications as applied to the borax deposit at Boron, California. D, 1969, University of California, Riverside. 228 p.

Raabe, Bruce A. Fixed ammonium nitrogen in shales. M, 1983, University of Texas, Arlington. 51 p.

Raabe, John A. Bedrock geology of the Allagash Lake-Caucomgomoc Lake map area, West central Maine. D, 1977, Boston University. 222 p.

Raabe, John A. Structural relationships in the northwestern Kingfield Quadrangle, Maine. M, 1969, Boston University. 47 p.

Raabe, Kenneth Charles. Mineralogy and geochemistry of gold-silver veins at the Hock Hocking Mine, Alma, Colorado. M, 1983, Purdue University. 201 p.

Raabe, Robert G. Structure and petrography of the Bullock Canyon, Buehman Canyon area, Pima County, Arizona. M, 1959, University of Arizona.

Raad, Awni Tewfiq Saleh. Genetic development of youthful soils derived from dolomitic materials in the Blue Springs drainage basin (near Guelph, Ontario). D, 1970, University of Guelph.

Raasch, Albert C., Jr. The Sunniland oil field of Collier County, Florida. M, 1954, Florida State University.

Raasch, Gilbert O. The (Permian) Wellington Formation in Oklahoma. D, 1947, University of Wisconsin-Madison.

Raba, Carl Franz, Jr. The static and dynamic response of a miniature friction pile in remolded clay. D, 1968, Texas A&M University. 174 p.

Rabah, N. A. Stratigraphy and microfacies analysis of the Berino Formation (Atokan-DesMoines), Vinton Canyon, El Paso County, Texas. M, 1976, University of Texas at El Paso.

Rabbio, Salvatore Frank. Ecological and taphonomic gradients in storm disturbed bryozoan communities of the Kope Formation (Cincinnatian Series, Upper Ordovician), Cincinnati Arch region. M, 1988, Michigan State University. 219 p.

Rabbitt, John Charles. Anthophyllite and its occurrence in southwestern Montana. D, 1947, Harvard University. 145 p.

Rabbitt, John Charles. The Ostracoda of the Mississippian Big Snowy Group in Montana. M, 1937, Montana College of Mineral Science & Technology. 57 p.

Rabchevsky, George A. Contributions to the geology of the Adams Run Anticline, Hardy County, West Virginia. M, 1963, George Washington University.

Rabchevsky, George A. The feasibility of detecting subsurface coal fires in Wyoming and Montana from the ground, on aerial photography and on satellite imagery. D, 1972, George Washington University. 229 p.

Rabek, Karen Elaine. Late Pleistocene tephrochronology of the western Gulf of Mexico. M, 1983, University of Georgia. 46 p.

Rabenhorst, Martin Capell. Genesis of soils and carbonate enriched horizons in a climo-sequence developed over Cretaceous limestone in central and West Texas. D, 1983, Texas A&M University. 267 p.

Raber, Ellen. Zircons from diamond-bearing kimberlites; oxide reactions, fission track dating and a mineral inclusion study. M, 1978, University of Massachusetts. 90 p.

Rabie, Farida Hamed. Mineral weathering and distribution in a series of salt-affected soils. D, 1967, University of California, Davis. 126 p.

Rabii, H. A. An investigation of the utility of Landsat 2 MSS data to the fire-danger rating area, and forest fuel analysis with Crater Lake National Park, Oregon. D, 1979, Oregon State University. 434 p.

Rabinowitz, D. Daniel. Environmental tritium as a hydrometeorologic tool in the Roswell Basin, New Mexico. D, 1972, New Mexico Institute of Mining and Technology. 268 p.

Rabinowitz, D. Daniel. Forced exchange of tritiated water with natural clays. M, 1969, New Mexico Institute of Mining and Technology. 72 p.

Rabinowitz, Michael Bruce. Lead contamination of the biosphere by human activity; a stable isotope study. D, 1974, University of California, Los Angeles.

Rabinowitz, Philip David. The continental margin of the Northwest Atlantic Ocean; a geophysic. D, 1973, Columbia University. 181 p.

Raby, Andrew G. Interpreted depositional environments of the Salt Wash and lower members of the Morrison Formation, Grand County, Utah. M, 1982, New Mexico Institute of Mining and Technology. 113 p.

Race, Charles Dana. Hydrogeochemistry and sources of chloride and sulfate in the Verdigris River basin in Oklahoma. M, 1985, Oklahoma State University. 167 p.

Race, Kelley Ann Clinton. Geochemical analysis and environmental reconstruction of a Devonian lake basin in northern Scotland. M, 1985, Oklahoma State University. 186 p.

Race, Ronald Williams. Incipient metamorphism in the Ohanapecosh Formation Washington. M, 1969, University of Washington. 37 p.

Racey, Jan Stewart. Conodont biostratigraphy of the Redwall Limestone of east-central Arizona. M, 1974, Arizona State University. 199 p.

Rach, Nina Marie. Tectonics, structure, and sedimentation of the Lake Victoria basin, East Africa. M, 1988, Duke University. 116 p.

Rachele, L. D. Palynology of the Legler lignite. D, 1974, New York University. 94 p.

Rachou, John F. Tertiary stratigraphy of the Rattlesnake Hills, central Wyoming. M, 1951, University of Wyoming. 70 p.

Racicot, Denis. Pétrographie et géochimie du Pluton de la Rivière Barlow, Chibougamau, Québec. M, 1988, Universite du Quebec a Chicoutimi. 115 p.

Rackley, David Holland. Hydroids of the pelagic Sargassum community of the Gulf Stream and Sargasso Sea. M, 1974, College of William and Mary.

Rackley, Ruffin I. Geology of Bean Mountain area. M, 1951, University of Tennessee, Knoxville. 77 p.

Rackoff, J. S. The osteology of Sterropterygion (Crossopterygii; Osteolepidae) and the origin of tetrapod locomotion. D, 1976, Yale University. 304 p.

Rad, Nader Shafii *see* Shafii Rad, Nader

Radabaugh, Robert Eugene. Dundee and Rogers City limestones (Devonian) of the northern part of the Southern Peninsula of Michigan. M, 1937, University of Michigan.

Radabaugh, Robert Eugene. Middle Devonian Rogers City Limestone and its gastropod fauna. D, 1942, University of Michigan.

Radack, Phyllis. Wave- and fluvial-dominated deltaic deposits of the Cerro Huerta and Canon del Tule formations (Upper Cretaceous), northeastern Mexico. M, 1986, University of Houston.

Radain, A. A. M. Petrogenesis of some peralkaline and non-peralkaline post-tectonic granites in the Arabian Shield, Kingdom of Saudi Arabia. D, 1978, University of Western Ontario. 247 p.

Radcliffe, Dennis. Mineralogical studies on safflorite-loellingite. D, 1966, Queen's University. 122 p.

Radcliffe, Dennis. The geology of the Birch Portage beryl pegmatite deposit, Saskatchewan. M, 1964, University of Alberta. 113 p.

Rader, Eugene Kenton. Ostracodes of the Middle Devonian Needmore Shale of Northwest Virginia. M, 1962, University of Virginia. 87 p.

Rader, Herbert L. Geology of Valley Forge Park and vicinity, Valley Forge, Pennsylvania. M, 1951, Lehigh University.

Radford, Robert M. Pleistocene barrier island formation, Myrtle Beach, South Carolina. M, 1975, University of South Carolina. 31 p.

Radford, Wilbur Edward. Relocation of earthquakes in the Lake Sinclair reservoir area. M, 1988, Georgia Institute of Technology. 56 p.

Radi, Mahmoud Diab. The downstream channel form change of streams on different geological settings. D, 1982, University of Oklahoma. 242 p.

Radinsky, Leonard B. Origin and early evolution of North American Tapiroidea. D, 1962, Yale University.

Radisi, John. A report on the geology of the upper and lower Turkeyfoot Townships, Confluence Quadrangle, Pennsylvania. M, 1954, University of Pittsburgh.

Radke, Bruce M. Carbonate sedimentation in tidal and epeiric environments and diagenetic overprints; the Ninmaroo Formation (Upper Cambrian-Lower Ordovician), central Australia. D, 1978, Rensselaer Polytechnic Institute. 254 p.

Radke, Frank, Jr. Iron-titanium oxide minerals in quartz microsyenite (Tertiary), Jeff Davis County, Texas. M, 1967, University of Kansas. 29 p.

Radle, Nancy. Vegetation history and lake-level changes at a saline lake in northeastern South Dakota. M, 1982, University of Minnesota, Minneapolis. 126 p.

Radler, Dollie. The Bartlesville Sand in Osage County. M, 1922, University of Oklahoma.

Radlick, Thomas. Depositional subenvironments of trona and associated lithologies in the Wilkins Peak Member of the Green River Formation, southwestern Wyoming. M, 1984, Wayne State University. 223 p.

Radloff, David Lee. Wildland classification with multivariate analysis and remote sensing techniques. D, 1983, Colorado State University. 116 p.

Radomsky, Patrick M. Dynamic crystallization experiments on magnesian olivine-rich and pyroxene-olivine chondrule composition. M, 1988, Rutgers, The State University, New Brunswick. 96 p.

Radon, Stanley F. The Medina Group (Lower Silurian) of western New York, with emphasis on the "White Facies" of the Grimsby Sandstone. M, 1987, Kent State University, Kent. 170 p.

Radovich, Barbara Jean. Leaking modes of Love waves and seismogram interpretation. M, 1973, Rice University. 39 p.

Radrikrjengkrai, P. Petrographic investigation of hydrothermal alteration in the Quartz Creek, Middle

Fork, Snoqualmie River, King County, Washington. M, 1971, University of Washington.

Radsbaugh, John Wesley. Glacial geology of St. Louis City and county (Missouri). M, 1929, Washington University. 56 p.

Radtke, Arthur S. The relationship of Arizona mining districts to major geologic structures and formations. M, 1960, University of Minnesota, Minneapolis.

Radtke, Arthur Sears. Minor elements in iron ores from the Western United States. D, 1965, Stanford University. 360 p.

Radville, Mark E. Ice regime of the Cold River, New Hampshire. M, 1981, Boston University. 123 p.

Radwan, A. M. Wave-current interactions in water of variable depth. D, 1975, North Carolina State University. 76 p.

Rady, Paul M. Structure and petrology of the Groat Mountain area, North Cascades, Washington. M, 1980, Western Washington University. 132 p.

Raede, Deborah Lynn. Major ion and trace metal contamination of groundwater at the Riverside Sanitary Landfill. M, 1986, San Diego State University. 176 p.

Raedeke, Linda Dismore. Petrogenesis of the Stillwater Complex, Montana. D, 1982, University of Washington. 212 p.

Raedeke, Linda Dismore. Stratigraphy and petrology of the Stillwater Complex, Montana. M, 1979, University of Washington. 89 p.

Raeissi-Ardakani, Ezatollah. Level basin irrigation design with water constraints. D, 1982, Colorado State University. 184 p.

Raeside, Robert P. Structure, metamorphism and migmatization of the Scrip Range, Mica Creek, British Columbia. D, 1982, University of Calgary. 204 p.

Raeside, Robert Pollock. A reinvestigation of the Ile Bizard kimberlite, and its ultramafic xenolith suite, Montreal. M, 1978, Queen's University. 87 p.

Rafaels, Maris. Gravity modeling of Central American subduction zones. M, 1977, Rutgers, The State University, New Brunswick. 37 p.

Rafalko, Leonard Gervus. Uncertainty in estimating breakthrough curves for groundwater monitoring studies. M, 1988, University of Virginia. 96 p.

Rafalowski, Mary Beth. Sedimentary geology of the Late Cambrian Honey Creek and Fort Sill formations as exposed in the Slick Hills of southwestern Oklahoma. M, 1984, Oklahoma State University. 147 p.

Rafalska-Bloch, Janina. Diagenesis and catagenesis of marine kerogen precursors. D, 1987, University of Oklahoma. 303 p.

Rafipour, Bijan. Phase and attenuation studies of seismic waves using phase-matched filtering technique. D, 1981, Southern Methodist University. 271 p.

Rafipour, Bijan. The petrology and structure of the Smoothing-iron Mountain area, northwestern Llano County, central Texas. M, 1976, Southern Methodist University. 68 p.

Rafle, M. A. Bathymetry of Recent marine Ostracoda off the coasts of Mississippi and Alabama. M, 1973, Washington University. 34 p.

Rafle, M. A. Paleobiogeography of Paleocene through early Eocene calcareous nannoplankton assemblages from the Pacific Ocean. D, 1977, Washington University. 138 p.

Raftery, Peter John. Marine geology and geologic evolution of the northern Channel Islands Platform west of San Miguel Island. M, 1984, California State University, Northridge. 118 p.

Rafuse, Bruce Elwood. Bathymetry of conodonts from the Bird Spring Formation of Arrow Canyon, Nevada. M, 1973, Washington State University. 60 p.

Ragan, Donal M. Geology of Butte Valley, Inyo County, California. M, 1953, University of Southern California.

Ragan, Donal MacKenzie. The geology of Twin Sisters Dunite in the northern Cascades, Washington. D, 1961, University of Washington. 88 p.

Ragan, Jerry Michael. Late Cretaceous to early Eocene foraminiferal zonation from a well in the Gulf of Mexico. M, 1981, Washington University. 35 p.

Ragan, Virginia M. Geothermometry and organic matter-mineral link for the mineralization at the Prescott zinc deposit and adjacent country rocks of Linn County, Kansas. M, 1987, University of Missouri, Kansas City. 96 p.

Ragan, Wendell J. Brachiopoda and Mollusca from the Burgner Formation in Southwest Missouri. D, 1959, University of Missouri, Columbia. 238 p.

Raghavan, Rajagopal. The instability of liquid-liquid interfaces and model systems for emulsion formation in porous media. D, 1970, Stanford University. 127 p.

Raghu, D. Finite element analysis of the behavior of nonlinear soil continua including dilatancy. D, 1975, [Texas Tech University]. 144 p.

Ragland, Betty Catherine Sims. Gravity investigation of a portion of the Ouachita central zone, southeastern Oklahoma and western Arkansas. M, 1988, University of Texas at Dallas. 80 p.

Ragland, Deborah Ann. Sedimentary geology of the Ordovician Cool Creek Formation as it is exposed in the Wichita Mountains of southwestern Oklahoma. M, 1983, Oklahoma State University. 171 p.

Ragland, Kenneth E. Structural analysis of the Packsaddle Schist (late Precambrian), Sandy Mountain area, Llano County, Texas. M, 1975, Northeast Louisiana University.

Ragland, Paul C. Chemical, radiometric, and mineralogic trend surfaces within the Enchanted Rock Batholith, Llano, and Gillespie counties, Texas. D, 1962, Rice University. 80 p.

Ragland, Paul C. Geochemical and petrological studies of the Lost Creek Gneiss, Mason and McCullough counties, Texas. M, 1961, Rice University. 99 p.

Ragle, Richard Charles. The geology of the Fort Riley Military Reservation and vicinity (Kansas). M, 1937, Colorado College.

Ragle, Richard H. Investigations in the formation of lake ice in a temperate climate. M, 1958, Dartmouth College.

Ragnarsdottir, Kristin Vala. Experimental and theoretical investigation of quartz and corundum solubilities and their application to phase equilibria and mass transfer in the geothermal system, Svartsengi, Iceland. D, 1984, [Michigan State University]. 128 p.

Rago, Frank Thomas. The Brassfield Formation of southern Indiana. M, 1952, Indiana University, Bloomington. 32 p.

Ragone, Stephen Edward. The hydrolysis and polymerization processes of iron (III) in aqueous solution. D, 1968, Pennsylvania State University, University Park. 98 p.

Ragsdale, James Allan. Petrology of Miocene Oakville Formation, Texas coastal plain. M, 1960, University of Texas, Austin.

Rahaim, Stephen David. The aminostratigraphy of the Sankaty Sand, Nantucket Island, Massachusetts. M, 1987, University of Delaware. 131 p.

Raham, Gerald Orr. Geology of the big ledge zinc deposit, British Columbia (Canada). M, 1967, University of Calgary. 90 p.

Rahim, Zillur. Proppant transport down a three-dimensional planar fracture. D, 1988, Texas A&M University. 215 p.

Rahimi, Hassan. Comparison of direct and indirect methods for determining the coefficient of permeability of clays. D, 1977, Oklahoma State University. 153 p.

Rahimi, Saeed. Solubility of $CaCO_3$ in calcite and calcareous soils. D, 1986, Washington State University. 62 p.

Rahm, David Allen. Geology of the main drainage tunnel, Boston, Massachusetts. D, 1960, Harvard University.

Rahman, A. Interrelationships, involving sensitivity, thixotropy and their relationship with electrokinetic phenomena in some Hawaiian soils. D, 1974, University of Hawaii. 482 p.

Rahman, A. Some electrokinetic phenomena in three Hawaiian soils. M, 1969, University of Hawaii. 157 p.

Rahman, Abdel Fattah Mostafa Abdel *see* Abdel Rahman, Abdel Fattah Mostafa

Rahman, Ata Ur. Lithofacies, porosity and log response of the lower San Andres Formation in the Palo Duro Basin. M, 1983, Texas Tech University. 116 p.

Rahman, Ata Ur. Lithostratigraphy of the subsurface Silurian rocks in western Kentucky. D, 1987, Texas Tech University. 181 p.

Rahman, John L. Geothermal studies of drill holes in northern Illinois and the Hueco Tanks region, Otero County, New Mexico. M, 1983, University of Texas at El Paso. 68 p.

Rahman, M. Shamimur. Analyses for wave-induced liquefaction in relation to off-shore construction. D, 1977, University of California, Berkeley. 203 p.

Rahman, Mohammed Golzar. A constitutive model for creep of a frozen sand. D, 1988, University of Manitoba.

Rahman, Mostafa A. Abdel *see* Abdel Rahman, Mostafa A.

Rahman, Yousuf H. Geology of the Wellsville-Calcite area; Chaffee and Fremont counties, Colorado. D, 1954, Colorado School of Mines. 146 p.

Rahmani, Riyadh A. Sedimentology and petrology of the Cedar District Formation, late Cretaceous, S.W. British Columbia. M, 1970, University of British Columbia.

Rahmani, Riyadh Abdul-Rahim. Heavy mineral analysis of Upper Cretaceous and Paleocene sandstones in Alberta and adjacent areas of Saskatchewan. D, 1973, University of Alberta. 239 p.

Rahmanian, Victor D. Deltaic sedimentation and structure of the Fox Hills and Laramie formations, upper Cretaceous, southeast of Boulder, Colorado. M, 1975, Colorado School of Mines. 83 p.

Rahmanian, Victor David K. Stratigraphy and sedimentology of the Upper Devonian Catskill and uppermost Trimmers Rock formations in central Pennsylvania. D, 1979, Pennsylvania State University, University Park. 396 p.

Rahmatian, Mansour. Petrography and facies of the Buda Limestone, Gomez Quadrangle, Trans-Pecos, Texas. M, 1983, University of Southwestern Louisiana. 88 p.

Rahn, Jerry Everett. The geology of the Meyers Cove area, Lemhi County, Idaho. M, 1979, University of Idaho. 111 p.

Rahn, Perry Hendricks. The inselbergs of southwestern Arizona. D, 1965, Pennsylvania State University, University Park. 158 p.

Rahn, William R., Jr. The role of Spartina alterniflora in the transfer of mercury in a salt marsh environment. M, 1973, Georgia Institute of Technology. 61 p.

Rahsman, Robert G., II. Stream pattern changes on Rock Creek near Roberts, Montana. M, 1973, Southern Illinois University, Carbondale. 61 p.

Rai, C. B. Electrical and elastic properties of basalts and ultramafic rocks as function of saturation, pressure and temperature. D, 1977, University of Hawaii. 155 p.

Rai, Dhanpat. Stratigraphy and genesis of soils from volcanic ash in the Blue Mountains of eastern Oregon. D, 1971, Oregon State University. 136 p.

Rai, Vijai Narain. Geology of a portion of the Nightingale and Truckee ranges, Washoe and Pershing counties, Nevada. M, 1968, University of Nevada. 45 p.

Rai, Vijai Narain. Pennsylvanian brachiopods of Nevada. D, 1972, University of Nevada. 135 p.

Raible, Clarence J. Geology of the Liberty Hill area, Crawford County, Arkansas. M, 1959, University of Arkansas, Fayetteville.

Raible, Leonard J. Minor structures of the Boston Mountain Monocline in T. 10, 11, and 12 N., R. 30 W., Crawford County, Arkansas. M, 1954, University of Arkansas, Fayetteville.

Raidl, Robert F. Petrology of the Hillman Tonalite in east-central Minnesota. M, 1973, Northern Illinois University. 39 p.

Raij, Bernardo Van *see* Van Raij, Bernardo

Raikes, Susan Ann. I, Regional variations in upper mantle compressional velocities beneath southern California; II, Post-shock temperatures; their experimental determination, calculation, and implications. D, 1978, California Institute of Technology. 314 p.

Railsback, Loren Bruce. Diagenetic history of the carbonate members of the Dennis Formation (Missourian, Upper Pennsylvanian) in Iowa, Missouri, and Kansas. M, 1983, University of Iowa. 129 p.

Railsback, Rickard Reed. Geologic hazards in the Pine Springs Canyon area, Guadalupe Mountains National Park. M, 1976, Texas Tech University. 75 p.

Railton, J. B. Vegetational and climatic history of southwestern Nova Scotia in relation to a South Mountain ice cap. D, 1973, Dalhousie University.

Rainbird, R. H. Sedimentology and geochemistry of the Firstbrook Member of the Gowganda Formation in the eastern Cobalt Basin, Ontario. M, 1985, Carleton University. 157 p.

Raine, Frank Frederick. Properties and uses of talc and soapstone of the Pacific Northwest; Part two. M, 1936, University of Washington. 81 p.

Raine, John Wesley, III. Cyclic sedimentation of the upper Allegheny Series in the Kittanning-Freeport area. M, 1953, University of Pittsburgh.

Raines, Gary L. Evaluation of multiband photography for rock discrimination. D, 1974, Colorado School of Mines. 86 p.

Raines, Gary L. Geology of the Sargents area, Gunnison and Saguache counties, Colorado. M, 1971, Colorado School of Mines. 55 p.

Raines, Robert B. Geology of Kimes Mountain area, Crawford County, Arkansas. M, 1959, University of Arkansas, Fayetteville.

Raines, W. A. Sub-tidal oscillations in Monterey harbor (California). M, 1967, United States Naval Academy.

Rainey, Clifford S. Differences in intensity of subsidiary reflections in plagioclase feldspar. M, 1977, University of California, Berkeley. 54 p.

Rainey, Mary Tindal. The Quaternary stratigraphy of the North Sulphur River, Delta County, Texas. M, 1974, Southern Methodist University. 28 p.

Rains, George Edward. Paleocene silcrete beds in the San Juan Basin. M, 1981, University of Arizona. 81 p.

Rainsberry, Lois Eileen. Pollen and spores of the Lower Cretaceous Blairmore Group in Saskatchewan. M, 1962, University of Saskatchewan. 85 p.

Rainville, George D. Ore petrology and nickel-manganese variations in olivines of mafic and ultramafic rocks with emphasis on the Harriman Peridotite, Knox County, Maine. D, 1976, Boston University. 310 p.

Rainville, Serge. Recommandations d'un programme d'exploration minière sur la proprieté de "Les mines Selbaie" au Québec. M, 1986, Ecole Polytechnique. 245 p.

Rainville, Serge. Recommandations d'un programme d'exploration minière sur la propriété de "les Mines Selbaie" au Québec. M, 1985, Ecole Polytechnique. 245 p.

Rainwater, Edward H. and Osborn, Elburt F. Geology of the Calumet, Colorado mining district. M, 1934, Northwestern University.

Rainwater, Kenneth Alvis. Solute transport in porous media; theoretical and experimental modeling of cation exchange. D, 1985, University of Texas, Austin. 245 p.

Raione, Richard Paul. Petrographic characteristization of the upper Elkorn number two seam of eastern Kentucky. M, 1983, University of Kentucky. 181 p.

Rais, Samira. Analysis of the flow close to the outer bank of a meander bend. D, 1985, Colorado State University. 303 p.

Raish, Dan. The role of tidal deformation in the origin of the Lunar Grid System. M, 1969, University of Washington. 44 p.

Raish, Henry Dean Eugene. Petrology of a limestone bank in the Winchell Formation (Upper Pennsylvanian) of Wise County, Texas. M, 1964, Texas Christian University.

Raisz, Erwin J. The illustration of geologic papers. M, 1924, Columbia University, Teachers College.

Raisz, Erwin J. The scenery of Mt. Desert Island; its origin and development. D, 1929, Columbia University, Teachers College.

Raitz, Charles Henry. The geology of a portion of southwestern Mora County, New Mexico. M, 1951, University of Iowa. 94 p.

Rajagopal, R. S. On the structural characterization of rocks. D, 1976, University of California, Davis. 281 p.

Rajagopal, Rangaswamy. Sewage disposal and water supply alternatives on the basis of water quality and economic criteria. D, 1974, University of Michigan.

Rajagopalan, Natasayyer. Stratigraphy, micropaleontology and paleontology of the Upper Cretaceous and lower Tertiary formation of Pondicherry, South India. D, 1963, Rutgers, The State University, New Brunswick. 208 p.

Rajamani, V. Aspects of the crystal chemistry of nickel and cobalt. D, 1974, SUNY at Stony Brook. 215 p.

Rajasekaran, Konnur C. Mineralogy and petrology of nepheline syenite in Mont Saint Hilaire, Quebec (Canada). D, 1968, McGill University.

Rakai, R. J. Crystal structure of spessartine and andradite at elevated temperatures. M, 1975, University of British Columbia.

Raker, Gary. Laboratory simulated diagenesis of pelitic sediments. M, 1982, University of Missouri, Columbia.

Raker, Sarah L. Chemistry of groundwater in tuffaceous rocks, central Nevada. M, 1987, University of Nevada. 111 p.

Raksaskulwong, Manop. Combined electrical studies in the southern Hueco Bolson, El Paso County, Texas. M, 1985, University of Texas at El Paso.

Raleigh, Cecil Baring. Fabrics of naturally and experimentally deformed olivine. D, 1963, University of California, Los Angeles. 105 p.

Raleigh, Cecil Baring. Structure and petrology of a part of the Orocopia Schists (Precambrian, California). M, 1959, Pomona College.

Raleigh, Robert Eugene, Jr. Nannofossil dating of Cordilleran suspect terranes of southwestern Oregon. M, 1984, Ohio University, Athens. 79 p.

Rall, Elizabeth Pretzer. Pennsylvanian and Lower Permian geology of Sutton and Schleicher counties, Texas. D, 1956, University of Illinois, Urbana. 140 p.

Rall, Raymond Wallace. Ostracods from the Depauporate Zone of the Maquoketa Shale. M, 1951, University of Illinois, Urbana.

Rall, Robert D. Stratigraphy, sedimentology and paleotopography of the Lower Jurassic in the Gilby-Medicine River fields, Alberta. M, 1980, University of Calgary. 142 p.

Ralph, Elizabeth Kennedy. Geophysical implications of radiocarbon measurements. D, 1973, University of Pennsylvania.

Ralston, Dale R. Impact of legal constraints on ground-water development in Idaho. D, 1974, University of Idaho. 131 p.

Ralston, Dale R. Influence of water well design on neutron logging. M, 1967, University of Arizona.

Ralston, Edward Charles. Geology and mineralization of a part of the Nelson Range, Inyo County, California. M, 1984, University of Nevada. 177 p.

Ralston, June Kathleen. A study of the diatomites from a portion of the type area of the Ellensburg Formation of south central Washington. M, 1984, University of Washington. 112 p.

Ram, A. The identification and interpretation of upper mantle travel time branches from slowness measurements made on data recorded at the Gauribidanur (India) and Yellowknife (Canada) seismic arrays. D, 1976, University of Western Ontario.

Ramaekers, P. P. J. A study of dental variability in early Wasatchian Phenacolemur (Paramomyidae, Primates). D, 1975, University of Toronto.

Ramage, Joseph Robert. Lithofacies, regional stratigraphy, and depositional systems of the Clear Fork Group (Permian) Palo Duro Basin, Texas Panhandle. M, 1987, University of Texas, Austin. 242 p.

Ramalingaswamy, V. M. Stratiform mineralization and origin of vein deposits, Bunker Hill Mine, Coeur d'Alene, Idaho. M, 1975, University of Washington. 46 p.

Raman, Athipet Bashyam. Elastic-plastic transition tests on various rock types. M, 1962, Michigan State University. 52 p.

Raman, Swaminathan V. Petrogeny of granites and associated rocks at Marblehead, Massachusetts. M, 1973, Boston University. 53 p.

Raman, Swaminathan Venkat. Petrology and geochemistry of anorthosite and associated rocks, Honey Brook, Pennsylvania. D, 1979, Rutgers, The State University, New Brunswick. 148 p.

Ramanantoandro, Ramanantsoa. A magnetic survey of the southern Socorro mountains, New Mexico. M, 1965, New Mexico Institute of Mining and Technology. 38 p.

Ramanantoandro, Ramanantsoa. Elastic anisotropy in dunites. D, 1971, University of Washington. 78 p.

Ramanjaneya, Gundarlahalli Sankarasetty. Consolidation characteristics of a varved clay. D, 1969, University of Connecticut. 115 p.

Ramanlal, Kirti Kumar. Velocity determination from multireceiver full-waveform acoustic-logging data. M, 1987, Texas A&M University. 112 p.

Ramarathnam, Sethurama. Geology and petrology of the southern portion of the Laramie anorthosite mass, Albany County, Wyoming. D, 1962, Colorado School of Mines. 131 p.

Ramay, Charles Lee. Clarification of Desmoinesian stratigraphy in the Pleasant Hill Syncline of the Criner Hills. M, 1957, University of Oklahoma. 50 p.

Rambaldi, E. R. Distribution of elements among feldspars, micas, and epidote in some metamorphic rocks (Precambrian) near Bancroft, Ontario. D, 1970, University of Ottawa. 136 p.

Ramberg, E. M. A field and laboratory study of Quaternary deposits in the Harestua region, southern Norway. M, 1978, University of Wyoming. 146 p.

Rambler, Mitchell Bruce. Ultraviolet irradiation of bacteria under anaerobic conditions; implications for Prephanerozoic evolution. D, 1980, Boston University. 131 p.

Rambo, Charles Edward. Sedimentary analysis of the Pennsylvanian System of southernmost Sangre de Cristo Mountains, New Mexico. M, 1960, University of Oklahoma. 218 p.

Rambo, Daniel J. Structural geology of the Clear Creek and Gros Ventre River area, Teton and Sublette counties, Wyoming. M, 1983, Miami University (Ohio). 43 p.

Rambo, George Daniel. The Mescalero, Moore, and east Caprock oil fields, Lea County, New Mexico. M, 1956, University of Oklahoma. 53 p.

Ramboseck, August F. Geology and ore deposits of the Golden Sunlight Mine and vicinity. M, 1946, Montana College of Mineral Science & Technology. 28 p.

Ramelli, Alan Ray. Late Quaternary tectonic activity of the Meers Fault, Southwest Oklahoma. M, 1988, University of Nevada. 123 p.

Ramelli, William P. Changes in beach behavior brought about by the construction of groins and jetties along the Ventura Coast (California). M, 1966, [University of California, Santa Barbara].

Ramer, Alan Rutledge. Petrography of a portion of the Josephine Peridotite Sheet (Jurassic), Josephine County, Oregon. M, 1967, University of Oregon. 121 p.

Ramey, Everett H. An analytical system for super-wide-angle photography with analytical plotter, AP/C. M, 1965, Ohio State University.

Ramiah, B. K. Time effects on the consolidation properties of clays. D, 1959, Purdue University.

Ramirez Berrera, Andreas. Gravity survey in the Morrison area, Jefferson County, Colorado. M, 1954, Colorado School of Mines. 70 p.

Ramirez Lopez, Miguel. Stratigraphy of the Mississippian System, Las Animas Arch, Colorado. M, 1973, Colorado School of Mines. 89 p.

Ramirez Munoz, Jose. Regional geology in the Opodepe mining area, Sonora, Mexico. M, 1979, University of Arizona. 98 p.

Ramirez P., Raul. A map representation theory for the evaluation of digital exchange formats. D, 1988, Ohio State University. 353 p.

Ramirez Rojas, Armando J. Geochemistry of mine effluents in the Front Range mineral belt of Colorado. M, 1976, Colorado School of Mines. 112 p.

Ramirez Serafinoff, Rafael Esteban de la Cruz. Stratigraphy of the Tertiary of the middle Magdalena Basin (Colombia), central and northern parts. M, 1988, University of Texas, Austin. 119 p.

Ramirez, Abelardo Luis. Preliminary engineering geology study for the proposed Portugues diversion tunnel, Ponce, Puerto Rico. M, 1979, Purdue University. 133 p.

Ramirez, Edward. Modelling geomagnetic polarity transitions. M, 1974, University of Miami. 92 p.

Ramirez, Guillermo. Chemical characterization of volcanic glass from central Mexico and its application to archeological studies. M, 1976, University of New Orleans. 115 p.

Ramirez, Jesus Emillio. An experimental investigaton on the nature and origin of microseisms at Saint Louis, Missouri. D, 1939, St. Louis University.

Ramirez, Jesus Emillio. Epicenters of earthquakes in the Republic of Colombia 1916-26, and their relation to the geologic structure. M, 1931, St. Louis University.

Ramirez, Jose R. Geology of the north part of the San Antonio Mountains, State of Sonora, Mexico. M, 1965, University of Arizona.

Ramirez, Octavio. Petrology and structure of the Precambrian metaigneous sequence in the Savage Run Creek area, Carbon County, Wyoming. M, 1971, Colorado State University. 149 p.

Ramirez, Vincent Rex. Geology of the San Andreas Fault at Tejon Pass, California. M, 1984, University of California, Santa Barbara. 256 p.

Ramirez-Rivera, Jaime. Process, prediction and measurement of soil-loss from watersheds. M, 1970, [Colorado State University].

Ramirez-Rojas, Armando Jose. Geochemistry of nutrient elements in water and sediment of the Tuy River basin, Venezuela. D, 1988, Pennsylvania State University, University Park. 277 p.

Ramondetta, Paul John. Heavy metals distribution in Jamaica Bay sediments. M, 1974, Brooklyn College (CUNY).

Ramos, Olga Nanet Arancibia see Arancibia Ramos, Olga Nanet

Ramos-Martinez, Luis. Program "MATCH", for depth matching of logs. M, 1986, Stanford University.

Ramp, E. R. Geologic controls on the regional distribution of dolines in eastern Iowa. M, 1977, Iowa State University of Science and Technology.

Ramp, Lenin. Heavy detrital minerals of the South Umpqua River drainage, Oregon. M, 1953, University of Oregon. 67 p.

Rampal, K. K. Filtering prediction and interpolation in photogrammetry. D, 1976, Ohio State University. 207 p.

Rampertaap, Autar. An experimental study on fracture propagation in brittle-ductile rocks in rift zones and in volcanic centers. M, 1980, Brooklyn College (CUNY).

Rampino, M. R. Quaternary history of South-central Long Island, New York. D, 1978, Columbia University, Teachers College. 755 p.

Rampton, Vernon Neil. Pleistocene geology of the Snag-Klutlan area, southwestern Yukon, Canada. D, 1969, University of Minnesota, Minneapolis. 279 p.

Ramsay, Colin Robert. Metamorphism and gold mineralization of Archean metasediments near Yellowknife, Northwest Territories. D, 1973, University of Alberta. 285 p.

Ramsayer, George Ralph. Experimental and theoretical study of deltaic sedimentation. D, 1974, University of Rochester. 152 p.

Ramsayer, George Ralph. Geological application of adaptive pattern recognition. M, 1971, University of Rochester. 63 p.

Ramsdell, Jack D. Recrystallization of cold-swaged commercially-pure vanadium. M, 1958, University of Nevada - Mackay School of Mines. 90 p.

Ramsdell, Lewis Stephen. The crystal structure of some metallic sulfides. D, 1925, University of Michigan.

Ramsdell, Robert C. A review of the stratigraphy of the Late Cretaceous and earliest Tertiary formations in New Jersey with a re-study of the synonomy of the contained invertebrate fossil forms. M, 1948, Rutgers, The State University, New Brunswick. 418 p.

Ramsden, John. Numerical methods in fabric analysis. D, 1975, University of Alberta. 434 p.

Ramsden, John. Till fabric studies in the Edmonton area, Alberta, with special emphasis on methodology (Pleistocene). M, 1970, University of Alberta. 205 p.

Ramsden, Todd Wallace. The structural geology and paleomagnetism of the sheeted dike complex in the Mitsero-Arakapas area, Troodos Ophiolite, Cyprus. M, 1987, University of California, Davis. 114 p.

Ramseier, Frederic Neil. Geochemical exploration of the Elkhorn mining district and Tizer Basin, Jefferson County, Montana. M, 1963, Montana College of Mineral Science & Technology. 65 p.

Ramsey, Albert Frank. Prime agricultural lands of Louisiana; identification, evaluation, and analysis of transition to nonagricultural uses. D, 1981, Louisiana State University. 212 p.

Ramsey, Bruce L. The physical and petrographic characteristics of formcoke produced experimentally from lignite and sub-bituminous coal. M, 1974, University of North Dakota. 67 p.

Ramsey, Christina U. Observations of changes with location and depth in some physical and chemical properties related to the manganese and iron content in Lake Chesdin, (Chesterfield county) Virginia. M, 1970, Virginia State University. 66 p.

Ramsey, Dean A. Lithostratigraphy and petrography of the Atoka Formation in the St. Paul area, Northwest Arkansas. M, 1983, University of Arkansas, Fayetteville. 90 p.

Ramsey, Elijah William, III. Remote sensing of water quality indicators in a lacustrine environment studied using optical modelling of water and atmospheric constituents and aircraft multispectral scanner imagery. D, 1988, University of South Carolina. 200 p.

Ramsey, Elmer W. Geology of the southern portion of Bolar Anticline, Highland County, Virginia. M, 1950, University of Virginia. 114 p.

Ramsey, Jacqueline M. Criteria for evaluation of abandoned oil fields. M, 1978, University of Texas, Arlington. 46 p.

Ramsey, James E. An investigation of selected stratigraphy and fossil sites in Penn Township, Westmoreland County, PA. M, 1966, Indiana University of Pennsylvania. 39 p.

Ramsey, John William, Jr. Perdiz Conglomerate, Presidio County, Texas. M, 1961, University of Texas, Austin.

Ramsey, Kelvin Wheeler. Stratigraphy and sedimentology of a late Pliocene intertidal to fluvial transgressive deposit; Bacons Castle Formation, upper York-James Peninsula. D, 1988, University of Delaware. 418 p.

Ramsey, Margaret. A study of the fossil fresh-water diatoms of Oregon. M, 1929, University of Oregon. 58 p.

Ramsey, Nancy Jo. Upper Emsian-upper Givetian conodonts from the Columbus and Delaware limestones (middle Devonian) and lower Olentangy shale (upper Devonian) of central Ohio. M, 1969, Ohio State University.

Ramsey, Robert M. An empirical tensor representation for the stress-dependence of magnetic susceptibility. M, 1981, Pennsylvania State University, University Park. 46 p.

Ramsey, Rodney Dean. Geology of the Big Goose Canyon area, Sheridan County, Wyoming. M, 1955, University of Wyoming. 111 p.

Ramspott, Lawrence D. Geology of the Eighteenmile Peak area and petrology of the Beaverhead Pluton, Idaho-Montana. D, 1962, Pennsylvania State University, University Park. 237 p.

Ramulu, Uddanapalli Subbarayappa Sree see Sree Ramulu, Uddanapalli Subbarayappa

Ranahan, Fedro Sigmundo Zazueta see Zazueta Ranahan, Fedro Sigmundo

Ranalli, Giorgio. Rheological properties of rocks as inferred from the study of earthquake aftershock sequences. D, 1970, University of Illinois, Urbana. 173 p.

Rance, Hugh. Superior-Churchill structural boundary, Wabowden, Manitoba. D, 1966, University of Western Ontario.

Rand, Wendell Phillips. An experimental study of some problems in the recovery of crude oil from loose sands by means of alkaline floods. M, 1929, University of Wisconsin-Madison.

Rand, Wendell Phillips. Generation of oil in rocks by shearing pressures; further studies of effects of heat on oil shales. D, 1931, University of Wisconsin-Madison.

Rand, William Whitehill. The geology of Santa Cruz Island, California. D, 1934, University of California, Berkeley. 192 p.

Randall, A. Henry, III. Storm-dominated shelf sedimentation marginal to a delta; an example from the Devonian of southwestern Virginia and eastern West Virginia. M, 1984, Virginia Polytechnic Institute and State University. 174 p.

Randall, Allan Dow. Glacial geology and groundwater possibilities in southern LaSalle and eastern Putnam counties, Illinois. M, 1955, University of Illinois, Urbana.

Randall, Arthur G. Areal geology of the Pine Cliff area (Chalk Creek) Summit County, Utah. M, 1952, University of Utah. 43 p.

Randall, Bruce Loyal. The analysis of four new oil and gas property valuation parameters; particularly, as applied to the acquisition of oil and gas properties. M, 1988, University of Oklahoma. 244 p.

Randall, Duane Chilton. The fauna of the Auburn chert. M, 1934, University of Missouri, Columbia.

Randall, Gaither M. Ground water resources of Northwest Dane County (Wisconsin). M, 1954, University of Wisconsin-Madison.

Randall, George. Bouguer gravity modeling of reservoir and underlying basement, Cerro Prieto geothermal field, Baja California, Mexico. M, 1984, California State University, Long Beach. 77 p.

Randall, J. H. Hydrogeology and water resources of middle Kirkland Creek basin, Yavapai County, Arizona. M, 1974, University of Arizona.

Randall, Jeffery Hunt. Halocarbons in ground water; Tucson, Arizona. D, 1983, University of Arizona. 159 p.

Randall, John A. The geology and ore deposits of Upper Mayflower Gulch, Summit County, Colorado. M, 1958, University of Colorado.

Randall, John W. Environment of deposition and paleoecology of the Cave Hill Member, Kinkaid Formation (upper Mississippian) in southern Illinois. M, 1970, Southern Illinois University, Carbondale. 151 p.

Randall, Karl Gordon. Identification of seismic reflections from the basement. M, 1985, Purdue University. 146 p.

Randall, Michael John. The radiation pattern of the long-period P-wave pulse; a measure of strain at the focus of deep earthquakes. D, 1968, University of California, Los Angeles. 107 p.

Randall, Richard G. Geology of the Salt Springs area, Death Valley, California, and its bearing on early Mesozoic regional tectonics. M, 1975, San Jose State University. 62 p.

Randall, Walter. An analysis of the subsurface structure and stratigraphy of the Salton Sea geothermal anomaly, Imperial Valley, California. D, 1974, University of California, Riverside. 159 p.

Randall, William Arthur, Jr. Delineation of a basaltic dike and a reconnaissance survey for banded iron formation in the Wolverine Basin, Gravelly Range, Montana, using differential gravity and magnetic fields. M, 1988, Wright State University. 70 p.

Randau, Paul Clemens. Principles of systematics, the system and phylogeny of Paleozoic ammonoids; Part I, The theory of phylogenetic systematics. M, 1962, University of Iowa. 422 p.

Randazzo, Anthony F. Petrography and stratigraphy of the Carolina Slate Belt, Union County, North Carolina. D, 1968, University of North Carolina, Chapel Hill. 79 p.

Randazzo, Anthony F. The stratigraphy of the Wadesboro Triassic basin in North and South Carolina. M, 1965, University of North Carolina, Chapel Hill. 52 p.

Randazzo, Peter Joseph. The influence of river bend morphology on the flow dynamics of gravel bed rivers. M, 1988, SUNY at Binghamton. 86 p.

Randazzo, Santi. Seismic model experiments using the solid physical modeling system. M, 1988, University of Houston.

Randell, David Howard. Liquefaction potential in coastal areas in and around Long Beach, California. M, 1983, California State University, Long Beach. 250 p.

Randolph, Eldred O. A study of the physiography of the Isle of Palms. M, 1915, University of North Carolina, Chapel Hill. 16 p.

Randolph, Ellis Edwin. Lithostratigraphy and subsurface study of the Chaetetes bearing lower Strawn Formation (Pennsylvanian), Gaines County, Texas. M, 1974, University of Oklahoma. 148 p.

Randolph, John. Influence of NEPA on Corps of Engineers water planning in California. D, 1976, Stanford University. 311 p.

Randolph, Robert Lee. Paleontology of the Swasey Limestone, Drum Mountains, west-central Utah. M, 1973, University of Utah. 73 p.

Raney, James H. The geology of a portion of the Shell Knob, Missouri, Quadrangle. M, 1929, University of Iowa.

Raney, Jay A. Geology of the Elk Creek-Stonyford area, northern California. D, 1976, University of Texas, Austin. 193 p.

Ranganathan, Vishnu. Aspects of subsurface brine formation in passive continental margin evaporite basins. D, 1988, Louisiana State University. 194 p.

Ranganathan, Vishnu. Significance of abundant K-feldspar in potassium-rich Cambrian shales of the Appalachian Basin. M, 1980, University of Cincinnati. 68 p.

Ranganayaki, Rambabu Pothireddy. Generalized thin sheet approximation for magnetotelluric modelling. D, 1978, Massachusetts Institute of Technology. 204 p.

Rangel, Hamilton Duncan. Geologic evolution of Fazenda Cedro paleosubmarine canyon; Espirito Santo Basin, Brazil. D, 1984, University of Texas, Austin. 238 p.

Rangel, Jorge Enrique. The effect of stratigraphy and clay mineralogy on the settlement characteristics of reclaimed mined land. M, 1979, Texas A&M University. 109 p.

Ranger, Julie Ann. Implications of a petrographic analysis of sandstone clasts within the Dunnage Formation, Newfoundland. M, 1988, SUNY at Buffalo. 86 p.

Ranger, Michael Joseph. The stratigraphy and depositional environment of the Lower Ordovician Bell Island and Wabana groups, Conception Bay, Newfoundland. M, 1978, Memorial University of Newfoundland. 216 p.

Rankin, D. S. Heat flow-heat production study in Nova Scotia. D, 1974, Dalhousie University.

Rankin, David. A theoretical and experimental study of structures in the magneto-telluric field. D, 1960, University of Alberta. 104 p.

Rankin, David Karl. Holocene geologic history of the Clatsop Plains foredune ridge complex. M, 1983, Portland State University. 189 p.

Rankin, Douglas Whiting. Bedrock geology of the Katahdin-Traveler area, Maine. D, 1961, Harvard University.

Rankin, James Scott. Geology of the south-central part of the Arvonia Syncline, central Piedmont, Virginia. M, 1979, University of Kentucky. 54 p.

Rankin, L. D. Stratigraphy of the B subzone Wentworth Formation, Windsor Group. M, 1974, Acadia University.

Rankin, Leslie Darrell. The geology of the Portage Lakes-Caribou area, northern New Brunswick. M, 1981, University of New Brunswick. 220 p.

Rankin, Mary L. Statistical analysis of the P/4 enamel pattern of Equus (Asinus) and Equus (Plesippus) from Sheridan County and Broadwater Quarries, Nebraska. M, 1984, Bowling Green State University. 48 p.

Rankin, Peter Watson. Mylonitic fabric development through the east flank of the Bitterroot Dome, Montana. M, 1982, University of Montana. 80 p.

Rankin, Wilbur D'Arcy. Geology of a portion of the Santa Ana Mountains, Orange County, California. M, 1928, University of California, Berkeley. 81 p.

Rankin, William E. Reconnaissance gravity surveying for drift-filled valleys in the Mercer Quadrangle, Pennsylvania. M, 1969, Pennsylvania State University, University Park. 49 p.

Rankis, Linda Victoria. Thermal maturation in microfossils from the Oread Formation (Pennsylvanian), Midcontinent, U.S.A. M, 1988, University of Nebraska, Lincoln. 180 p.

Rannefeld, James W. The stony corals of Enmedio Reef (Recent) off Veracruz, Mexico (Gulf of Mexico). M, 1972, Texas A&M University.

Ranney, Richard Willard. Soil forming processes in glossoboralfs of west-central Wisconsin. D, 1966, University of Wisconsin-Madison. 123 p.

Rannie, E. H. The mean and distribution of hydraulic conductivity in the Midwest Lake area, northern Saskatchewan. M, 1981, University of Waterloo. 179 p.

Rannie, W. F. An approach to the prediction of suspended sediment rating curves. D, 1976, University of Toronto.

Ransford, Gary Allen. Effects of accretion on the initial thermal state of the Moon. D, 1978, University of California, Los Angeles. 197 p.

Ransom, Michel Doyle. Genetic processes in seasonally wet soils on the Illinoian till plain, southwestern Ohio. D, 1984, Ohio State University. 380 p.

Ransome, Alfred L. Descriptive geology and ores of the Blind Spring Hill mining district, Mono County, California. M, 1937, Stanford University. 69 p.

Ransome, Frederick Leslie. Great Valley of California; a criticism of the theory of isostasy. D, 1896, University of California, Berkeley. 57 p.

Ranson, William Albrecht. Anorthosites of diverse magma types in the Puttuaaluk Lake area, Nain Complex, Labrador. D, 1979, University of Massachusetts. 101 p.

Ranson, William Albrecht. Geology and petrology of portions of the Zacatecas and Guadelupe quadrangles, Zacatecas, Mexico. M, 1975, University of New Orleans.

Ranspot, Henry W. Geology and uranium deposits of the Indian Creek area, Gunnison and Saguache counties, Colorado. M, 1958, University of Colorado.

Ranta, Donald E. Geology, alteration, and mineralization of the Winfield (LaPlata) District, Chaffee County, Colorado. D, 1974, Colorado School of Mines. 261 p.

Ranta, Donald Eli. Supergene enrichment at the Betty O'Neal Mine, Lander County, Nevada. M, 1967, University of Nevada. 47 p.

Ranta, Reino A. Contact effects of basic magmas on rhyolite. M, 1939, University of Minnesota, Minneapolis. 47 p.

Rantala, Raymond Henry. A study of the insoluble residues of the Whitewood Limestone, central Black Hills of South Dakota. M, 1947, University of Minnesota, Minneapolis. 48 p.

Rao Divi, Sri Ramachandra. Structural analysis of Grenville rocks near Bancroft, Ontario, Canada. D, 1972, University of Ottawa. 178 p.

Rao, B. V. Parameswara. A study of the effect of point defects on electrokinetic phenomena as they control the floatability of calcium fluoride. D, 1967, Pennsylvania State University, University Park. 116 p.

Rao, Chelluri. Distribution of rare-earth elements in fluorapatites from the Oka carbonatite complex, Quebec, Canada. M, 1969, University of Pittsburgh.

Rao, Kunduri Viswa Sundara. Paleomagnetism of the Ordovician redbeds of Bell island, Newfoundland. M, 1970, Memorial University of Newfoundland. 201 p.

Rao, M. Kesava. Abrasion of selected rocks. M, 1972, University of Kansas.

Rao, Manam Venkata Panduranga. Effects of effluent and influent seepage on the hydrodynamic forces acting on an idealized noncohesive sediment particle. D, 1969, Utah State University. 147 p.

Rao, Meera. Studies on model systems of possible prebiotic significance; Part I, Synthesis of phosphatidylcholine and phosphatidylethanolamine under possible primitive Earth conditions; Part II, The adsorptive and catalytic properties of clays. D, 1981, [University of Houston]. 201 p.

Rao, Nacharaju Manohar. Sedimentary sources and processes in the Bay of Bengal. M, 1978, Brooklyn College (CUNY).

Rao, Narasinga Bandiatmakur. A river system environmental modeling and simulation methodology. D, 1981, University of Iowa. 181 p.

Rao, Prasada C. Applications of computer techniques to the petrographic study of oolitic environments, Ste. Genevieve limestone (Mississippian), southern Illinois and eastern Missouri. D, 1970, University of Illinois, Urbana. 111 p.

Rao, S. K. Prediction of settlement in landfills for foundation design purposes. D, 1974, West Virginia University. 300 p.

Rao, Shankaranarayana R. N. An experimental study of subpressure in a freezing soil system. D, 1964, Rutgers, The State University, New Brunswick. 257 p.

Rao, Srinivas Thanner. The Bryant Mountain section of the Ancha Sill, Arizona; a multiple pulse injection. M, 1973, Brooklyn College (CUNY).

Rao, Talur Seshagiri. The influence of ionic strength and ion-pair formation on Na-Ca and Na-Mg ion-exchange equilibria. D, 1967, University of California, Riverside. 101 p.

Raoof-Malayeri, Mehdi. Attenuation of high frequency earthquake surface waves in South America. M, 1984, St. Louis University.

Raphael, Constantine Nicholas. Geomorphology and archeology, northwest Peloponnesos, Greece. D, 1968, Louisiana State University. 130 p.

Rapp, David William. A comparison of the type sections of the Jefferson City, Cotter, and Powell formations. M, 1956, University of Missouri, Columbia.

Rapp, George Robert, Jr. Geochemistry and mineralogy of the zoisite-epidote group. D, 1960, Pennsylvania State University, University Park. 88 p.

Rapp, John S. The geology and mineralization of the Iron Creek area, Custer County, South Dakota. M, 1970, South Dakota School of Mines & Technology.

Rapp, Keith Burleigh. Groundwater recharge in the Trinity Aquifer, central Texas. M, 1986, Baylor University. 132 p.

Rapp, Richard H. The orthometric height. M, 1961, Ohio State University.

Rapp, Richard Henry. The prediction of point and mean gravity anomalies through the use of a digital computer. D, 1964, Ohio State University. 189 p.

Rapp, Robert Paul. An experimental investigation of the solubility and dissolution kinetics of monazite and its implications for the thorium and rare earth chemistry of felsic magmas. M, 1985, Rensselaer Polytechnic Institute. 99 p.

Rappaport, Paul Aaron. Sediment flora of the western Barents Sea. M, 1986, University of Maine. 248 p.

Rappaport, Stephen M. The identification of effluents from rubber vulcanization. D, 1974, University of North Carolina, Chapel Hill. 138 p.

Rappenecker, Caspar. The regional and economic geography of Jamaica. D, 1936, Cornell University.

Rappeport, Melvyn Lewis. Studies of tidally-dominated shallow marine bed forms; lower Cook Trough, Cook Inlet, Alaska. D, 1982, Stanford University. 336 p.

Rapport, Eric J. Stratigraphic and structural controls on fluorspar mineralization in northern Valle Las Norias, Coahuila, Mexico. M, 1983, Texas A&M University. 79 p.

Rapson-McGugan, June E. Lithology and petrography of transitional Jurassic-Cretaceous clastic rocks, southern Rocky Mountains. M, 1963, University of Calgary. 75 p.

Rapstine, Inge Frances. Geology and geophysics of the Devonian Prairie Evaporite interval, Folsum Coulee area, Sheridan County, Montana. M, 1980, West Texas State University. 84 p.

Raring, Andrew Michael. Conodont biostratigraphy of the Chazy Group (Lower Middle Ordovician), Champlain Valley, New York and Vermont. D, 1972, Lehigh University. 163 p.

Rasbury, Sidney A. Depositional environment and diagenesis of Vicksburg sandstones, Tabasco Field, Hidalgo County, Texas. M, 1986, Tulane University. 155 p.

Raschilla, Stephen Nicholas. Analysis of cleavage in the Dunlap Formation, Pilot Mountains, West Central Nevada. M, 1980, Texas Christian University.

Rascoe, Bailey. Geology of the Fox Creek quadrangle, Cumberland County, Tennessee. M, 1951, Vanderbilt University.

Rashak, Edward P. Organic geochemistry of Woodford and Green River shales using hydrous pyrolysis. M, 1985, University of Texas at Dallas. 137 p.

Rashid, Muhammad Abdur. Palynology of the Bostwick Member of the Lake Murray Formation

(Pennsylvanian) of southern Oklahoma. M, 1968, University of Oklahoma. 122 p.

Rashid-Noah, Augustine Bundu. Designing subsurface drainage systems to avoid excessive drainage of sands. D, 1981, McGill University.

Rasho, John L. Effective rates of evaporation and chemical properties of the precipitates. M, 1969, University of New Mexico. 81 p.

Rashrash, Salem M. A quantitative hydrological study of Murzuq and Al Hamadah Al Hamra basins, Libya. M, 1984, Ohio University, Athens. 607 p.

Rashrash, Salem Mohamed. Fracture characteristics and their relationships to producing zones in deep wells, Raft River geothermal area. D, 1988, University of Idaho. 171 p.

Rask, Dale Hugo. Upper Cretaceous rocks in the Frenchman river valley (southwest Saskatchewan, Canada). M, 1969, University of Saskatchewan. 86 p.

Rask, James Harold. Manganese oxide mineralogy; transmission electron microscopy studies. D, 1988, Arizona State University. 279 p.

Raskin, Greg Steven. Iterative modelling; a new approach to the inversion of 1-D seismograms. M, 1987, Texas A&M University. 115 p.

Rasmussen, Clayton R. Gold and diamond placer concentrates of French Equatorial Africa. M, 1935, University of Minnesota, Minneapolis. 29 p.

Rasmussen, Donald L. Geology and mammalian paleontology of the Oligocene-Miocene Cabbage Patch Formation, central-western Montana. D, 1977, University of Kansas. 794 p.

Rasmussen, Donald Linden. Late Cenozoic geology of the Cabbage Patch area, Granite and Powell counties, Montana. M, 1969, University of Montana. 188 p.

Rasmussen, Gerald E. Desmoinesian (Pennsylvanian) gastropods from the Lonsdale Limestone of north-central Illinois. M, 1960, University of Wisconsin-Madison.

Rasmussen, John A. Fission track geochronology on a pegmatite from Embreeville, Pennsylvania. M, 1969, Rensselaer Polytechnic Institute. 32 p.

Rasmussen, Karen Hasine. Sources and natural removal processes for some gaseous atmospheric pollutants. M, 1974, Pennsylvania State University, University Park. 121 p.

Rasmussen, Keith A. Tectonic and climatic controls on alluvial fan sedimentation of the Cutler Formation (Permo-Pennsylvanian) in southwestern Colorado. M, 1982, New Mexico State University, Las Cruces. 117 p.

Rasmussen, Kenneth A. An ecologic and taphonomic analysis of submarine cave communities; Salt River Canyon, St. Croix, U.S.V.I. M, 1983, University of Rochester. 148 p.

Rasmussen, Michael G. Depositional environments of siliceous laminites in the Laney Member of the Green River Formation, Sublette County, Wyoming. M, 1983, Loma Linda University. 69 p.

Rasmussen, Noel Fredrich. The Mississippian, Devonian and Silurian systems in the sub-surface of Dallas County, Iowa. M, 1956, University of Nebraska, Lincoln.

Rasmussen, Patricia Elizabeth. Selenium in the environment; the determination of total selenium by G.F.A.A. and two alkyselenides by gas chromatograph – A.A. M, 1987, University of Toronto.

Rasmussen, Todd Christian. Fluid flow and solute transport through three-dimensional networks of variably saturated discrete fractures. D, 1988, University of Arizona. 328 p.

Rasmussen, William Charles. The probable error of sampling sediments for heavy mineral study. M, 1939, University of Chicago. 41 p.

Rasmussen, William Charles, Sr. Geology and hydrology of the "bays and basins" of Delaware. D, 1958, Bryn Mawr College. 206 p.

Rasor, Charles A. Mineralogy and petrography of the Tombstone mining district, Arizona. D, 1937, University of Arizona.

Rasor, Charles Alfred. Silver mineralization at the Sunshine Mine, Coeur d'Alene district, Idaho. M, 1934, University of Idaho. 17 p.

Rasor, J. P. Prospecting, developing, and mining bentonite deposits. M, 1946, University of Missouri, Rolla.

Rasor, R. W. The calculation of proppant transport in vertical hydraulic fractures using finite difference techniques. M, 1978, Texas A&M University.

Rasrikriengkrai, Piyamit. Petrographic investigation of hydrothermal alteration in the Quartz Creek, Middle Fork, Snoqualmie River, King County, Washington. M, 1971, University of Washington. 86 p.

Rassam, Ghassan Noel. Studies on certain glasses in the system $PbO-Al_2O_3-SiO_2$. M, 1963, Miami University (Ohio). 91 p.

Rassam, Ghassan Noel. Studies on the Platteville Formation (Middle Ordovician) of Minnesota, Iowa, and Wisconsin. D, 1967, University of Minnesota, Minneapolis. 173 p.

Rassios, Anne Ewing. Geology and evolution of the magmatic rocks of the Vourinos Ophiolite, northern Greece. D, 1981, University of California, Davis. 708 p.

Rassman, Barbara. A sedimentologic comparison of two dissimilar Ordovician edrioasteroid pavements (hardgrounds). M, 1981, University of Cincinnati. 224 p.

Rastegar, Iraj. A detailed study of the diabase dikes in the middle Haddam Quadrangle, Connecticut. M, 1972, Wesleyan University. 87 p.

Ratanalert, Pirmpoon. Sulfur studies with some Chernozemic soils. M, 1973, University of Alberta. 94 p.

Ratananaka, C. Community water supplies for rural areas of northeastern Thailand. M, 1973, University of Arizona.

Ratchford, Timothy Daniel. Red and gray sands of the Lower Cretaceous found in the interior Mississippi salt basin with emphasis on differential porosity and permeability. M, 1987, Northeast Louisiana University. 92 p.

Ratcliff, Gene A. Surface structure on the east flank of the Nemaha Anticline in Northeast Pottawatomie County, Kansas. M, 1957, Kansas State University. 33 p.

Ratcliff, Marvin W. Geology and mineralogy of altered zones in the Red River area, New Mexico. M, 1962, New Mexico Institute of Mining and Technology. 85 p.

Ratcliffe, Nicholas Morley. Bedrock geology of the Great Barrington area, Massachusetts. D, 1965, Pennsylvania State University, University Park. 269 p.

Rath, Bruce A. Stratigraphy and diagenesis of the Ervay Member of the Park City and Goose Egg formations (Permian), eastern Wind River basin, Wyoming. M, 1982, Colorado College.

Rath, David L. A study of middle to late Quaternary sediments in a karst trap. M, 1975, University of Missouri, Rolla.

Rath, Otto. A study in the integration of geophysics and subsurface geology, Kaufman County, Texas. M, 1956, [University of Oklahoma].

Rath, Ulrich E. G. A base metal occurrence in the Wollaston lake belt (Precambrian) (Saskatchewan, Canada). M, 1969, [University of Alberta].

Rathbun, Fred Charles. Abnormal pressures and conductivity anomaly, northern Green River basin, Wyoming. M, 1969, University of Tulsa. 25 p.

Rathburn, Anthony Earl. A paleoecological study of late Pleistocene ostracodes from the southern Champlain Sea. M, 1984, University of Vermont. 200 p.

Rathke, William Wilson. Structural relations of Taconic slices, southern Berlin Valley, New York. M, 1978, Cornell University.

Rathnayake, Rathnayake M. D. A combined analytical/numerical model for simulation of flow and mass transport in groundwater. D, 1985, Colorado State University. 175 p.

Ratigan, Joe Lawrence. A statistical fracture mechanics approach to the strength of brittle rock. D, 1981, University of California, Berkeley. 97 p.

Ratliff, James R. Geology of a portion of Cove Creek Township, Washington County, Arkansas. M, 1959, University of Arkansas, Fayetteville.

Ratliff, Jeffrey Allan. The Illinoian and pre-Illinoian glacial deposits of southeastern Indiana and southwestern Ohio. M, 1983, Miami University (Ohio). 105 p.

Ratliff, Larry Eugene. Environmental stratigraphy of the middle Ordovician carbonates in the Luttrell Belt, Union and Grainger counties, East Tennessee. D, 1974, University of Tennessee, Knoxville. 148 p.

Ratliff, Larry Eugene. Micropaleontologic study of the limestones in the Pennington Formation (Mississippian) in eastern Tennessee. M, 1968, University of Tennessee, Knoxville. 40 p.

Ratliff, Robert A. Deformation history of the Bullbreen Group, central western Spitsbergen. M, 1985, Wayne State University. 194 p.

Ratliff, Terry Wayne. An analysis of the randomness of soil compactors in the Mini-Sosie seismic system. M, 1987, Wright State University. 82 p.

Ratmiroff, Gregor N. De *see* De Ratmiroff, Gregor N.

Ratte, Charles A. Genesis of the Kinsman quartz monzonite, New Hampshire. M, 1955, Dartmouth College. 70 p.

Ratte, Charles Arthur. Rock alteration and ore genesis in the Iron Springs-Pinto mining district, Iron County, Utah. D, 1963, University of Arizona. 222 p.

Ratte, James Clifford. Bedrock geology of the Skitchewaug Mountain area, Claremont Quadrangle, New Hampshire-Vermont. M, 1952, Dartmouth College. 56 p.

Ratterman, Nancy G. Mineral reactions in tuffaceous sediments of the Green River Formation, Wyoming. M, 1980, University of Wyoming. 100 p.

Rau, Jon Llewellyn. Upper Devonian and Lower Mississippian Leptodesma from southwestern Pennsylvania. M, 1955, University of Cincinnati. 94 p.

Rau, Jon Llewelyn. Stratigraphy and paleontology of the Three Forks Formation (Upper Devonian) in south-western Montana. D, 1959, Yale University.

Rau, Robert L. Sedimentology of the Upper Cretaceous Winthrop Sandstone, northeastern Cascade Range, Washington. M, 1987, Eastern Washington University. 197 p.

Rau, Theresa. Stratigraphy of the Mississippian Madison Group of the Whitney Canyon-Carter Creek Field of southwest Wyoming. M, 1982, Kent State University, Kent. 156 p.

Rau, Weldon Willis. Foraminifera from the Oligocene Lincoln Formation in the Porter area, Washington. M, 1946, University of Iowa. 104 p.

Rau, Weldon Willis. Tertiary foraminifera from the Willapa River valley of Southwest Washington. D, 1950, University of Iowa. 138 p.

Raubvogel, David R. Petrology of the Middle Jurassic Twin Creek Limestone, Lincoln and Sublette counties, southwestern Wyoming. M, 1984, Utah State University. 187 p.

Rauch, Henry William. The effects of lithology and other hydrogeologic factors on the development of solution porosity in the middle Ordovician carbonates of central Pennsylvania. D, 1972, Pennsylvania State University, University Park. 530 p.

Rauch, Peter C. rtiary welded tuffs of the Ryan Spring area, Needle Range, Beaver County, Utah. M, 1975, University of Missouri, Rolla.

Rauch, William E. Sedimentary petrology, depositional environment, and tectonic implications of the upper Eocene Quimper Sandstone and Marrowstone Shale, northeastern Olympic Peninsula, Washington. M, 1985, Western Washington University. 102 p.

Raudsepp, John J. Lithology and altitude in Gaspé Peninsula, P.Q. (Canada). M, 1967, McGill University.

Raudsepp, Mati. Evaluation of amphibole synthesis and product characterization. D, 1985, University of Manitoba.

Raudsepp, Mati. Petrology and emplacement of a differentiated subvolcanic mafic sill complex in the early Precambrian Favourable Lake volcanic complex, northwestern Ontario. M, 1978, University of Manitoba. 115 p.

Rauenzahn, Kim Ann. The effects of mixed quartz-montmorillonite gouge on the frictional sliding of Tennessee Sandstone. M, 1985, Texas A&M University. 108 p.

Raup, David Malcolm. Classification and morphological variation in the genus Dendraster. D, 1957, Harvard University.

Raup, Omer Beaver. Clay mineralogy of the Pennsylvanian red beds and associated rocks Flanking the Ancestral Front Range of central Colorado. D, 1962, University of Colorado. 117 p.

Raup, Robert Bruce. Relationships between zinc ores and gossan at Hanover, New Mexico. M, 1952, University of Michigan.

Rauschkolb, Michael Howard. The geology of the Northwest Orebody, Twin Buttes Mine, Pima County, Arizona. M, 1983, University of Arizona. 174 p.

Rautman, Alison Eunice. Archaeological geology of the Henauhof Northwest Site, Germany. M, 1983, University of Michigan.

Rautman, Christopher A. Depositional environments of the "lower Sundance" Formation (upper Jurassic) of the eastern Wyoming region. D, 1976, University of Wisconsin-Madison. 88 p.

Rautman, Christopher Arthur. The Denali Fault system in the Dick Creek-Well Creek area, central Alaska Range, Alaska. M, 1974, University of Wisconsin-Madison. 141 p.

Rauzi, Steven L. Structural geology of eastern part of James Peak Quadrangle and western part of Sharp Mountain Quadrangle, Utah. M, 1979, Utah State University. 73 p.

Rava, Barry. Mariner 10 color-ratio data and the surface of Mercury. M, 1980, University of Pittsburgh.

Ravat, Dhananjay Narendra. Magnetic investigations in the St. Louis Arm of the New Madrid rift complex. M, 1984, Purdue University. 102 p.

Raveck, Karen L. An investigation of nuclear detonations and solid Earth tides as possible triggering mechanisms for earthquakes in central California during the time period 1969 through 1970. M, 1973, University of Wisconsin-Milwaukee.

Raven, K. G. The effect of sample size on the stress-permeability relationship of natural fractures. M, 1980, University of Waterloo.

Ravenhurst, Casey Edward. An isotopically and thermochronologically constrained model for lead-zinc and barium mineralization related to Carboniferous basin evolution in Nova Scotia, Canada. D, 1987, Dalhousie University. 251 p.

Ravenhurst, Casey Edward. Utility of digitally merged Seasat-A SAR, Landsat MSS, and magnetic field data sets for mapping lithology and structure in a vegetated terrain. M, 1980, Pennsylvania State University, University Park. 132 p.

Ravenhurst, William Richard. Microcomputer modelling of fixed-loop time-domain EM systems. M, 1986, University of Western Ontario. 215 p.

Ravenscroft, Arthur William. The geology of Big Bald Mountain, White Pine County, Nevada. M, 1974, San Diego State University.

Ravenscroft, John H. Mesothermal-hypothermal mineralization and genetically related wall rock alteration in Bismarck District, Madison County, Montana. M, 1970, University of Wisconsin-Milwaukee. 113 p.

Ravicz, Louis G. Enrichment of silver ores. M, 1915, University of Minnesota, Minneapolis. 48 p.

Ravina, Amnon Nathan. The effect of capillarity on the dimensional changes and the durability of saturated sedimentary rocks. M, 1986, University of Windsor. 166 p.

Raviola, F. P. Metamorphism, plutonism and deformation in the pateros; Alta Lake region, north-central Washington. M, 1988, San Jose State University. 181 p.

Ravn, Robert Lee. Stratigraphy, petrography and depositional history of the Hertha Formation (Upper Pennsylvanian), Midcontinent North America. D, 1981, University of Iowa. 247 p.

Ravn, Robert Lee. The palynology and paleoecology of a lower Cherokee (Pennsylvanian) coal from Wapello County, Iowa. M, 1977, University of Iowa. 254 p.

Ravneberg, N. M. Petrographic and aeromagnetic map study, central and western Maine. M, 1954, Columbia University, Teachers College.

Rawi, Yeha Tawfeq Al *see* Al Rawi, Yeha Tawfeq

Rawles, William Post. A study of the shapes of zircon crystals in igneous rocks. D, 1930, University of Wisconsin-Milwaukee.

Rawles, William Post. Some experimental studies of certain phases of sedimentation. M, 1926, Indiana University, Bloomington.

Rawlings, Gary Don. Identification and measurement of the major chemical forms of mercury in urban atmospheres. D, 1974, Texas A&M University. 190 p.

Rawlins, David M. Geology of the Buffalo Gap area, Custer and Fall River counties, South Dakota. M, 1978, South Dakota School of Mines & Technology.

Rawlinson, George Harmon. Studies of Pleistocene geology of northern Campbell County, Kentucky. M, 1952, University of Cincinnati. 27 p.

Rawlinson, Stuart. Paleoenvironment of depositions, paleocurrent directions, and the provenance of Tertiary deposits along Kachemak Bay, Kenai Peninsula, Alaska. M, 1979, University of Alaska, Fairbanks. 162 p.

Rawls, Vernon C., Jr. Geochemical prospecting in wallrock adjacent to vein deposits, Rosiclare, Illinois. M, 1957, Michigan State University. 82 p.

Rawls, W. J. Analysis of the shallow subsurface flow process in the Georgia coastal plain. D, 1976, Georgia Institute of Technology. 150 p.

Rawson, Donald Eugene. Geology of the Tecolate Hills area, Lincoln County, New Mexico. M, 1957, University of New Mexico. 77 p.

Rawson, Richard Ray. Geology of the southern part of the Spanish Fork Quadrangle, Utah. M, 1957, Brigham Young University. 33 p.

Rawson, Richard Ray. Petrographic facies analysis of the Ray Member, Kibbey Formation (Upper Mississippian), Williston Basin and central Montana. D, 1966, University of Wisconsin-Madison. 92 p.

Rawson, Shirley Ann. Regional metamorphism of rodingites and related rocks from the north-central Klamath Mountains, California. D, 1984, University of Oregon. 252 p.

Ray, Amal Kumar. Three-dimensional ray-theoretical traveltime modeling of COCORP data from southwestern Oklahoma. M, 1984, University of Texas at Dallas. 167 p.

Ray, Bradley Stephen. The holdfast of Scyphocrinites; a bioecologic, morphologic, and biostratigraphic study of Camarocrinus (= Lobolithus) in the Hunton Group of central Oklahoma. M, 1985, University of Texas, Arlington. 129 p.

Ray, Clayton Edward. The oryzomyine rodents of the Antillean subregion (Pleistocene-Recent). D, 1962, Harvard University.

Ray, David R. Geology of the Precambrian Red Bluff granite complex, Fusselman Canyon area, Franklin Mountains, El Paso County, Texas. M, 1982, University of Texas at El Paso.

Ray, Dipak Kumer. Geochemistry and petrology of the Mount Trident andesites, Katmai National Monument, Alaska. D, 1967, University of Alaska, Fairbanks. 213 p.

Ray, Earl S. Diagenesis of sandstones from the Douglas Creek Member of the Green River Formation (Eocene) at Red Wash Field, Uintah County, Utah. M, 1985, Texas A&M University. 86 p.

Ray, George Dale. Areal geology of the Farris Quadrangle, Pushmataha and Atoka counties, Oklahoma. M, 1960, University of Oklahoma. 87 p.

Ray, Glenn Lamar. An ion microprobe study of trace element partitioning between clinopyroxene and liquid in the diopside $(CaMgSi_2O_6)$-albite $(Na-AlSi_3O_8)$-anorthite $(CaAl_2Si_2O_8)$ system. D, 1981, Massachusetts Institute of Technology. 142 p.

Ray, Hubert Roy. Origin and treatment of soils. M, 1912, University of North Carolina, Chapel Hill. 14 p.

Ray, James Allen. Holcombe Branch and Democrat dunite bodies, Madison and Buncombe counties, North Carolina. M, 1962, North Carolina State University. 97 p.

Ray, James C. Genesis of the ore deposits of the Butte District, Montana. D, 1929, [Stanford University]. 87 p.

Ray, James C. The covellite zone; paragenesis of the ore minerals in the Butte District, Montana. M, 1915, [Stanford University].

Ray, Jimmie Dell. Stratigraphy and structure of the Cinnamon Mountain area, Gallatin County, Montana. M, 1967, Oregon State University. 156 p.

Ray, John T. Petrology of the Cenozoic igneous rocks of the Tinton District, Black Hills, South Dakota-Wyoming. M, 1979, University of North Dakota. 122 p.

Ray, Johnny. Stratigraphy of the Moran and Putnam formations (Albany Group, Permian System), Shackleford and Callahan counties, Texas. M, 1968, Baylor University. 230 p.

Ray, K. B. Calcareous concretions from the Gulf of Panama. M, 1974, University of Wisconsin-Milwaukee.

Ray, Leon Nicholas. Econometric models of domestic water consumption in the Tucson Metropolitan area. M, 1972, University of Arizona.

Ray, Louis Lamy. Certain minor features of valley glaciers and valley glaciation. M, 1932, Washington University. 174 p.

Ray, Louis Lamy, Jr. Geomorphology and Quaternary chronology of northeastern Colorado. D, 1938, Harvard University.

Ray, Phillip T. Age determinations of intrusives from the Wisconsin Range, Horlick Mountains (Antarctica) by the rubidium-strontium whole rock isochrom method. M, 1973, Ohio State University.

Ray, Pulak K. Variability and genesis of sedimentary structures of Mississippi River bar (Plaquemine Point, Louisiana). D, 1972, Louisiana State University.

Ray, Richard G. Geology and ore deposits of the Willow Creek mining district, south central Alaska. D, 1950, The Johns Hopkins University.

Ray, Richard G. Igneous and metamorphic rocks of the northern half of the Pawtucket Quadrangle, Rhode Island. M, 1943, Brown University.

Ray, Richard Paul. Changes in shear modulus and damping in cohesionless soils due to repeated loading. D, 1984, University of Michigan. 433 p.

Ray, Robert R. Seismic stratigraphic interpretation of the Fort Union Formation, western Wind River basin, Wyoming. M, 1983, Colorado School of Mines. 132 p.

Ray, Satyabrata. Mineralogy of the Jacksonburg Formation in eastern Pennsylvania and western New Jersey. D, 1957, Lehigh University. 124 p.

Ray, T. M. Determination and significance of geologic factors contributing to the September 13-14, 1973, flash flooding in Harrison County, Mississippi. M, 1975, University of Southern Mississippi.

Ray, Walter Barclay. The crystal structure of zunyite. D, 1956, California Institute of Technology. 194 p.

Ray, Wendy L. The origin and distribution of oolite zones in the Ste. Genevieve Limestone (Valmeyeran), Posey County, Indiana. M, 1986, Indiana University, Bloomington. 95 p.

Raybuck, M. S. Hydrogeology of a bedrock aquifer in the park terrace area, Binghamton, New York. M, 1982, SUNY at Binghamton. 49 p.

Raychaudhuri, Bimalendu. Studies of amphibolites and constituent hornblendes from an area of progressive metamorphism near Lead, South Dakota. D, 1960, California Institute of Technology. 209 p.

Raychaudhuri, Sunilkumar. Trace elements in the sulphide deposits of the Chibougamau District, Quebec. D, 1960, McGill University.

Raydon, Gerald Thomas. Geology of the northeastern Chiricahua Mountains, Arizona. M, 1953, University of California, Berkeley. 141 p.

Rayl, Robert Lee. A study of Warrensburg Sandstone. M, 1952, University of Missouri, Columbia.

Raymahashay, Bikash Chandra. Hot springs and hydrothermal alteration in the Paint Pot Hill area, Yellowstone Park (Wyoming). D, 1967, Harvard University.

Raymer, John Herbert. The Deese Group (Middle Pennsylvanian) of the Ardmore Basin, southern Oklahoma. M, 1987, University of Oklahoma. 289 p.

Raymond, Anne. Peat taphonomy of Recent mangrove peats and Upper Carboniferous coal-ball peats. D, 1983, University of Chicago. 293 p.

Raymond, Charles Forest. Flow in a transverse section of Athabasca glacier, Alberta, Canada. D, 1969, California Institute of Technology. 331 p.

Raymond, Dorothy Echols. Upper Ordovician conodonts in Alabama. M, 1973, University of Alabama.

Raymond, L. C. Major structural and physiographic features of Oregon. M, 1932, Massachusetts Institute of Technology. 63 p.

Raymond, Larry C. Structural geology of the Oxford Peak area, Bannock Range (Bannock, Franklin, and Oneida counties), Idaho. M, 1971, Utah State University. 48 p.

Raymond, Loren A. The stratigraphy and structural geology of the northern Lone Tree creek and southern Tracy quadrangles (San Joaquin County), California. M, 1969, San Jose State University. 143 p.

Raymond, Loren Arthur. Franciscan geology of the Mount Oso area, California. D, 1973, University of California, Davis. 185 p.

Raymond, Lynda. Sédimentologie et paléomilieu d'un talus avant-récifal; membre de l'Anse à la Loutre de la Formation d'Indian Point, Port-Daniel, Gaspésie. M, 1986, Universite Laval. 55 p.

Raymond, Martin Snider. The physical stratigraphy of the upper Mesozoic sediments of a portion of Glenn and Tehama counties, California. M, 1958, University of California, Berkeley. 63 p.

Raymond, Paul Cletus. Arenaceous foraminifera from the Osgood (Middle Silurian) Formation, Indiana. M, 1955, Indiana University, Bloomington. 27 p.

Raymond, Percy E. A Tropidoleptus faunule at Canadaigua Lake, New York. D, 1905, Yale University.

Raymond, Rhonda Karen. A model for the rapid flow of spherical granular material. D, 1985, Clarkson University. 160 p.

Raymond, Richard Brian. The paleolimnology of Bull Run Lake; disruption and stability in a natural system. D, 1983, Portland State University. 128 p.

Raymond, Richard H. Paleontology of a portion of the Nolichucky Shale and Maynardville Limestone (Cambrian) of the Powell Quadrangle, Tennessee. M, 1959, University of Tennessee, Knoxville. 58 p.

Raymond, Robert, Jr. Evolution of tidal flat, shallow subtidal taxonomic assemblages, Early Cambrian, western Nevada. D, 1975, University of California, Santa Cruz. 137 p.

Raymond, Scott. Chemical changes associated with burial diagenesis of Gulf Coast pelitic sediments. M, 1974, University of Missouri, Columbia.

Raymond, William F. A geologic investigation of the offshore sands and reefs on Broward County, Florida. M, 1972, Florida State University.

Raymond, William H. A study of the fauna of the Whitewater Formation at Indian Creek, Butler County, Ohio. M, 1960, Miami University (Ohio). 111 p.

Raymondi, Michael Joseph. The relationship of northeastern Ohio peat deposits to their surrounding surficial materials. M, 1981, Kent State University, Kent. 199 p.

Raymundo, M. E. The properties of the black earths of Hawaii. D, 1965, University of Hawaii. 115 p.

Rayne, Todd William. Geomorphology and Quaternary history of the Yellowstone River valley between Livingston and Big Timber, Montana. M, 1982, University of Wisconsin-Madison. 90 p.

Raynolds, Mary Vera. Geochemistry of fluids in the Cerro Colorado porphyry copper deposit, Panama. D, 1983, Harvard University. 235 p.

Raynolds, R. G. H. Satellite remote sensing of the McDermitt Caldera, Nevada-Oregon. M, 1976, Stanford University. 42 p.

Raynolds, Robert Gregory Honshu. The Plio-Pleistocene structural and stratigraphic evolution of the eastern Potwar Plateau, Pakistan. D, 1980, Dartmouth College. 265 p.

Rayome Goldblatt, Rosann E. A sedimentological study of some glaciofluvial deposits in the Binghamton, N.Y., region. M, 1987, SUNY at Binghamton. 69 p.

Raza, Saiyid and Papadakis, James. Geology of the Bare Hill and Montreal River areas, Keweenaw County, Michigan. M, 1954, Michigan Technological University. 83 p.

Razem, Allan C. Chert replacement mechanisms in a crinoidal biostrome; Fort Payne Formation (Mississippian), Tennessee. M, 1976, University of South Florida, Tampa. 107 p.

Razum, Brynne Anne. Mineralogy and geochemistry of lake deposits from Devlin's Park, Boulder County, Colorado. M, 1979, University of Colorado.

Rea, Campbell C. The erosion of Siletz Spit, Oregon. M, 1975, Oregon State University. 105 p.

Rea, David K. Stratigraphy of the red chert-pebble conglomerate in the Earp Formation (Upper Pennsylvanian and Permian), southeastern Arizona. M, 1967, University of Arizona.

Rea, David K. Tectonics of the East Pacific Rise, 5° to 12°S. D, 1975, Oregon State University. 139 p.

Rea, Raymond A. Influence of regolith properties on migration of septic tank effluent. M, 1977, University of South Florida, Tampa. 92 p.

Rea, Ronald G. Investigation of gannisters in Southeast Kentucky. M, 1986, Eastern Kentucky University. 58 p.

Reaber, Douglas W. Factors controlling carbonate mineralization at Laguan Mormona, Baja California Norte, Mexico. M, 1986, San Diego State University. 114 p.

Read, Barry Steven. Bright leaf and red clay; family farming and soil conservation in Piedmont North Carolina. D, 1988, University of North Carolina, Chapel Hill. 366 p.

Read, Burton Charles. Lower Cambrian stratigraphy of Pelly Mountains, central Yukon Territory. M, 1976, University of Calgary. 146 p.

Read, David L. Oxygen isotope composition of the 3,800 m.y. old Isua gneiss of Southwest Greenland. M, 1976, Northern Illinois University. 47 p.

Read, Edward Wade. The Honesdale Sandstone a key horizon for the Pocono Plateau in northeastern Pennsylvania. M, 1953, Lehigh University.

Read, John J. The geology and ore genesis of the gold quartz veins at the Big Hurrah Mine, Seward Peninsula, Alaska. M, 1985, Washington State University. 153 p.

Read, John Russell Lee. Slope reliability analyses using the principle of maximum entropy. D, 1987, Purdue University. 166 p.

Read, Mason Kent. Geology and economic resources of Birds Quadrangle, Illinois-Indiana. M, 1916, University of Illinois, Chicago.

Read, Peter Burland. Petrology and structure of Poplar Creek map-area, British Columbia. D, 1966, University of California, Berkeley. 195 p.

Read, Peter Burland. The geology of the Fraser Valley between Hope and Emory Creek. M, 1960, University of British Columbia.

Read, Robert Olcott. Geology of the Williams Creek area, Huerfano Quadrangle, Huerfano County, Colorado. M, 1956, University of Michigan.

Read, Steven Edward. Concentration of heavy minerals in braided channels; the effect of convergent flow. M, 1982, University of Illinois, Chicago.

Read, William. Environmental significance of a small deposit in the Permian Lueders Formation of Baylor County, Texas, containing terrestrial vertebrates and plants in association with marine invertebrates. D, 1942, University of Chicago. 52 p.

Read, William Harold. Geology of the Lillis Ranch Quadrangle. M, 1962, University of California, Berkeley. 87 p.

Readdy, Leigh A. Economic geology and photogeology of the Tsumeb area, Southwest Africa. M, 1972, University of Arizona.

Reade, Ernest Herbert, Jr. The geology of a portion of Newton and Walton counties, Georgia. M, 1960, Emory University. 65 p.

Reade, Harold Leslie, Jr. Stratigraphy and paleontology of the Monte Cristo Limestone, Goodsprings Quadrangle, Nevada. M, 1962, University of Southern California.

Reader, John Malcolm. A study of folds in the Womble Shale (Ordovician), Lake Ouachita, Arkansas. M, 1985, University of Missouri, Columbia. 88 p.

Reading, David John Richard. The geology and isotope geochemistry of the ankerite units, Dome Mine, Timmins. M, 1982, University of Waterloo. 178 p.

Ready, Jeffery A. The geology of the igneous rocks of Bear Butte (Meade County), South Dakota. M, 1968, University of Nebraska, Lincoln.

Reagan, Albert B. Contributions to the geology of the Navajo country, Arizona, with notes on the archaeology. D, 1924, Stanford University. 411 p.

Reagan, Albert B. Geology of northeastern Monroe County. M, 1904, Indiana University, Bloomington.

Reagan, Jeffrey F. Defining parameters in gravity exploration for groundwater. M, 1974, Michigan State University. 96 p.

Reagan, Mark Kenyon. Turrialba Volcano, Costa Rica; magmatism at the southeast terminus of the Central American Arc. D, 1987, University of California, Santa Cruz. 223 p.

Reagan, Mary. Interpretation of line-of-sight gravity data from the Pioneer Venus orbiter. M, 1987, Purdue University. 116 p.

Reagan, R. L. A finite-difference study of subterranean cavity detection and seismic tomography. D, 1978, University of Missouri, Rolla. 242 p.

Reagan, Wiley S. Relic volcanic glass deposits in the Ordovician Bogfork Chert of the Ouachita fold belt, southeastern Oklahoma. M, 1978, Northeast Louisiana University.

Reager, Richard David. Cryoplanation terraces of interior and western Alaska. D, 1975, Arizona State University. 326 p.

Real, Charles R. Local richter magnitude based on total signal direction from vertical short-period seismographs in the Southern California region. M, 1972, University of Southern California.

Realini, Michael J. Petrology and sedimentology of an Oligocene reef tract, southwestern Puerto Rico. M, 1978, Northern Illinois University. 125 p.

Ream, Lanny Ray. Economic geology of the Silver Creek mining district, Snohomish County, Washington. M, 1972, Washington State University. 59 p.

Reamer, Sharon Kae. Improvements to the Fourier gravity inversion method with applications. M, 1986, University of Texas at Dallas. 123 p.

Reams, Max W. Cave sediments and the geomorphic history of the Ozarks. D, 1968, Washington University. 167 p.

Reams, Max W. Some experimental evidence for a vadose origin of foibe (domepits). M, 1963, University of Kansas. 115 p.

Reamsbottom, Stanley Baily. Geology and metamorphism of the Mount Breakenridge area, Harrison Lake, British Columbia. D, 1974, University of British Columbia.

Reamsbottom, Stanley Baily. The geology of the Mount Breakenridge area, Harrison Lake, British Columbia. M, 1972, University of British Columbia.

Reardon, Eric John. Thermodynamic properties of some sulfate, carbonate and bicarbonate ion pairs. D, 1974, Pennsylvania State University, University Park. 93 p.

Reardon, Jeffry. Depositional environments of Anaheim Bay salt marsh, Seal Beach, California. M, 1981, California State University, Long Beach. 223 p.

Reasenberg, Paul Allen. Seismic near field of an air gun measured by a wide band accelerometer system. M, 1971, Massachusetts Institute of Technology. 66 p.

Reaser, Donald Frederick. The geology of the Ferris Quadrangle, Dallas and Ellis counties, Texas. M, 1958, Southern Methodist University. 20 p.

Reaser, Donald R. Geology of Cieneguilla area, Chihuahua and Texas. D, 1974, University of Texas, Austin. 397 p.

Reasoner, Melton Aaron. The late Quaternary lacustrine record from the upper Cataract Brook valley, Yoho National Park, British Columbia. M, 1988, University of Alberta. 215 p.

Reaves, Christopher Madison. The migration of iron and sulfur during the early diagenesis of marine sediments. D, 1984, Yale University. 439 p.

Reavis, Betty Hill. The geologic background of English poetry. M, 1938, George Washington University.

Reazin, David G. Ice-contact sedimentation of the Packerton Moraine, north-central Indiana. M, 1985, Indiana University, Bloomington. 275 p.

Rebagay, Teofila Velasco. The determination of zirconium and hafnium in meteorites and terrestrial materials by activation analysis. D, 1969, [University of Kentucky]. 185 p.

Rebbert, Carolyn Rose. Biotite oxidation; an experimental and thermodynamic approach. M, 1986, Virginia Polytechnic Institute and State University.

Reber, Jennifer Joy. Correlation and biomarker characterization of Woodford-type oil and source rock, Aylesworth Field, Marshall County, Oklahoma. M, 1988, University of Tulsa. 96 p.

Reber, Louis E. The relation between ore deposition, rock alteration, and magmatic differentiation at Morenci, Arizona. D, 1916, Yale University.

Reber, Spencer J. Stratigraphy and structure of the south-central and northern Beaver Dam Mountains, Washington County, Utah. M, 1951, Brigham Young University. 68 p.

Rebertus, Donald Gene. Sedimentary analysis of artesian spring sands, (Pleistocene and Modern), Meade County, Kansas. D, 1972, University of Iowa. 92 p.

Rebertus, Russell A. Occurrence and distribution of kaolin and gibbsite in Hapludults and Dystrochrepts formed from mica gneiss and schist in North Carolina. D, 1984, North Carolina State University. 227 p.

Rebholz, Irma. The highway pattern of metropolitan St. Louis. M, 1939, Washington University. 118 p.

Reblin, Michael Thomas. Regional gravity survey of the Dominican Republic. M, 1973, University of Utah. 124 p.

Rebne, Claudia A. A seismic stratigraphic study of the Muddy Formation, Bell Creek Field, Montana. M, 1985, Colorado School of Mines. 97 p.

Rebuck, Ernest Charles. Evaluation of unconfined aquifer parameters using a successive line relaxation finite difference model. D, 1972, University of Arizona.

Rebull, Peter Mario. Mechanical model analysis of clay behavior. D, 1967, Rensselaer Polytechnic Institute. 109 p.

Recca, Steven I. Fine-grained, millimeter-sized objects in ordinary chondrites and their relation to chondrules and matrix. M, 1988, University of New Mexico. 199 p.

Reches, Z. The development of monoclines. D, 1977, Stanford University. 250 p.

Rechtian, Richard Douglas. Amplitude independent static hysteresis damping as a model for earth materials. D, 1964, Washington University. 61 p.

Rechtien, Richard Douglas. Pseudo-gravity as derived from the vertical magnetic intensity. M, 1959, Washington University. 16 p.

Reck, Brian Harrison. Deformation and metamorphism in the southwestern Narragansett Basin and their relationship to granite intrusion. M, 1985, University of Texas, Austin. 76 p.

Reck, Donald F., Jr. The Lower Cretaceous stratigraphic sequence in the Chihuahua Trough, Presidio County, Texas. M, 1980, Sul Ross State University.

Reckendorf, Frank Fred. Drainage phenomena as related to a tundra regime in northern Alaska. M, 1964, Iowa State University of Science and Technology.

Reckendorf, Frank Fred. Techniques for identifying flood plains in Oregon. D, 1973, Oregon State University. 344 p.

Recks, Elizabeth Helen. A comparison of Cambro-Ordovician and modern sedimentary prisms in the New England area. M, 1969, Massachusetts Institute of Technology. 38 p.

Recny, C. J. Assessment of economic scale of operations in mine planning. M, 1978, Stanford University. 107 p.

Record, Richard Storey. Paleoenvironmental analysis of coastal marsh deposits in the Aguja Formation, Late Cretaceous, Trans-Pecos, Texas. M, 1988, Colorado School of Mines.

Record, Walter Ross, Jr. A subsurface study of the Simpson Group of rocks of Garvin County and adjacent area, Oklahoma. M, 1948, University of Oklahoma. 71 p.

Rector, Glasco Windrom. Paleontology and stratigraphy of a well core from Ashtabula, Ohio. M, 1950, University of Michigan.

Rector, Richard James. Geology of the east half of the Swauk Creek mining district. M, 1962, University of Washington. 73 p.

Rector, Roger J. A structural analysis of the San Antonio Formation, Rice Lake area, Manitoba. M, 1966, University of Manitoba.

Rector, Roger Joseph. An experimental investigation into the influence of environmental and fabric parameters on the deformational behavior of sandstones. D, 1970, McMaster University. 190 p.

Rector, Sharon. Fabric analysis of deformed Proterozoic metasedimentary rocks, western Wedel Jarlsberg Land, Spitsbergen. M, 1987, University of Wisconsin-Madison. 98 p.

Rector, William Kenna. The general geology of Moffat's Creek area in Augusta and Rockbridge counties, Virginia. M, 1958, University of Virginia. 98 p.

Rector, Willis Edward. Porosity and permeability studies in carbonates. M, 1958, University of Michigan.

Reddell, Donald Lee. Dispersion in groundwater flow systems. D, 1969, Colorado State University. 238 p.

Redden, Jack Allison. Geology of the Fourmile pegmatite area, Custer County, South Dakota. D, 1956, Harvard University.

Redden, Martha J. The crystal structure of darapskite. M, 1968, Massachusetts Institute of Technology. 63 p.

Reddi, Lakshmi Narayana. Probabilistic analysis of groundwater levels in hillside slopes. D, 1988, Ohio State University. 141 p.

Reddig, Ransom P. Application of the Rayleigh-FFT technique to topographic corrections in magnetotellurics. M, 1984, San Diego State University. 188 p.

Reddin, Nancy Jean. Investigations of the upper and lower Shelbyville tills in southwestern Ohio. M, 1981, Miami University (Ohio). 117 p.

Redding, Carter Eugene, Jr. Late Mesozoic rhyolitic activity in the Southwest Pacific; geochronological results. M, 1976, Case Western Reserve University.

Reddy, Chemicala Janardhan. Landsat lineament analysis and gas production from the Devonian shales of southeastern Ohio. M, 1984, Ohio University, Athens. 131 p.

Reddy, George R. Geology of the Queen Lake domes near Malaga, Eddy County, New Mexico. M, 1961, University of New Mexico. 84 p.

Reddy, Indupuru Kota. A magnetotelluric study of resistivity anistropy. M, 1968, University of Alberta. 75 p.

Reddy, Indupuru Kota. Magnetotelluric sounding in central Alberta. D, 1970, University of Alberta. 162 p.

Reddy, Kevin M. The use of Rayleigh waves in near-surface geologic studies. M, 1975, University of Wisconsin-Milwaukee.

Reddy, M. Rajendra. The petrology of a granite-olivine diabase contact, NE Alderney, Channel islands. M, 1969, Brooklyn College (CUNY).

Reddy, Nagendra Peesary. Characterization of coal breakage by continuous miners. D, 1988, West Virginia University. 195 p.

Reddy, Raja Palpunuri. Theoretical model to predict the growth and swelling behavior of clay on hydration. M, 1980, University of Oklahoma. 88 p.

Reddy, Ramesh Kumar T. A generalized model for evaluating area-potential in a mineral exploration program; a case study of silver exploration in parts of Idaho and Montana. D, 1987, University of Georgia. 119 p.

Reddy, Varakantham S. A study of stress distribution and measurement of axial strain around a pilot hole during overcoring. M, 1966, University of Missouri, Rolla.

Reddy, Vavula Srinivas. Crustal structure in New Mexico based on Project Gnome and microearthquake data. M, 1966, New Mexico Institute of Mining and Technology. 45 p.

Redfern, R. A. Environmental geology of the Marquam Hill area. M, 1973, Portland State University. 109 p.

Redfern, Richard Robert. Geology and uranium deposits on the Prince claims, Elko County, Nevada. M, 1978, University of California, Los Angeles.

Redfield, John Stowe. Differentiation of the Bokchito Formation, Love County, Oklahoma. M, 1929, University of Oklahoma. 36 p.

Redfield, Robert Crim. A study of the lower portion of the Boquillas Formation, Brewster County, Texas. M, 1940, University of Texas, Austin.

Redfield, Thomas F. Structural geology of part of the Leonia Quadrangle in Northeast Idaho. M, 1987, Western Washington University. 156 p.

Reding, Lynn Marie. North American Plate stress modeling; a finite element analysis. M, 1984, University of Arizona. 111 p.

Redman, Earl C. Geology of the Wyoming Hills, Mt. McKinley National Park, Alaska. M, 1974, University of Alaska, Fairbanks. 61 p.

Redman, F. H. Potassium availability in soils from the semi-arid areas of Oahu. M, 1958, University of Hawaii. 49 p.

Redman, Kenneth G. The Ryan oil and gas pool of Rush and Pawnee counties, west-central Kansas. M, 1947, University of Kansas. 47 p.

Redman, Robert H. Post-Mississippian geology of Love County, Oklahoma. M, 1964, University of Oklahoma. 73 p.

Redmond, Brian. The utility of Fourier estimates of grain shape in sedimentological studies. M, 1969, Michigan State University. 145 p.

Redmond, Brian Thomas. Sedimentary processes and products; an amber-bearing turbidite complex in the northern Dominican Republic. D, 1982, Rensselaer Polytechnic Institute. 456 p.

Redmond, Charles David. Paleontology and stratigraphy of the Lomita Formation. M, 1936, University of California, Los Angeles.

Redmond, Charles Edward. Genesis, morphology and classification of some till derived Chernozems of eastern North Dakota. D, 1964, Michigan State University. 330 p.

Redmond, John Charles. Effect of simulated overburden pressure on the resistivity, porosity, and permeability of selected sandstones. D, 1962, Pennsylvania State University, University Park. 149 p.

Redmond, John Lynn. Paleogeology of Paris Basin. M, 1961, Stanford University.

Redmond, John Lynn. Structural analysis of the Blue Canyon Formation (Mississippian), Sierra Nevada, Placer County, California. D, 1966, University of Oregon. 197 p.

Redmond, John Lynn. Tunnel route geology of the American River project, California. M, 1962, Stanford University.

Redmond, Roy J. An inferred crustal velocity structure of eastern Pennsylvania-northern New Jersey from inversion of wide-angle reflections. M, 1982, Lehigh University. 97 p.

Redpath, Bruce Beckwith. The seismic determination of glacial thickness. M, 1965, McGill University.

Redwine, James C. Soxhlet extraction of fly ash to determine the behavior of constituents during aqueous leaching. M, 1982, University of Georgia.

Redwine, Lowell Edwin and Varney, Frederick Merrill. Development of a coring instrument for submarine geological investigations. M, 1937, University of California, Los Angeles.

Redwine, Lowell Edwin. The tertiary Princeton submarine valley system beneath the Sacramento Valley, California. D, 1972, University of California, Los Angeles. 900 p.

Reeber, Robert Richard. Low temperature thermal expansion of Wurtzite-phases of IIB-IVB compounds. D, 1968, Ohio State University. 154 p.

Reeburgh, William Scott. Measurements of gases in sediments. D, 1967, The Johns Hopkins University. 102 p.

Reece, Dennis E. A study of leaching of metals from sediments and ores and the formation of acid mine water in the Bunker Hill Mine. M, 1974, University of Idaho. 117 p.

Reed, Alan A. Stratigraphy, depositional environments and sedimentary petrology of the Lower Jurassic East Berlin Formation, central Connecticut. M, 1976, University of Massachusetts. 175 p.

Reed, Billy Kirk. Geology of the Pre-Atokan unconformity of portions of Love and Carter counties, Oklahoma. M, 1957, University of Oklahoma. 59 p.

Reed, Bruce. Geology of the Lake Peters area, northeastern Brooks Range, Alaska. D, 1966, Harvard University.

Reed, Bruce Loring. Structural geology and ore deposits of the Bannockburn Basin, Lardeau area, British Columbia. M, 1958, Washington State University. 49 p.

Reed, Byram E., Jr. Petrology of the Dutchtown Formation (Ordovician), southeast Missouri. M, 1968, University of Missouri, Columbia.

Reed, Charles M. Insoluble residue studies of the (Silurian) Lockport Dolomite of New York State. M, 1936, University of Rochester. 61 p.

Reed, Darrell. Carbonate petrography and depositional environment of the Lodgepole Formation in southeastern Idaho. M, 1980, Eastern Washington University. 55 p.

Reed, David Allen. Core structure constraints derived from SKS and SKKS observations. M, 1974, Massachusetts Institute of Technology. 102 p.

Reed, Donald Lawrence. Structure and stratigraphy of the eastern Sunda Forearc, Indonesia; geologic consequences of arc-continent collision. D, 1985, University of California, San Diego. 261 p.

Reed, Ethbert F. The petrology of the Red River mining district. M, 1923, Columbia University, Teachers College.

Reed, Eugene Clifton. A study of the Emporia-Dover interval in Nebraska. M, 1933, University of Nebraska, Lincoln.

Reed, George Bruce. Application of kinematical geodesy for determining the short wave length components of the gravity field by satellite gradiometry. D, 1973, Ohio State University. 173 p.

Reed, Jack C. Economic development of a mineral deposit of southeastern Alaska. M, 1969, University of Arizona.

Reed, Jack Morce. Sediments of the Magdelena Group in the Creston Range of northwestern San Miguel County, New Mexico. M, 1958, Mississippi State University. 95 p.

Reed, James Bradly. Sedimentology of some Triassic-Middle Jurassic (?) siliciclastic flysch deposits, Northwest Baja California Norte, Mexico. M, 1985, San Diego State University. 265 p.

Reed, James Courtney. A microfacies analysis of a Virgilian algal bioherm, Hueco mountains, Texas. M, 1969, Texas Tech University. 90 p.

Reed, James Edward. A fluid inclusion study of the Tintic District, Utah. M, 1981, University of Missouri, Rolla. 83 p.

Reed, James Patrick. Factors influencing the decomposition of plant material in marine sediments; a bioenergetic model and laboratory microcosm approach. D, 1981, North Carolina State University. 76 p.

Reed, James Robert. Geology and geochemistry of the Proterozoic age Alder Group, central Mozatzal Mountains, Arizona. M, 1988, New Mexico Institute of Mining and Technology. 211 p.

Reed, James Stalford. An investigation of the constitution of transition metal ions in a spinel matrix. D, 1965, Alfred University. 132 p.

Reed, John Calvin. Geology of the Potsdam Quadrangle, New York. D, 1930, Princeton University. 158 p.

Reed, John Calvin. The geology of the Catoctin Formation near Luray, Virginia. D, 1954, The Johns Hopkins University.

Reed, John Daniel. The geology, geochemistry, and hydrology of four hot spring areas along the South Fork Payette River, between Lowman and Banks, Boise County, Idaho. M, 1986, Washington State University. 112 p.

Reed, Jon Edward. Enhancement/isolation wavenumber filtering of potential field data. M, 1980, Purdue University. 205 p.

Reed, Juliet Carrington. Sedimentary petrology of outcrops of the Upper Cretaceous Englishtown Formation of New Jersey. M, 1956, Bryn Mawr College. 129 p.

Reed, Kenneth John. Biometric study of the embryonic apparatus of selected species of the genus Lepidocyclina (Foraminifera). D, 1965, Cornell University. 114 p.

Reed, Kenneth John. Mid-Tertiary smaller foraminifera from a bore at Heywood, Victoria, Australia. M, 1964, Cornell University.

Reed, Lanny Joe. The development of preferred orientations in a system of rigid particles in a viscous matrix as the matrix is deformed by pure shear and simple shear. D, 1973, University of Tulsa. 97 p.

Reed, Leslie D. A geophysical investigation of groundwater supply, Morongo Indian Reservation, Riverside County, California. M, 1971, University of California, Riverside. 81 p.

Reed, Louis Calvin. Geology of the Midlothian Quadrangle, Ellis County, Texas. M, 1958, Southern Methodist University. 26 p.

Reed, Mark Hudson. Calculations of hydrothermal metasomatism and ore deposition in submarine volcanic rocks with special reference to the West Shasta District, California. D, 1977, University of California, Berkeley. 195 p.

Reed, Mary Catherine. A petrographic analysis of a gabbro dike at Everton, New York. M, 1956, Smith College. 77 p.

Reed, Norman H. A comparative study of the types of gold deposits. M, 1916, Columbia University, Teachers College.

Reed, Phillip Lewis. The petrology of Western Bluff Springs Granite and the Elkahatchee Quartz Diorite Gneiss, Tallapoosa and Clay counties, Alabama. M, 1980, Memphis State University.

Reed, Ralph D. Tertiary petrology in the Coalinga District, California. D, 1924, Stanford University. 63 p.

Reed, Richard K. Precambrian geology of the central Nacimiento Mountains, Sandoval County, New Mexico. M, 1971, University of New Mexico. 116 p.

Reed, Richard K. Structure and petrography of the Fraquita Peak area, Santa Cruz County, Arizona. M, 1966, University of Arizona.

Reed, Richard W. Computerized filing system of construction aggregates and the availability of construction aggregates in Ada County. M, 1976, University of Idaho. 82 p.

Reed, Robert C. Copper mineralization in Animikie sediments of the Marquette Range, Marquette County, Michigan. M, 1965, Michigan State University. 55 p.

Reed, Robert G. Structure and stratigraphy of the Alisitos Formation (early Cretaceous), Arroyo San Juan de Dios, east of El Rosario, Baja California (Mexico). M, 1967, University of San Diego.

Reed, Robert Marion. The analysis and fusibility of the ash from certain Washington coals. M, 1930, University of Washington. 32 p.

Reed, Roy Edwin, II. Paleoenvironmental analysis of biohermal facies, Mississippian Lake Valley Formation, northern Sacramento Mountains, New Mexico. M, 1982, Texas A&M University. 182 p.

Reed, Thomas B. Digital image processing and analysis techniques for SeaMARC II side scan sonar imagery. D, 1987, University of Hawaii.

Reed, Thomas Willis. Trace element distribution and alteration study of the Copper Cities Deposit, Arizona. M, 1975, Oklahoma State University. 62 p.

Reed, Timm L. Heavy metal concentrations in the sediments of a portion of the East Florida continental shelf. M, 1976, Florida Institute of Technology.

Reed, Timothy. Sequence of thrusting in the Northwest Montana thrust belt. M, 1982, Purdue University. 209 p.

Reed, V. Stephen. Stratigraphy and depositional environment of the upper Precambrian Hakatai Shale, Grand Canyon, Arizona. M, 1976, Northern Arizona University. 163 p.

Reed, Walter Edwin. Stratigraphic controls on the composition of petroleum and sedimentary organic matter. D, 1972, University of California, Berkeley. 296 p.

Reed, Walter Edwin. The geology of part of the southern Cherry Creek Mountains, Nevada. M, 1962, University of California, Berkeley. 57 p.

Reede, Roger John. Geology of southern Nelson County, North Dakota. D, 1972, University of North Dakota. 147 p.

Reeder, Jean Dolan. Nitrogen transformations in revegetated coal spoils. D, 1981, Colorado State University. 85 p.

Reeder, John William. Experimental studies of the effects of ephemeral stream-flow depths on infiltration. D, 1981, Stanford University. 193 p.

Reeder, Louis Robert. The feasibility of underground liquid waste disposal in northeastern Oklahoma. M, 1971, University of Tulsa. 94 p.

Reeder, Richard James. Phase transformations in dolomite. D, 1980, University of California, Berkeley. 163 p.

Reeder, Robert T. Feasibility and cash flow analysis of the Centennial Development Company's Larson coal lease property, Carbon County, Utah. M, 1976, Colorado School of Mines. 154 p.

Reeder, Sarah Jane. Studies in the fossil flora of the (Eocene) Wilcox Formation near Mansfield, Louisiana. M, 1928, Tulane University.

Reeder, William Glase. A review of Tertiary rodents of the family Heteromyidae. D, 1957, University of Michigan. 644 p.

Reeds, Chester A. The stratigraphy of the (Silurian-Devonian) Hunton Formation with introductory chapters on the physiography and structure of the Arbuckle Mountains, Oklahoma. D, 1910, Yale University.

Reedy, Milton Frank, Jr. Sections in the Eocene of the Gulf Coastal Plain of Texas and Louisiana. M, 1939, University of Texas, Austin.

Reef, John W. The Unity Reservoir rhyodacite tuff-breccia and associated volcanic rocks, Barker County, Oregon. M, 1983, Washington State University. 128 p.

Reel, David A. Geothermal gradients and heat flow in Florida. M, 1970, University of South Florida, Tampa. 66 p.

Reel, Ted Wesley. The evolution of geologic knowledge of the Cretaceous area of central Texas with an annotated bibliography of geologic work of pertinence to the area. M, 1960, Baylor University. 175 p.

Reel, Ted Wesley. The excavation and preparation of two fossilized whales (upper Eocene, from Jasper County, Mississippi). D, 1972, University of Southern Mississippi.

Reep, Thomas W. Van de *see* Van de Reep, Thomas W.

Rees, Christopher John. Metamorphism in the Canadian Shield of northern Saskatchewan. M, 1980, University of Regina. 136 p.

Rees, Christopher John. The Inermontane-Omineca Belt boundary in the Quesnel Lake area, east-central British Columbia; tectonic implications based on geology, structure and paleomagnetism. D, 1987, Carleton University. 421 p.

Rees, Delbert Clyde. Geology and diatremes of Desert Mountain, Utah. M, 1971, University of Utah. 57 p.

Rees, Kathleen Ann. A quantitative analysis of a physically based predictive model of sediment yield for small mountainous watersheds in Southern California. M, 1988, University of California, Los Angeles. 147 p.

Rees, Margaret N. A fault-controlled trough through a carbonate platform, Middle Cambrian House Range embayment, Utah and Nevada. D, 1984, University of Kansas. 207 p.

Rees, Margaret N. Paleoenvironments of the Upper Cambrian Johns Wash Limestone, House Range, Utah. M, 1975, University of Kansas.

Rees, Rhys Willis. Evolutionary trends of some Lower Devonian faunal stocks. M, 1949, Pennsylvania State University, University Park. 45 p.

Rees, Terry F. Interaction of neptunium (V) with selected naturally occurring organic ligands and a surface water fulvic acid. D, 1982, Colorado School of Mines. 108 p.

Rees, Todd Howard. Microstructural deformation of polycrystalline halite in transmitted light. M, 1984, Michigan Technological University. 128 p.

Reese, Dale O. Sedimentology of the Hardinsburg Formation in southeastern Illinois and northwestern Kentucky. M, 1959, University of Kansas. 77 p.

Reese, Donald Leon. Areal geology of the Spavinaw Lake area, Delaware County, Oklahoma. M, 1963, University of Oklahoma. 67 p.

Reese, James. Geological aspects of the Recent sediments of the Suisun Bay complex, San Francisco Bay, California. M, 1964, San Jose State University.

Reese, Joseph F. Stratigraphy and structure of the middle Proterozoic Belt Supergroup, southwestern Whitefish Range, Montana. M, 1988, University of Wisconsin-Madison. 190 p.

Reese, Joseph L. The size, shape, extent and continuity of the coal field at Madrid, Iowa. M, 1975, Iowa State University of Science and Technology.

Reese, Nathan Mark. Cenozoic tectonic history of the Ruby Mountains and adjacent areas, northeastern Nevada; constraints from radiometric dating and seismic reflection profiles. M, 1986, Southern Methodist University. 88 p.

Reese, Robert J. Salt kinematics of southern Bienville Parish, Louisiana. M, 1977, Louisiana State University.

Reese, Robert Lester. Computed solid-vapor equilibria in multicomponent systems. M, 1973, Arizona State University. 43 p.

Reese, Ronald S. Stratigraphy of the Entrada Sandstone and Todilto Limestone (Jurassic), north-central New Mexico. M, 1984, Colorado School of Mines. 182 p.

Reese, Stuart O. Effect of paleotopography and organic matter on the mineralogy and chemistry of sediments exposed during a Pennsylvanian diastem, Hazard coal zone, Leslie County, Kentucky. M, 1986, University of Tennessee, Knoxville. 115 p.

Reese, Thomas J. Electronic surveying with the laser. M, 1965, Ohio State University.

Reeside, J. B. Jr. The Helderberg and Tonoloway formations of central Pennsylvania, with special reference to the Siluro-Devonian boundary. D, 1915, The Johns Hopkins University.

Reesman, Arthur Lee. A study of clay mineral dissolution. D, 1966, University of Missouri, Rolla. 228 p.

Reesman, Arthur Lee. Dissolved minor elements in the hydrolysis of sixty-seven silicate rocks and minerals. M, 1961, University of Missouri, Columbia.

Reesman, Richard H. A rubidium-strontium isotopic investigation of the possibility of dating hydrothermal mineral deposits. D, 1968, Massachusetts Institute of Technology. 149 p.

Reesor, John Elgin. The White Creek Batholith and its geological environment in Dewar Creek map-area, B.C. D, 1952, Princeton University. 124 p.

Reesor, S. N. Thermal properties of bentonite-sand and kaolinite-sand mixtures. M, 1982, University of Waterloo. 136 p.

Reetz, Gene Rene. A hydrologic appraisal of Arizona's ground water as affected by the state code. M, 1969, University of Arizona.

Reeve, Donna M. Seismic detection of Mississippian Frobisher-Alida facies, Wiley and Mouse River Park fields, Bottineau and Renville counties, Williston Basin, North Dakota. M, 1986, Colorado School of Mines. 87 p.

Reeve, Edward John. Geochemistry of the Golding-Keene Pegmatite and adjacent rocks. D, 1972, University of Toronto.

Reeve, Edward John. Petrology and mineralogy of a gabbroic intrusion in Pardee Township near Port Arthur, Ontario (Canada). M, 1969, University of Wisconsin-Milwaukee.

Reeve, Richard L. Stratigraphy and petroleum geology of the Oriskany Sandstone and Lower and Middle Devonian strata in eastern and central Coshocton

County, Ohio. M, 1983, Kent State University, Kent. 162 p.

Reeve, Scott Cleveland. Magnetic reversal sequence in the upper portion of the Chinle Formation (Upper Triassic), Montoya, New Mexico. M, 1971, University of Texas, Austin.

Reeve, Scott Cleveland. Paleomagnetic studies of sedimentary rocks of Cambrian and Triassic age. D, 1975, University of Texas at Dallas. 425 p.

Reeve, William. Bedrock geology of the Blue Hills, Kitsap County, Washington. M, 1979, Colorado School of Mines. 58 p.

Reeves, Anita L. Geologic influences in human environment. M, 1967, Baylor University. 115 p.

Reeves, Corwin C. Phosphoria Formation, Johnson and Natrona counties, Wyoming. M, 1957, University of Oklahoma. 75 p.

Reeves, Corwin C., Jr. Some geomorphological, structural, and stratigraphic aspects of the Pliocene and Pleistocene sediments of the southern High Plains. D, 1970, Texas Tech University. 188 p.

Reeves, Donald W. Geodetic problems in the extension of horizontal and vertical control for the mapping of a new country. M, 1964, Ohio State University.

Reeves, Elaine Louise. A taxonomic revision of the gastropod family Nassariidae. M, 1964, Stanford University. 31 p.

Reeves, Frank. A discussion of the absence of water in certain petroleum-bearing strata of the Appalachian oil fields. D, 1916, The Johns Hopkins University.

Reeves, James Elmo. Some Pentremites of the Golconda Limestone (Mississippian) in Indiana. M, 1939, Indiana University, Bloomington. 29 p.

Reeves, James J. Determination of rear abutment zones by SH channel wave velocity. D, 1984, Colorado School of Mines. 245 p.

Reeves, James Ray. Ore deposits of the Battle Mountain-Eureka Mineral Belt, Lander County, Nevada. M, 1970, Texas Tech University. 71 p.

Reeves, John Edward. The Ghost Range basic intrusion, District of Cochrane, Ontario. M, 1950, McMaster University. 60 p.

Reeves, John Robert. Oil shales of Indiana. D, 1923, Indiana State University.

Reeves, John Robert. Preliminary report on oil shales of Indiana. M, 1921, Indiana State University. 46 p.

Reeves, Keith D. Hydrothermal alteration in the central Davis Mountains, Jeff Davis County, Texas; physical structure, chemical evolution and precious metals deposition potential of a fossil geothermal system. M, 1987, Sul Ross State University. 256 p.

Reeves, Richard Wayne. Modern channel entrenchment in the coastal ranges of central and southern California. D, 1970, University of California, Los Angeles. 294 p.

Reeves, Richard Wayne. Modifications of drainage in the El Segundo Sand Hills of coastal Southern California. M, 1964, University of California, Los Angeles. 139 p.

Reeves, Robert G. Geochemical and remote sensing exploration of the Upper Cement Creek area, Gunnison County, Colorado. M, 1979, University of Kentucky. 66 p.

Reeves, Robert Grier. Geology and mineral resources of the Monlevade and Rio Piracicaba quadrangles, Minas Gerais, Brazil. D, 1965, Stanford University. 192 p.

Reeves, Thomas Kenneth, Jr. El Progreso Quadrangle, El Progreso, Guatemala. M, 1967, Rice University. 87 p.

Reeves, Thomas Leslie. The Fairmount Formation (Upper Ordovician) on the eastern flank of the Cincinnati Arch and its Bryozoa fauna. M, 1966, University of Kentucky. 103 p.

Refai, Refai Taher. Geology and reservoir characteristics of the Red River Formation in Harding County, South Dakota. M, 1982, South Dakota School of Mines & Technology.

Refolo, Perry J. The geology of the McCullough Range, from Black Mountain to Henderson, Clark County, Nevada. M, 1988, Montclair State College. 57 p.

Regalbuto, David Philip. An isotopic investigation of groundwater in Leelanau County, Michigan. M, 1987, Michigan State University. 54 p.

Regan, Donald R. An aerial photogrammetric survey of long-term shoreline changes, southern Rhode Island coast. M, 1976, University of Rhode Island.

Regan, Janice. Numerical studies of propagation of L_g waves across ocean continent boundaries using the representation theorem. D, 1987, California Institute of Technology. 247 p.

Regan, Louis J., Jr. Origin of the Eocene sands of the Coalinga District, California. D, 1943, California Institute of Technology. 73 p.

Regan, Louis J., Jr. The composition, texture, structure, and probable origin of the (Eocene) "Gatchell" sand (California). M, 1941, California Institute of Technology. 51 p.

Regan, Peter T. Stratigraphy and facies relations of Deerparkian formations in southeastern New York. M, 1982, Rensselaer Polytechnic Institute. 73 p.

Regan, Robert D. Geophysical examination of Meteor Crater (Arizona). M, 1967, Boston College.

Regan, Robert David. The application of spectral methods in the analysis and interpretation of gravity data; a critical study. D, 1973, Michigan State University. 124 p.

Regan, Terence R. Determining the hydrogeologic parameters of individual aquifers in a multiaquifer system. M, 1987, University of New Hampshire. 100 p.

Regel, Bernard. Elastic properties of the "dimension" stone unit (Mississippian Salem Limestone), southern Indiana. M, 1979, Bowling Green State University. 73 p.

Reger, J. P. Mass movement on spoil outslopes of contour surface-mines, North-central West Virginia. D, 1977, West Virginia University. 262 p.

Reger, Richard D. Recent glacial history of Gulkana and College glaciers, central Alaska Range, Alaska. M, 1964, University of Alaska, Fairbanks. 75 p.

Reggio, Jose de Jesus Gomez *see* Gomez Reggio, Jose de Jesus

Regis, Andrew J. Geochemistry of the Green River Formation in the Bridger Basin, Wyoming. M, 1958, University of Utah. 52 p.

Register, Joseph K., Jr. Rb-Sr and related studies of the Salado Formation, southeastern New Mexico. M, 1979, University of New Mexico. 119 p.

Register, Marcia E. Geochemistry and geochronology of the Harding Pegmatite, Taos County, New Mexico. M, 1979, University of New Mexico. 145 p.

Regli, Robert. Petrography, porosity, and depositional environments of the Burro Canyon Formation and Dakota Sandstone of southwest Colorado. M, 1982, Bowling Green State University. 128 p.

Regnier, Jerome Philippe Mathieu. Cenozoic geology in the vicinity of Carlin, Nevada. D, 1958, Columbia University, Teachers College. 112 p.

Regnier, Jerome Philippe Mathieu. Mineralogy and paragenesis of the eastern part of the Elliston mining district, Montana. M, 1951, Montana College of Mineral Science & Technology. 38 p.

Regout, Robertus. Contribution to the correlation of the Colorado Group (Cretaceous) in eastern Wyoming and north-central Colorado. M, 1951, University of Colorado.

Regueiro S., Jose. Detection of multimode seam waves by vertical arrays. D, 1984, Colorado School of Mines. 180 p.

Rehder, Timothy Ray. Stratigraphy, sedimentology, and petrography of the Picuris Formation in Ranchos de Taos and Tres Ritos quadrangles, north-central New Mexico. M, 1986, Southern Methodist University. 110 p.

Reheis, M. J. Source, transportation, and deposition of debris on Arapaho Glacier, Front Range, Colorado. M, 1974, University of Colorado.

Reheis, Marith Cady. Chronologic and climatic control on soil development, northern Bighorn Basin, Wyoming and Montana. D, 1984, University of Colorado. 346 p.

Rehfuss, Isidore L. The role of anamorphism in the production of ores. M, 1913, University of Wisconsin-Milwaukee.

Rehkemper, Leonard James. Petrology of Springerville area, Apache County, Arizona. M, 1956, University of Texas, Austin.

Rehkemper, Leonard James. Sedimentology of Holocene estuarine deposits, Galveston Bay, Texas. D, 1969, Rice University. 218 p.

Rehm, B. W. Field dispersivity measurements in a shallow gravel aquifer. M, 1977, University of Waterloo.

Rehm, John M., Jr. Landslide potential in the Atlantic Highlands of New Jersey. M, 1978, Rutgers, The State University, New Brunswick. 35 p.

Rehmer, Judith. Petrology of the Esopus Shale; (Lower Devonian) New York and adjacent states. D, 1976, Harvard University.

Rehn, Edgar Ernest. Onondaga Group of parts of West Virginia and Virginia. M, 1942, Ohio State University.

Rehn, Warren M. Petrology and geochemistry of the Blue Joint area, southwestern Montana. M, 1983, Colorado School of Mines. 203 p.

Rehrig, William Allen. Fracturing and its effects in molybdenum mineralization at Queste, New Mexico. D, 1969, University of Arizona. 291 p.

Reich, Matthew A. Depositional and diagenetic history of a carbonate unit within the lower member of the Honaker Trail Formation (Pennsylvanian), San Juan Mountains, Colorado. M, 1986, Colorado School of Mines. 249 p.

Reichard, Eric George. Reconstruction of water availability along the San Luis Rey River in Southern California. M, 1982, Stanford University. 146 p.

Reichard, Eric George. The influence of hydrologic factors on the benefits of groundwater management in areas of irrigated agriculture. D, 1985, Stanford University. 138 p.

Reichard, James E. Fracture studies, lineament analysis and geology in the Temple Hill, Freedon and Gamaliel 7.5 quadrangles, south-central Kentucky. M, 1984, University of Toledo. 163 p.

Reiche, Parry. The geology of the Lucia Quadrangle, California. D, 1934, University of California, Berkeley. 186 p.

Reichelderfer, Jan L. Microfacies, diagenesis and porosity development in the Salem Limestone (Middle Mississippian), southern Illinois, U.S.A. M, 1985, University of Illinois, Urbana. 95 p.

Reichenbach, Ingrid G. An ensialic marginal basin in Wopmay Orogen, northwestern Canadian Shield. M, 1987, Carleton University. 129 p.

Reichenbach, Mary Elizabeth. Lithofacies analysis for the Lower Cretaceous glauconitic sandstone in the Medicine River area, central Alberta. M, 1981, University of Calgary. 208 p.

Reichert, Randall Lee. The geology of Burnt Ridge Quadrangle and vicinity, Ravalli County, Montana. M, 1983, Western Michigan University.

Reichert, Stanley O. Geology of the Golden-Green Mountain area, Jefferson County, Colorado. D, 1953, [Colorado School of Mines]. 136 p.

Reichert, William H. Bibliography and index of geology and mineral resources of Washington. M, 1958, University of Washington. 721 p.

Reichhard-Barends, Enrique. Quantification of risk in mineral exploration. M, 1984, Queen's University. 180 p.

Reichhoff, Colin Lee. Geology of a lower Proterozoic volcaniclastic sequence near Wausau, Marathon

County, Wisconsin. M, 1986, University of Minnesota, Duluth. 131 p.

Reichhoff, Jayne A. Two Keweenawan basaltic dike swarms in the Duluth area, Minnesota. M, 1987, University of Minnesota, Duluth. 168 p.

Reichle, Michael S. A seismological study of the Gulf of California; sonobuoy and teleseismic observations, and tectonic implications. D, 1975, University of California, San Diego. 273 p.

Reichlin, R. L. The crystal chemistry of orthogermanates. M, 1978, SUNY at Stony Brook.

Reid, Alan B. A palaeomagnetic study at 1800 million years in Canada. D, 1972, University of Alberta. 237 p.

Reid, Alastair Milne, II. Biostratigraphy of Naco Formation (Pennsylvanian) in south-central Arizona. D, 1968, University of Arizona. 322 p.

Reid, Alastair Milne, II. Stratigraphy and paleontology of the Naco Formation (Pennsylvanian-Permian) in the southern Dripping Spring Mountains, near Winkelman, Gila County, Arizona. M, 1966, University of Arizona.

Reid, Allan R. An attempt to localize kimberlite source areas for Venezuelan diamonds from stratigraphy and analysis of diamond mineral inclusions. D, 1976, Colorado School of Mines. 121 p.

Reid, Allan Robert. Bedrock geology of Cheam Ridge area, Chillwack region, British Columbia. M, 1968, University of Washington. 49 p.

Reid, Archibald M. Petrology of the Mount Megantic igneous complex, southern Quebec. M, 1961, University of Western Ontario. 87 p.

Reid, Archibald McMillan. Characteristics of orthopyroxenes from enstatite achondrites. D, 1964, University of Pittsburgh.

Reid, Brian C. Structural analysis of the southwest flank of the Wind River Uplift, Green River Lakes area, Sublette County, Wyoming. M, 1983, Miami University (Ohio). 67 p.

Reid, Chase S. The utility of microwave remote sensing techniques for soil moisture profile determination. M, 1978, University of Arkansas, Fayetteville.

Reid, D. F. The near-surface distribution of radium in the Gulf of Mexico and Caribbean Sea; temporal and spatial variability and hydrographic relationships. D, 1979, Texas A&M University. 226 p.

Reid, D. F. The Quaternary geology of the Lake Johanna region, west-central Minnesota. M, 1974, University of Minnesota, Duluth.

Reid, Fredrick Samuel, Jr. Microfacies and diagenesis of the Viola Limestone (Ordovician) of South Central Oklahoma. M, 1980, Southern Methodist University. 106 p.

Reid, Gary Carl. Literature evaluation of induced groundwater tracers, field tracer techniques, and hydrodynamic dispersion values in porous media. M, 1981, Texas Tech University. 102 p.

Reid, George O. The relationships among slope processes, soils, bedrock, and topography in an area of Franciscan terrane in Marin County, California. M, 1978, San Jose State University. 90 p.

Reid, Ian. The Rivera Plate; a study in seismology and plate tectonics. D, 1976, University of California, San Diego. 305 p.

Reid, J. Barry. A subsurface study of the Monteagle Limestone; Northeast Tennessee. M, 1981, Memphis State University. 80 p.

Reid, Jeffery P. The nature of immiscible displacement in carbonate reservoir rocks. M, 1980, University of Calgary. 215 p.

Reid, Jeffrey Clinton. Hazel Formation, Culberson and Hudspeth counties, Texas. M, 1974, University of Texas, Austin.

Reid, Jeffrey Clinton. Metallogeny of the tungsten deposits of the northeastern Brazil scheelite district, and applied exploration bedrock geochemistry (lithogeochemistry); Volume I and II. D, 1981, University of Georgia. 466 p.

Reid, John. The character of the sediments in the Saco Valley, North Conway Quadrangle, New Hampshire. M, 1952, Syracuse University.

Reid, John B., Jr. The roles of lherzolite and garnet pyroxenite in the constitution of the upper mantle. D, 1970, Massachusetts Institute of Technology. 149 p.

Reid, John Reynolds, Jr. Geology of Burt Lake, Cheboygan County, Michigan. M, 1957, University of Michigan.

Reid, John Reynolds, Jr. Structural glaciology of an ice layer in a firn fold, Camp Michigan, Bay of Whales, Ross Ice Shelf, Antarctica. D, 1961, University of Michigan. 129 p.

Reid, Joseph Hugh. Lead and zinc deposits in the sedimentary rocks of East Tennessee and Southwest Virginia. M, 1930, University of Missouri, Rolla.

Reid, Leslie Margaret. Sediment production from gravel-surfaced forest roads, Clearwater Basin, Washington. M, 1981, University of Washington. 247 p.

Reid, Mark E. Modeling variably-saturated groundwater flow in layered surficial deposits. D, 1988, University of California, Santa Cruz.

Reid, Mary Ruth. Chemical stratification of the crust; isotope, trace element, and major element constraints from crustally contaminated lavas and lower crustal xenoliths. D, 1987, Massachusetts Institute of Technology. 324 p.

Reid, Reginald E. Geologic hazards in a portion of East Flagstaff, Coconino County, Arizona. M, 1975, Northern Arizona University. 120 p.

Reid, Robert R. Some Devonian sections in southeastern Arizona, and their correlation. M, 1928, University of Arizona.

Reid, Rolland R. Crystalline rocks of the northern Tobacco Root Mountains, Madison County, Montana. D, 1959, University of Washington. 179 p.

Reid, Rolland R. Petrography and petrology of the rocks in the Fish Lake area, South Eastern Wallowa Mts., Oregon. M, 1953, University of Washington. 118 p.

Reid, Ruth Pamela. Apatite in a glacial lake. M, 1979, University of British Columbia.

Reid, Ruth Pamela. The facies and evolution of an Upper Triassic reef complex in northern Canada. D, 1985, University of Miami. 468 p.

Reid, S. A. Microfauna and zonation of Naco Formation (Pennsylvanian) from selected sections in South-central Arizona. M, 1973, University of Arizona.

Reid, Sarah R. Petrology, paleodepositional environment, and paleoecology of the Lower Devonian Ross Formation, west-central Tennessee. M, 1983, University of Tennessee, Knoxville. 143 p.

Reid, Stephen Anthony. Depositional environments of the Vaqueros Formation along upper Sespe Creek, Ventura County, California. M, 1979, California State University, Northridge. 129 p.

Reid, Steven Graham. Geology and ore deposits of the Mines Plomosas District, Chihuahua, Mexico. M, 1972, University of Texas at El Paso.

Reid, Steven K. Preliminary assessment of cementation and dolomitization of the Meagher Formation (Middle Cambrian), Southwest Montana. M, 1986, University of Idaho. 68 p.

Reid, William McCormick. Active faults in Houston, Texas. D, 1973, University of Texas, Austin. 146 p.

Reid, William McCormick. Geology and fracture patterns of west central Burnet County, Texas. M, 1968, University of Texas, Austin.

Reid, William Thomas. Clastic limestone in the upper Eagle Ford Shale, Dallas County, Texas. M, 1952, Southern Methodist University. 14 p.

Reid-Green, John Douglas. Determination of roundness and mineralogical maturity of quartz grains from six environments. M, 1983, Northeast Louisiana University. 83 p.

Reidel, Stephen Paul. The geology of the Serpent Mound cryptoexplosion structure. M, 1972, University of Cincinnati. 150 p.

Reidel, Stephen Paul. The stratigraphy and petrogenesis of the Grande Ronde Basalt in the lower Salmon and adjacent Snake River canyons. D, 1978, Washington State University. 415 p.

Reidenouer, David Raymond. The relationship between sulfur distribution and paleotopography in three selected coal seams of western Pennsylvania. M, 1966, Pennsylvania State University, University Park. 196 p.

Reider, Eugene R. Ostracoda of the Florena Shale of Kansas. M, 1952, University of Nebraska, Lincoln.

Reider, Richard Gary. A geomorphological and pedological interpretation of the White River Plateau, Colorado. D, 1971, University of Nebraska, Lincoln. 233 p.

Reidy, Denis. Three-dimensional modeling and interpretation of Bouguer gravity anomalies of plutonic intrusions in the northern Mississippi Embayment. D, 1986, St. Louis University. 282 p.

Reif, Douglas M. Depositional environments and paleogeography of the red bed members, Moenkopi Formation, Nevada. M, 1978, Arizona State University. 263 p.

Reif, Henry Ernest, Jr. Some factors effecting sediment composition of very coarse and granule size fractions; Saint Francis River, Southeast Missouri. M, 1975, Southern Illinois University, Carbondale. 79 p.

Reifenstein, Mark. The hydrochemistry of ground waters occurring in the Carbondale Group, Southwest Indiana. M, 1980, Indiana University, Bloomington. 123 p.

Reifenstuhl, Rocky. A geologic and geophysical study of the Goddard Hot Springs area, Baronof Island, S. E. Alaska. M, 1983, University of Alaska, Fairbanks. 113 p.

Reighard, Kenneth Frederick. A portion of the Rome Fault of Northwest Georgia. M, 1963, Emory University. 65 p.

Reihman, Mary Ann. Petrified pteridosperm foliage from Iowa coal balls. D, 1977, University of Iowa. 160 p.

Reijenstein d'Acierno, Carlos Enrique. Stratigraphy and economic geology of uranium-bearing sediments in the Poison Spider District, Natrona County, Wyoming. M, 1969, University of Missouri, Rolla.

Reik, Barry. The Tertiary stratigraphy of Clay County, Florida with emphasis on the Hawthorne Formation. M, 1981, Florida State University.

Reik, Gerhard Albert. Deformation modes in Caledonia gypsum (Silurian, Ontario), under confining pressure and constant strain. M, 1968, University of Toronto.

Reik, Gerhard Albert. Joints, microfractures and residual strain in Cardium Siltstone, South Ram River area, Alberta; a field and experimental investigation of factors that contribute to fracture porosity and permeability in sedimentary rock. D, 1973, University of Toronto.

Reike, Herman H., III. Rock mechanics applied to the solution of slope stability in the Santa Monica (California) Slates (Triassic). M, 1965, University of Southern California.

Reilinger, Robert Eric. Vertical crustal movements in seismically active areas of the United States from leveling observations. D, 1978, Cornell University. 162 p.

Reilkoff, Brian Rory. Processing reflection seismic data from the Athabasca Basin. M, 1982, University of Saskatchewan. 230 p.

Reilly, Brian Arthur. Structural analysis of the Paint Lake deformation zone, northern Ontario. M, 1988, Brock University. 189 p.

Reilly, Edgar Milton, Jr. Origins and distributions of North American birds. D, 1954, Cornell University. 329 p.

Reilly, George Alexander. An estimate of the composition of part of the Canadian Shield in northwestern Ontario. M, 1965, McMaster University. 91 p.

Reilly, George Alexander. Partitioning of Mn and Co between ZnS and FeS_2 as a function of temperature. D, 1977, University of Michigan. 161 p.

Reilly, James Francis, II. Rb/Sr geochronology of the granitic intrusives of the Hobbs Coast region, Marie Byrd Land, West Antarctica. M, 1987, University of Texas at Dallas. 76 p.

Reilly, James Patrick. Station position determination from correlated satellite observations in the NGS/DOD (BC-4) worldwide network. D, 1974, Ohio State University.

Reilly, Joseph Michael. A geologic and potential field investigation of the central Virginia Piedmont. M, 1980, Virginia Polytechnic Institute and State University.

Reilly, Maurine Brigid. The stratigraphy and petrography of the middle member of the Luning Formation, West-central Nevada. M, 1979, Texas Christian University. 45 p.

Reilly, Mercedes Catherine. The metamorphism of Precambrian diabase dikes in southeastern Pennsylvania (Montgomery County, Pennsylvania). M, 1969, Bryn Mawr College. 17 p.

Reilly, Stephen Moran. Biochemical systematics and evolution of the eastern North American newts, genus Notophthalmus. D, 1986, Southern Illinois University, Carbondale. 86 p.

Reilly, Thomas Eugene. Analysis of saltwater upconing beneath a pumping well. D, 1986, Polytechnic University. 181 p.

Reim, Kenneth Maurice. Washington salines as a ceramic flux. M, 1951, University of Washington. 65 p.

Reimann, Martha Campbell. Ground-water resources of Fulton County, Ohio. M, 1979, University of Toledo. 82 p.

Reimchen, Theodore Frederick Harold. Pleistocene mammals from the Saskatchewan gravels in Alberta, Canada. M, 1968, University of Alberta. 99 p.

Reimer, G. Michael. Fission-track geochronology; method for tectonic interpretation of apatite studies with examples from the central and southern Alps (Tertiary). D, 1972, University of Pennsylvania.

Reimer, Gerda Elise. The sedimentology and stratigraphy of the southern basin of glacial Lake Passaic, New Jersey. M, 1984, Rutgers, The State University, New Brunswick. 204 p.

Reimer, James Denis. A laboratory and numerical study of crystalline rock-water interaction. M, 1980, University of Waterloo.

Reimer, L. J. Petrology of Republican, Smoky Hill, and Kansas River sands near Junction City, Kansas. M, 1975, Kansas State University. 118 p.

Reimer, Louis R. Stratigraphy, paleohydrology and uranium deposits of Church Rock Quadrangle, McKinley County, New Mexico. M, 1969, Colorado School of Mines. 254 p.

Reimers, Clare Elizabeth. Sedimentary organic matter; distribution and alteration processes in the coastal upwelling region off Peru. D, 1982, Oregon State University. 219 p.

Reimers, David D. Coccoliths in the surface sediments of the Louisiana continental shelf (Pleistocene-Recent, Gulf of Mexico). M, 1972, Texas A&M University.

Reimers, David D. The calcareous nannoplankton of the Midway Group (Paleocene) of Alabama. D, 1976, Tulane University. 154 p.

Reimers, Richard F. Geology, collapse mechanisms and prediction of collapsible soils in El Llano, New Mexico. M, 1986, New Mexico Institute of Mining and Technology. 166 p.

Reimnitz, Erk. Late Quaternary history and sedimentation of the Copper River Delta and vicinity, Alaska. D, 1966, University of California, San Diego. 188 p.

Reinarts, Mary Susan. Depositional environment of Upper Cretaceous Lewis sandstones, Sand Wash Basin, Colorado. M, 1981, Texas A&M University. 91 p.

Reinbold, Grover. Mineralogy of (Mississippian) Degonia, (Pennsylvanian) Pounds, and (Pennsylvanian) Boskydell sandstones (Illinois). M, 1961, Southern Illinois University, Carbondale. 59 p.

Reinbold, Mark Lester. Late Devonian conodont biostratigraphy, Las Vegas Range, Clark County, Nevada. M, 1977, University of Illinois, Urbana.

Reinecke, Kurt M. Structural geology of Smith-Weasel creeks area, Lewis and Clark County, Montana. M, 1984, Kansas State University. 97 p.

Reinecke, Leopold. A contribution to the petrography of the Ithaca Drift (New York). M, 1909, Cornell University.

Reinecke, Leopold. The geology and ore deposits of the Beaverdell map area, British Columbia. D, 1914, Yale University.

Reinelt, Erhard Rudolph. The propagation of anomalous sound from large explosions. D, 1964, University of Alberta. 220 p.

Reiner, Steven Roy. Some effects of water discharge through the Mount Enterprise fault zone on surface water composition in southern Rusk County, Texas. M, 1988, Stephen F. Austin State University. 153 p.

Reinert, S. L. Environment of deposition and diagenetic history of upper Dakota Group rocks, Denver Basin. M, 1975, University of Missouri, Columbia.

Reinertsen, David Louis. Small spores from the Summit Coal of Boone County, Missouri. M, 1953, University of Missouri, Columbia.

Reinertsen, Robert Wessley. Radium content of a core sample taken from East Sound. M, 1947, University of Washington. 33 p.

Reinhard, Mahlon J. A., III. Study of the Ste. Genevieve Formation at selected localities in southern Illinois and western Kentucky. M, 1964, University of Missouri, Rolla.

Reinhardt, Edward Wade. Petrology of the Numabin Bay intrusions, Reindeer Lake, Saskatchewan. M, 1962, University of Saskatchewan. 96 p.

Reinhardt, Edward Wade. Phase relations of cordierite, garnet, biotite, and hypersthene in highhigh-grade pelitic gneisses of the Gananoque area, Ontario. D, 1965, Queen's University. 218 p.

Reinhardt, Juergen. Stratigraphy, sedimentology and Cambro-Ordovician paleogeography of the Frederick Valley, Maryland. D, 1973, The Johns Hopkins University.

Reinhart, Phillip Wingate. Geology of the South Mountain oil field, Ventura County, California. M, 1928, Stanford University. 62 p.

Reinhart, Phillip Wingate. Tertiary Arcidae of the Pacific slope of North America. D, 1933, Stanford University. 293 p.

Reinhart, Roy H. A review of the Sirenia and the Desmostylia. D, 1952, University of California, Berkeley. 321 p.

Reinhart, Roy H. Halianassa ossivalensis; a sirenian from the Pliocene Bone Valley Formation, Florida. M, 1947, University of Chicago. 25 p.

Reinhart, Wilbur Allen. Harmonic analysis of telluric contents in the period range of 6 to 120 seconds. M, 1962, University of Utah. 119 p.

Reinhart, William Robert. The nature and origin of local and regional variations in manganese nodules in the western Pacific. D, 1979, Washington State University. 144 p.

Reinholtz, Philip N. Distribution, petrology and depositional environment of "Bush City Shoestring Sandstone" and "Centerville Lagonda Sandstone" in Cherokee Group (Middle Pennsylvanian), southeastern Kansas. M, 1982, University of Iowa. 180 p.

Reining, Joseph Bradley. Variation of seismic velocity through the crust in Hardeman County, Texas. M, 1977, University of Texas at Dallas. 90 p.

Reinink-Smith, Linda M. The mineralogy, geochemistry and origin of bentonite partings in the Eocene Skookumchuck Formation, Centralia Mine, southwestern Washington. M, 1982, Western Washington University. 179 p.

Reinke, Charles Austin, Jr. Mechanical analysis of glacial till from the Marsailles, Iroquois, and Chatsworth moraines in Illinois and Indiana. M, 1954, University of Michigan.

Reinke, Robert Edward. Surface wave propagation in the Tularosa and Jornada del Muerto basins, South central New Mexico. D, 1978, Southern Methodist University. 148 p.

Reinking, Robert Louis. Geochemical prospecting for fluorspar in southern Illinois, using associated heavy metals. M, 1965, University of Illinois, Urbana.

Reinking, Robert Louis. Geochemistry of the Iron Mountain magnetite deposit, Albany County, Wyoming. D, 1967, University of Illinois, Urbana. 107 p.

Reinold, Marvin L. Coal resources of eastern Logan and southern Johnson counties (Arkansas). M, 1953, University of Arkansas, Fayetteville.

Reinsbakken, Arne. Detailed geological mapping and interpretation of the Grand Forks-Eholt area (Precambrian-Eocene), Boundary District, British Columbia. M, 1971, University of British Columbia.

Reinsch, Thomas Glynn. Genesis and lateral variability of sand- and silt-mantled soils in north central Oklahoma. D, 1982, Oklahoma State University. 66 p.

Reinson, Gerald Edward. The Middle Devonian Lower Prairie evaporite formation of central Saskatchewan. M, 1970, University of Saskatchewan. 117 p.

Reinthal, Carol Ann Armstrong. A stable isotope, petrographic, and fluid inclusion investigation of the Pioche mining district, Nevada. M, 1983, University of Wisconsin-Madison. 166 p.

Reinthal, William Arthur. Geochemical evolution of precious metal mineralization in the Cracker Creek District of the Blue Mountains, northeastern Oregon. D, 1986, University of Wisconsin-Madison. 165 p.

Reis, James Martin. Lithostratigraphy of Cambro-Ordovician rocks and radionuclide analysis of associated waters, northeastern Oklahoma. M, 1984, University of Oklahoma. 83 p.

Reisberg, Laurie Ceil. The isotopic and geochemical systematics of the Ronda ultramafic complex of southern Spain. D, 1988, Columbia University, Teachers College. 248 p.

Reiser, Allan R. Occurrence, paragenesis, and microscopic features of certain ores of the San Francisco mining district, Utah. M, 1934, University of Utah. 59 p.

Reiser, Samuel G. A petrographic study of the Patula Arkose as exposed in Sierra de la Gavia, Coahuila, Mexico. M, 1964, Louisiana State University.

Reiser, Wendel. The application of the Stremme-Ostendorff method of soil mapping to an area near Stuttgart, Germany. M, 1952, Columbia University, Teachers College.

Reishus, Mark. The Newcastle Formation (Lower Cretaceous) in the Williston Basin of North Dakota. M, 1967, University of North Dakota. 139 p.

Reising, Gayle Angela. A paleoenvironmental study based on Ostracoda of the Byram Formation (Oligocene) in central Mississippi. M, 1986, Northeast Louisiana University. 190 p.

Reiskind, Jeremy. Paleontology and stratigraphy of the Niobrara Formation (Upper Cretaceous) of eastern North Dakota with emphasis on the calcareous nannoplankton. D, 1986, University of North Dakota. 477 p.

Reisland, Jack N. Lithology of the Upper Cretaceous Frontier Formation, Dubois and Cumberland Gap areas, Wyoming. M, 1958, Miami University (Ohio). 83 p.

Reiswig, Kenneth N. Palynological differences between the Chuckanut and Huntingdon formations, northwestern Washington. M, 1982, Western Washington University. 61 p.

Reisz, A. Colbert. Electromagnetic probing of the lunar interior in the frequency region from one kilohertz to ten kilohertz. M, 1970, Massachusetts Institute of Technology. 37 p.

Reisz, R. Petrolacosaurus kansensis Lane, the oldest known diapsid reptile. D, 1975, McGill University.

Reitenbach, Jacob Andrew. Variation among Oligocene equids from the Big Badlands, South Dakota. M, 1975, South Dakota School of Mines & Technology.

Reiter, Bruce E. Controls on lead-zinc skarn mineralization, Iron Cap Mine area, Aravaipa District, Graham County, Arizona. M, 1980, Arizona State University. 46 p.

Reiter, David Ernest. Geology of Alamo Hueco and Dog Mountains, Hidalgo County, New Mexico. M, 1980, University of New Mexico. 100 p.

Reiter, Frederick Howard. Geology of a portion of the Gallatin Range in southwestern Montana. M, 1950, University of Michigan. 42 p.

Reiter, Jessie Oscar. Geology of Hitchcock Field, Galveston County, Texas. M, 1958, University of Houston.

Reiter, Leon. An investigation into the time term method in refraction seismology. M, 1968, University of Michigan.

Reiter, Leon. Rayleigh wave attenuation in the 15 to 50 second period range and regional models of the intrinsic attenuation of shear waves in the crust and uppermost mantle of North America. D, 1971, University of Michigan.

Reiter, Marshall Allan. Terrestrial heat flow and thermal conductivity in southwestern Virginia. D, 1969, Virginia Polytechnic Institute and State University.

Reiter, Martin. Seasonal variations in intertidal foraminifera of Santa Monica Bay, California. M, 1957, University of Southern California.

Reiter, Michael Anthony. The relationship between the development of benthic stream algae and the hydrodynamics associated with rough substrates. D, 1988, University of Virginia. 147 p.

Reith, Howard Cartnick. A study of earth science teachers and practices in North Dakota for the academic year 1968-1969. D, 1969, University of North Dakota. 68 p.

Reitz, Alison. The geology and petrology of the northern San Emigdio plutonic complex, San Emigdio Mountains, Southern California. M, 1986, University of California, Santa Barbara. 80 p.

Reitz, Bruce K. Evolution of Tertiary plutonic and volcanic rocks near Ravenna, Granite County, Montana. M, 1980, Kansas State University. 90 p.

Reitz, Donald D. Analysis of joint patterns in the Pikes Peak and Cripple Creek Granite of Rock Creek, El Paso County, Colorado. M, 1971, Louisiana State University.

Reitzel, John S. Studies of heat-flow at sea. D, 1961, Harvard University.

Rejas, Angel. Geology of the Cerros de Amado area, Socorro County, New Mexico. M, 1965, New Mexico Institute of Mining and Technology.

Rejhon, G. The limestone conglomerate of the Quebec Group. M, 1957, McGill University.

Rejon, G. A study of the Ordovician conglomerates near Matane, Quebec. M, 1957, McGill University.

Reker, Carl Caspar. A stratigraphic study of the Morrison and Cloverly formations in a part of central Wyoming. M, 1947, University of Missouri, Columbia.

Relf, Carolyn Diane. The formation and alteration history of a series of shear zones, Mirage Islands, Yellowknife Bay, Northwest Territories. M, 1988, Memorial University of Newfoundland. 350 p.

Reller, Gregory Joseph. Structure and petrology of the Deer Peaks area, western North Cascades, Washington. M, 1986, Western Washington University. 106 2 plates p.

Relly, B. H. A method for determining the solubility of sulphides. M, 1957, McGill University.

Relly, B. H. The geology of Buchans Mine, Newfoundland. D, 1960, McGill University.

Remenyi, Miklos Tamas. Geology of the Texas Canyon area, Los Angeles County, California. M, 1966, University of California, Los Angeles.

Remick, David Brear. A study of the producing formations in the oil fields of the United States east of the Rocky Mountains. M, 1942, University of Texas, Austin.

Remick, Jerome H., III. Sedimentation and stratigraphy of the Ross Mountain area, New Quebec. M, 1953, Michigan Technological University. 95 p.

Remington, Donald Bryce. Precambrian rocks of the Whistlers Mountain trail map-area, Jasper. M, 1960, University of Alberta. 54 p.

Remington, Newell C. A history of the gilsonite industry. M, 1958, University of Utah. 338 p.

Remondi, Benjamin William. Using the Global Positioning System (GPS) phase observable for relative geodesy; modelng, processing, and results. D, 1984, University of Texas, Austin. 324 p.

Remsen, Walter E., Jr. Geology of the San Nicolas Bank, southern California continental borderland. M, 1982, California State University, Northridge. 135 p.

Remson, Irwin. Hydrologic studies at Seabrook, New Jersey. D, 1954, Columbia University, Teachers College. 171 p.

Remy, Robert Reginald. Diagenesis of the mid-Middle Park Formation, central Grand County, Colorado. M, 1984, University of Colorado. 87 p.

Remy, Victor Felix Barua *see* Barua Remy, Victor Felix

Remz, Stuart R. The composition variations in some marine lutites. M, 1971, University of Rochester. 39 p.

Ren, Xiaofen. Healed microfracture orientations in granites from the Basin and Range Province, western Utah and eastern Nevada and their relationship to paleostresses. M, 1988, Brigham Young University. 20 p.

Renard, Vincent Paul Augustine. Virgin Bank; correlation of magnetism and gravity with geology. D, 1967, Rice University. 174 p.

Renaud, Charles Benham. Geology of China Draw area, Culberson and Reeves counties, Texas. M, 1950, University of Texas, Austin.

Renault, Jacques Roland. The geological conditions of molybdenite deposition as deduced from textural analysis. D, 1964, University of Toronto. 215 p.

Renault, Jaques R. The growth pressure of fibrous sodium chloride. M, 1960, New Mexico Institute of Mining and Technology. 59 p.

Renbarger, K. Scott. A crustal structure study of South America. M, 1984, University of Texas at El Paso.

Rencz, A. N. The relationship between heavy metals in the soil and their accumulation in various organs and plants growing in the Arctic. D, 1978, University of New Brunswick.

Renda, Christine A. Groundwater contamination in parts of the Denver-Arapahoe disposal site. M, 1984, Colorado State University. 106 p.

Render, Francis William. Groundwater development and electric and digital modelling of the upper carbonate aquifer, metropolitan Winnipeg area, Manitoba. M, 1970, University of Manitoba.

Renders, Peter Joseph Norbert. Aqueous phase compositions in equilibrium with the assemblage quartz-kaolinite-beryl at elevated temperatures. M, 1985, University of Toronto.

Rendigs, R. Slope erosion and sediment contribution within the Willow Creek watershed. M, 1977, University of Waterloo.

Rendina, Michael A. Provenance, petrography and diagenesis of some selected sandstones and mudrocks of the Dunkard Group (Upper Pennsylvanian-Permian) in Ohio, West Virginia and Pennsylvania. M, 1985, Miami University (Ohio). 143 p.

Rendon-Herrero, Oswald. A method for the prediction of washload (soil erosion) in certain small watersheds. D, 1972, Virginia Polytechnic Institute and State University. 104 p.

Rendu, Jean-Michel Marie. Some applications of statistics to decision making in mineral exploration. D, 1971, University of Colorado. 276 p.

Rene Rodriguez E. *see* Rodriguez E. Rene

Rene, Raymond Morgan. Magnetotelluric fields with H-polarization; effects of topography, anistropy, relief on a highly resistive or conductive substrate, and lateral variation in conductivity. D, 1973, University of Utah. 304 p.

Reneau, Steven Lee. Depositional and erosional history of hollows; application to landslide location and frequency, long-term erosion rates, and the effects of climatic change. D, 1988, University of California, Berkeley. 339 p.

Reneer, Bernal. Development of the Wheat (Cherry Canyon) oil field, Loving County, Texas. M, 1985, Baylor University. 74 p.

Renfro, Arthur R. Depositional history of the Elmont Limestone. M, 1963, Kansas State University. 68 p.

Renfro, Harold Bell. Lower Pennsylvanian sediments of northeastern Oklahoma. D, 1947, University of Wisconsin-Madison.

Renfro, Kenneth McDonald. The petrographic character of the Bostwick Member of the Dornick Hills Formation in the Ardmore Basin, Oklahoma. M, 1950, University of Oklahoma. 56 p.

Renfro, Millicent Aloyse. A stratigraphic study of the Graham Formation, Upper Pennsylvanian, of Jack County, Texas. M, 1942, Texas Christian University. 83 p.

Renfro, William C. Radioecology of ^{65}Zn in an arm of the Columbia River estuary. D, 1968, Oregon State University. 88 p.

Renfroe, Charles Albert. The subsurface Mississippian rocks of the northwestern Anadarko Basin. M, 1948, University of Oklahoma. 28 p.

Renick, B. Coleman. The correlation of the (Pennsylvanian) lower Allegheny-Pottsville section in western Pennsylvania. D, 1922, University of Chicago. 33 p.

Renick, Howard, Jr. Magnetotelluric investigations in the area of the Tobacco Root mountains, southwestern Montana; southwestern Montana; and southern Illinois, Indiana, and Ohio. D, 1969, Indiana University, Bloomington. 103 p.

Renier, Joseph Maurice. High-angle fault trends of the Idaho-Wyoming Overthrust Belt salient. M, 1982, Idaho State University. 99 p.

Renk, R. R. Naturally occurring organic compounds found in Hyrum Reservoir, Utah. D, 1977, Utah State University. 172 p.

Renke, Daniel Felix. Geology of a part of the Newbury Park Quadrangle, Ventura County, California. M, 1957, University of California, Los Angeles.

Renken, Paul. The geology of the Montgomery Creek mining district and the R & S Mine area, Benton Quadrangle, California-Nevada. M, 1980, University of Nevada. 47 p.

Renken, Robert A. A hydrologic budget and regression model study for the Four Mile-Seven Mile Creek Basin, Ohio. M, 1976, Miami University (Ohio). 181 p.

Renkin, Miriam L. Age, depth, and residual depth anomalies in the North Pacific; implications for thermal models of the lithosphere and upper mantle. M, 1986, University of Texas, Austin. 118 p.

Renne, Paul De *see* De Renne, Paul

Renne, Paul Randall. Permian to Jurassic tectonic evolution of the eastern Klamath Mountains, California. D, 1987, University of California, Berkeley. 132 p.

Rennebaum, Thomas D. The geology and geochemistry of the North Shore Group of volcanic rocks, northeastern Minnesota. M, 1978, Bowling Green State University. 101 p.

Renner, James M. Placement and paleogeographic significance of Permian-Triassic boundary, central and Southeast Wyoming. M, 1988, University of Wyoming. 135 p.

Renner, Joel L. The petrology of the contact rocks of the Duluth complex (Precambrian), Dunka river area,

Minnesota. M, 1969, University of Minnesota, Minneapolis. 81 p.

Renner, Rebecca. The geology and geochemistry of the Issineru Formation, northern Guyana. M, 1986, Cornell University. 252 p.

Renner, Richard Eugene. A study of the deposition of the Glen Rose Limestone in Parker County, Texas. M, 1961, Texas Christian University. 47 p.

Renney, Kenneth Michael. The Miocene Temblor flora of west-central California. M, 1972, University of California, San Diego.

Rennick, Walter Lee. A model of lead isotope evolution in the Earth's crust and upper mantle (less systematics and better isotope evolution). M, 1973, University of California, Santa Barbara.

Rennie, Douglas Paul. Late Pleistocene alpine glacial deposits in the Pine Forest Range, Nevada. M, 1987, University of Nevada. 54 p.

Renny, Edward. Glacial water levels in the Quinsigamond River valley, Massachusetts. M, 1957, Clark University.

Reno, Duane Hugh. Weathering at the Precambrian Paleozoic contact along the east side of the Front Range, Colorado. M, 1939, University of Colorado.

Rensberger, John Marshall. A new genus of platanistid cetacean from the Miocene of California. M, 1961, University of California, Berkeley. 75 p.

Rensberger, John Marshall. Entoptychine gophers of the John Day Formation, Oregon. D, 1967, University of California, Berkeley. 297 p.

Rensburg, Willem Cornelius Janse Van *see* Van Rensburg, Willem Cornelius Janse

Renshaw, Ernest Wilroy. Pennsylvanian sediments in Northwest Georgia. M, 1951, Emory University. 66 p.

Renshaw, James L. Precambrian geology of the Thompson Peak area, Santa Fe County, New Mexico. M, 1984, University of New Mexico. 197 p.

Rensick, David Gene. The response of montmorillonite to elevated temperature and pressure. M, 1971, University of Oklahoma. 47 p.

Renton, John Johnson. An experimental investigation of sandstone lithification. D, 1966, West Virginia University.

Renton, John Johnston. Crystal growth as a factor in sandstone weathering. M, 1959, West Virginia University.

Renwick, Gregory K. Sea-floor spreading and the evolution of the continental margins of Atlantic Canada. M, 1973, Dalhousie University.

Renwick, Patricia Louise. Paleoecology of a floodplain lake in the Durham Sub-basin of the Deep River basin (Late Triassic), North Carolina. M, 1988, Emory University. 324 p.

Renwick, W. H. Short term sediment movements and related channel changes; the Piceance Basin, Colorado. D, 1979, Clark University. 178 p.

Reny, J. J. P. Gilles. Textural and chemical variations in a diabase dike, St.-Pierre de Wakefield, Québec. M, 1984, University of Ottawa. 138 p.

Renz, Genelle Winona. The distribution and ecology of Radiolaria in the central Pacific; plankton and surface sediments. D, 1973, University of California, San Diego. 262 p.

Renzetti, Bert Lionel. Geology and petrogenesis at Chuquicamata, Chile. D, 1957, Indiana University, Bloomington. 71 p.

Renzetti, Bert Lionel. Geology of the Scranton Mine area, Tooele County, Utah. M, 1952, Indiana University, Bloomington. 32 p.

Renzetti, Phyllis A. Some foraminifera from the (Upper Cretaceous) Prairie Bluff Chalk of Mississippi. M, 1950, Columbia University, Teachers College.

Renzetti, Phyllis Jean. Fauna of the Threeforks Shale (Devonian) of southwestern Montana. D, 1961, Indiana University, Bloomington. 342 p.

Renzulli, Michael J. Analysis of asbestos in drinking water using stain enhancement and the direct collec-

tion method. M, 1988, Rutgers, The State University, Newark. 111 p.

Repasky, Ted R. Magnetotelluric profile of the Jacobsville Sandstone. M, 1984, Michigan Technological University. 132 p.

Repecka, Albert Lee. Geology of the type Toro Formation (Lower Cretaceous), San Luis Obispo County, California. M, 1940, University of California, Berkeley. 50 p.

Repeta, Daniel James. Transformations of carotenoids in the oceanic water column. D, 1982, Woods Hole Oceanographic Institution. 241 p.

Repetski, J. E. Conodonts from the El Paso Group (lower Ordovician) of West Texas. D, 1975, University of Missouri, Columbia. 251 p.

Repetski, John E. Conodonts of the Dutchtown Formation of Missouri. M, 1973, University of Missouri, Columbia.

Repetti, William C. New values for some of the discontinuities in the Earth. D, 1928, St. Louis University.

Repetto, F. L. Geology of the Cojitambo area, province of Canar, Republic of Ecuador, South America. M, 1975, University of Wyoming. 84 p.

Repic, Randall L. Economic effects of coal mine subsidence in selected area of southwestern Indiana. M, 1987, Indiana State University. 84 p.

Repine, Thomas Edward, Jr. Examination of Upper Mississippian-Lower Pennsylvanian thickening sequences in Lewis County, West Virginia. M, 1981, Indiana University of Pennsylvania. 80 p.

Reppe, Calvin Clark. The geology of Devils Kitchen Quadrangle; Big Horn County, Wyoming. M, 1981, Iowa State University of Science and Technology. 78 p.

Requarth, Jeffrey S. The evolution of Guelph (Silurian) Dolomite multistory reefs, White Rock Quarry, Clay Center, Ohio. M, 1978, Bowling Green State University. 221 p.

Requejo, Adolfo G. Geochemistry of biogenic alkenes in estuarine sediments. D, 1983, University of Rhode Island. 253 p.

Requist, Norris N. A study of the Madison Limestone in and around the Bighorn Basin, Wyoming. M, 1949, University of Kansas. 81 p.

Reshkin, Mark. Geomorphic history of the Jefferson Basin, Jefferson, Madison and Silverbow counties, Montana. D, 1963, Indiana University, Bloomington. 146 p.

Reshkin, Mark. Subsurface geology of the Chester Series in the Tri-county region, Pike, Gibson, and Warrick counties, Indiana. M, 1958, Indiana University, Bloomington. 35 p.

Resig, Johanna Martha. Ecology of foraminifera of Santa Cruz Basin, California. M, 1956, University of Southern California.

Resio, Donald Thomas. An integrated model of storm-generated waves and surges. D, 1974, University of Virginia. 270 p.

Resler, Ray Chester. Comparison of theoretical and observed regional gravity anomalies over the Wichita-Amarillo system and adjacent features, southwestern Oklahoma and the Texas Panhandle. M, 1955, University of Utah. 31 p.

Resley, William E. Ultrasonic determination of elastic properties of the olivine, $(Mg, Fe)_2SiO_4$, solid solution series. M, 1979, Pennsylvania State University, University Park. 38 p.

Resnick, Alan J. Petrological analysis of the Butler Hill-Breadtray Granite Pluton, St. Francois Mountains, southeastern Missouri. M, 1984, University of Toledo. 169 p.

Reso, A. Geology of the June Canyon region, Toquima Range, central Nevada. M, 1955, Columbia University, Teachers College.

Reso, Anthony. The geology of the Pahranagat Range, Lincoln County, Nevada. D, 1960, Rice University. 345 p.

Resser, Kurt Douglas. Petrography, diagenesis and depositional environments of the Codell Sandstone and Juana Lopez members of the Carlile Shale (Upper Cretaceous), south-central Colorado. M, 1976, University of Nebraska, Lincoln.

Ressetar, Robert M. Major and minor element geochemistry of the Holyoke Basalt, Connecticut Triassic Basin. M, 1976, University of South Carolina.

Ressetar, Robert M. Paleomagnetism, ages and compositions of Phanerozoic dike swarms, Egyptian Eastern Desert. D, 1979, University of South Carolina. 106 p.

Ressmeyer, Paul F. Biostratigraphy and depositional environments of the Hampton Formation, Lower Mississippian (Kinderhookian) of central Iowa. M, 1983, University of Iowa. 171 p.

Restrepo Mejia, Jorge I. A surface and ground water model for the conjunctive use of a stream-aquifer system. D, 1987, Colorado State University. 242 p.

Restrepo, Jorge J. Geology and mineral deposits of the Rio Hondo-Red River divide, Taos County, New Mexico. M, 1972, Colorado School of Mines. 132 p.

Restrepo, Juan F. A geochemical investigation of Pleistocene to Recent calc-alkaline volcanism in western Panama. M, 1987, University of South Florida, Tampa. 103 p.

Retelle, Michael James. Glacial geology and Quaternary marine and lacustrine stratigraphy of the Robeson Channel area, northeastern Ellesmere Island, N.W.T., Canada. D, 1985, University of Massachusetts. 243 p.

Retelle, Michael James. Surficial geology of the southern half of the Bernardston Quadrangle, Massachusetts and Vermont. M, 1979, University of Massachusetts. 98 p.

Retherford, Robert Morse. Late Quaternary geologic environments and their relation to archaeological studies in the Bella Bella-Bella Coola Region of the British Columbia Coast. M, 1972, University of Colorado.

Rettenmaier, Karl Albert. Provenance and genesis of the Mesquite Flat Breccia; Superstition volcanic field, Arizona. M, 1984, Arizona State University. 175 p.

Rettew, David Mark. Paleoenvironmental study of the Chaffee Group (Upper Devonian) in North-east Gunnison County, Colorado. M, 1978, University of Kentucky. 85 p.

Rettger, Robert E. Soft rock deformation. M, 1923, University of Wisconsin-Milwaukee.

Rettke, R. C. Clay mineralogy and clay mineral distribution patterns in Dakota Group sediments, northern Denver Basin, eastern Colorado and western Nebraska. D, 1976, Case Western Reserve University. 147 p.

Rettke, Robert Clark. Some clay mineralogy; gross lithofacies relationships in Upper Devonian off-delta sediments of western and west-central New York. M, 1967, SUNY at Buffalo. 18 p.

Retty, Joseph Arlington. Geology of the township of Gaboury and Blondeau, Temiskaming County, Quebec. D, 1931, Princeton University. 64 p.

Reusch, Douglas N. The New World Island Complex and its relationships to nearby formations, north-central Newfoundland. M, 1983, Memorial University of Newfoundland. 248 p.

Reusing, Stephen P. Geology of the Graves Mountain area, Lincoln and Wilkes counties, Georgia. M, 1979, University of Georgia.

Reuss, Robert James. An investigation of the use of seismic methods to track meteorological disturbances over Lake Superior and to locate strain releases taking place in the Houghton-Hancock areas (Michigan). M, 1962, Michigan Technological University. 61 p.

Reuss, Robert Lester. Geology and petrology of the Wilson Park area, Fremont County, Colorado. D, 1970, University of Michigan. 184 p.

Reuss, Robert Lester. Mineralogy and petrology of the Gem Park carbonatites (Precambrian), Custer and Fremont counties, Colorado. M, 1967, University of Michigan.

Reuter, Stephen G. A gravity study of the Escalante Desert region, southwestern Utah. M, 1981, University of Texas at El Paso.

Reutinger, Charles Anton. The Pleistocene geology of the Thornville Quadrangle and the western portion of the Zanesville Quadrangle. M, 1941, Ohio State University.

Reutter, David. Sources and distribution patterns of the late Pleistocene-Holocene sands on the central Texas-Louisiana continental shelf. M, 1985, Texas A&M University.

Revel, Humbert S. A comparative study of meanders. M, 1956, New York University.

Revell, Stephen. Recent pattern changes on Rock Creek in southern Montana and their relationship to the threshold concept. M, 1974, Southern Illinois University, Carbondale. 71 p.

Revell, Steve. Heavy mineral content of some Pleistocene sands of Leon and Wakulla counties, Florida. M, 1958, Florida State University.

Revelle, Roger. Marine bottom samples collected in the Pacific Ocean by the Carnegie on its seventh cruise. D, 1936, University of California, Berkeley.

Revere, Althea. A survey of the application of the electron microscope to the study of clay minerals. M, 1944, Columbia University, Teachers College.

Reverman, Karla M. Mechanisms for neomorphism of the Pleistocene Anastasia Formation, Florida. M, 1985, Duke University. 58 p.

Reves, William Dickenson, Jr. The clay minerals of the North Carolina coastal plain. M, 1956, University of North Carolina, Chapel Hill. 46 p.

Revetta, Frank A. The till petrology of the Tazewell drifts of west-central Indiana. M, 1957, Indiana University, Bloomington.

Revetta, Frank Alexander. A regional gravity survey of New York and eastern Pennsylvania. D, 1970, University of Rochester. 447 p.

Revilla, Charles E. The geology of the Duffield-Stickleyville area, Virginia. M, 1952, University of Virginia. 105 p.

Revock, K. L. An investigation of nuclear detonations and solid Earth tides as possible triggering mechanisms for earthquakes in central California during the time period 1969 through 1970. M, 1974, University of Wisconsin-Milwaukee.

Revol, Jacques. Laboratory magnetic observations related to earthquake prediction. M, 1977, University of California, Santa Barbara.

Revol, Jacques. Magnetization of polycrystalline magnetite and rocks under stress, and implications for tectonomagnetism. D, 1979, University of California, Santa Barbara. 330 p.

Rex, Robert Walter. Quartz in sediments of Central and North Pacific Basin. D, 1958, University of California, Los Angeles. 110 p.

Rex-Pelkey, Ilene. The biostratigraphy and paleoecology of the western flanks of the Eugene Island Block 208 Salt Dome, Plio-Pleistocene section, offshore, Louisiana. M, 1985, University of Southwestern Louisiana. 187 p.

Rexford, E. P. Mineralization of the Andover and Sulphur Hill mines near Andover, New Jersey. M, 1927, Massachusetts Institute of Technology. 69 p.

Rexroad, Carl Buckner. An investigation of the Pennsylvanian channel-form sandstones of Boone County, Missouri. M, 1950, University of Missouri, Columbia.

Rexroad, Carl Buckner. Conodonts from the type Chester, Illinois. D, 1955, University of Iowa. 71 p.

Rexroad, Richard L. Stratigraphy, sedimentary petrology and depositional environments of tillite in the upper Precambrian Mount Rogers Formation, Virginia. M, 1978, Louisiana State University.

Rey, N. A. C. Textural and geothermometric aspects of sodium distribution in feldspars from two areas in the Grenville Province, Quebec. M, 1978, University of Ottawa. 141 p.

Rey, Walter T. Casquino see Casquino Rey, Walter T.

Reyer, Robert Winslow. Virgil cyclothems of the Holder Formation in Beeman and Indian Wells canyons in the Sacramento Mountains, northwest of Alamogordo, New Mexico. M, 1958, University of Wisconsin-Madison. 65 p.

Reyes C., Ignacio A. Geology and uraniferous mineralization in the Sierra de Coneto, Durango, Mexico. M, 1985, University of Texas at El Paso.

Reyes, A. Guitron de los see Guitron de los Reyes, A.

Reyes, Benjamin Panganiban. Relations of permeability to particle size of homogeneous and heterogeneous sands and gravels. M, 1964, University of Idaho. 44 p.

Reyes, Carlos. Dynamic decomposition of a seismic trace with application to time-varying filtering. M, 1985, University of Houston.

Reyes-Garces, Rafael Armando. The geology of Libya and its oil fields. M, 1967, University of California, Los Angeles.

Reyes-Navarro, Jaime. Pyrite in coal; its forms and distribution as related to the environments of coal deposition in three selected coals from western Pennsylvania. M, 1976, Pennsylvania State University, University Park. 141 p.

Reyner, Millard Lester. Geology of the Tidal Wave mining district, Madison County, Montana. M, 1947, Montana College of Mineral Science & Technology. 27 p.

Reynolds, Charles B. Geology of the Hagan-La Madera area, Sandoval County, New Mexico. M, 1954, University of New Mexico. 82 p.

Reynolds, Daniel M. Origin and formation of clastic plugs and dikes of Union County, New Mexico. M, 1979, Wichita State University. 22 p.

Reynolds, David E. The discharge of South Carolina streams as it relates to link magnitude. M, 1972, University of South Carolina. 59 p.

Reynolds, David James. Structural and dimensional repetition in continental rifts. M, 1984, Duke University. 175 p.

Reynolds, David Johnson. Foreland deformation of the Precambrian Streerwitz Thrust in Allamoore, Texas. M, 1988, University of Texas at Dallas. 172 p.

Reynolds, Douglas Wade. Geology and porosity of a Devonian reservoir in the Wilfred gas storage field, Sullivan County, Indiana. M, 1987, Indiana University, Bloomington. 106 p.

Reynolds, Douglas Wade. Subsurface geology of the eastern part of Spencer County, Indiana. M, 1965, Indiana University, Bloomington. 52 p.

Reynolds, Edward B. A three-dimensional model study of acoustic wave propagation in a borehole through fractured and non-fractured media. M, 1968, Colorado School of Mines. 109 p.

Reynolds, J. Rex. Geology of southwestern Leavenworth County, Kansas. M, 1957, University of Kansas. 78 p.

Reynolds, J. W. Mafic and ultramafic rocks near Goodwater, Alabama; Coosa County, Alabama. M, 1973, University of Alabama.

Reynolds, James Howard, III. Chronology of Neogene tectonics in the Central Andes (27°-33°S) of western Argentina based on the magnetic polarity stratigraphy of foreland basin sediments. D, 1987, Dartmouth College. 353 p.

Reynolds, James Howard, III. Tertiary volcanic stratigraphy of northern Central America. M, 1977, Dartmouth College. 89 p.

Reynolds, Jeffrey. The petrology and geochemistry of basalts from the Mariana Trough. M, 1982, Cornell University. 79 p.

Reynolds, John Lawrence. A linear programming application to the two-dimensional inverse gravity problem. M, 1978, Indiana University, Bloomington. 193 p.

Reynolds, John Maurice. Sedimentary study of the Ocate-Entrada sandstone in Mora and San Miguel

counties, New Mexico. M, 1959, Mississippi State University. 61 p.

Reynolds, Martin Bruce. Pleistocene molluscan faunas of the Humboldt deposit, Ross County, Ohio. M, 1958, Ohio State University.

Reynolds, Merrill Johnson. The Decorah Formation of east-central Missouri. M, 1952, Washington University. 75 p.

Reynolds, Michael Anthony. A study of the age of the elements. D, 1970, University of Arkansas, Fayetteville. 62 p.

Reynolds, Mitchell William. Stratigraphy and structural geology of Titus and Titanothere canyons area, Death Valley (Inyo county), California. D, 1969, University of California, Berkeley. 310 p.

Reynolds, Nelly A. Trenton Formation (Ordovician). M, 1896, University of Kentucky.

Reynolds, Peter Herbert. A lead isotope study of ores and adjacent rocks. D, 1967, University of British Columbia.

Reynolds, R. L. Paleomagnetism of the Yellowstone tuffs and their associated air-fall ashes. D, 1975, University of Colorado. 300 p.

Reynolds, Richard Alan. Cenozoic cornutellid biostratigraphy and paleooceanography from Deep Sea Drilling Project core 77B of Leg 9 (Eastern Equatorial Pacific). M, 1978, Rice University. 88 p.

Reynolds, Richard Alan. Neogene radiolarian biostratigraphy and paleoceanography of the Northwest Pacific. D, 1979, Rice University. 218 p.

Reynolds, Richard L. Paleomagnetism of Pleistocene volcanic air-fall ashes; correlation to sources as a means of dating Pleistocene deposits. M, 1970, University of Colorado.

Reynolds, Robert Coltart, Jr. The petrology of some Archean rocks from the Kragero District of South Norway. D, 1955, Washington University. 171 p.

Reynolds, Robert Ramon. A comparative study of the quartz pebbles in the LaFayette gravel and the Caseyville Conglomerate of southern Illinois. M, 1942, University of Illinois, Urbana.

Reynolds, Robin Raible. Mesozoic lithofacies and their tectono-stratigraphic controls; eastern cratonic margin of the Neuquen back-arc basin, Argentina. M, 1987, North Carolina State University. 100 p.

Reynolds, S. J. Styles of deformation in windows and slide blocks of the Roberts Mountains thrust belt, central Nevada. M, 1977, University of Arizona. 166 p.

Reynolds, Sargent Thurber. Geology of the northern half of the Banack Quadrangle, Beaverhead County, Montana. M, 1962, University of California, Berkeley. 54 p.

Reynolds, Shawn Arvin. Depositional development and fluvial architecture of the Narrabeen Group, Illawarra District, Sydney Basin, Australia. M, 1988, University of Texas, Austin. 157 p.

Reynolds, Stephen James. Geology and geochronology of the South Mountains, central Arizona. D, 1982, University of Arizona. 240 p.

Reynolds, Stephen Kempster. Chemical and mineralogical nature of phosphate nodules from the Upper Cretaceous of north central Texas. M, 1988, University of Texas, Arlington. 92 p.

Reynolds, Suzanne. The fabrics of deep-sea detrital muds and mudstones; a scanning electron microscope study. D, 1988, University of Southern California.

Reynolds, Theodore James. Variations in hydrothermal fluid characteristics through time at the Santa Rita porphyry copper deposit, New Mexico. M, 1980, University of Arizona. 52 p.

Reynolds, Thomas B. Geology and regional tectonic significance of Blanca Peak area, Huerfano County, Colorado. M, 1986, Wichita State University. 165 p.

Reynolds, Thomas Emmett. New fossil mammals from the lowermost Paleocene of New Mexico. D, 1935, [St. Louis University].

Reynolds, W. D. Column studies of strontium and cesium transport through a granular geologic porous medium. M, 1978, University of Waterloo.

Reynolds, William Daniel. The Guelph Permeameter method for in situ measurement of field-saturated hydraulic conductivity and matrix flux potential. D, 1987, University of Guelph.

Reynolds, William Francis. Geology of South Palo Pinto Creek area, Eastland County, Texas. M, 1953, University of Texas, Austin.

Reynolds, William Jerome. Beach dynamics and the societal response to beach erosion at Marco Island, Florida. D, 1982, Rutgers, The State University, New Brunswick. 346 p.

Reynolds, William Kennedy. Distribution and paleoecology of benthic foraminifera in the Tallahatta Formation in Alabama. M, 1981, Auburn University. 177 p.

Reynolds, William Roger. Formation of cristobalite, zeolite and clay minerals in the Paleocene and lower Eocene of Alabama. D, 1966, Florida State University. 322 p.

Reynolds, William Roger. The lithostratigraphy and clay mineralogy of the Tampa-Hawthorn Sequence of Peninsular Florida. M, 1962, Florida State University.

Rezabek, Dale Henry. The chemical behavior of heavy metals at the water-sediment interface of selected streams in Maine based on ternary partitioning diagrams. M, 1988, Michigan State University. 130 p.

Rezaie, Nasser M. The hydrogeology of the Boone-St. Joe Aquifer of Benton County, Arkansas. M, 1979, University of Arkansas, Fayetteville.

Rezaie, Patricia Ann Jameson. Palynological correlation of upper Chesterian and Morrowan strata of Northwest Arkansas with their western European equivalents. M, 1980, University of Arkansas, Fayetteville.

Rezak, Richard. Stromatolites of the Belt Series in Glacier National Park and vicinity, Montana. D, 1957, Syracuse University. 94 p.

Rezak, Richard. The Summit Park structure, Jefferson County, Missouri. M, 1949, Washington University. 43 p.

Rezayat, Mohsen. Direct boundary integral equation methods for a class of earth-structure interaction problems. D, 1985, University of Kentucky. 219 p.

Rezende, Servulo Batista de *see* de Rezende, Servulo Batista

Rezigh, A. A. Two-dimensional front tracking model for microcomputers. M, 1988, University of Oklahoma. 100 p.

Rhea, Keith Pendleton. Geology of the Red Mountain area, Lake and Mendocino counties, California. M, 1966, University of California, Berkeley. 54 p.

Rheams, Karen F. The petrography and structure of the Mitchell Dam Amphibolite and the surrounding Higgins Ferry Formation, Chilton and Coosa counties, Alabama. M, 1982, University of Alabama. 124 p.

Rhee, Robert Weston Von *see* Von Rhee, Robert Weston

Rhett, Douglas William. Phase relationships and petrogenetic environment of Precambrian granites of the New Jersey Highlands. D, 1975, Rutgers, The State University, New Brunswick. 166 p.

Rhindress, Richard C. Structure of the Martinsburg Formation-Schoehary Ridge Syncline. M, 1966, Lehigh University.

Rhine, Janet L. Sedimentological and geomorphological reconstruction of the late Pleistocene Athabasca fan-delta, Northeast Alberta. M, 1984, University of Calgary.

Rhinehart, Julie. Lithofacies and paleoenvironments of Guelph-Lockport Group (Middle Silurian) subsurface of western Pennsylvania. M, 1979, SUNY, College at Fredonia. 109 p.

Rhoades, David Alan. The geology of the Perry Formation (Upper Devonian, Maine). M, 1963, University of Maine. 112 p.

Rhoades, James David. Interstratification tendencies of trioctahedral vermiculite and biotite produced by ion exchange equilibrations. D, 1966, University of California, Riverside. 178 p.

Rhoades, Matthew J. A structural analysis of a portion of the Montana fold and thrust belt near Canyon Ferry, Lewis and Clark County, Montana. M, 1984, Washington State University. 123 p.

Rhoades, Rendell. Evolution of the crayfish genus *Orconectes* section limosus. D, 1960, Ohio State University. 81 p.

Rhoads, Bruce Lane. Process and response in desert mountain fluvial systems. D, 1986, Arizona State University. 309 p.

Rhoads, G. H., Jr. Determination of aquifer parameters from well tides. M, 1976, Virginia Polytechnic Institute and State University.

Rhoads, Holly. Facies relationships of the Ingleside Formation in northern Colorado and southeastern Wyoming. M, 1987, Colorado School of Mines. 98 p.

Rhoads, Ray William. Mississippian rocks of northeastern Osage County, Oklahoma. M, 1968, University of Oklahoma. 87 p.

Rhode, David Ronald. History and petroleum potential of the Mississippian Leadville Formation in the western San Juan Basin and adjoining Four Corners platform area, New Mexico and Arizona. M, 1982, University of Colorado. 82 p.

Rhodehammel, Edward Charles. An interpretation of the pre-Pleistocene geomorphology of a portion of the Saginaw Lowland (Michigan). M, 1951, Michigan State University. 163 p.

Rhodenbaugh, Edward Franklin. Washington lignites and their utilization. M, 1915, University of Washington. 46 p.

Rhodes, Brady P. Kinematics, metamorphism, and tectonic history of the Spokane Dome mylonitic zone, southern Priest River Complex, northeastern Washington and northern Idaho. D, 1984, University of Montana. 142 p.

Rhodes, Brady P. Structure of the east flank of Kettle Dome, Ferry and Stevens counties, Washington. M, 1980, University of Washington. 97 p.

Rhodes, Dallas D. Geomorphology of two high-mountain streams, Lake County, Colorado. D, 1973, Syracuse University.

Rhodes, E. G. Three-dimensional analysis by refraction seismic methods of coastal sand bodies on the northeastern Massachusetts coast. M, 1971, University of Massachusetts.

Rhodes, Flora Lee. A petrologic analysis of the norite zone, Stillwater Complex, Montana. M, 1970, Case Western Reserve University. 148 p.

Rhodes, Howard Startup. The Mississippian stratigraphy of southern Alberta, Canada. M, 1957, Brigham Young University. 67 p.

Rhodes, James A. Stratigraphy and origin of the Pennsylvanian-Permian rocks of the Huerfano Park Quadrangle, Colorado. D, 1964, University of Michigan. 205 p.

Rhodes, James A. Stratigraphy and structural geology of the Buckley Mountain area, south central Wasatch Mountains, Utah. M, 1955, Brigham Young University. 57 p.

Rhodes, John Arthur. The depth interpretation of gravity data for drift-filled valleys in Erie County, Pennsylvania. M, 1980, Pennsylvania State University, University Park. 67 p.

Rhodes, Jonathan J. A reconnaissance of hydrologic nitrate transport in an undisturbed watershed near Lake Tahoe. M, 1985, University of Nevada. 254 p.

Rhodes, M. L. A paleomagnetic study of Precambrian rocks from the Anti-Atlas region, Morocco. M, 1976, Ohio State University. 128 p.

Rhodes, Mary Louise. The physical geology of an area near Humansville, Polk County, Missouri. M, 1939, University of Missouri, Columbia.

Rhodes, Randolph A. Characteristics of regraded coal spoils. M, 1981, Colorado State University. 136 p.

Rhodes, Rene George. Geology of the East Kirkland area (Ontario). M, 1940, Cornell University.

Rhodes, Richard L., Jr. The geology of the southwestern part of the Vernon, New York 7-1/2 minute Quadrangle. M, 1960, Syracuse University.

Rhodes, Richard Sanders, II. Mammalian paleoecology of the Farmdalian Craigmile and the Woodfordian Waubonsie local faunas, southwestern Iowa. D, 1982, University of Iowa. 131 p.

Rhodes, Rodney Charles. Volcanic rocks associated with the western part of the Mogollon plateau volcano-tectonic complex, southwestern New Mexico. D, 1970, University of New Mexico. 145 p.

Rial M., Jose Antonio. I, The Caracas, Venezuela earthquake of 1967; a multiple source event; II, Seismic waves at the epicenter's antipode. D, 1979, California Institute of Technology. 125 p.

Rial M., Jose Antonio. Seismic wave transmission across the Caribbean Plate; high attenuation on concave side of Lesser Antilles island arc. M, 1976, University of Michigan.

Ribaudo, Anthony J. The effect of manganese on the enstatite diopside solvus. M, 1979, Brooklyn College (CUNY).

Ribbe, Paul Hubert. An X-ray and optical investigation of the peristerite plagioclases. M, 1958, University of Wisconsin-Madison. 54 p.

Ribe, Neil Marshall. Towards a dynamic model of intraplate volcanism. D, 1981, University of Chicago. 210 p.

Ribeiro, Julio C. Mesozoic tectonic history of the Reconcavo Basin, Brazil. M, 1979, Syracuse University.

Riber, Joshua I. The behavior of thallium with respect to potassium and rubidium in coexisting micas and feldspars. M, 1967, Rice University. 25 p.

Ricart y Menendez, Fernando Osacar. The diagenesis of the axial corallite of Acropora cervicornis. D, 1977, Rensselaer Polytechnic Institute. 119 p.

Ricci, Armando T. Subsurface geology related to petroleum accumulation in Reno County, Kansas. M, 1955, Kansas State University. 49 p.

Ricci, Margaret Philleo. Geology, structure, and gravity of the northern margin of the Smartville Complex, northern Sierra Nevada Foothills, California. M, 1983, University of California, Davis. 130 p.

Ricciardi, Karen. A petrographic and chemical study of axinite and scapolite occurrences in the White Oaks mining district, Lincoln County, New Mexico. M, 1978, University of New Mexico. 128 p.

Riccio, Joseph F. Morphology of the Lower Cambrian Mesonacidae of the southern Marble Mountains, California. M, 1950, University of Southern California.

Riccio, Joseph Frank. Recent and upper Pleistocene sediments of the southwestern portion of Los Angeles County, California. D, 1965, University of Southern California.

Riccio, Luca Michelangelo. Stratigraphy and petrology of the peridotite-gabbro component of the western Newfoundland ophiolites. D, 1976, University of Western Ontario. 265 p.

Riccio, Luca Michelangelo. The Betts Cove ophiolite (Ordovician), Newfoundland. M, 1972, University of Western Ontario.

Rice, Benjamin John. Major crustal lineaments and the Rome Trough in West Virginia. M, 1983, Pennsylvania State University, University Park. 58 p.

Rice, Craig W. Aspects of the biogeochemistry of manganese in the photic zone. M, 1983, University of Hawaii at Manoa. 58 p.

Rice, Donald L. Trace element chemistry of aging marine detritus derived from coastal macrophytes. D, 1979, Georgia Institute of Technology. 159 p.

Rice, Dudley Dennison. Stratigraphy of Chinchaga and older Paleozoic formations (Devonian) of the Great Slave Lake area, southern Northwest Territories and northern Alberta (Canada). M, 1967, University of Alberta. 144 p.

Rice, Emery van Daell and Adams, M. Ian. High level silts and gravels. M, 1960, Bryn Mawr College. 27 p.

Rice, Glenn S. Silurian-Devonian stratigraphy of southern Madison, southern Garrard, and eastern Lincoln counties. M, 1962, University of Kentucky. 66 p.

Rice, H. J. Physiography of Cazenovia Valley (New York). D, 1881, Syracuse University.

Rice, H. M. Anthony. A (Pliocene) San Diego fauna in the Newhall Quadrangle, California. D, 1934, California Institute of Technology. 9 p.

Rice, H. M. Anthony. The geology and economic geology of the Cranbrook District, British Columbia. D, 1934, California Institute of Technology. 228 p.

Rice, Harington Molesworth Anthony. A survey of the general and economic geology of the east and south contact of the West Kootenay Batholith. M, 1931, University of British Columbia.

Rice, Jack Morris. A preliminary report on the phase equilibria and chemistry of metamorphosed impure dolomites, Ross Lake, Washington. M, 1972, University of Washington. 29 p.

Rice, Jack Morris. Progressive metamorphism of impure dolomite and mineral equilibria in the system CaO-MgO-SiO2-K2O-Al2O3-CO2-H2O; Part I, the Marysville aureole, Montana; Part II; The northern portion of the Boulder Batholith, Montana. D, 1975, University of Washington. 185 p.

Rice, James A. Studies on humus; I, Statistical studies on the elemental composition of humus; II, The humin fraction of humus. D, 1986, Colorado School of Mines. 315 p.

Rice, James E. Conodont biostratigraphy of the lower Horquilla Limestone (lower Derryan through lower Desmoinesian) in the New Well Peak section, Big Hatchet Mountains, Hidalgo County, New Mexico. M, 1985, New Mexico Institute of Mining and Technology. 191 p.

Rice, John Albert. Controls on silver mineralization in the Creede Formation, Creede, Colorado. M, 1984, Colorado State University. 150 p.

Rice, John B., Jr. Spatial and temporal landslide distribution and hazard evaluation analyzed by photogeologic mapping and relative dating techniques, Salt River Range, Wyoming. M, 1987, Utah State University. 129 p.

Rice, Jonathan Aaron. Sedimentology, provenance, and tectonic significance of the basal conglomerate of the Rainbow Gardens Member of the Miocene Horse Spring Formation, Lake Mead area, southeastern Nevada. M, 1987, University of Nevada, Las Vegas. 171 p.

Rice, Karen C. Hydrogeology and hydrochemistry of springs in Mantua Valley and vicinity, north-central Utah. M, 1987, Utah State University. 113 p.

Rice, Lee R. The treatment of iron and acid stream contamination in the Black Hills of South Dakota. M, 1970, South Dakota School of Mines & Technology.

Rice, M. Craig. Stylolitization mechanisms, permeability barriers and pressure solution generated cements; examples from the Lockport Formation (Middle Silurian) of Dundas, Ontario, and Nisku Formation (Upper Devonian) of central Alberta. M, 1986, McMaster University. 316 p.

Rice, Marion. Notes on the petrography of the Isle of Pines. M, 1918, Columbia University, Teachers College.

Rice, Phillip. The Bermuda Seamount; an investigation into the magnetic properties and the $^{40}Ar/^{39}Ar$ radiometric age of selected hydrothermally altered submarine flows. M, 1977, Dalhousie University.

Rice, R. M. Application of discriminant function analysis to geologic problems. M, 1973, University of Southern California. 122 p.

Rice, Randolph James. The hydraulic geometry of the lower portion of the Sunwapta River valley train, Jasper National Park, Alberta. M, 1979, University of Alberta. 160 p.

Rice, Randolph James. The sedimentology and petrology of the Okse Bay Group (Middle and Upper Devonian) on S.W. Ellesmere Island and North Kent Island in the Canadian Arctic Arhipelago. D, 1987, McMaster University. 769 p.

Rice, Raymond H. A study of the suspended load of the Trinity River, Texas. M, 1967, Rice University. 47 p.

Rice, Raymond Martin. Storm runoff from chaparral watersheds (San Dimas experimental forest), Southern California. D, 1970, Colorado State University. 162 p.

Rice, Roger F. Environments of deposition in the Mesaverde Formation (Upper Cretaceous), central Wyoming. M, 1970, University of Missouri, Columbia.

Rice, S. B. Diffusion and olivine growth in "FAMOUS" glasses. M, 1978, SUNY at Stony Brook.

Rice, Thomas John, Jr. Mineralogical transformations in soils derived from mafic rocks in the North Carolina Piedmont. D, 1981, North Carolina State University. 235 p.

Rice, Thomas L. Estimates of maximum past overburden for the Pierre Shale, Hayes area, South Dakota. M, 1987, Colorado School of Mines. 110 p.

Rice, W. Ralph. Water quality variation in Arkansas streams as influenced by the environment. M, 1976, University of Arkansas, Fayetteville.

Rice, William A. Geology of the Blind River-Spragge area, north shore of Lake Huron (Ontario). D, 1940, Yale University.

Rice, William David. Conodont zonation of the Battleship Wash Formation, late Mississippian, Arrow Canyon Range, Clark County, Nevada. M, 1971, University of Illinois, Urbana.

Rice, William Forrester. The systematics and biostratigraphy of the Brachiopoda of the Decorah Shale at St. Paul, Minnesota. M, 1985, University of Minnesota, Minneapolis. 142 p.

Rice, William N. The Darwinian theory of the origin of species. D, 1867, Yale University.

Rice, Winfield L. Paleolithic classifications. M, 1913, Columbia University, Teachers College.

Rich, Charles Clayton. Late Cenozoic geology of the lower Manawatu Valley, New Zealand. D, 1959, Harvard University.

Rich, D. W. Porosity in oolitic limestones. D, 1980, University of Illinois, Urbana. 195 p.

Rich, Douglas H. Economic geology of the Silvertip Basin, Sunlight mining region, Park County, Wyoming. M, 1974, Miami University (Ohio). 70 p.

Rich, Ernest I. Geology of the Wilbur Springs Quadrangle, Colusa and Lake counties, California. D, 1968, Stanford University. 122 p.

Rich, Ernest I. The geology of a portion of the area north of the Sandia Mountains, New Mexico. M, 1953, University of California, Los Angeles.

Rich, Frederick James. The origin and development of tree islands in the Okefenokee Swamp, as determined by peat petrography and pollen stratigraphy. D, 1979, Pennsylvania State University, University Park. 321 p.

Rich, Gretchen Remington. Heavy metal analysis of particulates in groundwater from South Dayton, Ohio, using X-ray fluorescence. M, 1986, Wright State University. 146 p.

Rich, John Lyon. Studies in the physiography of semi-arid regions. D, 1911, Cornell University.

Rich, John Lyon. Types of marginal glacial drainage in the Finger Lake region (New York). M, 1907, Cornell University.

Rich, Mark. Geology of the southern portion of the Pancake Summit Quadrangle, Nevada. M, 1956, University of Southern California.

Rich, Mark. Stratigraphic section and fusulinids of the Bird Spring Formation near Lee Canyon, Clark County, Nevada. D, 1959, University of Illinois, Urbana. 188 p.

Rich, Michael A. Foraminifera of the Graneros and Greenhorn formations (Upper Cretaceous) from one exposure near Sioux City, Iowa. M, 1975, Iowa State University of Science and Technology.

Rich, Patricia Vickers. Description of a new species of Neophrontops (family: Accipitridae) from the lower Pliocene of the Gypaetinae. M, 1969, Columbia University. 106 p.

Rich, Patricia Vickers. The history of Australia's non-passeriform birds; Part I, Antarctic dispersal routes, plate tectonics, and the origin of Australia's non-passeriform avifauna; Part II, The Dromornithidae, a family of large, extinct ground birds endemic to Australia, systematics and phylogenetic considerations. D, 1973, Columbia University. 1049 p.

Rich, Robert Alan. Fluid inclusions in metamorphosed Paleozoic rocks of eastern Vermont. D, 1975, Harvard University. 299 p.

Rich, Thomas B. Growth-form analysis and paleoecology of the corals of the Lake Ordovician through Mid-Silurian Fish Haven and Laketown formations, Bear River Range, North-central Utah. M, 1981, Utah State University. 209 p.

Rich, Thomas Hewitt. Deltatheridia, Carnivora, and Condylartha from the Paris Basin lower Eocene. M, 1967, University of California, Berkeley. 134 p.

Rich, Thomas Hewitt. Origin and history of the Erinaceinae and Brachyericinae (Mammalia: Insectivora) in North America. D, 1973, Columbia University. 327 p.

Richard, Benjamin Hinchcliffe. Geologic history of the intermontane basins of the Jefferson Island Quadrangle, Montana. D, 1966, Indiana University, Bloomington. 73 p.

Richard, G. A. Temporal and environmental variations in sediment accretion rates at Flax Pond, a Long Island salt marsh. M, 1976, SUNY at Stony Brook.

Richard, J. K. Characterization of a bedrock aquifer, Harpswell, Maine. M, 1976, Ohio State University. 145 p.

Richard, Linda R. Geochemical discrimination of the peraluminous Devonian-Carboniferous granitoids of Nova Scotia and Morocco. M, 1988, Dalhousie University. 346 p.

Richard, Paul Francois. A computer analysis of the flow of water and nutrients in agricultural soils as affected by subsurface drainage. D, 1988, University of British Columbia.

Richard, Pierre. La Méthode d'addition symbolique et la définition de la structure cristalline de l'ekanite de St-Hilaire, P.Q. D, 1971, Ecole Polytechnique.

Richard, Stephen Miller. Mesozoic shear zones, cooling, and Tertiary unroofing of the Harquahala Mountains, west-central Arizona. D, 1988, University of California, Santa Barbara. 312 p.

Richard, Stephen Miller. Structure and stratigraphy of the southern Little Harquahala Mountains, La Paz County, Arizona. M, 1983, University of Arizona. 154 p.

Richards, Adrian Frank. Geology, volcanology, and bathymetry of Isla San Benedicto, Mexico. D, 1957, University of California, Los Angeles. 225 p.

Richards, Alexander Moreno. Gravity analysis of the Tonto and Payson basins, central Arizona. M, 1987, Northern Arizona University. 162 p.

Richards, Arthur. Geology of the Kremmling area, Grand County, Colorado. D, 1941, University of Michigan.

Richards, Barry Charles. Uppermost Devonian and Lower Carboniferous stratigraphy, sedimentation, and diagenesis, southwestern district of Mackenzie and southeastern Yukon Territory. D, 1983, University of Kansas. 434 p.

Richards, Carrol A. Geology of a part of the Funeral Mountains, Death Valley National Monument, California. M, 1957, University of Southern California.

Richards, David Barton. Engineering geology of the proposed Straight Creek Tunnel, Clear Creek and Summit counties, Colorado. M, 1963, Colorado School of Mines. 131 p.

Richards, Edith Jean. Geology of the Fall Creek area, Snake River Range, Wyoming. M, 1948, University of Michigan.

Richards, Esther English. Fossil and Recent Turritellidae of the Pacific Coast of North America. M, 1920, University of California, Berkeley. 62 p.

Richards, Gene Edward. An investigation of the sedimentary processes between Scott Point, and the Marblehead Light, Catawba and Danbury townships, Ottawa County, Ohio. M, 1957, Ohio State University.

Richards, George L. Geology of the (Miocene) Santa Margarita Formation, San Luis Obispo County, California. M, 1933, Stanford University. 182 p.

Richards, Gordon Gwyn. Geology of the Ox Lake Cu-Mo porphyry deposit. M, 1974, University of British Columbia.

Richards, H. Glenn. Fenestrate Bryozoa of the Virgilian and Wolfcampian of Kansas. M, 1955, University of Kansas. 129 p.

Richards, James Anthony. Depositional history of the Sunniland Limestone (Lower Cretaceous) Racoon Point Field, Collier County, Florida. M, 1987, University of New Orleans. 175 p.

Richards, James K. The accuracies of stations in the super first-order triangulation at Cape Canaveral with respect to the density of geodimeter baselines in the net. M, 1962, Ohio State University.

Richards, Jeremy Peter. A fluid inclusion and stable isotope study of Keweenawan fissure-vein hosted copper sulphide mineralisation, Mamainse Point, Ontario. M, 1986, University of Toronto.

Richards, John Charles. Geology and ore deposits of the Pennsylvania Mine, Three Forks, Montana. M, 1947, Montana College of Mineral Science & Technology. 39 p.

Richards, Kenneth A. The engineering geology and relative stability of Mount Adams and parts of Walnut Hills and Columbia Parkway, Cincinnati, Ohio. M, 1982, University of Cincinnati. 111 p.

Richards, Kenneth C. Structural complications involving the McConnell Thrust near its southern termination, east of Beehive Mountain, southern Alberta foothills. M, 1987, University of Calgary. 141 p.

Richards, Mark Alan. Dynamical models for the Earth's geoid. D, 1986, California Institute of Technology. 329 p.

Richards, Matthew E. Subsurface geology of the Santa Clara Avenue oil field and the Las Posas area, Ventura Basin, California. M, 1986, Oregon State University. 58 p.

Richards, Michael L. A study of electrical conductivity in the earth near Peru. D, 1970, University of California, San Diego.

Richards, Nancy A. Structure in the Precambrian and Paleozoic rocks at Saint John, New Brunswick. M, 1971, University of New Brunswick. 73 p.

Richards, Paul Granston. A contribution to the theory of high frequency elastic waves, with applications to the shadow boundary of the Earth's core. D, 1970, California Institute of Technology. 290 p.

Richards, Paul William. Structural geology of the Crazy Mountain Syncline-Beartooth Mountain border east of Livingston, Montana. D, 1952, Cornell University. 81 p.

Richards, Peter Alan Leslie. The polarographic speciation of iron in a natural groundwater environment. M, 1982, Queen's University. 155 p.

Richards, R. Peter. Paleoecology of the brachiopod species of the Richmond group (upper Ordovician), southwestern Indiana and southeastern Ohio. D, 1970, University of Chicago. 244 p.

Richards, Susan Staben. A hydrogeologic study of South Russell and adjacent areas. M, 1981, Kent State University, Kent. 202 p.

Richards, Tom. Plutonic rocks (late Cretaceous–Miocene) between Hope and the 49th Parallel, southern British Columbia. D, 1971, University of British Columbia.

Richards, Trenton Hubert. Mineralogy and morphology of the Gruta Cuatro Palmas, Coahuila, Mexico. M, 1986, West Texas State University. 101 p.

Richardson, Albert L. Geology of the Mayoworth region, Johnson County, Wyoming. M, 1950, University of Wyoming. 59 p.

Richardson, Archie M. In-situ mechanical characterization of jointed crystalline rock. D, 1986, Colorado School of Mines. 543 p.

Richardson, Catherine Kessler. The solubility of fluorite in hydrothermal solutions. D, 1977, Harvard University.

Richardson, Clinton Preston. Mechanics of pore water uptake by overland flow. D, 1987, University of Kansas. 310 p.

Richardson, Darlene S. The origin of iron-rich layers in sediments of the western equatorial Atlantic Ocean. D, 1974, Columbia University. 213 p.

Richardson, David Bruce. Hydrographical description of the Ft. Pierce Inlet and the adjoining section of the Indian River. M, 1977, Florida Institute of Technology.

Richardson, David Newton. Relative durability of shale. D, 1984, University of Missouri, Rolla. 232 p.

Richardson, Donald A. Three dimensional geodesy and the solution of geometric problems. M, 1965, Ohio State University.

Richardson, Douglas Burton. The Indian Lands Study; an example of the application of geographic research to the analysis of complex energy and environmental policy issues. D, 1980, Michigan State University. 397 p.

Richardson, E. S. Opening of the Red Sea with two poles of rotation. M, 1976, University of Miami. 21 p.

Richardson, Eugene Stanley, Jr. A Middle Ordovician and some Lower Devonian conulariids, with two orthoceratids, from central Pennsylvania. M, 1942, Pennsylvania State University, University Park. 57 p.

Richardson, Eugene Stanley, Jr. Some fossil invertebrates of the Pennsylvanian Mazon Creek Coastal Plain, Illinois. D, 1954, Princeton University. 84 p.

Richardson, Everett Ellsworth. Geology of northern Boracho Quadrangle, Culberson County, Texas. M, 1950, University of Texas, Austin.

Richardson, G. B. A study of the red beds of the Black Hills of South Dakota and Wyoming. D, 1901, The Johns Hopkins University.

Richardson, G. N. The seismic design of reinforced earth walls. D, 1976, University of California, Los Angeles. 770 p.

Richardson, G. T. Environmental geology applied to highway site selection, West Lafayette, Indiana. M, 1976, Purdue University.

Richardson, George L. Geology and ore deposits of the Landusky mining district, Phillips County, Montana. M, 1973, University of Arizona. 64 p.

Richardson, Hibbard Ellsworth. The geology of the Sweet Home Petrified Forest. M, 1950, University of Oregon. 44 p.

Richardson, James Bushnell, III. The preceramic sequence and Pleistocene and post-Pleistocene climatic change in northwestern Peru. D, 1969, University of Illinois, Urbana. 329 p.

Richardson, Jean Madeline. Genesis of the East Kemptville greisen-hosted tin deposit, Davis Lake Complex, southwestern Nova Scotia, Canada. D, 1988, Carleton University.

Richardson, Jean Madeline. Geology and geochemistry of the East Kemptville greisen-hosted tin deposit, Yarmouth County, southwestern Nova Scotia. M, 1983, University of Toronto.

Richardson, Jennifer Lynn. An analysis of the sedimentary geology of the Jurassic Ralston Creek Formation as it is exposed in the vicinity of Canon City, Colorado. M, 1987, Oklahoma State University. 147 p.

Richardson, Jimmie Larry. Areal geology of western Washita County, Oklahoma. M, 1970, University of Oklahoma. 67 p.

Richardson, Jimmy David. Stratigraphy and depositional environment of the Wolfe City Formation (Upper Cretaceous), Northeast Texas. M, 1972, University of Texas, Arlington. 152 p.

Richardson, Keith Allan. The radioactivity, sites of alpha emitters, and radioactive disequilibrium in the Conway Granite of New Hampshire. D, 1963, Rice University. 90 p.

Richardson, Keith Allan. The thorium, uranium, and zirconium concentration in bauxites and their relationship to bauxite genesis. M, 1959, Rice University. 59 p.

Richardson, Larie Kenneth. Geology of the Alabama Hills, California. M, 1975, University of Nevada. 146 p.

Richardson, Larry J. A subsurface analysis of underground gas storage project in the Bagly-Hernson area of west central Iowa. M, 1977, Wichita State University. 73 p.

Richardson, Mark. Tectonic and stratigraphic evolution of the Neogene Gulf of Suez and northern Red Sea Rift; depositional environment and hydrocarbon source potential of evaporites. D, 1988, University of Rhode Island. 495 p.

Richardson, Mary Josephine. Composition and characteristics of particles in the ocean; evidence for present day resuspension. D, 1980, Woods Hole Oceanographic Institution. 237 p.

Richardson, Nathaniel R., Jr. Analysis of the magnetic anomaly map of Indiana. D, 1978, Purdue University. 180 p.

Richardson, Nathaniel Reginal, Jr. Geologic interpretation of selected anomalies within the Monrovia Quadrangle, Liberia, using an integration of geophysical methods. M, 1973, Michigan State University. 63 p.

Richardson, Paul W. Adsorption of copper on quartz. D, 1955, Massachusetts Institute of Technology. 62 p.

Richardson, Paul William. The Lost Horse Intrusives, Copper Mountain, British Columbia. M, 1950, University of British Columbia.

Richardson, Ralph O., Jr. A petrographic study of the Murphy marble and talc deposits at Murphy, North Carolina. M, 1973, University of South Carolina.

Richardson, Randall Miller. Intraplate stress and the driving mechanism for plate tectonics. D, 1978, Massachusetts Institute of Technology. 5 p.

Richardson, Raymond Moseley. Sedimentation and shore processes at Bolivar Peninsula, Galveston County, Texas. M, 1948, University of Texas, Austin.

Richardson, Remond W. Shoreline physiography of the Santa Lucia Mountains (California). M, 1924, Stanford University. 87 p.

Richardson, Rondald E. Petrography of Precambrian iron formation, Pembina County, North Dakota. M, 1975, University of North Dakota. 82 p.

Richardson, Stephen Hilary. Evolution of enriched mantle from derivative basalt, peridotite and diamond inclusion geochemistry. D, 1984, Massachusetts Institute of Technology. 191 p.

Richardson, Stephen Vance; Essere, Eric J. and Kesler, Stephen E. Origin and geochemistry of the Chapada Cu-Au deposit, Goias, Brazil; a metamorphosed wall rock porphyry copper deposit. M, 1984, University of Michigan.

Richardson, Steven M. Some volcanic rock types in a portion of Hull, Massachusetts. M, 1970, Boston University. 53 p.

Richardson, Steven McAfee. Fe-Mg exchange among garnet, cordierite, and biotite during retrograde metamorphism. D, 1975, Harvard University.

Richardson, Steven Michael. Stratigraphy and depositional environments of a marine-nonmarine Plio-Pleistocene sequence, western Salton Trough, California. M, 1984, San Diego State University. 112 p.

Richardson, Stuart. Physical-chemical and radiation properties of mountain streams. M, 1969, Colorado State University. 84 p.

Richardson, William S. Forecasting beach erosion along the oceanic coastlines of the northeast and mid-Atlantic states. M, 1977, College of William and Mary.

Richers, David Matthew. A hydrogeochemical study of the Corbin Sandstone Member of the Pennsylvanian Lee Formation in eastern Kentucky for uranium and base metal mineralization. D, 1980, University of Kentucky. 153 p.

Richers, David Matthew. The geochemistry of the Stillwater igneous complex, Montana. M, 1976, University of Kentucky. 110 p.

Richey, Charles Irwin. Canning Ridge Intrusion, Culberson County, Texas. M, 1961, University of Texas, Austin.

Richey, James M. The sedimentary environment of the Beach Swamp, and shoals of Cape Romano, Florida. M, 1961, Florida State University.

Richey, Jeffrey Edward. Phosphorus dynamics in Castle Lake, California. D, 1974, University of California, Davis. 162 p.

Richey, Joanna Sloane. Effects of urbanization on a lowland stream in western Washington. D, 1982, University of Washington. 248 p.

Richey, King Arthur. The marine invertebrate fauna of the Orinda Formation, and its stratigraphic relation to the Nannippus tehonensis Zone. M, 1939, University of California, Berkeley. 45 p.

Richey, Oleta May. Foraminifera of the Webberville Formation in Texas. M, 1928, University of Texas, Austin.

Richey, Ronald Glenn. Dolomitization in the North Adams oil field, Arenac County, Michigan. M, 1980, Michigan State University. 80 p.

Richey, Scott R. Geologic setting of the Blue Jacket volcanogenic, massive sulfide deposit, west-central Idaho. M, 1987, Colorado School of Mines. 126 p.

Richgels, Henry J. Geophysical reconnaissance in the Wisconsin part of Lake Superior Geosyncline. M, 1960, University of Wisconsin-Madison.

Richins, William D. Earthquake swarm near Denio, Nevada, February to April, 1973. M, 1974, University of Nevada. 57 p.

Richman, Lance Ramon. Sedimentary analysis of pre-Missoula gravels in the southeastern part of the Pasco Basin, Washington. M, 1981, Washington State University. 82 p.

Richmann, Debra L. Rb-Sr ages of the Red Mountain and Big Branch gneisses, Llano Uplift, Central Texas. M, 1977, University of Texas, Austin.

Richmond, Douglas P. Precambrian geology of Lake Plateau, Beartooth Mountains, Montana. M, 1987, Montana State University. 43 p.

Richmond, Gerald M. Quaternary stratigraphy of the La Sal Mountains, Utah. D, 1955, University of Colorado.

Richmond, James F. Petrology and structure of the San Bernardino Mountains north of Big Bear Lake, California. D, 1955, Stanford University. 145 p.

Richmond, James Frank. Geology of Burruel Ridge, northwestern Santa Ana Mountains, California. M, 1950, Pomona College.

Richmond, James Frank. Geology of Lucerne Valley Quadrangle (California). D, 1953, [Stanford University].

Richmond, R. N. Origin of magnetic anomalies over the northwestern portion of the Hawaiian Ridge. M, 1968, University of Hawaii. 64 p.

Richmond, Wallace Everett, Jr. Crystal chemistry of the phosphates, arsenates, and vanadates of the type $A_2XO_4(Z)$. D, 1939, Harvard University.

Richmond, William C. Grain size, organic carbon and trace metal dispersal patterns in Mugu Lagoon. M, 1976, California State University, Northridge. 114 p.

Richmond, William O. Paleozoic stratigraphy and sedimentation of the Slave Point Formation, southern Northwest Territories and northern Alberta. D, 1965, Stanford University. 565 p.

Richter, Alan. A volumetric analysis of Holocene sediments underlying present Delaware salt marshes inundated by Delaware Bay tides. M, 1974, University of Delaware.

Richter, Bernd Christian. Geochemical and hydrogeological characteristics of salt spring and shallow subsurface brines in the Rolling Plains of Texas and Southwest Oklahoma. M, 1983, University of Texas, Austin. 148 p.

Richter, Brian. Precipitation analysis for Piceance Basin, northwestern Colorado. M, 1982, Colorado State University. 187 p.

Richter, Brian E. Stratigraphy of the Frontier Sandstone (Upper Cretaceous), Vernal area, Uintah County, Utah. M, 1983, Colorado School of Mines. 86 p.

Richter, D. H. Mineralogy and origin of the Michipicoten iron formation. M, 1952, Queen's University.

Richter, Daniel deBoucherville, Jr. Prescribed fire; effects on water quality and nutrient cycling in forested watersheds in the Santee experimental forest in South Carolina. D, 1980, Duke University. 209 p.

Richter, Dennis Max. Periglacial features in the central Great Smoky Mountains. D, 1973, University of Georgia.

Richter, Frank M. Dynamical models for sea floor spreading. D, 1972, University of Chicago. 152 p.

Richter, Goetz M. Micro-lysimeter and field study of water and chemical movement through soil. M, 1984, University of California, Riverside. 161 p.

Richter, Henry Robert, Jr. Ground water resources in the part of Canyonlands National Park east of the Colorado River and contiguous Bureau of Land Management lands, Utah. M, 1980, University of Wyoming. 90 p.

Richter, Howard L. Response of plagioclase grain shape to metamorphism. M, 1970, Michigan State University. 111 p.

Richter, James B. and Breitenwischer, Robert. Geology of the northern half of the Marquand Quadrangle, Madison and Bollinger counties, Missouri. M, 1953, University of Michigan.

Richter, Joseph Gustav. Hydrogeology of a buried valley and adjacent aquifers in Southwest Pottawatomie County, Kansas. M, 1988, Kansas State University. 185 p.

Richter, Karen June. The propagation of Rayleigh surface waves through a nonhomogeneous region. D, 1987, University of Wisconsin-Madison. 173 p.

Richter, Michael A. A digital analysis of flexural waves propagating in freshwater lake ice. M, 1976, University of Wisconsin-Milwaukee.

Richter, Robert W. Areal geology of the Creta area, Jackson County, Oklahoma. M, 1960, University of Oklahoma. 126 p.

Richtmyer, Allan G. A subsurface study of the Middle Ordovician sequence in Morrow County, Ohio. M, 1965, Michigan State University. 92 p.

Riciputi, Lee Remo. Archean granulite metamorphism in the Nordlandet and Tasiusarsuaq terranes, southern West Greenland. M, 1987, University of Wisconsin-Madison. 70 p.

Rickard, Hilton L. A petrographic study of some typical igneous and metamorphic rocks of the Virginia Piedmont and Blue Ridge. M, 1937, University of Virginia. 79 p.

Rickard, Lawrence Vroman. Stratigraphy of the Upper Silurian Cobleskill, Bertie and Brayman formations of New York State. M, 1953, University of Rochester. 178 p.

Rickard, Lawrence Vroman. The Lower Devonian Helderbergian Series of central New York, a stratigraphic and paleoecological study. D, 1955, Cornell University.

Ricker, Karl Edwin. Quaternary geology in the southern Ogilvie Ranges, Yukon Territory and an investigation of morphological, periglacial, pedological and botanical criteria for possible use in the chronology of

morainal sequences. M, 1968, University of British Columbia.

Rickerich, Steven F. Sedimentology, stratigraphy, and structure of the Kittery Formation in the Portsmouth, New Hampshire, area. M, 1983, University of New Hampshire. 115 p.

Rickertsen, Mark Andrew. Chemical evaluation of source areas and flow paths for groundwaters in the vicinity of Cedar Bog, Champaign County, Ohio. M, 1988, Wright State University. 180 p.

Rickett, R. L. The action of oxygen and hydrogen sulphide upon iron chromium alloys at high temperature. D, 1933, University of Michigan.

Ricketts, B. D. Sedimentology and stratigraphy of eastern and central Belcher Islands, Northwest Territories. D, 1979, Carleton University. 314 p.

Ricketts, Brian M. A hydrogeochemical profile analysis of groundwater discharge zones within Cedar Bog, Champaign County, Ohio. M, 1987, Wright State University. 159 p.

Ricketts, Edward W. and Whaley, Keith Ray. Structure and stratigraphy of the Oak Ridge Fault-Santa Susana Fault intersection, Ventura Basin, California. M, 1975, Ohio University, Athens. 81 p.

Ricketts, James Edward. The geology of the Blockhouse Quadrangle, Blount County, Tennessee. M, 1942, University of Tennessee, Knoxville. 60 p.

Rickles, Sue E. Ecology, taxonomy and distribution of Holocene reefal ostracods, Veracruz, Mexico. M, 1975, University of Nebraska, Lincoln.

Rickman, Douglas L. A thermochemical study of the ore deposits of the Milliken Mine, New Lead Belt, Missouri. D, 1981, University of Missouri, Rolla. 336 p.

Rickman, Douglas Lee. Origin of celestite (strontium sulfate) ores in the Southwestern United States and northern Mexico. M, 1977, New Mexico Institute of Mining and Technology. 75 p.

Ricks, Cynthia L. Flood history and sedimentation at the mouth of Redwood Creek, Humboldt County, California. M, 1984, Oregon State University. 167 p.

Rico, B. S. Some geologic and exploration characteristics of porphyry copper deposits in a volcanic environment, Sonora, Mexico. M, 1975, University of Arizona.

Ricou, Michel L. Review of statistical techniques and their application to the earth sciences. M, 1979, Northeastern Illinois University.

Ricoult, Daniel Louis. Point defect-controlled creep and electron diffraction study of grain boundaries in olivine. D, 1984, Cornell University. 206 p.

Ricoy, Jose Ulises. Depositional systems in the Sparta Formation (Eocene) Gulf Coast basin of Texas. M, 1976, University of Texas, Austin.

Ridd, Merril Kay. A new land form map for Utah. M, 1960, University of Utah. 66 p.

Riddell, J. E. The geology of the Buffalo Ankerite gold mines, Ltd. (Porcupine District, Ontario). M, 1936, McGill University.

Riddell, John Evans. Wall rock alteration around base metal sulphide deposits of northwestern Quebec. D, 1953, McGill University.

Riddell, Walter. Some aspects of the rock bursts in the Kirkland Lake gold mines at Kirkland Lake, Ontario, Canada. M, 1936, New Mexico Institute of Mining and Technology. 77 p.

Riddle, J. M. Optimum mine environmental planning. D, 1977, Virginia Polytechnic Institute and State University. 196 p.

Rideg, Peter. Geology and structure of a portion of the Serra do Mar in eastern Sao Paulo, Brazil. D, 1974, SUNY at Binghamton. 156 p.

Ridenour, James. Depositional environments of the late Pleistocene American Falls Formation, southeastern Idaho. M, 1971, Idaho State University. 81 p.

Rider, Jonathan Richards. Bryozoa of the Sahul Shelf, northwest Australia. M, 1974, University of California, Berkeley. 362 p.

Ridge, John C. The surficial geology of the Great Valley section of the Ridge and Valley Province in eastern Northampton County, Pennsylvania, and Warren County, New Jersey. M, 1983, Lehigh University.

Ridge, John Charles. The Quaternary glacial and paleomagnetic record of the West Canada Creek and western Mohawk valleys of central New York. D, 1985, Syracuse University. 556 p.

Ridge, John D. A textural study of the beach sands of Cedar Point, Ohio, Lake Erie. M, 1932, University of Chicago. 45 p.

Ridge, John D. The genesis of the tri-state zinc and lead ores. D, 1935, University of Chicago. 69 p.

Ridgeway, David C. Geology of the Blaney, South Carolina, quadrangle. M, 1960, University of South Carolina. 66 p.

Ridgley, Jennie L. Chemical and mineral variation in the Lake Owen mafic complex, Albany County, Wyoming. M, 1972, University of Wyoming. 122 p.

Ridgley, Neill H. Precambrian rocks in the Blackhall Mountain area, Carbon County, Wyoming. M, 1971, University of Wyoming. 54 p.

Ridgway, Eric R. Mineralogy of the Steel Creek manganese deposit, Olympic Peninsula, Washington. M, 1986, Eastern Washington University. 70 p.

Ridgway, Jeffrey. Preparation and interpretation of a revised Magsat satellite magnetic anomaly map over South America. M, 1984, Purdue University.

Ridgway, Kenneth D. Geomorphic controls on facies transitions in the lower Mansfield Formation, southwestern Indiana. M, 1986, Indiana University, Bloomington. 145 p.

Ridgway, Lucille. Some laboratory methods in advanced mineralogy. M, 1925, Northwestern University.

Ridgway, Robert H. The geology of the principal manganese ore deposits of the world. M, 1934, George Washington University.

Ridland, G. Carmen. Mineralogy of the Negus and Con mines, Yellowknife, Northwest Territories. D, 1939, Princeton University. 32 p.

Ridler, Roland Hartley. Petrographic study of Crow Lake ultrabasic sill, Keewatin Volcanic Belt, northwestern Ontario, Canada. M, 1966, University of Toronto.

Ridler, Roland Hartley. The relationship of mineralization of volcanic stratigraphy in the Kirkland lake area, Ontario (Canada). D, 1969, University of Wisconsin-Madison. 168 p.

Ridley, Albert Paul. Devonian and Mississippian sedimentation and stratigraphy of the Mazourka Canyon area, Inyo Mountains, Inyo County, California. M, 1971, San Jose State University. 78 p.

Ridley, Kevin J. D. The non reproducibility of chemical analyses on reference rock standards and an evaluation of the cause of the problem. M, 1975, University of Windsor. 100 p.

Ridley, Susanne Larkin. Sedimentology and diagenesis of Pleistocene and Holocene limestones; an investigation of diagenetic alteration by an evaporated sea water brine, Salt Cay, Turks and Caicos islands, British West Indies. M, 1986, University of Calgary. 183 p.

Ridley, Wade Clark. Geology of the northwest corner of the Smithville Quadrangle, Bastrop County, Texas. M, 1955, University of Texas, Austin.

Ridlon, James Barr. Bathymetry and structure of San Clemente island, California, and tectonic implications for the Southern California continental borderland. D, 1970, Oregon State University. 246 p.

Ridlon, James Barr. Stratigraphy and paleontology of the Belden Formation (Pennsylvanian) in part of west central Colorado; Pitkin and Eagle counties. M, 1954, University of Colorado.

Rieb, Sidney L. Structural geology of the Nemaha Ridge in Kansas. M, 1954, Kansas State University. 40 p.

Riebessel, Ulf. Sinking and sedimentation characteristics of a diatom winter/spring bloom. M, 1987, University of Rhode Island.

Rieck, Richard Louis. The glacial geomorphology of an interlobate area in Southeast Michigan; relationships between landforms, sediments, and bedrock. D, 1976, Michigan State University. 216 p.

Riecken, Charles Christopher. Petrology of the Striped Rock Granite (Precambrian) and surrounding rocks, Grayson County, Virginia. D, 1966, Virginia Polytechnic Institute and State University. 212 p.

Riecker, Robert Edward. Geologic interpretation of hydrocarbon fluorescence. D, 1961, University of Colorado.

Rieckin, Charles Christopher. The thermoluminescent characteristics of Cenozoic calcite from Florida in relation to geologic occurrence. M, 1960, University of Florida. 57 p.

Riedell, Karl Brock. Geology and porphyry copper mineralization of the Fawn Peak intrusive complex, Methow Valley, Washington. M, 1979, University of Washington. 52 p.

Rieder, Milan. A study of natural and synthetic lithium-iron micas. D, 1968, The Johns Hopkins University. 272 p.

Riedesel, Mark Alan. Seismic moment tensor recovery at low frequencies. D, 1985, University of California, San Diego. 280 p.

Riediger, Cynthia Louise. Sedimentology and tectonic history of the Eureka Sound and Beaufort formations, southern Ellesmere Island, Arctic, Canada. M, 1985, University of British Columbia. 160 p.

Riedl, G. W. Geology of the eastern portion of Shirley Basin, Albany and Carbon counties, Wyoming. M, 1959, University of Wyoming. 49 p.

Riedle, Lisa Ann. Groundwater accounting for the Indian River Farms Water Management District of southeastern Florida. D, 1988, University of Alabama. 398 p.

Rieg, Louis Eugene. A petrographic and X-ray study of the dolomite distribution in certain Cambro-Ordovician in limestones of central and south-central Pennsylvania. D, 1958, University of Pittsburgh.

Riegel, Walter Leonard. Palynology of environments of peat formation (Recent) in southwestern Florida. D, 1965, Pennsylvania State University, University Park. 198 p.

Rieger, Samuel. Development of the A2 horizon in soils of the Palouse area. D, 1952, Washington State University. 107 p.

Riegler, Paul William. An aerial photogrammetric study of erosion and accretion, Boston Harbor islands, Massachusetts. M, 1981, University of Rhode Island.

Riehl, A. M. A water quality diffusion model by the finite element method for a continuous plume in a bounded channel. D, 1979, University of Louisville. 140 p.

Riehl, William George. Structural analysis in determination of the origin of the Canaan Mountain Formation (Cambrian (?), Connecticut). M, 1972, University of Wisconsin-Madison.

Riehle, James Donald. A study of scapolite in part of the Humboldt igneous complex, Nevada. M, 1969, Northwestern University.

Riehle, James Donald. Theoretical compaction profiles in ash-flow tuffs. D, 1970, Northwestern University. 191 p.

Rieke, Herman Henry, III. Compaction of Argillaceous sediments (20-500,000 PSI). D, 1970, University of Southern California.

Rieke, Herman Henry, III. Rock mechanics applied to the solution of slope stability problems in the Santa Monica slates. M, 1965, University of Southern California.

Riekels, Lynda M. Preferred orientation in a quartz mylonite from the Moine Thrust, Scotland. M, 1973, University of Illinois, Chicago.

Rieken, E. R. Computer generated fault surface determinations from earthquake foci; Washington State,

1969-1983. M, 1985, Washington State University. 126 p.

Riel, Stanley Joseph. A basal Oligocene local fauna from McCarty's Mountain, southwestern Montana. M, 1963, University of Montana. 74 p.

Riely, Samuel Leander. The petrology of the upper and lower block underclays of Clay, Owen, and Parke counties, Indiana. M, 1953, Indiana University, Bloomington. 40 p.

Ries, E. G., Jr. Distributary channel networks; field measurement, analysis and simulation modeling. D, 1975, University of California, Los Angeles. 416 p.

Ries, Edward Richard. A study of the Sasakwa Limestone Member of the Holdenville Formation. M, 1943, University of Oklahoma. 45 p.

Ries, Edward Richard. The geology of Okfuskee County, Oklahoma. D, 1951, University of Oklahoma. 213 p.

Ries, Gaila Vawn. Distribution and petrography of calcrete zones, southern High Plains, New Mexico and Texas. M, 1981, Texas Tech University. 158 p.

Ries, Heinrich. Monoclinic pyroxenes of New York State. D, 1896, Columbia University, Teachers College.

Ries, Minette Lillian. Fauna of the Glen Rose. M, 1929, University of Texas, Austin.

Riese, Arthur Carl. Adsorption of radium and thorium onto quartz and kaolinite; a comparison of solution/surface equilibria models. D, 1982, Colorado School of Mines. 292 p.

Riese, Ronald W. Precambrian geology of the southern part of the Rincon range (New Mexico). M, 1969, New Mexico Institute of Mining and Technology. 183 p.

Riese, Walter Charles. Geology and geochemistry of the Mount Taylor uranium deposit, Valencia County, New Mexico. M, 1977, University of New Mexico. 119 p.

Riese, Walter Charles. The Mount Taylor uranium deposit, San Mateo, New Mexico. D, 1980, University of New Mexico. 643 p.

Rieser, Robert B. Structural study of the Oak Ridge Fault between South Mountain and Wiley Canyon, Ventura County, California. M, 1976, Ohio University, Athens. 93 p.

Riesmeyer, William Duncan. Precambrian geology and ore deposits of the Pecos mining district, San Miguel and Santa Fe counties, New Mexico. M, 1978, University of New Mexico. 215 p.

Riess, C. Maurine. A thermal maturation study of Cretaceous petroleum source rocks in Presidio County, Texas. M, 1984, University of Texas at El Paso.

Riess, Stephani Kay. Structural geometry of the Charleston thrust fault, central Wasatch Mountains, Utah. M, 1985, University of Utah. 73 p.

Riestenberg, Mary M. Anchoring of thin colluvium on hillslopes in Cincinnati by roots of sugar maple and white ash. D, 1987, University of Cincinnati. 366 p.

Riestenberg, Mary M. The effect of woody vegetation on stabilizing slopes in the Cincinnati area, Ohio. M, 1981, University of Cincinnati. 79 p.

Riester, Debra. Relation of nearshore and shelf patterns of quartz sand shape variation to provenance and dynamics. M, 1980, University of South Carolina.

Rietman, Jan David. Remanent magnetization of the late Yakima Basalt (upper Miocene and lower Pliocene), Washington State. D, 1966, Stanford University. 87 p.

Rife, David Leroy. Barite fluid inclusion geothermometry, Cartersville mining district, Northwest Georgia. M, 1969, University of Tennessee, Knoxville. 67 p.

Riffelmacher, Wallace Edwin. Microspores of the lower Kittanning coal in its type area. M, 1952, Pennsylvania State University, University Park. 57 p.

Riffenburg, Harry Bucholz. Chemical character and alteration in ground waters of the northern Great Plains area. D, 1924, American University.

Rigbey, Stephen J. The effect of sorbed water on expansivity and durability of rock aggregates. M, 1980, University of Windsor. 171 p.

Rigby, F. A. Pressure ridge generated internal wave wakes at the base of the mixed layer in the Arctic Ocean. M, 1976, University of Washington. 35 p.

Rigby, J. K., Jr. Swain Quarry of the Fort Union Formation, middle Paleocene (Torrejonian), Carbon County, Wyoming; geologic setting and mammalian fauna. D, 1976, Columbia University, Teachers College. 305 p.

Rigby, J. Keith. Paleoecology of the (Permian) Delaware Mountain Group, Guadalupe Mountain area, Texas and New Mexico. D, 1952, Columbia University, Teachers College.

Rigby, J. Keith. Stratigraphy and structure of the Paleozoic rocks in the Selma Hills, Utah County, Utah. M, 1949, Brigham Young University. 108 p.

Rigby, James Gordon. The sedimentology, mineralogy and depositional environment of a sequence of Quaternary catastrophic flood-derived lacustrine turbidites near Spokane, Washington. M, 1982, University of Idaho. 132 p.

Rigby, M. A. Natural inventories as a foundation for an outdoor laboratory; a type example; Eidson Lake and Fairy quadrangles, Hamilton County, Texas. M, 1976, Baylor University.

Rigden, Sally Miranda. The determination of the equation of state of molten silicates at high pressures using shock-wave techniques. D, 1986, California Institute of Technology. 146 p.

Rigert, James Aloysius. Uniaxial and controlled-lateral strain tests on selected sedimentary rocks. D, 1980, Texas A&M University. 417 p.

Rigg, Charles Gordon. Paleontology and stratigraphy of a well penetrating Devonian, Silurian, and Ordovician rocks from Weare Township, Oceana County, Michigan. M, 1955, University of Michigan.

Rigg, David Michael. Relationships between structure and gold mineralization in Campbell Red Lake and Dickenson mines, Red Lake District, Ontario. M, 1980, Queen's University. 153 p.

Rigg, Robert Mader. Mineralization and rock alterations in the Nederland District, Colorado. M, 1937, University of Michigan.

Riggenbach, D. K. The effects of dredging on the stability of an ebb-tidal delta, Lynnhaven Inlet, Virginia. M, 1976, Old Dominion University. 112 p.

Riggert, Virginia Leigh. The geometry and mechanical development of the Heart Mountain thrust stack, Exshaw, Alberta. M, 1983, University of Calgary. 175 p.

Riggi, Salvador Anthony, Jr. A two dimensional solute transport model for pollutant flux in the vicinity of a landfill, Dayton, Ohio. M, 1983, Wright State University. 142 p.

Riggins, Earl Michael. The temporal-morphologic analysis of three Upper Devonian brachiopods from north-central Iowa. M, 1977, Pennsylvania State University, University Park. 178 p.

Riggins, Michael. The viscoelastic characterization of marine sediment in large-scale simple shear. D, 1981, Texas A&M University. 127 p.

Riggle, Mark Robbins. An environmental geologic study of Wright State University. M, 1981, Wright State University. 332 p.

Riggs, C. O. Shallow foundations on dense silty loessial soils. D, 1978, University of Missouri, Rolla. 277 p.

Riggs, Elliot Arthur. The Pennsylvanian Fusulinidae of the western Kentucky coal field. M, 1953, University of Wisconsin-Madison.

Riggs, Elliott Arthur. Fusulinids of the Keeler Canyon Formation, Inyo County, California. D, 1962, University of Illinois, Urbana.

Riggs, K. A., Jr. Pleistocene geology and soils in southern Iowa. D, 1956, Iowa State University of Science and Technology.

Riggs, Karl A., Jr. A facies study in the Gulf Coast Tertiary of Texas. M, 1952, Michigan State University. 32 p.

Riggs, Kenton Nile. A study of spontaneous potentials associated with Indiana coals. M, 1959, Indiana University, Bloomington. 39 p.

Riggs, Nancy Rosalind. Stratigraphy, structure, and mineralization of the Pajarito Mountains, Santa Cruz County, Arizona. M, 1985, University of Arizona. 102 p.

Riggs, Richard M. The (Ordovician) Chazyan section near South Hero, Vermont. M, 1947, Columbia University, Teachers College.

Riggs, Stanley. Strontium calcite of dolomitized limestone. M, 1962, Dartmouth College. 27 p.

Riggs, Stanley Robert. Phosphorite stratigraphy, sedimentation, and petrology of the Noralyn Mine, central Florida phosphate district. D, 1967, University of Montana. 267 p.

Riggsbee, Wade H. Use of chemical correlation techniques to determine groundwater flow patterns and aquifer characteristics in Bethel Township, Ohio. M, 1974, Wright State University. 178 p.

Rightmire, Craig Turner. A radiocarbon study of the age and origin of caliche deposits. M, 1967, University of Texas, Austin.

Rightmire, George Philip. Recent and subfossil human cranial variation from southern Africa; a re-assessment by multivariate statistical techniques. D, 1969, University of Wisconsin-Madison. 160 p.

Riglin, L. D. The Perpetual Landslide, Summerland, British Columbia. M, 1977, University of British Columbia.

Riglos, Mario Suarez see Suarez Riglos, Mario

Rigney, Harold William. The Middle Mississippian formations of North America. M, 1933, University of Chicago. 145 p.

Rigney, Harold William. The morphology of the skull of a young Galesaurus planiceps and related forms. D, 1937, University of Chicago. 71 p.

Rigo, Richard J. Middle and Upper Cambrian stratigraphy in the autochthon and allochthon of northern Utah. M, 1967, Brigham Young University.

Rigoti, Augustinho. Reduction of ambiguity in geoelectric models using multiple data sets. M, 1985, McGill University. 129 p.

Rigotti, Peter A. Aftershocks. M, 1972, Brown University.

Rigotti, Peter A. The paleomagnetism of the Palisade Sill and the development of the ARM correction method of paleointensity determination. D, 1976, University of Pittsburgh. 258 p.

Rigsby, George P. Glaciological studies in the St. Elias Range, Canada. M, 1950, California Institute of Technology. 31 p.

Rigsby, George P. Studies of crystal fabrics and structures in glaciers. D, 1953, California Institute of Technology. 51 p.

Rihani, Rushdi F. Geochemistry of Holocene salt marsh deposits in the vicinity of Sapelo Island, Georgia, U.S.A. D, 1971, University of Georgia. 223 p.

Rihani, Rushdi F. Variation in the mineralogy of heavy minerals on the south shore of Long Island, New York. D, 1968, University of Georgia.

Rihani, Rushdi Freih. Variation in the mineralogy of Recent marine sediments on the south shore of Long Island (New York). M, 1967, Wayne State University.

Rijo, Luiz. Modeling of electric and electromagnetic data. D, 1977, University of Utah. 257 p.

Rikard, Michael Wayne. Hydrologic and vegetative relationships of the Congaree Swamp National Monument. D, 1988, Clemson University. 113 p.

Rike, William M. A petrographic study of the upper sandstone units of the Sundance Formation (upper Jurassic) in the Dubois area, Fremont County, Wyoming. M, 1973, Miami University (Ohio). 85 p.

Riley, Alton O'Neil. Geology of the Doe Creek sandy limestone. M, 1961, University of Oklahoma. 70 p.

Riley, C. J. A gravity survey of the Kenora area, Ontario. M, 1965, University of Manitoba.

Riley, C. O. Geochemical analysis of clay seams from the La Posta quartz diorite, La Posta, San Diego County, California. M, 1978, San Diego State University.

Riley, Charles Marshall. A petrographic study of the Gunflint Formation near Loon Lake, Ontario. M, 1948, University of Minnesota, Minneapolis. 61 p.

Riley, Charles Marshall. The relations of the physical and chemical properties to the bloating of clays. D, 1950, University of Minnesota, Minneapolis. 53 p.

Riley, Christopher. The geology of the Rocky Mountains. M, 1929, University of British Columbia.

Riley, Christopher. The granite-porphyries of Great Bear Lake, Canada. D, 1934, University of Chicago. 58 p.

Riley, D. C. Wave equation synthesis and inversion of diffracted multiple seismic reflections. D, 1975, Stanford University. 108 p.

Riley, Francis Stevenson. Tiltmeter to measure surface subsidence around a pumping artesian well. M, 1960, Stanford University.

Riley, G. C. Geology of the Cumberland Sound area, Baffin Island (Northwest Territories). D, 1952, McGill University.

Riley, G. C. The bedrock geology of Makkovik and its relations to the Aillik and Kaipokok series (Precambrian, Labrador). M, 1951, McGill University.

Riley, Greg. Sedimentology and stratigraphy of a limestone pebble conglomerate at the base of the Late Triassic Chinle Formation, southern Nevada. M, 1987, Southern Illinois University, Carbondale. 180 p.

Riley, James J. The effects of saline solutions on the mineral composition of plants. D, 1968, University of Arizona.

Riley, James Lemuel. A regional study of the Lower Cretaceous Viking Formation of Alberta and Saskatchewan. M, 1955, Southern Methodist University. 50 p.

Riley, Jill K. Cation substitution in norsethite, $MgBa(CO_3)_2$. M, 1975, Miami University (Ohio). 15 p.

Riley, John A. Acid water implications for mine abandonment, Coeur d'Alene mining district, Idaho. M, 1985, University of Idaho. 148 p.

Riley, John M. Seismology–collection–interpretation. M, 1980, Emporia State University.

Riley, John Paul. Application of an electronic analog computer to the problems of river basin hydrology. D, 1967, Utah State University. 200 p.

Riley, Leonard B. Geology and ore deposits of Sierra Mojada, Coahuila, Mexico. D, 1936, Yale University.

Riley, Louise Anderson. A study of meteorites; a catalogue of South Carolina meteorites; descriptions and classification of meteorite specimens in the Geology Museum, University of South Carolina. M, 1963, University of South Carolina.

Riley, Michael James. A dynamic lake water quality model for the evaluation of lake treatment alternatives. D, 1988, University of Minnesota, Minneapolis. 277 p.

Riley, N. Allen. Structural petrology of the (Precambrian) Baraboo Quartzite (Wisconsin). D, 1947, University of Chicago. 22 p.

Riley, Paul E. Geology of an area along the Colorado River, T. 3 S., R. 85 W., Eagle County, Colorado. M, 1950, University of Colorado.

Riley, R. A. The character and origin of the Steeprock Buckshot (iron-ore from the Steeprock Mine at Atikokan, Ontario). M, 1969, Queen's University. 189 p.

Riley, Robert. Stratigraphic facies analysis of the upper Santa Fe Group, Fort Hancock and Camp Rice Formations, far West Texas and south-central New Mexico. M, 1984, University of Texas at El Paso.

Riley, Roger Ray. Pacific Northwest clays and shales for possible use lightweight aggregates. M, 1952, University of Washington. 72 p.

Riley, Ronald A. Stratigraphic and environmental significance of Silurian rocks, northwest Ohio. M, 1980, Bowling Green State University. 146 p.

Riley, Thomas Andrew. The petrogenetic evolution of a Late Jurassic island arc; the Rogue Formation, Klamath Mountains, Oregon. M, 1988, Stanford University. 40 p.

Rimal, Durga Nath. Mineralogy of rose muscovite and lepidolite from the Harding Pegmatite, Taos County, New Mexico. D, 1962, University of New Mexico. 94 p.

Rimando, Philip M. Design and implementation of filters for potential field data. M, 1987, University of Texas at El Paso.

Rimkus, Arvid J. A study of weathering in the Carajas region, Brazil. M, 1987, University of Western Ontario. 81 p.

Rimkus, Wayne Vincent. Development of novel fluorine combustion procedures for elemental coal analysis. D, 1986, Loyola University. 202 p.

Rimmer, Susan Margaret. Lateral and vertical variability in petrography and mineralogy of the lower Kittanning Seam, western Pennsylvania and eastern Ohio. D, 1985, Pennsylvania State University, University Park. 502 p.

Rimsaite, Yadvyga. Trace element study of coal seams, Cape Breton, Nova Scotia. M, 1953, Queen's University. 102 p.

Rimsnider, Donald Orin. Texture and mineralogy of beach and dune sands, southeastern shore of Lake Michigan. M, 1959, University of Illinois, Urbana.

Rimstidt, Daniel L. Water quality of an active strip-mining area in Greene County, Indiana. M, 1977, Indiana University, Bloomington. 31 p.

Rimstidt, James Donald. The kinetics of silica-water reactions. D, 1979, Pennsylvania State University, University Park. 148 p.

Rinaldi, Gianfrumco Giuseppe L. Intramolecular distribution of stable carbon isotopes in metabolic products. D, 1974, Indiana University, Bloomington. 44 p.

Rinaldi, Romano. Crystallography and chemistry of Li-Rb-Cs bearing micas from the Tanco (Chemalloy) pegmatite, Bernic lake, Manitoba. M, 1970, University of Manitoba.

Rinaldo-Lee, Marjory Beach. Hydrogeology and computer model of the Bass Lake area, Saint Croix County, Wisconsin. M, 1978, University of Wisconsin-Madison.

Rindsberg, Andrew K. Ichnology and paleoecology of the Sequatchie and Red Mountain formations (Ordovician-Silurian), Georgia-Tennessee. M, 1983, University of Georgia.

Rindsberg, Andrew Kinney. Distribution and preservation of biogenic sedimentary structures in the deep sea. D, 1986, Colorado School of Mines. 300 p.

Rine, James Marshall. Depositional environments and Holocene reconstruction of an argillaceous mud belt, Surinam, South America. D, 1980, University of Miami. 222 p.

Rinehart, Eric John. The use of microearthquakes to map an extensive magma body in the Socorro, New Mexico area. M, 1977, New Mexico Institute of Mining and Technology.

Rinehart, Jon. Petrography and mineralogy of some carbonatites of the Avon (Missouri) area. M, 1973, University of Nebraska, Lincoln.

Rinehart, Verrill Joanne. Investigation of twelve earthquakes off the Oregon and Northern California coasts. M, 1964, Oregon State University. 41 p.

Rinehart, Wilburn Allan. Harmonic analysis of telluric currents in the period range of 6-120 seconds. M, 1962, University of Utah. 119 p.

Ring, M. J. Magmatism and its effect on regional metamorphism; a geophysical approach. M, 1974, SUNY at Buffalo. 94 p.

Ringe, Louis Don. Geomorphology of the Palouse Hills, southeastern Washington. D, 1968, Washington State University. 73 p.

Ringe, Louis Don. Stratigraphy of the Triassic system in the Monkman Lake area, British Columbia. M, 1957, University of Idaho. 34 p.

Ringena, Delbert Bearne. The geology of the southeast quarter of Elsberry Quadrangle, Missouri. M, 1949, University of Iowa. 104 p.

Ringham, Kevin Lee. Stable isotope systematics of heavy oil deposits, Lower Cretaceous, Alberta, and Saskatchewan. M, 1986, University of Alberta. 93 p.

Ringle, John Edward. Clay mineralogy of some Paleozoic clastic sedimentary rocks of the Central Appalachians. M, 1966, American University. 44 p.

Ringler, R. W. Sphalerite geobarometry at the Calloway Mine, Ducktown, Tennessee. M, 1975, University of Illinois, Urbana. 39 p.

Ringo, William Pryor, Jr. A study of cementation and inherent crushing strength of sandstone. M, 1951, University of Kentucky. 70 p.

Ringrose, Charles D. A geochemical survey of stream sediments of the Piceance Creek basin, Colorado. M, 1977, Colorado School of Mines. 100 p.

Rinie, Robert J. Geology of the upper Morrowan strata (lower Pennsylvanian) in the Piney Creek area (north-central Arkansas). M, 1971, University of Arkansas, Fayetteville.

Rinne, Richard W. Geology of the Duke Point-Kulleet Bay area, Vancouver Island, B.C. M, 1973, Oregon State University. 63 p.

Rinowski, Robert D. Helms Formation, Franklin Mountains and Bishop Cap Hills, Texas and New Mexico. M, 1981, University of Texas at El Paso.

Rintala, William E., Jr. The electrical resistivity of near-surface bedrock units in northeastern Ohio; a laboratory study. M, 1980, Kent State University, Kent. 93 p.

Riordan, Carol J. Signal noise and MTF analysis of land cover change detection using Landsat data. M, 1980, Colorado State University. 127 p.

Riordan, E. J. Development of a drainage and flood control management system for urbanizing communities. D, 1978, Colorado State University. 231 p.

Riordon, Peter H. Geology of the Thetford-Black Lake District of Quebec with special reference to the asbestos deposits. D, 1952, McGill University.

Riordon, Peter H. The geology of a section of Beauchastel Township, Quebec. M, 1938, McGill University.

Rios, Juan H. Mineralogy and petrofabrics of the Quarryville-Octoraro area, Lancaster County, Pennsylvania. M, 1966, Franklin and Marshall College. 123 p.

Rios, Julio C. Vertical adjustment of a super-long aerial triangulation. M, 1961, Ohio State University.

Rios, Michael. Trace analysis of F^{19} in rocks by neutron and proton irradiation. M, 1967, Louisiana State University.

Rios, Nelson Guillermo. An exploratory model for the prediction of future recoverable reserves in the Seminole region of Oklahoma. M, 1972, University of Oklahoma. 91 p.

Rios, R. A. A non-linear programming model for evaluating water supply policies in the Texas coastal zone. D, 1975, University of Texas, Austin. 123 p.

Rioux, Robert Lester. A study of topography on a granitic terrain. M, 1955, University of Illinois, Urbana.

Rioux, Robert Lester. Geology of the Spence Kane area, Big Horn County, Wyoming. D, 1958, University of Illinois, Urbana. 270 p.

Ripley, Anneliese A. Paleoenvironmental interpretation of Tertiary carbonates in western Montana. M, 1987, University of Montana. 99 p.

Ripley, David P. Clogging in simulated glacial aquifers due to artificial recharge. M, 1972, University of Illinois, Chicago.

Ripley, Edward M. The ore petrology and structural geology of the Deer Lake mafic-ultramafic complex,

Efie, Itasca County, Minnesota. M, 1973, University of Minnesota, Minneapolis.

Ripley, Edward Michael. Mineralogic, fluid inclusion, and stable isotope studies of the stratabound copper deposits at the Raul Mine, Peru, South America. D, 1976, Pennsylvania State University, University Park. 163 p.

Ripley, William F. Geology of the Dutch area, Grainger and Claiborne counties, Tennessee. M, 1962, University of Tennessee, Knoxville. 32 p.

Rippee, David Scott. Geology along a cross-section through the frontal Ouachita Mountains in Pittsburg, Atoka, and Pushmataha counties, Oklahoma. M, 1981, University of Oklahoma. 90 p.

Ripple, Alfred Louis. Geology of Muldoon area, Fayette County, Texas. M, 1951, University of Texas, Austin.

Ripple, Charles D. Some physical responses to shear in montmorillonite-water systems. D, 1965, California State University, Bakersfield.

Ripy, Bruce Johnson. Effects of offshore petroleum production platforms on the mineralogy of suspended sediments, with emphasis on the clay mineralogy of a bottom water nepheloid layer, inner continental shelf, Grand Isle, Louisiana, Aug. 1972 to Jan. 1974. M, 1974, University of Florida. 85 p.

Ririe, George Todd. A loess terrace method for returning land to agricultural use while strip mining. M, 1976, University of Iowa. 66 p.

Ririe, George Todd. Precambrian mineralization and tectonic framework of Fremont County, Colorado. D, 1981, University of Iowa. 224 p.

Risatti, J. B., Jr. Geochemical and microbial aspects of Volo Bog, Lake County, Illinois. D, 1978, University of Illinois, Urbana. 106 p.

Risatti, James B. An investigation of the nannofossils in the upper Bluffport Marl, Ripley Formation and lower Prairie Bluff Chalk, (Upper Cretaceous), Oktibbeha County, Mississippi. M, 1973, Mississippi State University. 167 p.

Risbud, Subhash H. Metastability and crystallization studies in the silica-alumina system. D, 1976, University of California, Berkeley. 128 p.

Risch, David Lawrence. Magnetic anomalies northeast of Shatsky Plateau. M, 1982, Texas A&M University. 73 p.

Rishel, John Curtis. The geology of a portion of the Eve Mills zinc procpect, Monroe County, Tennessee. M, 1974, University of Tennessee, Knoxville. 74 p.

Rising, Brandt Albert. Phase relations among pyrite, marcasite and pyrrhotite below 300°C. D, 1973, Pennsylvania State University, University Park. 192 p.

Risk, Michael J. Shallow-water ripple marks (Recent) at Pinery Park, Lake Huron, Ontario. M, 1967, University of Western Ontario. 146 p.

Risley, David E. Paleotopography and areal stratigraphy of the sub-Pennsylvanian erosion surface, Daviess County, Indiana. M, 1986, Indiana University, Bloomington. 46 p.

Risley, George A. Biogeochemistry of the lower shale interval in simple Pennsylvanian cyclothems (Iowa). D, 1951, University of Wisconsin-Madison.

Risley, Lawrence J. Evidence for transitional facies in the earliest Upper Devonian of Southcentral New York State. M, 1978, SUNY, College at Oneonta. 116 p.

Risley, R. G. Jr. The structural geology of Byron-Garland anticlines, Park and Big Horn counties, Wyoming. M, 1961, University of Wyoming. 62 p.

Rismeyer, Neil W. Geology of the Precambrian Farmington Complex, Bountiful Peak, Morgan and Davis counties, Utah. M, 1981, University of Minnesota, Duluth.

Risner, Jeffrey Keith. A quantitative study of the McArthur water well field, Vinton County, Ohio. M, 1982, Ohio University, Athens. 278 p.

Rison, William. Isotopic studies of the rare gases in igneous rocks; implications for the mantle and atmosphere. D, 1980, University of California, Berkeley. 110 p.

Risser, Dennis W. Baseline study of surface waters prior to strip mining, Wilson Site, Knox County, Indiana. M, 1977, Indiana University, Bloomington. 114 p.

Risser, Jeffrey Allen. Phosphate buffering behavior of soils estimated from retention-release measurements. D, 1982, Pennsylvania State University, University Park. 87 p.

Ristau, Donn Albert. Shallow-water Demospongiae of north-central California; taxonomy and systematics, distribution and fossilization potential. D, 1977, University of California, Davis. 179 p.

Ristau, Donn Albert. Somewhere in the sea site a sediment shrouded sponge; the functional morphology and adaptive strategy of Polymastia Pachymastia Delaubenfels (Demospongia). M, 1973, University of California, Davis. 61 p.

Ristorcelli, Steven Joseph. Geology of the eastern Smith Lake ore trend, McKinley County, New Mexico, western Nuclear, Reno. M, 1980, University of New Mexico. 71 p.

Ristow, Walter W. Geographic studies of Orangeburg County, South Carolina. D, 1937, Clark University.

Ristvet, B. L. Reverse weathering reactions within Recent nearshore marine sediments, Kaneohe Bay, Oahu. D, 1977, Northwestern University. 314 p.

Ritch, Kurt D. Structural geomorphic relationships in the southeastern Edwards Plateau. M, 1984, Baylor University. 111 p.

Ritchey, Joseph L. Origin of divergent magmas at Crater Lake, Oregon. D, 1979, University of Oregon. 209 p.

Ritchey, Joseph Landon. Significance of piezobirefringence around minerals included in pyrope from diatremes, Four Corners area, Arizona. M, 1968, University of California, Los Angeles.

Ritchie, Alexander Webb. Geology of part of Las Tablas Quadrangle, Rio Arriba County, New Mexico. M, 1969, University of Texas, Austin.

Ritchie, Alexander Webb. Geology of the San Juan Sacatepequez Quadrangle, Guatemala, Central America. D, 1975, University of Texas, Austin. 151 p.

Ritchie, Beatrice. Tholeiitic lavas from the western Cascade Range, Oregon. M, 1987, University of Oregon. 94 p.

Ritchie, Eric Lee. Geochemical correlation of Tertiary volcanic ash, Idaho and Wyoming. M, 1981, University of New Orleans. 62 p.

Ritchie, James Graham. The thermal maturity of the Codell Sandstone-Carlile Shale interval (Cretaceous) in part of the Denver Basin, Colorado. M, 1985, University of Colorado. 230 p.

Ritchie, Paul Michael. A study of the copper-nickel-zinc deposit of Bird River Mines County, Ltd, southeastern Manitoba. M, 1973, University of Manitoba.

Ritchie, William Douglas. The Kneehills Tuff. M, 1957, University of Alberta. 66 p.

Ritger, Scott D. Methane-derived authigenic carbonates formed by subduction-induced pore water expulsion along the Oregon/Washington margin. M, 1985, Lehigh University. 66 p.

Ritland, J. H. Fossil forests of the Specimen Creek area, Yellowstone National Park, Montana. M, 1968, Andrews University. 62 p.

Rito, Robert F. De see De Rito, Robert F.

Rittenhouse, Gordon. A study of varve clays of northwestern Ontario. M, 1933, University of Chicago. 30 p.

Rittenhouse, Gordon. Geology of a portion of Savant Lake area, Ontario. D, 1935, University of Chicago. 65 p.

Ritter, Charles J. Color center development in natural quartz. M, 1962, Massachusetts Institute of Technology. 52 p.

Ritter, Charles John. Trace elements of gold-bearing quartz veins of the Lamaque Mine, Bourlamaque,

P.Q., Canada. D, 1971, University of Michigan. 226 p.

Ritter, Christine. Shapes and surface textures of quartz sand grains from glacial deposits; effects of source and transport. M, 1987, Texas A&M University. 62 p.

Ritter, Dale Franklin. Terrace development along the front of the Beartooth Mountains, southern Montana. D, 1964, Princeton University. 187 p.

Ritter, George S. Geology of the Blacksburg area, Montgomery County, Virginia. M, 1969, Virginia Polytechnic Institute and State University.

Ritter, John B. The response of alluvial fan systems to late Quaternary climatic change and local base-level change, eastern Mojave Desert, California. M, 1986, University of New Mexico. 193 p.

Ritter, John Ernest. The relation of seatrocks to the environment of deposition of back-barrier coals. M, 1981, Emory University. 120 p.

Ritter, John R. and Wolff, Roger G. Channel sandstones (Oligocene) of the eastern section of the Big Badlands of South Dakota. M, 1958, South Dakota School of Mines & Technology.

Ritter, John Robert. Beach dune ridges of San Diego County, California. M, 1964, University of California, San Diego.

Ritter, Michael H. The use of tritium for confirming areas of groundwater recharge, Meridian Township, Michigan. M, 1980, Michigan State University. 144 p.

Ritter, R. M. Anomalous deformation of the Dolgeville Facies, central New York State. M, 1983, SUNY at Buffalo. 87 p.

Ritter, Scott Myers. Permian and Triassic conodont evolution; rapid evolution of the Early Permian Sweetognathus lineage in the central and western United States and stasis in Middle Triassic Neogondolella at Fossil Hill, Humboldt Range, Nevada. D, 1986, University of Wisconsin-Madison. 292 p.

Ritter, Steven Paul. Igneous geology of the central portion of the Sierra Hechiceros, Chihuahua and Coahuila, Mexico. M, 1987, West Texas State University. 263 p.

Ritterbush, Linda Anita. Paleobiology of agnostid trilobites; functional morphology, growth, and evolution. M, 1981, California State University, Northridge. 84 p.

Rittersbacher, David J. Facies relationships of the Tensleep Sandstone and Minnelusa Formation, western Powder River basin, Johnson County, Wyoming. M, 1985, Colorado School of Mines. 188 p.

Rittgers, Fred Henry. A geographical survey of Blount County, Tennessee. M, 1941, University of Tennessee, Knoxville.

Rittschof, William F. Coastal geologic engineering parameters relative to docking designs, Little Sand Bay, Apostle Islands National lake shore park, Wisconsin. M, 1976, University of Rhode Island.

Ritzi, Robert William, Jr. The hydrogeologic setting and water resources of Vashon and Maury islands, King County, Washington. M, 1983, Wright State University. 115 p.

Ritzma, Howard R. Geology along the southwest flank of the Sierra Madre, Carbon County, Wyoming. M, 1949, University of Wyoming. 77 p.

Ritzwoller, Michael Herman. Magnetic anomalies over Antarctica and the surrounding oceans measured by Magsat. M, 1982, University of Wisconsin-Madison. 146 p.

Ritzwoller, Michael Herman. Observational constraints on the large scale aspherical structure of the deep Earth. D, 1987, University of California, San Diego. 109 p.

Riva, John F. Allochthonous Ordovician-Silurian cherts, argillites and volcanic rocks on Knoll Mountain, Elko County, Nevada. D, 1962, Columbia University, Teachers College. 141 p.

Riva, John F. Geology of a portion of the Diamond Range, White Pine County, Nevada. M, 1957, University of Nevada - Mackay School of Mines. 50 p.

Riva, Joseph Peter, Jr. Geology of the Sheep Creek-middle Cottonwood Creek area, Fremont County, Wyoming. M, 1959, University of Wyoming. 82 p.

Rivard, Benoit. Petrochemistry of a layered Archean magma chamber and its relation to models of basalt evolution. M, 1986, University of Toronto.

Rivas, Charlie, Jr. Hydrocarbon entrapment within porosity zones (P1, P2, & P3) of the Permian San Andres Formation, Tom Tom Field, Chaves County, New Mexico. M, 1986, West Texas State University. 128 p.

Rivera, Jorge Enrique Lugo. Biostratigraphy and foraminifera of the Neogene Coatzacoalcos Formation, in the Isthmian Salt Basin, southeastern Mexico. M, 1985, University of Texas, Austin. 157 p.

Rivera, Louis George. Upper Cretaceous foraminifera from the Budden Canyon Formation, northwestern Sacramento Valley, California. M, 1977, University of Washington. 86 p.

Rivera, Pedro A. Marin see Marin Rivera, Pedro A.

Rivera, Robert A. Groundwater supplies for mining operations. M, 1965, University of California, Berkeley. 189 p.

Riverin, Gerald. Wall-rock alteration at the Millenbach Mine, Noranda, Quebec. D, 1977, Queen's University. 256 p.

Riveroll, Gustavo Calderón see Calderón Riveroll, Gustavo

Rivers, C. Structures and textures of metamorphic rocks, Ompah area, Grenville Province, Ontario. D, 1976, University of Ottawa. 281 p.

Rivers, Mark Lloyd. Ultrasonic studies of silicate liquids. D, 1985, University of California, Berkeley. 223 p.

Rivers, Thomas D. Subsurface geology of Willisville area, Nevada County, Arkansas. M, 1959, University of Arkansas, Fayetteville.

Rives, John S. Paleoenvironmental analysis of the Joachim Formation (Middle Ordovician) of northern Arkansas. M, 1977, Louisiana State University.

Rivières, Jean des see des Rivières, Jean

Rix, Cecil Charles. Geology of Chinati Peak Quadrangle, Trans-Pecos Texas (R 528). D, 1953, University of Texas, Austin.

Rix, Cecil Charles. Petrography, northern Davis Mountains, Trans-Pecos, Texas. M, 1951, University of Texas, Austin.

Rixon, A. J. Effects of heavy applications of lime to soils derived from volcanic ash on the humid Hilo and Hamakua coasts, Island of Hawaii. D, 1962, University of Hawaii. 148 p.

Rizo, Jaime A. Geology of the Gypsum–Dotsero area, Eagle County, Colorado. M, 1971, Colorado School of Mines. 94 p.

Rizvi, S. Ali Ibne Hamid. Comparative physiography of the lower Ganges and lower Mississippi valleys. D, 1955, Louisiana State University. 286 p.

Rizvi, Saiyed Mohammed Naseer. The petrology and petrography of the Seguine Formation, Wilcox Group, Bastrop County, Texas. M, 1958, University of Texas, Austin.

Rizvi, Syed Saghir Ahmed. Investigation of water supply depletion in the upper Colorado River basin. D, 1967, University of Colorado. 177 p.

Rizzo, Jean G. Comparative study of world-wide jadeites and nephrites. M, 1982, Montclair State College. 183 p.

Rizzo, William Martin. The community metabolism and nutrient dynamics of a shoal sediment in a temperate estuary, with emphasis on temporal scales of variability. D, 1986, College of William and Mary. 124 p.

Roach, Carl H. A petrographic study of the Hartshorne Sandstone in southeastern Oklahoma. M, 1955, University of Oklahoma. 88 p.

Roach, Dan. Barite bodies west of Hemlo, Ontario; petrofabric and geochemical study. M, 1987, University of Ottawa. 136 p.

Roach, David M. The determination of refractive index distributions for oceanic particulates. D, 1975, Oregon State University. 173 p.

Roach, Samuel. Sediments of the Pecos River in New Mexico. M, 1939, Texas Tech University. 41 p.

Roach, W. R. Methodology for the estimation and evaluation of nonpoint pollution loading from watersheds in Oklahoma. D, 1977, University of Oklahoma. 208 p.

Roadifer, Jack E. A subsurface study of the (Cretaceous) Dakota Sandstone in South Dakota. M, 1962, South Dakota School of Mines & Technology.

Roadifer, Jack Ellsworth. Stratigraphy of the Petrified Forest National Park, Arizona. D, 1966, University of Arizona. 182 p.

Roales, Paul A. Ore and monzonite dike relations, Monitor Mine, Coeur d'Alene District, Idaho. M, 1973, University of Idaho. 142 p.

Roark, Clayton R. Left-lateral wrench faulting along the Amarillo Uplift and its significance in basement exploration, Carson County, Texas. M, 1985, University of Kansas. 96 p.

Roark, Louis. An oil field of Oklahoma. M, 1921, Indiana University, Bloomington.

Roark, Norris Wilson. Geology of Texas. M, 1921, University of Chicago. 146 p.

Roark, Philip W. Watershed management in the underdeveloped countries. M, 1968, [Colorado State University].

Roark, Robert C. Stratigraphy and microfacies of the Pennsylvanian System, North Franklin Mountains, Dona Ana County, New Mexico. M, 1986, University of Texas at El Paso.

Robb, George L. The geology of northern Perry Park, Douglas County, Colorado. M, 1949, Colorado School of Mines. 62 p.

Robb, Gregory Alan. Sedimentology and diagenesis of the Viking Formation, Garrington oil field, south-central Alberta. M, 1985, University of Alberta. 302 p.

Robb, James M. Structure of continental margin geology between Cape Rhir and Cape Sim, Morocco. M, 1970, University of Rhode Island.

Robb, Marion Glenn. Areal geology of the eastern Mount Ida area, Montgomery County, Arkansas. M, 1960, University of Oklahoma. 77 p.

Robberson, Thomas Allen. Stratigraphy and uranium potential of Virgilian through Leonardian strata in parts of Comanche, Cotton, and Tilman counties, Oklahoma and Wichita County, Texas. M, 1980, Oklahoma State University. 153 p.

Robbins, Barry Phillip. The Imperial Formation, northeastern Mackenzie Mountains, Northwest Territories. M, 1960, University of Alberta. 108 p.

Robbins, Carl Richard. The minerals of clay size in some limestones. M, 1951, University of Missouri, Columbia.

Robbins, Charles Henry. Geology of the Lantern Property near the Scossa mining district, Pershing County, Nevada. M, 1985, University of Nevada. 89 p.

Robbins, David B. Metamorphism and structure of the Encampment Creek area, British Columbia. M, 1976, University of Calgary. 171 p.

Robbins, Donald A. Late Cenozoic gravel deposits in the vicinity of Lewiston, Idaho. M, 1984, Washington State University. 56 p.

Robbins, Edward M., Jr. A contribution to the Upper Cretaceous-Lower Tertiary stratigraphy of the southeastern part of the Wind River basin, Wyoming. M, 1949, Colorado School of Mines. 76 p.

Robbins, Eleanora Iberall. "Fossil Lake Danville"; the paleoecology of Late Triassic ecosystem on the North Carolina-Virginia border. D, 1982, Pennsylvania State University, University Park. 464 p.

Robbins, Elizabeth A. Laramide structural geometry of the northwestern Beartooth Mountains, Montana. M, 1987, Colorado State University. 129 p.

Robbins, Gary Alan. Determining dispersion parameters to predict ground-water contamination. D, 1983, Texas A&M University. 238 p.

Robbins, Gary Alan. Radiogenic argon diffusion in muscovite under hydrothermal conditions. M, 1972, Brown University.

Robbins, Gerald Duane, Jr. Geology of the Yale Southwest Quadrangle, Payne County, Oklahoma. M, 1979, Oklahoma State University. 78 p.

Robbins, Harold Weston. The Pleistocene geology of Portales Valley, Roosevelt County, New Mexico and certain adjacent areas. M, 1941, University of Nebraska, Lincoln.

Robbins, James L. A seismic interpretation of the Zuercher Pool, Harper County, Kansas. M, 1968, Wichita State University. 78 p.

Robbins, Kevin Douglas. Simulated climate data inputs for DRAINMOD. D, 1988, North Carolina State University. 226 p.

Robbins, Lisa Louise. Morphologic variability and protein isolation and characterization of Recent planktonic foraminifera. D, 1987, University of Miami. 322 p.

Robbins, Peter. Recognition of post-glacial shorelines of Keuka Lake, central New York State. M, 1971, Virginia State University. 56 p.

Robbins, Stephen L. Gravity and magnetic data in the vicinity of the Calaveras, Hayward, and Silver creek faults, (near San Jose, Santa Clara County). M, 1969, San Jose State University. 33 p.

Robbins, William H. Foraminiferal ecology of Biloxi Bay, Mississippi. M, 1961, University of Missouri, Columbia.

Robbs, Edward E. The petrology and paleoenvironments of the Pennington Formation (Upper Mississippian), east central Tennessee and Northwest Georgia. M, 1980, Memphis State University.

Robelen, Peter G. The petrology of the Honeybrook anorthosite (PreCambrian-Grenville), Chester County, Pennsylvania. M, 1968, Bryn Mawr College. 30 p.

Roberson, Dana Shumard. The paleoecology, distribution and significance of circular Edwards Limestone (Cretaceous) bioherms in central Texas. M, 1971, Baylor University. 80 p.

Roberson, Gary D. The geomorphology of the Canadian River basin. M, 1972, Baylor University. 156 p.

Roberson, Herman Ellis. Petrology of Tertiary bentonites of Texas. D, 1959, University of Illinois, Urbana. 87 p.

Roberson, Herman Ellis. Petrology of the Carrizo and Marquez formations, Leon County, Texas. M, 1957, University of Texas, Austin.

Roberson, Lindon B. Application of the Wenner method for tracing shallow unfaulted strata in the Gulf Coast province. M, 1960, Texas A&M University. 49 p.

Roberson, Michel I. Continuous seismic profiler survey of Oceanographer, Gilbert, and Lydonia submarine canyons, Georges Bank. M, 1964, University of Kansas. 37 p.

Robert, Francois. Etude du mode de mise en place des veines aurifères de la Mine Sigma, Val d'Or, Québec. D, 1983, Ecole Polytechnique. 294 p.

Robert, François. Pétrographie et pétrochimie des roches encaissantes du gîte de Zn-Cu-Az de Manitou-Barvue, Val d'Or, Québec. M, 1980, Ecole Polytechnique. 208 p.

Robert, Jean Louis. Transition "Grenville-Keewatin" dans la région du Lac Témiscamingue. D, 1963, Universite Laval. 243 p.

Robert, Jean-Marc. Analyses des donées géologiques et géotechniques et appréciation des inclinomètres mis en place dans le secteur de St. Jean Vianney. M, 1974, Ecole Polytechnique.

Robert, Jean-Paul. Stratigraphic variations around Destourbes-Robion area, Castellane, Basses-Alpes, France. M, 1956, University of Texas, Austin.

Robert, Lance Christian. Structural geology and geometries of the Denton Duplex along the frontal Blue Ridge, near Hartford, Tennessee. M, 1987, University of Tennessee, Knoxville. 147 p.

Robert, Pierre Camille. Evaluation of some remote sensing techniques for soil and crop management. D, 1982, University of Minnesota, Minneapolis. 131 p.

Robert, Ray, Jr. Petrology and tectonic evolution of the Bowers Supergroup, northern Victoria Land, Antarctica. M, 1987, Western Washington University. 116 p.

Robert, Richard A. Thermodynamic properties of $CaMg(CO_3)_2$, $Mg_3Si_4O_{10}(OH)_2$, and $Ca_2Mg_5Si_8O_{22}(OH)_2$ from 12° to 300°K. D, 1957, University of Chicago.

Roberts, Albert Eugene. A petrographic study of the intrusive at Mary's Peak, Benton County, Oregon. M, 1949, University of Oregon. 70 p.

Roberts, Alice Lynn. Simulating flow and advective-dispersive transport in stochastically-generated fracture networks. M, 1984, University of Waterloo. 214 p.

Roberts, Andrew Clifford. A mineralogical investigation of alstonite, $BaCa(CO_3)_2$. M, 1976, Queen's University. 59 p.

Roberts, Arthur Cecil Batt. Preceramic occupations along the north shore of Lake Ontario. D, 1982, York University.

Roberts, Barry L. Paleoclimate in the Southwestern United States; a computer model. M, 1985, Kent State University, Kent. 189 p.

Roberts, C. K. The topography of the mid-Pacific mountains. M, 1968, United States Naval Academy.

Roberts, Carl Nelson. Geology of the Dallas Quadrangle. M, 1953, Southern Methodist University. 33 p.

Roberts, Carroll Norton. The Osterby Pegmatite, Dalarna, Sweden. M, 1949, Indiana University, Bloomington. 30 p.

Roberts, Charles L. The micropaleontology, biostratigraphy and paleoenvironment of the Kingshill Marl, St. Croix, U.S. Virgin Islands. M, 1972, Ohio University, Athens. 258 p.

Roberts, Charles Thomas. Laminated black shale-chert cyclicity in the Woodford Formation (Upper Devonian of southern Mid-continent). M, 1988, University of Texas at Dallas. 85 p.

Roberts, Clarence Everett. Preliminary study of the concretions in the (Devonian) Romney Shale, Highland County, Virginia. M, 1960, University of Virginia. 173 p.

Roberts, Clay and Hoffman, R. N. The occurrence and metallurgy of copper. M, 1913, University of Kansas.

Roberts, Colleen T. Cenozoic evolution of the northwestern Honey Lake basin, Lassen County, California. M, 1984, Colorado School of Mines. 159 p.

Roberts, Dar Alexander. Discrimination of vegetation anomalies associated with hydrocarbon microseeps using multi-date image subtraction, Railroad Valley, Nevada. M, 1986, Stanford University. 133 p.

Roberts, David. Form and function of the primate scapula. D, 1973, Yale University.

Roberts, David Blair. Relationships between lithology and microfossils in the lower Tertiary of northeastern Utah. M, 1953, University of Minnesota, Minneapolis. 51 p.

Roberts, David Chapin. Permocarboniferous flora of north-central Texas. M, 1952, Boston University.

Roberts, Ellis E. Geochemical and geobotanical prospecting for barium and copper. D, 1949, Stanford University. 83 p.

Roberts, Ellis E. Geology of the Alston District, Houghton and Baraga counties, Michigan. M, 1940, California Institute of Technology. 47 p.

Roberts, Eve. Geology of the southern part of the Wayan West Quadrangle, Caribou County, Idaho. M, 1982, San Jose State University. 117 p.

Roberts, Francis H. Ultramafic rocks along the Precambrian axis of southeastern Pennsylvania. M, 1968, Bryn Mawr College. 24 p.

Roberts, Francis H. Ultramafic rocks along the Precambrian axis of southeastern Pennsylvania "extension of Master's thesis". D, 1969, Bryn Mawr College. 42 p.

Roberts, Harry Heil. A paleoecological study of a lower Allegheny shale (Pennsylvanian) in eastern Ohio. M, 1966, Louisiana State University.

Roberts, Harry Heil. Recent carbonate sedimentation, North Sound, Grand Cayman island, British West Indies. D, 1969, Louisiana State University. 118 p.

Roberts, James C. Some remarks on convection motion when the stress is not a linear function of strain rate. D, 1972, Princeton University. 126 p.

Roberts, James Morgan, Sr. X-ray diffraction and chemical techniques for quantitative soil clay mineral analysis. D, 1974, Pennsylvania State University, University Park. 86 p.

Roberts, James W. Stratigraphy, sedimentology, and structure of the Swauk Formation along Tronsen Ridge, central Cascades, Washington. M, 1985, Washington State University. 188 p.

Roberts, Janis L. Hydrothermal alteration and fluid inclusion studies of the Cazaderos epithermal prospect, Zacatecas, Mexico. M, 1988, Dartmouth College. 107 p.

Roberts, Jeannette E. Agglutinated foraminifera from the Devonian of Wisconsin. M, 1972, University of Wisconsin-Milwaukee.

Roberts, Jeffrey B. The petrology and paleoenvironments of the Mississippian Big Clifty Formation in South-central Tennessee; implications for regional paleogeography. M, 1980, Memphis State University.

Roberts, John. Post-rift sediments of the United States middle Atlantic margin and speculations regarding tectonics in the source area. M, 1988, University of Delaware. 157 p.

Roberts, John Calvin. Late Cenozoic volcanic stratigraphy of the Swan Valley Graben between Palisades Dam and Pine Creek, Bonneville Co., Idaho. M, 1981, Idaho State University. 58 p.

Roberts, John F. Ostracoda from the Sundance (Jurassic) Formation of central Wyoming. M, 1934, University of Missouri, Columbia.

Roberts, John Lenox. Stratigraphy of the lower Kittanning cycle in northwestern Clearfield County, central Pennsylvania. M, 1959, Pennsylvania State University, University Park. 73 p.

Roberts, Joseph. Phosphates, their distribution, occurrence, genesis, and character. M, 1915, The Johns Hopkins University.

Roberts, Joseph K. The Triassic of northern Virginia. D, 1922, The Johns Hopkins University.

Roberts, Keith. The Precambrian geology of the Oliver—Spalding lakes region, northern Saskatchewan. M, 1979, University of Regina. 134 p.

Roberts, Larry E. Organic geochemistry of the Devonian-Mississippian Chattanooga Shale and Cumberland Plateau oils in Tennessee; a source to oil correlation study. M, 1987, University of Texas, Arlington. 226 p.

Roberts, Lillian. Solar cyclicity in the lacustrine Green River Formation (Eocene, Wyoming). M, 1988, University of Southern California.

Roberts, Malissa A. A petrographic and geochemical analysis of the Sparta granitic complex, Hancock County, Georgia. M, 1983, University of Georgia.

Roberts, Marion S. Upper Claiborne foraminifera of the "Saline Bayou" or "St. Maurice" type locality. M, 1934, Louisiana State University.

Roberts, Michael Anderson, Jr. Structure of Summit Dome, Chestnut Ridge, Pennsylvania. M, 1973, University of South Carolina.

Roberts, Michael J. Subsurface geology of the Spar Mountain Member of the Ste. Genevieve Limestone, Hamilton County, Illinois. M, 1984, University of Cincinnati. 231 p.

Roberts, Michael Taylor. Geology of the Blue Diamond breccia (Tertiary), Clark County, Nevada. M, 1968, Pennsylvania State University, University Park. 95 p.

Roberts, Michael Taylor. The stratigraphy and depositional environments of the lower part of the Crystal Spring Formation, Death Valley, California. D, 1974, Pennsylvania State University, University Park. 368 p.

Roberts, Paula. Geophysical investigation of the possible northeast extension of the New Madrid rift complex; gravity. M, 1984, Purdue University. 86 p.

Roberts, Philip Kenneth. Stratigraphy of the Green River Formation (Eocene), Uinta Basin, Utah. D, 1964, University of Utah. 286 p.

Roberts, R. B. Gravity studies of buried valleys north of Green Lakes State Park, central New York. M, 1978, Syracuse University.

Roberts, Ralph Jackson. Geology of the Antler Peak Quadrangle, Nevada. D, 1949, Yale University.

Roberts, Ralph Jackson. The petrography and ore deposits of the Dixie District, Idaho. M, 1938, University of Washington. 61 p.

Roberts, Robert G. Geology and geochemistry of Mattagami Lake Mine, Galinee Township, Quebec. D, 1966, McGill University.

Roberts, Sarah Ann. The geologic history of the post-Croatan (Pleistocene) sediments at the Lee Creek Mine, Beaufort County, North Carolina. M, 1981, East Carolina University. 76 p.

Roberts, Sheila M. Depositional environments and diagenesis of Permian scaphopod-bellerophontacean gastropod-bearing beds in southwestern Montana. M, 1983, University of Montana. 176 p.

Roberts, Stephen M. Subsurface facies model for a transgressive shallow-water, transitional clastic-carbonate environment; Martha Brae delta-reef complex, Falmouth, Jamaica. M, 1986, University of Oklahoma. 80 p.

Roberts, Steven Arland. Early hydrothermal alteration and mineralization in the Butte District, Montana. D, 1975, Harvard University.

Roberts, Thomas Adolph. Transitional lower delta plain-upper delta plain sediments in Middle Pennsylvanian coal-bearing strata, eastern Wartburg Basin, Tennessee. M, 1978, University of Tennessee, Knoxville. 272 p.

Roberts, Thomas G. Fusulinidae in upper Paleozoic of Peru. D, 1949, Columbia University, Teachers College.

Roberts, William. Sediments and suspended particulate matter of the Patuxent River drainage basin, Maryland; Dec. 68–July 69. D, 1971, George Washington University.

Roberts, William B. A study of river terraces of the Chattahoochee River between Chattahoochee, Florida and Fort Gaines, Georgia. M, 1958, Florida State University. 47 p.

Roberts, William B. Geology of a part of the Rosamond Hills area, Kern County, California. M, 1951, California Institute of Technology. 39 p.

Roberts, William Stephan. Regionalized feasibility study of cold weather earthwork. M, 1976, Purdue University. 190 p.

Robertson, Benjamin Telfer. Stratigraphic setting of some new and rare phosphate minerals in the Yukon Territory. M, 1980, University of Saskatchewan. 218 p.

Robertson, Billy Gene. Stratigraphy of the Williams Canyon Formation, Colorado. M, 1957, University of Oklahoma. 74 p.

Robertson, Bruce Edward. The paleoecology of the Tinton Formation (Upper Cretaceous), New Jersey coastal plain. M, 1972, Rutgers, The State University, New Brunswick. 114 p.

Robertson, C. K. Natural rates of methane production and their significance to carbon cycling in two small lakes. D, 1978, University of Michigan. 192 p.

Robertson, Cameron P. A reconnaissance survey of Kaladar Township, Lennox-Addington County, Ontario. M, 1940, Northwestern University.

Robertson, Charles E. The geology of the Galena Quadrangle, Missouri. M, 1961, University of Missouri, Columbia.

Robertson, Christopher Alan. Petrology, sedimentology, and structure of the Chuckanut Formation, Coal Mountain, Skagit County, Washington. M, 1981, University of Washington. 41 p.

Robertson, Craig. Potential impact of subsurface irrigation return flow on a portion of the Milk River and Milk River Aquifer in southern Alberta. M, 1988, University of Alberta. 173 p.

Robertson, D. The Temagami metadiorite sill; petrology and geochemistry. M, 1977, Laurentian University, Sudbury.

Robertson, Daniel Edward. Skarn mineralization at Iron Mountain, New Mexico. M, 1986, Arizona State University. 77 p.

Robertson, David John. Effects of clearcutting on lotic invertebrate communities in the central Appalachian Mountains. D, 1981, University of Pittsburgh. 225 p.

Robertson, David Kenneth. Ground-water availability in southern New Jersey; a model approach to estimation. D, 1973, Rutgers, The State University, New Brunswick. 199 p.

Robertson, David Knox. Isotope analysis for lead and sulphur from the Great Slave Lake area. M, 1966, University of Alberta. 66 p.

Robertson, David S. The petrology of the Kisseynew Gneiss of the Batty Lake area, Manitoba. D, 1953, Columbia University, Teachers College.

Robertson, Dennis. A study of eight samples of Leptomeryx (Mammalia, Artiodactylia) from the Oligocene of South Dakota. M, 1988, South Dakota School of Mines & Technology.

Robertson, Donelson Anthony. Petrographic analysis of the Pennsylvanian sandstones of Perry County, Kentucky. M, 1951, University of Illinois, Urbana.

Robertson, Douglas Scott. Geodetic and astrometric measurements with very-long-baseline interferometry. D, 1975, Massachusetts Institute of Technology. 187 p.

Robertson, E. Parameterization of a vertical heat transfer coefficient for use in an energy budget model for the estimation of lake evaporation. M, 1980, University of Waterloo.

Robertson, Eddie B. Pollen and spores as stratigraphic indices to the North Dakota Paleocene. D, 1975, University of Minnesota, Minneapolis. 243 p.

Robertson, Eugene C. An experimental study of flow and fracture in rocks. D, 1952, Harvard University.

Robertson, Florence. A critical study of the variations in the Earth's magnetic field in West Texas. M, 1936, St. Louis University.

Robertson, Florence. A mathematical analysis of numerical integration and its application to electromagnetic seismograms. D, 1945, St. Louis University.

Robertson, Forbes Smith. Geology and mineral deposits, Elliston Mining District, Powell County, Montana. D, 1956, University of Washington. 332 p.

Robertson, Forbes Smith. The igneous geology of the eastern Ironton and western Fredericktown quadrangles, Missouri. M, 1940, Washington University.

Robertson, Frederick N. Hexavalent chromium in the ground water in Paradise Valley, Maricopa County, Arizona. M, 1975, Arizona State University. 121 p.

Robertson, George C., III. Surficial deposits and geologic history, northern Bear Lake Valley, Idaho. M, 1978, Utah State University. 91 p.

Robertson, George K. The geology of the Santa Monica Mountains in the vicinity of Topanga Canyon, Los Angeles County, California. M, 1932, University of Southern California.

Robertson, George McAfee. The Tremataspidae (Silurian, Estonia). D, 1938, Yale University.

Robertson, Herbert Chapman, Jr. The petrology of the Lower Fry sandstone reservoir, South Crewes Field, Runnels County, Texas. M, 1959, Southern Methodist University. 38 p.

Robertson, J. B. The improvement of the physical and chemical properties of the dark magnesium Clay soils of the Hawaiian Islands. M, 1951, University of Hawaii. 69 p.

Robertson, J. D. Col. Sellers Mine, Leadville, Colorado. M, 1886, Washington University.

Robertson, J. D. Geophysical studies on the Ross ice shelf, Antarctica. D, 1975, University of Wisconsin-Madison. 227 p.

Robertson, J. O., Jr. Hydration of clays; effect of various organic and inorganic ions and electrochemical treatment of the swelling of various clays. D, 1975, University of Southern California.

Robertson, James Alexander. The general geology of part of the Blind River area. M, 1961, Queen's University. 432 p.

Robertson, James Alexander. The general geology of part of the Blind River area (Ontario). M, 1960, Queen's University.

Robertson, James Douglas. A seismic study of the structure and metamorphism of firn in western Antarctica. M, 1972, University of Wisconsin-Madison. 54 p.

Robertson, James H. Glacial to interglacial oceanographic changes in the Northwest Pacific, including a continuous record of the last 400,000 years. D, 1975, Columbia University. 355 p.

Robertson, James Magruder. Crystal growth from boiling solutions; an experimental study. M, 1968, University of Michigan.

Robertson, James Magruder. Geology and mineralogy of some copper sulfide deposits, Keweenaw County, Michigan. D, 1972, University of Michigan. 125 p.

Robertson, John Andrew. Environmental geology of the Chandler Quadrangle, Maricopa County, Arizona, Part II. M, 1986, Arizona State University. 102 p.

Robertson, John K. Experimental studies on rocks from the Deboullie stock, northern Maine; including phase relations in the water deficient environment. D, 1970, University of Chicago. 145 p.

Robertson, John Louis. Reflection seismic correlation and interpretation marginal to the Central Basin Platform, West Texas-New Mexico. M, 1961, University of Oklahoma. 182 p.

Robertson, Kenneth Scott. Stratigraphy, depositional environments, petrology, diagenesis, and hydrocarbon maturation related to the Red Fork Sandstone in north-central Oklahoma. M, 1983, Oklahoma State University. 137 p.

Robertson, Leo L. Geographical changes resulting from oil development in Oklahoma City and vicinity. M, 1937, University of Oklahoma. 71 p.

Robertson, Lloyd B. Geology of the Ingleside area northwest of Fort Collins (Larimer County), Colorado. M, 1950, Colorado School of Mines. 90 p.

Robertson, Marina. Engineering geology of northeastern Crow Canyon, Santa Ana Mountains, California. M, 1984, California State University, Long Beach. 101 p.

Robertson, Mary Spotswood. Petrography of the sodarich White Creek Stock, Fresno County, California. M, 1940, University of California, Berkeley. 49 p.

Robertson, P. Blyth. Petrography of the bedrock and breccia erratics in the region of Lac Couture, Quebec (Canada). M, 1965, Pennsylvania State University, University Park. 112 p.

Robertson, Paul. Strike-slip faulting in the vicinity of Trinidad and Tobago, West Indies, and its tectonic significance. M, 1986, University of Houston.

Robertson, Percival. Some problems of the middle Mississippi River during the Pleistocene time. D, 1936, Washington University. 143 p.

Robertson, Peter Kay. In-situ testing of soil with emphasis on its application to liquefaction assessment. D, 1983, University of British Columbia.

Robertson, Ray Henry. Glacial factors in the development of streams and their whitewater reaches on the Precambrian shield, Northern Saskatchewan. M, 1978, University of Saskatchewan. 241 p.

Robertson, Richard A. Precambrian rocks of the Windmill Islands, Budd Coast, Antarctica. M, 1961, University of Wyoming. 56 p.

Robertson, Richard Douglas. An early Tertiary pelagic microfauna from the Coast Range, western Washington County, Oregon. M, 1972, University of Oregon. 100 p.

Robertson, Roland Secrest. Chinle stratigraphy of St. Johns vicinity, Apache County, Arizona. M, 1956, University of Texas, Austin.

Robertson, Stephen Wood. Computer modeling of simultaneous garnet resorption and MnO diffusion, Llano County, Texas. M, 1987, University of Texas, Austin. 160 p.

Robertson, W. D. A description of the unconsolidated geological materials of the Grand River valley in the Waterloo area. M, 1977, University of Waterloo.

Robertson, William L. The hydrogeology of Streetsboro, Portage County, Ohio. M, 1983, Kent State University, Kent. 143 p.

Robeson, John Maxwell, Jr. A study of fossil pollen as found in peat from the Dismal Swamp area of Virginia and North Carolina. M, 1928, University of Virginia.

Robichaud, Stephen R. Palynomorphs from the Tipton Shale Member of the Green River formation, Southwest Wyoming. M, 1979, Louisiana State University.

Robie, Richard A. Thermodynamic properties of $CaMg(CO_3)_2$, $Mg_3Si_4O_{10}(OH)$, and $Ca_2Mg_5Si_4O_{22}(OH)_2$ from 12° to 300° K. D, 1957, University of Chicago. 107 p.

Robigou, Veronique. Metamorphism of the Cambrian Campito Formation, White-Inyo Mountains area, eastern California. M, 1984, University of California, Los Angeles. 101 p.

Robillard, Jacques. Etude des roches plutoniques mafiques du Mont Royal (Quèbec, Canada). M, 1968, Universite de Montreal.

Robin, Michel J. L. Cation-exchange during horizontal infiltration into soil. M, 1980, University of Guelph.

Robin, Pierre Yves. Mechanical significance of a fracture pattern in Precambrian metavolcanics, Tudor Twp., Ontario (Canada). M, 1968, University of Toronto.

Robin, Pierre-Yves François. Equilibrium of stressed solids with respect to phase changes, and its geological applications. D, 1974, Massachusetts Institute of Technology. 196 p.

Robinette, Michael Joseph. Ground water flow systems in lower Dry Valley, Caribou County, Idaho. M, 1977, University of Idaho. 21 p.

Robins, Alfred Raymond. Geology of the Alum Rock Canyon sector, Diablo Range, California. M, 194?, Stanford University.

Robinson, Alexander Maguire. The measurement of transition probabilities of atomic neon. D, 1966, University of British Columbia.

Robinson, Andres J., Jr. Subsurface geology of northeastern Young County, Texas. M, 1970, University of Oklahoma. 123 p.

Robinson, Andrew G. The origin and modification of a uraninite-bearing placer, Elliot Lake, Ontario. M, 1982, University of Toronto.

Robinson, B. A. Insoluble residues from the upper Monte Cristo Limestone and the lower Bird Spring Group (Carboniferous) at Mountain Springs, Clark County, Nevada. M, 1973, University of Southern California. 91 p.

Robinson, B. Spence. Petrology of the Presidio Formation and equivalent units; Presidio County, Texas. M, 1978, University of Houston.

Robinson, Bob R. Stratigraphy and environment of deposition of Member 9 of the Rawls Formation, Presidio County, Texas. M, 1976, University of Houston.

Robinson, Bob Russell. Geology of the D Cross Mountain Quadrangle, Socorro and Catron counties, New Mexico. D, 1981, University of Texas at El Paso. 225 p.

Robinson, Bobby Brick. Geology of the Holser Canyon area, Ventura County, California. M, 1956, University of California, Los Angeles.

Robinson, Bret A. Laboratory investigation of the phenomenon of stream-bed armoring in a degrading channel. M, 1986, Southern Illinois University, Carbondale. 93 p.

Robinson, Brian William. Studies on the Echo Bay silver deposit (Precambrian), Northwest Territories, Canada. D, 1971, University of Alberta. 229 p.

Robinson, Carl Francis. Stratigraphy and structural geology of Ahtanum Ridge, Yakima, Washington. M, 1966, University of Washington. 35 p.

Robinson, Charles Sheerwood. Geology of ore deposits of the Whitepine area, Tomichi mining district, Gunnison County, Colorado. D, 1955, University of Colorado.

Robinson, Clair W. Origin and erosion of soils. M, 1915, Yale University.

Robinson, Daniel O. The mineral composition of the colloidal clay fractions of some Arizona soils. M, 1947, University of Arizona.

Robinson, Donald James. Interpretation of gravity anomaly data from the Aravaipa Valley area, Graham and Pinal counties, Arizona. M, 1976, University of Arizona.

Robinson, Donald James. Stratigraphic relationships, geochemistry and genesis of the Redstone volcanic-hosted nickel deposit, Timmins, Ontario. D, 1982, University of Western Ontario. 370 p.

Robinson, Edwin G. Environment and genesis of apatite and mica deposits of West Portland Township, Quebec. D, 1951, University of Toronto.

Robinson, Edwin Simons. Geological structure of the Transantarctic Mountains and adjacent ice covered areas, Antarctica. D, 1964, University of Wisconsin-Madison. 319 p.

Robinson, Edwin Simons. Seismic refraction studies on the Ross Ice Shelf, Antarctica. M, 1957, University of Michigan.

Robinson, Elizabeth M. The effect of a shallow low viscosity zone on mantle convection and its expression at the surface of the Earth. D, 1987, Woods Hole Oceanographic Institution. 320 p.

Robinson, Enders. Predictive decomposition of time series with applications to seismic exploration. D, 1954, Massachusetts Institute of Technology. 281 p.

Robinson, Ernest Guy. A correlation of the Mesozoic strata of southwestern Alberta, Canada, and Montana. M, 1925, Cornell University. 30 p.

Robinson, Ernest Guy. Environment and genesis of apatite and mica deposits of West Portland Township, Quebec. D, 1953, University of Toronto.

Robinson, Gene D., Jr. An investigation of the fission tract method of geologic dating and its application to Blue Ridge muscovites. M, 1968, University of Tennessee, Knoxville. 43 p.

Robinson, Gene D., Jr. Partitioning of certain trace metals in soils in the Southeastern United States. D, 1978, University of Georgia. 292 p.

Robinson, George Calver. Geology of the southern half of the Millerstown, Pennsylvania, 15 minute topographic quadrangle. M, 1958, West Virginia University.

Robinson, George M. Hydrogeology of buried valleys in Geauga County, Ohio. M, 1972, Kent State University, Kent. 70 p.

Robinson, George W. The occurrence of rare earth elements in zircon. D, 1979, Queen's University. 155 p.

Robinson, Gerald B., Jr. Stratigraphy and Leonardian (Permian) fusulinid paleontology in central Prequop

Mountains, Elko County, Nevada. M, 1961, Brigham Young University. 145 p.

Robinson, Gershon DuVall. The geology of the Daulton Quadrangle, California. M, 1941, University of California, Berkeley. 95 p.

Robinson, Gilpin Rile, Jr. Bedrock geology of the Nashua River area, Massachusetts-New Hampshire. D, 1979, Harvard University.

Robinson, H. R. A study of the genus Baculites in the (Upper Cretaceous) Bearpaw Formation of Western Canada. M, 1942, McGill University.

Robinson, Harry R. Physiographic types as shown by topographic maps. M, 1933, University of Missouri, Columbia.

Robinson, Henry H. Geology of San Francisco Mountain and vicinity, Arizona. D, 1903, Yale University.

Robinson, J. E. A study of the cobalt sediments (Precambrian) of Dasserat Township, Quebec. M, 1951, McGill University.

Robinson, J. W. Reconnaissance geology of the northern Vizcaino Peninsula, Baja California Sur, Mexico. M, 1975, San Diego State University.

Robinson, James. Co-carbonization of selected American coals wth commercial additives. M, 1981, University of Pittsburgh.

Robinson, James E. Geology and mineralogy of the San Luis barite prospect, Belize, Central America. M, 1978, Colorado School of Mines. 119 p.

Robinson, James Holt. Geology of the northwest portion of the Waucoba Springs Quadrangle, Inyo County, California. M, 1964, California State University, Los Angeles.

Robinson, James Parker. Petrology and petrochemistry of granitic intrusives of the Cima Dome-Southern Ivanpah Mountains area, southeastern California. M, 1979, University of Southern California.

Robinson, James Richard. The ground water resources of the Laramie area, Albany County, Wyoming. M, 1956, University of Wyoming. 80 p.

Robinson, James Varney. The geochemistry and petrography of crustal xenoliths from La Olivina, Mexico. M, 1988, University of California, Santa Cruz.

Robinson, James William. A laboratory and numerical investigation of solute transport in discontinuous fracture systems. M, 1987, Memorial University of Newfoundland. 177 p.

Robinson, John Charles. HRVA; a velocity analysis technique for seismic data. D, 1969, University of Missouri, Rolla.

Robinson, Joseph Edward. Analysis by spatial filtering of some intermediate scale structures in southern Alberta (Canada). D, 1968, University of Alberta. 193 p.

Robinson, Keith. The crystal structures of zircon, clinohumite and the hornblendes; a determination of polyhedral distortion and order-disorder. D, 1971, Virginia Polytechnic Institute and State University. 143 p.

Robinson, Larry. Visible region absorption spectra of selected metamorphic and igneous amphiboles. M, 1966, University of Alaska, Fairbanks. 47 p.

Robinson, Lee. Geology of the Arlington Formation, Butte Lake area, Plumas County, California. M, 1975, University of California, Davis. 77 p.

Robinson, Lewis C., III. The geology and mineral resources of the Streator Quadrangle, Illinois. D, 1935, University of Chicago. 175 p.

Robinson, Louis H. Bases for including geology in the secondary curriculum. M, 1958, California State University, Chico. 52 p.

Robinson, M. C. Geology and gold deposits of the Sheep Creek mining camp, British Columbia. M, 1949, Queen's University. 102 p.

Robinson, Malcolm Campbell. The geologic setting and relationships of ore deposits in the West Kootenay District, B.C. D, 1951, Princeton University. 99 p.

Robinson, Mark and Patton, Thomas L. Bedrock geology, geochemistry and geophysics of Brooks Mountain, Seward Peninsula, Alaska. M, 1975, University of Alaska, Fairbanks. 106 p.

Robinson, Michael. A deltaic-back barrier model for the formation of the Beckley Coal Seam in southern West Virginia. M, 1975, University of South Carolina.

Robinson, Nelson M., Jr. Shallow crustal structure offshore of the neovolcanic zone in northern Iceland. M, 1975, University of Georgia.

Robinson, P. C. Geology and evolution of the Manitouwadge migmatite belt, Ontario, Canada. D, 1979, University of Western Ontario. 368 p.

Robinson, Paul David. An X-ray study of sideronatrite. M, 1963, Southern Illinois University, Carbondale. 12 p.

Robinson, Paul T. and Kasabach, Haig Frederick. A detailed study of the Old Baldy thrust fault, Huerfano and Costilla counties, Colorado. M, 1959, University of Michigan.

Robinson, Paul Thorton. The Cenozoic stratigraphy and structure of the central part of the Silver Peak Range, Esmeralda County, Nevada. D, 1964, University of California, Berkeley. 107 p.

Robinson, Peter. Gneiss domes of the Orange area, Massachusetts and New Hampshire. D, 1964, Harvard University.

Robinson, Peter. The fossil mammals of the Huerfano Formation (Eocene) of Colorado. D, 1960, Yale University.

Robinson, R. E. The molybdenite occurrence at the Bain property, Masham Township, Quebec. M, 1962, University of Toronto.

Robinson, R. F. The geology of the Orland property, Beauchastel Township, Province of Quebec. M, 1938, McGill University.

Robinson, R. L. Economic comparison of mineral exploration and acquisition; ore reserve replacement and growth strategies. M, 1985, Queen's University.

Robinson, Richard C. Sedimentology of beach ridge and nearshore deposits, pluvial Lake Cochise, southeastern Arizona. M, 1965, University of Arizona.

Robinson, Richard Dudley. A petrologic and structural study of a meta-igneous complex in the core of the Catoctin-Blue Ridge anticlinorium in northern Virginia. M, 1973, American University. 111 p.

Robinson, Richard N. The geology of the Precambrian basement of southern Ontario. M, 1982, University of Windsor. 145 p.

Robinson, Richard W. Ore mineralogy and fluid inclusion study of the southern Amethyst vein system, Creede Mining District, Colorado. M, 1981, New Mexico Institute of Mining and Technology. 85 p.

Robinson, Robert Blair. Structural geology and structural origin of the veins of the Coeur Mine, Coeur d'Alene mining district, Shoshone County, Idaho. M, 1980, University of Idaho. 98 p.

Robinson, Robert Bradley. Cambrian stratigraphy of southeastern Wyoming. M, 1957, University of Nebraska, Lincoln.

Robinson, Rosalie M. Tertiary volcanic rocks of the central Davis Mountains, Jeff Davis County, Texas. M, 1982, University of Texas at El Paso.

Robinson, Rosalind. The foraminifera of a deep-sea core from the Ross Sea, Antarctica. M, 1952, Smith College. 104 p.

Robinson, Russell. Interpretation of an aeromagnetic map of the Twin Sisters Dunite region, Washington. M, 1967, Stanford University.

Robinson, Russell Dow. San Juan and Paradox basins. M, 1958, Stanford University.

Robinson, Russell, Jr. Shear wave velocity in the Earth's mantle. D, 1971, Stanford University. 108 p.

Robinson, S. C. The lead-antimony-sulfur system, mineralogy and mineral synthesis. D, 1947, Queen's University. 242 p.

Robinson, S. D. Structural geology of Paleozoic rocks, Lake Massawippi area, Eastern Townships, Quebec. M, 1974, University of Ottawa. 140 p.

Robinson, Sara L. Geologic analysis of the Manti Landslide, Manti, Utah. M, 1977, [Colorado State University].

Robinson, Scott E. Geochemistry and petrology of the Lake Vermilion Formation, Ely greenstone belt, northeastern Minnesota. M, 1985, University of North Dakota. 181 p.

Robinson, Vincent A. Stratigraphy of the Lance and Fort Union formations in south central Wyoming. M, 1954, University of Nebraska, Lincoln.

Robinson, Vincent H. Geology of the road building materials of the New York State Thruway. M, 1955, New York University.

Robinson, Wilber I. The relationship of the Tetracoralla to the Hexacoralla. D, 1916, Yale University.

Robinson, William Conrad. Petrography and depositional environments of the Cretaceous (Cenomanian) Buda Limestone, northern Coahuila, Mexico. M, 1982, University of Texas, Arlington. 156 p.

Robinson, William G. The Flavrian Lake map area and the structural geology of the surrounding district (Quebec). D, 1941, McGill University.

Robinson, William G. The geology of a section of Mount Royal near the new building of the University of Montreal (Quebec). M, 1938, McGill University.

Robinson, William H. The study of surface and ground-water interactions in two watersheds. M, 1986, Indiana University, Bloomington. 107 p.

Robinson-Cook, Sylveen E. Wolframite; an infrared examination of transparency controls and fluid inclusions. M, 1986, New Mexico Institute of Mining and Technology. 120 p.

Robison, Brad Alan. Stratigraphy and petrology of some Mesozoic rocks in western Arizona. M, 1979, University of Arizona. 138 p.

Robison, Edward Clark. Geochemistry of lamprophyric rocks of the eastern Ouachita Mountains, Arkansas. M, 1976, University of Arkansas, Fayetteville.

Robison, James B. Maximum present value calculations with digital computers. M, 1965, University of Missouri, Rolla.

Robison, James Holt. Geology of the northwest portion of the Waucoba Spring Quadrangle, Inyo Mountains, California. M, 1964, University of California, Los Angeles.

Robison, Mary. Potassium/argon geochronology of the Piedmont/Blue Ridge boundary region near Lynchburg, Virginia. D, 1978, University of Pittsburgh.

Robison, Richard Ashby. Late Middle Cambrian faunas from the Wheeler and Marjum formations of western Utah. D, 1962, University of Texas, Austin.

Robison, Richard Ashby. Some Dresbachian and Franconian trilobites of western Utah. M, 1960, Brigham Young University. 59 p.

Robison, Robert M. The surficial geology and neotectonics of Hansel Valley, Box Elder County, Utah. M, 1986, Utah State University. 120 p.

Robison, Steven F. Paleocene (Puerean-Torregonian) Mammalian faunas of the North Horn Formation, central Utah. M, 1986, Brigham Young University. 133 p.

Robison, Vaughn David. Organic geochemical characterization of the Late Cretaceous-early Tertiary transgressive sequence found in the Duwi and Dakhla formations, Egypt. D, 1986, University of Oklahoma. 190 p.

Robitaille, René. Etude hydrogéologique du bassin versant du Lac Laflamme. M, 1985, Universite Laval. 184 p.

Robitshek, Melvin F. The surface approach to the subsurface study of the Spergen Formation in eastern Missouri. M, 1941, Washington University. 132 p.

Robl, Thomas L. Factors affecting the state of saturation of the waters of Puckett Spring Creek with respect to calcite. M, 1974, University of Kentucky. 60 p.

Robl, Thomas L. Factors controlling the geochemistry of vadose and stream waters in a carbonate terrain. D, 1977, University of Kentucky. 216 p.

Robles, Jorge de la Torre *see* de la Torre Robles, Jorge

Roblesky, Robert F. Upper Silurian (late Ludlovian) to Upper Devonian (Frasnian?) stratigraphy and depositional history of the Vendom Fiord region, Ellesmere Island, Canadian Arctic. M, 1979, University of Calgary. 230 p.

Robocker, J. E. The stratigraphy and petrography of the Ellis Group (Jurassic), west-central Montana. M, 1985, Montana College of Mineral Science & Technology. 125 p.

Robold, Edward Lynn. Use of remote sensing in the analysis of patterns formed by rock alteration products in the Cerbat Mountains, Arizona. M, 1981, University of Missouri, Rolla. 47 p.

Robords, Alan Carman. Terraces of the glacial Grand Valley. M, 1980, Michigan State University. 63 p.

Robotham, Hugh Beresford. Evaluation of alternative water resources management systems for the Sonoita Creek watershed. M, 1979, University of Arizona. 154 p.

Robson, G. M. Sudbury breccia. M, 1940, Queen's University. 47 p.

Robson, Homer L. The system MgSO$_4$-H$_2$O from 68° to 240°. D, 1925, University of California, Berkeley. 21 p.

Robson, James M. An optical study of the magmatic rocks of Butte, Montana. M, 1972, Montana College of Mineral Science & Technology. 71 p.

Robson, Richard Michael. Motion problems involved with gravity measurements at sea. M, 1962, University of Utah. 166 p.

Robson, Walter M. On the possibilities of the use of Doppler radar for the measurement of long geodetic lines. M, 1960, Ohio State University.

Robyn, Elisa S. A description of the Miocene Tranquillon Volcanics and a comparison with the Miocene Obispo Tuff. M, 1980, University of California, Santa Barbara.

Robyn, Thomas Lynn. Geology and petrology of the Strawberry Volcanics, NE Oregon. D, 1977, University of Oregon. 197 p.

Roca, Henri Joseph, III. The Oligocene foraminiferal biostratigraphy of the South Atlantic, 30° South latitude. M, 1982, Washington University. 46 p.

Roca, R. Luis. Engineering geology and relative slope stability of the Inyan Kara Hogback, Rapid City, South Dakota. M, 1981, South Dakota School of Mines & Technology. 144 p.

Roca-Ramisa, Luis. Simulation of groundwater contaminant transport in an unconfined aquifer; a systematic approach to parameter estimation. D, 1984, University of Missouri, Rolla. 255 p.

Roche, James Edward. Petrography and environmental study of middle Devonian reefs (lower Michigan from Alpena to Petosky), Michigan. D, 1969, University of Illinois, Urbana. 85 p.

Roche, James Edward. Petrography of selected Chesterian carbonates (Visean–Namurian) from the type area in southwestern Illinois. M, 1967, University of Illinois, Urbana.

Roche, Michael B. Quantitative paleoecology applied to coccoliths; paleooceanography of 0 to 127,000 YBP in the North Atlantic. M, 1972, Queens College (CUNY). 30 p.

Roché, Richard Louis. Stratigraphic and geochemical evolution of the Glass Buttes Complex, Oregon. M, 1987, Portland State University. 99 p.

Rocheleau, Michel. Sédimentologie de sédiments grossiers dans une séquence de flysch, l'Islet, Québec. M, 1971, Universite de Montreal.

Rocheleau, Michel. Séquence de structures sédimentaires dans un ensemble conglomérate-grès. M, 1973, Universite de Montreal.

Rochester, Eugene Wallace, Jr. Potable water availability on long ocean islands. D, 1970, North Carolina State University. 73 p.

Rochester, Haydon. Late-glacial and postglacial diatom assemblages of Berry Pond, Massachusetts, in relation to watershed ecosystem development. D, 1979, Indiana University, Bloomington. 76 p.

Rochette, Elizabeth A. Chemical weathering in the West Glacier Lake drainage basin, Snowy Range, Wyoming; implications for future acid deposition. M, 1987, University of Wyoming. 138 p.

Rochette, Francois Jules. Hydrogeological study of "Ruisseau des Eaux Volees" experimental (drainage) basin (Laurentian Provincial Park, Quebec). M, 1972, University of Western Ontario. 90 p.

Rochna, David A. Physiography and sedimentology of the "Outer Banks" between Cape Lookout and Ocracoke Inlet, North Carolina. M, 1961, University of Kansas. 91 p.

Rock, K, N. Legal, political, and institutional aspects of the Federal Geothermal Leasing Program. M, 1977, Stanford University. 63 p.

Rockaway, John Dobbling, Jr. Trend-surface analysis of ground water fluctuations. D, 1968, Purdue University. 176 p.

Rocken, Christian. The Global Positioning System; a new tool for tectonic studies. D, 1988, University of Colorado. 281 p.

Rockett, Thomas John. Phase relations in the systems silica, boron oxide-silica, and sodium oxide-boron oxide-silica. D, 1963, Ohio State University. 111 p.

Rockett, Thomas John. The Lewis Overthrust and associated minor structures (Montana). M, 1958, Boston College.

Rockey, David L. Geology of the eastern Vanocker laccolith area, Meade County, South Dakota. M, 1974, South Dakota School of Mines & Technology.

Rockie, John D. A glossary of terms in the fields of gravimetric and celestial geodesy. M, 1963, Ohio State University.

Rockingham, Christopher John. Metamorphism and metal zoning of the Hood River - 41 massive sulfide deposit, Slave structural province, N.W.T. M, 1979, University of Western Ontario. 125 p.

Rockwell, Charles. A mineral analysis of the sediments of the south shore of Long Island, New York. M, 1957, University of Oklahoma. 75 p.

Rockwell, Charles. Recent sedimentation in Great South Bay, Long Island, New York. D, 1974, Cornell University.

Rockwell, Thomas Kent. Soil chronology, geology, and neotectonics of the north central Ventura Basin, California. D, 1984, University of California, Santa Barbara. 522 p.

Rockwood, Dwight Nelson. Oil and gas occurrence in Gratiot County, Michigan. M, 1938, University of Michigan.

Rockwood, Walter G. Natural gas possibilities of the Oriskany Group in southeastern New York. M, 1956, Dartmouth College. 41 p.

Rodabaugh, Gary Lee. An investigation into processes contributing to voluntary exposure of Michigan anglers to contaminated waterways and contaminated fish. D, 1987, Michigan State University. 299 p.

Rodda, Peter Ulisse. Geology and paleontology of a portion of Shasta County, California. D, 1960, University of California, Los Angeles.

Roddick, J. C. Geochronology of the Tulameen and Hedley complexes (Mesozoic), British Columbia. M, 1970, Queen's University.

Roddick, James Archibald. Some features of the geology of the North Vancouver area (Canada). M, 1950, California Institute of Technology. 47 p.

Roddick, James Archibald. The plutonic rocks in the Vancouver North-Coquitlam area, in the southern Coast Mountains of British Columbia. D, 1955, University of Washington. 331 p.

Roddick, Susan L. A study of the marl deposits at Dry Lake, near Marbank, Ontario. M, 1970, Queen's University. 124 p.

Roddie, W. G. Relationship of the (Devonian) Cooking Lake Member of the Leduc D$_3$ Member bioherms in central Alberta. M, 1953, [University of Illinois, Urbana].

Roddy, David John. The Paleozoic crater at Flynn Creek, Tennessee. D, 1966, California Institute of Technology. 344 p.

Roddy, David John. Theory and procedure of the differential thermal analysis method and its application to hornblende. M, 1957, Miami University (Ohio). 147 p.

Rodeick, Craig A. The origin, distribution and depositional history of gravel deposits on the Beaufort Sea continental shelf, Alaska. M, 1975, San Jose State University. 87 p.

Roden, M. K. Rare earth elements distribution and strontium isotope data from the Gem Park igneous complex, Colorado. M, 1977, Kansas State University.

Roden, Michael F. Field geology and petrology of the Minette Diatreme at Buell Park, Apache County, Arizona. M, 1977, University of Texas, Austin.

Roden, Michael Frank. Geochemistry of the Earth's mantle, Nunivak Island, Alaska, and other areas; evidence from xenolith studies. D, 1982, Massachusetts Institute of Technology. 413 p.

Roden, Rocky Ray. Late Cenozoic seismic stratigraphy and structure of the northern Gulf of Alaska. M, 1980, Texas A&M University.

Rodenbaugh, Karl Hase. Groundwater monitoring program development, implementation, and evaluation; presentation of a case study at a solid waste landfill. D, 1987, University of California, Los Angeles. 232 p.

Rodenbeck, Sue Anita. Merging Pleistocene lithostratigraphy with geotechnical and hydrogeologic data; examples from eastern Wisconsin. M, 1988, University of Wisconsin-Madison. 286 p.

Roder, Dennis Lee. The petrology and chemistry of the Round Lake intrusion, Northwest Wisconsin. M, 1973, University of Wisconsin-Madison. 115 p.

Roderick, Gilbert Leroy. Water vapor-sodium montmorillonite interaction. D, 1965, Iowa State University of Science and Technology. 136 p.

Rodgers, Benjamin Kirby. Ostracoda and environmental interpretation of the Hurricane Lentil (middle Eocene), Texas (Leon and Houston counties). M, 1967, University of Texas, Austin.

Rodgers, David Walter. Thermal and structural evolution of the southern Deep Creek Range, west central Utah and east central Nevada. D, 1987, Stanford University. 185 p.

Rodgers, Donald Alvis. Deformation, stress accumulation, and secondary faulting in the vicinity of the Transverse Ranges of southern California. D, 1975, Brown University. 198 p.

Rodgers, Donald Alvis. Displacements and strains associated with a bend in a strike-slip fault. M, 1969, Brown University.

Rodgers, Dorothy Lynne. Possible effects of sea level rise on Kailua, Oahu and implications of Pacific atolls. M, 1988, University of Hawaii. 100 p.

Rodgers, Jack Pinknea. Sedimentary petrology of the Carrizo sand outcrop east of Bastrop, Bastrop County, Texas. M, 1947, University of Texas, Austin.

Rodgers, John. Geology and mineral deposits of Bumpass Cove, Unicoi and Washington counties, Tennessee. D, 1944, Yale University.

Rodgers, John. Stratigraphy and structure in the upper Champlain Valley. M, 1937, Cornell University.

Rodgers, Michael Robert. Geological and geochemical investigations along the northwestern extension of the Tyrone-Mount Union Lineament in the plateau province of northwestern Pennsylvania. M, 1981, University of Pittsburgh. 178 p.

Rodgers, Robert William. The stratigraphy of the Glen Rose Formation, (Cretaceous) type area, central Texas. M, 1965, Baylor University. 96 p.

Rodgers, T. Deane. A mineral analysis of the surface soils of Lubbock County, Texas. M, 1942, Texas Tech University. 64 p.

Rodgers, William Howard. Celestite from the Tonoloway Formation (Upper Silurian) at Hayfield, Frederick County, Virginia. M, 1965, University of Virginia. 97 p.

Rodiguez, Eileen. Quaternary geologic history of the San Juan nearshore area from Punta las Marias to Punta Cangrejos, Puerto Rico. M, 1987, University of South Florida, Tampa.

Rodin, Evald Maurice. The stratigraphy and structure of the Hillsboro oil field, Arkansas. M, 1949, University of Iowa. 84 p.

Rodine, J. D. Analysis of the mobilization of debris flows. D, 1975, Stanford University. 235 p.

Rodolfo, Kelvin Schmidt. Marine geology of the Andaman Basin, northeast Indian Ocean. D, 1967, University of Southern California.

Rodolfo, Kelvin Schmidt. Suspended sediment in Southern California waters. M, 1964, University of Southern California.

Rodphothong, Somphong. The induced polarization method in geophysics prospecting. M, 1971, New Mexico Institute of Mining and Technology.

Rodrigue, Raymond Fredrick. Attenuation of nitrate by sorption from solution onto montmorillonite clay. D, 1969, University of Southern California. 180 p.

Rodrigues, Cyril Gerard I. Benthonic foraminiferal associations of the lower St. Lawrence Estuary; a quantitative study. M, 1976, Carleton University. 146 p.

Rodrigues, Cyril Gerard I. Holocene microfauna and paleoceanography of the Gulf of St. Lawrence. D, 1981, Carleton University. 352 p.

Rodrigues, Francisco Soland de Oliveira. Evaluation of some chemical methods for predicting the availability of organic soil nitrogen. D, 1981, Mississippi State University. 102 p.

Rodriguez de la Garza, Fernando Javier. Unsteady state pressure behavior of a reservoir with a well intersected by a partially penetrating finite-conductivity vertical fracture. D, 1983, Stanford University. 183 p.

Rodriguez E. Rene. Friction and the mechanism of shallow focus earthquakes. D, 1976, St. Louis University. 269 p.

Rodriguez Gonzalez, Argenis. A petrographic study of relations between cementation and fracture porosity in fine-grained sandstones. M, 1977, University of Toronto.

Rodriguez Gonzalez, Argenis. Sedimentology of the Miocene Oficina Formation in the Cerro Negro area, Orinoco oil sands, Venezuela. D, 1986, University of Toronto.

Rodriguez Molina, Carlos. Engineering geology and relative stability of parts of Avondale, Cincinnati, Ohio. M, 1983, University of Cincinnati. 86 p.

Rodriguez Perez, C. E. Analysis of an underground opening in jointed rock. D, 1980, University of Illinois, Urbana. 226 p.

Rodriguez, Adolfo Chavez see Chavez Rodriguez, Adolfo

Rodriguez, Enrique Levy. Economic geology of the sulphur deposits of Sulphurdale, Utah. M, 1960, University of Utah. 74 p.

Rodriguez, Jaime Alberto. Exploration model for the coal-bearing Saginaw Formation; Michigan Basin. M, 1975, Michigan State University. 72 p.

Rodriguez, Joaquin. Invertebrate fauna of the Golconda Formation (middle Chester) of Indiana, western Kentucky, and southern Illinois. D, 1960, Indiana University, Bloomington. 259 p.

Rodriguez, Joaquin. Paleontology of the Mississippian formation of Knox County, Ohio. M, 1957, Ohio State University.

Rodriguez, Luis R. Pliocene marine ostracodes from the Cabo Blanco area, north-central Venezuela, South America. M, 1965, University of Kansas. 80 p.

Rodriguez, Maria C. Gravity investigation of crustal structures in Southwest Florida. M, 1984, University of South Florida, Tampa. 69 p.

Rodriguez, N. S. Computer-aided interpretation of seismic refraction data. M, 1976, Colorado School of Mines. 114 p.

Rodriguez, Rafael W. Origin, evolution, and morphology of the shoal Escollo de Arenas, Vieques, Puerto Rico, and its potential as a sand resource. M, 1979, University of North Carolina, Chapel Hill. 71 p.

Rodriguez, Simon E. Sedimentary facies of the Sokoman iron-bearing formation (Precambrian), Elross Creek area, Labrador, Newfoundland. M, 1968, Queen's University. 138 p.

Rodriguez-Amaya, C. A decomposed aquifer model suitable for management. D, 1976, Colorado State University. 117 p.

Rodriquez, Carlos Jose. Petrology and diagenesis of Pleistocene limestones, northeastern Yucatan Peninsula, Mexico. M, 1982, University of New Orleans. 80 p.

Rodriquez, Raphael T. Geology of the metamorphic terrane in the Acatlan area, Puebla, Mexico. M, 1970, Cornell University.

Rodvang, S. J. Geochemistry of the weathered zone of a fractured clayey deposit in southwestern Ontario. M, 1987, University of Waterloo. 177 p.

Rodwan, John Charles. Stratigraphic and sedimentologic analysis of the Middle Devonian Filer Sandstone. M, 1986, Western Michigan University.

Roe, Gene Vincent. An acoustic method for identifying sand fabric and liquefaction potential. D, 1981, University of New Hampshire. 151 p.

Roe, Glenn D. Rubidium-strontium analyses of ultramafic rocks and the origin of peridotites. D, 1965, Massachusetts Institute of Technology.

Roe, Glenn Dana. Mineralogy and some trace element studies of East Texas iron ores. M, 1961, Texas Christian University. 62 p.

Roe, Joseph Thomas. Geology of the Mount Fleecer area. M, 1948, Montana College of Mineral Science & Technology. 40 p.

Roe, Kevin K. Dating insular phosphorite; experimental results and preliminary evaluation. M, 1983, Florida State University.

Roe, Leon Mergeson, II. Sedimentary environments of the Java Group (upper Devonian); a three dimensional study. D, 1975, University of Rochester. 153 p.

Roe, Leon Mergeson, II. The origin and classification of pillow structures at Irondequoit Bay, Rochester, New York. M, 1972, University of Rochester. 32 p.

Roe, Newton Charles. Geology of the eastern Potato Hills, Pushmataha and Latimer counties, Oklahoma. M, 1955, University of Oklahoma. 63 p.

Roe, Robert Ralph. Geology of the Squaw Peak porphyry copper-molybdenum deposit, Yavapai County, Arizona. M, 1976, University of Arizona.

Roe, Walter B. Clay-veins in Illinois Springfield (Number five) coal. M, 1934, Northwestern University.

Roebroek, Edward John. Inversion of the lunar travel time data. M, 1974, University of Alberta. 101 p.

Roebuck, Sheila Joan. Predicting groundwater flows to underground coal mines in western Pennsylvania. M, 1980, Pennsylvania State University, University Park. 119 p.

Roecken, Christian. Acoustic emission in Westerly Granite. M, 1982, University of Colorado. 151 p.

Roecker, Steven William. Seismicity and tectonics of the Pamir Hindu Kush region of Central Asia. D, 1981, Massachusetts Institute of Technology. 297 p.

Roed, Murray A. Mid-Devonian productellid and chonetid brachiopods from northern Canada. M, 1961, University of Saskatchewan. 68 p.

Roed, Murray Anderson. Surficial geology of the Edson-Hinton area (Pleistocene), Alberta (Canada). D, 1968, University of Alberta. 200 p.

Roedder, Edwin W. Clinozoisite pseudomorphs after scapolite from Hillburn, New York. M, 1947, Columbia University, Teachers College.

Roedder, Edwin W. The system K_2O-MgO-SiO_2. D, 1950, Columbia University, Teachers College.

Roeder, Joseph Harrison. The symmetry of the folding of the Broad Top Synclinorium. M, 1949, University of Pittsburgh.

Roeder, Peter L. Phase relations in the Mg_2SiO_4-$CaAl_2Si_2O_8$-FeO-Fe_2O_3-SiO_2 systems and their bearing on crystallization of basaltic magma. D, 1960, Pennsylvania State University, University Park. 111 p.

Roegiers, Jean-Claude. The development and evaluation of a field method for in-situ stress determination using hydraulic fracturing. D, 1974, University of Minnesota, Minneapolis. 284 p.

Roegner, Harold F., Jr. Oil and gas reservoir influence functions. M, 1971, University of Oklahoma. 127 p.

Roehl, Perry O. The Paleozoic geosyncline of the Klamath Mountains, California. D, 1955, University of Wisconsin-Madison.

Roehler, H. W. Geology of Bates Hole, Wyoming. M, 1958, University of Wyoming. 58 p.

Roehrs, Robert C. A study of isolated Upper Devonian outcrops in central Missouri. M, 1958, University of Missouri, Columbia.

Roellig, Harold Frederick. The Osteoglossidae, fossil and Recent. D, 1967, Columbia University. 228 p.

Roeloffs, Evelyn Anne. Effects of pore pressure variations in critically stressed sandstones. M, 1978, University of Wisconsin-Madison.

Roeloffs, Evelyn Anne. Elasticity of saturated porous rocks; laboratory measurements and a crack problem. D, 1982, University of Wisconsin-Madison. 171 p.

Roelofs, Nicolas Henry. Thermal rearrangements of aromatic hydrocarbons. D, 1987, [University of Nevada]. 166 p.

Roemer, Cletus D. Petrology of the Davis Shale of the Potosi and Edge Hill Quadrangle. M, 1941, St. Louis University.

Roemer, Lamar B. Geology and geophysics of the Beata Ridge, Caribbean. M, 1973, Texas A&M University.

Roemer, Lamar B. Structure and stratigraphy of late Quaternary deposits on the outer Louisiana shelf. D, 1976, Texas A&M University. 268 p.

Roemermann, Donald Gregory. The subsurface Sellersburg Limestone of southern Indiana. M, 1954, Indiana University, Bloomington. 44 p.

Roemmel, Janet Sue. A petrographic and economic evaluation of the White Cow Intrusion, Little Rocky Mountains, Montana. M, 1982, University of Montana. 51 p.

Roen, John Brandt. The geology of the Lynn Window, Tuscarora Mountains, Eureka County, Nevada. M, 1962, University of California, Los Angeles.

Roeper, Timothy R. A geotechnical and hydrogeological investigation of two landfills in Portage County, Ohio. M, 1988, Kent State University, Kent. 193 p.

Roepke, Harlan H. The Nisswa Lake marl deposit, Crow Wing County, Minnesota. M, 1958, University of Minnesota, Minneapolis. 60 p.

Roepke, Harlan Hugh. Petrology of carbonate units in the Canyon group (Missourian series), central Texas. D, 1970, University of Texas, Austin. 371 p.

Roepke, Timothy J. Stratigraphy and microfacies analysis of the Pennsylvanian System, Robledo Mountains, Dona Ana County, New Mexico. M, 1984, University of Texas at El Paso.

Roeser, Edgar Waldemar. A Lower Mississippian (Kinderhookian-Osagian) crinoid fauna from the Cuyahoga Formation of northeastern Ohio. M, 1986, University of Cincinnati. 322 p.

Roeske, Sarah Melissa. Metamorphic and tectonic history of the Raspberry Schist, Kodiak Islands, Alaska. D, 1988, University of California, Santa Cruz. 219 p.

Roesler, Max. Geology of the iron-ore deposits of the Firmeza District, Oriente Province, Cuba. D, 1916, Yale University.

Roesler, Max. Some garnet rims in anorthosite rock from the Adirondacks (New York). M, 1915, Columbia University, Teachers College.

Roesler, Toby Albert. The structure and stratigraphy of a part of the Tepechitlan Quadrangle, Zacatecas, Mexico. M, 1987, University of New Orleans.

Roethel, Frank Joseph. The interactions of stabilized power plant coal wastes with the marine environment. D, 1981, SUNY at Stony Brook. 376 p.

Rofe, Rafael. Subsurface geology of the Dunnigan Hills gas field, Yolo County, California. M, 1960, Stanford University. 28 p.

Roffman, Haia. Natural and experimental weathering of basalts, southcentral New Mexico. D, 1972, New Mexico Institute of Mining and Technology.

Rofheart, Douglas H. Depositional environment, diagenesis and petrophysical properties of the upper Bluejacket Sandstone (Desmoinesian), KB Field, Allen County, Kansas. M, 1987, University of Kansas. 174 p.

Rog, Andre Mark. The geology and petrogenesis of the Volcano Cerro Grande, San Pedro, Nayarit, Mexico. M, 1983, University of New Orleans. 98 p.

Rogan, Allan Douglas. Laboratory analysis of some of the rock properties of low permeability oil and gas reservoirs. M, 1961, University of Saskatchewan. 84 p.

Rogatz, Henry. The geology and stratigraphy of the Texas Panhandle oil and gas fields. D, 1939, Columbia University, Teachers College.

Rogatz, Henry. The ostracode fauna of the Arroyo Formation, Wichita-Albany Group of the Permian in Tom Green County, Texas. M, 1932, Columbia University, Teachers College.

Rogers, Albert Mitchell, Jr. The effect of a dipping layer on P wave transmission. D, 1970, St. Louis University. 118 p.

Rogers, Allen S. The physical behavior and geologic control of radon in mountain streams. D, 1954, University of Utah. 58 p.

Rogers, Austin Flint. Crystallographic studies; (A) The morphology of certain organic compounds, (B) The calcites of the New Jersey Trap region, (C) New graphical methods. D, 1902, Columbia University, Teachers College. 36 p.

Rogers, Betty Ross. A structural problem in the Leyden Argillite, Bernardston, Massachusetts. M, 1941, Mount Holyoke College. 69 p.

Rogers, Caroline Sutherland. The response of a coral reef to sedimentation. D, 1977, University of Florida. 196 p.

Rogers, Cassandra T. The influence of petrographic factors on the strength and deformational behaviour of some quartz-rich sandstones. M, 1982, University of Windsor. 172 p.

Rogers, Charles William. Structural geology of the Round Rock Quadrangle, Williamson County, Texas. M, 1963, University of Texas, Austin.

Rogers, Christopher A. The effect of de-icing agents on water adsorption phenomena in rock aggregates. M, 1977, University of Windsor. 122 p.

Rogers, Daniel T. Petrology of the Middle Cambrian Langston and Ute formations in southeastern Idaho. M, 1987, Utah State University. 208 p.

Rogers, David K. Environmental geology of northern Carson City, Nevada. M, 1975, University of Nevada. 133 p.

Rogers, David M. Geomorphology and sedimentology of the Hancock County ridge complex, coastal Mississippi. M, 1984, Mississippi State University. 52 p.

Rogers, David P. A petrographic reconnaissance of granitic rocks in the Biscotasing area, Ontario. M, 1961, University of Toronto.

Rogers, Dennis J. Sandstone petrography of the Mesaverde Group of northwestern Colorado. M, 1975, Kent State University, Kent. 118 p.

Rogers, Frederick S. Tylothyris (Brachiopoda) from the Traverse Group (Devonian) of Michigan. M, 1984, University of Massachusetts. 89 p.

Rogers, G. C. The study of a micro-earthquake swarm. M, 1972, University of Hawaii.

Rogers, Gaillard S. Geology of the Cortlandt Series and its emery deposits (New York). D, 1911, Columbia University, Teachers College.

Rogers, Garry Colin. Seismotectonics of British Columbia. D, 1983, University of British Columbia. 247 p.

Rogers, Gary D. The structure and stratigraphy of the Proterozoic Hecla Hoek succession near Orvinfjellet, Wedel Jarlsberg Land, Spitsbergen. M, 1985, University of Wisconsin-Madison. 135 p.

Rogers, Jack A., Jr. Kimberlite exploration, Red Feather area, and petrology of the Chicken Park diatreme, northern Colorado. M, 1985, Colorado State University. 211 p.

Rogers, James E., Jr. Silurian and Lower Devonian stratigraphy and paleobasin development; Illinois Basin, Central United States. D, 1972, University of Illinois, Urbana. 144 p.

Rogers, James E., Jr. Upper Silurian and lower Devonian stratigraphy of the Illinois Basin, central United States. M, 1970, University of Illinois, Urbana. 69 p.

Rogers, James Edwin. Mineralogy of Oxford serpentine deposit, Llano County, Texas. M, 1961, University of Texas, Austin.

Rogers, James Joseph. Design of a system for predicting effects of vegetation manipulation of water yield in the Salt-Verde Basin. D, 1973, University of Arizona.

Rogers, James Joseph. Mathematical system theory and the ecosystem concept, an approach to modelling watershed behavior. M, 1971, University of Arizona.

Rogers, James Kenneth. Landslides in Kentucky. M, 1929, University of Cincinnati. 96 p.

Rogers, James Kenneth. The geology of Highland County, Ohio. D, 1933, University of Cincinnati. 236 p.

Rogers, James Palmer. Stratigraphy and paleontology of a core penetrating Upper Silurian and Middle Devonian strata in Lorain County, Ohio. M, 1953, University of Michigan.

Rogers, James Samuel. The hydraulic conductivity; water content relationship during non-steady flow in a vertical sand column. D, 1969, University of Illinois, Urbana. 65 p.

Rogers, Jerry Rowland. Seasonal streamflow generation for rivers with highly variable flow. D, 1970, Northwestern University.

Rogers, Jesse Armstead. Foraminifera from the (Eocene) Yazoo Clay of the Jackson Formation near Shubuta, Mississippi. M, 1936, Texas Tech University. 51 p.

Rogers, John C. Depositional environments and paleoecology of two quarry sites in the Middle Cambrian Marjum and Wheeler formations, House Range, Utah. M, 1984, Brigham Young University. 115 p.

Rogers, John Hiram, Jr. Geology of a portion of 17 ledge, Homestake Mine, Lead, South Dakota. M, 1984, University of Oregon. 135 p.

Rogers, John James William. Absorption of uranium by peat. M, 1952, University of Minnesota, Minneapolis.

Rogers, John James William. Textural studies in igneous rocks near Twenty-Nine Palms, California. D, 1955, California Institute of Technology. 258 p.

Rogers, John L. An investigation into the environment of production for the silt-size particles of loess. M, 1979, University of Houston.

Rogers, John Robert. Deep water sediments of northwestern Lake Huron. M, 1959, University of Illinois, Urbana. 36 p.

Rogers, Jonathan David. The genesis, properties and significance of fracturing in Colorado Plateau sandstones. D, 1982, University of California, Berkeley. 724 p.

Rogers, Kenneth Joseph. Relationships between velate structure and brood pouch in beyrichiid ostracods. M, 1957, University of Michigan.

Rogers, Leah Lucille. A cross sectional groundwater flow model of the Sheffield Illinois low level radioactive waste site. M, 1979, University of Illinois, Urbana. 73 p.

Rogers, Lewis F. The petrology-mineralogy of six Georgia kaolins. D, 1979, University of Georgia. 246 p.

Rogers, Lowell Thompson. Petrology of Grindstone Creek and Garner formations, Erath and Eastland counties, Texas. M, 1960, University of Texas, Austin.

Rogers, Luther Franklin, Jr. Arkosic conglomerate beds in Mason, Menard, and Kimble counties, Texas. M, 1955, Texas A&M University. 66 p.

Rogers, Margaret Anne Christie. Stratigraphy and structure of the Fredericksburg division (lower Cretaceous), northwest quarter lake Travis Quadrangle, Travis and Williamson counties, Texas. M, 1969, University of Texas, Austin.

Rogers, Marion Alan. Carbohydrates in plants and sediments from two Minnesota lakes. M, 1962, University of Minnesota, Minneapolis. 164 p.

Rogers, Marion Alan. Organic geochemistry of some Devonian black shales from eastern North America; carbohydrates. D, 1965, University of Minnesota, Minneapolis.

Rogers, Mark Arleigh. Petrology, geochemistry and significance of ultramafic inclusions in lavas from Guadalupe Island; an alkali basalt volcano in the eastern Pacific Ocean. M, 1981, Washington University. 162 p.

Rogers, Maynard. The Pleistocene geology of the London and the South Charleston quadrangles, and the eastern part of the Springfield Quadrangle, Ohio. M, 1938, Ohio State University.

Rogers, Melissa J. B. The determination of Q_{Lg} and Q_c as a function of frequency in the crust of Virginia and its environs. M, 1988, Virginia Polytechnic Institute and State University. 80 p.

Rogers, Patrick C. Origin of stratiform barite deposits, in the northern Shoshone Range and Osgood Mountains, northern Nevada. M, 1985, University of Idaho. 114 p.

Rogers, Philip R. Roof fall study of the Harrisburg (No. 5) Coal in portions of Saline and Williamson counties, Illinois. M, 1981, Southern Illinois University, Carbondale. 66 p.

Rogers, R. D. Copper mineralization in Pennsylvanian-Permian rocks of the Tonto Rim segment of the Mogollon Rim in central Arizona. M, 1977, University of Arizona. 65 p.

Rogers, Ralph D. Structural and geochemical evolution of a mineralized volcanic vent at Cerro de Pasco, Peru. D, 1983, University of Arizona. 142 p.

Rogers, Reginald Douglas, Jr. A petrographic study of the (Upper Ordovician) Bald Eagle, Juniata, and (Lower Silurian) Tuscarora formations in central Pennsylvania. M, 1939, Cornell University.

Rogers, Robert B. Electrical resistivity monitoring of fly ash leachate in a glacial aquifer. M, 1979, University of Wisconsin-Milwaukee. 123 p.

Rogers, Robert Errett. Geologic mapping of the Greybull-Basin area, Wyoming as interpreted from aerial photographs. M, 1948, University of Illinois, Urbana. 47 p.

Rogers, Robert John. A numerical model for simulating pedogenesis in semiarid regions. D, 1980, University of Utah. 296 p.

Rogers, Robert K. Element distribution in the Rio Tuba nickeliferous laterite deposit, Philippines. M, 1973, Michigan Technological University. 81 p.

Rogers, Scott W. Sediments of a seagrass bed in Anclote Anchorage, Tarpon Springs, Florida. M, 1977, University of South Florida, St. Petersburg.

Rogers, Thomas Hardin. Stratigraphy and structure of the southern Swisshelm Mountains, Cochise County, Arizona. M, 1957, University of California, Berkeley. 46 p.

Rogers, Thornwell. The geology of the Copper Chief Mine area, Yavapai County, Arizona. M, 1978, Northern Arizona University. 102 p.

Rogers, Timothy Joseph. Sedimentation and contemporaneous structures in part of the western Kentucky coal field. M, 1985, University of Kentucky.

Rogers, W. J. Sodium-calcium ion exchange on clay minerals at moderate to high ionic strengths. D, 1979, University of Tennessee, Knoxville. 142 p.

Rogers, Wilbur Frank. A lithologic study of the Dakota Group in certain deep wells in Nebraska. M, 1941, University of Nebraska, Lincoln.

Rogers, Wiley Samuel, III. Middle Ordovician stratigraphy of the Red Mountain area, Alabama. D, 1960, University of North Carolina, Chapel Hill. 272 p.

Rogers, Wiley Wamuel. The crystallographic and chemical examination of the crystal forms of titanite. M, 1951, Emory University. 54 p.

Rogers, William Brokaw. Depositional environments of the Skinner Ranch and Hess formations (lower Permian), Glass Mountains, West Texas. D, 1972, University of Texas, Austin. 551 p.

Rogers, William Donald. A faunal study of the Otterville Limestone Member of the Dornick Hills Formation in the Ardmore Basin. M, 1948, University of Oklahoma. 53 p.

Rogers, William John, Jr. The effects of subducting slabs on seismic amplitudes and travel-times from forearc and outer rise earthquakes. M, 1982, Michigan State University. 113 p.

Rogers, William Patrick. A geological and geophysical study of the central Puget Sound lowland. D, 1970, University of Washington. 123 p.

Rogers, William Patrick. General geology of Medicine Mountain area, Bighorn Mountains, Wyoming. M, 1958, University of Iowa. 94 p.

Roggensack, Kurt. Morphology, distribution, and chemistry of shield volcanoes of the central Trans-Mexican Volcanic Belt. M, 1988, Dartmouth College. 148 p.

Roggenthen, William Michael. Paleomagnetism of Late Cretaceous-early Tertiary European pelagic limestones. D, 1980, Princeton University. 98 p.

Roghani, Foad. Stratigraphy and geologic history of the Manitou Formation in the Garden Park area, Fremont and Teller counties, Colorado. M, 1975, Colorado School of Mines. 93 p.

Roglans-Ribas, Jordi. Disposal of spent nuclear fuel and high-level waste; design and technical/economic analysis. D, 1987, Iowa State University of Science and Technology. 269 p.

Rognerud, Walter N. Geophysical investigation of a gravity anomaly in the central portions of Rockland and Rousseau quadrangles, Michigan. M, 1974, Michigan Technological University. 207 p.

Rogozen, M. B. Coal slurry pipelines; the water issues. D, 1978, University of California, Los Angeles. 284 p.

Rohani-Najafabadi, Behzad. A nonlinear elastic-viscoplastic constitutive relationship for earth materials. D, 1970, Texas A&M University. 161 p.

Rohatgi, N. K. Fate of trace metals upon ocean disposal of wastewater effluents. D, 1975, University of Southern California.

Rohaus, Donna Marie. Structural analysis of the Siluro-Devonian cover in Pendleton County, West Virginia. M, 1987, West Virginia University. 134 p.

Rohay, Alan Charles. Crust and mantle structure of the North Cascades Range, Washington. D, 1982, University of Washington. 163 p.

Rohl, Arthur N. Alteration and mineralization in the Uravan mineral belt, Colorado. D, 1985, Columbia University, Teachers College. 168 p.

Rohl, Arthur N. Asphaltite-metallic sulfide association, La Bajada, New Mexico. M, 1959, Columbia University, Teachers College.

Rohr, David M. Geology of the Lovers Leap area, China Mountain Quadrangle, California. M, 1972, Oregon State University. 95 p.

Rohr, David. M. Stratigraphy, structure, and early Paleozoic Gastropoda of the Callahan area, Klamath Mountains, California. D, 1978, Oregon State University. 316 p.

Rohr, Gene M. Marine Upper Jurassic microfauna and stratigraphy of north-central Wyoming (Bighorn Basin). M, 1965, University of Iowa. 80 p.

Rohr, Kristin Marie Michener. A study of the seismic structure of upper oceanic crust using wide-angle reflections. D, 1983, Massachusetts Institute of Technology. 210 p.

Rohr, Steven Anthony. Stratigraphic analysis of the Prairie du Chien Group, Lower Peninsula, Michigan. M, 1985, Michigan State University. 151 p.

Rohrbacher, Timothy J. Selected trace elements, copper, and iron distributions in the Copper Harbor conglomerate (Precambrian), White Pine, Michigan. M, 1968, University of Toledo. 111 p.

Rohrbacher-Carls, M. R. Some sublethal effects of oiled sediment on a population of the marine amphipod Pontoporeia femorata (Kroeyar, 1842). M, 1978, Dalhousie University.

Rohrback, B. G. Analysis of low molecular weight products generated by thermal decomposition of organic matter in Recent sedimentary environments. D, 1979, University of California, Los Angeles. 208 p.

Rohrbacker, Robert G. Geology of the Temporal Gulch-Mansfield Canyon area, Santa Cruz County, Arizona. M, 1964, University of Arizona.

Rohrbough, Claude A. Subsurface geology of northwestern Kansas. M, 1956, Kansas State University. 50 p.

Rohrbough, R. D. and Koehr, J. E. Daily and quasi-weekly beach profile changes at Monterey, California. M, 1964, United States Naval Academy.

Rohrer, William Lyman Murphy see Murphy Rohrer, William Lyman

Rohrs, David Tullar. A light stable isotope study of the Roosevelt Hot Springs thermal area, southwestern Utah. M, 1980, University of Utah. 87 p.

Rohtert, William R. Paragenesis of the Thumbum uranium deposit, San Bernardino Mountains, California. M, 1981, University of Colorado. 149 p.

Roig, J. H. Use of heavy minerals as tracers of sand transport of the Santa Barbara-Oxnard shelf; Santa Barbara Channel, California. M, 1976, University of Southern California.

Roig, Lisa C. A Kalman filter application for the design of water quality monitoring programs. M, 1984, Colorado State University. 153 p.

Rojanasoonthon, Santhad. Morphology and genesis of gray podzolic soils in Thailand. D, 1972, Oregon State University. 226 p.

Rojas, Armando J. Ramirez see Ramirez Rojas, Armando J.

Rojas, Gloria G. Foraminifera from the Viche Formation, early-middle Miocene, northwestern Ecuador. M, 1980, University of California, Los Angeles.

Rojas, Isabel de see de Rojas, Isabel

Rojas, Leyla Amparo. Mineralogical influences on boron adsorption in soils from Colombia. D, 1988, University of California, Riverside. 129 p.

Rojas-Gonzalez, Luis Fernando. Analysis of laterally loaded drilled pier foundations. D, 1988, University of Pittsburgh. 218 p.

Rojstaczer, Stuart Alan. Moisture movement through layered soils of highly contrasting texture. M, 1981, University of Illinois, Urbana. 86 p.

Rojstaczer, Stuart Alan. The response of the water level in a well to atmospheric loading and Earth tides; theory and applications. D, 1988, Stanford University. 225 p.

Rokach, Allen. Ice-rafted sediment in the Bellingshausen Basin of the Southern Ocean; a climatic indicator in Antarctic deep-sea cores. M, 1973, Queens College (CUNY). 110 p.

Rokke, Stephen R. A study of the buried reefs in the Northwest Garza area, Garza County, Texas, based on seismograms and electric logs. M, 1957, Texas A&M University.

Rokke, Stephen Richard. Reef analysis on seismograms from northeast Garza County, Texas. M, 1958, Texas A&M University. 31 p.

Roland, George Warren. Phase relations and geologic application of the system Ag-As-S. D, 1966, Lehigh University. 203 p.

Roland, John L. A paleomagnetic age investigation of pre-Salmon Springs Drift Pleistocene deposits in the southern Puget Lowland, Washington. M, 1983, Western Washington University. 93 p.

Rold, John W. Structure and Pre-Pennsylvanian stratigraphy of the Wellsville area, Fremont County, Colorado. M, 1950, University of Colorado.

Roldan-Quintana, J. The geology and mineralization of the San Felipe area, East-central Sonora, Mexico. M, 1976, University of Iowa. 120 p.

Rolf, E. Gerald. The geology of the Anderson area, Grimes County, Texas. M, 1958, Texas A&M University. 120 p.

Rolfe, Martha Deette. A study of Illinois physiography. M, 1904, University of Illinois, Urbana.

Rolin, John. The refinement of the structure of $MgSO_4 \cdot 5H_2O$. M, 1966, University of Pittsburgh.

Roll, Margaret A. Effects of Acadian kyanite-zone metamorphism on relict granulite-facies assemblages, Mt. Mineral Formation, Pelham Dome, Massachusetts. M, 1987, University of Massachusetts. 257 p.

Roller, Julie Ann. Geometric and kinematic analysis of the Proterozoic upper Alder Group and the Slate Creek movement zone, central Mazatzal Mountains, central Arizona. M, 1987, Northern Arizona University. 105 p.

Rolling, David M. Low resistive Pleistocene pay sands; a case study, Ship Shoal 290-291 Field. M, 1979, Tulane University.

Rollins, Anthony. Geology of the Bachelor Mountain area, Linn and Marion counties, Oregon. M, 1976, Oregon State University. 83 p.

Rollins, David D. Shoreline and river processes at the site of the proposed Salmon River breakwater, Lake Ontario. M, 1980, SUNY, College at Fredonia. 111 p.

Rollins, Francis O'Rourke. Flexure of the lithosphere beneath the lower portion of the Devonian Catskill Delta. M, 1985, University of North Carolina, Chapel Hill. 61 p.

Rollins, Harold B. Gastropods from the Lower Mississippian Wassonville Limestone in southeastern Iowa. M, 1963, University of Wisconsin-Madison.

Rollins, Harold Bert. The phylogeny and functional morphology of the Knightitinae, Carinaropsinae, and Praematuratripidae (Gastropoda, Bellerophontacea). D, 1967, Columbia University. 188 p.

Rollins, John Flett. Stratigraphy and structure of the Goajira Peninsula, northwestern Venezuela and northeastern Colombia. D, 1960, University of Nebraska, Lincoln. 366 p.

Rollins, John Flett. Stratigraphy and structure of the Sheep Mountain area, in North Park, Jackson County, Colorado. M, 1952, University of Nebraska, Lincoln.

Rollins, Kyle Morris. The influence of buildings on potential liquefaction damage. D, 1987, University of California, Berkeley. 274 p.

Rolnick, L. S. Stability of gypsum and anhydrite in the geologic environment. D, 1954, Massachusetts Institute of Technology. 152 p.

Roloff, Glaucio. Strength of low and variable charge soils. D, 1988, University of Minnesota, Minneapolis. 118 p.

Rolph, Alan Lindsay. Structure and stratigraphy around the Elizabeth Mine, Vermont. M, 1982, University of Cincinnati. 228 p.

Roma-Hernandez, Mauricio. The Providence Island Formation in the Northern Appalachian region; a Lower-Middle Ordovician analogue to recent arid-semiarid tidal-flat carbonates in the Persian Gulf Trucial Coast. M, 1987, SUNY at Albany. 209 p.

Romain, Samuel Joseph St. *see* St. Romain, Samuel Joseph

Roman, Juan A. Deliz. Shallow seismic refraction survey, Gallatin County, Illinois. M, 1983, Southern Illinois University, Carbondale. 113 p.

Roman, Luis Alberto. A subsurface study of rocks of the Morrow Series of the Pennsylvanian System in Cimarron County, Oklahoma. M, 1973, Texas Christian University.

Roman, Ronald J. Manganese dioxide-sulfuric acid oxidation of molybdenite. M, 1964, New Mexico Institute of Mining and Technology. 55 p.

Romanak, Martin. Sedimentology and depositional environments of the basement sands, West Texas. M, 1988, University of Texas, Arlington. 143 p.

Romanelli, R. Environmental history of sand and gravel deposits of the Champlain Sea in the Gatineau Valley, Quebec. M, 1976, University of Ottawa. 119 p.

Romani-Cardenas, J. A. Fredy. Dependence of stability of slopes on initiation and progression of failure. D, 1970, Purdue University. 203 p.

Romanik, Peter B. Delineation of a karst groundwater basin in Sinking Valley, Pulaski County, Kentucky. M, 1986, Eastern Kentucky University. 85 p.

Romano, Joseph Vincent. Feldspar-grain size relations in the Mazomanie and Reno members of the Franconia Formation (Upper Cambrian), eastern Minnesota. M, 1982, University of Minnesota, Minneapolis. 94 p.

Romano, R. R. A study of selected heavy metals in the Grand Calumet River-Indiana Harbor canal system. D, 1976, Purdue University. 221 p.

Romans, Robert Charles. Palynology of some upper Cretaceous coals of Black Mesa, Arizona. D, 1969, University of Arizona. 193 p.

Rombauer, Alfred B. Placer mining at Pauline Placer, Clear Creek Co., Colorado. M, 1886, Washington University.

Rombauer, Alfred B. and Sauer, J. H. Report of Rosborough coal mine, Rosborough, Ill. M, 1889, Washington University.

Romberger, Samuel Bergstresser. Solubility of copper in aqueous sulfide solutions coexisting with covellite from 25 degrees C to 200 degrees C, with geologic applications. D, 1968, Pennsylvania State University, University Park. 216 p.

Romer, Henry Severyn De *see* De Romer, Henry Severyn

Romero, E. Optimization of steam drive processes by geometric programming. M, 1974, Stanford University. 178 p.

Romero, Gonzalo Cruz. Ion-exchange and calcium carbonate solubility in Calcareous soils. D, 1968, University of California, Riverside. 98 p.

Romero, John C. Geologic control of ground water in the Kiowa-Wolf-Comanche Creek area in central Adams County, Colorado. M, 1965, [Colorado State University].

Romey, William D. Geology of a part of the Etna Quadrangle, Siskiyou County, California. D, 1963, University of California, Berkeley. 96 p.

Romick, Jay D. The igneous petrology and geochemistry of northern Akutan Island, Alaska. M, 1982, University of Alaska, Fairbanks. 150 p.

Romig, Phillip R., Jr. Secular strains in NE Denver (Colorado), 1968. D, 1969, Colorado School of Mines. 60 p.

Romig, Woodfred Edward. The Latouche caving system. M, 1927, University of Nevada - Mackay School of Mines. 53 p.

Romine, Karen Kay. Neogene paleoceanography of the equatorial and North Pacific Ocean. D, 1983, University of Rhode Island. 302 p.

Rominger, Joseph F. Interrelationships of the geological and soil mechanics properties of the Lake Agassiz sediments. D, 1950, Northwestern University.

Rominger, Joseph F. Orientation analysis of fine-grained clastic sediments. M, 1943, Northwestern University.

Romito, Anthony A. Slope stability analysis of surface mine spoil banks. M, 1976, Kent State University, Kent. 38 p.

Romkens, Mathias Joseph Marie. Migration of mineral particles in ice with a temperature gradient. D, 1969, Cornell University. 120 p.

Romney, Carl Frederick, Jr. The Dixie Valley-Fairview Peak earthquakes of December 16, 1954. D, 1956, University of California, Berkeley. 75 p.

Ron, Enrique Castillo *see* Castillo Ron, Enrique

Rona, Peter Arnold. A seismic and sedimentological investigation of the continental terrace, continental rise, and abyssal plain of Cape Hatteras, North Carolina. D, 1967, Yale University. 346 p.

Ronai, Lili E. Recent reef-living foraminifera of Barbados, British West Indies. M, 1955, New York University.

Ronai, Peter. Brackish-water foraminifera of the New York Bight. M, 1952, New York University.

Ronalder, Nina Lynn Walker. Geologic investigation of the Austin-Taylor contact, Northeast Texas. M, 1982, University of Texas, Arlington. 196 p.

Ronan, Thomas E. Structural and paleoecological aspects of a modern marine soft-sediment community; an experimental field study. D, 1975, University of California, Davis. 220 p.

Ronbeck, Arthur C. The paragenesis of magnetite in sulphide ores. M, 1938, University of Minnesota, Minneapolis. 37 p.

Ronca, Luciano Bruno. Peel study of the Shawnee Group (Pennsylvanian). M, 1959, University of Kansas. 129 p.

Ronca, Luciano Bruno. Thermoluminescence as a paleoclimatological tool. D, 1963, University of Kansas. 100 p.

Ronen, Joshua Mordechai. Multichannel inversion in reflection seismology. D, 1986, Stanford University. 72 p.

Rones, Morris. A litho-stratigraphic, petrographic and chemical investigation of the lower Middle Ordovician carbonate rocks in central Pennsylvania. D, 1955, Pennsylvania State University, University Park. 345 p.

Rongitsch, Brian A. Relationship of the McGrath Gneiss and the Thomson Formation in the vicinity of Denhem, Pine County, Minnesota. M, 1971, Northern Illinois University. 55 p.

Ronkos, Charles Joseph. Geology, alteration, and mineralization in the pyroclastic and sedimentary deposits of the Bretz-Aurora Basin, McDermitt Caldera, Nevada-Oregon. M, 1981, University of Nevada. 92 p.

Ronnei, David M. Geohydrologic evaluation of a proposed coal-ash disposal site near Mandan, North Dakota. M, 1987, University of North Dakota. 206 p.

Roob, Christine K. Metamorphism in the Wabigoon Subprovince in the vicinity of Vermilion Bay and Sioux Lookout, Ontario. M, 1987, University of North Dakota. 118 p.

Roof, Raymond Bradley, Jr. The crystal structure of ferric acetylacetonate, an application of Fourier series to crystal structure analysis. D, 1955, University of Michigan.

Roohi, Mansoor. Topanga Formation in the Malibu Lake area. M, 1969, University of California, Los Angeles.

Rooke, Steven. Computer analysis of geologic and geochemical data of the Fort Cady borate prospect. M, 1982, University of Arizona. 147 p.

Rooney, Lawrence Fredrich. A stratigraphic study of the Permian formations of part of southwestern Montana. D, 1956, Indiana University, Bloomington. 111 p.

Rooney, Rosalia Eugenia (Rey). Relations between spontaneous potential and detailed petrology in three

drill holes. M, 1956, Indiana University, Bloomington. 31 p.

Rooney, Sean T. Subglacial geology of ice stream B, West Antarctica. D, 1988, University of Wisconsin-Madison. 188 p.

Rooney, Thomas P. Influence of Al/Si disordering on the infrared absorption spectra of cordierite. M, 1962, Columbia University, Teachers College.

Rooney, Thomas Peter. Mineralogical study of North Carolina phosphorite. D, 1965, Columbia University. 142 p.

Rooney, William Stephen, Jr. Endolithic organisms from the Arlington Reef complex, Great Barrier Reef, Australia; their external morphology, distribution, and geologic significance. M, 1971, Duke University. 73 p.

Roosendaal, Dan J. Van *see* Van Roosendaal, Dan J.

Root, Forrest Keith. Structure, petrology and mineralogy of pre-Beltian metamorphic rocks of the Pony-Sappington area, Madison County, Montana. D, 1965, Indiana University, Bloomington. 184 p.

Root, Kevin Gordon. Geology of the Delphine Creek area, southeastern British Columbia; implications for the Proterozoic and Paleozoic development of the Cordilleran divergent margin. D, 1987, University of Calgary. 446 p.

Root, Michael R. Computer applications to petroleum exploration, Osage County, Oklahoma. M, 1978, University of Alaska, Fairbanks. 148 p.

Root, Ralph R. Computerized terrain mapping of Yellowstone National Park. D, 1974, Colorado State University. 255 p.

Root, Robert Lee. Geology of the Smith and Morehouse-Hayden Fork area, Utah. M, 1952, University of Utah. 58 p.

Root, Robert William, Jr. A comparison of deterministic and geostatistical modeling methods as applied to numerical simulation of ground-water flow at the Savannah River Plant, South Carolina. D, 1987, Pennsylvania State University, University Park. 568 p.

Root, Ruth Eva. Globigerina pachyderma and globigerina quinqueloba in Late Cenozoic sediment of the Arctic Ocean. M, 1972, University of Wisconsin-Madison.

Root, Samuel. The Paleozoic geology of the Canal Flats area, British Columbia. M, 1954, University of Manitoba.

Root, Samuel I. Geology of Knox County, Ohio. D, 1958, Ohio State University. 330 p.

Root, Stephen Allen. The paleoecology of the Weno and Pawpaw formations (Lower Cretaceous) of north central Texas. M, 1975, University of Texas, Arlington. 114 p.

Root, Towner B. The glacial geology of the lower Illinois River valley. D, 1935, University of Chicago. 313 p.

Rooth, Guy Harlan. Biostratigraphy and paleoecology of the Coaledo and Bastendorff formations, southwestern Oregon. D, 1974, Oregon State University. 270 p.

Roots, Charles Frederick. Regional tectonic setting and evolution of the late Proterozoic Mount Harper volcanic complex, Ogilvie Mountains, Yukon. D, 1988, Carleton University. 127 p.

Roots, Ernest Frederick. Geology and mineral deposits of the Aiken Lake Map-area, British Columbia. D, 1949, Princeton University. 627 p.

Roots, Ernest Frederick. Geology of the Aiken Lake Map area, British Columbia. M, 1947, University of British Columbia.

Rooyani, Firouz. Characteristics and genesis of certain soils in the southern foothills of central Alborz, Iran. D, 1980, Utah State University. 131 p.

Rooze, Tom W. Field study of repetitive longshore bars, Truro, Massachusetts Bay. M, 1986, Massachusetts Institute of Technology. 83 p.

Roper, Douglas C. Mesoscopic fabric of Babbs Knobs area (Pulaski Thrust, East Tennessee). M, 1977, University of Tennessee, Knoxville. 83 p.

Roper, Edith D. Paleoecology of thr coral reefs and stromatoporocil faunas of the Callaway Lithofacies, Cedar Valley Formation, Boone and Callaway counties, Missouri. M, 1979, University of Missouri, Columbia.

Roper, Michael William. Geology of the Kelsey copper-molybdenum property, Okanogan County, Washington. M, 1973, Montana State University. 97 p.

Roper, Paul James. Geology of the Tamassee, Satolah and Cashiers quadrangles, Oconee County, South Carolina. D, 1970, University of North Carolina, Chapel Hill. 113 p.

Roper, Paul James. The paleoecology of a Middle Devonian stromatoporoid bioherm in southwestern Ontario. M, 1964, University of Nebraska, Lincoln.

Ropes, Leverett H. Proposal for instrumentation and study of the Gallatin seismic-sensitive well. M, 1963, Montana State University. 62 p.

Roquemore, G. The Cenozoic history of the Coso Mountains as determined by tuffaceous lacustrine deposits. M, 1977, California State University, Fresno.

Roquemore, Glenn Raymond. Active faults and associated tectonic stress in the Coso Range, California. D, 1981, University of Nevada. 231 p.

Roquemore, Sam Kendall. Clinoptilolite occurrence in the Tallahatta Formation (middle Eocene) of Southeast Mississippi. M, 1984, University of Mississippi. 106 p.

Roqueplo-Brouillet, C. Seismic stratigraphy of the eastern continental shelf of the Weddell Sea, Antarctica. M, 1983, Rice University. 132 p.

Rorick, Andrew Hammond. Sediment dispersal patterns and provenance of the Marshall Formation (Mississippian) in the Michigan Basin; a petrographic analysis. M, 1983, University of Toledo. 130 p.

Rorty, Melitta. A geostatistical analysis of soil hydraulic properties on an uplifted marine terrane. M, 1985, University of California, Santa Cruz.

Rosa, Andre L. R. Extraction of elastic parameters using seismic reflection amplitude. M, 1976, University of Houston.

Rosa, Eugene. Multiple deformations and minor structures of major rock units in the Lynchburg Quadrangle, Virginia. M, 1977, University of Kentucky. 92 p.

Rosa, Felipe. Mineralogy and petrogenesis study of the Stockdale Intrusive, Riley County, Kansas. M, 1966, Kansas State University. 77 p.

Rosa, Hermenegildo. Research problem. M, 1958, Stanford University.

Rosa, Hermenegildo. Structural interpretation of Mata oil field and stratigraphy of Reconcavo Basin. M, 1959, Stanford University.

Rosa, Joao Willy Correa. A global study on phase velocity, group velocity and attenuation of Rayleigh waves in the period range 20 to 100 seconds. D, 1987, Massachusetts Institute of Technology. 859 p.

Rosado, Roberto Victor. Devonian stratigraphy of south central New Mexico and far West Texas. M, 1970, University of Texas at El Paso.

Rosales, Anibal Marcia L. A methodology for quantifying pedogenesis. D, 1981, Cornell University. 262 p.

Rosato, V. J. Peruvian deep-sea sediments; evidence for continental accretion. M, 1974, Oregon State University. 93 p.

Rosauer, Mark Steven. Seismic and acoustic ship signatures during winter navigation on the St. Mary's River. M, 1980, University of Wisconsin-Milwaukee. 152 p.

Rosborough, George Walton. Satellite orbit perturbations due to the geopotential. D, 1986, University of Texas, Austin. 169 p.

Roscoe, Bradley Albert. Applicability of factor analysis to the determination of mineral matter in coal. D, 1981, University of Illinois, Urbana. 195 p.

Roscoe, Michael. Conodont biostratigraphy and facies relationships of the Lower Middle Ordovician strata in the upper Lake Champlain Valley. M, 1973, Ohio State University.

Roscoe, S. M. A silver-lead-zinc deposit in the Slocan District, British Columbia. M, 1948, Queen's University. 72 p.

Roscoe, Stuart M. Dilation diagrams, their application to vein-type ore deposits. D, 1951, Stanford University. 79 p.

Roscoe, William Edwin. Experimental deformation of natural chalcopyrite at temperatures up to 300°C. D, 1973, McGill University. 163 p.

Roscoe, William Edwin. Geology of the Caribou base metal deposit, Bathurst, New Brunswick, (Canada). M, 1969, McGill University. 120 p.

Rose, Arthur William. Trace elements in sulfide minerals from Central mining district, New Mexico, and the Bingham mining district, Utah. D, 1958, California Institute of Technology. 264 p.

Rose, Bruce. Geology of Savana District, British Columbia. D, 1913, Yale University.

Rose, Charles Cleland. Seismic moment, stress drop, strain energy, dislocation radius, and location of seismic acoustical emissions associated with a high alpine snowpack at Berthoud Pass, Colorado. M, 1981, Montana State University. 72 p.

Rose, Cindy L. Stratigraphy of the Lower Cretaceous in the Sabine Uplift area, Texas-Louisiana. M, 1981, Baylor University. 141 p.

Rose, E. R. Geological study of the iron deposits of eastern Ontario and western Quebec. D, 1952, Queen's University.

Rose, E. R. Geology of the Mings Bight area, (Newfoundland). M, 1945, Queen's University. 87 p.

Rose, E. R. The iron deposits of Eastern Canada and their origin. D, 1954, Queen's University. 142 p.

Rose, Edward K. Analysis of logarithmic spiral shaped shorelines of western Lake Erie. M, 1978, University of Toledo. 82 p.

Rose, Edwin R. Elements in sphene. M, 1959, Massachusetts Institute of Technology. 80 p.

Rose, Gregory Lloyd. A geophysical study in Beavercreek Township, Ohio. M, 1978, Wright State University. 70 p.

Rose, Hugh. Tidal gravity measurements at Ann Arbor, Michigan. M, 1952, University of Michigan.

Rose, James P. The engineering geology of the eastern Menominee River valley area of Milwaukee, Wisconsin. M, 1978, University of Wisconsin-Milwaukee.

Rose, Jeannette Noel. The fossils and rocks of eastern Iowa; a half-billion years of Iowa history. M, 1966, University of Iowa. 234 p.

Rose, John Creighton. The establishment of a reference gravity standard in North America. D, 1955, University of Wisconsin-Madison. 97 p.

Rose, John Kerr. A brief survey of national policies on federal land ownership, with special reference to studies conducted by committees of the Congress of Commissions of the Executive Branch of the Federal Government legislative reference series 1956. M, 1958, University of Michigan.

Rose, John Kerr. Transportation by pipe line in the United States. M, 1931, Indiana University, Bloomington. 32 p.

Rose, K. C. The Shannonville and Ameliasburg Pre-Cambrian inliers. M, 1947, Queen's University. 51 p.

Rose, K. D. The Clarkforkian land-mammal "age" and mammalian faunal composition across the Paleocene-Eocene boundary. D, 1979, University of Michigan. 628 p.

Rose, Lawrence. The petrology and heavy mineral facies analysis of the Binnewater Sandstone (Silurian), Ulster County, New York. M, 1968, Brooklyn College (CUNY).

Rose, M. W. Sedimentology of Estero la Cholla, northwest coast of Sonora, Mexico. M, 1975, University of Arizona.

Rose, Nicholas De see De Rose, Nicholas

Rose, Nicholas Martin. Prehnite-epidote phase relations in the Nordre Aputiteq and Kruuse Fjord intrusions, East Greenland. M, 1985, Stanford University. 57 p.

Rose, Peter Robert. Carboniferous stratigraphy of the Hall area, San Saba County, Texas. M, 1959, University of Texas, Austin.

Rose, Peter Robert. Edwards Formation (Lower Cretaceous), surface and subsurface, central Texas. D, 1968, University of Texas, Austin. 387 p.

Rose, R. Burton. An inductive study of mineral conductivity as a function of frequency. M, 1939, University of California, Los Angeles.

Rose, Raymond R. Investigation of dissolved constituent concentrations and assessment of chemical equilibrium controls in the saturated zone of two coal surface mining spoil banks. D, 1983, University of Tennessee, Knoxville. 390 p.

Rose, Robert Leon. Geology of the May Lake area, Yosemite National Park, California. D, 1957, University of California, Berkeley. 224 p.

Rose, Robert Leon. Tertiary volcanic domes of the Jackson Quadrangle, California. M, 1949, University of California, Berkeley. 88 p.

Rosé, Robert Rowland. The stratigraphy and structure of the southern Madison Range, Madison and Gallatin counties, Montana. M, 1967, Oregon State University. 173 p.

Rose, Seth E. The heavy metal adsorption characteristics of Hawthorne Formation sediments. M, 1981, University of Florida. 122 p.

Rose, Seth Edward. Dissolved oxygen systematics in the Tucson Basin aquifer, Arizona. D, 1987, University of Arizona. 185 p.

Rose, Stuart M. Bedrock geology of Woonsocket Basin area; Rhode Island and Massachusetts. M, 1985, Boston University. 160 p.

Rose, Susan Humphrey. Depositional environment and paleogeography of the Lysite Member, Wind River Formation (lower Eocene, central Wyoming). M, 1988, University of Iowa. 100 p.

Rose, Timothy Patrick. A model for the origin of the chemically zoned Ammonia Tanks Member of the Timber Mountain Tuff due to crystal fractionation. M, 1988, Michigan State University. 72 p.

Rose, William D., Jr. Geology of the northern part of the Camp Austin Quadrangle, Morgan County, Tennessee. M, 1953, Vanderbilt University.

Rose, William Ingersoll, Jr. The geology of the Santiaguito volcanic dome, Guatemala. D, 1970, Dartmouth College. 253 p.

Roseboom, Eugene Holloway, Jr. Geology of part of the granitic complex south of the Felch Mountain Range, Michigan. M, 1951, Ohio State University.

Rosebsloom, Eugene Holloway, Jr. Phase relations in the arsenic-rich portion of the system Co-Ni-Fe-As. D, 1958, Harvard University.

Rosell, Ramon Antonio. Some reactions of manganese with montmorillonite and soil organic matter. D, 1967, University of California, Berkeley. 195 p.

Rosen, Carol J. Karst geomorphology of the Door Peninsula, Wisconsin. M, 1984, University of Wisconsin-Milwaukee. 119 p.

Rosen, Douglas S. Beach and nearshore sedimentation on Caladesi Island State Park, Pinellas County, Florida. M, 1976, University of South Florida, Tampa. 114 p.

Rosen, G. E. A. Von see Von Rosen, G. E. A.

Rosen, Lawrence Collinger. The sedimentology and petrology of the early Proterozoic Gowganda Formation around Gowganda-Elk Lake, Ontario, Canada. M, 1985, University of Minnesota, Duluth. 170 p.

Rosen, Louis. Iron phosphates in New Jersey greensands. M, 1975, University of Delaware.

Rosen, Michael Elliott. The development of adsorption/thermal desorption for the determination of trace levels of volatile organic compounds in groundwater. D, 1988, Oregon Graduate Institute of Science and Technology. 348 p.

Rosen, Michael Robert. Petrographic and chemical evidence for a modern, brackish water dolomite; Chesapeake Bay, Maryland. M, 1984, University of Rochester. 73 p.

Rosen, Norman C. Comparison of mineralogical characteristics of Wisconsin glacial deposits, Highland-Fayette County area and Hocking Valley, Ohio. M, 1964, Ohio State University.

Rosen, Norman Charles. Heavy minerals of the Citronelle Formation (Pliocene) of the Gulf Coastal Plain. D, 1968, Louisiana State University. 197 p.

Rosen, P. S. The morphology and processes of the Virginia Chesapeake Bay shoreline. M, 1976, College of William and Mary.

Rosen, Peter S. Evolution and processes of Coatue Beach, Nantucket Island, Massachusetts; a cuspate shoreline. M, 1972, University of Massachusetts. 203 p.

Rosen, Sherman Jay. Stratigraphic distribution of red coloration in post-Michigan age rocks (Michigan). M, 1967, University of Michigan.

Rosen-Spence, A. F. De *see* De Rosen-Spence, A. F.

Rosenbauer, R. J. Silica diagenesis in Recent diatomaceous sediment from the Guaymas Basin in the Gulf of California. M, 1975, University of Southern California.

Rosenbaum, James. Theory of warping of Alaskan crustal plate at time of the Alaskan earthquake, 1964. M, 1971, Stanford University.

Rosenbaum, Joseph Griffin. Comparison of the paleomagnetic records in sediments from two late Pleistocene lake basins in Colorado. D, 1980, University of Colorado. 159 p.

Rosenberg, Ernest A. Geology and petrology of the northern Wenatchee Ridge area, northern Cascades, Washington. M, 1961, University of Washington. 109 p.

Rosenberg, Gary David. Patterned growth of the bivalve Chione undatella Sowerby relative to the environment. D, 1972, University of California, Los Angeles. 236 p.

Rosenberg, Louis J. and Coulson, Francis M. The formation and detailed description of a portion of Wind Cave (South Dakota). M, 1958, South Dakota School of Mines & Technology.

Rosenberg, Murray J. Temporal variability of beach profiles, Charlestown Beach, Rhode Island. M, 1985, University of Rhode Island.

Rosenberg, Philip E. Subsolidus studies in the system CaCO3-MgCO3FeCO3-MnCO3. D, 1960, Pennsylvania State University, University Park. 146 p.

Rosenberg, Robert Steven. The paleoecology of the late Wisconsinan Eagle Point local fauna, Clinton County, Iowa. M, 1983, University of Iowa. 70 p.

Rosenberger, Eric J. The relationship of Huronian sediments to associated igneous rocks in sections 22 and 23, T47N, R26W, Marquette County, Michigan. M, 1961, Michigan State University. 59 p.

Rosenberry, Donald Orville. Factors contributing to the formation of transient water-table mounds on the outflow side of a seepage lake, Williams Lake, central Minnesota. M, 1985, University of Minnesota, Minneapolis. 127 p.

Rosenblum, Mark B. Early diagenetic sheet crack cements of the Guadalupian (Permian) Shelf, Yates and Tansill formations, New Mexico (USA). M, 1984, University of Wisconsin-Madison. 95 p.

Rosenboom, Arthur Kenneth. Geology of the southern portion of the Pounding Mill Quadrangle, Tazewell County, Virginia. M, 1975, North Carolina State University. 101 p.

Rosencrans, Richard D. Crustal block tectonics; application to the Lake Erie-Maryland "block" in Eastern United States. M, 1983, Pennsylvania State University, University Park. 67 p.

Rosencrantz, E. J. The geology of the northern part of North Arm Massif, Bay of Islands ophiolite complex, Newfoundland; with application to upper oceanic crust lithology, structure, and genesis. D, 1980, SUNY at Albany. 378 p.

Rosendahl, B. T. Geological and geophysical studies of the Canton Trough. M, 1972, University of Hawaii. 120 p.

Rosendahl, Bruce R. Evolution of oceanic crust. D, 1976, University of California, San Diego. 153 p.

Rosene, Richard K. Micropaleontology of the Bearpaw Formation (Cretaceous), southwestern Alberta Foothills. M, 1972, University of Alberta. 133 p.

Rosenfarb, J. L. Effect of fabric on the directional shear strength of a kaolin clay. D, 1975, Northwestern University. 186 p.

Rosenfeld, Charles Louis. A quantitative investigation of relations among certain morpho-climatic parameters of drainage basins. D, 1973, University of Pittsburgh.

Rosenfeld, George Albert. Origin of oriented lakes, Arctic Coastal Plain, Alaska. M, 1958, Iowa State University of Science and Technology.

Rosenfeld, J. K. Nitrogen diagenesis in nearshore anoxic sediments. D, 1977, Yale University. 200 p.

Rosenfeld, John Lang. Geology of the southern part of the Chester Dome, Vermont. D, 1954, Harvard University.

Rosenfeld, Joshua Henry. Geology of the western Sierra de Santa Cruz, Guatemala, Central America; an ophiolite sequence. D, 1981, SUNY at Binghamton. 313 p.

Rosenfeld, Melvin Arthur. Petrographic variation in the Oriskany "sandstone complex". D, 1953, Pennsylvania State University, University Park. 220 p.

Rosenfeld, Melvin Arthur. Porosity; Part I, A survey of the problem; Part II, Some statistical techniques applied to porosity data. M, 1950, Pennsylvania State University, University Park. 203 p.

Rosenfeld, Sigmund Judith. A study of the Pleistocene shore lines between the Altamaha and the Savannah rivers in Georgia. M, 1955, Emory University. 51 p.

Rosengreen, Theodore Ernest. Surficial geology of the Maple Valley and Hobart quadrangles, Washington. M, 1965, University of Washington. 71 p.

Rosengreen, Theodore Ernest. The glacial geology of Highland County, Ohio. D, 1970, Ohio State University. 198 p.

Rosenkrans, Robert Russell. Correlation studies of the central and south central Pa bentonite occurrences. M, 1933, Pennsylvania State University, University Park. 108 p.

Rosenkrans, Robert Russell. The geographic distribution and stratigraphic occurrence of Ordovician altered volcanic materials in eastern North America. D, 1936, Princeton University. 370 p.

Rosenlund, Gene C. Geology and mineralization of the Cumberland Pass area, Gunnison County, Colorado. M, 1984, Colorado State University. 126 p.

Rosenmeier, Frederick Joseph. Stratigraphy and structure of the Table Mountain; Mission Peak area in the Wenatchee Mountains, central Washington. M, 1968, University of Washington. 44 p.

Rosenquest, Darl. Slope stability and landslide hazards in the Squaw Run area watershed, Allegheny County, PA. M, 1978, University of Pittsburgh.

Rosenshein, Arthur N. Stratigraphy of the Connelly Formation and associated Deerparkian strata (lower Devonia) in southeastern New York. M, 1969, SUNY at Buffalo.

Rosenshein, Joseph Samuel. Geohydrology of Pleistocene deposits and sustained yield of principal Pleistocene aquifer, Lake County, Indiana. D, 1967, University of Illinois, Urbana. 100 p.

Rosenstein, E. S. The Chazyan and Black Riveran formations of the Cornwall area, south-east Ontario; a paleoenvironmental reconstruction and diagenetic study. M, 1973, Queen's University. 106 p.

Rosenthal, David Bruce. Distribution of crust in the deep eastern Gulf of Mexico. M, 1987, University of Texas, Austin. 149 p.

Rosenthal, Lorne Richard Phillip. The stratigraphy, sedimentology and petrography of the Upper Creta-

ceous Wapiabi and Belly River formations in southwestern Alberta. M, 1984, McMaster University. 250 p.

Rosenthal, Robert John. A surface wave study of the South China Subplate. M, 1977, University of Southern California.

Rosentraub, M. S. Coastal policy development and self-evaluating agencies; information utilization and the South Coast Regional Commission. D, 1975, University of Southern California.

Rosenzweig, Abraham. A chemical, optical, and genetic study of hornblendes of southeastern Pennsylvania and Delaware. D, 1950, Bryn Mawr College. 37 p.

Rosewitz, Lura Ellen. Subsurface analysis, "Cherokee" Group (Desmoinesian), portions of Lincoln, Pottawatomie, Seminole, and Okfuskee counties, Oklahoma. M, 1981, University of Oklahoma. 57 p.

Roshardt, Mary Ann. Paleocurrent and textural analyses of the Saint Peter Sandstone (Ordovician) in a portion of south-central Wisconsin. M, 1965, University of Wisconsin-Madison.

Rosholt, John Nicholas. Evaluation of the Pa^{231}/U, Th^{230}/U method for Pleistocene dating. M, 1961, University of Miami. 37 p.

Rosholt, John Nicholas, Jr. Uranium in sediments. D, 1963, [Miami University (Ohio)]. 223 p.

Roshong, Carolyn Grace. The depositional environment and diagenetic history of the Chester "J" Limestone in portions of Dewey and Major counties, Oklahoma. M, 1986, University of Oklahoma. 250 p.

Rosowitz, Donald W. Palynology and paleoecology of the Riverton coal bed (Desmoinesian, Pennsylvanian) in southeastern Kansas. M, 1982, Wichita State University. 137 p.

Rospenda, Robert E. Ordovician stratigraphy of Wilcox-Sunwapta Pass area, Alberta, Canada. M, 1964, University of Wisconsin-Madison.

Ross, Alan. Diffusion of CO_2 in a sodium-aluminosilicate melt. M, 1979, Rensselaer Polytechnic Institute. 44 p.

Ross, Alex R. and St. John, Jack W. Geology of the northern Wyoming Range, Wyoming. M, 1950, University of Michigan.

Ross, Alex R. Pleistocene and Recent sediments in western Lake Erie. D, 1953, University of Michigan.

Ross, Arthur Henry, Jr. Geology of the Saltville Fault between Saltville and Broadford, Virginia. M, 1966, Virginia Polytechnic Institute and State University.

Ross, Barbara. Interpretation of maximum entropy derived dispersion curves from northern Alabama. M, 1984, Georgia Institute of Technology. 72 p.

Ross, Brian. A petrologic study of the Barden Peak Peridotite, Duluth Complex. M, 1985, University of Minnesota, Minneapolis.

Ross, Bruce C. Depositional environments of a Chesterian (Upper Mississippian) sandstone body in Breckinridge County, Kentucky. M, 1988, Eastern Kentucky University. 57 p.

Ross, Bruce E. The Pleistocene history of the San Francisco Bay along the Southern Crossing. M, 1977, San Jose State University. 121 p.

Ross, Charles A. The type Wolfcamp Series (Permian) Glass Mountains, Texas. D, 1959, Yale University.

Ross, Charles Richard, II. Phase relations in the omphacitic pyroxenes. D, 1986, University of California, Los Angeles. 265 p.

Ross, Christopher George Arthur. Evaluation of hydraulic and hydrogeochemical parameters of a deep fracture groundwater system in the area of Cuyamaca-Julian, San Diego County, California. M, 1987, San Diego State University. 147 p.

Ross, Clarence Samuel. The correlation of the Edgewood and Sexton Creek formations of the Alexandran Series in Missouri, Illinois, Iowa, and eastern Wisconsin. M, 1915, University of Illinois, Chicago.

Ross, Clarence Samuel. The differentiation and contact metamorphism of the Snowbank Lake Syenite in the Vermilion iron-bearing region of Minnesota. D, 1919, University of Illinois, Chicago.

Ross, Daniel. Mineral chemistry and petrogenesis of cumulate dunites from Blow Me Down Mountains, Newfoundland. M, 1988, University of Houston.

Ross, David A. The geochemical significance of the depositional environments of the bioherms in the Upper Pennsylvanian Bern Limestone in eastern Kansas. M, 1960, University of Kansas. 90 p.

Ross, David Alexander. The sediments and structure of the northern Middle America Trench. D, 1965, University of California, San Diego. 230 p.

Ross, David Ian. The distribution of gold and other elements in the uranium conglomerates of Elliott Lake, Canada. M, 1981, McMaster University. 104 p.

Ross, Donald C. The geology of the north one-half of the Elsberry Quadrangle, Missouri. M, 1949, University of Iowa. 197 p.

Ross, Donald C. The igneous and metamorphic rocks of a portion of Sequoia National Park, California. D, 1952, University of California, Los Angeles.

Ross, Edmund. Report on a portion of the Soda Springs mining district in Bernalillo County, New Mexico. M, 1909, [University of New Mexico].

Ross, Frederick W. Sedimentary structures and animal-sediment relationships, Old Tampa Bay, Florida. M, 1975, University of South Florida, Tampa. 129 p.

Ross, Gerald Marckres. Geology and depositional history of the Hornby Bay Group, Proterozoic, Northwest Territories, Canada. D, 1984, Carleton University. 344 p.

Ross, Gerhard John. Characterization of a montmorillonite in a northern Michigan podzol. D, 1965, [University of Michigan]. 112 p.

Ross, Grant Arrett. Case studies of soil stability problems resulting from earthquakes. D, 1968, University of California, Berkeley. 232 p.

Ross, Howard Persing. Detailed electrical surveys in the Triassic basin, North Chester County, Pennsylvania. M, 1963, Pennsylvania State University, University Park. 109 p.

Ross, Howard Persing. In situ determination of the remanent magnetic vector of two-dimensional tabular bodies. D, 1965, Pennsylvania State University, University Park. 116 p.

Ross, J. S. Uranium deposits of the Goldfields area, Saskatchewan. D, 1951, University of Toronto.

Ross, James. Recurrent species associations and species diversity of cytheracean ostracodes in the upper Austin and lower Taylor groups (Campanian, Upper Cretaceous) of Travis County, Texas. M, 1983, University of Houston.

Ross, James J. Photogeology as an exploration tool at Cerro de Pasco, Peru. M, 1970, Stanford University.

Ross, James Robert Holland. Archean nickel sulphide mineralization Lunnonshoot, Kambalda, Western Australia. D, 1974, University of California, Berkeley. 283 p.

Ross, Jeffrey Allen. ^{40}Ar/^{39}Ar dating of polymetamorphic amphibolite, blueschist, and eclogite of the Franciscan Complex, California. M, 1987, SUNY at Stony Brook. 137 p.

Ross, John Sawyer. Geology of central Payne County, Oklahoma. M, 1972, Oklahoma State University. 87 p.

Ross, Joseph Ray. A three-dimensional facies analysis of the canyon alluvial fill sequence, Chaco Canyon, New Mexico. M, 1978, University of New Mexico. 86 p.

Ross, L. C. A Quaternary stratigraphic cross-section through Kitchener-Waterloo. M, 1986, University of Waterloo. 54 p.

Ross, Landon T. The genus Crepidula in North America; its taxonomy, phylogeny, and distribution. M, 1965, Florida State University.

Ross, Landon T., Jr. Vermetid reefs and coastal development in Southwest Florida. D, 1965, Florida State University.

Ross, Malcolm. The crystallography, crystal structure, and crystal chemistry of various minerals and compounds belonging to the torbernite mineral group. D, 1962, Harvard University.

Ross, Malcolm Ingham. MAGANOM; a computer program for the modeling and interpretation of marine magnetic anomalies with an example from the Cayman Trough, Northwest Caribbean Sea. M, 1987, University of Texas at Dallas. 117 p.

Ross, Mark A. Stratigraphy of the Tamaulipas Limestone, northeastern Mexico. M, 1979, University of Texas, Arlington. 96 p.

Ross, Mark Allen. Vertical structure of estuarine fine sediment suspensions. D, 1988, University of Florida. 188 p.

Ross, Martin E. Quantitative petrography of Precambrian mafic dikes in the Bald Mountain area, Bighorn Mountains, Wyoming. M, 1970, Kent State University, Kent. 96 p.

Ross, Martin Edward. Stratigraphy, structure, and petrology of Columbia River Basalt in a portion of the Grande Ronde River - Blue Mountains area of Oregon and Washington. D, 1978, University of Idaho. 407 p.

Ross, Martin Nicholas. Tidal dissipation, thermal and dynamic evolution, and internal structure of several outer planet satellites, Mercury, and the Earth-Moon system. D, 1988, University of California, Los Angeles. 283 p.

Ross, Mary Harvey. Source and correlation of the (Lower Ordovician) Deepkill conglomerates (New York). M, 1948, Cornell University.

Ross, Mary Harvey. The Tabulata of the Hamilton Group (Middle Devonian of New York); Part I, The Favositidae. D, 1951, Cornell University.

Ross, Michael Lee. Petrology and geochemistry of alkalic rocks from the Sawmill Mountain area, Brewster County, Texas; evidence for multiple source regions. M, 1987, Texas Christian University. 154 p.

Ross, Nancy Lee. A thermochemical and lattice vibrational study of high pressure phase transitions in silicates and germanates. D, 1985, Arizona State University. 270 p.

Ross, Ralph B. Structures in the Conemaugh Formation near Bakerstown Station, Pennsylvania. M, 1933, University of Pittsburgh.

Ross, Reuben J., Jr. The stratigraphy of the (Lower Ordovician) Garden City Formation in northeastern Utah and its trilobite fauna. D, 1948, Yale University.

Ross, Richard Bush. A Jurassic Sawtooth fauna from southwestern Montana. M, 1950, University of Michigan. 33 p.

Ross, Robert Motague. The geology of the northwestern part of the Hermitage Quadrangle, Tennessee, with special reference to the stratigraphy and paleontology of the Hermitage Formation. M, 1932, Vanderbilt University.

Ross, Roland Case. Fossil geese of the McKittrick asphalt deposit. M, 1932, California Institute of Technology. 19 p.

Ross, Roy M., Jr. A quantitative sedimentary analysis of the Middle Devonian Traverse Group in the Michigan Basin. M, 1957, Michigan State University. 52 p.

Ross, Sheila Lynn. Origin and diagenesis of Mississippian carbonate buildups, Quanah Field, Hardman County, Texas. M, 1981, Texas A&M University. 142 p.

Ross, Stewart H. The geology of the Lac Deschenes map area, Quebec. D, 1951, Syracuse University.

Ross, Sylvia Yvonne Hall. Contributions to the geohydrology of Moscow Basin, Latah County, Idaho. M, 1965, University of Idaho. 116 p.

Ross, Theodore William. A stratigraphic and petrologic study of the Reelsville Limestone (lower Chester) in south-central Indiana. M, 1934, Indiana University, Bloomington.

Ross, Theodore William. Granitic pegmatites in Washington. D, 1969, Washington State University. 217 p.

Ross, Thomas Paul. Subsurface sequence of Eocene foraminifera, Simi Valley, California. M, 1959, University of California, Berkeley. 68 p.

Ross, Virginia F. An X-ray investigation of some low temperature solid phases of the systems Cu_2S-Sb_2S_3, Cu_2S-As_2S, Cu_2S-FeS, and Cu_2S-CuS. D, 1953, Massachusetts Institute of Technology. 77 p.

Ross, Wayne A. Subsurface study of eastern portion of Monroe Uplift, Mississippi. M, 1954, University of Wisconsin-Madison.

Ross, William C. The tectonic relationship of the Alaska Peninsula to the North American Cordillera; a plate tectonic and palynologic synthesis. M, 1979, Louisiana State University.

Ross, William Michael. Oil pollution as a developing international problem; a study of the Puget Sound and Strait of Georgia regions of Washington and British Columbia. D, 1972, University of Washington. 273 p.

Ross, William R. Depositional conditions and Recent changes in the Holocene West Pascagoula River delta. M, 1980, Memphis State University.

Ross, Wyn Charles. A two-dimensional simulation of tritium transport in the vadose zone at the Nevada Test Site. M, 1990, University of Nevada. 81 p.

Rossbach, Thomas J. Lithostratigraphy and molluscan biostratigraphy of the River Bend Formation at the Martin Marietta Quarry, New Bern, North Carolina. M, 1987, University of North Carolina, Chapel Hill. 173 p.

Rossbacher, Lisa Ann. Geomorphic studies of Mars. D, 1983, Princeton University. 262 p.

Rossell, Dean M. Alteration of the Deer Lake Peridotite in the vicinity of the Ropes Mine, Marquette County, Michigan. M, 1983, Michigan Technological University. 83 p.

Rossello, Pierina Onorina Blanch. An Eocene foraminiferal faunule from Contra Costa County, California. M, 1939, University of California, Berkeley. 174 p.

Rossen, Christine. Sedimentology of the Brushy Canyon Formation (Permian, early Guadalupian) in the onlap area, Guadalupe Mountains, West Texas. M, 1985, University of Wisconsin-Madison. 314 p.

Rossetter, R. J. Geology of the San Luis and San Lorenzo Island groups, Gulf of California. M, 1973, [University of California, San Diego].

Rossi, Dennis A. Depositional environment of the La Rose Formation, Miocene, Maracaibo Basin, Venezuela, as determined by SP curve characteristics. M, 1977, Wright State University. 105 p.

Rossi, Mario Eduardo. Impact of spatial clustering on geostatistical analysis. M, 1988, Stanford University. 101 p.

Rossi, Randall Steven. Land use and vegetation change in the oak woodland-savanna of northern San Luis Obispo County, California (1774-1978). D, 1979, University of California, Berkeley. 345 p.

Rossin, Edgar L. Petrography of a contact metamorphic deposit near Hanover, New Mexico. M, 1924, Columbia University, Teachers College.

Rossinsky, Victor. Sedimentation and Holocene history of the Loxahatchee River estuary, Jupiter, Florida. M, 1984, University of Miami. 242 p.

Rossiter, J. R. Interferometry depth sounding on the Athabasca Glacier (Alberta, Canada); development of the interferometry technique for lunar exploration. M, 1971, University of Toronto.

Rossiter, J. R. Interpretation of radio interferometry depth sounding, with emphasis on random scattering from temperate glaciers and the lunar surface. D, 1977, University of Toronto.

Rossmann, Ronald. Lake Michigan ferromanganese nodules. D, 1973, University of Michigan. 166 p.

Rossmiller, R. L. Land and water resource planning using goal programming. D, 1979, Iowa State University of Science and Technology. 1360 p.

Rosso, Weymar Allen. The ecology of the South Fork of the Saline River as affected by surface mine wastes. D, 1975, Southern Illinois University, Carbondale. 145 p.

Rossow, Joerg. A seismic investigation of The Geysers-Clear Lake geothermal area, Lake County, California, utilizing compressional- and shear-wave sources. M, 1982, Colorado School of Mines. 361 p.

Rossow, Joerg. An investigation of deconvolution techniques for transient electromagnetic records. D, 1987, Colorado School of Mines. 140 p.

Rostoker, Mendel David. Geology of the Canso Group in the Maritime Provinces of Canada. D, 1960, Boston University. 147 p.

Roszkowski, George Antoni. A petrochemical study of the Deer Lakes piemontite zone, eastern Sierra Nevada, California. M, 1984, San Diego State University. 64 p.

Rotan, Cleone M. A paleontological study of some marine invertebrates from the Upper Pennsylvanian Canyon Series of north central Texas. M, 1951, Smith College. 93 p.

Rotan, Pat Malone. Preferred orientation of plagioclase in basic rocks, Raggedy Mountains, southwestern Oklahoma. M, 1960, University of Oklahoma. 62 p.

Rotan, R. A. Organic residue study of lower Paleozoic rocks. M, 1952, University of Massachusetts.

Rotert, J. Field measurement of hydraulic conductivity in Hawaii oxisols. M, 1977, University of Hawaii. 77 p.

Roth, Barry. Late Cenozoic marine invertebrates from Northwest California and Southwest Oregon. D, 1979, University of California, Berkeley. 803 p.

Roth, Charles Barron. Ferrous-ferric ratio and cec changes with removal of ferruginous colloidal coatings from weathered micaceous vermiculites and soils. D, 1969, University of Wisconsin-Madison. 99 p.

Roth, Eldon Sherwood. Geology in the junior college. D, 1969, University of Southern California.

Roth, Eldon Sherwood. Landslides between Santa Monica and Point Dume. M, 1959, University of Southern California.

Roth, Frances Ann. Implications of stratigraphic completeness analysis for magnetic polarity stratigraphic studies. M, 1985, University of Arizona. 67 p.

Roth, Horst. A structural study of the Sutton Mountains, Quebec. D, 1965, McGill University.

Roth, James G. Delineation of the Railroad Valley flow system using a deuterium-calibrated groundwater model. M, 1988, University of Nevada. 156 p.

Roth, James Richard. The equilibrium between labile and crystalline phosphates in soils. D, 1968, Cornell University. 229 p.

Roth, Janet. Late Pleistocene vertebrate fauna from Edisto Island, South Carolina. M, 1979, University of Georgia.

Roth, Jim Craig. Geology of a portion of the southern Santa Ana Mountains, Orange County, California. M, 1958, University of California, Los Angeles.

Roth, John E. Porosity evolution of the Pahasapa (Madison) Limestone at Jewel Cave National Monument, Custer, South Dakota. M, 1977, South Dakota School of Mines & Technology.

Roth, John N. A gravitational investigation of fracture zones in Devonian rocks in portions of Arenac and Bay counties, Michigan. M, 1965, Michigan State University. 89 p.

Roth, Kingsley William. Stratigraphy of the pre-Frontier Cretaceous rocks in the Rattlesnake Hills, southwestern Natrona County, Wyo. M, 1955, University of Wyoming. 102 p.

Roth, Mark. Evaporitic environments and their relationship to porosity of associated carbonates in Williston Basin (Mississippian). M, 1984, Queens College (CUNY). 180 p.

Roth, Richard A. Landslide suspectibility in San Mateo County, California. M, 1982, Stanford University. 87 p.

Roth, Robert Ingersoll. Geology of central part of Mount Vernon quadrangle. M, 1926, University of Washington. 65 p.

Roth, Robert Leroy. Moisture movement from a point source. D, 1983, University of Arizona. 159 p.

Roth, Robert Sidney. Correlation of Pennsylvanian strata in the Hartville Uplift, Wyoming and the southern Black Hills, South Dakota. M, 1950, University of Illinois, Urbana.

Roth, Robert Sidney. The structure of montmorillonite in relation to the occurrence and properties of certain bentonites. D, 1951, University of Illinois, Urbana.

Roth, Susan Viola. Depositional environment of the Upper Cretaceous Tuscaloosa reservoir sandstones (Gulfian) in Profit Island Field, East Baton Rouge Parish, Louisiana. M, 1981, Texas A&M University. 110 p.

Rothammer, Christine Marie. A detailed shoreline monitoring grain size and total carbon analysis of three beach sections before and after Hurricane Alicia, Galveston Island, Texas. M, 1984, Texas Christian University.

Rothbard, David Rod. Diagenesis of the Lamotte Sandstone in the Southeast Missouri lead district. D, 1982, University of Missouri, Columbia. 251 p.

Rothe, George Henry, III. Earthquake swarms in the Columbia River basalts. D, 1978, University of Washington. 181 p.

Rothe, George Henry, III. Geophysical investigation of a diabase dike. M, 1973, Georgia Institute of Technology. 78 p.

Rothenberger, Jay Anderson. Geology of Goshen Pass and environs in Rockbridge County, Virginia. M, 1959, University of Virginia. 91 p.

Rotherham, D. C. A study of some radioactive granites and pegmatites in northern Saskatchewan. M, 1955, University of Saskatchewan. 51 p.

Rothermel, Samuel Royden. Hydrogeological and hydrochemical investigation of springs and wells in Hays and Comal counties, Texas. M, 1985, University of Arkansas, Fayetteville.

Rothman, Daniel Harris. Large near-surface anomalies, seismic reflection data, and simulated annealing. D, 1986, Stanford University. 107 p.

Rothman, Edward M. The petrology of the Berea Sandstone (Early Mississippian) of south-central Ohio and a portion of northern Kentucky. M, 1978, Miami University (Ohio). 105 p.

Rothman, Robert L. Crack distribution under uniaxial load and associated changes in seismic velocities prior to failure. D, 1975, Pennsylvania State University, University Park. 143 p.

Rothman, Robert L. Model studies for focal-depth determination at near-source stations. M, 1964, Pennsylvania State University, University Park. 88 p.

Rothrock, Edgar Paul. The geology of Cimarron County, Oklahoma. D, 1922, University of Chicago. 110 p.

Roths, Pamela J. Geologic mapping and structural analysis of the Caddo Mountains in western Montgomery County, Arkansas. M, 1988, Southern Illinois University, Carbondale. 73 p.

Rothschild, Donald Isador. Subsurface geology of Erath County, Texas. M, 1955, University of Texas, Austin.

Rothschild, Edward Robert. Hydrogeology and contaminant transport modeling of the Central Sand Plain, Wisconsin. M, 1982, University of Wisconsin-Madison. 215 p.

Rothwell, Bret. Geology of th Mackay 2SW and parts of the Harry Canyon (Hailey 1NE), Mackay 2NW, and Standhope Peak (Hailey 1SE) quadrangles. M, 1974, University of Wisconsin-Milwaukee. M,

Rothwell, Sally Ann. Continent-vergent shearing resulting from marginal basin collapse, southernmost Andes. M, 1987, University of Texas, Austin. 90 p.

Rotondi, Paul L. The petrology of the Trondequoit Limestone Formation (Middle Silurian) of western

New York State and Rockway, Ontario. M, 1986, Miami University (Ohio). 139 p.

Rotstein, Yair. Geophysical investigations in southeastern Mojave Desert, California. D, 1974, University of California, Riverside. 187 p.

Rott, Edward H., Jr. The ore deposits of the Gold Circle District, Nevada. M, 1930, University of California, Berkeley. 39 p.

Rottenfusser, Brian Albert. Petrology and depositional environments of the Watt Mountain Formation, northern Alberta. M, 1974, University of Calgary. 120 p.

Rotter, Harold A. A mechanical and petrographic analysis of the sandstones of the (Pennsylvanian) Savanna Formation in Pittsburg County, Oklahoma. M, 1956, [University of Oklahoma].

Rotter, Richard Joseph. Pb-210 and Pu-239, 240 in nearshore Gulf of Mexico sediments. M, 1985, Texas A&M University. 157 p.

Rottman, Marcia Louise Gaines. Euthecosomatous pteropods and planktonic foraminifera in Southeast Asia marine waters; species associations and distribution in sediments. D, 1977, University of Colorado. 252 p.

Rottmann, Carmen Juanita Farr. Physical parameters and interrelationships of modern beach sands, Pleistocene terrace sands, and Eocene sandstones from Cape Arago, Oregon; a study combining the evolution of grain morphology in the zone of surf action and local aspects of present and past depositional environments. D, 1970, University of Oregon. 238 p.

Rottweiler, Kurt A. Sedimentary structures in Percebu Bay and vicinity, Baja California, Mexico. M, 1966, University of Southern California.

Rotunno, Richard. Internal gravity waves in the nocturnal planetary boundary layer. D, 1976, Princeton University.

Roubanis, Aristidis Savvas. Geology of the Santa Ynez Fault, Gaviota Pass, Point Conception area, Santa Barbara County, California. M, 1962, University of California, Los Angeles.

Roueche, William Lee III. The Pleistocene-Holocene history of the iguanid lizard Urosaurus ornatus in the El Paso southwest. M, 1971, University of Texas at El Paso.

Rouhani, M. Subsurface geologic study of Dallas County. M, 1976, University of Iowa. 37 p.

Rouhani, Shahrokh. Optimal data collection in random fields. D, 1983, Harvard University. 189 p.

Rouleau, A. Analyse structurale du Groupe de Rosaire et des roches connexes de la région de Thetford Mines. M, 1976, Universite Laval.

Rouleau, Alain. Statistical characterization and numerical simulation of a fracture system; application to groundwater flow in the Stripa Granite. D, 1984, University of Waterloo. 416 p.

Roulidis, Christos Z. Petrographic study of the volcanic tuffs of Crandall Basin, Park County, Wyomimng. M, 1959, Wayne State University.

Roulier, Michael Henry. Evaluation of transient flow measurements on intact cores as a means of approximating the hydraulic conductivity of unsaturated soil under field conditions. D, 1972, University of California, Riverside. 50 p.

Roulo, David L. Comparative study of the composition of the macrozooplankton populations of two ecologically different areas of the Indian River during the summer months. M, 1977, Florida Institute of Technology.

Roulston, B. V. Stratigraphy and sedimentology of the lower Chaleur Group in Gaspe, Quebec. M, 1976, University of New Brunswick.

Roulston, John S. The petrology and structure of the Lanesville-Bayview area, Gloucester, Massachusetts. M, 1983, University of Kentucky. 153 p.

Round, Edna M. The Carboniferous flora of Rhode Island and its probable correlation. D, 1920, Brown University.

Rounds, Thomas Richard, Jr. Dinoflagellate biostratigraphy and organic-walled phytoplankton cyst paleoecology of the Demopolis-Ripley transition interval from the Upper Cretaceous Selma Group of Mississippi and Alabama. M, 1982, Virginia Polytechnic Institute and State University. 612 p.

Roundtree, Robert L. Subsurface geology of Rich Valley and southeast Rich Valley oil fields, Grant County, Oklahoma. M, 1961, University of Oklahoma. 61 p.

Rountree, John H. Petroleum geology of the Arbuckle Group, southern Osage and Pawnee counties, Oklahoma. M, 1980, Oklahoma State University. 71 p.

Rourke, Gerald F. Magnetic studies on the Ellesmere ice shelf. M, 1960, Boston College.

Rouse, George E. The trace element mineralogy of some metalliferous shales. D, 1968, Colorado School of Mines. 84 p.

Rouse, Glen E. Paleobotanical analysis of fossil plant remains associated with Canadian Lower Cretaceous coal measures. D, 1955, Ohio State University.

Rouse, John Thomas. Geology of the Valley area, Park County, Wyoming. D, 1932, Princeton University. 149 p.

Rouse, John Thomas. On the petrographic character and origin of the tuff-like rocks of Whitestone Mountain, Washington. M, 1930, University of Cincinnati. 58 p.

Rouse, Roland Carl. Structural and crystal chemical relations in the mineral senaite Pb(Ti, Fe, Mn, Mg)$_{21}$O$_{38}$. M, 1967, University of Michigan.

Rouse, Roland Carl. The crystal chemistry of some lead-oxygen compounds. D, 1972, University of Michigan. 107 p.

Rousell, D. H. The Blairmore Formation of southern Saskatchewan. M, 1956, University of British Columbia.

Rousell, D. H. The nature of the south boundary of the Nelson River gneiss zone (Proterozoic), (Manitoba). M, 1963, University of Manitoba.

Rousell, Donald H. The petrology of Archaean and Proterozoic rocks at Cross Lake, Manitoba and the effects of the Hudsonian Orogeny. D, 1965, University of Manitoba.

Roush, James Manfred. Marine geology of the western extension of the Transverse Ranges; Point Conception to Point Arguello. M, 1983, California State University, Northridge. 151 p.

Roush, Kathleen Ann. Depositional environments of the Eocene Domengine Formation near Coalinga, Fresno County, California. M, 1986, California State University, Northridge. 85 p.

Roush, Robert C. Sediment textures and internal structures; a comparison between central Oregon continental shelf sediments and adjacent coastal sediments. M, 1970, Oregon State University. 75 p.

Roush, Ted L. Effects of temperature on remotely sensed mafic mineral absorption features. M, 1984, University of Hawaii at Manoa. 129 p.

Roush, Ted L. The characterization of the spectral reflectance of mafic silicates, hydrated silicates, and hydrated silicate-water ice mixtures in the 0.6 to 4.5 m wavelength region and applications to planetary science. D, 1987, University of Hawaii.

Roush, Thomas L. Depositional environment of Lower Pennsylvanian and Upper Mississippian detrital sediments in Makanda Quadrangle, Union County, southern Illinois. M, 1972, Southern Illinois University, Carbondale. 66 p.

Roush, Tod Wayne. Geology of Sierra Hermosa Quadrangle (southern half) Zacatecas and San Luis Potosi, Mexico. M, 1981, University of Texas, Arlington. 114 p.

Rousseau, Joseph P. Groundwater hydrology of South Table Mountain, Jefferson County, Colorado. M, 1980, Colorado School of Mines. 238 p.

Rousseau, Normand. Migration des éléments métalliques dans un massif argileux naturel. M, 1986, Universite Laval. 229 p.

Routh, Darcia Layne. Conodont biostratigraphy of the Moorefield and lower Hindsville formations (Upper Mississippian) of the eastern Oklahoma Ozarks. M, 1981, University of Iowa. 37 p.

Routson, Ronald Chester. Illite solubility. D, 1970, Washington State University. 123 p.

Roux, Frederick Holmes Le *see* Le Roux, Frederick Holmes

Roux, P. H. Definition and monitoring of an industrial ground-water pollution problem. M, 1978, Queens College (CUNY). 65 p.

Roux, Wilfred Francois, Jr. Stratigraphy of upper Midway and lower Wilcox groups, west-central Alabama and east-central Mississippi. D, 1958, University of Texas, Austin. 271 p.

Rove, Olaf N. Petrology of Norwegian clays. M, 1925, University of Wisconsin-Madison.

Rove, Olaf N. Some physical characteristics of certain limestone ore horizons. D, 1939, Massachusetts Institute of Technology. 100 p.

Rovetta, Mark Rino. Microfracture growth during the high-temperature creep of peridotite due to the presence of CO$_2$ vapor. D, 1984, University of Washington. 139 p.

Rovey, C. E. Numerical model of flow in a stream-aquifer system. D, 1974, Colorado State University. 195 p.

Rovey, Charles W. Computer modelling of the interaction between Lake Michigan and the dolomite aquifer at Mequon, Wisconsin. D, 1983, University of Wisconsin-Milwaukee. 271 p.

Rowan, Andrew Thomas. Bank erosion, bank stability, and channel migration in salt marsh tidal channels. M, 1988, Rutgers, The State University, New Brunswick. 168 p.

Rowan, Charles David V. The nature and characteristics of lineaments mapped from satellite and aerial imagery in an area of south-central Colorado bounded by 105°00' to 105°30' west longitude to 38°15' to 38°52'30" north latitude. D, 1986, University of Nebraska, Lincoln. 379 p.

Rowan, Dana E. The glacial and periglacial geology of Spitsbergen, Svalbard. M, 1981, Arizona State University. 116 p.

Rowan, Lawrence Calvin. Geology of the Purgatory Mountain area, Botetourt County, Virginia. M, 1957, University of Virginia. 139 p.

Rowan, Lawrence Calvin. Structural analysis of the Quad-Wyoming Line Creeks area, Beartooth Mountains, Montana-Wyoming. D, 1964, University of Cincinnati. 462 p.

Rowan, M. Elizabeth Anderson. A computer model of the aquifer system at the University of Nebraska Mead Agricultural Field Laboratory. M, 1986, University of Nebraska, Lincoln. 72 p.

Rowden, Robert D. Depositional processes and sedimentological analyses of the Spring Canyon Member of the Blackhawk Formation (Campanian), Carbon County, Utah. M, 1985, University of Iowa. 120 p.

Rowe, Andrew Jackson. Subsurface geology, Burleson County, Texas. M, 1951, University of Texas, Austin.

Rowe, David W. Structural and petrologic history of northeastern Tobago, West Indies; a partial cross-section through a composite oceanic arc complex. M, 1987, University of Wyoming. 165 p.

Rowe, Dean E. An isopachous and structural map study of the Allegan area of southwestern Michigan. M, 1951, Wayne State University.

Rowe, Gilbert T. A study of the deep water benthos of the northwestern Gulf of Mexico. M, 1966, Texas A&M University.

Rowe, Jesse Perry. Deposits of volcanic ash of Montana. M, 1903, University of Nebraska, Lincoln.

Rowe, Jesse Perry. Montana coal and lignite deposits. D, 1906, University of Nebraska, Lincoln.

Rowe, R. B. The paleo-Devonian formations of Maryland. D, 1900, The Johns Hopkins University.

Rowe, Robert Burton. Petrology of the Richardson Deposit, Wilberforce, Ontario, Canada. D, 1951, University of Wisconsin-Madison.

Rowe, Robert Burton. The nickel-copper deposits at Lynn Lake, Manitoba. M, 1948, University of Toronto.

Rowe, Roger G. Geology and sulfide mineralization of the Clear Lake area (Precambrian), Vermillion District, Minnesota. M, 1971, Bowling Green State University. 79 p.

Rowe, Royle Carlton. Description and correlative evidence of the Brachiopoda and other faunal members of the Montana Madison Limestone. M, 1927, University of Montana. 98 p.

Rowe, Ruth. The geographic saga of an Ozark family. M, 1939, Washington University. 70 p.

Rowe, Timothy. Osteological diagnosis of Mammalia, L. 1758, and its relationship to extinct Synapsida. D, 1986, University of California, Berkeley. 465 p.

Rowe, William D., Jr. The geology and geochemistry of the Goodes Creek and Hawkins Branch tin occurrences in the inner Piedmont Belt, North Carolina. D, 1987, University of Texas at Dallas. 443 p.

Rowe, Winthrop A. Geology of the south-central Pueblo Mountains, Oregon–Nevada. M, 1971, Oregon State University. 81 p.

Rowekamp, Edward Terry. Size, shape and lithology of gravel (Recent) along the Knik River, Alaska. M, 1968, University of Colorado.

Rowell, Bruce Fenton. Paleoecology of the Doniphan Shale (upper Pennsylvanian) in the northern midcontinent region. D, 1972, University of Nebraska, Lincoln. 201 p.

Rowell, Bruce Fenton. Paleoecology of the Jackson Park shale (upper Pennsylvanian in the northern midcontinent region. M, 1970, University of Nebraska, Lincoln.

Rowell, Eleanor M. The structural geology of the limestones of the plateau front near Syracuse, New York. M, 1937, Syracuse University.

Rowell, Thomas David, III. The foraminifera and Ostracoda of the Anahuac Formation from a deep well in Matagorda County, Texas. M, 1958, University of Oklahoma. 198 p.

Rowell, William Frank. Platinum group elements and gold in the Wanapitei nipissing-type intrusion, northeastern Ontario. M, 1984, University of Western Ontario. 86 p.

Rowett, Charles Llewellyn, Jr. Biostratigraphic interpretation and coral fauna of the Wapanucka Formation of Oklahoma. D, 1962, University of Oklahoma. 306 p.

Rowett, Charles Llewellyn, Jr. Petrographic description of the Wapanucka Limestone at Limestone Gap, Atoka County, Oklahoma. M, 1959, Tulane University. 77 p.

Rowland, Bret. Sedimentary analyses of the Glendon; Marianna Limestones (Oligocene) sequence. M, 1973, Memphis State University.

Rowland, David. Kink band folding in the Green Pond Syncline, northern New Jersey and southern New York. D, 1978, University of South Carolina. 111 p.

Rowland, David Andrew. Pressure build up and drawdown behavior in undersaturated reservoirs of discontinuous permeability. D, 1969, Stanford University. 116 p.

Rowland, J. F. An X-ray study of gold and silver tellurides. M, 1950, Queen's University. 94 p.

Rowland, Mark R. Hydrologic study of the Silurian-Devonian aquifer of the upper Auglaize River basin of Ohio. M, 1969, University of Toledo. 106 p.

Rowland, Richard A. A petrotectonic analysis of cleavage in otherwise unmetamorphosed sediments (Virginia). D, 1938, Cornell University.

Rowland, Richard Atwell. Geology of the Angels Rest area. M, 1934, University of Cincinnati. 85 p.

Rowland, Richard Ernest. Geology of the Grouse Creek area, South Fork Mountains, California. M, 1966, University of California, Los Angeles.

Rowland, Robert William. Paleontology of the San Diego Formation (Pliocene) in northwestern Baja California, Mexico. M, 1968, University of California, Davis. 61 p.

Rowland, Scott K. The flow character of Hawaiian basalt lava. D, 1987, University of Hawaii.

Rowland, Stephen Mark. Environmental stratigraphy of the Lower Member of the Poleta Formation (Lower Cambrian), Esmeralda County, Nevada. D, 1978, University of California, Santa Cruz. 124 p.

Rowland, Tom Lee. Mississippian rocks in the subsurface of the Kingfisher-Guthrie area, Oklahoma. M, 1958, University of Oklahoma. 76 p.

Rowland, Tommy Lee. Lithostratigraphy and carbonate petrology of the Morrow Formation (Pennsylvanian), Braggs-Cookson area, northeastern Oklahoma. D, 1970, University of Oklahoma. 408 p.

Rowles, Lisa Dianne. Deformational history of the Hampton Creek Canyon area, northern Snake Range, Nevada. M, 1982, Stanford University. 80 p.

Rowlett, H. E. Seismic and tectonic studies of plate boundaries; Mid-Atlantic Ridge at 37°N, Oceanographer fracture zone and margin of the Philippine Sea Plate. D, 1978, Columbia University, Teachers College. 153 p.

Rowley, David B. Complex structure and stratigraphy of lower slices of the Taconic Allochthon near middle Granville, New York. M, 1980, SUNY at Albany. 258 p.

Rowley, David Ballantyne. Operation of the Wilson cycle in western New England during the early Paleozoic; with emphasis on the stratigraphy, structure, and emplacement of the Taconic allochthons. D, 1983, SUNY at Albany. 693 p.

Rowley, Eric Alfred. A mineralogical and petrological study of the Cape Porcupine area, Guysborough County, Nova Scotia. M, 1956, Dalhousie University.

Rowley, Peter DeWitt. Geology of the southern Sevier Plateau, Utah. D, 1968, University of Texas, Austin. 385 p.

Rowley, R. Blaine. Geology and mineral deposits of the Lodi Hills, Nye County, Utah. M, 1980, Brigham Young University.

Rowlinson, Norman R. Structural geology of the Carter Lake area, Larimer County, Colorado. M, 1957, University of Colorado.

Rowntree, Rowan A. Morphological change in a California estuary; sedimentation and marsh invasion at Bolinas Lagoon. D, 1973, University of California, Berkeley. 271 p.

Rowser, Edwin M. A study of the Silurian beds of northern Cedar County, Iowa. M, 1929, University of Iowa. 81 p.

Rowser, Edwin M. The Gower Formation of Iowa and its echinoderm fauna. D, 1932, University of Iowa. 188 p.

Rowshandel, Badiollah. A plasticity model for crustal rocks with application to strike-slip fault deformation. D, 1983, Northwestern University. 226 p.

Roxburgh, Kenneth R. A theory for the generation of "Intervals of pulsations of diminishing period". D, 1970, University of British Columbia.

Roxlo, Katherine Spencer. Uranium geochemistry of the Roosevelt Hot Springs thermal area, Utah. M, 1980, University of Utah. 62 p.

Roy, Amalendu. Investigations on the use of shaped electrodes for surface exploration of minerals. M, 1952, Colorado School of Mines. 47 p.

Roy, Amitava. Evolution of the igneous activity in southeastern Costa Rica and southwestern Panama from the middle Tertiary to the Recent. D, 1988, Louisiana State University. 330 p.

Roy, Andre Gerald. Optimality and its relationship to the hydraulic and angular geometry of rivers and lungs. D, 1982, SUNY at Buffalo. 333 p.

Roy, Bimal Chandra. Compression of equally spaced digital elevation model (DEM) data. D, 1987, Ohio State University. 268 p.

Roy, Chalmer John. The origin and significance of the chert in the zinc-lead district of Missouri, Kansas, Oklahoma. D, 1936, Harvard University.

Roy, Chalmer John. The origin of the ironstone nodules in the Benton shales, Alberta, Canada. M, 1930, University of Missouri, Columbia.

Roy, Charles. Géologie de la mine d'or Kiena. M, 1983, Ecole Polytechnique. 200 p.

Roy, David C. The Silurian of northeastern Aroostook County, Maine. D, 1970, Massachusetts Institute of Technology. 484 p.

Roy, David T. The vertebrate fauna of the Horton Bluffs. M, 1971, Acadia University.

Roy, Della Martin. Phase equilibria in the system MgO-Al2O3-H2O and in quaternary systems derived by the addition of SiO2, CO2, and NO2. D, 1952, Pennsylvania State University, University Park. 89 p.

Roy, Della Martin. Phase relations and structural phenomena in the fluoride model systems LiF-BeF2 and NaF-BeF2. M, 1949, Pennsylvania State University, University Park. 50 p.

Roy, Denis. Etude de la fracturation dans la partie ouest de la structure du Lac Manicouagan. M, 1969, Universite de Montreal.

Roy, Denis W. Origin and evolution of the Charlevoix cryptoexplosion structure. D, 1979, Princeton University. 528 p.

Roy, Donald Hilaire. Petrography of the upper Sundance Formation on the flanks of the Big Horn Mountains, Wyoming. M, 1960, University of Illinois, Urbana.

Roy, Edward C. Pleistocene non-marine Mollusca of northeastern Wisconsin. D, 1964, Ohio State University.

Roy, Emery Bernard. Geology of the Italian Mountain intrusives and associated Pb-Ag replacement deposits, Crested Butte and Taylor Park quadrangles, Gunnison County, Colorado. M, 1973, Oklahoma State University. 161 p.

Roy, Jean. Contribution à l'étude de la méthode E.M.V. M, 1974, Ecole Polytechnique.

Roy, Jean L. Electrical methods in mineral well logging. D, 1984, McGill University. 151 p.

Roy, Kenneth James. Stratigraphic analysis and environmental reconstruction of the boundary member of the Charlie Lake Formation (Triassic), northeastern British Columbia. D, 1968, Northwestern University. 190 p.

Roy, Malcom Bernard. Arenaceous foraminifera of the Lansing Group of Late Pennsylvanian Age (Missourian) from northeastern Kansas. M, 1966, University of Kansas. 61 p.

Roy, R. H. Insoluble residue studies of the Smackover Limestone Formation well cores of the Cadium #1 F.B. Smith, Rains County, Texas. M, 1977, East Texas State University.

Roy, Robert F. Heat flow measurements in the United States. D, 1963, Harvard University.

Roy, Sharat Kumar. The characteristics and origin of the Frankonia Sandstone of Leland, Sauk County, Wisconsin. M, 1924, University of Illinois, Chicago.

Roy, Sharat Kumar. The Upper Ordovician fauna of Frobisher Bay, Baffin Island. D, 1941, University of Chicago. 212 p.

Roy, Stephen Donald. Computer simulation model of coastal erosion on Lake Michigan. D, 1987, University of Illinois, Chicago. 395 p.

Roy, Stephen Donald. Evaluation of partial melting models of the origin of some Australian basalts; trace element evidence. M, 1975, Massachusetts Institute of Technology. 144 p.

Roy, William R. Glacial chronology of the South Boulder Valley, Tobacco Root Range, Montana. M, 1980, Indiana University, Bloomington. 163 p.

Roy, William Robert. On competitive adsorption of oxyanions by soils. D, 1985, University of Illinois, Urbana. 163 p.

Royall, P. Daniel. Late-Quaternary paleoecology and paleoenvironments of the western lowlands, Southeast Missouri. M, 1988, University of Tennessee, Knoxville. 181 p.

Roybal, Gretchen Hoffman. Facies relationships in a patch reef of the upper Mural Limestone in southeastern Arizona. M, 1979, University of Arizona. 76 p.

Royden, Leigh Handy. The evolution of the intra-Carpathian basins and their relationship to the Carpathian mountain system. D, 1982, Massachusetts Institute of Technology. 256 p.

Royo, Gilberto Rafael. Environment of deposition of the Yowlumne Sandstone; internal morphology and rock properties, Kern County, California. M, 1986, Texas A&M University. 167 p.

Royse, Chester F., Jr. Sediments of Willapa submarine canyon. M, 1964, University of Washington. 86 p.

Royse, Chester Franklin, Jr. A stratigraphic and sedimentologic analysis of the Tongue River and Sentinel Butte formations (Paleocene), western North Dakota. D, 1968, University of North Dakota. 338 p.

Royse, F., Jr. Geology of the Pine Creek Pass area, Big Hole Mountains, Teton and Bonneville counties, Idaho. M, 1957, University of Wyoming. 105 p.

Royse, Susan E. Soil geochemical study of the altered zones associated with the Gooseberry Mine area, Storey County, Nevada. M, 1986, University of Nevada. 105 p.

Rozacky, Wendy Johnson see Johnson Rozacky, Wendy

Rozanski, George. Microfauna of the (Upper Mississippian) Glen Dean Limestone (Illinois, Indiana, Kentucky). M, 1939, Columbia University, Teachers College.

Rozas Elqueta, Eduardo del see del Rozas Elqueta, Eduardo

Rozelle, John W. Lead and molybdenum dispersion in an arid environment; Sonora, Mexico. M, 1978, Colorado School of Mines. 140 p.

Rozelle, Richard Kent. Feldspar content, current lineations, and bedding characteristics of the Sites Formation, northern Yolo County, California. M, 1962, University of California, Berkeley. 57 p.

Rozen, Robert W. the geology of the Elberton East Quadrangle. M, 1978, University of Georgia.

Rozendal, Roger Anthony. The Upper Cretaceous rocks of Columbia County, Arkansas. M, 1957, University of Minnesota, Minneapolis. 55 p.

Rozilo, Paul John. Volcanics of the State Bridge area, Eagle County, Colorado. M, 1964, Case Western Reserve University.

Rozov, Wendy Cara. A detailed gravity study of the northern Newark Basin, Rockland County, New York. M, 1984, Lehigh University. 82 p.

Rubalcaba, Jose Ramirez. Geology of the north part of the San Antonio Mountains, State of Sonora, Mexico. M, 1965, University of Arizona.

Rubarts, William Eugene. The Boyle-Duffin-New Albany relationships in northern Casey and western Lincoln counties. M, 1959, University of Kentucky. 53 p.

Rubel, Daniel Nicholas. Geology of the Independence area, Sweet Grass and Park counties, Montana. D, 1964, University of Michigan. 192 p.

Rubel, Daniel Nicholas. Tertiary volcanic rocks of the Cooke City-Pilot Creek area, Montana, Wyoming. M, 1959, Wayne State University. 51 p.

Rubenstone, James L. Geology and geochemistry of early Tertiary submarine volcanic rocks of the Aleutian Islands, and their bearing on the development of the Aleutian Island arc. D, 1984, Cornell University. 367 p.

Ruberti, James A. Mineralogy and textural properties of clay rocks in the upper Sewickley sandstone and shale member, and the Benwood-Arnoldsburg Limestone interval, middle Monongahela Group (Pennsylvanian), Morgan, Muskingum, and Noble counties, Ohio. M, 1981, University of Akron. 119 p.

Rubin, A. E. Magmatic fractionation in mesosiderites. M, 1979, University of Illinois, Chicago.

Rubin, Alan Edward. Petrology and origin of brecciated chondritic meteorites. D, 1982, University of New Mexico. 220 p.

Rubin, Allan Mattathias. Dike propagation and crustal deformation in volcanic rift zones. D, 1988, Stanford University. 222 p.

Rubin, Charles M. Carbonate petrology across the top of the Ptychaspid Biomere, Survey Peak Formation, Alberta, Canada. M, 1980, University of Montana. 96 p.

Rubin, David M. Compositional variations in light mineral fractions of beach and nearshore lake sands from southern and eastern Lake Ontario. M, 1972, University of Rochester. 17 p.

Rubin, David M. Depositional environments and diagenesis of the Whitehall Formation (Cambro-Ordovician) of eastern New York and adjacent southwestern Vermont. D, 1975, Rensselaer Polytechnic Institute. 128 p.

Rubin, Jeffrey Neil. Mineralogy and ore genesis at the San Martin Mine, Zacatecas, Mexico. M, 1986, University of Texas, Austin. 98 p.

Rubin, Meyer. A radiocarbon chronology of glacial events during Wisconsin time. D, 1956, University of Chicago. 78 p.

Rubincam, David Parry. The early history of the lunar inclination. D, 1974, University of Maryland.

Rubins, Charles Curtis. Structural, stratigraphic and petrologic relations of rocks south of the Barth Island layered intrusion, Labrador. D, 1973, Syracuse University.

Rubinstein, Jacobo. Domestic water conservation; a component of long term water resources planning. D, 1982, Stanford University. 285 p.

Rubio-Montoya, David. Determination of the erodibility factor of soils and lignite spoils. D, 1981, Texas A&M University. 91 p.

Rubner, David Paul. The petrology, stratigraphy, paleoecology of the Laketown dolomite in east central Nevada. M, 1969, Northern Illinois University. 101 p.

Rubright, Richard D. Ore deposits of the Boulder Falls mining area, Boulder County, Colorado. M, 1941, University of Colorado.

Rubury, Eric Alan. The petrology of the central New York ultraalkaline dikes. M, 1981, University of Rochester. 82 p.

Ruby, Christopher. Morphology and sedimentation of the northern Gulf of Alaska. M, 1978, University of South Carolina.

Ruby, Christopher Houston. Clastic facies and stratigraphy of a rapidly retreating cuspate foreland, Cape Romain, South Carolina. D, 1981, University of South Carolina. 218 p.

Ruch-Hirzel, Mary Lou. General geology and log of the western half of the Ohio Turnpike. M, 1970, Bowling Green State University. 79 p.

Rucker, James B. Paleoecological analysis of Bryozoa of Venezuela-British Guiana shelf sediments. D, 1966, Louisiana State University.

Rucker, James Bivin. The relationship of trace elements in Crassostrea virginica to salinity of the habitat. M, 1961, University of Missouri, Columbia.

Rucker, Paul Douglas. Fluid inclusion and $\delta^{18}O$ study of the precious metal-bearing veins of the Wheaton River District, Yukon. M, 1988, University of Alberta. 120 p.

Ruckman, David W. Geologic land-use mapping of a part of Brazos County. M, 1978, Texas A&M University. 129 p.

Ruckman, John Hamilton. Faunal succession of the Coalinga east side field, Fresno County, California. M, 1914, University of California, Berkeley. 74 p.

Ruckmick, John Christian. Ultramafic intrusives and associated magnetite deposits at Union Bay, Southeast Alaska. D, 1957, California Institute of Technology. 145 p.

Rud, John Orlin. The geology of the southwest quarter of the Bone Mountain Quadrangle, Oregon. M, 1971, University of Oregon. 73 p.

Rudat, Juhani. Quaternary evolution of the San Pedro margin, California. M, 1980, California State University, Northridge. 137 p.

Rudd, Lawrence P. An analysis of the hillslope-floodplain boundary. M, 1980, University of Denver.

Rudd, Neilson. Some geologic factors bearing on the magnetization of the Onondaga Limestone. M, 1955, University of Minnesota, Minneapolis. 39 p.

Rudd, Robert Dean. Glacial deposits of the Yorkville, Illinois, Quadrangle. D, 1954, Northwestern University. 224 p.

Rudder, James, Jr. The significance of organic complexing in the mobility of iron in the KL Landfill leachate plume, Kalamazoo, Michigan. M, 1988, Western Michigan University.

Ruddiman, William Fitzhugh. Planktonic foraminifera of the subtropical north Atlantic gyre (vortex). D, 1969, Columbia University. 292 p.

Rude, LaVerne C. Surficial geology of northern half of Griggs County, North Dakota. M, 1966, University of North Dakota. 128 p.

Rude, Lawrence Culver. A study of the imperfect ditch method for rigid culverts. D, 1979, University of Virginia. 207 p.

Rudek, Evelyn Anne. Petrology of ultramafic and gabbroic inclusions in basaltic rocks of Kahoolawe Island, Hawaii. M, 1988, North Carolina State University. 51 p.

Rudell, Marjorie. Glacial gravels, Manhasset Formation of Long Island, as found in the pits of Goodwin Gallagher, Roslyn, New York. M, 1935, Columbia University, Teachers College.

Ruden, Stuart Michael. Correlation of late Cenozoic rhyolitic tuffs in Power and Bannock counties, southeastern Idaho. M, 1979, Idaho State University. 72 p.

Ruder, Michal Ellen. Interpretation and modeling of regional crustal structure of Southeastern United States using raw and filtered conventional and satellite gravity and magnetic data. D, 1986, Pennsylvania State University, University Park. 308 p.

Rudesill, Roger C. Some aspects of the use of native materials for art ceramics. M, 1978, University of Iowa. 107 p.

Rudin, Cyril. Pelagic foraminifera in long range correlations. M, 1953, New York University.

Rudin-Rodriguez, Fernando M. Interoceanic geodetic leveling in Costa Rica. M, 1961, Ohio State University.

Rudine, S. F. Geology and depositional environments of the Permian rocks, northern Del Norte Mountains, Brewster County, Texas. M, 1988, Sul Ross State University.

Rudkin, G. Thomas. Depositional environments and sand body morphologies of the Muddy Sandstone, west flank Big Horn Basin, Park County, Wyoming. M, 1986, University of Wyoming. 99 p.

Rudloff, Gregory A. Sedimentology and stratigraphy of Wisconsinan deposits, Lake Michigan bluffs, northern Illinois. M, 1988, University of Illinois, Chicago.

Rudman, Albert Julius. A study of seismic reflections from the surface of the basement complex in Indiana. D, 1963, Indiana University, Bloomington. 168 p.

Rudman, Albert Julius. A study of the electrical resistivity and seismic velocity of Indiana coals. M, 1954, Indiana University, Bloomington. 29 p.

Rudmann, Joseph Emmett. The zonal distribution and mineralogy of nickel in saprolitic ore from Nickel Mountain, Riddle, Oregon. M, 1970, Case Western Reserve University. 72 p.

Rudnick, Barbara J. A study of the geology and exploration of a pyrrhotite deposit in Precambrian units, Cuttingsville, VT. M, 1986, University of New Brunswick. 212 p.

Rudnick, David Thornton. Seasonality of community structure and carbon flow in Narragansett Bay sediments. D, 1984, University of Rhode Island. 334 p.

Rudnick, Jon. The nature of the rocks of the Mid-Atlantic Ridge as indicated by underwater photographs. M, 1968, Columbia University. 30 p.

Rudnick, Roberta L. Petrography, geochemistry and tectonic affinities of meta-igneous rocks from the Precambrian Carrizo Mountain Group, Van Horn, Texas. M, 1983, Sul Ross State University. 117 p.

Rudnyansky, Albert Julius. A study of the electrical resistivity and seismic velocity of Indiana coals. M, 1954, Indiana University, Bloomington. 29 p.

Rudolph, D. L. A quasi three-dimensional finite element model for steady-state analysis of multiaquifer systems. M, 1985, University of Waterloo. 100 p.

Rudolph, Joseph. The nationalization of mineral resources, its forms and stages of development in various countries. M, 1928, University of Wisconsin-Milwaukee.

Rudolph, Kurt W. Diagenesis of back-reef carbonates; an example from the Capitan Complex. M, 1978, University of Texas, Austin.

Rudser, Ralph Jay. Geology and geothermal potential of Susanville, Lassen County, California. M, 1979, University of California, Davis. 62 p.

Rudy, Donald James. Mars; high resolution VLA observations at wavelengths of 2 and 6cm and derived properties. D, 1987, California Institute of Technology. 156 p.

Rudy, Harold R. Permo-Carboniferous stratigraphy of the Banff-Jasper area, Alberta. M, 1958, University of Alberta. 66 p.

Rudy, Richard J. Structural damage from shale expansion at Kansas City, Missouri. M, 1983, University of Missouri, Kansas City. 89 p.

Rudy, Samuel. A study of the electrolytic oxidation of cinnabar ore for the recovery of mercury. M, 1972, University of Nevada. 42 p.

Rue, Edward E. Geology of the Carter Lake region north-west of Berthoud (Larimer County), Colorado. M, 1949, Colorado School of Mines. 64 p.

Rueb, Ronald A. Lineament analysis of the Black Hills, South Dakota and Wyoming. M, 1984, South Dakota School of Mines & Technology.

Ruebelmann, Kerry L. Geology and petrology of Archean migmatites and gneisses, Mount Baldy area, Horseshoe Lake Quadrangle, Wind River Mountains, Wyoming. M, 1988, Idaho State University. 119 p.

Rueda, J. E. G. Exploration and development at the La Negra Mine, Maconi, Queretaro, Mexico. M, 1975, University of Arizona.

Rueda, Jose E. Gaytan see Gaytan Rueda, Jose E.

Ruede, George M. A mapping and field study of the Unkpapa Sandstone in the Black Hills, South Dakota. M, 1951, University of Nebraska, Lincoln.

Ruedisili, Lon Chester. Stratigraphy and paleontology of the Mississippian bioherms in the northern part of the Sacramento Mountains, New Mexico. D, 1968, University of Wisconsin-Madison. 188 p.

Ruedisili, Lon Chester. The stratigraphy and paleontology, Permian Taku Group, Windy Arm, Tagish Lake, Yukon Territory. M, 1965, University of Wisconsin-Madison.

Rueger, Bruce Francis. Geology and palynology of the Paradox Formation (Desmoinesian), southeastern Utah. M, 1984, University of Colorado. 192 p.

Ruehr, B. B. Geology of the Devil's Gate area, Albany and Carbon counties, Wyoming. M, 1961, University of Wyoming. 48 p.

Ruendal, Aime Pamela. Petrology of sillimanite-grade metapelites in the Headquarters area, northern Idaho. M, 1987, University of Oregon. 111 p.

Ruetschilling, Richard L. Structure and stratigraphy of the San Ysidro Quadrangle, Sandoval County, New Mexico. M, 1973, University of New Mexico. 79 p.

Ruetz, Joseph William. Paleocene sedimentation in the northern Santa Lucia Range. M, 1977, Stanford University. 104 p.

Ruez, Paul H. A field and laboratory study of millerite and related Ni-bearing minerals. M, 1973, Miami University (Ohio). 63 p.

Ruff, Arthur W. The geology and ore deposits of the Indiana Mine area, Pima County, Arizona. M, 1951, University of Arizona.

Ruff, Barbara L. A Hemphillian vertebrate fauna from Antelope County, Nebraska. M, 1976, University of Georgia.

Ruff, Larry John. I, Great earthquakes and seismic coupling at subduction zones; II, The structure of the lowermost mantle determined by short period P-wave amplitudes. D, 1982, California Institute of Technology. 215 p.

Ruffel, Alice Veronica. The distribution of trace elements in crude oils using proton induced X-ray emission analysis and wavelengh-dispersive X-ray fluorescence. M, 1986, University of Oklahoma. 132 p.

Ruffin, Isiah Washington. Characterization and genesis of Geary soils in southeastern Nebraska. D, 1971, University of Nebraska, Lincoln. 127 p.

Ruffin, James H. Palynology of the Tebo Coal (Pennsylvanian) of Oklahoma. M, 1961, University of Oklahoma. 124 p.

Ruffman, Alan S. Crustal seismic studies in Hudson Bay, Canada. M, 1966, Dalhousie University. 127 p.

Rugg, Edwin Stanton. Geology of the Carter Mine, Gunnison County, Colorado. M, 1956, Colorado School of Mines. 54 p.

Ruggiero, J. G. Seismic risk criteria for New York City and surroundings. M, 1976, City College (CUNY).

Ruggiero, Robert Winslow. Depositional history and performance of a Bell Canyon Sandstone reservoir, Ford-Geraldine Field, West Texas. M, 1985, University of Texas, Austin. 242 p.

Rugh, Alex L. Temporal and spatial variations in surface sediment near dredge spoils, New London, Connecticut. M, 1977, Lehigh University.

Ruhe, Roberet Victory. The geology of Shelby County. M, 1948, Iowa State University of Science and Technology.

Ruhe, Robert Victory. Reclassification and correlation of the glacial drifts of northwestern Iowa and adjacent areas. D, 1950, University of Iowa. 124 p.

Ruhle, James L. The Mount Laurel and Wenonah sands of New Jersey. M, 1960, University of Massachusetts. 134 p.

Ruhlman, Fred Lee and Wade, Franklin Russell. A study of certain factors influencing the flow of hydrocarbons through reservoir sands. M, 1941, University of Southern California.

Ruisaard, Chris Ivan. Stratigraphy of the Miocene Alverson Formation, Imperial County, California. M, 1979, San Diego State University.

Ruisch, Edeltraud. Late Cenozoic size variation of the foraminifer Orbulina universa in the North Pacific Ocean. M, 1980, Rutgers, The State University, Newark. 101 p.

Ruitenberg, Arie Anne. Potential ore mineralization and alteration at the Mount Pleasant tin prospect, Charlotte County, New Brunswick. M, 1963, University of New Brunswick.

Ruiz Calzada, Carlos Edgardo. Development of a mathematical model for pesticide and sediment transport in the Iowa River. D, 1987, University of Iowa. 199 p.

Ruiz Castellanos, Mario. Rubidium - strontium geochronology of the Oaxaca and Acatlan metamorphic areas of southern Mexico. D, 1979, University of Texas at Dallas. 188 p.

Ruiz de la Pena Horcasitas, Gerardo. Structural controls on the mineralization of El Arco copper deposit, Baja California Norte, Mexico. M, 1979, Colorado School of Mines. 98 p.

Ruiz, Carlos Soto *see* Soto Ruiz, Carlos

Ruiz, Joaquin. Geology and geochemistry of fluorite ore deposits and associated rocks in northern Mexico. D, 1983, University of Michigan. 215 p.

Ruiz, Patricio. Probabilistic study of the behavior of structures during earthquakes. D, 1969, University of California, Berkeley. 99 p.

Ruiz-Elizondo, Jesus. A study of the safe yield and replenishment conditions for the Yucaipa Basin area, California. M, 1954, California Institute of Technology. 91 p.

Ruiz-Elizondo, Jesus. Geology of the St. Francis area, Los Angeles County (California). M, 1953, California Institute of Technology. 54 p.

Ruiz-Menacho, Carmen Maria. Subsolidus studies of the system Li2O-Al2O3-SiO2-H2O. M, 1959, Pennsylvania State University, University Park. 50 p.

Rukas, Justin M. The Sabine Group of Natchitoches Parish, Louisiana. M, 1939, Louisiana State University.

Rukavina, Norman Andrew. Mineral ratios in hydraulic sizes as indicators of nearshore sediment source. M, 1961, University of Western Ontario. 69 p.

Rukavina, Norman Andrew. Particle orientation in turbidites; theory and experiment. D, 1965, University of Rochester. 57 p.

Ruland, W. W. Fracture depths and active groundwater flow in a weathered clayey till in Lambton County, Ontario. M, 1988, University of Waterloo. 62 p.

Rule, Audrey Catherine. Refinement of the crystal structures of phengite-2M₁ and ferroan clinochlore; derivation of the six simple mica polytypes; determination of the stacking arrangement of the Tordal lepidolite and symmetry drawings of the seventeen unique plane groups. D, 1985, University of Wisconsin-Madison. 286 p.

Ruley, Eugene E. The geology of a portion of the Foothills Belt of the southwest side of the Canon City Embayment, Fremont County, Colorado. M, 1952, Colorado School of Mines. 66 p.

Rulli, Vernon G. Stratigraphy and taxonomy of the cryptostome bryozoan genus Sulcoretepora. M, 1973, Wayne State University.

Rulon, Jennifer. The development of multiple seepage faces along heterogeneous hillsides. D, 1984, University of British Columbia. 158 p.

Rumbaugh, James Orville, III. Effect of fracture permeability on radon-222 concentration in ground water of the Reading Prong, Pennsylvania. M, 1983, Pennsylvania State University, University Park. 110 p.

Rumble, Douglas, III. Stratigraphic structural and petrologic studies in the Mount Cube area, New Hampshire, Vermont. D, 1969, Harvard University.

Rundell, Bruce M. Depositional relationship between carbonate and clastic environments of the Early Permian Laborcita Formation near Tularosa, N.M. M, 1982, New Mexico Institute of Mining and Technology. 130 p.

Rundle, J. B. Anelastic processes in the strike slip faulting; application to the San Francisco earthquake of 1906. D, 1976, University of California, Los Angeles. 224 p.

Rundquist, L. A. A classification and analysis of natural rivers. D, 1975, Colorado State University. 404 p.

Runge, Erwin John, Jr. An analysis of the microfauna of the Kiddville layer of the Devonian Boyle Limestone, Marion County, Kentucky. M, 1959, Miami University (Ohio). 86 p.

Runge, Erwin John, Jr. Continental shelf sediments, Columbia River to Cape Blanco, Oregon. D, 1966, Oregon State University. 143 p.

Runkel, Anthony Charles. Geology and vertebrate paleontology of the Smith River basin, Montana. M, 1986, University of Montana. 80 p.

Runkel, Anthony Charles. Stratigraphy, sedimentology, and vertebrate paleontology of Eocene rocks, Big Bend region, Texas. D, 1988, University of Texas, Austin. 310 p.

Runkle, Dita Elisabeth. Geology and geochronometry of the Coast Plutonic Complex adjacent to Douglas,

Sue and Loretta channels, British Columbia. M, 1979, University of British Columbia.

Runnalls, N. D. Mineralogical variation in the Mackenzie Island batholith, Red Lake. M, 1933, Queen's University. 68 p.

Runnalls, R. J. Gravity modeling, Red Lake region, N.W. Ontario. M, 1978, University of Toronto.

Runne, Marjorie E. The nepheline syenites of Wausau, Wisconsin. M, 1938, Northwestern University.

Runnells, Donald DeMar. The copper deposits of Ruby Creek, Cosmos Hills, Alaska. D, 1963, Harvard University. 310 p.

Runnels, Tyson D. Application of geostatistical analysis to uranium mining, Shirley Basin, Wyoming. M, 1983, Colorado School of Mines. 136 p.

Runner, Delmar G. The geology and utilization of non-metallic deposits for use in highway construction. M, 1933, George Washington University. 64 p.

Runner, Joseph J. The Pre-Cambrian geology of the Nemo District, Black Hills, South Dakota, with special reference to a Pre-Cambrian unconformity. D, 1923, University of Chicago. 52 p.

Runyon, Cassandra J. Stratigraphy and erosion of the banks along the Osage River, Missouri. M, 1984, Southern Illinois University, Carbondale. 90 p.

Runyon, David M. Structure, stratigraphy, and tectonic history of the Indianola Quadrangle, central Utah. M, 1977, Brigham Young University. 82 p.

Runyon, Gary A. The distribution and variation of sulfide constituents of the Great Smoky Group, Ducktown area, Tennessee, with respect to metamorphic grade. M, 1983, University of Tennessee, Knoxville. 115 p.

Runyon, H. Everett. Chemical analyses of some oilwell waters of Russell, Ellis and Trego counties. M, 1936, Fort Hays State University. 34 p.

Runyon, Stephen Lane. A stratigraphic analysis of the Traverse Group of Michigan. M, 1976, Michigan State University. 86 p.

Ruof, Mark Anthony. A geotechnical investigation of selected coal mine waste embankments of east-central Ohio. M, 1987, Kent State University, Kent. 204 p.

Ruotsala, Albert Peter. A study of available chemical analyses of igneous rocks of Minnesota. M, 1955, University of Minnesota, Minneapolis. 70 p.

Ruotsala, Albert Peter. Some factors affecting the formation of anorthite in the solid state. D, 1962, University of Illinois, Urbana. 61 p.

Rupert, Frank. Biogeography of the Recent deepwater benthic foraminifera of the Mediterranean Sea. M, 1980, Florida State University.

Rupert, Michael G. Structure and stratigraphy of the Klondike Hills, southwestern New Mexico. M, 1986, New Mexico State University, Las Cruces. 138 p.

Rupke, Nicolaas Adrianus. Geologic studies of an early and middle Eocene flysch formation, southwestern Pyrenees, Spain. D, 1972, Princeton University. 378 p.

Rupnik, John J. The geology of the Wiley Canyon area, Oak Ridge Anticline, T. 3-4 N, R. 18-19 W, Ventura County, California. M, 1941, California Institute of Technology. 22 p.

Rupp, John Andrew. Tertiary rhyolite dikes and plutons of the northern Little Belt Mountains, Montana. M, 1980, Eastern Washington University. 136 p.

Ruppel, Carolyn Denise. Thermal-modelling of extensional tectonics. M, 1986, Massachusetts Institute of Technology. 116 p.

Ruppel, Edward Thompson. Geology of the Limestone Hills, Broadwater County, Montana. M, 1950, University of Wyoming. 104 p.

Ruppel, Stephen C. Conodont biostratigraphy and correlation of the Fort Payne Chert and Tuscumbia Limestone (Mississippian) at selected sites in northwestern Alabama. M, 1971, University of Florida. 74 p.

Ruppel, Stephen C. The stratigraphy, carbonate petrology, and depositional environments of the Chickamauga Group (Middle Ordovician) of northern

East Tennessee. D, 1979, University of Tennessee, Knoxville. 231 p.

Ruppert, Leslie F. Cathodoluminescent quartz grains in the Upper Freeport coal bed, west-central Pennsylvania; an indicator of detrital influx. M, 1988, George Washington University. 73 p.

Rush, Francis Eugene. Petrography and physical properties of some Devonian limestones of Iowa. M, 1957, Iowa State University of Science and Technology.

Rush, James D. Rare earth elements in wallrock and ore zone biotites and hornblendes, Ore Knob, North Carolina. M, 1973, University of North Carolina, Chapel Hill. 48 p.

Rush, Randy J. Geophysical and geotechnical investigations near Dupee Shaft, Hancock Mine, Hancock, Michigan. M, 1984, Michigan Technological University. 189 p.

Rush, Richard W. Primary structures of the (Ordovician) lower Chazy at South Hero, Vermont. M, 1948, Columbia University, Teachers College.

Rush, Richard W. Silurian rocks of western Millard County, Utah. D, 1954, Columbia University, Teachers College.

Rush, Thomas Dudley. The geology of southwestern Boone County, Missouri. M, 1950, University of Missouri, Columbia.

Rushin, Carol Jo. Interpretive and paleontologic values of Natural Trap Cave, Bighorn Mountains, Wyoming. M, 1973, University of Montana. 97 p.

Rushing, Emmett O., III. Geochemical alteration of the ground water in the Carrizo Sand of northwestern Anderson County, Texas. M, 1987, Stephen F. Austin State University. 96 p.

Rushing, Jodi A. Contact metamorphism of Paleozoic rocks near Stronghold Canyon, Dragoon Mountain, Arizona. M, 1978, University of Arizona.

Rushing, Robert S. Nick Springs and West Nick Springs fields, Union County, Arkansas. M, 1956, University of Arkansas, Fayetteville.

Rushing, V. Detection of potential clay deposits in the Jackson Purchase region of western Kentucky utilizing numerical analysis of Landsat thematic mapper data. M, 1988, Murray State University. 88 p.

Rushton, Betty Toombs. Wetland reclamation by accelerating succession. D, 1988, University of Florida. 267 p.

Rusling, Lee Judson, Jr. Structure, stratigraphy, and source of sediments of the Great Meadow Hill Syncline, Narragansett Basin, Massachusetts. M, 1961, Brown University.

Rusmore, Margaret Elizabeth. Geology and tectonic significance of the Upper Triassic Cadwallader Group and its bounding faults, southwestern British Columbia. D, 1985, University of Washington. 174 p.

Rusnak, Gene Alexander. The orientation of sand grains under conditions of unidirectional fluid flow. D, 1955, University of Chicago. 55 p.

Russ, Carol Alice. Contact metamorphism of mafic schist, Laramie anorthosite complex, Morton Pass, Wyoming. M, 1984, SUNY at Stony Brook. 311 p.

Russ, David P. Geology of the West Virginia part of the Princeton, West Virginia, Virginia Quadrangle. M, 1969, West Virginia University.

Russ, David Perry. The Quaternary geomorphology of the lower Red River valley, Louisiana. D, 1975, Pennsylvania State University, University Park. 426 p.

Russel, Ronald P. Geology of the Bayou Middle Fork Field, Claiborne Parish, Louisiana. M, 1979, Louisiana Tech University.

Russell, Alice E. Pebbles in the glacial till of eastern North Dakota. M, 1950, University of North Dakota. 47 p.

Russell, Billy Joe, Jr. Depositional and diagenetic history of Bodcaw Sand, Cotton Valley Group (Upper Jurassic), Longwood Field, Caddo Parish, Louisiana. M, 1983, Stephen F. Austin State University. 140 p.

Russell, Branch James. Pre-Tertiary paleogeography and tectonic history of the Jackson Mountains, northwestern Nevada. D, 1981, Northwestern University. 265 p.

Russell, Charles Eugene. Hydrogeologic investigations of flow in fractured tuffs, Rainier Mesa, Nevada Test Site. M, 1987, University of Nevada, Las Vegas. 154 p.

Russell, Charles W. Geology of the central portion of the Little Rocky Mountains, Phillips County, Montana. M, 1984, University of Idaho. 92 p.

Russell, Charles William. Crystallization history of the Banks Complex; implications for middle crustal evolution in Cordilleran batholithic terranes. D, 1988, University of Washington. 226 p.

Russell, Dale Alan. An early Cenozoic mammalian fauna from Togwotee Pass, Wyoming. M, 1960, University of California, Berkeley. 61 p.

Russell, Dale Alan. The skull of American mosasaurs. D, 1963, Columbia University, Teachers College. 413 p.

Russell, David Ray. Multi-channel processing of dispersed surface waves. D, 1987, St. Louis University. 162 p.

Russell, Dearl T. Surface geology of the Robbers Cave-Lodi area, Latimer County, Oklahoma. M, 1958, University of Oklahoma. 105 p.

Russell, Don Eugene. Clarendonian (Pliocene) fauna of Junatura, Oregon. M, 1956, University of California, Berkeley. 95 p.

Russell, Donald Arthur. Velocity-gradient relationships for water-saturated porous media. D, 1968, Purdue University. 155 p.

Russell, Edgar Ernest. Measurement of the optical constants of sapphire and quartz in the far infrared with the asymmetric Fourier transform method. D, 1966, Ohio State University. 119 p.

Russell, Edmund Louis, Jr. Stratigraphy and intraformational structures from well cores of the Upper Devonian of the Bradford oil field. M, 1951, Pennsylvania State University, University Park. 135 p.

Russell, Edward F. Geology of a portion of the Lemhi Range near Gilmore, Idaho. M, 1974, Eastern Washington University. 58 p.

Russell, Ernest Everett. Stratigraphy of the outcropping Cretaceous beds below the McNairy Sand in Tennessee. D, 1965, University of Tennessee, Knoxville. 197 p.

Russell, Ernest Everett. The vertebrate fossils of the Upper Cretaceous formations in Mississippi. M, 1955, Mississippi State University. 59 p.

Russell, Eugene. Petrology of the Tallohatte Formation (Eocene) in Grenada County, Mississippi. M, 1972, University of Mississippi.

Russell, Eugene Merle. Geography of an airway; St. Louis, Kansas City, Omaha. M, 1929, Washington University.

Russell, G. A. Discussion of the Grenville crystalline limestone in the vicinity of Kingston, Ontario. M, 1935, Queen's University. 53 p.

Russell, Gail S. U-Pb, Rb-Sr, and K-Ar isotopic studies bearing on the tectonic development of the southernmost Appalachian Orogen, Alabama. D, 1978, Florida State University. 209 p.

Russell, George A. A discussion of the relation of differential stress to crystal dissolution and growth. M, 1934, University of Minnesota, Minneapolis. 39 p.

Russell, James D. Determination of Ca^{18} by neutron activation. M, 1960, Massachusetts Institute of Technology. 107 p.

Russell, James Kelly. Petrology of Diamond Craters, S.E. Oregon. D, 1984, University of Calgary. 156 p.

Russell, James Kelly. The petrogenesis of the Thompson nickel belt gneisses, Paint Lake, Manitoba. M, 1980, University of Calgary. 199 p.

Russell, Jimmie Norton. Geology of Hannibal area, northwestern Erath County, Texas. M, 1954, University of Texas, Austin.

Russell, John L. Cemented slope and terrace deposits of Cenozoic age in western Marion County, Kansas. M, 1967, Kansas State University. 63 p.

Russell, John Lysle. Comparison of two late Paleozoic red shales of the Midcontinent region. D, 1974, University of Nebraska, Lincoln.

Russell, John Phillip. Strontium, cesium, iodine, and barium determinations in samples by ion-induced X-ray fluorescence. D, 1973, University of Florida. 119 p.

Russell, Karen L. Coral facies and diagenesis of a Pleistocene patch reef; Ambergris, Belize. M, 1986, Texas A&M University.

Russell, Kenneth Lloyd. Clay mineral origin and distribution on Astoria Fan. M, 1967, Oregon State University. 40 p.

Russell, Kenneth Lloyd. Geochemistry and halmyrolysis of clay minerals, Rio Ameca, Mexico. D, 1969, Princeton University. 66 p.

Russell, Larry Lee. Chemical aspects of ground water recharge with wastewaters. D, 1976, University of California, Berkeley. 425 p.

Russell, Laura M. Petrogenesis of a migmatite, Penobscot County, Maine; ultrametamorphism or intrusion?. M, 1984, Virginia Polytechnic Institute and State University. 290 p.

Russell, Lee Robin. Tectonic character of the Melones fault zone, western Sierra Nevada, California. D, 1977, Texas Tech University. 258 p.

Russell, Lee Robin. The structural and metamorphic history of the Oakhurst Roof Pendant, Mariposa and Madera counties, California. M, 1972, Texas Tech University. 70 p.

Russell, Loris Shano. Paleontology and stratigraphy of the uppermost Cretaceous and lower Tertiary formations of Alberta, Canada. D, 1930, Princeton University.

Russell, Mary Ellen. West Coast Naticidae. M, 1933, University of Washington. 53 p.

Russell, Merlin D. Foraminiferal biostratigraphy of the Pollocksville and Haywood Landing members of the Belgrade Formation, North Carolina coastal plain. M, 1987, University of North Carolina, Chapel Hill. 303 p.

Russell, Orville Ray. Geology of the Hominy area, Osage County, Oklahoma. M, 1955, University of Oklahoma. 84 p.

Russell, Perry Wooten. The Point Fermin submarine fan; a small late middle Miocene age fan within the Monterey Formation. M, 1988, California State University, Northridge. 82 p.

Russell, Richard Dana. The (Pliocene) Tehama Formation of Northern California. D, 1932, University of California, Berkeley. 133 p.

Russell, Richard Doncaster. The age of the Earth from studies of the radioactive decay of uranium, thorium and potassium. D, 1954, University of Toronto.

Russell, Richard Joel. Basin range structure and geomorphology of the Warner Range, northeastern California. D, 1926, University of California, Berkeley. 496 p.

Russell, Richard Verner. Basalts of Allen Ranch area, Uvalde County, Texas. M, 1965, University of Texas, Austin.

Russell, Rick Harold. Geology of the Blue Ouachita mountain area, Frontal Ouachita mountains, Arkansas. M, 1969, Northern Illinois University. 105 p.

Russell, Robert Guy. Geology of the Cedar Mountain Quadrangle, eastern Oregon. M, 1961, University of Oregon. 41 p.

Russell, Robert O. A temperature study of solid state transformation of selected aluminum silicates to mullite. M, 1965, Miami University (Ohio). 84 p.

Russell, Robert T. The geology of the pegmatites at Glastonbury, Connecticut. M, 1939, Northwestern University.

Russell, Robert Thayer. The geology of the Poncha fluorspar district, Chaffee County, Colorado. D, 1950, University of Cincinnati. 67 p.

Russell, Ronald Paul. Geology of the Bayou Middle Fork Field, Claiborne Parish, Louisiana. M, 1979, Louisiana Tech University.

Russell, Scott Lewis. Growth, morphology, habit and habitat of selected brachiopod and mollusc species from the Mead Peak Member of the Phosphoria Formation, Permian, northeastern Utah-southeastern Idaho-southwestern Wyoming. M, 1980, Utah State University. 155 p.

Russell, Stephen John. Stratigraphy and structure of Mesozoic metavolcanic rocks in the vicinity of Mt. Dana, Yosemite National Park, California. M, 1976, California State University, Fresno.

Russell, Steven Duffy. Seismic structure of the Galapagos spreading center at 86°W and the occurrence of a crustal low-velocity zone. M, 1979, Duke University. 171 p.

Russell, Suzanne J. Physical, chemical, and petrographic properties affecting the skid resistance of carbonate aggregates in Illinois Class III bituminous concrete pavements. M, 1975, University of Illinois, Urbana. 43 p.

Russell, Suzanne Jeannette. Petrography and depositional environment of the Herrin (No. 6) Seam in central, eastern and northwestern Illinois. D, 1983, Pennsylvania State University, University Park. 490 p.

Russell, Timothy Gray. Geology of Powell Valley in the LaFollette Quadrangle, Campbell County, Tennessee. M, 1966, University of Tennessee, Knoxville. 68 p.

Russell, William E. The (Pennsylvanian) Hill Creek Beds and the Meek Bend Limestone of the Lazy Bend Formation, Parker County, Texas. M, 1953, Texas Christian University.

Russell, William John. Some physical and chemical processes affecting magmas; Part I, A geochemical and petrological investigation of the trough bands of the Skaergaard Intrusion, East Greenland; Part II, Magma withdrawal from zoned magma bodies and related conduit mixing. M, 1987, University of Oregon. 114 p.

Russell, William L. Structural and stratigraphic problems in the Cretaceous System of western South Dakota. D, 1927, Yale University.

Russell, William L. The great Triassic fault of southern Connecticut and its structural and stratigraphic relations. M, 1922, Yale University.

Russo, Amelia Gloria. The Ostracoda fauna of the (Devonian) Arkona Shale, Arkona, Ontario. M, 1936, Columbia University, Teachers College.

Russo, Anthony F. A geological consideration of the proposed Broadview Heights-Brecksville sanitary landfill site. M, 1976, University of Akron. 55 p.

Russo, Grace-Louise M. Relief map of the Mohorovicic discontinuity with accompanying text. M, 1962, New York University.

Russo, Joseph F. Influence of framework grain composition on sandstone diagenesis, Lobo Formation (Tertiary), southern New Mexico. M, 1987, University of Texas at El Paso.

Russon, Michael P. Geology, depositional environments, and coal resources of the Helper 7.5′ Quadrangle, Carbon County, Utah. M, 1984, Brigham Young University. 168 p.

Rust, Aaron B. Petrography, chemistry, and mineralogy of selected weathering products, Long Binh, South Vietnam. M, 1974, University of South Florida, Tampa. 42 p.

Rust, Claude Charles. Conodonts of the Martinsburg Formation (Ordovician) of southwestern Virginia. D, 1968, Ohio State University. 199 p.

Rust, Derek John. Geologic, geomorphic and structural analysis of Quaternary tectonic behavior; San Andreas fault zone in the Transverse Ranges north of Los Angeles, California. D, 1985, University of California, Santa Barbara. 559 p.

Rust, George W. Colloidal primary ores at Cornwall Mines, southeastern Missouri. D, 1935, University of Chicago. 28 p.

Rust, James Edward. Amplitude spectra of rayleigh waves from an earthquake swarm in the Gulf of California. M, 1971, University of Michigan.

Rust, Lee D. Temporal and spatial variations in topography and surface texture on a tidal flat in the Great Bay Estuary, New Hampshire. M, 1980, University of New Hampshire. 75 p.

Rust, Richard Reynolds. Estimation of percolation from landfill final covers based on extreme climatic events. D, 1986, Texas A&M University. 284 p.

Rutan, Debra. The petrology and depositional environments of the Pennsylvanian Lawrence Formatin in eastern Kansas. M, 1980, University of Kansas. 126 p.

Rutford, Robert Hoxie. The glacial geology and geomorphology of the Ellsworth Mountains, West Antarctica. D, 1969, University of Minnesota, Minneapolis. 374 p.

Ruth, John Helms. Assessment of streambank erosion along the North Fork Flathead River, northwestern Montana. M, 1988, Montana State University. 99 p.

Ruth, John William. The genus Siphonalia of the Pacific Coast Tertiary. M, 1937, University of California, Berkeley. 51 p.

Ruth, Joseph Frank. Barium-lanthanum ratios and the petrogenesis of arc volcanics. M, 1979, University of Arizona. 137 p.

Rutherford, David W. A geochemical baseline survey of selected toxic elements in surficial materials in the vicinity of the Paraho demonstration oil shale processing plant, Anvil Points, Garfield County, Colorado. M, 1979, Colorado School of Mines. 153 p.

Rutherford, Gary J. Diagenesis of upper Miocene sediment; Terrebonne Trough, southern Terrebonne and Lafourche parishes, Louisiana. M, 1988, University of New Orleans.

Rutherford, Homer Morgan. Interpretation of reflection seismograms. M, 1933, University of Pittsburgh.

Rutherford, Malcolm John. An experimental study of biotite phase equilibria. M, 1968, The Johns Hopkins University. 328 p.

Rutherford, Malcolm John. Geothermometry of liquid inclusions in quartz, Coronation Mine, Flin Flon area, Saskatchewan. M, 1963, University of Saskatchewan. 40 p.

Rutherford, Mark S. Depositional environments and areal distribution of updip lower Tuscaloosa "Stringer" Member sandstones in portions of Amite and Wilkinson counties, Mississippi. M, 1988, University of Southwestern Louisiana. 148 p.

Rutherford, Ralph L. Geology of the Saunders Creek and Nordegg coal areas, Alberta, Canada. D, 1923, University of Wisconsin-Madison.

Ruths, Mark Allen. The reference-correction method for improving accuracy in the seismic location of trapped coal miners. M, 1977, Pennsylvania State University, University Park. 141 p.

Rutka, Margaret A. The sedimentology and petrography of the Whirlpool Sandstone (Lower Silurian) in outcrop and the subsurface in southern Ontario and upper New York State. M, 1986, McMaster University. 355 p.

Rutland, Carolyn. Carbonate phase petrology of carbonatites. M, 1979, University of Texas, Austin.

Rutland, Carolyn. Geochemistry of the Elkhorn Mountains volcanics, southwestern Montana; implications for the early evolution of a volcanic-plutonic complex. D, 1985, Michigan State University. 108 p.

Rutledge, Donald W. Contact phenomena of the Pokiok Granite, New Brunswick. M, 1954, University of New Brunswick.

Rutledge, Elliott Moye. Loess in Ohio; composition in relation to several local rivers. D, 1969, Ohio State University. 208 p.

Rutledge, Floyd Wayne. Cenozoic history of Springerville area, Apache County, Arizona. M, 1956, University of Texas, Austin.

Rutledge, Henry Mitchell. The Boyle Formation of southern Boyle County, Kentucky. M, 1957, University of Kentucky. 36 p.

Rutledge, J. J. The Clinton iron ores of Stone Valley, Huntingdon County, Penn. D, 1904, The Johns Hopkins University.

Rutledge, James Raymond. A study of fluid migration in porous media by stereoscopic radiographic techniques. M, 1966, Brigham Young University.

Rutledge, James Thomas. A shallow seismic refraction survey over a late Quaternary fault scarp west of the Santa Rita Mountains, Arizona. M, 1984, University of Arizona. 93 p.

Rutledge, Richard Boyden. A petrographic study of some mineral eutectics. M, 1921, University of Missouri, Columbia.

Rutledge, Richard Boyden. The geology of Lawrence County and Barry County, Missouri. D, 1924, University of Missouri, Columbia.

Rutstein, Martin Stuart. The partitioning of iron between natural and synthetic wollastonite, clinopyroxene and garnet. D, 1969, Brown University. 144 p.

Ruttan, George Douglas. The development of a gneiss zone in the Flin Flon area (Manitoba). M, 1936, University of Manitoba.

Rutter, Nathaniel. Surficial geology of the Banff area, Alberta. D, 1966, University of Alberta. 105 p.

Rutter, Nathaniel W. Foliation and other structures of the Gulkana Glacier, central Alaska Range, Alaska. M, 1962, University of Alaska, Fairbanks. 51 p.

Ruvalcaba-Ruiz, Delfino. Geology, alteration and fluid inclusions of the Santa Elena and Santo Nino fissures. M, 1980, Colorado State University. 98 p.

Ruvalcaba-Ruiz, Delfino Concepcion. Geology and origin of the Aquila iron deposit in southwestern Michoacan, Mexico. D, 1983, Colorado State University. 171 p.

Ruzicka, Joseph Frederick. Geology of the Santiago Shafter mines area, Kettle Falls, Stevens County, Washington. M, 1967, Washington State University. 67 p.

Ruzyla, Kenneth. Geomorphic features and processes on the south shore of Fire Island, New York. M, 1971, SUNY at Binghamton. 77 p.

Ruzyla, Kenneth. The relationship of diagenesis to porosity development and pore geometry in the Red River Formation (Upper Ordovician), Cabin Creek Field, Montana. D, 1980, Rensselaer Polytechnic Institute. 200 p.

Ryall, Alan S., Jr. P waves of the Hegben Lake, Montana earthquake of August 18, 1959. D, 1962, University of California, Berkeley. 76 p.

Ryall, Patrick J. C. A comparison between natural and laboratory oxidation of titanomagnetites in pillow lavas. D, 1975, Dalhousie University. 175 p.

Ryals, Gary N. Ground-water resources of the Arcadia Minden area, Louisiana. M, 1977, Northwestern State University. 60 p.

Ryan, Arthur Bruce. Progressive structural reworking of the Uivak gneisses, Jerusalem Harbour, northern Labrador. M, 1977, Memorial University of Newfoundland. 230 p.

Ryan, Barry Desmond. Structural geology and Rb-Sr geochronology of the Anarchist Mountain area, southcentral British Columbia. D, 1974, University of British Columbia.

Ryan, Dale Edward. Quaternary stratigraphy of the lower Mud Brook Basin, Northampton Township, Summit County, Ohio. M, 1980, University of Akron. 140 p.

Ryan, David. Paleomagnetism of Paisano Volcano, Texas. M, 1988, Texas A&M University. 153 p.

Ryan, David C. Foraminiferal biostratigraphy and paleoecology of the Cody Shale, Upper Slide Lake, Jackson Hole, Wyoming. M, 1981, University of Wyoming. 116 p.

Ryan, Dennis J. Geologic structure and petrography of the southern part of Cliff Walk, Newport, Rhode Island. M, 1952, Brown University.

Ryan, Edmond P. A summary of the engineering properties of subsurface soils in the Albuquerque area. D, 1974, University of New Mexico. 395 p.

Ryan, Edward McNeill. Conodonts from the Hardin Sandstone of Tennessee. D, 1943, University of British Columbia.

Ryan, Elizabeth B. Liquidus phase relations in part of the system Mg_2SiO_4-Fe_2SiO_4-$CaMgSi_2O_6$-$CaFeSi_2O_6$-SiO_2-$KAlSi_3O_8$ along the nickel-nickel oxide buffer curve. M, 1987, University of Texas at El Paso.

Ryan, John Arthur. The distribution of seismic wave energy at a free surface. D, 1959, Pennsylvania State University, University Park. 86 p.

Ryan, John Donald. The petrology of the diabase intrusions of the Delaware River valley. M, 1948, Lehigh University.

Ryan, John Donald. The sediments of Chesapeake Bay. D, 1952, The Johns Hopkins University.

Ryan, John F. Upper Devonian sandstones in the Arrow Canyon Range, Clark County, Nevada. M, 1972, University of Illinois, Urbana. 60 p.

Ryan, John Joseph, Jr. The (Ordovician) Martinsburg Formation of Northwest Virginia. M, 1963, University of Virginia. 83 p.

Ryan, Kim Kathleen. The effect of included phases on the growth of plagioclase porphyroblasts. M, 1973, Michigan State University. 41 p.

Ryan, M. C. An investigation of nitrogen compounds in the groundwater in the Valley of Mexico. M, 1987, University of Waterloo. 110 p.

Ryan, Michael P. High-temperature mechanical properties of basalt. D, 1979, Pennsylvania State University, University Park. 593 p.

Ryan, Michael Patrick. Structural analysis of the Red Fork-Powder River area, Johnson County, Wyoming. M, 1986, University of Iowa. 166 p.

Ryan, Michael Patrick. Textural adjustments in regional metamorphism. M, 1973, Michigan State University. 60 p.

Ryan, Nancy Joan. The chemical and anatomical compositions of coal precursors. D, 1985, Pennsylvania State University, University Park. 286 p.

Ryan, Patrick Joseph. A vertical intensity magnetic study of the western part of the Arbuckle Mountains. M, 1976, University of Oklahoma. 51 p.

Ryan, R. J. The paleontology and paleoecology of the Gays River Formation in Nova Scotia. M, 1978, Acadia University.

Ryan, Richard C. and Corey, Ronald Stewart. Geology of the Slide Mountain area, Huerfano and Costilla counties, Colorado. M, 1960, University of Michigan.

Ryan, Robert. A thickness study of the Ogallala Group in south-central Nebraska. M, 1959, University of Nebraska, Lincoln.

Ryan, Robert N., Jr. Benthic foraminiferal assemblages of the Austin Chalk and Taylor Marl in Dallas County, Texas. M, 1988, Tulane University.

Ryan, Roger M. The role of focal length of the photogrammetric camera in acutance and resolution. M, 1966, Ohio State University.

Ryan, Ruth M. Chemical, isotopic, and petrographic study of the sulfides in the Duluth Complex cloud zone. M, 1984, Indiana University, Bloomington. 88 p.

Ryan, Scott S. Description and paragenetic interpretation of quartz-carbonate-barite veins of the Hartford Basin. M, 1986, University of Connecticut. 132 p.

Ryan, Timothy Harold. Geochemistry of the basaltic rocks, Fishers Peak Mesa, Colorado. M, 1982, Iowa State University of Science and Technology. 90 p.

Ryan, Wallace and Powell, William I., Jr. Establishing high order control points for the Ohio State University campus area. M, 1960, Ohio State University.

Ryan, William Alexander, Jr. The conodonts from the Jefferson City Formation (Lower Ordovician) of Missouri. M, 1940, University of Missouri, Columbia.

Ryan, William Bradley Frear. The floor of the Mediterranean Sea. D, 1971, Columbia University. 404 p.

Ryan, William M. Geology of the Bluefield, West Virginia Quadrangle. M, 1969, West Virginia University.

Ryan, William P., Jr. Provenance of the Norphlet Sandstone, northern Gulf Coast. M, 1986, University of New Orleans. 136 p.

Ryason, Daniel John. The stratigraphy and structure of the Pipestone Canyon area in north central Washington. M, 1959, University of Washington. 45 p.

Rybarczyk, Sandra M. A hydrogeochemical study of three geothermal areas in Arizona, New Mexico and Texas. M, 1982, New Mexico State University, Las Cruces. 129 p.

Ryberg, George Ernest. The geology of the Jicarilla Mountains, Lincoln County, New Mexico. M, 1968, University of New Mexico. 95 p.

Ryberg, Paul Thomas. Lithofacies and depositional environments of the Coaledo Formation, Coos County, Oregon. M, 1978, University of Oregon. 159 p.

Ryberg, Paul Thomas. Sedimentation, structure and tectonics of the Umpqua Group (Paleocene the early Eocene), southwestern Oregon. D, 1984, University of Arizona. 422 p.

Rydelek, Paul Anthony. Observations of long-period motions of the Earth at the South Pole. D, 1983, University of California, Los Angeles. 123 p.

Rydell, Harold Stanford. An investigation of the rare earth element distribution in apatite as a function of the unit cell size. M, 1964, Florida State University.

Rydell, Harold Stanford. The implications of uranium isotope distribution associated with the Floridan aquifer of north Florida. D, 1969, Florida State University. 127 p.

Ryden, Bonnie L. Stepp. Structural analysis of the Llao Rock dacite (Pleistocene to Recent), Crater Lake (southwestern), Oregon. M, 1968, University of Oregon. 60 p.

Ryder, Albert. Geology of the southern Ruby Mountains, White Pine County, Nevada. M, 1982, San Diego State University. 129 p.

Ryder, Graham. A rationale for the origins of massif anorthosites. D, 1974, Michigan State University. 66 p.

Ryder, Henry L. The placement of the staurolite and sillimanite isograds in the Hill City and Mt. Rushmore quadrangles, South Dakota. M, 1978, University of Toledo. 82 p.

Ryder, June Margaret. Alluvial fans of post-glacial environments within British Columbia. D, 1970, University of British Columbia.

Ryder, Robert Thomas. The Beaverhead Formation; a late Cretaceous-Paleocene syntectonic deposit in southwestern Montana and east-central Idaho. D, 1968, Pennsylvania State University, University Park. 187 p.

Ryding, John. Characterization of a static trapping technique for the analysis of soil and groundwater contamination. M, 1985, Colorado School of Mines. 84 p.

Rye, Danny Michael. The stable and lead isotopes of parts of the northern Black Hills (South Dakota); age and origin of the Homestake (Precambrian) and surrounding (Tertiary) ore bodies. D, 1972, University of Minnesota, Minneapolis. 130 p.

Rye, Kenneth Alan. Geology and geochemistry of the Hoyle Pond Deposit, Timmins, Ontario. M, 1987, University of Western Ontario.

Rye, Raymond T., III. A paleoenvironmental study of some marine Pleistocene locations in southern Maryland. M, 1971, George Washington University.

Rye, Robert Orph. The carbon, hydrogen, and oxygen isotopic composition of the hydrothermal fluids responsible for the lead-zinc deposits at Providencia,

Zacatecas, Mexico. D, 1965, Princeton University. 108 p.

Ryer, T. A. Patterns of sedimentation and environmental reconstruction of the western margin of the Interior Cretaceous Seaway, Coalville and Rockport areas, Utah. D, 1975, Yale University. 219 p.

Ryerson, Charles Curtis. Models for calculating daily changes in soil frost depth in the midwestern United States. D, 1977, Southern Illinois University, Carbondale. 228 p.

Ryerson, Frederick J. Homogeneous and heterogeneous equilibria in silicate melts and meta-pelitic rocks. D, 1979, Brown University. 263 p.

Rylaarsdam, Katharine Worcester. Life histories and abundance patterns of some common Caribbean reef corals. D, 1981, The Johns Hopkins University. 142 p.

Ryland, Robert R. Relationship of thrust, torque, and rate of penetration in rotary drilling in brittle materials. M, 1954, Ohio State University.

Ryland, Stephen Lane. A gravity and magnetic study of the Galapagos Islands. M, 1971, University of Missouri, Columbia.

Ryley, Charles Christopher. Multielement taxonomy, biostratigraphy, and paleoecology of Late Triassic conodonts from the Mamonia Complex, southwestern Cyprus. M, 1987, Memorial University of Newfoundland. 191 p.

Rymer, Michael J. Stratigraphy of the Cache Formation (Pliocene and Pleistocene) in Clear Lake Basin, Lake County, California. M, 1978, San Jose State University. 99 p.

Rymer, Rodney Keith. Mineralogy of the Fire Clay coal seam and the related roof, floor, and tonstein rocks. M, 1981, University of Kentucky. 72 p.

Rynearson, Sylvester. Optimum spacing for oil wells. M, 1930, University of Pittsburgh.

Rynn, John M. W. Seismotectonics of the Arthur's Pass region, South Island, New Zealand and regional variations in t_s/t_p. D, 1976, Columbia University. 293 p.

Ryswyk, Albert Leonard Van *see* Van Ryswyk, Albert Leonard

Ryswyk, Roy J. Van *see* Van Ryswyk, Roy J.

Ryther, Thomas E. Geology of the Willow Creek area, Fremont County, Wyoming. M, 1956, University of Kansas. 133 p.

Rytuba, J. J. Mutual solubilities of pyrite, pyrrhotite, quartz, and gold in aqueous NaCl solutions from 200° to 500°C, and 500 to 1500 bars, and genesis of the Cortez gold deposit, Nevada. D, 1977, Stanford University. 148 p.

Ryu, Jisoo. Low frequency electromagnetic scattering. D, 1971, University of California, Berkeley. 188 p.

Ryu, Jisoo. Seismic ray theory and its applications. M, 1967, University of Minnesota, Minneapolis. 92 p.

Ryznar, Gerald John. Sulphur isotope investigation of the Quemont ore deposit (Quebec). M, 1965, University of Alberta. 48 p.

S., Eduardo Garcia *see* Garcia S., Eduardo

S., Jesus M. Castillo *see* Castillo S., Jesus M.

S., Jose Regueiro *see* Regueiro S., Jose

Saad, A. A. Hydrologic simulation in a semi-arid region. D, 1978, Georgia Institute of Technology. 259 p.

Saad, Afif Hani. Application of digital computer to magnetic data interpretation using surface integral method. M, 1962, University of Missouri, Rolla.

Saad, Afif Hani. Magnetic properties of ultramafic rocks from Red Mountain (Santa Clara and Stanislaus counties), California. D, 1968, Stanford University. 59 p.

Saad, Kamal Farid. Nonsteady flow toward wells which partially penetrate thick artesian aquifers. M, 1960, New Mexico Institute of Mining and Technology. 56 p.

Saadallah, Adnan A. Nature and lateral variation of host rock limestone in the Tri-State zinc-lead district. M, 1965, University of Missouri, Rolla.

Saadeghvazari, Mohamad Ala. Inelastic response of R/C bridges under horizontal and vertical earthquake motions. D, 1988, University of Illinois, Urbana. 290 p.

Saari, Kari Heikki Olavi. Analysis of plastic deformations (squeezing) of seams intersecting tunnels and shafts in rock. D, 1982, University of California, Berkeley. 183 p.

Sabag, Shahé Fares. The geochemistry and petrology of granitoids at Meggisi Lake, N.W. Ontario. M, 1979, University of Toronto.

Sabaka, Terence J. Harmonic analysis of satellite Doppler acceleration data over the Southeast Pacific Ocean. M, 1986, University of Akron. 217 p.

Sabatka, Edward Frank. Structural geology of the White River Oligocene in northeastern Sioux County, Nebraska. M, 1953, University of Nebraska, Lincoln.

Sabbagh, Suzanne Kathleen Boram. A preliminary study of fossils as stratigraphic indicators (1022-1820). D, 1964, University of Wisconsin-Madison. 225 p.

Sabel, Joseph M. The sedimentology of the Spearfish Formation. M, 1981, South Dakota School of Mines & Technology. 70 p.

Sabelin, Tatiana. Trace element and Sr isotopic geochemistry of the Keweenawan volcanics of Minnesota. M, 1979, University of Minnesota, Minneapolis. 47 p.

Sabels, Bruno Erich. Late Cenozoic volcanism in the San Francisco volcanic field and adjacent areas in north-central Arizona. D, 1960, University of Arizona.

Sabet, Mohamed A. Gravity and magnetic investigation west of Michigan Hill, South Park, Colorado. M, 1963, Colorado School of Mines. 75 p.

Sabet, Mohamed A. Gravity anomalies associated with the Elkhora Fault, South Park, Colorado. D, 1966, Colorado School of Mines. 166 p.

Sabin, Andrew. Modification of the genetic theory for the Friedensville zinc deposit through petrologic observations. M, 1985, University of Pittsburgh.

Sabins, Floyd F., Jr. Geology of the Cochise Head and western part of the Vanar Quadrangle, Arizona. D, 1955, Yale University.

Sabisky, Matthew Andrew. Finite strain, ductile flow and folding in central Raft River Mountains, northwestern Utah. M, 1985, University of Utah. 69 p.

Sable, Edward George. Geology of the Romanzof Mountains, Brooks Range, northeastern Alaska. D, 1965, University of Michigan. 270 p.

Sable, Edward George. Preliminary report on sedimentary and metamorphic rocks in part of the Romanzof Mountains, Brooks Range, northeastern Alaska. M, 1959, University of Michigan.

Sables, Bruno Erich. Late Cenozoic volcanism in the San Francisco volcanic field and adjacent areas in north-central Arizona. D, 1960, University of Arizona.

Sablock, Jeanette M. The geology of the Boat Mountain, West Hebgen Ridge area, Gallatin County, Montana. M, 1985, University of Idaho. 96 p.

Saboia, L. A. De *see* De Saboia, L. A.

Sabol, Donald Edwin, Jr. Geology of the Thermopolis Lineament, southern Bighorn Basin, Wyoming. M, 1985, University of Iowa. 104 p.

Sabol, George V. River dispersion; a skewed distribution. D, 1974, Colorado State University. 110 p.

Sabol, Joseph W. The geology of the Porterfield Quarry area, Smyth County, Virginia. M, 1958, Virginia Polytechnic Institute and State University.

Saboski, E. M. Physiological ecology of Hawaiian, marine, psammolittoral diatoms. D, 1976, University of Hawaii. 220 p.

Sabourin, Raymond. Estimation géostatistique du soufre dans une mine charbon au Cap Breton. M, 1975, Ecole Polytechnique.

Sabourin, Robert Joseph Edmond. Geology of the Bristol-Masham area, Pontiac and Gatineau counties, Quebec. D, 1955, Universite Laval.

Sabourin, Robert Joseph Edmond. The Meach Lake, Quebec, pseudo-conglomerate and associated phenomena. M, 1952, Universite Laval.

Sabtan, Abdullah A. Determination of longitudinal and lateral sediment distribution in reservoirs. D, 1988, South Dakota School of Mines & Technology.

Sacco, Paul Augustus. Upper Jurassic-Lower Cretaceous foraminiferal biostratigraphy, paleoecology, and paleobiogeography of the COST B-2 well. M, 1980, Rutgers, The State University, New Brunswick. 47 p.

Saccocia, Peter James. Sulfidation of oxide facies iron formation; implications for gold mineralization in the Glen Township Formation, Aitkin County, east-central Minnesota. M, 1987, University of Minnesota, Minneapolis. 93 p.

Sachdev, Sham Lal. New methods for the determination of trace quantities of vanadium. D, 1966, Louisiana State University. 82 p.

Sachdev, Suresh C. Phase relations in the lead-tin-antimony sulfide system and syntheses of cylindrite and franckeite. D, 1974, Miami University (Ohio). 148 p.

Sachdev, Suresh C. Size analysis of the pan fraction and mineral analysis of the sediments of the Muscamoot bay, Michigan. M, 1969, Wayne State University.

Sachi-Kocher, Afsar. Mineralogical and geochemical evolution of siliceous sediments in the Havallah Sequence, Nevada; possible geochemical indicators of depositional environment. M, 1987, Queens College (CUNY). 98 p.

Sachs, Harvey Maurice. Quantitative radiolarian-based paleo-oceanography in late Pleistocene subarctic Pacific sediments. D, 1973, Brown University.

Sachs, Jules. A biostratigraphic analysis of the (Eocene) Zeuglodon-bearing bed, Jackson Formation, Cocoa Post Office, Alabama. M, 1955, New York University.

Sachs, Jules Barry. Calcareous nannofossils of the Aftonian Shale (Pleistocene), Louisiana continental shelf. D, 1970, Tulane University. 132 p.

Sachs, Kelvin Norman. Stratigraphic and geographic distribution of some ostracodes from the Middle Devonian Moscow Formation in west-central New York. M, 1956, University of Rochester. 80 p.

Sachs, Kelvin Norman, Jr. Revision of the American Lepidocyclinas. D, 1960, Cornell University. 158 p.

Sachs, Peter L. The stratigraphy and structure of the southeast part of Oneida, New York, 7-1/2 minute Quadrangle. M, 1959, Syracuse University.

Sachs, Scott Donald. Magnetic differentiation of bottom current circulation in the Argentine Basin, South Atlantic Ocean. M, 1987, University of Texas, Arlington. 55 p.

Sack, Dorothy Irene. Studies of G.K. Gilbert, Lake Bonneville chronology reconstructions, and the Quaternary geology of Tule Valley, west-central Utah. D, 1988, University of Utah. 138 p.

Sack, Richard Olmstead. Studies of mafic granulites. D, 1979, Harvard University.

Sack, William R. Geometric analysis and kinematics of folding associated with overthrusting; Blue Ridge Province, Tennessee. M, 1988, Michigan State University. 158 p.

Sacker, Joshua. Use of hydrogeology to map groundwater barriers and flow, Lytle Creek fan area, upper Santa Ana River valley, San Bernardino County, California. M, 1988, University of Southern California.

Sacket, William Malcolm. Ionium-uranium ratios in marine deposited calcium carbonates and related materials. D, 1958, Washington University. 106 p.

Sackett, Duane H. Geology of the Gordon Park area, Fremont County, Colorado. M, 1961, University of Kansas. 38 p.

Sackheim, Margo J. Regional stratigraphy of the Louark (Upper Jurassic) Norphlet, Smackover, and Buckner formations in Rankin and northern Simpson counties, Mississippi. M, 1985, Tulane University. 165 p.

Sackreiter, Donald K. Latest Cenozoic stratigraphy of Lake Sakakawea area, Northeast Mercer County, North Dakota. M, 1973, University of North Dakota. 66 p.

Sackreiter, Donald K. Quaternary geology of the southern part of the Grand Forks and Bemidji quadrangles. D, 1975, University of North Dakota. 175 p.

Sacks, Paul Eric. Stratigraphy and structural geology of part of the southern Portneuf Range, southwestern Idaho. M, 1984, Bryn Mawr College. 86 p.

Sacre, Jeffrey Allen. Engineering properties of meander belt deposits of the lower Mississippi River valley. M, 1986, University of Missouri, Rolla. 159 p.

Sacris, Eduardo Milan. The diffusion of hydrogen in some liquid metals. D, 1969, Stanford University. 117 p.

Sacrison, W. R. A study of the Jurassic Unkpapa Sandstone of the Black Hills region, western South Dakota and eastern Wyoming. M, 1958, University of Wyoming. 78 p.

Sadd, James Lester. Petrology and geochemistry of the Buckhorn Asphalt (Desmoinesian) Arbuckle Mountains, Oklahoma. D, 1986, University of South Carolina. 168 p.

Sadd, James Lester. Sediment transport in a fringing reef, Cane Bay, St. Croix, United States Virgin Islands. M, 1980, University of Texas, Austin.

Sadeghi, Ali Mir Mohamad *see* Mir Mohamad Sadeghi, Ali

Sadeghi, Ali Mohammad. Use of stream sediments in geochemical exploration for carbonatite and uranium in central Arkansas. M, 1988, University of Arkansas, Fayetteville.

Sadeghi, Ali Reza. Structural geology of the Willard Peak areas, north-central Wasatch Mountains, Utah. M, 1973, University of Utah. 64 p.

Sadeghi, Mohammad-Ali. Development of a new method for the removal of hydrocarbon contaminants from bituminous sands. D, 1987, University of Southern California.

Sadeghipour, Jamshid. Parameter identification of groundwater aquifer models; a generalized least squares approach. D, 1984, University of California, Los Angeles. 118 p.

Sader, Steven Alan. Remote sensing data applications for the inventory and monitoring of renewable natural resources in Costa Rica. D, 1981, University of Idaho. 163 p.

Sadighi, Soleman. Study of the Tertiary rocks of the north-east portion of Byron Hot Spring Quadrangle, California. M, 1960, Stanford University.

Sadler, H. E. Nares Strait. D, 1975, Dalhousie University.

Sadler, J. F. A detailed study of Onwatin Formation. M, 1958, Queen's University. 155 p.

Sadler, James William. A study of the insoluble residues of the dolomites and limestones of the Canadian and Ozarkian systems in central Pennsylvania. M, 1933, Cornell University.

Sadler, R. Kumbe. Structure and stratigraphy of the Little Sheep Creek area, Beaverhead County, Montana. M, 1981, Oregon State University. 294 p.

Sadlick, Walter. Biostratigraphy of the Chainman Formation (Carboniferous), eastern Nevada and western Utah. D, 1965, University of Utah. 249 p.

Sadlick, Walter. Mississippian-Pennsylvanian boundary in northeastern Utah. M, 1955, University of Utah. 77 p.

Sado, Edward Vincent. The Quaternary stratigraphy and history of the Lucan map area, southwestern Ontario. M, 1980, University of Waterloo.

Sadowski, Raymond M. Clay-organic interactions. M, 1988, Colorado School of Mines. 208 p.

Saeed, El Tayeb M. Ground water appraisal of the Gash River basin in Kassala, Kassala Province, Sudan. M, 1968, University of Arizona.

Saeger, William Eldon. Geologic and subsurface investigation of the St. Louis, Missouri metropolitan area. M, 1975, Washington University. 264 p.

Saenger, Robert Craig. A seismic refraction survey of Marion County, Indiana. M, 1958, Indiana University, Bloomington. 19 p.

Saenz, Guadalupe. Geochemical exploration for petroleum in a marshy area; examination and statistical analysis of C_1-C_7 hydrocarbons in near surface samples taken over producing and barren structures. M, 1987, University of Texas at El Paso.

Saether, Ola Magne. Depositional history of the Devil's Kitchen sandstones and conglomerates in the Ardmore Basin, southern Oklahoma. M, 1976, University of Oklahoma. 95 p.

Saether, Ola Magne. The geochemistry of fluorine in Green River oil shale and oil-shale leachates. D, 1980, University of Colorado. 251 p.

Safai, Nader M. Simulation of saturated and unsaturated deformable porous media. D, 1978, Princeton University. 495 p.

Saffer, Parke E. A preliminary investigation of river and beach samples collected in the states of Florida, Georgia, and Alabama. M, 1955, Florida State University. 59 p.

Safford, Frederick Bargar. Origin of the Edwardsville Siltstone Member of the Muldraugh Formation (Mississippian) of south-central Indiana. M, 1975, University of Cincinnati. 91 p.

Safford, Wilbur L. Geology of the Phoenix-Rollinsville area, Boulder and Gilpin counties, Colorado. M, 1951, University of Colorado.

Safonov, Anatole Ivanovitch. Orogeny of the Urals. M, 1937, Cornell University.

Sagar, Budhi. Calibration and validation of aquifer models. D, 1973, University of Arizona. 191 p.

Sagasta, Paul Frederick. Seismic investigations of shallow-subsurface carbonates in the Bahama Islands. M, 1984, University of Texas, Austin. 165 p.

Sage, Cynthia. Environmental hazards as a basis for land-use planning in a rural portion of the Santa Barbara coastal area, California. M, 1972, University of California, Santa Barbara.

Sage, Janet D. Variable water pressure metamorphic assemblages in the Meguma Group, Nova Scotia. M, 1984, Virginia Polytechnic Institute and State University. 361 p.

Sage, Nathaniel McLean, Jr. Structural geology of the Antigonish-Pomquet area, Antigonish County, Nova Scotia. M, 1951, Massachusetts Institute of Technology. 84 p.

Sage, Nathaniel McLean, Jr. Windsor Group (Mississippian) stratigraphy in the Antigonish and Mahone Bay areas, Nova Scotia. D, 1953, Massachusetts Institute of Technology. 191 p.

Sage, Orrin G., Jr. Geology of the eastern portion of the "Chico" formations (lower(?) and upper Cretaceous), Simi Hills, California. M, 1971, University of California, Santa Barbara.

Sage, Orrin G., Jr. Paleocene geography of Southern California. M, 1973, Ohio University, Athens. 250 p.

Sage, Orrin, Jr. Paleocene geography of Southern California. D, 1973, University of California, Santa Barbara.

Sage, Ronald. Geology and mineralogy of the Cripple Creek syenite stock, Teller County, Colorado. M, 1966, Colorado School of Mines. 236 p.

Sage, Ronald Parker. Alkalic rock complexes - carbonatites of northern Ontario and their economic potential. D, 1986, Carleton University. 337 p.

Sager, William Warren. Seamount paleomagnetism and Pacific Plate tectonics. D, 1983, University of Hawaii. 489 p.

Saggar, Surinder Kumar. Studies of sulfur in relationship to carbon and nitrogen in soil organic matter. D, 1981, University of Saskatchewan.

Sagher, A. Availability of soil runoff phosphorus to algae. D, 1976, University of Wisconsin-Madison. 192 p.

Sagoci, H. F. Forced torsional oscillations of an elastic half space. D, 1944, Massachusetts Institute of Technology. 35 p.

Sagoe, Kweku-Mensah Olakunle. The significance of population breaks in the log-probability grain-size distributions of sand. M, 1975, University of Tulsa. 78 p.

Sagstad, Steven R. Hydrogeologic analysis of the southern Rathdrum Prairie area, Idaho. M, 1977, University of Idaho. 96 p.

Sague, Virginia Muessig. Migmatite occurrences in the Wissahickon Formation and related rocks, Pennsylvania-Delaware Piedmont. M, 1978, Bryn Mawr College. 91 p.

Saha, A. K. Mode of emplacement of some granitic plutons in southeastern Ontario. D, 1957, University of Toronto.

Saha, A. K. Studies of the mode of emplacement of some granitic plutons in Hastings County, Ontario. D, 1958, University of Toronto.

Saha, Prasenjit K. Subsolidus studies in the system $NaAlSiO_4$-$NaAlSi_3O_8$-H_2O. D, 1959, Pennsylvania State University, University Park. 151 p.

Sahagian, Dork. Sublithospheric upwelling distribution and its implications regarding hot spots and mantle convection. M, 1981, Rutgers, The State University, New Brunswick. 39 p.

Sahagian, Dork L. Epeirogeny and eustatic sea level changes as inferred from Cretaceous shoreline deposits. D, 1987, University of Chicago. 182 p.

Sahai, S. K. Thermal cracking of alkane (C_{21}) and thermal conductivity studies of carbonates and sandstones. M, 1976, University of Wyoming. 54 p.

Sahakian, Armen S. A petrographic and structural study of a portion of the Palmer Gneiss area, Marquette District, Michigan. M, 1959, Michigan State University. 65 p.

Sahakian, Armen Souren. Paleocurrent study of the Potsdam Sandstone (Cambrian) of New York and Quebec. D, 1964, Harvard University.

Sahanhaya, Sait. Areal geology of Toyah Lake area, Reeves County, Texas (Sa 19). M, 1952, University of Texas, Austin.

Sahimi, Muhammad. Transport and dispersion in porous media and related aspects of petroleum recovery. D, 1984, University of Minnesota, Duluth. 648 p.

Sahin, E. Boron and borates. M, 1978, Stanford University. 100 p.

Sahinen Uuno, Mathias. Mining districts of Montana. M, 1935, Montana College of Mineral Science & Technology. 109 p.

Sahl, Howard Leroy. Geology of the Soap Creek Dome, Big Horn County, Montana. M, 1952, University of Nebraska, Lincoln. 46 p.

Sahl, Lauren Elizabeth. Suspended sediment on the upper Texas continental shelf. D, 1984, Texas A&M University. 105 p.

Sahni, Ashok. The vertebrate fauna of the Judith River Formation (Upper Cretaceous), (southern) Montana. D, 1968, University of Minnesota, Minneapolis. 267 p.

Sahu, Basanta Kumar. Environments of deposition from the size analysis of clastic sediments. D, 1962, University of Wisconsin-Madison. 106 p.

Sai, Joseph Obodai. Effect of capillary hysteresis and spatial variability of hydraulic conductivity on two-dimensional hillslope soil moisture flow. D, 1982, Texas A&M University. 120 p.

Saia, Robert. The geology of the Wurtsboro, New York Quadrangle. M, 1966, New York University.

Said, Faraj. Sedimentation history of the Paleozoic rocks of the Ghadames Basin in Libyan Arab Republic. M, 1974, University of South Carolina.

Said, Rushdi. Foraminifera of the northern Red Sea. D, 1950, Harvard University.

Saidi, Ali Mohammad. A new treatment of the diffusivity equation. D, 1961, Stanford University. 75 p.

Saidji, Mohamed. Chitinozoan fauna of the Upper Ordovician Maquoketa Formation (Wisconsin and Iowa). M, 1968, University of Wisconsin-Madison.

Saif, Hakeem Thamir. Factors influencing base saturation and CA/Mg ratios in soils of southeastern Ohio. D, 1986, Ohio State University. 334 p.

Saif, S. I. Identification, correlation and origin of the Key Anacon-Brunswick mines ore horizon, Bathurst, New Brunswick. D, 1978, University of New Brunswick.

Saikia, Chandan Kumar. Waveform modeling of eastern North American earthquakes using short-distance recordings. D, 1985, St. Louis University. 248 p.

Sailor, Richard Vance. Attenuation of low frequency seismic energy. D, 1979, Harvard University.

Sain, Herman Andrew. Fossil stream bed of the Little River, Tennessee. M, 1972, Wright State University. 100 p.

Saines, Marvin. Geology of the College Corner, Ohio-Indiana area with a chapter on ground-water resources. M, 1966, Miami University (Ohio). 130 p.

Saines, Marvin. Hydrogeology and hydrogeochemistry of part of South Hadley, Massachusetts. D, 1973, University of Massachusetts. 252 p.

Saines, Steven James. Hydrogeochemical well water reconnaissance in Orange County, Indiana. M, 1983, Indiana University, Bloomington. 293 p.

Sainey, Timothy J. Petrography and origin of the ore deposits at Benson Mines, New York. M, 1973, Ohio University, Athens. 85 p.

Saing, S. Clarification of the nature of the kaolin materials in Hawaiian soils. M, 1964, University of Hawaii.

Sainsbury, Cleo L. Geology of the Mt. Olds-Clark Peak area, Juneau vicinity, Alaska. M, 1953, University of Colorado.

Sainsbury, Cleo Ladell. Geology and ore deposits of the central York Mountains, western Seward Peninsula, Alaska. D, 1965, Stanford University. 255 p.

Saint Clair, Charles S. Geologic reconnaissance of the Agua Fria River area, central Arizona. M, 1957, University of Arizona.

Saint Clair, Charles Spencer. The classification of minerals; some representative mineral systems from Agricola to Werner. D, 1965, University of Oklahoma. 298 p.

Saint Clair, Donald W. The laboratory study of limestones with reference to the Silurian-Devonian contact in the Syracuse region (New York). M, 1935, Syracuse University.

Saint Clair, Gregory M. Geology of the Seydell oil pool. M, 1985, Emporia State University.

Saint John, Ruth Nimmo. Replacement vs. impregnation in petrified wood. M, 1925, Cornell University.

Saint, Prem Kishor. Effects of landfill disposal of chemical wastes on groundwater quality. D, 1973, University of Minnesota, Minneapolis.

Saint-Amant, Marcel. Les paramètres de la conduction électrique non lineaire des roches. D, 1976, Ecole Polytechnique.

Saint-Amant, Marcel Michel Yvon. Frequency and temperature dependence of dielectric properties of some common rocks. M, 1968, Massachusetts Institute of Technology. 133 p.

Saisasong, Atapon. Study of distribution of residual oil saturation by statistical methods. M, 1981, University of Oklahoma. 75 p.

Saito, Akira. Development of transient electromagnetic physical modeling for geophysical exploration. D, 1984, Colorado School of Mines. 161 p.

Saitta, Bertoni Sandro. Bluejacket Formation (Pennsylvanian); a subsurface study in northeastern Oklahoma. M, 1968, University of Tulsa. 142 p.

Sakakeeny, Stephen Anthony. Development and application of the Thornwaite climatic water-balance in

New Hampshire. M, 1986, University of New Hampshire. 157 p.

Sakalowsky, P. P., Jr. Relationships between dynamic nearshore processes and beach changes within the inter-tidal zone of Napatree Beach, Rhode Island. D, 1972, Indiana State University. 116 p.

Sakamoto-Arnold, Carole. An observational study of cadmium and other constituents in the Connecticut River estuary. M, 1983, University of Connecticut. 76 p.

Sakdejayont, Kiet. A study of Poisson's ratio and V_p/V_s ratio in the Rio Grande Rift. M, 1974, New Mexico Institute of Mining and Technology.

Sakhan, Kousoum South. Simulation of the dynamics of erodible streams. D, 1972, Utah State University.

Sakkaf, Ali. Authigenic minerals of some Devonian limestones near Syracuse, New York. M, 1962, Syracuse University.

Sakoda, E. T. The marine geology and sedimentology of Hawaii Kai, Kuapa Pond and adjacent Maunalua Bay. M, 1975, University of Hawaii. 71 p.

Sakowski, Henry A. Variability in Gumbelina from the (Cretaceous) Navarro Formation of Texas. M, 1951, University of Nebraska, Lincoln.

Sakrison, Herbert C. Studies of the host rocks of the Lake Dufault Mine, Quebec (Canada). D, 1967, McGill University. 138 p.

Salahi, Dirgham Rida. Upper Cretaceous and Tertiary ostracoda from the Zelten area, Libya. M, 1964, George Washington University.

Salahuddin, Qazi. Sedimentology and sedimentary petrology, part of the lower Wilcox sediments, Hurricane Creek Field, Allen Parish, Louisiana. M, 1985, University of Southwestern Louisiana. 110 p.

Salameh, Hassan Ramadan. A comparative geomorphic analysis of alluvial fans derived from granitic watersheds. D, 1974, University of California, Los Angeles. 258 p.

Salami, Mohammad Reza. Constitutive modelling of concrete and rocks under multiaxial compressive loadings. D, 1986, University of Arizona. 441 p.

Salami, Moshudi Babajide. Biology, morphology, and phylogenetic relationships of Trochammina cf. T. quadriloba Hoglund, an agglutinating foraminifer. M, 1974, University of California, Berkeley. 74 p.

Salami, Satari Olatunde. A reconnaissance study of the ground-water flow systems in the central part of the Lewiston-Clarkston area, Idaho-Washington. M, 1978, University of Idaho. 105 p.

Salas, Carlos J. Braided fluvial architecture within a rapidly subsiding basin; the Pennsylvanian Cumberland Group southwest of Sand River, Nova Scotia. M, 1986, University of Ottawa. 332 p.

Salas, Guillermo Armando. Areal geology and petrology of the igneous rocks, Santa Ana Quadrangle, Sonora, Mexico. M, 1968, University of Oklahoma. 118 p.

Salas, Guillermo Armando. Contact metamorphism in Sierra de Mapimi, Durango, Mexico. D, 1971, Stanford University. 220 p.

Salas, Luis Armando Acurero *see* Acurero Salas, Luis Armando

Salata, Frank Vincent Michael La *see* La Salata, Frank Vincent Michael

Salaymeh, Saleem Rushdi. Distribution of uranium and plutonium isotopes in the environment. D, 1987, University of Arkansas, Fayetteville. 224 p.

Salaymeh, Talab A. Petrography and geochemistry of selected oil shales from central Jordan. M, 1984, Southern Illinois University, Carbondale. 68 p.

Salcedo, Marco Antonio. Identification of frost susceptible aggregates and their use in concrete or bituminous pavements. D, 1984, Purdue University. 194 p.

Sale, Clarence. Geology along the Clear Fork of the Trinity River southwest of Fort Worth, Texas, including Benbrook Lake. M, 1957, Texas Christian University.

Sale, Hershel E. An environmental study of the Anahuac Formation (Oligocene, subsurface) in the upper Gulf Coast region of Texas. M, 1959, University of Houston.

Sale, Michael John. Optimization techniques for instream flow allocations. D, 1981, University of Illinois, Urbana. 175 p.

Saleeb, Atef Fathy. Constitutive models for soils in landslides. D, 1981, Purdue University. 511 p.

Saleeby, Jason Brian. Structure, petrology and geochronology of the Kings-Kaweah mafic-ultramafic belt, southwestern Sierra Nevada foothills, California. D, 1975, University of California, Santa Barbara.

Saleem, Zubair A. Optimal utilization of water resources of a complex overdrawn basin in a semiarid irrigated area. D, 1969, New Mexico Institute of Mining and Technology. 120 p.

Saleh, Ali. The field evaluation of continuous and intermittent water application on different soil textures, as related to variations of soil physical properties. D, 1987, Utah State University. 129 p.

Saleh, Hamed Hussein. Effect of salinity, sodium adsorption ratio and depth of water table on soil salinization under cropping and fallowing conditions. D, 1980, Iowa State University of Science and Technology. 121 p.

Saleh, Saad T. Borehole stability in the Williston Basin; the Four Eyes Field case study. M, 1982, Colorado School of Mines. 125 p.

Saleh, Saad Turky. An investigation of the effect of core length on oil recovery by micellar displacement in porous media. D, 1987, Colorado School of Mines. 540 p.

Salehi, Iraj A. Report on the aeromagnetometer survey of Smith Valley (Nevada). M, 1966, Stanford University.

Saleknejad, Hossein. Some empirical correlations among some physical properties of mine rocks. M, 1967, University of Utah. 58 p.

Salem, A. M. Behavior of dredged materials in diked containment areas. D, 1975, Northwestern University. 257 p.

Salem, Bruce B. Interactive CRT display of Bouguer gravity models in computer-assisted instruction in geology. M, 1973, Stanford University.

Salem, Mohamed Halim. A hydraulic system analysis of the ground-water resources of the Western Desert, UAR (Egypt). D, 1965, University of Arizona.

Salem, Mohamed Rafik Ibrahim. Sedimentologic study of Eocene deposits and their hydrocarbon prospects in the area between El Bahariya Oasis and the Mediterranean littorals of the Western Desert of Egypt. D, 1973, University of Texas, Austin. 362 p.

Salem, Mohammad Zarif. Characteristics, genesis and classification of some soils of Afghanistan and a study of ant pedoturbation in a Wisconsin forest soil. D, 1969, University of Wisconsin-Madison. 170 p.

Salem, Mostafa. Fossil otoliths of some lower Cenozoic perciform fishes of the Gulf Coast. M, 1971, University of Missouri, Rolla.

Salem, Mostafa Juma. Geology of the Al-Khums area, northwestern Libya, with emphasis on the stratigraphy and biostratigraphy of the Al-Khums Formation. D, 1977, University of Missouri, Rolla. 268 p.

Salem-Mehemed, Salem S. The petroleum geology of the Khalifa, Samah fields (concession 59) and Beda Field (concession 47) in the southwestern portion of Sirte Basin, Libya. M, 1978, Ohio University, Athens. 63 p.

Salerno, Catherine M. Walpole bounds on the effective mechanical properties of multiphase aggregates. M, 1985, Rensselaer Polytechnic Institute. 117 p.

Sales, John Keith. Structural analysis of the Basin Range Province in terms of wrench faulting. D, 1966, University of Nevada - Mackay School of Mines. 178 p.

Salgado, Peter G. A kinetic study of the reaction between graphite and hydrogen. D, 1958, West Virginia University.

Salih, Hammed Mohammad. Soil, location, and climatic factors influencing available phosphorus level with depth in subsoil horizons of Iowa soils. D, 1980, Iowa State University of Science and Technology. 248 p.

Salisbury, Beverly J. Petrology and diagenesis of the La Casita Formation in Chorro and Cortinas canyons, northeastern Mexico. M, 1982, University of New Orleans. 55 p.

Salisbury, Gerald P. Structural geology of Elk Basin anticline, Park County, Wyoming and Carbon County, Montana. M, 1948, University of Wyoming. 47 p.

Salisbury, James P. Fluvial sedimentology and associated bone distribution in the northwestern corner (Steveville region), Dinosaur Provincial Park, Alberta, Canada. M, 1985, Wichita State University. 184 p.

Salisbury, John William, Jr. Geology and mineral resources of the northwest quarter of the Cohutta Mountain Quadrangle, Wisconsin. D, 1959, Yale University. 95 p.

Salisbury, Matthew Harold. A seismic deformation within the oceanic lithosphere. M, 1971, University of Washington.

Salisbury, Matthew Harold. Investigation of seismic velocities in the Bay of Islands, Newfoundland, ophiolite complex for comparison with oceanic seismic structure. D, 1974, University of Washington. 147 p.

Salisbury, Richard A. Jurassic stratigraphy of the southern Two-Thirds of North Dakota. M, 1966, University of North Dakota. 124 p.

Sallak, Sulieman. The conodont fauna of the Decorah Formation of Missouri. M, 1971, Washington University. 36 p.

Salleh, Mustapha Haji Mohd. The correlation of regional geophysical data of the conterminous United States. M, 1985, Purdue University. 195 p.

Sallenger, A. H., Jr. Mechanics of beach cusp formation. D, 1974, University of Virginia. 230 p.

Saller, Arthur Henry. Depositional setting of post-Antler Pennsylvanian strata in north-central Nevada. M, 1980, Stanford University. 118 p.

Saller, Arthur Henry. Diagenesis of Cenozoic limestones on Enewetak Atoll. D, 1984, Louisiana State University. 384 p.

Salloum, John Duane. Land disposal of newsprint mill effluents. D, 1973, McGill University.

Salman, Diab Salmon Ahmad. Geology of the Triangle Quadrangle area, Owyhee County, Idaho. M, 1972, University of Idaho. 65 p.

Salman, T. Adsorption of hexyl mercaptan on synthetic sphalerite. D, 1965, McGill University.

Salmassy, Vladimir Baroyant. Numerical models for the mechanical properties of saturated claywater systems. D, 1974, University of Utah. 103 p.

Salmon, Bette Christine. The experimental deformation of galena natural aggregates at moderate confining pressures from 25°C to 400°C. M, 1972, University of Michigan.

Salmon, Eleanor S. A molluscan faunule from the (Cretaceous) Pierre Formation in eastern Montana. M, 1934, Columbia University, Teachers College. 56 p.

Salmon, Eleanor S. Mohawkian Rafinesquinae (Ordovician). D, 1942, Columbia University, Teachers College.

Salomon, N. L. Stratigraphy of glacial deposits along the south shore of Lake Ontario, N.Y. M, 1976, Syracuse University.

Salomon, Ralph A. Calcareous nannoplankton of the Tupelo Tongue (Upper Cretaceous), Lee County, Mississippi. M, 1984, Mississippi State University. 194 p.

Salotti, Charles A. and Harley, William Frank. Geology of the upper Maes Creek area, Wet Mountains, Colorado. M, 1955, University of Michigan.

Salotti, Charles Anthony. Geology and petrology of the Cotopaxi-Howard area, Fremont County, Colorado. D, 1961, University of Michigan. 294 p.

Salpas, Peter Andrew. A geochemical study of anorthosite from the Stillwater Complex, Montana. D, 1985, Washington University. 193 p.

Salstrom, Philip L. Areal geology and stratigraphy of the Reedsburg Quadrangle, Wisconsin. M, 1962, University of Wisconsin-Madison.

Salt, Kenneth Julian. Archean geology of the Spanish Peaks area, southwestern Montana. M, 1987, Montana State University. 81 p.

Salter, P. F. Metal accumulation rates in sediments of the Mid-Atlantic Ridge near 37°N. M, 1977, Texas A&M University.

Salter, Robert Joel. Primary sedimentary structures, petrography and paleoenvironments of the Gartra Member of the Chinle Formation (Upper Triassic) in northern Utah. M, 1966, University of Utah. 78 p.

Salter, Timothy L. Crystallization of rhyolite lava flows from Hidalgo County, New Mexico; a study of texture and composition. M, 1982, University of Kansas. 111 p.

Salter, Timothy L. Texture and composition of a residual lateritic bauxite from Saline County, Arkansas. D, 1988, Indiana University, Bloomington. 276 p.

Saltman, David. The Stone Mountain Fault and the occurrence of radioactive minerals near Ela Mills, Carter County, Tennessee. M, 1965, University of Tennessee, Knoxville. 125 p.

Salton, George H. Some petrogenic, metamorphic and structural aspects in the history of Ontario gold deposits. M, 1930, University of Wisconsin-Madison.

Saltzberg, Edward R. Relation between rainfall and runoff in the Furnace Brook Basin and vicinity, Quincy, Mass. M, 1974, Boston University. 71 p.

Saltzer, Charles E. An investigation of the seismicity of the Aleutian-Alaskan Arc. M, 1972, University of Wisconsin-Milwaukee.

Saltzer, Sarah Dawn. Applications of computers to balanced cross-sections. M, 1986, Massachusetts Institute of Technology. 101 p.

Saltzman, D. The Stone Mountain Fault and the occurrence of radioactive minerals near Elk Mills, Carter County, Tennessee. M, 1965, University of Tennessee, Knoxville.

Saltzman, Eric. The mechanism of sulfate aerosol formation; chemical and sulfur isotopic evidence. M, 1983, University of Miami. 4 p.

Saltzman, Uzi. Rock quality determination for large-size stone used in protective blankets. D, 1975, Purdue University. 264 p.

Saluja, Sundar Singh. Study of the mechanism of rock failure under the action of explosives. D, 1963, University of Wisconsin-Madison. 186 p.

Salvador, Amos. Stratigraphy of the Chejende area, western Venezuela. D, 1950, Stanford University. 184 p.

Salvador, Phillip. A geomorphic study of rhyolite domes in the San Francisco volcanic field, northern Arizona. M, 1975, Stanford University. 59 p.

Salver, Henry Arthur. A stratigraphic and petrologic study of the Mississippian Loyalhanna Limestone of western Pennsylvania. M, 1962, University of Pittsburgh.

Salvino, John Francis. Light and heavy mineral diagenesis in the Cambrian Munising Formation. M, 1986, Michigan State University. 140 p.

Salway, A. A. Statistical estimation and prediction of avalanche activity from meteorological data for the Rogers Pass area of British Columbia. D, 1976, University of British Columbia.

Salyapongse, Sirot. Petrography and petrology of granitic rocks of the Llano Uplift, Texas. M, 1978, University of Texas, Austin.

Sama-Nupa-Win, Bancha. Geology of the Florence gas field, Vermillion Parish, Louisiana. M, 1975, Lehigh University. 43 p.

Samadi, Suleiman Afif. Hydraulic characteristics and flow prediction in high-gradient rough channels. D, 1981, Utah State University. 227 p.

Samai, Mahdi. Landsat-1 photointerpretation of geologic features in the Black Hills, South Dakota. M, 1976, South Dakota School of Mines & Technology.

Samanez, Wilfredo Ballon. Sedimentary petrology of the Lagonda Formation (Desmoinesian) in Saint Louis County and vicinity, Missouri. M, 1968, Washington University. 66 p.

Samarasinghe, Ananda Mahinda. Consolidation of soils predicted by finite and small strain theories. D, 1983, University of Kentucky. 193 p.

Sambol, M. Evidence of selection pressure in *Agerostrea mesenterica* (Bivalvia, Mollusca) in the Navesink Formation (upper Cretaceous) of New Jersey. M, 1974, Queens College (CUNY). 94 p.

Sambol, Melvin. Inter-demic selection differences in a Cretaceous oyster. D, 1985, The Johns Hopkins University. 437 p.

Samiento-Alarcon, A. Experimental study of pebble abrasion. M, 1945, University of Chicago.

Samii, Cyrus. The geology of the Flat Rock oil field, Upton County, Texas. M, 1960, University of Arizona.

Samman, Nabil Fahmi. Sedimentation and stratigraphy of the Rome Formation in East Tennessee. D, 1975, University of Tennessee, Knoxville. 337 p.

Sammel, Edward A. Compressibility of some granular sediments subjected to light static loading. M, 1959, Massachusetts Institute of Technology. 58 p.

Sammis, Catherine G. Mammalian osteo-archaeology of the Mitchell site, (39DV2), Mitchell, South Dakota. M, 1978, University of Iowa. 51 p.

Sammis, Charles George. Seismological applications of lattice theory. D, 1971, California Institute of Technology. 259 p.

Sammis, Neil C. Petrology of the Devonian Wapsipinicon Formation from a core and reference exposures, central-eastern Iowa. M, 1978, University of Iowa. 232 p.

Sammis, Theodore Wallace. Channel transmission losses in small watersheds. M, 1972, University of Arizona.

Sampara, Michel Ulrich. Etude comportement des roches au gel et dégel. M, 1972, Ecole Polytechnique. 91 p.

Sampattavanija, Suvit. Geologic relations and depositional environment of the Chupadera iron deposits, Torrance County, New Mexico. M, 1972, New Mexico Institute of Mining and Technology.

Samper Calvete, Francisco Javier. Statistical methods of analyzing hydrochemical, isotopic, and hydrological data from regional aquifers. D, 1986, University of Arizona. 556 p.

Sample, Charles H. The ostracods of the (Pennsylvanian) East Mountain Shale (Texas). M, 1932, Columbia University, Teachers College.

Sample, James Clifford. Structure, tectonics, and sedimentology of the Kodiak Formation, Kodiak and adjacent islands, Alaska. D, 1986, University of California, Santa Cruz. 173 p.

Sample, Milton David. The Meridian Sand (Eocene) of Alabama and Mississippi; a stratigraphic and sedimentary investigation. M, 1968, University of Alabama.

Sample, Raymond Dewey. Geology of the Lakewood region, Boulder County, Colorado. M, 1942, University of Colorado.

Sampson, Daniel Edward. A field and geochemical study of the Inyo volcanic chain, eastern California. M, 1986, University of California, Santa Cruz.

Sampson, Edward. The ferruginous chert formations of Notre Dame Bay, Newfoundland. D, 1920, Princeton University.

Sampson, Edward W. A petrographic description of a section of the Pennsylvanian-Permian. M, 1929, University of Oklahoma. 42 p.

Sampson, Geoffrey Alexander. Petrology of some Grenville volcanic and pelitic rocks from near Madoc, southeastern Ontario. D, 1972, University of Toronto.

Sams, David Bruce. Uranium/lead zircon geochronology, petrology, and structural geology of the crystalline rocks of the southernmost Sierra Nevada and Tehachapi Mountains, Kern County, California. D, 1986, California Institute of Technology. 354 p.

Sams, Richard Houston. Geology of the Charlie Canyon area, Northwest Los Angeles County, California. M, 1964, University of California, Los Angeles.

Samsel, Howard S. Geology of the southeast quarter of the Cross Mountain Quadrangle, Kern County, California. M, 1952, University of California, Los Angeles.

Samsel, W. A. A study of the longitudinal distribution of velocity in the upper Whippany River, New Jersey. M, 1973, Rutgers, The State University, New Brunswick. 63 p.

Samsela, John J., Jr. Sedimentology of the Leadville Limestone (Mississippian) and the Chaffee Group (Upper Devonian), Chaffee, Fremont, and Saguache counties, Colorado. M, 1980, Colorado School of Mines. 168 p.

Samson, Claire. Recording the Kapuskasing pilot reflection survey with refraction instruments; a feasibility study. M, 1986, University of Toronto.

Samson, J. C. Deep resistivity measurements in the Fraser Valley, British Columbia. M, 1968, University of British Columbia.

Samson, Michael R. Tortonian Archaeomonadaceae Krensdorf, Austria. M, 1976, University of Rhode Island.

Samson, Scott Douglas. Chemistry, mineralogy, and correlation of Ordovician bentonites. M, 1986, University of Minnesota, Minneapolis. 158 p.

Samuels, G. A mixing budget for the Strait of Georgia, British Columbia. M, 1979, University of British Columbia.

Samuels, Neil D. The sedimentology and petrology of the Brallier Formation; Upper Devonian turbidite slope facies of the Central Appalachians. M, 1979, University of Cincinnati. 153 p.

Samuelson, Alan Conrad. Finite element analyses of initial elastic deformation of vertically deformed crustal blocks. D, 1972, Pennsylvania State University, University Park. 260 p.

Samuelson, Donald R. Petrology and structure of Precambrian crystalline and Tertiary igneous rocks, Manhattan District, Larimer County, Colorado. M, 1971, Colorado State University. 101 p.

Samuelson, Kiff James. Geology of a section of the Northwest Tobacco Root Mountains, Madison County, Southwest Montana. M, 1984, Western Michigan University. 58 p.

San Filipo, William Anthony. Electromagnetic modeling and data acquisition for exploration geophysics. D, 1984, University of Utah. 204 p.

San Filipo, William Anthony. Three-dimensional offline refraction and wide angle reflection survey of the Wind River uplift seismic survey using a stationary array. M, 1978, University of Texas at Dallas. 81 p.

San Juan, Francisco Claudio. A study of uranium and thorium isotopes in ground waters and solids of two uranium mines, South Texas. D, 1982, Florida State University. 270 p.

San Juan, Francisco Claudio, Jr. Geochemical exploration study of an iron oxide deposit near Sparta, Georgia. M, 1977, Florida State University.

San Vicente, Napoleon Otero. First experimental, artificial massive recharge of aquifers project in Mexico, Laguna Seca, Coahuila, Mexico. M, 1981, SUNY, College at Fredonia. 60 p.

Sanborn, Albert F. Geology and paleontology of a part of the Big Bend Quadrangle, Shasta County, California. D, 1952, Stanford University. 98 p.

Sanborn, John D. Paleocurrents of the Vanport Limestone (Pennsylvania) in Lawrence County, Pennsylvania. M, 1973, Slippery Rock University. 44 p.

Sanborn, Mary Margaret. The role of brittle-ductile shear in the formation of gold-bearing quartz-carbonate veins in the west carbonate zone of the Cochenous Williams gold Mine, Red Lake, Ontario. M, 1987, University of Toronto.

Sanborn, Paul Thomas. Ferro-humic Podzols of coastal British Columbia; aspects of genesis and chemistry. D, 1987, University of British Columbia.

Sancar, Mustafa Sitki. Rock typing in shaly-sand. M, 1974, University of Tulsa. 134 p.

Sancetta, Constance A. Oceanography of the North Pacific during the last 18,000 years derived from fossil diatoms. D, 1977, Oregon State University. 121 p.

Sancetta, Constance Antonina. Climatic record of the past 130,000 years in North Atlantic deep-sea core V23-82; correlation with the terrestrial record. M, 1973, Brown University.

Sánchez Zamora, Osvaldo. Crustal structure and thermal gradients of the northern Gulf of California determined using spectral analysis of magnetic anomalies. D, 1988, Oregon State University. 127 p.

Sanchez Zamora, Oswaldo. Gravity and structure of the Pacific continental margin of central Mexico. M, 1981, Oregon State University. 39 p.

Sanchez, Arthur Ledda. Chemical speciation and adsorption behavior of plutonium in natural waters. D, 1983, University of Washington. 191 p.

Sanchez, B. V. Rotational dynamics of mathematical models of the nonrigid Earth. D, 1975, University of Texas, Austin. 250 p.

Sanchez, Ricardo Jose Padilla y see Padilla y Sanchez, Ricardo Jose

Sanchez, V. Mendoza see Mendoza Sanchez, V.

Sanchez, Victor M. Computer analysis of production decline curves. M, 1968, University of Oklahoma. 71 p.

Sanchez-Barreda, Luis Antonio. Geologic evolution of the continental margin of the Gulf of Tehuantepec in southwestern Mexico. D, 1981, University of Texas, Austin. 269 p.

Sanchez-Barreda, Luis Antonio. Sedimentology of Laguna Potosi and environs, State of Guerrero, Mexico. M, 1976, Rice University. 141 p.

Sánchez-Salinero, Ignacio. Analytical investigation of seismic methods used for engineering applications. D, 1987, University of Texas, Austin. 450 p.

Sancio, Rodolfo Traostino. Analysis of the stability of slopes in weathered rocks. D, 1979, University of California, Berkeley. 228 p.

Sancton, Joel A. The East Berlin Formation of Massachusetts; a Triassic alluvial fan complex. M, 1970, University of Massachusetts. 132 p.

Sand, Leonard Bertram. Mineralogy and petrology of the residual kaolins of the southern Appalachian region. D, 1952, Pennsylvania State University, University Park. 129 p.

Sandback, John Elmer. Investigation of metallurgical treatment of the dry gold-silver ore from the Golden Age Mine, Pioneerville, Idaho. M, 1924, University of Idaho. 40 p.

Sandberg, Adolph Engelbrekt. A cross section across the Keweenawan lavas at Duluth, Minnesota. D, 1937, University of Cincinnati. 65 p.

Sandberg, Adolph Engelbrekt. A mineralographic study of lead-zinc-silver ores from the Sugar Loaf District, Colorado. M, 1931, University of Cincinnati. 38 p.

Sandberg, Brian S. Shoreline recession; past, present, and future, Lake Sakakawea, North Dakota. M, 1986, University of North Dakota. 185 p.

Sandberg, Edward C. Gold quartz veins of the Julian District, California. M, 1929, California Institute of Technology.

Sandberg, Philip A. A microfaunal study of Tamiahua Lagoon, Veracruz, Mexico. M, 1961, Louisiana State University.

Sandberg, Stewart Kim. Controlled-source audiomagnetotellurics in geothermal exploration. M, 1980, University of Utah. 85 p.

Sandefur, Bennett T. The geology and paragenesis of the nickel ores of the Cuniptau Mine, Goward, Nipissing District, Ontario. D, 1943, University of Chicago. 64 p.

Sandefur, Bennett Toy. The amygdaloidal basalts of Damascus, Virginia. M, 1930, University of Kentucky. 47 p.

Sandefur, Craig A. Paleocurrent analysis of the Cretaceous Mitchell Formation, north-central Oregon. M, 1986, Loma Linda University. 80 p.

Sandeman, Hamish A. I. A field, petrographical and geochemical investigation of the Kennack Gneiss, Lizard Peninsula, southwestern England. M, 1988, Memorial University of Newfoundland. 361 p.

Sander, Edgar Anthony. Sub-surface geology of Red Willow and Hitchcock counties, Nebraska. M, 1965, Kansas State University.

Sander, John Egan. An analysis of the Permo-Carboniferous "red beds" of Michigan. M, 1959, Michigan State University. 50 p.

Sander, John Egan. Bog-watershed relationships utilizing electric analog modeling. D, 1971, Michigan State University. 225 p.

Sander, Mark VanDyke. Creation of the Stanford III porphyry copper model; an application of conditional simulation. M, 1981, Stanford University. 77 p.

Sander, Mark VanDyke. Epithermal gold-silver mineralization, wall-rock alteration, and geochemical evolution of hydrothermal fluids in the ash-flow at Round Mountain, Nevada. D, 1988, Stanford University. 309 p.

Sander, Nestor John. A monograph of the species of the lamellibranch genus Mulinia. M, 1938, University of California, Berkeley. 84 p.

Sander, Paul Martin. Depositional environment and taphonomy of some fossil vertebrate occurrences in Lower-Permian redbeds. M, 1984, University of Texas, Austin.

Sander, Stephan. The geometry of rifting in Lake Tanganyika, East Africa. M, 1986, Duke University. 44 p.

Sanderman, L. A. and Utterback, C. L. Radium content of some inshore bottom samples in the Pacific Northwest. D, 1943, University of Washington. 5 p.

Sanders, Anthony W. Petrology and stratigraphic relations of the Renfro Member of the Borden Formation (Mississippian) in southeast-central Kentucky. M, 1979, Eastern Kentucky University. 39 p.

Sanders, Christopher O'Neill. I, Seismotectonics of the San Jacinto fault zone and the Anza seismic gap; II, Imaging the shallow crust in volcanic areas with earthquake shear waves. D, 1987, California Institute of Technology. 188 p.

Sanders, Clarence Whitney, Jr. The composite stock of Snowbank Lake, Minnesota. M, 1926, University of Minnesota, Minneapolis. 62 p.

Sanders, David E. Massive calcite in the Vulture Mountains near Wickenburg, Maricopa County, Arizona. M, 1974, Northern Arizona University. 106 p.

Sanders, David P. Resistivity methods applied to pollution detection in a crystalline bedrock aquifer at Little Compton, Rhode Island. M, 1983, University of Rhode Island.

Sanders, David Thomas. Mineral resources of the Sevier River drainage, central Utah. M, 1962, Utah State University. 60 p.

Sanders, Donald T. Sandstones of the (Pennsylvanian) Douglas and Pedee groups as possible groundwater reservoirs in northeastern Kansas. M, 1957, University of Kansas. 47 p.

Sanders, Frank Stanley. An investigation of carbon flux in the sediment of Castle Lake, California. D, 1976, University of California, Davis. 131 p.

Sanders, George F., Jr. Geology of the Buffalo Peaks, Park and Chaffee counties, Colorado. M, 1975, Colorado School of Mines. 62 p.

Sanders, John Essington. Geology of the Pressmen's Home area, Hawkins and Grainger counties, Tennessee. D, 1953, Yale University.

Sanders, John Warren, Jr. The geology of the New Bloomfield area, Callaway County, Missouri. M, 1948, University of Missouri, Columbia.

Sanders, Laura Lourdes. Geochemistry and paleotemperatures of formation waters from the Lower Silurian "Clinton" Formation, eastern Ohio. D, 1986, Kent State University, Kent. 141 p.

Sanders, Malcolm Keith. Sand dunes of Bailey County, Texas. M, 1951, Texas Tech University. 43 p.

Sanders, Norman K. Recent interaction of man and physical environment in coastal Ventura County, California. M, 1964, University of California, Los Angeles. 84 p.

Sanders, Peter A. Geology of the Kingston mining district, Black Range, New Mexico. M, 1986, New Mexico State University, Las Cruces. 96 p.

Sanders, Richard Bryan. Sedimentology and isotope geochemistry of the Upper Cretaceous Ernst Member of the Boquillas Formation, Big Bend National Park, Texas. M, 1988, Texas Tech University. 170 p.

Sanders, Richard Pat. Petrology of the syenitic intrusive complex, Pleasant Mountain, Maine. D, 1971, University of Illinois, Urbana. 156 p.

Sanders, Richard Pat. The petrology of the Archean metadolerite pluton and some small Tertiary dikes in the Goose Lake-Mount Zimmer area, Park County, Montana. M, 1968, Northern Illinois University. 81 p.

Sanders, Robert. Stratigraphy and structure of the Thacher and Olney members of the (Lower Devonian) Manlius Formation (New York). M, 1956, Syracuse University.

Sanders, Robert Bruce. Conodonts of the Beaverfoot-Brisco Formation. M, 1962, University of Iowa. 115 p.

Sanders, Robert Bruce. Palynological investigation of Desmoinesian and Missourian strata, Elk City area, Oklahoma. D, 1967, University of Oklahoma. 175 p.

Sanders, Thomas P. Correlations and variations of some clay minerals and elements in the New Albany Shale, Kentucky. M, 1972, Eastern Kentucky University. 45 p.

Sanderson, Deborah Anne. Structural geometry and deformational history of the Highwood-Elbow area, Rocky Mountains, Alberta. M, 1987, University of Calgary. 149 p.

Sanderson, Dewey Dennis. Magnetic minerals and properties of the Melrose Stock (Early Cretaceous, eastern Nevada). D, 1972, Michigan State University. 218 p.

Sanderson, George Albert, Jr. Studies of magnesium in foraminiferal tests. D, 1954, University of Wisconsin-Madison.

Sanderson, I. D. Sedimentology and paleoenvironments of the Mount Watson Formation, upper Precambrian Uinta Mountain Group, Utah. D, 1978, University of Colorado. 161 p.

Sanderson, Ivan D. Sedimentary structures and their environmental significance in the Navajo Sandstone, San Rafael Swell, Utah. M, 1974, Brigham Young University. 246 p.

Sanderson, James O. G. Analysis of Cretaceous and Tertiary sediments and their stratigraphical relationships. M, 1924, [University of Alberta].

Sanderson, James O. G. The geology along the Red Deer River, Alberta. D, 1928, University of Toronto.

Sanderson, Robert Michael. Prediction equations for building vibrations. D, 1969, Virginia Polytechnic Institute and State University. 97 p.

Sandford, William Edward. Detailed three-dimensional structure of the deep crust in the U.S. Cordillera in north-central Washington based on COCORP data. M, 1986, Cornell University. 55 p.

Sandhu, I. S. Optimization of Manning's roughness coefficients for overland flow using a finite-element method. D, 1979, Virginia Polytechnic Institute and State University. 187 p.

Sandidge, John R. A study of the brachiopod Kingena wacoensis. M, 1922, Vanderbilt University.

Sandidge, John R. The foraminifera of the Ripley Formation in Alabama. D, 1928, The Johns Hopkins University.

Sandifer, Donald Ford. The Balcones fault zone north of Manchaca (Texas) (Sa 56). M, 1935, University of Texas, Austin.

Sandler, Carol. The recrystallization of calcite. M, 1974, Brooklyn College (CUNY).

Sandlin, Gary L. Petrology and lithostratigraphy of the Goodland Formation (lower Cretaceous) in north-central Texas. M, 1973, University of Texas, Arlington. 113 p.

Sandlin, Gary Stuart. Subsurface geology of Graham County, Kansas. M, 1957, Kansas State University. 44 p.

Sandlin, Larry F. Geology of the Greenland Sandstone, Winslow Formation (Pennsylvanian), Madison County, Arkansas. M, 1968, University of Kansas.

Sandlin, Walter Lee, Jr. A facies study of the Red Mountain Formation (Silurian) of Northwest Georgia. M, 1960, Emory University. 78 p.

Sandness, Gerald Allyn. A numerical evaluation of the Helmholtz integral in acoustic scattering. D, 1973, University of Wisconsin-Madison.

Sando, Thomas W. Trace elements in Hercynian granitic rocks of the southeastern Piedmont, U.S.A. M, 1979, University of North Carolina, Chapel Hill. 91 p.

Sando, William J. The stratigraphy and paleontology of the west belt of the (Lower Ordovician) Beekmantown Group in Maryland. D, 1953, The Johns Hopkins University.

Sandomirisky, Peter. Geology of the Henderson and Conley talc mines, Madoc, Ontario. M, 1954, University of Western Ontario.

Sandor, Jonathan Andrew. Soils at prehistoric agricultural terracing sites in New Mexico. D, 1983, University of California, Berkeley. 337 p.

Sandoval, Deig-Nevy. Fallout of uranium and plutonium from recent volcanic eruptions. D, 1984, University of Arkansas, Fayetteville. 131 p.

Sandoval, Jose C. Mineralization and ore deposition at the Tosca Mine, Santander, Colombia. M, 1941, Columbia University, Teachers College.

Sandoval, Mario. The behavior of fresh water-salt water interface in coastal aquifers in the Philippines. M, 1964, University of Arizona.

Sandrock, George Stephen. Petrology and paleoenvironment of a key stratigraphic unit at the New Market zinc mine, New Market, Tennessee. M, 1973, University of Tennessee, Knoxville. 54 p.

Sands, C. D. Mineralogy and diagenesis in Deep Sea Drilling Project cores, Sites 146 and 323. M, 1976, University of Wyoming. 60 p.

Sands, David R. Morphology, hydraulics and sediment transport patterns at Slocum River embayment; South Dartmouth, Massachusetts. M, 1984, Boston University. 176 p.

Sandstrom, Mark William. Hydrocarbons in surface sediments from the Alaska and Southern California continental shelves. M, 1978, University of California, Los Angeles.

Sandstrom, Melissa. Stratigraphy and structure of Chochal Limestone, El Tapon Canyon area, Cuilco Quadrangle, Guatemala. M, 1978, University of Pittsburgh.

Sandusky, Clinton Leroy. Sedimentology of Estero Marua, Sonora, Mexico. M, 1969, University of Arizona.

Sandvik, Peter Olaf. Metal distribution in ore deposits of central Alaska. D, 1964, Stanford University. 201 p.

Sandwell, David Thomas. Thermal isostasy; spreading ridges, fracture zones, and thermal swells. D, 1981, University of California, Los Angeles. 223 p.

Sandy, John Jr. Petrology and photogeology of the Lone Mountain intrusive, Esmeralda County, Nevada. M, 1965, Tulane University. 48 p.

Sanford, Allan Robert. Sec. 1, An analytical and experimental study of some simple geologic structures; Sec. 2, Gravity survey of a part of the Raymond and San Gabriel basins, Southern California. D, 1958, California Institute of Technology. 142 p.

Sanford, John Theron. The Ordovician and Silurian stratigraphy of portions of Monroe and Wayne counties, New York. D, 1930, Princeton University. 101 p.

Sanford, Richard Frederick. Regional metamorphism of ultramafic bodies and their contact zones, western New England. D, 1978, Harvard University.

Sanford, Robert Bailey, Jr. Grain size reconnaissance of the Virginia-North Carolina inner shelf; analysis by settling technique. M, 1970, Old Dominion University. 84 p.

Sanford, Steven J. Fission track annealing in wells M-94 and T-366 at the Cerro Prieto geothermal field, BC, Mexico, as a means of dating thermal events. M, 1981, University of California, Riverside. 105 p.

Sanford, Steven Ray. The geology and geochemistry of the spar and related deposits, Mineral County, Montana. M, 1972, Washington State University. 98 p.

Sanford, Thomas B. The measurement and interpretation of motional electric fields in the sea. D, 1967, Massachusetts Institute of Technology. 161 p.

Sanford, Thomas Herbert, Jr. Geology and ground water resources of the Huntsville area, Alabama. M, 1961, University of Alabama.

Sanford, Ward Earl. Assessing the potential for calcite dissolution in coastal saltwater mixing zones. D, 1987, Pennsylvania State University, University Park. 119 p.

Sanford, Wendell Glenn. The families of tetracorals. M, 1938, Yale University.

Sanford, William Casey. An analysis of depositional environments and diagenesis of Tertiary volcanic sediment deposits of the northeastern Lago de Atitlán region, Guatemala. M, 1979, Texas Tech University. 150 p.

Sangameshwar, Salem Ramachandra Rao. Trace element and ore mineralogy of the Osborne lake mine, Manitoba, (Canada). M, 1968, University of Saskatchewan. 65 p.

Sangameshwar, Salem Ramachandra Rao. Trace element and sulphur isotope geochemistry of the sulphide deposits of the Flin Flon and Snow Lake areas (Precambrian) of Saskatchewan and Manitoba. D, 1972, University of Saskatchewan. 177 p.

Sanger, Daniel B. Determination of post-Pleistocene depositional environments of Little Lake, San Salvador Island, Bahamas, using ostracode microfauna. M, 1983, University of Akron. 97 p.

Sanger, Gary Edward. An analysis of circular surface features near New Baltimore, Ohio, using remotely sensed data. M, 1980, University of Akron. 87 p.

Sangines, Eugenia Maria. Tin bearing granites from the Cordillera Real, Bolivia. M, 1985, San Diego State University. 174 p.

Sangree, John Brewster, Jr. Silurian of northern Indiana. D, 1960, Northwestern University. 190 p.

Sangrey, Dwight Abram. The behavior of soils subjected to repeated loading. D, 1968, Cornell University. 364 p.

Sangster, A. L. Metallogeny of base metal, gold and iron deposits of the the Grenville Province of southeastern Ontario. D, 1970, Queen's University. 356 p.

Sangster, D. F. Thermochemical studies of certain iron minerals. M, 1961, McGill University.

Sangster, Donald Frederick. The contact metasomatic magnetite deposits of southwestern BC. D, 1964, University of British Columbia. 403 p.

Sangtian, C. The nature of argillic horizons in Hawaiian Ultisols. M, 1969, University of Hawaii.

Sanguanruang, S. S. Water quality aspects of coal transportation by slurry pipeline. D, 1977, University of Arkansas, Fayetteville. 336 p.

Sanjeevareddi, Buggana S. A theoretical and experimental investigation of the earth resistivity method as applied to dipping strata. M, 1936, Colorado School of Mines. 25 p.

Sanjines, Raul. Ore controls and formation of the ore-bearing structures in the Idarado mines, San Miguel and Ouray counties, Colorado. M, 1967, University of North Dakota. 62 p.

Sanker-Narayan, P. Mathematical methods in the interpretation of magnetic data. D, 1961, University of Wisconsin-Madison. 241 p.

Sankey, Scott Norman. A history of trace metal pollution in the Tia Juana River estuary, San Diego County, California. M, 1980, San Diego State University.

Sanner, Wayne K. Tectonic significance of Early Miocene basin formation in the Box Canyon area, Mojave Desert. M, 1985, University of North Carolina, Chapel Hill. 67 p.

Sans, John Rudolfs. Mafic mineral chemistry and oxygen isotope composition of the Cloud Pass Batholith, Washington. D, 1983, University of Chicago. 376 p.

Sans, Roger Stephen. Origin of Devonian rock units in the southern Fish Creek Range, Nye County, Nevada. M, 1986, Oregon State University. 68 p.

Sanschagrin, Roland. Lower and Middle Ordovician of the St. Lawrence Lowlands; stratigraphy and historical geology. M, 1951, University of British Columbia.

Sanschagrin, Yves. Etudes des variations latérales et verticales des faciès dans les coulées de basaltes tholeiitiques du Groupe de Kinojévis, canton d'Aiguebelle, Abitibi. M, 1981, Universite du Quebec a Chicoutimi. 114 p.

Sansfacon, Robert. Etude structurale, stratigraphique et économique des formations sédimentaires et volcaniques de la région du Lac Bousquet, Comté d'Abitibi, Canton de Bousquet, Québec. M, 1984, Universite du Quebec a Chicoutimi. 163 p.

Sansome, Constance Jefferson. Minnesota geology; a laboratory and field manual for a first course in geology with an elaboration on the Ordovician stratigraphy of SOGN Quadrangle, Goodhue County. M, 1972, University of Minnesota, Minneapolis. 575 p.

Sansome, Constance Jefferson. Origin and configuration of the present-day land surface, Goodhue County, Minnesota. D, 1986, Oregon State University. 144 p.

Sansone, Francis Joseph. Volatile fatty acid cycling in anoxic coastal sediments. D, 1980, University of North Carolina, Chapel Hill. 92 p.

Sant, Jan Franklin Van see Van Sant, Jan Franklin

Sant, Mary Jane Van see Van Sant, Mary Jane

Santa, Rick J. Subsurface geology, diagenesis, and depositional environment of the Renault Formation, Hamilton County, Illionois. M, 1985, University of Cincinnati. 66 p.

Santamaria, Francisco Jose. Geochemistry and geochronology of the igneous rocks of the Venezuelan coast ranges and southern Caribbean Islands and their relation to tectonic evolution. D, 1972, Rice University. 112 p.

Santamaria, Stephen V. Petrographic analysis of the upper Glen Rose Formation (Lower Cretaceous), Alabama Ferry Field, Leon County, Texas. M, 1986, Stephen F. Austin State University. 123 p.

Santana, Derli Prudente. Soil formation in a toposequence of Oxisols from Patos de Minas region, Minas Gerais State, Brazil. D, 1984, Purdue University. 140 p.

Santangelo, Mark A. Magnetic survey over the Section 32-35 area of the Felch District, central Dickinson County, Michigan. M, 1987, Bowling Green State University. 99 p.

Santas, Photenos. Effects of soil clay content, cow manure, lead and water upon roadside soil communities. D, 1988, George Washington University. 257 p.

Santi, Paul Michael. The kinematics of debris flow transport down a canyon. M, 1988, Texas A&M University. 85 p.

Santiago, Donald Jose. A gravity and magnetic study of the Medford Anomaly, north-central Oklahoma. M, 1979, University of Oklahoma. 105 p.

Santillan Cruz, V. H. Finite-difference techniques in digital computer modeling of groundwater systems. M, 1977, University of Arizona.

Santillan, Hector Manuel. A study of the Eagle Ford Shale-Austin Chalk contact in Dallas County, Texas. M, 1957, Southern Methodist University.

Santini, Ronald J. The stratigraphy and petroleum geology of the "Newburg" porous carbonate zone of the Lockport Group, (Middle Silurian), in the subsurface of Summit County, northeast Ohio. M, 1981, Kent State University, Kent. 112 p.

Santis, J. E. De *see* De Santis, J. E.

Santo, L. T. Soil water hysteresis in the inter-aggregate voids of two Hawaiian Oxisols. M, 1974, University of Hawaii.

Santogrossi, Patricia A. Environmental analysis of a lithostrotionid biostrome, Yellowpine Limestone, Arrow Canyon, Clark County, Nevada. M, 1977, University of Illinois, Urbana. 75 p.

Santos, Claudio A. F. Faria *see* Faria Santos, Claudio A. F.

Santos, Edson R. Dos *see* Dos Santos, Edson R.

Santos, Elmer S.; Boydston, Donald and Hamil, Brenton M. Geology of the east central portion of the Huerfano Quadrangle, Huerfano County, Colorado. M, 1954, University of Michigan.

Santos, Emidio Gil. Disaggregation modeling of hydrologic time series. D, 1983, Colorado State University. 291 p.

Santos, John Joseph, Jr. Geochemical survey in the Silver, Arkansas, area utilizing computer analysis. M, 1972, Louisiana Tech University.

Santos, Mauro Carneiro. Pedogenesis of Luvisolic soils in east-central Saskatchewan. D, 1984, University of Saskatchewan.

Santos, Michele M. Evaluation of rock mass properties and abutment stability at the Gillis Falls dam site, Carroll County, Maryland. M, 1988, Colorado School of Mines. 157 p.

Santos, Rebecca de Regla. A continuum mechanical formulation of the equations governing wave propagation in hydrocarbon reservoirs. D, 1986, University of Calgary. 190 p.

Santos, Rebecca R. An attempt towards an interpretation of a continuous crustal seismic refraction survey in Manitoba. M, 1976, University of Manitoba.

Santos, Vanessa Anne. The geology of the Union de Tula area, Jalisco, Mexico. M, 1983, University of Kentucky. 90 p.

Santos-Ynigo, Luis Marcial. Geology, structure, and origin of the nickeliferous laterites of Nonoc Island, Surigao, Philippines. D, 1960, Stanford University. 296 p.

Sanvordenker, Viola Chang. Optical observation by transmitted light of the tetragonal to cubic phase transition in (ba, Sr)TiO₃ ceramics. D, 1963, University of Michigan.

Sanyal, Subir Kumar. The effect of temperature on electrical resistivity and capillary pressure behavior of porous media. D, 1971, Stanford University. 133 p.

Saparoea, C. M. G. Van den Berg van *see* Van den Berg van Saparoea, C. M. G.

Saperstone, Herb I. Sedimentology and paleotectonic setting of the Pennsylvanian Quadrant Sandstone, Southwest Montana. M, 1986, Colorado State University. 178 p.

Saplaco, S. R. Synthesis of the water budget on a semiarid watershed. D, 1977, University of Arizona. 78 p.

Saporito, Mark. Chemical and mineral studies of a core from Lake Patzcuaro, Mexico. M, 1975, University of Minnesota, Minneapolis. 55 p.

Sappenfield, Luther Weidner. A magnetic survey of the Adams County cryptovolcanic structure. M, 1950, University of Cincinnati. 27 p.

Sapper, Samuel E. Geology and geophysics of skarn deposits in the Berg-MacDougall area, south-central Alaska. M, 1982, Northeastern Illinois University. 110 p.

Sappington, Eric Jon. Heavy metal content in stream sediments adjacent to the Wright County, Missouri, sanitary landfill. M, 1987, Southwest Missouri State University. 54 p.

Saptarshi, Vidyadhar Chintamen. Geology of the south flank of the Wildwood Anticline near Dillon, Summit County, Colorado. M, 1946, Colorado School of Mines. 39 p.

Saraby, Fereydoon and Fotouchi, Manuchehre. Geology of the Dunkleberg District, Drummond Quadrangle, Montana. M, 1958, Michigan Technological University. 95 p.

Saracino, Anthony M. Sedimentology and petrology of the Forst Union Formation, North Knobs UCG site, Wyoming. M, 1984, Colorado State University. 166 p.

Saraf, Deoki Nandan. Measurement of fluid saturations by nuclear magnetic resonance and its application to three-phase relative permeability studies. D, 1966, University of California, Berkeley. 106 p.

Saragoni, Gustavo Rodolfo. Nonstationary aaalysis and simulation of earthquake ground motions. D, 1972, University of California, Los Angeles.

Sarapuu, Erich. The underground electrocarbonization and gasification of mineral fuels (T975). D, 1951, University of Missouri, Rolla.

Sarda, Gouri Saukar. Serpentine deposits of Easton, Pennsylvania, and Phillipsburg, New Jersey. M, 1950, Columbia University, Teachers College.

Sardi, Otto. The petrology and structure of the Precambrian complex of the Warm Spring Creek area, from South Fork to Wildcat Creek, Fremont County, Wyoming. M, 1961, Miami University (Ohio). 81 p.

Sardi, Otto. The yttria-stabilized isometric phase of zirconia. D, 1968, Indiana University, Bloomington. 68 p.

Sares, Steven William. Hydrologic and geomorphic development of a low relief evaporite karst drainage basin, southeastern New Mexico. M, 1984, University of New Mexico. 123 p.

Sarewitz, Daniel R. Geology of a part of the Heavens Gate Quadrangle, Seven Devils Mountains, western Idaho. M, 1983, Oregon State University. 144 p.

Sarewitz, Daniel R. The geologic evolution of western Mindoro Island, Philippines. D, 1986, Cornell University. 292 p.

Sarg, Frederick. Depositional environments of the Ames Limestone (Pennsylvanian) in the Pittsburgh area. M, 1971, University of Pittsburgh.

Sarg, J. F. Sedimentology of the carbonate-evaporite facies transition of the Seven Rivers Formation (Guadalupian, Permian) in Southeast New Mexico. D, 1976, University of Wisconsin-Madison. 341 p.

Sargent, Colleen G. Intertonguing Kayenta Formation and Navajo Sandstone (Lower Jurassic) in northeastern Arizona; analysis of fluvial-aeolian processes. M, 1984, Northern Arizona University. 190 p.

Sargent, Elwood Cather. The (Cretaceous) Woodbine Formation, its subsurface structure, extent, and characteristics (East Texas) (Sa73). M, 1930, University of Texas, Austin.

Sargent, John D. Geology of the Windsor Locks Quadrangle, Connecticut. M, 1949, SUNY at Buffalo.

Sargent, Kenneth Aaron. Chemical and isotopic investigation of stratigraphic and tectonic dolomites in the Arbuckle Group, Arbuckle Mountains, south central Oklahoma. D, 1974, University of Oklahoma. 183 p.

Sargent, Kenneth Aaron. Geology and petrology of selected tectonic dolomite areas in the Arbuckle group (upper Cambrian and lower Ordovician), Arbuckle mountains, south-central Oklahoma. M, 1969, University of Oklahoma. 85 p.

Sargent, Kenneth Albert. Alkaline rocks of the Nemogosenda Lake area, Ontario. M, 1957, University of Iowa. 84 p.

Sargent, Kenneth Albert. Allanite occurrence in the Horn area, Bighorn Mountains, Wyoming. D, 1960, University of Iowa. 136 p.

Sargent, M. W. Biostratigraphy of the upper Black River algal bioherms in the vicinity of Kingston, Ontario. M, 1970, Queen's University. 176 p.

Sargent, Melville Wayne. Depositional patterns in the Upper Cambrian Lyell Formtion, southern Canadian Rockies. D, 1975, University of Calgary. 261 p.

Sargent, Michael L. Relationships of iron titanium oxides and other accessory minerals to the development of the Laramie range anorthosite complex (Precambrian), Wyoming. M, 1970, University of Illinois, Urbana. 48 p.

Sargent, Robert Edward. Geology of the New Bullion Mine area, Tooele County, Utah. M, 1953, Indiana University, Bloomington. 49 p.

Sargent, T. H. E. Geology of the Bedwell River-Drinkwater Creek area, British Columbia. D, 1942, Massachusetts Institute of Technology. 142 p.

Sariahmed, Abdelwaheb. Synthesis of sequences of summer thunderstorm volumes for the Atterbury Watershed in the Tucson (Arizona) area. M, 1969, University of Arizona.

Saric, James A. The hydrogeological effects of abandoned underground coal mines, Muddy, Illinois. M, 1987, Northern Illinois University. 106 p.

Sarkar, Gautam Prasad. Late Cenozoic low-temperature near-surface tungsten-bearing manganese oxide cromanechite veins; Blackie Mine, FRA Cristobal Mountains, Sierra County and selected localities in Socorro, Grant, and Hidalgo counties, southwestern New Mexico. D, 1985, University of New Mexico. 194 p.

Sarkar, P. K. Petrology and geochemistry of the White Rock metavolcanic suite, Yarmouth, Nova Scotia. D, 1978, Dalhousie University.

Sarkesian, Arthur C. A petrofabric analyses of the Monkton quartzite (lower Cambrian) in west central Vermont. M, 1970, University of Vermont.

Sarkissian, Volga der *see* der Sarkissian, Volga

Sarle, Clifton J. A new eurypterid fauna. M, 1903, University of Rochester.

Sarle, Clifton J. The (Lower Silurian) Medina Formation and fauna of New York. D, 1906, Yale University.

Sarle, Laura L. Processes and resulting morphology of sand deposits within Beaufort Inlet, Carteret County, North Carolina. M, 1977, Duke University. 152 p.

Sarles, John E. The (Pennsylvanian) upper Seminole and lower Francis formations with emphasis on the DeNay Limestone (Oklahoma). M, 1941, University of Oklahoma. 25 p.

Sarmah, Suryya Kanta. Attenuation of compressional waves in the Earth's mantle. D, 1967, Oregon State University. 80 p.

Sarmiento, J. L. A study of mixing in the deep sea based on STD, radon-222, and radium-228 measurements. D, 1978, Columbia University, Teachers College. 289 p.

Sarmiento, Raul. Dip moveout by migration of radial trace profiles from shot records. M, 1987, University of Houston.

Sarmiento-Soto, Roberto. Geology and iron ore resources of the Paz de Rio region, Boyaca, Colombia. D, 1946, University of California, Berkeley. 181 p.

Sarna-Wojcicki, Andrei M. Correlation of late Cenozoic pyroclastic deposits in the central Coast Ranges of California. D, 1971, University of California, Berkeley. 172 p.

Sarnecki, Joseph Charles. Nannofossil biostratigraphy of the lower Pierre Formation in south-central South Dakota. M, 1974, University of South Dakota. 52 p.

Sarniak, Terry Michael. The sedimentological and structural analysis of Vinini and Valmy formations

(Ordovician), North-central Nevada. M, 1979, Wayne State University.

Saroop, Hayman Cecil. Sedimentology, paleoecology and diagenesis of middle and upper Eocene carbonate shoreline sequences, Crystal River, Florida. M, 1974, University of Florida. 165 p.

Sarpi, Ernesto. Emplacement of the flysch sequence in the Cilento area, southern Apennines, Italy. D, 1967, Columbia University. 194 p.

Sarris, Nelson James. Contributions to the stratigraphy and lithology of the Triassic sedimentary rocks in the Connecticut Valley of Massachusetts. M, 1955, University of Massachusetts. 41 p.

Sarsfield, L. J. Properties of cadmium complexes and their effect on toxicity to a biological system. D, 1976, University of Michigan. 169 p.

Sartain, Maxwell Roland. Subsurface geology of the Short Junction and West More Field area, Cleveland County, Oklahoma. M, 1957, University of Oklahoma. 59 p.

Sartell, Jonathan Floyd. A remote sensing characterization and comparison of the Semipalatinsk area of the Soviet Union and the geologically analogous Mount Katahdin region of north-central Maine. M, 1982, Pennsylvania State University, University Park. 165 p.

Sartin, Austin A., Jr. The geology of Starr Hill Quadrangle, Washington County, Arkansas. M, 1966, University of Arkansas, Fayetteville.

Sartin, Austin Albert, Jr. Paleoecology of the Palo Pinto Limestone (Pennsylvanian), north-central Texas. D, 1972, Southern Methodist University. 115 p.

Sartin, John Philip. A cross-sectional study of the oil producing rocks of Desmoinesian age (Pennsylvanian) in northeastern Oklahoma. M, 1958, University of Oklahoma. 91 p.

Saruk, Bertrand Alexander. Calibration of an electromagnetic seismometer. M, 1975, University of Alberta. 78 p.

Sarvas, P. The structure and magnetic fabric of the Quetico metasedimentary rocks in the Calm Lake-Peach Lake area, near Atikokan, northwestern Ontario. M, 1988, Lakehead University.

Sarver, Timothy John. Geochemical analysis of selected Eocene carbonate rocks of Peninsular Florida. M, 1978, University of Florida. 77 p.

Sarwar, A. K. M. Determination of the acoustical impedance of a layered medium by the Gopinath-Sondhi integral equation. D, 1983, Indiana University, Bloomington. 209 p.

Sarwar, Ghulam. A study of textures, element distribution, and base metal diffusion in the central Kentucky mineral district. M, 1977, University of Cincinnati. 142 p.

Sarwar, Ghulam. Geology of the Bela Ophiolites in the Wayaro area, Las Bela District, south central Pakistan. D, 1981, University of Cincinnati. 423 p.

Sarwar, Golam. Depositional model for the Middle Devonian Mahantango Formation of south-central Pennsylvania. M, 1984, SUNY at Stony Brook. 251 p.

Sarzenski, Darci José. Reservoir geology of the Miranga 1 Sandstone, Miranga Field, Recôncavo Basin, Brazil. M, 1982, University of Texas, Austin. 227 p.

Sasala, Connie S. Deformation zones and structural mechanics affecting Precambrian rocks, Gros Ventre Range, Wyoming. M, 1987, Miami University (Ohio).

Sasowsky, Ira D. Geomorphic significance of longitudinal stream profiles in fluviokarsts. M, 1988, Pennsylvania State University, University Park. 167 p.

Sass, Bruce M. Stability relationships of illite in solutions between 25° and 250°C. M, 1984, Washington State University. 131 p.

Sass, Daniel B. Some aspects of the paleontology, stratigraphy, and sedimentation of the Corry Sandstone of northwestern Pennsylvania. D, 1959, University of Cincinnati. 289 p.

Sass, Daniel Benjamin. Paleoecology and stratigraphy of the (Upper Devonian) Genundewa Limestone of western New York. M, 1951, University of Rochester. 113 p.

Sass, John H. Evaluation of "in situ" methods of measurements of thermal conductivity of rocks. M, 1961, University of Western Ontario.

Sassano, Giampaola. The nature and origin of the uranium mineralization (Precambrian) at the Fay Mine, Eldorado, Saskatchewan, Canada. D, 1972, University of Alberta. 305 p.

Sasscer, Reverdy G. Microscopic study of sediments. M, 1926, University of North Carolina, Chapel Hill. 21 p.

Sasse, Jerome B. A paleontologic study of the Temblor middle Miocene of Kern County, California. M, 1927, University of California, Berkeley. 174 p.

Sassen, Roger. Early diagenesis of fatty acids in mangrove-associated sediments, St. Croix, U. S. Virgin Islands. D, 1975, Lehigh University. 137 p.

Sassen, Roger. Fatty acid transformations in surface sediments of a New Jersey salt marsh. M, 1972, Lehigh University. 60 p.

Sasser, Walter B., III. Petrography of some Tennessee aggregates related to their engineering properties. M, 1973, Memphis State University.

Sasseville, D. R. Present and historic geochemical relationships in four Maine lakes. M, 1974, University of Maine. 67 p.

Sasso, Jane A. Palynology of the clastic units of the lower tongue of the Breathitt Formation (lower Pennsylvanian) of eastern Kentucky. M, 1976, Eastern Kentucky University. 93 p.

Sassos, Michael. Geochemistry as an exploration guide for roll-type uranium deposits in South Texas. M, 1984, Queens College (CUNY). 102 p.

Sastrosoedirdjo, Djoko W. A study on the mapping and surveying system in Indonesia. M, 1966, Ohio State University.

Sastrowihardjo, Sutadi. The soil conservation program in relation to watershed rehabilitation in Indonesia. M, 1988, [Colorado State University].

Satchell, Loretta Simmonds. Patterns of disturbance and vegetation change in the Miocene Succor Creek flora of Oregon-Idaho. D, 1983, Michigan State University. 155 p.

Sateesha, Malalur K. Analysis of data from a deep hole stress measurements device. D, 1974, South Dakota School of Mines & Technology.

Sater, G. S. Geology of the McQuat-Gauvin area, Mistassini Territory and Roberval electoral district, Quebec. M, 1958, McGill University.

Sathiyakumar, Neelakandan. Secondary recovery of groundwater by air injection; finite element model. D, 1987, Texas Tech University. 147 p.

Satin, Lowell Robert. Geology of the Bradford area of Huerfano Park, Huerfano County, Colorado. M, 1955, University of Michigan.

Satkin, Richard L. A geothermal resource evaluation at Castle Hot Springs, Arizona. M, 1981, Arizona State University. 147 p.

Satman, Abdurrahman. In-situ combustion models for the steam plateau and for fieldwide oil recovery. D, 1979, Stanford University. 112 p.

Sato, H. H. Interpretation of index properties of the unified classification system for Hawaiian soils. M, 1971, University of Hawaii.

Sato, Motoaki. The oxidation of sulfide ore bodies with special reference to self-potentials. D, 1959, University of Minnesota, Minneapolis. 335 p.

Satorius-Fox, Marsha R. Paleoecological analysis of micromammals from the Schmidt site, a Central Plains tradition village in Howard County, Nebraska. M, 1982, University of Iowa. 87 p.

Satoskar, Vijay Vishnu. Petrology, magnetic properties, and stratigraphy of the Elkhorn Mountains volcanic rocks (Late Cretaceous) of southern Jefferson County, Montana. D, 1971, Indiana University, Bloomington. 75 p.

Satterfield, Ira Robert. The bedrock geology of the Cobden Quadrangle (Illinois). M, 1965, Southern Illinois University, Carbondale. 159 p.

Satterfield, Joe. The geology of a portion of the Trap Mountains, Arkansas. M, 1982, University of Missouri, Columbia.

Satterfield, Will McSwain. Late Quaternary sedimentology and evolution of the continental slope (long 94°-95°), Northwest Gulf of Mexico. M, 1988, University of Texas, Austin. 121 p.

Satterthwait, D. F. Paleobiology and paleoecology of middle Cambrian algae from western North America. D, 1976, University of California, Los Angeles. 134 p.

Satterthwaite, Laurence Cyrus. The geology of the east-central part of Promontory Butte Quadrangle, Arizona. M, 1951, University of Iowa. 184 p.

Saturni, Ben A. Depositional history of the middle Valmeyeran Ramp Creek and Harrodsburg limestones of Southwest Indiana. M, 1985, Indiana University, Bloomington. 107 p.

Sauber, Jeanne. Temporal and characteristics of earthquakes preceding the larger events (M_z z) in Lake Jocassee, SC. M, 1979, University of South Carolina.

Sauchyn, David John. Open rock basins and debris fans in the Kananaskis area, southern Canadian Rocky Moutains. D, 1984, University of Waterloo. 321 p.

Saucier, Alva Eugene. The Morrison (upper Jurassic) and related formations in the Gallup, New Mexico region. M, 1967, University of New Mexico. 106 p.

Saucier, Roger Thomas. Recent geomorphic history of the Pontchartrain Basin, Louisiana. D, 1968, Louisiana State University. 189 p.

Sauck, William August. A laboratory study of induced electrical polarization in selected anamalous rock types. M, 1969, University of Arizona.

Sauck, William August. Compilation and preliminary interpretation of the Arizona aeromagnetic map. D, 1972, University of Arizona.

Sauer, Emil Karl. The application of geotechnical principles in road design problems. D, 1967, University of California, Berkeley. 345 p.

Sauer, Frank Joseph. The (Upper Triassic) Springdale Sandstone in the Zion Canyon region, southwestern Utah. M, 1956, University of Nebraska, Lincoln.

Sauer, J. H. and Rombauer, Alfred B. Report of Rosborough coal mine, Rosborough, Ill. M, 1889, Washington University.

Sauer, R. Tayler. A metamorphosed stratiform alteration zone as footwall to massive sulfide, Mineral District, Virginia. M, 1983, University of Western Ontario. 196 p.

Sauer, T. C., Jr. Volatile liquid hydrocarbons in the marine environment. D, 1978, Texas A&M University. 333 p.

Saueracker, Paul Robert. Solution features in Southeast Kansas. M, 1966, University of Kansas. 34 p.

Saul, John M. Fauna of the Accraian Series (Devonian of Ghana). M, 1960, Massachusetts Institute of Technology. 57 p.

Saul, LouElla Rankin. Senonian (Cretaceous) mollusks fron Chico Creek (California). M, 1959, University of California, Los Angeles.

Saul, Michael T. The steady-state equilibrium of Government Creek, Union County, Illinois. M, 1986, Southern Illinois University, Carbondale. 68 p.

Saul, Richard Brant. The geology of the southwest quarter of the Waucoba Mountain Quadrangle, California. M, 1959, University of California, Los Angeles.

Saul, William L. Brown dolomite zone of the Lehigh Acres Formation, Lower Cretaceous, South Florida. M, 1987, University of Southwestern Louisiana. 89 p.

Saull, Vincent A. Some aspects of chemical energy in geology. D, 1952, Massachusetts Institute of Technology. 167 p.

Saulnier, George J. Ground water resources and geomorphology of the Pass Creek Basin area, Albany and

Carbon counties, Wyoming. M, 1968, University of Wyoming. 91 p.

Saulnier, George J., Jr. Genesis of the saline waters of the Green River Formation, Piceance Basin, northwestern Colorado. D, 1978, University of Nevada. 130 p.

Saulnier, Henry Siddell. The paleopalynology of the (Paleocene) Fort Union coals of Red Lodge, Montana. M, 1950, University of Massachusetts. 62 p.

Sauls, Brian D. The geology, microstructures, and small-scale structures in the vicinity of Upper Cataract Lake, Gore Range, Colorado. M, 1982, Wright State University. 49 p.

Saum, Nicholas Mather. Detailed mineralogy in sediments from five south Atlantic deep-sea cores of differing stratigraphic ages. D, 1970, University of Missouri, Columbia. 188 p.

Saum, Nicholas Mather. Geochemical prospecting in the south-western Wisconsin zinc-lead district. M, 1962, University of Missouri, Columbia.

Saumsiegle, W. J. Stability and local effects of an offshore sand storage mound, Dam Neck disposal site, Virginia inner continental shelf. M, 1976, Old Dominion University. 92 p.

Saun, Richard Van *see* Van Saun, Richard

Saunders, Bruce W. Geology of the Wharton area, Madison County, Arkansas. M, 1967, University of Arkansas, Fayetteville.

Saunders, Cynthia Margaret. Controls of mineralization in the Betts Cove Ophiolite. M, 1985, Memorial University of Newfoundland. 200 p.

Saunders, David W. DeGraaff-Runter's model Earth anomalies. M, 1963, Ohio State University.

Saunders, Denise Marie. Geology of the Park, Utah west vein, Park City District, Utah. M, 1984, Colorado State University. 137 p.

Saunders, Jack McLeod. Lithology and insoluble residues of a core from the (Lower Devonian) Detroit River Group. M, 1949, University of Michigan.

Saunders, James Alexander. Petrochemistry, volcanic stratigraphy, and economic geology of the Waconichi Formation in Scott Township, Chibougamau, Quebec. M, 1978, University of Georgia.

Saunders, James Alexander. Petrology, mineralogy, and geochemistry of representative gold telluride ores from Colorado. D, 1986, Colorado School of Mines. 171 p.

Saunders, Jeffrey J. The distribution and taxonomy of Mammuthus in Arizona. M, 1970, University of Arizona.

Saunders, Jeffrey John. Late Pleistocene vertebrates of the western Ozark Highlands, Missouri. D, 1975, University of Arizona. 400 p.

Saunders, Margaret Janet. Mineralogy and metamorphism of the Valdez Group, Kenai Peninsula, Alaska. M, 1978, University of California, Los Angeles.

Saunders, Margaret M. Geologic and isotopic investigation of the South Willow Creek gold prospect, Madison County, Montana. M, 1985, Indiana University, Bloomington. 71 p.

Saunders, Ronald Stephen. The geology of the Southern Highlands of the Moon. D, 1970, Brown University. 158 p.

Saunders, Ronald Stephen. The use of hystrichospheres in the interpretation of environments of deposition of some Lower and Middle Silurian rocks of the north-central Appalachians. M, 1968, Brown University.

Saunders, William Bruce. Upper Mississippian (Chesterian) ammonoids from the Imo and Rhoda Creek formations, Arkansas and Oklahoma. D, 1971, University of Iowa. 184 p.

Saunderson, Carol Patricia. An investigation of the crystal structure of anhydrous lithium acetate. M, 1960, University of Manitoba.

Saunderson, H. C. Eskerine sedimentation; an analysis of hypotheses and an empirical test. D, 1974, University of Toronto.

Saur, William F. Computer simulation of solute transport in groundwater at a hazardous waste disposal site, Pullman, Washington. M, 1988, Washington State University. 61 p.

Sauter, Allan Wayne. Studies of the upper oceanic floor using ocean bottom seismometers. D, 1987, University of California, San Diego. 114 p.

Sauve, Jeffrey Allen. Near surface velocity reconstruction and diving wave tomography; Erawan Field, Gulf of Thailand. M, 1988, University of Texas, Austin. 174 p.

Sauve, Judith Ann. The Middle Ordovician of northeastern Mississippi; implications for the Black Warrior Basin. M, 1981, Vanderbilt University. 113 p.

Sauve, Pierre. Sedimentation and volcanic history of a part of the "Labrador Trough". M, 1953, Queen's University. 124 p.

Sauve, Pierre. The geology of the East half of the Gerido Lake area, New Quebec, Canada. D, 1957, The Johns Hopkins University.

Sauveplane, Claude Marcel. Analytical modelling of transient flow in wells in complex aquifer systems; characterization and productivity of these systems. D, 1987, University of Alberta. 295 p.

Savage, Bruce A. Petrology of the Ison Creek kimberlite, Kentucky. M, 1985, University of Cincinnati. 109 p.

Savage, Carleton N. The drumlins of the Waterloo-Watertown quadrangles, Wisconsin. M, 1940, Northwestern University.

Savage, Donald Elvin. Late Cenozoic vertebrates of the San Francisco Bay region (California). D, 1949, University of California, Berkeley. 142 p.

Savage, Donald Elvin. The Optima fauna, middle Pliocene, from Texas County, Oklahoma. M, 1939, University of Oklahoma. 127 p.

Savage, E. Lynn. A detailed study of Triassic sediments at Granton Quarry, Bergen County, New Jersey. M, 1955, New York University.

Savage, E. Lynn. The Triassic sediments of Rockland County, New York. D, 1967, Rutgers, The State University, New Brunswick. 193 p.

Savage, George Richard. Geology of the southwest quarter of the Chunky, Mississippi Quadrangle. M, 1965, Mississippi State University. 93 p.

Savage, Godfrey Hamilton. The design and analysis of a submerged, buoyant, anchored pipeline for transportation of natural gas through the deep ocean. D, 1970, Stanford University. 288 p.

Savage, James C. Wave propagation in a continuously stratified fluid. D, 1957, California Institute of Technology. 79 p.

Savage, James W. Cross sections of Mississippi River at the government locks and dams. M, 1938, University of Minnesota, Minneapolis. 42 p.

Savage, Kevin M. Petrographic studies of quartz-rich South American modern sands. M, 1986, University of Cincinnati. 152 p.

Savage, Martha Kane. Aftershocks of an M_L = 4.2 earthquake on the Island of Hawaii. M, 1984, University of Wisconsin-Madison. 61 p.

Savage, Martha Kane. Spectral properties of Hawaiian microearthquakes; source, site, and attenuation effects. D, 1987, University of Wisconsin-Madison. 223 p.

Savage, Rebbeca Jo. Modes of longshore variability in the development of a bar-trough morphology. M, 1988, College of William and Mary. 78 p.

Savage, Thomas Edmund. Detailed section of the Devonian formation in Johnson County, Iowa. M, 1898, University of Iowa. 71 p.

Savage, Thomas Edmund. The stratigraphy of the lower Paleozoic formations in southwestern Illinois. D, 1909, Yale University.

Savage, William S. Solution, transportation, and precipitation of manganese. D, 1935, University of Toronto.

Savage, William Underwood. Earthquake probability models; recurrence curves, aftershocks, and clusters. D, 1975, University of Nevada. 145 p.

Savage, William Z. Application of plastic flow analysis to drumlin formation (Pleistocene). M, 1968, Syracuse University.

Savage, William Z. Stress and displacement fields in stably folded rock layers. D, 1974, Texas A&M University. 207 p.

Savanick, George Adrian. An investigation of jarosite found in some radioactive lignites. M, 1961, Pennsylvania State University, University Park. 61 p.

Savard, Martine. Pétrographie en cathodoluminescence et évolution diagenetique d'une plate-forme calcaire, Silurien supérieur, Gaspésie, Québec. M, 1986, Universite Laval. 59 p.

Savard, Wilfred L. The sediments of Block Island Sound. M, 1966, University of Rhode Island.

Savarese, Joseph G. Hydrogeologic evaluation of a continental crystalline bedrock aquifer; Tiverton, RI. M, 1987, University of Rhode Island. 148 p.

Savarese, Michael L. Faunal and lithologic cyclicity in the Centerfield Member (Middle Devonian Hamilton Group) of western New York; a reinterpretation of depositional history. M, 1984, University of Rochester. 100 p.

Savci, Gultekin. Structural and metamorphic geology of the subophiolitic dynamothermal metamorphic sole and peridotite tectonites, Blow Me Down Massif, Newfoundland- Canada; tectonic implications for subduction and obduction. D, 1988, University of Houston. 379 p.

Savelle, J. M. Sedimentary and faunal facies of an Upper Silurian marine succession near Creswell Bay, Somerset Island, Northwest Territories. M, 1978, University of Ottawa. 134 p.

Savely, James P. Orientation and engineering properties of jointing in the Sierrita Pit, Arizona. M, 1972, University of Arizona.

Savely, James Palmer. Probabilistic analysis of fractured rock masses. D, 1987, University of Arizona. 340 p.

Savic, Milos. Study of seismic velocities using a minimum error zero-offset traveltime method. M, 1986, University of Houston.

Savin, Samuel Marvin. Oxygen and hydrogen isotope ratios in sedimentary rocks and minerals. D, 1967, California Institute of Technology. 228 p.

Savina, Mary Elizabeth. Studies in bedrock lithology and the nature of downslope movement. D, 1982, University of California, Berkeley. 297 p.

Savinelli, Peter. Australian mineral production and policy. M, 1983, University of Texas, Austin.

Savino, John Michael. The nature of long-period (20 to 130 sec) earth noise and importance of a pronounced noise minimum to detection of seismic events. D, 1971, University of Colorado. 146 p.

Savoy, Donald DeCoursey. Sedimentary processes along the Lake Erie shore at Magee Marsh, Ohio. M, 1956, Ohio State University.

Savoy, Lauret Edith. Seacliff erosion and the direction of longshore transport along portions of the Northern California Coast. M, 1983, University of California, Santa Cruz.

Savrda, Charles Edward. Development and evaluation of a trace fossil model for the reconstruction of paleo-oxygenation in marine environments. D, 1986, University of Southern California.

Savre, Wayland Carlyle. Stratigraphy, igneous rocks, mineral deposits and structure of a portion of the southwestern Bear Lodge Mountains, Crook County, Wyoming. M, 1954, University of Iowa. 260 p.

Savula, N. A. Light mineral petrology of sediments from Santa Monica and San Pedro bays, California continental borderland. M, 1978, University of Southern California.

Sawatsky, Don L. Tectonic style along the Elkhorn Thrust, eastern South Park and western Front Range,

Park County, Colorado. D, 1967, Colorado School of Mines. 206 p.

Sawatsky, L. H. The Lloydminster Field, Saskatchewan. M, 1958, University of Saskatchewan. 66 p.

Sawatzky, Don L. Tectonic style along the Elkhorn Thrust, eastern South Park and western Front Range, Park County, Colorado. D, 1967, Colorado School of Mines.

Sawford, Clayton. Sedimentary facies of the Black River Group (Ordovician), Ottawa Valley, Canada. D, 1972, Carleton University.

Sawford, Edward C. Stratigraphy and sedimentation of the Carson Creek Reef (Devonian), Alberta (Canada). M, 1967, Queen's University. 130 p.

Sawhill, Gary S. The effect of the spray irrigation of secondary treated effluent on the vegetation, soils and groundwater quality in a New Jersey pine barrens habitat. D, 1977, Rutgers, The State University, New Brunswick. 183 p.

Sawicki, David A. A structural and petrographic evaluation of the Thing Valley Lineament, San Diego County, California. M, 1978, San Diego State University.

Sawin, Robert Scott. Paleoenvironmental interpretation of a Virgilian (Pennsylvanian) stromatolite biostrome in northeastern Kansas. M, 1977, Kansas State University.

Sawiuk, Myron J. Geologic and petrochemical investigations of stratabound uranium mineralization, Karpinka Lake, Saskatchewan. M, 1984, McGill University. 172 p.

Sawka, Wayne Nickolas. Vertical and horizontal fractionation of the Tinemaha Granodiorite, John Muir Wilderness and Kings Canyon National Park, central Sierra Nevada, California. M, 1981, University of California, Los Angeles. 86 p.

Sawkins, Frederick J. Lead-zinc ore deposition in the light of fluid inclusion studies, Providencia, Zacatecas, Mexico. D, 1963, Princeton University. 144 p.

Sawlan, Jeffrey J. Early diagenetic remobilization of some transition metals in hemipelagic sediments. D, 1982, University of Washington. 251 p.

Sawlan, Michael Gary. Petrogenesis of late Cenozoic volcanic rocks from Baja California Sur, Mexico. D, 1986, University of California, Santa Cruz.

Sawtelle, E. Rossiter, Jr. The origin of Berea Sandstone (Mississippian) in the Michigan Basin. M, 1958, Michigan State University. 72 p.

Sawyer, Byrd Fanita Wall. Gold and silver rushes of Nevada, 1900-1910. D, 1931, University of California, Berkeley. 172 p.

Sawyer, Dale Stewart. Thermal evolution of the northern United States Atlantic continental margin. D, 1982, Massachusetts Institute of Technology. 264 p.

Sawyer, David Andrew. Late Cretaceous caldera volcanism and porphyry copper mineralization at Silver Bell, Pima County, Arizona; geology, petrology, and geochemistry. D, 1987, University of California, Santa Barbara. 400 p.

Sawyer, Edward William. The origin and formation of migmatites in the Archean Quetico metasedimentary belt near Kashabowie, Ontario, Canada. D, 1984, University of Toronto.

Sawyer, J. B. Paul. Porphyries of the Bathurst area, New Brunswick. M, 1957, University of Western Ontario. 124 p.

Sawyer, J. F. The occurrence of smaller foraminifera in the (Permian) Florena Shale (Kansas). M, 1954, Columbia University, Teachers College.

Sawyer, Kenneth Charles, III. Preservational patterns of organic material in a carbonate shelf environment, Southwest Puerto Rico. M, 1980, University of Oklahoma. 161 p.

Sawyer, Michael B. Geology of the Whitsett Quadrangle, Atascosa, Live Oak, and McMullen counties, Texas. M, 1980, Colorado School of Mines. 94 p.

Sawyer, Noah Gus, Jr. The geology of the Big Ridge Park Quadrangle, Knox, Union, and Anderson coun-

ties, Tennessee. M, 1947, University of Tennessee, Knoxville. 48 p.

Sawyer, Robert Frank. Gravity and ground magnetic surveys of the Thermo Hot Springs KGRA region, Beaver County, Utah. M, 1977, University of Utah. 142 p.

Sawyer, Robert Knowlton. Late Quaternary mineralogy and paleoenvironmental reconstruction of a freshwater peat-forming basin in the Florida Everglades. M, 1982, University of Florida. 147 p.

Sawyerr, Olumuyiwa Akinnade. Subsurface stratigraphic analysis of N.E. Verden–Dutton area, Caddo and Grady counties, Oklahoma. M, 1971, University of Oklahoma. 58 p.

Sax, Norman Alfred. Geology of the East Sugarloaf area, Boulder County, Colorado. M, 1948, University of Colorado.

Sax, Robert L. Simulation of density distributions from gravitational data with application to mapping a glacier. D, 1960, Massachusetts Institute of Technology. 168 p.

Saxby, Donald B. Geologic mapping of Sequatchie County, Tennessee, from aerial photographs (Sa97). M, 1947, University of Illinois, Urbana. 13 p.

Saxby, Donald William. Sampling problems and hydraulic factors related to the dispersion of scheelite in drainage sediments, Clea Property, Yukon Territory. M, 1985, University of British Columbia. 151 p.

Saxe, Joseph Archibald. Petrography of contact zones at Elkhorn, Idaho. M, 1910, Columbia University, Teachers College.

Saxena, Ram S. Petrology of some Allegheny Pennsylvanian rocks in eastern Ohio and western Pennsylvania. D, 1971, Louisiana State University. 200 p.

Saxon, Fred Chalmers. Water-resource evaluation of Morgan Valley, Morgan County, Utah. M, 1972, University of Utah. 118 p.

Saxton, Deborah C. Quaternary and environmental geology of northern Lafayette Parish, Louisiana. M, 1986, University of Southwestern Louisiana. 121 p.

Sayala, D. Regional geochemical exploration for mineral deposits in the Appalachians of central and northwestern Virginia. D, 1979, George Washington University. 374 p.

Sayala, Dasharatham. A geochemical study of distribution of elements in ores and limestones from New Jersey zinc mines, Hanover, New Mexico. M, 1972, University of New Mexico. 285 p.

Sayao, Otavio de Sampaio Ferraz Jardim. Beach profiles and littoral sand transport. D, 1982, Queen's University.

Saydam, A. Sacit. Evaluation of some of the electrode arrays used in induced polarization and resistivity surveying for their relative responses. M, 1975, University of Calgary. 165 p.

Sayed, Sayed Mourad. Expansion of long cylindrical cavities in nonlinear dilatant media. D, 1982, Duke University. 1982 p.

Sayeed, Usman Ahmed. Petrology and structure of the Kern Mountains plutonic complex, White Pine County, Nevada and Juab County, Utah. D, 1973, University of Nebraska, Lincoln.

Sayeed, Usman Ahmed. The tectonic setting of the Colchester plutons, Southwest Arm, Green Bay, Newfoundland. M, 1970, Memorial University of Newfoundland. 76 p.

Sayek, T. F. Pore pressure in isotropically and anisotropically consolidated clays. D, 1975, Wayne State University. 230 p.

Sayer, Christina J. An electrochemical study at 25°C of the leaching behaviour of gold and silver in inorganic and organic solutions. M, 1986, University of Alberta. 107 p.

Sayer, Suzanne. An integrated study of the Blue Hills porphyry and related units; Quincy and Milton, Massachusetts. M, 1974, Massachusetts Institute of Technology. 146 p.

Sayers, Frank Edward. The geology of a portion of the Manhattan Quadrangle, Horseshoe Hills area, Gal-

latin County, Montana. M, 1962, University of California, Berkeley. 67 p.

Sayili, Alaittin. Petrology and diagenesis of fan-delta deposits in the middle Minturn Formation (Penn), McCoy area, Eagle County, Colorado. M, 1985, University of Colorado.

Saylan, Serif. Seismic analysis of three-dimensional soil-structure interaction system with a rectangular base. D, 1982, George Washington University. 140 p.

Sayles, Frederick Livermore. Coastal sedimentation; Point San Pedro to Miramontes Point, California. M, 1965, University of California, Berkeley. 104 p.

Saylor, Timothy E. The geology of the Pearl-Pioneerville area, Idaho. M, 1967, Case Western Reserve University.

Saylor, Weldon Wayne. Late Paleozoic and early Mesozoic stratigraphy of the eastern Canon City Embayment, Colorado. M, 1955, University of Oklahoma. 104 p.

Sayman, Ali. Determination of an earthquake fault area and rupture velocity from the spectra of long period P waves. M, 1970, St. Louis University.

Sayre, Albert Nelson. The fauna of the (Pennsylvanian) Drum Limestone of Kansas and western Missouri. D, 1928, University of Chicago. 284 p.

Sayre, Robert Lewis. Geology and mineral deposits of the western portion, Church Hills, Millard County, Utah. M, 1971, University of Utah. 74 p.

Sayre, Ted M. Pore water extraction by triaxial compression from unsaturated tuff, Yucca Mountain, Nevada. M, 1985, Colorado School of Mines. 138 p.

Sayre, William W. Dispersion of mass in open-channel flow. D, 1967, Colorado State University. 112 p.

Sayyab, Abdullah Shakir. Cretaceous Ostracoda from the Persian Gulf area. D, 1956, Iowa State University of Science and Technology. 155 p.

Sayyab, Abdullah Shakir. Petrology of some Recent marine sediments off the North Carolina coast. M, 1954, Indiana University, Bloomington. 48 p.

Sayyah, Taha Ahmed. Geochronological studies of the Kinsley Stock, Nevada, and the Raft River Range, Utah. D, 1965, University of Utah. 112 p.

Sbag, Shahe Fares. The geochemistry and petrology of granitoids at Meggisi Lkae, N.W. Ontario. M, 1979, University of Toronto.

Sbar, Marc Lewis. Contemporary compressive stress and seismicity in eastern North America; an example of intra-plate tectonics. D, 1972, Columbia University. 120 p.

Sbeta, Ali M. Geology and petrography of Hosselokus Limestone, Shasta County, California. M, 1970, Stanford University.

Sblendorio Levy, J. S. The northeastern Atlantic Polar Front during the last glacial cycle; a floral investigation. M, 1975, Queens College (CUNY). 72 p.

Scadden, Raymond William. Analysis of Maumee III and Whittlesey Beach ridge sediments in Seneca County, Ohio. M, 1964, Bowling Green State University. 36 p.

Scafe, Donald W. A comprehensive study of six sections of the Deer Creek Limestone in Nebraska, Missouri and Kansas. M, 1963, University of Kansas. 31 p.

Scafe, Donald William. A clay mineral investigation of six cores from the Gulf of Mexico. D, 1968, Texas A&M University. 90 p.

Scaife, Norman Caldwell. Geology of the Camp Air West area, Mason County, TX. M, 1957, Texas A&M University.

Scales, Anthony Scott. Geochemical and petrologic characterization of the Glamorgan coal seam and implications for paleoenvironment reconstruction in southeastern Pike County, Kentucky. M, 1987, University of Tennessee, Knoxville. 116 p.

Scales, Bert. Geology of the Applebush Hill area, south Antelope Valley, Nye County, Nevada. M, 1961, University of Nevada. 50 p.

Schaff, Ross Gilbert. The geology of the Little Bigelow Mountain Quadrangle, Maine. D, 1963, Boston University. 162 p.

Schaff, Schuyler C. The 1968 Adel, Oregon, earthquake swarm. M, 1976, University of Nevada. 63 p.

Schaffel, Simon. Reconstruction of late-glacial and postglacial events in Long Island Sound, New York. D, 1971, New York University.

Schaffel, Simon. The geology of Breakneck Ridge and vicinity, Duchess and Putnam counties, southeastern New York. M, 1957, Rutgers, The State University, New Brunswick. 109 p.

Schafran, Gary Charles. The influence of ground water inputs on the porewater and sediment chemistry of a dilute, acidic lake. D, 1988, Syracuse University. 280 p.

Schafroth, Don Wallace. Structure and stratigraphy of the Cretaceous rocks south of the Empire Mountains, Pima and Santa Cruz counties, Arizona. D, 1965, University of Arizona. 194 p.

Schaftenaar, Carl Howard. Preferred orientation of calcite in deep-sea carbonates and its relationship to acoustic anisotropy. M, 1982, Texas A&M University. 57 p.

Schaftenaar, Wendy Elizabeth. Uranium occurrence in igneous rocks of the central Davis Mountains, West Texas. M, 1982, Texas A&M University. 110 p.

Schaik, Edward J. Van *see* Van Schaik, Edward J.

Schaik, Jacob Cornelis Van *see* Van Schaik, Jacob Cornelis

Schaiowitz, Michael. The petrology of a multiple basic sill in the Three Forks, Sappinton, and Lodgepole formations, Madison County, Montana. M, 1964, Indiana University, Bloomington. 41 p.

Schairer, John F. The mineralogy and paragenesis of the pegmatite at Collins Hill, Portland, Connecticut. M, 1926, Yale University.

Schake, Celia May LaLonde *see* LaLonde Schake, Celia May

Schake, Wayne Eugene. Carboniferous stratigraphy of the Wallace Creek area, San Saba County, Texas. M, 1961, University of Texas, Austin.

Schalck, Diane Kate. Geology, petrology, and alteration of the Sorrel Spring syenitic complex, Custer County, Idaho. M, 1987, University of Idaho. 112 p.

Schaleman, Harry James. The sand and gravel industry of the Cincinnati area. M, 1953, University of Cincinnati. 112 p.

Schalk, Marshall. A textural study of certain New England beaches. D, 1936, Harvard University.

Schall, James Douglas. Two-dimensional investigation of shear flow turbulence in open channel flows. D, 1983, Colorado State University. 258 p.

Schalla, Robert Allen. Paleozoic stratigraphy of the southern Mahogany Hills, Eureka County, Nevada. M, 1978, Oregon State University. 118 p.

Schaller, Phillip H. Applications of two-dimensional vidicon photometry; Venus. M, 1973, Massachusetts Institute of Technology. 113 p.

Schalles, J. F. Comparative limnology and ecosystem analysis of Carolina Bay ponds on the upper coastal plain of South Carolina. D, 1979, Emory University. 290 p.

Schallhorn, Janis K. Geochemistry of coal-fly ash leachates and their interaction with glacial sediment. M, 1982, University of Wisconsin-Milwaukee. 180 p.

Schamel, Steven. Eocene subduction in central Liguria, Italy. D, 1974, Yale University.

Schanck, J. W. Determination of petrographic and insoluble residue content of some Tennessee limestones in relation to their engineering properties. M, 1976, Memphis State University.

Schandl, Eva S. The feldspar mineralogy of the Sudbury Complex. M, 1982, McGill University. 147 p.

Schandle, Thomas M. Investigation of the Birch-Murnagham equation of state as a method for determining zero pressure elastic constants for the Earth's lower mantle. M, 1977, University of Minnesota, Duluth.

Schapiro, Norman. Petrographic studies of lower Kittanning (Pennsylvanian, Pennsylvania). D, 1955, West Virginia University.

Schapiro, Norman. The geology of a part of the Quinnesec greenstone complex, Dickinson County, Michigan. M, 1952, Ohio State University.

Scharf, David W. A Miocene mammalian fauna from Sucker Creek, southeastern Oregon. M, 1932, California Institute of Technology. 34 p.

Scharnberger, Charles Kirby. Plate tectonics and paleomagnetism of Antarctica. D, 1971, Washington University. 118 p.

Scharon, Harry Leroy. Application of geophysical methods in the Fredricktown lead district, Madison County, Missouri. D, 1946, The Johns Hopkins University.

Schasse, Henry William. The geology and mineral deposits of Jacks Mountain, in the Mount Union and Butler Knob 7 1/2-minute quadrangles, central Pennsylvania. M, 1978, Pennsylvania State University, University Park. 175 p.

Schatz, Barry Allen. Depositional environments of the upper Fountain and Ingleside formations between Lyons and Loveland, Colorado. M, 1986, University of Colorado. 118 p.

Schatz, Clifford Eugene. Source and characteristics of the tsunami observed along the coast of the Pacific Northwest on March 28, 1964. M, 1965, Oregon State University. 39 p.

Schatz, Frank Lee. Insoluble residues of the Silurian of southern Indiana. M, 1950, Miami University (Ohio). 49 p.

Schatz, John F. Thermal conductivity of earth materials at high temperatures. D, 1971, Massachusetts Institute of Technology. 199 p.

Schatzinger, R. A. Later Eocene (Uintan) lizards from the Greater San Diego area, California. M, 1975, San Diego State University.

Schatzinger, Richard Allen. Depositional environments and diagenesis of the eastern portion of the Horseshoe Atoll, West Texas. D, 1987, University of Texas, Austin. 171 p.

Schau, Mikkel Paul. Geology of the upper Triassic Nicola group in south central British Columbia. D, 1969, University of British Columbia.

Schaub, William J. Metamorphism and garnet zoning in the Blue Ridge near Chunky Gal Mountain, southwestern North Carolina. M, 1983, University of Kentucky. 94 p.

Schauble, Carl Eugene. Factors influencing copper availability in selected North Carolina coastal plain soils. D, 1969, North Carolina State University. 144 p.

Schaubs, Michael Paul. Geology and mineral deposits of the Bohemia mining district, Lane County, Oregon. M, 1978, Oregon State University. 135 p.

Schavran, Gabrielle. Structural features of a Laramide fold and thrust belt, east flank of the Sangre de Cristo Range, Colorado. M, 1984, Colorado School of Mines. 67 p.

Scheckler, Stephen Edward. Evolutionary trends in primitive Progymnosperms. M, 1970, Cornell University.

Schedl, Andrew D. The petrogenesis of the granites at Five Springs, Bighorn Mountains, Wyoming. M, 1979, University of Iowa. 176 p.

Schedl, Andrew David. The role of the brittle-ductile transition in the mechanics of detachment and motion of crystalline nappes. D, 1986, University of Michigan. 356 p.

Scheele, R. A. Earth science teaching units on geology, coal, oil, and their role in the generation of electrical energy. M, 1975, Virginia State University.

Scheerens, Clark L. An investigation of some geochemical variations in three Silurian dolomites from Northwest Ohio. M, 1979, University of Toledo. 75 p.

Scheerer, Paul Ervin. A petrological study of the textural features of the Nonewaug Granite (Connecticut). M, 1958, University of Wisconsin-Madison.

Scheetz, Barry Earl. The effects of order/disorder on the vibrational spectra of minerals. D, 1976, Pennsylvania State University, University Park. 210 p.

Scheetz, Barry Earl. Vibrational spectra of selected melilite minerals (Pennsylvania). M, 1972, Pennsylvania State University, University Park. 96 p.

Scheevel, Jay R. Soft-sediment and hard-rock deformation in the Chinle Formation, northeastern Arizona. M, 1983, Texas A&M University. 152 p.

Scheffe, Gregory L. Depositional environment and petrographic image analysis of porosity development of a Middle Pennsylvanian series hydrocarbon reservoir in the midcontinent region. M, 1986, Wichita State University. 202 p.

Scheffe, Richard Donald. Reacidification modeling and dose calculation procedures for calcium carbonate treated lakes. D, 1987, Clarkson University. 361 p.

Scheffler, Joanna March. A petrologic and tectonic comparison of the Hell's Canyon area, Oregon-Idaho, and Vancouver Island, British Columbia. M, 1983, Washington State University. 98 p.

Scheffler, Peter K. Geology and geophysics of the epicentral area of the August 2, 1974 McCormick South Carolina earthquake. M, 1975, University of South Carolina. 65 p.

Scheibach, Robert Bruce. Geothermal occurrences in Truckee Meadows, Washoe County, Nevada. M, 1975, University of Nevada. 74 p.

Scheid, Vernon E. Clay resources of Latah County, Idaho. M, 1940, University of Idaho. 76 p.

Scheid, Vernon Edward. Excelsior high-alumina clay deposit, Spokane County, Washington; 2 volumes. D, 1946, The Johns Hopkins University.

Scheidegger, Kenneth Fred. Temperatures and compositions of magmas ascending beneath actively spreading mid-ocean ridges. D, 1973, Oregon State University. 143 p.

Scheidle, Diana Lynn. Plagioclase zoning and compositional variation in anorthosite I and II along the Contact Mountain traverse, Stillwater Complex, Montana. M, 1983, Stanford University. 122 p.

Scheidt, Ronald C. Relation between natural radioactivity in sediment and potential heavy mineral enrichment on the Washington continental shelf. M, 1975, Oregon State University. 64 p.

Scheier, Nicholas William. Stochastic, temporal influences on contaminant transport in shallow, phreatic aquifers. D, 1983, University of Waterloo. 114 p.

Scheiern, Milton Ralph. A study of species of the ostracod genus Dizygopleura from the Middle Devonian Traverse Group of Michigan. M, 1953, University of Michigan.

Scheihing, Mark H. A paleoenvironmental analysis of the Shumway Cyclothem (Virgilian), Effingham County, Illinois. M, 1978, University of Illinois, Urbana. 184 p.

Scheihing, Mark Henry. Aspects of land plant biostratonomy in Upper Carboniferous deltas of Euramerica. D, 1982, University of Pennsylvania. 375 p.

Scheimer, James Francis. Experimental study of seismic scattering by a penny-shaped crack. D, 1978, Massachusetts Institute of Technology. 155 p.

Scheiner, Jonathan Edward. Crustal attenuation and velocity structure at the San Andreas fault zone in Central California. D, 1987, University of California, Berkeley. 206 p.

Scheldt, John Christian. Petrology of the Catahoula sandstones in East Texas. M, 1976, University of New Orleans. 58 p.

Scheliga, John Thomas, Jr. Geology and water resources of Warner Basin, San Diego County, CA. M, 1963, University of Southern California.

Schell, Bill Joseph. The stratigraphy and depositional history of the Verdigris-Higginsville interval in northeastern Oklahoma. M, 1955, University of Tulsa. 223 p.

Schell, Elmer Morris. Geology of the Marietta Mine and adjacent area, Elkhorn Mountains, Broadwater County, Montana. M, 1961, University of Montana. 61 p.

Schell, Roy T., III. Geologic and petrographic study of the arenaceous facies of the Batesville Formation in Independence County, Arkansas. M, 1971, Northeast Louisiana University.

Schell, William Willkomm. Foraminifera of the Eagle Ford Shale in the type area, Dallas and Tarrant counties. M, 1952, Southern Methodist University. 86 p.

Schellenberg, S. L. Groundwater flow and nonreactive tracer motion in heterogeneous statistically anisotropic porous media. M, 1987, University of Waterloo. 113 p.

Schellhorn, Robert Wayne. Bouguer gravity anomalies and crustal structure of northern Mexico. M, 1987, University of Texas at Dallas. 167 p.

Schellinger, David Kenneth. Curie depth determinations in the High Plateaus, Utah. M, 1972, University of Utah. 176 p.

Schemehorn, Neil R. Sedimentation study of New Albany Shale, Rockford Limestone, and New Providence Shale in Indiana. M, 1956, Indiana University, Bloomington. 39 p.

Schemel, Mart P. Small spore assemblages of mid-Pennsylvanian coals of West Virginia. D, 1957, West Virginia University.

Schenck, B. E. High precision geodesy applied to crustal studies. M, 1978, University of Hawaii.

Schenck, Barbara Jane. A mineralogical and petrographical study of the pegmatites in the vicinity of Lithia, Massachusetts. M, 1953, Smith College. 112 p.

Schenck, Hubert G. Marine Oligocene of Oregon. D, 1926, University of California, Berkeley. 370 p.

Schenck, Hubert Gregory. A preliminary report of the geology of the Eugene Quadrangle, Lane and Linn counties, Oregon. M, 1923, University of Oregon. 104 p.

Schenewerk, Philip Andrew. The accuracy of pulsed neutron capture logs for residual oil saturation determinations. D, 1981, University of Oklahoma. 95 p.

Schenk, Christopher Joseph. Analysis of stratification produced by eolian sand ripples. M, 1982, University of Michigan.

Schenk, Edward T. The stratigraphy and paleontology of the Triassic of the Suplee region of central Oregon. M, 1931, University of Oregon. 53 p.

Schenk, Paul Edward. The environment of cyclic sedimentation and the paleoecology of the Altamont Formation (Desmoinesian) of Iowa, Missouri, Kansas and northeastern Oklahoma. D, 1963, University of Wisconsin-Madison. 136 p.

Schenk, Paul Edward. The Gowganda (Precambrian) sediments at the south end of Timagami Lake, Ontario. M, 1961, University of Wisconsin-Madison.

Schenk, Paul M. The crustal tectonics and history of Europa; a structural, morphological, and comparative analysis. M, 1983, Northern Illinois University. 94 p.

Schenk, Paul Michael. Impact craters on icy satellites as probes of stratigraphy, tectonic history, impactor sources and crustal properties. D, 1988, Washington University. 208 p.

Schenker, Albert Rudolph, Jr. Particle-size distribution of late Cenozoic gravels in an arid region piedmont, Gila Mountains, Arizona. M, 1977, University of Arizona. 118 p.

Schenning, James W. Origin of hummocky glaciogenic sediment near Burlington, southeastern Wisconsin. M, 1987, Northern Illinois University. 146 p.

Schepis, Eugene Louis. Geology of eastern Douglas County, Georgia. M, 1952, Emory University. 56 p.

Scherb, Ivan Victor. An investigation of the Mill Creek earthquake of October, 1935. M, 1936, California Institute of Technology. 25 p.

Scherer, Oliver J. The relation of precipitation to ground water replenishment in Nebraska. M, 1937, University of Nebraska, Lincoln.

Scherer, Wolfgang. A mathematical model for the differential subsidence of intra-cratonic basins. D, 1973, Northwestern University.

Scherer, Wolfgang. Application of Markov chains to cyclical sedimentation in the Officina Formation (Oligocene), eastern Venezuela. M, 1968, Northwestern University.

Scherffius, William Edward. Zinc-lead-fluorite mineralization of the Shenandoah valley, north-central Virgina. M, 1969, Cornell University.

Scherkenbach, Daryl A. Potassium and rubidium metasomatism related to mineralization at the Julcani District, Peru. M, 1978, Michigan Technological University. 68 p.

Scherkenbach, Daryl Andrew. Geologic, mineralogic, fluid inclusion and geochemical studies of the mineralized breccias at Cumobabi, Sonora, Mexico. D, 1982, University of Minnesota, Minneapolis. 240 p.

Schern, Edward Michael. The occurrence of copper in clays in altered plagioclase phenocrysts. M, 1975, Arizona State University. 66 p.

Scherzer, Howard Jay. Glacial stratigraphy of St. Clair County, Michigan. M, 1978, University of Michigan.

Scherzer, Jolie L. Subsurface structure and stratigraphy of Rochester, N.Y. M, 1983, University of Rochester.

Schetter, William Cameron. The Precambrian surface of Idaho, Montana, North Dakota, South Dakota, and Wyoming. M, 1962, University of Oregon. 113 p.

Schettig, William J. Petrology and structure of Precambrian crystalline rocks and stratigraphy of post-Precambrian sedimentary rocks of the Upper Grace Creek area, Larimer County, Colorado. M, 1974, Colorado State University. 115 p.

Scheu, Steven Ray. A gravity and magnetic investigation east of Preston, Idaho. M, 1985, Idaho State University. 76 p.

Scheubel, Frank R. The geology and mineralization of the San Martin de Bolanos mining district, Jalisco, Mexico. M, 1983, University of Texas at El Paso. 173 p.

Scheuer, Tim Ellis. The recovery of subsurface reflectivity and impedance structure from reflection seismograms. M, 1981, University of British Columbia.

Scheufler, John H. Areal and structural geology of Beartooth Butte, Wyoming. M, 1954, Wayne State University. 26 p.

Scheufler, Lowell W. The geology of the Darrtown-McGonigle area, Butler County, Ohio. M, 1956, Miami University (Ohio). 58 p.

Schewe, John H. The variation of the angle of repose of loose granular materials as a function of grain density, roundness, and sphericity. M, 1979, Louisiana State University.

Schey, N. D. Penecontemporaneous, conjugate fold zones in McCree River Formation, Wyoming. M, 1976, University of Minnesota, Minneapolis.

Schiappa, Christopher. Petrology and geochemistry of the Coronaca Pluton, Greenwood County, South Carolina. M, 1988, East Carolina University. 73 p.

Schick, Charles W. Unconformity-type uranium mineralization at the Groveland Iron Mine, Dickinson County, Michigan. M, 1985, Bowling Green State University. 89 p.

Schick, Robert Bryant. Geology of the Morgan-Henefer area, Morgan and Summit counties, Utah. M, 1955, University of Utah. 54 p.

Schiebel, Lawrence Glenn. Lithology and depositional environments of a Cambrian section at Camp Creek, Montana. M, 1981, University of Idaho. 104 p.

Schieber, Juergen. The relationship between basin evolution and genesis of stratiform sulfide horizons in mid-Proterozoic sediments of central Montana (Belt Supergroup). D, 1985, University of Oregon. 811 p.

Schiebout, Judith Ann. Sedimentology of the Paleocene Black Peaks Formation, western Tornillo Flat, Big Bend National Park, Texas. M, 1970, University of Texas, Austin.

Schiebout, Judith Ann. Vertebrate paleontology and paleoecology of Paleocene Black Peaks Formation, Big Bend National Park, Texas. D, 1973, University of Texas, Austin. 264 p.

Schieck, David Ernest. The x-ray study of the clay minerals of sediments from two New York swamps. M, 1972, University of Michigan.

Schiefelbein, Debra Ruth Jessica. Carbonate lithofacies and depositional environments of the Upper Ordovician-Lower Silurian Hanson Creek Formation, central Nevada. M, 1984, University of Wisconsin-Milwaukee. 129 p.

Schiel, Kathryn A. The Dewey Lake Formation; end stage deposit of a peripheral foreland basin. M, 1988, University of Texas at El Paso.

Schiering, Mark Harrison. Petrology and geophysics of the Dark Ridge ultramafic body, Jackson County, North Carolina. M, 1979, Kent State University, Kent. 42 p.

Schierling, Eldon J. Origin of a structurally positive area in Gove County, Kansas. M, 1968, Wichita State University. 41 p.

Schierow, Linda-Jo. An evaluation of the Great Lakes nearshore index of environmental quality. D, 1983, University of Wisconsin-Madison. 439 p.

Schiesser, Clarence Frederick. The stratigraphy and the areal geology of the Stoddard Quadrangle, Wisconsin. M, 1948, University of Wisconsin-Madison. 174 p.

Schiferle, Jane Carol. Hydrologic conditions of floodplain sedimentation along the upper Missisquoi River, northern Vermont. M, 1988, Wright State University. 74 p.

Schiff, Anshel Judd. Some problems in engineering seismology. D, 1967, Purdue University. 96 p.

Schiff, Sherry Line. Acid neutralization in sediments of freshwater lakes. D, 1986, Columbia University, Teachers College. 354 p.

Schiffelbein, Paul Arthur. Stable isotope systematics in Pleistocene deep-sea sediment records. D, 1984, University of California, San Diego. 245 p.

Schiffman, Arnold. A study of the swash-surf energy system. M, 1965, University of Southern California.

Schiffman, P. Synthesis and stability relations of pumpellyite. D, 1978, Stanford University. 205 p.

Schiffman, Robert A. Development of activation analysis method for Hg determination in ppb range. M, 1973, New Mexico Institute of Mining and Technology.

Schiffries, Craig Mason. Magmatic and hydrothermal processes in layered intrusions. D, 1988, Harvard University. 257 p.

Schildt, Timothy Allen. Paleoecology and paleoenvironment of the Boggy Formation in the Franks Graben. M, 1981, Texas Christian University. 157 p.

Schile, C. A. Sedimentology of the "El Gallo Formation" (upper Cretaceous), El Rosario, Baja California, Mexico. M, 1974, San Diego State University.

Schile, Charles A. Sedimentology of the El Gallo Formation, El Rosario, Baja California, Mexico. M, 1973, San Diego State University.

Schilizzi, Paul P. G. Monitoring and prediction of surface movements above underground mines in the Eastern U.S. coalfields. D, 1987, Virginia Polytechnic Institute and State University. 255 p.

Schiller, Edward Alexander. Mineralogy and geology of the Guysborough Area, Nova Scotia, Canada. D, 1963, University of Utah. 199 p.

Schiller, Edward Alexander. Petrography and petrology of the andalusite schists in the Port Felix area, Nova Scotia, Canada. M, 1959, Michigan State University. 50 p.

Schillereff, Herbert Scott. Relationship among rock groups within and beneath the Humber Arm Allochthon at Fox Island River, western Newfoundland.

M, 1980, Memorial University of Newfoundland. 165 p.

Schillhahn, Ernest O. A restudy of the Hillsboro Sandstone of Highland County. M, 1929, Ohio State University.

Schilling, Frederick A., Jr. The Upper Cretaceous stratigraphy of the Pacheco Pass Quadrangle, California. D, 1962, Stanford University. 153 p.

Schilling, Jean-Guy E. Rare earth fractionation in Hawaiian volcanic rocks. D, 1966, Massachusetts Institute of Technology. 403 p.

Schilling, John Harold. The Questa molybdenum mine, Taos County, New Mexico. M, 1952, New Mexico Institute of Mining and Technology. 73 p.

Schilling, Karl H. New species from the San Lorenzo and Monterey series of the Emigdio region, California. M, 1918, University of California, Berkeley. 30 p.

Schilling, Keith Edwin. Aspects of pre-Illinoian glaciation in Southwest Iowa. M, 1988, Iowa State University of Science and Technology. 128 p.

Schilly, Michael McKernan. Interpretation of crustal seismic refraction and reflection profiles from Yellowstone and the eastern Snake River plain. M, 1979, University of Utah. 170 p.

Schilt, Frank Steve. Studies of continental and oceanic crust from seismic, geodetic, and magnetic observations. D, 1981, Cornell University. 206 p.

Schiltz, Debra Lynn. An evaluation of the uranium potential of the Catahoula Formation in parts of Jasper, Tyler, Polk and Angelina counties. M, 1981, Stephen F. Austin State University. 127 p.

Schimann, Karl. Geology of the Wakeham Bay area, eastern end of the Cape Smith Belt, New Quebec. D, 1978, University of Alberta. 426 p.

Schimel, David Steven. Nutrient and organic matter dynamics in grasslands; effects of fire and erosion. D, 1982, Colorado State University. 80 p.

Schimmelmann, Arndt. Stable isotopic studies on chitin. D, 1985, University of California, Los Angeles. 223 p.

Schimschal, Ulrich. An application of the deconvolution technique to model gravitational fields. M, 1972, Colorado School of Mines. 40 p.

Schimschal, Ulrich. Frequency domain and the domain solutions in the magneto-telluric method. D, 1977, Colorado School of Mines. 94 p.

Schindel, David Edward. Paleoecology and systematics of the Pennsylvanian Gastropoda of north-central Texas; an investigation into biotic responses to changing physical conditions. D, 1979, Harvard University.

Schinderle, Denis W. Comparative analysis of regional joint patterns with lineaments interpreted from Landsat multispectral imagery; northeastern Indiana. M, 1983, Indiana State University. 122 p.

Schindler, Jack Frederic. The geology of a portion of the Augusta Quadrangle, St. Charles County, Missouri. M, 1951, University of Missouri, Columbia.

Schindler, John Henry. Study of the drainage problems of a certain mine. M, 1933, University of Pittsburgh.

Schindler, John Norman. Rhenium and osmium in some Canadian ores by neutron activation analysis. D, 1975, McMaster University. 254 p.

Schindler, Kris L. Resistivity log-sonic log cross plots applied to subsurface carbonate facies analysis (Jeffersonville and North Vernon limestones, northern Clay Co., Indiana). M, 1982, Ball State University. 86 p.

Schindler, Norman R. Geology of the Waite-Ackerman-Montgomery Property, Duprat and Dufresnoy townships, Quebec. M, 1933, McGill University.

Schindler, Norman R. Igneous rocks of Duprat Lake and Rouyn Lake areas, Quebec. D, 1934, McGill University.

Schindler, Stanley Fred. Geology of the White Lake Hills, Utah. M, 1952, Brigham Young University. 66 p.

Schink, David Regier. The measurement of dissolved silicon-32 in sea water. D, 1963, University of California, San Diego.

Schink, Ernest Allen. Geology of the Red-Chris porphyry copper deposit, northwestern British Columbia. M, 1977, Queen's University. 211 p.

Schipper, Louis B., III. Behavior of major and trace elements in contact aureoles, northeastern Washington. M, 1979, Eastern Washington University. 65 p.

Schipper, Mark Raymond. A ground-water management model for the Washita River alluvial aquifer in Roger Mills and Custer counties, Oklahoma. M, 1983, Oklahoma State University. 147 p.

Schipper, Warren Bailey. The Tendoy Copper Queen Mine. M, 1955, University of Idaho. 38 p.

Schirmer, Tad William. Sequential thrusting beneath the Willard thrust fault, Wasatch Mountains, Ogden, Utah. M, 1985, Utah State University. 199 p.

Schissel, Donald J. Fenitization in the Heather Ann Complex, Thompson Falls, Montana. M, 1981, University of Montana. 77 p.

Schlaefer, Jill T. The surface geology of the Honor Rancho area of the east Ventura and western Soledad basins, California. M, 1978, Ohio University, Athens. 86 p.

Schlaikjer, Erich M. A new species of horse from the White River Formation of South Dakota. M, 1931, Columbia University, Teachers College.

Schlaikjer, Erich M. Contributions to the stratigraphy and paleontology of Goshen Hole area, Wyoming. D, 1935, Columbia University, Teachers College.

Schlain, Mildred Rachel. A description of the postcranial anatomy of a Miocene amphicyonid; Carnivora, Mammalia. M, 1980, University of Massachusetts. 233 p.

Schlaks, Francine D. Abundances of trace elements in iron disulfides in the Benevides roll-front uranium deposit, Webb County, Texas. M, 1978, Colorado School of Mines. 90 p.

Schlanger, Seymour O. Petrology of the limestones of Guam. D, 1959, The Johns Hopkins University.

Schlanger, Seymour Oscar. Stratigraphy and petrology of the Vincentown Formation in New Jersey. M, 1951, Rutgers, The State University, New Brunswick. 117 p.

Schlaudt, Charles McCammon. Phase equilibria and crystal chemistry of cement and refractory phases in the system $CaO-MgO-Al_2O_3-Fe_2O_3-CaF_2-P_2O_5-SiO_2$. D, 1964, Pennsylvania State University, University Park. 147 p.

Schlaudt, Charles McCammon. Temperatures of mineral associations with quartz from liquid inclusions. M, 1960, University of Texas, Austin.

Schlee, John Stevens. Geology of the Mutau Flat area, Ventura County, California. M, 1952, University of California, Los Angeles.

Schlee, John Stevens. Sedimentological analysis of the upland gravels of southern Maryland. D, 1956, The Johns Hopkins University.

Schlegel, Helen E. Stenstrom. A detailed paleontologic study of the microfauna of the Vanport Interval (Pennsylvanian), in southern Stark and northern Tuscarawas counties, Ohio. M, 1981, University of Akron. 98 p.

Schleh, Edward E. Toroweap and Kaibab Formations in a part of Parashant Canyon, Mohave County, Arizona. M, 1960, University of Kansas.

Schleh, Edward Eugene. Upper Devonian to Middle Pennsylvanian discontinuity-bounded sequences in a part of the Cordilleran region. D, 1963, University of Washington. 128 p.

Schlehuber, Michael Jude. Use of water level and hydrochemistry to map groundwater flow and subsurface geology in San Jacinto Valley, California. M, 1987, University of California, Riverside. 193 p.

Schleicher, David Lawrence. Emplacement mechanism of the Miraleste Tuff (Middle Miocene), Palos Verdes Hills, California. D, 1965, Pennsylvania State University, University Park. 72 p.

Schleifer, Stanley. Instrumentation for a geological model study. M, 1970, Brooklyn College (CUNY).

Schleiss, Wolfgang A. A study of vein mineralization and wall rock alteration at the Delaware Mine, Keweenaw County, Michigan. M, 1986, Michigan Technological University. 86 p.

Schlemm, L. G. W. Geology of the Lake Rowan Property (Red Lake District, Ontario). M, 1939, McGill University.

Schlemmer, Frederick C. Concentration of particulate matter in the eastern Gulf of Mexico; an indicator of surface circulation patterns. M, 1971, University of South Florida, St. Petersburg. 82 p.

Schlenker, J. L. A phenomenological treatment of thermal expansion in crystals of the lower symmetry classes and the crystal structures of $CaCoSi_2O_6$ and $CaNiSi_2O_6$. D, 1976, Virginia Polytechnic Institute and State University. 204 p.

Schlesinger, Benjamin. Development of a procedure for forecasting long-range environmental impacts. D, 1975, Stanford University. 150 p.

Schlicker, Herbert G. Columbia River Basalt in relation to stratigraphy of Northwest Oregon. M, 1954, Oregon State University. 93 p.

Schlicten, Harold Carl von. Geologic evidence on the antiquity of man in America. M, 1941, University of Cincinnati. 62 p.

Schlinger, Charles Martin. The magnetic petrology of the deep crust and the interpretation of regional magnetic anomalies. D, 1983, The Johns Hopkins University. 256 p.

Schlipp, Wayne Richard. Paleoenvironmental analysis based on the microfossils of the Lower Permian Elephant Canyon Formation, Canyonlands National Park, Utah. M, 1988, University of Wisconsin-Milwaukee. 180 p.

Schlocker, Julius. Geology of the Wilbur Springs District, California. M, 1941, University of California, Berkeley. 49 p.

Schloderer, John Peter. Geology and kinematic analysis of deformation in the Redington Pass area, Pima County, Arizona. M, 1974, University of Arizona.

Schloerb, Frederic Peter. Radio interferometric investigations of Saturn's rings at 3.71- and 1.30-cm wavelengths. D, 1978, California Institute of Technology. 215 p.

Schlorholtz, Michael William. Terrestrial heat flow in southeastern Ohio. M, 1979, Kent State University, Kent. 77 p.

Schloss, Jeffrey A. Changes in the potentiometric surface of the deep aquifer in the Joplin area, Missouri, 1900-present. M, 1986, Southwest Missouri State University.

Schlosser, Isaac Joseph. Effects of perturbations by agricultural land use on structure and function of stream ecosystems. D, 1981, University of Illinois, Urbana. 222 p.

Schlue, John William. Anisotropy of the upper mantle of the Pacific Basin. D, 1975, University of California, Los Angeles. 60 p.

Schlueter, James C. Geology of the upper Ojai-Timber Canyon area, Ventura County, California. M, 1976, Ohio University, Athens. 76 p.

Schluger, Paul Randolph. Perry Formation, St. Andrews Peninsula, St. Andrews, New Brunswick, Canada. M, 197?, University of Pennsylvania.

Schluger, Paul Randolph. Sedimentology of the Perry Formation (Upper Devonian), New Brunswick, Canada, and Maine, U.S.A. D, 1972, University of Illinois, Urbana. 152 p.

Schmachtenberg, William F. Geological implications of a study of an Upper Cretaceous epicontinental seaway fauna. D, 1983, University of Chicago. 245 p.

Schmalfuss, Bradford Roger. Post-Linda Vista faulting in a portion of the La Mesa and National City quadrangles, San Diego, California. M, 1981, San Diego State University.

Schmaltz, Lloyd John. Pebble lithology of Nebraskan and Kansas tills in north-central Missouri. D, 1959, University of Missouri, Columbia. 118 p.

Schmaltz, Lloyd John. Physiographic history of some Tertiary conglomerates along part of the northeast flank of the Wind River Mountains, Wyoming. M, 1956, University of Missouri, Columbia.

Schmalz, J. P., Jr. A statistical study of core analysis data. M, 1950, University of Oklahoma. 75 p.

Schmalz, Robert Fowler. A technique for quantitative modal analysis by X-ray diffraction, and its application to modern sediments of the Peru-Chile Trench. D, 1958, Harvard University.

Schmedeman, Otto C. Geology of the Isabella ore-body of Isabella, Tennessee. M, 1931, University of Wisconsin-Madison.

Schmedeman, Otto C. Interrelated problems of volcanic activity and ore genesis. D, 1937, Harvard University.

Schmedtje, Lucille C. Marine planation of southwestern New England. M, 1936, Smith College. 89 p.

Schmela, Ronald J. Geophysical and geological analysis of a fault-line linearity in the lower Clackamas River area (Clackamas County, Oregon). M, 1971, Portland State University. 113 p.

Schmid, Karl J. The geology and vein mineralization of Cedar Valley, east central Hualapai Mountains, Mohave County, Arizona. M, 1983, Northern Arizona University. 136 p.

Schmidley, Eric B. The sedimentology, paleogeography and tectonic setting of the Pennsylvanian Massillon Sandstone in east-central Ohio. M, 1987, University of Akron. 193 p.

Schmidt, Alan James. The nitrogen-phosphorus hydrochemistry of Las Vegas Wash. M, 1979, University of Illinois, Urbana. 78 p.

Schmidt, Allan Thomas. Petrology of the Lilesville granite batholith aureole, Anson and Richmond counties, North Carolina. M, 1972, University of Florida. 113 p.

Schmidt, Bennetta Lee. A study of X-ray diffraction and thermodynamic properties of synthetic, binary nepheline-kalsilite crystalline solutions. M, 1982, Pennsylvania State University, University Park. 101 p.

Schmidt, Birger. Settlements and ground movements associated with tunneling in soils. D, 1969, University of Illinois, Urbana. 234 p.

Schmidt, Bruno M. A study of a thousand-foot core from the Tonkawa oil field, Oklahoma. M, 1925, Yale University.

Schmidt, Christopher John. An analysis of folding and faulting in the northern Tobacco Root Mountains, Southwest Montana. D, 1975, Indiana University, Bloomington. 480 p.

Schmidt, Dwight Lyman. Petrography of Idaho Batholith, Valley Co., Idaho. M, 1957, University of Washington. 110 p.

Schmidt, Dwight Lyman. Quaternary geology of Bellevue area in Blaine and Camas counties, Idaho. D, 1961, University of Washington. 135 p.

Schmidt, Earl Albert. An investigation of dimensional grain orientation in a sandstone. M, 1958, Michigan State University. 46 p.

Schmidt, Eberhard A. Geology of the Mineral Mountain Quadrangle, Pinal County, Arizona. M, 1967, University of Arizona.

Schmidt, Eberhard Adalbert. A structural investigation of the northern Tortilla Mountains, Pinal County, Arizona. D, 1971, University of Arizona. 370 p.

Schmidt, Edward John. Water quality impact of non-point source contaminants in small tidal rivers. D, 1981, University of New Hampshire. 460 p.

Schmidt, Ehud Jeruham. Nuclear magnetic resonance and the broadband acoustic response of porous rocks. D, 1987, Stanford University. 198 p.

Schmidt, Eugene Karl. Plate tectonics, volcanic petrology, and ore formation in the Santa Rosalia area, Baja California, Mexico. M, 1975, University of Arizona.

Schmidt, Gene W. Interstitial water composition and geochemistry of deep Gulf Coast shales and sands. M, 1971, University of Tulsa. 121 p.

Schmidt, Gerry Lee. Alternative structure and process models in earth sciences education. D, 1975, University of South Carolina.

Schmidt, Gerry Lee. The Price Formation (Mississippian), southwestern Virginia and southeastern West Virginia. M, 1973, University of South Carolina.

Schmidt, Gregory Thomas. Geology of the northern Sierra el Encinals, Sonora, Mexico. M, 1978, Northern Arizona University. 80 p.

Schmidt, H. G. The relations of amplitudes of conventional seismic reflection records to the stratigraphy of the Shaunavon Formation (Jurassic) of southwestern Saskatchewan(Canada). M, 1967, University of Manitoba.

Schmidt, Harold A. The convergence of the Altamont and Pawnee limestones in Northeast Oklahoma. M, 1959, University of Kansas. 115 p.

Schmidt, James Scott. Geophysical basis and cartography of the complete Bouguer gravity anomaly map of Arizona. M, 1976, University of Arizona.

Schmidt, Jeanine Marie. Geology and geochemistry of the Arctic Prospect, Ambler District, Alaska. D, 1984, Stanford University. 314 p.

Schmidt, Jeanine Marie. Volcanogenic massive sulfides at Campo Seco (Calaveras County), California. M, 1978, Stanford University. 124 p.

Schmidt, John Christian, III. Geomorphology of alluvial sand deposits, Colorado River, Grand Canyon National Park, Arizona. D, 1987, The Johns Hopkins University. 217 p.

Schmidt, Kenneth D. The distribution of boron in the ground water of the Arvin-Caliente creek area, Kern County, California. M, 1969, University of Arizona.

Schmidt, Kenneth Dale. The distribution of nitrate in groundwater in the Fresno–Clovis metropolitan area, San Joaquin Valley, California. D, 1971, University of Arizona. 348 p.

Schmidt, Lane T. A comparison of classification techniques using Landsat thematic mapper and multi-spectral scanner data, for landcover classification of a portion of Calloway and Graves counties, Kentucky. M, 1984, Murray State University. 57 p.

Schmidt, Marcia. Sedimentology, petrology, and stratigraphy of the Quartz Arenite Member of the Harkless Formation, White-Inyo mountains, California. M, 1977, University of California, Santa Barbara.

Schmidt, Mark F. A geothermal study of Ohio with heat flow measurement in Norton Township, Summit County, Ohio. M, 1984, Kent State University, Kent. 75 p.

Schmidt, Mark Thomas. Petrology of metasedimentary rocks from Taylor Valley, South Victoria Land, Antarctica. M, 1986, Kent State University, Kent. 110 p.

Schmidt, Martin Leo. A geotechnical and hydrogeological investigation of waste-water treatment sludges and river sand to be used as sanitary landfill caps. D, 1985, Kent State University, Kent. 284 p.

Schmidt, Otto Mackenty. San Ramon Sandstone (Oligocene) in the Pacheco Syncline, California. M, 1958, Stanford University. 53 p.

Schmidt, Paul Gerhard. The geology of the Jarilla Mountains, Otero County, New Mexico. M, 1962, University of Minnesota, Minneapolis. 134 p.

Schmidt, Peter B. Geology of the State Bride area (Eagle County), Colorado. M, 1961, University of Colorado.

Schmidt, Richard Arthur. Microscopic extraterrestrial particles; Part I, Antarctic Peninsula traverse, 1961-62. D, 1963, University of Wisconsin-Madison. 193 p.

Schmidt, Richard Arthur. Temperatures of mineral formation in the Miami-Picher zinc-lead district (Oklahoma-Kansas) as indicated by liquid inclusions. M, 1959, University of Wisconsin-Madison.

Schmidt, Richard Carl. Adsorption of copper, lead, and zinc on some rock forming minerals and its effect on lake sediments. D, 1956, McGill University.

Schmidt, Richard Carl. Dispersion of copper, lead and zinc from mineralized zones in an area of moderate relief as indicated by soils and plants. M, 1955, McGill University.

Schmidt, Robert G. The subsurface geology of Freedom Township in the Baraboo iron-bearing district of Wisconsin. M, 1951, University of Wisconsin-Madison.

Schmidt, Robert G. Volcanic rocks of Saipan, Mariana Islands (Pacific Ocean). D, 1953, Harvard University.

Schmidt, Ronald G. Joint patterns in relation to local tear faults in the central foothills of Alberta. M, 1955, Columbia University, Teachers College.

Schmidt, Ronald G. Joint patterns in relation to regional and local structure in the central foothills belt of the Rocky Mountains of Alberta. D, 1957, University of Cincinnati. 185 p.

Schmidt, Ronald Roy. Planktonic foraminifera from the lower Tertiary of California. D, 1970, University of California, Los Angeles. 364 p.

Schmidt, Ruth A. M. Miocene Ostracoda from Yorktown Formation, Virginia. M, 1939, Columbia University, Teachers College.

Schmidt, Ruth A. M. Ostracoda from the Upper Cretaceous and lower Eocene of Maryland, Delaware, and Virginia. D, 1948, Columbia University, Teachers College.

Schmidt, Stephen L. Geology of the south part of Chuckanut Mountain, a structural and petrologic study. M, 1972, Western Washington University. 51 p.

Schmidt, Thomas G. Precambrian metavolcanic rocks and associated volcanogenic mineral deposits of the Fletcher Park and Green Mountains areas, Sierra Madre, Wyoming. M, 1983, University of Wyoming. 113 p.

Schmidt, Timothy Germer. Geology of the Embar Field, Andrews County, Texas. M, 1947, University of Texas, Austin.

Schmidt, Victor Edward. Varves in the Finger Lakes region of New York State. D, 1946, Cornell University.

Schmidt, Walter. A paleoenvironmental study of the Twiggs Clay, upper Eocene, of Georgia using fossil microorganisms. M, 1977, Florida State University.

Schmidt, Walter. Neogene stratigraphy and geologic history, Apalachicola Embayment, Florida. D, 1987, Florida State University. 250 p.

Schmidt, William Jay. Structure of the northern Sierra Nevada, California. D, 1985, Rice University. 272 p.

Schmidt, William Jay. Structure of the Oxbow area, Oregon and Idaho. M, 1980, Rice University. 61 p.

Schmidt, Winfried. Palaeobiology and carbonate petrology of part of the Hughes Creek Shale in northeastern Kansas. M, 1974, Kansas State University. 388 p.

Schmidt-Fonseca, Susan. An evaluation of the Albemarle County Runoff Control Ordinance. M, 1980, University of Virginia. 67 p.

Schmidtke, Klaus-Dieter. Determination of performance measure uncertainty in groundwater. D, 1985, University of Waterloo. 174 p.

Schmieg, Robert Eugene. Palynology of the Cabaniss coals in Henry County, Missouri. M, 1959, University of Missouri, Columbia.

Schmierer, Kurt E. Cretaceous remagnetization of the Cambro-Ordovician Bowers Supergroup, northern Victoria Land, Antarctica. M, 1987, Western Washington University. 122 p.

Schmiermund, Ronald Lee. Geology and geochemistry of uranium deposits near Penn Haven Junction, Carbon County, Pennsylvania. M, 1977, Pennsylvania State University, University Park. 153 p.

Schmincke, Hans-Ulrich. Petrology, paleocurrents and stratigraphy of the Ellenburg Formation and interbedded flows of the Yakima Basalt in south-central Washington. D, 1964, The Johns Hopkins University. 532 p.

Schmitt, Clifford T. Early cementation of periplatform sediment in Northwest Providence Channel, Bahamas. M, 1987, Miami University (Ohio). 56 p.

Schmitt, Dennis A. Topographic shadow analysis for geologic interpretation, south-central Idaho. M, 1978, University of Wisconsin-Milwaukee.

Schmitt, Douglas Ray. I, Applications of double-exposure holography to the measurement of in situ stress and the elastic moduli of rock from boreholes; II, Shock temperature measurements in fused quartz and crystalline NaCl to 35 GPa. D, 1987, California Institute of Technology. 187 p.

Schmitt, George Theodore. A petrographic investigation of the relationship of deposition of sediments in a group of eskers related to the Charlotte till plain (Michigan). M, 1949, Michigan State University. 52 p.

Schmitt, George Theodore. Regional stratigraphic analysis of the Middle and Upper marine Jurassic in the Northern Rocky Mountains-Great Plains. D, 1952, Northwestern University.

Schmitt, Harrison Ashley. An experimental investigation as to the possible commercial production of potash from Minnesota rocks. M, 1922, University of Minnesota, Minneapolis. 30 p.

Schmitt, Harrison Ashley. Contributions to the geology and ore deposits of southern Chihuahua and northern Durango (Mexico). D, 1926, University of Minnesota, Minneapolis. 91 p.

Schmitt, Harrison Hagan. Petrology and structure of the Eiksundsaal eclogite complex, Hareidland, Sunnmore, Norway. D, 1964, Harvard University.

Schmitt, James G. Description and interpretation of silicified skeletal material from the Park City Formation (Permian) of Wyoming. M, 1979, University of Wyoming. 84 p.

Schmitt, James G. Origin and sedimentary tectonics of Upper Cretaceous Frontier Formation conglomerates in the Wyoming-Idaho-Utah thrust belt. D, 1982, University of Wyoming. 239 p.

Schmitt, Leonard J., Jr. A lithic description of the Cutler Formation (Permian) in Big Indian Wash, Utah. M, 1964, Columbia University, Teachers College.

Schmitt, Leonard Joseph, Jr. Uranium and copper mineralization in the Big Indian Wash-Lisbon Valley mining area, southeastern Utah. D, 1968, Columbia University. 193 p.

Schmitt, Thomas J. Upper Mesozoic magnetostratigraphy. D, 1976, University of South Carolina. 65 p.

Schmitt, V. L. Petrology of the arcuate hornblende gabbro dike complex, Shadow Mountains, western San Bernardino County, southern California. M, 1975, University of Southern California.

Schmittle, John M. Brittle deformation and cataclasis on the southwest flank of the Llano Uplift, Mason County, Texas. M, 1987, Texas A&M University. 141 p.

Schmitz, Christopher. Geology of the Black Pearl Mine area, Yavapai County, Arizona. M, 1987, Arizona State University. 155 p.

Schmitz, Emmett Richard. Stream piracy and glacial diversion of Little Missouri River, North Dakota. M, 1955, University of North Dakota. 39 p.

Schmitz, Larry. The general geology of Ha Ha Tonka State Park and surrounding area near Camdenton, Missouri. M, 1982, University of Missouri, Columbia.

Schmitz, Richard Joseph. The geology of the Festus, Missouri and Selma, Illinois-Missouri quadrangles. M, 1966, University of Missouri, Rolla.

Schmoker, James William. Interpretation of aeromagnetic and gravity data from the San Francisco mountains vicinity, southwestern Utah. D, 1969, Virginia Polytechnic Institute and State University. 236 p.

Schmoll, H. R. Upper Mississippian quartzites in the White Pine Range, east central Nevada. M, 1955, Columbia University, Teachers College.

Schmus, William Randolph. The geochronology of the Blind River-Bruce mines area, Ontario, Canada. D, 1964, University of California, Los Angeles. 112 p.

Schnabel, Per Borre. Effects of local geology and distance from source of earthquake ground motions. D, 1973, University of California, Berkeley.

Schnabel, Robert W. Petrographic study of the (Precambrian) rocks exposed along Moose River between Lyons Falls and Fowlerville, New York. M, 1955, SUNY at Buffalo.

Schnable, Jon Edwin. The evolution and development of part of the northwest Florida coast. D, 1966, Florida State University.

Schnacke, Arthur W., Jr. Glacial geology and stratigraphy of southeastern Pierce and southwestern Benson counties, North Dakota. M, 1982, University of North Dakota. 140 p.

Schnaible, Dean R. The geology of the southwestern part of the Valsetz Quadrangle, Oregon. M, 1958, University of Oregon. 100 p.

Schnake, Carol J. A thermodynamic appraisal of mineral equilibria in siliceous dolomitic limestones during contact metamorphism. M, 1977, New Mexico Institute of Mining and Technology. 204 p.

Schnake, Carol Jeanne. Calculation of the chemical and thermodynamic consequences of differences between fluid and geostatic pressure in hydrothermal systems. D, 1981, University of California, Berkeley. 89 p.

Schnake, David W. Conditions of formation of the iron-bearing skarns at Lone Mountain, Lincoln County, New Mexico. M, 1977, New Mexico Institute of Mining and Technology. 88 p.

Schnapp, Madeline. Seismic investigation of the San Francisco volcanic field in Arizona. M, 1976, Stanford University.

Schnapp, Michael Gordon. Estimating the Q of low angular order split free oscillations of the Earth. M, 1982, University of Colorado. 74 p.

Schneck, M. C. Post-Horizon A sedimentology of the region between the Sohm and Hatteras abyssal plains. M, 1974, Queens College (CUNY). 99 p.

Schneck, William M. Lithostratigraphy of the McCoy Creek Group and Prospect Mountain Quartzite (upper Proterozoic and Lower Cambrian), Egan and Cherry Creek ranges, White Pine County, Nevada. M, 1986, Eastern Washington University. 109 p.

Schneeberger, Dale. Depositional environment of the Santiago Formation, Santa Ana Mountains, Orange County, California. M, 1984, California State University, Long Beach. 105 p.

Schneeflock, R. D. Possible origins of the Livingston fault zone; Sumter County, Alabama. M, 1972, University of Alabama.

Schneer, Cecil Jack. Polymorphism in one dimension. D, 1954, Cornell University.

Schneider, Allan Frank. Fauna of the Trenton Limestone near Waddle, central Pennsylvania. M, 1951, Pennsylvania State University, University Park. 115 p.

Schneider, Allan Frank. Pleistocene geology of part of central Minnesota. D, 1957, University of Minnesota, Minneapolis. 244 p.

Schneider, Arland David. Numerical analysis of the formation of groundwater mounds in layered media. D, 1976, University of California, Davis. 147 p.

Schneider, Dieter. The complex strain approximation in space and time applied to the kinematical analysis of relative horizontal crustal movements. D, 1982, University of New Brunswick.

Schneider, E. J. Surficial geology of the Grand Junction-Fruita area, Mesa County, Colorado. M, 1975, [Colorado State University].

Schneider, Eric Davis. Downslope and across-slope sedimentation as observed in the westernmost north Atlantic. D, 1970, Columbia University. 301 p.

Schneider, Eric Davis. The sediments of the Caicos Outer Ridge, north of the Bohama Islands, Atlantic Ocean. M, 1965, Columbia University. 42 p.

Schneider, Gary B. Cenozoic geology of the Madison Bluffs area, Gallatin County, Montana. M, 1970, Montana State University. 61 p.

Schneider, Gary Bradley. Petrology of the Pahasapa (Madison) Limestone of the northeastern Black Hills of South Dakota. D, 1973, South Dakota School of Mines & Technology.

Schneider, Harvey I. Trace element, mineralogical, and size distributions in suspended material samples from selected rivers in eastern Kansas. M, 1976, University of Kansas. 81 p.

Schneider, Harvey Ira. Mineral, chemical and textural variations along s-surfaces in selected outcrops of amphibolite, Georgia and Alabama Piedmont. D, 1980, Florida State University. 122 p.

Schneider, Howard John. Paleobiochemistry of algae. D, 1969, [University of Houston]. 190 p.

Schneider, Hyrum. A study of glauconite. D, 1926, University of Wisconsin-Madison.

Schneider, Hyrum. Mid-Tertiary deformation of western North America. M, 1911, University of Wisconsin-Madison.

Schneider, Jay A. Evolutionary ecology of post-Paleozoic crinoids. M, 1988, University of Cincinnati. 185 p.

Schneider, Jill Leslie. Structure, petrology, and genesis of the Jack Flats breccia pipe, Eagle County, Colorado. M, 1980, University of Colorado.

Schneider, John Frederick. Application of Omega navigation to timing and positioning of seismograph arrays. M, 1981, University of Wisconsin-Madison. 118 p.

Schneider, John Frederick. The intermediate-depth microearthquakes of the Bucaramanga Nest, Colombia. D, 1984, University of Wisconsin-Madison. 251 p.

Schneider, Kari J. Applications of photoacoustic microscopy to changes in the thermal properties of selected Lower Kittanning Seam coals with rank. M, 1986, Southern Illinois University, Carbondale. 117 p.

Schneider, Marie Diane. Subsidence, compaction, and thermal history of sediments in the northern North Sea. M, 1982, Massachusetts Institute of Technology. 71 p.

Schneider, Mark Edward. Major element compositions of silicates dissolved in super-critical fluids at mantle conditions; implications for mantle metasomatism. M, 1982, Pennsylvania State University, University Park. 50 p.

Schneider, Michael C. The Hardinsburg Formation, Upper Mississippian, of the Iron oil pool, White County, Illinois. M, 1956, Miami University (Ohio). 47 p.

Schneider, Michael Charles. Early Tertiary continental sediments of central and south-central Utah. D, 1967, Brigham Young University.

Schneider, Nicholas McCord. Sodium in Io's extended atmosphere. D, 1988, University of Arizona. 219 p.

Schneider, Nicholas P. Morphology of the Madison Range fault scarp, Southwest Montana; implications for fault history of segmentation. M, 1985, Miami University (Ohio). 118 p.

Schneider, P. F. Notes on the geology of the Onondaga Valley (New York). M, 1893, Syracuse University.

Schneider, R. C. Petrography of some carbonate rocks at Ferguson and LeGrand, Iowa. M, 1954, Iowa State University of Science and Technology.

Schneider, Raymond H. Debris slides and related flood damage resulting from Hurricane Camille, 19-20 August, and subsequent storm, 5-6 September 1969 in the Spring Creek drainage basin, Greenbrier County, West Virginia. D, 1973, University of Tennessee, Knoxville. 131 p.

Schneider, Raymond H. Microfauna of the Bachelor Formation and associated beds, southwestern Missouri. M, 1961, University of Missouri, Columbia.

Schneider, Richard C. Stratigraphy and depositional environments of the Mississippian rocks, Garnet Range-Bearmouth area, Granite County, Montana. M, 1988, Oregon State University. 186 p.

Schneider, Rita. Analysis of the benthonic foraminiferal thanatocoenoses from piston cores taken in Lac de Tunis, Tunisia. M, 1977, Duke University. 194 p.

Schneider, Robert V. The vertical distribution of uranium, thorium, and potassium in the Canadian Shield, Sudbury, Ontario, Canada. M, 1985, University of Texas at El Paso.

Schneider, Ronald L. Analysis of sedimentary structures in a modern bay environment; Maumee Bay, Ohio. M, 1975, University of Toledo. 99 p.

Schneider, Stephen Jay. A subsurface geologic study of the Bay Saint Elaine oil field in South Terrebonne Parish, Louisiana. M, 1958, Washington University. 33 p.

Schneider, Thomas. Geology of Dillahunty Quadrangle, Culberson County, Texas. M, 1951, University of Texas, Austin.

Schneider, William A. Investigation of the radiative contribution to the thermal conductivity in sodium chloride, dunite, and fused quartz. D, 1961, Massachusetts Institute of Technology. 177 p.

Schneider, William A., Jr. Dipmeter analysis by a Monte Carlo technique. M, 1987, Colorado School of Mines. 112 p.

Schneider, William D. Petrographic sources of the Brookline Member of the Roxbury Conglomerate. M, 1975, Boston University. 56 p.

Schneider, William Joseph. Analysis of the densification of reclaimed surface mined land. M, 1977, Texas A&M University. 125 p.

Schneiderman, Jill Stephanie. Ascutney Mountain revisited; petrology of the igneous complex and included breccia xenoliths. D, 1987, Harvard University. 315 p.

Schneidermann, Nahum. Selective dissolution of Recent coccoliths in the Atlantic Ocean. D, 1972, University of Illinois, Urbana. 175 p.

Schnetzler, Charles C. The composition and origin of tektites. D, 1962, Massachusetts Institute of Technology. 259 p.

Schneyer, Joel David. Geology of the Leppy Hills area, southern Silver Island Mountains, near Wendover, Utah. M, 1984, University of Texas, Austin. 57 p.

Schnieders, B. R. The geology and sulphide-facies iron-formations and associated rocks in the Lower Steel River-Little Steel Lake area, Terrace Bay, Ontario. M, 1987, Lakehead University.

Schnitker, Detmar Friedrich. Distribution of foraminifera on the North Carolina continental shelf. D, 1967, University of Illinois, Urbana. 169 p.

Schnitker, Detmar Friedrich. Upper Miocene foraminifera near Grimesland, Pitt County, North Carolina. M, 1966, University of North Carolina, Chapel Hill. 163 p.

Schnoebelen, Douglas J. Depositional environments of the lower Crooked Fork Group near Wartburg, Morgan County, Tennessee. M, 1983, University of Tennessee, Knoxville. 243 p.

Schnurr, Paul Eugene. Subsurface Simpson Group, Ward, Crane, Upton and Reagan counties, West Texas. M, 1955, University of Texas, Austin.

Schoch, Robert Milton. Systematics, functional morphology and macroevolution of the extinct Mammalian order Taeniodonta. D, 1983, Yale University. 733 p.

Schock, Michael Reed. A feasibility study of the silver sulfide ion-selective electrode as a geochemical exploration tool. M, 1978, Michigan State University. 115 p.

Schock, Robert Norman. Geology of the Pleistocene sediments, Troy North Quadrangle, New York. M, 1963, Rensselaer Polytechnic Institute.

Schock, Susan C. A method for dating Quaternary basalts in Hawaii. M, 1981, Michigan State University. 87 p.

Schock, William Wallace, Jr. Relations between stream sediment and source rock mineralogy, Piney Creek basin, western Wyoming. M, 1975, University of Wyoming. 92 p.

Schock, William Wallace, Jr. Stratigraphy and paleontology of the lower Dinwoody Formation, and its relation to the Permian-Triassic boundary in western Wyoming, southeastern Idaho and southwestern Montana. D, 1982, University of Wyoming. 250 p.

Schoell, John. The hydrogeology of the Skunk River regolith aquifer supplying Ames, Iowa. M, 1967, Iowa State University of Science and Technology.

Schoen, R. Geological report on the Copper Hill mineral district, Carbon County, Wyoming. M, 1953, University of Wyoming. 41 p.

Schoen, Robert. Petrology of iron-bearing rocks of the Clinton Group (Silurian) in New York State. D, 1963, Harvard University.

Schoenberg, Michael. The structure and stratigraphy of the Adirondack Lowlands near Gouverneur, New York. M, 1974, Cornell University.

Schoenborn, William Anthony. The structural geology of the Gainesville Quadrangle, Northeast Georgia. M, 1984, Miami University (Ohio). 165 p.

Schoendaller, Karen Sue. An evaluation of crustal contamination in the Cenozoic flood basalts of northwestern Colorado using petrography, strontium isotopes and trace elements. M, 1988, Colorado School of Mines. 135 p.

Schoenfeld, Mark Jean. Quaternary geology of the Burnt Fork area, Uinta mountains, Summit County, Utah. M, 1969, University of Wyoming. 80 p.

Schoenfeld, Perry C. Upper Cambrian stratigraphy of the northwestern Wind River Mountains in the Green River Lake area, Wyoming. M, 1940, Texas A&M University. 49 p.

Schoenike, Howard. Geology and exploration of the nickel silicate deposit at Riddle, Oregon. M, 1955, University of Wisconsin-Madison. 74 p.

Schoenlaub, Robert A. Equilibrium studies in the system of montecellite, glauchochroite and calcium fayalite. D, 1934, Ohio State University.

Schoenwald, Carolyn Paulette. The origin of the Waterbury Dome migmatite. M, 1971, University of Wisconsin-Madison. 72 p.

Schoewe, Walter Henry. Origin and history of extinct Lake Calvin. D, 1920, University of Iowa. 233 p.

Schoff, Stuart L. Geology of the Cedar Hills, Utah. D, 1937, Ohio State University.

Schoff, Stuart L. Oolites in the Mani Formation of central Utah. M, 1931, Ohio State University.

Schofield, Donald W. Surface features, and geological analysis of selected cores from Elizabeth Reef, Australia. M, 1980, University of Northern Colorado.

Schofield, James Dean. A gravity and magnetic investigation of the eastern portion of the Centennial Valley, Beaverhead County, Montana. M, 1980, Montana College of Mineral Science & Technology. 94 p.

Schofield, Neil Aubrey. The Porgera gold deposit, Papua New Guinea; a geostatistical study of underground ore reserves. M, 1988, Stanford University. 220 p.

Schofield, Richard Edward. Petrology of the Dundonald komatiites, Dundonald, Ontario. D, 1982, Rutgers, The State University, New Brunswick. 285 p.

Schofield, Richard Edward. The petrology and geochemistry of the Cienega Falls diabase sill, Salt River Canyon area, Gila County, Arizona. M, 1976, University of Arizona.

Schofield, Stuart James. Geology of East Kootenay, British Columbia, with special reference to the origin of granite in sills. D, 1912, Massachusetts Institute of Technology. 141 p.

Schold, Gary Paul. Analysis of an indigenous foraminiferal biocenosis from Buttonwood Sound, Florida Bay. M, 1977, Duke University. 89 p.

Scholl, Carol J. A structural study of potassium feldspar from Siscowit granite, New York-Connecticut. M, 1970, Miami University (Ohio). 48 p.

Scholl, David William. Geology and surrounding Recent marine sediments of Anacapa Island. M, 1959, University of Southern California.

Scholl, David William. Modern coastal swamp sedimentation, southwestern Florida. D, 1962, Stanford University. 139 p.

Scholl, Layne A. Effects of selected parameters in frequency-wavenumber migration. M, 1982, Indiana University, Bloomington. 206 p.

Scholl, Milton Richard, Jr. The geology of the Boracho Quadrangle, Culberson and Jeff Davis counties, Texas. M, 1948, University of Texas, Austin.

Scholle, Peter Allen. Geologic studies of the upper Cretaceous Monte Antola Formation, northern Apennines, Italy. D, 1970, Princeton University. 191 p.

Scholten, Arnold Gerhard. The reaction of phosphate with mineral surfaces and iron oxide gels. D, 1965, University of Wisconsin-Madison. 254 p.

Scholten, Robert. Geology of a part of the Beaverhead Mountains and the Nicholia Creek basin, Beaverhead, Montana and Clark County, Idaho. M, 1948, University of Michigan. 70 p.

Scholten, Robert. Geology of the Lima Peaks area, Beaverhead County, Montana, and Clark County, Idaho. D, 1950, University of Michigan. 377 p.

Scholtz, Judith Fessenden. Geology of the Woods Canyon drainage basin, Coconino County, Arizona. M, 1968, Northern Arizona University. 40 p.

Scholz, Christopher Alfred. Sediment distribution and sedimentation rates in the western arm of Lake Superior using 3.5 kHz seismic reflection profiles and ^{210}Pb geochronology. M, 1985, University of Minnesota, Duluth. 129 p.

Scholz, Christopher H. Microfracturing of rock in compression. D, 1967, Massachusetts Institute of Technology. 177 p.

Scholz, Sally A. The distribution of Recent foraminifera in Hawk Channel, Florida. M, 1962, University of Wisconsin-Madison.

Schombell, Leonard Frederick. Preliminary report on the geology of the Soledad Quadrangle, Monterey County, California. M, 1939, University of California, Berkeley. 49 p.

Schondorf, Amy von see von Schondorf, Amy

Schoner, Amy Elizabeth. The sedimentology and sedimentary petrology of the Lower Silurian Clinch Sandstone at Bean's Gap, Grainger County, Tennessee. M, 1985, University of Tennessee, Knoxville. 227 p.

Schonfeldt, Hilmar Von see Von Schonfeldt, Hilmar

Schoof, Craig Crandall. Geophysical input for seismic hazard analysis. D, 1985, Stanford University. 201 p.

Schooler, Owen Edwin. Pennsylvanian strata of the North River area, Madison County, Iowa. M, 1955, University of Nebraska, Lincoln.

Schooler, Richard A. Interpretation of rock and vapor phase relations in the Ruby Mountain volcanic complex, Chaffee County, Colorado. M, 1982, Bowling Green State University. 104 p.

Schooley, Jeannie Victoria. A study of the mineralogy of lower Cretaceous Mannville Group oil sand deposits, Alberta and West central Saskatchewan. M, 1975, University of Calgary. 134 p.

Schoon, Robert A. A geology of the Witten Quadrangle, South Dakota. M, 1958, University of South Dakota. 80 p.

Schoonmaker, Jane E. Magnesian calcite - seawater reactions; solubility and recrystallization behavior. D, 1981, Northwestern University. 277 p.

Schoonover, Floyd Eldon. The igneous rocks of the Fort Sill Reservation, Oklahoma. M, 1948, University of Oklahoma. 123 p.

Schoonover, Lois Margaret. A stratigraphic study of the mollusks of the Calvert and Choptank formations of southern Maryland. D, 1940, Bryn Mawr College. 303 p.

Schoonover, Lois Margaret. The Eocene crassatellas of the Atlantic and Gulf Coast provinces. M, 1936, Cornell University.

Schopf, Paul S. Numerical models of the oceanic heat transport. D, 1978, Princeton University.

Schopf, Thomas Joseph Morton. Conodonts of the Trenton Group (Ordovician) in New York, southern Ontario, and Quebec. D, 1964, Ohio State University. 169 p.

Schorn, Howard E. Revision of the fossil species of Mahonia from North America. M, 1966, California State University, Bakersfield.

Schornick, James C., Jr. Uranium and thorium isotope geochemistry in ferromanganese concretions from the Southern Ocean. D, 1972, Florida State University.

Schorr, Gregory Thomas. Study of seismic reflection data over Virginia Mesozoic basins. M, 1986, Virginia Polytechnic Institute and State University.

Schorsch, Laurie Jane. The Orleansville Turbidite, southern Balearic Basin, western Mediterranean Sea. M, 1980, Duke University. 165 p.

Schosinsky, G. E. Development of water-powered rainfall recorders. M, 1973, University of Arizona.

Schot, Eric H. The diagenetic origin of the dolomites, chert, and pyrite in the Jefferson City Formation. M, 1963, University of Missouri, Rolla.

Schowalter, Timothy T. Geology of part of the Creston range, Mora County, New Mexico. M, 1969, University of New Mexico. 70 p.

Schrader, David L. Holocene sedimentation in a low energy microtidal estuary, St. Lucie River, Florida (U.S.A.). M, 1984, University of South Florida, Tampa. 132 p.

Schrader, Edward J., Jr. A geochemical study of trace elements in sediment from a fluvial system heavily influenced by mining of coal. M, 1975, University of Tennessee, Knoxville. 86 p.

Schrader, Edward Leon, Jr. Weathering characeristics of molybdenum ore deposits with respect to the chemical and mineralogical aspects of Fe-Mo oxides. D, 1977, Duke University. 272 p.

Schraeder, Robert Lynn. Ablation of Ice Island Arlis II, 1961 (drifting ice station) (Alaska). M, 1968, University of Alaska, Fairbanks. 59 p.

Schrager, Gene. The feasibility of using tree-ring chronologies to reconstruct streamflow records for the Pemigewasset River, New Hampshire. M, 1987, University of New Hampshire. 91 p.

Schramke, Janet Ann. The kinetics of metamorphic hydration-dehydration reactions. D, 1984, Pennsylvania State University, University Park. 227 p.

Schramm, Donna J. Geology of the Muldoon Canyon-Muldoon Creek area, south-central Idaho. M, 1978, University of Wisconsin-Milwaukee.

Schramm, Eck Frank. The growth of the cement industry in Nebraska. M, 1908, University of Nebraska, Lincoln.

Schramm, Linda Sue. The role of volatiles in the cyclic Bandelier Tuffs, Jemez Mountains, New Mexico. M, 1982, Florida State University.

Schramm, Martin William, Jr. Geology of the northwestern part of the Boyertown Quadrangle, Berks County, Pennsylvania. M, 1955, University of Pittsburgh.

Schramm, Martin William, Jr. Paleogeologic and quantitative lithofacies analysis of the Simpson Group, Oklahoma. D, 1963, University of Oklahoma. 234 p.

Schramm, William H. Geology of the Sykes Spring area, Big Horn County, Wyoming. M, 1984, University of Southwestern Louisiana. 148 p.

Schrank, Joseph A. Formation of Red Sea lithified layers as studied by cathodoluminescence; Deep-Sea Drilling Project Leg 23B. M, 1975, Rensselaer Polytechnic Institute. 111 p.

Schrantz, Jonathan K. Regional gravity survey and modeling of the Rome Trough in southwestern West Virginia. M, 1984, Wright State University. 57 p.

Schraubstadtler, R. T. Concentration of Southwest Missouri ore. M, 1893, Washington State University.

Schrecongost, Milford Martin. An analysis of Recent sediments near the mouth of the Sandusky River, Ohio. M, 1963, Bowling Green State University. 29 p.

Schreiber, Anne Marie. A comparison of methods of block coal description. M, 1985, University of Kentucky. 91 p.

Schreiber, B. Charlotte. An analysis of three deep-sea cores from the north Atlantic. M, 1966, Rutgers, The State University, New Brunswick. 90 p.

Schreiber, B. Charlotte. Upper Miocene (Messinian) evaporite deposits of the Mediterranean Basin and their depositional environments. D, 1975, Rensselaer Polytechnic Institute. 499 p.

Schreiber, Berta L. Early to middle Miocene sea level fluctuations, biostratigraphy, and paleoecology of the Maryland subsurface (Calvert Formation). M, 1984, Rutgers, The State University, New Brunswick. 115 p.

Schreiber, Hans W. The geology of and the occurrence of uranium in the Nelson Pond area of the West Point Quadrangle, Putnam County, New York. M, 1958, Columbia University, Teachers College.

Schreiber, Joseph Frederick, Jr. Sedimentary record in Great Salt Lake, Utah. D, 1958, University of Utah. 110 p.

Schreiber, Richard Lee. A multivariate statistical analysis of the microcrinoid Allagecrinus sculptus from Oklahoma and New Mexico. M, 1971, University of Michigan.

Schreiber, Sue Anne. Geology of the Nelson Butte area, South-central Cascade Range, Washington. M, 1981, University of Washington. 81 p.

Schreier, H. Chemical terrain variability; a geomorphological approach using numerical and remote sensing techniques. D, 1976, University of British Columbia.

Schreifels, W. A. Some phase relations in the system CaO-MgO-iron oxide-SiO$_2$ in air and CO$_2$. M, 1974, Queens College (CUNY). 91 p.

Schreifels, Walter Arthur. Liquid-solid equilibria involving spinel, ilmenite and ferropseudobrookite in the system iron oxide - Al$_2$O$_3$-TiO$_2$ with implications for lunar and terrestrial petrology. D, 1976, Pennsylvania State University, University Park. 127 p.

Schreiner, Bryan T. Quaternary geology of the Precambrian shield south of 58° North latitude, Saskatchewan. M, 1980, University of Saskatchewan. 186 p.

Schremp, Lee A. Late Tertiary rodents from the Catamarca Province, Argentina; the Field Museum collections. M, 1984, Loma Linda University. 169 p.

Schreuder, Kenneth M. Predicting gypsum-anhydrite equilibria in the subsurface; a thermodynamic model. M, 1986, South Dakota School of Mines & Technology.

Schreuder, P. J. The determination of aquifer anisotropy by transmissivity tensor analysis. M, 1974, University of Arizona. unpaginated p.

Schriber, Craig Norman. Structural geology in the Northport area, Stevens County, Washington. M, 1981, Washington State University. 99 p.

Schriener, Alexander, Jr. Geology and mineralization of the north part of the Washougal mining district, Skamania County, Washington. M, 1979, Oregon State University. 135 p.

Schriver, George H. Petrochemistry of Precambrian greenstones and granodiorites in southeastern Oneida County, Wisconsin. M, 1973, University of Wisconsin-Milwaukee.

Schrock, Robert Rakes and Cumings, Edgar Roscoe. The geology of the Silurian rocks of northern Indiana. D, 1928, Indiana University, Bloomington. 293 p.

Schrock, Ronald L. Mixed-layered clays of the Triassic Chinle Formation, northern Arizona and southern Utah. M, 1975, Dartmouth College. 79 p.

Schroder, Charles H. Trace fossils of the Upper Cretaceous-lower Tertiary (formerly Tuscaloosa Formation) and basal Jackson Group, east-central Georgia. M, 1979, University of Georgia.

Schroder, Claudia J. Deep-water arenaceous foraminifera in the Northwest Atlantic Ocean. D, 1986, Dalhousie University. 287 p.

Schroder, Richard A. The effects of dissolution kinetics on the propagation of secondary porosity in carbonate aquifers; a laboratory simulation. M, 1987, Eastern Kentucky University. 104 p.

Schrodt, Augusta Kay. Depositional environments, provenance, and vertebrate paleontology of the Eocene-Oligocene Baca Formation, Catron County, New Mexico. M, 1980, Louisiana State University.

Schrodt, Joseph Keith. Temperature and axial strain rate effects on micromechanical behavior in triaxially compressed marbles. D, 1983, University of Illinois, Urbana. 166 p.

Schroeder, David Alan. Geology of the central Ouachitas in part of Pushmataha County, Oklahoma. M, 1975, Northern Illinois University. 63 p.

Schroeder, David Alan. Geology of the Granby and Strawberry Lake 7 1/2' quadrangles, Grand County, Colorado. D, 1984, University of Colorado. 448 p.

Schroeder, Eugene Robert. The geology of a portion of the Hermitage Quadrangle, Missouri. M, 1950, University of Iowa. 86 p.

Schroeder, F. W. A geophysical investigation of the Oceanographer fracture zone and the Mid-Atlantic Ridge in the vicinity of 35° north. D, 1977, Columbia University, Teachers College. 471 p.

Schroeder, Gerald. Effect of applied pressures on the radon characteristics of an underground mine environment. D, 1965, Massachusetts Institute of Technology. 154 p.

Schroeder, James E. Geology of a portion (Cretaceous or Jurassic) of the Ensenada Quadrangle, Baja California, Mexico. M, 1967, University of San Diego.

Schroeder, Johannes Herbert. Experimental dissolution of calcium, magnesium, and strontium from Recent biogenic carbonates; a model of diagenesis. D, 1968, George Washington University. 87 p.

Schroeder, Kim Erik. Structure and stratigraphy related to gold mineralization at the Howie Mine, south-central North Carolina. M, 1987, Ohio University, Athens. 201 p.

Schroeder, M. Richard. Sedimentology and petrography of a distributary channel complex in the Aguja Formation (late Campanian), Big Bend National Park, Texas. M, 1988, Texas Tech University. 172 p.

Schroeder, Mark E. Design criteria and economic evaluation of longwall-mining development schemes. M, 1988, Pennsylvania State University, University Park. 311 p.

Schroeder, Marvin L. Arenaceous foraminifera of the Shawnee Group of the Upper Pennsylvanian in southeastern Kansas. M, 1961, University of Kansas. 48 p.

Schroeder, Melvin Carroll. Pre-Cambrian stratigraphy of the Bead Lake District, Pend Oreille County, Washington. D, 1953, Washington State University. 92 p.

Schroeder, Melvin Carroll. The genesis of the Turk magnesite deposits of Stevens County, Washington. M, 1947, Washington State University. 19 p.

Schroeder, Paul A. Pyrite and associated authigenic heavy minerals of the continental slope off the Eastern United States. M, 1981, University of South Florida, St. Petersburg. 98 p.

Schroeder, Richard John. Dickite and kaolinite in Pennsylvanian rocks of southeast Kansas. M, 1967, University of Iowa. 83 p.

Schroeder, Robert John. Stratigraphy and structure of the southern two-thirds of the Glenrock Quadrangle, Converse County, Wyoming. M, 1953, University of Iowa. 153 p.

Schroeder, Roy A. Kinetics, mechanism and geochemical applications of amino acid racemization of various fossils. D, 1974, University of California, San Diego. 293 p.

Schroeder, Thomas Francis. Accuracy of sediment size analysis by pipette method. M, 1959, Washington University. 64 p.

Schroeder, Thomas Francis. Preparation and evaluation of submicron whiskers for possible use as elongated single domain particles. D, 1962, Washington State University.

Schroeder, Thomas Francis. Preparation and evaluation of submicron whiskers for possible use as singlar domain particles. D, 1959, Washington University. 19 p.

Schroeder, Tom Scot. Determination of the immediate source areas and probable sediment transport pathways of New Jersey beach sands. M, 1982, Lehigh University. 137 p.

Schroeder, Warren Lee. Liquefaction of saturated sands. D, 1967, University of Colorado. 129 p.

Schroeder, Wayne E. Lithologic study of glacial sediments in southeastern South Dakota. M, 1978, University of South Dakota. 165 p.

Schroeder, William Floyd. The empirical age-depth relation and depth anomalies in the Pacific Ocean basin. M, 1983, Purdue University. 56 p.

Schroedl, Alan Robert. The Archaic of the northern Colorado Plateau. D, 1976, University of Utah. 122 p.

Schroeter, Charles. Stratigraphy and structure of the Juncal Camp - Santa Ynez Fault sliver, southeastern Santa Barbara County, California. M, 1972, University of California, Santa Barbara.

Schroeter, Thomas Gordon. Geology of the Nippers Harbour area, Newfoundland (late Proterozoic to Devonian). M, 1971, University of Western Ontario. 93 p.

Schroth, Brian K. Water chemistry reconnaissance and geochemical modeling in the Meadow Valley Wash area, southern Nevada. M, 1987, University of Nevada. 97 p.

Schroth, Charles Lorenz. Analysis and prediction of the properties of western Samoa soils. D, 1970, University of Hawaii. 280 p.

Schroth, Eugene Howard. Some aspects of the power resources of Canada with special reference to coal. M, 1936, University of Illinois, Urbana.

Schrott, Robert Otto. Conodonts of the Aplington Formation (Devonian) in north-central Iowa. M, 1959, University of Nebraska, Lincoln.

Schrott, Robert Otto. Paleoecology and stratigraphy of the Lecompton Megacyclothem (Late Pennsylvanian) in the northern Midcontinent region. D, 1966, University of Nebraska, Lincoln. 367 p.

Schrull, Jeffrey Lee. Two dimensional spectral/statistical analysis of marine magnetic data; implications for depth-to-magnetic source. M, 1987, Texas A&M University. 103 p.

Schrunk, Verne K. Surficial geology of a part of the northeast flank of the Bridger Range, Montana. M, 1976, Montana State University. 132 p.

Schryver, Robert F. Geology of the Mound House area, Ormsby and Lyon counties, Nevada. M, 1961, University of Nevada - Mackay School of Mines. 47 p.

Schubbe, Dennis Lee. Hydrogeology of the Spearhead Lake area, Hubbard County, Minnesota. M, 1988, South Dakota School of Mines & Technology.

Schubel, Jerry Robert. Suspended sediment of the northern Chesapeake Bay. D, 1968, The Johns Hopkins University. 287 p.

Schubert, Carl Eric. Seafloor structure and tectonics east of the northern Lesser Antilles islands. D, 1977, University of Miami. 226 p.

Schubert, Carlos. Geology of the Barinitas-Santo Domingo region, southeastern Venezuelan Andes. D, 1967, Rice University. 155 p.

Schubert, Carlos. Grain size distributions of various materials in an abrasion mill. M, 1963, Rice University. 95 p.

Schubert, Christopher E. Kohler. A computer analysis of the coal resources of Clay County, Indiana. M, 1985, Indiana University, Bloomington. 110 p.

Schubert, Jeffrey Paul. A hydrogeologic and numerical simulation feasibility study of connector well dewatering of underground coal mines, Madera, Pennsylvania. M, 1978, Pennsylvania State University, University Park. 325 p.

Schuberth, Christopher John. Geology of the Port Jervis South New York-New Jersey-Pennsylvania Quadrangle. M, 1960, New York University. 44 p.

Schubring, Selma L. A statistical study of lead and zinc mining in Wisconsin. D, 1920, University of Wisconsin-Madison.

Schuett, Edwin Clarence, Jr. Petrography and water-mineral relationships of two Quaternary fills in eastern Nebraska. M, 1964, University of Nebraska, Lincoln.

Schuette, Gretchen. Recent marine diatom taphocoenoses off Peru and off Southwest Africa; reflection of coastal upwelling. D, 1980, Oregon State University. 115 p.

Schug, David Lynn. Neotectonics of the western reach of the Agua Blanca Fault, Baja California, Mexico. M, 1987, San Diego State University. 125 p.

Schuh, Henry Allen. The geology of Lodi Township, Athens County, Ohio. M, 1953, Ohio State University.

Schuh, Mary Louise. Geology of the Avakutakh Iron Formation, Labrador. M, 1981, Northern Illinois University. 192 p.

Schuiling, William T. Calcite-barite chrysocolla mineralization, Sacramento Mountains, California. M, 1978, University of Arizona.

Schuldt, Walter Carl. Cambrian strata of northeastern Iowa. D, 1940, University of Iowa.

Schule, J. W. Anisotropy of the upper mantle of the Pacific Basin. D, 1975, University of California, Los Angeles. 141 p.

Schulenberg, John Theodore. Cenozoic stratigraphy of Rim Rock country, Trans-Pecos, Texas. M, 1958, University of Texas, Austin.

Schulingkamp, Warren John, II. Petrology of the Ordovician Swan Peak Formation, southeastern Idaho and north-central Utah. M, 1972, Utah State University. 118 p.

Schull, H. W. Geology of the Port Jervis South Quadrangle, New York-New Jersey-Pennsylvania. M, 1959, New York University.

Schull, Herman Walker, III. Geology of the south wall of the South Line Creek Canyon, Beartooth Mountains, Montana-Wyoming. M, 1959, Columbia University. 123 p.

Schull, Herman Walter, III. X-ray pole figures of pyrrhotite. D, 1971, Columbia University. 123 p.

Schuller, Rudolph M. Use of the cupric ion-selective electrode as a geochemical exploration tool. M, 1976, Wright State University. 69 p.

Schulman, Edmund. History of precipitation and runoff in the Colorado Basin as indicated by tree-rings. D, 1944, Harvard University.

Schulman, Melvyn M. Sedimentary environments of a nearshore microfacies of the medial Ordovician Coboconk Limestone in southeastern Ontario. M, 1971, Boston University. 29 p.

Schult, Frederick Roy. Plate motions and tectonics of plate margins. D, 1986, Stanford University. 133 p.

Schult, Mark Frederick. Correlation between uranium levels, phosphate, and biostratigraphy in the Central Florida phosphate mining district. M, 1988, University of Florida. 76 p.

Schulte, Frank J. Groundwater of the Spiritwood Lake area, Stutsman County, North Dakota. D, 1972, University of North Dakota. 341 p.

Schulte, Frank J. Quantitative geomorphology and hydrology of the Cass River Basin, New Zealand. M, 1971, University of North Dakota. 684 p.

Schulte, George S. The Cottage Grove and Noxie sandstones (Pennsylvania; "Layton") in south-central Kansas. M, 1958, University of Kansas. 112 p.

Schulte, John Joseph. The bedrock geology of Knox County, Nebraska. M, 1952, University of Nebraska, Lincoln.

Schultejann, Patricia Ann. Late Mesozoic and Cenozoic tectonic history of south central California. D, 1984, University of California, San Diego. 229 p.

Schulten, Cathy S. Environmental geology of the Tempe Quadrangle, Maricopa County, Arizona; Part I. M, 1979, Arizona State University. 101 p.

Schultheiss, Norbert H. Petrography, source and origin of the Cadomin conglomerate (Cretaceous) between the North Saskatchewan and Athabasca rivers, Alberta. M, 1970, McGill University.

Schultink, Gerhardus. Statistical sampling of agricultural and natural resources data; a comparison of aerial sampling and area-frame sampling techniques with special reference to lesser developed countries. D, 1980, Michigan State University. 176 p.

Schultz, Adam. On the electrical heterogeneity of the Earth's interior; a global study of mid-mantle conductivity. D, 1985, University of Washington. 245 p.

Schultz, Albert West. Sedimentology and petrology of the Fountain Formation near Canon City, Colorado. D, 1986, Indiana University, Bloomington. 184 p.

Schultz, Alfred Reginald. The underground water supply of Wisconsin, northern Illinois, and the Northern Peninsula of Michigan. D, 1905, University of Chicago.

Schultz, Arthur H. and Jurko, Robert Clarence. Stratigraphy and paleontology of a core from Kent County, Ontario, Canada. M, 1953, University of Michigan.

Schultz, Arthur Philip. Broken-formations of the Pulaski thrust sheet near Pulaski, Virginia. D, 1983, Virginia Polytechnic Institute and State University. 111 p.

Schultz, Arthur Philip. Deformation associated with Pulaski over-thrusting in the Price Mountain and East Radford windows, Montgomery County, Southwest Virginia. M, 1979, Virginia Polytechnic Institute and State University.

Schultz, Charles Bertrand. Oreodonts from the Marsland and Sheep Creek Formations with notes on the Miocene stratigraphy of Nebraska. D, 1941, University of Nebraska, Lincoln.

Schultz, Charles Bertrand. The Pleistocene mammals of Nebraska, with additional notes concerning the association of artifacts and extinct mammals in Nebraska. M, 1933, University of Nebraska, Lincoln.

Schultz, D. M. Source, formation, and composition of suspended lipoidal material in Narragansett Bay, Rhode Island. D, 1974, University of Rhode Island. 219 p.

Schultz, Douglas J. Crystalline silica in mudrocks. M, 1975, University of Oklahoma. 48 p.

Schultz, Douglas J. Fourier analysis of quartz shape during pro-grade regional metamorphism. D, 1978, University of South Carolina. 92 p.

Schultz, Frederick J. An evaluation of techniques of cluster analysis applied to the stratigraphy of a Caribbean deep-sea core. M, 1979, University of Georgia.

Schultz, Gerald E. The relationship of Nevada mining districts to major geologic structures and formations. M, 1960, University of Minnesota, Minneapolis. 74 p.

Schultz, Gerald Edward. The geology and paleontology of a late Pleistocene basin in Southwest Kansas. D, 1966, University of Michigan. 123 p.

Schultz, Jerald David. Geomorphology, sedimentology, and Quaternary eolian history of the west-central

San Juan Basin, northwest New Mexico. M, 1980, University of New Mexico. 257 p.

Schultz, John R. A late Quaternary mammal fauna from the tar seeps of McKittrick, California. D, 1937, California Institute of Technology. 201 p.

Schultz, John R. Geology of the White's Point outfall sewer tunnel, California. D, 1937, California Institute of Technology. 43 p.

Schultz, John R. The chert of the Niagara series in the Chicago area. M, 1933, Northwestern University.

Schultz, Karin L. Paleomagnetism of Jurassic plutons in the central Klamath Mountains, southern Oregon and northern California. M, 1983, Oregon State University. 153 p.

Schultz, Lane D. A land-use study in the third dimension. M, 1972, Lehigh University.

Schultz, Lane D. The stratigraphy of the Trimmers Rock Formation in northeastern Pennsylvania. D, 1974, Lehigh University. 167 p.

Schultz, Leonard Gene. Potash metasomatism in the La Plata Mountains, Colorado. M, 1952, University of Illinois, Urbana.

Schultz, Leonard Gene. The petrology of underclays. D, 1954, University of Illinois, Urbana.

Schultz, Mark Gulliver. Recharge through some upland surface glacial drift in Portage County, Ohio. M, 1973, Kent State University, Kent. 83 p.

Schultz, Norbert. Two- and three-dimensional inversions of magnetic anomalies in the MARK area (Mid-Atlantic Ridge 23°N). M, 1988, University of Rhode Island.

Schultz, P. S. Velocity estimation by wave front synthesis. D, 1976, Stanford University. 177 p.

Schultz, Richard Allen. Martian global tectonics. M, 1982, Arizona State University. 113 p.

Schultz, Richard Allen. Mechanics of curved strike-slip faults. D, 1987, Purdue University. 142 p.

Schultz, Richard B. Depositional environment and petrology of the Iowa Point Shale Member, Topeka Limestone Formation, (Shawnee Group, Upper Pennsylvanian) in eastern Kansas. M, 1988, Wichita State University. 151 p.

Schultz, Roger Stephen. Geology of northwestern Newton and southwestern Walton counties, Georgia. M, 1961, Emory University. 46 p.

Schultz, T. R. Trichlorofluoromethane as a ground-water tracer for finite-state models. D, 1979, University of Arizona. 213 p.

Schultz, Thomas Allan. A study of bedrock topography of Franklin Township by a correlation of resistivity and seismic surveys. M, 1979, Kent State University, Kent. 208 p.

Schultz, Thomas R. Concentration and distribution of selected trace metals in the Maumee River basin, Ohio, Indiana, and Michigan. M, 1972, Ohio State University.

Schultz, Thomas R. Linear regression analysis of modified Mercalli intensity variations with distance for three seismic regions of the United States and Eastern Canada. M, 1974, Pennsylvania State University, University Park. 87 p.

Schultz, William R. A petrographic analysis of some highway aggregates in northern Virginia. M, 1952, University of Virginia. 61 p.

Schultz-Ela, Daniel Dennett. Strain patterns and deformation history of the Vermilion District, northeastern Minnesota. D, 1988, University of Minnesota, Minneapolis. 381 p.

Schulz, Klaus Jurgen. The petrology and geochemistry of Archean volcanics, western Vermilion District, northwestern Minnesota. D, 1977, University of Minnesota, Minneapolis. 366 p.

Schulz, Klaus Jurgen. The petrology and structural relations of some ultramafic bodies in the early Precambrian Newton Lake Formation, Vermilion District, northeastern Minnesota. M, 1974, University of Minnesota, Minneapolis.

Schulz, Michael Gerhard. The quantification of soil mass movements and their relationship to bedrock ge-

ology in the Bull Run Watershed, Multnomah and Clackamas counties, Oregon. M, 1981, Oregon State University. 170 p.

Schulze, Daniel J. Garnet pyroxenites and partially melted metamorphic eclogites from the Sullivan Buttes latite xenolith suite, Chino Valley, Arizona. M, 1977, Queen's University. 72 p.

Schulze, Daniel James. The petrology of ultramafic xenoliths from the Hamilton Branch kimberlite pipe, Elliott County, Kentucky. D, 1982, University of Texas at Dallas. 132 p.

Schulze, Jack D. The Pre-Dakota geology of the northern half of the Canon City Embayment, Colorado. M, 1954, University of Oklahoma. 163 p.

Schumacher, Ann Louise. Gravity survey and subsurface investigation of the Van Buren Quadrangle, Arkansas and Oklahoma. M, 1979, University of Arkansas, Fayetteville.

Schumacher, Dietmar. Conodont biostratigraphy and paleoenvironments of middle-upper Devonian boundary beds, Missouri. D, 1972, University of Missouri, Columbia. 140 p.

Schumacher, Dietmar. Conodonts and biostratigraphy of the Middle and Upper Devonian of Wisconsin. M, 1967, University of Wisconsin-Madison.

Schumacher, Gregory A. Sedimentology, biostratinomy, and paleoecology of Cincinnatian (Upper Ordovician) Iocrinus subcrassus crinoid (Echinodermata) assemblages. M, 1984, University of Cincinnati. 210 p.

Schumacher, James David. A study of near-bottom currents in North Carolina shelf waters. D, 1974, University of North Carolina, Chapel Hill. 134 p.

Schumacher, John C. Geology of Precambrian and Paleozoic rocks on part of western Casper Mountain, Natrona County, Wyoming. M, 1979, University of Akron. 81 p.

Schumacher, John Charles. Stratigraphic, geochemical, and petrologic studies of the Ammonoosuc Volcanics, north-central Massachusetts and southwestern New Hampshire. D, 1983, University of Massachusetts. 250 p.

Schumacher, John G. Chemical and isotopic investigations of crude oils in some Paleozoic reservoirs; west-central Kansas. M, 1988, Kansas State University. 187 p.

Schumacher, Madelyn. Depositional environment of the Bartlesville Sandstone, LaHarpe Field, Allen County, Kansas. M, 1976, Texas A&M University. 82 p.

Schumacher, Mark J. Site evaluation of the Town of Harmony Landfill. M, 1988, SUNY, College at Fredonia. 135 p.

Schumacher, Otto L. Geology and ore deposits of the Southwest Nacimiento Range, Sandoval County, New Mexico. M, 1972, University of New Mexico. 79 p.

Schumaker, Robert Clarke. Geology of the Pawlett Quadrangle, Vermont. D, 1960, Cornell University.

Schumaker, Robert D. Regional study of Kansas Permian evaporite formations. M, 1966, Wichita State University. 87 p.

Schuman, Richard L. Paleogeomorphology of the sub-Mesozoic and sub-Cretaceous unconformities in Kansas. M, 1963, University of Kansas. 38 p.

Schumann, John R. Description and geologic history of Cenozoic gravels in northern Nevada. M, 1962, Bowling Green State University. 33 p.

Schumann, Paul L. Metamorphic history of the Alto Allochthon and petrology of the adjacent Chauga Belt, Ayersville Quadrangle, Georgia. M, 1988, University of Tennessee, Knoxville. 95 p.

Schumm, Stanley A. Evolution of drainage systems and slopes in badlands at Perth Amboy, New Jersey. D, 1954, Columbia University, Teachers College.

Schupbach, Martin Albert. Comparison of slope and basinal sediments of a marginal cratonic basin (Pedregosa Basin, New Mexico) and a marginal geosynclinal basin (southern border of Piemontais Geosyncline, Bernina Nappe, Switzerland). D, 1974, Rice University. 126 p.

Schupp, Robert Donald. A study of the cobble beach cusps along Santa Monica Bay, California. M, 1955, University of Southern California.

Schuraytz, Benjamin Charles. Geochemical gradients in the Topopah Spring Member of the Paintbrush Tuff; evidence for eruption across a magmatic interface. D, 1988, Michigan State University. 146 p.

Schurer, Victoria Christine. Structural geology of the Kirkman-Diamond Creek Formation, West-central Oquirrh Mountains, Utah. M, 1979, University of Utah. 46 p.

Schussler, Sherryl Ann. Paleogene and Franciscan-Complex stratigraphy, southwestern San Rafael Mountains, Santa Barbara County, California. M, 1981, University of California, Santa Barbara. 235 p.

Schuster, David Conway. The nature and origin of the late Precambrian Gwna Melange, North Wales, United Kingdom. D, 1980, University of Illinois, Urbana. 389 p.

Schuster, Gerard Thomas. Seismic studies of crustal structure in Nicaragua and Costa Rica. M, 1977, University of Houston.

Schuster, Gerard Thomas. Some boundary integral equation methods and their application to seismic exploration. D, 1984, Columbia University, Teachers College. 148 p.

Schuster, J. E. Distribution of copper and the platinum group in mafic rocks of the Sierra Madre, Carbon County, Wyoming. M, 1972, University of Wyoming. 116 p.

Schuster, Robert L. A study of chert and shale gravel in concrete. D, 1961, Purdue University.

Schuster, Robert Lee. The glacial geology of Pickaway County, Ohio. M, 1952, Ohio State University.

Schutter, Stephen Richard. Petrology, clay mineralogy, paleontology, and depositional environments of four Missourian (Upper Pennsylvanian) shales of Midcontinent and Illinois basins (Volumes I-III). D, 1983, University of Iowa. 1208 p.

Schutter, Stephen Richard. The petrology and depositional environment of the Glenwood Shale in southeastern Minnesota and northeastern Iowa. M, 1978, University of Iowa. 149 p.

Schutts, Larry Davis. A resetting of paleoremanence in low-grade metavolcanics. M, 1974, Massachusetts Institute of Technology. 88 p.

Schutz, Donald F. Global variation in the chemical compositions of basalt and andesite. M, 1958, Rice University. 90 p.

Schutz, Joseph Leroy. Geology of precious metal mineralization in the Baldy Mountain District, Beaverhead County, Montana. M, 1986, Colorado State University. 190 p.

Schuver, Henry John. Modeling volcano morphology. M, 1988, Arizona State University. 120 p.

Schuyler, Andrew. Sedimentology and palaeoecology of miospores from the Middle to Upper Devonian Oneonta Formation; part of the Catskill Magnafacies, New York State. M, 1987, Pennsylvania State University, University Park. 132 p.

Schuyler, Donald Richard, II. Some mechanical properties of lanthanum metal. M, 1962, University of Nevada - Mackay School of Mines. 40 p.

Schuyler, Jeffrey N. An investigation into the effect of aqueous chemical environments on hydraulic fracture morphology and fracture strength of a quartz arenite. M, 1982, Indiana University, Bloomington. 125 p.

Schuyler, Sharon Stowe. The effects of acid precipitation on two soils of Pennsylvania. M, 1987, Pennsylvania State University, University Park. 220 p.

Schuyler, T. Kent. Petrology of the sub-layer along the north range of the Sudbury Irruptive. M, 1985, Rutgers, The State University, New Brunswick. 142 p.

Schuyler-Rossie, Christine. The use of refracted shear waves in groundwater exploration; a field test. M, 1985, Colorado School of Mines. 53 p.

Schwab, Anne Marie. The geomorphology and archaeological geology of the Bais Anthropological Project,

Phase II, Negros Oriental, Philippines. M, 1983, University of Michigan.

Schwab, Frederic L. An analysis of Phanerozoic sedimentation through time, western Wyoming. D, 1968, Harvard University.

Schwab, Frederick L. Geology of the Quosatana Butte area, Curry County, Southwest Oregon. M, 1963, University of Wisconsin-Madison. 114 p.

Schwab, Jean Ann. Ultrasonic determination of the elastic properties of fayalite as a function of pressure and temperature. M, 1983, Pennsylvania State University, University Park. 82 p.

Schwab, Joseph Alan. Depositional environments and diagenesis of the upper Smackover Formation, Hampstead County, Arkansas. M, 1978, Texas Tech University. 70 p.

Schwab, Joseph Patrick. Effective stress model for static and cyclic loading of granular soils. D, 1987, Syracuse University. 350 p.

Schwab, Karl W. The relation of conodonts to early vertebrates. M, 1963, University of Arizona.

Schwab, William C. The structural interpretation of two aeromagnetic lineaments in Rhode Island. M, 1976, University of Rhode Island.

Schwab, William Charles. Slope instability on the Alsek prodelta, Gulf of Alaska. D, 1985, Duke University. 490 p.

Schwade, Irving T. Geology of northern South Park, Colorado. M, 1935, Northwestern University.

Schwalbaum, William Jesse. Hydrogeology and geochemistry of the old Amherst landfill, Amherst, Massachusetts. M, 1983, University of Massachusetts. 166 p.

Schwaller, Mathew R. Remote sensing for geobotanical prospecting. D, 1983, University of Michigan. 127 p.

Schwandt, Craig Stuart. The petrogenesis of composite mafic-felsic dikes from the Vermilion granitic complex, NE Minnesota. M, 1988, University of Missouri, Columbia. 97 p.

Schwans, Peter. Stratal packages at the subsiding margin of the Cretaceous foreland basin, Utah. D, 1988, Ohio State University. 543 p.

Schwarberg, T. M. The geology of Muddy Mountain, south-east Natrona County, Wyoming. M, 1959, University of Wyoming.

Schwarbert, T. M., Jr. Geology of Muddy Mountain, Southeast Natrona County, Wyoming. M, 1959, University of Wyoming. 103 p.

Schwarcz, Henry Philip. I, Geology of the Winchester-Hemet area, Riverside County, California; II, Geochemical investigations of an arkosic quartzite of the Winchester-Hemet area, California. D, 1960, California Institute of Technology. 427 p.

Schwartz, Arnold Edward. Strength of rock. D, 1963, Georgia Institute of Technology.

Schwartz, Arthur H. Facies mosaic of the upper Yates and lower Tansill formations (Upper Permian), Rattlesnake Canyon, Guadalupe Mountains, New Mexico. M, 1981, University of Wisconsin-Madison.

Schwartz, D. P. Geology of the Zacapa Quadrangle and vicinity, Guatemala, Central America. D, 1976, SUNY at Binghamton. 252 p.

Schwartz, Dale Ann. Species-abundance distributions; potential applicability to fossil "communities". M, 1977, University of Rochester.

Schwartz, Daniel Evan. Sedimentology of the braided-to-meandering transition zone of the Red River, Oklahoma and Texas. D, 1978, University of Texas at Dallas. 394 p.

Schwartz, David. Geologic history of Elkhorn Slough, Monterey County, California. M, 1984, San Jose State University. 102 p.

Schwartz, David P. The geology of the Punta Guanajibo-Cuchilla Sabana Alta area, western Puerto Rico. M, 1970, Queens College (CUNY). 73 p.

Schwartz, Deborah Ellen. Nature and origin of the Banda seafloor, eastern Indonesia. M, 1985, University of California, Santa Cruz.

Schwartz, Dirk Anson. The deposition and diagenesis of the Bluell Zone, upper Mission Canyon Formation (Mississippian), Flaxton Field, Burke County, North Dakota. M, 1987, University of North Dakota. 246 p.

Schwartz, Franklin Walter. Digital simulation of hydrochemical patterns in regional groundwater flow. D, 1972, University of Illinois, Urbana. 74 p.

Schwartz, Franklin Walter. Geohydrology and hydrogeochemistry of groundwater; streamflow systems in the Wilson Creek experimental watershed, Manitoba. M, 1970, University of Manitoba.

Schwartz, George M. Earth movement in North America at the close of the Cretaceous. M, 1916, University of Wisconsin-Madison.

Schwartz, George M. The contrast in the effect of granite and gabbro intrusions on the Ely Greenstone. D, 1923, University of Minnesota, Minneapolis. 70 p.

Schwartz, Hilde Lisa. Paleoecology of the late Cenozoic fishes from the Turkana Basin, northern Kenya. D, 1983, University of California, Santa Cruz. 305 p.

Schwartz, James. The geology of oceanographic, Gilbert Lydonia submarine canyon south of Georges Bank, off northeastern U.S. M, 1965, University of Rhode Island.

Schwartz, Karen Ann. Core stratigraphy and petrography of the Reno Member of the Lone Rock Formation, at a dam site, Vernon County, western Wisconsin. M, 1979, University of Wisconsin-Madison.

Schwartz, Kenneth Bruce. Mossbauer spectroscopy and crystal chemistry of natural Fe-Ti garnets. M, 1977, Massachusetts Institute of Technology. 91 p.

Schwartz, Kenneth Bruce. Structure, crystal chemistry, and electronic interactions in metallic and semiconducting mixed-valence ternary platinum oxides; $M_xPt_3O_4$ and MPt_3O_6. D, 1982, SUNY at Stony Brook. 255 p.

Schwartz, Martha. Computer comparison of dedrifting versus tare removal in precision gravity surveys. M, 1984, California State University, Long Beach. 123 p.

Schwartz, Maurice. Field and laboratory study of sea level rise as a cause of shore erosion. M, 1964, Columbia University, Teachers College.

Schwartz, Maurice Leo. Beach profile translation; scale of shore erosion. D, 1966, Columbia University. 125 p.

Schwartz, Michael. Mass transport in calcium silicates. D, 1968, Rensselaer Polytechnic Institute. 206 p.

Schwartz, P. H. Analysis and performance of hydraulic sandfill levees. D, 1976, University of Iowa. 290 p.

Schwartz, Robert Karl. Stratigraphic and petrographic analysis of the Lower Cretaceous Blackleaf Formation, southwestern Montana. D, 1972, Indiana University, Bloomington. 268 p.

Schwartz, Roland J. Detailed geological reconnaissance of the central Tortilla Mountains, Pinal County, Arizona. M, 1954, University of Arizona.

Schwartz, Susan Ynid. A paleomagnetic study of thrust sheet rotations in response to foreland impingement in the Wyoming-Idaho overthrust belt. M, 1983, University of Michigan.

Schwartz, Susan Ynid. Fault zone heterogeneity and earthquake occurrence in subduction zones. D, 1988, University of Michigan. 170 p.

Schwartzlow, Carl R. The origin of the dolomite and the granularity of the Chouteau Formation. D, 1932, University of Missouri, Columbia.

Schwartzman, David William. Excess argon in minerals from the Stillwater Complex, (Precambrian, Montana). M, 1966, Brown University. 38 p.

Schwartzman, David William. Excess argon in the Stillwater Complex (Precambrian, Montana) and the problem of mantle-crustal degassing. D, 1971, Brown University. 120 p.

Schwarz, Charles Fredrick. Conceptual ecologic modeling in regional environmental management and land planning; a case study of Lake Tahoe water color transparency. D, 1982, University of California, Berkeley. 508 p.

Schwarz, Charles G. The paleomagnetism of Mesozoic sedimentary, volcanic, and plutonic rocks in the Methow Valley, eastern Okanogan County, Washington. M, 1985, Western Washington University. 144 p.

Schwarz, David L. Geology of the Lower Permian Dry Mountain Trough, Buck Mountain, Limestone Peak, and Secret Canyon areas, east-central Nevada. M, 1987, Joint program, Idaho State Univ. and Boise State Univ. 149 p.

Schwarz, Frederick P., Jr. Geology and ore deposits of Minnie Gulch, San Juan County, Colorado. D, 1967, Colorado School of Mines.

Schwarz, Mary E. Bowers. Cephalopods of the Barnett Formation, central Texas. M, 1975, University of Texas, Austin.

Schwarz, Ursula Agnes Maria. Chlorofluorocarbon (freon) 11 and 12 measurements and their application in multi-parameter water mass tracing in the coastal seas of southern British Columbia. D, 1988, University of Victoria. 186 p.

Schwarzbach, Theodore Jeremiah. Geology of the Beulah area, Pueblo County, Colorado. M, 1961, University of Texas, Austin.

Schwarze, David Martin. Geology of the Lava Hot Springs Area, Idaho. M, 1959, University of Idaho. 43 p.

Schwarzer, Rudolph R., Jr. Order-disorder relationships in plagioclase feldspars from the Adirondack Anorthosite Complex [Precambrian], New York. M, 1966, Rensselaer Polytechnic Institute. 78 p.

Schwarzer, Rudolph Reynolds. The concentration and distribution of uranium in coexisting pegmatite feldspars by the fission track method. D, 1969, Rensselaer Polytechnic Institute. 140 p.

Schwarzer, Theresa F. A theoretical treatment of the non-freezing behavior of certain argillaceous dolomites. M, 1966, Rensselaer Polytechnic Institute. 132 p.

Schwarzer, Theresa Frances. The distribution of major and trace elements in coexisting feldspar pairs from selected pegmatites in New York State. D, 1969, Rensselaer Polytechnic Institute. 245 p.

Schwarzman, Elisabeth C. Geology of Alum Rock Park, San Jose, California. M, 1969, Stanford University.

Schwarzwalder, Robert Nathan, Jr. Systematics and early evolution of the Platanaceae. D, 1986, Indiana University, Bloomington. 198 p.

Schweers, Frederick Paul. Some Lower Pennsylvanian foraminifers of Spencer County, Indiana. M, 1940, Indiana University, Bloomington. 25 p.

Schwegal, Steven R. Holocene foraminifera and Ostracoda from a New Jersey barrier island complex. M, 1981, University of Nebraska, Lincoln.

Schweger, Charles E. Pollen analyses of Iola Bay and paleoecology of the Two Creeks Interval (Pleistocene, eastern Wisconsin). M, 1966, University of Wisconsin-Madison.

Schweger, Charles Earl. Late Quaternary paleoecology of the Onion Portage region, northwestern Alaska. D, 1976, University of Alberta. 183 p.

Schweickert, Richard Allan. Shallow-level intrusions in the eastern Sierra Nevada. D, 1972, Stanford University. 85 p.

Schweig, Eugene Sidney, III. Late Cenozoic stratigraphy and tectonics of the Darwin Plateau, Inyo County, California. M, 1982, Stanford University. 85 p.

Schweig, Eugene Sidney, III. Neogene tectonics and paleogeography of the southwestern Great Basin, California. D, 1985, Stanford University. 229 p.

Schweinfurth, Mark Fred. Stratigraphy and paleoecology of the highest strata exposed at Cincinnati, Ohio. M, 1958, University of Cincinnati. 114 p.

Schweinfurth, Stanley Paul. Regional devolatilization of Devonian oil shales of eastern Pennsylvania (a preliminary study). M, 1953, University of Cincinnati. 41 p.

Schweitzer, Janet. Cleavage development in dolomite, Elbrook Formation, Pulaski thrust sheet, Southwest Virginia. M, 1984, Virginia Polytechnic Institute and State University. 111 p.

Schweitzer, Peter Neil. Ontogeny and heterochrony in the ostracode Cavellina Coryell from Lower Permian rocks in Kansas. M, 1986, University of Kansas. 43 p.

Schweller, William J. Chile Trench; extensional rupture of oceanic crust and the influence of tectonics on sediment distribution. M, 1976, Oregon State University. 90 p.

Schweller, William John. Origin and emplacement history of the Zambales Ophiolite, Luzon, Philippines. D, 1982, Cornell University. 152 p.

Schwellnus, Jurgen Erdmann Gotthilf. Ore controls in deposits of the Knob Lake area, Labrador Trough. D, 1957, Queen's University. 156 p.

Schwellnus, Jurgen Erdmann Gotthilf. Sedimentary and structural control of basal reef goldfields, Orange Free State, South Africa. M, 1953, Queen's University. 87 p.

Schwendeman, J. F. Trace element zoning in galena. M, 1975, University of Kentucky. 152 p.

Schwendinger, William W. A detailed resistivity survey with reference to the cause of the magnetic anomaly immediately southeast of St. Charles, in St. Louis County, Missouri. M, 1950, St. Louis University.

Schwengber, Janet Ruth. Agricultural districts and environmental land use decisions in Delaware County, New York. D, 1981, State University of New York, College of Environmental Science and Forestry. 201 p.

Schwenn, Mary Bernadette. Creep of olivine during hot-pressing. M, 1977, Massachusetts Institute of Technology. 105 p.

Schwert, D. P. Paleoentomological analyses of two postglacial sites in eastern North America. D, 1978, University of Waterloo.

Schwessinger, William T. Spinel-silicate equilibria in the system MgO-FeO-Fe_2O_3-Al_2O_3-Cr_2O_3-SiO_2 and comparisons with mafic igneous rocks. M, 1984, Pennsylvania State University, University Park. 128 p.

Schwetz, Diane L. Radioisotope geochronology of sediments in Green Bay, Lake Michigan. M, 1982, University of Wisconsin-Milwaukee. 161 p.

Schweyen, Timothy. The sedimentology and paleohydrology of Waldsea Lake, Saskatchewan, an ectogenic meromictic saline lake. M, 1984, University of Manitoba.

Schwietering, Joseph Francis. Devonian shales of Ohio and their eastern equivalents. D, 1970, Ohio State University.

Schwimmer, Barbara A. Stratigraphic distribution of brachiopods and pelecypods in the Upper Devonian (Famennian) Chagrin Shale in the Cuyahoga River valley, northeastern Ohio. M, 1988, Kent State University, Kent. 134 p.

Schwimmer, David R. The growth and allometry of the trilobite Phacops rana. M, 1969, SUNY at Buffalo. 44 p.

Schwimmer, David Richard. The middle-Cambrian biostratigraphy of Montana and Wyoming. D, 1973, SUNY at Stony Brook. 465 p.

Schwimmer, Peter M. Paleomagnetism of the Eocene volcanics in northeastern Washington, and implications for the tectonic evolution of the Pacific Northwest. M, 1981, Western Washington University. 93 p.

Schwimmer, Reed Andrew. Geochemistry and mineralogy of four ultramafic complexes in the Ketchikan area, southeastern Alaska. M, 1986, Bryn Mawr College. 47 p.

Schwind, Joseph J. Von see Von Schwind, Joseph J.

Schwind, Joseph John. Converted waves, PS type, resulting from large explosions. M, 1960, University of Utah. 124 p.

Schwind, R. F. Conodonts from the Hillsdale Limestone (Upper Mississippian) in the Greendale Syncline, Smyth and Washington counties, Virginia. M, 1967, Virginia Polytechnic Institute and State University.

Schwindinger, Kathleen Rose. Petrogenesis of olivine aggregates from the 1959 eruption of Kilauea Ski; synneusis and magma mixing. D, 1987, University of Chicago. 213 p.

Schwitter, Michael. Geometry changes of glaciers in relation to the time scales for adjustment to glacier change. M, 1988, University of Washington. 57 p.

Schwochow, S. D. Surficial geology of the Eastlake Quadrangle, Adams County, Colorado. M, 1972, Colorado School of Mines. 152 p.

Schymiczek, Herman Bodo. Benthic foraminifera and paleobathymetry of the Eocene Llajas Formation, southwestern Santa Susana Mountains, California. M, 1983, California State University, Northridge. 86 p.

Sciacca, Thomas P., Jr. An electron microscopic study of the aggregation of synthetic pitchblende. M, 1956, Columbia University, Teachers College.

Sciarrillo, Joanne R. Pliocene-Pleistocene biostratigraphy of a deep-sea piston core from the Equatorial Pacific. M, 1976, Rutgers, The State University, Newark. 64 p.

Scibek, John C. Clay mineral associatons in the turbidity maximum of the Delaware Estuary at low flow. M, 1982, University of Delaware, College of Marine Studies. 105 p.

Sciple, Larry. Depositional and diagenetic environments of Fort Payne Mounds (Mississippian), northern Cumberland Plateau, Tennessee. M, 1981, Vanderbilt University. 107 p.

Sclar, Charles B. The Preston Gabbro and the associated metamorphic gneisses, New London County, Connecticut. D, 1950, Yale University.

Sclosser, Thomas N. Surficial geology, western Lake Nipigon area, Ontario. M, 1983, University of Wisconsin-Milwaukee. 171 p.

Scoates, Reginald Francis Jon. The distribution of copper and nickel and related platinum group metals in orebodies at Gordon Lake, Ontario. M, 1963, University of Manitoba.

Scoates, Reginald Francis Jon. Ultramafic rocks and associated copper-nickel sulphide ores, Gordon Lake, Ontario. D, 1972, University of Manitoba.

Scobey, Ellis Hurlbut. Sedimentary studies of the Wapsipinicon Formation in Iowa. D, 1938, University of Iowa. 173 p.

Scobey, Ellis Hurlbut. The Alexandrian problem in Iowa. M, 1935, University of Iowa. 39 p.

Scobey, Warren Barrett. Origin of the (Devonian) "Helderberg" limestones in the Syracuse (New York) area. M, 1940, Syracuse University.

Scofield, D. H. Fold forms in the Monterey and Santa Margarita formations along Lopez Canyon, Tar Spring Ridge (7 1/2') Quadrangle, San Luis Obispo County, California. M, 1975, Stanford University. 61 p.

Scofield, Jane Ann Rutledge. Pollen analysis of the late Wisconsin sediments from the Willcox Basin, Arizona. M, 1973, Queens College (CUNY).

Scofield, Nancy L. C. Mineral chemistry applied to interrelated albitization, pumpellyitization, and native copper redistribution in some Portage Lake basalts, Michigan. D, 1976, Michigan Technological University. 192 p.

Scofield, Nancy Lou. Vertical variation in the layered series, Raggedy Mountain Gabbro Group (Precambrian), Kiowa County, Oklahoma. M, 1968, University of Oklahoma. 155 p.

Scolaro, Reginald Joseph. Paleoecology of the Bryozoa of the Chipola Formation (Miocene, lower), Clarksville area, Florida. D, 1968, Tulane University. 256 p.

Scolaro, Reginald Joseph. Some Florida upper Miocene Bryozoa. M, 1964, University of Florida. 103 p.

Scopel, Louis Joseph. The volcanic history of Jackson Hole, Wyoming. M, 1949, Wayne State University.

Scorer, John D. A survey of reservoir rock compaction, and its influence on well flow tests. M, 1971, Stanford University. 156 p.

Scotese, Christopher R. The assembly of Pangea; middle and late Paleozoic paleomagnetic results from North America. D, 1985, University of Chicago. 339 p.

Scotese, Thomas Richard. Generic siting and design of mined caverns for disposal of low-level radioactive wastes. M, 1981, University of Arizona. 317 p.

Scotford, David M. Structure of the Sugar Loaf Mountain area, Maryland, as a key to Piedmont stratigraphy. D, 1950, The Johns Hopkins University.

Scott, A. An examination of the effects of weathering on engineering design in southern Ontario shales. M, 1981, University of Waterloo. 117 p.

Scott, A. R. Organic and inorganic chemistry, oil source rock correlation, and diagenetic history of the Permian Spraberry Formation. M, 1988, Sul Ross State University.

Scott, Alan Johnson. Late Devonian and Early Mississippian conodont faunas of the Upper Mississippi Valley. D, 1958, University of Illinois, Urbana. 218 p.

Scott, Allan J. The Muddy Creek Formation; depositional environment, provenance, and tectonic significance in the western Lake Mead area, Nevada and Arizona. M, 1988, University of Nevada, Las Vegas. 114 p.

Scott, Andrew William. Structure and stratigraphy of the lower Wilcox, Lobo Trend, in central Webb County, Texas. M, 1985, Texas A&I University. 60 p.

Scott, Barry. The diorite complex beneath the Sullivan orebody and its associated alterations, Kimberly, British Columbia. M, 1954, Queen's University. 131 p.

Scott, Billy. Using infrared spectrophotometry in mineralogical and petrological analysis. M, 1969, SUNY at Buffalo.

Scott, Blaine Pierce. Trace copper and zinc in the Coronation Mine. M, 1963, University of Saskatchewan. 72 p.

Scott, Charles Thomas. Design of optimal two-stage multiresource surveys. D, 1981, University of Minnesota, Minneapolis. 153 p.

Scott, D. R. Marine benthic communities of the Reading Limestone (Upper Pennsylvanian), Atchison County, Kansas. M, 1973, Kansas State University. 135 p.

Scott, Darcy Lon. Geology of the Wasootch Creek map area. M, 1959, University of British Columbia.

Scott, Darcy Lon. Stratigraphy of the lower Rocky Mountain Supergroup in the southern Canadian Rocky Mountains. D, 1964, University of British Columbia. 271 p.

Scott, David. Investigation of a surface resistivity technique for the estimation of hydraulic conductivity in stratified drift aquifers. M, 1980, University of Connecticut. 98 p.

Scott, David B. Distributions and population dynamics of marsh-estuarine foraminifera with applications to relocating Holocene sea-level. D, 1977, Dalhousie University. 207 p.

Scott, David B. Recent benthonic foraminifera from Samish and Padilla bays, Washington. M, 1973, Western Washington University. 63 p.

Scott, David Holcomb. The geology of the southern Pancake Range and Lunar Crater volcanic field, Nye County, Nevada. D, 1969, University of California, Los Angeles. 209 p.

Scott, David Kendall. The use of the United States public land surveys as part of a general system of horizontal control. M, 1954, Ohio State University.

Scott, David Russell. Magmons; solitary waves arising in the buoyant ascent of magma by porous flow through a viscously deformable matrix. D, 1987, California Institute of Technology. 107 p.

Scott, Deborah L. Modern processes in a continental rift lake; an interpretation of 28 kHz seismic profiles from Lake Malawi, East Africa. M, 1988, Duke University. 82 p.

Scott, Diane M. A gravity survey of Hamilton County, Ohio, for the purpose of delineating buried valleys. M, 1985, Wright State University. 91 p.

Scott, Douglas Lindsay. A seismic study of the Haughton Crater, Devon Island, Canadian Arctic. M, 1988, University of Saskatchewan. 147 p.

Scott, Earl Harold. Digital processing of seismic reflection data for purposes of a comparison, Harlem Township, Delaware County, Ohio. M, 1984, Wright State University. 91 p.

Scott, Edith Elizabeth Bohrer. The genus Protosalvinia from Kentucky, U.S.A. D, 1980, George Washington University. 139 p.

Scott, Erwin Ralph. Structural control of ore deposition in the Upper Mississippian Valley lead-zinc district. M, 1934, Northwestern University.

Scott, Fenton J. Wall-rock alteration and ore deposition at the Needle Mountain copper deposits, Gaspe, Quebec. M, 1950, University of New Brunswick.

Scott, Gary Robert. Geology, petrology, Sr isotopes and paleomagnetism of the Hackberry Mountain volcanic center. M, 1974, University of Texas at Dallas. 71 p.

Scott, Gary Robert. Paleomagnetism of Carboniferous and Triassic strata from cratonic North America. D, 1975, University of Texas at Dallas. 167 p.

Scott, Gayle. The Duck Creek Formation of North Texas. M, 1920, Texas Christian University. 60 p.

Scott, George L., Jr. Areal geology of portions of Beckham, Greer, Kiowa, and Washita counties, Oklahoma. M, 1955, University of Oklahoma. 108 p.

Scott, George Norman. Temperature equilibration in boreholes; a statistical approach. M, 1982, University of Michigan.

Scott, Gerald Lee. Permian sedimentary frame work of the Four Corners region. D, 1960, University of Wisconsin-Madison. 137 p.

Scott, Gerald Lee. Some stratigraphic studies of the Atoka Formation in east-central Oklahoma. M, 1957, University of Wisconsin-Madison. 44 p.

Scott, Gertrude Murray. A study of certain faunal assemblages of the Meaford Formation (Upper Ordovician) from exposures by the Credit River and Streetsville, Ontario, Peel County. M, 1965, University of Toronto.

Scott, Graham Howard. The textural interpretation and genetic significance of some Canadian massive sulphide deposits. M, 1970, SUNY at Buffalo. 127 p.

Scott, H. S. Geology of the cryolite deposit at Ivigtut, Greenland. D, 1951, University of Toronto.

Scott, Halley Mering. Chouteau Formation of central Missouri. M, 1914, University of Missouri, Columbia.

Scott, Harold. Pleistocene Ostracoda from Illinois. D, 1953, [University of Illinois, Urbana].

Scott, Harold B. A statistical analysis of the heavy minerals of the offshore islands of the Mississippi Sound from Dauphin Island to Cat Island. M, 1960, Mississippi State University. 47 p.

Scott, Harold William. Studies of the Quadrant Group of Montana and Wyoming. D, 1935, University of Chicago. 100 p.

Scott, Harold William. The fauna of the Galena Limestone of northwestern Illinois. M, 1931, University of Illinois, Chicago.

Scott, Harold William. Zoological relationships of the conodonts. D, 1935, University of Chicago.

Scott, Henry Kenneth. Manganese deposits of Rolette County, North Dakota. M, 1935, McMaster University.

Scott, Horace A. The geology of the claims of the Superior and Boston Copper Company. M, 1918, Northwestern University.

Scott, Irving Day. Uber interessante Amerikanische pyritkrystalle. M, 1907, University of Michigan.

Scott, J. D. Crystal structure of synthetic Sn_2S_3 and the naturally occurring tin sulphide mineral, franckeite. D, 1970, Queen's University. 149 p.

Scott, James Alan Bryson. Upper Cretaceous foraminifera of the Haslam Qualicum and Trent River formations, Vancouver Island, British Columbia. D, 1974, University of Calgary. 174 p.

Scott, James B. Structure of the ore deposits at Santa Barbara, Chihuahua, Mexico. M, 1959, University of Nevada - Mackay School of Mines. 33 p.

Scott, James Douglas. Studies in the ternary system PbS-Bi_2S_3-Ag_2S. M, 1966, University of Toronto.

Scott, James W. The base of the Cambrian in southeastern Newfoundland. M, 1952, University of Wisconsin-Madison.

Scott, Jerry Douglas. Subsurface stratigraphic analysis, "Cherokee" Group (Pennsylvanian) northern Noble County, Oklahoma. M, 1970, University of Oklahoma. 56 p.

Scott, John Stanley. Surficial geology of the Elbow-Outlook area, Saskatchewan, Canada. D, 1960, University of Illinois, Urbana. 133 p.

Scott, Kenneth Charles. Hydrogeologic and geophysical analysis of selected diatremes in the Hopi Buttes area, Arizona. M, 1975, Northern Arizona University. 129 p.

Scott, Kenneth Eugene. Fauna and age of Leadville Limestone (Mississippian) in part of west-central Colorado; Pitkin and Eagle counties. M, 1954, University of Colorado.

Scott, Kevin McMillan. Depositional dynamics of a Cretaceous flysch sequence, Patagonian Andes, southern Chile. D, 1964, University of Wisconsin-Madison. 163 p.

Scott, Kevin McMillan. Geology of the Waucoba Springs area, Inyo Mountains, California. M, 1960, University of California, Los Angeles.

Scott, Kriston H. Laboratory simulation of the diagenesis of Gulf Coast Tertiary shales. M, 1981, University of Missouri, Columbia.

Scott, Lewis P., IV. Tectonic history of the Kentucky River fault system in eastern Kentucky during the Silurian and Devonian. M, 1978, Eastern Kentucky University. 59 p.

Scott, Margaret Ann. "Grass-roots" seismic safety program for citizens of Southeast Missouri. M, 1986, Southeast Missouri State University. 71 p.

Scott, Martha Lyles Richter. Distribution of clay minerals on the British Honduras Shelf. D, 1966, Rice University. 116 p.

Scott, Mary Woods. Annotated bibliography of the geology of North Dakota, 1806-1959. M, 1972, University of North Dakota. 133 p.

Scott, Norman. Modern vs. ancient braided stream deposits; a comparison between simulated sedimentary deposits and the Ivishak Formation of the Prudhoe Bay Field, Alaska. M, 1986, Stanford University. 103 p.

Scott, Norman Jackson, Jr. A zoogeographic analysis of the snakes of Costa Rica. D, 1969, University of Southern California. 405 p.

Scott, Owen A. Geology of the northeastern quarter of the Columbus Quadrangle, Mississippi. M, 1965, Mississippi State University. 55 p.

Scott, Patricia Frances. Applications of the Kirchhoff-Helmholtz integral to problems in body wave seismology. D, 1985, California Institute of Technology. 187 p.

Scott, Philip. Geochemistry and petrography of the Salmon River lead deposit, Cape Breton Island, Nova Scotia. M, 1980, Acadia University. 111 p.

Scott, Phyllis Wilk. Hydrogeological-structural analysis of the Woody Mountain Well Field area with geo-

Scott, Horace A. The geology of the claims of the Superior and Boston Copper Company. M, 1918, Northwestern University.

physical interpretations, Coconino County, Arizona. M, 1974, Northern Arizona University. 79 p.

Scott, R. W. Geology of Rabbit Ears Pass area, Jackson and Grand counties, Colorado. M, 1961, University of Wyoming. 73 p.

Scott, Ralph Carter, Jr. The geomorphic significance of debris avalanching in the Appalachian Blue Ridge Mountains. D, 1972, University of Georgia. 194 p.

Scott, Richard A. Fossil fruits and seeds from the Eocene Clarno Formation of Oregon. D, 1953, University of Michigan. 111 p.

Scott, Richard M. Examination of selected Holocene-Pleistocene beach environments, Southeast Atlantic Coast. M, 1976, University of Georgia.

Scott, Robert Blackburn, III. The Tertiary geology and ignimbrite petrology of the Grant Range, east-central Nevada. D, 1965, Rice University. 166 p.

Scott, Robert Brown. Tertiary foraminifera from the Oficina Field, Anzoategui, Venezuela. M, 1951, University of Illinois, Urbana.

Scott, Robert Earl. Radiation damage in rock-forming minerals. M, 1977, Massachusetts Institute of Technology. 195 p.

Scott, Robert J. Lithologic control of faunas in the Middle Ordovician Platteville Formation of Wisconsin. M, 1962, University of California, Los Angeles.

Scott, Robert Karl. Stratigraphy and depositional environments of a Neogene playa-lake system, China Ranch Beds, near Death Valley, California. M, 1985, Pennsylvania State University, University Park. 249 p.

Scott, Robert W. Paleontology and paleoecology of the Kiowa Formation (Lower Cretaceous) in Kansas. D, 1967, University of Kansas. 308 p.

Scott, Sally C. Ontogeny of Rhodocrinus douglassi var. serpens. M, 1962, University of Wisconsin-Madison.

Scott, Stephen M. The reprocessing and extended interpretation of seismic reflection data recorded over the Hayesville-Fries thrust sheet in southwestern North Carolina. M, 1987, Virginia Polytechnic Institute and State University.

Scott, Steven Donald. Silver mineralization in number 13 vein system, Siscoe Metals of Ontario, Gowganda. M, 1964, University of Western Ontario. 119 p.

Scott, Steven Donald. Stoichiometry and phase changes in zinc sulfide. D, 1968, Pennsylvania State University, University Park. 165 p.

Scott, Susan A. Trace element study of sulphides from the Temagami Mine, Ontario (Canada). M, 1969, McGill University. 148 p.

Scott, Theodore. Sand movement by waves. M, 1954, University of California, Berkeley. 69 p.

Scott, Thomas Dwayne, Jr. A geochemical and isotopic study of the Garber-Wellington Aquifer, Cleveland County, Oklahoma. M, 1988, University of Oklahoma. 122 p.

Scott, Thomas M. The clay mineralogy of the Silurian Crab Orchard Formation of east-central Kentucky. M, 1973, Eastern Kentucky University. 67 p.

Scott, Thomas Melvin. The lithostratigraphy of the Hawthorn Group (Miocene) of Florida. D, 1986, Florida State University. 478 p.

Scott, Thurman Eugene, Jr. A strain analysis of the Marquenas Quartzite and contact relationships of Ortega-Vadito groups. M, 1980, University of Texas at Dallas. 177 p.

Scott, Troy Calvin. Investigation of the Murray Landfill; an uncontrolled hazardous waste site. M, 1985, University of Nevada. 98 p.

Scott, W. D. Fluvial sedimentation characteristics of base level altered segments of the Holland River. M, 1975, University of Waterloo.

Scott, W. F. Differential thermal analysis of some minerals of the chlorite group. M, 1947, University of Utah. 31 p.

Scott, W. J. Experimental measurement of induced polarization of some synthetic metalliferous samples at low current densities. M, 1965, University of Toronto.

Scott, Waldemar F. Some engineering geological aspects of the Sydenham-Harrowsmith area, Ontario. M, 1971, Queen's University. 171 p.

Scott, Willard Frank. Regional physical stratigraphy of the Triassic in a part of the Eastern Cordillera. D, 1954, University of Washington. 142 p.

Scott, William A. Contact metamorphism at Britannia Mines. M, 1929, University of Wisconsin-Madison.

Scott, William E. Petrographic characteristics of late Quaternary tephra from Mauna Kea Volcano, Hawaii. M, 1971, University of Washington.

Scott, William E. Quaternary glacial and volcanic environments, Metolius River area, Oregon. D, 1974, Washington University. 95 p.

Scott, William Henry. Evolution of folds in the metamorphic rocks of western Dovrefjell, Norway (parts I and II). D, 1967, Yale University. 243 p.

Scott, William James. Phase angle measurements in the induced polarization method of geophysical prospecting. D, 1972, McGill University.

Scott, William P. A volcanic hosted gold occurrence in Marathon County, Wisconsin. M, 1988, University of Wisconsin-Madison. 95 p.

Scotto, Susan. U.S.-Mexican cooperation in the prevention of marine oil pollution. M, 1982, University of Delaware, College of Marine Studies. 158 p.

Scougale, John Douglas. The stratigraphy of the Hale Formation in Madison, Newton, Boone, and Carroll counties, Arkansas. M, 1954, University of Arkansas, Fayetteville.

Scowen, P. A. H. Re-equilibration of chromite from Kilauea Iki lava lake, Hawaii. M, 1986, Queen's University. 119 p.

Scoyoc, G. E. van see van Scoyoc, G. E.

Scranton, Mary Isabelle. The marine geochemistry of methane. D, 1977, Massachusetts Institute of Technology. 251 p.

Scratch, Richard Boyd. Geologic, structural, fluid inclusion and oxygen isotope study of the Lake George antimony deposit, southern New Brunswick. D, 1981, University of Western Ontario. 195 p.

Screaton, Elizabeth J. Fluid flow in the Barbados Ridge complex; a model of dewatering in the toe of the complex. M, 1988, University of California, Santa Cruz.

Scribbins, Brian Thomas. Exotic inclusions from the South Range sublayer, Sudbury. M, 1978, University of Toronto.

Scrimshire, Elven Rick. Hydrocarbon potential of the Pennsylvanian Tensleep Formation in central Big Horn County, Wyoming. M, 1985, West Texas State University. 65 p.

Scrivens, Paul R. Analysis of sonobuoy profiles from Tongue of the Ocean and Exuma Sound, Bahamas. M, 1983, University of Delaware, College of Marine Studies. 164 p.

Scrivner, Clarence Leland. Morphology, mineralogy and chemistry of the Lebanon silt loam. D, 1960, University of Missouri, Columbia. 175 p.

Scrivner, James V. The geology of Mine Ridge, Baker County, Oregon. M, 1983, Washington State University. 114 p.

Scroggs, Doyle L. Bedrock stratigraphy in Black Squirrel Creek Basin, El Paso County, Colorado. M, 1971, Colorado State University. 82 p.

Scrudato, Ronald John. Geology and petrology of the Diamond Joe Quarry, Arkansas. M, 1964, Tulane University. 32 p.

Scrudato, Ronald John. Kaolin and associated sediments of east-central Georgia. D, 1969, University of North Carolina, Chapel Hill. 97 p.

Scruton, Philip Challacombe. Marine geology of the Near Islands shelf, Alaska. D, 1953, University of California, Los Angeles.

Scruton, Philip Challacombe. The petrography and environment of deposition of the Warner, Little Cabin, and Hartsthorne sandstones of northeastern Oklahoma. M, 1949, University of Tulsa. 77 p.

Scudder, Phyllis Jeanne. Post-Mississippian topography and areal stratigraphy of Vanderburgh County, Indiana. M, 1958, Indiana University, Bloomington. 44 p.

Scudder, Ronald Jay. Geology of Cheesequake State Park, New Jersey. M, 1955, Rutgers, The State University, New Brunswick. 83 p.

Scuderi, Louis Anthony. A dendroclimatic and geomorphic investigation of late-Holocene glaciation, southern Sierra Nevada, California. D, 1984, University of California, Los Angeles. 247 p.

Scuderi, Louis Anthony. Environment influences on cirque form and location, southern Sierra Nevada, California. M, 1978, University of California, Los Angeles. 60 p.

Scull, Berton James. A further study of the igneous rocks in the Granite-Lugert area, Oklahoma. M, 1947, University of Oklahoma. 65 p.

Scull, Berton James. Origin and occurrence of barite in Arkansas. D, 1956, University of Oklahoma. 208 p.

Scully, Elizabeth Mary. Silvermine Granite; a differentiated ring pluton of the St. Francois Mountains Batholith, southeastern Missouri. M, 1978, University of Kansas. 45 p.

Sea, Frédéric. Analyse par ordinateur des linéaments structuraux de l'Ile d'Anticosti, Québec. M, 1987, Ecole Polytechnique. 190 p.

Seaber, Paul R. A detailed study of a portion of the Ordovician and Silurian of the Rochester Gorge (New York). M, 1957, University of Rochester. 187 p.

Seaber, Paul Robert. Variations in the chemical character of the water in the Englishtown Formation, New Jersey. D, 1962, University of Illinois, Urbana. 74 p.

Seaberg, John Karl. Geohydrologic interpretation of glacial geology near Williams Lake, central Minnesota, with emphasis on lake-groundwater interaction. M, 1985, University of Minnesota, Minneapolis. 141 p.

Seager, George F. Geology and ore deposits of the Iron Mountain and Little Backbone districts, Shasta County, California. D, 1936, Yale University.

Seager, Oramel Ainsworth. Some contact effects of the Duluth Gabbro on the Gunflint iron-bearing formation. M, 1929, Northwestern University.

Seager, William R. Geology of the Jarilla Mountains, Tularosa Basin, New Mexico. M, 1961, University of New Mexico. 80 p.

Seager, William Ralph. Geology of the Bunkerville section of the Virgin Mountains, Nevada and Arizona. D, 1966, University of Arizona. 118 p.

Seagle, E. P. Types of cavities in New England pegmatites. M, 1954, Columbia University, Teachers College.

Seagle, S. M. Field investigation of epiphytic algal populations from a spring fed aquatic environment in Texas with ancillary laboratory studies. D, 1977, University of Texas, Austin. 276 p.

Seal, Robert R., II. The Lake George W-Mo stockwork deposit, southwestern New Brunswick. M, 1984, Queen's University. 202 p.

Seal, Thomas Lee. Pre-Grenville dehydration metamorphism in the Adirondack Mountains, New York; evidence from pelitic and semi-pelitic metasediments. M, 1986, SUNY at Stony Brook. 84 p.

Seale, Gary L. Relationship of possible Silurian reef trend to middle Paleozoic stratigraphy and structure of the southern Illinois Basin of western Kentucky. M, 1981, University of Kentucky. 72 p.

Seale, John David. Depositional environments and diagenesis of Upper Pennsylvanian Marchand sandstones on south, east, and northeast flanks of the Anadarko Basin. M, 1980, Oklahoma State University. 74 p.

Seals, Mary J. Lithostratigraphic and depositional framework near-surface Upper Pennsylvanian and Lower Permian strata, southern Brazos Valley, north-central Texas. M, 1966, Baylor University. 128 p.

Seals, Wilburn Hale. A study of the insoluble residues of the Cap Mountain Formation of central Texas. M, 1939, University of Texas, Austin.

Sealy, Brian E. Geology of the Voca-North area, McCulloch County, Texas. M, 1963, Texas A&M University.

Sealy, George, Jr. Geologic history of Painthorse Quadrangle, Culberson County, Texas. M, 1953, University of Texas, Austin.

Sealy, James E., Jr. A mathematical model for river delta development and its application to the New Brazos River delta at Freeport, Texas. M, 1974, Texas A&M University. 130 p.

Seaman, David Martin. Minerals and mineral deposits of the San Juan region (San Juan, San Miguel, and Ouray counties), Colorado. M, 1934, University of Colorado.

Seaman, Jane Marie. Evaluation of local and regional groundwater contamination at the Athens County Landfill. M, 1984, Ohio University, Athens. 190 p.

Seaman, Sheila J. Geology and petrogenesis of ash-flow tuffs and rhyolitic lavas associated with the Gila Cliff Dwellings basin, Bursum Caldera Complex, southwestern New Mexico. D, 1988, University of New Mexico. 170 p.

Seaman, Sheila June. Geology and ore potential of the Jupiter Canyon region, Baboquivari Mountains, Arizona. M, 1983, University of Arizona. 99 p.

Seamount, Daniel T. Well log analysis of the hydrothermal altered sediments of Cerro Prieto geothermal field, Baja California, Mexico. M, 1981, University of California, Riverside. 165 p.

Seanor, Clinton E. Migmatites and mafic boudins of the Gannett Peak area, Sublette and Fremont counties, Wyoming. M, 1988, Colorado State University. 174 p.

Seara, J. L. Developments in electrical prospecting methods. M, 1977, University of Western Ontario.

Search, Marshall Allen. The Winthrop Sandstone, Methow Valley, Washington. M, 1931, Washington State University. 35 p.

Searight, Thomas Kay. Geology of the Humansville Quadrangle, Missouri. M, 1952, University of Missouri, Columbia.

Searight, Thomas Kay. Post-Cheltenham Pennsylvanian stratigraphy of the Columbia-Hannibal region, Missouri. D, 1959, University of Illinois, Urbana. 282 p.

Searight, Walter Vernon. A correlation of the uppermost Cambrian strata of the Upper Mississippi Valley. M, 1924, University of Iowa. 108 p.

Searight, Walter Vernon. The geology of the Beardstown Quadrangle, Illinois. D, 1927, University of Iowa. 192 p.

Searle, Clark Wellington. Magnetic resonance of pure and doped crystals of hematite. D, 1965, University of Minnesota, Minneapolis. 132 p.

Searls, Craig Allen. The piezomagnetic effect; limits on detectability using a combined vector and scalar magnetometer array. D, 1981, University of California, Los Angeles. 309 p.

Sears, Charles Edward, Jr. Hydrothermal alteration and mineralization in the Climax molybdenum deposit, Climax (Summit County), Colorado. D, 1953, Colorado School of Mines. 56 p.

Sears, Charles Edward, Jr. Petrography of the Blue Ridge hematite. M, 1935, Virginia Polytechnic Institute and State University.

Sears, David Hume. Geology of the Pantano Hill area, Pima County, Arizona. M, 1939, University of Arizona.

Sears, David, II. Geology of the Pantano Hill area, Pima County, Arizona. M, 1959, University of Arizona.

Sears, Frederick Mark. Analysis of microcracks in dry polycrystalline NaCl by ultrasonic signal processing. D, 1980, University of California, Berkeley. 364 p.

Sears, James W. Structural geology of the Precambrian Grand Canyon Series, Arizona. M, 1973, University of Wyoming. 112 p.

Sears, James Walter. Tectonic contrasts between the infrastructure and suprastructure of the Columbian Orogen, Albert Peak area, western Selkirk Mountain, British Columbia. D, 1979, Queen's University. 154 p.

Sears, Joseph McHutchon. The Hollis Basin, southwestern Oklahoma. M, 1951, University of Oklahoma. 71 p.

Sears, Julian D. Conditions of sedimentation in the Conemaugh Formation of West Virginia. D, 1919, The Johns Hopkins University.

Sears, Peter J. Mercury in the base metal and gold ores of the province of Quebec, (Canada). M, 1969, Universite Laval.

Sears, Philip C. Evolution of Platt shoals, northern North Carolina shelf; interferences from areal geology. M, 1973, Old Dominion University. 82 p.

Sears, R. Bonner. Selective modification of sand size alluvium, Colorado River, Texas. M, 1978, University of Texas, Austin.

Sears, Richard Sherwood. Geology of the area between Canjilon Mesa and Abiquiu, Rio Arriba County, New Mexico. M, 1953, University of New Mexico. 71 p.

Sears, Stephen O'Reilly. Facies interpretations and diagenetic modifications of the Sunniland Limestone (Lower Cretaceous), South Florida; a thin section and x-ray diffraction analysis. M, 1972, University of Florida. 80 p.

Sears, Stephen O'reilly. Inorganic and isotopic geochemistry of the unsaturated zone in a carbonate terrane. D, 1976, Pennsylvania State University, University Park. 244 p.

Sears, William Arthur, Jr. Geology of the Deer Creek-Smith Creek area, Converse and Natrona counties, Wyoming. M, 1949, University of Wyoming. 56 p.

Seashore, Robert Holmes. Geology of the region about Belton in northwestern Montana. M, 1924, University of Iowa. 75 p.

Seastrom, Wesley C. Structural geology of the Canyon Spring Quadrangle, Riverside County, California. M, 1953, University of Southern California.

Seaton, William Joseph. Foraminiferal biostratigraphy and paleoecology of the Aquia Formation near Hanover, Virginia. M, 1981, Virginia Polytechnic Institute and State University. 103 p.

Seaver, Daniel Badger. Quaternary evolution and deformation of the San Emigdio Mountains and their alluvial fans, Transverse Ranges, California. M, 1986, University of California, Santa Barbara. 116 p.

Seavey, Donald Barker. Influence of small-scale chemical variation of sediment on possible metamorphic assemblages. M, 1971, Michigan State University. 55 p.

Seay, Christopher S. Chemical and petrological characteristics of the intrusive rocks of the Quitman Mountains, Texas. M, 1973, Texas A&M University. 110 p.

Seay, John G., Jr. The geology of the northern Piceance Creek Basin, northwestern Colorado. M, 1970, Ohio University, Athens. 124 p.

Sebenik, Paul Gregory. Physicochemical transformations of sewage effluent releases in an ephemeral stream channel. M, 1975, University of Arizona.

Seborowski, K. Damian. The composition and origin of the Beemerville carbonatite, Sussex County, New Jersey. M, 1982, Rutgers, The State University, Newark. 58 p.

Sebring, Louie, Jr. The Slick-Wilcox Field, DeWitt and Goliad counties, Texas. M, 1947, University of Texas, Austin.

Secco, Richard Andrew. Electrical, thermal, and thermoelectrical properties of solid and liquid iron at high pressures. D, 1988, University of Western Ontario.

Seccombe, Philip Kenneth. Sulphur isotope and trace element geochemistry of sulphide mineralisation in the Birch-Uchi greenstone belt, northwestern Ontario. D, 1973, University of Manitoba.

Secor, Dana M. The paleontology of a portion of the Devonian of the Natural Bridge region of the Virginia. M, 1932, Columbia University, Teachers College.

Secor, Donald Terry, Jr. Geology of the Central Spring Mountains, Nevada. D, 1963, Stanford University. 197 p.

Secor, Donald Terry, Jr. Joints in the Camels Hump area, Vermont. M, 1959, Cornell University.

Secord, David J. A generalized ray theory synthetic seismogram program. M, 1979, Dalhousie University. 177 p.

Secord, Theresa Karen. The geology and geochemistry of the Ore Hill zinc-lead-copper deposit, Warren, New Hampshire. M, 1984, University of Wisconsin-Madison. 99 p.

Secrist, Mark H. The zinc deposits of East Tennessee. D, 1923, The Johns Hopkins University.

Sedam, Alan Charles. Alcova Limestone of the Big Horn Basin, Wyoming. M, 1957, Ohio State University.

Seddon, George. Middle Paleozoic conodonts from the Llano Uplift, central Texas. D, 1965, University of Minnesota, Minneapolis. 463 p.

Sedenquist, Daniel Frederic. Petrography and diagenesis of volcanogenic sediments from the San Augustin Plains, western New Mexico; an example of zeolite authigenesis in a saline, alkaline environment. M, 1986, University of Colorado. 116 p.

Sedgeley, David R. A paleomagnetic study of some welded tuffs in central Arizona. M, 1976, Arizona State University. 110 p.

Sediqi, Atiqullah. A sedimentological and geochemical study of the Bigfork Chert in the Ouachita Mountains and the Viola Limestone in the Arbuckle Mountains. D, 1985, University of Oklahoma. 168 p.

Sedivy, Robert Alan. K-Ar relationships in a Cambrian shale as a function of burial depth. M, 1979, Georgia Institute of Technology. 146 p.

Sedlacek, Wanda Jane. Lateral velocity variations and the occurrence of geopressure in Brazoria County, Texas. M, 1981, University of Texas, Austin. 161 p.

Sedlock, Richard Louis. Lithology and structure of the Morgan Lewis area, Barbados, West Indies. M, 1982, University of California, Santa Barbara. 173 p.

Sedlock, Richard Louis. Lithology, petrology, structure, and tectonics of blueschists and associated rocks, west-central Baja California, Mexico. D, 1988, Stanford University. 249 p.

Sedore, Jacquelin. Model study of sorting by debris flow. M, 1966, Stanford University.

See, Bennie E. Palynology of the Mississippian Marshall Sandstone and Coldwater Shale of Michigan. M, 1980, Bowling Green State University. 66 p.

See, J. Melville. Origin and distribution of molybdenum in the vicinity of the Glove Mine, Santa Cruz County, Arizona. M, 1964, University of Arizona.

See, Thomas. Apollo 16 impact melt-splashes. M, 1985, University of Houston.

Seeburger, Donald Alan. Studies of natural fractures, fault zone permeability, and a pore space-permeability model. D, 1981, Stanford University. 243 p.

Seed, Raymond Boulton. Soil-structure interaction effects of compaction-induced stresses and deflections. D, 1983, University of California, Berkeley. 473 p.

Seedorf, Douglas Christopher. Upper Cretaceous through Eocene stratigraphy of the southern Ventura Basin, California. M, 1983, Oregon State University. 53 p.

Seedorff, Charles Eric. Geology of the Royston porphyry copper prospect, Nye and Esmeralda counties, Nevada. M, 1981, Stanford University. 103 p.

Seedorff, Charles Eric. Henderson porphyry molybdenum deposit; cyclic alteration-mineralization and geo-

chemical evolution of topaz- and magnetite-bearing assemblages. D, 1987, Stanford University. 460 p.

Seefeldt, David R. and Glerup, Melvin O. Stream channels of the Scenic Member of the Brule Formation (Oligocene), western Big Badlands, South Dakota. M, 1958, South Dakota School of Mines & Technology.

Seeger, Charles Ronald. A geophysical and geological investigation of the Jeptha Knob structure, Shelby County, Kentucky. D, 1966, University of Pittsburgh. 152 p.

Seeger, Charles Ronald. An examination of the geological history of the Washington D.C. area with particular reference to sedimentary tectonics. M, 1958, George Washington University.

Seegmiller, Ben Lorin. Correlation and detection of wave velocity and stress. D, 1969, University of Utah. 116 p.

Seegmiller, Ben Lorin. Strength of modified short column. M, 1966, University of Utah. 140 p.

Seekatz, Jeffrey G. Stratigraphic and structural features of part of the Sigsbee Escarpment, northwestern Gulf of Mexico. M, 1978, University of Texas, Austin.

Seekins, L. C. Lateral variations in the upper mantle beneath the Philippine Sea; a surface wave dispersion study. M, 1975, University of Southern California.

Seekins, William C. Groundwater resources of the upper Sweetwater River Basin. M, 1974, San Diego State University.

Seel, Charles P. Two mineralization types of western Mexico. M, 1950, Columbia University, Teachers College.

Seeland, David Arthur. Paleocurrents of the late Precambrian to Early Ordovician (Basal Sauk) transgressive clastics of the Western and Northern United States, with a review of the stratigraphy. D, 1968, University of Utah. 276 p.

Seeland, David Arthur. Stratigraphy and cross-bedding of the Aladdin Sandstone, Black Hills, South Dakota. M, 1961, University of Minnesota, Minneapolis. 124 p.

Seelen, Mark Allan. Tidal periodicity of and temperature effect on shell growth increments in subtidal and intertidal populations of the bivalve Mercenaria mercenaria. M, 1981, Indiana University, Bloomington. 103 p.

Seeley, William Oran. Geology of the southeastern quarter of the Dixonville Quadrangle, Oregon. M, 1974, University of Oregon. 77 p.

Seeling, Alan. The Shannon Sandstone, a further look at the environment of deposition at Heldt Draw, Wyoming. M, 1977, University of Colorado.

Seeling, R. R. Rb/Sr whole-rock isotopic geochemistry and petrology of the tonalitic-trondhjemitic phase of the Archean quartzo-feldspathic gneiss, Sacred Heart-North Redwood area, Minnesota River valley. M, 1977, University of Minnesota, Duluth.

Seely, Diana M. Estimation of ground water recharge in New Hampshire. M, 1984, University of New Hampshire. 110 p.

Seely, Donald Randolph. Geology of the Talihina area, Pushmataha, Latimer, and LeFlore counties, Oklahoma. M, 1955, University of Oklahoma. 77 p.

Seely, Donald Randolph. Structure and stratigraphy of the Rich Mountain area, Oklahoma and Arkansas. D, 1962, University of Oklahoma. 264 p.

Seely, Mark Richard. Depositional environments and dolomitization-silicification diagenesis of Warsaw-Salem-St. Louis sequence (Meramecian) Southwest St. Louis County, Missouri. M, 1985, Washington University. 168 p.

Seereeram, Devo. Plasticity theory for granular media. D, 1986, University of Florida. 325 p.

Seery, Virginia B. Glaciers of the islands of the Arctic Ocean. M, 1934, Cornell University.

Seevers, William J. Stratigraphy of the Stotler Limestone (Virgilian) between the Kansas River and Neosho River valleys, Kansas. M, 1960, University of Kansas. 149 p.

Seewald, Clyde Ray. Sedimentology of the Whitsett Formation (upper Eocene), south-central Texas. M, 1966, University of Texas, Austin.

Seewald, Jeffrey Steven. Na and Ca metasomatism during hydrothermal basalt alteration; an experimental and theoretical study. M, 1987, University of Minnesota, Minneapolis. 127 p.

Seewald, Kenneth Oscar. Stratigraphy of the Austin Chalk, McLennan County, Texas. M, 1959, Baylor University. 44 p.

Seff, Philip. Gastropods and paleoclimatology of the late Pleistocene of the Republican River valley, Nebraska. M, 1952, University of Nebraska, Lincoln.

Seff, Philip. Stratigraphic geology and depositional environments of the 111 Branch area, Graham County, Arizona. D, 1962, University of Arizona. 216 p.

Segal, Lucille B. A regional study of Florida. M, 1937, Columbia University, Teachers College.

Segal, Ronald Henry. A study of certain cardiocarpalean ovules from the American Carboniferous. D, 1966, University of Kansas. 71 p.

Segal, Rose J. Field evidence for the presence of secondary structures in the (Precambrian) Inwood Limestone in the type locality (New York). M, 1934, Columbia University, Teachers College.

Segall, Julius. The origin and occurrence of certain crystallographic intergrowths. M, 1915, University of Minnesota, Minneapolis. 13 p.

Segall, Paul. The development of joints and faults in granitic rocks. D, 1981, Stanford University. 244 p.

Segar, Robert L. A gravity and magnetic investigation along eastern flank of the Ozark Uplift (Missouri). M, 1965, Washington University. 163 p.

Segeler, Marie-Louise. A Miocene ostracode fauna from southeast Victoria, Australia. M, 1967, New York University.

Seggern, D. H. Von *see* Von Seggern, D. H.

Seggern, David Henry von *see* von Seggern, David Henry

Seglund, James Arnold. Geology of part of the Tendoy Mountains, near Red Rock, Beaverhead County, Montana. M, 1949, University of Michigan. 41 p.

Sego, P. D. Magnetic investigations east of Malaita Island. M, 1969, University of Hawaii. 37 p.

Segovia Nerhot, Antonio Valentin. The geology of Planch L-12 (Peralonso-Medina area) of the geologic map of Colombia. D, 1963, Pennsylvania State University, University Park. 263 p.

Segretto, Peter Ssalvatore. The relationship between urbanization and streamflow in the Hillsborough River basin, 1940-1970. D, 1975, University of Florida. 266 p.

Seguin, Maurice. Geology of Knob Lake Ridge, Schefferville, Quebec. M, 1963, McGill University.

Seguin, Maurice. Phase relations in the Fe-C-O-S-(H2O) systems. D, 1965, McGill University.

Seguinot-Barbosa, Jose. Coastal modificaton and land transformation in the San Juan Bay area; Puerto Rico. D, 1983, Louisiana State University. 320 p.

Sehayek, Lily. Unsaturated/saturated flow of liquids through deformable soils; numerical solution and applications to hazardous waste landfills, lagoon leaks, and accidental spills. D, 1987, Rutgers, The State University, New Brunswick. 628 p.

Sehgal, M. M. Neutron well logging in basalt hydrogeology (type area–Oahu, Hawaii). M, 1974, University of Hawaii. 108 p.

Sehnke, Errol Douglas. Gravitational gliding structures, Horse Creek area, Teton County, Wyoming. M, 1969, University of Michigan.

Seibel, Erwin. Shore erosion at selected sites along lakes Michigan and Huron. D, 1972, University of Michigan. 190 p.

Seibel, Geoffrey C. A geochemical approach to the dolomitization of the Epler Formation, Berks County, Pa. M, 1982, Lehigh University. 96 p.

Seibert, Barre Alan. Study of the response to partial demagnetization in an alternating magnetic field of rock samples collected from the Palisades diabase near Alpine, New Jersey. M, 1967, Stanford University.

Seibert, Jeffrey Lynn. Analysis of the "Clinton" Sandstone and its hydrocarbon potential in south-central Mahoning County, Ohio. M, 1987, University of Akron. 124 p.

Seibert, W. E., Jr. Metamorphic geology of the Forest Hill District, Nova Scotia. M, 1949, Massachusetts Institute of Technology. 41 p.

Seidel, Robert Eugene. Deflation basin stratigraphy; southwestern North Dakota. D, 1986, University of North Dakota. 235 p.

Seidell, Barbara Castens. The anatomy of a modern marine siliciclastic sabkha in a rift valley setting; northwest Gulf of California tidal flats, Baja California, Mexico. D, 1984, The Johns Hopkins University. 405 p.

Seidelman, Paul J. Seismically-induced geologic hazards in San Mateo, California. M, 1975, San Jose State University. 65 p.

Seidemann, D. E. K-Ar and Ar^{40}/Ar^{39} dating of deep-sea igneous rocks. D, 1976, Yale University. 206 p.

Seiden, Hyman. Geology of Las Llajas Canyon, Ventura County, California. M, 1972, University of California, Los Angeles.

Seidensticker, C. Michael. Geochemistry and regional correlation of pre-Tertiary volcanic rocks in west-central Nevada. M, 1983, Rice University. 96 p.

Seiders, Victor Mann. Geology of central Miranda, Venezuela. D, 1963, Princeton University. 304 p.

Seidl, Richard F. Geology and uranium evaluation of the Dry Valley tuffs, Washoe County, Nevada. M, 1982, University of Nevada. 83 p.

Seidman, Peter. A study of the accuracy with which U.S. Geological Survey 71/2 minute topographic maps depict stream channel networks for small drainage basins along the Catskill Escarpment near Kingston, New York. M, 1976, Brooklyn College (CUNY).

Seifert, Douglas J. Petrology and stratigraphy of the middle Monongahela Group (Pennsylvanian) in northern Morgan and western Noble counties, Ohio. M, 1982, University of Akron. 65 p.

Seifert, Gregory G. Hydrogeochemistry of coal mine spoil piles of various ages in Macon County, Missouri. M, 1982, University of Missouri, Columbia.

Seifert, Karl Earl. The genesis of plagioclase twinning in the Nonewaug Granite. D, 1963, University of Wisconsin-Madison. 105 p.

Seigel, Harold O. Theoretical and experimental investigations into the applications of the phenomenon of overvoltage to geophysical prospecting. D, 1948, University of Toronto.

Seigler, William C. A study of the cooling history of a pluton in the Finaly Mountains, Hudspeth County, Texas. M, 1987, University of Texas at El Paso.

Seigley, Lynette Sue. Origin of cherts in the Burlington Limestone (lower Middle Mississippian) of southeastern Iowa and western Illinois. M, 1987, University of Iowa. 119 p.

Seik, Lawrence Michael. Stratigraphy and structure of Pike and southern Canton townships, Stark County, Ohio. M, 1959, Ohio State University.

Seiler, Charles D. Correlation of alluvial lenses along the Kansas River between Topeka and Ogden, Kansas. M, 1951, Kansas State University. 109 p.

Seiler, Frank Carl. Foraminifera of the Selma Chalk, Oktibbeha County, Mississippi. M, 1939, Mississippi State University. 77 p.

Seiler, Robert C. The petroleum geology and the environment of deposition of the Bradford Third sandstone (Devonian), Five Mile field, Cattaraugus County, New York. M, 1970, SUNY at Buffalo. 47 p.

Seiner, Maureen Bernice. Fusulinidae of the Laborcita Formation (Pennsylvanian and Permian), Sacramento

Mountains, New Mexico. M, 1967, Southern Methodist University. 74 p.

Seiple, Willard R. Engineering geology of expansive soils. M, 1968, University of Southern California.

Seithlheko, Edwin M. Studies of mean particle size and mineralogy of sands along selected transects on the Llano Estacado. M, 1975, Texas Tech University. 69 p.

Seitz, Harold R. The effect of a landfill on a hydrogeologic environment. M, 1971, University of Idaho. 118 p.

Seitz, James F. Investigation of type locality, Astoria Formation. M, 1948, University of Washington. 63 p.

Seitz, John N. A diagenetic study of Miocene carbonates from Indonesia utilizing cathodoluminescence. M, 1975, Rensselaer Polytechnic Institute. 112 p.

Sekel, David M. The distribution of Paleocene and Eocene calcareous nannofossils in the coastal plain of Virginia. M, 1980, Ohio University, Athens. 69 p.

Selby, Andrew C. The glacial geology of Darke County. M, 1978, Ohio State University.

Selby, Curt McKee. Petrologic and structural mechanisms of the lower pelitic schist on the western flank of the Woodstock Dome, Howard County, Maryland. M, 1978, West Virginia University.

Selby, Douglas Allen. Effects of cadmium bioaccumulation on community structure in artificial hardwater streams. D, 1983, Utah State University. 204 p.

Selby, Jonathan B. Depositional environments and petroleum potential, second Wall Creek Interval, Frontier Formation, Johnson and Natrona counties, Wyoming. M, 1983, Colorado School of Mines. 111 p.

Selden, Robert W. Techniques of microradiography using X-ray diffraction equipment. M, 1973, Wright State University. 71 p.

Self, Daniel Eugene. Holocene carbonate reef-flat and beachrock cementation, Enewetak Atoll, Marshall Islands. M, 1984, University of Texas, Austin. 143 p.

Self, Edward Moss. Geology of the Livermore area, Larimer County, Colorado. M, 1952, University of Kansas. 78 p.

Self, George William, Jr. Correlation between engineering and mineralogical properties of expansive soils at the Dallas–Fort Worth regional airport (Texas). M, 1971, Texas Christian University.

Self, Gregory Alan. Growth movement of the Citronelle Dome, and its genetic relationship to other salt structures in the southern Mississippi Salt Basin. M, 1981, University of Florida. 136 p.

Self, Robert P. Petrology of Holocene sediments in the Rio Nautla drainage basin and the adjacent beaches, Veracruz, Mexico. D, 1971, Rice University. 265 p.

Self, Robert Patrick. Petrology of the Duncan Sandstone (Permian) of south-central Oklahoma. M, 1966, University of Oklahoma. 133 p.

Self, Selden R. A restudy of the type localities of the lower Washita. M, 1928, Texas Christian University. 71 p.

Selfridge, George C. An X-ray and optical investigation of the serpentine minerals. D, 1936, Columbia University, Teachers College.

Selfridge, George C., Jr. A petrographic study of the Pre-Cambrian crystalline rocks of the "highlands" in the Schunemunk Quadrangle, southeastern New York. M, 1932, Columbia University, Teachers College.

Selim, Mohammed Abdel-Moniem. Determination of the irreducible minimum water saturation of porous solids by an evaporation method. M, 1952, University of Texas, Austin.

Selinger, Keith A. Secondary oil recovery in the central unit of the Bisti oil field, San Juan County, New Mexico. M, 1964, University of Arizona.

Selk, Bruce W. The manganese-enriched sediments of the Blanco Trough; evidence for hydrothermal activity in a fracture zone. M, 1978, Oregon State University. 137 p.

Selk, Donald Clair. Crustal and upper-mantle structures in Utah as determined by gravity profiles. M, 1976, University of Utah. 81 p.

Selkregg, Kevin R. Relationship between the chemical composition, optical properties, unit cell dimensions, and structural state of the mineral cordierite. M, 1979, Virginia Polytechnic Institute and State University.

Sell, James D. Bedding replacement deposit of the Magma Mine, Superior, Arizona. M, 1961, University of Arizona.

Sellards, Elias H. A study of some Paleozoic plants and insects. D, 1903, Yale University.

Sellars, Barbara D. The barite occurrence amd stratigraphy of the Siksikpuk Formation in Atigun River gorge, Brooks Range, Alaska. M, 1981, University of Alaska, Fairbanks. 108 p.

Sellars, Robert T., Jr. Subsurface structure and stratigraphy of the Kent Bayou-Turtle Bayou-North Turtle Bayou Complex, Terrebonne Parish, Louisiana. M, 1961, Tulane University. 27 p.

Selleck, Bruce Warren. Heavy mineral analysis of some lake and shore sands of southern Lake Ontario. M, 1972, University of Rochester. 23 p.

Selleck, Bruce Warren. Paleoenvironments and petrography of the Potsdam Sandstone, Theresa Formation and Ogdensburg Dolomite (U. Camb. - L. Ord.) of the southwestern St. Lawrence Valley, New York. D, 1975, University of Rochester. 163 p.

Selleck, William Lewis. Use of aerial photographs to delineate expansive soils in the San Francisco Bay area. M, 1972, Stanford University.

Sellers, David Henry Aikins. Late Paleozoic faunas of north-central Yukon Territory. M, 1960, University of Iowa. 128 p.

Sellers, George August. Hydrothermal experiments on the thermal stability of amino substances in sediments. D, 1966, California Institute of Technology. 152 p.

Sellers, J. E. Investigation of the Estes Park Meteorite; Larimer County, Colorado. M, 1928, University of Colorado.

Sellers, Robert T., Jr. Geology of the Mena-Board camp quadrangles, Polk County, Arkansas. D, 1966, Tulane University. 128 p.

Sellman, Paul V. Flow and ablation of Gulkana Glacier, central Alaska Range, Alaska. M, 1962, University of Alaska, Fairbanks. 36 p.

Sellmer, H. W. Geology and petrogenesis of the Serb Creek Intrusive Complex near Smithers, British Columbia. M, 1966, University of British Columbia.

Selman, Michael Lamar. OCS oil and gas development; impact on the coastal zone community. M, 1982, University of Virginia. 65 p.

Selmser, C. B. The petrology of part of Mount Royal near Cote des Neiges Village (Quebec). M, 1939, McGill University.

Seltin, Richard J. Review of the family Captorhinidae. D, 1956, University of Chicago. 77 p.

Seltzer, Geoffrey Owen. Glacial history and climatic change in the central Peruvian Andes. M, 1987, University of Minnesota, Minneapolis. 92 p.

Selverstone, Jane Elizabeth. Petrologic and fluid inclusion study of granulite xenoliths, Pali-Aike volcanic field, Chile. M, 1981, University of Colorado. 151 p.

Selverstone, Jane Elizabeth. Pressure-temperature-time constraints on metamorphism and tectonism in the Tauern Window, Eastern Alps. D, 1985, Massachusetts Institute of Technology. 279 p.

Selvius, Douglas Brian. Lithostratigraphy and algal-foraminiferal biostratigraphy of the Cupido Formation, Lower Cretaceous, in Bustamante Canyon and Potrero Garcia, Northeast Mexico. M, 1982, University of Michigan.

Selywn, Stephen. Effects of highly permeable layers on the strength of saturated sands. M, 1971, New York University.

Selznick, Martin Richard. Campanian foraminiferal paleoecology of Northeast Texas. M, 1984, University of Texas at Dallas. 239 p.

Seme-Abomo, Richard. A theoretical study of the amplitude variation with offset method in offshore Cameroon seismic exploration. M, 1985, Pennsylvania State University, University Park. 62 p.

Semet, Michel Pierre. Stability relations and crystal chemistry of the amphibole magnesiohastingsite, $NaCa_2Mg_4Fe^{3+}Si_6Al_2O_{22}(OH)_2$. D, 1972, University of California, Los Angeles.

Semken, Holmes Alford, Jr. Stratigraphy and paleontology of the (Pleistocene) McPherson Equus Beds (Sandahl Local Fauna), McPherson County, Kansas. D, 1965, University of Michigan. 107 p.

Semken, Holmes Alfred, Jr. Fossil vertebrates from Longhorn Cavern, Burnet County, Texas. M, 1960, University of Texas, Austin.

Semken, Steven Christian. A neodymium and strontium isotopic study of late Cenozoic basaltic volcanism in the southwestern Basin and Range Province. M, 1984, University of California, Los Angeles. 68 p.

Semkin, Raymond Garry. A limnogeochemical study of Sudbury area lakes. M, 1975, McMaster University. 248 p.

Semler, Charles E., Jr. A solid state study of the system $BaO\text{-}SiO_2$ and a portion of the system $BaO\text{-}Al_2O_3\text{-}SiO_2$. M, 1965, Miami University (Ohio). 104 p.

Semler, Charles Edward, Jr. Studies in the ternary system barium oxide-silica; the system celsian-silica-corundum. D, 1968, Ohio State University. 96 p.

Semmens, Dave. Three dimensional gravity modeling techniques with application to the Ennis geothermal area. M, 1987, Montana College of Mineral Science & Technology. 251 p.

Semmes, Douglas R. The geology of the San Juan District, Puerto Rico. D, 1917, Columbia University, Teachers College.

Sempels, Jean-Marie. Mathematical simulation of crystal nucleation, growth and the formation of textures. D, 1978, University of Ottawa. 154 p.

Sempere, Jean-Christophe. Occurrence and evolution of overlapping spreading centers. D, 1986, University of California, Santa Barbara. 238 p.

Semprini, Lewis. Modeling and field studies of radon-222 in geothermal reservoirs. D, 1986, Stanford University. 347 p.

Semrau Lago, R. Bearing capacity of circular footings on a two layered cohesive subsoil. D, 1979, Northwestern University. 217 p.

Semtner, Albert J., Jr. A numerical investigation of Arctic Ocean circulation. D, 1973, Princeton University.

Sen Gupta, Mrinal Kanti. A travel time study of P-waves using deep-focus earthquakes. M, 1972, Massachusetts Institute of Technology. 84 p.

Sen Gupta, Mrinal Kanti. The structure of the Earth's mantle from body wave observations. D, 1975, Massachusetts Institute of Technology. 542 p.

Sen Gupta, Pradip Kumar. The orientation of aliphatic amine cations on vermiculite. D, 1964, Washington University. 115 p.

Sen, Gautam. Part 1, Petrology of the ultramafic xenoliths on the Koolau Shield, Oahu, Hawaii; Part 2, Liquidus phase relations on the join anorthite-forsterite-silica at 10 kbar. D, 1981, [University of Texas at Dallas]. 204 p.

Sen, Mrinal K. Kirchhoff-Helmholtz synthetic seismograms. D, 1987, University of Hawaii.

Sen, Rajan. Dynamic analysis of buried structures. D, 1984, SUNY at Buffalo. 297 p.

Sen, Sisir Kumar. Potash content of natural plagioclases and the origin of antiperthites. D, 1957, University of Chicago. 16 p.

Send, Uwe. Heat and flow response to upwelling relaxations. D, 1988, University of California, San Diego.

Sendlein, Lyle Vernon Archie. Geology of the Sperry Mine, Des Moines County, Iowa. D, 1964, Iowa State University of Science and Technology. 83 p.

Sendlein, Lyle Vernon Archie. Vibrational study of soils. M, 1960, Washington University. 24 p.

Sene, Siki. A shallow seismic reflection experiment in Avra Valley, Pima County, Arizona. M, 1983, University of Arizona. 80 p.

Sener, C. An endochronic nonlinear inelastic constitutive law for cohesionless soils subjected to dynamic loading. D, 1979, Northwestern University. 332 p.

Senft, W. H., II. Concentrations of intracellular nutrients and rates of photosynthesis in algae. D, 1977, University of Minnesota, Minneapolis. 100 p.

Senftle, Frank E. Radioactive studies applied to certain geological problems. D, 1948, University of Toronto.

Senftle, Joseph Thomas. Relationships between coal constitution, thermoplastic properties and liquefaction behavior of coals and vitrinite concentrates from the lower Kittanning seam. D, 1981, Pennsylvania State University, University Park. 240 p.

Sengebush, Robert M. The geology and tectonic history of the Fourth of July Creek area, White Cloud Peaks, Custer County, Idaho. M, 1984, University of Montana. 79 p.

Senger, Rainer Klaus. Hydrogeology of Barton Springs, Austin, Texas. M, 1983, University of Texas, Austin. 119 p.

Sengor, Ali Mehmet Celal. Geometry and kinematics of continental deformation in zones of collision; examples from Central Europe and eastern Mediterranean. M, 1979, SUNY at Albany. 126 p.

Sengör, Ali Mehmet Celâl. The geology of the Albula Pass area, eastern Switzerland in its Tethyan setting; Palaeo-Tethyan factor in Neo-Tethyan opening. D, 1982, SUNY at Albany. 511 p.

Sengul, Mustafa. Numerical solution of heat conduction with phase change in cylindrical systems. D, 1977, Stanford University. 211 p.

Sengupta, Arijeet. Stratigraphic analysis of the Trenton Limestone of Michigan, Indiana, Illinois, Ohio and Wisconsin. M, 1984, University of Michigan.

Sengupta, Somnath. Recent carbonate sedimentation in Florida Bay; a study to define major sub-environments. M, 1985, Wichita State University. 135 p.

Seni, Steven John. Genetic stratigraphy of the Dockum Group (Triassic), Palo Duro Canyon, Panhandle, Texas. M, 1978, University of Texas, Austin.

Senich, Donald. X-ray diffraction and adsorption isotherm studies of the calcium montmorillonite-H_2O system. D, 1966, Iowa State University of Science and Technology. 202 p.

Senich, Michael A. Macrofossil distribution in the Stanton Limestone (Upper Pennsylvanian) in eastern Kansas. D, 1978, University of Iowa. 301 p.

Senich, Michael A. Relation of biotic assemblages to lithofacies in Stanton Limestone (upper Pennsylvanian), southeastern Kansas. M, 1975, University of Iowa. 198 p.

Sensintaffar, Jack L. Pliocene rhinoceroses from the Rhinoceros Hill Quarry, Kansas. M, 1952, University of Kansas. 93 p.

Senstius, Maurits Wilhelm. Studies on weathering and soil formation in tropical high altitudes. D, 1928, University of Michigan.

Senter, Lance E. Geology and porphyry molybdenum mineralization of the Turnley Ridge Stock in the Elkhorn mining district, Jefferson County, Montana. M, 1976, Eastern Washington University. 77 p.

Sentner, David A. The distribution of iron, chromium, and aluminum among coexisting phases, as illustrated by the phase assemblage spinel, mullite, silica, sesquioxide, and liquid in the system iron oxide-aluminum oxide-silica-chromium oxide at various oxygen pressures. M, 1980, Pennsylvania State University, University Park.

Senturk, H. A. Resistance to flow in sand-bed channels. D, 1976, Colorado State University. 194 p.

Senyaki, Abu Lwangq. Clay mineralogy of poorly drained soils developed from serpentinite rocks. D, 1977, University of California, Davis. 94 p.

Seo, Haeyoun. Permian non-fusulinid foraminifera from the northern Yukon Territory. M, 1977, University of Saskatchewan.

Seo, Jung Hoon. Detecting abnormal geopressure using seismic reflectivity. M, 1978, University of Texas, Austin.

Sepassi, Bahman. Removal of lateral inconsistency from seismic data by means of consistent strata. M, 1978, University of Texas, Austin.

Sepehr, Keyvan. Non-linear and time-dependent finite element modelling of underground excavations with special reference to induced seismicity in potash mining. D, 1988, University of Manitoba.

Sepehr, Mansour. Field condition hydrogeologic modeling for identification of salinity sources in a stream-aquifer system. D, 1984, Utah State University. 229 p.

Sepkoski, Joseph John, Jr. Dresbachian (Upper Cambrian) stratigraphy in Montana, Wyoming, and South Dakota. D, 1977, Harvard University.

Sequeira, Jose F., Jr. The depositional environments, diagenetic history, and porosity development of the upper Smackover Member of Eustace Field, Henderson County, Texas. M, 1987, Texas A&M University. 170 p.

Serafinoff, Rafael Esteban de la Cruz Ramirez see Ramirez Serafinoff, Rafael Esteban de la Cruz

Seraphim, Robert H. Some aspects of the geochemistry of fluorine. D, 1951, Massachusetts Institute of Technology. 109 p.

Seraphim, Robert Henry. A gold specularite deposit, Unuk River. M, 1948, University of British Columbia.

Serbeck, John W. Foraminiferal paleoecology of the Sage Breaks Shale, Centennial Valley, Wyoming. M, 1981, University of Wyoming. 138 p.

Sereda, Ivan Theodore. A crustal reflection, expanding-spread study in Southeast Saskatchewan and Southwest Manitoba. M, 1978, University of Saskatchewan. 352 p.

Serencsits, Colleen McCabe. Potassium-argon age determinations for some alkaline ring dike complexes of the Southeastern Desert of Egypt. M, 1977, University of Pennsylvania.

Serenko, Thomas J. Stratigraphy and palynology of the Tullock Formation of the Fort Union Group (Paleocene), McCone County, Montana. M, 1988, Colorado School of Mines. 67 p.

Sereno, Paul Callistus. The ornithischian dinosaur Psittacosaurus from the Lower Cretaceous of Asia and the relationships of the Ceratopsia. D, 1987, Columbia University, Teachers College. 572 p.

Sereno, Thomas John, Jr. The propagation of high frequency seismic energy through oceanic lithosphere. D, 1986, University of California, San Diego. 188 p.

Serfes, Michael Edward. The Wolverine Creek fault system; a newly recognized terrane boundary in the western Chugach Mountains of southern Alaska. M, 1984, Lehigh University. 67 p.

Sergeant, Richard E. Stratigraphic correlations of selected Alleghenian coals of the Princess Reserve District, Kentucky. M, 1979, Eastern Kentucky University. 63 p.

Sergoulopoulos, Alexandros. Shallow seismic reflection applied to St. Joseph's Church gas field, Suffield Township, Portage County, Ohio. M, 1988, Kent State University, Kent. 67 p.

Serim, Hakki Erdem. Structural geology along the Whiteoak Mountain Fault, Meigs County, Tennessee. M, 1965, University of Tennessee, Knoxville. 35 p.

Serlin, Bruce Steven. An Early Cretaceous fossil flora from Northwest Texas; its composition and implications. D, 1980, University of Texas, Austin. 206 p.

Sermer, Tamara. The behaviour of barium sulfate precipitate based membrane electrodes. M, 1972, University of Toronto.

Sermon, Thomas Croxford. A null current indicator. M, 1940, Michigan Technological University.

Serna, Carlos J. An investigation of groundwater contamination due to oil field brines in Portage County, Ohio. M, 1986, Miami University (Ohio). 98 p.

Serna-Isaza, Mario J. Geology and geochemistry of Calico Peak, Dolores County, Colorado. M, 1971, Colorado School of Mines. 88 p.

Serpa, Laura Fern. Detailed gravity and aeromagnetic surveys in the Black Rock Desert area, Utah. M, 1980, University of Utah. 210 p.

Serpa, Laura Fern. Structural analyses of extensional terranes in northeastern Kansas and Death Valley, California, from COCORP deep seismic reflection data. D, 1986, Cornell University. 172 p.

Serra, Kelsen Valente. Well testing for solution gas drive reservoirs. D, 1988, University of Tulsa. 286 p.

Serra, S. Structure and stratigraphy of pre-Silurian rocks in West-central Somerset County, Maine. M, 1973, Syracuse University.

Serra, Sandro. Styles of deformation in the ramp regions of overthrust faults. D, 1978, Texas A&M University. 133 p.

Serrag, Salaheddin Ali. An assessment of erosion processes as simulated using artificial rainfall; southwest conditions. D, 1987, New Mexico State University, Las Cruces. 173 p.

Serrano, Sergio Enrique. Analysis of stochastic groundwater flow problems in Sobolev space. D, 1985, University of Waterloo. 163 p.

Serreze, Mark C. Topoclimatic investigations of a small, sub-polar ice cap with implications for glacierization. M, 1985, University of Massachusetts. 201 p.

Serson, Paul H. The development of a universal airborne magnetometer. D, 1951, University of Toronto.

Servilla, Thomas. Unconformity at the base of the Raritan Formation in Middlesex County, New Jersey. M, 1960, Rutgers, The State University, New Brunswick. 83 p.

Serviss, Curtis Raymond, Jr. The use of rutile-ilmenite concentration ratios from X-ray diffraction on panned concentrates as an exploration method for rutile in the Piney River-Roseland titanium district, Virginia. M, 1979, University of Missouri, Rolla.

Serviss, Fred L. F. Trinidad coal field of Colorado; Las Animas County. M, 1922, Colorado School of Mines. 80 p.

Servos, Gary Gordon. A gravitational investigation of Niagaran reefs in southeastern Michigan. D, 1965, Michigan State University. 144 p.

Servos, Mark Roy. Fate and bioavailability of polychlorinated dibenzo-p-dioxins in aquatic environments. D, 1988, University of Manitoba.

Sessinghaus, Gustavus. The geology of an area in Litchfield County, Connecticut. M, 1899, University of Wisconsin-Madison.

Sestak, Andrew Aloysius. A laboratory investigation of the dearth of granular material in the 4-2 mm size range. M, 1952, University of Illinois, Urbana.

Sestak, Helen Maria. Stratigraphy and depositional environments of part of the Pennsylvanian Pottsville Formation in the Black Warrior Basin, Alabama and Mississippi. M, 1984, University of Alabama. 197 p.

Sestini, Julian. Miocene stratigraphy and depositional environments, Red Sea coast, Sudan. D, 1962, University of Chicago. 159 p.

Sethuraman, K. Petrology of Grenville metavolcanic rocks in the Bishop Corners-Donaldson area, Ontario. D, 1970, Carleton University. 160 p.

Setlock, G. H. Dissolved oxygen and nutrient distributions in interstitial waters of abyssal marine sediments; early diagenesis in manganese nodule localities from the eastern equatorial Pacific Ocean. D, 1979, University of Wisconsin-Madison. 504 p.

Setlow, Loren W. Dune reddening; a scanning electron microscope study of quartz and heavy mineral sand grains. M, 1972, Florida State University.

Seto, H. G. Synthesis and characterization of stable kaolin intercalation complexes; a possible means of valorization and diversification of kaolin clay minerals. D, 1977, University of Illinois, Urbana. 243 p.

Setoguchi, Takeshi. The Cedar Ridge local fauna (late Oligocene), Badwater Creek area, central Wyoming; paleontology and geology of the site. D, 1977, Texas Tech University. 204 p.

Setoguchi, Takeshi. The late Eocene marsupials and insectivores from the Tepee Trail Formation, Badwater, Wyoming. M, 1973, Texas Tech University. 109 p.

Setra, Abdelghani. Stratigraphy and microfacies analysis of the Mississippian Las Cruces Formation (Osage-Meramec), Vinton Canyon, El Paso County, Texas. M, 1976, University of Texas at El Paso.

Setterfield, Thomas Neal. Nature and significance of the McDougall-Despina fault set, Noranda, Quebec. M, 1984, University of Western Ontario. 148 p.

Settle, Alberta L. Feldspars as recorders of pegmatite petrogenesis; Bob Ingersoll and Peerless pegmatites, Black Hills, South Dakota. M, 1985, South Dakota School of Mines & Technology.

Settle, Mark F. A statistical analysis of volcanic activity at Stromboli, Italy. M, 1973, Massachusetts Institute of Technology. 82 p.

Settle, Mark F. P. Studies of impact cratering processes. D, 1979, Brown University. 189 p.

Settlemyre, Julius L., III. Chemical and suspended sediment budgets for a tidal creek, Charleston Harbor, S.C. D, 1975, University of South Carolina. 132 p.

Settlemyre, Julius L., III. Contrasting characteristics and trends along a linear complex of submarine volcanoes and ridges near the Reykjanes Ridge. M, 1972, University of South Carolina.

Settles, Patricia Leigh. The geochemistry of radioactive ground water within the Cambrian-Ordovician system of central Missouri. M, 1988, University of Missouri, Columbia.

Setty, M. G. Anantha Padmanabha. The paleontological and ecological study of fresh-water diatoms of Lake Bonneville Basin, Utah. D, 1963, University of Utah. 147 p.

Setudehnia, Ata-Ollah. Structure and stratigraphy of Big Rock Creek Formation, Valyermo, California. M, 1964, University of California, Riverside. 52 p.

Setzler, Robert Eric. Comanchean (Lower Cretaceous) and lower Dockum (Triassic) stratigraphy of the King Mountain area, Upton and Crane counties, Texas. M, 1988, University of Texas of the Permian Basin. 233 p.

Seur, Linda Perkins Le see Le Seur, Linda Perkins

Seuthé, Chantal. La Photographie multispectrale appliquée à l'etude structurale et pétrographique d'un secteur situé au nord-ouest de Rouyn-Noranda, Québec. M, 1981, Universite de Montreal. 114 p.

Seutter, Andrew Edward, III. Depositional environments and diagenetic history of Member B (Upper Ordovician) of the Mountain Springs Formation, southern Nevada and southeastern California. M, 1986, San Diego State University. 145 p.

Sever, C. K. Structural geology of the Sheephead Mountain area, Carbon County, Wyoming. M, 1975, University of Wyoming. 105 p.

Sever, Julia Rebecca. The organic geochemistry of hydrocarbons in coastal environments. D, 1970, University of Texas, Austin. 155 p.

Severance, Robert W. A dependence of the circulation on the surface waters of the Gulf of Mexico upon the horizontal distribution of surface temperatures. M, 1968, Florida State University.

Severin, Kenneth Paul. Functional morphology of benthic foraminifera. D, 1987, University of California, Davis. 220 p.

Severn, W. P. Stratigraphy of late Upper Cretaceous, Paleocene, and early Eocene deposits within the east flank of the Rock Springs Uplift, Sweetwater County, Wyoming. M, 1959, University of Wyoming. 90 p.

Severson, George D. The petrography of the Missouri Mountain Shale of Arkansas and Oklahoma. M, 1963, University of Tulsa. 105 p.

Severson, John Louis. A comparison of the Madison Group (Mississippian) with its subsurface equivalents in central Montana. D, 1952, University of Wisconsin-Madison. 87 p.

Severson, M. J. Petrology and sedimentation of early Precambrian graywackes in the eastern Vermilion District, northeastern Minnesota. M, 1978, University of Minnesota, Duluth.

Severson, Roger H. Depositional environments, facies relationships, and coal occurrence in Carboniferous sediments of the Narragansett Basin. M, 1981, University of Rhode Island.

Severy, Charles Lamb. Subsurface stratigraphy of the Chesterian Series, Southwest Kansas. M, 1975, University of Colorado.

Severy, Niles Ian. Calibration and installation of tiltmeters with a discussion of some tilt observations. M, 1973, University of Colorado.

Sevian, Walter Andrew. Nonlinear dispersion of groundwater disturbances. D, 1970, SUNY at Stony Brook. 49 p.

Sevigny, James H. Geochemistry and petrology of amphibolites, granites, and metasedimentary rocks, Monashee Mountains, southeastern Canadian Cordillera. D, 1988, University of Calgary. 149 p.

Sevigny, James H. Structure and petrology of the Tomyhoi Peak area, north Cascade Range, Washington. M, 1983, Western Washington University. 203 p.

Sevillano, Arturo Cabreros. Secondary dispersion of copper, molybdenum, tungsten and nickel in Mount Nungkok area, Saba (Pliocene), Malaysia. M, 1969, University of New Brunswick.

Sevon, William David. Geology of the Marindahl Quadrangle, South Dakota. M, 1958, University of South Dakota. 98 p.

Sevon, William David, III. Stratigraphy of the Ogallala Group of a portion of southwestern South Dakota. D, 1961, University of Illinois, Urbana. 101 p.

Sewall, Angela Jean. Structure and geochemistry of the upper plate of the Saddle Island Detachment, Lake Mead, Nevada. M, 1988, University of Nevada, Las Vegas. 81 p.

Seward, Allan E. The areal geology of the southern portion of the Reipetown Quadrangle, Nevada. M, 1962, University of Southern California.

Seward, Clay L. Geology of the Jordan Gap Quadrangle. M, 1950, Texas A&M University. 72 p.

Seward, Clay L. Petroleum prospects of a part of the Marfa Basin, Texas. D, 1953, Texas A&M University. 23 p.

Seward, Paul A. Sphalerite mineralization in the Viola Limestone (Ordovician) of south-central Kansas. M, 1982, Wichita State University. 96 p.

Sewell, Charles Robertson. Igneous petrology of Candelaria area, Presidio County, Texas. M, 1955, University of Texas, Austin.

Sewell, H. James. Petrology and diagenesis of the Grimsby and Thorold sandstones (Silurian), Ontario and New York; implications for the origin of quartz arenites. M, 1986, Bowling Green State University. 99 p.

Sewell, John Michael. Geochemical characteristics of surface and ground water from various Carboniferous paleoenvironments in northeastern Kentucky. M, 1976, University of South Carolina.

Sexauer, Mae Lynn. The geology and origin of the pyrophyllite deposits in southwestern Granville County, North Carolina. M, 1983, University of North Carolina, Chapel Hill. 85 p.

Sexsmith, Suzanne L. Depositional environments of the Salt Wash Member of the Morrison Formation in Grand and Emery counties, Utah. M, 1980, New Mexico Institute of Mining and Technology. 165 p.

Sexton, Alan J. Geology of the Sporting Mountain area, southeastern Cape Breton Island, Nova Scotia. M, 1988, Acadia University. 215 p.

Sexton, James V. The ostracode genus Cytherelloidea. M, 1950, Louisiana State University.

Sexton, John Lloyd. Ellipticity of Rayleigh waves recorded in the Midwest. D, 1973, Indiana University, Bloomington. 381 p.

Sexton, Thomas F. Seismic study of various (Pleistocene) glacial deposits (Massachusetts). M, 1951, Boston College.

Sexton, Walter Jerome. Morphology and sediment character of mesotidal shoreline depositional environments. D, 1987, University of South Carolina. 314 p.

Sexton, William T. Selected forest soil-parent material relationships in the Clearwater and Nezperce national forests. D, 1986, University of Idaho. 86 p.

Seyb, S. M. Paleomagnetic study of oriented piston cores from the Northwest Pacific. M, 1977, University of Hawaii. 88 p.

Seyedghasemipour, Seyedjavad. Petroleum resource estimation in a a partially explored region with a sequential land release scheme. D, 1988, North Carolina State University. 211 p.

Seyfert, Carl Keenan, Jr. Geology of the Sawyers Bar area, Klamath Mountains, northern California. D, 1965, Stanford University. 227 p.

Seyfried, Conrad Kent. Concretions as indicators of the compaction of the Ohio black shale, southern Ohio. M, 1953, University of Cincinnati. 57 p.

Seyfried, W. E., Jr. Seawater-basalt interaction from 25°-300°C and 1-500 bars; implications for the origin of submarine metal-bearing hydrothermal solutions and regulation of ocean chemistry. D, 1977, University of Southern California.

Seyfried, William E., Jr. The redox stability of glauconite. M, 1974, Louisiana State University.

Seyler, Beverly. The Whirlpool Sandstone (Medina Group) in outcrop and the subsurface; a description and interpretation of environment of deposition. M, 1981, SUNY, College at Fredonia. 95 p.

Seyler, Douglas J. Middle Ordovician of the Michigan Basin. M, 1974, Michigan State University. 31 p.

Seymour, B. O. Geology of the talc mine at East Johnson, Vermont. M, 1977, Queens College (CUNY). 172 p.

Seymour, Catherine. Some foraminifera from the (Eocene) Penon Sandstone, Matanzas Province, Cuba. M, 1946, Columbia University, Teachers College.

Seymour, David. Geologic features of the Mount Guyot area, Summit County, Colorado. M, 1962, Colorado School of Mines. 102 p.

Seymour, Frank F. Gravity of the area surrounding the Blue Mountains Seismological Observatory (Oregon). M, 1965, Southern Methodist University. 27 p.

Seymour, Kevin Lloyd. The Felinae (Mammalia:Felidae) from the late Pleistocene tar seeps at Talara, Peru, with a critical examination of the fossil and Recent Felinae of North and South America. M, 1983, University of Toronto. 303 p.

Seymour, Richard Scott. Petrology and geochemistry of the Coffeepot Stock, N.E. Nevada; a record of crystallization history and hydrothermal fluid migrations. M, 1980, University of Oregon. 237 p.

Seyrafian, A. Stratigraphy, petrology, and depositional analysis of the Wolfcampian, Shafer Limestone Member, of the Elephant Canyon Formation, East-central Utah. M, 1975, Fort Hays State University.

Seyrafian, Ali. The Humboldt Fault; Pottawatomie County, Kansas, Precambrian to present. M, 1978, University of Kansas. 44 p.

Seyyedian-Choobi, M. Motion of the surface of a layered elastic half space produced by a buried dislocation pulse. D, 1976, University of Illinois, Urbana. 158 p.

Sgambat, Jeffrey Peter. Hydrogeology, ground-water quality and waste-water management; Nassau and Suffolk counties, New York. M, 1978, Pennsylvania State University, University Park. 120 p.

Sgouras, John D. Stratigraphy and conodont biostratigraphy of the Silurian St. Clair-Lafferty limestones (undifferentiated), Searcy and Stone counties, Arkansas. M, 1979, University of New Orleans.

Shaak, Graig Dennis. Species diversity and community structure of the Brush Creek Member, Conemaugh Group (Pennsylvanian) of the Appalachian Basin (of western Pennsylvania). D, 1972, University of Pittsburgh. 133 p.

Shaar, Edwin Willis, Jr. Mineralogy of selected world soil samples, with implications regarding the abrasion/corrosion potential of environmental dust on military ordnance and a hypothesis for the Southeast Asia problem. M, 1973, United States Naval Academy.

Shaarawy, Mostafa A. Razik. Quantitative geophysical study of the Prue Sand reservoir (Pennsylvanian) in a portion of Lincoln County, Oklahoma. M, 1962, University of Oklahoma. 145 p.

Shaath, Samir Khali. Quantitative analysis for Jalu-Augila Project. M, 1976, Ohio University, Athens. 170 p.

Shackelford, Charles Duane. Diffusion of inorganic chemical wastes in compacted clay. D, 1988, University of Texas, Austin. 474 p.

Shackelford, T. J. Structural geology of the Rawhide Mountains, Mohave County, Arizona. D, 1976, University of Southern California.

Shackleton, James Stephen. A petrographic study of the granitic intrusives of the Middle Foster Lake area, northern Saskatchewan. M, 1957, University of Saskatchewan. 41 p.

Shaddrick, David R. Metaconglomerates in the east central Black Hills (South Dakota). M, 1971, South Dakota School of Mines & Technology.

Shade, Harry D. The geology of an area from Moon Lake to Union Peak, Fremont County, Wyoming. M, 1959, Miami University (Ohio). 100 p.

Shade, John W. Mineral equilibria investigations in the Cincinnatian Series shales of Ohio, Indiana, and Kentucky. M, 1963, Miami University (Ohio). 61 p.

Shade, John William. Hydrolysis equilibria in the system $K_2O-Al_2O_3-SiO_2-H_2O$. D, 1968, Pennsylvania State University, University Park. 182 p.

Shadid, Omar Shakir Abdu Samara. Hydrogeologic aspects of potential wastewater reuse areas near Idaho Falls—Blackfoot, Idaho. M, 1971, University of Idaho. 87 p.

Shadix, Shirley J. Geology, taphonomy and metapodial analysis of a Pleistocene horse quarry, Hartley County, Texas. M, 1975, West Texas State University. 69 p.

Shadle, Harry Wallace. The petrography of the Pocono Formation...?. M, 1957, Pennsylvania State University, University Park. 52 p.

Shadul, Shadul A. The Bijou Creek damsites and reservoirs of Adams and Arapahoe counties, Colorado. M, 1971, Colorado School of Mines. 114 p.

Shaefer, George. Shelf sediments off San Clemente island, California. M, 1970, University of Southern California.

Shaeffer, Stanley B. Geological analysis of petroleum production. M, 1942, Colorado School of Mines. 55 p.

Shafer, Daivd Scott. Late-Quaternary paleoecologic, geomorphic, and paleoclimatic history of the Flat Laurel Gap, Blue Ridge Mountains, North Carolina. M, 1984, University of Tennessee, Knoxville. 148 p.

Shafer, David Clark. Petrology and depositional environment of the Beck Spring Dolomite, southern Death Valley region, California; 1 volume. M, 1983, University of California, Davis.

Shafer, Elena Camilli. Trivalent-pentravalent substitutions in silica structures. M, 1955, Pennsylvania State University, University Park. 54 p.

Shafer, Harry E., Jr. A petrographic study of the Oriskany Sandstone (Lower Devonian) in the Campbell Creek oil and gas field, Kanawha County, West Virginia. M, 1963, West Virginia University.

Shafer, J. M. An interactive river basin water management model; synthesis and application. D, 1979, Colorado State University. 262 p.

Shafer, Martin Merrill. Biogeochemistry and cycling of water column particulates in southern Lake Michigan. D, 1988, University of Wisconsin-Madison. 447 p.

Shaffer, Bernard Leroy. Microfloral successions in Permian evaporites. M, 1961, University of Missouri, Columbia.

Shaffer, Bernard LeRoy. Palynology of the Michigan "red beds". D, 1969, Michigan State University. 261 p.

Shaffer, Mark E. Geology and geochemistry of the Consolation Creek area, Puruni, Northwest Guyana. M, 1969, Colorado School of Mines. 147 p.

Shaffer, Marvin James. A transition state physico-chemical model predicting nitrification rates in soil-water systems. D, 1972, University of Arizona.

Shaffer, Marvin James. Prediction of nitrogen movement from alkaline soils. M, 1970, University of Arizona.

Shaffer, N. R. Regional distribution of $^{87}Sr/^{86}Sr$ ratios and mixing of sediments in the Ross Sea, Antarctica. M, 1974, Ohio State University.

Shaffer, Paul Raymond. Erosion surfaces of the Southern Appalachians. D, 1945, Ohio State University.

Shaffer, Paul Raymond. Pleistocene geology of the Fostoria, Ohio, Quadrangle. M, 1937, Ohio State University.

Shaffer, Ronald von see von Shaffer, Ronald

Shaffer, William Leroy. Geology of the Hogback Mountain area, northern Big Belt Mountains, Montana. M, 1971, University of New Mexico. 66 p.

Shafii Rad, Nader. Static and dynamic behavior of cemented sands. D, 1983, Stanford University. 345 p.

Shafiq, Moayad Abdulla. Foraminifera and environmental interpretation of the Weches Formation (middle Eocene), eastern Texas. M, 1969, University of Texas, Austin.

Shafiqullah, Muhammad. Geochronology of Cretaceous-Tertiary boundary, Alberta, Canada. M, 1963, University of Alberta. 65 p.

Shafiqullah, Muhammad. The diffusion characteristics of argon in nepheline and the K/Ar geochronology of the Oka carbonatite complex, (Oka, Quebec) and its aureole. D, 1969, Carleton University. 192 p.

Shafiuddin, Mohammed. Spray River Formation near Banff and Cadomin. M, 1960, University of Alberta. 114 p.

Shafor, Kenneth W. The geology of the Trenton Quadrangle, Butler County, Ohio. M, 1959, Miami University (Ohio). 101 p.

Shagam, Reginald. Geology of central Aragua, Venezuela. D, 1956, Princeton University. 145 p.

Shah, Balkumar P. Crystallization of silica gel in the presence of lithium salts. M, 1970, Michigan State University. 53 p.

Shah, Balkumar Prataprai. Evaluation of natural aggregates in Kalamazoo County and vicinity, Michigan. D, 1971, Michigan State University. 193 p.

Shah, Dasharathlal K. Some Middle Devonian stromatoporoids from Esterhazy, Saskatchewan. M, 1966, McGill University.

Shah, Safdar Ali. Comparative ore reserve estimates of the Van Stone Mine. M, 1969, University of Idaho. 56 p.

Shah, Syed Mohammad Ibrahim. Stratigraphy and paleobotany of the Weiser area, Idaho. M, 1966, University of Idaho. 191 p.

Shah, Syed Mohammad Ibrahim. Stratigraphy and paleobotany of Weiser area (Washington County), Idaho. D, 1968, University of Idaho. 166 p.

Shahabi, Mohammad Ali. Analysis of the distribution of petroleum deposits in Vigo County, Indiana, through computer processing of Landsat multispectral scanner data. M, 1979, Indiana State University. 87 p.

Shahbazi, Mohsen. Analytic techniques for determining groundwater flow fields. D, 1967, University of California, Berkeley. 147 p.

Shaheen, Elias J. Crustal structure and tectonics of the Middle East. M, 1977, University of Texas at El Paso.

Shahghasemi, Ebrahim. Simulation modeling of erosion processes on small agricultural watersheds. D, 1980, Iowa State University of Science and Technology. 286 p.

Shahriar, Mostafa. Application of iterative inversion techniques in crustal seismology. D, 1987, University of Alberta. 107 p.

Shaikh, Alaeddin. Surface erosion of compacted clays. D, 1986, Colorado State University. 145 p.

Shaikh, Zafar Mohammed. Geology and some aspects of the paleontology of the Ordovician Long Point Formation, Port au Port Peninsula, Newfoundland. M, 1971, Memorial University of Newfoundland. 165 p.

Shakal, A. F. Tectonic strain relief as a function of earthquake magnitude and distance. M, 1972, University of Wisconsin-Milwaukee.

Shakal, Anthony Frank. Analysis and modelling of the effects of the source and medium on strong motion. D, 1980, Massachusetts Institute of Technology. 264 p.

Shakel, Douglas Wilson. The geology of layered gneisses in part of the Santa Catalina Forerange, Pima County, Arizona. M, 1974, University of Arizona.

Shakoor, Abdul. Evaluation of methods for predicting durability characteristics of argillaceous carbonate aggregates for highway pavements. D, 1982, Purdue University. 252 p.

Shakshuki, Mokhtar. Focal mechanism determination using (SV/P)z amplitude ratios in Anna, Ohio area. M, 1987, Purdue University. 91 p.

Shalaby, Hany. Analyse des structures géométriques et interprétation rhéologique d'un slump cambrien dans le flysch de Saint-Jean–Port-Joli, Appalaches du Québec. M, 1977, Universite de Montreal.

Shalaby, Hany. Microfaciès algaire de la plate-forme du Saint-Laurent (Ordovicien moyen). D, 1981, Universite de Montreal. 328 p.

Shalaby, Mohamed A. E. A. Prospects and determinants for national and regional development in Zambia; the second national development plan and beyond. D, 1973, University of Pittsburgh. 549 p.

Shaler, Millard K. Coal of the Creek Nation, Indian territory. M, 1904, University of Kansas.

Shaller, Philip J. The influence of carbon dioxide gas on the emanation of sulfur gases from oxidizing pyrite. M, 1985, Montana College of Mineral Science & Technology. 162 p.

Shallom, Lizzie J. Pleistocene molluscan fauna of the La Blanc Deposits Matapedia County, Quebec, Canada. M, 1965, Ohio State University.

Shambaugh, John Scott. Structure of the Black Gap area, Brewster County, Texas. M, 1951, University of Texas, Austin.

Shamel, Charles Harmonas. Geology in the law. D, 1907, Columbia University, Teachers College.

Shamlian, Robert. Petrology of the White Mountain volcanic field, Inyo and Mono counties, California. M, 1979, University of California, Santa Barbara.

Shan, Frederick C. Chazy Group (Ordovician) trilobites. D, 1965, Harvard University.

Shanabrook, David Clark. A geophysical and geological study of the basement complex along the Peshekee River, Marquette County, northern Michigan. M, 1978, Michigan State University. 84 p.

Shanahan, Edward. Pollutional study of the San Pablo Bay. M, 1974, Stanford University.

Shanahan, Peter. Ground water model used to sense subsurface structure in an alluvial basin near Tombstone, Arizona. M, 1974, Stanford University.

Shane, John Denis, Sr. The palynology, biostratigraphy and paleoecology of the Umiat Delta Complex (middle Albian-early Cenomanian), North Slope, Alaska. D, 1984, Arizona State University. 375 p.

Shaner, Linda Ann. Reconnaissance study of metal sulfide deposition in tidal flat and sabkha-like environments, Gulf of California, Sonora, Mexico. M, 1982, University of Arizona. 80 p.

Shanholtz, Wendell H. Ordovician limestones in the vicinity of Hoges Store, Giles County, Virginia. M, 1956, Virginia Polytechnic Institute and State University.

Shank, Douglas Carl. Environmental geology in the Phoenix Mountains, Maricopa County, Arizona. M, 1973, Arizona State University. 39 p.

Shank, John C. A detailed magnetic survey in the Triassic Basin, North Chester County, Pennsylvania. M, 1961, Pennsylvania State University, University Park. 74 p.

Shank, Robert D. A seismic linear slip interface computer model for the detection of SV-shear reflections and tube wave generation from large fractures. M, 1986, Colorado School of Mines. 201 p.

Shank, Stephen Everett. The Devonian stratigraphy of Ward Mountain, Nevada. M, 1957, University of California, Berkeley.

Shankar, Nilakantan Jothi. Influence of tidal inlets on salinity and related phenomena in estuaries. D, 1970, University of Texas, Austin. 163 p.

Shankle, John Dyer, III. The "Flippen" Sandstone of parts of Taylor and Callahan counties, Texas. M, 1958, University of Oklahoma. 49 p.

Shanklin, Bob. Stratigraphic correlations between eastern Kentucky gas wells by means of insoluble residues. M, 1939, University of Cincinnati. 11 p.

Shanklin, Robert Elstone. An improved approach to crystal symmetry and the derivation and description of the thirty-two crystal classes by means of the stereographic projection and group theory. D, 1971, Ohio State University. 262 p.

Shanks, Jack Lee. Petrology of the Saint Joe Limestone in its type area, northcentral Arkansas. M, 1976, University of Arkansas, Fayetteville.

Shanks, W. C., III. Geochemical and sulfur isotope study of Red Sea geothermal systems. D, 1976, University of Southern California.

Shanks, Wayne C., III. Experimental study of manganese and iron migration in marine sediments. M, 1971, Louisiana State University.

Shanks, William Scott. A structural analysis of the Cedar Lake Dome, northwestern Ontario. M, 1986, University of Toronto.

Shanley, F. E. A geologic and economic analysis of the disseminated gold investment alternative. M, 1977, Stanford University. 254 p.

Shanley, Gerard E. The hydrography of the Oyster River Estuary (New Hampshire). M, 1972, University of New Hampshire. 89 p.

Shanley, Keith W. Stratigraphy and depositional model, upper Mission Canyon Formation (Mississippian), Northeast Williston Basin, North Dakota. M, 1983, Colorado School of Mines. 172 p.

Shanmugam, Ganapathy. A petrographic study of Simpson Group (Ordovician) sandstones, southern Oklahoma. M, 1972, Ohio University, Athens. 85 p.

Shanmugam, Ganapathy. The stratigraphy, sedimentology, and tectonics of the Middle Ordovician Sevier Shale basin in East Tennessee. D, 1978, University of Tennessee, Knoxville. 222 p.

Shannon, Charles William. The iron ore deposits of Indiana. M, 1907, Indiana University, Bloomington.

Shannon, Dennis L. Comparison of the foraminiferal faunas on both sides of the Tampa-Suwannee contact (Oligocene–Miocene) in Florida. M, 1967, Florida State University.

Shannon, E. H. M. Clay mineralogy and geochemistry of the Cherokee Group (Middle Pennsylvanian), Allen County, Kansas. D, 1978, Syracuse University. 220 p.

Shannon, Ellen Carol. Tissotiidae from Peru. M, 1950, University of Illinois, Urbana.

Shannon, Eugene Himie. Differentiation and alteration of the Caldwell Sill, Ontario. M, 1973, Michigan State University. 63 p.

Shannon, James R. Geology of the Mount Aetna cauldron complex, Sawatch Range, Colorado. D, 1988, Colorado School of Mines. 434 p.

Shannon, James R. Igneous petrology, geochemistry and fission track ages of a portion of the Baguio mineral district, northern Luzon, Philippines. M, 1979, Colorado School of Mines. 173 p.

Shannon, John P. Reconnaissance geology of the Howe Peak area, Butte County, Idaho. M, 1960, Northwestern University.

Shannon, John Philip, Jr. Hunton Group and related strata in Oklahoma. D, 1961, Northwestern University. 96 p.

Shannon, Kathleen Marie. Hydrogeochemical groundwater reconnaissance in Lawrence County, Indiana. M, 1986, Indiana University, Bloomington. 200 p.

Shannon, Lee T. Stratigraphy of the Blair Formation, an Upper Cretaceous slope and basin deposit, Rock Springs Uplift, Sweetwater, County, Wyoming. M, 1983, Colorado School of Mines. 149 p.

Shannon, Lynn Carlton. Lithofacies and structural features of Middle Pennsylvanian strata in southwestern Parker County, Texas. M, 1964, Texas Christian University.

Shannon, Margarita C. The relationship between strain and magnetic susceptibility anisotropy in mylonitized Henderson Gneiss. M, 1988, University of Massachusetts. 55 p.

Shannon, Patrick Joseph. The geology of the Pawhuska area, Osage County, Oklahoma. M, 1954, University of Oklahoma. 98 p.

Shannon, Samuel W. Selected Alabama mosasaurs. M, 1975, University of Alabama.

Shannon, Spencer Sweet, Jr. Geochemical reconnaissance of the Vienna District, Blaine and Camas counties, Idaho. D, 1963, University of Idaho. 158 p.

Shannon, W. J. A study to determine the feasibility of resource recovery from the solid waste generated in a specific planning area. D, 1976, University of Northern Colorado. 168 p.

Shannon, William M. Lithogeochemical characterization of intrusive rocks comprising the Quitman-Sierra Blanca igneous complex, Hudspeth County, Texas. M, 1986, University of Texas at El Paso.

Shapiro, Alan M. A similarity model of nonlinear convection. D, 1987, The Johns Hopkins University. 131 p.

Shapiro, Allen Marc. Fractured porous media; equation development and parameter identification. D, 1981, Princeton University. 387 p.

Shapiro, Arthur A. Acropora palmata; skeletal strength of a hermatypic coral, and its adaptive significance. M, 1980, Brooklyn College (CUNY).

Shapiro, Bruce E. Chemistry and petrography of the Navajoe Mountains basalt spilites of southern Oklahoma. M, 1981, University of Texas, Arlington. 116 p.

Shapiro, Earl A. Determination of optimum sample size for paleoecological investigations of the basal part of the Choptank Formation (Miocene), Maryland. M, 1973, Pennsylvania State University, University Park. 111 p.

Shapiro, Earl A. Evolution and natural selection in the Miocene genus Chesapecten (Mollusca; Bivalvia). D, 1977, Bryn Mawr College. 322 p.

Shapiro, Howard E. Production of lightweight concrete aggregate from Pacific Northwest clays and shales. M, 1952, University of Washington. 64 p.

Shapiro, Lewis Harold. Structural geology of the Black Hills region (South Dakota and Wyoming) and implications for the origin of the uplifts of the middle Rocky Mountain Province. D, 1971, University of Minnesota, Minneapolis. 343 p.

Shapiro, Neal. Deep seabed mining; an international seabed authority or private property rights?. M, 1982, University of Delaware, College of Marine Studies. 161 p.

Shapley, Robert A. The subsurface structure and stratigraphy related to petroleum accumulation in Pawnee County, Kansas. M, 1956, Kansas State University. 50 p.

Shappel, Maple D. The cleavage of ionic minerals; the crystal structure of bixbyite and the c-modification of the sexquioxides. D, 1933, California Institute of Technology. 48 p.

Shappell, Maple P. The cleavage of ionic minerals; the crystal structure of bixbyite and the C-modification of the sexquioxides. D, 1933, California Institute of Technology.

Shappirio, Joel Rez. Geology and petrology of the Tallahassee Creek area, Fremont County, Colorado. D, 1963, University of Michigan. 282 p.

Shapre, Roger Dale. Petrology, geochemistry, and metamorphic history of the Henson Creek Dunite, Mitchell County, North Carolina. M, 1979, University of Georgia.

Sharan, S. K. Earthquake response of dam-reservoir-foundation systems. D, 1978, University of Waterloo.

Sharangpani, Shirish C. A trace element study of barite in the Central Kentucky mineral district. M, 1982, University of Kentucky. 75 p.

Sharar, Taleb M. Abu *see* Abu Sharar, Taleb M.

Sharata, Salem Muftah. Geologic factors controlling hydrocarbon occurrence in the upper Minnelusa Formation in northeastern Wyoming. M, 1982, South Dakota School of Mines & Technology. 65 p.

Shareck, Andre. Etude géochimique des komatiites de Spinifex Ridge, Comté de La Motte, Québec. M, 1983, Universite du Quebec a Chicoutimi. 104 p.

Shareghi, Ehsan A. Morphology and characteristics of zircons in the Elkahatchee quartz diorite gneiss in northern Alabama Piedmont. M, 1981, Memphis State University. 95 p.

Sharga, Paul J. Petrological and structural history of the lineated garnetiferous gneiss Gore Mountain, New York. M, 1986, Lehigh University.

Sharief, Farooq Abdulsattar M. Depositional environment and regional significance of the Sakaka Sandstone, northwestern Saudi Arabia. M, 1974, Rice University. 50 p.

Sharief, Farooq Abdulsattar M. Sedimentary facies of the Jilh Formation, Saudi Arabia; and regional paleostratigraphy and tectonic evolution of the Middle East during the Middle Triassic Period. D, 1977, Rice University. 219 p.

Shariff, Asghar Jahani. Geology and paleoenvironment of the Bunker gas field, Solano County, California. M, 1983, California State University, Northridge. 48 p.

Sharma, Bijon. Interpretation of gravity data due to faults and dikes. D, 1968, McGill University.

Sharma, Bijon. The vertical gradient of gravity in gravitational interpretation. M, 1964, University of Minnesota, Minneapolis.

Sharma, Deepak Kumar. Kinetics of oil bank formation. D, 1987, University of California, Berkeley. 374 p.

Sharma, Ghanshyam Datta. Geology of the Peters Field, Saint Clair County, Michigan. D, 1961, University of Michigan. 156 p.

Sharma, Kamal N. M. Structural analysis of the Piscatosin Synform, Baskatong Reservoir (E) maparea, Quebec. D, 1969, Queen's University. 208 p.

Sharma, M. L. Influence of soil structure on water retention, water movement and thermodynamic properties of absorbed water. D, 1966, University of Hawaii.

Sharma, Mukul M. Transport of particulate suspensions in porous media. D, 1985, University of Southern California.

Sharma, Shatish K. A study of the basement conditions of the Norman dam site area, Cleveland County, Oklahoma. M, 1966, University of Oklahoma. 74 p.

Sharma, Sunil. A generalized seismic environment for soil structure interaction. D, 1986, Purdue University. 240 p.

Sharman, George F. The plate tectonic evolution of the Gulf of California. D, 1976, University of California, San Diego. 137 p.

Sharni, Dan. Earth gravity models based on distinct 5° by 5° values of topography and Moho. D, 1966, Ohio State University. 236 p.

Sharp, Byron J. Mineralization in Little Cottonwood Canyon, Utah. D, 1955, University of Utah. 79 p.

Sharp, Byron J. The mineralogy of the (Tertiary) Fox Clay deposit (Utah County), Utah. M, 1949, University of Utah. 28 p.

Sharp, Everett Ray. Petrography and petrology of the Southgate Member, Eden Group, southwestern Ohio. M, 1957, Ohio State University.

Sharp, George Carter, Jr. Stratigraphy and structure of the Greenstone Mountain area, Beaverhead County, Montana. M, 1970, Oregon State University. 121 p.

Sharp, Gerald L. Geochemistry and petrogenesis of the Diana Complex, Northwest Adirondacks, New York. M, 1987, University of Rochester. 142 p.

Sharp, Gregory A. Solute transport in fractal heterogeneous porous media. M, 1988, University of Nevada. 71 p.

Sharp, Henry S. The physical history of the Connecticut shoreline. D, 1929, Columbia University, Teachers College.

Sharp, James E. Structural geology and shaft construction. M, 1969, University of Arizona.

Sharp, James Wilfrid George. The analysis of pumping test data from artesian aquifers. M, 1966, University of Saskatchewan. 141 p.

Sharp, James William. West-central Utah; palinspastically restored sections constrained by Cocorp seismic reflecion data. M, 1984, Cornell University. 60 p.

Sharp, John. Gravity and magnetic investigations of the Alfred and Tatnic intrusive complexes, York County, Maine. M, 1976, University of New Hampshire. 120 p.

Sharp, John B. Water chemistry of the Ouachita Mountain springs excluding hot springs, Arkansas. M, 1980, University of Arkansas, Fayetteville.

Sharp, John L. Petrology and diagenetic history of the lower Atoka (Pennsylvanian) Spiro Sandstone in northwestern Arkansas and the Arkoma Basin of eastern Oklahoma. M, 1984, University of Arkansas, Fayetteville.

Sharp, John Malcolm, Jr. An investigation of energy transport in thick sequences of compacting sediments. D, 1974, University of Illinois, Urbana. 151 p.

Sharp, John V. A. The uranium deposits in the (Jurassic) Morrison Formation, Church Rock area, Grant's District, New Mexico. M, 1955, Columbia University, Teachers College.

Sharp, John Van Alstyne. Unconformities within basal marine Cretaceous rocks of the Piceance Basin, Colorado. D, 1963, University of Colorado. 189 p.

Sharp, Kenneth Denver. Paleontology and stratigraphy of the Upper Cretaceous rocks between Mount Taylor and the Rio Puerco. M, 1953, University of New Mexico. 93 p.

Sharp, Kevan Denton. Development of a model for probabilistic slope stability analysis and application to a tailings dam. D, 1981, Utah State University. 219 p.

Sharp, Robert Jay. The geology, geochemistry, and sulfur isotopes of the Anyox massive sulfide deposits. M, 1980, University of Alberta. 211 p.

Sharp, Robert Phillip. Cenozoic geology of the Ruby-East Humboldt Range, northeastern Nevada. D, 1938, Harvard University.

Sharp, Robert Phillip. Geology of the Ravenna Quadrangle, Los Angeles County, California. M, 1935, California Institute of Technology. 81 p.

Sharp, Robert R., Jr.   A geological engineering evaluation of an underground nuclear test site. D, 1972, University of Arizona.

Sharp, Robert R., Jr.   Some magnetic properties of a part of Pikes Peak iron deposit, Maricopa County, Arizona. M, 1962, University of Arizona.

Sharp, Robert Victor.   Geology of the San Jacinto Fault Zone in the Peninsular Ranges of southern California. D, 1965, California Institute of Technology. 207 p.

Sharp, Samuel.   Boundary element methods in quasistatic thermoelasticity with applications in rock mechanics. D, 1982, University of Minnesota, Minneapolis. 356 p.

Sharp, Warren Douglas.   Structure, petrology, and geochronology of a part of the central Sierra Nevada Foothills metamorphic belt, California. D, 1984, University of California, Berkeley. 206 p.

Sharp, Willard Edwin.   The escape of the Earth's atmosphere. M, 1960, University of California, Los Angeles.

Sharp, Willard Edwin.   The system $CaO-CO_2-H_2O$ in the two phase region calcite + aqueous solution and its application to the origin of quartz-calcite veins. D, 1964, University of California, Los Angeles. 178 p.

Sharp, William McMillan.   The structural geology of the Ruth Hope and Silversmith mines. M, 1950, University of British Columbia.

Sharp, William Wheeler.   Butler Clay (Wilcox Group), Bastrop County, Texas. M, 1951, University of Texas, Austin.

Sharp, Zachary David.   Metamorphism and oxygen isotope geochemistry of the northern Wind River Range, Wyoming. D, 1988, University of Michigan. 203 p.

Sharpe, Charles F. S.   Geology of the vicinity of Whiskey and Muddy gaps, Wyoming. M, 1931, Columbia University, Teachers College.

Sharpe, Charles F. S.   Landslides and related phenomena; a study of mass-movements of soil and rock. D, 1938, Columbia University, Teachers College.

Sharpe, Charles Lee.   Sedimentological interpretation of Tertiary carbonate rocks from west-central Florida. M, 1980, University of Florida. 170 p.

Sharpe, D. R.   Mudflows in the San Juan Mountains, Colorado; flow constraints, frequency and erosional effectiveness. M, 1974, University of Colorado.

Sharpe, H. N.   A thermal history model for the Earth with parameterized convection. D, 1978, University of Toronto.

Sharpe, Joanna L.   Geochemistry of the Cargill carbonatite complex, Kapuskasing, Ontario. M, 1987, Carleton University. 73 p.

Sharpe, John I.   The Welsford igneous complex. M, 1958, University of New Brunswick.

Sharpe, Karen Broderick.   Subsurface geology and hydrocarbon accumulation, West Flank Oil Springs oil field, Magoffin County, Kentucky. M, 1983, University of New Orleans. 87 p.

Sharpe, Lois Kremer.   Paragenesis of the pegmatites of southern Jackson and Macon counties, North Carolina. D, 1942, Northwestern University.

Sharpe, R. D.   Petrology, geochemistry, and metamorphic history of the Henson Creek Dunite, Mitchell County, North Carolina. M, 1979, University of Georgia.

Sharpe, Robert James.   A petrological study of some Archean metamorphic rocks from Miminiska Lake area, northwestern Ontario. M, 1979, University of Toronto.

Sharpe, Selina.   Deep sea deposits in continental areas. M, 1896, University of California, Berkeley. 13 p.

Sharps, Seymour L.   Geology of King Mountain, Routt County, Colorado, and correlation of Pre-Mancos (Cretaceous), Post-Weber (Permian) formations, northwestern Colorado. M, 1956, University of Colorado.

Sharps, Seymour Lytton.   Geology of Pagoda Quadrangle, northwestern Colorado; Moffat, Routt, Rio Blanco, and Garfield counties. D, 1962, University of Colorado. 364 p.

Sharpton, Virgil Le Roy.   Analysis of topography and implications for the tectonic evolution of the Moon and Venus; Part I, Lunar substructure, mare ridges and the tectonic evolution of mare basins; Part II, A comparison of the regional slope charcteristics of Venus and Earth. D, 1985, Brown University. 187 p.

Sharrett, Janice Beechwood.   A comparison of measured and predicted soil loss in Albemarle County, Virginia. M, 1981, University of Virginia. 61 p.

Sharry, John.   The geology of the western Tehachapi Mountains, California. D, 1981, Massachusetts Institute of Technology. 215 p.

Sharry, John.   Thermodynamics of phase transitions in Mg_2SiO_4. M, 1977, University of Rochester. 51 p.

Shaser, Joseph Leroy.   Anomalous carbonate and granitic erratics along the west flank of the Bighorn Mountains, Wyoming. M, 1978, Iowa State University of Science and Technology.

Shatoury, Hamad Mohamed El see El Shatoury, Hamad Mohamed

Shattuck, G. B.   The Shoal Creek fauna; a contribution to Cretaceous invertebrate paleontology (Texas). D, 1897, The Johns Hopkins University.

Shatzer, D. C.   Conodont biostratigraphy of the Fremont and Priest Canyon formations (Upper Ordovician) at Kerber Creek, Saguache County, Colorado. M, 1976, Ohio State University. 141 p.

Shaub, Benjamin Martin.   A unique feldspar deposit near DeKalb Junction, St. Lawrence County, New York. M, 1928, Cornell University.

Shaub, Benjamin Martin.   Banding in filled fissure veins. D, 1929, Cornell University.

Shaub, Francis Jean.   Interpretation of a gravity profile across the Gettysburg Triassic basin. M, 1975, Pennsylvania State University, University Park. 65 p.

Shaughnessy, Anna Catarina.   Remagnetization of the Eocene oceanic formation on Barbados, West Indies. M, 1980, Massachusetts Institute of Technology. 31 p.

Shaver, Kenneth C.   Dacites of the northern Gallatin and western Beartooth ranges, Montana. M, 1974, Montana State University. 109 p.

Shaver, Kenneth Charles.   Geology of the Rustic 7 1/2' quadrangle, northern Front Range, Colorado. D, 1980, University of Colorado. 209 p.

Shaver, R. B.   Nitrate enrichment of ground water in Shelby Township, Michigan. M, 1975, Wayne State University.

Shaver, Robert Harold.   Pegmatites in the Georges Mills area, New Hampshire. M, 1949, University of Illinois, Urbana.

Shaver, Robert Harold.   The morphology, ontogeny, and classification of the Ostracod families Bairdiidae, Cypridae, Cytherellidae, and Healdiidae. D, 1951, University of Illinois, Urbana.

Shaver, Stephen Allen.   The Hall (Nevada moly) molybdenum deposit, Nye County, Nevada; geology, alteration, mineralization, and geochemical dispersion. D, 1984, Stanford University. 560 p.

Shaw, Alan B.   Stratigraphy and structure of the St. Albans area, Vermont. D, 1949, Harvard University.

Shaw, Allen Vaughan.   The preferential acceptance of certain ions into the ferrite spinel lattice. M, 1965, Michigan State University. 53 p.

Shaw, B. R.   Quantitative lithostratigraphic correlation of digitized borehole-log records; upper Glen Rose Formation, Northeast Texas. D, 1978, Syracuse University. 180 p.

Shaw, Brian Robert.   Geology of the Albion-Scipio Trend, southern Michigan. M, 1975, University of Michigan.

Shaw, Carl G.   Strain analysis of Laramide deformation, Bighorn Mountains, Wyoming. M, 1986, Iowa State University of Science and Technology. 51 p.

Shaw, Charles E., Jr.   Geology and petrochemistry of the Milford area, Massachusetts. D, 1967, Brown University. 141 p.

Shaw, Charles E., Jr.   Stratigraphy and structural geology of the Sylacagua area, Alabama. M, 1961, Brown University.

Shaw, Charles McEwen.   An investigation of some chemical reactions involved in the genesis of metamorphic rocks. D, 1956, University of California, Berkeley. 91 p.

Shaw, Christopher William.   Late Pleistocene and Holocene geologic history of the Taylor-Hilgard portion of the southern Madison Range, southwestern Montana. M, 1988, University of Idaho. 181 p.

Shaw, Cynthia A.   Eocene and Oligocene silicoflagellate biostratigraphy for DSDP Leg 71, Sites 511 and 512, South Atlantic Ocean. M, 1981, University of Georgia.

Shaw, Daniel B.   A study of the variations in the mineralogical composition of shales. M, 1959, University of Houston.

Shaw, David Andrew.   Structural setting of the Adamant Pluton, northern Selkirk Mountains, British Columbia. D, 1980, Carleton University. 184 p.

Shaw, David Montgomery.   Thermodynamic studies of paleoclimates. D, 1969, Columbia University. 138 p.

Shaw, David William.   Determination of global kinetics of coal volatiles combustion. D, 1988, Ohio State University. 149 p.

Shaw, Denis M.   The geochemistry of thallium. D, 1951, University of Chicago. 76 p.

Shaw, Don Wayne.   Mutual adsorption of clay minerals and colloidal hydrous chromium oxide; an electron microscopy investigation. D, 1965, [Baylor University]. 130 p.

Shaw, Donald H.   Effect of taxation on the base metals industry of Mexico. M, 1960, Massachusetts Institute of Technology. 103 p.

Shaw, Earl Bennett.   The evolution of agricultural practices in Jasper County, Iowa. M, 1929, Washington University. 188 p.

Shaw, Ernest W.   Guelph and Eramosa formations (Silurian) of the Ontario Peninsula. D, 1937, University of Toronto.

Shaw, Frederick C.   Chazy Group trilobites. D, 1965, Harvard University.

Shaw, Frederick Carleton.   Richmondian (Upper Ordovician) Rafinesquinae, Cincinnati, Ohio. M, 1960, University of Cincinnati. 156 p.

Shaw, G.   The geology and petrography of Viewmount Avenue, Westmount, Quebec. M, 1934, McGill University.

Shaw, George Hamill, III.   Operating principles and potential applications of Barnes' hydrothermal rocking autoclave. M, 1969, University of Washington.

Shaw, George Hamill, III.   The effect of solid-solid phase changes on seismic velocities. D, 1971, University of Washington. 73 p.

Shaw, Henry Francis, III.   Sm-Nd and Rb-Sr isotopic systematics of tektites and other impactites, Appalachian mafic rocks, and marine carbonates and phosphates. D, 1984, California Institute of Technology. 296 p.

Shaw, Herbert Richard.   Mineralogical studies in the Bunker Hill Mine, Idaho. D, 1959, University of California, Berkeley. 182 p.

Shaw, Jerry C.   Bacterial fouling of a model core system. M, 1982, University of Calgary. 134 p.

Shaw, Jimmy E.   An investigation of ground-water contamination by oil field brine disposal in Morrow and Delaware counties, Ohio. M, 1966, Ohio State University.

Shaw, John Damon.   Late Pleistocene paleontology of Orcas, Shaw, Lopez and San Juan islands of the San Juan archipelago. M, 1972, University of Washington. 60 p.

Shaw, John Holmes.   Petrography and petrology of some intrusive bodies in the southern Willamette Valley, Oregon. M, 1964, University of Oregon. 77 p.

Shaw, Jonathan E.   Anionic nutrients in ground water at a spray irrigation site, Tampa, Florida. M, 1980, University of South Florida, Tampa. 57 p.

Shaw, Martha Jane. Sediment volume changes in the vicinity of Oceanside Harbor. M, 1986, San Diego State University. 199 p.

Shaw, Nancy Sue. Ostracoda from the DeQueen Limestone. M, 1961, Louisiana State University.

Shaw, Neil B. Biostratigraphy of the Cowlitz Formation in the upper Nehalem River basin, Northwest Oregon. M, 1986, Portland State University. 110 p.

Shaw, Nolan Gail. Cheilostomata of the Gulfian Cretaceous of southwestern Arkansas. D, 1967, Louisiana State University.

Shaw, Nolan Gail. Geology of the Benbrook Quadrangle, Tarrant County, Texas. M, 1956, Southern Methodist University. 38 p.

Shaw, Peter Robert. Waveform inversion of explosion seismology data. D, 1983, University of California, San Diego. 211 p.

Shaw, Richard Frank, Jr. A subsurface study of the post-Morrowan Series of the Pennsylvanian System of Township 2 South, Range 3 West, Carter County, Oklahoma. M, 1954, University of Oklahoma. 77 p.

Shaw, Richard Michael. Mobile ground magnetometer survey of a portion of the Southern Peninsula of Michigan. M, 1971, Michigan State University. 84 p.

Shaw, Ronald L. Mineralogy and petrology of a Precambrian interbedded chert and carbonate section, Baraga Basin, Baraga County, Michigan. M, 1974, Bowling Green State University. 232 p.

Shaw, Stephen Lynn. Geology and landuse capability of the Castle Hills Quadrangle, Bexar County, Texas. M, 1974, University of Texas, Austin.

Shaw, Thomas Howard. Lithostratigraphy and conodont biostratigraphy of the Newala Limestone in its type area, south-central Alabama. M, 1987, Emory University. 550 p.

Shaw, Timothy Jean. The early diagenesis of transition metals in nearshore sediments. D, 1988, University of California, San Diego. 180 p.

Shaw, William S. The Cumberland Basin of deposition. D, 1951, Massachusetts Institute of Technology. 193 p.

Shawa, Monzer S. Depositional pattern within the upper Cretaceous, southwest of Coalinga (Fresno County), California. M, 1970, University of San Diego.

Shawe, Daniel Reeves. Heavy detrital minerals in stream sands of the eastern Sierra Nevada between Leevining and Independence, California. D, 1953, Stanford University. 200 p.

Shawesh, Othman Mohamed. Vegetation types of semi-arid rangelands in northwestern Libya, North Africa. D, 1981, University of Wyoming. 240 p.

Shayani, Sohrab. Leakage from reservoirs on limestone terraines. M, 1972, University of Illinois, Urbana. 149 p.

Shayes, F. P. Oil and gas report on North Louisiana and South Arkansas. M, 1961, Louisiana State University.

Shayes, Frederick Pine. Oil and gas report on North Louisiana and South Arkansas. M, 1925, University of Missouri, Columbia.

Shazly, Hanssan El *see* El Shazly, Hanssan

Shea, Damian. Solubility and precipitation of copper-sulfide. D, 1985, University of Maryland. 266 p.

Shea, Frank Stoddard. Observations on the Walton Barytes Deposit, Walton, Hants County, Nova Scotia. M, 1958, Dalhousie University. 68 p.

Shea, Gregory Thomas Francis. A petrologic study of basalts from the Magic Mountain hydrothermal area, southern Explorer Ridge, Northeast Pacific Ocean. M, 1987, University of British Columbia. 90 p.

Shea, James F., Jr. Geology of the Ware Quadrangle, Massachusetts. M, 1953, University of Massachusetts. 50 p.

Shea, James H. Stratigraphy of the Lower Ordovician New Richmond Sandstone in the Upper Mississippi Valley. M, 1960, University of Wisconsin-Madison.

Shea, James Herbert. Petrology and stratigraphy of sediments from southern and central Lake Michigan. D, 1964, University of Illinois, Urbana. 99 p.

Shea, John V. Oil, its origin and production. M, 1947, Bates College.

Shea, Michael Curtis. Geology and highway location considerations in the Orofino–Kamiah–Nez Perce area (northwestern), Idaho. M, 1970, University of Idaho. 88 p.

Shea, Michael E. Uranium nuclide migration at Marysvale, Utah; natural analog for radioactive waste isolation. M, 1982, University of California, Riverside. 123 p.

Shea, Robert Michael. Bimodal volcanism in the Northeast Basin and Range; petrology of the Wildcat Hills, Box Elder County, Utah. M, 1985, University of Utah. 79 p.

Shea, Valois Ruth. Geology of the White Raven Mine, Ward mining district, Boulder County, Colorado. M, 1988, University of Colorado. 141 p.

Sheafer, William Lesley, II. A study of cyclical sedimentation of the lower Conemaugh Group in the Northeast Pittsburgh area. M, 1950, University of Pittsburgh.

Sheaffer, Sandra. Geographic distribution, abundance, and classification of fish teeth in some Quaternary deep-sea sediments. M, 1979, University of Delaware.

Sheahan, Joseph W. Delineation of landslides in West Virginia by use of a portable proton magnetometer. M, 1977, University of Toledo. 89 p.

Shearer, Carroll Dean. Geology of the igneous intrusions of the northern Valle Las Norias, Coahuila, Mexico. M, 1980, West Texas State University. 98 p.

Shearer, Charles. Geochemical and geological investigation of the Pawtuckaway Mountain plutonic complex, Rockingham County, New Hampshire. M, 1976, University of New Hampshire. 78 p.

Shearer, Charles Kenneth. Petrography, mineral chemistry, and geochemistry of the Hardwick Tonalite and associated igneous rocks, central Massachusetts. D, 1983, University of Massachusetts. 265 p.

Shearer, Charles Raymond. Terrestrial heat flow studies in eastern Arizona. D, 1979, New Mexico Institute of Mining and Technology. 184 p.

Shearer, Clement Fletcher. Regional flood analyses and land use planning in an area of inadequate data; Santa Cruz County, California. D, 1978, University of California, Santa Cruz.

Shearer, David. Modern and early Holocene Arctic deltas, Melville Island, N.W.T., Canada. M, 1975, University of South Carolina.

Shearer, Eugene Merle. Geology of Red Dirt Creek area, Eagle County, Colorado. M, 1951, University of Colorado.

Shearer, Gerald Brian. Chromium in plants, soil, stream sediments, and water from the Illinois River basin, southwestern Oregon. M, 1972, Ohio State University.

Shearer, Harold K. Rock weathering and laterization in Brazil. M, 1914, University of Wisconsin-Madison.

Shearer, James Moxley. Detailed grain size analysis of Recent marine sediments and post glacial history of Port au Port Bay, West Newfoundland. M, 1973, Memorial University of Newfoundland. 234 p.

Shearer, Jay Nevin. Structural geology of eastern part of the Malad Summit Quadrangle, Idaho. M, 1975, Utah State University. 82 p.

Shearer, Peter Marston. Anisotropy in the oceanic lithosphere; the Ngendei seismic refraction experiment. D, 1986, University of California, San Diego. 142 p.

Shearer, Ralph Duward. A cross section through the Triassic basin. M, 1927, University of North Carolina, Chapel Hill. 23 p.

Shearer-Fullerton, Amanda. The petrology and mineralogy of Tertiary(?) olivine trachyte in the Harrington Peak Quadrangle, southeastern Idaho. M, 1985, Utah State University. 62 p.

Sheasby, Norman Michael. Depositional patterns of the upper Devonian Waterways Formation, Swan Hills area, Alberta. M, 1971, University of Calgary. 69 p.

Sheatsley, Larry Lee. Pleistocene molluscan faunas of the Aultman Deposit, Stark County, Ohio. M, 1960, Ohio State University.

Shebl, Mamdouh Abdel-Aal. The relation of sandstone diagenesis to petrology and depth; the Mt. Simon, St. Peter, and Aux Vases sandstones, subsurface, Illinois Basin. M, 1985, Southern Illinois University, Carbondale. 161 p.

Shedd, Solon S. Geography and geology of Washington. M, 1907, [Stanford University]. 116 p.

Shedd, Solon S. The clays of the State of Washington; their geology, mineralogy, and technology. D, 1910, Stanford University. 355 p.

Shedlock, Kaye M. Structure and tectonics of North China. D, 1986, Massachusetts Institute of Technology. 194 p.

Shedlock, Robert John. Mineralogy, petrology, and geochemistry of a fassaite-bearing tactite near Helena, Montana. M, 1975, University of Michigan. 58 p.

Sheedlo, Mark Kenneth. Structural geology of the northern Snowcrest Range, Beaverhead and Madison counties, Montana. M, 1984, Western Michigan University. 132 p.

Sheedy, Katherine A. The petrogenesis of the banded gneiss of the Wilmington North Quadrangle, Delaware. M, 1975, University of Delaware.

Sheehan, Jack Richard. Structure and stratigraphy of northwestern Contra Costa County, California. M, 1956, University of California, Berkeley. 61 p.

Sheehan, Leo. Geology and uranium deposits of portions of the Mountain City and Rowland 15-minute quadrangles, Elko County, Nevada. M, 1978, University of Idaho. 90 p.

Sheehan, Mark Charles. The postglacial vegetational history of the Argolid Peninsula, Greece. D, 1979, Indiana University, Bloomington. 84 p.

Sheehan, Michael Anthony. Ichnology, depositional environment, and paleoecology of the upper Pennington Formation (Upper Mississippian), Dougherty Gap, Walker County, Georgia. M, 1988, University of Georgia. 211 p.

Sheehan, Peter Michael. Silurian Brachiopoda, community ecology, and stratigraphic geology in western Utah and eastern Nevada, with a section on late Ordovician stratigraphy. D, 1971, University of California, Berkeley. 538 p.

Sheehan, Peter Michael. Siluro-Devonian brachiopods from Chihuahua, Mexico. M, 1967, University of California, Berkeley. 114 p.

Sheely, Milton Jerome, Jr. Pb-210 geochronology of Recent sediments from the Tijuana River estuary, San Diego County, California. M, 1979, San Diego State University.

Sheely, William S. A study of the heavy minerals of the Dakota Sandstone members and the Jackpile Member of the Morrison Formation in the southeastern San Juan Basin, New Mexico. M, 1977, Bowling Green State University. 139 p.

Sheets, M. Meredith. Contributions to the geology of the Cascade Mountains in the vicinity of Mount Hood. M, 1932, University of Oregon. 71 p.

Shefchik, William T. The Paleozoic geology of the Potaman Peak, Ziegler Basin, Bowery Creek, Bowery Peak, Herd Lake, Galena Peak, Ryan Peak, and Meridian Peak quadrangles, Custer County, Idaho. M, 1977, University of Wisconsin-Milwaukee.

Sheffels, Barbara Moths. Structural constraints on crustal shortening in the Bolivian Andes. D, 1988, Massachusetts Institute of Technology. 170 p.

Sheffer, Bernard Douglas. Geology of the western Alcova area. M, 1951, University of Wyoming. 74 p.

Sheffer, Herman Weaver. An investigation of some germanium occurrences in southwestern New Mexico. M, 1963, New Mexico Institute of Mining and Technology. 68 p.

Sheffey, Renata. Geology of the Centreville Quarry, Fairfax County, Virginia. M, 1988, Virginia State University. 106 p.

Sheffield, Tatum M. Petrography and geochemistry of the West Potrillo basalt field, Dona Ana County, New Mexico. M, 1981, University of Texas at El Paso.

Sheffy, M. V. A study of Myricaceae from Eocene sediments of southeastern North America. D, 1972, Indiana University, Bloomington.

Shefloe, Allyn Carlyle. Metallurgical investigation of antimony-gold ore from Valley County, Idaho. M, 1941, University of Idaho. 31 p.

Shegelski, Roy Jan. Geology and mineralogy of the Terra silver mine, Camsell River, Northwest Territories. M, 1973, University of Toronto.

Shegelski, Roy Jan. Stratigraphy and geochemistry of Archean iron formations in the Sturgeon Lake-Savant Lake greenstone terrain, N.W. Ontario. D, 1978, University of Toronto.

Shegewi, Omar M. Stratigraphy, sedimentary petrology, and depositional environments of parts of the Chickamauga Supergroup (Middle Ordovician), Walker County, Northwest Georgia. M, 1985, Emory University. 119 p.

Shehabi, G. Mineral study and determination of depositional environment of Red River channel sediment. M, 1975, East Texas State University.

Shehata, William Makram. Geohydrology of Mount Vernon Canyon area, Jefferson County, Colorado. D, 1971, Colorado School of Mines. 164 p.

Sheikh Ali, Khadim S. Stratigraphic and paleoenvironmental study of the St. Louis Limestone (Mississippian), and subjacent beds of the Salem Limestone, Gulletts Creek section, Lawrence County, Indiana. M, 1974, Indiana University, Bloomington. 78 p.

Sheikh, Abdul Mannan. Geology and ore deposits of Las Culias tungsten district, Pima County, Arizona. M, 1966, University of Arizona.

Sheikh, Abdul Mannan. Mineralogical studies of the Sunshine Mine, Kellogg, Idaho. D, 1976, University of Idaho. 148 p.

Sheikh, Abdul Razzak. Geological engineering study of T. 16 N., R. 5 W., Kingfisher County, Oklahoma. M, 1965, University of Oklahoma. 52 p.

Sheikh, Izaz A. The effect of convergence on entrance losses in pumped wells. M, 1965, University of Arizona.

Sheikh-ol-Eslami, Bahman. An analysis of "clay-adsorbed' water using nuclear magnetic resonance spectroscopy. D, 1967, University of California, Davis. 110 p.

Shekarchi, Ebraham. Heavy accessory minerals of the Pennsylvanian formation of Walden Ridge, Tennessee. M, 1951, University of Tennessee, Knoxville. 49 p.

Shelburne, Kevin Lee. Estimation of parameters of two mathematical models of surface runoff. M, 1976, New Mexico Institute of Mining and Technology.

Shelburne, Orville B. Some stratigraphic studies of the (Lower Cretaceous) Kiamichi Formation in central Texas. M, 1956, University of Wisconsin-Madison.

Shelburne, Orville Berlin, Jr. Geology of the Boktukola Syncline area of the Ouachita Mountains of Oklahoma. D, 1959, University of Wisconsin-Madison. 155 p.

Shelby, Cader Alverd. Heavy minerals in the Wellborn Formation, Lee and Burleson counties, Texas. M, 1962, University of Texas, Austin.

Shelby, Debra L. An occurrence of dolomite in the Lower Ordovician Arbuckle Group in the Slick Hills of southwestern Oklahoma. M, 19??, Texas Christian University.

Shelby, Lynn. A comparison of linears mapped from digitally processed Landsat data to faults depicted on geologic maps; the fluorspar district, western Ken-

tucky and southern Illinois. M, 1982, Murray State University. 88 p.

Shelby, Phillip R. Depositional history of the St. Joe-Boone Formations in northern Arkansas. M, 1986, University of Arkansas, Fayetteville.

Shelby, Thomas Hall, Jr. The geology of an area lying along Barton Creek southwest of Austin, Texas. M, 1934, University of Texas, Austin.

Shelden, Arthur William. Geology of the NW 15-minute Ural Quadrangle, Lincoln County, Montana. M, 1961, Montana State University. 65 p.

Shelden, Francis D. Stratigraphy and sedimentary petrology of the Silurian formations of central Manitoulin Island, Ontario. M, 1961, Wayne State University.

Sheldon, D. H. The geology of a portion of the Solstice Quadrangle, Los Angeles County. M, 1932, University of Southern California.

Sheldon, Elisabeth Shepard. Continuity and change in plant usage from the Mississippian to the historic period. D, 1982, University of Alabama. 144 p.

Sheldon, Estelle L. Geographic influences upon the history and economic development of North Dakota. M, 1926, University of Wisconsin-Madison.

Sheldon, Lyndon L. Analysis of formulas for the change of vertical-deflection components. M, 1961, Ohio State University.

Sheldon, Pearl Gertrude. The Atlantic slope Arcas. D, 1911, Cornell University.

Sheldon, Pearl Gertrude. The development of the genus Arca in America. M, 1909, Cornell University.

Sheldon, Richard Porter. Physical stratigraphy of the Phosphoria Formation in northwestern Wyoming. D, 1956, Stanford University. 105 p.

Sheldon, Wichita F. Heavy mineral study of some of the sands of the KMA oil field, Wichita and Archer counties, Texas. M, 1940, Texas Tech University. 83 p.

Sheldon, William Knowles. Geology of Lund area, Travis County. M, 1949, University of Texas, Austin.

Shelefka, Michael A. The origin of Cedar Butte; an evolved tholeiitic construct at the center of the eastern Snake River plain, Idaho. M, 1981, SUNY at Buffalo. 54 p.

Shell, Jesse Allen. Stratigraphy, petrology, and sedimentology of the Lower Devonian sandstones of southwestern Virginia. M, 1984, University of North Carolina, Chapel Hill. 149 p.

Shellebarger, Jeffrey. Stratigraphy, structure, and metamorphism in portions of the Jacks Gap and Coosa Bald 7 1/2' quadrangles, Georgia. M, 1980, University of Georgia.

Shelley, Geoffrey K. Stratigraphy and depositional environment of the Lower Permian Putnam Formation, and Lost Creek and Hords Creek beds of the Admiral Formation in north-central Texas. M, 1984, University of Texas, Arlington. 205 p.

Shelley, Joanne Ross. The geology of the western San Juan Islands, Washington State. M, 1971, University of Washington. 64 p.

Shellhorn, Mark A. The role of crystal contamination at the Butcher Ridge igneous complex, Antarctica. M, 1982, New Mexico Institute of Mining and Technology. 169 p.

Shelnutt, John P. Intrusive breccias of the Julcani Silver District, Peru. M, 1980, Michigan Technological University. 56 p.

Shelp, Gene Sidney. The distribution and dispersion of gold in glacial till associated with gold mineralization in the Canadian Shield. M, 1985, Queen's University. 155 p.

Shelton, David Camp. Thickness of alluvium and evaluation of aggregate resources in the Lower Cache La Poudre River Valley, Colorado, by electrical resistivity techniques. M, 1972, University of Colorado.

Shelton, Dean H. The geology of a portion of the Solstice Quadrangle, Los Angeles County (California). M, 1932, University of Southern California.

Shelton, Deborah H. The geochemistry and petrogenesis of gabbroic rocks from Attu Island, Aleutian Islands, Alaska. M, 1986, Cornell University. 178 p.

Shelton, Glenmore Lorraine. Experimental deformation of single phase and polyphase crustal rocks at high pressures and temperatures. D, 1981, Brown University. 155 p.

Shelton, John S. The Miocene Glendora Volcanics in eastern Los Angeles County, California. D, 1947, Pomona College.

Shelton, John Wayne. Studies of the Strawn-Canyon boundary, Pennsylvanian, in north-central Texas. D, 1953, University of Illinois, Urbana.

Shelton, John Wayne. The Mesaverde Group, Cretaceous, in the northwest part of the San Juan Basin, New Mexico and Colorado. M, 1951, University of Illinois, Urbana. 80 p.

Shelton, Kevin Louis. Evolution of the porphyry copper and skarn deposits at Murdochville, Gaspe Peninsula, Quebec; a geochemical, stable isotopic, and fluid inclusion study. D, 1982, Yale University. 286 p.

Shelton, Marlyn Lyle. A multiple-storage model for the uniformly flowing Deschutes River. D, 1973, Southern Illinois University, Carbondale. 107 p.

Shelton, Theodore Brian. Decomposition of oil pollutants in natural bottom sediments (of New Jersey rivers). D, 1972, Rutgers, The State University, New Brunswick. 154 p.

Shelton-V, Bert J. Geology and petroleum prospects of Darien, southeastern Panama. M, 1952, Oregon State University. 62 p.

Shelvey, Stephanie Anne. The stratigraphy and environment of deposition of productive Wilcox clays in west central Freestone and Southeast Limestone counties, Texas. M, 1986, Texas A&M University.

Shemeliuk, Edward Michael. A practical application and evaluation of the OPH-2 electromagnetic method. M, 1973, University of Manitoba.

Shemeliuk, Virginia A. B. Compositional variations in an iron-rich sequence of carbonate and siliceous sediments, Turquetil Lake area, Northwest Territories. M, 1976, University of Manitoba.

Shen, Elizabeth Jean. A sediment-profile camera study of the sediment-water interface in Milwaukee Harbor, Wisconsin. M, 1984, University of Wisconsin-Milwaukee. 159 p.

Shen, Glen T. Lead and cadmium geochemistry of corals; reconstruction of historic perturbations in the upper ocean. D, 1986, Massachusetts Institute of Technology. 233 p.

Shen, P.-Y. Dynamics of the liquid outer core of the Earth. D, 1975, University of Western Ontario.

Shen, Pouyan. Observation and thermodynamic calculations of stoichiometry and stability fields of wustite at various temperature and pressure conditions; their implications to the Earth. D, 1982, Cornell University. 211 p.

Shenberger, David M. Gold solubility in aqueous sulfide solutions. M, 1985, Pennsylvania State University, University Park. 102 p.

Sheng, Cheng Chun. The geology of the Amco Lake, Eurnet Creek and Wreck Lake, Coppermine River area, NWT. M, 1958, University of British Columbia.

Sheng, Jopan. Multiphase immiscible flow through porous media. D, 1986, Virginia Polytechnic Institute and State University. 134 p.

Sheng, Tse Cheng. Development of a landslide classification for mountain watersheds of Taiwan, China. M, 1965, Colorado State University. 206 p.

Sheng, Zhengzhi. A new mechanism of recoil fractionation of uranium isotopes. D, 1987, University of Arkansas, Fayetteville. 283 p.

Shenkel, Claude W., Jr. A lithologic definition of the Hermosa Formation; Pennsylvanian, Colorado, Utah. D, 1952, University of Colorado.

Shenker, Alan Edward. Glacial, periglacial, and cryopedogenic development of arctic-alpine terrains in the C-26 sector of the Juneau Icefield, Atlin Provincial Park, N.W. British Columbia, Canada. M, 1979, University of Idaho. 184 p.

Shenon, Philip J. Geology and ore deposits of the Bannack District, Montana. D, 1926, University of Minnesota, Minneapolis. 48 p.

Shenon, Philip John. A metallurgical study of the Bay Horse Mine ore with notes on the geology (near Huntington, Oregon). M, 1924, University of Idaho. 82 p.

Shenton, Edward. A study of the foraminifera and sediments of Matagorda Bay, Texas. M, 1957, Texas A&M University.

Shepard, Brian K. Petrography and environments of deposition of the Carter Sandstone (Mississippian) in the Black Warrior Basin of Alabama and Mississippi. M, 1979, University of Alabama.

Shepard, Francis P. The structure and stratigraphy of the Rocky Mountain Trench from Gateway to Golden (British Columbia). D, 1922, University of Chicago. 23 p.

Shepard, J. Scott. Modeling water contents of Kentucky soils. D, 1982, University of Kentucky. 122 p.

Shepard, James R. Correlation of soil test data with airphoto patterns for trafficability study. M, 1951, Purdue University.

Shepard, John Bixby, Jr. The geology of part of the San Gabriel fault zone, Los Angeles and Ventura counties, California. M, 1961, University of California, Los Angeles.

Shepard, John Lynn. Sedimentology of a Pennsylvanian, coal-bearing, cratonic delta; upper Tradewater Formation, western Kentucky. M, 1980, University of Illinois, Urbana. 170 p.

Shepard, Mark D. Geology and ore deposits of the Hermosa mining district, Sierra County, New Mexico. M, 1984, University of Texas at El Paso.

Shepard, Nancy. The geothermal history of the Plateau Province of northern Alabama. M, 1985, Tulane University. 75 p.

Shepard, Norman. The petrology and mineralogy of the Cross Lake area (Quebec). D, 1960, University of Toronto.

Shepard, P. J. Geology and mineralization of the southern Silver Star Stock, Washougal mining district, Skamania County, Washington. M, 1980, Oregon State University. 113 p.

Shepard, Timothy Mark. Geology of the Rosillos Mountains, Trans-Pecos, Texas. M, 1982, Texas Christian University. 136 p.

Shepeck, Anthony W. Characterization of the ore host rock and hydrothermal system at the Ropes gold mine, Ishpeming, Michigan. M, 1985, Michigan Technological University. 140 p.

Shephard, L. E. Geotechnical properties and their relation to geologic processes in South Pass outer continental shelf lease area blocks 28, 47 and 48, offshore Louisiana. M, 1977, Texas A&M University.

Shephard, Les Edward. Geotechnical properties of select convergent margin sediments. D, 1981, Texas A&M University. 183 p.

Shepheard, William Wayne. Depositional environment of the upper Cretaceous Bearpaw-Edmonton transition zone, Drumheller "Badlands", Alberta, (Canada). M, 1969, University of Calgary. 127 p.

Shepherd, Ashley. A study of the magnetic anomalies in the eastern Gulf of Mexico. M, 1983, University of Houston.

Shepherd, Charles Edward. Three dimensional experimental study of acoustic-elastic wave scattering and diffraction. D, 1981, Columbia University, Teachers College. 80 p.

Shepherd, Glenn Lincoln. Geology of the Whitaker Peak-Canton Canyon area, Southern California. M, 1960, University of California, Los Angeles.

Shepherd, H. E. Distribution of sulfur in the Palouse silt loam as influenced by four sulfur containing fertilizers. M, 1974, University of Idaho. 93 p.

Shepherd, Jackson Howard. A study of certain rocks of the California Lake map area, northern Manitoba. M, 1954, University of Manitoba.

Shepherd, Norman. Petrography and mineralogy of the Cross Lake area, Ungava, New Quebec. D, 1960, University of Toronto.

Shepherd, Norman. The mineralogy of a metamorphosed iron formation from Payne Bay, Ungava. M, 1957, University of Toronto.

Shepherd, Rodney D. The sedimentology of the Upper Cretaceous Horsethief Formation, Lewis and Clark County, Montana. M, 1977, University of Missouri, Columbia. 93 p.

Shepherd, Russell G. A model study of river incision. M, 1972, [Colorado State University].

Shepherd, Russell G. Geomorphic operation, evolution, and equilibria, Sand Creek Watershed, Llano region, central Texas. D, 1975, University of Texas, Austin. 224 p.

Shepherd, T. A. A methodology for evaluating uranium tailings management alternatives. D, 1979, Colorado State University. 450 p.

Shepley, Susan I. Taxonomy and biostratigraphy of Paleogene age sediments from cores in the Northwest Atlantic and Caribbean using diatoms. M, 1982, University of Delaware. 164 p.

Sheppard, Don G. Geology of the northwest quarter of Booneville Quadrangle, Logan and Franklin counties, Arkansas, with special emphasis on ancient depositional environments. M, 1978, Northeast Louisiana University.

Sheppard, John Selwyn. The influence of geothermal temperature gradients upon vegetation patterns in Yellowstone National Park. D, 1971, Colorado State University. 170 p.

Sheppard, Richard Abner. Petrology of the Simcoe Mountains volcanic area, Washington. D, 1960, The Johns Hopkins University. 153 p.

Sheppard, Simon Mark Foster. Carbon and oxygen isotope studies in marbles. D, 1966, McMaster University. 185 p.

Sheppard, Stephen Charles. Plant phosphorus requirements and soil phosphorus reactions as influenced by temperature. D, 1982, University of Manitoba.

Shepperd, Jane Elizabeth. Development of a salt marsh on the Fraser Delta at Boundary Bay, British Columbia, Canada. M, 1981, University of British Columbia. 99 p.

Shepps, Vincent Chester. Correlation of the tills of northeastern Ohio by size analysis. M, 1952, University of Illinois, Urbana.

Shepps, Vincent Chester. The glacial geology of a part of northwestern Pennsylvania. D, 1955, University of Illinois, Urbana.

Sher, Mohammad Tahir. Fluoride contents of synthetic 10 hydrated kaolinites. M, 1986, SUNY at Buffalo. 56 p.

Sherburne, Roger Wayne. Crust-mantle structure in continental South America and its relation to sea floor spreading. D, 1975, Pennsylvania State University, University Park. 153 p.

Sherer, Richard Lowell. Nephrite deposits of the Granite, the Seminoe and the Laramie mountains (Wyoming). D, 1970, University of Wyoming. 194 p.

Sheridan, David S. Permian, Triassic, and Jurassic stratigraphy of the McCoy area (Routt, Grand, and Eagle counties), west-central Colorado. M, 1950, University of Colorado.

Sheridan, Douglas M. Geology of the High Climb Pegmatite, Custer County, South Dakota. M, 1951, University of Minnesota, Minneapolis. 76 p.

Sheridan, James T. Stratigraphic and paleontologic evidence at Pillar Point, San Mateo County for late Tertiary movement on the San Andreas Fault. M, 1972, San Francisco State University.

Sheridan, John Francis. A methodology for determining flash flood forecasting systems for a community. D, 1977, Oklahoma State University. 153 p.

Sheridan, John Joseph. Modeling geologic noise for MAD applications. M, 1976, Naval Postgraduate School.

Sheridan, John Thomas. Paleontology and stratigraphy of known outcrops of the (Lower Ordovician) "Holston" Formation in North Georgia. M, 1951, Emory University. 55 p.

Sheridan, Michael Francis. The mineralogy and petrology of the Bishop Tuff (Pleistocene of California). D, 1965, Stanford University. 193 p.

Sheridan, Robert E. Seaward extension of the Canadian Appalachians. D, 1968, Columbia University. 170 p.

Sheridan, Robert E. Seismic-refraction measurements of the continental margin east of Florida. M, 1965, Columbia University. 73 p.

Sheridan, Sally Wright. Remote sensing of annual grasslands for geologic information; a literature analysis. M, 1982, Stanford University. 197 p.

Sherif, Alamin Abdalla. Precambrian geology of a portion of the Mason Quadrangle, Mason County, Texas. M, 1984, University of Texas of the Permian Basin. 50 p.

Sherif, Nasreddin. Modal analysis of heavy minerals by X-ray diffraction and textural studies of New Jersey beach sands. M, 1971, University of Toledo. 87 p.

Sheriff, Akbar. Origin and distribution of terrigenous detritus in the late early Permian of the central Cordilleran Miogeosyncline. M, 1975, San Jose State University. 69 p.

Sheriff, Robert Edward. A curved crystal reflection X-ray spectrometer. M, 1947, Ohio State University.

Sheriff, Robert Edward. The measurement of nuclear gyromagnetic ratios. D, 1950, Ohio State University.

Sheriff, Steven D. Paleomagnetic studies in Wyoming; the Leucite Hills volcanic field, the Rattlesnake Hill volcanic field, and the Green River Formation. D, 1981, University of Wyoming.

Sheriff, Steven D. Paleomagnetism of the San Juan volcanic field, southwestern Colorado. M, 1976, Western Washington University. 113 p.

Sherkarchi, Ebraham. The geology of the Flag-Pond Quadrangle, Tennessee-North Carolina. D, 1959, University of Tennessee, Knoxville. 140 p.

Sherlock, Sean M. Structure and stratigraphy of early Proterozoic rocks, McDonald Mountain-Breadpan Mountain area, northern Sierra Anchas, Gila County, Arizona. M, 1986, Northern Arizona University. 81 p.

Sherman, David Michael. Electronic structures of iron and manganese oxides with applications to their mineralogy. D, 1984, Massachusetts Institute of Technology. 230 p.

Sherman, Don Kerry. Upper Eocene biostratigraphy of the8967077 Snow Creek area, northeastern Olympic Peninsula, Washington. M, 1960, University of Washington. 116 p.

Sherman, Douglas B. Utilization of paired cords in sedimentological studies of the Mozambique Channel in the southwest Indian Ocean. M, 1966, University of Southern California.

Sherman, Douglas Joel. Longshore currents; a stress balance approach. D, 1983, University of Toronto.

Sherman, Frank B., Jr. Petrology of the Lyons Sandstone (Permian) (Northern Colorado). M, 1968, Colorado State University. 54 p.

Sherman, James E. Hartshorne heavy minerals, Franklin, Sebastian, Scott counties, Arkansas. M, 1955, University of Arkansas, Fayetteville.

Sherman, John W. Post-Pleistocene diatom assemblages in New England lake sediments. D, 1976, University of Delaware. 330 p.

Sherman, John W. Post-Pleistocene diatom stratigraphy in cores from Lake Champlain, Vermont. M, 1972, University of Vermont.

Sherman, Peter Edward. A gravity study of the Narragansett Basin region, Massachusetts and Rhode Island. M, 1980, Boston College.

Sherman, Sharon F. Aluminum silicate polymorph relationships in the Pennsylvania Piedmont. M, 1973, University of Delaware.

Sherman, Thomas Floyd. Sedimentology of the Upper Cretaceous Ericson Formation, Rock Springs Uplift, Wyoming. M, 1983, Colorado State University. 207 p.

Shero, B. R. An interpretation of temporal and spatial variations in the abundance of diatom taxa in sediments from Yellowstone Lake, Wyoming. D, 1977, University of Wyoming. 178 p.

Sherr, Margot S. An analysis of the economics of petroleum extraction on the continental shelf of eastern China. M, 1985, Stanford University. 134 p.

Sherrer, P. L. Cataclastic rocks of the Rawhide Mountains, West-central Arizona. M, 1976, San Diego State University.

Sherriff, Barbara Lucy. Relationships between MAS NMR and mineral structure. D, 1988, McMaster University. 214 p.

Sherrill, John F., III. Pyrite and other forms of sulfur in Recent sediments at three sites in South Louisiana. M, 1987, University of Southwestern Louisiana. 104 p.

Sherrill, Richard E. Symmetry of Northern Appalachian Foreland folds. D, 1933, Cornell University.

Sherrill, Richard Ellis. Post-Mississippian folds and faults in north central Oklahoma. M, 1928, Cornell University.

Sherrill, Timothy Wilson. Evaluation of the role of environmental contamination in the microbial degradation of polyaromatic hydrocarbons. D, 1982, University of Tennessee, Knoxville. 96 p.

Sherrod, David R. Geology, petrology, and volcanic history of a portion of the Cascade Range between latitudes 43°-44°N, central Oregon, U.S.A. D, 1986, University of California, Santa Barbara. 320 p.

Sherrod, Neil R. Glacial geology of western Sheridan County, North Dakota. M, 1963, University of North Dakota. 91 p.

Sherrod, Thomas D. Lithofacies analysis and diagenesis of deep water lower Atoka Sandstone (Pennsylvanian), Perry County, Arkansas. M, 1986, University of Arkansas, Fayetteville.

Sherry, Paul F. Geohydrology and refraction seismology. M, 1956, Boston College.

Shervais, John W. Petrology and structure of the Alpine lherzolite massif at Balmuccia, Italy. D, 1979, University of California, Santa Barbara.

Sherwin, Jo-Ann Major Koch. A theoretical study of decollement folding with application to the Plateau Province of the central Appalachians. D, 1972, Brown University. 162 p.

Sherwonit, William Edward. A petrographic study of the Catalina Gneiss in the forerange of the Santa Catalina Mountains, Arizona. M, 1974, University of Arizona.

Sherwood, Elizabeth Schneider. Study of a Precambrian terrane in central Wisconsin near Pittsville. M, 1976, University of Wisconsin-Madison.

Sherwood, Herbert Gordon. The quantitative mineralogy of the forty-five Canadian base metal sulphide ore deposits. D, 1967, University of Manitoba.

Sherwood, Herbert Gordon. The significance of Cu, Pb, Zn, Au, and Ag distribution in Canadian base metal deposits. M, 1964, University of Manitoba.

Sherwood, Kirk W. Geologic structure along the Hines Creek Fault west of the Wood River, north central Alaska Range. M, 1973, University of Wisconsin-Madison.

Sherwood, Kirk W. Stratigraphy, metamorphic geology, and structural geology of the central Alaska Range, Alaska. D, 1979, University of Wisconsin-Madison. 949 p.

Sherwood, Ronald W. Coccolithoporid biostratigraphy and systematics of the middle Eocene Weches Formation of Texas. D, 1972, Washington University. 201 p.

Sherwood, William Cullen. Structure of the Jacksonburg Formation in Northampton and Leigh counties, Pennsylvania. D, 1961, Lehigh University. 143 p.

Sherwood, William Cullen. The petrography of some Cambrian and Ordovician limestones occurring in Virginia. M, 1958, University of Virginia. 67 p.

Sherzer, William Hittell. Geological report on Monroe County (Michigan). D, 1901, University of Michigan.

Sheta, Mohamed Aly. Dynamic analysis of shallow and pile foundations. D, 1981, University of Western Ontario.

Sheth, Kertikant R. Investigations of the feasibility of transporting ore in open-flumes. M, 1964, University of Missouri, Rolla.

Sheth, Madhusudan. A heavy mineral study of Pleistocene and Holocene sediments near Nome, Alaska. M, 1971, San Jose State University. 83 p.

Sheth, Pranlal Girdharlal. Shaped charges; their mechanism and applications in mining. D, 1953, University of Missouri, Columbia.

Shettel, Don Landis, Jr. Experimental determination of oxygen isotopic fractionation between H_2O and hydrous silicate melts. D, 1978, Pennsylvania State University, University Park. 116 p.

Shettel, Don Landis, Jr. The solubility of quartz in supercritical H_2O-CO_2 fluids. M, 1974, Pennsylvania State University, University Park. 52 p.

Shettigara, K. V. Electrical resistivity investigation of the Schofield high-level water body, Oahu, Hawaii. D, 1985, University of Hawaii.

Shetty, Y. G. Chromatographic estimation of sulfur gases in Hawaiian volcanic gases; sampling investigations. M, 1965, University of Hawaii. 50 p.

Sheu, D. D. Phosphate content as a possible paleosalinity indicator of carbonate depositional environments; observation in a core from lower McKnight Formation, Comanchean (Cretaceous), South Texas. M, 1977, University of Texas, Arlington. 63 p.

Sheu, Der-Duen. The geochemistry of Orca Basin sediments. D, 1983, Texas A&M University. 147 p.

Sheu, Jiun-Chyuan. Applications and limitations of the spectral-analysis-of-surface-waves method. D, 1987, University of Texas, Austin. 305 p.

Sheu, Nien-Jen Wang. Calcareous nannofossils of the Brownstown Marl and Gober Chalk (Upper Cretaceous) of the upper Austin Group, Fannin County, Northeast Texas. M, 1982, University of Texas, Arlington. 91 p.

Sheu, Wa-Ye. Modeling of stress-strain-strength behavior of a clay under cyclic loading. D, 1984, University of Colorado. 358 p.

Shevenell, Thomas Cortland. The effect of rain on intertidal estuarine sediment transport. D, 1986, University of New Hampshire. 214 p.

Shew, Roger D. Geology of the area between Fayetteville and Elizabethtown, North Carolina. M, 1979, University of North Carolina, Chapel Hill. 114 p.

Shewbridge, Scott Edward. The influence of reinforcement properties on the strength and deformation characteristics of a reinforced sand. D, 1987, University of California, Berkeley. 195 p.

Shewmake, Dan W. The geology of the Spirits Creek area. M, 1963, University of Arkansas, Fayetteville.

Shewmaker, Sherman Nelson. Variations in stream channel process-form relationships in environmentally distinctive regions; Midwest versus Great Plains. D, 1982, Indiana University, Bloomington. 269 p.

Shewman, Frederick Charles. Phosphate content of Great Basin (Utah) ground waters and methods for appraising their contamination potential. D, 1971, Utah State University.

Shewman, Robert W. Phase relations in the pentlandite region of the Fe-Ni-S system. M, 1966, McGill University.

Shi, Chung-Shin. Electrical conductivity of silica in the presence of water and methanol vapor. D, 1972, Stanford University. 109 p.

Shi, Nungjane Carl. Reverse sediment transport induced by amplitude modulated waves. D, 1983, University of Washington. 121 p.

Shi, Yaolin. Pore pressure, temperature and deformation in accretionary complexes; a coupled study on the Barbados Ridge complex. D, 1986, University of California, Berkeley. 409 p.

Shibata, Tsugio. Petrology of the Oceanographer fracture zone (35°N35°W). D, 1976, SUNY at Albany. 129 p.

Shideler, Gerald Lee. Pennsylvanian sedimentational patterns of the Michigan Basin. M, 1965, University of Illinois, Urbana.

Shideler, Gerald Lee. Petrography and provenance of the Johns Valley Boulders (pre-Mississippian), Mountains, southeastern Oklahoma and southwestern Arkansas. D, 1968, University of Wisconsin-Madison. 153 p.

Shideler, James Henry, Jr. The geology of the Silver Creek area, northern Cascades, Washington. M, 1965, University of Washington. 94 p.

Shideler, William Henry. The evolution of North American spirifers. D, 1910, Cornell University.

Shieh, Chih-Shin. Adsorption of dissolved vanadium onto particulate matter in seawater. D, 1988, Florida Institute of Technology. 119 p.

Shieh, Chiou-Fen. Polarization analysis of complex seismic wave field. D, 1988, St. Louis University. 134 p.

Shieh, Yuch-ning. Oxygen, carbon, and hydrogen isotopes studies of contact metamorphism. D, 1969, California Institute of Technology. 355 p.

Shiekh, Khalid. Use of magnetic reversal timelines to reconstruct the Miocene landscape near Chinji Village, Pakistan. M, 1984, Dartmouth College. 103 p.

Shields, Hilbert Nathaniel. Comparative geology and geochemistry with respect to precious metal mineralization of selected California Coast Range mercury mining districts. M, 1983, University of Nevada. 122 p.

Shields, James C., II. Geology and ore deposits of the Dives and Gold Ridge groups, Dos Cabezas, Arizona. M, 1940, University of Arizona.

Shields, John F. Structural and stratigraphic controls on lower Atokan gas production, Johnson County, Arkansas. M, 1986, University of Arkansas, Fayetteville.

Shields, Kermit E. Structure of the northwestern margin of the San Fernando Valley, Los Angeles County, California. M, 1977, Ohio University, Athens. 82 p.

Shields, Martin L. Outcrop to basin stratigraphy and structure of the Glen Rose Limestone, central Texas. M, 1984, Baylor University. 191 p.

Shields, R. C. The Magador zinc deposits, Barrante Township, Quebec. M, 1955, McGill University.

Shields, Richard H. Depositional environments of the sediments in Marsh Valley near Arimo, Idaho. M, 1978, Idaho State University. 68 p.

Shields, Robert McC. The $Rb^{87}-Sr^{87}$ age of stony meteorites. D, 1965, Massachusetts Institute of Technology. 365 p.

Shier, Daniel Edward. Vermetid reefs and coastal development in Southwest Florida. D, 1965, Florida State University. 152 p.

Shiever, John Wayne. Hydrothermal synthesis and crystallographic measurements of the fayalite-forsterite series. M, 1967, Texas Christian University. 38 p.

Shiferaw, Alemu. The Softestad iron ore deposit, Southern Norway. M, 1973, University of Alberta. 111 p.

Shifflett, Frances Elaine. Foraminifera of the (lower Eocene) Aquia Formation at the type locality, Virginia and southern Maryland. D, 1948, The Johns Hopkins University.

Shifflett, Howard Richard. Geomorphology of the Kaskaskia River, Illinois. D, 1973, Washington University. 180 p.

Shifflett, Howard Richard. The geography of the Ogden Valley (Utah). M, 1960, University of Utah. 102 p.

Shigley, James Edwin. Phosphate minerals in granitic pegmatites; a study of primary and secondary phosphates from the Stewart Pegmatite, Pala, California. D, 1982, Stanford University. 558 p.

Shih, Chi-Yu. The rare Earth geochemistry of oceanic igneous rocks. D, 1972, Columbia University. 151 p.

Shih, Chung-Chi. Inversion of marine gravity data. M, 1982, Texas A&M University.

Shih, Ernest Hsiao Hsin. Multispectral reflectance and image textural signatures of arid alluvial geomorphic surfaces in the Castle Dome Mountains and piedmont, southwestern Arizona. M, 1982, University of Arizona. 287 p.

Shih, John Shai-Fu. The nature and origin of fine-scale sea-floor relief. D, 1980, Massachusetts Institute of Technology. 222 p.

Shih, Keh-gong. On the reduction and interpretation of ocean-floor temperature and heat flow data. D, 1968, Oregon State University. 103 p.

Shih, T-M. Physical properties of Gaspe skarn. M, 1965, McGill University.

Shih, Tai-Chang. Marine magnetic anomalies from the western and northern Philippine Sea; implications for the evolution of marginal basins. D, 1978, University of Texas, Austin. 190 p.

Shih, Xiao Rung. Propagation of seismic waves in oceanic crust. M, 1985, Texas Tech University. 60 p.

Shikoh, Mirza M. The Paleogene strata of western Pakistan. M, 1956, University of Missouri, Columbia.

Shiller, Alan Mark. The geochemistry of particulate major elements in the Santa Barbara Basin and observations on the calcium carbonate-carbon dioxide system in the ocean. D, 1982, University of California, San Diego. 214 p.

Shiller, Gerald I. Suspension in selected streams of the Southern California Bight in the Flood of 1969. M, 1972, University of Southern California.

Shillibeer, Harry A. The ages of rocks by the measurement of radiogenic argon. M, 1953, University of Toronto.

Shillibeer, Harry A. The potassium-argon method of age determinations. D, 1955, University of Toronto.

Shilts, William W. A laboratory study of late Pleistocene sediments in the Jay Peak, Irasburg, and Memphramagog quadrangles, Vermont. M, 1965, Miami University (Ohio). 90 p.

Shilts, William Weimer. Pleistocene geology of the Lac Megantic region, southeastern Quebec, Canada. D, 1970, Syracuse University. 227 p.

Shim, J. H. Distribution and taxonomy of planktonic marine diatoms in the Strait of Georgia, B.C. D, 1976, University of British Columbia.

Shimamoto, Toshihiko. Effects of fault-gouge on the frictional properties of rocks; an experimental study. D, 1977, Texas A&M University. 213 p.

Shimazaki, Yoshohiko. Mineralogy of basic carbonate minerals of copper and zinc. D, 1957, Stanford University. 93 p.

Shimeall, Clark M. The Weber Sandstone (Pennsylvanian and Permian) of Daggett County, Utah. M, 1968, Bowling Green State University. 74 p.

Shimer, J. A. Spectrographic analysis of New England granites and pegmatites. D, 1942, Massachusetts Institute of Technology. 77 p.

Shimer, J. A. Structural geology of the Norfolk Basin (Massachusetts). M, 1939, Massachusetts Institute of Technology. 54 p.

Shimko, Kenneth Andrew. A study on structures in two valley glaciers in northern Sweden by observations and computer modeling. M, 1987, University of Minnesota, Minneapolis. 155 p.

Shimoyama, Akira. Catalytic transformation of n-fatty acids to n-alkanes in relation to petroleum genesis. D, 1971, Washington University. 114 p.

Shin, Chang-soo. Nonlinear elastic wave inversion by blocky parameterization. D, 1988, University of Tulsa. 123 p.

Shin, Charng-Jeng. Dynamic soil-structure interaction. D, 1987, University of Colorado. 271 p.

Shin, Myong Sup. Hawaii; a study of images and anti-images with a special focus on its dry lands. D, 1979, University of Minnesota, Minneapolis. 309 p.

Shin, Tzay-Chyn Tony. Lg and coda wave studies of eastern Canada. D, 1985, St. Louis University. 203 p.

Shine, Brendan F. Engineering geology of the Battle Mountain landslide south of Minturn, Colorado. M, 1985, Colorado School of Mines. 125 p.

Shinn, Mike Reed. Structural geology of the Brentwood-St. Paul area, Northwest Arkansas. M, 1979, University of Arkansas, Fayetteville.

Shinn, Richard. A study of climate sensitivity to the size and configuration of large continental ice sheets. M, 1988, University of Miami. 155 p.

Shinn, Rory K. Petrology and environment of deposition of the Collier Shale (Lower Ordovician), western Garland and eastern Montgomery counties, Arkansas. M, 1988, University of New Orleans.

Shinohara, Kiyoshi. Calculation and use of steam/water relative permeabilities in geothermal reservoirs. D, 1978, Stanford University. 192 p.

Shinol, John Henry. Seismic analysis of a complex structure in Stephens County, Oklahoma. M, 1988, University of Oklahoma. 101 p.

Shipley, Raymond Dale. Local depositional trends of "Cherokee" sandstones, Payne County, Oklahoma. M, 1975, Oklahoma State University. 48 p.

Shipley, Susan. Erosional modification of the downwind tephra lobe of the 18 May 1980 eruption of Mount Saint Helens, Washington. M, 1983, University of Washington. 73 p.

Shipley, Thomas Howard. Echo characteristics, reflector horizons and geology of the western central North Atlantic. D, 1975, Rice University. 194 p.

Shipley, Webster E., III. Geology, petrology and geochemistry of the Mountain Pine Ridge Batholith, Belize, Central America. M, 1978, Colorado School of Mines. 90 p.

Shipman, Wayne D. Saltwater-bearing aquifers at the periphery of Narragansett Bay; geoelectric and geohydrologic characteristics. M, 1978, University of Rhode Island.

Shipp, B. G. Geology of an area east of Bates Hole, Carbon and Albany counties, Wyoming. M, 1959, University of Wyoming. 69 p.

Shipp, Craig. Morphology and sedimentology of a single-barred nearshore system, eastern Long Island, New York. M, 1980, University of South Carolina.

Shipp, Thomas C. A simplified technique for determining phosphorus loading in lakes and applications using shallow ground water samples. M, 1975, Virginia State University. 60 p.

Shipton, Washburne D. Geology of the Sparta Quadrangle, Wisconsin. M, 1916, University of Iowa. 81 p.

Shirasuna, Takeshi. Finite element analyses on cohesive soil behavior due to advanced shield tunneling. D, 1985, Virginia Polytechnic Institute and State University. 520 p.

Shirazi, G. A. Salinity status of some selected Hawaiian soils. M, 1966, University of Hawaii. 83 p.

Shirey, Burrell Peter. F-salts of the Salina Group of the Michigan Basin. M, 1983, Michigan State University. 108 p.

Shirey, S. B. Sulfides and sulfur content of the Kiglapait layered intrusion, Labrador. M, 1975, University of Massachusetts. 76 p.

Shirey, Steven Bottome. The origin of Archean crust in the Rainy Lake area, Ontario. D, 1984, SUNY at Stony Brook. 411 p.

Shirley, Brooke Howard. Geology of a portion of Big Horn County, Montana. M, 1941, University of Iowa. 54 p.

Shirley, David Noyes. Magmatic differentiation in partially molten systems; applications to the Moon and the Palisades Sill. D, 1986, University of California, Los Angeles. 202 p.

Shirley, Dennis H. Geochemical facies analysis of the Surprise Canyon Formation in Fern Glen channelway, central Grand Canyon, Arizona. M, 1988, Northern Arizona University. 240 p.

Shirley, John E. Structural geology north of the junction of Bear Creek with East Fork Wind River, Fremont County, Wyoming. M, 1979, Miami University (Ohio). 55 p.

Shirley, Richard H., Jr. Petrography and prediction of reservoir rock properties in the Sussex Sandstone, Powder River basin, Wyoming. M, 1977, Texas A&M University. 91 p.

Shirley, Steven Hayden. Stratigraphic and structural subsurface analysis of pre-Missourian rocks in Northwest Norman area, Cleveland and McClain counties, Oklahoma. M, 1986, University of Oklahoma. 98 p.

Shishkevish, Leo. Sakesaria cotteri and its variants (Saudi Arabia). M, 1954, New York University.

Shishkevish, Leo J. The distribution of foraminifera in cores from the northern Scotia Arc, Scotia Sea. D, 1964, New York University. 159 p.

Shive, Peter Northrop. The effect of internal stress on remanent magnetic properties of nickel. D, 1968, Stanford University. 51 p.

Shiveler, Donna Jean. Sedimentary petrology, depositional environment and diagenesis of the Cambrian St. Charles Formation, southeastern Idaho. M, 1986, University of Idaho. 82 p.

Shively, C. Effects of chemicals on the properties of Gulf Coast drilling muds. M, 1939, Louisiana State University.

Shively, Margaret V. Reconnaissance geology of the genesis and development of soils in the upper Palouse River area of Idaho. M, 1977, University of Idaho. 45 p.

Shiver, Richard Steven. The geology of the Heath Springs Quadrangle, Heath Springs, South Carolina. M, 1974, North Carolina State University. 39 p.

Shklanka, Roman. Repeated metamorphism and deformation of evaporite-bearing sediments, Little Maria Mountains, California. D, 1963, Stanford University. 156 p.

Shklanka, Roman. The petrogenesis of the Porter Lake Dome, Porter-Blackstone Lakes area, northern Saskatchewan. M, 1957, University of Saskatchewan. 100 p.

Shlemon, R. J. Geology of Red Spring Anticline, Hot Springs County, Wyoming. M, 1959, University of Wyoming. 73 p.

Shlemon, Roy James. Landform-soil relationships in northern Sacramento County, California. D, 1967, University of California, Berkeley. 335 p.

Shlien, Seymour. Automatic classification of seismic detections from large-aperture seismic arrays. D, 1972, Massachusetts Institute of Technology. 133 p.

Shmitka, Richard Otto. Geology of the eastern portion of Lion Canyon Quadrangle, Ventura County, California. M, 1970, University of California, Davis. 86 p.

Shoberg, Thomas Gilford. Collisional accretion of the terrestrial planets. M, 1982, Washington University. 132 p.

Shockey, Philip Nelson. Northern portion; Burning Springs Anticline (West Virginia). M, 1954, West Virginia University.

Shockey, Philip Nelson. Reconnaissance geology of the Leesburg Quadrangle, Lemhi County, Idaho. D, 1957, Cornell University.

Shoemaker, Abbott Hall and Somers, George. The geology of the El Tiro Mine, Silver Bell, Arizona. M, 1924, University of Arizona.

Shoemaker, Craig Alan. A relocation of earthquakes in the area of the Gorda Rise. M, 1984, Southern Illinois University, Carbondale. 111 p.

Shoemaker, Eugene M. Impact mechanics at Meteor Crater, Arizona. D, 1959, Princeton University. 55 p.

Shoemaker, Eugene M. Petrology of the (Precambrian) Hopewell Series in the Ojo Caliente District of

New Mexico. M, 1948, California Institute of Technology. 66 p.

Shoemaker, Helen E. Cuestas of the North and Baltic seas. M, 1925, University of Wisconsin-Madison.

Shoemaker, James Scovell. The suppression of multiple reflections on vertical seismic profiles. M, 1983, Pennsylvania State University, University Park. 125 p.

Shoemaker, Phillip Wayne. Depositional environment and reservoir characteristics of the lower Vicksburg Sandstones, East McAllen Ranch Field, Hidalgo County, Texas. M, 1978, Texas A&M University. 128 p.

Shoemaker, Richard Walter. Geology of the Shirley and Seminoe Mountain area, Carbon County, Wyoming. M, 1936, University of Wyoming. 65 p.

Shoemaker, Robert Earl. Fossil leaves of the Hell Creek and Tullock formations of eastern Montana. M, 1964, University of Montana. 118 p.

Shoemaker, Robert Earl. Pollen and spores of the Judith River Formation (Campanian), central Montana. D, 1969, University of Minnesota, Minneapolis. 262 p.

Shoemaker, William A. The geology of the Raymond Canyon area, Sublette Range, Lincoln County, Wyoming. M, 1987, Idaho State University. 55 p.

Shoffner, David A. Geology of Reinecker Ridge-Bald Hill area, South Park, Park County, Colorado. M, 1974, Colorado School of Mines. 82 p.

Shoja-Taheri, Jafar. Geophysical investigations along Forest highway 16, Houghton and Ontonagon counties, Michigan. M, 1970, Michigan Technological University. 73 p.

Shoja-Taheri, Jafar. Seismological studies of strong motion records. D, 1977, University of California, Berkeley. 288 p.

Shokes, R. F. Rate-dependent distributions of lead-210 and interstitial sulfate in sediments of the Mississippi River delta. D, 1976, Texas A&M University. 133 p.

Sholes, Mark A. Stratigraphy and petrography of the Arkansas Novaculite of Arkansas and Oklahoma. D, 1978, University of Texas, Austin. 312 p.

Sholes, Mark Allen. The stratigraphic (Paleozoic), and structural geology of Henry County, Iowa. M, 1967, University of Iowa. 88 p.

Sholin, Michael Hugh. Landslide hazards in The Dalles, Wasco County, Oregon. M, 1982, Oregon State University. 47 p.

Sholkovitz, Edward R. The chemical and physical oceanography and interstitial water chemistry of the Santa Barbara basin. D, 1973, University of California, San Diego.

Sholl, Vinton Hubbard. The stratigraphy of the Templeton Member of the Upper Cretaceous Woodbine Formation in eastern Denton County, Texas. M, 1956, Southern Methodist University. 27 p.

Shomaker, John Wayne. Geology of the southern portion of the Sandia granite (Precambrian), Sandia Mountains, Bernalillo County, New Mexico. M, 1965, University of New Mexico. 80 p.

Shomali, Bahman Saghatchian. Ostracodes from the shales of the Lingle Formation (Middle Devonian) of southern Illinois. M, 1970, Southern Illinois University, Carbondale. 65 p.

Shomo, Sarah J. Geology of a buried Triassic - Jurassic basin, coastal plain, Virginia. M, 1982, University of North Carolina, Chapel Hill. 57 p.

Shonfelt, John P. Geologic heterogeneities of a Chesteran sandstone reservoir, Kinney-Lower Chester Field, Stevens and Seward counties, Kansas, and their effect on hydrocarbon production. M, 1988, Wichita State University. 206 p.

Shook, Ellen L. Foraminifera of the (Oligocene) Byram calcareous marl, Vicksburg, Mississippi. M, 1936, University of Mississippi.

Shope, Linda S. Historical shoreline changes and modern sediment transport, Bogue Inlet to New River

Inlet, North Carolina. M, 1980, Bowling Green State University. 189 p.

Shope, Steven B. Regional groundwater flow and contamination transport in the vicinity of the Tolend Road landfill, Dover, New Hampshire. M, 1986, University of New Hampshire. 136 p.

Shope, Steven Michael. Electromagnetic coal seam tomography. D, 1987, Pennsylvania State University, University Park. 308 p.

Shor, Alexander N. Bottom currents and abyssal sedimentation processes south of Iceland. D, 1979, Woods Hole Oceanographic Institution. 246 p.

Shor, George G., Jr. Crustal structure and reflections from the Mohorovicic discontinuity in Southern California. D, 1954, California Institute of Technology. 158 p.

Shorb, William Murray. Stratigraphy, facies analysis and depositional environments of the Moenkopi Formation (Lower Triassic), Washington County, Utah, and Clark and Lincoln counties, Nevada. M, 1983, Duke University. 205 p.

Shore, Jesse Paul. Extrusion of Yule Marble. D, 1977, University of California, Berkeley. 152 p.

Shore, Lawrence R. Uranium concentration variations in stream water in response to changing stream discharge from paired watersheds in western Montana. M, 1980, University of Montana. 118 p.

Shore, Michael J. Prediction of the frequency domain response of explosive sources with comparison to the BOXCAR event (Nevada). M, 1972, Pennsylvania State University, University Park. 53 p.

Shore, Michael James. Short period P-wave attenuation in the middle and lower mantle of the Earth. D, 1983, The Johns Hopkins University. 202 p.

Shore, Patrick John. Quantitative analysis in structural geology and in the determination of thermodynamic properties of minerals. D, 1985, Southern Methodist University. 262 p.

Shore, Rochelle C. The conodont fauna and paleoecology of the lower Chambersburg Formation (Middle Ordovician) of south-central Pennsylvania. M, 1971, University of Missouri, Columbia.

Shorey, Edwin F. Geology of part of southern Morrow County, Northeast Oregon. M, 1976, Oregon State University. 131 p.

Shorey, Mark David. Estimates of extension in the North Sea Central Graben from analysis of high quality seismic data. M, 1987, University of Texas, Austin. 135 p.

Short, Allan McIlroy. A chemical and optical study of piedmontite from Shadow Lake, Madera County, California. M, 1933, Cornell University.

Short, Andrew Damien. Beach dynamics and nearshore morphology of the Alaskan Arctic coast. D, 1973, Louisiana State University.

Short, B. L. A geologic and petrographic study of the Ferris-Haggerty mining area, Carbon County, Wyoming. M, 1958, University of Wyoming. 138 p.

Short, David George. Application of electrical resistivity methods to the detection of subsurface solution cavities underlying highways. M, 1981, University of Florida. 163 p.

Short, Maxwell Naylor. Deep-level chalcocite at Superior, Arizona, and at Butte, Montana. D, 1923, Harvard University. 174 p.

Short, Michael Ray. A study of various properties of sediments occurring as bed material in selected reaches of the Kentucky River. M, 1973, University of Kentucky. 141 p.

Short, Michael Ray. Petrology of the Pennington and Lee formations of northeastern Kentucky and the Sharon Conglomerate of southeastern Ohio. D, 1978, University of Cincinnati. 216 p.

Short, Nicholas Martin. The Ste. Genevieve Formation at its type locality. M, 1954, Washington University. 124 p.

Short, Nicholas N. Behavior of trace elements in rock weathering and soil formation. D, 1958, Massachusetts Institute of Technology. 218 p.

Short, S. K. Holocene palynology in Labrador-Ungava; climatic history and culture change on the central coast. D, 1978, University of Colorado. 244 p.

Short, William R., Jr. Geology of the Santa Teresa Hills, Santa Clara County, California. M, 1987, California State University, Hayward. 112 p.

Shortridge, Barbara Gimla. Map lettering as a quantitative symbol; a preliminary investigation. D, 1977, University of Kansas. 206 p.

Shortridge, Charles Glen. The geological relationships of water loss and gain problems on Battle Creek near Hermosa, South Dakota. M, 1954, South Dakota School of Mines & Technology.

Shortt, Eric R. Source characterization of the October 30, 1983, Narman-Horasan earthquake. M, 1985, Massachusetts Institute of Technology. 82 p.

Shortt, Trevor Alan Leslie. Investigations in shallow reflection seismology. M, 1988, University of Western Ontario. 267 p.

Shotts, T. W. and Ellis, H. A. The investigation of Osage City, Kansas, clays. M, 1911, [University of Kansas].

Shotwell, James D. Sandstone classification. M, 1976, Brooklyn College (CUNY).

Shotwell, Jesse Arnold. A Hemphillian mammalian fauna from McKay Reservoir, Oregon. D, 1953, University of California, Berkeley. 150 p.

Shotwell, L. Brad. Morphology of zircons from Precambrian gneiss in the southern Bighorn Mountains, Wyoming. M, 1973, Kent State University, Kent. 71 p.

Shotyk, William. The inorganic geochemistry of peats and the physical chemistry of waters from some Sphagnum bogs (Volume I and II). D, 1987, University of Western Ontario.

Shou, Philip M. Y. Kinetics and mechanisms of several amino acid diagenetic reactions in aqueous solutions and in fossils. D, 1979, University of California, San Diego. 178 p.

Shouldice, James Robert. Geology of a part of St. Helena Quadrangle, California. M, 1947, University of California, Berkeley. 54 p.

Shoup, Robert Charles. Correlation of Landsat lineaments with geologic structures, north-central Oklahoma. M, 1980, University of Oklahoma. 123 p.

Shourd, Melvin Lee. A paleoecological study of the Decorah subgroup (Middle Ordovician) in the middle Mississippi Valley. D, 1972, Washington University. 222 p.

Shover, Edward Franklin. Petrology of upper Paleozoic clays and shales of north central Texas. D, 1961, University of Illinois, Urbana. 194 p.

Shovic, Henry Folke. Genesis, classification, and variability of the Kitsap Series, Washington. D, 1979, Washington State University. 157 p.

Showalter, Donald Lee. Composition trends in australites, impact glasses, and associated natural materials by activation analysis. D, 1970, [University of Kentucky]. 188 p.

Showalter, James Aswell. Simulation of seismic data and residual migration. M, 1986, University of New Orleans. 77 p.

Showers, Kate Barger. Assessment of the land use potential of Ha Makhopo, Lesotho, southern Africa; a holistic approach to agricultural evaluation. D, 1982, Cornell University. 333 p.

Showers, William J. Isotopic trends of calcareous plankton across the Equatorial Pacific high productivity zone. D, 1982, University of Hawaii. 282 p.

Showers, William James. Biometry and biology of an Antarctic foraminiferid Rosalina globularis. M, 1978, University of California, Davis. 93 p.

Shows, Thaddeus N. Geology of the Burnsville Quadrangle, Tishomingo Alcorn and Prentiss counties, Mississippi. M, 1963, University of Tennessee, Knoxville. 33 p.

Shrake, Tom. Geology and hydrothermal alteration of the Pan disseminated gold occurrence, White Pine County, Nevada. M, 1984, University of Idaho. 75 p.

Shrestha, Rajendra K. Seismic studies across the Crater Island and West Newfoundland grabens, Box Elder County, Utah. M, 1986, University of Oklahoma. 185 p.

Shrestha, Ramesh Lal. Local geodetic networks adjustments in three dimensions. D, 1983, University of Wisconsin-Madison. 280 p.

Shreve, Ronald Lee. Geology and mechanics of the Blackhawk Rockslide, Lucerne Valley, California. D, 1959, California Institute of Technology. 79 p.

Shride, Andrew Fletcher. Some aspects of younger Precambrian geology in southern Arizona. D, 1961, University of Arizona. 294 p.

Shrier, Tracy. Hydrothermal transport and deposition of uranium in granitic host rocks. M, 1980, University of Utah. 70 p.

Shriram, Calcutta R. Geological significance of abnormal pressures in deep wells of South Louisiana (Miocene). M, 1967, University of Tulsa. 117 p.

Shrivastava, Jai Nandan. Certain trace element distribution in the Searchlight, Nevada quartz monzonite, Clark County, Nevada. M, 1961, University of Missouri, Rolla.

Shroba, Cynthia Susan. Brachiopod biostratigraphy and paleoenvironments across the Chesterian/Morrowan (Mississippian/Pennsylvanian) boundary at Arrow Canyon, Clark County, Nevada. M, 1988, University of Illinois, Urbana. 132 p.

Shroba, R. R. Soil development in Quaternary tills, rockglacier deposits, and taluses, southern and central Rocky Mountains. D, 1977, University of Colorado. 449 p.

Shrock, Robert Rakes. A report on the Mississippian geology of the United States. M, 1926, Indiana University, Bloomington.

Shrock, Robert Rakes. The stratigraphy and paleontology of the West Franklin (Somerville) Limestone. M, 1926, Indiana University, Bloomington. 76 p.

Shrode, Raymond Scott. Temperature of formation of some bedded Illinois fluorite crystals. M, 1950, University of Illinois, Urbana.

Shroder, John F., Jr. Stratigraphic and tectonic history of the Moncton Group (Mississippian) of nonmarine red beds of New Brunswick, Canada. M, 1963, University of Massachusetts. 81 p.

Shroder, John Ford, Jr. Landslides of Utah. D, 1967, University of Utah. 337 p.

Shropshire, Kenneth Lee. The Chinle and Jelm formations (Triassic) of north-central Colorado. D, 1974, University of Colorado.

Shu Kay Joseph Lee. Multilayer gravity inversion using Fourier transforms. M, 1977, University of Alberta. 144 p.

Shuart, Susan Rae. Equilibria in the system Fe-Cr-Ti-O, with special emphasis on spinel, ilmenite, and pseudobrookite phases at 1300 degrees C and 1200 degrees C and an oxygen pressure of 10^{-10} atmosphere. M, 1975, Pennsylvania State University, University Park.

Shubak, Kenneth Arnold. New species of Exogyra from the Cretaceous Indidura Formation of Mexico. M, 1960, University of Michigan.

Shubat, Michael Andrew. Stratigraphy, petrochemistry, petrography, and structural geology of the Columbia River Basalt in the Minam-Wallowa River area, Northeast Oregon. M, 1979, Washington State University. 156 p.

Shuchman, Robert A. Quantification of SAR signatures of shallow water ocean topography. D, 1982, University of Michigan.

Shuck, Edward L. A seismic survey of the Ralston area, Jefferson County, Colorado. M, 1976, Colorado School of Mines. 45 p.

Shuck, Gordon R. Filtration properties of Pacific Northwest diatomite. M, 1952, University of Washington. 51 p.

Shudofsky, Gordon N. Source mechanisms and focal depths of East African earthquakes and Rayleigh wave phase velocities in Africa. D, 1984, Princeton University. 191 p.

Shuff, Sheldon G. A mineralogical study of Cincinnatian Series limestones. M, 1974, Miami University (Ohio). 67 p.

Shufflebarger, Thomas Edwin, Jr. The geology of the Lower Cambrian metasediments in the Poor Mountain area of Floyd, Montgomery, and Roanoke counties, Virginia. M, 1953, Virginia Polytechnic Institute and State University.

Shuford, Marlene Eloise. Biostratigraphy and sedimentation rates of the Taiwan continental slope. M, 1977, University of Oregon. 126 p.

Shukis, Paul S. A Petrographic study of sand sized Sediments from the upper Saint Francis River, Missouri. M, 1972, Southern Illinois University, Carbondale. 63 p.

Shukla, Narendra R. Sedimentary petrology of the Ripley Formation (Upper Cretaceous) in Oktibbeha County, Mississippi. M, 1972, Mississippi State University. 170 p.

Shukla, Surendra Shanker. Chemistry of inorganic phosphorus in lake sediments. D, 1972, University of Wisconsin-Madison.

Shukla, Vijai. Dolomitization in the Lockport Formation (Middle Silurian), in New York. D, 1980, Rensselaer Polytechnic Institute. 192 p.

Shuler, Edward Hooper. Geology of a portion of the San Timoteo Canyon Badlands near Beaumont, California. M, 1952, University of Southern California.

Shuler, Ellis William. The geology of the Walker Mountain overthrust block in southwestern Virginia. D, 1915, Harvard University.

Shuleski, Paul J. Seismic fault motion and SV screening by shallow magma bodies in the vicinity, Socorro, New Mexico. M, 1977, New Mexico Institute of Mining and Technology.

Shulhof, William Peter. Mineralogy of the Lone Eagle uranium-bearing deposits, Boulder Batholith, Jefferson Co., Mont. M, 1955, Pennsylvania State University, University Park. 85 p.

Shulhof, William Peter. Relationships between uranium and some other trace elements in pyrite, galena, and sphalerite from vein deposits. D, 1960, Pennsylvania State University, University Park. 108 p.

Shulik, Stephen J. The paleomagnetism of the Carboniferous strata in the northern Appalachian Basin with applications toward magnetostratigraphy. D, 1979, University of Pittsburgh. 305 p.

Shulkaew, Pitak. Petrography and origin of the lower Cotton Valley sandstones in Kildare Field, Cass County, East Texas. M, 1983, Northeast Louisiana University. 82 p.

Shull, Roger Don. Radioactivity transport in water; simulation of sustained releases to selected river environments. D, 1968, University of Texas, Austin. 254 p.

Shullaw, Byron L. The geology of Magnolia mining district; Boulder County, Colorado. M, 1951, University of Colorado.

Shulman, Chaim. Stratigraphic analysis of the Cherokee Group (Pennsylvanian) in adjacent portions of Lincoln, Logan and Oklahoma counties, Oklahoma. M, 1965, University of Tulsa. 30 p.

Shultz, Albert W. Sedimentology of conglomerates of debris flow and water flow origin in the Permo-Pennsylvanian Cutler and Fountain formations of Colorado. M, 1980, Indiana University, Bloomington. 97 p.

Shultz, Charles High. Petrology of Mount Washburn, Yellowstone National Park, Wyoming. D, 1962, Ohio State University. 225 p.

Shultz, David James. Stable carbon isotope variations in organic and inorganic carbon reservoirs in the Fenholloway River estuary and the Mississippi River estuary. D, 1974, Florida State University.

Shultz, Julianna M. Mid-Tertiary volcanic rocks of the Timberwolf Mountain area, south-central Cascades, Washington. M, 1988, Western Washington University. 145 p.

Shum, Che-Kwan. Altimeter methods in satellite geodesy. D, 1982, University of Texas, Austin. 414 p.

Shumac, Karen May. Composition of alluvial sands in a small, high relief drainage basin in the South Mountains of North Carolina. M, 1983, North Carolina State University. 38 p.

Shumaker, Robert Clarke. Geology of the Pawlet Quadrangle, Vermont. D, 1960, Cornell University. 109 p.

Shumaker, Robert Clarke. Till texture variations and the Pleistocene deposits of the Union Springs and Scipio quadrangles, Cayuga County, New York. M, 1957, Cornell University.

Shuman, Christopher A. Fracture studies and in situ permeability testing with borehole packers. M, 1987, Pennsylvania State University, University Park. 185 p.

Shuman, Fred Leon, Jr. A similitude study of soil strength instruments. D, 1966, Iowa State University of Science and Technology. 189 p.

Shumard, C. Brent. Palynology of a lacustrine sinkhole facies, and the geologic history of a (late Pleistocene?) basin in Clark County, southwestern Kansas. M, 1974, Wichita State University. 155 p.

Shump, Kenneth W. Streamflow and water quality effects of groundwater discharge to Steamboat Creek, Nevada. M, 1985, University of Nevada. 101 p.

Shumway, Dinah O'Sullivan. Geochemistry of the Ammonoosuc Volcanics; tectonic setting of massive sulfide ore deposits. M, 1984, Dartmouth College. 102 p.

Shumway, George Alfred, Jr. Sedimentary copper, Tatamagouche area, Nova Scotia. M, 1951, Massachusetts Institute of Technology. 84 p.

Shumway, George Alfred, Jr. Sound speed and absorption studies of marine sediments by a resonance method. D, 1958, University of California, Los Angeles. 141 p.

Shumway, Ramon Dwight. Geology of the Lime Hill area, Asotin County, Washington. M, 1960, Washington State University. 54 p.

Shunik, Thomas W. Conodont biostratigraphy of some middle and upper Paleozoic limestones, central Idaho. M, 1974, University of Wisconsin-Milwaukee.

Shupack, Benjamin. Some foraminifera from Long Island and New York Harbor. M, 1932, Columbia University, Teachers College.

Shupack, Dora. Foraminifera of the Lincoln Tunnel (New York). M, 1939, Columbia University, Teachers College.

Shupe, Mark G. A three-dimensional simulation of the groundwater flow system of south-central Florida. M, 1986, Ohio University, Athens. 557 p.

Shure, Loren. Modern mathematical methods in geomagnetism. D, 1982, University of California, San Diego. 150 p.

Shurig, Donald G. Power augers and Earth resistivity units as supplements to drilling machines. M, 1957, Purdue University.

Shurnas, Marshall Kenneth. The stratigraphy and micropaleontology of a well in western Florida. M, 1949, University of Missouri, Kansas City.

Shurr, George W. Marine cycles in the lower Montana Group, Montana and South Dakota. D, 1975, University of Montana. 310 p.

Shurr, George W. Paleocene tectonics in south-central Montana. M, 1967, Northwestern University. 118 p.

Shuster, Evan Thomas. Seasonal variations in carbonate spring water chemistry related to ground water flow. M, 1970, Pennsylvania State University, University Park. 148 p.

Shuster, Mark William. The origin and sedimentary evolution of the northern Green River basin, western Wyoming. D, 1986, University of Wyoming. 356 p.

Shuster, Robert Duncan. Chemistry and petrography of caldera-related ash flows, St. Francois Mountains, Missouri; Grassy Mountain Ignimbrite and Taum Sauk Rhyolite. M, 1978, University of Kansas. 45 p.

Shuster, Robert Duncan. Geochemical and isotopic evidence for the petrogenesis of the northeastern Idaho Batholith, Bitterroot Range, Montana and Idaho. D, 1985, University of Kansas. 93 p.

Shuter, Kelli A. Surface and subsurface hydrologic processes in Big Meadows, Rocky Mountain National Park, Colorado. M, 1988, Colorado School of Mines. 136 p.

Shy, Timothy Laurence. Statistical analysis of controls on the thickness of three Breathitt Formation coal seams; Knott, Perry, and Breathitt counties, Kentucky. M, 1985, University of Kentucky. 207 p.

Shykind, Edwin B. Quantitative studies in geomorphology; subaerial and submarine erosional environments. D, 1956, University of Chicago. 73 p.

Shyu, Chuen T. Numerical analysis of critical field functions for thermal convection in vertical or quasivertical Darcy flow slabs. D, 1979, Oregon State University. 190 p.

Shyu, Jinn-Hwa. The blockage of surface waves by longer waves and currents. D, 1986, The Johns Hopkins University. 147 p.

Siah, Shu-Sheng Jonathan. Two-dimensional shear flow of a granular material. D, 1983, Clarkson University. 172 p.

Sial, Alcides Nobregq. Petrology and tectonic significance of the post-Paleozoic basaltic rocks of Northeast Brazil. D, 1974, University of California, Davis. 403 p.

Siami, Mehdi. Arsenic profiles in sediments and sedimentation processes along the slope of a lake basin. D, 1981, Michigan State University. 60 p.

Siapno, William David. Geology of part of northern Sangre de Cristo Mountains (Fremont County), Colorado. M, 1953, University of Colorado.

Sibbett, Bruce Scott. Geology of the northeast part of the Loon Creek mining district, Custer County, Idaho. M, 1976, University of Idaho. 111 p.

Sibiya, Victor. The surface geochemistry of the Tletletsi area, southeastern Botswana. M, 1982, University of Manitoba.

Sibley, David Michael. The structural fabric of the Rockmart Slate and its relation to the timing of orogenesis in the Valley and Ridge Province of Northwest Georgia. M, 1983, Auburn University. 109 p.

Sibley, Duncan Fawcett. Intergranular pressure solution in the Tuscarora Orthoquartzite. D, 1975, University of Oklahoma. 69 p.

Sibley, Duncan Fawcett. Marine diagenesis of carbonate sediments, Bonaire, Netherlands Antilles. M, 1971, Rutgers, The State University, New Brunswick. 40 p.

Sibley, Michael J. Geology of the Magruder Mine-Chambers Prospect area, Lincoln and Wilkes counties, Georgia. M, 1982, University of Georgia.

Sibley, S. F. U and Th distributions in meteorites in relation to the Pu^{244} problem. M, 1974, Washington University. 111 p.

Sibley, Thomas Howard. The use of X-ray fluorescent spectroscopy for monitoring heavy metals in aquatic environments. D, 1976, University of California, Davis. 158 p.

Sibol, Matthew Steven. Network locational testing and velocity variations in central Virginia. M, 1982, Virginia Polytechnic Institute and State University. 124 p.

Sibray, Steven Sherman. Mineralogy, petrology, and geochemistry of some lavas from Kohala Volcano, Hawaii. M, 1977, University of New Mexico. 113 p.

Sibul, U. Auriferous quartz float tracing in the King Bay area, Sturgeon Lake, northwestern Ontario. M, 1963, University of Toronto.

Sibul, Ulo. Groundwater investigation in the vicinity of Morden, Manitoba (Canada). M, 1968, University of Saskatchewan. 160 p.

Sichko, Michael J. Structural and petrological study of (The Second Watchung) basaltic flow (upper Triassic) near Pluckemin, New Jersey. M, 1970, Brooklyn College (CUNY).

Sickafoose, Donald Kim. Ecostratigraphy of the lower Middle Ordovician and paleoecology of Porterfieldian and Wildernessian biotas at Evans Ferry, Tennessee. M, 1979, University of Tennessee, Knoxville. 185 p.

Sicking, Charles John. Sampling requirements for reflection seismograms in geophysical data acquisition. D, 1980, University of Texas, Austin. 227 p.

Sickles, James Michael. Geology and geologic hazards of the Kenwood-Glen Ellen area, eastern Sonoma County, California. M, 1974, University of California, Davis. 103 p.

Sicks, G. C. The kinetics of silica dissolution from volcanic glass in the marine environment. M, 1975, University of Hawaii. 80 p.

Siclen, Dewitt Clinton Van *see* Van Siclen, Dewitt Clinton

Sidder, Gary Brian. Metallization and alteration of the Monterrosas Mine, Ica, Peru. M, 1981, University of Oregon. 109 p.

Sidder, Gray Brian. Ore genesis at the Monterrosas Deposit in the Coastal Batholith, Ica, Peru. D, 1985, Oregon State University. 221 p.

Siddhanta, Sushil Kumar. The nature of clay minerals in some Wyoming bentonites and the estimation of the montmorillonite content by differential thermal analysis; a critical study. M, 1956, University of Illinois, Urbana.

Siddharthan, Rajaratnam. A two-dimensional non-linear static and dynamic response analysis of soil structures. D, 1984, University of British Columbia.

Siddiqi, Farhat Hussain. Strength evaluation of cohesionless soils with oversize particles. D, 1984, University of California, Davis. 180 p.

Siddiqi, Khalid Omar. An economic evaluation and feasibility study of deep drilling operations in the United States. D, 1981, Louisiana Tech University. 169 p.

Siddiqui, Sayeed Ahmed. Dispersion analysis of seismic data. M, 1971, University of Tulsa. 112 p.

Siddiqui, Shamsul Hasan. Hydrogeologic factors influencing well yields and aquifer hydraulic properties of folded and faulted carbonate rocks in central Pennsylvania. D, 1969, Pennsylvania State University, University Park. 568 p.

Siddiqui, Wasit Ahmed. A study of the relationship between basement faulting and consequent structures in the overlying sediments. M, 1965, New Mexico Institute of Mining and Technology. 42 p.

Sides, James Ronald. A study of the emplacement of a shallow granite batholith; the St. Francois Mountains, Missouri. D, 1978, University of Kansas. 135 p.

Sides, James Wesley. The geology of the central Butte Mountains, White Pine County, Nevada. D, 1966, Stanford University. 225 p.

Sidharta, Ananta Sigit. Prediction of creep failure of excavations in overconsolidated clay with pore pressure dissipation. D, 1985, Colorado State University. 271 p.

Sidhu, Surain Singh. Structural and magnetic nature of crystalline pyrrhotite. D, 1937, University of Pittsburgh.

Sidjabat, Mulia M. Oxidation of hydrogen sulfide in sea water. M, 1967, [University of Miami].

Sidle, William C. Geology of North Craters of the Moon National Monument, Idaho. M, 1979, Portland State University. 64 p.

Sidler, Aubrey Gene. The origin of heavy minerals in the Boise Basin, Idaho. M, 1957, University of Idaho. 62 p.

Sidner, B. R. Late Pleistocene geologic history of the outer continental shelf and upper continental slope, Northwest Gulf of Mexico. D, 1977, Texas A&M University. 145 p.

Sidner, Bruce Robert. Foraminiferal evidence of late Quaternary sea level fluctuations from the west flower garden bank. M, 1973, Texas A&M University.

Sidwell, Raymond. The Colorado Series of south central Wyoming. D, 1928, University of Iowa. 192 p.

Sidwell, Raymond. The stratigraphy and fauna of the Colorado Series in Laramie Basin, Wyoming. M, 1926, University of Iowa. 78 p.

Siebels, Charles Joseph. Petrology, clay mineralogy and conodont distribution of the Cherryvale Formation, Upper Pennsylvanian, Midcontinent. M, 1981, University of Iowa. 164 p.

Siebenmann, Kathleen F. Potential for groundwater contamination by phenols and PAH adsorbed onto Mount St. Helens ash/sediments. M, 1982, Washington State University. 36 p.

Sieber, Kurt DeLance. Preparation and characterization of ABO$_4$ type tungstates and molybdates. D, 1983, Brown University. 89 p.

Siebert, Harry L. A new depositional mechanism to produce cross-lamination. M, 1961, University of Alabama.

Siebert, R. M. The stability of MgHCO$^+_3$ and MgCO$_3$ ion-pairs from 10°C to 90°C. D, 1974, University of Missouri, Columbia. 117 p.

Siebert, Robert M. The solubility of dolomite below 100°C. M, 1970, University of Missouri, Columbia.

Sieck, Herman C. Gravity investigation of the Monterey-Salinas area, California. M, 1964, Stanford University.

Siedlecki, Mary. Trace-element analysis of the clay-silt fraction of the Pungo River Formation, North Carolina. M, 1983, North Carolina State University. 128 p.

Siefert, August Carl. Studies on the hydration of clays. D, 1942, Pennsylvania State University, University Park. 118 p.

Siefken, David Lee. Selected aquifer pump tests, southeastern Hillsborough and southwestern Polk counties, Florida. M, 1974, University of Florida. 73 p.

Siegal, Barry Steven. Crater morphology; an indicator of origin?. D, 1973, Pennsylvania State University, University Park. 165 p.

Siegal, Barry Steven. Effect of grain size, "regolith" thickness, gas pressure, and duration of gas streaming on the morphology of fluidization craters. M, 1971, Pennsylvania State University, University Park. 180 p.

Siegal, Sherman M. Subsurface geology of the upper Tradewater and lower Carbondale groups in south-central Saline County, Illinois. M, 1979, Wright State University. 70 p.

Siegel, David. Stratigraphy of the Putnam Peak Basalt and correlation to the Lovejoy Formation, California. M, 1988, California State University, Hayward. 119 p.

Siegel, Donald I. Quartzite genesis in the upper Johnnie Formation [Precambrian, southern Death Valley, California]. M, 1971, Pennsylvania State University, University Park. 141 p.

Siegel, Donald Ira. Hydrogeochemistry and kinetics of silicate weathering in a gabbroic watershed; Filson Creek, northeastern Minnesota. D, 1981, University of Minnesota, Minneapolis. 289 p.

Siegel, Frederic H. The significance of trace elements in the study of Recent and Pleistocene corals and related sediments of the Florida reef area. M, 1959, University of Kansas.

Siegel, Frederic Richard. Synthesis and geologic origin of sedimentary dolomites. D, 1961, University of Kansas. 96 p.

Siegel, John Alan. Point defects and the electrical properties of apatite. D, 1970, Princeton University. 180 p.

Siegel, Malcolm Dean. Studies of the mineralogy, chemical composition, textures and distribution of manganese nodules at a site in the north equatorial Pacific Ocean. D, 1981, Harvard University.

Siegel, Randall David. Paleoclimatic significance of D/H and ^{13}C/^{12}C ratios in Pleistocene and Holocene wood. M, 1983, University of Arizona. 105 p.

Siegelberg, Alan I. An investigation of relative X-ray diffraction intensities of olivine. M, 1977, Brooklyn College (CUNY).

Siegert, Rudolf B. Sedimentary parameters of Upper Barataria Bay, Louisiana. M, 1961, Texas A&M University.

Siegfried, Joshua Floyd. Geology of the Colorado River from Moab, Utah, to the inflow of the Green River. M, 1927, University of Utah. 74 p.

Siegfried, Robert T. Stratigraphy and chronology of raised marine terraces, Bay View Ridge, Skagit County, Washington. M, 1978, Western Washington University. 52 p.

Siegfried, Robert Wayne. Differential strain analysis; application to shock induced microfractures. D, 1977, Massachusetts Institute of Technology. 158 p.

Siegfus, Stanley. A geological study of the Boise Basin. M, 1923, University of Idaho. 44 p.

Siegfus, Stanley Spencer. A reconnaissance of the Promontory Point mining district, Utah. M, 1924, University of Utah. 165 p.

Siegler, Violet Bernice. Hoback Formation (Paleocene) of western Wyoming. M, 1946, University of Michigan.

Siegmann, James M. Clay mineralogy and burial diagenesis of the Atoka Formation (Pennsylvanian), southeastern Oklahoma. M, 1979, University of Texas, Austin.

Siegmann, Sheryl A. Stratigraphy and depositional environment of the Late Mississippian Monroe Canyon Formation in Southeast Idaho and northern Utah. M, 1984, University of Idaho. 93 p.

Siegrist, Henry Galt, Jr. Multivariate statistical study of two sandstones. D, 1961, Pennsylvania State University, University Park. 166 p.

Siegrist, Henry Galt, Jr. Petrology of graphite-bearing rocks of Chester County, Pennsylvania. M, 1959, Pennsylvania State University, University Park. 113 p.

Sieh, K. E. A study of Holocene displacement history along the south-central reach of the San Andreas Fault. D, 1977, Stanford University. 243 p.

Sieja, Donald Michael. Clay mineralogy of glacial Lake Missoula varves, Missoula County, Montana. M, 1959, University of Montana. 50 p.

Siemens, Allen G. The areal geology of the Tenkiller Ferry area, Sequoyah and Muskogee counties, Oklahoma. M, 1950, University of Oklahoma. 57 p.

Siemens, Michael A. Subsurface geology of the Simpson Group (Ordovician), Sumner County, Kansas. M, 1985, Wichita State University. 99 p.

Siemers, Charles Troy. Stratigraphy, paleoecology, and environmental analysis of upper part of Dakota Formation (Cretaceous), central Kansas. D, 1971, Indiana University, Bloomington. 333 p.

Siemers, William Terry. Stratigraphy and petrology of Mississippian, Pennsylvanian, and Permian rocks in the Magdalena area, Socorro County, New Mexico. M, 1973, New Mexico Institute of Mining and Technology. 133 p.

Siemiatkowska, Krystyna. Fenitic breccias in the Sudbury area (Ontario). M, 1972, McGill University. 89 p.

Siems, Barbara Ann. Surface to subsurface correlation of Columbia River Basalt using geophysical data, in parts of Adams and Franklin counties, Washington. M, 1973, Washington State University. 53 p.

Siems, Peter L. Volcanic and economic geology of the Rosita Hills and Silver Cliffs districts, Custer County, Colorado. D, 1967, Colorado School of Mines. 222 p.

Sienko, Dennis Alan. Crustal structure of south-central Pennsylvania determined from wide-angle reflections and refractions. M, 1982, Pennsylvania State University, University Park. 127 p.

Sierraalta, Carlos Alberto Lujan *see* Lujan Sierraalta, Carlos Alberto

Sierveld, Fred Grove. Geology of a part of Pattiway Ridge, Kern and Ventura counties, California. M, 1957, University of California, Los Angeles.

Siesser, William G. Geology of the Cuilco Quadrangle, Guatemala (west central). M, 1967, Louisiana State University.

Siever, Raymond. Mississippian-Pennsylvanian unconformity in southern Illinois. D, 1950, University of Chicago. 71 p.

Siever, Raymond. Structure and key beds of the Pennsylvanian System in Richland County, Illinois. M, 1947, University of Chicago.

Sieverding, Jayne Louise. Stratigraphy and environmental analysis of the Osgood Formation (Niagaran), southeastern Indiana. M, 1981, Indiana University, Bloomington. 176 p.

Sigalove, Joel J. Carbon-14 content and origin of caliche. M, 1969, University of Arizona.

Sigleo, A. M. C. Organic and inorganic geochemistry of the petrification of wood. D, 1977, University of Arizona. 88 p.

Sigleo, Anne M. C. Trace-element geochemistry of southwestern turquoise. M, 1970, University of New Mexico. 92 p.

Sigmund, James Martin. Geology of a Miocene rhyodacite lava flow, southern Tushar Mountains, Utah. M, 1979, Kent State University, Kent. 35 p.

Signor, Carl Wilson, Jr. Hypogene zoning in metalliferous deposits. M, 1955, University of Michigan.

Signor, Philip W., III. Species richness in the Phanerozoic; an investigation of sampling effects. M, 197?, University of Rochester. 36 p.

Signor, Philip White, III. Shell form and function in turritelliform gastropods. D, 1982, The Johns Hopkins University.

Signore, Alan G. Del *see* Del Signore, Alan G.

Sigsby, Robert J. Scoria of North Dakota. D, 1966, University of North Dakota. 245 p.

Sigsby, Robert M. A study of the pyritic-carbonaceous slate from the Sherwood Mine, Iron River, Michigan. M, 1960, Michigan Technological University. 124 p.

Sigurdson, David R. Mineral paragenesis and fluid inclusion thermometry at four western U. S. tungsten deposits. D, 1974, University of California, Riverside. 233 p.

Sikander, A. H. Structural analysis of the lower Paleozoic rocks of western Gaspé (Quebec). D, 1967, University of Ottawa. 134 p.

Sikes, C. S. Calcification of Cladophora glomerata. D, 1976, University of Wisconsin-Madison. 135 p.

Sikich, S. W. Stratigraphy of the Upper Triassic Stanaker formations of the eastern Uinta Mountain area, northeastern Utah and northwestern Colorado. M, 1960, University of Wyoming. 119 p.

Sikka, Desh B. A radiometric survey of Redwater Oilfield, Alberta, Canada. D, 1959, McGill University.

Sikka, Desh B. Sedimentation in Raritan River and Raritan Bay (New Jersey). M, 1954, New York University.

Sikka, Deshbandhu. A radiometric survey of Redwater oil field, Alberta. D, 1960, McGill University.

Sikkila, Kevin M. A structural analysis of Proterozoic metasediments, northern Falls River, Baraga County, Michigan. M, 1987, Michigan Technological University. 103 p.

Sikkink, Pamela Gayla Lindell. Depositional environments and biostratigraphy of the Lower Triassic Thaynes Formation, southwestern Montana. M, 1984, University of Montana. 161 p.

Sikora, Gerald S. Groundwater quality of a carbonate aquifer in the Bellevue, Ohio area. M, 1975, University of Toledo. 74 p.

Sikora, Paul J. Quantitative analysis of foraminifera from the Kemp Clay, north-central Texas. M, 1984, University of Texas, Austin. 229 p.

Sikorski, Peter Edwin. Remote sensing detection and geological interpretation of the eastern Snake River

plain area, Idaho-Montana. M, 1986, Texas Tech University. 59 p.

Silber, Jay Brian. A tectonic reconstruction of the Ural Mountains, U.S.S.R. M, 1982, Michigan State University. 134 p.

Silberling, Norman John. Pre-Tertiary stratigraphy and Upper Triassic paleontology of the Union District, Shoshone Mountains, Nevada. D, 1957, Stanford University. 263 p.

Silberman, Louis. A sedimentological study of the Gulf beaches of Sanibel and Captiva islands, Florida. M, 1979, Florida State University.

Silberman, Miler Louis. Isothermal compression of two iron-silicon alloys to 290 kilobars at 23°C. M, 1967, University of Rochester. 72 p.

Silberman, Miles Louis. Time of ore deposition in epithermal veins of western Nevada and eastern California as measured by K-Ar isotopic ages on primary adularia and alunite. D, 1971, University of Rochester. 149 p.

Siler, Jerry C. Correlations of subsurface Upper Cambrian and Lower Ordovician strata in Ellis County, Kansas, by insoluble residues. M, 1958, University of Kansas. 63 p.

Silfer, Jeffrey A. Facies and diagenesis of the Hartford Limestone (Topeka Limestone Formation, Upper Pennsylvanian Series) algal-mound complex in east-central Kansas. M, 1986, Wichita State University. 168 p.

Silgado, Enrique F. The partition of amplitudes for an SV-wave by Zoeppritz's method. M, 1944, California Institute of Technology. 22 p.

Silins, Lauma. Strontium and yttrium contents of fluorites from the Minerva No. 1 Mine, Cave-in-Rock, Illinois. D, 1974, Boston University. 215 p.

Silitonga, Parahum H. Geology of part of the Kittridge Springs Quadrangle, Elko County, Nevada. M, 1974, Colorado School of Mines. 88 p.

Silk, Ernest S. The geology and ore deposits of Hamilton, Nevada. M, 1931, Yale University.

Silka, Lyle Ramsay. Hydrogeochemistry of the Washita River alluvium in Caddo and Grady counties, Oklahoma. M, 1975, Oklahoma State University. 98 p.

Silker, Ted H. Surface geology-soil-site relationships in western Gulf Coastal Plain and inland areas. D, 1974, Oklahoma State University. 103 p.

Sill, G. T. Chemical reactions at the surface and in the atmosphere of Venus. D, 1976, University of Arizona. 207 p.

Sill, William Dudley. The rhynchosaurs of South America. D, 1969, Harvard University.

Sill, William Robert. Study of electrode noise. M, 1963, Massachusetts Institute of Technology. 100 p.

Siller, Thomas Joseph. Seismic response of tiedback retaining walls. D, 1988, Carnegie-Mellon University. 119 p.

Silliman, Alan Holt. A study of stress drops and apparent stresses of microearthquakes beneath Adak Canyon of the central Aleutian island arc. M, 1985, University of Colorado.

Silliman, Stephen Edward Joseph. Stochastic analysis of high-permeability paths in the subsurface. D, 1986, University of Arizona. 195 p.

Silling, Rose Mary. A gravity study of the Chiwaukum Graben, Washington. M, 1979, University of Washington. 100 p.

Silman, J. F. B. Structural control of pitchblende-bearing fractures at Nesbitt-Labine uranium mine, Saskatchewan. M, 1954, Queen's University.

Silman, Jack Forrest Banning. The stabilities of some oxidized copper minerals in aqueous solutions at 25°C and 1 atmosphere total pressure. D, 1959, Harvard University.

Silva, G. L. R. De *see* De Silva, G. L. R.

Silva, Joao Batista Correa da *see* da Silva, Joao Batista Correa

Silva, Juan Jose Cervantes *see* Cervantes Silva, Juan Jose

Silva, L. R. Landslides in the upper Kings Creek and Deer Creek watersheds, Castle Rock State Park, California. M, 1976, San Jose State University. 100 p.

Silva, Luis R. Two-layer master curves for electromagnetic soundings. M, 1969, Colorado School of Mines. 177 p.

Silva, Maria Augusta Martins da see da Silva, Maria Augusta Martins

Silva, Walter Joseph, Jr. Wave propagation in anelastic media with applications to seismology. D, 1978, University of California, Berkeley. 166 p.

Silva, Zenaide Carvalho Goncales da see da Silva, Zenaide Carvalho Goncales

Silver, Burr Arthur. North American mid-Jurassic through mid-Cretaceous stratigraphic patterns of Colorado Plateau, Rocky Mountains and Great Plains. D, 1966, University of Washington. 88 p.

Silver, Burr Arthur. The Bluebonnet Member, Lake Waco Formation (Upper Cretaceous), central Texas, a lagoonal deposit. M, 1963, Baylor University. 48 p.

Silver, Caswell. Geology and ore deposits of the Dunmore mine, Ouray County, Colorado. M, 1946, University of New Mexico. 55 p.

Silver, Douglas Balfour. The distribution of tungsten in limestone contact environments, Silver Bell Mine, Dos Cabezas Mountains, Arizona. M, 1980, University of Arizona. 58 p.

Silver, Eli Alfred. Structure of the continental margin (off northern California), north of the Gorda escarpment. D, 1969, University of California, San Diego. 136 p.

Silver, Leon Theodore. Paragenesis of the ores of Poughkeepsie Gulch, San Juan Mountains, Colorado. M, 1948, University of New Mexico. 81 p.

Silver, Leon Theodore. The structure and petrology of the Johnny Lyon Hills area, Cochise County, Arizona. D, 1955, California Institute of Technology. 407 p.

Silver, Lynn Alison. Water in silicate glasses. D, 1988, California Institute of Technology. 319 p.

Silver, Paul Gordon. Optimal estimation of scalar seismic moment. D, 1982, University of California, San Diego. 251 p.

Silver, Ronald. Crystal chemistry of anhydrous beryllium silicates. M, 1973, Miami University (Ohio). 40 p.

Silver, Wendy Ilene. Gravity investigation of the deep structure of the northern Overthrust Belt, Idaho and Wyoming. M, 1979, Texas A&M University. 95 p.

Silverberg, B. A. Selected studies in the fine structure of characean cells. D, 1974, University of Toronto.

Silverberg, David Scott. Structure and petrology of the Whitechuck Mountain - Mount Pugh area, north Cascade Range, Washington. M, 1985, Western Washington University. 173 3 plates p.

Silverberg, Norman. Reconnaissance of the upper continental slope off Sable Island bank, Nova Scotia. M, 1965, Dalhousie University. 79 p.

Silverberg, Norman. Sedimentology of the surface sediments of the East Siberian and Laptev seas. D, 1972, University of Washington. 184 p.

Silverman, Alan N. Geochemical and biogeochemical studies in the Hansonburg mining district, New Mexico. M, 1975, University of Missouri, Rolla.

Silverman, Arnold J. Structural terminations of vein type ore deposits; Part 1, The Sydney Mine, Pine Creek area, Coeur d'Alene mining region, Idaho. M, 1958, Columbia University, Teachers College.

Silverman, Arnold Joel. Investigation of the parameters controlling base metal dispersion in selected ore deposits. D, 1963, Columbia University, Teachers College. 231 p.

Silverman, Eugene Norton. X-ray diffraction study of orientation in Chattanooga Shale. M, 1955, Pennsylvania State University, University Park. 72 p.

Silverman, Franklin M. The Lambertville Sill, New Jersey. M, 1973, Brooklyn College (CUNY).

Silverman, Marc S. Structural geology and interpretation of trace element distribution in soils overlying a fault zone in the northern Shenandoah Valley, Jefferson County, West Virginia. M, 1975, University of Toledo. 202 p.

Silverman, Maxwell. The possibility of contamination in underwater piston core samples. M, 1956, University of Illinois, Urbana.

Silverman, Sol Robert. The isotopic composition of oxygen in natural silicates. D, 1950, University of Chicago. 52 p.

Silverman, Stephen J. The Cortland-Beemerville magmatic trend across the Central Appalachians. M, 1987, University of Rochester.

Silvers, Eric Richard. Episodic deposition of the Late Pennsylvanian Hermosa Formation in southeastern Utah. M, 1987, University of Nebraska, Lincoln. 92 p.

Silversides, David A. Petrology and molybdenite mineralization of the Lucky Ship Igneous Complex (ranges between Lower Jurassic and Early Tertiary) (west central British Columbia). M, 1968, University of Manitoba.

Silverstein, Hank. The New Haven Arkose of Connecticut (Triassic). M, 1974, Brooklyn College (CUNY).

Silverston, Elliot. The stable channel as shaped to flow and sediment. D, 1981, University of Arizona. 238 p.

Silverstone, Brahm S. An application of the linear inversion technique for finding the solution of surface structures using PL leaking modes. M, 1973, University of Toronto.

Silverts, Leif E. Application of the dye dilution technique to periodic summations of streamflow from mountain watersheds. M, 1967, [Colorado State University].

Silvestri, V. Performance of sensitive clays under variable stresses. D, 1974, McGill University.

Silvia, M. T. Deconvolution of geophysical time series. D, 1977, Northeastern University. 302 p.

Simandl, George. Geology and geochemistry of talc deposits in the Madoc area. M, 1985, Carleton University. 153 p.

Simard, Alain. Stratigraphie et volcanisme dans la partie orientale de la bande volcano-sedimentaire archeenne Frotet-Evans, Quebec. D, 1987, Universite de Montreal.

Simard, G. Natural isotopes and groundwater flow systems in the Eaton River basin. M, 1973, University of Waterloo.

Simigian, Sandra Lynn. An analysis of crystallographic orientation and grain shapes of pyrrhotite from Ducktown, Tennessee. M, 1984, University of Western Ontario. 201 p.

Simila, Gerald Wayne. Seismic velocity structure and associated tectonics of Northern California. D, 1980, University of California, Berkeley. 186 p.

Simkin, Thomas Edward. The picritic sills of northwest Trotternish, Isle of Skye, Scotland. D, 1965, Princeton University. 139 p.

Simkover, Elizabeth Gail. A groundwater flow model of the aquifer intercommunication area, Hanford Site, Washington. M, 1986, Portland State University. 120 p.

Simmen, Louis E. Petrography of the Blue Syenite of Granite Mountain, Pulaski County, Arkansas. M, 1955, University of Arkansas, Fayetteville.

Simmerman, Graham Hanson, Jr. Coal bearing strata of the middle and upper Kanawha Formation (Middle Pennsylvanian Series), Boone, Logan and Raleigh counties, southern West Virginia. M, 1986, North Carolina State University. 93 p.

Simmers, Rick J. Hydrologic analysis of Granger and western Bath townships. M, 1985, University of Akron. 185 p.

Simmon, Douglas E. Statistical evaluation of slug aquifer evaluation methods. M, 1983, University of South Florida, Tampa. 119 p.

Simmonds, Robert Tobin. The structure and stratigraphy of the southeastern portion of the Canastota (New York) Quadrangle. M, 1958, Syracuse University.

Simmonds, Robert Tobin. Wisconsinan geochronological determinations based upon amine rates. D, 1962, University of Illinois, Urbana. 39 p.

Simmons, Ardyth M. The geology of Mount Hope, a volcanic dome in the Colorado Plateau-Basin and Range transition zone, Arizona. M, 1986, SUNY at Buffalo. 156 p.

Simmons, Benjamin Titus. Geology of the Bristol gas field, New York. M, 1936, University of Rochester. 79 p.

Simmons, Dale L. Effects of urbanization on the base flow of south shore streams on Long Island, New York. M, 1979, SUNY at Binghamton. 531 p.

Simmons, David. The non-theropsid reptiles of the Lufeng Basin, Yunnan, China. D, 1959, University of Chicago.

Simmons, David W. Geomorphology and sedimentology of humid-temperate alluvial fans along the west flank of the Blue Ridge Mountains, Shenandoah Valley, Virginia. M, 1988, Southern Illinois University, Carbondale. 107 p.

Simmons, E. C. Origins of four anorthosite suites. D, 1976, SUNY at Stony Brook. 214 p.

Simmons, George Clarke. The Russell Ranch Formation. M, 1950, Washington State University. 26 p.

Simmons, George Mills. Geology of a portion of the Hope Bay Greenstone belt; District of Mackenzie, Northwest Territories, Canada. M, 1969, University of Iowa. 100 p.

Simmons, Gregory R. Depositional environments and diagenesis of the Atkins Sandstone, Pennsylvanian to Permian, northeastern Midland Basin. M, 1982, Oklahoma State University. 65 p.

Simmons, J. A. Kent. The biogeochemistry of the Devonshire lens, Bermuda. M, 1983, University of New Hampshire. 86 p.

Simmons, James Andre Kent. Major and minor ion geochemistry of groundwaters from Bermuda. D, 1987, University of New Hampshire. 262 p.

Simmons, James Layton, Jr. Traveltime inversion for a 3-D near surface velocity model. M, 1987, University of Texas, Austin. 288 p.

Simmons, K. R. K-Ar age dating of plagioclase from Precambrian mafic dikes. M, 1975, SUNY at Stony Brook.

Simmons, Kent J. A. The biogeochemistry of the Devonshire Lens, Bermuda. M, 1983, University of New Hampshire. 86 p.

Simmons, M. M. U. S. Environmental Protection Agency policy development for selected issues in municipal wastewater treatment. D, 1977, University of California, Los Angeles. 156 p.

Simmons, Marlys Gail. Contact metamorphism at the Cameron Creek Laccolith, Grant County, New Mexico. M, 1984, Northeast Louisiana University. 58 p.

Simmons, Marvin Gene. Gravity survey in northern New York. D, 1962, Harvard University.

Simmons, Marvin Gene. The photo-extinction method for the measurement of silt-sized particles. M, 1958, Southern Methodist University. 41 p.

Simmons, Michael L. Stratigraphy and paleoenvironments of Thetis, Kuper, and adjacent islands, B.C. M, 1973, Oregon State University. 114 p.

Simmons, Noel G. Structural analysis of the Valley and Ridge extension of the Parsons Lineament. M, 1983, Virginia Polytechnic Institute and State University. 106 p.

Simmons, Richard G. Magnetic susceptibility study of the Poston Butte porphyry copper deposit, Pinal County, Arizona. M, 1974, Arizona State University. 56 p.

Simmons, Stuart Frank. Fluid inclusion and alteration studies of the Washington Mine, Sonora, Mexico. M, 1982, University of Minnesota, Minneapolis. 77 p.

Simmons, Stuart Frank. Physio-chemical nature of the mineralizing solutions for the St. Niño Vein; results from fluid inclusion, deuterium, oxygen and helium studies in the Fresnillo District, Zacatecas, Mexico.

D, 1987, University of Minnesota, Minneapolis. 254 p.

Simmons, Thomas Paul. A comparison of two different well designs through the use of specific electrical conductance. M, 1988, Wright State University. 88 p.

Simmons, Vernon P. Investigation of the 1 kHz sound absorption in sea water. D, 1975, University of California, San Diego. 185 p.

Simmons, Weldon A., Jr. Stratigraphy and sedimentation of the Paleozoic rocks in the Maya Mountains, British Honduras. M, 1972, Louisiana State University.

Simmons, William Alexander. Stratigraphy and depositional environments of the Middle Cambrian Maryville Limestone (Conasauga Group) near Thorn Hill, Tennessee. M, 1984, University of Tennessee, Knoxville. 276 p.

Simmons, William Bruce Jr. Mineralogy, petrology, and trace element geochemistry of the South Platte granite-pegmatite system. D, 1973, University of Michigan. 292 p.

Simmons, William Bruce, Jr. Mineralogy of south Georgia and North Carolina phosphorite. M, 1968, University of Georgia. 77 p.

Simmons, Woodrow W. The geology of the Cleveland Mine area, Gila County, Arizona. M, 1938, University of Arizona.

Simms, Frederick Eugene. Geology of the Morgantown north, West Virginia-Pennsylvania, 7.5 minute topographic quadrangle. M, 1960, West Virginia University.

Simms, Frederick Eugene, Jr. The igneous petrology, geochemistry and structural geology of part of the northern Crazy Mountains, Montana. D, 1966, University of Cincinnati. 339 p.

Simms, John J. A study of the bedrock valleys of the Kansas and Missouri rivers in the vicinity of Kansas City. M, 1975, University of Kansas. 106 p.

Simms, Richard W. Geology of the Rayado area, Colfax County, New Mexico. M, 1965, University of New Mexico. 90 p.

Simnacher, Faroy. Stratigraphy, depositional environments and paleontology of the Cathedral Bluffs tongue of the Wasatch Formation (Eocene), Parnell Creek area, Sweetwater County, Wyoming. M, 1970, University of Wyoming. 102 p.

Simon, Andrew Dorsey. Development of a coupled areal to cross-sectional model for the simulation of stratified reservoirs. D, 1980, University of Missouri, Rolla. 140 p.

Simon, Andrew L. Gravity flow toward horizontal collector wells. D, 1962, Purdue University.

Simon, D. E. Physical and chemical changes related to epigenetic dolomitization. D, 1972, Iowa State University of Science and Technology.

Simon, David Eugene. The partition of calcium among cementing compounds in aging highway concrete. M, 1968, Iowa State University of Science and Technology.

Simon, Donald Bruce. Geology of the Silver Hill area, Socorro County, New Mexico. M, 1974, New Mexico Institute of Mining and Technology. 101 p.

Simon, F. O. The distribution of chromium and tungsten in the rocks and minerals from the Southern California Batholith. D, 1972, University of Maryland. 130 p.

Simon, Henry Francis. Near-infrared and Mössbauer study of cation site occupancies in tourmalines. M, 1973, Massachusetts Institute of Technology. 67 p.

Simon, Jack Aaron. Correlation studies of Upper Pennsylvanian rocks in southwest-central Illinois. M, 1946, University of Illinois, Urbana.

Simon, John Frederick. High resolution spectroscopy of quadrupole features on Jupiter, Saturn and Neptune, and HD dipole absorption on Uranus. M, 1987, Washington University. 91 p.

Simon, Nancy Shoemaker. Nitrogen cycling in Potomac River transition zone shallow water sediments. D, 1984, [American University]. 202 p.

Simon, Peter R. Climatic significance of nonmarine mollusks from the Illinoian Stage of the Pleistocene from southwestern Kansas and northwestern Oklahoma. M, 1977, Kent State University, Kent. 55 p.

Simon, Robert B. Processing and evaluation of seismic reflection data obtained using the Betsy Seisgun source for an energy source study in Greene County, Ohio. M, 1984, Wright State University. 97 p.

Simon, Ruth B. Geology of the Camp Creek area, northern Skamania County, Washington. M, 1972, University of Washington. 33 p.

Simon, Steven B. Lunar regolith breccias and soils; comparative petrology, chemistry and evolution. D, 1988, South Dakota School of Mines & Technology.

Simon, Steven Bruce. Petrography, petrology and origin of chondrules in the Allende Meteorite. M, 1980, University of Massachusetts. 97 p.

Simonberg, Elliott Mark. The origin and development of case-hardening in the northeastern Spring mountains, Clark County, Nevada. M, 1969, Pennsylvania State University, University Park. 100 p.

Simonds, Charles Henry. Quartz latite and associated altered sandstone along the Big Pine Fault, Ventura County, California. M, 1967, Stanford University.

Simonds, Charles Henry. Recrystallization and grain growth of quartz and hematite in a metamorphic gradient, Negaunee Iron Formation, Upper Peninsula, Michigan. D, 1971, University of Illinois, Urbana. 94 p.

Simonds, Charles Henry, III. Structure and contact metamorphism of calc-silicate rocks, Inyo Mountains, California. M, 1969, University of Illinois, Urbana.

Simonds, Francis M. The occurrence and manufacture of aluminum, nickel, and cobalt; chlorination of gold ores. D, 1889, Columbia University, Teachers College.

Simoneau, Pierre. Pétrographie, sédimentologie et analyse de faciès de la Formation de Daubrée, Chapais, Québec. M, 1987, Universite du Quebec a Chicoutimi. 475 p.

Simonetti, Antonio. The geochronology of Acadian plutonism in southeastern Quebec. M, 1988, McGill University. 168 p.

Simoni, Tully R., Jr. Geology of the Loma Prieta area, Santa Clara and Santa Cruz counties, California. M, 1974, San Jose State University. 75 p.

Simonis, Edvardas Karolis. Geology of part of the Oden Quadrangle, Ouachita mountains, Arkansas. M, 1968, Northern Illinois University. 116 p.

Simons, Daryl B. Theory and design of stable channels in alluvial materials. D, 1957, Colorado State University. 394 p.

Simons, Elwyn Laverne. The Paleocene Pantodonta and their allies. D, 1956, Princeton University. 310 p.

Simons, Frank S. Geology and ore deposits of the Zimapan mining district, Mexico. D, 1951, Stanford University. 273 p.

Simons, Jeffrey H. Detailed mapping of the Danbury Gneiss and adjacent rocks, Sauratown Mountains, North Carolina. M, 1982, University of North Carolina, Chapel Hill. 141 p.

Simons, Merton Eugene. Insoluble residues of the (Devonian) Traverse Group in the Charlevoix-Petoskey area, Michigan. M, 1949, Wayne State University.

Simons, Merton Eugene. Insoluble residues of the Devonian Traverse Group, Michigan. D, 1953, University of North Carolina, Chapel Hill. 118 p.

Simons, Philip Yale. Polymorphism of TiO_2 minerals. D, 1967, Pennsylvania State University, University Park. 134 p.

Simons, Robert Keith. Analysis of bank protection using a probability of motion approach. D, 1986, Colorado State University. 191 p.

Simonsen, August Henry. Wreford fenestrates; important bryozoans in a lower Permian megacyclothem in Kansas, Oklahoma, and Nebraska. D, 1977, Pennsylvania State University, University Park. 131 p.

Simonson, Bruce Miller. Sedimentology of Precambrian iron-formations with special reference to the Sokoman Formation and associated deposits of northeastern Canada. D, 1982, The Johns Hopkins University. 475 p.

Simonson, Eric Robb. The Tertiary volcanic rocks of the Mount Daniel area, central Washington Cascades. M, 1981, University of Washington. 81 p.

Simonson, John C. B. The stratigraphy, carbonate petrology, depositional environments, and paleoecology of the Middle Ordovician Moccasin Formation near Thorn Hill, Tennessee. M, 1985, University of Tennessee, Knoxville. 279 p.

Simonson, Russell Ray. Conglomerates of the Sespe and Topanga formations of Dry Canyon Quadrangle, Santa Monica Mountains, California. M, 1936, University of California, Los Angeles.

Simony, Philip Steven. Origin of the Apsley Paragneiss. M, 1960, McMaster University. 79 p.

Simpkins, Tim H. Geology and geochemistry of the Aguachile Mountain fluorspar-beryllium mining district, northern Coahuila, Mexico. M, 1983, Sul Ross State University. 137 p.

Simpkins, W. W. Surficial geology and geomorphology of Forest County, Wisconsin. M, 1979, University of Wisconsin-Madison.

Simpson, Altus L. Physiography of the Canyon Spring Quadrangle, Riverside County, California. M, 1958, University of Southern California.

Simpson, C. Leon. Hydrologic investigation of a sanitary landfill at Middletown, Ohio. M, 1973, Ohio State University.

Simpson, Dale Rodekohr. Geology of the Ramona Pegmatites, San Diego County, California. D, 1960, California Institute of Technology. 197 p.

Simpson, David F. Geology of the Ambler 4B extension of the Smucker volcanogenic massive sulfide deposit, Ambler District, Alaska. M, 1983, University of Alaska, Fairbanks. 135 p.

Simpson, David Gordon. A centrifuge model study of the evolution of laccolithic intrusions. M, 1980, Queen's University. 199 p.

Simpson, David H. A study of certain mineral deposits at the headwaters of the Spillmacheen River, British Columbia. D, 1952, McGill University.

Simpson, David Hope. Petrogenesis of the silicate minerals associated with copper ores, Gaspe, Quebec. M, 1941, McGill University.

Simpson, David Hope. The stratigraphy, structure, and mineral deposits at the headwaters of Spillamacheen River, British Columbia. D, 1951, McGill University.

Simpson, David Paul. Hydraulics of two small gravelly tidal inlets. M, 1976, University of Washington. 142 p.

Simpson, David William. A record seismic system for refraction studies at sea. M, 1968, Dalhousie University. 77 p.

Simpson, Donald H. Modeling of bedload transport and hydraulic roughness in mountain streams. M, 1978, Colorado State University. 93 p.

Simpson, E. S. Buried preglacial channels in the Albany-Schenectady area in New York. M, 1945, Columbia University, Teachers College.

Simpson, Edward Cannon. Geology of the Elizabeth Lake Quadrangle (Calfornia). D, 1933, University of California, Berkeley. 166 p.

Simpson, Edward L. The geometry and structure of interdune deposits at White Sands National Monument, New Mexico. M, 1983, University of Nebraska, Lincoln.

Simpson, Edward Leonard. Facies analysis and tectonic implications of the Cambrian Chilhowe Group, central and southern Virginia. D, 1987, Virginia Polytechnic Institute and State University.

Simpson, Elizabeth J. A photogrammetric survey of backbarrier accretion on the Rhode Island barrier beaches. M, 1977, University of Rhode Island.

Simpson, Eugene S. Traverse dispersion in liquid flow through porous media. D, 1960, Columbia University, Teachers College.

Simpson, Evanna Lois. Mineralogy and geochemistry of an ocellar minette sill, northern New Brunswick. M, 1980, University of New Brunswick.

Simpson, Frederick Muir. The mineralogy of pollucite and beryl from the Tanco Pegmatite at Bernic Lake, Manitoba. M, 1974, University of Manitoba.

Simpson, Garey L. Shoal migration and dune encroachment, North Harbor, Moss Landing, California. M, 1978, San Jose State University. 48 p.

Simpson, George Gaylord. Mesozoic mammals. D, 1926, Yale University.

Simpson, Harry James, Jr. Closed basin lakes as a tool in geochemistry. D, 1970, Columbia University. 325 p.

Simpson, Howard E., Jr. Geology of an area about Yankton, South Dakota. D, 1953, Yale University.

Simpson, Howard Edwin, Jr. The Pleistocene geology of Garrison Quadrangle, North Dakota. M, 1942, University of Illinois, Urbana.

Simpson, Howard Muncie. Palynology and the vertical sedimentary profile of Missourian strata, Tulsa County, Oklahoma. M, 1969, University of Tulsa. 64 p.

Simpson, I. D., Jr. Geology of the Strang area, Mayes County, Oklahoma. M, 1951, University of Oklahoma. 89 p.

Simpson, Jack Ezelle. Studies of hydrolysis effects upon soil colloids. M, 1939, Louisiana State University.

Simpson, James William. Potassium argon dating of volcanic rocks from Mogollon plateau area of New Mexico and Arizona. M, 1970, University of Toronto.

Simpson, Jimmie D. An insoluble residue study of the Comanche Peak and Edwards limestones (lower Cretaceous), Bosque and western McLennan counties, Texas. M, 1967, Texas A&M University. 67 p.

Simpson, Jo-anna R. A field and laboratory study of a section of the Williamsburg Granodiorite (Carboniferous?) and pegmatites in Haydenville, Massachusetts. M, 1966, Smith College. 52 p.

Simpson, John Page, III. The geology of Carmel Bay, California. M, 1972, United States Naval Academy.

Simpson, Kenneth M. The minerals of the Ellenville zinc mine. M, 1906, Columbia University, Teachers College.

Simpson, Kenneth Reed. Petrology and depositional environment of the Upper Devonian Grandview Dolomite, Custer County, Idaho. M, 1983, University of Idaho. 109 p.

Simpson, Larry Clark. Paleontology of the Garber Formation (Lower Permian), Tillman County, Oklahoma. M, 1976, University of Oklahoma. 216 p.

Simpson, Leon. Preliminary hydrogeologic investigation of a sanitary landfill at Middletown, Ohio. M, 1973, Ohio State University.

Simpson, Lloyd William, Jr. A foraminiferal and sedimentary analysis of the Vicksburg Group (Oligocene) in Warren and Hinds counties, Mississippi. M, 1965, Mississippi State University. 165 p.

Simpson, M. S., Jr. Statistical approaches to certain problems in geophysics. D, 1953, Massachusetts Institute of Technology. 142 p.

Simpson, Michael. Structural analysis of the Kings Mountain Belt, Kings Mountain, N.C. M, 1977, University of South Carolina.

Simpson, Peter Robert. Quantitative measurements of the optical properties of opaque minerals. M, 1965, University of New Brunswick.

Simpson, R. de *see* de Simpson, R.

Simpson, R. W., Jr. A gravimetric and structural analysis of the Shelburne Falls Dome, Massachusetts. M, 1974, University of Maine. 58 p.

Simpson, R. W., Jr. Paleomagnetic evidence for tectonic rotation of the Oregon Coast Range. D, 1977, Stanford University. 167 p.

Simpson, Robert A. Analysis of a large-scale disturbance in the mesosphere. D, 1962, University of Chicago. 102 p.

Simpson, Ronald D. Permian Brachiopoda from far West Texas and southwestern New Mexico. D, 1984, University of Texas at El Paso. 429 p.

Simpson, Stephen J. Structure and petrology of the Rhodes Peak Cauldron, Montana and Idaho. M, 1985, University of Montana. 105 p.

Simpson, Thomas A. General geology and structural features of the Birmingham red iron ore district, Alabama. M, 1959, University of Alabama.

Simpson, Thomas Mason. The geology and hydrothermal alteration of the Sulphurets Deposits, Northwest British Columbia. M, 1983, University of Idaho. 99 p.

Sims, Dewey Leroy. Subsurface studies of the West Franklin Limestone and Trivoli Sandstone in Wayne County, Illinois. M, 1957, University of Illinois, Urbana.

Sims, Donald R., Jr. Variation in grain shape and surface texture of fine quartz sands in the South Texas eolian sands sheet. M, 1984, Texas A&M University.

Sims, Francis R. Bedrock geology of the Omak Quadrangle and the northeast portion of the Ruky Quadrangle, Okanogan County, Washington. M, 1984, University of Idaho. 93 p.

Sims, Frank Chanberg. Geology of the west end of the Laramie Range, Natrona County, Wyoming. M, 1948, University of Wyoming. 99 p.

Sims, James Rae. Retention of boron by layer silicates, sesquioxides and soil materials. D, 1966, University of California, Riverside. 112 p.

Sims, James Thomas. Phosphorus adsorption studies with an acid, aluminous soil and aluminum hydroxide suspensions. D, 1981, Michigan State University. 91 p.

Sims, John David. A sedimentary petrographic study of the upper Fort Union Group, northern Crazy Mountains, Montana. M, 1964, University of Cincinnati. 40 p.

Sims, John David. Geology and sedimentology of the Livingston Group (Upper Cretaceous and Paleocene), northern Crazy Mountains, (Meager, Park, and Gallatin counties), Montana. D, 1967, Northwestern University. 212 p.

Sims, Michael S. The systematics, paleoecology, and paleobiogeography of Gonioloboceras (Ammonoidea) from the Middle and Upper Pennsylvanian of North America. M, 1987, Ohio University, Athens. 108 p.

Sims, Paul Kibler. Geology of the Dover magnetite district, New Jersey. D, 1951, Princeton University. 226 p.

Sims, Richard Carlton. Depositional environment and regional significance of the Upper Cambrian Candland Shale Member of the Orr Formation in the House Range, western Utah. M, 1985, University of Kansas. 47 p.

Sims, Richard M. A manual for operations and maintenance of a seismic station at Slippery Rock State College. M, 1975, Slippery Rock University. 181 p.

Sims, Samuel John. Geology of part of the Santa Rosa Mountains, Riverside County, California. D, 1961, Stanford University. 112 p.

Sims, Samuel John. Geology of the Sandy Mountain area, Llano County, Texas. M, 1957, University of Texas, Austin.

Sims, W. A. Sorption of copper, lead and zinc on American Petroleum Institute reference clays K-4, M-23 and M-25. M, 1959, McGill University.

Sims, William Eldon. Methods of magnetotelluric analysis. D, 1969, University of Texas, Austin. 95 p.

Sims, William R. Facies analysis of carbonate rocks on the southern end of San Salvador Island, Bahamas. M, 1987, University of Akron. 135 p.

Sinanuwong, S. Cation exchange equilibria in irrigated tropical soils. D, 1972, University of Hawaii. 206 p.

Sinclair, A. J. Mine evaluation; the Erzberg Molybdenite Property, Greenland. M, 1958, University of Toronto.

Sinclair, Alastair James. A lead isotope study of mineral deposits in the Kootenay Arc. D, 1984, University of British Columbia.

Sinclair, Brian J. Estimation of grout absorption in fractured rock foundations. D, 1972, University of Illinois, Urbana. 316 p.

Sinclair, Calvin R. The stratigraphy and structure of the Mineola-Danville area, Montgomery County, Missouri. M, 1956, University of Missouri, Columbia.

Sinclair, George W. The biology of the Conularida. D, 1949, McGill University.

Sinclair, Horace A. Apatite fission track analysis of the Triassic Portland Formation, Connecticut River valley. M, 1981, Rensselaer Polytechnic Institute. 64 p.

Sinclair, John Taylor, Jr. Subsurface geology of the Sentous zone development, Englewood oil field, California. M, 1943, University of Southern California.

Sinclair, Joseph H. Geology and physiography of the San Juancito District, Honduras. M, 1912, University of Rochester. 121 p.

Sinclair, Patricia Drew. Carbonate and bicarbonate complexes of lead in hydrothermal solutions from 25° to 150°C, with geologic applications. D, 1973, University of Wisconsin-Madison.

Sinclair, R. D. Digital analysis of response of the groundwater reservoir to long term pumping, Welland area, Ontario. M, 1974, University of Waterloo.

Sinclair, Richard H. Surface resistivity methods for the delineation of a coal refuse bank fire. M, 1982, Kent State University, Kent. 136 p.

Sinclair, Sheryl Lynn. A field study of the nearshore groundwater system of Mono Lake. M, 1988, University of California, Santa Cruz.

Sinclair, Steven Whitney. Analysis of macroscopic fractures on Teton Anticline, northwestern Montana. M, 1980, Texas A&M University.

Sinclair, William. Geology of the Upper East Fork of the San Juan River area, Mineral and Archuleta counties; Colorado. M, 1963, Colorado School of Mines. 133 p.

Sinclair, William D. Geology of the No. 5 zone, Horne Mine, Noranda, Quebec, Canada. M, 1970, University of Wisconsin-Madison.

Sinclair, William David. The solubility of copper and copper sulfides in aqueous chloride solutions from 25° to 250°C, with geologic applications. D, 1973, University of Wisconsin-Madison.

Sinclair, William J. The exploration of the Potter Creek Cave. D, 1904, University of California, Berkeley. 27 p.

Sine, Franklin Arthur, Jr. An experimental study on the incorporation of Zn_{2+} in calcite at low-temperature. M, 1983, Lehigh University.

Sinex, Scott A. A chemical and mineralogical study of the saprolites of the Blue Ridge of Ashe County, North Carolina. M, 1975, Miami University (Ohio). 55 p.

Sinex, Scott Alden. Trace element geochemistry of modern sediments from Chesapeake Bay. D, 1981, University of Maryland. 203 p.

Singamsetti, Surya Rao. Diffusion of sediment in a submerged jet. D, 1965, Iowa State University of Science and Technology. 106 p.

Singdahlsen, Donald Scott. Structural geology of the Swift Reservoir culmination, Sawtooth Range, Montana. M, 1986, Montana State University. 124 p.

Singer, Bradley Sherwood. Petrology and geochemistry of Polvadera Group rocks of La Grulla Plateau, Northwest Jemez volcanic field, New Mexico; evidence favoring evolution by assimilation-fractional crystallization. M, 1985, University of New Mexico. 149 p.

Singer, Donald Allen. Multivariate statistical analysis of the unit regional value of mineral resources (Pennsylvania). D, 1971, Pennsylvania State University, University Park. 210 p.

Singer, Donald Allen. Petrographic investigation of three cored wells of the Salt Wash Member of the Morrison Formation, Montrose County, Colorado. M, 1968, Pennsylvania State University, University Park. 112 p.

Singer, Harvey A. Heat transport by steady state plumes with strongly temperature dependent viscosity. D, 1986, The Johns Hopkins University. 212 p.

Singer, Jeff. Adsorption of phosphorus on silica gel as influenced by treatment with trimethylchlorosilane; to develop a suitable diluent for fine textured soils. M, 1976, Rensselaer Polytechnic Institute. 57 p.

Singer, Jill Karen. A study in the hydrodynamics of sediment transport. M, 1982, Rice University. 151 p.

Singer, Jill Karen. Terrigenous, biogenic, and volcaniclastic sedimentation patterns of the Bransfield Strait and bays of the northern Antarctic Peninsula; implications for Quaternary glacial history. D, 1987, Rice University. 358 p.

Singer, Karen M. Evaluating thermal rock conductivity of sandstone from aggregate using the divided bar. M, 1987, Kent State University, Kent. 110 p.

Singer, Michael P. A computer simulation model of the growth and desiccation of pluvial lakes in the Southwestern United States; Volume 1, Description of the model. M, 1985, Kent State University, Kent. 436 p.

Singer, Robert Bennett. The composition of the Martian dark regions; observations and analysis. D, 1980, Massachusetts Institute of Technology. 301 p.

Singer, Stephen H. Geology of the west half of the Disautel 15′ Quadrangle, Okanogan County, Washington. M, 1984, University of Idaho. 101 p.

Singewald, Joseph T., Jr. The iron ores of Maryland in the Piedmont and Appalachian regions. D, 1909, The Johns Hopkins University.

Singewald, Quenton D. Igneous rocks from the Andes of central Peru. D, 1926, The Johns Hopkins University.

Singh, Awtar. Creep phenomena in soils. D, 1966, University of California, Berkeley. 200 p.

Singh, B. R. Studies on nitrogen transformation and nitrate adsorption in soils. M, 1968, University of Hawaii.

Singh, Chaitanya. Palynology of the Mannville Group (Lower Cretaceous), central Alberta. M, 1964, University of Alberta. 443 p.

Singh, Chatra Ket. Petrology of the Signal Hill and Blackhead formations (Proterozoic), Avalon peninsula, Newfoundland, (Canada). M, 1969, Memorial University of Newfoundland. 96 p.

Singh, Gambhir. Investigation of the physical properties of reservoir rocks by electric well logging in Graham County, Kansas. M, 1965, Kansas State University. 74 p.

Singh, Harbhajan. Construction and application of a water quality model for the upper Blackfoot River basin in the Caribou National Forest, Idaho. D, 1979, University of Idaho. 240 p.

Singh, Harinder, Jr. Diabase intrusions of a portion of the Durham Triassic Basin, North Carolina. M, 1963, University of North Carolina, Chapel Hill. 23 p.

Singh, Harnek. Remote sensing for detecting riverbed sediments and computer mapping the time trends in the south fork of the Salmon River. D, 1981, University of Idaho. 131 p.

Singh, Jogeshwar Preet. The influence of seismic source directivity on strong ground motions. D, 1981, University of California, Berkeley. 195 p.

Singh, Krishna Kumar. Comparison of the Buckley-Leverett technique with a numerical simulation model containing heterogeneity and capillary pressure. M, 1970, University of Missouri, Rolla.

Singh, Maghar. Well-log analysis of the Milk River (Campanian) strata of southwestern Saskatchewan. M, 1982, University of Windsor. 167 p.

Singh, Ram Narain. An approach to genesis of Pine Point mineralization, Northwest Territories, Canada (Devonian). M, 1967, University of Alberta. 129 p.

Singh, Raman J. Conodonts from the Bellevue Limestone Member of the McMillan Formation (Upper Ordovician) of Cincinnati, Ohio. M, 1966, University of Cincinnati. 98 p.

Singh, Raman J. Trepostomatous bryozoan fauna and stratigraphy of the Bellevue Limestone, Upper Ordovician, in the tri-state area of Ohio, Indiana and Kentucky. D, 1971, University of Cincinnati. 188 p.

Singh, S. Undisturbed sampling and cyclic load testing of sands. D, 1979, University of California, Berkeley. 138 p.

Singh, Shri K. Transient response of a permeable conducting sphere embedded in a conducting infinite space. D, 1971, Columbia University. 124 p.

Singh, Sudarshan. A seismic model study of methods to minimize surface wave interference in reflection seismology. M, 1972, University of Minnesota, Minneapolis.

Singh, Sudarshan. Regionalization of crustal Q in the continental United States. D, 1981, St. Louis University. 203 p.

Singh, Sudesh Kumar. Petrological and mineralogical studies of the Joan Lake agpaitic complex, central Labrador. D, 1972, University of Ottawa. 179 p.

Singh, Surendra. Statistical analysis of microearthquakes of the Socorro region (New Mexico). D, 1970, New Mexico Institute of Mining and Technology. 156 p.

Singh, Vijay. Computer calculated analysis of subsurface formation porosities in central New York State. D, 1983, Syracuse University. 243 p.

Singh, Vijay. Recognition of subsurface subtle features using spatial analysis techniques in the Santa Cruz Basin (Bolivia). M, 1979, Syracuse University.

Singh, Yash Pal. Finite element analyses of cellular cofferdams. D, 1987, Virginia Polytechnic Institute and State University. 259 p.

Singh, Yogendra L. and Pradham, Bi-swa M. Geology of the area between Virden and Red Rock, Hidalgo and Grant counties, New Mexico. M, 1960, University of New Mexico. 75 p.

Singhroy, V. H. Sedimentation and erosion studies in Wilson Creek watershed, Manitoba, with particular reference to shale bank retreat. M, 1977, University of Manitoba.

Single, Erwin Leroy. Contorted strata of the Lower Mississippian rocks in Pike and Ross counties, Ohio. M, 1956, University of Cincinnati. 112 p.

Singler, Caroline Susan. Carbonate petrology and conodont biostratigraphy of a Mississippian carbonate unit, East Pahranagat Range, Lincoln County, Nevada. M, 1987, Washington State University. 107 p.

Singler, Charles R. Sedimentology of the White River group (Oligocene) in northwestern Nebraska. D, 1969, University of Nebraska, Lincoln. 161 p.

Singler, Charles Richard. Stratigraphy of the Pleasanton Group (Pennsylvanian) in the northern Midcontinent. M, 1965, University of Nebraska, Lincoln.

Singler, James Charles. The effect of varying the parameters of vane shear tests on marine sediments. M, 1971, United States Naval Academy.

Singletary, Henry McLean. The geology of the Mebane Quadrangle, North Carolina. M, 1972, North Carolina State University. 66 p.

Singleton, G. A. Weathering in a soil chronosequence. D, 1979, University of British Columbia.

Singleton, Paul C. Nature of interlayer material in silicate clays of selected Oregon soils. D, 1966, Oregon State University. 84 p.

Singleton, Scott Wayne. High resolution seismic and paleomagnetic study of the structure and sedimentation of Sweet and Phleger banks, northern Gulf of Mexico. M, 1988, Texas A&M University. 131 p.

Sinh Do Lam see Do Lam Sinh

Sinha, Akhaury K. The isotopic evolution of common lead and its bearing on the age of the Earth. D, 1969, University of California, Santa Barbara.

Sinha, Bhudeo Narayan. Geochemistry of lead, zinc, copper, cobalt, nickel, cadmium, iron and manganese in surface waters and stream sediments, St. Francis River drainage basin, southeastern Missouri. D, 1980, University of Missouri, Rolla. 312 p.

Sinha, Bhudeo Narayan. Precambrian volcanic rocks of Stegall and Mule mountains, Carter and Shannon counties, Missouri. M, 1976, University of Missouri, Rolla.

Sinha, Bindeshwari Narain. Geology of the region around Parkdale, Fremont County, Colorado. D, 1951, Colorado School of Mines. 129 p.

Sinha, Evelyn Zepel. Geomorphology of the lower Coastal Plain from the Savannah River area, Georgia, to the Roanoke River area, North Carolina. D, 1959, University of North Carolina, Chapel Hill. 141 p.

Sinha, M. N. Geochemistry and petrology of mafic dikes of Mackenzie and Sudbury swarms. M, 1983, University of Waterloo. 188 p.

Sinibaldi, Alfredo Salvador Galvez see Galvez Sinibaldi, Alfredo Salvador

Sinna, Ravindra P. Petrology of volcanic rocks of North mountain, Nova Scotia. D, 1970, Dalhousie University. 154 p.

Sinnett, Donald L. Subsurface stratigraphy and structure of the Garnett East Quadrangle, Anderson County, Kansas. M, 1985, Emporia State University.

Sinno, Yehia Ahmed. Crustal structure of the southern Rio Grande Rift determined from seismic refraction, surface wave dispersion, and gravity data. D, 1984, University of Texas at El Paso. 195 p.

Sinnock, Scott. Geomorphology of the Uncompahgre Plateau and Grand Valley, western Colorado. D, 1978, Purdue University. 201 p.

Sinnokrot, Ali Amin. The effect of temperature on oil-water capillary pressure curves of limestones and sandstones. D, 1969, Stanford University. 85 p.

Sinnott, Allen. A textural study of certain beaches south of San Francisco. M, 1941, Stanford University. 61 p.

Sint Jan, Michel Leopold Van see Van Sint Jan, Michel Leopold

Sinton, John Blatnik. Suppression of long-decay multiples by predictive deconvolution. M, 1977, University of Texas at Dallas. 99 p.

Sinton, John M. A study of granitization in an area in northern Saskatchewan. M, 1971, University of Oregon. 64 p.

Sinton, Peter O. Three-dimensional, steady-state, finite-difference model of the ground-water flow system in the Death Valley ground-water basin, Nevada-California. M, 1987, Colorado School of Mines. 145 p.

Siok, Jerome P., III. Geologic history of the Siksikpuk Formation on the Endicott Mountains and Picnic Creek allochthons, north central Brooks Range, Alaska. M, 1985, University of Alaska, Fairbanks. 253 p.

Siok, William J. Total water storage capacity of the Fall River-Lakota Sandstones, and of the Dakota Formation East of the Zero Skull Creek Shale Line in South Dakota. M, 1973, South Dakota School of Mines & Technology.

Sipe, Dwight Randy. Depositional environments and origin of bounding surfaces in the Ingleside Formation, Livermore area, Colorado. M, 1984, University of Wyoming. 99 p.

Sipiera, P. P., Jr. Devitrification studies on chemical compositions corresponding to Ca-Al-rich inclusions in the Allende Meteorite. M, 1975, Northeastern Illinois University.

Sipkin, Stuart A. Constraints on Earth structure determined from observations of multiple ScS. D, 1979, University of California, San Diego. 237 p.

Sipling, Philip Jay. A kinetic study on the microcline-sanidine transformation. M, 1971, Brown University.

Sipling, Philip Jay. Coherent exsolution in the sanidine-analbite series. D, 1975, Brown University. 93 p.

Sippel, Katharine N. Depositional and tectonic setting of the Ephraim Formation, southeastern Idaho and western Wyoming. M, 1982, University of Wyoming. 157 p.

Sipperly, David William. Tectonic history of the Sierra del Alambre, northern Chihuahua, Mexico. M, 1967, University of Texas, Austin.

Siragusa, Giorgio. Structural analysis of a mylonite from the Trout Lake area, Lac la Ronge region, Saskatchewan, Canada. M, 1969, University of Toronto.

Siraki, E. S. Amplitude variations of short-period teleseismic P waves. D, 1975, University of Connecticut. 129 p.

Siratovich, Edmund Norman. The Upper Cretaceous rocks of Union County, Arkansas. M, 1960, University of Minnesota, Duluth.

Sircar, Jayanta Kumar. Computer aided watershed segmentation for spatially distributed hydrologic modeling. D, 1986, University of Maryland. 534 p.

Sirey, Cordella R. Estimation of porosity/depth or density/depth curves. M, 1984, Stanford University.

Siribhakdi, Kanchit. Evidence for Cambrian pyroclastic vulcanism preserved in the Rome Formation near Oak Ridge, Anderson County, Tennessee. M, 1976, University of Tennessee, Knoxville. 51 p.

Siriunas, John Michael. Primary trace element dispersion in the stratigraphic horizon containing an Archean massive sulphide ore body; Wilroy Mines Ltd. No. 4 zone, Manitouwadge, Ontario. M, 1979, University of Toronto.

Siriwardane, Hema Jayalath. Nonlinear soil-structure interaction analysis of one-, two-, and three-dimensional problems using finite element method. D, 1980, Virginia Polytechnic Institute and State University. 358 p.

Sirkin, Gerald L. The petrology and petrography of the Triassic-Jurassic sandstone, Currie (Elko county), Nevada. M, 1970, University of Nebraska, Lincoln.

Sirkin, Leslie A. Late Pleistocene palynology and chronology of western Island and eastern Staten Island, New York. D, 1965, New York University.

Sirkin, Leslie Arthur. The (Upper Devonian) Oneonta-upper Ithaca transition in central New York. M, 1957, Cornell University.

Sirles, Phil C. Shear-wave velocity and attenuation analysis of liquefiable soils in the south Truckee Meadows, Washoe County, Nevada. M, 1987, University of Nevada. 157 p.

Sirois, Brenda. Application of a modular three-dimensional finite difference ground-water flow model to a glacial valley fill stream-aquifer system in the Rockaway drainage basin, NJ. M, 1986, Lehigh University.

Siroky, Francis. Geochemical variation of basalts from the Bay of Islands Ophiolite, Newfoundland, Canada. M, 1983, University of Houston.

Siroonian, Harold Ara. Distribution of lithium in the granitic rocks of the Preissac-Lacorne area, Quebec. M, 1958, McMaster University. 66 p.

Sirota, Thomas. A petrographic analysis of coke containing weathered coal. M, 1982, Southern Illinois University, Carbondale. 88 p.

Sirrine, George Keith. Geology of the Springerville-Saint Johns area, Apache County, Arizona. D, 1958, University of Texas, Austin. 273 p.

Sirrine, George Keith. Geology of Warm Springs Mountain, Goshen, Utah. M, 1953, Brigham Young University. 83 p.

Sirvas, Ernesto. Petrography and mineralization of the East Pima Project. M, 1957, Indiana University, Bloomington. 51 p.

Sisk, Steve W. Water resources in glacial till (Pleistocene), Worth County, Missouri. M, 1972, University of Missouri, Columbia.

Siskind, David Eugene. Seismic model study of refraction arrivals in a three-layered structure. M, 1966, Pennsylvania State University, University Park. 38 p.

Siskind, David Eugene. The pressurization and failure of model underground openings. D, 1971, Pennsylvania State University, University Park. 117 p.

Sisler, John Joseph. Geology of the southwest portion of the Laurel Quadrangle, Santa Cruz County, California. M, 1960, Stanford University.

Sisler, John Joseph. Paleogeological study of southwestern Oklahoma. M, 1959, Stanford University.

Sisson, Thomas Winslow. Sedimentary characteristics of the airfall deposit produced by the major pyroclastic surge of May 18, 1980, at Mount St. Helens, Washington. M, 1982, University of California, Santa Barbara. 145 p.

Sisson, Virginia Baker. Contact metamorphism and fluid evolution associated with the intrusion of the Ponder Pluton, Coast plutonic complex, British Columbia, Canada. D, 1985, Princeton University. 345 p.

Sitar, N. Behavior of slopes in weakly cemented soils under static and dynamic loading. D, 1979, Stanford University. 183 p.

Sitar, N. Optimization of dewatering schemes. M, 1975, Stanford University. 60 p.

Sites, R. Geology and structural analysis of the Smoke Holes (Pendleton County, West Virginia). M, 1971, West Virginia University.

Sites, R. S. Structural analysis of the Petersburg Lineament, central Appalachians. D, 1978, West Virginia University. 434 p.

Sitler, Gary Wilson. Depositional environment and petrography of the Mississippian Chainman and Diamond Peak formations, east-central Nevada. M, 1982, University of Idaho. 146 p.

Sitler, Guy F. An investigation of the petrographic composition of the Chilton Coal, Kanawha age, West Virginia, by a reflected light method. M, 1956, University of Pittsburgh.

Sitler, Robert Francis. Glacial geology of a part of western Pennsylvania. D, 1957, University of Illinois, Urbana. 150 p.

Sitler, Robert Francis. Petrography of the Wisconsin tills of northeastern Ohio and northwestern Pennsylvania. M, 1955, University of Illinois, Urbana.

Sitterly, Preston. Paleoecological analysis of the Meadow Marble (Middle Ordovician) in Blount County, Tennessee. M, 1976, University of Tennessee, Knoxville. 162 p.

Siudyla, E. A. A hydrogeologic investigation of aromatic hydrocarbons in the aquifer supplying Ames, Iowa. M, 1975, Iowa State University of Science and Technology.

Sivaborvorn, Vichai. Re-study of hydrocarbon distribution around the Hilbig Oil Field, Bastrop County, Texas. M, 1974, University of Texas, Austin.

Sivakugan, Nagaratnam. Anisotropy and stress path effects in clays. D, 1987, Purdue University. 222 p.

Sivakumar, Ramamurthy. Small-scale crater tests in concrete and sand. M, 1986, New Mexico Institute of Mining and Technology. 209 p.

Sivarajasingham, Sivasupramaniam. Weathering and soil formation on ultrabasic and basic rocks under humid tropical conditions. D, 1961, Cornell University. 247 p.

Sivaraman, Tirupattur V. A potassium-argon age study of the basement complex of Mauritania. M, 1976, Florida State University.

Sivaraman, Tirupattur V. Geochronology of the Precambrians of Rajasthan, India. D, 1983, Florida State University. 148 p.

Sivenas, Prokopios. Crystallization history of the Highland Valley porphyry copper deposit, British Columbia. M, 1976, McMaster University. 164 p.

Sivenas, Prokopis. Aspects of electrochemistry applied to the study of mississippi valley type ore deposits. D, 1981, University of Toronto.

Sivils, David J. Geology of northern Sierra de Palomas, Chihuahua, Mexico. M, 1988, University of Texas at El Paso.

Sivon, Paul A. Stratigraphy and paleontology of the Maquoketa Group (Upper Ordovician) at Wequiock Creek, eastern Wisconsin. M, 1979, University of Wisconsin-Milwaukee. 76 p.

Siwiec, Steven F. The multivariate rotation method of shape analysis as applied to pebbles; a case study from the Jackson Hole-Gros Ventre River area, Teton County, Wyoming. M, 1986, Lehigh University.

Six, Don Eldon. Subsurface geology of the Mounts area, Gibson County, Indiana. M, 1951, Indiana University, Bloomington. 24 p.

Six, Ray L. The Reagan Sandstone. M, 1929, University of Oklahoma. 101 p.

Sixt, Karen C. Temperature of deformation and minimum amount of transport in the Bitterroot mylonite zone, Bitterroot Range, Montana. M, 1988, University of Montana. 69 p.

Sixt, Shirley Claire Smith. Depositional environments, diagenesis and stratigraphy of the Gilmore City Formation (Mississippian) near Humboldt, north-central Iowa. M, 1983, University of Iowa. 164 p.

Sixta, David P. Comparison and analysis of downgoing waveforms from land seismic sources. M, 1982, Colorado School of Mines. 451 p.

Siy, Suzan Elizabeth. Geochemical and petrographic study of phosphate nodules of the Woodford Shale (Upper Devonian-Lower Mississippian) of southern Oklahoma. M, 1988, Texas Tech University. 172 p.

Siyam, Youssuf Mustafa. Accuracy of earthwork calculations from digital elevation data. D, 1981, University of Illinois, Urbana. 285 p.

Size, William Bachtrup. Petrology of the Red Hill syenitic complex, Moultonboro, New Hampshire. D, 1971, University of Illinois, Urbana. 134 p.

Size, William Bachtrup. The petrology of the Goose Lake Tertiary intrusive complex, Park County, Montana. M, 1967, Northern Illinois University. 96 p.

Sizgoric, Martha. Gas transfer of molybdenum sulphide. M, 1969, McGill University. 68 p.

Skalbeck, John D. Paleomagnetism of the Early Triassic Koipato Group, western Nevada, and its tectonic implications. M, 1985, Western Washington University. 206 p.

Skaller, P. M. G. Plant colonization and soil development in the Jamesville limestone quarry. D, 1978, State University of New York, College of Environmental Science and Forestry. 227 p.

Skalnik, Petr. A study of the Upper Maquoketa fossil assemblages of northwestern Illinois. M, 1973, University of Illinois, Chicago.

Skapinsky, Stanley Alfred. The geology of the Kingfield Quadrangle, Maine. D, 1961, Boston University. 266 p.

Skarie, Richard Luther. Hydrologic, geochemical, and mineralogical aspects of saline soils along road ditches in the Red River valley. D, 1986, North Dakota State University. 106 p.

Skean, Donald Minter. Sediments in northern Core Sound, North Carolina. M, 1959, University of North Carolina, Chapel Hill. 42 p.

Skeels, Dorr Covell. Structural geology of the Trail Creek Canyon Mt. area, Montana. D, 1936, Princeton University. 113 p.

Skeels, Margaret Anne. The mastodons and mammoths of Michigan. M, 1961, University of Michigan.

Skeeters, Warren Ware. The migration and accumulation of petroleum and natural gas. M, 1942, Colorado School of Mines. 106 p.

Skehan, James W. Geology of the Wilmington, Vermont, area. D, 1953, Harvard University.

Skelly, Lawrence. The Imogene oil field, Atascosa County, Texas. M, 1947, University of Texas, Austin.

Skelly, Michael F. The geology of the Moapa Peak area, southern Mormon Mountains, Clark and Lincoln counties, Nevada. M, 1987, Northern Arizona University. 150 p.

Skelly, Raymond Lee. Sedimentology of a fluvial-to-marine transition; the Chilhowee Group (Lower Cambrian), Southeast Tennessee. M, 1987, University of Tennessee, Knoxville. 211 p.

Skelly, William A. An investigation of the strength and structural properties of mine pillars in the Pocahontas No. 3 Bed, Wyoming and McDowell counties, West Virginia. M, 1977, Colorado School of Mines. 229 p.

Skerky, Barbara Blanche. Submarine lithification on the Blake Plateau (Southeast Atlantic continental margin). M, 1972, Duke University. 82 p.

Skerlec, Grant M. Geology of the Acarigua area, Venezuela. D, 1979, Princeton University. 301 p.

Skeryanc, Anthony J. Texture, fabric, and composition of fine-grained terrigenous sediments from the Graneros Member of the Mancos Shale, San Juan Basin, New Mexico. M, 1977, University of New Mexico. 137 p.

Sketchley, Dale Albert. The nature of carbonate alteration in basalt at Erickson gold mine, Cassiar, north-central British Columbia. M, 1986, University of British Columbia. 129 p.

Skewes, Milka Alexandra. Petrology of the early formed hydrothermal veins within the central potassic alteration zone of the Los Pelambres porphyry copper deposit, Chile. M, 1984, University of Colorado. 103 p.

Skibicky, Taras V. Use of matched filters to form an additive array in electromagnetic sounding. M, 1982, University of Wisconsin-Madison. 77 p.

Skibo, D. N. Trace-element fractionation and transport in the Earth's mantle. M, 1966, University of Western Ontario.

Skibo, Donald Nicholas. Europium geochemistry; trace element partitioning in silicate-molten salt systems. D, 1976, Massachusetts Institute of Technology. 539 p.

Skidmore, Charles M. A geophysical and tectonic study of the Colima Graben, Mexico. M, 1988, University of New Orleans.

Skidmore, Wilfred Brian. The geology of the Gastonquay-Mourier area, Gaspe Peninsula. D, 1959, Princeton University. 186 p.

Skiles, David Glenn. Petrology of the Chico Ridge limestone bank (upper Pennsylvanian), north-central Texas. M, 1973, University of Texas, Arlington.

Skiles, J. W., III. Human population and water usage in the lower basin of the San Joaquin Valley, California. D, 1977, University of California, Irvine. 290 p.

Skillman, Margaret W. Intrusives of central Minnesota. D, 1946, University of Minnesota, Minneapolis.

Skillman, Margaret W. Some silicic intrusives of eastern central Minnesota. M, 1945, University of Minnesota, Minneapolis. 96 p.

Skinner, Brian J. Thermal expansions of selected isometric minerals, as determined by X-ray measurements. D, 1955, Harvard University.

Skinner, Hubert C. Foraminifera from Arkadelphia Marl exposures near Hope, Arkansas. D, 1954, University of Oklahoma. 257 p.

Skinner, Hubert C. Ostracoda from Arkadelphia Marl exposures near Hope, Arkansas. M, 1953, University of Oklahoma. 70 p.

Skinner, Jeffrey. Paleostress analysis of the Greenbrier Group (Mississippian) Monroe County, West Virginia. M, 1979, University of Toledo. 66 p.

Skinner, John Russo. An experimental investigation of magnetic susceptibility in weak steady fields. M, 1978, University of Oklahoma. 102 p.

Skinner, Orion L. Foraminiferal biostratigraphy and correlation of the lower part of the Hilliard Shale and equivalents, southwestern Wyoming and northeastern Utah. M, 1982, University of Wyoming. 178 p.

Skinner, Ralph. A study of some intrusive rocks and replacement phenomena in the Salmon Arm area, British Columbia (Shuswap Terraine, Precambrian). M, 1961, McGill University.

Skinner, Ralph. Geology of the Tetagouche Group, Bathurst, New Brunswick. D, 1956, McGill University.

Skinner, Robert G. Quaternary stratigraphy of Moose River Basin, Ontario, Canada. D, 1971, University of Washington. 91 p.

Skinner, Robert G. Quaternary stratigraphy of the Moose river basin, Ontario, Canada. M, 1970, University of Washington.

Skinner, Roland B. The geology of the Luttrell area, Union and Grainger counties, Tennessee. M, 1961, University of Tennessee, Knoxville. 38 p.

Skinner, Walter Swart. The Tully Formation of eastern Pennsylvania. M, 1948, Lehigh University.

Skinner, William Robert. Geologic evolution of the Beartooth Mountains, Montana and Wyoming; Part 8, Ultramafic rocks in the Highline Trail Lakes area, Wyoming. D, 1966, Columbia University. 90 p.

Skipp, Betty A. Geology of the Dead Horse Creek area northeast of Salida (Chaffee County), Colorado. M, 1956, University of Colorado.

Skipp, Betty Ann Lindberg. Contraction and extension faults in the southern Beaverhead Mountains, Idaho and Montana. D, 1985, University of Colorado. 170 p.

Skippen, George Barber. A study of the distribution of palladium in igneous rocks. M, 1963, McMaster University. 89 p.

Skippen, George Barber. An experimental study of the metamorphism of siliceous carbonate rocks. D, 1967, The Johns Hopkins University. 320 p.

Skipper, Keith. Depositional mechanics of atypical turbidites, Cloridorme Formation, Gaspe, Quebec. M, 1970, McMaster University. 137 p.

Skirrow, Roger G. Silicification in a semiconformable alteration zone below the Chisel Lake massive sulphide deposit, Manitoba. M, 1987, Carleton University. 94 p.

Skirvin, Raymond Taylor. The underground course of the Sante Fe River near High Springs, Florida. M, 1962, University of Florida. 55 p.

Sklar, Fred Hal. Water budget, benthological characterization, and simulation of aquatic material flows in a Louisiana freshwater swamp. D, 1983, Louisiana State University. 296 p.

Sklar, Maurice. Petrology of the volcanic rocks of the region around Boulder Dam (Nevada, Arizona). M, 1938, California Institute of Technology.

Sklar, Paul Jeffrey. Petrologic, petrographic and paleotectonic investigation of Precambrian mafic intrusives in eastern South Dakota. M, 1982, University of Iowa. 98 p.

Sklarew, D. S. Analysis of kerogen in Precambrian stromatolites. D, 1978, University of Arizona. 131 p.

Sklash, M. G. A conceptual model of watershed response to rainfall developed through the use of oxygen-18 as a natural tracer. M, 1975, University of Waterloo.

Sklash, M. G. The role of groundwater in storm and snowmelt runoff generation. D, 1978, University of Waterloo.

Sklenar, Scott. Genesis of an Eocene lake system within the Washakie Basin of southwestern Wyoming. M, 1982, San Jose State University. 89 p.

Sklenar, Walter Martin. Leeside erosion rate in relation to slipface bedform stability. M, 1980, Syracuse University.

Skoch, Edwin James. Seasonal changes in phosphate, iron, and carbon occurring in the bottom sediments, near Rattlesnake Island, in western Lake Erie, 1966 to 1968. D, 1968, Ohio State University. 55 p.

Skogstrom, H. Clifford. Paleontological and paleoecological studies of the Pelecypoda of the Fox Hills Formation. M, 1959, University of South Dakota. 72 p.

Skokan, C. K. A time-domain electromagnetic survey of the East Rift Zone, Kilauea Volcano, Hawaii. D, 1975, Colorado School of Mines. 62 p.

Skokan, Catherine A. King. Time-domain electromagnetic coupling. M, 1971, Colorado School of Mines. 62 p.

Skolasky, Robert A. The use of color infrared and panchromatic aerial photography for drainage density and analysis and soil mottling studies of Onion Creek terraces, Travis and Hays counties, Texas. M, 1978, University of Texas, Austin.

Skolnick, Herbert. The lithology and stratigraphy of the Tokio Formation of McCurtain County, Oklahoma. M, 1949, University of Oklahoma. 35 p.

Skolnick, Herbert. The stratigraphy and paleontology of a part of the Lower Cretaceous rocks of the Black Hills area. D, 1952, University of Iowa. 142 p.

Skopec, Robert A. Organic geochemistry of the Alligator River complex peat, North Carolina. M, 1983, Kent State University, Kent. 153 p.

Skorpen, Allan J. Magnetic profiles across the Aleutian Trench and Ridge. M, 1968, Oregon State University. 43 p.

Skotnicki, Michael Charles. Stratigraphy and depositional environments of the Upper Cretaceous Blufftown Formation, eastern Gulf Coastal Plain of Alabama. M, 1985, Auburn University. 162 p.

Skov, Niels Aage. Factors influencing the salinity difference between the North Atlantic and North Pacific oceans. M, 1965, Oregon State University. 44 p.

Skov, Niels Aage. The ice cover of the Greenland Sea; an evaluation of oceanographic and meteorological causes for year-to-year variations. D, 1968, Oregon State University. 88 p.

Skow, Donald Lester. An approach to increasing the resolution of the seismic reflection method in delineating salt-dome flanks. D, 1971, Washington University. 213 p.

Skrivan, J. A. Application of the Sagar method for the solution of the inverse problem in ground-water hydrology. M, 1975, University of Arizona.

Skrzyniecki, Alan Francis. Geology and geochemistry of the Sheep Rock mineralized norite, Albany County, Wyoming. D, 1973, University of Illinois, Urbana. 115 p.

Skrzyniecki, Alan Francis. Geology of the Sheep Rock mineralized norite, Albany County, Wyoming. M, 1970, University of Illinois, Urbana. 71 p.

Skrzyniecki, Randal G. Geochemistry, stratigraphy and petrography of some Ohio Middle Devonian limestones. M, 1972, University of Toledo. 101 p.

Skulski, Thomas. The tectonic and magmatic evolution of the central segment of the Archean La Grande greenstone belt, central Quebec. M, 1986, University of Toronto.

Skurla, Steven J. Geology of the Sturgill Peak area, Washington County, Idaho. M, 1974, Oregon State University. 98 p.

Skvarla, John J. The Ostracoda of the Silurian Brownsport Formation of western Tennessee. M, 1958, Miami University (Ohio). 144 p.

Skwara, Theresa. Late Pleistocene vertebrate fauna from Riddell site near Saskatoon, Canada. M, 1978, University of Saskatchewan. 198 p.

Skyllingstad, Paul E. Depositional environments and diagenetic history of the Virgin Member of the Moenkopi Formation. M, 1977, Arizona State University. 146 p.

Slack, Harold A. Field measurement of the radioactivity of rocks. D, 1952, University of Toronto.

Slack, Howard A. The measurement of absolute radioactivity of rocks using a Geiger counter. D, 1951, University of Toronto.

Slack, J. F. Hypogene zoning and multistage vein mineralization in the Lake City area, western San Juan Mountains, Colorado. D, 1976, Stanford University. 397 p.

Slack, John F. Structure, petrology, and ore deposits of the Indian Springs (Delano Mountains) region, Elko County, Nevada. M, 1972, Miami University (Ohio). 159 p.

Slack, Paul B. Structural geology of the northeast part of the Rio Puerco Fault zone, Sandoval County, New Mexico. M, 1973, University of New Mexico. 74 p.

Slade, M. Lyle. Pennsylvanian and Permian fusulinids of the Ferguson Mountain area, Elko County, Nevada. M, 1961, Brigham Young University. 92 p.

Slade, R. C. Hydrogeologic investigation of Carpinteria ground water basin, Santa Barbara County, California. M, 1975, University of Southern California.

Slaght, W. H. A petrographic study of the Copper Cliff offset in the Sudbury District (Ontario). M, 1951, McGill University.

Slagle, E. S. The paleontology and paleoecology of the Hillsdale Limestone (Mississippian, Meramecian), Washington County, Virginia. M, 1978, East Carolina University. 176 p.

Slagle, Letha P. Depositional systems and structures of the middle Eocene Domengine-Yokut Sandstone, Vallecitos, California. M, 1979, Stanford University. 59 p.

Slagley, Scott A. Petrography and stratigraphy of the Ninemile Formation (Lower Ordovician) of central Nevada. M, 1984, University of Missouri, Columbia. 97 p.

Slaine, D. D. Geophysical mapping of subsurface contaminants. M, 1983, University of Waterloo. 192 p.

Slama, Don C. Revision of the bryozoan genus Rhombopora Meek. M, 1952, University of Nebraska, Lincoln.

Slankis, J. A. Magnetization of the Kapiko Iron Formation (Precambrian), Thunder Bay District, Ontario. M, 1966, University of Western Ontario.

Slankis, John Aris. Telluric and magnetotelluric surveys at 8 Hz. D, 1970, McGill University.

Slapp, Kevin P. Heavy mineral analysis of selected loess samples from Franklin and Catahoula parishes, Louisiana. M, 1987, Northeast Louisiana University. 63 p.

Slate, Houston Leale. Petroleum geology of the Taloga-Custer City area, Dewey and Custer counties, Oklahoma. M, 1962, University of Oklahoma. 58 p.

Slater, David H. The stratigraphy and paleoecology of the Tamiami Formation in Hendry County, Florida. M, 1978, Florida State University.

Slater, Jennifer Margaret. On denitrification in coastal marine sediments. D, 1986, SUNY at Stony Brook. 146 p.

Slater, Jock R. Some relationships between deformation, mineralization and ore genesis at the Fay Mine, Eldorado, Saskatchewan. M, 1982, University of Alberta. 122 p.

Slater, L. E. A multi-wavelength distance-measuring instrument for geophysical experiments. D, 1975, University of Washington. 137 p.

Slater, Mical N. ology and mineral deposits of the western Cuddy Mountain District, western Idaho. M, 1969, Oregon State University. 82 p.

Slater, Richard A. Sedimentary environment in Suisun Bay, California (San Fransisco Bay area). M, 1965, University of Southern California.

Slator, Dorothy Stevenson. Sandstone diagenesis and its variation with deltaic depositional environments, Upper Cretaceous, southern Rio Escondido basin, Coahuila, Mexico. M, 1980, University of Texas, Austin.

Slatt, Roger Malcolm. Nature and distribution of sediments in the Norris Glacier outwash area, upper Taku Inlet, southeastern Alaska. M, 1967, University of Alaska, Fairbanks. 45 p.

Slatt, Roger Malcolm. Sedimentological and geochemical aspects of sediment and water from Ten Alaskan valley glaciers. D, 1970, University of Alaska, Fairbanks. 134 p.

Slatten, Mark H. The Windmill Limestone at Wenban Peak, southern Cortez Mountains, Nevada. M, 1978, University of California, Riverside. 170 p.

Slaughter, Arthur Edwin. The stratigraphic value of certain cryptostomatous Bryozoa of the Traverse Formation, Middle Devonian age, of Michigan. M, 1950, Michigan State University. 37 p.

Slaughter, Cecil Bryan. Chemistry of the near-surface groundwater, Great Salt Plains, Alfalfa County, Oklahoma. M, 1988, Iowa State University of Science and Technology. 42 p.

Slaughter, John. Experimental determination and thermodynamic calculation of equilibria in the system CaO-MgO-SiO$_2$-H$_2$O-CO$_2$. D, 1976, Pennsylvania State University, University Park. 115 p.

Slaughter, John. Genesis of the stratiform copper deposits of the Catskill Formation (middle and upper Devonian and lower Mississippian) in northeastern Pennsylvania. M, 1970, Rutgers, The State University, New Brunswick. 67 p.

Slaughter, Maynard. A study of mineralogical and other properties of selected raw and fired fireclays. M, 1957, University of Missouri, Columbia.

Slaughter, Maynard. The crystal structure of aluminum tetroxycarbide. D, 1962, University of Pittsburgh.

Slaughter, Thad A. Lithology and strontium distribution of the De Queen Formation at the main Highland gypsum quarry, Highland, Arkansas. M, 1985, Stephen F. Austin State University. 95 p.

Slaughter, Turbit Henry. Shore erosion in tidewater Maryland. M, 1949, The Johns Hopkins University.

Slavens, Margaret Dever. Clays in industry in Ohio. M, 1928, University of North Carolina, Chapel Hill. 42 p.

Slavik, Harold Joseph, Jr. Reconnaissance geology of the Cayuse Point 7.5 minute quadrangle, Elmore County, Idaho. M, 1987, University of Idaho. 107 p.

Slavin, E. J. Process and mechanism of stream bank failures along Brown's River, Vermont. M, 1977, University of Vermont.

Slawinski, Michael A. Investigation of inhomogeneous body waves in an elastic/anelastic medium. M, 1988, University of Calgary. 82 p.

Slawson, Chester Baker. The thermo-optical properties of heulandite. D, 1925, University of Michigan.

Slawson, Guenton Cyril, Jr. Water quality in the lower Colorado River system and effects of reservoirs. M, 1972, University of Arizona.

Slawson, William Francis. Lead in potassium feldspars from Basin and Range monzonites. D, 1958, University of Utah. 82 p.

Slaydon, Robert Earl, Jr. Hydrothermal synthesis and analysis of some olivine group minerals. M, 1961, Texas Christian University. 43 p.

Slaymaker, Susan Clark. A plate tectonics model for the Southern Appalachians. D, 1974, University of North Carolina, Chapel Hill. 69 p.

Slayton, David F. Field evidence for shale membrane filtration of groundwater, south-central Michigan. M, 1982, Michigan State University. 80 p.

Slebir, Edward Joseph. Geology of North Cement Creek area, Gunnison County, Colorado. M, 1957, Colorado School of Mines. 93 p.

Slechta, John J. Glauconite and co-existing clay of the Clayton Formation, southern Illinois. M, 1974, Southern Illinois University, Carbondale. 66 p.

Slechta, Marc W. Sedimentation patterns and processes at Bass River inlet. M, 1982, Boston University. 154 p.

Sledz, James John. Computer model for determining fracture porosity and permeability in the Conasauga Group, Oak Ridge National Laboratory, Tennessee. M, 1980, University of Tennessee, Knoxville. 139 p.

Sledz, Janine Gajda. Petrologic, mineralogic, and ion exchange characteristics of the Rome Formation and Pumpkin Valley Shale on the Oak Ridge National Laboratory Reservation, Oak Ridge, Tennessee. M, 1980, University of Tennessee, Knoxville. 105 p.

Sleeman, Lyle H., Jr. Petrography and petrology of the igneous intrusive in Woodson County, Kansas. M, 1959, Kansas State University. 40 p.

Sleeman, Lyle Herman, Jr. The petrology and sedimentation history of the Callaway Formation in central and northeastern Missouri. D, 1964, University of Missouri, Columbia. 269 p.

Sleep, Norman Harvey. Deep structure and geophysical processes beneath island arcs. D, 1973, Massachusetts Institute of Technology. 274 p.

Sleep, Norman Harvey. Topography and tectonics of the intersections of fracture zone with central rifts. M, 1969, Massachusetts Institute of Technology. 18 p.

Sleeper, James Lockert, Jr. An investigation of the possible relationship between certain physiographic features and subsurface structure in Lubbock County, Texas. M, 1981, University of Texas at Dallas. 23 p.

Sleeper, James Lockert, Jr. Investigation of the possible relationship between certain physiographic features and subsurface structure in Lubbock County, Texas. M, 1941, Texas Tech University. 23 p.

Sleight, Vergil G. The geology of Ogishkemuncie Lake and vicinity (Minnesota). D, 1933, Northwestern University.

Slemmons, David Burton. Geology of the Sonora Pass region. D, 1953, University of California, Berkeley. 222 p.

Slentz, Loren Williams. Tertiary Salt Lake Group in the Great Salt Lake basin. D, 1955, University of Utah. 59 p.

Slessor, David K. Holocene history of New London Bay, Prince Edward Island (Canada). M, 1972, Queen's University. 104 p.

Slewitzke, Edward B. A revision of the species of Hexagonaria in the Cedar Valley Limestone. M, 1961, University of Kansas. 53 p.

Slifko-Welch, Christine M. The Harrison Avenue landslide, Pierre, South Dakota. M, 1981, South Dakota School of Mines & Technology. 106 p.

Sliger, Kenneth Leon. Geology of the lower James River area, Mason County, Texas. M, 1957, Texas A&M University.

Slingerland, Rudy Lynn. Processes, responses, and resulting stratigraphic sequences of barrier island tidal inlets as deduced from Assawoman Inlet, Virginia. D, 1977, Pennsylvania State University, University Park. 456 p.

Slingerland, Rudy Lynn. Transportation and hydraulic equivalence relationships of light and heavy minerals in sands. M, 1973, Pennsylvania State University, University Park. 115 p.

Slingluff, Frank Peter. Sedimentation and shore processes of southwestern Galveston Island, Galveston City, Texas. M, 1948, University of Texas, Austin.

Slipp, R. M. Areal geology of the Marymac map-area, New Quebec. D, 1954, McGill University.

Slipp, R. M. Base metal deposits in the "Labrador Trough" between Lake Harveng and Lac Aulneau, New Quebec. D, 1957, McGill University.

Slipp, R. M. The geology of the Round Pond map area, Newfoundland. M, 1952, McGill University.

Sliter, William Volk. Upper Cretaceous foraminifera from Southern California and northwestern Baja California, Mexico. D, 1966, University of California, Los Angeles. 554 p.

Slitor, Truman Wentworth. The petrology, petrography and structural geology of the Precambrian complex of the Torrey Creek area, Fremont County, Wyoming. M, 1969, Miami University (Ohio). 106 p.

Sliva, Thomas W. Lower Caseyville (Lower Pennsylvanian) depositional environments; Union and Johnson counties, Illinois. M, 1972, Southern Illinois University, Carbondale. 47 p.

Slivitsky, Anne. Coupe structurale à travers la partie sud-ouest des Appalaches du Québec. M, 1983, Universite Laval. 21 p.

Sloan, Doris. Ecostratigraphic study of Sangamon sediments beneath central San Francisco Bay. D, 1981, University of California, Berkeley. 338 p.

Sloan, Doris. Middle Ordovician environments of deposition, Mazourka Canyon, Inyo County, California. M, 1975, University of California, Berkeley. 125 p.

Sloan, Heather. Bathymetry and magnetics of an overlapping spreading center at 16°20′N on the East Pacific Rise with photoelastic modelling of curved overlapping cracks. M, 1986, University of California, Santa Barbara. 76 p.

Sloan, James. Cenozoic organic carbon deposition in the deep sea. M, 1985, University of Miami. 197 p.

Sloan, John F. A comparative application of the methods of Stokes and Hirvonen to the computation of the undulation of the geoid. M, 1961, Ohio State University.

Sloan, Jon Roger. Radiolarians of the North Philippine Sea; their biostratigraphy, preservation, and paleoecology. D, 1981, University of California, Davis. 154 p.

Sloan, Kenneth W. Distribution and ecology of Ostracoda and foraminifera in the Bennett Shale. M, 1963, Kansas State University. 98 p.

Sloan, Robert E. The paleoecology of the Pennsylvanian marine shales of Palo Pinto County, Texas. D, 1953, University of Chicago. 16 p.

Sloanaker, Charles Jasper. The geomorphology and Pleistocene drainage history of the Turkey Creek basin, Dickinson County, Kansas. M, 1950, University of Kansas. 70 p.

Sloane, Bryan Jennings. A stratigraphic study of the Cotton Valley Bodcaw Sand and its relation to pre-Cotton Valley structure. M, 1957, Louisiana State University.

Slocki, Stanley Francis. The physical stratigraphy of the Georgetown Formation equivalents in Tarrant, Denton, and Cooke counties. M, 1957, Texas Christian University. 17 p.

Slocomb, J. P. An analysis of the relationship between community structure and ecosystem function. D, 1979, Virginia Polytechnic Institute and State University. 225 p.

Slocum, Gilbert. Gypsite of Toyah Quadrangle, Reeves County, Texas. M, 1951, University of Texas, Austin.

Slocum, R. C. A study of the post-Boone outliers of eastern Mayes, southern Delaware, and northern Adair counties, Oklahoma. M, 1953, University of Oklahoma. 91 p.

Slodowski, Thomas Raymond. Geology of the Yauco area, Puerto Rico. D, 1956, Princeton University. 177 p.

Slone, George T. Conodont paleoecology of the Renfro Member of the Borden Formation and Newman Limestone of eastern Kentucky. M, 1975, Eastern Kentucky University. 54 p.

Slorp, L. H. Relationships of species composition in a forest community to topographic variation; an example from the Hoosier Hills. D, 1977, University of Illinois, Urbana. 144 p.

Slosek, Jean. The spatial distribution of morbidity and mortality caused by earthquakes. M, 1986, University of Massachusetts. 189 p.

Sloss, Laurence L. Devonian rugose corals from the Traverse Beds of Michigan. D, 1937, University of Chicago. 87 p.

Sloss, Peter William. Coastal processes under hurricane action; numerical simulation of a free-boundary shoreline. D, 1972, Rice University. 139 p.

Slosson, James E. Lithofacies and sedimentary-paleogeographic analysis of the Los Angeles Repetto Basin. D, 1958, University of Southern California.

Slosson, James Edward. Sedimentation in area of Diversion Dam, Figueredo Wash, New Mexico. M, 1950, University of Southern California.

Slover, Susan M. Fining upward sequences in the lower Mt. Shields Formation, middle Proterozoic Belt Supergroup, west-central Montana. M, 1982, University of Montana. 55 p.

Slovinsky, Raymond LeRoy. Mineralogical variation of Wyoming bentonites and its significance. D, 1958, University of Illinois, Urbana. 126 p.

Slow, Evan S. Quartz grain surface textures and carbonate cements in sandstones. M, 1985, Bowling Green State University. 127 p.

Slowey, Austin Henry. The effect of wind on beach erosion on the Outer Banks at Pea Island, North Carolina. M, 1971, North Carolina State University. 43 p.

Slowey, James Frank, Jr. Studies on the distribution of copper, manganese and zinc in the ocean using neutron activation analysis. D, 1966, Texas A&M University. 115 p.

Slucher, Ernie R. Sedimentation patterns and tectonic controls of Early to Middle Pennsylvanian rocks, south-central-eastern Kentucky. M, 1986, Eastern Kentucky University. 98 p.

Slusarski, Mark Leo. Photogeologic fracture traces and lineaments in the Wartburg Basin section of the Cumberland Plateau physiographic subprovince, Tennessee. M, 1979, University of Tennessee, Knoxville. 72 p.

Slyker, Robert G. Geologic and geophysical reconnaissance of the Valle de San Felipe region, Baja California, Mexico. M, 1970, University of San Diego.

Smaglik, Suzanne M. Petrogenesis and tectonic implications for Archean mafic and ultramafic magmas in the Elmer's Rock greenstone belt, Laramie Range, Wyoming. M, 1987, Colorado School of Mines. 126 p.

Smajovic, I. Rusty-weathering paragneisses of the (Precambrian) Grenville Province (Ontario). M, 1960, McGill University.

Smaldone, John R. Aquifer assessment of Groveland, Massachusetts. M, 1984, Boston University. 80 p.

Smale, Gordon R. A field and petrographic study of the Freda Formation along the Montreal River, Gogebic County, Michigan. M, 1958, Michigan State University. 49 p.

Smale, Timothy. Soft sedimentation deformation in the Southern Ridge Basin, Transverse Ranges, California. M, 1978, University of California, Santa Barbara.

Small, Benjamin A., III. Inventory and mathematical definition of estuarine meanders in coastal Louisiana. M, 1977, Louisiana State University.

Small, G. G. Groundwater recharge and quality transformations during the initiation and management of a new stabilization lagoon. M, 1973, University of Arizona.

Small, John, Jr. Stratigraphy of Southwest Ecuador and Ancon oil field studies. D, 1962, University of Colorado. 212 p.

Small, Steven B. The Al:Ti:Fe ratio; a valuable tool in lunar basalt mineralogy and petrology. M, 1976, Brooklyn College (CUNY).

Small, Thomas Wayne. The spatial distribution of hillslope form and process in selected Driftless Area watersheds, southwestern Wisconsin. D, 1973, University of Wisconsin-Madison.

Small, William David. Cordilleran geochronology deduced from hydrothermal leads. D, 1970, University of British Columbia. 97 p.

Smalley, Richard Curtis. An isotopic and geochemical investigation of the hydrogeologic and geothermal systems in the Safford Basin, Arizona. M, 1983, University of Arizona. 85 p.

Smalley, Robert G. Trace elements and geologic structure of some hot springs of Yellowstone National Park. D, 1948, University of Chicago. 119 p.

Smalley, Robert, Jr. Two earthquake studies; 1, Seismicity of the Argentine Andean foreland, and 2, A renormalization group approach to earthquake mechanics. D, 1988, Cornell University. 188 p.

Smallwood, Alan Robert. An abrasion hardness classification for sandstone. M, 1970, University of Illinois, Urbana. 127 p.

Smallwood, James C. The geology of the Hamilton-Sevenmile area, Butler County, Ohio. M, 1958, Miami University (Ohio). 105 p.

Smallwood, Kenneth Keith. Geology of the Wasa mining area, Granite County, Montana. M, 1956, University of Montana. 51 p.

Smart, Burton. Geologic study and mapping of Pennsylvanian rocks in western Missouri. M, 1957, University of Iowa. 123 p.

Smart, Eugene. Linear high-resolution frequency wavenumber analysis. D, 1976, Southern Methodist University. 76 p.

Smart, Miles Millard, III. Stream-watershed relationships in the Missouri Ozark Plateau Province. D, 1980, University of Missouri, Columbia. 182 p.

Smartt, Richard A. Late Pleistocene and Recent Microtus from southcentral and southwestern New Mexico. M, 1972, University of Texas at El Paso.

Smath, Richard A. Discriminating relationships among basic lithologies and engineering parameters obtained from the point-load and slake durability tests. M, 1983, Eastern Kentucky University. 68 p.

Smathers, Nancy Preas. The paleomagnetism of Jurassic and Triassic limestones in the upper Austroalpine unit and the tectonic implications. M, 1987, University of Florida. 189 p.

Smedes, Harry Wynn. Geology of art of the northern Wallowa Mountains, Oregon. D, 1959, University of Washington. 217 p.

Smedley, Gary Lee. The design and theory of a seismometer. M, 1966, University of Kansas.

Smedley, Harold Orian. A study of the (Pennsylvanian) Oread Formation in Nebraska. M, 1933, University of Nebraska, Lincoln.

Smedley, Jack Elwood. The stratigraphy of part of the Lemhi Mountains, Idaho. M, 1948, University of Idaho. 34 p.

Smeds, Russell Clarence. The geology of the northern half of the White Cross Quadrangle, North Carolina. M, 1972, University of North Carolina, Chapel Hill. 61 p.

Smee, Barry Warren. Laboratory and field evidence in support of electrogeochemically enhanced ionic diffusion through glaciolacustrine sediment. D, 1983, University of New Brunswick.

Smelik, Eugene Alan. An X-ray diffraction study of displacive phase transitions in terrestrial tridymite. M, 1987, University of North Carolina, Chapel Hill. 161 p.

Smelley, Randal Keith. Sedimentary zonation of the Ogallala Aquifer. M, 1980, Texas Tech University. 65 p.

Smerchanski, Mark Gerald. The geology of the Scotia gold property number two. M, 1938, Virginia Polytechnic Institute and State University.

Smethie, W. M., Jr. An investigation of vertical mixing rates in fjords using naturally occurring radon-222 and salinity as tracers. D, 1979, University of Washington. 247 p.

Smiley, Charles Jack. A preliminary report on the Ellensburg flora of Washington with special reference to the genus Paulownia. M, 1954, University of California, Berkeley. 45 p.

Smiley, Charles Jack. The Ellensburg and Selah floras of central Washington. D, 1960, University of California, Berkeley. 278 p.

Smiraldo, Mark S. Lithology, porosity development, and silica cement source of the "Clinton" Formation in eastern Ohio. M, 1985, University of Akron. 132 p.

Smiser, Jerome Standley. A study of the echinoid fragments in the Cretaceous rocks of Texas. D, 1931, Princeton University.

Smiser, Jerome Standley. A study of the fragments of some representative Texas Cretaceous echinoids. M, 1929, Texas Christian University. 89 p.

Smit, David E. Pennsylvanian conodonts from the Ardmore Basin, southern Oklahoma (Lake Murray Golf Course formations). M, 1967, University of Iowa. 52 p.

Smit, David Ernst. Stratigraphy and sedimentary petrology of the Cambrian and lower Ordovician shelf facies of western Newfoundland. D, 1971, University of Iowa. 192 p.

Smit, Johannes Hendricus. Sedimentology, metamorphism, and structure of the LaGorce Formation, LaGorce Mountains, Upper Scott Glacier area, Antarctica. M, 1981, Arizona State University. 83 p.

Smith, Abigail Marion. Rates of production and accumulation of carbonate sediments on a temperate pocket beach, Gulf of Maine. M, 1984, Massachusetts Institute of Technology. 72 p.

Smith, Alan Barrett. Paleoecology of the Trent Formation (lower Miocene, North Carolina). M, 1958, University of Michigan.

Smith, Alan D. Isotopic and geochemical studies of Terrane I, south-central British Columbia. D, 1986, University of Alberta. 212 p.

Smith, Alan Gilbert. Structure and stratigraphy of the Northwest Whitefish Range, Lincoln County, Montana. D, 1963, Princeton University. 151 p.

Smith, Alan Lewis. Petrology of Quaternary basic lavas of California and a note on sphene-perovskite paragenesis. D, 1969, University of California, Berkeley. 140 p.

Smith, Alan Robert. Techniques for obtaining fabric data from coarse clastic sediments. M, 1968, Brigham Young University.

Smith, Alan Robert. The petrology and geochemistry of the lower zone of the Mulcahy Gabbro, northwestern Ontario. M, 1987, University of Western Ontario. 174 p.

Smith, Albert Turner. The stress distribution beneath island arcs. M, 1971, Massachusetts Institute of Technology. 104 p.

Smith, Albert Turner. Time-dependent strain accumulation and release at island arcs; implications for the 1946 Nankaido earthquake. D, 1975, Massachusetts Institute of Technology. 292 p.

Smith, Albert W. A survey of the glacial landforms in the headwaters of the Snake River, Summit County, Colorado. M, 1949, University of Colorado.

Smith, Alexander. Structural petrology; Crestmore, California. D, 1947, California Institute of Technology.

Smith, Alexander. The geochemistry and paragenesis of the ores of the Cactus Mine, Kern County, California. D, 1947, California Institute of Technology.

Smith, Alexander. The structure of the eastern belt of the Cordillera in Canada. M, 1933, University of British Columbia.

Smith, Alison Jean. The taxonomy and paleoecology of the Holocene freshwater Ostracoda of Pickerel Lake, South Dakota. M, 1987, University of Delaware. 244 p.

Smith, Allan Conrad, Jr. In situ stresses and small anticlinal features in eastern North America. M, 1977, Cornell University.

Smith, Althea Page. Petrogenesis and structure in the Mt. Chocorua area, New Hampshire. D, 1940, Harvard University.

Smith, Alvin H. Areal geology of Elk City area, Beckham and Roger Mills counties, Oklahoma. M, 1964, University of Oklahoma. 64 p.

Smith, Anne Lauren. Correlation of Monterey Formation organic geochemistry to lithology, Santa Maria-Santa Barbara basins, California. M, 1985, University of Wyoming. 261 p.

Smith, Arthur Edward. Petrology of the tourmaline-bearing layered granitic rocks from the Black Hills of South Dakota. M, 1960, University of Missouri, Columbia.

Smith, Arthur Rankin. The modal composition of the major intrusions of the Yosemite Valley. M, 1958, University of California, Berkeley. 65 p.

Smith, Arthur Tremaine. Geology of redbed Cu-U occurrences in the Upper Devonian Catskill Fm., Pennsylvania. D, 1983, Pennsylvania State University, University Park. 234 p.

Smith, Arthur Tremaine. Stratigraphic and sedimentologic controls for copper and uranium in red-beds of the Upper Devonian Catskill Formation in Pennsylvania. M, 1980, Pennsylvania State University, University Park. 216 p.

Smith, Arthur Young. Experimental investigation of some textures of massive sulphide ores. M, 1961, Queen's University. 178 p.

Smith, Avery Edward. A subsurface study of the Rochester oil field, Gibson County, Indiana and Wabash County, Illinois. M, 1950, University of Oklahoma. 83 p.

Smith, B. G. Geology of the Adair area, Gaines and Terry counties, Texas. M, 1959, University of Tulsa. 99 p.

Smith, Barbara J. Effects of Permian evaporite dissolution in Kansas. M, 1976, Kansas State University. 107 p.

Smith, Barry Samuel. Occurrence and quality of ground water in the Oxford-Morning Sun area, Ohio. M, 1982, Miami University (Ohio). 85 p.

Smith, Bennett Lawrence. The Grenville geology of southeastern Ontario. D, 1954, Syracuse University. 205 p.

Smith, Bernice Young. Lower Tertiary foraminifera from Contra Costa County, California. M, 1951, University of California, Berkeley. 178 p.

Smith, Bill Ross. Mineralogy of selected soils on the lower coastal plain of North Carolina. D, 1970, North Carolina State University. 164 p.

Smith, Bradley Earl. The geology and mineral resources of Palmer and northern Wesley townships, Washington County, Ohio. M, 1960, Ohio University, Athens. 164 p.

Smith, Bradley Keller. Plastic deformation of garnets; mechanical behavior and associated microstructures. D, 1982, University of California, Berkeley. 208 p.

Smith, Bradley Wayne. A new torsional shear-wave generator and its geophysical application. M, 1981, North Carolina State University. 37 p.

Smith, Brian Alan. Upper Cretaceous stratigraphy and the mod-Cenomanian unconformity of east-central Mexico. D, 1986, University of Texas, Austin. 247 p.

Smith, Brian Mitchell. Oxygen- and strontium-isotopic studies of the Skye intrusive complex, Northwest Scotland. D, 1981, Brown University. 82 p.

Smith, Bruce C. Investigation of spectral methods of depth estimation with application to aeromagnetic anomalies in Southwest Georgia. M, 1981, Indiana University, Bloomington. 306 p.

Smith, Bruce D. Geologic and geophysical investigation of an area of Precambrian rocks, central Laramie Range, Albany County, Wyoming. M, 1967, University of Wyoming. 118 p.

Smith, Bruce Dyfrig. Interpretation of electromagnetic field measurements. D, 1975, University of Utah. 244 p.

Smith, Bruce Edward. A stratigraphic investigation around a landfill near Tremont City, Ohio, using seismic refraction and electrical resistivity. M, 1986, Wright State University. 100 p.

Smith, Bruce L. Geology of the southeastern portion of the Inchelium Quadrangle, Stevens County, Washington. M, 1982, Eastern Washington University. 60 p.

Smith, C. J. Geology and some mineral deposits of San Diego and Imperial counties, California. M, 1923, University of Minnesota, Duluth.

Smith, C. L. Effects of man upon the geomorphology of the Rampart Range, Colorado. D, 1977, University of Georgia. 247 p.

Smith, Cale Clinton. The stratigraphy of the Oread Limestone of an area near the Kansas-Oklahoma state line. M, 1938, University of Iowa. 55 p.

Smith, Cameron Outcalt. Hydrocarbon exploration in the North Sea and adjacent basins. M, 1975, Pennsylvania State University, University Park. 149 p.

Smith, Carl C. Facies analysis and depositional history of Antlers sands (Cretaceous), Callahan Divide area, Texas. M, 1972, Louisiana State University.

Smith, Casey C. Underground hydrofracturing in-situ stress measurements near the Keweenaw Fault in Upper Michigan. M, 1980, Michigan Technological University. 128 p.

Smith, Cassius Crowell. The Minerva Mine and survey methods employed in connection with the same at Atlanta, Idaho. M, 1927, University of Nevada - Mackay School of Mines. 15 p.

Smith, Cathlee. The relationship of old Mississippi River channels to the origin of Reelfoot Lake, Tennessee. M, 1981, Southern Illinois University, Carbondale. 92 p.

Smith, Charles Andrew Francis, III. Diversity patterns and their interpretability as stochastic variables. M, 1976, University of Rochester. 84 p.

Smith, Charles Culberson. Calcareous nannoplankton and stratigraphy of the upper Eagle Ford and lower Austin formations, Texas. D, 1973, University of Texas at Dallas. 336 p.

Smith, Charles Culberson. Foraminifera, paleoecology, and biostratigraphy of the Paleocene "Ostrea thirsae beds", Nanafalia Formation, west-central Alabama. M, 1967, University of Houston.

Smith, Charles David. Cation affinity for diagenetic materials in the Hensel Sandstone (Cretaceous), Gillespie County, Texas. M, 1988, Stephen F. Austin State University. 136 p.

Smith, Charles Edward. The Genesee sub-stage; its characters and distribution. M, 1902, Cornell University.

Smith, Charles G., Jr. Geohydrology of the shallow aquifers of Baton Rouge, Louisiana. M, 1969, Louisiana State University.

Smith, Charles I. Stratigraphy of the upper Austin in the vicinity of Dallas, Texas. M, 1955, Louisiana State University.

Smith, Charles Isaac. Physical stratigraphy and facies analysis, Lower Cretaceous formations, northern Coahuila, Mexico. D, 1966, University of Michigan. 217 p.

Smith, Charles Randy. Provenance and depositional environments of the La Casita Formation, Sierra Madre Oriental, Southwest of Monterrey, northeastern Mexico. M, 1987, University of New Orleans. 143 p.

Smith, Chester Martin, Jr. Distribution of alpha tracks in auto radiographs of some uraniferous base metal sulfide minerals. M, 1959, Pennsylvania State University, University Park. 58 p.

Smith, Chester Martin, Jr. Quantitative petrographic comparison of the Bradford Third and Lewis Runs sands. D, 1964, Pennsylvania State University, University Park. 121 p.

Smith, Christian S. On the electrical evaluation of three southern New Mexico geothermal areas. M, 1977, University of New Mexico. 113 p.

Smith, Christopher R. Hydrogeology of the Carbondale and Raccoon Creek groups, Pennsylvanian System, Vigo, Clay and Sullivan counties, Indiana. M, 1983, Indiana University, Bloomington. 93 p.

Smith, Christy Harvey L. Sedimentology of the Late Cretaceous (Santonian - Maestrichtian) Tres Pasos Formation, Ultima Esperanza District, southern Chile. M, 1977, University of Wisconsin-Madison.

Smith, Cindy Lynn. Natural ground water systems and associated freshwater carbonate deposits in southern Oklahoma. M, 1984, Oklahoma State University. 153 p.

Smith, Clay T. Geology and ore deposits of the northeast quarter of the Seiad Quadrangle, California. M, 1940, California Institute of Technology. 49 p.

Smith, Clay T. The biostratigraphy of Glyceramis Vegatchii in California. D, 1943, California Institute of Technology. 40 p.

Smith, Clay T. The origin of some chromite deposits in the Pacific Coast region. D, 1943, California Institute of Technology. 86 p.

Smith, Cleon Verl. Geology of the North Canyon area, southern Wasatch Mountains, Utah. M, 1956, Brigham Young University. 32 p.

Smith, Clyde Louis. Geology of eastern Mount Bennett Hills, Camas, Gooding, and Lincoln counties, Idaho. D, 1966, University of Idaho. 129 p.

Smith, Clyde Louis. Stratigraphy of the Red Mountain Formation (Lower Pennsylvanian?) of northwestern Washington. M, 1961, University of British Columbia. 96 p.

Smith, Clyde Moffett. The relations of the Turor Diorite and the adjoining amphibolites. M, 1928, University of Illinois, Chicago.

Smith, Cole L. Chemical controls on weathering and trace metal distribution at Teels Marsh, Nevada. D, 1974, University of Wyoming. 108 p.

Smith, Cole L. Design and testing of a downhole continuous wave generator. M, 1968, University of Missouri, Rolla.

Smith, Constant Ann. Echinoderm systematics and paleoecology, Lexington Limestone and parts of the Clays Ferry Formation (Middle Ordovician) central Kentucky. M, 1986, University of Kentucky. 99 p.

Smith, Corilss M., Jr. Public relations of the minerals industries. M, 1980, University of Nevada. 72 p.

Smith, Craig Anthony. Clay size mineral variations and their relationship to natural gas migration and accumulation in a section of the Forbes Formation, Sacramento Valley, California. M, 1986, San Diego State University. 102 p.

Smith, Craig B. Kimberlites and mantle derived xenoliths at Iron Mountain, Wyoming. M, 1977, Colorado State University. 229 p.

Smith, D. E. A study of the silica in some Iowa diatoms. M, 1970, Iowa State University of Science and Technology.

Smith, D. L. Modern and fossil diatoms of Porters Lake, Nova Scotia; a postglacial history with special reference to marine intrusions. M, 1984, Dalhousie University. 166 p.

Smith, D. P. The effects of urbanization on water chemistry and sediment geochemistry in parts of the Grand River basin, southern Ontario. M, 1974, University of Waterloo.

Smith, Dan Howard. Origin and development of beach cusps at Monterey Bay, California. M, 1973, United States Naval Academy.

Smith, Dana K. A comparison of Sattlegger and Raymap migration algorithms. M, 1987, [Southern Methodist University].

Smith, Daniel. Interactive seismic processing parameter selection. M, 1986, University of Houston.

Smith, Daniel T. A gravity investigation of northern Alaska with emphasis on the Wiseman Quadrangle and adjacent Dalton Highway. M, 1986, University of Alaska, Fairbanks. 116 p.

Smith, David A. Hydrodynamic flow in Lower Cretaceous Muddy Sandstone, Rozet Field, Powder River basin, Wyoming. M, 1983, Texas A&M University.

Smith, David A. Lead-zinc mineralization in the Ponca-Boxley area, Arkansas. M, 1978, University of Arkansas, Fayetteville.

Smith, David A. Relationship between the flow regime and nitrogen species in a waste-sludge disposal pond, Hillsborough County, Florida. M, 1981, University of South Florida, Tampa. 73 p.

Smith, David B. Leachability of uranium and other elements from freshly erupted volcanic ash and water-soluble material on aerosols collected within volcanic eruption clouds. D, 1980, Colorado School of Mines. 182 p.

Smith, David Burl. Physical erosion and denudation rates in Cartwright Basin and vicinity, Williamson County, Tennessee. M, 1972, Vanderbilt University.

Smith, David Duane. Development of a method to calculated sand failure conditions. M, 1979, Stanford University.

Smith, David Dwyer. The geomorphology of part of the San Francisco Peninsula, California. D, 1960, Stanford University. 433 p.

Smith, David G. Pleistocene geology and geomorphology of the San Pedro River valley, Cochise County, Arizona. M, 1963, University of Arizona.

Smith, David K. Mesozoic geology northwest of Holcomb Valley, San Bernardino Mountains, California. M, 1982, University of California, Riverside. 87 p.

Smith, David L. Analysis of the 15° finite-difference wave equation; applications to seismic migration. M, 1983, Indiana University, Bloomington. 169 p.

Smith, David Lawrence. First order drainage basin morphology in a portion of the Sierra Nevada, California. D, 1966, University of Oregon. 183 p.

Smith, David M. A diatom abundance stratigraphy and dispersed ash tephrochronology for the South Atlantic sector of the Southern Ocean. M, 1982, University of Georgia.

Smith, David P. Paleontology and paleoecology of the basal New Providence Shale (Osagian; Mississippian) at Paris Landing, Tennessee. M, 1978, Indiana University, Bloomington. 173 p.

Smith, David Robertson. Quartz grain surface microtextures from sediments of the Santa Barbara littoral cells, California. M, 1979, University of Southern California.

Smith, David S. Geology of the eastern half of the Tecpan Guatemala Quadrangle, Guatemala, Central America. M, 1981, University of South Florida, Tampa.

Smith, Deane Kingsley, Jr. The crystal structure of uranophane $Ca(H_2O)(UO_2)(SiO_4)_2·3H_2O$. D, 1956, University of Minnesota, Minneapolis. 45 p.

Smith, Deborah Kay. The statistics of seamount populations in the Pacific Ocean. D, 1985, University of California, San Diego. 232 p.

Smith, Denver Jeter. Miocene foraminifera of the Harang sediments of southern Louisiana. M, 1948, Louisiana State University.

Smith, Derald Glen. Aggradation and channel braiding in the North Saskatchewan River, Alberta, Canada. D, 1973, The Johns Hopkins University.

Smith, Diane M. Inventory and assessment of the disposal of coal slurry and mine drainage precipitate wastes into underground coal mines in West Virginia. M, 1987, West Virginia University. 404 p.

Smith, Diane Ruth. The mineralogy and phase chemistry of silicic tephras erupted from Mount St. Helens volcano, Washington. M, 1980, Rice University. 158 p.

Smith, Diane Ruth. The petrology and geochemistry of High Cascade volcanics in southern Washington; Mount St. Helens Volcano and the Indian Heaven basalt field. D, 1984, Rice University. 423 p.

Smith, Donald Allen. Hydrogeology in the vicinity of Juliaetta, Idaho. M, 1984, University of Idaho. 134 p.

Smith, Donald Eugene. Application of a combined seismic refraction and reflection technique to delineate coal in southeastern Ohio. M, 1982, Ohio University, Athens. 144 p.

Smith, Donald L. Geology of northeast McCurtain and southeast LeFlore counties, Oklahoma. M, 1967, University of Wisconsin-Madison.

Smith, Donald L. Late Paleozoic reef structure and sedimentation of the Greenland area, Washington County, Arkansas. M, 1962, University of Arkansas, Fayetteville.

Smith, Donald Laurence. Stratigraphy and carbonate petrology of the Mississippian Lodgepole Formation in central Montana. D, 1972, University of Montana. 143 p.

Smith, Donald Leigh. A lithologic study of the (Ordovician) Stony Mountain and (Silurian) Stonewall Formations in southern Manitoba. M, 1963, University of Manitoba.

Smith, Donald Leith. The Tippecanoe Sequence (Ordovician) in Western North America. D, 1966, University of Washington. 83 p.

Smith, Donnie Fay. Petrography of the Newala Limestone (Lower Ordovician) in the vicinity of Shelby and Bibb counties, Alabama. M, 1975, University of New Orleans.

Smith, Dorian Glenys Whitney. Lower Devonian bentonites from Gaspe, P.Q. M, 1960, University of Alberta. 158 p.

Smith, Douglas. Mineralogy and petrology of an olivine diabase sill complex and associated unusually potassic granophyres, Sierra Ancha, central Arizona. D, 1969, California Institute of Technology. 345 p.

Smith, Douglas Lee. The vertical distribution of heat production and heat flow in northwestern Mexico. D, 1972, University of Minnesota, Minneapolis. 189 p.

Smith, Douglas Michael. Methane diffusion and desorption in coal. D, 1982, [University of New Mexico]. 206 p.

Smith, Duane D. Geology of the northeast quarter of Carrizo Mountain Quadrangle. M, 1962, University of Southern California.

Smith, Duncan Ross. Physical and chemical studies of the sialic volcanic breccias, Kakagi Lake, Ontario. M, 1971, McMaster University. 79 p.

Smith, Earl Winston. Subsurface geology of eastern Kay County, Oklahoma, and southern Cowley County, Kansas. M, 1954, University of Oklahoma. 59 p.

Smith, Edgar E. N. Metamorphism of the Contact Lake area, Saskatchewan. M, 1949, Northwestern University.

Smith, Edgar Ernest Norval. Structure, wall rock alteration, and ore deposits at Martin Lake, Saskatchewan. D, 1952, Harvard University.

Smith, Edgar M. Exploration for a buried valley by resistivity and thermal probe surveys. M, 1971, Miami University (Ohio). 101 p.

Smith, Edward James, Jr. Stratigraphic relationship of the Jackson and Catahoula formations in Brazos County, Texas. M, 1942, Texas A&M University.

Smith, Edward Thornton, Jr. Geology of the Gideon Area, Cherokee County, Oklahoma. M, 1952, University of Oklahoma. 117 p.

Smith, Elizabeth J. Paleoecologic aspects of modern macroinvertebrate communities of southern Laguna Madre, Texas. M, 1985, Stephen F. Austin State University. 77 p.

Smith, Eric C. The nature of mercury anomalies at the New Calumet mines area, Quebec. M, 1971, McGill University.

Smith, Ernest Marshall, Jr. Fractures in the Carolina Slates and nearby areas in North Carolina. M, 1951, University of North Carolina, Chapel Hill. 25 p.

Smith, Essie Alma. A quantitative study of the variation of Pentremites conoideus. M, 1904, Indiana University, Bloomington.

Smith, Ethan Timothy. Mathematical models for environmental quality management. D, 1974, Rutgers, The State University, New Brunswick. 380 p.

Smith, Eugene Irwin. Comparison of selected lunar and terrestrial volcanic domes. D, 1970, University of New Mexico. 200 p.

Smith, Eugene Irwin. Criteria for the determination of flow direction in volcanic rocks. M, 1968, University of New Mexico. 118 p.

Smith, Eugene L. Geology and geochemical soil survey of the Tunnell Ranch Mine area, (Santa Barbara County), California. M, 1969, San Jose State University. 62 p.

Smith, Everett Newman. Late Quaternary vegetational history at Cupola Pond, Ozark National Scenic Riverways, southeastern Missouri. M, 1984, University of Tennessee, Knoxville. 115 p.

Smith, F. Gordon. Experiments on the transportation and deposition of sulphides in alkaline sulphide solutions. M, 1939, University of Manitoba.

Smith, Foster D., Jr. Cyclic sedimentation, Oficina Formation (Oligocene and lower Miocene), Venezuela. D, 1960, New York University.

Smith, Francis deSales. A fluid inclusion study of the Burlington Limestone (Mississippian), southeastern

Iowa and western Illinois. M, 1984, SUNY at Stony Brook. 210 p.

Smith, Frank C. Geology, mineralization, and exploration potential of the McGhee Peak area, San Simon mining district, Hidalgo County, New Mexico. M, 1987, University of New Mexico. 176 p.

Smith, Frederick E. Some of the foraminifera of the Vicksburg (Oligocene) Group in Louisiana. M, 1934, Louisiana State University.

Smith, Frederick J. Mineralization of the Boss Bixby anomaly, Iron and Dent counties, Missouri. M, 1968, University of Missouri, Rolla.

Smith, Freeman M. A volumetric-threshold infiltration model. D, 1971, Colorado State University.

Smith, G. P. Geology and structure study of the Meguma Series (Ordovician) in the Bedford area. M, 1951, Acadia University.

Smith, Garon Corder. Quantification of metal ion complexation in multiligand mixtures. D, 1983, Colorado School of Mines. 687 p.

Smith, Gary Allen. Stratigraphy, sedimentology, and petrology of Neogene rocks in the Deschutes Basin, central Oregon; a record of continental-margin volcanism and its influence on fluvial sedimentation in an arc-adjacent basin. D, 1986, Oregon State University. 467 p.

Smith, Gary B. Some effects of sewage discharge to the marine environment. D, 1974, University of California, San Diego. 351 p.

Smith, Gary E. Depositional systems and facies control of copper mineralization - San Angelo Formation (Permian), north Texas. M, 1974, University of Texas, Austin.

Smith, Gary Parker. Stratigraphy and paleontology of the Lower Devonian sequence, Southwest Ellesmere Island, Canadian Arctic Archipelago. D, 1984, McGill University. 513 p.

Smith, Geoffrey Wayne. Surficial geology of the Shuswap River drainage, British Columbia, (Canada). D, 1969, Ohio State University. 207 p.

Smith, George E. The geology and ore deposits of the Mowry Mine area, Santa Cruz County, Arizona. M, 1956, University of Arizona.

Smith, George E., III. Lithostratigraphic relationships of coastal plain units in Lexington County and adjacent areas, South Carolina. M, 1979, University of South Carolina.

Smith, George F., Jr. Basal Windsor rocks (Upper Mississippian) of Antigonish County, Nova Scotia. M, 1956, Massachusetts Institute of Technology. 169 p.

Smith, George I. Geology of the Cache Creek region. M, 1951, California Institute of Technology. 72 p.

Smith, George Irving. Geology and petrology of the Lava Mountains, San Bernardino County, California. D, 1956, California Institute of Technology. 230 p.

Smith, George O. The geology of the Fox Islands, Maine. D, 1896, The Johns Hopkins University.

Smith, George Taylor. Sedimentary petrology of the Callville Limestone and Pakoon Formation (Pennsylvanian-Permian) at Iceberg Canyon, Clark County, Nevada. M, 1972, Memphis State University.

Smith, George Wendell. The Ferron Point and Genshaw formations in Cheboygan and western Presque Isle counties, Michigan. M, 1942, Michigan State University. 90 p.

Smith, Gerald Nelson. The Permian vertebrates of Oklahoma. M, 1927, University of Oklahoma. 46 p.

Smith, Gerald Ray. Distribution and evolution of the North American fishes of the subgenus Pantosteus. D, 1965, University of Michigan. 368 p.

Smith, Gilbert B. Gravity and magnetic based computer modeling of an area in west central Louisiana. M, 1984, University of Southwestern Louisiana.

Smith, Glen N. Mean gravity anomaly prediction from terrestrial gravity data and satellite altimetry data. D, 1974, Ohio State University. 150 p.

Smith, Glenn Allen. A palynological investigation of Eocene localities in Northern California. M, 1987, University of California, Davis. 130 p.

Smith, Glenn Scott. Seismic velocity studies of liquid-saturated sand. M, 1957, University of Utah. 37 p.

Smith, Gordon Egbert. A focal mechanism study using both P-wave first motions and S-wave polarization angles. M, 1980, Georgia Institute of Technology. 128 p.

Smith, Gordon McNeal, IV. An investigation of iron removal processes in the Merrimack River estuary, Massachusetts. M, 1982, University of New Hampshire. 85 p.

Smith, Grant McKay. Geology of Southwest Lake Mountain, Utah. M, 1951, Brigham Young University. 40 p.

Smith, Grant Sackett. Paleoenvironmental reconstruction of Eocene fossil soils from the Clarno Formation in eastern Oregon. M, 1988, University of Oregon. 167 p.

Smith, Gregory Alan. In situ orientation and calibration of three-component downhole seismometers. M, 1988, Texas A&M University. 92 p.

Smith, Gregory John. Finite element pre-processor for hydrogeologic modelling. M, 1984, University of Alberta. 204 p.

Smith, Gregory Ogden. Stratigraphy and sedimentology of upper Cretaceous and upper Miocene red bed conglomerates and associated shallow marine rocks in the Sierra Madre Range, northern Santa Barbara County, California. M, 1985, University of California, Santa Barbara. 115 p.

Smith, Gregory Paul. Interpretation of imagery lineations, Logan Creek area, Southeast Missouri. M, 1973, University of Missouri, Rolla.

Smith, Gregory Warren. Stratigraphy, sedimentology, and petrology of the Cambria Slab, San Luis Obispo County, California. M, 1978, University of New Mexico. 123 p.

Smith, Guy Michael. Some applications of magnetic resonance spectroscopy to rock magnetism. D, 1981, University of Washington. 270 p.

Smith, Guy William. Molt stages of Geisina unispinosa n. sp. from southwestern Colorado. M, 1953, University of Illinois, Urbana.

Smith, Hampton. Origin of some of the siliceous Miocene rocks of California. D, 1934, California Institute of Technology. 105 p.

Smith, Hampton. The stratigraphic position of some of the diatomite horizons in the Los Angeles Basin. D, 1934, California Institute of Technology. 3 p.

Smith, Harold T. U. The Tertiary and Quaternary geology of the Abiquiu Quadrangle, New Mexico. D, 1936, Harvard University.

Smith, Harris Theodore. A statistical model for determining the sediment yields from urban and rural landscapes along Breakneck Creek, Portage and Stark counties, Ohio. M, 1974, Kent State University, Kent. 60 p.

Smith, Harry Dean. An experimental study of the diffusion of Na, K, and Rb in magmatic silicate liquids. D, 1973, University of Oregon. 207 p.

Smith, Harry Lee. Cretaceous stratigraphy of Carrizo drainage basin, Apache County, Arizona, and Catron and Valencia counties, New Mexico. M, 1956, University of Texas, Austin.

Smith, Helen V. The fossil flora of Rockville, Oregon. M, 1932, University of Oregon. 44 p.

Smith, Henry Carl, Jr. Sedimentary fabrics in unconsolidated sands. M, 1952, Michigan State University. 30 p.

Smith, Henry H. Development of a methodology for integration of water from several limited sources in the Caribbean islands. D, 1985, Colorado State University. 177 p.

Smith, Homer J. Fauna of the Buckhorn Asphalt. D, 1935, University of Chicago. 68 p.

Smith, Hugh Preston. Foraminifera of the Wagonwheel Formation. M, 1950, University of California, Berkeley. 93 p.

Smith, Hugh Preston. The Thaynes Formation (lower Triassic) of the Moenkopi group (lower and middle Triassic), north-central Utah. D, 1969, University of Utah. 378 p.

Smith, Ian S. Petrographic investigation of a granitic pegmatite (Grenville province) in the Bancroft area (Saranac Uranium mines Ltd., northeast of Troy Hill), Ontario, Canada. M, 1969, Slippery Rock University. 74 p.

Smith, Isabel Fothergill. Anorthosite in the Piedmont Province of Pa. D, 1923, Bryn Mawr College. 42 p.

Smith, J. B. A finite element model for liquid waste movement in a two-dimensional nonhomogeneous groundwater system. M, 1973, University of Waterloo.

Smith, J. Fred, Jr. Geology of the Devil Ridge area, Hudspeth County, Texas. D, 1939, Harvard University.

Smith, J. L. A stochastic analysis of steady-state groundwater flow in a bounded domain. D, 1978, University of British Columbia.

Smith, J. R. Areal geology of the McKenzie Township (south half), Chibougamau region, 1 inch to 1000 feet (Quebec). D, 1953, [Princeton University].

Smith, J. R. Geology of Montauban-Les Mines mineralized area, Quebec. M, 1950, Universite Laval.

Smith, J. R. Petrography of the Frontier Formation in the southeastern Bighorn Basin, Hot Springs and Washakie counties, Wyoming. M, 1957, University of Wyoming. 62 p.

Smith, James Allen. Insoluble residue study of Pennsylvanian strata exposed in San Juan Canyon, San Juan County, Utah. M, 1957, University of New Mexico. 84 p.

Smith, James August. The Moodys Branch Marl of Mississippi. M, 1952, Mississippi State University. 117 p.

Smith, James B. Preliminary investigations of the association of organic material and carbon dioxide with sedimentary particles. D, 1961, Texas A&M University.

Smith, James C. The depositional environment of the Mariah Hill coal seam (Dubois County, Indiana) and its correlation with coal property parameters. M, 1985, University of Cincinnati. 71 p.

Smith, James D. The dynamics of sand waves and sand ridges. D, 1968, University of Chicago. 78 p.

Smith, James Douglas. Depositional environments of the Tertiary Colton and Basal Green River formations in Emma Par, Utah. M, 1986, Brigham Young University. 174 p.

Smith, James Dungan. A study of shell orientation by tidal currents in Barnstable Harbor, Massachusetts. M, 1963, Brown University.

Smith, James George. An integrated geophysical study of the Grenville Front in Lake Huron. M, 1988, Purdue University. 122 p.

Smith, James Gordon, II. Petrology of the southern Pine Forest Range, Humboldt County, Nevada. D, 1966, Stanford University. 150 p.

Smith, James K. Ostracoda of the Prairie Bluff chalk (Cretaceous) and the Pine Barren member of the Clayton Formation (Paleocene), Lowndes County, Alabama. M, 1969, University of Alabama.

Smith, James Mitchell. A study of the Wisconsin glaciation in southeastern Indiana and southwestern Ohio. M, 1949, Miami University (Ohio). 78 p.

Smith, James Richard. Independent yields from the photofission of [232]Th. D, 1986, New Mexico State University, Las Cruces. 150 p.

Smith, James Robert. Low temperature plagioclases. D, 1954, Princeton University. 73 p.

Smith, James Ronald. Shallow igneous intrusion; Covington, Tennessee, as known from gravity and magnetic data. M, 1974, Vanderbilt University.

Smith, James T. Ground water resources of Big Blue and Kansas River valleys from Manhattan to Wamego, Kansas. M, 1959, Kansas State University. 74 p.

Smith, James Thomas. An interpretation of gravity anomalies of northeastern Utah. M, 1973, University of Utah. 74 p.

Smith, James Wiliam. The Saltville fault near Mooresburg, Tennessee. D, 1968, University of Tennessee, Knoxville. 149 p.

Smith, James William. Geology of an area along the Cartersville Fault near Fairmount, Georgia. M, 1959, Emory University. 41 p.

Smith, Jan G. The geology of the Clear Creek area, Montana-Idaho. M, 1961, Pennsylvania State University, University Park. 75 p.

Smith, Jan H. Tensile strength symmetries of polycrystalline freshwater ice. M, 1979, University of Wisconsin-Milwaukee. 78 p.

Smith, Jane Elizabeth Inch. Ostracods from the Middle Devonian Traverse Group of Emmet and Charlevoix counties of Michigan. D, 1959, University of Michigan. 224 p.

Smith, Janet Yvonne. On the fate of dissolved aromatic hydrocarbons in groundwater flow systems; experimental determination of absorption from water on quartz aquifer materials and computer simulated solute transport with dispersion from instantaneous point sources in confined aquifers. M, 1986, University of Minnesota, Minneapolis. 200 p.

Smith, Jeffrey William. The petrology of the Mississippian Redwall Limestone in northern Yavapai County, Arizona. M, 1974, Northern Arizona University. 88 p.

Smith, Jeffry A. The petrology of the Solsville and Pecksport members of the Marcellus Formation (middle Devonian) and the Ashokan and Plattekill members of the Skaneateles Formation (middle Devonian) in the Catskill Front, southeastern New York state. M, 1970, Rensselaer Polytechnic Institute. 45 p.

Smith, Jennifer Margaret. Biodeposition by and paleoecological significance of the ribbed mussel Geukensia demissa in a salt marsh, Sapelo Island, Georgia. D, 1983, University of Georgia. 187 p.

Smith, Jennifer Margaret. P, T, and relative timing of metamorphism in the aureole around the Anvil Batholith, south central Yukon. M, 1988, University of Alberta. 121 p.

Smith, Jeremy Torquil. Rapid inversion of multi-dimensional magnetotelluric data. D, 1988, University of Washington. 160 p.

Smith, Jerry D. The Blake event; a brief late Pleistocene geomagnetic reversal. M, 1969, Columbia University. 26 p.

Smith, Jerry P. Petrography of Brite Ignimbrite (Tertiary Vieja Group), Trans-Pecos Texas. M, 1967, Kansas State University.

Smith, Joe Earl. Areal geologic map of southeastern Williamson County, Texas. M, 1949, University of Texas, Austin.

Smith, John Alan. Tidal fluctuations of the Florida current (Summer). M, 1968, [University of Miami].

Smith, John C. Coarse clay-fine clay ratios across the Mesozoic-Cenozoic contact in Falls, Milam, and Travis counties, Texas. M, 1966, Texas A&M University. 78 p.

Smith, John C. Structural geology and mineralogy of Keymet Mines Limited, Petit Rocher Nord, Gloucester County, New Brunswick. M, 1954, University of New Brunswick.

Smith, John E. A real and stratigraphic geology of the Waukee, Iowa Quadrangle. M, 1911, Iowa State University of Science and Technology.

Smith, John Livingstone. Petrology, mineralogy, and chemistry of the Tobacco Root Batholith, Madison County, Montana. D, 1970, Indiana University, Bloomington. 164 p.

Smith, John Millard. The mineralogy of some glacial lake clays. M, 1956, Indiana University, Bloomington. 32 p.

Smith, John Peter. Foraminifera of the (Upper Cretaceous) Taylor and Navarro formations (Texas). M, 1931, Texas Christian University. 118 p.

Smith, Joseph Blake and Potter, Lloyd Dean. Some uranium mines with production in the Black Hills of Wyoming and South Dakota. M, 1958, South Dakota School of Mines & Technology.

Smith, Joseph Thurston, Jr. Areal geology of Candelaria area, Presidio County, Trans-Pecos, Texas. M, 1956, University of Texas, Austin.

Smith, Joseph Walter. The Kessler Limestone of Northwest Arkansas. M, 1951, University of Arkansas, Fayetteville.

Smith, Joyceanne. A study of fixtures, structures and fabrics of two examples of alpine peridotite. M, 1978, University of Toronto.

Smith, Judith Marilyn. The bedrock geology of the vicinity of Torrington, Connecticut. M, 1960, University of Wisconsin-Madison.

Smith, K. E. A study of the silica in some Iowa diatoms. M, 1970, Iowa State University of Science and Technology.

Smith, Karl J. A gravity and tectonic study of the southwestern portion of the Ouachita System. M, 1986, University of Texas at El Paso.

Smith, Kathleen S. Adsorption of copper and lead onto goethite as a function of pH, ionic strength, and metal and total carbonate concentrations. M, 1986, Colorado School of Mines. 117 p.

Smith, Kelsey Anne. Normal faulting in an extensional domain; constraints from seismic reflection interpretation and modeling. M, 1984, University of Utah. 165 p.

Smith, Kenneth G. A standard for grading topographic texture. M, 1949, Columbia University, Teachers College.

Smith, Kenneth G. Chemical equilibrium in some metamorphosed iron-rich sediments (Ordovician). M, 1969, University of Kentucky. 116 p.

Smith, Kenneth G. Erosional processes and landforms in Badlands National Monument, South Dakota. D, 1953, Columbia University, Teachers College.

Smith, Kent W. Structure and petrology of Tertiary volcanic rocks near Etna, Utah. M, 1980, Utah State University. 83 p.

Smith, Kevin. Petrology and origin of Precambrian metamorphic rocks in the eastern Ruby Mountains, southwestern Montana. M, 1980, University of Montana. 84 p.

Smith, L. B., Jr. Submarine geomorphology and sedimentation patterns of the Gyre intraslope basin, Northwest Gulf of Mexico. M, 1975, Texas A&M University.

Smith, L. C. Stratigraphy of Red Mountain Formation (Lower Pennsylvanian?) of northwestern Washington. M, 1961, University of British Columbia.

Smith, Larry. The interaction of an ice sheet with harbor structures in McKinley Marina, Milwaukee, Wisconsin. M, 1981, University of Wisconsin-Milwaukee. 112 p.

Smith, Larry B. Geology and uranium geochemistry of the western margin of the Thirtynine Mile volcanic field, Park, Chaffee and Fremont counties, Colorado. M, 1982, Colorado School of Mines. 355 p.

Smith, Larry Noel. Basin analysis of the lower Eocene San Juan Formation, San Juan Basin, New Mexico and Colorado. D, 1988, University of New Mexico. 166 p.

Smith, Larry S. Paleoenvironments of the upper Entrada Sandstone and the Curtis formations on the west flank of the San Rafael Swell, Emery County, Utah. M, 1976, Brigham Young University.

Smith, Larry Warren. Depositional setting and stratigraphy of the Tuscaloosa Formation, central Alabama to west-central Georgia. M, 1984, Auburn University. 139 p.

Smith, Laura Ann. Micritization by boring, infilling and primary micritic carbonate in modern and ancient

carbonate allochems. M, 1985, University of Illinois, Urbana. 50 p.

Smith, Laurence Lowe. The French Creek magnetite deposits of southeastern Penn. D, 1924, The Johns Hopkins University.

Smith, Laurence Lowe. The stratigraphy, structure and physiography of the Delaware Valley from the water gap to Kintnersville, Pennsylvania. M, 1919, Lafayette College.

Smith, Laurence Lynwood. Physiography of South Carolina. M, 1933, University of South Carolina. 43 p.

Smith, Lawrence Noel. Late Cenozoic fluvial evolution in the northern Chaco River drainage basin, northwestern New Mexico. M, 1983, University of New Mexico. 209 p.

Smith, Lawrence Ralph. Niagaran (Middle Silurian) reef trends of southcentral Michigan. M, 1977, University of Michigan.

Smith, Lawson Mottley. Geomorphic development of alluvial fans in the Yazoo Basin, northwestern Mississippi. D, 1983, University of Illinois, Urbana. 260 p.

Smith, LeBrun N. The Eocene stratigraphy of central South Carolina. M, 1957, University of South Carolina. 67 p.

Smith, Lee Anderson. Biostratigraphy of late Tertiary and Quaternary subsurface deposits of southern Louisiana. D, 1962, Stanford University. 177 p.

Smith, Lee Anderson. Recognition of Plio-Miocene environments from megafaunas in well cuttings of southern Louisiana. M, 1959, Texas Christian University. 75 p.

Smith, LeRoy W. An examination of the petrographic types of the Ely Greenstone (Precambrian) occurring in the area of Twin Lakes, Minnesota. M, 1968, Michigan State University. 57 p.

Smith, Lester B., Jr. Submarine geomorphology and sedimentation patterns of the Tyre intraslope basin, Northwest Gulf of Mexico. M, 1965, Texas A&M University.

Smith, Lewis A. The geology of the Commonwealth Mine. M, 1927, University of Arizona.

Smith, Lewis B. An investigation of the Lake City District, Colorado, with mill for treatment of ore from Cleveland Prospect. M, 1915, University of Kansas.

Smith, Linda. Lithostratigraphic correlation, petrology and paleoenvironment of the Ordovician Simpson Group, Golden Trend area, south-central Oklahoma. M, 1988, University of Southern California.

Smith, Linda F. The crystal structure of clinoptilolite. D, 1981, Colorado School of Mines. 211 p.

Smith, Lloyd Beecher. Peculiar jointage caused by a peridotite dike in Fayette County, Pennsylvania. M, 1913, Pennsylvania State University, University Park.

Smith, M. K. Filter theory of linear operators with seismic applications. D, 1954, Massachusetts Institute of Technology. 94 p.

Smith, Mackey. Stratigraphy and chronology of the Tacoma area, Washington. M, 1972, Western Washington University. 38 p.

Smith, Mahlon. Geology of the Matson-Defiance area in St. Charles County, Missouri. M, 1958, Washington University. 58 p.

Smith, March E. Geology of the northern half of Mt. Pleasant Quadrangle, Arkansas, with emphasis on the economic aspects of the St. Peter Sandstone. M, 1976, Northeast Louisiana University.

Smith, Marguerite Adelle. Queenston Formation (Upper Ordovician) of western New York. M, 1938, University of Rochester. 141 p.

Smith, Marian McNally. The effect of sand surface texture of the primary recovery of bitumen from the Athabasca Oil Sands. D, 1986, University of South Carolina. 127 p.

Smith, Marie Theresa. Recurring shoaling-upward sequences of a carbonate tidal flat, Cool Creek Formation (Lower Ordovician), Arbuckle Mountains, Oklahoma. M, 1980, University of Wisconsin-Madison.

Smith, Mark A. Analysis of gravity data from the Verde Valley, Yavapai County, Arizona. M, 1984, Northern Arizona University. 59 p.

Smith, Mark E. A method for predicting the impact of urban growth on flood expectancy in southeastern New England. D, 1978, Colorado State University. 10 p.

Smith, Mark Francis. Rare-earth elements in the Adirondack Tupper-Saranac Mangerite. M, 1976, Iowa State University of Science and Technology.

Smith, Mark Newton. Depositional environment and diagenesis of the upper member of the Smackover Formation (Upper Jurassic), Keoun Creek Field, Columbia and Lafayette counties, Arkansas. M, 1986, Northeast Louisiana University. 73 p.

Smith, Martin L. The normal modes of a rotating, elliptical Earth. D, 1974, Princeton University. 155 p.

Smith, Marvin Lyle. Ostracodes from Middle Devonian Ledyard, Wanakah, and Tichenor shales of western New York. M, 1952, University of Rochester. 126 p.

Smith, Mary Ann Hrivnak. Estimation of Venus wind velocities from high-resolution infrared spectra. D, 1978, University of Chicago. 122 p.

Smith, Matthew Clay. Measurement and prediction of herbicide transport into shallow groundwater. D, 1988, University of Florida. 214 p.

Smith, Maurice Harold. Structure contour map of the pre-Pennsylvanian surface in Illinois (Sm621). M, 1941, University of Illinois, Urbana. 27 p.

Smith, Melvin Owen. Stratigraphy and structure of the Middle Ordovician rocks in the vicinity of Calhoun, McMinn County, Tennessee. D, 1976, University of Tennessee, Knoxville. 155 p.

Smith, Melvin Owen. The problem of leaking farm ponds and their relation to stratigraphic zones and other factors in the inner Blue Grass region of Kentucky. M, 1964, University of Kentucky. 67 p.

Smith, Merlin C., Jr. An investigation into feasible methods of mean evaluation determination for geodetic applications. M, 1963, Ohio State University.

Smith, Merritt B. Ground water in the Livermore Valley, California. M, 1934, Stanford University. 83 p.

Smith, Michael. Lower Tertiary planktonic biostratigraphy, California Aqueduct section, northwestern Kern County, California. M, 1974, California State University, Fresno.

Smith, Michael A. Geology and trace element geochemistry of the Fort Davis Mountains, Trans-Pecos, Texas. D, 1975, University of Texas, Austin. 255 p.

Smith, Michael A. Selectivity of adventitious test material by Reophax curtus (foraminifera). M, 1969, University of Kansas. 26 p.

Smith, Michael Forrest. Geochemical constraints on alteration in the Galapagos fossil hydrothermal system. M, 1987, University of Florida. 200 p.

Smith, Michael J. A gravity survey of northeastern Jackson County, Illinois. M, 1975, Southern Illinois University, Carbondale. 57 p.

Smith, Michael John. Marine geology of the northwestern Weddell Sea and adjacent coastal fjords and bays; implications for glacial history. M, 1985, Rice University. 157 p.

Smith, Michael L. Coastal warm spring systems along northeastern Baja California. M, 1978, San Diego State University.

Smith, Michael Roy. Geology and mineralization of the southeastern Gillis Range, Mineral County, Nevada. M, 1981, University of Nevada. 103 p.

Smith, Michael William. Evaluating ground-water recharge potential in the Valley and Ridge region of the Central Appalachians using a digital-overlay technique. M, 1986, Pennsylvania State University, University Park. 289 p.

Smith, Michael William. Factors affecting the distribution of permafrost Mackenzie Delta N.W.T. D, 1973, University of British Columbia.

Smith, Michael William. Short term environmental effects of dredging between Stingaree Cove and Big Pasture Bayou, Bolivar Peninsula, Galveston County, Texas. M, 1972, University of Texas, Arlington. 157 p.

Smith, Moira T. Structure and petrology of the Grandy Ridge-Lake Shannon area, North Cascades, Washington. M, 1986, Western Washington University. 156 2 plates p.

Smith, Morland E. Element distribution between co-existing feldspars in high-grade metamorphic rocks. D, 1966, Queen's University. 154 p.

Smith, Muriel C. The Long Island Sound sub-bottom topography in the area between 73°00'W and 73°30'W. M, 1963, Columbia University, Teachers College.

Smith, N. S. Burden-rock stiffness and its effect on fragmentation in bench blasting. D, 1976, University of Missouri, Rolla. 164 p.

Smith, Natasha D. Muscular and ligamentary attachments in the feet of Tertiary and Recent horses, with special reference to evolutionary development. M, 1934, University of California, Berkeley. 15 p.

Smith, Nathaniel Greene. Stratigraphy, petrography and geochemistry of Upper Devonian black shales, Gataga District, north-central British Columbia. M, 1985, University of Texas, Austin. 108 p.

Smith, Neal Johnstone. The geology of a part of the Carbona Quadrangle, California. M, 1936, University of California, Berkeley. 45 p.

Smith, Ned Myron. Applied sedimentology of the Salem Limestone. D, 1962, Indiana University, Bloomington. 127 p.

Smith, Ned Myron. Salem Limestone in the Bedford-Bloomington quarry belt. M, 1955, Indiana University, Bloomington. 118 p.

Smith, Norman Dwight. A stratigraphic and sedimentologic analysis of some Lower and Middle Silurian clastic rocks of the north-central Appalachians. D, 1967, Brown University. 204 p.

Smith, Norman H. Santa Clara gravels along Corte de Madera Creek. M, 1931, Stanford University.

Smith, Norman Walter. Engineering geology of Athens, south and east central areas. M, 1974, Ohio University, Athens. 43 p.

Smith, Ollie L., Jr. The geology of the Concord Quadrangle, Knox, Blount and Loudon counties, Tennessee. M, 1950, University of Tennessee, Knoxville. 50 p.

Smith, P. K. The structural geology of the southeastern part of the Kentville area (2IH/2E), Nova Scotia. M, 1976, Acadia University.

Smith, P. M. Geological setting, timing and controls of gold mineralization at the Duport Deposit, Shoal Lake, Ontario. M, 1987, University of Waterloo. 316 p.

Smith, Pamela Louise. Tectono-stratigraphic significance of the Oceanic Formation for the development of the central western Barbados Ridge. M, 1987, Cornell University. 164 p.

Smith, Patricia Gould. Stratigraphy and sedimentology of Devonian and Mississippian strata flanking the Homestake shear zone, northeastern Sawatch Uplift, Eagle County, Colorado. M, 1987, Colorado School of Mines. 132 p.

Smith, Patrick E. Rb-Sr whole rock and U-Pb zircon geochronology of the Michipicoten greenstone belt, Wawa area, northwestern Ontario. M, 1981, University of Windsor. 112 p.

Smith, Patrick Edmund. Uranium-thorium-lead geochronology of Archean rocks of the eastern Superior Province and application of initial lead and hafnium isotope ratios to greenstone belt. D, 1988, University of Toronto.

Smith, Patsy J. Recent foraminifera of Central America; quantitative and qualitative analysis of the subfamily Bolivinae. M, 1958, Pomona College.

Smith, Paul L. Biostratigraphy of the Snowshoe Formation (Jurassic) in the Izee area, Grant County, Oregon. M, 1976, Portland State University. 213 p.

Smith, Paul Laurence. Biostratigraphy and ammonoid fauna of the Lower Jurassic (Sinemurian, Pliensbachian and lowest Toarcian) of eastern Oregon and western Nevada. D, 1981, McMaster University. 368 p.

Smith, Peter H. Stratigraphy and structure of a part of the Hopewell area, Nova Scotia with emphasis on the Horton Group. M, 1962, Northwestern University.

Smith, Peter Henderson. The structure and petrology of the Basler-Eau Claire granite complex, District of Mackenzie, N.W.T., Canada. D, 1966, Northwestern University. 277 p.

Smith, Philip A. "Depositional edge" versus "outcrop edge"; a study of the coastal plain formations of New Jersey. M, 1971, Columbia University. 45 p.

Smith, Philip Alson. Discrimination and provenance of tills in northwestern Connecticut. D, 1985, McGill University. 324 p.

Smith, Philip C. Carbon-14 dates on a new sea level high stand from Marie-Galante, F.W.I. M, 1976, Brooklyn College (CUNY).

Smith, Philip Gerard. A hydrogeological survey of Belchertown, Massachusetts. M, 1975, University of Massachusetts. 68 p.

Smith, Phillip Sidney. The copper sulphide deposits of Orange County, Vermont. D, 1904, Harvard University.

Smith, R. H. Development of a dynamic flood routing model for small meandering rivers. D, 1978, University of Missouri, Rolla. 179 p.

Smith, R. S. Sedimentology of the deltas of the Moose and Diligent rivers, Minas Basin, Bay of Fundy, Nova Scotia. M, 1969, University of Pennsylvania.

Smith, Ralph Emerson. Studies of the Weno Formation north of the Brazos River. M, 1939, Texas Christian University. 77 p.

Smith, Ralph Ernest. Geology of the Mill Creek area, San Bernardino County, California. M, 1960, University of California, Los Angeles.

Smith, Randall. Gravity and magnetic anomaly studies in West Texas. M, 1981, Purdue University. 116 p.

Smith, Randall Blain. Geology of the Harper Ranch Group (Carboniferous-Permian) and Nicola Group (Upper Triassic) northeast of Kamloops, British Columbia. M, 1979, University of British Columbia.

Smith, Randall Blain. Sedimentology and tectonics of a Miocene collision complex and overlying late orogenic clastic strata, Buton Island, eastern Indonesia. D, 1982, University of California, Santa Cruz.

Smith, Randall William. Effects of imperfections on the magnetic properties of Fe_2O_3. D, 1968, University of Pittsburgh. 165 p.

Smith, Randall William. Reversible susceptibility and remanent magnetization of ferromagnetic minerals. M, 1965, University of Pittsburgh.

Smith, Raymond I. The geology of the northwest part of Snow Peak Quadrangle, Oregon. M, 1958, Oregon State University. 93 p.

Smith, Raymond James. Geology of portions of the Humphreys and Sylmar quadrangles, Los Angeles County, California. M, 1948, California Institute of Technology. 52 p.

Smith, Raymond James. The geology of the Los Teques-Cua region, Venezuela. D, 1951, Princeton University. 170 p.

Smith, Raymond Newton. Gennaeocrinus variabilis, a new crinoid species from the Middle Devonian Bell Shale of Michigan. M, 1961, University of Michigan.

Smith, Raymond Newton. Musculature and muscle scars of Chlamydotheca Arcuata (Sars) and Cypridopsis Vidua (O. F. Muller) (Ostracoda; Cyprididae). D, 1964, University of Michigan.

Smith, Rebecca Ann Pope. Geochemistry of volcanic ash in the Salt Lake Group, Bonneville Basin, Utah, Idaho, and Nevada. M, 1975, University of Utah. 93 p.

Smith, Richard Cronan. Drifting heat sources in a viscous fluid; a model for continental drift. D, 1972, University of California, Los Angeles. 111 p.

Smith, Richard Elbridge. Bottom sediments of western Lake Superior. M, 1960, University of Illinois, Urbana.

Smith, Richard Elbridge. Petrographic properties influencing porosity and permeability in the carbonate-quartz system as represented by the Gatesburg Formation. D, 1966, Pennsylvania State University, University Park. 220 p.

Smith, Richard J. The effect of damming on trace metal distribution in the Tennessee River-Four Loudon Lake system of East Tennessee. M, 1975, University of Tennessee, Knoxville. 49 p.

Smith, Richard L. Diagenesis of Miocene sandstones, South Louisiana. M, 1984, Texas A&M University. 153 p.

Smith, Richard M. Geology of the Buda-Kyle area, Hays County, Texas. M, 1978, University of Texas, Austin.

Smith, Richard P. Structure and petrology of Spanish Peaks dikes, south central Colorado. D, 1975, University of Colorado. 270 p.

Smith, Richard Paul. Structural geology of the Precambrian rocks of the Panorama Peak area, Larimer County, Colorado. M, 1968, University of Colorado.

Smith, Richard V. Provenance of mid-Atlantic continental margin sediments from the Cost B-2 test well. M, 1980, University of Delaware.

Smith, Richard Wellington. The geology and origin of the brown and blue phosphate rock deposits. M, 1926, Cornell University.

Smith, Ricky. The flora of the roof shales of the lower Hartshorne Coal, Scott County, Arkansas. M, 1987, Northeast Louisiana University. 57 p.

Smith, Riley Semour, Jr. A study of the Chinle-Shinarump beds in the Leupp-Holbrook area, Arizona. D, 1957, University of Arizona. 204 p.

Smith, Riley Seymour, Jr. Igneous rocks of the Snyder Lake area. M, 1951, University of Tulsa. 59 p.

Smith, Rita Monahan. Numerical bias in correlation; carbonate modal data, stratigraphic thickness data, and geochemical data. M, 1977, University of Houston.

Smith, Robert B. Geology of the Monte Cristo area, Bear River Range, Utah. M, 1965, Utah State University. 77 p.

Smith, Robert Baer. A regional gravity survey of western and central Montana. D, 1967, University of Utah. 193 p.

Smith, Robert Cameron. Mineralogic and fluid-inclusion studies of epithermal gold-quartz veins in the Oatman District, northwestern Arizona. M, 1984, University of Arizona. 233 p.

Smith, Robert Charles, II. Geochemistry of Triassic diabase from southeastern Pennsylvania. D, 1973, Pennsylvania State University, University Park. 262 p.

Smith, Robert E. Petrogenesis of clastic plugs, Union County, New Mexico. M, 1975, West Texas State University. 38 p.

Smith, Robert Earl. A lattice analogy for the solution of some nonlinear stress problems. D, 1965, University of Texas, Austin. 145 p.

Smith, Robert Hamilton. Geology of the Rich Patch Valley area, Alleghany County, Virginia. M, 1955, University of Virginia. 143 p.

Smith, Robert Hendell. Fauna of the Whites Creek Miocene, Walton County, Florida. M, 1940, Louisiana State University.

Smith, Robert Kay. Mineralogy and petrology of Silurian bioherms of eastern Iowa. M, 1967, University of Iowa. 124 p.

Smith, Robert Kay. The mineralogy and petrology of the contact metamorphic aureole around the Alta Stock, (Salt Lake County) Utah. D, 1972, University of Iowa. 215 p.

Smith, Robert Martin. Geotechnical properties of glacial lake, Great Falls sediments in the Missouri River Canyon, Montana. M, 1975, University of Idaho. 122 p.

Smith, Robert Ryland. The geology of the southeast quarter of the Montgomery City Quadrangle, Missouri. M, 1953, University of Missouri, Columbia.

Smith, Robert W. Aqueous chemistry of molybdenum at elevated temperatures and pressures with applications to porphyry molybdenum deposits. D, 1983, New Mexico Institute of Mining and Technology. 311 p.

Smith, Roberta Katharine. Glacio marine foraminifera of British Columbia and southeastern Alaska. D, 1966, University of British Columbia.

Smith, Roberta Lynn. The meteoric overprint on marine phreatic sediments; the late Pleistocene of South Florida. M, 1987, Duke University. 84 p.

Smith, Rodney S. Regional study of joints in the northern Piceance Basin, Northwest Colorado. M, 1980, Colorado School of Mines. 126 p.

Smith, Roger Dayton. Heat flow studies within the Socorro Mt. thermal anomaly. M, 1974, New Mexico Institute of Mining and Technology.

Smith, Roger F. Structural and metamorphic evolution of Proterozoic rocks in the northern Taos Range, Taos County, New Mexico. M, 1988, University of New Mexico. 176 p.

Smith, Roger Norman. Heat flow of the western Snake River plain. M, 1980, Washington State University. 111 p.

Smith, Roger Stanley Uhr. Late-Quaternary fluvial and tectonic history of Panamint Valley, Inyo and San Bernadino counties, California. D, 1976, California Institute of Technology. 314 p.

Smith, Roger Stanley Uhr. Migration and wind regime of small barchan dunes within the Algodones dune chain, Imperial County, California. M, 1970, University of Arizona.

Smith, Roger W. Sedimentological analyses of deep sea sediment cores from the South Atlantic Ocean. M, 1978, Queen's University. 199 p.

Smith, Ronald D. Paleontology of the Columbiana Shale near Corsica, Pennsylvania. M, 1968, Pennsylvania State University, University Park. 164 p.

Smith, Ronald Ellis. A study of consolidation of cohesive soils under constant rates of strain. D, 1968, North Carolina State University. 159 p.

Smith, Ross Wilbert. The state of Al (III) in aqueous solution and adsorption of hydrolysis products on Al_2O_3. D, 1969, Stanford University. 194 p.

Smith, Rossman William. Thermal dynamics at the earth-air interface; the implications for remote sensing of the geologic environment. M, 1969, Stanford University. 58 p.

Smith, Roy E. Taxonomy and ecology of Atrypella spp. and the Atrypella community. M, 1973, Oregon State University. 105 p.

Smith, Roy Edward. Lower Devonian (Lochkovian) brachiopods paleoecology and biostratigraphy of the Canadian Arctic Islands. D, 1976, Oregon State University. 362 p.

Smith, Roy George Gerhard. Rapid and simple methods for the determination of trace amounts of As^{+3}, As^{+5}, and total arsenic in geological, biological and environmental samples using atomic absorption spectrophotometry. M, 1977, University of Toronto.

Smith, Russell. Correlation of the producing sands in the Richburg area, Allegany County, New York. M, 1953, University of Michigan.

Smith, Russell. The geology of the Redwing area, Huerfano County, Colorado. D, 1961, University of Michigan. 156 p.

Smith, Russell Clarence. Hydrology and stratigraphy of several Cypress domes and the surrounding area of Alachua County, Florida. M, 1975, University of Florida. 87 p.

Smith, Samuel Joseph. Susceptibility of interlayer potassium in illites to exchange. D, 1967, Iowa State University of Science and Technology. 115 p.

Smith, Samuel Milton. Mineralogy and trace element study of the manganese oxide deposits, Burgin Mine,

east Tintic mining district, Utah County, Utah. M, 1971, Brigham Young University. 122 p.

Smith, Sandra Leslie Rhyne. Predictive geomorphic criteria for recognizing the presence and behavior of faults in the Atlantic Coastal Plain. M, 1983, North Carolina State University. 58 p.

Smith, Scot Earle. Application of remote sensing techniques to the study of the impacts of the Aswan High Dam. D, 1982, University of Michigan. 240 p.

Smith, Scott Raymond. Deposition and diagenesis of the upper Mount Head Formation (Loomis, Marston and Carnarvon members), Plateau Mountain, Alberta. M, 1978, University of Manitoba.

Smith, Sharon Lynne. Distribution of Recent foraminifera in lower Florida Bay. M, 1964, University of Wisconsin-Madison.

Smith, Shea C. A biogeochemical reconnaissance survey of Chandalar Quadrangle, Alaska. M, 1977, Colorado School of Mines. 269 p.

Smith, Shelby W. A sedimentary study of the Jurassic of Quay County, New Mexico. M, 1951, Texas Tech University. 55 p.

Smith, Simon Andrew. The sedimentology of the late Precambrian Recontre Formation, Fortune Bay, Newfoundland. M, 1983, Memorial University of Newfoundland. 191 p.

Smith, Stephen. An environmental model for Saint Joe (Lower Mississippian) carbonate mounds (Southwest Missouri and adjacent areas in Oklahoma and Arkansas). M, 1966, Northwestern University.

Smith, Stephen C. Characteristics of underclays and roof rocks associated with the Springfield (V) Coal in southwestern Indiana. M, 1984, Indiana University, Bloomington. 39 p.

Smith, Stephen M. Carbonate petrology and depositional environments of carbonate buildups in the Devonian Guilmette Formation near White Horse Pass, Elko, Nevada. M, 1984, Brigham Young University. 139 p.

Smith, Stephen Pritchard. Studies of noble gases in meteorites and in the Earth. D, 1979, California Institute of Technology. 425 p.

Smith, Stephen Vaughan. Calcium carbonate budget of the southern California continental borderland. D, 1970, University of Hawaii. 174 p.

Smith, Steven Don. A theoretical study of the flow of slightly compressible non-Newtonian fluids in eccentric annuli. M, 1988, University of Oklahoma. 109 p.

Smith, Stewart. Precambrian geology of the Jawbone Mountain area, Rio Arriba County, NM. M, 1986, New Mexico Institute of Mining and Technology. 99 p.

Smith, Stewart Wilson. An investigation of the Earth's free oscillations. D, 1961, California Institute of Technology. 80 p.

Smith, Susan. Geochemistry and petrology of residual mantle and plutonic rocks from the Lewis Hills Massif, Bay of Islands and coastal complex ophiolites, Newfoundland, Canada. M, 1985, University of Houston.

Smith, Tad Monnett. Dolomitization of the Devonian Jefferson Formation in south-central Montana; evidence for multiple dolomitization and recrystallization events in cyclic platform carbonates. M, 1987, Washington State University. 107 p.

Smith, Ted J. Evaluation of geochemical exploration methods. M, 1983, University of Michigan.

Smith, Ted J. Genetic models at the Hollinger-McIntyre gold deposits, Timmins, Ontario. M, 1983, University of Michigan.

Smith, Terence A. Mineralization in Tertiary porphyries of the Black Hills, with special reference to the Homestake Mine. M, 1935, University of Minnesota, Minneapolis. 52 p.

Smith, Terence Robert. Conservation principles and stability in the evolution of drainage systems. D, 1971, The Johns Hopkins University. 101 p.

Smith, Terrance Lee. The relationship between filtered gravity anomalies and tectonic provinces in northwestern Oklahoma. M, 1984, University of Missouri, Columbia. 67 p.

Smith, Terri Lynn. Geologic and volcanic development of a near-ridge seamount and new evidence for the diversity of origin of seamount hyaloclastites; results from integrated Alvin, Alvin/Angus and laboratory study. M, 1987, Washington University. 377 p.

Smith, Theodore Lee. The geology of the Antimony area, Garfield and Piute County areas. M, 1957, University of Utah. 39 p.

Smith, Thomas Andrew. A seismic refraction investigation of the Manson disturbed area (northwest Iowa). M, 1971, Iowa State University of Science and Technology.

Smith, Thomas Andrew. Three-dimensional Kirchhoff methods of acoustic modeling and migration. D, 1981, University of Houston. 229 p.

Smith, Thomas Edward. An aeromagnetic investigation of the Dixie Valley-Carson Sink area, Nevada. M, 1965, Stanford University. 58 p.

Smith, Thomas Edward. Geology, economic geochemistry, and placer gold resources of the western Clearwater mountains, east central Alaska. D, 1971, University of Nevada - Mackay School of Mines. 497 p.

Smith, Thomas G. Paleoecology of upper Eocene formations in Louisiana, Mississippi and Alabama. M, 1972, Northeast Louisiana University.

Smith, Thomas Joseph, III. The influence of grazing by greater snow geese (Anser caerulescens atlantica) on belowground production, detritus availability, and inorganic soil nitrogen in three North Carolina salt marshes. M, 1979, University of Virginia. 66 p.

Smith, Thomas N. Stratigraphy and sedimentation of the Onion Peak area, Clatsop County, Oregon. M, 1975, Oregon State University. 190 p.

Smith, Timothy Ben. Gravity study of the Fumarole Butte area, Juab and Millard counties, Utah. M, 1974, University of Utah. 54 p.

Smith, Timothy Ellis. Paleomagnetism of the lower Mesozoic diabase and arkose of Connecticut and Maryland. D, 1976, Ohio State University. 334 p.

Smith, Trevor David. Pressuremeter design method for single piles subjected to static lateral load (Volumes I and II). D, 1983, Texas A&M University. 441 p.

Smith, Verl Leon. Hypogene alteration at the Esperanza Mine, Pima County, Arizona. M, 1975, University of Arizona.

Smith, Victor G. Origin of cone-in-cone structures in Upper Devonian and Lower Mississippian shales of north-central Ohio. M, 1982, Bowling Green State University. 144 p.

Smith, Victor Mackusick. Geology of the upper Castaic Creek region, Los Angeles County, California. M, 1951, University of California, Los Angeles.

Smith, Virginia H. A review of peneplain interpretations of the Appalachian Plateau. M, 1938, Columbia University, Teachers College.

Smith, Walter J. Cenozoic stratigraphy near Redington, Pima County, Arizona. M, 1966, University of Arizona.

Smith, Walter Lorane. Stratigraphic correlation of southeastern Utah. M, 1951, Brigham Young University. 52 p.

Smith, Walter R. The Upper Cretaceous invertebrate fauna of Alaska. D, 1924, The Johns Hopkins University.

Smith, Ward C. Geology of the Caribou Stock in the Front Range, Colorado. D, 1936, Yale University.

Smith, Warren DuPre. Development of Scaphites. M, 1904, [Stanford University].

Smith, Warren DuPree. The coal deposits of Batan Island. D, 1908, University of Wisconsin-Madison.

Smith, Warren Slocum. Geology of the Skykomish Basin, Washington. D, 1916, Columbia University, Teachers College. 70 p.

Smith, Warren Slocum. The geology and mineral resources of northeastern King County, Washington.

M, 1913, Columbia University, Teachers College. 35 p.

Smith, Warwick Denison. A finite element study of the effects of structural irregularities on body wave propagation. D, 1975, University of California, Berkeley. 162 p.

Smith, William Algene, Jr. An integrated study of the seismotectonics of the Bowman seismic zone, South Carolina. D, 1987, University of South Carolina. 397 p.

Smith, William Calhoun. The geology of some Pleistocene deposits and their engineering properties. D, 1961, University of Illinois, Urbana. 222 p.

Smith, William Calhoun. The petrography of the Bushberg Sandstone of Missouri. M, 1940, University of Missouri, Columbia.

Smith, William D. Composition and depositional environment of the Albert Formation oil shales, New Brunswick. M, 1985, Dalhousie University. 286 p.

Smith, William Everett. The petrology of some Lower Mississippian carbonate rocks of North Alabama. M, 1961, University of Alabama.

Smith, William Gill. Sedimentary environments and environmental change in the peat-forming area of south Florida. D, 1968, Pennsylvania State University, University Park. 264 p.

Smith, William H. A geological and geochemical investigation of mineralization on Hill 3560, 40 Mile District, Alaska. M, 1968, University of Alaska, Fairbanks. 48 p.

Smith, William H. A geophysical investigation of offshore Kudat Peninsula, northern Borneo. M, 1975, Colorado School of Mines. 51 p.

Smith, William Henking. Geology of Newport Township, Washington County. M, 1948, Ohio State University.

Smith, William Henry. A sedimentary study of the Purgatoire Formation of Quay County, New Mexico. M, 1951, Texas Tech University. 53 p.

Smith, William Horn. A revision of the family Kloedenellidae. M, 1950, University of Illinois, Urbana.

Smith, William Howard. An analysis of Kope Formation clay mineralogy with respect to landslide activity in Hamilton County, Ohio. M, 1980, Wright State University. 178 p.

Smith, William Kerrison. Long-term highwall stability in the northeastern Powder River basin, Wyoming and Montana. D, 1980, University of Arizona. 124 p.

Smith, William LaRue. The geology of the Conasauga Formation (Cambrian) in the vicinity of Ranger, Georgia. M, 1958, Emory University. 27 p.

Smith, William Lee. The nature and origin of skarn. M, 1951, Rutgers, The State University, New Brunswick. 155 p.

Smith, William Oliver. Freshwater ostracodes from the Upper Pennsylvanian and Lower Permian of Ohio. M, 1949, University of Iowa. 33 p.

Smith, William Sidney Tangier. The geology of Santa Catalina Island. D, 1897, University of California, Berkeley. 71 p.

Smith, William Theodore. Geology of part of the Tendoy-Medicine Lodge area, Beaverhead County, Montana. M, 1948, University of Michigan. 50 p.

Smith, William Thomas. An application of foraminiferal ecology, vicinity of Horn Island, Mississippi. M, 1958, University of Missouri, Columbia.

Smith-Evernden, Roberta. Glacio-marine Foraminifera of British Columbia and southeast Alaska. D, 1967, University of British Columbia.

Smitheringale, W. V. The manganese occurrences of the Maritime Provinces, Canada. D, 1928, Massachusetts Institute of Technology. 277 p.

Smitheringale, William G. Isotopic composition of sulphur and the Triassic igneous rocks of Eastern United States. D, 1962, Massachusetts Institute of Technology. 241 p.

Smitheringale, William Vickers. Antimony. M, 1925, University of British Columbia.

Smitherman, Eugene Alston. A reservoir study of the Canyon Reef limestone field, Scurry County, Texas. M, 1952, University of Texas, Austin.

Smitherman, James R. Geology of the Stibnite roof pendant, Valley County, Idaho. M, 1985, University of Idaho. 62 p.

Smithers, Ronald Milton. Tellurium isotope fractionation in nature and in the laboratory. M, 1965, University of Alberta. 101 p.

Smithson, Scott B. The geology of the southeastern Leucite Hills, Sweetwater County, Wyoming. M, 1959, University of Wyoming. 94 p.

Smithwick, Jack Allison. Subsurface correlation of the Wilcox Group in Louisiana. M, 1954, Louisiana State University.

Smithwick, Maureen. A global climatic model based on Permian paleogeography. M, 1976, University of South Carolina.

Smits, James Robert. Textural variation and chemistry of Recent intergranular submarine cemented and organically bound oolites, Schooner Cays, Bahamas. M, 1982, University of Texas, Austin. 189 p.

Smol, John Paul. Postglacial changes in fossil algal assemblages from three Canadian lakes. D, 1982, Queen's University.

Smolak, George. An investigation of the metallurgical treatment of a low grade silver ore from the Mineral Mine at Mineral, Idaho. M, 1924, University of Idaho. 51 p.

Smoliga, John A. The Udall Mine, Vermont, a stratiform massive sulfide deposit of submarine exhalative origin. M, 1985, Western Connecticut State University.

Smolski, Chester E. Physiography of glacial marine deposits at Ipswich, Massachusetts. M, 1953, Clark University.

Smookler, Stanley. A continuously recording interferometer in the calibration of strainmeters. M, 1968, Colorado School of Mines. 194 p.

Smoor, Peter Bernard. Dimensional grain orientation studies of turbidite graywackes. M, 1960, McMaster University. 97 p.

Smoor, Peter Bernard. Hydrochemical facies study of ground water in the Tucson Basin (Pima County, Arizona). D, 1967, University of Arizona. 282 p.

Smoot, Edith L. Phloem anatomy and phylogeny of selected Carboniferous ferns and pteridosperms. D, 1983, Ohio State University. 190 p.

Smoot, John Leach. Hydrogeology and mathematical model of ground-water flow in the Pullman-Moscow region, Washington and Idaho. M, 1987, University of Idaho. 118 p.

Smoot, Joseph P. Sedimentology of a saline closed basin; the Wilkins Peak Member, Green River Formation (Eocene), Wyoming. D, 1978, The Johns Hopkins University. 331 p.

Smoot, Thomas W. The structure and petrography of the Titicus Mountain-Scott Ridge area, Fairfield County, Connecticut, and Westchester County, New York. M, 1956, Miami University (Ohio). 52 p.

Smoot, Thomas William. Clay mineralogy of the pre-Pennsylvanian sandstones and shales of the Illinois Basin. D, 1959, University of Illinois, Urbana. 120 p.

Smoot, Virginia Ellen B. Conodonts from isolated Devonian outcrops. M, 1958, University of Missouri, Columbia.

Smosna, Richard Allan. Tungsten Gap Chert member, BSc Formation, Bird Spring group (upper Mississippian, Pennsylvanian, and lower Permian), Clark County, Nevada. M, 1970, University of Illinois, Urbana. 52 p.

Smosna, Richard Allan. Upper Pennsylvanian-Lower Permian stratigraphy of southern and eastern Nevada. D, 1973, University of Illinois, Urbana. 109 p.

Smouse, Deforrest. High-temperature, high-pressure investigation of cobalt metasilicate and the synthesis of cubic boron nitride. M, 1965, Brigham Young University. 120 p.

Smunt, Frank Michael. Stratigraphy and petrography of the Dutch Creek Sandstone, Union and Alexander counties, Illinois. M, 1964, Southern Illinois University, Carbondale. 58 p.

Smyers, Larry F. Magnetic anomalies over a portion of the Granite Mountains, central Wyoming. M, 1979, University of North Dakota. 102 p.

Smyers, Norman B. Clastic intrusions in the Moreno shale (upper Cretaceous and Paleocene?), east flank-Panoche hills, (southern) California. M, 1970, University of San Diego.

Smyk, Mark C. Geology of Archean interflow sedimentary rocks and their relationship to Ag mineralization in the Cobalt area, Ontario. M, 1987, Carleton University. 87 p.

Smykowski, Anthony Steve. Dam safety; statistics of dam failures and acoustic emissions from seepage and piping. M, 1981, University of Idaho. 95 p.

Smyth, Andrew H. Groundwater and land subsidence investigations of the Hueco Bolson in New Mexico, Texas and Mexico. M, 1986, New Mexico State University, Las Cruces. 109 p.

Smyth, Ann Lindsay. Pedogenesis and diagenesis of the Olive Hill Clay Bed, Breathitt Formation (Carboniferous), northeastern Kentucky. M, 1984, University of Cincinnati. 283 p.

Smyth, Charles H., Jr. A third occurrence of peridotite in central New York. D, 1892, Columbia University, Teachers College.

Smyth, D. J. A. Hydrogeological and geochemical studies above the water table in an inactive uranium tailings impoundment near Elliot Lake, Ontario. M, 1981, University of Waterloo.

Smyth, John Thomas. Seismic data processing, Trenton reflector analysis and general geologic interpretation in Harlem Township, Delaware County, Ohio. M, 1985, Wright State University. 62 p.

Smyth, Joseph. High temperature single-crystal studies on low-calcium pyroxenes. D, 1970, University of Chicago. 91 p.

Smyth, Pauline. Fusulinidae in the Pennsylvanian of Ohio. M, 1951, Ohio State University.

Smyth, Walter Ronald. The stratigraphy and structure of the southern part of the Hare Bay Allochthon, N.W. Newfoundland. D, 1973, Memorial University of Newfoundland. 250 p.

Smyth, William David. Hölmboe waves. M, 1986, University of Toronto.

Smythe, Donald DeCou. The iron ore deposits of Iron Mountain, Socorro County, New Mexico. M, 1921, Cornell University.

Smythe, Frank Ward, Jr. Variations in the volcanic and nonvolcanic components of subpolar North Atlantic ice-rafted sediment. M, 1982, Memphis State University.

Smythe, William David. The detectability of clathrate hydrates in the outer solar system. D, 1979, University of California, Los Angeles. 116 p.

Snavely, D. S. Effects of channelization of the Eel River, Indiana. M, 1975, Washington University. 105 p.

Snavely, Parke D. Geology of the coastal area between Cape Kiwanda and Cape Foulweather, Oregon. M, 1952, University of California, Los Angeles.

Snavely, Parke Detweiler, III. Depositional and diagenetic history of the Thebes Formation (lower Eocene), Egypt, and implications for early Red Sea tectonism. D, 1984, University of California, Santa Cruz. 750 p.

Snavely, Richard K. Two-dimensional seismic modeling; a cross-sectional view of the Bay Marchand salt dome. M, 1988, University of New Orleans.

Snead, David Edward. Creep rupture of saturated undisturbed clays. D, 1970, University of British Columbia.

Snead, Robert G. The study of skeletal calcarenites in the vicinity of the Arca Reef Group, Campeche Bank, Mexico. M, 1964, Texas A&M University.

Snead, Robert Garland. Microflora diagnosis of the Cretaceous-Tertiary boundary, central Alberta (Canada). D, 1968, University of Alberta. 216 p.

Snead, Rodman Eldredge. Physical geography of the Las Bela coastal plain, West Pakistan. D, 1963, Louisiana State University. 211 p.

Snedden, John William. Origin and sedimentary characteristics of discrete sand beds in modern sediments of the central Texas continental shelf. D, 1985, Louisiana State University. 267 p.

Snedden, John William. Stratigraphy and depositional environment of the San Miguel lignite deposit, northern McMullen and southeastern Atascosa counties, Texas. M, 1979, Texas A&M University. 162 p.

Snedden, Loring B. Stratigraphy and micropaleontology of the (Miocene) Rincon Formation of California. M, 1931, Stanford University. 40 p.

Snee, Lawrence Warren. Emplacement and cooling of the Pioneer Batholith, southwestern Montana. D, 1982, Ohio State University. 339 p.

Snee, Lawrence Warren. Petrography, K-Ar ages and field relations of the igneous rocks of part of the Pioneer Batholith, southwestern Montana. M, 1978, Ohio State University. 110 p.

Sneed, David Richard. Reservoir rock-property calculations from thin section measurements. M, 1988, Texas A&M University.

Sneed, Edmund David. Roundness and sphericity of Colorado River pebbles. M, 1955, University of Texas, Austin.

Sneed, Henry Eugene. Stratigraphic control of the Paint Creek-Bethel oil accumulation in the Maud North Consolidated Area, Wabash County, Illinois. M, 1950, University of Illinois, Urbana.

Sneeringer, Margaret Riggs. The geochemistry of co-existing glass, inclusions in basalts dredged from the West Indian Ocean triple junction. M, 1979, Massachusetts Institute of Technology. 154 p.

Sneeringer, Margaret Riggs. The geochemistry of co-existing glass, phenocrysts, and glass inclusions in basalts dredged from the West Indian Ocean triple junction. M, 1976, Massachusetts Institute of Technology. 154 p.

Sneeringer, Mark Albert. Strontium and samarium diffusion in diopside. D, 1981, Massachusetts Institute of Technology. 277 p.

Sneh, Amihai. Sedimentary environments of the northern gulfs of the Red Sea. D, 1978, Rensselaer Polytechnic Institute. 106 p.

Sneh, Amihai. Stratigraphy of the Lower Cretaceous system of northern Israel, with special emphasis on depositional environments and economic significance. M, 1974, Rensselaer Polytechnic Institute. 68 p.

Sneider, John Scott. Depositional environment of the Caballos Formation, San Francisco Field, Neiva subbasin, upper Magdalena Valley, Colombia. M, 1988, Texas A&M University. 157 p.

Sneider, Robert Morton. Petrology of the Croton Falls mafic complex, southeastern New York. D, 1962, University of Wisconsin-Madison. 138 p.

Snelgrove, Alfred Kitchener. Geology and ore deposits of Betts Cove-Tilt Cove area, Notre Dame Bay, Nfld. D, 1930, Princeton University.

Snell, N. S. Flexure of a viscoelastic lithosphere. M, 1975, Northwestern University.

Snellenburg, J. W. A chemical and petrographic study of the chondrules in the unequilibrated ordinary chondrites, Semarkona and Krymka. D, 1978, SUNY at Stony Brook. 187 p.

Snellenburg, Jonathan W. A computer model of the cation distribution in orthopyroxene. M, 1972, Dartmouth College. 56 p.

Snelling, Walter O. The old copper mine at Bristol, Connecticut. M, 1906, Yale University.

Snelson, Sigmund. Geology in northern Ruby Mountains and eastern Humboldt Range, Elko County, northeastern Nevada. D, 1957, University of Washington. 214 p.

Snelson, Sigmund. Geology of southern Pequop Mountains, Elko County, northeastern Nevada. M, 1955, University of Washington.

Snetsinger, Kenneth George. Late Tertiary and Quaternary history of the lower Arroyo Seco area, Monterey County, California. M, 1962, Stanford University. 36 p.

Snetsinger, Kenneth George. Petrology and mineralogy of metamorphic and intrusive rocks, northwest part of the Bass Lake 15' Quadrangle, Madera and Mariposa counties, California. D, 1966, Stanford University. 209 p.

Sneyd, Deana S. Deformation of ordinary chondritic meteorites. M, 1987, University of Tennessee, Knoxville. 106 p.

Snider, David W. Mineralogic relationships within the gradational zone between the Siamo Formation (Precambrian) and the Negaunee iron formation (Precambrian), sections 7 and 8, T. 47 N., R. 26 W., Marquette County, Michigan. M, 1972, Bowling Green State University. 80 p.

Snider, Frederic G. Analysis of magnetic and chemical data from Mesozoic diabase dikes of the Appalachians, with implications for the presence of a Triassic hotspot in the Carolinas. M, 1975, Wesleyan University. 61 p.

Snider, Henry Irwin. Stratigraphy and associated tectonics of the Upper Permian Castile-Salado-Rustler Evaporite Complex, Delaware Basin, West Texas and Southeast New Mexico. D, 1966, University of New Mexico. 196 p.

Snider, Henry Irwin. Subsurface geology and the Mount Simon (Upper Cambrian); Precambrian contact in southeastern Minnesota. M, 1967, University of Minnesota, Minneapolis. 46 p.

Snider, John Luther. Geology of Seven Heart Quadrangle, Culberson County, Texas. M, 1955, University of Texas, Austin.

Snider, Luther Crocker. Soil survey of Daviess County, Indiana. M, 1909, Indiana University, Bloomington.

Snider, Luther Crocker. The geology and paleontology of the Mississippian rocks of northwestern Oklahoma. D, 1915, University of Chicago.

Snider, Lynn Vivion. Integrated multidisciplinary lineament analysis of the Ouachita Mountains of Arkansas and Oklahoma. M, 1984, University of Arkansas, Fayetteville. 124 p.

Sniedovich, Moshe. On the theory and modeling of dynamic programming with applications in reservoir operation. M, 1976, University of Arizona.

Sniegocki, Richard Ted. The geology of the Powell Station Quadrangle, Knox and Anderson counties, Tennessee. M, 1949, University of Tennessee, Knoxville. 63 p.

Sniffen, Jane. The geology of Florida; a guide to sources of information. M, 1967, University of Florida. 74 p.

Snipes, David Strange. Stratigraphy and sedimentation of the Middendorf Formation between the Lynches River, South Carolina, and the Ocmulgee River, Georgia. D, 1965, University of North Carolina, Chapel Hill. 140 p.

Snitbhan, N. Plasticity solutions for slopes in anisotropic, inhomogeneous soil. D, 1975, [Lehigh University]. 181 p.

Snively, Norman Ray. Genesis of the migmatites and associated Pre-cambrian formations near Bergen, Colorado Front Range. D, 1948, University of Michigan.

Snodgrass, Donald Blaine. A study of Lake Michigan bottom sediments. M, 1952, University of Illinois, Urbana.

Snodgrass, Elvis Dean. The geology of the Church area, Adair County, Oklahoma. M, 1952, University of Oklahoma. 55 p.

Snodgrass, Thomas W. The geology of the southeastern quarter of the Richland Center Quadrangle, Wisconsin. M, 1955, University of Wisconsin-Madison.

Snoke, Arthur Wilmot. The petrology and structure of the Preston Peak area, Del Norte and Siskiyou counties, California. D, 1972, Stanford University. 274 p.

Snook, James Donald. Geology of the north half of the Arden 7 1/2-minute quadrangle, Stevens County, Washington. M, 1981, University of Idaho. 51 p.

Snook, James R. Geology of the Bald Mountain area, Richmond Quadrangle, Oregon. M, 1957, Oregon State University. 69 p.

Snook, James Ronald. Petrology of parts of the Tonasket and Omak Lake quadrangles, north central Washington. D, 1962, University of Washington. 158 p.

Snorrason, Arni. Analysis of multivariate stochastic hydrological systems using transfer function-noise models. D, 1983, University of Illinois, Urbana. 123 p.

Snow Boles, Jennifer Lee. Aqueous thermal degradation of naturally occurring aromatic acids and the synthetic chelating agent disodium EDTA. D, 1986, Princeton University. 150 p.

Snow, Charles Bruce. Stratigraphy of basal sandstones in the Green River Formation (Eocene, lower and middle), northwest Piceance Basin, Rio Blanco County, Colorado. M, 1969, Colorado School of Mines. 108 p.

Snow, Dale L. Study of the "sand" deposit between Manhattan and Wamego, Kansas. M, 1963, Kansas State University. 66 p.

Snow, David Tunison. A parallel plate model of fractured permeable media. D, 1965, University of California, Berkeley. 331 p.

Snow, David Tunison. The geology of the northeast corner of Alameda County and adjacent portions of Contra Costa County, California. M, 1957, University of California, Berkeley. 188 p.

Snow, Eleanour Anne. Selected studies in mineral kinetics. D, 1987, Brown University. 180 p.

Snow, Geoffrey Greacen. Mineralogy and geology of the Dolly Varden Mountains, Elko County, Nevada. D, 1964, University of Utah. 187 p.

Snow, Geoffrey Greacen. Plutonic rocks of the Sonoma Quadrangle, Nevada. M, 1960, University of Utah. 44 p.

Snow, John Humphrey. Study of structural and tectonic patterns in south-central Utah as interpreted from gravity and aeromagnetic data. M, 1978, University of Utah. 245 p.

Snow, Jonathan E. Isotopic investigations in two ancient basaltic complexes and their implications for the evolution of the crust and mantle. M, 1986, University of Rochester. 75 p.

Snow, Lester A. The copper industry and water in Arizona. M, 1976, University of Arizona.

Snow, P. D. Mathematical modeling of phosphorus exchange between sediments and overlying water in shallow eutrophic lakes. D, 1976, University of Massachusetts. 289 p.

Snow, Phillip D. Lake Erie; hydroxy-apatite saturation and plankton concentrations. M, 1968, Syracuse University.

Snow, Randall J. The depositional environment of the Late Carboniferous, coal-bearing upper Thorburn Member of the Stellarton Group, Pictou Coalfield, New Glasgow, Nova Scotia. M, 1988, Acadia University. 257 p.

Snow, Robin Scott. Mathematical modelling of longitudinal profile adjustment in alluvial streams. D, 1983, Pennsylvania State University, University Park. 249 p.

Snow, Roland B. Equilibrium studies in the system FeO-Al₂O₃-SiO₂. D, 1936, Ohio State University.

Snow, William Eugene. The ore deposits of Alaska and the Yukon. M, 1936, University of British Columbia.

Snowden, Jesse Otho. Geologic and chemical environment of Biloxi Bay, Mississippi. M, 1961, University of Missouri, Columbia.

Snowden, Jesse Otho, Jr. Petrology of Mississippi loess. D, 1966, University of Missouri, Columbia. 217 p.

Snowden, John M. An investigation of practical solutions of large systems of normal equations (geodetic science). M, 1966, Ohio State University.

Snowdon, Lloyd R. Organic geochemistry of the Upper Cretaceous/Tertiary delta complexes of the Beaufort-Mackenzie sedimentary basin, northern Canada. D, 1978, Rice University. 137 p.

Snyder, Barry L. Stratigraphy of the middle part of the Council Grove Group (Early Permian) of Nebraska and Kansas. M, 1968, University of Nebraska, Lincoln.

Snyder, David Bufton. Foreland crustal geometries in the Andes of Argentina and the Zagros of Iran from seismic reflection and gravity data. D, 1987, Cornell University. 208 p.

Snyder, Don Otis. Geology of the Empire Abo Field, Eddy County, New Mexico. M, 1962, University of New Mexico. 77 p.

Snyder, Don Otis. Stratigraphic analysis of the Baca Formation (Eocene(?)), west central New Mexico. D, 1971, University of New Mexico. 160 p.

Snyder, Donald D., III. A gravity survey of South Park, Colorado. D, 1968, Colorado School of Mines. 168 p.

Snyder, Donald L. Palynology and geophotometry of the Middle Cretaceous rocks in Ellsworth and Russell counties, Kansas. M, 1963, Kansas State University. 65 p.

Snyder, Edward M. Taxonomy and biostratigraphy of the Bryozoa of the Gerster Formation (Permian), northeastern Nevada. M, 1976, Eastern Washington University. 171 p.

Snyder, Edward McKinley. Taxonomy, functional morphology and paleoecology of the Fenestellidae and Polyporiidae (Fenestelloidea, Bryozoa) of the Warsaw Formation (Valmeyeran, Mississippian) of the Mississippi Valley. D, 1984, University of Illinois, Urbana. 802 p.

Snyder, Frank George. Petrofabric studies in the Wausau section of the Wisconsin Batholith. D, 1947, University of Wisconsin-Madison.

Snyder, Frank S. A spatial and temporal analysis of the Sleeping Bear Dunes complex, Michigan (a contribution to the geomorphology of perched dunes in humid continental regions). D, 1985, University of Pittsburgh. 221 p.

Snyder, Geoffrey William. Size distributions of suspended and bottom sediments in and around Quinault submarine canyon; implications for modern sediment transport and accumulation off the Washington coast. M, 1984, Lehigh University. 90 p.

Snyder, George L. Some considerations of Aleutian lavas with special emphasis on those of Little Sitkin Island. M, 1952, Dartmouth College.

Snyder, Gregory A. A Nd and Sr isotopic and trace element study of a mafic-ultramafic dike swarm, Laramie Range, Wyoming; evidence for late Archean-early Proterozoic rifting of the Wyoming Craton. M, 1986, Colorado School of Mines. 135 p.

Snyder, Hollice Andrew. A gravity and magnetic study of the Skalkaho pyroxenite-syenite complex, western Montana. M, 1986, University of Montana. 40 p.

Snyder, J. M. Petrography and depositional environment of Upper Pennsylvanian sandstones, subsurface of Fisher County, Texas. M, 1977, University of Texas, Arlington. 144 p.

Snyder, James D. Plastic flowage of salt in mines at Hutchinson and Lyons, Kansas. M, 1960, University of Kansas. 38 p.

Snyder, Janet Greer. A metamorphic and structural study of the Port Simpson area, British Columbia. M, 1980, Bryn Mawr College. 88 p.

Snyder, Jay. Heat flow in the southern Mesilla Basin, Dona Ana County, New Mexico, with an analysis of the east Potrillo geothermal system. M, 1986, New Mexico State University, Las Cruces. 252 p.

Snyder, Jeremy. Life habits of diminutive bivalve molluscs in the Maquoketa Formation (upper Ordovician) (midwestern U.S.). M, 1970, Northwestern University.

Snyder, John Frank. A gravity study of Rowan County, North Carolina. M, 1963, University of North Carolina, Chapel Hill. 37 p.

Snyder, John Frank. Petrography and minor element distribution in wall rocks from the Mascot-Jefferson City zinc district. D, 1975, University of Tennessee, Knoxville. 209 p.

Snyder, John Lemoyne. Compositional variations in garnet. M, 1953, Dartmouth College. 62 p.

Snyder, John LeMoyne. Geochemical study of the Duluth Lopolith. D, 1957, Northwestern University. 190 p.

Snyder, Kenneth Dele. Geochemical orientation survey in the Seven Devils mining district [Idaho]. M, 1973, University of Idaho. 111 p.

Snyder, Kent Everett. Pedogenesis and landscape development in the Salamanca Re-entrant, southwestern New York. D, 1988, Cornell University. 345 p.

Snyder, Kossouth. A sulfur isotope study of the pyrite to pyrrhotite conversion. M, 1988, Indiana University, Bloomington. 33 p.

Snyder, Linden E. Report on the Rolling Hills Country Club landslide and nearby slopes, South Table Mountain, Jefferson County, Colorado. M, 1977, Colorado School of Mines. 137 p.

Snyder, Michael Robert. A quantitative evaluation of the bioerosion potential of modern marine endolithic algae and fungi. M, 1982, Duke University. 51 p.

Snyder, Randy William. The development and application of Fourier transform infrared spectroscopic techniques to the characterization of coal and oil shale. D, 1982, Pennsylvania State University, University Park. 163 p.

Snyder, Robert D., Jr. A seismic reflection investigation near Whitehall, Montana. M, 1978, Montana College of Mineral Science & Technology. 72 p.

Snyder, Robert Dean. Geology of the east-central rectangle of the Red Mesa Quadrangle, La Plata County, Colorado. M, 1952, University of Illinois, Urbana.

Snyder, Robert H. Geology and geochemistry of the Skootamatta syenite and surrounding rocks, southeastern Ontario, Canada. M, 1978, SUNY at Buffalo. 69 p.

Snyder, Ronald Wayne. The stratigraphy of the Roubidoux and Gasconade Formation of northern Arkansas. M, 1976, University of Arkansas, Fayetteville.

Snyder, Scott William. Distribution of planktonic foraminifera in surface sediments of the Gulf of Mexico. D, 1974, Tulane University.

Snyder, Victor Abram. Theoretical aspects and measurement of tensile strength in unsaturated soils. D, 1980, Cornell University. 103 p.

Snyder, W. S. Origin and exploration for ore deposits in upper Paleozoic chert-greenstone complexes of northern Nevada. D, 1977, Stanford University. 169 p.

Snydsman, W. E. Simultaneous inversion of seismic travel-times and amplitudes. D, 1979, University of Washington. 102 p.

Soal, Norman. The identification and study of sulphide silver-bearing minerals by selective iridescent filming. M, 1939, Montana College of Mineral Science & Technology. 57 p.

Soar, Linda Katherine. Paleoecology of phylloid algal mud mounds, Honaker Trail Formation (Pennsylvanian), Southwest Colorado. M, 1984, University of Texas, Austin. 69 p.

Soares, E. F. A deterministic-stochastic model for sediment storage in large reservoirs. D, 1975, University of Waterloo.

Sobba, Donald H. Paleoecology and biostratigraphy of the Shubuta Clay Member of the Yazoo Clay in Mississippi and Alabama. M, 1976, Northeast Louisiana University.

Sobel, Lloyd S. Sedimentology of the Blackbird mining district, Lemhi County, Idaho. M, 1982, University of Cincinnati. 235 p.

Sobel, Phyllis A. The phase P'dP' as a means for determining upper mantle structure. D, 1978, University of Minnesota, Minneapolis. 142 p.

Sobhanie, Mohammad Eghbal. Numerical study of three different soil material models. D, 1986, University of Akron. 253 p.

Sobocinski, Robert Walter, Jr. Characterization of bedload transport in a gravel-bottomed stream by modeling the distribution of transport of cesium-137. M, 1987, University of Utah. 106 p.

Sobol, Joseph Walter. The petrology of Grenville Marbles in the vicinity of Bancroft, Ontario. M, 1973, University of Michigan.

Socci, Anthony D. Paleophosphate determinations and paleohydrographic reconstructions using a statistically-derived isotopic model. D, 1982, Florida State University. 97 p.

Socci, Anthony D. The sedimentary petrology of Bridger C in the southern Green River basin, southern Wyoming. M, 1978, University of Wisconsin-Milwaukee.

Socha, Betty Jean. The glacial geology of the Baraboo area, Wisconsin, and application of remote sensing to mapping surficial geology. M, 1984, University of Wisconsin-Madison. 154 p.

Socolow, Arthur Abraham. Genesis of ore deposits at Ducktown, Tennessee. M, 1947, Columbia University, Teachers College.

Socolow, Arthur Abraham. Geology of the Irwin District of Colorado. D, 1955, Columbia University, Teachers College. 113 p.

Socratous, G. A model for optimal development of the water resources of a semi-arid region. D, 1976, University of Pennsylvania. 402 p.

Sodbinow, E. S. Some seismic studies in central and eastern Tennessee. M, 1977, Virginia Polytechnic Institute and State University.

Soderberg, Roger Kenneth. Gravity and tectonic study of the Rough Creek fault zone and related features. M, 1976, University of Kentucky. 36 p.

Soderblom, Laurence Albert. The distribution and ages of regional lithologies in the lunar Maria. D, 1970, California Institute of Technology. 148 p.

Soderman, Jarmo Georg William. Microscopic investigation of Mississippian crinoidal limestones, Stobo, Indiana. M, 1960, University of Illinois, Urbana.

Soderman, Jarmo Georg William. Petrography of algal bioherms in the Burnt Bluff Group (Silurian), Wisconsin. D, 1962, University of Illinois, Urbana. 114 p.

Soderman, Kristopher Lorne. The behaviour of geotextile reinforced embankments. D, 1987, University of Western Ontario.

Soderstrom, Glen S. Till-fabric analysis of the Farmdale Drift of northwestern Illinois. M, 1958, University of Missouri, Columbia.

Soeder, Daniel J. Bioerosion in the rocky intertidal zone at Missouri Key, Florida. M, 1978, Bowling Green State University. 107 p.

Soegaard, J. K. Sedimentological constraints on Precambrian crustal evolution in northern New Mexico. D, 1984, Virginia Polytechnic Institute and State University.

Soegaard, Kristian. Sedimentological constraints on Precambrian crustal evolution in northern New Mexico. D, 1984, Virginia Polytechnic Institute and State University. 193 p.

Soeller, Stephen Anton. Quaternary and environmental geology of Lemmon Valley, Nevada. M, 1978, University of Nevada. 70 p.

Soemarso, C. The effects of production rates and some reservoir parameters on recovery in a strong water drive gas reservoir. M, 1978, Texas A&M University.

Soens, David Dale. Stratigraphy and sedimentology of the Tombigbee Sand Member, Eutaw Formation (Cretaceous-Campanian Stage), of northeastern Mississippi. M, 1984, University of Alabama. 193 p.

Soeria-Atmadja, Rubini. Petrology of the granophyre of Isle au Haut, Maine. D, 1963, University of Illinois, Urbana. 73 p.

Soeripto, Ruskamto. Structural framework of the western Williston Basin, Northeast Montana and adjacent areas. M, 1986, Colorado School of Mines. 80 p.

Soesilo, Dwisuryo Indroyono. A multi-level remote sensing technique for regional geologic investigation in tropical areas through a study of the Sukabumi and Lembang regions, West Java, Indonesia. D, 1987, University of Iowa. 251 p.

Sofranko, Ronald A. Seismic interpretation and geologic analysis of Harlem Township, Delaware County, Ohio, to determine potential gas production. M, 1985, Wright State University. 107 p.

Sofranoff, Stephanie E. Geology, alteration, and mineralization of the carbonate mining district and surrounding area, Lawrence County, South Dakota. M, 1979, South Dakota School of Mines & Technology.

Sohl, Norman Frederick. The gastropod fauna of the Ripley and Owl Creek formations. M, 1951, University of Illinois, Urbana.

Sohl, Norman Frederick. The gastropods of the Late Cretaceous Ripley, Owl Creek, and Prairie Bluff formations. D, 1954, University of Illinois, Urbana. 551 p.

Sohn, Israel G. Ostracoda from the Mauch Chunk (Mississippian of West Virginia). M, 1938, Columbia University, Teachers College.

Sohn, Joonik. Crack propagation in earth embankment subjected to fault movement. D, 1988, University of California, Davis. 305 p.

Sohnge, Paul Gerhard. The geology of the Messina copper mines and adjoining country (South Africa). D, 1944, Harvard University.

Soholt, Donald Eugene. Metamorphic facies of the Groveland Mine, northern Michigan. M, 1964, University of Minnesota, Minneapolis. 147 p.

Sohon, Robert S. Ostracodes of the Glenerie Limestone at Catskill, New York. M, 1960, Pennsylvania State University, University Park. 133 p.

Soileau, John Millard. Evaluation of free iron oxide distribution and source and significance of red and yellow coloration in certain well-drained coastal plain soils. D, 1962, North Carolina State University. 138 p.

Soileau, Lyndon Sewell. Analysis of fine-grained turbidites in intraslope basins of the northwestern Gulf of Mexico. M, 1987, University of New Orleans. 81 p.

Soister, Paul E. Geology of Santa Ana Mesa and adjoining areas, Sandoval County, New Mexico. M, 1952, University of New Mexico. 126 p.

Soja, Constance Meredith. Paleontologic, paleoecologic, and sedimentologic studies of Lower Devonian facies, Kasaan and Wadleigh islands, southeastern Alaska. D, 1985, University of Oregon. 682 p.

Sokol, Daniel. The hydrogeology of the San Francisquito Creek basin, San Mateo and Santa Clara counties, California. D, 1964, Stanford University. 274 p.

Sokolow, B. B. An environmental assessment of the San Joaquin nuclear project; a case study of nuclear power plant siting in California. D, 1977, University of California, Los Angeles. 110 p.

Sokolowski, T. J. Elastic constants and computer programs for calculating the elastic parameters of materials subject to modest pressures or temperatures. M, 1970, University of Hawaii. 159 p.

Sokolsky, George E.; Guyton, J. Stephen and Hutton, John R. Geology of the Devil's Hole area, Custer and Huerfano counties, Colorado. M, 1960, University of Michigan.

Sol, Ayhan. Chemical and petrographic variations across transverse profiles of four early Mesozoic diabase dikes from North Carolina. M, 1987, Florida State University. 227 p.

Solak, Mustafa Remzi. Magnetic survey in the Iron King Mine area, Jackson Mountains, Nevada. M, 1962, University of Utah. 66 p.

Solanki, Jawahirlal J. Surface wave investigations of the crust and upper mantle of the western United States. D, 1974, University of Pittsburgh. 131 p.

Solano, Marcelo. 3D seismic study in northern Peru. M, 1986, Colorado School of Mines. 120 p.

Solano-Borrego, Ariel E. Microseismicity on the Gorda Ridge. M, 1982, Oregon State University. 76 p.

Solano-Borrego, Ariel Enrique. Crustal structure and seismicity of the Gorda Ridge. D, 1985, Oregon State University. 185 p.

Solano-Rico, Baltazar. Some geologic and exploration characteristics of porphyry copper deposits in a volcanic environment, Sonora, Mexico. M, 1975, University of Arizona.

Solar, Carlos W. Del see Del Solar, Carlos W.

Solari, Allison Jarvis. The Tertiary formations of the Copperopolis Quadrangle (California). M, 1936, University of California, Berkeley. 66 p.

Solbczyk, Stanley Michael. Implications of variations in observed Earth-tide measurements at the Granite Mountain records vault, Salt Lake County, Utah. M, 1977, University of Utah. 88 p.

Solberg, Peter Harvey. Structural relations between the Shuswap and Cahe Creek complexes near Kalamalka Lake, southern British Columbia. M, 1976, University of British Columbia.

Solberg, Roger A. Petrology and petrofabrics of the Felch metronite quarry, Dickinson County, Michigan. M, 1958, Michigan State University. 52 p.

Solberg, Teresa Christine. Fe oxidation and weathering studies of Antarctic and SNC meteorites. M, 1987, Massachusetts Institute of Technology. 96 p.

Soldate, Albert Mills, Jr. Propagation of transient waves on non-uniform bathymetry. D, 1987, University of Miami. 138 p.

Solebello, Louis P., Jr. A geochemical investigation of volcanic rocks from the islands of Evvia, Skyros, and Eustratios; north central Aegean Sea, Greece. M, 1988, University of South Florida, Tampa. 103 p.

Soler, T. Secular motion of the pole and global plate tectonics. D, 1977, Ohio State University. 223 p.

Soles, James A. A study of the open system, high pressure-temperature experimentation, transportation and deposition of some sulphides. D, 1960, McGill University.

Soles, James Albert. Further studies on the hydrothermal stability of quartz. M, 1954, University of British Columbia.

Solheim, Larry Peter. An axially symmetric, constant viscosity, mantle-convection model. M, 1986, University of Toronto.

Solie, Diana Nelson. The Middle Fork plutonic complex; a plutonic association of coeval peralkaline and metaluminous magmas in the north-central Alaska Range. D, 1988, Virginia Polytechnic Institute and State University. 278 p.

Solien, M. A. Conodont biostratigraphy of the Lower Triassic Thaynes Formation. M, 1975, University of Wisconsin-Madison.

Solien, Mark Aldon. Conodont biostratigraphy of the Lower Triassic Thaynes Formation, Utah. M, 1976, University of Wisconsin-Madison. 94 p.

Soliman, Afifi H. Accuracy of graphical adjustment of an aero-levelling strip triangulation. M, 1962, Ohio State University.

Soliman, Hosny El-Desouky Ahmed. Stratigraphy and sedimentology of Lower Cretaceous Sykes Mountain Formation, Bighorn Basin, Wyoming. D, 1988, Iowa State University of Science and Technology. 173 p.

Soliman, Mohamed. Numerical modeling of thermal recovery processes. D, 1979, Stanford University. 121 p.

Soliman, Soliman Mahmoud. General geology of the Isabel-Eylar area, California, and petrology of Franciscan Sandstones. D, 1958, Stanford University. 174 p.

Solis-Iriarte, Raul Fernando. Late Tertiary and Quaternary depositional system in the subsurface of central coastal Texas. D, 1980, University of Texas, Austin. 323 p.

Solis-Iriarte, Raul Fernando. Subsurface geology (Paleozoic) of the central cart of the Fort Worth Basin, Texas. M, 1972, University of Texas, Austin.

Soller, David. The stratigraphy and clay mineralogy of Pleistocene deposits of the Lewisburg Quadrangle, southwestern Ohio. M, 1978, Miami University (Ohio). 128 p.

Soller, David Rugh. The Quaternary history and stratigraphy of the Cape Fear River valley, North Carolina. D, 1984, George Washington University. 225 p.

Sollid, Sherman A. Surficial geology of the Porcupine drainage basin, Gallatin County, southwest Montana. M, 1973, Montana State University. 100 p.

Solliday, James Richard. Biostratigraphy of the lower Neuse River estuary, Craven and Pamlico counties, North Carolina. M, 1962, University of Houston.

Solmonson, Donald W. The relationship between rheological behavior and chemical composition of a Wyoming bentonite. M, 1961, South Dakota School of Mines & Technology.

Solohub, J. T. A test of grain-size distribution statistics as indicators of depositional environments, at Grand Beach, Manitoba (Canada). M, 1967, University of Manitoba.

Solomah, Ahmed Gabr. High level radioactive waste management. D, 1980, North Carolina State University. 109 p.

Solomon, Barry J. Geology and oil shale resources near Elko, Nevada. M, 1979, San Jose State University. 142 p.

Solomon, Douglas Kip. Seasonal variations of carbon dioxide in a montane soil. M, 1985, University of Utah. 82 p.

Solomon, George Cleve. Kinetics of interaction between biotite adamellite, granitic gneiss and aqueous sodium chloride solutions. M, 1978, Pennsylvania State University, University Park. 109 p.

Solomon, Hayden S. A study of the water quality of the Farmington River, Connecticut. M, 1984, Boston University. 204 p.

Solomon, Peter J. Sulfur isotope and textural studies of the ores at Balmat, New York and Mt. Isa, Queensland (Australia). D, 1963, Harvard University.

Solomon, Sean C. Seismic-wave attenuation and the state of the upper mantle. D, 1971, Massachusetts Institute of Technology. 321 p.

Solomon, Steven M. Sedimentology and fossil-fuel potential of the Upper Carboniferous Barachois Group, western Newfoundland. M, 1987, Memorial University of Newfoundland. 256 p.

Solomons, Eugenie. An investigation of the mechanism of formation for small subsiding basins. M, 1985, University of North Carolina, Chapel Hill. 37 p.

Solonyka, Edward Richard. Mink Narrows copper deposit, Lake Athapapuskow, Manitoba. M, 1974, University of Manitoba.

Solounias, N. The Turolian fauna from the island of Samos, Greece with special emphasis on the hyaenids and the bovids. D, 1979, University of Colorado. 465 p.

Solow, Andrew R. The multivariate gaussian approach in geostatistics. M, 1980, Stanford University. 56 p.

Soloyanis, Susan Constance. Paleomagnetic properties of some New England tills. D, 1978, University of Massachusetts. 88 p.

Solter, Donald D. Distribution of rubidium, strontium, zirconium and iron of Porphyry Mountain and age of the Silver Plume Granite, Jamestown, Colorado. M, 1966, Ohio State University.

Solyom, Val. A study of selected dissolved minerals in water samples from drilled water wells located in representative portions of the DeKalb and Sycamore

quadrangles. M, 1964, Northern Illinois University. 35 p.

Somasekhara, Kananur V. Megafauna and paleoecology of two shale units in the Haney and Menard formations (upper Mississippian) in southern Illinois. M, 1970, Southern Illinois University, Carbondale. 87 p.

Somasundarum, Sujithan. Constitutive modelling for anisotropic hardening behavior with applications to cohesionless soils. D, 1986, University of Arizona. 339 p.

Somayajula, Chavali Rama. Paleomagnetic studies of volcanic rocks in Ireland, Newfoundland and Labrador. M, 1969, Memorial University of Newfoundland. 197 p.

Somers, George and Shoemaker, Abbott Hall. The geology of the El Tiro Mine, Silver Bell, Arizona. M, 1924, University of Arizona.

Somers, George Brooks. Correlations of the anomalies of vertical intensity of the Earth's magnetic field, with the regional geology of North America (T521). D, 1930, Colorado School of Mines. 314 p.

Somers, George Henry. Petrogenesis of the White Lake Pluton. M, 1984, University of Western Ontario. 154 p.

Somers, Lee Bert Hamill. Bathymetry of the western African continental margin; senegal to Ivory Coast. D, 1969, University of Michigan. 78 p.

Somers, Lee Bert Hamill. Small scale variation in nearshore sediments, vicinity of Sleeping Bear Point, Lake Michigan. M, 1965, University of Illinois, Urbana.

Somers, Ransom Evarts. Geology of the Burro Mountains copper district, New Mexico. D, 1915, Cornell University.

Somerville, Malcolm Robert. Elasticity, constitution and temperature of Earth's lower mantle. D, 1977, University of California, Berkeley. 134 p.

Somerville, Paul Graham. P coda evidence for a layer of anomalous velocity in the upper crust beneath Leduc, Alberta. M, 1972, University of British Columbia.

Somerville, Paul Graham. Time domain determination of earthquake fault parameters from short-period P-waves. D, 1976, University of British Columbia.

Sommarstrom, S. J. The land-water interface of inland lakes and its implications for ecological planning. D, 1976, University of Michigan. 240 p.

Sommer, Hulda Henrietta. On the question of dispersion in the first preliminary seismic waves. D, 1930, University of California, Berkeley. 94 p.

Sommer, Jesse Valentine. Structural geology and metamorphic petrology of the northern Sierra Estrella, Maricopa County, Arizona. M, 1982, Arizona State University. 127 p.

Sommer, Michael Anthony, II. Analysis and interpretation of the gases released from various sites in rocks and minerals. D, 1974, University of Tulsa. 309 p.

Sommer, Michael Anthony, II. Quantitative analysis of the volatile components of selected basalts and ultrabasic-Ultramafic rocks. M, 1972, University of Tulsa. 96 p.

Sommer, Nicholas A. Folds and cleavage in the Womble Shale (Lower and Middle Ordovician) of Arkansas. M, 1967, University of Missouri, Columbia.

Sommer, Nicholas Anthony. Structural analysis of the Blakely Mountain area, Arkansas. D, 1971, University of Missouri, Columbia. 146 p.

Sommer, Sheldon. Mineralogy and geochemistry of Recent carbonate sediments. M, 1964, Texas A&M University.

Sommer, Sheldon Emanuel. Characterization and application of cathodoluminescence from manganese activated carbonate minerals. D, 1969, Pennsylvania State University, University Park. 124 p.

Sommerfeld, Richard. The heats of formation of zoisite, muscovite, and anorthite and thermochemical calculations on some minerals involving these minerals. D, 1965, University of Chicago. 94 p.

Sommers, David Arthur. Geology of the Ural Northeast 15 minute Quadrangle, Lincoln County, New York. M, 1961, University of Rochester. 52 p.

Sommers, David Arthur. Stratigraphy and structure of a portion of the Bob Marshall wilderness area, northwestern Montana. D, 1966, University of Massachusetts. 234 p.

Somogyi, F. Dewatering and drainage of red mud tailings. D, 1976, University of Michigan. 314 p.

Sompongse, Duangporn. The role of wetland soils in nitrogen and phosphorus removal from agricultural drainage water. D, 1982, University of Florida. 166 p.

Son, Kang-Hyee. Interpretation of electromagnetic dipole-dipole frequency sounding data over a vertically stratified Earth. D, 1985, North Carolina State University. 161 p.

Sonaike, Susanna Yetunde. The relative stabilities of metal-organic complexes in Recent lake sediment. M, 1975, Michigan State University. 25 p.

Sonder, Leslie Jean. Thermal and mechanical models of continental deformation. D, 1986, Harvard University. 177 p.

Sonderegger, John L., II. A photogeologic and structural study of a limestone terrane with emphasis on fractures affecting ground water occurrence, (Limestone county, Alabama). M, 1969, University of Alabama.

Sonderegger, John L., II. A preliminary investigation of the dissolution kinetics of strontianite and witherite. D, 1974, New Mexico Institute of Mining and Technology. 100 p.

Sondergaard, Jon N. Stratigraphy and petrology of the Nooksack Group in the Glacier Creek-Skyline Divide area, North Cascades, Washington. M, 1979, Western Washington University. 103 p.

Sondergeld, Carl H. An experimental study of two-phase hydrothermal convection in a porous medium with applications to geological problems. D, 1977, Cornell University. 115 p.

Sondergeld, Carl H. Interpretation of elastic properties derived from polycrystals of geophysical interest. M, 1973, Queens College (CUNY). 164 p.

Sonderman, Frank James. Geology of the Weepah mining district, Esmeralda County, Nevada. M, 1971, University of Nevada. 85 p.

Sondheim, Mark Weiss. Numerical assessment of soil properties in relation to classification and genesis. D, 1982, University of British Columbia.

Sonerholm, Paul Arthur. Normative mineral distributions in Utah Lake sediments; a statistical analysis. M, 1974, Brigham Young University. 117 p.

Song, Byong-Mu. A study of fundamental engineering characteristics of Recent and Pleistocene marine sediments of Sabine Pass area, Gulf of Mexico. D, 1967, Texas A&M University. 74 p.

Songer, Nathan L. Seepage velocities in stress-relief fractures in the eastern Kentucky coal field. M, 1987, Eastern Kentucky University. 87 p.

Songsirikul, Benja. Associations and survivorship patterns of late Paleozoic brachiopods. M, 1978, University of Kansas. 116 p.

Sonido, Ernesto P. Gravity survey for chromite over rugged topography, Coto ore body, Masinloc area, Zimbales Prov. Luzon Island, Philippines. D, 1963, Washington University. 63 p.

Sonido, Ernesto P. Gravity survey of the Golden-Morrison-Denver area, Colorado. M, 1959, Colorado School of Mines. 47 p.

Sonnad, Jagadeesh Ramana. Application of digital filtering in the study of the geological structure under the LASA. M, 1972, Texas A&M University. 150 p.

Sonnad, Jagadeesh Ramanna. Apparent attenuation associated with seismic wave interference in two-component cyclic stratification. D, 1980, Texas A&M University. 122 p.

Sonnamaker, Charles P. Subsurface geology of southeastern Throckmorton County, Texas. M, 1959, University of Oklahoma. 61 p.

Sonnefield, Robert D. Geology of Northwest Jackson County with special emphasis on the Caseyville Formation. M, 1981, Southern Illinois University, Carbondale. 85 p.

Sonneman, Howard S. Geology of the Boney Mountain area, Santa Monica Mountains, California. M, 1956, University of California, Los Angeles.

Sonnenberg, Frank Payler. The Greenbrier Formation in eastern Kentucky. M, 1949, University of Cincinnati. 21 p.

Sonnenberg, Stephen A. Tectonics, sedimentation, and petroleum potential, northern Denver Basin, Colorado, Wyoming, and Nebraska. D, 1981, Colorado School of Mines. 215 p.

Sonnenberg, Stephen Arnold. Properties of natural gas reservoirs in Cotton Valley sandstones, Northwest Louisiana. M, 1975, Texas A&M University. 125 p.

Sonnevil, Ronald Alan. Evolution of skarn at Monte Cristo, Nevada. M, 1979, Stanford University. 81 p.

Sonneville, Joseph Leonardus Johannes De see De Sonneville, Joseph Leonardus Johannes

Sontag, Karen D. Delineation of a coal burn edge with seismic refraction and magnetics. M, 1984, Wright State University. 55 p.

Sontag, Richard Joseph. Regional gravity survey of parts of Beaver, Millard, Piute, and Sevier counties, Utah. M, 1965, University of Utah. 30 p.

Soo, Kwong Yin. A geochemical and petrological study of volcanic rocks in the Beardmore-Geraldton Archaean greenstone belt, northwestern Ontario. M, 1988, Brock University. 149 p.

Soo, Sweanum. Studies of plain and reinforced frozen soil structures. D, 1984, Michigan State University. 283 p.

Sood, Manmohan Kumar. Melting relations of selected alkaline rocks under water vapor and controlled oxygen pressure. M, 1968, University of Western Ontario. 116 p.

Sood, Manmohan Kumar. Phase relations in the system diopside-albite leucite and its implications to leucite-bearing rocks. D, 1969, University of Western Ontario. 135 p.

Sooky, Attilla A. The flow through a meander-flood plain geometry. D, 1964, Purdue University.

Soonawala, N. M. Diffusion of radon-222 in overburden and its application to uranium exploration. D, 1976, McGill University.

Soonthornsaratul, Chekchanok. Petrology and diagenetic history of Queen City sandstones (Eocene), Mestena Grande Field, Jim Hogg County, Texas. M, 1988, Texas A&I University. 171 p.

Soper, Donald Arthur. Application of wrench-fault tectonics in the Carolina slate belt of south central North Carolina. M, 1974, University of Florida. 54 p.

Soper, Edgar K. The buried rock surface and pre-glacial river valleys of Minneapolis and vicinity. M, 1914, University of Minnesota, Minneapolis. 31 p.

Soper, Edgar K. The origin, occurrence, and uses of Minnesota peat. D, 1916, University of Minnesota, Minneapolis.

Soper, Elmer Gail. Geology of a portion of the Timber Quadrangle, Oregon. M, 1974, University of Oregon. 102 p.

Soper, Harland. Geology of the thermal water area and deposits north of Dubois, Wyoming. M, 1942, Texas Tech University. 80 p.

Sopher, Stephen R. Palaeomagnetic study of the Sudbury Irruptive. M, 1962, University of British Columbia.

Sopher, Stephen R. Paleomagnetic study of the Sudbury Irruptive (Ontario). M, 1961, Carleton University. 93 p.

Sophocleous, Marios Andreou. Analysis of heat and water transport in unsaturated-saturated porous media. D, 1978, University of Alberta. 271 p.

Sopkin, Sandra Meryl. Ultrasonic determination of the elastic properties of single-crystal fayalite, Fe$_2$SiO$_4$.

M, 1982, Pennsylvania State University, University Park. 58 p.

Sopp, George P. Geology of the Montana Mine area, Empire Mountains, Arizona. M, 1940, University of Arizona.

Sopuck, Vladimir Joseph. A lithogeochemical approach in the search for areas of felsic volcanic rocks associated with mineralization in the Canadian Shield. D, 1977, Queen's University. 296 p.

Sopuck, Vladimir Joseph. Structural geology in the Birch Lake-Uchi Lake metavolcanic-metasedimentary belt, Mitchell Township, Ontario. M, 1971, University of Manitoba.

Soranno, Michael Andrew. Effects of urban growth on stream flow regimes on Long Island, New York. D, 1981, Indiana University, Bloomington. 177 p.

Sorauf, Christine M. A numerical study of caustics and polarizations produced by seismic waves interacting with three-dimensional geological structures. M, 1987, University of North Carolina, Chapel Hill. 54 p.

Sorauf, James Edward. Devonian stratigraphy of eastern Jasper Park, Alberta, Canada. M, 1955, University of Wisconsin-Madison.

Sorauf, James Edward. Structural geology and stratigraphy of the Whitmore area, Mohave County, Arizona. D, 1962, University of Kansas. 398 p.

Sorbara, James Paul. The geology and geochemistry of the Innerring Lake area, Back River Complex, Northwest Territories. M, 1979, University of Toronto.

Soregaroli, Arthur Earl. Geology of the Boss Mountain Mine (Central), British Columbia. D, 1968, University of British Columbia.

Soregaroli, Arthur Earl. Geology of the McKim Creek area, Lemhi County, Idaho. M, 1961, University of Idaho. 53 p.

Sorem, Ronald Keith. Origin of manganese deposits of Busuanga Island, Philippines. D, 1958, University of Wisconsin-Madison.

Sorensen, Arthur. Petrographic and geochemical investigation of wallrock alteration, Silver Summit Mine, Coeur d'Alene mining dist., Shoshone Co., Idaho. M, 1957, University of Washington. 211 p.

Sorensen, Charles Elliott. A study of active processes affecting grain-size and chemical distribution in three selected basins of Florida Bay, Florida. M, 1985, Wichita State University. 146 p.

Sorensen, Gary Frank. Petrogenesis of the Lewis Peak-Elk Peak area, Judith Mountains, Fergus County, Montana. M, 1985, University of Montana. 109 p.

Sorensen, Mark Randall. The Queen of Bronze copper deposit, southwestern Oregon; an example of sub-sea floor massive sulfide mineralization. M, 1983, University of Oregon. 205 p.

Sorensen, Martin L. Petrology of some sandstones from the eastern Olympic Peninsula, Washington. M, 1971, San Jose State University. 57 p.

Sorensen, Paul Arnold. Sedimentology and sedimentary petrology of a Paleogene basin near Laguna San Ignacio, Estado de Baja California Sur, Mexico. M, 1983, University of California, Santa Barbara. 123 p.

Sorensen, Sorena Svea. Petrology of basement rocks of the California continental borderland and the Los Angeles Basin. D, 1984, University of California, Los Angeles. 447 p.

Sorenson, Curtis James. Interrelationships between soils and climate and between paleosols and paleoclimates; forest/tundra ecotone, north central Canada. D, 1973, University of Wisconsin-Madison.

Sorenson, Eric R. The vegetation and water chemistry of two patterned fens in northern Maine. M, 1986, University of Maine. 89 p.

Sorenson, G. E., Jr. Geology of the Thousand Springs Valley area, Madison and Teton counties, Idaho. M, 1961, University of Wyoming. 95 p.

Sorenson, Raymond P. Clay mineralogy of suspended-load transport, Guadalupe River, Texas. M, 1975, University of Texas, Austin.

Sorenson, Robert Eugene. The geology and ore deposits of the South Mountain mining district, Owyhee County, Idaho. M, 1927, University of Idaho. 22 p.

Sorenson, Seval C. The igneous intrusives of the South Cuyuna iron range and their relation to the iron ore bodies. M, 1930, University of Minnesota, Minneapolis. 49 p.

Sorey, M. L. Numerical modeling of liquid geothermal systems. D, 1976, University of California, Berkeley. 76 p.

Sorgenfrei, Harold. Gas production from the New Albany Shale. M, 1952, Indiana University, Bloomington. 26 p.

Sorkness, H. O. Geology of soil making and agricultural geology. M, 1901, University of Minnesota, Minneapolis.

Soroka, Leonard Gregory. Modern Bermuda lichenoporids, their skeletal morphology, variability, and reefal ecology. D, 1977, Pennsylvania State University, University Park. 206 p.

Soroka, William L. The effect of low temperature oxidation on magnetic directions within Atlantic Ocean pillow basalts. M, 1981, Michigan Technological University. 93 p.

Soronen, George Charles. A study of Welleria aftonensis Warthin from the (Middle Devonian) Gravel Point Formation of Michigan. M, 1956, University of Michigan.

Soroos, R. L. Determination of hydraulic conductivity of some Oahu aquifers with step-drawdown test data. M, 1973, University of Hawaii. 239 p.

Sorrel, Frank D. Areal geology of the Quinlan area, Woodward County, Oklahoma. M, 1961, University of Oklahoma. 79 p.

Sorrell, Charles A. Metamorphism and metasomatism of the Inwood Marble, Poundridge area, New York. M, 1958, Miami University (Ohio). 68 p.

Sorrell, Charles Arnold. Solid state formation of barium, strontium, and lead feldspars in clay-sulfate mixtures. D, 1961, University of Illinois, Urbana. 80 p.

Sorrells, Gordon Guthrey. Studies in seismology. D, 1971, Southern Methodist University.

Sorrells, Gordon Guthrey. Variations of longitudinal wave velocity in low porosity rocks. M, 1962, Southern Methodist University. 73 p.

Sorrentino, Anthony Vincent. Yorktown cerioporids from Colerain Beach; globular cyclostome byrozoans in Pliocene-?Miocene shelly sands along the Chowan River in northeastern North Carolina. D, 1977, Pennsylvania State University, University Park. 185 p.

Sorrwar, Gholam. Middle Devonian ostracods from the Ferron Point Formation, Cheboygan and Presque Isle counties, Michigan. M, 1961, Michigan State University. 68 p.

Sorrwar, Gholam. Upper Cretaceous and Tertiary ostracod faunas from Kohat District of West Pakistan. D, 1970, Michigan State University. 232 p.

Soske, Joshua Lawrence. Theory of magnetic methods of applied geophysics with applications to the San Andreas Fault. D, 1935, California Institute of Technology. 110 p.

Soto Ruiz, Carlos. Chemical composition of deep subsurface waters of the Anadarko Basin. M, 1974, University of Tulsa. 56 p.

Soto, A. E. Engineering geology and relative stability of part of Ladera, San Mateo County, California. M, 1975, Stanford University. 80 p.

Soto-Vargas, M. Fernando. Study of the effects on dissolved-solids content ground water due to phreatophytes. M, 1963, New Mexico Institute of Mining and Technology.

Sotonoff, M. L. X-ray diffraction analysis and particle size distribution of the Francis Creek Shale of northern and central Illinois. M, 1976, Northeastern Illinois University.

Sottek, T. C. Geology of the Deadman Mountain and Whitewood Anticline area, Meade-Lawrence counties, South Dakota. M, 1959, South Dakota School of Mines & Technology.

Souaya, Fernand Joseph. A study of the microfossils in some Miocene well sections and outcrops in Egypt. M, 1948, University of Texas, Austin.

Soucek, Charles H. Structure and stratigraphy of Sumner County, Kansas, related to petroleum accumulation. M, 1959, Kansas State University. 64 p.

Souch, Bertram Elford. Investigation of Great Bear Lake minerals with appendix on a microscopic technique for the determination of opaque minerals. M, 1933, University of Alberta. 149 p.

Soucie, Gordon E. A lithochemical study of metasomatic alteration in the Temagami greenstone belt, N.E. Ontario. M, 1979, Laurentian University, Sudbury.

Soucy, René Modélisation du comportement de l'aquifére du Cap-de-la-Madeleine. M, 1986, Universite Laval. 71 p.

Souder, Karl Cameron. The hydrochemistry of thermal waters of southeastern Idaho, western Wyoming and northeastern Utah. M, 1985, University of Idaho. 139 p.

Souders, Robert H. Angle of emergence of seismic P waves and its variation with frequency. M, 1967, Oregon State University. 56 p.

Souders, Robert Patton. Petrology of the reefoid limestones of the Kettle Falls area, Stevens County, Washington. M, 1967, Washington State University. 53 p.

Souflis, Constantinos L. Seismic damage analysis of earth retaining structures and natural slopes. D, 1985, Rensselaer Polytechnic Institute. 265 p.

Soukup, James J. Hydrogeologic investigation of a municipal landfill in Broome County, New York. M, 1988, University of Massachusetts. 157 p.

Soukup, William G. A computer model and environmental analysis of the Rotterdam Aquifer, Schenectady County, New York. M, 1978, Ohio University, Athens. 85 p.

Soule, Charles H. Tectonic geomorphology of Big Chino Fault, Yavapai County, Arizona. M, 1978, University of Arizona.

Soule, James McGovern. Structural geology of the northern Animas Mountains, Hidalgo County, New Mexico. M, 1971, University of New Mexico. 47 p.

Soule, Kenneth Dana. The igneous geology of Bear Mountain and vicinity, Wichita Mountains, Oklahoma. M, 1951, University of Oklahoma. 51 p.

Soule, Mary Alice. Evaluation of reclamation of land strip-mined for coal in Crawford County, Kansas, using remote sensing techniques. M, 1974, University of Kansas. 45 p.

Soule, Ralph P. A gravity survey and analysis of the Republic graben of northeastern Washington. M, 1976, Western Washington University. 72 p.

Sousa Carvalho Martins, Verónica E. de *see* de Sousa Carvalho Martins, Verónica E.

Sousa, Francis Xavier. Geology of the Middlemarch Mine and vicinity, central Dragoon Mountains, Cochise County, Arizona. M, 1980, University of Arizona. 107 p.

Sousa, W. P. Disturbance and ecological succession in marine intertidal boulder fields. D, 1977, [University of California, Santa Barbara]. 240 p.

Souster, W. E. Characteristics of thin loess deposits in the Swift Current map area (72J). M, 1973, University of Saskatchewan.

Souter, James Edwin. Environmental mapping of the Oread and Lecompton megacycles of the Shawnee Group (Upper Pennsylvanian of the Midcontinent). M, 1966, University of Illinois, Urbana.

South, Bernard Carl. Mineralogy and stable isotope geochemistry of the Devonian Iron Mountain massive sulfide deposit, Shasta County, California. M, 1985, University of California, Davis. 141 p.

South, David Long. Laboratory studies of fluid flow through borehole seals. D, 1983, University of Arizona. 298 p.

South, David Long. Sulphide zoning at the Lakeshore copper deposit, Pinal County, Arizona. M, 1972, University of Arizona.

South, Mark Veeder. Stratigraphy, depositional environment, petrology and diagenetic character of the Morrow reservoir sands, Southwest Canton Field, Blaine and Dewey counties, Oklahoma. M, 1983, Oklahoma State University. 179 p.

Southard, Alvin Reid. Classification and genesis of soils high in black shale. D, 1963, Cornell University. 120 p.

Southard, John B. Turbulence and momentum transport in flow between concentric rotating cylinders. D, 1966, Harvard University.

Southard, Lloyd Colman. The diatoms of a Washington peat bog. M, 1931, University of Washington. 45 p.

Souther, Jack Gordon. The geology of Terrace area, Coast District, British Columbia. D, 1956, Princeton University. 131 p.

Southerland, Elizabeth. A continuous simulation modeling approach to nonpoint pollution management. D, 1982, Virginia Polytechnic Institute and State University. 165 p.

Southernwood, Renee. Late Cretaceous limestone clast conglomerates of Honduras. M, 1986, University of Texas at Dallas. 299 p.

Southhard, Randal Jay. Subsoil blocky structure formation in North Carolina coastal plain soils. D, 1983, North Carolina State University. 170 p.

Southwick, David Leroy. Mafic and ultramfic rocks of the Ingalls-Peshastin area, Washington, and their geologic setting. D, 1962, The Johns Hopkins University.

Southwick, P. Resistivity survey for glass sand in Hantsport, Nova Scotia. M, 1950, Massachusetts Institute of Technology.

Southwick, Peter Frederick. Velocity of shear waves through unconsolidated materials. D, 1952, Massachusetts Institute of Technology. 63 p.

Southwick, Robert S. Geology of the south-central part of the Schell Peaks Quadrangle, Nevada. M, 1962, University of Southern California.

Southwick, Stanley Harpham. Inorganic constituents of crude oil. D, 1951, Massachusetts Institute of Technology. 113 p.

Southwick, Thomas S. Geology of a portion of the Santa Ana Mountains (California). M, 1929, California Institute of Technology. 45 p.

Southworth, Dennis. Geology of the Goodnews Bay ultramafic complex. M, 1986, University of Alaska, Fairbanks. 115 p.

Southworth, J. H. Laboratory and in situ measurement of sound velocity and mass physical properties of some unconsolidated marine sediment cores. M, 1971, University of Hawaii. 132 p.

Souto, Antonio Pedro Dias. A bottom gravity survey of Carmel Bay, California. M, 1973, Naval Postgraduate School.

Souto, Antonio Pedro Dias *see* Dias Souto, Antonio Pedro

Souto-Maior Filho, Joel. Temperature studies in a case study of the applications of thermal remote sensing. M, 1972, University of Wisconsin-Madison.

Souza, Euler Magno de *see* de Souza, Euler Magno

Souza, Jairo M. Transmission of seismic energy through the Brazilian Parana Basin layered basalt stack. M, 1982, University of Texas, Austin.

Souza, Jairo Marcondes de *see* de Souza, Jairo Marcondes

Souza, Manuel Edward. Physiography and structure of the Oregon Coast Range Province. M, 1927, University of Oregon. 50 p.

Souza, Murilo M. Wavelet extraction parameters of homomorphic deconvolution. M, 1976, University of Houston.

Sowayan, Abdulrahman M. Seismic studies of the bedrock valley system in northwestern Illinois. M, 1969, Northern Illinois University. 67 p.

Sowers, George M. Structure and petrology of the Pre-Cambrian granites near Red Lodge, Montana. M, 1944, The Johns Hopkins University. 62 p.

Sowers, Janet Marie. Pedogenic calcretes of the Kyle Canyon alluvial fan, southern Nevada; morphology and development. D, 1985, University of California, Berkeley. 184 p.

Sowers, John William. The prediction of landslides typical of eastern Ohio. M, 1975, Ohio University, Athens. 133 p.

Sowers, Nancy R. Geologic factors affecting the geohydrology of the Tongue River valley, Sheridan County, Wyoming. M, 1979, University of Wyoming. 112 p.

Sowers, Todd. Stable isotopic composition of occluded gases in ice cores. M, 1987, University of Rhode Island.

Soydemir, Cetin. Long term stability of cohesive soil media. D, 1967, Princeton University. 307 p.

Soyupak, S. Modifications to sanitary landfill leachate organic matter migrating through soil. D, 1979, University of Waterloo.

Sozanski, A. G. Geochemistry of ferromanganese oxide concretions and associated sediments and bottom waters from Shebandowan Lakes, Ontario, Northwest Territories. M, 1974, University of Ottawa. 139 p.

Spackman, William Jr. The flora of the (Eocene) Brandon Lignite; geological aspects and a comparison of the flora with its modern equivalents (Vermont). D, 1949, Harvard University.

Spader, David H. Early diagenesis of carbonate sediments in a supra-tidal environment, Bonaire, the Netherlands Antilles. M, 1983, University of New Hampshire. 90 p.

Spahn, Ronald A., Jr. Geology of the southwestern Horseshoe Hills area, Gallatin County, Montana. M, 1967, Montana State University. 95 p.

Spahr, Cynthia L. Economic geology of the Midnight, Black Knight, and associated magnetite claims, Lone Mountain, Lincoln County, New Mexico. M, 1983, New Mexico Institute of Mining and Technology. 196 p.

Spaid-Reitz, Malia K. Petrogenesis of a bimodal assemblage of alkali-basalt and rhyolitic ignimbrite, Gravelly Range, Southwest Montana. M, 1980, Kansas State University. 92 p.

Spain, David. A study of the gravel of the middle and upper Cumberland River. M, 1940, Vanderbilt University.

Spain, Ernest. An occurrence of Pleistocene clay near Indian Mound, Stewart County, Tennessee. M, 1933, Vanderbilt University.

Spalding, James Simon. Phosphate exploration and property evaluation in southeastern Idaho, illustrated by the Dry Valley area. M, 1974, Utah State University. 132 p.

Spalding, Robert W. Minor movements of the Earth's crust during the Pennsylvanian period in the Laramie Basin region. M, 1928, University of Wyoming. 64 p.

Spalding, Roy Follansbee. The contemporary geochemistry of uranium in the Gulf of Mexico distributive province. D, 1972, Texas A&M University. 268 p.

Spalding, Thomas D. Subsurface stratigraphy and depositional environments of the Middle Mississippian St. Louis Formation, Monteagle Limestone and adjacent formations in north-central Tennessee. M, 1982, University of Kentucky. 162 p.

Spall, Henry Roger. Thermal demagnetization of some Spitzbergen dolerites. M, 1968, Southern Methodist University. 23 p.

Spalvins, Karlis. The stratigraphy of the Conasauga Group in the vicinity of Adairsville Quadrangle, Georgia. M, 1967, Emory University. 62 p.

Spane, Frank A., Jr. Evaluation of factors influencing the inorganic water-quality regimen of Carson River, Carson Valley, Nevada-California. D, 1977, University of Nevada - Mackay School of Mines. 205 p.

Spane, Frank A., Jr. Hydrogeologic studies near Akron, Colorado. M, 1971, Colorado State University. 104 p.

Spang, John Harvey. A structural study across a transition of radiometric age dates in the Gallatin Canyon area, Gallatin County, Montana. M, 1967, Brown University. 82 p.

Spang, John Harvey. Mechanics of folding of the Becraft Limestone (Helderberg Group, lower Devonian) on Becraft Mountain, Columbia County, New York. D, 1971, Brown University.

Spangenberg, Norma E. Partitioning and hydrologic process models in watershed yield models. M, 1969, [Colorado State University].

Spangenberg, Norman E. A physically-based approach to watershed partitioning. D, 1976, Colorado State University. 93 p.

Spangler, D. R. A characteristic approach to ground motion determination. D, 1977, University of Illinois, Urbana. 115 p.

Spangler, Daniel Patrick. A geophysical study of the hydrogeology of the Walnut-Gulch experimental watershed, Tombstone, Arizona. D, 1969, University of Arizona. 148 p.

Spangler, Daniel Patrick. Engineering uses of geology as applied by the Virginia Department of Highways. M, 1964, University of Virginia. 135 p.

Spangler, John Franklin. The geology of the Graveston Quadrangle, Union and Knox counties, Tennessee. M, 1949, University of Tennessee, Knoxville. 68 p.

Spangler, Lawrence E. Karst hydrogeology of northern Fayette and southern Scott counties, Kentucky. M, 1982, University of Kentucky. 103 p.

Spangler, Walter Blue. The geology of the Chester Series in an area in southern Orange County, Indiana. M, 1940, Indiana University, Bloomington. 18 p.

Spanglet, M. Whole-rock Rb-Sr isochron data for the Lowerre and Poughquag quartzites; a possible case of provenance ages in psammitic rocks. M, 1974, Queens College (CUNY). 124 p.

Spanjers, Raymond Peter. Lineament and fracture analysis of three molybdenite-bearing granitic plutons in the eastern Piedmont of North Carolina. M, 1983, North Carolina State University. 42 p.

Spanos, Thomas James Timothy. A stability analysis of thermo-mechanical flow in the Earth's mantle. D, 1977, University of Alberta. 121 p.

Spanski, Gregory Thomas. Chromite of the Twin Sisters Mountains, Washington. M, 1963, University of Illinois, Urbana. 41 p.

Spanski, Gregory Thomas. Geology and geochemistry of the Trail Creek kyanite deposit, Albany County, Wyoming. D, 1966, University of Illinois, Urbana. 122 p.

Sparacin, W. G. Predicting field settlements of compressible soils due to complex loading. D, 1975, Cornell University. 206 p.

Sparenberg, George Russell. The Paleozoic formation south of the San Saba River in McCulloch County, Texas. M, 1932, University of Texas, Austin.

Spariosu, Dann Jack. Paleomagnetic investigations of northern Appalachian terrane history during the middle to late Paleozoic. D, 1984, Columbia University, Teachers College. 300 p.

Sparkes, Ann Katherine. Petrology of the East Piute Pluton, southeastern California. M, 1981, Vanderbilt University. 119 p.

Sparks, Billy Joe. The geology of the Marmaton Group of north-western Rogers County, Oklahoma. M, 1955, University of Oklahoma. 66 p.

Sparks, Dale D. Orbitoid foraminifera of the California Eocene. M, 1924, Stanford University. 38 p.

Sparks, Dennis Michael. Microfloral zonation and correlation of some lower Tertiary rocks in southwest

Washington and some conclusions concerning the paleoecology of the flora. D, 1967, Michigan State University. 214 p.

Sparks, Diane K. Epizoans on the Devonian brachiopod, Paraspirifer bownockeri (Stewart). M, 1978, Bowling Green State University. 38 p.

Sparks, Thomas Norton. Sedimentary structures in Hatteras abyssal plain sediments. M, 1979, Duke University. 123 p.

Sparks, William E. The paleoecology and biostratigraphy of uppermost Eocene and lowermost Oligocene strata of Little Stave Creek, Alabama. M, 1967, Rutgers, The State University, New Brunswick. 104 p.

Sparlin, Mark Alan. Crustal structure of the eastern Snake River plain determined from ray-trace modeling of seismic refraction data. M, 1981, Purdue University.

Sparling, Dale R. Occurrence of Prasopora in a core from the Northville area, Michigan. M, 1956, Wayne State University.

Sparling, Dale Richard. Geology of Ottawa County, Ohio. D, 1965, Ohio State University. 298 p.

Sparling, Kirk Darren. Depositional history and hydrocarbon potential of the lower Douglas "A" Sandstone (Upper Pennsylvanian) in a portion of Hemphill County, Texas. M, 1988, West Texas State University. 103 p.

Spasari, John V. Shoreline processes and sediment response related to the origin of beach cusps on Whidbey and Fidalgo Islands, Washington. M, 1978, Western Washington University. 107 p.

Spasojevic, Miodrag P. Numerical simulation of two-dimensional (plan view) unsteady water and sediment movement in natural watercourses. D, 1988, University of Iowa. 243 p.

Spat, A. G. Iron formations and associated rocks in the Mount Wright area, Quebec. M, 1959, McGill University.

Spatz, David Moore. Genetic, spectral, and Landsat thematic mapper imagery; relationships between desert varnish and Tertiary volcanic host rocks, southern Nevada. D, 1988, University of Nevada. 364 p.

Spatz, David Moore. Geology and alteration-mineralization zoning of the Pine Flat porphyry copper occurrence, Yavapai County, Arizona. M, 1974, University of Arizona.

Spatz, Jeffrey Michael. Structure of the post-Caledonian sequence on Midterhuken Peninsula, Spitsbergen. M, 1981, University of Wisconsin-Madison.

Spatz, Paige Herzon. The Cretaceous sequence on Midterhuken Peninsula, Spitsbergen. M, 1983, University of Wisconsin-Madison. 144 p.

Spaulding, Karen Lee. Petrology and geochemistry of mid-Tertiary volcanic rocks, La Perla, Chihuahua, Mexico. M, 1985, University of Oklahoma. 83 p.

Spaulding, Walter Geoffrey. Pollen analysis of fossil dung of Ovis canadensis from southern Nevada. M, 1974, University of Arizona.

Spaulding, Walter Geoffrey. The late Quaternary vegetation of a southern Nevada mountain range. D, 1981, University of Arizona. 307 p.

Spaulding, Walter Miles. Salt domes; their nature, origin, and composition. M, 1952, University of Michigan.

Spaven, Harvey Robert. Granite tectonics in part of Eden Township, Sudbury District, Ontario. M, 1966, McMaster University. 80 p.

Spaw, Joan Mussleer. Vertical distribution, ecology and preservation of Recent polycystine Radiolaria of the North Atlantic Ocean (southern Sargasso Sea region). D, 1979, Rice University. 250 p.

Spaw, Joan Mussler. Paleo-environments of Middle Pennsylvanian Chaetetes lithotopes, Texas and New Mexico. M, 1977, Rice University. 80 p.

Spaw, Richard Hoencke. Late Pleistocene stratigraphy and geologic development of Cozumel Island, Quintana Roo, Mexico. M, 1977, Rice University. 143 p.

Spayne, Robert William. A unified concept of drumlins based on an investigation of eastern Massachusetts drumlins. D, 1972, Clark University.

Spear, D. B. Explosion - collapse hypothesis for the origin of Crater Rings, Elmore County, Idaho. M, 1975, SUNY at Buffalo. 31 p.

Spear, D. B. The geology and volcanic history of the Big Southern Butte-East Butte area, eastern Snake River plain, Idaho. D, 1979, SUNY at Buffalo. 127 p.

Spear, F. S. Phase equilibria and mineral chemistry of a hydrothermally synthesized amphibolite. D, 1976, University of California, Los Angeles. 192 p.

Spear, James W. Environments of deposition of the Fort Thompson Formation (Pleistocene, southern Florida), based on the microfauna. M, 1974, University of Akron. 162 p.

Spear, John H. A study of the (Mississippian) Short Creek Oolite, Ottawa County, Oklahoma. M, 1951, University of Oklahoma.

Spear, Ray William. The history of high-elevation vegetation in the White Mountains of New Hampshire. D, 1981, University of Minnesota, Minneapolis. 257 p.

Spear, Steven G. Geologic mapping of erosional susceptibility. M, 1971, University of Southern California.

Spear, Steven Gregory. Distribution and relative age of selected landforms in Death Valley National Monument, California and Nevada. D, 1986, University of California, Los Angeles. 310 p.

Spearing, Darwin Robert. Stratigraphy, sedimentation and tectonic history of the Paleocene-Eocene Hoback Formation of western Wyoming. D, 1969, University of Michigan. 205 p.

Spearman, Charles. Notes on the stratigraphy of an area in the Catskill region of eastern New York. M, 1912, Columbia University, Teachers College.

Spearnak, Mark. Selected soil water ion concentrations of a mountain watershed. M, 1977, Colorado State University. 169 p.

Spears, David B. Strain in Ordovician and Devonian shales from the Massanutten Synclinorium; implications for fold development and tectonic history. M, 1983, Virginia Polytechnic Institute and State University. 70 p.

Spechler, Rick M. Chemical character of water in the upper part of the Floridan Aquifer, near Tarpon Springs, Florida. M, 1983, University of South Florida, Tampa. 95 p.

Specht, Daniel J. A new solubility model and its application to the Mississippi Valley lead deposits. M, 1983, California State University, Hayward. 108 p.

Specht, Glenwood W. Geometry and reservoir characteristics of turbidite sand bodies in Castaic Junction field, (Los Angeles, California). M, 1969, University of Southern California.

Specht, Ralph W. Petrologic and stratigraphic analysis of bioclastic limestones in the Sundance Formation; Upper Jurassic of South-central Wyoming. M, 1978, University of Iowa. 120 p.

Specht, Thomas David. Architecture of the Malawi Rift, East Africa. M, 1987, Duke University. 56 p.

Specht, Thomas Henry. A subsurface study of the sandstone in the Big Clifty Formation in southwestern Indiana. M, 1985, Indiana University, Bloomington. 92 p.

Specter, Robert Michael. Effects of pore structure on mixing in stable, single-phase miscible displacements. M, 1984, New Mexico Institute of Mining and Technology. 110 p.

Spector, Allan. Spectral analysis of aeromagnetic data. D, 1968, University of Toronto.

Spector, I. H. Manganese deposits in the Riding mountain (upper Cretaceous) and Stonewall (Ordovician) formations in southern Manitoba. M, 1941, University of Manitoba.

Speden, Ian Gordon. Paleozoology of Lamellibranchia from the type area of Fox Hills Formation (upper Cretaceous, Maestrichtian), South Dakota. D, 1965, Yale University. 551 p.

Speed, Bert Lewis. A sedimentary study of the Yeso Formation of central and northern New Mexico. M, 1958, Texas Tech University. 143 p.

Speed, Robert Clarke. Scapolitized gabbroic complex, West Humboldt Range, Nevada. D, 1962, Stanford University. 255 p.

Speelman, J. D. Megaspore palynology and paleoecology, Foremost Formation (Upper Cretaceous), southeastern Alberta. M, 1978, University of Calgary. 114 p.

Speer, J. A. The stratigraphy and metamorphism of the Snyder Group, Labrador. D, 1976, Virginia Polytechnic Institute and State University. 226 p.

Speer, John A. Plagioclases from the Kiglapait Intrusion, Labrador. M, 1973, Virginia Polytechnic Institute and State University.

Speer, John H. A study of the Short Creek Oolite, Ottawa County, Oklahoma. M, 1951, University of Oklahoma. 50 p.

Speer, Kevin George. The influence of geothermal sources on deep ocean temperature, salinity, and flow fields. D, 1988, Massachusetts Institute of Technology. 146 p.

Speer, Michael Carr. Magnesium vapor solubility and interactions in liquid iron-base alloys. D, 1970, Stanford University. 141 p.

Speer, Paul Edward. Tidal distortion in shallow estuaries. D, 1984, Woods Hole Oceanographic Institution. 210 p.

Speer, Roberta D. Fossil bison remains from the Rex Rodgers Site, Briscoe County, Texas. M, 1975, West Texas State University. 109 p.

Speer, Stephen William. Abo Formation (Early Permian), Sacramento Mountains, New Mexico; a dry alluvial fan and associated basin-fill. M, 1983, University of Texas, Austin. 129 p.

Speer, W. E. Geology of the McDermitt Mine area, Humboldt County, Nevada. M, 1977, University of Arizona. 65 p.

Speers, E. C. Age relations and origin of Sudbury breccias, Ontario. D, 1956, Queen's University. 393 p.

Speers, E. C. Mineralization of the O'Sullivan Lake area (Ontario). M, 1951, University of Toronto.

Speers, E. C. The breccias of the Sudbury area (Ontario). D, 1955, Queen's University.

Speice, Charles. Pine Field, Cedar Creek Anticline, eastern Montana. M, 1957, South Dakota School of Mines & Technology.

Speidel, David Harold. Element distribution among coexisting phases in the system MgO-FeO-Fe₂O₃-SiO₂-TiO₂ as a function of temperature, oxygen fugacity and bulk composition. D, 1964, Pennsylvania State University, University Park. 142 p.

Speidel, W. C. Nearshore sediment distribution at San Onofre, California. M, 1973, [University of California, San Diego].

Speirs, David. A resistivity survey north of Kingston, Ontario. M, 1970, Queen's University. 82 p.

Speliotis, Dionysios Elias. A cryomagnetic study of iron oxides. D, 1961, University of Minnesota, Minneapolis. 133 p.

Spell, Terry Lee. Geochemistry of Valle Grande Member ring fracture rhyolites, Valles Caldera, N.M. M, 1987, New Mexico Institute of Mining and Technology. 213 p.

Spellman, James Wheeler. The geology of a gypsum deposit in Northwest Dawes County, Nebraska. M, 1976, University of Nebraska, Lincoln.

Spellman, Jane L. Simulations of ground water flow at a dredged disposal site, coastal plain, North Carolina. M, 1986, University of Arkansas, Fayetteville.

Spelman, A. R. Geology of the area between Bed Rick Creek and the west fork of Labonte Creek, Converse County, Wyoming. M, 1959, University of Wyoming. 81 p.

Spelman, Allen Rathjen. Stratigraphy of Lower Ordovician Nittany Dolomite in central Pennsylvania. D, 1965, Pennsylvania State University, University Park. 419 p.

Spence, George D. Gravity and seismic studies in the southern Rocky Mountain trench. M, 1976, University of British Columbia.

Spence, George Daniel. Seismic structure across the active subduction zone of western Canada. D, 1984, University of British Columbia.

Spence, Jeffrey Gordon. Geology of the Mineral Hill interlayered amphibolite-augen gneiss complex, Lemhi County, Idaho. M, 1984, University of Idaho. 240 p.

Spence, John Alen. Possible ore forming fluids from sedimentary accumulations. M, 1966, McGill University.

Spence, William Henry. Relict plagioclase phenocrysts from metavolcanic rocks mantling a granite dome and the petrology of the associated rocks in the Grenville Province of southeastern Ontario (Canada). D, 1968, Rutgers, The State University, New Brunswick. 75 p.

Spence, William John. Crustal and upper mantle P-wave velocity heterogeneity and the problem of earthquake location. D, 1973, Pennsylvania State University, University Park. 190 p.

Spencer, Alexander B. Geology of the basic rocks of the eastern portion of the Raggedy Mountains, southwestern Oklahoma. M, 1961, University of Oklahoma. 46 p.

Spencer, Alexander Burke. Alkalic igneous rocks of Uvalde County, Texas. D, 1966, University of Texas, Austin. 189 p.

Spencer, Alice Whitham. Evaporite facies related to reservoir geology, Seven Rivers Formation (Permian), Yates Field, Texas. M, 1987, University of Texas, Austin. 125 p.

Spencer, Arthur C. The geology of Massanutten Mountain in Virginia. D, 1896, The Johns Hopkins University.

Spencer, Barry Craig. Stratiform magnetite-ilmenite deposits of the McClure Mountain-Iron Mountain ultramafic-alkalic complex, Fremont County, Colorado. M, 1978, University of Michigan.

Spencer, Charles G. Reliability of an electrical resistivity method for measuring soil moisture. M, 1982, University of Missouri, Kansas City. 195 p.

Spencer, Charles Grason. The Loogootee Gas Field. M, 1939, Indiana University, Bloomington. 19 p.

Spencer, Charles Winthrop. A petrographic study of the underclay of the Number 6 coal in Illinois. M, 1955, University of Illinois, Urbana.

Spencer, Clyde H., Jr. Statistical correlation of platinum metals with heavy minerals in stream sediments of the Klamath Mountains, California. M, 1971, San Jose State University. 62 p.

Spencer, Edgar W. Geological evolution of the Beartooth Mountains, Montana and Wyoming; Part II, Fracture patterns. D, 1960, Columbia University, Teachers College.

Spencer, Frank Darwyn. Geology in the vicinity of Santa Ynez Canyon, Santa Monica Mountains. M, 1932, University of Southern California.

Spencer, J. W., Jr. Ultrasonic velocities in rocks under crustal conditions. D, 1975, Stanford University. 88 p.

Spencer, James R. Distribution and petrology of clinoptilolite of the Tallahatta Formation (middle Eocene) of Clarke County, Alabama. M, 1983, University of Mississippi. 139 p.

Spencer, Jean. Geologic factors influencing mutation. M, 1964, Baylor University. 136 p.

Spencer, Jeffrey Allen. The effect of hurricanes on sediment accumulation; Graveline Bay, Mississippi. M, 1982, University of New Orleans. 152 p.

Spencer, Jesse Garvin. Geology of the northern portion of the Panther Creek Quadrangle, Mason County, Texas. M, 1988, University of Texas of the Permian Basin. 66 p.

Spencer, JoEllen Page. A comparison of the effects of analysis techniques and computer systems in remote sensing technology and a reference data collection technique. D, 1981, University of Alaska, Fairbanks. 207 p.

Spencer, John Michael. A statistical study of problematical microfossils from the Ordovician Maquoketa Formation of eastern Iowa. M, 1974, Iowa State University of Science and Technology.

Spencer, John Robert. The surfaces of Europa, Ganymede, and Callisto; an investigation using Voyager IRIS thermal infrared spectra. D, 1987, University of Arizona. 228 p.

Spencer, Jon Eric. Geology and geochronology of the Avawatz Mountains, San Bernardino County, California. D, 1981, Massachusetts Institute of Technology. 183 p.

Spencer, Khalil Joseph. Isotopic, major and trace element constraints on the sources of granites in an 1800 Ma old igneous complex near St. Cloud, Minnesota. D, 1987, SUNY at Stony Brook. 266 p.

Spencer, Maria Frances. Petrographic description of the Whitehorse and other Permian and Pennsylvanian sandstones of Oklahoma. M, 1930, University of Oklahoma. 62 p.

Spencer, Mary Jo. Trace metals in seawater; chelation capacities, conditional stability constants, and water sampler evaluations. D, 1984, University of Miami. 116 p.

Spencer, Patrick K. Stratigraphy, lithology, and depositional environment of the Black Prince Formation, southeastern Arizona and southwestern New Mexico. M, 1980, Western Washington University. 51 p.

Spencer, Patrick Kevin. Lower Tertiary biostratigraphy and paleoecology of the Quilcene-Discovery Bay area, northeast Olympic Peninsula, Washington. D, 1984, University of Washington. 173 p.

Spencer, Randall Scott. Multivariant analysis of variance of Neochonetes granulifer (Owen); its implication with respect to evolution, correlation, and time-stratigraphy. D, 1968, University of Kansas. 128 p.

Spencer, Randall Scott. Spiriferacea and Punctospiracea of the Pennsylvanian System of Kansas. M, 1962, University of Kansas. 105 p.

Spencer, Ronald J. Silicate and carbonate sediment-water relationships in Walker Lake, Nevada. M, 1977, University of Nevada. 98 p.

Spencer, Ronald James. The geochemical evolution of Great Salt Lake, Utah. D, 1983, The Johns Hopkins University. 329 p.

Spencer, S. M. Ordovician sequence at Gap Mountain (Giles County), Virginia. M, 1968, Virginia Polytechnic Institute and State University.

Spencer, Sherman Glenn. The geology and petrology of the Almond Pluton, Randolph County, Alabama. M, 1973, Memphis State University.

Spencer, Sue Ann. Groundwater movement in the Paleozoic rocks and impact of petroleum production on water levels in the southwestern Bighorn Basin, Wyoming. M, 1986, University of Wyoming. 165 p.

Spencer, Terry Warren. Studies in the acoustic pulse propagation. D, 1956, California Institute of Technology. 112 p.

Spengler, Charles Joseph. The upper three-phase region in a portion of the system KalSi₃O₈-SiO₂-H₂O at water pressures from two to seven kilobars. D, 1965, Pennsylvania State University, University Park. 178 p.

Spengler, Thomas J. The volcanic stratigraphy and petrology of Rock Pile Mountain Wilderness Area and adjacent area to the south, St. Francois Mountains, SE Missouri. M, 1987, University of Toledo. 207 p.

Speno, Leo Anthony. The Tyler Formation of central Montana. M, 1958, University of Colorado. 79 p.

Spera, F. J. Aspects of the thermodynamic and transport behavior of basic magmas. D, 1977, University of California, Berkeley. 184 p.

Sperandio, R. J. The geology of the Brookfield Diorite, New Milford Quadrangle, Connecticut. M, 1974, George Washington University.

Sperandio, Robert J. The petrology and structure of the Hualapai Flat area, Washoe, Pershing, and Humboldt counties, Nevada. D, 1978, Colorado School of Mines. 154 p.

Speranza, Angelo. Sedimentation and diagenesis of the Pekisko Formation (Mississippian), Canyon Creek, Alberta. M, 1984, University of Waterloo. 278 p.

Sperb, Grace. An invertebrate fauna from the (Triassic) lower Star Peak Formation of the West Humboldt Range, Nevada. M, 1937, Columbia University, Teachers College.

Sperber, Christine Martina. Structural breaks in the accretionary complex from the western Gulf of Alaska; subduction of the trailing edge of the Yakutat Terrane and the transition from tectonic erosion to accretion. M, 1987, Stanford University.

Sperling, Tedd F. A study of velocity anisotropy with respect to horizontal receivers. M, 1984, Michigan State University. 100 p.

Sperr, Jay T. Xenoliths of the Leucite Hills volcanic rocks, Sweetwater County, Wyoming. M, 1985, University of Wyoming. 57 p.

Sperrazza, Joseph Thomas. Reef foraminifera of Raroia Atoll, French Oceania (Pacific Ocean). M, 1954, New York University.

Sperrazza, Josephine. The Cretaceous and Tertiary foraminifera of eastern, central and western Sicily. M, 1958, New York University.

Sperry, Steven W. The Flagstaff Formation; depositional environment and paleoecology of clastic depositional Salina, Utah. M, 1980, Brigham Young University.

Spesshardt, Scott Alan. Late Devonian-Early Mississippian phosphorite-bearing shales, Arbuckle Mountain region, south-central Oklahoma. M, 1985, Texas Tech University. 109 p.

Spetseris, Jerry. Depositional model of the Queen City Formation (Eocene) in the Hagist Ranch Field, Duval and McMullen counties, Texas; delineation of lithofacies using seismic-stratigraphic modeling. M, 1986, University of Houston.

Spetzler, Harmut A. W. Part 1; the effect of temperature and partial melting on velocity and attenuation in a simple binary system; pt. 2; effect of temperature and pressure on elastic properties of polycrystalline and single crystal MgO. D, 1969, California Institute of Technology. 246 p.

Spevack, B. Z. A geochemical study of interbedded gneisses and amphibolites at the Shelburne Falls dome, northwestern Massachusetts. M, 1979, SUNY at Binghamton. 127 p.

Speyer, Patricia M. Alkaline earth metal chemistry in Adirondack Lake sediments. M, 1985, University of Rochester. 66 p.

Speyer, Stephen E. Trilobite clustering in the Hamilton Group of New York State; behavioral paleobiology of Middle Devonian trilobites. M, 1983, University of Rochester. 113 p.

Speyer, Stephen Eric. Taphonomy and paleoecology of Middle Devonian (Hamilton Group) trilobite assemblages. D, 1985, University of Rochester. 248 p.

Speziale, Mario H. Correlations of hydrocarbons in shales and oil, Peoria Field, Colorado. M, 1972, Colorado School of Mines. 68 p.

Sphar, Joe D. The origin of the Wreford Chert (Wolfcampian, Permian) of Kansas. M, 1965, Wichita State University. 86 p.

Spice, John Overstreet. Geology of northwest part of Dry Devil Quadrangle, Val Verde, Texas. M, 1954, University of Texas, Austin.

Spicer, R. C. Glaciers in the Olympic Mountains, Washington; present distribution and recent variations. M, 1986, University of Washington. 158 p.

Spicer, Stanley R. Stratigraphy and paleontology of the eastern portion of the Jones Sink, Meade County, Kansas. M, 1975, Kent State University, Kent. 69 p.

Spicker, Fredrick A. The geology, alteration, and mineralization of Parkview Mountain Grand and Jackson counties, Colorado. M, 1973, Colorado State University. 109 p.

Spicola, John. Asymmetry of the A-B-C model with regard to the evolution of Dog Island, Florida. M, 1984, Florida State University.

Spiegel, Zane. Hydraulics of certain stream-connected aquifer systems. D, 1962, New Mexico Institute of Mining and Technology. 107 p.

Spiegelberg, Frederick, III. Stratigraphy of northern Sierra de Ventana, Municipio de Ojinaga, Chihuahua, Mexico. M, 1961, University of Texas, Austin.

Spieglan, Mark Joseph. A finite element model of subduction angle determination and its implications for circulation in the mantle. D, 1983, University of Chicago. 106 p.

Spieker, Andrew Maute. Hydrogeologic aspects of an analog model study of the Fairfield-New Baltimore area, Ohio. D, 1965, Stanford University. 167 p.

Spieker, Edmund M. The molluscan fauna of the Zorritos Formation of northern Peru. D, 1921, The Johns Hopkins University.

Spies, Annette. Estuarine microplankton ecology; an experimental approach. D, 1984, University of British Columbia.

Spies, David C. Possible source beds where reservoir rocks have epigenetic porosity. M, 1951, University of Michigan.

Spies, William Andrew. Lateral and vertical distribution of seven major and minor elements in the No. 11 Seam of the Western Kentucky coal field of the Eastern Interior Basin. M, 1977, University of Kentucky. 71 p.

Spieth, Mary Ann. Two detailed seismic studies in central California; Part I, Earthquake clustering and crustal structure studies of the San Andreas Fault near San Juan Bautista; Part II, Seismic velocity structure along the Sierra foothills near Oroville, California. D, 1981, Stanford University. 174 p.

Spigai, Joseph J. A study of the Rome Formation in the Valley and Ridge Province of East Tennessee. M, 1963, University of Tennessee, Knoxville. 179 p.

Spigai, Joseph John. Marine geology of the continental margin off southern Oregon. D, 1971, Oregon State University. 214 p.

Spigner, Benjamin Cantrell. Hydrogeology of Mississippian aquifers on the southeast flank of the Birmingham Anticline, Jefferson County, Alabama. M, 1975, University of Alabama.

Spiker, Carlisle Titus, Jr. Geology of northwestern Fluvanna County, Virginia. M, 1961, University of Virginia. 55 p.

Spikes, Clayton Henry. A gravimetric survey of the Santa Cruz-Ano Nuevo Point continental shelf and adjacent coastline. M, 1973, United States Naval Academy.

Spilde, Michael N. Tantalum-niobium mineralization as a recorder of pegmatite evolution; Bob Ingersoll and Tin Mountain pegmatites, Black Hills, South Dakota. M, 1987, South Dakota School of Mines & Technology.

Spilhaus, Athelstan F. Study of some optical variables of sea water. M, 1960, Massachusetts Institute of Technology. 29 p.

Spiller, J. Evolution of turritellid gastropods from the Miocene and Pliocene of the Atlantic Coastal Plain. D, 1977, SUNY at Stony Brook. 222 p.

Spiller, Jason. The geology of part of the Olds Ferry Quadrangle, Oregon. M, 1958, University of Oregon. 84 p.

Spiller, Reginal Wayne. Univariate and multivariate analysis of water well yields as related to fracture traces and lineaments in the Martinsburg Shale in eastern Pennsylvania. M, 1979, Pennsylvania State University, University Park. 231 p.

Spillman, T. An analysis of the glacial features of eastern Morrow and western Knox counties, Ohio, using Landsat MSS data. M, 1986, Murray State University. 68 p.

Spindel, Sylvia Frances. Microfossils of the upper part of the Fort Union Formation, southeastern Montana. M, 1974, University of Montana. 51 p.

Spindler, Richard. A gravity interpretation of structural features in the Cajon Pass area, San Bernardino County, California. M, 1988, California State University, Long Beach. 60 p.

Spindler, William M., III. Structure and stratigraphy of a small Plio-Pleistocene depocenter, Louisiana continental shelf. M, 1948, University of Texas, Austin.

Spink, Walter John. Stratigraphy and structure of the Paleozoic rocks of northwestern New Jersey. D, 1967, Rutgers, The State University, New Brunswick. 311 p.

Spink, Walter John. Structure of the Cambro-Ordovician rocks of Sussex County, New Jersey. M, 1963, Rutgers, The State University, New Brunswick. 114 p.

Spinler, Paul. An analysis of the mineral industries of the republics of China, the Philippines, and Korea, the Kingdom of Thailand, and New Zealand. M, 1984, University of Texas, Austin.

Spinney, Edward E. The preliminary surficial geology of the Newburyport West Quadrangle, Massachusetts. M, 1975, Boston University. 51 p.

Spinnler, Gerard Eugene. HRTEM study of antigorite, pyroxene-serpentine reactions, and chlorite. D, 1985, Arizona State University. 249 p.

Spinnler, Gerard Eugene. The stability of wolframite-type phases in ternary WO_3-bearing systems. M, 1979, Miami University (Ohio). 67 p.

Spino, D. Petrology of the igneous rocks of the New Bay and Loon Bay areas, Newfoundland. M, 1953, McGill University.

Spinosa, Claude. The Permian ammonoid families Paraceltitidae and Xenodiscidae. M, 1965, University of Iowa. 62 p.

Spinosa, Claude. The Xenoidiscidae, Permian otoceratacean ammonoids. D, 1968, University of Iowa. 148 p.

Spirakis, Charles Stanley. Kinetics of sulfate reduction, metastable sulfur, and the genesis of hydrothermal uraninite deposits, mississippi valley lead-zinc deposits, barite deposits and epithermal base and precious metal deposits. D, 1983, University of Colorado. 96 p.

Spirakis, Charles Stanley. The sub-Kaskaskia unconformity of the Illinois Basin. M, 1974, Northwestern University.

Spirn, Regin V. Rare-earth distributions in the marine environment. D, 1966, Massachusetts Institute of Technology. 165 p.

Spiro, David Alan. Urea leaching in a sandy ooil. M, 1972, University of Arizona.

Spiroff, Kiril. Geological observations of the Block P Mine, Hughesville, Montana. M, 1934, Michigan Technological University. 22 p.

Spittler, Thomas E. Volcanic petrology and stratigraphy of nonmarine strata, Orocopia Mountains; their bearing on Neogene slip on the San Andreas Fault, southern California. M, 1974, University of California, Riverside. 115 p.

Spitz, A. H. A petrofabric analysis of ductile shear zones of the Dana Hill metagabbro and the Deiana Gneiss, Northwest Adirondack Mountains, New York. M, 1985, SUNY at Binghamton. 28 p.

Spitz, Dan Spencer. Effect of porosity and mineralogy on the deformation of St. Peter Sandstone. M, 1972, Iowa State University of Science and Technology.

Spitz, Guy. Etude pétrographique et pétrochimique des roches volcaniques autour du gisement de Louvem. M, 1973, Ecole Polytechnique. 73 p.

Spitz, Herbert M. Subsurface geology of the southeastern Cuyama Valley, southern Coast Ranges, California. M, 1986, Oregon State University. 83 p.

Spitzer, Jeanette. Correlation of Pleistocene glacial deposits in northwestern Hamilton County, Ohio. M, 1979, Miami University (Ohio). 119 p.

Spitznas, Roger L. A study of pothole erosion along Little Kimshew Creek, Butte County, California. M, 1950, University of Missouri, Columbia.

Spivack, Arthur J. Boron isotope geochemistry. D, 1986, Massachusetts Institute of Technology. 184 p.

Spivack, Joe. The interpretation of geothermal gradients. M, 1934, University of Manitoba.

Spivak, Joseph. The system NaAlSiO₄-CaSiO₃-Na₂SiO₃. D, 1942, University of Chicago. 82 p.

Spivey, Robert Charles. Geology of northeastern Jack County, Texas. M, 1936, University of Iowa. 58 p.

Spivey, Robert Charles. The ostracodes of the Maquoketa Shale of Iowa. D, 1938, University of Iowa. 41 p.

Spock, Leslie E. Computations of geologic time. M, 1924, Columbia University, Teachers College.

Spock, Leslie E. Geological reconnaissance of parts of Grand, Jackson, and Larimer counties, Colorado. D, 1929, Columbia University, Teachers College.

Spoelhof, Robert William. Pennsylvanian stratigraphy and tectonics in the Lime Creek-Molas Lake area, San Juan County, Colorado. D, 1974, Colorado School of Mines. 193 p.

Spoelhof, Robert William. Structure and stratigraphy of portions of the Meridian Peak and Herd Peak quadrangles, Custer County, south-central Idaho. M, 1972, Colorado School of Mines. 86 p.

Spoerl, Carol Lynn. Modeling of porosity, permeability, and microstructure of the St. Peter Sandstone. M, 1983, University of Wisconsin-Milwaukee. 127 p.

Spohn, Thomas. Mineralogy of the residual soils which overlie a talc deposit in the State Line Serpentine, Pennsylvania. M, 1981, Rutgers, The State University, Newark. 72 p.

Spoley, Robert J. The bedrock geology of the Oxford Quadrangle, Butler-Preble counties, Ohio. M, 1967, Miami University (Ohio). 140 p.

Spoljaric, Nenad. Pleistocene sedimentology; middletown-Odessa area (New Castle County), Delaware. D, 1970, Bryn Mawr College. 193 p.

Spooner, Charles M. Sr⁸⁷/Sr⁸⁶ ratios, K/Rb ratios, and rare-earth element abundances in pyroxene granulites (Precambrian, World Wide). D, 1969, Massachusetts Institute of Technology.

Spooner, Harry V., Jr. The subsurface Wilcox Group of southwestern Jefferson and northern Adams counties, Mississippi. M, 1955, University of Oklahoma. 70 p.

Spooner, Ian Stewart. Applied sedimentological evaluation of a glacial deposit, Joyceville, Ontario. M, 1988, Queen's University. 221 p.

Spooner, Jill A. Field and laboratory study of fracture characteristics as a function of bed curvature in folded dolomites, Sawtooth Mountains, Montana. M, 1984, University of Oklahoma. 135 p.

Sporer, Peggy. Lithostratigraphy and biostratigraphy of the Lower Ordovician Pogonip Ridge core, White Pine Range, Nevada. M, 1982, University of Missouri, Columbia.

Spotti, Adler E. Some geological descriptions of the roof strata and associated structures of the Herrin (No. 6) coal bed in the Staunton-Gillespie region, Illinois. M, 1941, University of Illinois, Urbana. 53 p.

Spotts, John Hugh. Heavy minerals of some granitic rocks of Central California. D, 1959, Stanford University. 111 p.

Spotts, John Hugh. The absorption of infrared radiation by some silicate minerals. M, 1951, University of Missouri, Columbia.

Spradlin, Charles Buckner. Geology of the Beavers Bend State Park area, McCurtain County, Oklahoma. M, 1959, University of Oklahoma. 105 p.

Spradlin, Ernest J. Stratigraphy of the Tertiary volcanic rocks, Joyita Hills area, Socorro County, New Mexico. M, 1976, University of New Mexico. 73 p.

Spradlin, Scott Dunbar. Miocene fluvial systems; Southeast Texas. M, 1980, University of Texas, Austin.

Sprague, Anthony Ross Grafton. Depositional environment and petrology of the lower member of the Pennsylvanian Atoka Formation, Ouachita Mountains, Arkansas and Oklahoma. D, 1985, University of Texas at Dallas. 553 p.

Sprague, Charles Warren. Engineering geology of the Gainesville urban area, Alachua County, FL. M, 1983, University of Florida. 145 p.

Sprague, Douglas W. Geology and economic significance of Pleistocene channel and terrace deposits of the San Diego mainland shelf, California. M, 1971, University of Southern California.

Sprague, Edward K. Aminostratigraphy of relict salt marsh deposits (late Holocene), St. Catherine Island, Georgia. M, 1987, University of Georgia. 94 p.

Sprague, Gloria Davis. Paleoenvironments of the Main Street Limestone and its equivalents (Lower Cretaceous) of central Texas. M, 1982, Stephen F. Austin State University. 94 p.

Spraitzar, Ronald F. Investigations of authigenic clays in sandstones. M, 1977, Miami University (Ohio). 93 p.

Spraker, Larry A. An analysis of unsupervised classification algorithms using a geographic information system approach. M, 1987, Indiana State University. 56 p.

Spratt, Deborah Anne. Finite strain and deformation mechanisms in carbonate thrust sheets. D, 1987, The Johns Hopkins University. 233 p.

Spratt, Joseph Grant. Some heavy mineral investigations of the Turner Valley and Gas Field, Alberta. M, 1928, University of Manitoba.

Spratt, Ray Steven. Measurements of secular variations in gravity. D, 1981, [University of California, San Diego]. 132 p.

Spray, Karen L. Impact of surface-water and groundwater withdrawal on discharge and channel morphology along the Arkansas River, Lakin to Dodge City, Kansas. M, 1986, University of Kansas. 102 p.

Sprecher, Terry Ann. Wallrock alteration, vein structure, and preliminary fluid-inclusion studies, Gooseberry Mine, Storey County, Nevada. M, 1985, University of Nevada. 93 p.

Spreen, Christian August. A history of placer gold mining in Oregon, 1850-1870. M, 1939, University of Oregon. 117 p.

Spreiter, Terry Anne. Factors affecting gully formation and distribution in coastal San Mateo County, California. M, 1979, Stanford University. 58 p.

Spreng, Alfred Carl. Cyclic sedimentation in the (Mississippian) Banff Formation, Alberta, Canada. D, 1950, University of Wisconsin-Madison.

Spreng, Alfred Carl. Mississippian cyclic sedimentation, Wapiti Lake area, British Columbia. M, 1948, University of Kansas.

Spreng, W. Carl. Upper Devonian and Lower Mississippian strata on the flanks of the western Uinta Mountains, Utah. M, 1978, Brigham Young University.

Sprenke, Kenneth Fredrick. An application of the p-coda spectral ratio method to crustal structure in central Alberta. M, 1972, University of Alberta. 101 p.

Spresser, Ralph G. Flow of water through sand, transitional between Darcy and turbulent flow. M, 1963, Stanford University.

Spring, S. and Andersen, D. A geologic investigation of the Ferguson Pegmatite (South Dakota). M, 1957, South Dakota School of Mines & Technology.

Springer, D. G. T. Paragenesis in pegmatites in southeastern Manitoba. D, 1951, University of Toronto.

Springer, Dale Ann. Community gradients in the Martinsburg Formation (Ordovician), southwestern Virginia. D, 1982, Virginia Polytechnic Institute and State University. 326 p.

Springer, G. D. Petrography and ores of the Hermiston-McCauley property, Strathy Township, Ontario. M, 1942, University of Toronto.

Springer, George H. The structure and age relationships of rocks of the east central margin of the Narragansett Basin (Rhode Island-Massachusetts). M, 1940, Brown University.

Springer, James E. Structural analysis of the southeastern Livermore Basin, California. M, 1983, San Jose State University. 178 p.

Springer, Robert Kenneth. Geology of the Pine Hill intrusive complex, El Dorado County, California. D, 1971, University of California, Davis. 362 p.

Springer, Robert Kenneth. The Pine Hill intrusive complex (Jurassic), El Dorado County, California. M, 1968, University of California, Davis. 99 p.

Springfield, Charles Winston, Jr. Hybrid stress finite elements for three-dimensional, linear elastic stress and fracture analysis of nearly or precisely incompressible materials. D, 1983, Georgia Institute of Technology. 246 p.

Sprinkel, Douglas A. Structural geology of the Cutler Dam Quadrangle and northern part of Honeyville Quadrangle, Utah. M, 1976, Utah State University. 58 p.

Sprinkle, C. L. A study of factors controlling the chemical quality of water in Cartwright Creek basin, Williamson County, Tennessee. M, 1974, Vanderbilt University.

Sprinkle, James Thomas. Morphology and evolution of the blastozoan echinoderms (Paleozoic). D, 1971, Harvard University.

Sprinsky, William Harold. Gridding of satellite photographs. M, 1966, Ohio State University.

Sprouble, John C. A study of the (Ordovician) Cobourg Formation (Ontario). D, 1935, University of Toronto.

Sproule, W. R. Control of ore deposition; Con, Rycon and and Negus mines. M, 1952, Queen's University. 78 p.

Sprouse, Dan P. The feasibility of entering into a uranium exploration and development program. M, 1976, Colorado School of Mines. 206 p.

Sprouse, Donald West. Subsurface Upper Mississippian rocks of West Virginia. M, 1954, University of Illinois, Urbana.

Sprowl, Donald Richard. The paleomagnetic record from Elk Lake, MN, and its implications. D, 1985, University of Minnesota, Minneapolis. 156 p.

Spruill, Richard K. The volcanic geology of the Rancho Peñas Azules area, Chihuahua, Mexico. M, 1976, East Carolina University. 99 p.

Spruill, Richard Kent. Petrology and geochemistry of Cretaceous to Oligocene volcanic rocks from the Calera-del Nido Range, Chihuahua, Mexico. D, 1981, University of North Carolina, Chapel Hill. 106 p.

Spruit, Jeffry Dean. The paleoecology of carbonaceous (algal) material in the Middle Devonian Rockport Quarry Limestone of the northeastern portion of Michigan's Lower Peninsula. M, 1981, Western Michigan University. 98 p.

Sprunt, E. C. S. Pressure solution, cementation, and cathodoluminescence with emphasis on quartz. D, 1977, Stanford University. 135 p.

Sprunt, Eve S. Scanning electron microscope study of cracks and pores in crystalline rocks. M, 1973, Massachusetts Institute of Technology. 75 p.

Sprunt, Hugh Hamilton. Bed aggradation experiments in a small recirculating flume employing 40 micron silt with special reference to turbidity current depositional structures. M, 1972, Massachusetts Institute of Technology. 160 p.

Spry, Paul Graeme. The synthesis, stability, origin and exploration significance of zincian spinels. D, 1984, University of Toronto.

Spudich, Paul A. Oceanic crustal studies using waveform analysis and shear waves. D, 1979, University of California, San Diego.

Spudich, Paul A. The response of the ocean bottom to explosive sound. M, 1975, University of California, San Diego.

Spudis, Paul D. The geology of the lunar multi-ring basins. D, 1982, Arizona State University. 291 p.

Spulber, Susan Jane Dixon. Origin of granitic rocks in oceanic crust; an experimental study. D, 1981, Brown University. 114 p.

Spurgeon, Paul A. A Pennsylvanian compression-impression flora from eastern Kentucky; fossil plants from strata above Grannies Branch and Rocky Branch of Goose Creek, Clay County, Kentucky. M, 1980, Eastern Kentucky University. 51 p.

Spurney, John C. Geology of the Iron Peak Intrusion, Iron County, Utah. M, 1984, Kent State University, Kent. 83 p.

Spurr, Malcolm R. Reef bioerosion and sediment production by excavating sponges; families Clionidae and Adociidae. M, 1975, Louisiana State University.

Spuy, Peter M. Van der *see* Van der Spuy, Peter M.

Spycher, G. Extractable forms of Al and Fe in acid western Oregon soils. M, 1973, Oregon State University. 84 p.

Spycher, Nicolas François. Boiling and acidification in epithermal systems; numerical modeling of transport and deposition of base, precious and volatile metals. D, 1987, University of Oregon. 253 p.

Spydell, D. Randall. Recent patterns of sedimentation in Lake Powell, Arizona-Utah. M, 1976, Dartmouth College. 73 p.

Spyrakos, Constantine Christoforos. Dynamic response of strip foundations by the time domain BEM-FEM method. D, 1984, University of Minnesota, Duluth. 267 p.

Squair, Cheryl Anne. Surface karst on Grand Cayman Island, British West Indies. M, 1988, University of Alberta. 119 p.

Squarer, David. An analysis of relationships between flow conditions and statistical measures of bed configurations in straight and curved alluvial channels. D, 1968, University of Iowa. 173 p.

Squier, Lyman Radley. A study of deformations in selected rockfill and earth dams. D, 1967, University of Illinois, Urbana. 453 p.

Squiller, Samuel F. The geochemistry of franklinite and associated minerals from the Sterling Hill zinc deposit, Sussex County, New Jersey. M, 1976, Lehigh University. 231 p.

Squires, Donald Fleming. Middle Pennsylvanian pleurotomariid gastropods from St. Louis, Missouri. M, 1952, University of Kansas. 92 p.

Squires, Donald Fleming. The Cretaceous and Tertiary corals of New Zealand. D, 1955, Cornell University. 269 p.

Squires, G. H. An experimental investigation of the consolidation and shear strength characteristics of a Seattle clay. M, 1985, University of Washington. 127 p.

Squires, Henry Dayton. Areal and structural geology of Saint George map area. D, 1927, University of Wisconsin-Madison.

Squires, Jean M. Rhomboporoid bryozoans from the Pennsylvanian of eastern Kansas. M, 1952, University of Kansas. 35 p.

Squires, L. E. Algal response to a thermal effluent; study of a power station on the Provo River, Utah, USA. D, 1977, Brigham Young University. 64 p.

Squires, Livia J. Two algorithms for the complex cepstrum with application to wavelet extraction and phase unwrapping. M, 1986, University of Houston.

Squires, Richard Lane. Burial environment, diagenesis, mineralogy, and Mg and Sr contents of skeletal carbonates in the Buckhorn Asphalt of middle Pennsylvanian age, Arbuckle Mountains, Oklahoma. D, 1973, California Institute of Technology. 226 p.

Squires, Richard Lane. Origin of reeflike masses in the upper member of the San Andres Formation (Lower and Upper Permian), central Guadalupe Mountains, Eddy County, New Mexico. M, 1968, University of New Mexico. 124 p.

Squires, Stewart G. Seismic interpretation of Permian salt dissolution features, northeastern Colorado. M, 1986, Colorado School of Mines. 54 p.

Squyres, John Benjamin. Origin and depositional environment of uranium deposits of the Grants region (Valencia county) New Mexico. D, 1970, Stanford University. 242 p.

Squyres, Steven Weldon. The morphology and evolution of Ganymede and Callisto. D, 1981, Cornell University. 353 p.

Sramek, Steven F. Study of the non-point source pollutant concentrations in the runoff from a natural submerged marsh and scrub pine/palmetto area of Kennedy Space Center, Florida. M, 1976, Florida Institute of Technology.

Sree Ramulu, Uddanapalli Subbarayappa. Phosphorus reactions with soils high in iron oxides. D, 1966, University of California, Riverside. 122 p.

Sriburi, Thavivongse. Flood damage reduction by stream network control. D, 1983, Colorado State University. 233 p.

Sridharan, Asuri. Some studies on the strength of partly saturated clays. D, 1968, Purdue University. 198 p.

Sridharan, Nagalaxmi. Aqueous environmental chemistry of phosphorus in lower Green Bay, Wisconsin. D, 1972, University of Wisconsin-Madison.

Sriisraporn, Somchai. Deep structure of the Southern Oklahoma Aulacogen. M, 1977, University of Oklahoma. 68 p.

Srimal, Neptune. Geology and oxygen, strontium and ^{40}Ar/^{39}Ar isotopic study of India-Asia collision in the Ladakh and Karakoram Himalaya, Northwest India. D, 1986, University of Rochester. 288 p.

Srinivasan, Vajapeyam S. The mechanics of flat bed flow and occurrence of bed forms in alluvial channels. D, 1969, University of Waterloo.

Srinivasan, Venkatraman. Anisotropic swelling characteristics of compacted clay. D, 1970, Oklahoma State University. 181 p.

Sriramadas, Aluru. Geology of the Manchester Quadrangle, New Hampshire. D, 1955, Harvard University.

Sriruang, Somsakdi. A study of rock facture induced by dynamic tensile stress and its applications to fracture mechanics. M, 1972, Michigan Technological University. 165 p.

Srivastava, G. H. Optical and digital processing of geological surfaces in Kansas. D, 1975, Syracuse University. 326 p.

Srivastava, Praveen. Structural and tectonic evolution of the Kumaon Himalayas based on balanced cross-sections and a study of deformation mechanisms along fault zones. M, 1988, University of Rochester.

Srivastava, Prem N. Environmental studies, breccias of the Ancient Wall reef complex (Devonian), Alberta. M, 1970, McGill University. 88 p.

Srivastava, Rae Mohan. A non-ergodic framework for variograms and covariance functions. M, 1987, Stanford University. 113 p.

Srivastava, Satish Kumar. Angiosperm microflora of the Edmonton Formation (Cretaceous), Alberta, Canada. D, 1968, University of Alberta. 378 p.

Srivastava, Satish Kumar. Palynology of Late Cretaceous mammal-beds, Scollard, Alberta. M, 1965, University of Alberta. 129 p.

Srivastava, Surat Prasa. An investigation of the magnetotelluric method for determining subsurface resistivities. D, 1962, University of British Columbia.

Srogi, Elizabeth Lee Ann. The petrogenesis of the igneous and metamorphic rocks in the Wilmington Complex, Pennsylvania-Delaware Piedmont. D, 1988, University of Pennsylvania. 645 p.

St-Arnaud, L. Numerical modelling of leachate collection under sanitary landfills. M, 1987, University of Waterloo. 71 p.

St-Hilaire, Camil. Cartographie électromagnétique aéroportée. M, 1975, Ecole Polytechnique.

St-Onge, Marc Robert. Metamorphic conditions of the low-pressure internal zone of North-central Wopmay Orogen, Northwest Territories, Canada. D, 1981, Queen's University. 240 p.

St. Amant, Marcel M. Y. Frequency and temperature dependence of dielectric properties of some common rocks. M, 1968, Massachusetts Institute of Technology. 135 p.

St. Aubin, Thomas E. A geochemical study of the extrusive volcanic formations (Tertiary) along Little Sunlight Creek and the Geers Point–Black Mountain area, Park County, Wyoming. M, 1971, Wayne State University.

St. Clair, Ann Elizabeth. Quality of water in the Edwards Aquifer, central Travis County, Texas. M, 1979, University of Texas, Austin.

St. Clair, James H. Geodetic application of occultations observed at one station. M, 1961, Ohio State University.

St. Clair, Stuart. Ore deposits produced by magmatic segregation, with special reference to nickel ores of the Sudbury District, Ontario. M, 1913, University of Iowa. 49 p.

St. Germain, Louis Charles. Depositional dynamics, Brushy Canyon Formation (Lower Permian), Delaware Basin, western Texas. M, 1966, Texas Tech University. 119 p.

St. Jean, Joseph. A Middle Pennsylvanian foraminiferal fauna from Dubois County, Indiana. M, 1953, Indiana University, Bloomington. 139 p.

St. Jean, Joseph. Middle Devonian Stromatoporoidea from Indiana, Kentucky, and Ohio. D, 1956, Indiana University, Bloomington. 412 p.

St. John, Billy Eugene. Geology of the Tecolote Point area, Rio Arriba County, New Mexico. M, 1960, University of Texas, Austin.

St. John, Billy Eugene. Structural geology of Black Gap area, Brewster County, Texas. D, 1965, University of Texas, Austin.

St. John, Jack W. and Ross, Alex R. Geology of the northern Wyoming Range, Wyoming. M, 1950, University of Michigan.

St. John, Jack W. The Cool Creek Formation, an example of a Lower Ordovician peritidal carbonate deposit from the Arbuckle Mountains, Murray County, Oklahoma. M, 1979, University of Texas at Dallas. 130 p.

St. Jorre, Louise de *see* de St. Jorre, Louise

St. Julien, Pierre. Les Appalaches; pétrographie, stratigraphie, et tectonique. D, 1963, Universite Laval.

St. Lawrence, William. Ultrasonic emissions in snow, Gallatin County, Montana. M, 1973, Montana State University. 36 p.

St. Louis, Robert Michael. Geochemistry of the platinum group elements in the Tulameen ultramafic complex, British Columbia. M, 1984, University of Alberta. 143 p.

St. Romain, Samuel Joseph. Stratigraphy and foraminiferal paleontology of the upper Vicksburg-lower Chickasawhay (Oligocene) in Mississippi and western Alabama. M, 1980, University of New Orleans. 121 p.

Staab, Robert F. Analysis of gravity and magnetic surveys transecting monoclinical flexures in Yavapai County, Arizona. M, 1978, Northern Arizona University. 87 p.

Staal, Cees R. van *see* van Staal, Cees R.

Staargaard, Christiaan Frederik. Lithogeochemical features associated with volcanogenic massive sulphide mineralization in the Sturgeon Lake area of Ontario, Canada. M, 1981, Queen's University. 200 p.

Staatz, Mortimer H. Geology of the Quartz Creek Pegmatite, Gunnison County, Colorado. D, 1952, Columbia University, Teachers College.

Staatz, Mortimer H. and Norton, James J. The Precambrian geology of the Los Pinos Range, New Mexico. M, 1942, Northwestern University.

Staay, Robert Van der *see* Van der Staay, Robert

Stablein, Newton Kingman. Petrogenesis of microcline megacrysts in the New York canyon pluton, Stillwater range, Nevada. M, 1970, Northwestern University.

Stablein, Newton Kingman, III. Nature and origin of feldspars of the Tunnel City Group, central and southern Wisconsin. D, 1974, University of British Columbia.

Stacey, John Sydney. A method of ratio recording for lead isotopes in mass spectrometry. D, 1962, University of British Columbia.

Stacey, Karl. Petroleum and gas in the economy of Oklahoma. D, 1955, Clark University.

Stach, Robert L. Mineralogy and stratigraphy of the Niobrara Formation, South Dakota. M, 1976, University of South Dakota. 53 p.

Stackelberg, Paul E. The role of intergranular pressure solution in the diagenesis of the St. Peter Sandstone along a traverse of the Illinois Basin. M, 1987, University of Missouri, Columbia. 113 p.

Stacy, Ann L. Geology of the area around the Langmuir laboratory, Magdalena mountains, Socorro County, New Mexico. M, 1968, New Mexico Institute of Mining and Technology. 69 p.

Stacy, Curtis Clyde, Jr. Calhoun County geology (Mississippi). M, 1952, University of Mississippi.

Stacy, Howard Elwell. Invertebrate paleontology and paleoecology of the later Pleistocene of the Lower Medicine Creek Valley, Nebraska. M, 1949, University of Nebraska, Lincoln.

Stacy, Howard Elwell. The Lower Cretaceous microfauna from Trinidad and adjacent areas. D, 1966, University of Michigan. 256 p.

Stacy, Maurice Cyrus. Stratigraphy and paleontology of the Windsor Group (Upper Mississippian) in parts of Cape Breton Island, Nova Scotia. D, 1952, Massachusetts Institute of Technology. 191 p.

Stacy, Robert R. A study of the Winsor Formation of Upper Jurassic age near type section. M, 1957, University of Nebraska, Lincoln.

Stadelman, D. K. Carbonatization in the Barber-Larder lake area (Ontario). M, 1939, University of Toronto.

Stadler, Carl Albert. The geology of the Goldbelt Spring area, northern Panamint Range, Inyo County, California. M, 1968, University of Oregon. 78 p.

Stadnik, Paul M. Trace element characterization of pyrite in the Michigamme Formation, Upper Peninsula, Michigan. M, 1983, Michigan Technological University. 145 p.

Stadum, Carol J. The development and analysis of a paleontological park in the Pecten Reef of the Monterey Formation, Orange County, California. M, 1982, California State University, Long Beach. 113 p.

Staff, George McDonald. The causes of structural and morphological changes of faunal communities within cyclic carbonate and clastic units of the Georgetown Formation (Comanchean-Lower Cretaceous) of Texas. M, 1978, Texas A&M University.

Staff, George McDonald. The nature of information loss in the paleoecological reconstruction of benthic macrofaunal communities using faunal assemblages from the Recent Texas coastal environment. D, 1983, Texas A&M University. 225 p.

Staffeld, Byron Clifford, Jr. Sand-shale ratios of the part of the Pennsylvanian in Richland County, Illinois. M, 1954, University of Illinois, Urbana.

Stafford, Donald Bennett. Development and evaluation of a procedure for using aerial photographs to conduct a survey of coastal erosion. D, 1968, North Carolina State University. 230 p.

Stafford, Gerald Maner. A study of the geologic sections in northern Travis County, Texas. M, 1930, University of Texas, Austin.

Stafford, Howard Straus. A regional geographical study of Guano Valley, a section of the Basin Range area of southeastern Oregon. M, 1935, University of Oregon. 68 p.

Stafford, John Michael. An evaluation of the carbonate cements and their diagenesis on selected banks,

outer continental shelf; northern Gulf of Mexico. M, 1982, Texas A&M University. 78 p.

Stafford, Lester Earl. The geology of the Stillwell area, Adair County, Oklahoma. M, 1951, University of Oklahoma. 66 p.

Stafford, Mark R. Hydrogeology, groundwater chemistry, and resistivity of a contaminated shallow aquifer system in southern Bond County, Illinois. M, 1987, Southern Illinois University, Carbondale. 167 p.

Stafford, Thomas F., Jr. Features of jointing in the inner Bluegrass of Kentucky. M, 1962, University of Kentucky. 72 p.

Stafford, Thomas Wier, Jr. Quaternary stratigraphy, geochronology, and carbon isotope geology of alluvial deposits in the Texas Panhandle. D, 1984, University of Arizona. 180 p.

Stafford, Wilbur L. Geology of the Phoenix-Rollinsville area, Boulder and Gilpin counties, Colorado. M, 1951, University of Colorado.

Stageman, J. Christopher. Depositional facies and provenance of lower Paleozoic sandstones of the Bliss, El Paso, and Montoya formations, southern New Mexico and West Texas. M, 1987, New Mexico State University, Las Cruces. 101 p.

Stager, Jay Curt. Environmental changes at Lake Cheshi, Zambia, since 40,000 years B.P. with additional information from Lake Victoria, East Africa. D, 1984, Duke University. 213 p.

Stagg, Julie Wenger. Petrology of the basal conglomerate of the Triassic Pekin Formation, Sanford Subbasin, North Carolina. M, 1984, University of North Carolina, Chapel Hill. 79 p.

Staggs, James Otis. The Curry Field, Ouachita County, Arkansas. M, 1953, University of Arkansas, Fayetteville.

Stagner, Howard Ralph. Geology of the Carter Lake area (Larimer County), Colorado. M, 1933, University of Colorado.

Stagner, Willbur Lowell. Paleogeography of the Uinta Basin during Uinta B time. M, 1939, University of Colorado.

Staheli, Albert C. Surface beach ridge sagging as an indicator to subsurface limestone solution, central peninsular Florida. M, 1969, University of North Carolina, Chapel Hill. 44 p.

Staheli, Albert Clifford. Topographic criteria for recognition of a threshold of erosion by the Laurentide ice sheet. D, 1971, University of North Carolina, Chapel Hill. 82 p.

Stahl, Lloyd E. The marine geology of Tampa bay (Florida). M, 1969, Florida State University.

Stahl, Stephen David. Pre-Cenozoic structural geology and tectonic history of the Sonoma Range, north-central Nevada. D, 1987, Northwestern University. 301 p.

Stahnke, Clyde Raymond. The genesis of a chrono-climo-sequence of mollisols in west-central Oklahoma. D, 1968, Oklahoma State University. 124 p.

Stainbrook, Don J. Heat flow in the Boss deposit of Missouri. M, 1966, University of Missouri, Rolla.

Stainbrook, Merrill Addison. Geology and paleontology of the Cedar Valley zones at Brandon, Iowa. M, 1922, University of Iowa. 91 p.

Stainbrook, Merrill Addison. The Brachiopoda of the Cedar Valley beds of the Iowa Devonian. D, 1927, University of Iowa. 231 p.

Stakes, Debra S. Submarine hydrothermal systems; variations in mineralogy, chemistry, temperatures and the alteration of oceanic layer II. D, 1979, Oregon State University. 189 p.

Staley, George G. The geology of the War Eagle Quadrangle, Benton County, Arkansas. M, 1962, University of Arkansas, Fayetteville.

Staley, Jack W. Deformational textures of the Day Book Dunite. M, 1979, University of Cincinnati. 108 p.

Staley, Walter Goodwin, Jr. Application of the two-feldspar geothermometer. D, 1962, Washington University. 39 p.

Staley, Walter Goodwin, Jr. Mechanism of reaction between corundum and cristobalite in the subsolidus region of the system $Al_2O_3SiO_2$. D, 1968, Pennsylvania State University, University Park. 142 p.

Stalker, Archibald M. A study of erosion surfaces in the southern part of the Eastern Townships of Quebec. M, 1948, McGill University.

Stalker, Archibald M. The geology of the Red Deer area, Alberta, with particular reference to the geomorphology and water supply. D, 1950, McGill University.

Stall, Albert M. Geology of Kent Bayou and Turtle Bayou oil and gas fields, Terrebonne Parish, Louisiana. M, 1957, University of Oklahoma. 35 p.

Stallard, Mary Loah. Albitization and zeolitization in Triassic/Jurassic volcanogenic rocks, Hokonui Hills, New Zealand. M, 1986, University of California, Santa Barbara. 151 p.

Stallard, Paul C. Geology of the Pine Mountain area of Pike County, Kentucky. M, 1961, University of Kentucky. 73 p.

Stallard, Robert Forster. Major element geochemistry of the Amazon River system. D, 1980, Massachusetts Institute of Technology. 366 p.

Stallman, Georgia. Mineralogy and geochemistry of the Echo Park Alluvium, Colorado. M, 1981, University of Missouri, Columbia.

Stalmach, Daniel Miles. Dispersion and excitation of leaking modes in layered media. D, 1972, Rice University. 142 p.

Stam, Alan C. The Teklanika Formation (Paleocene), Mount Fellows area, central Alaska Range, Alaska. M, 1980, University of Wisconsin-Madison.

Stam, Beert. Quantitative analysis of Middle and Late Jurassic foraminifera from Portugal and its implications for the Grand Banks of Newfoundland. D, 1985, Dalhousie University. 343 p.

Stam, Marianne. Morphology of Martian impact basins; modification styles and tectonic setting. M, 1985, University of Massachusetts. 90 p.

Stamatakos, John A. Paleomagnetism of Eocene plutonic rocks, Matanuska Valley, Alaska. M, 1988, Lehigh University. 70 p.

Stamatelopoulou-Seymour, Karen Catherine. Metamorphosed volcanogenic Pb-Zn deposits at Montauban, Quebec. M, 1975, McGill University. 230 p.

Stamatelopoulou-Seymour, Karen Catherine. Volcanic petrogenesis in the Lac Guyer greenstone belt, James Bay area, Quebec. D, 1982, McGill University. 307 p.

Stamm, John Francis. Geology at the intersection of the Death Valley and Garlock fault zones, southern Death Valley, California. M, 1981, Pennsylvania State University, University Park. 123 p.

Stamm, Robert G. Conodont biostratigraphy of the Scott Peak Formation (Upper Mississippian) of south-central Idaho. M, 1985, University of Idaho. 112 p.

Stanbrow, Gregory E., Jr. A subsurface study of the "Cherokee" Group, Grant County, and a portion of Alfalfa County, Oklahoma. M, 1960, University of Oklahoma. 34 p.

Stancel, Steven George. Alluvial deposition in the Mauch Chunk-Pottsville transition zone (Upper Mississippian-Lower Pennsylvanian) in East central Pennsylvania. M, 1980, SUNY at Binghamton. 93 p.

Stancliffe, Richard John. Vertically stacked barrier island systems, Sego Sandstone (Campanian), Northwest Colorado. M, 1984, University of Texas, Austin. 156 p.

Stancioff, Andrew Simeon. Geology and geomorphology of the Lost River area, Hardy County, West Virginia. M, 1964, George Washington University.

Stanczyk, Dennis T. Effects of development on barrier island evolution; Bogue Banks, North Carolina. M, 1975, Duke University.

Standen, Allan Richard. Mineralization characteristics of the Scotia-Vanderbilt vein system, Silverton, Colorado. M, 1987, University of Texas, Austin. 128 p.

Stander, Edward. The stratigraphy and structural geology of the Twillingate region. M, 1984, Memorial University of Newfoundland. 138 p.

Stander, Thomas. Structural nature of the Humboldt fault zone in northeastern Nemaha County, Kansas. M, 1981, University of Kansas. 23 p.

Standhardt, Barbara R. Vertebrate paleontology of the Cretaceous/Tertiary transition of Big Bend National Park, Texas. D, 1986, Louisiana State University. 320 p.

Standing, Keith F. A computer method for interpreting magnetic anomalies over dike-like structures. M, 1971, University of Manitoba.

Standish, Richard Perkins. Structural geology and metamorphism of the Mallard-Larkins Peak area, Clearwater and Shoshone counties, Idaho. M, 1973, University of Idaho. 84 p.

Standlee, Larry Aaven. Geology of the northern Sierra Nevada basement rocks, Quincy-Downieville area, California. D, 1978, Rice University. 221 p.

Standlee, Larry Aven. Structure and petrography of the middle Feather River ultramafic body, Plumas County, California. M, 1976, Rice University. 77 p.

Standridge, Mark C. Size frequency distribution and composition of chemically altered non-transported saprolitized source rocks. M, 1983, University of Georgia.

Stanford, Loudon Roberts. Glacial geology of the upper South Fork, Payette River, south-central Idaho. M, 1982, University of Idaho. 83 p.

Stanford, Scott Daniel. Stratigraphy and structure of till and outwash in drumlins near Waukesha, Wisconsin. M, 1982, University of Wisconsin-Madison. 131 p.

Stanforth, Robert Rhodes. Adsorption kinetics and isotopic exchange of phosphate on goethite. D, 1981, University of Wisconsin-Madison. 285 p.

Stang, Peter M. Butyl tin compounds in the sediment of San Diego Bay, California. M, 1985, San Diego State University. 61 p.

Stangl, David William. Impact of surface coal mining and reclamation on the hydrogeology at Iowa Coal Project Demonstration Mine No. 1, Mahaska County, Iowa. M, 1978, Iowa State University of Science and Technology.

Stangl, Frank J., Jr. A rudistid reef in the Edwards Limestone, near Belton, Texas. M, 1927, Texas Christian University. 80 p.

Stanin, Frederick Theodore. Armalcolite/ilmenite in lunar basalts; mineral chemistry, paragenesis, and origin of textures. M, 1979, University of Tennessee, Knoxville. 58 p.

Stanke, Faith Alane. Fluvial geomorphology of two contrasting tributaries of the Vermilion River, east-central Illinois. M, 1988, University of Illinois, Urbana. 131 p.

Stanker, Larry Henry. Upper Pleistocene calcareous nannofossils from three Caribbean cores. M, 1973, University of Illinois, Urbana.

Stanley, Alan David. Relation between secondary structures in Athabasca Glacier and laboratory deformed ice (Japan Park, Alberta, Canada). D, 1966, University of British Columbia.

Stanley, Alan David. The geology of Pioneer Gold Mine, Lillooet Mining Division, BC. M, 1960, University of British Columbia.

Stanley, Alice Roberta. Reconnaissance investigation of three areas in Montana for hydrogeologic suitability to host a hazardous waste disposal facility. M, 1988, Montana State University. 164 p.

Stanley, Barbara (Vis). The use of certain remote sensing techniques to study aquifers in Douglas and Franklin counties, Kansas. M, 1981, University of Kansas. 41 p.

Stanley, Charles Bernard. Kinematic implications of footwall structures. M, 1983, Virginia Polytechnic Institute and State University. 71 p.

Stanley, Clifford R. The geology and geochemistry of the Daisy Creek Prospect, a stratabound copper-silver occurrence in western Montana. M, 1984, University of British Columbia. 277 p.

Stanley, Clifford Read. Comparison of data classification procedures in applied geochemistry using Monte Carlo simulation. D, 1988, University of British Columbia.

Stanley, Daniel G. L. Lower Tertiary foraminifera of the Vidono Shale near Puerto La Cruz, Venezuela. M, 1958, Brown University.

Stanley, Daniel R. Geochemical investigation of some Archean banded iron formations from Southwest Montana. M, 1988, University of Florida. 136 p.

Stanley, Donald Alvora. A preliminary investigation of the low-sodium portion of the system BeO-Na$_2$O-SiO$_2$-H$_2$O. D, 1966, University of Utah. 76 p.

Stanley, Edith Matilda. Biogeography and evolution of "bipolar" Radiolaria. D, 1981, University of California, Davis. 134 p.

Stanley, Edward Alexander. Some Mississippian conodonts from the high resistivity of the Nancy Watson No. 1 Well in Northeast Mississippi. M, 1956, Pennsylvania State University, University Park. 73 p.

Stanley, Edward Alexander. Upper Cretaceous and lower Tertiary sporomorphae from northwestern South Dakota. D, 1960, Pennsylvania State University, University Park. 358 p.

Stanley, Everett Michael. A two-dimensional time-dependent numerical model investigation of the coastal sea circulation around the Chesapeake Bay entrance. D, 1976, College of William and Mary.

Stanley, Frederick C. A critical study of the composition of hornblende. D, 1905, Yale University.

Stanley, G. A. An investigation for armour stone for use in a tidal barrier in the Bay of Fundy region. M, 1979, Acadia University. 300 p.

Stanley, G. D., Jr. Triassic coral buildups of western North America. D, 1977, University of Kansas. 293 p.

Stanley, George D., Jr. Environments and communities of the Carters Limestone (Ordovician) in central Tennessee. M, 1972, Memphis State University.

Stanley, George Mahon. Abandoned strands of Isle Royale and northeastern Lake Superior. D, 1932, University of Michigan.

Stanley, George Mahon. Origin of laccoliths with special reference to the theory of William H. Hobbs. M, 1930, University of Michigan.

Stanley, James Theodore. A study in sedimentation of parts of the Petersburg and Dugger formations in the Winslow area, Indiana. M, 1952, Miami University (Ohio). 44 p.

Stanley, Jennifer Sue. Sediment sorption studies of five common volatile organic compounds. M, 1988, Wright State University. 128 p.

Stanley, Kenneth Oliver. Tectonic and sedimentologic history of the lower Jurassic Sunrise and Dunlap formations, west-central Nevada. D, 1969, University of Wisconsin-Madison. 112 p.

Stanley, Kenneth Oliver. The structural history of the Clearwater Fault, northwest Los Angeles County, California. M, 1966, University of California, Los Angeles.

Stanley, Kirk William. Economic geology of the Ahtell-Slana District, Alaska. M, 1958, Montana College of Mineral Science & Technology. 87 p.

Stanley, Larry Gerald. A eugeosynclinal orthoquartzite facies in the eastern Slate Belt rocks of Nash County, North Carolina. M, 1978, North Carolina State University. 81 p.

Stanley, Richard Graham. A shoaling-upward carbonate sequence in the Dogger (Middle Jurassic) of the Central High Atlas of Morocco. M, 1976, Rice University. 177 p.

Stanley, Richard Graham. Middle Tertiary sedimentation and tectonics of the La Honda Basin, central California. D, 1984, University of California, Santa Cruz. 485 p.

Stanley, Richard J. The relationship between groundwater transmissibility and fracture occurrence in the Sharon Conglomerate (Pennsylvanian) of Portage County, Ohio. M, 1973, Kent State University, Kent. 53 p.

Stanley, Roderick G. Trace fossils and the environments of deposition of the Caseyville and Abbott formations (Lower Pennsylvanian), of southern Illinois. M, 1980, Southern Illinois University, Carbondale. 110 p.

Stanley, Rolfe S. Metamorphic stratigraphy and structural geometry in the Collinsville Quadrangle, Connecticut. D, 1962, Yale University.

Stanley, Roy A. Geophysical exploration for ground water and bedrock topography near Wauseon, Ohio. M, 1987, University of Toledo. 275 p.

Stanley, Sarah G. Conodont micropaleontology and biostratigraphy of the Silverwood Limestone Member, Universal Limestone Member, and Salt Creek Limestone Lens (Desmoinesian). M, 1982, Ball State University. 77 p.

Stanley, Steven A. Surface characteristics of geologic materials and their relation to topography, Halloran Springs area, California through computer processing of multispectral scanner data. M, 1980, Indiana State University. 56 p.

Stanley, Steven Mitchell. Relationship between shell form and mode of life among bivalve Mollusca. D, 1968, Yale University.

Stanley, Thomas M. Stratigraphy, ichnology, and paleoichnology of the Deadwood Formation (Upper Cambrian - Lower Ordovician), northern Black Hills, South Dakota. M, 1984, Kent State University, Kent. 224 p.

Stanley, William Dal. An integrated geophysical study related to ground water conditions in Cache Valley, Utah and Idaho. D, 1971, University of Utah. 173 p.

Stann, Leon Kruk. Well spacing and its effect on ultimate recovery. M, 1947, University of Pittsburgh.

Stanonis, Francis L. Geology of the Bell's Ferry and Pond River oil pools (eastern interior basin, Hopkins and McLean counties, Kentucky). M, 1956, University of Kentucky. 46 p.

Stanonis, Frank L. The petrology of the Chipmunk Sand and its relationship to reservoir properties. D, 1958, Pennsylvania State University, University Park. 151 p.

Stanonis, Frank L., III. Depositional and diagenetic history and reservoir properties of the upper Harrodsburg-Salem-lower St. Louis formations; subsurface, North central Kentucky. M, 1981, Vanderbilt University. 110 p.

Stanonis, Gregory L. Geologic study of the Livingston Limestone to locate possible quarry sites in Clark County, Illinois. M, 1977, Southern Illinois University, Carbondale. 68 p.

Stansberry, Gerald Francis. The Viking Formation, central Alberta. M, 1957, University of Alberta. 124 p.

Stanton, George D. Factors influencing porosity and permeability, Wilcox Group (Eocene), Karnes County, Texas. M, 1977, University of Texas, Austin.

Stanton, James Clifford. Crustal structure of the central High Plains of Texas from Rayleigh wave dispersion. M, 1972, Texas Tech University. 51 p.

Stanton, Linda Janice. Early and middle Llandovery (Silurian) brachiopods from the Bowling Green Dolomite (Edgewood Group), northeastern Missouri. M, 1988, University of Wisconsin-Milwaukee. 138 p.

Stanton, M. S. Heavy accessory mineral study applied to local Precambrian correlation. M, 1941, Queen's University. 90 p.

Stanton, Michael. A numerical study of strain heating in a model of continental lithospheric extension. M, 1986, SUNY at Buffalo. 55 p.

Stanton, Michael D. Rock glaciers in the Three Apostles area, central Sawatch Range, Colorado. M, 1980, University of Colorado at Colorado Springs. 98 p.

Stanton, Peter T. Sedimentology, diagenesis and reservoir potential of the Pennsylvanian Tyler Formation, central Montana. M, 1986, University of Colorado at Colorado Springs. 126 p.

Stanton, Robert Guy. The paleoenvironment of the Summerville Formation on the west side of the San Rafael Swell, Emery County, Utah. M, 1975, Brigham Young University.

Stanton, Robert James, Jr. Paleoecology of the upper Miocene Castaic Formation, Los Angeles County, California. D, 1960, California Institute of Technology. 361 p.

Stanton, Roberta Anne. Genesis of uranium-copper occurrence 74-1E, Baker Lake, Northwest Territories. M, 1979, University of Western Ontario. 167 p.

Stanton, W. Layton, Jr. Geology of the Adelaide Quadrangle, California. D, 1931, California Institute of Technology. 138 p.

Stanturf, John Alvin, IV. Effects of added nitrogen on trees and soil of deciduous forests in southern New York. D, 1983, Cornell University. 153 p.

Stanzel, Theodore Edward. The occurrence and economic analysis of sulphur reserves on Otapan Dome, Isthmian saline basin, Mexico. M, 1972, [University of Tulsa].

Stapanian, Maritza Irene. Induced fission track measurements of carbonaceous chondrite Th/U ratios and Th/U microdistributions in Allende inclusions. D, 1981, California Institute of Technology. 290 p.

Staples, Bruce Allen. The nature and content of certain trace elements in selected galenas. M, 1964, University of Arizona.

Staples, Lloyd W. Mineral determination by microchemical methods. D, 1936, Stanford University. 155 p.

Staples, Lloyd William and Cook, Charles Wilford. Microscopic investigation of molybdenite ore from Climax, Colorado. D, 1913, University of Michigan.

Staples, M. E. Stratigraphy and depositional environments of the Goodland Formation in Texas. M, 1977, Baylor University. 84 p.

Stapleton, Richard Pierce. Ultrastructures of foraminifera as shown by electron microscopy. D, 1969, University of Southern California.

Staplin, Frank Lyons. Lithology and micropaleontology of the Marble Falls Formation, Texas. M, 1950, University of Texas, Austin.

Staplin, Frank Lyons. Pleistocene Ostracoda of Illinois. D, 1953, University of Illinois, Urbana.

Stapor, Francis W., Jr. Stratigraphy of the Todilto Formation (Upper Jurassic) in the Ghost Ranch area, New Mexico. M, 1968, University of Wisconsin-Madison.

Stapor, Frank W., Jr. Coastal sand budgets and Holocene beach ridge plain development, Northwest Florida. D, 1973, Florida State University. 235 p.

Stapp, Wilford Lee. The geology of T.X.L. oil field, Ector County, Texas. M, 1946, University of Texas, Austin.

Star, Clarence. A paleocurrent study of the outer conglomerate along the Montreal River, Gogebic County, Michigan. M, 1959, Michigan State University. 51 p.

Star, Ira. Gas reserves in the Medina Group of northwestern Pennsylvania as related to fracture porosity and stratigraphic control. M, 1983, Pennsylvania State University, University Park. 100 p.

Starcher, Robert Warren. A constructional morphological analysis of the fenestrate colony meshwork. D, 1987, Rutgers, The State University, New Brunswick. 255 p.

Starcher, Robert Warren. Effects of simulated growth parameters on bryozoan colony form. M, 1982, Michigan State University. 133 p.

Starich, Patrick. South-central United States magnetic anomaly. M, 1984, Purdue University. 76 p.

Stark, Charles Edwin, Jr. A sedimentary study of the Skiatook Group in portions of Okfuskee and Seminole counties, Oklahoma. M, 1959, University of Oklahoma. 57 p.

Stark, Howard Everett. Geology and paleontology of the northern Whittier Hills, California. M, 1949, Pomona College.

Stark, Jack H. Geology of the Mena Mine with emphasis on the mineralogy and paragenesis of the Mena Mine deposit. M, 1979, Colorado State University. 177 p.

Stark, James Roland. Surficial geology and groundwater geology of the Babbitt-Kawishiw area, northeastern Minnesota with planning implications. M, 1977, University of Wisconsin-Madison.

Stark, Jessie B. An analysis of some physical characteristics of the glacial tills from northeastern Illinois. M, 1945, Northwestern University.

Stark, John Thomas. The geology of the Kekequabic Lake area, northeastern Minnesota. D, 1927, University of Chicago. 222 p.

Stark, John Thomas. The metamorphism of the upper Huronian sediments by the Keweenawan gabbro and Logan diabase sills in the Gunflint Lake District of the Lake Superior region. M, 1922, Northwestern University.

Stark, Michael Paul. The geology of the Winterset and Saint Charles quadrangles, Madison County, Iowa. M, 1973, Iowa State University of Science and Technology.

Stark, Norman Paul. Areal geology of the Upton region, Summit County, Utah. M, 1953, University of Utah. 39 p.

Stark, Philip Bradford. Travel time inversion; inference and regularization. D, 1986, University of California, San Diego. 120 p.

Stark, Philip Herald. A subsurface stratigraphic study of the Pennsylvanian System in the Marietta Basin of south-central Oklahoma and north-central Texas. M, 1961, University of Wisconsin-Madison.

Stark, Philip Herald. Distribution and significance of foraminifera in the Atoka Formation in the central Ouachita Mountains of Oklahoma. D, 1963, University of Wisconsin-Madison. 144 p.

Stark, Timothy D. Mechanisms of strength loss in stiff clays. D, 1987, Virginia Polytechnic Institute and State University. 308 p.

Stark, Tracy Joseph. Information extraction from deep water seismic reflection data; LASE Line 2. D, 1986, University of Texas, Austin. 510 p.

Stark, William James and Mullineaux, Donald R. The glacial geology of the City of Seattle. M, 1950, Washington State University. 89 p.

Starke, George Wesley. Persistent lithologic horizons of the Prairie du Chien Formation from the type section eastward to the crest of the Wisconsin Arch. M, 1949, University of Wisconsin-Madison. 29 p.

Starke, John Metcalf, Jr. Areal geology of northeastern Cherokee County, Oklahoma. M, 1960, University of Oklahoma. 107 p.

Starkey, Caldwell. A late Upper Cretaceous invertebrate fauna of southeastern Wyoming. M, 1939, University of Wyoming. 24 p.

Starkey, Kimberly J. Geology, petrography, chemistry and petrogenesis of the Nannies Peak intrusive complex, Elko County, Nevada. M, 1987, University of Wyoming. 72 p.

Starkey, Michael J. The hydrogeology of the Hoffman Road solid waste disposal facility, Toledo, Ohio. M, 1985, University of Toledo. 214 p.

Starkey, Robert James. Low-grade metamorphism of the Karmutsen volcanics, Vancouver Island, British Columbia. D, 1988, University of Wyoming. 177 p.

Starkey, Sarah J. $Cu_3^{2+} Fe_4^{3+} (VO_4)_6$, a new sublimate mineral from the fumaroles of Izalco Volcano, El Salvador. M, 1986, Miami University (Ohio). 61 p.

Starks, Michael J. Mixing-zone diagenesis in a carbonate aquifer, west-central Florida. M, 1986, University of South Florida, Tampa. 98 p.

Starks, T. L. Succession of algae on surface mined lands in western North Dakota. D, 1979, [University of North Dakota]. 280 p.

Starlin, Leigh Ann. The effects of Mount St. Helens ash co-disposal with solid waste on leachate character. M, 1986, Washington State University. 48 p.

Starmer, Robert John. Precambrian geology of the upper Big Goose area, Bighorn Mountains, Wyoming. M, 1969, University of Cincinnati. 102 p.

Starmer, Robert John. The distribution and geochemistry of the Big Timber dike swarm, Crazy Mountains, Montana. D, 1972, University of Cincinnati. 90 p.

Starn, Jon Jeffrey. A three-dimensional ground water flow model of a glacial outwash terrace, The Plains, Ohio. M, 1987, Ohio University, Athens. 197 p.

Starnes, Jasper Leon. A study of the Cretaceous-Tertiary contact in western Caldwell County, Texas. M, 1948, University of Texas, Austin.

Starquist, Virginia L. The stratigraphy and structural geology of the central portion of the Mount Tom and East Mountain ridges (Massachusetts). M, 1943, Smith College. 49 p.

Starr, Charles Comfort. The Mars Mine. M, 1902, Columbia University, Teachers College.

Starr, James Patrick. Geology and petrology of Ruminahui Volcano, Ecuador. M, 1984, University of Oregon. 98 p.

Starr, Mark Andrew Michael. Water and nutrient budgets of Liberty Lake Marsh. M, 1984, Washington State University. 77 p.

Starr, Robert Brewster. Geology of the Twin Peaks Mine, Lemhi County, Idaho. M, 1955, Cornell University.

Starr, Robert Charles. An investigation of the role of labile organic carbon in denitrification in shallow sandy aquifers. D, 1988, University of Waterloo. 148 p.

Starr, Robert Charles. Parametric study and laboratory investigation of reactive solute transport in fractured porous media. M, 1984, University of Waterloo. 61 p.

Starr, Stephen G. Statistical analysis and interpretation of selected sands from two exploratory tests near Bogra, East Pakistan. M, 1964, Wayne State University.

Stasiuk, Lavern D. Thermal maturation and organic petrology of Mesozoic strata of southern Saskatchewan. M, 1988, University of Regina. 178 p.

Stasko, Lawrence E. Trace fossils of the Middle Ordovician Platteville Formation (McGregor Member) in Southwest Wisconsin. M, 1974, University of Wisconsin-Madison. 106 p.

Staten, Walter T. A chemical study of the silicate minerals of the Great Gossan Lead and surrounding rocks in south-western Virginia. M, 1976, Virginia Polytechnic Institute and State University.

Statham, Kenneth Francis. Stratigraphy and reservoir properties, Devonian Kee Scarp Formation, Norman Wells, Northwest Territories, (Canada). M, 1969, University of Saskatchewan. 130 p.

Stathis, George John. Geology of the southern portion of Spanish Springs Valley Quadrangle, Nevada. M, 1960, University of Nevada. 83 p.

Statkewicz, Edmund. Geology of the northern half of the Otisco Valley Quadrangle (7 1/2 minute series) New York. M, 1956, Syracuse University.

Statler, Anthony Trabue. Geology of the Thompsons Station area, Spring Hill Quadrangle, Williamson County, Tennessee. M, 1951, Vanderbilt University.

Statom, Richard A. Surface geology of the Halltown Quadrangle, Franklin County, Alabama. M, 1988, Mississippi State University. 103 p.

Staub, Ann Marie. Geology of the Picture Rock Hills Quadrangle, southwestern Keg Mountains, Juab County, Utah. M, 1975, University of Utah. 87 p.

Staub, Harrison L. and Laurie, Archibald M. Some studies of multiphase relative permeability in consolidated California oil sands as determined by the capillary pressure displacement method. M, 1950, University of Southern California.

Staub, James Rodney. Tectonically controlled distribution of thick mineable bodies of the Beckley Seam coal in southern West Virginia. D, 1985, University of South Carolina. 139 p.

Staub, James Rodney. The Snuggedy Swamp of South Carolina; a modern back-barrier coal-forming environment. M, 1977, University of South Carolina. 40 p.

Staub, William Praed. A geophysical survey of the De Soto area, Missouri. M, 1960, Washington University. 49 p.

Staub, William Praed. Seismic refraction, a technique for subsurface investigation in Iowa. D, 1969, Iowa State University of Science and Technology. 152 p.

Stauber, D. A. Crustal structure in the Battle Mountain heat flow high in northern Nevada from seismic refraction profiles and Rayleigh wave phase velocities. D, 1980, Stanford University. 330 p.

Stauble, Donald Keith. The interaction of swash and sediment on the backshore in the presence and absence of barriers. D, 1979, University of Virginia. 127 p.

Staubo, John Peder. The Viking Formation (Cretaceous) in southeastern Saskatchewan, a tidal current deposit. M, 1970, University of Manitoba.

Stauder, William Vincent. The plane of polarization of S-waves as related to the mechanism at the focus of an earthquake. D, 1959, University of California, Berkeley. 212 p.

Stauffer, Clinton Raymond. The Devonian limestones of central Ohio. M, 1906, Ohio State University.

Stauffer, Clinton Raymond. The relationship of the Middle Devonian faunas of Ohio. D, 1909, University of Chicago. 201 p.

Stauffer, John Edward. Geology of an area west of Wolcott, Eagle County, Colorado. M, 1953, University of Colorado.

Stauffer, Karl Walter. Quantitative petrographic study of Paleozoic carbonate rocks, Caballo Mountains, New Mexico. D, 1961, Stanford University. 202 p.

Stauffer, Melvyn Roy. The geology of the Ski Lodge Road map-area, Jasper, Alberta. M, 1961, University of Alberta. 62 p.

Stauffer, Peter Hermann. Sedimentation of lower Tertiary marine deposits, Santa Ynez Mountains, California. D, 1965, Stanford University. 231 p.

Stauffer, Thomas B. The distribution of dissolved fatty acids in the James River estuary and adjacent ocean water. M, 1969, College of William and Mary.

Stauffer, Thomas Bennett. Sorption of nonpolar organics on minerals and aquifer materials. D, 1987, College of William and Mary. 121 p.

Stauffer, Truman Parker, Sr. Occupance and use of underground space in the greater Kansas City area. D, 1972, University of Nebraska, Lincoln. 128 p.

Stauft, P. The Becher oilfield, Southwest Ontario. M, 1952, University of Toronto.

Staursky, Geoffrey N. The petrology of selected limestone beds of the Bull Fork Formation (Cincinnatian Series), Caesar Creek Lake spillway, Warren County, Ohio. M, 1981, Miami University (Ohio). 154 p.

Stauss, Lynne D. Anomalous paleomagnetic directions in Eocene dikes of central Washington. M, 1982, Western Washington University. 101 p.

Stavert, Larry W. A geochemical reconnaissance investigation of Mount Baker andesites. M, 1971, Western Washington University. 60 p.

Stavnes, Sandra A. A preliminary study of the subsurface temperature distribution in Kansas and its relationship to the geology. M, 1982, University of Kansas. 311 p.

Stea, Rudolph R. The properties, correlation and interpretation of Pleistocene sediments in central Nova Scotia. M, 1982, Dalhousie University. 215 p.

Stead, Frederick Lee. Foraminifera of the Glen Rose Formation, Lower Cretaceous, of central Texas. M, 1950, University of Texas, Austin.

Steadman, David William. Fossil birds, reptiles, and mammals from Isla Floreana, Galapagos Archipelago. D, 1982, University of Arizona. 254 p.

Steadman, Edward N. Palynology of the Hagel lignite bed and associated strata, Sentinel Butte Formation (Paleocene), in central North Dakota. M, 1985, University of North Dakota. 241 p.

Stearley, Ralph Francis. Bioerosion by intertidal endolithic invertebrates; Puerto Peñasco, Sonora, Mexico. M, 1988, University of Utah. 114 p.

Stearn, Colin W. Stratigraphy and coral faunas of the Silurian of southern Manitoba. D, 1952, Yale University.

Stearn, Noel Hudson. The use of the dip needle as a geological instrument. D, 1926, University of Wisconsin-Madison.

Stearns, Bruce G. Petrography and depositional environments of a lignite from Lauderdale County, Tennessee. M, 1984, South Dakota School of Mines & Technology.

Stearns, Carola Hill. Late Cretaceous-early Tertiary paleomagnetism of Aruba and Bonaire (Netherlands Leeward Antilles). M, 1981, University of Michigan.

Stearns, Carola Hill. Middle to late Paleozoic paleomagnetic results from the Appalachian-Caledonian and North Greenland fold belts. D, 1988, University of Michigan. 128 p.

Stearns, Charles E. Tertiary and Quaternary geology of the Galisteo-Tonque area, New Mexico. D, 1950, Harvard University.

Stearns, Charles Edward. Geology of the eastern part of Mount Jura, California. M, 1962, University of Oregon. 75 p.

Stearns, David W. Drape folds over uplifted basement blocks with emphasis on the Wyoming Province. D, 1970, Texas A&M University. 130 p.

Stearns, David W. Igneous geology of a portion of the Galena District, Lawrence County, South Dakota. M, 1955, South Dakota School of Mines & Technology.

Stearns, Donald L. Geology of the Permo-Carboniferous strata of western Washington County, Ohio. M, 1957, Miami University (Ohio). 54 p.

Stearns, H. T. The geology of Kau District, Hawaii. D, 1926, George Washington University. 216 p.

Stearns, Margaret Dorothy. Investigations as to the origin of the manganese nodules found in oceanic depths. M, 1930, [University of Michigan].

Stearns, Margaret Dorothy. The petrology of the Marshall Formation of Michigan. D, 1933, University of Michigan.

Stearns, Richard Gordon. Pennsylvanian geology of the Crab Orchard Mountain area, Tennessee. D, 1953, Northwestern University.

Stearns, Richard Gordon. Pennsylvanian stratigraphy of the Keith Springs area, Franklin County, Tennessee. M, 1949, Vanderbilt University.

Stearns, Stanley W. Disseminated epithermal precious metals in the Santa Fe District, Mineral County, Nevada. M, 1982, Stanford University. 109 p.

Stearns, Steven Vincent. Incipient diagenesis of sediments from the Pigmy Basin, northern Gulf of Mexico. M, 1986, Texas A&M University.

Stebbins, Jonathan Farwell. Enthalpies and heat capacities of silicate liquids; calorimetric measurements and thermodynamic calculations. D, 1983, University of California, Berkeley. 157 p.

Stebbins, Robert H. Field description of the Percy Islands ultramafic complex, southeastern Alaska. M, 1957, Columbia University, Teachers College.

Stebnisky, Richard J. Hydraulic conductivity as a function of depositional environments on a barrier island, Anclote Key, Pinellas and Pasco counties, Florida. M, 1987, University of South Florida, Tampa. 95 p.

Steck, Lee Karl. Q structure beneath the Mono Craters-Long Valley region. M, 1986, University of California, Santa Barbara. 34 p.

Steckler, Michael Stuart. The thermal and mechanical evolution of atlantic-type continental margins. D, 1981, Columbia University, Teachers College. 266 p.

Steckley, Robert Cecil. Fabric analysis of joints and microfractures in central and eastern Marathon County, Wisconsin. M, 1970, University of Wisconsin-Milwaukee.

Stecyk, Amy N. Sedimentology and stratigraphic correlations of Lower Pennsylvanian strata exposed along Roaring Creek, Parke County, Indiana. M, 1985, University of Illinois, Urbana. 121 p.

Steder, Robert M. A field and petrographic study of the Millie Pit and adjacent area, Dickinson County, Michigan. M, 1958, Michigan State University. 45 p.

Stedischulte, Victor Cyril. The Japanese earthquake of March 29, 1928, and the problem of depth of focus. D, 1932, University of California, Berkeley. 104 p.

Stedman, Thomas Gentry. Astogeny of fenestrate bryozoans and their potential use in biostratigraphy. M, 1982, Southern Illinois University, Carbondale. 154 p.

Steece, Fred V. The geology of the Canton, South Dakota-Iowa, Quadrangle. M, 1957, University of South Dakota. 91 p.

Steed, Douglas A. Geology of the Sterling Quadrangle, Sanpete County, Utah. M, 1979, Brigham Young University.

Steed, Michael O. Micropaleontology and paleoecology of the Glendon Limestone. M, 1973, Northeast Louisiana University.

Steed, Robert W. The depositional environment and stratigraphy of the Mississippian Mission Canyon Formation, east-central Williston Basin, North Dakota. M, 1983, Duke University. 163 p.

Steefel, Carl Iver. Porphyry mineralization in the Elkhorn District, Montana. M, 1982, University of Colorado. 143 p.

Steeg, Carl Ver *see* Ver Steeg, Carl

Steeg, David James Ver *see* Ver Steeg, David James

Steeg, Karl Van *see* Van Steeg, Karl

Steege, Lauren. The Milbank Granite. M, 1957, South Dakota School of Mines & Technology.

Steel, O. J. A theoretical and experimental investigation of pressure solution. M, 1988, Lakehead University.

Steel, Warren George. Dikes of the Durham Triassic basin near Chapel Hill, North Carolina. M, 1949, University of North Carolina, Chapel Hill. 26 p.

Steele, Anthony D. Characteristics of faults as determined with the VLF receiver. M, 1982, University of Nebraska, Lincoln. 215 p.

Steele, David R. Physical stratigraphy and petrology of the Cretaceous Sierra Madre Limestone, west-central Chiapas, Mexico. M, 1982, University of Texas, Arlington. 174 p.

Steele, Elizabeth Anne. Depositional environments of the sediments in southern Marsh Valley, Bannock County, Idaho. M, 1980, Idaho State University. 58 p.

Steele, Frederick Abbott. The crystal structure of tricalcium aluminate. D, 1928, Pennsylvania State University, University Park. 24 p.

Steele, George Alexander, III. Stratigraphy and depositional history of Bogue Banks, North Carolina. M, 1980, Duke University. 201 p.

Steele, Grant. Stratigraphic interpretation of the Pennsylvanian-Permian systems of the eastern Great Basin (Nevada-Utah). D, 1959, University of Washington. 294 p.

Steele, Ian McKay. Electron microprobe and X-ray diffraction study of natural idocrase. D, 1971, University of Illinois, Urbana. 81 p.

Steele, John Davis. The relationship of $^{87}Sr/^{86}Sr$ ratios of contacted lithologies to the ratios in stream waters of the Scioto River basin, Ohio. M, 1973, Wright State University. 77 p.

Steele, John Lisle. A heat flow study in the Turtle Lake Quadrangle, Washington. M, 1975, Southern Methodist University.

Steele, John Lisle. Subsurface temperatures from surface heat flow data using equivalent plane sources. D, 1978, Southern Methodist University. 186 p.

Steele, Kenneth Franklin. Chemical variations parallel and perpendicular to strike in two Mesozoic dolerite dikes, North Carolina and South Carolina. D, 1971, University of North Carolina, Chapel Hill. 202 p.

Steele, Kenneth George. Utilizing glacial geology in uranium exploration, Dismal Lakes, Northwest Territories. M, 1985, University of Alberta. 180 p.

Steele, Kenneth Kane. Stratigraphy, depositional environment and petrology of the Battle Spring Formation in the Green Mountain area, central Wyoming. M, 1985, University of Wyoming. 144 p.

Steele, M. Molluscan fauna of the Ordovician Chaumont Formation near Braeside, Ontario. M, 1966, University of Ottawa. 85 p.

Steele, Sarah Ellen. Fold-fault complexes in the Appalachian Bend between Spruce Creek and Williamsburg, PA. M, 1986, Pennsylvania State University, University Park. 110 p.

Steele, Timothy Doak. Seasonal variations in chemical quality of surface water in the Pesvadero Creek watershed, San Mateo County, California. D, 1968, Stanford University. 196 p.

Steele, W. C. Quaternary stream terraces in the northwestern Sacramento Valley, Glenn, Tehama, and Shasta counties, California. D, 1979, Stanford University. 270 p.

Steele, William Kenneth. The paleomagnetism of the Iztaccihuatl volcano, Mexico. D, 1970, Case Western Reserve University. 291 p.

Steele-Petrovich, Helen Miriam. Stratigraphy and paleoenvironments of Middle Ordovician carbonate rocks, Ottawa Valley Canada. D, 1984, Yale University. 494 p.

Steemson, Gregory Hugh. A theoretical evaluation of the magnetic induced polarization method. M, 1982, University of Utah. 77 p.

Steen, Virginia Carol. Effects of weathering environment on the refractive index of pumice glass from Glacier Peak, Washington and Mount Mazama (Crater Lake), Oregon. M, 1965, Washington State University. 147 p.

Steen-McIntyre, Virginia C. Collection, preparation, petrographic description and approximate dating of tephra (volcanic ash). D, 1977, University of Idaho. 167 p.

Steenland, Nelson C. Deflection of the vertical in the Bahama Islands. M, 1947, Columbia University, Teachers College.

Steenland, Nelson C. Interpretation of aeromagnetic maps. D, 1951, Columbia University, Teachers College.

Steensma, Gilein. Attenuation of coda waves in central Alaska. M, 1985, University of Alaska, Fairbanks. 97 p.

Steenstrup, Susan Jeanette. Morphology and genesis of central peaks, central pits, and central pitted peaks in Martian craters. M, 1986, University of Massachusetts. 165 p.

Steeples, D. W. Teleseismic P-delays in geothermal exploration with application to Long Valley, California. D, 1975, Stanford University. 237 p.

Steeples, Donald Wallace. Resistivity methods in prospecting for ground water. M, 1970, Kansas State University. 48 p.

Steer, Bradley Laurance. Paleohydrology of the Eocene Ballena Gravels, San Diego County, California. M, 1980, San Diego State University.

Steere, Eugene A. Glacial moraines of Montana. M, 1895, University of Wisconsin-Madison.

Steeves, Michael Albert. The petrology of the metavolcanic rocks of the Rusty Lake area, Manitoba. M, 1972, University of Manitoba.

Steeves, Robin Roy. Estimation of gravity tilt response to atmospheric phenomena at the Fredericton tiltmetric station using a least squares response method. D, 1981, University of New Brunswick.

Stefani, Joseph Paul. Applications of teleseismic body waves to shallow earth structure. D, 1984, Stanford University. 144 p.

Stefaniak, Gary John. The depositional history of the Point Pleasant Member of the Cynthiana Formation in northern Kentucky. M, 1984, Western Michigan University.

Stefansson, Karl. Petrography and petrofabrics of gneiss domes near Baltimore. M, 1943, The Johns Hopkins University.

Steffen, Lyle J. Intraformational folding in the Minnekahta Limestone (Lower Permian) of the Fanny Peak area, Weston County, Wyoming. M, 1972, South Dakota School of Mines & Technology.

Steffen, Robert W. A redescription of the Cripple Creek Granite, Cripple Creek, Colorado. M, 1964, University of Kansas. 62 p.

Steffens, Gary Scott. Analysis of opaque minerals in the Vanoss Group (Pennsylvanian), Oklahoma. M, 1980, University of Oklahoma. 106 p.

Steffens, Thomas J. A mineralogical and chemical study of the Milwaukee Harbor sediments. M, 1978, University of Wisconsin-Milwaukee.

Steffensen, Carl K. Diagenetic history and the evolution of porosity in the Cotton Valley Limestone, Teague Townsite Field, Freestone County, Texas. M, 1982, Texas A&M University. 134 p.

Steffes, Darren A. A systematic wettability study of selected Alberta oil reservoirs. M, 1988, University of Calgary. 225 p.

Steflik, Martin. Electrical resistivity characteristics of water-bearing fractures in the North Georgia Piedmont. M, 1988, University of Georgia. 43 p.

Stegall, Melton J. A sedimentary analysis of the Dakota Sandstone of the Las Vegas Basin, San Miguel County, New Mexico. M, 1957, Mississippi State University. 55 p.

Stegen, Ralph J. Geology and geochemistry of the Smuggler Mine Ag-Pb-Zn-Cu-Ba mantos deposits, Aspen, Colorado. M, 1988, Colorado State University. 223 p.

Stehli, Francis G. The Brachiopoda of the (Permian) lower Bone Spring, Sierra Diablo, western Texas. D, 1953, Columbia University, Teachers College.

Stehm, Mark. Isolation and evaluation of subsurface geologic structures through spatial analysis of Paleozoic units in New York State. M, 1981, Syracuse University.

Stehman, C. F. Planktonic foraminifera from the Mid-Atlantic Ridge at 45° north. D, 1975, Dalhousie University.

Stehman, Charles F. Historical implications derived from core E-6282, Northeast Providence Channel, Bahamas. M, 1969, Duke University. 97 p.

Steidl, P. A. A petrographic study of the depositional environment and diagenetic history of the upper Knox Group (Lower Ordovician) in Southwest Virginia. M, 1978, Vanderbilt University.

Steidl, Peter F. A petrographic study of sand-size sediments from the Middle St. Francis River, southeastern Missouri. M, 1974, Southern Illinois University, Carbondale. 71 p.

Steidtmann, Edward. A graphic comparison of the alteration of igneous rocks by weathering with their alteration by hot solution. M, 1907, University of Wisconsin-Madison.

Steidtmann, Edward. The evolution of limestone and dolomite. D, 1910, University of Wisconsin-Madison.

Steidtmann, James. The geochemical cycle of iodine in the Hubbard Brook watershed, New Hampshire. M, 1962, Dartmouth College. 31 p.

Steidtmann, James Richard. Sedimentation, stratigraphy and tectonic history of the early Eocene Pass Peak Formation, central-western Wyoming. D, 1968, University of Michigan. 169 p.

Steiffer, R. John. Geology of the Oakvale, West Virginia area. M, 1971, University of Akron. 28 p.

Steigert, Frederick. Seismicity of the Southern Appalachians seismic zone in Alabama. M, 1982, Georgia Institute of Technology. 145 p.

Steigerwald, Celia H. Bibliographic and petrochemical data bases for molybdenite deposits. M, 1983, Eastern Washington University. 709 p.

Steim, Joseph Michael. The very-broad-band seismograph. D, 1986, Harvard University. 341 p.

Stein, Carol Ann. Part I, Heat transfer, seismicity and intraplate deformation in the central Indian Ocean; Part II, The transition between the Sheba Ridge and Owen Basin; rifting of old oceanic lithosphere. D, 1984, Columbia University, Teachers College. 156 p.

Stein, Carol Lynn. Dissolution of diatoms and diagenesis in siliceous sediments. D, 1977, Harvard University.

Stein, Harlan L. Geology of the Cochiti mining district, Jemez Mountains, New Mexico. M, 1983, University of New Mexico. 122 p.

Stein, Holly Jayne. A lead, strontium, and sulfur isotope study of Laramide-Tertiary intrusions and mineralization in the Colorado mineral belt with emphasis on climax-type porphyry molybdenum systems plus a summary of other newly acquired isotopic and rare earth element data. D, 1985, University of North Carolina, Chapel Hill. 502 p.

Stein, Holly Jean. Evidence for intertidal-supratidal facies control of strataform ore at Magmont Mine, Viburnum Trend, Southeast Missouri. M, 1978, University of North Carolina, Chapel Hill. 49 p.

Stein, Jean. An elaboration of two methods to investigate unfrozen water movement in a snow-soil environment. D, 1985, University of Alaska, Fairbanks. 310 p.

Stein, Jeffrey Allen. Upper Cretaceous (Campanian-Maestrichtian) dinoflagellate cysts from the Great Valley Group, Central California. D, 1983, Stanford University. 505 p.

Stein, Ronald John. Numerical models of faulting. M, 1978, University of Oklahoma. 74 p.

Stein, Ross Simmon. Contemporary and Quaternary deformation in the Transverse Ranges of Southern California. D, 1980, Stanford University. 130 p.

Stein, Seth Auram. I, Seismological study of the Ninetyeast and Chagos-Laccadive ridges, Indian Ocean; II, Models for asymmetric and oblique spreading at midocean ridges; III, Attenuation studies using split normal modes. D, 1978, California Institute of Technology. 215 p.

Stein, Walter T. Mammals from archaeological sites, Point of Pines, Arizona. M, 1962, University of Arizona.

Stein, Walter William, Jr. Geology of Ramsey Quadrangle, Culberson and Reeves counties, Texas. M, 1952, University of Texas, Austin.

Stein, William Earl, Jr. Reinvestigation of the Iridopteridinae of Arnold from the Middle Devonian of New York and Virginia. D, 1980, University of Michigan. 247 p.

Steinbach, Robert C. Geology of the Azure area, Grand County, Colorado. M, 1956, University of Colorado.

Steinberg, D. J. The equilibrium, non-global, self-gravitating, loading pole tide; its contributions to the Chandler wobble period and a comparason with the actual pole tide. M, 1984, SUNY at Binghamton. 1984 p.

Steinberg, Roger T. Geology and petrology of the Lost Creek barite mine, Union County, Tennessee. M, 1981, University of Tennessee, Knoxville. 165 p.

Steinberg, Spencer Martin. Part I; The marine organic chemistry of α-keto acids and oxalic acid; Part II; The chemical decomposition reactions of proteins in calcareous fossils and their effect on amino acid based geochronological methods. D, 1982, University of California, San Diego. 422 p.

Steinborn, Terry L. Particle-size effects in energy dispersive X-ray fluorescence analysis. D, 1976, University of New Mexico. 71 p.

Steinborn, Terry Laurence. Trace element geochemistry of several volcanic centers in the High Cascades. M, 1972, University of Oregon. 115 p.

Steineck, Paul. Microfauna and stratigraphy of Monmouth County, New Jersey, offshore borings. M, 1966, New York University.

Steineck, Paul Lewis. Paleoecologic and systematic analysis of foraminifera from the Eocene-Miocene Montpelier and Lower Coastal groups, Jamaica, West Indies. D, 1973, Louisiana State University.

Steinemann, Christopher F. A gravity study of the Middlesboro cryptoexplosive structure. M, 1980, University of Kentucky. 58 p.

Steinen, Randolph P. Diagenetic modification of some Pleistocene limestones from the subsurface of Barbados, West Indies. D, 1973, Brown University. 152 p.

Steinen, Randolph P. Stratigraphy of the middle and upper Miocene Barstow Formation, San Bernardino County, California. M, 1966, University of California, Riverside. 150 p.

Steiner, David Robert. On the velocity resolution of deep seismic reflection surveys. M, 1984, Cornell University. 27 p.

Steiner, Jeffrey Carl. An experimental study of the assemblage alkali feldspar-liquid-quartz in the system $NaAlSi_3O_8$-$KAlSi_3O_8$-SiO_2-H_2O at 4000 bars. D, 1970, Stanford University. 108 p.

Steiner, Johann. Lower Miette rocks at Jasper, Alberta. M, 1962, University of Alberta. 65 p.

Steiner, Mark A. Petrology of sandstone from the Bullion Creek and Sentinel Butte formations (Paleocene), Little Missouri Badlands, North Dakota. M, 1978, University of North Dakota. 153 p.

Steiner, Maureen B. Contributions to paleomagnetism. D, 1974, University of Texas at Dallas. 196 p.

Steiner, Maureen Bernice. Fusulinidae of the Laborcita Formation, Sacramento Mountains, New Mexico. M, 1967, Southern Methodist University. 74 p.

Steiner, Richard James. The terraces along the Cedar River from Cedar Rapids to Moscow, Iowa. M, 1961, University of Iowa. 61 p.

Steiner, Robert L. Evaluation of geochemical methods of prospecting for manganese in northeastern Tennessee. M, 1958, University of Tennessee, Knoxville. 55 p.

Steiner, Roland Christian. Short-term forecasting of municipal water use. D, 1984, The Johns Hopkins University. 139 p.

Steiner, Waldon W. An interpretation of size analysis of a beach sand, Sterling State Park, Monroe County, Michigan. M, 1952, Wayne State University.

Steiner, Werner. Soil washing processes for the removal of hydrophobic organic compounds from soils and sediments. D, 1988, New York University. 157 p.

Steinfurth, Carl. The role of pre-existing grain surfaces in recrystallization of the Mississippian Bayport Limestone (Michigan). M, 1972, Michigan State University. 48 p.

Steingraber, Walter A. Crystallite size determination of organically precipitated aragonite by X-ray diffraction line broadening techniques and its paleoecological implications. M, 1969, University of South Florida, Tampa. 83 p.

Steinhardt, Christoph K. Structure and stratigraphy of West Haven, Vermont. M, 1983, SUNY at Albany. 167 p.

Steinhart, John Shannon. Explosion studies of continental structure. D, 1961, University of Wisconsin-Madison. 479 p.

Steinhilber, Patricia Mary. Fate of sewage sludge derived zinc relative to soil fractions and plant utilization. D, 1981, University of Georgia. 79 p.

Steinhilper, F. A. Particulate organic matter in Kaneohe Bay, Oahu, Hawaii. M, 1970, University of Hawaii. 53 p.

Steinhoff, Raymond Okley. Geology of the Lavon area, Collin County, Texas. M, 1948, Southern Methodist University. 51 p.

Steinhoff, Raymond Okley. Structural analysis of certain oil and gas fields in South Louisiana Abbeville-Perry-South Perry structures, South Louisiana. D, 1965, Texas A&M University.

Steininger, Roger Claude. Geology of the Kingsley mining district (Elko County, Nevada). M, 1966, Brigham Young University.

Steininger, Roger Claude. Trace elements in sphalerite, galena, pyrite, and chalcopyrite from molybdenum and non-molybdenum systems. D, 1986, Colorado State University. 338 p.

Steinkamp, Allen L. A study of the changes in the potentiometric surface 1930-1986, Springfield area, Missouri. M, 1987, Southwest Missouri State University. 123 p.

Steinke, Theodore R. An evaluation of map design and map reading using eye movement recordings. D, 1979, University of Kansas. 335 p.

Steinker, Don C. Preliminary report on the Recent foraminifera of southern Florida. M, 1961, University of Kansas.

Steinker, Donald Cooper. Foraminifera of the rocky intertidal zone of the central California coast with emphasis on the biology of Rosalina columbiensis (Cushman). D, 1969, University of California, Berkeley. 277 p.

Steinker, Paula Dziak. Shallow-water foraminifera, Jewfish Cay, Bahamas. M, 1973, Bowling Green State University. 99 p.

Steinkraus, William E. A petrographic and petrologic analysis of a Cambrian and Devonian section at Mingus Mountain, Arizona. M, 1961, Michigan State University. 143 p.

Steinman, Dale Marie P. Cretaceous benthonic foraminifera of the Eagle Ford Formation [Texas]. M, 1974, University of Idaho. 63 p.

Steinmetz, J. C. The character, identification, and ultrastructure of selected serpulid (annelid polychaete) tubes from South Florida and the Bahamas. M, 1975, University of Illinois, Urbana. 96 p.

Steinmetz, John. A biostratigraphic and biogeographic analysis of the global distribution of calcareous nannofossils from the early and middle Miocene. D, 1978, University of Miami. 246 p.

Steinmetz, Richard. Size and shape of quartzose pebbles from three New Jersey gravels. M, 1957, Pennsylvania State University, University Park. 116 p.

Steinmetz, Richard. Wasatch sandstones in the eastern Green River basin. D, 1962, Northwestern University. 207 p.

Steinpress, Martin Garth. Neogene stratigraphy and structure of the Dixon area, Espanola Basin, northcentral New Mexico. M, 1980, University of New Mexico. 127 p.

Steinthorsson, Sigurdur. The oxide mineralogy, initial oxidation state, and deuteric alteration in some Precambrian diabase dike swarms in Canada. D, 1974, Princeton University. 224 p.

Stejer, Francis Adrien, Jr. The geology and ore deposition of the Bonanza Mine, Emery (Zosel) mining district, Powell County, Montana. M, 1948, Montana College of Mineral Science & Technology. 39 p.

Stekl, Peter J. A hydrogeologic investigation of two small watersheds of the eastern Quabbin Basin. M, 1985, University of Massachusetts. 169 p.

Stelck, Charles R. Cenomanian-Albian foraminifera of Western Canada. D, 1950, Stanford University. 207 p.

Stelck, Charles Richard. Geology of Pouce Coupe River area, Alberta-British Columbia. M, 1941, University of Alberta. 159 p.

Stell, Mary Jane Armitage. Chemical evolution of waters in the Lansing-Kansas City groups in western Kansas. M, 1988, Kansas State University. 97 p.

Stella, Mark Phillip. Paleocommunities of the Weches Formation, (middle Eocene), East Texas. M, 1986, Stephen F. Austin State University. 117 p.

Stella, Pamela J. Geology of the northern Baie Verte Peninsula, Newfoundland, Canada. M, 1987, SUNY at Albany. 107 p.

Stellas, Michael James. The origin and development of rhombic beach rills. M, 1975, Rutgers, The State University, Newark. 55 p.

Steller, David DeLong. Geology of the Crystal Falls Creek area, Custer County, Colorado. M, 1963, University of Michigan.

Steller, Dorothy L. The epifanual elements on the Brachiopoda of the Silica Formation (Middle Devonian, Ohio). M, 1965, Bowling Green State University. 79 p.

Stelljes, Von D. A gravity investigation of Jefferson County, Kansas. M, 1964, University of Kansas. 36 p.

Stellman, Terry Allen. Seismic anistropy; fracture detection. M, 1987, Ohio University, Athens. 260 p.

Steltenpohl, Mark Gregory. The structural and metamorphic history of Skanland, North Norway, and its significance for tectonics in Scandinavia. D, 1985, University of North Carolina, Chapel Hill. 181 p.

Steltenpohl, Mark Gregory. The structure and stratigraphy of the Ofoten Synform, North Norway. M, 1983, University of Alabama. 119 p.

Stelter, L. H. Inventory, food habits, and trace element levels of selected fauna of Colorado's oil shale region. D, 1979, Colorado State University. 109 p.

Stelzer, William T. A subsurface study of the Middle Ordovician sequence in Ohio. M, 1966, Michigan State University. 121 p.

Stembridge, James Edward, Jr. Shoreline changes and physiographic hazards on the Oregon coast. D, 1975, University of Oregon. 202 p.

Stemen, Kim S. Glacial stratigraphy of portions of Lincoln and Knox counties, Maine. M, 1979, Ohio University, Athens.

Stemle, Steven Von see Von Stemle, Steven

Stempvoort, Dale van see van Stempvoort, Dale

Stenberg, Carroll Dean. Stratigraphy and structure of the Roscoe Quadrangle, Missouri. M, 1953, University of Iowa. 82 p.

Stene, L. P. Holocene and present alluvial investigations, Porcupine Hills, southwestern Alberta. D, 1976, University of Western Ontario.

Steneck, Robert Steven. Adaptive trends in the ecology and evolution of crustose coralline algae (Rhodophyta, Corallinaceae). D, 1982, The Johns Hopkins University. 262 p.

Stengel, K. C. Onset of convection in a variable viscosity fluid. M, 1977, University of Washington. 179 p.

Stenger, William J. Flash heating of pulverized coal through the plastic range. M, 1951, West Virginia University.

Stengl, Gerald Edward. Simpson Group, Andrews County, Texas. M, 1954, University of Texas, Austin.

Stengle, Eugene Henry. The effect of shot depth on the generation of seismic energy. M, 1954, Pennsylvania State University, University Park. 76 p.

Stennett, Albert J. Micropaleontology of part of the Washita Group, Kent Quadrangle, Texas. M, 1956, Texas Tech University. 127 p.

Stensaas, Larry. Paleontology and stratigraphy of the Joana Limestone (Mississippian) at Ward Mountain, Nevada. M, 1957, University of California, Berkeley. 103 p.

Stensland, Donald E. Geology of part of the northern half of the Bend Quadrangle, Jefferson and Deschutes counties, Oregon. M, 1970, Oregon State University. 118 p.

Stensland, Robert Dean. A tectonic analysis of the lithologic associations in the Frontier Formation (Upper Cretaceous), Big Horn County, Wyoming. M, 1965, Iowa State University of Science and Technology.

Stensrud, Howard Lewis. Geology of the Lake Owens mafic complex, Albany County, Wyoming. M, 1963, University of Wyoming. 46 p.

Stensrud, Howard Lewis. Trace and minor element geochemistry of Precambrian muscovites of northern

New Mexico. D, 1970, University of Washington. 130 p.

Stenstrom, Richard. Some geochemical aspects of diagenesis in a shale. D, 1964, University of Chicago. 109 p.

Stenzel, Sheila R. Stratigraphy and sedimentology of the Upper Cambrian Wonewoc Formation in the Baraboo and Kickapoo river valleys, Wisconsin. M, 1983, University of Wisconsin-Madison. 235 p.

Stepanek, Bryan Earl. Diagenesis of paleokarst collapse breccias; the Ordovician El Paso and Montoya groups, West Texas. M, 1984, University of Michigan.

Stepanek, John G. Geological interpretation of an aeromagnetic anomaly near Randalia, northeastern Iowa. M, 1978, University of Iowa. 113 p.

Stephanatos, Basilis Nikolaos. Mathematical modelling of the movement of heat, water, and solutes in two-dimensional, saturated-unsaturated, surface water-groundwater flow systems. D, 1987, University of Illinois, Urbana. 325 p.

Stephen, J. N. Upper Devonian organic reef at Golden Spike, Alberta, Canada. M, 1952, University of Wyoming. 96 p.

Stephen, Michael F. Sedimentary aspects of the New River Delta, Salton Sea, Imperial County, California. M, 1972, University of Southern California.

Stephen, Michael Frederick. Effects of seawall construction on beach and inlet morphology and dynamics at Caxambas Pass, Florida. D, 1981, University of South Carolina. 208 p.

Stephen, Walter Mitchell. The Rainbow Mine, Montana. M, 1911, Columbia University, Teachers College.

Stephens, Christopher Douglas. A short-term microearthquake survey in the central New Hebrides island arc; closing the gap. M, 1978, Cornell University.

Stephens, D. B. Analysis of constant head borehole infiltration tests in the vadose zone. D, 1979, University of Arizona. 384 p.

Stephens, Daniel Guy. Distribution of life and death assemblages of cheilostome Bryozoa in a tidal creek system. M, 1971, University of South Carolina. 64 p.

Stephens, Daniel Guy. Sedimentary and faunal analyses of the Santee River estuaries. D, 1973, University of South Carolina.

Stephens, E., Jr. Intensity distribution study and fault plane solution for the Cape Mendocino, California, earthquake of October 11, 1956. M, 1958, University of Wyoming. 54 p.

Stephens, Edward Harrison. Structure of the Rolesville Batholith and adjacent metamorphic terranes in the east-central Piedmont, North Carolina; a geophysical perspective. M, 1988, North Carolina State University. 166 p.

Stephens, Edward Vernon. The subsurface geology of the Black Warrior Basin in Mississippi. M, 1958, University of Oklahoma. 40 p.

Stephens, George. Hydraulic fracturing theory for conditions of thermal stress. M, 1981, Pennsylvania State University, University Park.

Stephens, George C. Stratigraphy and structure of a portion of the basal Silurian clastics in eastern Pennsylvania. M, 1969, George Washington University.

Stephens, George Christopher. The deology of the Salal Creek pluton, southwestern British Columbia, Canada. D, 1972, Lehigh University. 241 p.

Stephens, Hal Grant. The geology of the Gore area, Warren County, Missouri, and petrography of the rocks. M, 1941, University of Missouri, Columbia.

Stephens, Harold Criss. The stratigraphy and sedimentology of the late Precambrian Boston Bay Group in the vicinity of Squantum, Massachusetts. M, 1987, Ohio University, Athens. 118 p.

Stephens, J. J.; Haas, James J. and Kaarsberg, J. T. Geology of the Poison Canyon area, Huerfano County, Colorado. M, 1956, University of Michigan.

Stephens, Jack E. Engineering materials and problems of the New England-Maritime Provinces. M, 1955, Purdue University.

Stephens, James D. Hydrothermal alteration of the United States mines, West Mountain (Bingham) District, Utah. D, 1960, University of Utah. 83 p.

Stephens, James Gilbert. A preliminary study of the basal sandy zone of the Ohio-Chattanooga-New Albany black shale. M, 1953, University of Cincinnati. 47 p.

Stephens, Jerry Clair. The genetic significance of pebble morphology. M, 1961, University of Colorado.

Stephens, Jim. The Alma Sandstones of the Atoka Formation; a stratigraphic investigation in the Arkoma Basin, Arkansas. M, 1985, University of Arkansas, Fayetteville.

Stephens, John James, III. Stratigraphy and paleontology of a late Pleistocene basin, Harper County, Oklahoma. D, 1959, University of Michigan. 73 p.

Stephens, L. E. An interpretation of gravity and magnetic anomalies in the Kingston area, Ontario. M, 1970, Queen's University. 87 p.

Stephens, Mark Randall. The shape of ground water contamination zones induced by solid waste disposal sites located on alluvial floodplains. M, 1974, Iowa State University of Science and Technology.

Stephens, Maynard Moody. Microscopic effects of carbon arc beam on the polished surfaces of minerals. D, 1934, University of Minnesota, Minneapolis. 62 p.

Stephens, Maynard Moody. The effect of light on polished surfaces of silver minerals. M, 1931, University of Minnesota, Minneapolis. 25 p.

Stephens, Michael Anthony. Brachiopod genus Lepidocyclus Wang and its discrimination from Rhynchotrema Hall. M, 1966, University of Cincinnati. 139 p.

Stephens, Michael P. Ion exchange selectivity of clay under conditions simulating subsurface environments. D, 1974, University of Illinois, Urbana. 91 p.

Stephens, Randall A. Depositional history and diagenesis of the upper Mission Canyon and lower Charles formations (Mississippian), Billings County, North Dakota. M, 1986, University of North Dakota. 236 p.

Stephens, Raymond Weathers, Jr. Stratigraphy and Ostracoda of the Ripley Formation of western Georgia. D, 1960, Louisiana State University. 85 p.

Stephens, Robert Leck. The effect of water content variation and freeze-thaw cycles on the resilient modulus of a silty highway subgrade soil. M, 1980, University of Idaho. 117 p.

Stephens, Robert Monroe. Factors in secondary recovery of petroleum in water-flooding in Illinois. M, 1949, University of Illinois, Urbana. 61 p.

Stephens, Robert W. Fine particle flotation in a differential pressure system. M, 1958, University of Nevada - Mackay School of Mines. 73 p.

Stephens, Thomas C. A fracture analysis in portions of Pendleton, Grant, Hardy, Randolph and Tucker counties, West Virginia. M, 1984, University of Toledo. 129 p.

Stephens, William. Reconnaissance geology of Late Jurassic-Early Cretaceous rocks and Precambrian overthrusts in the Basomari-La Lanina area, north central Sonora, Mexico. M, 1988, University of Pittsburgh.

Stephenson, David A. Stratigraphy of the Entrada-Preuss and Curtis-Stump formations in southwestern Wyoming and the Uinta Mountains of Utah. M, 1961, Washington State University. 96 p.

Stephenson, David Arthur. Hydrogeology of glacial deposits associated with buried Mahomet Bedrock Valley of east-central Illinois. D, 1965, University of Illinois, Urbana. 156 p.

Stephenson, Edgar L. The geology of the Youngstown region. M, 1933, Ohio State University.

Stephenson, Eugene Austin. Studies in hydrothermal alteration; I. D, 1915, University of Chicago. 19 p.

Stephenson, Gordon R. Stratigraphy of the Thaynes Formation in southeastern Idaho. M, 1964, Washington State University. 119 p.

Stephenson, Hubert Kirk. Contributions to the mineralogy of chromite based on the chromite deposit of Casper Mt., Wyoming. D, 1941, Princeton University. 82 p.

Stephenson, James Todd. Petrology, lithofacies and depositional analysis of the Boyle Dolomite, Middle Devonian in east-central Kentucky. M, 1979, University of Cincinnati. 205 p.

Stephenson, John Francis. Gold deposits of the Rice Lake—Beresford Lake area, southeastern Manitoba. D, 1972, University of Manitoba.

Stephenson, John P. Paleoenvironmental analysis of the Lenoir Limestone of southern Knox County, Tennessee. M, 1973, University of Tennessee, Knoxville. 99 p.

Stephenson, Larry Gene. Pedee and Douglas groups (Pennsylvanian) of southeastern Nebraska and adjacent region. M, 1958, University of Nebraska, Lincoln.

Stephenson, Lloyd W. The Mesozoic deposits of the coastal plain of North America. D, 1907, The Johns Hopkins University.

Stephenson, R. Stresses developed at an aseismic continental margin. M, 1977, Carleton University. 124 p.

Stephenson, Robert C. Titaniferous magnetite deposits of the Lake Sanford area, New York. D, 1943, The Johns Hopkins University.

Stephenson, Thomas Edwin. Sources of error on the decrepitation method of study of liquid inclusions. D, 1952, University of Wisconsin-Madison.

Stephey, Anne. A cold vapor technique for measuring mercury concentrations in rocks with an application using sediments from Cachuma Creek, Santa Barbara County, California. M, 1977, University of California, Santa Barbara.

Stepp, Jesse Carl. An investigation of earthquake risk in the Puget Sound area by use of the Type 1 distribution of largest extremes. D, 1971, Pennsylvania State University, University Park. 131 p.

Stepp, Jesse Carl. Regional gravity of parts of Tooele and Box Elder counties, Utah, and Elko County, Nevada. M, 1961, University of Utah. 40 p.

Steppe, Michael C. Pikes Peak Granite rampart range site, CO. M, 1981, Colorado State University. 111 p.

Steppuhn, Harold W. A system for detecting fluorescent tracers in streamflow. D, 1970, Colorado State University. 206 p.

Stepusin, Susan M. Vertical variations in the mineralogical and chemical composition of the underclay of the Herrin (No. 6) Coal in southwestern Illinois. M, 1978, University of Illinois, Urbana. 68 p.

Sterenberg, Velma Zwaantje. Structural and metamorphic history of the East Cleaver Lake Zone, Hackett River, N.W.T. M, 1984, Queen's University. 210 p.

Stergiopoulos, Apostolo Basil. Geophysical crustal studies off the Southwest Greenland margin. M, 1984, Dalhousie University.

Stergiopoulos, Stergios. An experimental study of inertial waves in a fluid contained in a rotating cylindrical cavity during spin-up from rest. D, 1982, York University.

Steritz, John William. The southern termination of the Hosgri fault zone, offshore south-central California. M, 1986, University of California, Santa Barbara. 78 p.

Sterling, Donald Lowell. Model studies on the usefulness of Gauss' mechanical quadrature formula as a sample design. M, 1960, Michigan Technological University. 137 p.

Sterling, Richard P. Behavior of synthetic and biogenic amorphous silica in laboratory-simulated diagenesis. M, 1968, University of Houston.

Sterling, Richard P. Stream channel response to reduced irrigation return flow sediment loads. M, 1983, University of Idaho. 113 p.

Sterling, Stephen Charles. The geology of the inner basin margin, Newport Beach to Dana Point, Orange County, California. M, 1982, California State University, Northridge. 88 p.

Stern, Charles R. Melting relations of gabbro-tonalite-granite-red clay with H_2O at 30kb; the implications for melting in subduction zones. D, 1973, University of Chicago. 217 p.

Stern, Laura A. Mineralogy and geochemical evolution of the Little Three layered pegmatite-aplite intrusive, Ramon, CA. M, 1985, Stanford University. 62 p.

Stern, Robert J. Late Precambrian ensimatic volcanism in the central Eastern Desert of Egypt. D, 1979, University of California, San Diego. 228 p.

Stern, Thomas Whital. Sedimentation and shore processes on the northeastern portion of Galveston Island, Texas. M, 1948, University of Texas, Austin.

Sternbach, Charles A. Shoreline erosion processes and properties unique to cold regions; an in-depth literature review. M, 1981, Rensselaer Polytechnic Institute. 335 p.

Sternbach, Charles Alan. Deep-burial diagenesis and dolomitization in the Hunton Group carbonate rocks (Upper Ordovician to Lower Devonian) in the Anadarko Basin of Oklahoma and Texas. D, 1984, Rensselaer Polytechnic Institute. 144 p.

Sternbach, Linda Raine. Carbonate facies and diagenesis of the Cambrian-Ordovician shelf and slope-margin (the Knox Group and Conasauga Formation), Appalachian fold belt, Alabama. M, 1984, Rensselaer Polytechnic Institute. 165 p.

Sternberg, Ben K. Controlled source electromagnetic soundings of the crust in northern Wisconsin. M, 1974, University of Wisconsin-Madison. 50 p.

Sternberg, Ben K. Electrical resistivity structure of the crust in the southern extension of the Canadian Shield. D, 1977, University of Wisconsin-Madison. 496 p.

Sternberg, Charles William. Cranial foramina and certain other skull openings in some members of the family Canidae. M, 1941, University of Chicago. 21 p.

Sternberg, Hilgard O'Reilly. A contribution to the geomorphology of the False River area, Louisiana. D, 1956, Louisiana State University. 181 p.

Sternberg, Raymond Martin. The cranial morphology of the Devonian crossopterygian Eusthenopteron. D, 1940, University of Toronto.

Sternberg, Richard Walter. Observations of boundary layer flow in a tidal current. D, 1965, University of Washington. 71 p.

Sternberg, Richard Walter. Recent sediments in Bellingham Bay, Washington. M, 1961, University of Washington. 65 p.

Sternberg, Robert S. An archaeomagnetic paleointensity study using some Indian potsherds from Snaketown, AZ. M, 1977, University of Arizona.

Sternberg, Robert Saul. Archaeomagnetic secular variation of direction and paleointensity in the American Southwest. D, 1982, University of Arizona. 338 p.

Sterne, Edward J. Clay mineralogy and carbon-nitrogen geochemistry of the Lik and Competition Creek stratiform Zn-Pb-Ag base metal deposits, Delong Mountains, northern Alaska. M, 1981, Dartmouth College. 156 p.

Sterner, Steven Michael. Calcite precipitation and thermal evaluation of the Cerro Prieto geothermal system; a fluid inclusion and stable isotope study. M, 1985, Pennsylvania State University, University Park. 67 p.

Sternfeld, Jeffrey N. Petrology, hydrothermal mineralogy, stable isotope geochemistry and fluid inclusion geothermometry of selected wells in The Geysers geothermal resource area, Sonoma County, California. M, 1981, University of California, Riverside. 202 p.

Sternin, Jay E. Subsurface geology of Geary and Morris counties, Kansas. M, 1961, Kansas State University. 62 p.

Sternlof, Kurt Richard. Structural style and kinematic history of the active Panamint-Saline extensional system, Inyo County, California. M, 1988, Massachusetts Institute of Technology. 40 p.

Sterr, Horst Michael. The seismo-tectonic history and morphological evolution of late Quaternary fault scarps in southwestern Utah. D, 1980, University of Colorado. 286 p.

Sterrett, Robert John. Factors and mechanics of bluff erosion on Wisconsin's Great Lakes shorelines. D, 1980, University of Wisconsin-Madison. 389 p.

Sterrett, Robert John. The geology and hydrogeology of University Bay, Madison, Wisconsin. M, 1975, University of Wisconsin-Madison.

Stesky, Robert Michael. The magnetism of mid-Atlantic ridge sediments near 45°N. M, 1970, University of Toronto.

Stesky, Robert Michael. The mechanical behavior of faulted rock at high temperature and pressure. D, 1975, Massachusetts Institute of Technology. 197 p.

Stettes, Steven S. Changes in the potentiometric surface in Franklin County, Missouri; 1931-1980. M, 1986, Southwest Missouri State University. 63 p.

Stetz, Donna Jane. Geomorphic analysis of a portion of southeastern Wisconsin using Landsat imagery. M, 1978, University of Wisconsin-Madison. 69 p.

Steuber, Alan M. An analysis of the lateral distribution of trace elements in the galenas from the Fredericktown, Missouri, mines. M, 1961, University of Washington.

Steubing, William C. Notes on the geology of Helvetia, Pima County, Arizona. M, 1912, Columbia University, Teachers College.

Steuer, Fred. Geology of the McCarthy Mountain area, Beaverhead and Madison counties, Montana. M, 1956, University of Utah. 70 p.

Steuer, Mark R. Structural and depositional history of Mesozoic allochthonous rocks in the eastern Garfield Hills, West Central Nevada. M, 1978, Texas Christian University. 36 p.

Steuer, Mark Raymond. Structural geology of the Salem area, New York. D, 1983, Bryn Mawr College. 161 p.

Steuerwald, Bradley A. Faunal, sedimentary and magnetic investigations of Arctic Ocean bottom cores. M, 1969, University of Wisconsin-Madison.

Steuerwald, John B. Petrofabric analysis of the Knox Dolomite and Holston Marble in the Mascot-Jefferson City zinc district, Tennessee. M, 1957, University of Tennessee, Knoxville. 50 p.

Steven, Thomas A. Metamorphism and the origin of the granitic rocks in the Northgate District, Colorado. D, 1950, University of California, Los Angeles.

Stevens, Anne E. Earthquake mechanism studies with S waves. D, 1965, University of Western Ontario.

Stevens, Anthony John. A ground magnetic survey of Marion County, Florida. M, 1981, University of Florida. 103 p.

Stevens, Calvin H. Stratigraphy and paleontology of the McCoy (Routt and Eagle counties), Colorado area. M, 1958, University of Colorado.

Stevens, Calvin Howes. Paleoecology and stratigraphy of pre-Kaibab Permian rocks in the Ely Basin, Nevada and Utah. D, 1963, University of Southern California. 262 p.

Stevens, Carolyn C. A provenance study of the Tertiary sandstones in the Healy and Lignite Creek coal basins, Alaska. M, 1971, University of Alaska, Fairbanks. 122 p.

Stevens, D. Scott. Precious metal deposits associated with alkaline rocks, North American Cordillera south of 41 degrees N.; computer exercises in pattern, recognition, and prediction. M, 1984, Eastern Washington University. 66 p.

Stevens, Dale John. The high Uintas; a landform analysis. D, 1969, University of California, Los Angeles. 201 p.

Stevens, David Lee. Geology and ore deposits of the Antelope mining district, Pershing County, Nevada. M, 1971, University of Nevada. 89 p.

Stevens, David W. A survey of the factors which affect mining of the Lower Mississippian coals in Montgomery County, Virginia. M, 1959, Virginia Polytechnic Institute and State University.

Stevens, Donald L. Geology and geochemistry of the Denali prospect, Clearwater Mountains, Alaska. D, 1971, University of Alaska, Fairbanks. 81 p.

Stevens, Douglas N. Geology and geochemistry of the Buckskin Joe Mine and vicinity, Buckskin-Mosquito mining district, Park County, Colorado. D, 1965, Colorado School of Mines. 97 p.

Stevens, Edward H. The geology of the Sheep Mountain remnant of the Heart Mountain thrust sheet, Park County, Wyoming. D, 1938, University of Chicago. 55 p.

Stevens, George R. Nature and distribution of S-planes in Maryland and southern Pennsylvania. D, 1959, The Johns Hopkins University.

Stevens, Gordon M. A hydrogeologic study of selected bedrock wells near Oneonta, New York. M, 1988, SUNY, College at Oneonta. 187 p.

Stevens, J. R. An investigation of the sources of error in the potassium-argon method of geologic time age determinations of minerals and rocks. M, 1955, University of Toronto.

Stevens, James Bowie. Geology of the Castolon area, Big Bend National Park, Brewster County, Texas. D, 1969, University of Texas, Austin. 163 p.

Stevens, James Bowie. Geology of the Meade County State Park area, Kansas. M, 1963, University of Michigan.

Stevens, James Crosby. Ground-water geology of Hovey area, Brewster and Pecos counties, Texas. M, 1957, University of Texas, Austin.

Stevens, James Crosby. Paleosedimentary environments of Jurassic lower Sundance Formation, northern Denver Basin using cluster analysis. D, 1974, University of Colorado.

Stevens, Jeffry Lowell. Seismic stress relaxation phenomena in an inhomogeneously prestressed medium. D, 1980, University of Colorado. 288 p.

Stevens, Kirk. Fluid inclusion and geological studies on the Zn-Pb-Cu vein system and Lemieux Dome, Gaspé, Québec. M, 1986, University of Toronto.

Stevens, Marian Merrill. Application of remote sensing to the assessment of surface characteristics of selected Mojave Desert playas for military purposes. D, 1988, University of Missouri, Rolla. 218 p.

Stevens, Mark Gerald. Geology of the Iron Dyke Mine, Homestead, Oregon. M, 1981, University of Utah. 78 p.

Stevens, Neil E. The petrology of a hornblende andesite-granodiorite composite dike, Granite Falls, Minnesota. M, 1974, University of Missouri, Columbia.

Stevens, Nelson Pierce. A comparative study of the Mt. Holyoke diabase and the dike at Dry Brook (Massachusetts). M, 1937, University of Massachusetts.

Stevens, R. K. Lower Paleozoic evolution of West Newfoundland. D, 1976, University of Western Ontario. 286 p.

Stevens, Richard K. Igneous petrology of Paisano Mine area, Brewster County, Texas. M, 1974, West Texas State University.

Stevens, Robert Keith. Geology of the Humber Arm area, West Newfoundland. M, 1965, Memorial University of Newfoundland. 121 p.

Stevens, Robert Louis. Petrology of the Gilmore City Formation (Miss.), Iowa. M, 1959, Iowa State University of Science and Technology.

Stevens, Robert Paul. Paragenesis of the mineral in the Einstein vein, Madison County, Missouri. M, 1958, University of Missouri, Rolla.

Stevens, Samuel S. The stratigraphy and structure of the western half of the Jamesville 7-1/2 minute Quadrangle, New York. M, 1958, Syracuse University.

Stevens, Stanley S. Petrogenesis of a tonstein in the Appalachian bituminous basin. M, 1979, Eastern Kentucky University. 83 p.

Stevens, Waldo Eugene. Igneous petrology and geomorphology of the Huehuetenango Quadrangle, Guatemala, (Central America). M, 1969, Louisiana State University.

Stevens, William Walter, Jr. Age of the sandstone outliers of eastern Washington and Benton counties, Arkansas. M, 1948, University of Oklahoma. 71 p.

Stevenson, Andrew J. Molecular absorption interference by iron on zinc analyses using atomic absorption spectroscopy. M, 1974, Stanford University.

Stevenson, C. D. A study of currents in southern Monterey bay (California). M, 1964, United States Naval Academy.

Stevenson, David Lloyd. General geology of the east flank of the Big Horn Mountains south-southwest of Buffalo, Wyoming. M, 1956, University of Iowa. 46 p.

Stevenson, Donald A. Precursory Ts/Tp of Lake Jocassee earthquakes. M, 1977, University of South Carolina.

Stevenson, Douglas Roy. Geologic and groundwater investigations in the Marmot Creek Experimental Basin, Alberta (Quaternary). M, 1967, University of Alberta. 111 p.

Stevenson, Frank Vincent. Devonian formations of New Mexico. M, 1941, University of Chicago. 43 p.

Stevenson, Frederick. Response of the Black Mountain, South Africa, sulfide deposit to various geophysical techniques and implications for exploration of similar deposits. M, 1985, University of Arizona. 152 p.

Stevenson, Gene M. Stratigraphy of the Dox Formation, Precambrian, Grand Canyon, Arizona. M, 1973, Northern Arizona University. 225 p.

Stevenson, Ira M. Geology of the Truro map-area, Colchester and Hants counties, Nova Scotia. D, 1954, McGill University.

Stevenson, Ira M. Structure and petrology of the barite deposit at Bookfield, Colchester County, Nova Scotia. M, 1951, McGill University.

Stevenson, J. S. Mineralization at the Eustis Mine, Eustis, Quebec. D, 1934, Massachusetts Institute of Technology. 186 p.

Stevenson, Jack C. A vertical magnetic intensity study of the Grand Rapids anomaly. M, 1964, Michigan State University. 30 p.

Stevenson, Jeffrey D. Solid veined pyrobitumen from the Panel Mine in the Elliot Lake uranium district, Ontario. M, 1987, Bowling Green State University. 65 p.

Stevenson, Merritt Raymond. Subsurface currents off the Oregon coast. D, 1966, Oregon State University. 140 p.

Stevenson, R. J. A mafic layered intrusion of Keweenawan age near Finland, Lake County, Minnesota. M, 1974, University of Minnesota, Duluth.

Stevenson, Ralph G., Jr. A study of the La Lande, New Mexico, Yonozu, Japan, and Glorieta Mountain, New Mexico, meteorites. M, 1950, University of New Mexico. 45 p.

Stevenson, Ralph Girard, Jr. Mineralogy, petrology, and geochemistry of a complex mineral association near Pony, southwestern Montana. D, 1965, Indiana University, Bloomington. 130 p.

Stevenson, Raymond H. Areal geology of the northwest portion of Canadian County, Oklahoma. M, 1958, University of Oklahoma. 74 p.

Stevenson, Robert E. The Cretaceous stratigraphy of the southern Santa Ana Mountains, California. M, 1948, University of California, Los Angeles.

Stevenson, Robert E. The marshlands at Newport Bay, California. D, 1955, University of Southern California.

Stevenson, Robert Evans. Petrology of the Ringold in the Palouse area, Washington. M, 1942, Washington State University. 43 p.

Stevenson, Robert Evans. Stratigraphy and structural geology of central New York. D, 1950, Lehigh University. 184 p.

Stevenson, Robert James. Amphiboles at the base of the Duluth Complex, Minnesota. D, 1982, University of Minnesota, Minneapolis. 174 p.

Stevenson, Robert Louis. The effects of a supplementary programmed textbook on transfer of training to particular geologic interpretations; an experimental study. D, 1969, Southern Illinois University, Carbondale. 183 p.

Stevenson, Ross Kelley. Implications of amazonite to sulfide-silicate equilibria. M, 1985, McGill University. 311 p.

Stevenson, Wilbur Lloyd. Pennsylvanian conodonts of Illinois. M, 1955, University of Illinois, Urbana.

Stever, Rex Hale. Geologic structure of the Pennsylvanian rocks, Tecolote Mountain area, New Mexico. M, 1951, Texas Tech University. 41 p.

Stever, Richard Clay. Depositional and diagenetic framework of the Lower Permian Chase Group, southern Hugoton Embayment. M, 1987, University of Oklahoma. 198 p.

Steward, Hugh Leighton. Paleocurrent study of sandstones in the (Pennsylvanian) Canyon Group, Palo Pinto and Eastland counties, Texas. M, 1960, Southern Methodist University. 21 p.

Steward, Robert G. Light attenuation measurements of suspended particulate matter; northeastern Gulf of Mexico. M, 1981, University of South Florida, St. Petersburg.

Stewart, Alastair James. Structural evolution of the White Range Nappe, central Australia. D, 1971, Yale University. 397 p.

Stewart, Charles A. Coals of Whatcom County, Washington; their geology, character, and methods employed in mining them. M, 1923, Washington State University. 108 p.

Stewart, Charles A. The geology and ore deposits of the Silverbell mining district, Arizona. D, 1912, Columbia University, Teachers College.

Stewart, Charles A. The magnetic belts of Putnam County, New York. M, 1907, Columbia University, Teachers College.

Stewart, Collin L. Rock mass response to longwall mining of a thick coal seam utilizing shield-type supports. M, 1977, Colorado School of Mines. 384 p.

Stewart, Craig. Interchange of water between the Missouri River and the surrounding alluvial aquifer. M, 1982, University of Missouri, Columbia.

Stewart, Daniel Russell. Comparison of rock mass classification systems; Mt. Axtell Tunnel, Gunnison County, Colorado. M, 1981, University of Arizona. 136 p.

Stewart, Dave. Geology of the northern half of the Baker Peak Quadrangle, Blaine and Camas counties, Idaho. M, 1987, Idaho State University. 113 p.

Stewart, David Benjamin. Rapakivi granite of the Deer Isle region, Maine. D, 1956, Harvard University.

Stewart, David Mark. The vibrational modes of spheroidal cavities in elastic solids. D, 1971, University of Missouri, Rolla.

Stewart, David Perry. The surface geology and Pleistocene history of the Watertown and Sackets Harbor quadrangles, New York. D, 1954, Syracuse University.

Stewart, David Perry. The surface geology of Wexford County, Michigan. M, 1948, Michigan State University. 32 p.

Stewart, Dennis D. The geology and petrology of the Lake George syenite stock, Park and Teller counties, Colorado. M, 1964, Colorado School of Mines. 115 p.

Stewart, Dion Carlyle. A petrographic, chemical and experimental study of kaersutite occurrences at Dish Hill, California, with implications for volatiles in the upper mantle. D, 1980, Pennsylvania State University, University Park. 98 p.

Stewart, Dion Carlyle. Crystal clots in calc-alkaline andesites as breakdown products of high-Al amphiboles. M, 1975, Pennsylvania State University, University Park.

Stewart, Donald G. General geology, channeling and uranium mineralization, Triassic Shinarump Conglomerate, Circle Cliffs area, Utah. M, 1956, Brigham Young University. 38 p.

Stewart, Douglas Bruce. A survey of the differential thermal analysis method and its application to geological problems. M, 1953, Miami University (Ohio). 24 p.

Stewart, Duncan. Geology and petrography of the Antarctic continent. D, 1933, University of Michigan.

Stewart, Duncan. The nature and occurrence of hydrous and anhydrous products of metamorphism at Manton, Rhode Island. M, 1930, Brown University.

Stewart, E. B. A study of the lead-zinc mineralization at Jubilee, Victoria County, Nova Scotia. M, 1978, Acadia University.

Stewart, Edwin Mack. Upper Pennsylvanian rugose corals from the Possum Kingdom area of Palo Pinto County, Texas. M, 1961, Texas Christian University. 69 p.

Stewart, Francis Jr. A map of portions of Muskogee and McIntosh counties, Oklahoma, with special reference to the Inola Limestone and Secor Coal. M, 1949, University of Oklahoma. 81 p.

Stewart, Gary C. The geology of the Parguera area and insular shelf, Southwest Puerto Rico. M, 1985, University of Oklahoma. 137 p.

Stewart, Gary F. A study of fracture and drainage patterns in northern Cleveland County, Oklahoma. M, 1962, University of Oklahoma. 64 p.

Stewart, Gary Franklin. A basis for prediction of denudation and erosion rates in central Kansas. D, 1973, University of Kansas. 215 p.

Stewart, George. Geology of the southwesternmost Whipple Mountains, San Bernardino County, California. M, 1988, University of Southern California.

Stewart, Glen William. Subsurface conditions in and around Syracuse, New York. M, 1937, Syracuse University.

Stewart, Gordon Selbie. Complexity of rupture propagation in large earthquakes in relation to tectonic environment. D, 1982, California Institute of Technology. 256 p.

Stewart, Grace Anne. The fauna of the (Lower Devonian) Little Saline Limestone in Ste. Genevieve County, Missouri. D, 1923, University of Chicago. 66 p.

Stewart, Gregory Lee. Experimental impact abrasion and its implications for eolian studies of Earth and Mars. M, 1983, Arizona State University. 300 p.

Stewart, Harris B., Jr. Sediments and the environment of deposition in a coastal lagoon. D, 1956, University of California, Los Angeles. 355 p.

Stewart, Harry Edward. Photogeology and basin configuration of Mare Smythii region of the Moon. M, 1974, Northwestern State University. 39 p.

Stewart, I. E. Physical geography, geology and economic minerals of Canada. M, 1915, McMaster University.

Stewart, James Conrad. Geology of the Morningstar Mine area, Greaterville mining district, Pima County, Arizona. M, 1971, University of Arizona.

Stewart, James Pirtle, Jr. An experimental investigation into the uplift capacity of drilled shaft foundations in cohesionless soils. D, 1981, Cornell University. 427 p.

Stewart, James S. The geology of the Disturbed Belt of southwestern Alberta. D, 1916, Yale University.

Stewart, Joe Dean. Taxonomy, paleoecology, and stratigraphy of the halecostome-inoceramid associations of the North American Upper Cretaceous epicontinental seaways. D, 1984, University of Kansas. 207 p.

Stewart, John Arden. Change in social and cognitive structures during a scientific revolution; plate tectonics and geology. D, 1979, University of Wisconsin-Madison. 599 p.

Stewart, John Conyngham. The stratigraphy, physiography, and structural geology of the Dryhead-Garvin Basin, south central Montana. D, 1957, Princeton University. 160 p.

Stewart, John Harris. Stratigraphy and origin of the Chinle Formation (Upper Triassic) on the Colorado Plateau. D, 1961, Stanford University. 247 p.

Stewart, John M. Deep-sea fan deposits of the Blakely Formation (Lower-Middle Ordovician) of the Benton Uplift, Ouachita Mountains, Arkansas. M, 1983, University of Missouri, Columbia.

Stewart, John McG. Cored sediments of the Mid-Atlantic Ridge near 45° N. M, 1972, Dalhousie University.

Stewart, John Patrick. Petrology and geochemistry of the intrusive spatially associated with the Logtung W-Mo Prospect, south-central Yukon Territory. M, 1983, University of Toronto.

Stewart, John Rolland. Secondary dolomitization of the Trenton Limestone in the Northville oil field (Michigan). M, 1957, University of Michigan.

Stewart, John W. Lithologic investigation of the Schroyer Limestone in Pottawatomie, Riley, and Geary counties, Kansas. M, 1963, Kansas State University. 54 p.

Stewart, Katherine C. and Stewart, Roscoe E. Local relationships of the Mollusca of the Wildcat coast section, Humboldt County, California. M, 1935, University of Southern California.

Stewart, Keith John. The geology of the Austin Brook, No. 1 sulphide ore body, Bathurst, Gloucester County, New Brunswick. M, 1954, University of New Brunswick.

Stewart, Kevin George. A study of compressional and extensional structures in the Northern Apennine fold-and-thrust belt. D, 1987, University of California, Berkeley. 184 p.

Stewart, Lincoln A. A study of the Ajo copper ore minerals. M, 1933, University of Arizona.

Stewart, Lincoln Adair. The petrology of the Prospect porphyritic gneiss of Connecticut. D, 1935, Columbia University, Teachers College.

Stewart, Lisa Maureen. Strain release along oceanic transform faults. D, 1983, Yale University. 293 p.

Stewart, Lloyd Lincoln. A chemical analysis of axinite found near Five Lakes, Placer County, California. M, 1916, University of California, Berkeley. 7 p.

Stewart, M. T. An integrated geologic, hydrologic, and geophysical investigation of drift aquifers, western Outagamie County, Wisconsin. D, 1976, University of Wisconsin-Madison. 183 p.

Stewart, Mark Thurston. Pre-Woodfordian drifts of north-central Wisconsin. M, 1973, University of Wisconsin-Madison. 82 p.

Stewart, Michael. Hydrogeology of the upper part of the Mesaverde Group, Williams Fork Mountains, Routt and Moffatt counties, Colorado. M, 1983, Colorado School of Mines. 210 p.

Stewart, Michael K. Stable isotopes in raindrops and lakes; laboratory and natural studies. D, 1974, University of Pennsylvania.

Stewart, Moyle Duanne. Stratigraphy and paleontology of the lower Oquirrh Formation of Provo Canyon in the south central Wasatch Mountains, Utah. M, 1950, Brigham Young University.

Stewart, Nanna Beth Bolling. The freeze-thaw resistance of some Pennsylvania gravel aggregates. M, 1970, Pennsylvania State University, University Park. 81 p.

Stewart, Otis Floyd. Origin and occurrence of vermiculite at Tigerville, South Carolina. M, 1949, University of South Carolina.

Stewart, Peter William. Geology, geochemistry, and geochronology and genesis of granitoid clasts in breccia-conglomerates, MacLean extension orebody, Buchans, Newfoundland. M, 1985, Memorial University of Newfoundland. 327 p.

Stewart, Rae Alden. Aerodist controlled photography for topographic mapping. D, 1973, Ohio State University.

Stewart, Ralph B. Gabb's California Cretaceous and Tertiary type lamellibranchs. D, 1928, The Johns Hopkins University.

Stewart, Randall. Stratigraphy and petrology of the Alamo Creek Basalt, Big Bend National Park, Brewster County, Texas. M, 1984, University of Houston.

Stewart, Richard. Recent sedimentary history of St. Joseph Bay, Florida. M, 1962, Florida State University.

Stewart, Richard John. Petrology, metamorphism and structural relations of graywackes in the Western Olympic peninsula, Washington. D, 1970, Stanford University. 123 p.

Stewart, Richard Lee. An evaluation of grain size, shape, and roundness parameters in determining depositional environment in Pleistocene sediments from Newport, Oregon. M, 1967, University of Oregon. 77 p.

Stewart, Robert. The geology of the Benjamin River Deposit, New Brunswick. M, 1979, Carleton University. 125 p.

Stewart, Robert A. The glacial geology of Wildhorse Canyon, Custer County, Idaho. M, 1977, Lehigh University. 102 p.

Stewart, Robert Arthur. Glacial and glaciolacustrine sedimentation in Lake Maumee III near Port Stanley, southwestern Ontario. D, 1982, University of Western Ontario. 350 p.

Stewart, Robert Donald. Radar investigation of the Cote Blanche salt dome. M, 1974, Texas A&M University.

Stewart, Robert H. Some Hamilton ostracodes of western New York State. M, 1946, Northwestern University.

Stewart, Robert R. Vertical seismic profiling; the one-dimensional forward and inverse problems. D, 1983, Massachusetts Institute of Technology. 246 p.

Stewart, Robert W. The reef limestones of the North Snyder oil field, Scurry County, Texas. D, 1951, Massachusetts Institute of Technology. 215 p.

Stewart, Robert W. The use of polished sections to study features of sedimentary rocks. M, 1946, Massachusetts Institute of Technology. 27 p.

Stewart, Roger Malcolm. Shock wave compression and the Earth's core. D, 1970, University of California, Berkeley. 219 p.

Stewart, Roscoe E. and Stewart, Katherine C. Local relationships of the Mollusca of the Wildcat coast section, Humboldt County, California. M, 1935, University of Southern California.

Stewart, Samuel Woods. Gravity surveys of Ogden Valley, Weber County, Utah. M, 1956, University of Utah. 23 p.

Stewart, Samuel Woods. Seismic ray theory applied to refraction surveys of the Earth's crust in Missouri. D, 1966, St. Louis University. 208 p.

Stewart, Scott. Morphology and systematics of 3-dimensionally preserved Late Ordovician graptolites from Anticosti Island, Quebec. M, 1988, SUNY at Buffalo. 225 p.

Stewart, T. Lori. Carbonate petrology and sedimentology of the Miocene Pungo River Formation, Onslow Bay, North Carolina continental shelf. M, 1985, East Carolina University. 184 p.

Stewart, Thomas George. Deglacial-marine sediments from Clements Markham Inlet, Ellesmere Island, Northwest Territories, Canada. D, 1988, University of Alberta. 229 p.

Stewart, W. Douglas. Pennsylvanian shallow marine and aeolian sediments, Tyrwhitt, Storelk and Tobermory formations of southeastern British Columbia. M, 1978, McMaster University. 277 p.

Stewart, Walter Alan. Structure and oil possibilities of the west flank of the Denver Basin (Jefferson County), north-central Colorado. D, 1953, Colorado School of Mines. 121 p.

Stibbe, Ehud. A field method for the determination of the hydraulic conductivity of various soil layers in a profile with the aid of isolated undisturbed monoliths. D, 1965, Ohio State University. 136 p.

Stice, Gary Dennis. Geology and petrology of the Manu'a Islands, American Samoa. D, 1966, University of Hawaii. 191 p.

Sticha, Jill Marie. The stratigraphy and sedimentation of the Turner Sandy Member of the Carlile Shale, western South Dakota. M, 1981, Northern Illinois University. 112 p.

Sticht, John H. H. Geomorphology and glacial geology along the Alaska Highway in Yukon Territory and Alaska. D, 1952, Harvard University.

Stichtenoth, Craig W. Endoliths in fossils from the Waynesville Formation (Upper Ordovician) of southeastern Indiana and northern Kentucky. M, 1977, Bowling Green State University. 60 p.

Stickel, John Frederick, Jr. The igneous geology of the Bonne Terre Quadrangle. M, 1949, Washington University. 95 p.

Sticker, Edwin E. Geology and reservoir analysis of the Lapeyrouse Field, Terrebonne Parish, Louisiana. M, 1979, University of New Orleans.

Stickney, James Francis. Investigation of Recent movement along the Rough Creek fault system in Webster and McLean counties, Kentucky. M, 1985, Eastern Kentucky University. 47 p.

Stickney, Michael C. Seismic and gravity studies of faulting in the Kalispell Valley, Northwest Montana. M, 1980, University of Montana. 82 p.

Stickney, Roger B. Sedimentology, stratigraphy, and structure of the Late Cretaceous rocks of Mayne and Samuel islands, British Columbia. M, 1977, Oregon State University. 226 p.

Stickney, Webster Fairbanks. Vectorial modification of the shape of quartz grains by abrasion. M, 1955, University of Tennessee, Knoxville. 24 p.

Stiebel, W. H. An investigation of the occurrence and origin of nitrate in groundwater in the Swifts Brook watershed. M, 1977, University of Waterloo.

Stiegler, James Harold. The morphology and genesis of certain Virginia soils with deep silty horizons. D, 1971, Virginia Polytechnic Institute and State University. 159 p.

Stieglitz, Ronald Dennis. Scanning electron microscopy of the fine fraction of recent carbonate sediments from Bimini, Bahamas. D, 1970, University of Illinois, Urbana. 100 p.

Stieglitz, Ronald Dennis. Sedimentary structures in Canadian through Givetian (Ordovician, Middle Devonian) rocks, Arrow Canyon Range, Clark County, Nevada. M, 1967, University of Illinois, Urbana.

Stierman, D. J. A study of stress-dependent velocity variations in situ. D, 1977, Stanford University. 208 p.

Stieve, Alice Leutung. Petrologic variation of the granulites and related gneisses of Pine Mountain Terrane, Georgia. M, 1984, Emory University. 107 p.

Stifel, Peter Beekman. Geology of the Terrace and Hogup Mountains, Box Elder County, Utah. D, 1964, University of Utah. 248 p.

Stifler, James Fairman. Finite element modelling of torsional eigenvibrations of the Earth. D, 1979, University of California, Berkeley. 96 p.

Stiles, Aden Edmund. A study of the Cretaceous-Tertiary contact in Bastrop and Travis counties, Texas, south of the Colorado River. M, 1931, University of Texas, Austin.

Stiles, Craig A. Geology and alteration of the west fork of Mayfield Creek area, Custer County, Idaho. M, 1976, University of Idaho. 138 p.

Stiles, Newell Thrift. Mass property relationships of sediments from the Hatteras abyssal plain. M, 1966, American University. 107 p.

Stiles, Walter W. A subsurface study of Sligo Field in Bossier Parish, Louisiana. M, 1976, Northwestern State University. 156 p.

Still, Arthur Rood. Uranium at copper cities and other porphyry copper deposits, Miami District, Arizona. D, 1962, Harvard University.

Stillings, Lisa L. Iron and manganese removal within a wetland constructed for the treatment of acid mine drainage. M, 1988, Kent State University, Kent. 98 p.

Stillman, Francis Benedict. A reconnaissance of the Wasatch Front between Alpine and American Fork canyons, Utah County, Utah. M, 1927, Cornell University.

Stillman, Neil Warren. A gravity study of the Wolcott, gas storage field in northwestern Indiana. M, 1977, Purdue University. 118 p.

Stillwell, Martha. Depositional environments and diagenesis of Cisco Carbonates, Roosevelt County, New Mexico. M, 1982, Texas Tech University. 109 p.

Stilson, Willam P. Seasonal changes of water quality in a reservoir containing acid mine drainage. M, 1969, West Virginia University.

Stilwell, D. P. A grain size study of the Loup river system. M, 1978, University of Nebraska, Lincoln.

Stilwell, Jeffrey Darl. The biostratigraphy of early Tertiary macroinvertebrates from the La Meseta Formation, Seymour Island, Antarctic Peninsula. M, 1988, Purdue University. 485 p.

Stilwell, Randy. Petrology and stratigraphy of the Triassic La Boca Formation, Sierra Madre Oriental, northeastern Mexico. M, 1980, University of New Orleans. 151 p.

Stimac, James Alan. Volcanic rocks and ore deposits of the Cusihuiriachic-Cuauhtemoc area, Chihuahua, Mexico. M, 1983, University of Texas, Austin. 178 p.

Stimac, John P. Geology of the Stockade Mountain 15′ Quadrangle, Malheur and Harney counties, Oregon. M, 1988, Fort Hays State University. 79 p.

Stimson, Eric Jordan. Geology and metamorphic petrology of the Elkhorn Ridge area, northeastern Oregon. M, 1980, University of Oregon. 123 p.

Stimson, James Roy. Control and distribution of porosity in the Red River "C" laminated member at the Brush Lake Field. M, 1985, Montana State University. 76 p.

Stinchcomb, Bruce L. The paleontology and biostratigraphy of the Lower Ordovician Gasconade Formation of Missouri. D, 1978, University of Missouri, Rolla. 193 p.

Stinchcomb, Bruce Leonard. New Mollusca of the Potosi and Eminence formations (Cambrian). M, 1965, Washington University. 53 p.

Stindl, Heribert. Clay mineralogy of the Cottonwood Limestone (Permian) near Manhattan, Kansas. M, 1966, Kansas State University. 87 p.

Stindl, Heribert. Electron microprobe study of diabase-granite contact zones in composite dikes, Mount Desert island, Maine. D, 1971, University of Illinois, Urbana. 86 p.

Stine, Alan D. Sedimentology of the Early Cretaceous lower sandstone member of the Thermopolis Shale, southwestern Montana. M, 1986, Montana State University. 62 p.

Stine, C. M. Stratigraphy of the Ohanapecosh Formation north of Hamilton Buttes, south-central Washington. M, 1987, Portland State University. 92 p.

Stine, Joseph G. Geology of southern Muskogee County, Oklahoma. M, 1958, University of Oklahoma. 89 p.

Stine, Scott William. Mono Lake; the past 4,000 years. D, 1987, University of California, Berkeley. 732 p.

Stingelin, Ronald W. Phyto-geologic relationships in the vicinity of Bethlehem, Pennsylvania. M, 1959, Lehigh University.

Stingelin, Ronald Werner. Late-Glacial and Postglacial vegetational history in the north-central Appalachian region. D, 1965, Pennsylvania State University, University Park. 217 p.

Stinnett, James William, Jr. A geochemical study of calc-alkaline andesites and associated volcanic rocks from the Mogollon-Datil Field, southwestern New Mexico. D, 1980, Miami University (Ohio). 271 p.

Stinnett, Landy A. An experimental approach of determining the influence of fabric and confining pressure on the mechanical properties of rocks. M, 1963, South Dakota School of Mines & Technology.

Stinson, James E., II. Engineering geology criteria for dredged material disposal in upper Laguna Madre, Texas. M, 1977, Texas A&M University. 93 p.

Stinson, Kerry James. Analysis of geomagnetic depth sounding data. M, 1981, University of British Columbia.

Stinson, Melvin C. Mineralogy of the heavy minerals from some placers of central Idaho. M, 1950, University of Idaho. 61 p.

Stinson, William Dank. Geology of northeast part of the Cushing Quadrangle, southern Rusk County, Texas. M, 1952, University of Texas, Austin.

Stipp, S. L. Evaluation of an equilibrium speciation model for the aqueous iron-sulfate system at 25°C. M, 1983, University of Waterloo. 106 p.

Stipp, Thomas F. The Eocene foraminifera of the Marysville buttes, Sutter County, California. M, 1926, Stanford University. 70 p.

Stirewalt, Gerry Lewis. Structural analysis of the Sauratown Mountains anticlinorium and the Brevard lithologic zone, North Carolina. D, 1970, University of North Carolina, Chapel Hill. 146 p.

Stirling, John Alexander Robert. Crystallization of a felsic dyke. M, 1979, University of New Brunswick.

Stirling, William Alex. Local earthquake ground motion scaling for Southeast Missouri. M, 1977, St. Louis University.

Stirton, Ruben Arthur. A phylogeny of North American Equidae with observations on the development of teeth. D, 1940, University of California, Berkeley. 88 p.

Stites, Robert L. Stratigraphy of the Upper Cretaceous Frontier Sandstone, North Park Basin, Jackson County, Colorado. M, 1986, Colorado School of Mines. 125 p.

Stith, David Allen. Petrology of the Hennesey Shale (Permian), Wichita Mountain area, Oklahoma. M, 1968, University of Oklahoma. 113 p.

Stitt, James Harry. Carboniferous stratigraphy of the Bend area, San Saba County, Texas. M, 1964, University of Texas, Austin.

Stitt, James Harry. Late Cambrian and earliest Ordovician trilobites, Timbered Hills and lower Arbuckle groups, western Arbuckle Mountains, Murray County, Oklahoma. D, 1968, University of Texas, Austin.

Stitt, Leonard Timothy. Geology of the Ventura and Soledad basins in the vicinity of Castaic, Los Angeles County, California. M, 1981, Oregon State University. 124 p.

Stix, John. Volcanic facies and geochemistry of Archean lava flows and pyroclastic rocks near Kenora, Ontario, Canada. M, 1985, University of Toronto.

Stoakes, Franklin A. Depositional history and economic potential of lower and middle Devonian (Gedinnian-Eifelian) sediments of the Moose River basin of northern Ontario. M, 1975, University of Windsor. 121 p.

Stoakes, Franklin Arthur. Sea level control of carbonate-shale deposition during progradational basin-filling; the Upper Devonian Duvernay and Ireton formations of Alberta, Canada. D, 1979, University of Calgary. 346 p.

Stobbe, Helen R. Petrology of volcanic rocks of northeastern New Mexico. D, 1947, Columbia University, Teachers College.

Stobbe, Helen R. The geology of the island of Jamaica, British West Indies. M, 1931, Columbia University, Teachers College.

Stobbe, William Joseph. Petrology and stratigraphy of the Dodds Creek Member, Galesburg Formation (Upper Pennsylvanian) of southeastern Kansas. M, 1971, Wichita State University. 76 p.

Stock, Carl. Upper Devonian stromatoporoides of north-central Iowa. M, 1974, SUNY at Binghamton. 123 p.

Stock, Carl W. Upper Silurian (Pridoli) Stromatoporoidea of New York. D, 1977, University of North Carolina, Chapel Hill. 154 p.

Stock, Chester. The Pleistocene fauna of Hawver Cave. D, 1917, University of California, Berkeley. 515 p.

Stock, Joann Miriam. Kinematic constraints on the evolution of the Gulf of California extension province, northeastern Baja California, Mexico. D, 1988, Massachusetts Institute of Technology. 128 p.

Stock, Joann Miriam. Uncertainties in the relative positions of the Australia, Antarctica, Lord Howe, and Pacific plates during the Tertiary. M, 1981, Massachusetts Institute of Technology. 106 p.

Stock, Mark D. The geohydrology of the shallow aquifers in the vicinity of Old Woman Anticline, Niobrara County, Wyoming. M, 1981, University of Wyoming. 87 p.

Stockar, David V. Contemporary tectonics of the Lancaster, Pennsylvania, seismic zone. M, 1986, Pennsylvania State University, University Park. 226 p.

Stockard, Donald P. Location of earthquake foci in Hawaii. M, 1967, Dartmouth College. 76 p.

Stockdale, Paris Buell. Stylolites, their nature and origin. M, 1921, Indiana University, Bloomington. 97 p.

Stockdale, Paris Buell. The Borden (Knobstone) rocks of southern Indiana. D, 1930, Indiana University, Bloomington. 330 p.

Stockdale, Richard G. A geologic study of the chemical quality of Medicine Lake (South Dakota). M, 1971, University of South Dakota. 59 p.

Stocken, C. G. Petrographic methods of determining the source of clastic sediments. M, 1950, McGill University.

Stocker, George Robert. Surface to subsurface correlation of the Madison Group (Mississippian) in eastern Montana and adjacent areas based on insoluble residues. M, 1954, University of Kansas. 298 p.

Stocker, Richard Louis. The acoustic properties of copper-lead and copper-silver partial melts. D, 1973, Yale University.

Stockey, R. A. Morphology and reproductive biology of Cerro Cuadrado fossil conifers. D, 1977, Ohio State University. 149 p.

Stockhausen, Edward Joseph. Gamma radiation in geopressured shale. M, 1981, University of Florida. 96 p.

Stocking, Hobart Ebey. Examination of a drill core through the caprock of the Boling Salt Dome, Texas. D, 1949, University of Chicago. 74 p.

Stocking, Hobert Ebey. The process of analcitization in the Terlingua, Texas, region. M, 1936, The Johns Hopkins University.

Stockmal, Glen Stanley. Modeling of large-scale accretionary wedge and thin-skinned thrust-and-fold belt mechanics. D, 1984, Brown University. 283 p.

Stockmal, Glen Stanley. Structural geology of the northern termination of the Lewis Thrust, Front Ranges, southern Canadian Rocky Mountains. M, 1979, University of Calgary. 174 p.

Stockman, Harlan Wheelock. Noble metals in the Ronda and Josephine peridotites. D, 1982, Massachusetts Institute of Technology. 236 p.

Stockton, Charles W. Hydrogeology of Upper Box Elder Valley, Larimer County, Colorado. M, 1965, Colorado State University. 157 p.

Stockton, Charles Wayne and Stoeck, Penelope L. The feasibility of augmenting hydrologic records using tree-ring data. D, 1971, University of Arizona.

Stockton, Marjorie Moore. Geology of the gabbroic rocks in southern Cooperton Quadrangle and northern Odetta Quadrangle, Oklahoma. M, 1984, University of Texas, Arlington. 83 p.

Stockton, Scott L. Feasibility of using petroleum seismic methods to resolve thin beds in salt. M, 1976, Colorado School of Mines. 80 p.

Stockwell, Clifford Howard. Pegmatite dykes and associated rocks of southeastern Manitoba and adjacent portions of Ontario. D, 1930, University of Wisconsin-Madison.

Stoddard, Andrew. Mathematical model of oxygen depletion in the New York Bight; an analysis of physical, biological, and chemical factors in 1975 and 1976. D, 1983, University of Washington. 364 p.

Stoddard, E. F. Granulite facies metamorphism in the Colton-Rainbow Falls area, Northwest Adirondacks, New York. D, 1976, University of California, Los Angeles. 308 p.

Stoddard, Paul V. A petrographic and geochemical analysis of the Zana Granite and Kowaliga augen gneiss; northern Piedmont, Alabama. M, 1983, Memphis State University.

Stoddard, Robert Russell. Geology of a portion of south-central Archuleta County, Colorado. M, 1957, Colorado School of Mines. 178 p.

Stodghill, Allan M. Resistivity investigation of the Coastal Ridge Aquifer hydrostatigraphy, Martin County, Florida. M, 1983, University of South Florida, Tampa. 273 p.

Stodt, John Allan. Application of the audiomagnetotelluric and magnetotelluric methods in the Marysville Montana geothermal area. M, 1976, University of Utah. 151 p.

Stodt, John Allan. Noise analysis for conventional and remote reference magnetotelluric data. D, 1983, University of Utah. 230 p.

Stoeck, Penelope L. Petrology and contact effects of the Hatfield Pluton of Belchertown Tonalite (Devonian) in the Whately–Northampton area, western Massachusetts. M, 1971, University of Massachusetts. 83 p.

Stoeck, Penelope L. and Stockton, Charles Wayne. The feasibility of augmenting hydrologic records using tree-ring data. D, 1971, University of Arizona.

Stoeckinger, William T. Geology of the McQuady oil pool, Breckinridge County, Kentucky. M, 1957, University of Kentucky. 32 p.

Stoelting, P. K. The concept of an esker, esker form, and esker form system in eastern Wisconsin. D, 1978, University of Wisconsin-Milwaukee. 563 p.

Stoermer, Eugene Filmore. Post-Pleistocene diatoms from Lake West Okoboji, Iowa. D, 1963, Iowa State University of Science and Technology. 233 p.

Stoertz, G. E. Oxidized lead-zinc ores of the Shoshone Mines, Tecopa, California. M, 1955, Columbia University, Teachers College.

Stoertz, Mary W. Evaluation of groundwater recharge in the Central Sand Plain of Wisconsin. M, 1985, University of Wisconsin-Madison. 159 p.

Stoeser, Douglas. Geology of a portion of the Great Pond Quadrangle, Maine. M, 1966, University of Maine. 88 p.

Stoeser, Douglas Benjamin. Mafic and ultramafic xenoliths of cumulus origin, San Francisco Volcanic Field, Arizona. D, 1973, University of Oregon. 260 p.

Stoessell, Ronald K. Experimental study of multi-cation diffusion in an artificial quartz sandstone. M, 1974, Louisiana State University.

Stoessell, Ronald Keith. Geochemical studies of two magnesium silicates, sepiolite and kerolite. D, 1977, University of California, Berkeley. 122 p.

Stoesz, LeRoy Warren. A sedimentation study of early upper Pleistocene intertill deposits in northwestern Lancaster County, Nebraska. M, 1949, University of Nebraska, Lincoln.

Stoever, Edward Carl, Jr. Geology of the Pass Creek area, Huerfano and Costilla counties, Colorado. D, 1959, University of Michigan. 114 p.

Stoever, Edward Carl, Jr. and Austin, Ward Hunting. Reconnaissance geology of the south flank of Cinnamon Mountain, Gallatin County, Montana. M, 1950, [University of Michigan]. 102 p.

Stoffa, P. L. The application of homomorphic deconvolution to shallow water marine seismology. D, 1974, Columbia University. 169 p.

Stoffel, Keith L. Glacial geology of the southern Flathead Valley, Montana. M, 1980, University of Montana. 149 p.

Stoffer, Philip Ward. The geology of the northwestern half of Whiskey Mountain near Dubois, Fremont County, Wyoming. M, 1983, Miami University (Ohio). 84 p.

Stoffey, Philip Stephen. Transmissibility of the Little Blue River basin in southeastern Nebraska. M, 1968, University of Nebraska, Lincoln.

Stoffregen, Roger Eben. Genesis of acid-sulfate alteration and Au-Cu-Ag mineralization at Summitville, Colorado (including sections on supergene alteration and clay mineralogy of the deposit). D, 1985, University of California, Berkeley. 216 p.

Stoffregen, Roger Eben. Intrusives on the eastern edge of the Coast Range Batholith near Terrace, British Columbia. M, 1978, Bryn Mawr College. 70 p.

Stoffyn, M. A. The fate of dissolved aluminum within the oceans. D, 1979, Northwestern University. 167 p.

Stohl, Frances Virginia. Fluxing experiments on the possible stability ranges of andalusite, sillimanite and kyanite and jadeite at one atmosphere. M, 1969, Ohio State University.

Stohl, Francis Virginia. The crystal chemistry of the uranyl silicate minerals. D, 1974, Pennsylvania State University, University Park. 124 p.

Stohr, Christopher J. Delineation of sinkholes and the topographic effects on multispectral response. M, 1974, Purdue University. 132 p.

Stoiber, G. A. Streptelasmatina of the Centerfield Limestone (Middle Devonian) of western New York. M, 1977, SUNY at Buffalo. 132 p.

Stoiber, R. E. Genetic significance of minor elements in sphalerite. D, 1937, Massachusetts Institute of Technology. 129 p.

Stoker, Carol R. Vertical structure and convective dynamics of the equatorial region on Jupiter. D, 1983, University of Colorado. 218 p.

Stokes, Christopher P. The effects of titanium and chromium on the enstatite-diopside solvus. M, 1983, Brooklyn College (CUNY).

Stokes, William Lee. Lithology and stratigraphy of the Red Plateau, Emery County, Utah. M, 1938, Brigham Young University. 50 p.

Stokes, William Lee. Stratigraphy of the Morrison Formation and related deposits in and adjacent to the Colorado Plateau. D, 1941, Princeton University. 168 p.

Stokesbury, Walter Allen. A faunal study of the Dallas area in an attempt to determine the faunal horizon or horizons present and to correlate with other Oregon and California horizons. M, 1933, Oregon State University. 74 p.

Stokke, Per R. Holocene and present day sedimentation on Nitinat deep sea fan, Northeast Pacific Ocean. D, 1976, Lehigh University. 241 p.

Stokley, John Allen. Geology of the Rose Run (Brassfield) iron ores of Bath County, Kentucky. M, 1948, University of Kentucky. 35 p.

Stokowski, Steven J., Jr. Sedimentary depositional environment of the Ordovician St. Peter Formation, and distribution and origin of jarosite and soluble components at the Clayton silica mine, Clayton, Iowa. M, 1982, South Dakota School of Mines & Technology. 116 p.

Stolar, John, Jr. Megaspores and the Devonian-Mississippian boundary along Route 322, Centre County, Pennsylvania. D, 1978, Pennsylvania State University, University Park. 121 p.

Stoley, Aaron Kenneth. A glacial outwash study in South Dakota. M, 1956, University of Nebraska, Lincoln.

Stolfus, Martha A. Supposed marine terraces of southeastern Connecticut. M, 1923, Columbia University, Teachers College.

Stoll, Sarah Johanna. A foraminiferal study of the Kara Sea north of 76 degrees north latitude. M, 1968, University of Wisconsin-Madison.

Stoll, W. C. Geology and geochemistry associations of beryllium. D, 1942, Massachusetts Institute of Technology. 86 p.

Stoll, W. C. Relations of structure to mineral deposition at the Independence Mine, Alaska. M, 1941, Massachusetts Institute of Technology.

Stollar, Richard Lloyd. Geology and some engineering properties of near surface Pennsylvanian shales in northeastern Ohio. M, 1976, Kent State University, Kent. 52 p.

Stollar, Robert L. Applications of geophysics in hydrogeology, Boxelder Creek valley, (Adams and Weld counties) Colorado. M, 1969, Colorado State University. 151 p.

Stolle, James Michael. Stratigraphy of the lower Tertiary and Upper Cretaceous (?) continental strata in the Canyon Range, Juab County, Utah. M, 1977, Brigham Young University.

Stollenwerk, Kenneth G. Removal of nickel and chromium from solution by sediment in a southeastern Kentucky stream. M, 1974, University of Kentucky. 77 p.

Stollenwerk, Kenneth George. Geochemistry of leachate from retorted and unretorted Colorado oil shale. D, 1980, University of Colorado. 236 p.

Stolper, Edward Manin. Igneous petrology of differentiated meteorites. D, 1979, Harvard University.

Stolte, Christian. Geometry of the Wadati-Benioff zone and crustal structure of the Middle American Trench in southern Mexico from microearthquakes. M, 1986, University of California, Santa Cruz.

Stolz, Harry P. and Winckel, E. E. Subsurface correlation of a portion of the southwest flank of the Long Beach oil field, with particular reference to legal proceedings involving trespass deviational drilling. M, 1939, University of Southern California.

Stolzenberg, J. E. A stochastic model for the transport of a trace chemical in a regional environment. D, 1975, University of Wisconsin-Madison. 263 p.

Stolzman, Robert A. A paleomagnetic and rock magnetic study of selected granitic rocks of Rhode Island. M, 1978, University of Rhode Island.

Stommel, Harrison Edfred. Seismic investigations in the Golden-Denver area (Colorado) (T729). D, 1951, Colorado School of Mines. 83 p.

Stone, B. D. The Quaternary geology of the Plainfield and Jewett City quadrangles, central eastern Connecticut. D, 1974, The Johns Hopkins University. 237 p.

Stone, Byron D. Deglaciation events in part of the Manchester South 7.5 (Minute) Quadrangle, south central New Hampshire. M, 1971, University of Vermont.

Stone, C. Basalts from Hawaii Geothermal Project Well-A. M, 1977, University of Hawaii.

Stone, Catherine S. The relationship of plant species distributions to flood frequency of the Willimantic River, Connecticut. M, 1973, Boston University. 442 p.

Stone, Charles David. Ferromagnetic resonance intensity (Is); a rapid method for determining the mode of origin of lunar glass beads. M, 1982, University of Tennessee, Knoxville. 89 p.

Stone, Charles G. Geology of T. 12 N., R. 21 W., Johnson County, Arkansas. M, 1957, University of Arkansas, Fayetteville.

Stone, Christopher Talbott. The hydrogeology of a small esker-aquifer in central Maine. M, 1986, University of Maine. 114 p.

Stone, Craig A. Experimental and theoretical characterization of effective interactions near ^{132}Sn. D, 1987, University of Maryland. 278 p.

Stone, David J. The geology of the upper Dunleith Formation (Prosser Member, Galena Formation) of Middle Ordovician age in southeastern Minnesota. M, 1980, University of Minnesota, Duluth.

Stone, David Michael Raymond. Model studies on the stability of reinforced mine backfill. D, 1985, Queen's University.

Stone, Denise M. Paleoenvironment and secondary porosity of the upper Jackfork Sandstone, central Arkansas. M, 1981, Memphis State University. 84 p.

Stone, Denver Cedrill. The Sydney Lake fault zone in Ontario and Manitoba, Canada. D, 1982, University of Toronto.

Stone, Donald B. The Moosehead Plateau in eastern New England. M, 1942, Union College. 38 p.

Stone, Donald Sherwood. Origin and significance of the breccias along the northwestern side of Lake Champlain. M, 1951, Cornell University.

Stone, Dwayne David. Desmoinesian conodonts from Utah, Colorado and Iowa. D, 1968, University of Utah. 259 p.

Stone, Gary Calvin. Paleoecology and diagenesis of the Avon Park Formation, Gulf Hammock Wildlife Management Area, Levy County, Florida. M, 1975, University of Florida. 128 p.

Stone, George Thomas. Petrology of upper Cenozoic basalts of the western Snake River plain, Idaho. D, 1967, University of Colorado. 417 p.

Stone, Gerald L. Bighorn (Ordovician) conodonts from Wyoming. M, 1957, Iowa State University of Science and Technology.

Stone, Gerald Leslie. Bighorn conodonts from Wyoming. M, 1958, University of Iowa. 60 p.

Stone, Irving Charles. A study of some Fairfax County, Virginia soils. M, 1961, George Washington University.

Stone, Irving Charles, Jr. Geochemistry and mineralogy of continental shelf sediments off the South Carolina coast. D, 1967, George Washington University. 112 p.

Stone, James A. A lithostratigraphic correlation of post Eocene pre-Quaternary rocks of central Florida. M, 1960, Florida State University.

Stone, James M. Petrology of the granitic pluton at Oak Creek, Fremont County, Colorado. M, 1988, Kansas State University. 139 p.

Stone, Jerome. The geology of the Elliston Phosphate District, Powell County, Montana. M, 1952, University of Montana. 52 p.

Stone, John B. The geology of Kilauea (Hawaii). D, 1926, Yale University. 60 p.

Stone, John Elmer. Insoluble residues of the Madison Group from five sections on the eastern flanks of the Big Horn Mountains, Wyoming. M, 1958, University of Illinois, Urbana. 23 p.

Stone, John Elmer. Pleistocene geology of Clark County, northeastern Missouri. D, 1960, University of Illinois, Urbana. 206 p.

Stone, John Grover, II. Ore genesis in the Naica District, Chihuahua, Mexico. D, 1959, Stanford University. 123 p.

Stone, Joseph Fred. Palynology of the Almond Formation (upper Cretaceous), Rock Springs uplift, Wyoming. D, 1971, Michigan State University. 190 p.

Stone, Joseph Fred. Palynology of the Fort Belknap Coal (Pennsylvanian) of north-central Texas. M, 1966, Texas Christian University.

Stone, Kenneth Coy. Measurement and simulation of soil water status under field conditions. D, 1987, University of Florida. 167 p.

Stone, Larry J. Van *see* Van Stone, Larry J.

Stone, Paul. Stratigraphy, depositional history, and paleogeographic significance of Pennsylvanian and Permian rocks in the Owens Valley-Death Valley region, California (Volumes I and II). D, 1985, Stanford University. 470 p.

Stone, Ralph Arthur. Waulsortian-type bioherms of Mississippian age, central Bridger Range, Montana. M, 1971, University of Wisconsin-Madison. 108 p.

Stone, Randolph. Hydrogeology of the Nishnabotna River Basin. D, 1971, Iowa State University of Science and Technology. 184 p.

Stone, Randolph. The geometry and origin of the oolitic sandstone bodies of the Sundance Formation (Upper Jurassic) in the Big Horn Basin, Wyoming. M, 1967, Iowa State University of Science and Technology.

Stone, Richard B. A quantitative study of benthic fauna in lower Chesapeake Bay with emphasis on animal-sediment relationships. M, 1963, College of William and Mary.

Stone, Richard O'Neil. A geologic investigation of playa lakes. D, 1955, University of Southern California.

Stone, Richard O'Neill. A sedimentary study and classification of playa lakes. M, 1953, University of Southern California.

Stone, Robert. Groundwater geology, geochemistry, and hydrology of the southeastern San Joaquin Valley, California. D, 1955, University of California, Los Angeles.

Stone, Solon W. Structure and stratigraphy of the Milton area, Vermont. D, 1951, Harvard University.

Stone, Solon W. The origin of the (Silurian) Manlius Group of limestones in central New York. M, 1940, Syracuse University.

Stone, Thomas A. Geologic mapping of the Uana-Bendego area, Bahia State, Brazil, using Landsat digital data. M, 1982, Dartmouth College. 1117 p.

Stone, Timothy Storer. The quality of glacial sand and gravel resources as related to environmental conditions and landforms in the Binghamton region, New York. M, 1981, SUNY at Binghamton. 100 p.

Stone, W. L. An assessment of alternative seawater intrusion control strategies for the Oxnard Plain of Ventura County, California. D, 1978, University of California, Los Angeles. 188 p.

Stone, Wilbur Alan. Relation of the stratigraphy and geomorphology to the occurrence of ground water in northeastern Jefferson County, Kansas. M, 1965, University of Kansas. 84 p.

Stone, William Burgess, Jr. Mineralogic and textural dispersal patterns within the Permian Post Oak Formation of southwestern Oklahoma. M, 1977, University of Oklahoma. 117 p.

Stone, William D. Geology of the Black Mountain area, Apple Valley Quadrangle, California. M, 1964, University of California, Riverside. 127 p.

Stone, William Edward. Nature and significance of metamorphism in gold concentration, Bousquet Township, Abitibi greenstone belt, Northwest Quebec. D, 1988, University of Western Ontario. 441 p.

Stone, William J. Stratigraphy and sedimentary history of middle Cenozoic (Oligocene and Miocene) deposits in North Dakota. D, 1973, University of North Dakota. 217 p.

Stone, William J. Stratigraphy of the Minnelusa Formation (Pennsylvanian-Permian) along the western and northern flanks of the Black Hills, Wyoming and South Dakota. M, 1969, Kent State University, Kent. 285 p.

Stone, William Leroy. Mesozoic sedimentary facies and tectonic framework of the northern Tindouf Basin; North Africa. M, 1977, North Carolina State University. 82 p.

Stonebraker, Jack Douglas. Anisotropy of magnetic susceptibility studies of a limited metamorphic suite. M, 1969, Florida State University.

Stonebraker, Jack Douglas. The potassium-argon geochronology of the Brevard Fault Zone, southern Appalachians. D, 1973, Florida State University.

Stoneburner, Richard Kelty. Subsurface study of the Cherokee Group on the western flank of the Central Kansas Uplift in portions of Trego, Ellis, Rush and Ness counties, Kansas. M, 1982, Wichita State University. 67 p.

Stoneburner, Roger W. Q. Mississippian stratigraphy and coral zones of the Wapiti Lake area, British Columbia, Canada. M, 1948, University of Kansas. 86 p.

Stonecipher, Sharon A. Origin, distribution and diagenesis of deep-sea phillipsite and clinoptilolite. D, 1977, University of California, San Diego. 238 p.

Stonehouse, Harold B. Trace elements and their association with mineralization. D, 1952, University of Toronto.

Stonehouse, James Mrcus. Movement of mineralizing fluids, Bonanza mining district, Nicaragua. M, 1976, Dartmouth College. 66 p.

Stoneman, Deborah Ann. Structural geology of the Plomosa Pass area, northern Plomosa Mountains, La Paz county, Arizona. M, 1985, University of Arizona. 99 p.

Stonerook, William H. Mining nitrate at Oficina Pedro de Valdivia, Chile. M, 1932, New Mexico Institute of Mining and Technology. 24 p.

Stonestreet, Albert Lee. A study of lacustrine deposits in the valleys of some of the major tributaries to the Hocking River, Athens County, Ohio. M, 1965, Ohio University, Athens. 66 p.

Stonestrom, David Arthur. Co-determination and comparisons of hysteresis-affected, parametric functions of unsaturated flow; water-content dependence of matric pressure, air-trapping, and fluid permeabilities in a non-swelling soil. D, 1987, Stanford University. 307 p.

Stooke, Philip John. Cartography of non-spherical worlds. D, 1988, University of Victoria. 169 p.

Stookey, Donald Graham. Stratigraphy of the Des Moines Series of southeastern Iowa. D, 1935, University of Iowa. 135 p.

Stookey, Donald Graham. The Pennsylvanian outliers of Iowa. M, 1932, University of Iowa.

Stoops, Sheryl Lynn. Identification and occurrence of the clay minerals in the reservoir sandstones of the Grimsby Formation (Medina Group-Lower Silurian), Genesee County, New York. M, 1986, SUNY at Buffalo. 98 p.

Stopen, Lynne E. Geometry and deformation history of mylonitic rocks and silicified zones along the Mesozoic Connecticut Valley border fault, western Massachusetts. M, 1988, University of Massachusetts. 176 p.

Stopper, Robert Francis. Geology southeast of Page Mill Road, Palo Alto Quadrangle. M, 194?, Stanford University.

Storch, Robert H. Stratigraphy of the Chugwater Formation in the area adjacent to Dubois, Wyoming. M, 1950, Miami University (Ohio). 39 p.

Storch, Sara Glen Power. Geology of the Blue River mining district, Linn and Lane counties, Oregon. M, 1978, Oregon State University. 70 p.

Storer, John Edgar, III. The Wood Mountain fauna; an upper Miocene mammalian assemblage from southern Saskatchewan. D, 1971, University of Toronto.

Storey, Lester Oscar. Geology of a portion of Bannock County, Idaho. M, 1959, University of Idaho. 58 p.

Stork, Allen Louis. Silicic magmatism in an island Arc, Fiji, Southwest Pacific; implications for continental growth. D, 1984, University of California, Santa Cruz. 288 p.

Stork, Christof. Ray trace tomographic velocity analysis of surface seismic reflection data. D, 1988, California Institute of Technology. 664 p.

Storm, Allen Bruce. Stratigraphy and petrology of the Grassy Mountain Formation (Hemphillian, middle Pliocene), Malheur County, Oregon. M, 1975, University of Oregon. 63 p.

Storm, David Russell. A predictive method for assessing the impact of maintenance dredging in an estuary. D, 1973, University of California, Davis. 153 p.

Storm, Evelyn V. A study of a diminutive fauna from the Marcellus Formation (Middle Devonian-Erian) from sites in Albany County, New York, and Sussex County, New Jersey. M, 1985, Montclair State College. 27 p.

Storm, Linda M. Major and trace metal concentration of Dietz No. 2 and Rosebud coal via direct reading emission spectrometry. M, 1978, Montana College of Mineral Science & Technology. 135 p.

Storm, Paul Jennings. A petrographic study of the Merchantville Clay of Camden and Burlington counties, New Jersey, and its stratigraphic significance. D, 1930, University of Pennsylvania. 26 p.

Storm, Paul Jennings. Evidences of sedimentation rhythms in the (Lower Cambrian to Middle Ordovician) Shenandoah Limestone near Ivy Rock, Montgomery County, Pennsylvania. M, 1926, University of Pennsylvania.

Storm, Paul Vissing. Channel deposits of the Salt Wash Member, Morrison Formation, Arizona. M, 1955, Iowa State University of Science and Technology.

Stormer, John Charles. The volcanic petrology of northeastern New Mexico. D, 1971, University of California, Berkeley. 126 p.

Storms, Erik. Hydrogeologic investigation of Williamsburg, Massachusetts. M, 1987, University of Massachusetts. 109 p.

Storr, John Frederick. Ecology and oceanography of the coral reef tract, Abaco Island, Bahamas. D, 1955, Cornell University.

Storrer, James. The occurrence, origin, and distribution of tungsten. M, 1914, Cornell University.

Storrs, Glenn William. A review of occurrences of the Plesiosauria (Reptilia:Sauropterygia) in Texas with description of new material. M, 1981, University of Texas, Austin. 226 p.

Storrs, Glenn William. Anatomy and relationships of Corosaurus alcovensis (Reptilia; Nothosauria) and the Triassic Alcova Limestone of Wyoming. D, 1986, Yale University. 392 p.

Storrs, Lucius S. The Rocky Mountain coal fields. M, 1904, University of Nebraska, Lincoln.

Storti, Frank W. The geochemistry of ^{210}Pb in the Southeastern, United States estuarine system. M, 1980, Georgia Institute of Technology. 42 p.

Stott, Donald Franklin. The Alberta Group and equivalent rocks, Rocky Mountain foothills, Alberta. D, 1958, Princeton University. 748 p.

Stott, Donald Franklin. The Jurassic stratigraphy of Manitoba. M, 1954, University of Manitoba.

Stott, G. M. The sequence of tectonic events in the Radcliff Gneiss terrain of the Grenville Province, Radcliff Township, Ontario. M, 1977, University of Waterloo.

Stott, George F. Mining and milling of complex lead and zinc ores in the Park City District, Utah. M, 1916, University of Utah. 64 p.

Stott, Gregory Myles. A structural analysis of the central part of the Archean Shebandowan greenstone belt and a crescent-shaped granitoid pluton, northwestern Ontario. D, 1986, University of Toronto.

Stott, Laurence Richard. A gravity survey of a late Cenozoic graben in the Wyoming-Idaho thrust belt. M, 1974, University of Michigan.

Stottlemyer, Laura Lee. Variations in spatial responses along the Virginia barrier islands. M, 1978, University of Virginia. 35 p.

Stottlemyre, James Arthur. An investigation of sand and sandstone permeability degradation at elevated temperature and pressure. D, 1981, University of Washington. 168 p.

Stottlemyre, James Arthur. Experimental deformation of the granodioritic material composing the Mount Stuart Batholith in the vicinity of Stevens Pass. M, 1973, University of Washington. 53 p.

Stottmann, Walter. The feasibility of integrated ground and surface water use in humid regions. D, 1974, Pennsylvania State University, University Park. 226 p.

Stotts, John Louis. Structure and stratigraphy of the northwestern Indio Hills, Riverside County, California. M, 1965, University of California, Riverside. 208 p.

Stotz, Tina M. Permeability of peat to selected toxic leachates. M, 1987, San Jose State University. 138 p.

Stouder, Ralph Eugene. A geological, stratigraphical, and structural study of the Daviess County oil fields, lying principally in Veale, Barr, and Reeve townships, Daviess County, Indiana. M, 1927, Indiana University, Bloomington. 52 p.

Stoudt, David L. Depositional environments of lower Cretaceous Muddy sandstones, Recluse area, Campbell County, Wyoming. M, 1974, Texas A&M University. 118 p.

Stoufer, R. N. The effect of organic carbon on the concentrations of iron and hydrogen sulfide in ground water. M, 1975, University of Missouri, Columbia.

Stouffer, Stephen Gerald. Landslides in the Coal Hill area, Kane County, Utah. M, 1964, University of Utah. 102 p.

Stouge, Svend Sandbergh. Conodonts of the Table Head Formation (Middle Ordovician), western Newfoundland. M, 1980, Memorial University of Newfoundland. 413 p.

Stough, Joan B. A statistical study of some Pennsylvanian faunas near Wellsville (Chaffee and Fremont counties), Colorado. M, 1957, University of Colorado.

Stoughton, Dean. Interpretation of seismic reflection data from the San Luis Valley, South-central Colorado. M, 1977, Colorado School of Mines. 100 p.

Stout, Earl Douglas. The Prentice area, Terry and Yoakum counties, Texas. M, 1954, Texas Tech University. 61 p.

Stout, James H. Bedrock geology between Rainy Creek and Denali Fault, eastern Alaska Range, Alaska. M, 1965, University of Alaska, Fairbanks. 75 p.

Stout, James Harry. Geology of the Fyresvatn-Nisservatn area, Telemark, Norway (Precambrian). D, 1970, Harvard University.

Stout, Jerry Dale. Structure and origin of Terlingua Uplift, Brewster and Presidio counties, Texas. M, 1979, West Texas State University. 79 p.

Stout, John L. Geology of the Siloam area, Pueblo County, Colorado. M, 1959, Colorado School of Mines. 143 p.

Stout, Koehler Sheridan. Geology and mines of the Ogden Mountain mining district, Powell County, Montana. M, 1949, Montana College of Mineral Science & Technology. 57 p.

Stout, Martin Lindy. Geology of a part of the south-central Cascade Mountains of Washington. D, 1959, University of Washington. 183 p.

Stout, Martin Lindy. Geology of the southwestern portion of the Mt. Stuart quadrangle, Washington. M, 1957, University of Washington. 115 p.

Stout, Mavis Z. Mineralogy and petrology of Quaternary lavas, Snake River Plain, Idaho and the cation distribution in natural titanomagnetites. M, 1975, University of Calgary. 150 p.

Stout, Paul Michael. Calcium carbonate sedimentation on the northeast insular shelf of Puerto Rico. M, 1979, Duke University. 107 p.

Stout, Paul Michael. Chemical diagenesis of pelagic biogenic sediments from the Equatorial Pacific. D, 1985, University of California, San Diego. 238 p.

Stout, Scott Alan. A microscopic investigation of the fate of secondary xylem during peatification and the early stages of coal formation. M, 1985, Pennsylvania State University, University Park. 309 p.

Stout, Scott Alan. Tracing the microscopical and chemical origin of huminitic macerals in coal. D, 1988, Pennsylvania State University, University Park. 351 p.

Stout, Thompson Mylan. A stratigraphic study of the Oligocene rodents in the Nebraska State Museum. M, 1937, University of Nebraska, Lincoln.

Stoutamire, Steve. A new middle Miocene vertebrate fauna from the Florida Panhandle. M, 1975, Texas Tech University. 131 p.

Stovall, James Curl. Pleistocene geology and physiography of the Wallowa Mountains. M, 1929, University of Oregon. 115 p.

Stovall, John Willis. A study of the stratigraphy and fauna of the Cannon Formation in the vicinity of Nashville, Tennessee. M, 1927, Vanderbilt University.

Stovall, John Willis. Geology of the Cimarron River valley in Cimarron County, Oklahoma. D, 1938, University of Chicago. 60 p.

Stovall, Robert L. The relationships between gravity anomalies and the geology of the Blue Ridge Province near Floyd, Virginia. M, 1984, Virginia Polytechnic Institute and State University. 64 p.

Stovall, Robert M. Cavity detection by surface wave distortion, a finite difference model. M, 1978, University of Missouri, Rolla.

Stover, Bruce King. Debris-flow origin of high-level sloping surfaces on the northern flanks of Battlement Mesa, and surficial geology of parts of the North Mamm Peak, Rifle, and Rulison quadrangles, Garfield County, Colorado. M, 1984, University of Colorado. 75 p.

Stover, Lewis E. Pt. I, Stratigraphy and paleontology of the (Devonian) Moscow Formation (Hamilton) in central and western New York; Pt II, Ostracoda from the (Middle Devonian) Windom Shale in western New York. D, 1956, University of Rochester. 160 p.

Stover, Mark K. A comparative analysis of three modes of operation of a soil compactor as an energy source for seismic reflection exploration. M, 1986, Wright State University. 82 p.

Stover, Stewart L. Micropaleontology and paleoecology of the Gasport and Moody's Branch formations (upper Eocene) in southwestern Alabama. M, 1967, Northeast Louisiana University.

Stow, Dorrik, A. V. Late Quaternary stratigraphy and sedimentation on the Nova Scotian outer continental margin. D, 1977, Dalhousie University. 360 p.

Stow, Marcellus H. An occurrence of (Devonian) Oriskany Sandstone with celestite cement. M, 1927, Cornell University.

Stow, Marcellus H. Contribution to the petrography of the Oriskany Sandstone. D, 1931, Cornell University.

Stow, Stephen Harrington. A radiometric and chemical study of the binary Fitzwilliam Granite (Late Carboniferous) of New Hampshire. M, 1965, Rice University. 79 p.

Stow, Stephen Harrington. The distribution of elements among the coexisting phases of the Precambrian meta-sediments of West Texas. D, 1966, Rice University. 121 p.

Stowell, Harold Hilton. Sphalerite geobarometry in metamorphic rocks and the tectonic history of the Coast Ranges near Holkham Bay, southeastern Alaska. D, 1987, Princeton University. 376 p.

Stowers, Douglas I. The Prairie Bluff Formation in the Buena Vista Quadrangle, Chickasaw County, Mississippi. M, 1961, Mississippi State University. 69 p.

Stowers, Robert Earl, II. Stratigraphy and geochronology of Pleistocene carbonates; Sandy Point area, southern San Salvador Island, Bahamas. M, 1988, Mississippi State University. 103 p.

Stoyer, Charles Hayes. Numerical solutions of the response of a two-dimensional Earth to an oscillating magnetic dipole source with application to a ground-water field study. D, 1974, Pennsylvania State University, University Park. 211 p.

Straccia, Joseph Robert. Stratigraphy and structure of the upper Wilcox Group (lower Eocene) in the Rosita gas field, Duval County, Texas. M, 1980, Texas A&M University. 107 p.

Strachan, Betsy M. The biostratigraphy and paleoecology of Miocene benthic Foraminiferida from the Salina Basin, State of Veracruz, Mexico. M, 1986, Tulane University. 167 p.

Strachan, Clyde G. The problem of the correlation of the Mendota and Black Earth dolomites. M, 1927, University of Wisconsin-Madison.

Strachan, Donald G. Stratigraphy of the Jarilla Mountains, Otero County, New Mexico. M, 1977, New Mexico Institute of Mining and Technology. 135 p.

Stracher, Glenn B. Structure and petrology of Precambrian rocks in zones of high ductile strain, Southeast Adirondack Mountains of New York State. M, 1986, University of Nebraska, Lincoln. 74 p.

Strack, Kurt-Martin. A method for the determination of the thermal conductivity of tight sandstones using a variable state approach. M, 1981, Colorado School of Mines. 162 p.

Strader, Harold L. The petroleum resources of Wyoming. M, 1927, Columbia University, Teachers College.

Stradley, Ann Chalmers. Hydrology and subsurface geology of the Mississippian Madison Group and potential water use for agriculture and industry, northwestern Montana plains. D, 1981, University of Montana. 238 p.

Strahl, Erwin O. The relationship among selected minerals, trace elements, and organic constituents of several black shales. D, 1958, Pennsylvania State University, University Park. 155 p.

Strahler, Arthur N. Hypotheses of stream development in the folded Appalachians of Pennsylvania. D, 1944, Columbia University, Teachers College.

Strahler, Arthur N. Landslides of the Vermilion and Echo cliffs, northern Arizona. M, 1940, Columbia University, Teachers College.

Strain, Lamar Asal, Jr. Eocene and Oligocene biostratigraphy of the Twin Rivers area, Clallam County, Washington. M, 1964, University of Washington. 131 p.

Strain, William Samuel. Blancan mammalian fauna and Pleistocene formations, Hudspeth County, Texas. D, 1964, University of Texas, Austin. 174 p.

Strain, William Samuel. The Pleistocene geology of part of the Washita River valley, Grady County, Oklahoma. M, 1937, University of Oklahoma. 102 p.

Strait, Melissa Marie. Chemical variations in enstatite achondrites. D, 1983, Arizona State University. 185 p.

Straka, Joseph John. Conodonts from the Kinderhookian Series Washington County, Iowa. M, 1966, University of Iowa. 139 p.

Straka, Joseph John, II. Age and correlation of the Goddard (Mississippian) and Springer (Mississippian-Pennsylvanian) formations in southern Oklahoma as determined by conodonts. D, 1969, University of Iowa. 283 p.

Straley, Harrison W., III. Field and laboratory studies of nonconcentric folding. D, 1939, University of Chicago.

Straley, Harrison Wilson, III. Some geomagnetic traverses in the Appalachian Mountains. D, 1938, University of North Carolina, Chapel Hill. 22 p.

Strand, Carl Ludvig. Pre-1900 earthquakes of Baja California and San Diego County. M, 1980, San Diego State University.

Strand, Edwin H. Experiments on the separation of copper minerals by solution. M, 1930, University of Minnesota, Minneapolis. 15 p.

Strand, Jesse Richard. Structural geology of pre-Tertiary rocks in southeastern Washington and adjacent portions of Idaho. M, 1949, Washington State University. 25 p.

Strand, Robert Leroy. Geology of a portion of the Blue Nose Mountain Quadrangle and adjacent areas, Plumas and Sierra counties, California. M, 1972, University of California, Davis. 86 p.

Strand, Rudolph G. A study of the effects of regional metamorphism on calcareous concretions. M, 1958, Northwestern University.

Strandberg, Carl H. The use of USGS land use/land cover maps to determine regional nonpoint source water pollution potential. M, 1980, San Jose State University. 106 p.

Strange, Elizabeth Allison. The geochemistry of the Barber mafic complex, Rowan County, North Carolina. M, 1983, University of Tennessee, Knoxville. 91 p.

Strange, Louis C. Geology of the (Lower Cretaceous) Patuxent Sandstone of northeastern Virginia. M, 1934, University of Virginia. 112 p.

Strange, Nettie S. A census study of foraminifera in the lower Taylor Group, upper Cretaceous, Travis and Williamson counties, Texas. M, 1975, University of Texas, Austin.

Strange, William D., Jr. Dolomitization in the Ellenburger Group, Val Verde Basin, West Texas, and its relationship to brecciation. M, 1988, Stephen F. Austin State University. 74 p.

Strangways, H. F. The washing of bituminous coal with notes on special experiments on certain Nova Scotian coals. M, 1908, McGill University.

Stransky, Terry. The petrology of the eastern funnel of the Cortlandt Complex. M, 1976, Brooklyn College (CUNY).

Strasen, James L. Nearshore sedimentary dynamics in western Lake Michigan at Terry Andrae-Kohler Park, Wisconsin. M, 1978, University of Wisconsin-Milwaukee.

Strassberg, Morton D. The geology of the southwest portion of the Montgomery City Quadrangle, Missouri. M, 1953, University of Missouri, Columbia.

Strathearn, Gary Edward. Paleoecologic analyses of cyclically (seasonally) banded middle Proterozoic cyanobacterial communities. D, 1984, University of California, Los Angeles. 295 p.

Strathouse, Elizabeth Cerates. Late Cenozoic fluvial sediments in Barton Flats and evidence for orogenesis, San Bernardino Mountains, San Bernardino County, Southern California. M, 1983, University of California, Riverside. 132 p.

Strathouse, Scott Mitchell. Genesis and mineralogy of soils on an alluvial fan, western Mojave Desert, California. D, 1982, University of California, Riverside. 180 p.

Strathouse, Scott Mitchell. Nitrogen geochemistry and clay mineralogy in two drainage basins of the eastern Diablo Range, Central California. M, 1978, University of California, Riverside. 89 p.

Stratton, Alan Jerome. The thermally-driven fluid-loop dynamo; field reversal, extinction, and regeneration. D, 1974, Stanford University. 146 p.

Stratton, Everett Franklin. Magnetic exploration for iron ores. D, 1933, Harvard University.

Stratton, J. Lynn. The stratigraphic and structural relationships on the central Kansas uplift. M, 1942, University of Oklahoma. 42 p.

Stratton, James Forrest. Studies of Polypora from the Speed Member, North Vernon Limestone (Eifelian, middle Devonian) in southern Indiana. D, 1975, Indiana University, Bloomington. 238 p.

Straube, Elsie Joan. The Florissant Basin, St. Louis County. M, 1933, Washington University. 108 p.

Straughan, George M. Geology of the Big Lime Formation (middle Mississippian) in eastern Kentucky. M, 1941, University of Kentucky. 71 p.

Strausberg, Sanford Irvin. An analysis of local buff and blue glacial deposits. M, 1955, University of Michigan.

Strauss, Robert C. Variations in plagioclase zoning in response to an evolving physicochemical environment; applications to the interpretation of crystallization processes in the Caribou Mountain Pluton, California. M, 1983, University of Arizona. 90 p.

Strauss, Robert G. Structure in the vicinity of the C-JD-7 mining area, Paradox Valley, Montrose County, Colorado. M, 1982, Michigan Technological University. 90 p.

Strauss, Ruth Ann. Trace elements in northern Virginia Middle Ordovician carbonates; implications for diagenesis. M, 1988, Old Dominion University. 111 p.

Strautman, Sabina Y. Determining dispersion coefficients from a landfill leachate plume. M, 1984, Texas A&M University. 50 p.

Stravers, Jay Anthony. Glacial geology of outer Meta Incognita Peninsula, southern Baffin Island, Arctic Canada. D, 1986, University of Colorado. 263 p.

Stravers, Loreen K. Stegeman. Palynology and deglaciation history of the central Labrador-Ungava Peninsula. M, 1981, University of Colorado. 171 p.

Straw, William Thomas. Geomorphology, hydrogeology, and economic geology of the Ohio River Valley, Mauckport to Cannelton, Indiana. D, 1968, Indiana University, Bloomington. 182 p.

Straw, William Thomas. Stratigraphy of the Marble Hill facies and contiguous strata of the Waynesville Formation (Richmondian) in southeastern Indiana and north-central Kentucky. M, 1960, Indiana University, Bloomington. 50 p.

Strawn, Mary Baker. The geography of Marsh Valley, Utah. M, 1964, University of Utah. 111 p.

Strawson, Frederick MacLeod, Jr. The geology of the Permian Carbon Ridge Formation, East-central Nevada. M, 1981, University of Nevada. 140 p.

Strayer, Geoffrey Ben. A remote sensing investigation of porphyry copper mineralization in Pinal and Gila counties, Arizona. M, 1987, Purdue University. 99 p.

Strayton, James P. Microstructural analysis of the Massanutten Sandstone, northwest Virginia. M, 1987, Bowling Green State University. 92 p.

Strean, Bernard Max, Jr. Breakdown and erosion of sandstone bedrock and the subsequent breakdown and movements of fragments produced in portions of Pomona Quadrangle, Illinois. M, 1963, Southern Illinois University, Carbondale. 51 p.

Streb, Lawrence Lambert. The use of the scanning electron microscope to observe conductivity phenomena in selected rocks. M, 1983, Pennsylvania State University, University Park. 72 p.

Strebeck, John William. Structure of the Precambrian basement in the Ozark Plateau as inferred from gravity and remote sensing data. M, 1982, Washington University. 149 p.

Strecker, Manfred Reinhard. Late Cenozoic landscape development, the Santa Maria Valley, Northwest Argentina. D, 1986, Cornell University. 277 p.

Street, Billy Andres. Geology of the Paleozoic sediments in the Washakie Park area, Wyoming. M, 1951, University of Missouri, Columbia.

Street, James S. Significance of variations in till constitution in the Rush Creek area, New York. M, 1963, Syracuse University.

Street, James Stewart. Glacial geology of the eastern and southern portions of the Tug Hill Plateau, New York. D, 1966, Syracuse University. 247 p.

Street, Michael O. Textural attributes and reef development of the Pleistocene Key Largo Limestone, South Florida. M, 1978, Bowling Green State University. 44 p.

Street, Peter J. Trilobite zones in the Murray Range, Pine Pass map area, British Columbia. M, 1967, University of British Columbia.

Street, Ronald Leon. Contemporary earthquake mechanics in central United States. D, 1976, St. Louis University. 271 p.

Street-Martin, Leah V. The chemical composition of the Shuksan Metamorphic Suite in the Gee Point-Finney Creek area, North Cascades, Washington. M, 1981, Western Washington University. 76 p.

Streeter, Michael Edward. The geology of the southern Bull Mountain area, Jefferson County, Montana. M, 1983, Western Michigan University. 89 p.

Streeter, Sereno Stephen. Foraminiferal distribution in the sediments of the Great Bahama Bank (Andros Lobe). D, 1963, Columbia University, Teachers College. 233 p.

Streeton, Dwight Harold. The geology of the Prairie Evaporite Formation (Middle Devonian) of the Yorkton area of Saskatchewan (Canada). M, 1967, University of Saskatchewan. 81 p.

Streeton, Eric Grant. The middle Devonian Winnipegosis Formation of west-central Saskatchewan. M, 1971, University of Saskatchewan. 142 p.

Streib, Donald L. Geology of the Wheeling, West Virginia-Ohio 7.5 minute topographic quadrangle. M, 1969, West Virginia University.

Streib, Donald Lamar. Analysis of benzene and chloroform soluble organic compounds of coal and their relationships to stratigraphy and paleoenvironments. D, 1972, West Virginia University. 167 p.

Streit, Lori Ann. Development of quantitative analytical methods for secondary ion mass spectrometry. D, 1987, Arizona State University. 228 p.

Streits, Robert. Subsurface stratigraphy and hydrology of the Rillito Creek-Tanque Verde Wash area, Tucson, Arizona. M, 1962, University of Arizona.

Streitz, Andrew Ryan. The application of shallow reflection seismic investigations to the delineation of buried drift features. M, 1988, University of Minnesota, Minneapolis. 80 p.

Strelitz, Richard A. Source processes of three complex deep focus earthquakes. D, 1977, Princeton University. 180 p.

Strete, Ralph F. The stratigraphy of the Dodge's Creek section, Butler County, Ohio. M, 1931, Miami University (Ohio). 35 p.

Stricker, Gary D. Geology of the Four Bear Mountain area, Park County, Wyoming. M, 1965, Wayne State University.

Stricker, Gary Dale. Carbonate microfacies of the Pogonip Group (lower Ordovician), Arrow Canyon Range, Clark County, Nevada. D, 1973, University of Illinois, Urbana. 76 p.

Strickland, Douglas K. Depositional environments, community structure, and paleotectonics of Leavick Tarn Dolomite Member, Manitou Formation (lower Ordovician), northern Canon City Embayment, Colorado. M, 1975, University of Wisconsin-Milwaukee.

Strickland, Frank G. Structure, stratigraphy and economic geology of Goat Ridge, Lune County, New Mexico. M, 1980, University of Texas at El Paso.

Strickland, Kenton E. The geology of Stobie and Marconi townships (Precambrian), District of Sudbury, Ontario, Canada. M, 1971, Bowling Green State University. 108 p.

Strickland, L. D. A cold-flow mining study of a vortex coal gasifier. D, 1973, West Virginia University.

Strickland, Matthew O. The areal distribution and depositional setting of the McLouth Sandstone in Jefferson and Leavenworth counties, Kansas. M, 1987, University of Southwestern Louisiana. 85 p.

Strickler, David L. Paleomagnetism of three Late Cretaceous granitic plutons, North Cascades, Washington. M, 1982, Western Washington University. 99 p.

Strickler, Donald Ward. The 1:5 and defect spinels in the system Li₂O-Fe₂O₃-Al₂O₃. M, 1959, Pennsylvania State University, University Park. 54 p.

Strickler, William John. Stratigraphy and sedimentation of the upper part of the Amsden Formation (Pennsylvanian), Tendoy Mountains, (Beaverhead County) Montana. M, 1972, University of Montana. 96 p.

Stricklin, Claude R. Geophysical survey of the Lemei Rock-Steamboat Mountain area, Washington. M, 1975, University of Puget Sound. 23 p.

Stricklin, Fred Lee, Jr. Pleistocene terraces along the Brazos and Wichita Rivers, central and northern Texas. D, 1953, Louisiana State University.

Strider, Mark. Sandstone diagenesis of the Vermego Formation and Trinidad Sandstone, Cimmaron, New Mexico area. M, 1980, Tulane University.

Striegl, Robert G. Exchange and transport of 14-carbon dioxide in the unsaturated zone. D, 1988, University of Wisconsin-Madison. 175 p.

Strife, Stuart C. Diagenetic fabrics of Coeymans (Lower Devonian) reef carbonates of central New York. M, 1977, Rensselaer Polytechnic Institute. 111 p.

Stringer, C. Pleas, Jr. Subsurface geology of western Payne County, Oklahoma. M, 1956, University of Oklahoma. 56 p.

Stringer, Gary L. A study of the upper Eocene otoliths and related fauna of the Yazoo Clay in Caldwell Parish, Louisiana. M, 1977, Northeast Louisiana University.

Stringer, Richard S. Geology of the Krebs Group, Inola area, Rogers and Mayes counties, Oklahoma. M, 1959, University of Oklahoma. 63 p.

Stringfield, Victor Timothy. The structural geology of parts of St. Louis and St. Louis County, Missouri. M, 1927, Washington University. 91 p.

Stringham, Bronson. Mineralization in the West Tintic mining district (Utah). D, 1942, Columbia University, Teachers College.

Strobel, Calvin Jerome. Model studies of geothermal fluids production from consolidated porous media. M, 1973, Stanford University. 105 p.

Strobel, Guye. Palaeomagnetic analyses of the Leaf Rapids area in Manitoba. M, 1988, University of Manitoba. 245 p.

Strobel, John Stuart. A quantitative study of attenuation under Southern California. M, 1985, University of North Carolina, Chapel Hill. 100 p.

Strobel, Robert J. Stratigraphy and structure of the Paleozoic rocks in the Rush Creek drainage, northern Ritter Range Pendant, California. M, 1986, University of Nevada. 123 p.

Strobell, John Dixon. Geology of the Carrizo Mountain area, Arizona. D, 1956, Yale University.

Strobl, Rudolph S. Stratigraphy of the Glauconitic Member (Mannville Group), Medicine River Field and adjacent areas, south-central Alberta. M, 1986, University of Alberta. 175 p.

Stroebel, Kenneth H. Loss of recharge to the aquifer of the Kent City well field as a result of siltation in Breakneck Creek, Portage County. M, 1983, Kent State University, Kent. 171 p.

Strogonoff, Robert Francis. Common Minerals in northwestern Ohio and their geologic occurrence. M, 1966, Bowling Green State University. 67 p.

Stroh, James M. A preliminary report on the San Quintin volcanic field, Baja, Calif., Mexico. M, 1971, University of Washington.

Stroh, James Michael. Latest Cenozoic volcanism, Baja California Norte, Mexico; solubility of alumina in orthopyroxene plus spinel as a geobarometer in complex systems; application to spinel-bearing alpine-type peridotite. D, 1975, University of Washington. 200 p.

Stroh, Patricia Tucker. Chemical and statistical studies of sialic rocks of the western Dharwar Craton, southern India. D, 1986, University of North Carolina, Chapel Hill. 165 p.

Strojan, C. L. The ecological impact of zinc smelter pollution on forest soil communities. D, 1975, Rutgers, The State University, New Brunswick. 102 p.

Strom, Richard N. Phosphorus fractionation in estuarine and marsh sediments. D, 1976, University of Delaware. 164 p.

Strom, Richard N. Sediment distribution in southwestern Delaware Bay. M, 1972, University of Delaware.

Strom, Robert Gregson. Stratigraphy of the southwest corner of the Mindego Quadrangle (California). M, 1957, Stanford University. 58 p.

Stromberg, Peter A. Landslide problems related to housing development in central California. M, 1967, San Francisco State University.

Stromdahl, A. W. An application of the Stanford Watershed Model IV; a hydrologic simulation study of three drainage basins in Lake County, Illinois. M, 1975, Northeastern Illinois University.

Strommel, H. E. Seismic investigation in the Golden-Denver area. D, 1951, Colorado School of Mines.

Stromquist, Albert W., Jr. Geometry and growth of grabens, lower Red Lake Canyon area, Canyonlands National Park, Utah. M, 1976, University of Massachusetts. 118 p.

Stromquist, Barbara Haworth Adams. The relationship between paleotemperatures, burial depth and diagenesis in the Lower Cretaceous sandstones of the Peace River, Spirit River and Gething formations of Northwest Alberta. M, 1985, University of Calgary. 134 p.

Stronach, J. A. Observational and modelling studies of the Fraser River plume. D, 1977, University of British Columbia.

Stronach, Nicholas John. Sedimentology and paleoecology of a shale basin; the Fernie Formation of the southern Rocky Mountains. D, 1981, University of Calgary. 398 p.

Strong, Catherine C. Depositional environment of Hosston sandstones (Lower Cretaceous), Bogalusa Field, Washington Parish, Louisiana. M, 1983, Texas A&M University. 107 p.

Strong, Ceylon Perseus, Jr. Geology of the Pe Ell-Doty area, Washington. M, 1967, University of Washington. 150 p.

Strong, Ceylon Perseus, Jr. Physical and biostratigraphic relations of the Colorado group (lower and upper Cretaceous) in west-central Montana. D, 1969, University of Washington. 176 p.

Strong, Cyrus. A textural study of beach and river sediments along the Texas Gulf Coast. M, 1959, Rice University. 48 p.

Strong, Daniel McSpadden. Subsurface geology of Craig, Mayes, and eastern Nowata and Rogers counties, Oklahoma. M, 1961, University of Oklahoma. 227 p.

Strong, David F. A study of apatite and related phosphates in the system CaO-MgO-P₂O₅-H₂O. M, 1967, Lehigh University.

Strong, Despina. Vanadium and nickel complexes in the Alberta oil sands. D, 1986, Washington State University. 184 p.

Strong, Percy George. The sedimentology and depositional history of the St. Roch Formation near St. Jean-Port-Joli, Quebec. M, 1978, McMaster University. 128 p.

Strong, Richard M. Subsurface geology of Barber County, Kansas. M, 1960, Kansas State University. 75 p.

Strong, Robert H. Thermal maturity of Franciscan and related strata in southern Humboldt County, N. California. M, 1986, University of Missouri, Columbia.

Strong, Walter Morrill. Structural geology of Pilot Knob area, Travis County, Texas. M, 1957, University of Texas, Austin.

Strong, Wayne L. Pre- and postsettlement palynology of southern Alberta. M, 1975, University of Calgary.

Strongin, Oscar. Geology and ore deposits of Apache Hills and northern Sierra Rica, Hidalgo County, New Mexico. D, 1958, Columbia University, Teachers College. 299 p.

Strongin, Oscar. Reconnaissance of the geology and ore deposits of the Apache Hills and Sierra Rica, New Mexico. M, 1953, Columbia University, Teachers College.

Stross, Richard Anthony. Lateral permeability variations in the thin water table aquifer adjacent to four cypress ponds in Florida. M, 1983, University of Florida.

Strothmann, Frederick Henry. Conodonts from the Kimmswick of eastern Missouri. M, 1940, University of Missouri, Columbia.

Stroud, Carlos R. Geology of townships 16 and 17 North, Range 9 West, Izard County, Arkansas. M, 1958, University of Arkansas, Fayetteville.

Stroud, Charles Brasher. Marine worms of the Texas Cretaceous. M, 1931, Texas Christian University. 53 p.

Stroud, Paul W. Oxidation characteristics of petroleum residual oils and asphalts (T553). D, 1934, Colorado School of Mines. 76 p.

Stroud, Raymond Brown. The areal distribution of radioactivity in the Potash Sulphur Springs Complex. M, 1951, University of Arkansas, Fayetteville.

Stroup, Janet B. Geologic investigations in the Cayman Trough and the nature of the plutonic foundation of the oceanic crust. M, 1982, SUNY at Albany. 189 p.

Strowd, William Bruce. A review of the upper Cenozoic stratigraphy overlying the Columbia River Basalt Group in western Idaho. M, 1981, University of Idaho. 124 p.

Struble, Richard Allen. A petrographic study of the Columbus and Delaware formations in northern Ohio. M, 1952, Ohio State University.

Struble, Richard Allen. Soil mapping and subsurface profiling for highway design utilizing geological, geophysical, airphoto, and direct subsurface exploration techniques. D, 1966, Ohio State University. 279 p.

Struby, William D. Structural geology of the Red Mountain area, Larimer County, Colorado. M, 1957, University of Colorado.

Struck, Rodney Grant. Permeability and connectivity characteristics of discontinuous fracture networks. M, 1985, San Diego State University. 184 p.

Struhsacker, Debra Winter. Mixed basalt-rhyolite assemblages in Yellowstone National Park; the petrogenetic significance of magma mixing. M, 1978, University of Montana. 112 p.

Struhsacker, Eric M. Geothermal systems of the Corwin Springs-Gardiner area, Montana; possible structural and lithologic controls. M, 1976, Montana State University. 93 p.

Struhsaker, James Frederick. Source and distribution of shell along Matagorda Peninsula, Texas. M, 1976, University of Michigan.

Struik, Lambertus Cornelis. Geology of the Barkerville-Cariboo River area, central British Columbia. D, 1980, University of Calgary. 335 p.

Strum, Stuart. Lithofacies and depositional environments of the Raton Formation (Upper Cretaceous-Paleocene) of northeastern New Mexico. M, 1985, North Carolina State University. 81 p.

Strumpher, Phillipus J. Gully erosion in Dry Creek, Nebraska. M, 1983, Colorado State University. 97 p.

Strung, J. A magnetotellurics survey across North America. M, 1971, University of Toronto.

Strunk, Kevin Lee. Structural relationships of the Cottage Grove and Shawneetown fault systems near Equality, Illinois, as inferred from geophysical data. M, 1984, Southern Illinois University, Carbondale. 89 p.

Strunk, Paul M. The subsurface geology of the Greenwood field area located in Morton County, Kansas and Baca County, Colorado. M, 1958, Kansas State University. 38 p.

Strunk, William L. The Galena Formation of the Root River valley, Minnesota. M, 1921, University of Minnesota, Minneapolis. 35 p.

Struthers, Joseph. On the study of slags from lead furnaces, with the object of producing liquidation or crust effect. D, 1895, Columbia University, Teachers College.

Struthers, Parke H., Jr. The geology of the Cazenovia, New York 7-1/2 minute Quadrangle. M, 1958, Syracuse University.

Struthers, Robert Allen. Sedimentation in the St. Johns River estuary in the vicinty of Orange Park, Florida. M, 1981, University of Florida. 111 p.

Strutz, Timothy Arthur. A pre-Pennsylvanian paleogeologic study of Michigan. M, 1978, Michigan State University. 68 p.

Strybel, Daniel Z. A descriptive study and correlation of four intrusives on the southern margin of the south-eastern Snake River plain near Pocatello, Idaho. M, 1984, Wayne State University. 102 p.

Stryhas, Bart Andrew. Progressive refolding in high strain regimes; an application to the Maggia Nappe, Ticino, Switzerland and the Lamoille Canyon Nappe, Ruby Mountains, Nevada. D, 1988, Washington State University. 188 p.

Stryhas, Bart Andrew. Structural analysis of Five Lakes Butte, Northeast Idaho. M, 1985, University of Idaho. 93 p.

Stuart, Alfred Wright. A detailed petrographic study of the Paleozoic sediments in the area of Fairmount, Georgia. M, 1956, Emory University. 33 p.

Stuart, Charles J. The stratigraphy, sedimentology, and tectonic implications of the San Onofre Breccia, southern California. D, 1975, University of California, Santa Barbara. 339 p.

Stuart, Charles Juhami. Metamorphism in the central Flint Creek Range, Montana. M, 1966, University of Montana. 103 p.

Stuart, Charles K. Temperate water carbonate deposition on the beaches of California; Recent sediments, ancient analogs, and depositional model. M, 1982, Rensselaer Polytechnic Institute. 132 p.

Stuart, David J. Gravity survey over parts of the Olympic Peninsula and Puget Trough, Washington. M, 1960, Stanford University.

Stuart, Edmund J. Distribution of uranium and associated elements in the Soldier Meadow Tuff, northwestern Nevada. M, 1979, Michigan Technological University. 74 p.

Stuart, James Edward. Stratigraphy and structure of a portion of southwestern New York Butte Quadrangle, Inyo County, CA. M, 1976, San Jose State University. 178 p.

Stuart, John Alexander. The crystal chemistry and dehydration of the clinoptilolite - heulandite zeolite series. D, 1985, Arizona State University. 222 p.

Stuart, John William. Structural relations between dickite, metadickite, and their high temperature phases. M, 1958, Washington University. 44 p.

Stuart, Robert J. Bryozoa from the Lower Permian limestone in the vicinity of the Sunflower Reservoir, Elko County, Nevada. M, 1962, Bowling Green State University. 33 p.

Stuart, Roy Armstrong. Geology of the Kemano-Tahtsa area, British Columbia. D, 1956, Princeton University. 88 p.

Stuart, Roy Armstrong. Petrofabrics of the (Ordovician) Wonah Quartzite, Stanford Range, British Columbia. M, 1952, Dartmouth College. 54 p.

Stuart, Tom Jeffrey. The effects of freshet turbidity on selected aspects of the biogeochemistry and the trophic status of Flathead Lake, Montana, U.S.A. D, 1983, University of North Texas. 244 p.

Stuart, William D. Evaporite deposition in a layered sea; a wind-driven dynamical model. D, 1971, Northwestern University.

Stuart, William D. Interpretation of Bermuda gravity measurements in terms of pedestal structure and carbonate stratigraphy. M, 1969, Northwestern University.

Stuart, William J., Jr. Stratigraphy of the Green River Formation west of the Rock Springs Uplift, Sweetwater County, Wyoming. M, 1964, University of Wyoming. 50 p.

Stuart-Alexander, Desiree Elizabeth. Contrasting deformation of Paleozoic and Mesozoic rocks near Sierra City, northern Sierra Nevada, California. D, 1967, Stanford University. 120 p.

Stubbins, J. B. Goodwood iron deposit. M, 1950, Queen's University.

Stubblefield, William Lynn. Genesis and modification of the sand ridges; inner and middle New Jersey shelf, U.S.A. D, 1980, Texas A&M University. 261 p.

Stubblefield, William Lynn. Petrographic and geochemical examination of the Ordovician Oneota Dolomite in the building stone districts of southeastern Minnesota. M, 1971, University of Iowa. 154 p.

Stubbs, Gale Susan. Geology of a contact zone; Dolly Varden Mountains, Elko County, Nevada. M, 1984, University of Colorado. 86 p.

Stubbs, J. L., Jr. Kinematic analyses of discfolds in Devonian Millboro Formation in the central-southern Appalachian junction zone. M, 1977, Virginia Polytechnic Institute and State University.

Stubbs, James E. Textural characteristic of some sands of the Permian Delaware Mountain Group of Texas. M, 1952, Columbia University, Teachers College.

Stubenrauch, Alan L. Seismicity of the Nevada Test Site; April 1, 1973, to October 1, 1975. M, 1977, University of Wisconsin-Milwaukee.

Stucchi, D. J. Seiches in the north west arm of Halifax Harbour. M, 1975, Dalhousie University.

Stuckey, George H. Ground-water resources of Putnam County, Ohio. M, 1988, University of Toledo. 144 p.

Stuckey, Jasper L. Notes on the Triassic east of Chapel Hill. M, 1920, University of North Carolina, Chapel Hill. 7 p.

Stuckless, John Shearing. The geology of the volcanic sequence associated with the Black Mesa Caldera, Arizona. M, 1969, Arizona State University. 76 p.

Stuckless, John Shearing. The petrology and petrography of the volcanic sequence associated with the Superstition Caldera, Superstition Mountains, Arizona. D, 1971, Stanford University. 113 p.

Stucky, Jasper L. The pyrophyllite deposits of the Deep River region of North Carolina. D, 1924, Cornell University.

Stude, Jerry R. Permian scolecodonts of the Fort Riley Limestone; fauna, stratigraphy and paleoecology. M, 1961, Wichita State University. 100 p.

Studebaker, Irving G. Structure and stratigraphy of the Helmet Peak area, Pima County, Arizona. M, 1960, University of Arizona.

Studebaker, Irving G. The effect of simulated geologic features on rock mass properties. D, 1977, University of Arizona.

Studemeister, Paul Alexander. An Archean felsic stock with peripheral gold and copper occurrences, Abotossaway Township District of Algoma, Ontario. D, 1982, University of Western Ontario. 501 p.

Studley, Clarence K. The Chico fauna of the type locality. M, 1912, University of California, Berkeley. 36 p.

Studley, Gregory Wayne. Structural geology and quantitative geomorphology for several selected areas in Blackfoot Mountain Range, southeastern Idaho. M, 1981, Idaho State University. 103 p.

Studley, Kermit. The distribution of heat producing elements in the Mount Waldo Pluton, south-central Maine. M, 1983, SUNY at Buffalo. 55 p.

Studlick, J. R. J. An analysis of orphaned strip-mines in Addison Quadrangle, Gallia County, Ohio. M, 1977, Ohio State University. 753 p.

Studt, Charles W. Geology of the Breckenridge area of Missouri with particular reference to oil and gas possibilities. M, 1921, Washington University.

Stueber, Alan Michael. A geochemical study of ultramafic rocks. D, 1965, University of California, San Diego. 200 p.

Stueber, Alan Michael. An analysis of the lateral distribution of trace elements in galenas from the Fredericktown, Missouri Mine. M, 1961, Washington University. 34 p.

Stuenitz, Holger. Reaction mechanisms, continuous and discontinuous compositional zonation in different amphiboles from a polymetamorphic metagabbro. M, 1987, Indiana University, Bloomington. 66 p.

Stugard, Frederick, Jr. Pegmatites of the Middleton area, Connecticut. D, 1950, Yale University.

Stuhr, Steven Walter. Geology of the Round Lake Intrusion, Sawyer County, Wisconsin. M, 1976, University of Wisconsin-Madison.

Stukas, Vidas. Plagioclase release patterns; a high resolution ^{40}Ar-^{39}Ar study. D, 1977, Dalhousie University. 141 p.

Stukel, Donald Joseph, II. Ichnology and paleoenvironmental analysis of the Late Devonian (Famennian) Chagrin Shale of northeast Ohio. M, 1987, Kent State University, Kent. 95 p.

Stukey, Arthur Herbert, Jr. Stratigraphic relations of Pennsylvanian-Permian strata, Manzanita Mountains, New Mexico. M, 1968, University of New Mexico. 64 p.

Stukhart, George, Jr. The theory of geodetic observations. D, 1968, Ohio State University.

Stull, Robert John. Petrology of graywackes from the Nooksack Group. M, 1966, University of Washington. 20 p.

Stull, Robert John. The geochemistry of the southeastern portion of the Golden Horn Batholith, (post-lower Cretaceous), northern Cascades, Washington. D, 1969, University of Washington. 127 p.

Stults, Arthur Carl. Foraminifera of the Kiamichi Formation of the Texas Panhandle. M, 1935, Texas Tech University. 56 p.

Stumm, Erwin Charles. New species of Devonian corals from the Eureka District, Nevada. M, 1933, George Washington University. 34 p.

Stumm, Erwin Charles. Part I, The lower Middle Devonian tetracorals of the Nevada Limestone; Part II, Upper Middle Devonian rugose corals of the Nevada Limestone. D, 1936, Princeton University.

Stump, Arthur Darrell. The microstructure of marine sediments. D, 1964, Oregon State University. 114 p.

Stump, Brian Williams. Investigation of seismic sources by the linear inversion of seismograms. D, 1979, University of California, Berkeley. 273 p.

Stump, Daniel. A hypothesis for sink development above solution-mine brine cavities in the Detroit area. M, 1980, University of Illinois, Urbana. 106 p.

Stump, E. On the late Precambrian-early Paleozoic metavolcanic and metasedimentary rocks of the Queen Maud Mountains, Antarctica, and a comparison with rocks of similar age from southern Africa. D, 1976, Ohio State University. 274 p.

Stump, James Duffield. Sedimentology of a carbonate tidal inlet in the lower Florida Keys. M, 1984, University of South Florida, Tampa. 113 p.

Stump, Richard Webster. Properties of silt deposits in Matanuska Valley, Alaska. M, 1956, Iowa State University of Science and Technology.

Stump, T. E. Stratigraphy and paleontology of the Imperial Formation in the western Colorado Desert. M, 1972, [University of California, San Diego].

Stump, Thomas Edward and Farmer, Jack Dewayne. Studies in the form, function, development and evolution of the Bryozoa. D, 1978, University of California, Davis. 232 p.

Stump, Thomas Edward. The evolutionary biogeography of the West Mexican Pectinidae (Mollusca; Bivalvia). D, 1979, University of California, Davis. 520 p.

Stumpe, Kim Michael. The sedimentology of the Teapot Sandstone, Southwest Powder River basin, Wyoming. M, 1983, Colorado State University. 164 p.

Stumpenhaus, Cathie L. Phase relations in the system CdO-MgO-SiO$_2$ and comparison with the system CaO-MgO-SiO$_2$. M, 1976, Miami University (Ohio). 55 p.

Stumpf, Gary A. Comparison of hydrologic data from mined and unmined areas in the Pennsylvanian age Carbondale Group of Daviess, Pike, and Gibson counties, Indiana. M, 1982, Indiana University, Bloomington. 157 p.

Stumpf, H. G. Movement of cyclonic eddies in the western Sargasso Sea. M, 1973, University of Southern California. 128 p.

Stumpf, Richard Paul. Analysis of suspended sediment distributions in the surface waters of Delaware Bay using remote sensing of optical properties. D, 1984, University of Delaware, College of Marine Studies. 170 p.

Stupak, William A. The petrography and petrochemistry of the Eocene volcanics, Rattlesnake Hills, central

Wyoming. M, 1984, University of New Brunswick. 338 p.

Sturchio, Neil Colrick. Geology, petrology, and geochronology of the metamorphic rocks of Meatiq Dome, central Eastern Desert, Egypt. D, 1983, Washington University. 218 p.

Sturdavant, Charles D. Sedimentary environments and structure of the Cretaceous rocks of Saturna and Tumbo islands, British Columbia. M, 1976, Oregon State University. 195 p.

Sturdevant, James A. Petrography of the Olalla Stock, Okanagan Mountains, British Columbia. M, 1963, University of New Mexico. 84 p.

Sturdevant, James Anton. Relationships between fracturing, hydrothermal zoning, and copper mineralization at the Big Bug Pluton, Big Bug mining district, Yavapai County, Arizona. D, 1975, Pennsylvania State University, University Park. 299 p.

Sturdivant, Ann Elizabeth. Uptake of ammonium by potassium-bearing silicates in the Guaymas Basin hydrothermal system, Gulf of California. M, 1988, University of California, Riverside. 78 p.

Sture, S. Strain-softening behavior of geologic materials and its effect on structural response. D, 1976, University of Colorado. 359 p.

Sturgeon, David A. and Ballou, William D. The formation and detailed description of a portion of Wind Cave, South Dakota. M, 1958, South Dakota School of Mines & Technology.

Sturgeon, Ernest Sidney. The Gasport dolomite in the vicinity of Hamilton, Ontario. M, 1955, McMaster University.

Sturgeon, Myron T. A contribution to the (Pennsylvanian) Allegheny fauna of eastern Ohio. D, 1936, Ohio State University.

Sturgeon, Myron T. The stratigraphy and paleontology of the middle portion of the Allegheny Formation of the Lisbon Ohio Quadrangle. M, 1933, Ohio State University.

Sturgess, Steven W. Structural geology of the Alum Fork area, Benton Uplift, Arkansas. M, 1986, University of Missouri, Columbia. 125 p.

Sturgis, Douglas S. Origin of sedimentary structures and stratigraphic variations in a sand pit in the Oak Openings Sand Belt, Lucas County, Ohio. M, 1985, Bowling Green State University. 110 p.

Sturgul, John Roman. Effects of surface irregularities on the underground stress field. D, 1967, University of Illinois, Urbana. 121 p.

Sturm, David H. Morphogenetic and phylogenetic relationships in monopleurid and radiolitid rudists (Mollusca, Bivalvia, Hippuritacea) in the Comanche Cretaceous of central Texas. M, 1976, Louisiana State University.

Sturm, Edward. A sedimentary study of the river terraces of Chugwater Creek, Laramie and Platte counties, Wyoming. M, 1950, University of Minnesota, Minneapolis. 67 p.

Sturm, Edward. Mineralogy and petrology of the Newark Group sediments of New Jersey. D, 1956, Rutgers, The State University, New Brunswick. 219 p.

Sturm, Frederick Henry. General geology of some replacement monazite deposits in Lemhi County, Idaho. M, 1954, University of Idaho. 64 p.

Sturm, J. F. The measurement of potassium 40 in minerals for radioactive age determinations. M, 1954, University of Toronto.

Sturm, John J. Conodont biostratigraphy and paleoecology of the Hegler and Pinery Members, Bell Canyon Formation (Permian) in the Delaware Basin of West Texas. M, 1975, Texas Tech University. 68 p.

Sturm, Stephen D. Depositional environments and sandstone diagenesis in the Tyler Formation (Pennsylvanian), southwestern North Dakota. M, 1982, University of North Dakota. 238 p.

Sturman, Bozidar D. Mineralogy and petrology of Hab Mine, Saskatchewan. M, 1971, Queen's University. 90 p.

Sturnick, Mark A. Metamorphic petrology, geothermo-barometry and geochronology of the eastern Kigluaik Mountains, Seward Peninsula, Alaska. M, 1984, University of Alaska, Fairbanks. 175 p.

Sturr, H. D., Jr. Continental shelf waves over a continental slope. M, 1969, United States Naval Academy.

Sturz, A. A. Selenium contamination in well water, Ramona Municipal Water District, Ramona, California. M, 1976, San Diego State University.

Stutzman, Paul E. Non-marine and marine influenced Herrin Coal and associated roof shales from southern Illinois; a mineralogical and statistical analysis. M, 1983, Southern Illinois University, Carbondale. 103 p.

Styan, William Bruce. The sedimentology, petrography and geochemistry of some Fraser Delta peat deposits. M, 1981, University of British Columbia. 188 p.

Styles, Thomas Richard. Holocene and late Pleistocene geology of Napoleon Hollow archaeological site in lower Illinois River valley. M, 1984, University of Illinois, Urbana. 276 p.

Stylianopoulos, L. C. Airphoto study and mapping of south-central Indiana sandstone-shale-limestone soil minerals. M, 1955, Purdue University.

Styzen, Michael J. Late Eocene foraminiferal systematics, biostratigraphy and paleoecology of DSDP Hole 267B, Southeast Indian Ocean. M, 1980, Northern Illinois University. 154 p.

Su, Bo-Chin. Fluid inclusion geothermometry of fluorite, Star Range, Utah. M, 1976, University of Utah. 63 p.

Su, Chen-Bin. Analytic theorems and applications to synthetic seismogram analysis. M, 1978, University of Missouri, Rolla.

Su, Chong-G. Hydraulic functions of soils from physical experiments and their applications. D, 1976, Oregon State University. 130 p.

Su, Fu-Chen. Seismic effects of faulting in coal seams; numerical modeling. D, 1976, Colorado School of Mines. 252 p.

Su, Ho-Jeen. Heat transfer in porous media with fluid phase changes. D, 1981, University of California, Berkeley. 108 p.

Su, Hon-Hsieh. A study on flow through porous medium. D, 1968, Michigan State University. 132 p.

Su, Sergio S. Philippines seismicity and structures. M, 1957, Boston College.

Su, Sergio S. The use of leaking modes in seismogram interpretation and in crust-mantle studies. D, 1965, Columbia University. 32 p.

Su, Shu-Chun. Alkali feldspars; ordering, composition and optical properties. D, 1986, Virginia Polytechnic Institute and State University. 141 p.

Su, Wen Huane. Development of ultrasonic techniques for the measurement of in-situ stresses. D, 1982, West Virginia University. 154 p.

Suarez Riglos, Mario. Some Devonian fossils from the state of Piaui, Brazil. M, 1967, University of Cincinnati. 121 p.

Suarez, Donald Louis. Heavy metals in waters and soil associated with several Pennsylvania landfills. D, 1974, Pennsylvania State University, University Park. 222 p.

Suarez, Gerardo. Seismicity, tectonics, and surface wave propagation in the Central Andes. D, 1982, Massachusetts Institute of Technology. 260 p.

Suárez, Gerardo. Seismicity, tectonics, and surface wave propagation in the Central Andes. D, 1983, Massachusetts Institute of Technology. 260 p.

Suarez, Max J. An evaluation of the astronomical theory of the ice ages. D, 1976, Princeton University. 118 p.

Suayah, Ismail B. Geochemistry, chronology and petrogenesis of the Wadi Yebigue Pluton, central Tibisti Massif, Libya. M, 1984, University of North Carolina, Chapel Hill. 82 p.

Subagio, Hardjosubroto. Volcanic ash influence and characterization of soils on the eastern plain of the Barisan Range between Sitiung-Kotabaru and Bangko, Sumatra. D, 1988, North Carolina State University. 309 p.

Subbarao, Eleswarapu Chinna. A study of the fundamental properties of Puget Sound glacial clays. M, 1953, University of Washington. 74 p.

Subbarayudu, G. V. The Rb-Sr isotopic composition and the origin of the Laramie anorthosite-mangerite complex, Laramie Range, Wyoming. D, 1975, SUNY at Buffalo. 118 p.

Subhas, Tella. Structural analysis of the Musquash area, Saint John County, New Brunswick (Canada). M, 1970, University of New Brunswick.

Sublette, William R. Stability analysis of wedge type rock slope failures. M, 1976, University of Arizona.

Subramanian, Vaidyanatha. Mechanisms of fixation of the trace metals manganese and nickel by ferric hydroxide. D, 1973, Northwestern University.

Subranmaniam, Anantharama Parameswara. Mineralogy and petrology of the Sittampundi Complex, Salem District, Madras State, India. D, 1952, Princeton University. 164 p.

Suchecki, Robert K. Sedimentology, petrology and structural geology of the Cow Head Klippe; Broom Point, St. Paul's Inlet, and Black Brook, western Newfoundland. M, 1975, University of Massachusetts. 180 p.

Suchecki, Robert Kenneth. Sedimentary history and diagenesis of volcanogenic rocks, Upper Jurassic and Lower Cretaceous Great Valley Sequence, northern California. D, 1980, University of Texas, Austin. 314 p.

Suchit, O. H. Magnetometric-resistivity studies. M, 1975, University of Toronto.

Suchomel, Diane Marie. Paleoecology and petrology of Pipe Creek Jr. Reed (Niagaran-Cayugan), Grant County, Indiana. M, 1975, Indiana University, Bloomington. 38 p.

Suchomel, T. J. Geology and mineralogy of the Harding Pegmatite, Taos County, New Mexico. M, 1976, University of Illinois, Urbana. 88 p.

Suchsland, Reinhard John. Quantitative analysis of the benthic foraminifera of San Pedro Basin, California. M, 1979, University of Southern California.

Suczek, C. A. Tectonic relations of the Harmony Formation, northern Nevada. D, 1977, Stanford University. 108 p.

Suda, Robert U. The Morton Gneiss, Morton, Minnesota. M, 1975, Northern Illinois University. 36 p.

Sudano, Peter L. The mineralogy of fine-grained sediment in the New Jersey nearshore region; implications for sediment sources and dispersal patterns. M, 1982, Lehigh University. 131 p.

Sudar, Susan A. Subsurface and hydrogeologic investigation of the existing operation and proposed expansion of South Side Sanitary Landfill, Marion County, Indiana. M, 1987, Purdue University. 148 p.

Suddhiprakarn, Chairat. Wave propagation in heterogeneous media. D, 1984, University of Texas, Austin. 203 p.

Sudicky, E. A. Field observation of nonuniform dispersivity in an unconfined sandy aquifer. M, 1979, University of Waterloo.

Sudicky, Edward Allan. An advection-diffusion theory of contaminant transport for stratified porous media. D, 1983, University of Waterloo. 203 p.

Suek, D. H. Biostratigraphy and carbonate petrology of a Devonian (Frasnian) biostrome near Mountain Pass, southeastern California. M, 1975, California State University, Fresno.

Suekawa, Harry S. Study of the Kennecott Copper Corp., Great Salt Lake Authority Mill tailings tests. M, 1968, University of Utah. 41 p.

Suemnicht, G. A. The geology of the Canada del Oro headwaters, Santa Catalina Mountains, Arizona. M, 1977, University of Arizona. 108 p.

Suen, Chi-Yeung John. Geochemistry of peridotites and associated mafic rocks, Ronda Ultramafic Complex, Spain. D, 1978, Massachusetts Institute of Technology. 283 p.

Sues, Hans-Dieter. Advanced mammal-like reptiles from the Early Jurassic of Arizona. D, 1984, Harvard University. 315 p.

Sues, Hans-Dieter. The anatomy and relationships of Stegoceras validus (Reptilia: Ornithischia) from the Judith River Formation of Alberta. M, 1977, University of Alberta. 183 p.

Suess, Erwin. Calcium carbonate interaction with organic compounds. D, 1968, Lehigh University. 162 p.

Suess, Erwin. Opal-phytoliths. M, 1966, Kansas State University. 77 p.

Suess, Irma L. Problem of post-glacial differential uplift. M, 1932, University of Rochester. 82 p.

Suess, Steven Tyler. Some effects of gravitational tides on a rotating fluid. D, 1969, University of California, Los Angeles. 197 p.

Suffel, George G. The dolomites of western Oklahoma. D, 1929, Stanford University. 218 p.

Sufi, Arshad Hussain. Temperature effects on oil-water relative permeabilities for unconsolidated sands. D, 1982, Stanford University. 283 p.

Sugai, Susan Frances. Processes controlling trace metal and nutrient geochemistry in two Southeast Alaskan fjords. D, 1985, University of Alaska, Fairbanks. 156 p.

Sugarman, Peter J. The geological interpretation of gravity anomalies in the vicinity of Raritan Bay, New Jersey and New York. M, 1981, University of Delaware. 135 p.

Suggs, James De Shae. Sediments and topography of the western Gulf of Mexico. M, 1967, Texas Tech University. 72 p.

Sugihara, Teruo. Nitrogen dynamics in a lagoon development and an adjacent salt marsh. D, 1981, Rutgers, The State University, New Brunswick. 352 p.

Sugumaran, Vijayan. High pressure water jet fragmentation of frozen ground. M, 1985, University of Alaska, Fairbanks. 134 p.

Suh, Jung Hee. Earth tides; surface strain measurements in NE Denver (Colorado). M, 1971, Colorado School of Mines. 38 p.

Suh, Mancheol. A seismic study of an impact feature in Cass County, Michigan. M, 1985, Western Michigan University.

Suhayda, Joseph Nicolas. Experimental study of the shoaling transformation of waves on a sloping bottom. D, 1972, University of California, San Diego.

Suhm, Raymond Walter. Geology of the northern part of the Batesville manganese district, Arkansas. M, 1963, Southern Illinois University, Carbondale. 63 p.

Suhm, Raymond Walter. Stratigraphy of the Everton Formation (early medial Ordovician) along the Buffalo-White River traverse, northern Arkansas. D, 1970, University of Nebraska, Lincoln. 502 p.

Suhr, David Olaf. A geological and geochemical study for copper and zinc in the Mackay area, Custer County, Idaho. M, 1964, University of Idaho. 56 p.

Sukhajintanakan, Warawan. Solubilization and mobilization of silica in relation to marine sediments. M, 1976, Pennsylvania State University, University Park. 145 p.

Sukup, J. W. Cretaceous planktonic foraminifera from Salzgitter-Salder, West Germany. M, 1978, University of Wyoming. 56 p.

Sulanowski, Jacek Kazimierz. Field study of relationship between organic matter and sedimentary particles. D, 1978, University of Chicago. 96 p.

Sulanowski, Jacek S. K. Shell structure of the Polyplacophora (Mollusca). M, 1972, Wayne State University.

Sulaym-an, Sulaym-an Mahm-ud. General geology of the Isabel-Eylar area, California, and petrology of

Franciscan sandstones. D, 1958, Stanford University. 137 p.

Sule, P. O. A magnetotelluric investigation of Pemberton area, British Columbia. M, 1976, University of British Columbia.

Suleiman, Abdunnur S. Integrated geophysical and subsurface studies of tectonic features in northeastern New Mexico and adjacent regions. M, 1984, University of Texas at El Paso.

Suleiman, Ibrahim Sharif. Gravity and heat flow studies in the Sirte Basin, Libya. D, 1985, University of Texas at El Paso. 200 p.

Sulek, John A. A new method and technique for studying small foraminifera in random thin-sections. M, 1954, New York University.

Sulenski, Robert J. The paleoenvironments and fossil communities of the Bellvale sandstone (middle Devonian) and Skunnemunk conglomerate (middle Devonian?) at Schunemunk mountain, New York. M, 1969, Queens College (CUNY). 138 p.

Sulik, John F. Stratigraphy and structure of the Montosa Canyon area, Santa Cruz County, Arizona. M, 1957, University of Arizona.

Sulima, J. H. Ecology of an epeiric sea deposit; Decorah Formation (middle Ordovician), upper Mississippi Valley. D, 1975, Northwestern University. 288 p.

Sulima, John H. Lower Jurassic stratigraphy in Coal canyon, West Humboldt range, Nevada. M, 1970, Northwestern University.

Sulkoske, William C. Cretaceous microplankton assemblages from the Albian to Campanian of Wyoming. M, 1975, University of Arizona.

Sullins, Charles Jefferson. Red Hills intrusion, northern Hudspeth County, Texas. M, 1971, University of Texas, Austin.

Sullivan, A. M. Orthophosphate adsorption by iron oxide complexes in lakeland soil profiles of Lexington County, South Carolina. M, 1974, University of South Carolina. 100 p.

Sullivan, Amy E. Middle and late Wisconsinan paleoecology of western Illinois and northcentral Iowa. M, 1986, University of Iowa. 144 p.

Sullivan, Barbara A. Multichannel seismic investigation of the Beata Ridge. M, 19??, Texas A&M University.

Sullivan, Brian G. Palynostratigraphy of dinoflagellate (Pyrrhophyta) assemblages from the Pebble shale unit (Hauterivian-Barremian) and Torok Formation (Albian) from selected Arctic Alaska test wells. M, 1986, Idaho State University. 164 p.

Sullivan, Catherine E. Uranium and other trace element geochemistry of the Hopi Buttes volcanic province, northeastern Arizona. M, 1978, University of New Mexico. 82 p.

Sullivan, Dan Allen, Jr. A paleocurrent study of Upper Mississippian and Lower Pennsylvanian rocks in the frontal Ouachita Mountains and Arkansas Valley. D, 1966, Washington University. 134 p.

Sullivan, David S. Sediment transport patterns at Winthrop Beach, Massachusetts. M, 1982, Boston University. 144 p.

Sullivan, Edward A. Multiple reflections in marine exploration. M, 1965, [Boston University].

Sullivan, Eileen M. Petrology of Upper Mississippian Denmar Formation, Greenbriar Group, northern Randolph County, West Virginia. M, 1985, University of North Carolina, Chapel Hill. 100 p.

Sullivan, Eleanor R. A correlation study of soil properties. M, 1957, Boston College.

Sullivan, Frank Rogers. Foraminifera from the type section of the San Lorenzo Formation, Santa Cruz County, California. M, 1956, University of California, Berkeley. 140 p.

Sullivan, Frank Rogers. Lower Tertiary nannoplankton from the California coast ranges; Part 1, Paleocene; Part 2, Eocene. D, 1964, University of California, Berkeley. 74 p.

Sullivan, J. S. Paleozoic stratigraphy as an ore control of hydrothermal, base-metal, sulfide deposits in New Mexico. D, 1973, University of Missouri, Rolla. 168 p.

Sullivan, James G. Chemistry and structural state of plagioclase feldspars from a differentiated dolerite dike. M, 1972, University of North Carolina, Chapel Hill. 51 p.

Sullivan, Jeffery Alan. Non-parametric estimation of spatial distributions. D, 1984, Stanford University. 368 p.

Sullivan, Jeffery Alan. The relationship of metallogenic zones and local geological features to lode gold orebodies, central Sierra Nevada foothills, California. M, 1980, University of Arizona. 89 p.

Sullivan, Jerry W. Some chemical and mineralogical aspects of plutonic rocks from the North Arm Mountain Massif, Bay of Islands Ophiolite, Newfoundland. M, 1981, SUNY at Albany. 149 p.

Sullivan, John Denis. Late Pleistocene-Holocene transgressive barrier island sequence; evidence for a fluctuating sea level, Hilton Head Island area, South Carolina. M, 1988, Georgia State University. 145 p.

Sullivan, John J. Environment of deposition and reservoir properties of Teapot sandstones, (Upper Cretaceous), Well Draw Field, Converse County, Wyoming. M, 1982, Texas A&M University. 176 p.

Sullivan, John S., Jr. Sedimentation and carbonate petrology of the Westerville limestone member (Pennsylvanian), Raytown, Missouri. M, 1969, University of Missouri, Columbia.

Sullivan, John W. Geology and mineral resources of the Port au Port area, Newfoundland. D, 1940, Yale University.

Sullivan, Joseph Edward. Geomorphic effectiveness of a high-magnitude rare flood in central Texas. M, 1983, University of Texas, Austin. 214 p.

Sullivan, Karen Louise. Organic facies variation of the Woodford Shale in western Oklahoma. M, 1983, University of Oklahoma. 101 p.

Sullivan, Kathryn D. The structure and evolution of the Newfoundland Basin, offshore eastern Canada. D, 1978, Dalhousie University. 225 p.

Sullivan, Keith Barry. Sandstone and shale diagenesis of the Frio Formation (Oligocene), Texas Gulf Coast; a close look at sandstone/shale contacts. M, 1988, University of Texas, Austin. 242 p.

Sullivan, Kevin James. Petrography, diagenetic history, and development of porosity in the Richfield Member of the Lower Middle Devonian Lucas Formation, northeast Isabella County, central Michigan Basin. M, 1986, Western Michigan University.

Sullivan, Michael Francis. Prestack Kirchhoff inversion and modelling in 2.5 dimensions. D, 1987, Colorado School of Mines. 121 p.

Sullivan, Michael P. The Granville Facies; Middle Ordovician barrier-beach hydrocarbon reservoirs in south-central Kentucky. M, 1983, University of Cincinnati. 150 p.

Sullivan, Michael Parnell. A management model for Lake Texana and the Lavaca-Navidad River basin based on the freshwater inflow needs of the Lavaca-Tres Palacios Estuary. D, 1986, University of Texas, Austin. 476 p.

Sullivan, Michael W. Geology and geochemistry of the Burlington Mine, Jamestown District, Boulder County, Colorado. M, 1973, University of Colorado.

Sullivan, Patrick Jay. Naturally-occurring nitrate pollution in the soils of the western San Joaquin Valley, California. D, 1978, University of California, Riverside. 131 p.

Sullivan, Robert Michael. Lower vertebrates from Swain Quarry "Fort Union Formation", middle Paleocene (Torrejonian), Carbon County, Wyoming. D, 1980, Michigan State University. 56 p.

Sullivan, Stephen Bradley. An investigation of flow in two groundwater basins in the inner Bluegrass karst region, Kentucky, using fluorescent dyes. M, 1983, University of Kentucky. 84 p.

Sullivan, Stephen J., Jr. A seismo-geological study of the Andean region. M, 1954, Boston College.

Sullivan, Thomas C. Diagenetic history of the Smackover Formation (Jurassic) of Meriwether Lake Field, Lafayette County, Arkansas. M, 1985, University of Arkansas, Fayetteville. 97 p.

Sullivan, William C. A telluric current study of the Rio Grande Valley near Belen, New Mexico. M, 1960, New Mexico Institute of Mining and Technology. 37 p.

Sullwold, Harold H. Geology of a portion of the San Joaquin Hills, Orange County, California. M, 1940, University of California, Los Angeles.

Sullwold, Harold H. The Tarzana Fan, a deep submarine delta of late Miocene age, Los Angeles County, California. D, 1958, University of California, Los Angeles.

Sultan, Ghazi Hashim. Downstream variation of grain size and mineralogy of sand of the modern Brazos River, Texas. M, 1964, University of Texas, Austin.

Sultan, Mohamed Ibrahim. Geology, petrology, and geochemistry of a Younger Granite pluton, central East Desert of Egypt; importance of mixing. D, 1984, Washington University. 228 p.

Suman, Daniel Oscar. Agricultural burning in Panama and Central America; burning parameters and the coastal sedimentary record. D, 1983, University of California, San Diego. 172 p.

Sumarac, Dragoslav. Self-consistent model for the brittle response of solids. D, 1987, University of Illinois, Chicago. 144 p.

Sumartojo, Jojok. A study of the mineralogy and geochemistry of the Tindelpina (upper Proterozoic), Adelaide Geosyncline, South Australia. D, 1974, University of Cincinnati. 159 p.

Sumartojo, Jojok. Precambrian rocks in Kentucky. M, 1966, University of Kentucky. 41 p.

Sumida, Stuart Shigeo. Alternation of neural spine height in Permo-Carboniferous tetrapods and a reappraisal of primitive modes of terrestrial locomotion. D, 1987, University of California, Los Angeles. 374 p.

Summa, Lori Louise. Use of tephra layers occurring in both altered and unaltered states to study diagenetic processes. D, 1986, University of California, Davis. 180 p.

Summer, Neil Steven. Maturation, diagenesis and diagenetic processes in sediments underlying thick volcanic strata, Oregon. M, 1987, University of California, Davis. 87 p.

Summer, Rebecca M. Mined land reclamation capability, Colorado County, Texas. M, 1975, University of Texas, Austin.

Summer, Rebecca Mae. Alpine soil erodibility of Trail Ridge, Rocky Mountain National Park, Colorado. D, 1980, University of Colorado. 200 p.

Summerford, H. Edgar. Geology of a portion of the St. Xavier Quadrangle, Montana. M, 1941, University of Iowa. 57 p.

Summers, Chester. Hydrological investigation in karst terrain utilizing geophysical and geochemical methods. M, 1985, University of Arkansas, Fayetteville.

Summers, Donna M. The petrology of two late Pliocene bentonites and the enclosing rocks, Upper Siwalik Subgroup, northern Punjab and southwestern Kashmir, Pakistan. D, 1978, Dartmouth College. 237 p.

Summers, Karen Varley. Palagonite and pillow basalts of the Columbia River Group. M, 1975, Washington State University. 99 p.

Summers, Robert Michael. The design of dredged spoil containment basins using mass sediment properties. D, 1981, The Johns Hopkins University. 275 p.

Summers, William Kelly. A study of the subsurface lithology of the St. Louis Limestone in south central Indiana. M, 1957, Indiana University, Bloomington. 35 p.

Summerson, Charles Henry. Some Pennsylvanian faunas of Tennessee, eastern Kentucky, and West Virginia. D, 1942, University of Illinois, Urbana.

Summerson, Charles Henry. The cardinal process of the Productidae. M, 1940, University of Illinois, Urbana. 20 p.

Sumner, John R. Tectonic significance of geophysical investigations in southwestern Arizona and northwestern Sonora, Mexico, at the head of the Gulf of California. D, 1971, Stanford University. 91 p.

Sumner, John Stewart. Geophysical studies of the Waterloo Range, Wisconsin. D, 1956, University of Wisconsin-Madison. 80 p.

Sumner, Richard Lee. The calcium-barium-magnesium carbonate system at room temperature. M, 1977, Southern Illinois University, Carbondale. 44 p.

Sumner, Roger Dean. Attenuation of earthquake generate P waves along the western flank of the Andes. D, 1965, University of Wisconsin-Madison.

Sumner, Wendolyn R. Structural geology and tectonic setting of the Cherry Creek metamorphic suite, southern Madison Range. M, 1988, Colorado State University. 138 p.

Sumsion, R. S. Stratigraphy and fusulinid paleontology of Permian exposures in the vicinity of Eureka, Nevada. M, 1974, San Jose State University. 127 p.

Sun, Albert Yen. Structure and stratigraphy of the Berzeliustinden area, Wedel Jarlsberg Land and Torell Land, Spitsbergen. M, 1980, University of Wisconsin-Madison.

Sun, Albert Yen. Structure and tectonic evolution of the Precambrian amphibolite-gneiss complex east of Eau Claire, Wisconsin. D, 1984, University of Wisconsin-Madison. 382 p.

Sun, Chin-Hong. COASTAL; a distributed hydrologic simulation model for lower coastal plain watersheds in Georgia. D, 1985, University of Georgia. 229 p.

Sun, Kwang-Hua David. The soil mechanics and clay mineralogy of glacial tills in a portion of southwestern Ohio. M, 1975, Miami University (Ohio). 51 p.

Sun, Min. Sr isotopic study of ultramafic nodules from Neogene alkaline lavas of British Columbia, Canada, and Josephine Peridotite, southwestern Oregon, U.S.A. M, 1985, University of British Columbia. 133 p.

Sun, Ming-Shan. A petrographic study of the Eocene Jackson Group of Mississippi and adjacent areas. D, 1950, Louisiana State University.

Sun, Ming-Shan. Some feldspars of Black Hills pegmatites (South Dakota). M, 1947, University of Chicago. 26 p.

Sun, Robert Jencheu. Numerical simulation and imaging of seismic wavefields. D, 1988, University of Texas at Dallas. 226 p.

Sun, Shine-Soon. Lead isotope studies of young volcanic rocks from oceanic islands, mid-ocean ridges, and island arcs. D, 1974, Columbia University. 139 p.

Sun, Stanley S. S. Fission track study of the Cheney Pond titaniferous iron ore deposit, Tahawus, New York. D, 1971, Washington University. 134 p.

Sund, J. Olaf. Origin of New Brunswick gypsum deposits. M, 1958, University of New Brunswick.

Sunda, Laxman Singh. Use of spent sulphite liquor for stabilization of fracture rock and sealing off water-bearing formations. M, 1964, University of Missouri, Rolla.

Sundaram, Panchanatham Naga. Water pressure and resistivity changes during stick-slip and stable sliding in direct shear of rock surfaces. D, 1978, University of California, Berkeley. 424 p.

Sundberg, Frederick Allen. Upper Cambrian paleobiology and depositional environments of the lower Nopah Formation, California and Nevada. M, 1979, San Diego State University.

Sunde, Robert Lynn. Geologic controls on oil and gas occurrence in the Giddings Field, Lee and Burleson counties, Texas. M, 1981, University of Wisconsin-Milwaukee. 91 p.

Sundeen, Curtis R. The structure of laccoliths. M, 1939, University of Minnesota, Minneapolis. 35 p.

Sundeen, Daniel Alvin. Petrology and geochemistry of the Haverhill 15' Quadrangle, southeastern New Hampshire. D, 1970, Indiana University, Bloomington. 211 p.

Sundeen, Kerry D. Assessment of surface water quality on an Appalachian watershed. M, 1979, Colorado State University. 109 p.

Sundeen, Stanley Paul. A textural study of a portion of the critical zone of the Bushveld Complex (Precambrian, Union of South Africa). M, 1965, University of Wisconsin-Madison.

Sundeen, Stanley Paul. The petrology of a magnetite rich portion of the Negaunee Iron-Formation in the southeast part of the Marquette Range, Michigan. D, 1968, University of Wisconsin-Madison. 102 p.

Sundeen, Stanley Wilford. A petrographic study of the basic dikes of the Saganaga and Snowbank Lake intrusives, and a general review of the literature on lamprophyres. D, 1937, University of Minnesota, Minneapolis. 140 p.

Sundelius, Harold Wesley. Occurrence and origin of the Peg Claims spodumene pegmatites, Knox County, Maine. D, 1959, University of Wisconsin-Madison. 121 p.

Sundell, Kent A. The geology of the North Fork of Owl Creek area, Absaroka Range, Hot Springs County, Wyoming. M, 1980, University of Wyoming. 168 p.

Sundell, Kent Allan. The Castle Rocks Chaos; a gigantic Eocene landslide-debris flow within the southeastern Absaroka Range, Wyoming. D, 1985, University of California, Santa Barbara. 383 p.

Sunderman, Harvey Cofer. Relationship of the minor structures to the major structures in the Lynchburg area, Virginia. M, 1947, University of Kentucky. 52 p.

Sunderman, Harvey Cofer. The "Martic Overthrust" in the Lynchburg area, Virginia. D, 1951, University of Wisconsin-Madison.

Sunderman, Jack. Geology of the Carbonate Mountain area, Sangre de Cristo Range, Colorado. M, 1956, University of Michigan.

Sunderman, Jack Allen. Mineral deposits at the Mississippian-Pennsylvanian unconformity in southwestern Indiana. D, 1963, Indiana University, Bloomington. 107 p.

Sundharovat, Swai. Geology of the Parkdale area, Fremont County, Colorado. M, 1956, University of Colorado.

Sundheimer, Glenn Robert. Short-term (twelve-year) seismicity as a predictor of long-term seismic activity. M, 1975, Pennsylvania State University, University Park. 76 p.

Sundin, Philip James. Investigation of siliceous intervals within the Ogallala Formation from analysis of subsurface core material. M, 1974, Texas Tech University. 63 p.

Sundquist, Eric Thorsten. Carbon dioxide in the oceans; some effects on sea water and carbonate sediments. D, 1979, Harvard University.

Sundvik, Michael Todd. Relationship of acoustic basement relief to seafloor spreading in the western North Atlantic Basin and East Pacific Rise. D, 1986, University of Rhode Island. 282 p.

Suneby, Lena B. Biostratigraphy of the Upper Triassic - Lower Jurassic Heiberg Formation, eastern Sverdrup Basin, Arctic Canada. M, 1984, University of Calgary. 245 p.

Suneson, Mark A. Reconstruction of paleoenvironments and assemblages in a Pennsylvanian (Desmoinesian) chaetetes-bearing limestone from southwestern Colorado. M, 1984, University of Texas, Austin.

Suneson, Neil H. The geology of the northern portion of the Superstition-Superior volcanic field, Arizona. M, 1976, Arizona State University. 123 p.

Suneson, Neil Hedner. The origin of bimodal volcanism, West-central Arizona. D, 1980, University of California, Santa Barbara. 326 p.

Sung, Chien-Min. The nature of the olivine→spinel transition in the Mg_2SiO_4-Fe_2SiO_4 system and its geophysical implications. D, 1976, Massachusetts Institute of Technology. 337 p.

Sung, Jen-Chun. Analysis of crack-tip fields in a viscoplastic material. D, 1987, Northwestern University. 144 p.

Sung, Roger. Plane wave decomposition with application to amplitude-offset analyses. M, 1985, University of Houston.

Sung, Wonmo. Development, testing and application of a multi-well numerical coal seam degasification simulator. D, 1987, Pennsylvania State University, University Park. 239 p.

Sungy, Eugene D. The nature of the Heart Mountain Fault in the vicinity of Dead Indian Hill, Park County, Wyoming. M, 1977, Texas A&M University. 86 p.

Sunwall, Mark T. Strontium isotopic compositions of brine, oil, and reservoir rock from petroleum fields of southeastern Ohio. M, 1976, Miami University (Ohio). 92 p.

Sunzeri, Christine Cresswell. The lean brown land; a study of the relationship between landform and plant ecology in the Black Rock Desert. M, 1975, University of California, Santa Cruz.

Suo, Lisheng. Hydraulic transients in rock-bored tunnels. D, 1988, University of Michigan. 121 p.

Suparman, Agus. A study of sedimentary structures in the Lower Carboniferous Horton Group of western Cape Breton Island, Nova Scotia. M, 1964, University of Ottawa. 75 p.

Suphasin, Chai. The subsurface geology of the Grimsby "Clinton" sandstone of Trumbull County, northeastern Ohio. M, 1979, Kent State University, Kent. 59 p.

Supina, Richard D. A geological-geophysical groundwater study for Negaunee Township, Marquette County, Michigan. M, 1974, Michigan Technological University.

Supko, Peter Richard. A quantitative X-ray diffraction method of the mineralogical analysis of carbonate sediments from the Tongue of the Ocean, Bahamas. M, 1963, University of Miami. 158 p.

Supko, Peter Richard. Deposition and diagenetic features in subsurface Bahamian rocks (San Salvador). D, 1970, University of Miami. 179 p.

Supkow, Donald. Stratigraphy and structure of the Spider Lake Formation, Churchill and Spider Lake quadrangles, Maine. M, 1965, University of Maine. 108 p.

Supkow, Donald James. Subsurface heat flow as a means for determining aquifer characteristics in the Tucson Basin, Pima County, Arizona. D, 1971, University of Arizona.

Suppe, John Edward. Franciscan (Jurassic-Cretaceous) geology of the Leech Lake mountain-Anthony Peak region, Northeastern Coast ranges, California. D, 1969, Yale University. 129 p.

Supplee, Jeffrey A. Geology of the Kings Mountain gold mine, Kings Mountain, North Carolina. M, 1986, University of North Carolina, Chapel Hill. 140 p.

Sura, Michael Anthony. Application of F-K migration to delineate structural and stratigraphic traps. M, 1986, University of New Orleans. 82 p.

Sura, Suzanne Hopkins. Conodonts from the Everton Formation (Early Middle Ordovician), Newton, Carroll, and Boone counties, northwestern Arkansas. M, 1987, University of New Orleans. 147 p.

Surblis, Benjamin. Dispersion of nickel in the environment from plating waste at Mogadore, Ohio. M, 1971, University of Akron. 33 p.

Surdam, Ronald Clarence. Low-grade metamorphism of the Karmutsen Group (Triassic) Buttle Lake area, Vancouver Island, British Columbia. D, 1967, University of California, Los Angeles. 336 p.

Surgenor, J. W. Lacustrine sediments as indicators of fluctuations of Riukojietna Glacier; Lappland, Sweden. M, 1978, University of Maine. 64 p.

Surkan, Alvin J. Electromagnetic models and magnetic fields of simple and composite conductors. D, 1959, University of Western Ontario.

Surles, M. A., Jr. Stratigraphy of the Eagle Ford Group (Upper Cretaceous) and its source-rock potential in the East Texas Basin. M, 1986, Baylor University. 219 p.

Surles, Terri L. Chemical and thermal variations accompanying formation of garnet skarns near Patagonia, AZ. M, 1978, University of Arizona.

Suro Perez, Vinicio. Indicator kriging based on principal component analysis. M, 1988, Stanford University. 168 p.

Surovell, Elizabeth Jean. Climatic influence on slope ratios in areas of dendritic drainage. M, 1968, Columbia University. 38 p.

Suryanarayana, Bhamidipaty. Mechanics of degradation and aggradation in a laboratory flume. D, 1969, Colorado State University. 112 p.

Suryanto, Untung. Stratigraphy and petroleum geology of the J Sandstone in portions of Boulder, Larimer, and Weld counties, Colorado. M, 1979, Colorado School of Mines. 173 p.

Susak, Nicholas John. Spectra, thermodynamics, and molecular chemistry of some divalent transition metal chloro-complexes in hydrothermal solution to 300°C. D, 1981, Princeton University.

Suska, Maria Magdalena. Mid-Devonian Elk Point Group and Cambrian rocks of north-central Alberta. M, 1963, University of Alberta. 95 p.

Susko, John M. Causes and implications of a linear magnetic anomaly in the Gilmanton and Penacook quadrangles, New Hampshire. M, 1980, Ohio University, Athens. 86 p.

Susman, Kenneth R. Post-Miocene subsurface stratigraphy of Shackleford Banks, Carteret County, North Carolina. M, 1975, Duke University. 85 p.

Susong, Bruce Irvin. Subsurface studies of Pennsylvanian sandstones on the western side of the Illinois Basin. M, 1955, University of Illinois, Urbana.

Susong, David Dunbar. Structure of an earthquake rupture segment boundary in the Lost River fault zone, Idaho; implications for rupture propagation during the 1983 Borah Peak earthquake. M, 1987, University of Utah. 52 p.

Sussko, Roger J. Sedimentology of the siliciclastic to carbonate transition on the Southwest Florida inner shelf. M, 1988, University of South Florida, Tampa. 83 p.

Sussman, David. The geology of Apoyo Caldera, Nicaragua. M, 1982, Dartmouth College. 166 p.

Sussman, Jeffery A. Investigation of the physical and chemical character of zircons from uraniferous Precambrian conglomerates as a possible exploration guide. M, 1984, South Dakota School of Mines & Technology.

Susuki, Takeo. Stratigraphic paleontology of the Topanga Formation at the type locality, Santa Monica Mountains, California. M, 1951, University of California, Los Angeles. 85 p.

Suszkowski, D. J. Sedimentology of Newark Bay, New Jersey; an urban estuarine bay. D, 1978, University of Delaware, College of Marine Studies. 237 p.

Sutch, Patricia Leigh. Historic seismicity of Honduras, 1539-1978. M, 1979, Stanford University. 87 p.

Sutcliffe, John Russell. A statistical analysis of the minerals in the Zone of Orbitolina, Trans-Pecos, Texas. M, 1961, Texas Tech University. 138 p.

Sutcliffe, Richard Harry. The petrology, mineral chemistry and tectonics of Proterozoic rift-related igneous rocks at Lake Nipigon, Ontario. D, 1986, University of Western Ontario. 325 p.

Suter, David R. The bifoliate cryptostome Bryozoa (Ectoprocta) from the Middle Ordovician Lincolnshire Limestone, Rockingham County, Virginia. M, 1973, Virginia Polytechnic Institute and State University.

Suter, John Robert. Concentration, distribution, and behavior of heavy metals in Recent sediments, Corpus Christi ship channel inner harbor. M, 1980, University of Texas, Austin.

Suter, John Robert. Late Quaternary facies and sea level history, Southwest Louisiana continental shelf. D, 1986, Louisiana State University. 347 p.

Suter, Tammy D. A comparison of the Lower Silurian clastic units in the Great Valley Province of Virginia and the Valley and Ridge Province of Virginia and West Virginia. M, 1987, Bowling Green State University. 125 p.

Suthakorn, Phairat. Paragenesis of the Blanchard Deposit, Bingham, Llansanburg, Socorro, New Mexico. M, 1977, New Mexico Institute of Mining and Technology.

Suthard, James A. Stratigraphy and paleontology in Fish Lake Valley, Esmeralda County, Nevada. M, 1965, University of California, Riverside. 108 p.

Sutherland, Berry. Silt-sized heavy minerals of some Texas Gulf Coast Tertiary shales. M, 1968, University of Houston.

Sutherland, D. B. Gravity investigations in the Ottawa-Bonnechere Graben [Ontario]. M, 1954, University of Toronto.

Sutherland, Garry Neil. Sedimentology of the Upper Cretaceous Frenchman Formation in the Frenchman River valley, Saskatchewan. M, 1977, University of Saskatchewan. 144 p.

Sutherland, George Donald. The microfauna of the lower Neogastroplites Zone (Cretaceous), northeastern British Columbia. M, 1971, University of Alberta. 100 p.

Sutherland, J. Clark. The clays of Orange and Riverside counties, Southern California. M, 1930, California Institute of Technology. 128 p.

Sutherland, Jane Louise. Stratigraphy and sedimentology of the Upper Cambrian Lone Rock Formation, western Wisconsin; focus on the Reno Member. M, 1986, University of Wisconsin-Madison. 81 p.

Sutherland, Jeffrey Clark. Mineral-water equilibrium, Great Lakes; aluminosilicates. D, 1968, Syracuse University. 114 p.

Sutherland, John W., Jr. Petroleum geology of the Tuscaloosa Group, North Louisiana. M, 1978, Louisiana State University.

Sutherland, Mortimer Y., Jr. A comparative study of the Virginia granites. M, 1935, University of Virginia. 65 p.

Sutherland, Susan M. The petrography and environmental interpretation of the Benwood Limestone, Monongahela Group (Pennsylvanian), in Marshall County, West Virginia and Belmont County, Ohio. M, 1986, Miami University (Ohio). 81 p.

Sutherland-Brown, Atholl. The structure and stratigraphy of the Antler Creek area, British Columbia. D, 1954, Princeton University. 142 p.

Sutley, David E. A petrographic and paleoenvironmental study of the Jemison Chert, Chilton County, Alabama. M, 1977, University of Alabama.

Sutley, William Christopher. Structure and seismic stratigraphy of post-Ouachita Paleozoic to Jurassic sediments of southeastern Arkansas, northeastern Louisiana, and west-central Mississippi. M, 1988, Stephen F. Austin State University. 161 p.

Sutphen, C. F. The petrology of a Triassic diabase intrusion near Frederick, Maryland. M, 1975, Temple University.

Sutphin, Hoyt Baldwin. Occurrence and structural control of collapse features on the southern Marble Plateau, Coconino County, Arizona. M, 1986, Northern Arizona University. 139 p.

Sutter, John Frederick. Application of the K/Ar method to the dating of cataclastically deformed rocks. D, 1970, Rice University. 129 p.

Sutter, John Frederick. Geochronology of major thrusts, southern Great Basin, California. M, 1968, Rice University. 32 p.

Sutterlin, Peter G. The (Early Silurian) Thorold Sandstone in the Niagara Peninsula, Ontario, Canada. M, 1954, McMaster University.

Sutterlin, Peter George. Uppermost Devonian (Post Woodbend) studies in southern Alberta area. D, 1958, Northwestern University. 128 p.

Suttner, Lee Joseph. Analysis of the Upper Jurassic-Lower Cretaceous Morrison and Kootenai formations, southern Montana. D, 1966, University of Wisconsin-Madison. 132 p.

Suttner, Lee Joseph. Geology of Brillion Ridge, east-central Wisconsin. M, 1958, University of Wisconsin-Madison.

Sutton, Arle Herbert. Geology of the southern part of the Dawson Springs Quadrangle, Kentucky. D, 1927, University of Chicago. 138 p.

Sutton, C. The solubility of aromatic hydrocarbons and the geochemistry of hydrocarbons in the eastern Gulf of Mexico. D, 1974, Florida State University. 210 p.

Sutton, Donald Grant. The Middle Devonian of southern Indiana with special emphasis upon the Geneva Formation. M, 1936, Washington University. 89 p.

Sutton, Eral Maurice. Channel filling of the Hardinsburg Formation, Gibson County, Indiana. M, 1954, Miami University (Ohio). 21 p.

Sutton, George H. Physical analysis of deep sea sediments. D, 1958, Columbia University, Teachers College.

Sutton, George H. Seismic refraction measurements in the Atlantic Ocean. M, 1953, Columbia University, Teachers College.

Sutton, J. S. and Jones, D. L. Geology of Panoche Valley Quadrangle (California). M, 1953, [Stanford University].

Sutton, John F. Mammals of the Anceney local fauna (late Miocene) of Montana. D, 1977, Texas Tech University. 257 p.

Sutton, Kenneth George. The geology of Mount Jefferson. M, 1974, University of Oregon. 120 p.

Sutton, Margot Jean. Stratigraphy and sedimentology of the upper Marmaton Group and the Pleasanton Group (Pennsylvanian) of southeastern Kansas. M, 1985, University of Iowa. 138 p.

Sutton, P. A. Field injection test to study the migration of selected organics in an unconfined sandy aquifer. M, 1982, University of Western Ontario.

Sutton, Robert George. Stratigraphy and structure of the Batavia Quadrangle, New York. M, 1950, University of Rochester. 141 p.

Sutton, Robert George. The stratigraphy of the Naples Group (Devonian) in western New York. D, 1956, The Johns Hopkins University.

Sutton, Sally Jo. The development of slaty cleavage at Ocoee Gorge, Tennessee. D, 1987, University of Cincinnati. 294 p.

Sutton, Stanley Matthew, Jr. Urban fluvial geometamorphosis. M, 1980, University of Texas, Austin.

Sutton, Stephen Roy. Thermoluminescence dating study of shock-metamorphosed rocks from Meteor Crater, Arizona. D, 1984, Washington University. 214 p.

Sutton, Stephen T. Spherical harmonic representation of the gravitational potential of discrete mass elements, with application to estimating the scale and distribution of heterogeneity within the Earth. M, 1986, University of Michigan. 41 p.

Sutton, Thomas C. Geology of the Virginia Horn area (Mesabi Range, Minnesota). M, 1963, University of Minnesota, Minneapolis. 97 p.

Sutton, Thomas Culver. Relationship between metamorphism and geologic structure along the Great Smoky fault system, Parksville Quadrangle, Polk and Bradley counties, Tennessee. D, 1971, University of Tennessee, Knoxville. 198 p.

Sutton, Victoria A. Toxic metal concentration and distribution in soils of four abandoned landfill sites, Springfield, Missouri. M, 1981, Southwest Missouri State University.

Sutton, W. R. Contact action of granite gneiss in Ontario. M, 1933, Queen's University. 59 p.

Sutton, Walter John. Geology of the Copper Rand Mine, Chibougamau, Quebec. M, 1959, University of Michigan.

Sutton, Willard Holmes. A study of the mineral constitution and ceramic properties of some shales from Pennsylvania. D, 1957, Pennsylvania State University, University Park. 134 p.

Suva, Melinda S. Cannon see Cannon Suva, Melinda S.

Suwanasing, Akanit. Geology and nickeliferous laterite deposits of Ban Tha Kradan Nok Quadrangle, Prachinburi Province, eastern Thailand. M, 1972, Lehigh University. 128 p.

Suydam, James David. Sedimentology, provenance, and paleotectonic significance of the Cretaceous Newark Canyon Formation, Cortez Mountains, Nevada. M, 1988, Montana State University. 90 p.

Suydam, John R. In-situ seismic velocity measurements in the Berkeley Pit, Butte, Montana. M, 1972, Montana College of Mineral Science & Technology. 134 p.

Suydam, R. B. Overthrusting in the South Labarge Creek area, Lincoln and Sublette counties, Wyoming. M, 1963, University of Wyoming. 88 p.

Suyenaga, W. An investigation of two models of the Doppler recordings associated with earthquake Rayleigh waves. M, 1973, University of Hawaii. 33 p.

Suyenaga, W. Earth deformation in response to surface loading; application to the formation of the Hawaiian Ridge. D, 1977, University of Hawaii. 47 p.

Suzuki, C. K. The determination of bound water in some Hawaiian soils by the Karl Fischer reagent. M, 1952, University of Hawaii. 48 p.

Sveinsdottir, Edda Lilja. Geological factors controlling the difference in compressive strength of basalts from Iceland. M, 1984, Queen's University. 84 p.

Svendsen, Augie Eugene. A microlithological study of the Plattsmouth, Beil and Ervine Creek members (Upper Pennsylvanian) in Southeast Nebraska and adjacent areas. M, 1961, University of Nebraska, Lincoln.

Svendsen, Mark T. Water management strategies and practices at the tertiary level; three Philippine irrigation systems. D, 1983, Cornell University. 311 p.

Svendsen, Robert Frederick Jr. Optical radiation from shock-compressed materials. D, 1988, California Institute of Technology. 405 p.

Svensson, Cynthia T. Studies of copper in various types of Washington estuaries. M, 1972, University of Washington.

Sverdlove, Marc Selig. Planktonic foraminiferal ecology of the eastern equatorial Pacific Ocean; including a paleoceanographic reconstruction of the Panama Basin for the last 320,000 years. D, 1983, University of Cincinnati. 357 p.

Sverdlove, Mark Selig. Stratigraphy and suggested phylogeny of Deflandrea vestita (Brideaux) new comb. and Deflandrea echinoidea Cookson and Eisenack. M, 1974, Queens College (CUNY). 156 p.

Sverdrup, Keith Allen. Seismotectonic studies in the Pacific Ocean basin. D, 1981, University of California, San Diego. 459 p.

Sverjensky, Dimitri Alexander. The origin of a mississippi valley-type deposit in the Viburnum Trend, Southeast Missouri. D, 1980, Yale University. 156 p.

Sveter, Owen D. An investigation of the contact between the Oakville and Catahoula formations (Miocene) in Grimes County, Texas. M, 1969, Texas A&M University. 69 p.

Svetlich, William G. Martinsburg limestones (Ordovician) in the Lebanon area, Pennsylvania. M, 1953, Wayne State University.

Svoboda, Joseph Otto. Gravity and magnetic interpretation of portions of the Plum River fault zone in East-central Iowa. M, 1980, University of Iowa. 167 p.

Svoboda, Mark Scott. The depositional and petrographic analysis of the Diamond Peak Formation in western White Pine County, Nevada. M, 1988, University of Nevada. 187 p.

Svoboda, Richard Frank. A detailed sedimentation and petrographic analysis of the Monroe Creek Formation of the Pine Ridge area of Northwest Nebraska. M, 1950, University of Nebraska, Lincoln.

Swade, John W. Conodont distribution, paleoecology, and preliminary biostratigraphy of the upper Cherokee and Marmaton groups (upper Desmoinesian, Middle Pennsylvanian) from two cores in south-central Iowa. M, 1982, University of Iowa. 118 p.

Swadley, W. C. Petrology in the Christmas Mountains Gabbro, Brewster County, Texas. M, 1958, University of Texas, Austin.

Swager, Dennis Ray. Stratigraphy of the Upper Devonian-Lower Mississippian shale sequence in the eastern Kentucky outcrop belts. M, 1978, University of Kentucky. 116 p.

Swagor, Nick S. The Cardium Conglomerate of the Carrot Creek Field, central Alberta. M, 1975, University of Calgary. 151 p.

Swain, Albert M. A history of fire and vegetation in northeastern Minnesota as recorded in lake sediments. D, 1975, University of Minnesota, Minneapolis. 103 p.

Swain, B. W. Fort Union Formation, west flank of the Sierra Madre, Carbon County, Wyoming. M, 1957, University of Wyoming. 132 p.

Swain, Edward Balcom. The paucity of blue-green algae in meromictic Brownie Lake; iron limitation or heavy metal toxicity?. D, 1984, University of Minnesota, Duluth. 90 p.

Swain, Frederick Morrill, Jr. Some faunas from the Onondaga Formation in central Pennsylvania and West Virginia. M, 1939, Pennsylvania State University, University Park. 48 p.

Swain, Frederick Morrill, Jr. Stratigraphy of the Cotton Valley Beds of the northern Gulf Coastal Plain. D, 1943, University of Kansas.

Swain, Patricia Breckenridge. Upper Triassic radiolarians from the Brooks Range, Alaska. M, 1981, University of California, Los Angeles. 174 p.

Swain, Patricia C. The development of some bogs in eastern Minnesota. D, 1979, University of Minnesota, Minneapolis. 258 p.

Swain, Robert L. Block caving at Braden Copper Company's Teniente Mine. M, 1958, University of Nevada - Mackay School of Mines. 29 p.

Swain, Walter C. Coal spoil characteristics and natural revegetation on a western Washington strip mine; an investigation ten years after mining. M, 1975, University of Washington. 115 p.

Swainbank, Ian G. Isotope composition of lead. D, 1967, Columbia University. 39 p.

Swainbank, Richard Charles. The geochemistry and petrology of eclogitic rocks near Fairbanks, Alaska. D, 1971, University of Alaska, Fairbanks. 155 p.

Swales, David L. Petrology and sedimentation of the Big Injun and Squaw sandstones, Granny's Creek Field, West Virginia. M, 1988, West Virginia University. 128 p.

Swales, William E. Mauch Chunk (Pennsylvanian) series in northern West Virginia and Southwest Pennsylvania. M, 1951, West Virginia University.

Swan, Arthur Graham. The differential thermal analysis of certain carbonate minerals. M, 1953, University of Alberta. 74 p.

Swan, F. H. Pleistocene and Holocene deposits of the lower Cache la Poudre River basin, Colorado. D, 1975, The Johns Hopkins University. 232 p.

Swan, Monte Morgan. The Stockton Pass Fault; an element of the Texas Lineament. M, 1976, University of Arizona.

Swan, Victor LaMarr. The petrogenesis of the Mount Baker volcanics, Washington. D, 1980, Washington State University. 630 p.

Swanberg, Chandler Alfred. A gravity survey over the central part of the Okanogan range, Washington. M, 1968, Southern Methodist University. 68 p.

Swanberg, Chandler Alfred. The measurement, areal distribution, and geophysical significance of heat generation in the Idaho Batholith and adjacent areas in eastern Oregon and western Montana. D, 1971, Southern Methodist University. 174 p.

Swanger, Henry Jay. Surface waves in strong ground motion with applications to offshore environments. D, 1981, Stanford University. 155 p.

Swank, Willard J. Structural history of the Canyon Range Thrust, central Utah. M, 1978, Ohio State University.

Swann, David Henry. The Favosites alpenensis-lineage in the Middle Devonian Traverse Group in Michigan. D, 1940, University of Michigan.

Swann, Gordon Alfred. Structure and petrology of Precambrian rocks in the Cache La Poudre-Rist Canyon area, Larimer County, Colorado. D, 1962, University of Colorado. 119 p.

Swanson, Andrew. Environmental controls of size and shape of Sowerbyella in the Trenton Group (Ordovician of New York State). M, 1985, Queens College (CUNY). 99 p.

Swanson, Clarence Otto. The genesis of the Texada Island magnetite deposits. D, 1924, University of Wisconsin-Madison.

Swanson, Clarence Otto. The replacement at the Homestake Mine. M, 1922, University of British Columbia.

Swanson, Cleo R. Stratigraphy and structure of Fish Creek area, Sublette County, Wyoming. M, 1952, University of Iowa. 109 p.

Swanson, David Karl. Properties of peatlands in relation to environmental factors in Minnesota. D, 1988, University of Minnesota, Minneapolis. 231 p.

Swanson, Donald Alan. The middle and late Cenozoic volcanic rock of the Tieton River area, south-central Washington. D, 1965, The Johns Hopkins University. 334 p.

Swanson, Donald K. A comparative study of ab initio generated geometries for first and second row atom oxide molecules with corresponding geometries in solids. M, 1980, Virginia Polytechnic Institute and State University.

Swanson, Donald Keith. High-temperature crystal chemical formalisms applied to $K_2Si^{VI}Si_3^{IV}O_9$ and $NaGaSi_3O_8$. D, 1986, SUNY at Stony Brook. 125 p.

Swanson, Earl H. Archaeological studies in the Vantage region of the Columbia Plateau, northwestern America. D, 1956, University of Washington. 243 p.

Swanson, Eric Craig. Bonneville flood deposits along the Snake River near Lewiston, Idaho. M, 1984, Washington State University. 74 p.

Swanson, Eric Rice. Petrology and volcanic stratigraphy of the Durango area, Durango, Mexico. M, 1974, University of Texas, Austin.

Swanson, Eric Rice. Reconnaissance geology of the Tomochic-Ocampo area, Sierra Madre Occidental, Chihuahua, Mexico. D, 1977, University of Texas, Austin. 154 p.

Swanson, Frederick John. Morphogenesis and shape sorting of coarse sediment in the Elk River, southwestern Oregon. D, 1972, University of Oregon. 148 p.

Swanson, Harold A. Hydrothermal alteration of feldspar. M, 1933, University of Minnesota, Minneapolis. 31 p.

Swanson, J. Craig. A three dimensional numerical model system of coastal circulation and water quality. D, 1987, University of Rhode Island. 157 p.

Swanson, James Walter. Geology of the southern portion of West Mountain, Utah. M, 1952, Brigham Young University. 69 p.

Swanson, Karen Anne. The effect of dissolved catechol on the dissolution of amorphous silica in seawater. D, 1988, Pennsylvania State University, University Park. 166 p.

Swanson, L. Craig. Hydrogeology at the site of a proposed uranium subgrade disposal pit in the Gas Hills uranium mining district, Fremont County, Wyoming. M, 1984, University of Idaho. 116 p.

Swanson, Mark H. Granite platforms of the western Wichita Mountains, Oklahoma. M, 1987, University of Toledo. 252 p.

Swanson, Mark Thomas. Geochemistry of sphalerite concentrates from the Balmat-Edwards zinc ores, St. Lawrence County, New York. M, 1979, Lehigh University. 171 p.

Swanson, Mark Thomas. The structure and tectonics of Mesozoic dike swarms in eastern New England. D, 1982, SUNY at Albany. 410 p.

Swanson, Mitchell Lee. Soil piping and gully erosion along coastal San Mateo County, California. M, 1983, University of California, Santa Cruz.

Swanson, Peter Lee. Stress corrosion cracking in Westerly Granite; an examination by the double torsion technique. M, 1980, University of Colorado.

Swanson, Peter Lee. Subcritical fracture propagation in rocks; an examination using the methods of fracture mechanics and non-destructive testing. D, 1984, University of Colorado. 282 p.

Swanson, Richard Wayne. Geologic subsurface study of the Cherokee Group in southeastern Nebraska. M, 1957, University of Nebraska, Lincoln.

Swanson, Rodney Duane. A stratigraphic-geochemical study of the Troutdale Formation and Sandy River Mudstone in the Portland Basin and lower Columbia River gorge. M, 1986, Portland State University. 103 2 plates a.

Swanson, Roger Glenn. Geology of a portion of the Las Vegas, New Mexico Quadrangle. M, 1950, University of Iowa. 145 p.

Swanson, Roger Warren. A petrographic study of small dikes cutting diabase sills near Finland, Minnesota. M, 1937, University of Minnesota, Minneapolis. 60 p.

Swanson, Roy Ivar. The extent of the Clare Dolomite, a marker horizon of Michigan. M, 1955, Michigan State University. 81 p.

Swanson, Ruth A. Structure, diversity, and homogeneity of benthic epifaunal communities, Buzzards Bay, Massachusetts. M, 1971, Wayne State University.

Swanson, Samuel Edward. Mineralogy and petrology of the Rocklin pluton (Jurassic), Placer and Sacramento counties, California. M, 1970, University of California, Davis. 86 p.

Swanson, Samuel Edward. Phase equilibria and crystal growth in granodioritic and related systems with H_2O and H_2O+CO_2. D, 1974, Stanford University. 128 p.

Swanson, Sherman Roger. Infiltration, soil erosion, nitrogen loss and soil profile characteristics of Oregon lands occupied by three subspecies of Artemisia tridentata. D, 1983, Oregon State University. 140 p.

Swanson, Stephen Robert. Development of constitutive equations for rocks. D, 1970, University of Utah. 140 p.

Swanson, Steven Brian. Two examples of secondary alterations associated with mid-ocean ridges. M, 1975, Texas A&M University. 87 p.

Swanson, Steven Carl. Sedimentology and provenance of the South Park Member of the Kingston Peak Formation, Panamint Range, California. M, 1982, University of California, Los Angeles. 162 p.

Swanson, Teresa H. A crystal structure analysis of lepidolite $2M_1$. M, 1980, University of Wisconsin-Madison.

Swanson, Vernon Emanuel. The stratigraphy of the (Upper Devonian or Mississippian) Hardin, Chattanooga, and (Lower Mississippian) Maury formations in the western Highland Rim area of Tennessee. M, 1950, Columbia University, Teachers College.

Swanson, Wallace A. Source of soil minerals in Labette County, Kansas. M, 1953, Kansas State University. 59 p.

Swanson, William Alfred. Seismicity and crustal structure of the Pamir-Alai region in Soviet Central Asia. M, 1988, Indiana University, Bloomington. 172 p.

Swanston, Douglas N. Geology and slope failure in the Maybeso Valley, Prince of Wales Island, Alaska. D, 1967, Michigan State University. 206 p.

Swanston, Douglas Neil. Glacial geology of the northwest portion of Hancock County, Ohio. M, 1962, Bowling Green State University. 36 p.

Swapp, Susan Mathilda. Metamorphism and mass transfer in calcareous concretions from the Black Hills of South Dakota. D, 1982, Yale University. 214 p.

Swarbrick, James C. Geology of the Sheep Mountain area and vicinity, Mitchell Quadrangle, Oregon. M, 1953, Oregon State University. 100 p.

Sward, Cynthia A. The distribution of selected trace metals in scleractinian coral skeletal material. M, 1969, Florida State University.

Swarts, S. W. A conservation and land use plan for the eastern Mojave Desert, California. D, 1975, University of California, Los Angeles. 170 p.

Swartz, C. K. The Columbus Formation of central-northern Ohio. D, 1904, The Johns Hopkins University.

Swartz, Charles C. The Devono-Carbonic section of northwestern Pennsylvania. M, 1904, Columbia University, Teachers College.

Swartz, Daniel Herbert. The "red beds" of Michigan. M, 1951, University of Michigan.

Swartz, Daniel Herbert, III. Techniques for mapping petroleum reservoirs in the Hunton Group and the "second Wilcox Sand", in parts of Lincoln, Logan, and Payne counties, Oklahoma. M, 1987, Oklahoma State University. 69 p.

Swartz, Frank M. The Helderbergian of West Virginia and Virginia. D, 1926, The Johns Hopkins University.

Swartz, Joel H. The age and stratigraphy of the Chattanooga black shale of Tennessee. D, 1923, The Johns Hopkins University.

Swartzlow, Carl Robert. The origin and occurrence of oolites in the (Devonian) Sylamore Formation in central Boone County, Missouri. M, 1929, University of Missouri, Columbia.

Swartzlow, Carl Robert. The origin of the dolomite and the granularity of the (Mississippian) Chouteau Formation. D, 1932, University of Missouri, Columbia.

Swarzenski, Wolfgang V. Erosion surfaces in northwestern Maine. D, 1954, Boston University. 177 p.

Swauger, David A. Geology and structure of the Green River Lakes area, Sublette County, Wyoming. M, 1982, University of Wyoming. 90 p.

Swayne, L. E. Pleistocene geology of the Digby area, Nova Scotia. M, 1952, Acadia University.

Swayze, Gregg A. Tectonic and metamorphic history of the Proterozoic Red Creek Terrane, northeastern Uinta Mountains, Utah. M, 1985, Colorado School of Mines. 95 p.

Swe, Win. Stratigraphy and structure of late Mesozoic rocks south and southeast of Clear Lake, California. D, 1968, Stanford University. 100 p.

Swe, Win. Structural and stratigraphic relationships along the northwestern border of the Wadesboro Basin of North Carolina. M, 1963, University of North Carolina, Chapel Hill. 64 p.

Swearingen, Ted L. The significance of ground water in the Houghton Lake drainage basin. M, 1973, Michigan State University. 76 p.

Sweazy, Christopher L. A relationship between gravity and the erosional highs on the Knox unconformity in Morrow County, Ohio. M, 1987, Wright State University. 78 p.

Sweeney, Brian Philip. Liquefaction evaluation using a miniature cone penetrometer and a large scale calibration chamber. D, 1987, Stanford University. 299 p.

Sweeney, George Le Jeune. A geologic reconnaissance of the Whitefish Range, Flathead and Lincoln counties, Montana. M, 1955, University of Montana. 48 p.

Sweeney, Gerald Thornton. Geology of Copper Basin, Custer County, Idaho. M, 1957, University of Idaho. 66 p.

Sweeney, J. A. Geology/urban planning interface. M, 1975, San Diego State University.

Sweeney, J. J. A study of acoustic velocity and dielectric permittivity anisotropy in relation to finite strain in deformed rock. D, 1980, University of Illinois, Urbana. 233 p.

Sweeney, J. S. A study of the life forms in shallow water deposits in Lower Cretaceous (Texas). M, 1918, Texas Christian University.

Sweeney, John Francis, Jr. Detailed gravity investigation of the shapes of granitic intrusives, south-central Maine, and implications regarding their mode of emplacement. D, 1972, SUNY at Buffalo. 69 p.

Sweeney, John Francis, Jr. Some aspects of the Wisconsin glacial stage in the Springville, New York (Erie county) quadrangle. M, 1969, SUNY at Buffalo. 64 p.

Sweeney, Mary Jo. Geochemistry of garnets from the North Ore Shoot, Bingham District, Utah. M, 1980, University of Utah. 154 p.

Sweeney, Robert Eugene. Pyritization during diagenesis of marine sediments. D, 1972, University of California, Los Angeles. 200 p.

Sweeney, S. The nature and distribution of loess in Canada; a preliminary investigation. M, 1985, University of Waterloo. 487 p.

Sweet, Arnold Lawrence. A stochastic model for predicting the subsidence of granular media. D, 1964, Purdue University. 111 p.

Sweet, Arthur R. The taxonomy, evolution, and stratigraphic value of azolla and azollopsis in the Upper Cretaceous and Early Tertiary. D, 1972, University of Calgary. 403 p.

Sweet, John M.; Freeman, Leroy Bradford and Tillman, Chauncey. Geology of the Henrys Lake Mountains, Fremont County, Idaho and Madison and Gallatin counties, Montana. M, 1950, University of Michigan. 83 p.

Sweet, Lois Bigelow. Rock gorges associated with the outlet of Cayuga Lake (New York). M, 1934, Cornell University.

Sweet, Michael Louis. A paleoenvironmental analysis of the late Missourian (Pennsylvanian) Bonpas Limestone Member, Mattoon Formation, east-central Illinois. M, 1983, University of Illinois, Urbana. 63 p.

Sweet, Randolph. Evaluation of the ground water resources of Ballabgarh and Palwal Tehsils, Haryana State, India. M, 1972, University of Oregon. 90 p.

Sweet, Rebecca Gail. A sedimentary analysis of the Upper Jurassic Morrison Formation as it is exposed in the vicinity of Canon City, Colorado. M, 1984, Oklahoma State University. 210 p.

Sweet, Walter Clarence. Geology of the southern portion of Manitou Park, Colorado. M, 1952, University of Iowa. 92 p.

Sweet, Walter Clarence. The Harding and Fremont formations of Colorado and their faunas. D, 1954, Iowa State University of Science and Technology.

Sweet, William Edward, Jr. Electrical resistivity logging in unconsolidated sediments (TAMU-SC-72-205) (Recent, Gulf of Mexico). D, 1972, Texas A&M University.

Sweet, William Edward, Jr. Geology of the Katemcy–Voca area, Mason and McCulloch counties, Texas. M, 1957, Texas A&M University.

Sweetkind, Donald Steven. Fission track studies of the Nelson Batholith, British Columbia, and the Idaho Batholith, Idaho. M, 1986, Southern Methodist University. 92 p.

Sweetland, Theodore Wessley. Ecological aspects of soil management and practices at a forest tree nursery located on the coastal plain, Virginia. M, 1979, Virginia State University. 73 p.

Sweetman, Edwin A. Geological report of Cuyuna iron ore district property. M, 1917, University of Minnesota, Minneapolis. 11 p.

Sweezy, John L. The role of trace elements in argentiferous galenas. M, 1979, University of Texas, Arlington. 72 p.

Sweide, Alan P. Synecology and faunal succession of the Upper Mississippian Great Blue Limestone, Bear River Range and Wellsville Mountain, North-central Utah. M, 1977, Utah State University. 152 p.

Sweigard, Richard Joseph. A procedure for environmental site planning and postmining land use planning for surface-minable land; development, analysis, and application. D, 1984, Pennsylvania State University, University Park. 317 p.

Sweigard, Richard Joseph. The role of bending in block-glide landslides and overthrust faults; an analytical model. M, 1979, Pennsylvania State University, University Park. 144 p.

Swem, Charles Edward. The Fox Hills Formation (Upper Cretaceous) of the Rocky Mountain region, with special reference to the Pikes Peak region; Colorado. M, 1936, University of Colorado.

Swemba, Michael J. The relationship between the cell dimensions, trace element analysis and color of northwest Ohio fluorites. M, 1974, University of Toledo. 43 p.

Swensen, Arthur Jaren. Anisoceratidae and Hamitidae (Ammonoidea) from the Cretaceous of Texas and Utah. M, 1962, Brigham Young University. 82 p.

Swenson, Alan L. An evaluation of ERTS-1 and U-2 imagery on the Tensleep Fault and southern Bighorn Basin, Wyoming. M, 1974, University of Iowa. 70 p.

Swenson, Alan Lee. Mechanics of Laramide deformation along the east flank of the Laramie Range, Wyoming and Front Range, Colorado. D, 1980, University of Colorado. 199 p.

Swenson, David Howard. Geochemistry of three Cascade volcanoes. M, 1973, New Mexico Institute of Mining and Technology. 101 p.

Swenson, David R. The Pennsylvanian Madera Formation, near Jemez Springs, New Mexico. M, 1977, University of New Mexico. 172 p.

Swenson, Donald Bruce. The heavy mineral assemblages of the Chadron and Brule formations in northwestern Nebraska. M, 1959, University of Nebraska, Lincoln.

Swenson, Erick M. Electromagnetic measurements of tidal transport. M, 1978, University of New Hampshire. 58 p.

Swenson, Frank Albert. Geology of the northwest flank of the Gros Ventre Mountains, Wyoming. D, 1942, University of Iowa. 100 p.

Swenson, Frank Albert. Sedimentation near junction of Maquoketa and Mississippi rivers. M, 1940, University of Iowa. 20 p.

Swenson, Guy Andrew, III. The ground water hydrology of Jacumba Valley, California and Baja California. M, 1981, San Diego State University.

Swesnik, Robert Malcolm. Geology of the West Edmond oil fields, central Oklahoma. M, 1945, University of Oklahoma. 65 p.

Swetland, Paul J. Lipid geochemistry of Delaware salt marsh environments. M, 1975, University of Delaware.

Swetnam, Monte N. Geology of the Pelton Creek area, Albany and Carbon counties. M, 1961, University of Wyoming. 80 p.

Swett, Earl R., Jr. The surface expression of the Zeandale Dome. M, 1959, Kansas State University. 59 p.

Swett, Keene. Petrology and paragenesis of the Manitou Formation (Ordovician) along the Front Range of Colorado; El Paso County and vicinity. M, 1961, University of Colorado.

Swick, Leo Emmett. The Pleistocene geology of the Bellefontaine and East Liberty quadrangles, Ohio. M, 1941, Ohio State University.

Swick, Norman Eugene. The Berne Conglomerate in Licking County and part of Fairfield County, Ohio. M, 1956, Ohio State University.

Swick, Steven E. Depositional environment and areal distribution of upper Textularia "W" sands in Terrebonne and Lafourche parishes, Louisiana. M, 1987, University of Southwestern Louisiana. 109 p.

Swiderski, Donald L. An analysis of patterns of homoplasy in freshwater snails of the family Planordidae (Pulmonata: Basommatophora). M, 1985, Michigan State University. 153 p.

Swiderski, Helena S. Artificial thermoluminescence behavior of carbonate host rocks in the vicinity of zinc-lead and fluorite-zinc-lead deposits and its potential as an exploration tool. D, 1976, Columbia University. 58 p.

Swiderski, Helena S. Scheelite-quartz mineralization in the older Precambrian Big Bug Group of the Yavapai Series in the Ora Flame area, Prescott National Forest, Yavapai County, Arizona. M, 1973, Columbia University. 235 p.

Swift, Charles M., Jr. A magnetotelluric investigation of an electrical conductivity anomaly in the southwestern United States. D, 1967, Massachusetts Institute of Technology. 222 p.

Swift, Donald Josiah Palmer. Origin of the Cretaceous Peedee Formation of the Carolina Coastal Plain. D, 1964, University of North Carolina, Chapel Hill. 151 p.

Swift, Douglas Baldi. Lithofacies of the Cuchillo Formation, southern Sierra de Juarez, Chihuahua, Mexico. M, 1973, University of Texas at El Paso.

Swift, Ellsworth Rowley. Study of the Morrison Formation and related strata, north-central New Mexico. M, 1956, University of New Mexico. 79 p.

Swift, Peter N. Precambrian metavolcanic rocks and associated volcanogenic mineral deposits of the southwestern Sierra Madre, Wyoming. M, 1982, University of Wyoming. 61 p.

Swift, Peter Norton. Early Proterozoic turbidite deposition and melange deformation, southeastern Arizona. D, 1987, University of Arizona. 161 p.

Swift, Robert N. A study of the effects of tidal current of suspended matter at the mouth of Delaware bay. M, 1970, Millersville University.

Swift, Roy Erwin. The characteristics and uses of Pacific Northwest diatomite. M, 1940, University of Washington. 164 p.

Swift, Stephen A. Holocene accumulation rates of pelagic sediment components in the Panama Basin, eastern equatorial Pacific. M, 1976, Oregon State University. 91 p.

Swift, Stephen Atherton. Cenozoic geology of the continental slope and rise off western Nova Scotia. D, 1985, Massachusetts Institute of Technology. 188 p.

Swift, William Arnold. Subsurface foraminiferal zones in the Charlotte Field, Atascosa County, Texas. M, 1951, University of Missouri, Columbia.

Swiger, Rual Bower. A paper on the "cup-coral" member of the Glenn Formation of Oklahoma. M, 1925, University of Oklahoma. 36 p.

Swihart, George H. Boron isotopic composition of boron minerals and tracer applications. D, 1987, University of Chicago. 160 p.

Swihart, George Hammond. The geochemistry of phyllosilicate weathering in the Conway Granite, New Hampshire. M, 1981, University of Iowa. 65 p.

Swilley, Gerald Kirk. Fourier grain shape analysis of sediments from the continental shelf off the coast of northern Oregon. M, 1986, Wichita State University. 109 p.

Swinamer, Ralph Terrance. The geomorphology, petrography, geochemistry and petrogenesis of the volcanic rocks in the Sierra del Chichinautzin, Mexico. M, 1986, Queen's University. 212 p.

Swinchatt, Jonathan Phillip. The significance of textural variation and skeletal breakdown in some Recent carbonate sediments. D, 1963, Harvard University.

Swinchatt, Peter F. E. Alteration and mineralization features at the Thornburg Mine, Saguache County, Colorado. M, 1956, Columbia University, Teachers College.

Swinden, Harold Scott. Chemical sedimentation associated with mid-Paleozoic volcanism in central Newfoundland. M, 1977, Memorial University of Newfoundland. 234 p.

Swinden, Harold Scott. Ordovician volcanism and mineralization in the Wild Bight Group, central Newfoundland; a geological, petrological, geochemical and isotopic study. D, 1988, Memorial University of Newfoundland. 452 p.

Swindler, James P. Sedimentology of the low energy coastal region between the Crystal and Withlacoochee rivers; Florida West Coast. M, 1973, University of Florida. 97 p.

Swineford, Ada. Fabric of a gravel talus. M, 1942, University of Chicago. 41 p.

Swineford, Ada. Petrology of outcropping post-Wolfcampian Permian rocks in Kansas. D, 1954, Pennsylvania State University, University Park. 317 p.

Swinehart, J. B., II. Cenozoic geology of the North Platte River valley, Morrill and Garden counties, Nebraska. M, 1979, University of Nebraska, Lincoln.

Swinehart, Thomas W. The Leith sandstone lentil (Dunkard Group, Pennsylvanian-Permian) of Washington County, Ohio and Pleasants County, West Virginia. M, 1969, Miami University (Ohio). 83 p.

Swinford, Edward M. Geology of the Peebles Quadrangle, Adams County, Ohio. M, 1983, Eastern Kentucky University. 104 p.

Swingen, Regina Anne. Evaluation of the St. Peter Sandstone and Joachim Dolomite for compressed air energy storage, with emphasis on thermal properties. M, 1981, University of Wisconsin-Milwaukee. 154 p.

Swingle, George D. Petrography of the Chilhowee Group, near Walland, Tennessee. M, 1949, University of Tennessee, Knoxville. 35 p.

Swingle, George D. Stratigraphy and structure of the Cleveland area, Tennessee. D, 1955, University of Wisconsin-Madison.

Swinnerton, Allyn Coates. Geology of a portion of the Castleton Quadrangle (Vermont). D, 1922, Harvard University.

Swinney, C. M.; Daviess, S. N. and Kellum, L. B. Geology and oil possibilities of the southwestern part of the Wide Bay Anticline, Alaska. M, 1945, University of Michigan.

Swinney, Chauncey Melvin. The geology and petrology of the Sonora Quadrangle, California. D, 1949, Stanford University. 90 p.

Swinzow, George K. Diastrophism in the light of thermal oscillation. D, 1957, Boston University. 182 p.

Swires, Charles Jonathan Jr. Source area of Mauch Chunk sediments in the Hurricane Ridge Syncline in southeastern West Virginia. M, 1972, University of Akron. 41 p.

Swirydczuk, Krystyna. Sedimentology of the Pliocene Glenns Ferry oolite and its stratigraphic setting in the western Snake River plain. D, 1980, University of Michigan. 257 p.

Swirydczuk, Krystyna. Tephra stratigraphy of sedimentary rocks associated with the Glenns Ferry Formation, western Snake River plain. M, 1977, University of Michigan.

Swisher, Carl Celso, III. Stratigraphy and biostratigraphy of the eastern portion of Wildcat Ridge, western Nebraska. M, 1982, University of Nebraska, Lincoln. 172 p.

Swisher, Marian Margaret. The determination of cardinal optical parameters of biaxial crystals by the orientation compensation method. M, 1969, University of Cincinnati. 25 p.

Switek, John. Trace metal distribution in rocks, stream sediments and soils of the Deep Creek basin, Pottawatomie County, Kansas. M, 1977, Kansas State University. 60 p.

Switek, Michael John. Stratigraphic section of the Elkhorn Ridge Argillite, Elkhorn Ridge area, Baker County, Oregon. M, 1967, University of Oregon. 92 p.

Switzer, George. Glaucophane schists of the central California Coast Ranges. D, 1942, Harvard University.

Swoboda, Allen Ray. Thermodynamics of cation exchange in montmorillonite clay. D, 1967, Texas A&M University. 69 p.

Swolfs, Henri Samuel. Influence of pore-fluid chemistry and temperature on fracture of sandstone under confining pressure. D, 1971, Texas A&M University.

Swolfs, Henri Samuel. Seismic refraction studies in the southwestern Gulf of Mexico. M, 1967, Texas A&M University. 42 p.

Sword, Charles Hege, Jr. Tomographic determination of interval velocities from reflection seismic data; the method of controlled directional reception. D, 1987, Stanford University. 112 p.

Swulius, Thomas. Oxygen isotope geochemistry of an early Archean Iron formation Isua, West Greenland. M, 1976, Northern Illinois University. 62 p.

Syder, J. F. Gravity study of Rowan County, North Carolina. M, 1963, University of North Carolina, Chapel Hill.

Sydnor, Robert H. Geology of the northeast border of the San Jacinto Pluton, Palm Springs, California. M, 1975, University of California, Riverside. 121 p.

Sydora, Larry John. Effects of the downgoing lithosphere. M, 1977, University of Alberta. 86 p.

Sydow, Marc Wolfgang. Seismicity and crustal structure of north-central Arizona. M, 1987, Northern Arizona University. 92 p.

Syed, Atiq Ahmad. Effects of the earthquake source mechanism on P wave magnitude determination. D, 1971, St. Louis University. 167 p.

Sygusch, Jurgen E. Refinement of CsTi(SO4)2, 12H2O crystal structure. M, 1969, McGill University. 39 p.

Sykes, Charles Ronald. Geology of the southeastern part of Fork Ridge Quadrangle, Claiborne County, Tennessee. M, 1968, University of Tennessee, Knoxville. 66 p.

Sykes, Donald Windsor. Mid-Devonian smooth spiriferoids from the Northwest Territories. M, 1962, University of Saskatchewan. 112 p.

Sykes, Julia Ann. High-grade metamorphism of iron-rich bodies in Archean gneiss, Wind River Mountains, Wyoming. M, 1985, University of Minnesota, Duluth. 131 p.

Sykes, Lynn R. An experimental study of compressional velocities in deep sea sediments. M, 1960, Massachusetts Institute of Technology. 81 p.

Sykes, Lynn R. The propagation of short-period seismic surface waves across oceanic areas. D, 1964, Columbia University, Teachers College.

Sykes, Martha Lynn. Ascent of granitic magma; constraints from thermodynamics and phase equilibria. D, 1986, Arizona State University. 312 p.

Sykes, Martha Lynn. Hydrous mineral stability as a function of fluid composition; a biotite melting experiment and a model for melting curves. M, 1979, Arizona State University. 121 p.

Sykes, Robert E. Seismic survey for determining geological control of ground water of the northwestern Jornado del Muerto, T. 4S., R. 3E, Socorro County, New Mexico. M, 1954, New Mexico Institute of Mining and Technology. 19 p.

Sykora, Jerry J. Stratigraphy, sedimentology and diagenesis-porosity relationships of the subsurface, Neocomian Parsons Group, Mackenzie Delta area, Northwest Territories, Canada. M, 1984, Carleton University. 167 p.

Sylla, N'Fanly. Une méthode d'interprétation des anomalies magnétiques par les moindres carrés. M, 1969, Ecole Polytechnique. 51 p.

Sylvester, Arthur Gibbs. Geology of the Vradel granitic pluton, Telemark, Norway. M, 1963, University of California, Los Angeles.

Sylvester, Arthur Gibbs. Structural and metamorphic petrology of the contact aureole of Papoose Flat Pluton, Inyo Mountains, California. D, 1966, University of California, Los Angeles. 206 p.

Sylvester, George H. Depositional environment of diamictites in the Headquarters Formation, Medicine Bow Mountains, southeastern Wyoming. M, 1973, University of Wyoming. 92 p.

Sylvester, James F. Assimilation of limestone by quartz monzonite magma, Ferber mining district, Nevada. M, 1950, University of Utah. 23 p.

Sylvester, Kathleen M. Paleontological and stratigraphical study of the Racine Dolomite (Middle Silurian) at Racine, Wisconsin. M, 1977, University of Wisconsin-Milwaukee.

Sylvester, Kenneth Albert. Ground water in upper Dry Valley and Little Long Valley, Caribou County, Idaho. M, 1975, University of Idaho. 97 p.

Sylvester, Kevin John. Geophysical investigations of the hydrogeology of the Goldendale-Centerville areas, Washington. M, 1978, Washington State University. 160 p.

Sylvester, Paul Joseph. Geology, petrology, and tectonic setting of the mafic rocks of the 1480 MA old granite-rhyolite terrane of Missouri, USA; (Volumes one and two). D, 1984, Washington University. 815 p.

Sylvester, Robert Kilbur. Scolecodonts from central Missouri. M, 1957, University of Missouri, Columbia.

Sylvester, Steven J. The petrology and geochemistry of a small zoned pegmatite (Permian); Middletown District, Connecticut. M, 1971, Franklin and Marshall College.

Sylvestre, M. A finite element model for salt water upconing and its application to the Magdalen Islands Aquifer. M, 1974, University of Waterloo.

Sylvia, Dennis Ashton. Depositional, diagenetic, and subsidence history of the Redwall Limestone, Grand Canyon National Park, Arizona. M, 1985, University of Arizona. 243 p.

Sylwester, Richard E. The determination of active fault zones in Puget Sound, Washington by means of continuous seismic profiling. M, 1971, University of Washington. 88 p.

Symborski, Mark Andrew. Testing the Modified Universal Soil Loss Equation (MUSLE) on an urbanizing Piedmont watershed in Maryland. M, 1988, University of Maryland.

Syme, A. M. Glacial features in the vicinity of Knob Lake, Labrador. M, 1951, McGill University.

Syme, Eric Charles. Petrogenesis of the boundary intrusions, Flin Flon area, Saskatchewan-Manitoba. M, 1975, University of Saskatchewan. 169 p.

Symecko, Ronald Edward. Glacil geology of the Orchard Park, New York Quadrangle. M, 1967, SUNY at Buffalo. 64 p.

Symes, James Leo, III. Geochemical behavior and distribution of copper and iron in marine waters. D, 1983, University of Rhode Island. 272 p.

Symmes, K. H. The biogeochemistry of copper in Indian Lake, Worcester, Massachusetts; a lake treated annually with copper sulfate. D, 1976, University of Massachusetts. 236 p.

Symonds, Paul S. Measurement of flow and pressure in fluids under pulsating conditions. M, 1941, Cornell University.

Symonds, Robert B. Transport and enrichment of elements in high temperature gases at Merapi Volcano, Indonesia. M, 1985, Michigan Technological University. 102 p.

Symons, David Thorburn Arthur. Paleomagnetic studies of Lake Superior iron ore deposits. D, 1965, University of Toronto.

Synowiec, Karen A. Stratigraphy of the Neda Formation in eastern Wisconsin, northern Illinois and eastern Iowa. M, 1981, University of Wisconsin-Milwaukee. 115 p.

Sypniewski, Bruce F. Trace element dispersion in groundwater from carbonate aquifers in northwestern Sandusky County, Ohio. M, 1982, Kent State University, Kent. 84 p.

Syriopoulou, Dimitra. Two-dimensional modeling of contaminant plumes in advection-dominated groundwater transport. D, 1987, [Vanderbilt University]. 181 p.

Syrjamaki, Robert M. The Prairie du Chien Group of the Michigan Basin. M, 1977, Michigan State University. 140 p.

Syverson, Kent Maurice. The glacial geology of the Kettle interlobate moraine region, Washington County, Wisconsin. M, 1988, University of Wisconsin-Madison. 123 p.

Syverson, Tim L. History and origin of debris torrents in the Smith Creek drainage, Whatcom County, Washington. M, 1984, Western Washington University. 84 p.

Syvert, Raymond J. Iterative analysis of the leakage coefficient for three aquifers. M, 1973, University of Toledo. 176 p.

Syvitski, J. P. M. Sedimentological advances concerning the flocculation and zooplankton pelletization of suspended sediment in Howe Sound, British Columbia; a fjord receiving glacial meltwater. D, 1978, University of British Columbia.

Szabelak, Stanley A. Colluvial slope stability estimate and driven anchor field experiment in a portion of Glenwood Canyon, Colorado. M, 1984, Colorado School of Mines. 180 p.

Szabo, Barney Julius. Distribution of barium in sea water in the Antillean-Caribbean region. M, 1966, [University of Miami].

Szabo, Ernest. Pennsylvanian paleotectonics of the Paradox region in parts of Utah, Arizona, New Mexico, and Colorado. D, 1968, University of New Mexico. 144 p.

Szabo, Ernest. Stratigraphy and paleontology of the Carboniferous rocks of the Cedro Canyon area, Manzanita Mountains, Bernalillo County, New Mexico. M, 1953, University of New Mexico. 137 p.

Szabo, John Paul. The Quaternary history of the lower part of Pioneer Creek basin, Cedar and Jones counties, Iowa. D, 1975, University of Iowa. 173 p.

Szabo, Michael Wallace. Quaternary geology, Alabama River basin, Alabama. M, 1972, University of Alabama. 63 p.

Szabo, N. L. Dispersion of indicators by glacial transportation at Mount Pleasant. D, 1975, University of New Brunswick.

Szabo, Nicholas. Sampling procedure in glacial contrast features selected sampling studies for grain size analysis in consolidated deposits. M, 1970, University of Connecticut. 75 p.

Szak, Caroline. The nature and timing of late Quaternary events in eastern Long Island Sound. M, 1987, University of Rhode Island. 83 p.

Szatai, John Endre. The geology of parts of the Redrock Mountain, Warm Spring, Violin Canyon, and Red Mountain quadrangles, Los Angeles County, California. D, 1961, University of Southern California. 313 p.

Szavits-Nossan, Vlasta. Intrinsic time behavior of cohesive soils during consolidation. D, 1988, University of Colorado. 447 p.

Szczepanowski, Stanley Peter. Aspects of groundwater flow in lake sediments, East Twin Lake, Portage County, Ohio. M, 1976, Kent State University, Kent. 71 p.

Szecsody, James Edward. Use of major ion chemistry and environmental isotopes to delineate subsurface

flow in Eagle Valley, Nevada. M, 1982, University of Nevada. 195 p.

Szekeley, Francisco L. The environmental cost of techno-economic development; copper production in Cananea, Sonora, Mexico. D, 1974, Washington University. 278 p.

Szekely, Thomas S. The geology of the Puquina-Omate area, southwestern Peru. D, 1963, Harvard University.

Szerbiak, Robert Bruce. Accurate seismic phase-velocities from interfering surface-waves using homomorphic deconvolution. M, 1981, Texas A&M University.

Szerdy, Frank Steven. Flow slide failures associated with low level vibrations. D, 1985, University of California, Berkeley. 294 p.

Szetu, Sui-Shing. Geochemistry of the formation of diamonds. D, 1954, University of Toronto.

Szewczyk, Z. J. A gravity study of an Archean granitic area north of Ignace, Ontario. M, 1974, University of Toronto.

Szigeti, George Joseph. Sedimentology and paleontology of the Upper Jurassic Unkpapa Sandstone and Morrison Formation, East flank of the Black Hills, South Dakota. M, 1979, South Dakota School of Mines & Technology.

Szmuc, Eugene Joseph. Stratigraphy and paleontology of the Cuyahoga Formation of northern Ohio. D, 1957, Ohio State University. 644 p.

Szmuc, Eugene Joseph. Stratigraphy of the post-Berea rocks in northwestern Trumbull and southeastern Ashtabula counties, Ohio. M, 1953, Ohio State University.

Szpakowski, James. The cascaded process applied to finite-difference migration; methods and seismic applications. M, 1987, Indiana University, Bloomington. 182 p.

Szpakowski, Sally. The use of algorithm LSQR in inverting simulated travel times for a refraction fan-shot geotomography experiment. M, 1987, Indiana University, Bloomington. 99 p.

Szumigala, David. Geology and geochemistry of the Tin Creek zinc-lead-silver skarn prospects, Farewell mining district, southwestern Alaska. M, 1986, University of Alaska, Fairbanks. 144 p.

Szustakowski, Robert James. Chemistry and evolution of formation waters in western New York. M, 1987, Syracuse University. 78 p.

Szuwalski, Daniel Robert. The petrology of the ultramafic rocks in the footwall Levack Complex, Fraser area, Sudbury, Canada (possible source of the Sudbury sublayer ultramafic xenoliths). M, 1984, Rutgers, The State University, New Brunswick. 116 p.

Szydlik, Stephen J. Sedimentary processes on the continental slope around Cape Hatteras. M, 1980, University of South Florida, St. Petersburg. 158 p.

Szymanski, Daniel. Probing beneath the Blue Ridge-Piedmont overthrust through the Grandfather Mountain Window. M, 1987, Purdue University. 238 p.

Szymanski, Maciej Boleslaw. A frictional theory of plasticity for granular materials. D, 1981, Carleton University.

Szymanski, William N. Petrology of the Paoli-Beaver Bend and Reelsville-Beech Creek Limestone members of the Newman Limestone (upper Mississippian) in East-central Kentucky. M, 1975, Eastern Kentucky University. 82 p.

T'an Hsi Chou see Hsi Chou T'an

Ta Ni Pe see Ni Ta Pe

Tabachnick, Rachel. Evolving entities in fossil populations; a morphometric analysis of Miocene planktonic foraminifera (Globorotalia). D, 1988, University of Michigan. 216 p.

Tabachnick, Rachel Ann. Morphologic variation in Miocene Globigerina bulloides d'Orbigny, Newport Bay, California. M, 1981, University of California, Los Angeles. 120 p.

Taban, Osman. Stratigraphy, lithology, depositional and diagenetic environments of the Antelope Valley Limestone at the Antelope Range and Martin Ridge section in central Nevada. M, 1986, California State University, Long Beach. 180 p.

Tabatabai, Habibollah. Centrifugal modeling of underground structures subjected to blast loading. D, 1987, University of Florida. 312 p.

Tabatabai, Mehran. Petrography and geochemistry of the McArthur Township area, Ontario. M, 1979, Brock University. 141 p.

Tabatabaie-Raissi, Mansour. The flexible volume method for dynamic soil-structure interaction analysis. D, 1982, University of California, Berkeley. 298 p.

Tabba, M. M. Risk analysis of slope stability with special reference to Canadian sensitive clays. D, 1979, McGill University.

Tabbert, Robert Leland. Pennsylvanian fusulinids from the Ardmore Basin (Oklahoma). M, 1954, University of Wisconsin-Madison.

Tabbi-Anneni, Abdelhafid. Subsurface geology of south-central Calcasieu Parish, Louisiana. M, 1975, University of Southwestern Louisiana. 82 p.

Tabbutt, Kenneth Dean. Fission track chronology of foreland basins in the eastern Andes; magmatic and tectonic implications. M, 1986, Dartmouth College. 100 p.

Taber, Arthur P. An investigation of structure in the Upper Mississippi Valley. M, 1930, University of Minnesota, Minneapolis. 20 p.

Taber, David R. Upper Jurassic stratigraphy and paleogeography of the North Celtic Sea basin, offshore Ireland. M, 19??, University of Arkansas, Fayetteville.

Taber, Edward C., III. Holocene estuarine and marine sediments in the Galveston Bay (Texas) and adjacent shelf areas. M, 1971, University of Houston.

Taber, Edward Carroll. Report on San Pedro Point area and supplementary report. M, 1938, Stanford University.

Taber, James Tobert. Seismicity of geothermal areas, Iceland, El Salvador, Imperial Valley, California, New Mexico, Alaska, Hawaii, New Zealand. M, 1972, New Mexico Institute of Mining and Technology.

Taber, Joseph John, Jr. Crustal structure and seismicity of the Washington continental margin. D, 1983, University of Washington. 159 p.

Taber, Robert William. The megafauna of the Vanport Limestone of Ohio. M, 1951, University of Missouri, Columbia.

Taber, Stephen. Geology of the gold belt in the James River basin, Virginia. D, 1913, University of Virginia. 271 p.

Tabesh, Elahe. Radioactive anomalies exploration in the vicinity of Nishabour turquoise mine, northeastern Iran. M, 1979, Syracuse University.

Tabet, David Elias. Structure and petrology of the Mellen igneous intrusive complex near Mellen, Wisconsin. M, 1974, University of Wisconsin-Madison. 58 p.

Tabidian, Mohamad Ali. Stream/aquifer relationships along the Big Blue River near Beatrice, Nebraska. D, 1987, University of Nebraska, Lincoln. 300 p.

Tabios, Guillermo Quesada, III. Nonlinear stochastic modeling of river dissolved oxygen. D, 1984, Colorado State University. 289 p.

Tabor, John Raymond. Deformational and metamorphic history of Archean rocks in the Rainy Lake District, northern Minnesota. D, 1988, University of Minnesota, Minneapolis. 262 p.

Tabor, John Raymond. Nature and sequence of deformation in the southwestern limb of the Kingston orocline. M, 1985, University of Illinois, Urbana. 87 p.

Tabor, Lawrence LaVerne. Geology of the Crater sulphur deposits, Inyo County, California. M, 1935, University of California, Berkeley. 93 p.

Tabor, Norman Richard. Ostracods of the family Hollinidae from the (Middle Devonian) Genshaw Formation of Michigan. M, 1951, University of Michigan.

Tabor, Richard L. A study of total hardness, magnesium, and calcium concentration at various depths in Wasco Lake (Illinois). M, 1964, Northern Illinois University.

Tabor, Rowland Whitney. Crystalline geology of area south of Cascade Pass, northern Cascades. D, 1961, University of Washington. 205 p.

Tabor, Rowland Whitney. Structure and petrology of Magic-Formidable region, northern Cascades, Washington. M, 1958, University of Washington. 81 p.

Tabora, Oscar. Surface geology of the Weston Quadrangle, Alabama. M, 1985, Mississippi State University. 94 p.

Tabrez, A. Sedimentary aspects of late Quaternary deposits in the Lake Nipissing area. M, 1983, Lakehead University.

Tabrum, Alan Robert. A contribution to the mammalian paleontology of the Ogallala Group of South-central South Dakota. M, 1981, South Dakota School of Mines & Technology. 408 p.

Tacker, Allen B. Sedimentology of the Lagonda Formation (Pennsylvanian), north-central Missouri. M, 1970, University of Missouri, Columbia.

Tacker, Robert Christopher. The partitioning of molybdenum between magnetite and a synthetic rhyolitic melt. M, 1985, University of Maryland.

Tadkod, Mohamed-Ali. Petrography and petrochemistry of the rock types from Hell Roaring Lakes area, Montana. M, 1976, University of Cincinnati. 107 p.

Taer, Andrew D. Fracture patterns and intragranular strain in the Massanutten Sandstone, northwest Virginia. M, 1985, Bowling Green State University. 44 p.

Taft, A. G. A geotechnical evaluation of ash residue from solid waste incineration. M, 1988, Kent State University, Kent. 148 p.

Taft, J. L., III. Phosphorus cycling in the plankton of Chesapeake Bay. D, 1974, The Johns Hopkins University. 204 p.

Taft, William H. Studies of the Ogallala Sands in south-central South Dakota. M, 1958, University of South Dakota. 52 p.

Taft, William Harrison. Geology of an area in the South Quien Sabe Quadrangle, California. M, 1957, Stanford University.

Taft, William Harrison. Pre-Mississippian to pre-Permian paleogeography of south-central Oklahoma. M, 1959, Stanford University.

Taft, William Harrison. Unconsolidated carbonate sediments of Florida Bay, Florida. D, 1962, Stanford University. 70 p.

Tafuri, William Joseph. A geochemical study of the barite deposits of Mineral County, Nevada. M, 1973, University of Nevada. 69 p.

Tafuri, William Joseph. Geology and geochemistry of the Mercur mining district, Tooele County, Utah. D, 1987, University of Utah. 194 p.

Tag, Peter Harrison. Studies of the rheology and thermal behavior of strike-slip faults. M, 1979, Cornell University.

Taggart, Bruce E. Net shore-drift of Kitsap County, Washington. M, 1984, Western Washington University. 95 p.

Taggart, James Nash. Geology of Mt. Powell Quadrangle, Colorado. D, 1962, Harvard University.

Taggart, James Nash. Problems in correlation of terraces along the Trinity River in Dallas County, Texas. M, 1953, Southern Methodist University. 21 p.

Taggart, Joseph Edgar, Jr. Abundances of Ca, Fe, Mg, Cu, Mn, and Ni in minerals separated from ultramafic rocks. M, 1969, Miami University (Ohio). 52 p.

Taggart, R. E. Palynology and paleoecology of the Miocene Sucker Creek flora from the Oregon-Idaho boundary. D, 1971, Michigan State University.

Taggart, Ralph E. A preliminary analysis of the microfossils of the Bridge Creek Shale (Oligocene), Oregon, with special emphasis on the palynology of the deposit. M, 1967, Ohio University, Athens. 90 p.

Tagudin, Jill Ellen. A correlation between fault vergence, fault spacing and sediment type in the North Panama thrust belt. M, 1988, University of California, Santa Cruz.

Tague, Glenn Charles. Post-glacial geology of the Grand Marais Embayment, Berrien County, Michigan. D, 1942, University of Michigan.

Taha, Rozlan Mohammad. Single well model for field studies of sandstone matrix acidizing. D, 1987, University of Texas, Austin. 165 p.

Tahir, Abdul Haleem. Engineering properties of loess in the United States. D, 1971, Catholic University of America. 227 p.

Tahiri, Mohamed. Kinetics and mechanisms of the solubilization of the organic matter of a Moroccan oil shale in a toluene solution at moderate temperatures. D, 1988, University of Oklahoma. 232 p.

Tahmazian, Garabed A. Petrography of the (Upper and Middle Cambrian) Troy and (Upper and Middle Devonian) Martin clastics sequence at Roosevelt Dam, Gila County, Arizona. M, 1965, University of Arizona.

Tai, K. C. Analysis and synthesis of flood control measures. D, 1975, Colorado State University. 288 p.

Taigbenu, Akpofure Efemena. A new boundary element formulation applied to unsteady aquifer problems. D, 1985, Cornell University. 210 p.

Tailleur, Irvin L. Ore deposits of the Clayton area, Custer County, Idaho. M, 1948, Cornell University.

Tainter, Patrick A. Investigation of stratigraphic and paleostructural controls on hydrocarbon migration and entrapment in Cretaceous D and J sandstones of the Denver Basin. M, 1982, University of Colorado. 235 p.

Taioli, Fabio. Laboratory evaluation of waveguides for acoustic emission/microseismic monitoring. M, 1987, Pennsylvania State University, University Park. 144 p.

Taira, A. Design and calibration of a photo-extinction settling tube for grain size analysis. D, 1976, University of Texas at Dallas. 365 p.

Taira, H. A study of the Io/Th (Th-230/Th-232) chronology of some marine sediments using solid-state detectors. M, 1967, University of Hawaii. 39 p.

Tait, Donald Burkholder. The system diopside-forsterite-anorthite. M, 1949, Pennsylvania State University, University Park. 33 p.

Tait, James Fulton. The effects of seawalls on beaches. M, 1988, University of California, Santa Cruz.

Tait, Larry. The character of organic matter and the partitioning of trace and rare earth elements in black shales; Blondeau Formation, Chibougamau, Quebec. M, 1987, Universite du Quebec a Chicoutimi. 140 p.

Tait, Sandra Elizabeth. Breccias of Mount Pleasant tin deposit, New Brunswick. M, 1965, McGill University. 89 p.

Takacs, Michael James. The oxidation of chromium by manganese oxide; the nature and controls of the reaction. M, 1988, Michigan State University. 271 p.

Takagi, Robert Shigern. Construction and operation of a Cartesian diver apparatus for weighing ostracods. M, 1960, University of Michigan.

Takahashi, Kozo. Vertical flux, ecology and dissolution of Radiolaria in tropical oceans; implications for the silica cycle. D, 1981, Woods Hole Oceanographic Institution. 461 p.

Takahashi, Taro. Supergene alteration of zinc and lead deposits. D, 1957, Columbia University, Teachers College.

Takayanagi, Kazufumi. Heavy metals in anoxic and oxic sediments of the Indian River near Melbourne, Florida. M, 1978, Florida Institute of Technology.

Takayanagi, Kazufumi. The marine geochemistry of selenium. D, 1982, Old Dominion University. 131 p.

Take, W. F. Study of the (Precambrian) Wolf nepheline belt, Renfrew County, Ontario (Canada). M, 1954, Dalhousie University.

Takeshita, Toru. Texture development and plastic anisotropy in deformed polycrystals; theories, experiments and geological implications. D, 1987, University of California, Berkeley. 211 p.

Takeuchi, Shozaburo. Variation of elasticity in water-saturated rocks near the melting temperature of ice. M, 1971, Massachusetts Institute of Technology. 69 p.

Takigiku, Ray. Isotopic and molecular indicators of origins of organic compounds in sediments. D, 1987, Indiana University, Bloomington. 248 p.

Talanda, Jean. Fission-track geochronology and geothermometry of Late Cretaceous-Early Tertiary epizonal plutons, in and adjacent to the Lombard thrust fault, southwestern Montana. M, 1988, Western Michigan University.

Talay, Theodore A. Usage and limitations of characteristic vector analysis of remote sensing multispectral data for the identification and quantification of water quality parameters. D, 1981, Old Dominion University. 394 p.

Talbett, Michael Steven. Acid mine drainage and the acid drainage index. D, 1981, Indiana State University. 127 p.

Talbot, Curtis L. Modern evaporite origin at Howell and Soap Hole Lakes, Albany County, Wyoming. M, 1967, University of Wyoming. 56 p.

Talbot, James Paul. Grain morphology and origins of fine ash deposits from the 180 A.D. Taupo eruption, central North Island, New Zealand. M, 1988, University of Texas, Arlington. 262 p.

Talbot, Mignon. Contributions to a revision of the (Lower Devonian) Helderbergian fauna of New York. D, 1904, Yale University.

Talbot, Robert Walter. Atmospheric fluxes and geochemistries of stable Pb, Pb-210, and Po-210 in Crystal Lake, Wisconsin. D, 1981, University of Wisconsin-Madison. 274 p.

Talbot, W. Robert. Sedimentology, depositional environments and paleomagnetic reconnaissance of the Upper Cretaceous Hell Creek Formation, northeastern Butte County, South Dakota. M, 1985, South Dakota School of Mines & Technology.

Talbott, Lyle W. The De Beque canyon slide, Mesa County, Colorado. M, 1969, Colorado State University. 128 p.

Talbott, William Charles. The geology of the Smolan Pool area, Saline County, Kansas. M, 1954, University of Oklahoma. 56 p.

Taleb, Mohammed. Structure and subsurface geology of northwestern margin of Sirte Basin in Libyan Arab Republic. M, 1977, University of South Carolina.

Taleb, T. Sediment properties and environment of deposition of the Minnelusa Formation (Permo-Pennsylvanian), northern Black Hills, South Dakota and Wyoming. M, 1971, University of Missouri, Rolla.

Talerico, Frank. An assessment of analytical methods and their biases in the determination of major elements in rock analysis. M, 1978, University of Windsor. 204 p.

Taliaferro, Lindsay C., III. Seasonal changes in the partitioning of heavy metals in a southwestern Ohio soil. M, 1987, Miami University (Ohio). 70 p.

Taliaferro, Nicholas Lloyd. The manganese deposits of the Sierra Nevada of California. D, 1920, University of California, Berkeley.

Talkiewicz, Joseph M. A petrographic analysis of the Martinsburg Formation, north-central New Jersey; an insight into the tectonic setting. M, 1986, SUNY at Buffalo. 85 p.

Talkington, Raymond. The petrology and geochemistry of the Harriman Lherzolite, Union, Maine. M, 1976, University of New Hampshire. 80 p.

Talkington, Raymond Willis. The geology, petrology and petrogenesis of the White Hills Peridotite, St. Anthony Complex, northwestern Newfoundland. D, 1981, Memorial University of Newfoundland. 292 p.

Tallamraju, R. K. M. R. Inference of stratospheric minor constituents from satellite limb radiant intensity measurements. D, 1975, University of Michigan. 163 p.

Tallan, M. E. Systematics and biostratigraphy of the early Hemphillian Bemis local fauna, Ellis County, Kansas. M, 1978, Fort Hays State University. 36 p.

Talley, Gilbert Arthur. A laboratory study of stream sorting and deposition. M, 1934, University of Iowa. 47 p.

Talley, James Bishop. The subsurface geology of northeastern Noble County, Oklahoma. M, 1955, University of Oklahoma. 58 p.

Talley, John H. Sedimentology of a coastal plain core. M, 1974, Franklin and Marshall College. 96 p.

Talley, Kieth L. Descriptive geology of the Redwood Mountain Outlier of the South Fork Mountain Schist, northern Coast Ranges, California. M, 1976, Southern Methodist University. 86 p.

Tallman, Ann M. The glacial and periglacial geomorphology of the Fourth of July Creek valley, Atlin region, Cassiar District, northwestern British Columbia. D, 1975, Michigan State University. 178 p.

Tallon, Walter Adam. The geology of Eniirikku Island, Bikini Atoll, Marshall Islands. M, 1955, University of Pittsburgh.

Tally, Taz. The effects of geology and large organic debris on stream channel morphology and process from streams flowing through old growth redwood forests in northwestern California. D, 1980, University of California, Santa Barbara.

Tallyn, Lee Ann K. Scabland mounds of the Cheney Quadrangle, Spokane County, Washington. M, 1981, Eastern Washington University. 94 p.

Tallyn, Robert Bernard. Petrology and stratigraphy of graywackes in the northwestern Olympic Mountains, Washington. M, 1972, University of Washington. 36 p.

Talmadge, Thomas White. Structure and stratigraphy of Taconic Hills near Old Chatham, New York. M, 1956, New York University.

Talmage, Sterling Booth. Quantitative standards for color and hardness of ore minerals. D, 1925, Harvard University.

Talmage, Sterling Booth. Some lead and zinc occurrences in eastern Pennsylvania. M, 1923, Lehigh University.

Talman, Stephen James. Feldspar dissolution and chemical weathering of a granodiorite with emphasis on the REE. M, 1987, University of Western Ontario. 174 p.

Talpey, James G. Geochemical and structural evolution of Archean gneisses at South Pass, Wyoming. M, 1984, University of Rochester. 102 p.

Talukdar, Suhas Chandra. Implications of petrological study of Paleozoic-Mesozoic volcanic rocks of Mineral County, Nevada. D, 1973, Rice University. 196 p.

Talwani, Manik. Gravity anomalies in the Bahamas and their interpretation. D, 1960, Columbia University, Teachers College. 90 p.

Talwani, Pradeep. Seismic wave velocities in unconsolidated materials and some lunar implications. D, 1973, Stanford University. 114 p.

Tam, Kwok F. Dielectric property measurements of rocks in the VHF-UHF region. D, 1974, Texas A&M University. 141 p.

Tam, S. S. Mechanisms and spatial patterns of erosion and instability in the Joe's River basin, Barbados. D, 1975, McGill University.

Tama, K. Charge, colloidal and structural stability interrelationships for selected Hawaiian soils. M, 1975, University of Hawaii. 135 p.

Tamburi, Alfred J. Geology and the water resource system of the Indus Plains. D, 1974, Colorado State University. 380 p.

Tamburi, Alfred J. Rock creep and environment. M, 1971, Colorado State University. 94 p.

Tamburro, Edie T. Geochemistry of the Sykesville Diamictite in central Maryland; an investigation of clast-matrix activity during metamorphism, with tectonic implications. M, 1986, Virginia Polytechnic Institute and State University.

Tamesis, Emmanuel Valerio. Cretaceous stratigraphy and sedimentation in the Avenal Ridge--Reef Ridge area, Fresno and Kings counties, California. D, 1966, Stanford University. 233 p.

Tamesis, Emmanuel Valerio. Pre-Pennsylvanian and post-Mississippian paleogeology of Texas and a lithologic and isopachous analysis of the Mississippian of Texas. M, 1955, Stanford University.

Tamimi, Y. N. Ammonium fixation in Hawaiian soils. D, 1964, University of Hawaii. 86 p.

Tamm, Allan H. An examination of the stratigraphy of the lower sand member of the lower Oakville Formation with special reference to the localization of uranium mineralization. M, 1977, Brooklyn College (CUNY).

Tamm, Lucille C. Electron petrography of bent augite crystals from the Triassic traps of the Mid-Atlantic states. M, 1973, Pennsylvania State University, University Park.

Tampoe, Tara J. Geochemical constraints on the future of agriculture in Sri Lanka. M, 1988, University of Western Ontario. 309 p.

Tamura, Allen Y. Origin, nature and significance of the residual mounds on Danby Dry lake, San Bernardino County, California. M, 1969, University of Southern California.

Tan, D. Y. Finite element analysis and design of chemically stabilized tunnels. D, 1977, Stanford University. 219 p.

Tan, Ek-Khoo. Stability of soil slopes. D, 1947, Columbia University, Teachers College.

Tan, Francis C. Carbon and oxygen isotope studies of the Great Estuarine series (Jurassic) of Scotland. D, 1969, Pennsylvania State University, University Park. 205 p.

Tan, Francis C. Paleotemperature studies L5W3= Li-Ping on Ordovician rocks. D, 1965, McGill University.

Tan, Hua Hui. Analysis and computer program for soil-structure-fluid interaction. D, 1988, University of California, Berkeley. 225 p.

Tan, J. T. Late Miocene to early Pliocene foraminifera from eastern and southeastern Viti Levu, Fiji. M, 1975, Carleton University.

Tan, J. Tony. Late Triassic-Jurassic dinoflagellate biostratigraphy, western Arctic Canada. D, 1979, University of Calgary. 296 p.

Tan, Kuo-Yu. The silicate mineralogy and petrology of the Elberton Granite. M, 1983, University of Georgia.

Tan, Li-Ping. The metamorphism of Taiwan Miocene coals. M, 1963, Columbia University, Teachers College.

Tan, Li-Ping. The New York pegmatite deposits. D, 1965, Columbia University. 185 p.

Tan, Robert Yaunchan. Estimation and optimization of the serviceability of underground water transmission network systems under seismic risk. D, 1980, Columbia University, Teachers College. 78 p.

Tan, Siang Swie. An analysis of structural lineaments on Vancouver Island, British Columbia. M, 1972, University of California, Los Angeles.

Tanabe, M. J. Selected physical properties of various soil and mediums as influenced by different compaction levels. M, 1972, University of Hawaii. 68 p.

Tanaka, Harry Harumi. Geology of the Gomez Park area, Jeff Davis County, Texas. M, 1948, University of Texas, Austin.

Tanaka, Kenneth Lloyd. Geology of the Olympus Mons region of Mars. D, 1983, University of California, Santa Barbara. 294 p.

Tanaka, Roderick Taira. Trace element indicators of uranium-concentrating processes in Elliot Lake uraniferous conglomerate. M, 1983, University of Toronto.

Tanck, Glen Steven. A paleoenvironmental interpretation of the Upper Cambrian Galesville Sandstone of South-central and West-central Wisconsin. M, 1977, University of Wisconsin-Madison.

Tanczyk, Elizabeth. Original strike and latitude of dykes and their bearing on the state of stress in the lithosphere. M, 1979, Carleton University. 84 p.

Tanenbaum, Ronald J. Geological engineering survey of the Tucson Basin, Pima County, Arizona. M, 1972, University of Arizona.

Tanenbaum, Ronald Joel. Recommendations for the geotechnical analysis and design of shallow excavations with vertical or near-vertical walls. D, 1975, Texas A&M University. 188 p.

Tang Kong, V. William. The effect of wettability and pore geometry on three phase fluid displacement in porous media. M, 1988, University of Calgary. 111 p.

Tang, Alan M. The processing and interpretation of long wavelength aeromagnetic anomalies. M, 1980, University of Manitoba. 117 p.

Tang, Alice C. Studies in the organic halogen compounds in sea water. M, 1956, Massachusetts Institute of Technology.

Tang, C. H.-W. A multi parametric lake model. D, 1975, University of Washington. 165 p.

Tang, David Hsin-Ying. Development of design criteria for mechanical roof bolting in underground coal mines. D, 1984, West Virginia University. 275 p.

Tang, Huey. A gravity survey of the New Baltimore area, Ohio. M, 1981, University of Akron. 90 p.

Tang, Hwei-Feng Nadine. Pressure relationship between sodium chloride and ruby by X-ray diffraction technique and ruby fluorescence methods. M, 1975, University of Rochester. 47 p.

Tang, James I. S. Studies of liquid effluents from synthetic fuel conversion processes. D, 1981, University of Southern California.

Tang, Ming. Isotopically anomalous Ne and Xe in meteorites and their carriers, SiC, and diamond. D, 1987, University of Chicago. 200 p.

Tang, Patrick S. Late Quaternary stratigraphy of Mullach brook and adjacent areas, Cape Breton, Nova Scotia. M, 1970, Acadia University.

Tang, Pei-Tau. Model study of sanitary landfill in North Carolina. M, 1980, North Carolina State University. 93 p.

Tang, Rex Wai Yuen. Geothermal exploration by telluric currents in the Klamath Falls area, Oregon. M, 1974, Oregon State University. 86 p.

Tang, Roger. Ocean bottom friction study using hydrodynamic modelling and SEASAT-ALT data. M, 1985, University of Manitoba.

Tang, Su. Paleozoic lichenoporid-like bryozoans. M, 1988, Pennsylvania State University, University Park. 152 p.

Tang, Wen Jian. Satellite magnetic signatures of oceanic hot spots; interpretation of MAGSAT anomalies over Iceland and Hawaii regions. M, 1985, University of Iowa. 140 p.

Tangchawal, Sanga. Engineering properties and clay mineralogy of Illinois coal-mine roof shales and underclays. D, 1988, University of Missouri, Rolla. 248 p.

Tanger, John Carroll, IV. Calculation of the standard partial molal thermodynamic properties of aqueous ions and electrolytes at high pressures and temperatures. D, 1986, University of California, Berkeley. 139 p.

Tanguay, Marc G. Etudes des pyroxènes monocliniques et diopside d'Oka. M, 1963, Ecole Polytechnique. 156 p.

Tanguay, Marc Gilles. Aerial photography and multi-spectral remote sensing for engineering soils mapping. D, 1969, Purdue University. 362 p.

Tanimoto, Toshiro. Coupling and attenuation of torsional modes in a heterogeneous Earth. D, 1982, University of California, Berkeley. 87 p.

Tanis, James Iran. Isostasy and crustal thickness in Utah and adjacent areas as revealed by gravity profiles. D, 1962, University of Utah. 178 p.

Tank, Nihat. Lithostratigraphic analysis of the Kinderhookian and Osagean series in the continental interior of the United States of America. M, 1966, University of Wisconsin-Madison. 58 p.

Tank, Ronald Warren. Clay mineralogy of some lower Tertiary (Paleogene) sediments from Denmark. D, 1962, Indiana University, Bloomington. 97 p.

Tank, Ronald Warren. The (Jurassic) Morrison Formation of the Black Hills area, South Dakota and Wyoming. M, 1955, University of Wisconsin-Madison.

Tanksley, David Arthur. Oil migration studies. M, 1953, Miami University (Ohio). 33 p.

Tanner, George F. The Mt. Carmel Fault and associated features in south-central Indiana. M, 1985, Indiana University, Bloomington. 66 p.

Tanner, James G. Gravity anomalies in the Gaspe Peninsula, Quebec. M, 1958, University of Western Ontario. 120 p.

Tanner, James Henry. Subsurface geology of the Lower Permian, Palo Duro Basin, Texas. M, 1957, Texas Tech University. 149 p.

Tanner, James Thomas. The determination of antimony in natural materials by neutron activation. D, 1966, [University of Kentucky].

Tanner, Jeffrey. Chemical and isotopic evidence of the origin of hydrocarbons and source potential of rocks from the Vicksburg and Jackson formations of Slick Ranch area, Starr County, Texas. M, 1987, University of Houston.

Tanner, Joseph Jarratt. Geology of the Castle Mountain area, Montana. D, 1949, Princeton University. 179 p.

Tanner, Stephen Bruce. Experimental kinetic study of the reaction; calcite+quartz = wollastonite+carbon dioxide, at geologically-relevant P-T conditions. M, 1984, Pennsylvania State University, University Park. 89 p.

Tanner, Vasco Myron. Deltas of the Lake Bonneville. M, 1920, University of Utah. 44 p.

Tanner, William Francis, Jr. A study of the characteristics and sedimentation of certain sand dunes in Lynn, Lamb, and Bailey counties. M, 1939, Texas Tech University. 71 p.

Tanner, William Francis, Jr. The geology of Seminole County, Oklahoma. D, 1953, University of Oklahoma. 297 p.

Tanner, William Roger. A fossil flora from the Beartooth Butte Formation of northern Wyoming; microform. D, 1983, Southern Illinois University, Carbondale. 222 p.

Tanoli, Saifullah Khan. A study of the mechanics of fracture closure. M, 1982, University of Toronto.

Tanoli, Saifullah Khan. Stratigraphy, sedimentology, and ichnology of the Cambrian-Ordovician Saint John Group, southern New Brunswick, Canada. D, 1987, University of New Brunswick.

Tansey, Vivian Ouray. Paleontological study of the (Devonian) Helderberg and Oriskany formations of Ste. Genevieve County, Missouri. D, 1921, University of Chicago.

Tanski, Joseph James. Episodic bluff erosion on the north shore of Long Island, New York. M, 1981, SUNY at Stony Brook.

Tanski, Stephen A. Provenance study of the Middle Ordovician sandstones of New York and western Vermont. M, 1984, SUNY at Albany. 112 p.

Tansley, Wilfred. Geology and petrology of the Tobacco Root Mountains, Montana. D, 1933, University of Chicago. 104 p.

Tansley, Wilfred. The proportion of quartz to feldspar in graphic granite. M, 1931, University of Chicago. 24 p.

Tantikom, Supachai. An analysis of stress distribution around piles. D, 1988, University of Alabama. 134 p.

Tanton, Thomas L. The relative importance of meteoric and magmatic waters in the deposition of certain primary ores. D, 1915, University of Wisconsin-Madison.

Tao, Shu. Migration and translocation of topically applied copper, lead, zinc, cadmium and nickel in soil profiles during accelerated leaching. D, 1984, University of Kansas. 340 p.

Taplin, A. C. A sheared and altered pendant in the Cassiar Batholith, headwaters of Stikine River, British Columbia. M, 1954, University of British Columbia.

Taplin, Arthur Cyril. A sheared and altered pendant in the Cassiar Batholith, headwaters of the Stikine River, BC. M, 1951, University of British Columbia.

Tapp, Gayle Standridge. Computer modeling of fracture pattern in a single layer fold using the finite element method. M, 1977, University of Oklahoma. 83 p.

Tapp, James Bryan. Breccias and megabreccias of the Arbuckle Mountains, Southern Oklahoma Aulacogen, Oklahoma. M, 1978, University of Oklahoma. 126 p.

Tapp, James Bryan. Relationships of rock cleavage fabrics to incremental and accumulated strain in a portion of the Blue Ridge, Virginia. D, 1983, University of Oklahoma. 322 p.

Tappan, Helen Nina. A microfaunal study of the Grayson Formation (Lower Cretaceous) of northern Texas and southern Oklahoma. M, 1938, University of Oklahoma. 142 p.

Tappe, John. The chemistry, petrology and structure of the Big Timber Igneous Complex, Crazy Mountains, Montana. D, 1966, University of Cincinnati. 177 p.

Tapper, Charles Joseph. Geochemistry of greenstone and amphibolite in the eastern Black Hills, South Dakota. M, 1984, South Dakota School of Mines & Technology.

Tapper, Wilfred Bonno. Geology of Greene County, Iowa. M, 1938, University of Iowa. 59 p.

Tappmeyer, D. M. The application of orbital imagery in the interpretation of the geology of Norwood area, Washakie County, southern Bighorn Basin, Wyoming. M, 1974, University of Iowa. 65 p.

Tapscott, Christopher Robert. The evolution of the Indian Ocean triple junction and the finite rotation problem. D, 1979, Woods Hole Oceanographic Institution. 210 p.

Tara, Muriel Elizabeth. The geology and petrography of Little Presque Isle, Marquette County, Michigan. M, 1950, Michigan State University. 31 p.

Tarabzouni, Mohamed Ahmed. Computer-enhanced Landsat images for ground water exploration in the northern Arabian Shield; Ha'il test site. D, 1981, University of Tennessee, Knoxville. 279 p.

Tarache, Crisalida. Stratigraphy and uranium potential of the La Quinta Formation of Jurassic age, North-central Tachira, Venezuela. M, 1980, Colorado School of Mines. 103 p.

Taranik, James V. Stratigraphic and structural evolution of Breckenridge area, central Colorado. D, 1974, Colorado School of Mines. 221 p.

Tarantolo, P. J., Jr. Electromagnetic probing of salt with high frequency radar. D, 1978, Texas A&M University. 303 p.

Taras, Brian Daniel. Sr, Nd and Pb isotope and trace element geochemistry of the New England seamount chain. M, 1984, Massachusetts Institute of Technology. 91 p.

Tarazona, Carlos. An investigation of selected methods for the discrimination of natural and explosive seismic sources. M, 1975, University of Wisconsin-Milwaukee.

Tarbell, Eleanor. The (Devonian) Antrim-(Mississippian) Ellsworth and Coldwater formations of Michigan. M, 1945, University of Michigan.

Tarbox, David L. Environmental geology of portions of the townships of Bristol and Monkton, Vermont; a case study for landuse and planning. D, 1972, Cornell University.

Tarbutton, R. J. Petrography of some selected Upper Cretaceous Selma Group sediments in parts of Humphreys, Sharkey, and Yazoo counties, Mississippi. M, 1979, University of Southern Mississippi.

Tarduno, John Anthony. Cretaceous absolute motion of Pacific oceanic rises; linking the continental and oceanic records through paleomagnetic analysis. D, 1987, Stanford University. 241 p.

Targgart, F. Arthur. Geological field trip in the Phoenix area (Arizona). M, 1967, Virginia State University. 29 p.

Tariki, Abdulla Homoud. Geology of Saudi Arabia. M, 1947, University of Texas, Austin.

Tariq, Syed Mohammad. A study of the behavior of layered reservoirs with wellbore storage and skin effect. D, 1978, Stanford University. 161 p.

Tarkington, Daniel K. Paleoecology of fusulinids from the upper Hughes Creek Shale (Lower Permian) of east-central Kansas. M, 1977, University of Kansas. 76 p.

Tarkoy, P. J. Rock hardness index properties and geotechnical parameters for predicting tunnel boring machine performance. D, 1975, University of Illinois, Urbana. 347 p.

Tarkoy, Peter J. Lithostratigraphy and petrography of the Upper Cambrian Maynardville Formation within the Hunter Valley Fault Belt in east Tennessee. M, 1967, University of Tennessee, Knoxville. 99 p.

Tarleton, William Addison. The geology of the southern portion of the Hightown Anticline, Monterey Quadrangle, Highland County, Virginia. M, 1948, University of Virginia. 89 p.

Tarman, D. W. Application of dynamic analysis of mineral grains to some structural problems in the Empire Mountains, Pima County, Arizona. D, 1975, University of Arizona. 198 p.

Tarman, Donald. Analysis of electrical resistivity measurements of shallow deposits. M, 1967, Iowa State University of Science and Technology.

Tarr, Arthur Charles. The dispersion of Rayleigh waves in the crust and upper mantle under the western north Atlantic Ocean, Gulf of Mexico, and Caribbean Sea. D, 1968, University of Pittsburgh. 155 p.

Tarr, Karen M. Critique of a groundwater solute transport model of radionuclide migration from an uranium mill. M, 1988, University of Idaho. 112 p.

Tarr, William Arthur. The barite deposits of Missouri and the geology of the Barite District, Missouri. D, 1916, University of Chicago. 111 p.

Tarrant, G. A. Taxonomy, biostratigraphy and paleoecology of Late Ordovician conodonts from southern Ontario. M, 1977, University of Waterloo.

Tarshis, Andrew Lorie. Electron microscopic examination of zeolitization and cooling devitrification in silicic volcanic rocks. D, 1982, University of California, Santa Cruz. 257 p.

Tartamella, Natale John. The foraminifera of the Saratoga Chalk. M, 1951, University of Kansas.

Taruvinga, Peter Rangarirai. Geomorphic cartography; the Canadian perspective; an assessment of selected landform mapping approaches in Canada. D, 1981, University of Waterloo.

Tary, Alexander. Correlation of seismic velocities and some mechanical properties of unconsolidated sediments. M, 1969, University of California, Riverside. 93 p.

Tarzi, J. G. Weathering of mica minerals in selected Ontario soils. D, 1977, University of Guelph.

Tasch, Paul. Causes and paleoecological significance of dwarfed fossil marine invertebrates. D, 1952, University of Iowa. 246 p.

Tasch, Paul. Fauna of the Upper Cambrian Warrior Formation of central Pennsylvania. M, 1950, Pennsylvania State University, University Park. 47 p.

Tascin, Mustafa T. Exploration for a geothermal system in the Lualualei Valley, Oahu, Hawaii. M, 1975, Colorado School of Mines. 87 p.

Tasillo, Anne Marie. Paleomagnetism of the Blake River Group, Abitibi greenstone belt, Ontario. M, 1980, University of Toronto.

Taskey, Ronald D. Relationships of clay mineralogy to landscape stability in western Oregon. D, 1978, Oregon State University. 223 p.

Tasse, Normand. Sédimentologie d'une bande de roches pyroclastiques archéennes de la région de Rouyn-Noranda, Quebec. M, 1976, Universite de Montreal.

Tasse, Normand. Sèdimentologie du flysch à Helminthoides de la Nappe du Parpaillon, Embrunais-Ubaye, Hautes-Alpes, France. D, 1982, McGill University. 226 p.

Tassell, Jay Van *see* Van Tassell, Jay

Tassone, Jeffrey Allen. The Cunningham Sandstone; a canyon fan deposit. M, 1977, University of Tulsa. 65 p.

Tassos, Stavros T. Holocene sediments and some oceanographic parameters of Kalloni Bay, Lesvos, Greece. M, 1975, University of Minnesota, Minneapolis. 159 p.

Tatalovic, Radmilo. Numerical and physical modeling studies of the resolution of thin beds for thickness calculation. M, 1985, University of Houston.

Tatar, Jhumar Mal. The Dunvegan Sandstone of the type area. M, 1964, University of Alberta. 54 p.

Tatar, Philip J. Analysis of south-central Nevada Earth strain measurements and the potential influence of Earth tides on microearthquakes recorded at Groom Mine, Nevada. M, 1974, University of Wisconsin-Milwaukee.

Tate, William Lewis. The depositional environment, petrology, diagenesis, and petroleum geology of the Red Fork Sandstone in central Oklahoma. M, 1985, Oklahoma State University. 189 p.

Taterka, Bruce D. Bedrock geology of Central Park, New York City. M, 1987, University of Massachusetts. 111 p.

Tatham, Robert H. Geologic inferences from seismic surface wave analysis of small magnitude earthquakes and V_p/V_s as a direct hydrocarbon indicator. D, 1975, Columbia University. 143 p.

Tatlock, Derek Bruce. Stratigraphy and paleontology of a well core through Upper and Middle Silurian strata in Oakland County, Michigan. M, 1955, University of Michigan.

Tatman, James B. The chemistry and mineralogy of ultramafic nodules from Salt Lake (Oahu), Hawaii. M, 1971, University of Washington. 58 p.

Tator, Benjamin Almon. Piedmont interstream surfaces of the Colorado Springs region. D, 1948, Louisiana State University.

Tator, Benjamin Almon. Smaller Miocene Mollusca (Atlantic and Gulf coasts). M, 1941, Louisiana State University.

Tatro, James O. Ostracoda of the Upper Cretaceous Marlbrook Marl of Arkansas. M, 1961, University of Kansas. 91 p.

Tattam, Charles M. Application of the electrical resistivity method of geophysical prospecting to problems of underground water. D, 1932, Colorado School of Mines. 79 p.

Tattersall, Ian. Crania and dentitions of archaeolemurinae (lemuroidea, primates) (Madagascar). D, 1971, Yale University.

Tatum, Emmett Perry, Jr. The ranges and distribution of several guide types of foraminifera in the Gulf Coast. M, 1928, Louisiana State University.

Tatum, T. E., Jr. Shallow geologic features of the upper continental slope, northwestern Gulf of Mexico. M, 1977, Texas A&M University.

Taubeneck, William Harris. Bald Mountain Batholith, Elkhorn Mountains, Oregon. D, 1955, Columbia University, Teachers College.

Taubeneck, William Harris. Geology of the northeast corner of the Dayville Quadrangle, Oregon. M, 1950, Oregon State University. 154 p.

Taublieb, Edward J. Recent sediments in eastern Lake Erie. M, 1969, SUNY at Buffalo. 33 p.

Taucher, Leonard Max. Geology of the Cookstove Basin area, Big Horn County, Wyoming. M, 1953, University of Wyoming. 88 p.

Taucher, Paul J. Reprocessing and interpretation of high resolution seismic reflection data from a uranium exploration prospect in the Granite Mountains area of central Wyoming. M, 1988, University of Wyoming. 136 p.

Taucher, Susan E. Gray. Distribution and paleoecology of benthic foraminifera from the Upper Cretaceous lower Hilliard Shale, Blazon Gap, southwestern Wyoming. M, 1986, University of Wyoming. 195 p.

Tauchid, Mohamad. The geochemistry of molybdenum in the Bathurst District, (New Brunswick). M, 1966, University of Ottawa. 113 p.

Taufen, Paul M. A stream drainage geochemical survey of the Goodnews and Hagemeister Island quadrangles; comparison of sample types. M, 1976, Colorado School of Mines. 97 p.

Tausch, Earl Harry. Natural variations in the isotope abundance ratios of molybdenum. D, 1974, Case Western Reserve University.

Tauvers, Peter Rolfs. Structure sections through the Marathon Basin, Trans-Pecos Texas; implications for basement-influenced deformation. D, 1988, University of Texas, Austin. 207 p.

Tauxe, Lisa. Rock magnetism and paleomagnetism of Miocene fluvial sediments in northern Pakistan. D, 1983, Columbia University, Teachers College. 144 p.

Tavares, Ricardo A. Semi-quantitative analysis of Jamaican bauxites. M, 1976, Brooklyn College (CUNY).

Tavassoli, Abolghasem. Soil nutrient availability during reclamation of salt-affected soils. D, 1980, University of Arizona. 138 p.

Tavolaro, John F. Sedimentology of the upper East River; an urban tidal strait. M, 1986, Queens College (CUNY). 218 p.

Tavora, Flavio J. Hydrogeochemical studies on selected mineralized areas in Colorado. M, 1971, Colorado School of Mines. 199 p.

Tawashi, A. M. H. El *see* El Tawashi, A. M. H.

Taweel, Michael Elias, Jr. The geology of the Eagle Rock and Highland Park area, Los Angeles County, California. M, 1963, University of Southern California.

Tawfik, F. M. Water quality within East Portland terraces. M, 1974, Portland State University. 71 p.

Tawfiq, Kamal Sulaiman. Effect of time and anisotropy on dynamic properties of cohesive soils. D, 1987, University of Maryland. 303 p.

Taxer, Karlheinz J. The crystal structure of rhodizite. M, 1966, Massachusetts Institute of Technology. 88 p.

Taylor, A. H. Carbonate stratigraphy and petrology; Robb Lake zinc-lead property, northeastern British Columbia. M, 1977, Carleton University. 7616 p.

Taylor, Alan Bruce. Geology of the Lake Massawippi region, Quebec Appalachians. M, 1983, University of Western Ontario. 163 p.

Taylor, Alisa J. The mammalian fauna from the mid-Irvingtonian Fyllan Cave local fauna, Travis County, Texas. M, 1982, University of Texas, Austin.

Taylor, Allan Beowulf. Mineral transformations and geochemical mass balance of a disturbed forested watershed. M, 1988, Michigan State University. 132 p.

Taylor, Allan Maurice. Crystal chemical relations in the P zeolite group. D, 1962, Pennsylvania State University, University Park. 240 p.

Taylor, Amy E. A biostratigraphic investigation of the Columbus Limestone at Marblehead, Ohio. M, 1983, University of Toledo. 133 p.

Taylor, Andrew M. Geohydrologic investigations in the Mesilla Valley, New Mexico. M, 1967, New Mexico State University, Las Cruces. 130 p.

Taylor, Andrew M. Stratigraphy and sedimentary environments of the Lower Cretaceous in portions of Las Animas, Otero, and Bent counties, Colorado. D, 1974, Colorado School of Mines. 211 p.

Taylor, B. Origin and significance of C-O-H fluids in the formation of Ca-Fe-Si skarn, Osgood Mountains, Humboldt County, Nevada. D, 1976, Stanford University. 306 p.

Taylor, Bobbey Ben. Geologic aspects of surface and ground water in North Louisiana. M, 1964, Louisiana Tech University.

Taylor, Brian. On the tectonic evolution of marginal basins in northern Melanesia and the South China Sea. D, 1982, Columbia University, Teachers College. 200 p.

Taylor, Brian Burke. Prediction of settlement of strip foundations on layered soils. D, 1983, University of Waterloo. 265 p.

Taylor, Bruce. Heat flow studies and geothermal exploration in western Trans-Pecos Texas. D, 1981, University of Texas at El Paso. 339 p.

Taylor, Bruce E. Precambrian geology of the Byers Peak area, central Front Range, Colorado. M, 1971, Colorado School of Mines. 118 p.

Taylor, Carl Alvin, Jr. Geology and gravity of the northern end of the Inner Piedmont and Charlotte Belt, Davie County Mesozoic basin, and surrounding area, central North Carolina Piedmont. M, 1982, University of North Carolina, Chapel Hill. 142 p.

Taylor, Carl H. Sulfur gases produced by the decomposition of sulfide minerals; application to metallic minerals exploration. M, 1981, University of Michigan.

Taylor, Charles Mosser. Mineralogical applications of the electron beam microprobe in the study of a gold silver deposit, Knob Hill Mine, Republic, Washington. D, 1968, Stanford University. 280 p.

Taylor, Colin Hubert. Sediment discharge from the Eaton River basin (Quebec) during spring runoff. D, 1972, McGill University.

Taylor, D. G. Biostratigraphy of the type Weberg Member, Snowshoe Formation, Grant County, Oregon. M, 1977, Portland State University. 183 p.

Taylor, D. J. Reflection seismic study in the Laramie anorthosite complex, southern Laramie Range, Wyoming. M, 1978, University of Wyoming. 170 p.

Taylor, Daniel T. Geology and mineralization of the Atlanta mining district and adjacent areas, Elmore County, Idaho. M, 1986, University of Idaho. 145 p.

Taylor, David Gene. Jurassic (Bajocian) ammonite biostratigraphy and macroinvertebrate paleoecology of the Snowshoe Formation, East-central Oregon. D, 1981, University of California, Berkeley. 340 p.

Taylor, David N. Bone histology of fossil reptiles. M, 1941, University of California, Berkeley. 317 p.

Taylor, David O. A new shale in the Chicago area and its stratigraphic position. M, 1930, Northwestern University.

Taylor, David R. Sedimentology of the Moosebar Tongue and bounding strata, Lower Cretaceous Blairmore Group, south-central foothills, Alberta. M, 1981, McMaster University. 239 p.

Taylor, David W. A., Jr. The structure of the crust and upper mantle near Midway Island. M, 1983, New Mexico State University, Las Cruces. 86 p.

Taylor, David Winship. Survey of the paleo-obiogeographic relationships of North American angiosperms from the Upper Cretaceous and Paleogene. D, 1987, University of Connecticut. 338 p.

Taylor, David Wyatt Aiken, Jr. Source studies over a wide range in earthquake magnitude. D, 1988, Virginia Polytechnic Institute and State University. 106 p.

Taylor, Dennis Ritch. Stratigraphy of the Gomez Park area, Trans-Pecos, Texas. M, 1952, University of Texas, Austin.

Taylor, Denver Walter. Stratigraphy and fauna of (Devonian) Detroit River Group (Michigan, Ontario, Ohio). M, 1951, University of Michigan.

Taylor, Don Ray. Geology, Gholson and Aquilla quadrangles, McLennan and Hill counties, Texas. M, 1962, Baylor University. 82 p.

Taylor, Donald Alfred. Detailed stratigraphy of the Edmonton District. M, 1934, University of Alberta. 85 p.

Taylor, Donald Richard. Faulting as related to lead-zinc mineralization, west end of the Milliken Mine, Reynolds County, Missouri. M, 1983, University of Missouri, Rolla. 82 p.

Taylor, Donald S. Paleoecology of the Choctawhatchee deposits at Jackson Bluff, Florida. M, 1962, University of Houston.

Taylor, Dorothy Ann. The geology of the Gunnison Plateau Front in the vicinity of Wales, Utah. M, 1948, Ohio State University.

Taylor, Doug. Utility of Landsat SIR-A and SIR-B imagery for geologic and hydrologic analysis in northern Africa. M, 1988, University of Arkansas, Fayetteville.

Taylor, Douglas William. Carbonate petrology and depositional environments of the limestone member of the Carmel Formation near Carmel Junction, Kane County, Utah. M, 1981, Brigham Young University.

Taylor, Dwight Willard. Late Cenozoic paleoecology and molluscan faunas of the High Plains. D, 1957, University of California, Berkeley. 351 p.

Taylor, Dwight Willard. Some late Cenozoic molluscan faunas from Kansas and Nebraska. M, 1954, University of California, Berkeley. 188 p.

Taylor, Earle Frederick. General geology of the Glendevey area, Colorado. M, 1936, University of Iowa. 49 p.

Taylor, Edward Drummond. Crystallographic studies. M, 1940, Universite Laval.

Taylor, Edward M. Geology of the Clarno Basin, Mitchell Quadrangle, Oregon. M, 1960, Oregon State University. 173 p.

Taylor, Edward Morgan. Recent vulcanism between Three Fingered Jack and North Sister, Oregon Cascade Range. D, 1967, Washington State University. 84 p.

Taylor, Elliott. Oceanic sedimentation and geotechnical stratigraphy; hemipelagic carbonates and red clays. D, 1984, Texas A&M University. 158 p.

Taylor, Emily Mahealani. Impact of time and climate on Quaternary soils in the Yucca Mountain area of the Nevada Test Site. M, 1986, University of Colorado. 217 p.

Taylor, F. C. A petrographic study of the gabbro-diabase series of the coast of Labrador, south of Hamilton Inlet. M, 1951, McGill University.

Taylor, F. C. Petrology of the Serpentine Belt, Matheson District (Ontario, Canada). D, 1953, McGill University.

Taylor, F. C. The petrology of serpentine bodies in the Matheson District, Ontario. D, 1955, McGill University.

Taylor, Frederick W. Depositional environment of the Corbin Sandstone Member of the Lee Formation (Lower Pennsylvanian) of eastern Kentucky. M, 1977, Eastern Kentucky University. 63 p.

Taylor, Frederick Wiley, Jr. Quaternary tectonic and sea-level history, Tonga and Fiji, Southwest Pacific. D, 1978, Cornell University. 366 p.

Taylor, G. Jeffrey. Electron microprobe study of the metallic minerals in ordinary chondrites. D, 1970, Rice University. 133 p.

Taylor, G. Jeffrey. On the thermal history of chondrites. M, 1968, Rice University. 16 p.

Taylor, G. Lynn. Analysis of iron in layer silicates by Mossbauer spectroscopy. M, 1967, Michigan Technological University. 88 p.

Taylor, Garvin Lawrence. Differentiation and correlation of Black Hills Precambrian granites. D, 1934, University of Iowa. 36 p.

Taylor, Garvin Lawrence. Geology of the Southwest Kansas gas area. M, 1931, University of Iowa. 69 p.

Taylor, Gary Allen. Upper Ordovician graptolites of the Cincinnati, Ohio area. M, 1974, University of Cincinnati. 144 p.

Taylor, Gary Edward. Depositional environments of the Eocene through Oligocene Sespe Formation in the northern Simi Valley area, Ventura County, Southern California. M, 1984, California State University, Northridge. 74 p.

Taylor, George F. Geology of the Merced Hills with a section on the radioactivity of the oils and waters (California). M, 1931, California Institute of Technology. 28 p.

Taylor, George F. Quaternary fault structure of the Bishop region, east-central California. D, 1933, California Institute of Technology. 216 p.

Taylor, George Johnston. A detailed gravity survey of Alachua County, Florida, with structural and magnetic interpretation. M, 1975, University of Florida. 82 p.

Taylor, Gerald Lynn. Stratigraphy, sedimentology and sulphide mineralization of the Kona Dolomite (Lower Middle Precambrian, Marquette Trough, Michigan) (Parts I-III). D, 1972, Michigan Technological University. 112 p.

Taylor, Gilbert D. The geology of the Limon area of Costa Rica. D, 1975, Louisiana State University. 147 p.

Taylor, Gilbert D., Jr. Conodonts from the Mansfield and Brazil formations (Morrowan) of the Illinois Basin. M, 1971, University of Illinois, Urbana. 75 p.

Taylor, Gordon Cosmos. The geology of the Island of Margarita, Venezuela. D, 1960, Princeton University. 144 p.

Taylor, Gregson William. Structural, sedimentological, and petrological setting of the Lang-Halsted Sequence and Duncan Peak Chert, lower Shoo Fly Complex, northern Sierra Nevada, California. M, 1986, San Diego State University. 110 p.

Taylor, Harold R. Hydrogeology of an Iowa City sanitary landfill; a preliminary study. M, 1972, University of Iowa. 74 p.

Taylor, Harry L. Geology and mineralogy of Coldstream copper mines. M, 1957, University of Minnesota, Minneapolis. 49 p.

Taylor, Henry B. Study of some ores from Austin, Nevada. M, 1912, Columbia University, Teachers College.

Taylor, Henry Clyde. Melting relations in the system $MgO-Al_2O_3-SiO_2$ at 15 kilobars. M, 1971, Southern Methodist University. 92 p.

Taylor, Henry Clyde. Silicate mineral chemistry of selected Apollo 15, 16 and 17 soils. D, 1973, University of Texas at Dallas. 272 p.

Taylor, Henry G. Geology of a portion of the Kensington Quadrangle, Northwest Georgia. M, 1951, Emory University. 122 p.

Taylor, Hugh Pettingill, Jr. O^{18}/O^{16} ratios in coexisting minerals of igneous and metamorphic rocks. D, 1959, California Institute of Technology. 171 p.

Taylor, Ian Edward. Middle and upper Keweenawan siliciclastics of the Lake Superior Basin. D, 1987, McMaster University. 319 p.

Taylor, Ira Daniel. Pennsylvanian stratigraphy of Daviess County, Indiana. M, 1952, Indiana University, Bloomington. 29 p.

Taylor, J. R. Structural geology of the Sickle and Wasekwan groups, Mynarski Lakes, Manitoba. M, 1978, University of Manitoba. 60 p.

Taylor, J. W. Paleozoic fossils of a geological section in Wyoming. M, 1897, University of Wyoming.

Taylor, Jack Allen. The Lower Pennsylvanian Primrose Sandstone of Oklahoma. M, 1951, University of Oklahoma. 140 p.

Taylor, James Barton. The geology of the Indian Springs region, Clark County, Nevada. M, 1963, University of California, Los Angeles.

Taylor, James C. Milton. Petrography of some Ordovician, Silurian, and Mississippian limestones as related to their engineering properties. M, 1975, Memphis State University.

Taylor, James Carlton. Geology of Camp Pendleton area, San Diego County (California). M, 1950, Pomona College.

Taylor, James Carlton. The petrology of some subsurface Miocene sediments from the Brea oil field and vicinity, Southern California. D, 1953, Pomona College.

Taylor, James Grover V. Distribution of hydrocarbon fluids and their compositions in volatile oil reservoirs during depletion. D, 1966, Stanford University. 101 p.

Taylor, James Michael. Geology of the Drum Mountains; Millard and Juab counties, Utah. M, 1979, Brigham Young University.

Taylor, James Michael. Geology of the Sterling Quadrangle, Sanpete County, Uath. M, 1979, Brigham Young University.

Taylor, James Rulie. The geology of east-central Boone County, Missouri. M, 1950, University of Missouri, Columbia.

Taylor, Jane M. Pore space reduction in sandstones. M, 1949, University of Cincinnati. 28 p.

Taylor, John Dallas. Geology of the Elkins Quadrangle, Washington County, Arkansas. M, 1964, University of Arkansas, Fayetteville.

Taylor, John Dallas. Upper Mississippian (Chesteran) ammonoids from the Pitkin Formation in Arkansas. D, 1972, University of Iowa. 146 p.

Taylor, John Felix. Biostratigraphy and litho-stratigraphy of the Cambrian-Ordovician boundary interval, Texas and Oklahoma, USA. D, 1984, University of Missouri, Columbia. 248 p.

Taylor, Johnnie D. Photo-linear analysis as an exploration technique for part of the Chandalar and Wiseman quadrangles, Brooks Range, Alaska. M, 1974, University of Arizona.

Taylor, Joseph K. Geology of the Nevada scheelite mine, Mineral County, Nevada. M, 1982, University of Nevada. 94 p.

Taylor, Julie Marie. The composition of island arc detritus from Southwest Japan; provenance and tectonic history. M, 1982, Michigan State University. 42 p.

Taylor, Karen S. Lithologies and distribution of clasts in the Elephant Moraine, Allan Hills, South Victoria Land, Antarctica. M, 1988, Kent State University, Kent. 121 p.

Taylor, Keith R. Structural and stratigraphic analysis of Precambrian rocks, south-central Ladron Mountains, Socorro County, New Mexico. M, 1986, University of New Mexico. 166 p.

Taylor, Kendrick C., Jr. Sonic logging at Dye-3, Greenland. M, 1982, University of Wisconsin-Madison. 64 p.

Taylor, Kendrick Cashman. Application of borehole geophysical methods to shallow groundwater investigations. D, 1987, University of Nevada. 194 p.

Taylor, L. B. Well interference in the South Silica Field. M, 1951, University of Missouri, Rolla.

Taylor, LaRon. Biogeochemical exploration for Cu, Pb, and Zn mineral deposits, using juniper and sagebrush, Dugway Range, Utah. M, 1975, Brigham Young University. 85 p.

Taylor, Larry E. Bedrock geology and its influence on groundwater resources in the Hedgesville and Williamsport 7 1/2 minute Quadrangle, Berkeley County, West Virginia. M, 1974, University of Toledo. 82 p.

Taylor, Lawrence A. The system Ag-Fe-S; phase equilibria and geologic applications. D, 1968, Lehigh University. 186 p.

Taylor, Lawrence D. Ice petrofabric studies of Angiussaq Lake, Northwest Greenland. M, 1958, Dartmouth College.

Taylor, Lawrence Dow. Ice structures, Burroughs Glacier, Southeast Alaska. D, 1962, Ohio State University. 227 p.

Taylor, Lisa Gale. Geological and geochemical evolution of enargite-bearing veins at the Togo Mine, Central City mining district, Colorado. M, 1985, Michigan Technological University. 73 p.

Taylor, Louis. The conodonts and age of the Welden Limestone. M, 1941, University of Missouri, Columbia.

Taylor, Louis Henry. Geochronology of the Torrejonian sediments, Nacimiento Formation, San Juan Basin, New Mexico. M, 1977, University of Arizona.

Taylor, Louis Henry. Review of Torrejonian mammals from the San Juan Basin, New Mexico. D, 1984, University of Arizona. 571 p.

Taylor, M. X-ray radial distribution studies of silicate mineral glasses. D, 1978, Stanford University. 223 p.

Taylor, M. W. Mineral reactions in the tuffaceous sediments at Teels Marsh, Nevada. M, 1978, University of Wyoming. 90 p.

Taylor, Mark. Fluid inclusion evidence for fluid mixing, Mascot-Jefferson City zinc district, Tennessee. M, 1982, University of Michigan.

Taylor, Melvin Hall. Siluro-Devonian strata in central Kansas. D, 1947, Columbia University, Teachers College.

Taylor, Michael Evan. Biostratigraphy of the upper Cambrian (upper Franconian-Tempealeauan stages) in the Central Great Basin, Nevada and Utah. D, 1971, University of California, Berkeley. 428 p.

Taylor, Michael Evan. The Lower Devonian Water Canyon Formation of northeastern Utah. M, 1963, Utah State University. 63 p.

Taylor, Michael Francis. A survey of soil freezing on the east side of the Sierra Nevada. M, 1969, University of Nevada. 187 p.

Taylor, Omer James. Correlation of volcanic rocks in Santa Cruz County, Arizona. M, 1960, University of Arizona.

Taylor, P. T. Effect of sample size on determination of thermal conductivity by modified hot-wire method. M, 1962, Pennsylvania State University, University Park. 52 p.

Taylor, Pamela Sue. Precipitation of calcium-magnesium carbonates. M, 1977, Wright State University. 59 p.

Taylor, Patrick Timothy. Interpretation of the heat flow pattern from the Sumatra Trench. D, 1965, Stanford University. 69 p.

Taylor, Paul Scott. Mineral variations in the silver veins of Guanajuato, Mexico. D, 1971, Dartmouth College. 139 p.

Taylor, Paul Scott. Soluble material on volcanic ash. M, 1969, Dartmouth College. 77 p.

Taylor, Ralph E. Origin of the cap rock of Louisiana salt domes. D, 1938, Louisiana State University.

Taylor, Randall E. Sulfur forms, trace elements, and mineralogy of the Danville Coal Member (VII), Dugger Formation (Pennsylvanian) in southwestern Indiana. M, 1986, Indiana University, Bloomington. 87 p.

Taylor, Raymond John, Jr. The relation of forest vegetation to soils and geology in the Gulf Coastal Plain in Oklahoma. D, 1967, University of Oklahoma. 57 p.

Taylor, Read. Stratigraphic and structural analysis and geologic data processing of remote sensing imagery in the Nipomo, eastern Santa Lucia Mountain area. M, 1988, University of Southern California.

Taylor, Richard Bartlette. Investigations on the Duluth Gabbro. M, 1953, University of Minnesota, Minneapolis. 54 p.

Taylor, Richard Bartlette. Petrology and petrography of the Duluth gabbro complex near Duluth, Minnesota. D, 1955, University of Minnesota, Minneapolis. 108 p.

Taylor, Richard David. Geology of the South Drew Field, Ouachita Parish, Louisiana. M, 1979, Louisiana Tech University.

Taylor, Richard H. Planktonic foraminiferal biostratigraphy of the Demopolis Formation (Campanian/Maastrichtian) in Lowndes and Oktibbeha counties, Mississippi. M, 1985, Mississippi State University. 134 p.

Taylor, Richard Spence. A study of some high-latitude patterned-ground features. D, 1956, University of Minnesota, Minneapolis. 299 p.

Taylor, Richard Spence. Geologic reconnaissance of the middle Back River region, District of Keewatin, N.W.T., Canada. M, 1955, University of Minnesota, Minneapolis. 85 p.

Taylor, Ricky Joe. Petrography of the layered series, Saddle Mountain Quadrangle, eastern Wichita Mountains, Oklahoma. M, 1978, University of Texas, Arlington. 66 p.

Taylor, Robert Clark. The geology of the Foraker area, Osage County, Oklahoma. M, 1953, University of Oklahoma. 108 p.

Taylor, Robert F. Areal geology of the Brownfield Quadrangle, Illinois. M, 1967, Southern Illinois University, Carbondale. 103 p.

Taylor, Robert Joseph. Manganese geochemistry in Galveston Bay sediment. D, 1987, Texas A&M University. 275 p.

Taylor, Robert Kruse. Geology of the central portion of the Rumsey Quadrangle, California. M, 1955, University of California, Berkeley. 78 p.

Taylor, Robert Warren. Remote determinations of Earth structure from relative event analysis with applications to the Mid-Atlantic Ridge. D, 1972, Pennsylvania State University, University Park. 200 p.

Taylor, Robert Wesley. An experimental study of the system FeO-Fe$_2$O$_3$-TiO$_2$ and its bearing on mineralogical problems. D, 1961, Pennsylvania State University, University Park. 94 p.

Taylor, Robert Wesley. Phase equilibria in the system FeO-Fe$_2$O$_3$-TiO$_2$. M, 1958, University of California, Berkeley. 63 p.

Taylor, Roger Loren. Cenozoic volcanism, block faulting, and erosion in the northern White Mountains, Nevada. M, 1965, University of California, Berkeley. 99 p.

Taylor, Ronald S. The "Warsaw"-Salem formations in Taylor and Green counties of Kentucky. M, 1962, University of Kentucky. 53 p.

Taylor, Ronald Shearer. Paleoecology of ostracodes from the Luman Tongue and Tipton Member (early Eocene) of the Green River Formation, Wyoming. D, 1972, University of Kansas. 96 p.

Taylor, Roy Owen. The stratigraphy and structure of the northeast quarter of the Fordland Quadrangle, Missouri. M, 1958, University of Missouri, Columbia.

Taylor, Russell N. Geology of the Cuba Sandstone. M, 1950, Lehigh University.

Taylor, S. J. Carbonate petrology of the Lower Permian Brown Dolomite, Carson County, Texas. M, 1978, West Texas State University. 56 p.

Taylor, Samuel Guy, Jr. Gravity investigation of the southern San Francisco Bay area, California. D, 1957, Stanford University. 159 p.

Taylor, Scott H. Pleistocene to Recent benthic foraminifera from DSDP Site 207, Lord Howe Rise, Tasman Sea. M, 1985, University of California, Los Angeles. 126 p.

Taylor, Sharon I. A simple inversion method to aid in the interpretation of borehole TDEM data. M, 1985, Queen's University. 105 p.

Taylor, Sheryl M. The geochemistry and petrology of Hamilton Mounds, Wisconsin. M, 1983, Northern Illinois University. 74 p.

Taylor, Sidney William. Geology of Marystown map sheet (E/2), Burin Peninsula, southeastern Newfoundland. M, 1976, Memorial University of Newfoundland. 165 p.

Taylor, Simon D. A fluid inclusion study of the Wheal Remfry Breccia, Cornwall, U.K. M, 1982, University of Windsor. 144 p.

Taylor, Stanley D. The solid solution relationship among anglesite (PbSO₄), barite (BaSO₄), and celestite (SrSO₄). M, 1962, Miami University (Ohio). 78 p.

Taylor, Stephen Bernard. Stratigraphy, sedimentology, and paleogeography of the Swauk Formation in the Liberty area, central Cascades, Washington. M, 1985, Washington State University. 199 p.

Taylor, Steven. The mineral industries of Latin America. M, 1983, University of Texas, Austin.

Taylor, Steven Renold. Crust and upper mantle structure of the northeastern United States. D, 1980, Massachusetts Institute of Technology. 288 p.

Taylor, Stuart Ross. Geochemistry of some New Zealand igneous and metamorphic rocks. D, 1954, Indiana University, Bloomington. 94 p.

Taylor, Teresa Ann. Effects of temperature and particle size on a model for estimating the rheological parameters of mudflows. M, 1984, Washington State University. 84 p.

Taylor, Terry Dean. Interpretation of large-scale cross-strata in a borehole; a computer simulation model. M, 1986, Pennsylvania State University, University Park. 171 p.

Taylor, Terry Lee. The basalt stratigraphy and structure of the Saddle Mountains of south-central Washington. M, 1976, Washington State University. 116 p.

Taylor, Theodore Warren. A petrological and geochemical study of the O.K. copper-molybdenum deposit, Beaver County, Utah. M, 1983, Lehigh University.

Taylor, Thomas Garrett. The silica refractories of Pennsylvania. M, 1922, Pennsylvania State University, University Park.

Taylor, Thomas Raymond. Petrographic and geochemical characteristics of dolomite types and the origin of ferroan dolomite in the Trenton Formation, Michigan Basin. D, 1982, Michigan State University. 75 p.

Taylor, Thomas Raymond. The origin of composite dikes at Mount Desert Island, Maine; an example of coexisting acidic and basic magmas. M, 1979, Michigan State University. 72 p.

Taylor, Vernon A. An experimental study of some replacement processes. D, 1965, Florida State University. 109 p.

Taylor, Vernon A. Geology of the Columbia North quadrangle. M, 1949, University of South Carolina. 27 p.

Taylor, W. L. W. Copper-nickel sulphide deposits of the Werner Lake, Ontario and Bird River, Manitoba areas. M, 1950, McGill University. 65 p.

Taylor, Wallace K. Study of structural relationship of the Riley County, Kansas, intrusions to the Abilene Arch. M, 1950, Kansas State University. 25 p.

Taylor, Waller Eugene. The Cypress (Jackson) Formation of Warrick, Spencer, Perry, Dubois and Crawford counties, Indiana. M, 1951, Indiana University, Bloomington. 20 p.

Taylor, Wanda Jean. Superposition of thin-skinned normal faulting on Sevier Orogenic Belt thrusts, northern Mormon Mountains, Lincoln County, Nevada. M, 1984, Syracuse University.

Taylor, Warren LeRoy. The geology of Ihman East oil field. M, 1949, University of Illinois, Urbana.

Taylor, William Harlan. Studies on Recent ostracodes in relation to the taxonomy of fossil ostracodes. M, 1931, Stanford University. 38 p.

Taylor, William R. Geology and geochemistry of a uranium-rich area in southwestern Newfoundland. M, 1971, Memorial University of Newfoundland. 60 p.

Tays, Gerald. The stratigraphy, structure, and areal geology of parts of the Waterville and Skowhegan quadrangles, Maine. M, 1965, University of Maine. 88 p.

Tazelaar, James Fulton. A geological and geophysical survey of the Stony Point, Virginia, iron-copper deposit. M, 1958, University of Virginia. 53 p.

Tchakerian, Vatche Panos. The Pismo coastal dune complex, California; geomorphology and textural relationships. M, 1983, University of California, Los Angeles. 107 p.

Tchinda, Fidele. Trough cross-stratification geometry and methods for determination of paleoflow direction. M, 1988, Pennsylvania State University, University Park. 141 p.

Tchombe, Laurence Puande. Hydraulic interpretation of grain-size distributions; River South Esk, Glen Clova, Scotland. M, 1981, SUNY at Binghamton. 73 p.

Teagle, John. The stratigraphy of the Permian and Triassic formation in Guadalupe County, New Mexico. M, 1932, University of Texas, Austin.

Teague, Alan Gaither. Focal mechanisms for eastern Tennessee earthquakes, 1981-1983. M, 1985, Virginia Polytechnic Institute and State University.

Teague, Lisa Shomura. Diagenesis of the Grande Ronde basalt formation, Pasco Basin, Washington. M, 1980, University of California, Berkeley. 105 p.

Teague, Richard Darnell. Depositional environment and diagenesis of the Viola Group (Ordovician), Criner Hills and Marietta Basin, southern Oklahoma. M, 1985, Oklahoma State University. 117 p.

Teal, Lewis W. Geology and petrology of the Chispa Mountain Quadrangle and vicinity, Culberson and Jeff Davis counties, Texas. M, 1979, University of Texas at El Paso.

Teal, Philip Rae. Stratigraphy, sedimentology, volcanology and development of the Archean Manitou Group, northwestern Ontario, Canada. D, 1979, McMaster University. 291 p.

Teal, Suzanne Elizabeth. The geology and petrology of the Firesand River carbonatite complex, northwestern Ontario. M, 1979, McMaster University. 102 p.

Tearpock, Daniel John. Structural analysis of the Wissahickon Formation along the northern section of the Wissahickon Creek valley, Philadelphia, Pa. M, 1977, Temple University.

Teas, Livingstone Pierson. The relation of sphalerite to other sulphides in ores. M, 1917, Cornell University.

Tebbutt, Gordon E. Lithogenesis of a distinctive Permian carbonate fabric, Big Horn, and Washakie counties, Wyoming. M, 1964, University of Wyoming. 81 p.

Tebbutt, Gordon Edward. Diagenesis of Pleistocene limestone on Ambergris Cay, British Honduras. D, 1967, Rice University. 138 p.

Tebedge, Sleshi. Fossil Suidae from the middle Awash Valley, Afar Depression, Ethiopia. M, 1980, University of Texas, Austin.

Tebedge, Sleshi. Paleontology and paleoecology of the Pleistocene mammalian fauna of Dark Canyon Cave, Eddy County, New Mexico. D, 1988, University of Texas, Austin. 301 p.

Tedder, Kenneth H. Slope stability in the North Saskatchewan River valley. M, 1986, University of Alberta. 271 p.

Tedesco, Lawrence. Artificial compaction of modern lime mud from southern Florida. D, 1981, University of Missouri, Columbia. 147 p.

Tedesco, Steve A. Selected trace elements in the rocks above the strippable coal seams in southern Illinois. M, 1980, Southern Illinois University, Carbondale. 192 p.

Tedford, Frederick J. Depositional environment of upper Wilcox sandstones, Northeast Thompsonville Field, Jim Hogg and Webb counties, Texas. M, 1977, Texas A&M University. 97 p.

Tedford, Richard Hall. The fossil Macropodidae from Lake Menindee, New South Wales. D, 1960, University of California, Berkeley. 165 p.

Tedlie, William Donald. Acid rocks associated an intrusive complex, Coppermine River Area. M, 1960, University of British Columbia.

Tedrahn, David C. Rayleigh wave dispersion and its relation to porosity. M, 1980, University of Wisconsin-Milwaukee.

Tedrick, Patricia Ann. Direction of longshore drift, littoral current velocities, and sand migration along the Pinellas County, Florida coastline from 1925 to 1970, with prediction of future trends. M, 1972, University of South Florida, Tampa. 115 p.

Tedrick, Robert Lowell. Ordovician geology of the Prophet River map-area, British Columbia. M, 1962, University of Alberta. 30 p.

Tedrow, Harvey Louis. Report on the Metates Mining Company. M, 1922, University of Missouri, Rolla.

Tee, Deebari Porobe. Three-dimensional finite element modeling of groundwater flow. D, 1983, Oklahoma State University. 183 p.

Tee, K.-T. Tide-induced residual current in Minas Channel and Minas Basin. D, 1975, Dalhousie University.

Teel, Derrick Brehm. Sedimentology and tectonic significance of the Copper Basin Formation in the eastern Whipple Mountains, San Bernardino County, California. M, 1983, San Diego State University. 128 p.

Teepen, Kristina L. Sulfide mineralization in Washington Basin, Idaho. M, 1985, Kent State University, Kent. 84 p.

Teerman, Stanley C. A petrographic study of coals from the Hanna coal field, Wyoming. M, 1983, Southern Illinois University, Carbondale. 167 p.

Teeter, James Wallace. The distribution of Recent marine ostracodes from British Honduras. D, 1966, Rice University. 221 p.

Teeter, James Wallis. The ostracod fauna of the Eramosa Member of the Lockport Formation (Middle Silurian). M, 1962, McMaster University. 89 p.

Teeter, Kathy. A neutron powder diffraction study of ordering in Fe-Mn olivines. M, 1988, McMaster University. 109 p.

Teflian, Samuel. A study of the Vanport Limestone in the Oak Hill area, southern Ohio. M, 1952, Ohio State University.

Tegland, Nellie May. Fauna of the type Blakely, upper Oligocene of Washington, with special reference to the genera of Galeodea. D, 1930, University of California, Berkeley. 158 p.

Tegland, Nellie May. Fossil fauna from Restoration Point, Bainbridge Island. M, 1924, University of Washington. 32 p.

Tehan, Robert E. The stratigraphy and sedimentary petrology of the Pitkin (Chesterian) Limestone, Washington and Crawford counties, Arkansas. M, 1977, University of Arkansas, Fayetteville.

Teichert, John A. Geology of the southern Stansbury Range, Tooele County, Utah. M, 1958, University of Utah. 79 p.

Teichmann, Friedrich. The geochemistry of chlorites as a function of metamorphic grade, mineral assemblage and bulk composition in the Rangeley and Rumford areas, western Maine. M, 1988, University of Maine. 175 p.

Teichmann, Warren James. The Alcova Limestone of east-central Wyoming. M, 1956, Ohio State University.

Teifke, Robert H. Comparative petrology of uppermost Cretaceous and Tertiary (non-marine) rocks exposed along the McLeod River and its tributaries, Alberta (Canada). M, 1972, University of Kansas. 98 p.

Teir, Lennart T. Rhyolitic epithermal mineralization of the Boulder Batholith, Montana. M, 1941, Oregon State University. 83 p.

Teis, Maurice R. Pennsylvanian ostracods from the Brownwood Shale of Wise County, Texas. M, 1931, University of Chicago. 50 p.

Teissere, Ronald Franklin. Paleomagnetic investigation of the San Marcos Gabbro, Southern California. M, 1968, University of California, Riverside. 57 p.

Teitsma, A. Fission xenon dating. D, 1975, McMaster University.

Teitz, Martin W. Late Proterozoic Yellowhead and Astoria carbonate platforms, southwest of Jasper, Alberta. M, 1985, McGill University. 107 p.

Tejada, Jorge A. Portugal *see* Portugal Tejada, Jorge A.

Tejirian, Haig G. Trend analysis of aeromagnetic data in the Caledonia area of southern New Brunswick. M, 1974, University of New Brunswick.

Tekbali, Ali Omar. Aspects of palynostratigraphy in western Libya Silurian-Lower Cretaceous; Part 1, Paleozoic palynology; Part 2, Mesozoic palynology. M, 1987, University of California, Davis. 504 p.

Tekiner, Yasar. Determination of the parameters needed for some statistical methods of well logging (2 volumes; vol. 2, Appendix and data). M, 1970, Colorado School of Mines.

Tekneci, Zeki. Extractive process for copper. M, 1977, Stanford University. 125 p.

Tekverk, Raymond W. Glacial-marine clays in the Norridgewook area, Maine. M, 1974, Boston University. 69 p.

Teleki, Paul Geza. Differentiation of materials formerly assigned to the Alachua Formation (Pliocene, northern Florida). M, 1966, University of Florida. 101 p.

Teleki, Paul Geza. Measurement of boundary shear in oscillating flow in presence of roughness. D, 1970, Louisiana State University. 213 p.

Tella, Subhas. Microstructures and preferred orientations of quartz in tectonites of different metamorphic grade. D, 1980, McGill University. 163 p.

Telle, Whitney R. Geology of the Lay Dam Formation, Chilton County, Alabama. M, 1983, University of Alabama. 93 p.

Tellefsen, Mark James. The preparation and characterization of several oxide systems with the spinel and olivine structures. D, 1983, Brown University. 106 p.

Teller, Jacob Abe. Early evolution of planet Earth. D, 1973, Yeshiva University. 87 p.

Teller, James T. The glacial geology of Clinton County, Ohio. M, 1964, Ohio State University.

Teller, James Tobias. Early Pleistocene glaciation and drainage of southwest Ohio, southeast Indiana and northern Kentucky. D, 1970, University of Cincinnati. 115 p.

Tellez, J. R. An analysis of the effects of target attitude and texture on the shuttle multispectral infrared radiometer, Sierra Vieja. M, 1988, Sul Ross State University.

Tellier, Anthony Harrison. Geothermal waters of Arizona. M, 1973, Arizona State University. 30 p.

Tellier, Kathleen E. Calculated one-dimensional X-ray diffraction patterns of illite/smectite as fundamental particle aggregates; implications for the structure of illite/smectite and for the smectite-to-illite transition. D, 1988, Dartmouth College. 159 p.

Tellier, Kathleen E. Clay mineralogy in the upper Miocene Monterey and Etchegoin formations, southeastern Lost Hills oil field, San Joaquin Basin, California. M, 1985, Dartmouth College. 64 p.

Telljohann, Eric P. Authigenic clays of the Cretaceous Muddy Sandstone, Powder River basin, Wyoming. M, 1986, Bowling Green State University. 46 p.

Teme, So-Ngo Clifford. Physical modelling in rock slope stability evaluations. D, 1982, Purdue University. 458 p.

Temeng, Kwaku Ofori. Pressure distributions in asymmetric circular systems. M, 1982, Stanford University. 58 p.

Tempelman Kluit, Dirk Jacob. Geology of the Haggart Creek-Dublin Gulch area, Mayo District, Yukon. M, 1964, University of British Columbia.

Tempelman-Kluit, Dirk J. The stratigraphy and structure of the Keno Hill Quartzite (Precambrian), Yukon. D, 1966, McGill University.

Temple, Carol I. Preliminary investigation; gabbro complexes of New England (Nahant, Massachusetts; Sugarloaf, Maine; and Pierce Pond, Maine. M, 1967, Boston University. 22 p.

Temple, D. M. G. Geology of the hydrothermal field at 26°N, Mid-Atlantic Ridge; interpretations from ocean-bottom photography. M, 1977, Texas A&M University.

Temple, Dennis Charles. Mount Ogden granite, (Precambrian), (west side of Wasatch range, east of Ogden, Utah). M, 1969, University of Utah. 30 p.

Temple, James M. Permian salt dissolution related to alkaline lake basins, southern High Plains, Texas. M, 1986, Texas Tech University. 80 p.

Temple, Peter G. Geology of Bathurst Island group, District of Franklin, Northwest Territories (Canada). D, 1965, Princeton University. 206 p.

Temple, Vernon J., Jr. Structural relations along the western end of the Arrowhead Fault, Muddy Mountains, Nevada. M, 1977, Texas A&M University. 92 p.

Temples, Tommy Joe. Stable isotope variations in foraminifera from eastern Caribbean cores; stratigraphic implications. M, 1978, University of Georgia.

Templeton, Bonnie Carolyn. The fruits and seeds of the Rancho La Brea Pleistocene deposits. D, 1964, Oregon State University. 224 p.

Templeton, Eugene C. The geology and stratigraphy of the San Jose Quadrangle, California. M, 1912, Stanford University. 84 p.

Templeton, George Daniel, III. Trace metal-organic matter interactions during early diagenesis in anoxic estuarine sediments. D, 1980, University of New Hampshire. 304 p.

Templeton, Justus Stevens. The geology of part of the Woosung Quadrangle. D, 1940, University of Illinois, Urbana.

Templeton, Terry R. A study of geophysical methods for shallow subsurface investigations in the Memphis, Tennessee area. M, 1974, Memphis State University.

Temudom, Ladda. Mechanical properties of Eastern gas shale and geological factors affecting these properties. M, 1980, Michigan Technological University. 84 p.

ten Brink, Marilyn Rae Buchholtz. Radioisotope mobility across the sediment water interface in the deep sea. D, 1987, Columbia University, Teachers College. 472 p.

Ten Brink, Norman Wayne. Holocene deleveling and glacial history between Sondre Stromfjord and the Greenland ice sheet, West Greenland. D, 1972, University of Washington. 191 p.

Ten Brink, Norman Wayne. Pleistocene geology of the Stillwater drainage and Beartooth Mountains near Nye, Montana. M, 1968, Franklin and Marshall College. 183 p.

Ten Brink, Uri S. Lithospheric flexure and Hawaiian volcanism; a multichannel seismic perspective. D, 1986, Columbia University, Teachers College. 225 p.

Ten Eyck, James R. An evaluation of Fourier analysis in determining depositional processes from grain-size distributions. M, 1980, Indiana University, Bloomington. 138 p.

Ten Have, Lewis Earl. Relationship of dolomite/limestone ratios to the structure and producing zones of the West Branch oil field, Ogemaw County, Michigan. M, 1979, Michigan State University. 100 p.

Tench, Robert Norman. The paleontology of some loess-like deposits of Story County, Iowa. M, 1955, Iowa State University of Science and Technology.

Tendall, Bruce Alan. Mineralogy and petrology of Precambrian ultramafic bodies from the Tobacco Root Mountains, Madison County, Montana. M, 1978, Indiana University, Bloomington. 127 p.

Teng, Hai-Chuan. Marble deposits and the marble industry of the Knoxville area. M, 1948, University of Tennessee, Knoxville. 148 p.

Teng, Hau Chong. A lithogeochemical study of the St. Lawrence Granite, Newfoundland. M, 1974, Memorial University of Newfoundland. 194 p.

Teng, Louis Suh-Yui. Seismic stratigraphic study of the California continental borderland basins; structure, stratigraphy, and sedimentation. D, 1985, University of Southern California.

Teng, Ray Tsao Dah. Determination of osmium isotopes in meteoritic and crustal samples with accelerator mass spectrometry. M, 1986, University of Rochester. 104 p.

Teng, Ta-Liang. Body-wave and earthquake source studies. D, 1966, California Institute of Technology. 205 p.

Teng, William Ling. Remote sensing for landforms and soils in the arid Southwest United States. D, 1984, Cornell University. 404 p.

Tenharmsel, Ronald L. Investigation of late Tertiary to Recent movement along the bounding faults of the Shearer Graben within the Kentucky River fault system in southern Clark County, Kentucky. M, 1982, Eastern Kentucky University. 58 p.

Tennant, Harold Ellsworth. The Columbia Basin project. M, 1937, University of Washington. 78 p.

Tennant, Steven Hunter. Lithostratigraphy and depositional environments of the upper Dornick Hills Group (Lower Pennsylvanian) in the northern part of the Ardmore Basin, Oklahoma. M, 1981, University of Oklahoma. 291 p.

Tenney, Christopher M. Facies analysis of the Kindblade Formation, upper Arbuckle Group, southern Oklahoma. M, 1984, University of Oklahoma. 110 p.

Tennissen, Anthony. The distribution, occurrence, and origin of the gypsum in the (Silurian) Camillus Formation of the Syracuse East Quadrangle, New York. M, 1952, Syracuse University.

Tennissen, Anthony Cornelius. Mineralogy, petrography and ceramic properties of lower Cabaniss underclays in western Missouri. D, 1963, University of Missouri, Columbia. 201 p.

Tenny, Ralph Emil. Trace elements in the (Mississippian) Windsor limestones of Cape Breton Island (Nova Scotia, Canada). M, 1951, Massachusetts Institute of Technology. 37 p.

Tennyson, L. C. Subsurface hydrology and ion solution chemistry related to effluent disposal sites in the Missouri Ozarks. D, 1977, University of Missouri, Columbia. 195 p.

Tennyson, Marilyn Elizabeth. Petrology of Jurassic-Cretaceous volcanic sandstones, North Cascade Range, Washington and British Columbia. M, 1972, University of Washington. 31 p.

Tennyson, Marilyn Elizabeth. Stratigraphy, structure, and tectonic setting of Jurassic and Cretaceous sedimentary rocks in the west-central Methow-Pasayten area, northeastern Cascade Range, Washington and British Columbia. D, 1974, University of Washington. 112 p.

Tenorio, P. A. Hydrologic-data-network design concepts pertinent to ground-water resources inventory in the Hawaiian Islands. M, 1969, University of Hawaii. 118 p.

Teoman, Mahmet S. Groundwater contamination at Acton, Massachusetts. M, 1981, Boston University. 115 p.

Tepedino, Victor. A mineralogical study of sulphide ores from the Kidd Creek Mine, Timmins, Ontario, Canada. M, 1969, Michigan Technological University. 69 p.

Tepper, Dorothy H. Hydrogeologic setting and geochemistry of residual periglacial Pleistocene seawater in wells in Maine. M, 1980, University of Maine. 126 p.

Tepper, Jeffrey Hamilton. Petrology of the Chilliwack composite batholith, Mt. Sefrit area, North Cascades,

Washington. M, 1985, University of Washington. 102 p.

Tepperman, Mark. The stratigraphy of some Late Devonian sandstones in Braxton County, West Virginia. M, 1978, West Virginia University.

Terada, K. Determination of the mass ratio of the isotopes of carbon and hydrogen present in volcanic effluvia. M, 1954, University of Hawaii.

Terchunian, Aram V. Hen and Chickens Shoal, Delaware; evolution of a modern nearshore marine feature. M, 1984, University of Delaware, College of Marine Studies.

Terefenko, Robert. Variations in sodium concentration and hydrogeology of Farm River watershed, in Blue Hills region, near Braintree, Massachusetts. M, 1987, Boston University. 287 p.

Terich, Thomas Anthony. Bayocean Spit, Tillamook, Oregon; early economic development and erosion history. D, 1974, Oregon State University. 145 p.

Terlecky, Peter Michael Jr. The origin, stratigraphy and post depositional history of a late Pleistocene marl deposit near Caledonia, (northwestern) New York. D, 1969, University of Rochester. 161 p.

Terlecky, Peter Michael, Jr. The nature and distribution of oolites on the Atlantic continental shelf of the southeastern United States. M, 1967, Duke University. 46 p.

Terman, Marc L. Depositional and diagenetic history of the basal carbonate unit of the Upper Jurassic Todilto Formation, northwest New Mexico and southwest Colorado. M, 1984, Bowling Green State University. 119 p.

Terpening, John Nathan. Geology of part of the eastern Santa Monica Mountains. M, 1951, University of California, Los Angeles.

Terpstra, Paul. A computer simulation of the temporal development of geothermal energy systems with implications for the Salton Sea geothermal field of California. M, 1980, University of Illinois, Chicago.

Terracciano, Stephen Alan. Utility of ^{222}Rn as a tracer of ground water flow in near-shore sediments. M, 1986, SUNY at Stony Brook. 94 p.

Terranova, Thomas F. Multivariate analysis of geological, hydrological, and soil mechanical controls on slope stability in central Virginia. M, 1987, Southern Illinois University, Carbondale. 113 p.

Terrel, Ronald L. A classification and glossary of land forms and parent materials. M, 1961, Purdue University.

Terrell, Bruce C. The stratigraphy and structure of the Bethel-Boktukola synclinal area, Ouachita Mountains, Oklahoma. M, 1975, Northern Illinois University. 47 p.

Terrell, Don Michael. Trend and genesis of the Pennsylvanian Elgin Sandstone in the western part of northeastern Oklahoma. M, 1972, Oklahoma State University. 79 p.

Terrell, Forrest M. Lateral facies and paleoecology of Permian Elephant Canyon Formation, Grand Canyon, Utah. M, 1972, Brigham Young University. 44 p.

Terrell, John H. Separation of zircon and biotite from bentonite for absolute dating and possibilities for dating certain West Texas bentonites. M, 1960, Rice University. 44 p.

Terres, Donald Albert. Thermal water systems of the western Transverse Ranges, California. M, 1984, University of California, Santa Barbara. 126 p.

Terres, Richard Ralph. Paleomagnetism and tectonics of the central and eastern Transverse Ranges, Southern California. D, 1984, University of California, Santa Barbara. 364 p.

Terriere, Robert Theodore. Geology of Grosvenor Quadrangle, Texas, and petrology of some of its Pennsylvanian limestones. D, 1960, University of Texas, Austin. 207 p.

Terriere, Robert Theodore. The Mississippian sediments of the Bedford Quadrangle region. M, 1951, Pennsylvania State University, University Park. 103 p.

Terrill, Arthur C. The ores of Comstock Lode (Nevada). M, 1914, Columbia University, Teachers College.

Terry, Ann H. The geology of the Whipple Mountain Fault (Precambrian), southeastern California. M, 1972, [University of California, San Diego].

Terry, David Brian. Late Wisconsinan proglacial and ice-marginal sedimentation in the Susquehanna Valley, near Windsor, New York. M, 1985, SUNY at Binghamton. 120 p.

Terry, Ira James. The stratigraphic position of an Upper Cretaceous sand in Clay and Chickasaw counties, Mississippi. M, 1957, Mississippi State University. 21 p.

Terry, Judith Shoemaker. A description of a marine molluscan faunule from the Lobo Formation (Paleocene) of Fresno County, California. M, 1964, Stanford University. 48 p.

Terry, Judith Shoemaker. Taxonomy and distribution in space and time of the marine gastropod genera Argobuccinum, Fusitriton, and Priene (Family Cymatiidae). D, 1968, Stanford University. 212 p.

Terry, Orlyn Lee. The stratigraphy and paleontology of the Otter Formation, Montana. M, 1953, Washington State University. 98 p.

Terry, Owen W. Prospecting with a view toward mining operation. M, 1937, New Mexico Institute of Mining and Technology. 9 p.

Terry, Richard Dean. Bibliography of marine geology and oceanography, California coast. M, 1956, University of Southern California.

Terry, Richard Dean. Continental slopes of the world. D, 1965, University of Southern California. 930 p.

Terry, Steven H. Devonian-lowermost Mississippian lithostratigraphy and conodont biostratigraphy of the Batesville District, northeastern Arkansas. M, 1981, University of Arkansas, Fayetteville. 100 p.

Tertz, James. Geology of the Cottontown Quadrangle, Sumner County, Tennessee. M, 1956, Vanderbilt University.

Teruta, Yuko. A study of the distribution of some noble metals in the Merensky Horizon, Bushveld igneous complex. M, 1974, McMaster University. 137 p.

Terwilliger, F. Wells. The glacial geology and ground water resources of Van Buren County, Michigan. M, 1952, Michigan State University. 104 p.

Terwilliger, Valery Jane. Mechanical effects of chaparral disturbances on soil slip patterns in the Transverse Ranges of Southern California. D, 1988, University of California, Los Angeles. 218 p.

Terzakis, George N. Geomorphology and geology of Cape Cod; past, present and future. M, 1977, Pennsylvania State University, University Park. 288 p.

Teseneer, Ronald Lee. Selected trace elements in pyrite from the North Carolina slate belt. M, 1978, North Carolina State University. 60 p.

Teshima, Janet Marie. Iron meteorite parent bodies; inferences based on inclusion mineralogy and composition. M, 1983, Arizona State University. 91 p.

Teskey, D. J. Design of a semi-automated three-dimensional interpretation system for potential field data. D, 1978, McGill University.

Teskey, Maurice Forgie. The nature and origin of coking coals. D, 1934, University of Toronto.

Tesky, D. J. Computations of apparent resistivity profiles by numerical methods. M, 1968, University of Toronto.

Tesmer, Irving Howard. Stratigraphy and paleontology of the (Upper Devonian) lower Canadaway of Chautauqua County, New York. M, 1948, SUNY at Buffalo.

Tesmer, Irving Howard. Stratigraphy and paleontology of the Cherry Creek Quadrangle, New York. D, 1954, Syracuse University. 333 p.

Tesoriero, Anthony John. Distribution of trace water around brine leaks in the Avery Island salt mine; implications for natural migration of water in salt. M, 1985, Arizona State University. 102 p.

Tessari, Oscar Jose. Model ages and applied whole rock geochemistry of silver-lead-zinc veins, Keno Hill-Galena Hill mining camp, Yukon Territory. M, 1979, University of British Columbia.

Tessier, G. Robert. Pétrologie d'une partie de la Formation de Charny près de. M, 1950, Universite Laval.

Tessier, Gérard. Paleomagnetism of late Wisconsin lake sediments of southeastern Quebec. M, 1983, McGill University. 236 p.

Tessman, J. S. Investigation and modeling of the interaction of H^+, Cu^{2+}, and Cd^{2+} ions with fulvic acid. M, 1987, University of Waterloo. 140 p.

Tessman, Norman. Fossil sharks of Florida. M, 1969, University of Florida. 132 p.

Test, Thomas Alvin. Sediments derived from the weathering of a primordial anorthositic crust. M, 1976, Michigan State University. 58 p.

Testa, Stephen Michael. Geochemistry of Mesozoic dolerites from Liberia, Africa and Spitsbergen. M, 1978, California State University, Northridge. 112 p.

Testarmata, Margaret M. Magnetostratigraphy of the Eocene-Oligocene Vieja Group, Trans-Pecos, Texas. M, 1978, University of Texas, Austin.

Tester, Allen Crawford. The Dakota Stage of the type locality. D, 1929, University of Wisconsin-Madison.

Tester, Allen Crawford. The Pennsylvanian of the San Juan Basin. M, 1921, University of Kansas.

Teti, M. J. The surface water geochemistry of Deckers Creek drainage basin, West Virginia. M, 1976, West Virginia University. 76 p.

Tetreault, Andre R. Bedrock geology of the western half of the Westhampton Quadrangle, Massachusetts. M, 1959, University of Massachusetts. 95 p.

Tetrick, Martha Jane. The palynology of the Graneros Shale in five counties in Kansas. M, 1979, Wichita State University. 125 p.

Tetrick, Paul Roderick. Glacial geology of Oberon Quadrangle, North Dakota. M, 1947, University of Cincinnati. 58 p.

Tettenhorst, Rodney Tampa. Inhomogeneities in minerals of the montmorillonite group. M, 1957, Washington University. 62 p.

Tettenhorst, Rodney Tampa. Structural aspects of some montmorillonite organic reactions. D, 1960, University of Illinois, Urbana.

Tettleton, Burvon B. Subsurface geology of the Altus-Tipton area, Jackson and Tillman counties, southwestern Oklahoma. M, 1958, University of Oklahoma. 44 p.

Tetzlaff, Daniel Matias. A simulation model of clastic sedimentary processes. D, 1987, Stanford University. 367 p.

Teufel, Lawrence William. An experimental study of hydraulic fracture propagation in layered rock. D, 1979, Texas A&M University. 102 p.

Teufel, Lawrence William. The measurement of contact areas and temperature during frictional sliding of Tennessee Sandstone. M, 1976, Texas A&M University. 64 p.

Teufel, Michael Richard. Application of a long baseline bistatic acoustic sounder to the study of temperature inversions near the Earth's surface. M, 1975, Pennsylvania State University, University Park. 75 p.

Tew, John H. Big Cottonwood Canyon and its importance as a water source. M, 1971, University of Utah. 107 p.

Tew, Katherine Hine. A quantitative geomorphological investigation of the pre-Cretaceous erosion surface beneath the inner coastal plain of North Carolina. M, 1981, North Carolina State University. 40 p.

Tewhey, John. The petrology and structure of the crystalline rocks in the Irmo Quadrangle, South Carolina. M, 1968, University of South Carolina.

Tewhey, John D. The controls of biotite-cordierite-chlorite-garnet equilibria in the contact aureole of the Cupsuptic Pluton, West-central Maine and the two-phase region in the CaO-SiO$_2$ system; experimental data and thermodynamic analysis. D, 1975, Brown University. 160 p.

Tewksbury, Barbara Jarvis. Polyphase deformation and contact relationships of the Precambrian Uncompahgre Formation, Needle Mountains, southwestern Colorado. D, 1981, University of Colorado. 475 p.

Textoris, Daniel Andrew. Petrography and evolution of Niagaran reefs, Indiana. D, 1963, University of Illinois, Urbana.

Textoris, Daniel Andrew. Stratigraphy of the Green River Formation in the Bridger Basin, Wyoming. M, 1960, Ohio State University.

Textoris, Steven D. Stratigraphy and depositional history of late Pleistocene Key Largo patch reefs underlying Florida Bay. M, 1988, Wichita State University. 165 p.

Thacker, Joseph L., Jr. A study of the subsurface stratigraphy of the upper Cambrian in western Missouri. M, 1974, University of Missouri, Rolla.

Thacker, Mark Sloan. Geology and geochemistry of early-Proterozoic supracrustal rocks from the northern Sangre de Cristo Mountains and adjacent area, central Colorado. M, 1988, New Mexico Institute of Mining and Technology. 129 p.

Thackrey, Edmund Lee. Geology of Sheep Mountain, Fremont County, Wyoming. M, 1937, University of Missouri, Columbia.

Thacpaw, Saw Clarence. Geology of the Ruby Star Ranch area, Twin Buttes mining district, Pima County, Arizona. M, 1960, University of Arizona.

Thaden, Robert Emerson. The Porcupine Mountain "red rock" (Michigan). M, 1950, Michigan State University. 103 p.

Thaeler, John D. Carbonate cement stratigraphy and diagenesis of the Salem Formation in central Kentucky. M, 1979, University of Cincinnati. 119 p.

Thakur, Tukrel Radhakishin. Statistical models of river meanders. D, 1970, University of Illinois, Urbana. 163 p.

Thalman, Katherine L. A Pleistocene lagoon and its modern analogue, San Salvador, Bahamas. M, 1983, University of Akron. 166 p.

Thames, Clement Beal, Jr. Geology of the Yearlinghead Mountain area, Llano County, Texas. M, 1957, University of Texas, Austin.

Thamm, John Kenneth. Geology of part of the Seven Mountains District of central Pennsylvania. M, 1956, Pennsylvania State University, University Park. 169 p.

Thangsuphanich, Ittichai. Regional gravity survey of the southern Mineral Mountains, Beaver County, Utah. M, 1976, University of Utah. 38 p.

Thapa, Khagendra. Detection of critical points; the first step to automatic line generalization. D, 1987, Ohio State University. 198 p.

Thapar, Mangat R. Propagation of Rayleigh waves along perturbed boundaries. D, 1968, University of Western Ontario.

Tharalson, Darryl Bruce. Heavy minerals in Recent alluvium along the eastern flank of the Front Range, Golden to Canon City, Colorado. M, 1966, University of Colorado.

Tharalson, Darryl Bruce. Permian ammonoid family Perrinitidae. D, 1973, University of Iowa. 146 p.

Tharin, James Cotter. Glacial geology of the Calgary, Alberta area. D, 1960, University of Illinois, Urbana.

Tharin, James Cotter. Textural studies of the Wisconsin tills of northwestern Pennsylvania. M, 1958, University of Illinois, Urbana.

Tharp, Marie. Subsurface studies of the (Lower Devonian) Detroit River series (Michigan, Ontario, Ohio). M, 1945, University of Michigan.

Tharp, Paul A. The probable role of faults and fissures in oil and gas accumulation. M, 1925, University of Oklahoma. 36 p.

Tharp, Thomas M. Numerical model study of subduction and the deformation of the oceanic lithosphere. D, 1978, University of Wisconsin-Madison. 217 p.

Tharp, Timothy C. Subsurface geology and paleogeography of the lower Ste. Genevieve Limestone in Hamilton County, Illinois. M, 1983, University of Cincinnati. 93 p.

Tharp, Tommy Lee. Aspects of Leon River drainage history with implications to other central Texas streams. M, 1988, Baylor University. 260 p.

Tharpe, L. W. Fluvial sediments of the Rio Achiguate and its tributaries, Guatemala. M, 1976, University of Missouri, Columbia.

Thasan, Pat. Particulate matter in northwestern Arkansas rains. M, 1987, University of Arkansas, Fayetteville.

Thatcher, Richard Whitfield. The bedrock topography of a portion of the City of St. Louis, Missouri. M, 1927, Washington University. 51 p.

Thatcher, Robert James. Correlation of structural history with reservoir fluid distribution in the Lindsborg Pool, Kansas. M, 1961, University of Oklahoma. 53 p.

Thatcher, Russell N. The technique of X-ray diffraction pattern measurement. M, 1938, Columbia University, Teachers College.

Thatcher, Wayne Raymond. Surface wave propagation and source studies, Gulf of California. D, 1972, California Institute of Technology. 147 p.

Thayer, Charles Walter. Marine paleoecology of the Upper Devonian Genesee Group of New York. D, 1972, Yale University. 240 p.

Thayer, David W. Pennsylvanian lepospondyl amphibians from the Swisshelm Mountains, Cochise County, Arizona. M, 1973, University of Arizona.

Thayer, Donald D. A study of ground motion caused by air explosions, sonic booms, and thunder. M, 1964, New Mexico Institute of Mining and Technology. 61 p.

Thayer, James Bliss. Geology of the Steer Creek area northeast of Salida, Colorado. M, 1959, University of Colorado. 79 p.

Thayer, Paul Arthur. Geology of the Dan River and Davie County Triassic basins, North Carolina. D, 1967, University of North Carolina, Chapel Hill. 178 p.

Thayer, Richard E. Telluric-magnetotelluric investigations of regional geothermal processes in Iceland. D, 1975, Brown University. 291 p.

Thayer, Thomas P. The general geology of the North Santiam River section of the Oregon Cascades. D, 1934, California Institute of Technology. 92 p.

Thayer, Thomas P. and Howland, Arthur L. The geology of Gabamichigami Lake, Minnesota. M, 1931, Northwestern University.

Thayer, Thomas P. The stratigraphy and paleontology of the Salem Hills, Oregon. D, 1934, California Institute of Technology. 7 p.

Thayer, Valerie Lynn. Diatoms in Lake Superior sediments; distribution, stratigraphy, and taxonomy. M, 1981, University of Minnesota, Duluth.

Thede, Ray John. The petrology of the type Thomaston (Reynolds Bridge) granite (Connecticut). M, 1958, University of Wisconsin-Madison.

Theilig, Elizabeth Eilene. Formation of pressure ridges and emplacement of compound basaltic lava flows. D, 1986, Arizona State University. 214 p.

Theilig, Elizabeth Eilene. Plains and channels in the Lunae Planum; Chryse Planitia region of Mars. M, 1979, Arizona State University. 59 p.

Theiling, Stanley Cecil. The Pleistocene fauna of Lost River sink, Iowa County, Wisconsin. M, 1973, University of Iowa. 66 p.

Thein, Maung. A petrologic study of the Permo-Pennsylvanian red beds of central Colorado with special reference to the development of red color. M, 1963, Northwestern University.

Thein, Maung. A petrological study of the Upper Cambrian rocks in parts of Wisconsin and Minnesota. D, 1966, Northwestern University. 270 p.

Thein, Myint Lwin. Chester (Upper Mississippian) Gastropoda of the upper Mississippi and lower Ohio valleys. D, 1965, University of Chicago. 293 p.

Theis, Charles Vernon. Geology of Henderson County, Kentucky. D, 1929, University of Cincinnati. 215 p.

Theis, Nicholas James. Comparative mineralogy of the Quirke No. 1 and New Quirke mines, Rio Algom Mines Limited, Elliot Lake, Ontario. M, 1973, Queen's University. 61 p.

Theis, Nicholas James. Uranium-bearing and associated minerals in their geochemical and sedimentological context, Elliot Lake, Ontario. D, 1976, Queen's University. 158 p.

Theis, Sidney Wayne. Multifrequency remote sensing of soil moisture. D, 1982, Texas A&M University. 150 p.

Theis, William P. Stratigraphy and petrology of the Antigua Formation (Oligocene); Antigua, British West Indies. M, 1980, Northern Illinois University. 221 p.

Theiss, Mary Elizabeth. The fauna of the Ames Limestone in the Pittsburgh Quadrangle. M, 1940, University of Pittsburgh.

Theiss, Richard M. Environment of deposition of Woodbine and Eagleford sandstones, Leon, Houston, and Madison counties, Texas. M, 1983, Texas A&M University.

Thelen, Paul and Knopf, Adolph. Sketch of the geology of Mineral King, California. M, 1905, University of California, Berkeley. 35 p.

Theodore, Theodore George. Structure and petrology of the gneisses and mylonites at Coyote Mountain, Borrego Springs, California. D, 1967, University of California, Los Angeles. 351 p.

Theodosis, Steven Daniel. Metamorphism of the Randville Dolomite in the Felch mountain range, Michigan. M, 1948, Northwestern University.

Theodosis, Steven Daniel. The geology of the Melrose area, Beaverhead and Silver Bow counties, Montana. D, 1956, Indiana University, Bloomington. 118 p.

Theokritoff, Sergius. Hetero-epitaxial growth of large crystals of a metastable phase; germania (Quartz). M, 1970, Pennsylvania State University, University Park. 48 p.

Theoret, Dennis R. Compositional zoning and ilmenite inclusions in porphyroblasts from staurolite grade rocks, Black Hills, South Dakota. M, 1986, University of Akron. 99 p.

Theriault, François. Lithofacies, diagenesis and related reservoir properties of the Upper Devonian Grosmont Formation, northern Alberta. M, 1984, University of Calgary. 207 p.

Therkelsen, Edward R. An investigation of various methods of second order longitude determinations. M, 1965, Ohio State University.

Therrien, Pierre. Présentation d'un modèle numérique eulérien-lagrangien destiné à la simulation de cas de contamination sans réaction chimique dans les eaux souterraines. M, 1986, Universite Laval. 169 p.

Theyab, A. and Al-Mishwt, A. T. Geology, mineralogy, and petrochemistry of Al-Halgah Pluton, At-Taif, Saudi Arabia. D, 1977, University of Wisconsin-Madison. 312 p.

Theyer, Fritz. Late Neogene paleomagnetic and planktonic zonation, southeastern Indian Ocean-Tasman Basin. D, 1972, University of Southern California. 249 p.

Thiagalingam, K. Effect of temperature and biological control chemicals on nitrogen transformation in Hawaiian soils. M, 1967, University of Hawaii. 60 p.

Thibault, Newman William. Celestite and associated minerals from geodes near Chittenango Falls, New York. M, 1934, Syracuse University.

Thibault, Newman William. Morphological and structural crystallography and optical properties of silicon carbide. D, 1943, University of Michigan.

Thibault, Yves. Géologie et pétrologie des gabbros et des dykes de la région de Phini, complexe ophiolitique de Troodos, Chypre. M, 1987, Universite Laval. 219 p.

Thibodaux, Bernadette L. Sedimentological comparison between Somerero Key and Loo Key, Florida. M, 1972, Louisiana State University.

Thibodeau-Jordan, Dawne Marie. Paleomagnetic study of Upper Triassic carbonate rocks from northwestern Sicily. M, 1981, Wright State University. 86 p.

Thibodeaux, Jerry Lee. Inner shelf hydrography, Peard Bay, Alaska. M, 1980, Louisiana State University.

Thieben, Scott E. Geochemistry and petrography of the basaltic and carbonate rocks, Suregei-Asille volcanic district, East Turkana, Kenya. M, 1980, Iowa State University of Science and Technology.

Thiede, D. S. Geological implications of variation in heavy metal content in the Elk Point evaporite sequence, Saskatchewan, Canada. M, 1975, University of Wisconsin-Madison.

Thiede, D. S. The genesis of metalliferous brines from evaporites; a study based upon the Middle Devonian Elk Point Group of Canada. D, 1978, University of Wisconsin-Madison. 223 p.

Thiel, George Alfred. The manganese minerals; their identification and paragenesis, with special reference to the manganese ores of the Cuyuna Range. D, 1923, University of Minnesota, Minneapolis.

Thiel, Paul Thomas. Mineralogical variations in a diabase sill near Nimrod, Granite County, Montana. M, 1961, University of Montana. 66 p.

Thieling, Stanley Cecil. The Pleistocene fauna of Lost River sink, Iowa County, Wisconsin. M, 1973, University of Iowa.

Thieme, Martin Alan. Computer simulation of scraper-pusher behavior during early open pit mining operations. D, 1968, University of Missouri, Rolla.

Thienprasert, Ammuayachai. Seismic refraction interpretation by computer technique. M, 1974, New Mexico Institute of Mining and Technology.

Thies, Barry Peter. Structural configuration of Horse Creek Canyon, Hoback Range, Wyoming. M, 1974, University of Michigan.

Thies, Jennifer L. Analysis of crinoid communities and their associated lithofacies within the Fort Payne Formation (Lower Mississippian) in the vicinity of Burkesville, Kentucky (Cumberland County). M, 1987, University of Cincinnati. 171 p.

Thies, K. J. The recognition of environmental differences by means of mortality patterns and growth rates of Pycnodonte convexa (Ostreidae) in the Navesink Formation (Cretaceous, New Jersey). M, 1976, Queens College (CUNY). 68 p.

Thies, Roland Kendall. Laboratory investigation of the accuracy and consistency of depth determination by earth resistivity methods. M, 1932, University of Colorado.

Thiesmeyer, Lincoln R. The plutonic rocks of northwestern Fauquier County, Virginia. D, 1937, Harvard University.

Thiesse, Mark F. Geology of Paleozoic basinal rocks in the northern Fox Range; Washoe County, Nevada. M, 1988, University of Nevada. 86 p.

Thiessen, Richard Leigh. Theoretical and computer assisted studies in tectonics, structural geology and isotope dating. D, 1980, SUNY at Albany. 500 p.

Thill, Richard E. Petrofabric correlations with compressional wave velocity variation. M, 1967, University of Minnesota, Minneapolis. 124 p.

Thiruvathukal, John V. A regional gravity study of basement and crustal structures in the Southern Peninsula of Michigan. M, 1963, Michigan State University. 55 p.

Thiruvathukal, John Varkey. Regional gravity of Oregon. D, 1968, Oregon State University. 92 p.

Thoburn, Thomas C. Least squares polynomial fitting of gravity data, a case history study over the Garber oil field, Garfield County, Oklahoma. M, 1977, Wright State University. 93 p.

Thole, Ronald Henry. Joint analysis of the magnetic belt of Stevens County, Washington. M, 1970, Washington State University. 89 p.

Thom, Bruce Graham. Coastal and fluvial landforms, Horry and Marion counties, South Carolina. D, 1967, Louisiana State University. 112 p.

Thom, Emma N. An outline of the geology of New Zealand. M, 1930, George Washington University.

Thom, M. Geomorphic and structural studies in Hispaniola using Landsat, Skylab and low-altitude aerial photographs. M, 1976, George Washington University.

Thom, William T., Jr. Problems of the Cretaceous-Eocene boundary in Montana and the Dakotas. D, 1917, The Johns Hopkins University. 189 p.

Thomann, William F. Igneous and metamorphic petrology of the Honey Brook Uplands in the Elverson, Pottstown and Phoenixville 7 1/2' quadrangles, southeastern Pennsylvania. M, 1977, Bryn Mawr College. 65 p.

Thomann, William Frederick. Petrology and geochemistry of the Precambrian Thunderbird Formation, Franklin Mountains, El Paso County, Texas. D, 1980, University of Texas at El Paso. 184 p.

Thomas, A. Volcanic stratigraphy of the Izok Lake greenstone belt, District of Mackenzie, N.W.T. M, 1978, University of Western Ontario. 110 p.

Thomas, Abram O. Echinoderms of the Iowa Devonian. D, 1923, University of Chicago. 183 p.

Thomas, Abram Owen. A comparison and revision of the fauna of the Lime Creek and Independence shales. M, 1909, University of Iowa.

Thomas, Andrew R. The Porters Creek Formation (Paleocene) of the northern Mississippi Embayment; clay mineral variations and depositional interpretation. M, 1981, Indiana University, Bloomington. 100 p.

Thomas, Andrew Russell. Gravity investigation of selected alpine ultramafic bodies in western North Carolina. M, 1982, Kent State University, Kent. 96 p.

Thomas, Anne Valerie. A petrological and fluid inclusion study of the Tanco Pegmatite, S.E. Manitoba. M, 1985, University of Toronto.

Thomas, B. L. Nitrate contamination in the groundwater of the Annapolis-Corwallis Valley, N. S. M, 1974, University of Waterloo.

Thomas, Barbara R. Composition and cell parameters of authigenic albite from the Manlius and Coeymans formations (Lower Devonian) of eastern New York. M, 1968, University of Kansas. 61 p.

Thomas, Blakemore Ewing. Geology and ore deposits of the Walapai District, Mohave County, Arizona. D, 1949, California Institute of Technology. 187 p.

Thomas, Byron K. Structural geology and stratigraphy of Sugar Loaf Anticlinorium and adjacent Piedmont area, Maryland. D, 1953, The Johns Hopkins University.

Thomas, Carol Varner. The foraminifera of the Cretaceous-Tertiary contact in Clay County, Mississippi. M, 1960, Mississippi State University. 45 p.

Thomas, Carroll Morgan. Origin of the Permian pisolites, Guadalupe Mountains, southern New Mexico and West Texas. M, 1964, Texas Tech University. 116 p.

Thomas, Charles Ward, Jr. Lithology and paleontology of Antarctic Ocean bottom sediments. M, 1953, Washington University. 47 p.

Thomas, Chester Ward, Jr. Structural reconnaissance of the Byrnesville area, Jefferson County, Missouri. M, 1956, Washington University. 49 p.

Thomas, Clark L. The geology of the Derby area in the Denver Basin of Colorado; Adams and Arapahoe counties. M, 1942, Colorado School of Mines. 66 p.

Thomas, Clifford Ward, Jr. The rock aggregate and ground-water resources of Keokuk County, Iowa. M, 1959, University of Iowa. 210 p.

Thomas, D. The isotopic profile of gases from the summit and flank of Kilauea Volcano. D, 1977, University of Hawaii. 196 p.

Thomas, Dale Edmund. A new American species of Calamopitys from the Devonian. D, 1932, Cornell University.

Thomas, David A. Middle Atokan stratigraphy of the southern Arkoma Basin. M, 1983, University of Arkansas, Fayetteville.

Thomas, David Erben. Village land use in Northeast Thailand; predicting the effects of development policy on village use of wildlands. D, 1988, University of California, Berkeley. 210 p.

Thomas, David J. The Tertiary geology and systematic paleontology (phylum mollusca) of the Guajira Peninsula, Colombia, South America. D, 1972, SUNY at Binghamton. 147 p.

Thomas, David James. Distribution, geological controls and genesis of uraniferous pegmatites in the Cree Lake zone of northern Saskatchewan. M, 1983, University of Regina. 213 p.

Thomas, Dennis R. The morphology of Venus; relief and texture considerations for imaging radar. M, 1982, University of Arkansas, Fayetteville. 145 p.

Thomas, Edward F. Characteristics of shelf deposition controlled by carbonate platform margin configuration; Devonian Tor Limestone, central Nevada. M, 1987, University of Nevada, Las Vegas. 73 p.

Thomas, Edwin S. Landslide forms and their origin in the middle Coast Ranges. M, 1939, University of California, Berkeley. 198 p.

Thomas, Emil Paul. The geology of the Claiborne (Eocene) Group of Mississippi as far north as Grenada County. D, 1942, Louisiana State University.

Thomas, Erich. Biostratigraphy of the Bilk Limestone (Permian), northwestern Nevada. M, 1972, Western Washington University. 109 p.

Thomas, Everett R. Paleoecology of the Pawnee Formation (Pennsylvanian, Des Moinesian) of Missouri. M, 1963, University of Wisconsin-Madison.

Thomas, F. C. Lower Scotian Slope benthic foraminiferal faunas past and present, with taxonomic outline. M, 1985, Dalhousie University. 199 p.

Thomas, Faiz N. Ecology and distribution of Foraminifera of a traverse in the northwestern Gulf of Mexico. M, 1966, University of Missouri, Rolla.

Thomas, George Ligon. Petrography of the Catahoula Formation in Texas. M, 1960, University of Texas, Austin.

Thomas, George Martz. The geology of the northeast third of the Ritter Quadrangle, Oregon. M, 1956, University of Oregon. 60 p.

Thomas, George McConnell. A comparison of the paleomagnetic character of some varves and tills from the Connecticut Valley. M, 1984, University of Massachusetts. 136 p.

Thomas, George W. Investigation of dusellite. M, 1929, Colorado School of Mines. 54 p.

Thomas, Gerald M. Structural geology of the Badger Pass Area, Southwest Montana. M, 1981, University of Montana. 58 p.

Thomas, Gerald William. Stratigraphic paleontology of the Morgan Formation, Uinta Mountains and vicinity. M, 1958, Washington State University. 137 p.

Thomas, Gilbert Edward. The South Flat and related formations in the northern part of the Gunnison Plateau, Utah. M, 1960, Ohio State University.

Thomas, Glenn H. Geology of the Indian Springs Quadrangle, Tooele and Juab counties, Utah. M, 1958, Brigham Young University. 35 p.

Thomas, Gordon Wallace. A theoretical analysis of forward combustion in petroleum reservoirs. D, 1963, Stanford University. 110 p.

Thomas, Harold Eugene. Geology of Cedar City and Parowan valleys, Iron County, Utah. D, 1947, University of Chicago. 82 p.

Thomas, Harry E. Late Pleistocene planation surfaces of the Paleozoic region of Arkansas. M, 1958, University of Arkansas, Fayetteville.

Thomas, Harry George. Correlation of the Madison Group of the Bighorn Mountains and the Powder

River basin area. M, 1953, Northwestern University. 72 p.

Thomas, Herman Hoit. Trace element contamination in tholeiitic basalts and a garnet peridotite as determined by an acid leaching technique. D, 1973, University of Pennsylvania.

Thomas, Horace Davis. Geology and deformation of the Centennial Valley, Wyoming. M, 1928, University of Wyoming. 46 p.

Thomas, Horace Davis. Phosphoria (Permian) and Dinwoody (Triassic) tongues in lower Chugwater of central and southeastern Wyoming. D, 1934, Columbia University, Teachers College.

Thomas, Hugo F. Differential thermal analysis of clay from the mudstones of a lower Cincinnatian Series core, Butler County, Ohio. M, 1959, Miami University (Ohio). 44 p.

Thomas, Hugo Frederick. Late-glacial sedimentation near Burlington, Vermont. D, 1964, University of Missouri, Columbia.

Thomas, J. E. Multi-channel estimate of the seismic wavelet. M, 1976, University of Houston.

Thomas, James M. Water chemistry and flow system of the Early Pennsylvanian Mansfield Formation aquifer system in Clay County, Indiana. M, 1980, Indiana University, Bloomington. 76 p.

Thomas, Jimmy N. Determination and refinement of the crystal structure of quenstedtite. M, 1973, Southern Illinois University, Carbondale. 84 p.

Thomas, John A. Geology of the Cloquet area, northeastern Minnesota. M, 1959, University of Minnesota, Minneapolis. 65 p.

Thomas, John Alroy. The geology and ore deposits of the Tascuela area, Sierrita Mountains, Pima County, Arizona. D, 1966, [University of Michigan].

Thomas, John B. A detailed study of the Pleistocene geology of a portion of Preble County, southwestern Ohio. M, 1966, Miami University (Ohio). 119 p.

Thomas, John B. The stratigraphy, sedimentology, and petrology of the early Tertiary Claron Formation in the Red Canyon area of the south-central High Plateaus, Utah. M, 1985, Kent State University, Kent. 135 p.

Thomas, John Byron. Mineralogic dispersal patterns in the Vanoss Formation, south-central Oklahoma. D, 1973, University of Oklahoma. 162 p.

Thomas, John J. A detailed structural analysis of the minor structures of the Williamstown area, Massachusetts. M, 1965, Northwestern University.

Thomas, John Jenks. The detection of low angle thrust fault surfaces by mesostructural analysis as applied in western Vermont. D, 1968, University of Kansas. 89 p.

Thomas, John Moore. Potential deep reservoir beds in western Pennsylvania. M, 1955, University of Pittsburgh.

Thomas, John Neil. Geology of the Inter-State 244 Interchange, Kirkwood, Missouri. M, 1965, University of Missouri, Rolla.

Thomas, June Marie. Correlation and petrogenesis of the Miocene volcanic rocks in the San Emigdio and San Juan Bautista areas, California. M, 1986, California State University, Northridge. 84 p.

Thomas, Karen Kay. Paleozoic stratigraphy and structure of a part of the northwestern Sulphur Springs Range, Eureka County, Nevada. M, 1985, University of California, Riverside. 79 p.

Thomas, Kimberly Jaye. Deformation and metamorphism of the central Narragansett Basin of Rhode Island. M, 1981, University of Texas, Austin. 96 p.

Thomas, Kimberly Kodidek. The origin of sillimanite in Essex, Connecticut. M, 1984, Indiana University, Bloomington. 101 p.

Thomas, Laurence E. Subsurface geology of the Criner area, McClain County, Oklahoma. M, 1962, University of Oklahoma. 63 p.

Thomas, Lee A. The petrology and paleoenvironments of the Mississippian Big Clifty Formation in East

Tennessee. M, 1981, Memphis State University. 84 p.

Thomas, Leo Almor. Devonian-Mississippian formation of Southeast Iowa. D, 1948, University of Missouri, Columbia.

Thomas, Leo Almor. Foraminifera of the Crockett and Stone City formations of Texas. M, 1942, University of Missouri, Columbia.

Thomas, Leonard C. Stratigraphy and structure of the pre-Cambrian rocks of the southeastern Black Hills, South Dakota. D, 1932, University of Iowa. 78 p.

Thomas, Leonard C. The Precambian Cherry Creek Series of the Madison Valley of southwestern Montana. M, 1928, University of Iowa. 89 p.

Thomas, Lewis F. A geographic study of Greene County, Missouri. M, 1917, University of Missouri, Columbia.

Thomas, Lloyd. Petrology and stratigraphy of the Greenbrier Limestone (Upper Mississippian) in Wayne, Mingo and Logan counties, West Virginia. M, 1967, West Virginia University.

Thomas, M. W. Structural geology of a gneissic terrain, Laurie Lake, northern Manitoba. M, 1980, University of Manitoba. 59 p.

Thomas, Mark A. Petrology and diagenesis of the Osagean Series, western Sedgwick Basin, Kansas. M, 1982, University of Missouri, Columbia.

Thomas, Mark B. Depth-porosity relationships in the Viking and Cardium formations of central Alberta. M, 1977, University of Calgary. 147 p.

Thomas, Mark H. Stratigraphy of the Dakota and Burro Canyon formations near Gunnison, Colorado. M, 1981, Colorado School of Mines. 76 p.

Thomas, Michael B. Structural geology along part of the Blackfoot Fault system near Potomac, Missoula County, Montana. M, 1987, University of Montana. 52 p.

Thomas, Michael Robert. A conceptual framework for ecosystem planning and management. D, 1980, Michigan State University. 165 p.

Thomas, Murrell Dee. The petrography and origin of the sandstone members of the Wind River Formation in the Wind River basin, Wyoming. M, 1942, University of Missouri, Columbia.

Thomas, Peter B. The nature of gold deposits in shear zones in the Cordova Gabbro, Grenville Province, Ontario. M, 1985, University of Ottawa. 154 p.

Thomas, Peter C. Probable rejuvenation of the piedmont in the James River drainage basin. M, 1973, University of North Carolina, Chapel Hill. 66 p.

Thomas, Peter Chew. The morphology of Phobos and Deimos. D, 1978, Cornell University. 286 p.

Thomas, R. The rates of denudation on Mont St. Hilaire, Quebec. M, 1974, McGill University. 164 p.

Thomas, R. N. A north-south subsurface correlation of Silurian formations in eastern Kentucky. M, 1940, [University of Kentucky].

Thomas, R. P. The nature of the observed motion of shear waves of the Nevada earthquake of 1932. M, 1949, University of California, Berkeley. 54 p.

Thomas, R. Scott. Structural and strain analysis of the Crystal Creek area, Gros Ventre Mountains, Wyoming. M, 1986, Miami University (Ohio). 107 p.

Thomas, Robert Brinley. Geochemical exploration using heavy minerals in the San Luis Valley, Colorado. M, 1986, University of Akron. 87 p.

Thomas, Robert Curtiss. Paleontology and carbonate petrology across the marjumiid-pterocephaliid biomere boundary, southwestern Montana. M, 1987, University of Montana. 147 p.

Thomas, Robert D., Jr. Geology of the Pleasant Mountain syenite complex, southwestern Maine. M, 1974, Wesleyan University.

Thomas, Roger David Keen. Functional morphology, ecology and evolution in the genus Glycymeris (Bivalvia, Cretaceous–Recent). D, 1971, Harvard University.

Thomas, Ronny G. The geomorphic evolution of the Pecos River system (Texas, New Mexico). M, 1971, Baylor University. 90 p.

Thomas, Susan. Generation of interval traveltime profiles from resistivity logs using lithologic correlation. M, 1984, Stanford University.

Thomas, Terry Robert. Environmental planning for geothermal energy resource exploration, development, and utilization. D, 1982, University of California, Los Angeles. 126 p.

Thomas, Trevor James. Geology of the Buffalo Valley Prospect, Lander County, Nevada. M, 1985, University of Nevada. 118 p.

Thomas, W. M. Stability relations of the amphibole hastingsite, $NaCa_2Fe^{2+}_4Fe^{3+}Si_6Al_2O_{22}(OH)_2$. D, 1979, University of California, Los Angeles. 196 p.

Thomas, Walter L. Geology and ore deposits of the Rosemont area, Pima County, Arizona. M, 1931, University of Arizona.

Thomas, Warren Baxter. Palynology of the Ripley lignitic clays, Clay County, Mississippi. M, 1960, Mississippi State University. 65 p.

Thomas, Wayne Barker. The structural geology of Lower Ogden Canyon, (Weber County), Utah. M, 1940, University of Utah. 43 p.

Thomas, William A. Upper Mississippian stratigraphy of southwestern Virginia, southern West Virginia, and eastern Kentucky. D, 1960, Virginia Polytechnic Institute and State University.

Thomas, William Andrew. Stratigraphy of the Devonian Chaffee Formation of northeastern Gunnison County, Colorado. M, 1957, University of Kentucky. 84 p.

Thomas, William Avery. A geological and stratigraphic study of the Hazelton oil field, lying principally in Washington Township, Gibson County, Indiana. M, 1924, Indiana University, Bloomington. 66 p.

Thomas, William Dennis. Dikes of the Clear Creek area, Wasatch Plateau, Utah. M, 1976, Utah State University. 73 p.

Thomason, Robert Edward. Volcanic stratigraphy and epithermal mineralization of the Delamar silver mine, Owyhee County, Idaho. M, 1983, Oregon State University. 111 p.

Thomason, Thomas J. Polar motion. M, 1968, Ohio State University.

Thomasson, Maurice R. The geology of the Big Springs area, Missouri. M, 1953, University of Missouri, Columbia.

Thomasson, Maurice Ray. Late Paleozoic stratigraphy and paleotectonics of central and eastern Idaho. D, 1959, University of Wisconsin-Madison. 288 p.

Thomerson, Jamie Edward. Micropaleontology of Orbitolina Zone, Trans-Pecos, Texas. M, 1961, Texas Tech University. 118 p.

Thomes, Margaret S. Textural and petrographic analyses of certain marine clays and shales with special reference to the Decorah Shale of Minnesota. M, 1937, University of Minnesota, Minneapolis. 45 p.

Thomlinson, A. G. Devonian fauna of the Cordilleran region of Canada. M, 1953, University of British Columbia.

Thomlinson, Arnold Gordon. Upper Devonian corals of the Canadian Cordilleran region. M, 1954, University of British Columbia.

Thompson, Allan McMaster. Sedimentologic and geochemical studies of the Bald Eagle-Juniata color boundary (Upper Ordovician), central Pennsylvania. D, 1968, Brown University. 282 p.

Thompson, Alyce D. Nitrogen fixation in Virginia salt marshes and the effects of chronic oil pollution on nitrogen fixation in the Mobjack Bay marshes. M, 1977, College of William and Mary.

Thompson, Amy Gale. Syn-metamorphic intrusions in the Observation Peak area, Klamath Mountains, southern Oregon. M, 1988, Texas Tech University. 95 p.

Thompson, Andrew A. Tidal work on Marguerite Bay, Antarctica. M, 1948, Columbia University, Teachers College.

Thompson, Arthur P. On the relation of pyrrhotite to chalcopyrite and other sulphides. M, 1912, Columbia University, Teachers College.

Thompson, Braden Jay. The geology of the upper Proterozoic Scout Mountain Member, Pocatello Formation, Garden Creek Gap, Bannock Range, southeastern Idaho. M, 1982, Idaho State University. 41 p.

Thompson, Bruce A. Petrography of the Upper Carboniferous-Permian sandstones of the southern half of the Dunkard Basin. M, 1963, Miami University (Ohio). 115 p.

Thompson, Bruce Gregory. Crustal electrical conductivity studies in the Georgia Piedmont. D, 1982, Cornell University. 135 p.

Thompson, C. B. H. The distribution of carbohydrates and their associations with metal ions in selected Grey soils. M, 1973, University of British Columbia.

Thompson, C. Sheldon. Determination of the composition of plagioclase feldspars by means of infrared spectroscopy. D, 1967, University of Utah. 52 p.

Thompson, Calvin John. The geology of the northern third of the Susanville Quadrangle, Oregon. M, 1956, University of Oregon. 90 p.

Thompson, Candace M. Characterization of the heavy metal content of coal-bearing stratigraphy from the New River basin, Tennessee; heavy metals released by static leaching of the Big Mary coal sequence. M, 1977, University of Tennessee, Knoxville. 53 p.

Thompson, Cheryl I. The role of ice as an agent of erosion and deposition of an estuarine tidal flat. M, 1977, University of New Hampshire. 64 p.

Thompson, Craig Dickerson. Geology of Bassam Park, Chaffee County, Colorado. M, 1957, University of Colorado.

Thompson, D. A. Petroleum geology and sedimentary petrology of the First Sandstone, Five Mile Field, Cattaraugus County, New York. M, 1975, SUNY at Buffalo. 83 p.

Thompson, D. E. Application of fluid-instability analysis to glacier flow. D, 1976, University of California, Los Angeles. 111 p.

Thompson, Dale Richard. Astogenetic study of some trepostomatous Bryozoa in the Waynesville Formation of the Richmond Group. M, 1963, Miami University (Ohio). 138 p.

Thompson, Dan A. Mineralogy and origin of the "medicine earth" sulfates in the Bucatunna clay, central Mississippi. M, 1978, University of Mississippi. 91 p.

Thompson, Daniel Lee. Stratigraphy and petrography of the Muzquiz Formation (Upper Cretaceous) Sabinas coal basin, northeastern Mexico. M, 1981, University of New Orleans. 61 p.

Thompson, Danny Lee. The nature of anorthosite; country rock interaction during granulite facies metamorphism; and example from the Whitestone Anorthosite. M, 1983, McMaster University. 312 p.

Thompson, David. The geology of the Evening Star tin mine and surrounding region, San Bernandino County, California. M, 1978, University of Missouri, Rolla.

Thompson, David Grosh. A study of the Pleistocene deposits and water-bearing strata of the Hardinville and Sumner quadrangles, Illinois. M, 1913, University of Illinois, Chicago.

Thompson, Debora B. Hydrogeology of the Niagara Dolomite aquifer at Wind Point, Wisconsin, and its interaction with Lake Michigan. M, 1981, University of Wisconsin-Milwaukee. 108 p.

Thompson, Diana M. Atoka Group (Lower-Middle Pennsylvanian), northern Fort Worth Basin, Texas; terrigenous depositional systems, diagenesis, reservoir distribution and quality. M, 1982, University of Texas, Austin.

Thompson, Donald E. Paleoecology of the San Diego marine Pleistocene. M, 1967, San Diego State University.

Thompson, Donald Eugene. Paleoecology of the Pamlico Formation, Saint Mary's County, Maryland. D, 1972, Rutgers, The State University, New Brunswick. 179 p.

Thompson, Donald John. A paleoecologic and biostratigraphic analysis of late Neogene sediments in the Gulf Coast. D, 1973, Washington University. 190 p.

Thompson, Donald Merrell. An analysis of the anomalous pediment in Boulder Valley, Jefferson County, Montana. D, 1982, Indiana State University. 91 p.

Thompson, E. C. Some possible guides to ore in the Red Lake Camp (Ontario). M, 1960, University of Toronto.

Thompson, E. M. 911 years of microparticle deposition at the South Pole; climatic interpretation. D, 1979, Ohio State University. 218 p.

Thompson, Ellen M. Investigation of the nature of the microparticles from the Byrd Station, Antarctica, deep ice core and their relationship to the climate of that period. M, 1975, Ohio State University.

Thompson, Esther H. Morphology and taxonomy of Cyclonema hall (Gastropoda), upper Ordovician, Cincinnatian province. M, 1969, Miami University (Ohio). 148 p.

Thompson, Frederick Henry. Hydrology of the glacial section along the Manistee river, Kalkaska, Grand Traverse, Wexford, and Missaukee counties, Michigan. M, 1969, Michigan Technological University. 54 p.

Thompson, Gary Gene. Paleoecology of palynomorphs in the Mancos shale, southwestern Colorado. D, 1969, Michigan State University. 200 p.

Thompson, Gary L. Investigation of gravity and magnetic anomalies near the Clarendon-Linden structure, western New York. M, 1979, University of Rhode Island.

Thompson, George A., Jr. Autoradiography of minerals. M, 1942, Massachusetts Institute of Technology. 50 p.

Thompson, George A., Jr. Structural geology of the Terlingua quicksilver district, Texas. D, 1949, Stanford University. 79 p.

Thompson, George David. Petrography of a portion of the marine banks of the Oolagah Limestone (Pennsylvanian) Tulsa County, Oklahoma. M, 1967, University of Tulsa. 173 p.

Thompson, George Fayette. Geology and paleontology of Johnson County, Iowa. M, 1898, University of Iowa. 144 p.

Thompson, George M. A magnetic study of fault-diabase dike relationships in the Cedar Creek Community, (South Carolina). M, 1968, University of South Carolina.

Thompson, Gerald Leon. A study of part of the preCambrian granite-greenstone complex in southeastern Florence County, and northwestern Marinette County, Wisconsin. M, 1955, Ohio State University.

Thompson, Glenn Michael. Trichlorofluoromethane, a new hydrologic tool for tracing and dating ground water. D, 1976, Indiana University, Bloomington. 93 p.

Thompson, Glenn Michael. Uranium series dating of stalagmites from Blanchard Springs Caverns, Stone County, Arkansas. M, 1973, Memphis State University.

Thompson, Graham R. Investigations of potassium-rubidium ratios in carbonate rocks of problematical origin. M, 1964, Dartmouth College. 24 p.

Thompson, Graham R. The crystal chemistry of glauconite; an explanation for low radiometric ages from glauconites. D, 1971, Case Western Reserve University. 157 p.

Thompson, Gregory L. Changes occurring in coal during in situ gasification. M, 1980, Iowa State University of Science and Technology.

Thompson, Harold Reid. Geomorphology of Pangnirtung Pass, Baffin Island, Northwest Territories (Canada). D, 1954, McGill University.

Thompson, Henry Dewey. Geomorphology of the Hudson Gorge in the highlands (New York). D, 1935, New York University.

Thompson, Henry Travis. A study of the physiographic history of swamp lands in relation to the problem of their drainage. M, 1928, University of North Carolina, Chapel Hill. 49 p.

Thompson, Ida Louise. Biological rhythms of shell growth and valve movement in bivalves. D, 1972, University of Chicago. 242 p.

Thompson, J. B., Jr. A gneiss dome in Southeast Vermont. D, 1950, Massachusetts Institute of Technology. 214 p.

Thompson, J. P. Stratigraphy and geochemistry of the Scots Bay Formation, Nova Scotia. M, 1974, Acadia University.

Thompson, James D., Jr. The coastal upwelling cycle on a beta-plane; hydrodynamics and thermodynamics. D, 1974, Florida State University. 154 p.

Thompson, James H. Precambrian geology of the Emigrant Canyon area, Panamint Range, California. M, 1963, University of Southern California.

Thompson, James R. Depositional environments of the Upper Ordovician Sequatchie and Silurian Rockwood formations of upper East Tennessee. M, 1984, University of Tennessee, Knoxville. 199 p.

Thompson, James R. The structural relations of the Negaunee-Ishpeming District of the Marquette Iron Range, Lake Superior. M, 1892, University of Wisconsin-Madison.

Thompson, John. Hydrogeologic characteristics of glaciolacustrine sediments. M, 1987, University of Rhode Island. 153 p.

Thompson, John David. Morphometric analysis of bedrock-controlled drainage basins in glaciated terrain of Summit and Portage counties, Ohio. M, 1980, University of Akron. 69 p.

Thompson, John Francis Hugh. The geology of the Vakkerlien nickel deposit, Kvikne, Norway. M, 1978, University of Toronto.

Thompson, John Francis Hugh. The intrusion and crystallization of gabbros, central Maine, and genesis of their associated sulfides. D, 1982, University of Toronto.

Thompson, John H. The molybdenum industry at Climax, Colorado. M, 1943, University of Colorado.

Thompson, John N. The Deforest Creek Landslide and sediment transport in Deer Creek, Skagit County, Washington. M, 1988, Western Washington University. 80 p.

Thompson, John Peters. The geology of a part of St. Louis County, Missouri. M, 1928, Washington University. 95 p.

Thompson, John Robert., Jr. Geology of Wet Beaver Creek canyon, Yavapai County, Arizona. M, 1968, Northern Arizona University. 69 p.

Thompson, John W., III. Ground-water resistivity investigation on Cape Hatteras Island, North Carolina outer banks. M, 1980, Miami University (Ohio). 125 p.

Thompson, Jon Louis. A paleontologic study of Rose's and Reid's Bluff, St. Marys River, Nassau County, Florida. M, 1962, University of Florida. 87 p.

Thompson, Joyce Ann. Molluscan biostratigraphy and physical stratigraphy of the Miocene Astoria(?) Formation in western Washington. M, 1978, University of Washington. 59 p.

Thompson, Keith F. Redistribution of potassium in synthetic micas analogous to illite. D, 1966, Massachusetts Institute of Technology. 201 p.

Thompson, Keith Goodwin. Stratigraphy and petrology of the Hamblin-Cleopatra Volcano, Clark County, Nevada. M, 1985, University of Texas, Austin. 306 p.

Thompson, Keith S. Prediction of water level declines in the Casper Aquifer in the vicinity of Laramie, Wyoming. M, 1979, University of Wyoming. 81 p.

Thompson, Kenneth. Mineral deposits of the Deep Creek Mountains, Tooele and Juab counties, Utah. D, 1970, University of Utah. 350 p.

Thompson, L. G. Microparticles, ice sheets and climate. D, 1976, Ohio State University. 216 p.

Thompson, Laird Berry. Distribution of living benthonic foraminifera near Isla de los Estados, Tierra del Fuego, Argentina. M, 1974, University of California, Davis. 70 p.

Thompson, Laird Berry. The Campanian shelf in Northeast Texas. D, 1983, University of Texas at Dallas. 204 p.

Thompson, Leon Garfield. Edges as specifiers of image quality. D, 1972, Ohio State University. 169 p.

Thompson, Lonnie G. Analysis of the concentration of microparticles in an ice core from Byrd Station, Antarctica. M, 1973, Ohio State University.

Thompson, Loren Edward, Jr. A sedimentational study of the Marietta Sequence (Dunkard) sandstones from Marietta, Ohio to Freeport, West Virginia. M, 1963, Ohio University, Athens. 105 p.

Thompson, Lucas G. An investigation of rock alteration in the Flin Flon ore-body (Manitoba). M, 1924, University of Manitoba.

Thompson, M. Gary. Depositional environments of the Maxon Formation (Lower Cretaceous), Marathon region, West Texas. M, 1977, University of Texas, Austin.

Thompson, Marcus Luther. The Cretaceous-Eocene contact in Mississippi. M, 1933, University of Iowa. 51 p.

Thompson, Marcus Luther. The fusulinids of the Des Moines Series of Iowa. D, 1934, University of Iowa. 99 p.

Thompson, Mark E. Geology of the Thompson Creek-Slate Creek area, South-central Idaho. M, 1977, University of Wisconsin-Milwaukee.

Thompson, Mark Edward. Microfeatures on quartz surfaces produced by aeolian abrasion. M, 1982, Arizona State University. 70 p.

Thompson, Mary E. Contributions to the mineralogy of uranium and vanadium, and to the geochemistry of carbonates and sea water. D, 1964, Harvard University.

Thompson, Nils Wilder. The geology of the Fairport Quadrangle, Ellis and Russell counties, Kansas. M, 1988, Fort Hays State University. 63 p.

Thompson, P. C. School site nature centers in environmental education. M, 1977, Northeastern Illinois University.

Thompson, P. J. Stratigraphy and structure of Shattuck Ridge, Bakersfield and Waterville, Vermont. M, 1975, University of Vermont.

Thompson, Peter. Speleochronology and late Pleistocene climates inferred O, C, H, U and Th isotopic abundances in speleothems. D, 1973, McMaster University. 352 p.

Thompson, Peter Hamilton. Stratigraphy, structure and metamorphism of the Flinton Group in the Bishop Corners-Madoc area, Grenville Province, eastern Ontario. D, 1972, Carleton University. 268 p.

Thompson, Peter James. Stratigraphy, structure, and metamorphism in the Monadnock Quadrangle, New Hampshire. D, 1985, University of Massachusetts.

Thompson, Peter Robert. Planktonic foraminiferal biostratigraphy of Oligocene deep-sea piston cores (middle latitude Atlantic and Pacific oceans). M, 1972, Rutgers, The State University, New Brunswick. 73 p.

Thompson, R. M. Descriptive mineralogy of the tellurides. D, 1947, University of Toronto.

Thompson, Randolph Charles. The development of fractures in the Mesozoic volcanic rocks adjacent to the Sierrita porphyry copper deposit, Pima County, Arizona. M, 1981, University of Arizona. 85 p.

Thompson, Raymond M. Stratigraphic sections of pre-Cody Upper Cretaceous rocks in central Wyoming. M, 1950, University of Wyoming. 21 p.

Thompson, Richard J. The structure and stratigraphy of a portion of the southern edge of the Dead River basin, Marquette County, Michigan. M, 1961, Michigan State University. 63 p.

Thompson, Richard Lee. The geology of the middle one-third of the Glide Quadrangle, Douglas County, Oregon. M, 1968, University of Oregon. 57 p.

Thompson, Richard Rogers. Lithostratigraphy of Middle Ordovician Salona and Coburn formations of the Trenton Group in central Pennsylvania. D, 1961, Pennsylvania State University, University Park. 297 p.

Thompson, Robert I. Geology of the Akolkolex River area near Revelstoke, British Columbia. D, 1972, Queen's University. 125 p.

Thompson, Robert M. Morphogenesis of Blowout Canyon debris deposit, Mt. Baird Idaho-Wyoming 7 1/2' Quadrangle. M, 1978, Idaho State University. 54 p.

Thompson, Robert Mitchell. Minor elements in sphalerite and some associated minerals. M, 1943, University of British Columbia.

Thompson, Robert S. Late Pleistocene and Holocene packrat middens from Smith Creek Canyon, White Pine County, Nevada. M, 1978, University of Arizona.

Thompson, Robert Stephen. Late Pleistocene and Holocene environments in the Great Basin. D, 1984, University of Arizona. 275 p.

Thompson, Robert Wayne. Tidal flat sedimentation on the Colorado River delta, northwestern Gulf of California. D, 1965, University of California, San Diego.

Thompson, Rodney. Sedimentation of the Yukon Delta, Alaska. M, 1984, University of Houston.

Thompson, Roger B., Jr. Landfill influenced degradation of stream water. M, 1975, University of Vermont.

Thompson, Sam, III. Geology of the southern part of the Fra Cristobal Range, Sierra County, New Mexico. M, 1955, University of New Mexico. 75 p.

Thompson, Shannon Wesley. Carbonate complexation and adsorption of rare elements in seawater. M, 1987, University of South Florida, St. Petersburg.

Thompson, Sheree A. Stratigraphy of the Upper Jurassic rocks of the Sabine Uplift area, Texas-Louisiana. M, 1981, Baylor University. 173 p.

Thompson, Stanley D. Estimation of regional trends in gravity using gram orthogonal polynomials. M, 1968, University of Missouri, Rolla.

Thompson, Stephen C. Depositional environments and history of the Winnipeg Group (Ordovician), Williston Basin, North Dakota. M, 1984, University of North Dakota. 210 p.

Thompson, Stephen L. Stratigraphic and structural relationships governing Mississippian gas occurrence in Mecosta County and the Falmouth gas field in Missaukee County, Michigan. M, 1963, Michigan State University. 74 p.

Thompson, Stewart N. Sea-level rise along the Maine coast during the last 3,000 years. M, 1973, University of Maine. 78 p.

Thompson, Susan Lewis. Ferron Sandstone Member of the Mancos Shale; a Turonian mixed-energy deltaic system. M, 1985, University of Texas, Austin. 165 p.

Thompson, Thomas Dick, III. Adsorption of purines, pyrimidines and nucleosides on montmorillonites and illites. D, 1969, Pennsylvania State University, University Park. 86 p.

Thompson, Thomas Dick, III. Complexes of water and some simple n-alkyl compounds with a synthetic montmorillonite. M, 1965, Pennsylvania State University, University Park. 68 p.

Thompson, Thomas L. Conodonts of the Meramecian Stage (Mississippian) from the subsurface of western Kansas. M, 1962, University of Kansas. 65 p.

Thompson, Thomas Luman. Stratigraphy, tectonics, structure, and gravity in the Rocky Mountain Trench area, southeastern British Columbia, Canada. D, 1962, Stanford University. 280 p.

Thompson, Thomas Luther. Conodonts from the Meramecian Stage (Upper Mississippian) of Kansas. D, 1965, University of Iowa. 212 p.

Thompson, Thomas Marvin. Geology of the Clifton Forge area, Alleghany-Botetourt counties, Virginia. M, 1955, University of Virginia. 109 p.

Thompson, Timothy James. Outer-fan depositional lobes of the lower to middle Eocene Juncal Formation, southern San Rafael Mountains, California. M, 1987, University of California, Santa Barbara. 96 p.

Thompson, Todd A. Limestone-clast conglomerates in the Early Cretaceous foreland basin in southwestern Montana; origin and significance. M, 1984, Indiana University, Bloomington. 73 p.

Thompson, Todd Alan. Sedimentology, internal architecture and depositional history of the Indiana Dunes National Lakeshore and State Park. D, 1987, Indiana University, Bloomington. 129 p.

Thompson, Tommy B. Geology of the Sierra Blanca, Lincoln and Otero counties, New Mexico. D, 1966, University of New Mexico. 146 p.

Thompson, Tommy B. The geology of the South Mountain area, Bernalillo, Sandoval, and Santa Fe counties, New Mexico. M, 1963, University of New Mexico. 69 p.

Thompson, Troy Richard. Fracturing in the Wasatch fault zone; implications for fluid flow, rock mass strength and earthquake processes. M, 1988, University of Utah. 186 p.

Thompson, Walter H. Mineralization and metamorphism of the Pend d'Oreille-Salmo area [Idaho]. M, 1952, University of Toronto.

Thompson, Warren Charles. Geological oceanography of the Atchafalaya Bay area, Louisiana. D, 1954, Texas A&M University.

Thompson, Warren F. Cathodoluminescence colors and textures of metamorphic rocks of the Black Hills, South Dakota. M, 1983, University of Akron. 94 p.

Thompson, Warren O. Original structure of beaches. D, 1934, Stanford University. 227 p.

Thompson, Will Francis. The regional morphological character of New England mountains. D, 1960, Clark University. 262 p.

Thompson, William Allen. Mineralogy of tin, tungsten, and tantalum deposits of the Harney Peak region (South Dakota). M, 1930, Iowa State University of Science and Technology.

Thompson, William B. A study of the magneto-telluric method of prospecting. M, 1958, Massachusetts Institute of Technology. 54 p.

Thompson, William David. Stratigraphy of black shale facies of Green River Formation (Eocene), Uinta Basin, Utah. M, 1971, University of Utah. 61 p.

Thompson, William E. The concentration of selected elements in brines of Perry County, Ohio. M, 1973, Ohio State University.

Thompson, William Hayes. Crustal structure near the Iceland Research Drilling Project borehole from a seismic refraction survey. M, 1980, University of Washington. 67 p.

Thompson, William Joseph. Structure of Spring Mountain Mesozoic formations, Fremont County, Wyoming. M, 1952, Miami University (Ohio). 27 p.

Thompson, William Ross. Hydrogeology of the Jocko Valley, west-central Montana. M, 1988, University of Montana. 120 p.

Thompson, Willis H., Jr. The conodonts of the Platteville Formation of southeastern Minnesota. M, 1959, University of Minnesota, Minneapolis. 168 p.

Thompson, Woodrow B. The drainage and glacial history of the Still River Valley, southwestern Connecticut. M, 1971, University of Vermont.

Thompson, Woodrow Burr. The Quaternary geology of the Danbury-New Milford area, Connecticut. D, 1975, Ohio State University. 193 p.

Thompstone, Robert Marshal. Topics in hydrological time series modelling. D, 1983, University of Waterloo. 227 p.

Thoms, John A. The type upper Marietta Sandstone (Dunkard Series) of southeastern Ohio. M, 1956, Miami University (Ohio). 75 p.

Thoms, Richard Edwin. Biostratigraphy of the Umpqua Formation (lower to middle Eocene), southwest Oregon. D, 1965, University of California, Berkeley. 235 p.

Thoms, Richard Edwin. The geology and Eocene biostratigraphy of southern Quimper Peninsula area, Washington. M, 1959, University of Washington. 102 p.

Thomsen, Anton Gaarde. A watershed information system. D, 1980, Colorado State University. 176 p.

Thomsen, Bert W. Surface-water supply for the city of Williams, Coconino County, Arizona. M, 1968, University of Arizona.

Thomsen, Harry L. Textural variations of the (Lower Mississippian) Berea Sandstone of northern Ohio. M, 1934, Oberlin College.

Thomsen, K. O. Hydrogeology for urban planning, Kane County, Illinois, U.S.A. M, 1974, Northeastern Illinois University.

Thomsen, Leon. On the fourth-order anharmonic equation of state of solids. D, 1969, Columbia University. 172 p.

Thomsen, Mark Andrew. Petrology of the mudrocks of the Dunkard Group (Upper Pennsylvanian-Permian), northern Dunkard Basin, West Virginia and Pennsylvania. M, 1980, Miami University (Ohio). 136 p.

Thomson, Alan Frank. Petrology and cementation of (Devonian) Oriskany Sandstone. M, 1954, West Virginia University.

Thomson, Alan Frank. Petrology of the Silurian quartzites and conglomerate in New Jersey. D, 1957, Rutgers, The State University, New Brunswick. 457 p.

Thomson, D. B. A study of the combination of terrestrial and satellite geodetic networks. D, 1976, University of New Brunswick.

Thomson, David James. Analytical and laboratory model studies of electromagnetic induction within the Earth. D, 1973, University of Victoria. 217 p.

Thomson, Ellis. I, Quantitative microscopic analysis; II, A qualitative and quantitative determination of the ores of Cobalt, Ontario. D, 1929, Harvard University.

Thomson, G. Colloids in groundwater; field and laboratory experiments and a literature review. M, 1986, University of Waterloo. 131 p.

Thomson, James Alan. Fossil communities of the Skaneateles Formation (Middle Devonian, Hamilton Group) in central New York. M, 1978, Syracuse University.

Thomson, James E. The nature and origin of the nepheline syenites and related alkali syenites of Coldwell, Ontario. D, 1932, University of Wisconsin-Madison.

Thomson, John Pretiss. Genesis of the Swauk placer gold, Swauk mining district, Washington. M, 1932, Washington State University. 76 p.

Thomson, Kenneth. Geology of the Sugarloaf fault block, South Franklin Mountains, El Paso County, Texas. M, 1974, University of Texas at El Paso.

Thomson, Kenneth Clair. Mineral deposits of the Deep Creek mountains, Tooele and Juab counties, Utah. D, 1970, University of Utah. 483 p.

Thomson, Ker C. A study in visco-elastic wave propagation by photoiscollasticity. D, 1965, Colorado School of Mines. 274 p.

Thomson, Margaret C. Biofacies and paleoecology of Upper Carboniferous fossil plant assemblages. M, 1978, University of Pennsylvania.

Thomson, Margaret Lee. The Crixas gold deposit, Brazil; metamorphism, metasomatism and gold mineralization. D, 1987, University of Western Ontario. 345 p.

Thomson, Richard Edward. Theoretical studies of the circulation of the Subarctic Pacific region and the generation of Kelvin type waves by atmospheric disturbances. D, 1971, University of British Columbia.

Thomson, Robert. A study of the nickel intrusive, Sudbury, Ontario. D, 1935, University of Chicago. 51 p.

Thomson, Robert Charles. Lower to Middle Jurassic (Pliensbachian to Bajocian) stratigraphy and Pliensbachian ammonite fauna of the northern Spatsizi area, north-central British Columbia. M, 1985, University of British Columbia. 183 p.

Thomson, Robert D. Mining of mineral aggregates in urban areas. D, 1980, University of Pittsburgh. 180 p.

Thomson, William A., III. A geophysical investigation of a possibly Recent fault in southwestern Oklahoma. M, 1986, University of Oklahoma. 113 p.

Thonis, Michael. Correlations between manganese and copper ore deposits and their relationship to plate tectonics. M, 1974, Massachusetts Institute of Technology. 110 p.

Thor, Devin. Depositional environment and paleogeographic setting of the Santa Margarita Formation, Ventura Co., California. M, 1977, California State University, Northridge. 144 p.

Thorarinsson FreyrFreyr Thorarinsson

Thorarinsson Freyr *see* Freyr Thorarinsson

Thorbjarnardottir, Bergthora. Relative earthquake location by cross-correlation of waveforms; application to preshock-mainshock-aftershock. M, 1985, University of Utah. 98 p.

Thoreson, Ronald F. Reconnaissance geology of the western Smokey, Ceibo Grande, and Ceibo Chico rivers, Belize, Central America. M, 1980, University of Idaho. 71 p.

Thorgrimsson, S. Suitability of glacial and glaciofluvial soils from the Vestfjordur Peninsula, Iceland, in embankment dam site. M, 1978, University of Arizona.

Thorkelson, Derek John. Volcanic stratigraphy and petrology of the Mid-Cretaceous Spences Bridge Group near Kingsvale, southwestern British Columbia. M, 1986, University of British Columbia. 119 p.

Thorkelson, John M., Jr. Depositional environments of select shelfal sands associated with the Permian Basin of West Texas and southeastern New Mexico. M, 1983, University of Southwestern Louisiana. 68 p.

Thorleifson, Allan James. A marine deep seismic sounding survey over Winona Basin. M, 1978, University of British Columbia.

Thorleifson, James Tracy. A modified stepwise fluorination procedure for the oxygen isotopic analysis of hydrous silica. M, 1984, Arizona State University. 71 p.

Thorleifson, L. Harvey. The eastern outlets of Lake Agassiz. M, 1983, University of Manitoba.

Thormahlen, David J. Geology of the northwest one-quarter of the Prineville Quadrangle, central Oregon. M, 1984, Oregon State University. 106 p.

Thorman, Charles Hadley. Geology of Wood Hills, Elko Co., Nev. M, 1960, University of Washington. 49 p.

Thorman, Charles Hadley. Structure and stratigraphy of the Wood Hills and a portion of the northern Pequop Mtns., Elko Co., Nev. D, 1962, University of Washington. 218 p.

Thorn, Kevin Arthur. NMR structural investigations of aquatic humic substances. D, 1984, University of Arizona. 209 p.

Thornber, Carl Richard. Factors controlling the ferric-ferrous ratio in basaltic liquids. M, 1977, Queen's University. 101 p.

Thornburg, Robert. Stratigraphic relationships of the Cambrian Knox and overlying Middle Ordovician strata. M, 1979, University of Pittsburgh.

Thornburg, Todd Mark. Sedimentary basins of the Peru continental margin; structure, stratigraphy, and Cenozoic tectonics from 6°S to 16°S latitude. M, 1981, Oregon State University. 60 p.

Thornburg, Todd Mark. Sedimentation in the Chile Trench. D, 1985, Oregon State University. 182 p.

Thornburgh, Hubert Robert. Geology and ore deposits of the St. John's quicksilver mine, Solano County, California. M, 1923, University of California, Berkeley. 125 p.

Thornburn, Thomas Hampton. The effect of certain wetting agents on the water intake and erodibility of soils. D, 1941, Michigan State University. 49 p.

Thornbury, William David. Glacial geology of southern and south-central Indiana. D, 1936, Indiana University, Bloomington. 204 p.

Thornbury, William David. Glaciation on the east side of the Colorado Front Range between James Peak and Longs Peak; Boulder County. M, 1928, University of Colorado.

Thorndale, C. William. Washington's Green River Coal Company; 1880-1930. M, 1965, University of Washington. 168 p.

Thorndike, Alan. Structure of flow fields. D, 1978, University of Washington. 90 p.

Thorndycraft, R. B. An investigation of silicate ores; beneficiation and X-ray analytical procedures. M, 1974, University of Washington. 101 p.

Thorne, Julian Arthur. Studies in stratology; the physics of stratigraphy. D, 1985, Columbia University, Teachers College. 530 p.

Thorne, P. An ecosystem assessment technique for environmental impact statements. M, 1974, University of Arizona.

Thornhill, Ronald Roger. Distribution of clay in Recent sand. M, 1978, University of New Orleans. 129 p.

Thornhill, Stephen Alan. The Austin Chalk and its petroleum potential; south-central Texas. M, 1982, Baylor University. 161 p.

Thorniley, B. K. Experiments with copper minerals at high temperatures. M, 1962, McGill University.

Thornton, Charles P. The geology of the Mount Jackson Quadrangle, Virginia. D, 1953, Yale University.

Thornton, Edward Clifford. Anorthosite-gabbro-granophyre relationships, Mount Sheridan area, Oklahoma. M, 1975, Rice University. 65 p.

Thornton, Edward Clifford. Experimental and theoretical modeling of sediment-seawater hydrothermal interaction at constant temperature and in a thermal gradient; implications for the diagenesis and metamorphism of marine clay and the subseabed disposal of nuclear waste. D, 1983, University of Minnesota, Minneapolis. 201 p.

Thornton, James P. Stratigraphy and petrology of sand units near the base of the Atoka Formation, Crawford County, Arkansas. M, 1968, University of Arkansas, Fayetteville.

Thornton, Jeffrey A. The distribution of reactive silicate in the Piscataqua River estuary of New Hampshire-Maine. M, 1976, University of New Hampshire. 65 p.

Thornton, Patricia Ann. Stratigraphy, structural relation, petrology and chemical correlation of the Kilgore Complex in south eastern Idaho. M, 1988, Wayne State University.

Thornton, Robert C. Geology of the northern half of the Ellington Quadrangle, Missouri. M, 1963, University of Missouri, Rolla.

Thornton, Scott Ellis. Holocene stratigraphy and sedimentary processes in Santa Barbara Basin; influence of tectonics, ocean circulation, climate and mass movement. D, 1981, University of Southern California.

Thornton, Scott Ellis. The Holocene evolution of a coastal lagoon, Lake Tunis, Tunisia. M, 1976, Duke University. 151 p.

Thornton, Wayne D. Mississippian rocks in the subsurface of Alfalfa and parts of Woods and Grant counties, northwestern Oklahoma. M, 1958, University of Oklahoma. 71 p.

Thornton, Wendy Barclay. Distortion of stream channel responses in a semiarid environment; Sand Creek, El Paso County, Colorado. M, 1985, University of Colorado at Colorado Springs. 117 p.

Thornton, William Mynn, Jr. New processes for the analytical separation of titanium from iron, aluminum, and phosphoric acid. D, 1914, Yale University.

Thornton, William S. Minimum disposal depth investigation of southern Coffey and northern Woodson counties, Kansas. M, 1985, Emporia State University. 23 p.

Thoroman, Marilyn. K-Ar geochronology of the Dokhan Volcanics of the central Eastern Desert, Red Sea Hills, Egypt. M, 1985, Georgia Institute of Technology. 72 p.

Thorp, Eldon M. Calcareous shallow-water marine deposits of Florida and the Bahamas. D, 1934, University of California, Berkeley. 129 p.

Thorpe, Douglas G. Mineralogy and petrology of Precambrian metavolanic rocks, Squaw Peak, Phoenix, Arizona. M, 1980, Arizona State University. 96 p.

Thorpe, L. R. The taxonomy and paleoecology of the B subzone productids from the Windsor Group in the western part of the Minas Sub-basin. M, 1974, Acadia University.

Thorpe, Malcolm R. The geology of the Abajo Mountains, San Juan County, Utah. D, 1916, Yale University.

Thorpe, Ralph Irving. The controls of hypogene sulphide zoning, Rossland, British Columbia. D, 1967, University of Wisconsin-Madison. 141 p.

Thorpe, Ralph Irving. The radioactive mineralogy of the ore conglomerates at Panel Mine, Blind River, Ontario. M, 1963, Queen's University. 148 p.

Thorsen, Carl Elmer. Stratigraphy and Ostracoda of the Brownstown and Tokio formations; Southwest Arkansas. D, 1959, Louisiana State University. 164 p.

Thorsen, Carl P. E. The attitude of the Pleistocene depositional surfaces in the northern portions of East Feliciana and West Feliciana parishes, Louisiana and the adjoining portions of Mississippi. M, 1954, Louisiana State University.

Thorsen, Gerald Wayne. The geology of an amphibolite unit in northern Stevens County and Pend Oreille County, Washington. M, 1966, Washington State University. 71 p.

Thorson, E. F. Rocks and rock alteration in part of the Malartic area (Quebec). M, 1940, McGill University.

Thorson, Jeffrey R. Velocity stack and slant slack inversion methods. D, 1984, Stanford University. 157 p.

Thorson, Jon Peer. Igneous petrology of the Oatman District, Mohave County, Arizona. D, 1971, University of California, Santa Barbara. 233 p.

Thorson, Robert M. The Late Quaternary history of the Dry Creek area, central Alaska. M, 1975, University of Alaska, Fairbanks. 85 p.

Thorson, Robert Mark. Isostatic effects of the last glaciation in the Puget Lowland, Washington. D, 1979, University of Washington. 154 p.

Thorstad, Linda Elaine. The Upper Triassic "Kutcho Formation", Cassiar Mountains, north-central British Columbia. M, 1983, University of British Columbia. 271 p.

Thorsteinsson, Erik. Upper Mississippian Visean ammonoids of the Peel River, Yukon. M, 1972, University of Iowa. 103 p.

Thorsteinsson, Raymond. Geology of Cornwallis and Little Cornwallis islands, Arctic Archipelago, Northwest Territories. D, 1955, University of Kansas. 247 p.

Thorsteinsson, Thorsteinn. Preliminary report on the geology of Colgrove Butte, Hettinger County, North Dakota. M, 1949, University of North Dakota. 31 p.

Thorstenson, Donald Carl. Equilibrium distribution of small organic molecules in natural waters. D, 1969, Northwestern University. 159 p.

Thorup, Richard Russell. The stratigraphy of the (Miocene) Vaqueros Formation at its type locality, Monterey County, California. M, 1942, Stanford University. 70 p.

Thrailkill, John Vernon. A speleological investigation of Fulford Cave, Eagle County, Colorado. M, 1955, University of Colorado.

Thrailkill, John Vernon. Studies in the excavation of limestone caves and the deposition of speleothems; Part 1, Chemical and hydrologic factors in the excavation of limestone caves; Part 2, Water chemistry and carbonate speleothem relationships in Carlsbad Caverns, New Mexico. D, 1965, Princeton University. 206 p.

Thrall, F.G. Geotechnical significance of poorly crystalline soils derived from volcanic ash. D, 1981, Oregon State University. 445 p.

Thrasher, Glenn P. Median valley crustal structure and sea floor spreading at the Gorda Ridge, 42°N latitude. M, 1978, Oregon State University. 67 p.

Thrasher, Lawrence C. Macrofossils and biostratigraphy of the Bakken Formation (Devonian and Mississippian) in western North Dakota. M, 1985, University of North Dakota. 292 p.

Thrasher, Ronald E. A petrographic study and reconnaissance geologic map of the Tepee Trail Formation and a post-Tepee Trail volcanic deposit in the Brooks Lake Creek Falls area, Fremont County, Wyoming. M, 1959, Miami University (Ohio). 83 p.

Threadgold, Ian Malcolm. The crystal structures of hellyerite and nacrite. D, 1963, University of Wisconsin-Madison. 107 p.

Threet, Richard Lowell. Geology and the origin of the Lake Ontario Basin. M, 1949, University of Illinois, Urbana.

Threet, Richard Lowell. Geology of Red Hills area, Iron Co., Utah. D, 1952, University of Washington. 107 p.

Threinen, David T. Stratigraphy and petrography of the Mayes Group of northern Arkansas. M, 1961, Northwestern University.

Threlkeld, William Earl. Geology of the stockwork molybdenite occurrence, Jamestown, Colorado. M, 1982, University of Western Ontario. 109 p.

Thresher, John E., Jr. Stratigraphy, structure, and metamorphism in the Richmond area (Chittenden county), (Vermont). M, 1970, University of Vermont.

Thrivikramaji, K. P. Sedimentology and paleohydraulics of Pleistocene Syracuse channels, Onondaga County, New York. D, 1977, Syracuse University. 200 p.

Throckmorton, Michael. Petrology of the Castle Lake peridotite-gabbro mass, eastern Klamath Mountains, California. M, 1978, University of California, Santa Barbara.

Thronton, Edward Bennett. Internal density currents generated in a density stratified reservoir during withdrawal. M, 1966, Oregon State University. 70 p.

Throop, Allen H. The nature and origin of black chrysocolla at the Inspiration Mine (Inspiration), Arizona. M, 1970, Arizona State University. 56 p.

Throop, Robert Neblett. Geology of the northern Sierra Vieja Mountains, Trans-Pecos, Texas. M, 1949, University of Texas, Austin.

Thrupp, Gordon Alan. The paleomagnetism of Paleogene lava flows on the Alaska Peninsula, and the tectonics of southern Alaska. D, 1987, University of California, Santa Cruz. 261 p.

Thudium, C. L. The geological development of Lake Superior. M, 1978, Northeastern Illinois University.

Thum, James Arthur. A seismic and gravity analysis of the subsurface geology in Fairfield County, Ohio. M, 1983, Wright State University. 85 p.

Thumtrakul, Wilaiwan. The carbon dioxide system and its role on trace metal chemistry in the marine environment. D, 1984, University of Rhode Island. 163 p.

Thune, Howard Willis. Mineralogy of the ore deposits of the western portion of the Little Eightmile mining district, Lemhi County, Idaho. M, 1941, University of Idaho. 49 p.

Thunell, R. C. Late Cenozoic biostratigraphy and paleoceanography of the Mediterranean Sea. D, 1978, University of Rhode Island. 363 p.

Thurber, Clifford H. Earth structure and earthquake locations in the Coyote Lake area, Central California. D, 1981, Massachusetts Institute of Technology. 332 p.

Thurber, David Lawrence. Natural variation in the ratio U^{234}/U^{238} investigation of the potential of U^{234} for Pleistocene chronology. D, 1963, Columbia University, Teachers College. 172 p.

Thurber, James E. Petrology and Cu-Mo mineralization of the Kennedy Stock, East Range, Pershing County, Nevada. M, 1982, Colorado State University. 114 p.

Thurber, Judson B. General and economic geology of the West Kootenay Batholith with special reference to the Cranbrook area. M, 1937, University of British Columbia.

Thuren, John B. A fluidic controlled seismic source. D, 1969, Colorado School of Mines. 85 p.

Thurlow, Ernest Emmanuel. Geology and ore deposits of the Lower Hot Springs mining district, Madison County, Montana. M, 1941, Montana College of Mineral Science & Technology. 85 p.

Thurlow, Ernest Huntington. The water quality and bottom sediment characteristics of New Jersey lagoon developments. D, 1974, Rutgers, The State University, New Brunswick. 344 p.

Thurlow, John Geoffrey. Geology, ore deposits and applied rock geochemistry of the Buchans Group, Newfoundland. D, 1981, Memorial University of Newfoundland. 305 p.

Thurlow, John Geoffrey. Lithogeochemical studies in the vicinity of the Buchans massive sulphide deposits, central Newfoundland. M, 1973, Memorial University of Newfoundland. 171 p.

Thurman, Charles P. Origin and stratigraphic relations of the Ohio River Formation (Paleocene), Tip Top (Mississippian), and Mooretown Formation (Upper Mississippian) of southern Indiana and west central Kentucky. M, 1973, University of Kentucky. 41 p.

Thurman, E. M. Isolation, characterization, and geochemical significance of humic substances from ground water. D, 1979, University of Colorado. 230 p.

Thurman, Franklin A. The Ostracoda of the (Paleocene) Kincaid Formation (Texas). M, 1938, University of Chicago. 44 p.

Thurmer, G. Sidney. Analysis of palynomorph succession in Pennsylvanian strata, Morgan County, Tennessee. M, 1976, University of Tennessee, Knoxville. 50 p.

Thurmond, Carol J. The geology of the Chennault 7 1/2' Quadrangle, Georgia-South Carolina, and its significance to the problem of the Charlotte Belt-Carolina Slate Belt boundary. M, 1979, University of Georgia.

Thurmond, John Tydings. Lower vertebrates and paleoecology of the Trinity group (lower Cretaceous) in north central Texas. D, 1969, Southern Methodist University. 128 p.

Thurmond, John Tydings. Quaternary deposits of the east fork of the Trinity River, northcentral Texas. M, 1967, Southern Methodist University. 74 p.

Thurmond, Valerie. The carbonate system in Ca-Mg-Na-Cl brines up to 6.0 molal NaCl at 25°C. M, 1982, University of Miami. 129 p.

Thurn, Richard L. Petrofabric investigation of the southern portion of the Bear Mountain Dome, Black Hills, South Dakota. M, 1968, South Dakota School of Mines & Technology.

Thurrell, Robert F., III. Anthophyllite-cummingtonite assemblages in the Clinton Quadrangle (Connecticut). M, 1970, University of Rochester.

Thurston, Peter Bouck. Geochemistry and provenance of Archean metasedimentary rocks in the southwestern Beartooth Mountains. M, 1986, Montana State University. 74 p.

Thurston, Phillips Cole. Petrography of the Baltimore Gneiss (Precambrian) at Glen Mills, Chester County, Pennsylvania. M, 1965, Bryn Mawr College. 37 p.

Thurston, Phillips Cole. The volcanology and trace element geochemistry of cyclical volcanism in the Archean Confederation Lake area, northwestern Ontario. D, 1981, University of Western Ontario. 554 p.

Thurston, William Richardson. Geology of manganese deposit at Clinton Point, New Jersey. M, 1943, Columbia University, Teachers College.

Thurston, William Richardson. Pegmatites of the Crystal Mountain District, Larimer County, Colorado. D, 1953, Columbia University, Teachers College.

Thurwachter, Jeffrey E. Sedimentology of Neogene basin-fill deposits, lower Tornillo Creek area, Big Bend National Park, Texas. M, 1984, University of Texas, Austin. 87 p.

Thusu, Bindraban. Palynology of the Dunvegan Formation (Cretaceous) in type area, on Peace River, Alberta (Canada). M, 1968, University of Alberta. 176 p.

Thwaites, Fredrik T. Geology of the southeast quarter of the Cross Plains Quadrangle, Dane County, Wisconsin. M, 1908, University of Wisconsin-Madison.

Tiab, Djebbar. Analysis of multiple-sealing-fault systems and closed rectangular reservoirs by type curve matching. D, 1976, University of Oklahoma. 253 p.

Tibbetts, Benton L. Subsurface geology of Norton County, Kansas. M, 1958, Kansas State University. 55 p.

Tibbs, Nicholas Howard. Background concentrations of copper, lead, and zinc in streams of the "new lead belt", (southeast portion of Missouri from Ellington, to Viburnum, Missouri). M, 1969, University of Missouri, Rolla.

Tibbs, Nicholas Howard. Wall rock geochemistry of the Chester Vein, Sunshine Mine, Kellogg, Idaho. D, 1972, University of Missouri, Rolla. 120 p.

Tibby, Richard B. Results of oceanographic investigations off the California coast, 1939-1940. D, 1944, [University of California, San Diego].

Ticken, Edward John. Geology of the inner basin margin; Dana Point to San Onofre, California. M, 1983, California State University, Northridge. 85 p.

Tickner, Bruce. Stratigraphic studies and microfacies analysis, of the Jurassic succession, Malone Mountains, West Texas. M, 1987, University of Texas at Dallas. 204 p.

Tidball, Ronald Richard. A study of soil development on dated pumice deposits from Mount Mazama, Oregon. D, 1965, University of California, Berkeley. 251 p.

Tidwell, William D. A Lower Pennsylvanian flora from Utah and its stratigraphic significance. D, 1966, Michigan State University. 200 p.

Tidwell, William D. Early Pennsylvanian flora from the Manning Canyon Shale, Utah. M, 1962, Brigham Young University. 101 p.

Tidy, Enrique Finch. Petrology of leucite-bearing volcanic plugs, Deep Spring valley, California. M, 1972, University of California, Berkeley. 84 p.

Tie, An. On scattering of seismic waves by a spherical obstacle. D, 1987, Georgia Institute of Technology. 141 p.

Tiedemann, Herbert Allen. The geology of the Postoak Window, Roane County, Tennessee. M, 1956, University of Tennessee, Knoxville. 59 p.

Tieh, Thomas Ta-pin. Heavy mineral assemblages in some Tertiary sediments near Stanford University, California. D, 1965, Stanford University. 193 p.

Tieje, Arthur Jerrold. The Cambrian sediments of the Bighorn Mountains. D, 1920, University of Minnesota, Minneapolis. 186 p.

Tieman, David J. Deformation textures in barite and fluorite of the Del Rio District and their implications as to the relative timing of deformation and mineralization. M, 1978, University of Tennessee, Knoxville. 84 p.

Tieman, Joannes Jacobus. A theory concerning the behavior of waves in inhomogeneous media. M, 1980, University of Calgary. 128 p.

Tiemann, Theodore D. The effect of hydrogen ion concentration on the flotation of some southwestern Wisconsin zinc ores and of a certain pyrrhotitic gold ore. D, 1933, University of Wisconsin-Madison.

Tien, Jean H. Investigation on seismic wave forms. D, 1972, Colorado School of Mines. 106 p.

Tien, Pei-Lin. Distribution and character of water-insoluble residues in Hutchison Salt (Wellington Formation, Permian), Sedgwick County, Kansas. M, 1965, University of Kansas. 64 p.

Tien, Pei-Lin. Petrology and mineralogy of underclays in Cherokee Group (middle Pennsylvanian), Kansas. D, 1973, University of Kansas. 117 p.

Tien, Yu Bun. Site response and seismicity in the Seattle area. D, 1970, University of Washington. 195 p.

Tierney, Michael T. Temperature and pressure effects on the chemical durability of borosilicate simulated nuclear waste glass. M, 1983, Rensselaer Polytechnic Institute. 59 p.

Tietbohl, Douglass Ralph. Structure and stratigraphy of the Hayden Creek area, Lemhi Range, east-central Idaho. M, 1981, Pennsylvania State University, University Park. 121 p.

Tiety, P. J. Holocene sedimentation and evolution of the Great Lake area, North Carolina. M, 1981, University of North Carolina, Chapel Hill.

Tietz, Frederic A. The Adena oil field area, Morgan and Adams counties, Colorado. M, 1956, University of Colorado.

Tietz, Paul G. Holocene sedimentation and evolution of the Great Lake area, North Carolina. M, 1981, University of North Carolina, Chapel Hill. 82 p.

Tiezzi, Lawrence James. Petrogenesis of basalts from the Mid-Atlantic Ridge, 26°N. D, 1982, Texas A&M University. 128 p.

Tiezzi, Lawrence James. Petrology of a dredged cumulate-textured gabbroic complex from the Mid-Atlantic Ridge, latitude 26°N. M, 1977, Texas A&M University. 117 p.

Tiezzi, Pamela Anne. Petrography and diagenesis of the Mississippian Lodgepole Formation, south-central Montana. M, 1984, University of Texas, Austin. 181 p.

Tiffin, Donald Lloyd. Continuous seismic reflection profiling in the Strait of Georgia, British Columbia (Canada). D, 1970, University of British Columbia.

Tighe, Edward J. A study of the coal deposits of South America. M, 1942, Catholic University of America. 105 p.

Tight, George William. The origin and development of the Ohio River. D, 1902, University of Chicago.

Tihor, Sharon Louise. The mineralogical composition of the carbonate rocks of the Kirkland Lake-Larder Lake gold camp. M, 1978, McMaster University. 93 p.

Tikrity, Sammi Sherif. Subsurface geology of part of Pennsylvanian System, southwest quarter of Wise County, Texas. M, 1964, University of Texas, Austin.

Tikrity, Sammi Sherif. Tectonic genesis of the Ozark Uplift. D, 1968, Washington University. 196 p.

Tilander, Nathaniel G. Application of the generalized inverse technique to the estimation of residual statics corrections. M, 1978, Indiana University, Bloomington. 234 p.

Tilden, Jean Ellen. Certain trace elements among coexisting sulfide minerals in selected ore deposits of the southeastern United States. M, 1971, North Carolina State University. 43 p.

Tilford, Maxwell Joseph. Structural analysis of the southern and western Greenhorn Range, Madison County, Montana, and magnetic beneficiation of Montana talc cores. M, 1978, Indiana University, Bloomington. 181 p.

Tilford, Norman Ross. Engineering geology-Stewart Mountain dam site (Arizona). M, 1966, Arizona State University. 42 p.

Tiliouine, Boualem. Nonstationary analysis and simulation of seismic signals. D, 1983, Stanford University. 265 p.

Tilke, Peter Gerhard. Caledonian structure, metamorphism, geochronology, and tectonics of the Sitas-Singis area, Sweden. D, 1987, Massachusetts Institute of Technology. 295 p.

Till, Alison Berna. Crystalline rocks of the Kigluaik Mountains, Seward Peninsula, Alaska. M, 1980, University of Washington.

Till, Henry Anthony. A computer model for three-dimensional simulation of thermal discharges into rivers. D, 1973, University of Missouri, Rolla.

Tiller, Kevin George. Specific soprtion of some heavy metal cations by pure minerals and soil clays. D, 1961, Cornell University.

Tilley, Barbara J. Sedimentology and clay mineralogy of the glauconitic sandstone, Suffield heavy oil sands, southeastern Alberta. M, 1982, University of Alberta. 127 p.

Tilley, Barbara Jean. Diagenesis and porewater evolution in Cretaceous sedimentary rocks of the Alberta deep basin. D, 1988, University of Alberta. 222 p.

Tilley, Craig W. Geology of the Spanish Creek basin area, Madison and Gallatin counties, Montana. M, 1976, Montana State University. 111 p.

Tilley-Grogan, Carol. Oil retention of unconsolidated deposits. M, 1978, University of Minnesota, Duluth.

Tilling, Robert Ingersoll. Batholith emplacement and contact-metamorphism in the Paipote-Tierre Amarilla area, Atacama Province, Chile. D, 1963, Yale University.

Tillman, Arthur G. The Mississippi Gorge, successive adjustments to the environment, La Crosse, Wisconsin, to Winona, Minnesota. D, 1928, University of Wisconsin-Madison.

Tillman, Chauncey; Freeman, Leroy Bradford and Sweet, John M. Geology of the Henrys Lake Mountains, Fremont County, Idaho and Madison and Gallatin counties, Montana. M, 1950, University of Michigan. 83 p.

Tillman, Chauncey Glenn. Stratigraphy and brachiopod fauna of the (Silurian) Osgood Formation, Laurel Limestone, and Waldron Shale of southeastern Indiana. D, 1962, Harvard University.

Tillman, J. Edward. Structure and petrology of the Sugarloaf gabbroic massif, Franklin County, western Maine. M, 1973, Syracuse University.

Tillman, Jack Louis. Geology of the Tiawah area, Rogers and Mayes counties, Oklahoma. M, 1952, University of Oklahoma. 65 p.

Tillman, John Robert. A study of the Middle Devonian species of the genera Alveolites and Planalveolites from the Traverse Group of Michigan. M, 1958, University of Michigan.

Tillman, John Robert. Variation in species of *Mucrospirifer* from Middle Devonian rocks of Michigan, Ontario and Ohio. D, 1962, University of Michigan. 43 p.

Tillman, Joseph W. Post-Pliocene displacement history of the Kentucky River Fault in northwest Madison and southeast Jessamine counties, Kentucky. M, 1985, Eastern Kentucky University. 62 p.

Tillman, Peter Douglas. Geochemistry of the Blue Hills igneous complex, Massachusetts; a statistical study. M, 1977, University of North Carolina, Chapel Hill. 121 p.

Tillman, Roderick Whitbeck. Petrology and paleoenvironments of the Robinson Member, Minturn Formation (Pennsylvanian), Eagle Basin (Routt, Grand, and Eagle counties), Colorado. D, 1967, University of Colorado. 278 p.

Tillman, Roderick Whitbeck. The stratigraphy and areal geology of the Terryville Quadrangle, Wisconsin. M, 1960, University of Wisconsin-Madison.

Tillman, Stephen Edward. A sonic method for petrographic analysis. M, 1974, Michigan State University. 7 p.

Tillotson, Harold Harman. The Viola Limestone of the Arbuckle Mountains of southern Oklahoma. M, 1924, University of Oklahoma. 22 p.

Tills, Linda Ann. Suspended sediment transport in four Midwestern rivers. M, 1977, University of Illinois, Urbana.

Tilson, Seymour. Geology of the Mohonk Lake Quadrangle, New York. M, 1957, New York University.

Tilton, John Littlefield. The Pleistocene deposits in Warren County, Iowa. D, 1910, University of Chicago. 42 p.

Tilton, Terry L. Middle Devonian scolecodonts of the Silica Formation, (northwestern Ohio). M, 1969, Bowling Green State University. 50 p.

Timbel, Ned R. Structural geology of the Indian Point Quadrangle, Fremont County, Wyoming. M, 1978, Miami University (Ohio). 71 p.

Timco, G. W. J. High pressure dielectric properties of perovskite ferroelectrics related to the Earth's mantle. D, 1977, University of Western Ontario.

Timco, G. W. J. High pressure hysteresis loop studies of virtually clamped crystals of ferroelectric triglycine sulphate and Rochelle salt. M, 1974, University of Western Ontario.

Timken, Hye Kyung Cho. Solid-state nuclear magnetic resonance studies of quadrupolar nuclei in inorganic systems. D, 1987, University of Illinois, Urbana. 213 p.

Timko, Donald Joseph. Geology of portions of the Boyertown and Allentown West Quadrangle, Berks County, Pennsylvania. M, 1956, University of Pittsburgh.

Timlin, Dennis James. Evaluating the effect of soil erosion on soil productivity through simulation modeling and geostatistical analysis. D, 1987, Cornell University. 174 p.

Timm, Bert Clifford. The geology of the southern Cornudas Mountains, Texas, and New Mexico. M, 1941, University of Texas, Austin.

Timm, John Jay. Age and significance of Paleozoic sedimentary rocks in the southern River Mountains, Clark County, Nevada. M, 1985, University of Nevada, Las Vegas. 62 p.

Timm, Ronald W. Precambrian geology of the Broadwater River area, Beartooth Mountains, Montana. M, 1979, Northern Illinois University. 110 p.

Timm, Susan. The structure and stratigraphy of the Columbia River Basalt in the Hood River valley, Oregon. M, 1979, Portland State University. 60 p.

Timmer, Robert Scott. Geology and sedimentary copper deposits in the western part of the Jarosa and Seven Springs quadrangles, Rio Arriba and Sandoval counties, New Mexico. M, 1976, University of New Mexico. 151 p.

Timmerman, David Harold. Deformation characteristics of sand subjected to anisotropic cyclic dynamic loading. D, 1969, Michigan State University. 118 p.

Timmins, Edwin Allen. Paleomagnetism and geochronology of the Bird River greenstone belt. M, 1985, University of Windsor. 130 p.

Timmons, Dale M. Stratigraphy, lithofacies and depositional environment of the Cowlitz Formation, T. 4 and 5N., R. 5W. Northwest Oregon. M, 1981, Portland State University. 89 p.

Timoney, Kevin Patrick. A geobotanical investigation of the subarctic forest-tundra of the Northwest Territories. D, 1988, University of Alberta. 292 p.

Timothy, Mary. The effect of glaciation on the river systems of Wisconsin. M, 1922, University of Wisconsin-Madison.

Timroth, Ron A. Geology and petrology of the South Dome Pluton, Park County, Colorado. M, 1958, University of Colorado.

Timson, Glenn H. Geochemistry of hydrothermal alteration and related precious metal mineralization, Wire Patch porphyry molybdenum prospect, eastern Breckenridge mining district, Summit County, Colorado. M, 1977, Ohio State University. 113 p.

Tindall, Terry Allen. The movement of salts through soil as affected by subsoil structure. D, 1983, Oklahoma State University. 48 p.

Tindell, William Norman. Cyclical sedimentation of the middle Conemaugh Series in the Pittsburgh area. M, 1950, University of Pittsburgh.

Ting, Cheng. Application of the theory of plasticity to drape folds. D, 19??, Texas A&M University.

Ting, Chuen Pu. The mineralogy of the Perry Farm Shale. M, 1950, University of Missouri, Columbia.

Ting, Francis Ta-Chuan. The petrology of the lower Kittanning Coal (Pennsylvanian) in western Pennsylvania. D, 1967, Pennsylvania State University, University Park. 147 p.

Ting, San Chen-Shin. Integral equation modeling of 3-D magnetotelluric response. M, 1980, University of Utah. 54 p.

Tingey, John Craig. Palynology of the Lower Cretaceous Bear River Formation in the Overthrust Belt of southwestern Wyoming. D, 1978, Michigan State University. 167 p.

Tingle, Tracy Neil. Experiments and observations bearing on the solubility and diffusivity of carbon in olivine. D, 1987, University of California, Davis. 148 p.

Tingle, Woodrow Wilson. Geology of the clays of Twiggs County, Georgia. M, 1957, University of North Carolina, Chapel Hill. 71 p.

Tingley, Icyl Cathryn. Geology, petrology, and tungsten mineralization of the southern Lodi Hills, Nye County, Nevada. M, 1986, University of Nevada. 327 p.

Tingley, Joseph V. Guides to exploration in the Sierra Nevada tungsten province. M, 1963, University of Nevada. 72 p.

Tinkel, Anthony Robert. Natural gas seeps in the northern Gulf of Mexico. M, 1973, Texas A&M University.

Tinker, Clarence N.; Hulstrand, Richard F. and Wendt, Roy L. Geology of the Turkey Creek-Williams Creek area, Huerfano County, Colorado. M, 1955, University of Michigan.

Tinker, John R., Jr. Rates of hillslope lowering in the Badlands of North Dakota. D, 1970, University of North Dakota. 110 p.

Tinker, John R., Jr. Time and soil development on lateral moraines, Martin River glacier, south-central Alaska. M, 1967, University of North Dakota. 99 p.

Tinker, Scott W. Lithostratigraphy and biostratigraphy of the Aptian La Peña Formation, Northeast Mexico and South Texas. M, 1985, University of Michigan. 63 p.

Tinker, Wesley R. A genetic classification of the beaches of the Ohio shoreline of Lake Erie. M, 1959, Ohio State University.

Tinkham, Daniel J. Resistance of high-gradient of streams in New Hampshire. M, 1986, University of New Hampshire. 127 p.

Tinkle, Anthony R. A geophysical and geological investigation of the Yucatan Basin and its margins. D, 19??, Texas A&M University.

Tinkle, Anthony Robert. Sediment velocity structure of the Yucatan Basin, Caribbean Sea from conventional CDP seismic data. D, 1981, Texas A&M University. 392 p.

Tinklepaugh, Betty M. A chemical, statistical and structural analysis of secondary dolomitization in the Rogers City-Dundee Formation of the central Michigan Basin. D, 1957, Michigan State University. 126 p.

Tinklepaugh, Betty M. A sedimentary, petrographic, and statistical study of certain glacial clays of northern Michigan as an aid in correlation. M, 1955, Michigan State University. 94 p.

Tinl, Teresa J. Aluminum translocation in shallow Norwegian Inceptisols; a response to acid precipitation, Project RAIN, Sogndal, Norway. M, 1986, Northern Arizona University. 278 p.

Tinney, James Craig. Trading quality for quantity; an assessment of salinity contamination generated by groundwater conservation policy in the Tucson Basin. D, 1986, University of Arizona. 245 p.

Tinoco, Fernando Heriberto. Shear strength of granular materials. D, 1968, Iowa State University of Science and Technology. 185 p.

Tinsley, J. C. Quaternary geology of northern Salinas Valley, Monterey County, California. D, 1975, Stanford University. 257 p.

Tint, Maung Thaw. Microfossils and petrology of the Mississippian limestones, northern Stansbury Range, Tooele County, Utah. M, 1963, University of Utah. various pagination p.

Tintera, John J. The identification and interpretation of karst features in the Bellevue-Castalia region of Ohio. M, 1980, Bowling Green State University. 63 p.

Tinucci, John Paul. Modeling of rock mass discontinuities for deformation analysis. D, 1985, University of California, Berkeley. 592 p.

Tiong, Siong-Sui. Driving mechanisms of plate tectonics; the forward problem. M, 1980, Carleton University. 116 p.

Tiphane, M. Pershing Township map-area (Quebec). M, 1947, McGill University.

Tiphane, M. Petrography of Brome and Shefford mountains (southern Quebec). D, 1953, McGill University.

Tipnis, Ravindra Shalad. Biostratigraphy and paleontology of Late Cambrian to late Middle Ordovician conodonts from southwestern District of Mackenzie, Northwest Territories. D, 1978, University of Alberta. 473 p.

Tipper, Howard Watson. Revision of the Hazelton and Takla groups of central British Columbia. D, 1954, Washington State University. 105 p.

Tippett, C. R. A structural and stratigraphic cross-section through the Selkirk fan axis, Selkirk Mountains, southeastern British Columbia. M, 1976, Carleton University. 176 p.

Tippett, Clinton Raymond. A geological cross-section through the southern margin of the Foxe fold belt, Baffin Island, Arctic Canada, and its relevance to the tectonic evolution of the northeastern Churchill Province. D, 1980, Queen's University. 490 p.

Tippett, Michael Charles. The geology of the Copper Basin ore deposits, Lander County, Nevada. M, 1967, University of Nevada. 31 p.

Tippie, Frank E. The subsurface stratigraphy of the (Mississippian) lower Chester formations and the upper Ste. Genevieve members in southwestern Illinois. M, 1945, University of Chicago. 57 p.

Tippie, Mark William. Transformation of sparse frequency-domain electromagnetic data to the time domain. M, 1984, University of Utah. 57 p.

Tippit, Phyllis Russell. The biostratigraphy and taxonomy of Mesozoic Radiolaria from the Samail Ophiolite and Hawasina Complex, Oman. D, 1981, University of Texas at Dallas. 396 p.

Tipple, Gregory L. Clay mineralogy and Atterberg limits on the Taylor Group in the vicinity of Austin, Texas. M, 1975, University of Texas, Austin.

Tipton, Ann. Landslide in the Diablo Range near Patterson, California. M, 1967, Stanford University.

Tipton, Ann. Oligocene faunas and biochronology in the subsurface, Southwest San Joaquin Valley, California. M, 1970, University of California, Berkeley. 81 p.

Tipton, Merlin J. Geology of the Akron Quadrangle, South Dakota-Iowa. M, 1958, University of South Dakota. 52 p.

Tipton, Ronald M. Evaluation of the Iowa Pore Index test and expected durability factor for predicting the frost durability of selected carbonate rocks. M, 1986, University of Toledo. 280 p.

Tipton, William Everett. Geologic structure of northern Hurd Draw and southern Foster quadrangles,

Culberson County, Texas. M, 1951, University of Texas, Austin.

Tirey, Homer Luvois, Jr. The geology of the southeastern quarter of the Frisco, Texas, Quadrangle. M, 1950, University of Oklahoma. 79 p.

Tirey, Martha Margaret. Displacements required during multiple drape folding along the Northwest Bighorn Mountain Front, Wyoming. M, 1978, Texas A&M University. 61 p.

Tirrell, Ann Louise. Conodont biostratigraphy of Member A (Lower to lower Middle Ordovician) Mountain Springs Formation, southern Great Basin. M, 1985, San Diego State University. 180 p.

Tirsch, Franklin Steven. River basin water quality monitoring network design. D, 1983, University of Massachusetts. 236 p.

Tischendorf, Wilheim. An evaluation of infiltration theory and methods. M, 1963, Colorado State University. 91 p.

Tischler, Herbert. Devonian and Mississippian of Rest Spring area, California. M, 1955, University of California, Berkeley. 77 p.

Tischler, Herbert. The Pennsylvanian and Permian stratigraphy of the Huerfano Park area, Colorado. D, 1961, University of Michigan. 226 p.

Tisdale, Ernest Edward. The geology of the Heart Butte Quadrangle, North Dakota. M, 1940, Texas A&M University. 71 p.

Tisdale, Ronald Marion. Field characteristics and recognition of upper delta plain sediments, Vowell Mountain and Cross Mountain formations, Windrock Quadrangle, Tennessee. M, 1974, University of Tennessee, Knoxville. 128 p.

Tisdale, Todd Street. Estimating nonpoint source pollution; an evaluation of methods for forecasting reservoir sedimentation and eutrophication. M, 1980, University of Virginia. 90 p.

Tisdel, Fred Walter. Primary distribution of sedimentary particle size, with reference to Rosin's law of crushing. M, 1940, University of Chicago. 35 p.

Tisin, Abdulmehdi Bektash. Design and analysis of an optimum multichannel deconvolution filter based on normal moveout stretched wavelets. D, 1986, University of Pittsburgh. 96 p.

Tisoncik, D. D. The structure and petrology of a partial ring dike, South Pond, New Hampshire. M, 1974, University of Connecticut. 84 p.

Tissue, Jeffery Stephen. A paleoenvironmental analysis of the Middle Devonian sandstones in the Upper Mississippi Valley. M, 1977, University of Illinois, Urbana.

Tisza, Stephen Thomas. A subsurface study of the Tar Springs Formation (Mississippian), Union-Webster-Henderson counties, Kentucky. M, 1958, University of Illinois, Urbana. 37 p.

Titaro, Dino. The uranium-iron oxide mineral association in the Burma Lake Pegmatite, Sept-Iles, Quebec. M, 1980, University of Western Ontario. 177 p.

Titcomb, Jane. The Devonian section at Bisbee, Arizona. M, 1932, University of Minnesota, Minneapolis. 24 p.

Titley, Spencer R. Silication as an ore control, Linchburg Mine, Socorro County, New Mexico. D, 1958, University of Arizona.

Titman, G. D. Interspecific competition for resources; an experimental and theoretical study. D, 1976, University of Michigan. 73 p.

Tittlebaum, M. E. Investigation of leachate heavy metal and organic carbon content stabilization through leachate recirculation. D, 1979, University of Louisville. 167 p.

Titus, Charles A. O. Factors controlling the karst progress in Monroe County, Illinois. M, 1976, Southern Illinois University, Carbondale. 123 p.

Titus, Frank Bethel, Jr. Late Tertiary and Quaternary hydrogeology of Estancia Basin, central New Mexico. D, 1969, University of New Mexico. 179 p.

Titus, Frank Bethel, Jr. Recent unconsolidated sediments between the northern Keys and the mainland in Florida. M, 1958, University of Illinois, Urbana.

Titus, Robert C. A nearshore facies of the Lockatong Formation (upper Triassic) of northeast New Jersey and its implications on the environment of deposition of the Lockatong sedimentary cycles. M, 1971, Boston University. 37 p.

Titus, Robert Charles. Fossil communities and paleoecology of the medial Ordovician Kings Falls and Sugar River limestones (Trenton Group) of northwestern and central New York. D, 1974, Boston University.

Titus, Russell Gerard. A study of the physical and chemical variations in the garnet group from the unique orebodies at Franklin, and at Sterling Hill in Ogdensburg, Sussex County, New Jersey. M, 1986, Montclair State College. 66 p.

Titus, Willard Sidney, III. Development and application of some quantitative stratigraphic techniques to the Coos Bay Coalfield, a Tertiary fluvio-deltaic complex in southwestern Oregon. M, 1987, Portland State University. 113 p.

Tituskin, Susan E. Rare earth element distribution in the Negaunee Iron Formation, Marquette District, Michigan. M, 1983, Michigan State University. 99 p.

Tivey, Maurice Anthony. The central anomaly magnetic high; its source and implications for ocean crust construction and evolution. D, 1988, University of Washington. 131 p.

Tiwari, Suresh Chandra. Source and behavior of potassium in Cecil soils. D, 1969, University of Georgia. 135 p.

Tiyamani, Chanchai. Evaluation of design criteria for sediment detention basins in surface mining. D, 1983, Virginia Polytechnic Institute and State University. 348 p.

Tizzard, Paul G. Viking deposition in the Suffield area, Alberta. M, 1974, University of Alberta. 126 p.

Tlapek, John William. Genetic significance of maximum stability temperatures of certain three-layer clay minerals in selected soil samples. M, 1962, University of Missouri, Columbia.

Tleel, Jack Wadie. Surface geology of Dammam Dome, Eastern Province, Saudi Arabia. M, 1972, Texas Christian University. 85 p.

Tobar, Alvaro Barra. Stratigraphy and structure of the El Salvador-Potrerillos region, Atacama, Chile. D, 1977, University of California, Berkeley. 117 p.

Tobey, Eugene Francis. Geologic and petrologic relationships between the Thirtynine Mile volcanic field and the Cripple Creek volcanic center (central Colorado). M, 1969, New Mexico Institute of Mining and Technology. 61 p.

Tobey, Eugene Francis. Geology of the Bull Valley intrusive-extrusive complex and genesis of the associated iron deposits. D, 1976, University of Oregon. 244 p.

Tobey, William H. The Cripple Creek mining district (Colorado). M, 1902, University of Kansas.

Tobias, Steven Martin. The mixed phase reflection seismic wavelet; its determination and extraction on the complex z plane. M, 1977, Pennsylvania State University, University Park. 123 p.

Tobias, Theodore. Suspension and sediment stratification in some clay-water systems. M, 1966, Michigan Technological University. 86 p.

Tobiassen, Richard Torre. Selected trace element analyses of whole rock and separated phosphate grains from the Miocene Pungo River Formation, North Carolina. M, 1982, University of North Carolina, Chapel Hill. 66 p.

Tobin, Don Graybille. Microscopic criteria for defining macroscopic modes of deformation in experimentally deformed oolitic limestone. D, 1966, Columbia University. 102 p.

Tobin, Don Grayville. Subsurface structure from reflective seismology in Clinton, Fayette, Highland, Pike, and Ross counties, Ohio. M, 1961, Ohio State University.

Tobin, Rick Curtis. A model for cyclic deposition in the Cincinnatian Series of southwestern Ohio, northern Kentucky, and southeastern Indiana. D, 1982, University of Cincinnati. 500 p.

Tobin, Rick Curtis. Sedimentology of the Fairview Formation, Miamitown Shale, and Bellevue Limestone (Upper Ordovician, Cincinnatian Series) of southwestern Ohio and northern Kentucky. M, 1980, University of Cincinnati. 144 p.

Tobisch, Mary Kathryn. Part 1, Late Cenozoic geology of the central Motagua Valley, Guatemala; Part II, Uplift rates, deformation, and neotectonics of Holocene marine terraces from Point Delgada to Cape Mendocino, California. D, 1986, University of California, Santa Cruz. 568 p.

Tobisch, Othmar Tardin. Geology of the Crane Flat-Pilot Peak area, Yosemite District, California. M, 1960, University of California, Berkeley. 89 p.

Tobison, Norman Murray. Brachiopod faunas of the Pennsylvanian Coal City and Myrick Station limestones of Iowa and Missouri. M, 1950, University of Wisconsin-Madison. 35 p.

Tobola, David Philip. Determination of reservoir permeability from repeated induction logging. D, 1988, Texas A&M University. 224 p.

Tocher, Don. Seismic velocities and structure in Northern California and Nevada. D, 1956, University of California, Berkeley. 120 p.

Tochtenhagen, Mark S. Computer enhancement of three problem seismic lines in Steuben County, New York. M, 1988, Wright State University. 79 p.

Todd, Brian Jeremy. Iceberg scouring on Saglek Bank, northern Labrador Shelf. M, 1984, Dalhousie University. 176 p.

Todd, Brian Jeremy. Offset margins of the Canadian eastern continental shelf; crustal structure and thermal evolution. D, 1988, Dalhousie University. 227 p.

Todd, Charles Payson. Isolation, display and interpretation of offset dependent phenomena in seismic reflection data using offset to depth (ODR) range partial stacking. M, 1986, University of Texas, Austin. 235 p.

Todd, Eric Donald. Seismicity of the Bath County, Virginia, locale. M, 1982, Virginia Polytechnic Institute and State University. 100 p.

Todd, Harry Wayne. Area geology of the Moyers Quadrangle, Pushmataha County, Oklahoma. M, 1960, University of Oklahoma. 72 p.

Todd, James Forrest. The aquatic geochemistry of the particle-reactive radionuclides. D, 1984, Old Dominion University. 213 p.

Todd, James Hodkins. A contribution to the study of Pleistocene history of the Upper Mississippi River. D, 1942, University of Minnesota, Minneapolis.

Todd, Jean P. Survey of the near-shore deposits of Lake Michigan adjacent to Northwestern University. M, 1937, Northwestern University.

Todd, Margaret Ruth. Glacial geology of Hamma Hamma Valley and its relation to glacial history of Puget Sound basin. M, 1939, University of Washington. 48 p.

Todd, Raymond C. An investigation of the seismic wave propagation properties of a thin unsaturated layer as a wave guide. M, 1971, Michigan State University. 116 p.

Todd, Robert G. Spore analysis of the "Alvis" coal of western Missouri. M, 1957, University of Missouri, Columbia.

Todd, Stanley G. Bedrock geology of the southern part of Tom Miner Basin, Park and Gallatin counties, Montana. M, 1969, Montana State University. 63 p.

Todd, Stanley Glenn. The geology and mineral deposits of the Spirit Pluton and its metamorphic aureole, Stevens and Pend Oreille counties, Washington. D, 1973, Washington State University. 153 p.

Todd, Terrence P. Effect of cracks on elastic properties of low porosity rocks. D, 1973, Massachusetts Institute of Technology. 337 p.

Todd, Thomas Waterman. Areal petrology of the Sacajawea and Amsden formations and the Tensleep Sandstone, Big Horn Basin, Wyoming. D, 1959, University of Texas, Austin. 426 p.

Todd, Thomas Waterman. Comparative petrology of the Carrizo and Newby sandstones, Bastrop County, Texas. M, 1956, University of California, Davis.

Todd, Victoria Roy. Structure and petrology of metamorphosed rocks in central Grouse Creek Mountains, Box Elder County, Utah. D, 1973, Stanford University. 373 p.

Todd, Wallace. Typical lake deposits of the Great Basin. M, 1931, Stanford University. 96 p.

Todd, William C. A gravity, magnetic and seismic investigation of the Providencia Mine vicinity, New Almaden, California. M, 1976, San Jose State University. 59 p.

Todd-Brown, William Edward, Jr. Statistical analysis of dolomite outcrop fracture distributions to evaluate well core sized samples (Sawtooth Mountains, Montana). M, 1983, University of Oklahoma. 165 p.

Toder, Daniel R. A study of minerals found in the Franklin-Sterling Hill area, Sussex County, New Jersey. M, 1981, Montclair State College. 127 p.

Todoeschuck, John Peter. Non-linear seismic attenuation in the Earth as applied to the free oscillations. D, 1985, McGill University. 178 p.

Todoeschuck, John Peter. The compressibility of the Earth's core and the anti-dynamo theorems. M, 1979, Memorial University of Newfoundland. 73 p.

Toelle, Brian Edward. Structural geology of the Persimmon Gap area of Big Bend National Park, Brewster County, Texas, based on remote sensing and field methods. M, 1981, Stephen F. Austin State University. 109 p.

Toepelman, Walter Carl. The geology of a portion of the Slim Buttes region of northwestern South Dakota, with special reference to unusual structural features due to slumping. D, 1925, University of Chicago. 77 p.

Toeppe, Victor Francis. Structures in Little Horse Creek area, Fremont County, Wyoming. M, 1952, Miami University (Ohio). 21 p.

Toft, Paul Bernard. Magsat anomalies over the Man Shield of the West African Craton in relation to magnetization and evolution of the lithosphere. D, 1988, University of Massachusetts. 239 p.

Toggweiler, John Robert. 1, A multi-tracer study of the abyssal water column of the deep Bering Sea, including sediment interactions; II, A six zone regionalized model for bomb radiotracers and CO_2 in the upper kilometer of the Pacific Ocean. D, 1983, Columbia University, Teachers College. 425 p.

Togola, N'Golo. Pétrographie du faciès grésoconglomératique de la Formation de Duparquet, d'âge archéen, région de Rouyn, Abitibi, Québec. M, 1986, Universite Laval. 70 p.

Toha, Franciscus Xaverius. Wave-induced response of poro-elastic offshore foundations. D, 1983, University of Wisconsin-Madison. 283 p.

Tohill, Bruce Owen. Stratigraphy and sedimentary structures of the Crow Mountain Member, Chugwater Formation (Triassic and Permian) of southeast Big Horn Basin, Wyoming. M, 1965, University of Nebraska, Lincoln.

Toit, Charl du *see* du Toit, Charl

Tokar, Ronald Albert. Dispersion of chromium in the environment from plating waste at Mogadore, Ohio. M, 1971, University of Akron. 31 p.

Tokarsky, Orest. Geology and groundwater resources (Quaternary) of the Grimshaw area, Alberta (Canada). M, 1967, University of Alberta. 218 p.

Tokarz, Marek T. Synthesis and physico-chemical properties of aluminosilicates; 1, Sodalite-type aluminosilicates obtained through low-temperature hydrothermal transformation of kaolinite-type minerals; II, Cross-linked smectites. D, 1985, University of Utah. 200 p.

Tokhais, Ali. A quantitative study of the Wasia well field (Saudi Arabia). M, 1982, Ohio University, Athens. 413 p.

Tokoro, Atsuo. Sigma phase formation in binary alloys of the transition elements. M, 1959, University of Nevada - Mackay School of Mines. 52 p.

Toksoz, M. Nafi. Velocities of long period surface waves and microseisms and their use in structural studies; I, Mantle Love and mantle Rayleigh waves and the structure of the Earth's upper mantle; II, Microseisms and their application to seismic exploration. D, 1963, California Institute of Technology. 90 p.

Tokunaga, Tetsu Keith. The temperature dependence of gas diffusivities in porous media. D, 1986, University of California, Berkeley. 149 p.

Tolan, Terry Leo. The stratigraphic relationships of the Columbia River Basalt Group in the lower Columbia River Gorge of Oregon and Washington. M, 1982, Portland State University. 151 p.

Tolbert, Gene Edward. Structure and ore deposits of the Raposos Mine, Minas Gerais, Brazil. D, 1962, Harvard University.

Tolbert, James N. The sources, pathways, and sinks of Cr, Mn, Fe, Co, Ni, and Cu in near surface sabkha sediments; Laguna Madre Flats, Texas. M, 1985, Michigan State University. 154 p.

Tolderlund, Douglas Stanley. Seasonal distributional patterns of planktonic foraminifera at five ocean stations in the western North Atlantic. D, 1969, Columbia University. 210 p.

Toldi, John L. Velocity analysis without picking. D, 1985, Stanford University. 118 p.

Tole, Mwakio Peter. Factors controlling the kinetics of silicate-water interactions. D, 1982, Pennsylvania State University, University Park. 234 p.

Tole, Peter Mwakio. The uranium content of zircons from the Catskill Formation, eastern Pennsylvania. M, 1979, Pennsylvania State University, University Park. 155 p.

Toler, Henry Miles. Limestone reservoirs of Kentucky. M, 1929, University of Illinois, Chicago.

Toler, Larry Gene. Weathering of a quartz monzonite near Lolo Hot Springs, Missoula County, Montana. M, 1959, University of Montana. 44 p.

Tolgay, Mitat Yumnu. A sedimentary study of the Harding Formation near Canon City, Colorado. M, 1952, University of Oklahoma. 78 p.

Tolle, Timothy V. Watershed and climate influences on flood frequency distributions in the Willamette River basin. D, 1979, Oregon State University. 167 p.

Tollefson, Everett Harold. Problems in briquetting North Dakota lignite. M, 1925, University of Minnesota, Minneapolis. 21 p.

Tollefson, Linda Joyce Sindelar. Paleoenvironmental analysis of the Kokomo and Kenneth limestone members of the Salina Formation in the vicinity of Logansport, Indiana. M, 1978, University of Illinois, Urbana. 173 p.

Tollefson, Oscar W. Geology of central Middle Park, Colorado. D, 1956, University of Colorado. 167 p.

Tollefson, Oscar W. Structure of the foothills region from Gregory Canyon to the southern Boulder County line, Colorado. M, 1942, University of Colorado.

Tolley, M. A. Hydrothermal alteration of ash-flow tuffs in the Indian Peak District of southwestern Utah. M, 1971, University of Missouri, Rolla.

Tollo, Richard. Contact metamorphism in the Weymouth Formaton, Nahant, Massachusetts. M, 1976, University of New Hampshire. 56 p.

Tollo, Richard Paul. Petrography and mineral chemistry of ultramafic and related inclusions from the Orapa A/K-1 kimberlite pipe, Botswana. D, 1982, University of Massachusetts. 218 p.

Tolman, A. L. The hydrogeology of the White Clay Lake area, Shawano County, Wisconsin. M, 1975, University of Wisconsin-Madison.

Tolman, Carl. The geology and ore deposits of the Chilco Lake area (Canada). M, 1925, Yale University.

Tolman, Carl. The geology of the Big Eddy Lake area, Sudbury District, Ontario. D, 1927, Yale University.

Tolman, Davis Nichols. Linear analysis from remote sensor data; Arkansas Valley. M, 1979, University of Arkansas, Fayetteville.

Tolman, Frank Bronson. Recent littoral foraminifera of the Central California coast. M, 1933, Stanford University.

Tolman, Richard Robbins. A guide to the geology of the central Wasatch Mountains for teachers in secondary schools. M, 1964, University of Utah. 145 p.

Tolson, Patrick M. Geology of the Seaside-Young's River Falls area, Clatsop County, Oregon. M, 1976, Oregon State University. 191 p.

Tolson, Ralph Bradley. Structure, stratigraphy, tectonic evolution and petroleum source potential of the Hope Basin, southern Chukchi Sea, Alaska. D, 1987, Stanford University. 294 p.

Tolstoy, Ivan. Punched card calculation of suostatic and topographic reductions. M, 1947, Columbia University, Teachers College.

Tolstoy, Ivan. The T-phase of shallow-focus earthquakes. D, 1950, Columbia University, Teachers College.

Tolunay, Aykut. A sedimentological study of the Jacobsville (Michigan) sandstone (Precambrian or Cambrian). M, 1970, Michigan Technological University. 58 p.

Tom, Gene Francis. Some heavy mineral placers of the Idaho Batholith north of Salmon River, Idaho. M, 1950, University of Illinois, Urbana.

Tomajczyk, Charles F. Modulation transfer function and the role of focal length of a photogrammetric camera in image generation. M, 1966, Ohio State University.

Tomasello, Richard S. Boussinesq approximation. M, 1973, Florida Institute of Technology.

Tomassetti, John A. Geology of Lockport (Silurian) rocks at the Ohio Lime Company Quarry, Woodville, Ohio. M, 1981, Bowling Green State University. 72 p.

Tomastik, Thomas E. Geology of the southern part of Buck Mountain, White Pine County, Nevada. M, 1981, Ohio University, Athens. 240 p.

Tombale, Akolang Russia. An interpretation of lithogeochemical data of the Guichon Creek Batholith. M, 1984, University of British Columbia. 141 p.

Tombaugh, Karen. Biostratigraphy of the Permian Shedhorn Sandstone and Ervay and Franson Members of the Park City Formation, southeastern Gros Ventre Mountains, Wyoming. M, 1973, University of Wyoming. 103 p.

Tomczyk, Ted. Analyses of the thermal regimes of the Valles Caldera and Yellowstone Caldera by downward continuation of temperature gradient data and of the Snake River plain by employing a thermo-mechanical model. M, 1987, Purdue University. 95 p.

Tome, R. F. Studies on the seismoelectric effect. M, 1975, University of Toronto.

Tomes, Reynold Joseph. Environmental and diagenetic history of the "Upper" Warner Sandstone, SW Vernon County, Missouri. M, 1986, Colorado State University. 208 p.

Tomida, Yukimitsu. Dragonian mammals and Paleocene magnetostratigraphy, North Horn Formation, central Utah. M, 1978, University of Arizona.

Tomida, Yukimitsu. Small mammal fossils and correlation of continental deposits, Safford and Duncan basins, Arizona. D, 1985, University of Arizona. 269 p.

Tomik, John C. Surficial geology for land use planning in the Jamestown, New York, area. M, 1982, SUNY, College at Fredonia. 52 p.

Tomikel, John. Land surface in Erie County, Pennsylvania. M, 1962, Syracuse University.

Tomkus, Mindaugas. Zinc mineralization at Balmat, N.Y. M, 1976, Brooklyn College (CUNY).

Tomlinson, Charles W. The origin of red beds; a study of the conditions of origin of the Permo-Carboniferous and Triassic red beds of the Western United States. M, 1914, University of Wisconsin-Madison.

Tomlinson, Charles Weldon. The middle Paleozoic stratigraphy of the Central Rocky Mountain region. D, 1916, University of Chicago. 58 p.

Tomlinson, Robin Gayle. Shuttle imaging radar; recognition of lithologic and structural characteristics. M, 1986, Purdue University. 132 p.

Tompkin, J. M. Aggregate resources within Routt National Forest. M, 1976, Colorado State University. 213 p.

Tompkins, K. E. Evaluation of the ground water resources of southern Suffolk County, L.I., New York. M, 1975, SUNY at Binghamton. 70 p.

Tompkins, Linda Anne. The Koidu kimberlite complex, Sierra Leone, West Africa. M, 1983, University of Massachusetts. 340 p.

Tompkins, Robert E. A study of the relationship between the clay mineralogy and selected geotechnical properties in Norfolk Canyon. M, 1978, University of South Florida, Tampa. 94 p.

Tompkins, Robert Eugene. Origin and occurrence of selected worldwide calcrete deposits. D, 1980, Texas A&M University. 156 p.

Tompson, Willard D. Geology of the northern part of the Cherry Creek metamorphics, Madison County, Montana. M, 1959, Montana State University. 53 p.

Tomson, Janice. Early-middle Pleistocene paleo-oceanography of the Southwest Atlantic sector of the Southern Ocean. M, 1985, San Jose State University. 44 p.

Tomten, David Charles. Geothermometry and geobarometry of metamorphosed Belt Series, northwest of the Idaho Batholith, Idaho. M, 1985, University of Utah. 78 p.

Tondu, R. Joe. Depositional systems of the Upper Cretaceous Mancos and Mesaverde groups, Axial Basin region, northwestern Colorado. M, 1976, University of Texas, Austin.

Toney, Jennifer Diana. Historical sediment storage and pollen stratigraphy in the Connecticut River estuary. M, 1987, Wesleyan University. 168 p.

Toney, Jimmie C. A statistical analysis of Orbitolina in Trans-Pecos, Texas. M, 1962, [Texas Tech University].

Tong, Chiun Shing. Dispersion and salt water intrusion in non-homogeneous porous media. D, 1971, University of Wisconsin-Madison. 284 p.

Tong, James A. Upper Miocene and Pliocene mollusks from northern Venezuela. D, 1930, The Johns Hopkins University.

Tong, Y. Engineering behaviour of cohesionless soils. D, 1975, University of Waterloo.

Tongtaow, Chalermkiat. Wave propagation along a cylindrical borehole in a transversely isotropic medium. D, 1982, Colorado School of Mines. 308 p.

Tonking, William Harry. Geology of the Puertecito Quadrangle, Socorro County, New Mexico. D, 1953, Princeton University. 159 p.

Tonnsen, John J. Stratigraphy, petrography and environment of deposition of the Frontier Formation on the western margin of the Crazy Mountain Basin, South-central Montana. M, 1975, Montana State University. 160 p.

Tonroy, Lucky Less. Pollution and underground water in Levelland, Texas, and vicinity. M, 1957, Texas Tech University. 41 p.

Tonti, Edmond Charles. Certain disconformities and lithology of the Vicksburg Stage of southeastern United States. D, 1955, Louisiana State University. 123 p.

Toogood, David J. Repeated folding in the Precambrian basement of the Laramie Mountains, Albany and Platte counties, Wyoming. M, 1967, University of Wyoming. 74 p.

Toohey, Loren Milton. Equidae of the upper part of the Marsland Formation. M, 1950, University of Nebraska, Lincoln.

Toohey, Loren Milton. Oligocene and early Miocene nimravids of North America. D, 1953, Princeton University. 121 p.

Tooker, Edwin Wilson. Barite deposit near Buckmanville, Pennsylvania. M, 1949, Lehigh University.

Tooker, Edwin Wilson. Thermal transformations in some layer silicate minerals. D, 1952, University of Illinois, Urbana.

Tooley, Richard D. The role of geothermal gradients on the motion of underground fluids. D, 1958, Massachusetts Institute of Technology. 70 p.

Toombs, Ralph Belmore. Some geologic factors relating to the laboratory examination of Recent sediments. M, 1953, University of British Columbia.

Toomey, Donald Francis. Lateral homogeneity in a "middle-limestone member" (Leavenworth) of a Kansas Pennsylvanian megacyclothem. D, 1964, Rice University. 238 p.

Toomey, Donald Francis. Paleontology and stratigraphy of the Carboniferous rocks of the Placitas region, northern Sandia Mountains, Sandoval County, New Mexico. M, 1953, University of New Mexico. 192 p.

Toomey, Douglas Ray. The tectonics and three-dimensional structure of spreading centers; microearthquake studies and tomographic inversions. D, 1987, Woods Hole Oceanographic Institution. 210 p.

Toomey, William J. A stratigraphic analysis of the Wasatch Formation, western Powder River basin, Wyoming. M, 1977, Wright State University. 57 p.

Toots, Heinrich A. Paleoecological studies on the Mesaverde Formation in the Laramie Basin. M, 1962, University of Wyoming. 112 p.

Toots, Heinrich A. Reconstruction of continental environments; the Oligocene Wyoming. D, 1965, University of Wyoming. 176 p.

Toppan, Frederick Willcox. The geology of Maine. M, 1932, Union College. 142 p.

Topper, Ralf E. Fine structure of the Benioff zone beneath the Central Aleutian Arc. M, 1978, University of Colorado.

Toppozada, Tousson Mohamed Roushdy. Seismic investigation of crustal structure and upper mantle velocity in the state of New Mexico and vicinity. D, 1974, New Mexico Institute of Mining and Technology. 152 p.

Toprak, Selami. Petrographic characterizations of coal in the Kozlu and Kilic formations (Westphalian A), Zonguldak, Turkey. M, 1984, University of Pittsburgh.

Torabian, Ali. Movement of phenolic compounds in ground water. D, 1988, Oklahoma State University. 144 p.

Toran, Laura Ellen. Sulfate contamination in groundwater near an abandoned mine; hydrogeochemical modeling, microbiology and isotope geochemistry. D, 1986, University of Wisconsin-Madison. 227 p.

Torcoletti, Paul James. A remote sensing and ground-based spectral study of the geobotanical relationships on Mt. Moosilauke, New Hampshire. M, 1988, Dartmouth College. 230 p.

Torgersen, T. L. Limnologic studies using the tritium-helium-3 tracer pair; a survey evaluation of the method. D, 1977, Columbia University, Teachers College. 238 p.

Torguson, William R., Jr. Morphology and taxonomy of the Ostracoda assemblage from the Benwood and Arnoldsburg members of the Monongahela Group (Pennsylvanian) in Morgan and Noble counties, Ohio. M, 1977, University of Akron. 65 p.

Torkelson, Bruce Emil. Gas adsorption on crushed minerals and rocks. M, 1977, University of Tulsa. 58 p.

Torlucci, Joseph, Jr. The distribution of heavy metal concentrations in sediment surrounding a sanitary landfill in the Hackensack Meadowlands, New Jersey.

M, 1982, Rutgers, The State University, Newark. 129 p.

Tormey, Brian B. Geomorphology of the Falls stretch of the Potomac River. D, 1980, Pennsylvania State University, University Park. 359 p.

Torney, Barbara Calhoon. Depositional and diagenetic history of the Fort Atkinson Member of the Maquoketa Formation (Upper Ordovician) in northeastern Iowa. M, 1983, University of Iowa. 146 p.

Toron, Praphat. Ground water in the Boise area, Idaho. M, 1964, University of Idaho. 48 p.

Torrance, James Kenneth. Hydraulic conductivity and pressure as limitations to the rate of frost heaving. D, 1968, Cornell University. 83 p.

Torrance, Jeffrey Gordon. Structural geology of evaporite piercement walls and adjacent synclines, Axel Heiberg Island, Canadian Arctic Archipelago. M, 1986, University of Toronto.

Torre Robles, Jorge de la *see* de la Torre Robles, Jorge

Torres, Adrian R. Landslides at Bolinas, Marin County, California. M, 1975, San Jose State University. 123 p.

Torres, Alfredo Arriola *see* Arriola Torres, Alfredo

Torres, Linda M. Radium-226 in plankton on the West Florida shelf. M, 1988, University of South Florida, St. Petersburg. 49 p.

Torres, Max Antonio. Depositional facies and Markov analysis of Pennsylvanian strata, Lookout Mountain, Northwest Georgia. M, 1987, Georgia State University. 215 p.

Torres, Pedro Leon. Gravity and magnetic study of normally and reversely polarized intrusions, Menominee County, Michigan. M, 1976, Michigan State University. 115 p.

Torres, Wilson Fabian Jaramillo *see* Jaramillo Torres, Wilson Fabian

Torres-Robles, Rafael. Phase equilibria in the system: n-butane, 1-butene, 1,3.butadiene, acetonitrile, water (120-160 F). D, 1980, University of Kansas. 275 p.

Torres-Roldan, Victor. Summary of Eocene stratigraphy at the base of Jim Mountain, north fork of the Shoshone River, northwestern Wyoming; Stratigraphy of the Eocene Willwood, Aycross, and Wapiti formations along the north fork of the Shoshone River, north-central Wyoming. M, 1983, University of Michigan.

Torresan, Michael E. Fabric and its relation to sedimentological and geotechnical properties of near-surface sediment in Shelikof Strait, Alaska. M, 1984, San Jose State University. 220 p.

Torrey, Victor Hugo, III. Some effects of rate of loading, method of loading, and applied total stress path on the critical void ratio of a fine uniform sand. D, 1981, Texas A&M University. 244 p.

Torries, Thomas F. Geology of the northeast quarter of the West Point, Mississippi, Quadrangle and related bentonites. M, 1963, Mississippi State University. 62 p.

Torrini, Rudolph Edward, Jr. The structural and stratigraphic history of the Oceanic Beds of Barbados; implications for forearc basin/structural high tectonics. D, 1985, Northwestern University. 233 p.

Tortorelli, Robert Louis. A methodology to assess the impact of a changing flood plain determination on an ungaged urban basin. D, 1981, Oklahoma State University. 188 p.

Tortosa, Delio J. The geology of the Cenex Mine, Beaverlodge, Saskatchewan. M, 1983, University of Saskatchewan. 141 p.

Tosaya, Carol Ann. Acoustical properties of clay-bearing rocks. D, 1982, Stanford University. 145 p.

Toscano, Marguerite Ann. Vertical sedimentary sequences of Delaware's Holocene prograding spit and transgressive barrier based on sedimentary discriminant analysis. M, 1986, University of Delaware. 144 p.

Tosdal, Richard Mark. Mesozoic rock units along the Late Cretaceous Mule Mountains thrust system, southeastern California and southwestern Arizona. D, 1988, University of California, Santa Barbara. 394 p.

Tosdal, Richard Mark. The timing of the geomorphic and tectonic evolution of the southernmost Peruvian Andes. M, 1978, Queen's University. 136 p.

Toskos, Theodoros. A structural and gravity transect along the New Jersey Highlands and adjacent Valley and Ridge, in northern New Jersey. M, 1984, Rutgers, The State University, New Brunswick. 156 p.

Toste, A. P. The sterol molecule; its analysis and utility as a chemotaxonomic marker and a fine geochemical probe into Earth's past. D, 1976, University of California, Berkeley. 467 p.

Toth, D. J. Organic and inorganic reactions in near shore and deep sea sediments. D, 1976, Northwestern University. 170 p.

Toth, John C. Facies of Paleocene alluvial plains, Powder River basin, Montana. M, 1982, North Carolina State University. 73 p.

Toth, John R. Deposition of submarine hydrothermal manganese and iron, and evidence for hydrothermal input of volatile elements to the ocean. M, 1977, Oregon State University. 79 p.

Toth, Margo Irene. Petrology, geochemistry, and origin of the Red Mountain ultramafic body near Seldovia, Alaska. M, 1979, University of Colorado.

Toth, Margo Irene. Structure, petrochemistry, and origin of the Bear Creek and Paradise plutons, Bitterroot Lobe of the Idaho Batholith. D, 1983, University of Colorado. 337 p.

Totten, David K. Biostratigraphy of some Late Paleozoic sediments in the Naco Hills area near Bisbee, Arizona. M, 1972, University of Arizona.

Totten, Matthew Wayne. Grain size distribution of quartz as a function of distance from shoreline, Blaine Formation (Permian), western Oklahoma. M, 1979, University of Oklahoma. 77 p.

Totten, Stanley Martin. Glacial geology of Richland County, Ohio. D, 1962, University of Illinois, Urbana. 153 p.

Totten, Stanley Martin. Quartz-feldspar ratios of tills in northeastern Ohio. M, 1960, University of Illinois, Urbana.

Totten, William B. Subsurface stratigraphy of the Ste. Genevieve Limestone (Valmeyeran) and associated units, Owen County, Indiana. M, 1985, Indiana University, Bloomington. 109 p.

Toub, Joyce Silverstein. The petrology and mineralogy of the Adirondack highland metasediments, Ausable Forks Quadrangle, New York. M, 1973, Rensselaer Polytechnic Institute. 62 p.

Toufigh, Mohammad Mohsen. Behavior of unsaturated soil and its influence on soil-soil interaction at an interface. D, 1987, University of Arizona. 432 p.

Toukan, Ziad R. Cation exchange properties of mine tailings. M, 1971, University of Idaho. 48 p.

Toulmin, Lyman D., Jr. Carbon ratio and moisture content of Pennsylvanian coals in Alabama and their relation to geologic structure and possible petroleum occurrence. M, 1934, University of Alabama.

Toulmin, Lyman D., Jr. The Salt Mountain Limestone of Alabama. D, 1939, Princeton University. 392 p.

Toulmin, Priestley, III. Bedrock geology of the Salem area, Massachusetts. D, 1959, Harvard University.

Toulmin, Priestley, III. Petrography and petrology of Rito Alto Stock, Custer and Saguache counties, Colorado. M, 1953, University of Colorado.

Toung, Kouang Shu. Geology of the Poorman Mine and adjacent area, Boulder County, Colorado. M, 1947, University of Colorado.

Touqan, Omar I. Hydrological and mechanical characteristics of soil in the area of salt deposits, northwest Phoenix, Arizona. M, 1970, University of Arizona.

Tour, Timothy Earle La *see* La Tour, Timothy Earle

Tourek, Thomas James. The depositional environments and sediment accumulation models for the upper Silurian Wills Creek Shale and Tonolaway

Limestone, central Appalachians. D, 1970, The Johns Hopkins University. 393 p.

Touring, Roscoe Manville. Structure and stratigraphy of the La Honda and San Gregorio quadrangles, San Mateo County, California. D, 1959, Stanford University. 294 p.

Tousley, Robert M. The influence of sulphate-reducing bacteria in secondary enrichment of copper ores. M, 1928, University of Minnesota, Minneapolis. 33 p.

Toussaint, C. R. A method for the determination of regional values associated with the assessment of environmental impacts. D, 1975, University of Oklahoma. 219 p.

Touwaide, Marcel E. A study of the genesis of the Boleo copper deposit. D, 1927, Stanford University. 121 p.

Touysinhthiphonexay, Kimball C. N. A statistical analysis of discontinuous arroyos in Northwest New Mexico, U.S.A.; topographic, lithologic, and vegetational controls. D, 1987, Pennsylvania State University, University Park. 79 p.

Touysinhthiphonexay, Kimball Cooke Nettleton. The effect of strip mining on stream morphology and behavior with emphasis on twenty-nine small watersheds in central Pennsylvania. M, 1982, Pennsylvania State University, University Park. 129 p.

Touysinhthiphonexay, Yen. Oxygen isotope fractionation between quartz and magnetite; an experimental determination at 600°C and 5 Kb. M, 1981, Pennsylvania State University, University Park. 71 p.

Tovall, Walter Massey. Some aspects of the geology of the Milk River and Pakowki formations (southern Alberta). D, 1956, University of Toronto.

Tovell, Walter Massey. Geology of the nodular shale of the middle and upper Miocene of the western Los Angeles Basin (California). M, 1942, California Institute of Technology. 39 p.

Tovell, Walter Massey. Some aspects of the geology of the (Upper Cretaceous) Milk River and Pakowki formations (southern Alberta). D, 1956, University of Toronto.

Towe, Kenneth McCarn. Directional features and source of sediments of the Narragansett Basin, Rhode Island and Massachusetts. M, 1959, Brown University.

Towe, Kenneth McCarn. Lateral variations in clay mineralogy across major facies boundaries in the Middle Devonian (Ludlowville), New York. D, 1961, University of Illinois, Urbana. 81 p.

Toweh, Solomon H. Landsat image analysis of linears and lineaments in the tri-state district, Missouri-Oklahoma-Kansas. M, 1978, University of Missouri, Rolla.

Towell, David Garrett. Rare-earth distribution in some rocks and associated minerals of the batholith of Southern California. D, 1963, Massachusetts Institute of Technology. 179 p.

Tower, Chris. Hydrogeology of a possible geothermal system near Deeth, Nevada. M, 1982, Colorado School of Mines. 101 p.

Tower, Deborah A. Heavy metal distribution in the sediments of the waters near the Kennedy Space Center. M, 1975, Florida Institute of Technology.

Tower, Dennis B. Geology of the central Pueblo Mountains, Harney County, Oregon. M, 1972, Oregon State University. 96 p.

Towle, Guy N. Stress effects on acoustic velocities of rocks. D, 1978, Colorado School of Mines. 235 p.

Towles, Henry Clay, Jr. A study in integration of geology and geophysics. M, 1951, Southern Methodist University. 20 p.

Townley, Paul Joseph. Preliminary investigation for underground storage of pipeline gas in the Bruer and Flora pools, Mist gas field, Columbia County, Oregon. M, 1985, Portland State University. 147 p.

Towns, Danny Joe. Distribution, depositional environment, and reservoir properties of the Pennsylvanian Cottage Grove Sandstone, South Gage Field, Oklahoma. M, 1978, Oklahoma State University. 98 p.

Townsend, Frank Charles. The influence of sesquioxides on some physicochemical and engineering properties of a lateritic soil. D, 1970, Oklahoma State University. 165 p.

Townsend, James Robert. Geology of a portion of the Newhall Quadrangle, Los Angeles County, California. M, 1940, University of California, Los Angeles.

Townsend, Margaret Anne. Hydrogeology and health effects of a sewage irrigation site, Kerrville, Texas. M, 1981, University of Texas, Austin. 165 p.

Townsend, Peter H. A study of heavy mineral dispersal from the Ausable and Lamoille rivers (Lake Champlain). M, 1970, University of Vermont.

Townsend, Roger Neal. The depositional environment and diagenesis of the Upper Smackover Member in Womack Hill Field, Clarke and Choctaw counties, Alabama. M, 1986, Northeast Louisiana University. 97 p.

Townsend, T. E. Discrimination of alteration in the Crooks Gap, Wyoming uranium district using laboratory and Landsat spectral reflectance data. M, 1978, Stanford University. 84 p.

Townsend, Timothy Elwood. Discrimination of iron alteration minerals in remote sensing data. D, 1984, Stanford University. 262 p.

Towse, Donald Frederick. The (Cretaceous) Frontier Formation in the Southwest Powder River basin, Wyoming. D, 1951, Massachusetts Institute of Technology. 163 p.

Toy, Billy Reynolds. Subsurface stratigraphy and larger foraminifera of Sari Singh #1, Pakistan. M, 1959, University of Illinois, Urbana.

Tozer, Edward T. Uppermost Cretaceous and Paleocene non-marine molluscan faunas of southwestern Alberta. D, 1952, University of Toronto.

Tozer, Warren Wilson. The history of gold mining in the Swauk, Peshastin, and Cle Elum mining districts of the Wenatchee Mountains, 1853-1899. M, 1965, Washington State University. 118 p.

Trabant, Dennis Carlyle. Diagenesis of the seasonal snow cover of interior Alaska. M, 1970, University of Alaska, Fairbanks. 48 p.

Trabant, P. K. Submarine geomorphology and geology of the Mississippi River delta front. D, 1978, Texas A&M University. 131 p.

Trabant, Peter K. Consolidation characteristics and related geotechnical properties of sediments retrieved by the Glomar Challenger from the Gulf of Mexico. M, 1972, Texas A&M University.

Trabelsi, Ali M. Deposition and diagenesis of the Fort Terrett Formation, near Junction, Texas. M, 1984, Texas Tech University. 131 p.

Tracey, Joshua I., Jr. Subsurface geology of Bikini Atoll, Marshall Islands. D, 1950, Yale University.

Tracy, Bradford M. The characterization of the chemical and trace element content of selected oil shales of Ohio. M, 1983, University of Toledo. 242 p.

Tracy, Paul W. Fluoride-calcium interactions in calcareous soil systems. M, 1982, University of Idaho. 65 p.

Tracy, R. W. The plutonic geology of the northern portion of the Giant Forest Quadrangle, California. M, 1968, University of Hawaii.

Tracy, Robert J. High grade metamorphic reactions and partial melting in pelitic schist, Quabbin Reservoir area, Massachusetts. D, 1975, University of Massachusetts. 133 p.

Tracy, Robert James. The petrology of the eastern end of the Cortlandt complex (age undetermined), Westchester County, New York. M, 1970, Brown University.

Tracy, William Craig. Structure and stratigraphy of the central White Pine Range, East-central Nevada. M, 1980, California State University, Long Beach. 66 p.

Trafton, Burke O. Experimental hydrothermal alteration of a quartz monzonite porphyry. M, 1963, Pennsylvania State University, University Park. 72 p.

Traganza, Eugene Dewees. Dynamics of the carbon dioxide system on the Great Bahama Bank. D, 1966, [University of Miami]. 239 p.

Trail, Robert Bruce. A field trip through glacial deposits in eastern Connecticut. M, 1967, Virginia State University. 40 p.

Traill, R. J. Crystal structure of high temperature albite. D, 1953, Queen's University. 86 p.

Traill, R. J. Synthesis and X-ray study of uranium sulphate minerals. M, 1951, Queen's University. 76 p.

Train, Fiona F. A study of the maps and the computerised indexing of the Institute of Geological Sciences map collection. M, 1981, City College (CUNY). 67 p.

Trainer, David Woolsey. Intrusive andesites and associated rocks, Kekequabic Lake, northeastern Minnesota. M, 1923, Northwestern University.

Trainer, David Woolsey, Jr. Wisconsin molding sands, their occurrence and properties. D, 1932, Cornell University.

Trainer, Frank Wilson. Eocene foraminifera from Virginia. M, 1948, University of Virginia. 57 p.

Trainer, Frank Wilson. Geology and ground-water resources of the Matanuska Valley agricultural area, Alaska. D, 1953, Harvard University.

Tralli, David Marcelo. Lateral variations in mantle P velocity for a tectonically regionalized Earth. D, 1986, University of California, Berkeley. 163 p.

Trammell, John W. Geology of the Cumberland Pass area, Gunnison County, Colorado. M, 1961, University of Colorado.

Trammell, John W. Strata-bound copper mineralization in the Empire Formation and Ravalli Group, Belt Supergroup, Northwest Montana. M, 1975, University of Washington. 70 p.

Trancynger, Thomas C. Mineral paragenesis of the Magmont ores, Viburnum Trend, Southeast Missouri. M, 1975, University of Missouri, Rolla.

Trangmar, Bruce Blair. Spatial variability of soil properties in Sitiung, West Sumatra, Indonesia. D, 1984, University of Hawaii. 354 p.

Tranter, Charles Enoch, Jr. Correlation of gravel deposits in the La Plata River valley, La Plata County, Colorado. M, 1954, University of Illinois, Urbana. 85 p.

Trapasso, Linda. The geology of the Torbrook Syncline, Kings and Annapolis counties, Nova Scotia. M, 1979, Acadia University.

Trapnell, Don Edward. Areal geology of southwestern Canadian County, Oklahoma. M, 1961, University of Oklahoma. 63 p.

Trapp, Harold R. Heavy minerals of the Arkansas River from Buena Vista, Colorado to Wichita, Kansas. M, 1966, Wichita State University. 195 p.

Trapp, Henry, Jr. The Noix Oolite of Pike County, Missouri. M, 1950, Washington University. 163 p.

Trapp, John Siegfried. A geochemical study of the Maquoketa Formation (upper Ordovician) in Pike County, Missouri. M, 1969, University of Missouri, Rolla.

Trapp, John Siegfried. Trend surface analysis as an aid in exploration for Mississippi Valley type ore deposits (Missouri). D, 1972, University of Missouri, Rolla. 171 p.

Trask, C. B. Mineralogy, texture, and longshore transport of beach sand, eastern shore of Lake Ontario. D, 1976, Syracuse University. 317 p.

Trask, Charles Brian. Roundness, sphericity, and form of pocket-beach gravels (Recent), Mount Desert Island, Hancock County, Maine. M, 1972, University of Texas, Austin.

Trask, Newell J., Jr. Geology of the Buford area, Rio Blanco County, Colorado. M, 1956, University of Colorado.

Trask, Newell J., Jr. Stratigraphy and structure in the Vernon-Chesterfield area, Massachusetts, New Hampshire, Vermont. D, 1964, Harvard University.

Trask, Parker Davies. A study of the fauna and stratigraphy of the Briones Formation of middle California. M, 1920, University of California, Berkeley. 82 p.

Trask, Parker Davies. The geology of the Point Sur Quadrangle, California. D, 1923, University of California, Berkeley. 181 p.

Trask, Richard P. A study of estuarine bottom stress. M, 1979, University of New Hampshire. 56 p.

Traub, John H. Petrographic analysis and environmental interpretation of the Strodes Creek Member (Upper Ordovician) of the Lexington Limestone. M, 1982, Eastern Kentucky University. 65 p.

Trauba, W. C. Petrography of pre-Tertiary rocks of the Blue Mountains, Umatilla County, Northeast Oregon. M, 1975, Oregon State University. 171 p.

Traut, Marc W. Sedimentology and Recent sedimentary history, San Felipe area, Baja California, Mexico. M, 1977, Arizona State University. 139 p.

Trautman, Ray Love. A study and description of some sediments from Southwestern United States. M, 1932, University of Kentucky. 85 p.

Trautman, T. A. Stratigraphy and petrology of Tertiary clastic sediments, Seymour Island, Antarctica. M, 1976, Ohio State University. 170 p.

Trautmann, Charles H. Engineering geology of Franciscan melange terrane in the Red Hill area, Petaluma, California. M, 1976, Stanford University. 74 p.

Trautmann, Charles Home. Behavior of pipe in dry sand under lateral and uplift loading. D, 1983, Cornell University. 328 p.

Travers, Ian C. An environmental analysis of Charlottetown Harbour and Cardigan Bay, Prince Edward Island. M, 1975, Queen's University. 166 p.

Travers, Mark Aaron. Fracture toughness testing of the sandstone associated with the Dugger Formation overlying the No. 5 Coal in Indiana. M, 1986, Purdue University. 150 p.

Travers, William Brailsford. Geology of the Newell Creek area, Boulder Creek, California. M, 1959, Stanford University. 44 p.

Travers, William Brailsford. The Late Eocene transgression and associated deformation in part of the northern Apennine Mountains, Italy (Part I.); a zone of Miocene subduction in the northern Apennine Mountains, Italy (Part II.). D, 1972, Princeton University. 275 p.

Traverse, Alfred F., Jr. The pollen and spores of the (Eocene) Brandon Lignite; a coal in Vermont of lower Tertiary age. D, 1951, Harvard University.

Travis, Abe. Geology of Oklahoma County. M, 1930, University of Oklahoma. 46 p.

Travis, Charles. Pyrite from Cornwall, Lebanon County, Pennsylvania. D, 1906, University of Pennsylvania.

Travis, Deborah Sue. Chronostratigraphy, depositional rates, continental margin progradation, and growth-fault dynamics within the Tertiary wedge, San Marcos Arch, Northwest Gulf of Mexico. M, 1988, University of Texas, Austin. 125 p.

Travis, Everett Joyce. Areal geology of Harral Quadrangle, Culberson County, Texas. M, 1951, University of Texas, Austin.

Travis, Fred Lee. The stratigraphy of the upper Marmaton Group of Rogers County, Oklahoma. M, 1942, University of Iowa. 83 p.

Travis, Jack Watson. Paleoenvironmental study of the Helderberg Group of West Virginia and Virginia. D, 1971, Michigan State University. 150 p.

Travis, Jack Watson. Stratigraphic and petrographic study of the (Silurian) McKenzie Formation in West Virginia. M, 1962, West Virginia University.

Travis, Lynne S. The depositional environments and reservoir characteristics of the upper Morrow "A" Sandstone, Postle Field, Texas County, Oklahoma. M, 1987, Texas A&M University. 91 p.

Travis, Patricia Ann Asiala. An analysis of Pleistocene sediments in an aquifer recharge area, Kalamazoo, Michigan. D, 1966, Michigan State University. 143 p.

Travis, Paul Leonard, Jr. Geology of the area near the north end of Summer Lake, Lake County, Oregon. M, 1977, University of Oregon. 95 p.

Travis, Russell Burton. The geology of the Sebastopol Quadrangle, California. D, 1951, University of California, Berkeley. 145 p.

Travis, Steven L. Uranium mineralization in Jim Wells County, Texas. M, 1981, Wichita State University. 50 p.

Traxler, J. Douglas. Geology of the east central Santa Monica Mountains. M, 1948, University of California, Los Angeles.

Traylor, Charles Tim. The evaluation of a methodology to measure manual digitizing error in cartographic data bases. D, 1979, University of Kansas. 122 p.

Traylor, Henry Grady. Geology of a portion of the Kensington Quadrangle, Northwest Georgia. M, 1951, University of Iowa. 122 p.

Treadgold, Galen E. Modelling and interpretation of the oceanic crustal structure north of the Puerto Rico Trench. M, 1985, University of Texas, Austin.

Treadway, Jeffrey A. Shallow seismic study of a fault scarp near Borah Peak, Idaho. M, 1987, University of Kansas. 80 p.

Treadway, Keith Richard. The petrology of glacial tills from Tippecanoe County, Indiana, and adjacent areas. M, 1958, Indiana University, Bloomington. 59 p.

Treadwell, Carol Jane. Geomorphic evidence for contemporary uplift, northeastern Adirondack Mountains, New York. M, 1988, SUNY at Binghamton. 170 p.

Treadwell, D. D. The influence of gravity, prestress, compressibility, and layering on soil resistance to static penetration. D, 1976, University of California, Berkeley. 227 p.

Treadwell, Robert Cuthrell. Nature of the Moody's Branch-Cockfield contact in Sabine Parish, Louisiana, and adjacent areas. M, 1951, Louisiana State University.

Treadwell, Robert Cuthrell. Sedimentology and ecology of southeast coastal Louisiana. D, 1955, Louisiana State University.

Treasher, Ray C. Geology of the Pullman quadrangle (Washington). M, 1925, Washington State University. 74 p.

Treat, Cheryl Lee. Metamorphic and structural history of the Hicks Butte inlier, south-central Cascades, Washington. M, 1987, San Jose State University. 93 p.

Trebing, Harry Evan. Chromite compositional variation in the Twin Sisters Dunite, Washington State. M, 1988, Michigan State University. 74 p.

Treckman, John F. Petrography of the Upper Cretaceous Terry and Hygiene sandstones in the Denver Basin; Colorado. M, 1960, University of Colorado.

Trees, Charles Connett. Suspended solids in a basin of the lagoonal system of Florida. M, 1977, Florida Institute of Technology.

Treesh, Michael Irvin. Depositional environments of the Salina Group (upper Silurian) in New York State. D, 1973, Rensselaer Polytechnic Institute. 127 p.

Trefethen, Helen B. Studies of origin of eskers in Maine. M, 1934, University of Wisconsin-Madison.

Trefethen, Joseph M. The Lincoln Sill, southeastern Maine. D, 1935, University of Wisconsin-Madison.

Trefethen, Joseph Muzzy. The origin of Georges Bank. M, 1932, University of Illinois, Chicago.

Trefry, J. H., III. The transport of heavy metals by the Mississippi River and their fate in the Gulf of Mexico. D, 1977, Texas A&M University. 236 p.

Trefz, Richard Joseph. An analysis of small-scale structures in the Late Precambrian pelitic schists of the Moinian System of Anglesey, United Kingdom. M, 1973, University of Illinois, Urbana. 127 p.

Tregaskis, Scott W. Geological and geochemical studies of the Woodbury zinc and lead occurrences, Bedford County, Pennsylvania. M, 1979, Pennsylvania State University, University Park. 174 p.

Tréhu, Anne Martine. Seismicity and structure of the Orozco transform fault. D, 1982, Massachusetts Institute of Technology. 370 p.

Trehu, Anne Martine. Seismicity and structure of the Orozco transform fault from ocean bottom seismic observations. D, 1982, Woods Hole Oceanographic Institution. 370 p.

Treiman, Allan Harvey. Precambrian geology of the Ojo Caliente Quadrangle, Rio Arriba and Taos counties, New Mexico. M, 1978, Stanford University. 93 p.

Treiman, Allan Harvey. The OKA carbonatite complex, Quebec; aspects of carbonatite petrogenesis. D, 1982, University of Michigan. 182 p.

Treitel, S. Wavelet model seismic noise. M, 1955, Massachusetts Institute of Technology. 32 p.

Treitel, Sven. On the dissipation of seismic energy from source to surface. D, 1958, Massachusetts Institute of Technology. 174 p.

Trekell, Rex Elroy. Predicting formation conditions of natural gas hydrates at elevated pressures. M, 1965, University of Oklahoma. 78 p.

Trela, John Joseph. Soil formation on Tertiary landsurfaces of the New Jersey coastal plain. D, 1984, Rutgers, The State University, New Brunswick. 620 p.

Treloar, Anne Marie. The conodonts of the Ordovician Maquoketa Formation in Iowa. M, 1957, University of Iowa. 73 p.

Treloar, Nathan A. Viscous remanent magnetization acquisition at lower crustal temperatures. M, 1985, University of Wyoming. 58 p.

Tremaine, John W. A restudy of structural and stratigraphic relations in western New Hampshire. M, 1957, Dartmouth College. 81 p.

Tremba, Edward Louis. Isotope geochemistry of strontium in carbonate and evaporite rocks of marine origin. D, 1973, Ohio State University.

Trembath, Gerald. The compositional variation of staurolite in the area of the Anderson Mine, Snow Lake, Manitoba. M, 1986, University of Manitoba.

Trembath, Lowell Thomas. A study of the potassium feldspars in some igneous and metamorphic rocks from the Moak-Thompson map area, Manitoba. M, 1961, University of Manitoba.

Trembath, Lowell Thomas. Refinement of the fayalite structure. D, 1964, Queen's University. 80 p.

Tremblay, André. Etude du contrôle structural de la minéralisation dans la "zone du toit" de la mine Copper Rand à Chibougamau. M, 1981, Universite du Quebec a Chicoutimi. 129 p.

Tremblay, François. Etude de reconnaissance en géochimie isotopique de l'oxygène et de l'hydrogène; application à quelques minéralisations de la région de Chibougamau, Québec, Canada. M, 1987, Universite du Quebec a Chicoutimi. 99 p.

Tremblay, Guy. Etude des déformations du métaquartzite de la Galette, du Comté Charlevoix. M, 1985, Universite du Quebec a Chicoutimi. 128 p.

Tremblay, Léo-Paul. Contribution à l'étude des feldspaths des roches du Mont Tremblant et de celles du Pine Hill (Quebec). M, 1944, Universite Laval.

Tremblay, Leo-Paul. Geology of the Lacorne-Barraute area, Abitibi County, Quebec. D, 1947, University of Toronto.

Tremblay, Michel. Etude de l'hétérogénéité du gisement de kaolin de Château-Richer et de son incidence sur les propriétés des mousses d'argile. M, 1987, Universite du Quebec a Chicoutimi. 149 p.

Tremblay, Mosseau. Geology of the Williamson diamond mine, Mwadue, Tanganyika. D, 1956, McGill University.

Tremblay, P.-R. The optimized management of the Cap-de-la-Madeleine Aquifer by means of a digital model. M, 1975, University of Waterloo.

Tremblay, Pierrette. Mineralogy and geochemistry of the radioactive pegmatites of the Mont-Laurier area, Quebec. M, 1974, Queen's University. 133 p.

Trembley, Susan B. Environments of deposition of the Plio-Pleistocene Fort Denaud Member, Caloosahatchee Formation, southern Florida, based on ostracod faunas. M, 1974, University of Akron. 175 p.

Trembly, Jeffrey Allen. Hydrothermally altered basalts from the Mariana Trough. M, 1982, University of Arizona. 42 p.

Trembly, Lynn Dale. Primary seismic waves near explosions. M, 1965, Oregon State University. 58 p.

Trembly, Lynn Dale. Seismic source characteristics from explosion generated Pwaves. D, 1968, Oregon State University. 137 p.

Trembly, W. A., Jr. Depositional environments of an upper portion of the Modelo Formation, Los Angeles County, California. M, 1987, California State University, Northridge. 126 p.

Tremper, Lauren Roy. Lithofacies and stratigraphic analysis of the Salina Group of the "North Slope" of the Michigan Basin. M, 1973, University of Michigan.

Trench, Elaine. Wetlands and flood discharge in the Taunton River watershed. M, 1980, Boston University. 188 p.

Trench, N. R. The geomorphology and paleoenvironmental history of the Lake City landslide complex, Southwest Colorado. M, 1978, University of Colorado. 149 p.

Trenchard, Walter H. An environmental study of the subsurface Miocene of Matagorda County, Texas. M, 1961, Texas A&M University. 225 p.

Trengove, Stanley Albin. Hydrothermal oxidation of manganese minerals. D, 1934, University of Minnesota, Minneapolis. 22 p.

Trenholm, L. S. Geology of the Amm gold mine, Cadillac Township, Quebec. M, 1939, McGill University.

Trent, Dee Dexter. Structural geology of the upper Rock Creek area, Inyo County, California, and its relation to the regional structure of the Sierra Nevada. D, 1973, University of Arizona. 206 p.

Trent, Donald Eugene. High pressure synthesis and influence of pressure on cation distribution in 2-3 and 2-4 spinels. D, 1970, Ohio State University. 80 p.

Trent, William Richard. Some aspects of the petrology and geochemistry of selected freshwater carbonates. M, 1978, Oklahoma State University. 121 p.

Trentham, Robert Craig. Leaching of uranium from felsic volcanics and volcanoclastics; model, experimental studies and analysis of sites. D, 1981, University of Texas at El Paso. 227 p.

Trepagnier, Albert James. The lower Frio subsurface geology of the Bayou Mallet - Lewisburg - Opelousas, Louisiana area. M, 1970, University of Southwestern Louisiana.

Trepasso, Linda. The geology of the Torbrook Syncline, Kings and Annapolis counties, Nova Scotia. M, 1979, Acadia University.

Trescott, Peter Chapin. An investigation of the groundwater resources of the Annapolis-Cornwallis Valley, Nova Scotia. D, 1967, University of Illinois, Urbana. 312 p.

Trescott, Peter Chapin. The influence of preceding sedimentary patterns on the thickness of a Pennsylvanian coal. M, 1964, University of Illinois, Urbana.

Tress, David Edward Von *see* Von Tress, David Edward

Tresselt, Peter. Recent beach and coastal dune sands at Pismo Beach, California. M, 1960, University of California, Los Angeles.

Tresslar, R. C. The living benthonic foraminiferal fauna of the West Flower Garden Bank coral reef and biostrome. M, 1974, Texas A&M University.

Tressler, Richard Ernest. Crystal chemistry and physical properties of some sulfospinels and their polymorphs. D, 1967, Pennsylvania State University, University Park. 135 p.

Trethewey, Ben Clifford. Relationships of certain structural elements to iron ore concentrations in a portion of the Cary Mine, Hurley, Wisconsin. M, 1958, Michigan State University. 32 p.

Trettin, Hans Peter. Geology of the Fraser River valley between Lillooet and Big Bar Creek. D, 1960, University of British Columbia.

Trettin, Hans Peter. Silver Cup Mine, Lardeau; regional framework and structural ore control. M, 1957, University of British Columbia.

Trevena, A. S. Depositional models and the Shinarump Member and the Sonsela Sandstone Bed of the Chinle Formation, northeastern Arizona and northwestern New Mexico. M, 1975, University of Arizona.

Trevena, A. S. Studies in sandstone petrology; origin of the Precambrian Mazatzal Quartzite and provenance of detrital feldspar. D, 1979, University of Utah. 401 p.

Trever, Paula Fern. Geology of the Gardner Mountain area, Happy Valley Quadrangle, Cochise County, Arizona. M, 1983, University of Arizona. 130 p.

Treves, Samuel B. General geology of the Seafoam mining district, Custer County, Idaho. M, 1953, University of Idaho. 50 p.

Treves, Samuel Blain. Geology of the Carney Lake Complex, Dickinson County, Michigan. D, 1959, Ohio State University. 196 p.

Trevino, Ramon H., III. Facies and depositional environments on the Boquillas Formation, Upper Cretaceous of Southwest Texas. M, 1988, University of Texas, Arlington. 120 p.

Trew, James R. Slope development in the San Dimas experimental forest, California. M, 1956, Brown University.

Treworgy, Janis Driver. Stratigraphy and depositional settings of the Chesterian (Mississippian) Fraileys/Big Clifty and Haney formations in the Illinois Basin. D, 1985, University of Illinois, Urbana. 202 p.

Trexler, Bryson D., Jr. The hydrogeology of acid production in a lead-zinc mine. D, 1975, University of Idaho. 164 p.

Trexler, Bryson Douglas, Jr. Petrography, porosity, and hydraulic conductivity of the Castle Hayne aquifer, Castle Hayne, North Carolina. M, 1974, North Carolina State University. 65 p.

Trexler, David William. Stratigraphy and structure of the Coalville area, northeastern Utah. D, 1955, The Johns Hopkins University.

Trexler, Dennis Thomas. Geology of the northwest quarter of the Cecilville Quadrangle, Siskiyou County, California. M, 1966, University of Southern California.

Trexler, James Hugh, Jr. Depositional environment of the basal Atoka Formation in northeastern Oklahoma, based on biogenic structure and faunal evidence. M, 1976, University of Oklahoma. 77 p.

Trexler, James Hugh, Jr. Stratigraphy, sedimentology and tectonic significance of the Upper Cretaceous Virginian Ridge Formation, Methow Basin, Washington; implications for tectonic history of the North Cascades. D, 1984, University of Washington. 172 p.

Trexler, John Peter. Geology of Godfrey Ridge, near Stroudsburg, Pennsylvania. M, 1953, Lehigh University.

Trexler, John Peter. The geology of the Klingerstown, Valley View, and Lykens quadrangles, southern anthracite field, Pennsylvania (V. 1-3). D, 1964, University of Michigan. 749 p.

Treybig, Lucille Evelyn. Subsurface geology of the Luling fault zone. M, 1942, University of Texas, Austin.

Triana, Rebecca. The Panhandle Field; a case study. M, 1986, University of Texas, Austin.

Trible, Marla. Ground magnetometer survey in the Valley of Ten Thousand Smokes, Alaska. M, 1972, University of Alaska, Fairbanks. 310 p.

Trice, E. L. Jack, III. Conodont biostratigraphy and stratigraphic relationships of the Strawn Group (Pennsylvanian), Colorado and Brazos River valleys, central Texas. M, 1984, Baylor University. 140 p.

Trice, Edwin Leslie, Jr. Geology of Lagunitas Lake area, Rio Arriba County, New Mexico. M, 1957, University of Texas, Austin.

Triegel, Elly Kirsten. Changes in cadmium and nickel adsorption over time in a periodically submerged soil. D, 1984, University of Tennessee, Knoxville. 164 p.

Trifunac, Mihailo Dimitrije. Investigation of strong earthquake ground motion. D, 1969, California Institute of Technology. 153 p.

Trigger, James Kendall. Geology of the south-central part of the Sitkum Quadrangle, Coos County, Oregon. M, 1966, University of Oregon. 79 p.

Trimble, Allen Ben. Seismicity and contemporary tectonics of the Yellowstone Park-Hebgen Lake region. M, 1973, University of Utah. 66 p.

Trimble, David Charlton. Styles of sedimentation in the Chilhowee Group in the southern part of the Del Rio District. M, 1985, University of Kentucky. 136 p.

Trimble, Deborah. The application of uranium-thorium systematics to rocks from the Lassen Dome field. M, 1984, San Jose State University. 94 p.

Trimble, James K. The geology of the Kenosha Pass area and vicinity, Park County, Colorado. M, 1960, Colorado School of Mines. 107 p.

Trimble, Larry Merc. Geology and ore deposits of the San Rafael River mining area, Emery County, Utah. M, 1977, University of Utah. 284 p.

Trimble, Stanley Wayne. A geographic analysis of erosive land use on the southern Piedmont, 1700-1970. D, 1973, University of Georgia.

Tringale, Philip Thomas. Soil identification in-situ using an acoustic cone penetrometer. D, 1983, University of California, Berkeley. 427 p.

Tripathi, Vijay Shankar. Geochemical prospecting on Jasper Ridge. M, 1976, Stanford University. 145 p.

Tripathi, Vijay Shankar. Uranium(VI) transport modeling; geochemical data and submodels. D, 1984, Stanford University. 319 p.

Triplehorn, Don Murray. The petrology of glauconite. D, 1961, University of Illinois, Urbana.

Tripp, Alan Craig. Multidimensional electromagnetic modeling. D, 1982, University of Utah. 191 p.

Tripp, Allan C. Electromagnetic and Schlumberger resistivity sounding in the Roosevelt Hot Springs known geothermal resource area. M, 1977, University of Utah. 110 p.

Tripp, Angela M. Lower Mississippian Fort Payne Formation of southern Illinois. M, 1981, Northern Illinois University. 67 p.

Tripp, Eugene C. The geology of the northern half of the Pancake Summit Quadrangle, Nevada. M, 1957, University of Southern California.

Tripp, Russell Maurice. Classification of organic shales based on thermographic analysis. D, 1948, Massachusetts Institute of Technology. 97 p.

Tripp, Steven Edward. Rubidium-strontium and uranium-lead geochronology of the northeastern border zone of the Idaho Batholith, Bitterroot Range, Montana. M, 1976, Western Michigan University. 73 p.

Tripp, William James. Density and distribution of macroborers of the modern Glossifungites ichnofacies; St. Catherines Island, Georgia. M, 1984, University of Georgia. 164 p.

Trippet, William A., III. A physical inventory of Somervell County. M, 1973, Baylor University. 231 p.

Trippi, Michael Herbert. Chemical fingerprinting, clay mineralogy and heavy minerals of several Champlainian K-bentonites of north-central and eastern New York. M, 1986, University of Cincinnati. 157 p.

Tristan, Elorrieta Nimio Juvenal. Rb-Sr dates of crystalline rocks from southern Israel. M, 1980, University of Texas at Dallas. 38 p.

Trites, Albert F. Differential thermal analysis of goethite and lepidocrocite. M, 1948, Columbia University, Teachers College.

Tritschler, Charles W. The nature and origin of the drainage pattern in Hamilton Township, Warren County, Ohio. M, 1956, University of Cincinnati. 42 p.

Trivedi, Harshadrai P. Mineralogy of Recent sediments from selected cores along the southeast coast of Puerto Rico. M, 1968, Texas A&M University. 72 p.

Trivedi, Nikhilesh Chandrakant. Electrokinetic behavior of asbestos minerals as a function of aging time in water. M, 1970, University of Nevada. 70 p.

Trocki, Linda Katherine. Ages of uranium mineralization and lead loss in the Key Lake Deposit, northern Saskatchewan, Canada. M, 1983, Pennsylvania State University, University Park. 38 p.

Troell, Arthur R. Mississippian bioherms of southwestern Missouri and northwestern Arkansas. M, 1960, University of Missouri, Rolla.

Troell, Arthur Richard, Jr. Sedimentary facies of the Toronto Limestone, lower limestone member of the Oread Megacyclothem (Virgilian) of Kansas. D, 1965, Rice University. 334 p.

Troelsen, Johannes C. Geology of the Bonne Bay-Trout River area, Newfoundland. D, 1947, Yale University. 323 p.

Troendle, C. A. A variable source area model for stormflow prediction on first order forested watersheds. D, 1979, University of Georgia. 124 p.

Troensegaard, Kingdon W., II. Development of cleavage in a portion of the Bays Mountain Synclinorium, Greene, Cocke, Hamblen, and Jefferson counties, Tennessee. M, 1965, University of Tennessee, Knoxville. 39 p.

Trojan, Michael. Diagenesis of the Upper Jurassic Taylor Sandstone and its effect on well log response, Terryville Field, Lincoln Parish, Louisiana. M, 1984, Texas A&M University.

Trojan, William R. Devonian stratigraphy and depositional environments of the northern Antelope Range, Eureka County, Nevada. M, 1979, Oregon State University. 134 p.

Trojer, Felix J. Crystallographic study of calcium. D, 1969, Massachusetts Institute of Technology. 133 p.

Trojer, Felix J. The refinement of the structure of sulvanite, Cu_3US_4. M, 1966, Massachusetts Institute of Technology. 49 p.

Trollope, Frederick Hamilton. A lower microfauna of the Loon River Formation, northern Alberta. M, 1951, University of Alberta. 96 p.

Trombetta, Michael J. Evolution of the Eocene Avon volcanic complex, Powell County, Montana. M, 1987, Montana State University. 112 p.

Trombley, Thomas J. A radio echo-sounding survey of Athabasca Glacier, Alberta, Canada. M, 1986, University of New Hampshire. 64 p.

Trommer, Jeff. Effect of sediment texture and fabric on hydraulic conductivity of beach sediments, Florida. M, 1987, University of South Florida, Tampa. 76 p.

Tromp, Paul L. Stratigraphy and depositional environments of the "Leo Sands" of the Minnelusa Formation, Wyoming and South Dakota. M, 1981, University of Wyoming. 69 p.

Troncoso, J. H. In situ impulse test for determination of soil shear modulus as a function of strain. D, 1975, University of Illinois, Urbana. 344 p.

Trone, Paul Max. Textural and mineralogical characteristics of altered Grande Ronde Basalt, northeastern Oregon; a natural analog for a nuclear waste repository in basalt. M, 1987, Portland State University. 152 p.

Tronnes, Reidar Gjermund. The incorporation of Ti in phlogopite in a simplified, synthetic system; a potential geothermobarometer for upper mantle and lower crustal rocks. D, 1985, University of Western Ontario. 133 p.

Troop, Andrew John. The geology of the Ogama-Rockland gold mine (southeastern Manitoba). M, 1950, University of Manitoba.

Troop, Douglas Grant. The petrology and geochemistry of Ordovician banded iron formations and associated rocks at the Flat Landing Brook massive sulphide deposit, northern New Brunswick. M, 1984, University of Toronto.

Troschinetz, John. Stratigraphic relationships, chemical correlation and possible source of the tuff of Spencer and Heise of the eastern Snake River plain, in Idaho. M, 1983, Wayne State University. 148 p.

Troseth, Frank Paton. Radial transmission factors for spherical waves in elastic media. D, 1954, Colorado School of Mines. 78 p.

Trost, Lawrence C. The economic geology of the Dairy Farm and Valley View mine areas, Placer County, California. M, 1962, University of California, Berkeley. 80 p.

Trost, Paul B. Effect of humic-acid-type organics on the secondary dispersion of mercury. D, 1970, Colorado School of Mines. 162 p.

Trostel, Everett G. Production gas-oil ratio as a function of the distribution of hydrocarbons in the underground reservoir. M, 1940, University of Southern California.

Troster, John Gooch. Recent marine sedimentation in Halfmoon Bay, California. M, 1949, Stanford University. 61 p.

Trott, Barbara G. Gravity study of the Golden Spike and Westerose South reefs, Alberta. M, 1981, University of Calgary. 98 p.

Trotter, Charles Leonard. Palyno-botanical and stratigraphic studies of three lignite, drill cores (Paleocene) from Harding County, South Dakota. D, 1963, Pennsylvania State University, University Park. 346 p.

Trotter, John Francis. Geology of the Nowood Creek-Ten Sleep area, Washakie County, Wyoming. M, 1954, University of Wyoming. 79 p.

Trotter, Sherry F. Algal-bank complex in the upper Utopia Limestone (Late Pennsylvanian), Sumner County, Kansas. M, 1981, Wichita State University. 57 p.

Trottier, Jacques. Synthèse métallogénique des dépôts sulfureux volcanogènes de la ceinture ophiolitique des Apalaches du Québec, région de l'Estrie. D, 1988, Ecole Polytechnique. 201 p.

Troughton, G. H. Stratigraphy of the Vizcaino Peninsula near Asuncion Bay, Territorio de Baja California, Mexico. M, 1974, San Diego State University.

Troup, Arthur George. Geochemical investigations of ferromanganese concretions from three Canadian lakes. M, 1969, McMaster University. 82 p.

Troup, Bruce Neil. The interaction of iron with phosphate, carbonate and sulfide in Chesapeake Bay interstitial waters; a thermodynamic interpretation. D, 1974, The Johns Hopkins University.

Troup, W. R. Metasomatic wall rock alteration of the Mattagami Lake Mine. M, 1975, University of Waterloo.

Trout, Laurence Emory. The geology and paleontology of the Simpson Formation. M, 1913, University of Oklahoma. 218 p.

Trout, Michael Lynn. Origin of bromide-rich brines in southern Arkansas. M, 1974, University of Missouri, Columbia.

Troutman, Thomas William. The King's Canyon Lineament, central California. M, 1979, University of Nevada. 93 p.

Troutt, William Richard. An occurrence of large scale, inactive, sorted patterned ground south of the glacial border in central Pennsylvania. M, 1971, Pennsylvania State University, University Park. 71 p.

Trow, James William. Early Mesozoic sinking of the Appalachia-Ouachita tract. M, 1945, University of Chicago. 63 p.

Trow, James William. The (Precambrian) Sturgeon Quartzite of the Menominee District, Michigan. D, 1948, University of Chicago. 60 p.

Trowbridge, Arthur Carleton. The geology of the Owena Valley, California region with special reference to the terrestrial deposits. D, 1911, University of Chicago.

Trowbridge, Raymond Maxwell. Pleistocene glacial deposits and pre-glacial drainage pattern of northwestern Missouri. D, 1938, University of Missouri, Columbia.

Trowbridge, Raymond Maxwell. The physical composition of the Chugwater Formation. M, 1930, University of Missouri, Columbia.

Troxel, Bennie Wyatt. Geology of the northwestern part of the Shadow Mountains, western San Bernardino County, California. M, 1958, University of California, Los Angeles.

Troxell, Edward L. The terraces of Kellogg Ravine. M, 1911, Northwestern University.

Troxell, Edward L. The vertebrate fossils of Rock Creek, Texas. D, 1914, Yale University.

Troyanowski, Larry D. A subsurface study of the Eloi Bay Field, St. Bernard Parish, Louisiana. M, 1982, University of Southwestern Louisiana. 52 p.

Troyer, David Robert. Petrography and paleoenvironments of the Chazyan Rockcliffe Formation of the Ottawa-Hull area. M, 1979, Carleton University. 200 p.

Troyer, Max L. Geology of the Lander (Hudson) and Plunkett anticlines and vicinity, Fremont County, Wyoming. M, 1951, University of Wyoming. 68 p.

Truax, Stephen, III. Foraminiferal distribution and paleoecology of the Weno Formation (Lower Cretaceous) along Sycamore Creek, Tarrant County, Texas. M, 1980, University of Southwestern Louisiana. 129 p.

Truccano, Norman D. Ratio classification of forested areas lying above petroleum deposits in western Illinois utilizing thematic mapper. M, 1986, Indiana State University. 86 p.

Truckle, Daniel M. Geology of the Calvert Hill area, Beaverhead County, Montana. M, 1988, Montana College of Mineral Science & Technology. 94 p.

Trudeau, Douglas A. Hydrogeologic investigation of the Littlefield Springs. M, 1979, University of Nevada. 136 p.

Trudeau, M. A study of the distribution of diffracted intensity in reciprocal space for franckeite-type minerals. M, 1975, McGill University. 56 p.

Trudeau, P. N. Ecology of barrier beaches in South central Maine (Popham Beach State Park, Reid State Park, and Small Point Beach). D, 1979, University of Massachusetts. 412 p.

Trudeau, Yvon. Pétrographie et géochimie des roches du secteur environnant pour la mine Bruneau, Chibougamau, Quebec. M, 1983, Universite du Quebec a Chicoutimi. 134 p.

Trudel, Pierre. Le volcanisme archéen et la géologie structurale de la région de Clericy, Abitibi, Québec. D, 1980, Ecole Polytechnique.

Trudel, Pierre. Le Volcanisme Archéen et la géologie structurale de la région de Clericy, Abitibi, Québec. D, 1979, Universite de Montreal.

Trudel, Pierre. Pétrographie et pétrologie de la partie nord de la ceinture métavolcanique de Rouyn-Noranda. M, 1975, Ecole Polytechnique.

Trudell, Laurence G. and Evans, Stewart Thompson. Geology of the South Veta Creek area, Huerfano Quadrangle, Huerfano County, Colorado. M, 1958, University of Michigan.

Trudell, Mark Russell. Factors affecting the occurrence and rate of denitrification in shallow groundwater flow systems. M, 1980, University of Waterloo.

Trudick, Lee S. The biogeochemistry of trace element uptake in plants and soils as an indicator of fluorspar deposits. M, 1979, Kent State University, Kent. 89 p.

Trudnak, Gerald S. Paleotemperature analysis of upper Eocene - lower Oligocene sediments of Alabama. M, 1978, Northeast Louisiana University.

Trudu, Alfonso Giacomo. Petrology, structure and origin of the K-1 scheelite orebody, Westfeld Sector, Felbertal, Austria. M, 1984, Queen's University. 383 p.

True, D. G. Undrained vertical penetration into ocean bottom soils. D, 1976, University of California, Berkeley. 226 p.

Trueblood, Peter Martin. Explanatory text for a geologic map of an area north of Quseir, Egypt. M, 1981, Bryn Mawr College. 72 p.

Trueman, David Lawrence. Petrological, structural and magnetic studies of a layered basic intrusion; Bird River Sill (Precambrian), Manitoba. M, 1971, University of Manitoba.

Trueman, David Lawrence. Stratigraphy, structure and metamorphic petrology of the Archean greenstone belt at Bird River, Manitoba. D, 1980, University of Manitoba.

Trueman, Edward Albert George. Minor elements in rocks and minerals from the Copper Mountain mining district, British Columbia. M, 1973, Queen's University. 114 p.

Trueman, Joseph Douglas. Igneous rocks of the Blue Hills Complex. M, 1909, Massachusetts Institute of Technology. 73 p.

Trueman, Joseph Douglas. The value of certain criteria for the determination of the origin of foliated crystalline rocks. D, 1911, University of Wisconsin-Madison.

Truesdell, Alfred Hemingway. The study of natural glasses through their behavior as membrane electrodes. D, 1962, Harvard University.

Truesdell, Page Ernest. Mammoth Cave. M, 1939, University of Cincinnati. 66 p.

Truesdell, W. H. Dikes of the St. Louis River District. M, 1906, University of Minnesota, Minneapolis.

Truettner, Laura Elizabeth. Mineral weathering and sources of alkalinity in two Adirondack lake watersheds. M, 1984, University of Massachusetts. 147 p.

Truettner, Walter James, Jr. Stratigraphy and paleontology of a core from Silurian and Devonian strata of Lorain County, Ohio. M, 1954, University of Michigan.

Truex, John Neithardt. Geology of the northern part of the Santa Monica Mountains between Beverly Glen and Laurel Canyon Boulevard. M, 1950, University of California, Los Angeles.

Truitt, Duane J. The Eastern Idaho Environmental Survey. M, 1986, University of Idaho. 62 p.

Truitt, Paul Bettman. The geology of the Spencer oil pool, Posey County, Indiana. M, 1951, University of Cincinnati. 33 p.

Trujillo, Alan P. Sedimentology of the Scanlan Conglomerate Member of the Pioneer Formation (Proterozoic) in the Pleasant Valley area, Gila County, Arizona. M, 1984, Northern Arizona University. 242 p.

Trujillo, Ernest Floyd. Upper Cretaceous foraminifera from near Redding, Shasta County, California. M, 1958, University of California, Los Angeles.

Trujillo, Mario R. Petrology and facies analysis of the Drinkard Sandy Member (lower Leonard), in the central Drinkard Unit No. 431 well, Lea County, New Mexico. M, 1983, University of Texas at El Paso. 121 p.

Truman, W. E., III. Geology of the Blue Ridge Front near Riner, Virginia. M, 1976, Virginia Polytechnic Institute and State University.

Trumbly, Nancy Irene. Chemical equilibria between Holocene and Pleistocene intertidal/shallow subtidal carbonate rocks and associated interstitial waters, Discovery Bay, Jamaica. M, 1984, University of Oklahoma. 135 p.

Trumbly, Philip Nelson. Stratigraphy and sedimentation of the Cretaceous Brown Mountain Sandstone near Coalinga, California. M, 1983, San Diego State University. 155 p.

Trumbo, D. B. The Parsons Lineament, Tucker County, West Virginia. M, 1976, West Virginia University. 81 p.

Trumbull, Robert Bruce. Petrology of flecked gneisses in the northern Wet Mountains, Fremont County, Colorado. M, 1984, University of New Mexico. 93 p.

Trumburro, Edie. Geochemistry of the Sykesville Diamictite in central Maryland; an investigation of clast-matrix activity during metamorphism, with tectonic implications. M, 1986, Virginia Polytechnic Institute and State University.

Trummel, John E. Creep testing of samples of sylvinite, carnallite, halite, and tachyhydrite, Sergipe Basin, east-central Brazil. M, 1974, Stanford University.

Trummel, John E. Engineering geology of part of Los Altos Hills and Stanford University land, northern Santa Clara County, California. M, 1974, Stanford University.

Trump, George William. Geology of the northeastern part of the Boyertown Quadrangle, Berks County, Pennsylvania. M, 1954, University of Pittsburgh.

Truschel, Anthony D. A reservoir-routing model calibration method relating storage elements to basin geomorphology for peak runoff prediction from extreme summer storm events in ungaged arid watersheds. M, 1983, University of Nevada. 140 p.

Truscott, Frederick Wilson, Jr. Insoluble residue zones in the Bonneterre Formation. M, 1951, Washington University. 52 p.

Truscott, Marilyn Gail. Diatoms of the Sturgeon Lake marl (recent), Saskatchewan, (Canada). M, 1969, University of Saskatchewan. 153 p.

Truscott, Marilyn Gail. Petrology and geochemistry of igneous rocks of East Butte, Sweetgrass Hills, Montana. D, 1975, University of Saskatchewan. 174 p.

Trusler, James R. Analysis of folding and its influence on the formation of nickel sulfide deposits in La Motte Township, Quebec. M, 1972, Michigan Technological University. 45 p.

Trussell, Devin Brian. Measurement of V.L.F. fields. D, 1977, University of California, Berkeley. 118 p.

Truxell, Robert Eugene. Chonetid brachiopods of the Florena Shale (Permian) of Kansas. M, 1952, University of Nebraska, Lincoln.

Truxes, Lee Sayles. The geology of the Silurian Red Mountain Formation from Taylor Ridge to Horn Mountain (Georgia). M, 1956, Emory University. 50 p.

Trygstad, Joyce C. The petrology of Mesozoic dolerite dikes in southern New Hampshire and Maine. M, 1979, University of North Carolina, Chapel Hill. 132 p.

Tryon, Lansing E. Techniques of natural carbon-14 determination. M, 1952, Columbia University, Teachers College.

Trzcienski, Walter Edward Jr. Staurolite paragenesis and metamorphic reactions in pelitic rocks of the Whetstone Lake area, southeastern Ontario. D, 1971, McGill University. 150 p.

Tsai, Cheng Yun. Description of zeolites from certain localities in the Eocene basalts of western Washington. M, 1933, University of Washington. 28 p.

Tsai, Ching-Chang James. Limitation of marine seismic profiling for deep crustal reflections and reduction of water bottom multiples and scattered noise from the rough basaltic basement. D, 1981, University of Texas, Austin. 262 p.

Tsai, Chong-Shien. An improved solution procedure to the fluid-structure interaction problem as applied to the dam-reservoir system. D, 1987, SUNY at Buffalo. 383 p.

Tsai, Helen M. H. Mineralogical and geochemical investigations of mineral inclusions in diamond, kimberlite and associated rocks. D, 1978, Purdue University. 202 p.

Tsai, Jaime Irong. Three-dimensional behavior of re-molded overconsolidated clay. D, 1985, University of California, Los Angeles. 328 p.

Tsai, Kuang-Jung. A hydromechanical erosion model for surface-mined areas. D, 1983, Montana State University. 133 p.

Tsai, Louis Loung-Yie. Characterization of the Pittsburgh No. 8 Coal from three localities in southeastern Ohio. M, 1980, University of Toledo. 226 p.

Tsai, Louis Loung-Yie. Reflectance indicatrix of vitrinite; its orientation and application. D, 1985, West Virginia University. 66 p.

Tsai, Meng-Chin. Role of tin catalysts in the hydroliquefaction of coal. D, 1981, SUNY at Buffalo. 192 p.

Tsai, Nien Chien. Influence of local geology on earthquake ground motion. D, 1969, California Institute of Technology. 206 p.

Tsai, Yi-Ben. Determination of focal depths of earthquakes in the mid-oceanic ridges from amplitude spectra of surface waves. D, 1969, Massachusetts Institute of Technology. 144 p.

Tsao, A. C. ERGIS data bank for land and resources utilization. D, 1975, University of Wisconsin-Madison. 210 p.

Tsao, Tze-Tzong. Stability analysis of cohesionless dune sand slope resulting from mining. M, 1983, University of Michigan.

Tsau, Jau-Ping. Resistivity soundings tested by sample measurements employing circular soundings to search for faults on the Tiptonville Dome. M, 1979, Vanderbilt University.

Tsay, Siuh-Chun. An experimental investigation on the use of a caustic-nitrogen injection process in a waterflooded oil reservoir. D, 1982, University of Oklahoma. 130 p.

Tschirhart, Rochie. Fracture-trace analysis to increase the probability of locating groundwater for Murray County, Georgia. M, 1984, Georgia Institute of Technology. 56 p.

Tschirhart, Sid C. Velocity and logarithmic decrement in multiphase media. M, 1978, University of Texas at El Paso.

Tschopp, D. G. Distribution of heavy minerals as related to mean grain size and sorting in upper Cambrian sandstones of western Wisconsin. M, 1975, Northern Illinois University.

Tschupp, Edward Walter. Modeling the High Plains Aquifer, South Dakota; refinement of model hydrologic data and grid size. M, 1987, Colorado School of Mines. 105 p.

Tse, Eric Wai Keung. Application of critical state soil mechanics to electric K_0 stepped blade. D, 1988, Iowa State University of Science and Technology. 224 p.

Tse, Simon Tak-Chan. Mechanics of crustal strike-slip earthquakes in relation to frictional constitutive response. D, 1986, Harvard University. 180 p.

Tsegay, Tekleab. Sedimentary geology of the Reagan Formation (Upper Cambrian) of the Blue Creek Canyon, Slick Hills, Southwest Oklahoma. M, 1983, Oklahoma State University. 90 p.

Tseng, K. H. A new model for the crust in the vicinity of Vancouver Island (British Columbia). M, 1968, University of British Columbia.

Tsenn, Michael C. Computer simulation of hydrodynamic flow in Lower Cretaceous Muddy Sandstone, Bell Creek Field, Montana. M, 1983, Texas A&M University. 186 p.

Tsentas, Constantine I. Endolithic infestation within carbonate substrates "planted" on the St. Croix shelf. M, 1974, Duke University. 90 p.

Tsernoglou, Demetrius. Contribution to the study of the structure and microbiology of marine and freshwater sediments. M, 1962, Dalhousie University.

Tshudy, Dale. Macruran decapods, and their epibionts, from the Lopez de Bertodano Formation (Late Cretaceous), Seymour Island, Antarctica. M, 1987, Kent State University, Kent. 82 p.

Tsikos, George. Fracture analysis and scale modelling in a centrifugally-induced diapiric (three-dimensional) strain field. M, 1987, Queen's University. 210 p.

Tsiris, Vassilios L. Organic geochemistry and thermal history of the uppermost Morrow Shale (Lower Pennsylvanian) in the Anadarko Basin, Oklahoma. M, 1983, University of Oklahoma. 163 p.

Tsiza, Stephen Thomas. A subsurface study of the Tar Springs Formation, Union-Webster-Henderson counties, Kentucky. M, 1958, University of Illinois, Chicago.

Tso, J. L. Sulfidation of synthetic biotites. M, 1977, Virginia Polytechnic Institute and State University.

Tso, Jonathan Lee. Geology of the Ashe Formation between Fries and Galax, Virginia. D, 1987, Virginia Polytechnic Institute and State University. 393 p.

Tsou, J.-L. Study of CO_2 released from stored deep sea sediments. M, 1974, Columbia University. 27 p.

Tsou, Po. Relation of drill stem test data to the producibility of formations below 13,000 feet in Anadarko Basin, western Oklahoma. M, 1967, University of Oklahoma. 83 p.

Tsubota, K. The frequency and the time domain responses of a buried two-dimensional inhomogeneity. D, 1979, Colorado School of Mines. 166 p.

Tsuchiya, C. Liquefaction of fully and partially saturated sands. D, 1977, University of Washington. 142 p.

Tsui, Ping-Sheng. Spatial and temporal variabilities of subsurface drainage in irrigated agriculture. D, 1985, University of Nevada - Mackay School of Mines. 165 p.

Tsui, Po Chow. Deformation, ground subsidence, and slope movements along the Salt River Escarpment in Wood Buffalo National Park. M, 1982, University of Alberta. 158 p.

Tsui, Po Chow. Geotechnical investigations of glaciotectonic deformation in central and southern Alberta. D, 1987, University of Alberta. 452 p.

Tsui, Tien Fung. Laser microprobe analysis of fluid inclusions in quartz. D, 1976, Harvard University.

Tsuji, G. Y. Measurement and evaluation of soil water transmission coefficients in some Hawaiian Latosols. M, 1967, University of Hawaii. 88 p.

Tsuji, Karl Sei. Silver mineralization of the El Tigre Mine and volcanic resurgence in the Chiricahua Mountains, Cochise County, Arizona. M, 1984, University of Arizona. 140 p.

Tsutsui, Bruce Osamu. Storm generated, episodic sediment movements off Kahe Point, Oahu, Hawaii. M, 1985, University of Hawaii at Manoa. 79 p.

Tu, Christopher King-Woo. Flow in unconfined aquifers analyzed by finite element method. D, 1975, University of California, Davis. 157 p.

Tu, Hsin-yuan. The glass content of Alaskan soil. M, 1959, American University. 48 p.

Tu, Jizheng. Simulation of a haulage dispatching system in an open pit mine. D, 1984, University of Utah. 193 p.

Tu, John Siuming. Solution of general two-dimensional inverse heat conduction problems and one-dimensional inverse melting problems. D, 1988, Michigan State University. 145 p.

Tu, Kwang-Chi. The hydrothermal synthesis of Mg-mica and Mg-chlorite. D, 1949, University of Minnesota, Minneapolis. 69 p.

Tuach, John. Structural and stratigraphic setting of the Ming and other sulphide deposits in the Rambler area, Newfoundland. M, 1975, Memorial University of Newfoundland. 128 p.

Tuach, John. The Ackley high-silica magmatic/metallogenic system and associated post-tectonic granites, Southeast Newfoundland. D, 1988, Memorial University of Newfoundland. 455 p.

Tuan-Sarif, Tuan Besar Bin. The influence of organic inhibitors on the crystallization of gypsum. M, 1983, Iowa State University of Science and Technology. 107 p.

Tubb, John Beaufort, Jr. Environmental study of stages within the Brereton Cyclothem of Illinois in the eastern and part of the Western Interior coal basins. M, 1961, University of Illinois, Urbana.

Tubb, John Beaufort, Jr. Regional study of the limestones within the Brereton Cyclothem (Pennsylvanian). D, 1963, University of Illinois, Urbana. 119 p.

Tubbs, Donald W. Measurement of slow soil movements. M, 1971, University of Washington. 67 p.

Tubbs, Donald Willis. Causes, mechanisms and prediction of landsliding in Seattle. D, 1975, University of Washington. 88 p.

Tubbs, Robert E., Jr. Depositional history of the White Rim Sandstone, Wayne and Fairfield counties, Utah. M, 1984, Texas A&M University. 124 p.

Tubman, Kenneth M. Full waveform acoustic logs in radially layered boreholes. D, 1984, Massachusetts Institute of Technology. 250 p.

Tubman, Michael W. The interaction of surface waves and a linear viscoelastic sea floor. M, 1978, Louisiana State University.

Tucholke, Brian Edward. The history of sedimentation and abyssal circulation on the Greater Antilles outer ridge. D, 1973, Massachusetts Institute of Technology. 314 p.

Tuck, D. Richard, Jr. Major environmental variables affecting grain size distribution in the shoaling-wave zone under storm conditions at Virginia Beach, Virginia. M, 1969, College of William and Mary.

Tuck, Ralph. The geology and ore deposits of the Blue River mining district (Oregon). M, 1927, University of Oregon. 60 p.

Tuck, Ralph. The geology and origin of a lead-zinc deposit at Geneva Lake, Ontario. D, 1930, Cornell University.

Tucker, Annette B. Fossil decapod crustaceans from the lower Tertiary of the Prince William Sound region, Gulf of Alaska. M, 1988, Kent State University, Kent. 109 p.

Tucker, Brian E. Source mechanisms of aftershocks of the 1971 San Fernando, California, earthquake. D, 1975, University of California, San Diego. 245 p.

Tucker, Bruce C. Groundwater seepage prediction for bedrock excavations in the Kingston area, Kingston, Ontario, Canada. M, 1987, Queen's University. 118 p.

Tucker, Charles E. Present and proposed locations for underground storage of natural gas in West Virginia. M, 1949, West Virginia University.

Tucker, Charles Odell. A petrographic comparison of the (Permian) lower Spraberry and the Dean siltstones of the northern Midland Basin of West Texas. M, 1955, Texas Tech University. 65 p.

Tucker, Charlie Alexander, Jr. The geology of Miller Cove, Blount County, Tennessee. M, 1951, University of Tennessee, Knoxville. 46 p.

Tucker, Christopher M. The glacial geomorphology of west-central Newfoundland; Halls Bay to the Topsails. M, 1973, Memorial University of Newfoundland. 131 p.

Tucker, Daniel R. Abundances of Na, Mn, Cr, Sc, and Co in the Holcombe Branch, North Carolina, ultramafic intrusion. M, 1968, Miami University (Ohio). 74 p.

Tucker, Daniel R. Stratigraphy and structure of Precambrian Y (Belt?) metasedimentary and associated rocks, Goldstone Mountain Quadrangle, Lemhi County, Idaho, and Beaverhead County, Montana. D, 1975, Miami University (Ohio). 221 p.

Tucker, Delos Raymond. Subsurface Lower Cretaceous stratigraphy, central Texas. D, 1962, University of Texas, Austin.

Tucker, Elizabeth R. Geology and structure of the Brothers fault zone in the central part of the Millican SE Quadrangle, Deschutes County, Oregon. M, 1976, Oregon State University. 88 p.

Tucker, Eva. Mechanical and compositional analysis of sands in the low portion of Mill Creek valley, Cincinnati, Ohio. M, 1962, University of Cincinnati. 87 p.

Tucker, Glenn Jefferson. Land application of an industrial sludge and its effect on heavy metal relationships in a soil-plant system. D, 1980, University of Kansas. 161 p.

Tucker, Glennda B. Morphologic parameters of Mount Mazama and Glacier Peak tephras; a scanning-electron microscope study. M, 1977, University of Washington. 108 p.

Tucker, Helen Ione. The Atlantic and Gulf Coast Tertiary Pectinidae of the United States. D, 1937, Cornell University.

Tucker, Helen Ione. The Cenozoic pectens of the east coast of the United States. M, 1928, Cornell University.

Tucker, James D. Detailed geochemical survey of a Sn-Mo anomaly in the southern Wah Wah Mountains, southwestern Utah. M, 1980, Colorado School of Mines. 131 p.

Tucker, James William. Structural analysis on geologic history of the Cedar Fouche area, Lake Ouachita, Arkansas. M, 1980, Texas A&M University. 114 p.

Tucker, Leroy Maddy. Geology of the Scipio Quadrangle, Utah. D, 1954, Ohio State University.

Tucker, Mark. A geological interpretation of magnetic anomalies in the Sugarloaf Mountain area, Maryland. M, 1983, University of Pittsburgh.

Tucker, Mary L. Geochemical correlation of crude oils and source rocks in Black Mesa Basin, northeastern Arizona. M, 1983, Northern Arizona University. 80 p.

Tucker, Melinda R. Morphology and taxonomy of the class Cyclocystoidea (Echinodermata). M, 1968, Miami University (Ohio). 121 p.

Tucker, Norman Alvi. The relationship of the subsurface geology to the petroleum accumulation in Harvey County, Kansas. M, 1956, Kansas State University. 41 p.

Tucker, R. D. Bedrock geology of the Barre area, central Massachusetts. M, 1977, University of Massachusetts. 132 p.

Tucker, R. Lee. Distribution and facies of the Hardinsburg Formation (Upper Mississippian) in southern Illinois. M, 1968, Southern Illinois University, Carbondale. 60 p.

Tucker, R. W. The sedimentary environment of an Arctic lagoon. M, 1973, University of Alaska, Fairbanks. 119 p.

Tucker, Robert David. Geology of Vestranden west of Trondheim, south-central Norway. D, 1985, Yale University. 347 p.

Tucker, Robert E. A geochemical study of St. John, U. S. Virgin Islands. D, 1987, Colorado School of Mines. 405 p.

Tucker, Robert Scott. A reconnaissance study of metamorphism in the Big Maria Mountains, Riverside County, California. M, 1980, San Diego State University.

Tucker, Robert William. Differentiation of sedimentary environments by statistical methods. D, 1977, Washington State University. 243 p.

Tucker, Rodney John. The megafauna of the Decaturville chert zone of Tennessee. M, 1958, University of Missouri, Columbia.

Tucker, William Motier. Effect of the Wisconsin glacier on the Whitewater Valley of Indiana. M, 1909, Indiana University, Bloomington.

Tucker, William Motier. The hydrology of Indiana. D, 1916, Indiana University, Bloomington. 146 p.

Tuckey, Michael Edward. Global biogeography, biostratigraphy and evolutionary patterns of Ordovician and Silurian Bryozoa. D, 1988, Michigan State University. 179 p.

Tuckey, Michael Edward. Sexual dimorphism in the Devonian trilobite Phacops rana. M, 1975, Michigan State University. 90 p.

Tudor, Daniel Strain. A geophysical study of the Kentland disturbed area (northwestern Indiana). D, 1971, Indiana University, Bloomington. 116 p.

Tudor, Daniel Strain. A seismic refraction survey of Johnson County, Indiana. M, 1957, Indiana University, Bloomington. 31 p.

Tudor, Matthew Sanford. Geology of the west-central flank of the Laramie Range, Albany County, Wyoming. M, 1952, University of Wyoming. 51 p.

Tuefel, Lawrence William. The measurement of contact areas and temperature during frictional sliding of Tennessee sandstone. M, 1976, Texas A&M University.

Tuffnell, Pamela Anne. Triarthrus biostratigraphy of the Whitby Formation (Upper Ordovician) of southern Ontario. M, 1986, University of Toronto.

Tuffy, F. Chert in the Ordovician of southern Quebec. M, 1955, McGill University.

Tuftee, Kelly Krenz. Interpretation of the gravity and magnetic fields of the southwest part of the Iron Mountain Quadrangle, Wisconsin. M, 1981, Northern Illinois University. 78 p.

Tufts, Susan. An X-ray diffraction study of opaline material. M, 1971, University of Virginia. 69 p.

Tugal, Halil. Acoustic identification of marine sediments by stochastic methods. D, 1977, University of New Hampshire. 160 p.

Tuke, M. F. The lower Paleozoic rocks and klippen of the Pistolet Bay area northern Newfoundland. D, 1966, University of Ottawa. 142 p.

Tuksal, Ilker. Seismicity of the North Anatolian fault system in the domain of space, time and magnitude. M, 1976, St. Louis University.

Tula, Alex. Structure and petrology of the Pacific Ridge area, Colusa and Lake counties, California. M, 1978, Rice University. 65 p.

Tull, James Franklin. The geology and structure of Vestvaagoey in Lofoten, North Norway. D, 1973, Rice University. 191 p.

Tuller, Burl A. Motion on the S phase of earthquakes. D, 1955, University of California, Berkeley. 129 p.

Tuller, Jack N. Relationship of runoff, flow duration and recharge to basin characteristic in central Ohio. M, 1975, Ohio State University.

Tullio, Lee Dolores di *see* di Tullio, Lee Dolores

Tullis, Edward Langdon. The composition and origin of certain commercial clays of northern Idaho. M, 1932, University of Idaho. 27 p.

Tullis, Edward Langdon. The geology and petrography of Latah County, Idaho. D, 1940, University of Chicago. 218 p.

Tullis, Julia Ann. Preferred orientations in experimentally deformed quartzites. D, 1971, University of California, Los Angeles. 362 p.

Tullis, Terry Edson. Experimental development of preferred orientation of mica during recrystallization. D, 1971, University of California, Los Angeles. 278 p.

Tully, Deborah G. A statistical approach to the palynology of the Taylor-Copland Coal, Knott County, Kentucky. M, 1984, Eastern Kentucky University. 83 p.

Tully, John P. Oceanography of Alberni Inlet. D, 1948, University of Washington. 329 p.

Tully, Stephen Anthony. Lithofacies and diagenetic relationships within the San Andres Formation, Quay and De Baca counties, New Mexico. M, 1979, Texas Tech University. 68 p.

Tulucu, K. Application of the multilevel approach to the management of land and water resources in agricultural production. D, 1975, Utah State University. 214 p.

Tuman, Vladimir S. Elastic energy propagation in medium under variable stresses. D, 1964, Stanford University. 90 p.

Tumialan, Pedro H. Sulfide mineralogy and fabrics of the iron ores at Benson mines (Lewis County), New York. M, 1968, University of Missouri, Rolla.

Tuminas, Alvydas. Structural and stratigraphic relations in the Grass Valley-Colfax area of the northern Sierra Nevada foothills, California. D, 1983, University of California, Davis. 442 p.

Tuncer, E. R. Engineering behavior and classification of lateritic soils in relation to soil genesis. D, 1976, Iowa State University of Science and Technology. 137 p.

Tunell, George. The oxidation of disseminated copper ores in altered porphyry. D, 1930, Harvard University.

Tung, H.-S. Impacts of contour coal mining on streamflow; a case study of the New River watershed, Tennessee. D, 1975, University of Tennessee, Knoxville. 138 p.

Tung, J. P.-Y. The surface wave study of crustal and upper mantle structures of mainland China. D, 1975, University of Southern California.

Tung, Ping Ya James. A study of the wall rocks and their relationship to the ore body at the Calloway Mine, Ducktown, Tennessee. M, 1968, University of Tennessee, Knoxville. 67 p.

Tunnell, Felix Maxwell. Geology of Antelope area, Culberson and Reeves counties, Texas. M, 1952, University of Texas, Austin.

Tuozzolo, Peter A. Holocene foraminiferal distribution in New York and vicinity. M, 1962, New York University.

Tupas, Mateo H. The significance of mineral relationships in the Upper Mississippi Valley lead-zinc ores. D, 1950, University of Wisconsin-Madison.

Tupper, William M. Geology of Burnt Hill tungsten mine, York County, New Brunswick. M, 1955, University of New Brunswick.

Tupper, William M. Sulphur isotope ratios and the origin of the sulfide deposits in the Bathurst area of northern New Brunswick. D, 1959, Massachusetts Institute of Technology. 184 p.

Tur, Stephen Martin. Depositional environments and stratigraphic relationships of the Upper Cretaceous Pierre Shale, Trinidad Sandstone and Vermejo Formation of the Raton Basin, Trinidad, Colorado. M, 1979, Texas Tech University. 61 p.

Turbak, Abdulaziz S. Kinematic-wave volume balance models applied to surface irrigation systems. D, 1983, Colorado State University. 189 p.

Turbeville, Bruce N. Mineralogy, facies analysis, and interpretation of felsic tephra deposits of the Puye Formation, Jemez Mountains, New Mexico. M, 1986, University of Texas, Arlington. 428 p.

Turco, Caroline Ann. The conodont fauna of the Upper Ordovician Southgate Member, southwestern Ohio. M, 1957, Ohio State University.

Turco, Kevin. Calcareous nannofossils from the lower Tertiary of North Carolina. M, 1979, Ohio University, Athens.

Turcotte, Frederick Thomas. Epicenters and focal depths in the Hollister region in Central California. D, 1964, University of California, Berkeley. 124 p.

Turek, Andrew. Geology of Lake Cinch Mines Limited, Uranium City, Saskatchewan. M, 1962, University of Alberta. 85 p.

Turek, Frank. Environmental geology of the Apache Junction Quadrangle, Arizona. M, 1975, Arizona State University. 58 p.

Turek, Jeffery Lee. Analysis by simulation of the disposition of nuclear fuel waste. D, 1980, Virginia Polytechnic Institute and State University. 251 p.

Turekian, Karl. Strontium content of limestones and fossils. M, 1951, Columbia University, Teachers College.

Turekian, Karl K. The geochemistry of strontium. D, 1958, Columbia University, Teachers College.

Turgut, Suleyman. The kinetics of the silica transformations and their applications to natural processes. M, 1974, Northwestern University.

Turinetti, James D. Computer analysis of multispectral image densities to classify land-uses. M, 1972, Ohio State University.

Turjoman, A. M. The behavior of lead as a migrating pollutant in six Saudi Arabian soils. D, 1978, University of Arizona. 162 p.

Turk, John T. A study of diffusion in clay-water systems by chemical and electrical methods. D, 1976, University of California, San Diego. 127 p.

Turk, Leland Jan. Hydrogeology of the Bonneville Salt flats, northwestern Utah. D, 1969, Stanford University. 328 p.

Turk, Lon B. Geology of the Carr City and Maud pools of Seminole and Pottawatomie counties, Oklahoma. M, 1933, University of Wisconsin-Madison.

Türkarslan, Muharrem. Near-surface seismic properties for Permian rock formations at selected sites in Oklahoma. M, 1979, University of Oklahoma. 78 p.

Turkeli, Arif. Analysis of a continuous gravity profile from the Madeira abyssal plain through the Strait of Gibraltar to the Alboran Basin. M, 1963, Columbia University, Teachers College.

Turker, Yucel. Short-term variation of runoff-rainfall ratios in Nova Scotia international hydrologic decade watersheds. M, 1969, Dalhousie University.

Turkopp, John. The stratigraphy and areal geology of Flint Ridge. M, 1915, Ohio State University.

Turley, Charles H. Petrology of several rhyolites from western Juab and Millard counties, Utah. M, 1979, University of Utah. 72 p.

Turley, Mitchell Reed. Upper Devonian sediments of the Bedford Quadrangle. M, 1952, Pennsylvania State University, University Park. 99 p.

Turmelle, John M. Stratigraphic and paleocurrent analysis of the Burro Canyon Formation and the Dakota Sandstone, northeastern San Juan Basin area, Colorado and New Mexico. M, 1979, Bowling Green State University. 75 p.

Turmelle, Thomas Jeffrey. Lithostratigraphy and depositional environments of the Mayes Formation (Mississippian) in Adair County, Oklahoma. M, 1982, University of Oklahoma. 144 p.

Turnbull, Lawrence Stur Levant, Jr. Determination of seismic source parameters using far-field surface wave spectra. D, 1976, Pennsylvania State University, University Park. 453 p.

Turnbull, Lionel Graham. Pleochroic haloes of the uranium series in biotite. M, 1933, Dalhousie University.

Turnbull, R. W. Engineering geology of the Eden Canyon area, near Castro Valley, Alameda County, California. M, 1976, Stanford University. 107 p.

Turneaure, Frederick S. The tin deposits of Llallagua, Bolivia, and their genetic significance. D, 1933, Harvard University.

Turner, Alan Keith Finlay. The formation, distribution, and engineering uses of sand and gravel in parts of York and Ontario counties, Province of Ontario, Canada. M, 1964, Columbia University, Teachers College.

Turner, B. W. Mineral distribution within the sediments of Pearl Harbor. M, 1975, University of Hawaii. 93 p.

Turner, Barbara Lee. Study of activity on a fault in Los Altos Hills, Santa Clara County, California. M, 1971, Stanford University.

Turner, Brian Buddington. Configuration and petrogenesis of a geologically critical area in the (Precambrian) of southeastern Adirondack Mountains, New York. D, 1967, University of Kansas. 156 p.

Turner, Christine E. Facies of the Toroweap Formation, Marble Canyon, Arizona. M, 1974, Northern Arizona University. 120 p.

Turner, Colin C. Archaean sedimentation; alluvial fan and turbidite deposits, Little Vermilion Lake, northwestern Ontario. M, 1972, McMaster University. 211 p.

Turner, D. J. An introduction to minor element studies in the Noranda District (Quebec). M, 1961, University of Toronto.

Turner, Daniel Stoughton. Heavy accessory mineral and radioactivity studies of the igneous rocks in the

Wausau area (Wisconsin). D, 1948, University of Wisconsin-Madison.

Turner, David Raiford. Petrologic and fluid inclusion study of zinc skarns of the Buckhorn area, Central Mining District, Grand County, New Mexico. M, 1985, University of Utah. 110 p.

Turner, Dennis L. Geology and economic geology of the Juniper Hill-Sycamore Creek area, Reading Mountain Quadrangle, Grant County, New Mexico. M, 1978, Colorado School of Mines. 152 p.

Turner, Donald A. Diagenetic patterns of surface and subsurface samples from the bioherm facies of the Edgecliff Member of the Onondaga Formation (Middle Devonian) of New York State. M, 1977, Rensselaer Polytechnic Institute. 118 p.

Turner, Donald H. Inner-nearshore topography and sediment distribution in western Lake Michigan at Terry Andrae-Kohler State Park, Wisconsin. M, 1972, University of Wisconsin-Milwaukee.

Turner, Donald Lloyd. Potassium argon dates concerning the Tertiary foraminiferal time scale and San Andreas Fault displacement. D, 1968, University of California, Berkeley. 99 p.

Turner, E. J. The use of shape as a nominal variable on multipattern dot maps. D, 1977, University of Washington. 264 p.

Turner, Francis Earl. Geology of the Quail Lake region (Mojave Desert, California). M, 1928, California Institute of Technology. 23 p.

Turner, Francis Earl. The stratigraphy and molluscan faunas of the Eocene of western Oregon. D, 1934, University of California, Berkeley. 273 p.

Turner, Frank Paul. Stratigraphy and structure of Provo Canyon between Bridal Veil Falls and South Fork. M, 1952, Brigham Young University. 46 p.

Turner, Gerry H. Energy partitioning in high-velocity-impact cratering. D, 1969, University of Utah. 133 p.

Turner, Gerry H. Projectile effects and subsurface disturbances in high-velocity-impact cratering in lead. M, 1960, University of Utah. 81 p.

Turner, Gregory Larkin. Textures of the iron-titanium of the minerals of the Laramie Range, Wyoming. M, 1947, University of Chicago. 31 p.

Turner, Homer G. Terranes of Greensboro, Vermont. M, 1914, Syracuse University.

Turner, Irving L. Geology along the Saltville Fault, Philadelphia Quadrangle, Loudon and Monroe counties, Tennessee. M, 1960, University of Tennessee, Knoxville. 57 p.

Turner, J. D. Heavy accessory minerals. M, 1933, Queen's University. 28 p.

Turner, J. M. Response of illite and chlorite in the Mid-Ordovician Utica Shale of northern New York State to the maximum temperatures generated during burial diagenesis. M, 1985, SUNY at Binghamton. 33 p.

Turner, James. Yalobusha County geology (Mississippi). M, 1952, University of Mississippi.

Turner, James R. Depositional environment and reservoir characteristics of Frio Sandstone, McCampbell deep field, Aransas County, Texas. M, 1977, Texas A&M University. 141 p.

Turner, Jeffrey S. Geochemical investigations into the formation of siliceous coatings of the St. Peter Sandstone. M, 1985, University of Missouri, Columbia. 76 p.

Turner, John Charles. The structural geology of the southeast Rainbow mountain area, central Alaska range, Alaska. M, 1969, University of Wisconsin-Madison.

Turner, John Patrick. Experimental analysis of drilled shaft foundations subjected to repeated axial loads under drained conditions; (Volumes I and II). D, 1986, Cornell University. 427 p.

Turner, Joseph Brian. Petrography and geochemistry of basalts from the Bullpen Lake sequence, northern Sierra Nevada, California; a test of a working hypothesis. M, 1986, San Diego State University. 105 p.

Turner, Joseph Gresham. Comparison of mechanical properties of four types of sandstone described by Ferm's classification. M, 1982, University of Kentucky. 68 p.

Turner, Kenneth Harold. Self-potential surveys of a reclaimed coal strip mine, Clearfield County, Pennsylvania. M, 1982, Pennsylvania State University, University Park. 89 p.

Turner, Kent, Jr. A determination of the sulfur isotopic signature of an ore-forming fluid from the Sierrita porphyry copper deposit, Pima County, Arizona. M, 1983, University of Arizona. 93 p.

Turner, L. A. A comparison of metal zoning with age determination and gravity data in Canada. M, 1961, University of Toronto.

Turner, Laurie Jeanne. Form of aluminum (III) in dilute aqueous solution. M, 1984, McMaster University. 108 p.

Turner, Laurie Jeanne. Sorption of sulfate on iron oxide minerals; hematite and goethite. D, 1988, McMaster University. 158 p.

Turner, Lawrence D. A study of the relationship between dissolved organic matter and metals in streams of the Republic area, northeastern Washington. M, 1981, Eastern Washington University. 67 p.

Turner, Mary Ann. A faunal assemblage from the lower Ash Hollow Formation (Neogene) of southern Nebraska. M, 1972, University of Nebraska, Lincoln.

Turner, Mortimer Darling. Geology of the San Sebastian area of northwestern Puerto Rico and the tectonic development of Puerto Rico and surrounding areas. D, 1972, University of Kansas. 109 p.

Turner, Mortimer Darling. The Ione Formation of the Buena Vista area, Amador County, California. M, 1954, University of California, Berkeley. 180 p.

Turner, Neil L. Geology of the Ruth Quadrangle, White Pine County, Nevada. M, 1963, University of Southern California.

Turner, Neil Lee. Carboniferous stratigraphy of western San Saba County, Texas. D, 1970, University of Texas, Austin. 436 p.

Turner, P. J. Seismic reflection survey in Lake Champlain. M, 1977, University of Vermont.

Turner, Patricia Ann. The sulfide mineralogy and the behavior of S and certain transition elements in the Skaergaard Intrusion, East Greenland. M, 1986, Dartmouth College. 81 p.

Turner, Philip Ambrose. Sedimentation in the Upper Cretaceous of east-central Georgia. M, 1959, Cornell University. 30 p.

Turner, R. R. A mass balance approach to understanding the effect of an impoundment on the export of selected solutes and particulate matter from a watershed. D, 1976, Florida State University. 295 p.

Turner, Ralph R. The significance of color banding in the upper layer of Kara sea (Arctic Ocean) sediments. M, 1970, Florida State University.

Turner, Robert D. Cenozoic pisolitic limestone and caliche near San Angelo (Tom Green and Concho counties), West Texas. M, 1966, University of Houston.

Turner, Robert J. The effects of a mid-foreshore groundwater effluent zone on tidal-cycle sediment distribution. M, 1987, Western Washington University. 29 p.

Turner, Robert John Whitlock. The genesis of stratiform lead-zinc deposits, Jason Property, Macmillan Pass, Yukon. D, 1987, Stanford University. 219 p.

Turner, Robert John Whitlock. The geology of the east-central Tobin Range, Nevada. M, 1982, Stanford University. 113 p.

Turner, Robert Spilman. Biogeochemistry of trace elements in the McDonalds Branch watershed, New Jersey Pine Barrens. D, 1983, University of Pennsylvania. 333 p.

Turner, Ronald Fredric. The paleoecologic and paleobiogeographic implications of the Maestrichtian Cheilostomata (Bryozoa) of the Navesink Formation.

D, 1973, Rutgers, The State University, New Brunswick. 371 p.

Turner, Roosevelt. Kinetic studies of acid dissolution of montmorillonite and kaolinite. D, 1964, University of California, Davis. 95 p.

Turner, Ryan David. Miocene folding and faulting of an evolving volcanic center in the Castle Mountains, southeastern California and southern Nevada. M, 1985, University of North Carolina, Chapel Hill. 56 p.

Turner, Shirley. A structural study of tunnel manganese oxides by high-resolution transmission electron microscopy. D, 1982, Arizona State University. 239 p.

Turner, Shirley. High resolution transmission electron microscopy of some manganese oxide and silicate minerals. M, 1978, Arizona State University. 163 p.

Turner, Thomas Edward. Geology of Great Sitkin Island Volcano, Aleutian Islands. M, 1947, University of Washington. 36 p.

Turner, Thomas R. The geology of the Northern Complex near Herman, Michigan. M, 1973, Michigan Technological University. 72 p.

Turner, W. N. The North Carolina Sandhills. D, 1949, University of North Carolina, Chapel Hill.

Turner, Warren H. The luminescence of some framework silicates. M, 1967, Washington University. 126 p.

Turner, William. Effective confining pressure and fluid discharge along fractures. M, 1973, University of South Carolina.

Turner, William B. A lithologic and faunal study of the Arnheim Formation in Casey, Washington, Nelson, and Trimble counties, Kentucky. M, 1964, University of Kentucky. 59 p.

Turner, William L. The geology of the Vesta 7 1/2′ Quadrangle, Georgia. M, 1987, University of Georgia. 204 p.

Turner, William Louis. Geology of the Eagle Ford Quadrangle, Dallas County, Texas. M, 1950, Southern Methodist University. 29 p.

Turner, William Morrow. Geology of the Polis–Kathikas area, Cyprus. D, 1971, University of New Mexico. 430 p.

Turner, William Morrow. Heavy mineral distribution in stream gravels in the Sandy Springs, Maryland and Kensington, Maryland 7 1/2′ Quadrangle. M, 1965, Pennsylvania State University, University Park. 111 p.

Turner, William Newton. Quarries of Knox County, Tennessee. M, 1931, University of Tennessee, Knoxville. 159 p.

Turner, William S. A geochemical analysis of Upper Silurian dolomites of northwest Ohio. M, 1977, University of Toledo. 82 p.

Turner-Peterson, Christine Elizabeth. Sedimentology of the Westwater Canyon and Brushy Basin members, Upper Jurassic Morrison Formation, Colorado Plateau, and relationship to uranium mineralization. D, 1987, University of Colorado. 184 p.

Turnmire, J. B. Simulated discharge from small watersheds before, during, and after disturbance. D, 1987, University of Tennessee, Knoxville. 173 p.

Turnock, Allan Charles. The analysis of aluminum and sodium in igneous rocks by induced radioactivity. M, 1956, University of Manitoba.

Turnock, Allan Charles. The stability and phase relationships of FeAl spinels and iron chlorites. D, 1960, The Johns Hopkins University.

Turpening, Roger Munson. A linear mode filter for seismic waves. D, 1966, University of Michigan. 31 p.

Turpening, Roger Munson. Some observations of P_n spectra from the New Madrid earthquake of February 2, 1962. M, 1963, University of Michigan.

Turpening, Walter Ray. An approximate solution for air-coupled Rayleigh waves propagating across a vertical boundary. M, 1972, Michigan State University. 50 p.

Turpin, Bernard. An experimental study of the self-potential associated with graphitic bodies. M, 1960, Colorado School of Mines. 35 p.

Turpin, Evelyn M. An analysis of some physical characteristics of the glacial tills from northeastern Illinois. M, 1945, Northwestern University.

Turpin, Robert David. Evaluation of photogrammetric techniques for engineering measurements. D, 1957, Ohio State University.

Turrill, Sheldon Lee. The petrology of the Gilboy and Uniontown sandstones in Washington and Athens counties, Ohio. M, 1960, Ohio University, Athens. 116 p.

Turrin, Brent David. Thermal history and estimated emplacement depths of plutons within the Tower Peak Quadrangle, central Sierra Nevada, California, from disparities in K-Ar mineral ages. M, 1984, Stanford University. 42 p.

Turyn, May E. Crustal structure of New England using short period Rayleigh wave dispersion; a theoretical and empirical study. M, 1967, Boston College.

Tuschall, John Richard, Jr. Heavy metal complexation with naturally occurring organic ligands in wetland ecosystems. D, 1981, University of Florida. 212 p.

Tuthill, Jonathan Dale. The propagation of Stoneley waves. M, 1980, University of Washington. 103 p.

Tuthill, Rosalind L. The hydrothermal behavior of basalts in their melting range at 5 kilobars. M, 1968, Pennsylvania State University, University Park. 170 p.

Tuthill, Samuel James. A comparison of the late Wisconsinan molluscan fauna of the Missouri Coteau District (North Dakota) with a modern Alaskan analogue. D, 1969, University of North Dakota. 234 p.

Tuthill, Samuel James. Mollusks from Wisconsinan (Pleistocene) ice-contact sediments of the Missouri Coteau in central North Dakota. M, 1963, University of North Dakota. 102 p.

Tuthill, Schuyler K. The geology of the Nebo Quadrangle in Missouri. M, 1953, University of Missouri, Columbia.

Tuttle, Arthur L. Leaching of a silver ore from the Gipsey Mine, New Mexico. M, 1893, Washington University.

Tuttle, Frances. Textural and mineralogical analysis of late Wisconsin glacial and post glacial deposits from Suffolk County, New York. M, 1943, Smith College. 56 p.

Tuttle, Jesse L., Jr. Foraminifera of the Brownstone Formation (Cretaceous) of southwestern Arkansas. M, 1964, University of Oklahoma. 138 p.

Tuttle, Lawrence Elliott. Geography of the Sanpete County oasis, Utah. M, 1948, Brigham Young University. 95 p.

Tuttle, Martitia Powell. Earthquake potential of the San Gregorio-Hosgri fault zone, California. M, 1985, University of California, Santa Cruz.

Tuttle, Orville Frank. Petrographic interpretation of the Ordovician-Silurian boundary problem in central Pennsylvania. M, 1940, Pennsylvania State University, University Park. 71 p.

Tuttle, Orville Frank. Structural petrology of planes of liquid inclusions. D, 1949, Massachusetts Institute of Technology. 93 p.

Tuttle, Russell Howard. A study of the chimpanzee hand with comments on hominoid evolution. D, 1965, University of California, Berkeley. 248 p.

Tuttle, Sherwood D. The Quaternary geology of the coastal region of New Hampshire. D, 1953, Harvard University.

Tuttle, Sherwood Dodge. The geomorphology of Pelton basin, Glacier Peak quadrangle, Washington. M, 1941, Washington State University. 17 p.

Tütüncü, Azra Nur. The effects of saturation salinity, temperature and viscosity on wave attenuation and velocity. M, 1983, Stanford University. 117 p.

Tuyl, Francis M. Van *see* Van Tuyl, Francis M.

Tuyl, Francis Maurice Van *see* Van Tuyl, Francis Maurice

Tuysuz, Necati. Heavy mineral distribution in the basal Deadwood Formation; a possible geochemical tool for Precambrian mineralization in the northern Black Hills, South Dakota. M, 1987, South Dakota School of Mines & Technology.

Tuzinski, Patrick A. Rare-alkali ion halos surrounding the Bob Ingersoll lithium bearing pegmatite mine, Keystone, Black Hills, South Dakota. M, 1983, Kent State University, Kent. 121 p.

Tvrdik, Timothy N. Petrology of the Precambrian basement rocks of the Matlock Project cores; northwestern Iowa. M, 1983, University of Iowa. 113 p.

Twardy, Stanley A. Systematic descriptions of the Yorktown Gastropoda fauna (Virginia). M, 1936, University of Virginia. 138 p.

Tway, Linda Elaine. Geologic applications of Late Pennsylvanian ichthyoliths from the Midcontinent region. D, 1982, University of Oklahoma. 333 p.

Tway, Linda Elaine. Pennsylvanian ichthyoliths from the Shawnee Group of eastern Kansas. M, 1977, University of Oklahoma. 136 p.

Tweddale, John B. The relationship of discharge to hydrochemical and sulfur isotope variations in spring waters of south-central Indiana. M, 1987, Indiana University, Bloomington. 165 p.

Twedt, Thomas J. Simulation of the physical subsystem of a mountain stream ecosystem by digital computer. D, 1975, Utah State University. 130 p.

Tweedy, Norma A. An x-ray powder examination of some pegmatite minerals from southwestern Manitoba. M, 1962, University of Manitoba.

Twell, Buddy H. Elastic wave propagation in highway pavements. D, 1967, Texas A&M University.

Twell, Charles F. Relation of petroleum accumulation to clay minerals in some Lansing and Kansas City Group (Pennsylvanian) cores of western Kansas. M, 1964, University of Kansas. 55 p.

Twenhofel, William H. Geology, stratigraphy, and physiography of Anticosti Island. D, 1912, Yale University.

Twenhofel, William S. Geology of the Alaska-Juneau lode system, Alaska. D, 1952, University of Wisconsin-Madison.

Twenter, Floyd Robert. Relation of texture and mineral composition to topographic expression of some Colorado Plateau sandstones. M, 1956, University of Missouri, Columbia.

Tweto, Ogden Linne. Petrography of the Precambrian and Paleozoic rocks of Montana and Yellowstone National Park. M, 1937, University of Montana. 245 p.

Tweto, Ogden Linne. Pre-cambrian and Laramide geology of the Vasquez Mountains, Colorado. D, 1947, University of Michigan.

Twichell, David Cushman, Jr. Erosion of the Florida Escarpment; eastern Gulf of Mexico. D, 1988, University of Rhode Island. 186 p.

Twigg, Daniel Bruce. The stratigraphy and structural geology of the Corning Quadrangle, New York. M, 1961, University of Rochester. 44 p.

Twigg, Robert W. Geology of the Cumberland Quadrangle (West Virginia). M, 1957, West Virginia University.

Twinem, J. C. Geologic and flotation study of an oxidized Cripple Creek gold ore (Colorado). M, 1934, Massachusetts Institute of Technology. 70 p.

Twining, John Theodore. Pelecypods of the middle Eocene Stone City Beds of Texas; Part II. M, 1954, University of Texas, Austin.

Twiss, Elizabeth S. Po 210 in sediments off the Washington coast. M, 1974, University of Washington.

Twiss, Page Charles. Geology of Van Horn Mountains, Trans-Pecos, Texas. D, 1959, University of Texas, Austin. 274 p.

Twiss, Page Charles. The non-carbonate mineralogy of some Permian and Pennsylvanian limestone. M, 1955, Kansas State University. 83 p.

Twiss, Robert J. A micromorphic theory of mixtures with application to elastic polycrystalline solids. D, 1970, Princeton University. 104 p.

Twiss, Stuart Nelson. Stratigraphy of Saddle Mountains. M, 1933, Washington State University. 35 p.

Twitchell, M. W. The Cenozoic Cassiduloidea of the U.S. D, 1905, The Johns Hopkins University.

Twitchell, Paul F. Solar-terrestrial relationships. M, 1962, Boston College.

Twombly, Gregory L. The nature of non-recoverable strain in experimentally deformed Indiana Limestone and Yule Marble. M, 1980, University of Calgary. 158 p.

Twomey, Arthur Allen. Seismic refraction studies in surficial materials of Victoria Land, Antarctica. M, 1968, University of Wisconsin-Madison.

Tworo, A. The nature and origin of lead-zinc mineralization, Middle Silurian dolomites, southern Ontario. M, 1985, University of Waterloo. 275 p.

Twyman, James DeWitt. Igneous geology of the diorite-quartz monzonite, Loco Mountain Stock (western half), Crazy Mountains, Montana. M, 1979, University of Cincinnati. 135 p.

Twyman, James DeWitt. The generation, crystallization, and differentiation of carbonatite magmas; evidence from the Argor and Cargill complexes, Ontario. D, 1983, University of Toronto.

Twyman, Terry Rayno. Major and trace element redistribution during metamorphism; geochemical characterization of absorbtion and mass transfer processes. D, 1983, University of Toronto.

Twyman, Terry Rayno. Mineralogy, texture and chemistry of the serpentinites and serpentinized ultramafic rocks of the Canyon Mountain Complex, John Day, Oregon. M, 1979, University of Cincinnati. 128 p.

Tyburczy, James Albert. High pressure electrical conductivity of naturally-occurring silicate melts. D, 1983, University of Oregon. 214 p.

Tychsen, Paul Charles. A sedimentation study of the Brule Formation in Northwest Nebraska. D, 1954, University of Nebraska, Lincoln. 241 p.

Tychsen, Paul Charles. Geology and ground-water hydrology of the Heart River Irrigation Project and the Dickinson area, North Dakota. M, 1949, University of Nebraska, Lincoln.

Tydlaska, LeRoy Jerome. Geology of Palm Valley Quadrangle, Williamson County, Texas. M, 1951, University of Texas, Austin.

Tye, Rennie V. Sedimentary petrology of some Kansas areas. M, 1946, Kansas State University. 43 p.

Tye, Robert S. Non-marine Atchafalaya deltas; processes and products of interdistributary basin alluviation, south-central Louisiana. D, 1986, Louisiana State University. 437 p.

Tygesen, Jeffery Dean. Settlement monitor for mine waste embankments. M, 1984, University of Utah. 106 p.

Tyler, D. A. Horizontal crustal movement in the San Francisco Bay area. D, 1976, University of Wisconsin-Madison. 200 p.

Tyler, D. B. Macrosystems ecology; a metastructure for environmental studies. D, 1976, University of California, Berkeley. 366 p.

Tyler, David Lynn. Stratigraphy and structure of the late Precambrian-Early Cambrian clastic metasedimentary rocks of the Baldwin Lake area, San Bernardino Mountains, California. M, 1976, Rice University. 37 p.

Tyler, Henry Johnson. Contribution to the petrology of the (Silurian) Thorold Sandstone (New York). M, 1940, Cornell University.

Tyler, J. Gary. Subsurface geology and depositional systems of the Upper Mississippian–Lower Pennsylvanian Bayport and Saginaw formations, central Michigan Basin. M, 1980, Wayne State University.

Tyler, Jeremy Guy Anthony. Daily effect of waves on an exposed, natural beach in Brevard County, Florida. M, 1972, Florida Institute of Technology.

Tyler, John H. Geology and mineral resources of the Abingdon area, Washington County, Virginia. M, 1960, Virginia Polytechnic Institute and State University.

Tyler, John Howard. Petrology, fauna, and paleoecology of the type Four Mile Dam Limestone, Alpena County, Michigan. D, 1963, University of Michigan. 162 p.

Tyler, Noel. Jurassic depositional history and vanadium-uranium deposits, Slick Rock District, Colorado Plateau. D, 1981, Colorado State University. 285 p.

Tyler, Ronald D. A morphological comparison of lunar and terrestrial lava flows. M, 1972, South Dakota School of Mines & Technology.

Tyler, Ronald Douglas. Chloride metasomatism in the southern part of the Pierrepont Quadrangle, Adirondack Mountains, New York. D, 1978, SUNY at Binghamton. 518 p.

Tyler, Stanley A. A study of sediments from the coasts of North Carolina and Florida. M, 1929, University of Wisconsin-Madison.

Tyler, Stanley A. The heavy minerals of the (Ordovician) St. Peter Sandstone in Wisconsin. D, 1935, University of Wisconsin-Madison.

Tyler, Stanley Roy. Geology of the Hudson Quadrangle. M, 1956, University of Minnesota, Minneapolis. 195 p.

Tyler, Theodore F. The Petrology of the Dotsero Formation (central northwestern Colorado). M, 1972, Colorado State University. 137 p.

Tynan, Eugene Joseph. Silicoflagellates of the Calvert Formation (Miocene) of Maryland. M, 1956, University of Massachusetts. 42 p.

Tynan, Eugene Joseph. The order Hystrichosphaerida. D, 1962, University of Oklahoma. 586 p.

Tynan, Mark Christian. A new group of corals and other microfossils (echinoderms, sponges, crustaceans, foraminifers, molluscs, brachiopods, problematica) from the Early Cambrian Deep Spring, Campito, and Poleta formations, White-Inyo Mountains, California. D, 1981, University of Iowa. 170 p.

Tynan, Mark Christian. Conodont biostratigraphy of the Mississippian Chainman Formation, western Millard County, Utah. M, 1977, University of Iowa. 87 p.

Tyne, Arthur M. Van *see* Van Tyne, Arthur M.

Tyner, Clarence Graham. Geology of the Abbott and Peoria quadrangles, Hill County, Texas. M, 1964, Baylor University. 111 p.

Tyner, Grace Nell. Field geology, petrology, and trace element geochemistry of the Sullivan Buttes Latite, Yavapai County, Arizona. M, 1979, University of Texas, Austin.

Tyner, Grace Nell. Geology and petrogenesis of the Sullivan Buttes Latite, Yavapai County, Arizona; field and geochemical evidence. D, 1984, University of Texas, Austin. 286 p.

Tyraskis, Panagiotis A. New techniques in the analysis of geophysical data modelled as a multichannel autoregressive random process. D, 1983, McGill University.

Tyrrell, Miles Edward. Refractory uses of Pacific Northwest chromites. M, 1946, University of Washington. 133 p.

Tyrrell, Willis Woodbury, Jr. Geology of the Whetstone Mountain area, Cochise and Pima counties, Arizona. D, 1957, Yale University. 330 p.

Tysdal, Russell Gene. Geology of a part of the north end of the Gallatin Range, Gallatin County, Montana. M, 1966, Montana State University. 95 p.

Tysdal, Russell Gene. Geology of the north end of the Ruby range, southwestern Montana. D, 1970, University of Montana. 187 p.

Tyson, Alfred Knox. The source of the water along the Balcones fault escarpment. M, 1924, University of Texas, Austin.

Tyson, R. Michael. The mineralogy and petrology of the Partridge River Troctolite in the Babbitt-Hoyt Lakes region of the Duluth Complex, northeastern Minnesota. D, 1979, Miami University (Ohio). 179 p.

Tyson, Robert Michael. Hornfelsed basalts in the Duluth Complex. M, 1976, Cornell University.

Tyson, Rogert G., Jr. Sediments of a FLorida Bay basin. M, 1981, University of South Florida, Tampa. 89 p.

Tzeng, Rong-Fung. Migration of vertical seismic profiles by time reversal extrapolation and ray tracing. M, 1984, University of Houston.

Tzeng, Shih-Ying. Low-grade metamorphism in East-central Puerto Rico. D, 1976, University of Pittsburgh. 172 p.

Tzeng, Wen Shyr. Investigation of SV to P wave amplitude ratio for determining focal mechanism. M, 1982, Georgia Institute of Technology. 119 p.

Tziavos, Christos C. Sedimentology, ecology, and paleogeography of the Sperchios Valley and Maliakos Gulf, Greece. M, 1977, University of Delaware.

Tzong, Tsair-Jyh. Hybrid modelling of soil-structure interaction in layered media. D, 1983, University of California, Berkeley. 142 p.

U., Alberto Lobo Guerrero *see* Lobo Guerrero U., Alberto

Uan-On, Thanakorn. Nonlinear reservoir analysis with risk-benefit objectives. D, 1983, University of California, Davis. 145 p.

Uba, Humphreys Douglas. Hydrothermal reactions of Wyoming bentonite in the presence of mixed salts and their effects on the rheology of bentonite fluids. M, 1988, Texas Tech University. 78 p.

Ubani, Ephraim Agbawo. The well performance of naturally fractured lenticular sand reservoirs. D, 1985, University of Oklahoma. 154 p.

Uber, Harvey A. The characteristics of zircon as a means of identifying the origin of gneisses and schists. M, 1917, University of Wisconsin-Madison.

Ubiera Castro, Antonio Amilcar. The occurrence and properties of hydroxy-interlayered silicate clays in some soils of the Dominican Republic. D, 1987, North Carolina State University. 396 p.

Ucakuwun, Elias Kerukaba. The pegmatites and granitoid rocks of the Dryden area, northwestern Ontario. M, 1981, University of Manitoba. 149 p.

Ucar, R. Decoupled explosive charge effects on blasting performance. M, 1975, University of Missouri, Rolla.

Uchida, A. Water quality modelling of mine acid drainage. D, 1979, SUNY at Buffalo. 228 p.

Uchio, Takayasu. Ecology of living benthonic foraminifera from the San Diego, California area. M, 1957, University of California, Los Angeles. 248 p.

Uchupi, Elazar. Continental margin from San Francisco, California, to Cedros Island, Baja California. D, 1962, University of Southern California.

Uchupi, Elazar. Submarine geology of the Santa Rosa-Cortes Ridge. M, 1954, University of Southern California.

Uchytil, Steven James. Functional morphologic studies of certain brachiopods from the Upper Devonian (Frasnian) of Iowa and a reevaluation of the feeding mechanism in richthofeniacean brachiopods. M, 1979, University of California, Davis. 80 p.

Udaka, T. Analysis of response of large embankments to traveling base motions. D, 1975, University of California, Berkeley. 351 p.

Udaloy, Anne Greenough. Arsenic mobilization in response to the draining and filling of the reservoir at Milltown, Montana. M, 1988, University of Montana. 140 p.

Udayashankar, K. V. Depositional environment, petrology, and diagenesis of Red Fork Sandstone in central Dewey County, Oklahoma. M, 1985, Oklahoma State University. 188 p.

Udegbunam, Emmanuel Onyekwelu. Migration of natural gas from storage reservoirs. D, 1983, University of Michigan. 195 p.

Udell, Stewart. Contact metamorphism in certain ore deposits of the Seven Devils District, Idaho. M, 1928, University of Idaho. 47 p.

Udias, Agustin Vallina. A numerical method for focal mechanism determination using S wave data. M, 1964, St. Louis University.

Udo, Eno J. Zinc adsorption and desorption in calcareous soils. D, 1968, University of Arizona. 105 p.

Udwadia, Firdaus Erach. Investigations of earthquake and microtremor ground motions. D, 1972, California Institute of Technology. 148 p.

Ueckert, John Fant. Interpretation of the Lewisville Sequence, Van oil field, Van Zandt County, Texas. M, 1981, Texas A&M University. 93 p.

Ueda, H. A comparison of a ferromagnetic core coil and an air core coil as the sensor head of the induction magnetometer. M, 1975, University of British Columbia.

Uehara, G. The nature and properties of the soils of the red and black complex of the Hawaiian Islands. M, 1956, University of Hawaii. 44 p.

Ueng, Charles Wen-Long. The fractionation of carbon and oxygen isotopes in a banded iron formation, Marquette District, northern Michigan. M, 1980, Northern Illinois University.

Ueng, Wen-Long Charles. The early Proterozoic tectonic history of south-central Lake Superior region. D, 1986, Stanford University. 128 p.

Ugaz, Oscar Guillermo. An experimental and numerical study of impact driving of open-ended pipe piles in dense saturated sand. D, 1988, [University of Houston]. 159 p.

Ugland, Richard Olav. An investigation of paleomagnetic secular variation in the sediments of Yellowstone Lake and Jackson Lake, Wyoming. M, 1976, University of Utah. 67 p.

Uglow, William L. A study of methods of mine valuation and assessments, with special reference to the zinc mines of Southwest Wisconsin. D, 1914, University of Wisconsin-Madison.

Uglow, William L. On the origin of the secondary silicate zones at the contacts of intrusives with limestones. M, 1912, University of Wisconsin-Madison.

Ugolini, Fiorenzo C. Soil development on the red beds of New Jersey. D, 1960, Rutgers, The State University, New Brunswick. 117 p.

Ugrinic, George M. The geology and ground-water resources of Bernalillo County, New Mexico. M, 1950, University of New Mexico. 80 p.

Uhl, Ben Forrest. Igneous rocks of the Arbuckle Mountains. M, 1932, University of Oklahoma. 54 p.

Uhl, David A. Conodont distribution and paleoenvironmental significance in the Deer Creek Cyclothem (Upper Pennsylvanian) in southeastern Nebraska. M, 1981, University of Nebraska, Lincoln.

Uhl, V. W., Jr. The occurrence of ground water in the Satpura region of central India. M, 1976, University of Arizona.

Uhlir, David Mason. Sedimentology of the Sundance Formation, northern Wyoming. D, 1987, Iowa State University of Science and Technology. 119 p.

Uhrhammer, R. Shear wave velocity structure in the Earth from differential shear wave measurements. D, 1977, University of California, Berkeley. 180 p.

Uhri, Duane C. The correlation of Cincinnatian environments by spectrographic analyses of associated fossils. M, 1954, Michigan State University. 54 p.

Uhrich, William G. Seismic hazards in Idaho. M, 1986, University of Idaho. 95 p.

Uhrig, Leonard F. Structural study of a portion of the Lang and Humphreys quadrangles, Los Angeles County, California. M, 1936, California Institute of Technology. 42 p.

Uhrin, David C. The feasibility of producing a photo-interpretation map of landslide prone areas in eastern Washington suitable for land-use planning. M, 1973, Ohio State University.

Ujueta, Guillermo. Structural geology of the western flank of the Elkins Valley Anticline, Elkins area,

West Virginia. M, 1964, Pennsylvania State University, University Park. 48 p.

Ukaji, K. Analysis of soil-foundation-structure interaction during earthquakes. D, 1975, Stanford University. 267 p.

Ukayli, Mustafa Ahmad. Hydrogeology and digital modeling of the buried-valley aquifer and the Silurian-Devonian carbonate aquifer in the Scioto River basin, Ohio. D, 1978, Ohio State University. 287 p.

Ukazim, Emenike Otuonyeadike. Kirchhoff integral applied to two-dimensional seismic modeling of a growth fault in the Niger Delta region. D, 1980, Texas A&M University. 167 p.

Ulbrich, Horstpeter Herberto Gustavo Jose. I, Contact aureoles around granites in the Blanco Mountain Quadrangle, White Mountains, Calif.; II, Crystallographic studies on scapolites; III, An examination of enthalpies of minerals in the system SiO_2-Al_2O_3-K_2O-H_2O. D, 1973, University of California, Berkeley. 325 p.

Ulbrich, Mable Norma Costas. Systematics of eucrites and howardite meteorites and a petrographic study of representative individual eucrites. M, 1971, University of California, Berkeley.

Ullman, William John. The sedimentary geochemistry of iodine and bromine. D, 1982, University of Chicago. 335 p.

Ullmann-Beck, Gary. Chemistry composition, structure and associations of individual coal macerals. M, 1982, Purdue University. 117 p.

Ullmer, Edwin. Molybdenum trace analysis of certain phreatophytes as a biogeochemical prospecting method in the sedimentary basins of southern Arizona. M, 1975, University of Arizona.

Ullrich, C. R. An experimental study of the time-rate of swelling. D, 1975, University of Illinois, Urbana. 366 p.

Ulmer, Gene Carlton. Oxidation-reduction reactions and equilibrium phase relations at 1300°C at oxygen pressures from O.21 to 10^{-14} atmospheres for the spinel solid solution series $FeCr_2O_4$-$MgCr_2O_4$ and $FeAl_1O_4$-$MgAl_2O_4$. D, 1964, Pennsylvania State University, University Park. 166 p.

Ulmer, James H. Quaternary stratigraphy of the Lake Sakakawea area, McLean County, North Dakota. M, 1973, University of North Dakota. 57 p.

Ulmer, Joseph Walter, Jr. An examination of the Bokkeveld Devonian fauna of South Africa. M, 1958, University of Cincinnati. 140 p.

Ulmo, George J., Jr. Geology of portions of the Cony Mountain, Sweetwater Needles, Sweetwater Gap, and Christina Lake quadrangles, Wyoming. M, 1979, University of Missouri, Columbia.

Ulmschneider, Rene Joseph. Exchangeable cations and paleosalinity; a test using shales of the Appalachian Basin. M, 1978, University of Cincinnati. 118 p.

Ulrich, Barbara Carol. The Eocene foraminiferal biostratigraphy of the Atlantic Coastal Plain of New Jersey. M, 1976, Rutgers, The State University, New Brunswick. 72 p.

Ulrich, George Erwin. Petrology and structure of the Porcupine Mountain area, Summit County, Colorado. D, 1963, University of Colorado. 205 p.

Ulrich, Gilbert Wayne. The mechanics of central peak formation in shock wave cratering events. D, 1974, Massachusetts Institute of Technology. 170 p.

Ulrich, Suzanne Danner. Formation of a platinum-rich beach placer deposit, Goodnews Bay, Alaska. M, 1984, University of Texas, Austin. 179 p.

Ulrick, James Steven. The relevance of Hack's stream gradient index to a study of Dry Creek, Yolo County, California. M, 1981, University of California, Davis. 99 p.

Ulriksen, Carlos E. Regional geology, geochronology and metallogeny of the coastal cordillera of Chile between 20°30′ and 26° south. M, 1979, Dalhousie University. 221 p.

Ulrych, Tadeusz J. The preparation of lead tetramethyl for mass spectrometer analysis. M, 1960, University of British Columbia.

Ulrych, Tadeusz Jan. Gas source mass spectrometry of trace leads from Sudbury, Ontario. D, 1962, University of British Columbia.

Ulteig, J. R. Upper Niagaran and Cayugan stratigraphy of northeastern Ohio and adjacent areas. M, 1963, University of Wyoming. 112 p.

Ulugur, Mustafa Elmas. Fluvial physiography as a factor in basin response. D, 1969, Colorado State University. 168 p.

Um, Junho. A fast algorithm for two-point seismic ray tracing. M, 1986, SUNY at Stony Brook. 82 p.

Umar, P. A. Mineral resource potential; Rouyn-Noranda region, Quebec. D, 1978, McGill University. 305 p.

Umari, A. M. A. Identification of aquifer dispersivities in two dimensional, transient, groundwater contaminant transport; an optimization approach. D, 1977, Cornell University. 133 p.

Umbach, Elmer Dean. Geology of the Sage Creek area, Carbon County, Wyoming. M, 1948, University of Wyoming. 36 p.

Umbach, Paul Henry. History of the geographical and geological explorations in Wyoming prior to 1900. M, 1935, University of Wyoming. 62 p.

Umhoefer, Paul John. Age and structure of a deformed Triassic sequence in the upper Wood River valley, central Alaska Range, Alaska. M, 1979, University of Wisconsin-Madison.

Umphress, A. M. Lithostratigraphic zonation of the Ellenburger Group with the Belco Hickman No. 13 well, Barnhart Field, Reagan Uplift, Texas. M, 1977, University of Texas, Arlington. 156 p.

Umpleby, Joseph Bertram. Drainage history of Warm Springs Creek, central Montana. M, 1909, University of Utah. 37 p.

Umpleby, Joseph Bertram. Geology and ore deposits of the Republic mining district (Washington). D, 1910, University of Chicago. 65 p.

Umpleby, Stuart Standish. Faulting, accumulation and fluid distribution in Ramsey Pool, Payne County, Oklahoma. M, 1954, University of Oklahoma. 34 p.

Umshler, Dennis B. Source of the Evan's Mound obsidian. M, 1975, New Mexico Institute of Mining and Technology. 38 p.

Unash, Cora Louise. The stratigraphy and fauna of the Sundance Formation of the northern Black Hills. M, 1925, University of Iowa. 75 p.

Underberg, Gregory L. Revisions to a geomorphic events simulation model for a hypothetical nuclear waste disposal site in the Columbia Plateau, Washington. M, 1983, Kent State University, Kent. 176 p.

Underhill, Douglas H. Structural studies in the Romanet Lake-Dumphy Lake area, east side Labrador Trough, P.Q. (Canada). M, 1967, McGill University.

Underhill, Douglas Henry. The nature of deformation in experimentally deformed calcite-cemented sandstones. D, 1972, McMaster University. 198 p.

Underhill, James. Areal geology of Lower Clear Creek (Jefferson County), Colorado. D, 1905, University of Colorado.

Underwood, J. K. The stratigraphic distribution of mercury in lake sediments from Halifax County, Nova Scotia. M, 1974, Dalhousie University.

Underwood, James O. The Emporia Formation in southeastern Kansas. M, 1964, Wichita State University. 146 p.

Underwood, James Ross, Jr. Geology of Carrizo Valley, Apache County, Arizona. M, 1956, University of Texas, Austin.

Underwood, James Ross, Jr. Geology of Eagle Mountains and vicinity, Trans-Pecos, Texas. D, 1962, University of Texas, Austin. 604 p.

Underwood, Joan Elizabeth. Evaluation of ground water pollution potential from surface impoundments in northern Idaho and eastern Washington. M, 1981, University of Idaho. 180 p.

Underwood, Mark Roland. Stratigraphy and depositional history of the Bourbon Flags (Upper Pennsylvanian), an enigmatic unit in Linn and Bourbon counties in east-central Kansas. M, 1984, University of Iowa. 111 p.

Underwood, Michael Bruce. Franciscan and related rocks of southern Humboldt County, Northern California Coast Ranges; analysis of structure, tectonics, sedimentary petrology, paleogeography, depositional history, and thermal maturity. D, 1984, Cornell University. 376 p.

Underwood, Prescott, Jr. Geology of the southwestern Big Snowy Mountains, Montana. M, 1955, University of Kansas. 57 p.

Underwood, W. D. Determination of provinience of potsherds by heavy mineral analysis. M, 1977, SUNY at Buffalo. 50 p.

Unfer, Louis, Jr. Study of factors influencing the rank of Wyoming coals. M, 1951, University of Wyoming. 54 p.

Ung, M. K. Geology and economic appraisal of graphite deposits in Wake County. M, 1965, North Carolina State University.

Ungate, Carol A. Neogene-Quaternary basin, fill history of the Pahsimeroi Valley, Idaho. M, 1988, Idaho State University. 105 p.

Unger, Henry E. Geology of the Union copper deposit, Gold Hill District, central North Carolina. M, 1982, University of North Carolina, Chapel Hill. 86 p.

Unger, John D. Melting of granite under effective confining pressure. M, 1967, Massachusetts Institute of Technology. 42 p.

Unger, John Duey. The microearthquake activity of Mount Rainier, Washington. D, 1969, Dartmouth College. 179 p.

Unger, Michael Allen. Investigation of tributyltin; water/sediment interactions. D, 1988, College of William and Mary. 90 p.

Ungrady, Timothy Edward. Stratigraphic analysis of the Piney Point Formation of Maryland. M, 1980, Rutgers, The State University, New Brunswick. 60 p.

Unites, Dennis F. Environmental geology for planning, Estes Park area, Colorado; a pilot study in the crystalline uplands. M, 1974, Colorado State University. 129 p.

UnKauf, John Cameron. Physical stratigraphy of the Hoxbar Group, Ardmore Basin, southern Oklahoma. M, 1970, Tulane University. 168 p.

Unklesbay, Athel Glyde. Geology of southwestern Guernsey County, Ohio. M, 1940, University of Iowa. 87 p.

Unklesbay, Athel Glyde. The siphuncle of late Paleozoic ammonoids. D, 1942, University of Iowa. 87 p.

Unni, C. K. Chlorine and bromine degassing during submarine and subaerial volcanism. D, 1976, University of Rhode Island. 292 p.

Unrau, J. D. A detailed field and laboratory investigation of compressional wave velocities through natural and artificial soils at permafrost temperatures. M, 1985, University of Waterloo. 263 p.

Unruh, Daniel M. The U-Pb age of L chondrites determined by terrestrial Pb contamination corrections. M, 1980, Colorado School of Mines. 54 p.

Unruh, Mark Eugene. Geochemistry and petrology of Triassic granitoids of the Sierra Nevada Batholith, California and Nevada. M, 1985, California State University, Northridge. 94 p.

Unterman, Billie Ruple. A study of the Wildcat Fault in the Berkeley Hills, California. M, 1935, University of California, Berkeley. 154 p.

Unuiboje, Felix E. Palynology of the fire clay coal bed of eastern Kentucky. M, 1987, Eastern Kentucky University. 34 p.

Unver, Olcay Ismail Hakki. Simulation and optimization of real-time operations of multireservoir systems under flooding conditions. D, 1987, University of Texas, Austin. 328 p.

Uotila, Urho Antti Kalevi. Investigations on the gravity field and shape of the earth. D, 1959, Ohio State University. 185 p.

Up de Graff, Jaye Ellen. Gravity study of the northern boundary of the western Transverse Ranges, California. M, 1984, University of California, Santa Barbara. 171 p.

Upadhyay, Hansa Datt. Geology of the Gullbridge copper deposit, central Newfoundland. M, 1970, Memorial University of Newfoundland. 134 p.

Upadhyay, Hansa Datt. The Betts Cove ophiolite and related rocks of the Snooks Arm Group, Newfoundland. D, 1973, Memorial University of Newfoundland. 224 p.

Upchurch, Clyde Neil. Geology of the southwest quarter of Troy, North Carolina Quadrangle. M, 1968, North Carolina State University. 90 p.

Upchurch, Michael Lee. Petrology of the Eocene Castle Hayne Limestone at Ideal Cement Quarry, New Hanover County, North Carolina. M, 1973, University of North Carolina, Chapel Hill. 97 p.

Upchurch, Sam B. Sediment-fauna relationships in the Wayne Group (Silurian) of middle Tennessee. M, 1966, Northwestern University.

Upchurch, Sam Bayliss. Sedimentation on the Bermuda platform. D, 1970, Northwestern University. 243 p.

Updegraff, Nancy A. Pelecypoda of the Brush Creek (Pennsylvanian) of Ohio. M, 1969, Bowling Green State University. 101 p.

Updegraff, Richard Alan. The Quaternary history of the Goose Lake Channel area, Clinton and Jackson counties, Iowa. M, 1981, University of Iowa. 155 p.

Updike, Randall Groves. Glacial geology of the San Francisco peaks, Coconino County, Arizona. M, 1969, Arizona State University. 154 p.

Updike, Randall Groves. The geology of the San Francisco Peaks, Arizona. D, 1977, Arizona State University. 423 p.

Upham, Gregory A. A study of trace metals found in sediment in a stream before and during strip mining of coal in the Cumberland Plateau of eastern Tennessee. M, 1975, University of Tennessee, Knoxville. 145 p.

Upham, Sidney H., Jr. A study of the possibility of locating transverse positions in shoestring sands through sedimentary analysis of modern offshore bars. M, 1950, Columbia University, Teachers College.

Uphoff, Thomas L. Subsurface stratigraphy and structure of the Mesilla and Hueco bolsons, El Paso region, Texas and New Mexico. M, 1978, University of Texas at El Paso.

Upitis, Uldis. The Rapitan Formation (late Precambrian), Yukon and Northwest Territories. M, 1966, McGill University.

Upp, Charles S. A gravity survey of the Moffat, Eisenhower and Johnson tunnels in the Front Range of Colorado. M, 1985, Colorado School of Mines. 139 p.

Upp, Jerry E. The Permian section of southeastern Nebraska. M, 1932, University of Nebraska, Lincoln.

Upp, Robert Rexford. Holocene activity on the Maacama Fault, Mendocino County, California. D, 1982, Stanford University. 196 p.

Upperco, Jesse R. Stratigraphy and areal geology of Rose Township, Carroll County, Ohio. M, 1961, Ohio State University.

Uppuluri, Venkata Rao. A stratigraphic and compositional study of basalts of the Columbia River Group near Prineville, central Oregon. M, 1973, University of Oregon. 87 p.

Upshaw, Charles Francis. Palynology of the Frontier Formation, northwestern Wind River basin, Wyoming. D, 1959, University of Missouri, Columbia. 483 p.

Upshaw, Charles Francis. The age of the Chickasawhay Limestone in Mississippi. M, 1953, Mississippi State University. 104 p.

Upshaw, Laurel P. The genus Sphenodiscus in the Upper Cretaceous of Mississippi. M, 1954, Mississippi State University. 135 p.

Upson, Joseph E. II. Tertiary geology and geomorphology of Culebra Reentrant, southern Colorado. D, 1938, Harvard University.

Upson, M. E. The Ostracoda of the Permian System of Nebraska. M, 1933, University of Nebraska, Lincoln.

Upson, Roberta Hastings. Geologic significance of the Latah flora of northern Nez Perce County, Idaho. M, 1940, University of Idaho. 74 p.

Upson, Susan Adelaide. Lacustrine sediment variations as indicators of late Holocene climatic fluctuations, Arapaho Cirque, Colorado Front Range. M, 1980, Idaho State University. 57 p.

Urban, James Bartel. Microfossils of the Woodford Shale (Devonian) of Oklahoma. M, 1960, University of Oklahoma. 77 p.

Urban, James Bartel. Palynology of the mineral coal (Pennsylvanian) of Oklahoma and Kansas. D, 1962, University of Oklahoma. 158 p.

Urban, Logan L. Palynology of the Drywood and Bluejacket coals (Pennsylvanian) of Oklahoma. M, 1965, University of Oklahoma. 92 p.

Urban, Noel Richard. The nature and origins of acidity in bogs. D, 1987, University of Minnesota, Duluth. 418 p.

Urban, Scott D. The mineralogy and petrology of uranium bearing veins in the Antionoli Claims area, Silver Bow County, Montana. M, 1977, Bowling Green State University. 137 p.

Urban, Stuart D. The application of stratigraphic models to revegetation conditions for Gulf Coast surface lignite mines. M, 1984, Texas A&M University. 181 p.

Urban, Thomas Charles. Terrestrial heat flow in middle Atlantic States. D, 1970, University of Rochester. 398 p.

Urbanczyk, Kevin M. Petrogenesis of the Rawls Formation, Bofecillos Mountains, Trans-Pecos, Texas. M, 1987, Sul Ross State University. 156 p.

Urbanec, Don Alan. Stream terraces and related deposits in the Austin area, Texas. M, 1963, University of Texas, Austin.

Urbani, Franco. Petrology of the (Precambrian) igneous and metaigneous rocks of the Almont area, Gunnison County, (Colorado). M, 1972, University of Kentucky. 224 p.

Urbani, Franco. Phase equilibria and spatial extent of chemical equilibration of magmatite rocks from Colorado, U.S.A., and Venezuela. D, 1975, University of Kentucky. 417 p.

Urbanowicz, Karla Louise. The geology and petrology of the Chetco Complex, Klamath Mountains, southwestern Oregon. M, 1986, University of California, Davis. 131 p.

Uresk, Jack R. Sedimentary environment of the Cretaceous Ferron Sandstone near Caineville, Utah. M, 1978, Brigham Young University.

Urian, Brett A. The subsurface stratigraphy, structure, and petroleum geology of the Clinton section (Lower Silurian) in Wayne County, Ohio. M, 1986, Kent State University, Kent. 96 p.

Urick, R. J. The determination of sound velocity in core samples. M, 1939, California Institute of Technology. 21 p.

Urish, D. W. A study of the theoretical and practical determination of hydrogeological parameters in glacial outwash sands by surface geoelectrics. D, 1978, University of Rhode Island. 336 p.

Urish-McLatchie, Carol Lynn. Microfloral borers in Recent Caribbean scleractinian corals. M, 1976, Rice University. 103 p.

Urosevic, Milovan. Some effects of an anisotropic medium on P and SW waves; a physical modeling study. M, 1985, University of Houston.

Urquhart, Glen. Magnetite deposits of the Savage River, Tasmania. M, 1965, McGill University.

Urquhart, Joanne. Depth of emplacement and cooling history of the Skaergaard Intrusion, East Greenland as revealed by fission track dates. M, 1986, Dartmouth College. 58 p.

Urquhart, Scott Allen. A magnetotelluric investigation of Newberry Volcano, Oregon. M, 1988, University of Oregon. 75 p.

Urquhart, William Edward S. Geological evaluation of a magnetic survey in the Matagami region of Quebec. D, 1988, University of Toronto.

Urrea, Victor Hugo Noguera see Noguera Urrea, Victor Hugo

Urschel, Stephen F. Paleo-depth of burial of Lower Ordovician Beekmantown Group carbonates in New York State. M, 1984, Rensselaer Polytechnic Institute. 88 p.

Usbug, Enis. The geology of the western end of the Baraboo Syncline (Precambrian–central southern Wisconsin). M, 1968, University of Wisconsin-Madison.

Usdansky, Steven Ira. Topologic properties of c-component, (c+4) phase petrogenetic grids with applications to silica and metamorphic rocks in the Gold Creek area, Gunnison County, Colorado. D, 1981, University of Minnesota, Minneapolis. 170 p.

Ushah, Abdurrazag. Application of Hilbert transform in geophysics. M, 1986, University of Manitoba.

Usher, John Leslie. The stratigrpahy and paleontology of the Upper Cretaceous rocks of Vancouver Island, British Columbia. D, 1950, McGill University.

Usher, S. J. One dimensional analytical solution, two layer aquitard. M, 1986, University of Waterloo.

Usiriprisan, Chamroon. Geology of the Woodville Hills Intrusive body, Lawrence County, South Dakota. M, 1979, South Dakota School of Mines & Technology.

Usmani, Tariq U. Seismic stratigraphic analysis of the southwestern abyssal Gulf of Mexico. M, 1980, [University of Houston].

Usselman, Thomas Michael. The liquidus relations at high pressure in the iron-rich sulfur-poor portion of the Fe-Ni-S system and their application to the core of the Earth. D, 1973, Lehigh University. 153 p.

Ussler, William, III. Experimental and theoretical studies of mixed magmas. D, 1988, University of North Carolina, Chapel Hill. 147 p.

Ussler, William, III. The rare earth element geochemistry of the Cranberry magnetite ores, Avery County, North Carolina. M, 1980, University of North Carolina, Chapel Hill. 108 p.

Usunoff, Eduardo Jorge. Factors affecting the movement and distribution of fluoride in aquifers. D, 1988, University of Arizona. 369 p.

Uszynski, Bruce J. Stratigraphic framework and depositional systems; upper Bloyd and lower Atoka (Pennsylvanian) strata, Arkoma Basin of central Arkansas. M, 1982, University of Arkansas, Fayetteville.

Utech, Nancy. Geochemical and petrographic analyses of travertine precipitating waters and associated travertine deposits, Arbuckle Mountains, Oklahoma. M, 1988, University of Houston.

Utgaard, John Edward. Fenestrate bryozoans from the Glen Dean Limestone (Middle Chester) of southern Indiana. M, 1961, Indiana University, Bloomington. 50 p.

Utgaard, John Edward. Trepostomatous bryozoan fauna of the upper part of the Whitewater Formation (Cincinnatian) of eastern Indiana and western Ohio. D, 1963, Indiana University, Bloomington. 202 p.

Uthe, R. E. Assessment of soil conductance and Ph in exploration geochemistry for selected mining areas of New Brunswick, Canada. D, 1978, University of New Brunswick.

Uthe, Richard E. Geochemistry of amphibolites from the southern Bighorn mountains, Wyoming (Precambrian). M, 1970, Kent State University, Kent. 79 p.

Uthman, William. Hydrogeology of the Hamilton North and Corvallis quadrangles, Bitterroot Valley, southwestern Montana. M, 1988, University of Montana. 232 p.

Utine, Mehmet T. Extraction of minor elements from todorokite (MnO_2) and associated solid state transformations. D, 1974, Stanford University.

Utley, Kenneth W. Stratigraphy of the Pliocene Quiburis Formation near Mammoth, Arizona. M, 1980, Arizona State University. 178 p.

Utseth, Rolf Halvor. Numerical simulation of gas injection in oil reservoirs. D, 1980, University of Texas, Austin. 182 p.

Uttamo, Wutti. Application of Landsat imagery for regional geologic and geomorphic mapping, northern Mississippi. M, 1979, University of Mississippi. 80 p.

Utter, Gordon Stanley. A structural and sedimentary study of the Washington Field, St. Laudry Parish, Louisiana. M, 1959, Michigan State University. 117 p.

Utterback, C. L. and Sanderman, L. A. Radium content of some inshore bottom samples in the Pacific Northwest. D, 1943, University of Washington. 5 p.

Utterback, Donald Desmond. A study of outcropping bituminous limestones and sandstones with reference to porosity and to the origin and migration of petroleum. D, 1936, University of Illinois, Urbana.

Utterback, Donald Desmond. The Silurian limestones of northeastern Illinois. M, 1932, University of Illinois, Chicago.

Utterback, W. C. The geology and mineral deposits of Eden Valley-Saddle Peaks and vicinity, southeastern Coos and northeastern Curry counties, Oregon. M, 1973, Oregon State University. 81 p.

Utting, John. Geology of the Codroy Valley, southwestern Newfoundland, including results of a preliminary palynological investigation. M, 1965, Memorial University of Newfoundland. 89 p.

Utting, M. G. The generation of stormflow on a glaciated hillslope in coastal British Columbia. M, 1978, University of British Columbia.

Uttley, J. S. The stratigraphy of the Maxville Group of Ohio and correlative strata in adjacent areas. D, 1974, Ohio State University. 269 p.

Utzig, Esther W. Some geographic aspects of agriculture in Champaign County (Illinois) (Utl). M, 1926, University of Illinois, Urbana.

Uuno, Mathias Sahinen see Sahinen Uuno, Mathias

Uutala, Allen J. Paleolimnological assessment of the effects of lake acidification on Chironomidae (Diptera) assemblages in the Adirondack region of New York. D, 1987, SUNY at Buffalo. 163 p.

Uy, Dominador C. Relationships among the geology, rock temperature and ore mineralogy at the Magma Mine, Superior (Pinal county), Arizona. M, 1965, New Mexico Institute of Mining and Technology. 50 p.

Uyeno, Thomas Tadashi. Conodonts of the Hull Member, Ottawa Formation (Middle Ordovician), of the Ottawa-Hull District, Canada. M, 1963, University of Iowa. 275 p.

Uyeno, Thomas Tadashi. Conodonts of the Waterways Formation (Upper Devonian) of northwestern and central Alberta (Canada). D, 1966, University of Iowa. 238 p.

Uygur, Kadir. Diagenesis and reservoir qualities of the Jurassic Navajo (Nugget) Sandstone in Utah and southwestern Wyoming. D, 1983, University of Utah. 339 p.

Uygur, Kadir. Hydraulic and petrographic characteristics of the Navajo Sandstone in southern Utah. M, 1980, University of Utah. 134 p.

Uytana, Veronica Feliciano. Physiochemical characteristics during potassic alteration of the porphyry copper deposit at Ajo, Arizona. M, 1983, University of Arizona. 96 p.

Uzochukwu, Godfrey A. Properties, genesis, and classification of soils on two geomorphic surfaces in the North Platte River valley in western Nebraska. D, 1983, University of Nebraska, Lincoln. 131 p.

Uzoigewe, Andrew Chukudebelu. Emulsion rheology and flow through synthetic porous media. D, 1970, Stanford University. 202 p.

Uzuakpunwa, Anene B. Petrology and structure of the western Massachusetts ultramafic belt. D, 1976, Boston University. 181 p.

Uzuakpunwa, Anene Benedict. Structural studies of the Gander and Davidsville groups in the Carmanville-Ladle Cove areas, Newfoundland. M, 1973, Memorial University of Newfoundland. 136 p.

V., Jesus Orangel Pereira see Pereira V., Jesus Orangel

Vaag, Myra Kathleen. Stratigraphy of the Permian Rainvalley Formation, southeastern Arizona. M, 1984, University of Arizona. 135 p.

Vacher, Henry Leonard. Cambrian section near Hunters, northeastern Washington. M, 1969, Northwestern University. 76 p.

Vacher, Henry Leonard. Late Pleistocene sea-level history; bermuda evidence. D, 1971, Northwestern University.

Vaden, David W. Petrology and diagenesis of the Short Creek Oolite Member of the Boone Formation, Northeast Oklahoma. M, 1987, Oklahoma State University. 197 p.

Vadnais, Raymond R. Quantitative terrain factors as related to soil parent materials and their engineering classification. D, 1965, University of Illinois, Urbana. 166 p.

Vadnal, John Louis. A numerical model for steady flow in meandering alluvial channels. D, 1984, University of Iowa. 152 p.

Vafidis, Antonios. Supercomputer finite difference methods for seismic wave propagation. D, 1988, University of Alberta. 164 p.

Vagners, Uldis Janis. Lithologic relationship of till to carbonate bedrock in southern Ontario. M, 1966, University of Western Ontario. 156 p.

Vagners, Uldis Janis. Mineral distribution in tills (Pleistocene) in central and southern Ontario. D, 1970, University of Western Ontario. 279 p.

Vagt, G. Oliver. The tectonic significance of some intrusive rocks in the Kenora-Westhawk lake area, Ontario. M, 1968, University of Manitoba.

Vagt, Peter John. Characterization of a landfill-derived contaminant plume in glacial and bedrock aquifers, Dupage County, Illinois. D, 1987, Northern Illinois University. 355 p.

Vagt, Peter John. Vertical and horizontal stability of streams in northern Illinois. M, 1982, Northern Illinois University. 139 p.

Vagt, William Arthur. Petrology of Monadnock Mt., Vt. M, 1976, Massachusetts Institute of Technology. 68 p.

Vagvolgyi, Andrew Louis. Palynology of type McMurray Formation. M, 1964, University of Alberta. 132 p.

Vahrenkamp, Volker C. Constraints on the formation of platform dolomite; a geochemical study of late Tertiary dolomite from Little Bahama Bank, Bahamas. D, 1988, University of Miami. 434 p.

Vahrenkamp, Volker Christian. Processes and environment of deposition of the late Pleistocene recessional Fort Wayne Moraine. M, 1983, University of Michigan.

Vaid, Yoginder, P. Comparative behaviour of an undisturbed clay under triaxial and plane strain conditions. D, 1971, University of British Columbia.

Vaiden, Robert Clifford. Biostratigraphy and paleoenvironment of Morrowan (Zone 20) Brachiopoda, Bird Spring Group, Arrow Canyon, Clark County, Nevada. M, 1985, University of Illinois, Urbana. 153 p.

Vail, John Randolph. Geology of the Racing River area, British Columbia. M, 1957, University of British Columbia.

Vail, Peter Robbins. Stratigraphy and lithofacies of Upper Mississippian rocks in the Cumberland Plateau. D, 1959, Northwestern University. 207 p.

Vail, Peter Robbins. The igneous and metamorphic complex of the East Boulder River area, Montana. M, 1955, Northwestern University. 68 p.

Vail, Ronald Grant. Origin of the paha topography in the Garden Plain upland, Whiteside County, Illinois. M, 1969, Northern Illinois University. 47 p.

Vail, Ruth Staron. The relationship of gneiss composition to dynamic history as shown by analysis of variance of trend surface analysis, Ragged Top area, Albany County, Wyoming. M, 1968, University of Illinois, Urbana. 56 p.

Vail, Scott G. Geology and geochemistry of the Oregon Mountain area, southwestern Oregon and northern California. D, 1977, Oregon State University. 159 p.

Vairavamurthy, Appathurai. Studies on the biosynthesis of dimethylsulfide and β-dimethylsulfoniopropionate by a coccolithophorid alga, Hymenomonas carterae. M, 1984, Florida State University.

Vaitl, Jonathan David. Geology and structure of a northern Sierra Nevada foothills melange and adjacent areas, Butte County, California. M, 1980, University of California, Davis. 93 p.

Valachi, Laszlo Zoltan. The petrology and petrography of the Norris Peridotite in Union County, Tennessee. M, 1963, University of Tennessee, Knoxville. 57 p.

Valasek, David W. Stratigraphy of the Ohio Creek Member of the Williams Fork Formation, Piceance Creek Gap to Rifle Gap, Garfield and Rio Blanco counties, Colorado. M, 1986, Colorado School of Mines. 126 p.

Valasek, Paul A. Processing and interpretation of seismic reflection data from two Cordilleran metamorphic core complexes; the Ruby Mountains-East Humboldt Range, NV, and the Picacho Mountains, AZ. M, 1987, University of Wyoming. 260 p.

Valastro, Salvatore, Jr. A new technique for the radiocarbon dating of mortar. M, 1975, University of Texas, Austin.

Valderrama, Rafael. The Skinner Sandstone Zone in central Oklahoma. M, 1974, University of Tulsa. 63 p.

Valdes, Carlos M. Analysis of the Petatlan aftershocks; numbers, energy release, and asperities. M, 1983, University of Wisconsin-Madison. 61 p.

Valdes, Fernando, Jr. Magnetic and gravity investigations of ultramafic rocks near Camp Meeker, California. M, 1971, San Jose State University. 116 p.

Valdes, Gustavo E. Geology of Jones Hill Quadrangle, South Park, Park County, Colorado. M, 1967, Colorado School of Mines. 86 p.

Valdés, José de Jesús. Bisti oil field, New Mexico; an evaluation of proposed microseepage effects on surface carbonate cementation and ferromanganese content of soils and plants. M, 1983, University of Massachusetts. 136 p.

Valdovinos, Dennis L. Petrography of the lamprophyres in the eastern Ouachitas, Arkansas. M, 1967, University of Arkansas, Fayetteville.

Valenca, Joel Gomes. Geology, petrography and petrogenesis of some alkaline igneous complexes of Rio de Janeiro State, Brazil. D, 1980, University of Western Ontario. 248 p.

Valencia, M. J. Tertiary and Quaternary sediments of the Ontong Java Plateau area. D, 1972, University of Hawaii. 196 p.

Valencia, Mark John. Electron microscopic investigation into the possibilities of refined taxonomic identifications of fossil pollen. M, 1968, University of Texas, Austin.

Valencia, Shirley M. The origin, distribution and engineering characteristics of surficial material in the San Gabriel Valley, California. M, 1966, University of Southern California.

Valente, Jose T. Study of the relations between subsidence, drawdown, and lithology in the Houston-Galveston area. M, 1976, University of Texas, Austin.

Valenti, Gerard L. Structural geology of the Laketown Quadrangle, Rich County, Utah. M, 1980, University of Wyoming. 107 p.

Valentine, Grant M. Origin of banding in an intrusive rock east of Buckley, Washington. M, 1943, Washington State University. 42 p.

Valentine, Gregory Allen. Field and theoretical aspects of explosive volcanic transport processes. D, 1988, University of California, Santa Barbara. 228 p.

Valentine, James William. Paleoecologic molluscan geography of the Californian Pleistocene. D, 1958, University of California, Los Angeles.

Valentine, James William. Pleistocene paleontology of a portion of the northwest coast of Baja California, Mexico. M, 1954, University of California, Los Angeles.

Valentine, Michael James. Structure and tectonics of the southern Gebel Duwi area, Queseir region, Eastern Desert of Egypt. M, 1985, University of Massachusetts. 141 p.

Valentine, Page. Holocene Ostracode zoogeography of the Cape Hatteras region, and climatic implication of a late Pleistocene Ostracode assemblage of the Norfolk, Virginia area. M, 1970, George Washington University.

Valentine, Page Climenson. Zoogeography of the Holocene Ostracoda off western North America and paleoclimatic implications. D, 1973, University of California, Davis. 123 p.

Valentine, Robert Miles. A subsurface geological study of the Redfield gas storage area. M, 1960, University of Iowa. 68 p.

Valentine, Wilbur G. Geology of the Cananea Mountains, Sonora, Mexico. D, 1936, Columbia University, Teachers College.

Valentine, Wilbur G. The nature and composition of tetrahedrite. M, 1926, University of Rochester. 105 p.

Valentino, Albert J. Magnetite-franklinite-pyrophanite intergrowths of the Sterling Hill zinc deposit, Sussex County, New Jersey; an analytical and experimental study. M, 1983, Lehigh University. 89 p.

Valia, Hardarshan S. Miocene sedimentation and stratigraphy of the Delaware coastal plain. D, 1976, Boston University. 269 p.

Valia, Hardarshan Singh. Petrology of the uppermost Miocene, Kent County, Delaware. M, 1971, Bryn Mawr College. 63 p.

Valin, Reed Van see Van Valin, Reed

Valizaden-Alavi, Hedayatollah. Pattern analysis of benthic boundary layer momentum and sediment transport. D, 1983, Ohio State University. 430 p.

Valkenburg, Nicholas. Water quality and mineral solubilities in surface and ground water; Oak Openings community, Lucas County, Ohio. M, 1973, University of Toledo. 46 p.

Valkenburgh, Alvin Van see Van Valkenburgh, Alvin

Valkenburgh, Blaire van see van Valkenburgh, Blaire

Vallance, James W. Late Quaternary volcanic stratigraphy on the southwestern flank of Mount Adams Volcano, Washington. M, 1986, University of Colorado. 122 p.

Valle, Placido D. La see La Valle, Placido D.

Valle, R. S. Della see Della Valle, R. S.

Valle, Raul del see del Valle, Raul

Valle, Richard Saverio Della see Della Valle, Richard Saverio

Valleau, Douglas Nelson. Nature and distribution of near-surface suspended particulate matter contributing to the turbidity stress-field through a coral reef tract of the Florida Keys. M, 1977, University of Florida. 89 p.

Vallee, Marc-Alex. Scale models on transient response of multiple conductors. M, 1981, University of Toronto.

Vallejo, L. E. Mechanics of the stability and development of the Great Lakes coastal bluffs. D, 1977, [University of Michigan]. 261 p.

Valles, Peter Kerry. Sand composition of the Santa Clara River, California; implications for tectonic setting. M, 1985, University of California, Los Angeles. 126 p.

Valletta, Robert Michael. Structures and phase equilibria of binary rare earth metal systems. D, 1959, Iowa State University of Science and Technology. 88 p.

Valley, John William. Calc-silicate reactions in Grenville Marble, Adirondack Mountains, New York. M, 1977, University of Michigan.

Valley, John Williams. The role of fluids during metamorphism of marbles and associated rocks in the Adirondack Mountains, New York. D, 1980, University of Michigan. 257 p.

Valliant, Robert Irwin. The geology, stratigraphic relationships and genesis of the Bousquet gold deposit, Northwest Quebec. D, 1981, University of Western Ontario. 325 p.

Vallier, Tracy Lowell. The geology of part of the Snake River Canyon and adjacent areas in northeastern Oregon and western Idaho. D, 1967, Oregon State University. 267 p.

Vallieres, Andre. Relations stratigraphiques et structurales du Super Groupe de Québec dans la région de Sainte-Malachie. M, 1971, Universite de Montreal.

Vallières, André Stratigraphie et structure de l'Orogène Taconique de la région de Rivière-du-Loup, Québec. D, 1984, Universite Laval. 316 p.

Valls, Hannia Azuola see Azuola Valls, Hannia

Valocchi, Albert Joseph. Transport of ion-exchanging solutes during groundwater recharge. D, 1981, Stanford University. 263 p.

Valusek, Jay E. Biostratigraphy and depositional environments of the lower member of the Antelope Valley Limestone and correlations (Lower and Middle Ordovician), central and eastern Nevada and western Utah. M, 1984, Colorado School of Mines. 187 p.

Vamosi, Sendor. Theoretical examination of the higher order theodolites. M, 1961, Ohio State University.

Van Allen, Bruce R. Hydrothermal iron ore and related alterations in volcanic rocks of La Perla, Chihuahua, Mexico. M, 1978, University of Texas, Austin.

Van Alstine, David Ralph. Apparent polar wandering with respect to North America since the late Precambrian. D, 1979, California Institute of Technology. 368 p.

Van Alstine, James Bruce. Paleontology of brackishwater faunas in two tongues of the Cannonball Formation (Paleocene, Danian), Slope and Golden Valley counties, southwestern North Dakota. M, 1974, University of North Dakota. 101 p.

Van Alstine, James Bruce. Postglacial ostracod distribution and paleoecology, Devils Lake basin, northeastern North Dakota. D, 1980, University of North Dakota. 159 p.

Van Alstine, Ralph Erkstine. Geology and mineral deposits of the St. Lawrence area, Burin Peninsula, Newfoundland. D, 1944, Princeton University.

Van Alstine, Ralph Erkstine. Geology of the Shawkey gold mines, Shawkey, Quebec. M, 1938, Northwestern University.

Van Altena, Peter James. Some Keweenawan sediments of Ontonagon County, Michigan. M, 1951, Michigan Technological University. 47 p.

Van Amringe, John Howard. Geology of a part of the western San Emigdio Mountains, California. M, 1957, University of California, Los Angeles.

Van Arsdale, B. E., Jr. Geology of the area east of Rye, Pueblo County, Colorado. M, 1952, Colorado School of Mines. 68 p.

Van Atta, Robert Otis. Sedimentary petrology of some Tertiary formations, upper Nehalem River basin, Oregon. D, 1971, Oregon State University. 245 p.

Van Beek, Johannes Laurens. Rhythmic patterns of beach topography. D, 1973, Louisiana State University.

Van Beek, Johannes Laurens. The relationship between waves and the changes of the sub-aerial beach profile. M, 1965, Louisiana State University.

Van Beever, Hank G. The significance of the distribution of clasts within the Great Pond Esker and adjacent till (Hancock County, Maine). M, 1971, University of Maine. 63 p.

Van Berkel, Gary Joseph. The role of kerogen in the origin and evolution of nickel and vanadyl geoporphyrins. D, 1987, Washington State University. 246 p.

Van Besien, Alphonse Camille. A photoelastic study of the stresses around and underground opening and its support system. M, 1969, University of Missouri, Rolla.

Van Beuren, Victor Vignot. Stratigraphy of the Sunbury Shale (Lower Mississippian) of the central Appalachian Basin. M, 1980, University of Cincinnati. 150 p.

Van Beveren, Oscar F. Geology and ore deposits of the Logan Mine, Boulder County, Colorado. M, 1932, University of Colorado.

Van Biersel, Thomas P. V. Hydrogeology and chemistry of an oil-field brine plume within a shallow aquifer system in southern Bond County, Illinois. M, 1985, Southern Illinois University, Carbondale. 162 p.

Van Blaricom, Richard. Induced polarization and resistivity modeling of dikes and other selected structural features. M, 1971, University of Arizona.

van Bosse, Jacqueline Y. Metamorphism and alteration in the aureole of the McGerrigle Mountains Pluton, Gaspé, Quebec. M, 1986, University of Toronto.

Van Breemen, Otto. Thermally induced relocation of strontium and rubidium in a granodiorite. M, 1965, University of Alberta. 34 p.

Van Buren, Mark. Seismic stratigraphy of a modern carbonate slope; northern Little Bahama Bank. M, 1984, San Jose State University. 115 p.

Van Buren, Wayne Martin. Geochemistry of syenite in Arkansas. M, 1977, University of Arkansas, Fayetteville.

Van Burkalow, Anastasia. Angle of repose and angle of sliding friction; an experimental study. D, 1945, Columbia University, Teachers College.

Van Burkalow, Anastasia. Methods of valley floor widening. M, 1933, Columbia University, Teachers College.

Van Buskirk, Donald Robert. Slope condition related to aspect along highway sideslopes in northeast Ohio. M, 1978, Kent State University, Kent. 67 p.

Van Buskirk, Steven C. Marine algal-bank development of the Lawrence Formation in Chautauqua County, Kansas. M, 1986, Wichita State University. 146 p.

Van Camp, Quentin Walter. Geology of the Big Mountains area, Santa Susana and Simi quadrangles, Ventura County, California. M, 1959, University of California, Los Angeles.

van Camp, Scott Gregory. Geochronology and geochemistry of cataclastic rocks from the Linville Falls Fault, North Carolina. M, 1982, University of North Carolina, Chapel Hill. 40 p.

Van Cott, Harrison C. Microtextural sequences in rocks and artificial crystallizations. M, 1963, Columbia University, Teachers College.

Van Coutren, Lewis Anderson. A subsurface study of the Aux Vases Formation in portions of Pike and Dubois counties, Indiana. M, 1960, Indiana University, Bloomington. 33 p.

Van Couvering, John Anthony. Geology of the Chilcoot Quadrangle, Plumas and Lassen counties, California. M, 1962, University of California, Los Angeles.

Van Couvering, Martin and Allen, Harry B. Geology of the Devils Den District, northwestern Kern County, California. M, 1941, University of California, Los Angeles.

Van Dalen, Stephen Craig. Depositional systems and natural resources of the middle Eocene Yegua Formation, South and central Texas Coastal Plain. M, 1981, University of Texas, Austin. 132 p.

Van Dam, Dale A. Basin morphology and depositional environments of the lower Mohnian (upper Miocene) La Vida Member of the Puente Formation, Los Angeles Basin, California. M, 1985, University of Utah. 173 p.

Van Dam, George Henry. A late Pleistocene herpetofauna from Baker Bluff Cave, Sullivan County, Tennessee. M, 1976, Michigan State University. 31 p.

van de Graaff, Fredric R. Upper Cretaceous stratigraphy of the central part of Utah. M, 1962, Utah State University. 107 p.

Van de Graaff, Fredric Ray. Depositional environments and petrology of the Castlegate sandstone (Cretaceous), east-central Utah. D, 1969, University of Missouri, Columbia. 135 p.

Van de Kamp, Peter Cornelius. Geochemistry and classification of amphibolites and related rocks. M, 1964, McMaster University. 106 p.

van de Poll, Henk Wouter. Carboniferous volcanic and sedimentary rocks of the lower Shin Creek area, Sunbury County, New Brunswick. M, 1963, University of New Brunswick.

Van de Reep, Thomas W. Deposition and diagenesis of the Mississippian Midale Beds in the Tatagwa-Neptune-Bromhead areas of southeastern Saskatchewan. M, 1987, University of Saskatchewan. 281 p.

Van de Verg, Philip E. Seismic refraction investigation of the Dunes thermal anomaly, Imperial Valley, California. M, 1976, University of California, Riverside. 81 p.

Van de Voorde, Barbara Wiley. A geophysical study of a Precambrian crustal boundary in Minnesota. M, 1980, Northern Illinois University. 43 p.

Van Dellen, Kenneth J. The Port Huron Moraine and associated glacial features, St. Clair and Sanilac counties, Michigan. M, 1978, University of Michigan.

Van den Berg van Saparoea, C. M. G. Complexation of copper in natural waters; determination of complexing capacities and conditional stability constants in natural waters by MnO_2, and the implications for phytoplankton toxicity. D, 1979, McMaster University. 260 p.

Van Den Berg, Alexander Nicolaas. Anion distribution among coexisting minerals of Grenville marbles. M, 1975, University of Michigan.

Van Den Berg, Jacob. Pennsylvanian stratigraphy above the Shoal Creek Limestone in Illinois. M, 1956, University of Illinois, Urbana.

Van den Berge, Johannes Christian. The paleo-geomorphological features of the Lovejoy Formation (Eocene, upper or Oligocene, lower), in the Sacramento Valley, California. M, 1968, University of California, Davis.

Van den Heuvel, Peter. Petrography of the Boone Formation, Northwest Arkansas. M, 1979, University of Arkansas, Fayetteville.

Van Denburgh, Alber Stevens. Analysis and comparison of waters from four serpentine and quartz diorite drainage areas. M, 1959, Stanford University.

Van Der Flier, Eileen. Geochemistry and sedimentology of two Cretaceous coal deposits in Canada. D, 1985, University of Western Ontario. 302 p.

Van der Hoeven, G. A. Factors influencing the use of plant material to develop a functional regional landscape in the Great Plains. D, 1977, Virginia Polytechnic Institute and State University. 334 p.

van der Laan, Sieger Robbert. An experimental study of boninite genesis. D, 1987, University of Illinois, Chicago. 118 p.

Van der Leeden, Fritz. The groundwater resources of Westchester County, New York. M, 1962, New York University.

van der Leeden, John. Stratigraphy, structure and metamorphism in the northern Selkirk Mountains

southwest of Argonaut Mountain, southeastern British Columbia. M, 1976, Carleton University. 105 p.

van der Loop-Avery, Mary Louise. The stratigraphy and depositional environment of the Pueblo Formation (Lower Permian) north central Texas. M, 1983, University of Texas, Arlington. 258 p.

van der Meyden, Hendrik Jan. Petrology of gabbroic xenoliths from the summit cone of Mauna Kea Volcano, Hawaii. M, 1985, North Carolina State University. 69 p.

Van der Plank, Adrian. Petrology and geochemistry of some diamond drill cores from the Saskatchewan potash deposits. M, 1963, University of Wisconsin-Madison.

Van Der Pluijm, Bernardus Adrianus. Geology and microstructures of eastern New World Island, Newfoundland, and implications for the Northern Appalachians. D, 1984, University of New Brunswick.

Van der Poel, Washinton I., III. A reconnaissance of the late Tertiary and Quaternary geology, geomorphology and contemporary surface hydrology of the Rattlesnake Creek watershed, Missoula County, Montana. M, 1979, University of Montana. 85 p.

Van der Spuy, Peter M. Geology and geochemical investigations of geophysical anomalies, Sierra Rica, Hidalgo County, New Mexico. M, 1970, Colorado School of Mines. 156 p.

Van der Staay, Robert. Soil surface area affects on microbial denitrification. M, 1974, University of California, Riverside.

Van der Ven, Paulus Hendrikus. Seismic stratigraphy and depositional systems of northeastern Santos Basin, offshore southeastern Brazil. M, 1983, University of Texas, Austin. 151 p.

Van Deusen, John Ernest, III. Mapping geothermal anomalies in the Klamath Falls Oregon region with gravity and aeromagnetic data. M, 1978, University of Oregon. 116 p.

Van Devender, Thomas Roger. Late Pleistocene plants and animals of the Sonoran Desert; a survey of ancient packrat middens in southeastern Arizona. D, 1973, University of Arizona.

Van Deventer, Bruce Robert. Petrology of the Moenkopi Formation (Early Triassic), Uinta Mountain area, northeastern Utah. M, 1974, University of Utah. 99 p.

Van Deventer, James Bartlett. Cuyahoga Falls water survey and evaluation. M, 1971, University of Akron. 34 p.

Van Diver, Bradford Babbitt. Geology of the Calumet mining district, Chaffee County, Colorado. M, 1958, University of Colorado.

Van Diver, Bradford Babbitt. Petrology of the metamorphic rocks, Wenatchee Ridge area, central northern Cascades, Washington. D, 1964, University of Washington. 140 p.

Van Donk, Jan. The oxygen isotope record in deep sea sediments. D, 1970, Columbia University. 39 p.

Van Driel, James Nicholas. Geologic input in a computer land analysis system. M, 1974, Iowa State University of Science and Technology.

Van Duym, Dirk Peter. The mineralogy of the Grassy Creek and Saverton formations. M, 1954, University of Missouri, Rolla.

Van Dyke, Lindell Howard. A study of the structure of a portion of the Gros Ventre Hoback Ranges of western Wyoming. M, 1948, University of Michigan.

Van Dyke, R. J. Geology and depositional environment of the reservoir sandstone, Kincaid oil field, Anderson County, Kansas. M, 1976, University of Kansas. 80 p.

Van Eck, Orville J. Some methods of determining porosity and permeability. M, 1953, University of Michigan.

Van Eeckhout, Edward M. The effect of moisture on the mechanical properties of coal mine shales. D, 1974, University of Minnesota, Minneapolis. 176 p.

van Es, Harold Mathijs. Field-scale water relations for an eroded Hapludult. D, 1988, North Carolina State University. 171 p.

Van Esbroeck, Guillaume. A new teaching diagram for igneous rocks. M, 1923, Columbia University, Teachers College.

van Everdingen, David Allard. Hydrogeological study of a surficial sand and gravel unit in South Kitchener, Ontario. M, 1984, University of Waterloo. 100 p.

Van Fleet, John M. Geomorphology of the Box Canyon drainage basin, Santa Rita Mountains, Pima County, Arizona. M, 1973, University of Arizona.

Van Fossen, John Doan. A study of the Rafinesquinae of the middle Maysville (Upper Ordovician), Cincinnati, Ohio. M, 1951, University of Cincinnati. 98 p.

van Gelder, Susan M. Geology of the Bald Rock granitic batholith, South Carolina. M, 1980, University of Tennessee, Knoxville. 70 p.

Van Gilder, Harold R. The (Permian) Double Mountain Group of Oklahoma. M, 1931, Yale University.

Van Gilder, Kerry L. The manganese orebody at the Three Kids Mine, Clark County, Nevada. M, 1963, University of Nevada - Mackay School of Mines. 94 p.

Van Gundy, Clarence Edgar. The relations of the Upper Cretaceous and (Paleocene) Martinez formations in the northern part of the Adelaide Quadrangle (California). M, 1934, University of California, Berkeley. 60 p.

Van Harryok, Harry Jerrold. Subsolidus equilibria in the system Fe$_3$O$_4$-Mn$_3$O$_4$. M, 1956, Pennsylvania State University, University Park. 33 p.

Van Hart, Dirk. The physical stratigraphy of the Saluda and Whitewater formations (Cincinnatian Series), southeastern Indiana. M, 1966, Miami University (Ohio). 142 p.

Van Havern, Bruce P. Soil water phenomena of a shortgrass prairie site. M, 1974, Colorado State University. 186 p.

Van Hecke, Michael Clement. Sulfation kinetics of the Bayer process residue and manganese sea nodules. D, 1972, Stanford University. 99 p.

van Heerden, Ivor Llewellyn. Deltaic sedimentation in eastern Atchafalaya Bay, Louisiana. D, 1983, Louisiana State University. 161 p.

van Heerden, Ivor Llewellyn. Sedimentary responses during flood and non-flood conditions, new Atchafalaya Delta, Louisiana. M, 1980, Louisiana State University.

Van Heerden, Willem Maartens. Splash erosion as affected by the angle of incidence of raindrop impact. D, 1964, Purdue University. 160 p.

Van Hees, Edmond Harry Peter. Auriferous ankerite vein genesis in the Aunor Mine, Timmins, Ontario. M, 1979, University of Western Ontario. 138 p.

Van Heeswijk, Marijke. Shallow crustal structure of the caldera of Axial Seamount, Juan de Fuca Ridge. M, 1987, Oregon State University. 80 p.

Van Hessert, Christian. The effect of oxygen fugacity on the melting of shale. M, 1971, University of Minnesota, Minneapolis.

Van Hise, Charles R. Crystalline rocks of the Wisconsin Valley. M, 1882, University of Wisconsin-Madison.

Van Hise, Charles R. The Penokee iron-bearing series of northern Wisconsin and Michigan. D, 1892, University of Wisconsin-Madison.

van Hissenhoven, René Method and application for locating earthquakes in a two-dimensional velocity-depth model. D, 1988, University of Wisconsin-Madison. 207 p.

Van Hoeven, William, Jr. Organic geochemistry. D, 1969, University of California, Berkeley. 243 p.

Van Hook, Harry Jerrold. The ternary system Ag$_2$S-Bi$_2$S$_3$-PbS. D, 1959, Pennsylvania State University, University Park. 105 p.

Van Horn, Clifford Layne. Effects of a Silurian reef on Mississippian and Pennsylvanian sediments. M, 1956, University of Illinois, Urbana.

Van Horn, John E. Seismotectonic study of the southwestern Lake Superior region utilizing the Upper Michigan-northern Wisconsin seismic network. M, 1980, Michigan Technological University. 146 p.

Van Horn, Michael D. Stratigraphy of the Almond Formation, east-central flank of the Rock Springs Uplift, Sweetwater County, Wyoming. M, 1979, Colorado School of Mines. 179 p.

Van Horn, Robert Gary. Chert in the Columbus and Delaware limestones. M, 1972, Ohio State University.

Van Horn, Stephen R. Compositional zoning of garnets within a progressive regional metamorphic terrane, Blackerby Ridge, Juneau, Alaska. M, 1983, University of Missouri, Columbia.

Van Horn, William Lewis. Late Cenozoic beds in the upper Safford Valley, Graham County, Arizona. M, 1957, University of Arizona.

Van Horn, William Lewis. Trace metal uptake by (Recent) sediments of the upper James Estuary, Virginia. D, 1972, University of Virginia. 103 p.

Van Houten, Franklyn Bosworth. Part I, Stratigraphy of the Willwood and Tatman formations in northeastern Wyoming; Part II, Early Eocene mammalian faunas of North America. D, 1941, Princeton University. 167 p.

Van Ingen, L. B., III. Structural geology and tectonics; Troublesome Creek valley area, Carbon, Wyoming. M, 1978, University of Wyoming. 76 p.

Van Ingen, Robert. Selenium in the Solbec Deposit, Quebec. M, 1962, Dartmouth College. 67 p.

Van Kauwenbergh, James B. Diagenetic study of the carbonate rocks of the Island of San Salvador, Bahamas. M, 1985, University of Akron. 34 p.

Van Keuren, Lewis Karl, III. Spacial-diurnal variations of the ambient seawater chemistry of a Holocene ooid shoal, Browns Cay, Bahamas. M, 1987, University of Oklahoma. 86 p.

Van Klavaren, Richard William. Hydraulic erosion resistance of thawing soils. D, 1987, Washington State University. 218 p.

Van Kooten, Gerald K. An ultrapotassic basaltic suite from the central Sierra Nevada, California; a study of the mineralogy, petrology, geochemistry and isotopic composition. D, 1980, University of California, Santa Barbara. 112 p.

Van Kooten, Gerald K. Olivine zoning and petrology of a crater in the San Francisco volcanic field, Arizona. M, 1975, Arizona State University. 82 p.

van Kranendonk, Martin Julian. On the origin of thin meta-anorthositic units in the southern Central Gneiss Belt of Ontario, Grenville structural province. M, 1987, University of Toronto.

van Leeuwen, Wim. Computer assisted land evaluation for waste disposal in eastern Monroe County, Michigan. M, 1975, Michigan State University. 51 p.

Van Lewen, Melvin C. The geology of St. Ignace Island, Ontario, and a correlation of the Keweenawan Series of the Lake Superior region. M, 1957, Massachusetts Institute of Technology.

Van Lieu, Junius A. A study of the Chattanooga Shale. M, 1953, University of Missouri, Columbia.

Van Liew, Michael Wayne. Dynamic simulation of water and sediment discharge from agricultural watersheds. D, 1983, Washington State University. 349 p.

Van Loan, Paul Rose. An x-ray crystal structure determination of aenigmatite. D, 1968, McGill University. 130 p.

Van Loan, Paul Rose. An X-ray study of two triclinic minerals. M, 1958, University of Toronto.

Van Lopik, Jack Richard. Recent geology and geomorphic history of central coastal Louisiana. D, 1955, Louisiana State University.

Van Luik, A. E. J. Equilibrium chemistry of heavy metals in concentrated electrolyte solution. D, 1978, Utah State University. 165 p.

Van Maness, L., Jr. A quantitative geomorphic study of stream valley symmetry in the Eden Shale-Outer

Blue Grass area of northern Kentucky. M, 1977, Indiana State University. 122 p.

van Middelaar, Wilhelmus T. Alteration and mica chemistry of the granitoid associated with the Can-Tung scheelite skarn, Tungsten, Northwest Territories, Canada. D, 1988, University of Georgia. 261 p.

Van Nest, Julieann. Holocene stratigraphy and geomorphic history of the Buchanan drainage; a small tributary to the South Skunk River near Ames, Story County, Iowa. M, 1987, University of Iowa. 168 p.

Van Nieuwenhuise, Donald S. Benthic molluscan communities and environmental study of the Cape Romain tidal inlet complex, South Carolina. M, 1977, University of Houston.

Van Nieuwenhuise, Donald S. Stratigraphy and ostracode biostratigraphy of the Black Mingo Formation, South Carolina. D, 1978, University of South Carolina. 108 p.

Van Nieuwenhuise, Robert. Well level fluctuations related to the occurrence of earthquakes at Lake Jocassee, South Carolina. M, 1979, University of South Carolina.

van Nieuwenhuyse, Ulrich Eric. Geology and geochemistry of the Pyrola massive sulfide deposit, Admiralty Island, Southeast Alaska. M, 1984, University of Arizona. 170 p.

van Noort, Peter John. A preliminary investigation of hydrocarbon contamination, Alliance, Nebraska. M, 1988, University of Nebraska, Lincoln. 175 p.

Van Nostrand, Amy K. The geology of Solheimasandur, South Iceland and comparison with other areas modified by catastrophic flooding. M, 1981, University of Georgia.

van Nostrand, Timothy Stuart. Geothermometry-geobarometry and ^{40}Ar/^{39}Ar incremental release dating in the Sandwich Bay area, Grenville Province, eastern Labrador. M, 1987, Memorial University of Newfoundland. 203 p.

van Oss, Hendrik G. Trace element-organic carbon correlation analysis of the Mancos Shale (Upper Cretaceous). M, 1978, Dartmouth College. 108 p.

Van Peteghem, James Karl. Studies pertaining to the mineral chemistry of sodalite, nosean and hauyne. M, 1961, McMaster University. 147 p.

Van Raij, Bernardo. Electrochemical properties of some Brazilian soils. D, 1971, Cornell University.

Van Rensburg, Willem Cornelius Janse. Copper mineralization in the carbonate members and phoscorite, Phalaborwa, South Africa. D, 1965, University of Wisconsin-Madison. 121 p.

Van Roosendaal, Dan J. An analysis of rock structures and strain in cleaved pelitic rocks, east branch of the Huron River, Baraga County, Michigan. M, 1985, Michigan Technological University. 82 p.

Van Ryswyk, Albert Leonard. Forest and Alpine soils of south-central British Columbia. D, 1969, Washington State University. 189 p.

Van Ryswyk, Roy J. Rock bolt testing at the Churchill Falls Power Project, Labrador, Canada. M, 1972, University of Illinois, Urbana. 100 p.

Van Sant, Jan Franklin. Pennsylvanian fusulinids from Whiskey Canyon, New Mexico. M, 1958, University of Kansas. 152 p.

Van Sant, Jan Franklin. The Crawfordsville crinoid fauna of Indiana. D, 1963, University of Kansas. 483 p.

Van Sant, Mary Jane. Applications of the piezoelectric crystal detector in environmental analysis. D, 1984, University of New Orleans. 163 p.

van Saparoea, C. M. G. Van den Berg *see* Van den Berg van Saparoea, C. M. G.

Van Saun, Richard. An evaluation of the use of pore pressure coefficients in the analysis of rapid drawdown in earth slopes. D, 1985, University of Texas, Austin. 196 p.

Van Schaik, Edward J. Geology of selected areas of Cu Zn mineralization of Searcy County, Arkansas. M, 1975, University of Arkansas, Fayetteville.

Van Schaik, Jacob Cornelis. Ion diffusion in clay-water systems. D, 1964, Colorado State University. 74 p.

van Scoyoc, G. E. Surface and structural properties of palygorskite. D, 1976, Purdue University. 94 p.

Van Siclen, Dewitt Clinton. Fossil organic reefs of a West Texas area. D, 1951, Princeton University.

Van Sint Jan, Michel Leopold. Ground and lining behavior of shallow underground rock chambers for the Washington, D.C. subway. D, 1982, University of Illinois, Urbana. 308 p.

van Staal, Cees R. The structure and metamorphism of the Brunswick Mines area, Bathurst, New Brunswick, Canada. D, 1985, University of New Brunswick.

Van Steeg, Karl. Drainage changes in northeastern Washington. M, 1923, University of Chicago.

Van Steeg, Karl. Wind gaps and water gaps of the Northern Appalachians, their characteristics and significance. D, 1930, Columbia University, Teachers College.

van Stempvoort, Dale. Chazy Group, carbonate sedimentology and diagenesis; southern Quebec. M, 1985, McGill University. 201 p.

Van Stone, Larry J. Measurement and prediction of anisotropic thermal conductivity. M, 1985, South Dakota School of Mines & Technology.

Van Tassell, Jay. Deposition of a carbonate turbidite on the Silver abyssal plain. D, 1979, Duke University. 172 p.

Van Tuyl, Francis M. The origin of dolomite. D, 1915, Columbia University, Teachers College.

Van Tuyl, Francis Maurice. Geology of the Keokuk Beds of the central Mississippi Valley. M, 1912, University of Iowa. 121 p.

Van Tyne, Arthur M. Petrology of the Euclid Avenue peridotite intrusive (Syracuse, New York). M, 1958, Syracuse University.

Van Valin, Reed. An investigation of methods for the selective removal and characterization of trace metals in sediments. M, 1980, University of Miami. 142 p.

Van Valkenburgh, Alvin. Geology of the Poorman Dike, Boulder County, Colorado. M, 1938, University of Colorado.

van Valkenburgh, Blaire. A morphological analysis of ecological separation within past and present predator guilds. D, 1984, The Johns Hopkins University. 299 p.

Van Vesien, Alphonse C. A photoelastic study of the stresses around an underground opening and its support system. M, 1969, University of Missouri, Rolla.

Van Vlack, L. H. Interfacial energies of some metallic-nonmetallic systems and their relation to microstructure. D, 1950, University of Chicago. 81 p.

Van Vleet, E. S. Diagenesis of hydrocarbons, fatty acids, and isoprenoid alcohols in marine sediments. D, 1978, University of Rhode Island. 253 p.

Van Vloten, Roger. A contribution to the statistical study of joints with application in Cortlandt Complex (New York). M, 1950, Columbia University, Teachers College.

Van Voast, Wayne Adams. General geology and geomorphology of the Emigrant Gulch-Mill Creek area, Park County, Montana. M, 1964, Montana State University. 65 p.

van Voorhis, David. Seismic interpretation of the Wind River Mountains. M, 1982, Texas A&M University. 113 p.

Van Voorhis, Gerald D. A method for computing the magnetization of two-dimensional structures. M, 1964, Michigan Technological University. 37 p.

Van Wagoner, John Charles. Lower and Middle Pennsylvanian rocks of the northern Sacramento Mountains; a study of contemporaneous carbonate and siliciclastic deposition in an active tectonic setting. D, 1977, Rice University. 204 p.

Van Wagoner, John Charles. Paleoenvironmental analysis of the lower Middle Pennsylvanian deep water sediments, Fort Worth Basin, central Texas. M, 1976, Rice University. 89 p.

van Wagoner, Steven Lewis. Geology of a part of the Zaca Lake Quadrangle, Santa Barbara County, California. M, 1981, California State University, Northridge. 143 p.

Van Walsum, N. An alternating direction Galerkin technique for simulation of transient density-dependent flow in groundwater transport. M, 1987, University of Waterloo. 101 p.

Van Ward, Roland. The geological and geographic writings of Thomas Jefferson. M, 1938, University of Virginia. 219 p.

Van West, Olaf. Geology of the San Benito Islands and southwestern part of Cedros Island, Baja California. M, 1958, Pomona College.

Van Wie, William Arthur. Sedimentology of alluvial fans in southern Nevada. D, 1976, University of Cincinnati. 249 p.

Van Wie, William Arthur. The petrology of the Floyds Knob Glauconite. M, 1971, University of Cincinnati. 90 p.

Van Wyckhouse, Roger J. A study of test borings from the Pleistocene of the southeastern Michigan glacial lake plain. M, 1966, Wayne State University.

Van Wyk, Stefanus J. Selected aspects of Holocene sedimentation, Pensacola area, Florida. M, 1973, Colorado School of Mines. 211 p.

Van Zant, Kent Lee. Late- and postglacial vegetational history of northern Iowa. D, 1976, University of Iowa. 123 p.

Van Zant, Kent Lee. Pleistocene stratigraphy in a portion of Northeast Iowa (Kansan, Tazewell and Cary tills). M, 1973, University of Iowa. 52 p.

van Zyl, D. J. A. Seepage erosion of geotechnical structures subjected to confined flow; a probabilistic design approach. D, 1979, Purdue University. 261 p.

VanArsdale, Roy Burbank. Geology of Strawberry Valley and regional implications. D, 1979, University of Utah. 80 p.

VanArsdale, Roy Burbank. The influence of caliche on the geometry of arroyos near Buckeye, Arizona. M, 1974, University of Cincinnati. 69 p.

VanArsdall, David E. Lithostratigraphy of the Conasauga Group within the Hunter Valley and Copper Creek strike belts, north eastern Tennessee. M, 1974, Eastern Kentucky University. 66 p.

Vanbuskirk, John Reed. Investigation of reservoir conditions of lower Deese Sandstones (Pennsylvanian) for a flood project in the North Alma Pool, Stephens County, Oklahoma. M, 1960, University of Oklahoma. 165 p.

Vance, Dana Joslyn. Relocations of some seismic events in the Puget Sound area 1951–1968. M, 1971, University of Washington. 56 p.

Vance, David B. The geology and geochemistry of three hot spring systems in the Shoup geothermal area, Lemhi County, Idaho. M, 1986, Washington State University. 153 p.

Vance, G. F., Jr. Holocene sediments of Mission Bay, Texas. M, 1975, University of Houston.

Vance, Joseph Alan. The geology of the Sauk River area in the northern Cascades of Washington. D, 1957, University of Washington. 312 p.

Vance, Kenneth Raymond. The petrology of a Pennsylvanian cyclothem. M, 1953, Indiana University, Bloomington. 40 p.

Vance, Randall Blaine. Geology of the NW 1/4 of the Wallace 15′ Quadrangle, Shoshone County, Idaho. M, 1981, University of Idaho. 103 p.

Vance, Richard. Geology of the Hardt Creek-Tonto Creek area, Gila County, Arizona. M, 1983, Northern Arizona University. 99 p.

Vance, Robert Kelly. Geology and geochemistry of the Gunnison intrusive complex, Gunnison County, Colorado. M, 1984, University of Kentucky. 260 p.

Vandall, Thomas Andrew. Paleomagnetism of the Michipicoten and Gamitigama greenstone belts of the Wawa Subprovince, Ontario. M, 1986, University of Windsor. 117 p.

Vandamme, Luc Michel Pierre. A three-dimensional displacement discontinuity model for the analysis of hydraulically propagated fractures. D, 1986, University of Toronto.

Vande Kamp, Brad Douglas. The geology of the Smith Canyon area, Okanogan County, Washington. M, 1986, Eastern Washington University. 46 1 plate p.

Vandell, Terry Delores. Analysis of the hydrogeology of the Phosphoria Formation in lower Dry Valley, Caribou County, Idaho. M, 1978, University of Idaho. 225 p.

Vandemoer, Catherine. The hydrogeochemistry of recharge processes and implications for water management in Southwestern United States. D, 1988, University of Arizona. 186 p.

Vander Hoof, Vertress Lawrence. A skeleton of Aelurodon haydeni from the later Tertiary of Nevada. M, 1932, University of California, Berkeley.

Vander Horck, Mark Patrick. Diagenesis in the Sioux Quartzite. M, 1984, University of Minnesota, Duluth. 101 p.

Vander Kooi, Verna. Paleoenvironmental history of eastern Florida Bay based on foraminiferida. M, 1977, University of Akron. 172 p.

Vander Ley, J. W. Petrology of the Hampton Formation at Eagle City, Iowa. M, 1962, Iowa State University of Science and Technology.

Vander Schaaff, Bertis J. III. Petrology and mineralogy of the "Old Workings area", El Salvador, Chile. M, 1965, University of Missouri, Rolla.

Vander, Pyl. Physiographic development of the Lake Quinsigamond Valley (Massachusetts). M, 1953, Clark University.

VanderBrug, G. J. Linear feature detection and mapping. D, 1977, University of Maryland. 263 p.

Vanderburgh, Sandy. Geomorphology and sedimentology of the Holocene Slave River delta, Northwest Territories. M, 1987, University of Calgary.

Vandergon, Mark. Turbidite depositional environment of the Upper Cretaceous to Tertiary Canning Formation and related deposits, Arctic National Wildlife Refuge, Alaska. M, 1988, University of Alaska, Fairbanks.

Vanderhill, James B. Geology and paleontology of the Patrick Buttes, Sioux County, Nebraska, and Coshen County, Wyoming. M, 1980, University of Nebraska, Lincoln.

Vanderhill, James Burke. Lithostratigraphy, vertebrate paleontology, and magnetostratigraphy of Plio-Pleistocene sediments in the Mesilla Basin, New Mexico. D, 1986, University of Texas, Austin. 330 p.

VanderHoof, Vertress Lawrence. A study of the Miocene sirenian Desmostylus. D, 1935, University of California, Berkeley. 156 p.

Vanderhurst, William Lee. The Santa Clara Formation and orogenesis of Monte Bello Ridge, Northwest Santa Clara County, California. M, 1982, San Jose State University. 114 p.

Vanderpoel, Frank. The nitrogen compounds of cellulose; the deposit of infusorial earth near Drakesville, New Jersey. D, 1894, Columbia University, Teachers College.

Vanderpool, Harold Claude. A preliminary study of the principle groups in southwestern Arkansas, southeastern Oklahoma, and northwestern Tennessee. M, 1919, [University of Oklahoma].

Vanderpool, Harold Claude. A preliminary study of the Trinity Group in southwestern Arkansas, southeastern Oklahoma, and northern Texas. M, 1928, University of Oklahoma. 31 p.

Vanderpool, N. L. A mathematical model of landform development in the Canyonlands region of southeastern Utah. D, 1979, Stanford University. 166 p.

Vanderpool, Robert E. Areal geology of the Featherston area, Pittsburgh County, Oklahoma. M, 1959, University of Oklahoma. 56 p.

Vandersluis, George D. Solid state crystallization and polymorphism in the systems CuO-SiO₂ and CuO-SiO₂-Al₂O₃ with CuO-O-Cu₂O substitution. M, 1965, Miami University (Ohio). 111 p.

Vandervoort, Dirk Sheridan. Sedimentology, provenance, and tectonic implications of the Cretaceous Newark Canyon Formation, east-central Nevada. M, 1987, Montana State University. 145 p.

Vanderwal, K. S. Compositional and textural variations in the Vashon till and underlying drifts in the northern and central Puget Lowland, Washington. M, 1985, University of Washington. 110 p.

Vanderwilt, John W. I, The nature of polished surfaces; abrasion instead of "amorphous flow"; II, Improvements in the polishing of ores. D, 1927, Harvard University.

Vanderwood, Timothy B. Strontium isotope systematics in the Hamersley Range; theories of origin of banded iron formations and their significance to atmospheric history. M, 1977, Florida State University.

VanderWood, Timothy Blake. The development of the Abitibi greenstone belt; evidence from ion-microprobe-determines lead isotope ratios in zircon. D, 1983, University of Chicago. 172 p.

Vandike, James E. Hydrogeology of the North Fork River basin above Tecumseh, Missouri. M, 1979, South Dakota School of Mines & Technology.

Vandine, Douglas F. Geotechnical and geological engineering study of Drynoch landslide, British Columbia. M, 1974, Queen's University. 122 p.

Vandiver-Powell, Lorraine. The structure, stratigraphy and correlation of the Grande Ronde Basalts on Tygh Ridge, Wasco County, Oregon. M, 1978, University of Idaho. 57 p.

Vandivier, John Carl, III. Generation of vertical and radial seismograms in an elastic layer overlying an elastic halfspace. M, 1986, Indiana University, Bloomington. 139 p.

Vandor, H. The use of environmental isotopes to study infiltration and mixing in the unsaturated zone. M, 1977, University of Waterloo.

VanDorpe, Paul E. The lithologic comparison of lower Cherokee (Pennsylvanian) rocks from three cores in southeastern Iowa. M, 1980, Wayne State University.

VanDorston, Philip L. Environmental analysis of the Swan Peak Formation (middle Ordovician) in the Bear River range, north-central Utah and southeastern Idaho. M, 1969, Utah State University. 126 p.

Vandrevu, B. R. An analysis of gravity anomalies over Hawaiian volcanoes. M, 1970, University of Hawaii. 55 p.

Vane, Gregg Allen. A seismic reflection investigation of the Jefferson River basin, Jefferson, Madison, and Silver Bow counties, Montana. M, 1972, Indiana University, Bloomington. 39 p.

Vanek, James R. Seismic investigation of a positive gravity anomaly (Buchanan County, Iowa). M, 1968, University of Iowa. 49 p.

VanGundy, Robert D. Upper Cretaceous foraminifera from a Peedee Formation outcrop, Scuffleton, North Carolina. M, 1983, University of North Carolina, Chapel Hill. 100 p.

Vaniman, David Timothy. Studies on the Godani Granodiorite, Northern Nigeria; petrology, chemistry and mineralogy. D, 1976, University of California, Santa Cruz.

Vaninetti, Gerald Eugene. Coal stratigraphy of the John Henry Member of the Straight Cliffs Formation, Kaiparowits Plateau, Utah. M, 1979, University of Utah. 274 p.

Vanko, David Alan. Petrology of the Humboldt Lopolith, N.W. Nevada, with emphasis on marialitic scapolitization and the synthesis of marialitic scapolite. D, 1982, Northwestern University. 249 p.

Vann, Richard Pickard. Paleontology of the Upper Cretaceous rocks of Chaco Canyon, New Mexico. M, 1931, University of New Mexico. 64 p.

Vanstrum, Vincent B. The zonation and paleoecology of the Calasahatchee Formation based on Ostracoda. M, 1961, Florida State University.

VanWagner, Elmer, III. An integrated investigation of the Bowling Green Fault using multispectral reflectance, potential field, seismic, and well log data sets. M, 1988, Bowling Green State University. 171 p.

VanWormer, James D. Solid earth tides as a triggering mechanism for earthquakes. M, 1967, University of Nevada. 69 p.

Vanzant, James Harvey. The geology of the Thomas and Hubbard oil pools of Kay County, Oklahoma. M, 1926, University of Oklahoma. 13 p.

Vaos, Stephanos Pantelis. The application of scanning electron microscopy to the study of kaolin with emphasis on the kaolins of Georgia and South Carolina. D, 1980, Florida State University. 244 p.

Vaos, Stephanos Pantelis. The unit cell dimensions of some apatites in relation to their chemical composition. M, 1966, Florida State University.

Varadhi, S. N. Foundation response to soil transmitted loads. D, 1977, Illinois Institute of Technology. 217 p.

Varady, Ella H. Physiographic aspects of the Gettysburg campaign. M, 1926, Columbia University, Teachers College.

Varchol, Douglas J. A reflection seismic survey of the Vema fracture zone on the Mid-Atlantic Ridge. M, 1980, Pennsylvania State University, University Park.

Varela, Francisco Ernesto, Jr. Tourmaline in the Cananea District, Sonora, Mexico; 1 volume. M, 1972, University of California, Berkeley.

Varga, Lisa Louise. Dolomitization of the Brassfield Formation (Lower Silurian) in Adams County, Ohio. M, 1981, Western Michigan University. 74 p.

Varga, Robert Joseph. Stratigraphy and superposed deformation of a Paleozoic and Mesozoic sedimentary sequence in the Harquahala Mountains, Arizona. M, 1976, University of Arizona.

Varga, Robert Joseph. Structural and tectonic evolution of the early Paleozoic Shoo Fly Formation, northern Sierra Nevada Range, California. D, 1980, University of California, Davis. 248 p.

Vargas D., Francisco H. Geology of the Cotopaxi Inlier on the northern trend of the Sangre de Cristo Range, Fremont County, Colorado. M, 1960, Colorado School of Mines. 84 p.

Vargas, A. Correlation by trace elements of the Hudson River Shale in southeastern New York and the Martinsburg Formation in northwestern New Jersey and eastern Pennsylvania. M, 1976, Queens College (CUNY). 82 p.

Vargas, H. Rodrigo. Photogeologic map of the Jelm Mountain Quadrangle, Albany County, Wyoming, prepared from color infrared imagery. M, 1973, University of Wyoming. 76 p.

Vargas, John F. Deterministic deconvolution model for high resolution seismic data in Kansas. M, 1984, University of Kansas. 82 p.

Vargas, Jose Eusebio. Petrogenic study of the Black Hand Member of the Cuyahoga Formation (Mississippian) in central Ohio. M, 1975, Ohio University, Athens. 151 p.

Vargas-Semprun, D. On the stochastic modeling of daily streamflows. D, 1977, Colorado State University. 215 p.

Vargo, Elena Fisher. Some geologic considerations for the site selection of solid-waste disposal areas in North Georgia. M, 1982, Emory University. 115 p.

Vargo, Jan M. Structural geology of a portion of the eastern Rand Mountains, Kern and San Bernardino counties, California. M, 1972, University of Southern California.

Varlashkin, Charlotte M. Comparative petrology of intra- and extra-macrofossil sediment within Pleistocene to Cretaceous strata of the North Carolina coastal plain. M, 1986, East Carolina University. 105 p.

Varley, Christopher John. The sedimentology and diagenesis of the Cadomin Formation, Elmworth area, northwestern Alberta. M, 1982, University of Calgary. 173 p.

Varma, Madan Mohan. Seismicity of the eastern half of the United States (exclusive of New England). D, 1975, Indiana University, Bloomington. 176 p.

Varnell, Ronnie J. Geology of the Hat Top Mountain Quadrangle, Grant and Luna counties, New Mexico. M, 1976, University of Texas at El Paso.

Varner, Vicki. A cost surface model for preliminary evaluation of selected physical factors associated with residential site development in south-central Greene County, Missouri. M, 1987, Southwest Missouri State University.

Varney, Frederick Merrill and Redwine, Lowell Edwin. Development of a coring instrument for submarine geological investigations. M, 1937, University of California, Los Angeles.

Varney, Peter J. Depositional environment of the Mineral Fork (Precambrian), Wasatch Mountains, Utah. M, 1972, University of Utah. 134 p.

Varnum, Nick C. Application of geographic information systems techniques to assess natural hazards in the east-central Sierra Nevada. M, 1987, University of Nevada. 62 p.

Varrin, Robert Douglas. A pre-Cretaceous channel in the Plainsboro, N. J., area as determined by seismic-refraction measurements. M, 1957, Princeton University. 35 p.

Varrin, Robert Douglas. Model analysis of unsteady flow to multiaquifer wells. D, 1968, University of Delaware, College of Marine Studies. 139 p.

Varsek, John L. Seismic characteristics of the Seisgun source. M, 1984, University of Calgary. 143 p.

Varvaro, Gasper Gus. A geologic report on the Mamou Field, Evangeline Parish, Louisiana. M, 1949, Louisiana State University.

Varvaro, Gasper Gus. Geology of Evangeline and Saint Landry parishes. D, 1958, Louisiana State University.

Vasco, Donald Wyman. Inversion of static displacement of the Earth's surface. D, 1986, University of California, Berkeley. 115 p.

Vasconcelos, J. J. Optimization of a regional water resource quality management system. D, 1976, University of California, Berkeley. 383 p.

Vasconcelos, Jose Aluizio de see de Vasconcelos, Jose Aluizio

Vasconcelos, Paulo M. Gold geochemistry in a semi-arid weathering environment; a case study of the Fazenda Brasileiro Deposit, Bahia, Brazil. M, 1987, University of Texas, Austin. 254 p.

Vasilakos, Nikolaos Petrou. Coal desulfurization by selective chlorinolysis. D, 1981, California Institute of Technology. 160 p.

Vasilkovs, Irene N. $^{87}Sr/^{86}Sr$ and geochemical variations in the Pennsylvanian and Permian nonmarine and marine limestones of southeastern Ohio, northwestern West Virginia, and southwestern Pennsylvania. M, 1987, Miami University (Ohio).

Vaskey, Gordon Thomas. Carboniferous facies and some Pennsylvanian brachiopods, western Prince of Wales Island, southeastern Alaska. M, 1982, University of Oregon. 255 p.

Vasquez Perez, Adalberto. Economic geology of the Alamos mining district, Sonora, Mexico. M, 1975, University of Arizona.

Vasquez, Enrique Eduardo. The Oca Fault of northern South America. M, 1971, University of Tulsa. 90 p.

Vasquez, Julio C. Temperature of calcite deposition in the Pine Point lead-zinc deposits, Northwest Territories. M, 1968, Queen's University. 115 p.

Vasquez, L. The Cananea copper district, Sonora, Mexico. M, 1955, Columbia University, Teachers College.

Vasquez-Herrera, Andres R. The behavior of undrained contractive sand and its effect on seismic liquefaction flow failures of earth structures. D, 1988, Rensselaer Polytechnic Institute. 545 p.

Vassallo, Carol Frances. Geomorphic history and sediment dynamics of a dredged inlet on a developed shoreline; Townsends Inlet, New Jersey. M, 1988, Rutgers, The State University, New Brunswick. 144 p.

Vassallo, Kathryn L. Bottom reflectivity, microphysiography, and their indications for near-bottom sedimentation processes, western North Atlantic. M, 1983, SUNY at Buffalo. 58 p.

Vassaro, John Joseph. A mathematical model of Chincoteague Bay, Virginia. M, 1976, College of William and Mary.

Vassiliou, Marios S. The energy release in earthquakes, and subduction zone seismicity and stress in slabs. D, 1983, California Institute of Technology. 351 p.

Vastan, Andrew Charles. A numerical study of the tsunami response at an island. D, 1967, Texas A&M University.

Vatis, Martin D. Space triangulation adjustment. M, 1962, Ohio State University.

Vaughan, F. R. The origin and diagenesis of the Arroyo Penasco collapse breccia. M, 1978, SUNY at Stony Brook.

Vaughan, Francis Edward. Geology of San Bernardino Mountains north of San Gorgonio Pass (California). D, 1918, University of California, Berkeley.

Vaughan, Francis Edward. Geology of the San Bernardino Mountains north of San Gorgonio Pass. M, 1916, University of California, Berkeley. 71 p.

Vaughan, Hague Hingston. Prehistoric disturbance of vegetation in the area of Lake Yaxha, Peten, Guatemala. D, 1979, University of Florida. 176 p.

Vaughan, Henry. The Brainard Outlier, a southerly occurrence of the Rensselaer Plateau (New York). M, 1934, Union College. 29 p.

Vaughan, J. L., Jr. Biostratigraphic analysis of Mallory's (1959) Ulatisian benthic foraminiferal stage (Eocene of California). M, 1976, University of Southern California. 85 p.

Vaughan, M. T. Elasticity and crystal structure in aluminosilicates and pyroxenes. D, 1979, SUNY at Stony Brook. 145 p.

Vaughan, Michael J. Geology of the Lewisville West Quadrangle, Denton County, Texas. M, 1973, University of Texas, Arlington. 62 p.

Vaughan, Patrick Robison. Alluvial stratigraphy and neotectonics along the Elsinore Fault at Agua Tibia Mountain, California. M, 1987, University of Colorado. 182 p.

Vaughan, Peter James. Acoustic emissions associated with the development of creep instability in experimentally deformed dunite. M, 1973, University of California, Los Angeles. 141 p.

Vaughan, Peter James. Deformation mechanisms in the olivine and spinel phases of magnesium germanate and applications to the Earth's mantle. D, 1979, University of California, Santa Cruz. 345 p.

Vaughan, R. J. Disturbance in a stressed environment; bank vegetation of the Gulf Intracoastal Waterway at Cedar Lakes, Texas. D, 1977, Texas A&M University. 124 p.

Vaughan, R. Lee, Jr. Diagenetic effects on reservoir development in the Upper Jurassic Norphlet Formation, Mobile and Baldwin counties and offshore Alabama. M, 1985, University of Alabama. 143 p.

Vaughan, Thomas Wayland. The Eocene and lower Oligocene coral faunas of the United States, with descriptions of a few doubtfully Cretaceous species. D, 1903, Harvard University.

Vaughan, W. S. The anorthosite at Oka, Quebec. M, 1962, McGill University.

Vaughn, Danny M. Paleohydrology and geomorphology of selected reaches of the upper Wabash Valley, Indiana. D, 1984, Indiana State University. 225 p.

Vaughn, David E. W. The crystallization ranges of the Spruce Pine and Harding pegmatites. M, 1963, Pennsylvania State University, University Park. 61 p.

Vaughn, Elliott Benson. Three-dimensional velocity variation across the San Andreas fault zone near Parkfield, California. M, 1979, University of Texas at Dallas.

Vaughn, Patty Hollyfield. Mesozoic sedimentary rock features resulting from volume movements required in drape folds at corners of basement blocks; Casper Mountain area, Wyoming. M, 1976, Texas A&M University.

Vaughn, Peter P. The Permian reptile Araeoscelis restudied. D, 1954, Harvard University.

Vaughn, Rodney Lynn. Sedimentology of the Dakota Formation (Cretaceous), Uinta Mountains, northeastern Utah. M, 1973, University of Utah. 86 p.

Vaughn, William James, Jr. Geology of the west fork of the Madison River area, Montana. M, 1948, University of Michigan. 43 p.

Vaught, Richmond Murphy. Development of engineering geologic performance standards for land-use regulation in Sabine Pass, Texas. M, 1982, Texas A&M University. 134 p.

Vause, James E., Jr. Submarine geomorphic and sedimentological investigations off part of the Florida Panhandle coast. M, 1957, Florida State University.

Vaux, Walter Gregson. The motion and collection of suspended particles within streambed gravel. D, 1968, University of Minnesota, Minneapolis. 142 p.

Vavra, Charles Lee. Provenance and alteration of the Triassic Fremouw and Falla formations, central Transantarctic Mountains, Antarctica. D, 1982, Ohio State University. 193 p.

vay, Joseph C. De *see* De vay, Joseph C.

Vaz, Jesus Eduardo. Radiation damage in crystalline calcium carbonate. M, 1965, University of Kansas. 65 p.

Vaz, Jesus Eduardo. Thermoluminescence-metamictization reactions in zircon. D, 1969, George Washington University. 70 p.

Vazin, Hassan. Finite element analysis of soil compaction resulting from vibratory tillage. D, 1982, University of Tennessee, Knoxville. 83 p.

Vaziri-Zanjani, Hans Hamid. Nonlinear temperature and consolidation analysis of gassy soils. D, 1986, University of British Columbia.

Vazzana, Michael Eugene. Stratigraphy, sedimentary petrology and basin evolution of the Abiquiu Formation, north-central New Mexico. M, 1980, University of New Mexico. 115 p.

Veal, Harry Kaufman. Subsurface study of South Forrest County and adjacent area, Mississippi. M, 1956, University of Oklahoma. 73 p.

Veatch, Maurice D. Post-Permian geology and ground water resources of Jefferson County, Nebraska. M, 1963, Kansas State University.

Veatch, Maurice Deyo. Ground-water occurrence, movement and hydrochemistry within a complex stratigraphic framework, Jefferson County, Nebraska. D, 1969, Stanford University. 288 p.

Veblen, David Rodli. Triple- and mixed-chain biopyriboles from Chester, Vermont. D, 1976, Harvard University.

Vecchioli, John. Pre-Cambrian rocks in the Jenny Jump Mountain area. M, 1957, Rutgers, The State University, New Brunswick. 41 p.

Vedagiri, Velpari. Strength and durability of basalt fiber and basalt-fiber cement composites. D, 1987, Washington State University. 236 p.

Vedder, John G. The Eocene and Paleocene of the Northwest Santa Ana Mountains (Southern California). M, 1950, Pomona College.

Vedder, Laurel Kathleen. Stratigraphic relationship between the Late Jurassic Canelo Hills Volcanics and the Glance Conglomerate, southeastern Arizona. M, 1984, University of Arizona. 129 p.

Vedros, Stephen G. The Red Oak Sandstone; a submarine fan deposit. M, 1976, University of Tulsa. 61 p.

Veeh, Hans Herbert. A petrographic study of the cementation in the Lyons Sandstone (Permian, Colorado). M, 1959, University of Colorado.

Veeh, Hans Herbert. Thorium-230/uranium-238 and uranium-234/uranium-238 ages of elevated Pleistocene coral reefs and their geological implications. D, 1965, University of California, San Diego. 103 p.

Veen, Cynthia A. Gravity anomalies and their structural implications for the southern Oregon Cascade Mountains and adjoining Basin and Range Province. M, 1982, Oregon State University. 86 p.

Veen, Cynthia Ann. A geophysical definition of a Klamath Falls graben fault. M, 1979, Portland State University. 74 p.

Veesaert, Marlin Joseph. Geology of parts of Creve Coeur and Chesterfield Quadrangle. M, 1952, Washington University.

Vega, Luis Alfonso. The alteration and mineralization of the Alum Wash Prospect, Mohave County, Arizona. M, 1984, University of Arizona. 57 p.

Vehrs, Robert Alan. On an occurrence of "quartzite" in the southern complex near Palmer, Marquette County, Michigan. M, 1959, Michigan State University. 76 p.

Vehrs, T. I. Tectonic, petrologic and stratigraphic analysis of the Bigelow Range, Stratton and Little Bigelow Mountain quadrangles, northwestern Maine. D, 1975, Syracuse University. 243 p.

Veilleux, B. M. Structure and stratigraphy of the Sherbrooke Series (Upper Cambrian) in the Memphremagog area (Quebec). M, 1949, McGill University.

Veinus, Julia. The Benbolt crinoids (Ordovician) of southwest Virginia and northeast Tennessee. M, 1969, Syracuse University.

Veis, George. Geodetic application of observations of the Moon, artificial satellites, and rockets. D, 1958, Ohio State University.

Veitch, John D. The transportation of ore metals by amino acids. D, 1972, Lehigh University. 171 p.

Veith, Karl F. A geophysical study of a portion of the Midcontinent gravity high. M, 1966, University of Minnesota, Minneapolis. 56 p.

Veith, Karl Fredrick. The relationship of island arc seismicity to plate tectonics (parts I and II). D, 1974, Southern Methodist University. 29 p.

Velbel, Danita Brandt *see* Brandt Velbel, Danita

Velbel, Michael Anthony. Mineral transformations during rock weathering, and geochemical mass-balances in forested watersheds of the Southern Appalachians. D, 1984, Yale University. 189 p.

Velde, Bruce Dietrich. The significance of the polytypic form of illite in sedimentary Paleozoic rocks. D, 1962, University of Montana. 89 p.

Velden, Trude Vander. Mineralogical and petrological investigation of dark rims in low petrologic types of ordinary and carbonaceous chondrites. M, 1979, University of Houston.

Veldhuis, Jerry H. A geophysical and geological analysis of the Salt Flat Basin of Trans-Pecos Texas and southern New Mexico. M, 1980, University of Texas at El Paso.

Veldhuyzen, Hendrik. Sources and transport of late Quaternary sediments, Karlsefni Trough, Labrador Shelf. M, 1981, McGill University. 311 p.

Velho, Luis Rousset. The diffusion and solubility of oxygen in platinum and platinum-nickel alloys. D, 1971, Stanford University. 182 p.

Velinsky, David Jay. The geochemistry of selenium and sulfur in a coastal salt marsh. D, 1987, Old Dominion University. 211 p.

Vella, Alfred J. Ether-derived alkanes and their role as bacterial biomarkers in sediments. D, 1984, Colorado School of Mines. 375 p.

Vellutini, David. Organic maturation and source rock potential of Mesozoic and Tertiary strata, Queen Charlotte Islands, British Columbia. M, 1988, University of British Columbia. 248 p.

Vemuri, Ramesam. The chemistry and mineralogy of >2μ size fraction of non-marine cyclothems (Dunkard Group-Upper Pennsylvanian–Permian in Ohio, U.S.A.). D, 1968, McMaster University. 139 p.

Ven, Paulus Hendrikus Van der *see* Van der Ven, Paulus Hendrikus

Venchiarutti, Daniel A. The origin of mafic inclusions in topaz rhyolites from Spor Mountain, Utah; evidence for magma mixing. M, 1987, University of Iowa. 83 p.

Vendetti, Michael J. Sedimentology of the Rocky Ridge Sandstone (Upper Cretaceous), Cheyenne Basin, Colorado. M, 1985, Colorado State University. 165 p.

Venditti, Anthony R. Petrochemistry of Precambrian rocks in southeastern Oneida County, Wisconsin. M, 1973, University of Wisconsin-Milwaukee.

Vendl, Lawrence J. Relationship between water discharge and the chemical constituents of Illinois streams. M, 1978, Northern Illinois University. 177 p.

Vendl, M. A. Sedimentation in glacier-fed Peyto Lake, Alberta. M, 1978, University of Illinois, Chicago.

Venecia, Kathryn Elizabeth Wolberg-de *see* Wolberg-de Venecia, Kathryn Elizabeth

Venkatakrishnan, Ramesh. A remote sensing based study of the structural evolution of the Coeur d'Alene mining district, Idaho. D, 1982, University of Idaho. 175 p.

Venkataraman, Sundaram. Clay minerals and discontinuities in nine southern Ontario soil profiles. D, 1971, University of Guelph.

Venkatesan, Thandalalai R. Systematics of potassium-argon and argon40-argon39 dating of the terrestrial sample 132022 from the Fiskenaesset Complex in West Greenland, lunar soils 75081, 71501, 12033 and a lunar rock sample 67915. D, 1976, University of Minnesota, Minneapolis. 119 p.

Venkatesh, Eswarahalli S. Application of petroleum data system to oil field evaluations. M, 1980, University of Oklahoma. 71 p.

Venkateswaran, G. P. Phase relations in the system nepheline (NaAlSiO₄)-leucite(KAlSi₂O₆)-akermanite(Ca₂MgSi₂- O₇) and their petrological significance to potash and soda-rich undersaturated alkaline lavas. M, 1972, University of Western Ontario. 113 p.

Venkateswaran, Ravi T. Upper Glen Rose depositional environment (Lower Cretaceous), Travis County, Texas. M, 1977, Wayne State University. 118 p.

Venkitasubramanyan, Calicut S. Large scale superimposed folds in Precambrian of the Actinolite-Kaladar area, southeastern Ontario. D, 1969, Queen's University. 224 p.

Vennum, Walter Robert. Petrology of the Castle Crags Pluton, Shasta and Siskiyou counties, California. D, 1971, Stanford University. 160 p.

Venour, E. R. The Swan River Formation (Cretaceous) in Manitoba. M, 1958, University of Manitoba.

Venso, Nolan J. Paleoenvironmental interpretation of the lower part of the Cool Creek Formation by means of stromatolite morphotypes and lithofacies. M, 1975, Northeast Louisiana University.

Venter, Ronald H. McConnell Thrust and associated structures at Mount Yamnuska, Alberta, Canada. M, 1973, University of Calgary. 118 p.

Venters, Edwards. Determination of the non-durable constituents of Indiana gravels. M, 1951, Purdue University.

Venticenque, Salvatore M. Paleoenvironment of the Marshalltown Formation (Upper Cretaceous) in New Jersey. M, 1972, Brooklyn College (CUNY).

Vento, Frank. The geology and geomorphology of the Rench Site, Peoria County, Illinois. M, 1982, University of Pittsburgh.

Vento, Frank John. The geology and geoarchaeology of the Bay Springs rockshelters, Tishomingo County,

Mississippi. D, 1986, University of Pittsburgh. 485 p.

Ventress, William Pynchon Stewart. Stratigraphy and megafossils of the Frisco Formation. M, 1958, University of Oklahoma. 144 p.

Venturoli, Karen A. An evolution of the ground-water resources of Lucas County, Ohio. M, 1978, University of Toledo. 133 p.

Venys, James Joseph. Creek County, Oklahoma; a faulted structure of the Viola Limestone. M, 1981, Wright State University. 94 p.

Venzke, Carl Peter. The effects of suburban development on the ground water systems of Cedarbury and Grafton townships, Wisconsin. M, 1974, University of Wisconsin-Milwaukee.

Ver Steeg, Carl. Drainage changes in northeastern Washington. M, 1926, University of Chicago. 116 p.

Ver Steeg, David James. Electric analog model of the regolith aquifer supplying Ames, Iowa. M, 1968, Iowa State University of Science and Technology.

Vera, Elpidio de la Cruz. The relation between composition and physical properties of some Philippine clays. M, 1953, University of Illinois, Urbana.

Vera, Ramon H. Geochronology of some Precambrian rocks of the Blue Ridge area, Fremont County, Colorado. M, 1973, University of Kansas. 45 p.

Verastegui-Mackee, Pedro. The pelecypod genus Venericardia in the Paleocene and Eocene of western North America. M, 1953, Stanford University. 112 p.

Verbeek, Earl Raymond. Structural evolution of the Somport area, West-central Pyrenees, France and Spain. D, 1975, Pennsylvania State University, University Park. 197 p.

Verbeek, Karen Jane Wenrich. The mechanism of emplacement of the Marble Mountain (White Horse Hills) Laccolith (Tertiary-Pliocene, Flagstaff, Arizona). M, 1971, Pennsylvania State University, University Park. 112 p.

Verbeek, Karen Jane Wenrich. Trace and major element chemistry and the petrogenesis of lavas from the upper portion of San Francisco Mountain, Arizona. D, 1975, Pennsylvania State University, University Park. 209 p.

Vercoutere, Thomas. Sedimentation across the oxygen minimum zone on the continental slope offshore central California. M, 1984, San Jose State University. 129 p.

Verdejo, Cecilia. Sediment types and sources of heavy minerals, Half Moon Bay marine terrace. M, 1963, Stanford University.

Vere, Victor Kurt. The biostratigraphy, evolution and paleoautecology of Cryptolithus (Trilobita) in New York State and western Vermont. D, 1972, Syracuse University. 136 p.

Verg, Philip E. Van de *see* Van de Verg, Philip E.

Verge, M. J. A three-dimensional saturated-unsaturated groundwater flow model for practical applications. D, 1975, University of Waterloo.

Vergo, Norma. Wallrock alteration at the Bulldog Mountain Mine, Creede mining district, Colorado. M, 1986, University of Illinois, Urbana. 88 p.

Vergos, Spiros. Seismic and electrical definition of sand-gravel deposits beneath clay in Essex County, Ontario. M, 1979, University of Windsor. 171 p.

VerHoeve, Mark W. The petrology and reservoir character of the Entrada Sandstone (Jurassic) Durango, Colorado. M, 1982, University of Texas, Austin.

Verhoeven, Cornelius Simon. Possibility of determining restricted reef conditions from the areal distribution of the widely distributed contemporaneous detrital and lagunal facies. M, 1948, Michigan State University. 41 p.

Verhoogen, Jean. Geology of Mount Saint Helens, Washington. D, 1936, Stanford University. 115 p.

Verhulst, Albert T. The petrology of three sandstone bodies in the Kansas City group (Upper Pennsylvanian) of northeastern Kansas and northwestern Missouri. M, 1970, Wichita State University. 117 p.

Verhulst, Galen G. Core analyses for hydrocarbon production potential determination in Jackson and Cass counties, Missouri. M, 1986, University of Missouri, Kansas City. 242 p.

Verish, Nicholas Paul. Reservoir trends, depositional environments, and petroleum geology of "Cherokee" sandstones in Tll-13N, R4-5E, central Oklahoma. M, 1978, Oklahoma State University. 69 p.

Verkouteren, R. Michael. Trace element characterization of forsterite chondrites and meteorites of similar redox state. D, 1984, Purdue University. 189 p.

Verleun, Leo Johannes. Base metal supply potential in northern and southern Canada; a comparative economic study. M, 1984, Queen's University. 198 p.

Verly, Georges. Etude géostatistique d'un gisement d'uranium de type roll front du Wyoming. M, 1980, Ecole Polytechnique. 209 p.

Verm, Richard Wayne. Interactive image processing of synthetic seismic wavefronts. D, 1983, University of Houston. 134 p.

Verma, Ambika Prasad. Gravity anomalies and basement elevations in the midcontinent. M, 1971, Michigan State University. 95 p.

Verma, Harish Mitter. Late Jurassic ammonites and stratigraphy of Sierra Catorce, San Luis Potosi, Mexico. D, 1972, McMaster University. 259 p.

Verma, Harish Mitter. Upper Triassic Eumorphotis and Meleagrinella (Bivalvia) from British Columbia. M, 1968, McMaster University. 131 p.

Verma, Pramod Kumar. Contributions to the geology and petrology of Nahant and Weymouth, Massachusetts. D, 1973, Harvard University.

Verma, Raj Kumar. Elasticity of several high density crystals. D, 1960, Harvard University.

Verma, Rameshwar Dayal. Physical analysis of the outflow from an unconfined aquifer. D, 1969, Cornell University. 238 p.

Vermeulen, Mark V. Data processing and interpretation of seismic refraction data using the generalized reciprocal method. M, 1988, Wright State University. 208 p.

Vermillion, Peter A. Environment of deposition of Marginulina Zone sands, northern Vermilion Parish, Louisiana. M, 1987, University of Southwestern Louisiana. 67 p.

Verner, Frederick Carr. A study of selected gravity and magnetic anomalies and their relationship to seismicity in the eastern Midcontinent. M, 1985, Purdue University. 92 p.

Vernon, Gail Franklin, Jr. Depositional environment and paleoecologic setting of a Cretaceous oyster biostrome. M, 1973, Texas Christian University. 54 p.

Vernon, James Hayes. Three-dimensional finite element modeling for geologic implications in forced folding. D, 1987, University of Oklahoma. 293 p.

Vernon, James Wesley. Geology of the Douglas Mine, Idaho. M, 1953, University of California, Berkeley. 78 p.

Vernon, James Wesley. Shelf sediment transport system (along the Southern California coast). D, 1966, University of Southern California. 142 p.

Vernon, Peter David. Tills of the Lethbridge area, Alberta; their stratigraphy, fabric, and composition. M, 1962, Carleton University. 137 p.

Vernon, Robert Orion. Geology of the McCalla area, Alabama. M, 1937, [University of Iowa].

Vernon, Robert Orion. The geology of Holmes and Washington counties, Florida. D, 1941, Louisiana State University.

Vernon, Roger Clay. Geology of the Crozet-Pasture Fence Mountain area, Albemarle County, Virginia. M, 1952, University of Virginia. 99 p.

Vernon-Chamberlain, Valerie Elaine. The geochemistry and geochronology of the Malton Gneiss complex, British Columbia. D, 1983, University of Alberta. 212 p.

Vernour, E. R. The Swan River Formation in Manitoba. M, 1957, University of Manitoba.

Verpaelst, Pierre. Géochimie et géochronologie des roches granitiques et paragnéissiques, région de la rivière Eastmain inférieure, Province du Lac Supérieur. M, 1977, Universite de Montreal.

Verplanck, Emily Pierce. Temporal variations in volume and geochemistry of volcanism in the western Cascades, Oregon. M, 1985, Oregon State University. 115 p.

Verplanck, Philip L. A field and geochemical study of the boundary between the Nanga Parbat-Haramosh Massif and the Ladakh Arc terrane, northern Pakistan. M, 1987, Oregon State University. 136 p.

VerPlanck, William E., Jr. Geology of a gypsum deposit in the Little Maria Mountains, Riverside County, California. M, 1950, Stanford University.

VerPloeg, Alan James. Microscopic analysis of modes of deformation in experimentally deformed adirondack anorthosite. M, 1973, University of Iowa.

Verrall, Peter. Geology of the Horseshoe Hills area, Montana. D, 1955, Princeton University. 371 p.

Verrastro, Robert T. Structure, stratigraphy, and hydrodynamics of northwestern Vermilion Parish, Louisiana. M, 1982, University of Southwestern Louisiana. 148 p.

Verrillo, Dan E. Geology and petrography of the Tertiary volcanic rocks in the Southwest Eagle Mountains, Hudspeth County, Texas. M, 1979, University of Texas at El Paso.

Verross, Victoria Ann. Supratidal and intertidal shell deposits in a back-barrier environment, Wassaw Sound, Chatham County, Georgia. M, 1980, University of Texas, Austin.

Verry, Elon Sanford. Water quality dynamics in shallow water impounds of north central Minnesota. D, 1983, Colorado State University. 166 p.

Verschoor, Karin van Romondt. Paleobotany of the Tertiary McAbee Beds (middle Eocene), McAbee, British Columbia. M, 1974, University of Calgary. 127 p.

Verseput, Timothy Dean. Structure of the South Mountain State Park area, Burke County, North Carolina. M, 1980, Southern Illinois University, Carbondale. 62 p.

Versfelt, Joseph W. Relationships between Precambrian basement structures and the Cenozoic rift architecture of the Malawi (Nyasa) rift zone, East Africa. M, 1988, Duke University. 68 p.

Versfelt, Porter LaRoy, Jr. Richmond sections in Ripley County, Indiana. M, 1953, Miami University (Ohio). 48 p.

Versic, Ronald James. The role of molecular clustering in the growth of crystals and the theory of liquids. D, 1969, Ohio State University.

Vertiz, Salvador Ortiz *see* Ortiz Vertiz, Salvador

Vertucci, Frank Anthony. The reflectance and fluorescence properties of Adirondack mountain region lakes applied to the remote sensing of lake chemistry. D, 1988, Cornell University. 389 p.

Verville, George Julius. Pennsylvanian and Permian stratigraphy of Elk County, Kansas. D, 1952, University of Wisconsin-Madison.

Vervoort, Jeffrey D. Petrology and geochemistry of the Archean rocks of the Jap Lake area, N.E. Minnesota. M, 1987, University of Minnesota, Duluth. 193 p.

Verwiebe, Walter August. The Devono-Carboniferous boundary rocks of Ohio, Pennsylvania and New York. D, 1918, Cornell University.

Vesely, Leon Robert. The pre-Cambrian geology of the upper Squaw Creek area, Black Hills, South Dakota. M, 1932, University of Iowa. 57 p.

Veseth, Michael K. Paleomagnetics of mid-Tertiary and younger volcanic rocks of the Kofa Mountains, southwestern Arizona. M, 1985, San Diego State University. 168 p.

Vesey, Brian K. Stratigraphic and sedimentological history of the Oligocene upper Frio Camerina embayment area in southwestern Louisiana. M, 1984, University of Southwestern Louisiana.

Vesey, Jamsie Roberts. A petrologic study of three cores from the Rodessa Formation, Bossier Parish, Louisiana. M, 1987, University of Southwestern Louisiana. 109 p.

Vesien, Alphonse C. Van *see* Van Vesien, Alphonse C.

Veska, Eric. A hypothesis for the geochemical anomaly in the North Creek watershed. M, 1978, Brock University. 152 p.

Veska, Eric. Origin and subsurface migration of radio-nuclides from waste rock at an abandoned uranium mine near Bancroft, Ontario. D, 1983, University of Waterloo. 386 p.

Vespucci, Paul Daniel. Petrology and geochemistry of the late Cenozoic volcanic rocks of the Dominican Republic. D, 1988, George Washington University. 303 p.

Vespucci, Paul Daniel. The depositional history of Mosquito Lagoon. M, 1974, University of Florida. 142 p.

Vessal, Ali. Application of geophysical techniques to coal mine planning. M, 1981, Southern Illinois University, Carbondale. 83 p.

Vessell, Richard. Hydrology, morphology, and sedimentology of the Rio Guacalate, volcanic highlands, Guatemala. M, 1977, University of Missouri, Columbia.

Vessell, Richard K. Recent and ancient volcaniclastic sedimentation on an active continental margin. D, 1979, Texas Tech University. 125 p.

Vest, Ernest Louis, Jr. Paleontology and stratigraphy of the Ordovician limestones in Chattanooga Valley, Georgia. M, 1952, Emory University. 127 p.

Vest, Harry Arthur. Structure of Sierra del Porvenir, Chihuahua, Mexico. M, 1959, University of Texas, Austin.

Vest, Jimmy Thomas. Morrowan strata of the Greers Ferry Reservoir area. M, 1962, University of Arkansas, Fayetteville.

Vest, William C. Geology of the Clinch River area, Union, Claiborne and Grainger counties, Tennessee. M, 1963, University of Tennessee, Knoxville. 39 p.

Vestal, Franklin E. Remnants of ancient peneplains in and adjacent to the Rocky Mountain System of the United States. M, 1921, University of Chicago. 86 p.

Vestal, Jack Herring. The Smackover limestone formation of southern Arkansas. M, 1948, University of Oklahoma. 52 p.

Vetorino, Robert Morris. A contribution to the drainage problem in northwestern Boone County, Kentucky. M, 1954, University of Cincinnati. 31 p.

Vetter, James R., Jr. Rotation of remanence through tectonic fabric development; a paleomagnetic and magnetic susceptibility anisotropy study of the Waynesboro Formation, south central Pennsylvania. M, 1987, Lehigh University. 136 p.

Vetter, Mark. Hartford and Deerfield basins framework mineralogies; independent evidence for provenance, current indicators and tectonic history. M, 1988, Wright State University. 71 p.

Vetter, Scott Keith. Geochemistry of basaltic rocks from the southeastern Brazilian margin; evidence for enriched and depleted mantle sources during continental rifting. M, 1984, North Carolina State University. 126 p.

Vezzoli, Gary C. A theoretical comparison of surface wave dispersion in a low velocity saturated layer and in a high velocity floating ice sheet. M, 1966, [Boston University].

Vhay, John S. Geology of a part of the Beartooth Mountain Front near Nye, Montana. D, 1934, Princeton University. 112 p.

Via, Edwin K. Geology of the Wolf Creek-Piney Ridge area. M, 1962, Virginia Polytechnic Institute and State University.

Via, William Noel. Geochemistry of lunar soil formation. M, 1976, University of Tennessee, Knoxville. 77 p.

Vian, Richard W. Investigations in the bismuth carbonate group. M, 1959, Miami University (Ohio). 19 p.

Vian, Richard Wright. Geology of the Devils Hole area, Fremont County, Colorado. D, 1965, University of Michigan. 189 p.

Viana, R. T. Incorporation of power-law fluid effects in the presence of leak-off in the Perkins and Kern model of vertical fracture propagation. M, 1981, Colorado School of Mines. 58 p.

Viani, Chris William. Stratigraphy and mineralogy of tills in Knox County, Ohio. M, 1986, University of Akron. 98 p.

Viard, James Philip. Description of grain-size distribution curves from the Platte River system. M, 1977, Texas Christian University. 23 p.

Viau, Christian Alain. Diagenesis, sedimentology and structure of the Swan Hills Formation, Swan Hills Field, central Alberta, Canada. D, 1986, University of Calgary. 574 p.

Vibetti, Ndoba Joseph. A study of deep fluid circulation in the Troodos Ophiolite, Cyprus. D, 1987, University of Western Ontario. 239 p.

Vicars, Robert Glenn. The sedimentology of the upper Harrison Formation in the Agate Fossil Beds National Monument area. M, 1979, Texas Christian University. 44 p.

Vice, Daniel Hoy. The geology and petrography of the Babocomari Ranch area, Santa Cruz-Cochise counties, Arizona. M, 1974, Arizona State University. 152 p.

Vice, Mari Ann. Depositional environments and diagenesis in an interval of the Mission Canyon Limestone (Madison Group, Mississippian), south-central Montana and northern Wyoming. M, 1988, Southern Illinois University, Carbondale. 149 p.

Vicencio, Raul. Models for the morphology and morphogenesis of the ammonoid shell. D, 1973, McMaster University. 116 p.

Vicente Vidal Lorandi, V. M. Studies of marine hydrothermal activity in a coastal environment-Punta Banda, Baja California Norte, Mexico- and its geochemical implications for modelling marine hydrothermal processes in the ocean. D, 1978, University of California, San Diego. 246 p.

Vicente, Ernesto Edgardo. Pore water pressure increase in loose saturated sand at level sites during three directional earthquake loading. D, 1983, Rensselaer Polytechnic Institute. 272 p.

Vicente, Napoleon Otero San *see* San Vicente, Napoleon Otero

Vicenzi, Edward. The geology, petrology, and petrogenesis of Isla Marchena, Galapagos Archipelago. M, 1985, University of Oregon. 96 p.

Vichit, Pongsak. Origin of corundum in basalt. M, 1975, New Mexico Institute of Mining and Technology. 140 p.

Vick, Alphonso Roscoe. Some pollen profiles from the coastal plain of North Carolina. D, 1961, Syracuse University. 99 p.

Vick, William Edward. Sedimentary structures of the Ogallala of Lubbock County, Texas. M, 1950, Texas Tech University. 36 p.

Vickers, Michael A. Paleontology of the Blufftown Formation (Upper Cretaceous) Chattahoochee River region, Georgia-Alabama. M, 1967, Florida State University.

Vickers, William Ward. North Palisade Glacier, Sierra Nevada; a theory on the extent of sub-surface ice. M, 1956, University of California, Los Angeles. 93 p.

Vickery, Ann Marie. Physical constraints on the origin of shergottites, nakhlites, and chassignites. D, 1984, SUNY at Stony Brook. 249 p.

Vickery, Frederick P. Structural dynamics of the Livermore region (California). D, 1924, Stanford University. 70 p.

Vickery, Frederick P. The physiography of the Santa Cruz Quadrangle (California). M, 1919, Stanford University. 109 p.

Vickery, Ward Rollin. Eastborough oil field, Sedgwick County, Kansas. M, 1932, Wichita State University. 60 p.

Vickrey, Earl Wayne. The topography and geology of the Griffey Creek valley. M, 1913, Indiana University, Bloomington.

Victor, Iris. Wolframite deposit in the Burnt Hill Brook area, New Brunswick, Canada. M, 1956, Columbia University, Teachers College.

Victor, Linda. Structures of the continental margin of Central America from northern Nicaragua to northern Panama. M, 1976, Oregon State University. 76 p.

Victor, R. The taxonomy and distribution of freshwater ostracods (Crustacea; Ostracoda) of Malaysia, Indonesia and the Philippines. D, 1979, University of Waterloo.

Vidal Lorandi, Francisco Vicente. Part I, The metabolism of arsenic in marine bacteria and yeast; Part II, Stable isotopes of helium, nitrogen and carbon in the geothermal gases of the subaerial and submarine hydrothermal systems of the Ensenada Quadrangle in Baja California Norte, Mexico; Part III, Life at high temperatures in the sea; thermophilic marine bacteria isolated from submarine hot springs, coastal seawater and heat exchangers of seawater cooled power plants. D, 1980, University of California, San Diego. 125 p.

Vidal Lorandi, V. M. Vicente *see* Vicente Vidal Lorandi, V. M.

Vidal, Francisco Suarez. Jurassic stratigraphy, depositional environment and paleogeography on the east flank of the Tamaulipas Paleopeninsula. M, 1984, San Diego State University. 138 p.

Vidal, Jose Rabasso. Geology of an upper Paleozoic sequence in north-central Canelo Hills, Santa Cruz County, Arizona. M, 1971, University of Arizona.

Vidale, John Emilio. Application of two-dimensional finite-difference wave simulation to earthquakes, earth structure, and seismic hazard. D, 1987, California Institute of Technology. 158 p.

Vidale, Rosemary Jacobson. Calc-silicate bands and metasomatism in a chemical gradient. D, 1968, Yale University. 105 p.

Videgar, Frank D. A geochemical study of the Tertiary volcanic rocks of northwestern Washington. M, 1975, Western Washington University. 83 p.

Videlock, Shari Lynn. The stratigraphy and sedimentation of Cluett Key, Florida Bay. M, 1983, University of Connecticut. 172 p.

Videtich, Patricia Ellen. Origin, marine diagenesis, and early fresh-water diagenesis of limestones and dolomites (Tertiary-Recent); stable isotopic, electron microprobe, and petrographic studies. D, 1982, Brown University. 297 p.

Vidrine, Dana Marie. Geochemistry and petrology of the Cold Springs Breccia Formation, Wichita Mountains, Oklahoma. M, 1983, University of Missouri, Rolla. 103 p.

Vieaux, Don George. A foraminiferal micro fauna of the Denton Formation in the vicinity of Denison, Grayson County, Texas. M, 1939, University of Oklahoma. 89 p.

Viecelli, James Anthony. Generation of Rayleigh waves by underground nuclear explosions; an examination of the effect of spall impact and site configuration. D, 1973, University of California, Davis. 72 p.

Vieira, Mario E. C. Time-series study of sanding in Ventura Harbor, California. M, 1974, Naval Postgraduate School.

Vieira, Sidney Rosa. Geostatistical analyses of some agronomical observations. D, 1981, University of California, Davis. 261 p.

Viekman, Bruce. Secondary circulation in the bottom boundary layer over sedimentary furrows. M, 1988, University of Rhode Island.

Viele, George Washington. The geology of the Flat Creek area, Lewis and Clark County, Montana. D, 1966, University of Utah. 213 p.

Vierbuchen, Richard C., Jr. The tectonics of northeastern Venezuela and the southeastern Caribbean Sea. D, 1979, Princeton University. 175 p.

Vierma, Luis F. Correlation of crude oils with source rocks in a portion of the Maracaibo Basin, Venezuela. M, 1984, Indiana University, Bloomington. 159 p.

Vietti, Barbara Tomes. The geohydrology of the Black Butte and Canyon Creek areas, Bighorn Mountains, Wyoming. M, 1977, University of Wyoming. 55 p.

Vietti, John S. Structural geology of the Ryckman Creek Anticline area, Lincoln and Uinta counties, Wyoming. M, 1974, University of Wyoming. 106 p.

Vig, Reuben J. Geology of the unconsolidated deposits of Lake County, Indiana. M, 1963, University of North Dakota. 68 p.

Viglino, Janet Atkinson. Hydrogen isotope exchange between aluminous chlorite and water. M, 1985, Southern Methodist University. 97 p.

Vigoren, LaVerne and Arcilise, Casper. Uraniferous siltstone of the Lonesome Pete No. 2 Claim, South Cave Hills, Harding County, South Dakota. M, 1957, South Dakota School of Mines & Technology.

Vigrass, Laurence William. Geology of the Suplee area, Crook, Grant, and Harney counties, Oregon. D, 1961, Stanford University. 317 p.

Vigrass, Laurence William. Jurassic stratigraphy of southern Saskatchewan. M, 1952, University of Saskatchewan. 74 p.

Vikre, P. G. Geology and silver mineralization of the Rochester District, Pershing County, Nevada. D, 1978, Stanford University. 471 p.

Viksne, Andris. Seismic velocity studies of synthetic and natural rock cores. M, 1957, University of Utah. 35 p.

Vilas, Faith. Spectral reflectance curves of the Planet Mercury. M, 1975, Massachusetts Institute of Technology. 70 p.

Viletto, John, Jr. Channel morphology in northwestern Pennsylvania. D, 1968, Pennsylvania State University, University Park. 246 p.

Vilinskas, Peter. Diffusion of sodium ions in sodalite. D, 1964, University of Connecticut. 105 p.

Vilks, G. Quantitative analysis of foraminifera in the Bras d'Or Lakes. M, 1966, Dalhousie University.

Vilks, Peter. Copper sorption on kaolinite. D, 1985, McMaster University. 307 p.

Vilks, Peter. Li distribution between chlorite and albite in a common vapor phase. M, 1981, McMaster University. 95 p.

Villaescusa Cordova, Ernesto. Slope stability analysis at La Caridad Mine, Nacozari, Sonora, Mexico. M, 1987, Colorado School of Mines. 138 p.

Villalobos, Hector. Engineering geology reconnaissance of five damsites and reservoir study areas, Similkameen River, Okanogan County, Washington. M, 1982, San Jose State University. 136 p.

Villamayor, Faustino Paysan. A model for determining soil contrast and its application to five different soil orders. D, 1987, Oregon State University. 344 p.

Villamil, R. J., Jr. Studies on the moisture regime of the Vergennes soil. D, 1978, University of Vermont. 168 p.

Villanueva, Enrique Osmar Jacome *see* Jacome Villanueva, Enrique Osmar

Villar, James Walter. Metamorphic petrology of the Animikie Series in the Republic Trough area, Marquette County, Michigan. D, 1965, Michigan State University. 185 p.

Villar, James Walter. Petrology and petrofabrics at the Newton Falls Pit, Star Lake, New York. M, 1956, Michigan State University. 70 p.

Villard, D. J. Factors affecting the distribution of uranium in stream sediments, Newcastle area, New Brunswick. M, 1972, University of New Brunswick.

Villarroel, Patricio. Mineral assemblages and their stabilities in the Montana-Argentine vein, Telluride (San Miguel County), Colorado. M, 1970, Michigan Technological University. 58 p.

Villars, Paul Emile. A preliminary investigation of the use of radio waves in geophysical reconnaissance. M, 1952, St. Louis University.

Villas, Cathleen Anna. The Holocene evolution of the Acheloos River delta, northwestern Greece; associated environments, geomorphology and microfossils. M, 1984, University of Delaware. 222 p.

Villas, Raimundo Netuno Nobre. Fracture analysis, hydrodynamic properties and mineral abundance in the altered igneous wall rocks of the Mayflower Mine, Park City District, Utah. D, 1975, University of Utah. 271 p.

Villaume, James F. Geochemistry of some Pre-Cambrian ultra mafic rocks. M, 1973, Pennsylvania State University, University Park.

Villeneuve, Michael E. Pb isotopes as evidence for an early Proterozoic source for garnet granites of Wopmay Orogen, N.W.T. M, 1988, Washington University. 136 p.

Villet, Willem Christiaan Bouwer. Acoustic emissions during the static penetration of soils. D, 1981, University of California, Berkeley. 411 p.

Villien, Alain. Central Utah deformation belt. D, 1984, University of Colorado. 362 p.

Villiers, Johan Pieter Roos de *see* de Villiers, Johan Pieter Roos

Villiers, Johanne De *see* De Villiers, Johanne

Vincelette, Richard Roy. Structural geology of the Mt. Stirling Quadrangle, Nevada, and related scale-model experiments. D, 1964, Stanford University. 141 p.

Vincent, Bruce Douglas. Markov analysis of the Blairmore Group (Lower Cretaceous), Alberta Foothills. M, 1977, University of Alberta. 282 p.

Vincent, Charles. Storms and erosion, Cape Hatteras, North Carolina. M, 1971, University of Virginia. 49 p.

Vincent, Charles Linwood. Quantification of shoreline meandering. D, 1973, University of Virginia. 113 p.

Vincent, Douglas Anderson. Evaporites and associated clastics within the Sumner Group (Lower Permian) of central and western Kansas. M, 1965, University of Kansas. 63 p.

Vincent, Edith S. Oceanography and late Quaternary planktonic foraminifera, southwestern Indian Ocean. D, 1972, University of Southern California. 367 p.

Vincent, Frank S. A paleoecological study of the lower Miocene in northern St. Mary Parish, Louisiana, based on foraminifera from wells on the north flank of the Jeanerette gas field. M, 1975, Louisiana State University.

Vincent, Harold R. Metamorphosed mafic dykes within the Cranberry Gneiss and their relation to amphiobolite of the Ashe Formation, North Carolina-Tennessee. M, 1981, Eastern Kentucky University. 70 p.

Vincent, J. S. Melting phenomena in rocks of the anorthosite suite. M, 1963, McGill University.

Vincent, Jerry William. Lithofacies and biofacies of the Haney limestone (Mississippian), Illinois, Indiana, and Kentucky. D, 1971, Texas A&M University. 168 p.

Vincent, Judy Ann. Petrology and geochemistry of syenites of Sawtooth Mountain, Davis Mountains, Jeff Davis County, Texas. M, 1988, Texas Tech University. 98 p.

Vincent, Kenneth C. The application of radioactive tracers in the study of adsorption and self-diffusion in certain minerals. D, 1943, Massachusetts Institute of Technology. 108 p.

Vincent, Robert J. Structural geology and petrofabrics of the Reading Prong near Reading, Pennsylvania. M, 1967, Lehigh University.

Vincent, Robert Keller. A thermal infrared ratio imaging method for mapping compositional variations among silicate rock types. D, 1973, University of Michigan.

Vincent, Samir Ambrose. Sedimentation of Keyser Limestone (Upper Silurian and Lower Devonian), (Keyser, West Virginia). M, 1966, American University. 37 p.

Vincent, W. F. Ecophysiological studies on the aphotic phytoplankton of Lake Tahoe, California-Nevada. D, 1977, University of California, Davis. 250 p.

Vincenzo, Theresa E. De see De Vincenzo, Theresa E.

Vinckier, Thomas Alan. Hydrogeology of the Dakota Group aquifer with emphasis on the radium-226 content of its contained ground water, Canon City Embayment, Fremont and Pueblo counties, Colorado. M, 1978, University of Colorado.

Vinet, Marshall Justin. Geology of Sierra Baluartes and Sierra de Pajaros Azules, Coahuila, Mexico. M, 1975, University of New Orleans.

Vineyard, Jerry Daniel. Origin and development of Cave Spring, Shannon County, Missouri. M, 1963, University of Missouri, Columbia.

Vineyard, William Lawton. The geologic and economic aspects of water flooding the Ikemire-Henry Leases, Main Pool, Crawford County, Illinois. M, 1950, University of Illinois, Urbana.

Vining, M. R. The Spuzzum Pluton northwest of Hope, B.C. M, 1977, University of British Columbia.

Vinje, S. P. Archean geology of an area between Knife Lake and Kekekabic Lake, eastern Vermilion District, northeastern Minnesota. M, 1978, University of Minnesota, Duluth.

Vink, Gregory Evans. Continent rifting and plate tectonic reconstructions, with applications to the Norwegian-Greenland Sea. D, 1983, Princeton University. 138 p.

Vinopal, Robert J. Effect of shape sorting on the grain volume of carbonate sands and gravels. M, 1976, Kent State University, Kent. 90 p.

Vinopal, Robert James. Petrology of the Upper Devonian clastic sequence in Jackson and Lincoln counties, West Virginia, and Wise County, Virginia. D, 1981, West Virginia University. 252 p.

Vinson, George Larry. The geology of the Cross Mountain Anticline, Moffatt County, Colorado. M, 1955, Texas A&M University. 123 p.

Vinson, Thomas Edward. A paleomagnetic study of the Upper Cretaceous to early Eocene Eureka Sound Formation, Strathcona Fiord, Ellesmere Island, Canada. M, 1981, University of Wisconsin-Milwaukee. 125 p.

Vinton, R. P. Gravity studies in uranium districts, southern Powder River Basin, Wyoming. M, 1976, University of Wyoming. 58 p.

Violette, John La see La Violette, John

Viret, Marc. Relocation study of Virginia earthquakes (1959-1981) using the joint hypocenter determination and joint epicenter determination methods. M, 1982, Virginia Polytechnic Institute and State University. 112 p.

Virgin, William Wallace, Jr. The structure and petrography of the Concord Granite in the Concord area, New Hampshire. D, 1964, Lehigh University. 191 p.

Virta, Robert Lee. An evaluation of the adequacy of morphological data for determining the carcinogenicity of minerals. M, 1988, University of Maine.

Viscio, Paul James. Petrology of TiO$_2$-polymorph-bearing vein deposits adjacent to the Magnet Cove intrusion. M, 1981, Washington University. 93 p.

Visco, C. The geomorphic effects of off road vehicles on the beach, Fire Island, N.Y. M, 1977, SUNY at Binghamton. 73 p.

Visconti, Robert Vincent. Paleozoic stratigraphy and structure of the Dry Creek area, Elko and Eureka counties, Nevada. M, 1983, Oregon State University. 67 p.

Visconty, Greg. Rock slope stability studies in Siskiyou National Forest. M, 1988, Portland State University. 91 p.

Visentin, G. Aspects of the inorganic amorphous system of Humo ferric podzols of the lower mainland of British Columbia. M, 1974, University of British Columbia.

Visger, Frank J. The geology of the Uncle Jess molybdenite deposit, Custer County, Idaho. M, 1974, Eastern Washington University. 30 p.

Visher, Glenn S. Differentiation in two diabase sills in Northeast Minnesota. M, 1956, Northwestern University.

Visher, Glenn S. Geology of the Moxie Pluton, west central Maine. D, 1960, Northwestern University. 143 p.

Visher, Peggy M. Sedimentology and three dimensional facies relations within a tidally-influenced carboniferous deltas; the Big Clifty Formation, Sulphur, Indiana. M, 1980, Indiana University, Bloomington. 152 p.

Visocky, Adrian P. Sand model experiments for wells penetrating two artesian aquifers. M, 1964, New Mexico Institute of Mining and Technology. 64 p.

Visser, Alex Theo. An evaluation of unpaved road performance and maintenance. D, 1981, University of Texas, Austin. 354 p.

Visser, John. Sedimentology and taphonomy of a Styracosaurus bonebed in the Late Cretaceous Judith River Formation, Dinosaur Provincial Park, Alberta. M, 1986, University of Calgary. 150 p.

Visser, W. Experimental investigation of silicate liquid immiscibility in the system K$_2$O-FeO-Al$_2$O$_3$-SiO$_2$-TiO$_2$-P$_2$O$_5$. D, 1979, University of Illinois, Chicago. 141 p.

Viste, Daniel Ralph. A water resource study of eastern Bath Township, Greene County, Ohio. M, 1975, Wright State University. 84 p.

Visvanathan, Thellur Rangswamy. Analysis of earthquakes in relation to variations in Earth's rotation rate. D, 1973, University of South Carolina.

Viswanathan, Subramanian. A petrological and geochemical study of granitic and metamorphic rocks in and adjacent to western part of the Giants Range Batholith, northeastern Minnesota. D, 1971, University of Minnesota, Minneapolis.

Vita, Charles Ludwig. Arctic route geotechnical characterization and analysis; a systems approach. D, 1985, University of Washington. 251 p.

Vitali, Rino. Surficial mapping in Old Mystic Quadrangle (Connecticut). Geology at the Upper Skungamuung River basin. M, 1969, University of Connecticut. 43 p.

Vitaliano, Charles J. Contact metamorphism at Rye Patch, Nevada. D, 1944, Columbia University, Teachers College.

Vitaliano, Charles J. Tungsten deposits of the Snake Range, White Pine County, Nevada. M, 1938, Columbia University, Teachers College.

Vitanage, Piyadas. Study of zircon types in the Ceylon Pre-cambrian complex. D, 1957, University of Chicago. 11 p.

Vitani, Nicholas M. Conodont biostratigraphy of the lower Strawn Group (Pennsylvanian); evaluation of structure and stratigraphic relationships, Colorado River valley, Texas. M, 1987, Baylor University. 89 p.

Vitayasupakorn, Vichai. Development of an electroosmotic field test for evaluation of consolidation parameters of soils. D, 1986, University of Washington. 175 p.

Vitcenda, John Frederick. Correlation of the Charles Formation (Mississippian) from central Montana into the Bighorn Mountains, Wyoming. M, 1958, University of Wisconsin-Madison. 61 p.

Vitek, John Dennis. The mounds of south-central Colorado; an investigation of geographic and geomorphic characteristics. D, 1973, University of Iowa. 229 p.

Vito, Steven A. De see De Vito, Steven A.

Vitorello, Icaro. Heat flow and radiogenic heat production in Brazil with implications for thermal evolution of continents. D, 1978, University of Michigan. 153 p.

Vitorello, Icaro. Paleomagnetic studies of late Pleistocene and Holocene sediments from Lake Michigan cores. M, 1975, University of Michigan.

Vitousek, Peter Morrison. The regulation of element concentrations in mountain streams in the northeast-

ern United States. D, 1975, Dartmouth College. 98 p.

Vitt, Alfred Weston. The (Pliocene) Pinole Tuff east of San Francisco Bay, California. M, 1936, University of California, Berkeley. 70 p.

Vittorio, Louis F. Paleomagnetism of Late Cretaceous sedimentary rocks, Matanuska and Boulder Creek valleys, south central Alaska. M, 1988, Lehigh University. 36 p.

Vitz, Howard Engeler. Some Cretaceous ostracodes from North and South Carolina. M, 1939, University of North Carolina, Chapel Hill. 41 p.

Viveiros, John J. Cenozoic tectonics of the Great Salt Lake for seismic reflection data. M, 1986, University of Utah. 81 p.

Viveros, José G. Study of P-waves through layered and fault geometries, using the two-dimensional model techniques. M, 1968, Rice University. 70 p.

Vivian, Gary J. The geology of the Blackdome epithermal deposit, B.C. M, 1988, University of Alberta. 203 p.

Vizgirda, Joana Marija. Dynamic properties of carbonates and applications to cratering processes. D, 1982, California Institute of Technology. 213 p.

Vlack, L. H. Van see Van Vlack, L. H.

Vlam, Heber Adolf. Petrology of Lake Bonneville, gravels (Pleistocene), Salt Lake County, Utah. M, 1963, University of Utah. 58 p.

Vleet, E. S. Van see Van Vleet, E. S.

Vliek, P. J. Effects of metamorphism and structure on aeromagnetic anomalies over the Carolina slate belt near Roxboro, North Carolina. M, 1979, Virginia Polytechnic Institute and State University.

Vloten, Roger Van see Van Vloten, Roger

Voast, Wayne Adams Van see Van Voast, Wayne Adams

Vocke, Robert Donald, Jr. Petrogenetic modelling in an Archean gneiss terrain, Saglek, northern Labrador. D, 1983, SUNY at Stony Brook. 294 p.

Vodrazka, Walter C. Engineering and industrial applications of Selma chalk. M, 1962, Mississippi State University. 127 p.

Voegeli, David Afred. The origin of composite dike rocks from the North Eastern Desert and Sinai, Egypt. M, 1985, University of Texas at Dallas. 165 p.

Voelger, Klaus. Cenozoic deposits in the southern foothills of the Santa Catalina Mountains near Tucson, Arizona. M, 1953, University of Arizona.

Voelker, George Edmund. An analysis of data obtained from vane shear tests of Recent marine sediment. M, 1973, United States Naval Academy.

Vogel, Donald A. Surface mining and ground water quality in eastern Rush Creek basin, Perry County, Ohio. M, 1985, Ohio University, Athens. 150 p.

Vogel, Irene D. Bryozoans of the Toroweap Formation, Dry Lake Range and North Muddy Mountains, Clark County, Nevada. M, 1976, Eastern Washington University. 75 p.

Vogel, James William. Late Quaternary sedimentary facies of the southern Sierra Leone and Liberian continental shelf and upper slope, Northwest Africa. D, 1982, University of Rhode Island. 375 p.

Vogel, James William. The geology of southernmost Juab Valley and adjacent highlands, Juab County, Utah. M, 1957, Ohio State University.

Vogel, John David. Geology and ore deposits of the Klondike Ridge area, Colorado. D, 1960, Stanford University. 290 p.

Vogel, Karen L. The petrology of pelitic rocks of the Bugtown Formation, Black Hills, South Dakota. M, 1985, Kent State University, Kent. 106 p.

Vogel, Kenneth Daniel. Deformation in the lower Great Valley Sequence; the Paskenta fault zone of Northern California. M, 1985, University of Texas, Austin. 130 p.

Vogel, Peter Nicholas. Carbonate island hydrology and solution conduit genesis; San Salvador Island,

Bahamas. M, 1988, Mississippi State University. 143 p.

Vogel, Richard Mark. Development and comparison of models for drought simulation. M, 1979, University of Virginia. 89 p.

Vogel, Richard Mark. The variability of reservoir storage estimates. D, 1985, Cornell University. 216 p.

Vogel, Thomas A. The sedimentary origin of the Housatonic Highlands gneiss complex in the Cornwall area (Connecticut). M, 1961, University of Wisconsin-Madison.

Vogel, Thomas Adolph. The petrogenic significance of plagioclase twinning. D, 1963, University of Wisconsin-Madison. 107 p.

Vogelpohl, Sidney. Mineralogy and geochemistry of manganese oxide mineralization in the eastern half of the West-Central Manganese District of Arkansas. M, 1977, University of Arkansas, Fayetteville.

Vogfjord, Kristin S. The Meckering earthquake of October 14, 1968; a possible downward propagating rupture. M, 1986, Pennsylvania State University, University Park. 50 p.

Vogl, Eric G. Chemical effects of selected trace-metals from sanitary land-fill leachates on ground water quality. M, 1981, University of Kansas. 86 p.

Vogler, David L. Dissolution of quartz, cristobalite, chalcedonies, and opal-cristobalites in dilute organic acids at room temperature and its implication to weathering. M, 1971, University of South Florida, Tampa. 85 p.

Vogler, Herbert A., III. Major and trace element geochemistry of the Laguna del Perro area playa-bolson complex, Torrance County, New Mexico. M, 1983, University of New Mexico. 247 p.

Vogt, Beverly Frobenius. The stratigraphy and structure of the Columbia River Basalt Group in the Bull Run watershed, Multnomah and Clackamas counties, Oregon. M, 1981, Portland State University. 151 p.

Vogt, Eric. A geochemical study of the Mount Kinabalu Batholith, Malaysia. M, 1988, University of Illinois, Chicago. 108 p.

Vogt, Jay Nathan. Dolomitization and anhydrite diagenesis of the San Andres (Permian) Formation, Gaines County, Texas. M, 1986, University of Texas, Austin. 203 p.

Vogt, Mary Cameron. Study of a deep well in Neptune Township, New Jersey. M, 1946, Bryn Mawr College. 54 p.

Vogt, Peter Richard. A reconnaissance geophysical survey of the North, Norwegian, Greenland, Kara and Barents seas and the Arctic Ocean. D, 1968, University of Wisconsin-Madison. 211 p.

Vogt, Peter Richard. Interpretations of magnetic anomalies over the Mid-Atlantic Ridge between 42°N and 47°N. M, 1965, University of Wisconsin-Madison.

Vogt, Robert R. The fauna of the (Devonian) Cedar Valley Limestone near Buffalo, Iowa. M, 1932, Iowa State University of Science and Technology.

Vogt, Timothy J. The petrology, mineralogy and chemistry of the Cora Mine; a hydrothermal lead-silver deposit, Black Hills, South Dakota. M, 1984, South Dakota School of Mines & Technology.

Voight, Barry. Structural studies in west-central Vermont. D, 1965, Columbia University. 298 p.

Voight, David J. A petrochemical and magnetic study of the volcanic greenstones in northwestern Marathon County, Wisconsin. M, 1970, University of Wisconsin-Milwaukee.

Voight, David Scott. The effects of bottom current erosion on sediment diagenesis; investigation of ice-rafted debris and manganese micronodule occurrences in the Southern Ocean. M, 1980, University of Wisconsin-Madison.

Voight, Donald Edward. The solubility of Al$_2$SiO$_5$ in the system KAlSi$_3$O$_8$-SiO$_2$-H$_2$O at 2 Kbar, and its implications for melt speciation. M, 1983, Pennsylvania State University, University Park. 33 p.

Voight, Kenneth. Geochemical study of Leon Mountain analcite syenogabbro, Brewster County, Texas. M, 1985, University of Houston.

Voigt, Virginia. Texture of limestones. M, 1939, University of Minnesota, Minneapolis. 110 p.

Voit, Roland L. Petrology and mineralogy of Tertiary(?) volcanic rocks west and southwest of Kelton (Box Elder Co.), Utah. M, 1985, Utah State University. 90 p.

Vokes, Emily Hoskins. Cenozoic Muricinae of the western Atlantic region. D, 1967, Tulane University. 494 p.

Vokes, Emily Hoskins. The gastropod genus Murex s. s. in the Cenozoic of the western Atlantic region. M, 1962, Tulane University. 66 p.

Vokes, Harold Ernest. Middle Eocene molluscan faunas of the Vallecitos and Coalinga areas. D, 1935, University of California, Berkeley. 707 p.

Volchok, Herbert L. The ionium method of age determination. D, 1955, Columbia University, Teachers College.

Volchok, Herbert L. Thick source alpha count of some representative deep sea cores. M, 1951, Columbia University, Teachers College.

Volckmann, Richard Peter. Geology of the Crestone Peak area, Sangre de Cristo Range, Colorado. D, 1965, University of Michigan. 148 p.

Volckmann, Richard Peter and Lasca, Norman Paul. Geology of the Music Pass area in the Sangre de Cristo Range, Colorado. M, 1961, University of Michigan.

Voldseth, Nels Edward. Geology of the Cottonwood Canyon area, Bighorn Mountains, Wyoming. M, 1973, University of Iowa. 100 p.

Volk, J. A. Structural analysis and kinematic interpretation of rocks in the southern portion of the Okanogan gneiss dome, north-central Washington. M, 1986, Washington State University. 156 p.

Volk, Joseph Anthony. Theory and design of electronic circuits applicable to the measurement of Earth motion. D, 1950, St. Louis University.

Volk, Karen Wagner. Preliminary paleomagnetic results for a rossville-type diabase dike system in southeastern Pennsylvania. M, 1974, Pennsylvania State University, University Park.

Volk, Karen Wagner. The paleomagnetism of Mesozoic diabase and the deformational history of southeastern Pennsylvania. D, 1977, Pennsylvania State University, University Park. 164 p.

Volk, Norman J. The nature of potash fixation in soils, and the isolation and identification of a potash silicate formed. D, 1932, University of Wisconsin-Madison.

Volk, Tyler. Multi-property modeling of the marine biosphere in relation to global climate and carbon cycles. D, 1984, New York University. 368 p.

Volkert, David G. Stratigraphy and petrology of the Colton Formation (Eocene), Gunnison Plateau, central Utah. M, 1980, Northern Illinois University. 132 p.

Volkert, Richard Allen. A determinative study of the structural state and composition of alkali feldspars from pegmatites along Route 15, Morris and Sussex counties, New Jersey. M, 1984, Montclair State College. 114 p.

Volkmann, Robert G. Geology of the Mackay 3NE Quadrangle, Custer, Blaine, and Butte counties, Idaho. M, 1972, University of Wisconsin-Milwaukee.

Vollendorf, William Charles. A microscopic study of the silver-zinc ores of the Pulacayo District, Bolivia. M, 1955, University of Michigan.

Vollmer, Frederick W. Structural studies of the Ordovician flysch and melange in Albany County, New York. M, 1981, SUNY at Albany. 151 p.

Vollmer, Frederick Wolfer. A structural study of the Grovudal fold-nappe, northern Dovrefjell, central Norway. D, 1985, University of Minnesota, Minneapolis. 260 p.

Vollo, N. B. The geology of the Henderson copper deposit, Chibougamau region, Quebec. M, 1959, McGill University.

Volman, Kathleen Cushman. Paleoenvironmental implications of botanical data from Meadowcroft Rockshelter, Pennsylvania. D, 1981, Texas A&M University. 236 p.

Volpe, Alan Max. Petrogenesis and Sr-Nd isotopic geochemistry of basalts from Western Pacific backarc basins and Precambrian mafic amphibolites and diorites in the Delhi Supergroup. D, 1988, University of California, San Diego. 269 p.

Volpi, Mary. A sedimentary analysis of carbonate sands in the Florida Keys; beach vs. subtidal sediments. M, 1985, Lehigh University. 124 p.

Volpi, Richard Wayne. The influence of Pennsylvanian bedrock on the composition of pre-Woodfordian tills in Columbiana County, Ohio. M, 1987, University of Akron. 122 p.

Voltz, Charles Frederick. Paleomagnetism of the Upper Ordovician Maquoketa Shale and Lower Silurian Dolomite in eastern Wisconsin. M, 1983, University of Wisconsin-Milwaukee. 245 p.

Volz, Gary Arlen. Seismic investigation of the River Falls Basin in western Wisconsin and part of the Saint Croix Horst (Upper Cambrian) in southeastern Minnesota. M, 1968, University of Minnesota, Minneapolis. 140 p.

Volz, Marilyn. The vein minerals of the Connecticut Valley Basalt in Massachusetts. M, 1955, Smith College. 106 p.

Volz, Steven Alan. Preliminary report on a late Pleistocene death-trap fauna from Monroe Co., Indiana. M, 1977, Indiana University, Bloomington. 31 p.

Volz, William Richard. Travel time perturbations in the crust and upper mantle in the Southeast. M, 1979, Georgia Institute of Technology. 198 p.

Von Almen, William F. Palynology of selected coals of northeastern and north central Missouri. M, 1959, University of Missouri, Columbia.

Von Almen, William Frederick. Palynomorphs of the Woodford shale of south central Oklahoma with observations on their significance in zonation and paleoecology. D, 1970, Michigan State University. 222 p.

Von Bargen, David J. Geology, geochemistry and mineralogy of Box Butte County, Nebraska. M, 1977, Purdue University. 66 p.

Von Bargen, David J. The silver-antimony-mercury system and the mineralogy of the Black Hawk District, New Mexico. D, 1979, Purdue University. 226 p.

von Bargen, Nikolaus. Permeabilities, electrical conductivities and interfacial energies of partially molten systems at textural equilibrium. D, 1986, University of Oregon. 126 p.

von Baumgaertner, I. Some new processes of beach dynamics, Robert Moses State Park, Fire Island, New York. D, 1977, Columbia University, Teachers College. 249 p.

Von Bergen, Donald. Microfacies, depositional environments and diagenesis of Atokan carbonates, Delaware Basin, Reeves County, Texas, U.S.A. M, 1985, University of Illinois, Urbana. 101 p.

Von Bergen, Donald. Natural and experimentally-simulated stylolitic porosity in carbonate rocks. D, 1988, University of Illinois, Urbana. 130 p.

Von Bitter, Peter H. Environmental control of conodont distribution in the Shawnee Group (Upper Pennsylvanian) of eastern Kansas. D, 1972, University of Kansas. 402 p.

von Bitter, Peter Hans. Echinoderms as guide fossils in the correlation of the Windsor Group (Mississippian) subzones of the Minas Sub-basin (Nova Scotia). M, 1966, Acadia University.

von Borstal, B. E. The physical behavior of oil in sandy beaches, McNabs Island, Nova Scotia. M, 1974, Dalhousie University.

von Breymann, Marta T. Magnesium in hemipelagic environments; surface reactions in the sediment-pore water system. D, 1988, Oregon State University. 245 p.

Von Damm, Karen Louise. Chemistry of submarine hydrothermal solutions at 21 North, East Pacific Rise and Guaymas Basin, Gulf of California. D, 1984, Massachusetts Institute of Technology. 240 p.

Von Demfange, W. C., Jr. An investigation of the vertical distribution of sulfur forms in surface mine spoils, Henry County, Missouri. M, 1974, University of Missouri, Rolla.

von der Hoya, H. Austin, II. A reflection seismic and magnetic investigation of the Tela Basin; northern offshore Honduras. M, 1986, Southern Methodist University. 112 p.

von der Osten, Erimar Alfred. Age and correlation of the Barranquin Formation of northeastern Venezuela. D, 1956, Stanford University. 195 p.

Von Dohlen, Edward Lee. Application of applied geochemical methods to the search for podiform chromite concentrations at Cedars ultramafic body near Cazadero, California. M, 1977, Stanford University. 90 p.

Von Engeln, Oscar Dierich. Phenomena associated with glacier drainage and wastage, with especial reference to observations in the Yakutat Bay region, Alaska. D, 1911, Cornell University.

Von Estoff, Fritz E. The (Eocene and Oligocene) Kreyenhagen Shale at the type locality (California). M, 1929, Stanford University. 106 p.

von Feldt, Ann Elizabeth. The taxonomic and phylogenetic affinities of the species of Globotruncanita Reiss. M, 1984, University of Texas at Dallas. 161 p.

von Frese, Ralph Robert Benedict. Magnetic exploration of historical Midwestern archaeological sites as exemplified by a survey of Ft. Ouiatenon (12-T-9). M, 1978, Purdue University. 66 p.

von Frese, Ralph Robert Benedikt. Lithospheric interpretation and modeling of satellite elevation gravity and magnetic anomaly data. D, 1980, Purdue University. 180 p.

Von Herzen, Richard Pierre. Pacific Ocean floor heat flow measurements, their interpretation and geophysical implications. D, 1960, University of California, Los Angeles. 119 p.

Von Holdt, Laura Lynn. Foraminifera of the Storm King Mountain Shale Member, Mancos Shale, western Colorado. M, 1982, University of Colorado. 135 p.

Von Horn, Robert. Chert in the Columbus and Delaware Limestones (Devonian, Ohio). M, 1972, Ohio State University.

von Huene, Roland Ernest. Structural geology and gravimetry of Indian Wells Valley, southeastern California. D, 1960, University of California, Los Angeles.

von Maluski, Barbara Janine. The processing of seismic reflection data acquired with the Wacker source and the Mini-Sosie method. M, 1984, Wright State University. 120 p.

von Metzsch, Ernst Hans. Decision analysis in mineral exploration applied to porphyry copper deposits. D, 1976, Harvard University.

Von Rhee, Robert Weston. Model of deposition of Batestown Till in east-central Illinois. M, 1977, University of Illinois, Urbana.

Von Rosen, G. E. A. Pasco Gneiss breccia (Mesozoic?). M, 1966, University of British Columbia.

von Schondorf, Amy. Sedimentary facies and paleohydraulics of an ice-contact glacial outwash plain, Germany Flats, New Jersey. M, 1987, Rutgers, The State University, New Brunswick. 122 p.

Von Schonfeldt, Hilmar. An experimental study of open-hole hydraulic fracturing as a stress measurement method with particular emphasis on field tests. D, 1970, University of Minnesota, Minneapolis.

Von Schwind, Joseph J. Characteristics of gravity waves of permanent form. D, 1968, Texas A&M University.

Von Seggern, D. H. Electromagnetic mapping of Hawaiian lava tubes. M, 1967, University of Hawaii. 41 p.

von Seggern, David Henry. Investigation of seismic precursors before major earthquakes and of the state of stress on fault planes. D, 1982, Pennsylvania State University, University Park. 352 p.

von Shaffer, Ronald. Geologic and geophysical analysis of adjacent portions of Hancock and Wood counties, Ohio. M, 1982, Wright State University. 90 p.

Von Stemle, Steven. The effects of tributary mixing on the heavy mineral composition of sediment derived from the Beartooth Mountains, south-central Montana. M, 1985, Southern Illinois University, Carbondale. 82 p.

Von Tress, David Edward. Some Waynesville and Liberty Bryozoa from Versailles, Ripley County, Indiana. M, 1954, Indiana University, Bloomington. 61 p.

Vonarx, Clifford E. Geology of Horse Shoe Curve type; Devonian to Lower Pennsylvanian. M, 1979, Pennsylvania State University, University Park. 66 p.

Vonder Haar, Stephen P. Evaporites and algal mats at Laguna Mormona, Pacific coast, Baja California, Mexico. D, 1976, University of Southern California.

Vonder Haar, Stephen P. Semi-arid coastal evaporite environment at Laguna Mormona, Pacific Coast, Baja California, Mexico. M, 1972, University of Southern California.

Vonder Linden, Karl. An analysis of the Portuguese Bend landslide, Palos Verdes Hills, California. D, 1972, Stanford University. 405 p.

Vonderharr, Jerry. Sedimentology and paleoecology of the DeChelly Sandstone (Permian) of northeastern Arizona. M, 1986, Northern Arizona University. 137 p.

Vonderohe, Alan Paul. Photogrammetric systems for the analysis of planar displacements in soil. D, 1981, University of Illinois, Urbana. 262 p.

Vondra, Carl Frank. The stratigraphy of the Chadron Formation in northwestern Nebraska. M, 1958, University of Nebraska, Lincoln.

Vondra, Carl Frank. The stratigraphy of the Gering Formation in the Wildcat Ridge in western Nebraska. D, 1962, University of Nebraska, Lincoln.

Voner, Frederick Ronald. Age and origin of granitic rocks in the Farmington Quadrangle, Maine. M, 1980, Miami University (Ohio). 107 p.

Voner, Frederick Ronald. Crustal evolution of the Hopedale Block, Labrador, Canada. D, 1985, Miami University (Ohio). 240 p.

vonGonten, Glenn. Sedimentology of the Brazos River Formation (Desmoinesian) of north-central Texas. M, 1985, University of Texas, Arlington. 258 p.

Vonheeder, Ellis R. Origin and development of coastal landforms at Point Francis, Washington. M, 1972, Western Washington University. 92 p.

Vonhof, Jan Albert. Tertiary gravels and sands in southern Saskatchewan. M, 1965, University of Saskatchewan. 99 p.

Vonhof, Jan Albert. Tertiary gravels and sands in the Canadian Great plains. D, 1969, University of Saskatchewan. 279 p.

Voogd, Beatrice De *see* De Voogd, Beatrice

Vookerding, Clifford C. A study of some chemical and biological properties of metaphosphates and their utilization as soil fertilizers. D, 1943, Cornell University.

Voorde, Barbara Wiley Van de *see* Van de Voorde, Barbara Wiley

Voorhees, Brent J. Stratigraphy and facies of the Lower Carmel Formation (Middle Jurassic), southwestern Utah. M, 1978, Northern Arizona University. 161 p.

Voorhees, David H. The stratigraphy and sedimentary petrography of the Oriskany Sandstone (Lower Devonian) in central and eastern New York State. M, 1982, Rensselaer Polytechnic Institute. 112 p.

Voorhees, Gerald E. Upper Cretaceous stratigraphy and overthrusting in the Deadman, Blind Bull and Horse Creek area, Lincoln County, Wyoming. M, 1964, University of Wyoming. 88 p.

Voorhees, John K. Physical properties of selected soils in the Norridgewock, Maine, Quadrangle. M, 1959, Columbia University, Teachers College.

Voorhies, Coerte Van. Magnetic location of Earth's core-mantle boundary and estimates of the adjacent fluid motion. D, 1984, University of Colorado. 367 p.

Voorhies, Michael R. Taphonomy and population dynamics of an early Pliocene vertebrate fauna, Knox County, Nebraska. D, 1966, University of Wyoming. 197 p.

Voorhis, David van *see* van Voorhis, David

Voorhis, Gerald D. Van *see* Van Voorhis, Gerald D.

Vopni, Lorne Kenelm. Stratigraphy of the Horn Plateau Formation; a middle Devonian reef, Northwestern Territories, (Canada). M, 1969, University of Alberta. 115 p.

Voran, Roxie Lynn. Fossil assemblages, stratigraphy, and depositional environments of the Crouse Limestone (Lower Permian) in north central Kansas. M, 1977, Kansas State University. 208 p.

Vorauer, A. G. A geomechanical investigation of the weathered zone at three sites in southwestern Ontario. M, 1988, University of Waterloo. 112 p.

Vorhis, Robert Carson. Paleozoic stratigraphy of Preble County, Ohio. M, 1941, University of Iowa. 53 p.

Voris, Richard Hensler. Geology of a southwest portion of the Ignacio Quadrangle, La Plata County, Colorado. M, 1952, University of Illinois, Urbana.

Vormelker, Joel David. Geology of the High Springs Quadrangle, Florida. M, 1966, University of Florida. 57 p.

Vormelker, R. S. Vertical distribution of foraminifera, Upper Chalk Member of the Austin Formation, northern Ellis County, Texas. M, 1962, Southern Methodist University. 58 p.

Vorwald, Gary R. Paleontology and paleoecology of the upper Wheeler Formation (Late Middle Cambrian), Drum Mountains, west-central Utah. M, 1984, University of Kansas. 176 p.

Vos, Richard G. Destructive deltaic sedimentation; an example from the upper Paleozoic of southern Morocco. D, 1975, University of South Carolina. 105 p.

Vosburg, David Lee. Geology of the Burbank-Shidler area, Osage County, Oklahoma. M, 1954, University of Oklahoma. 110 p.

Vosburg, David Lee. Permian subsurface evaporites in the Anadarko Basin of the western Oklahoma-Texas Panhandle region. D, 1963, University of Oklahoma. 137 p.

Voshinin, Natalie. Foraminifera of the Manasquan Formation in New Jersey. M, 1955, Rutgers, The State University, New Brunswick. 166 p.

Voskuil, Walter H. A regional study in the depletion of soil phosphorus. M, 1922, University of Wisconsin-Madison.

Voskuil, Walter H. The economic geography of northern Price County, Wisconsin. D, 1924, University of Wisconsin-Madison.

Voss, J. An empirical method for the determination of gravity terrain corrections. M, 1974, University of Hawaii. 74 p.

Voss, James D. Geology of ancient blanket and channel deposits (Pennsylvanian) exposed near Wamego, Kansas. M, 1972, Kansas State University. 104 p.

Voss, James T. Burlington and Keokuk formations of the Upper Mississippi Valley area. M, 1963, St. Louis University.

Voss, Patrick Charles. A plan for development of the new DeKalb County Forest Preserve. M, 1974, Northern Illinois University. 52 p.

Voss, R. L. The characteristics and genesis of the Akaka and Hilo soils of the Island of Hawaii. M, 1970, University of Hawaii. 94 p.

Voss-Roberts, Kevin David. Silenis (Ostracoda, Metacopina) from the Silurian of Gotland; morphology, ontogeny, and stratigraphic distribution. M, 1987, Arizona State University. 147 p.

Vossler, Donald A. On the influence of terrestrial heat flow anomalies on the temperature microstructure at the ocean floor. M, 1968, Oregon State University. 72 p.

Vossler, Donald Alan. Anisotropic media and the determination of subsurface velocity by the use of surface seismic reflection data. D, 1971, Virginia Polytechnic Institute and State University.

Vossler, Shawna M. Ichnology of the Cardium Formation (Pembina area). M, 1988, University of Alberta. 288 p.

Votaw, Robert B. Conodont biostratigraphy of the Black River Group (Middle Ordovician) and equivalent rocks of the eastern midcontinent, North America. D, 1972, Ohio State University.

Voto, Richard H. De *see* De Voto, Richard H.

Voultsos, Mark. Phase equilibrium studies in the system NaAlSiO₄-CaMgSi₂O₆-SiO₂-iron oxide at variable oxygen fugacities, and some petrologic implications. D, 1972, Pennsylvania State University, University Park. 38 p.

Vowell, Bobby Gene. Sedimentation survey of Lake Ellsworth, Caddo and Comanche counties, Oklahoma. M, 1969, Oklahoma State University. 38 p.

Vozoff, K. Quantitative analysis of earth resistivity data. D, 1956, Massachusetts Institute of Technology. 74 p.

Vozoff, Keeva. Gravity investigation in north-central Pennsylvania. M, 1951, Pennsylvania State University, University Park. 52 p.

Vralsted, David A. Distribution of shallow ice-bordered sediments, Harrison Bay, Alaska. M, 1986, University of Alaska, Fairbanks. 134 p.

Vrba, Sheryl L. Precambrian geology of the Cleator area, Yavapai County, Arizona. M, 1980, Northern Arizona University. 96 p.

Vredenbrugh, Larry Dale. Sulfur isotopic investigation of petroleum, Wind River Basin, Wyoming. D, 1968, University of Washington. 107 p.

Vredenburgh, Larry Dale. Reactivity and expansion phenomena as related to physical properties of carbonate rocks. M, 1964, Iowa State University of Science and Technology.

Vredevoogd, James J. Substitution mechanism of chromium in clinopyroxene. M, 1974, University of Illinois, Chicago.

Vreeken, Willem Jaap. Geomorphic regimen of small watersheds in loess, Tama County, Iowa. D, 1972, Iowa State University of Science and Technology. 327 p.

Vreeland, John Howard. Gravity anomalies and geology of the Jenny Jump Mountain area, New Jersey. M, 1965, Princeton University. 69 p.

Vrh, Steven J. Petrology of the Grayson sandstone member of the Lee Formation, northeastern Kentucky. M, 1985, Bowling Green State University. 86 p.

Vries, George A. De *see* De Vries, George A.

Vries, Janet L. de *see* de Vries, Janet L.

Vries, Thomas John de *see* de Vries, Thomas John

Vrolijk, Peter John. Experimental study of sand transport and deposition in high-velocity surge. M, 1981, Massachusetts Institute of Technology. 90 p.

Vrolijk, Peter John. Paleohydrogeology and fluid evolution of the Kodiak accretion complex, Alaska. D, 1987, University of California, Santa Cruz. 250 p.

Vu, Xuan-Lan. Géologie de la Mine d'Or Belmoral, Val d'Or, Québec. M, 1985, Ecole Polytechnique. 71 p.

Vucetic, Mladen. Pore pressure buildup and liquefaction at level sandy sites during earthquakes. D, 1986, Rensselaer Polytechnic Institute. 665 p.

Vuich, John S. A geologic reconnaissance and mineral evaluation, Wheeler Wash area, Hualapai Mountains,

Mohave County, Arizona. M, 1974, University of Arizona.

Vujnich, Joseph William. The role and implications of ecological planning in making land use decisions for the Busiek State Forest. M, 1983, Southwest Missouri State University. 146 p.

Vuke, Susan M. Depositional environments of the Cretaceous Thermopolis, Muddy and Mowry formations, southern Madison and Gallatin ranges, Montana. M, 1982, University of Montana. 141 p.

Vukovich, John William. Hydro-stratigraphic units of the surficial deposits of east-central Illinois. M, 1967, University of Illinois, Urbana.

Vyas, Y. K. Erosion of dredged material islands due to waves and currents. D, 1977, Texas A&M University. 202 p.

Vyhmeister, Gerald Erwin. Palynological correlation of the Kentucky No. 12 Coal. D, 1981, Loma Linda University. 160 p.

Vyles, Charles E., III. Geology of Indian Creek area, Franklin and Johnson counties, Arkansas. M, 1966, University of Arkansas, Fayetteville.

Waag, Charles J. A report on the geology of a portion of the coal fields of Pierce County, Washington. M, 1958, University of Pittsburgh.

Waag, Charles Joseph. Structural geology of the Mount Bigelow, Bear Wallow, Mount Lemmon area, Santa Catalina Mtns., Arizona. D, 1968, University of Arizona. 220 p.

Waage, Karl M. Fire clay deposits of eastern Fremont, western Pueblo and adjacent counties, Colorado. D, 1946, Princeton University. 171 p.

Waanders, Gerald Lee. Palynology of the Monmouth Group (Maastrichtian) from Monmouth County, New Jersey, U.S.A. D, 1974, Michigan State University. 204 p.

Waboso, Chijoke Ezekiel. The delineation of non-layered anomalous velocity zones by a combination of fan shooting and least squares analysis. M, 1977, University of Western Ontario.

Waboso, Chijoke Ezekiel. The solubility of rare gases in silicate melts; implications for K-Ar dating, Earth-atmosphere evolution and Earth degassing processes. D, 1980, University of Western Ontario.

Wach, Phillip Hanby. Geology of the west-central part of the Malad Range (Oneida County), Idaho. M, 1967, Utah State University. 71 p.

Wachs, Daniel. Petrology and depositional history of limestones in the Franciscan Formation of California. D, 1973, University of California, Santa Cruz.

Wachs, Thomas C. Modification and application of remote sensing techniques as applied to mineral exploration in the Pilbara region of Western Australia. M, 1975, University of Kentucky. 105 p.

Wachs, W. C., Jr. The Atlas lands. M, 1931, [University of Cincinnati].

Wachtell, Douglas Lowell. Sedimentology and tectonic setting of conglomerate in Paleocene North Horn Formation, Spanish Fork Canyon and Mount Nebo areas, central Utah. M, 1988, University of Utah. 133 p.

Wachter, Bruce George. Rapid fresh and altered rock-Analysis for exploration reconnaissance; infrared absorption applications in the Monitor District, California. D, 1971, Stanford University. 115 p.

Wachter, Jack P. Geochemical prospecting for mercury in the Terlingua quicksilver district, Texas. M, 1974, Kent State University, Kent. 56 p.

Wacker, Herbert James. The stratigraphy and structure of Cretaceous rocks in north-central Sierra de Juarez, Chihuahua, Mexico. M, 1972, University of Texas at El Paso.

Wacker, John Frederick. Composition of noble gases in the Abee Meteorite, and the origin of the enstatite chondrites. D, 1982, University of Arizona. 176 p.

Wackwitz, Linda K. Regional tectonic systems of the Pacific Northwest delineated from ERTS-1 imagery. M, 1975, University of Montana. 62 p.

Waddel, Claudia True. Study of the interrelationships among chemical and petrographic variables of United States coals. M, 1978, Pennsylvania State University, University Park. 240 p.

Waddell, Courtney. Geology and economic resources of the Shelburn, Fairbanks, Hutton, and Pimento quadrangles. D, 1952, Indiana University, Bloomington. 63 p.

Waddell, Courtney. Relationship of Pennsylvanian and Devonian structures in west-central Sullivan County, Indiana. M, 1949, Indiana University, Bloomington. 13 p.

Waddell, Dwight Ernest. Statistical analysis of the fusulinid genera Fusulinella, Fusulina, Wedekindellina?, and Triticites in the Ardmore Basin. D, 1964, University of Oklahoma. 278 p.

Waddell, Dwight Ernest. The Atokan-Desmoinesian contact in the Ardmore Basin, Oklahoma, as defined by fusulinids. M, 1959, University of Oklahoma. 95 p.

Waddell, Evans. The dynamics of swash and its implication to beach response. D, 1973, Louisiana State University.

Waddell, Richard K., Jr. Environmental geology of the Helotes Quadrangle, Bexar County, Texas. M, 1977, University of Texas, Austin.

Waddell, Richard Kent. Patoka oil field, Marion County, Illinois. M, 1941, University of Texas, Austin.

Waddell, Richard Kent, Jr. Evaluation of a surficial application of limestone and flue dust in the abatement of acidic drainage; Jonathan Run drainage basin at Interstate 80, Centre County, Pennsylvania. D, 1978, Pennsylvania State University, University Park. 317 p.

Waddell, William Henry. Cadotte and Paddy members of the Peace River Formation, northwestern Alberta. M, 1957, University of Saskatchewan. 69 p.

Waddell-Sheets, Carol. Microprobe determination of the rate of calcium migration in clay-cement systems using archeological materials. M, 1983, SUNY at Buffalo. 36 p.

Waddington, Dennis Howson. Foliation and mineral lineation in the Moon River Synform, Grenville structural province, Ontario. M, 1973, University of Toronto.

Waddington, Edwin Donald. Accurate modelling of glacier flow. D, 1982, University of British Columbia.

Waddington, Edwin Donald. Numerical seismograms by the Cagniard-de Hoop method for core diffraction problems. M, 1973, University of Alberta. 160 p.

Waddington, J. B. The Upper Paleozoic brachiopod subfamily spiriferellinae from the Canadian arctic, and its significance for paleogeography, paleoclimatology and continental drift. M, 1972, University of Toronto.

Waddington, Jean C. B. Vegetational changes associated with settlement and land-clearance in Minnesota over the last 125 years; a comparison of historical and sedimentary records. D, 1978, University of Minnesota, Minneapolis. 188 p.

Wade, Bruce Jerome. The petrography, diagenesis and depositional setting of the Pennsylvanian Cottage Grove Sandstone in Dewey, Ellis, Roger Mills, and Woodward counties, Oklahoma. M, 1987, Oklahoma State University. 175 p.

Wade, Don Earl. Is Fredericksburg Group a practical unit in Trans-Pecos Texas?. M, 1954, University of Texas, Austin.

Wade, Edward J. Reduction of mineral matter in coal throughout a preparation plant. M, 1977, Southern Illinois University, Carbondale. 59 p.

Wade, Franklin Alton. Some contributions to the geology, glaciology, and geography of Antarctica. D, 1937, The Johns Hopkins University.

Wade, Franklin Russell and Ruhlman, Fred Lee. A study of certain factors influencing the flow of hydrocarbons through reservoir sands. M, 1941, University of Southern California.

Wade, George David. Structural geology of a portion of the Alabama tin belt, Coosa County, Alabama. M, 1986, University of Alabama. 77 p.

Wade, Jay Alan. Conodont paleontology of the Velpen Member, Linton Formation (Desmoinesian), Parke County, Indiana. M, 1978, Indiana University, Bloomington. 146 p.

Wade, Kenneth. Uranium in situ solution mining and groundwater quality at Grover test site, Weld Cty., CO. M, 1981, Colorado State University. 91 p.

Wade, M. J. Organophosphorus pesticides in the marine environment; their transport and fate. D, 1979, University of Rhode Island. 187 p.

Wade, T. L. Sedimentary geochemistry of hydrocarbons from Narragansett Bay, Rhode Island; incorporation, distribution and fate. D, 1978, University of Rhode Island. 107 p.

Wade, W. J. X-ray diffraction studies on the weathering of pyrophyllite deposits. M, 1979, Vanderbilt University.

Wade, W. Michael. Geology of the northern part of the Cooper Mountain Batholith, north-central Cascades, Washington. M, 1988, San Jose State University. 88 p.

Wadekamper, Donald. Electrical porcelains formulated from Pacific Northwest raw materials with special effects of quartz grain size. M, 1968, University of Washington. 42 p.

Wadell, Hakon A. Volume, shape, and roundness of rock particles. D, 1932, University of Chicago. 122 p.

Wadell, James Sanders. Sedimentation and stratigraphy of the Verde Formation (Pliocene), Yavapai County, Arizona. M, 1972, Arizona State University. 110 p.

Wadhwa, Nand P. Suspended sediment load studies of the Brazos River between Waco and Richmond, Texas. M, 1961, Texas A&M University.

Wadleigh, Moire Anne. Marine geochemical cycle of strontium. M, 1982, University of Ottawa. 187 p.

Wadsworth, Albert Hodges, Jr. The lower Colorado River, Texas. M, 1941, University of Texas, Austin.

Wadsworth, Donald V. Approximate integration methods applied to wave propagation. D, 1958, Massachusetts Institute of Technology. 128 p.

Wadsworth, Joseph Rogers. Transport mechanisms operating on a recurved spit, Tawas Point, Michigan. M, 1975, University of Michigan.

Wadsworth, Joseph Rogers, Jr. Geomorphic characteristics of tidal drainage networks in the Duplin River system, Sapelo Island, Georgia. D, 1980, University of Georgia. 283 p.

Wadsworth, Marshman Edward. On the classification of rocks. D, 1879, Harvard University.

Wadsworth, William B. Petrogenesis of a quartz diorite pluton near Pembine, Wisconsin. M, 1962, Northwestern University.

Wadsworth, William Bingham. Quantitative petrology of epizonal and mesozonal catazonal granitic plutons. D, 1966, Northwestern University. 315 p.

Waechter, Noel B. Hydrology, morphology, and sedimentology of an ephemeral braided stream; Prairie Dog Town fork of the Red River, Texas panhandle. M, 1972, University of Texas, Austin.

Waegli, Jerome A. Geology and mineralization of the Uncle Sam Vein and surrounding area, San Juan County, Colorado. M, 1979, Colorado School of Mines. 127 p.

Waesche, Hugh Henry. The areal geology of the Blacksburg region. M, 1935, Virginia Polytechnic Institute and State University.

Wafer, James Oscar. An electric log study of structure, thickness and permeability of the Aux Vases Formation, Mt. Vernon, Illinois area. M, 1955, University of Illinois, Urbana.

Wagener, Henry Dickerson. Areal modal variation in the Farrington igneous complex, Chatham and Orange counties, North Carolina. M, 1964, University of North Carolina, Chapel Hill. 50 p.

Wagener, Henry Dickerson. Petrology of the adamellites, granites and related metamorphic rocks of the Winnsboro Quadrangle, South Carolina. D, 1970, University of North Carolina, Chapel Hill. 92 p.

Wagenhofer, Paul Joseph. Analysis of volatiles in primary and secondary fluid inclusions of minerals from the Pasto Bueno tungsten-base ore deposit, northern Peru. M, 1974, University of Tulsa. 91 p.

Wagenhoffer, Albert J. The biostratigraphy of the lower Helderbergian formations (Lower Devonian) as exposed along Wallpack Ridge, Sussex County, New Jersey. M, 1977, Montclair State College. 44 p.

Waggoner, Eugene Benjamin. The nature of the schist basement in the western part of the Los Angeles Basin. M, 1939, University of California, Los Angeles.

Waggoner, Gail Louise. Sedimentary analysis of gravel deposits in the vicinity of Clarkston, Washington. M, 1981, Washington State University. 107 p.

Waggoner, James Allen. Unconsolidated shelf sediments in the area of Scripps and La Jolla submarine canyons. M, 1979, San Diego State University.

Waggoner, Raymond Russell. Environmental geology problems of Pyramid Lake basin. M, 1975, University of Nevada. 95 p.

Waggoner, Thomas D. A method of differentiating carbonate and silicate facies of the Negaunee Iron Formation (Precambrian, northern Michigan) by spectrochemical analysis. M, 1967, Michigan State University. 55 p.

Wagner, Annette. Debris reworking processes and environments on Trident and Castnor glaciers, Alaska Range, Alaska. M, 1986, Lehigh University.

Wagner, Brian Jeffrey. A statistical methodology for estimating transport parameters for aquifers and streams; applications to one-dimensional systems. M, 1985, Stanford University. 85 p.

Wagner, Brian Jeffrey. Optimal groundwater quality management under uncertainty. D, 1988, Stanford University. 230 p.

Wagner, Carol Daily. Evolution among some clypeasteroid echinoids. D, 1970, University of California, Berkeley. 271 p.

Wagner, Carroll M. The San Lorenzo Series of the San Emigdio region, California. M, 1918, University of California, Berkeley. 34 p.

Wagner, Chancellor Philip. Geology of the Lyons area (Boulder County), Colorado. M, 1940, University of Colorado.

Wagner, Charles Gregory. A geophysical study of the Cave Creek basin, Maricopa County, Arizona. M, 1979, University of Arizona. 61 p.

Wagner, D. B. Lower Permian paleogeography and fusulinid paleontology, northeastern Nevada and western Utah. M, 1975, San Jose State University. 254 p.

Wagner, Dallas M. Deposition, diagenesis, and porosity distribution of North Russell (Devonian) Field, Gaines County, Texas. M, 1988, Texas Tech University. 123 p.

Wagner, Dana Bernadine. Environmental history of central San Francisco Bay with emphasis on foraminiferal paleontology and clay mineralogy. D, 1978, University of California, Berkeley. 274 p.

Wagner, Daniel P. Acid sulfate weathering in upland soils of the Maryland coastal plain. D, 1982, University of Maryland. 187 p.

Wagner, David L. Mesozoic geology of the Walter Springs area, Napa County, California. M, 1975, San Jose State University. 68 p.

Wagner, Donald E. Statistical decision theory applied to the focal mechanisms of Peruvian earthquakes. D, 1972, St. Louis University.

Wagner, Frances Joan Estelle. Paleontology and stratigraphy of the marine Pleistocene deposits of southwestern British Columbia. D, 1954, Stanford University. 141 p.

Wagner, Frederick John, Jr. The geology and mineral relationships of the Cornwall copper mine area in Ste. Genevieve County, Missouri. M, 1954, Washington University. 68 p.

Wagner, George H. Trace elements in the sediments of the Buffalo River, Arkansas. M, 1974, University of Arkansas, Fayetteville.

Wagner, George Robert. Sedimentation study of a lower Pennsylvanian conglomerate in Martin County, Indiana. M, 1956, Indiana University, Bloomington. 44 p.

Wagner, Gregory S. Waveform inversion for five African earthquakes and tectonic implications for continental deformation. M, 1986, Pennsylvania State University, University Park. 57 p.

Wagner, Harry Arthur, III. Stratigraphy and environmental interpretation of the Cuchillo, Penigno, Lagrima, and Finlay formations, Lower Cretaceous, Juarez Mountains, Chihuahua, Mexico. M, 1975, University of Texas at El Paso.

Wagner, Hugh McKinlay. A new species of Pliotaxidea (Mustelidae, Carnivora) with a discussion of the evolution of the Nearctic Taxidiinae. M, 1974, University of California, Berkeley. 59 p.

Wagner, Hugh McKinlay. Geochronology of the Mehrten Formation in Stanislaus County, California. D, 1981, University of California, Riverside. 385 p.

Wagner, James Kendall. Stratigraphy of the Gilmore City Formation of north-central Iowa. M, 1962, Iowa State University of Science and Technology.

Wagner, Jeffrey Karl. Reflection spectroscopy of stony meteorites in the vacuum ultraviolet spectral region. D, 1980, University of Pittsburgh. 232 p.

Wagner, Jesse Ross. Late Cenozoic history of the Coast Ranges east of San Francisco Bay. D, 1978, University of California, Berkeley. 161 p.

Wagner, John Paul. Geology of Sawtooth Ridge Quadrangle, Kern County, California. M, 1951, University of California, Berkeley. 132 p.

Wagner, Joseph J. Geology of the Haycock Mountain 7.5 minute quadrangle, western Garfield County, Utah. M, 1984, Kent State University, Kent. 68 p.

Wagner, Karma L. Hydrocarbon production potential of Mississippian rocks in Cowley County, Kansas. M, 1984, Wichita State University. 132 p.

Wagner, Keith Brian. Petrology and diagenesis of the Smackover Formation (Upper Jurassic); Blackjack Creek Field, Florida. M, 1978, Duke University. 94 p.

Wagner, Lawrence Henry. Origin of some Recent tractional sedimentary structures in the Rio Puerco near Bernardo, Socorro County, New Mexico. M, 1963, University of New Mexico. 64 p.

Wagner, Mary Emma. Granulite facies rocks near Wayne, Pennsylvania. M, 1966, Bryn Mawr College. 34 p.

Wagner, Mary Emma. Metamorphism of the Precambrian Baltimore gneiss in southeastern Pennsylvania. D, 1972, Bryn Mawr College. 89 p.

Wagner, Norman Spencer. The structural geology of part of the Pocono Plateau (Pennsylvania). M, 1933, Cornell University.

Wagner, Oscar Emil. Fauna of the Silurian rocks of northern Illinois. M, 1929, University of Illinois, Chicago.

Wagner, Oscar Emil. The paleontology and stratigraphy of the Kaibab Limestone. D, 1932, University of Illinois, Chicago.

Wagner, Paul Anthony. Geochronology of the Ameralik dykes at Isua, West Greenland. M, 1982, University of Alberta. 92 p.

Wagner, Paul David. Geochemical characterization of meteoric diagenesis in limestone; development and applications. D, 1983, Brown University. 391 p.

Wagner, Philip Lee. A study of Beyrichia tuberculata (Kloden), a Silurian ostracod from the glacial drift of northern Germany. M, 1955, University of Michigan.

Wagner, Richard A. Geology for environmental planning in the Franklin Quadrangle, southwestern Ohio. M, 1974, Miami University (Ohio). 59 p.

Wagner, Richard A., Jr. Sand modeling of crustal extension. M, 1985, Massachusetts Institute of Technology. 54 p.

Wagner, Richard J. Algal reefs in early Paleozoic strata of Southeast Missouri. M, 1961, St. Louis University.

Wagner, Richard Joseph. Stratigraphic and structural controls and genesis of barite deposits in Washington County, Missouri. D, 1973, University of Michigan. 337 p.

Wagner, Robert David. The condensation and stability of chemically heterogeneous systems. M, 1979, Arizona State University. 87 p.

Wagner, Sheila L. Effect of Laramide folding on previously folded Precambrian metamorphic rocks, Madison County, Montana. M, 1967, Indiana University, Bloomington. 27 p.

Wagner, Walter Richard. Life and death assemblages of Zone 10 of the Calvert Formation, Calvert County, Maryland. M, 1955, Bryn Mawr College. 59 p.

Wagner, Walther O. Reservoir study of Cretaceous Woodbine Fields in the northern East Texas Basin. M, 1987, Baylor University. 139 p.

Wagner, Warren R. Geology and ore deposits of the Rocky Bar District, Idaho. M, 1939, University of Idaho. 32 p.

Wagner, Warren R. The geology of part of the south slope of the St. Joe Mountains, Shoshone County, Idaho. D, 1947, The Johns Hopkins University.

Wagner, Wayne R. Geology of the Chibex gold deposit, Chibougamau, Quebec. M, 1978, Universite du Quebec a Chicoutimi. 105 p.

Wagner, William Philip. Correlation of Rocky Mountain and Laurentide glacial chronologies in southwestern Alberta, Canada. D, 1966, University of Michigan. 169 p.

Wagner, William Phillip. Snow facies on the Kaskawaulsh Glacier, Yukon Territory, Canada. M, 1963, University of Michigan.

Wagoner, Jeffrey L. Stratigraphy and sedimentation of the Pleistocene Brawley and Borrego formations in the San Felipe Hills area, Imperial Valley, California, U.S.A. M, 1977, University of California, Riverside. 128 p.

Wagoner, John Charles Van see Van Wagoner, John Charles

Wagoner, Steven Lewis van see van Wagoner, Steven Lewis

Wagstaff, Donald Allan. The geology of the SW 1/4 of the Meyers Cove Quadrangle, Lemhi County, Idaho. M, 1979, University of Idaho. 101 p.

Wagstaff, Melvin D., Jr. Three-dimensional velocity structure of the lithosphere beneath the Hanford Array in Washington State. M, 1986, University of North Carolina, Chapel Hill. 119 p.

Wagstaff, Ronald A. Analysis of the dynamic forces exerted in deep-sea sediments during penetration by a probe. M, 1966, [University of Miami].

Wahab, Mahmoud M. Abdel see Abdel Wahab, Mahmoud M.

Wahab, Nasar Ahamad. Analysis of water resources in Safford, Arizona. M, 1972, University of Arizona.

Wahab, Osman Abdel. Aplites and pegmatites in certain productive and barren North American Laramide and mid-Tertiary intrusions. D, 1974, University of Arizona. 246 p.

Wahab, Osman Abdel. The origin of the Kutum lead-zinc deposits, Darfur Province, Republic Sudan. M, 1968, University of Arizona.

Waheb, M. A. Plastic deformation of rocks. M, 1953, University of Missouri, Rolla.

Waheed, Abdul. Sedimentology of the coal bearing Bearpaw-Horseshoe Canyon Formation (Upper Cretaceous), Drumheller area, Alberta, Canada. M, 1983, University of Toronto.

Wahid, Ibrahim Abdel see Abdel Wahid, Ibrahim

Wahl, Carl Christian. A regional color study of subsurface Dakota Group samples from certain deep wells in Nebraska, including a summary list of Nebraska deep wells. M, 1948, University of Nebraska, Lincoln.

Wahl, David E., Jr. Geology of the El Salto Strip, Durango, Mexico. M, 1973, University of Texas, Austin.

Wahl, David Edwin, Jr. Mid-Tertiary volcanic geology in parts of Greenlee County, Arizona, Grant and Hidalgo counties, New Mexico. D, 1980, Arizona State University. 144 p.

Wahl, Floyd Michael. A petrographic study of the underclay of the No. 5 coal in Illinois. M, 1957, University of Illinois, Urbana.

Wahl, Floyd Michael. Reaction at elevated temperatures as investigated by continuous X-ray diffraction. D, 1958, University of Illinois, Urbana. 94 p.

Wahl, Harry Albert. An electrolytic model study of the effects of horizontal fractures on well productivity. M, 1956, University of Oklahoma. 100 p.

Wahl, Harry Albert. Sand movement in horizontal fractures. D, 1963, University of Oklahoma. 140 p.

Wahl, John Lesslie. Rock geochemical exploration at the Heath Steele and Key Anacon deposits, New Brunswick. D, 1978, University of New Brunswick.

Wahl, Ronald Richard. Interpretation of gravity data from the Carson Sink area, Nevada. M, 1965, Stanford University.

Wahl, William G. The Canica-Cawatose map area (Quebec). D, 1947, McGill University.

Wahlert, John Howard. The cranial foramina of protrogomorphous and sciuromorphous rodents; an anatomical and phylogenetic study. D, 1972, Harvard University.

Wahli, Catherine. The electromigration of copper and sulfate through a porous medium as a potential method for ground water remediation. M, 1988, University of Colorado. 157 p.

Wahlig, Barry Glenn. Transport of suspended matter through rock formations. D, 1980, Georgia Institute of Technology. 179 p.

Wahlman, Gregory Paul. Middle and Upper Ordovician Monoplacophora and bellerophontacean Gastropoda of the Cincinnati Arch region. D, 1985, University of Cincinnati. 815 p.

Wahlman, Gregory Paul. Stratigraphy, structure, paleontology, and paleoecology of the Silurian reef at Montpelier, Indiana. M, 1974, Indiana University, Bloomington. 71 p.

Wahlstedt, Warren J. Geology of the Milsap Creek and Temple Canyon areas, Fremont County, Colorado. M, 1964, University of Kansas. 49 p.

Wahlstrom, Ernest E. The geology of west-central Boulder County, Colorado. D, 1939, Harvard University.

Wahlstrom, Ernest Eugene. Geology of the Lake Albion region, Boulder County, Colorado. M, 1933, University of Colorado.

Wahnon, Ethel. Multivariate analysis of geochemical lake water and sediment data from northeastern Saskatchewan; application for uranium exploration. M, 1983, Ecole Polytechnique. 99 p.

Wahr, J. M. The tidal motions of a rotating, elliptical, elastic and oceanless Earth. D, 1979, University of Colorado. 227 p.

Wahrhaftig, Clyde A. Geology of the southeast portion of Mt. Lowe Quadrangle (California). M, 1941, California Institute of Technology.

Wahrhaftig, Clyde A. Quaternary geology of the Nenana River and adjacent parts of the Alaska Range, Alaska. D, 1953, Harvard University.

Wahrhaftig, Leon. Insoluble residues of the Wabaunsee Group limestones of the Upper Pennsylvanian in northeastern Kansas. M, 1952, University of Kansas. 129 p.

Wai, Raymond Sheung-Che. Rock behavior under elevated temperatures. D, 1981, University of Western Ontario.

Wai, U. Thit. Reservoir properties from cores compared with well logs, Taglu gas field, Canada. M, 1975, University of Tulsa. 103 p.

Waines, H. R. Stromatoporoids from the upper Abitibi Limestone. M, 1956, University of Toronto.

Waines, R. H. A study of stromatoporoids from northern Ontario. M, 1956, University of Toronto.

Waines, Russell Hamilton. Devonian stromatoporoids of Nevada. D, 1965, University of California, Berkeley. 505 p.

Wainner, Kenneth Fred. A petrologic study of certain Upper Pennsylvanian sandstones of the Kansas River valley, Kansas. M, 1939, University of Nebraska, Lincoln.

Wainright, Elizabeth Jane. Calculation of isostatic gravity anomalies and isostatic geoid heights using two-dimensional filtering; implications for structure in subduction zones. D, 1983, Texas A&M University. 97 p.

Wainwright, John Ernest Nolan. Morphology and taxonomy of some Middle Silurian Ostracoda. D, 1959, University of Illinois, Urbana. 161 p.

Wainwright, John Ernest Nolan. Some late Tertiary Ostracoda from the Makran Coast, Pakistan. M, 1957, University of Illinois, Urbana.

Wainwright, Kenneth E. An interpretation of the Big Springs, Kansas, gravity anomaly. M, 1960, University of Kansas. 44 p.

Wainwright, Laurin L. The geology of Greenland-Prairie Grove area, Washington County, Arkansas. M, 1961, University of Arkansas, Fayetteville.

Waisgerber, William. Later Mesozoic stratigraphy of the Jim Robertson Ranch area, central Oregon. M, 1956, Oregon State University. 82 p.

Waisley, Sandra L. Petrography and depositional environment of the Cuesta del Cura Formation, upper Albian-lower Cenomanian, northeastern Mexico. M, 1978, University of Texas, Austin.

Waisman, Dave. Geology and mineralization of the Black Pine Mine, Granite County, Montana. M, 1985, University of Montana. 66 p.

Waite, Burt A. Environmental geology of the Huntington Valley, Vermont. M, 1971, University of Vermont.

Waite, Herbert A. A study of the geology and ground water resources of Keith County, Nebraska. M, 1937, University of Nebraska, Lincoln.

Waite, Lowell E. Biostratigraphy and paleoenvironmental analysis of the Sierra Madre Limestone (Cretaceous), Chiapas, southern Mexico. D, 1983, University of Texas, Arlington. 192 p.

Waite, Harold. The Silurian of the Kearsarge area California. M, 1953, University of California, Berkeley. 0 p.

Waite, Verdi V. A review of the Paleozoic stratigraphy of Oklahoma. M, 1917, University of Oklahoma. 120 p.

Waitt, Mary Cooper. Desert dunes of the Kermit Sandhills, Winkler County, Texas. M, 1969, University of Texas, Austin.

Waitt, Richard Brown Jr. Geomorphology and glacial geology of the Methow drainage basin, eastern North Cascade Range, Washington. D, 1972, University of Washington. 154 p.

Waitt, Richard Brown, Jr. Ignimbrites of the Sierra Madre Occidental between Durango and Mazatlan, Mexico. M, 1970, University of Texas, Austin.

Waitzenegger, Bernard. Pétrographie et géochimie des épontes quartzodioritiques de la mine d'or Ferderber (Belmoral), Val d'Or, Québec. M, 1986, Ecole Polytechnique. 152 p.

Wakefield, Lawrence Woodbury. Geology of the Boetcher Ridge-Sheep Mountain-Delanos Butte area, North Park (Jackson County), Colorado. M, 1952, University of Colorado.

Wakeham, Stuart Glenwood. The geochemistry of hydrocarbons in Lake Washington. D, 1976, University of Washington. 192 p.

Wakeham, Susan Elizabeth. Petrochemical patterns in young pillow basalts dredged from Juan De Fuca and Gorda ridges. M, 1978, Oregon State University. 95 p.

Wakeland, M. E., Jr. Provenance and dispersal patterns of fine-grained sediments in Long Island Sound. D, 1977, University of Connecticut. 262 p.

Wakeley, John Raymond. The nature and origin of a chert-rich beach sand, Rodeo Cove, Marin Peninsula, California. M, 1968, University of California, Berkeley. 91 p.

Wakeley, Lillian Donley. Petrology of the Upper Nounan-Worm Creek Sequence, Upper Cambrian Nounan and St. Charles formations, Southeast Idaho. M, 1975, Utah State University. 127 p.

Wakely, William James. Fossil gastropods from the Pleistocene terrace-fills of the Elkhorn River valley, Nebraska. M, 1955, University of Nebraska, Lincoln.

Wakelyn, Brian D. Petrology of the Smackover Formation (Jurassic), Perry and Stone counties, Mississippi. M, 1976, Tulane University.

Wakeman, James Fisher. The structure, stratigraphy, and gas storage potentials of an area near West Point, Indiana. M, 1957, Indiana University, Bloomington. 50 p.

Wakhungu, Judi W. The micropaleontology of the Jurassic-Cretaceous boundary at Cape Espichel, Portugal. M, 1986, Acadia University. 174 p.

Walasko, James Thadius. Petrology of a Permo-Carboniferous section, northern Jasper National Park. M, 1962, University of Alberta. 123 p.

Walawender, Michael J. Petrogenesis of the Bear Mountain amphibolites (southwestern Black Hills, west of Custer, South Dakota). M, 1967, South Dakota School of Mines & Technology.

Walawender, Michael John. A study of the Charlevoix structure, Quebec, Canada. D, 1972, Pennsylvania State University, University Park. 93 p.

Walbridge, Stephen Rorick. Optimal well field designs for the rehabilitation of the Gloucester special waste site near Ottawa, Ontario, Canada. M, 1985, Stanford University. 153 p.

Walch, Andrew F. A quantitative study of some Platteville (Ordovician) gastropods. M, 1959, University of Wisconsin-Madison.

Walch, C. A. Recent abyssal benthic foraminifera from the eastern equatorial Pacific. M, 1978, University of Southern California.

Walchessen, Anne Aurelia. The geology and ore deposits of a portion of the Tyndall mining district, Santa Cruz County, Arizona. M, 1983, University of Arizona. 89 p.

Walck, Marianne Carol. Teleseismic array analysis of upper mantle compressional velocity structure. D, 1984, California Institute of Technology. 243 p.

Walcott, Albert J. A study of some factors influencing crystal habit. D, 1926, University of Michigan.

Wald, Alan R. The impact of truck traffic and road maintenance on suspended-sediment yield from a 14' standard forest road. M, 1975, University of Washington. 38 p.

Wald, S. D. Biometric characterization of Schizodus (Bivalvia) shells from Lander, Wyoming. M, 1974, Columbia University. 32 p.

Waldbaum, David R. Calorimetric investigation of the alkali feldspars. D, 1966, Harvard University.

Waldbaum, David R. Structural and thermodynamic properties of silver iodide. M, 1960, Massachusetts Institute of Technology. 61 p.

Waldbaum, Raymond. Geology of the Portuguese Mountain area, Nye County, Nevada. M, 1970, University of San Diego.

Walden, Kyle Douglas. A stratigraphic and structural study of Coal Mine Basin, Idaho-Oregon. M, 1986, Michigan State University. 70 p.

Walden, Ronald L. The origin of schist and amphibolite bodies within the Groveland iron mine, central Dickinson County, northern Michigan. M, 1984, Bowling Green State University. 117 p.

Walder, Joseph Scott. Coupling between fluid flow and deformation in porous crustal rocks. D, 1984, Stanford University. 264 p.

Walder, Joseph Scott. Field and theoretical studies of subglacial hydrology. M, 1980, Stanford University. 88 p.

Waldichuk, Michael. Physical oceanography of the Strait of Georgia, British Columbia. D, 1955, University of Washington. 275 p.

Waldman, Estella. Time significance of gamma-ray peaks; an example from the Helderberg Group (Lower Devonian), New York. M, 1979, Rensselaer Polytechnic Institute. 28 p.

Waldo, Allen Worcester. Identification of the copper ore minerals by means of X-ray powder diffraction patterns. D, 1934, Harvard University.

Waldo, Allen Worcester. The Lonsdale Limestone and its fauna in Illinois. M, 1928, University of Illinois, Chicago.

Waldo, Jaunell Jean. A survey of the engineering geologic information needs of land use planners. M, 1986, Washington State University. 137 4 plates p.

Waldon, Colin E. Chemistry and mineralogy of Morgan and Omaha saline lakes, Natrona County, Wyoming. M, 1971, University of Wyoming. 63 p.

Waldram, Robert James. Conodonts from the Cooper Limestone (Middle Devonian) of Missouri. M, 1942, University of Missouri, Columbia.

Waldrip, David Bennett. The effect of sanitary landfills on water quality in southern Indiana. D, 1975, Indiana University, Bloomington. 159 p.

Waldron, David Anthony. Structural characteristics of a subducting oceanic plate off western Canada. M, 1982, University of British Columbia. 131 p.

Waldron, John Francis. Reconnaissance geology and ground water study of a part of Socorro County, New Mexico. D, 1957, Stanford University. 302 p.

Waldron, Kim Ann. High pressure contact metamorphism in the aureole of the Cortlandt Complex, southeastern New York. D, 1986, Yale University. 142 p.

Waldron, Richard L. Hydrothermal alteration of the Gamma Ridge rocks, on Glacier Peak, and their relation to hot spring activity. M, 1986, Western Washington University. 57 p.

Waldron, Robert P. A seasonal ecologic study of foraminifera from Timbalier Bay, Louisiana. M, 1958, Louisiana State University.

Waldroop, William W. A petroscopic study of factors affecting reservoir rock characteristics. M, 1964, University of Oklahoma. 47 p.

Waldrop, Ann Lyneve Chapman. The study of minerals with the formula type $ABXO_4(Z)$ with special attention to the crystal structures of the triplite-triploidite group. D, 1970, Massachusetts Institute of Technology. 136 p.

Waldrop, Henry A. Arapahoe Glacier, Boulder County, Colorado. M, 1962, University of Colorado.

Waldschmidt, William A. Examination of some lead ores from the Coeur d'Alene District, Idaho. M, 1922, Massachusetts Institute of Technology. 49 p.

Waldschmidt, William Albert. Table Mountains and associated igneous rock near Golden (Jefferson County), Colorado. D, 1938, University of Colorado.

Walen, Michael B. Petrogenesis of the granitic rocks of part of the upper Granite Gorge, Grand Canyon, Arizona. M, 1973, Western Washington University. 92 p.

Walen, Sarah Kimball. Hydrogeology and artificial recharge potential of Holden, Massachusetts. M, 1983, University of Massachusetts. 138 p.

Walia, Daman S. Phase relations in the $PbS-Bi_2S_3-Sb_2S_3-As_2S_3$ system. D, 1977, Miami University (Ohio). 179 p.

Walia, Daman S. Phase relations in the systems; $PbS-As_2S_3$ and $PbS-As_2S_3-Bi_2S_3$. M, 1972, Miami University (Ohio). 40 p.

Walk, Hugh Gerard. Stratigraphy of the Leitchfield Group, Kentucky. M, 1948, University of Illinois, Urbana.

Walka, Joseph August. The geography of the Grays Summit Saddle. M, 1936, Washington University. 64 p.

Walker, A. T., III. Trace element variation in the volcanic rocks of the Adak and Umnak Islands of the Aleutian Arc. M, 1974, Columbia University. 39 p.

Walker, Alfred Thomas, III. The Preston Gabbro of southeastern Connecticut; geochemistry and history of formation. D, 1983, Lehigh University. 208 p.

Walker, Alta. Geological evaluation of remote sensing imagery of the Mesabi Range, Minn. M, 1971, University of Minnesota, Minneapolis.

Walker, Alta S. Inert gas investigations of five Apollo 11 and 12 breccias and of an Apollo 17 soil sample. D, 1977, Rice University. 146 p.

Walker, Ann Leslie. Comparison of anomalously low vitrinite reflectance values with other thermal maturation indices in and near the Playa del Rey Oilfield, California. M, 1982, University of Washington. 190 p.

Walker, April Rubens. Agriculture and nonrenewable resources; the case of irrigation on the Ogallala Aquifer. M, 1988, West Virginia University. 88 p.

Walker, Bruce Dawson. Soils of the Truelove Lowland and vicinity, Devon Island, Northwest Territories. M, 1976, University of Alberta.

Walker, Bruce M. Petrogenesis of oceanic granites from the Aves Ridge in the Caribbean Basin. M, 1972, Michigan State University. 7 p.

Walker, Bruce Michael. The origin of quartz-feldspar, graphic intergrowths. D, 1975, Michigan State University. 65 p.

Walker, Carl Hampton. Comparative study of the Newark of North Carolina and New England. M, 1923, University of North Carolina, Chapel Hill. 25 p.

Walker, Cecil Lester. Geology of the Cumberland Reservoir-Little Muddy Creek area, Lincoln and Uinta counties, Wyoming. M, 1943, University of Wyoming. 210 p.

Walker, Charles Stephen. The stratigraphy and depositional environment of the upper Proterozoic and Cambrian Brigham Group near Mink Creek, Idaho. M, 1983, Idaho State University. 55 p.

Walker, Charles William. False cap rock overlying Gulf Coast salt domes; analysis and origin. M, 1968, University of Mississippi.

Walker, Charles William. The nature and origin of caprock overlying Gulf Coast salt dome. D, 1972, Louisiana State University. 269 p.

Walker, Clifford A. Stratigraphy and environmental facies of the Glen Rose Formation, Coryell and Lampasas counties, Texas. M, 1973, Baylor University. 119 p.

Walker, Constance M. Stratigraphy, sedimentation and mineralization of the Vinini Formation, Elko County, Nevada. M, 1985, Colorado School of Mines. 108 p.

Walker, D. A. A study of the northwestern Pacific upper mantle. M, 1965, University of Hawaii.

Walker, D. A. Oceanic mantle phases recorded on hydrophones and seismographs in the northwestern Pacific at distances between 7° and 40°. D, 1971, University of Hawaii. 65 p.

Walker, Danny Norbert. Studies on the late Pleistocene mammalian fauna of Wyoming. D, 1986, University of Wyoming. 223 p.

Walker, David. Experimental petrology of lunar basalts. D, 1972, Harvard University.

Walker, Douglas J. An experimental study of wind ripples. M, 1981, University of Waterloo.

Walker, Edward Bullock, III. Stratigraphic relations of the Boston Basin sediments (Massachusetts). M, 1947, Massachusetts Institute of Technology. 65 p.

Walker, Eugene H. I, Erosion surfaces and the uplift of the Andes in the vicinity of Llallagua, Bolivia; II, The primary mineralization of the eight principal disseminated-type copper deposits of the Western United States. D, 1948, Harvard University.

Walker, Frank Haff. Boyle Limestone of the southeastern Knob Belt. M, 1950, University of Kentucky. 37 p.

Walker, George Ernest. Geology and ground water of Amargosa Valley, Nevada and California. M, 1963, University of Oklahoma. 67 p.

Walker, George Pinckney, III. Subsurface geologic reconnaissance of Kerr Basin and adjacent areas, south-central Texas. M, 1967, University of Texas, Austin.

Walker, George W. A study of the sedimentary petrography of the (Eocene) Sierra Blanca Limestone and associated rocks, Santa Barbara County, California. M, 1948, Stanford University. 36 p.

Walker, Gerald Grant. Transient electromagnetics for permafrost. D, 1988, University of Alaska, Fairbanks. 308 p.

Walker, Giles E. Intrusive relations of the batholith of Southern California near Bonsall, California. M, 1959, University of Arizona.

Walker, Graham Thomas. High plains depressions in eastern Colorado; distribution, classification, and genesis. D, 1985, University of Denver. 297 p.

Walker, Ian Robert. Late-Quaternary palaeoecology of Chironomidae (Diptera; Insecta) from lake sediments in British Columbia. D, 1988, Simon Fraser University. 204 p.

Walker, Ian Robert. Geology and ground water resources of the Wagner Quadrangle, South Dakota. M, 1963, University of South Dakota. 225 p.

Walker, J. L. Pedogenesis of some highly ferruginous formations in Hawaii. D, 1962, University of Hawaii. 406 p.

Walker, J. P. Sedimentology of the Medina and Clinton groups (lower and middle Silurian) of western and central New York. M, 1975, University of Illinois, Urbana. 105 p.

Walker, J. W. R. Origin of granophyre and alteration of gabbro and andesites, Godfrey Township, Ontario. M, 1954, Queen's University. 95 p.

Walker, Jack R. Correlation of the basal Mesozoic from the outcrop to the subsurface in Northeast Texas. M, 1959, Texas Christian University.

Walker, James Allen. Geochemical constraints on the petrogenesis of behind-the-front volcanism in Central America. M, 1980, Rutgers, The State University, New Brunswick. 124 p.

Walker, James Allen. Volcanic rocks from Nejapa and Granada cinder cone alignments, Nicaragua, Central America. D, 1982, Rutgers, The State University, New Brunswick. 143 p.

Walker, James D. Tectonics and dispersal of Hayner Basin sediments (Miocene), San Diego Mountain, Dona Ana County, south-central New Mexico. M, 1986, New Mexico State University, Las Cruces. 49 p.

Walker, James Douglas. An experimental study of wind ripples. M, 1981, Massachusetts Institute of Technology. 145 p.

Walker, James Douglas. Permo-Triassic paleogeography and tectonics of the Southwestern United States. D, 1985, Massachusetts Institute of Technology. 224 p.

Walker, James K. The geology and hydrogeology of a portion of the Great Miami Valley Aquifer south of Dayton, Ohio. M, 1983, Wright State University. 110 p.

Walker, James Steven. Asbestos and the asbestiform habit of minerals. M, 1981, University of Minnesota, Minneapolis. 164 p.

Walker, Jeffrey Ross. Pliocene stratigraphy of the Klamath River Gorge, Oregon. M, 1983, Dartmouth College. 126 p.

Walker, Jeffrey Ross. Structural and compositional aspects of low-grade metamorphic chlorite. D, 1987, Dartmouth College. 62 p.

Walker, Jerry C. A gravity and magnetic study of south-central Fayette County, Ohio. M, 1979, Wright State University. 105 p.

Walker, Jimmy Roy. The geomorphic evolution of the Southern High Plains. M, 1977, Baylor University. 123 p.

Walker, Joe Dudgeon, Jr. Subsurface geology of Fayette County, Texas. M, 1955, University of Texas, Austin.

Walker, John. The capabilities and limitations of the Oneonta, N. Y. municipal water supply. M, 1985, SUNY, College at Oneonta. 52 p.

Walker, John Anthony, Jr. Stratigraphy and facies relationships of selected roadcuts in Tuscarawas, Guernsey, and Coshocton counties, Ohio. M, 1975, Ohio University, Athens. 225 p.

Walker, John Fortune. Geology and mineral deposits of Windermere map area, British Columbia. D, 1924, Princeton University.

Walker, John Weldon. Stratigraphy and insoluble residues of the Inkster Junction core, Wayne County, Michigan. M, 1953, University of Michigan.

Walker, John William Richard. Geology of the Jackfish-Middleton area, District of Thunder Bay, Ontario, Canada. D, 1961, University of California, Los Angeles.

Walker, Keith F. The geology of the northern half of Rockwell Quadrangle, Texas. M, 1948, University of Oklahoma. 34 p.

Walker, Kenneth Russell. The stratigraphy and petrography of the Price-Pocono Formation in a portion of southwestern Virginia. M, 1964, University of North Carolina, Chapel Hill. 123 p.

Walker, Kenneth Russell. The stratigraphy, environmental sedimentology, and community ecology of the middle Ordovician Black River group of New York State. D, 1969, Yale University. 368 p.

Walker, Laurence G. Stratigraphy of the Ordovician Martinsburg Formation in southwestern Virginia. D, 1967, Harvard University.

Walker, Laurence Graves. Geology of the Mt. Hope area, Garden Valley Quadrangle, Nevada. M, 1962, University of California, Los Angeles.

Walker, Lawrence Price. Depositional environments, petrography and diagenesis of the Middle to Upper Devonian Misener Sandstone in north-central Oklahoma. M, 1986, Oklahoma State University. 97 p.

Walker, Lydia P. Growth rates and mortality factors of selected Miocene gastropods (Saint Mary-Yorktown Formation) of the Virginia coastal plains, Virginia. M, 1972, Virginia State University. 132 p.

Walker, Marilyn Drew. Vegetation and floristics of pingos, central Arctic coastal plain, Alaska. D, 1987, University of Colorado. 432 p.

Walker, Michael. Late Wisconsin ice in the Morley Flats area of the Bow Valley and adjacent areas of the Kananaskis Valley, Alberta. M, 1971, University of Calgary.

Walker, Monte Eugene. Petrology of the limestones of the Renault Formation (Mississippian System) from southwestern Illinois and southeastern Missouri. M, 1985, Southern Illinois University, Carbondale. 122 p.

Walker, Nan Delene. Physical responses of southern Florida and northern Bahama lagoon waters to severe cold air outbreaks and effects on hermatypic coral reefs. M, 1982, Louisiana State University. 114 p.

Walker, Nancy Denning. Remote sensing analysis of southern Walker Lane. M, 1986, University of Nevada. 117 p.

Walker, Nicholas Warren. U/Pb geochronologic and petrologic studies in the Blue Mountains Terrane, northeastern Oregon and westernmost-central Idaho; implications for pre-Tertiary tectonic evolution. D, 1986, University of California, Santa Barbara. 224 p.

Walker, Nola Constance. A study of the application of biogeochemistry in the Key Lake area, Saskatchewan. M, 1980, University of Regina. 169 p.

Walker, Patricia Ellen Grove. A regional study of the diagenetic and geochemical character of the Pennsylvanian Morrow Formation, Anadarko Basin, Oklahoma. M, 1986, Oklahoma State University. 156 p.

Walker, Patrick M. Geologic hazard analysis of suburban development on alluvial deposits in the Basin and Range Province. M, 1981, University of Nevada. 163 p.

Walker, Philip Caleb. The forest sequence of the Hartstown bog area [Pennsylvania]. D, 1958, University of Pittsburgh. 96 p.

Walker, Richard John. The origin of the Tin Mountain Pegmatite, Black Hills, South Dakota. D, 1984, SUNY at Stony Brook. 400 p.

Walker, Richard Randall. The geology and uranium deposits of Proterozoic rocks, Simpson Islands, Northwest Territories. M, 1977, University of Alberta. 193 p.

Walker, Robert Dean. Structural petrology at the Crystal Falls Municipal Dam, Iron County, Michigan. M, 1956, Michigan State University. 51 p.

Walker, Robert Edwin. A statistical study of the evolution of Mesolobus mesolobus. M, 1952, University of Wisconsin-Madison.

Walker, Robert F. Geomorphology of the southern part of Big Smoky Valley, Esmeralda County, Nevada. M, 1966, University of Massachusetts. 101 p.

Walker, Robert J. The mineralogy and distribution of phyllosilicates associated with the Henderson molybdenite deposit, Clear Creek County, Colorado. M, 1982, Colorado School of Mines. 64 p.

Walker, Robert Keith. Structural geology and distribution of Lower and Middle Pennsylvanian sandstones in adjacent portions of Lincoln, Okfuskee, Seminole, and Pottawatomie counties, Oklahoma. M, 1982, Oklahoma State University. 56 p.

Walker, Russell Allen. Geology of the Filley Quadrangle, Cedar County, Missouri. M, 1961, University of Missouri, Columbia.

Walker, Scott Donald. Mesozoic tectonic evolution of the Twin Buttes Mine area, Pima County, Arizona; implicatons for a regional tectonic control of ore deposits in the Pima mining district. M, 1982, University of Arizona. 140 p.

Walker, Stephen Dade. Generalized linear inversion using tau-p forward modeling. M, 1987, University of Texas of the Permian Basin. 76 p.

Walker, Stephen David. Geology of the auriferous Chadbourne Breccia, Noranda, Quebec. M, 1981, University of Western Ontario. 97 p.

Walker, Steven T. Sedimentary structures of the western Florida slope and eastern Mississippi cone; distribution and geological implications. M, 1984, University of South Florida, St. Petersburg.

Walker, Steven W. H. High-intensity magnetic separation; beneficiation of bauxites to produce refractory and chemical grades. M, 1978, Indiana University, Bloomington. 69 p.

Walker, Susan Claire. The nature and origin of bentonites and potassium bentonites, northwestern Montana. M, 1987, University of Montana. 33 p.

Walker, Theodore Roscoe. The petrology of the Upper Cambrian sandstones of central Texas. D, 1952, University of Wisconsin-Madison.

Walker, Thomas F. The geology of the Elliston area, western Montana. M, 1967, University of North Dakota. 136 p.

Walker, Thomas Franklin. Stratigraphy and depositional environments of the Morrison and Kootenai formations in the Great Falls area; central Montana. D, 1974, University of Montana. 195 p.

Walker, Thomas Henry. The geology of the northwest quarter of the Ironton Quadrangle. M, 1942, Washington University.

Walker, Thomas Wiley. A study of cyclic aspects of origin and development of limestone caverns. M, 1952, University of Houston.

Walker, Timothy Dane. The evolution of taxonomic diversity in an adaptive mosaic. D, 1984, University of California, Santa Barbara. 156 p.

Walker, Valerie A. Depositional environments and petroleum potential of upper Jelm and Sundance formations, northern Laramie Basin, Wyoming. M, 1982, Colorado College.

Walker, Valerie Ann. Relationships among several breccia pipes and a lead-silver vein in the Copper Creek mining district, Pinal County, Arizona. M, 1979, University of Arizona. 163 p.

Walker, W. Archean magmatism and metallogeny; geological sequences in some Archean orefields. D, 1976, University of Ottawa. 385 p.

Walker, W. W., Jr. Some analytical methods applied to lake water quality problems. D, 1977, Harvard University. 557 p.

Walker, Warren. The geology of Township 15 North, Range 27 West, Madison County, Northwest Arkansas. M, 1951, University of Arkansas, Fayetteville.

Walker, Wayne T., Jr. Petrology and mineralization of the C-1 Zone, Sweetwater Mine, Viburnum Trend, Southeast Missouri. M, 1976, University of Missouri, Rolla.

Walker, William B. Petrography of the Pearlette Volcanic Ash (Pleistocene) in southeastern Nebraska. M, 1967, University of Nebraska, Lincoln.

Walker, William Joseph. The state and solubility of cadmium as related to xenotic inorganic phases generated homogeneously in soils. D, 1985, University of California, Davis. 151 p.

Walker, William Peter, Jr. Geology of a part of the Owyhee Mountains south of Silver City, Idaho. M, 1965, University of Idaho. 59 p.

Walker, Willis Lavern. Earth dams; geotechnical considerations in design and construction. D, 1984, Oklahoma State University. 315 p.

Walkey, Clifton. Geochemistry and structural setting of a geothermal spring located north of the Washington-Oregon border proximate to the Snake River. M, 1984, Washington State University.

Walko, George R. Precambrian geology of the northeastern Kinikinik Quadrangle, Northern Front range, Colorado. M, 1969, SUNY at Buffalo. 64 p.

Wall, David Joseph. Hydrologic time series model choice. D, 1981, University of Pittsburgh. 228 p.

Wall, Duncan Arthur. Dinoflagellate cysts and acritarchs from California Current surface sediments. D, 1986, University of Saskatchewan. 304 p.

Wall, Ellen R. The occurrence of staurolite and its implications for polymetamorphism in the Mt. Washington area, New Hampshire. M, 1988, University of Maine. 124 p.

Wall, Frederick M. The clay mineralogy of continental slope and rise sediments off the Eastern United States. M, 1981, University of South Florida, St. Petersburg. 81 p.

Wall, James Ray. The age of the Gosport Sand in Alabama. M, 1952, Mississippi State University. 62 p.

Wall, James Roy. Geology and regional relationships of the Sierra del Fraile, Nuevo Leon, Mexico. D, 1961, Louisiana State University. 142 p.

Wall, John Hallet. Jurassic microfaunas of Saskatchewan (Canada). D, 1957, University of Missouri, Columbia.

Wall, John Hallett. Cenomanian-Turonian foraminifera from Kaskapau Formation, Peace River area, Western Canada. M, 1951, University of Alberta. 134 p.

Wall, John Hallett. Jurassic microfaunas from Saskatchewan, Western Canada. D, 1958, University of Missouri, Columbia. 448 p.

Wall, Leeman Jack. Early predictions of oil recovery from a graphical solution of the straight line material balance. M, 1963, University of Oklahoma. 39 p.

Wall, Linda Sue. Distribution of northern Channel Islands foraminifera and its relation to sediments. M, 1979, University of Southern California.

Wall, Robert Ecki. Geophysical investigations in the central basin of Lake Erie. D, 1965, Columbia University. 73 p.

Wall, William Patrick. Systematics, phylogeny, and functional morphology of the Amynodontidae (Perissodactyla; Rhinocerotoidea). D, 1981, University of Massachusetts. 320 p.

Wallace, A. R. Geology and ore deposits, Kennedy mining district, Pershing County, Nevada. M, 1977, University of Colorado.

Wallace, Alan Joseph. The Lamotte Sandstone in the region of the Farmington Anticline. M, 1938, Washington University. 106 p.

Wallace, Alan Ryon. Alteration and vein mineralization, Schwartzwalder uranium deposit, Front Range, Colorado. D, 1983, Oregon State University. 172 p.

Wallace, Andy B. Geology and mineral deposits of the Pyramid District, southern Washoe County, Nevada. D, 1975, University of Nevada. 197 p.

Wallace, Andy Bert. Mineral deposits of the Indio Mountains, Hudspeth County, Texas. M, 1972, University of Texas at El Paso.

Wallace, Blanche. Petrology and geochemistry of the Wahoo Creek Formation, Stone Mountain Quadrangle, Georgia. M, 1981, Georgia Institute of Technology. 148 p.

Wallace, Charles Michael. Relationship of pore size to texture in some carbonate rocks. M, 1962, Iowa State University of Science and Technology.

Wallace, Chester A. A basin analysis of the Upper Precambrian Uinta Mountain group, Utah. D, 1972, University of California, Santa Barbara.

Wallace, Cloyd Russell. The Productinae of the Chouteau Limestone. M, 1930, University of Missouri, Columbia.

Wallace, Debbie L. Regionalizing water quality; the Province model. M, 1980, Northern Illinois University. 113 p.

Wallace, Derek. Geothermal measurements on the Sohm abyssal plain, Northwest Atlantic. M, 1985, [Dalhousie University]. 183 p.

Wallace, Don Lee. Subsurface geology of the Chitwood area, Grady County, Oklahoma. M, 1953, University of Oklahoma. 69 p.

Wallace, Donna. Evolution of the drainage in a portion of the Bighorn Basin near Shell, Wyoming. M, 1971, Iowa State University of Science and Technology.

Wallace, Dorothy M. Piedmont terraces in Pennsylvania and Maryland. M, 1935, Columbia University, Teachers College.

Wallace, G. T., Jr. Particulate matter in surface seawater; sources, chemical composition, and affinity for the sea-air interface. D, 1976, University of Rhode Island. 303 p.

Wallace, Garnet Cecil Grant. Deposition and diagenesis of the upper Red River Formation (upper Fort Garry Member) and Stony Mountain Formation (Upper Ordovician) north and west of Winnipeg, Manitoba. M, 1979, University of Manitoba. 170 p.

Wallace, Graeme M. B. Petrogenesis of the McGerrigle plutonic complex, Gaspe, Quebec. M, 1988, McGill University. 296 p.

Wallace, Henry. Differentiation trends in Shesheeb Bay section of Osler volcanics (Precambrian, Ontario). M, 1972, University of Toronto.

Wallace, James. Lithium uptake in burial diagenesis. M, 1974, Kent State University, Kent.

Wallace, James Alan. The geology of the Aufeas Mine at Silver Creek, BC. M, 1942, University of British Columbia.

Wallace, James Hay. A structure refinement of six metamorphic cordierites. M, 1978, University of California, Berkeley. 50 p.

Wallace, James Ray. Mineralogy and textures of certain manganese deposits in the Hampton-Shady Valley districts of Northeast Tennessee. M, 1973, University of Tennessee, Knoxville. 95 p.

Wallace, Jane House. A petrographical and petrological study of the Anorthosite, Grenville sediments, and Diabase at Mountain Pond, Franklin County, New York. M, 1948, Smith College. 105 p.

Wallace, Kenneth C. Depositional history of the Codell Sandstone (Upper Cretaceous) in Kansas. M, 1979, Fort Hays State University. 51 p.

Wallace, Kevin John. Clay petrology of the Bucatunna Formation. M, 1984, University of New Orleans. 264 p.

Wallace, Kirk D. Estimation of source parameters for micro-earthquakes near the Sleep Hollow oil field, Red Willow County, Nebraska. M, 1986, University of Kansas. 96 p.

Wallace, Maurice H. Geochemical investigations in the Tri-State zinc and lead mining district. M, 1942, University of Kansas. 72 p.

Wallace, Peter Ian. Strain analysis; Upper Manitou Lake, northwestern Ontario. M, 1975, McMaster University. 126 p.

Wallace, Pollack Austin. Geology and development of the West Edmond oil pool. M, 1935, University of Oklahoma. 37 p.

Wallace, R. Scott. Quantification of net shore-drift rates in Puget Sound and the Strait of Juan de Fuca, Washington. M, 1987, Western Washington University. 58 p.

Wallace, Richard Warren. A finite-element, planar-flow model of Camas Prairie, Idaho. D, 1972, University of Idaho. 91 p.

Wallace, Richard Warren. Initial erosional effects on Cary drift plain, central Iowa. M, 1961, Iowa State University of Science and Technology.

Wallace, Robert Duncan. Evaluation of possible detachment faulting west of the San Andreas Fault, southern Santa Rosa Mountains, California. M, 1982, San Diego State University. 77 p.

Wallace, Robert Earl. A Miocene mammalian fauna from Beatty Buttes, Oregon. D, 1946, California Institute of Technology.

Wallace, Robert Earl. A portion of the San Andreas Rift in Southern California. D, 1946, California Institute of Technology. 113 p.

Wallace, Robert Earl. Volcanic tuff beds of the (Miocene) Mint Canyon Formation (Southern California). M, 1940, California Institute of Technology. 73 p.

Wallace, Robert J. The paleoecology of the Browns Town and Montpelier limestones (Oligocene-Miocene) of Jamaica. M, 1970, Northern Illinois University. 88 p.

Wallace, Robert James. A reconnaissance of the sedimentology and ecology of Glovers Reef atoll, Belize (British Honduras). D, 1975, Princeton University. 150 p.

Wallace, Robert Manning. Stratigraphy and structure of a part of the Canada del Oro District, Santa Catalina Mountains, Pinal County, Arizona. M, 1951, University of Arizona.

Wallace, Roberts Manning. Structure of the northern end of the Santa Catalina Mountains, Arizona. D, 1955, University of Arizona. 93 p.

Wallace, Ronald G. Late Cenozoic mass movement along part of the west edge of Wasatch Plateau, Utah. M, 1964, Ohio State University.

Wallace, Ronald Gary. Types and rates of alpine mass movement, west edge of Boulder County, Colorado Front Range. D, 1967, Ohio State University. 240 p.

Wallace, Ronald James. The origin and diagenesis of the phosphate deposit in the middle Miocene Hawthorn Formation in Northeast Chatham County, Georgia. M, 1980, University of Kansas. 71 p.

Wallace, Ronald Louis. Gravity survey of the Kent Quadrangle, Portage County, Ohio. M, 1978, Kent State University, Kent. 132 p.

Wallace, Stewart Raynor. Geology of part of the Tendoy Mountains, Beaverhead County, Montana. M, 1948, University of Michigan. 56 p.

Wallace, Stewart Raynor. The petrology of the Judith Mountains, Fergus County, Montana. D, 1951, University of Michigan. 245 p.

Wallace, Terry Charles, Jr. Long period regional body waves. D, 1983, California Institute of Technology. 186 p.

Wallace, Wesley K. Bedrock geology of the Ross Lake fault zone in the Skymo Creek area, North Cascades

National Park, Washington. M, 1976, University of Washington. 111 p.

Wallace, Wesley K. Structure and petrology of a portion of a regional thrust zone in the central Chugach Mountains, Alaska. D, 1981, University of Washington. 253 p.

Wallace, William. Investigation of Darcy's law for low hydraulic gradients. M, 1962, Stanford University.

Wallace, William E., Jr. A study of deep-seated domes of South Louisiana. D, 1943, Louisiana State University.

Wallace, William E., Jr. Foraminifera of the Jackson Eocene of Danville Landing, Ouachita River, Catahoula Parish, Louisiana. M, 1932, Louisiana State University.

Wallace, William John, Jr. The development of the chlorinity-salinity concept in oceanography. D, 1971, Oregon State University. 359 p.

Wallach, Joseph. Origin of steeply inclined fractures in central and western New York. M, 1968, Syracuse University.

Wallach, Joseph Leonard. The metamorphism and structural geology of the Hinchinbrooke Gneiss and its age relationship to metasedimentary and metavolcanic rocks of the Grenville Group. D, 1973, Queen's University. 141 p.

Wallem, Daniel B. Environmental, diagenetic, and source rock analysis of the Bear River Formation, western Wyoming. M, 1981, University of Wyoming. 101 p.

Waller, Harold E., Jr. The geology of the Paymaster and Olivette mining areas, Pima County, Arizona. M, 1960, University of Arizona.

Waller, Harry O'Neal. Foraminiferal biofacies off the South China Coast. M, 1958, University of Southern California.

Waller, J. O. Influence of geology on the water resources of the upper Roanoke River basin, Virginia. D, 1976, Virginia Polytechnic Institute and State University. 561 p.

Waller, James Allen. Vertical seismic profile observations of microearthquake wave attenuation, Oroville, California. M, 1984, University of California, Santa Barbara. 112 p.

Waller, James O. Influence of oxide variation on properties of artificial and natural basaltic glasses. M, 1967, Virginia Polytechnic Institute and State University.

Waller, Louis Raymond. Water quality standards as determined for the Big Muddy River in southern Illinois through the process of model development. D, 1973, Southern Illinois University, Carbondale. 200 p.

Waller, Michael Reginald. Lake beds in the Chiricahua National Monument. M, 1952, University of Tulsa. 96 p.

Waller, Philip Charles. Study of some Cambrian sediments from Jasper Park, Alberta. M, 1959, University of Alberta. 128 p.

Waller, Roger M. Water sediment ejections of the 1964 Alaska earthquake. M, 1969, University of Arizona.

Waller, Stephen F. The Cumberland Formation and Leipers Limestone (Upper Ordovician) in the south-central Appalachian Basin; stratigraphy, petrology and depositional environment. M, 1983, University of Cincinnati. 243 p.

Waller, Thomas H. Lower Cisco (upper Pennsylvanian) sedimentary framework and facies-depositional model for the eastern shelf, north-central Texas. D, 1974, University of Kansas. 150 p.

Waller, Thomas Howell. The stratigraphy of the Graham and Thrifty formations, Cisco Group (Pennsylvanian), southeastern Stephens County, Texas. M, 1966, Baylor University. 350 p.

Waller, Thomas R. A comparative study of the Pitkin (Mississippian) and Hale (Pennsylvanian) faunas of northeastern Oklahoma. M, 1962, University of Wisconsin-Madison.

Waller, Thomas Richard. Late Cenozoic evolution of the Aequipecten gibbus Stock. D, 1966, Columbia University. 291 p.

Walles, Frank E. Niagara pinnacle reefs of South central Michigan. M, 1980, Michigan State University. 90 p.

Walley, David Stephen. Component magnetization of the iron formation and deposits at the Moose Mountain Mine, Capreol, Ontario. M, 1980, University of Windsor. 137 p.

Wallick, Brian P. Sedimentology of the Bullion Creek and Sentinel Butte formations (Paleocene) in a part of southern McKenzie County, North Dakota. M, 1984, University of North Dakota. 245 p.

Wallick, Edward Israel. A tritium tracer study of the Florida Aquifer in the Big Bend area of Florida. M, 1969, Florida State University.

Wallick, Edward Israel. Isotopic and chemical considerations in radiocarbon dating of groundwater within the arid Tucson Basin, Arizona. D, 1973, University of Arizona.

Wallin, Charles Stanton. Consolidation and shear strength of north-central Gulf of Mexico sediments. M, 1968, University of Washington. 83 p.

Wallin, E. Timothy. Stratigraphy and paleoenvironments of the Engle coal field, Sierra County, New Mexico. M, 1983, New Mexico Institute of Mining and Technology. 127 p.

Wallin, S. R. The sediments in the head of Carmel submarine canyon (California). M, 1968, United States Naval Academy.

Wallinga, Mark A. Petrographic and stratigraphic analysis of the Atoka Formation, northern Franklin and northwestern Johnson counties, Arkansas. M, 1986, University of Arkansas, Fayetteville.

Wallis, B. Franklin. The geology and economic value of the Wapanucka Limestone of Oklahoma. D, 1915, The Johns Hopkins University.

Wallis, James R. A factor analysis of soil erosion and stream sedimentation in Northern California. D, 1965, University of California, Berkeley. 141 p.

Wallis, P. M. Sources, transportation, and utilization of dissolved organic matter in groundwater and streams. D, 1978, University of Waterloo.

Wallis, Paul Francis. In situ determination of rock mass quality and discontinuity frequencies by acoustic borehole logging techniques. M, 1980, University of Saskatchewan. 123 p.

Wallis, Thomas Irvin. Stratigraphy of the Ordovician Maravillas Formation. M, 1958, Texas Tech University. 53 p.

Wallmann, Peter Caswell. Mechanical models for correlation of ring-fracture eruptions at Pantelleria, Strait of Sicily, with glacial sea-level drawdown. M, 1986, Stanford University. 34 p.

Walls, Billy. Geology of Bell Gin Quadrangle, Williamson County, Texas. M, 1950, University of Texas, Austin.

Walls, Ian A. Provenance of the Jurassic Norphlet Formation in Southwest Alabama. M, 1985, University of Alabama. 193 p.

Walls, James G. Holston Marble at Asbury, Tennessee. M, 1930, University of Tennessee, Knoxville. 72 p.

Walls, James Gray. The Dolly Varden phase of the Holston Marble. D, 1946, University of North Carolina, Chapel Hill. 47 p.

Walls, Joel Dan. Effects of pore pressure, confining pressure and partial saturation on permeability of sandstones. D, 1983, Stanford University. 125 p.

Walls, R. A. Cementation history and porosity development, Golden Spike reef complex (Devonian), Alberta. D, 1978, McGill University. 226 p.

Walls, Richard A. Dissolution kinetics of strontium and magnesium in carbonate skeletal material. M, 1973, University of North Carolina, Chapel Hill. 68 p.

Walls, Thomas. Stratigraphic correlation in the Eocene of Ventura County, California, using trace elements in

clay fractions. M, 1985, California State University, Long Beach. 154 p.

Walmsley, M. E. Soil-water chemistry relationships and characterization of the physical environment - intermittent permafrost zone, MacKenzie Valley, N.W.T. M, 1974, University of British Columbia.

Waln, Kirk Alexander. Palynology of the Lincoln Creek Formation (Oligocene), Grays Harbor basin, western Washington. M, 1985, University of Iowa. 212 p.

Walniuk, Daria M. Microscopic textures, lattice orientations and deformation mechanisms of deformed quartz in metasediments from western Spitsbergen. M, 1983, Wayne State University. 114 p.

Waloweek, W. Plant microfossil studies of three coal seams near Joggins, Nova Scotia. M, 1952, University of Massachusetts.

Walper, Jack Louis. Geology fo the Coban-Purulha area, Alta Verapaz, Guatemala. D, 1957, University of Texas, Austin.

Walper, Jack Louis. Geology of the Coban-Purulha area, Alta Verapaz, Guatemala. D, 1958, University of Texas, Austin. 149 p.

Walper, Jack Louis. Igneous rocks of the Cold Springs area, Wichita Mountains, Oklahoma. M, 1949, University of Oklahoma. 71 p.

Walpole, Robert Leonard. Microfacies study of the Rundle Group, Front Ranges, central Alberta. D, 1961, University of Illinois, Urbana. 111 p.

Walraven, David. A model for the formation of the Earth's core. M, 1976, Rice University. 53 p.

Walrond, Grantley Wainright. Minerals policy development strategy in Guyana. D, 1979, University of Alberta. 418 p.

Walrond, Henry. Geology of the upper Santa Ynez Valley area, Santa Barbara, California. M, 1952, University of Southern California.

Walsh, Carol Ann. Rates of reaction of covellite and blaubleibender covellite with ferric iron at pH 2.0. M, 1984, Virginia Polytechnic Institute and State University. 49 p.

Walsh, Daniel Charles. Geophysical and hydrogeological investigation of the Easthampton and Northampton Landfills in Hampshire County, Massachusetts. M, 1987, University of Massachusetts. 336 p.

Walsh, Daniel Hallaron. An investigation of the direction of initial motions and some characteristics of waves from Mexican earthquakes, as recorded at St. Louis and Florissant, Missouri. M, 1951, St. Louis University.

Walsh, Daniel Hallaron. An observational study of the origin of short period microseisms near Saint Louis, Missouri. D, 1953, St. Louis University.

Walsh, David Vernon. Geological and geochemical relationships of the Desert Mountain igneous complex rocks. M, 1987, University of Iowa. 64 p.

Walsh, Deborah Anne. An assessment of the mineral resources of the United Kingdom and Republic of Ireland. M, 1979, Pennsylvania State University, University Park. 134 p.

Walsh, Don. The Mississippi River outflow, its seasonal variation and its surface characteristics. D, 1968, Texas A&M University.

Walsh, Ian David. Resuspension and the rebound process; implications of sediment trap studies in the northern Pacific. M, 1986, Oregon State University. 108 p.

Walsh, J. Andrew. Sedimentology and late Holocene evolution of the Lubec Embayment. M, 1988, University of Maine. 434 p.

Walsh, J. E. Disposal and utilization of hydraulically dredged lake sediments in limited containment areas. D, 1977, University of Massachusetts. 283 p.

Walsh, James F. Geology and geochemistry of the Pamour #1 Mine, Timmins, Ontario, Canada; fluid immiscibility in the H_2O-CO_2-CH_4 system as a control on ore deposition. M, 1986, University of Michigan. 60 p.

Walsh, James Paul Jr. Drainage basins and areas of past flooding in the Windsor area, Colorado. M, 1973, University of Colorado.

Walsh, M. H. A study of the Ordovician basal Bighorn Sandstone of the central and northern Big Horn Mountains, Wyoming. M, 1957, University of Wyoming. 81 p.

Walsh, Marcus Whitley. Stratigraphy and petrography of the Sligo Formation (Cretaceous) of Marion and Harrison counties, Texas. M, 1968, University of Tulsa. 86 p.

Walsh, Mark Patrick. Geochemical flow modeling. D, 1983, University of Texas, Austin. 526 p.

Walsh, P. R. The distribution and potential sources and fluxes of arsenic in the atmosphere. D, 1978, University of Rhode Island. 301 p.

Walsh, R. P. Geology of the Spokane silver and lead mine. M, 1928, University of Minnesota, Minneapolis. 35 p.

Walsh, Robert G., Sr. The Keller Plan in college introductory physical geology; a comparison with the conventional teaching method. D, 1975, University of North Dakota. 80 p.

Walsh, Stephen J. An investigation into the comparative utility of color infrared aerial photography and LANDSAT data for detailed surface cover type mapping within Crater Lake National Park, Oregon. D, 1978, Oregon State University. 356 p.

Walsh, Thelma Helaine. Glaciation of the Taylor Basin area, Madison Range, southwestern Montana. D, 1976, University of Idaho. 244 p.

Walsh, Thelma Helaine. Quaternary geology of the east portion of West Fork Basin, Gallatin County, Montana. M, 1971, Montana State University. 83 p.

Walsh, Thomas F. The proboscideans of the Black Hawk Ranch fauna, Mount Diablo, California. M, 1959, University of California, Berkeley. 83 p.

Walsh, Timothy John. Texture, mineralogy and wave energy relations in intertidal sediments in a closed system, San Miguel Island, California. M, 1979, University of California, Los Angeles.

Walsh, W. P. Spectrum analysis in seismology. D, 1954, Massachusetts Institute of Technology. 165 p.

Walsh, William Egan, Jr. The use of surface wave techniques for verification of dynamic rigidity measurements in a kaolinite-water artificial sediment. M, 1971, United States Naval Academy.

Walston, Gerald M. Bedrock topography and glacial drift thickness of Macomb County and the south half of Saint Clair County, Michigan. M, 1967, Wayne State University.

Walsum, N. Van see Van Walsum, N.

Walter, David R. Conodont biostratigraphy of the Mississippian rocks of northwestern Arizona. M, 1976, Arizona State University. 185 p.

Walter, Edward Joseph. An extension of the method of numerical integration of seismograms to include a third integration. D, 1944, St. Louis University.

Walter, Edward Joseph. Earthquake wave velocities and travel times, and crustal structure south of Saint Louis. M, 1940, St. Louis University.

Walter, Gary R. The sedimentology and morphologic development of a small artificial channel. M, 1974, University of Missouri, Columbia.

Walter, Gary Robert. The effects of molecular diffusion on groundwater solute transport through fractured tuff. D, 1985, University of Arizona. 200 p.

Walter, Henry Glenn. The Dinwoody Formation of western Wyoming. M, 1931, University of Missouri, Columbia.

Walter, John R. Interpretation of paleogeography and depositional environment for the Ames and Brush Creek marine horizons, Conemaugh Formation (Pennsylvanian) in northeastern Kentucky. M, 1979, Eastern Kentucky University. 67 p.

Walter, John William. An experimental study on the application of subsurface hydraulic mining in tar sands. M, 1982, Stanford University. 87 p.

Walter, John William. Application of subsurface hydraulic mining in tar sands. M, 1981, Stanford University.

Walter, Joseph Charles, Jr. Paleontology of Rustler Formation, Culberson County, Texas. M, 1951, University of Texas, Austin.

Walter, Kenneth Gaines. A study of the pegmatites of the Stone Mountain-Lithonia-Panola Shoals area (Georgia). M, 1958, Emory University. 61 p.

Walter, Laurie Rianne. Limb adaptations in kannemeyeriid dicynodonts. D, 1985, Yale University. 276 p.

Walter, Lawrence E., Jr. Geology of the Marigold area, Fremont and Teller counties, Colorado. M, 1959, University of Kansas. 63 p.

Walter, Louis S. Cleavage tendencies in beta-quartz. M, 1955, University of Tennessee, Knoxville. 76 p.

Walter, Louis Simon. Pressure-temperature univariant equilibria of some reactions in the system, CaO-MgO-SiO_2-CO_2. D, 1960, Pennsylvania State University, University Park. 96 p.

Walter, Lynn M. The effects of phosphate on the dissolution kinetics of biogenic magnesium calcites. M, 1978, Louisiana State University.

Walter, Lynn Marie. The dissolution kinetics of shallow water carbonate grain types; effects of mineralogy, microstructure, and solution chemistry. D, 1983, University of Miami. 347 p.

Walter, Otto T. Trilobites of Iowa and some related Paleozoic forms. D, 1923, University of Iowa.

Walter, R. Bryan. Depositional environment, petrology, diagenesis and petroleum geology of the Cottage Grove Sandstone, North Concho Field, Canadian County, Oklahoma. M, 1985, Oklahoma State University. 156 p.

Walter, Robert Curtis. The volcanic history of the Hadar early-man site and the surrounding Afar region of Ethiopia. D, 1980, Case Western Reserve University.

Walter, Rudolph J. The construction of Caballo Dam in New Mexico. M, 1940, New Mexico Institute of Mining and Technology.

Walters, Charles Philip. Energy of the Earth's rotation applied to the deformation of Southern California. D, 1957, Cornell University.

Walters, Charles Phillip. Study of local concretions. M, 1938, Kansas State University. 80 p.

Walters, Clifford Carol. Organic geochemistry of the 3,800 million year old metasediments from Isua, Greenland. D, 1981, University of Maryland. 314 p.

Walters, Diana. Aluminum; a global composite analysis of a commodity. M, 1986, University of Texas, Austin.

Walters, Donald Lee. The pre-Woodford subcrop and its relationship to an overlying detrital lithofacies in Northeast Marshall and Southwest Johnston counties, Oklahoma. M, 1958, University of Oklahoma. 37 p.

Walters, Donna Lynn. Three-dimensional seismic stratigraphic study of Minnelusa Formation, Powder River basin, Campbell County, Wyoming. M, 1988, Texas A&M University. 98 p.

Walters, Gerard Michael. The subsurface stratigraphy and petroleum geology of the "Clinton" Sandstone (Lower Silurian), northeast Ohio. M, 1980, Kent State University, Kent. 230 p.

Walters, James Carter. Origin and paleoclimatic significance of fossil periglacial phenomena in central and northern New Jersey. D, 1975, Rutgers, The State University, New Brunswick. 138 p.

Walters, James Ettore. Glacial geology of Countess of Warwick Sound, Baffin Island, Arctic Canada. M, 1988, University of Colorado. 110 p.

Walters, James K. The depositional and diagenetic history of the Chappel Formation, Conley Field, Hardeman County, Texas. M, 1984, Texas A&M University. 125 p.

Walters, Jon K. Seismic reflection investigation in the Great Salt Lake Desert, Utah. M, 1984, University of Oklahoma. 109 p.

Walters, Kenneth Allan. The geology of the Baldy Mountain area, Colfax County, New Mexico. M, 1979, Northern Arizona University. 87 p.

Walters, Kenneth Lamont. Microfacies relationships of the Mississippian Midale Carbonate of the Glen Ewenfield, southeastern Saskatchewan. M, 1983, University of Regina. 169 p.

Walters, Lawrence Albert. Major-and trace-element studies of selected pegmatites and their wall rocks in Rumford District, Maine. D, 1965, University of Wisconsin-Madison. 104 p.

Walters, Lester J., Jr. Bound halogens in sediments. D, 1968, Massachusetts Institute of Technology. 233 p.

Walters, Mark A. Geological and geochemical evidence for a possible concealed mineral deposit near the Old Hadley mining district, Luna County, New Mexico. M, 1972, Stanford University. 63 p.

Walters, Mathias Joseph. Regional variation in grain size in the Aux Vases (Mississippian) Sandstone. M, 1958, University of Illinois, Urbana.

Walters, R. F. Geology of the Independence Mountain area, North Park, Colorado. M, 1953, University of Wyoming. 65 p.

Walters, Randy R. Structural geometries, fabrics, and stratigraphic relationships in the Cades Cove region, Great Smoky Mountains National Park, Tennessee. M, 1988, University of Tennessee, Knoxville. 155 p.

Walters, Richard Francis. Stratigraphic study of the Brive Basin, Southwest France. D, 1957, Stanford University. 218 p.

Walters, Robert Derek. Seismic stratigraphy and salt tectonics of Plio-Pleistocene deposits, continental slope and Upper Mississippi fan, North Gulf of Mexico. M, 1985, University of Texas, Austin. 204 p.

Walters, Robert F. Buried Precambrian hills in the Kraft-Prusa District, northeastern Barton County, central Kansas. D, 1945, The Johns Hopkins University.

Walters, Robert Fred. Contributions to the glacial geology of Monroe County, New York. M, 1938, University of Rochester. 88 p.

Walters, Robert L. Geologic study of collapses in the American Rock Crusher Mine, Wyandotte County, Kansas. M, 1984, University of Kansas. 213 p.

Walters, Roy A. Deep circulation in a fjord type estuary. M, 1973, University of Washington.

Walters, Stephen. A study of basic intrusives from the Peter Lake Complex. M, 1982, University of Regina. 172 p.

Walters, Sylvia Jene. Sedimentology and depositional environment of the Saluda Formation (Upper Ordovician) on Highway 421, Jefferson County, Indiana. M, 1988, University of Cincinnati. 183 p.

Walthall, Bennie H. Peripheral gulf rifting in Northeast Texas. M, 1963, University of Tulsa. 59 p.

Walthall, Bennie Harrell. Stratigraphy and structure, part of Athens Plateau, southern Ouachitas, Arkansas. D, 1966, Columbia University. 104 p.

Walther, Deborah Stacy. Mining on federal lands and a proposal for change. M, 1987, University of California, Los Angeles. 182 p.

Walther, John Victor. Thermodynamic analysis of equilibrium phase relations among minerals and aqueous solutions in metasomatic processes. D, 1978, University of California, Berkeley. 138 p.

Walthier, Thomas N. Geology and mineral deposits of the area between Corner Brook and Stephenville, western Newfoundland. D, 1948, Columbia University, Teachers College.

Walthier, Thomas N. Geology of the Felknor Mine area, Jefferson County, Tennessee. M, 1945, Columbia University, Teachers College.

Walthour, Steven Douglas. Sedimentology and petrology of limestones in the Womble Shale, Ouachita Mountains, Arkansas. M, 1983, University of Arkansas, Fayetteville.

Waltman, M. R. Petrology of the ore horizon and ore paragenesis at the Portis Mine, Franklin County,

Waltman, Reid Martin. Stratigraphy and purification of New Jersey glass sand with emphasis on beneficiation of limonitic (nugget) sand by magnetic separation. M, 1948, Rutgers, The State University, New Brunswick. 144 p.

North Carolina. M, 1985, University of North Carolina, Chapel Hill.

Waltman, William John. The stratigraphy and genesis of pre-Wisconsinan soils in the Allegheny Plateau. D, 1985, Pennsylvania State University, University Park. 377 p.

Walton, Anne Helene. Magnetostratigraphy and the ages of Bridgerian and Uintan faunas in the lower and middle members of the Devil's Graveyard Formation, Trans-Pecos, Texas. M, 1986, University of Texas, Austin. 135 p.

Walton, Anthony Warrick. Clay mineralogy of the upper Jackson Group (upper Eocene) and the Catahoula Formation (Miocene and Oligocene), east-central Texas. M, 1968, University of Texas, Austin.

Walton, Anthony Warrick. Sedimentary petrology and zeolitic diagenesis of the Vieja Group (Eocene-Oligocene), Presidio County, Texas. D, 1972, University of Texas, Austin. 277 p.

Walton, David S. The geology of Slater Park, Routt County, and the stratigraphy of the Lewis Shale (Upper Cretaceous), northern Routt County, Colorado. M, 1963, Colorado School of Mines. 179 p.

Walton, Frank Dennis. Sedimentary dynamics under tidal influences, Big Grass island, Taylor County, Florida. M, 1970, Florida State University.

Walton, Godfrey John. The petrography of a fluorite-wollastonite skarn in regionally metamorphosed greenschist rocks of the garnet zone, Lake Township, Ontario. M, 1978, Queen's University. 100 p.

Walton, John C. Sediment characteristics and processes of beaches and bars, Presque Isle, Erie, Pennsylvania. M, 1978, SUNY, College at Fredonia. 157 p.

Walton, John Calvin. Redox equilibria in a freshwater lake. M, 1981, University of Virginia. 104 p.

Walton, Lori A. Geology and geochemistry of the Venus Au-Ag-Pb-An vein deposit, Yukon Territory. M, 1987, University of Alberta. 113 p.

Walton, Matthew S., Jr. Magnetic exploration of the nickel-copper deposits of Bohemia Basin, southeastern Alaska. M, 1947, Columbia University, Teachers College.

Walton, Matthew S., Jr. The Blashke Island ultrabasic complex with notes on related areas in Southeast Alaska. D, 1953, Columbia University, Teachers College.

Walton, Otis Raymond. Particle-dynamics modeling of geological materials. D, 1980, University of California, Davis. 239 p.

Walton, Paul Talmadge. Geology of the Cretaceous of the Uinta Basin, Utah. D, 1942, Massachusetts Institute of Technology. 120 p.

Walton, Paul Talmadge. The origin of a gold placer deposit in Sublette County, Wyoming. M, 1940, University of Utah. 28 p.

Walton, Robert G. Lithofacies study of Missourian rocks (Pennsylvanian) in northwestern Kansas and adjacent areas. M, 1960, University of Kansas. 53 p.

Walton, Todd Leon, Jr. Littoral sand transport on beaches. D, 1979, University of Florida. 322 p.

Walton, Wahnes. Generic descriptions and suture patterns of the Cretaceous ammonites of Texas. M, 1941, University of Texas, Austin.

Walton, Wayne J. A., Jr. A study of crystallization from water of compounds in the system $^+Fe^{++}$-Cu^{++}-SO_4^+. M, 1968, University of Arizona.

Walton, Wayne J. A., Jr. Phase equilibrium in the system $Mg_3(PO_4)_2$-$Ca_3(PO_4)_2$-Mg_2SiO_4 (Farringtonite-Whitlockite-Forsterite) at one atmosphere pressure. D, 1972, Ohio State University.

Walton, William L. Geology of the Carlos-East area, Grimes County, Texas. M, 1959, Texas A&M University. 97 p.

Walton, William R. Ecology of living benthonic foraminifera, Todos Santos Bay, Baja California. D, 1954, University of California, Los Angeles. 221 p.

Waltz, James Patterson, II. An analysis of selected landslides in Alameda and Contra Costa counties, California. D, 1967, Stanford University. 208 p.

Waltz, Michael David. The evolution of shallowing-upwards reef to oolite sequences at the leeward margin of Caicos Platform, B.W.I. M, 1988, University of Miami. 95 p.

Waltz, Stephen Ray. Evaluation of shapes of quartz silt grains as provenance in central Michigan. D, 1972, Michigan State University. 39 p.

Walz, Clarence M. The ore deposits of India. M, 1921, University of Minnesota, Minneapolis. 90 p.

Walz, David Henry. Sewage renovation and surface-water quality, Lakeway resort community, Travis County, Texas. M, 1974, University of Texas, Austin.

Walz, David M. Copper deposits associated with clastic plugs, Black Mesa District, Oklahoma. M, 1982, University of Nebraska, Lincoln. 68 p.

Wampler, J. Marion. An isotopic study of lead in sedimentary pyrite. D, 1963, Columbia University, Teachers College.

Wamser, Robert Charles. Barium in some common igneous and metamorphic rocks. M, 1977, Northern Illinois University. 60 p.

Wan, Chun Yan. Petrography and chemistry of inclusions in trachybasalts of the Silver Peak region, Esmeralda Co., Nevada. M, 1974, University of California, Riverside. 104 p.

Wan, Hsien-Ming. The lizardite-nepouite and the kerolite-pimelite series of minerals. M, 1975, Pennsylvania State University, University Park. 109 p.

Wan, N. M. B. Nik *see* Nik Wan, N. M. B.

Wana-Etyem, Charles. Static and dynamic water content-pressure head relations of porous media. D, 1982, Colorado State University. 155 p.

Wanab, Osman Abdel. The origin of the Kutum lead-zinc deposits, Darfur Province, Republic of Sudan. M, 1968, University of Arizona.

Wanakule, Nisai. A model for determining optimal pumping and recharge of large-scale aquifers. D, 1984, University of Texas, Austin. 301 p.

Wanamaker, Barbara Jo. The kinetics of crack healing and the chemical and mechanical re-equilibration of fluid inclusions in San Carlos olivine. D, 1986, Princeton University. 232 p.

Wandke, Alfred D., Jr. Investigations relating to the origin of chalcocite. D, 1950, Harvard University.

Wandtke, Alfred. The geology of the Portsmouth Basin, Maine and New Hampshire. D, 1917, Harvard University.

Wandzilak, Michael. Crystal symmetry and elastic constants. M, 1969, Massachusetts Institute of Technology. 24 p.

Wanek, Leo James. Geology of an area east of Wolcott, Eagle County, Colorado. M, 1953, University of Colorado.

Wanenmacher, Joseph M. The Paleozoic strata of the Baraboo region, Wisconsin. D, 1932, University of Wisconsin-Madison.

Wanenmacher, Joseph M. The solubility of $Na_2Si_3O_7$ and $Ca_3Si_2O_7$ in akermanite. M, 1924, University of Wisconsin-Madison.

Wang Wei *see* Wei Wang

Wang, Albert Min-Hao. Comparison of linear and nonlinear acoustic probing of rock salt. M, 19??, Texas A&M University.

Wang, Ben. Development of a 2-D large-scale micellar/polymer simulator. D, 1982, University of Texas, Austin. 356 p.

Wang, Chang. Genetic studies of Lansing and Langford soils (central New York). D, 1971, Cornell University. 175 p.

Wang, Chen Yu. Numerical modeling of subduction in a fluid of constant viscosity. D, 1973, Rice University. 108 p.

Wang, Chi-fung. Elastic wave propagation in homogeneous transversely isotropic medium with symmetry axis parallel to the free surface. M, 1973, Massachusetts Institute of Technology. 89 p.

Wang, Chi-Yuen. The figure of the Earth as obtained from satellite data and its geophysical implications. D, 1964, Harvard University.

Wang, Chien-Ying. Wave theory for seismogram synthesis. D, 1981, St. Louis University. 258 p.

Wang, Ching-Pi. A mathematical model of a withdrawal system for contaminated groundwater. M, 1983, University of Idaho. 148 p.

Wang, Chu C. Geology of the Ta Ching Shan Range and its coal fields, Suiyuan, China. M, 1930, University of Wisconsin-Madison.

Wang, Chuching. A three-dimensional finite element model coupled with parameter identification for aquifer solute transport. D, 1986, University of California, Los Angeles. 139 p.

Wang, Chung Yu. The Ell Mine; project of a drift mine in California. M, 1904, Columbia University, Teachers College.

Wang, Chung-Ho. $^{18}O/^{16}O$ ratios in marine diatoms as paleoclimate and paleoceanography indicators for the late Quaternary. D, 1984, University of Hawaii.

Wang, D.-P. Coastal water response to the variable wind-theory and coastal upwelling experiment. D, 1975, University of Miami. 183 p.

Wang, Feng-Hui. Recent sediments in Puget Sound and portions of Washington Sound and Lake Washington. D, 1955, University of Washington. 160 p.

Wang, Fun-Den. The strength alteration of rocks and glass in liquid environments. D, 1966, University of Illinois, Urbana. 88 p.

Wang, Hau-Ran. Magnetic modeling of total field anomalies; application to estimation of basalt thickness near Reardan, Washington. M, 1980, Eastern Washington University. 48 p.

Wang, Hau-Ran. Magnetic modelling of the Columbia River basalts. M, 1979, University of Washington.

Wang, Herbert Fan. Elasticity of some mantle crystal structures. D, 1971, Massachusetts Institute of Technology. 211 p.

Wang, Huei-Yuin Wen. A seismological investigation of Tibetan Plateau; attenuation, velocity structure, and seismic source process. D, 1983, St. Louis University. 257 p.

Wang, Jason. Sedimentology of Norian (Late Triassic) cherts and carbonates from the Peninsular terrane; Puale Bay, Alaska Peninsula. M, 1987, Syracuse University. 124 p.

Wang, Jeen-Hwa. The propagation of Rayleigh waves in laterally heterogeneous media. D, 1982, SUNY at Binghamton. 235 p.

Wang, Jia Shung. Thermal alterations of potassium exchangeability in micaceous mineral particles of different size. D, 1981, Iowa State University of Science and Technology. 157 p.

Wang, Kia-Kang. The geology of Ouachita Parish, Louisiana. D, 1951, Louisiana State University.

Wang, Kong. Emplacement analysis of the Starvation Flat quartz monzonite pluton, northeastern Washington; by using the anisotropy of magnetic susceptibility method. M, 1982, Eastern Washington University. 34 p.

Wang, Kung Teh. Statistical study of suspension load of the Trinity River (Texas). M, 1967, Rice University. 71 p.

Wang, Kung-Ping. Controlling factors in the future development of the Chinese coal industry. D, 1947, Columbia University, Teachers College.

Wang, Min-Chao. Catalytic role of selected soil minerals in the abiotic formation of humic substances and the associated reactions. D, 1987, University of Saskatchewan.

Wang, Ming Kuang. Formation of goethite and hematite at 70°C. D, 1978, Rutgers, The State University, New Brunswick. 73 p.

Wang, Ning-Wu. Laboratory study of the diffusion controlled contact polarization curve (CPC) technique. M, 1983, University of Utah. 77 p.

Wang, Peter. Estimation of elastic parameters using seismic amplitude versus offset information. M, 1986, University of Houston.

Wang, Shi-Chen. Tectonic implications of global seismicity. D, 1982, Stanford University. 76 p.

Wang, Shih-Hsien. Study of coals by diffuse reflectance Fourier transform infrared spectrometry. D, 1984, Ohio University. 202 p.

Wang, Shin. Sphalerite pole figure analysis and metamorphic textures, Matagami Lake Mine, Quebec, Canada. D, 1973, Columbia University. 174 p.

Wang, Simon Yaou-Dong. Computer assisted tomography and its application in the study of multiphase flow through porous media. D, 1983, [Princeton University]. 376 p.

Wang, Sou-Yung. Velocity-independent processing of seismic data. D, 1988, [University of Houston]. 238 p.

Wang, Tzupo. Endochronic theory for the mechanical behavior of cohesionless soil under static loading. D, 1980, University of Iowa. 165 p.

Wang, Wei-Ming. Leachate transport modeling. D, 1985, Oklahoma State University. 147 p.

Wang, Wei-Yeong. The heat capacity of coal chars. D, 1982, City College (CUNY). 168 p.

Wang, Wen Chiang. Flow characteristics over alluvial bedforms. D, 1984, Colorado State University. 318 p.

Wang, Wha-ching. Texture, composition and geological significance of shelf sediments of Taiwan. M, 1974, University of Oregon. 116 p.

Wang, Xi-Shuo. A contribution to the interpretation of magnetovariation fields. D, 1987, University of Alberta. 217 p.

Wang, Xiaomin. A study on the equilibrium grossular + clinochlore = 3 diopside + 2 spinel + 4 H_2O. M, 1986, University of British Columbia. 100 p.

Wang, Yeh David. Investigation of constitutive relations for weakly cemented sands. D, 1986, University of California, Berkeley. 309 p.

Wang, Yinchang T. The formation of the oxidized ores of zinc from the sulphide. D, 1915, Columbia University, Teachers College.

Wang, Yongjia. Numerical model for computing time-dependent displacements and stresses in rock mechanics. D, 1983, University of Minnesota, Minneapolis. 238 p.

Wang, Yuan. Geology and coal metamorphism of dacite in the Chuifen gold mine, northern Taiwan. M, 1963, Pennsylvania State University, University Park. 113 p.

Wang, Yuan Sung. The contamination and bioconcentration of aldrin, dieldrin and endrin in lower lakes at Rocky Mountain Arsenal. D, 1988, Colorado State University. 139 p.

Wang, Yun Chung. Mylonitization and deformational history in the vicinity of the Cold Springs fault zone, Haywood County, North Carolina. M, 1979, Wayne State University.

Wang, Yun Fei. Geological and geophysical studies of the Gilson Mountains and vicinity, Juab County, Utah. D, 1970, University of Utah. 196 p.

Wang, Yung-Liang. Local hypocenter determination in linearly varying layers applied to earthquakes in the Denver area. D, 1965, Colorado School of Mines. 211 p.

Wang, Yunshuen. Petrology and mineralogy of Tertiary(?) volcanic rocks of Snowville area, Utah, and Tertiary-Quaternary(?) volcanic rocks of Table Mountain and Holbrook areas, Idaho. M, 1985, Utah State University. 83 p.

Wangensteen, Martin Walter. The hydrogeological impact of heavy metal-laden ash landfilling at the WLSSD Landfill, Rice Lake, Minnesota. M, 1988, University of Minnesota, Duluth. 212 p.

Wanger, D. B. Environmental history of central San Francisco Bay with emphasis on foraminiferal paleontology and clay mineralogy. D, 1978, University of California, Berkeley.

Wanger, Johnny P. Relationships among uranium, thorium, and other elements in igneous rock series from the Carolina Piedmont. M, 1972, University of North Carolina, Chapel Hill. 64 p.

Wanklyn, Robert Paul. Stratigraphy and depositional environments of the Ostracode Member of the McMurray Formation (Lower Cretaceous; late Aptian-early Albian) in west-central Alberta. M, 1985, University of Colorado.

Wanless, Harold. Sediments of Biscayne Bay; distribution and depositional history. M, 1967, University of Miami. 259 p.

Wanless, Harold Rogers. Cambrian of the Grand Canyon; a re-evaluation of the depositional environment. D, 1973, The Johns Hopkins University.

Wanless, Harold Rollin. The lithology and stratigraphy of the White River Beds of South Dakota. D, 1923, Princeton University.

Wannamaker, Philip Ein. Three-dimensional magnetotelluric interpretation. D, 1983, University of Utah. 241 p.

Wanner, Henry Eckert. Some additional faunal remains from the Trias of York County, Pennsylvania. D, 1926, University of Pennsylvania.

Wanninkhof, Richard Hendrik. Gas exchange across the air-water interface determined with man made and natural traces. D, 1986, Columbia University, Teachers College. 306 p.

Wanslow, Julie Lenore. Stratigraphy and depositional framework of the Red Oak Sandstone, Atoka Formation, in the Arkoma Basin of Oklahoma. M, 19??, University of Arkansas, Fayetteville.

Wantland, Dart. Geophysical surveys of three placer areas in Colorado. M, 1935, Colorado School of Mines. 136 p.

Wantland, Kenneth Franklin. Recent benthonic Foraminifera of the British Honduras shelf. D, 1967, Rice University. 326 p.

Wantland, Kenneth Franklin. Recent foraminifera of Trinity Bay, Texas. M, 1964, Rice University. 64 p.

Wanty, Duane Allen. The spatial variation of several hydrogeochemical parameters within the Apple Creek drainage basin, North Dakota. M, 1984, Oklahoma State University. 105 p.

Wanty, Richard B. Geochemistry of vanadium in an epigenetic sandstone-hosted vanadium-uranium deposit, Henry Basin, Utah. D, 1986, Colorado School of Mines. 198 p.

Wanyeki, Simon. Well field design for the Merti Beds Aquifer, North Eastern Province, Kenya. M, 1979, Ohio University, Athens. 110 p.

Wanzong, Walter F. A study of a proposed randomness test for mineral samples. M, 1964, Michigan Technological University. 56 p.

Wappler, John H. Quartz latite dikes of the Louisiana State University Geology Camp area (El Paso County, Colorado). M, 1963, Louisiana State University.

Waraksa, Irene Rosalina. The petrology and structural geology of the Precambrian igneous rocks near Wallace, Chester County, Pennsylvania. M, 1952, Bryn Mawr College. 17 p.

Warburton, David Lewis. Moessbauer effect studies of olivines. D, 1978, University of Chicago. 177 p.

Warburton, Wayne L. Hydrology and copper budget of Torch Lake, Houghton County, Michigan. M, 1987, Michigan Technological University. 74 p.

Ward, A. Wesley, Jr. Windforms and wind trends on Mars; an evaluation of Martian surficial geology from Mariner 9 and Viking spacecraft television images. D, 1978, University of Washington. 201 p.

Ward, Albert Noll. A petrographic and stratigraphic study of the Avant Limestone Member of the Iola Formation in northeastern Oklahoma. M, 1964, University of Tulsa. 172 p.

Ward, Albert Noll, Jr. Pre-Pennsylvanian stratigraphy of the San Juan Mountain area in southwestern Colorado. D, 1966, University of Colorado. 204 p.

Ward, Alexander Wesley, Jr. Petrology and chemistry of the Huntzinger Flow, Columbia River Basalt, Washington. M, 1975, University of Washington. 133 p.

Ward, Alfred H. and McKnight, Edwin Thor. Geology of Snohomish Quadrangle. M, 1925, University of Washington. 95 p.

Ward, Andrew David. Characterizing infiltration through reconstructed surface mine profiles. D, 1981, [University of Kentucky]. 554 p.

Ward, Barbara L. Late Quaternary foraminifera from elevated deposits of the Capes Royds - Barne area, Ross Island, Antarctica. M, 1979, Northern Illinois University. 205 p.

Ward, Christopher Allan. Structural geology and tectonic history of Paleozoic rocks in the Sierra de las Montillas, east-central Chihuahua, Mexico. M, 1977, Texas Christian University.

Ward, Daniel Lee. Geology of area immediately west of Georgetown, Williamson County, Texas. M, 1950, University of Texas, Austin.

Ward, David Barry. Rb-Sr dating techniques applied to a metamorphosed Proterozoic terrane in the southern Sangre de Cristo Mountains, north-central New Mexico. M, 1986, University of New Mexico. 152 p.

Ward, Dederick C., III. The geology of the Bald Mountain area, Boulder County, Colorado. M, 1958, University of Colorado.

Ward, Dwight Edward. Geology of the southern end of the Medicine Bow Mountains (Jackson and Larimer counties), Colorado. D, 1959, University of Colorado. 200 p.

Ward, Fred Darrell. The subsurface geology of Noble County, Oklahoma. M, 1958, University of Oklahoma. 70 p.

Ward, Freeman. Geology of the New Haven-Branford region (Connecticut). D, 1908, Yale University.

Ward, G. Seasonal variation in composition of the Red river (Manitoba). M, 1926, University of Manitoba.

Ward, George William. A chemical and optical study of the black tourmalines. D, 1928, University of Minnesota, Minneapolis. 52 p.

Ward, J. A. Stratigraphy, depositional environments and diagenesis of the El Doctor Platform, Queretaro, Mexico. D, 1979, SUNY at Binghamton. 184 p.

Ward, J. Harold Edgar. The geology of the Haden area, Shackelford and Callahan counties, Texas. M, 1940, University of Texas, Austin.

Ward, James G. Petrology of the Rapakivi Granite of the Great Wass Pluton (Late Devonian), Washington County, Maine. D, 1972, University of Illinois, Urbana.

Ward, Jeffrey Kost. Geomorphic effects of pyroclastic volcanism; a deterministic modeling approach. D, 1973, University of Iowa. 183 p.

Ward, Jerome Vincent. Lower Cretaceous angiospermic pollen from the Cheyenne and Kiowa formations (Albian) of Kansas, U.S.A. D, 1983, Arizona State University. 255 p.

Ward, JoAnn. A comparative analysis of the hydraulic parameters of urbanized and non-urbanized drainage basins in southwestern Michigan. M, 1978, Southern Illinois University, Carbondale. 180 p.

Ward, John F. Preliminary study of the foraminifera of the Marianna Limestone in Alabama. M, 1958, University of Alabama.

Ward, John Robert. A study of the joint patterns in gently dipping sedimentary rocks of south-central Kansas. M, 1964, Wichita State University. 126 p.

Ward, Larry Guy. Physical and sedimentological processes in a salt marsh tidal channel; Kiawah Island, South Carolina. D, 1978, University of South Carolina. 191 p.

Ward, Larry Guy. The morphology and hydrology of the Skeidara and Skaftafellsa river distributaries, Skeidararsandur, Iceland. M, 1974, University of South Carolina.

Ward, Lauck. Stratigraphic revision of the middle Eocene-Oligocene and lower Miocene Atlantic Coastal Plain of North Carolina. M, 1977, University of South Carolina.

Ward, Lauck Walton. Chronostratigraphy and molluscan biostratigraphy of the Miocene; middle Atlantic Coastal Plain of North America. D, 1980, University of South Carolina. 169 p.

Ward, Marsha Jane. The glacial history of early Lake Saginaw. M, 1979, Michigan State University. 51 p.

Ward, Michael B. The volcanic geology of the Castle Hot Springs area, Yavapai County, Arizona. M, 1977, Arizona State University. 74 p.

Ward, Paul T. Stratigraphy of the Gallatin Group (Cambrian) of northwestern Wyoming. M, 1972, University of Missouri, Columbia.

Ward, Peter Douglas. Stratigraphy of Upper Cretaceous rocks on Orcas, Waldron, and Sucia Islands. M, 1973, University of Washington. 44 p.

Ward, Peter Douglas. Stratigraphy, paleoecology and functional morphology of heteromorph ammonites of the Upper Cretaceous Nanaimo Group, B.C. and Washington. D, 1976, McMaster University. 194 p.

Ward, Peter Langdon. I, A new interpretation of the geology of Iceland; a detailed study of a boundary between lithospheric plates; II, Microearthquakes, swarms , and the geothermal areas of Iceland. D, 1970, Columbia University, Teachers College. 162 p.

Ward, Peter Langdon. Volcanic and seismic activity in Katmai National Monument, Alaska. M, 1967, Columbia University. 78 p.

Ward, R. M. P-wave travel time residuals at Alq. M, 1974, New Mexico Institute of Mining and Technology.

Ward, Richard Brendan. Sedimentology, diagenesis and provenance of the slope channel deposits exposed within the upper siliceous member of the Monterey Formation exposed west of Gaviota Beach, California. M, 1984, Stanford University. 101 p.

Ward, Richard C. Geology of the Huronian Supergroup (Precambrian) in the Shiner Syncline, Vernon Township, Sudbury District, Ontario. M, 1972, Bowling Green State University. 150 p.

Ward, Richard F. The geology of the (Precambrian) Wissahickon Formation in Delaware. M, 1956, New York University.

Ward, Richard Floyd. Petrology and metamorphism of the Wilmington Complex, Delaware and adjacent Pennsylvania and Maryland. D, 1958, Bryn Mawr College. 103 p.

Ward, Robert Alan. Geology of the northern half of the Reipetown Quadrangle, Nevada. M, 1962, University of Southern California.

Ward, Robert Lee. Optimal potentiometric surface modification for groundwater contaminant management. D, 1988, University of Arkansas, Fayetteville. 245 p.

Ward, Roland Van *see* Van Ward, Roland

Ward, Ronald Arthur. A study of the limestone nodules in the Eden Shales. M, 1966, University of Cincinnati. 123 p.

Ward, Ronald W. Synthesis of teleseismic P-waves from sources near transition zones. D, 1971, Massachusetts Institute of Technology. 208 p.

Ward, Rosalyn Julia. Isolated bioherms in the Tansill Formation (Guadalupian), Dark Canyon, Guadalupe Mountains, southeastern New Mexico. M, 1988, Texas Christian University. 54 p.

Ward, S. H. A theoretical and experimental study of the electromagnetic method of geophysical prospecting. D, 1952, University of Toronto.

Ward, Steven N. Two studies of long period body waves; upper mantle reflected and converted phases; ringing P waves and submarine faulting. D, 1978, Princeton University. 121 p.

Ward, Suzanne. Environment of deposition and diagenesis of the Stuart City Limestone (upper Edwards), Sperry Field, Karnes County, Texas. M, 1979, Texas A&M University. 76 p.

Ward, T. J. Factor of safety approach to landslide potential delineation. D, 1976, Colorado State University. 128 p.

Ward, Theresia A. The depositional environment of the Upper Jurassic Salt Wash Member of the Morrison Formation, Slick Rock District, San Miguel County, Colorado. M, 1981, New Mexico Institute of Mining and Technology. 182 p.

Ward, Timothy James. Relationship of basin characteristics to selected water chemistry parameters in upper Carson River Basin. M, 1973, University of Nevada. 102 p.

Ward, W. E., II. Jointing in a selected area of the Warrior coal field, Alabama. M, 1977, University of Alabama.

Ward, Wendi I. Lithofacies and depositional environments of a portion of the Stones River Formation in Etowah and Dekalb counties, Northeast Alabama. M, 1983, University of Alabama. 267 p.

Ward, William Cruse. Diagenesis of Quaternary eolianites of NE Quintana Roo, Mexico. D, 1970, Rice University. 243 p.

Ward, William Cruse. Geology of the Bartons Creek area, Bastrop and Fayette counties, Texas. M, 1957, University of Texas, Austin.

Ward, William D., Jr. Sedimentology of the Minnelusa Formation, Dark Canyon, Rapid City, South Dakota. M, 1979, South Dakota School of Mines & Technology.

Ward, William James. The Colorado Group in southwestern Saskatchewan. M, 1962, University of Saskatchewan. 59 p.

Ward, William O. The America Mine landslide, Almaden Quicksilver Park, Santa Clara County, California. M, 1985, San Jose State University. 94 p.

Ward, William Paul. Nearshore sedimentation in lower Lake Huron. M, 1971, Wayne State University.

Ward, William Roger. I, The formation of planetesimals; II, Tidal friction and generalized Cassini's Laws in the solar system. D, 1973, California Institute of Technology. 94 p.

Wardani, Sayed A. El *see* El Wardani, Sayed A.

Warde, John Maxwell. Geology and clays of the Kootenai Formation of Montana; clay industry of Montana. M, 1937, Montana College of Mineral Science & Technology. 84 p.

Wardell, Henry Russel. The geology of the Potts Canyon mining area near Superior, Arizona. M, 1941, University of Arizona.

Wardlaw, B. R. The biostratigraphy and paleoecology of the Gerster Formation (upper Permian) in Nevada and Utah. D, 1975, Case Western Reserve University. 247 p.

Wardle, R. J. The stratigraphy and tectonics of the Greenhead Group; its relationship to Hadrynian and Paleozoic rocks, southern New Brunswick. D, 1978, University of New Brunswick.

Wardle, William Clyde. Eocene foraminifera from the Lucia Shale. M, 1957, University of California, Berkeley. 89 p.

Wardrop, Richard T. Source of anomalous barium in groundwater supplies of Indiana County, Pennsylvania. M, 1988, Pennsylvania State University, University Park. 401 p.

Ware, Don Westmont. Stratigraphy and depositional environments of the Roberts Mountains Formation in southern Nevada. M, 1979, San Diego State University.

Ware, Douglas C. Amphibolite complex of the Blackhall Mountain area, southeastern Sierra Madre, Wyoming. M, 1982, University of Wyoming. 71 p.

Ware, George Hunter. Induced electrical polarization and ground water studies. M, 1966, University of California, Berkeley. 96 p.

Ware, George Hunter. Theoretical and field investigations of telluric currents and induced polarization. D, 1974, University of California, Berkeley. 96 p.

Ware, Glen Chase, Jr. The geology of a portion of the Mecca Hills, Riverside County, California. M, 1958, University of California, Los Angeles.

Ware, Herbert Earl, Jr. Surface and shallow subsurface investigation of the Senora Formation of northeastern Oklahoma. M, 1954, University of Oklahoma. 89 p.

Ware, Jerry Allen. Scattering and inverse-scattering problems in a continuously varying elastic medium. D, 1969, Massachusetts Institute of Technology. 115 p.

Ware, John McKee. The paleontology and stratigraphy of the Frontier Formation on the east side of the Wind River Mountains, Wyoming. M, 1927, University of Missouri, Columbia.

Ware, Michael James. Depositional history and stratigraphy of the Aphebian Tamarack River Formation and the Paleohelikian Sims Formation, western Labrador. M, 1983, Memorial University of Newfoundland. 119 p.

Ware, Thomas III. Sparks Pool, Lincoln County, Oklahoma. M, 1951, University of Missouri, Columbia.

Wareham, Stephen I. Geology and petroleum potential of the Hay Creek Anticline, north-central Oregon. M, 1986, Loma Linda University. 65 p.

Waren, Kirk Bernon. Fracture controlled erosional processes and groundwater flow in the Niagara Group carbonates of southwest Ohio. M, 1988, Wright State University. 101 p.

Wares, Roy M. Petrology of part of the Michikamau anorthosite intrusion, Labrador. M, 1971, Queen's University. 138 p.

Waresback, Damon B. The Puye Formation, New Mexico; analysis of a continental rift-filling, volcaniclastic alluvial-fan sequence. M, 1986, University of Texas, Arlington. 269 p.

Warfield, Robert George. Stratigraphy and structure of the northeast quarter of the Richwoods Quadrangle, Missouri. M, 1953, University of Iowa. 104 p.

Warford, A. L. Methanogenesis in polluted and naturally anoxic marine sediments. D, 1977, University of California, Los Angeles. 217 p.

Warford, Andrew Craig. Mantle convection at marginal stability. M, 1979, Rice University. 63 p.

Warg, Jamison B. A palynological study of shales and "coals" of a Devonian-Mississippian transition zone, central Pennsylvania. M, 1972, Pennsylvania State University, University Park.

Wargo, Joseph G. Geology of a portion of the Coyote-Quinlan Complex, Pima County, Arizona. M, 1954, University of Arizona.

Wargo, Joseph George. The geology of the Schoolhouse Mountain Quadrangle, Grants County, New Mexico. D, 1959, University of Arizona. 248 p.

Waring, C. Joseph. Science, religions and origins, an innovative course for clarifying issues in science and religion. D, 1979, University of South Carolina. 64 p.

Waring, Clarence A. Stratigraphic and faunal relations of the (Paleocene) Martinez to the (Upper Cretaceous) Chico and (Eocene) Tejon of Southern California. M, 1914, [Stanford University].

Waring, Juliana. Depositional environments of the lower Cretaceous Muddy Sandstone, southeastern Montana. D, 1975, Texas A&M University. 210 p.

Waring, Juliana. Paleoecology of the Snyder Creek Formation (Upper Devonian) of central Missouri. M, 1972, University of Missouri, Columbia.

Waring, Marcus H. Geology of the Windsor Group (Mississippian) reference section, Newport Landing (Avondale), Hants County, Nova Scotia. M, 1967, Acadia University.

Waring, Robert Gordon. Geology of the phosphatic shale member of the Park City Formation in central and north central Utah. M, 1952, Brigham Young University. 34 p.

Waring, Ronald Anthony. Sedimentology of the Bradore Formation, southern Labrador, Newfoundland. M, 1975, Memorial University of Newfoundland. 143 p.

Warinner, J. Ernest. Adsorption of radionuclides on clay minerals. M, 1962, College of William and Mary.

Waritay, Lanfia T. S. Deposition and diagenesis of Silurian carbonates and sulfates, northern Michigan Basin. M, 1984, Texas Tech University. 89 p.

Warith, Mostafa Mohamed Abdel *see* Abdel Warith, Mostafa Mohamed

Wark, David Austin. Geology of the mid-Tertiary volcanic terrane at Buenos Aires, Chihuahua, Mexico. M, 1983, University of Texas, Austin. 156 p.

Warlow, Joseph Charles. Petrography and trace metal chemistry of intrusion breccias, eastern Breckenridge mining district, Summit County, Colorado. M, 1978, Ohio State University.

Warman, James Clark. Some aspects of the (Permian) Washington Coal (West Virginia). M, 1952, West Virginia University.

Warmath, Alex T. The sedimentary petrology and lithofacies of the Pitkin Formation in western Madison and eastern Washington counties, Arkansas. M, 1976, University of Arkansas, Fayetteville.

Warmbrodt, R. E. Sedimentology of the Jefferson City Formation, Franklin County, Missouri. M, 1975, University of Missouri, Columbia.

Warme, John Edward. Paleoecological aspects of the Recent ecology of Mugu Lagoon, California. D, 1966, University of California, Los Angeles. 458 p.

Warmkessel, Carl Andrew. Geology in the vicinity of Fordwick, Virginia. D, 1951, Lehigh University.

Warmkessel, Carl Andrew. Geology of Brown Ridge near Craigsville, Virginia. M, 1942, Lehigh University.

Warn, G. Frederick. Silurian stratigraphic relationships and Brassfield correlations. M, 1941, Northwestern University.

Warn, John Michael. The disparid crinoid Superfamilies Homocrinacea (Ordovician-Silurian) and Cincinnaticrinacea (Ordovician). D, 1974, University of Cincinnati. 296 p.

Warn, John Michael. Variation of the Cincinnatian (Upper Ordovician) crinoid Heterocrinus heterodactylus Hall; ontogeny, regeneration, and pathology. M, 1971, University of Cincinnati. 127 p.

Warncke, Darryl Dean. Investigation of the components involved in the diffusion of zinc in soil. D, 1970, Purdue University. 140 p.

Warne, Andrew G. Stratigraphic analysis of the Upper Devonian Greenland Gap Group and Lockhaven Formation near the Allegheny Front of central Pennsylvania. M, 1986, Rutgers, The State University, New Brunswick. 219 p.

Warner, Albert J. The description and origin of the clastic dikes associated with Sheep Mountain Anticline in the Big Horn Basin, Wyoming. M, 1968, Iowa State University of Science and Technology.

Warner, Albert J. Upper Niagaran and lower Cayugan stratigraphy and depositional environments of the central Appalachian Basin. D, 1978, University of Iowa. 212 p.

Warner, Ambrose Deidriche. The Richland gas field of Richland Parish, Louisiana. M, 1931, Louisiana State University.

Warner, Barry G. Late Quaternary paleoecology of eastern Graham Island, Queen Charlotte Islands, British Columbia, Canada. D, 1984, Simon Fraser University. 190 p.

Warner, David John. Fistuliporid bryozoans of the Wreford Megacyclothem (Lower Permian) in Kansas. M, 1972, Pennsylvania State University, University Park. 63 p.

Warner, Don L. Stratigraphy of Mancos-Mesa Verde (Cretaceous) inter-tonguing, Southeast Piceance Basin, Colorado, and geology of a portion of the Grand Hogback, Garfield County, Colorado. M, 1961, Colorado School of Mines. 170 p.

Warner, Don Lee. An analysis of the influence of physical-chemical factors upon the consolidation of fine-grained clastic sediments. D, 1964, University of California, Berkeley. 149 p.

Warner, Earl, Jr. The conodont fauna of the Upper Ordovician Economy Member, southwestern Ohio. M, 1956, Ohio State University.

Warner, Edward Mark. Petrology and structural geology of igneous and metamorphic rocks, west side of Eureka Valley, Inyo Mountains, California. M, 1971, University of California, Los Angeles.

Warner, Esther Ruth. A study of the (Upper Mississippian) Chester Series of west-central Indiana. M, 1944, Indiana University, Bloomington. 21 p.

Warner, Frederic Kent. Investigation of caving conditions and water resources potential of abandoned iron mines near Hurley and Montreal, Wisconsin. M, 1980, University of Wisconsin-Madison.

Warner, James Brian. Variscan fabrics and structural geometry of the Mispec-Cape Spencer region, Saint John, New Brunswick. M, 1985, Ohio University, Athens. 94 p.

Warner, James Walter. Finite element 2-D transport model of groundwater restoration for in situ solution mining of uranium. D, 1981, Colorado State University. 336 p.

Warner, Jeffrey L. The geology of the Buckfield Quadrangle, Maine. D, 1967, Harvard University.

Warner, Julian Dean. Geology and mineralization of the Blue Rock Mine, northeastern Rincon Mountains, Pima County, Arizona. M, 1982, University of Arizona. 131 p.

Warner, Lawrence Allen. Structure and petrology of the southern Edsel Ford Ranges, Antarctica. D, 1942, The Johns Hopkins University.

Warner, Mark James. Chlorofluoromethanes F-11 and F-12; their solubilities in water and seawater and studies of their distributions in the South Atlantic and North Pacific oceans. D, 1988, University of California, San Diego.

Warner, Marvin Eugene and Frugoni, James John. A magnetic study of selected intrusives in Jefferson, Madison and Gallatin counties, Montana. M, 1958, Indiana University, Bloomington. 54 p.

Warner, Maurice Armond. The origin of the Rex chert. D, 1956, University of Wisconsin-Madison. 94 p.

Warner, Maurice Lee. Environmental impact analysis; an examination of three methodologies. D, 1973, University of Wisconsin-Madison.

Warner, Mont Marcellus. Correlation study of the Mesozoic stratigraphy of Utah and the adjacent portions of surrounding states. M, 1949, Brigham Young University. 120 p.

Warner, Mont Marcellus. Sedimentation of the Duchesne River Formation, Uinta Basin, Utah. D, 1963, University of Iowa. 339 p.

Warner, Paul Freeman. A study of the present bedrock surface of the east half of Township 44 N., Range 2E., state of Illinois. M, 1968, Northern Illinois University.

Warner, Philip Edmund. Regional study of the "Planulina interval"; a productive zone in the lower Miocene sediments of Texas and Louisiana. M, 1982, University of Massachusetts. 185 p.

Warner, Ralph Hartwin. Structural geology of Carboniferous rocks near Marble Falls, Burnet County, Texas. M, 1961, University of Texas, Austin.

Warner, Richard Charles. Cost effective erosion and sediment control systems. D, 1982, Clemson University. 370 p.

Warner, Richard Dudley. Experimental investigations in the system CaO-MgO-SiO₂-H₂O. D, 1971, Stanford University. 178 p.

Warner, Robert Adolph. Studies in the effects of temperature, pressure, capillarity, and electricity as re-lated to oil migration. M, 1954, Miami University (Ohio). 39 p.

Warner, Robert O. Groundwater geology of the Provo-Orem, Utah area. M, 1951, Brigham Young University. 54 p.

Warner, Ronald L. Petrographic and chemical correlation of eight hyalo-rhyolite intrusives of southeastern Idaho. M, 1970, Wayne State University. 119 p.

Warner, Scott David. Modeling the aqueous geochemical evolution of ground water within the Grande Ronde Basalt, Columbia Plateau, Washington. M, 1986, Indiana University, Bloomington. 242 p.

Warner, William Crim. Some Pennsylvanian foraminifera from the Lenapah Limestone of northeastern Oklahoma. M, 1939, University of Missouri, Columbia.

Warning, Karl R. Transgressive-regressive deposits of Difunta Group (Upper Cretaceous-Paleocene), Parras Basin, northeastern Mexico. M, 1977, University of Texas, Austin.

Warnke, Detlef Andaeas. A geologic study of the Halloran Hills, central Mojave Desert, California. D, 1965, University of Southern California. 291 p.

Warnock, Frank B. The Pleistocene mammal fauna and associated artifacts of the San Pedro Springs, Pecos County, Texas. D, 1972, University of Southern Mississippi.

Warnock, George F. Geology of Johobob Mines Limited, Yukon Territory, Canada. M, 1963, University of Arizona.

Warren, Albert David. Ecology of foraminifera of the Buras-Scofield Bayou region, Southeast Louisiana. M, 1954, Louisiana State University.

Warren, Bruce A. Topographic influences on the path of the Gulf Stream. D, 1962, Massachusetts Institute of Technology. 107 p.

Warren, Charles Hyde. Investigations in mineralogy and crystallography including a description of four new minerals from Franklin, New Jersey. D, 1899, Yale University.

Warren, Charles R. Hood River Conglomerate (Miocene or Pliocene) in southern central Washington. D, 1939, Yale University. 282 p.

Warren, David H. Seismic refraction measurements in the Atlantic Ocean northeast of Recife, Brazil. M, 1956, Columbia University, Teachers College.

Warren, Deborah R. Diagenesis of North Cliff phosphate deposits, San Andres Island, Colombia. M, 1984, Tulane University.

Warren, Elbert Clay. A geologic report of the Beck's Mill Quadrangle, Washington County, Indiana. M, 1951, Indiana University, Bloomington. 28 p.

Warren, Elmer John. The bedrock topography of the Keweenaw Penninsula, Michigan. D, 1982, Michigan Technological University. 213 p.

Warren, Jack Roland. A study of magnetic anomalies associated with ultrabasic dikes in the western Kentucky fluorspar district. M, 1955, Indiana University, Bloomington. 89 p.

Warren, James Wolfe. Growth zones in the skeleton of recent and fossil vertebrates. D, 1963, University of California, Los Angeles. 292 p.

Warren, John Stanley. Dinoflagellates and acritarchs from the Upper Jurassic and Lower Cretaceous rocks on the west side of the Sacramento Valley, California. D, 1967, Stanford University. 497 p.

Warren, K. R. Thesis on a petrological study of rocks of the Springhill Mine area and their relation to violent stress relief. M, 1958, Technical University of Nova Scotia.

Warren, Kenneth M., Jr. The Mill Creek Flora, Roca Shale, Wabaunsee County, Kansas. M, 1969, Kansas State University. 49 p.

Warren, P. H. Geochemical studies of pristine, nonmare lunar rocks, and models of nonmare rock genesis. D, 1979, University of California, Los Angeles. 216 p.

Warren, Percival S. The geology of the Banff area (Alberta, Canada). D, 1924, University of Toronto.

Warren, Ray Noble. Geology of a portion of the Stump Hills in southern Shelby County and northern Chilton County, Alabama. M, 1969, Tulane University. 120 p.

Warren, Richard G. Characterization of the lower crust-upper mantle of the Engle Basin, Rio Grande Rift, from a petrochemical and field geologic study of basalts and their inclusions. M, 1978, University of New Mexico. 156 p.

Warren, Roy Kenneth. Magnetic investigation of the Copper Range area of Michigan. M, 1960, University of Washington.

Warren, Thomas Ernest. Structural and metamorphic history of Grenville Province tectonites (Precambrian) in central Dryden Township, Ontario (Canada). M, 1967, University of Wisconsin-Madison.

Warren, Tom Hillary. Stratigraphy and sedimentation of the Pennsylvanian-Permian Fountain Formation, Fremont County, Colorado. M, 1960, University of Oklahoma. 225 p.

Warren, Victoria L. Petrography and environmental history of the Vienna Limestone Member (upper Chesterian) of the southwestern Indiana outcrop. M, 1985, Indiana University, Bloomington. 108 p.

Warren, Virginia Ada. Conodonts and fish remains from the Ardmore cyclothem of Iowa. M, 1948, University of Wisconsin-Madison. 53 p.

Warren, Walter Cyrus. Age of certain andesites in the Mount Aix, Washington, Quadrangle. M, 1933, Washington State University. 24 p.

Warrick, R. A. Volcano hazard in the United States; a research assessment. D, 1975, University of Colorado. 197 p.

Warring, Juliana. Depositional environments of the Lower Cretaceous muddy sandstone, southeastern Montana. D, 1975, Texas A&M University. 245 p.

Warringer, Ben, IV. The Geiger counter. M, 1940, Virginia Polytechnic Institute and State University.

Warrner, Charles Joseph. Subsurface stratigraphy of the Berea and Cussewago sandstones in eastern Ohio. M, 1978, Kent State University, Kent. 65 p.

Warry, Norman David. Formation of ferric phosphate minerals and adsorption of phosphate on amorphous iron oxide. M, 1972, McMaster University. 111 p.

Warshauer, Steven M. A preliminary investigation of Ostracoda from the Kope Formation (Upper Ordovician). M, 1969, University of Cincinnati. 85 p.

Warshauer, Steven M. The taxonomy, ontogeny, biostratigraphy and paleoecology of the Edenian (Upper Ordovician) ostracods of the Ohio Valley. D, 1973, University of Cincinnati. 222 p.

Warshaw, Charlotte Marsh. The mineralogy of glauconite. D, 1957, Pennsylvania State University, University Park. 167 p.

Warshaw, Israel. Phase equilibrium and crystal chemical relationships in rare earth system. D, 1961, Pennsylvania State University, University Park. 107 p.

Warshaw, Israel. Studies in the quaternary system $MgO-Cr_2O_5-Al_2O_3-SiO_2$. M, 1953, Pennsylvania State University, University Park. 56 p.

Warsi, Waris Ejaz Khan. Convergence tectonics of the Peru Trench; 8°S-15°S latitude. D, 1983, Texas A&M University. 182 p.

Warsi, Waris Ejaz Khan. Plate tectonics and the Himalayan Orogeny; a modelling study based on gravity data. M, 1976, Massachusetts Institute of Technology. 66 p.

Warso, Miriam Rebecca. The application of absolute dating techniques to distal glaciolacustrine sediments in a high energy proglacial lake in southeastern British Columbia. M, 1985, University of California, Los Angeles. 143 p.

Warter, Janet Lee Kirchner. Palynology of a lignite of lower Eocene (Wilcox) age from Kemper County, Mississippi. D, 1965, Louisiana State University. 213 p.

Warthen, Robert Carl. Bedrock geology of the northwest quarter of the Dongola Quadrangle, Illinois. M, 1962, Southern Illinois University, Carbondale. 71 p.

Warthin, Aldred S., Jr. Micropaleontology of the (Pennsylvanian) Wetumka, Wewoka, and Holdenville formations (Oklahoma). D, 1931, Columbia University, Teachers College.

Wartman, Brad L. Stratigraphy of the Inyan Kara Formation (Lower Cretaceous) in the vicinity of the Nesson Anticline, northwestern North Dakota. M, 1983, University of North Dakota. 126 p.

Warwick, David B. The response of the Kentucky River drainage basin to a lowering of base level control. M, 1985, Western Michigan University.

Warwick, John Jules. Modeling nitrogen transformations in stream environments. D, 1983, Pennsylvania State University, University Park. 223 p.

Warwick, Peter Delawet. Depositional environments and petrology of the Felix Coal interval (Eocene), Powder River basin. D, 1985, University of Kentucky. 349 p.

Warwick, Peter Delawet. The geology of some lignite-bearing fluvial deposits (Paleocene), southwestern North Dakota. M, 1982, North Carolina State University. 116 p.

Warwick, W. F. Man and the Bay of Quinte, Lake Ontario; 2800 years of cultural influence, with special reference to the Chironomidae (Diptera), sedimentation and eutrophication. D, 1978, University of Manitoba.

Warzeski, E. Robert. Growth history and sedimentary dynamics of Caesars Creek Bank. M, 1976, University of Miami. 198 p.

Warzeski, Edward Robert. Facies patterns and diagenesis of a Lower Cretaceous carbonate shelf; northeastern Sonora and southeastern Arizona. D, 1983, SUNY at Binghamton. 401 p.

Waschitz, Martin. The organic geochemistry of nearshore sediments, New York Bight apex. M, 1980, Brooklyn College (CUNY).

Wasem, Richard. The Pendleton Formation. M, 1942, Louisiana State University.

Wash, Thelma H. Quaternary geology of the east portion of West Fork Basin, Gallatin County, Montana. M, 1971, Montana State University.

Washburn, Alan T. Early Pennsylvanian crinoids from the south-central Wasatch Mountains of central Utah. M, 1968, Brigham Young University.

Washburn, Albert L. Reconnaissance geology of portions of Victoria Island and immediately adjacent regions, Arctic Canada. D, 1942, Yale University.

Washburn, Edward Davis III. The geology of part of the Hollister Quadrangle, San Benito and Santa Clara counties, California. M, 1946, University of California, Berkeley. 42 p.

Washburn, George Robert. Geology of the Manti Canyon area, central Utah. M, 1948, Ohio State University.

Washburn, Judy. Deposition, diagenesis, and porosity relationships of Mississippian Chappel Limestone, Shackelford County, Texas. M, 1978, Texas Tech University. 90 p.

Washburn, Robert Henry. Geology of the Granite Bluff Complex, Dickinson County, Michigan. M, 1961, University of Nebraska, Lincoln.

Washburn, Robert Henry. Structure and Paleozoic stratigraphy of the Toiyabe Range, southern Lander County, Nevada. D, 1966, Columbia University. 94 p.

Washburne, Catharine Lorena. Thermodynamics and speciation of lead in seawater. M, 1981, University of Delaware, College of Marine Studies. 105 p.

Washburne, James C. Parameterization of spectral induced polarization data and laboratory and in situ spectral induced polarization measurements; West Shasta copper-zinc district, Shasta, CA. M, 1982, Colorado School of Mines. 443 p.

Washington, Paul A. Structural analysis of an area near Middlebury, Vermont. M, 1981, SUNY at Albany. 77 p.

Washington, Paul Allyn. Mechanics of thrust fault formation. D, 1987, University of Connecticut. 199 p.

Washken, Edward. Plastic deformation and recrystallization of mineral aggregates. D, 1946, Massachusetts Institute of Technology. 189 p.

Wasilewski, Peter. Correspondence between magnetic and textural changes in titanomagnetites and basaltic rocks. D, 1969, University of Pittsburgh.

Wasilewski, Peter Joseph. Magnetic properties of rocks collected in eastern Ellsworth Land, Antarctica. M, 1965, George Washington University.

Waskom, John D. Quartz grain roundness as an indicator of depositional environments of part of the coast of Panhandle Florida. M, 1957, Florida State University.

Waskom, John Dennis. Geology and geophysics of the Lilesville granite batholith North Carolina. D, 1970, University of North Carolina, Chapel Hill. 78 p.

Waslenchuk, D. G. The distribution and transport of heavy metals in bed sediments of the Ottawa River. M, 1975, University of Ottawa. 184 p.

Waslenchuk, Dennis G. The geochemistry of arsenic in the continental shelf environment. D, 1977, Georgia Institute of Technology. 69 p.

Wasowski, Janusz J. Sedimentary and tectonic history of the northeastern portion of the Dunnage Melange and surrounding units, Newfoundland. M, 1983, SUNY at Buffalo. 86 p.

Wasowski, Janusz Josef. Geology and plate tectonic significance of rock units in the New Bay-Bay of Exploits area, north-central Newfoundland. D, 1986, SUNY at Buffalo. 219 p.

Wassel, Raymond Anthony. Effects of acid mine drainage on heterotrophic bacterial communities in a freshwater impoundment. M, 1982, University of Virginia. 119 p.

Wassenaar, Leonard I. Geochemical and paleoecological investigation using marine molluscs of late Quaternary marine submergences, Quebec, Ontario, British Columbia. M, 1986, Brock University. 174 p.

Wasserburg, Gerald Joseph. The branching ratio of potassium 40 in the dating of geologic material. D, 1954, University of Chicago.

Wasserman, Gilbert. Magnetic survey of the Staten Island Serpentine (New York). M, 1956, New York University.

Wasson, Edward Bassett. The geology of a part of the Metz Quadrangle, San Beniot and Monterey counties, California. M, 1948, Stanford University. 54 p.

Wasson, Isabel B. Sub-Trenton (Ordovician) formations in Ohio. M, 1934, Columbia University, Teachers College.

Wasteneys, Richard Alan. Headquarters Granite of the Wichita Mountains, Oklahoma. M, 1962, University of Oklahoma. 48 p.

Wasuwanich, Pipob. Models of basalt petrogenesis; a study of lower Keweenawan diabase dikes and middle Keweenawan Portage Lake Lavas, Michigan. M, 1979, Michigan State University. 71 p.

Waszczak, Ronald F. Ecology and distribution of Recent plant dwelling foraminifera off Big Pine Key, Florida. M, 1978, Bowling Green State University. 159 p.

Watanabe, Roy Yoshinobu. Geology of the Waugh Lake metasedimentary complex, northeastern Alberta. M, 1961, University of Alberta. 89 p.

Watanabe, Roy Yoshinobu. Petrology of cataclastic rocks of northeastern Alberta. D, 1966, University of Alberta. 219 p.

Waterfall, Louis N. A contribution to the paleontology of the Fernando Group of Ventura County. M, 1927, University of California, Berkeley. 23 p.

Waterman, Arthur Stephen. Conodont biostratigraphy, paleontology, and paleoecology of the Trenton and Lexington limestones in southeastern Indiana. M, 1975, Indiana University, Bloomington. 65 p.

Waterman, Glenn C. Gold deposits of the Eagle Mountain area, Potaro District, British Guiana. M, 1950, [Stanford University].

Waterman, Herbert Douglas. Geology of the Ashcroft area, Pitkin County, Colorado. M, 1955, Colorado School of Mines.

Waterman, Willis D. Factors determining the colors of red and green shales. M, 1951, Kansas State University. 66 p.

Waters, Aaron C. Geology of the southern half of the Chelan Quadrangle, Washington. D, 1930, Yale University. 256 p.

Waters, Aaron C. Structural and petrographic study, Glass Buttes, Lake County, Oregon. M, 1927, University of Washington. 43 p.

Waters, Arnold E., Jr. The placers of the rampart of Hot Springs districts, Alaska. D, 1933, The Johns Hopkins University.

Waters, Barbara Tihen. Osteology of diprotodon (Marsupialia, Mammalia). M, 1969, University of California, Berkeley. 302 p.

Waters, Brent Blakely. Sedimentology and paleogeography of the Upper Cambrian Waterfowl Formation, southern Canadian Rockies. M, 1986, University of Calgary. 216 p.

Waters, Douglas L. Depositional cycles and coral distribution, Mission Canyon and Charles formations, Madison Group (Mississippian), Williston Basin, North Dakota. M, 1984, University of North Dakota. 173 p.

Waters, James A. The Springer Member of the Glenn Formation of the Ardmore Quadrangle. M, 1925, University of Oklahoma. 44 p.

Waters, Jeffrey Phillip. A geophysical and geochemical investigation of selected collapse features on the Coconino Plateau in northern Arizona. M, 1988, Northern Arizona University. 112 p.

Waters, John Nelson. Oligocene foraminifera from Church Creek, Santa Lucia Mountain, California. M, 1963, University of California, Berkeley. 108 p.

Waters, Johnny Arlton. Shape analysis of a population of the blastoid genus Pentremites. M, 1975, Indiana University, Bloomington. 65 p.

Waters, Johnny Arlton. The paleontology and paleoecology of the lower Bangor Limestone (Chesterian, Mississippian) in northwestern Alabama. D, 1978, Indiana University, Bloomington. 193 p.

Waters, Kenneth Montelle, Jr. A geological report upon the Holland area, Dubois County, Indiana. M, 1950, Indiana University, Bloomington. 14 p.

Waters, Laura A. Correlation of upper Eocene and lower Oligocene strata in Mississippi and Alabama. M, 1983, University of Alabama. 179 p.

Waters, Michael Richard. Lake Cahuilla; late Quaternary lacustrine history of the Salton Trough, California. M, 1980, University of Arizona. 74 p.

Waters, Michael Richard. The late Quaternary geology and archaeology of Whitewater Draw, southeastern Arizona. D, 1983, University of Arizona. 125 p.

Waters, Peter Mackenzie. Geostatistics of the Estevan Coal seam in the Boundary Dam Mine, southern Saskatchewan. M, 1984, University of Alberta. 234 p.

Waters, Robert R. The glacial geomorphology of the Pekislo Creek-Happy Valley area, Alberta. M, 1975, University of Calgary.

Waters, Roger Kenneth. Earthquake doublets and mixed magnitude populations in earthquake prediction. M, 1984, Pennsylvania State University, University Park. 146 p.

Waters, Roger Kenneth, III. Lunar paleomagnetic intensity determination using anhysteretic remanent magnetization. M, 1978, University of Oklahoma. 41 p.

Waters, Ronald Hobart. The effect of porosity on shearing resistance and thermal conductivity for amorphous soils in vacuum. D, 1967, Texas A&M University. 175 p.

Waters, Susan Alice. Factors controlling the generation of acid mine drainage in bituminous surface coal mines. M, 1981, Pennsylvania State University, University Park. 247 p.

Waters, Virginia J. Foraminiferal paleoecology and biostratigraphy of the Pungo River Formation, southern Onslow Bay, North Carolina continental shelf. M, 1983, East Carolina University. 186 p.

Wathen, John B. Factors affecting levels of Rn-222 in wells drilled into two-mica granites in Maine and New Hampshire. M, 1986, University of New Hampshire. 97 p.

Watkins, David H. Metamorphism of iron formations on the Melville Peninsula, North West Territories. M, 1973, Carleton University. 103 p.

Watkins, David Kibler. Calcareous nannofossil paleobiogeography in the Cretaceous Greenhorn Sea. D, 1984, Florida State University. 119 p.

Watkins, Elizabeth Ann. Vertical grain size progressions as an aid in interpreting the depositional environments of the Queen City Formation (Eocene), East Texas. M, 1987, Rice University. 287 p.

Watkins, Guyton Hampton. Ammonium aluminosilicates; the examination of a mechanism for the high temperature condensation of ammonia in circumplanetary subnebulae. M, 1981, Massachusetts Institute of Technology. 55 p.

Watkins, Guyton Hampton, Jr. The consequences of cometary and asteroidal impacts on the volatile inventories of the terrestrial planets. D, 1983, Massachusetts Institute of Technology. 216 p.

Watkins, Henry Vaughan, Jr. A subsurface study of the lower Tuscaloosa Formation (Cretaceous) in southern Mississippi. M, 1962, University of Oklahoma. 86 p.

Watkins, Irvine Cabell. Geology of a portion of the gold-pyrite belt in the vicinity of Tabscott, Goochland County, Virginia. M, 1931, University of Virginia. 121 p.

Watkins, Jackie Lloyd. Geology of the Cedar Hill Quadrangle, Dallas and Ellis counties, Texas. M, 1954, Southern Methodist University. 12 p.

Watkins, Jackie Lloyd. Middle Devonian auloporid corals from the Traverse Group of Michigan. D, 1958, University of Michigan.

Watkins, James G. Foraminiferal ecology around the Orange County, California ocean outfall. M, 1959, University of Southern California.

Watkins, Jeffery Alan. Tunisian (North Africa) beach sands. M, 1971, Duke University. 65 p.

Watkins, Joe Henry. Origin of the phosphate deposits of the southeastern states. D, 1942, University of North Carolina, Chapel Hill. 182 p.

Watkins, Joe Henry. Origin of the phosphates of South Carolina. M, 1937, University of North Carolina, Chapel Hill. 54 p.

Watkins, Joel Smith, Jr. Gravity and magnetism of the Ouachita structural belt in central Texas. D, 1961, University of Texas, Austin.

Watkins, John Joseph. The geology of the Corbet Cu-Zn deposit and the environment of ore deposition in the Central Noranda area. M, 1980, Queen's University. 130 p.

Watkins, John M. Depositional environment of Upper Cretaceous Woodbine sandstones, Kurten Field, Brazos County, Texas. M, 1982, Texas A&M University. 48 p.

Watkins, John Phillip. A ground magnetic and subsurface investigation of the Van Buren Quadrangle, Arkansas and Oklahoma. M, 1978, University of Arkansas, Fayetteville.

Watkins, Kenneth N. Clay mineralogy of some Permian and Pennsylvanian limestones. M, 1957, Kansas State University. 64 p.

Watkins, Michael L. Joint analysis and diagenetic studies in the Upper Rockdale Run Formation (Lower Ordovician) in the Great Valley of West Virginia. M, 1986, University of Toledo. 127 p.

Watkins, Norman David. Studies in paleomagnetism. M, 1961, University of Alberta. 139 p.

Watkins, Rodney Martin. A report on the carboniferous system between Hirz Mountain and Kabyai Creek, Shasta County, Northern California. M, 1972, University of California, Berkeley. 161 p.

Watkins, Russell Allen. The Cretaceous section at Lanigan in south-central Saskatchewan. M, 1978, University of Saskatchewan. 102 p.

Watkins, Thomas A. The geology of the Copper House, Copper Mountain, and Parashant breccia pipes; western Grand Canyon, Mohave County, Arizona. M, 1975, Colorado School of Mines. 83 p.

Watkins, William A. A preliminary study of kerogen and kerogen shales. M, 1923, University of Oklahoma. 53 p.

Watkins, William Merle, II. Use of a finned tube for temperature control in absorption. M, 1951, West Virginia University.

Watkinson, David Hugh. Melting relationships in parts of the system $Na_2O-K_2O-CaO-Al_2O_3-SiO_2-CO_2-H_2O$ with applications to carbonate and alkalic rocks. D, 1965, Pennsylvania State University, University Park. 203 p.

Watkinson, David Hugh. Petrochemistry of the mafic-rich rocks, Lac des Mille Lacs area, northwestern Ontario. M, 1963, McMaster University. 75 p.

Watney, Willard Lynn. Major and minor element distributions in some Pennsylvanian shales of Iowa; their paleoenvironmental interpretations. M, 1972, Iowa State University of Science and Technology.

Watney, Willard Lynn. Origin of four Upper Pennsylvanian (Missourian) cyclothems in the subsurface of western Kansas; application to search for accumulation of petroleum. D, 1985, University of Kansas. 630 p.

Watso, David Charles. The effect of tectonic subsidence on sedimentation processes during deposition of four Late Cambrian formations, upper Middle West, United States of America. M, 1988, University of Illinois, Urbana. 264 p.

Watson, Alicia Tyler. An appraisal of the mineral resources of New Zealand. M, 1977, Pennsylvania State University, University Park. 129 p.

Watson, Barry N. Effects of the August 17, 1959 earthquake and subsequent quaking upon the thermal features of Yellowstone National Park. M, 1959, University of Arizona.

Watson, Barry Norton. Structure and petrology of the eastern portion of the Silver Bell Mountains, Pima County, Arizona. D, 1964, University of Arizona. 255 p.

Watson, Bruce F. The tectonic evolution of Kamchatka Peninsula, USSR. M, 1985, Michigan State University. 202 p.

Watson, Chester Conrad. An assessment of the Lower Mississippi River below Natchez, Mississippi. D, 1982, Colorado State University. 175 p.

Watson, Christopher Lex. Relationships between liquid surface tension, contact angle and infiltration for porous media. D, 1969, University of California, Riverside. 50 p.

Watson, Donald Whitman. Geology and structural evolution of the Geco Massive sulfide deposit at Manitouwadge, northwestern Ontario, Canada. D, 1970, University of Michigan. 304 p.

Watson, Donald Whitman. Ore localization at the Holden Mine, Chelan County, Washington. M, 1957, Washington State University. 43 p.

Watson, Douglas William. Middle Devonian salt formations of Alberta. M, 1965, University of Saskatchewan. 62 p.

Watson, Edward Bruce. Experimental studies bearing on the nature of silicate melts and their role in trace element geochemistry. D, 1976, Massachusetts Institute of Technology. 171 p.

Watson, Edward H. The pegmatites of Maryland, with discussion of associated phenomena. D, 1929, The Johns Hopkins University.

Watson, Elizabeth Anne. Stratigraphic occurrences of Discocyclina in the Eocene of California. M, 1941, Stanford University. 65 p.

Watson, Fred Somervill, Jr. Fullerton South Ellenburger Field, Andrews County, Texas. M, 1951, University of Texas, Austin.

Watson, Gordon Peter. Alteration zone geo-chemistry, Lake George antimony deposit. M, 1981, University of New Brunswick.

Watson, Gordon Peter. Ore types and fluid regimes; Macassa gold mine, Kirkland Lake. D, 1984, University of Western Ontario. 341 p.

Watson, Ian A. Structure of the Rattlesnake area, west-central Montana. M, 1984, University of Montana. 55 p.

Watson, James Knox. Stratigraphic correlation of the Niagaran-lower Salina units. M, 1972, University of Michigan.

Watson, Jerry A. Ferruginous layers in sediments from the Gulf of Mexico. M, 1968, Texas A&M University.

Watson, Jerry Palmer. Reservoir simulation model studies of the Postle Morrow "A" Sandstone. M, 1974, University of Oklahoma. 44 p.

Watson, Jeter Marvin. Optimization of pipeline siting resulting from oil and gas development, offshore Virginia. M, 1977, University of Virginia. 68 p.

Watson, John. Geology of the Bear Den Mountain area, Montgomery County, Arkansas. M, 1959, University of Oklahoma. 63 p.

Watson, John E. An economic examination of the Cecil-Argo claim group, Clear Creek County, Colorado. M, 1976, Colorado School of Mines. 74 p.

Watson, John Gaul. The Lower Carboniferous of the Diamond Peak area, Nevada. M, 1939, Cornell University.

Watson, Joseph Quealy. Stratigraphy and depositional environment of the Wilcox lignite deposit south of Hallsville, Harrison County, Texas. M, 1979, Texas A&M University. 224 p.

Watson, Kenneth. I, The thermal conductivity measurements of selected silicate powders in vacuum of 150-350 degrees K; II, An interpretation of the Moon's eclipse and lunation cooling as observed through the Earth's atmosphere from 8-14 microns. D, 1964, California Institute of Technology. 152 p.

Watson, Kenneth DePencer. Geology and mineral deposits of the Baie Verte-Mings Bight area, Newfoundland. D, 1940, Princeton University. 48 p.

Watson, Kenneth Wayne. Stochastic modeling of the initial distribution of surface applied water and dissolved solutes. D, 1983, [University of Kentucky]. 152 p.

Watson, Louis Camille. A comparison of WRENS evapotranspiration estimates to actual losses in north central New Mexico. M, 1983, Colorado State University. 90 p.

Watson, Lowell Brent. Petrologic sedimentary analysis of the basal Cretaceous of Parker and Hood counties, Texas. M, 1961, Texas Christian University. 123 p.

Watson, Michael Guy. Stratigraphy and environment of deposits of the State Quarry Limestone, Johnson County, Iowa. M, 1974, University of Iowa. 140 p.

Watson, Patricia Helen Wanless. Genesis and zoning of silver-gold veins in the Beaverdell area, South-central British Columbia. M, 1981, University of British Columbia. 156 p.

Watson, Phillip Charles. Dendrochronologic reconstruction of water levels for Pyramid Lake, Nevada, 1745 to 1904 A.D. M, 1977, University of Nevada. 124 p.

Watson, R. I. Microtopography of quartz grain surface. M, 1972, University of Toronto.

Watson, Ralph Mayhew, Jr. Stratigraphy and petrography of some sandstones of the uppermost Edmonton and lowermost Paskapoo formations (Cretaceous) of west-central Alberta, Canada. M, 1965, Syracuse University.

Watson, Randall O. An insoluble residue study of the Mississippian and Pennsylvanian limestone of the western Boston Mountains, Arkansas. M, 1959, University of Arkansas, Fayetteville.

Watson, Richard A. Landslides on the east flank of the Chuska Mountains, New Mexico. M, 1959, University of Minnesota, Minneapolis. 45 p.

Watson, Richard Clovis. An analysis of federal laws and regulations affecting mineral location on public land. M, 1980, University of Arizona. 79 p.

Watson, Richard L. Origin of shell beaches, Padre Island, Texas. M, 1968, University of Texas, Austin.

Watson, Richard L. The relationship between littoral drift rate and the longshore component of wave energy flux. D, 1975, University of Texas, Austin. 119 p.

Watson, Robert Brian Fraser. Experiments on ultramafic rocks and volatiles at high temperatures and pressures. M, 1964, University of British Columbia.

Watson, Robert J. Interpretation of resistivity measurements in electrical prospecting. D, 1933, University of Colorado.

Watson, Robert Joseph. Ground vibrations from a periodic source. D, 1958, Pennsylvania State University, University Park. 111 p.

Watson, Robert L. Pleistocene stratigraphy across a portion of the Hartwell Moraine in southwestern Ohio. M, 1973, Miami University (Ohio). 133 p.

Watson, Robert William. The effect of the steep temperature gradient on relative permeability measurements. D, 1987, Pennsylvania State University, University Park. 172 p.

Watson, S. Michelle. The Boulder Batholith as a source for the Elkhorn Mountains Volcanics, southeast quarter of the Deerlodge 15' Quadrangle, southwestern Montana. M, 1986, University of Montana. 100 p.

Watson, Samuel Eugene, Jr. The geology of a part of the Healdsburg Quadrangle, Sonoma County, northern Coast Ranges, California. M, 1941, University of California, Berkeley.

Watson, Simon Timothy. Conodonts from a core of the Nita and Goldwyer formations (Lower-Middle Ordovician) of the Canning Basin, Western Australia. M, 1987, Memorial University of Newfoundland. 171 p.

Watson, Stuart T. Petrography of some South Louisiana subsurface Tertiary rocks. D, 1965, Louisiana State University. 148 p.

Watson, Stuart Tucker. Pennsylvanian strata of the Middle River area of Madison County, Iowa. M, 1955, University of Nebraska, Lincoln.

Watson, Thomas. Leaking mode propagation in layered elastic media. D, 1970, University of Pittsburgh.

Watson, Thomas. The hydraulic characteristics of massive crystalline rock formations in the metropolitan Atlanta area, Georgia. M, 1987, Georgia Institute of Technology. 109 p.

Watson, Thomas Leonard. Some higher levels in the post-glacial development of the Finger Lakes of New York State. D, 1897, Cornell University.

Watson, Vicki Jean. Seasonal phosphorus dynamics in a stratified lake ecosystem; an evaluation of external and internal loading and control. D, 1981, University of Wisconsin-Madison. 297 p.

Watson, William Gorom. Inhomogeneities of the Ramsey Member of the Permian Bell Canyon Formation, Geralding Ford Field, Culberson and Reeves counties, Texas. M, 1974, University of Texas, Arlington.

Watson, William W. A quantitative faunal investigation of the upper five feet of the middle Devonian Wanakah shale at Cazenovia creek, (Erie county, New York). M, 1970, SUNY at Buffalo. 87 p.

Watson-White, M. A petrological study of acid volcanic rocks in part of the Aillik Series, Labrador. M, 1976, McGill University. 92 p.

Watt, Peter J. Measurement of Fe^{+3}/Fe^{+2} ratios in natural titanomagnetites using the Mossbauer effect. M, 1972, Dalhousie University.

Watt, Terry L. Some applications of robust statistical procedures to problems in seismic signal processing. D, 1984, University of Tulsa. 170 p.

Watt, Walton Delbert. A tracer study of the phosphorus cycle in sea water. M, 1962, Dalhousie University.

Wattayakorn, Gullaya. On the occurrence and origins of hopanoids in the Chesapeake Bay. D, 1983, College of William and Mary. 178 p.

Wattenbarger, Robert Allen. Effects of turbulence, wellbore damage, wellbore storage, and vertical fractures on gas well testing. D, 1969, Stanford University. 139 p.

Watters, Thomas Robert. Compressional deformation and tectonic evolution of the Tharsis region of Mars. D, 1985, George Washington University. 267 p.

Watters, Thomas Robert. The aubrites; their origin and relationship to enstatite chondrites. M, 1979, Bryn Mawr College. 45 p.

Watterson, Karen. Earth science microcomputer software; an overview and a contribution. M, 1984, Virginia State University. 129 p.

Watthanachan, Suwit. Origin of saline deposits in the Khorat Plateau, Thailand. M, 1964, University of Alabama.

Wattrus, Nigel James. Two-dimensional velocity anomaly reconstruction by seismic tomography. D, 1984, University of Minnesota, Minneapolis. 245 p.

Watts, Bradice C. The Precambrian geology of the Bear Creek area, Medicine Bow Mountains, Wyoming. M, 1971, Bowling Green State University. 59 p.

Watts, Chester Frederick. Development, testing, and evaluation of a microcomputer system for rapid collection and analysis of geological structure data related to rock slope stability. D, 1983, Purdue University. 409 p.

Watts, D. R. Upper Keweenawan and lower Paleozoic paleomagnetism of the North American Craton. D, 1979, University of Michigan. 270 p.

Watts, Doyle Robin. A paleomagnetic study of four Mesozoic diabase dike swarms of the Southern Appalachian Mountains. M, 1975, Ohio State University.

Watts, G. P. Methods for measuring thermal conductivity of earth materials with application to anisotropy in basalt and heat flux through lake sediments. M, 1975, University of Hawaii at Manoa. 103 p.

Watts, Keith Fred. Evolution of a carbonate slope facies along a South Tethyan continental margin; the Mesozoic Sumeini Group and the Qumayrah Facies of the Muti Formation, Oman. D, 1985, University of California, Santa Cruz.

Watts, Kenneth Robert. Assessment of Landsat imagery for the investigation of fracturing in an unconfined chert and carbonate aquifer. M, 1977, Oklahoma State University. 85 p.

Watts, Linda Jean. Kettle hole and sandur development at the margin of the Casement Glacier, Glacier Bay National Park, Alaska. M, 1988, University of Nebraska, Lincoln. 121 p.

Watts, Raymond Douglas. Magnetotelluric fields over round structures. D, 1972, University of Toronto.

Watts, Royal J. Petrology of the (Pennsylvanian) Homewood Sandstone in Kanawha and Mineral counties, West Virginia. M, 1959, West Virginia University.

Watts, Stephen H. A study of the Santee Till (middle Pleistocene) in northeastern Nebraska. M, 1971, University of Nebraska, Lincoln.

Watts, Terrance Roger. Grain size variations in the beach sands of Long Point, Lake Erie. M, 1962, McMaster University. 105 p.

Watwood, Mary Elizabeth. Environmental parameters regulating sulfur retention in a variety of forest soils. D, 1987, University of Georgia. 152 p.

Waugh, John Russell, II. Late tectonism in the Big Bend region of Trans-Pecos, Texas. M, 1977, Southern Methodist University. 44 p.

Wauters, John Ferdinand. The Sparta Aquifer, northern Brazos County, Texas. M, 1956, Texas A&M University. 51 p.

Wavra, Craig Scott. Structural geology and petrology of the SW 1/4 Mahoney Butte Quadrangle, Blaine County, Idaho. M, 1985, University of Idaho. 127 p.

Wavrek, David A. Chemical and petrographic characteristics of the Upper Silurian (Cayugan) carbonates and evaporites, Tymochtee Formation, Ottawa County, Ohio. M, 1985, University of Toledo. 208 p.

Wawersik, Wolfgang R. Detailed analysis of rock failure in laboratory compression tests. D, 1968, University of Minnesota, Minneapolis. 200 p.

Way, Dana Clark. A reconnaissance study of granitoid plutonism in southwestern Yukon Territory. M, 1977, Queen's University. 187 p.

Way, Douglas Stewart. Drainage density; a function of rock properties, relief, and climate. D, 1982, Clark University. 260 p.

Way, Harold G. The Silurian of Manitoulin Island, Ontario. D, 1936, University of Toronto.

Way, Helen Sue Kincaid. Structural study of the Hunton Lime of the Wilzetta Field, T12-13N, R5E, Lincoln County, Oklahoma, pertaining to the exploration for hydrocarbons. M, 1983, Oklahoma State University. 40 p.

Way, J. H., Jr. Petrology and sedimentation of a section of the Carboniferous sedimentary rocks at Joggins, Nova Scotia; a vertical profile. M, 1967, University of Pennsylvania.

Way, John H., Jr. Deposition environmental analysis, Middle and Upper Devonian sedimentary rocks, Catskill Mountain area, New York. D, 1972, Rensselaer Polytechnic Institute. 125 p.

Way, Shao Chih. The study of ground water movement with the use of computer model at the Socorro Grant, New Mexico. M, 1971, New Mexico Institute of Mining and Technology.

Way, Shao-Chih. Methods for the determination of three-dimensional formation permeability with field examples from in-situ uranium ore bodies. D, 1980, University of Wyoming. 228 p.

Waychison, M. Mineralogy and petrology of the Tichegami Group amphibolites, Mistassini Territory, Quebec. M, 1976, McGill University. 246 p.

Waychunas, Glen Alfred. Mossbauer, X-ray, optical and chemical study of cation arrangements and defect association in Fm3m solid solutions in the system periclase-wustite-lithium ferrite. D, 1979, University of California, Los Angeles. 479 p.

Wayland, John Rex. Geology of the Baum Limestone of southern Oklahoma. M, 1954, Southern Methodist University. 25 p.

Wayland, Russell Gibson. A mineralogical study of Black Hills cummingtonite. M, 1935, University of Minnesota, Minneapolis. 31 p.

Wayland, Russell Gibson. Geology of the Juneau region, Alaska, with special reference to the Alaska Juneau ore body. D, 1939, University of Minnesota, Minneapolis. 92 p.

Waylett, Annette Shelton. The Antarctic fantasy; suspension of sovereignty and the potential for offshore mineral development. M, 1987, University of Idaho. 68 p.

Wayman, Cooper Harry. Measurements of dissociation pressures of hydrous minerals using thermistors. D, 1959, Michigan State University. 128 p.

Wayman, Cooper Harry. Petrography of basic electric-furnace slags. M, 1954, University of Pittsburgh.

Waymire, Kelly Sue. Geochemistry of the regionally metamorphosed mafic rocks of the Standing Pond Volcanics, Vermont. M, 1984, University of Kentucky. 156 p.

Wayne, Christopher J. Sea and marsh grasses; their effect on wave energy and near-shore sand transport. M, 1975, Florida State University.

Wayne, David Matthew. Electron microprobe analysis of rare-earth-element-bearing phases from the White Cloud Pegmatite, South Platte District, Jefferson County, Colorado. M, 1986, University of New Orleans. 122 p.

Wayne, William John. The glacial geology of Wabash County, Indiana. M, 1950, Indiana University, Bloomington. 107 p.

Wayne, William John. Thickness of drift and bedrock physiography of Indiana north of the Wisconsin glacial boundary. D, 1952, Indiana University, Bloomington. 91 p.

Waywanko, Andrea O. Sedimentary and geophysical well log analysis of the Clearwater Formation (Lower Cretaceous), Cold Lake, Alberta. M, 1984, University of Alberta. 163 p.

Weagle, Lawrence Townsend. Geology of the Darby Creek valley adjacent to Bryn Mawr, Pennsylvania. M, 1941, Bryn Mawr College. 25 p.

Weakliem, John Herbert. The petrological evolution of garnet-rich meta-igneous bodies in the southeastern Adirondack Mountains, New York. M, 1984, Lehigh University.

Weakly, Edward Cletus. The geology and groundwater resources of Polk County, Nebraska. M, 1966, University of Nebraska, Lincoln.

Weand, B. L. The chemical limnology of Lake Bonney, Antarctica with emphasis on trace metals and nutrients. D, 1976, Virginia Polytechnic Institute and State University. 200 p.

Weant, George Edward, III. Airborne particulate production from feldspar processing (western North Carolina). M, 1973, North Carolina State University. 43 p.

Weart, Richard Claude. Geology of the northern flank of the Wind River Mountains, Fremont County, Wyoming. M, 1948, Syracuse University.

Weart, Richard Claude. Pennsylvanian and Permian fusulinids of western Montana and central Idaho. D, 1950, University of Illinois, Urbana. 164 p.

Weart, Wendell Duane. Determination of the seismic characteristics of desert alluvium. D, 1961, University of Wisconsin-Madison. 138 p.

Weary, David John. Dinoflagellate paleoecology and biostratigraphy of the middle Eocene Tallahatta and Lisbon formations from the Baldwin County, Alabama, core. M, 1988, Virginia Polytechnic Institute and State University.

Weatherill, Philip Mathew. Seismic stratigraphy and tectonic evolution of the Stord Basin, northern North Sea. M, 1988, University of Texas, Austin. 147 p.

Weatherington, Julie B. A geologic and land-use development survey of Blendon and Plain townships, Franklin County, Ohio. M, 1978, Ohio State University.

Weathers, Gerald. A geological investigation of an area in Hickory and Polk counties, Missouri. M, 1950, Washington University. 144 p.

Weathers, L. Michael. Geology for land and ground water planning in Putnam County, Indiana. M, 1975, DePauw University.

Weathers, Maura Susan. Geology, petrography and origin of josephinite. M, 1976, Cornell University.

Weathers, Maura Susan. Mantle-derived iron-nickel alloys. D, 1978, Cornell University. 173 p.

Weaver, Benjamin Franklin. Reconnaissance geology and K-Ar geochronology of the Trigo Mountains detachment terrane, Yuma County, Arizona. M, 1982, San Diego State University. 119 p.

Weaver, Charles Edward. Mineralogy and petrology of some Paleozoic clays from central Pennsylvania. D, 1952, Pennsylvania State University, University Park. 145 p.

Weaver, Charles Edwin. Geology of the Napa Quadrangle, California. D, 1907, University of California, Berkeley. 38 p.

Weaver, Charles Edwin. Petrography and petrology of rocks near the Quadrant-Phosphoria boundary in S.W. Montana. M, 1950, Pennsylvania State University, University Park. 150 p.

Weaver, Clark L. Geology of the Blue Mountain Quadrangle, Beaver and Iron counties, Utah. M, 1979, Brigham Young University.

Weaver, Craig Steven. Seismic events on Cascade volcanoes. D, 1976, University of Washington. 151 p.

Weaver, David F. Seismological determination of crustal thickness in southern Alberta. M, 1963, University of Alberta. 132 p.

Weaver, Dennis W. A gravity survey of the buried Teays Valley in northern Mercer County, Ohio. M, 1984, Bowling Green State University. 176 p.

Weaver, Donald W. Eocene foraminifera from west of Refugio Pass, California. M, 1957, University of California, Berkeley. 106 p.

Weaver, Donald W. The paleontology and stratigraphy of the Gaviota Formation, Santa Barbara County, California. D, 1960, University of California, Berkeley. 244 p.

Weaver, Erich. Miocene stratigraphy, silica diagenesis and paleomagnetism of Santa Rosa Island. M, 1985, University of California, Santa Barbara. 144 p.

Weaver, Faye Janet. Source rock studies of natural seep oils near Parsons Pond on the west coast of Newfoundland. M, 1988, Memorial University of Newfoundland. 178 p.

Weaver, Fred M. Late Miocene and Pliocene radiolarian paleobiogeography and biostratigraphy of the Southern Ocean. D, 1976, Florida State University. 183 p.

Weaver, Fred M. Pliocene paleoclimatic and paleoglacial history of East Antarctica recorded in deep sea piston cores. M, 1973, Florida State University.

Weaver, G. D. Environmental hazards of oil-shale development. D, 1973, The Johns Hopkins University. 401 p.

Weaver, J. Geothermal temperature measurements using oscillators. M, 1967, University of Toronto.

Weaver, J. F. The sorption of cadmium on silica and kaolin in the presence of some organic and inorganic ligands. D, 1979, University of Maryland. 319 p.

Weaver, J. Scott. Calculation of the B-V relation for the B1 phase of NaCl up to 300 kilobars at 25°C. M, 1970, University of Rochester.

Weaver, John Christian B. Miocene foraminifera of the Graves Creek section of the Monterey Formation, San Luis Obispo County, California. M, 1986, University of California, Davis. 145 p.

Weaver, John Ferry. Petrology of the Memesagamesing Lake Complex. M, 1954, University of Cincinnati. 49 p.

Weaver, John Scott. Equation of state of NaCl, MgO and stishovite. D, 1971, University of Rochester. 173 p.

Weaver, Kenneth Newcomer. The geology of the Hanover area, York County, Pennsylvania. D, 1955, The Johns Hopkins University.

Weaver, M. S. The structural evolution of North central Texas. M, 1977, Baylor University. 117 p.

Weaver, Oscar David, Jr. The geology of the Balcones fault zone south from Water Park, Texas, to State Highway Twenty-Nine. M, 1947, University of Texas, Austin.

Weaver, Oscar Dee, Jr. The geology of Hughes County, Oklahoma. D, 1952, University of Oklahoma. 201 p.

Weaver, Richard. Geology interpretation of the Ruby Star Ranch area, Twin Buttes mining district, Pima County, Arizona. M, 1965, University of Arizona.

Weaver, Robert Michael. Montmorillonite genesis in soils as influenced by the activities of monosilicic acid and various cations in the matrix solution. D, 1970, University of Wisconsin-Madison. 261 p.

Weaver, Ronald Lee. Aspects of the radiation budget related to fast ice decay, Broughton Island, Baffin Island, N.W.T. M, 1976, University of Colorado.

Weaver, Sarah C. The hydrogeochemistry of the principal aquifers in Las Vegas Valley, Nevada, and the chemical effects of artificial recharge of Colorado River water. M, 1982, University of Nevada. 159 p.

Weaver, Sidney Mark. Finite strain study of the Collier Formation in the Benton Uplift of western Arkansas. M, 1985, Southern Illinois University, Carbondale. 86 p.

Weaver, Stephen George. The geology of the Bovard mining district, Gabbs Valley Range, Mineral County, Nevada. M, 1982, University of Nevada. 85 p.

Weaver, Stephen George. The Patagonian Batholith at 48°S latitude, Chile; implications for the petrological and geochemical evolution of calc-alkaline batholiths. D, 1988, Colorado School of Mines. 115 p.

Weaver, Tamie R. Groundwater geochemistry and radionuclide activity in the Cambrian-Ordovician sandstone aquifer of Dodge and Fond du Lac counties, Wisconsin. M, 1988, University of Wisconsin-Madison. 111 p.

Weaver, Tamie Renee. Diagenesis in the Trenton Group (Ordovician; New York). M, 1984, Cornell University. 78 p.

Weaver, Thomas Adrian. The opacities of transparent materials as a function of temperature and wave length, and their geophysical implications. D, 1973, University of Chicago. 118 p.

Weaver, Timothy Otis. Earthquake analysis of soil-structure systems utilizing Fourier transform techniques. D, 1980, University of Virginia. 110 p.

Weaver, Trinchitella Marianne. Late Quaternary paleoclimatic and paleoglacial history of the southeastern Pacific Subantarctic and Antarctic region; analysis of a glacial cycle. M, 1979, Florida State University.

Weaver, William E. Experimental study of alluvial fans. D, 1984, Colorado State University. 538 p.

Weaver, William Ray. Upper Eocene foraminifera from the southwestern Santa Ynez Mountains, California. M, 1956, University of California, Berkeley. 83 p.

Weaverling, Paul Harrison. Early Paleozoic tectonic and sedimentary evolution of the Reelfoot-Rough Creek rift system, Midcontinent, U. S. M, 1987, University of Missouri, Columbia. 116 p.

Webb, Anthony J. Electrochemical determination of crystallization oxygen fugacities in Monteregian lamprophyre dikes (Quebec). M, 1970, McGill University. 35 p.

Webb, Chris Cynthia. Deposition, diagenesis and porosity relationships of the Odom Limestone. M, 1978, Texas Tech University. 72 p.

Webb, David B. The osteology of Camelops. M, 1961, University of California, Berkeley. 89 p.

Webb, David Knowlton, Jr. A clay mineral study of some bentonites from the Sheridan, Wyoming area. M, 1959, University of Illinois, Urbana.

Webb, David Knowlton, Jr. Vertical variations in the clay mineralogy of sandstone, shale and underclay members of Pennsylvanian cyclothems. D, 1961, University of Illinois, Urbana. 111 p.

Webb, David Ralph. The relationship of ore to structural features of the Campbell shear zone and hanging wall stratigraphy in the Con Mine, Yellowknife, Northwest Territories. M, 1983, Queen's University. 181 p.

Webb, Douglas R. Conemaugh fossil invertebrates from eastern Ohio; Gastropoda. M, 1972, Ohio University, Athens. 147 p.

Webb, Edwin James. The Devonian of West Texas and southeastern New Mexico. M, 1946, University of Pittsburgh.

Webb, Elmer James. Cambrian sedimentation and structural evolution of the Rome Trough in Kentucky. D, 1980, University of Cincinnati. 98 p.

Webb, Frank S. Surface geology of the Eufaula-Texanna area, McIntosh and Pittsburg counties, Oklahoma. M, 1957, University of Oklahoma. 60 p.

Webb, Fred. Geology of the Middle Ordovician limestones in the Rich Valley area, Smyth County, Virginia. M, 1959, Virginia Polytechnic Institute and State University.

Webb, Fred, Jr. Geology of the Big Walker Mountain-Crockett Cove area, Bland, Pulaski and Wythe counties, Virginia. D, 1965, Virginia Polytechnic Institute and State University. 252 p.

Webb, G. R. Benthic ecology and taphonomy of a Bay of Fundy rocky subtidal community, with particular reference to the articulate brachiopod Terebratulina septentrionalis (Couthouy). M, 1976, University of New Brunswick.

Webb, Gregory Edward. Coral fauna and carbonate mound development, Pitkin Formation (Chesterian), North America. M, 1984, University of Oklahoma. 267 p.

Webb, Gregory W. Middle Ordovician stratigraphy in eastern Nevada and western Utah. D, 1954, Columbia University, Teachers College.

Webb, Gregory W. Stratigraphy of the Ordovician quartzites of the Great Basin. M, 1949, Columbia University, Teachers College.

Webb, James Edward. Reconnaissance survey of the (Precambrian to Pennsylvanian) Talladega Series in parts of Polk and Haralson counties, Georgia. M, 1957, Cornell University. 35 p.

Webb, James Franklin. Geology of the San Diego margin, Carlsbad to La Jolla, San Diego County, California. M, 1984, California State University, Northridge. 188 p.

Webb, James H. The geology of Green Forest Township, Carroll County, Arkansas. M, 1961, University of Arkansas, Fayetteville.

Webb, James Sutton. The subsurface stratigraphy of the Laurel Dolomite (Silurian) in west-central Kentucky. M, 1984, University of Kentucky. 97 p.

Webb, James W. Allegheny sedimentary geology in vicinity of Ashland, Kentucky. D, 1963, Louisiana State University.

Webb, John Benwell. Stratigraphy and structure of the Foothills belt, western Alberta, between Highwood and Berland rivers. M, 1930, University of Manitoba.

Webb, John Charles. Petrology and potential sources of uranium in Tertiary rocks, Logan County, northeastern Colorado. M, 1980, University of Colorado.

Webb, John Hanor. The geology of an area southwest of Douglas, Converse County, Wyoming. M, 1941, University of Oklahoma. 94 p.

Webb, John Purcell. A seismological study of the tectonics of a portion of the Southwest Pacific. D, 1954, St. Louis University.

Webb, John W. The wearing properties of mineral aggregates in highway surfaces. M, 1970, University of Virginia. 84 p.

Webb, Kathryn Wenthe. Depositional sub-environments of Santa Rosa Island, Florida. M, 1985, University of Alabama. 105 p.

Webb, L. E. and Glenn, W. H. An investigation of longshore currents at Moss Landing (Monterey County) California. M, 1966, United States Naval Academy.

Webb, Lyndall. The evaluation of cementation mechanisms in kaolinite. M, 1974, Georgia Institute of Technology. 41 p.

Webb, M. D. Pathways of CO_2, O_2, CO and CH_4 in water flowing over a coral reef, Kaneohe Bay, Oahu, Hawaii. M, 1977, University of Hawaii. 107 p.

Webb, Philip K. Areal geology of the Cavanal Syncline. M, 1958, University of Oklahoma. 113 p.

Webb, R. J. Localization of prismatic orebody by stratigraphic/structural controls, South Trend, Pine Point lead-zinc property, Northwest Territories, Can. M, 1986, University of Waterloo. 245 p.

Webb, Robert Howard. Late Holocene flooding on the Escalante River, south-central Utah. D, 1985, University of Arizona. 245 p.

Webb, Robert Howard. The effects of controlled motorcycle traffic on a Mojave Desert soil. M, 1980, Stanford University. 87 p.

Webb, Robert T. Petrography, structure, and mineralization of the Meadow Creek area, Chelan County, Washington. M, 1957, University of Arizona. 51 p.

Webb, Robert Wallace. Geology of a portion of the southern Sierra Nevada of California; the northern Kernville Quadrangle. D, 1937, California Institute of Technology.

Webb, Robert Wallace. Paleontology of the Pleistocene of Point Loma, San Diego County, California. D, 1937, California Institute of Technology.

Webb, Robert Wallace. The geology of eastern Sierra Pelona Ridge and vicinity in the southeastern part of the Elizabeth Lake Quadrangle, California. M, 1932, California Institute of Technology. 83 p.

Webb, Sam Nail. An occurrence of bentonite and its commercial possibilities in Houston County, Texas. M, 1942, University of Texas, Austin.

Webb, Spahr Chapman. Observations of seafloor pressure and electric field fluctuations. D, 1984, University of California, San Diego. 208 p.

Webb, Steve William, III. The interaction of ion-exchanged montmorillonite clays and polyethylene oxide polymers. D, 1984, University of Alabama. 206 p.

Webb, Steven Ray. A subsurface study of the Devonian Caballos Novaculite of the Tobosa Basin of West Texas. M, 1975, Texas Tech University. 67 p.

Webb, Timothy C. Depositional history of the Armstrong Brook-Pointe Rochette area, northern New Brunswick. M, 1983, University of New Brunswick. 100 p.

Webb, William Felton. Precambrian geology and ore deposits near Poland Junction, Yavapai County, Arizona. M, 1979, University of Arizona. 113 p.

Webb, William Martin. The geology of Mullet Lake (Michigan). M, 1964, University of Michigan.

Webber, B. S. A description of the Madrid, Iowa, coal field. M, 1926, Iowa State University of Science and Technology.

Webber, Benjamin N. The Upper Carboniferous stratigraphy of the Galiuro Mountains. M, 1925, University of Arizona.

Webber, George R. Age determinations of Massachusetts granites from radiogenic lead in zircon. D, 1955, Massachusetts Institute of Technology. 84 p.

Webber, George Roger. Spectrochemical analysis of the White Mountain magma series and some Finnish granites. M, 1952, McMaster University. 27 p.

Webber, Irma Eleanor Schmidt. Pleistocene woods from Carpinteria, California. M, 1927, University of California, Berkeley. 37 p.

Webber, Irma Eleanor Schmidt. Pleistocene woods from the coast of California. D, 1929, University of California, Berkeley.

Webber, Karen Louise. The Mammoth Mountain and Wason Park tuffs; magmatic evolution in the central San Juan volcanic field, southwestern Colorado. D, 1988, Rice University. 243 p.

Weber, Anthony J. An integrated geological/geophysical interpretation of the Pachuta Creek area (Clarke County, Mississippi). M, 1979, University of Southwestern Louisiana.

Weber, Eric F. A stochastic model and risk analysis of artesian well depth and well yield in the Fairbanks area, Alaska. M, 1986, University of Alaska, Fairbanks. 196 p.

Weber, F. Harold. The geology and mineral deposits of the Ord Mountains District, San Bernardino County, California. M, 1956, University of California, Los Angeles.

Weber, Gerald Eric. Geology of the fluvial deposits of the Colorado River Valley, central Texas. M, 1968, University of Texas, Austin.

Weber, Gerald Eric. Recurrence intervals and recency of faulting along the San Gregorio fault zone, San Mateo County, California. D, 1980, University of California, Santa Cruz. 277 p.

Weber, Hans-Peter. Stoichiometric and non-stoichiometric magnetites. A structural investigation by Mossbauer spectrometry. D, 1972, University of Chicago. 99 p.

Weber, J. N. E. The chemical and crystallochemical approach to the dolomite problem. D, 1962, University of Toronto.

Weber, James R. Structural geology of the northeastern flank of the Uinta Mountains, Moffat County, Colorado. M, 1971, Colorado School of Mines. 68 p.

Weber, Janine F. The biostratigraphy and paleogeography of the Lower Ordovician Ninemile Formation, Pogonip Group, central Nevada. M, 1983, Colorado School of Mines. 138 p.

Weber, Jo Ann T. Oscillatory compositional zoning in grossular-andradite garnets. M, 1985, University of New Mexico. 106 p.

Weber, John C. Bedrock geology and structural analysis in portions of the Cossatot Mountains, Athens Plateau and Mazarn Basin, Pike and Montgomery counties, Arkansas. M, 1986, Southern Illinois University, Carbondale. 112 p.

Weber, Jon N. E. The geochemistry of some graywackes and some shales. M, 1959, McMaster University. 149 p.

Weber, Kenneth A. Characterization of storm water runoff by numerical analysis. M, 1981, University of South Florida, Tampa. 77 p.

Weber, L. James. Environmental analysis of a Virgilian (Pennsylvanian) carbonate sequence within Rhodes Canyon, San Andres Mountains, New Mexico. M, 1983, New Mexico Institute of Mining and Technology. 150 p.

Weber, Lawrence C. Depositional environments of the Tate Member of the Ashlock Formation in east central Kentucky. M, 1974, Eastern Kentucky University. 85 p.

Weber, Lawrence James, Jr. Paleoenvironmental analysis and test of stratigraphic cyclicity in the Nolichucky Shale and Maynardville Limestone (Upper Cambrian) in central East Tennessee. D, 1988, University of Tennessee, Knoxville. 389 p.

Weber, Linda. Lithologic and organic source analysis; Stevens interval, Cal Canal Field, Kern County, California. M, 1985, Stanford University. 100 p.

Weber, Michael H. The Gaussian beam method in seismology. M, 1983, University of Toronto. 198 p.

Weber, Mitch W. The study of slope movements along Interstate 77 between Canton, Ohio, and Ripley, West Virginia, with special emphasis on rock falls and rock topples. M, 1985, Kent State University, Kent. 288 p.

Weber, Neil Victor. The relationships among linear trends, bedrock fractures, and linear drainage lines in the Oak Creek upland area of the Colorado Plateau, Arizona. D, 1972, Indiana State University. 169 p.

Weber, Paul Wesley. Geology and an analysis of the structure at Moss Beach, California. M, 1958, Stanford University.

Weber, Richard Elmo. Petrology and sedimentation of the upper Precambrian Sioux Quartzite, Minnesota, South Dakota and Iowa. M, 1981, University of Minnesota, Duluth.

Weber, Richard G. Fractures and shear vertical seismic profiling. M, 1986, Colorado School of Mines. 93 p.

Weber, Robert, II. The geology of the east-central portion of the Huachuca Mountains, Arizona. D, 1950, University of Arizona.

Weber, Scott James. Petrography and diagenesis of the McLish Formation, Simpson Group (Middle Ordovician), southeastern Anadarko Basin, Oklahoma. M, 1987, Oklahoma State University. 98 p.

Weber, Wilfred W. L. The geology of the Duverny area with particular reference to the granitic rocks. D, 1950, University of Toronto.

Weber, William Mark. Correlation of Pleistocene glaciation in the Bitterroot Range, Montana, with fluctuations of glacial Lake Missoula. D, 1971, University of Washington. 109 p.

Weber, William Mark. General geology and geomorphology of the Middle Creek area, Gallatin County, Montana. M, 1965, Montana State University. 86 p.

Weberg, Erik D. Structural geology of the Cretaceous rocks in Sun River Canyon, Teton County and Lewis and Clark County, Northwest Montana. M, 1986, Washington State University. 102 p.

Webernick, Nelson Ellsworth. Gypsum karst, Culberson County, Texas. M, 1952, University of Texas, Austin.

Webers, Gerald F. A study of the conodonts of the Dubuque Formation of Minnesota. M, 1961, University of Minnesota, Minneapolis. 50 p.

Webers, Gerald F. A study of the Middle and Upper Ordovician conodont faunas of Minnesota. D, 1964, University of Minnesota, Minneapolis. 176 p.

Webster, David Alexander. A study of the axial and pediment gravels in the eastern part of the Santo Domingo Basin, New Mexico. M, 1966, University of New Mexico. 114 p.

Webster, Elizabeth Archer. Partitioning of alkali elements, REE, Sc and Ba between silicic melts and aqueous fluids. M, 1981, Arizona State University. 119 p.

Webster, Gary D. Geology of the Canon City-Twin Mountain area, Fremont County, Colorado. M, 1959, University of Kansas. 99 p.

Webster, Gary Dean. Biostratigraphy of the pre-Des Moines part of the Bird Spring Formation, northern Clark and southern Lincoln counties, Nevada. D, 1966, University of California, Los Angeles. 268 p.

Webster, J. M. Stratigraphy and distribution of the Upper Cretaceous D Sandstone in a portion of West-central Denver Basin. M, 1976, University of Iowa. 61 p.

Webster, J. R. Analysis of potassium and calcium dynamics in stream ecosystems on three southern Appalachian watersheds of contrasting vegetation. D, 1975, University of Georgia. 244 p.

Webster, James D. Meteorologic influences on soil-gas release; as applied to mineral exploration and earthquake prediction. M, 1980, Colorado School of Mines. 151 p.

Webster, James Dale. Experimental study of a F-, Cl- and lithophile trace element-enriched vitrophyre from Spor Mountain, Utah. D, 1987, Arizona State University. 208 p.

Webster, John Robert. Petrochemistry and origin of the plagioclase gneisses of the Killingworth Dome, south-central Connecticut. M, 1985, Indiana University, Bloomington. 64 p.

Webster, Maude Martha. The weathering of soil minerals. M, 1927, University of North Carolina, Chapel Hill. 17 p.

Webster, Michael Stilson. Regional study of the Newcastle Formation of South Dakota. M, 1963, Michigan State University. 46 p.

Webster, Rick L. Analysis of petroleum source rocks of the Bakken Formation (Devonian and Mississippian) in North Dakota. M, 1982, University of North Dakota. 150 p.

Webster, Robert E. Structural geology of the Southwest Edwards Plateau area, Edwards, Kinney and Val Verde counties, Texas. M, 1978, University of Texas, Arlington. 197 p.

Webster, Russell N. Pleistocene deposition in the Monongahela River valley of northern West Virginia. M, 1949, West Virginia University.

Webster, Stephen Leroy. Interaction effects due to subsidence in multiple seam mining. M, 1983, Virginia Polytechnic Institute and State University. 130 p.

Webster, Terence A. Faulting and slumping during deposition of the Precambrian Prichard Formation, near Quinns, Montana; a model for sulfide deposition. M, 1981, University of Montana. 71 p.

Webster, Thomas Craig. Subsurface stratigraphy of Lower and Middle Mississippian formations of the southeastern Illinois Basin; West-central Kentucky. M, 1981, University of Kentucky. 177 p.

Webster, Thomas F. A description and an analysis of meanders in the Gulf Stream. D, 1961, Massachusetts Institute of Technology. 142 p.

Wechsler, Barry Andrew. Crystallographic studies of titanomagnetites and ilmenite. D, 1981, SUNY at Stony Brook. 181 p.

Wedderburn, Leslie Ansel. Karst hydrogeology of Nassau Valley, Jamaica. M, 1967, University of Illinois, Urbana.

Weddle, Thomas K. Petrology of Upper Triassic sandstones from the Hartford, Pomperaug, and Newark basins. M, 1979, University of Massachusetts.

Wedekind, Frank E. Geochemical survey of tungsten along the Young America Fault, Magdalena mining district, New Mexico. M, 1962, New Mexico Institute of Mining and Technology. 59 p.

Wedekind, James E. The stratigraphy, depositional environments, bryozoan paleoecology, and taphonomy of the middle Chickamauga Group near Thorn Hill, Grainger County, Tennessee. M, 1986, University of Tennessee, Knoxville. 179 p.

Wedel, Arthur Albert. Geologic structure of the Devonian strata of south central New York. D, 1930, Cornell University.

Wedemeyer, Richard C. Geochemistry and geochronology of the Sims Granite, Eastern Carolina Slate Belt, North Carolina. M, 1981, East Carolina University. 63 p.

Wedge, William Keith. Petrography, chemical composition and stratigraphic setting of the coals of North-central Missouri. D, 1973, University of Missouri, Rolla.

Wedge, William Keith. Petrology and petrography of the Croweburg Coal, north central Missouri. M, 1971, University of Missouri, Rolla.

Wedgeworth, Bruce Steven. Ita Mai Tai Guyot; a comparative geophysical study of western Pacific seamounts. M, 1985, University of Hawaii. 90 p.

Wedow, Helmuth Jr. A study of the stratigraphy and sedimentation of the (Upper Devonian) Genundewa Limestone in western New York. M, 1939, SUNY at Buffalo.

Wee, Pamela Sui Lian. The development of rhenium-187/osmium-187 system in the age dating of iron meteorites. M, 1986, University of Toronto.

Wee, Soo-Meen. Gravity modelling of the western Marquette area, Michigan. M, 1985, Michigan State University. 111 p.

Weech, Philip S. A review of groundwater in the Bahamas. M, 1982, Colorado State University. 121 p.

Weed, Charles Edward. The relationship between acid mine drainage and the benthic macroinvertebrate community structure in the Tioga River, Tioga County, Pennsylvania. D, 1971, Pennsylvania State University, University Park. 77 p.

Weed, Daniel Del. Implications of an eolian sandstone unit of the basal Morrison Formation, central Wyoming. M, 1988, Iowa State University of Science and Technology. 149 p.

Weed, Rebecca. Chronology of chemical and biological weathering of cold desert sandstones in the Dry Valleys, Antarctica. M, 1985, University of Maine. 142 p.

Weeden, Dennis Alvin. Geology of the Harper area, Malheur County, Oregon. M, 1963, University of Oregon. 94 p.

Weeden, Harmer Allen. Multiscale photo interpretation and mapping for highway constructors. D, 1965, Cornell University. 332 p.

Weedman, Suzanne Dallas. Depositional environment and petrography of the upper Freeport Limestone in Indiana and Armstrong counties, Pennsylvania. D, 1988, Pennsylvania State University, University Park. 395 p.

Weedman, Suzanne Dallas. Turbulent boundary layer interpretation of sand streaks and parting lineation. M, 1983, Pennsylvania State University, University Park. 132 p.

Weedmark, Ron Harold. An analysis of some fundamental properties of the slant stack. M, 1985, University of Calgary. 106 p.

Weege, Randall J. The stages of alteration of ilmenite. M, 1955, University of Wisconsin-Madison.

Weekes, Alan F. Lithofacies and paleoenvironments of a Silurian core, Sandusky, Ohio. M, 1981, SUNY, College at Fredonia. 87 p.

Weekes, David C. Geology and petrology of the Rainy Creek igneous complex near Libby, Montana. M, 1981, Montana College of Mineral Science & Technology. 56 p.

Weeks, Albert W. The hydraulic theory of oil migration and accumulation. M, 1924, University of Wisconsin-Madison.

Weeks, Albert William. Late Cenozoic deposits of the Texas coastal plain between the Brazos River and the Rio Grande. D, 1941, University of Texas, Austin.

Weeks, Gary C. Precambrian geology of the Boulder River area, Beartooth Mountains, Montana. M, 1980, University of Montana. 58 p.

Weeks, Herbert J. The origin of some metamorphic rocks from Jackson County, Wisconsin. M, 1921, University of Wisconsin-Madison.

Weeks, John David. Some aspects of frictional sliding at high normal stress. D, 1980, Stanford University. 181 p.

Weeks, L. J. The geology of southwestern Cape Breton Island. M, 1952, Queen's University. 168 p.

Weeks, Leslie Vernon. Water conductivity and diffusivity of soils measured under transient flow conditions. D, 1966, University of California, Riverside. 143 p.

Weeks, Lewis Austin. The tectonic significance of the major joints in the First Watchung Flow, New Jersey. M, 1950, Columbia University, Teachers College.

Weeks, Robin James. Paleomagnetic records of the geomagnetic field; reversal records and reversal stratigraphy. D, 1988, University of California, Santa Barbara. 98 p.

Weeks, Ross M. The relative ages of the chalcopyrite and the rhyolite dyke zone orebodies at Quemont Mine, Quebec. M, 1963, Dalhousie University. 142 p.

Weeks, Victor L. Gravity and magnetic investigations in the south-central part of the Ishpeming greenstone belt, Marquette County, MI. M, 1987, Michigan Technological University. 61 p.

Weeks, Wilford Frank. A thermochemical study of equilibrium relations during metamorphosis of siliceous rock. D, 1956, University of Chicago. 41 p.

Weeks, Wilford Frank. The structure and petrology of the Ute Mountains, Colorado. M, 1953, University of Illinois, Chicago.

Weems, R. E. Doswellia kaltenbachi; an unusual newly discovered reptile from the Upper Triassic of Virginia. D, 1978, George Washington University. 160 p.

Weems, Robert E. Geology of the Hanover Academy and Ashland quadrangles, Virginia. M, 1974, Virginia Polytechnic Institute and State University.

Weeraratne, Saroj Premasiri. Modelling of stress strain behavior of sand under rotational stress fields. D, 1988, University of Massachusetts. 213 p.

Weerasinghe, Asoka. Stratigraphy and paleontology of the Ordovician Long Point Formation (middle Ordovician?), Long Point, Port au Port peninsula, Newfoundland. M, 1970, Memorial University of Newfoundland. 171 p.

Wegemann, Carroll H. Some notes on river development in the vicinity of Danville, Illinois. M, 1907, University of Wisconsin-Madison.

Wegenast, W. G. The footwall material of "B" orebody, Steeprock Lake, Ontario. M, 1954, Queen's University. 79 p.

Wegert, Emily Landris. Cation exchange in pollucite. M, 1981, University of Toledo. 64 p.

Wegg, D. S., Jr. Bingham mining district, Utah. M, 1915, University of Utah. 224 p.

Wegner, Robert C. Gravity anomalies in east-central Pennsylvania. M, 1972, Lehigh University.

Wegner, Robert Carl. Gravity anomalies, crustal structure and heat flow. D, 1978, Rice University. 85 p.

Wegner, W. W. The rate of metamorphic reactions; an example from the system MgO-SiO$_2$-H$_2$O. D, 1978, University of California, Los Angeles. 176 p.

Wegner, Warren W. The Denali Fault (Hines Creek Stand) in northeastern Mount McKinley National Park, Alaska. M, 1972, University of Wisconsin-Madison.

Wegrzyn, Richard S. Sedimentary petrology of the Jordan Formation. M, 1973, Northern Illinois University. 148 p.

Wegweiser, Arthur Ervin. Ecology and distribution of some foraminifera from shallow waters on the continental shelf of the Northeastern United States. D, 1966, Washington University. 106 p.

Wehmeyer, Lisa Kathryn. Denitrification and nitrate movement in the shallow alluvial aquifer of the West Des Moines River, Palo Alto County, Iowa. M, 1988, University of Iowa. 129 p.

Wehmiller, John F. Amino acid diagenesis in fossil calcareous organisms. D, 1971, Columbia University. 329 p.

Wehn, David C. The systematics of trace-element partitioning between coexisting micas in staurolite-zone schists, Black Hills, South Dakota. M, 1987, Kent State University, Kent. 87 p.

Wehr, Frederick Lewis, II. Geology of the Lynchburg Group in the Culpeper and Rockfish River areas, Virginia. D, 1983, Virginia Polytechnic Institute and State University. 267 p.

Wehrenberg, John Patterson. Diabase dikes of Mount Desert Island, Maine. M, 1952, University of Illinois, Urbana. 49 p.

Wehrenberg, John Patterson. Diffusion of zinc through carbonate systems. D, 1956, University of Illinois, Urbana. 74 p.

Wehrfritz, Barbara D. The Rhame Bed (Slope Formation, Paleocene), a silcrete and deep weathering profile, in southwestern North Dakota. M, 1978, University of North Dakota. 158 p.

Wehrle, Mary E. Holocene vegetational history of Ritterbush Pond, Lamoille County, Vermont. M, 1983, Queens College (CUNY). 36 p.

Wehrman, Ken C. A study of the transition zone between the loess hills and Sand Hills in central Nebraska. M, 1961, University of Nebraska, Lincoln.

Wei Wang. Bio-sedimentary facies on the East Flower Garden Bank, Northwest Gulf of Mexico. M, 1981, Texas A&M University.

Wei, Chenkou. Heat and mass transfer in porous media in convective heating and/or mirowave heating. D, 1984, University of Minnesota, Duluth. 213 p.

Wei, Chi-Sheng. Fluid inclusion of geothermometry of Tri-State sphalerite. M, 1975, University of Missouri, Rolla.

Wei, Kuo-Yen. Tempo and mode of evolution in Neogene planktonic foraminifera; taxonomic and morphological evidence. D, 1987, University of Rhode Island. 418 p.

Weibel, Carl Pius. Stratigraphy, depositional history, and brachiopod paleontology of Virgilian strata of east-central Illinois. D, 1988, University of Illinois, Urbana. 233 p.

Weibel, Carl Pius. Systematics, biostratigraphy and paleoenvironment of Morrowan and Atokan syringoporid corals of the Bird Spring Group, Arrow Canyon Range, Clark County, Nevada. M, 1982, University of Illinois, Urbana. 286 p.

Weibel, William Lee. Depositional history and geology of the Cloudburst Formation near Mammoth, Arizona. M, 1981, University of Arizona.

Weiblen, Paul Willard. A funnel-shaped, gabbro-troctolite intrusion in the Duluth Complex, Lake County, Minnesota. D, 1965, University of Minnesota, Minneapolis. 169 p.

Weichert, Dieter Horst. Digital analysis of mass spectra. D, 1965, University of British Columbia.

Weichman, B. E. Stratigraphy of the Phosphoria Formation in the southern portion of the Wind River basin, Wyoming. M, 1958, University of Wyoming. 72 p.

Weichman, David A. Depositional environment, provenance and tectonic significance of the Hanna Formation in Pass Creek basin, Carbon County, Wyoming. M, 1988, University of Wyoming. 113 p.

Weide, D. L. Postglacial geomorphology and environments of the Warner Valley - Hart Mountain area, Oregon. D, 1974, University of California, Los Angeles. 311 p.

Weidemann, Donna Elizabeth. Oxygen and hydrogen isotope variations in mid-Tertiary intrusions, Gunnison County, Colorado. M, 1986, Rice University. 154 p.

Weidenheim, Jan Peter. The petrography, structure, and stratigraphy of Powell Buttes, Crook County, central Oregon. M, 1981, Oregon State University. 95 p.

Weidenschilling, Stuart John. Aspects of planetary formation. D, 1976, Massachusetts Institute of Technology. 124 p.

Weider, Mark F. Comparison of numerical and analytical solutions of groundwater flow for Coors' coal mine, Keensburg, Colorado. M, 1982, [Colorado State University].

Weider, Melvin I. The geology of the Beacon Hill-Colossal Cave area, Pima County, Arizona. M, 1957, University of Arizona.

Weidie, Alfred E., Jr. The stratigraphy and structure of the Parras Basin, Coahuila and Nuevo Leon, Mexico. D, 1961, Louisiana State University.

Weidler, Mark E. The pre-Pleistocene geology of Jefferson County, Nebraska. M, 1954, University of Nebraska, Lincoln.

Weidman, Robert McMaster. Geology of the King City Quadrangle, California. D, 1959, University of California, Berkeley. 229 p.

Weidman, Robert McMaster. Structural features of the Mississippian-Pennsylvanian contact in parts of Lawrence, Martin, and Orange counties, Indiana. M, 1949, Indiana University, Bloomington. 112 p.

Weidman, Samuel. The geology of the pre-Cambrian igneous rocks of the Fox River valley, Wisconsin. D, 1898, University of Wisconsin-Madison.

Weidner, Donald James. Rayleigh waves from mid-ocean ridge earthquakes; source and path effects. D, 1972, Massachusetts Institute of Technology. 256 p.

Weidner, Jerry R. X-ray spectrochemical investigations of the Cincinnatian Series shale of Ohio, Indiana, and Kentucky. M, 1962, Miami University (Ohio). 96 p.

Weidner, Jerry Raymond. Phase equilibria in a portion of the system Fe-C-O from 250 to 10,000 bars and 400°C to 1200°C and its petrologic significance. D, 1968, Pennsylvania State University, University Park. 171 p.

Weidner, Melvin I. Geology of the Beacon Hill-Colossal Cave area, Pima County, Arizona. M, 1957, University of Arizona.

Weidner, William E. Paleoecological interpretation of echinocarid arthropod assemblages in the Late Devonian (Famennian) Chagrin Shale, northeastern Ohio. M, 1983, Kent State University, Kent. 66 p.

Weigand, Peter Woolson. Major and trace element geochemistry of the Mesozoic dolerite dikes from eastern North America. D, 1970, University of North Carolina, Chapel Hill. 162 p.

Weigand, Peter Woolson. Structural control of metasomatism in the Albemarle area, N.C. M, 1969, University of North Carolina, Chapel Hill. 62 p.

Weigel, Robert D. Fossil vertebrates of Vero, Florida. D, 1958, University of Florida. 87 p.

Weigle, James. Effects of subnormal temperature on some physical characteristics of minerals; a preliminary investigation. M, 1948, Syracuse University.

Weihaupt, John George, Jr. A morphometric study of the floodplains of White River and Bayou Bartholomew, Arkansas and Louisiana. D, 1973, University of Wisconsin-Milwaukee.

Weil, Charles B., Jr.   A model for the distribution, dynamics, and evolution of Holocene sediments and morphologic features of Delaware Bay. D, 1976, University of Delaware. 429 p.

Weiland, Erick F.   Fission track dating applied to uranium mineralization. M, 1979, Colorado School of Mines. 74 p.

Weiland, Thomas Joseph.   Petrology of volcanic rocks in Lower Cretaceous formations of northeastern Puerto Rico. D, 1988, University of North Carolina, Chapel Hill. 205 p.

Weiland, Thomas Joseph.   The petrology, chemistry and evolution of the Aguacate Formation, Tilaran Cordillera and Aguacate Mountains, Costa Rica, Central America. M, 1984, University of North Carolina, Chapel Hill. 131 p.

Weilbacher, Carol A.   Interpretation of the laminations of the Oligocene Florissant Lake deposits, Colorado. M, 1963, University of New Mexico. 115 p.

Weiler, Roland R.   A study of the surface structure of marine sediments by gas adsorption and related techniques. D, 1965, Dalhousie University. 131 p.

Weill, Daniel Francis.   A petrologic study of the cyclic sediments in the Pennsylvanian of eastern Kansas, western Missouri. M, 1958, University of Illinois, Urbana.

Weill, Daniel Francis.   Relative stability of crystalline phases in the Al2O3-SiO2 system. D, 1962, University of California, Berkeley. 105 p.

Weimer, Douglas James.   Petrology of the mudstones and shales of the Dunkard Group (Pennsylvanian-Permian), western Dunkard Basin in southeastern Ohio and central West Virginia. M, 1980, Miami University (Ohio). 95 p.

Weimer, Paul.   The bedrock geology of the Ridgway area, Ouray County, Colorado. M, 1980, University of Colorado. 91 p.

Weimer, Robert H.   A geologic investigation of the Talladega Group in Polk and Haralson counties, Georgia. M, 1976, Florida State University.

Weimer, Robert Jay.   A spectrochemical correlation study of Paleozoic rocks in West Texas and New Mexico. D, 1953, Stanford University. 107 p.

Weimer, Robert Jay.   Geology of the Whiskey Gap-Muddy Gap area, Wyoming. M, 1949, University of Wyoming. 88 p.

Weinberg, David M.   Structural geology of Beaver Creek ore, Big Belt mountains, Montana. M, 1970, University of New Mexico. 75 p.

Weinberg, David Michael.   Two-dimensional kinematic analyses of selected aspects of folding in the Rocky Mountain foreland, and their geologic implications. D, 1977, Texas A&M University. 121 p.

Weinberg, Edgar Leon.   Deep water sediments of western Lake Huron. M, 1948, University of Illinois, Urbana.

Weinbrandt, Richard Mickey.   The effect of temperature on relative permeability. D, 1972, Stanford University. 96 p.

Weiner, John Louis.   Environmental study of stages within the Summum Cyclothem of Illinois in the east central United States. M, 1961, University of Illinois, Urbana.

Weiner, John Louis.   The Old Fort Point Formation (late Precambrian), Jasper, Alberta (Canada). D, 1966, University of Alberta. 188 p.

Weiner, Mitchell S.   Slope aspect effects on Landsat signatures in forested terrain. M, 1979, Colorado State University. 113 p.

Weiner, Stephen.   Aspects of the biochemistry of the organic matrix of extant and fossil molluscs. D, 1977, California Institute of Technology. 114 p.

Weiner, Stephen Paul.   Deposition and stratification of oblique dunes, South Padre Island, Texas. M, 1981, University of Texas, Austin. 127 p.

Weiner, William.   A petrographic and geochemical study of chert formation around Thornton Reef Complex, Thornton, Illinois. M, 1974, University of Illinois, Urbana.

Weingarten, Baruch.   Geochemical and clay-mineral characteristics of lake sediments from the Venezuelan Andes; modern climatic relations and paleoclimatic interpretation. D, 1988, University of Massachusetts. 229 p.

Weingarten, Baruch.   Tectonic and paleoclimatic significance of a late-Cenozoic Paleosol from the Central Andes, Venezuela. M, 1977, University of Pennsylvania.

Weingeist, Leo.   The ostracode genus Eucytherura and its species from the Cretaceous and Tertiary of the Gulf Coast. M, 1948, Louisiana State University.

Weinheimer, Amy L.   Radiolarian responses to the 1957-1958 and 1964 El Ninos and 1963 anti-El Nino and a search for similar events in the fossil record. M, 1985, Rice University. 162 p.

Weinheimer, Gerald Joseph.   The geology and geothermal potential of the upper Madison Valley between Wolf Creek and the Missouri Flats, Madison County, Montana. M, 1979, Montana State University. 108 p.

Weinheimer, Roger L.   Vertical sequences and diagenesis in Breathitt sandstones of eastern Kentucky. M, 1982, University of Cincinnati. 115 p.

Weinkauf, Ronald Albert.   The Columbia Basin Project, Washington; concept and reality, lessons for public policy. D, 1974, Oregon State University. 227 p.

Weinle, Arthur R.   Physical geology; a laboratory manual. M, 1973, Ohio State University.

Weinman, Barry L.   Geophysical, geochemical, and remote sensing studies of Pennsylvania's thermal springs. M, 1976, Pennsylvania State University, University Park.

Weinman, Zvi H.   Analysis of littoral transport; Cape Henry, Virginia to the Virginia-North Carolina border. M, 1971, Old Dominion University. 64 p.

Weinmeister, Marcus Paul.   Origin of upper Bell Canyon reservoir sandstones (Guadalupian), El Mar and Paduca fields, Southeast New Mexico and West Texas. M, 1978, Texas A&M University. 72 p.

Weinreb, Gary.   Hydrogeologic investigation of thermal contamination within the Wabash Valley outwash aquifer, West Lafayette, Indiana. M, 1987, Purdue University.

Weinstein, Robert P.   The sedimentology of a portion of the continental shelf off Georgia. M, 1972, Florida State University.

Weinstock, Kenneth J.   Geomorphic hazards and their relationships to the development of the coastal zone of Southern California. M, 1981, University of California, Los Angeles.

Weintraub, David Leon.   The properties of electrical porcelain bodies formulated from Pacific Northwest raw material. M, 1949, University of Washington. 58 p.

Weintraub, Gary S.   A potassium-argon age study of the Basement Complex of Uganda. M, 1971, Florida State University.

Weintraub, Jill.   An assessment of the susceptibility of two alpine watersheds to surface water acidification; Sierra Nevada, California. M, 1986, Indiana University, Bloomington. 186 p.

Weintraub, Judy Montoya.   Mineral paragenesis and wall rock alteration at the Ophir Mines, Tooele County, Utah. M, 1957, University of Utah. 44 p.

Weintritt, Donald J.   An analysis of clay minerals in Recent sediments from East Bay, Galveston. M, 1957, University of Houston.

Weinzierl, John F.   Subsurface geology of the north part of the Blackwell Field, Oklahoma. M, 1922, University of Oklahoma. 12 p.

Weir, G. M.   Inlet formation and washover processes at North Pond, eastern Lake Ontario. M, 1977, SUNY at Buffalo. 75 p.

Weir, Harvey C.   A gravity profile across Newfoundland. M, 1970, Memorial University of Newfoundland. 164 p.

Weir, L. Alison.   Thermal regimes of small basins, an investigation of the effects of intrabasinal conductive and advective heat transport. M, 1986, Pennsylvania State University, University Park. 74 p.

Weir, Marjorie Fraser.   The world distribution of the Devonian. M, 1934, University of Alberta. 103 p.

Weir, Robert H., Jr.   Mineralogic, fluid inclusion, and stable isotope studies of several gold mines in the Mother Lode, Tuolumne and Mariposa counties, California. M, 1986, Pennsylvania State University, University Park. 98 p.

Weir, William G.   Electrical resistivity survey of Sandy Hook, San Salvador, Bahamas. M, 1988, University of Akron. 144 p.

Weirauch, Douglas A., Jr.   The measurement of the monatomic sodium and potassium activities of selected feldspars by atomic absorption spectrophotometry. M, 1974, University of Wisconsin-Milwaukee.

Weirauch, Douglas Allan, Jr.   The kinetics of the pseudobrookite decomposition reaction and its application to problems in lunar petrology. D, 1978, Pennsylvania State University, University Park. 101 p.

Weirich, Frank.   Sedimentation processes in a high altitude proglacial lake in southeastern British Columbia. D, 1982, University of Toronto.

Weis, Brent R.   Foraminiferal assemblages associated with South Florida coral reefs. M, 1977, Bowling Green State University. 115 p.

Weis, George Franklin.   Cyclic sedimentation in the Pottsville Series of the Lawrence County area, Pennsylvania. M, 1956, University of Pittsburgh.

Weis, Lawrence A.   A petrochemical investigation of the Winona Quadrangle basalts, northern Michigan. M, 1974, University of Toledo. 82 p.

Weis, Leonard Walter.   Origin of the Tigerton Anorthosite (Wisconsin). D, 1965, University of Wisconsin-Madison. 103 p.

Weis, Michelle A.   Sedimentology and depositional environments of the Permian lower Cutler Group; Turk's Head, Canyonlands National Park, Utah. M, 1988, University of Utah. 229 p.

Weis, Paul L.   Fluid inclusions in certain minerals of the zoned pegmatites of the Black Hills, South Dakota, and their significance. D, 1952, University of Wisconsin-Madison.

Weisberg, Maurice M.   Economic history of tungsten mining in Boulder County, Colorado. M, 1942, University of Colorado.

Weisberg, Michael.   The mineralogy and petrology of the Khohar L3 Chondrite. M, 1984, Brooklyn College (CUNY).

Weisberg, R. H.   The non tidal flow in the Providence River of Narragansett Bay; a stochastic approach to estuarine circulation. D, 1975, University of Rhode Island. 136 p.

Weisblatt, Edward A.   Multispectral discrimination of deltaic environments. D, 1972, Louisiana State University.

Weisbord, Norman Ed.   Venezuelan Devonian fossils. M, 1928, Cornell University.

Weise, Bonnie Renee.   Wave-dominated deltaic systems of the Upper Cretaceous San Miguel Formation, Maverick Basin, South Texas. M, 1979, University of Texas, Austin.

Weise, James R.   Petrology and stratigraphy of the Benwood Limestone and Arnoldsburg Limestone members of the Monongahela Group (Pennsylvanian) in Muskingum and Noble counties, Ohio. M, 1978, University of Akron. 85 p.

Weise, James Richard.   Stratigraphy, lithofacies, and biofacies of the U-Bar Formation (Aptian-Albian) of the Big Hatchet Mountains, Hidalgo County, New Mexico. D, 1982, University of Texas at El Paso. 403 p.

Weise, John Herbert.   Geology and mineral resources of the Neenach Quadrangle, California. D, 1947, University of California, Los Angeles.

Weise, John Herbert.   Geology of a portion of the Santa Paula Quadrangle (California). M, 1941, University of California, Los Angeles.

Weisel, Clifford Paul. The atmospheric flux of elements from the ocean. D, 1981, University of Rhode Island. 184 p.

Weisenberg, Charles William. Petrology and structure of the Ivanpah Mountains area, California. M, 1973, Rice University. 60 p.

Weisenberg, Charles William. Structural development of the Red Hill portion of the Feather River ultramafic complex, Plumas County, California. D, 1979, Rice University. 181 p.

Weisenburger, Kenneth William. Reflection seismic data acquisition and processing for enhanced interpretation of high resolution objectives. M, 1985, Virginia Polytechnic Institute and State University.

Weisenfluh, Gerald Alan. An epizonal trondhjemite-quartz keratophyre complex near Calhoun Falls, South Carolina. M, 1977, University of South Carolina. 45 p.

Weisenfluh, Gerald Alan. Controls on deposition of the Pratt seam, Black Warrior Basin, Alabama. D, 1982, University of South Carolina. 87 p.

Weisenstein, Debra Kay. A two-dimensional global dispersion model applied to several halocarbons. M, 1983, Massachusetts Institute of Technology. 51 p.

Weiser, Jeanne. A method of dyeing as applied to certain plagioclase feldspars. M, 1948, Syracuse University.

Weiser, Robert Neal. Palynology of the Lexington Mystic Coals. M, 1960, University of Missouri, Columbia.

Weisgarber, Sherry Lee. The use of seismic refraction and resistivity surveys to determine the failure surface of the Stumpy Basin landslide in northeastern Ohio. M, 1983, Kent State University, Kent. 101 p.

Weishampel, David Bruce. The evolution of ornithopod jaw mechnics. D, 1981, University of Pennsylvania. 248 p.

Weishampel, David Bruce. The functional significance of narial crest development in the Lambeosaurinae (Reptilia-Ornithischia) with reference to Parasaurolophus. M, 1978, University of Toronto.

Weishar, L. L. An examination of shoaling wave parameters. M, 1976, College of William and Mary.

Weishar, Lee. Statistical study of process-response system on a tideless coast. D, 1982, Purdue University. 130 p.

Weisman, Melanie Custer. Geology of the Pine and northern Buckhead Mesa quadrangles, Mogollon Rim region, central Arizona. M, 1984, Northern Arizona University. 126 p.

Weismeyer, Albert L., Jr. Geology of the northern portions of the Seventeen Palms and Fonts Point quadrangles, Imperial and San Diego counties, California. M, 1968, University of Southern California.

Weiss, Alan E. Groundwater geochemistry, hydrogeology and geochemical modeling of the Housatonic River basin, southern Berkshire County, Massachusetts. M, 1984, University of Massachusetts. 107 p.

Weiss, Alan J. Analysis of paleoecologic and stratigraphic control of rocklandian (Medial Ordovician) conodont assemblages of northwestern New York and southeastern Ontario. M, 1972, Boston University. 95 p.

Weiss, Benjamin. Comparative study of some Vermont talc deposits. M, 1961, University of Michigan.

Weiss, Christopher Paul. A comparison of the thermal profiles of two adjacent thrust plates in western Montana. M, 1987, University of Montana. 76 p.

Weiss, Dennis. A study of four cores from the Haverstraw Bay, Tappan Zee Bay area of the Hudson River, New York. M, 1967, New York University.

Weiss, Dennis. Late Pleistocene stratigraphy and paleoecology of the lower Hudson River estuary. D, 1971, New York University.

Weiss, Edwin M. Political control of strategic minerals in Latin America. M, 1951, University of Wisconsin-Madison.

Weiss, Eric A. Environmental sedimentology of the Middle Ordovician and paleoecology of a portion of

the Witten Formation at Solway, Tennessee. M, 1981, University of Tennessee, Knoxville. 229 p.

Weiss, Eriol Joseph. Equilibrium in the system CaO-MnO-SiO₂. D, 1949, Ohio State University.

Weiss, Garrett D. Dissolved methane in the Columbia River Basalt groundwaters beneath the Hanford Site, Washington State. M, 1985, Indiana University, Bloomington. 106 p.

Weiss, Gayle C. A depositional analysis of the arkose member (Middle Proterozoic) of the Troy Quartzite in central Arizona. M, 1986, Northern Arizona University. 191 p.

Weiss, Judith Vera. Contact metamorphism in Van Artsdalen's Quarry near Feasterville, Pennsylvania. M, 1945, Bryn Mawr College. 40 p.

Weiss, Judith Verra. The Wissahickon Schist at Philadelphia, Pennsylvania. D, 1948, Bryn Mawr College. 68 p.

Weiss, Lawrence. Origin of the (Pleistocene) Gardiners Clay in eastern Long Island, New York. M, 1951, New York University.

Weiss, Malcolm Pickett. The stratigraphy and stratigraphic paleontology of the upper Middle Ordovician rocks of Fillmore County, Minnesota. D, 1953, University of Minnesota, Minneapolis.

Weiss, Martin. Ostracod fauna of the (Middle Devonian) Norway Point Formation of Michigan. M, 1951, University of Michigan.

Weiss, Martin. Ostracods of the family Hollinidae from the Middle Devonian formations of Michigan and adjacent areas. D, 1954, University of Michigan. 162 p.

Weiss, Paula. Fossil minosoid leaflets from the middle Eocene (Claiborne Formation), Graves County, Kentucky. M, 1980, Indiana University, Bloomington. 76 p.

Weiss, Ray F. Dissolved gases and total inorganic carbon in seawater; distribution, solubilities, and shipboard gas chromatography. D, 1970, University of California, San Diego.

Weiss, Richard B. Geology and preliminary evaluation of ground stability in part of Woodside Highlands, Portola Valley, California. M, 1970, Stanford University.

Weiss, Richard L. Trace elements in four "porphyry copper" deposits in the Southwestern United States. M, 1963, Columbia University, Teachers College.

Weiss, Richard Lawrence. Outcrops as guides to copper ore; four examples. D, 1965, Columbia University. 194 p.

Weiss, Richard Marion. The geology of Jackson Township, Jackson County, Ohio. M, 1951, Ohio State University.

Weiss, Steven I. Geologic and paleomagnetic studies at the Stonewall Mountain and Black Mountain volcanic centers, southern Nevada. M, 1987, University of Nevada. 67 p.

Weiss, W. G., Jr. The environmental chemistry of selenium in the San Bernard River. D, 1977, Texas A&M University. 125 p.

Weissberg, Byron Goodspeed. Geochemical and petrographic aspects of arsenic deposits. D, 1964, University of California, Los Angeles. 176 p.

Weisse, Patricia A. Correlation between the paleomagnetism and magnetic mineralogy of the Nahant Gabbro, Salem gabbro-diorite and the gabbro at Salem Neck, eastern Massachusetts. M, 1983, University of Massachusetts. 137 p.

Weissenberger, John Arthur William. Sedimentology and conodont biostratigraphy of the Upper Devonian Fairholme Group, west-central Alberta. D, 1988, University of Calgary. 266 p.

Weissenberger, Ken William. Lakeview uranium area, Lake County, Oregon; constraints on genetic modelling from a district-scale perspective. D, 1984, Stanford University. 367 p.

Weissenborn, Helen Frances. A study of the Columbia River basalts at Spokane, Washington, with a

comparison of the "Rimrock" and "Valley" flows. M, 1955, Smith College. 64 p.

Weissenborn, Paul R. The Precambrian geology of the western portion of the Rochford gold-mining district, Black Hills, South Dakota. M, 1987, South Dakota School of Mines & Technology.

Weissenburger, Ken William. Application of ground water chemistry to stratabound uranium exploration, western Wyoming. M, 1978, Stanford University. 79 p.

Weissinger, John Leonard. A geological investigation in west-central Union County (Mississippi). M, 1961, University of Mississippi.

Weissman, Arthur Bruce. Environmental change from the perspectives of aesthetics and geomorphology. D, 1982, The Johns Hopkins University. 183 p.

Weissman, M. The fabric and magnesium content in pore filling calcite in vugs from the Tonoloway Formation at Hively Gap, Pendleton County, West Virginia. M, 1976, George Washington University.

Weissman, Simha. The use of photogrammetric methods to investigate surface movement of the Antarctic ice sheet. M, 1964, Ohio State University.

Weissmann, Gary Stephen. Alluvial architecture of a sheet sandstone, Willwood Formation, Bighorn Basin, Wyoming. M, 1988, University of Colorado. 120 p.

Weissmann, Robert Charles. Petrography of some Iowa limestones. M, 1953, Iowa State University of Science and Technology.

Weist, William G., Jr. Paleontology and stratigraphy of Cambro-Ordovician rocks in the Colorado Springs area, vicinity of El Paso and Fremont counties, Colorado. M, 1956, University of Colorado.

Weisz, Reuben Nathan. A methodology for planning land use and engineering alternatives for floodplain management. D, 1973, University of Arizona.

Weitz, John H., Jr. A mesoscopic fabric analysis of the Tellico-Sevier Shale belt at northeastern Douglas Lake, Tennessee. M, 1983, University of Tennessee, Knoxville. 239 p.

Weitz, John Hills. The Mercer fire-clay in Clinton and Centre counties, Pennsylvania. D, 1954, Pennsylvania State University, University Park. 107 p.

Weitz, Joseph L. Geology of the Bay of Islands area, western Newfoundland. D, 1954, Yale University.

Weitz, Joseph Leonard. Geology of the Bay of Islands area, western Newfoundland. D, 1965, Yale University. 208 p.

Weitz, Thomas James. Geology and ore deposition at the I-10 Prospect, Cochise County, Arizona. M, 1976, University of Arizona.

Weitzman, M. J. A probabilistic model for predicting groundwater levels in the Greenbrook well field, Kitchener, Ontario. M, 1980, University of Waterloo.

Weixelman, Wesley D. Geology and the northeast portion of Des Arc Quadrangle, Iron and Madison counties, Missouri. M, 1959, University of Missouri, Rolla.

Welber, P. W. A petrologic study of compositionally zoned plagioclase from the Rocky Hill Stock, Tulare County, California. M, 1977, University of Arizona. 62 p.

Welbourn, Martha Lynne. Ecologically based forest policy analysis; fire management and land disposals in the Tanana River basin, Alaska. D, 1983, Cornell University. 248 p.

Welby, Charles W. Occurrence of the alkali metals in some modern sediments. D, 1952, Massachusetts Institute of Technology. 309 p.

Welby, Charles W. The geology of the central part of the La Panza Quadrangle, San Luis Obispo County, California. M, 1949, University of California, Berkeley. 107 p.

Welc-LePain, Joan L. The source rupture process of the Great Banda Sea earthquake of November 4, 1963. M, 1986, University of Michigan. 21 p.

Welch, Alan Herbert. Geothermal resources of the western Black Rock Desert, northwestern Nevada; hydrology and aqueous geochemistry. D, 1985, University of Nevada. 150 p.

Welch, Alan Herbert. Lead isotopic compositions of some eugeosynclinal sediments and their relation to the origin of Cascade andesites. M, 1974, University of California, Santa Barbara. 23 p.

Welch, Carl Martin. A preliminary report on the geology of the Mineral Hill area, Crook County, Wyoming. M, 1974, South Dakota School of Mines & Technology.

Welch, David Michael. Slope analysis and evolution on protected lacustrine bluffs. D, 1972, University of Western Ontario.

Welch, Fred., Jr. The geology of the Castle Rock area, Douglas County, Colorado. M, 1969, Colorado School of Mines.

Welch, James Robert. Petrology and development of algal banks in the Millersville Limestone Member (Bond Formation, Upper Pennsylvanian) of the Illinois Basin. M, 1975, Indiana University, Bloomington. 21 p.

Welch, Jane Marie. CoO-Al$_2$O$_3$-TiO$_2$ as a model for equilibria involving phases of pseudobrookite, ilmenite and spinel structures. M, 1974, Pennsylvania State University, University Park. 35 p.

Welch, Jane Marie. Phase relations in the iron oxide-chromium oxide-titanium oxide-aluminum oxide system at low oxygen pressures and their bearing on lunar petrogenesis. D, 1978, Pennsylvania State University, University Park. 135 p.

Welch, Jerome V. The influence of soil properties, water quality, and degree of water saturation on the movement of nickel in sludge-amended soil. D, 1979, University of California, Riverside. 215 p.

Welch, John Charles. The geology of the Argenta Stock, Beaverhead County, Montana. M, 1983, Purdue University. 74 p.

Welch, Paul M. A field experiment to improve perforation efficiency. M, 1987, Colorado School of Mines. 86 p.

Welch, Robert Gerald. Arenaceous foraminifera of the Linn and Zarah subgroups of Pennsylvanian age (Missourian) of eastern Kansas. M, 1964, University of Kansas. 48 p.

Welch, Robert Newman. A study of the description of two cores from the "corniferous" of western Powell County, Kentucky. M, 1937, University of Kentucky. 76 p.

Welch, Vorrin J. A structural and sedimentation study of the Shawnee Group in central Kansas and the possible relationship to oil producing areas. M, 1951, Kansas State University. 38 p.

Welday, Edward E. Geology of the San Guillermo Mountain area, California. M, 1960, Pomona College.

Welden, Charles Morris. A comparative petrographic study of the Kalkberg and Tioga bentonites (Devonian, New York, Pennsylvania and neighboring states). M, 1966, Brown University.

Welder, Frank A. Deltaic processes in Cubits Gap area, Plaquemines Parish, Louisiana. D, 1955, Louisiana State University. 255 p.

Welder, George Edward. Geology of the Basalt area, Eagle and Pitkin counties, Colorado. M, 1954, University of Colorado.

Weldon, Charles S. Jurassic sediments of the Las Vegas, New Mexico area. M, 1951, Texas Tech University. 64 p.

Weldon, John B., Jr. The geology of the Pasadena-Eagle Rock area, Los Angeles County, California. M, 1955, Pomona College.

Weldon, Ray James, II. The late Cenozoic geology of Cajon Pass; implications for tectonics and sedimentation along the San Andreas Fault. D, 1986, California Institute of Technology. 481 p.

Welford, Mark R. Quaternary terrace formation in the Little Salmon River basin. M, 1988, University of Idaho. 116 p.

Welhan, J. A. Feasibility of the oxygen-18 mass balance method in the calculation of the water balance of a small lake. M, 1974, University of Waterloo.

Welhan, John Andrew. Carbon and hydrogen gases in hydrothermal systems; the search for a mantle source. D, 1981, University of California, San Diego. 216 p.

Weliky, Karen. Clay-organic associations in marine sediments; carbon, nitrogen, and amino acids in the fine grained fractions. M, 1983, Oregon State University. 166 p.

Welker, D. B. A paleoclimatic study of three Southern Ocean deep-sea cores. D, 1978, Case Western Reserve University. 325 p.

Welker, Kenneth Kramer. Rock failure in deep mines; field and experimental studies. D, 1933, Harvard University.

Welker, Mary. Structure and deformation mechanisms along the Tonale Line in Italy. M, 1985, Texas A&M University. 127 p.

Welkie, Carol Jean Jigliotti. Geophysical-geological exploration for offshore sand and gravel, western Lake Michigan. D, 1980, University of Wisconsin-Madison. 188 p.

Well, Ralph Gordon. Some relationships in the system ZrO$_2$-FeO-SiO$_2$. D, 1961, University of Michigan.

Wellborn, Jewel E. F. Stratigraphy of the Mesaverde Formation, Mt. Gunnison coal property, Gunnison County, Colorado. M, 1982, Colorado School of Mines. 91 p.

Welleck, Mary N. Dikes of the Speedway region (New York). M, 1919, Columbia University, Teachers College.

Wellendorf, William George. Aeolian microtextures on quartz sand. M, 1979, Arizona State University. 89 p.

Weller, James Marvin. The geology of Edmonson County, Kentucky. D, 1927, University of Chicago. 125 p.

Weller, Roger Nelson. Geologic interpretation of the Northwest Shore of Oceanus Procellarum (Moon) based on photoenhancement of Orbiter 4 photographs. M, 1972, University of Arizona.

Weller, Stuart. Studies of the Paleozoic faunas of the interior continental basin of North America. D, 1901, Yale University.

Welles, Samuel P. Revision of elasmosaurid plesiosaurs. D, 1940, University of California, Berkeley. 78 p.

Wellington, Gerard Michael. The role of competition, niche diversification and predation on the structure and organization of a fringing coral reef in the Gulf of Panama. D, 1981, [University of California, Santa Barbara]. 196 p.

Wellman, Dean Castor. The insoluble residues of the Dundee and upper Monroe formations of central Michigan. M, 1936, Washington University. 72 p.

Wellman, Robert Persey. Nitrogen isotope fractionation in chemical and microbiological systems. D, 1969, University of Alberta. 184 p.

Wellman, Samuel S. Stratigraphy, petrology, and sedimentology of the nonmarine Honda Formation (Miocene), upper Magdalena Valley, Colombia. D, 1968, Princeton University. 262 p.

Wellman, Samuel Sidney. Stratigraphy of the lower Miocene Gering Formation, Pine Ridge area, northwestern Nebraska. M, 1964, University of Nebraska, Lincoln.

Wellman, Thomas Robert. The stability of sodalite in the system NaAlSi$_3$O$_8$-KAlSi$_3$O$_8$-NaAlSiO$_4$-NaCl-KCl-H$_2$O. D, 1969, Yale University. 165 p.

Wells, A. J. Middle Pennsylvanian fusulinids of the Naco Formation (Dripping Springs Mtns.) near Winkelman, Gila County, Arizona. M, 1965, University of Arizona.

Wells, Anke Marie Neumann. Evolution of a humid tropical landscape in Northcentral Costa Rica as deduced from geomorphic and pedogenic evidence. D, 1979, University of Kansas. 285 p.

Wells, Dana. Lower Middle Mississippian of southeastern West Virginia. D, 1949, Columbia University, Teachers College.

Wells, Dana. Notes on the paleontology of the Garrison Formation of Kansas. M, 1929, University of Kansas. 91 p.

Wells, David Rolfe. A study of textural parameters of the Central and South Texas barrier island beaches. M, 1976, University of Texas, Arlington. 179 p.

Wells, Donald Loren. Geology of the eastern San Felipe Hills, Imperial Valley, California; implications for wrench faulting in the southern San Jacinto fault zone. M, 1987, San Diego State University. 136 p.

Wells, Douglas E. Geology and petrochemistry of the Rock Branch Quadrangle, Elbert County, Georgia. M, 1983, University of Georgia.

Wells, Eddie N. Absorption bands in lunar glasses. D, 1977, University of Pittsburgh.

Wells, Francis Gerritt. The hydrothermal alteration of olivine. D, 1928, University of Minnesota, Minneapolis. 46 p.

Wells, Frederick J. On the separation of the annual and Chandler spectral components in astronomic latitudes and polar motion data. D, 1972, Brown University.

Wells, Frederick Joseph. Acoustic-gravity waves around an explosion source in the atmosphere. M, 1968, Brown University.

Wells, Gary Steven. A morphological chemical and petrological examination of Archean pillow basalts from the Abitibi greenstone belt. D, 1980, Queen's University. 281 p.

Wells, Ian. A spatial analysis methodology for predicting archaeological sites in Delaware and its potential application in remote sensing. M, 1981, University of Delaware, College of Marine Studies. 102 p.

Wells, J. M. The extraction and mass spectrometric analysis of trace quantities of lead. M, 1966, University of Toronto.

Wells, J. T. Shallow-water waves and fluid-mud dynamics, coast of Surinam, South America. D, 1977, Louisiana State University. 113 p.

Wells, James Aertsen. Pleistocene geology of Union County, Pa. M, 1960, Pennsylvania State University, University Park. 70 p.

Wells, James Alan. Environmental education topics in current geoscience texts. M, 1972, Indiana University of Pennsylvania. 40 p.

Wells, Jane Freeman. A study of thought relating to strip mining in the Commonwealth of Kentucky. D, 1973, Indiana University, Bloomington. 156 p.

Wells, John Cawse. Petrology and structure of the Crystal Lake area, Los Angeles County, California. M, 1938, California Institute of Technology. 31 p.

Wells, John D. A study of the Eskridge Shale. M, 1950, Kansas State University. 35 p.

Wells, John H. Placer examination principles and practice. M, 1970, University of Nevada. 155 p.

Wells, John Maurice. Spherical harmonic analysis of paleomagnetic data. D, 1969, University of California, Berkeley. 315 p.

Wells, John Thomas. Particle size distribution and small scale bed-forms on sand waves, Chesapeake Bay entrance, Virginia. M, 1973, Old Dominion University. 111 p.

Wells, John West. Corals of the Cretaceous of the Atlantic and Gulf coastal plains and Western Interior of the United States. D, 1933, Cornell University.

Wells, John West. Corals of the Trinity division of the Comanchean (Lower Cretaceous) of central Texas. M, 1930, Cornell University.

Wells, Karen. Detailed correlation of the Woods Run marine unit (Upper Pennsylvanian) in southwestern Pennsylvania. M, 1985, University of Pittsburgh.

Wells, Kathy M. The modern Burlington baymouth bar of Lake Ontario and its ancient Lake Iroquois analogues. M, 1976, University of Akron. 74 p.

Wells, Liss Eleanor. Holocene fluvial and shoreline history as a function of human and geologic factors in arid northern Peru. D, 1988, Stanford University. 396 p.

Wells, Neil Andrew. Marine and continental sedimentation in the early Cenozoic Kohat Basin and adjacent northwestern Indo-Pakistan. D, 1984, University of Michigan. 479 p.

Wells, P. D. A seismic reflection and refraction study near the Blue Ridge Thrust in Virginia. M, 1975, Virginia Polytechnic Institute and State University.

Wells, R. A. Structural analysis of Precambrian rocks in northeastern Badwater Quadrangle, Wyoming. M, 1975, University of Wyoming. 54 p.

Wells, Randall W. Investigation of carbonate intervals within the Ogallala Formation from analysis of subsurface core material. M, 1974, Texas Tech University. 77 p.

Wells, Ray Edward. Geology of the Drake Peak rhyolite complex and the surrounding area, Lake County, Oregon. M, 1975, University of Oregon. 130 p.

Wells, Ray Edward. Paleomagnetism and geology of Eocene volcanic rocks in Southwest Washington; constraints on mechanisms of rotation and their regional tectonic significance. D, 1982, University of California, Santa Cruz. 261 p.

Wells, Richard Bruce. Orthoquartzites of the Oquirrh Formation. M, 1963, Brigham Young University. 82 p.

Wells, Richard F. Fresh water invertebrate living traces of the Mississippi alluvial valley, near Baton Rouge, Louisiana. M, 1977, Louisiana State University.

Wells, Stephen Gene. A study of surficial processes and geomorphic history of a basin in the Sonoran Desert, southwestern Arizona. D, 1976, University of Cincinnati. 328 p.

Wells, Stephen Gene. Geomorphology of the sinkhole plain in the Pennyroyal Plateau of the central Kentucky karst. M, 1973, University of Cincinnati. 122 p.

Wells, Terry L. The hydrogeology of the Sharon Conglomerate (Pennsylvanian) in Geauga County, Ohio. M, 1970, Kent State University, Kent. 45 p.

Wellstead, Carl F. Fossil lizards of the Valentine Formation. M, 1977, University of Nebraska, Lincoln.

Wellstead, Carl Frederick. Taxonomic revision of the Permo-Carboniferous lepospondyl amphibian families Lysorophidae and Molgophidae. D, 1985, McGill University.

Welsch, Dennis G. Environmental geology of the McDowell Mountains area, Maricopa County, Arizona; Part II. M, 1977, Arizona State University. 68 p.

Welsh, Fred, Jr. Geology of the Castle Rock area, Douglas County, Colorado. M, 1969, Colorado School of Mines. 93 p.

Welsh, Gerald Merritt. Space and water for industry in Mill Creek valley, Cincinnatti. M, 1957, University of Cincinnati. 64 p.

Welsh, J. L. Petrology of the northwest margin of the Sacred Heart Pluton and adjacent Archaean gneisses from the Minnesota River valley; Redwood and Yellow Medicine counties, Minnesota. M, 1976, University of Minnesota, Duluth.

Welsh, James Lowell. Structure, petrology, and metamorphism of the Marble Mountains area, Siskiyou County, California. D, 1982, University of Wisconsin-Madison. 302 p.

Welsh, James P. The origin of the Day Creek Dolomite (Permian) of Kansas. M, 1966, Wichita State University. 67 p.

Welsh, James Patrick Jr. Patterns of compositional variation in some glaciofluvial sediments in the Lower Peninsula of Michigan. D, 1971, Michigan State University. 98 p.

Welsh, John Elliot. Geology of the Sheep Mountain-Delaney Butte area, North Park, Colorado. M, 1951, University of Wyoming. 56 p.

Welsh, John Elliott. Biostratigraphy of the Pennsylvanian and Permian systems in southern Nevada. D, 1959, University of Utah. 415 p.

Welsh, Joyce R. Lithofacies and diagenetic overprints within the lower San Andres Formation, Quay and Roosevelt counties, New Mexico. M, 1982, Texas Tech University. 99 p.

Welsh, Robert. Oriskany Sandstone; depositional environment and fracture porosity in Somerset County, Pennsylvania. M, 1984, University of Pittsburgh.

Welsh, Thomas M. Variations in telluric current power spectrums with local time and magnetic activity. M, 1967, Colorado School of Mines. 113 p.

Weltin, Timothy P. Effect of sediment sorting and variation of mean grain size of sand on shell abrasion. M, 1970, Wayne State University.

Welton, B. J. Late Cretaceous and Cenozoic Squalomorphii of the Northeast Pacific Ocean. D, 1979, University of California, Berkeley. 569 p.

Welty, Curtis Michael. Petrology and diagenesis of the Ojo Caliente Sandstone (Miocene?), Espanola Basin, New Mexico. M, 1988, University of Colorado. 118 p.

Weltz, Mark Allen. Observed and estimated (ERHYM-II model) water budgets for South Texas rangelands. D, 1987, Texas A&M University. 190 p.

Wemple, Edna Mary. New cestraciont teeth from the West-American Triassic. M, 1906, University of California, Berkeley. 26 p.

Wen, Cheng-Lee. A study of bolson fill thickness in the southern Rio Grande Rift, southern New Mexico, West Texas and northern Chihuahua. M, 1983, University of Texas at El Paso. 74 p.

Wen, Huei-Yuin. Love waves as an explosion earthquake discriminant. M, 1977, St. Louis University.

Wen, Jianping. Isotope geochemistry of the Oka carbonatite complex, Quebec. M, 1985, Carleton University.

Wen, Jing. Two approaches to analysis of seismic data from structurally complicated regions. M, 1984, University of Texas at Dallas. 118 p.

Wen, W. Electrokinetic behavior and flotation of oxidized coal. D, 1977, Pennsylvania State University, University Park. 343 p.

Wen-Jen Wu. Algebraic Kirchhoff-Trorey inversion. D, 1979, [University of Houston].

Wenban-Smith, Alan Kenneth. The petrology of part of the Glamorgan Gabbro, Ontario (Canada). M, 1967, University of Toronto.

Wenberg, Edwin Hugo. Insoluble residue studies of the Missouri and Virgil series of Iowa. D, 1941, University of Iowa. 165 p.

Wenberg, Edwin Hugo. The Paleozoic stratigraphy of Lorain County, Ohio. M, 1938, Oberlin College. 91 p.

Wendel, Clifford Arthur. The origin of various types of disseminated deposits of oxidized copper ore in northern Chile. M, 1965, Montana College of Mineral Science & Technology. 35 p.

Wendelin, R. Franz. A subsurface stratigraphic study from the top Wilcox to the top Vicksburg in central Louisiana. D, 1963, University of Pittsburgh.

Wendell, Clarence Adami. A microscopic study of the Butte vein minerals. M, 1935, Montana College of Mineral Science & Technology. 35 p.

Wendell, Daniel E. Geology, alteration, and geochemistry of the McGinness Hills area, Lander County, Nevada. M, 1985, University of Nevada. 123 p.

Wendell, William G. The structure and stratigraphy of the Virgin valley, McGee mountain area, Humboldt County, Nevada. M, 1970, Oregon State University. 130 p.

Wenden, Henry Edward. The influence of ionic diffusion on some properties of quartz. D, 1958, Harvard University.

Wender, Lawrence Edwin. Chemical and mineralogical evolution of the Cenozoic volcanics of the Marysvale, Utah area. M, 1976, University of Utah. 57 p.

Wendland, Wayne M. Dating (^{14}C) the temporal limits of climatic episodes during the Holocene. D, 1972, University of Wisconsin-Madison.

Wendlandt, Richard Frederick. Investigations bearing on the origins of potassic magmas; Part I, Melting relations in the system kalsilite-forsterite-silica-carbon dioxide to 30 kilobars, and Part II, Stability of phlogopite in natural spinel lherzolite and in the system K_2O-MgO-Al_2O_3-SiO_2-H_2O-CO_2 as a function of volatile composition at high pressures and high temperatures. D, 1978, Pennsylvania State University, University Park. 182 p.

Wendler, Arno Paul. A petrographical study of sands from middle Tertiary formations east of the Guadalupe River, Texas. D, 1934, University of Texas, Austin.

Wendler, Arno Paul. Heavy minerals of the Catahoula of Fayette County, Texas. M, 1932, University of Texas, Austin.

Wendt, C. J. Geology, alteration, and mineralization of the Batamonte Ranch area, northern Sonora, Mexico. M, 1977, University of Arizona. 110 p.

Wendt, Roy L.; Hulstrand, Richard F. and Tinker, Clarence N. Geology of the Turkey Creek-Williams Creek area, Huerfano County, Colorado. M, 1955, University of Michigan.

Wendte, John C. Facies analysis and stratigraphy of the Raton Formation, late Cretaceous-Paleocene, south central Colorado and north central New Mexico. M, 1970, University of Wisconsin-Madison.

Wendte, John Curtis. Sedimentation and diagenesis of the Cooking Lake platform and lower Leduc reef facies, upper Devonian, Redwater, Alberta. D, 1974, University of California, Santa Cruz. 237 p.

Wenger, Lloyd Miller, Jr. Solubility of crude oil and heavy petroleum distillation fractions in methane (with water present) at elevated temperatures and pressures as applied to the primary migration of petroleum. M, 1982, Idaho State University. 118 p.

Wenger, Lloyd Miller, Jr. Variations in organic geochemistry of anoxic-oxic black shale-carbonate sequences in the Pennsylvanian of the Midcontinent, U.S.A. D, 1987, Rice University. 620 p.

Wengerd, Sherman Alexander. Lithographic variation of the (Ordovician) Viola Limestone in south-central Oklahoma. D, 1947, Harvard University.

Wenk, Warren Joel. A seismic refraction model of the Hartford Basin in southern New England. M, 1983, University of Connecticut. 44 p.

Wenkam, Chiye. Late Quaternary changes in the oceanography of the eastern tropical Pacific. M, 1977, Oregon State University. 143 p.

Wennagel, Dale Anderson. Hydrogeologic contributions to ecosystem analysis; a different use for ground-water models. M, 1974, Oklahoma State University. 106 p.

Wennekens, Marcel Pat. Marine environment and macro-benthos of the waters of Puget Sound, San Juan Archipelago, southern Georgia Strait, and Strait of Juan de Fuca. D, 1959, University of Washington. 298 p.

Wenner, David Bruce. Hydrogen and oxygen isotopic studies of serpentinization of ultramafic rocks. D, 1971, California Institute of Technology. 400 p.

Wentworth, Carl Merrick, Jr. The Upper Cretaceous and lower Tertiary rocks of the Gualala area, northern Coast Ranges, California. D, 1967, Stanford University. 229 p.

Wentworth, Chester Keeler. Petrology and origin of the post-Miocene terrace gravels of the Middle Atlantic slope; a study of clastic sediments. D, 1923, University of Iowa. 119 p.

Wentworth, Chester Keeler. Quantitative studies of the shapes of pebbles. M, 1921, University of Iowa. 67 p.

Wentworth, David W. The Lower Ordovician in the eastern Mohawk Valley (New York). M, 1959, West Virginia University.

Wentworth, Sally Ann. An investigation of fine-grained micas with emphasis on their hydrous character. D, 1967, Pennsylvania State University, University Park. 98 p.

Wentworth, Susan Jane. Petrology of light-colored loose fragments from the Apollo 17 deep drill core, Nasa Johnson Space Center, Houston. M, 1980, University of New Mexico. 192 p.

Wentworth, T. R. The vegetation of limestone and granite soils in the mountains of southeastern Arizona. D, 1976, Cornell University. 1118 p.

Wenz, Kenneth P., Jr. Chemical stratigraphy of lavas exposed in the walls of two pit craters on Kilauea Volcano, Hawaii. M, 1988, University of Massachusetts. 106 p.

Wenzel, Robert John. An investigation of the mechanics of vertical sorting in permafrost soils. M, 1962, Northern Illinois University. 34 p.

Wenzel, Roger A. Effect of floods on selected sand bars; San Bernard and Brazos rivers, Austin County, Texas. M, 1975, University of Texas, Austin.

Wepfer, A. J. The asbestiform-fiber contamination of Lake Superior and the resulting potential health hazard; an interpretation of interlocking physical and human geographical systems. D, 1977, [Oklahoma State University]. 218 p.

Werbach, David. A study of the LaSalle Falls massive sulphide prospect and adjacent area, Florence County, Wisconsin. M, 1987, Northern Illinois University. 121 p.

Werblow, Jack. The limits of comprehensive environmental planning for policy development and decision making. D, 1974, University of Cincinnati. 197 p.

Werkheiser, William H. The hydrogeology of the Sayre-Waverly area, New York-Pennsylvania. M, 1987, University of Massachusetts. 147 p.

Werle, James L. Allard Stock, La Plata Mountains, Colorado; an enigmatic porphyry copper-precious metals deposit. M, 1983, Eastern Washington University. 60 p.

Werle, Kevin J. Global oceanic stratigraphy of the lower Miocene. M, 1982, Ohio University, Athens. 149 p.

Werme, Douglas R. Precambrian geology of the Defiance Uplift, Arizona. M, 1981, Northern Arizona University. 88 p.

Wermeyer, Raymond Alfred. Pleistocene deposits of the St. Louis area. M, 1955, Washington University. 96 p.

Wermiel, Dan E. Diagenetic history and dolomitization of Tamarora Sequence (Mississippian) carbonates of the Pedregosa Basin, southeastern Arizona, southwestern New Mexico, and northern Sonora and Chihuahua, Mexico. M, 1978, Arizona State University. 156 p.

Wermund, Edmund G., Jr. Glauconite in early Tertiary sediments of the Gulf coastal province. D, 1961, Louisiana State University.

Werner, Alan. Glacial geology of the McKinley River area, Alaska; with an evaluation of various relative age dating techniques. M, 1982, Southern Illinois University, Carbondale. 147 p.

Werner, Alan. Holocene glaciation and climatic change, Spitsbergen, Svalbard. D, 1988, University of Colorado. 325 p.

Werner, Eberhard Wolfgang. Origin of solution features; Cloverlik Valley, Pocahontas County, West Virginia (Paleocene Geology). M, 1972, Rutgers, The State University, New Brunswick. 58 p.

Werner, Harry Jay. A preliminary investigation of the Lovingston Gneiss and its related phases in the Lovingston Quadrangle of central Virginia. M, 1949, Washington University. 98 p.

Werner, Harry Jay. The geology of Humber Valley, Newfoundland. D, 1956, Syracuse University. 145 p.

Werner, James Edward. A porosity study of the Conococheague Dolomite near Leesport, Pennsylvania. M, 1951, University of Pittsburgh.

Werner, Marian Adair. Comparative study of shallow water sediments in the Barrow, Alaska area. M, 1959, Smith College. 67 p.

Werner, Matthew Lambert III. Petrofabric analysis of the Glarner Freiberg, Kanton Glarus, Switzerland. D, 1973, Pennsylvania State University, University Park. 58 p.

Werner, Michael Askam. Equilibria in cement paste-carbonate aggregate reactions. M, 1962, Iowa State University of Science and Technology.

Werner, Michael Robert. Superposed Mesozoic deformations, southeastern Inyo Mountains, California. M, 1979, California State University, Northridge. 69 p.

Werner, Sanford L. The chemical characteristics of the ground water in the San Gabriel Valley, California. M, 1965, University of Southern California.

Werner, Sarah R. The role of tellurium in hydrothermal gold transport. M, 1984, Pennsylvania State University, University Park.

Werner, Walter C. The Devonian coral faunas at the falls of the Ohio River (Kentucky, Indiana). D, 1930, George Washington University.

Werner, William G. Sedimentology of the Cutler Formation (Pennsylvanian-Permian) near Gateway, Colorado, and Fisher Towers, Utah. M, 1972, University of Missouri, Columbia.

Wernicke, Brian Philip. Processes of extensional tectonics. D, 1982, Massachusetts Institute of Technology. 170 p.

Wernicke, Rolf Stephan. Geology, petrography and geochemistry of the El Topo Pluton, Baja California Norte, Mexico. M, 1987, San Diego State University. 232 p.

Wernlund, Russell J. Biostratigraphy and paleoecology of Holothurian sclerites from the Pinery Member, Bell Canyon Formation (Permian) of the Delaware Basin of West Texas. M, 1977, Texas Tech University. 122 p.

Werrell, William L. Stream gaging by continuous injection of tracer elements. M, 1967, University of Arizona.

Werrell, William Lewis. Pennsylvanian ostracods and fusulinids of Tijeras and Cedro canyons, Bernalillo County, New Mexico. M, 1961, University of New Mexico. 102 p.

Wershaw, Robert Lawrence. Oxygen isotope fractionation in the system quartz-water. D, 1963, University of Texas, Austin. 104 p.

Wert, Richard T. A frictional model of a two-part unbounded ocean basin. M, 1968, Texas A&M University.

Werth, Lee Forrest. A comparison of remote sensing technique applications to wetland vegetation classification in Minnesota's 7-county metropolitan area. D, 1980, University of Minnesota, Minneapolis. 271 p.

Wertman, Ronald La Mar. Subsurface study of the Tar Springs Formation, Newburg and Calhoun quadrangles, Kentucky. M, 1958, University of Illinois, Urbana.

Wertz, Jacques B. Logarithmic pattern in river placer deposits. M, 1948, Columbia University, Teachers College.

Wertz, Jacques Bernard. Mechanism of erosion and deposition along channelways. D, 1962, University of Arizona. 622 p.

Wertz, James Claude. Geology of the Cottontown Quadrangle, Sumner County, Tennessee. M, 1956, Vanderbilt University.

Wertz, William Earl. The depositional environments and petrography of the Stirling Quartzite, Death Valley region, California and Nevada. D, 1983, Pennsylvania State University, University Park. 210 p.

Wescott, Eugene M. The effect of topography and geology on telluric currents. M, 1960, University of Alaska, Fairbanks. 60 p.

Wescott, William A. A model for shallow marine epeiric sea deposition; The Tar Springs Sandstone; (Upper Mississippian) in southern Illinois. M, 1976, Southern Illinois University, Carbondale. 141 p.

Wescott, William A. The Yallahs fan delta, southeastern Jamaica; a depositional model for active tectonic coastlines. D, 1979, Colorado State University. 189 p.

Wesendunk, Paul Robert. Lower Tertiary foraminifera from the southern Santa Cruz Mountains, California. M, 1956, University of California, Berkeley. 132 p.

Wesh, Richard Adams. Invertebrate faunas of the shales of the Upper Pennsylvanian Fresnal Group, Sacramento Mountains, New Mexico. M, 1955, University of Wisconsin-Madison.

Wesley, George Rutherford. Geology of the Kentucky River fault zone in Casey and a part of Lincoln counties, Kentucky. M, 1934, University of Kentucky. 54 p.

Wesley, Richard Hal. Geochemical study of a drill core. M, 1949, Washington University. 34 p.

Wesling, John R. Glacial geology and soil development of Winsor Creek drainage basin, southernmost Sangre de Cristo Mountains, New Mexico. M, 1988, University of New Mexico. 186 p.

Weslow, Vanessa Maria. Magnetotelluric profiling of the Peninsular Ranges in northern Baja California, Mexico. M, 1985, San Diego State University. 174 p.

Wesner, George Mack. The importance of water quality in water resources management. D, 1972, University of Colorado.

Wesnousky, Steven Glenn. Crustal deformation and earthquake risk in Japan. D, 1982, Columbia University, Teachers College. 243 p.

Wesolowski, David. Geochemistry of tungsten in scheelite deposits; the skarn ores at King Island, Tasmania. D, 1984, Pennsylvania State University, University Park. 465 p.

Wessel, Gregory R. Geology of the Big Pass Tungsten Mine, Beaver County, Utah. M, 1977, University of Missouri, Rolla.

Wessel, Gregory Ralph. Shallow stratigraphy, structure, and salt-related features, Yates oil field area, Pecos and Crockett counties, Texas. D, 1988, Colorado School of Mines. 144 p.

Wessel, James McCandless. paleocurrent analysis and petrographic study of Late Triassic Turners Falls Sandstone and Mount Toby Conglomerate in north central Massachusetts. M, 1969, University of Massachusetts. 157 p.

Wessel, James McCandless. Sedimentary petrology of the Springdale and Botwood formations, Central Mobile Belt, Newfoundland, Canada. D, 1975, University of Massachusetts. 216 p.

Wessell, J. M. Sedimentary history of Upper Triassic alluvial fan complexes in north-central Massachusetts. M, 1969, University of Massachusetts.

Wesselman, Henry Barnard, III. Pliocene micromammals from the lower Omo Valley, Ethiopia; systematics and paleoecology. D, 1982, University of California, Berkeley. 392 p.

Wesson, John Nolan. Post-Paleozoic stratigraphy, igneous petrography, economic geology, and geomorphology of the Nebaj Quadrangle, Guatemala, C.A. M, 1973, Louisiana State University.

Wesson, Robert L. Geophysical interpretation of the Carrizo Plain. M, 1968, Stanford University.

Wesson, Robert Laughlin. Seismic ray computations in laterally inhomogeneous crustal models. D, 1970, Stanford University. 100 p.

West, Alvin E. Surface geology of northeastern Lincoln County, Oklahoma. M, 1955, University of Oklahoma. 53 p.

West, Barbara J. Carbonate petrology and paleoenvironmental analysis of Pennsylvanian (?) and Permian limestone units in the Mission Argillite, Stevens County, Washington. M, 1976, Eastern Washington University. 46 p.

West, Christine M. Stratigraphy and depositional environments of the lower Sundance Formation, eastern Bighorn Basin, Wyoming. M, 1984, University of Wyoming. 94 p.

West, Cliff Merrell, Jr. Geology of the Middleway, West Virginia, quadrangle. M, 1963, West Virginia University.

West, David Lawrence. Geology of the Wilson Creek-Mill Creek fault zone; the north flank of the former Mill Creek basin, San Bernardino County, California. M, 1987, University of California, Riverside. 94 p.

West, David P., Jr. $^{40}Ar/^{39}Ar$ mineral ages from southwestern Maine; evidence for late Paleozoic metamorphism and Mesozoic faulting. M, 1988, University of Maine. 199 p.

West, Eldon S. Biostratigraphy and paleoecology of the lower Morrison Formation of Cimarron County, Oklahoma. M, 1978, Wichita State University. 61 p.

West, Gordon F. Theoretical studies for induction prospecting. M, 1957, University of Toronto.

West, Heidi DeEtte. An analysis of California's system of environmental regulation. D, 1982, University of California, Los Angeles. 99 p.

West, Howard Bruce. The origin and evolution of lavas from Haleakala Crater, Hawaii. D, 1988, Rice University. 360 p.

West, James M. Structure and ore-genesis; Little Deer Deposit, Whaleback Mine, Springdale, Newfoundland. M, 1972, Queen's University. 71 p.

West, Jerry R. First Thought Mountain, Stevens County, Washington; a volcano within a caldera. M, 1976, Eastern Washington University. 30 p.

West, John Wesley. Geology of the southwestern part of the Malibu Beach Quadrangle, Los Angeles County, California. M, 1955, University of California, Los Angeles.

West, Joseph Edward. Trace element geochemistry of the coarse crystalline ore-breccia dolomite in the Flat Gap Mine, Treadway, Tennessee. D, 1974, University of Tennessee, Knoxville. 49 p.

West, Joseph Edward and dWest, Joseph E. Trace element study of wallrocks adjacent to a mineralized breccia body in the New Market zinc mine, New Market, Tennessee. M, 1970, University of Tennessee, Knoxville. 91 p.

West, Kenneth Ernest. H_2O^{18}/H_2O^{16} variations in ice and snow in mountainous regions of Canada. D, 1972, University of Alberta. 181 p.

West, Larry Thomas. Genesis of soils and carbonate enriched horizons associated with soft limestones in central Texas. D, 1986, Texas A&M University. 271 p.

West, Mark Allen. The effect of diagenesis on enhanced recovery methods in Frio reservoir sandstones of the middle Texas Gulf Coast. M, 1981, Texas A&M University. 105 p.

West, Michael W. The Quaternary geology, reported surface faulting and seismicity along the east flank of the Gore Range, Summit County, Colorado. M, 1977, Colorado School of Mines. 209 p.

West, Olaf Van *see* Van West, Olaf

West, Philip J. Geology of the Stokes Canyon reservoir site. M, 1962, University of Southern California.

West, Philip W. The sorption of fluoride ion with special reference to fluoride removal from potable waters. M, 1936, University of North Dakota. 24 p.

West, Richard C. Borehole transient EM response of a three-dimensional fracture zone in a conductive half-space. M, 1986, University of Utah. 113 p.

West, Robert E. Analysis of gravity data from the Aora valley area, Pima County, Arizona. M, 1970, University of Arizona.

West, Robert Elmer. A regional bouguer gravity anomaly map (scale, 1:500,000) of Arizona. D, 1972, University of Arizona.

West, Ronald R. The brachiopod superfamily Orthotetacea of the Missourian Stage (Pennsylvanian) of Kansas. M, 1962, University of Kansas. 120 p.

West, Ronald Robert. Marine communities of a portion of the Wewoka Formation (Pennsylvanian) in Hughes County, Oklahoma. D, 1970, University of Oklahoma. 342 p.

West, Samuel W. Geology and ground water resources of the Dry Creek area, Cassia and Twin Falls counties, Idaho. M, 1956, University of Arizona.

West, Terry Ronald. Vibrational phenomena of loess. M, 1962, Washington University. 36 p.

West, Walter Scott. A mineral comparison of the lead and zinc deposits of southwestern Wisconsin with those of Tennessee. M, 1937, University of Tennessee, Knoxville. 92 p.

Westbrook, Marston, Jr. Measurement of bedload transport in a shallow marine environment. M, 1972, University of Washington.

Westbrook, Stephen Henry. Elemental distribution and variation in volcanic rocks from the western Carolina slate belt. M, 1977, North Carolina State University. 84 p.

Westcott, James Franklin. Differential thermal analysis of some Missouri clays. M, 1947, University of Missouri, Columbia.

Westen, Diane. Recognition criteria for young multiple surface ruptures along the Meers Fault in southwestern Oklahoma. M, 1985, Texas A&M University. 69 p.

Wester, L. L. Changing patterns of vegetation on the west side and south end of the San Joaquin Valley during historic time. D, 1975, University of California, Los Angeles. 203 p.

Westerdahl, Howard Ellsworth. A study of particulate organic ^{32}phosphorus in a simulated sediment-water system. D, 1973, University of Oklahoma. 189 p.

Westerfield, Michael J. Geology and magnetics of the Cerros Clemente-Llano Verde, northwestern Sonora, Mexico. M, 1988, University of Cincinnati. 124 p.

Westergaard, Edwin H. A new cornute species from the Ashgill of Northern Ireland. M, 1986, Baylor University. 193 p.

Westerholm, Allan Sixten. Stratigraphic study of the Middle Jurassic of Berry, Nevernais, Auxois, France. M, 1957, New York University.

Westerman, C. J. A petrogenetic study of the Guichon Creek Batholith (early Jurassic), British Columbia. M, 1970, University of British Columbia.

Westerman, Christopher John. Tectonic evolution of a part of the English River Subprovince, northwestern Ontario. D, 1978, McMaster University. 292 p.

Westerman, David Scott. Petrology of the Pocomoonshine gabbro-diorite, Big Lake Quadrangle, Maine. D, 1972, Lehigh University. 212 p.

Westerman, Julius D. A seismic refraction and reflection study in the Great Salt Lake Desert, Utah. M, 1986, University of Oklahoma. 261 p.

Westermann, Dale Thomas. The effect of selected potassium salts on the availability of soil manganese. D, 1969, Oregon State University. 141 p.

Western, Stephen Kent. The interdependence of gravity anomalies with topography in Chihuahua, Mexico, West Texas, and southern New Mexico. M, 1979, Texas Christian University. 114 p.

Western, Wayne H. A minerals availability program for the United States with a special emphasis on cobalt. M, 1985, University of Utah. 237 p.

Westervelt, Mary Lynn. An annotated bibliography with notes on the Eocene-Oligocene series of Texas and Louisiana. M, 1941, University of Oklahoma. 93 p.

Westervelt, Ralph Donaldson. An investigation of the sulphide mineralization at the Kootenay chief ore body, Bluebell Mine, British Columbia. M, 1960, Queen's University. 165 p.

Westervelt, T. N. Structural superposition in the Lake O'Hara region, Yoho and Kootenay national parks, British Columbia, Canada. D, 1979, University of Wyoming. 290 p.

Westfahl, Diane E. The crystal structures of raspite $PbWO_4$. M, 1977, Colorado School of Mines. 74 p.

Westfahl, Richard K. An unconfined compression testing machine for marine sediments. M, 1970, United States Naval Academy.

Westfall, Dwayne Gene. Effect of drying on aluminum and other extractable cations in some strongly acid soils. D, 1968, Washington State University. 171 p.

Westfall, Jonathan E. Geology and orientation geochemistry of a portion of the Carico Lake Quadrangle, Lander County, north-central Nevada. M, 1986, University of Idaho. 118 p.

Westgate, J. W. The appendicular skeleton of Aelurodon (Prohyaena) taxoides (Canidae); an analysis of functional morphology as related to paleoecologic role. M, 1978, University of Nebraska, Lincoln.

Westgate, James William. Biostratigraphic and paleoecologic implications of the first Eocene land mammal fauna from the North American coastal plain. D, 1988, University of Texas, Austin. 194 p.

Westgate, John Arthur. Surficial geology of the Foremost-Cypress Hills area. D, 1964, University of Alberta. 208 p.

Westgate, Lewis Gardner. The geology of the northern part of Jenny Jump Mountain in Warren County, New Jersey. D, 1896, Harvard University.

Westgate, Linda Marie. Variations in ratios of sulfur isotopes in the Herrin (No. 6) coal member of Illinois. M, 1983, University of Illinois, Urbana. 79 p.

Westhoff, David Edward. Application of kinematic wave theory to estimation of runoff from selected areas of Nevada. M, 1979, University of Nevada. 95 p.

Westhusing, James K. The geology of the northern third of the Sutherlin Quadrangle (Oregon). M, 1959, University of Oregon. 111 p.

Westjohn, David B. Finite strain in the Precambrian Kona Formation, Marquette County, Michigan. M, 1978, Michigan State University.

Westland, Anthony James. The deep earthquake of the Central Pacific, April 16, 1937. M, 1937, St. Louis University.

Westlund, Carlyle W. Abandoned oil and gas wells in Pennsylvania, their effect on the ground-water flow system, and plugging priority for ground-water protection. M, 1976, Pennsylvania State University, University Park. 81 p.

Westmoreland, Harry. A subsurface study of the Delhi area of northeastern Louisiana. M, 1947, University of Oklahoma. 50 p.

Weston, Ayla. Statistical examination of gravity anomalies over preglaciated terrains. M, 1965, Carleton University. 84 p.

Weston, D. P. Distribution of macrobenthic invertebrates on the NC continental shelf with consideration of sediment, hydrography and biogeography. D, 1983, College of William and Mary.

Weston, Loren Kinsman. Application of a chemical equilibrium model in the determination of pH of natural ground waters. M, 1972, University of Arizona.

Weston, Marion D. Illustrated key to the fossil flora of the upper Paleozoic coal measures of the United States; study of the Rhode Island species of Pecopteris, Aliopteris, and Mariopteris. D, 1917, Brown University.

Weston, Ray Franklin. Geology of part of Cushing Quadrangle, southwestern Rusk County, Texas. M, 1951, University of Texas, Austin.

Westover, James Donald. Chromatographic examination of gilsonite. D, 1966, Brigham Young University. 155 p.

Westphal, Jerome Anthony. Digital simulation of inorganic water quality of Tahoe-Truckee system, Nevada-California. D, 1973, University of Nevada. 220 p.

Westphall, Michael. History and anatomy of a ringed-reef complex, Belize, Central America. M, 1986, University of Miami. 225 p.

Westrich, H. R. Fluoride-hydroxyl exchange equilibria in several hydrous minerals. D, 1978, Arizona State University. 152 p.

Westrich, Henry R. An experimental determination of the solubility of molybdenum in pH-buffered KCl-HCl fluids. M, 1975, University of Cincinnati. 105 p.

Westrich, Joseph Theodore. The consequences and controls of bacterial sulfate reduction in marine sediments. D, 1983, Yale University. 564 p.

Westrop, Stephen Richard. Late Cambrian and earliest Ordovician trilobites, southern Canadian Rocky Mountains, Alberta. D, 1984, University of Toronto.

Westrop, Stephen Richard. Systematics and palaeoecology of Upper Ordovician trilobites from the Red River Formation (Selkirk Member) of southern Manitoba. M, 1980, University of Toronto.

Wetendorf, Fred H. Environment of deposition of parts of the Brereton, Jamestown, and Bankston cyclothems (Middle Pennsylvanian) of Williamson County, Illinois. M, 1967, Southern Illinois University, Carbondale. 89 p.

Wetherbee, Paul K. Fluid inclusions and Archean lode gold, Sturgeon Lake-Savant Lake, Northwest Ontario. M, 1988, University of Wisconsin-Madison. 75 p.

Wetherell, Anthony John. Multiple measurements of structure in a shrink-swell soil. M, 1967, University of California, Riverside. 53 p.

Wetherell, Clyde E. Geology of part of the southeastern Wallowa Mountains, northeastern Oregon. M, 1960, Oregon State University. 209 p.

Wetherell, Dennis Gene. Geology and ore genesis of the Sam Goosly copper-silver-antimony deposit, British Columbia. M, 1979, University of British Columbia.

Wethington, Lynette Diane. Structural relationships and chronologies of two chert to clastic rock successions, western Sierra Nevada, California. M, 1976, University of Texas at Dallas. 107 p.

Wethington, William Orville. A petrographic study of the Permian formations of Woods County, Oklahoma. M, 1932, University of Oklahoma. 57 p.

Wetlaufer, Pamela H. Geochemistry and mineralogy of the carbonates of the Goede mining district, Colorado. M, 1977, George Washington University.

Wetmiller, R. J. An earthquake swarm on the Queen Charlotte islands fracture zone (British Columbia). M, 1969, University of British Columbia.

Wetmore, Clinton C. Petrographic analysis of the Flint Ridge, flint of the upper Breathitt Formation (Pennsylvanian) in Breathitt and Magoffin counties in east-central Kentucky. M, 1978, Eastern Kentucky University. 65 p.

Wetmore, Karen Louise. Test strength, mobility, and functional morphology of benthic foraminifera. D, 1988, The Johns Hopkins University. 197 p.

Wetsel, Eldon R. The geology of Cass area, Watuluia Quadrangle, Franklin County, Arkansas. M, 1963, University of Arkansas, Fayetteville.

Wetter, Raymond Emil. A Cenomanian microfauna from upper Fort St. John strata. M, 1951, University of Alberta. 87 p.

Wetterauer, R. H. The Mina Deflection; a new interpretation based on the history of the Lower Jurassic Dunlap Formation, western Nevada. D, 1977, Northwestern University. 185 p.

Wetterauer, Richard H. The Humboldt Lopolith of western Nevada; a magnetic model. M, 1972, Northwestern University.

Wetterhall, Walter S. The groundwater resources of Chemung County, New York. M, 1959, Syracuse University.

Wetzel, John Hall. Upper Mississippian sediments of the Pomquet River and Knoydart areas, Nova Scotia. M, 1952, Massachusetts Institute of Technology. 89 p.

Wetzel, John M. Petrography of selected limestones of the Fairview Formation and Bellevue Member, Mc-Millan Formation (Cincinnatian Series) from southwestern Hamilton County, Ohio. M, 1968, Miami University (Ohio). 93 p.

Wetzel, N. Paleoecology and carbonate petrology of a Devonian (Frasnian) stromatoporoid reef, Mountain Springs Summit, Nevada. M, 1978, California State University, Fresno.

Wetzel, Richard Allen. Axisymmetric stress wave propagation in sand. D, 1969, Illinois Institute of Technology. 174 p.

Wetzel, Wayne Allen. Weathering and development of weathering residuals on Boulder Batholith, southwestern Montana. D, 1983, Simon Fraser University. 353 p.

Wetzstein, Eric E. Sedimentology and tectonic significance of a Cretaceous conglomerate in the eastern Klamath Mountains, California. M, 1986, University of Nebraska, Lincoln. 46 p.

Wexler, E. J. Ground-water flow and solute transport at a municipal landfill site on Long Island, New York. M, 1987, University of Waterloo. 175 p.

Weyenberg, Lynn Ellen. Depositional and diagenetic history of the Middle Ordovician carbonates of the Shenandoah Valley, northern Virginia. M, 1987, Old Dominion University. 182 p.

Weyer, Laura L. The application of optical diffraction analysis to the study of microfracture propagation and the contribution of microfractures to rock failure. M, 1987, University of Wisconsin-Milwaukee. 195 p.

Weymouth, Andrew Allen. The cretaceous stratigraphy and paleontology of Sucia Island, San Juan Group, Washington. M, 1928, University of Washington. 56 p.

Weynand, G. W. The resolution of two closely spaced wavelets in the presence of noise. M, 1977, Texas A&M University.

Whaite, Peter. An automated sample line for the preparation of O^{18}/O^{16} isotope analyses from water samples. M, 1982, University of British Columbia. 179 p.

Whalen, Joseph Bruce. Geology and geochemistry of molybdenite showings of the Ackley City Batholith, Fortune Bay, Newfoundland. M, 1976, Memorial University of Newfoundland. 267 p.

Whalen, Michael T. The carbonate petrology and paleoecology of Upper Triassic limestones of the Wallowa Terrane, Oregon and Idaho. M, 1985, University of Montana. 151 p.

Whalen, Patricia Ann. Ecology and significance of foraminiferal assemblages from western Andros Island, Great Bahama Bank. M, 1967, Columbia University. 34 p.

Whalen, Patricia Ann. Lower Jurassic radiolarian biostratigraphy of the Kunga Formation, Queen Charlotte Islands, British Columbia, and the San Hipolito Formation, Baja California Sur. D, 1985, University of Texas at Dallas. 440 p.

Whalen, Stephen C. The chemical limnology and limnetic primary of the Tongue River Reservoir, Montana. M, 1979, Montana State University. 205 p.

Whaley, Harry Max. The geology of a part of the Rock Creek Quadrangle on the north side of the San Gabriel Mountains, Los Angeles County, California. M, 1937, University of California, Los Angeles.

Whaley, Joseph Floyd. Some computational aids for shallow refraction seismograms. M, 1957, Indiana University, Bloomington. 61 p.

Whaley, Keith Ray and Ricketts, Edward W. Structure and stratigraphy of the Oak Ridge Fault-Santa Susana Fault intersection, Ventura Basin, California. M, 1975, Ohio University, Athens. 81 p.

Whaley, Peter Walter. A litho-genetic model for rocks of a lower deltaic plain sequence. D, 1969, Louisiana State University. 168 p.

Whaley, Peter Walter. A petrographic study of the Tyrone Limestone. M, 1964, University of Kentucky. 45 p.

Whalin, Robert Warren. The limit of applicability of linear wave refraction theory in a convergence zone. D, 1971, Texas A&M University. 175 p.

Whan, Wen J. The direct current resistivity modeling of thin insulators. D, 1979, Colorado School of Mines. 135 p.

Whang, Chen-Wen. The direct determination of arsenic and antimony in sea water by anodic stripping voltammetry. D, 1984, Queen's University.

Whang, Jooho. Migration of radioactive wastes from shallow land burial site under saturated and unsaturated conditions. D, 1986, Georgia Institute of Technology. 229 p.

Wharton, Amy Laura. Depositional environment of the Middle Pennsylvanian granite wash; Lambert 1, Hryhor, and Sundance fields, northern Palo Duro Basin, Oldham County, Texas. M, 1986, Texas A&M University.

Wharton, David Ian. Dinoflagellates from Middle Jurassic sediments of southern Alaska. D, 1988, Stanford University. 467 p.

Wharton, George B., Jr. Snow regimen of Mount Wrangell Caldera (Alaska). M, 1966, University of Alaska, Fairbanks. 63 p.

Wharton, H. M. A geological study of the York Harbour Mine, Humber District, Newfoundland. M, 1955, Columbia University, Teachers College.

Wharton, Jay Bigelow, Jr. Microfauna of the lower Jackson Formation, Montgomery, Louisiana. M, 1935, University of Oklahoma. 100 p.

Wharton, Mary E. Floristics and vegetation of the Devonian-Mississippian black-shale region of Kentucky. D, 1946, University of Michigan.

Wharton, Ralph E. The sub-Eden geology of the Denmark oil pool, Morrow County, Ohio. M, 1964, Ohio State University.

Wharton, Richard J. Some evidence on the validity of the exsolution model for the formation of antiperthites. M, 1972, Michigan State University. 16 p.

Wharton, Robert Andrew, Jr. Ecology of algal mats and their role in the formation of stromatolites in Antarctic dry valley lakes. D, 1982, Virginia Polytechnic Institute and State University. 131 p.

Whatley, Arthur F. The stratigraphy of the Glendon Limestone in the Vicksburg area, Mississippi (Glendon-Oligocene). M, 1950, Mississippi State University. 69 p.

Whatley, James Broughton. Geology and mineral resources of part of the University of Alabama lands and the adjoining areas in the vicinity of Woods Creek, Jefferson County, Alabama. M, 1949, University of Alabama.

Whealdon, Edwin Phillips. Origin and age of the redrock sandstone of Marion County, Iowa. M, 1937, University of Iowa. 20 p.

Wheatcraft, S. W. Buoyant plumes in density-stratified porous media; a laboratory model study of waste injection into a static environment. M, 1976, University of Hawaii. 113 p.

Wheatcraft, Suzanne Bragg. Hydrogeochemistry of the Sundre Aquifer, Minot, North Dakota. M, 1987, Oklahoma State University. 108 p.

Wheatcroft, Robert Arthur. Ichnology and major facies of the Ripley Formation and Providence Sand (Upper Cretaceous), Chattahoochee River valley. M, 1984, University of Georgia. 193 p.

Wheatfill, Edward Lewis. The possibilities of locating and developing deeper aquifers to augment the water well production for Long Beach, California. M, 1957, University of California, Los Angeles.

Wheatley, Kenneth Lewis. The sedimentology, stratigraphy and structural geology of the Carswell Formation, northern Saskatchewan. M, 1985, University of Saskatchewan. 176 p.

Wheatley, Todd L. Experimental thermal maturation of Type III organic matter; kinetics and catalysis. M, 1988, University of Missouri, Columbia.

Wheatley-Doyle, Michelle D. Paleocene ostracodes from the inner coastal plain of North Carolina. M, 1984, University of Delaware. 397 p.

Wheeler, A. Edward. Upper Cretaceous foraminifera of the Santa Ana Mountains, California. M, 1952, University of California, Los Angeles.

Wheeler, Alfred H. Natural air drying of diatomaceous earth. M, 1956, University of Nevada - Mackay School of Mines. 34 p.

Wheeler, Cary F. R. The mineralogy, petrology, geochemistry and petrogenesis of the Mount Poser gabbroic pluton, Southern California. M, 1979, University of Windsor. 129 p.

Wheeler, Charles Brown. Geology of the Skyline-Corte Madera area, San Mateo County, California. M, 1953, Stanford University.

Wheeler, Charles Thomas, Jr. and Bruder, Karl Fritz. Geology of the Greaser Creek area, west central Huerfano Park, Huerfano County, Colorado. M, 1955, University of Michigan.

Wheeler, David M. Stratigraphy and sedimentology of the Tensleep Sandstone, Southeast Bighorn Basin, Wyoming. M, 1986, Colorado School of Mines.

Wheeler, Dooley Peyton, Jr. Geology and mineralization in the vicinity of the Bondurant Mine, Mariposa County, California. M, 1941, University of California, Berkeley. 152 p.

Wheeler, Everett Pepperell. A study of some diabase dikes of the Labrador coast. D, 1930, Cornell University.

Wheeler, Everett Pepperell. Chemical and optical study of olivine from Monhegan Island, Maine. M, 1926, Cornell University.

Wheeler, Garland Edgar. Zonation of the Mississippian strata in the vicinity of Pigeon Mountain in Northwest Georgia. M, 1954, Emory University. 58 p.

Wheeler, Girard Emory. The west wall of the New England Triassic lowland, a study of Triassic border structure and geomorphology. D, 1937, Columbia University, Teachers College.

Wheeler, Gregory R. Geology of the Vinegar Hill area, Grant County, Oregon. D, 1976, University of Washington. 94 p.

Wheeler, Gregory R. Nuclear power plant siting in North Clallam County, Washington; a study of geologic considerations for power sites along the northern coast of the Olympic Peninsula. M, 1971, University of Washington. 54 p.

Wheeler, Harry E. Stratigraphy and paleontology of the (Permian) McCloud Limestone of Northern California. M, 1932, Stanford University. 71 p.

Wheeler, Harry E. The fauna and correlation of the McCloud Limestone of Northern California. D, 1934, Stanford University. 165 p.

Wheeler, J. O. Areal geology of plutonic rocks of the Whitehorse map-area, Yukon Territory (Canada). D, 1954, Columbia University, Teachers College.

Wheeler, James D. A pegmatitic magnetite deposit in northern St. Louis County, Minnesota. M, 1919, University of Minnesota, Minneapolis. 26 p.

Wheeler, James William. Jurassic and Cretaceous microplankton from the central Alborz Mountains, Iran. M, 1982, University of Saskatchewan. 337 p.

Wheeler, John Oliver. An examination of the analyses of igneous rocks of British Columbia. M, 1949, University of British Columbia.

Wheeler, John Oliver. Evolution and history of the Whitehorse Trough as illustrated by the geology of Whitehorse map-area, Yukon. D, 1956, Columbia University, Teachers College. 194 p.

Wheeler, Joseph Bowen. The geology of an area of five square miles northwest of Austin, Texas. M, 1934, University of Texas, Austin.

Wheeler, Joseph Orby. Geology of northeastern front of Davis Mountains, Trans-Pecos, Texas. M, 1956, University of Texas, Austin.

Wheeler, Karen Lynn. Maestrichtian shoreline sedimentation in northeastern Montana. M, 1983, University of Iowa. 155 p.

Wheeler, Mark Crawford. Oxygen and sulfur isotope geochemistry of sulfide oxidation during formation of acid mine drainage. M, 1984, University of California, Davis. 206 p.

Wheeler, Mark Thomas. Late Cenozoic gravel deposits of the Lower Snake River, Washington. M, 1980, Washington State University. 102 p.

Wheeler, Merlin L. Electric analog model study of the hydrology of the Saginaw Formation in the Lansing, Michigan area. M, 1967, Michigan State University. 74 p.

Wheeler, Merlin Leroy. Application of thermocouple psychrometers to field measurements of soil moisture potential. D, 1972, University of Arizona.

Wheeler, Richard Brian. Environmental trace element geochemistry of sediments of the Buccaneer offshore oil and gas field; factors controlling concentrations of trace elements in sediments. M, 1979, Rice University.

Wheeler, Richard F. Geology of the Sewing Machine Pass Quadrangle, central Wah Wah Range, Beaver County, Utah. M, 1980, Brigham Young University. 191 p.

Wheeler, Robert B. The stratigraphy and petrography of the Moyers Formation, Oklahoma and Arkansas. M, 1974, Northern Illinois University. 74 p.

Wheeler, Robert J. Water resource and hazard planning report for the Clark Fork River valley above Missoula, Missoula County, Montana. M, 1975, University of Montana. 69 p.

Wheeler, Robert Reid. Cambrian and Lower Ordovician of the Adirondack border. D, 1942, Harvard University.

Wheeler, Russell L. Folding history of the metamorphic rocks of eastern Dovrefjell, Norway. D, 1973, Princeton University. 233 p.

Wheeler, Walter H. A revision of the uintatheres. D, 1952, Yale University.

Wheeler, Walter H. and Dorr, John Adam, Jr. Geology of a part of the Ruby Basin, Madison County, Montana. M, 1948, University of Michigan.

Wheeler, William A. A geological and geophysical investigation of the area near Merkel, Taylor County, Texas. M, 1983, University of Texas at Dallas. 42 p.

Wheelock, Graham. The sedimentology and mineralization of the Archean Elsburg No. 5 reef on Vaal Reefs West Gold Mine, Transvaal, South Africa. M, 1987, University of Cincinnati. 104 p.

Wheelock, Thomas G. B. Jurassic ophiuroid echinoderms from Utah and Wyoming. M, 1971, University of Wyoming. 168 p.

Wheelright, Mona Yvonne. Preliminary palynology of some Utah and Wyoming coals. M, 1958, University of Utah. 132 p.

Whelan, Charlene J. Mineralogy of the Knox Group carbonates (Cambrian–Ordovician) from three wells in middle Tennessee. M, 1971, University of Tennessee, Knoxville. 51 p.

Whelan, Hugh Thomas More. Geostatistical estimation of the spatial distributions of porosity and percent clay in a Miocene Stevens Turbidite reservoir; Yowlumne Field, California. M, 1984, Stanford University. 139 p.

Whelan, James Arthur. A study of hisingerite and related silicates. D, 1959, University of Minnesota, Minneapolis. 166 p.

Whelan, James P. The comparative analysis of the mammal remains from three prehistoric Indian sites in Alameda County. M, 1970, San Francisco State University.

Whelan, Joseph F. Trace element and sulfur isotopic comparison of anhydrites from Belmat-Edwards, N.Y., with sedimentary and hydrothermal anhydrite. M, 1974, Pennsylvania State University, University Park. 99 p.

Whelan, Peter M. Geochemical and petrologic studies relating to the origin of realgar at Kramer, California. M, 1967, University of Wisconsin-Madison.

Whelan, Peter Michael. A geochemical and geochronological study of late Miocene to Recent volcanism, Fiji, Southwest Pacific. D, 1988, University of California, Santa Cruz.

Whelan, Thomas Joseph, Jr. Foraminiferal distribution in the Delaware Bay area. M, 1954, Rutgers, The State University, New Brunswick. 116 p.

Wherry, Edgar Theodore. Contributions to the mineralogy of the (Triassic) Newark Group in Pennsylvania. D, 1909, University of Pennsylvania.

Wherry, Stephen D. Mineralization of Granite Mountain, Lander County, Nevada. M, 1982, University of Texas at Dallas. 112 p.

Whetten, John T. Geology of St. Croix, U. S. Virgin Islands. D, 1961, Princeton University. 102 p.

Whetten, John Theodore. The geology of the central part of the Soldier Pass Quadrangle, Inyo County, California. M, 1959, University of California, Berkeley. 72 p.

Whiddon, Deborah Justice. Surface geology of Twin Island Field, Cameron Parish, Louisiana. M, 1982, University of New Orleans. 50 p.

Whigham, Linda. Stratigraphy of the Fredericksburg Group north of the Colorado River, Texas. M, 1980, Baylor University. 311 p.

Whiita, Richard A. Geologic field trip in the Lorain, Ohio area. M, 1968, Virginia State University. 50 p.

Whillans, I. M. Mass-balance and ice flow along the Byrd Station strain network, Antarctica. D, 1975, Ohio State University. 151 p.

Whipkey, Charles Evans. Provenance of lower Tertiary sandstones of the Powder River basin, Wyoming. M, 1988, North Carolina State University. 145 p.

Whipple, Arthur Paul. A gravity survey of the Babler State Park area, Missouri. M, 1958, St. Louis University.

Whipple, Earle Raymond. Quantitative Mossbauer spectra and chemistry of iron. D, 1974, Massachusetts Institute of Technology. 205 p.

Whipple, George Leslie. Larger foraminifera from the lower Miocene of Vitilevu, Fiji Islands. M, 1929, University of California, Berkeley. 22 p.

Whipple, James Wilburn. Depositional environment of the middle Proterozoic Spokane Formation-Empire Formation transition zone, West-central Montana. M, 1980, University of Colorado.

Whipple, Janice M. Glacial geology of the area between Little Falls and Richfield Springs, New York. D, 1969, Rensselaer Polytechnic Institute. 213 p.

Whipple, Ross. Radioactivity correlation in an ore deposit near Park City, Utah. M, 1949, University of Utah. 40 p.

Whippo, Robert. The optical mineralogy of certain Black Hills and Badlands rocks. M, 1960, South Dakota School of Mines & Technology.

Whisnant, Jack Summey. Geology of the southeastern quarter of the Marion 15-minute quadrangle, North Carolina. M, 1975, University of North Carolina, Chapel Hill. 50 p.

Whisonant, Robert Clyde. Stratigraphy and origin of the late Paleozoic Parkwood Formation in Jefferson, Shelby and Saint Clair counties, Alabama. M, 1965, Florida State University.

Whisonant, Robert Clyde. Stratigraphy and petrology of the basal Cambrian Chilhowee Group in central-eastern and southeastern Tennessee. D, 1967, Florida State University. 249 p.

Whistler, David P. Stratigraphy and small fossil vertebrates of the Ricardo Formation (Pliocene, lower), Kern County, California. D, 1969, University of California, Berkeley. 276 p.

Whistler, David Paul. A new Hemingfordian (middle Miocene) mammalian fauna from Boron, California and its stratigraphic implications within the western Mojave Desert. M, 1965, University of California, Riverside. 142 p.

Whitacre, Halford E. The geology of the Madera-Agua Caliente canyons area, southern Arizona. M, 1964, University of Arizona.

Whitacre, Thomas James. A geophysical investigation of the San Juan Belt and its relationship to mineralization at Cripple Creek, Colorado. M, 1986, Purdue University. 112 p.

Whitacre, Timothy Patrick. Study of mineral catalyst and organic components of several Ohio coals for a SRC II-process. M, 1981, University of Toledo. 191 p.

Whitaker, Doyle Gene. Stratigraphic study of some upper Strawn beds in Palo Pinto County, Texas. M, 1956, Texas Christian University. 64 p.

Whitaker, James T. Ultrasonic velocity measurements of Precambrian metamorphic rocks and their correlation with field measurements. M, 1966, Michigan State University. 101 p.

Whitaker, John Carroll. The geology of the Catoctin Mountain, Maryland and Virginia. D, 1953, The Johns Hopkins University.

Whitaker, John Henry. Geology of the Willow Creek area, Elko County, Nevada. M, 1985, Oregon State University. 116 p.

Whitaker, Kent Y. Depositional environments of the Bisher Dolomite Formation (Middle Silurian), Lewis County, Kentucky. M, 1987, Miami University (Ohio). 83 p.

Whitaker, Laura Rothenberg. The geology of the Saddleback Mountain area, Northwood Quadrangle, southeastern New Hampshire. M, 1983, University of New Hampshire. 137 p.

Whitaker, Richard M. Grain size distribution and depositional processes of the Casper Formation (Pennsylvanian-Permian), southern Laramie Basin, Wyoming. M, 1972, University of Wyoming. 66 p.

Whitaker, Sidney Hopkins. Geology of the Wood Mountain area (72-G), Saskatchewan. D, 1965, University of Illinois, Urbana. 150 p.

Whitaker, Stephen Taylor. The Flaxville Formation in the Scobey-Opheim area, northeastern Montana. M, 1980, University of Colorado. 164 p.

Whitbeck, Florence. Great Lakes-St. Lawrence deep waterway. M, 1921, University of Wisconsin-Madison.

Whitbeck, Luanne F. Computer simulation of erosion at the Western New York Nuclear Service Center, Cattaraugus County, New York. M, 1983, Rensselaer Polytechnic Institute. 45 p.

Whitcomb, Charles W. Sedimentology of Bonner Springs Formation (Pennsylvanian), southeastern Nebraska and adjacent regions. M, 1965, University of Nebraska, Lincoln.

Whitcomb, Harold A. Geology of the Morgan Mine area, Twin Buttes, Arizona. M, 1948, University of Arizona.

Whitcomb, James. Marine geophysical studies offshore - Newport, Oregon. M, 1965, Oregon State University. 51 p.

Whitcomb, James Hall. Part I, A study of the velocity structure of the Earth by the use of core phases; Part II, The 1971 San Fernando earthquake series focal mechanisms and tectonics. D, 1973, California Institute of Technology. 443 p.

Whitcomb, John Byington. A daily municipal wateruse model; case study comparing West Los Angeles, California and Fairfax County, Virginia. D, 1988, The Johns Hopkins University. 114 p.

Whitcomb, Lawrence. Correlation of Ordovician limestone at Salona Co., Penna. D, 1930, Princeton University.

Whitcomb, Natalie Jo. Effects of oil pollution on selected species of benthic foraminifera from the lower York River, Virginia. M, 1977, Duke University. 137 p.

White, A. F. Sodium and potassium coprecipitation in calcium carbonate. D, 1975, Northwestern University. 165 p.

White, Americus Frederic. Composition of the waters of Rockbridge County, Virginia and their relation to the geological formations. D, 1906, Washington & Lee University.

White, Arthur James. A 1982 restoration feasibility study of Lakes of the Four Seasons in northwestern Indiana. D, 1985, Ball State University. 238 p.

White, Bob O. Geology of West Mountain and northern portion of Long Ridge, Utah County, Utah. M, 1953, Brigham Young University. 187 p.

White, Bradford Stephen. Carbon dioxide, argon, and water in silicate liquids at high pressure. D, 1988, University of California, Los Angeles. 221 p.

White, Brian George. Structural analysis of a small area in the northeast border zone of the Idaho Batholith, Idaho. M, 1969, University of Montana. 51 p.

White, Brian George. Thrusting in the northwestern Montana overthrust belt. D, 1978, University of Montana. 103 p.

White, Brian Nelson. The isthmian link, antitropicality and American biogeography; relationships and distributional history of the Atherinopsinae (Pisces: Atherinidae). D, 1983, University of Southern California.

White, Carla A. Petrology and mineral chemistry of some Jan Mayen volcanics. M, 1979, SUNY at Albany. 197 p.

White, Charlene K. A strain analysis in three imbricate areas in the Moine Thrust, Scotland. M, 1988, Miami University (Ohio).

White, Charlotte Anne. The petrology and geochemistry of peralkaline granite and volcanic rocks near Davis Inlet, Labrador. M, 1982, Memorial University of Newfoundland. 209 p.

White, Christine Anne. Geology of Precambrian metasedimentary rocks of Lester Mountain, Colorado; a study of depositional environment, metamorphism and structure. M, 1987, University of Iowa. 178 p.

White, Christopher D. Propagation of cavities in unconsolidated sand. M, 1982, Stanford University.

White, Christopher David. Representation of heterogeneity for numerical reservoir simulation. D, 1987, Stanford University. 274 p.

White, Christopher Minot. Surficial geology and mode of deglaciation in the Wilmington Quadrangle, Vermont. M, 1978, University of Massachusetts. 212 p.

White, Craig Kenneth. Geology of the San Diego onshore-offshore area, southern California. M, 1969, University of Nevada. 90 p.

White, Craig M. Bedrock geology of the Eights coast, Ellsworth land, Antarctica. M, 1970, University of Wisconsin-Madison.

White, Craig McKibben. Geology and geochemistry of volcanic rocks in the Detroit area, western Cascade Range, Oregon. D, 1980, University of Oregon. 178 p.

White, Daniel Howard. Predicting fragmentation characteristics of a block caving orebody. M, 1977, University of Arizona.

White, Daniel J. Three-dimensional modeling of seismic structure in the Puget Sound region. M, 1988, Pennsylvania State University, University Park. 64 p.

White, David Archer. Stratigraphy and structure of the Mesabi Range, Minnesota. D, 1954, University of Minnesota, Minneapolis. 171 p.

White, David Dean. Bedrock geology and stratigraphy of the Late Cretaceous Nanaimo Group of the Maple Bay-Cowichan Bay area, British Columbia. M, 1983, Oregon State University. 69 p.

White, David L. A Rb-Sr isotopic geochronologic study of Precambrian intrusives of South-central New Mexico. D, 1977, Miami University (Ohio). 147 p.

White, David L. The geology of the Pittsburg Landing area, Snake River Canyon, Oregon and Idaho. M, 1972, Indiana State University. 98 p.

White, David Ross. Fluvial-dominated deltaic deposits of the marine Plio-Pleistocene lower Saugus Formation, northern Simi Valley, California. M, 1985, California State University, Northridge. 86 p.

White, Dixon N. Geology of the upper James River area, Mason County, TX. M, 1961, Texas A&M University. 84 p.

White, Donald E. Inorganic quality of water as related to environment in the lower Box Elder Creek Valley, Larimer County, Colorado. M, 1964, Colorado State University. 97 p.

White, Donald E. The molybdenite deposits of the Rencontre East area, Nfld. D, 1939, Princeton University.

White, Dossey Hurdle, Jr. Ostracods of the Moodys Branch Formation at Little Stave Creek, Clarke County, Alabama. M, 1964, University of Alabama.

White, Elijah, Jr. Upper Devonian Chemung and Brallier formations in outcrop and subsurface of southeastern West Virginia and adjacent Virginia. M, 1984, University of North Carolina, Chapel Hill. 59 p.

White, Elizabeth Loczi. Role of carbonate rocks in modifying extreme flow behavior. D, 1975, Pennsylvania State University, University Park. 176 p.

White, Ella Marie. Features of physical geology in Southern California. M, 1937, University of Southern California.

White, Eugene Wilbert. Chemical characterization of materials by X-ray spectroscopy. D, 1965, Pennsylvania State University, University Park. 198 p.

White, Eugene Wilbert. Uranium mineralization in some North and South Dakota lignites. M, 1958, Pennsylvania State University, University Park. 79 p.

White, French Robertson, Jr. Areal geology and possible structural relationships of the Muenster Arch, northwestern Cooke County, Texas. M, 1948, Texas Christian University. 33 p.

White, George Willard. The limestone caves and caverns of Ohio. M, 1925, Ohio State University.

White, George Willard. The Pleistocene geology of the region of the reentrant angle in glacial boundary in northcentral Ohio. D, 1933, Ohio State University.

White, Harold O., Jr. Feasibility of artificial recharge to the High Plains Aquifer, northwestern Oklahoma. M, 1984, Oklahoma State University. 113 p.

White, Harold Richard. Eocene non-marine Ostracoda of Wyoming. M, 1948, University of Illinois, Urbana.

White, Howard James. Stratigraphy of the lower member, Koobi Fora Formation, southern Karari Escarpment, East Turkana Basin, Kenya. M, 1976, Iowa State University of Science and Technology.

White, Howard James. The stratigraphy of the southern Pab Range, Pakistan. D, 1981, Iowa State University of Science and Technology. 177 p.

White, J. C. Determination of subsurface structure from gravity anomaly measurements (W584). M, 1952, University of Texas, Austin.

White, J. D. Phase relations on the join phlogopite-sodium phlogopite at two kilobars water pressure. M, 1974, University of Illinois, Chicago.

White, J. D. Technical and environmental aspects of oil shale processing. D, 1978, University of California, Los Angeles. 296 p.

White, J. R., Jr. The stratigraphy and depositional environment of the "Carter Sand" of Chester age, in the subsurface of Lamar and Fayette counties, Alabama. M, 1976, University of Alabama.

White, Jack C. Geology of a Mississippian reef complex near Huntsville, Arkansas. M, 1963, University of Arkansas, Fayetteville.

White, James A. L. An investigation of some long term output forecasting techniques for the mineral industry. D, 1960, Massachusetts Institute of Technology. 372 p.

White, James A. L. The use of the logistic curve in forecasting mineral production. M, 1958, Massachusetts Institute of Technology. 239 p.

White, James D. L. The Doyle Creek Formation (Seven Devils Group), Pittsburg Landing, Idaho; a study of intra-arc sedimentation. M, 1985, University of Missouri, Columbia.

White, James Edward Moseby. A review of the modern theories concerning the formation of the ore-forming fluid. M, 1951, University of Michigan.

White, James R. The study in fluvial sedimentation using settling velocity techniques. M, 1966, Pennsylvania State University, University Park. 81 p.

White, James Robert. A study of the Mississippian deposits in the subsurface of north-central Texas. M, 1948, Texas Christian University. 64 p.

White, James Robert. The geology of the Little Warm Spring Creek area, Fremont County, Wyoming. M, 1951, Miami University (Ohio). 45 p.

White, Joe R., Jr. Foraminiferida and paleoecology of the Oligocene Spring Marl in Mississippi. M, 1974, Northeast Louisiana University.

White, John Francis. Geology of a portion of the Tortilla Mountains, Pinal County, Arizona, emphasizing granitic xenoliths in diabase and associated high- and low-temperature feldspars. D, 1955, University of California, Berkeley. 188 p.

White, John Fullington. Depositional environments of Pennsylvanian Redoak Mountain Formation, northern Cumberland Plateau, Tennessee. M, 1975, University of Tennessee, Knoxville. 118 p.

White, John Lester. The development of fractures in the Harris Ranch Quartz Monzonite related to the Sierrita porphyry copper system, Pima County, Arizona. M, 1980, University of Arizona. 39 p.

White, John M., Jr. The Brushy Mountain structure, Sequoyah and Adair counties, Oklahoma. M, 1955, University of Oklahoma. 37 p.

White, John S., Jr. Plattnerite, a description of the species from natural crystals. M, 1966, University of Arizona.

White, John Weldon. Topaz-bearing pegmatites and gem topaz in the Llano Uplift, Texas. M, 1960, University of Texas, Austin.

White, Jon M. Effects of small temperature changes on measured apparent residual strain in selected rock cylinders. M, 1973, South Dakota School of Mines & Technology.

White, Jon M. Experimental stress analysis and rock mechanics research techniques applied to selected problems in engineering and earth science. D, 1975, South Dakota School of Mines & Technology. 222 p.

White, Joseph Clancy. A study of deformation within the Flinton Group conglomerates, southeastern Ontario. D, 1979, University of Western Ontario. 351 p.

White, Kenneth Lewis. Pedogenic chronology of the Santa Ana River terraces. D, 1976, University of California, Los Angeles. 167 p.

White, Larry Michael. Deposition and diagenesis of the Five Finger Carbonate, Council Grove Group, Hugoton Field, Kansas. M, 1981, Texas Tech University. 89 p.

White, Lee H. Petrology and depositional environment of the Severy Shale (Pennsylvanian) of southeastern Kansas and northeastern Oklahoma. M, 1987, Wichita State University. 330 p.

White, Lloyd Arthur. The geology of the Middle Ordovician of the Knoxville Belt, Bearden Quadrangle, Knox County, Tennessee. M, 1955, University of Tennessee, Knoxville. 41 p.

White, Marjorie Ann. Structural geology of the Irons Fork-North Fork creeks area, Lake Ouachita, Arkansas. M, 1980, Texas A&M University. 73 p.

White, Martha. The fluvial geomorphology of an ephemeral stream in southern Illinois. M, 1975, Southern Illinois University, Carbondale. 93 p.

White, Maurice Douglas. The clay mineralogy of the Claiborne Formation in Carroll and Weakley counties, Tennessee. M, 1985, Memphis State University. 44 p.

White, Max Gregg, Jr. Geology of the upper basin of North Fork Powder River, Bighorn Mountains, Wyoming. M, 1940, University of Iowa. 73 p.

White, Maynard P. Some index foraminifera of the Tampico Embayment area of Mexico. D, 1928, Columbia University, Teachers College.

White, Maynard P. Tertiary Mollusca of Chiapas and Tabasco (Mexico). M, 1923, Columbia University, Teachers College.

White, O. L. The application of soil consolidation tests to the determination of Wisconsin ice thickness in the Toronto region (Ontario). M, 1962, University of Toronto.

White, Owen L. The application of soil consolidation tests to the determination of Wisconsin ice thick-

nesses in the Toronto region. M, 1961, University of Toronto.

White, Owen Lister. Pleistocene geology of the Bolton area, Ontario. D, 1970, University of Illinois, Urbana. 305 p.

White, P. J. Geology of the Island Mountain area, Okanogan County, Washington. M, 1986, University of Washington. 80 p.

White, Paul Gary. Rock glaciers in the San Juan Mountains, Colorado. D, 1973, University of Denver.

White, Randal Ocee. The geology of the Schminski molybdenum deposit (Colville Indian Reservation), Ferry County, Washington. M, 1981, Washington State University. 103 p.

White, Rene. Paleomagnetism of the Tulare Formation from cores and surface exposures, west-central and southwestern San Joaquin Valley. M, 1987, California State University, Long Beach. 272 p.

White, Rex Harding, Jr. Petrology and depositional pattern in the upper Austin Group, Pilot Knob area, Travis County, Texas. M, 1960, University of Texas, Austin.

White, Richard E. Stratigraphy and structure of the Cedar Creek area of the Madison Range, Madison County, Montana. M, 1974, Oregon State University. 134 p.

White, Richard J. Sediment provenance and dispersal in Green Bay, Lake Michigan. M, 1982, University of Wisconsin-Milwaukee. 122 p.

White, Richard J. Southern Ocean silicoflagellate and ebridian biostratigraphy, the opening of the Drake Passage, and the Miocene of the Ross Sea, Antarctica. M, 1980, Northern Illinois University. 205 p.

White, Richard LeRoy. A study of resistivity logging problems by the use of wedge models. M, 1957, Indiana University, Bloomington. 70 p.

White, Richard William. Ultra-mafic inclusions in basaltic rocks from Hawaii. D, 1965, University of California, Berkeley. 160 p.

White, Robert C. Age of the (Miocene) "Modelo" in Haskell Canyon, easternmost Ventura Basin, California. M, 1947, California Institute of Technology. 35 p.

White, Robert J. Geochemistry of Frank Oster Lakes, Gooding County, Idaho. M, 1971, Idaho State University. 103 p.

White, Robert M. Noise level in a mountain gravity survey. M, 1976, University of Calgary. 95 p.

White, Robert Rankin. Water table response to barometric pressure changes; a laboratory investigation. M, 1971, University of Arizona.

White, Robin Shepard. The determination of seasonal variations in groundwater recharge by deuterium and oxygen-18 analysis for the Tucson Basin, Arizona. M, 1976, University of Arizona.

White, Roger Lee. An analysis of the manganese content of the Fort Payne Formation of northwestern Georgia and northeastern Alabama. M, 1988, University of Georgia. 100 p.

White, Ronald K. Foraminifera from two sections of the Cody Shale, central Wyoming. M, 1961, University of Wyoming. 84 p.

White, S. L. Clay mineralogy of the Wells Creek Dolomite, Stones River Group, and Nashville Group, in central Tennessee. M, 1974, Vanderbilt University.

White, Sidney E. A geologic investigation of the late Pleistocene history of the Volcano Popocatepetl, Mexico. D, 1951, Syracuse University.

White, Stanley F. Petrology of the Cenozoic igneous rocks of the Lytle Creek area, Bear Lodge Mountains, Wyoming. M, 1980, University of North Dakota. 69 p.

White, Stanton M. The (Pennsylvanian) Wayside Sand and Walter Johnson Siltstone of northeastern Oklahoma and southeastern Kansas. M, 1960, University of Rochester. 49 p.

White, Stanton Morse. The mineralogy and geochemistry of the sediments on the continental shelf off the

Washington-Oregon coast. D, 1968, University of Washington. 213 p.

White, Stephen Joseph. Uranium potential in the Antlers Formation south of the Belton-Tishomingo Uplift, southern Oklahoma. M, 1977, Oklahoma State University. 117 p.

White, Steven C. Coarse-grained sediment transport at a continental shelf site off the New Hampshire coast. M, 1975, University of New Hampshire. 60 p.

White, Theodore G. The (Ordovician) Black River, Trenton, and Utica formations in the Champlain Valley of New York and Vermont. D, 1899, Columbia University, Teachers College.

White, Thomas Edward. Characterization of fluid flow and wastewater transport processes within a wisconsin-type mound. D, 1986, Purdue University. 325 p.

White, Thomas Ellis. Nearshore sand transport. D, 1987, University of California, San Diego. 233 p.

White, Vincent Lee. Geology of Dallas Anticline, Fremont County, central Wyoming. M, 1951, University of Wyoming. 70 p.

White, W. M. Geochemistry of igneous rocks from the central North Atlantic; the Azores and the Mid-Atlantic Ridge. D, 1977, University of Rhode Island. 304 p.

White, Walter Stanley. Geology of the Pacoima-Little Tujunga area (California). M, 1937, California Institute of Technology. 59 p.

White, Walter Stanley. Structural geology of the Vermont portion of the Woodsville Quadrangle. D, 1946, Harvard University.

White, William A. A study of Montana black sands. M, 1934, Montana College of Mineral Science & Technology. 25 p.

White, William Alexander. A petrographic examination of some Atlantic coastal sands. M, 1931, University of North Carolina, Chapel Hill. 17 p.

White, William Alexander. The mineralogy of desert sands. D, 1938, University of North Carolina, Chapel Hill. 41 p.

White, William Arthur. The properties of clays. M, 1947, University of Illinois, Urbana.

White, William Arthur. Water absorption properties of homoionic clay minerals. D, 1955, University of Illinois, Urbana.

White, William Blaine. Phase relations in the system lead-oxygen. D, 1962, Pennsylvania State University, University Park. 157 p.

White, William Dennis. Effects of forest-fire devegetation on watershed geomorphology in Bandelier National Monument, New Mexico. M, 1981, University of New Mexico. 209 p.

White, William Harrison. Geology and ore deposition of Silbak Premier Mine. M, 1939, University of British Columbia.

White, William Harrison. The mechanism and environment in veins and lodes. D, 1942, University of Toronto.

White, William Robert. Miocene and Pliocene foraminifera from the vicinity of San Juan Capistrano, California. M, 1953, University of Southern California.

White, William Robert Hugh. The structure of the Earth's crust in the vicinity of Vancouver Island from seismic and gravity observations. D, 1962, University of British Columbia.

White, William Wesley, III. Paleontology and depositional environments of the Cambrian Wheeler Formation, Drum Mountains, west-central Utah. M, 1973, University of Utah. 135 p.

White, Willis Harkness. Plutonic rocks of the southern Seven Devils Mountains, Idaho. D, 1968, Oregon State University. 177 p.

White, Willis, H. Geology of the Picture Gorge Quadrangle, Oregon. M, 1964, Oregon State University. 154 p.

Whitebread, Donald Harvey. Geology of the Pitkin area, Gunnison County, Colorado. M, 1951, University of Colorado.

Whited, Joseph Michael. Petrology, provenance, and tectonic significance of Upper Cretaceous Ohio Creek Member, Williams Fork Formation, Piceance Creek basin, Colorado. M, 1987, University of Utah. 145 p.

Whiteford, William B. Age relations and provenance of upper Paleozoic sandstones in the Golconda Allochthon; the Schoonover Sequence, northern Independence Mountains, Elko Co., Nevada. M, 1984, Stanford University. 126 p.

Whitehead, Ann Waybright. Petrology of the Beaufort Formation (Paleocene), Lenoir - Craven county line, North Carolina. M, 1981, University of North Carolina, Chapel Hill. 52 p.

Whitehead, Jack Waters. The petrology of the Sanford Basin Triassic sediments, North Carolina. M, 1962, University of Missouri, Columbia.

Whitehead, Mark L. Geology and talc occurrences of the Benson Ranch, Beaverhead County, Montana. M, 1979, Montana College of Mineral Science & Technology. 53 p.

Whitehead, N. H., III. The stratigraphy, sedimentology, and conodont paleontology of the Floyds Knob Bed and Edwardsville Member of the Muldraugh Formation (Valmeyeran), southern Indiana and Northcentral Kentucky, parts 1 and 2. M, 1976, University of Illinois, Urbana. 443 p.

Whitehead, Neil Harwood, III. Lithostratigraphy of the Grainger-Price Formation, northeastern Tennessee and southwestern Virginia, and depositional history of the Lower Mississippian of the East central United States. D, 1979, University of North Carolina, Chapel Hill. 214 p.

Whitehead, Robert Edgar. Application of rock geochemistry to problems of mineral exploration and ore genesis at Heath Steele mines, New Brunswick. D, 1973, University of New Brunswick.

Whitehead, W. L. Veins of Chanarcillo, Chile. D, 1918, Massachusetts Institute of Technology. 89 p.

Whitehurst, B. B. Duration magnitude of eastern United States earthquakes at World Wide Standard Seismograph Network stations. M, 1977, Virginia Polytechnic Institute and State University.

Whitehurst, Thomas M. Geology of the Muddy Run area, central Bath County, Virginia. M, 1984, University of North Carolina, Chapel Hill. 131 p.

Whiteley, Karen R. Geology of a portion of the Hayward 7 1/2′ Quadrangle, Alameda County, California. M, 1978, San Jose State University. 133 p.

Whitely, Joy L. Sediment supply and bedload transport; a regression model for mountain streams. M, 1980, [Colorado State University].

Whiteman, Edith J. Regional physiography of the capitol district of New York State; Albany, Cohoes, Troy, and Schenectady quadrangles. M, 1930, Columbia University, Teachers College.

Whitesides, Virgil Stuart, Jr. Surface geology of the Stidham area, McIntosh County, Oklahoma. M, 1957, University of Oklahoma. 109 p.

Whitfield, John M. The relationship between the petrology and the thorium and uranium contents of some granite rocks. D, 1958, Rice University. 169 p.

Whitfield, Merrick S., Jr. Pine Island Formation (Cretaceous) of North-central Louisiana. M, 1962, Louisiana Tech University.

Whitford, Arley C. Tertiary fossil woods of Nebraska. M, 1916, University of Nebraska, Lincoln.

Whitford, Stanley D. A regional study of the Muddy Sandstone in the Wind River basin, Wyoming. M, 1959, Northwestern University.

Whitford-Stark, James Leslie. A comparison of the origin and evolution of a circular and an irregular lunar mare. D, 1980, Brown University. 372 p.

Whitham, K. Laboratory scintillation counters applied to some geophysical problems. D, 1951, University of Toronto.

Whiting, Brian M. The Gulf of Suez; a new interpretation based upon subsidence modelling. M, 1984, University of North Carolina, Chapel Hill. 48 p.

Whiting, Francis B. An investigation of the Good Hope Mine, Hedley, British Columbia. M, 1948, McGill University.

Whiting, Francis B. Use of biotite for strontium age measurements. D, 1951, Massachusetts Institute of Technology. 118 p.

Whiting, James Freeman. Acoustic velocities in poorly consolidated clastic materials. M, 1982, University of Wisconsin-Madison. 46 p.

Whiting, Jerry Max. Counterstressing rock for ground support. D, 1968, Stanford University. 148 p.

Whiting, L. R. A gravity survey of the eastern portion of the Ross Ice Shelf, Antarctica. M, 1975, University of Wisconsin-Madison.

Whiting, Lester Le Roy. Rosiclare Sandstone, Cooks Mills area, Coles and Douglas counties, Illinois. M, 1958, University of Illinois, Urbana.

Whiting, Phillip Howard. Depositional environment of Red Fork sandstones, deep Anadarko Basin, western Oklahoma. M, 1982, Texas A&M University. 82 p.

Whiting, William Martin. A subsurface study of the post-Knox unconformity and related rock units in Morrow County, Ohio. M, 1965, Michigan State University. 100 p.

Whiting-McBride, Celia Kathleen. Small laccoliths and feeder dikes of the northern Adel Mountain volcanics. M, 1977, University of Montana. 74 p.

Whitla, Raymond E. The geology and ore deposits of the Lost Lake District, Boulder County, Colorado. M, 1939, University of Kansas. 82 p.

Whitlatch, George Isaac. A report on field and laboratory investigations of some clays and shales from Brown, Owen, Morgan, Putnam, Monroe, Clay and Greene counties, introduced by a review of the technology of ceramics. M, 1929, Indiana University, Bloomington. 135 p.

Whitlatch, George Isaac. The clay resources of Indiana. D, 1932, Indiana University, Bloomington. 444 p.

Whitley, Donald Lee. A stratigraphic and sedimentologic analysis of Cretaceous rocks in Northwest Iowa. M, 1980, University of Iowa. 81 p.

Whitley, Richard L. A petrographic, mineralogical, and geochemical comparison of some Precambrian granites near Butternut, Wisconsin. M, 1966, Bowling Green State University. 43 p.

Whitlock, C. H., III. Fundamental analysis of the linear multiple regression technique for quantification of water quality parameters from remote sensing data. D, 1977, Old Dominion University. 184 p.

Whitlock, James D. Utilizing the National Uranium Resource Evaluation (NURE) program's geochemical data; a case study of Billings Quadrangle, Montana/Wyoming. M, 1979, Montana College of Mineral Science & Technology. 100 p.

Whitlock, William W. Geology of the Steele Butte Quadrangle, Garfield County, Utah. M, 1984, Brigham Young University. 165 p.

Whitlow, Sallie Ida. Study of the silver apparent complexation capacity of the Susquehanna and Chenango rivers. M, 1982, SUNY at Binghamton. 83 p.

Whitman, Alfred Russell. Genesis of the ores of the Cobalt District, Ontario, Canada. D, 1921, University of California, Berkeley.

Whitman, Alfred Russell. The vadose synthesis of pyrite. M, 1913, University of California, Berkeley. 20 p.

Whitman, C. M. A hypothetical mathematical model of the benthic algal mat in Lake Bonney, Antarctica. D, 1976, Virginia Polytechnic Institute and State University. 116 p.

Whitman, Harry M. Geology of the Pearis Mountain area, Virginia. M, 1964, Virginia Polytechnic Institute and State University.

Whitman, Jill M. Tectonic and bathymetric evolution of the Pacific Ocean basin since 74 Ma. M, 1981, University of Miami. 168 p.

Whitman, Rick R. Physical stratigraphy of upper Miocene to lower Pleistocene sediment from the Amerasian Basin, Arctic Ocean. M, 1977, University of Wisconsin-Madison.

Whitmarsh, James Hardin. The geology and topography of the Bloomington Waterworks catchment. M, 1914, Indiana University, Bloomington.

Whitmore, Duncan. Chromium-bearing micas. M, 1940, Queen's University. 64 p.

Whitmore, Duncan Richard Elmer. Proterozoic rocks of the Squaw Lake-Woollett Lake area, west central Labrador. D, 1943, Princeton University. 111 p.

Whitmore, Frank C., Jr. Cranial morphology of some early Tertiary Artiodactyla. D, 1942, Harvard University.

Whitmore, Frank Clifford, Jr. Upper Silurian ostracode faunas from Nearpass Quarries, New Jersey. M, 1939, Pennsylvania State University, University Park. 92 p.

Whitmore, Janet Lynn. The vertebrate paleontology of the Late Cretaceous (Lancian) localities in the Lance Formation, northern Niobrara County, Wyoming. M, 1988, South Dakota School of Mines & Technology.

Whitmore, John Day. A gravimeter survey of a portion of Monroe County, Illinois. M, 1958, St. Louis University.

Whitmore, Randall Paul. Fossil Chiroptera from the Winnfield salt dome, Winnfield, Louisiana. M, 1987, Northeast Louisiana University. 70 p.

Whitney, B. L. Campanian-Maestrichtian and Paleocene dinoflagellate and acritarch assemblages from the Maryland-Delaware coastal plain. D, 1976, Virginia Polytechnic Institute and State University. 349 p.

Whitney, C. G. The paragenesis of synthetic phyllosilicates on the talc-phlogopite join. D, 1979, University of Illinois, Urbana. 237 p.

Whitney, C. Gene. Clay mineralogy of early-stage weathering products of ultramafic bedrock in alpine zones of the North Cascades, Washington, and Klamath Mountains, California. M, 1975, Western Washington University. 57 p.

Whitney, Dan D. Characterization of the non-Darcy flow coefficient in propped hydraulic fractures. M, 1988, University of Oklahoma. 126 p.

Whitney, Francis Luther. Bibliography and index of North American Mesozoic invertebrates. D, 1928, Cornell University.

Whitney, Francis Luther. Fauna of the Buda Limestone (Lower Cretaceous, Texas). M, 1911, Cornell University.

Whitney, Fred L. The Boulder Oil Field, Boulder County, Colorado. M, 1956, University of Colorado.

Whitney, Harold Tichenor, Jr. Plastic movement of soft clay in a sheeted excavation. D, 1969, Northwestern University. 129 p.

Whitney, James A. Partial melting relationships of three granitic rocks (Southeast New England). M, 1969, Massachusetts Institute of Technology. 143 p.

Whitney, James Arthur. History of granodioritic and related magma systems; an experimental study. D, 1972, Stanford University. 206 p.

Whitney, John W. Geology of the Heusser Mountain Pluton, White Pine County, Nevada. M, 1971, University of Nebraska, Lincoln.

Whitney, John Wallis. Paleomagnetism and rock magnetism of the Cretaceous Black Peak Batholith, North Cascades, Washington. D, 1975, University of Washington. 145 p.

Whitney, John Wilber. An analysis of copper production, processing, and trade patterns, 1950-1972. D, 1976, Pennsylvania State University, University Park. 161 p.

Whitney, John William. The geology and geomorphology of the Helmand Basin, Afghanistan. D, 1984, The Johns Hopkins University. 239 p.

Whitney, Marion Isabelle. Fauna of the (Lower Cretaceous) Glen Rose Formation. M, 1931, University of Texas, Austin.

Whitney, Marion Isabelle. Fauna of the Glen Rose Formation. D, 1937, University of Texas, Austin.

Whitney, Mark S. Geology of the Cumo prospect, Boise County, Idaho. M, 1975, Colorado School of Mines. 84 p.

Whitney, Olive Therese. Seismic-stratigraphic study of Early Cretaceous rocks, western Beaufort Sea, Alaska. M, 1978, University of California, Santa Cruz.

Whitney, Philip R. Rubidium-strontium geochronology of argillaceous sediments. D, 1962, Massachusetts Institute of Technology. 166 p.

Whitney, Richard Lee. Stratigraphy and structure of the northeastern part of the Tucson Mountains. M, 1957, University of Arizona.

Whitney, Robert A. Structural-tectonic analysis of northern Dixie Valley, Nevada. M, 1980, University of Nevada. 65 p.

Whitney, Sam Weslie. The influences of alkaline steam injection on the physical properties of formation rock and proppants. M, 1988, University of Oklahoma. 199 p.

Whiton, Geoffrey Arthur. Fossil plants applied to the dating of the Hazelton Group. M, 1962, University of British Columbia.

Whitsett, Charles K. Paleomagnetism of the Catoctin Formation, Skyline Drive, Virginia. M, 1978, University of Cincinnati. 97 p.

Whitsett, Robert M. Gravity measurements and their structural implications for the continental margin of southern Peru. D, 1976, Oregon State University. 82 p.

Whitson, David Neale. Geochemical stratigraphy of the Dooley rhyolite breccia and Tertiary basalts in the Dooley Mountain Quadrangle, Oregon. M, 1988, Portland State University. 122 p.

Whittaker, Peter J. Geology of the East central Port Coldwell Complex from the Pic River to Red Sucker Cove. M, 1979, McMaster University. 246 p.

Whittaker, Peter James. Geology and petrogenesis of chromite and chrome spinel in alpine-type peridotites of the Cache Creek Group, British Columbia. D, 1983, Carleton University. 339 p.

Whittall, Kenneth Patrick. Exploring magnetotelluric nonuniqueness using inverse scattering methods. D, 1987, University of British Columbia.

Whittall, Kenneth Patrick. Inversion of magnetotelluric impedances from above young lithosphere. M, 1982, University of British Columbia. 89 p.

Whittecar, George Richard, Jr. Geomorphic history and Pleistocene stratigraphy of the Pecatonica River valley, Wisconsin and Illinois. D, 1979, University of Wisconsin-Madison. 211 p.

Whittecar, George Richard, Jr. The glacial geology of the Waukesha Drumlin Field, Waukesha County, Wisconsin. M, 1976, University of Wisconsin-Madison.

Whittemore, Arthur Snow, III. Co-mingling of magmas in the Belknap Mountains Complex, central New Hampshire. M, 1988, University of Vermont. 201 p.

Whittemore, Donald Osgood. The chemistry and mineralogy of ferric oxhydroxides precipitated in sulfate solutions. D, 1973, Pennsylvania State University, University Park. 159 p.

Whittemore, Osgood James, Jr. Basin refractories from Washington olivine. M, 1941, University of Washington. 66 p.

Whitten, Christopher James. Depositional environment of downdip Yegua (Eocene) Sandstone, Jackson County, Texas. M, 1988, Texas A&M University. 191 p.

Whitten, Harriet Virginia. Consideration of soils with reference to certain soils in the vicinity of Austin, Texas. M, 1900, University of Texas, Austin.

Whittier, Donald A. Character and distribution of mineralization associated with magnetite bodies northeast of Aquila, Michoacan, Mexico. M, 1964, University of Arizona.

Whittier, William Harrison. Investigation of the iron ore resources of the Northwest. M, 1917, University of Washington. 128 p.

Whittington, David. Geology of the Stone Hill massive sulfide copper mine, Cleburne and Randolph counties, Alabama. M, 1982, University of Alabama. 143 p.

Whittington, David. Mesozoic diabase dikes of North Carolina. D, 1988, Florida State University. 258 p.

Whittles, Arthur Brice Leroy. The precise measurement of mass spectrometer. M, 1960, University of British Columbia.

Whittles, Arthur Brice Leroy. Trace lead isotope studies with gas source mass spectrometry. D, 1964, University of British Columbia. 215 p.

Whitton, Elliott. A study of foraminifera in the Clayton Formation in Wilcox County, Alabama. M, 1963, Florida State University.

Whorton, Chester Deward. The geology of a portion of the Mineola Dome of Montgomery County, Missouri. M, 1926, University of Missouri, Columbia.

Whyatt, J. K. Geomechanics of the Caladay Shaft. M, 1986, University of Idaho. 195 p.

Whyman, L. O. The Garber oil and gas field, Oklahoma. M, 1921, University of Nebraska, Lincoln.

Whynot, John David. Mineralogy and early diagenesis of deep Gulf of Mexico basin sediments. D, 1986, Texas A&M University. 126 p.

Whyte, James Bernard. Primary dispersion in wall rocks of the A1 and A2 zones of the Selbaie Deposit, Quebec. M, 1984, Queen's University. 322 p.

Wiberg, Leanne. The Hico Structure; a possible astrobleme in north-central Texas, U.S.A. M, 1981, Texas Christian University.

Wiberg, Patricia Louise. Mechanics of bedload sediment transport. D, 1987, University of Washington. 132 p.

Wicander, Edwin Reed. Diversity and abundance fluctuations in a late Devonian-early Mississippian phytoplankton assemblage from Northeast Ohio, U.S.A. D, 1973, University of California, Los Angeles.

Wice, Richard B. Tertiary volcanic rocks in Southwest Montana and adjacent Idaho as possible source rocks for epigenetic stratabound uranium deposits. M, 1982, Western Washington University. 97 p.

Wicke, Heather Dawn. Feasibility of recharging the Avra Valley Aquifer with Santa Cruz River flood flows. M, 1978, University of Arizona.

Wicke, Heather Dawn. Risk assessment for water quality management. D, 1983, University of Michigan. 273 p.

Wickenden, Robert T. D. The Upper Cretaceous foraminifera of the Praine provinces (Canada). D, 1931, Harvard University.

Wicker, Karen M. Recent changes in physiography of the Buffalo Cove, Atchafalaya Basin, Louisiana. M, 1975, Louisiana State University.

Wicker, Russell Alan. An investigation of the subsurface of peninsular Florida using gravity modeling techniques. M, 1977, University of Florida. 86 p.

Wickham, J. T. Glacial geology of North-central and western Champaign County, Illinois. M, 1976, University of Illinois, Urbana. 83 p.

Wickham, John S. Structural geology of the western slope of the Blue Ridge near Front Royal, Virginia. D, 1969, The Johns Hopkins University.

Wickham, Susan Specht. The Tiskilwa Till Member, Wedron Formation; a regional study in northeastern Illinois. M, 1979, University of Illinois, Urbana. 229 p.

Wickland, David Charles. Deposition, diagenesis, and porosity evolution of the upper Smackover Formation (Jurassic), Edgewood Fields, Van Zandt County, Texas. M, 1981, Texas Tech University. 120 p.

Wicklein, Phillip. Geology of the nickeliferous soapstone deposits (Cretaceous) of Saline County, Arkansas. M, 1967, University of Missouri, Rolla.

Wickliff, Diana. The effects of rainstorm events on the water chemistry of a river with emphasis on the heavy metal content. M, 1987, University of Arkansas, Fayetteville.

Wicklund, Mark A. Geology and genesis of the rocks of the Graeber uranium area, Stevens County, Washington. M, 1984, Eastern Washington University. 160 p.

Wicks, F. J. DTA (differential thermal analysis) of sediments of the Lake Agassiz Basin in Metro. Winnipeg. M, 1965, University of Manitoba.

Wicks, John L. A regional study of joints in the Fall River Formation (Cretaceous), Black Hills of South Dakota and Wyoming. M, 1980, University of Toledo. 74 p.

Wickstrom, Alden Eugene. Geological and engineering properties of fine-grained sands of eastern Iowa. M, 1956, Iowa State University of Science and Technology.

Wickstrom, Charles W. Geology of the eastern half of the Big Ridge Quadrangle, North Carolina. M, 1979, University of New Orleans.

Wickstrom, Conrad Eugene. Blue-green algal and ostracod interactions in an Oregon hot spring. D, 1974, University of Oregon. 220 p.

Wickstrom, Lawrence H. Geology of a Miocene felsic tuff and overlying basalts, southwestern Tushar Mountains, Utah. M, 1982, Kent State University, Kent. 61 p.

Wickwire, Daniel William. Biostratigraphy of the Lodgepole Limestone, Samaria Mountain, southeastern Idaho. M, 1984, Washington State University. 147 p.

Wickwire, Grant T. Physical features of the Triassic lowland of the New Haven, Middletown, and Meriden quadrangles, in Connecticut. M, 1929, Yale University.

Wickwire, Joy McIntosh. A pilot study to evaluate effects of Blackwood Creek channel restoration work on suspended sediment yield. M, 1979, University of Nevada. 89 p.

Widdicombe, Roberta E. Quaternary surficial features in the Delaware Basin, Eddy and Lea counties, New Mexico. M, 1979, University of New Mexico. 178 p.

Wideman, Charles. Strain steps associated with earthquakes. M, 1967, Colorado School of Mines. 38 p.

Widhelm, Sally. Geologic and environmental analysis for development purposes of Burns Ranch area, San Mateo County, California. M, 1976, Stanford University. 75 p.

Widjaja, Hadi. Scale and time effects in hydraulic fracturing. D, 1983, University of California, Berkeley. 224 p.

Widmann, Roy K. Episodic and seasonal effects of rainfall on spring water chemistry in a limestone terrane. M, 1982, University of Arkansas, Fayetteville.

Widmayer, Margaret A. Depositional model of the sandstone beds in the Tongue River Member of the Fort Union Formation (Paleocene), Decker, Montana. M, 1977, Montana State University. 123 p.

Widmayer, Ronald Edward. Feldspar geothermometry of the Hell Canyon Pluton, Boulder Batholith, Montana. M, 1978, Michigan State University. 47 p.

Widmer, Kemble. The geology of the Hermitage Bay area, Nfld. D, 1951, Princeton University. 459 p.

Widmier, John Michael. Mesozoic stratigraphy of the west-central Klamath Province; a study of eugeosynclinal sedimentation. D, 1963, University of Wisconsin-Madison. 141 p.

Widmier, John Michael. Paleozoic rocks along lower North Fork Canyon, Independence Mountains, Elko County, Nevada. M, 1960, Columbia University, Teachers College.

Widness, Scott E. The low-temperature geothermal resource of the Moses Lake, Ritzville-Connell area, east-central Washington. M, 1986, Washington State University. 357 p.

Widom, Elisabeth. Synneusis of olivine crystals in the 1959-60 eruption of Kilauea Volcano. M, 1984, Cornell University. 43 p.

Wie, William Arthur Van see Van Wie, William Arthur

Wiebe, Robert Alan. Structure and petrology of Ventana Cones area, California. D, 1966, Stanford University. 112 p.

Wiebe, Robert Alan. The geology of Mount Pilchuck. M, 1963, University of Washington. 53 p.

Wiechmann, Ferdinand G. Fusion structures in meteorites. D, 1882, Columbia University, Teachers College.

Wiechmann, Mark Jerdone. Mineralogy, petrology and structure of the metamorphosed Pennsylvanian sediments in the Providence area, Rhode Island - Massachusetts. M, 1979, Indiana University, Bloomington. 220 p.

Wieckowski, Miriam Anna. The stratigraphy, structure, and environmental interpretation of the Albion Group (Lower Silurian) in Coshocton County, Ohio. M, 1986, Kent State University, Kent. 112 p.

Wieczorek, G. F. Landslide susceptibility evaluation in the Santa Cruz Range, San Mateo County, California. D, 1978, University of California, Berkeley. 223 p.

Wieczorek, William Frederick. A pedogeomorphic analysis of first-order drainage basins in the Allegheny High Plateau. D, 1988, SUNY at Buffalo. 157 p.

Wied, Otto J. Geology of the Encampment area, Carbon County, Wyo. M, 1960, University of Wyoming. 52 p.

Wiedenheft, Judy Ann. Geohydrology of South Lake Tahoe, California; implications for land-use planning and resource management. M, 1980, University of California, Davis. 105 p.

Wiedenhoeft, G. R. Structural geology of parts of the Metaline District, northeastern Washington. M, 1986, Washington State University. 115 p.

Wiedey, Lionel William. Notes on the Vaqueros and Temblor formations of the California Miocene with descriptions of new species. M, 1928, Stanford University. 97 p.

Wiedey, Lionel William. The Oligocene and Miocene marine molluscan faunas of the Pacific Coast and their correlations. D, 1929, Stanford University. 547 p.

Wiedie, Alfred E., Jr. The stratigraphy and structure of the Parras Basin, Coahuila and Nuevo Leon, Mexico. D, 1961, Louisiana State University.

Wiedlin, Matthew Paul. An evaluation of field capacity as a parameter for groundwater recharge estimates. M, 1986, San Diego State University. 160 p.

Wiedman, Lawrence Alan. Community paleoecological study of the Silica Shale equivalent of northeastern Indiana. M, 1982, Wright State University. 63 p.

Wiedmann, John Philip. Geology of the upper Tularcitos Creek-Cachagua Creek area, Jamesburg Quadrangle, California. M, 1964, Stanford University. 80 p.

Wiedmann, Sebastian Paul. Geology of Sierra del Fraile and vicinity, San Luis Potosi, Mexico. M, 1979, University of Texas, Arlington. 102 p.

Wieg, Paul Kenneth. Deposition and diagenesis of Knox Group dolomites, Northeast Tennessee. M, 1987, Duke University. 192 p.

Wiegand, J. Patrick. Dune morphology and sedimentology at Great Sand Dunes National Monument. M, 1977, Colorado State University. 178 p.

Wiegand, Jeffrey W. Variation in composition of three granitic stocks associated with ore deposits (Lincoln Co., Nevada; Zacatecas, Mexico). M, 1961, Columbia University, Teachers College.

Wiegel, William E. A heavy mineral study of contemporary sands derived from the Atoka Formation in the Boston Mountains and the Fourche Mountains, Arkansas. M, 1959, University of Arkansas, Fayetteville.

Wiegman, Ronald W. Late Cretaceous and early Tertiary stratigraphy of the Little Mountain area, Sweetwater County, Wyoming. M, 1964, University of Wyoming. 53 p.

Wieland, Edward Paul. The geology of Toponce Creek area, Caribou County, Idaho. M, 1977, Idaho State University. 30 p.

Wieland, George R. Osteology of some fossil turtles. D, 1900, Yale University.

Wielchowsky, Charles Carl. Criteria for distinguishing Pleistocene (?) alluvial terrace deposits from the Coker Formation in the Cottondale, Alabama area. M, 1975, University of Alabama.

Wielchowsky, Charles Carl. The geology of the Brevard Zone and adjacent terranes in Alabama. D, 1983, Rice University. 298 p.

Wieluns, Robert. The thermal behavior of wollastonite. D, 1969, Rutgers, The State University, New Brunswick. 105 p.

Wiener, Jacky M. Geologic modeling in the Lake Wauberg-Chacala Pond vicinity utilizing seismic refraction techniques. M, 1982, University of Florida. 185 p.

Wiener, Leonard S. Structural geology of the Lone Mountain area, Grainger and Claiborne counties, Tennessee. M, 1965, University of Tennessee, Knoxville. 80 p.

Wiener, Richard Witt. Archean layered gabbro-anorthosite complex, Tessiuyakh Bay, Labrador. M, 1976, University of Massachusetts. 44 p.

Wiener, Richard Witt. Stratigraphy, structural geology, and petrology of bedrock along the Adirondack Highlands; Northwest Lowlands boundary near Harrisville, New York. D, 1981, University of Massachusetts. 144 p.

Wiengeist, Leo. The Ostracoda genus Eucytherura and its species from the Cretaceous and Tertiary of the Gulf Coast. M, 1948, Louisiana State University.

Wiens, Douglas Alvin. Oceanic intraplate seismicity; implications for the rheology and tectonics of the oceanic lithosphere. D, 1985, Northwestern University. 235 p.

Wiens, Roger Craig. Laboratory shock emplacement of gases into basalt, and comparison with trapped gases in shergottite EETA 79001. D, 1988, University of Minnesota, Minneapolis. 187 p.

Wier, Charles Eugene. Correlation of the upper part of Pennsylvanian rocks in southwestern Indiana. D, 1955, Indiana University, Bloomington. 110 p.

Wier, Charles Eugene. Geology and mineral deposits of the Jasonville Quadrangle, Indiana. M, 1950, Indiana University, Bloomington. 44 p.

Wier, Donald Raymond. Solubility of hydroxylapatite and some phosphate rocks. D, 1968, Iowa State University of Science and Technology. 185 p.

Wier, Karen E. Amphibolites in the Wissahickon Formation of the Pennsylvania-Delaware Piedmont region. D, 1962, Bryn Mawr College. 55 p.

Wier, Stuart Kirkland. Observations of long period surface wave multipathing at the high-gain long-period seismograph stations. D, 1977, Princeton University. 184 p.

Wierman, Douglas A. The hydrogeology of the Cedarburg Bog area. M, 1979, University of Wisconsin-Milwaukee. 101 p.

Wiersma, Cynthia Leigh. Relative rates of reaction of pyrite and marcasite with ferric iron at low pH. M, 1982, Virginia Polytechnic Institute and State University. 55 p.

Wiese, John Herbert. Geology and mineral resources of the Neenach Quadrangle, California. D, 1947, University of California, Los Angeles.

Wiese, John Herbert. The geology of a portion of the Santa Paula Quadrangle, California. M, 1941, University of California, Los Angeles.

Wiese, Larry Bruce. Pollen variability of the Two Creeks Forest Bed. M, 1979, University of Wisconsin-Madison.

Wiese, Robert George, Jr. Petrology and geochemistry of a copper-bearing Precambrian shale, White Pine, Michigan. D, 1961, Harvard University.

Wiese, Wolfgang. Geological setting and surficial sediments of Fatty Basin, a shallow inlet on the west coast of Vancouver Island, British Columbia. M, 1971, University of British Columbia.

Wiesenburg, Denis Alan. Geochemistry of dissolved gases in the hypersaline Orca Basin. D, 1980, Texas A&M University. 282 p.

Wiesnet, Donald R. A study of the Ordovician sediments along Deer River, New York. M, 1951, SUNY at Buffalo.

Wiethe, John David. The environmental geology of the Maryville Quadrangle (Tennessee). M, 1972, University of Tennessee, Knoxville. 82 p.

Wietrzychowski, Joseph. The geology of the Rhinebeck Northwest Quadrangle, New York. M, 1961, New York University.

Wigand, Peter Ernest. Diamond Pond, Harney County, Oregon; Man and marsh in the eastern Oregon desert. D, 1985, Washington State University. 280 p.

Wigbels, Frank B. Relation of character of subsurface water to geological structure. M, 1942, University of Pittsburgh.

Wigger, Stephen Thomas. A gravity investigation of mountain flank thrusting and normal faulting, Madison and Tobacco Root ranges, Montana. M, 1985, Western Michigan University.

Wiggett, G. J. Ecology, ethology and biostratigraphy of early metazoan faunas in eastern California. D, 1975, University of California, Davis. 347 p.

Wiggin, Roger Clay. Depositional environments of the Cambrian Ignacio Formation and Devonian pre-Elbert conglomerate, San Juan Mountains, southwestern Colorado. M, 1987, University of Texas, Austin. 246 p.

Wiggins, Brian Douglas. Volcanism and sedimentation of a Late Jurassic to Early Cretaceous Franciscan terrane near Bosley Butte, southwestern Oregon. D, 1980, University of California, Berkeley. 322 p.

Wiggins, John Henry, Jr. The effects of variations in the time delay between detonations on quarry blast vibrations. M, 1955, St. Louis University.

Wiggins, Lovell B. A reconnaissance investigation of chalcopyrite-sphalerite relationships in the Cu-Fe-Zn-S system. M, 1974, Virginia Polytechnic Institute and State University.

Wiggins, Peter Nelson, III. Geology of the Ham Gossett oil field, Kaufman County, Texas. M, 1953, Southern Methodist University. 12 p.

Wiggins, Virgil D. Palynomorph fossils from the Goddard Formation (Mississippian) of southern Oklahoma. M, 1962, University of Oklahoma. 123 p.

Wiggins, William David. Petrography and depositional environments of the Hale Formation in Madison County, northwestern Arkansas. M, 1978, Tulane University. 83 p.

Wiggins, William David, III. Depositional history and microspar development in reducing pore water, Marble Falls Limestone (Pennsylvanian) and Barnett Shale (Mississippian), central Texas. D, 1982, University of Texas, Austin. 145 p.

Wigginton, William Barclay. Subsurface stratigraphy and structure of Mississippian sediments of the Odon East oil field, Daviess County, Indiana. M, 1958, Pennsylvania State University, University Park. 108 p.

Wigglesworth, Edward. The geology of Martha's Vineyard, Massachusetts. D, 1917, Harvard University.

Wigglesworth, John Bradley. Geology and geochemistry of the Almanor and Glidden barite deposits, Northern California. M, 1984, University of Nevada. 103 p.

Wiggs, Calvin R. Depositional environment and tectonic significance of the Permo-Triassic Lykins Formation, Golden-Morrison area, Jefferson County, Colorado. M, 1986, Colorado School of Mines. 170 p.

Wight, David Clayton. Aneurophytalean progymnosperms from the Middle Devonian Millboro Shale of southwestern Virginia. D, 1985, University of Michigan. 194 p.

Wightman • Wilde

Wightman, D. The sedimentology and paleotidal significance of a late Pleistocene raised beach, Advocate Harbour, Nova Scotia. M, 1975, Dalhousie University. 157 p.

Wightman, Daryl M. Late Pleistocene glaciofluvial and glaciomarine sediments on the north side of the Minas Basin, Nova Scotia. D, 1980, Dalhousie University. 426 p.

Wightman, John F. Vein silica paragenesis in Triassic basalt, near Centerville, Digby Neck, Nova Scotia. M, 1970, Acadia University.

Wightman, R. H. Caving procedure at the Crestmore limestone mine of the Riverside Cement Company, Riverside County, Calif. M, 1949, University of Missouri, Rolla.

Wightman, Robert Bradford. Geology of Valentine area, Jeff Davis County, Texas. M, 1953, University of Texas, Austin.

Wigington, Parker Jamison, Jr. Occurrence and accumulation of selected trace metals in soils of urban runoff control structures. D, 1981, Virginia Polytechnic Institute and State University. 358 p.

Wigington, Richard James Stephen. The age and orthid fauna of the lower Whittaker Formation in the southern Mackenzie Mountains, Northwest Territories. M, 1977, University of Western Ontario. 148 p.

Wiginton, Randal Lynn. Earth structure beneath the Gulf of Mexico indicated by Rayleigh-wave dispersion. M, 1971, Texas Tech University. 74 p.

Wigley, Cynthia R. A radiolarian analysis of the Monterey Formation; paleoceanographic reconstructions of the Neogene California Current system. D, 1985, Rice University. 480 p.

Wigley, Cynthia R. Radiolaria in the Holocene sediment of the Gulf of Mexico and the basins of Southern California; assemblage changes with water depth and eutrophism. M, 1982, Rice University. 140 p.

Wigley, Perry Braswell, Jr. Conodonts from lower Middle Ordovician rocks of Russell and Scott counties, Virginia. D, 1967, Virginia Polytechnic Institute and State University.

Wigley, Perry Braswell, Jr. Geology of the Hemp area, Fannin County, Georgia. M, 1965, University of Virginia.

Wigley, William C. The glacial geology of the Copper Basin, Custer County, Idaho; a morphologic and pedogenic approach. M, 1976, Lehigh University. 106 p.

Wigmosta, Mark Steven. Rheology and flow dynamics of the Toutle debris flows from Mt. St. Helens. M, 1983, University of Washington. 184 p.

Wiig, Stephen Victor. Stratigraphy and depositional environment of the Des Moinesian rocks in a portion of Mahaska County, Iowa. M, 1976, University of Nebraska, Lincoln.

Wijayaratne, Rammali Devlina. Sorption of organic pollutants on natural estuarine colloids. D, 1982, University of Maryland. 154 p.

Wijeyawickrema, Chandrasi. Allocation of ground water rights on the basis of farmers' consumptive water needs. D, 1986, University of Oklahoma. 260 p.

Wijeyesekera, Sunil David. The molecular and crystal orbitals of transition metal clusters and solid state compounds; their relationship to the structure and properties of marcasite and arsenopyrite, transition metal carbides and triosmium clusters. D, 1983, Cornell University. 233 p.

Wikander, Frederick Gerdes. The eastern North American geosyncline in the Proterozoic era. M, 1949, University of Pittsburgh.

Wikoff, Penny Marie. A zeta potential study of Idaho phosphoria. M, 1978, University of Idaho. 100 p.

Wilband, John Truax. Geochemistry and origin of the Wheatland sericite deposits, Platte County, Wyoming. D, 1966, University of Illinois, Urbana. 113 p.

Wilband, John Truax. The acid volcanic complex of the Bourinot Group, Cape Breton Island, Nova Scotia. M, 1962, University of New Brunswick.

Wilbanks, John Randall. Geology of the Fosdick mountains, Marie Byrd land, west Antarctica. D, 1969, Texas Tech University. 246 p.

Wilbanks, John Randall. Zircons from the Copper Flat Intrusion, Hillsboro, New Mexico. M, 1966, Texas Tech University. 41 p.

Wilber, Robert Jude. Late Quaternary history of a leeward carbonate bank margin; a chronostratigraphic approach. D, 1981, University of North Carolina, Chapel Hill. 277 p.

Wilber, Robert Jude. Petrology of submarine-lithified hardgrounds and lithoherms from the deep flank environment of Little Bahama Bank (northeastern Straits of Florida). M, 1976, Duke University. 241 p.

Wilbert, Louis Joseph, Jr. Faunal zones in the Pendleton Formation. M, 1942, Louisiana State University.

Wilbert, Louis Joseph, Jr. The Jacksonian stage in southeastern Arkansas. D, 1951, University of Kansas. 161 p.

Wilbert, William Pope. Geology of Sierra de la Paila; Coahuila, Mexico. D, 1976, Tulane University. 242 p.

Wilbert, William Pope. Stratigraphy of the Georgetown Formation, central Texas. M, 1963, University of Texas, Austin.

Wilbrand, John T. The acid volcanic complex of the (Cambrian) Bourinot Group, Cape Breton Island, Nova Scotia. M, 1962, University of New Brunswick.

Wilbur, David Truxton. The crystal structure and chemical constitution of calcite and aragonite. D, 1932, Cornell University.

Wilbur, Doris Marion. The physiography of Millers River, Massachusetts and New Hampshire. M, 1929, University of Iowa. 94 p.

Wilbur, Doris Marion. The Pleistocene and Recent history of the Randolph, New York area. D, 1931, University of Iowa. 174 p.

Wilbur, J. Scott. Shawangunk Mountains, New York zinc-lead-copper veins; fluid inclusion, geochemical and isotope studies. M, 1986, Eastern Washington University. 65 p.

Wilbur, Lyman. The nature and origin of Quaternary basaltic volcanism in the central Mojave Desert area, California. M, 1980, San Jose State University. 133 p.

Wilbur, Robert Olas. Petrographic analyses of the insoluble residue of the Permian, Chase, and Council Grove limestones with regard to the origin of chert. M, 1956, Kansas State University. 88 p.

Wilburn, David R. Isothermal compression of spinel (Fe$_2$SiO$_4$) up to 100 kbar with the gasketed diamond cell. M, 1975, University of Rochester. 52 p.

Wilcken, Phyllis D. The (Mississippian) Brazer Formation in the Beck Spur area, central Wasatch Mountains (Utah). M, 1936, University of Utah. 55 p.

Wilcock, Myrtle Lavinia. The Taconic Mountains of New England. M, 1931, Boston University.

Wilcock, Peter Richard. Bed-load transport of mixed-size sediments. D, 1987, Massachusetts Institute of Technology. 205 p.

Wilcox, Charles R. Pennsylvanian geology of Appanoose County, Iowa. M, 1941, Iowa State University of Science and Technology.

Wilcox, Douglas Abel. The effects of deicing salts on water chemistry and vegetation in Pinhook Bog, Indiana. D, 1982, Purdue University. 156 p.

Wilcox, Floyd B. Origin of interstitial porosity in the (Devonian) Oriskany Sandstone. M, 1957, West Virginia University.

Wilcox, Fred H. The Rattlesnake Jack Mine; a gold mine at Galena, South Dakota. M, 1923, University of Minnesota, Minneapolis. 50 p.

Wilcox, Glenn Avery. The terraces of the upper San Francisco Bay region, with an appendix on the geology of the Suisun-Potrero Hills, California. M, 1911, University of California, Berkeley. 194 p.

Wilcox, John T. A basaltic breccia plug on East Grants Ridge, Valencia County, New Mexico. M, 1962, Columbia University, Teachers College.

Wilcox, John Thomas. The Grants Ridge uranium area, New Mexico. D, 1964, Columbia University, Teachers College. 112 p.

Wilcox, L. E. An analysis of gravity prediction methods for continental areas. D, 1974, University of Hawaii. 284 p.

Wilcox, L. E. An evaluation of areal gravity-elevation relations using covariance and empirical techniques. M, 1971, University of Hawaii. 118 p.

Wilcox, Lee Warren. Ultrastructural and phylogenetic studies of three freshwater dinoflagellates. D, 1984, University of Wisconsin-Madison. 120 p.

Wilcox, Margret Schnaitman. An Upper Devonian flora from central New York State. D, 1967, Cornell University. 83 p.

Wilcox, Marion. The Tertiary Cancellaridae of western North America. M, 1926, University of California, Berkeley. 114 p.

Wilcox, Ralph E., Jr. The geology of the Guardarraya region in the Sierra de San Lorenzo, Durango, Mexico. M, 1975, University of Texas at El Paso.

Wilcox, Ray E. Contact relations between rhyolite and basalt on Gardiner River, Yellowstone Park, Wyoming. D, 1941, University of Wisconsin-Madison.

Wilcox, Robert. Evidence of Early Cretaceous age for the Bissett Formation, western Glass Mountains, Brewster and Pecos counties, West Texas; implications for regional stratigraphy, paleontology and tectonics. M, 198?, Sul Ross State University.

Wilcox, Ronald E. Metadolerite dike swarm in Bakersville-Roan Mountain area, North Carolina. D, 1959, Columbia University, Teachers College.

Wilcox, Ronald Erwin. Structural geology of the Garden of the Gods, Colorado. M, 1952, Iowa State University of Science and Technology.

Wilcoxon, J. A. Calcareous nannoplankton from tropical Indo-Pacific deep-sea cores. D, 1977, University of Southern California.

Wilcoxon, James A. Distribution of foraminifera off the South Atlantic coast of the United States. M, 1962, University of Southern California.

Wilczynski, Michael W. Trace element concentrations as a potential tool for stratigraphic correlation of Cretaceous bentonites in Wyoming. M, 1985, Wayne State University. 53 p.

Wild, Jack. Types of structures in southern Saskatchewan. M, 1960, University of Saskatchewan. 68 p.

Wild, Robert C. Sedimentary analysis of the Middle Devonian Sylvania Sandstone in the Michigan Basin. M, 1958, Michigan State University. 61 p.

Wild, Thomas. Mixing and hybridization contrasted magmas in the Tigalak intrusion, Nain anorthosite complex, Labrador. D, 1985, University of North Carolina, Chapel Hill. 191 p.

Wildanger, Edward George. Sinkhole-type subsidence over abandoned coal mines in St. David, Illinois. M, 1980, University of Illinois, Urbana. 88 p.

Wilde, Edith M. Stratigraphy and petrography of the Fox Hills Formation in the Cedar Creek Anticline area of eastern Montana. M, 1984, Montana College of Mineral Science & Technology. 249 p.

Wilde, Garner Lee. Stratigraphic study of fusulinids from the Pennsylvanian formations of Parker County, Texas. M, 1952, Texas Christian University. 45 p.

Wilde, Pat. Recent sediments of the Monterey Deep-Sea Fan, Monterey, California. D, 1965, Harvard University.

Wilde, Roger D. Geology of the Stratosphere Bowl-Hayward area, Pennington and Custer counties, South Dakota. M, 1964, South Dakota School of Mines & Technology.

Wilde, Walter Rowland. K-feldspar megacrysts of the Cathedral Peak quartz monzonite, Yosemite National Park, California; 1 volume. M, 1971, University of California, Berkeley.

Wildeman, Joseph W. Geology of a portion of the Waltersburg Quadrangle, Pope County, Illinois. M, 1976, Southern Illinois University, Carbondale. 57 p.

Wilder, D. G. Insular biogeography; afroalpine areas of East Africa. D, 1977, University of Colorado. 129 p.

Wilder, Frank Alonzo. The age and origin of the gypsum deposits of Webster County, Iowa. D, 1902, University of Chicago. 28 p.

Wilder, Graham. The Pleistocene stratigraphy of the Eaton South Quadrangle, Ohio. M, 1987, Miami University (Ohio). 85 p.

Wilder, Newell Morris. Mountain bumps in the Middlesboro Syncline. M, 1934, University of Kentucky. 47 p.

Wilder, Nicea Trindade. Late Paleozoic lycopodiaceous megaspores of Brazil. D, 1980, University of Arizona. 105 p.

Wilder, Steven V. The groundwater hydrology of the Cedarburg Bog. M, 1973, University of Wisconsin-Milwaukee.

Wilder, William. Hydrogeology of the alluvial aquifer at Dilkon, Arizona. M, 1981, Northern Arizona University. 198 p.

Wilderman, Candie Caplan. The floristic composition and distribution patterns of diatom assemblages in the Severn River estuary, Maryland. D, 1984, The Johns Hopkins University. 702 p.

Wildharber, Jimmie L. Suspended sediment over the continental shelf (Southern California). M, 1966, University of Southern California.

Wildman, Ernest Atkins. Researches on the structure of anthracene. D, 1922, University of Illinois, Urbana.

Wildman, Sally Ann. Tectonic subsidence and evolution of the Gulf of Suez. M, 1985, University of North Carolina, Chapel Hill. 77 p.

Wildman, William Edwards. Serpentinite weathering and clay mineral formation in some California clays. D, 1967, University of California, Davis. 167 p.

Wildrick, Linton Leigh. Geochemical equilibria in Pleistocene sediments of the southeast Puget Sound drainage basin. M, 1976, University of Washington. 78 p.

Wildy, Vernon L. Diatomaceous Earth deposits (Calvert Formation, Cenozoic Era) in Richmond, Virginia. M, 1972, Virginia State University. 61 p.

Wiles, Greg. Deglaciation of the southern Catskills. M, 1987, SUNY at Binghamton. 86 p.

Wiles, Terry David. Modelling of hydraulic fracture propagation in a discontinuous rock mass using the displacement discontinuity method. D, 1986, University of Toronto.

Wiles, William W. Pore concentration of the planktonic foraminifer, Globigerina eggeri, as an index to Quaternary climates. D, 1960, Columbia University, Teachers College. 124 p.

Wiley, Bruce Henry. Foraminifera of the upper Eocene type Skookumchuck Formation and the Eo-Oligocene Lincoln Creek Formation, Wabash Traverse, Centralia, Washington. M, 1979, University of Washington. 151 p.

Wiley, Dennis Roy. Petrology of bituminous sandstones in the Green River Formation, (Eocene) southeastern Uinta Basin, Utah. M, 1967, University of Utah. various pagination p.

Wiley, Kenneth George. Hydrogeological investigation examining the identification and cleanup of chromium contaminated groundwater in Richland Township, Kalamazoo County, Michigan. M, 1984, Wright State University. 162 p.

Wiley, Michael Alan. Correlation of geology with gravity and magnetic anomalies, Van Horn-Sierra Blanca region, Trans-Pecos Texas. D, 1970, University of Texas, Austin. 381 p.

Wiley, Michael Alan. Stratigraphy and structure of the Jackson Mountain-Tobin Wash area, Southwest Utah. M, 1963, University of Texas, Austin.

Wiley, Michael T. Basal deposits of the Uinta Mountain Group, Brown's Park, Utah; an upper Precambrian fanglomerate. M, 1984, Purdue University. 114 p.

Wiley, Robert W. An iterative seismic ray-migration scheme for complex structures. D, 1980, Colorado School of Mines. 61 p.

Wiley, Samuel Rogers. A study of the Cretaceous Tertiary contact in northeastern Caldwell County, Texas. M, 1948, University of Texas, Austin.

Wiley, Thomas James. Pliocene shallow-water sediment gravity flows at Moss Beach, San Mateo County, California. M, 1983, Stanford University. 57 p.

Wiley, William Eldon. The middle Devonian Watt Formation of the Pine Point area, Northwest Territories. M, 1970, University of Saskatchewan. 150 p.

Wilgus, Cheryl Kathleen. A stable isotope study of Permian and Triassic marine evaporite and carbonate rocks, Western Interior, U.S.A. D, 1981, University of Oregon. 109 p.

Wilgus, Wallace LaFetra. Heavy minerals of the Dresbach Sandstone of western Wisconsin. D, 1933, University of Wisconsin-Madison.

Wilgus, Wallace LaFetra. The petrology of the Dresbach Sandstone of western Wisconsin. M, 1930, Washington University. 37 p.

Wilhelm, Philip Arthur, Jr. A shear wave velocity profile of the Mojave Block from teleseismic P-waveforms. M, 1985, SUNY at Binghamton. 48 p.

Wilhelm, Rudolph. Determination of the sedimentary thickness of the Sonora Geosyncline by Rayleigh wave dispersion. M, 1972, University of Texas at El Paso.

Wilhelm, Steven John. Utilization of Landsat thematic mapper for a lithologic and structural analysis of the Slick Hills area, southwestern Oklahoma. M, 1987, Texas Christian University. 47 p.

Wilhelms, Don Edward. Geology of part of the Nopah and Resting Spring ranges, Inyo County, California. D, 1963, University of California, Los Angeles. 297 p.

Wilhelms, Don Edward. The geology of the eastern portion of the Spring Garden Quadrangle, Plumas County, California. M, 1958, University of California, Los Angeles.

Wilhelmy, Jean-Francois. Etude minéralogique de l'altération météorique d'un till Laurentien en milieu tempéré froid (Forêt Montmorency, Parc des Laurentides, Québec). M, 1983, Universite Laval.

Wiliams, Winston L. Petrography and microfacies of the Devonian Guilmette Formation in the Pequop Mountains, Elko County, Nevada. M, 1984, Brigham Young University. 186 p.

Wilie, Enid Evelyn. The cultural influence of the Balcones fault zone (Texas). M, 1940, University of Texas, Austin.

Wilk, Charles Kenneth. A petrographic and geochemical investigation of a brown glass from Williams, Arizona. M, 1979, Wright State University. 46 p.

Wilk, Grant B. Interpretation and geochemical analysis of "Clinton" brines from Mahoning County, Ohio. M, 1987, University of Akron. 73 p.

Wilke, Kurtis Merle. The geology of the Garnet-Coloma area, Garnet Range, Montana. M, 1986, Iowa State University of Science and Technology. 169 p.

Wilkening, Laurel Lynn. On the early history of meteorites; evidence from glasses, from fossil particle tracks and from the noble gases. D, 1970, [University of California, San Diego]. 166 p.

Wilkening, Lee. Conodont biostratigraphy near the Mississippian/Pennsylvanian boundary in the New Well Peak section, Big Hatchet Mountains, Hidalgo County, N.M. M, 1984, New Mexico Institute of Mining and Technology. 214 p.

Wilkening, Richard Matthew. The geology and hydrothermal alteration of the Bear Creek Butte area, Crook County, central Oregon. M, 1986, Portland State University. 129 p.

Wilkens, R. H. Windborne volcanic ash from DSDP sites 84, 154A, and 158. M, 1977, SUNY at Binghamton. 87 p.

Wilkens, Roy Henry. Seismic properties and petrofabrics of ultramafic and mafic rocks from the Red Mountain ophiolite, New Zealand. D, 1981, University of Washington. 159 p.

Wilkenson, R. P. Environments of deposition of the Norphlet Formation (Jurassic) in South Alabama. M, 1981, University of Alabama. 141 p.

Wilkerson, Albert S. Study of the chemical-optical relations of the olivine group of minerals. M, 1924, University of Wisconsin-Milwaukee.

Wilkerson, Albert Samuel. Geology and ore deposits of the Magnolia mining district, Boulder County, Colorado. D, 1938, University of Michigan.

Wilkerson, Amy. Eclogite remnants in Purdy Hill Quadrangle, Mason County, Texas; P-T implications for the Llano Uplift. M, 1987, University of Texas, Austin. 116 p.

Wilkerson, Gregg. Geology of the Batopilas mining district, Chihuahua, Mexico (Volumes I and II). D, 1983, University of Texas at El Paso. 970 p.

Wilkerson, William Louis. The geology of the southern part of the High Steens Mountains, Oregon. M, 1958, University of Oregon. 89 p.

Wilkes, Jerry. A model for the extent and direction of the Wadena Lobe in Minnesota; a field study. M, 1968, University of Minnesota, Minneapolis. 31 p.

Wilkie, Lorna Christine. The conodont fauna of the Upper Ordovician McMicken Member, southwestern Ohio. M, 1957, Ohio State University.

Wilkins, Gary. Late-Quaternary vegetational history at Jackson Pond, Larue County, Kentucky. M, 1985, University of Tennessee, Knoxville. 172 p.

Wilkins, Jerry I. Geology of the Sulphur Lick Field in south-central Kentucky. M, 1983, University of Kentucky. 112 p.

Wilkins, Joe. An induced-polarization survey at Meteor Crater, Arizona. M, 1974, University of Arizona.

Wilkins, Richard Llewellyn. Stratigraphy and structure of the Shady Dolomite (Lower Cambrian), Linville Falls region, northern McDowell County, North Carolina. M, 1966, North Carolina State University. 31 p.

Wilkinson, Bruce H. Paleoecology of the Permian Ervay Member of the Park City Formation in north-central Wyoming. M, 1967, University of Wyoming. 140 p.

Wilkinson, Bruce Harvey. Matagorda Island; the evolution of a Gulf Coast barrier complex. D, 1974, University of Texas, Austin.

Wilkinson, Michael. Petrology and alteration in the core of the Bear Lodge Tertiary intrusive complex, Bear Lodge Mountains, Crook County, Wyoming. M, 1982, University of North Dakota. 127 p.

Wilkinson, Peter H. Test wall ultrastructure and minor element composition of Fusulina from the Ely Limestone, east-central Nevada. M, 1983, University of Georgia.

Wilkinson, Rex. A comparative investigation of Crowley's Ridge gravels in Poinsett and Craighead counties, Arkansas. M, 1982, Memphis State University.

Wilkinson, Sarah E. The geology of the northeast quarter of the Silk Hope Quadrangle, Carolina slate belt, North Carolina. M, 1978, University of North Carolina, Chapel Hill. 56 p.

Wilkinson, Stephen J. Geology and sulphide mineralization of the marginal phases of the Coldwell Complex, northwestern Ontario. M, 1983, Carleton University. 129 p.

Wilkinson, Thomas Allen. The subsurface geology of the northern Healdton area, Carter, Stephens, and Jefferson counties, Oklahoma. M, 1955, University of Oklahoma. 72 p.

Wilkinson, William Donald. The petrography of the Clarno Formation of Oregon with special reference to the Mutton Mountains. D, 1932, University of Oregon. 87 p.

Wilkinson • Williams

Wilkinson, William Holbrook, Jr. Geology of the Tres Montosas-Cat Mountain area, Socorro County, New Mexico. M, 1976, New Mexico Institute of Mining and Technology. 158 p.

Wilkinson, William Holbrook, Jr. The distribution of alteration and mineralization assemblages of the Mineral Park Mine, Mohave County, Arizona. D, 1981, University of Arizona. 144 p.

Wilks, Maureen E. The geology of the Steep Rock Group, northwestern Ontario; a major Archaean unconformity and Archaean stromatolites. M, 1986, University of Saskatchewan. 212 p.

Will, Thomas Michael. Structural investigations on experimentally and naturally produced slickensides. M, 1987, SUNY at Albany. 156 p.

Willams, Neil D. The effects of elevated temperature on the engineering properties of seafloor sediments. D, 1982, University of California, Berkeley. 622 p.

Willand, T. N. Mineralogy of the clay-and silt-size fractions of the St. Peter Sandstone from the subsurface in northern Illinois. M, 1976, Northern Illinois University. 104 p.

Willard, Allen Dale. Surficial geology of Bear Lake Valley, Utah. M, 1959, Utah State University.

Willard, Bradford. The stratigraphy and palaeontology of the Delaware water gap (New Jersey, Pennsylvania). D, 1923, Harvard University.

Willard, Gates. The age and origin of the lead-zinc ores of the Mississippi Valley and eastern Tennessee. M, 1953, University of Michigan.

Willard, James S. Geology and sandstone petrography of the Puerto El Alamo area, northeastern Sonora, Mexico. M, 1988, University of Cincinnati. 250 p.

Willard, Jane M. Regional directions of ice flow along the southwestern margin of the Laurentide Ice Sheet as indicated by distribution of Sioux Quartzite erratics. M, 1980, University of Kansas. 87 p.

Willard, Max E. The mineralization at the Polaris Mine, Shoshone County, Idaho. M, 1938, University of Idaho. 21 p.

Willard, Parry Don. Tertiary igneous rocks of northeastern Cache Valley (Franklin County), Idaho. M, 1972, Utah State University. 54 p.

Willard, Paul D., Jr. The use of microchemistry in the determination of minerals. M, 1937, University of Minnesota, Minneapolis. 66 p.

Willard, Robert Jackson. The geology of the Kennebago Lake Quadrangle [Maine]. D, 1958, Boston University. 370 p.

Willcox, R. E., Jr. The geology of the Guardarraya region in the Sierra de San Lorenzo, Durango, Mexico. M, 1975, University of Texas at El Paso.

Willden, Charles Ronald. Geology of the Jackson Mountains, Humboldt County, Nevada. D, 1960, Stanford University. 146 p.

Willden, Charles Ronald. The nature of the igneous-sediment contact in the U.S. Mine, Bingham, Utah. M, 1952, University of Utah. 41 p.

Wille, Douglas Michael. The COCORP Surrency bright spot; possible evidence for fluid in the deep crust. M, 1986, Cornell University. 46 p.

Willemann, Raymond James. Deflection of spherical planetary lithospheres. D, 1986, Cornell University. 144 p.

Willes, Sidney Blaine. The mineral alteration products of Keetley Kamas area, Utah. M, 1962, Brigham Young University. 28 p.

Willett, Sean D. Spatial variation of temperature and thermal history of the Uinta Basin. D, 1988, University of Utah. 121 p.

Willette, P. D. A structural, stratigraphic and hydrodynamic analysis of the Paleozoic section of central New York State. D, 1979, Syracuse University. 138 p.

Willette, P. D. The feasibility of automatic classification of surficial deposits using digital MSS Landsat data. M, 1978, Syracuse University.

Willey, J. D. Physical chemistry of silica in seawater and marine sediments. D, 1975, Dalhousie University.

Willhour, Robert R. Structure and stratigraphy of a part of the Front-Laramie Range foothills in northern Colorado (Larimer County) and southern Wyoming. M, 1958, University of Colorado.

William, I. A. Some consideration on the fusibility and vitrifying temperatures of common clays. M, 1902, Iowa State University of Science and Technology.

William, Owen R. Hydraulic conductivity of mountain soils. M, 1976, [Colorado State University].

Williams, A. L. Heavy metal concentrations in the waters of the Springfield area, Missouri. M, 1973, University of Missouri, Rolla.

Williams, Adele. Double-ringed and multiringed basins on the Moon and Mercury; morphologic correlations with size and other parameters. M, 1982, University of Houston.

Williams, Alan Evan. Investigation of oxygen-18 depletion of igneous rocks and ancient meteoric-hydrothermal circulation in the Alamosa River Stock region, Colorado. D, 1980, Brown University. 284 p.

Williams, Albert Earl. Geology of the southeast quarter of the Chunky, Mississippi, Quadrangle. M, 1966, Mississippi State University. 105 p.

Williams, Anthony Brackett. A graphic display of plate tectonics in the Tethys Sea. M, 1972, Massachusetts Institute of Technology. 99 p.

Williams, Arthur James. Physiographic history of the "driftless area" of Iowa. D, 1923, University of Iowa. 116 p.

Williams, Arthur James. Physiographic studies in and around Dubuque, Iowa. M, 1914, University of Iowa. 52 p.

Williams, Barbara Jean Radovich. Source parameters for large earthquakes from high gain long period body wave spectra. D, 1977, University of Michigan. 280 p.

Williams, Barbara L. An investigation of the application of ^{29}Si magic angle spinning nuclear magnetic resonance in geology. M, 1984, Brock University. 166 p.

Williams, Barrett J. Glacial geology of south-central Kidder County, North Dakota. M, 1960, University of North Dakota. 91 p.

Williams, Billye Roan. A petrographic study of the sandstone members of the Simpson Group in western Garvin County, Oklahoma. M, 1957, University of Oklahoma. 65 p.

WIlliams, Bobby G. Landsat and geochemical investigation, Reagan and Crockett counties, Texas. M, 1985, University of Texas of the Permian Basin. 160 p.

Williams, C. T. Late Tertiary mammalian faunas of the San Francisco Bay region, California. D, 1976, University of California, Berkeley. 271 p.

Williams, Carol Alvis. Palynology of Upper Cretaceous and lower Tertiary strata from the northern Raton Basin, south-central Colorado. D, 1984, University of Colorado. 313 p.

Williams, Cassandra. Age and correlation of lens in the Gurabo Formation, Dominican Republic. M, 1979, Tulane University.

Williams, Charles Coburn. Comparison of internal and external morphologic features of some Kansas fusulinids. M, 1941, University of Kansas. 55 p.

Williams, Charles D. Recent landscape transformation in Northeast Louisiana. M, 1972, Louisiana State University.

Williams, Charles Dudley. Pre-Permian geology of the Pratt Anticline area in south central Kansas. M, 1968, Wichita State University. 116 p.

Williams, Charles E. Computer program for plotting earthquake hypocenters in subduction zones. M, 1973, New Mexico Institute of Mining and Technology.

Williams, Charles Enyart. The economic potential of the Lower Hartshorne coal on Pine Mountain, Heavener, Oklahoma. M, 1978, Oklahoma State University. 109 p.

Williams, Charles R. Geology of the Franconia region in New Hampshire. D, 1934, Harvard University.

Williams, Christopher. Geology of the Nomlaki Tuff and other silicic ashflows of the Tuscan Formation, Northern California. M, 1978, University of California, Santa Barbara.

Williams, Clifford Ralph. Alteration of the Study Butte intrusive, Terlingua, Brewster County, Texas. M, 1967, Texas Tech University. 85 p.

Williams, D. F. Planktonic foraminiferal paleoecology in deep-sea sediments of the Indian Ocean. D, 1976, University of Rhode Island. 297 p.

Williams, D. R. Riverbank stability at Ottawa. D, 1979, Queen's University.

Williams, D. Sam. Saltwater upconing potential in the Oscotillo-Coyote Wells Basin, Imperial County, California. M, 1986, San Diego State University. 128 p.

Williams, Daniel Bernhard. Structural and geochemical study of the South Sulphur asphalt deposits, Murray County, Oklahoma. M, 1983, University of Oklahoma. 163 p.

Williams, Daryll Wayne. The Anna, Ohio earthquake zone and the establishment of the Anna gravity network. M, 1976, University of Michigan.

Williams, David. Chromite deposits of Mountain Lake, Mistassini Territory, Quebec. M, 1965, Universite Laval.

Williams, David A. A general coal reflectance study of the Eastern Kentucky Coalfield. M, 1979, Eastern Kentucky University. 38 p.

Williams, David E. Wave propagation effects observed in aftershock waveforms of the January 9, 1982 Miramichi, New Brunswick earthquake. M, 1986, Pennsylvania State University, University Park. 72 p.

Williams, David L. The geochemical evolution of saline groundwater within a fresh water aquifer south of Oakes, North Dakota. M, 1984, University of North Dakota. 328 p.

Williams, David Lee. Heat loss and hydrothermal circulation due to sea-floor spreading. D, 1974, Massachusetts Institute of Technology. 139 p.

Williams, David M. Geology and paleoenvironment of the Zilpha Formation in North central Mississippi. M, 1980, Memphis State University.

Williams, David Richard. Regional structure and tectonics of Venus inferred from admittance analysis of gravity and topography. D, 1987, University of California, Los Angeles. 290 p.

Williams, Dennis E. Preliminary geohydrologic study of a portion of the Owens valley ground water reservoir (Inyo county, California). D, 1969, New Mexico Institute of Mining and Technology. 194 p.

Williams, Dennis E. Viscous-model study of groundwater flow in a wedge-shaped aquifer. M, 1965, New Mexico Institute of Mining and Technology. 59 p.

Williams, Dennis S. Ostracoda of the Yorktown Formation (Miocene), (Petersburg, Virginia). M, 1970, Virginia State University. 68 p.

Williams, Donald Lee. Interaction between landfill leachates and carbonate-derived residual soils. M, 1974, University of Missouri, Columbia.

Williams, Donald Roy. Clastic dikes and chalcedony veins in the White River badlands of Nebraska and South Dakota. M, 1962, University of Illinois, Urbana. 81 p.

Williams, Donna Jo. Mining-related and tectonic seismicity in the East Mountain area, Wasatch Plateau, central Utah. M, 1987, University of Utah. 130 p.

Williams, Duane H. A study of atmospheric space-charge density near the surface of the Earth. M, 1958, New Mexico Institute of Mining and Technology. 33 p.

Williams, Dwight Drue. An investigation of nitrogen contamination in the groundwater due to rural prac-

874

Bibliography of Geoscience Theses

tices in the Four-Mile Creek drainage basin of Preble County, Ohio. M, 1988, Wright State University. 161 p.

Williams, E. Joy. Precambrian plate tectonics; a geodynamic approach. M, 1986, Carleton University. 82 p.

Williams, Edmund Jay. Geomorphic features and history of the lower part of Logan Canyon, Utah. M, 1964, Utah State University. 64 p.

Williams, Edwin Philip. Geology of the Cardston area, Alberta, Canada. D, 1956, Harvard University.

Williams, Edwin Philip. Hanging wall quartzites, Sullivan Mine. M, 1942, University of British Columbia.

Williams, Ella Ruth Tews. Origin of the late-Paleozoic plutonic massifs of Morocco. M, 1975, Michigan State University. 34 p.

Williams, Ernest B. An investigation of chloride contamination in Tuscarawas and Muskingum River valleys, Ohio. M, 1973, Ohio State University.

Williams, Eugene Griffin. Stratigraphy of the Allegheny Series of the Clearfield Basin, Clearfield County, Pennsylvania. D, 1957, Pennsylvania State University, University Park. 568 p.

Williams, Eugene Griffin. The geology of the Cranberry Lake area, Maine. M, 1952, University of Illinois, Urbana.

Williams, Eula C. Tectonic influence on the Atoka shelf-to-basin transition Arkoma Basin and frontal Ouachitas, Oklahoma. M, 1984, University of Missouri, Columbia. 102 p.

Williams, F. J. Devonian stratigraphy of southern Saskatchewan. M, 1952, University of Saskatchewan. 58 p.

Williams, F. W. Hydrogeologic investigation of Boston and Northampton townships, Summit County, Ohio. M, 1983, Kent State University, Kent. 349 p.

Williams, Floyd Elmer. Geology of the North Selma Hills area, Utah County, Utah. M, 1951, Brigham Young University. 63 p.

Williams, Floyd J. Structural control of uranium deposits, Sierra Ancha region, Gila County, Arizona. D, 1959, Columbia University, Teachers College.

Williams, Floyd James. The Geology of the Stevens Mine, Clear Creek County, Colorado. M, 1951, Colorado School of Mines. 43 p.

Williams, Frank Brierley. Characterization of mixed-layer mica-montmorillonites. D, 1963, Washington University. 133 p.

Williams, Frank E. Fiords of southeastern Alaska. M, 1912, University of Wisconsin-Madison.

Williams, Frank E. Industrial development in the Schuylkill Valley. D, 1928, University of Wisconsin-Madison.

Williams, Frederick D. Water pollution control training; the educational role of the United States Environmental Protection Agency. D, 1979, University of Cincinnati. 204 p.

Williams, Frederick Enslow. Fusulinid fauna of the Naco Limestone, near Bisbee, Arizona. M, 1951, University of Illinois, Urbana.

Williams, Frederick M. G. Structural studies in the Rioux Quarry, Cowansville, Quebec. M, 1966, McGill University.

Williams, Garnett P. Some aspects of the eolian saltation load. D, 1963, University of Chicago. 96 p.

Williams, George A. The coal deposits and Cretaceous stratigraphy of the western part of the Black Mesa, Arizona. D, 1951, University of Arizona.

Williams, George K. An investigation of the iron deposits in the East River mountain district. M, 1958, Virginia Polytechnic Institute and State University.

Williams, George K. Geology and geochemistry of the sedimentary phosphate deposits of northern peninsular Florida. D, 1971, Florida State University.

Williams, George Quigley. Petrography and stratigraphy of the (Devonian) Portage Group in the Honeoye Quadrangle, New York. M, 1947, University of Rochester. 89 p.

Williams, George Quigley. The stratigraphy of the type Ithaca Formation (Upper Devonian, New York). D, 1951, Cornell University.

Williams, Gerald. Transition from ductile to brittle deformation at the head of the Deerfield Basin, Bernardston-Leyden area, Massachusetts. M, 1979, University of Massachusetts. 106 p.

Williams, Gordon Donald Clarence. The Mannville Group, central Alberta. D, 1960, University of Alberta. 106 p.

Williams, H. E. An analysis of precipitation patterns and trends in the North American desert region. D, 1979, Arizona State University. 344 p.

Williams, Harold. A petrographic study of the metamorphic rocks of the Chisel Lake area, northern Manitoba. D, 1961, University of Toronto.

Williams, Harold. Tilting igneous complex, Fogo District, Newfoundland. M, 1957, Memorial University of Newfoundland. 63 p.

Williams, Harold H. Fluorine, chlorine and bromine in carbonate rocks in relation to the chemical history of ocean water and dolmitization. D, 1969, McMaster University. 173 p.

Williams, Harold H. Some aspects of ion exchange in shales. M, 1967, University of Calgary. 182 p.

Williams, Harold L. The subsurface correlation of certain Pennsylvanian limestones between Oklahoma City and the vicinity of Tulsa. M, 1941, University of Oklahoma. 101 p.

Williams, Harriet L. Some problems concerning the gneisses in the vicinity of Ossining (New York). M, 1936, Columbia University, Teachers College.

Williams, Harry F. L. Sea-level change and delta growth; Fraser Delta, British Columbia. D, 1988, Simon Fraser University. 256 p.

Williams, Harry Franklin. Geology of the Kent area, Culberson County, Texas. M, 1949, University of Texas, Austin.

Williams, Higbee George. Geology and ore deposits of an area east of Warm Springs, Montana. M, 1951, Montana College of Mineral Science & Technology. 73 p.

Williams, Ian Stephen. An investigation of some aspects of geomagnetic reversals. D, 1982, University of California, Santa Barbara. 315 p.

Williams, Ira A. The comparative accuracy of the methods for determining the percentages of the several components of an igneous rock. M, 1936, Columbia University, Teachers College.

Williams, Jack Edward. The depositional environments of the Entrada Sandstone (Jurassic) from Beulah, Pueblo County, Colorado, to the Cimarron River canyon, Union County, New Mexico. M, 1987, Fort Hays State University. 57 p.

Williams, Jack Riley. A petrographic study of the subsurface Gallup Sandstone of San Juan County, New Mexico. M, 1956, Texas Tech University. 114 p.

Williams, James D. Variability study of carbonate rocks in Lost Burro Gap, Panamint Mountains, Death Valley, California. M, 1965, Pomona College.

Williams, James Dana. The petrography and differentiation of a composite sill from the San Rafael Swell region, Utah. M, 1983, Arizona State University. 123 p.

Williams, James Frank, III. Location of land areas Appropriate for the disposal of wastewaters, central Canyon County, Idaho. M, 1973, University of Idaho. 127 p.

Williams, James H. Classification of surficial materials. D, 1975, University of Missouri, Rolla. 376 p.

Williams, James Hadley. The geology of the New London area, Ralls County, Missouri. M, 1952, University of Missouri, Columbia.

Williams, James Jerome. Geology of part of the Orocopia Mountains, Riverside County, California. M, 1957, University of California, Los Angeles.

Williams, James Oliver. Reconnaissance geochemistry of the West Mountain area, west-central Idaho. M, 1977, University of Idaho. 238 p.

Williams, James Steele. The Louisiana Limestone of northeastern Missouri and its fauna. D, 1924, University of Missouri, Columbia.

Williams, James Steele. The variations in the Stropheodontas of the Snyder Creek Shale. M, 1922, University of Missouri, Columbia.

Williams, James Stewart. The fauna and stratigraphy of the (Upper Devonian) Tully Limestone of New York. D, 1932, George Washington University.

Williams, Jefferson Boone. An EMAP survey of the southern Wind River Overthrust, Wyoming. M, 1988, University of Texas, Austin. 170 p.

Williams, Jeffrey R. A geochemical investigation of the metasomatized ultramafic bodies in the Winston-Salem Quadrangle, northwestern North Carolina–southwestern Virginia. M, 1979, Miami University (Ohio). 147 p.

Williams, John G. Sedimentary petrology of the Kessler Limestone, Washington and Crawford counties, Arkansas. M, 1975, University of Arkansas, Fayetteville.

Williams, John Herbert. The hydrogeology of the Danville area, Pennsylvania. M, 1980, Pennsylvania State University, University Park. 251 p.

Williams, John Raynesford. Micro-fauna of the Weches Formation (Eocene) at Smithville, Texas. M, 1938, University of Illinois, Urbana.

Williams, John S. The relative contribution of local and regional atmospheric pollutants to lake sediments in northern New England. M, 1980, University of Maine. 59 p.

Williams, John Stuart and Frezon, Sherwood E. Cambrian stratigraphy and paleontology in the Teton Mountains, Teton County, Wyoming. M, 1963, University of Michigan.

Williams, John T. A theory of induced polarization for an idealized metal-solution interface. M, 1977, Colorado School of Mines. 108 p.

Williams, John Thomas. Silurian conodont biostratigraphy and depositional environments of the Hidden Valley Dolomite, Inyo County, California. M, 1980, San Diego State University.

Williams, John Wharton. Geomorphic history of Carmel valley and Monterey peninsula, California. D, 1970, Stanford University. 135 p.

Williams, Jonathan D. Provenance and depositional controls on the diagenesis of sandstones associated with the Laney Member of the Green River Formation, southwestern Wyoming. M, 1987, University of Wyoming. 74 p.

Williams, Joseph D. The petrology and petrography of sediments from the Sigsbee blanket, Yucatan Shelf, Mexico. M, 1963, Texas A&M University.

Williams, Karl Wendel. Depositional dynamics of the Queen Formation (Permian), New Mexico and Texas. M, 1967, Texas Tech University. 107 p.

Williams, Karl Wendel. Principles of cementation, environmental framework, and diagenesis of the Grayburg and Queen formations (Permian), New Mexico and Texas. D, 1969, Texas Tech University. 198 p.

Williams, Kathleen Marie. The Mount Diablo Ophiolite, Contra Costa County, California. M, 1983, San Jose State University. 156 p.

Williams, Kenneth E. The geology of the western part of Alachua County, Florida. M, 1974, University of Florida. 149 p.

Williams, Kerstin Margareta. Late Quaternary paleo-oceanography of the Baffin Bay region, based on diatoms. D, 1988, University of Colorado. 283 p.

Williams, Kerstin Margareta. Marine diatom assemblages from Baffin Bay and Davis Strait. M, 1984, University of Colorado. 111 p.

Williams, Larry D. Petrology and petrography of a section across the Bitterroot Lobe of the Idaho Batholith. D, 1977, University of Montana. 221 p.

Williams, Larry Dean. Neogeological landforms and neoglacial chronology of the Wallowa Mountains, northeastern Oregon. M, 1974, University of Massachusetts. 97 p.

Williams, Lawrence Darryl. Some factors influencing cirque glacierization on eastern Cumberland Peninsula, Baffin Island, Canada. M, 1972, University of Colorado.

Williams, Leonora May. A restudy of the (Lower Cretaceous) Comanche Peak Formation (Texas). M, 1929, Texas Christian University.

Williams, Loretta Ann. The sedimentational history of the Bear Gulch Limestone (Middle Carboniferous, central Montana); an explanation of "how them fish swam between them rocks". D, 1981, Princeton University. 251 p.

Williams, Lou Page. The physiography of the New-Kanawha river system (West Virginia, Ohio). M, 1935, University of Chicago.

Williams, Lou Page. The surface textures of sedimentary particles and grains. D, 1947, University of Chicago. 266 p.

Williams, Lyman O'Dell, Jr. Geology of the Crazy Woman Canyon-Powder River Pass-Tensleep Canyon area, Bighorn Mountains, Wyoming. D, 1962, University of Iowa. 325 p.

Williams, Marguerite Thomas. A study of the history of erosion in the Anacostia drainage basin (District of Columbia). D, 1942, Catholic University of America. 59 p.

Williams, Martin C. Magnetic properties of the Archean, Pikwitonei crustal cross-section, Manitoba, Canada. M, 1985, University of Wyoming. 83 p.

Williams, Mary Louise. Two geologic field trips in the Lynchburg Quadrangle, Virginia. M, 1971, Virginia State University. 40 p.

Williams, Matt B. Sedimentology and stratigraphy of the Queen Formation, Millard Field, Pecos County, Texas. M, 1984, Texas A&M University. 51 p.

Williams, Maurice. Bedrock geology of the Drakes Creek fault complex between Huntsville and Drakes Creek, Arkansas. M, 1965, University of Arkansas, Fayetteville.

Williams, Merton Y. Geology of the Arisaig-Antigonish District, Nova Scotia. D, 1912, Yale University.

Williams, Michael E. Origin of spiral coprolites from central Kansas. M, 1972, University of Kansas. 49 p.

Williams, Michael E. The "cladodont level" sharks of the Pennsylvanian black shales of central North America. D, 1979, University of Kansas. 279 p.

Williams, Michael J. Origin of recent carbonate sediments in the Blackwater River system, Ten Thousand Islands, Florida. M, 1982, University of South Florida, Tampa. 92 p.

Williams, Michael L. Stratigraphic, structural and metamorphic relationships in Proterozoic rocks from northern New Mexico. D, 1987, University of New Mexico. 138 p.

Williams, Michael Lloyd. Geology of the copper occurrence at Copper Hill, Picuris Mountains, New Mexico. M, 1982, University of Arizona. 98 p.

Williams, Michael M. Geology for land use planning, Sun Valley-Ketchum, Idaho. M, 1976, University of Idaho. 95 p.

Williams, Michael W. Distribution of foraminifera in Anclote Bay, Florida. M, 1977, Bowling Green State University. 93 p.

Williams, Mona S. Factors influencing the external morphology of stromatoporoids in the Upper Silurian (Pridoli) of central New York. M, 1987, University of Alabama. 87 p.

Williams, Myron T. Correlation and stratigraphy of certain beds near Harrison, Montana. M, 1928, University of Iowa. 130 p.

Williams, N. The formation of sedimentary-type stratiform sulfide deposits. D, 1976, Yale University. 353 p.

Williams, Nancy Susan. Stratigraphy and structure of the east-central Tendoy Range, southwestern Montana. M, 1984, University of North Carolina, Chapel Hill. 104 p.

Williams, Nelson Noel. Pollen analysis of two central Ohio bogs. D, 1962, Ohio State University. 72 p.

Williams, Norman Charles. Wall rock alteration, Mayflower Mine, Park City, Utah. D, 1952, Columbia University, Teachers College.

Williams, Pamela K. Middle Atoka depositional systems of the southwestern Arkoma Basin, Arkansas. M, 1985, University of Arkansas, Fayetteville. 53 p.

Williams, Paul. Plagioclase zoning profiles of a small pluton in Arizona. M, 1981, University of Arizona. 46 p.

Williams, Paul Lincoln. Glacial geology, Stanley Basin, Idaho. M, 1957, University of Washington. 67 p.

Williams, Paul Lincoln. Stratigraphy and petrography of the Quichapa Group (Oligocene?), southwestern Utah and southeastern Nevada. D, 1967, University of Washington. 139 p.

Williams, Paula S. Lithostratigraphy and depositional systems of the middle Atoka Formation, Arkoma Basin. M, 1983, University of Arkansas, Fayetteville. 88 p.

Williams, Peter Montague. Organic acids found in Pacific Ocean waters. D, 1960, University of California, Los Angeles. 74 p.

Williams, Philip Anthony. Geology along the south flank of the Owl Creek Mountains, Wyoming. M, 1941, University of Missouri, Columbia.

Williams, Quentin Christopher. Physics of high-pressure melts. D, 1988, University of California, Berkeley. 93 p.

Williams, R. Dave. Recent surface sediments of Redfish Lake Creek, Fishhook Creek, and Redfish Lake, Sawtooth Range, Custer County, Idaho. M, 1973, Idaho State University. 100 p.

Williams, R. M. Contact metamorphism of the (Precambrian or Cambrian) Ellsworth Schist at Bluehill, Maine. M, 1927, Massachusetts Institute of Technology. 48 p.

Williams, Richard. Miocene volcanism in the central Conejo Hills, Ventura County, California. M, 1977, University of California, Santa Barbara.

Williams, Richard C. The Shenandoah limestones of the Hagerstown Quadrangle. D, 1912, The Johns Hopkins University.

Williams, Richard E., Jr. Remote sensing techniques applied to mineral exploration in the heavily vegetated terrain of the Reading Prong of New York and New Jersey. M, 1979, Stanford University. 90 p.

Williams, Richard H. Provenance study of late Cenozoic sand and gravel deposits in northeastern Nebraska. M, 1984, University of Nebraska, Lincoln.

Williams, Richard John. Reaction constants in the system Fe-MgO-SiO$_2$-O$_2$, between 1300° and 900° at one atmosphere; theory experiment, and application. D, 1970, The Johns Hopkins University. 232 p.

Williams, Richard Llewellyn. A method to evaluate the economics of thermal recovery projects. M, 1980, Stanford University. 270 p.

Williams, Richard Lynn. Progressive failure of red Conemaugh shale. D, 1982, Ohio State University. 173 p.

Williams, Richard M. An economic analysis of low grade gold deposits. M, 1961, University of Toronto.

Williams, Richard S., Sr. and Freed, Robert Lowell. Geology of the Buck Mountain area, Costill County, Colorado. M, 1963, University of Michigan.

Williams, Richard Sugden, Jr. and Freed, Robert Lowell. Geology of the Buck Mountain area, Costilla County, Colorado. M, 1962, University of Michigan.

Williams, Richard Sugden, Jr. Geomorphology of a portion of the northern coastal plain of Puerto Rico. D, 1965, Pennsylvania State University, University Park. 223 p.

Williams, Richard T., II. The ocean tide beneath the Ross Ice Shelf. M, 1976, Virginia Polytechnic Institute and State University.

Williams, Richard Trudo. Microbial aspects of carbon cycling in peatlands. D, 1982, University of Minnesota, Minneapolis. 254 p.

Williams, Robert Bruce. A study of the faunal succession found in Cerro Gordo Member of the Lime Creek Formation (Upper Devonian) at Rockford and Bird Hills, north-central Iowa. M, 1967, Pennsylvania State University, University Park. 155 p.

Williams, Robert C. A faunule from the (Upper Cretaceous) Eagle Ford Formation of Texas. M, 1936, Columbia University, Teachers College.

Williams, Robert L. Late Niagaran and early Cayugan lithofacies of Waubuno Reef, Ontario (Canada). M, 1968, Ohio State University.

Williams, Robert Lee. Statistical symbols for maps; their design and relative values. D, 1957, Harvard University.

Williams, Roderick David. Structural and metamorphic geology of the Bass Lake area, northern Bitterroot Range, Ravalli County, Montana. M, 1976, University of Montana. 53 p.

Williams, Rodney King. The geology of Cape Disappointment Quadrangle and a portion of Fort Columbia Quadrangle, Washington. M, 1952, University of Oregon. 56 p.

Williams, Roger Bennett. Recent marine podocopid Ostracoda of Narragansett Bay, Rhode Island. M, 1966, University of Kansas.

Williams, Ronald Calvin. The mechanics of rock fill consolidation. M, 1963, Georgia Institute of Technology.

Williams, Ronald Leon. Stratification overturn and homogenization at West Branch Reservoir, Portage County, Ohio. M, 1972, University of Akron. 55 p.

Williams, Ronald O. Turbidity and sedimentation; Tanajib/Manifa coastal waters northwestern Arabian Gulf. M, 1984, San Jose State University. 136 p.

Williams, Ross Wood. Uranium and thorium decay series disequilibria in young volcanic rocks. D, 1988, University of California, Santa Cruz. 173 p.

Williams, Roy Edward. Shallow hydrogeology of glacial drifts in northeastern Illinois. D, 1966, University of Illinois, Urbana. 190 p.

Williams, Sheila R. Relationship of ground water chemistry to photolineaments in a karst aquifer. M, 1985, University of South Florida, Tampa. 138 p.

Williams, Sherilyn Coretta. The shear strength of gypsum single and polycrystals and its implications for petrofabric analysis. D, 1987, The Johns Hopkins University. 170 p.

Williams, Sidney A. A study of chlorastrolite. M, 1957, Michigan Technological University. 54 p.

Williams, Sidney Arthur. The mineralogy of the Mildren and Steppe mining districts, Pima County, Arizona. D, 1962, University of Arizona. 176 p.

Williams, Stanley Nichols. Geology and eruptive mechanisms of Masaya Caldera complex, Nicaragua. D, 1983, Dartmouth College. 169 p.

Williams, Stanley Nichols. The October 1902 eruption of Santa Maria Volcano, Guatemala. M, 1979, Dartmouth College. 137 p.

Williams, Stephen Loring. Paleoecology of the Monotis-bearing beds (upper Triassic), Hoyt Canyon, Clan Alpine Mountains, west-central Nevada. D, 1974, Stanford University. 132 p.

Williams, Stephen N. Stratigraphy, facies relationships, and depositional environments of Morowan and Atokan strata, eastern Arkoma Basin, Arkansas. M, 1986, University of Missouri, Columbia. 85 p.

Williams, Steven Hamilton. A comparative planetological study of particle speed and concentration during aeolian saltation. D, 1987, Arizona State University. 155 p.

Williams, Steven John. Aspects of barrier island and shallow shelf sedimentation; west of Pensacola, Florida. M, 1974, Colorado School of Mines. 161 p.

Williams, Stuart Charles. Stratigraphy, facies evolution, and diagenesis of late Cenozoic limestones and dolomites, Little Bahama Bank, Bahamas. D, 1985, University of Miami. 489 p.

Williams, Susan. Late Cenozoic volcanism in the Rio Grande Rift; trace element, strontium isotopic and neodymium isotopic geochemistry of the Taos Plateau Volcanics. D, 1984, University of Minnesota, Minneapolis. 211 p.

Williams, Terry Lynn. Environmental and reservoir mapping trends of the Lower Silurian Medina Group, northwestern Warren County, Pennsylvania. M, 1986, SUNY at Buffalo. 84 p.

Williams, Thomas Bowerman. Coking processes for Western Canada. M, 1913, Queen's University.

Williams, Thomas Bowerman. The Comox coal basin. D, 1925, University of Wisconsin-Madison.

Williams, Thomas Clifford. On the temperatures and stresses involved in regional metamorphism. M, 1969, University of Michigan.

Williams, Thomas Ellis. Correlation of insoluble residues in the Austin Chalk of southern Dallas County, Texas. M, 1957, Southern Methodist University. 15 p.

Williams, Thomas Ellis. Fusulinidae of the Hueco Limestone (Permian), Hueco Mountains, Texas. D, 1962, Yale University.

Williams, Thomas J. Use of the anomalous ratio k/p in the interpretation of gravimetric and magnetic data; a report. M, 1959, Stanford University.

Williams, Thomas R. Late Pleistocene lake level maxima and shoreline deformation in the Basin and Range Province, Western United States. M, 1982, Colorado State University. 59 p.

Williams, Thomas Roy. Geothermal potential in the Bearmouth area, Montana. M, 1975, University of Montana. 81 p.

Williams, Timothy Anderson. Sedimentary structures of the Torrey Sandstone north of San Diego, California. M, 1974, San Diego State University.

Williams, V. Eileen. Coal petrology of the Tulameen Coalfield, south central British Columbia. M, 1978, Western Washington University. 77 p.

Williams, Van Slyck. Glacial geology of the drainage basin of the Middle Fork of the Snoqualmie River. M, 1971, University of Washington. 45 p.

Williams, Van Slyck. Neotectonic implications of the alluvial record in the Sapta Kosi drainage basin, Nepalese Himalayas. D, 1977, University of Washington. 80 p.

Williams, Vera Eileen. Palynological study of the continental shelf sediments of the Labrador Sea. D, 1986, University of British Columbia.

Williams, Vernon Leslie. Subsurface geology of the Bayou Field, Carter County, Oklahoma. M, 1954, University of Oklahoma. 50 p.

Williams, W. W. Fundamental properties of five Iowa sands. M, 1953, Iowa State University of Science and Technology.

Williams, Wayne K. The petrology and depositional environment of the Bangor Limestone (Upper Mississippian) in Southeast Tennessee and Northwest Georgia. M, 1980, Memphis State University.

Williams, Wilbur S. Geology of the eastern half of the Engineer Mountain Quadrangle, San Juan and La Plata counties, Colorado. M, 1965, University of Colorado.

Williams, William C. Precambrian geology of the Dinwoody Lakes region, Wind River Range, Fremont County, Wyoming. M, 1980, University of Wyoming. 90 p.

Williams, William P. Preferred orientation of olivine in the Seiad Creek Peridotite, Siskiyou County, California. M, 1944, Northwestern University.

Williams, William Thomas. Ground-water flow along the total shoreline of Austin Lake and its environmental contribution to pollution of the lake. D, 1982, Western Michigan University. 130 p.

Williams-Jones, Anthony Eric. A field and theoretical study of the thermal metamorphism of Trenton Limestone in the aureole of Mount Royal, Quebec. D, 1973, Queen's University. 198 p.

Williamson, Alexander M. A detailed paleomagnetic study of certain Triassic formations along the Delaware River. M, 1962, Princeton University. 89 p.

Williamson, Anne L. Stratigraphy and structure of the Escalante Mine Formation, Iron County, Utah. M, 1988, New Mexico State University, Las Cruces. 52 p.

Williamson, Bonnie Marie. Formation of authigenic silicate minerals in Miocene volcaniclastic rocks, Boron, California. M, 1987, University of California, Santa Barbara. 89 p.

Williamson, Charles R. Depositional processes, diagenesis and reservoir properties of Permian deep-sea sandstones, Bell Canyon Formation, Texas-New Mexico. D, 1978, University of Texas, Austin. 276 p.

Williamson, Charles Ross. Carbonate petrology of the Green River Formation (Eocene), Uinta Basin, Utah and Colorado. M, 1972, University of Utah. 77 p.

Williamson, Don Alan. The geology of a portion of the eastern Cady Mountains, Mojave Desert, California. M, 1980, University of California, Riverside. 148 p.

Williamson, Eddie A. Petrology of Lower Ordovician Blakely and Crystal Mountain formations, Ouachita Mount Core, Arkansas. M, 1973, University of Missouri, Columbia.

Williamson, Edward F. Urban geology of Temple, Bell County, Texas. M, 1967, Baylor University. 53 p.

Williamson, J. R. The origin and occurrence of the chromite deposits of the Eastern Townships, Quebec. D, 1933, McGill University.

Williamson, J. W. The effect of wellbore storage and damage at the producing well on interference test analysis. M, 1977, Stanford University. 14 p.

Williamson, James A. Landslide susceptibility near Tomales Bay, California. M, 1975, San Francisco State University.

Williamson, John C. The petrology and petrography of the intrusive igneous rocks of the Las Vegas Quadrangle, New Mexico. M, 1936, Texas Tech University. 102 p.

Williamson, John David. Onshore-offshore sand transport on Del Monte Beach, California. M, 1972, United States Naval Academy.

Williamson, Lee Foster. A subsurface study of a channel filled with Hardinsburg Sand south of McLeansboro, Hamilton County, Illinois. M, 1956, University of Illinois, Urbana.

Williamson, Marie-Claude. The Cretaceous igneous province of the Sverdrup Basin, Canadian Arctic; field relations and petrochemical studies. D, 1988, Dalhousie University. 417 p.

Williamson, Marjorie. A study of ostracod fauna of the (Silurian) Waldron Shale, Flat Rock Creek, St. Paul, Indiana. M, 1934, Columbia University, Teachers College.

Williamson, Mark A. Benthic foraminiferal assemblages on the continental margin off Nova Scotia; a multivariate approach. D, 1983, Dalhousie University. 348 p.

Williamson, Mary Frances. Glacial physiography of Auburn, Massachusetts. M, 1942, Clark University.

Williamson, Michael. Compositional and mineralogical variations in deep sea ferromanganese nodules. M, 1974, University of Washington. 41 p.

Williamson, Norman L. Depositional environments of the Pocono Formation in southern West Virginia. D, 1974, West Virginia University. 171 p.

Williamson, Paul Bain. Foraminifera from the Arcadia Park section of the Eagle Ford Formation. M, 1950, Southern Methodist University. 58 p.

Williamson, Richard Edward. A photogrammetric map of the Ohio State University campus. M, 1957, Ohio State University.

Williamson, Robert L. Geology of the Paul-2 Sand, North Big Island Field, Rapides Parish, Louisiana. M, 1977, Louisiana Tech University.

Williamson, Terrence C. The paragenesis of pyrite-pyrrhotite-bearing black phyllites in the Waterville-Augusta area, Maine. M, 1968, University of Maine. 63 p.

Williamson, Thomas E. The taxonomy, osteology, functional morphology, taphonomy and paleobiology of early Cenozoic Meniscotherium (Mammalia, "Condylarthra"). M, 1988, University of New Mexico.

Williamson, Turner Franklin. Petrology of the lower Arroyo Penasco (Mississippian), Taos County, New Mexico. M, 1979, University of Texas, Austin.

Williamson, William Edward. Role of varying moments of inertia in Earth stresses. M, 1965, American University. 52 p.

Willian, Mark A. Sandstone petrology of the Cretaceous Lusardi and Cabrillo formations; evidence for deep dissection of the Peninsular Ranges magmatic arc by Late Cretaceous time. M, 1986, San Diego State University. 63 p.

Williard, John Earlton. Evolution and environment. M, 1936, George Washington University.

Willibey, Tom Dean. Changes in mineral composition with metamorphism of part of the Oreville Formation, Black Hills, South Dakota. M, 1975, South Dakota School of Mines & Technology.

Willie, Kelly Delon. Processing of seismic data for locating extensions of the Trenton gas zone, Harlem Township, Delaware County, Ohio. M, 1985, Wright State University. 100 p.

Willimon, Edward Lloyd. Quaternary gastropods and paleoecology of the Trinity River flood plain of Dallas County, Texas. M, 1970, Southern Methodist University. 89 p.

Willing, E. S., Jr. A field study in the relationship of Baltimore Gneiss, Cockeysville Marble and Wissahickon Schist in the vicinity of Marshallton, Pa. M, 1949, Bryn Mawr College. 14 p.

Willingham, Charles Richard. A gravity survey of the San Bernardino Valley, Southern California. M, 1968, University of California, Riverside. 96 p.

Willingham, Daniel L. Stratigraphy and sedimentology of the upper Santa Fe Group in the El Paso region, West Texas and south-central New Mexico. M, 1980, University of Texas at El Paso.

Willis, Brian James. Sedimentology of the upper Middle Devonian Catskill magnafacies at the Catskill Front, New York. M, 1987, SUNY at Binghamton. 140 p.

Willis, Clifford Leon. Geology of northeastern quarter Chiwaukum quadrangle, Washington. D, 1950, University of Washington. 158 p.

Willis, David Edwin. An investigation of seismic wave propagation in the eastern United States. D, 1968, University of Michigan. 93 p.

Willis, David Edwin. Observations on the seismometer plant resonance problem. M, 1957, University of Michigan.

Willis, David Grinnell. Analysis of deformation in sedimentary rocks with application to the Newport-Inglewood Uplift. D, 1954, Stanford University. 106 p.

Willis, Donald Kenyon. Textural comparison of insular and mainland shelf sediments, continental borderland, California. M, 1979, University of Southern California.

Willis, Gar Charles. The geology of the east central portion of the Paskenta Quadrangle, Tehama County, California. M, 1962, University of California, Berkeley. 58 p.

Willis, Gerald F. Geology of the Birch Creek molybdenite prospect, Beaverhead County, Montana. M, 1978, University of Montana. 74 p.

Willis, Grant C. Geology, depositional environment, and coal resources of the Sego Canyon 7 1/2 minute quadrangle, near Green River, east-central Utah. M, 1986, Brigham Young University. 208 p.

Willis, John Bruce. Geology of the Yellowjacket Anticline area, Rio Blanco County, Colorado. M, 1957, Colorado School of Mines. 91 p.

Willis, John W. The shallow structure of Tampa Bay. M, 1984, University of South Florida, St. Petersburg. 79 p.

Willis, Joseph P., Jr. Geology of the eastern portion of the Hurricane Quadrangle, Utah. M, 1961, University of Southern California.

Willis, Julie Barrott. Early Miocene bimodal volcanism, northern Wilson Creek Range, Lincoln County, Nevada; geochronology and petrology. M, 1985, Brigham Young University.

Willis, Mark Elliott. Seismic velocity and attenuation from full waveform acoustic logs. D, 1977, Massachusetts Institute of Technology. 357 p.

Willis, Paul Dewey. Geology of the (SNOW) area, Pushmataha County, Oklahoma. M, 1954, University of Oklahoma. 60 p.

Willis, Richard Porter. Geology of the Porcupine Creek area, Big Horn County, Wyoming. M, 1953, University of Wyoming. 64 p.

Willis, Robin. The physiography of the San Andreas Fault between the Pajaro Gap and the Cholame Plains (California). D, 1924, Stanford University. 385 p.

Willis, Robin. The physiography of the southern Santa Cruz Mountains (California). M, 1922, Stanford University. 92 p.

Willis, Ronald Porter. Upper Mississippian-Lower Pennsylvanian stratigraphy of central Montana and Williston Basin. D, 1958, University of Illinois, Urbana. 77 p.

Willis, William Haywood. Relationship of photo-lineaments to well yields and ground water chemistry in North central Benton County, Arkansas. M, 1978, University of Arkansas, Fayetteville.

Williston, Samuel W. Synopsis of the North American Syrphidae. D, 1885, Yale University.

Willman, Harold Bowen. An attempt to correlate the Pennsylvanian sandstones of the Havana Quadrangle by the composition and structure of their grains. M, 1928, University of Illinois, Chicago.

Willman, Harold Bowen. General geology and mineral resources of the Illinois Deep Waterway from Chicago to Peoria. D, 1931, University of Illinois, Chicago.

Willott, John. Analysis of modern vertical deformation in the western Transverse Ranges, California. M, 1972, University of California, Santa Barbara.

Willoughby, Janice Kay Sowards. Geology of Prichard Formation and Ravalli Group rocks in the SE 1/4 of the Kellogg 15' Quadrangle, Shoshone County, Idaho. M, 1986, University of Idaho. 155 p.

Willoughby, R. H. Paleontology and stratigraphy of the Shady Formation near Austinville, Virginia. M, 1977, Virginia Polytechnic Institute and State University.

Willoughby, William W. Geology and evaluation of the CB silver deposit, North Animas Mountains, Hidalgo County, New Mexico. M, 1985, University of Idaho. 146 p.

Wills, Bill Frank. Heavy minerals of Spring Creek drainage basin, Burnet County, Texas. M, 1951, University of Texas, Austin.

Wills, Christopher J. Structure and stratigraphy of the lower Orvindalen area, Wedel Jarlsberg Land, Spitsbergen. M, 1984, University of Wisconsin-Madison.

Wills, Donald L. The Sonora Sandstone (Mississippian) of west-central Illinois. D, 1971, University of Iowa. 73 p.

Wills, J. G. Geology of the Pine Mountain area, Natrona County, Wyoming. M, 1955, University of Wyoming. 65 p.

Wills, Robert H. A digital phase coded ground probing radar. D, 1987, Dartmouth College. 105 p.

Willson, Kenneth. Northern part of the Tow Creek Anticline, Routt County, Colorado. M, 1920, University of Colorado.

Willson, Robert G. Hydrogeology of the Shell Creek area, Mountrail County, North Dakota. M, 1967, University of North Dakota. 68 p.

Wilmar, Glenn C. Relation of diagenesis to lithofacies distribution, Nugget Sandstone, Washakie Basin, Wyoming. M, 1986, University of Texas at El Paso.

Wilmot, Barry R. The geology of the Lower Cretaceous Mannville Group, Edam, Saskatchewan. M, 1985, University of Calgary. 160 p.

Wilmoth, Benton M. Mineralogy of some (Pennsylvanian) Conemaugh limestones. M, 1956, West Virginia University.

Wilmott, Charles L., Jr. The Reagan Formation of the Wichita Mountains, Oklahoma. M, 1957, University of Oklahoma. 60 p.

Wilpot, Ralph H. The Paleozoic stratigraphy of the Los Pinos Range, New Mexico. M, 1942, Northwestern University.

Wilsey, William L. M. Depositional stratigraphy of the Cypress Sandstone and related formations of Hamilton County, Illinois. M, 1984, University of Cincinnati. 222 p.

Wilshire, Howard Gordon. The history of Tertiary volcanism near Ebbetts Pass, Alpine County, California. D, 1956, University of California, Berkeley. 126 p.

Wilshusen, John P. The structure and stratigraphy of the Little Elk Canyon area, Meade and Lawrence counties, South Dakota. M, 1963, South Dakota School of Mines & Technology.

Wilson, Alfred W. G. Physical geology of central Ontario. D, 1901, Harvard University.

Wilson, Alice Evelyn. The geology of the Cornwall District, Ontario. D, 1929, University of Chicago. 280 p.

Wilson, Andrew Moore. Response modeling of Holocene sediments within the shallow substrate of central Mississippi Sound. M, 1984, University of Mississippi. 89 p.

Wilson, Ann Catherine. Petrography of basalts. M, 1983, Queen's University. 232 p.

Wilson, Arthur Oliver. Effect of cupric sulfate solutions on the copper sulfide minerals. M, 1931, Montana College of Mineral Science & Technology. 43 p.

Wilson, Augustus O'Hara, Jr. Petrgenesis of parts of Black River age strata of the Chickamauga limestone Etowah and Dekalb counties, Alabama. D, 1971, University of North Carolina, Chapel Hill. 120 p.

Wilson, B. D. The origin of slaty cleavage and associated structures, Precambrian Lookout Schist, Medicine Bow Mountains, Wyoming. M, 1975, University of Wyoming. 87 p.

Wilson, B. T. Gold deposits in the vicinity of Straw Lake. M, 1936, Queen's University. 4 p.

Wilson, Barry James. Depositional environments and diagenesis of sandstone facies in the Aux Vases Formation (Mississippian), Illinois Basin. M, 1985, Southern Illinois University, Carbondale. 130 p.

Wilson, Bradley S. A sulphur isotope and structural study of the silver vein host rocks at Cobalt, Ontario. M, 1987, Carleton University. 156 p.

Wilson, Bruce A. An investigation of the groundwater ridging theory for large groundwater contributions to streams during storm runoff events. M, 1981, University of Windsor. 95 p.

Wilson, Bruce Craig. Gravitational sinking of oceanic lithosphere and the nature of orogeny. M, 1975, Queen's University. 55 p.

Wilson, Cathy Jean. Runoff and pore pressure development in hollows. D, 1988, University of California, Berkeley. 291 p.

Wilson, Charles William, Jr. Geology of the Muskogee-Whitfield area, Oklahoma. D, 1931, Princeton University. 61 p.

Wilson, Charles William, Jr. The stratigraphy and paleontology of the Carter's Creek Limestone. M, 1928, Vanderbilt University.

Wilson, Chester H. A detailed petrographic and chemical study of differentiation in Keweenawan lava flows of Michigan. D, 1975, Michigan State University. 75 p.

Wilson, Chester H. Petrology of the Algomah Mine area, Ontonagon County, Michigan. M, 1967, Michigan State University. 50 p.

Wilson, Clark L. The geology of the Black Forest Mine area, Spruce Mountain, Nevada. M, 1938, University of Arizona.

Wilson, Clark R. Meteorological excitation of the Earth's wobble. D, 1975, University of California, San Diego. 116 p.

Wilson, Clayton Hill. Depositional, structural, and thermal evolution of a Pleistocene oil-productive area; Texas-Louisiana continental shelf. M, 1985, University of Texas, Austin. 93 p.

Wilson, Clyde A. Ore controls of the San Xavier Mine, Pima County, Arizona. M, 1960, University of Arizona.

Wilson, Dail Adair. The orientation-compensation method for precise refractive index determination of doubly refracting substances. M, 1967, University of Cincinnati. 32 p.

Wilson, Daniel Allen. A seismic and gravity investigation of the North Boulder River and Jefferson River valleys, Madison and Jefferson counties, Montana. M, 1962, Indiana University, Bloomington. 43 p.

Wilson, Daniel L. The origin of serpentinites associated with the Shuksan Metamorphic Suite near Gee Point, Washington. M, 1978, Western Washington University. 64 p.

Wilson, David D. East-west seismic profile of McMurdo Sound, Antarctica. M, 1982, Northern Illinois University. 57 p.

Wilson, David Vernon. Geophysical investigation of the subsurface structure of Deep Springs Valley, California. M, 1975, University of California, Los Angeles.

Wilson, Deborah Crotty. Ocean tidal loading in Southeastern United States. M, 1978, Virginia Polytechnic Institute and State University.

Wilson, Donald Edward. Studies of the metal-ion binding characteristics of soluble, natural organic materials. D, 1974, University of Alaska, Fairbanks. 70 p.

Wilson, Douglas Hord. Petrography and paleoenvironment of the Upper Cretaceous Anacacho Formation in Uvalde and Kinney counties of Southwest Texas. M, 1982, University of Texas at Dallas. 211 p.

Wilson, Douglas Slade. Tectonic history of the Juan de Fuca Ridge. D, 1986, Stanford University. 195 p.

Wilson, Doyle C. Quartz and feldspar in mudrocks, Triassic Moenkopi Formation, southeastern Nevada. M, 1978, Arizona State University. 108 p.

Wilson, Duncan Campbell Ogden, Jr. Stratigraphy of Black Gap area, Brewster County, Texas. M, 1951, University of Texas, Austin.

Wilson, Earl O. The plasticity of finely ground minerals with water. D, 1935, Massachusetts Institute of Technology. 93 p.

Wilson, Edith Newton. Dolomitization of the Triassic Latemar buildup, Dolomites, northern Italy. D, 1988, The Johns Hopkins University. 285 p.

Wilson, Edmon D. A study of the effect of weathering, silt content, and depth of burial on physical properties of shale samples from north-central Texas. M, 1958, Texas A&M University. 58 p.

Wilson, Edward Carl. Corals and other significant fossils from the upper Paleozoic McCloud Limestone of northern California. D, 1967, University of California, Berkeley. 292 p.

Wilson, Edward Carl. The Pennsylvanian and Permian paleontology and stratigraphy of Ward Mountain, White Pine County, Nevada. M, 1960, University of California, Berkeley. 138 p.

Wilson, Edward Norman. Geology of the Shooks Gap Quadrangle, Missouri. M, 1947, Washington University. 82 p.

Wilson, Eldred Dewey. The Matzatzal Quartzite, a new Pre-Cambrian formation of central Arizona. M, 1922, University of Arizona.

Wilson, Eldred Dewey. The Pre-Cambrian Mazatzal revolution in central Arizona. D, 1937, Harvard University.

Wilson, Ellery H. Geological engineering study of the Coffintop Dam and reservoir site near Lyons, Boulder County, Colorado. M, 1977, Colorado School of Mines. 140 p.

Wilson, Eugene Jack. Foraminifera from the Gaviota Formation east of Gaviota Creek, California. M, 1949, University of California, Berkeley. 118 p.

Wilson, Everett E. Foraminifera from outcrops of the Pierre Shale (Upper Cretaceous) of North Dakota. M, 1958, University of North Dakota. 134 p.

Wilson, Forest Wayne. The origion of some mud pockets in Panola County, Texas. M, 1931, University of Texas, Austin.

Wilson, Forrest Raymond. Vinton and Edgerly salt domes, Calcasieu Parish, Louisiana; a gravity interpretation using least squares polynomial fitting and three-dimensional computer modeling. M, 1983, Wright State University. 128 p.

Wilson, Francis Smith. Foraminifera from the Pennsylvanian Wolf Mountain Shale near Graford, Texas. M, 1954, University of Tulsa. 57 p.

Wilson, Frank W. Barrier reefs of the Stanton Formation (Missourian) in southeastern Kansas. M, 1957, Kansas State University. 50 p.

Wilson, Frederic H. Arsenic and water, Pedro Dome-Cleary Summit area, Alaska. M, 1975, University of Alaska, Fairbanks. 91 p.

Wilson, Frederick A. Zircons in metasomatized amphibolites, Orchard Beach, Bronx, New York. M, 1970, Brooklyn College (CUNY).

Wilson, Frederick Albert. Geophysical and geologic studies in southern Mecklenburg County and vicinity, North Carolina and South Carolina. D, 1981, George Washington University.

Wilson, Frederick Henley. Late Mesozoic and Cenozoic tectonics and the age of porphyry copper prospects; Chignik and Sutwik Island quadrangles, Alaska Peninsula. D, 1980, Dartmouth College. 177 p.

Wilson, Gene Douglass. The instars and shell morphology of Ilyocypris scotti. M, 1954, University of Illinois, Urbana.

Wilson, Geoffrey Evans. Experimental and field studies of scheelite in tactite deposits of the Stormy Day Mine, Pershing County, Nevada. M, 1975, University of Nevada. 108 p.

Wilson, George Alexander. A petrographic description of a plutonic mass on Gambier Island, Howe Sound, British Columbia. M, 1951, University of British Columbia.

Wilson, George M. Geophysical investigations in the Humboldt River valley near Winnemucca, Nevada. M, 1960, University of Nevada. 30 p.

Wilson, George Miller. The stratigraphy of the (Pennsylvanian) McLeansboro Group of Vermilion and Edgar counties, Illinois (W693). M, 1944, University of Illinois, Urbana. 31 p.

Wilson, George Newton, Jr. Ostracodes of the Kiamichi Formation (Albian) in north-central Texas. M, 1973, University of Texas, Arlington. 66 p.

Wilson, George William. The geology of the southern half of Rockwall Quadrangle, Texas. M, 1948, University of Oklahoma. 40 p.

Wilson, Gilbert. The Pre-Cambrian trendlines. M, 1926, University of Wisconsin-Madison.

Wilson, Glenn Alan. Cassiterite solubility and metal chloride speciation in supercritical solutions. D, 1986, The Johns Hopkins University. 170 p.

Wilson, Glenn Venson. Characterization and modeling of the dynamics of ground-water mounds underneath septic tank-filter fields. D, 1986, University of Arkansas, Fayetteville. 161 p.

Wilson, Gregory Allen. The effects of subsurface dissolution of Permian salt on the deposition, stratigraphy and structure of the Ogallala Formation (late Miocene age) Northeast Potter County, Texas. M, 1988, West Texas State University. 107 p.

Wilson, Gregory C. Debris flows and alluvial fan development in the Appalachians. M, 1987, Southern Illinois University, Carbondale. 165 p.

Wilson, Gregory C. Structural geology of the southern part of the northern Animas Mountains, south-central Hidalgo County, New Mexico. M, 1986, New Mexico State University, Las Cruces. 115 p.

Wilson, Guilford, James, Jr. Geology of the Big Bend of the Llano River area, Mason County, Texas. M, 1957, Texas A&M University. 113 p.

Wilson, Guy D. Stratigraphy, depositional environments, and petrology of Wilcox Group coals in north-central Louisiana. M, 1988, Northeast Louisiana University. 143 p.

Wilson, H. D. and Hendry, N. W. Geology and quicksilver deposits of the Coso Hot Springs area, Inyo County, California. M, 1939, California Institute of Technology. 63 p.

Wilson, H. D. B. Geochemical studies of the epithermal deposits at Goldfield, Nevada. D, 1942, California Institute of Technology. 72 p.

Wilson, H. D. B. Stratigraphy of the Cretaceous and Eocene rocks of the Santa Monica Mountains (California). D, 1942, California Institute of Technology. 31 p.

Wilson, Herschell Thomas. Stratigraphy and paleontology of a well core through Silurian and Ordovician strata in Oakland County, Michigan. M, 1955, University of Michigan.

Wilson, Ivan Franklin. Geology of the San Benito Quadrangle, California. D, 1942, University of California, Berkeley. 167 p.

Wilson, J. N. The stratigraphy of the Lloydminster and adjacent areas (Alberta) with special emphasis on the lithology of the Lower Cretaceous. M, 1947, University of Alberta. 167 p.

Wilson, Jacqueline B. Stratigraphy of the (Upper Cretaceous) Sussex Sandstone, Powder River basin, Wyoming. M, 1951, University of Wyoming. 22 p.

Wilson, James A. Stratigraphy of the Wyandotte Limestone (Upper Pennsylvanian) in the Kansas River area. M, 1959, University of Kansas. 135 p.

Wilson, James G. Geophysical investigations of Perkins area, Michigan. M, 1952, Michigan Technological University. 20 p.

Wilson, James K. Investigation of the late Tertiary to Recent movement along the east bounding fault of the Shearer Graben within the Kentucky River fault system in southern Clark County, Kentucky. M, 1981, Eastern Kentucky University. 45 p.

Wilson, James Lee. Calcareous nannoplankton from stratigraphic sections of the Ocala Group (upper Eocene) and the Marianna Limestone (middle Oligocene) of northwestern Florida. M, 1971, University of Georgia.

Wilson, James Lee. The faunas and stratigraphic relations of the Gatesburg and Conococheague formations (Upper Cambrian) in the Central Appalachians. D, 1949, Yale University.

Wilson, James Lee. The trilobite fauna of the Camaraspis Zone in the basal Wilberns Limestone of Texas. M, 1944, University of Texas, Austin.

Wilson, James R. A source for the public water supply of the Town of Meyers Falls, Washington. M, 1938, Washington State University. 26 p.

Wilson, James Robert. An X-ray fluorescence study of trace element distributions in metamorphic rocks and potential economic mineralization near Huntdale, North Carolina. M, 1973, University of Tennessee, Knoxville. 55 p.

Wilson, James Robert. Glaciated dolomite karst in the Bear River Range, Utah. D, 1976, University of Utah. 136 p.

Wilson, James Stanley. Cambrian paleontology and stratigraphy of the Miller area, Esmeralda County, Nevada. M, 1961, University of California, Los Angeles.

Wilson, James Steven. Glaciological interpretation of the Donna boulder train; Cluff Lake, Saskatchewan, Canada. M, 1988, University of Saskatchewan. 102 p.

Wilson, James T. The South Atlantic earthquake of August 28, 1933. D, 1939, University of California, Berkeley. 48 p.

Wilson, Janine M. Sediment-water-biomass interaction of toxic metals in the western basin, Lake Erie. M, 1978, Bowling Green State University. 113 p.

Wilson, Jeffrey Kent. Influence of focal depth on the displacement spectra of earthquakes. M, 1983, Georgia Institute of Technology. 78 p.

Wilson, Jeffrey Leigh. Geology and engineering aspects of Boraxo Pit, Death Valley, California. M, 1976, University of Southern California.

Wilson, Jeffrey Warren. Fluid inclusion geochemistry of the Granisle and Bell copper porphyry. M, 1978, University of Toronto.

Wilson, Jerry C., Jr. Engineering properties of Southern California borderland sediments. M, 1970, University of Southern California.

Wilson, Jerry Calvert, Jr. Technology and economics assessment of developing an Arctic offshore petroleum area in Alaska (Chukchi Sea). D, 1983, University of California, Los Angeles. 217 p.

Wilson, John Allen. Physical modelling to assess the dynamic behavior of rock slopes. M, 1979, University of Arizona. 86 p.

Wilson, John Andrew. An interpretation of the skull of Buettneria with reference to the cartilage and soft parts (Triassic of Texas). D, 1940, University of Michigan.

Wilson, John Charles. Analysis of the observed earthquake response of a multiple span bridge. D, 1984, California Institute of Technology. 167 p.

Wilson, John Coe. Geology of the Alta Stock, Utah. D, 1961, California Institute of Technology. 236 p.

Wilson, John Coe. The Gypsum Spring Formation (Middle Jurassic) of western Wyoming. M, 1955, University of Kansas. 216 p.

Wilson, John E. Stratigraphy and structure of part of the Abiquiu embayment of the Rio Grande Rift, New Mexico. M, 1977, University of New Mexico. 90 p.

Wilson, John Ewing. Geology of Rustler Spring area, Culberson County, Texas. M, 1951, University of Texas, Austin.

Wilson, John H. The Pleistocene formations of Sankaty Head, Nantucket. M, 1905, Columbia University, Teachers College.

Wilson, John Joseph. Cretaceous stratigraphy of the Central Andes of Peru. D, 1961, Harvard University.

Wilson, John M. Conodont biostratigraphy and paleoecology of the Middle Ordovician (Chazyan) sequence in Ellett Valley, Virginia. M, 1977, Virginia Polytechnic Institute and State University.

Wilson, John M. Geohydrology of Skillman Basin on the western Highland rim, Tennessee. M, 1969, Vanderbilt University.

Wilson, John M. The geology of a Precambrian area near Rochford, South Dakota. M, 1951, South Dakota School of Mines & Technology.

Wilson, John O. Decomposition of litter of Spartina alterniflora in a salt marsh ecosystem; biochemical and geochemical studies. D, 1985, Boston University. 175 p.

Wilson, John R. Stratigraphic analysis of the Strawn Sequence in north central Texas. M, 1952, Northwestern University.

Wilson, John Randall. Geology, alteration, and mineralization of the Korn Kob Mine area, Pima County, Arizona. M, 1977, University of Arizona. 103 p.

Wilson, John T. Portage County revisited; an analysis of the occurrence of oil and gas in the Lower Silurian "Clinton" sandstone reservoir in Portage County, Ohio. M, 1988, Kent State University. 163 p.

Wilson, John Thomas. Geology of Seventeenmile Point, Old Dad Mountain Quadrangle, southeastern California. M, 1978, University of Southern California.

Wilson, John Tuzo. Geology of the Mill Creek-Stillwater area, Montana. D, 1936, Princeton University. 130 p.

Wilson, Joseph M. The gas fields of North Webster Parish, Louisiana. M, 1924, University of Missouri, Rolla.

Wilson, Joseph Raymond. Geology of the Price Lake area, Mason County, Washington. M, 1975, North Carolina State University. 79 p.

Wilson, Joseph Raymond. Structural development of the Kettle gneiss dome in the Boyds and Bangs Mountain quadrangles, Northeastern Washington. D, 1981, Washington State University. 156 p.

Wilson, Kenneth L. Eocene and related geology of a portion of the San Luis Rey and Encinitas quadrangles, San Diego County, California. M, 1972, University of California, Riverside. 135 p.

Wilson, Kevin M. Diagenesis of phosphorus in carbonate sediments from Bermuda. M, 1979, University of New Hampshire. 187 p.

Wilson, Kim Suzanne. The geology and epithermal silver mineralization of the Reymert Mine, Pinal County, Arizona. M, 1984, University of Arizona. 100 p.

Wilson, Laurence MacKenzie. Surficial glacial deposits of the Michigan-Saginaw lobes in the Grand Rapids area, Michigan, a study of relationships. M, 1955, Michigan State University. 86 p.

Wilson, Lee. Suspended sediment yield of United States rivers as a function of climate. D, 1971, University of Colorado. 416 p.

Wilson, Leslie Edward. Stratigraphy and paleontology of Garfield Co., Montana with notes on unconformity at base of Lance Formation. M, 1926, University of Washington. 76 p.

Wilson, Malcolm Alan. Palynology of three sections across the uppermost Cretaceous/Paleocene boundary in the Yukon Territory and District of Mackenzie, Canada. M, 1977, University of Saskatchewan. 232 p.

Wilson, Malcolm Alan. The climatic and vegetational history of the Postglacial in central Saskatchewan. D, 1981, University of Saskatchewan. 188 p.

Wilson, Malcolm E. The granodiorites and related rocks of southwestern Oregon. M, 1909, University of Iowa.

Wilson, Margaret T. Functional morphology of the genus Lyreidus. M, 1986, Kent State University, Kent. 76 p.

Wilson, Mark A. The geology and depositional history of the Mississippian Chappel Limestone mounds of Shackelford County, Texas. M, 1983, University of Texas, Arlington. 117 p.

Wilson, Mark Allan. The Chesterian and Morrowan environments and ecology of the western Bird Spring Basin (Nevada). D, 1982, University of California, Berkeley. 295 p.

Wilson, Mark Dale. The geology of the Upper Sixmile Canyon area, central Utah. M, 1949, Ohio State University.

Wilson, Mark Robert. Petrogenesis of fluids associated with unconformity-type uranium deposits in Saskatchewan, and with the Zortman-Landusky Au-Ag deposits in Montana. D, 1987, University of Saskatchewan. 158 p.

Wilson, Mark Vincent Hardman. Fossil fishes of the Tertiary of British Columbia. D, 1974, University of Toronto.

Wilson, Michael. Once upon a river; archaeology and geology of the Bow River valley at Calgary, Alberta, Canada. D, 1981, University of Calgary. 464 p.

Wilson, Michael Alan. Aluminum extraction and mineralogical solubility of Nason and Dothan soils with potassium salt solutions. D, 1983, Virginia Polytechnic Institute and State University. 137 p.

Wilson, Michael D. A structural and statistical study of migmatites in the Front Range, Colorado. M, 1965, Northwestern University.

Wilson, Michael David. The stratigraphy and origin of the Beaverhead Group in the Lima area, southwestern Montana. D, 1967, Northwestern University. 183 p.

Wilson, Michael L. Petrology and origin of Archean lithologies in the southern Tobacco Root and northern Ruby Ranges of southwestern Montana. M, 1981, University of Montana. 92 p.

Wilson, Michael P. Gravity studies in the vicinity of Walnut Creek. M, 1973, SUNY, College at Fredonia. 108 p.

Wilson, Michael Peter. Catastrophic discharge of Lake Warren in the Batavia-Genesee region. D, 1981, Syracuse University. 119 p.

Wilson, Michael Phillip. Groundwater contamination by the herbicide atrazine, Weld County, Colorado. D, 1986, Colorado State University. 208 p.

Wilson, Michael S. A point load test for aggregate durability. M, 1978, University of Calgary. 107 p.

Wilson, Monte Dale. Cretaceous stratigraphy of the southern Madison and Gallatin ranges, southwestern Montana. D, 1970, University of Idaho. 86 p.

Wilson, Morley Evans. Preliminary memoir on the Abitibi District, Pontiac County, Quebec. D, 1912, McGill University.

Wilson, N. F. Barrier island morphology at Rodanthe, North Carolina. M, 1973, University of Virginia.

Wilson, N. R. Review of the dressing works at the Einstein silver mine, with a project for treating the ore. M, 1879, Washington University.

Wilson, Nancy L. A study of the upper Winnipegosis in south-central Saskatchewan. M, 1985, University of Saskatchewan. 93 p.

Wilson, Nathaniel Carl. Structure and metamorphism of the Colebrooke Schist southeast of Langlois, southwestern Oregon. M, 1986, University of Oregon. 121 p.

Wilson, Neil T. Gravity study of subsurface geological structure in the central Appalachian Great Valley province, Virginia. M, 1978, SUNY at Buffalo. 58 p.

Wilson, Norman Albert. Calcareous nannoplankton from the Rosario Formation (upper Cretaceous) of southern California. M, 1970, University of California, Davis. 77 p.

Wilson, Norman L. A petrological study of the rocks of Mount Johnson, Quebec. M, 1933, McGill University.

Wilson, Norman L. An investigation of the metamorphism of the Orijarvi type with special reference to the zinc-lead deposits at Montauben-les-Mines, Province of Quebec. D, 1939, McGill University.

Wilson, Patricia McDowell. Petrology of the Cenozoic volcanic rocks of the basal Clarno Formation, central Oregon. M, 1973, Rice University. 47 p.

Wilson, Paula Nelson. Thermal and chemical evolution of hydrothermal fluids at the Ophir Hill Mine, Ophir District, Tooele County, Utah. M, 1986, University of Utah. 155 p.

Wilson, Peter G. An environmental-planning study, Wheeling, West Virginia–Ohio 7.5 minute topographical quadrangle. M, 1971, West Virginia University.

Wilson, Philip Roy. The geology and mineralogy of the Yreka copper property, Quatsino Sound, BC. M, 1955, University of British Columbia.

Wilson, Philo Calhoun. Geology of the northern part of the Allensville Quadrangle, Pennsylvania. M, 1950, Cornell University.

Wilson, Philo Calhoun. Pennsylvanian stratigraphy of the Powder River basin, Wyoming, and adjoining areas. D, 1954, Washington State University. 411 p.

Wilson, R. Wayne. Borehole television logging; a new tool for downhole evaluation in the Devonian shale. M, 1988, Eastern Kentucky University. 57 p.

Wilson, Raymond C., Jr. The structural geology of the Shadow Mountains area, San Bernardino County, California. M, 1966, Rice University. 45 p.

Wilson, Raymond Carl Jr. The mechanical properties of the shear zone of the Lewis overthrust, Glacier Park, Montana. D, 1970, Texas A&M University. 89 p.

Wilson, Raymond Edgar. Radium content of core samples taken from Hood Canal. D, 1942, University of Washington. 29 p.

Wilson, Richard D. Geology of the southeastern part of the Point Dume Quadrangle. M, 1955, University of California, Los Angeles.

Wilson, Richard Fairfield. The stratigraphy and sedimentology of the Kayenta and Moenave formations, Vermilion Cliffs region, Utah and Arizona. D, 1959, Stanford University. 401 p.

Wilson, Richard Leland. Systematic and faunal analysis of a lower Pliocene vertebrate assemblage from Trego County, Kansas. D, 1967, University of Michigan.

Wilson, Richard Leland. The Pleistocene vertebrates of Michigan. M, 1965, University of Michigan.

Wilson, Richard Shirl. Geochemical exploration reconnaissance in the Avery area, Idaho. M, 1963, University of Idaho. 81 p.

Wilson, Richard T., Jr. Statistical analysis of roof falls in eastern Kentucky coal mines. M, 1983, Eastern Kentucky University. 97 p.

Wilson, Richard W. Ground water investigations of a portion of the Kendrick Project, Natrona, Wyoming. M, 1951, University of Wyoming. 83 p.

Wilson, Robert E. A study of the chemical composition and structures of the silicate minerals to determine the effect of their silica and alumina content. M, 1939, University of Minnesota, Minneapolis. 165 p.

Wilson, Robert E. A study of the dispersion and flushing of water-borne materials in the northwest branch of Baltimore Harbor, (Maryland). M, 1970, The Johns Hopkins University.

Wilson, Robert F. Geology of the Scotian Shelf near Ingonish, Nova Scotia (Mississippian-Pennsylvanian). M, 1972, Dalhousie University. 89 p.

Wilson, Robert J. Bedrock geology of the Ossipee Lake area, New Hampshire. D, 1966, Harvard University.

Wilson, Robert Lake. Geology of Addison and Springfield townships, Gallia County, Ohio. M, 1950, University of Iowa. 89 p.

Wilson, Robert Lake. Pennsylvanian stratigraphy of the northern part of Sand Mountain, (Dekalb County), Alabama, (Dade County), Georgia and (Marion County), Tennessee. D, 1967, University of Tennessee, Knoxville. 110 p.

Wilson, Robert Lee. Stratigraphy and economic geology of the Chinle Formation, northeastern Arizona. D, 1956, University of Arizona. 332 p.

Wilson, Robert Rogers. A reconnaissance of the geology of the Adelaida Quadrangle, San Luis Obispo County, California, with special reference to the stratigraphy of the Miocene formations. M, 1930, Stanford University. 99 p.

Wilson, Robert Terry. Reconnaisance geology and petrology of the San Carlos area, Sonora, Mexico. M, 1978, Arizona State University. 107 p.

Wilson, Robert W. Pliocene rodents of western North America. D, 1936, California Institute of Technology. 138 p.

Wilson, Robert W. Rodents and lagomorphs of the (Pleistocene) Carpinteria asphalt (California). M, 1932, California Institute of Technology. 20 p.

Wilson, Robert W. The heavy accessory minerals of the Val Verde Tonalite (California). D, 1936, California Institute of Technology. 38 p.

Wilson, Roger Lenox. Variation in Pennsylvanian lophophyllidid corals. M, 1957, University of Illinois, Urbana.

Wilson, Roy Arthur. Geological and economic resources of Bridger Mountains, Montana. M, 1917, University of Montana.

Wilson, Roy Arthur. Geology and physiography of the Mission Range, Montana. D, 1921, University of Chicago. 107 p.

Wilson, Russell L. A study of artificially gravel packed fiberglass well screens. M, 1980, Ohio University, Athens. 76 p.

Wilson, Samuel M. Paleoecology of the Kenosha Shale (Upper Pennsylvanian) in the northern Midcontinent region. M, 1977, University of Nebraska, Lincoln.

Wilson, Sharon Lee. Structure and nature of the shallow basement beneath the Hardeman Basin, Oklahoma, as revealed by COCORP seismic reflection surveys. M, 1984, University of Wyoming. 117 p.

Wilson, Steven E. Molluscan bivalves of the Arkona and Silica formations (Devonian), Ontario and Ohio. M, 1973, Wayne State University.

Wilson, Steven M. Geology of the Fall Branch, Tennessee area. M, 1979, University of Tennessee, Knoxville. 80 p.

Wilson, Steven O'Hara. Pre-plutonic structure and stratigraphy of the Little Shay Mountain area, San Bernardino Mountains, San Bernardino County, California. M, 1985, University of California, Riverside. 194 p.

Wilson, T. Yates. The Rensselaer Grit problem (Lower Cambrian, New York). M, 1932, Union College. 77 p.

Wilson, Terry Jean. Stratigraphic and structural evolution of the Ultima Esperanza foreland fold-thrust belt, Patagonian Andes, southern Chile. D, 1983, Columbia University, Teachers College. 430 p.

Wilson, Thomas A. The geology near Mineral, California. M, 1961, University of California, Berkeley. 91 p.

Wilson, Thomas Carroll. Structure and stratigraphy of the Chestnut Ridge anticline near Uniontown, Pennsylvania. M, 1936, University of Pittsburgh.

Wilson, Thomas Carroll. The Burning Springs-Volcano Anticline, West Virginia. D, 1951, Cornell University.

Wilson, Thomas Hornor. Cross-strike structural discontinuities; tear faults and transfer zones in the Central Appalachians of West Virginia. D, 1980, West Virginia University. 265 p.

Wilson, Thomas M. Ground water yields and availability in fractured crystalline rocks, Wake County, North Carolina. M, 1982, North Carolina State University. 83 p.

Wilson, Thomas Vincent. Porosity reduction in a Cambrian quartz arenite; Galesville Sandstone, South central Wisconsin. M, 1977, Michigan State University. 77 p.

Wilson, Thomas Virgil. Prediction of baseflow for a Piedmont Watershed. D, 1972, North Carolina State University. 156 p.

Wilson, Todd Montgomery. Distribution and depositional environment of the Pennsylvanian Marchand Sandstone, Northwest Chickasha and Northwest Norge fields, Oklahoma. M, 1976, Oklahoma State University. 73 p.

Wilson, Tommie Claud. Numerical simulation of forward combustion in a radial system. D, 1970, University of Missouri, Rolla.

Wilson, V. V. The systematics and paleoecology of two late Pleistocene herpetofaunas from the Southeastern United States. D, 1975, Michigan State University. 74 p.

Wilson, Walter Byron. Study of clay slips in coal mines. M, 1914, University of Missouri, Columbia.

Wilson, Wendell Eugene. Trace element geochemistry and geochronology of early Precambrian granulite facies metamorphic rocks near Granite Falls in the Minnesota River valley. D, 1976, University of Minnesota, Minneapolis. 150 p.

Wilson, Wendell Eugene, Jr. The Bondoc mesosiderite; mineralogy and petrology of the metal nodules. M, 1972, Arizona State University. 74 p.

Wilson, Wesley Raphiel. Numerical calculation of electrical dispersion relations in rocks. M, 1977, University of Utah. 181 p.

Wilson, Wesley Raphiel. Thermal studies in a geothermal area. D, 1980, University of Utah. 156 p.

Wilson, Wilbur Dean. Subsurface Simpson Group, Crockett County, Texas. M, 1954, University of Texas, Austin.

Wilson, William Edward, III. The geology of the piedmont slopes in the Winnemucca area, Nevada. D, 1963, University of Illinois, Urbana. 227 p.

Wilson, William Feathergail. Sedimentary petrography and sedimentary structures of the Cambrian Hickory Sandstone Member, central Texas. M, 1962, University of Texas, Austin.

Wilson, William Harold. Petrology of the Wood River area, southern Absaroka Mountains, Park County, Wyoming. D, 1960, University of Utah. 150 p.

Wilson, William Harold. The (Pennsylvanian and Permian) Casper Formation in the Red Buttes area (Wyoming). M, 1950, University of Wyoming. 38 p.

Wilson, William Hugh. The distribution of copper in the mafic minerals of the Guichon Creek batholith. M, 1979, University of Toronto.

Wilson, William Jay. Subaqueous flow-markings in the Lower Mississippian strata of south-central Ohio and adjacent parts of Kentucky. M, 1950, University of Cincinnati. 37 p.

Wilson, William L. Lithologic and base level control of groundwater flow paths in the Garrison Chapel area, Indiana. M, 1985, Indiana State University. 211 p.

Wilson, William M. The stratigraphy of the Noonday Dolomite in the Clark Mountain thrust complex, San Bernardino County, California. M, 1975, Rice University. 49 p.

Wilson, Woodrow Pitkin. The geology of the south half of the Columbia Quadrangle, Boone County, Missouri. M, 1938, University of Missouri, Columbia.

Wilt, Jan C. Petrology and stratigraphy of the Colina limestone (Permian) in Cochise County, Arizona. M, 1969, University of Arizona.

Wilt, Michael J. An electrical survey of the Dunes geothermal anomaly and surrounding region, Imperial Valley, California. M, 1975, University of California, Riverside. 128 p.

Wiltenmuth, Kathleen. Probable source beds for the Woodbine oil (Upper Cretaceous) in East Texas. M, 1977, Tulane University.

Wilton, D. H. C. A genetic model for the Sustut copper deposit, North-central British Columbia. M, 1978, University of British Columbia.

Wilton, Derek Harold Clement. Metallogenic, tectonic and geochemical evolution of the Cape Ray fault zone with emphasis on electrum mineralization. D, 1984, Memorial University of Newfoundland. 618 p.

Wiltschko, David V. Mechanics of Appalachian Plateau structures. D, 1978, Brown University. 149 p.

Wiltse, Elliott Woodrow. Surface and subsurface study of the Southwest Davis oil field, sections 11 and 14, T 1 S, R 1 E, Murray County, Oklahoma. M, 1978, University of Oklahoma. 72 p.

Wiltse, Milton Adair, Jr. An electron microprobe investigation of a natural sulfide assemblage occurring at Balmat, New York. D, 1968, Indiana University, Bloomington. 352 p.

Wiltshire, John C. The origin and sedimentology of the Puna submarine canyon, Hawaii. D, 1983, University of Hawaii.

Wiman, S. K. Stratochronology and foraminiferal biostratigraphy of the Mio-Pliocene sedimentary sequence of central and northeastern Tunisia. D, 1976, University of Colorado. 184 p.

Wiman, W. David. Petrology of the Threemile Limestone (Lower Permian) in the Flint Hills of Kansas. M, 1966, Kansas State University.

Wimberg, William B. Analyses of liquid and gaseous hydrocarbons present in the New Albany Shale. M, 1979, Indiana University, Bloomington. 75 p.

Wimberley, C. Stanley. Marine sediments north of Scripps submarine canyon, La Jolla, California. M, 1954, University of Texas, Austin.

Wimberley, C. Stanley. Sediments of the Southern California mainland shelf. D, 1964, University of Southern California. 220 p.

Wimmer, Joseph L. Fracture occurrence in fresh water lake ice. M, 1977, University of Wisconsin-Milwaukee.

Win, Maung Soe. Geology and ore deposits of the Margaret Ann Mine, Silver Bow County, Montana. M, 1955, Montana College of Mineral Science & Technology. 72 p.

Win, Pe. Gravity survey of the Escalante Desert and vicinity, in Iron and Washington counties, Utah. M, 1980, University of Utah. 156 p.

Winans, Melissa Constance. Revision of North American fossil species of the genus Equus (Mammalia: Perissodactyla: Equidae). D, 1985, University of Texas, Austin. 282 p.

Winar, Richard Marion. The stratigraphy of the Molas Formation of southwestern Colorado. M, 1955, University of Illinois, Urbana.

Winbourn, Gary D. Geology, hydrology and hydrogeochemistry of the West Brush Creek development area, Beulah Mine, Mercer County, North Dakota. M, 1986, University of North Dakota. 178 p.

Winchell, Horace. The Honolulu Series (Pleistocene and Recent), Oahu, Territory of Hawaii. D, 1941, Harvard University. 215 p.

Winchell, Richard L. Relationship of the (Pennsylvanian) Lansing Group and the Tonganoxie (Stalnaker) Sandstone in south-central Kansas. M, 1957, University of Kansas. 76 p.

Winchell, Robert Eugene, Jr. X-ray study and synthesis of some copper-lead oxychlorides. D, 1963, Ohio State University. 245 p.

Winchester, Paul Drake. Carbonate diagenesis and cyclic sedimentation in the Wolfcampian Laborcita Formation of the Sacramento Mountains, New Mexico. D, 1976, Rice University. 176 p.

Winchester, Paul Drake. Geology of the Freeport (Texas) Rocks. M, 1971, Rice University. 310 p.

Winckel, E. E. and Stolz, Harry P. Subsurface correlation of a portion of the southwest flank of the Long Beach oil field, with particular reference to legal proceedings involving trespass deviational drilling. M, 1939, University of Southern California.

Winczewski, Laramie Martin. Paleocene coal-bearing sediments of the Williston Basin, North Dakota; an interaction between fluvial systems and an intracratonic basin. D, 1982, University of North Dakota. 298 p.

Wind, Frank. The stratigraphic zonation and paleoecology of conodonts of the Chase Group, lower Permian, of Kansas. M, 1973, Florida State University.

Wind, Frank H. Late Campanian and Maestrichtian calcareous nannoplankton biogeography and high-latitude biostratigraphy. D, 1979, Florida State University. 346 p.

Wind, Gerrit. Structural geology of the northern Dogtooth range, British Columbia (Canada). M, 1967, University of Calgary. 72 p.

Winder, Charles Gordon. Geology of the Mississippian terrain of the Truro-Green Oaks area, Nova Scotia, Canada. M, 1951, Cornell University.

Winder, Charles Gordon. Paleoecology and sedimentation of (Ordovician) Mohawkian limestones in south-central Ontario; a revision of Mohawkian and Cincinnatian stratigraphy. D, 1953, Cornell University.

Windfordner, John S. Geology and mineral resources of the University of Alabama property and adjoining areas in the vicinity of Rockhouse Creek, Tuscaloosa and Jefferson counties, Alabama. M, 1949, University of Alabama.

Windham, Steve R. Stratigraphy, paleontology, and structure of the Mississippian System in Ringgold Quadrangle, Georgia. M, 1956, Emory University. 92 p.

Windle, Delbert Leroy, Jr. Conemaugh cephalopods of Ohio. M, 1970, Ohio University, Athens. 139 p.

Windle, Delbert Leroy, Jr. Studies in Carboniferous nautiloids; cyrtocones and annulate orthocones. D, 1973, University of Iowa. 427 p.

Windom, Herbert L. Atmospheric dust records in glacial snow fields; implications to marine sedimentation. D, 1968, University of California, San Diego.

Windom, Kenneth Earl. The effect of reduced activity of anorthite on the reaction grossular + quartz = anorthite + wollastonite; a model for plagioclase in the Earth's lower crust and upper mantle. D, 1976, Pennsylvania State University, University Park. 36 p.

Windsor, J. G., Jr. Isolation and characterization of estuarine and marine sedimentary humic acids. D, 1976, College of William and Mary. 117 p.

Windsor, John Golay, Jr. Fatty acids and hydrocarbons in the surface waters of the York River. M, 1972, College of William and Mary.

Winegar, Robert Charles. The petrology of the Lost Creek (Powell County) Stock (late Cretaceous) and its relationship to the Mount Powell Batholith (late Cretaceous), Montana. M, 1971, University of Montana. 60 p.

Winegardner, Duane L. Hydrologic study of the Silurian aquifer of the Portage River Basin and adjacent Lake Erie tributary areas, Ohio. M, 1971, University of Toledo. 72 p.

Winester, Daniel. Basement features under the Southern Appalachians from gravity and magnetics. M, 1984, Georgia Institute of Technology. 104 p.

Wineteer, Craig Brian. Changes in fluvial style from the Lytle Formation (Lower Cretaceous) of Middle Park Basin, north-central Colorado. M, 1986, University of Colorado. 132 p.

Winfield, W. D. B. Volcanic, plutonic and Cu-Fe skarn rocks, Spout Lake, B.C. M, 1975, University of Western Ontario. 121 p.

Winfree, Keith Evan. Depositional environments of the St. Peter Sandstone of the upper Midwest. M, 1983, University of Wisconsin-Madison.

Winfrey, Elaine Clare. Mound growth, facies relationships, and diagenesis of Waulsortian carbonate mud mounds, Big Snowy Mountains, Montana. M, 1983, University of Iowa. 227 p.

Wing, Charles G. An on-the-bottom-sea gravimeter. D, 1966, Massachusetts Institute of Technology. 219 p.

Wing, Lawrence Alvin. Serpentine and asbestos in Maine. M, 1951, University of Maine. 56 p.

Wing, Margaret M. Genesis of the copper deposits of the United States. M, 1935, Columbia University, Teachers College.

Wing, Monta Eldo. Geology of the Lawrence area. M, 1921, University of Kansas. 112 p.

Wing, Monta Eldo. The Silurian Gastropoda of northeastern Illinois. D, 1923, University of Chicago. 111 p.

Wing, Richard Sherman. Structural analysis from radar imagery, eastern Panamanian isthmus. D, 1970, University of Kansas. 192 p.

Wing, Richard Sherman. The Frontier Formation of the Hanna Basin, Wyoming. M, 1950, University of Wisconsin-Madison. 79 p.

Wing, Robert Busch. The igneous rock types of the eastern half of the St. Francois Mountains. M, 1932, Washington University. 62 p.

Wing, Robert Claude. Prospective unit operation of oil and gas reservoirs in Pennsylvania. M, 1953, University of Pittsburgh.

Wing, Samuel James Courtney. Cementation and diagenesis in the Upper Cretaceous Frenchman Formation in the Frenchman River valley, Saskatchewan. M, 1984, University of Saskatchewan. 124 p.

Wing, Scott Louis. A study of paleoecology and paleobotany in the Willwood Formation (early Eocene, Wyoming). D, 1981, Yale University. 416 p.

Wingard, Norman Edward. A heavy mineral investigation of glacial tills in western New York. M, 1962, Michigan State University. 64 p.

Wingard, Norman Edward. Economic and petrographic evaluation of gravel resources in southern Michigan. D, 1969, Michigan State University. 124 p.

Wingard, Paul Sidney. Geology of the Castine--Blue Hill area, Maine. D, 1961, University of Illinois, Urbana. 147 p.

Wingard, Paul Sidney. Structural and petrographic analysis of the Colony Pegmatite, Alstead, New Hampshire. M, 1955, Miami University (Ohio). 42 p.

Wingate, Frederick Huston. Geology of part of northeastern Uinta County, Wyoming. M, 1961, University of Utah. 104 p.

Wingate, Frederick Huston. Palynology of the Denton Shale (lower Cretaceous) of southeastern Oklahoma. D, 1974, University of Oklahoma. 300 p.

Wingert, Everett Arvin. Potential role of optical data processing in geo-cartographic spatial analysis. D, 1973, University of Washington. 281 p.

Wingert, John Richard. Geology of the Bald Mountain area, Bighorn Mountains, Wyoming. M, 1958, University of Iowa. 110 p.

Wingerter, Jeffrey Hush. Depositional environment of the Revett Formation, Precambrian Belt Supergroup, northern Idaho and northwestern Montana. M, 1982, Eastern Washington University. 119 p.

Wingfield, Betsey C. A marine platform in southern New England; implications for the regional tectonic history. M, 1988, University of Connecticut. 52 p.

Wingo, James Raymond. Velocity and attenuation analysis of Gulf Coast sediments using vertical seismic profiling. M, 1981, Massachusetts Institute of Technology. 89 p.

Winkelbauer, Howard M. Experimental studies concerning solid solution, phase relations and geothermometry of wolframite. M, 1974, Miami University (Ohio). 83 p.

Winkelmaier, Joseph R. Ground water flow characteristics in fractured basalt in a zero order basin. M, 1987, University of Idaho. 128 p.

Winker, Charles David. Late Pleistocene fluvial-deltaic deposition, Texas coastal plain and shelf. M, 1979, University of Texas, Austin.

Winker, Charles David. Neogene stratigraphy of the Fish Creek-Vallecito section, Southern California; implications for early history of the northern Gulf of California and Colorado Delta. D, 1987, University of Arizona. 622 p.

Winker, Gregory James. Petrography and diagenesis of the Hosston Sandstone, Washington Parish, Louisiana. M, 1982, Texas A&M University.

Winkle, Candace J. The paleoenvironment of the Cedar Valley Limestone (Middle Devonian) in northwestern Illinois. M, 1979, University of Wisconsin-Milwaukee.

Winkler, Bruno Oscar. An amplifier suitable for use in applied seismology. D, 1933, Colorado School of Mines. 45 p.

Winkler, Dale Alvin. Paleoecology and taphonomy of an early Eocene mammalian fauna in the Clarks Fork Basin, northwestern Wyoming (U.S.A.). M, 1981, University of Michigan.

Winkler, Dale Alvin. Stratigraphy, vertebrate paleontology and depositional history of the Ogallala Group in Blanco and Yellowhouse canyons, northwestern Texas. D, 1985, University of Texas, Austin. 278 p.

Winkler, Fred E. Analysis of gravity data from the Aubrey Valley area, Coconino and Yavapai counties, Arizona. M, 1986, Northern Arizona University. 122 p.

Winkler, Gary R. Sedimentology and geomorphic significance of the Bishop conglomerate (Oligocene or Miocene) and the Browns Park Formation (Miocene?), eastern Uinta mountains, Utah, Colorado, and Wyoming. M, 1970, University of Utah. 115 p.

Winkler, Hans. Paleozoic geology of the Shovel Mountain area, Blanco Quadrangle, Texas. M, 1929, University of Texas, Austin.

Winkler, Hartmut A. Proposed apparatus and methods for a study of seismic energy transmission. M, 1951, New Mexico Institute of Mining and Technology.

Winkler, K. W. The effects of pore fluids and frictional sliding on seismic attenuation. D, 1979, Stanford University. 195 p.

Winkler, Marjorie Green. Late-glacial and Holocene environmental history of south-central Wisconsin; a study of upland and wetland ecosystems. D, 1985, University of Wisconsin-Madison. 292 p.

Winkler, Patrick Lynn. Source mechanisms of earthquakes associated with coal mines in eastern Utah. M, 1972, University of Utah. 75 p.

Winkler, Ron. Radionuclide geochronology in Beaver Lake, Wisconsin. M, 1984, University of Wisconsin-Milwaukee. 128 p.

Winkler, Virgil D. Stratigraphic geology of the (Upper Mississippian) Pennington of eastern Kentucky and east-central Tennessee. D, 1941, University of Illinois, Urbana. 84 p.

Winkler, Virgil Dean. Phanocrinus and related Mississippian Crinoidea. M, 1939, University of Illinois, Urbana.

Winland, Hubert Dale. Diagenesis of carbonate grains in marine and meteoric waters. D, 1971, Brown University.

Winland, Hubert Dale. Insoluble residue study in correlation of the Arbuckle Group in southern Oklahoma. M, 1955, University of Tulsa. 121 p.

Winn, Cathie Eileen. Vegetational history and geochronology of several sites in south southwestern Ontario with discussion on mastodon extinction in southern Ontario. M, 1978, Brock University. 374 p.

Winn, Peter Stewart. Structure and petrology of Precambrian rocks in the Black Canyon area, Yavapai County, Arizona. M, 1982, University of Utah. 101 p.

Winn, R. D., Jr. Late Mesozoic flysch of Tierra del Fuego and South Georgia Island; a sedimentologic approach to lithosphere plate restoration. D, 1975, University of Wisconsin-Madison. 172 p.

Winn, Robert. A sandstone body analysis and paleocurrent study of the Medicine Bow Formation, Hanna and Carbon basins, Wyoming. M, 1971, Lehigh University.

Winn, Robert Maurice. Clarification of lake water prior to artificial recharge by wells. M, 1960, Texas Tech University. 56 p.

Winn, Robert Maurice. Hydrogeology of the Albian Formation, Algeria. D, 1973, Texas Tech University. 126 p.

Winn, Ronald Frederick. Postglacial paleoecology and effects of European settlement on the environment of Lake Hunger and Lake Lisgar, southwestern Ontario. M, 1976, Brock University. 170 p.

Winn, Vernard. Geology of the Carrolton Quadrangle, Dallas and Denton counties, Texas. M, 1953, Southern Methodist University. 16 p.

Winn, William M. A study of the relation between discharge and suspended sediment in rivers and streams. D, 1984, The Johns Hopkins University. 631 p.

Winningham, Bruce R. Speciation and fate of heavy metals in a Xenia silt loam amended with milorganite (sewage sludge) in southwestern Ohio. M, 1988, Miami University (Ohio).

Winograd, Isaac J. Groundwater conditions and geology of Sunshine Valley and western Taos County, New Mexico. M, 1958, Columbia University, Teachers College.

Winograd, Issac Judah. Origin of major springs in the Amargosa Desert of Nevada and Death Valley, California. D, 1971, University of Arizona. 278 p.

Winslow, Donald Clarence. Laboratory study of lake ice. M, 1952, University of Michigan.

Winslow, John Durfee. The stratigraphy and bedrock topography of Portage County, Ohio, and their relation to ground-water resources. D, 1957, University of Illinois, Urbana. 121 p.

Winslow, Marcia Ring. Plant microfossils from Upper Devonian and Lower Mississippian rocks of Ohio. M, 1954, Ohio State University.

Winslow, Margaret Anne. Mesozoic and Cenozoic tectonics of the fold and thrust belt in southernmost South America and stratigraphic history of the Cordilleran margin of the Magallanes Basin. D, 1980, Columbia University, Teachers College. 409 p.

Winslow, Michael L. The paleoecology and stratigraphy of the Harpersville Formation, southwestern Stephens County, Texas. M, 1983, University of Texas, Arlington. 196 p.

Winslow, Nancy S. Timing and tectonic significance of Upper Jurassic and Lower Cretaceous nonmarine sediments, Bighorn Basin, Wyoming and Montana. M, 1986, University of Wyoming. 133 p.

Winsor, Henry. The paleogeography and paleoenvironments of the Middle Pennsylvanian part of the Wood River Formation in south-central Idaho. M, 1981, San Jose State University. 81 p.

Winsor, Mark F. Altitudinal distribution of granitic landforms in the south-central Sierra Nevada, California. D, 1984, University of Wisconsin-Milwaukee. 498 p.

Winsten, Miriam S. Chemical and mineralogical differences of marine and non-marine shales of the Dakota Sandstone and adjacent units, southwestern Colorado. M, 1982, Bowling Green State University. 124 p.

Winston, Donald, II. Geology of the Leon Creek area, Mason County, Texas; paleontology of the upper Wilberns Formation. M, 1957, University of Texas, Austin.

Winston, Donald, II. Stratigraphy and carbonate petrology of the Marble Falls Formation, Mason and Kimble counties, Texas. D, 1963, University of Texas, Austin. 477 p.

Winston, George Otis. An interpretation of structural conditions in the Griffin oil field. M, 1947, Indiana University, Bloomington. 16 p.

Winston, John G. Crystallographic and electrical properties of pyrrhotite grown by vapor transport reactions. M, 1972, University of Wisconsin-Milwaukee.

Winston, Michael R. Geology of the Silver Monument area, Sierra County, New Mexico. M, 1975, University of Texas at El Paso.

Winston, Richard Baury. Paleoecology and taphonomy of Middle Pennsylvanian-age coal-swamp plants; Herrin Coal-ball peat and coal, Peabody Camp 11 Mine, Herrin Coal, western Kentucky. D, 1986, University of Illinois, Urbana. 149 p.

Winter, Bryce. Isotopic investigation of carbonate fracture fills in the Monterey Formation, California. M, 1987, Arizona State University. 141 p.

Winter, Claud Victor. Geological structure of part of Boracho Quadrangle, Culberson County, Texas. M, 1951, University of Texas, Austin.

Winter, Gary Allan. The mineralogy and petrology of the carbonates and pyroxenoids from the manganese deposit near Bald Knob, North Carolina. M, 1977, University of Michigan.

Winter, Gerry Vernon. Ground water flow systems in the phosphate sequence, Caribou County, Idaho. M, 1979, University of Idaho. 120 p.

Winter, Johannes Antonius Franciscus. Fredericksburg and Washita strata (subsurface Lower Cretaceous), Southwest Texas. D, 1961, University of Texas, Austin. 194 p.

Winter, John Keith. The structure and crystal chemistry of high and low albite and the Al_2SiO_5 polymorphs at high temperatures. D, 1978, University of Washington. 137 p.

Winter, John Keith. Trace element concentrations in metaperidotites and associated gneisses compared to contained micas. M, 1971, University of Washington. 66 p.

Winter, Richard. Form and process of a gravel-bearing stream in western New York. M, 1974, SUNY, College at Fredonia. 79 p.

Winter, Thomas C. A pollen analysis of Kirchner Marsh, Dakota County, Minnesota. M, 1961, University of Minnesota, Minneapolis. 35 p.

Winter, Thomas C. The hydrologic setting of lakes in Minnesota and adjacent states with emphasis on the interaction of lakes and ground water. D, 1976, University of Minnesota, Minneapolis. 251 p.

Winterer, Edward Litton. Geology of southeastern Ventura Basin, Los Angeles County, California. D, 1954, University of California, Los Angeles.

Winterer, Joan I. Biostratigraphy of Bouse Formation; a Pliocene Gulf of California deposit in California, Arizona, and Nevada. M, 1975, California State University, Long Beach. 132 p.

Winterfeld, Gustav F. Geology and mammalian paleontology of the Fort Union Formation, eastern Rock Springs Uplift, Sweetwater County, Wyoming. M, 1979, University of Wyoming. 188 p.

Winterfeld, Gustav F. Laramide tectonism, deposition, and early Cenozoic stratigraphy of the northwestern Wind River basin and Washakie Range, Wyoming. D, 1986, University of Wyoming. 446 p.

Wintermute, Thomas Judson. Stratigraphy of the Tonoloway and Keyser limestones near Bedford, Pennsylvania. M, 1952, Pennsylvania State University, University Park. 61 p.

Winters, Alec T. Extensional faulting in the MARK area. M, 1986, Duke University. 68 p.

Winters, Allen S. Geology and ore deposits of the Castle Mountain mining district, Meagher County, Montana. M, 1965, Montana College of Mineral Science & Technology. 96 p.

Winters, Dermont. Pittsburgh Red Beds; stratigraphy and slope stability in Allegheny County, Pennsylvania. M, 1972, University of Pittsburgh.

Winters, H. H. The Pleistocene fauna of the Manix Beds in the Mojave Desert, California. M, 1954, California Institute of Technology. 68 p.

Winters, Harold Abraham. The late Pleistocene geomorphology of the Jamestown area, North Dakota. D, 1960, Northwestern University. 168 p.

Winters, Harry J., Jr. A programmed method for the computation of brine and solid. M, 1966, University of Arizona.

Winters, Jay Arthur. Comparative sandstone and mudrock diagenesis in the Jackfork Group, northern flank of the Broken Bow-Benton Anticline, southeastern Oklahoma. M, 1984, University of Oklahoma. 62 p.

Winters, Mark B. An investigation of fluid inclusions and geochemistry of ore formation in the Cedar Creek breccia pipe, North Santiam mining district, Oregon. M, 1985, Western Washington University. 104 p.

Winters, Martha Diane. A paleomagnetic study of Early Tertiary basalts in west Texas. M, 1968, Rice University. 80 p.

Winters, Stephen S. Supai Formation (Permian) of eastern Arizona. D, 1964, Columbia University, Teachers College.

Winters, Stephen Samuel. A Permian sequence in eastern Arizona. D, 1955, Columbia University, Teachers College.

Winters, Stephen Samuel. Stratigraphy of the Lower Permian in the Fort Apache Indian Reservation, Arizona. M, 1948, Columbia University, Teachers College.

Winters, Steven L. In situ retardation of trace organics in groundwater discharge to a sandy stream bed. M, 1984, University of Waterloo. 95 p.

Winters, Warren Jon. Stratigraphy and sedimentology of Paleogene arkosic and volcaniclastic strata, Johnson Creek-Chambers Creek area, southern Cascade Range, Washington. M, 1984, Portland State University. 160 p.

Wintringham, Neil Andrews. The geology of Dutchess County, New York. M, 1947, Cornell University.

Wintsch, Robert P. Mineral-solution equilibria and the tectonics of eastern Connecticut. D, 1975, Brown University. 101 p.

Wintz, Edward K. Geodetic field use of an optically visible Earth satellite. M, 1961, Ohio State University.

Wintz, Edward K. Geometric applications of artificial Earth satellite oberservations. D, 1965, Ohio State University.

Winzeler, Ted J. Petrology and evolution of Silurian reef and associated rocks, Buckland Quarry, Ohio. M, 1974, Bowling Green State University. 103 p.

Winzer, Stephen R. Metamorphism and chemical equilibrium in some rocks from the Kootenay Arc. D, 1973, University of Alberta. 335 p.

Wipf, Robert A. Effects of joint fillings on the mechanical behavior of rocks. M, 1974, University of Wisconsin-Milwaukee.

Wire, Jeremy Crosby. Geology of the Currant Creek District, Nye and White Pine counties, Nevada. M, 1961, University of California, Los Angeles.

Wireman, Michael. Nitrate pollution of groundwater in glacial sediments underlying a fertigated site in Kalamazoo County, Michigan. M, 1987, Western Michigan University.

Wirey, Gary Lee. A subsurface study of lower Chester rocks in the Union-Bowman Consolidated Pool, Pike and Gibson counties, Indiana. M, 1958, Indiana University, Bloomington. 41 p.

Wiringa, Leon Otis. The pre-Cambrian geology of Spokane District, Black Hills, South Dakota. M, 1932, University of Iowa. 102 p.

Wirojanagud, Prakob. Saline water upconing due to pumpage in unconfined aquifers. D, 1982, University of Texas, Austin. 229 p.

Wirojanagud, Wanpen. Particle size distributions in flocculent sedimentation. D, 1983, University of Texas, Austin. 272 p.

Wirth, Gerald Sheldon. The heterogeneous kinetics of silica dissolution in aqueous media. D, 1980, University of California, San Diego.

Wirth, Karl R. The geology and geochemistry of the Grao Papa Group, Serra dos Carajas, Para, Brazil. M, 1986, Cornell University. 284 p.

Wise, Diane. Paleozoic geology of the Dobbin Summit-Clear Creek area, Monitor Range, Nye County, Nevada. M, 1977, Oregon State University. 137 p.

Wise, Donald Underkofler. Tectonics and tectonic heredity in the southern Beartooth Mountains, Wyoming. D, 1957, Princeton University. 198 p.

Wise, Henry M. Geology and petrography of igneous intrusions of northern Hueco Mountains, El Paso and Hudspeth counties, Texas. M, 1977, University of Texas at El Paso.

Wise, James Charlton. Cambrian stratigraphy of the Sandy Post Office area, Blanco County, Texas. M, 1964, University of Texas, Austin.

Wise, Joseph Patrick. Geology and petrography of a portion of the Galice Formation, Babyfoot Lake area, Southwest Oregon. M, 1969, Idaho State University. 108 p.

Wise, Karen Ann Winterhoff. A petrographic analysis of the Strawn Limestone (Middle Pennsylvanian) within the Millican Field, Coke County, Texas. M, 1985, Stephen F. Austin State University. 105 p.

Wise, Michael Anthony. Geochemistry and crystal chemistry of Nb, Ta and Sn minerals from the Yellowknife pegmatite field, N.W.T. D, 1987, University of Manitoba.

Wise, Michael Terence. The Paleozoic biostratigraphy of the East Dobbin Creek area, northern Nye County, Nevada. M, 1977, University of Oregon. 156 p.

Wise, Richard Atlee. Eastern Front Range and foothills geology of central Jefferson County, north-central Colorado. D, 1952, Colorado School of Mines. 192 p.

Wise, Sherwood Willing, Jr. Scanning electron microscope study of Molluscan shell ultrastructures. D, 1970, University of Illinois, Urbana. 145 p.

Wise, Sherwood Willing, Jr. The ultrastructure of some selected mollusk shells. D, 1965, University of Illinois, Urbana.

Wise, Thomas W. Significance of solution cleavage in the Chapman Ridge Anticline, Tennessee. M, 1980, University of Tennessee, Knoxville. 101 p.

Wise, William S. Paragenesis of some impure marbles from the Santa Lucia Mountains, California. M, 1958, Stanford University.

Wise, William Stewart. The geology and mineralogy of the Wind River valley area, Washington and the stability of celadonite; 2 volumes. D, 1961, The Johns Hopkins University.

Wisehart, Richard McGhee. Paleoenvironmental analysis of the Bear Lake Formation (upper and middle Miocene), Alaska Peninsula, Alaska. M, 1971, University of California, Los Angeles.

Wiseman, Frederick Mathew. Paleoecology and the prehistoric Maya; a history of man-land relationships in the tropics. M, 1974, University of Arizona.

Wiseman, Frederick Matthew. Agricultural and historical ecology of the Lage region of Peten, Guatemala. D, 1978, University of Arizona.

Wisham, C. M., Jr. The geology of the Honolua area of West Maui, Hawaii. M, 1975, University of Hawaii at Manoa. 46 p.

Wishart, James Scotland. The chemical composition of chromite as related to its manner of origin. D, 1935, Princeton University.

Wishart, James Scotland. The geology of Disappointment Lake, Lake County, Minnesota. M, 1929, University of Rochester. 80 p.

Wisker, George E. Late Holocene evolution and stratigraphic units of Eatons Neck Basin, Long Island, New York. M, 1980, University of Delaware.

Wismer, Raymond J. The accessory minerals of the Magnet Cove igneous complex. M, 1937, University of Kansas. 49 p.

Wisniowiecki, Michael James. A Pennsylvanian paleomagnetic pole from the mineralized, Late Cambrian Bonne Terre Formation, Southeast Missouri. M, 1982, University of Michigan.

Wissig, George Conrad, Jr. Bedrock geology of the Ossining, New York, Quadrangle. D, 1970, Syracuse University. 230 p.

Wissinger, Diane E. Paleoecology and paleoenvironment of the Brereton Limestone (Pennsylvanian, Desmoinesian) in a portion of the Illinois Basin. M, 1987, Southern Illinois University, Carbondale. 100 p.

Wissler, Stanley G. Methods of deposition of sources of oil. M, 1923, Columbia University, Teachers College.

Wissler, Thomas Martin. Sandstone pore structure; a quantitative analysis of digital SEM images. D, 1987, Massachusetts Institute of Technology. 566 p.

Wissmann, Gerd. Marine geophysical studies offshore Cape Flattery, Washington, and the implications for regional tectonics. M, 1971, University of Washington. 96 p.

Wiswall, Charles Gilbert. Structural styles of the southern boundary of the Sapphire tectonic block, Anaconda-Pintlar Wilderness area, Montana. M, 1976, University of Montana. 62 p.

Wiswall, Charles Gilbert. Structure and petrography below the Bitterroot Dome, Idaho Batholith, near Paradise, Idaho. D, 1979, University of Montana. 129 p.

Witanachchi, Channa Devinda. Metamorphism and deformation in the Wissahickon Schist near Bryn Mawr, Pennsylvania. M, 1986, Bryn Mawr College. 82 p.

Witczak, Matthew Walter. A generalized investigation of selected highway design and construction factors by regional geomorphic units within the continental United States (volumes I and II). D, 1970, Purdue University. 420 p.

Witebsky, Susan Nadine. Paleontology and sedimentology of the Haymond boulder beds (Martin Ranch), Marathon Basin, Trans-Pecos, Texas. M, 1987, University of Texas, Austin. 121 p.

Witham, Roger. Geology and slope movement hazards of the Livermore Ranch area, Lake and Napa counties, California. M, 1981, San Jose State University. 184 p.

Witherbee, Kermit G. Environmental geology and computer mapping of the Oneonta Quadrangle, N.Y.; a study of land development limitations based on natural factors. M, 1979, SUNY, College at Oneonta. 87 p.

Withers, James Gordon, Jr. The suitability of activated carbon for removal of gum-forming hydrocarbons from synthesis gas. M, 1947, West Virginia University.

Withers, James Henry. Sliding resistance along discontinuities in rock masses. D, 1964, University of Illinois, Urbana. 133 p.

Withers, Katrina D. Grain shape variations in late Pleistocene and Holocene fluvial and shelf sands in the northwestern Gulf of Mexico, and the relationship of source and shelf paleogeography. M, 1984, Texas A&M University.

Withers, Robert John. Seismicity and stress determination at man-made lakes. D, 1977, University of Alberta. 241 p.

Withers, Robert Louis. Mineral deposits of the Northrim Mine and a brief enquiry into the genesis of veins of the (Ag, Bi, Ni, Co, As) type. M, 1979, University of Alberta. 271 p.

Withers, Ronald Carlton. A geologic study of the post-Meramec pre-Missouri sediments of Bradley area in central Oklahoma. M, 1960, University of Oklahoma. 60 p.

Witherspoon, A. J. Optical mineralogy of the rocks of Lassen Volcanic National Park. M, 1955, University of California, Berkeley. 52 p.

Witherspoon, James Mark. Groundwater resource analysis of the Arkansas Ouachita Mountains. M, 1979, Indiana State University. 61 p.

Witherspoon, Paul Adams, Jr. Studies on petroleum with the ultracentrifuge. D, 1957, University of Illinois, Urbana.

Witherspoon, Robert R. Geology of the western half of the Big Ridge Quadrangle, North Carolina. M, 1982, University of New Orleans. 80 p.

Witherspoon, William Dale. Structure of Blue Ridge thrust front, Tennessee, Southern Appalachians. D, 1981, University of Tennessee, Knoxville. 165 p.

Withington, Charles Francis. Geology of the Paradox Quadrangle, Montrose County, Colorado. M, 1949, University of Rochester. 87 p.

Withjack, Martha O. An analytical investigation of the mechanics of continental rifting. D, 1978, Brown University. 114 p.

Withrow, Jon R. A subsurface study of Tuscaloosa (Gulfian Cretaceous) formations in the area of Tensas Parish, Louisiana. M, 1963, University of Oklahoma. 66 p.

Withrow, Philip Charles. The subsurface geology of the Maysville area, Township 4 North, Range 2 West, Garvin County, Oklahoma. M, 1957, University of Oklahoma. 42 p.

Withstandley, D. W., III. Short-period microseisms in Northern California. M, 1952, University of California, Berkeley. 72 p.

Witinok, Patricia Mary. Distributed watershed and sedimentation model. M, 1979, University of Iowa. 246 p.

Witkind, Irving Jerome. Cretaceous from the middle Magdelana Valley, Colombia, South America. M, 1941, Columbia University, Teachers College.

Witkind, Irving Jerome. Geology and ore deposits of the Monument Valley area, Apache and Navajo counties, Arizona. D, 1956, University of Colorado. 284 p.

Witkowski, Marlene Ann. Characterization of suspended sediments in bottom nepheloid layers, Nova Scotian shelf, Canada. M, 1974, University of Wisconsin-Milwaukee.

Witkowski, Robert. The morphology and chemical composition of biotites from various geological origins. M, 1973, University of Pittsburgh.

Witkus, M. A. The petrology of the Winnsboro composite dike. M, 1973, University of South Carolina. 122 p.

Witmer, Daniel C. An investigation of the inclusion of the concept of organic evolution in junior high schools Earth Science. M, 1968, Millersville University.

Witmer, Richard Everett. Waveform analysis of geographic patterns recorded on visible and infrared imagery. D, 1967, University of Florida. 225 p.

Witmer, Roger J. Taxonomy and biostratigraphy of lower Tertiary dinoflagellate assemblages from the Atlantic Coastal Plain near Richmond, Virginia. M, 1975, Virginia Polytechnic Institute and State University.

Witmer, Roger J. Tertiary dinoflagellate, acritarch, and chlorophyte assemblages from the Oak Grove core, Virginia coastal plain. D, 1987, Virginia Polytechnic Institute and State University. 693 p.

Witner, Thomas W. Sedimentology of the lower Chowan River, North Carolina. M, 1984, University of North Carolina, Chapel Hill. 102 p.

Witous, John M. Infrared absorption trends of some trioctahedral micas. M, 1969, University of South Florida, Tampa. 76 p.

Witrock, R. B. Longevity and growth rate by direct ageing of the scallop Chesapecten nefrens from different biofacies of the Choptank Formation, Miocene of Maryland. M, 1978, Queens College (CUNY). 135 p.

Witschard, Maurice G. Contrasting structural style in siliclastic and carbonate lithologies of an offscraped sequence; the Peralta accretionary prism, Hispaniola. M, 1988, University of California, Santa Cruz.

Witt, William John. A geophysical interpretation of Jeptha Knob, Shelby County, Kentucky. M, 1951, University of Cincinnati. 36 p.

Wittbrodt, Paul Raymond. Paleomagnetism and petrology of St. Matthew Island, Bering Sea, Alaska. M, 1985, University of Alaska, Fairbanks. 210 p.

Witte, Duncan M. Geology of the massive sulfide deposits of the Silver Peak District, Douglas County, Oregon. M, 1977, Stanford University. 75 p.

Witte, Herman C. Geology of the Limekiln Canyon and Four Eyes Canyon areas, south-westernmost Montana. M, 1965, Pennsylvania State University, University Park. 85 p.

Witte, Ron W. The surficial geology and Woodfordian glaciation of a portion of the Kittatinny Valley and the New Jersey Highlands in Sussex County, New Jersey. M, 1988, Lehigh University. 281 p.

Witte, William K., Jr. Paleomagnetism and paleogeography of Cretaceous northern Alaska. M, 1982, University of Alaska, Fairbanks. 146 p.

Wittels, Mark C. A thermo-chemical analysis of some amphiboles. D, 1951, Massachusetts Institute of Technology. 127 p.

Wittels, Razel A. Tectonic overpressure in Franciscan terrains (Pacific coast). M, 1969, Massachusetts Institute of Technology. 25 p.

Witteman, John P. The computer simulation of phosphate removal from wastewater using lime. M, 1976, McMaster University. 120 p.

Witter, David N. Stratal architecture and volumetric distribution of facies tracts, upper Mission Canyon Formation (Mississippian), Williston Basin, North Dakota. M, 1988, Colorado School of Mines.

Witter, Donald Paul, Jr. Conodont biostratigraphy of the Upper Devonian in the Globe-Mammoth area, Arizona. M, 1976, University of Arizona.

Witter, Robert Allen. Radiochemical neutron activation analysis of zinc and antimony in Group IVA iron meteorites. M, 1978, University of California, Los Angeles.

Wittine, Arthur H. A study of a late Pleistocene lake in the Cuyahoga River valley, Summit County, Ohio. M, 1970, Kent State University, Kent. 64 p.

Wittke, James Henry. Geochemistry and isotope geology of basalts of the Arizona transition zone (Yavapai

County) and their tectonic significance. D, 1984, University of Texas, Austin. 255 p.

Wittman, Glenn Howard. Determination of the groundwater flow regime of a seepage lake, Silver Lake, Summit County, Ohio. M, 1979, Kent State University, Kent. 63 p.

Wittoesch, David F. Data processing of a seismic line in an area of thick glacial till using surface consistent automatic statics; Osceola County, Michigan. M, 1987, Wright State University. 69 p.

Wittpenn, Nancy Ann. Inversions of magnetic anomalies in the South Atlantic and quantification of observed variations in magnetization solutions. M, 1987, University of Miami. 151 p.

Wittreich, Curtis D. Methods of analysis in geological engineering. M, 1987, University of Idaho. 169 p.

Wittrup, Mark B. The origin of water leaks in Saskatchewan potash mines. M, 1988, University of Saskatchewan. 116 p.

Wittstrom, Martin D., Jr. Sedimentology of the Leadville Limestone (Mississippian), northeastern Gunnison County, Colorado. M, 1979, Colorado School of Mines. 159 p.

Wittur, Glen Eric. Domestic processing of mine output in Canada, with case studies on zinc and copper refining. D, 1974, Pennsylvania State University, University Park. 450 p.

Witty, John Edward. Classification and distribution of soils on Whiteface Mountain, Essex County, New York. D, 1968, Cornell University. 301 p.

Witucki, Gerald Stanislaus. A study of subsidence in the Wilmington oil field, California. M, 1960, University of California, Los Angeles.

Witzel, Frank Lewis. Guidebook and road logs for the geology of Dawei and northern Sioux counties, Nebraska. M, 1974, Chadron State College.

Witzel, William Thomas. The resources and industries of the New River drainage basin in West Virginia. M, 1958, University of Tennessee, Knoxville.

Witzke, Brian J. Echinoderms of the Hopkinton Dolomite (Lower Silurian), eastern Iowa. M, 1976, University of Iowa. 224 p.

Witzke, Brian J. Stratigraphy, depositional environments, and diagenesis of the eastern Iowa Silurian sequence. D, 1981, University of Iowa. 547 p.

Wixon, Roy Stephen. Systematics and biostratigraphy of brachiopods from the Chickamauga Group (Middle Ordovician) in the Alabama Valley and Ridge. M, 1984, University of Alabama. 157 p.

Wixted, James Bernard. Sedimentary aspects of the Livingston Conglomerate, southeastern Kentucky. M, 1977, University of Kentucky. 63 p.

Wiygul, Gary J. A subsurface study of the lower Tuscaloosa Formation at Olive Field, Pike and Amite counties, Mississippi. M, 1987, Northeast Louisiana University. 145 p.

Wnuk, Christopher. The paleoecology and growth habits of arborescent lycopods from an in situ swamp forest of Middle Pennsylvanian age (Bernice Basin, Sullivan County, Pennsylvania). D, 1984, University of Pennsylvania. 162 p.

Woakes, Michael Edward. Potassium-argon dating of mineralization at Butte, Montana. M, 1960, University of California, Berkeley. 42 p.

Wobber, Francis John. Eocene carbonates of Sind, West Pakistan. M, 1961, University of Illinois, Urbana.

Wobus, Reinhard Arthur. Petrology and structure of Precambrian rocks of the Puma Hills, Southern Front Range, Colorado. D, 1966, Stanford University. 146 p.

Wodzicki, Antoni. The distribution of some trace elements near contact aureoles in the Santa Rosa Range, Nevada. D, 1965, Stanford University. 94 p.

Woeber, A. Frederick. Elastic constants of selected rocks and minerals. M, 1961, Rensselaer Polytechnic Institute. 54 p.

Woeber, Arthur Frederick. Influence of layer parameters on leaking modes. D, 1969, Rice University. 80 p.

Woeller, R. M. Greenbrook well field management study 1981-1982. M, 1982, University of Western Ontario.

Woerner, Eric Gerard. A palynological investigation of the Mary Lee and Blue Creek coal seams in the northern Black Warrior Basin, Alabama. M, 1981, Auburn University. 115 p.

Woerns, Norbert M. Landscape geochemical investigations in the vicinity of a lead deposit near Snertingdal, southern Norway. M, 1976, Brock University. 194 p.

Woessner, William W. Hydrostratigraphy, flow system analyses and mining impact analyses of the coal bearing Fort Union Formation and related deposits, northern Cheyenne Reservation, Montana. D, 1978, University of Wisconsin-Madison. 208 p.

Woessner, William Wendling. Hydrogeology of four solid waste disposal sites in Alachua County, Florida. M, 1973, University of Florida. 155 p.

Wofford, Melissa K. Young. Skarn formation adjacent to the Whitehorn Stock, Chaffee County, Colorado. M, 1986, Texas Tech University. 72 p.

Wogsland, Karen L. The effect of urban storm water injection by Class V wells on the Missoula Aquifer, Missoula, Montana. M, 1988, University of Montana. 133 p.

Wohl, Ellen Eva. Northern Australian paleofloods as paleoclimatic indicators. D, 1988, University of Arizona. 291 p.

Wohlabaugh, Norman. Petrology of the Big Cusp Algal Dolomite; an informal member of the Kona Dolomite, Marquette, Michigan. M, 1980, Bowling Green State University. 167 p.

Wohlberg, Elwood Geron. Genesis of a meta-gabbroic sill in the Amisk Lake area, northern Saskatchewan. M, 1964, University of Saskatchewan.

Wohletz, Kenneth Harold. A model of pyroclastic surge. M, 1977, Arizona State University. 174 p.

Wohletz, Kenneth Harold. Explosive hydromagmatic volcanism. D, 1980, Arizona State University. 303 p.

Wohlford, Duane Dennis. Petrology and Structure of Precambrian rocks in the Cache La Poudre Canyon-Livermore area, Larimer County. D, 1965, University of Colorado. 147 p.

Woidneck, Robert Keith. Thermal waters of northwestern Baja California between Guadalupe and San Vicente. M, 1979, San Diego State University.

Wojciechowski, Walter Anthony. Gulf Coast salt domes; their distribution description and origin. M, 1952, University of Michigan.

Wojdak, P. J. Alteration at the Sam Goosly copper-silver deposit, British Columbia. M, 1974, McMaster University.

Wojniak, Wayne Stanley. A magnetotelluric sounding at the Tucson Magnetic and Seismological Observatory. M, 1979, University of Arizona. 194 p.

Wojtal, Aileen Marie. Trace element chemistry of quartz from quartz arenites. M, 1975, Indiana University, Bloomington. 60 p.

Wojtal, Steven F. Finite deformation in thrust sheets and their material properties. D, 1982, The Johns Hopkins University.

Wojtal, Steven Francis. Finite deformation in thrust sheets and their material properties. D, 1983, The Johns Hopkins University. 344 p.

Wolansky, Richard M. A gravity survey and the interpretation of gravity data of the Ocala National Forest area, Florida. M, 1973, University of South Florida, Tampa. 44 p.

Wolberg, Andrew Charles. Sedimentology of the Lower Cambrian Gog Group, British Columbia; an Early Cambrian tidal deposit. M, 1986, University of Alberta. 215 p.

Wolberg, Donald L. The mammalian paleontology of the late Paleocene (Tiffanian) Circle and Olive locali-

ties, McCone and Powder River counties, Montana. D, 1978, University of Minnesota, Minneapolis. 385 p.

Wolberg, Peter W. Structural geology of the Montpelier Canyon Quadrangle Bear Lake County, Idaho. M, 1983, University of Wyoming. 58 p.

Wolberg-de Venecia, Kathryn Elizabeth. Processes controlling the migration of a chromium contaminate plume through the application of a finite-difference solute-transport model. M, 1987, University of Toledo. 280 p.

Wolbrink, Mark A. Geology of the Mackay 3 NW Quadrangle, Idaho. M, 1970, University of Wisconsin-Milwaukee.

Wolcott, Albert. A study of some factors influencing crystal habit. D, 1926, University of Michigan.

Wolcott, Helen. The development of geologic thought and theory in the Connecticut Valley. M, 1934, Smith College. 70 p.

Wolcott, Henry Newton. A metallographic study of ores from Gilpin and Clear Creek counties, Colorado. M, 1916, Cornell University.

Wold, John Pearson. Geology and geophysics of the Poison Spider uranium prospect, Natrona County, Wyoming. M, 1978, Cornell University.

Wold, John Schiller. Interglacial consequents of central New York. M, 1939, Cornell University.

Wold, Richard John. Development of a digital recording magnetometer system and its application to the western Lake Superior region. D, 1966, University of Wisconsin-Madison. 275 p.

Wold, Ronald Odin. Composition and structural state variation of potassium and plagioclase feldspars from the Philipsburg Batholith (Cretaceous, Granite County), Montana. M, 1972, University of Montana. 95 p.

Wolde-Medhin, Bekele. Swell characteristics of expansive soils. D, 1980, University of Washington. 214 p.

Woldegabriel, Giday. Volcanotectonic history of the central sector of the main Ethiopian Rift; a geochronological, geochemical and petrological approach. D, 1988, Case Western Reserve University. 433 p.

Woldenburg, M. J. Stratigraphy of the Brazer Limestone of the Lost River Range, south central Idaho. M, 1958, University of Wisconsin-Madison.

Woldetensae, H. Comparative hydraulic conductivity studies in the lower Perch Lake basin. M, 1975, University of Waterloo.

Wolery, T. J. Some chemical aspects of hydrothermal processes at mid-oceanic ridges; a theoretical study; I, Basalt-sea water reaction and chemical cycling between the oceanic crust and the oceans; II, Calculation of chemical equilibrium between aqueous solutions and minerals. D, 1978, Northwestern University. 275 p.

Wolery, Thomas Jay. Vertical distribution of mercury, nickel, and chromium in Lake Erie sediments. M, 1973, Bowling Green State University. 194 p.

Wolf, Detlef Karl-Heinz. Dynamics of the continental lithosphere. D, 1985, University of Toronto.

Wolf, Douglas A. Identification of endmembers for magmas mixing in Little Sitkin Volcano, Alaska. M, 1987, SUNY at Albany. 201 p.

Wolf, Eric R. A detailed spectral reflectivity study of Copernicus (Moon) and its ejecta. M, 1970, Massachusetts Institute of Technology. 41 p.

Wolf, G. H. The fauna of the Lewis Formation. M, 1938, University of Wyoming. 42 p.

Wolf, George Henry. Ab initio theoretical predictions of high-pressure and high-temperature properties of lower mantle mineral phases. D, 1987, University of California, Berkeley. 164 p.

Wolf, Gerard V. Marine-bank development in the Oread Limestone (Shawnee Group, Upper Pennsylvanian) in southeastern Kansas and northern Oklahoma. M, 1984, Wichita State University. 104 p.

Wolf, James Kurt. Influence of landscape position on soil water, runoff, soil erosion and crop yield. D, 1987, North Dakota State University. 117 p.

Wolf, Michael Gene. Regional free-air gravity and tectonic observations in the United States. M, 1977, Northern Illinois University. 81 p.

Wolf, Michael Walter. Geology of the Sulphur Springs area, Fremont County, Wyoming. M, 1943, University of Missouri, Columbia.

Wolf, Stephen C. Current patterns and mass transport of clastic sediments in the near-shore regions of Monterey Bay (California). M, 1968, San Jose State University. 176 p.

Wolfanger, Louis A. The major soil divisions of the United States; a pedologic geographic survey. D, 1930, Columbia University, Teachers College.

Wolfbauer, Claudia A. Chemical petrology of non-marine carbonates in the western Bridger Basin, Wyoming. D, 1972, University of Wyoming. 146 p.

Wolfe, Caleb Wroe. Classification of minerals of the type A₃(XO₄)₂nH₂O. D, 1940, Harvard University.

Wolfe, David F. Noncombustible mineral matter in the Pawnee coal (Paleocene), Powder River County, Montana. M, 1969, Montana College of Mineral Science & Technology. 64 p.

Wolfe, Edward Winslow. Geology of Fairfield County, Ohio. D, 1961, Ohio State University. 352 p.

Wolfe, H. E. Optical and X-ray study of the low plagioclases. M, 1976, Virginia Polytechnic Institute and State University.

Wolfe, Jack A. Tertiary Juglandaceae of western North America. M, 1959, University of California, Berkeley. 111 p.

Wolfe, Jack Albert. Early Miocene floras of Northwest Oregon. D, 1960, University of California, Berkeley. 254 p.

Wolfe, Jack C. dt/dA measurement of short short period P waves in the upper mantle for the western North America using LASA. M, 1970, Massachusetts Institute of Technology. 55 p.

Wolfe, James Alvis. Forest soil characteristics as influenced by vegetation and bedrock in the spruce-fir zone of the Great Smoky Mountains. D, 1967, University of Tennessee, Knoxville. 193 p.

Wolfe, John A. Geology of the Masonville mining district, Larimer County, Colorado. M, 1953, Colorado School of Mines. 110 p.

Wolfe, Joseph Andrew. Fractures and oil. M, 1952, University of Michigan.

Wolfe, L. A. X-ray spectroscopic method for the analysis of manganese in marine sediments. M, 1969, University of Hawaii. 67 p.

Wolfe, Mitchell Dean. The relationship between forest management and landsliding in the Klamath Mountains of northwestern California. M, 1982, San Jose State University. 73 p.

Wolfe, Peter Edward. Subsequent topography and expression of structure, lithology and soil. D, 1941, Princeton University.

Wolfe, R. W. Polytypism of kaolin and pyrophyllite. D, 1974, SUNY at Buffalo. 107 p.

Wolfe, Robert Willard. Petrology and structure of the Precambrian and Tertiary crystalline rocks of the northern half of the Rustic Quadrangle, Mummy Range, Colorado. M, 1967, SUNY at Buffalo. 42 p.

Wolfe, Ronald A. A description of the geological deformations and some of their possible effects upon building construction at Indiana University of Pennsylvania. M, 1966, Indiana University of Pennsylvania. 34 p.

Wolfe, Stephen Howard. Geology and geochronology of the Manicouagan-Mushalagan lakes structure Grenville Province, Quebec, Canada. D, 1972, California Institute of Technology. 473 p.

Wolfe, Steven P. Au and Ag content of fresh Quaternary calc-alkalic andesites. M, 1984, Michigan Technological University. 120 p.

Wolfe, William John. Petrology, mineralogy and geochemistry of the Blue River ultramafic intrusion,

Cassiar District, British Columbia. D, 1967, Yale University. 195 p.

Wolfer, Donald H. Age and depth of ore folds. M, 1923, University of Minnesota, Minneapolis. 30 p.

Wolff, Breno. Microfacies, depositional environments, and diagenesis of the Amapá carbonates (Paleocene-middle Miocene), Foz do Amazonas Basin, offshore NE Brazil. D, 1984, University of Illinois, Urbana. 179 p.

Wolff, Ernest Nichols. Geology of the northern half of the Caviness Quadrangle, Oregon. D, 1965, University of Oregon. 200 p.

Wolff, Ernest Nichols. The geology of the upper Willow Creek-Cow Valley area of northern Malheur County, Oregon. M, 1959, University of Oregon. 93 p.

Wolff, John Elliot. The geology of Hoosac Mountain and adjacent territory (Massachusetts). D, 1889, Harvard University.

Wolff, John Marvin. The geochemical nature of an Archean plutonic-volcanic suite as exemplified by the Kakagi-Stephen lakes area, N.W. Ontario. M, 1977, McMaster University. 137 p.

Wolff, Manfred Paul. Deltaic sedimentation of the Middle Devonian Marcellus Formation in southeastern New York. D, 1967, Cornell University. 231 p.

Wolff, Manfred Paul. Stratigraphy and clay mineralogy of (Middle Devonian) lower Hamilton sedimentary rocks along the Catskill escarpment in southeastern New York. M, 1963, University of Rochester. 73 p.

Wolff, Martin. Geologically based fractured reservoir simulator. D, 1987, University of Texas, Austin. 436 p.

Wolff, Robert A. Ultramafic lenses within the Middle Ordovician Partridge Formation, Bronson Hill Anticlinorium, central Massachusetts. M, 1979, University of Massachusetts. 162 p.

Wolff, Roger G. and Ritter, John R. Channel sandstones (Oligocene) of the eastern section of the Big Badlands of South Dakota. M, 1958, South Dakota School of Mines & Technology.

Wolff, Roger Gene. Upper Keweenawan sedimentary rocks on Isle Royale, Lake Superior. M, 1969, University of Wisconsin-Madison.

Wolff, Roger Glen. Structural aspects of clay minerals using infrared absorption. D, 1961, University of Illinois, Urbana. 124 p.

Wolff, Roger Glen. The dearth of certain sizes of materials in sediments. M, 1960, University of Illinois, Urbana.

Wolff, Ronald Gilbert. Paleoecology of a late Pleistocene (Rancholabrean) vertebrate fauna from Rodeo, California. D, 1971, University of California, Berkeley. 136 p.

Wolffing, Craig L. Fluid inclusion studies in Precambrian rocks, Isua, West Greenland. M, 1978, Northern Illinois University. 101 p.

Wolfgram, Diane. Wall rock alteration and the localization of gold in the Homestake Mine, Lead, South Dakota. D, 1977, University of California, Berkeley. 139 p.

Wolfgram, Peter Arthur August. Development and application of a short-baseline electromagnetic exploration technique for the ocean floor. D, 1985, University of Toronto.

Wolford, John J. The geology of Owen County, north central Kentucky. D, 1932, The Johns Hopkins University.

Wolford, John J. The geology of the Oregonia-Ft. Ancient region, Warren County, Ohio. M, 1927, Ohio State University.

Wolfram, Katherine. Functional mechanics of jaw musculature and skeleton in the shark Notorhynchus, and its application to the fossil record. M, 1985, University of Nebraska, Lincoln.

Wolfson, Isobel K. A study of the tin mineralization and lithogeochemistry in the area of the Wedgeport

Pluton, southwestern Nova Scotia. M, 1983, Dalhousie University. 411 p.

Wolfson, Michael Stephen. Stratigraphy and paleoecology of the Checkerboard Formation (Pennsylvanian) of northeastern Oklahoma. M, 1963, University of Oklahoma. 121 p.

Wolfteich, Carl Martin. The use of vertical grain size progressions in establishing the depositional environment of ancient sand bodies. M, 1982, Rice University. 188 p.

Wolgemuth, Kenneth M. The marine geochemistry of radium and barium. D, 1972, Columbia University. 146 p.

Wolgemuth, Kenneth M. The vertical distribution of barium in the Pacific Ocean. M, 1968, Columbia University. 23 p.

Wolhuter, Louis E. The Opemisca Lake Pluton (Precambrian); a geochemical and petrological study. D, 1968, McGill University. 184 p.

Wolka, Kevin K. A stochastic analysis of pollutant movement in groundwater. D, 1986, Iowa State University of Science and Technology. 149 p.

Wolkdoff, Vladimir E. Pegmatite minerals in Noyes Mountain region, Maine. M, 1978, SUNY at Buffalo.

Wolkodoff, Vladimir E. Pegmatite minerals in the Noyes Mountain region, Maine. M, 1949, SUNY at Buffalo.

Wollard, George Prior. A report on the building and ornamental stones of Georgia. M, 1934, Georgia Institute of Technology. 150 p.

Wolle, Peter. The geology of the south half of the Tully, New York, Quadrangle. M, 1956, Syracuse University.

Wolleben, James Anthony. Biostratigraphy of the Ojinaga and San Carlos formations (Upper Cretaceous) of West Texas and northeastern Chihuahua (Mexico). D, 1966, University of Texas, Austin. 81 p.

Wollenberg, Harold. Earth materials for low-background radiation shielding. M, 1962, University of California, Berkeley. 136 p.

Woller, Kevin Lowell. A study of local earthquakes in Oklahoma recorded on 1-3 Hertz seismographs. M, 1978, University of Oklahoma. 110 p.

Woller, Neil M. Geology of the Willamette Pass area, Cascade Range, Oregon. M, 1986, Portland State University. 118 p.

Wollman, Constance Elizabeth. Fauna of the basal Welge Sandstone, Llano Uplift, Texas. M, 1952, University of Texas, Austin.

Wollschlager, Larry R. Hypostratotype (reference section) of lower Permian (Wolfcamp) shallow shelf carbonates in Hueco Mountains, El Paso and Hudspeth counties, Texas. M, 1975, University of Texas at El Paso.

Wolman, Markley G. The channel characteristics of Brandywine Creek, Pennsylvania. D, 1953, Harvard University.

Wolock, David Michael. Topographic and soil hydraulic control of flow paths and soil contact time; effects on surface water acidification. D, 1988, University of Virginia. 202 p.

Wolofsky, Leib. Geology of the Candego Mine, Gaspe North County, Quebec. M, 1955, McGill University.

Wolofsky, Leib. Hydrothermal experiments with variable pore pressure and shear stress in part of the MgO-SiO₂- H₂O system. D, 1957, McGill University.

Wolofsky, Leib. The geology of the Candego Mine, Gaspé North County, Quebec. M, 1954, McGill University.

Wolosin, Carl A. Seismic investigation and petroleum evaluation of central Lake Michigan. M, 1973, University of Wisconsin-Milwaukee.

Wolosz, Thomas Henry. Population variations in colonizing communities of Devonian patch reefs. M, 1977, Brooklyn College (CUNY).

Wolosz, Thomas Henry Matthew. Paleoecology, sedimentology, and massive favositid fauna of Roberts

Hill and Albrights reefs (Edgecliff Member, Onondaga Formation of New York). D, 1984, SUNY at Stony Brook. 414 p.

Woloszyn, Danuta. Ash flows of the Valley Springs Formation, Calaveras County, California. M, 1979, University of California, Berkeley. 75 p.

Wolske, Roxanne L. Paleomagnetic study of the Barron Quartzite of northwestern Wisconsin. M, 1985, University of Wisconsin-Milwaukee. 189 p.

Wolter, John A. The emerging discipline of cartography. D, 1975, University of Minnesota, Minneapolis. 357 p.

Wolterding, D. Facies-associated morphologic changes in the brachiopod Leptaena "rhomboidalis" (Wilckens). M, 1977, Queens College (CUNY). 93 p.

Wolters, Bernd. Seismicity and earthquake focal mechanisms and their tectonic implications in southern Central America and northwestern South America. M, 1983, Stanford University. 89 p.

Woltz, David. The chemistry of groundwaters in the Jemez area (New Mexico) and a magnetic survey of a potential source of magmatic fluids. M, 1973, University of New Mexico. 90 p.

Wolverson, Nancy Jean. Geology and hydrothermal alteration of the Palmetto Property, Esmeralda County, Nevada. M, 1987, University of Nevada. 82 p.

Womack, Bernard Anderson. Application of geophysical methods to exploration of the Carolina Bays. M, 1981, North Carolina State University. 63 p.

Womack, Stephen Hasie. Provenance of the Pottsville Formation in the Cahaba Syncline. M, 1983, University of Alabama. 130 p.

Womack, W. Raymond. Erosional history of Douglas Creek, northwestern Colorado. M, 1975, [Colorado State University].

Womer, M. B. A study of the ash rings of Split Butte, a maar crater of the South-central Snake River plain, Idaho. M, 1977, SUNY at Buffalo. 53 p.

Womochel, Daniel Robert. Taphonomy and paleoecology of the Slaton local fauna (Pleistocene, Texas). D, 1977, Texas Tech University. 148 p.

Won, Ihn Jae. Representation theorems for the electrodynamic diffraction problem and their applications. D, 1973, Columbia University.

Woncik, John. The Upper Cretaceous rocks of Lafayette County, Arkansas. M, 1951, University of Minnesota, Minneapolis. 41 p.

Wonder, James David. The origin of manganese-rich metasediments and their relationship to iron formation and base metal deposits, western Georgia Piedmont. M, 1987, Iowa State University of Science and Technology. 167 p.

Wonderley, Patricia Faith. The geology of the Wood's Mountain area in the Virginia Piedmont. M, 1981, University of Kentucky. 185 p.

Wones, David R. Phase relations of biotite. D, 1960, Massachusetts Institute of Technology. 157 p.

Wonfor, John Stephen. The Lower Cretaceous and Jurassic of the Lethbridge-Foremost area, Alberta, Canada. M, 1947, University of Oklahoma. 50 p.

Wong, A. S. Temperature investigation of coexisting dolomite and calcite in marble from Gatineau Park and surrounding area, Quebec. M, 1971, University of Ottawa. 77 p.

Wong, Albert H. Oxygen isotope study of the Lower Critical Zone of the central sector of the eastern Bushveld igneous complex, South Africa. M, 1982, University of Wisconsin-Madison. 109 p.

Wong, Anne B. The mineralogy, chemistry, and uranium distribution in the Spirit Pluton, northeastern Washington. M, 1978, Bowling Green State University. 113 p.

Wong, Chi Shing. The distribution of inorganic carbon in the eastern tropical Pacific Ocean. D, 1968, University of California, San Diego.

Wong, Daniel On-Cheong. Driveability and load transfer characteristics of vibro-driven piles. D, 1988, [University of Houston]. 389 p.

Wong, George T. F. The marine chemistry of iodates. M, 1973, Massachusetts Institute of Technology. 137 p.

Wong, George Tin Fuk. Dissolved inorganic and particulate iodine in the oceans. D, 1976, Massachusetts Institute of Technology. 272 p.

Wong, H. Donald. Subsurface geology of Fall River County, South Dakota. M, 1960, University of South Dakota. 91 p.

Wong, Henry Kwok-Hin. Petrology and provenance of the Eocene Wilcox Group, Northeast Texas. M, 1986, University of Texas, Austin. 139 p.

Wong, Her Yue. Clay mineralogy of lower Paleozoic rocks in Beavers Bend State Park, Ouachita Mountains, Oklahoma. M, 1964, University of Oklahoma. 48 p.

Wong, Her Yue. Clay petrology of the Atoka Formation (middle Pennsylvanian), eastern Oklahoma. D, 1969, University of Oklahoma. 107 p.

Wong, Ivan Gynmun. Site amplification of seismic shear waves in Salt Lake Valley, Utah. M, 1976, University of Utah. 63 p.

Wong, Jade Starr. The petrology and petrochemistry of the Gold Hill syenite-pyroxenite complex, Potlatch Quadrangle, Idaho. M, 1980, Washington State University. 104 p.

Wong, Joseph. Modelling in electromagnetic prospecting techniques. M, 1973, University of Toronto.

Wong, Joseph. Some theoretical and experimental aspects of the electrical and electromagnetic properties of geological materials. D, 1979, University of Toronto.

Wong, Kai Sin. Elasto-plastic finite element analyses of passive earth pressure tests. D, 1978, University of California, Berkeley. 373 p.

Wong, Ken. Effects of mineral taxation on a marginal resource. M, 1985, University of Utah. 164 p.

Wong, Kong-Cheng. Source modelling of Taiwan earthquakes from the wave forms of long-period P-waves. M, 1984, University of Colorado. 101 p.

Wong, Margaret S. Palynostratigraphic correlation of uppermost Devonian strata of medial South America (Bolivia and Paraguay), Brazil and North Africa. M, 1980, Rutgers, The State University, Newark. 104 p.

Wong, Pak K. Sedimentology and diagenesis of the Upper Devonian Kaybob stratigraphic reef. M, 1978, University of Calgary. 207 p.

Wong, Parkin. The mineral resources of China. M, 1914, Cornell University.

Wong, Paul Kwok-Ting. A determination of the velocity of the uppermost oceanic crust. M, 1981, University of Washington. 66 p.

Wong, Peter Kin. Numerical models of subduction dip angle with variable viscosity. M, 1981, Rice University. 54 p.

Wong, Poh-Poh. Beach changes and sand movement in low energy environments, West Coast, Barbados. D, 1971, McGill University.

Wong, Sam J. The paleoecology of the Keyser Limestone; a re-evaluation. M, 1985, Virginia Polytechnic Institute and State University.

Wong, Teng-Fong. Post-failure behavior of Westerly Granite at elevated temperatures. D, 1981, Massachusetts Institute of Technology. 169 p.

Wong, William Wai-Lun. Carbon isotope fractionation by marine phytoplankton. D, 1976, Texas A&M University. 126 p.

Wongsawat, S. Barometric and pump test determination of the characteristics of a carbonate aquifer in the vicinity of Norman Creek, Phelps County, Missouri. M, 1974, University of Missouri, Rolla.

Wongwiwat, Kraiwut. Gravity survey in southern end of Albuquerque-Belen Basin, Socorro County, New Mexico. M, 1970, New Mexico Institute of Mining and Technology.

Wonn, Philip M. Ecology and spatial distribution at the microenvironmental level of modern foraminifera in San Carlos Bay area, Ft. Myer, Florida. M, 1986, Bowling Green State University. 107 p.

Wonson-Liukkonen, Barbara. Geochemistry of two small lakes in northeastern Minnesota in relation to atmospheric inputs. M, 1987, University of Minnesota, Duluth. 145 p.

Woo, A. A. Desmond. An investigation of the methods of calibrating a seismometer. M, 1954, University of Toronto.

Woo, Ching-Chang. The Pre-Cambrian geology and amphibolites of the Nemo District, Black Hills, South Dakota. D, 1952, University of Chicago. 148 p.

Woo, Hyoseop. Sediment transport in hyperconcentrated flows. D, 1985, Colorado State University. 270 p.

Woo, Kyung Sik. Carbonate diagenesis and biostratigraphy of the Anahuac Formation at Damon Mound, Texas. M, 1982, Texas A&M University. 104 p.

Woo, Kyung Sik. Isotopic-textural-chemical studies of mid-Cretaceous limestones; implications for carbonate diagenesis and paleooceanography. D, 1986, University of Illinois, Urbana. 246 p.

Woock, Robert David. The quantitative mineralogy and mineral stratigraphy of the Late Devonian, Early Mississippian black shales of eastern Kentucky. M, 1980, University of Kentucky. 101 p.

Wood, Albert E. Evolution and relationships of the heteromyid rodents with new forms from the Tertiary of western North America. D, 1935, Columbia University, Teachers College.

Wood, Allan D. A geologic and mineralogical study of the Bethlehem copper property at Highland Valley, British Columbia. M, 1968, Oregon State University. 80 p.

Wood, Arthur Eugene. A study of Georgia clay chemically and spectroscopically. M, 1909, Vanderbilt University.

Wood, Barry R. Geomorphology of Elben cave-mouth deposits and associated karst phenomena (Roane county, Tennessee). M, 1969, University of Tennessee, Knoxville. 121 p.

Wood, Becky Leigh. Development of a structural framework from seismic reflection data. M, 1988, University of Texas, Austin. 102 p.

Wood, C. B. Stratigraphy and paleontology of the Bridger Formation northeast of Opal, Lincoln County, Wyoming. M, 1966, University of Wyoming. 118 p.

Wood, Charles A. A geophysical investigation of MacDougal Crater, Sonora, Mexico. M, 1972, University of Arizona.

Wood, Charles A. Morphometric studies of planetary landforms; impact craters and volcanoes. D, 1979, Brown University. 230 p.

Wood, Cynthia. A study of fluid inclusions in quartz veins associated with the Octoraro phyllite in southeastern Pennsylvania. M, 1976, Bryn Mawr College. 70 p.

Wood, Cynthia. Chemical and textural zoning in metamorphic garnets, Rangeley area, Maine. D, 1981, University of Wisconsin-Madison. 617 p.

Wood, David G. A study of stream transport of sand sized sediment, Battle Creek, Black Hills, South Dakota. M, 1970, University of Houston.

Wood, David Mahlon. Pattern and process in primary succession in high elevation habitats on Mount St. Helens. D, 1987, University of Washington. 124 p.

Wood, Douglas R., II. Geology at Timber Mountain Pass, northern Seaman Range, Nye County, Nevada. M, 1986, University of Texas at El Paso.

Wood, Edward Boyne. Morganfield South oil field, Union County, Kentucky. M, 1955, University of Kentucky. 40 p.

Wood, Ella L. The development of the modern concept of geography. D, 1927, University of Wisconsin-Madison.

Wood, Elvira. A revision of Troost's crinoids of Tennessee. M, 1908, Columbia University, Teachers College.

Wood, Elvira. The phylogeny of certain Cerithiidae. D, 1910, Columbia University, Teachers College.

Wood, Elwyn Devere. A study of the geochemistry of gold in the marine environment. D, 1971, University of Alaska, Fairbanks. 176 p.

Wood, Eric S. A kinematic-wave model for predicting subsurface stormflow in upland watersheds. M, 1986, University of New Hampshire. 244 p.

Wood, Francis W. Some characteristics of earthquake surface waves. D, 1951, Harvard University.

Wood, George V. A comparison of three quartzites. D, 1960, Pennsylvania State University, University Park. 172 p.

Wood, Gordon Daniel. Palynology and paleobotany of the Java and lowermost Canadaway formations, Upper Devonian (Senecan/Chautauquan), New York State. D, 1978, Michigan State University. 243 p.

Wood, Gordon Daniel, II. Observations on the morphology and distribution of Chitinozoa from the Silica Formation (middle Devonian) of northwestern Ohio. M, 1973, University of Michigan.

Wood, Harry Warren. A study of the correlation of the Mississippian of the Bloomington (Indiana) Quadrangle with the standard section of the Mississippi Valley. M, 1915, Indiana University, Bloomington.

Wood, Hiram Budd. Insoluble residues of the Bonneterre Dolomite. M, 1938, University of Missouri, Columbia.

Wood, Horace E., 2nd. Some early Tertiary rhinoceroses and hyracodonts. D, 1927, Columbia University, Teachers College.

Wood, J. D. The geology of the Castle Rock area, Grant, Harney and Malheur counties, Oregon. M, 1976, Portland State University. 123 p.

Wood, J. H. Barrier island field studies for high school students on the Outer Banks of North Carolina. M, 1974, Virginia State University.

Wood, J. R. Prediction of mineral solubilities in concentrated brines; a thermodynamic approach. D, 1972, The Johns Hopkins University. 131 p.

Wood, James L. Influence of repetitious freeze-thaw on structure and shear strength of Leda (Massena) Clay. D, 1976, Clarkson University. 181 p.

Wood, James Michael. Sedimentology of the Late Cretaceous Judith River Formation, "Cathedral" area, Dinosaur Provincial Park, Alberta. M, 1985, University of Calgary. 215 p.

Wood, John. The stratigraphy and sedimentation of the upper Huronian rocks in the Rawhide lake-Flack lake area, Ontario (Precambrian). M, 1970, University of Western Ontario. 237 p.

Wood, John A. The Chickamauga stratigraphy of a portion of Raccoon Valley, Union County, Tennessee. M, 1962, University of Tennessee, Knoxville. 40 p.

Wood, John Anderson. Internal pressures in freezing soils. D, 1988, Carleton University. 261 p.

Wood, John Edwin. Geology of southwestern Barren County, Kentucky. M, 1938, University of Iowa. 73 p.

Wood, John W. The geophysical and geological characteristics of fracture zones in the carbonate Floridan Aquifer. M, 1985, University of South Florida, Tampa. 72 p.

Wood, John William. A stratigraphic study of the Gallup Sandstone in San Juan County, New Mexico. M, 1956, Texas Tech University. 45 p.

Wood, John William. Geology of Apache Mountains, Trans-Pecos Texas. D, 1965, University of Texas, Austin. 265 p.

Wood, Joseph Miller. The flora of the lower Block coal of Greene County, Indiana. D, 1960, Indiana University, Bloomington. 197 p.

Wood, Karrie Champneys. Distribution of Recent benthic foraminifera on the Northern California continental shelf. M, 1984, University of Washington. 91 p.

Wood, Laura Fain. Geochemistry and petrogenesis of the Cuttingsville Intrusion, Vermont. M, 1984, University of North Carolina, Chapel Hill. 105 p.

Wood, Lawrence A. Surficial geology east of the Hudson River in the Hudson North and Stottville quadrangles, New York. M, 1980, Rensselaer Polytechnic Institute. 32 p.

Wood, Lawrence Charles. Reflection and refraction of body waves generated by a cylindrical cavity near an elastic interface. D, 1966, University of Utah. 239 p.

Wood, Leonard E. Geology of the Lower Hot Springs faulted area, Cement Creek, Gunnison County, Colorado. M, 1957, University of Kentucky. 58 p.

Wood, Leonard Eugene. Bottom sediments of Saginaw Bay, Michigan. D, 1958, Michigan State University. 325 p.

Wood, Lucile H. Prehistoric man in North America. M, 1942, Case Western Reserve University.

Wood, Lyman W. The geology of the road materials of southern Iowa. M, 1931, Iowa State University of Science and Technology.

Wood, Mabel Vivian. Geographic landscape of the northwest industrial district, Metropolitan St. Louis. M, 1936, Washington University. 162 p.

Wood, Marcus Irwin. The coordination and structural role of aluminum-oxide in immiscible silicate liquids in two systems; A, $(SiO_2.TiO_2.Al_2O_3.CaO.MgO.FeO)$; and B, $(SiO_2TiO_2.Al_2O_3.CaO.MgO.FeO.Na_2O.K_2O)$. D, 1980, Brown University. 118 p.

Wood, Maria Luisa. Sedimentology and architecture of Gilbert- and mouth bar-type fan deltas, Paradox Basin, Colorado. M, 1988, Colorado State University. 188 p.

Wood, Mary Connor. A petrographic study of the Confederate Limestone in the Ardmore Basin, Carter and Love counties, Oklahoma. M, 1952, University of Oklahoma. 75 p.

Wood, Michael Lee. Geological and geochemical studies of the Rode Ranch pegmatite area, Llano County, Texas. M, 1965, Texas Christian University.

Wood, Michael M. Metamorphic effects of the Leatherwood Quartz Diorite, Santa Catalina Mountains, Pima County, Arizona. M, 1963, University of Arizona.

Wood, Michael Manning. The crystal structures of ransomite $CuFe_2(SO_4)_4 \cdot 6H_2O$, and roemerite, $Fe^{2+}Fe^{3+}(SO_4)_4 \cdot 14H_2O$, and proposed classification for the transition metal sulfate hydrates. D, 1969, University of Arizona. 68 p.

Wood, Milton Darroll. The influence of ocean tidal loading on solid earth tides and tilts in the San Francisco bay region, California. D, 1969, Stanford University. 107 p.

Wood, P. A. Characteristics, comparisons, classification, and erodibility of some northern Alabama coal mine spoils. D, 1979, Auburn University. 221 p.

Wood, Paul A. A Miocene camel from Wellton, Yuma County, Arizona. M, 1956, University of Arizona.

Wood, Paul Alan. Pleistocene fauna from 111 Ranch area, Graham County, Arizona. D, 1962, University of Arizona. 134 p.

Wood, Peter Colin. The Hollinger-McIntyre gold-quartz vein system, Timmins, Ontario; geological characteristics, fluid properties and light stable isotope geochemistry. M, 1987, University of Toronto.

Wood, Raymond A. Converted wave reflections in exploration seismic records. M, 1978, University of Texas, Austin.

Wood, Robert H., II. Conodont distribution in facies of the Stanton Formation (Upper Pennsylvanian, Missourian) in southeastern Kansas. M, 1977, University of Iowa. 121 p.

Wood, Robert Staples. Stratigraphy of the (Lower Ordovician) Stonehenge Limestone in the northwestern part of the Shenandoah Valley, Virginia. M, 1962, University of Virginia. 84 p.

Wood, Roger L. Stratigraphy of the Zeandale Limestone (Upper Pennsylvanian) in Shawnee, Osage, and Lyon counties, Kansas. M, 1959, University of Kansas. 165 p.

Wood, Scott Alan. Some aspects of the physical chemistry of hydrothermal ore-forming solutions. D, 1985, Princeton University. 279 p.

Wood, Spencer Hoffman. Holocene stratigraphy and chronology of mountain meadows, Sierra Nevada, California. D, 1975, California Institute of Technology. 197 p.

Wood, Thomas R. The hydrogeology of the Wanapum Basalt, Creston study area, Lincoln County, Washington. M, 1987, Washington State University. 166 p.

Wood, Timothy Eldridge. Biological and chemical control of phosphorus cycling in a northern hardwood forest. D, 1980, Yale University. 215 p.

Wood, Warren W. Distribution and stratigraphic position of late Precambrian diabase dikes in parts of northern Michigan. M, 1962, Michigan State University. 57 p.

Wood, Warren W. Geochemistry of ground water of the Saginaw Formation in the upper Grand River basin, Michigan. D, 1969, Michigan State University. 104 p.

Wood, William. Transformation of breaking wave parameters over a submarine bar. D, 1971, Michigan State University. 210 p.

Wood, William H. The Cambrian and Devonian carbonate rocks at Yampai Cliffs, Mohave County, Arizona. D, 1955, University of Arizona.

Wood, William James. Areal geology of the Coalville vicinity, Summit County, Utah. M, 1953, University of Utah. various pagination p.

Woodard, Geoffrey Davidson. The Cenozoic succession of the West Colorado Desert, San Diego and Imperial counties, Southern California. D, 1963, University of California, Berkeley. 216 p.

Woodard, Henry H. The geology and paragenesis of the Lord Hill Pegmatite, Stoneham, Maine. M, 1949, Dartmouth College. 53 p.

Woodard, Henry H., Jr. Diffusion of chemical elements in some naturally occurring silicate inclusions. D, 1955, University of Chicago. 69 p.

Woodard, Jan N. Subsurface stratigraphy of the Strawn and Canyon groups of west-central Texas, Concho and Menard counties. M, 1983, Baylor University. 74 p.

Woodard, Thomas William. Geology of the Lookout Mountain area, northern Black Range, Sierra County, New Mexico. M, 1982, University of New Mexico. 95 p.

Woodas, Nicholas A. Some lithologic and stratigraphic aspects of selected outcrops of Miocene and Pliocene formations in North Carolina. M, 1965, University of North Carolina, Chapel Hill. 75 p.

Woodberry, Marjorie. Habitats of trilobites. M, 1939, The Johns Hopkins University.

Woodburne, Michael Osgood. The Alcoota fauna; an integrated geologic and paleontologic study. D, 1966, University of California, Berkeley. 534 p.

Woodburne, Michael Osgood. Upper Pliocene geology and vertebrate paleontology of part of Meade Basin, Kansas. M, 1960, University of Michigan.

Woodbury, Allan David. Simultaneous inversion of thermal and hydrogeologic data. D, 1987, University of British Columbia.

Woodbury, Homer Olwin. Structure of the Boulder Arch, Boulder County, Colorado. M, 1942, University of Colorado.

Woodbury, Jerry L. A petrographic study of the Aurora area, Wisconsin. M, 1962, University of Wisconsin-Madison.

Woodcock, Deborah. Use of wood-anatomical variables of bur oak (Quercus macrocarpa) in the reconstruction of climate. D, 1987, University of Nebraska, Lincoln. 114 p.

Woodcock, Edgar. Some new occurrences of minerals in California. M, 1918, University of California, Berkeley.

Woodcock, S. F. Crustal structure of the Tehuantepec Ridge and adjacent continental margins of southwestern Mexico and western Guatemala. M, 1976, Oregon State University. 52 p.

Woodell, Charles E. The Mississippian fauna of the Redwall Limestone near Jerome, Arizona. M, 1927, University of Arizona.

Wooden, Joseph Lovell. Geochemistry and Rb-Sr geochronology of Precambrian mafic dikes from the Beartooth, Ruby Range, and Tobacco Root Mountains, Montana. D, 1975, University of North Carolina, Chapel Hill. 229 p.

Woodfill, Robert D. A gravity investigation of Precambrian crystalline rocks, Squaw Rock area, Albany and Platte counties, Wyoming. M, 1968, University of Wyoming. 54 p.

Woodfill, Robert Dean. A geologic and petrographic investigation of a northern part of the Keetley volcanic field (Cenozoic), Summit and Wasatch counties, Utah. D, 1972, Purdue University. 168 p.

Woodford, Alfred Oswald. The San Onofre Breccia, its nature and origin. D, 1923, University of California, Berkeley. 81 p.

Woodhams, Richard L. and Newman, Karl Robert. Stratigraphy and paleontology of a core from Loraine County, Ohio. M, 1954, University of Michigan.

Woodhead, James A. The crystallographic and calorimetric effects of Al-Si distribution on the tetrahedral sites of melilite. D, 1977, Princeton University. 278 p.

Woodhouse, Bruce Alan. Geochemical indicators of petroleum migration in Red River Formation waters of South Dakota. M, 1979, South Dakota School of Mines & Technology.

Woodhouse, Elizabeth Gail. Model of stable geomorphic form for an urban stream, College Station, Texas. M, 1988, Texas A&M University. 126 p.

Woodhull, Patricia. Revision of the genus Prosserella. M, 1953, University of Michigan.

Woodland, Alan Butler. Halogens in biotite, sericite, and apatite in relation to alteration and mineralization in the vicinity of Mount Manitou, Bonanza mining district, Saguache County, Colorado. M, 1984, University of Oregon. 242 p.

Woodland, Bertram. Minor structures and their bearing on the major geologic structure of the Burke area. D, 1962, University of Chicago. 151 p.

Woodland, Roland Bert. Stratigraphic significance of Mississippian endothyroid foraminifera in central Utah. M, 1957, Brigham Young University. 73 p.

Woodley, Nancy Karen Fish. An investigation of landfill disposal of blast furnace slag from secondary lead smelters. D, 1984, University of Alabama. 213 p.

Woodman, James T. Availability of ground water, Coastal Bend region, Texas. M, 1975, University of Texas, Austin.

Woodman, Joseph Edmund. Geology of the Moose River gold district, Halifax County, Nova Scotia, together with the pre-Carboniferous history of the Meguma Series. D, 1902, Harvard University.

Woodman, Neal. A subarctic fauna from the late Wisconsinan Elkader Site, Clayton County, Iowa. M, 1982, University of Iowa. 56 p.

Woodmansee, Helen. Geology of the (Precambrian) Grenville Series of an area in the southern part of the Hammond Quadrangle, New York. M, 1950, Syracuse University.

Woodmansee, Walter. Geology of the igneous rocks and orthogneisses of an area in the southern part of the Hammond Quadrangle, New York. M, 1950, Syracuse University.

Woodring, S. M. Engineering soils mapping from multispectral remote sensing data using computer-assisted analysis. M, 1973, Purdue University.

Woodring, Wendell P. The Mollusca of the (Miocene) Bowden Beds of Jamaica. D, 1916, The Johns Hopkins University.

Woodrome, Larry S. Uranium; Trans-Pecos, Texas Tertiary intrusive and groundwater anomalies. M, 1980, University of Texas at El Paso.

Woodrow, Donald Lawrence. The paleoecology and paleontology of a portion of the Upper Devonian in south-central New York. M, 1960, University of Rochester. 104 p.

Woodrow, Donald Lawrence. Upper Devonian stratigraphy and sedimentation in Bradford and Susquehanna counties, Pennsylvania. D, 1965, University of Rochester. 165 p.

Woodruff, Charles M., Jr. The limits of deformation of the Howell structure, Lincoln County, Tennessee. M, 1968, Vanderbilt University.

Woodruff, Charles Marsh, Jr. Land-use limitations related to geology in the Lake Travis vicinity, Travis and Burnet counties, Texas. D, 1973, University of Texas, Austin.

Woodruff, Edwin C. Underground gas storage. M, 1954, University of Missouri, Columbia.

Woodruff, Elmer G. The geology of Cass County. M, 1904, University of Nebraska, Lincoln.

Woodruff, Fay. Deep sea benthic foraminiferal changes associated with the middle Miocene oxygen isotopic event, DSDP Site 289, equatorial Pacific. M, 1979, University of Southern California.

Woodruff, J. L. A photometric centrifuge for rapid size analysis of fine sediments. M, 1972, University of Hawaii. 42 p.

Woodruff, John Grunt. Areal and structural geology of the Wellsville Quadrangle, New York. D, 1936, University of Michigan.

Woodruff, John M. The depositional environment of the Garnett fossil locality, Rock Lake Shale Member of the Pennsylvanian Stanton Formation, Anderson County, Kansas. M, 1984, Duke University. 79 p.

Woodruff, Michael S. Strontium isotope compositions of mississippi valley-type deposits enclosed in Cambrian, Ordovician and Mississippian host carbonates. M, 1980, Miami University (Ohio). 69 p.

Woods, A. J. Marine terraces between Playa el Marron and Morro Santo Domingo, central Baja California, Mexico. D, 1978, University of California, Los Angeles. 205 p.

Woods, Arnold Martin. Trace fossils of the Whitsett Formation (upper Eocene) of Karnes County, Texas. M, 1981, University of Texas, Austin. 141 p.

Woods, Barry Bradford. Use of a multiprobe conductivity array in a groundwater contamination monitoring program. M, 1988, Wright State University. 89 p.

Woods, Delmer Maurice. Sedimentary study of the San Angelo Formation. M, 1947, Texas Tech University. 30 p.

Woods, Dennis V. A model study of the Crone borehole pulse electromagnetic (PEM) system. M, 1975, Queen's University. 329 p.

Woods, Diana M. The paleoecology of Crepidula (Gastropoda), James City Formation (Pleistocene), North Carolina. M, 1987, Tulane University. 119 p.

Woods, Earl Hazen. Geology of Long Valley, California. M, 1924, University of Iowa. 72 p.

Woods, Edmund Bert. Areal differentiation of slope forms as contrasted in semi-arid and humid climatic regimes. D, 1970, University of Iowa. 110 p.

Woods, Ella Jean. A study of the occurrence density and aquifers of some high fluoride ground waters of the Central Basin, Tennessee. M, 1977, Vanderbilt University.

Woods, Everett Kenneth. Some Devonian and Mississippian conodonts from northern Arkansas. M, 1955, University of Missouri, Columbia.

Woods, Henry Harper. Depositional and diagenetic history, upper Stuart City Formation (Cretaceous), South Texas. M, 1980, University of New Orleans. 122 p.

Woods, Madeline Marie. Seismic stratigraphy and fault activity of the inner San Pedro Shelf, California.

M, 1984, California State University, Northridge. 90 p.

Woods, Marion Marshall. Microfacies and depositional history of the San Jose Lentil, DiFunta Group (Upper Cretaceous-Tertiary), Nuevo Leon, Mexico. M, 1982, University of New Orleans. 80 p.

Woods, Mark Thomas. Shallow crustal structure as inferred from short period Rayleigh wave dispersion. M, 1986, St. Louis University.

Woods, Marvin O. Depositional subenvironments in a closed basin; the Shepard Formation (middle Proterozoic Belt Supergroup), southern Mission, Swan, and Lewis & Clark ranges, Montana. M, 1986, University of Montana. 215 p.

Woods, Michael Damian. The nature, origin, and significance of flow layering in a rhyolite flow, Wamsutta Formation, South Attleboro, Massachusetts. M, 1961, Brown University.

Woods, Michael J. Petrography and geochronology of basic and ultrabasic inclusions from kimberlites of Riley County, Kansas. M, 1970, Kansas State University. 96 p.

Woods, Michael J. Textural and geochemical features of the Highwood Mountains volcanics, central Montana. D, 1974, University of Montana. 121 p.

Woods, Paul Fredric. Primary productivity in Lake Koocanusa, Montana. D, 1979, University of Idaho. 112 p.

Woods, Raymond Douglas. Petrographic interpretation of some sections of the Carrizo Formation in central Texas. M, 1934, University of Texas, Austin.

Woods, Robert S. A correlation of gravity and magnetic anomalies in central Coffee County and western Anderson County, Kansas. M, 1978, Wichita State University. 51 p.

Woods, Terri Lee. Calculated solution-solid relations in the low temperature system CaO-MgO-FeO-CO$_2$-H$_2$O. D, 1988, University of South Florida, Tampa. 140 p.

Woods, Thomas F. Processing and interpretation of CDP, VSP and sonic log data acquired in the Rio Grande Rift, Dona Ana County, New Mexico. M, 1987, University of Wyoming. 236 p.

Woodside, Edward R. Foraminifera from the Cantua Sandstone Member of the Lodo Formation, San Benito County, California. M, 1958, University of California, Berkeley. 85 p.

Woodside, John M. Gravity anomalies in the eastern Mediterranean Sea. M, 1968, Massachusetts Institute of Technology. 92 p.

Woodson, Frederick Jennings. Lithologic and structural controls on karst landforms of the Mitchell Plain, Indiana, and Pennroyal Plateau, Kentucky. M, 1981, Indiana State University. 132 p.

Woodson, James C. Geology of the southeastern portion of the Snowball Quadrangle, Searcy County, Arkansas. M, 1967, University of Arkansas, Fayetteville.

Woodson, John Pierce. Areal geology of the southwest Pickens area, Pushmataha County, Oklahoma. M, 1964, University of Oklahoma. 81 p.

Woodson, Walter Browne, III. A bottom gravity survey of the continental shelf between Point Lobos and Point Sur, California. M, 1973, United States Naval Academy.

Woodster, Warren Scriver. Phosphate in the eastern North Pacific Ocean. D, 1972, University of California, Los Angeles. 83 p.

Woodsum, Glenn Craig. Chemical characterization of shallow groundwater at the Kennedy Space Center. M, 1974, Florida Institute of Technology.

Woodsworth, Glenn James. Interrelations of metamorphism, plutonism, and deformation in the Mt. Raleigh area, Coast Mountains, British Columbia. D, 1974, Princeton University. 250 p.

Woodward, Albert F. The geology of the Whitestone Mountain area, Washington. M, 1936, Stanford University. 30 p.

Woodward, C. W. D. The Newark Island layered intrusion. D, 1976, Syracuse University. 181 p.

Woodward, Charles W. D. Origin of garnets in pegmatites of the Grand Teton Range, Wyoming. M, 1972, Syracuse University.

Woodward, D. G. A hydrogeologic investigation of the Cache La Poudre River alluvium in the Windsor Triangle area, Colorado. M, 1975, University of Colorado.

Woodward, Harold Walter. Insoluble residues of the Devonian, Southern Rocky Mountains, Western Canada. D, 1953, University of Wisconsin-Madison.

Woodward, Harriette B. The geology of the Island of Haiti. M, 1929, Columbia University, Teachers College.

Woodward, Herbert P. Correlation tables of North American stratigraphy. M, 1936, Columbia University, Teachers College.

Woodward, Herbert P. Geology and mineral resources of the Roanoke area of Virginia. D, 1932, Columbia University, Teachers College.

Woodward, John Eylar. Geology of Killam deep test near Valentine, Jeff Davis County, Trans-Pecos, Texas. M, 1954, University of Texas, Austin.

Woodward, Lee Albert. Geology of central part of the Flathead Range, Montana. M, 1959, University of Montana. 45 p.

Woodward, Lee Albert. Structure and stratigraphy of the central Egan Range, White Pine County, Nevada. D, 1962, University of Washington. 145 p.

Woodward, Nicholas Brugger. Structural geometry of the Snake River Range, Idaho and Wyoming. D, 1981, The Johns Hopkins University. 401 p.

Woodward, Philip V. Regional evaluation of formation fluid salinity by spontaneous potential log, Ivishak Sandstone (Triassic), North Slope, Alaska. M, 1987, San Jose State University. 71 p.

Woodward, R. J. Sea-floor spreading during the past 10 million years on the East Pacific Rise between 35°S and 53°S, and the identification of short period pole reversal events. M, 1974, Texas A&M University.

Woodward, Stephen C. Paleocurrents of the Mississippian Ste. Genevieve Limestone and equivalents in the Eastern United States. M, 1983, University of Cincinnati. 276 p.

Woodward, Thomas C. Electromagnetic anomalies with scale models. M, 1950, Colorado School of Mines. 38 p.

Woodward, Thomas Canby. Geology of Deadman Butte area, Natrona County, Wyoming. D, 1955, University of Texas, Austin.

Woodward, Thomas Michael. Geology of the Lemitar Mountains, Socorro County, New Mexico. M, 1974, New Mexico Institute of Mining and Technology. 73 p.

Woodward, Truman P. A preliminary study of the sands and gravels of Louisiana. M, 1940, Louisiana State University.

Woodward, Warren M. Sedimentary study of the Deadwood Formation of the Black Hills, South Dakota. M, 1937, University of Minnesota, Minneapolis. 44 p.

Woodwell, Grant R. Fluid migration in an overthrust sequence of the Canadian Cordillera. D, 1985, Yale University. 266 p.

Woodworth-Lynas, Christopher M. T. Geology and structure of the Hare Bay Allochthon at Quirpon Island, northern Newfoundland. M, 1983, Memorial University of Newfoundland. 210 p.

Woodyard, Kenneth Eugene. Clays of St. Johns vicinity, Arizona and New Mexico. M, 1956, University of Texas, Austin.

Woodzick, Thomas L. Geophysical and remote-sensing characteristics of the Colorado-Wyoming kimberlite occurrences. D, 1987, Colorado State University. 322 p.

Woodzick, Thomas L. Predictive and partitive aspects of strain energy release in the Aleutian Islands, 1900-1970. M, 1972, University of Wisconsin-Milwaukee.

Woolard, Louis Eugene. Fluorescence analysis of pyritic uranium ores. M, 1951, Columbia University, Teachers College.

Woolard, Louis Eugene. Late Tertiary rhyolitic eruptions and uranium mineralization, Marysvale, Utah. D, 1955, Columbia University, Teachers College. 227 p.

Wooley, Jerry D. Relic evaporite deposits of the lower member of the Devonian-Mississippian Arkansas novaculite of the Ouachita Mountains, Arkansas. M, 1977, Northeast Louisiana University.

Wooley, William Leeman. Shallow cores from the subaerial portion of the Mississippi River delta. M, 1941, Louisiana State University.

Woolf, Theresa Skwara. Late Pleistocene vertebrate fauna from Riddell site near Saskatoon, Canada. M, 1978, University of Saskatchewan.

Woolford, Jean H. A genetic classification of rock island in streams. M, 1938, Columbia University, Teachers College.

Woollard, George Prior. Gravity anomalies and their relation to geologic structure. D, 1937, Princeton University. 106 p.

Woollen, Ian D. Structural framework, lithostratigraphy, and depositional environments of Upper Cretaceous sediments of eastern South Carolina. D, 1978, University of South Carolina. 297 p.

Woollen, Ian D. Structure, stratigraphy and environmental analysis of the Lower Carboniferous rocks near Azro. M, 1974, University of South Carolina.

Woollett, LeRoy Andrew. Geology of southern half of the China and Dillahunty quadrangles, Culberson and Reeves counties, Texas. M, 1951, University of Texas, Austin.

Woolley, J. J. Sedimentology of the Sespe and Vaqueros formations, Santa Rosa Island, California. M, 1978, San Diego State University.

Woolridge, Bruce Alan. Systematic identification of Cenozoic teleost scales in the northwestern Gulf Coastal region, with a paleoecological application. D, 1986, Southern Methodist University. 173 p.

Woolsey, Issac W. Geology of the Squaw-Marshall Creek area, Mason County, Texas. M, 1958, Texas A&M University.

Woolsey, James R., Jr. Neogene stratigraphy of the Georgia coast and inner continental shelf. D, 1977, University of Georgia. 244 p.

Woolsey, James R., Jr. The geology of Clarke County, Georgia. M, 1973, University of Georgia. 109 p.

Woolsey, Leonard L. A Rb-Sr geochronologic study of the Republic metamorphic node, Republic (Marquette County), Michigan. M, 1971, University of Kansas. 102 p.

Woolsey, Leonard Lee. Investigation of roof shales performance under variable environmental conditions. D, 1981, University of Missouri, Rolla. 144 p.

Woolsey, Thomas S. Physical modeling of diatreme emplacement. M, 1972, Colorado State University. 103 p.

Woolverton, D. G. Cu-Fe-S mineralization of the Sweetwater Mine, Reynolds County, Missouri. M, 1975, University of Missouri, Columbia.

Woolverton, Ralph S. The Camray discovery dyke and associated uranium deposits (Ontario). M, 1950, McGill University.

Woolverton, Ralph S. The Lumby Lake greenstone belt (British Columbia). D, 1954, McGill University.

Woolverton, Ralph S. The Lumby Lake greenstone belt, Thunder Bay District, Ontario. D, 1953, McGill University.

Wooten, M. W. Stratigraphy, structure, and metamorphism in portions of the Hayesville and Hiawassee 7.5 quadrangles, North Carolina. M, 1980, University of Georgia.

Wooten, Maria Jane. The (Mississippian) Coldwater Formation in the area of the type locality (Michigan). M, 1951, Wayne State University.

Wooten, Richard Mark. Stratigraphy, structure, and metamorphism in portions of the Hayesville and Hiawassee 7 1/2' quadrangles, North Carolina. M, 1980, University of Georgia.

Wootton, C. F. Pleistocene Mollusca of the Colon Deposit, St. Joseph County, Michigan. M, 1974, Ohio State University.

Wopat, Michael A. Part 1, The karst of southeastern Minnesota; Part 2, Methods for statistical analysis of polymodal two-dimensional orientation data. M, 1974, University of Wisconsin-Madison. 86 p.

Worayingyong, Kaweepoj. Analysis of one-dimensional vertical and radial consolidation by physical discrete element models. D, 1981, University of Texas, Austin. 212 p.

Worcester, Peter A. The reconnaissance geology of the southwest one-quarter of the Dundee Meadows Quadrangle, Fremont County, Wyoming and a petrologic study of the Wiggins Formation. M, 1967, Miami University (Ohio). 80 p.

Worcester, Peter A. The volcanic stratigraphy and petrography of the northern half of the Nacozari District, Sonora, Mexico. D, 1976, Miami University (Ohio). 242 p.

Worcester, Peter F. Reciprocal acoustic transmission in a mid-ocean environment. D, 1977, University of California, San Diego. 80 p.

Worcester, Philip G. The physiography of Colorado. D, 1924, University of Chicago. 339 p.

Worcester, Philip George. Geology of the Ward region, Boulder County, Colorado. M, 1911, University of Colorado.

Worden, John A. Pre-Desmoinesian isopachous and paleogeologic studies of the Amarillo-Hugoton area. M, 1959, University of Oklahoma. 85 p.

Work, David L. Depositional patterns of the Lewisville sandstones, northern Hawkins Field, Wood County, Texas. M, 1987, Texas A&M University. 141 p.

Work, Paul Murray. The stratigraphy and paleontology of the Minnelusa Formation of the southern Black Hills of South Dakota. M, 1931, University of Iowa. 189 p.

Work, Rebecca Diana. Depositional environment of Upper Devonian gas producing sandstones, Westmoreland County, southwestern Pennsylvania. M, 1988, Texas A&M University. 75 p.

Workman, Charles Edwin. Geology of the Sipe Springs area, Comanche County, Texas. M, 1961, University of Texas, Austin.

Workman, Lewis Edwin. Pleistocene series in vicinity of Thompson Reef (Thornton, Illinois). M, 1925, University of Chicago. 25 p.

Workman, Robert R., Jr. Foraminiferal assemblages of the nearshore inner continental shelf, Nags Head and Wilmington areas, North Carolina. M, 1981, East Carolina University. 161 p.

Workman, William Edward. Barite from the White River Formation (Oligocene) of northeastern Colorado, with emphasis on crystallography and geochemistry. M, 1964, University of Virginia. 124 p.

Workman, William Edward. Wollastonite in regionally metamorphosed rocks, Blount Mountain, Llano County, Texas. D, 1968, University of Texas, Austin. 173 p.

Workum, R. H. The stratigraphy of the Mesaverde Formation of the west flank of the Casper Arch, Natrona County, Wyoming. M, 1959, University of Wyoming. 70 p.

Worl, Ronald G. Superimposed deformations in Precambrian rocks near South Pass, Wyoming. M, 1963, University of Wyoming. 53 p.

Worl, Ronald G. Taconite and migmatite in the northern Wind River Mountains, Fremont, Sublette and Teton counties, Wyoming. D, 1968, University of Wyoming. 138 p.

Worland, Vincent Peter. Geologic and engineering aspects regarding the elimination of combined sewer discharge in Indianapolis, Indiana by underground conveyance and storage. M, 1979, Purdue University. 166 p.

Worley, George T. Geology of the Arcadia salt dome area, Bienville Parish, Louisiana. M, 1962, Louisiana State University.

Worley, John Cochran. The glades of Preston County, West Virginia. M, 1951, University of Pittsburgh.

Worley, Paul Lawrence Hill. Sedimentology and stratigraphy of the Tule Wash area, Mohave County, Arizona. M, 1979, Arizona State University. 99 p.

Wornardt, Walter William. Stratigraphic distribution of diatom floras from the "Mio-Pliocene" of California. D, 1963, University of California, Berkeley. 208 p.

Wornardt, Walter William, Jr. Stratigraphy, paleontology, and coral zonation of the Brazer Limestone (Mississippian), Lost River Range, Arco-Howe area, Idaho. M, 1958, University of Wisconsin-Madison.

Woronick, Robert Eugene. Burial diagenesis of the Lower Cretaceous Pearsall and lower Glen Rose formations, South Texas; a petrographic and geochemical study. M, 1985, University of Texas, Austin. 91 p.

Woronow, Alexander. A size-frequency study of large Martian craters. D, 1975, Harvard University.

Woronow, Alexander Nick. The coulomb-mohr criterion; an experimental study of its effectiveness as a predictor of rock failure. M, 1973, University of Houston.

Worrall, D. M. Structural geology of the Round Mountain area, Uinta County, Wyoming. M, 1975, University of Wyoming. 79 p.

Worrall, Dan M. Geology of the South Yolla Bolly area, northern California, and its tectonic implications. D, 1979, University of Texas, Austin. 292 p.

Worrall, John Griggs, III. Deposition and diagenesis of the Jurassic Smackover Formation, Hatter's Pond Field, SW Alabama. M, 1988, University of Texas, Austin. 237 p.

Worrel, Elizabeth Ann. Paleoecology and biostratigraphy of Wells 1, 2, and 4, Vermilion area, Block 265, offshore Louisiana. M, 1986, University of Texas, Austin. 89 p.

Worsley, Thomas. The sedimentology of Whites Creek Delta (Recent) in Watts Bar Lake (Roane and Rhea counties), Tennessee. M, 1967, University of Tennessee, Knoxville. 56 p.

Worsley, Thomas Raymond. The nature of the terminal Cretaceous event as evidenced by calcareous nanno-plankton extinctions in Alabama and other areas. D, 1970, University of Illinois, Urbana. 174 p.

Worstall, Robert Stewart. The subsurface geology of the Clinton Sandstone in Copley Township, Summit County and Sharon Township, Medina County, Ohio. M, 1986, Kent State University, Kent. 115 p.

Worstell, Paula Jane. Foraminifera of the Upper Greenhorn and Lower Carlile Formation (Upper Cretaceous), Colorado and Kansas. M, 1966, University of Colorado.

Worth, John Kirk. Stratigraphy of the lower Horton Bluff Formation (Mississippian), Wolfville, Kings County, Nova Scotia (Canada). M, 1969, Acadia University.

Wortham, Kenneth E. The effects of pollution on the benthic macroinvertebrates of Big Lick Creek, Indiana. D, 1974, Ball State University. 125 p.

Worthen, John Aldrich. Deposition, diagenesis, and porosity relationships of the lower San Andres Formation, Quay and Roosevelt counties, New Mexico. M, 1979, Texas Tech University. 83 p.

Worthington, David W. Ultrasonic studies of wave propagation through layered media. M, 1969, Virginia Polytechnic Institute and State University.

Worthington, J. E. A biogeochemical technique applied at the Shawangunk Mine. M, 1954, Columbia University, Teachers College.

Worthington, June. Structural analyses of the Calaveras Formation, Stanislaus River, California. M, 1978, California State University, Fresno.

Worthington, June. Structural analysis and metamorphic petrology of part of the Calaveras Formation, Stanislaus River, California. M, 1977, California State University, Fresno.

Worthington, Ralph E. Petrology of Middle Pennsylvanian (Desmoinesian) "Upper Bluejacket" Sandstone (Cherokee Group) of Bourbon, Crawford, and Cherokee counties, Kansas. M, 1982, University of Iowa. 108 p.

Wortman, Ann Aber. Origin and stratigraphic relations of the basal Bolsa Quartzite Conglomerate on Dos Cabezas Ridge, Cochise County, Arizona. M, 1984, University of Arizona. 61 p.

Wortman, Richard A. Environmental implications of surface water resource development in the middle Rio Grande drainage, New Mexico. M, 1971, University of New Mexico. 129 p.

Worzel, J. Lamar. Explosion sounds in shallow water. D, 1949, Columbia University, Teachers College.

Worzel, J. Lamar. Photography of the ocean bottom. M, 1948, Columbia University, Teachers College.

Wosick, Frederick D. Stratigraphy and paleontology of Upper Cretaceous Morden Member (Vermilion River Formation) in the outcrop area, northeastern North Dakota. M, 1977, University of North Dakota. 152 p.

Wosinski, J. F. The bedrock geology of the southern half of the Chepachet Quadrangle, Rhode Island. M, 1958, Brown University.

Wotorson, Cletus S. Geologic interpretation of the aeromagnetic anomalies in the Narragansett bay area and eastern Connecticut. M, 1968, Wesleyan University.

Wotruba, Nancy Jane. Fluid inclusion geothermometry of the ores of the Metaline mining district, Washington. M, 1978, Washington State University. 54 p.

Wotruba, Patrick Roy. Contact metamorphic effects of the Gem Stocks of sulfide mineral assemblages characteristic of the Coeur d'Alene mining district, Idaho; a sulfide phase equilibria study. M, 1983, University of Idaho. 125 p.

Woussen, Gérard. Les Monzonites du Mont-Royal (one of the Monteregian hills-alkaline intrusives) (Montreal, Quebec, Canada). M, 1969, Universite de Montreal.

Woussen, Gérard. Pétrologie du complexe igné de Brome. M, 1974, Universite de Montreal.

Wozab, David Hyrum. The chemical characteristics of the ground water in San Fernando Valley, California. M, 1952, University of Southern California.

Woznessensky, Boris. Structural analysis of the fracture pattern at White Pine Mine, Ontonagon County, Michigan. M, 1967, Michigan Technological University. 107 p.

Wozniak, Karl C. Geology of the northern part of the southeast Three Sisters Quadrangle, Oregon. M, 1982, Oregon State University. 98 p.

Wracher, David A. The geology and mineralization of the Peck mountain area, Hornet Quadrangle, Idaho. M, 1970, Oregon State University. 78 p.

Wraight, Joseph. Geography of Rockwoods Reserve. M, 1941, Washington University. 100 p.

Wrath, William Frederick. Marine sedimentation around Catalina and San Clemente islands. M, 1936, University of Illinois, Urbana.

Wray, Charles F. The geology of the northwest quarter of Ironside Mountain Quadrangle, Grant and Baker counties, Oregon. M, 1946, University of Rochester. 71 p.

Wray, Cloyd Field. Geology of the Calhoun Field. M, 1962, Louisiana Tech University.

Wray, Franklin C. Megascopic fauna of the (Upper Mississippian) Glen Dean Formation, Perry County, Missouri. M, 1934, University of Chicago. 59 p.

Wray, Irene. The (Devonian) Onondaga Limestone and its insoluble residues from sections of New York State. M, 1936, University of Rochester. 71 p.

Wray, James E. A field and petrographic study of the Ruby Creek area, Madison County, Montana. M, 1959, Michigan State University. 49 p.

Wray, John Lee. Mississippian Foraminifera from the central Appalachian region. D, 1956, University of Wisconsin-Madison. 70 p.

Wray, John Lee. The (Mississippian) Greenbrier Series in northern West Virginia and its correlates in southwestern Pennsylvania. M, 1951, West Virginia University.

Wray, William B., Jr. Geological and statistical review of structural and zoning problems, Butte District, Montana. D, 1972, University of California, Berkeley. 272 p.

Wrenn, John Harry. Dinocyst biostratigraphy of Seymour Island, Palmer Peninsula, Antarctica. D, 1982, Louisiana State University. 544 p.

Wrenn, John Harry Wycoff. Cenozoic subsurface micropaleontology and geology of eastern Taylor Valley. M, 1976, Northern Illinois University. 255 p.

Wright, Alexandra P. Shoreline and beach changes on Honeymoon Island, Pinellas County, Florida, 1967-1971. M, 1972, University of South Florida, Tampa. 48 p.

Wright, Alfred Edwin. A study of the Brazos River Sandstone and conglomerate of Parker and Palo Pinto counties, Texas. M, 1955, Texas Christian University. 65 p.

Wright, Andrew Clemmons. The subsurface structure of the Garber oil and gas field, Oklahoma. M, 1921, University of Oklahoma. 18 p.

Wright, Audrey Anne. Sediment distribution and depositional processes in the Lesser Antilles intraoceanic island arc, eastern Caribbean. D, 1983, University of California, Santa Cruz. 203 p.

Wright, Charles Edward. The Cretaceous rocks of the Qu'Appelle River valley in Saskatchewan. M, 1979, University of Saskatchewan. 167 p.

Wright, Charles Malcolm. Geology and origin of the pollucite-bearing Montgary Pegmatite, Manitoba. D, 1961, University of Wisconsin-Madison. 111 p.

Wright, Charles Malcolm. Pyrite zones in the hanging-wall of the Steep Rock ore area (Ontario). M, 1959, Queen's University. 138 p.

Wright, Cynthia Ann Roseman. Environmental study of the Liverpool Cyclothem of the Eastern Interior Basin and the Forest City Basin. M, 1963, University of Illinois, Urbana.

Wright, Cynthia Roseman. Environmental mapping of the beds of the Liverpool Cyclothem in the Illinois Basin and equivalent strata in the northern-mid-continent region. D, 1965, University of Illinois, Urbana. 100 p.

Wright, Daniel Frederick. Data integration and geochemical evaluation of Meguma Terrane, Nova Scotia, for gold mineralization. M, 1988, University of Ottawa. 82 p.

Wright, David Brian. Later Miocene Tayassuidae (Artiodactyla, Mammalia) of North America. M, 1983, University of Nebraska, Lincoln. 350 p.

Wright, David Craig. Stratigraphy of the (Ordovician) Chickamauga Limestone in the Kensington Quadrangle (Georgia). M, 1952, Emory University. 44 p.

Wright, David S. Coal deposits characterization by gamma-gamma density/percent dry ash relationships. M, 1984, Texas A&M University.

Wright, Dorothy Alden Davis. Belt of thrusts in Wyoming and Colorado. M, 1948, University of Michigan.

Wright, Elizabeth. Petrology and geochemistry of shield-building and post-erosional lava series of Samoa; implications for mantle heterogeneity and magma genesis. D, 1986, University of California, San Diego. 305 p.

Wright, Ellen Margrethe Marie Krogh. Stratification and paleocirculation patterns of the Upper Cretaceous

Western Interior Seaway of North America. D, 1986, Yale University. 135 p.

Wright, Ernest George. Petrology of granite molybdenite systems. M, 1979, Eastern Washington University. 111 p.

Wright, Frank J. Physiography of the upper James River basin in Virginia. D, 1918, Columbia University, Teachers College.

Wright, Frank M. The Pleistocene stratigraphy of the Farmersville and the northern part of the Middletown quadrangles, southwestern Ohio. M, 1970, Miami University (Ohio). 189 p.

Wright, Frank Myron, III. The Pleistocene and Recent geology of the Oneida-Rome District, New York. D, 1972, Syracuse University. 308 p.

Wright, Frederick F. The development and application of a fluorescent marking technique for tracing sand movements on beaches. M, 1961, Columbia University, Teachers College.

Wright, Frederick Fenning. The marine geology of San Miguel Gap off Point Conception, California. D, 1967, University of Southern California. 271 p.

Wright, G. M. The geology and ore deposits of the Hard Rock gold mine, Little Long Lac area, Ontario. M, 1947, Queen's University. 38 p.

Wright, Gordon R. An evaluation of specific ion electrodes for use in pollution studies. M, 1971, University of Toronto.

Wright, Grant MacL. Geology of the Ranji Lake and Ghost Lake areas, Northwest Territories, Canada. D, 1950, Yale University.

Wright, Harold D. Mineralogical study of certain Colorado and Ontario uraninite deposits. D, 1952, Columbia University, Teachers College.

Wright, Harold M. The ores of Copper Mountain, British Columbia. M, 1933, University of British Columbia.

Wright, Herbert Edgar, Jr. The Tertiary and Quaternary geology of the lower Rio Puerco area, New Mexico. D, 1943, Harvard University.

Wright, J. Seismic crustal studies in North-western Ontario. D, 1977, University of Toronto.

Wright, J. D. The age of some granite batholiths north of Lake Huron and the genetic relation of the arsenical gold ores to the Keweenawan granites. D, 1934, University of Toronto.

Wright, James Arthur. Geothermal investigations using in situ techniques. D, 1968, University of Toronto.

Wright, James Arthur. The measurement of heat flow through the Earth's crust near the surface. M, 1965, University of Toronto.

Wright, James Clifton. Genesis of the iron ore deposits at Benson Mines, St. Lawrence County, New York. M, 1949, Cornell University.

Wright, James E. Geology of the Carolina slate belt in the vicinity of Durham, North Carolina. M, 1974, Virginia Polytechnic Institute and State University.

Wright, James Earl. Geology and U-Pb geochronology of the western Paleozoic and Triassic subprovince, Klamath Mountains, Northern California. D, 1981, University of California, Santa Barbara.

Wright, James L. Numerical modeling of a magnetic induced polarization (M.I.P.) problem. M, 1975, Stanford University.

Wright, Janet Decker. Pedogenic horizons and associated opalized rhizoliths in the Ogallala Group of western Nebraska. M, 1987, University of Nebraska, Lincoln. 71 p.

Wright, Jean Davies. The type species of Spinocyrtia Fredericks and two new species of this brachiopod genus from the Middle Devonian Hamilton Group in the Thedford-Arkona region of southwestern Ontario. M, 1955, University of Michigan.

Wright, Jerome J. A textural and thickness study of the Peorian loess in Nebraska. M, 1947, University of Nebraska, Lincoln.

Wright, Jerome J. Petrology of the Devonian rocks in eastern Pima and Cochise counties, Arizona. D, 1964, University of Arizona. 202 p.

Wright, Jesse F. A study of the characteristics of the terrace deposits in southeastern Mississippi (terraces-Cenozoic). M, 1951, Mississippi State University. 98 p.

Wright, John Buel. Geology of the western Cooper Mountain area, Freemont County, Wyoming. M, 1956, University of Texas, Austin.

Wright, John Clinton, Jr. Geochemistry of heavy metals in the C horizon of a sandy loam soil in Pennsylvania. M, 1978, Pennsylvania State University, University Park. 174 p.

Wright, John Frank. The geology of the Brockville-Mallorytown map area (Ontario). D, 1923, University of Chicago. 144 p.

Wright, Judith. Rare earth element distributions in Recent and fossil apatite; implications for paleoceanography and stratigraphy. D, 1985, University of Oregon. 281 p.

Wright, Lauren A. An invertebrate assemblage from the (Miocene) "Modelo" Formation of Reynier Canyon, Los Angeles County, California. D, 1951, California Institute of Technology. 31 p.

Wright, Lauren A. Geology and origin of talc deposits of eastern California. D, 1951, California Institute of Technology. 181 p.

Wright, Lauren A. Geology of the Mint Canyon series and its relation to the "Modelo" Formation and to other adjacent formations, Los Angeles County, California. M, 1943, University of Southern California.

Wright, Leo Milfred. The micropaleontology of the Spaniard Limestone Member of northeastern Oklahoma. M, 1949, University of Tulsa. 69 p.

Wright, Leo Milfred. The paleontology of the Chester Series of southwestern Missouri. D, 1952, University of Missouri, Columbia.

Wright, Lynn Donelson. Circulation, effluent diffusion and sediment transport; mouth of South Pass, Mississippi River delta. D, 1970, Louisiana State University. 95 p.

Wright, Nancy Elin Peck. Compositional variation in the Stone Mountain Granite (Georgia). M, 1963, Emory University. 53 p.

Wright, P. J. Underfit meanders of the French Broad River, North Carolina. M, 1942, Columbia University, Teachers College.

Wright, Paul Randall. Mechanical analysis of sediments of the scablands of eastern Washington. M, 1932, University of Chicago. 32 p.

Wright, Phillip Michael. Geothermal gradient and regional heat flow in Utah. D, 1966, University of Utah. 199 p.

Wright, R. L. The geology of the Pioneer Ultramafite, Bralorne, British Columbia. M, 1974, University of British Columbia. 179 p.

Wright, R. M. Aspects of the geology of Tertiary limestones in West-central Jamaica, West Indies. D, 1976, Stanford University. 320 p.

Wright, Ramil Carter. Foraminiferal ecology in a back-reef environment, Molasses Reef, Florida. D, 1964, University of Illinois, Urbana. 124 p.

Wright, Ramil Carter. Petrology of the Gallatin Formation, east flank, Big Horn Mountains, Wyoming. M, 1962, University of Illinois, Urbana.

Wright, Reginald D. The Cooper Marl (Oligocene), Charleston County, South Carolina. M, 1973, Virginia State University. 40 p.

Wright, Richard E. Stratigraphic and tectonic interpretation of Oquirrh Formation, Stansbury Mountains, Utah. M, 1961, Brigham Young University. 166 p.

Wright, Richard Frederic. Forest fire; impact on the hydrology, chemistry, and sediments of small lakes in northeastern Minnesota. D, 1974, University of Minnesota, Minneapolis.

Wright, Richard Kyle. The water balance of a lichen tundra underlain by permafrost. D, 1980, McGill University.

Wright, Robert Harvey. Geology of central Marin County, California. D, 1982, University of California, Santa Cruz. 283 p.

Wright, Robert J. Dike and sill alteration, Santa Rita, New Mexico. D, 1947, Columbia University, Teachers College.

Wright, Robert J. Underfit meanders of the French Broad River, North Carolina. M, 1942, Columbia University, Teachers College.

Wright, Robert John. Molybdenum status of Texas soils; fertilization trials, anion exchange estimations, and correlation with soil chemical properties. D, 1982, Texas A&M University. 146 p.

Wright, Robert Paul. Geology of the San Francisco Quadrangle, Coahuila, Mexico. M, 1967, University of Michigan.

Wright, Robert Paul. The marine Jurassic of Wyoming and South Dakota; its paleoenvironments and paleobiogeography. D, 1971, University of Michigan. 257 p.

Wright, Robyn. Paleoenvironmental interpretation of the Upper Cretaceous Pt. Lookout Sandstone; implications for shoreline progradation and basin tectonic history, San Juan Basin, New Mexico. D, 1984, Rice University. 447 p.

Wright, Robyn. Sediment gravity transport on the Weddell Sea continental margin. M, 1980, Rice University. 96 p.

Wright, Roland Finley. Some Pennsylvanian and Lower Permian fusulinids from Clark County, Nevada. M, 1954, University of Illinois, Urbana.

Wright, Sally. Contribution to research in surface conductivity; hydrogeological and petrophysical applications. M, 1972, University of Minnesota, Minneapolis. 79 p.

Wright, Samuel Alexander. Mines as an alternative to shallow land burial of low-level nuclear wastes. M, 1981, University of Arizona. 184 p.

Wright, Sarah D. A structural, stratigraphic, and petrochemical study of Willow Creek Canyon, Quinn Canyon Range, Nye County, Nevada. M, 1987, University of North Carolina, Chapel Hill. 73 p.

Wright, Stephen F. Analysis of small scale structures developed during monoclinal folding; Biebel Monocline, Gunnison, Colorado. M, 1985, University of Minnesota, Minneapolis. 156 p.

Wright, Stephen F. Early Proterozoic deformational history of the Kiruna District, northern Sweden. D, 1988, University of Minnesota, Minneapolis. 237 p.

Wright, Stephen S. Seismic stratigraphy and depositional history of Holocene sediments on the central Texas Gulf Coast. M, 1980, University of Texas, Austin.

Wright, Steward A. Subsurface geology of the Midway Field, Lafayette County, Arkansas. M, 1967, University of Arkansas, Fayetteville.

Wright, Thomas Lee. The occurrence and genesis of graphite in the Mellen Gabbro, Ashland County, Wisconsin. M, 1956, University of Illinois, Urbana.

Wright, Thomas Llewellyn. The mineralogy and petrogenesis of the southern part of the Tatoosh Pluton, Mount Rainier National Park, Washington; 2 volumes. D, 1961, The Johns Hopkins University.

Wright, Thomas O. Sedimentary geochemistry of central Pamlico Sound, North Carolina. D, 1974, George Washington University. 303 p.

Wright, Thomas O. Sedimentation and geochemistry of surficial material, Ocracoke Island, Cape Hatteras, North Carolina. M, 1971, George Washington University.

Wright, W. L. Geology of the Cascade Mountain Goat, Mount Baker-Snoqualmie National Forest, Washington. M, 1977, Western Washington University.

Wright, William Allen. Skarn-formation at Pine Creek Mine; Bishop, California. D, 1973, University of California, Berkeley. 135 p.

Wright, William Herbert, III. Rock deformation in the Slate belt of west-central Vermont. D, 1970, University of Illinois, Urbana. 115 p.

Wright, William J. Geology of the New Ross map-area, with an introductory chapter on the gold-bearing series and the granites of southern Nova Scotia. D, 1915, Yale University.

Wright, Willis I. The composition and occurrence of garnets. D, 1937, University of Minnesota, Minneapolis. 75 p.

Wright-Clark, Judith. The geochemistry of conodont apatite; secular variations inferred by comparison with the cerium-iron system in the modern ocean. M, 1982, University of Oregon. 67 p.

Wright-Grassham, Anne C. Volcanic geology, mineralogy, and petrogenesis of the Discovery volcanic subprovince, southern Victoria Land, Australia. D, 1988, New Mexico Institute of Mining and Technology. 460 p.

Wrightson, Walter, Jr. Petrogenesis of the Lick Fork Ni-Co prospect, Floyd Co., Virginia. M, 1981, University of Tennessee, Knoxville. 112 p.

Wrightstone, Gregory. The stratigraphy and depositional environment of the Ravencliff Formation in McDowell and Wyoming counties, West Virginia. M, 1985, West Virginia University. 98 p.

Writt, Robert Joseph. The depositional environments of the Grimsby Formation, in the subsurface of central Lake Erie, Ontario. M, 1977, University of Windsor. 131 p.

Wroble, John Lee. Stratigraphy and sedimentation of the Popo Agie, Nugget, and Sundance formations in central Wyoming. M, 1953, University of Wyoming. 151 p.

Wroblewski, Frank G. Rationale for geology field trips in a high school science curriculum. M, 1977, Rensselaer Polytechnic Institute. 345 p.

Wrolstad, Keith H. Applications of source signature deconvolution to airgun seismic profiling and the measurement of attenuation from reflection seismograms. D, 1979, Oregon State University. 325 p.

Wrucke, Chester T. Geology of the Warm Springs area (Nevada). M, 1952, Stanford University.

Wrucke, Chester Theodore, Jr. Precambiran and Permian rocks in the vicinity of Warm Spring Canyon, Penamint Range, California. D, 1966, Stanford University. 215 p.

Wu Wen-Jen see Wen-Jen Wu

Wu, Arthur Han. On the failure load of soil anchors by limit analysis. D, 1981, George Washington University. 127 p.

Wu, Changsheng. A study of short period seismic noise. D, 1966, Rice University. 71 p.

Wu, Changsheng. Azimuthal variation in Pn velocity around the Gnome explosion. M, 1963, Rice University. 42 p.

Wu, Chia-Hsin. The maceral composition of Permian coals from the Parana Basin, Brazil as related to their environment of deposition. M, 1979, University of Toledo. 114 p.

Wu, Cho-Sen. Finite element analysis of fabric reinforced sand. D, 1987, University of Michigan. 355 p.

Wu, Dah Cheng. Clay mineralogy and geochemistry of the upper Flowerpot shale (Permian), (Major and Blaine counties, Oklahoma). D, 1969, University of Oklahoma. 115 p.

Wu, Dah Cheng. Mineralogy and chemistry of chlorite from the Anderson Talc Deposit, Saline County, Arkansas. M, 1966, University of Oklahoma. 62 p.

Wu, Francis Taming. I, Lower limit of the total energy of earthquakes and partitioning of energy among seismic waves; II, Reflected waves and crustal structures. D, 1966, California Institute of Technology. 251 p.

Wu, Guoping. Depositional mechanics and significance of the Eureka Quartzite in central Nevada. M, 1984, University of Nebraska, Lincoln.

Wu, H. C. A Moessbauer effect study of titanomaghemite. D, 1975, University of Wyoming. 55 p.

Wu, I-Pai. Hydrography of small watersheds in Indiana and hydrodynamics of overland flow. D, 1963, Purdue University.

Wu, I. Hsiung. Geochemistry of tetrahedrite-tennanite at Casapalca, Peru. D, 1975, Harvard University.

Wu, J. C. Inversion of traveltime data for seismic velocity structure in three dimensions. D, 1977, University of Washington. 71 p.

Wu, Jianjun. Analysis of the data of the 1984 Kapuskasing seismic experiment. M, 1987, University of Western Ontario. 189 p.

Wu, Jy-Shing. Development and application of a stormwater assessment model (SWAM). D, 1980, Rutgers, The State University, New Brunswick. 253 p.

Wu, Ming-Chee. The temperature and geometry influences on an underground opening in the frozen ground. M, 1985, University of Alaska, Fairbanks. 186 p.

Wu, Patrick Pak-Cheuk. The viscosity of the deep mantle. D, 1982, University of Toronto.

Wu, Ru-Chuan. Frequency domain computation of synthetic vertical seismic profiles. M, 1983, Texas A&M University.

Wu, Ru-Chuan. Modeling wave propagation in boreholes to determine rock properties. D, 1986, Texas A&M University. 135 p.

Wu, Ru-Shan. Seismic wave scattering and the small scale inhomogeneities in the lithosphere. D, 1984, Massachusetts Institute of Technology. 305 p.

Wu, S. S.-C. Mars synthetic topographic mapping. D, 1976, University of Arizona. 196 p.

Wu, Sheng Tung. Electromagnetic wave propagations in disrupted coal seam. M, 1988, Pennsylvania State University, University Park.

Wu, Shi-ming. Capillary effects on dynamic modulus of fine-grained cohesionless soils. D, 1983, University of Michigan. 226 p.

Wu, Tsai-Way. Geochemistry and petrogenesis of some granitoids in the Grenville Province of Ontario and their tectonic implications. D, 1984, University of Western Ontario. 623 p.

Wu, Tsai-Way. Structural, stratigraphic and geochemical studies of the Horwood Peninsula – Gander Bay area, Northeast Newfoundland. M, 1980, Brock University. 185 p.

Wu, Wei. An analysis of close seam interaction problems in the Appalachian coal fields. D, 1987, Virginia Polytechnic Institute and State University. 236 p.

Wu, Wen-Jen. Algebraic Kirchhoff-Trorey inversion. D, 1981, University of Houston. 92 p.

Wu, Wen-Jen. Seismic modeling source studies of stick-slip failure along a pre-existing fault in a stressed plate. M, 1978, SUNY at Binghamton. 61 p.

Wu, Y. H. Effect of roughness and its spatial variability on runoff hydrographs. D, 1978, Colorado State University. 189 p.

Wu, YeeMing Timothy. The detailed study of natural remanent magnetization in the Tatoosh granodiorite intrusion of Mount Rainier. D, 1974, University of Pittsburgh. 194 p.

Wuckert, Arthur Emil. Bioseries of the ostracode Ponderodictya in the Traverse Group of Michigan. M, 1950, Michigan State University. 38 p.

Wuellner, Dirck E. Tectonic evolution of the Marathon region with emphasis on mid-Carboniferous tectonism and sedimentation. M, 1985, University of Texas at El Paso.

Wuensch, Bernhardt John. The nature of the crystal structures of some sulfide minerals with substructures. D, 1963, Massachusetts Institute of Technology. 234 p.

Wuenschel, Paul Clarence. Gravity measurements and their interpretation in South America between latitudes 15° and 33° south. D, 1955, Columbia University, Teachers College. 228 p.

Wuerch, Helmuth Victor, III. Diagenesis and water chemistry of the Woodbine Group in the East Texas Basin. M, 1986, Utah State University. 129 p.

Wuestner, Charles E. R. The lithology and insoluble residue zones of the Bonneterre dolomite. M, 1952, Washington University. 55 p.

Wulf, George Richard. Geology of the Fannie Peak area, Weston County, Wyoming. M, 1955, South Dakota School of Mines & Technology.

Wulf, George Richard. Lower Cretaceous (Albian) rocks in northern Great Plains [U.S.]. D, 1959, University of Michigan. 311 p.

Wulff, Julie L. Ordination of paleocommunities of the Lodgepole Limestone. M, 1986, University of Illinois, Chicago. 146 p.

Wulkowicz, Gerald. Chloride balance of the Salt Creek Basin, Chicago metropolitan area. M, 1973, University of Illinois, Chicago.

Wunder, Susan Jean. Diagenetic features and inferred diagenetic processes in partially altered corals from the Key Largo Limestone (Pleistocene) south Florida. M, 1974, University of Illinois, Urbana.

Wunderman, Richard Lloyd. Amatitlán, an active resurgent caldera immediately south of Guatemala City, Guatemala. M, 1982, Michigan Technological University. 192 p.

Wunderman, Richard Lloyd. Crustal structure across the exposed axis of the Midcontinental Rift and adjacent flanks, based on magnetotelluric data, central Minnesota-Wisconsin; a case for crustal inhomogeneity and possible reactivation tectonics. D, 1988, Michigan Technological University. 224 p.

Wunsch, Carl I. Tidally driven Rossy waves. D, 1967, Massachusetts Institute of Technology. 147 p.

Wunsch, David Robert. Fluoride distribution and relation to chemical character of ground water in Northeast Ohio. M, 1982, University of Akron. 101 p.

Wuolo, Ray Wilbert. Laboratory studies of arsenic adsorption in alluvium contaminated with gold-mine tailings along Whitewood Creek, Black Hills, South Dakota. M, 1986, South Dakota School of Mines & Technology.

Wuorinen, Vilho. A methodology for mapping total risk in urban areas. D, 1979, University of Victoria. 346 p.

Wurdinger, Stephanie. Petrology and structure of Precambrian gneisses at Holcombe, Chippewa County, Wisconsin. M, 1980, University of Minnesota, Duluth.

Wurtsbaugh, Wayne Alden. Internal and external controls on plankton abundance in a large eutrophic lake; Clear Lake, California. D, 1983, University of California, Davis. 119 p.

Wust, Stephen Louis. Geology and structure of the Paradise Range, Mojave Desert, California. M, 1981, Stanford University. 69 p.

Wust, Stephen Louis. Tectonic development of the Pioneer structural complex, Pioneer Mountains, central Idaho. D, 1986, University of Arizona. 88 p.

Wuthrich, Dennis Richard. Fluid flow in the Barbados Ridge complex; permeability estimates and numerical simulations of fluid flow rates, flow directions, and pore pressures. M, 1988, University of California, Santa Cruz.

Wutke, William Bruce, Jr. A comparison of kriging, linear and bicubic spline interpolation techniques as applied to geomagnetic field data. M, 1979, University of California, Davis.

Wuttig, Frank J. Occurrence, distribution and movement of salts and moisture in permafrost near Fairbanks, Alaska. M, 1988, University of Alaska, Fairbanks. 120 p.

Wyatt, Danny J. Carbonate mud mounds from the Lower Ordovician Wah Wah Limestone on the Ibex area, western Millard County, western Utah. M, 1978, Brigham Young University. 44 p.

Wyatt, Frank K. Measurement of continuous strain; Pinon Flat Observatory. D, 1988, University of California, San Diego.

Wyatt, Glen Milton. Hydrogeology and geothermal potential of the Radersburg Valley, Broadwater County, Montana. M, 1984, Montana State University. 160 p.

Wyble, D. O. A study of the regional gravity of the southwestern end of the Iron River District, Iron River, Michigan. M, 1950, Michigan Technological University. 16 p.

Wyble, Donald O. The effect of pressure on conductivity, porosity and permeability of oil-bearing sandstones. D, 1958, Pennsylvania State University, University Park. 71 p.

Wychgram, Daniel C. Geology of the Hayden Pass-Orient Mine area, northern Sangre de Cristo Mountains, Colorado; a geologic remote sensing evaluation. M, 1972, Colorado School of Mines. 130 p.

Wyckhouse, Roger J. Van *see* Van Wyckhouse, Roger J.

Wyckoff, Barkley Sudduth. Geology of the east side of the Guffey volcanic center (Oligocene), Park County, Colorado. M, 1969, New Mexico Institute of Mining and Technology. 64 p.

Wyckoff, Dorothy. Geology of the Mt. Gausta region in Telemark, Norway. D, 1933, Bryn Mawr College. 72 p.

Wyder, John Ernest. Geophysical and geological study of surficial deposits near Frobisher, Saskatchewan (Canada). D, 1968, University of Saskatchewan. 196 p.

Wyder, John Ernest. Reconnaissance resistivity surveys in south-eastern Manitoba. M, 1964, University of Saskatchewan. 137 p.

Wyers, Gerard Paul. Petrogenesis of calc-alkaline and alkaline magmas from the southern and eastern Aegean Sea, Greece. D, 1987, Ohio State University. 381 p.

Wygant, G. T. Petrology and paleoecology of the Ledyard and Wanakah Shale members of Ludlowville Formation, Cazenovia Creek, western New York. M, 1985, SUNY at Binghamton. 75 p.

Wygant, Thomas G. Mineralogical changes and structures associated with the fault-dike system in the Empire Mine, Palmer, Michigan. M, 1969, Bowling Green State University. 68 p.

Wyk, Stefanus J. Van *see* Van Wyk, Stefanus J.

Wykes, E. R. The petrology of some crystalline rocks of the Perth Sheet, Ontario. M, 1931, McGill University.

Wyles, James C. Vitrinite reflectances of coaly materials and their implications in the New Baltimore area, Ohio. M, 1981, University of Akron. 70 p.

Wylie, Albert Sidney, Jr. Structural and metamorphic geology of the Falls Lake area, Wake County, North Carolina. M, 1984, North Carolina State University. 79 p.

Wylie, James Louis. Sedimentary history of the Sue Peaks Formation (Lower Cretaceous), Trans-Pecos region, Texas. M, 1987, Stephen F. Austin State University. 134 p.

Wylie, Jerry Austin. Sedimentary facies depositional environments and sea-level cycles of the Upper Cretaceous Mooreville Chalk (Campanian) west-central Alabama. M, 1987, Auburn University. 144 p.

Wylie, R. W. Resolution of seismic reflection. M, 1956, Massachusetts Institute of Technology. 71 p.

Wyman, Anne F.; Newcomb, Ester Hollis and Wyman, R. Geology of the northern Snake River Range, Idaho and Wyoming. M, 1949, University of Michigan.

Wyman, R.; Newcomb, Ester Hollis and Wyman, Anne F. Geology of the northern Snake River Range, Idaho and Wyoming. M, 1949, University of Michigan.

Wyman, Richard Vaughn. The relationship of ore exploration targets to regional structure in the Lake Mead metallogenic province. D, 1974, University of Arizona.

Wyman, Rosemary. Potential modelling of gravity and levelling data over Cerro Prieto geothermal field. M, 1983, California State University, Long Beach. 73 p.

Wymer, Richard Edward. The petrology of east central Casper Mountain, Wyoming. M, 1979, University of Akron. 95 p.

Wymore, Ivan F. Water requirements for stabilization of spent shale. D, 1974, Colorado State University. 152 p.

Wynn, Jeffrey Curran. Electromagnetic coupling in induced polarization. D, 1974, University of Arizona.

Wynn, Lester L. A study of the Simpson Group of south central Kansas. M, 1959, Wichita State University. 103 p.

Wynn, S. L. A comparison of data collection and analysis methods used to monitor impacts over time in a severely disturbed wetland. D, 1979, University of Wisconsin-Madison. 417 p.

Wynne, Milo E. Geology of a portion of Fremont County, Colorado. M, 1962, University of Kansas. 41 p.

Wynne, Paula Jane. A lithogeochemical study of the host rocks of the Strickland showing. M, 1983, Memorial University of Newfoundland. 313 p.

Wynne-Edwards, Hugh Robert. The structure and petrogenesis of the Westport concordant pluton in the Grenville, Ontario. M, 1956, Queen's University. 238 p.

Wynne-Edwards, Hugh Robert. The structure and petrology of the Grenville-type rocks in the Westport area, Ontario. D, 1959, Queen's University. 316 p.

Wys, Egbert Christiaan De *see* De Wys, Egbert Christiaan

Wys, J. Negus-de *see* Negus-de Wys, J.

Wysocki, Donald John. Characterization of the loamy surficial sediments in a type area of the Iowan erosion surface. D, 1983, Iowa State University of Science and Technology. 250 p.

Wysocki, Martin. Lithofacies and paleoenvironments, Lockport Formation, Allegany County, NY. M, 1980, SUNY, College at Fredonia. 70 p.

Wyss, Max. Observation and interpretation of tectonic strain release mechanisms. D, 1970, California Institute of Technology. 256 p.

Wyszynski, Joseph. Petrography and geochemistry of Tertiary igneous rocks, southern Quitman Mountains and vicinity, Hudspeth County, Texas. M, 1985, University of Texas, Arlington. 165 p.

Wytzes, Jetze. Development of a groundwater model for the Henry's Fork and Rigby Fan areas, upper Snake River basin, Idaho. D, 1980, University of Idaho. 196 p.

Xenophontos, Costas. Engineering geology of three powersites on the Chamouchouane River, Quebec. M, 1972, Queen's University. 189 p.

Xenophontos, Costas. Geology, petrology, and geochemistry of part of the Smartville Complex, northern Sierra Nevada Foothills, California. D, 1984, University of California, Davis. 446 p.

Xia, Chunshou. A microstructural study of mantle peridotites from the Mid-Atlantic Ridge, near 23°N. M, 1988, University of Houston.

Xia, Zong-Guo. Geology of the western boundary of the Taconic Allochthon near Troy and the anastomosing cleavage in the Taconic melange. M, 1983, SUNY at Albany. 189 p.

Xu, Song. Microcoercivity and bulk coercivity in multidomain materials. D, 1988, University of Washington. 170 p.

Xue, Xianyu. Geochemical and isotopic studies of ultramafic xenoliths from West Kettle River, southern British Columbia. M, 1988, University of Alberta. 116 p.

Xueping Ma *see* Ma Xueping

y Menendez, Fernando Osacar Ricart *see* Ricart y Menendez, Fernando Osacar

y Sanchez, Ricardo Jose Padilla *see* Padilla y Sanchez, Ricardo Jose

Yacoub, Nazieh K. Magnetic survey of the Jefferson area, Park County, Colorado. M, 1965, Colorado School of Mines. 30 p.

Yacoub, Naziek K. Attenuation of seismic surface waves and its regional variation in the Eurasian continental crust. D, 1976, St. Louis University. 204 p.

Yacoub, U. Al-Hajji. A quantitative hydrologic study of field "A", South Kuwait. M, 1976, Ohio University, Athens. 76 p.

Yadon, Douglas Mark. Field map and stability analysis of the Schneiderman landslide. M, 1975, Stanford University.

Yager, Milan King. Pennsylvanian geology of north-central Comanche County, Texas. M, 1961, University of Texas, Austin.

Yaghmour, Farouk Abdul-Khaleg. The relationship between historical growth and planned development growth of the Balqa-Amman region of Jordan. D, 1981, SUNY at Buffalo. 158 p.

Yaghubpur, A. Petrology and economic aspects of basic Precambrian basement rocks in Clay County, Iowa. M, 1976, University of Iowa. 185 p.

Yaghubpur, Abdolmajid. Preliminary geologic appraisal and economic aspects of the Precambrian basement of Iowa. D, 1979, University of Iowa. 294 p.

Yagishita, Koji. Mid- to Late Cretaceous sedimentation in the Queen Charlotte Islands, British Columbia; lithofacies, paleocurrent and petrographic analyses of sediments. D, 1985, University of Toronto.

Yagishita, Koji. Petrographic analyses of sandstones from the Tablones Formation (Upper Cretaceous) in northwestern Peru. M, 1977, Northern Illinois University. 106 p.

Yahney, Gordon K. Determination of the major sediment plumes in the western basin of Lake Erie. M, 1978, Bowling Green State University. 124 p.

Yaibuathes, N. An investigation of the tropical Histosols in Hawaii. D, 1971, University of Hawaii. 176 p.

Yake, Daniel Glen. Small dam safety; geotechnical aspects. M, 1983, University of Idaho. 173 p.

Yakzan, Azmi Mohd. Pollen analysis and diatom stratigraphy from Essex Bay marsh, Massachusetts. M, 1987, Boston University. 85 p.

Yaldezian, John George, II. Miocene geology of the east-central portion of the San Rafael Wilderness, Santa Barbara County, California. M, 1984, California State University, Northridge. 145 p.

Yale, David Paul. Network modelling of flow, storage and deformation in porous rocks. D, 1984, Stanford University. 182 p.

Yale, Fred Roger. Petrography of a cyclical sequence, upper Madera Limestone, Sandia Mountains, New Mexico. M, 1964, University of New Mexico. 81 p.

Yale, Leslie Berlincourt. An investigation of serpentinization and rodingitization. M, 1983, Stanford University. 113 p.

Yale, Mark William. Depositional environment and reservoirs morphology of Spraberry sandstones, Parks Field, Midland County, Texas. M, 1986, Texas A&M University.

Yalkovsky, Ralph. Relationship between paleotemperature and carbonate content in a deep sea core. D, 1956, University of Chicago. 43 p.

Yamada, John A. Plate motion and blueschist emplacement. M, 1978, Cornell University.

Yamamoto, Jaime. Rupture processes of some complex earthquakes in southern Mexico. D, 1978, St. Louis University. 203 p.

Yamamoto, S. Sedimentary geochemistry of carbonate-silicate cyclic sedimentation in deep sea. D, 1977, Syracuse University. 218 p.

Yamamoto, S. Sedimentary particles suspended in Asian marginal seas. M, 1974, Syracuse University.

Yamani, Mohamed Abdullah Abdu. Geology of the oolitic hematite of Wadi Fatima, Saudi Arabia, and the economics of its exploitation. D, 1968, Cornell University.

Yamani, Mohamed Abdullah Abdu. Spectrochemical and petrological study of oolitic hematite of Saudi Arabia. D, 1966, Cornell University.

Yamanka, T. Phlogopite-chlorite equilibrium relations. M, 1968, SUNY at Binghamton. 53 p.

Yamashiro, C. H. Potassium analysis of selected rocks by solid source mass spectrometry. M, 1965, University of Hawaii.

Yan, B. Source mechanism study of the 1982 New Brunswick earthquake sequence using a phase-matched-filtering method and a surface wave spectral amplitude-ratio method. M, 1985, Pennsylvania State University, University Park. 141 p.

Yan, Chun-Yeung. Indo-Australia Plate motion determined from Australia and Tasman Sea hotspot trails. M, 1988, University of Hawaii. 104 p.

Yanagida, R. Y. The crystal structure of several nitrogen-containing complexes of zeolite 4A. M, 1973, University of Hawaii. 66 p.

Yanase, Y. Isotopic studies in geological materials controlled heating experiments with the Ar^{40}/Ar^{39} dating technique. D, 1970, University of Toronto.

Yancey, Clyde L. Geology and elemental distribution of the Mississippian Pahasapa Limestone-Pennsylvanian Minnelusa Formation unconformity, southwestern Black Hills, South Dakota. M, 1978, South Dakota School of Mines & Technology.

Yancey, Elizabeth Stilphen. Early middle Ordovician marine benthic communities in southern Nevada and California, 1 volume. M, 1971, University of California, Berkeley.

Yancey, Thomas Erwin. Recent sediments of Monterey Bay, California. M, 1969, University of California, Berkeley. 145 p.

Yancheski, Tadeusz. Hydrology of the Hockessin area, Delaware. M, 1986, University of Delaware.

Yancy, Thomas E. Biostratigraphy, paleoecology and paleontology of the Arcturus Group (Permian; eastern Nevada and western Utah). D, 1971, University of California, Berkeley. 148 p.

Yandell, Lon R. Uranium and gold mineralization in the upper Ruby River basin, Madison County, southwestern Montana. M, 1983, Western Washington University. 107 p.

Yanex, Amade. Littoral processes and sediments of the inner continental shelf of the southern Bay of Campeche. M, 1968, Texas A&M University.

Yanez Pintado, Galo. Geology and geomorphology of the Roraima Group, southeastern Venezuela. D, 1984, Purdue University. 146 p.

Yanez-Correa, Amado. Littoral processes and nature of the sediments of the inner shelf off Terminos Lagoon, Campeche, Mexico. M, 1968, Texas A&M University.

Yanful, Ernest K. Geotechnical properties of clay soils in a brine environment. M, 1982, University of Windsor. 221 p.

Yanful, Ernest Kwesi. Heavy metal migration through clay below a domestic waste site. D, 1986, University of Western Ontario.

Yang, Albert In Che. Variations of natural radiocarbon during the last 11 millenia and geophysical mechanisms for producing them. D, 1971, University of Washington. 102 p.

Yang, Chieh-Hou. Elastic waves; P and S conversions in first arriving reflections and critical refractions. D, 1976, Colorado School of Mines. 242 p.

Yang, Chung-Tien. The magnetoelectric phenomenon reviewed and evaluated as an exploration tool to detect hydrocarbon accumulation. M, 1986, University of Texas at Dallas. 94 p.

Yang, Houng-Yi. Phase equilibria of the join akermanite-anorthite-forsterite in the system CaO-MgO-Al_2O_3-SiO_2 at atmospheric pressure. D, 1970, Ohio State University. 97 p.

Yang, Hsi Chi. 2-D analysis of U-frame structures in elastic-plastic soil medium. D, 1988, University of Florida. 292 p.

Yang, Jae Eui. Characterization of soil K availability by chemical and thermodynamic parameters. D, 1987, Montana State University. 184 p.

Yang, Jane-Fu Jeff. Role of lateral stress in slope stability of stiff overconsolidated clays and clayshales. D, 1987, Iowa State University of Science and Technology. 172 p.

Yang, Jih-Ping. Generalized velocity filters, shot-size calibration in the frequency domain, and seismotectonic study in the northeastern United States. D, 1984, Columbia University, Teachers College. 117 p.

Yang, Jinn-Chuang. Numerical simulation of bed evolution in multichannel river systems. D, 1986, University of Iowa. 205 p.

Yang, Jun-Yang Chen. Research to establish ecological standards for water resources developments. D, 1981, University of Oklahoma. 287 p.

Yang, Ker-Chie. The kinetics of coagulation of interacting polydisperse systems. D, 1976, Pennsylvania State University, University Park. 367 p.

Yang, King-Chih. The Bryozoa of the Greendale Member, Cynthiana Formation, Kentucky. M, 1948, Miami University (Ohio). 41 p.

Yang, Lien-Chu. The distribution and taxonomy of ostracodes and benthic foraminifers in late Pleistocene and Holocene sediments of the Troad (Biga Peninsula), Turkey. M, 1982, University of Delaware. 209 p.

Yang, Mai. Time dependent deformation and stress diffusion in the lithosphere. D, 1981, Massachusetts Institute of Technology. 291 p.

Yang, Mary Mei-ling. Raman spectroscopic investigations of hydrothermal solutions. D, 1988, Princeton University. 170 p.

Yang, Pe-Shen. Nonlinear finite element analysis of piles in integral abutment bridges. D, 1984, Iowa State University of Science and Technology. 241 p.

Yang, Qun. Upper Jurassic (upper Tithonian) Radiolaria from the Taman Formation, east-central Mexico. D, 1988, University of Texas at Dallas. 288 p.

Yang, S. J. Accurate RMS velocity determinations from CGP data by measuring residual NMO with an adaptive method. D, 1976, Texas A&M University. 197 p.

Yang, S. J. Velocity determination from velocity spectra. M, 1973, Texas A&M University.

Yang, Shih Te. Airphoto interpretation of drainage and soils of Fountain County, Indiana. M, 1947, Purdue University.

Yang, Shyue-rong. Major chemical relationships in the volcanic and intrusive rocks of southeastern Guatemala and their relationship to the tectonic model of an island arc. M, 1976, University of Texas, Arlington. 62 p.

Yang, Shyue-Rong Vincent. Petrological and geochemical approaches to the origin of the San Angelo-Flowerpot red beds (Permian) and their associated stratiform copper mineralizations in north central Texas and southwestern Oklahoma. D, 1985, University of Texas at Dallas. 318 p.

Yang, Tsu-Hsi. Oscillatory flow in a porous medium. D, 1967, University of Nebraska, Lincoln. 61 p.

Yang, Tsun-Yi. Molluscs of the Middle Devonian Traverse Group of Michigan. D, 1939, Yale University.

Yang, Tsung. Sand dispersion in a laboratory flume. D, 1968, Colorado State University. 177 p.

Yang, Tsung-Wen. A constitutive model for the mechanical properties of rocks. D, 1982, University of Missouri, Rolla. 141 p.

Yang, Wan-Fa. Volatilization, leaching, and degradation of petroleum oils in sand and soils systems. D, 1981, North Carolina State University. 196 p.

Yang, Wang Hong. High-resolution silicon-29, and sodium 23 nuclear magnetic resonance spectroscopic study of aluminum-silicon disordering in annealed albite and oligoclase. M, 1984, University of Illinois, Urbana. 74 p.

Yang, Wang-Hong Alex. Hydrothermal reaction of crystalline albite, sodium aluminosilicate glass, and a rhyolitic-composition glass with aqueous solutions; a solid-state study. D, 1988, University of Illinois, Urbana. 141 p.

Yang, Wei-Chong. Surf zone properties and on/offshore sediment transport. D, 1982, University of Delaware. 220 p.

Yang, Wei-Liang William. Mapping desert alluvial deposits near McCoy Mountains, southeastern California, using Landsat TM and Seasat SAR data. M, 1985, University of California, Los Angeles. 165 p.

Yang, Wen-Cai. New methods of potential field data processing in regional and mining geophysics. D, 1984, McGill University.

Yang, Xiangning. Evolutionary stasis in Neogene bivalve lineages; a morphometric evaluation. D, 1986, The Johns Hopkins University. 185 p.

Yang, Young Kyu. Microcomputer techniques for the creation and analysis of 7 1/2′ image map from Landsat MSS, RBV, and thematic mapper images; Volumes I and II. D, 1985, Texas A&M University. 429 p.

Yaniga, Paul M. Geochemistry and geology of the streams in the vicinity of the junction of the Nesquehoning Creek and the Leigh River. M, 1974, Lehigh University.

Yanizeski, George Michael. Phenomenological characteristics of the laminar flow of neutrally buoyant particles in a rectangular channel of high aspect ratio. D, 1968, Carnegie-Mellon University. 155 p.

Yannacci, Dawna S. The Buckskin Breccia; a block and ash-flow tuff of Oligocene age in the southwestern high plateaus of Utah. M, 1986, Kent State University, Kent. 107 p.

Yanoski, Mark A. Subsurface facies interpretation of the Minnelusa Formation in the Powder River basin, Wyoming. M, 1983, University of Missouri, Columbia.

Yao, Chia-Chi George. Geological structures from downward continuation of gravity anomalies. M, 1980, Texas A&M University.

Yao, Chien-Chang David. A preconcentration method and a new adsorbent for trace organic analysis in environmental samples. D, 1987, [University of Houston]. 127 p.

Yao, Marcel Abet *see* Abet Yao, Marcel

Yao, Neng-chun. Analysis of hydrographic techniques; a study of different effects of geostrophic current computations by using various hydrographic data processing methods. M, 1967, Oregon State University. 72 p.

Yaowanoiyothin, Winai. Petrology of the Black Brook granitic suite and associated gneiss, northeastern Cape Breton Highlands, Nova Scotia. M, 1988, Acadia University. 257 p.

Yapp, Crayton Jeffery. The variations and climatic significance of D/H ratios in the carbon-bound hydrogen of cellulose in trees. D, 1980, California Institute of Technology. 337 p.

Yarbrough, Ronald Edward. A geographical study of a micro-region in Appalachia; the Clover Fork River valley of Harlan County, Kentucky. D, 1972, University of Tennessee, Knoxville. 152 p.

Yarbrough, Ronald Edward. The common clay industries of East Tennessee. M, 1963, University of Tennessee, Knoxville. 85 p.

Yardley, D. H. The geology of an area at Kashabowie, Ontario, and the Coutchiching problem. M, 1947, Queen's University. 37 p.

Yardley, Donald H. The geology of the northern part of the Chalco Lake area, Northwest Territories, Canada. D, 1951, University of Minnesota, Minneapolis. 130 p.

Yarka, Paul James. Computer-aided hydrodynamic, stratigraphic, and structural analysis of the Paleozoic System in the Four Corners region. M, 1981, Syracuse University.

Yarlot, Mark. Sr:Ca, Mg:Ca, O^{18}:O^{16}, and C^{13}:C^{12} ratios of Neochonetes granulifer; implications for paleosalinity. M, 1982, University of Kansas. 139 p.

Yarnal, Brenton M. The sequential development of a rock glacier-like landform, Mount Assiniboine Provincial Park, British Columbia. M, 1979, University of Calgary.

Yarnal, Brenton Murray. The relationship between synoptic-scale atmospheric circulation and glacier mass balance in southwestern Canada. D, 1983, Simon Fraser University. 213 p.

Yarrow, G. R. Paleoecologic study of part of the Hughes Creek Shale (lower Permian) in north-central Kansas. M, 1974, Kansas State University. 247 p.

Yarter, William Vernon. Geology, geochemistry, alteration, and mass transfer in the Sol Prospect, a sub-economic porphyry copper-molybdenum deposit, Safford District, Graham County, Arizona. M, 1981, University of Arizona. 136 p.

Yarus, Jeffrey M. Bedrock identification by Fourier grain shape analysis of saprolite quartz; South Carolina Piedmont. M, 1976, University of South Carolina. 38 p.

Yarus, Jeffrey M. Shape variation of sand sized quartz in the eastern Gulf of Alaska and its hydrodynamic interpretation; Fourier grain shape analysis. D, 1978, University of South Carolina. 51 p.

Yarussi, Michael. Stratigraphic relationships of the Lower Ordovician sandstone within the Knox dolomite group. M, 1979, University of Pittsburgh.

Yarwood, Walter S. Colloids in iron ore sedimentation with special reference to a pisolitic hematite on Black Island, Manitoba. M, 1924, University of Manitoba.

Yasar, Tuncay. Ray parameters of shear waves recorded at LASA (Large Aperture Seismic Array in Montana). M, 1970, St. Louis University. 53 p.

Yasar, Tuncay. Study of seismic shear waves and upper mantle structure in the Western United States. D, 1974, St. Louis University.

Yassa, Guirguis Fahmy. Adjustment of aerial triangulation using orthogonal transformations. D, 1973, Cornell University. 138 p.

Yassin, Adel Taha. The vertical distribution of salts in a soil profile during the drainage process. D, 1986, Utah State University. 207 p.

Yassin, Yassin Yahya. An entropy formulation to model underground excavation induced disturbance in geomaterials. D, 1988, Colorado State University. 183 p.

Yasso, Warren E. Fluorescent coatings on coarse sediments; an integrated system. M, 1961, Columbia University, Teachers College.

Yasso, Warren E. Geometry and development of spit-bar shorelines at Horseshoe Cove, Sandy Hook, New Jersey. D, 1964, Columbia University, Teachers College.

Yasuda, Memorie. Geographic control of ocean circulation during the Late Cretaceous; comparison of results of an ocean general circulation model with oxygen isotope paleotemperatures. M, 1988, University of Southern California.

Yates, Ann Marie. X-ray fluorescence analysis of chondritic meteorites. D, 1966, Arizona State University. 91 p.

Yates, Arthur Berkeley. Structure of the Homestake ore body. D, 1931, Harvard University.

Yates, Douglas Morris. The solubility and stability relationships of muscovite. M, 1987, Washington State University. 122 p.

Yates, Eugene Adams, III. The geology and mineralization of the Park Canyon uranium prospect, Custer County, south-central Idaho. M, 1980, Washington State University. 146 p.

Yates, Harvey Emmons. A study of the insoluble residues of some Pennsylvanian and Permian limestones of central Texas (Y27). M, 1936, University of Texas, Austin.

Yates, Martin G. Geology and geochemistry of the Ag-Cu-Pb deposits in the Galena Mine, Coeur d'Alene District, Idaho. D, 1987, Indiana University, Bloomington. 220 p.

Yates, Martin G. Major and trace elements content of the Springfield V Coal in southwestern Indiana. M, 1984, Indiana University, Bloomington. 143 p.

Yates, Michael Timothy. Elastic anisotropy in rocks from the Stillwater Igneous Complex (Precambrian), Montana and the Tinaquillo Peridotite (Precambrian), Venezuela. D, 1968, Princeton University. 86 p.

Yates, Robert Giertz. Quicksilver deposits of the Terlingua District, Texas. D, 1961, Stanford University. 238 p.

Yates, Scott Raymond. Geostatistical methods for estimating soil properties. D, 1985, University of Arizona. 230 p.

Yatkola, Daniel Arthur. Middle Miocene environments of western Nebraska and adjacent states. M, 1972, University of Wyoming. 111 p.

Yatsevitch, Yuri. The surficial geology of the Gloverville area (Fulton county, New York). M, 1968, Rensselaer Polytechnic Institute.

Yau, Dah-Miin. Stratigraphy and micropaleontology of some Middle Cretaceous rocks in the North Carolina coastal plain. M, 1984, University of Delaware. 106 p.

Yau, Mary Wing-Chi. Studies of the Recent foraminifera of the Sunda Shelf and Java Sea. M, 1974, Rice University. 134 p.

Yau, Yu-Chyi Lancy. Microscopic studies of hydrothermally metamorphosed shales from the Salton Sea geothermal field, California, USA. D, 1986, University of Michigan. 174 p.

Yau, Yu-Chyi Lancy. Structure and phase relations in the system Li_2O - Fe_2O_3 - TiO_2. M, 1982, Miami University (Ohio). 50 p.

Yazdanian, Aminollah. Sustained-yield groundwater basin management; a goal-programming approach. D, 1986, University of Arkansas, Fayetteville. 143 p.

Yazicigil, H. Mathematical modelling and management of ground water contaminated by aromatic hydrocarbons in the aquifer supplying Ames, Iowa. M, 1977, Iowa State University of Science and Technology.

Yeager, John Conner. Stratigraphy of southern Sierra Pilares, Municipio de Ojinaga, Chihuahua, Mexico. M, 1960, University of Texas, Austin.

Yeager, Richard Neil. Geology, landsliding, and slope stability in the Little Sewickley Creek watershed, Allegheny County, Pennsylvania. M, 1981, University of Pittsburgh. 125 p.

Yeakel, Jesse David. The geochemistry and petrography of peats from the Okefenokee Swamp, Georgia, and the Everglades and coastal regions of southern Florida; a study of variable interrelationships and peat-type differences. D, 1981, Pennsylvania State University, University Park. 313 p.

Yeakel, Lloyd S. Geology of the ore body at Warwick, Pennsylvania, and vicinity. M, 1955, Northwestern University.

Yeakel, Lloyd Stanley. Paleocurrents and paleogeography of the Tuscarora Quartzite (Silurian; Pennsylvania, Maryland, West Virginia, Virginia). D, 1959, The Johns Hopkins University.

Yearsley, John R. Diffusion of salt and momentum in Lake Maracaibo, Venezuela. M, 1968, Massachusetts Institute of Technology. 104 p.

Yeates, David G. A reconstruction of late-glacial and postglacial events in New Haven Harbor, Connecticut; a high resolution subbottom seismic stratigraphic study. M, 1979, Rensselaer Polytechnic Institute. 170 p.

Yeatman, Robert Andrew. The structural integrity of locating sewage lagoons on an abandoned tailings pile. M, 1985, University of Idaho. 112 p.

Yeaton, Walter James. The geology of Zillah Quadrangle, Washington. M, 1923, University of Chicago. 88 p.

Yeats, Kenneth James. Geology and structure of the northern Dome Rock Mountains, La Paz County, Arizona. M, 1985, University of Arizona. 123 p.

Yeats, Robert S. Petrology and structure of Spruce Mountain area, Elko County, Nevada. M, 1956, University of Washington. 80 p.

Yeats, Robert Sheppard. Geology of Skykomish area, Cascade Mountains, Washington. D, 1958, University of Washington. 243 p.

Yeats, Robert Sheppard, Jr. Petrology and structure of the Mount Baring area, northern Cascades, Washington. M, 1956, University of Washington. 80 p.

Yeats, Vestal Liarly. The areal geology of the Moab 4 NW Quadrangle, Grand County, Utah. M, 1961, Texas Tech University. 96 p.

Yeatts, Daniel Solomon. Subsidence due to longwall mining; a geological study. M, 1987, West Virginia University. 172 p.

Yedlin, M. J. Disk ray theory in transversely isotropic media. D, 1978, University of British Columbia.

Yedlosky, Robert Joseph. Petrographic features of sandstones that affect their suitability for road material. M, 1960, West Virginia University.

Yee, Carlton S. Soil and hydrologic factors affecting stability of natural slopes in the Oregon Coast Range. D, 1975, Oregon State University. 204 p.

Yee, Eugene Chan. The magnetotelluric impedance tensor; its reconstruction and analysis. D, 1985, University of Saskatchewan.

Yeend, Warren E. Geology of the Cimarron Ridge-Cimarron Creek area, San Juan Mountains (Montrose, Ouray, and Gunnison counties), Colorado. M, 1961, University of Colorado.

Yeend, Warren Ernest. Quaternary geology of the Grand Mesa area, western Colorado. D, 1965, University of Wisconsin-Madison. 222 p.

Yegulalp, Tuncel Mustafa. A decision making problem in mineral exploration. D, 1968, Columbia University. 136 p.

Yeh, Betty. Mechanisms of the 1927 Lampoc earthquake from surface wave analysis. M, 1975, University of Washington. 91 p.

Yeh, Chih-Cheng. A petrographic study of the Morrison and Cloverly formations in a part of central Wyoming. M, 1948, University of Missouri, Columbia.

Yeh, H.-W. Oxygen isotope studies of ocean sediments during sedimentation and burial diagenesis. D, 1974, Case Western Reserve University. 147 p.

Yeh, Hund-Der. Models for open pit mine dewatering. D, 1982, University of Texas, Austin. 145 p.

Yeh, Long-Tsu. The distribution of the rare-earth elements in Silurian pelitic schists from northwestern Maine. M, 1973, Kansas State University.

Yeh, Pai-Tao. Application of airphoto interpretation techniques in determining runoff from a selected watershed. D, 1953, Purdue University.

Yeh, T. C. J. Effects of urbanization patterns and seasonal variations on runoff. M, 1978, University of Illinois, Chicago.

Yeh, William Wen-Gong. Moisture movement in a horizontal soil column under the influence of an applied pressure. D, 1967, Stanford University. 131 p.

Yeh, Yaw-Huei. Failure mechanisms of buried pipelines under fault movement and soil liquefaction. D, 1983, University of Oklahoma. 211 p.

Yeh, Yeong Tein. A new surface-wave inversion method for determining lateral variations in crust-mantle structure with application to China. D, 1979, Pennsylvania State University, University Park. 245 p.

Yehle, Lynn A. Solution-formed wedges and pipes (soils pendants) in deposits of Wisconsin age. M, 1954, University of Wisconsin-Madison.

Yeilding, Cindy Ann. Stratigraphy and sedimentary tectonics of the Upper Mississippian Greenbriar Group in eastern West Virginia. M, 1984, University of North Carolina, Chapel Hill. 117 p.

Yekini, Bourahm Bourahim. Interactions of orthophosphate with iron-oxyhydroxide minerals found in soils. D, 1980, University of Florida. 123 p.

Yeko, John D. Quantitative analysis of minerals by X-ray powder diffraction. M, 1980, Texas Tech University. 65 p.

Yelderman, J., Jr. Environmental atlas for the extraterritorial jurisdiction of Waco, Texas. M, 1977, Baylor University. 12 p.

Yelderman, Joe C., Jr. The relationship of geologic conditions to hydrologic responses in fluvial aquifers with emphasis on leaky artesian conditions and storage. D, 1983, University of Wisconsin-Madison. 279 p.

Yelin, T. S. The Seattle earthquake swarms and Puget Basin focal mechanisms and their tectonic implications. M, 1982, University of Washington. 96 p.

Yelken, Douglas Lynn. Depositional environment, petrology, and diagenesis of the Laverty-Hoover Sandstone in Beaver, Harper, Ellis, and Woodward counties, Oklahoma, and Lipscomb County, Texas. M, 1985, Oklahoma State University. 107 p.

Yellin, Samuel. Ice tectonics and icequake occurrence on Green Bay. M, 1979, University of Wisconsin-Milwaukee. 121 p.

Yelverton, Charles A. Geology of the southern portion of Old Dad Mountain Quadrangle, San Bernardino County, California. M, 1963, University of Southern California.

Yen, Chung-Cheng. A deterministic-probabilistic modeling approach applied to the Owens Valley groundwater basin. D, 1985, University of California, Irvine. 231 p.

Yen, Fu-Su. Correlation of tuff layers in the Green River Formation, Utah, using biotite compositions. M, 1974, University of Utah. 45 p.

Yen, Hsiang-Jen. Rigid block model for transient analysis of rock structures. D, 1982, Northwestern University. 130 p.

Yen, Luis. Paleomagnetic investigations of a gabbroic dike in the St. Francois Mountains, Missouri. M, 1962, Washington University. 55 p.

Yen, Tzuhua Edward. Velocity determination of the very shallow lunar crust. M, 1979, Texas A&M University.

Yen, Zora Meei-meei. Environmental and geological significance of Globorotalia inflata (d'Orbigny) (Eocene). M, 1971, University of Southern California.

Yenne, Keith Austin. The paleontology and stratigraphy of the Spergen Limestone in eastern Missouri. M, 1939, Washington University. 89 p.

Yeo, Gary Matthew. The Rapitan Group; relevance to the global association of late Proterozoic glaciation and iron-formation. D, 1984, University of Western Ontario. 603 p.

Yeo, Ross K. Animal-sediment relationships and the ecology of the intertidal mudflat environment, Minas Basin, Bay of Fundy, Nova Scotia. M, 1978, McMaster University. 396 p.

Yeo, Ross Kenneth. The stratigraphy and sedimentology of Upper Cretaceous sediments of southwestern California and Baja California, Mexico. D, 1982, Rice University. 622 p.

Yeomans, Bruce Wyatt. The geology and economic petrology of the Archean Newton Lake Formation, Boulder Bay area, St. Louis County, Minnesota. M, 1984, University of Minnesota, Duluth. 133 p.

Yerima, Bernard Palmer Kfuban. Soil genesis, phosphorus and micronutrients of selected Vertisols and associated Alfisols of northern Cameroon. D, 1986, Texas A&M University. 345 p.

Yerino, Lawrence N. Petrography of sands from the Peru-Chile fore-arc and adjacent areas. M, 1982, University of Cincinnati. 161 p.

Yerkes, Robert F. The geology of a portion of the Cajon Pass area, San Bernardino County, California. M, 1951, Pomona College.

Yersak, Thomas E. Gravity study of Staten Island and vicinity. M, 1977, Rutgers, The State University, New Brunswick. 48 p.

Yesberger, William Lloyd, Jr. Depositional environments of the Lamotte Sandstone in Southeast Missouri. M, 1982, University of Missouri, Columbia.

Yeskis, Douglas Jerome. Dehydroxylation studies of kaolinite, pyrophyllite, and muscovite using high-pressure differential thermal analysis. M, 1983, University of Illinois, Chicago.

Yett, Jan Reynolds. Eocene Foraminifera from the Olequa Creek Member of the Cowlitz Formation, southwestern Washington. M, 1979, University of Washington. 146 p.

Yettaw, Gordon A. Upper Cambrian and older rocks of the Security-Thalman No. 1 well, Berrien County, Michigan. M, 1967, Michigan State University. 72 p.

Yetter, Riley Glen. Stratigraphy (and structure) of a portion of the Sweetwater Uplift, Fremont County, Wyoming. M, 1954, University of Nebraska, Lincoln.

Yeung, Terence C. A study of magnetic anomalies in the Yucatan Basin. M, 1981, [University of Houston].

Yewisiak, Paul P. Conglomerate facies of the Culpeper Triassic Basin, Virginia. M, 1970, University of Virginia. 66 p.

Yi, Dar. Field investigation of the aftershocks from the October 1, 1987 Whittier Narrows earthquake. M, 1988, California State University, Long Beach. 67 p.

Yi, Zhixin. On the shallow hydrothermal regime of the Salton Sea geothermal field, California. M, 1987, University of California, Riverside. 265 p.

Yih, Liang Foo. Restudy of the specimens from Franklin Furnace, New Jersey, in the collection of Columbia University. M, 1922, Columbia University, Teachers College.

Yildiz, Mehmet. Structure and petrography of Black Rock, Apache County, Arizona. M, 1961, University of Arizona.

Yilmaz, H. Genesis of uranium deposits in Neogene sedimentary rocks, Menderes metamorphic massif, Turkey. D, 1979, University of Western Ontario. 221 p.

Yilmaz, O. Pre-stack partial migration. D, 1979, Stanford University. 126 p.

Yilmaz, Pinar Oya. Alakir Cay unit of the Antalya Complex (Turkey); an example of ocean floor obduction. M, 1978, Bryn Mawr College. 91 p.

Yilmaz, Pinar Oya. Geology of the Antalya Complex, SW Turkey. D, 1981, University of Texas, Austin.

Yim, Chik-Sing. Effects of transient foundation uplift on earthquake response of structures. D, 1983, University of California, Berkeley. 126 p.

Yim, Patrick C. Q. Some features of sea level fluctuations at island stations in the tropical Pacific. M, 1978, University of Hawaii.

Yim, T. B. Solid waste and sewage sludge management for the city of Seoul, Korea. D, 1975, University of Oklahoma. 252 p.

Yin, An. Geometry, kinematics, and a mechanical analysis of a strip of the Lewis Allochthon from Peril Peak to Bison Mountain, Glacier National Park, Montana. D, 1988, University of Southern California.

Yin, An. Geometry, kinematics, and mechanical analysis of a strip of the Lewis Allochthon from Peril Peak to Bison Mountain, Glacier National Park, Montana. M, 1988, University of California, Los Angeles.

Yin, David D. Microfacies analysis and depositional systems of Potrero Chico (Northeast Mexico). M, 1988, University of Texas at Dallas. 70 p.

Yin, Peigui. Generation and accumulation of hydrocarbons in the Gippsland Basin, southeastern Australia. D, 1988, University of Wyoming. 285 p.

Yinger, Mark Andrew. Geology of the eastern part of the Loon Creek mining district, Custer County, Idaho. M, 1976, University of Idaho. 103 p.

Yingling, Virginia Leigh. Timing of initiation of the Sevier Orogeny; Morrison and Cedar Mountain formations and Dakota Sandstone, east-central Utah. M, 1987, University of Wyoming. 169 p.

Yingst, Parke O. Upgrading western coking coal by conventional ore-dressing methods. D, 1960, Colorado School of Mines. 208 p.

Yip, Freddy F. Trace fossils of the Tallahatta Formation (Claiborne Group) in east-central Mississippi. M, 1981, Mississippi State University. 74 p.

Yiu, Shih-Kao. A measure of the energy flux represented by a seismogram. D, 1968, Pennsylvania State University, University Park. 173 p.

Yllarramendi, John A. Kavanagh see Kavanagh Yllarramendi, John A.

Yochelson, Ellis L. Permian Gastropoda of the Southwestern United States. D, 1955, Columbia University, Teachers College.

Yochelson, Ellis Leon. Gastropoda of the Middle Ordovician Peery and Murfreesboro formations of Virginia and Tennessee. M, 1950, University of Kansas. 143 p.

Yockum, Eric T. Petrography of the middle and upper Holder Formation (Upper Pennsylvanian), Sacramento Mountains, New Mexico. M, 1985, Texas Tech University. 92 p.

Yoder, Gary Eldon. Stratigraphy, structure, and paleoecology of a Silurian reef at Francesville, Indiana. M, 1982, Indiana University, Bloomington. 113 p.

Yoder, H. S., Jr. The stability relations of grossularite. D, 1948, Massachusetts Institute of Technology. 132 p.

Yoder, Nelson B. Microfacies analysis of a biogenetic bank, Lake Valley Formation (Osagian), Sacramento mountains, New Mexico. M, 1968, Texas Tech University. 96 p.

Yogendrakumar, Muthucumarasamy. Dynamic soil-structure interaction; theory and verification. D, 1988, University of British Columbia.

Yogodzinski, Gene M. The Deschutes Formation-High Cascade transition in the Whitewater River area, Jefferson County, Oregon. M, 1986, Oregon State University. 165 p.

Yoho, William Herbert. Geology of Audubon County, Iowa. M, 1939, University of Iowa. 136 p.

Yokley, John W. Geology of Horse Basin and Jerry Peak quadrangles, Custer County, Idaho. M, 1974, University of Wisconsin-Milwaukee.

Yokoyama, J. S. Soil-air-water relationship in Hawaiian soils. M, 1969, University of Hawaii. 81 p.

Yokoyama, Ken T. Geology of the Engadine Group (Silurian) in the subsurface of the northern peninsula of Michigan. M, 1981, Bowling Green State University. 84 p.

Yole, Raymond William. A faunal and stratigraphic study of upper Paleozoic rocks of Vancouver Island, British Columbia, Canada. D, 1965, University of British Columbia. 338 p.

Yolton, James S. The Dry Lake section of the Brazer Formation. M, 1943, Utah State University. 29 p.

Yomogida, Kiyoshi. Amplitude and phase variations of surface waves in a laterally heterogeneous earth. D, 1986, Massachusetts Institute of Technology. 227 p.

Yon, James W., Jr. The Hawthorn Formation (Miocene) between Chattahoochee and Ellaville, Florida. M, 1953, Florida State University.

Yonce, Joseph B. A study of suspended sediments in rivers with particular emphasis on the James River (Virginia) basin. M, 1970, Virginia State University. 85 p.

Yong, Kingston Cheng Wu. Geopressured-geothermal reservoir parameters of the Blessing Prospect area, Matagorda County, Texas. M, 1980, University of Texas, Austin.

Yong, Yan. Stochastic earthquake modeling and dynamic response analysis. D, 1987, University of Illinois, Urbana. 172 p.

Yonge, Charles J. Stable isotope studies of water extracted from speleothems. D, 1982, McMaster University. 298 p.

Yonk, Allen K. Age of mineralization of the Upper Mississippi Valley lead-zinc district. M, 1981, Northern Illinois University.

Yonkee, W. Adolph. Mineralogy and structural relationships of cleavage in the Twin Creek Formation within part of the Crawford thrust sheet in Wyoming and Idaho. M, 1983, University of Wyoming. 125 p.

Yonover, Robert. Laser decrepitation and analysis of fluid inclusions from the Meguma Complex, Nova Scotia; nature of the ore forming fluids. M, 1984, Florida State University.

Yont, D. R. Devonian Winnipegosis reefs in west central Saskatchewan. M, 1960, University of Saskatchewan. 55 p.

Yoo, Kyung Hak. Characteristics of nutrients and total solids in irrigation return flows. M, 1974, University of Idaho. 80 p.

Yoo, T.-S. A theory for vibratory compaction of soil. D, 1975, SUNY at Buffalo. 315 p.

Yoon, Kern Shin. Development of simple techniques for in-situ determination of rock deformation, with particular reference to specimen size and far-field stress effect. D, 1987, University of Wisconsin-Madison. 236 p.

Yoon, Kwi-Hyon. Synthetic seismogram modeling of fine oceanic crustal structure near Guadalupe Island, Mexico. M, 1987, University of Texas at Dallas. 93 p.

Yoon, Tai Nam. The Cambrian and lower Ordovician stratigraphy of the Saint John area, New Brunswick, (Canada). M, 1970, University of New Brunswick.

Yoon, Y. S. Identification of aquifer parameters in a large unconfined system; with finite elements and constrained nonlinear least squares optimization. D, 1976, University of California, Los Angeles. 79 p.

Yorath, Christopher J. The determination of sediment dispersal patterns by statistical and factor analyses, northeastern Scotian Shelf (Recent, Atlantic seacoast near Sable Island). D, 1967, Queen's University. 204 p.

Yorath, Christopher John. Micropaleontology of the Deer Bay Formation, Arctic Archipelago, Canada. M, 1962, University of Alberta. 94 p.

York, H. F. Geology of the Elk Mountain Anticline, North Park, Colorado. M, 1954, University of Wyoming. 77 p.

York, James Earl, III. Seismotectonics in intraplate and interplate regions; eastern North America, eastern Taiwan, China, and the New Hebrides. D, 1977, Cornell University. 162 p.

York, Linda Louise. Aminostratigraphy of Stetson Pit and Ponzer areas of North Carolina by Pleistocene mollusk analysis. M, 1985, University of Delaware. 188 p.

York, Terrell M. Geology of the Tok antimony mine, Tok, Alaska. M, 1980, Colorado School of Mines. 94 p.

Yorston, H. J. Geology of the south half of the Meramec Spring Quadrangle, Missouri. M, 1954, University of Missouri, Rolla.

Yose, Lyndon A. Autocyclic versus allocyclic controls on deposition of a mixed-clastic outer-platform-to-basin sequence; Middle to Upper Pennsylvanian of southeastern California. M, 1987, University of Wyoming. 79 p.

Yost, Carl R. Gravity survey of the Cheney Quadrangle, Washington. M, 1976, Eastern Washington University. 37 p.

Yost, Coyd Bickley, Jr. A magnetic profile across a part of northern West Virginia. M, 1949, West Virginia University.

Yost, Lawrence W. Integral migration of seismic sections taken over complex geologic structures. D, 1980, Colorado School of Mines. 111 p.

You, Yet-Cheng. Data acquisition and analysis system for the application of triaxial test and direct shear test in determining the residual shear strength of clays from landslide site. D, 1988, Oklahoma State University. 171 p.

Youash, Younathan Yousif. Correlation of gravity observations with the geology of southern Burnet County, Texas. M, 1961, University of Texas, Austin.

Youash, Younathan Yousif. Dynamic physical properties of rocks. M, 1964, University of Texas, Austin. 165 p.

Youash, Younathan Yousif. Experimental deformation of layered rocks. D, 1965, University of Texas, Austin. 206 p.

Youd, Thomas Leslie. The engineering properties of cohesionless materials during vibration. D, 1967, Iowa State University of Science and Technology. 101 p.

Youn, Oong. Effective deconvolution operator design by modification of input data. M, 1986, California State University, Long Beach. 229 p.

Youn, Sung Ho. Comparison of porosity and density values of shale from cores and well logs. M, 1974, University of Tulsa. 62 p.

Younce, Gordon Baldwin. Structure geology and stratigraphy of the Bonavista bay region, Newfoundland, (Canada). D, 1970, Cornell University. 214 p.

Young, Addison. Studies of gravity anomalies. M, 1926, Yale University.

Young, Brian T. The stability of selected road cuts along the Ohio River as influenced by valley stress relief joints. M, 1988, Kent State University, Kent. 164 p.

Young, Bruce Cash. The geomorphology of the Pinto Basin, southern California. D, 1968, University of California, Los Angeles. 315 p.

Young, C. T. Magnetotelluric measurements of conductivity anomalies in northern Wisonsin. D, 1977, University of Wisconsin-Madison. 165 p.

Young, Carl Michael. Morphological variation in middle Miocene Uvigerina from the northern Gulf of Mexico. M, 1986, University of New Orleans. 77 p.

Young, Carl R. Optimum vertical hydraulic fracturing for improving oil recovery using aluminum pellets. M, 1962, University of Oklahoma. 50 p.

Young, Chapman, III. Applications of dislocation theory to upper-mantle deformation. D, 1966, Stanford University. 154 p.

Young, Chapman, III. The geology north of White Cloud Canyon, Stillwater Range, Nevada. M, 1963, Stanford University. 66 p.

Young, Charles Robert. Paleontology, paleoecology, and sedimentology of the lower 25 feet of the Fort Riley Limestone. M, 1969, Wichita State University. 91 p.

Young, Chi Yuh. Attenuation of teleseismic P-waves within geothermal systems. D, 1979, University of Texas at Dallas. 192 p.

Young, Ching Ju Jennifer. Prediction and elimination of pressure generated ground noise on long period seismograms using optimum filters. M, 1976, Southern Methodist University. 35 p.

Young, Christopher J. Evidence for a shear velocity discontinuity in the lower mantle beneath India and the Indian Ocean. M, 1986, University of Michigan. 23 p.

Young, Dae Sik. Effect of peripheral top cuts on the strength of mine pillar models. M, 1968, University of Utah. 86 p.

Young, Dae Sik. Stress analysis in open-pit mines. D, 1972, University of Utah. 102 p.

Young, Daniel R. Molluscan paleontology of the Pliocene Peace Valley "Beds" and Ridge Route "Formation" (Ridge Basin Group), Ridge Basin, Southern California. M, 1980, University of North Dakota. 166 p.

Young, David K. Chemistry of Chesapeake Bay sediments. M, 1962, College of William and Mary.

Young, David Lucius. A kinematic interpretation of ductile deformation and shear-induced quartz c-axis fabric development in the Montevideo Gneiss near Granite Falls, S.W. Minnesota. M, 1987, University of Missouri, Columbia. 201 p.

Young, David Marion. The Cynthiana of the southwestern Bluegrass. M, 1931, University of Kentucky. 48 p.

Young, David Paul. The history of deformation and fluid phenomena in the top of the Wilderness Suite, Santa Catalina Mountains, Pima County, Arizona. M, 1988, University of Arizona. 144 p.

Young, David Ross. The distribution of cesium, rubidium and potassium in the quasi-marine ecosystem of the Salton Sea. D, 1970, University of California, San Diego.

Young, Davis A. Alkali metasomatism related to the Pinos Altos pegmatite body, Rio Arriba County, New Mexico. M, 1965, Pennsylvania State University, University Park. 72 p.

Young, Davis Alan. Petrology and structure of the west central New Jersey highlands. D, 1969, Brown University. 211 p.

Young, Durward Dudley. A magnetometer survey of an anomaly south of Valley Park, Missouri, in Saint Louis and Jefferson counties. M, 1955, St. Louis University.

Young, Edward J. A. A ground water problem in the north shore area, Nova Scotia. M, 1950, Massachusetts Institute of Technology. 55 p.

Young, Edward J. A. Trace element studies in sediments. D, 1954, Massachusetts Institute of Technology. 109 p.

Young, Ford. Hydrotungstite, a new mineral from Oruro, Bolivia. M, 1941, Columbia University, Teachers College.

Young, Francis Millard. A study of the heavy minerals of the streams of the West Fork of the Bitterroot River, Montana. M, 1965, Montana College of Mineral Science & Technology. 102 p.

Young, Frederick Griffin. Sedimentary cycles and facies in the correlation and interpretation of lower Cambrian rocks, east-central British Columbia. D, 1970, McGill University.

Young, Frederick P, Jr. Chronological distribution of the Pisces. M, 1930, Columbia University, Teachers College.

Young, Frederick P., Jr. Black River (Ordovician) stratigraphy and faunas (New York). D, 1943, Columbia University, Teachers College.

Young, Genevieve B. C. Stratigraphy and petrology of the Lower Cretaceous J Sandstone, Wattenberg Gas Field, Weld County, Colorado. M, 1987, Colorado School of Mines. 369 p.

Young, George A. Geology and petrology of Mount Yamaska, Province of Quebec. D, 1904, Yale University.

Young, George E. Geology of Billies Mountain Quadrangle, Utah County, Utah. M, 1975, Brigham Young University.

Young, George Husband. Crystal studies with certain nitrogen-substituted sulfonamides. D, 1936, Pennsylvania State University, University Park. 114 p.

Young, Gerald Clifton. A study of the physical stratigraphy in the northeast portion of the Paskenta Quadrangle, Tehama County, California. M, 1958, University of California, Berkeley. 159 p.

Young, Graham Arthur. Paleoecology, biostratigraphy, and systematics of Silurian tabulate corals from the Chaleur Bay region, Eastern Canada. D, 1988, University of New Brunswick. 520 p.

Young, Graham Arthur. Systematics and biostratigraphy of Middle Cambrian trilobites from the Cow Head Group, western Newfoundland. M, 1984, University of Toronto.

Young, Gregory B. Identification of seismic waves reflected from the top of the upper mantle low velocity zone. M, 1974, Purdue University. 100 p.

Young, H. L. Land and water acquisition in relation to present and potential land use change in South Park, Colorado. D, 1975, University of Colorado. 231 p.

Young, Harvey Ray. Petrology and heavy minerals of the Viking Formation, west-central Alberta. M, 1959, University of Alberta. 111 p.

Young, Harvey Ray. Petrology of the Virden Member of the Lodgepole Formation (Mississippian) in southwestern Manitoba. D, 1973, Queen's University. 385 p.

Young, I. A petrographic study of the rocks found in the vicinity of Tarrytown Quadrangle (New York). M, 1911, Columbia University, Teachers College.

Young, Ian Fairley. Structure of the western margin of the Queen Charlotte Basin, British Columbia. M, 1981, University of British Columbia. 380 p.

Young, J. Llewellyn. Glaciation in the Logan Quadrangle, Utah. M, 1939, Utah State University. 79 p.

Young, Jackson Smallwood. Limestone reservoirs of southern Oklahoma. M, 1929, University of Illinois, Urbana.

Young, James Lewis, Jr. A geological investigation of the causes of failure of concrete aggregate used in Kentucky highways. M, 1949, University of Kentucky. 38 p.

Young, James R. Saratoga; the bubbles of reputation and their implications for an embryonic rift system in the upper Hudson River valley. M, 1980, SUNY at Albany. 198 p.

Young, James Stanton. Paragenesis and geochemistry of the Granite gold district, Grant County, Oregon. M, 1979, Colorado State University. 115 p.

Young, Jay D. Underground storage of natural gas and spontaneous potential investigation of the Bistineau storage field, Louisiana. M, 1979, Northeast Louisiana University.

Young, Jay Marc. Geology of the nearshore continental shelf and coastal area, northern San Diego County, California. M, 1980, San Diego State University.

Young, Jean T. Pliocene - Pleistocene foraminifera from Timms Point, San Pedro and Bathhouse Beach, Santa Barbara, California. M, 1979, University of California, Los Angeles.

Young, Joe B. Deposition, diagenesis and distribution of an "Upper Silurian" dolostone reservoir, FK Devonian Field. M, 1984, Texas Tech University. 78 p.

Young, John A. Associations and origin of the limestone and steatite deposits of Rhode Island. M, 1934, Brown University.

Young, John A., Jr. Paleontology and stratigraphy of the Pennsylvanian strata near Taos, New Mexico. D, 1946, Harvard University.

Young, John C. Geology of the southern Lakeside Mountains (Tooele County, Utah). M, 1953, University of Utah. 90 p.

Young, John Cannon. Structure and stratigraphy in the north-central Schell Creek Range, eastern Nevada. D, 1959, Princeton University. 207 p.

Young, John D. Late Cenozoic geology of the lower Rio Puerco, Valencia and Socorro counties, New Mexico. M, 1982, New Mexico Institute of Mining and Technology. 126 p.

Young, John Walter. The relationship between lamprophyre dykes and ore deposits with special references to BC. M, 1948, University of British Columbia.

Young, K. P. A summary of the evidence of the Laramie revolution, Colorado and Wyoming. M, 1942, University of Wyoming. 74 p.

Young, Keith Preston. Stratigraphy and paleontology of the (Cretaceous) Frontier Formation, southern Montana. D, 1948, University of Wisconsin-Madison. 123 p.

Young, Kirby David. Fracture zone deformation of a Tertiary lava pile, north-central Iceland. M, 1983, Pennsylvania State University, University Park. 119 p.

Young, Lawrence Edward. A hydrogeologic investigation of the Brady Lake and portions of the adjacent Breakneck Creek drainages near Kent, Ohio. M, 1980, Kent State University, Kent. 139 p.

Young, Leonard Maurice. Dimensional grain orientation studies of Recent Canadian River sands. M, 1960, University of Oklahoma. 46 p.

Young, Leonard Maurice. Sedimentary petrology of the Marathon Formation (Lower Ordovician), Trans-

Pecos, Texas. D, 1968, University of Texas, Austin. 268 p.

Young, Malcolm G. Geology of a portion of the Sweetwater-Horse Greek drainages, Park County, Wyoming. M, 1966, Wayne State University.

Young, Maria Leigh. Petrology and origin of Archean rocks in the Armstead Anticline of Southwest Montana. M, 1982, University of Montana. 66 p.

Young, Michael A. A high resolution paleomagnetic study of Recent sediments from Long Lake, Hennepin County, Minnesota; correlation of paleomagnetism with paleoclimatic data. M, 1979, University of Minnesota, Minneapolis.

Young, Michael H. A comparative study to determine the effects of coal mining on the hydrology of small watersheds in southeastern Ohio. M, 1986, Ohio University, Athens. 500 p.

Young, Michael Steven. Willwood metaquarzite conglomerate in a southwestern portion of the Bighorn Basin, Wyoming. M, 1972, Iowa State University of Science and Technology.

Young, Peter C. Surface geology and soil geochemistry of the Buena Vista Mine area, Patagonia mountain, Santa Cruz County, Arizona. M, 1969, Colorado School of Mines. 119 p.

Young, Peter David. An active source electromagnetic method for probing the Earth's electrical conductivity structure beneath the ocean floor. D, 1981, University of California, San Diego. 255 p.

Young, Peter Frederick. Physical properties and microstructures of a selected portion of the titanium-vanadium-silicon alloy system. M, 1956, University of Nevada. 44 p.

Young, Rebecca H. Estimating the composition of illite and chlorite using experimental and theoretical techniques. M, 1984, Bowling Green State University. 114 p.

Young, Rechard Andrew. Cenozoic geology along the edge of the Colorado Plateau in northwestern Arizona. D, 1966, Washington University. 167 p.

Young, Rex James. Petrology of the Siesta Formation, Berkeley Hills, California. M, 1961, University of California, Berkeley. 71 p.

Young, Richard Wescoe. A geochemical investigation of dredging in Bayboro Harbor and the Port of St. Petersburg. M, 1984, University of South Florida, St. Petersburg.

Young, Robert Alexander. Deep-sea sedimentation on the northwestern African continental margin. M, 1972, Massachusetts Institute of Technology. 81 p.

Young, Robert Alexander. Flow and sediment properties influencing erosion of fine-grained marine sediments; sea floor and laboratory experiments. D, 1975, Massachusetts Institute of Technology. 202 p.

Young, Robert Brigham. Geology analysis of a part of northeastern Utah using E.R.T.S. multi-spectral imagery. M, 1984, Brigham Young University. 211 p.

Young, Robert C. Temperatures of late Precambrian metamorphism, Llano Uplift, Texas. M, 1976, Northeast Louisiana University.

Young, Robert Glenn. Stratigraphic relations in the Upper Cretaceous of the Book Cliffs, Utah-Colorado. D, 1952, Ohio State University. 201 p.

Young, Robert Spencer. A study of the industrial coals of Logan County, West Virginia. M, 1951, University of Virginia. 157 p.

Young, Robert Spencer. The geology of the Edinburg, Virginia-West Virginia Quadrangle. D, 1954, Cornell University.

Young, Robert Thomas. Significance of the magnesium/calcium ratio as related to structure in the Stony Lake oil field, Michigan. M, 1955, Michigan State University. 58 p.

Young, Robert Thomas. Subsurface geology of East Pauls Valley area, Garvin County, Oklahoma. M, 1960, University of Oklahoma. 46 p.

Young, Roger A. Stresses of thermal origin in the Moon. M, 1968, Stanford University.

Young, Ronald Earl. Geology and biostratigraphy of the Knappton, Washington area. M, 1966, University of Washington. 186 p.

Young, Ruth Hope. Triangular coiling in the Cephalopoda (Bend, Texas, and Clarita, Oklahoma). M, 1941, University of Chicago. 69 p.

Young, Seward D. Stratigraphy and petrology of the lower Atoka Formation in the Lee Creek area of Washington County, Arkansas. M, 1975, University of Arkansas, Fayetteville.

Young, Stephen Robert. Characterization of and parameters controlling small faults in naturally deformed, porous sandstones. M, 1982, Texas A&M University. 118 p.

Young, Steven Wilford. A size analysis of columnar burrow structures in selected exposures of Permian strat in southwestern Montana. M, 1973, Indiana University, Bloomington. 75 p.

Young, Steven Wilford. Petrography of Holocene fluvial sands derived from regional metamorphic source rocks. D, 1975, Indiana University, Bloomington. 144 p.

Young, Susan L. Structural history of the Jordan Creek area, northern Madison Range, Madison County, Montana. M, 1985, University of Texas, Austin.

Young, T. C. The dynamics of accumulation of phosphorus by the sediments of the lake system of the water quality management project at Michigan State University. D, 1977, Michigan State University. 96 p.

Young, Terence K. A seismic investigation of North and South Table Mountains near Golden, Jefferson County, Colorado. M, 1977, Colorado School of Mines. 54 p.

Young, Terence K. The application of generalized ray theory to the study of elastic wave propagation in the borehole environment. D, 1979, Colorado School of Mines. 118 p.

Young, Timothy J. Examination of fenestral structures and cements in the Tyrone Limestone (Middle Ordovician) in central Kentucky. M, 1986, Eastern Kentucky University. 69 p.

Young, Wilfred Ray. The origin and classification of bedding planes. M, 1950, Texas Tech University. 74 p.

Young, William L. The iron bearing formations of the Michipicoten area, Ontario. D, 1954, McGill University.

Younger, James Allen. Chromitites in the lower part of the anorthosite series, Bushveld Complex, Union of South Africa. M, 1958, University of Wisconsin-Madison.

Younger, Michael Alan. Petrography and depositional environments of the Cretaceous (Cenomanian) Eagle Mountains Formation, western Trans-Pecos, Texas. M, 1988, University of Texas, Arlington. 174 p.

Younger, Paul Lawrence. Barite travertine from southwestern Oklahoma and west-central Colorado. M, 1986, Oklahoma State University. 163 p.

Youngkin, Mark T. Late Cenozoic volcanism and tectonism of the Eagle Lake area, Lassen County, California. M, 1980, Colorado School of Mines. 106 p.

Youngquist, Walter Lewellyn. The cephalopod fauna of the White Pine Shale of Nevada. D, 1948, University of Iowa. 140 p.

Youngquist, Walter Lewellyn. Upper Devonian conodonts from the Independence Shale of Iowa. M, 1943, University of Iowa. 49 p.

Youngs, Robert Riggs. A three-dimensional effective stress model for cyclicly loaded granular soils. D, 1982, University of California, Berkeley. 322 p.

Youngs, Steven Wilcox. Geology and geochemistry of the Running Springs geothermal area, southern Bitterroot Lobe, Idaho Batholith. M, 1981, Washington State University. 125 p.

Younker, Jean Kay. Evaluation of the utility of two dimensional Fourier shape analysis for the study of ostracode carapaces. M, 1971, Michigan State University. 63 p.

Younker, Jean Lower. Analytical paleontology; patterns of taxonomic extinction. D, 1976, Michigan State University. 114 p.

Younker, Leland Wilbur. Process model for sedimentation under turbulent flow conditions. M, 1972, Michigan State University. 47 p.

Younker, Leland Wilbur. Role of mantle derived magmas in the production of crustal melts. D, 1974, Michigan State University. 72 p.

Younkin, Robert LeFevre. Spectrophotometry of the Moon, Mars and Uranus. D, 1970, University of California, Los Angeles. 188 p.

Younse, Gary A. The stratigraphy and phosphoritic rocks of the Robinson Canyon-Laureles Grade area. M, 1979, San Jose State University. 127 p.

Yount, James C. A neoglacial chronology for the Independence Pass area, Colorado using graph-theoretic classification methods. M, 1970, University of Colorado.

Youse, Arthur. A subsurface study of the gas producing zones of the Greenbrier Limestone (Mississippian) in southern West Virginia and eastern Kentucky. M, 1963, West Virginia University.

Yousef, Ali Abdulah. A surface wave dispersion study of the lithospheric structure of Africa. D, 1987, University of Texas at El Paso. 130 p.

Yousef, Ali Abdullah. A study of time residuals in the Socorro area for P_0 arrivals from mining explosions at Anata Rita, Tyrone, New Mexico, and Morenci, Arizona. M, 1977, New Mexico Institute of Mining and Technology.

Yousefpour, M. V. Geology and contact pyrometasomatic ore deposits at the Continental Mine, Fierro, New Mexico. M, 1977, Colorado School of Mines. 182 p.

Yousefpour, Mohammed Vali. Genesis of skarn polymetallic deposits in the Hanover-Fierro area and a geotechnical evaluation of pit slope stability at the Continental Mine, Fierro, New Mexico. D, 1979, Colorado School of Mines. 401 p.

Youshah, Bashir M. A geophysical study of the iron ore deposit of Wadi Shatti area (Libya). M, 1980, Ohio University, Athens. 195 p.

Yousif, Majeed H. Stability of oil/water interface during immiscible displacement in porous media. D, 1987, Colorado School of Mines. 140 p.

Yousif, Nesreen Bashir. Finite element analysis of some time dependent construction problems in geotechnical engineering. D, 1985, SUNY at Buffalo. 243 p.

Youssefnia, Iradj. Western North Atlantic Paleocene benthonic foraminiferal paleoecology. D, 1974, Rutgers, The State University, New Brunswick. 125 p.

Youssefnia, Iradj. X-ray analysis, geochemistry and description of the Vincentown microfossils (late Paleocene to early Eocene) (Burlington county, New Jersey). M, 1969, Brooklyn College (CUNY).

Ysalgue, Sarah E. An outline of the geomorphology of western Cuba. M, 1944, Columbia University, Teachers College.

Yu, Guey-Kuen. Regionalized shear velocity models beneath the Pacific Ocean from Love and Rayleigh wave dispersion. D, 1978, St. Louis University. 193 p.

Yu, Ho-Shing. A study of the bottom sediments of the Trinity River system, Tarrant County, Texas. M, 1974, Texas Christian University. 44 p.

Yu, Ho-Shing. Three aspects of sandstone diagenesis; compaction and cementation of quartz arenites and chemical changes in graywackes. D, 1979, University of Cincinnati. 135 p.

Yu, J. P. Applications of geochemistry in geothermal reservoir engineering. M, 1978, Stanford University. 27 p.

Yu, Jen Haur. Some thermal magnetic properties of baked clays. M, 1978, University of Oklahoma. 118 p.

Yu, Jianxin. Ostracode distribution within the Bush Creek Member (Missourian, Upper Pennsylvanian) of the northern American Appalachian Basin. M, 1986, University of Pittsburgh.

Yu, John Pingshun. Multi-criteria optimization model for thermal recovery processes. D, 1983, University of Oklahoma. 218 p.

Yu, Keun Bai. The hydrology of the Okefenokee Swamp watershed with emphasis on groundwater flow. D, 1986, University of Georgia. 242 p.

Yu, P. L. Microchemical determination of minerals. M, 1948, McGill University.

Yu, Shu-Cheng. Crystallographic study on two synthetic arsenic-sulfide crystals. M, 1971, University of Minnesota, Minneapolis.

Yu, Shu-cheng. The crystal chemistry of a Zn-Li silicate and defect substructures of augitic pyroxenes and their implications. D, 1976, Pennsylvania State University, University Park. 98 p.

Yu, Thiann-Ruey. A Rayleigh wave dispersion technique for geoexploration. D, 1974, McGill University.

Yu, Y-S. Physical properties of a Sigma Mine porphyry. M, 1965, McGill University.

Yu, Y. K. Groundwater pollution potential of confined land disposal of dredged material. D, 1979, University of Southern California.

Yuan, Ding-Wen. Relation of Magsat and gravity anomalies to the main tectonic provinces of South America. M, 1982, University of Pittsburgh.

Yuan, Georgia. The geomorphic development of an hydraulic mining site in Nevada County, California. M, 1979, Stanford University. 57 p.

Yuan, Jennwei. Sediments in the lower New York and Raritan bays. D, 1976, Lehigh University. 202 p.

Yuan, Li-Ping. The petrography of permeability; determination of relationships between thin section data and physical measurements. D, 1987, University of South Carolina. 134 p.

Yuan, Pao-Chiang. Three-dimensional finite element modeling of pollutant transport in aquifers. D, 1986, Oklahoma State University. 294 p.

Yuan, Peter B. Stratigraphy, sedimentology, and geologic evolution of eastern Terraba Trough, southwestern Costa Rica. D, 1984, Louisiana State University. 123 p.

Yuan, Peter Bee-Deh. Neocomian sedimentation of Sable Island area, Scotian Shelf, Eastern Canada. M, 1979, Carleton University. 173 p.

Yuan, Wen Lin. Removal of phosphate from water by aluminum. D, 1972, Rutgers, The State University, New Brunswick. 79 p.

Yudhbir. Engineering behaviour of heavily overconsolidated clays and clay shales with special reference to long-term stability. D, 1969, Cornell University. 171 p.

Yudovin, Susan Mary. Texture and mineralogy of heavy mineral enriched beach sand, Dockweiler State Beach, Southern California. M, 1979, University of California, Los Angeles.

Yue, G. K.-L. The fate of certain atmospheric trace gases and their interactions with the water cycle. D, 1975, SUNY at Albany. 157 p.

Yuen, C. M. Rock-structure time interaction in lined circular tunnels in high horizontal stress field. D, 1979, University of Western Ontario.

Yuen, Cheong-Yip. The clay mineralogy and chemical composition of sediments in Hadidi, northern Syria. M, 1979, University of Wisconsin-Milwaukee. 74 p.

Yuen, Chew F. Petroleum prospects of China. M, 1946, Columbia University, Teachers College.

Yuen, D. A. Some problems of local flows in the Earth's upper mantle. D, 1978, University of California, Los Angeles. 226 p.

Yuen, D. T.-C. A study of a reflection seismic survey from Lambton County, South western Ontario. M, 1974, University of Western Ontario.

Yuen, Dexter L. Analysis of five-spot tracer tests to determine reservoir layering. M, 1978, Stanford University.

Yukler, M. A. Analysis of error in groundwater modelling. D, 1976, University of Kansas. 194 p.

Yukler, M. A. The potential of multiband photography for revelation of geologic, hydrologic, and botanical data at Clinton Damsite, Douglas County, Kansas. M, 1974, University of Kansas. 79 p.

Yule, Alan. The Hope Brook gold deposit, Newfoundland, Canada; surface geology, representative lithochemistry, and styles of hydrothermal alteration. M, 1988, Dalhousie University. 249 p.

Yule, J. Douglas. Petrology and structure of south-central Tobago, West Indies; part of a displaced fragment of a Mesozoic oceanic island arc terrane. M, 1988, University of Wyoming. 226 p.

Yules, John A. Continuous seismic profiling studies of the President Roads area, Boston Harbor, Massachusetts. M, 1966, Massachusetts Institute of Technology. 32 p.

Yulke, Sandra Gay. Causes of coloration in blue and rose quartzes. M, 1977, Massachusetts Institute of Technology. 90 p.

Yun, S. Geology and skarn ore mineralization of the Yeonhwa-Ulchin zinc-lead mining district, southeastern Taebaegsan region, Korea. D, 1979, Stanford University. 409 p.

Yund, Richard Allen. The system Ni-As-S. D, 1960, University of Illinois, Urbana. 126 p.

Yungul, Sulhi. Magnetic survey of the San Gabriel Wash, Los Angeles County, California. M, 1944, California Institute of Technology. 24 p.

Yungul, Sulhi. Some uses of the spontaneous polarization method. M, 1945, California Institute of Technology. 70 p.

Yungul, Sulhi H. Gravity interpretation of shale sections, with specific reference to reefs, similar folds, salt domes, buried ridges, and faults. D, 1962, Texas A&M University. 81 p.

Yunker, Gerald G. Structural geology and petrology of the South Pass City area, Wyoming. M, 1979, University of Missouri, Columbia.

Yuras, Walter. Origin of the Lincoln fold belt, Lincoln County, New Mexico. M, 1977, New Mexico Institute of Mining and Technology. 97 p.

Yurdakul, Ali R. Investigation and determination of planimetric accuracy of photogrammetric maps plotted for cadastral purposes. M, 1961, Ohio State University.

Yuretich, Richard Francis. Sedimentology, geochemistry and geological significance of modern sediments in Lake Rudolf (Lake Turkana), eastern Rift Valley, Kenya. D, 1976, Princeton University. 322 p.

Yurewicz, D. A. Sedimentology, paleoecology, and diagenesis of the massive facies of the lower and middle Capitan Limestone (Permian), Guadalupe Mountains, New Mexico and West Texas. D, 1976, University of Wisconsin-Madison. 296 p.

Yurkas, George John. A petrographic study of the Elwen Sandstone of south-central Indiana. M, 1958, Indiana University, Bloomington. 53 p.

Yurkovich, Steven Peter. Cordierite; structural relations and phase equilibria. D, 1972, Brown University. 89 p.

Yurkovich, Steven Peter. The effects of neutron bombardment on the kinetics of the rhombic-clinoenstatite transition. M, 1968, Brown University.

Yusas, Michael Ray. Structural evolution of the Roosevelt Hot Springs geothermal reservoir. M, 1979, University of Utah. 62 p.

Yust, William W. A mesoscopic fabric analysis of a portion of the Tellico-Sevier Belt of East Tennessee. M, 1976, University of Tennessee, Knoxville. 106 p.

Zaadnoordijk, Willem Jan. Analytic elements for transient groundwater flow. D, 1988, University of Minnesota, Minneapolis. 150 p.

Zaaza, M. W. The depositional facies, diagenesis and reservoir heterogeneity of the upper San Andres Formation in West Seminole Field, Gaines County, Texas. D, 1978, University of Tulsa. 183 p.

Zaaza, Mahomoud Wafaie. Stratigraphic distribution of giant petroleum fields as controlled by worldwide correlative unconformities and onlap-offlap stratigraphic sequences. M, 1974, University of Tulsa. 126 p.

Zaback, Doreen Ann. Classification of crude oil based on stable carbon isotopes. M, 1987, University of Texas at Dallas. 209 p.

Zabawa, C. F. Flocculation in the turbidity maximum of northern Chesapeake Bay. D, 1978, University of South Carolina. 137 p.

Zabawa, Pamela Jean. Investigation of surficial structural geology of portions of Beckham, Custer, Roger Mills, and Washita counties, Oklahoma. M, 1976, University of Oklahoma. 98 p.

Zabel, Garrett Edward. A petrographic and chemical study of ternary feldspars. M, 1977, University of Houston.

Zablocki, Frank Stefan. A gravity study in the Deep River-Wadesboro Triassic basin of North Carolina. M, 1959, University of North Carolina, Chapel Hill. 44 p.

Zaborniak, Helen Mary. The brachiopod family Spiriferidae of the Lower Mississippian Lodgepole Formation of Manitoba. M, 1956, University of Michigan.

Zabowski, Darlene. Seasonal variations in the soil solutions of a subalpine spodosol collected using low-tension lysimetry and centrifugation. D, 1988, University of Washington. 120 p.

Zabriskie, Walter Edward. Geology of northern Davis Mountains, Jeff Davis County, Texas. M, 1951, University of Texas, Austin.

Zabriskie, Walter Edward. Petrology and petrography of Permian carbonate rocks, Arcturus Basin, Nevada and Utah. D, 1967, Brigham Young University.

Zacate, Michael Everett. The geology of northern Illinois. M, 1967, Northern Illinois University. 183 p.

Zachara, John M. A solution chemistry and electron spectroscopic study of zinc adsorption and precipitation on calcite. D, 1987, Washington State University. 189 p.

Zachariadis, R. G. P. Distribution of sedimentary structures along the Murray fracture zone. M, 1969, University of Hawaii. 55 p.

Zachariadis, R. G. P. The structure and tectonics of the Murray fracture zone west of the Hawaiian Ridge. D, 1973, University of Hawaii. 57 p.

Zachary, Alvin Leslie. Mineralogical and chemical study of the genesis of Miami, Russell, and Avonburg soils. D, 1966, Purdue University. 114 p.

Zachos, James C. Aspects of Late Cretaceous and Paleogene oceanic climate and productivity. D, 1988, University of Rhode Island. 565 p.

Zachos, Louis George. Stratigraphy and petrology of two shallow wells, Citrus and Levy counties, Florida. M, 1978, University of Florida. 105 p.

Zachry, Doy Lawrence, Jr. Carboniferous stratigraphy of the Chappel area, San Saba County, Texas. D, 1969, University of Texas, Austin. 396 p.

Zachry, Doy Lawrence, Jr. The lithology of the Fayetteville Black Shale, Washington County, Arkansas. M, 1964, University of Arkansas, Fayetteville.

Zack, Allen Lad. An investigation into the limits of gravinometric and seismic prospecting for subsurface solution openings. M, 1967, Vanderbilt University.

Zadina, William Louis. Structural variations of muscovite in porphyry copper systems. M, 1982, University of Arizona. 71 p.

Zadins, Zintars Z. Structure of the Northern Appalachian thrust belt at Cementon, New York. M, 1983, University of Rochester. 137 p.

Zadnik, Valentine Edward. Microfacies study of the Wabash Reef, Wabash, Indiana. M, 1958, University of Illinois, Urbana. 54 p.

Zadnik, Valentine Edward. Petrography of the Upper Cambrian dolomites of Warren County, New Jersey. D, 1960, University of Illinois, Urbana. 155 p.

Zaeff, Gene D. The occurrence and distribution of minerals in the Pittsburg No. 8 coal of southeastern Ohio. M, 1981, University of Toledo. 152 p.

Zaengle, Donald G. Provenance and depositional environment of the basal conglomerate of Moenave-Kayenta equivalent strata, Spring Mountains, southern Nevada. M, 1984, Southern Illinois University, Carbondale. 122 p.

Zagaar, Abdussalam. Petrography of the Oswego Limestone (Pennsylvanian) in Dewey County and parts of Custer and Ellis counties, Oklahoma. M, 1965, University of Tulsa. 98 p.

Zager, John P. The interaction of surface and ground water in Lower Nashotah Lake, Waukesha County, Wisconsin. M, 1981, University of Wisconsin-Milwaukee.

Zaghi, Nourollah. Application of hydraulic doublets to prevent intrusion of extraneous waters into potable aquifers. D, 1977, Stanford University. 121 p.

Zagorski, Theodore W. Depositional environments and diagenesis of the subsurface Tribes Hill Formation (Lower Ordovician) of the Mohawk Valley region, New York. M, 1981, Rensselaer Polytechnic Institute. 98 p.

Zahariev, G. K. The exponential tonnage-grade relationship and its potential use in mineral exploration. D, 1975, Columbia University. 121 p.

Zahary, Robert Gene. An ecological analysis of the floral and faunal assemblages on temperate artificial marine reefs at Santa Catalina Island, California. D, 1982, University of Southern California.

Zahavi, Abraham. Movement of bed material in a small reach of a perennial stream. D, 1981, Clark University. 181 p.

Zahedi, Jafar. Subsurface geology of the uppermost part of the Mesaverde Group of San Juan Basin. M, 1976, University of Texas, Arlington. 94 p.

Zaheha, Robert D. Holocene paleoenvironmental history of Six-Pack Pond, San Salvador Island, Bahamas. M, 1987, University of Akron. 100 p.

Zaher, Mohammad Abduz. A study of the clays (Carboniferous-Paleocene (?)) of the Salt range and Kala-Chitta hills, West Pakistan. M, 1969, Michigan Technological University. 74 p.

Zahn, Jack C. A gravity survey of the Serpent Mound area in southern Ohio. M, 1965, Ohio State University.

Zahn, Paul D. Stratigraphy and depositional environments of the Brigham Group, Cub River area, Franklin County, Idaho. M, 1987, Idaho State University. 196 p.

Zahony, Stephen G. Sulfide minerals of the Lebanon Dome area. M, 1966, Dartmouth College. 54 p.

Zahralban, Thomas A. Application of thermodynamics to the determination of appropriate mantle geotherms. M, 1982, Brooklyn College (CUNY). 45 p.

Zaikowski, A. Incorporation of noble gases during synthesis and equilibration of serpentine; implication for meteoritical noble gas abundances. D, 1977, SUNY at Stony Brook. 124 p.

Zainuddin, Syed Mohammad. Petrology of the grantic rocks in the vicinity of Republic Trough in the Upper Peninsula of Michigan. D, 1971, Michigan State University. 80 p.

Zainy, Muhammad A. Economic evaluation of the petroleum industry of Iraq. M, 1976, Colorado School of Mines. 131 p.

Zaitlin, Brian A. Sedimentology of the Pirate Cove, Fleurant and Bonaventure formations of the western Baie des Chaleurs area, Maritime Canada; a depositional and tectonic model. M, 1981, University of Ottawa. 197 p.

Zaitlin, Brian Allen. Sedimentology of the Cobequid Bay - Salmon River Estuary, Bay of Fundy, Canada. D, 1987, Queen's University. 391 p.

Zaitzeff, James Boris. Middle Ordovician Black River ostracods from a well core, Jackson County, Michigan. M, 1962, Michigan State University. 87 p.

Zaitzeff, James Boris. Taxonomic and stratigraphic significance of dinoflagellates and acritarchs of the Navarro Group (Maestrichtian) (Upper Cretaceous) from east central and southwest Texas. D, 1967, Michigan State University. 203 p.

Zajac, Ihor Stephan. The stratigraphy and mineralogy of the Sokoman Formation (Precambrian) in the Knob Lake area, Quebec and Newfoundland. D, 1972, University of Michigan.

Zajac, Ihor Stephen. The geology of Vulcan Ridge-Dewar Creek area. M, 1960, University of British Columbia.

Zajic, William E. Geology of the Dry Creek area in Fremont County, Wyoming. M, 1955, University of Kansas. 107 p.

Zakaria, Abdul Aziz. Approximating the fall of a water table in drained and undrained land with root extraction of water. D, 1987, Utah State University. 163 p.

Zakikhani, Mansour. Study of flow and mass transport in multilayered aquifers using boundary integral method. D, 1988, Georgia Institute of Technology. 213 p.

Zakir, Fawaz Abdul Rahman. Geology of the Ablah area, southern Hijaz Quadrangle, Kingdom of Saudi Arabia. M, 1972, South Dakota School of Mines & Technology.

Zakir, Fawaz Abdul Rahman. Preliminary study of the geology and tectonics of the Raghama Formation, Maqna area, Wadi As'Sirhan Quadrangle, Kingdom of Saudi Arabia. D, 1982, South Dakota School of Mines & Technology. 240 p.

Zakis, W. N. Geology of the east flank of the Bighorn Mountains near Dayton, Sheridan County, Wyoming. M, 1950, University of Wyoming. 137 p.

Zakrzewski, Allan G. Effects of radial, cyclical flow of heated, compressed air in Upper Cambrian sandstones of Northwest Illinois. M, 1983, University of Wisconsin-Milwaukee. 305 p.

Zakrzewski, Richard Jerome. A study of the primitive vole, Ogmodontomys from the Blancan of southwestern Kansas. M, 1965, University of Michigan.

Zakrzewski, Richard Jerome. The rodents from the Hagerman local fauna, upper Pliocene of (southwestern) Idaho. D, 1968, University of Michigan. 90 p.

Zakus, Paul D. Sedimentary petrography and stratigraphy of the Mississippian Whitewater Lake Member of southwestern Manitoba (Canada). M, 1967, University of Manitoba.

Zalan, Pedro V. Stratigraphy and petroleum potential at the Acarau and Piaui-Camocim sub-basins, Ceara Basin, offshore northeastern Brazil. M, 1983, Colorado School of Mines. 155 p.

Zalan, Pedro Victor. Tectonics and sedimentation of the Piaui-Camocim Sub-basin, Ceara Basin, offshore northeastern Brazil. D, 1984, Colorado School of Mines. 133 p.

Zalan, Thomas Anthony. The use of multi-energy-group neutron diffusion theory to numerically evaluate the relative utility of three dual-detector neutron porosity well logging tools. D, 1988, Colorado School of Mines. 159 p.

Zaleha, Michael James. The Hell Creek Formation (Maastrichtian), Glendive area, Montana; sedimentology, paleoenvironments, and provenance and their stratigraphic and taphonomic implications. M, 1988, Ohio University, Athens. 138 p.

Zalesny, Emil R. Foraminiferal ecology of Santa Monica Bay (California). M, 1956, University of Southern California.

Zalidis, George C. Experimental and theoretical analyses of capillary rise through porous media. D, 1988, Michigan State University. 127 p.

Zall, L. S. Photo-geology and remote sensing systems for locating ore deposits. D, 1976, Cornell University. 227 p.

Zalusky, Donald W. Holts Summit (Devonian) condonts from Missouri. M, 1958, University of Missouri, Columbia.

Zalusky, Donald W. Internal structure and biostratigraphy of some Rose Hill (middle Silurian) chitinozoans. D, 1976, University of Delaware. 288 p.

Zaman, M. M. Influence of interface behavior in dynamic soil-structure interaction problems. D, 1982, University of Arizona. 448 p.

Zaman, Mohammad Qamar. Efficient rates of oil and gas production through regulation. M, 1953, University of Texas, Austin.

Zamboras, Robert L. The geology and hydrothermal mineralization at the Molly property and vicinity, Snohomish County, Washington. M, 1979, Western Washington University. 88 p.

Zambrano, Elias. Geology of Cienega mining district, northwestern Yuma County, Arizona. M, 1965, University of Missouri, Rolla.

Zambresky, Liana. The calibration of a portable induction magnetometer system. M, 1977, University of British Columbia.

Zamel, Abdulla Z. Al. Study of Recent carbonate and evaporite sediments in the Sabkha, Kuwait, Persian Gulf. M, 1972, Lehigh University.

Zamora Guerrero, David Hipolito. Some useful digital techniques applied to remote sensing data. M, 1984, Stanford University. 189 p.

Zamora, Lucas Guillermo. Silurian rocks of the Permian basin region. M, 1968, University of Texas, Austin.

Zamora, Osvaldo Sánchez see Sánchez Zamora, Osvaldo

Zamora, Oswaldo Sanchez see Sanchez Zamora, Oswaldo

Zamore, Yale. Age of the Blue Ridge, southern Appalachians of North Carolina and Tennessee and degree of metamorphism required to fully homogenize strontium isotopes. M, 1975, Brooklyn College (CUNY).

Zampogna, Ralph V. Lithologic analysis and correlation of the Lowville Limestone from selected locations in Eastern United States. M, 1970, Indiana University of Pennsylvania. 63 p.

Zamzow, Craig Edward. Tertiary volcanics of the eastern Eagle Mountains, Hudspeth County, Texas. D, 1983, University of Texas at El Paso. 209 p.

Zanbak, C. Experimental evaluation of stability analysis methods for some rock slopes by a physical model. D, 1978, University of Illinois, Urbana. 156 p.

Zandell, Charles H. Microstratigraphy of the Rock Bluff Limestone, Virgilian of eastern Kansas. M, 1963, University of Kansas. 73 p.

Zandt, George. Study of three-dimensional heterogeneity beneath seismic arrays in central California and Yellowstone, Wyoming. D, 1978, Massachusetts Institute of Technology. 440 p.

Zaneveld, Jacques Ronald Victor. Optical and hydrographic observations of the Cromwell current between 92°00′ West and the Galapagos Islands. D, 1972, Oregon State University. 87 p.

Zanghi, Elizabeth M. A reconnaissance mineralogical and chemical study of interdistributary sediments from two borings taken in the Atchafalaya Basin. M, 1988, University of Southwestern Louisiana. 118 p.

Zannos, John A. Gasoline contamination of private wells in Westport, Massachusetts. M, 1988, Boston University. 106 p.

Zanoria, Elmer. The depositional and volcanological origin of the Diliman volcaniclastic formation, southwestern Luzon, Philippines. M, 1988, University of Illinois, Chicago. 163 p.

Zant, Kent Lee Van see Van Zant, Kent Lee

Zantop, Half Al. Trace element distribution in manganese oxides and iron oxides from the San Francisco manganese deposit, Jalisco, Mexico. D, 1969, Stanford University. 176 p.

Zapata, Raul Emilio. Soil erosion by overland flow with rainfall. D, 1987, University of Florida. 400 p.

Zapata, Vito Joseph. A theoretical analysis of viscous crossflow. D, 1981, University of Texas, Austin. 288 p.

Zapffe, Carl. The effects of a basic igneous intrusion on a Lake Superior iron-bearing formation. M, 1908, University of Wisconsin-Madison.

Zapico, N. M. Aquifer evaluation procedures; core acquisition and an assessment of slug tests. M, 1987, University of Waterloo. 195 p.

Zaporozec, A. Hydrogeologic aspects of ground-water management in Wisconsin. D, 1975, University of Wisconsin-Madison. 254 p.

Zapp, Alfred Dexter. Geology of the northeastern Cornudas Mountains, New Mexico. M, 1941, University of Texas, Austin.

Zappe, Steven Orvil. In situ seismic velocities of granitic rocks, Mojave Desert, California. M, 1979, University of California, Riverside. 135 p.

Zardeskas, Ralph Anthony. A bathymetric chart of Carmel Bay, California. M, 1971, United States Naval Academy.

Zareai, Amanollah. Bed forms and bed form friction in alluvial channel flow. D, 1983, University of California, Davis. 122 p.

Zarillo, Gary A. Cuspate shoreforms of West Passage, Narragansett Bay, Rhode Island. M, 1975, University of Rhode Island.

Zarillo, Gary A. Interrelation of hydrodynamics and sediment transport in a salt marsh estuary. D, 1979, University of Georgia. 236 p.

Zarins, Andrejs. Origins and geologic history of siliceous metacolloidal deposits, Cathedral Mountain Quadrangle, Brewster County, West Texas. M, 1977, University of Nebraska, Lincoln.

Zarkanellas, Antois J. New bottom sampling apparatus. M, 1973, Florida Institute of Technology.

Zarra, Lawrence. Biostratigraphy of planktonic foraminifera from Oligocene and lowermost Miocene rocks of the North Carolina coastal plain. M, 1988, University of Delaware. 260 p.

Zarrow, Lorraine. Structural relationships and geochemical investigation of the Lynn Volcanics, Pine Hill, North Boston Quadrangle, Mass. M, 1978, Massachusetts Institute of Technology. 184 p.

Zarth, R. J. The Quaternary geology of the Wrenshall and Frogner quadrangles, northeastern Minnesota. M, 1977, University of Minnesota, Duluth.

Zartman, Robert Eugene. A geochronological study of the Lone Grove Pluton from the Llano Uplift, Texas. D, 1963, California Institute of Technology. 142 p.

Zarzavatjian, Papken A. Detection of buried basement highs by air-photo drainage pattern analysis, Reynolds and Wayne counties, Missouri. M, 1957, University of Missouri, Rolla.

Zastrow, M. E. Stratigraphy and depositional environments of the Black Creek Formation along the Neuse River, North Carolina. M, 1982, University of North Carolina, Chapel Hill.

Zatezalo, Jo Lynn. A study of groundwater problems in an underground coal mine. M, 1982, Ohio University, Athens. 126 p.

Zatezalo, M. P. Hydrogeologic evolution of pollutant dispersion from municipal sewage lagoons located on a floodplain. M, 1977, University of Missouri, Columbia.

Zaturecky, John William. Cleavage and related structures in Precambrian rocks near Jasper, Alberta (Canada). M, 1968, University of Alberta. 169 p.

Zauderer, Jeffrey. A Neoglacial pollen record from Osgood Swamp, California. M, 1973, University of Arizona.

Zavada, Michael Stephan. Morphology, ultrastucture and evolutionary significance of monosulcate pollen. D, 1982, University of Connecticut. 225 p.

Zavesky, Richard Roy. Fluvial flow profiles based on optimization of energy loss distribution. D, 1982, Polytechnic University. 237 p.

Zavodni, Zavis. Physical testing study of the Cananea Mine rock (Cananea, Sonora, Mexico). M, 1969, University of Arizona.

Zavodni, Zavis Marian. Influence of geologic structure on slope stability in the Cananea mining district, Cananea, Mexico. D, 1971, University of Arizona. 113 p.

Zaw, Khin. The Cantung E-zone orebody, Tungsten, Northwest Territories; a major scheelite skarn deposit. M, 1976, Queen's University. 327 p.

Zawacki, Stephen John. The growth of non-stoichiometric apatite; a physiochemical study. D, 1988, SUNY at Buffalo. 328 p.

Zawislak, Ronald Lynn. Laramide deformation and problems of deep crustal reflection profiling encountered in the Wind River COCORP line, Wyoming. D, 1980, University of Wyoming. 234 p.

Zawistowski, Stanley J. Biostromes in the Rapid Member of the Cedar Valley Limestone (Devonian) in east-central Iowa. M, 1971, University of Iowa.

Zayachkivsky, B. Granitoids and rare-element pegmatites of the Georgia Lake area, northwestern Ontario. M, 1985, Lakehead University.

Zayatz, Mark. The economic geology and petrogenesis of the Blue Mountain nepheline syenite complex, Ontario, Canada. M, 1984, University of Alaska, Fairbanks. 119 p.

Zayed, Mostafa Mahmoud. The dynamic response of multi-story shear buildings with non-symmetrical cross section subjected to random ground motion. D, 1965, University of Colorado. 229 p.

Zazac, I. S. The geology of Vulcan Ridge-Dewar Creek area, British Columbia. M, 1957, University of British Columbia.

Zazou, Samiha Mahoud. A faunule from shale unit in the lower Ely Formation, west-central Utah. M, 1967, University of Utah. 66 p.

Zazueta Ranahan, Fedro Sigmundo. Simulation of agricultural drainage systems. D, 1982, Colorado State University. 245 p.

Zbinden, Elizabeth Anne. Structure and petrology of the dike swarm of Waianae Volcano. M, 1984, University of Hawaii at Manoa. 151 p.

Zbur, Richard Thomas. A geophysical investigation of Indian Wells Valleys, California. M, 1962, University of Utah. 150 p.

Zdanowicz, Ted. A. Stratigraphy and structure of the Horseshoe Gulch area, Etna and China Mountain quadrangles, California. M, 1972, Oregon State University. 88 p.

Zdepski, John Mark. Stratigraphy, mineralogy, and zonal relations of the Sun massive-sulfide deposit, Ambler District, Northwest Alaska. M, 1980, University of Alaska, Fairbanks. 93 p.

Zdinak, Andrew Patrick. Geology of the northwest portion of the Loyal Valley Quadrangle, Mason County, Texas. M, 1988, University of Texas of the Permian Basin. 112 p.

Zdzinski, Alexander Jules. Paleomagnetism and diagenesis of the Triassic Chugwater Group, Wyoming. M, 1985, University of Oklahoma. 149 p.

Zebal, George Patterson. The Upper Cretaceous paleontology and stratigraphy of the Simi Hills, Los Angeles and Ventura counties, California. M, 1943, California Institute of Technology. 51 p.

Zebarth, Bernard John. Saturated and unsaturated flow in a hummocky landscape in relation to topography and soil morphology. D, 1988, University of Saskatchewan.

Zecharias, Yemane Berhan. Geomorphic analysis of groundwater outflow from mountainous watersheds. D, 1984, Cornell University. 291 p.

Zeck, Wayne Anthony. Strain in a pair of multilayered en echelon folds. D, 1982, University of California, Los Angeles. 161 p.

Zee, G. T. Y. Sea level variation patterns in the Pacific Ocean. M, 1975, University of Hawaii. 47 p.

Zeff, Marjorie Lee. Microborings within carbonate substrates from the aphotic, deep-marine environment. M, 1977, Duke University. 115 p.

Zeff, Marjorie Lee. Tidal channel morphometry, flow, and sedimentation in a back-barrier salt-marsh; Avalon/Stone Harbor, New Jersey. D, 1987, Rutgers, The State University, New Brunswick. 149 p.

Zehner, Harold H. Ground water resources of Somerset, Carbondale, and Murphysboro Townships, Jackson County, Illinois. M, 1968, Southern Illinois University, Carbondale. 90 p.

Zehner, Richard E. Petrology, structure, and tectonic setting of the Hog Heaven Volcanics, and their relationship to mineralization. M, 1987, University of Montana. 139 p.

Zehr, Danny D. The mammals from an early Pliocene local fauna in Ellis County, Kansas. M, 1974, Fort Hays State University.

Zeidner, Martin Aaron. Some properties of endellite from Lawrence County, Missouri. M, 1950, University of Missouri, Columbia.

Zeigler, John Milton. Geology of the Blacktail area, Beaverhead County, Montana. D, 1954, Harvard University.

Zeigler, Lynn E. Sedimentary petrology and depositional environments of the Upper Ordovician Sequatchie and Lower Silurian Red Mountain Formation, northwestern Georgia. M, 1988, Emory University. 281 p.

Zeihen, Gregory D. Paragenetic relationships zoning, and mineralogy of at the Black Pine Mine, Granite County, Montana. M, 1985, University of Arizona. 103 p.

Zeihen, Lester Gregory. Some observations on the mineralogy of the chromite deposits of south-central Montana. M, 1937, Montana College of Mineral Science & Technology. 27 p.

Zeiner, Thomas C. Upper Sundance-lower Morrison sedimentology, Wyoming, and its stratigraphic implications. M, 1974, University of Missouri, Columbia.

Zeip, Vera Lydia. Geographic thinking in Ozark literature. M, 1939, Washington University.

Zeiss, Harvey S. Dinoflagellate cyst zonation of some Upper Jurassic and Lower Cretaceous strata penetrated by the Sun KR Panarctic Skybattle Bay Well, Sverdrup Basin, Arctic Archipelago, Canada. M, 1976, Pennsylvania State University, University Park. 37 p.

Zeitler, Peter Karl. The tectonic interpretation of fission track ages from the Himalayan ranges of northern Pakistan. M, 1980, Dartmouth College. 92 p.

Zeitler, Peter Karl. Uplift and cooling history of the NW Himalaya, northern Pakistan; evidence from fission-track and ^{40}Ar/^{39}Ar cooling ages. D, 1983, Dartmouth College. 266 p.

Zeitlin, Michael J. Variability and predictability of submarine ground-water flow into a coastal lagoon, Great South Bay, New York. M, 1980, SUNY at Stony Brook.

Zeizel, Arthur John. Groundwater geology of the shallow aquifers in DuPage County, Illinois. D, 1960, University of Illinois, Urbana. 130 p.

Zeizel, Eugene Paul. An evaluation of a variable water supply for Lovelock valley, Pershing and Churchill counties, Nevada. M, 1967, University of Nevada. 30 p.

Zekmi, Nadir. Subsurface geology of the Four Isle Field, Southeast Louisiana. M, 1980, University of Southwestern Louisiana. 50 p.

Zekri, Abdurrazzag Yusef. Interfacial tensions of surfactant mixtures against hydrocarbon liquids at elevated temperatures and pressures. D, 1982, University of Southern California.

Zekulin, Alexander Darius. Seismic velocity discontinuities in the Precambrian basement of southeastern Kentucky. M, 1985, University of Kentucky. 65 p.

Zelasko, Joseph Simon. An investigation of the influences of particle size, size gradation and particle

shape on the shear strength and packing behavior of quartziferous sands. D, 1966, Northwestern University. 272 p.

Zelazek, David Paul. Petrology of the Middle Cambrian Blacksmith Formation, southeastern Idaho and northernmost Utah. M, 1981, Utah State University. 132 p.

Zelenka, Brian. Distribution and interpretation of granitic uranium occurrences on the Vermejo Park Ranch, north central New Mexico. M, 1984, University of Alaska, Fairbanks. 133 p.

Zelewski, Gregg. Simultaneous estimation of interval velocity and structure. M, 1985, Colorado School of Mines. 175 p.

Zeliff, Clifford W. Subsurface analysis, "Cherokee" Group (Pennsylvanian), northern Kingfisher County, Oklahoma. M, 1975, University of Oklahoma. 53 p.

Zelinski, William P. Geologic evaluation of the Kelvin copper-molybdenum prospect, Pinal County, Arizona. M, 1973, New Mexico Institute of Mining and Technology. 72 p.

Zelinsky, Anne E. Geologic aspects of geothermal development in northern Dixie Valley, Nevada. M, 1980, University of Nevada. 102 p.

Zelios, Dan G. Petrographic analysis and diagenetic history of the upper Smackover Formation, Columbia County, Arkansas. M, 1986, University of Arkansas, Fayetteville.

Zell, Mary G. Brachiopoda and graphic correlation of the late Middle Cambrian Holm Dal Formation, Peary Land, North Greenland. M, 1986, University of Kansas. 82 p.

Zell, Paul. Paleoecology and stratigraphy of the Middle Devonian Moscow Formation in the Chenango Valley, New York. M, 1985, University of Pittsburgh.

Zell, R. A Mississippian coral assemblage from Darby Canyon, Teton County, Wyoming. M, 1959, University of Wyoming. 114 p.

Zelle, William C. The Chapin Mine and the Atlantic Mill. M, 1891, Washington University.

Zeller, Craig G. Structural geology of Teton Pass, Wyoming. M, 1981, University of Missouri, Columbia.

Zeller, Doris Eulalia Nadine. Endothyroid foraminiferal faunas from the Lower Carboniferous of England and Algeria. D, 1954, University of Wisconsin-Madison.

Zeller, Edward J. Endothyroid foraminifera from the Cordilleran Geosyncline. D, 1951, University of Wisconsin-Madison.

Zeller, Edward J. The stratigraphic significance of endothyroid Foraminifera. M, 1948, University of Kansas. 91 p.

Zeller, Howard Davis. The geology of the west-central portion of the Gunnison Plateau, Utah. M, 1949, Ohio State University.

Zeller, Robert Allen, Jr. The geology of the Big Hatchet Peak Quadrangle, Hidalgo County, New Mexico. D, 1958, University of California, Los Angeles.

Zeller, Robert Allen, Jr. The structural geology and mineralization of Sinking Valley, Pennsylvania. M, 1949, Pennsylvania State University, University Park. 64 p.

Zeller, Ronald P. Paleoecology of the Long Trail Shale Member of the Great Blue Limestone, Oquirrh Range, Utah. M, 1958, Brigham Young University. 36 p.

Zellmer, John Theodore. Environmental and engineering geology study of Bridgeport Valley, Mono County, California. M, 1977, University of Nevada. 211 p.

Zellmer, John Theodore. Recent deformation in the Saline Valley region, Inyo County, California. D, 1980, University of Nevada. 224 p.

Zellmer, Lauren Ann. Dissolution kinetics of crystalline and amorphous albite. M, 1986, Pennsylvania State University, University Park. 170 p.

Zellouf, Khemissi. Land and water resources evaluation for land use planning of the Saoura Valley, Algeria. M, 1980, Purdue University.

Zelonka, Frederick A. Frequency characteristics of seismic waves in shallow sediments produced by falling weight. M, 1959, University of Western Ontario.

Zelt, Frederick Bruce. Natural gamma-ray spectrometry, lithofacies, and depositional environments of selected Upper Cretaceous marine mudrocks, Western United States, including Tropic Shale and Tununk Member of Mancos Shale. D, 1985, Princeton University. 334 p.

Zelt, Karl-Heinz. Investigation of the cause of earthquakes in southeastern Tennessee and northern Georgia using focal mechanisms and models of crustal stress. D, 1988, Georgia Institute of Technology. 255 p.

Zeltner, Walter Anthony. Charge development at the goethite/water interface; effects of aggregation and carbonate adsorption. D, 1986, University of Wisconsin-Madison. 350 p.

Zelwer, Ruben. Some spectral characteristics of geomagnetic vibrations. M, 1965, University of California, Berkeley. 135 p.

Zelwer, Ruben. Spatial uniformity of geomagnetic micropulsations on the 0.001 Hz to 1 Hz frequency range and over distances of the order of 250 km. D, 1971, University of California, Berkeley. 237 p.

Zeman, A. J. Geotechnical properties of Lake Erie clays. M, 1976, McGill University. 133 p.

Zemansky, Gilbert Marek. Water quality regulation during construction of the Trans-Alaska oil pipeline system. D, 1983, University of Washington. 957 p.

Zemboski, Steven S. Depositional environments and facies distribution of the Millican carbonate buildup, Coke County, Texas. M, 1985, University of Texas, Arlington. 162 p.

Zemlicka, George. Source process study with the inclusion of the effects of near-source bathymetric structure for submarine events in the Gulf of California. M, 1988, University of Texas, Austin. 182 p.

Zemmels, Ivar. A study of the sediment composition and sedimentary geochemical processes in the vicinity of the Pacific-Antarctic Ridge. D, 1978, Florida State University. 349 p.

Zempolich, William G. Formation and diagenesis of Precambrian carbonate sediments as seen through petrographic and geochemical examination of the late Proterozoic Beck Spring Dolomite of eastern California. M, 1985, [University of Michigan].

Zen, E-an. Metamorphism of sedimentary rocks in the Castleton area, Vermont. D, 1955, Harvard University.

Zengeni, Teddy Godfrey. PKKP and the fine structure of the earth's core. D, 1970, Stanford University. 77 p.

Zenger, Donald Henry. Geology of the Cooperstown, New York, Quadrangle. M, 1959, Dartmouth College. 114 p.

Zenger, Donald Henry. Stratigraphy of the Lockport Formation (Silurian) in New York State. D, 1962, Cornell University. 327 p.

Zenone, Chester R. Glacio-hydrological parameters of the mass balance of Lemon Glacier, Juneau Icefield, Alaska, 1965-67. M, 1972, Michigan State University. 156 p.

Zenor, John Julian. Analysis of a focus log electrode in a non-homogeneous medium. D, 1968, University of Missouri, Rolla.

Zent, Aaron Patrick. Distribution and state of H_2O in the high latitude subsurface of Mars. M, 1985, University of Hawaii. 106 p.

Zentani, A. S. Gravity and magnetic investigations in portions of Benton, Newton and Warren counties, Indiana. M, 1975, Purdue University.

Zentilli, Marcos. Geological evolution and metallogenic relationships in the Andes of northern Chile between 26° and 29° South. D, 1974, Queen's University. 446 p.

Zentmeyer, Jan Penn. Petrology and diagenesis of the Lower Mississippian Price Formation, southwestern Virginia. M, 1985, Virginia Polytechnic Institute and State University.

Zeosky, Joseph E. Gravity and magnetic surveys of west-central Louisiana; implications for lignite exploration. M, 1982, University of Southwestern Louisiana. 183 p.

Zepeda, Ricardo L. Tectonic geomorphology of the Goleta-Santa Barbara area, California. M, 1987, University of California, Santa Barbara. 108 p.

Zeppieri, James Benjamin. Synchronous parallel evolution among three early Oligocene lineages of trissocyclid Radiolaria from the equatorial Pacific Ocean. M, 1982, Duke University. 85 p.

Zernitz, Emilie R. Drainage patterns and their significance. M, 1930, Columbia University, Teachers College.

Zerpa, Oswaldo. Soil information input to land use planning. M, 1983, [Colorado State University].

Zerrahn, Gregory Joseph. Tectonics and sedimentation associated with the Taconic Orogeny (Ordovician) of New York State. M, 1976, University of Arizona.

Zerva, Aspasia. Stochastic differential ground motion and structural response. D, 1986, University of Illinois, Urbana. 138 p.

Zervas, Chris Eugene. A finite element investigation of topographic variation at mid-ocean ridges spreading at the same rate. D, 1988, University of Washington. 190 p.

Zervas, Chris Eugene. A time term analysis of Pn velocities in Washington. M, 1984, University of Washington. 134 p.

Zerwick, Susan. The analysis of adsorbed gases from alteration minerals as a potential exploration tool for epithermal vein deposits. M, 1983, University of Minnesota, Minneapolis. 84 p.

Zetterlund, Dale. Mobility of uranium during devitrification and alteration of the Gillespie Tuff in Southwest New Mexico. M, 1982, University of Kansas. 124 p.

Zeuch, David Henry. The dislocation substructure of experimentally deformed synthetic dunite. D, 1980, University of California, Davis. 399 p.

Zeuss, Hilario. Geology of the Raton area, Colfax County, New Mexico. M, 1967, Colorado School of Mines. 108 p.

Zevallos, Raul A. Geology of the Acari iron mining district, Arequipa, Peru (South America). M, 1967, University of Missouri, Rolla.

Zevallos-Herrera, Francisco J. Stratigraphic trap interpretation of Los Organos oil field, northwestern Peru. M, 1958, Stanford University.

Zgambo, Thomas Patrick. Calcium silicates; glass content and hydration behavior. D, 1987, University of North Texas. 308 p.

Zhang, Da-chun. Nitrogen concentrations and isotopic compositions of some terrestrial rocks. D, 1988, University of Chicago. 157 p.

Zhang, Jiajun. Determination of source finiteness and depth of large earthquakes. D, 1988, California Institute of Technology. 219 p.

Zhang, Jiaxiang. Lithospheric flexure and deformation-induced gravity changes; effect of elastic compressibility and gravitation on a multi-layered, thick plate model. M, 1986, SUNY at Stony Brook. 91 p.

Zhang, Ke-ke. A study of buoyancy driven flows and magnetic field generation in rotating spherical shells. D, 1987, University of California, Los Angeles. 211 p.

Zhang, Peizhen. Rate, amount, and style of late Cenozoic deformation of southern Ningxia, northeastern margin of Tibetan Plateau. D, 1988, Massachusetts Institute of Technology. 258 p.

Zhang, Xiaomao. Fluid inclusion and stable isotope studies of the gold deposits in Okanagan Valley, British Columbia. M, 1986, University of Alberta. 111 p.

Zhang, Zhenzhong. Interaction of water and organic compounds with clay as determined from heat of immersion and heat of adsorption. D, 1988, Purdue University. 135 p.

Zhang, Zhimeng. Plate tectonics and high P/T metamorphic rocks of China. D, 1986, Stanford University. 210 p.

Zhao, Naiyu. Hydrogeochemistry of a Missouri River alluvial aquifer at McBaine, MO. M, 1986, University of Missouri, Columbia.

Zhao, Wu-Ling. Mechanical studies on the Tibetan Plateau, accretionary wedges, and continental rifts. D, 1985, Princeton University. 139 p.

Zhao, Xixi. A paleomagnetic study of Phanerozoic rock units from Eastern China. D, 1987, University of California, Santa Cruz. 993 p.

Zhao, Zhong Yan. Slaty cleavage and its formation mechanisms in metasediments of western Spitsbergen. M, 1982, Wayne State University. 104 p.

Zhao, Zhongliang. A laboratory study of borehole breakouts with emphasis on the pore pressure effect. M, 1988, University of Wisconsin-Madison. 76 p.

Zheng, Chunmiao. New solution and model for evaluation of groundwater pollution control. D, 1988, University of Wisconsin-Madison. 159 p.

Zheng, Hong. Effects of exchangeable cations of coagulation-flocculation and swelling behavior of smectites. M, 1988, Texas Tech University. 141 p.

Zheng, Hong. The crystal structures of two chlorites. M, 1987, University of Wisconsin-Madison. 69 p.

Zhong, Shaojun. An experimental investigation of "calcite and aragonite precipitation rates in seawater as a function of salinity" and its applications to some geological problems. M, 1988, McGill University. 117 p.

Zhong, William J. S. Compositional and structural changes in chlorite from some metamorphosed ultramafic bodies in North Carolina and Virginia. M, 1982, Miami University (Ohio). 73 p.

Zhou, Di. Adjustment of geochemical background data by multivariate statistics; application to NURE hydrogeochemical and stream sediment reconnaissance for uranium in the Hot Springs Quadrangle, South Dakota. D, 1984, University of Kansas. 358 p.

Zhou, Hua-wei. Prismatic method in solving the gravitational potential, with applications at Cerro Prieto geothermal field, northern Mexico. M, 1984, California State University, Long Beach. 87 p.

Zhou, Xianliang. Copper-complexing organic ligands in seawater; the influence of phytoplankton activity. D, 1988, Dalhousie University.

Zhou, Xinquan. Pressure balancing for the prevention of mine fire and a new computer program for this purpose. M, 1985, Michigan Technological University. 191 p.

Zhou, Yingxin. Designing for upper seam stability in multi-seam mining. D, 1988, Virginia Polytechnic Institute and State University. 216 p.

Zhu, Bingfu. Nuclear waste glass leaching in a simulated granite repository. D, 1987, University of Florida. 212 p.

Zhu, Tianfei. Deep structures beneath the Michigan Basin from reprocessed COCORP data; seismic ray theory and its application in reflection seismology. D, 1986, Cornell University. 203 p.

Ziagos, John Peter. Theoretical and empirical terrestrial heat flow studies. D, 1983, Southern Methodist University. 143 p.

Zick, A. D. Structural geology of the southeast corner of the Paron Quadrangle, Arkansas. M, 1977, University of Missouri, Columbia.

Zickus, Thomas Arunas. Landsat imagery as a tool in locating high yield wells in a crystalline rock area of the West-central Piedmont of North Carolina. M, 1981, North Carolina State University. 54 p.

Ziebarth, Harold Clarence. Micropaleontology and stratigraphy of the subsurface "Heath" Formation (Mississippian-Pennsylvanian) of western North Dakota. M, 1962, University of North Dakota. 145 p.

Ziebarth, Harold Clarence. The stratigraphy and economic potential of Permo-Pennsylvanian strata (Minnelusa Group) in southwestern North Dakota. D, 1972, University of North Dakota. 631 p.

Ziebell, Walter Richard. Minerals of the Idaho Batholith. M, 1949, University of Illinois, Urbana. 44 p.

Ziebell, Warren Gilbert. Interpretation of Pennsylvanian sedimentation from textural studies of strata at Superior, Arizona. M, 1955, University of Illinois, Urbana. 124 p.

Zieg, Gerald A. Stratigraphy, sedimentology, and diagenesis of the Precambrian upper Newland Limestone, central Montana. M, 1981, University of Montana. 182 p.

Ziegenfus, Robert Charles. Municipal natural resource inventories; an analysis and suggested methodology. D, 1980, Rutgers, The State University, New Brunswick. 264 p.

Zieglar, Donald Lowell. Late Paleozoic and early Mesozoic red beds in the Williston Basin of North Dakota and adjacent areas. D, 1959, Harvard University.

Ziegler, Carl Kirk. A numerical analysis of sediment transport in shallow water. D, 1986, [University of California, Santa Barbara]. 216 p.

Ziegler, Charles B. The structure and petrology of the Swift Creek area, western North Cascades, Washington. M, 1986, Western Washington University. 191 5 plates in.

Ziegler, Victor. Foothills structure in northern Colorado. D, 1917, University of Denver.

Ziegler, Victor. The Ravenswood grano-diorite. M, 1910, Columbia University, Teachers College.

Ziehlke, Daniel V. Structural geology of Sickel Outlier near Notigi Lake, Manitoba. M, 1973, University of Manitoba.

Zielinski, G. W. Thermal history of the Norwegian-Greenland Sea and its rifted continental margin. D, 1977, Columbia University, Teachers College. 166 p.

Zielinski, Gregory A. Paleoenvironmental implications of lacustrine sedimentation patterns in the Temple Lake valley, Wyoming. D, 1987, University of Massachusetts. 564 p.

Zielinski, Gregory Anthony. An analysis of the morphology, hydrology and climate of the Grays Lake drainage basin, Bonneville and Caribou counties, Idaho. M, 1980, Idaho State University. 73 p.

Zielinski, Robert A. Gough Island; evaluation of a fractional crystallization model and an experimental study of the partitioning of a rare element in the system diopside/water. D, 1972, Massachusetts Institute of Technology. 184 p.

Ziemba, Eugene Anthony. Microfacies studies of the Platteville Group. M, 1955, University of Illinois, Urbana. 69 p.

Ziemer, Robert Ruhl. Logging effects on soil moisture losses. D, 1978, Colorado State University. 146 p.

Ziemianski, Wayne P. Clay mineral changes associated with intensification of glaciation in the Norwegian-Greenland Sea. M, 1979, San Jose State University. 44 p.

Zientek, Michael Leslie. Petrogenesis of the basal zone of the Stillwater Complex, Montana. D, 1983, Stanford University. 249 p.

Zier, Steven Jonathan. Voltage as a function of clay flocculation. M, 1961, Tulane University. 145 p.

Zierenberg, Robert A. Recent seafloor metallogenesis; examples from the Atlantis II Deep, Red Sea and 21°N East Pacific Rise. D, 1983, University of Wisconsin-Madison. 231 p.

Zierer, Clifford Maynard. Geographic contrasts in the agriculture of southeastern Indiana. M, 1923, Indiana University, Bloomington.

Zigan, Steve Michael. Structure and stratigraphy of the La Porte ophiolitic sequence, northern Sierra Nevada, California. M, 1981, University of California, Davis. 100 p.

Zigich, Daniel K. Evaluating the effectiveness of Landsat data as a tool for locating buried pre-glacial valleys in eastern South Dakota. M, 1980, South Dakota School of Mines & Technology.

Zigmont, James H. Valley glaciation in the White Cloud Peaks, Custer County, Idaho. M, 1982, Lehigh University. 134 p.

Zilans, A. Quaternary geology of the Mackinac Basin, Lake Huron. M, 1985, University of Waterloo. 275 p.

Zilczer, Janet Ann. Adularia and orthoclase; X-ray diffraction and crystal structure analyses of diffuse streaking and aluminum/silicon ordering in natural potassium feldspars. D, 1981, George Washington University. 266 p.

Zilinski, Robert E., Jr. Geology of the central part of the Lucero Uplift, Valencia County, New Mexico. M, 1976, University of New Mexico. 69 p.

Zilka, Nicholas T. Geology of the German Peak area, Custer County, Idaho. M, 1969, Idaho State University. 92 p.

Zimbeck, Donald Allen. Gravity survey along northward-trending profiles across the boundary between the Basin and Range Province and Colorado Plateau. M, 1965, University of Utah. 111 p.

Zimbelman, David R. Geology of the Polaris 1SE Quadrangle, Beaverhead County, Montana. M, 1984, University of Colorado. 158 p.

Zimbelman, James Ray. Geologic interpretation of remote sensing data for the Martian volcano Ascraeus Mons. D, 1984, Arizona State University. 287 p.

Zimbrick, Grant David. The lithostratigraphy of the Morrowan Bloyd and McCully formations (Lower Pennsylvanian) in southern Adair and parts of Cherokee and Sequoyah counties, Oklahoma. M, 1978, University of Oklahoma. 182 p.

Zimdars, Marjorie Ann. Cepstrum analysis applied to the problem of multiple event discrimination. M, 1974, University of Wisconsin-Milwaukee.

Zimmer, Bonnie Jeanne. Nitrogen dynamics in the surface waters of the New Jersey Pine Barrens. D, 1981, Rutgers, The State University, New Brunswick. 313 p.

Zimmer, James A. The geology of the Chamberlain iron deposit, Roane County, Tennessee. M, 1964, University of Tennessee, Knoxville. 79 p.

Zimmer, Louis George. The stratigraphy and structure of the Middle Piney Creek area, Sublette and Lincoln counties, Wyoming. M, 1952, University of Iowa. 106 p.

Zimmer, Paul William. Phase petrology of Lyon Mountain magnesite deposits in the northeastern Adirondacks. M, 1947, Washington State University. 25 p.

Zimmer-Dauphinee, Susan A. A survey of 24 elements in North Dakota lignite (Fort Union Group, Paleocene) and possible geologic implications. M, 1983, University of North Dakota. 60 p.

Zimmerer, Sheilah Marie. A study of the englacial and subglacial hydrology of Storglaciären, northern Sweden. M, 1987, University of Minnesota, Minneapolis. 148 p.

Zimmerli, Edward Joseph. The Tioughnioga Valley section of the Ithaca Formation (Upper Devonian of New York). M, 1957, Cornell University.

Zimmerman, Charles C. The use of magnetic devices in oil exploration. M, 1927, University of Pittsburgh.

Zimmerman, Charles J. Geochemistry of andesites and related rocks, central-north Rio Grande Rift, New Mexico. M, 1979, University of New Mexico. 111 p.

Zimmerman, David W. Potential geologic hazards to the Salt Lake City refineries, Chevron Oil and Northwest Natural Gas pipelines, and possible ramifications to Pocatello, Idaho's energy supply and distribution system. M, 1987, Idaho State University. 89 p.

Zimmerman, Don Z. The geology of the Long Tom area, Lane County, Oregon. M, 1927, University of Oregon. 62 p.

Zimmerman, Donald A. Fusulinids of the (Pennsylvanian) Hermosa Formation of southwestern Colorado. M, 1951, University of Wisconsin-Madison.

Zimmerman, H. L. Bedrock geology of western Story County. M, 1952, Iowa State University of Science and Technology.

Zimmerman, Herman B. The size analysis of the sediments of Nauset Harbor, Cape Cod, Massachusetts. M, 1963, University of Massachusetts. 94 p.

Zimmerman, James Arthur. Pleistocene molluscan faunas of the Newell Lake deposit, Logan County, Ohio. M, 1958, Ohio State University.

Zimmerman, James Blaisdell. Jeff Conglomerate, northeastern Davis Mountains, Texas. M, 1950, University of Texas, Austin.

Zimmerman, James T. Geology of the Cove Creek area, Millard County and Beaver County, Utah. M, 1961, University of Utah. 91 p.

Zimmerman, Jay, Jr. Structure and petrology of rocks underlying the Vourinos Complex, northern Greece. D, 1968, Princeton University. 100 p.

Zimmerman, John, Jr. The origin of the Tuomey Sandstone (Tertiary), Fresno County, California. M, 1942, Stanford University. 71 p.

Zimmerman, Keith. Sedimentary petrology of the upper Jackfork Sandstones, Scott County, Arkansas. M, 1987, University of Arkansas, Fayetteville.

Zimmerman, Laurie S. Paleoecology of Permian bryozoan bioherms in the Glass Mountains, West Texas. M, 1985, Pennsylvania State University, University Park. 362 p.

Zimmerman, M. B. Long-run mineral supply; the case of coal in the United States. D, 1975, Massachusetts Institute of Technology. 212 p.

Zimmerman, Marc James. Aquatic biogeochemistry; a chemical equilibrium approach. D, 1980, University of Georgia. 101 p.

Zimmerman, Neil Jay. Wastewater spray transport in land application. D, 1980, University of North Carolina, Chapel Hill. 246 p.

Zimmerman, Peter J. Aspects of the depositional and diagenetic history of the Pleistocene Miami Oolite in the southern Florida Keys (Big Pine Key to Key West). M, 1985, Wichita State University. 134 p.

Zimmerman, Richard Albert. Geology of the Annapolis area, Iron County, Mo. M, 1959, University of Missouri, Rolla.

Zimmerman, Richard Albert. The origin of the bedded Arkansas barite deposits (with special reference to the genetic value of sedimentary features in the ore). D, 1964, University of Missouri, Rolla.

Zimmerman, Robert Wayne. The effect of pore structure on the pore and bulk compressibilities of consolidated sandstones. D, 1984, University of California, Berkeley. 122 p.

Zimmerman, Ronald K. Aspects of early Allegheny (Pennsylvanian) depositional environments in eastern Ohio. D, 1966, Louisiana State University. 183 p.

Zimmerman, Ronald K. X-ray diffraction technique for estimating quartz content of some Recent sediments. M, 1963, Louisiana State University.

Zimmerman, Thomas J. Recent and Pleistocene deposits of the Mississippi Delta platform. M, 1958, Louisiana State University.

Zimmerman, Tom Van. Petrology of a Morrison stratigraphic section in Fremont County, Wyoming. M, 1956, University of Missouri, Columbia.

Zimmermann, R. A. The interpretation of apatite fission track ages with an application to the study of uplift since the Cretaceous in eastern North America. D, 1977, University of Pennsylvania. 155 p.

Zimmermann, Regula Dorothea. Structural fabric and geology of the Golden Quadrangle, Idaho. M, 1982, University of Massachusetts. 106 p.

Zimpfer, Gerald Lee. Development of laboratory river channels. M, 1975, Colorado State University. 124 p.

Zimpfer, Gerald Lee. Hydrology and geomorphology of an experimental drainage basin. M, 1982, Colorado State University. 243 p.

Zindler, Gregory Alan. Geochemical processes in the Earth's mantle and the nature of crust-mantle interactions; evidence from studies of Nd and Sr isotope ratios in mantle-derived igneous rocks and lherzolite nodules. D, 1980, Massachusetts Institute of Technology. 263 p.

Zingula, Richard P. Cretaceous foraminifera from the Sacramento Valley, California. D, 1958, Louisiana State University.

Zink, Larry A. Local facies relations of the Dennis Formation (Kansas City Group, Pennsylvanian System), in northwestern Kansas. M, 1985, Wichita State University. 69 p.

Zink, Robert Miller. Certain aspects of the ecology of Venus and Mya at Morgan Bay and at Bunganuc, Maine; 2 volumes. M, 1952, University of Maine.

Zinkgraf, Joel P. A survey of ground motion resulting from forging operations. M, 1977, University of Wisconsin-Milwaukee.

Zinmeister, William J. Paleocene biostratigraphy of the Simi Hills, Ventura County, California. D, 1974, University of California, Riverside. 333 p.

Zinn, Justin. Petrography of the (Precambrian) Keweenawan lava flows of Michigan. M, 1930, Michigan Technological University. 26 p.

Zinn, Justin. The upper Huronian of the Marquette District of Michigan. D, 1933, University of Wisconsin-Madison.

Zinn, Lori A. Subsurface stratigraphy of the Ste. Genevieve Limestone (Meramecian) and relations to underlying Silurian reefs, Greene County, Indiana. M, 1983, Indiana University, Bloomington. 187 p.

Zinn, Robert Leonard. Cenozoic geology of Presidio area, Presidio County, Trans-Pecos, Texas. M, 1953, University of Texas, Austin.

Zinner, Ronald Eric. Geohydrology of the Rainy Creek igneous complex near Libby, Montana. M, 1983, University of Nevada. 118 p.

Zinser, Robert W. Stratigraphic distribution of the microfossils (exclusive of the Fusulinidae) in the Lansing Group in east-central Kansas. M, 1950, University of Kansas. 43 p.

Zinter, Glenn G. The geology and petrology of the Big Hole Canyon Pluton, Silverbow and Beaverhead counties, Montana. M, 1982, SUNY at Buffalo. 70 p.

Zinz, Barry Lynn. Environmental framework and diagenesis of the Yates Formation (Permian), Apache Mountains, Culberson County, Texas. M, 1971, Texas Tech University. 132 p.

Zinz, Barry Lynn. Environmental framework and diagenesis of the Yates Formation, Apache Mountains, Culberson County, Texas. M, 1977, University of Texas at Dallas. 132 p.

Ziony, Joseph Israel. Analysis of systematic jointing in part of the Monument Upwarp, southeastern Utah. D, 1966, University of California, Los Angeles. 152 p.

Ziony, Joseph Israel. Geology of the Abel Mountain area, Kern and Ventura counties, California. M, 1958, University of California, Los Angeles.

Zipf, R. Karl, Jr. The mechanics of fine fragment formation in coal. D, 1988, Pennsylvania State University, University Park. 302 p.

Zipp, Joel Frederick. Carbonate turbidites of the southern Blake Basin (Atlantic Ocean). M, 1972, University of Wisconsin-Madison.

Zipperer, Wayne C. Vegetation and landscape analysis of woodlots in central New York. D, 1987, State University of New York, College of Environmental Science and Forestry. 373 p.

Zippi, Pierre A. Calcium carbonate dissolution history for deep-sea cores adjacent Portugal. M, 1982, University of Georgia.

Zirino, Albert Rocco. Voltammetric measurement, speciation and distribution of zinc in ocean water. D, 1970, University of Washington. 205 p.

Zirkle, Robert Gale. Fusulinid fauna from the Naco Group in the Chiricahua Mountains near Portal, Cochise County, Arizona. M, 1952, University of Illinois, Urbana. 93 p.

Zirschky, John Herbert. Spatial analysis of hazardous waste data using geostatistics. D, 1984, Clemson University. 232 p.

Zitek, W. O. A study of inverse Q as a function of depth in western Washington. M, 1982, University of Washington. 87 p.

Zlotnik, Elias. Upper Cretaceous deep-sea fan and related lithofacies, San Diego, California; distribution and implications. M, 1981, San Diego State University.

Zmoda, Andrew J. Geochemistry of volcanic rocks from the Grandfather Mountain Window, northwestern North Carolina. M, 1987, University of North Carolina, Chapel Hill. 145 p.

Znidarcic, Dobroslav. Laboratory determination of consolidation properties of cohesive soil. D, 1982, University of Colorado. 185 p.

Zoback, M. D. High pressure deformation and fluid flow in sandstone, granite, and granular materials. D, 1975, Stanford University. 230 p.

Zoback, M. L. C. Mid-Miocene rifting in North-central Nevada; a detailed study of late Cenozoic deformation in the northern Basin and Range. D, 1978, Stanford University. 259 p.

Zoble, Jerry E. Stratigraphy of the Cretaceous Cloverly Formation and Crooked Creek Member of the Thermopolis Formation in the northeastern Big Horn Basin, Carbon County, Montana. M, 1957, University of Wyoming. 85 p.

Zodrow, Erwin Lorenz. Contribution to the informal theory of geological mineral sample, in two parts. D, 1973, University of Western Ontario. 321 p.

Zody, Steven P. Seismic refraction investigation of the shallow subsurface of the lower Rio Salado, northwest of San Acacia, N.M. M, 1988, New Mexico Institute of Mining and Technology. 80 p.

Zoerb, Richard M. Geology of the Elk Mountain area, Jackson County, Colorado. M, 1954, Colorado School of Mines. 158 p.

Zoerner, Frederick P. The geology of the central Elk Mountains, Colorado. M, 1974, University of Wyoming. 117 p.

Zoeten, Ruurdjan de *see* de Zoeten, Ruurdjan

Zogg, William D. Geology of the Colorado Gulch Turquoise Lake area, northern Sawatch Range, Lake County, Colorado. M, 1976, Colorado School of Mines. 187 p.

Zoghet, Mouine Fahed. Alpine surface soil movement. D, 1969, Colorado State University. 161 p.

Zoghi, Manoochehr. Stability of colluvial slopes during earthquakes. D, 1988, University of Cincinnati. 341 p.

Zohdy, Adel Abd El-Rahman. Earth resistivity and seismic refraction investigations in Santa Clara County, California. D, 1964, Stanford University. 142 p.

Zolensky, Michael Ewing. The crystal structure and twinning of meta-uranocircite. M, 1980, Pennsylvania State University, University Park. 79 p.

Zolensky, Michael Ewing. The structures and crystal chemistry of the autunite and meta-autunite mineral groups. D, 1983, Pennsylvania State University, University Park. 222 p.

Zolidis, Nancy Ritter. Restoration of a mined peatbog in Delafield Township, Waukesha County, Wisconsin; field and computer model studies of the hydrogeology and the growth of fen buckthorn (Rhamnus frangula). D, 1988, University of Wisconsin-Madison. 208 p.

Zoller, Lawrence J. Discussion of the geological department of an oil company. M, 1921, University of Missouri, Rolla.

Zollweg, James Edward. Seismic studies of the Reynolds County earthquake sequences. M, 1981, St. Louis University.

Zolnai, Andrew S. A regional cross-section across the Southern Province adjacent to Lake Huron, Ontario; the role of the Murray fault zone in the tectonism of the Southern Province. M, 1982, Queen's University. 94 p.

Zoltai, Tibor Z. The structure of coesite and the classification of tetrahedral structures. D, 1959, Massachusetts Institute of Technology. 158 p.

Zomorodi-Ardebili, Kaveh. Optimization of design and operation of artificial groundwater recharge facilities. D, 1988, Utah State University. 196 p.

Zones, Christe P. Petrographic and petrofabric study of the metamorphic rocks north of Carson City, Ormsby County, Nevada. M, 1958, University of Nevada. 80 p.

Zons, Frederick W. A new volumetric method for the determination of thorium in the presence of other rare earths, and its application to the analysis of monazite rock. D, 1911, Columbia University, Teachers College.

Zorich, Theodore M. Stream hydrograph by fluorescent dyes. D, 1966, Colorado State University. 92 p.

Zotto, Maria. A study of the dinoflagellate stratigraphy at Site 100, Deep Sea Drilling Project. M, 1986, Queens College (CUNY). 64 p.

Zou, Daihua. Numerical analysis of rock failure and laboratory study of the related acoustic emission. D, 1988, University of British Columbia.

Zoukaghe, Mimoun. Moisture diffusion and generated stresses in expansive soils by the boundary element methods. D, 1985, University of Missouri, Rolla. 134 p.

Zouki, Ashour Y. El *see* El Zouki, Ashour Y.

Zouwen, Dawn Elaine Vander. Structure and evolution of southern Okinawa Trough. M, 1984, Texas A&M University. 96 p.

Zrupko, M. M. Petrofabric analysis of late Precambrian-Cambrian quartzites from southeastern California. M, 1975, University of Southern California.

Zubari, Waleed Khalil. A numerical three-dimensional flow model for the Dammam Aquifer system; Bahrain and eastern Saudi Arabia. M, 1986, Ohio University, Athens. 409 p.

Zuber, Maria Theresa. Unstable deformation in layered media; application to planetary lithospheres. D, 1986, Brown University. 156 p.

Zuberi, Shafiq Ahmed. Subsurface structure of the Knox, Chazy, and Trenton formations in Morrow County, Ohio, as determined from well log data. M, 1987, Wright State University. 68 p.

Zuberi, Zaheer H. Differential reaction analysis (DRA); a possible new method of thermal analysis. M, 1975, University of Tennessee, Knoxville. 51 p.

Zubovic, Peter. Minor element content of coal from Illinois beds 5 and 6 and their correlatives in Indiana and western Kentucky. M, 1960, American University. 79 p.

Zucca, John Justin. The crustal structure of Kilauea and Mauna Loa volcanoes, Hawaii, from seismic refraction and gravity data. D, 1981, Stanford University. 136 p.

Zuccaro, Bruce. A trace element survey of surficial materials on Colorado oil shale tract C-b and vicinity, Rio Blanco County, Colorado. M, 1978, Colorado School of Mines. 130 p.

Zuch, Donald M., II. Lineament analysis, fracture studies and geology in the Sulphur Lick, Tompkinsville, and Gamaliel 7.5 minute quadrangles, south-central Kentucky. M, 1986, University of Toledo. 190 p.

Zucker, Sandy M. Petrology of some calc-silicate rocks from the Grenville Series, southeastern Adirondacks, New York. M, 1979, Brooklyn College (CUNY).

Zui, Yuval. A study of geo-electrical experiments. M, 1959, New York University.

Zuker, J. Stevens. Orientation and reconnaissance geochemical exploration surveys of the Monte Cristo Range and Pilot Mountains, Esmeralda and Mineral counties, Nevada. M, 1986, Colorado School of Mines. 103 p.

Zukoski, C. F. A thesis on the smelting of Missouri zinc ores. M, 1888, Washington University.

Zukoski, E. L. Treatment of a silver-lead ore. M, 1884, Washington University.

Zullig, James Joseph. Interaction of organic compounds with carbonate mineral surfaces in seawater and related solutions; Volume I and II. D, 1985, Texas A&M University. 364 p.

Zullo, Victor August. Cenozoic Balanomorphia of the Pacific Coast of North America. M, 1960, University of California, Berkeley. 147 p.

Zullo, Victor August. Classification and phylogeny of the Balanomorpha (Cirripedia). D, 1963, University of California, Berkeley. 372 p.

Zuloaga, Guillermo. Geology of the iron deposits of the Sierra de Imataca, Venezuela. D, 1930, Massachusetts Institute of Technology. 129 p.

Zumberge, J. E. The organic analyses and the development of the Vaal Reef carbon seams of the Witwatersrand gold deposits. D, 1976, University of Arizona. 127 p.

Zumberge, James H. The origin and classification of the lakes of Minnesota. D, 1950, University of Minnesota, Minneapolis. 138 p.

Zumberge, John Edward. Ozonolysis of the kerogen in the Transvaal stromatolitic limestone. M, 1973, University of Arizona.

Zumwalt, Gary Spencer. The effect of trophic resource stability on genetic complexity in Macoma (Bivalvia). D, 1984, University of California, Davis. 126 p.

Zumwalt, Gary Spencer. The functional morphology of the tropical brachiopod Thecidellina congregata, Cooper 1954. M, 1976, University of California, Davis. 135 p.

Zumwalt, Robert Wayne. Part I; quantitative gas-liquid chromatography of amino acid N-trifluoroacetyl n-butyl esters; Part II, A search for organics in hydrolysates of lunar fines; Part III, A search for amino acids in Apollo 11 and 12 lunar fines. D, 1971, University of Missouri, Columbia. 122 p.

Zuniga Izaguirre, M. A. Gravity and magnetic survey of the Sula Valley, Honduras, Central America. D, 1975, University of Texas, Austin. 171 p.

Zúñiga, Francisco Ramón. A study of earthquake source parameters and stress processes. D, 1987, University of Colorado. 178 p.

Zupan, Alan-Jon Wellward. Surficial sediments and sedimentary structures; middle ground, Padre Island, Texas. M, 1971, Texas A&M University.

Zuppann, C. W., Jr. Petrography and paleoenvironment of the Hermitage Formation in the western half of the Central Basin and in the western Highland Rim, Tennessee. M, 1974, Vanderbilt University.

Zuraff, Steven J. Three-dimensional computer modeling of basement topography based on gravity data in central southernmost Alabama. M, 1979, University of Southwestern Louisiana. 56 p.

Zuravel, David Lee. Wenlockian (Silurian) agglutinated foraminifera from the Wayne Formation, Tennessee. M, 1986, Texas Tech University. 130 p.

Zurawski, Ronald Philip. A reef community, its development in the Carters Limestone (Middle Ordovician) in Giles County, Tennessee. M, 1973, Vanderbilt University.

Zurbrigg, H. F. Study of nickeliferous pyrrhotites. M, 1933, Queen's University. 65 p.

Zurbuch, Jeffrey S. Ground water chemistry as an exploration tool for natural gas in southwestern West Virginia. M, 1988, West Virginia University. 221 p.

Zurcher, Hannes. A study of the crust in Puget Sound using a fixed seismic source. M, 1976, University of Washington. 62 p.

Zurcher, Hannes George. Spontaneous earthquake rupture simulated by a finite element code using variable fracture energy absorption. D, 1985, University of Washington. 265 p.

Zurinski, Stephanie Ann. Deposition and diagenesis of Empire Abo Field, Eddy County, New Mexico. M, 1979, Texas Tech University. 88 p.

Zuzo, P. L. Patterns of trace element distribution with altitude in the soils, litter, and native leaves of Mt. San Jacinto. M, 1976, University of Virginia. 81 p.

Zvanut, Frank Joseph. Purification of quartz sands and muscovite mica of the Pacific Northwest. M, 1933, University of Washington. 105 p.

Zvanut, Frank Joseph. Pyrochemical changes in Missouri halloysite. D, 1937, University of Missouri, Rolla.

Zvibleman, Barry. The feasibility of induced recharge for the city of Mequon. M, 1983, University of Wisconsin-Milwaukee. 110 p.

Zwanzig, H. Structural geology of the Long Lake area, Man. M, 1969, University of Manitoba.

Zwanzig, Herman V. Structural transitions between the foreland zone and the core zone of the Columbian Orogen, Selkirk Mountains, British Columbia. D, 1973, Queen's University. 158 p.

Zwart, David. Atlantic Coast exploration program, geological and economic. M, 1975, [University of Tulsa].

Zwart, Michael J. North American microtektites from Deep Sea Drilling Project cores. M, 1977, University of Delaware.

Zwart, Peter A. An investigation of the stratigraphic occurrence of Georgia tektites. M, 1978, University of Delaware.

Zwartendyk, Jan. A petrographic study of the Granite Wash in the Clear Hills area, Alberta. M, 1957, McGill University.

Zwashka, Mark R. Geological investigation of the Idaho gold district, Clay County, Alabama. M, 1986, Auburn University. 159 p.

Zweig, Julie. Processing and interpretation of seismic data in the northeastern Gulf of Alaska, offshore Yakutat. M, 1985, University of Houston.

Zweng, Paul L. Evolution of the Toquepala porphyry Cu-(Mo) deposit, Peru. M, 1984, Queen's University. 131 p.

Zwicker, Deborah L. Cambro-Ordovician sandstones of the northern Michigan Basin. M, 1983, Michigan State University. 53 p.

Zygarlicke, Christopher J. A mineralogical study of the Harmon lignite bed, Bullion Creek Formation (Paleocene) Bowman County, North Dakota. M, 1987, University of North Dakota. 187 p.

Zyl, D. J. A. van see van Zyl, D. J. A.

Zylstra, Elise. Molluscan associations of the Caicos Bank, British West Indies. M, 1985, University of North Carolina, Chapel Hill. 81 p.

Zymela, Steve. ESR dating of Pleistocene deposits. M, 1986, McMaster University. 118 p.

Zytner, Richard Gustav. Fate of perchloroethylene in unsaturated soil environments. D, 1988, [University of Windsor].

DEGREE RECIPIENTS AT EACH INSTITUTION

Acadia University
Wolfville, NS B0P 1X0

61 Master's

1920s: Haycock, M. H.

1930s: Hancock, L. T.

1950s: Cote, R. P.; Dunlop, W. B.; Grant, D. A.; Hudgins, A. D.; Johnson, C. G.; Loring, D. H.; MacNeill, R. H.; Oldale, H. R.; Purdy, C. A.; Smith, G. P.; Swayne, L. E.

1960s: Crowell, Gordon D.; Giles, Peter S.; Lewis, W. L.; von Bitter, Peter Hans; Waring, Marcus H.; Worth, John Kirk

1970s: Adams, Kenneth D.; Atkinson, Susan J.; Blakeney, R. S.; Boehner, R. C.; Buckley, D. W.; Chu, Peter H. T.; Davidson, D. D.; Doyle, Eibhlin; Durocher, A. C.; Freeman, Gary W.; Griffin, M. G.; Hill, J. D.; Holleman, M.; Irrinki, R. R.; Jones, B. E.; Liew, M. Y.-C.; MacDonald, Donald J.; McCulloch, Paul D.; Murphy, J. B.; O'Beirne, A. M.; Rankin, L. D.; Roy, David T.; Ryan, R. J.; Smith, P. K.; Stanley, G. A.; Stewart, E. B.; Tang, Patrick S.; Thompson, J. P.; Thorpe, L. R.; Trapasso, Linda; Trepasso, Linda; Wightman, John F.

1980s: Dennis, Frank Anthony Richard; Gardiner, John J.; Hossley, James Glenn; Hussain, Mahbub; Leybourne, Matthew Iain; Scott, Philip; Sexton, Alan J.; Snow, Randall J.; Wakhungu, Judi W.; Yaowanoiyothin, Winai

Adelphi University
Garden City, NY 11530

2 Master's, 1 Doctoral

1970s: Chiang, E.

1980s: Bakker, Allen; Furhmann, Mark

University of Akron
Akron, OH 44325

143 Master's, 5 Doctoral

1960s: Anthony, Gaylord Dean

1970s: Bain, Leslie Gay; Bickley, John A.; Bowyer, Robert C.; Brasaemle, Joan E.; Carney, Craig A.; Coburn, Jo Ann; Dannemiller, Gary Thomas; Dixon, Jeanette M.; Elliott, James Barry; Fantel, Richard J.; Foster, Robert A.; Frankovits, Nicholas D.; Franks, Bernard J.; Frlich, Waldo J.; Giuffria, Ruth; Gray, John D.; Haefka, Delbert John; Henning, Roger John; Henry, Scott Duray; Hodges, David Alvin; Houghton, LeRoy Kingsbury, III; Hummel, Judythe Ann; Keller, Donald Frederick; Keller, John David; Klingel, Eric John; Knox, John Harold; Landin, William; Lee, Raymond Frederick; Lorson, Richard C.; Manner, Barbara Marras; Marsek, Frank A.; Massaro, David A.; McKirgan, Bruce Stephen; Olson, Thomas L.; Quay, Paul C.; Russo, Anthony F.; Schumacher, John C.; Spear, James W.; Steiffer, R. John; Surblis, Benjamin; Swires, Charles Jonathan Jr.; Tokar, Ronald Albert; Torguson, William R., Jr.; Trembley, Susan B.; Van Deventer, James Bartlett; Vander Kooi, Verna; Weise, James R.; Wells, Kathy M.; Williams, Ronald Leon; Wymer, Richard Edward

1980s: Abrajano, Teofilo A., Jr.; Allahiari, Morteza; Alonso, Manuel; Angle, Michael Paul; Bahr, Tim J.; Barnett, Robert G.; Beck, William C.; Beyke, Robert J.; Billman, Thomas A., Jr.; Blauch, Matthew E.; Bomback, Ronald L.; Bonzo, Kevin M.; Bowman, Patricia A.; Bray, Thomas F., Jr.; Brocculeri, Thomas; Bruno, Pierre W.; Bruno, Sherrie L.; Burke, James Charles; Carnes, Lynne H.; Castillo, Paterno R.; Cheek, William M., Jr.; Chen, Cary Ching-chi; Clabaugh, Charles Donald; Clarke, Barbara G.; Colopietro, Margaret R.;

Corwin, Bert N.; Croley, Clifford W.; Crotty, Kevin J.; Dakoski, Andrea Marie; Daly, Philip J.; Delmastro, G. A.; Dicke-Burke, Collette; Donovan, Kevin; Eller, John August; Eshler, Lynn M.; Fashola, Ahmed B.; Fernandez, Réne L.; Florentino, Eugene; Gardner, Stephen P.; Garvey, John Thomas; Gaughan, Maryann; Gilbert, Deidre M.; Good, Charles Neil; Green, Scott Robert; Hansen, Lawrence; Heirendt, Kenneth M.; Heppard, Philip D.; Jean, Jiin-Shuh; Jones, Edward J.; Katzmark, Robert Raymond; Kendall, Robert Lee; Kesebir, Musa Mustafa; Knuth, Martin C.; Layton, Albert W., Jr.; Leelanitkul, S.; Lou, Ken-An; Mangun, Mark; Martinez, Ricardo D.; Mobasseri, Shahpur; Moore, Craig H.; Nanna, Richard F.; Nutt, William H.; Olver, Richard; Ospanik, Laddy Franklin; Pasternack, Stephen C.; Polasky, Mark Edward; Post, Richard Edward; Preston, Michael B.; Pringle, Patrick; Quick, Thomas J.; Ruberti, James A.; Ryan, Dale Edward; Sabaka, Terence J.; Sanger, Daniel B.; Sanger, Gary Edward; Schlegel, Helen E. Stenstrom; Schmidley, Eric B.; Seibert, Jeffrey Lynn; Seifert, Douglas J.; Simmers, Rick J.; Sims, William R.; Smiraldo, Mark S.; Sobhanie, Mohammad Eghbal; Tang, Huey; Thalman, Katherine L.; Theoret, Dennis R.; Thomas, Robert Brinley; Thompson, John David; Thompson, Warren F.; Van Kauwenbergh, James B.; Viani, Chris William; Volpi, Richard Wayne; Weir, William G.; Wilk, Grant B.; Wunsch, David Robert; Wyles, James C.; Zaheha, Robert D.

University of Alabama
Tuscaloosa, AL 35487

124 Master's, 11 Doctoral

1920s: Jones, Walter Bryan

1930s: Glass, Theodore Gunter; Toulmin, Lyman D., Jr.

1940s: Boswell, Ernest Harrison; Higgs, William Reginald; LaMoreaux, Phillip E.; Naff, John Davis; Whatley, James Broughton; Windfordner, John S.

1950s: Audesey, Joseph Louis, Jr.; Barksdale, Joe M.; Drennen, Charles William; Harris, James Zack; Hisey, William Murphy; Palmore, Robert Donald; Simpson, Thomas A.; Ward, John F.

1960s: Boyles, James McGregor; Britton, Thomas Abbot, Jr.; Brockman, George Frederic; Bryan, Jack Howard; Dayton, Frank Herbert, Jr.; Ehringer, Robert Ferris; Fleming, Randall J.; Harris, Eugene B., Jr.; Hyde, Luter Willis; Jinkins, Ronnie L.; Keeley, James Chester; Lile, Thomas Craig; Marsalis, W. E.; McNeal, James E.; McNutt, William Paul, Jr.; Moore, Donald B.; Neathery, Thornton Lee; Newman, Harry E., III; Newton, John Gordon; Sample, Milton David; Sanford, Thomas Herbert, Jr.; Siebert, Harry L.; Smith, James K.; Smith, William Everett; Sonderegger, John L., II; Watthanachan, Suwit; White, Dossey Hurdle, Jr.

1970s: Beavers, W. M.; Blake, Alan Brian; Bloss, Pamela; Chapman, Willie Eugene; Chuamthaisong, Charoen; Clark, John Robert; Dillon, Andrew Crawford, III; Dobbs, Danny M.; Drummond, S. E.; Files, Edgar James, Jr.; Gilbert, Oscar Edward, Jr.; Gilliam, W. B.; Hargett, W. G.; Hill, D. O.; Holler, D. P.; Hunter, Cindy Carothers; Kidd, Jack James; Masingill, John H., III; McWilliams, Richebourg Gaillard, Jr.; Moravec, George Frank; Muangnoicharoen, Nopadon; Naughton, Margaret M.; Oliver, Gary Earl; Price, R. C.; Raymond, Dorothy Echols; Reynolds, J. W.; Schneeflock, R. D.; Shannon, Samuel W.; Shepard, Brian K.; Spigner, Benjamin Cantrell; Sutley, David E.; Szabo, Michael Wallace; Ward, W. E., II; White, J. R., Jr.; Wielchowsky, Charles Carl

1980s: Al-Ansari, Jasem Mohammad; Ash, Nadim F.; Batchelder, Eric C.; Bearden, Bennett L.; By-

erly, Benjamin Edward; Carter, William W., Jr.; Chase, Duane D.; Cook, Terry Allen; Cunningham, Alan E.; Defant, Marc J.; di Giovanni, Marcel, Jr.; Engman, Mary Anne; Ferrill, Benjamin Arnold; Fisher, D. Ramsey; Hertig, Stephen Paul; Higginbotham, David R.; Hines, Robert Arthur, Jr.; Holmes, Ann Elizabeth; Holmes, James W.; Hooks, J. David; Hubbard, Perry, Jr.; Jenkins, Christine Maria; Katz, William Meyer; Laird, John W.; Lee, Alison Marian; Leverett, David Earl; Long, Aubrey Lamar; Miesfeldt, Mark Alan; Mink, Robert M.; Moffett, Tola Burton; Moss, Neil E.; Osborne, Walter Edward; Payton, J. Wayne; Pendexter, William Sands; Rheams, Karen F.; Riedle, Lisa Ann; Scanlon, Bridget R.; Sestak, Helen Maria; Sheldon, Elisabeth Shepard; Soens, David Dale; Steltenpohl, Mark Gregory; Tantikom, Supachai; Telle, Whitney R.; Vaughan, R. Lee, Jr.; Wade, George David; Walls, Ian A.; Ward, Wendi I.; Waters, Laura A.; Webb, Kathryn Wenthe; Webb, Steve William, III; Whittington, David; Wilkenson, R. P.; Williams, Mona S.; Wixon, Roy Stephen; Womack, Stephen Hasie; Woodley, Nancy Karen Fish

University of Alaska, Fairbanks
Fairbanks, AK 99775

158 Master's, 34 Doctoral

1960s: Bingham, Douglas K.; Blackwell, John Michael; Blake, J. Roger; Bond, Gerad C.; Britton, Joe M.; Brown, Jim McCaslin; Church, Richard E.; Davis, Thomas C., Jr.; Durfee, M. Charles; Furst, George A.; Furst, Martha Jean; Glavinovich, Paul S.; Hanson, Kenneth E.; Hanson, Larry Gene; Kienle, Juergen; Kreitner, Jerry D.; Matthews, John V., Jr.; Mayo, Lawrence Rulph; Moores, Eugene A.; Morrison, Donald Allen; Petocz, Ronald George; Phillips, Walter T., Jr.; Quinlan, Alician V.; Ray, Dipak Kumer; Reger, Richard D.; Robinson, Larry; Rutter, Nathaniel W.; Schraeder, Robert Lynn; Sellman, Paul V.; Slatt, Roger Malcolm; Smith, William H.; Stout, James H.; Wescott, Eugene M.; Wharton, George B., Jr.

1970s: Ager, Thomas A.; Amna, Kei; Ariey, Catherine A.; Barrett, Stephen A.; Biggar, Norma E.; Bingham, Douglas K.; Blodgett, Robert B.; Cameron, Christopher Paul; Carden, John R.; Chatterjee, Biswanath; Clardy, Bruce; Corbin, Samuel W.; Davies, John Norman; Davies, John Norman; Dean, Kenneson; Decker, John E.; Deininger, James W.; Estes, Steve A.; Fairchild, Drena K. T.; Gebhardt, Robert L.; Hackett, Steve W.; Heggie, D. T.; Hoffman, Barry L. P.; Holm, Bjarne; Huang, Paul; Jones, Brian K.; Kerin, L. John; Loder, Theodore C.; Madonna, James A.; Matteson, Charles; Metz, Paul; Metzner, Ron; Nelson, Gordon L.; Nelson, Richard V., Jr.; Nye, Christopher; Ordonez, José Luis; Packer, Duane Russell; Patton, Thomas L.; Peace, Jerry; Pearson, Christopher F.; Peek, Bradley C.; Prentki, R. T.; Rawlinson, Stuart; Redman, Earl C.; Robinson, Mark; Root, Michael R.; Slatt, Roger Malcolm; Stevens, Carolyn C.; Stevens, Donald L.; Swainbank, Richard Charles; Thorson, Robert M.; Trabant, Dennis Carlyle; Trible, Marla; Tucker, R. W.; Wilson, Donald Edward; Wilson, Frederic H.; Wood, Elwyn Devere

1980s: Adams, David D.; Agnew, James D.; Albanese, Mary; Allegro, Gayle L.; Alperin, Marc Jon; Anderson, Nancy Louise; Arce, Gary N.; Arunapuram, Sundararajan; Baker, Grant Cody; Bakke, Arne A.; Bender, Gary; Blum, Joel David; Bodnar, Dirk A.; Bond, James F.; Bracey, Dewey R.; Brown, Perry L.; Brown, William Gregor; Buckingham, Martin L.; Bundtzen, Thomas K.; Burton, Jeffrey P.; Calvin, James S.; Candler, Rudolph John, II; Carlson, Randall; Clarke, Theodore S.; Clautice, Karen H.; Clough, James G.; Coleman, Donald A.; Collett, Timothy S.; Cornwell, Jeffrey Clayton; Daley, E. Ellen; Dickey, Douglas B.;

Dilles, Peter Alden; Dilley, Thomas E.; East, Jennifer S.; Engle, Kathryn Yvonne; Estabrook, Charles Hershey; Ewy, Bradford James; Foley, Jeffrey Young; Fontana, Michael R.; Ford, Michael J.; Fox, Steven W.; Freeman, Curtis J.; George, Thomas H.; Glover, David Mark; Gobelman, Steven; Grebmeier, Jacqueline Mary; Hagen-Leveille, Janice; Hall, Mark H.; Hardy, Steven B.; Harris, Ronald Albert; Herzberg, Peter Jansen; Hogarty, Barry Jane; Hok, Charlotte I.; Hong, Gi Hoon; House, Gordon D.; Huber, Jeffrey A.; Izett, Glen Arthur; Knock, Douglas G.; Kucinski, Russell; Law, Kyin-kouk Hubert; Lockhart, Andrew; Lueck, Larry; Matava, Timothy; McCrum, Michael Arthur; McMillin, Steven; Mellor, Jack Conrad; Moore, Michael; Mortensen, Thomas William; Motyka, Roman John; Musgrave, David L.; Myers, Gregory L.; Panuska, Bruce C.; Panuska, Bruce C.; Paris, Chester E.; Peterson, John K.; Pujol, Jose M.; Reifenstuhl, Rocky; Romick, Jay D.; Sellars, Barbara D.; Simpson, David F.; Siok, Jerome P., III; Smith, Daniel T.; Southworth, Dennis; Spencer, JoEllen Paige; Steensma, Gilein; Stein, Jean; Sturnick, Mark A.; Sugai, Susan Frances; Sugumaran, Vijayan; Szumigala, David; Vandergon, Mark; Vralsted, David A.; Walker, Gerald Grant; Weber, Eric F.; Wittbrodt, Paul Raymond; Witte, William K., Jr.; Wu, Ming-Chee; Wuttig, Frank J.; Zayatz, Mark; Zdepski, John Mark; Zelenka, Brian

University of Alberta
Edmonton, AB   T6G 2E3

356 Master's, 131 Doctoral

1920s: Sanderson, James O. G.

1930s: Beach, H. Hamilton; Buckham, Alex Fraser; Crockford, Michael Bertram Bray; Hicks, H. S.; Howells, William Crompton; Lowther, George Kenneth; MacDonald, Roderick Dickson; Souch, Bertram Elford; Taylor, Donald Alfred; Weir, Marjorie Fraser

1940s: Andrichuk, John Michael; Carr, John Lawrence; Conybeare, Charles Eric Bruce; Crawford, Ian Douglas; Edie, Ralph William; Erdman, Oscar Alvin; Fox, Frederick Glenn; Kidd, Stuart James; Kunst, Henery; MacDonald, William Delbert; McKinnon, Frederick Allan; Stelck, Charles Richard; Wilson, J. N.

1950s: Antoniuk, Stephen Alexander; Bahan, Walter George; Bayrock, Luboslaw Antin; Berry, Andrew David; Beveridge, Alexander James; Borden, Robert Leslie; Brayton, Darryl Merritt; Brodie, David R.; Bullock, D. B.; Burwash, Ronald Allan; Byrne, Patrick James Sherwood; Campbell, Donald Lorne; Clow, William Henry Arthur; Colborne, Gerald Laverne; Duff, Denny Emerson; Farvolden, Robert N.; Fischbuch, Norman Robert; Greiner, Hugo Robert; Howard, Ronald Adrian; Hughes, George Muggah; Hunt, Graham Hugh; Jennings, Edward Wallace; Koch, Norris Gayle; Kryczka, Adam Alexander William; Lenz, Alfred Carl; Leslie, Gordon Anthony; Macrae, Leslie Blair; Martin, Leonard John; Mawdsley, James Cleugh; McMullen, Robert Michael; Mellon, George Barry; Moore, Wayne Ewing; Nielsen, Arne Rudolph; Nikiforuk, Zan Frank; Norris, Arnold Willy; Orr, John Barrie Bain; Patton, William John Hudson; Ritchie, William Douglas; Rudy, Harold R.; Stansberry, Gerald Francis; Swan, Arthur Graham; Trollope, Frederick Hamilton; Wall, John Hallett; Waller, Philip Charles; Wetter, Raymond Emil; Young, Harvey Ray

1960s: Adshead, John Douglas; Agarwal, Ram Gopal; Ahmad, Khwaja Gulzar; Akehurst, Alfred; Alpaslan, Tümer; Anan-Yorke, Rowland; Bielenstein, Hans Uwe; Bihl, Gerhard; Bilan, Larry Joseph; Bird, Gordon Winslow; Brown, Hugh Matiland; Bukhari, Syed Amir; Cameron, Bruce; Chandra, Nellutla Naveena; Chi, Byung I.; Cholach, Michael S.; Chrismas, Lawrence Philip; Clissold, Roger J.; Clowes, Ronald Martin; Clowes, Ronald Martin; Corneil, Barry David; Cruden, David Milne; Currie, Donald Varcoe; Davidson, Alexander, Jr.; Davies,

Edward Jullian Llewelyn; de la Cruz, Nga; Delorme, Denis Larry; Dodds, Christopher James; Eliuk, Leslie Samuel; Ellis, Robert Malcolm; Etherington, John Robert; Evans, Calvin Ralph; Evans, Thomas Lester; Fitzgerald, Edward Leo; Gabert, Gordon Michael; Germundson, Robert Kenneth; Given, Mary Michie; Goodrich, Laurel E.; Green, David Christopher; Griffiths, Reginald; Hanson, Larry; Hasegawa, Henery; Hemmings, Charles David; Hills, Leonard Vincent; Hjortenberg, Erik; Hunt, Graham Hugh; Husain, Sheikh Ansar; Jenik, Albert James; Johnston, Paul Francis; Jones, Gareth H. S.; Jull, Robert Kingsley; Kanasewich, Ernest Raymond; Kent, Donald Martin Joseph; Khamesra, Daulat Singh; Khan, Shahid; Kieller, Bernard John; Kirmani, Khalil-ullah; Klassen, Rudolph Waldemar; Kramers, John William; Kurtz, Ronald; Lafon, Guy Michel; Larson, Murray Lloyd; Leavitt, Eugene Millidge; Leech, Alice Payne; Lennox, D. H.; MacPherson, Donald Stuart; Maiklem, William Robert; Maureau, Gerrit T. F. R.; McClure, D.; McKay, Mary Winifred; Meistrell, Frank Joseph; Money, Peter L.; Montalbetti, James; Mueke, Gunter Kurt; Murthy, Gummuluru Satyanarayana; Nascimbene, Giovanni Giuseppe; Newland, Bernard Terence A.; Nielson, Grant Leroy; O'Brien, Donald Edward; O'Nions, Robert Keith; Oldenburg, Douglas; Peeples, Wayne; Pelzer, Ernest Edward; Peterman, Zell Edwin; Pfaff, Dieter; Platt, Ronald Lorne; Pronko, Peter Paul; Quon, Charles; Radcliffe, Dennis; Rankin, David; Rath, Ulrich E. G.; Reddy, Indupuru Kota; Reimchen, Theodore Frederick Harold; Reinelt, Erhard Rudolph; Remington, Donald Bryce; Rice, Dudley Dennison; Robbins, Barry Phillip; Robertson, David Knox; Robinson, Joseph Edward; Roed, Murray Anderson; Rutter, Nathaniel; Ryznar, Gerald John; Shafiqullah, Muhammad; Shafiuddin, Mohammed; Singh, Chaitanya; Singh, Ram Narain; Smith, Dorian Glenys Whitney; Smithers, Ronald Milton; Snead, Robert Garland; Srivastava, Satish Kumar; Srivastava, Satish Kumar; Stauffer, Melvyn Roy; Steiner, Johann; Stevenson, Douglas Roy; Suska, Maria Magdalena; Tatar, Jhumar Mal; Tedrick, Robert Lowell; Thusu, Bindraban; Tokarsky, Orest; Turek, Andrew; Vagvolgyi, Andrew Louis; Van Breemen, Otto; Vopni, Lorne Kenelm; Walasko, James Thadius; Watanabe, Roy Yoshinobu; Watanabe, Roy Yoshinobu; Watkins, Norman David; Weaver, David F.; Weiner, John Louis; Wellman, Robert Persey; Westgate, John Arthur; Williams, Gordon Donald Clarence; Yorath, Christopher John; Zaturecky, John William

1970s: Abou-Kassem, Jamal Hussein; Abraham, Amenti; Acham, Patrick A.; Achtman, Malcolm; Alabi, Adeniyi Oluremi; Alexander, Frederick John; Alpaslan, Tümer; Amajor, Levi Chukwuemeka; Anan-Yorke, Rowland; Anderson, Chester Washington, III; Au, Chong Ying Daniel; Aubut, Alan James; Badham, John Patrick Nicholas; Bannister, John Richard; Bates, Allan Clifford; Binda, Pier Luigi; Bloy, Graeme Richard; Boetzkes, Peter C.; Brame, Simon; Brown, R. W.; Brown, Rodney Stuart; Burnie, Stephen Wilbur; Cameron-Schimann, Monique; Camfield, P. Adrian; Cape, David Frank; Cartier, Gerard Leon Marcel; Chandra, Nellutla Naveena; Choi, A. P.; Churney, Garth; Costaschuk, Suzanne Mickey; Cox, John; Crowe, Allan; Davies, Linda M.; Davis, Donald Wayne; Day, L. Wayne; de Beer, Johannes Hendrik; Dickie, Geoffrey; Dickins, D. G.; Dodson, Peter John; Donaghy, Thomas James; Emerson, Donald; Farkas, Arpad; Flach, Peter Donald; Flint, David Warren; Ganley, David Charles; Gilliland, John Michael; Goble, Ronald James; Gold, Christopher Malcolm; Greig, John; Gudjurgis, Paul Joseph; Hall-Beyer, Barton MacNeill; Hall-Beyer, Mryka Christine; Harland, Rex; Haverslew, Roderick Edwin; Havskov, Jens; Heal, George Edward Newton; Hedinger, Adam Stefan; Hibbs, Roy Dean, Jr.; Hodgson, Geffrey David; Hoiles, Harley Harold Kristjan; Hron, Marie Petra; Hughes, John David; Hughes, Terence John; Hunt, Robert N.; Hussin, Ismail Bin; Iwuagwu, Chukwumaeze Julian; Jackson, Stewart Albert;

Kao, Dominique Wen; Kenyon, John Michael; Kesmarky, Susanna; Kilby, Ward Eldon; Kirkland, Kenneth John; Kloepfer, John Gerard; Knapik, Leonard Joseph; Kuo, Say Lee; Kuo, Say Lee; Kvill, D. R.; Lam, Hing-Lan; Lee, Shu Kay Joseph; Leung, Sydney Kwok-On; Lines, Laurence Richard; Lorberg, Eberhard F.; Mann, Gary Dale; Marks, Larry Wayne; Matthews, John V., Jr.; Mawer, Malcolm Frank; Moir, Rory Douglas; Morra, Franco Piero; Mozeson, Charles Edward; Narasimha Chary, K.; Naylor, Bruce Gordon; Nielson, Grant Leroy; Nielson, Peter Alfred; Nowak, Robert Lars; Oberg, Clayton John; Olade, Moses; Park, John Keith; Patterson, Terence Edward; Peto, Peter S.; Pinsent, Robert Hugh; Pratt, Ernest George; Rahmani, Riyadh Abdul-Rahim; Ramsay, Colin Robert; Ramsden, John; Ramsden, John; Ratanalert, Pirmpoon; Reddy, Indupuru Kota; Reid, Alan B.; Rice, Randolph James; Robinson, Brian William; Roebroek, Edward John; Rosene, Richard K.; Saruk, Bertrand Alexander; Sassano, Giampaola; Schimann, Karl; Schweger, Charles Earl; Shiferaw, Alemu; Shu Kay Joseph Lee; Sophocleous, Marios Andreou; Spanos, Thomas James Timothy; Sprenke, Kenneth Fredrick; Sues, Hans-Dieter; Sutherland, George Donald; Sydora, Larry John; Tipnis, Ravindra Shalad; Tizzard, Paul G.; Vincent, Bruce Douglas; Waddington, Edwin Donald; Walker, Bruce Dawson; Walker, Richard Randall; Walrond, Grantley Wainright; West, Kenneth Ernest; Winzer, Stephen R.; Withers, Robert John; Withers, Robert Louis

1980s: Adair, Robin N.; Aggarwal, Pradeep Kumar; Aggarwal, Pradeep Kumar; Amajor, Levi Chukwuemeka; Andriashek, Laurence Douglas; Ansdell, Kevin Michael; Apon, Johan Frederik; Arnott, Robert William Charles; Arnott, William Charles; Banks, Craig Stewart; Barta, Leslie Ann; Berndt, Kathleen A.; Blackadar, Donald William; Blackwell, Bonnie; Braco, Paulo, Jr.; Brearley, Mark; Brown, D.; Brown, Isobel Julia; Burtch, Steven Douglas; Burwash, Elizabeth Jean; Casey, John Joseph; Catto, Norman Rhoderick; Catto, Norman Rhoderick; Cavell, Patricia Anne; Chan, Gee Hung; Changkakoti, Amarendra; Chaplin, Catherine Elizabeth; Chatwin, Stephen C.; Collette, Ronald; Crowe, Allan S.; Dagenais, Georges Roman; de Matos, Milton Martins; de St. Jorre, Louise; Dean, Michael Edward; Dimitrakopoulos, Roussos-Georgios; Dingwell, Donald Bruce; Duke, Michael John Maclachlan; Dunn, John Todd; English, Paul James; Evans, David John Alexander; Evans, Stephen George; Fahner, Lewis George; Fedoruk, Richard Alexander; Fernando, Angelo Ransirimal; Fluet, Darrell Wayne; Fossey, Kenneth Wayne; Fox, Andrew J.; Freeman, James Thomas; Fyvie, Donald James; Gagnon, Lawrence Gregory; Ganley, David Charles; Gentzis, Thomas; Ghosh, Dipak Kumar; Gillen, K.; Giusti, Lorenzino; Goff, Stephen Patrick; Goodbody, Quentin Harald; Goodbody, Quentin Harald; Grant, Stacy Kent; Habib, Antoine Ghali Elia; Harding, Steven Craig; Heany, Heinrich Karl, Jr.; Hill, Kevin Charles; Hook, Simon John; Hughes, Rhys Leckie; Indraratna, Buddhima Nalin; Iwuagwu, Chukwumaeze Julian; Johnston, Paul Anthony Frederick; Johnston, Stephen Thomas; Kamenka, Louis Anthony D.; Kasperski, Kim Louise; Kirk, John S.; Kontak, Daniel Joseph; Krstic, Dragan; Kulig, John Joseph; Kvill, D. R.; Lantos, Julie Ann; Lee, Desmond Nyuk Hin; Lemmen, Donald Stanley; Leslie, Louise E.; Levson, Victor Mathew; Lhotka, Paul Gordon; Liverman, David Gordon Earl; Lockhart, Elizabeth Blair; MacDonald, Donald E.; Macrides, Costas; Magwood, James P. A.; Marks, Larry Wayne; McClymont, Gordon Lee; McCourt, George H.; McLellan, P. J. A.; Metcalfe, Paul; Miller, Richard George; Milne-Home, William Alexander; Montgomery, Keith; Morison, Stephen Robert; Mortensen, Paul Stuart; Negro, Arsenio, Jr.; Nentwich, Franz Werner; Ng, Kwok-Choi Samuel; Niemann, Knut Olaf; Nurkowski, John Ronald; Ophori, Duke Urhobo; Ortigoza Cruz, Felipe; Over, D. Jeffrey; Park, Darrell G.; Pate, Colin Roger; Paterson, Lorraine Ann; Phadke, Suhas; Polikar, Marcel; Power, M.; Proudfoot, David Nelson; Reasoner, Melton Aaron; Ringham, Kevin Lee; Robb, Gregory Alan; Robertson, Craig; Rucker, Paul Douglas; Sauveplane, Claude Marcel; Sayer, Christina J.; Shahriar, Mostafa; Sharp, Robert Jay; Slater, Jock R.; Smith, Alan D.; Smith, Gregory John; Smith, Jennifer Margaret; Squair, Cheryl Anne; St. Louis, Robert Michael; Steele, Kenneth George; Stewart, Thomas George; Strobl, Rudolph S.; Tedder, Kenneth H.; Tilley, Barbara J.; Tilley, Barbara Jean; Timoney, Kevin Patrick; Tsui, Po Chow; Tsui, Po Chow; Vafidis, Antonios; Vernon-Chamberlain, Valerie Elaine; Vivian, Gary J.; Vossler, Shawna M.; Wagner, Paul Anthony; Walton, Lori A.; Wang, Xi-Shuo; Waters, Peter Mackenzie; Waywanko, Andrea O.; Wolberg, Andrew Charles; Xue, Xianyu; Zhang, Xiaomao

Alfred University
Alfred, NY 14802

2 Doctoral

1960s: Levitt, Stephen Robert; Reed, James Stalford

American University
Washington, DC 20016

29 Master's, 5 Doctoral

1920s: Howard, Charles Spaulding; Jarvis, Clarence Sylvester; Riffenburg, Harry Bucholz

1930s: Allred, David Hammond; Erdreich, Emil; Foster, Margaret D.; Olmstead, Lewis Bertie

1950s: Tu, Hsin-yuan

1960s: Anderson, Arthur H.; Berberian, George Assadour; Bromery, Randolph Wilson; Brower, James Clinton; Butler, Edward Taylor; Casey, Donald Joseph; Dennis, Leonard Stanley; Edsall, Douglas Wayne; Eisenhard, Robert M.; Gallagher, James Frederick; Hickling, Nelson Lawson; Hyers, Merlyn Eugene; Kephart, William W.; Lander, John French; McDonald, Sister Mary Aquin; McNay, Lewis Morris; Norton, Annette H.; Oman, Charles Lee; Ringle, John Edward; Stiles, Newell Thrift; Vincent, Samir Ambrose; Williamson, William Edward; Zubovic, Peter

1970s: Monroe, Frederick Fales; Robinson, Richard Dudley

1980s: Simon, Nancy Shoemaker

Amherst College
Amherst, MA 01002

1 Master's

1980s: Barton, Colleen

Andrews University
Berrien Springs, MI 49104

2 Master's

1960s: Lugenbeal, M. P.; Ritland, J. H.

University of Arizona
Tucson, AZ 85721

822 Master's, 424 Doctoral

1910s: Cochran, H. Merle; Joseph-Haakevitch, Phincas E.

1920s: Brown, Ronald LaBern; Bruhn, Henry H.; Feiss, Julian William; Gordon, Ernest Rollin; Heineman, Robert E. S.; Lausen, Carl; Park, Charles Frederick, Jr.; Reid, Robert R.; Shoemaker, Abbott Hall; Smith, Lewis A.; Somers, George; Webber, Benjamin N.; Wilson, Eldred Dewey; Woodell, Charles E.

1930s: Alberding, Herbert; Alexis, Carl O.; Bishop, Ottey M.; Borland, Gerald, C.; Cook, Frederic S.; Dunham, Montgomery S.; Eckel, Edwin B.; Enlows, Harold E.; Entwhistle, Lawson P.; Galbraith, Frederic W., III; Gebhardt, Rudolph Carl; Gillingham, Thomas E.; Hernon, Robert Mann; Hernon, Robert Mann; Higdon, Charles E.; Hill, G. F.;

Lausen, Carl; Lee, Charles A.; Legge, John A.; Ormsby, Walter B.; Ornsby, Walter B.; Peterson, Nels P.; Peterson, Nels P.; Rasor, Charles A.; Sears, David Hume; Simmons, Woodrow W.; Stewart, Lincoln A.; Thomas, Walter L.; Wilson, Clark L.

1940s: Ageton, Robert William; Alexis, Carl O.; Campbell, Donald F.; Cederstrom, D. John; Feth, John H.; Hague, William C.; Harshbarger, John W.; Harshbarger, John W.; Harshman, Elbert N.; Hogue, William C.; Hoiles, Randolph Gerald; Houser, Frederick N.; Johnson, Vard H.; Jones, William R.; Kartchnor, Wayne E.; Kiersch, George A.; Kuhn, Truman H.; Loring, William B.; Marvin, Thomas Crockett; Mayuga, Manuel Nieva; Mayuga, Manuel Nieva; Moore, David Lafayette; Peng, Chi-jui; Popoff, Constantine C.; Price, William E., Jr.; Robinson, Daniel O.; Shields, James C., II; Sopp, George P.; Wardell, Henry Russel; Whitcomb, Harold A.

1950s: Acker, Clement J.; Akol, Halim; Anderson, Roger Y.; Anthony, John W.; Bailey, James Stuart; Banerjee, Anil K.; Banherjee, Anil K.; Bejnar, Waldomere; Belden, William A.; Bennett, Paul J.; Bissett, David H.; Blissenback, Erich; Bollin, Edgar M.; Brennan, Daniel J.; Brimhall, Willis H.; Britt, Terence L.; Brittain, Richard L.; Bromfield, Calvin S.; Browne, Jonathan F.; Bryant, Donald L.; Bryant, Donald L.; Bryner, Leonid; Burnette, Charles R.; Callahan, Joseph Thomas; Chew, Randall T.; Clark, Jackson L.; Colby, Robert E.; Cooley, Maurice E.; Coulson, Otis B.; Davis, Robert E.; Dickerman, Robert W.; Donald, Peter G.; Emigh, George Donald; Evensen, Charles G.; Faick, John Nicholas; Fries, Carl, Jr.; Gray, Robert S.; Grundy, Wilbur D.; Havenor, Kay Charles; Heindl, Leopold A.; Heyman, Arthur M.; Hillebrand, James R.; Hook, Donald L.; Howell, Paul William; Hughes, Paul W.; Imsiler, James B.; Jackson, Robert L.; Jones, Richard D.; Kerns, John R.; Kidwai, Zamir U.; Kinnison, John E.; Kurtz, William L.; Layton, Donald W.; LeMone, David V.; Loring, William Bacheller; Lovering, Tom G.; Ludden, Raymond W.; Lutton, Richard J.; MacKallor, Jules A.; McClymonds, Neal E.; Merrin, Seymour; Michel, Fred A., Jr.; Mirsky, Arthur; Moore, Richard T.; Neumeier, Donald P.; Paick, John N.; Papke, Keith G.; Peirce, Frederick Lowell; Peirce, Howard Wesley; Plafker, Lloyd; Raabe, Robert G.; Ruff, Arthur W.; Saint Clair, Charles S.; Schwartz, Roland J.; Sears, David, II; Smith, George E.; Smith, Riley Semour, Jr.; Sulik, John F.; Titley, Spencer R.; Van Horn, William Lewis; Voelger, Klaus; Walker, Giles E.; Wallace, Robert Manning; Wallace, Roberts Manning; Wargo, Joseph G.; Wargo, Joseph George; Watson, Barry N.; Webb, Robert T.; Weber, Robert, II; Weider, Melvin I.; Weidner, Melvin I.; West, Samuel W.; Whitney, Richard Lee; Williams, George A.; Wilson, Robert Lee; Wood, Paul A.; Wood, William H.

1960s: Abdullatif, Abdullatif A.; Abu Ajamieh, M. M.; Adam, David P.; Agasie, John M.; Agenbroad, Larry D.; Agenbroad, Larry Delmar; Anderson, Frank J.; Arif, Abdul H.; Arnold, Leavitt Clark; Ascencious, Alejandro C.; Assadi, Seid Mohamad; Balla, John Coleman; Barclay, C. S.; Barozzi, Rolando; Barter, Charles F.; Beane, Richard E.; Benites, Lois A.; Bennett, John N., Jr.; Bennett, Paul Joseph; Bhatti, Sabir A.; Bhuyan, Ganesh Ch; Bikerman, Michael; Bikerman, Michael; Binder, Alan B.; Bock, Charles Mitchell; Bohrer, Vorsila Laurene; Braun, Eric R.; Braun, Gerald E.; Breed, William J.; Briedis, John; Briscoe, James A.; Broderick, Jon P.; Brown, J. W.; Bruan, Gerald E.; Budo, Shoro; Burroughs, Richard Lee; Butler, William C.; Cantwell, Richard J.; Cetinay, Huseyin Turgut; Chaffee, Maurice A.; Chaffee, Maurice Ahlborn; Chambers, Arthur Edwin; Champney, Richard D.; Chaudhri, Ata-ur-Rehman; Cherkener, Douglas S.; Ciancanelli, Eugene V.; Clarke, Craig W.; Clay, Donald W.; Collier, J. Maurice; Conrad, Robert D.; Cooley, Keith R.; Coombs, Stanley L.; Cordes, Edwin H.; Cornelius, Kenneth D.; Crossley, Robert W.; Cruikshank, Dale Paul; Cruikshank,

Dale Paul; Cunningham, John Edward; Daniel, Herbert R.; Davenport, Ronald E.; Davis, Richard Warren; Denney, Phillip P.; DeWilliam, Patrick P.; Dickinson, Robert G.; Dickinson, Robert Gerald; Diery, Hassen D.; Dirks, Thomas N.; Dubin, David Joel; Dunlap, Richard E.; Durand, Harvey S.; Durand, Harvey S.; Dyer, Kenneth Lee; Echavez, Joaquin; Elston, Donald Parker; Erickson, Rolfe C.; Erickson, Rolfe Craig; Estes, Wayne Shelton; Ettinger, Leonard J.; Evensen, James M.; Evensen, James Mitchell; Everts, James Mitchell; Fair, Charles Leroy; Feldman, Arlen D.; Fergusson, William Blake; Frazier, Robert H.; Ganus, William J.; Gardner, Murray Curtis; Garza, Sergio; Gass, Harold; Gerrard, Thomas A.; Gilani, Maqsood Ali Shah; Gilman, Chandler R.; Gordon, Yoram; Gottesfeld, Allen S.; Grandi, Rolando Barozzi; Gray, Irving Bernard; Gray, Robert Stephen; Graybeal, Frederick Turner; Green, William D.; Greenstein, Gerald; Griswold, George Bullard; Gross, Michael P.; Gumble, Gordon E.; Halva, Carroll J.; Hammel, David J.; Hammer, Donald F.; Handverger, Paul A.; Hanson, Hiram Stanley; Harbour, Jerry; Hardas, Avinash Vishnu; Hartmann, William K.; Hatheway, Allen W.; Haxby, Ronald L.; Haynes, Caleb Vance, Jr.; Heatwole, David A.; Hedge, Carl E.; Hench, Stephen Wayne; Hevley, Richard Holmes; Hickey, John J.; Hite, John B.; Hoffman, Victor J.; Hoffman, Victor Joseph; Holliday, Este F.; Horton, John W.; Hostetter, Heber P., III; Hoyt, John W.; Huntoon, Peter W.; Ibrahim, Mohamed S.; Iles, Calvert D.; Ishag, Abudulla Bassan; Iskander, Wilson; Israelsen, C. Earl; Jemmett, Joe Paul; Jenney, William Willis, Jr.; Jinks, Jimmie E.; Johnson, Allen Harold; Jones, Neil Owen; Jorden, Roger M.; Karim, M. A.; Karkanis, Basily George; Kawar, Kamel A.; Kelley, Richard J., Jr.; Kennedy, Richard Ray; Kilbourne, Deane Earl; Kirkland, Larry A.; Klingmueller, Lothar M. L.; Kothavala, Rustam Zal; Kuck, Peter H.; Laidley, Richard Allan; Lang, William J.; Laughlin, A. William; Laughlin, Alexander William; Lee Moreno, Jose; Lee, Jose M.; Lee, L. Courtland; Leger, Arthur R.; Lehman, Norman E.; Lehr, Jay H.; Little, William; Livingston, Donald E.; Livingston, Donald Everett; Long, Austin; Lootens, Douglas Joseph; Lovejoy, Earl Mark Earl; Luepke, Gretchen; Lutton, Richard Joseph; Lyford, Forest Parsons; Lynch, Dean W.; Maddox, George E.; Maddox, George Edward; Marlowe, James I.; Mat, Bruce T.; Mathewson, Christopher C.; Mathias, William F.; Matlock, William Gerald; Matter, Philip, III; Mauger, Richard Leroy; Mawla, Haulam; May, Bruce Tipton; McColly, Robert A.; McCoy, Scott, Jr.; McCullough, Edgar Joseph, Jr.; McKenna, John J.; McLain, John P.; Mehdi, Purnendu K.; Mehringer, Peter J., Jr.; Merz, Joy J.; Metz, Robert; Meyer, William; Mickle, David Grant; Micklin, Richard F.; Mielke, James Edward; Miles, Charles Hammond; Miller, James B.; Miller, Wade E.; Min, Maung Myo; Moench, Allen Forbes; Montgomery, Errol Lee; Monzon, Felipe G.; Moore, Robert A.; Morin, George C.; Mullens, Rockne Lyle; Myhrman, Matts A.; Nasseredin, M. T.; Nations, Jack Dale; Nelson, Eleanor; Nelson, Frank J.; Nowatzki, Edward Alexander; Nye, Thomas Spencer; Olson, Harry J.; Olson, Harry J.; Osborne, Paul S.; Pantoja Alor, Jerjes; Pashley, Emil Frederick, Jr.; Passmore, Virginia L.; Patch, Susan; Payne, Charles Marshall; Peabody, David M.; Percious, Donald Joseph; Percious, Judith K.; Perry, David V.; Perry, David Vinson; Peterson, Dennis E.; Peterson, Richard C.; Peterson, Richard Charles; Pfirman, Richard S.; Phiancharoen, Charoen; Pilkington, Harold Dean; Pine, Gordon L.; Pine, Gordon Leroy; Pipkin, Bernard Wallace; Platt, Wallace S.; Plut, Frederick W.; Purdom, William Berlin; Quraishi, Raziuddin; Ralston, Dale R.; Ramirez, Jose R.; Ratte, Charles Arthur; Rea, David K.; Reed, Jack C.; Reed, Richard K.; Reetz, Gene Rene; Rehrig, William Allen; Reid, Alastair Milne, II; Reid, Alastair Milne, II; Riley, James J.; Roadifer, Jack Ellsworth; Robinson, Richard C.; Rohrbacker, Robert G.; Romans, Robert Charles; Rubalcaba, Jose Ramirez; Sabels, Bruno Erich; Sables, Bruno Erich; Saeed, El Tayeb M.; Salem, Mohamed Halim; Samii, Cyrus; Sandoval, Mario; Sandusky, Clinton Leroy; Sariahmed, Abdelwaheb; Sauck, William August; Schafroth, Don Wallace; Schmidt, Eberhard A.; Schmidt, Kenneth D.; Schwab, Karl W.; Seager, William Ralph; See, J. Melville; Seff, Philip; Selinger, Keith A.; Sell, James D.; Sharp, James E.; Sharp, Robert R., Jr.; Sheikh, Abdul Mannan; Sheikh, Izaz A.; Shride, Andrew Fletcher; Sigalove, Joel J.; Smith, David G.; Smith, Walter J.; Smoor, Peter Bernard; Spangler, Daniel Patrick; Staples, Bruce Allen; Stein, Walter T.; Streits, Robert; Studebaker, Irving G.; Tahmazian, Garabed A.; Taylor, Omer James; Thacpaw, Saw Clarence; Thomsen, Bert W.; Udo, Eno J.; Waag, Charles Joseph; Wahab, Osman Abdel; Waller, Harold E., Jr.; Waller, Roger M.; Walton, Wayne J. A., Jr.; Wanab, Osman Abdel; Warnock, George F.; Watson, Barry Norton; Weaver, Richard; Wells, A. J.; Werrell, William L.; Wertz, Jacques Bernard; Whitacre, Halford E.; White, John S., Jr.; Whittier, Donald A.; Williams, Sidney Arthur; Wilson, Clyde A.; Wilt, Jan C.; Winters, Harry J., Jr.; Wood, Michael M.; Wood, Michael Manning; Wood, Paul Alan; Wright, Jerome J.; Yildiz, Mehmet; Zavodni, Zavis

1970s: Abdullatif, A. A.; Abdulrazzak, M. J.; Aberra, G. B.; Abu-Obeid, Hamid A.; Abu-Taha, Mohammad F.; Adam, David Peter; Aguilar-Maldonado, A.; Ahrens, Gary Louis; Aiken, C. L. V.; Ajayi, O.; Al-Eryani, Mohamed Lotf; Al-Hadithi, A. H.; Al-Hadithi, A. H.; Aldovino, Lino Pineda; Allen, C. C.; Ames, Martha H.; Anderson, S. R.; Applebaum, S.; Arenson, John Dean; Arnold, L. D.; Arnold, Leavitt Clark, Jr.; Baldwin, Evelyn Joan; Ball, Andrew David; Balla, John C.; Bandurski, E. L.; Bangsainoi, S.; Barrie, K. A.; Bartos, Frances Maribel; Baskin, J. A.; Bejnar, Craig Russel; Belan, Ricky Allen; Bennett, Catheryn MacDonald; Bennett, K. C.; Berlanga-Galindo, E. R.; Beschta, R. L.; Bittson, Andrew George; Bladh, Katherine Laing; Bladh, Katherine Liana; Bladh, Kenneth Walter; Bladh, Kennith Walter; Blake, David William; Bodnar, R. J.; Bokhari, S. M. H.; Bolin, David Samuel; Bond, Frederick William; Boster, Mark Alan; Boster, Mark Alan; Boster, Ronald Stephen; Bostick, Kent Anthony; Bostock, C. A.; Boyd, Daniel T.; Boyer, D. G.; Boyer, Jeffrey Alan; Bradbeer, G. E.; Bradfish, Larry James; Bressler, S. L.; Brook, Doyle Kenneth, Jr.; Brown, Calvin C.; Brown, T. C., Jr.; Bryant, Jeffrey W.; Budden, R. T.; Butler, William Charles; Calderon-Riveroll, Gustavo; Call, Richard D.; Cameron, Donald Eugene; Campana, M. E.; Campana, M. E.; Cannon, Philip Jan; Capuano, Regina M.; Carlson, F. R.; Carrigan, Francis J.; Carroll, B. J.; Chaemsaithong, Kanchit; Chakarun, John Douglas; Champney, Richard Daniel; Chen, Y.; Choate, M. L.; Christie, Fritz Jay; Christman, Jerry L.; Clausen, George Samuel; Clay, Donald Wayne; Cleveland, Gaylord; Clyma, Wayne; Cole, R. C., Jr.; Coleman, Dennis D.; Colony, Wayne Edward; Conkey, Laura Elizabeth; Cook, E. R.; Corbett, Ronald K.; Craig, Richard Michael; Crowl, William James; Crump, Terry Richard; Cummings, Robert Adams; Curtin, George; Curtis, C. S.; Danielito, Tan Franco; Davis, Jerry Dean; Davis, Jerry Dean; de la Fuente Duch, M. F. F.; De la Fuente Duch, Mauricio Fernando; Dean, D. A.; DeCook, Kenneth James; Delaney, J. R.; Dewhurst, JoAnna; DeWitt, Ed; Dodge, Constance Nuss; Dolloff, Mary Helen; Douglas, A. V.; Douglas, Arthur Vern; Dove, Floyd Harvey; Dreier, J. E., Jr.; Duffield, C.; Duffy, D. M.; Dumeyer, John McMurray; Dunn, J. L.; Durning, William Perry; Duvick, Daniel Nelson; Earl, Thomas Alexander; Eastwood, Raymond Lester; Edmiston, Robert; Edson, G. M.; Edwards, Larry John; Eggleton, Richard Elton; Eliopulos, George J.; Ellis, R. B.; Engel, Michael Harris; Erwin, J. W.; Evans, Kenneth; Evans, Thomas J.; Everett, Lorne Gordon; Everett, Wayne Leonard; Farias-Garcia, Ramon; Fellows, Michael Lewis; Fernandez, J. A.; Figueroa, Julio Aguiles Pastor; Filho, J. M.; Fischer, J. N.; Flynn, Lawrence John; Foerster, Eugene Paul; Fogg, Graham Edwin, Jr.; Fouts, James A.; Franco, D. T.; Frank, Thomas Russell; Fredericksen, Rick Stewart; Frondorf, A. F.; Frost, Eric George; Gallaher, Bruce Morris; Galvez, J. A.; Ganus, William Joseph; Gass, T. E.; Gates, Joseph Spencer; Gaytan Rueda, Jose E.; Gentry, Donald W.; Gerlach, Terrence Melvin; Gisler, Patrick Michael; Goodoff, L. R.; Gordon, Yoram; Gottschalk, Richard Robert; Gould, J.; Graybeal, Frederick T.; Greenwood, David E.; Griepentrog, Thomas E.; Grimm, Joan P.; Grinshpan, Zvi; Guitron de los Reyes, A.; Guma'a, G. S.; Hackman, David B.; Hall, Denis Kane; Hammel, David J.; Hampf, Andrew W.; Hampton, N. F.; Harding, Lucy E.; Hargan, Bruce Alan; Hargis, D. R.; Harlan, Howard Marshall, II; Harlan, Julie de Azevedo; Harrison, Jessica A.; Hastings, David J.; Hathaway, Allen Wayne; Hazenbush, George Cordery; Hazzaa, Abdullah F.; Herzog, R. H., Jr.; Hillman, H. F.; Himes, Marshall D.; Hirt, William Carl; Hoelle, John Lowell; Holcomb, R. T.; Honey, James Gilbert; Howell, K. K.; Hsieh, Paul Anthony; Hulburt, Margery Ann; Hulse, Scott E.; Huntoon, Peter Wesley; Iberall, Eleanora R.; Ijirigho, B. T.; Ingersoll, D. L.; Irwin, Thomas D.; Jacobs, L. L., III; Jacobs, Louis Leo, III; Jacobsen, W. L.; Janbek, Tayseer T.; Jarroud, O. A.; Jensen, Louis; Johnston, I. M.; Jones, N. O.; Jones, P. L.; Jones, Richard Edwin; Jones, Rollin C.; Jones, William C.; Judge, Robert Michael; Kalthem, M. S.; Kao, Samuel E.; Kendorski, Francis S., III; Kessler, Edward Joseph; Kimball, D. B.; King, James Edward; Kiven, Charles Wilkinson; Klingmueller, Lothar Max Ludwig; Knapp, R.; Knapp, R. B.; Knight, Jerry Eugene; Knight, Louis Harold, Jr.; Knight, Louis Harold, Jr.; Koenig, Brian A.; Kreamer, D. K.; Kreis, Henry G.; Kresan, P. L.; Krewedl, Dieter Anton; Krzysztofowicz, Roman; Kuck, P. H.; Kunen, S. M.; Kunkel, James Robert; Kurupakorn, Somchai; Kypfer, Marvin Douglas; Ladd, T. W.; Laine, R. P.; Lammers, George Eber; Lane, Douglas L.; Laney, Robert Lee; Lange, Nixon Richard; Langlois, Joseph David; Larson, Peter Brennan; Laughon, Robert Bush; Layton, D. W.; Lee, Jose M.; Lee, M., Jose; Lehman, N. E.; Leonhart, L.; Lessman, James Lamont; Leventhal, Joel Stephen; Liming, Richard Brett; Liongson, Leonardo Quesada; Lipinski, Paul William; Lodewick, Richard Ballard; Loghry, James D.; Loidolt, Lawrence H.; Lovely, C. J.; Lowenfels, Harold Stuart; Lynch, Daniel James, II; Lynn, C. G.; Lytle, Jamie Laverne; Lytle-Webb, Jamie; Marjaniemi, Darwin Keith; Marsh, Bruce David; Marshak, R. Stephen; Masri, Fahad Isa; Massanat, Yousef M.; Massanat, Yousif M.; Mathewson, Christopher C.; Mathez, Edmond Albigese; Matis, John P.; Mayer, Larry; Mburu, S. G.; McCauley, Charles Anthony; McClure, R. K.; McDonald, John Harlan; McFadden, Leslie D.; Meader, N. M.; Meader, Sally Jo; Meschede, Louis Henry; Miller, Stanley Mark; Mills, W. C.; Mohon, J. P., Jr.; Mokhtar, Talal Ali; Montgomery, Errol Lee; Moores, Richard C., II; Morris, Marvin; Morse, J. G.; Mortimer, Robert E.; Mosesso, Michael Angelo; Moskowitz, B. M.; Mossesso, Michael Angelo; Muller, A. B.; Naseer, S. A.; Newsom, Horton E.; Nibler, Gerald John; Nicholas, D. E.; Nielson, Roger L.; O'Neil, Thomas J.; O'Rourke, Mary Kay; Oben-Nyarko, K.; Oben-Nyarko, K.; Oropeza, Romolo Marquez; Osterkamp, W. R.; Osterkamp, Waite R.; Packard, F. A.; Palacios-Velez, Enrique; Parker, Robert W.; Peebles, R. W.; Perez, A. V.; Peterson, Robert Howard; Phanartzis, Christos Apostolou; Phillips, A. M., III; Phillips, Fred Melville; Phillips, Mark Paul; Polanshek, D. H.; Popkin, Barney Paul; Potter, Steven C.; Potucek, Tony Lee; Powell, John S.; Preece, Richard Kellar, III; Price, William Evans; Puckett, James Carl, Jr.; Purves, W. J.; Qabazard, F. A.; Qahwash, Abdellatif Ahmad; Qahwash, Abdullatif A.; Quick, Jay Dudley; Ramirez Munoz, Jose; Randall, J. H.; Ratananaka, C.; Ray, Leon Nicholas; Readdy, Leigh A.; Rebuck, Ernest Charles; Reid, S. A.; Reynolds, S. J.; Richardson, George L.; Rico, B. S.; Robinson, Donald James; Robison, Brad Alan; Robotham, Hugh Beresford; Roe, Robert Ralph; Rogers, James Joseph; Rogers, James Joseph; Rogers, R. D.; Rose, M. W.; Roybal, Gretchen Hoffman; Rueda, J. E. G.; Rushing, Jodi A.; Ruth, Joseph Frank; Sagar, Budhi; Sammis, Theodore Wallace; Santillan Cruz, V. H.; Saplaco, S. R.; Sauck, William August; Saunders, Jeffrey J.; Saunders, Jeffrey John; Savely, James P.; Scarborough, Robert Bryan; Schenker, Albert Rudolph, Jr.; Schloderer, John Peter; Schmidt, Eberhard Adalbert; Schmidt, Eugene Karl; Schmidt, James Scott; Schmidt, Kenneth Dale; Schofield, Richard Edward; Schosinsky, G. E.; Schreuder, P. J.; Schuiling, William T.; Schultz, T. R.; Sebenik, Paul Gregory; Shaffer, Marvin James; Shaffer, Marvin James; Shakel, Douglas Wilson; Sharp, Robert R., Jr.; Sherwonit, William Edward; Sigleo, A. M. C.; Sill, G. T.; Sklarew, D. S.; Skrivan, J. A.; Slawson, Guenton Cyril, Jr.; Small, G. G.; Smith, Roger Stanley Uhr; Smith, Verl Leon; Sniedovich, Moshe; Snow, Lester A.; Solano-Rico, Baltazar; Soule, Charles H.; South, David Long; Spatz, David Moore; Spaulding, Walter Geoffrey; Speer, W. E.; Spiro, David Alan; Stephens, D. B.; Sternberg, Robert S.; Stewart, James Conrad; Stockton, Charles Wayne; Stoeck, Penelope L.; Studebaker, Irving G.; Sublette, William R.; Suemnicht, G. A.; Sulkoske, William C.; Supkow, Donald James; Surles, Terri L.; Swan, Monte Morgan; Tanenbaum, Ronald J.; Tarman, D. W.; Taylor, Johnnie D.; Taylor, Louis Henry; Thayer, David W.; Thompson, Robert S.; Thorgrimsson, S.; Thorne, P.; Tomida, Yukimitsu; Totten, David K.; Touqan, Omar I.; Trent, Dee Dexter; Trevena, A. S.; Turjoman, A. M.; Uhl, V. W., Jr.; Ullmer, Edwin; Van Blaricom, Richard; Van Devender, Thomas Roger; Van Fleet, John M.; Varga, Robert Joseph; Vasquez Perez, Adalberto; Vidal, Jose Rabasso; Vuich, John S.; Wagner, Charles Gregory; Wahab, Nasar Ahamad; Wahab, Osman Abdel; Walker, Valerie Ann; Wallick, Edward Israel; Webb, William Felton; Weisz, Reuben Nathan; Weitz, Thomas James; Welber, P. W.; Weller, Roger Nelson; Wendt, C. J.; West, Robert E.; West, Robert Elmer; Weston, Loren Kinsman; Wheeler, Merlin Leroy; White, Daniel Howard; White, Robert Rankin; White, Robin Shepard; Wicke, Heather Dawn; Wilkins, Joe; Wilson, John Allen; Wilson, John Randall; Winograd, Issac Judah; Wiseman, Frederick Mathew; Wiseman, Frederick Matthew; Witter, Donald Paul, Jr.; Wojniak, Wayne Stanley; Wood, Charles A.; Wu, S. S.-C.; Wyman, Richard Vaughn; Wynn, Jeffrey Curran; Zauderer, Jeffrey; Zavodni, Zavis Marian; Zerrahn, Gregory Joseph; Zumberge, J. E.; Zumberge, John Edward

1980s: Abe, Joseph Michael; Abedrabboh, Walid; Aboaziza, Abdelaziz Hassan; Adar, Eilon; Ajayi, Owolabi; Akman, Hulya Hayriye; Alawi, Adnan Jassim; Alfi, Abdulaziz Adnan Sharif; Ali, Molla Mohammad; Anderson, Phillip; Anderson, Rodney Scott; Anthony, Elizabeth Youngblood; Archibald, Lawrence Eben; Arias, Rojo Hector Manuel; Armin, Richard Alan; Ashouri, Ali Reza; Asmerom, Yemane; Azevedo, Luiz Otavio Roffe; Baafi, Ernest Yaw; Balcer, Richard Allen; Ballard, Stanton Neal; Bartolini, Claudio; Beard, Linda Sue; Benson, Gregory Scott; Binsariti, Abdalla Abdurazig; Boyd, John Whitney, IV; Brakenridge, George Robert; Brannon, Charles Andrew; Brikowski, Tom Harry; Butler, Edwin Farnham, Jr.; Bykerk-Kauffman, Ann; Calderone, Gary Jude; Capuano, Regina Marie; Carr, James Russell; Carr, James Russell; Carrera-Ramirez, Jesus; Carrigan, Francis John; Catlin, Steven Allen; Cervantes Silva, Juan Jose; Cervantes-Montoya, Jesus Alberto; Chadwick, Oliver Austin; Chavez Martinez, Mario Luis; Chavez Rodriguez, Adolfo; Chen, Hsien Wu; Cheng, Song-Lin; Clarke, Michael; Cleaveland, Malcolm Kent; Cobb, Steven Lloyd; Cochran, Ann; Cole, Kenneth Lee; Conkey, Laura Elizabeth; Cox, Billie Lea; Crespi, Jean Marie; Cropper, John Philip; Crumpler, Larry Steven; Cuevas Leree, Juan Antonio; Cunningham, Cindy Carolyn; Currier, Debra Ann; Davis, Stanley Graham; de Sousa Carvalho Martins, Verónica E.; di Tullio, Lee Dolores; Dietrich, Donald R.; Dietz, David Delbert; Douglas, Dean Alan; Drumm, Eric Corman; Eftekharzadeh, Shahriar; El Didy, Sherif Mohamed Ahmed; El Ghonemy, Hamdi Mohamed Riad; El-Haris, Mamdouh Khamis; Enders, Merritt Stephen; Engel, Michael Harris; Erjavec, James Laurence; Ethington, Edgar Francis; Ethridge, Loch Lee; Fabryka-Martin, June Taylor; Fall, Patricia Lynn; Faruque, M. O.; Ferguson, Robert Clark; Fischer, Anne Marie; Fishman, Kenneth Lawrence; Flaccus, Christopher Edward; Flanagan, Peter William; Flynn, Lawrence John; Fowler, Linda Leigh; Fuenkajorn, Kittitep; Galagoda, Herath Mahinda; Gardulski, Anne Frances; Gerla, Philip Joseph; Giudice, Philip Michael; Goodlin, Thomas Charles; Gordon, Julia Perry; Grajales-Nishimura, Jose Manuel; Gray, Matthew Dean; Green, Ronald Thomas; Grover, Jeffrey Alan; Gruszka, Thomas Peter; Gustin, Mae Sexauer; Hackman, David Brent; Hall, Dwight Lyman; Halvorson, Phyllis Heather Fett; Hambrick, Dixie Ann; Harding, Lucy Elizabeth; Harms, Tekla Ann; Harris, Jonathan O.; Haskin, Richard Allen; Hauck, Wayne Russell; Haverland, Raymond Louis; Haynes, Frederick Mitchell; Heichel, Kimberlee Sue; Heller, Paul Lewis; Hennessy, Joe Allen; Hess, Alison Anne; Hirschboeck, Katherine Kristin; Horvath, Emilio Hubert; Hostetler, Charles James; Hsieh, Paul Anthony; Huston, David Lowell; Ijirigho, Bruce Tajinere; Jacobs, Bonnie Fine; Jacobson, Elizabeth Ann; Jeffrey, Robert Graham, Jr.; Jeffrey, Robert Graham, Jr.; Johansen, Steven John; Johnson, James Wesley; Johnson, Lawrence Clinton; Jones, Jay W.; Joseph, Nancy Lee; Kaehler, Charles Alfred; Kanbergs, Karlis; Kanschat, Katherine Ann; Karnieli, Arnon; Kaufmann, Ronald Steven; Kistner, David John; Kluth, Charles Frederick; Knudsen, Harvey Peter; Knuepfer, Peter Louis Kruger; Koterba, Michael Taylor; Krantz, Robert Warren; Krantz, Robert Warren; Kreamer, David Kenneth; Lanier, William Paul; Lantz, Rik Earl; Lapham, Wayne Wright; Lawton, Timothy Frost; Leake, Martha Alan; Leavitt, Steven Warren; Lee, Han Yeang; Lee, Han Yeang; Lepry, Louis Anthony, Jr.; LeVeque, Richard Alan; Levine, Steven Joel; Lingrey, Steven Howard; Long, Keith Richard; Lopes, Vincente Lucio; Lynch, Daniel James, II; Lysonski, Joseph Charles; MacInnes, Scott Charles; Manske, Scott Lee; Mark, Roger Alan; Marozas, Dianne Catherine; Marozas, Dianne Catherine; Marrin, Donn Louis; Maus, Daniel Albert; May, Steven Robert; Mayer, Larry; Mazaris, George Michael; McAlaster, Penelope; McCord, Virgil Alexander Stuart; McFadden, Leslie David; McHargue, Lanny Ray; McNew, Gregory E.; Mead, Jim I.; Meko, David Michael; Menges, Christopher Martin; Merritts, Dorothy Jane; Meyer, Jeffrey Wayne; Monreal, Rogelio; Mooradian, Michael Minas; More, Syver Wakeman; Morehouse, Jeffrey Allen; Murphy, Ellyn Margaret; Myers, Bruce Eric; Myers, Genne Marie; Nagaraj, Benamanahalli Kempegowda; Nagi, David Michael; Naruk, Stephen John; Newsom, Horton Elwood; Nichols, Elizabeth Ann; Nnaji, Soronadi; Norris, James Richard; O'Rourke, Mary Kay; Oppenheimer, Joan Mary; Ottoni, Theophilo Benedicto Filho; Pardieck, Daniel Lee; Payawal, Pacifico Cruz; Phillips, Fred Melville; Piekenbrock, Joseph Robert; Pietenpol, David John; Piniazkiewicz, Robert J.; Plouff, Michael Thomas; Porto, Everaldo Rocha; Puls, Robert William; Quade, Jay; Rains, George Edward; Randall, Jeffery Hunt; Rasmussen, Todd Christian; Rauschkolb, Michael Howard; Reding, Lynn Marie; Reynolds, Stephen James; Reynolds, Theodore James; Richard, Stephen Miller; Riggs, Nancy Rosalind; Rogers, Ralph D.; Rooke, Steven; Rose, Seth Edward; Roth, Frances Ann; Roth, Robert Leroy; Rutledge, James Thomas; Ryberg, Paul Thomas; Salami, Mohammad Reza; Samper Calvete, Francisco Javier; Savely, James Palmer; Schneider, Nicholas McCord; Scotese, Thomas Richard; Seaman, Sheila June; Sene, Siki; Shaner, Linda Ann; Shih, Ernest Hsiao Hsin; Siegel, Randall David; Silliman, Stephen Edward Joseph; Silver, Douglas Balfour; Silverston, Elliot; Smalley, Richard Curtis; Smith,

Robert Cameron; Smith, William Kerrison; Somasundarum, Sujithan; Sousa, Francis Xavier; South, David Long; Spaulding, Walter Geoffrey; Spencer, John Robert; Stafford, Thomas Wier, Jr.; Steadman, David William; Sternberg, Robert Saul; Stevenson, Frederick; Stewart, Daniel Russell; Stoneman, Deborah Ann; Strauss, Robert C.; Sullivan, Jeffery Alan; Swift, Peter Norton; Sylvia, Dennis Ashton; Tavassoli, Abolghasem; Taylor, Louis Henry; Thompson, Randolph Charles; Thompson, Robert Stephen; Thorn, Kevin Arthur; Tinney, James Craig; Tomida, Yukimitsu; Toufigh, Mohammad Mohsen; Trembly, Jeffrey Allen; Trever, Paula Fern; Tsuji, Karl Sei; Turner, Kent, Jr.; Usunoff, Eduardo Jorge; Uytana, Veronica Feliciano; Vaag, Myra Kathleen; van Nieuwenhuyse, Ulrich Eric; Vandemoer, Catherine; Vedder, Laurel Kathleen; Vega, Luis Alfonso; Wacker, John Frederick; Walchessen, Anne Aurelia; Walker, Scott Donald; Walter, Gary Robert; Warner, Julian Dean; Waters, Michael Richard; Waters, Michael Richard; Watson, Richard Clovis; Webb, Robert Howard; Weibel, William Lee; White, John Lester; Wilder, Nicea Trindade; Wilkinson, William Holbrook, Jr.; Williams, Michael Lloyd; Williams, Paul; Wilson, Kim Suzanne; Winker, Charles David; Wohl, Ellen Eva; Wortman, Ann Aber; Wright, Samuel Alexander; Wust, Stephen Louis; Yarter, William Vernon; Yates, Scott Raymond; Yeats, Kenneth James; Young, David Paul; Zadina, William Louis; Zaman, M. M.; Zeihen, Gregory D.

Arizona State University
Tempe, AZ 85287

175 Master's, 71 Doctoral

1960s: Akers, J. P.; Avedisian, Gary Edward; Daggett, Larry Leon; Deal, Edmond Graham; Fodor, Ronald Victor; Gibson, Everett Kay, Jr.; Haimovitz, Allan; Hall, Kenneth McCoy; Hoyer, Marcus Conrad; Lowery, Carol Janette; Nava, David Francis; Newton, George Denson; Pederson, Edward Peter; Powell, Richard Conger; Stuckless, John Shearing; Tilford, Norman Ross; Updike, Randall Groves; Yates, Ann Marie

1970s: Agurkis, Edward N.; Anderson, Susan Leslie; Aylor, Joseph Garnett, Jr.; Bahlburg, William C.; Batchelder, George Lewis; Beaulieu, Patrick Leo; Bell, Elaine J.; Bell, John W.; Blazey, Edward Brice; Bliss, James D.; Brown, Ronald G.; Cameron, Suzanne P.; Christenson, Gary E.; Clary, Thomas A.; Clayton, Nancy Ann; Cloran, Courtenay A.; Conyers, Lawrence B.; Coons, William E., III; Cordy, Gail E.; Cousins, Noel Boyd; Cripe, Jerry Dale; Cripe, Jerry Dale; Daniel, Debra L.; Deslauriers, Edward Charles; Elvidge, Christopher David; Fielden, John R., III; Gordon, Arthur J.; Gray, Gary W.; Hillier, Mark R.; Holway, Jeffrey V.; Hsu, S. I.; Janders, David J.; Jennings, Mark D.; Johnpeer, Gary D.; Johnson, Steven B.; Kayler, Kyle L.; Kelsey, Graham Landers; Klosterman, Michael Joseph; Knock, Kent K.; Kokalis, Peter George; Lausten, Charles Dean; Lee, Gaylon Keith; Lewis, M. A.; Lin, H. P.; Little, Lynne A.; London, David; Maisano, Marilyn Dew; Malone, Gary Bruce; Matlick, Joseph S., III; McCurry, Wilson G.; Merrill, Robert Kimball; Merrill, Robert Kimball; Messenger, Jane A.; Norby, Rodney Dale; Peters, Dusty; Petersen, Lee Edward; Petersen, Lee Edward; Pickens, Caroline M.; Pope, Clifton Washington, Jr.; Racey, Jan Stewart; Reager, Richard David; Reese, Robert Lester; Reif, Douglas M.; Robertson, Frederick N.; Schern, Edward Michael; Schulten, Cathy S.; Sedgeley, David R.; Shank, Douglas Carl; Simmons, Richard G.; Skyllingstad, Paul E.; Suneson, Neil H.; Sykes, Martha Lynn; Tellier, Anthony Harrison; Theilig, Elizabeth Eilene; Throop, Allen H.; Traut, Marc W.; Turek, Frank; Turner, Shirley; Updike, Randall Groves; Van Kooten, Gerald K.; Vice, Daniel Hoy; Wadell, James Sanders; Wagner, Robert David; Walter, David R.; Ward, Michael B.; Wellendorf, William George; Welsch, Dennis G.; Wermiel, Dan E.;

Westrich, H. R.; Williams, H. E.; Wilson, Doyle C.; Wilson, Robert Terry; Wilson, Wendell Eugene, Jr.; Wohletz, Kenneth Harold; Worley, Paul Lawrence Hill

1980s: Abrahams, Jennifer Anne; Aden, Gary David; Albin, Edward Francis, Jr.; Amalfi, Frederick Anthony; Annis, David Robert; Bales, James; Barnett, Albert Prinnon; Beeler, Nick M.; Beeunas, Mark Anthony; Benson, Christopher Joseph; Bertka, Constance M.; Bikun, James V.; Blount, Howard Grady, II; Borg, Scott Gerald; Borg, Scott Gerald; Bougan, Shelley June; Bowers, Mark Thomas; Brady, Steven C.; Brittingham, Peter Lane; Bruck, Glenn Ralph; Burton, James Hutson; Canepa, Julie Ann; Capobianco, Christopher John; Caporuscio, Florie Andre; Carpenter, F. Owen; Charania, Equbalali Hassanali; Christiansen, Eric H.; Chung, Ming-Ping; Connolly, James Alexander; Copp, John Frederick; Correa, Brian Paul; Couch, Nathan Pierce, Jr.; Craddock, Robert Anthony; Davis, Gene Michael; Doorn, Stacy Seaman; Dunn, Dennis P.; Earl, Richard Allen; Eby, Henry Eckert; Eppler, Dean Bener; Esperanca, Sonia; Fan, Gary Guoyou; Feldman, Mark David; Ferguson, Kurt Mathew; Frank, Mark Steven; Greer, John Craig; Haimson, Marshall; Hargrove, Howard Ralph; Harmala, John Clifford; Heintz, Greta Marie; Hennessy, Joel; Herpfer, Marc Andreas; Horn, Marty; Hoyos-Patino, Fabian; Hsu, Tung; Hyers, Albert D.; Isagholian, Varush; Jagiello, Keith James; Jakobsson, Sigurdur; Jordan, Mark Steven; Kafura, Craig John; Katrinak, Karen Ann; Kenny, Ray; Kilbey, Thomas Ryan; Kite, William McDougall; Klobcar, Cheryl Louise; Komorowski, Jean-Christophe; Konstanty, Kevin Michael; Kortemeier, Curtis Paul; Kortemeier, Winifred Talbert; Lalko, Lynn-Edward; Lapham, Kathryn Elizabeth; Larson, Michael Kenneth; London, David; Lowry, Patrick H.; Manera, Paul Allen; Manley, Curtis Robert; Martel, Linda Marie Viglienzone; McCormick, Tamsin Cordner; McEwen, Alfred Sherman; McMillan, Paul Francis; Miller, Elizabeth Jean; Mohammadi, Hossein K.; Montz, Melissa Jean; Moyer, Thomas Carl; Moyer, Thomas Carl; Neder, Reinhard Bernhard Wilhelm; Neet, Kerrie Elise; O'Hara, Patrick Francis; Park, Debra Ann; Patera, Edward Smyth, Jr.; Peterfreund, Alan Richard; Post, Jeffrey Edward; Prowell, Sarah Eddings; Rask, James Harold; Reiter, Bruce E.; Rettenmaier, Karl Albert; Rhoads, Bruce Lane; Robertson, Daniel Edward; Robertson, John Andrew; Ross, Nancy Lee; Rowan, Dana E.; Satkin, Richard L.; Schmitz, Christopher; Schultz, Richard Allen; Schuver, Henry John; Shane, John Denis, Sr.; Smit, Johannes Hendricus; Sommer, Jesse Valentine; Spinnler, Gerard Eugene; Spudis, Paul D.; Stewart, Gregory Lee; Strait, Melissa Marie; Streit, Lori Ann; Stuart, John Alexander; Sykes, Martha Lynn; Teshima, Janet Marie; Tesoriero, Anthony John; Theilig, Elizabeth Eilene; Thompson, Mark Edward; Thorleifson, James Tracy; Thorpe, Douglas G.; Turner, Shirley; Utley, Kenneth W.; Voss-Roberts, Kevin David; Wahl, David Edwin, Jr.; Ward, Jerome Vincent; Webster, Elizabeth Archer; Webster, James Dale; Williams, James Dana; Williams, Steven Hamilton; Winter, Bryce; Wohletz, Kenneth Harold; Zimbelman, James Ray

University of Arkansas, Fayetteville
Fayetteville, AR 72701

318 Master's, 29 Doctoral

1930s: Payne, James Norman

1940s: Gosnell, George Edward

1950s: Al-Refai, Badir Hashim; Arnold, Joseph Jenk, Jr.; Ault, Ruey A.; Badir, Hashim Al-Refai; Baxter, James W.; Brown, Harry L.; Cammack, James; Case, James E.; Casey, John Eagle; Cavallas, Joe F.; Clements, Jake E., Jr.; Cole, Marshall F.; Connelly, Francis B.; Cook, Vance Oliver; Davis, James L.; DeJarnett, Presley J.; Downs, A. Fred; Downs, Harry F.; Eby, Thomas J., Jr.; Edmonson, Park Dale; Edwards, Dan Cabe; Flocks, Gerald

Walter; Freeman, Thomas, Jr.; Haddad, Richard; Hensley, Perry John; Herron, Ellis Doyle; Hille, Oscar Roy; Hollyfield, Charles E.; Hopkins, M. E.; Hudson, John G.; Jones, Eugene L.; Jonte, John Haworth; Knowles, Leonard Ivison; Knox, Burnal Ray; Kornhaus, James W.; Latta, James M.; Lines, William B.; Lines, William B.; McClellan, Thurman Ralph; McMahan, Troy; McRae, Edward Walton; Mussett, Jack D.; Oglesby, Gayle Arden; Phillips, Chase A.; Piles, Charles Foster; Pittman, William Gene; Pohlo, Ross H.; Powell, William E.; Raible, Clarence J.; Raible, Leonard J.; Raines, Robert B.; Ratliff, James R.; Reinold, Marvin L.; Rivers, Thomas D.; Rushing, Robert S.; Scanlan, Richard Scott; Scougale, John Douglas; Sherman, James E.; Simmen, Louis E.; Smith, Joseph Walter; Staggs, James Otis; Stone, Charles G.; Stroud, Carlos R.; Stroud, Raymond Brown; Thomas, Harry E.; Walker, Warren; Watson, Randall O.; Wiegel, William E.

1960s: Addington, James W.; Arrington, Jimmy L.; Barrett, Richard B.; Bennett, Robert; Bishop, William H.; Bogard, Donald Dale; Boutwell, Jerry; Carr, Leo C.; Cate, Paul David; Chinn, Alvin A.; Chittenden, David Morse, II; Cintron, John, Jr.; Clanton, Quinton David; Clardy, Benjamin F.; Clark, Robert Stephen; Coston, Wendell Ray; Crouch, Walter H., Jr.; Davis, Charles L.; Deal, James E.; Doria-Medina, Jorge; Gibbons, John F., III; Gilley, Jerry D.; Gunter, Bobby Dean; Hanford, Charles Robert; Hickcox, Charles Woodridge; Johnson, Ray T.; Kimbro, Charles; Lybarger, James H.; Manuel, Oliver Keith; Mapes, Royal H.; Mason, Richard Harper; McCaleb, James A.; McEntire, John A., III; McMoran, William Dalton; McNully, Claude V.; Metts, David; Mooney, Tom D.; Moore, Howard Earl; Moseley, Cecil Robert; Nix, Joe Franklin; Owens, Don Ray; Parcher, James Vernon; Parker, Patrick LeGrand; Patton, Delmar Keith; Pfiefer, James E., Jr.; Pryor, Stanley J.; Sartin, Austin A., Jr.; Saunders, Bruce W.; Shewmake, Dan W.; Smith, Donald L.; Staley, George G.; Taylor, John Dallas; Thornton, James P.; Valdovinos, Dennis L.; Vest, Jimmy Thomas; Vyles, Charles H., III; Wainwright, Laurin L.; Webb, James H.; Wetsel, Eldon R.; White, Jack C.; Williams, Maurice; Woodson, James C.; Wright, Steward A.; Zachry, Doy Lawrence, Jr.

1970s: Askeland, James Philip; Baker, Cathy; Beckman, Michael A.; Berry, Richard A.; Black, Robert B.; Branch, Jerry D.; Brooks, Richard O.; Cardneaux, Christopher A.; Carney, George; Chapman, John Gary; Coughlin, Terry Lee; Davis, Leo Carson; Dillard, Taylor W.; Edson, James E.; Ewald, H. K., III; Fouke, Michael A.; Gaines, Elizabeth; Glenn, E. Charlotte; Glenn, John M.; Grayson, Robert C., Jr.; Green, Stephen N.; Grubbs, Robert S.; Hall, Jeffrey D.; Hanson, Bradford C.; Heathcote, Richard C.; Heathcote, Susan Kay Hudson; Henderson, Joe D.; Holt, Larry E.; Hoover, Elwin C.; Howard, James Michael; Hughes, J. W.; Hunt, Michael C.; Hurley, Stephen C.; Iheme, Uzoma N.; Jackson, James B.; Keisler, Ronnie S.; Kelley, Hiram; Lamb, Garland Clayton; Liner, Jeffery Lynn; Liner, Robert T.; Lucas, Philip E.; Marks, Joel Harvey; McFarland, John D., III; McGee, Kenneth Ray; Melton, Richard W.; Morrison, James Douglas; Petersen, Cheryl L.; Peterson, Raymond Judd; Pittenger, Gary C.; Potter, Robert W., II; Puckette, William L.; Reid, Chase S.; Reynolds, Michael Anthony; Rezaie, Nasser M.; Rice, W. Ralph; Rinie, Robert J.; Robison, Edward Clark; Sanguanruang, S. S.; Schumacher, Ann Louise; Shanks, Jack Lee; Shinn, Mike Reed; Smith, David A.; Snyder, Ronald Wayne; Tehan, Robert E.; Tolman, Davis Nichols; Van Buren, Wayne Martin; Van den Heuvel, Peter; Van Schaik, Edward J.; Vogelpohl, Sidney; Wagner, George H.; Warmath, Alex T.; Watkins, John Phillip; Williams, John G.; Willis, William Haywood; Young, Seward D.

1980s: Adams, Mark David; Adamski, James Clifford; Amini, Mohamed Karim; Ashworth, Richard Allan; Bakhtiar, Steven Norouz; Ballenger, Benja-

min David; Barlow, Charles A.; Bath, Thomas Patrick; Beardsley, Reginald Huse; Berlau, Charles E.; Bevill, James Cecil, Jr.; Borengasser, Marcus X.; Boudra, Robert A.; Bridges, Lindell C.; Burroughs, Richard K.; Cains, William Tyson; Carr, John Loften, III; Cavendor, Philip N.; Chilton, Rosalie; Chitsazan, Manouchehr; Clark, B. Christopher; Coffey, William S.; Collar, Paul David; Cox, Randel T.; Cox, Timothy L.; Crabtree, Billy J.; Crowder, Rowley Keith; Dark, William M.; Delavan, Gerald; Demarcke, Janet A. S.; Dilday, T. F., III; Downs, John W.; Duncan, Robert C.; Duran, William Kent; Eddy, Paul Southworth, Jr.; Essien, Isang O.; Fisher, Champe Andrews, Jr.; Forsythe, Roger; Foshee, Ronald R.; Fritsche, Glen D.; Fritsche, Kenneth L.; Gandl, Lynnette Anne; Gaston, Andy; Gillespie, David R.; Goodman, Wyndal M.; Gray, Michael G.; Guimon, Robert Kyle; Haq, Munir Ul; Harris, Gary D.; Hartmetz, Christopher Pate; Hawkins, Wildon D.; Hogue, John; Hokett, Samuel Lee; Holland, Krista; Jefferies, Brenda K.; Johnston, Jay S.; Johnston, Joseph Eggleston; Keck, Bradly Dwight; Kilpatrick, John M.; Knight, Kenneth S.; Knight, Thomas B.; Kresse, Timothy M.; Kurrus, Andrew William, III; Lamb, Cynthia B.; Lamiotte, Louis; Laney, Stephen E.; Lang, Roy; Leding, Edward A., III; Lee, Hen-Chen; Liebe, William Mather; Liebelt, Michael F.; Liou, John Chwen-Haw; Lisle, Barbara; Lonsinger, Lu-Anne P.; Lundy, Gerald W.; Marsh, Jeffrey Robert; McBride, John Henry; McGilvery, Thomas A.; McGowan, Michael Francis; McMurrough, Hugh Kenneth; McQueen, Kay C.; Melvin, Judith L.; Miller, Marilyn Sue; Mir Mohamad Sadeghi, Ali; Mollison, Richard Allen; Moyer, Chris; Murdaugh, Daniel J.; Muse, Paul Stephen; Neely, Dorothy G.; Nooncaster, John R.; Parker, R. Jay; Parr, Deborah L. Jones; Peurifoy, Raymond E.; Pinkley, Victoria Elizabeth; Plafcan, Maria; Post, Eugene; Price, Charles Raines, Jr.; Quick, Ray A.; Ramsey, Dean A.; Rezaie, Patricia Ann Jameson; Rothermel, Samuel Royden; Sadeghi, Ali Mohammad; Salaymeh, Saleem Rushdi; Sandoval, Deig-Nevy; Sharp, John B.; Sharp, John L.; Shelby, Phillip R.; Sheng, Zhengzhi; Sherrod, Thomas D.; Shields, John F.; Snider, Lynn Vivion; Spellman, Jane L.; Stephens, Jim; Sullivan, Thomas C.; Summers, Chester; Taylor, Doug; Terry, Steven H.; Thasan, Pat; Thomas, David A.; Thomas, Dennis R.; Uszynski, Bruce J.; Wallinga, Mark A.; Walthour, Steven Douglas; Ward, Robert Lee; Wickliff, Diana; Widmann, Roy K.; Williams, Pamela K.; Williams, Paula S.; Wilson, Glenn Venson; Yazdanian, Aminollah; Zelios, Dan G.; Zimmerman, Keith

Auburn University
Auburn, AL 36849

27 Master's, 5 Doctoral

1960s: Bailey, Alvin Cornell

1970s: Wood, P. A.

1980s: Allen, Nancy Edwards; Arce, Rodolfo Gagarin; Barnett, Wayne Stephen; Bearce, Steven Craig; Brown, David Edward; Butts, Rayburn L.; Colberg, Mark Robert; Crawford, Joseph Edward; Esposito, Richard A., Jr.; Frinak, Timothy R.; Gibson, Michael Allen; Gray, Tony Douglas; Haas, Christopher Allen; Harrer, Joseph W., Jr.; Heinrich, Nathan Daniel; Hicks, Benjamin Keith; Hildick, Margaret E.; Hue, Nguyen Van; Johnson, Mark James; Karathanasis, Anastasios D.; McDaniel, Charles Russell, Jr.; Neal, William L.; Peebles, Mark Whitney; Reynolds, William Kennedy; Sibley, David Michael; Skotnicki, Michael Charles; Smith, Larry Warren; Woerner, Eric Gerard; Wylie, Jerry Austin; Zwashka, Mark R.

Ball State University
Muncie, IN 47306

22 Master's, 3 Doctoral

1960s: Judd, Robert William

1970s: Beerbower, D. C.; Davis, R. L.; Follis, M.; Harvey, Rose Mary; Maroney, David G.; Mourdock, R. E.; O'Gorman, John R.; Pentecost, David C.; Prowant, S. O.; Wortham, Kenneth E.

1980s: Baker, Robert J.; Behnami, Farhad; Bloemker, J. Mark; Burns, Danny E.; Fisher, David M.; Glasby, Virginia June; Johnston, David Kent; Killey, Myrna Marie; Madigosky, Stephen Robert; May, Suzette Kimball; Owens, Robert N.; Schindler, Kris L.; Stanley, Sarah G.; White, Arthur James

Bates College
Lewiston, ME 04240

1 Master's

1940s: Shea, John V.

Baylor University
Waco, TX 76798

164 Master's, 2 Doctoral

1950s: Bates, Allen Neal; Holloway, Harold Deen; Jameson, James Boyd; Seewald, Kenneth Oscar

1960s: Atlee, William Augustus; Beall, Arthur Oren, Jr.; Boone, Peter A.; Brown, Johnnie Boyd; Burket, John Maxwell; Chamness, Ralph S.; Chou, Chung-chi; Fandrich, Joe W.; Font, Robert G.; Fox, William Joseph; Frost, Jackie Glenn; Henningsen, Elmer Robert; Hopkins, Otho Neil; Jones, James Ogden; King, George Leslie, Jr.; Lewand, Raymond; Madani, Mastaneh; McGowen, Joseph H.; Moore, Thomas H.; Payne, William Ross; Proctor, Cleo V.; Ray, Johnny; Reel, Ted Wesley; Reeves, Anita L.; Rodgers, Robert William; Seals, Mary J.; Shaw, Don Wayne; Silver, Burr Arthur; Spencer, Jean; Taylor, Don Ray; Tyner, Clarence Graham; Waller, Thomas Howell; Williamson, Edward F.

1970s: Aguayo, Eduardo; Allen, Peter; Bain, James S.; Baldwin, Ellwood E.; Bammel, B. H.; Belcher, R. C.; Brown, Thomas E.; Byrd, C. Leon; Chandler, Carol Reisen; Clark, E., Jr.; Davis, Elizabeth R.; Davis, Keith William; Dobbins, Lloyd R.; Dreyer, Boyd V.; Epps, Lawrence W.; Flatt, C. D.; Howell, Gary D.; Hudson, Presley C.; Jackson, T. C.; Jessen, Margaret Lynn; Lambert, George; Leach, E. D.; Miller, M. A.; Mosteller, M. A.; Mudd, Wayne Adrian; Owen, M. T.; Perez, Olivia Ramoz; Pool, J. R.; Rigby, M. A.; Roberson, Dana Shumard; Roberson, Gary D.; Staples, M. S.; Thomas, Ronny G.; Trippet, William A., III; Walker, Clifford A.; Walker, Jimmy Roy; Weaver, M. S.; Yelderman, J., Jr.

1980s: Abdel Wahab, Mahmoud M.; Allen, Elan Anderson; Anderson, Brian D.; Anderson, L. Marlow; Atchley, Stacy C.; Barrett, Daniel Patrick; Beck, John H.; Bishop, Sue Lynn; Bochneak, Diane Lynn; Boylan, David Michael; Brothers, James; Brown, Thomas E.; Calavan, Charles W.; Campbell, Kevin; Cannata, Stan Lee; Carrillo, Victor; Cervenka, Robert E.; Charvat, William A.; Clark, Martha Ann; Craddock, Greg F.; Crass, David B.; Davidson, William T.; Dolliver, Paul N.; Duffin, Michael E.; Durler, David L.; Fassauer, Patti; Fletcher, Thomas E.; Ford, Michael E.; Frost, Richard W.; Gawloski, Ted; Giarratana, Ann Marie; Goldsberry, Stephen Lee; Granata, Glenn Walter; Green, Connie L.; Guillette, Brian R.; Gundersen, Thomas D.; Hancharik, Joan; Hawthorne, Hal W.; Hawthorne, J. Michael; Haycock, Susan; Hayward, Chris; Hightower, Jay Harold; Hild, Gregory Phillip; Hoadley, Carol R.; Homan, Kimberly Sue; Hoy, Steven M.; Jackson, Richard Lance; Jaffe, Daniel G.; Jamieson, William H., Jr.; Keyes, Steven Lynn; Kidman, Mark; Kingston, Jim; Koger, Curtis; Krystinik, Jon G.; Lemons, David Ray; Leuty, Joseph L.; Luginbill, Charles Philip; Martinez,

Norma; Matthews, Truitt F.; McFarland, Gregory J.; McKee, Bryce J.; McKnight, Cleavy L.; Meyerhoff, James C.; Meyerhoff, Lisa H.; Montgomery, James Alan; Morris, David G.; Nahm, Jay W.; Narramore, Rebecca Lynn; Oldani, Martin J.; Oleson, Nan E.; Onyegam, Emmanuel I.; Palladino, Deanna L.; Pettigrew, Robert J., Jr.; Pinkus, Joel R.; Pollard, Craig D.; Poorman, Stephen E.; Rapp, Keith Burleigh; Reneer, Bernal; Ritch, Kurt D.; Rose, Cindy L.; Shields, Martin L.; Surles, M. A., Jr.; Tharp, Tommy Lee; Thompson, Sheree A.; Thornhill, Stephen Alan; Trice, E. L. Jack, III; Vitani, Nicholas M.; Wagner, Walther O.; Westergaard, Edwin H.; Whigham, Linda; Woodard, Jan N.

Boise State University
Boise, ID   83725

6 Master's

1970s: Ireton, M. Frank

1980s: Beem, Leigh Ivan; Foss, Donald J.; Johnson, Rex J.; Lawrence, David C.; Schwarz, David L.

Boston College
Chestnut Hill, MA   02167

51 Master's, 3 Doctoral

1950s: Brooks, John E.; Burke, John Joseph; Costley, Wayne J.; Crosby, George S.; Crowley, Francis A., Jr.; Daly, Paul J.; Frey, John H.; Geilissee, P. J.; Gouin, Pierre; Graham, Richard H.; Holt, Richard J.; Iwin, Francis R., Jr.; LeBlanc, Gabriel; Murphy, Vincent J.; Nolan, George W.; Rockett, Thomas John; Sexton, Thomas F.; Sherry, Paul F.; Su, Sergio S.; Sullivan, Eleanor R.; Sullivan, Stephen J., Jr.

1960s: Boutilier, Robert Francis; Chin, Chen; Colombini, Victor Domenic; DeCaprariis, Pascal Peter; Fisk, Peter J.; Gore, Richard Z.; Hissenhoven, Rene Van, S. J.; Lacroix, A. V.; LeGarde, Charles N.; Levine, Edward Neil; Mao, Maurus Nai-Hsien; Ossing, Henry A.; Peerzada, Michael; Regan, Robert D.; Rourke, Gerald F.; Turyn, May E.; Twitchell, Paul F.

1970s: Banks, P. T., Jr.; Centorino, J. R.; Daly, Edward Joseph; Ingersoll, D. S.; Kolodny, Carole Renee; Kucinskas, D. P.; Li, T. M.; Martin, L. G.

1980s: Breunig, Peter A.; Buxton, Rebecca E.; Cicerone, Robert D.; Holden, Mark Kellogg; Huidobro N., Pablo; Klimkiewicz, George C.; McCague, John Joseph; Sherman, Peter Edward

Boston University
Boston, MA   02215

111 Master's, 63 Doctoral

1920s: Crane, Calista

1930s: Wilcock, Myrtle Lavinia

1950s: Borns, Harold William, Jr.; Cariani, Anthony Robert; Fellows, Ralph Sanborn; Hight, Richard Parker; Hutchinson, Robert O.; Moench, Robert H.; Roberts, David Chapin; Swarzenski, Wolfgang V.; Swinzow, George K.; Willard, Robert Jackson

1960s: Abu-Moustafa, Adel H.; Akeson, R. S.; Al-Hashimi, Abdul Razak K.; Benjamins, Janet; Berkebile, Charles Alan; Boutlier, R. F.; Breitling, William F.; Brewer, Thomas; Burke, Robert Francis; Chen, Chin; Davis, William Edwin, Jr.; Farnsworth, Ray Lothrop; Fischer, William L.; Friis, Karin L.; Frimpter, Michael Howard; Furlong, Ira Ellsworth; Gleba, Peter; Glidden, Philip Eugene; Handford, Lincoln S.; Hasan, Manzoor; Lyons, Paul Christopher; Mariano, Anthony Nick; Meszoely, Charles Aladar Maria; Miles, Paul R.; Mohsen, Lotfi A. F. M.; Morrison, Andrew D.; Murphy, John R.; Pettyjohn, Wayne Arvin; Raabe, John A.; Rostoker, Mendel David; Schaff, Ross Gilbert; Skapinsky, Stanley Alfred; Sullivan, Edward A.; Temple, Carol I.; Vezzoli, Gary C.

1970s: Arora, C. R.; Badawy, Assem M.; Boreske, John Robert, Jr.; Brewer, Thomas; Bukhari, Mohamed; Cabaniss, Gerry Henderson; Chasen, Edith A.; Defieux, R. J.; Dunn, Pete J.; Eby, George Nelson; Fay, Jan E.; Feldman, Lawrence; Fessenden, Franklin Wheeler; Frank, Wendy L.; Ghosh, Sudipta K.; Gore, Richard Z.; Hatzikostantis, Nicholas G.; Hellier, Nancy W.; Hlavin, William J.; Hlavin, William J.; Hoekzema, Robert B.; Housman, John J., Jr.; Huang, Chen-Feng; Hunnewell, Dorothy S.; Jeanne, Richard A.; Jones, J. R.; Jones, J. R.; Jordan, Richard J.; Kaktins, Uldis; Kamal, Rami A; Kaplan, W. A.; Kelly, Michael A.; Killius, D. R.; Kurz, Steven L.; Kurz, Steven L.; Langer, William H.; Lawlor, John Francis; Leonard, Jay E.; Lidback, M. M.; Mahoney, John J.; Mallio, William Joseph; Mangion, Stephen M.; Morgenstern, Lillian; Mudge, L. Taylor; Nellis, David A.; O'Brien, Arnold Leo; O'Neill, Margaret M.; Pealey, Annette; Raabe, John A.; Rainville, George D.; Raman, Swaminathan V.; Richardson, Steven M.; Saltzberg, Edward R.; Schneider, William D.; Schulman, Melvyn M.; Silins, Lauma; Spinney, Edward E.; Stone, Catherine S.; Tekverk, Raymond W.; Titus, Robert C.; Titus, Robert Charles; Uzuakpunwa, Anene B.; Valia, Hardarshan S.; Weiss, Alan J.

1980s: Adler, Robert C.; Attaway, Dorothy Claire; Ayer, China O.; Barker, Charles; Burnett, Linda S.; Caraco, Nina Marie; Chormann, Jeffrey; Coffin, Catherine; D'Amore, Denis W.; D'Amore, Denis W.; DeSimone, Leslie; Diaz, Eugenio G.; Dougherty, Percy H.; Duran, Philip B.; Edwards, Gerald B.; Faldetta, Sarah; Feinberg, Paul M.; Fish, Craig B.; Foley, Mary Kathryn; Freile, Deborah; Giblin, Anne Ellen; Graham, Robert L.; Hanna, Sara Ross; Hanson, Lindley Stuart; Hart, Steven W.; Hoagland, Matthew; Hoffman, Emily J.; Holland, William; Hudak, Larry J.; Ibrahim, Noor Azim; Jacobson, Carolyn; Kelly, Edward, Jr.; Kirchoff, Scharine; Lagace, Paul; Lang, Edwin F.; Lepzelter, Carol; Levin, Douglas R.; Lincoln, Jonathan; Magee, Andrew D.; McGlew, Peter J.; Mogekwu, Emmanuel; Moog, Polly Lu; Nelson, Karin E.; Newman, Ewa N.; Newman, Richard; Olney, Sylvie L.; Phippen, Peter D.; Pivetz, Bruce E.; Porter, Suzanne; Radville, Mark E.; Rambler, Mitchell Bruce; Rose, Stuart M.; Sands, David R.; Slechta, Marc W.; Smaldone, John R.; Solomon, Hayden S.; Sullivan, David S.; Teoman, Mahmet S.; Terefenko, Robert; Trench, Elaine; Wilson, John O.; Yakzan, Azmi Mohd; Zannos, John A.

Bowling Green State University
Bowling Green, OH   43403

222 Master's, 1 Doctoral

1960s: Anderson, Brooks D., II; Bailey, Thomas Corwin, III; Boyd, Cynthia Stiles; Britt, Claude J.; Brutvan, William J., Jr.; Bugh, James Edwin; Bush, Edward Allen, Jr.; Cochran, Michael D.; Daniels, Paul A., Jr.; Duncan, Dennis C.; Greb, Wayne S.; Grubb, James M.; Hadden, David R.; Jackson, James A.; Keller, Paul Henry; Krewedl, Dieter A.; Kussow, Roger G.; Lene, Gene Wilfred; Lin, Chong-Pin; Lindsey, David S.; Louden, Richard Owen; MacTavish, John Nickolas; Mayher, Andrea M.; Ottum, Margaret G.; Pauken, Robert J.; Peters, Thomas J.; Scadden, Raymond William; Schrecongost, Milford Martin; Schumann, John R.; Shimeall, Clark M.; Steller, Dorothy L.; Strogonoff, Robert Francis; Stuart, Robert J.; Swanston, Douglas Neil; Tilton, Terry L.; Updegraff, Nancy A.; Whitley, Richard L.; Wygant, Thomas G.

1970s: Andrews, Edward J.; Atwater, David E.; Bartlett, Wayne A.; Birak, Donald J.; Boyd, R. Michael; Bryan, Michael R.; Burns, Gregory K.; Busanus, James William; Campbell, Marie A.; Caruso, Joel W.; Cleneay, Charles A.; Cunningham, Chris P.; Cunningham, Frederick Franklin, Jr.; Cuzella, Jerome J.; Echelbarger, Michael J.; Eutsler, Robert L.; Evans, Elizabeth Lee; Ferrall, Kim Wallace; Floyd, Jack Curtis; Foust, Denny G.; George, Daniel T.; Grant, Keith; Graves, Lawrence

S.; Hackathorn, Merrianne; Haidarian, Mohammad Reza; Hoare, Thomas Bertram; Hoover, Jon R.; Hostenske, Dale J.; Huh, John Mun Suk; Jaycox, Robert Eugene; Johnson, David M.; Jolly, James G.; Keenan, Steven J.; Kindt, Eugene Anthony; Kostura, John R.; Kovacik, Thomas Louis; Lanz, Robert C.; Levinson, Andrea R.; Markowitz, Philip E.; McCullough, Warren D.; McGuire, Donn; McMaster, Larry; Meyer, James Arthur; Morro, Richard D.; Motten, Roger H., III; Newhart, Joseph A.; Nielsen, David M.; Nwankwo, Linus N.; Ochsenbein, Gary Dean; Overman, William C.; Pality, George D., Jr.; Peterson, Robert Michael; Pfeiffer, Dan E.; Przywara, Mark S.; Quick, Robert Carl, III; Regel, Bernard; Rennebaum, Thomas D.; Requarth, Jeffrey S.; Rowe, Roger G.; Ruch-Hirzel, Mary Lou; Shaw, Ronald L.; Sheely, William S.; Snider, David W.; Soeder, Daniel J.; Sparks, Diane K.; Steinker, Paula Dziak; Stichtenoth, Craig W.; Street, Michael O.; Strickland, Kenton E.; Turmelle, John M.; Urban, Scott D.; Ward, Richard C.; Waszczak, Ronald F.; Watts, Bradice C.; Weis, Brent R.; Williams, Michael W.; Wilson, Janine M.; Winzeler, Ted J.; Wolery, Thomas Jay; Wong, Anne B.; Yahney, Gordon K.

1980s: Andersen, Christine L.; Anderson, John R.; Anderson, Laurie C.; Balogh, Randy J.; Bambrick, Thomas C.; Barnett, Jarrall; Berger, Deborah J.; Billman, Robert B., III; Bogner, Kurt A.; Breckling, Robert; Brindle, Wendy D.; Butlien, Lawrence J.; Cade, Perry A.; Cassetta, Dominick M.; Christopher, Cranston C.; Clayton, Deborah A.; Click, David L.; Conklin, Michael W.; Connelly, Jeffrey B.; Crider, Val Joe; Cubberley, Alan J.; Cummins, Laura E.; David, Timothy L.; De Brock, Michael D.; Diedrich, Robin P.; Ditzel, Krista D.; Dragone, Annette M.; Duddy, Kathleen A.; Dyka, Mary Ann K.; Foreman, J. Lincoln; Fuehrer, David W.; Furman, Francis Chandler; Goss, William A.; Graber, Stuart M.; Grimm, Michael A.; Grube, Michael H.; Guy, Donald E., Jr.; Haefner, Ralph J.; Hagopian, John J.; Harbaugh, Jeffrey; Harrison, James E.; Heaney, Michael J., III; Henry, E. Marie; Hoffman, Robert A.; House, Valerie Hust; Hull, Douglas A.; Kappler, Jonathan; Kennedy, William Allen; Klotz, Jack A.; Krug, Donald J.; Lehle, Peter F.; Lewis, Eric S.; Lyons, Nancy M.; Mackovjak, Dennis; Maharidge, Allan D.; Maloney, Moira N.; Mancuso, Christina M.; Mecionis, Robert; Metzger, Robert J.; Miller, Donna L.; Miller, John J.; Miller, Nathaniel R.; Mohamad, Kamel B.; Motamedi, Saadi; Nickel, Brian K.; Noon, Patrick L.; Norman, Linda S.; Palmer, Catherine Grace; Peckins, Eric L.; Pennington, Richard L.; Polley, Mark R.; Potter, Christopher D.; Price, Pamela J.; Pugsley, Robert L.; Rankin, Mary L.; Regli, Robert; Riley, Ronald A.; Santangelo, Mark A.; Scanlan, Mark J.; Schick, Charles W.; Schooler, Richard A.; See, Bennie E.; Sewell, H. James; Shope, Linda S.; Slow, Evan S.; Smith, Victor G.; Stevenson, Jeffrey D.; Strayton, James P.; Sturgis, Douglas S.; Suter, Tammy D.; Taer, Andrew D.; Telljohann, Eric P.; Terman, Marc L.; Tintera, John J.; Tomassetti, John A.; VanWagner, Elmer, III; Vrh, Steven J.; Walden, Ronald L.; Weaver, Dennis W.; Winsten, Miriam S.; Wohlabaugh, Norman; Wonn, Philip M.; Yokoyama, Ken T.; Young, Rebecca H.

Brigham Young University
Provo, UT   84602

292 Master's, 20 Doctoral

1930s: Buss, Walter R.; Coffman, W. Elmo; Condon, David Delancey; Dennis, Eldon; Dixon, Howard B.; Harris, A. Wayne; Johnson, Vard Hayes; Stokes, William Lee

1940s: Bullock, Kenneth C.; Gwynn, Thomas Andrew; Mecham, Derral F.; Rigby, J. Keith; Tuttle, Lawrence Elliott; Warner, Mont Marcellus

1950s: Abbott, Ward Owen; Bentley, Craig B.; Boyden, Thomas A.; Brown, Ralph Sherman; Bullock, Reuben M. [Lynn]; Burge, Donald Lockwood; Calderwood, Keith W.; Clark, David L.;

Clark, Robert S.; Cline, Charles W.; Croft, Mack G.; Crosby, Gary Wayne; David, Briant LeRoy; Davis, Del E.; Davis, H. Clyde; Davis, Leland J.; Dearden, Melvin O.; Demars, Lorenzo C.; Elison, James H.; Erickson, Einar C.; Evans, Max Thomas; Foster, John M.; Franson, Oral M.; Gaines, Patrick W.; Gates, Robert W.; Gould, Wilburn James; Hamblin, William Kenneth; Harris, DeVerle; Harris, Harold Duane; Hebertson, Keith M.; Hebrew, Quey Chester; Henderson, Gerald V.; Hodgson, Robert A.; Hoffman, Floyd H.; Hosford, Gregory F.; Johns, Kenneth Herbert; Johnson, Kenneth Dee; Knight, Lester L.; Larsen, Norbert William; Livingston, Vaughn Edward, Jr.; Madsen, Jack William; Madsen, Russel A.; Maxfield, E. Blair; McFarland, Carl R.; McFarlane, James J.; Moulton, Floyd C.; Moyle, Richard W.; Murphy, Don R.; Nolan, Grace Margaret; Okerlund, Maeser D.; Olsen, Ben L.; Ornelas, Richard Henry; Peacock, C. Herschel; Perkins, Richard F.; Petersen, Herbert Neil; Petersen, Morris Smith; Peterson, Dallas O.; Peterson, Deverl J.; Peterson, Parley Royal; Pitcher, Grant Grow; Powell, Dean Keith; Prescott, Max W.; Price, Jack Rex; Rawson, Richard Ray; Reber, Spencer J.; Rhodes, Howard Startup; Rhodes, James A.; Schindler, Stanley Fred; Sirrine, George Keith; Smith, Cleon Verl; Smith, Grant McKay; Smith, Walter Lorane; Stewart, Donald G.; Stewart, Moyle Duanne; Swanson, James Walter; Thomas, Glenn H.; Turner, Frank Paul; Waring, Robert Gordon; Warner, Robert O.; White, Bob O.; Williams, Floyd Elmer; Woodland, Roland Bert; Zeller, Ronald P.

1960s: Baer, James Logan; Baer, James Logan; Bain, Roger J.; Beach, Gary A.; Berge, Charles William; Berge, John Stuart; Blake, John W.; Bodily, Norman Mark; Bordine, Burton W.; Brady, Michael J.; Bullock, Ladell R.; Burckle, Lloyd H.; Chamberlain, Kent C.; Embree, Glen F.; Foutz, Dell R.; Fowkes, Elliott Jay; Hanks, Keith Lynn; Hanks, Teddy L.; Hodgkinson, Kenneth A.; Howard, James Dolan; Jensen, Ronald Grant; John, Edward Charles; Jones, Alan M.; Kattleman, Donald Franklin; Kaufmann, Harold E.; Kennedy, Richard R.; Lufkin, John L.; Markland, Thomas Richard; Marshall, Frederick Charles; Mayou, Taylor Vinton; Mollazal, Yazdan; Nadjmabadi, Siavash; Pitcher, Max Grow; Prince, Donald; Rigo, Richard J.; Robinson, Gerald B., Jr.; Robison, Richard Ashby; Rutledge, James Raymond; Schneider, Michael Charles; Slade, M. Lyle; Smith, Alan Robert; Smouse, Deforrest; Steininger, Roger Claude; Swensen, Arthur Jaren; Tidwell, William D.; Washburn, Alan T.; Wells, Richard Bruce; Westover, James Donald; Willes, Sidney Blaine; Wright, Richard E.; Zabriskie, Walter Edward

1970s: Ahlborn, Robert C.; Alexander, David William; Anderson, Stephen Robert; Anderson, Thomas C.; Armstrong, Robert Morgan; Astin, Gary Kent; Bagshaw, Lawrence H.; Bagshaw, Rebecca Lillywhite; Ballou, Robert L.; Belnap, Dennis Wayne; Bingham, Clair C.; Black, Bruce Allan; Bohn, Ralph T.; Braithwaite, Lee Fred; Bushman, Arthur Vern; Campbell, Dennis R.; Chamberlain, Alan K.; Chesser, William La Grand; Chidsey, Thomas C., Jr.; Church, S. B.; Church, Stephen B.; Clark, Eugene E.; Cleavinger, Howard Ben, II; Cranor, John Ira; Cranor, John Ira; Davis, John Daniel; Demeter, Eugene J.; Derr, Michael E.; Desai, Jyotindra Ishwarbhai; Driggs, Allan F.; Dustin, J. D.; Gilland, James Kenneth; Hampton, George Lee, III; Hannum, Cheryl Ann; Harris, Daniel R.; Haugh, Galen Rudolph; Hawks, Ralph L.; Hogg, Norman Carroll; Hoggan, Roger D.; Hoggan, Roger D.; Holmes, Richard D.; Hoover, James David; James, Bruce Howard; Klopp, Helen C.; Ladle, Garth Harrison; Lambert, David James; Larsen, Norbert William; Lawyer, Gary Frank; Leedom, Stephen H.; Lewis, Paul Heywood; Lindsay, Robert F.; Loftsgaarden, Jan L.; Lohrengel, Carl Frederick; Lowrey, Ronald Ovel; Luke, Keith Joseph; Madsen, James Henry; Mahfoud, Robert F.; Maxfield, E. Blair; Melton, Robert A.; Merrill, Richard C.; Newman, Donald Hughes; Newman, Gary James; Nielson, R. LaRell; Nixon, Robert

Paul; Norris, Earl G.; Oliveira, Michael E.; Olsen, John Roger; Orgill, Jeffrey R.; Osborne, Steven D.; Ott, Valen D.; Palmer, Dennis Erwin; Palmer, Joel Oleen; Pape, Lance W.; Peterson, Allen R.; Piekarski, Leonard L.; Pierce, Carlos R.; Pinnell, Michael Lu; Runyon, David M.; Sanderson, Ivan D.; Smith, Larry S.; Smith, Samuel Milton; Sonerholm, Paul Arthur; Spreng, W. Carl; Squires, L. E.; Stanton, Robert Guy; Steed, Douglas A.; Stolle, James Michael; Taylor, James Michael; Taylor, James Michael; Taylor, LaRon; Terrell, Forrest M.; Uresk, Jack R.; Weaver, Clark L.; Wyatt, Danny J.; Young, George E.

1980s: Alexander, Velinda D. H.; Baclawski, Paul; Bashford, Howard H.; Brandley, Richard T.; Bunnell, Mark D.; Carroll, Richard E.; Chamberlain, Randy C.; Clayton, Robert W.; Coffin, Brian D.; Conner, John; Dattilo, Benjamin F.; Davis, Robert L.; Dean, Jim; Felt, Vince; George, Steven E.; Gosney, Terry C.; Greenhalgh, Brian R.; Hamblin, Alden H.; Hamblin, Russell D.; Hammond, Becky Jane; Hansen, Chris D.; Harris, E. Donald, Jr.; Hatch, Floyd; Haymond, Dan; Heaton, Timothy H.; Higgins, Janice M.; Hill, Richard Bruce; Holladay, John C.; Hunt, Charles B.; Hunt, Gregory L.; Hurst, Carolyn; Jenkins, David E.; Jensen, Mark E.; Jensen, Nolan R.; Jenson, John; Johansen, Jeffrey R.; Johnson, Brad T.; Johnson, Chris N.; Jones, Gregory L.; Jorgensen, Gregory J.; Keller, David R.; Kitzmiller, John Michael, III; Knowles, Steven Paul; Lambert, Ralph E.; Little, William W.; Martin, Glen Edward, II; Martorana, Anthony; Meibos, Lynn Clark; Millard, Alban Willis, Jr.; Morris, Stuart K.; Morton, Loren B.; Nethercott, Mark A.; Nielson, Dru R.; Oberhansley, Gary; Ren, Xiaofen; Robison, Steven F.; Rogers, John C.; Rowley, R. Blaine; Russon, Michael P.; Smith, James Douglas; Smith, Stephen M.; Sperry, Steven W.; Taylor, Douglas William; Wheeler, Richard F.; Whitlock, William W.; Wiliams, Winston L.; Willis, Grant C.; Willis, Julie Barrott; Young, Robert Brigham

University of British Columbia
Vancouver, BC V6T 2B4

423 Master's, 204 Doctoral

1920s: David, Newton Fraser Gordon; Emmons, Richard Conrad; Gillanders, Earle Burdette; Guernsey, Terrant Dickie; Jackson, Gerald Christopher Arden; Jones, William Alfred; Kania, Joseph Ernest Anthony; Lang, Arthur Hamilton; Logie, Russell Moore; Millward, Lewis G.; Osborne, Freleigh Fritz; Paterson, Philip G.; Pollock, J. R.; Price, Peter; Riley, Christopher; Smitheringale, William Vickers; Swanson, Clarence Otto

1930s: Armstrong, John Edward; Black, James Murray; Cummings, John Moss; David, Edwin Philip; Duffell, Stanley; Freshwater, Norman G. Morgan; Graham, Roy; Gray, John Gardiner; Johnson, J. R.; Killin, Alan Ferguson; Leckie, Phyllis Gilmour; Lord, Clifford Symington; McCammon, James William; McKecknie, Neil Douglas; Okulitch, Vladimir J.; Rice, Harington Molesworth Anthony; Smith, Alexander; Snow, William Eugene; Thurber, Judson B.; White, William Harrison; Wright, Harold M.

1940s: Allen, A. R.; Bacon, William Russell; Carbonneau, Come; Carlisle, Donald; Cheriton, Camon Glenn; Chierton, Camon G.; Deleen, John L.; Diebel, John Keith; Fyles, James Thomas; Gaul, R. F.; Gouin, Leon Oliver; Hoadley, John William; Hodgson, Alexander Goldie; Howatson, Charles Henry; Hughes, Richard David; Irish, Ernest J. W.; Irwin, Arthur Bonshaw; Johnston, William George; Joubin, Francis Renault; Kerr, Samuel Aubrey; King, Norma Louise; Lamb, John; Lee, James William; Leitch, Henry Cedric Browning; Little, Heward Wallace; Lunde, Magnus; MacDonald, Ralph Crawford; Maconachie, James Roy Alexander; Manifold, Albert Hedley; Mathews, William Henry; McEachern, Ronald Graham; McLellan, Robert Bryant; Morris, Arthur; Ney, Charles S.; Roots, Ernest Frederick; Ryan, Edward McNeill;

Seraphim, Robert Henry; Thompson, Robert Mitchell; Wallace, James Alan; Wheeler, John Oliver; Williams, Edwin Philip; Young, John Walter

1950s: Bacon, W. R.; Baker, Reginald Anthony; Bell, Gordon Lennox; Best, Raymond Victor; Bostock, Hewitt Hamilton; Burley, Brian John; Calkin, Parker Emerson; Carswell, Henry Thomas; Chown, Edward Holton MacPhail; Crabb, John Johnson; Fortescue, John Adrian Claude; Fortesque, John; Frebold, Fred; Frebold, Fridtjof Albert; Fyles, John Gladstone; Gabrielse, Hubert; Gale, Robert Earle; Gower, John Arthur; Greenwood, Hugh John; Greggs, Robert George; Grove, Edward Willis; Hall, Donald H.; Hansuld, John Alexander; Heddle, Duncan Walker; Hillhouse, Douglas Neil; Hoen, Ernest L. W. B.; Johnson, Ronald Dwight; Jones, William Charles; Kawase, Yoshio; Kierans, Martin Devalera; Lee, Randolph; Matheson, Marion Henderson; McDougall, James John; McEachern, R. G.; Menzies, Morris McCallum; Money, Peter Lawrence; Morris, Peter Gerald; Nelson, Samuel James; Padgham, William Albert; Papezik, V. S.; Papezik, Vladimir Stephen; Purdie, J. J.; Richardson, Paul William; Rousell, D. H.; Sanschagrin, Roland; Scott, Darcy Lon; Sharp, William McMillan; Sheng, Cheng Chun; Soles, James Albert; Taplin, A. C.; Taplin, Arthur Cyril; Thomlinson, A. G.; Thomlinson, Arnold Gordon; Toombs, Ralph Belmore; Trettin, Hans Peter; Vail, John Randolph; Wilson, George Alexander; Wilson, Philip Roy; Zazac, I. S.

1960s: Allard, Jean-Louis; Allen, Donald G.; Atkinson, Gerald; Basham, P. W.; Birnie, Thomas Alexander; Blenkinsop, John; Bottinga, Jan; Bright, Edward G.; Byrne, Peter Michael; Caner, Bernard; Cannon, Wayne Harry; Carson, David John Temple; Childs, John Frazer; Church, Barry Neil; Clement, Maurice James Young; Coates, James A.; Coode, Alan Melvill; Cos, Raymond Lee; Cox, Raymond Lee; Crossley, D. J.; Currie, Ralph Gordon; Davidson, Anthony; Davidson, Anthony; Davidson, Donald Alexander; Dawson, Robin Humphrey; Deas, A.; Dirom, G. E.; Dosso, Harry William; Drummond, Arthur Darryl; Farquharson, Robin Bruce; Fuchs, Jens Peter; Graham, John Donald; Greenhouse, John Phillips; Greggs, Robert George; Gurbuz, Behic M.; Hara, Elmer Hiroshi; Harvey, R. W.; Hasegawa, Henry S.; Hills, Leonard Vincent; Hopkins, William Stephen, Jr.; Hopkins, William Stephen, Jr.; Hyndman, R. D.; Ishii, H.; Jambor, John Leslie; Jensen, O. G.; Kalra, A. K.; Kanasewich, Ernest Raymond; Kerfoot, Denis Edward; Kirkham, Rodney Victor; Kollar, Frank; Koo, Ja Hak; Kwak, Teunis Adrianus Pieter; Lambert, A. I.; Lambert, Maurice Bernard; Lambert, Maurice Bernard; Livingstone, C.E.; Lumbers, Sydney Blake; Lund, John C.; Lund, John Casper; Mansinha, Lalatendu; McFadden, C. P.; McMillan, William John; McTaggart-Cowan, G. H.; Meldrum, R. D.; Michkofsky, R. N.; Monger, James William Heron; Monger, James William Heron; Montgomery, Joseph Hilton; Montgomery, Joseph Hilton; Ng, Tai-Ping; Nguyen, K. K.; Nishida, Atsuhiro; Northcote, Kenneth Eugene; Northcote, Kenneth Eugene; Osatenko, Myron John; Osborne, Willis Williams; Ostic, Ronald George; Panteleyev, Andrejs; Paterson, W. Stanley B.; Petrak, J. A.; Pickering, Dennison John; Piel, Kenneth Martin; Presto, Vittorio Annibale Guisepe; Read, Peter Burland; Reynolds, Peter Herbert; Ricker, Karl Edwin; Robinson, Alexander Maguire; Samson, J. C.; Sangster, Donald Frederick; Schau, Mikkel Paul; Scott, Darcy Lon; Sellmer, H. W.; Sinclair, Alastair James; Smith, Clyde Louis; Smith, L. C.; Smith, Roberta Katharine; Smith-Evernden, Roberta; Sopher, Stephen R.; Soregaroli, Arthur Earl; Srivastava, Surat Prasa; Stacey, John Sydney; Stanley, Alan David; Stanley, Alan David; Street, Peter J.; Tedlie, William Donald; Tempelman Kluit, Dirk Jacob; Trettin, Hans Peter; Tseng, K. H.; Ulrych, Tadeusz J.; Ulrych, Tadeusz Jan; Von Rosen, G. E. A.; Watson, Robert Brian Fraser; Weichert, Dieter Horst; Wetmiller, R. J.; White, William Robert Hugh; Whiton, Geoffrey Arthur; Whittles, Arthur Brice Leroy; Whittles, Arthur Brice Leroy; Yole, Raymond William; Zajac, Ihor Stephen

1970s: Ager, C. A.; Ager, Charles Arthur; Ahern, T. K.; Anderton, Lesley Jean; Annas, R. M.; Armstrong, W. P.; Ashley, G. M.; Athaide, D. J. A.; Barr, Sandra Marie; Barr, Sandra Marie; Bartholomew, Paul Richard; Belik, G. D.; Bell, L. M.; Beltagy, Ali Ibrahim; Bennett, G. T.; Berman, Robert G.; Bertrand, Aimee; Bertrand, Wayne Gerrard; Bevier, M. L.; Bhoojedhur, S.; Bihl, Gerhard; Birnie, D. J.; Blenkinsop, John; Bolduc, Pierre-Michel; Bourne, Douglas Randal; Brabac, Dragan; Bremner, James M.; Bremner, Trevor John; Broersma, K.; Brown, D. A.; Buckingham, W. R.; Buckley, J. R.; Buttrick, S. C.; Cann, Robert Michael; Cannon, Wayne Harry; Cargill, D. G.; Carlberg, R. G.; Carne, Robert Clifton; Carter, Lionel; Carter, N. C.; Cave, W. R.; Cawthorne, Nigel George; Chandra, Bhuvanesh; Cheng, J.-D.; Cheung, Henry P. Y.; Christie, James Stanley; Christopher, Peter Allen; Chronic, Felicie Jane; Church, M. A.; Clague, John Joseph; Classen, David Farley; Clayton, Robert W.; Cooper, Roger Brian; Cordes, R. E.; Cross, C. H.; Crossley, David John; Culbert, Richard Revis; Cumming, William B.; Daniel, P. E.; Davidson, L. W.; Davies, John C.; Dawson, Kenneth Murray; Ditson, G. M.; Dobell, P. E. R.; Doyle, P. J.; Doyle, Patrick Joseph; Dragert, Herb; Dragert, Herbert; Duffy, C. J.; Elliot, A. J. M.; Errington, J. C.; Fletcher, Christopher John Nield; Forsyth, D. A. G.; Francis, Donald Michael; Fraser, J. R.; Gell, W. A.; Gilbert, Robert; Godwin, C. I.; Goh, Rocque Tien-Lock; Green, D. R.; Green, N. L.; Green, Nathan Louis; Green, William Robert; Grette, J. F.; Grieve, D. A.; Hawes, R. A.; Hayles, J. G.; Hebda, R. J.; Hendershot, W. H.; Herring, Bernard Geoffrey; Herzer, Richard H.; Hicock, Stephen Robert; Hodge, Robert A.; Hodgins, D. O.; Hoffman, Stanley Joel; Hoffman, Stanley Joel; Hoffmann, Joseph Walter; Holmes, Andrew; Htoon, Myat; Jarvis, Gary Trevor; Jensen, Oliver George; Johnson, Ian Mayhew; Johnson, R. D.; Jose, Barrie Frederick; Jurkevics, Andrejs; Karvinen, William O.; King, D. R.; Kloosterman, Bruce; Knize, S.; Lager, G. A.; Lajoie, Jules Joseph; Leary, George Merilin; LeBlanc, M. J.; Lecheminant, Anthony Norman; LeCouteur, Peter Clifford; Lee, C. A.; Lee, K. W.; Lett, R. E. W.; Levy, Shlomo; Lines, L. R.; Linton, John Alexander; Littlejohn, Alastair Lewis; Livingstone, Kent W.; Loveless, Arthur John; Luternauer, John Leland; Lynch, S.; MacDonald, Alan Stratton; MacDonald, Robert D.; Malecek, S. J.; Maxwell, R. J.; Maynard, D. E.; McLaren, G. P.; McLeod, J. A.; Medford, G. A.; Mesard, Peter Morris; Miller, Hugh Gordon; Miller, J. K.; Miller, Jack H. L.; Misener, Donald James; Misener, Donald James; Mitchell, David Laurie; Mitchell, Gerald George; Mitchell, William Sutherland; Moon, W.; Moore, Dennis Patrick; Moore, Patrick Albert; Morganti, John Michael; Mortensen, James Kenneth; Morton, Penelope Cane; Nagel, Joe Jochen; Napoleoni, Jean-Gerard Pascal; Narod, B. B.; Narod, B. B.; Nelson, Joanne Lee; Nielsen, K. C.; Nowell, A. R. M.; Okulitch, Andrew V.; Olade, M. A. D.; Olson, R. A.; Oriel, William Michael; Orr, John F. W.; Ozard, John Malcolm; Panteleyev, A.; Pareja, German J.; Parrish, Randall Richardson; Paterson, Ian Arthur; Pharo, Christopher Howard; Pharo, Christopher Howard; Pigage, L. C.; Pigage, Lee Case; Price, Barry James; Rahmani, Riyadh A.; Rakai, R. J.; Reamsbottom, Stanley Baily; Reamsbottom, Stanley Baily; Reid, Ruth Pamela; Reinsbakken, Arne; Richards, Gordon Gwyn; Richards, Tom; Riglin, L. D.; Roxburgh, Kenneth R.; Runkle, Dita Elisabeth; Ryan, Barry Desmond; Ryder, June Margaret; Salway, A. A.; Samuels, G.; Schreier, H.; Shim, J. H.; Singleton, G. A.; Small, William David; Smith, J. L.; Smith, Michael William; Smith, Randall Blain; Snead, David Edward; Solberg, Peter Harvey; Somerville, Paul Graham; Somerville, Paul Graham; Spence, George D.; Stablein, Newton Kingman, III; Stronach, J. A.;

Sule, P. O.; Syvitski, J. P. M.; Tessari, Oscar Jose; Thompson, C. B. H.; Thomson, Richard Edward; Thorleifson, Allan James; Tiffin, Donald Lloyd; Ueda, H.; Utting, M. G.; Vaid, Yoginder, P.; Vining, M. R.; Visentin, G.; Walmsley, M. E.; Westerman, C. J.; Wetherell, Dennis Gene; Wiese, Wolfgang; Wilton, D. H. C.; Wright, R. L.; Yedlin, M. J.; Zambresky, Liana

1980s: Ahern, Timothy Keith; Allen, David Peter Beddome; Andrew, Anne; Andrew, Kathryn Pauline Elizabeth; Archambault, Marthe; Arthur, Andrew John; Au, Chong Ying Daniel; Barrett, Gary Edward; Bartholomew, Paul Richard; Berman, Robert Glenn; Bhatia, Shobha Krisna; Bird, David Neil; Bloodgood, Mary Anne; Bloom, Mark Stephen; Bradford, John Allan; Broatch, Jane Catherine; Brown, Derek Anthony; Buchanan, Peter; Burnett, Ronald Gordon; Bustin, Robert Marc; Cahn, Lorie S.; Campbell, Susan Wendy; Carmichael, Scott Matthew McKenzie; Carter, Elizabeth Sibbald; Carter, Matthew; Carye, Jeffrey Alyn; Champigny, Normand; Chern, Jin-Ching; Coenraads, Robert Raymond; Cook, Raymond Arnold; Cudrak, Constance Frances; Day, Stephen John; De Capitani, Christian Emile; de Mora, Stephen John; Denton, Alexander W. S.; Desloges, Joseph Robert; Devlin, Barry David; Donald, Roberta L.; Duncan, Ian James; Elsby, Darren C.; Engi, Jill Ellen; England, Lindy Alison; Erdman, Linda Ruth; Ewing, Thomas Edward; Fillipone, Jeffrey A.; Forbes, Donald Lawrence; Forman, Robert Douglas; Forster, Craig Burton; Forster, Douglas Burton; Francois, Roger; Friedman, Richard M.; Fullagar, Peter Kelsham; Gabites, Janet Elizabeth; Gallie, Thomas Muir, III; Gao, Zu-Cheng; Garven, Grant; Garwin, Steven Lee; Getsinger, Jennifer Suzanne; Goldin, Alan; Goldsmith, Locke B.; Gorzynski, George Arthur; Goutier, Françoise Melanie; Hammerstrom, Lyle Thomas; Hay, Alexander Edward; Henderson, Charles Murray; Hickson, Catherine Jean; Holbek, Peter Michael; Horn, James R.; Isachsen, Clark; Jamieson, Gordon Reginald; Jones, Francis Hugh Melvill; Jones, Ian Frederick; Juras, Stephen Joseph; Knight, John Bruce; Kwong, Yan-Tat John; Laidlaw, James Stuart; Leckie, Donald Gordon; Levy, Shlomo; Lewis, Peter D.; Link, Christine Marie; Lisowski, Michael; Logan, James Metcalfe; Losher, Albert Justin; Lyngberg, Erik; Mäder, Urs Karl; Malott, Mary Lou; Mase, Charles William; Massmann, Joel Warren; Maxwell, Michael George; McColl, Kathryn Margaret; McCollor, Douglas Clayton; McDonald, Bruce Walter Robert; McKenzie, Kathleen Jane; McPhail, Derry Campbell; Milford, John Calverley; Moffat, Ian William; Montgomery, John R.; Moon, David Earl; Morganti, John Michael; Nauss, Anne L.; Negussey, Dawit; Nixon, Graham Tom; Novak, Michael David; Pakalnis, Rimas C. Thomas; Parkinson, David Lamon; Parrish, Randall Richardson; Perkins, Ernest Henry; Perkins, Ernest Henry; Price, Michael Glyn; Richard, Paul Francois; Riediger, Cynthia Louise; Robertson, Peter Kay; Rogers, Garry Colin; Rulon, Jennifer; Sanborn, Paul Thomas; Saxby, Donald William; Scheuer, Tim Ellis; Shea, Gregory Thomas Francis; Shepperd, Jane Elizabeth; Siddharthan, Rajaratnam; Sinclair, Alastair James; Sketchley, Dale Albert; Sondheim, Mark Weiss; Spence, George Daniel; Spies, Annette; Stanley, Clifford R.; Stanley, Clifford Read; Stinson, Kerry James; Styan, William Bruce; Sun, Min; Thomson, Robert Charles; Thorkelson, Derek John; Thorstad, Linda Elaine; Tombale, Akolang Russia; Vaziri-Zanjani, Hans Hamid; Vellutini, David; Waddington, Edwin Donald; Waldron, David Anthony; Wang, Xiaomin; Watson, Patricia Helen Wanless; Whaite, Peter; Whittall, Kenneth Patrick; Whittall, Kenneth Patrick; Williams, Vera Eileen; Woodbury, Allan David; Yogendrakumar, Muthucumarasamy; Young, Ian Fairley; Zou, Daihua

Brock University
St. Catharines, ON L2S 3A1

37 Master's

1970s: Delaney, Garry D.; Fisher, James Edward; Ghorashi-Zadeh, Medhi; Gombos, Frances; Laporte, Pierre J.; McAtee, Christopher L.; Mostaghel, Mohammad Ali; Parkins, William George; Payne, Craig William Charles; Tabatabai, Mehran; Veska, Eric; Winn, Cathie Eileen; Winn, Ronald Frederick; Woerns, Norbert M.

1980s: Barnsley, John Anthony; Bazinet, J. Paul; Buck, Shane; Chaudhry, Mohammad Naveed Hayat; Chen, Chang-Sen; Cumming, Bradley R.; Davison, James Gregory; Dillon, David Lloyd; Kester, Stephen Joseph; Kresz, David; Leyland, James G.; Lorek, Edward G.; Maddalena, Albert L.; Mihychuk, Maryann; Morrison, Joan O.; Otto, Judith E.; Pastirik, George Paul; Podolak, Wilfred E.; Reilly, Brian Arthur; Soo, Kwong Yin; Wassenaar, Leonard I.; Williams, Barbara L.; Wu, Tsai-Way

Brooklyn College (CUNY)
Brooklyn, NY 11210

77 Master's, 1 Doctoral

1960s: Banino, George M.; Boreske, John R., Jr.; Deganello, Sergio; Fine, Charles D.; Reddy, M. Rajendra; Rose, Lawrence; Youssefnia, Iradj

1970s: Agostino, Patrick N.; Baiamonte, Matthew J.; Bassir, Franklin; Benimoff, Alan Irwin; Benitt, Theodore G.; Berent, Louis J.; Blackbeer, Lawrence E.; Bookman, Marcia; Chin, Danton J.; Connors, Stephen Dennis; Ejiaku, Sammuel A.; Eriksen, Harold W.; Falchook, Martin G.; Ferrero, Walter; Fishbein, Alan S.; Focone, Joseph A.; Geddes, Arthur; Henderson, Lawrence; Isuk, Edet E.; Jones, Walter D.; Kantrowitz, Ralph; Katz, Solomon Stuart; Kessler, Edgar M.; Kornfeld, Itzchak; Lieberman, Marcus; Lucania, John A.; Mazzullo, Salvatore J.; Montag, Rafael L.; Okulewicz, Steven C.; Olynyk, Max; Phillips, Leonard; Ramondetta, Paul John; Rao, Nacharaju Manohar; Rao, Srinivas Thanner; Ribaudo, Anthony J.; Sandler, Carol; Schleifer, Stanley; Seidman, Peter; Shotwell, James D.; Sichko, Michael J.; Siegelberg, Alan I.; Silverman, Franklin M.; Silverstein, Hank; Small, Steven B.; Smith, Philip C.; Stransky, Terry; Tamm, Allan H.; Tavares, Ricardo A.; Tomkus, Mindaugas; Venticenque, Salvatore M.; Wilson, Frederick A.; Wolosz, Thomas Henry; Zamore, Yale; Zucker, Sandy M.

1980s: Akhionbare, Monday; Castagna, John Patrick; Clifford, Edward; Geralnick, Alan; Hershkowitz, Zoltan; Hess, Lillian M.; Kazakos, George K.; Kurzius, Elizabeth C.; Moore, William A., Jr.; Palestino, Robert; Pillsbury, Stephen W.; Rampertaap, Autar; Shapiro, Arthur A.; Stokes, Christopher P.; Waschitz, Martin; Weisberg, Michael; Zahralban, Thomas A.

Brown University
Providence, RI 02912

99 Master's, 121 Doctoral

1910s: Hawkins, Alfred C.; Weston, Marion D.

1920s: Cleaves, Arthur B.; Gedney, E. K.; Haring, Alfred M., Jr.; Round, Edna M.

1930s: Beach, J. S.; Chace, Frederick M.; Drewett, Margaaret E.; Emery, Herbert M.; Hautau, Gordon H.; Killeen, Pemberton L.; Millington, Berton R.; Stewart, Duncan; Young, John A.

1940s: Lytton, Gwyn B.; Potter, Donald Brandreth; Ray, Richard G.; Springer, George H.

1950s: Acker, Richard C.; Boone, Gary McGregor; Caldwell, Dabney Withers; Carter, Carol S.; Collins, Glendon Elmer; Erwin, Robert B.; Fofonoff, Nick P.; Frost, Richard J.; Hall, Bradford A.; Kozak, Samuel J., Jr.; Lane, Bernard O.; Pollock, S. J.; Prescott, Glenn C.; Ryan, Dennis J.; Stanley,

Daniel G. L.; Towe, Kenneth McCarn; Trew, James R.; Wosinski, J. F.

1960s: Anderson, Edwin Joseph; Balsam, William Lando; Bonar, Kermit Mark; Brumbaugh, Richard L.; Buyce, Michael R.; Buzas, Martin Alexander; Cheng, Yung-Yu; Crowley, Donald Joe; Crowley, Donald Joe; Day, Howard Wilman; Feininger, Tomas; Feininger, Tomas; Fletcher, Raymond Charles; Fletcher, Raymond Charles; Foland, Kenneth Austin; Folchetti, John Robert; Forbes, Warren Clarence, Jr.; Hall, Henry Thompson; Harakal, J. E.; Harper, John David; Head, James William, III; Hecht, Kurt; Helenek, Henry Leon; Hofmann, Albrecht Werner; Hofmann, Albrecht Werner; Hohl, Julia Catherine Beaman; Jahn, Bor-ming; Kiesewetter, Carl Herman; Koch, Jo-Ann Major (Sherwin); Lamb, Robert Alvis; Lin, Jian-Shengzhong; Mahoney, John Wells, Jr.; McCallister, Robert Hood; McCave, Ian Nicholas; McGetchin, Thomas Richard; Mesolella, Kenneth Joseph; Murray, Daniel Patrick; Park, Hong Bong; Park, Yong Ahn; Parr, John Thomas; Pingitore, Nicholas Elias, Jr.; Poore, Richard Zell; Rodgers, Donald Alvis; Rusling, Lee Judson, Jr.; Rutstein, Martin Stuart; Saunders, Ronald Stephen; Schwartzman, David William; Shaw, Charles E., Jr.; Shaw, Charles E., Jr.; Smith, James Dungan; Smith, Norman Dwight; Spang, John Harvey; Thompson, Allan McMaster; Welden, Charles Morris; Wells, Frederick Joseph; Woods, Michael Damian; Young, Davis Alan; Yurkovich, Steven Peter

1970s: Alewine, Ralph Wilson, III; Allan, John R.; Arvidson, Raymond Ernst; Arvidson, Raymond Ernst; Aurelia, Michael Anthony, III; Balsam, William Lando; Benson, Larry V.; Bernabo, J. Christopher; Boersma, Anne; Boersma, Anne; Briskin, Madeleine; Cleary, M. P.; Crowley, Thomas John; Crowley, Thomas John; Day, Howard Wilman; Eby, David Eugene; Epstein, Claude Murray; Fairbanks, Richard G.; Flessa, Karl Walter; Foland, Kenneth Austin; Galloway, Walter Bruce; Gamper-Bravo, Martha Alicia; Gebelein, Conrad Dennis; Grillot, Larry Ray; Grillot, Larry Ray; Groshong, Richard Hughes, Jr.; Halley, Robert Bruce; Harris, William Howard; Harrison, Randolph S.; Hassler, H. Patricia; Hawke, Bernard R.; Helenek, Henry Leon; Husseini, Moujahed I.; Husseini, Sadad Ibrahim; Husseini, Sadad Ibrahim; Hutson, William H.; Jones, Kenneth L.; Jovanovich, Dushon Bogdan; Kasper, Robert B.; Kasper, Robert Basil; Kronenberg, Andreas K.; Laseski, Ruth Anne; Lin, Jian-Shengzhong; Lis, Michael Gregory; Lohmann, George P.; Luz, Boaz; Mahlburg, Suzanne E.; Martin, Henry Jerome; McCallister, Robert Hood; Murray, Daniel Patrick; Pingitore, Nicholas Elias, Jr.; Poore, Richard Zell; Rigotti, Peter A.; Robbins, Gary Alan; Rodgers, Donald Alvis; Ryerson, Frederick J.; Sachs, Harvey Maurice; Sancetta, Constance Antonina; Saunders, Ronald Stephen; Schwartzman, David William; Settle, Mark F. P.; Sherwin, Jo-Ann Major Koch; Sipling, Philip Jay; Sipling, Philip Jay; Spang, John Harvey; Steinen, Randolph P.; Tewhey, John D.; Thayer, Richard E.; Tracy, Robert James; Wells, Frederick J.; Wiltschko, David V.; Winland, Hubert Dale; Wintsch, Robert P.; Withjack, Martha O.; Wood, Charles A.; Yurkovich, Steven Peter

1980s: Belanger, Paul Edward; Brown, Alton Arthur; Burnell, James Russell, Jr.; Carroll, Michael Robert; Christoffersen, Roy Gray; Cintala, Mark John; Crowley, Julia Coolidge; Cullen, James Leo; Curry, William Baetzel; Dell'Angelo, Lisa Nicole; Dickenson, Melville Pierce, III; Dickinson, James Edward, Jr.; Dowsett, Harry James; Ellison, Adam James Gillmar; Eysteinsson, Hjalmar; Farver, John Richard; Fisher, Donald Myron; Garvin, James Brian; Gephart, John Wesley; Hale-Erlich, Wendy Susan; Heide, Kathleen Mae; Helfenstein, Paul; Humphrey, John Dean; Kuo, Ban-Yuan; Major, Richard Paul; Michaud, Marion Catharine; Murchie, Scott Lawrence; Nishimura, Clyde Edwin; O'Hara, Kieran D.; Overpeck, Jonathan Taylor; Pedersen, Jens R.; Peterson, Larry Curtis; Phipps

Morgan, William Jason; Prentice, Michael Lanman; Prince, Roger Allan; Sharpton, Virgil Le Roy; Shelton, Glenmore Lorraine; Sieber, Kurt DeLance; Smith, Brian Mitchell; Snow, Eleanour Anne; Spulber, Susan Jane Dixon; Stockmal, Glen Stanley; Tellefsen, Mark James; Videtich, Patricia Ellen; Wagner, Paul David; Whitford-Stark, James Leslie; Williams, Alan Evan; Wood, Marcus Irwin; Zuber, Maria Theresa

Bryn Mawr College
Bryn Mawr, PA 19010

71 Master's, 30 Doctoral

1900s: Bliss, Eleanora Frances

1910s: Bliss, Eleanora Frances; Jonas, Anna Isabel

1920s: Cobb, Margaret Cameron; Morningstar, Helen; Smith, Isabel Fothergill

1930s: Abbey, Marjorie Best; Armstrong, Elizabeth Jeanne; Armstrong, Jane Crozier; Auerbach, Pauline Dorothy; Benedict, Dorothy K.; Dedman, Kathryn K.; Kingsley, Louise; Meier, Adolph Ernest; Wyckoff, Dorothy

1940s: Albigese, Muriel; Bell, Jane; Bell, Jane; Boudreau, Cynthia Elizabeth; Cameron, Narcissa S.; Ch'ih, Chi Shang; Ch'ih, Chi Shang; Dorsey, Anna Laura; Hunt, E.; Klingsberg, Cyrus; Lutz, Katherine; Lutz, Katherine; Schoonover, Lois Margaret; Vogt, Mary Cameron; Weagle, Lawrence Townsend; Weiss, Judith Vera; Weiss, Judith Verra; Willing, E. S., Jr.

1950s: Clavan, Walter; Davis, Annin-Gray; Dike, Paul A.; Greenwood, Richard, III; Jansen, George James; Klosterman, Gregory; Norton, Dorita Anne; Rasmussen, William Charles, Sr.; Reed, Juliet Carrington; Rosenzweig, Abraham; Wagner, Walter Richard; Waraksa, Irene Rosalina; Ward, Richard Floyd

1960s: Adams, M. Ian; Bennett, Lee C., Jr.; Feden, Robert Henry; Hoskins, Donald Martin; Johannson, Thora Marian; Jordan, Robert R.; Jordan, Robert R.; Marsters, Beverly Ann; Montgomery, Sonya Paris; Mumby, Joyce Ione; Olmsted, Franklin Howard; Reilly, Mercedes Catherine; Rice, Emery van Daell; Robelen, Peter G.; Roberts, Francis H.; Roberts, Francis H.; Thurston, Phillips Cole; Wagner, Mary Emma; Wier, Karen E.

1970s: Amenta, Roddy Vincent; Coward, Robert Irvin; Demmon, Floyd Earl, III; Hart, Dabney Gardner; Hollis, Stephen Hall; Huntsman, John Robert; Huntsman, John Robert; Kuhlman, Robert; Mark, Lawrence Edward; Sague, Virginia Muessig; Shapiro, Earl A.; Spoljaric, Nenad; Stoffregen, Roger Eben; Thomann, William F.; Valia, Hardarshan Singh; Wagner, Mary Emma; Watters, Thomas Robert; Wood, Cynthia; Yilmaz, Pinar Oya

1980s: Bond, Paul Norman; Cook, Robert Davis; Coulter, David H.; Gray, Mary Elizabeth; Hefferan, Kevin Patrick; Horowitz, Marcie R.; Krage, Susan Marie; Lawler, Jeanne Passante; Murphy, Brendan E.; Petrie, John David; Pupa, Diane Marie; Sacks, Paul Eric; Schwimmer, Reed Andrew; Snyder, Janet Greer; Steuer, Mark Raymond; Trueblood, Peter Martin; Witanachchi, Channa Devinda

University of Calgary
Calgary, AB T2N 1N4

214 Master's, 45 Doctoral

1960s: Baxter, Sonny; Brunger, Allan G.; Clack, William J. F.; Deere, Raymond Edward; Dolphin, Dale Robert; Ellison, Albert H.; Gorveatt, Arnold Charles; Harrison, Randolph Stephen; Havard, Kenneth R.; Johnson, Charlie Ernest; Jones, Jonathan Wyn; Labute, Gary James; Langhus, Bruce Gunnar; Leask, Dennis M.; May, Ronald William; McPherson, Harold J.; Morgan, Allan V.; Naqvi, Ikram Husain; Netolitzky, Ronald Kort; Pheasant, David R.; Raham, Gerald Orr; Rapson-McGugan, June E.; Shepheard, William Wayne; Williams, Harold H.; Wind, Gerrit

1970s: Adeniji, Francis A.; Alley, Neville F.; Boreski, Charles V.; Borowski, Robert D.; Boydell, Anthony N.; Boydell, Anthony N.; Bradshaw, John Yates; Braman, Dennis Richard; Brown, Stephen Phillip; Bujak, Catherine A.; Bustin, R. Marc; Cauffman, Lewis B.; Chi, Byung I.; Christie, David L.; Cornett, Shawn; Corrigan, Anthony F.; Craw, David; Crawford, Frank D.; Cruickshank, Roy Douglas; Davis, Thomas L.; Day, David L.; DeVries, Christiaan D. S.; Dewis, Frederick John; Douglas, Thomas Richard; Downes, Hilary; Duford, James Matthew; Dunsmore, Hugh E.; Embry, Ashton Fox, III; Embry, Ashton Fox, III; Ethier, Valerie Mary Girling; Ferguson, Angus J.; Fiesinger, Donald William; Fox, Julian C.; Gallagher, Maureen T.; Glendinning, Gerald R.; Graham, T. Gordon; Grams, Bryan A.; Grawbarger, David J.; Gunther, Paul R.; Hanneman, Debbie L.; Haug, Jerry L.; Havard, Christina J.; Hawes, Richard John; Higgs, Roger Y.; Hill, Roderic P.; Hockley, Glenn D.; Hodgkinson, John; Holmes, Donald W.; Howell, James Douglas; Jackson, Lionel E.; James, David Paul; Jamison, William Richard; Johnson, Clifford D.; Jones, Jonathan Wyn; Keeler, Robert George; Klewchuk, Peter; Knitter, Clifford Charles; Krause, Frederico Fernando; Letts, Robert E.; Love, Michael A.; Macrae, John M.; Matt, C. Diane; McCormack, Martin R. H.; McIlreath, Ian Alexander; McLaren, Patrick; Mee, Cathleen E.; Meilliez, Francis; Miller, Bruce E.; Mitchell, Wendy J.; Mossop, Grant D.; Netterville, John A.; Nicoll, Larry D.; O'Connor, Michael J.; Ogunyomi, Olugbenga; Okon, Edem Effiong; Oliver, Elisabeth Margaret; Oyibo, Chamberlain Oruwari; Palonen, Pentti Arnold; Palonen, Pentti Arnold; Poulton, Terrence Patrick; Pratt, J. Lynn; Putnam, Peter Edward; Read, Burton Charles; Robbins, David B.; Roblesky, Robert F.; Rottenfusser, Brian Albert; Sargent, Melville Wayne; Saydam, A. Sacit; Schooley, Jeannie Victoria; Scott, James Alan Bryson; Sheasby, Norman Michael; Speelman, J. D.; Stoakes, Franklin Arthur; Stockmal, Glen Stanley; Stout, Mavis Z.; Strong, Wayne L.; Swagor, Nick S.; Sweet, Arthur R.; Tan, J. Tony; Thomas, Mark B.; Venter, Ronald H.; Verschoor, Karin van Romondt; Walker, Michael; Waters, Robert R.; White, Robert M.; Wilson, Michael S.; Wong, Pak K.; Yarnal, Brenton M.

1980s: Abercrombie, Hugh James; Anderson, Neil L.; Aulstead, Kathy Loree; Bays, Alan R.; Beattie, Edward T.; Beauchamp, Benoit; Bélanger-Davis, Colette E.; Blumstengel, Wayne; Boadu, Fred Kofi; Bradley, Cheryl; Braman, Dennis Richard; Brodylo, Leslie A.; Burden, Elliott T.; Calverley, Anne; Calvert, H. Thomas; Candido, Aladino; Carey, J. Anne; Celestino, Tarcisio Barreto; Chatellier, Jean-Yves; Cheadle, Scott Philip; Coflin, Kevin C.; Corbett, Cynthia Rena Ruth; Crowe, Gregory George; Cummins, Catharine; Currie, Lisel D.; Cutler, William Gerald; Dechesne, Roland George; Diegel, Scott G.; Doig, D. J.; Dougherty, Sean M.; Dove, Jane E.; du Toit, Charl; Dudley, Jon Steven; Dufresne, Denis; Dwyer, Mary Kathleen; Easley, Donald J.; Eaton, David W. S.; Ferri, Filippo; Fischer, Brian Frederick Gustav; Fitzpatrick, Mark; Freeman, Kimberley June; Freiholz, Ginette; Gardner, Howard David; Gilhooly, Murray Gordon; Gintautas, Peter Alan; Gremell, Paul E.; Griffith, L. A.; Gustafson, Catherine; Halwas, David Bruce; Hamill, Gary Bruce; Hearn, Deborah J.; Heise, Roy H.; Hugo, Ken J.; Hyslop, Kevin D.; Jin, Zhenkui; Kalstrom, Eric T.; Kay, Anthony Edward; Kirker, Jill Kathleen; Knox, Alexander Walter; Kosciusko, Kim Anne; Kostaschuk, Raymond A.; Kutluk, Hatice; Lang, Harold R.; Lee, Julie M.; Lee, Linda J.; Lefebvre, Rene; Locking, Tracy; Lortie, Johanne; Lowey, Grant William; Lowey, Grant William; Lowey, Jennifer Fortune; Lyatsky, Henry; MacDonald, Glen M.; Machemer, Steven Dean; Marion, Donat J.; Maurel, Laurie Elisabeth; McDonough, Michael Robert; Mihalynuk, Mitchell George; Moffat, Ian William; Morrison, Michael L.; Nicholls, Elizabeth L.; Oke, Christopher; Olsen-Heise, Katrina Edith Desa; Patterson, Judith Gay; Paukert, Gary William; Pell, Jennifer; Poley, Denise

F.; Putnam, Peter Edward; Quin, Andrew; Raeside, Robert P.; Rall, Robert D.; Reichenbach, Mary Elizabeth; Reid, Jeffery P.; Rhine, Janet L.; Richards, Kenneth C.; Ridley, Susanne Larkin; Riggert, Virginia Leigh; Root, Kevin Gordon; Russell, James Kelly; Russell, James Kelly; Sanderson, Deborah Anne; Santos, Rebecca de Regla; Sevigny, James H.; Shaw, Jerry C.; Slawinski, Michael A.; Steffes, Darren A.; Stromquist, Barbara Haworth Adams; Stronach, Nicholas John; Struik, Lambertus Cornelis; Suneby, Lena B.; Tang Kong, V. William; Theriault, François; Tieman, Joannes Jacobus; Trott, Barbara G.; Twombly, Gregory L.; Vanderburgh, Sandy; Varley, Christopher John; Varsek, John L.; Viau, Christian Alain; Visser, John; Waters, Brent Blakely; Weedmark, Ron Harold; Weissenberger, John Arthur William; Wilmot, Barry R.; Wilson, Michael; Wood, James Michael

University of California, Berkeley
Berkeley, CA 94720

425 Master's, 780 Doctoral

1890s: Anderson, Frank Marion; Fairbanks, Harold Wellman; Fairbanks, Harold Wellman; Fairbanks, Harold Wellman; Fairbanks, Harold Wellman; Fairbanks, Harold Wellman; Fairbanks, Harold Wellman; Fairbanks, Harold Wellman; Fairbanks, Harold Wellman; Fairbanks, Harold Wellman; Fairbanks, Harold Wellman; Fairbanks, Harold Wellman; Fairbanks, Harold Wellman; Fairbanks, Harold Wellman; Fairbanks, Harold Wellman; Fairbanks, Harold Wellman; Fairbanks, Harold Wellman; Fairbanks, Harold Wellman; Fairbanks, Harold Wellman; Fairbanks, Harold Wellman; Fisher, Grace Merriam; Louderback, George Davis; Manson, Marsden; Palache, Charles; Ransome, Frederick Leslie; Sharpe, Selina; Smith, William Sidney Tangier

1900s: Knopf, Adolph; Knopf, Adolph; Sinclair, William J.; Thelen, Paul; Weaver, Charles Edwin; Wemple, Edna Mary

1910s: Baker, Charles Laurence; Bradley, Walter W.; Buwalda, John Peter; Clark, Bruce Lawrence; Clark, Clifton Wirt; Davis, Elmer Fred; Davis, Elmer Fred; Dickerson, Roy Ernest; Durst, Fred M.; English, Walter A.; Kew, William Stephen Webster; Kew, William Stephen Webster; Larsen, Esper Signius; Martin, Bruce; Miller, Loye Holmes; Nomland, Jorgen O.; Nomland, Jorgen O.; Packard, Earl Leroy; Ruckman, John Hamilton; Schilling, Karl H.; Stewart, Lloyd Lincoln; Stock, Chester; Studley, Clarence K.; Vaughan, Francis Edward; Vaughan, Francis Edward; Wagner, Carroll M.; Whitman, Alfred Russell; Wilcox, Glenn Avery; Woodcock, Edgar

1920s: Abad, Leopoldo F.; Allen, Victor Thomas; Anderson, Charles Alfred; Bailey, Thomas Laval; Bailey, Thomas Laval; Byerly, Perry E.; Chan, Chun Young; Collins, Isabelle D.; Corey, William Henry; Crook, Theo Helsel; Edwards, Merwin Guy; Elftman, Herbert Oliver; Foshag, William Frederick; Frost, Frederick Hazard; Hanna, Marcus Albert; Howard, Hildegarde; Howard, Hildegarde; Hudson, Frank Samuel; Hulin, Carlton D.; Lee, Huyler Wells; Mason, Herbert Louis; Meserve, Clement Dann; Mitchell, George Dampier; Morse, Roy Robert; Moss, Frank Ambrose; Moss, Frank Ambrose; Nelson, Richard Newman; Nichols, William Marvin; Norton, Richard Drake; Oakeshott, Gordon Blaisdell; Pabst, Adolf R.; Palmer, Dorothy Bryant (Kemper); Pressler, Edward D.; Rankin, Wilbur D'Arcy; Richards, Esther English; Robson, Homer L.; Russell, Richard Joel; Sasse, Jerome B.; Schenck, Hubert G.; Taliaferro, Nicholas Lloyd; Thornburgh, Hubert Robert; Trask, Parker Davies; Trask, Parker Davies; Waterfall, Louis N.; Webber, Irma Eleanor Schmidt; Webber, Irma Eleanor Schmidt; Whipple, George Leslie; Whitman, Alfred Russell; Wilcox, Marion; Woodford, Alfred Oswald

1930s: Anderson, Howard T.; Andrews, Philip; Axelrod, Daniel I.; Bentson, Herdis; Boulger, Martha Lillian; Bramkamp, Richard A.; Bruff, Stephan

Cartland; Carder, Dean Samuel; Clark, Samuel Gilbert; Coats, Robert Roy; Condit, Carlton; Doell, Edward Charles; Dosch, Earl Fuller; Durham, John Wyatt; Durrell, Cordell; Dyk, Karl; Effinger, William Lloyd; Enders, Dean W.; Etherington, Thomas John; Fleming, Richard Howell; Gardescu, Ionel Ion; Gardner, Dion Lowell; Gee, Herbert Caran; Gilbert, Charles Merwin; Gilmore, Susan Potbury; Goudy, Clyde LeRoy; Gregory, Joseph Tracy; Hazzard, John Charles; Herold, C. Lathrop; Hesse, Curtis Julian; Jensen, Christian; Johnson, Francis Alfred; Johnson, Francis Alfred; Kesselli, John E.; Knox, Newton Booth; La Motte, Robert S.; MacDonald, Gordon Andrew; MacGinitie, Harry D.; Mason, Herbert Louis; McGrew, Paul Orman; Merriam, Charles W.; Mull, Bert Hathaway; Oliver, Elizabeth Sumner; Phillips, Ross M.; Rand, William Whitehill; Reiche, Parry; Revelle, Roger; Richey, King Arthur; Rossello, Pierina Onorina Blanch; Rott, Edward H., Jr.; Russell, Richard Dana; Ruth, John William; Sander, Nestor John; Sawyer, Byrd Fanita Wall; Schombell, Leonard Frederick; Simpson, Edward Cannon; Smith, Natasha D.; Smith, Neal Johnstone; Solari, Allison Jarvis; Sommer, Hulda Henrietta; Stedischulte, Victor Cyril; Tabor, Lawrence LaVerne; Tegland, Nellie May; Thomas, Edwin S.; Thorp, Eldon M.; Turner, Francis Earl; Unterman, Billie Ruple; Van Gundy, Clarence Edgar; Vander Hoof, Vertress Lawrence; VanderHoof, Vertress Lawrence; Vitt, Alfred Weston; Vokes, Harold Ernest; Wilson, James T.

1940s: Adkins, John Nathaniel; Allen, John Eliot; Angel, Loren Henry; Atkinson, James Ernest; Aydinoglu, Mustafa Ali; Bell, Gordon Leon; Bentson, Herdis; Borglin, Edward Kenneth; Boyd, Harold Alfred, Jr.; Brice, James Coble; Carter, William Horace; Chesterman, Charles Wesley; Christensen, Andrew Lee; Clark, Arthur Watts; Conrey, Bert Louis, Jr.; Crittenden, Max D., Jr.; Durham, John Wyatt; Erickson, Martin Richard; Fiedler, William Morris; Gardner, Robert A.; Goss, Charles Richard; Graham, Joseph John; Harding, John William, Jr.; Harman, John Warren; Harrington, William Cornell; Herkenham, Marjorie Watson; Hotz, Preston Enslow; Huang, Wei Ta; Huey, Arthur Sidney; Hurlbut, Elvin Millard, Jr.; Johnston, Stedwell; Kiilsgaard, Thor H.; Lambert, Earl Freeman; Leith, Carlton J.; MacDonald, James Reid; Macdonald, James Reid; Mallory, Virgil Standish; Mandra, York T.; Manning, George A.; Masson, Peter H.; Mathews, William Henry; McAndrews, Martin George; McIntyre, James R.; Merriam, Richard Holmes; Mielenz, Richard Childs; Minton, Morris Cresswell; Newton, Ralph James; Ogle, Burdette Adrian; Owens, John Snowden; Pabst, Marie Bertha; Parker, Pierre Edward; Patten, Philip R.; Peabody, Frank Elmer; Peabody, Frank Elmer; Peryam, Richard Calvin; Phillips, Irvine Lewis; Repecka, Albert Lee; Robertson, Mary Spotswood; Robinson, Gershon DuVall; Rose, Robert Leon; Sarmiento-Soto, Roberto; Savage, Donald Elvin; Schlocker, Julius; Shouldice, James Robert; Stirton, Ruben Arthur; Taylor, David N.; Thomas, R. P.; Washburn, Edward Davis III; Watson, Samuel Eugene, Jr.; Welby, Charles W.; Welles, Samuel P.; Wheeler, Dooley Peyton, Jr.; Wilson, Eugene Jack; Wilson, Ivan Franklin

1950s: Aarons, Bernard Louis; Addicott, Warren Oliver; Aho, Aaro Emil; Alexander, Charles S.; Alfors, John Theodore; Allison, Edwin Chester; Attridge, John; Back, William; Barr, Frank Theodore; Bauer, Francis Harry; Beal, Laurence Hastings; Benedict, Reba Ward; Blaisdell, Robert Clark; Booth, Charles Vincent; Borg, Iris Parnell; Bowen, Oliver Earle, Jr.; Bowers, Robert Alwyn; Bowers, Robert Alwyn; Boyd, Harold Alfred, Jr.; Brewer, William August, III; Brice, James Coble; Briggs, Louis Isaac; Brooks, Elwood Ralph; Browning, John Leverett; Burke, Willard F.; Cameron, John Baades; Cebull, Stanley Edward; Chandra, Deb Kumar; Christensen, Mark Newell; Churkin, Michael, Jr.; Cifelli, Richard; Classen, Willard John, Jr.; Collins, Donald Francher; Coogan, Alan

Hall; Cook, Theodore Davis; Cosgriff, John William; Cox, Allan Verne; Creely, Robert Scott; Crutcher, John Fulton; Crutchfield, William Henry, Jr.; Curtis, Garniss Hearfield; Darrow, Richard Lee; Day, David William, Jr.; De Bremaecker, Jean-Claude; De Noyer, John Milford; DeLeen, John L.; DeLise, Knoxie Carlton; Dempster, R. E.; Dickson, B. A.; Dixon, Helen Roberts; Doell, Richard Rayman; Dondanville, Richard Fred; Douglass, Robert Marshall; Doumani, George Iskandar; Downs, Theodore; Droullard, Emerson Keith; Duley, Dale Hamilton; Dunn, James Robert; Eaton, Jerry Paul; Eisenhardt, William Charles; Elliott, Douglas Howard; Emerson, William Keith; Engstrom, David Bert; Epis, Rudy Charles; Estes, Richard Dean; Evernden, Jack Foord; Fairchild, William W.; Farr, John Brent; Fields, R. W.; Finch, Warren Irvin; Fournier, Robert Orville; Frames, Donald Wayland; Freeman, Val LeRoy; Fulmer, Charles Virgil; Gastil, Russel Gordon; Gay, Thomas Edwards, Jr.; Gilbert, Francis Louisa; Giles, Eugene; Girard, Lewis V.; Glen, William; Glover, Joseph John Edmund; Goldberg, Irving; Goodwin, Joseph Grant; Gray, Jane; Gray, Raymond Franklin; Green, Morton; Greife, John Luverne; Grier, Albert William; Gutierrez, Dora; Hacker, Robert Norris; Halsey, Jonathan Horace; Ham, Cornelius Kimball; Hemley, John Julian; Herlyn, Henry Traver; Herrera, Leo John, Jr.; Higgins, Charles Graham, Jr.; Hodder, Robert William; Hollister, Victor F.; Hoover, Linn; Hornaday, Gordon Raymer; Hunt, John Prior; Isaacs, Kalman Nathan; Johnson, Robert Francis; Kilmer, Frank Hale; King, Truxton W.; Kirk, Mahlon V.; Klucking, Edward Paul; Langston, Wann, Jr.; Lerbekmo, John Franklin; Lindsey, E. H.; Lipson, Joseph I.; Loney, Sabra Osborn; Lutz, George C.; Lyon, Ronald James Pearson; Maddock, Marshall E.; Mallory, Virgil Standish; McKenna, Malcolm Carnegie; McLaughlin, Donald Hamilton, Jr.; Mero, John L.; Molander, Gene Emery; Murphy, Michael Joseph; Nelson, Rex W.; Nye, Thomas Spencer; Oakes, Millis Henry; Oberling, John James; Oberling, John James; Oestreich, Ernest Sebastian; Ogilvie, Thomas F.; Ogle, Burdette Adrian; Ortalda, Robert A.; Parker, Ronald Bruce; Pease, Maurice Henry, Jr.; Peikert, Ernest William; Pelletier, Willis Joseph; Pestana, Harold R.; Plafker, George; Pratt, Willis Layton, Jr.; Puffer, E. L.; Quaide, William Lee; Quaide, William Lee; Raydon, Gerald Thomas; Raymond, Martin Snider; Reinhart, Roy H.; Rogers, Thomas Hardin; Romney, Carl Frederick, Jr.; Rose, Robert Leon; Ross, Thomas Paul; Russell, Don Eugene; Scott, Theodore; Shank, Stephen Everett; Shaw, Charles McEwen; Shaw, Herbert Richard; Sheehan, Jack Richard; Shotwell, Jesse Arnold; Slemmons, David Burton; Smiley, Charles Jack; Smith, Arthur Rankin; Smith, Bernice Young; Smith, Hugh Preston; Snow, David Tunison; Stauder, William Vincent; Stensaas, Larry; Sullivan, Frank Rogers; Taylor, Dwight Willard; Taylor, Dwight Willard; Taylor, Robert Kruse; Taylor, Robert Wesley; Tischler, Herbert; Tocher, Don; Travis, Russell Burton; Tuller, Burl A.; Turner, Mortimer Darling; Vernon, James Wesley; Wagner, John Paul; Waite, Roy Harold; Walsh, Thomas F.; Wardle, William Clyde; Weaver, Donald W.; Weaver, William Ray; Weidman, Robert McMaster; Wesendunk, Paul Robert; Whetten, John Theodore; White, John Francis; Wilshire, Howard Gordon; Witherspoon, A. J.; Withstandley, D. W., III; Wolfe, Jack A.; Woodside, Edward R.; Young, Gerald Clifton

1960s: Abdel-aal, Farouk Mostafa; Abou-seida, Mohamed Mokhles; Adachi, Toshihisa; Adegoke, Sylvester; Al Rawi, Yeha Tawfeq; Allison, Edwin Chester; Allison, Richard Case; Applegate, Robert Lewis; Appuhn, Richard A.; Arden, Daniel Douglas; Arulanandan, Kandiah; Baird, Alexander Kennedy; Baker, William Laird; Barnes, Lawrence Gayle; Beaty, Chester B.; Belsky, Theodore; Best, Myron Gene; Bischoff, James Louden; Blake, Daniel Bryan; Blusson, Stewart Lynn; Bodmer, Rene; Bole, George Robert; Boyle, Michael William;

University of California, Berkeley

Bramble, Dennis M.; Braslau, David; Brewer, William August, III; Brodersen, Ray Arlyn; Brown, Edwin Hacker; Buffler, Richard Thurman; Burnett, John L.; Burns, Roger George; Byerly, John Robert; Campbell, David Lowell; Carmichael, Dugald Macaulay; Case, James Edward; Cavender, Wayne Sherrell; Chang, Chin-yung; Cherry, John A.; Chinn, William; Chopra, Anil Kumar; Christensen, Roberta Smith; Chu, Wen-kuan; Clague, John Joseph; Clemens, William Alvin, Jr.; Cochran, George Raymond; Coe, Robert Stephen; Collett, Raymond T.; Colwell, Jane; Cook, Frances Govean; Cook, Harry Edgar, III; Cosgriff, John William; Crawford, María Luisa Busé; Crawford, William Arthur; Cross, Ralph Herbert, III; Curry, Robert Rodney; Dailey, Donald Howard; Dalrymple, Gary Brent; Daniels, David Lee; Davidson, Lloyd Arthur; Davis, Gregory Arlen; Derr, John Sebring; Dews, Jon Robert; Dezfulian, Houshang; Dias, Carlos Alberto; Dibaj, Mostafa; Douglass, John Liddell; Druckerman, Daniel; Drummond, Arthur Darryl; Dudley, Priscilla Perkins; Eastwood, William Clifford; Essene, Eric John; Estes, Richard Dean; Ewoldsen, Hans Martin; Faustman, Walter Francis; Filson, John Roy; Firby, James Ronald; Firby, James Ronald; Firby, Jean Brower; Fleck, Robert Joseph; Foster, Merrill White; Fraser, Douglas Culton; Freeze, Roy Allan; Fuller, Brent D.; Galehouse, Jon Scott; Gangloff, Roland Anthony; Gardner, Robert Alexander; Ghent, Edward Dale; Gilbert, Neil Jay; Glickstein, Sara; Gluskoter, Harold Jay; Goldstein, Norman Edward; Goldstein, Norman Edward; Gonsalves, Ronald George; Goodman, Richard Edwin; Gray, Donald Harford; Grivetti, L. E.; Grommé, Charles Sherman; Hall, Minard Lane; Hall, Nelson Timothy; Haller, Charles Regis; Hamilton, Robert Morrison; Hansen, Richard Otto; Hassan, Mamdouh Abdel-Ghafoor; Haug, Patricia Ann; Hay, Edward Alexander; Helley, Edward John; Henderson, Gordon William; Hirst, Terence John; Hodgson, Charles Jonathan; Holdaway, Michael Jon; Hollander, Margaret; Holloway, Ralph Leslie, Jr.; Houston, William Newton; Hyndman, Donald William; Ismail, Farouk Taha Ahmed; Isselhardt, Courtney Francis; Jack, Robert Norman; Jahn, Melvin Edward; James, Gideon T.; Janda, Richard John; Jenkins, Sidney Ford; Jepsen, Anders Frede; Jewell, Thomas Ross; Johnson, Maureen G.; Johnston, Ian McKay; Kaar, Robert Frederick; Kadib, Abdel-Latif Abdullah; Kalkanis, George; Keevil, Norman Bell, Jr.; Keevil, Norman M.; Kern, John William; Kerrick, Derrill Maylon; Kiefer, Fred William, Jr.; Kilmer, Frank Hale; Kirsch, Stephen Augustine; Kishk, Fawzy Mohamed; Kistler, Ronald Wayne; Klikoff, Waldemar A., Jr.; Klucking, Edward Paul; Klucking, Edward Paul; Kumar, Ashok; Kuru, Durmus; Lajoie, Kenneth Robert; Langenheim, Virginia McCutcheon; Langston, Robert Burlison; Lee, Kenneth Lester; Leonardos, Othon Henry; Leung, Irene Sheung-Ying; Levi, Beatriz; Licari, Gerald Richard; Licari, Joan Perusse; Lin, Wu-Nan; Lindsay, Everett Harold, Jr.; Liu, Shih-Chi; Loney, Robert Ahlberg; Lorber, Harvey Raymond; MacClintock, Copeland; Mahony, John Daniel; Malhotra, Anil Kumar; Mann, Alan Eugene; Marchand, Dennis Eugene; Marcus, Leslie Floyd; Marzke, Mary Ronald Walpole; Masters, Bruce Allen; Mathur, Jagdish Narain; Mawby, John Evans; Mawby, John Evans; McAleer, Joseph Francis; McBirney, Alexander Robert; McCall, Mary A.; McCarthy, Eugene Desmond; McKay, David Stuart; McKee, Edwin H.; McNitt, James Raymond; Means, Winthrop Dickinson; Merrill, Ronald Thomas; Mesenbrink, Joseph H.; Meyers, William John; Mintz, Leigh Wayne; Mitchell, Edward D.; Moila, Richard James; Moore, Charles Aurelius, Jr.; Moore, Donald Bruce; Morrison, Huntley Frank; Mottern, Hugh Henry, Jr.; Nash, Douglas B.; Nations, Jack Dale; Neuman, Shlomo Peter; Nevin, Andrew Emmet; Niazi, Mansour; Nicholls, James Watson; Nielsen, Richard Leroy; Norrany, Iraj; Nowroozi, Ali Asghar; O'Brien, Douglas Patrick; Obert, Karl Richard; Obradovich, John Dinko; Pabst, Marie Bertha; Paduana, Joseph Anthony; Page, Norman J.; Pagenhart,

Thomas Harsha; Parry, Robert J.; Pepin, Robert Osborne; Phillips, Roger Jay; Plane, Michael Dudley; Podosek, Frank Anthony; Primmer, Stanley Russell; Read, Peter Burland; Read, William Harold; Reed, Walter Edwin; Rensberger, John Marshall; Rensberger, John Marshall; Reynolds, Mitchell William; Reynolds, Sargent Thurber; Rhea, Keith Pendleton; Rich, Thomas Hewitt; Rivera, Robert A.; Robinson, Paul Thorton; Romey, William D.; Rosell, Ramon Antonio; Ross, Grant Arrett; Rozelle, Richard Kent; Ruiz, Patricio; Russell, Dale Alan; Ryall, Alan S., Jr.; Saraf, Deoki Nandan; Sauer, Emil Karl; Sayers, Frank Edward; Sayles, Frederick Livermore; Shahbazi, Mohsen; Sheehan, Peter Michael; Shlemon, Roy James; Singh, Awtar; Smiley, Charles Jack; Smith, Alan Lewis; Snow, David Tunison; Steinker, Donald Cooper; Sullivan, Frank Rogers; Taylor, Roger Loren; Tedford, Richard Hall; Thoms, Richard Edwin; Tidball, Ronald Richard; Tobisch, Othmar Tardin; Trost, Lawrence C.; Turcotte, Frederick Thomas; Turner, Donald Lloyd; Tuttle, Russell Howard; Van Hoeven, William, Jr.; Waines, Russell Hamilton; Wakeley, John Raymond; Wallis, James R.; Ware, George Hunter; Warner, Don Lee; Waters, Barbara Tihen; Waters, John Nelson; Weaver, Donald W.; Webb, David B.; Weill, Daniel Francis; Wells, John Maurice; Whistler, David P.; White, Richard William; Willis, Gar Charles; Wilson, Edward Carl; Wilson, Edward Carl; Wilson, Thomas A.; Woakes, Michael Edward; Wolfe, Jack Albert; Wollenberg, Harold; Woodard, Geoffrey Davidson; Woodburne, Michael Osgood; Wornardt, Walter William; Yancey, Thomas Erwin; Young, Rex James; Zelwer, Ruben; Zullo, Victor August; Zullo, Victor August

1970s: Aagaard, Per; Ahern, Kevin Edward; Ahern, Kevin Edward; Al-Shawaf, Taha Daud; Altintas, Sabri; Alvarez-Bejar, Ramon; Amy, Gary Lee; Andrews, Edmund Daniel; Archibald, James David; Ayatollahi, M. S.; Bacon, Charles Robert; Bakbak, Mohamed Rida; Bakun, William Henry; Bandyopadhyay, Sunirmal; Banerjee, Nani Gopal; Barker, Richard M.; Barker, Robert Wadhams; Barnes, Lawrence Gayle; Behrman, Philip George; Berger, Ernst; Berry, David R.; Berta, Ann Alisa; Beyer, John Henry, Jr.; Bird, Dennis Keith; Borcherdt, Roger D.; Bracewell, L. W.; Bramble, Dennis Marley; Brett-Surman, Michael Keith; Brimhall, George H., Jr.; Brown, Francis Harold; Budge, David R.; Burns, Lary Kent; Bushee, Jonathan; Butler, Paul Ray; Campbell, Colin Robert; Cerling, Bettina; Cerling, Thure Edward; Chan, Kwok Chun; Chang, Ching Shung; Chen, Chao-Hsia; Clark, Samuel Harvey, Jr.; Coggon, John Henry; Cole, Stuart Loren; Cornelius, Scott Bedford; Corwin, Robert F.; Custer, Stephan Gregory; Daud, Badruddin Haroon; Delany, Joan Marie; Dengler, Lorinda Ann; Dewey, James William; Dey, Abhijit; Dickman, Steven Richard; Domning, Daryl Paul; Domning, Daryl Paul; Donnelly, J. M.; Drake, F. Lawrence; Drake, Robert Edward; Edwards, Stephen Walter; Eisenberg, Alfredo; Elias, Ghanem; Ellis, Richard Keller; Erskine, Mellville Cox, Jr.; Etayo-Serna, F.; Files, Frederic Grant; Fitzpatrick, Joan Juliana; Flowers, George Conrad; Frisch, Conny Jean; Fuller, Brent D.; Gale, J. E.; Gangloff, R. A.; Giovannetti, Dennis; Godoy, Estanislao Pirzio-Biroli; Goodin, Sarah Elizabeth; Gutierrez, J. A.; Haible, William Wilson; Hamati, R. E.; Hamilton, Robert Bruce; Hansen, William Walter; Hart, Earl W.; Heming, Robert Frederick; Heuze, Francois; Hildenbrand, T. G.; Hildreth, Edward Wesley; Hirschfield, Sue Ellen; Hittinger, M.; Hochstetler, Laurel Huggins; Hoffman, Kenneth Alan; Hohmann, Gerald Wayne; Hornaday, Gordon Raymer; Howard, Kenneth Leon, Jr.; Huffman, Othis Frank; Hutchison, John Howard; Irvine, Pamela Joe; Iwai, Katsuhiko; Jacobson, Mark Ivan; Jain, Birendra Kumar; Jaouni, Abdur-Rahim Khalil; Jaworski, Gary William; Jiracek, George R.; Jones, G. M.; Kalra, Ashok; Kane, Phillip S.; Kasim, A. G.; Kavazanjian, E., Jr.; Kendall, Ernest W.; Kharaka, Yousif Khoshu; Kojan, Eugene; Korentajer, Leonid; Krohn, I. M.; Lacerda, W. A.;

Lander, E. B.; Lander, E. B.; Lattanner, Alan V.; Laws, Richard Anthony; Lawson, Douglas Allan; Lee, Florence Ling; Lee, K. H.; Lee, Luther; Leppaluoto, David Alan; Leslie, Kenneth Campbell; Leu, L.-K.; Liaw, Alfred L.; Lim, Sheldon C. P.; Lin, Wu-Nan; Linn, George Willison; Lippmann, Marcelo E.; Lisle, Thomas E.; Litehiser, J. J., Jr.; Lourenco, Jose S.; Lowder, Garry George; Madden, Cary T.; Mahmood, Arshud; Majer, E.; Makdisi, Faiz Isbir; Marsh, Bruce David; Marshall, Larry G.; Mazzella, A. T.; McEdwards, Donald George; McNally, Karen Cook; Merino, Enrique; Meyer, Wallace Harold, Jr.; Miller, Wade E.; Minard, Claude Russell; Mojtahedi, Soheil; Mori, Kenji; Morris, P. A.; Munthe, Lynn Kathleen; Munthe, Lynn Kathleen; Murphy, Sean E.; Myer, Larry Richard; Namiq, Laith Ismail; Nash, William Purcell; Nasu, Mitsuru; Navolio, Michael Edward; Nelson, Clifford Melvin; Nelson, Stephen Allen; Nieuwenhuis, Carol Ann; Nord, Gordon Ludwig; Norris, Gary Martin; Novacek, Michael John; Olson, Peter Lee; Olson, Todd Rowland; Osborn, Gerald Davis; Osuch, Lawrence Theodore; Page, Robert Hull; Pape, Donald A.; Parkison, Gary Alden; Patet, Alix; Peppin, William Alan; Phillips, Franklin Jay; Plasil, Georg; Prestegaard, Karen; Price, Jonathan Greenway; Proffett, John Maddon, Jr.; Qamar, Anthony; Quigley, Donald Walker; Rahman, M. Shamimur; Rainey, Clifford S.; Reed, Mark Hudson; Reed, Walter Edwin; Rider, Jonathan Richards; Risbud, Subhash H.; Ross, James Robert Holland; Rossi, Randall Steven; Roth, Barry; Rowntree, Rowan A.; Russell, Larry Lee; Ryu, Jisoo; Salami, Moshudi Babajide; Sancio, Rodolfo Traostino; Sarna-Wojcicki, Andrei M.; Schnabel, Per Borre; Sheehan, Peter Michael; Shoja-Taheri, Jafar; Shore, Jesse Paul; Silva, Walter Joseph, Jr.; Singh, S.; Sloan, Doris; Smith, Warwick Denison; Somerville, Malcolm Robert; Sorey, M. L.; Spera, F. J.; Stewart, Roger Malcolm; Stifler, James Fairman; Stoessell, Ronald Keith; Stormer, John Charles; Stump, Brian Williams; Sundaram, Panchanatham Naga; Taylor, Michael Evan; Tidy, Enrique Finch; Tipton, Ann; Tobar, Alvaro Barra; Toste, A. P.; Treadwell, D. D.; True, D. G.; Trussell, Devin Brian; Tyler, D. B.; Udaka, T.; Uhrhammer, R.; Ulbrich, Horstpeter Herberto Gustavo Jose; Ulbrich, Mabel Norma Costas; Varela, Francisco Ernesto, Jr.; Vasconcelos, J. J.; Wagner, Carol Daily; Wagner, Dana Bernadine; Wagner, Hugh McKinlay; Wagner, Jesse Ross; Wallace, James Hay; Walther, John Victor; Wanger, D. B.; Ware, George Hunter; Watkins, Rodney Martin; Welton, B. J.; Wieczorek, G. F.; Wilde, Walter Rowland; Williams, C. T.; Wolff, Ronald Gilbert; Wolfgram, Diane; Woloszyn, Danuta; Wong, Kai Sin; Wray, William B., Jr.; Wright, William Allen; Yancey, Elizabeth Stilphen; Yancy, Thomas E.; Zelwer, Ruben

1980s: Abrahamson, Norman Alan; Adib, Mazen Elias; Ague, Daria Monica; Ague, Jay James; Ahn, Joonhong; Aiyesimoju, Kolawole Oluyomi; Allan, James Frederick; Alpers, Charles N.; Ansari, Gholam Reza; Araya Montoya, Rodrigo; Ayala, Luis; Baldauf, Jack Gerald; Bartel, David Clark; Bayo, Eduardo; Bell, Thomas Edward; Benson, Sally Merrick; Berc, Jeri Lynne; Bice, David Clifford; Bodvarsson, Gudmundur Svavar; Böhlke, John Karl Friedrich Paul; Bonaparte, Rudolph; Bowers, Teresa Suter; Boyce, Glenn Markland; Boyle, William John; Bozorgnia, Yousef; Brandon, Thomas Lyle; Brown, Ian Roderick; Brunsing, Thomas Peter; Bryant, Laurie Jean; Bryant, Samuel Morris; Buscheck, Thomas Alan; Carter, David Powell; Cawlfield, Jeff D.; Chalhoub, Michel Soto; Chan, Lap-yan; Chan, Lung Sang; Chavez, William Xavier, Jr.; Chen, Cheng-Hsing; Chen, Jen-Hwa; Chen, Jian-Chu; Chu, Chaw-Long; Chun, Kin-Yip; Clymer, Richard Wayne; Crempien, Jorge Laborie; Croteau, Paul; Cummins, Phillip; D'Orazio, Timothy Bruno; Dakessian, Suren; Darragh, Robert Bernard; Deino, Alan L.; Dey, Thomas Nathanel; Dingus, Lowell Wilson; Dones, Henry Cling, Jr.; Dorn, Geoffrey Alan; Dunn, Robert Jeffrey; Ed-

920 Bibliography of Geoscience Theses

wards, Stephen Walter; Eisenberg, Richard Alan; Elsworth, Derek; Endo, Howard; Erskine, Bradley Gene; Evans, Leonard Thomas; Evans, Mark David; Fenves, Gregory Louis; Fernandez, Ricardo; Fleming, Lorraine Nellita; Flexser, Steven; Foglia, Mauro Felice; Gauthier, Jacques Armand; Ghiorso, Mark Stefan; Gillerman, Virginia Sue; Goldstein, Susan Twyla; Goodwin, Laurel Bernice; Goodwin, Peter; Green, Sandra Lynn; Greenwald, Roy Fuld; Griffin, Patrick Maesa; Griffith, Michael Craig; Hale, George Robert; Hall, Robert Forrest, III; Hall, William Gordon; Hallager, William Sherman; Hansen, Roger Alan; Harden, Jennifer Willa; Harder, Leslie Frederick, Jr.; Harms, Richard William; Harper, Gregory Don; Hasenaka, Toshiaki; Hausback, Brian Peter; Hegedus, Andreas Gerhard; Heinz, Dion Larsen; Hirsch, Lee Mark; Hoang, Viet Thai; Huntsman, Scott Read; Huyck, Holly Louise; Irwin, James Joseph; Jackson, Kenneth James; James, Matthew Joseph; Javete, Donald Francis; Jeyapalan, Kanagasabai; Johnson, Jeffrey Alan; Johnson, Kenneth Allen; Johnson, Steven Carl; Ju, Jiann-Wen; Karasaki, Kenzi; Keavney, Joseph Michael; Kemeny, John McKenzie; Khalvati, Mehdi; Khamenehpour, Bahram; Kheradyar, Tara; Kim, Chang-Lak; Kirschbaum, Charles Louis; Kosco, Daniel Gregory; Kramer, Steven Lawrence; Kruge, Michael Anthony; Kuhn, Matthew Randell; Kuo, James Shaw-Han; Kyser, T. Kurtis; Labson, Victor Franklin; Lai, Cheng-Hsien; Lai, Shyh-Shiun; Lambe, Robert Noah; Lashof, Daniel Abram; Laws, Richard Anthony; Lee, Hei Yip; Lee, Richard Cacy; Lehre, Andre Kenneth; Lessard, Ghislain J. P.; Lettis, William Robert; Liao, Wen-Gen; Liu, Wen David; Long, Jane C. S.; Low, Bak Kong; Lucia, Patrick Chester; Luckman, Paul Gavin; Lugar, Gary Lance; Luhr, James Francis; Lung, Han-Chuan; Lux, Gayle Elizabeth; Mahood, Gail Ann; Mansour, Wahid Omar; Maqtadir, Abdul; Marron, Donna Carol; McClure, James Graham; McCormick, Michael; McKenzie, William Frank; McLaughlin, Keith Lynn; Meadows, Mark Allen; Medina-Melo, Francisco Jorge; Mehrotra, Vikram Pratap; Meike, Annemarie; Mejia, Lelio Hernan; Meyer, Herbert William; Montanari, Alessandro; Montellano, Marisol; Morgan, Thomas Glen; Mozley, Edward Clarence; Murphy, William Marshall; Murtha, Patricia Ellen; Myers-Bohlke, Brenda; Nesbitt, Elizabeth Anne; Ngunjiri, Philip Gichonge; Niebuhr, Walter Ward, II; Nolting, Richard Massie, III; O'Connell, Daniel Robert; Oppliger, Gary Lee; Palen, Walter Albert; Palmer, Stephen Philip; Paulsson, Bjorn Nils Patrick; Peiper, John Christopher; Perez, Francisco Luis; Perman, Roseanne Chambers; Peterson, John Edward, Jr.; Prestegaard, Karen Leah; Priestaf, Iris Gail; Pyles, Marvin Russell; Pyrak, Laura Jeanne; Ratigan, Joe Lawrence; Reeder, Richard James; Reneau, Steven Lee; Renne, Paul Randall; Rison, William; Rivers, Mark Lloyd; Rogers, Jonathan David; Rollins, Kyle Morris; Rowe, Timothy; Saari, Kari Heikki Olavi; Sandor, Jonathan Andrew; Savina, Mary Elizabeth; Scheiner, Jonathan Edward; Schnake, Carol Jeanne; Schwarz, Charles Fredrick; Sears, Frederick Mark; Seed, Raymond Boulton; Sharma, Deepak Kumar; Sharp, Warren Douglas; Shewbridge, Scott Edward; Shi, Yaolin; Simila, Gerald Wayne; Singh, Jogeshwar Preet; Sloan, Doris; Smith, Bradley Keller; Sowers, Janet Marie; Stebbins, Jonathan Farwell; Stewart, Kevin George; Stine, Scott William; Stoffregen, Roger Eben; Su, Ho-Jeen; Szerdy, Frank Steven; Tabatabaie-Raissi, Mansour; Takeshita, Toru; Tan, Hua Hui; Tanger, John Carroll, IV; Tanimoto, Toshiro; Taylor, Dan Gene; Teague, Lisa Shomura; Thomas, David Erben; Tinucci, John Paul; Tokunaga, Tetsu Keith; Tralli, David Marcelo; Tringale, Philip Thomas; Tzong, Tsair-Jyh; Vasco, Donald Wyman; Villet, Willem Christiaan Bouwer; Wang, Yeh David; Wesselman, Henry Barnard, III; Widjaja, Hadi; Wiggins, Brian Douglas; Willams, Neil D.; Williams, Quentin Christopher; Wilson, Cathy Jean; Wilson, Mark Allan; Wolf, George Henry; Yim, Chik-Sing; Youngs, Robert Riggs; Zimmerman, Robert Wayne

University of California, Davis
Davis, CA 95616

112 Master's, 142 Doctoral

1950s: Huble, Christoph W. H.; Todd, Thomas Waterman

1960s: Al-Khafif, Soud Mostafa; Aron, Gert; Atallah, Nicolas Jamil; Baldar, Nouri Amin; Brownell, James R.; Court, James Edward; Crawford, Kenneth Edgar; El-Domiaty, Awatif Mohammed; El-Nahal, Mohamed Abdelmonem M. H.; Esmaili, Houshang; Guitjens, Johannes Caspar; Harris, Dahl Le Roy; Hendricks, David M.; Howard, Ephraim Manasseh; Keller, Edward Anthony; Kimball, Royal Duane; Kimura, Hubert Satoshi; Kriz, George James; Labib, Tarik Mohamed; MacCracken, Michael Calvin; Miller, William Lawrence; Monroe, William Allen; Rabie, Farida Hamed; Rowland, Robert William; Sheikh-ol-Eslami, Bahman; Springer, Robert Kenneth; Turner, Roosevelt; Van den Berge, Johannes Christian; Wildman, William Edwards

1970s: Abrishamchi, Ahmad; Afshar, Abas; Alizadeh, Amin; Ariathurai, Chita Ranjan; Bachman, Steven Bruce; Basu, Asish Ranjam; Bedrossian, Trinda Lee; Berry, David T.; Bezore, Stephen Patrick; Blohm, Susan Jeanne; Blum, Justin Lawrence; Brizzolara, Donald William; Buer, Kill Yngvar; Busby, Linda Lucille; Calabro, Camille Elinore; Clark, Michael Sidney; Cole, Kenneth Arthur; Conger, Susan Jane Deutsch; D'Allura, Jad Alan; Daus, Steven Jean; Day, Sumner Daniel; De Laca, Ted Edwin; Dooley, Robert Lee; Ehrenberg, Stephan Neville; Enrico, Roy John; Erskian, Malcolm Gregory; Erskian, Malcolm Gregory; Farmer, Jack Dewayne; Finger, Kenneth L.; Fragaszy, Richard John; Gupta, Sumant K.; Hector, Scott T.; Herd, Howard Henry; Higgins, Chris Thomas; Hwang, Ralph Bang-Yen; Kandiah, Arumugam; Kato, Terence Tetsuo; Katopodes, Nikolas Demetrios; Kennedy, George Lindsay; Kimmel, Bruce Lee, II; Koenigs, Robert Louis; Kramer, Jerry Curtis; Krebs, William Nelson; Lovenburg, Mervin Frank; Macfarlane, Ian Charles; Martz, Paul Warren; Matsutsuyu, Bruce Akira; McCauley, Marvin Leon; McKeel, Daniel Royce; Miller, Edward James; Miyasaki, Brent; Moger, Seth R.; Murty, Vadali Venkata Narasimha; Naney, Michael Terrance; Neame, Peter Austin; Newhall, Christopher George; Nychas, Anastaios Emmanuel; O'Day, Michael Stephen; Paerl, Hans William; Parke, David L.; Rajagopal, R. S.; Raymond, Loren Arthur; Richey, Jeffrey Edward; Ristau, Donn Albert; Ristau, Donn Albert; Robinson, Lee; Ronan, Thomas E.; Rudser, Ralph Jay; Sanders, Frank Stanley; Schneider, Arland David; Senyaki, Abu Lwangq; Shmitka, Richard Otto; Showers, William James; Sial, Alcides Nobregq; Sibley, Thomas Howard; Sickles, James Michael; Springer, Robert Kenneth; Storm, David Russell; Strand, Robert Leroy; Stump, Thomas Edward; Stump, Thomas Edward; Swanson, Samuel Edward; Thompson, Laird Berry; Tu, Christopher King-Woo; Uchytil, Steven James; Valentine, Page Climenson; Viecelli, James Anthony; Vincent, W. F.; Wiggett, G. J.; Wilson, Norman Albert; Wutke, William Bruce, Jr.; Zumwalt, Gary Spencer

1980s: Abdulmumini, Salisu; Anandarajah, Annalingham; Anderson, Ruth Anne; Andresen, Arild; Angelakis, Andreas Nikolaos; Aniku, Jacob Robert Francis; Arora, Sushil Kumar; Arulmoli, Kandiah; Azrag, Elfadil Abd Elrahman; Bannan, David Bruce; Beard, James Sudler; Bedaiwy, Mohamed Naguib A.; Beiersdorfer, Raymond Emil; Benedict, Nathan Blair; Bernhard, Joan M.; Bertucci, Paul Frederick; Bettison, Lori Ann; Bobbitt, John Bailey; Brady, Roland Hamilton, III; Burnley, Pamela Carol; Busacca, Alan James; Busch, Robert Edward, Jr.; Butler, Paul Ray; Caparis, Petros P.; Casey, William Howard; Chen, Chin; Cohen, Andrew Scott; Cooper, Janice Hatchell; Cornett, Duane Charles; Cramer, Richard Stanley; Cray, Edward Joseph, Jr.; De vay, Joseph C.; Derewetzky, Aram Noah; Derstler, Kraig Lawrence; Deyo, Allen Elwin; Dienger, Jennifer Lynn; Dilek, Yildirim; Doehne, Eric Ferguson; Eddy, Carol Ann; Edelman, Steven Harold; Ehman, Kenneth Dean; Evangelou, Vasilios Petros; Finch, Michael O.; Foolad, Hamid Reza; Frampton, James Alan; Franks, Alvin LeRoy; Frantz, Gregory Alan; Gardner, Jamie Neal; Gates, Timothy Kevin; Gevirtzman, Debra Ann; Ghorbanzadeh-Rendi, Ali; Giaramita, Mario Joseph; Goldman, David Marc; Goncalves Ferreira, Alfredo Augusto Cunhal; Greene, Laurence Robert; Gronewold, Robert L.; Hacker, Bradley Russell; Haglund, John Louis; Hall, Susan Margaret; Hannah, Judith Louise; Hotton, Carol Louise; Howard, Jeffrey Kellogg; Huntington, Gordon Leland; Ingraham, Neil Layton; Jafroudi, Siamak; Jefferies, Paula Therese; Jenkins, Dennis Bruce; Kachanoski, Reginald Gary; Kelley, John Stewart; Kim, Yong Sik; Kusnick, Judith Elaine; Latham, Margie Ann Patterson; Lathan, Thomas Stanyer, Jr.; Lindsley-Griffin, Nancy; Madrid, Victor Manuel; Mazaheri, Seyed Ahmad; McCormick, John Murray; McMackin, Matthew Robert; Mechergui, Mohamed; Merriam, Martha; Miille, Michael James; Mirbagheri-Firoozabad, Seyed Ahmad; Moezzi, Bahman; Monsen, Susan Ann; Moses, Lynn Jane; Motumah, Linus Kiambati; Murray, Kent Stephen; Nater, Edward Arthur; Negrini, Robert Mark; Nelson, Thomas Arthur; Nour-el-Din, Mohamed Mohamed; Onken, Beth Renee; Onyejekwe, Okey Oseloka; Oskoorouchi, Ali Mohammad; Pandolfi, John Michael; Patrick, Brian E.; Paulsen, Steven George; Peterson, David Harland; Ramsden, Todd Wallace; Rassios, Anne Ewing; Ricci, Margaret Philleo; Severin, Kenneth Paul; Shafer, David Clark; Siddiqi, Farhat Hussain; Sloan, Jon Roger; Smith, Glenn Allen; Sohn, Joonik; South, Bernard Carl; Stanley, Edith Matilda; Summa, Lori Louise; Summer, Neil Steven; Tekbali, Ali Omar; Tingle, Tracy Neil; Tuminas, Alvydas; Uan-On, Thanakorn; Ulrick, James Steven; Urbanowicz, Karla Louise; Vaitl, Jonathan David; Varga, Robert Joseph; Vieira, Sidney Rosa; Walker, William Joseph; Walton, Otis Raymond; Weaver, John Christian B.; Wheeler, Mark Crawford; Wiedenheft, Judy Ann; Wurtsbaugh, Wayne Alden; Xenophontos, Costas; Zareai, Amanollah; Zeuch, David Henry; Zigan, Steve Michael; Zumwalt, Gary Spencer

University of California, Irvine
Irvine, CA 92717

1 Master's, 5 Doctoral

1970s: Christensen, E. R.; Skiles, J. W., III

1980s: Burtner, Don Reed; Hromadka, T. V., II; Mayer, Edward William; Yen, Chung-Cheng

University of California, Los Angeles
Los Angeles, CA 90024

407 Master's, 436 Doctoral

1930s: Dreyer, Frances Eaton; Dryer, F. E.; Gottsdanker, Eugene Nathan; Graham, David H.; Hopper, Richard H.; Irving, Earl Montgomery; Johnston, Robert L.; Kerr, Albert Ritz; MacDonald, Gordon Andrew; Redmond, Charles David; Redwine, Lowell Edwin; Rose, R. Burton; Simonson, Russell Ray; Varney, Frederich Merrill; Waggoner, Eugene Benjamin; Whaley, Harry Max

1940s: Adams, Charles E.; Allen, Harry B.; Arthur, Robert S.; Banks, John L., Jr.; Borax, Eugene; Conrad, Stanley D.; Cooper, Jack Charles; Creasey, Cyrus; Crowell, John Chambers; Daley, A. Cowles; Daviess, Steven Norman; Elam, Jack G.; Fawley, Allan Priest; Fine, Spencer F.; Ford, Waldo Emerson; Goldberg, James; Handin, John Walter; Hindman, James C.; Kelly, Robert Bowen; Kupfer, Donald Harry; Levorsen, R. D.; Loofbourow, John Stewart; Loversen, Robert I.; Maynard, Robert G.; McGill, John Thomas; Munk, Walter H.; Neuerburg, George Joseph; Norris, Robert N.; Oliver, Garnet W.; Poole, David M.; Stevenson, Robert E.; Sullwold, Harold H.; Townsend, James Robert; Traxler, J. Douglas; Van Couvering, Martin;

Weise, John Herbert; Weise, John Herbert; Wiese, John Herbert; Wiese, John Herbert

1950s: Adams, William L.; Aitken, James Drynan; Anderson, T. C. I.; Arnestad, Kenneth H.; Arntson, Ronald Hughes; Azmon, Emanuel; Badger, Robyn Lucas; Bailey, Harry P.; Bain, Roland John; Barosh, Patrick James; Bateman, Paul Charles; Beatie, Robert Lee; Bergen, Frederick Winfield; Birman, Joseph Harold; Bishop, William C.; Blanc, Robert Parmelee; Bloomfield, G. D.; Boden, Brian P.; Bradshaw, John Stratlii; Brady, Thomas James; Brinton, Edward; Brown, George Earl; Brown, Richard Shaw; Bush, Gordon L.; Butcher, William S.; Cameron, William Maxwell; Carman, Max F., Jr.; Carsola, A. J.; Carter, Neville Louis; Chauvel, Jean Paul; Christensen, Andrew Dougan; Cleveland, George B.; Cooney, Robert Lawrence; Corcoran, Eugene Francis; Cordova, Simon; Cowan, A. Gordon; Curray, Joseph Ross; Dickson, Frank W.; Dudley, Paul H., Jr.; Ehlig, Perry Lawrence; Ehrreich, Albert LeRoy; Elliott, John L.; Engel, Celeste Gilpin; Erickson, Harold Dean; Eschner, Stanford; Ewing, Gifford C.; Exum, Frank Allen; Faggioli, R. E.; Fantozzi, Joseph H.; Fantozzi, Joseph H.; Fisher, Robert Lloyd; Fitzgerald, Jennifer Kerry; Flynn, Dan Bruce; Folsom, Theodore Robert; Forbes, Ronald Frederick Scott; Frakes, Lawrence Austin; Fredericks, Robert Warren; Gamble, James Harold; Gillies, Warren Douglas; Gillou, Robert B.; Goldstein, Gilbert; Gould, James G.; Gross, David James; Guillou, Robert Barton; Guynes, George Eldridge; Hagen, Donald Wesley; Hall, Francis R.; Hamilton, Neil W.; Hamilton, Neil W.; Hamilton, Warren Bell; Hammond, Paul Ellsworth; Hansen, Don Allred; Hardey, Gordon Williams; Harding, Tod P.; Harris, Herbert; Harris, Walter Stephan; Hart, James Martin; Hartman, Donald Carl; Hazenbush, George C.; Heard, Hugh Corey; Hetherington, George Edward; Higgs, Donald V.; Hook, Joseph Frederick; Hsu, Kenneth Jinghwa; Hudson, Edward Wallace; Humphrey, Fred L.; Hunter, Hugh Edwards; Hurley, Robert Joseph; Inman, Douglas L.; Jennings, Charles Williams; Jennings, Robert Allen; Jestes, Edward Calvin; Johnson, Bradford Knowlton; Johnson, Bradford Knowlton; Keys, Scott Walter; Kiessling, Edmund; Kirkpatrick, John Curtis; Knutson, Carroll Field; Konigsmark, Theodore A.; Kovanick, Mark G.; Kovinick, Mark G.; Lamar, Donald Lee; Larson, Edwin Eric; Lian, Harold Maynard; Lindsay, Donald R.; Lloyd, George Perry, II; Lonsdale, Richard E.; Ludwick, John C.; Lung, Richard; Lyman, John; MacIvor, Keith Alan; Madsen, Stanley Harold; Martin, David Rolo; Maxell, Arthur Eugene; McCulloh, Thane Hubert; McCullough, Thomas Richard; McGill, John Thomas; McJennet, George Stanley; McMath, Vernon Everett; Merifield, Paul Milton; Miller, Clarence J.; Morrison, Robert Rex; Murphy, Michael Arthur; Nasu, Noriyuki; Natland, M. L.; Neuerburg, George Joseph; Newman, Peter Vincent; Newton, Robert Chaffer; Norris, Robert M.; Novotny, James R.; Oliver, Thomas Albert; Olson, Jerry Chipman; Orwig, Eugene R.; Orwig, Robert E.; Padick, Clement; Page, Gordon B.; Paige, Lennon Troy; Pasta, Dave; Pelline, Joseph Emmett; Perry, L.; Perry, LeRoy J.; Pittman, Edward Dale; Pollard, Dalton Leon; Pontius, David C.; Pritchard, D. W.; Proctor, Richard James; Renke, Daniel Felix; Rex, Robert Walter; Rich, Ernest I.; Richards, Adrian Frank; Robinson, Bobby Brick; Ross, Donald C.; Roth, Jim Craig; Samsel, Howard S.; Saul, LouElla Rankin; Saul, Richard Brant; Scanlin, Donald Gray; Schlee, John Stevens; Scruton, Philip Challacombe; Shumway, George Alfred, Jr.; Sierveld, Fred Grove; Smith, Victor Mackusick; Snavely, Parke D.; Sonneman, Howard S.; Steven, Thomas A.; Stewart, Harris B., Jr.; Stone, Robert; Sullwold, Harold H.; Susuki, Takeo; Terpening, John Nathan; Troxel, Bennie Wyatt; Truex, John Neithardt; Trujillo, Ernest Floyd; Uchio, Takayasu; Valentine, James William; Valentine, James William; Van Amringe, John Howard; Van Camp, Quentin Walter; Vickers, William Ward; Walton, William R.; Ware, Glen Chase, Jr.;

Weber, F. Harold; West, John Wesley; Wheatfill, Edward Lewis; Wheeler, A. Edward; Wilhelms, Don Edward; Williams, James Jerome; Wilson, Richard D.; Winterer, Edward Litton; Zeller, Robert Allen, Jr.; Ziony, Joseph Israel

1960s: Ahmed, Elhag Elhadi Ali; Akpati, Benjamin Nwaka; Alam el Din, Ibrahim Osman; Anderhalt, Robert Walter; Anderson, Robert Matthews; Asihene, Edmund Buahin; Atluri, Chandrasekera Rao; Azmon, Emanuel; Babcock, James Nissen; Baker, David Warren; Barrows, Allan Geer, Jr.; Bartow, James Alan; Bass, Ralph Oswald; Bates, Edmon Elkins, Jr.; Beheiry, Salah El Deen Abdalla; Berman, David S.; Bertholf, Harold Wyman; Beus, Stanley Spencer; Blackerby, Bruce Alfred; Blatt, Harvey; Blok, Jack H.; Blount, Charles Werner; Bowser, Carl James; Burton, William Dunn; Carter, Neville Louis; Carver, John Arthur; Castle, Robert Oliver; Chamberlain, Theodore K.; Champeny, Jon Duckett; Coates, Donald Allen; Colville, Patricia Ann; Corbato, Charles Edward; Crandall, Bradford G.; Dailey, Donald Howard; Davis, Briant Leroy; Dawson, James Clifford; deQuadros, Antonio Melicio; Dorsey, Ridgeley E.; Douglas, Robert Guy; Drake, Lon David; Duggan, Michael D.; Dutton, William George; Ehrreich, Albert LeRoy; Eidel, John James; El Din, Ibrahim Oslam Alam; Evans, James George; Fan, Pow-Foong; Fan, Powfoong; Fernandez, Alfred Peter; Fernow, Donald Lloyd; Filice, Alan Lewis; Frakes, Lawrence Austin; Fritsche, Albert Eugene; Gallick, Cyril M.; Gans, Roger Frederick; Getting, Ivan Craig; Gilbert, Murray Charles; Goldman, Don; Goldman, Harold B.; Golomb, Berl; Green, Harry Western, II; Harvill, Lee Lon; Heard, Hugh Corey; Hemborg, Thomas Harold; Hope, Roger Allen; Hsu, Liangchi; Iqbal, Mir Weseluddin Ahmed; Jackson, Everett Dale; Janke, Norman Charles; Jestes, Edward Calvin; Johnson, John Granville; Johnson, John Granville; Johnson, Raymond Larry; Kahle, James Edward; Kane, Henry Edward; Kern, John Philip; Killen, John Lippincott; Kim, Young Il; Kingsley, John; Kolodny, Yehoshua; Konigsberg, Richard Leonard; Kuhn, Michael William; Kuniyoshi, Shingi; La Mori, Phillip Noel; Lamar, Donald Lee; Larson, Allan Richard; Larson, Allan Richard; Laskowski, Edward Albin; Lawrence, Edmond Francis; Learned, Robert Eugene; Lee, William Hung Kan; Lee-Hu, Chin-Nan; Lipps, Jere Henry; Lloyd-Morris, Anthony Edward; Loeser, Cornelius James; Lumsden, William Watt, Jr.; MacDougall, Robert Earl; Matthews, Jerry Lee; McNeil, Mary Deligant; Meade, Robert Francis; Medaris, Levi Gordon, Jr.; Michael, Eugene Donald; Michels, Joseph William; Miller, Richard Harry; Morton, Douglas Maxwell; Mrowka, Jack Peter; Neder, Irving R.; Neumann, Else Rahnhild; Newton, Mark Shepard; Newton, Robert Chaffer; Nichols, Maynard Meldrim; Nickle, Neil LeRoy; Nili-Esfahani, Alireza; Nissenbaum, Arie; Olaechea, Julio M.; Ovenshine, Alexander Thomas; Palmer, Leonard Arthur; Pittman, Edward Dale; Player, Gary Farnsworth; Popenoe, Frank Wallace; Poyner, William Donald; Presley, Bobby Joe; Protzman, Donald LeRoy; Raleigh, Cecil Baring; Randall, Michael John; Reeves, Richard Wayne; Remenyi, Miklos Tamas; Reyes-Garces, Rafael Armando; Ritchey, Joseph Landon; Robison, James Holt; Rodda, Peter Ulisse; Roen, John Brandt; Roohi, Mansoor; Roubanis, Aristidis Savvas; Rowland, Richard Ernest; Sams, Richard Houston; Sanders, Norman K.; Schmus, William Randolph; Scott, David Holcomb; Scott, Kevin McMillan; Scott, Robert J.; Sharp, Willard Edwin; Sharp, Willard Edwin; Shepard, John Bixby, Jr.; Shepherd, Glenn Lincoln; Sliter, William Volk; Smith, Ralph Ernest; Stanley, Kenneth Oliver; Stevens, Dale John; Suess, Steven Tyler; Surdam, Ronald Clarence; Sylvester, Arthur Gibbs; Sylvester, Arthur Gibbs; Taylor, James Barton; Theodore, Theodore George; Tresselt, Peter; Van Couvering, John Anthony; Von Herzen, Richard Pierre; von Huene, Roland Ernest; Walker, John William Richard; Warme, John Edward; Warren, James Wolfe; Web-

ster, Gary Dean; Weissberg, Byron Goodspeed; Wilhelms, Don Edward; Williams, Peter Montague; Wilson, James Stanley; Wire, Jeremy Crosby; Witucki, Gerald Stanislaus; Young, Bruce Cash; Ziony, Joseph Israel

1970s: Adams, Herbert Gaston; Afiattalab, Firooz; Aguado, Edward; Ahmed, A. A.; Alpert, Stephen P.; Anderhalt, Robert W.; Annaki, M.; Appelbaum, B. S.; Asihene, Kwame ANane Buahin; Asquith, Donald Owen; Bacheller, John, III; Bachman, Steven Bruce; Bailey, Robert Gale; Balderman, Morris Aaron; Balogh, Richard Stephen; Band, Lawrence Ephram; Barron, John Arthur; Barrows, Katherine Jadwiga; Berger, Byron Roland; Bild, R. W.; Birchard, George Franklin; Biswas, Nirendra Nath; Blacic, James Donald; Blacic, Jan Marie; Booth, Michael Cameron; Brisbin, William Corbett; Brown, Amalie Jo; Budnik, Roy Theodore; Burkhard, N.; Burton, Vinston; Campbell, K. W.; Carey, Dwight L.; Carthew, John Arthur; Chambers, Donald D.; Chaney, R. C.; Chang, Fong-shun; Chriss, Terry Michael; Clarke, Anthony Orr; Claypool, George Edwin; Cline, Joel Dudley; Clymer, Richard W.; Colbath, George Kent; Cornell, William Crowninshield; Correa, Orlando Gonzales; Countryman, Robert Loren; Crawford, K. E.; Croft, Steven Kent; Cuong, Pham Giem; Damassa, Sarah Pierce; Demarest, Harold Hunt, Jr.; Dollinger, Gerald Lee; Donovan, Terrence John; Duncan, John Leslie, Jr.; Ehrenberg, Stephen Neville; Fairchild, T. R.; Farhat, J. S.; Finnerty, A. A.; Fouda, Ahmed Ali; Freeborn, W. P.; Frost, G. P.; Gallagher, B. J.; Garcia, M. O.; Gardner, David Allison; Garrett, P. M.; Gerdin, R. B.; Glover, Bernard K.; Goldhaber, Martin Bruce; Gonzales, Orlando Jose; Grudewicz, Eugene; Gustafson, William Ivor; Hallet, B.; Hallman, Leon Charles; Hanna, F. M.; Harris, Alan William; Haugh, Bruce Nilsson; Heikes, Kenneth Eugene; Heller, Jeff C.; Hildebrand-Mittlefehldt, Nurit; Hill, Merton H., III; Hill, Robert Lee; Holman, W. R.; Hood, L. L.; Hopkins, J. K.; Horning, Bryan Lee; Horodyski, Robert Joseph; Howard, Robert Bruce; Howe, Dennis Milton; Hoylman, Edward Wayne; Hradilek, P. J.; Hu, R. E. W.; Hurst, R. W.; Hutchinson, Charles F.; Jacobson, Sara Sue; Jensen, J. R.; Johnson, J. A.; Johnston, I. S.; Juda, Peter John; Kahane, S. W.; Kalil, E. K.; Kato, T. T.; Kettenring, Kenneth Norman; King, Thomas J.; Kirby, Stephen Homer; Kohl, Martin Sanford; Kolker, Oded; Kosiur, David Richard; Krishnan, T. K.; Kuniyoshi, Shingi; Laity, Julie Ellen; Lang, A. J.; Lastrico, Roberto Mario; Lee, Cherylene Alice; Leeds, Alan Robert; LeFever, R. D.; LeFever, Richard David; Leong, Eugene Yee; Licari, Gerald Richard; Lincoln, T. N.; Liou, Juhn-Guang; Lipshie, Steven Ross; Lister, Kenneth Henry; Liu, Kon-Kee; Luk, King-sing; Lustig, Lidia Diana; Macdonald, Robert, III; Mankiewicz, Paul Joseph; Marzolf, John E.; Mazzucchelli, Vincent George; McCoard, David; McCormick, J. W.; Miller, C. F.; Miller, David MacArthur; Miller, M. B. F.; Miller, R. H.; Mitchel, R.; Mittlefehldt, David Wayne; Moir, Gordon James; Moore, Johnnie Nathan; Moore, Johnnie Nathan; Moscati, A. F., Jr.; Moss, John Lawrence; Mount, Jack Douglas; Mrowka, Jack Peter; Myrick, A. C., Jr.; Nakanishi, K. K.; Neder, Irving Robert; Newman, Bradford Scott; Nyland, Edo; O'Keefe, J. D.; Oehler, Dorothy Zeller; Oehler, John Harlan; Pasqualetti, Martin J.; Payton, Patrick Herbert; Perez, Humberto Ramon; Peters, Kenneth Eric; Petersen, Robert M.; Petrowski, Nila Chari; Place, John Louis; Post, Robert Louis, Jr.; Post, Robert Louis, Jr.; Prior, Scott William; Putnam, Burleigh John, III; Rabinowitz, Michael Bruce; Ransford, Gary Allen; Redfern, Richard Robert; Redwine, Lowell Edwin; Reeves, Richard Wayne; Richardson, G. N.; Ries, E. G., Jr.; Rogozen, M. B.; Rohrback, B. G.; Rosenberg, Gary David; Rundle, J. B.; Salameh, Hassan Ramadan; Sandstrom, Mark William; Saragoni, Gustavo Rodolfo; Satterthwait, D. F.; Saunders, Margaret Janet; Schaal, Rand Brian; Schlue, John William; Schmidt, Ronald Roy;

Schule, J. W.; Scuderi, Louis Anthony; Seiden, Hyman; Semet, Michel Pierre; Simmons, M. M.; Smith, Richard Cronan; Smythe, William David; Sokolow, B. B.Sokolov, B. B.; Spear, F. S.; Stoddard, E. F.; Stone, W. L.; Swarts, S. W.; Sweeney, Robert Eugene; Tan, Siang Swie; Thomas, W. M.; Thompson, D. E.; Tullis, Julia Ann; Tullis, Terry Edson; Vaughan, Peter James; Walsh, Timothy John; Warford, A. L.; Warner, Edward Mark; Warren, P. H.; Waychunas, Glen Alfred; Wegner, W. W.; Weide, D. L.; Wester, L. L.; White, J. D.; White, Kenneth Lewis; Wicander, Edwin Reed; Wilson, David Vernon; Wisehart, Richard McGhee; Witter, Robert Allen; Woods, A. J.; Woodster, Warren Scriver; Yoon, Y. S.; Young, Jean T.; Younkin, Robert LeFevre; Yudovin, Susan Mary; Yuen, D. A.

1980s: Abelson, Robert Stephen; Adler, Lori Lynn; Anderhalt, Robert Walter; Anderson, Donna Schmidt; Apted, Michael John; Armstrong, Lisa Fellows; Askenaizer, Daniel Jay; Avanessian, Vahe; Bailey, Marcia Lynn; Band, Lawrence Ephram; Barbato, Lucia Sabina; Baumgardner, John Rudolph; Beck-von-Peccoz, Charles Morse, Jr.; Bilodeau, Bruce Joseph; Bloeser, Bonnie; Bolton, Edward Warren; Booth, Michael Cameron; Breslin, Patricia Anne; Brown, Amalie Jo; Brumbaugh, Robert Wayne; Bruner, William Michael; Cameron, Jeri Lynn; Carey, Dwight Lee; Carlson, William Douglas; Castens, Pamela G.; Cavazza, William; Chang, Syhhong; Chen, Chang; Chen, Tzerhong; Cheng, Ching-Chau Abe; Christensen, John Neil; Christensen, Philip Russel; Cloos, Mark Peter; Crews, Anita Lucille; Cynn, Hyunchae; Daniels, Jeffrey Irwin; de Lima, Edmilson Santos; Delany, Joan Marie; Ditteon, Richard P.; Doose, Paul Robin; Dorn, Ronald I.; Drake, Joan Elizabeth; Drake, Thomas George; Eganhouse, Robert Paul, Jr.; Ekas, Leslie Marie; El-Aghel, Asseddigh M.; Elachi, Charles; Ellsworth, Kirk Anthony; Erdem, Fahri; Ergas, Raymond Andrew; Farmer, Garland Langhorne; Fates, Dailey Gilbert; Finch, Christian Charles; Fishbein, Evan F.; Fitzgerald, Cathleen Marie; Ford, Leonard Neal, Jr.; Frishman, David; Fritts, Steven Grant; Gallup, Marc Richmond; Geijer, Theresa Anna Maria; Gergen, Leslie Dickson; Getz, Boyd Steven; Ghaly, Fatma M. Abd El Rahman; Glazner, Allen; Gonzalez, James M.; Goodge, John William; Green, Susan Molly; Grossman, Jeffrey N.; Grove, Marty; Hacker, Bradley Russell; Halderman, Tom Pepin; Halpern, Henry Ira; Hamann, Walter Edward; Hansen, Vicki Lynn; Hanson, Royce Brooks; Hazen, Richard Stewart; Hendrix, Eric Douglas; Hoag, Barbara Lillian; Hsu, Nien-Shieng; Huizinga, Bradley James; Hunsaker, Carolyn Thomas; Idiz, Erdem Fahri; Ingwell, Tim Harvey; Jacobson, Carl Ernest; Jenden, Peter Donald; Jennings, Kenneth Van Baker; Johnson, Jeffrey Alan; Johnston, David Dean; Jones, LaDon Carlos; Kallemeyn, Gregory William; Katz, Marvin; Kay, David William; Kettenring, Kenneth Norman, Jr.; Kettler, Richard M.; Kim, Moonkyum; Kirkgard, Mark Mitchell; Klein, Philip Alan; Knauer, Larry Craig; Koch, Philip Samuel; Krause, Robert Georg Fritz; Kretchmer, Andrea Gail; Kyte, Frank Thomas; Laity, Julie Ellen; Lane, Charles L.; Ledje, Hakan Karl; Lee, Jeffrey Alan; Lee, Kil Seong; Lewis, Donald Austin; Li, Yong-Gang; Liang, George Ching-Chi; Liao, Amy Hueymei; Licari, Joan Perusse; Lincoln, Beth Zigmont; Lipshie, Steven Ross; Liu, Chi-Ching; Loeher, Larry Leonard; Lu, Shan-tan; Lustig, Claire; Luth, Robert William; Macdonald, Calum; Mankiewicz, Carol; Mankiewicz, Paul Stephen; Marshall, Brian David; Marshall, Clare Philomena; Matzner, David Marc; Mazurek, Monica Ann; McCarthy, James M.; McCauley, Marlene Louise; McCurry, Michael Owen; McKone, Thomas Edward; Mendelson, Carl Victor; Milbauer, John A.; Mora, A. Roland; Morelan, Alexander Edward; Mortonson-Liedle, Judith D.; Mounzer, Maroun C.; Muchow, Charlotte I.; Mulligan, Kevin Reilley; Murrin, Theresa Eileen; Musselwhite, Donald Stanley; Nagata, Masato; Nakaki, David Kiyoshi; Needham, Edward Keith;

Nelson, Bruce Kert; Norris, Paula Jean; Ntiamoah-Agyakwa, Yaw; Nyberg, Albert Victor, Jr.; O'Connell, Rita Marie; O'Hirok, Linda Susan; Or, Arthur Chunchiu; Ord, Alison; Packer, Bonnie Marcia; Palmer, Francis Henry, Jr.; Parker, Erich Charles; Patterson, Roy Timothy; Peacock, Simon Muir; Perry, Frank Vinton; Pettis, Rani Hathaway; Phelps, Dorothy A.; Pickthorn, William Joseph; Pierce, Marcus Lacy; Plain, Donald Robert; Pleskot, Larry Kenneth; Polovina, Joseph Stanley; Rees, Kathleen Ann; Robigou, Veronique; Rodenbaugh, Karl Hase; Rojas, Gloria G.; Ross, Charles Richard, II; Ross, Martin Nicholas; Rydelek, Paul Anthony; Sadeghipour, Jamshid; Sandwell, David Thomas; Sawka, Wayne Nickolas; Schimmelmann, Arndt; Scuderi, Louis Anthony; Searls, Craig Allen; Semken, Steven Christian; Shirley, David Noyes; Sorensen, Sorena Svea; Spear, Steven Gregory; Strathearn, Gary Edward; Sumida, Stuart Shigeo; Swain, Patricia Breckenridge; Swanson, Steven Carl; Tabachnick, Rachel Ann; Taylor, Scott H.; Tchakerian, Vatche Panos; Terwilliger, Valery Jane; Thomas, Terry Robert; Tsai, Jaime Irong; Valles, Peter Kerry; Walther, Deborah Stacy; Wang, Chuching; Warso, Miriam Rebecca; Weinstock, Kenneth J.; West, Heidi DeEtte; White, Bradford Stephen; Williams, David Richard; Wilson, Jerry Calvert, Jr.; Yang, Wei-Liang William; Yin, An; Zeck, Wayne Anthony; Zhang, Ke-ke

University of California, Riverside
Riverside, CA 92521

146 Master's, 94 Doctoral

1960s: Arca, M.; Arora, Harpal Singh; Babcock, Elkanah Andrew; Beck, Myrl Emil, Jr.; Beyer, Larry Albert; Brown, Arthur R.; Butler, Godfrey Phillip; Doner, Harvey Ervin; Donovan, Terrence J.; Drake, Robert E.; Duggan, Dannie E.; Fett, John David; Ghaeni, Mohammad R.; Gibson, Ronald C.; Givens, Charles Ray; Grannell, Roswitha Barenberg; Griffin, Nancy Lindsley; Gronberg, Eric C.; Holzhey, Charles Steven; Hoover, Richard Alan; Hunsaker, Vaughn Edward; Jefferson, George Thomas; Johnson, Charles Jerome; Joshi, Martand Shipadrado; Lawrence, Edmond Francis; Le Roux, Frederick Holmes; Learned, Robert Eugene; Leon, Luis Alfredo; Leon, Luis Alfredo; Londono, Oscar Ospino; Martin, Neill W.; McLean, Gordon William; McNeal, Brian Lester; Meek, Burk D.; Muhammed, Shah; Mustafa, Mukhtar Ahmed; Nickerson, Theodore Dean; Norton, Denis Locklin; Ohrbom, Richard R.; Orr, William N.; Raab, Werner Joseph; Rao, Talur Seshagiri; Rhoades, James David; Romero, Gonzalo Cruz; Setudehnia, Ata-Ollah; Sims, James Rae; Sree Ramulu, Uddanapalli Subbarayappa; Steinen, Randolph P.; Stone, William D.; Stotts, John Louis; Suthard, James A.; Tary, Alexander; Teissere, Ronald Franklin; Watson, Christopher Lex; Weeks, Leslie Vernon; Wetherell, Anthony John; Whistler, David Paul; Willingham, Charles Richard

1970s: Aba-Husayn, Mansur Mohammed; Abudelgawad, Gilani M.; Anderson, Charles Peter; Appelt-Rossi, Herbert; Arthur, Michael A.; Barker, Charles Edward; Berendsen, Pieter; Bird, Dennis K.; Black, William E.; Bowden, Thomas D.; Bowman, Rudolf A.; Brem, Gerald F.; Cardenas-Cala, Rafael; Cheatum, Craig E.; Colman, Royce Luther; Dahleen, William; De Courten, Frank L.; de Oliveira Morais, Francisco Ilton; Dingus, Lowell W.; Dronyk, Michael P.; Dunham, John B.; Edwards, Lucy E.; Estrada, Jose Andres; Finney, Stanley C.; Freckman, John T.; Frith, Robert B.; Galindo-Griffith, Glenn; Garcia-Miragaya, J.; Gibali, Abdalla Sasi; Gilpin, Bernard E., IV; Giraldez-Cervera, Juan V.; Golz, David Jon; Goss, Ronald; Griffin, John Roy; Hadley, David Milton; Hill, Thomas G.; Hoagland, James R.; Humphreys, Eugene D.; Hunter, Robert D.; Jamieson, Iain M.; Keech, Dorothy Ann; Kendall, Carol; Kennedy, Michael Phipps; Kennedy, Patrick J.; Kinyali, Samuel M.; Knox, Richard D.; Kolesar, Peter Thomas, Jr.; Langenkamp, David F.; Long, John F.; Maas, John

P.; Maas, John P.; Magalhaes, Antonio F.; Margolis, Stanley; Matti, Jonathan C.; McGovney, James E.; McKibben, Michael Andersen; Mees, Ronald L.; Mehuys, Guy Robert; Miller, Susan T.; Minch, John Albert; Moran, Andrew I.; Morgan, Thomas G.; Moseley, Craig G.; Mueller, Wolfgang H. T.; Parcel, Rodney F.; Peck, Donald; Price, Leigh Charles; Randall, Walter; Reed, Leslie D.; Rotstein, Yair; Roulier, Michael Henry; Sigurdson, David R.; Slatten, Mark H.; Spittler, Thomas E.; Strathouse, Scott Mitchell; Sullivan, Patrick Jay; Sydnor, Robert H.; Van de Verg, Philip E.; Van der Staay, Robert; Wagoner, Jeffrey L.; Wan, Chun Yan; Welch, Jerome V.; Wilson, Kenneth L.; Wilt, Michael J.; Zappe, Steven Orvil; Zinmeister, William J.

1980s: Abu Sharar, Taleb M.; Aksoy, Rahmi; Aly, Saleh Mohamed; Amundson, Ronald Gene; Andes, Jerry Philip, Jr.; Ball, Nancy Beatrice; Baveye, Philippe; Biasi, Glenn Paul; Blackman, Thomas Donald; Boden, James R.; Bringhurst, Kelly Norman; Butters, Greg Lee; Carlton, Cleet Francis; Casey, Brian J.; Cavit, Douglas S.; Chambers, Jefferson K.; Chaney, Dan Scott; Charlet, Laurent; Cheevers, Craig Wallace; Choy-Manzanilla, Jose Enrique; Christiansen, David James, Jr.; Chung, Hung Tan; Coffman, Richard L.; Collins, John Bartlett; Cox, Brett Forrest; Damiata, Brian Neal; Demirer, Ali; Diallo, Mamadou Mouctar; Duecker, Gregory Thomas; El-Amamy, Mohaded Muftah; Emmerich, William Eugene; Evola, Gena M.; Foster, John Hugh; Garcia-Ocampo, Alvaro; Goldberg, Sabine Ruth; Hall, Donald Lewis, Jr.; Harrington, Robert Joseph; Helalia, Awad Mohamed Ahmed; Hesterberg, Dean L. R.; Hull, Carter Dean; Hwong, Tzer Jong; Ibrahim, Hassan Suliman; James, Barry; Johnson, Joseph A.; Johnston, Clifford; Jones, David Laurence; Kam, Marlene Ngit Sim; Kazi, Wallid M.; Kleijn, Willem Bastiaan; Kooser, Marilyn Ann; Lambert, Douglas Wade; LaPensee, Earl Francis; Lear, Janet Marie; LeClair, Joseph Paul; Lee, Dar-Yuan; LeFebvre, Ben Heywood; Lei, Hsiang-Yuan; Lilje, Anneliese; Liu, Wen-Cheh; Lukk, Michael E.; Mainzinger, Brent David; Marryott, Robert Allen; Martin, Jay John; May, Steven R.; Mehegan, James M.; Miller, Keith Ray; Miller, Michael J.; Moraes, Jose Francisco Valente; Morea, Michael Frank; Muramoto, Frank Shigeki; Neville, Scott L.; Oakes, Charles Steger; Owen, William Patrick; Patterson, Brooks Alan; Peryea, Francis Joseph; Phelan, Patrick John; Power, Jeanne Denise; Prenosil, Wolfgang Peter; Quinn, James P.; Richter, Goetz M.; Rojas, Leyla Amparo; Sanford, Steven J.; Schlehuber, Michael Jude; Seamount, Daniel T.; Shea, Michael E.; Smith, David K.; Sternfeld, Jeffrey N.; Strathouse, Elizabeth Cerates; Strathouse, Scott Mitchell; Sturdivant, Ann Elizabeth; Thomas, Karen Kay; Wagner, Hugh McKinlay; West, David Lawrence; Williamson, Don Alan; Wilson, Steven O'Hara; Yi, Zhixin

University of California, San Diego
La Jolla, CA 92093

29 Master's, 248 Doctoral

1940s: Tibby, Richard B.

1950s: El Wardani, Sayed A.; Inman, Douglas L.

1960s: Acosta, Marcial G.; Ayer, Nathan John, Jr.; Bada, Jeffrey Lee; Bailey, Stephen Milton; Banks, Norman Guy; Barnes, Steven Solomon; Beeson, Marvin Howard; Berger, Wolfgang Helmut; Blackman, Abner; Booker, John Ratcliffe; Bottinga, Yan; Calvert, Stephen Edward; Carpenter, Roy; Cohen, Lewis Hart; Condie, Kent Carl; d'Anglejan-Chatillon, Bruno F.; Daetwyler, Calvin Crowell, Jr.; Dahlen, Francis Anthony, Jr.; Dill, Robert Floyd; Dymond, Jack Roland; Earl, John Leslie; Elliott, William J.; Filloux, Jean Henri; Gibbs, Ronald John; Golik, Abraham; Heath, George Ross; Helmberger, Donald Vincent; Kling, Stanley Arba; Komar, Paul Douglas; Krause, Dale Curtiss; Kudo, Albert Masakiyo; Lankford, Robert Renninger; Larsen, Jimmy Carl; Luyendyk, Bruce Peter; Mac-

University of California, Santa Barbara

Kenzie, Glenn Staghan; McGeary, David Fitz Randolph; Meyer, Charles, Jr.; Miller, Jacquelin N.; Minch, John A.; Moore, Theodore Carlton, Jr.; Morris, Gerald Brooks; Normark, William R.; Parker, Frank Z.; Phillips, Richard Porter; Piper, David Zink; Pushkar, Paul Demitru; Reimnitz, Erk; Ritter, John Robert; Ross, David Alexander; Schink, David Regier; Silver, Eli Alfred; Stueber, Alan Michael; Thompson, Robert Wayne; Veeh, Hans Herbert; Windom, Herbert L.; Wong, Chi Shing

1970s: Adelseck, Charles G., Jr.; Agnew, Duncan C.; Andersen, R. L.; Anderson, Roger Neeson; Andreae, Meinrat O.; Archuleta, Ralph J.; Atwater, Tanya Maria; Aubrey, David G.; Barker, Terrance G.; Barnes, Ross Owen; Batiza, Rodey; Berger, Jonathan; Bibee, Leonard D.; Birkhahn, P. C.; Brecher, Aviva; Bruland, Kenneth W.; Buland, Raymond P.; Cain, W. F.; Cairns, James Lowell; Chase, Clement Grasham; Christensen, R. J.; Chung, Yu-Chia; Clague, David A.; Corliss, John Burt; Costa, Steven L.; Couture, Rex A.; Crane, Kathleen; Cromwell, John E.; Crouch, James K.; Day, Steven M.; de Kehoe, J. R., Jr.; Delaney, Joan; Delany, Anthony Charles; Dingler, John R.; Divis, Allan F.; Dixon, Timothy H.; Dratler, Jay, Jr.; Eells, John L.; Farrell, William E.; Fiadeiro, Manuel E.; Francheteau, Jean M.; Freyne, D. M.; Galloway, James N.; Garmany, Jan D.; Geehan, Gregory W.; Greenhouse, John Phillips; Greenslate, Jimmie L.; Grow, John Allen; Guza, Robert T.; Hall, R. E.; Hardy, L. R.; Hart, Michael; Hartzell, Stephen H.; Heinbokel, John F.; Herring, James R.; Hodson, Robert E.; Huestis, Stephen P.; James, A. H.; Jarrard, Richard D.; Johnson, David Ashby; Johnson, Leonard Evans; Johnson, Thomas C.; Karig, Daniel Edmund; Keene, John B.; Kelm, Donald L.; Kingery, F. A.; Klitgord, Kim D.; Kohl, C. P.; Kroopnick, Peter M.; Lam, Ronald Ka-Wei; Lange, Robherd Edward; Larson, Roger Lee; Lawver, L. A.; Lee, Cynthia L.; Lindstrom, Richard Mark; Linick, T. W.; Lonsdale, Peter Frank; Macdougall, John Douglas; Mayer, Larry A.; Mayo, A. L.; McDonald, James M.; McDuff, Russell E.; McNutt, Marcia K.; Nason, Robert Dohrmann; Natland, James H.; Nishimori, Richard K.; Nyman, Douglas Christian; Oldenburg, Douglas William; Orcutt, John A.; Parke, Michael E.; Parker, J. B.; Peltzer, Edward T., III; Perry, Mary J.; Pistek, Pavel; Poelchau, Harald S.; Reichle, Michael S.; Reid, Ian; Renney, Kenneth Michael; Renz, Genelle Winona; Richards, Michael L.; Rosendahl, Bruce R.; Rossetter, R. J.; Schroeder, Roy A.; Sharman, George F.; Sholkovitz, Edward R.; Shou, Philip M. Y.; Simmons, Vernon P.; Sipkin, Stuart A.; Smith, Gary B.; Speidel, W. C.; Spudich, Paul A.; Spudich, Paul A.; Stern, Robert J.; Stonecipher, Sharon A.; Stump, T. E.; Suhayda, Joseph Nicolas; Terry, Ann H.; Tucker, Brian E.; Turk, John T.; Vicente Vidal Lorandi, V. M.; Weiss, Ray F.; Wilkening, Laurel Lynn; Wilson, Clark R.; Worcester, Peter F.; Young, David Ross

1980s: Adair, Richard Glen; Anooshehpoor, Abdolrasool; Baker, Paul Arthur; Bales, Roger Curtis; Baltuck, Miriam; Beaudry, Desiree; Becker, Keir; Bloomer, Sherman Harrison; Bralower, Timothy James; Brienzo, Richard K.; Burdige, David Jay; Campbell, Andrew Craig; Carlson, Richard Walter; Chao, Benjamin Feng; Chelton, Dudley Boyd, Jr.; Cho, Byung C.; Constable, Catherine Gwen; Crane, Stephen Ernest; Creager, Kenneth Clark; Czajkowski, Jaroslaw; Druffel, Ellen Mary; Dunbar, Robert Bruce; Evans, Cynthia Ann; Fox, Richard Lyn; Francis, Robert Daniel; Gomberg, Joan Susan; Harvie, Charles Edmund; Haymon, Rachel Michal; Henry, Marilee; Hills, Scott Jean; Hinman, Nancy Wheeler; Hough, Susan Elizabeth; Jacobson, Randall Scott; Johnson, Richard Foster; Kastens, Kim Anne; Kent, Douglas Bernard; Kieckhefer, Robert Mariner; Kim, Kyung-Ryul; King, Jerry Leon; Kleinrock, Martin Charles; Lee, Dong Soo; Lee, Homa Jesse; Lerner-Lam, Arthur Lawrence; Lindberg, Craig Robert; Liu, Char-Shine; MacKenzie, Kevin Ralph; Mahoney, John Joseph; Marchisio, Giovanni, B.; Meisling, Kristian

E.; Mendez, Andres J.; Metzler, Christopher Virgil; Moammar, Mustafa Omar; Munguia-Orozco, Luis; Murrell, Michael Tildon; Nava Pichardo, Fidencio Alejandro; Newman, Sally; Ogg, James George; Olson, Allen Hiram; Park, Jeffrey John; Paull, Charles Kerr; Poreda, Robert Joseph; Price, Barbara Ann; Reed, Donald Lawrence; Riedesel, Mark Alan; Ritzwoller, Michael Herman; Sauter, Allan Wayne; Schiffelbein, Paul Arthur; Schultejann, Patricia Ann; Send, Uwe; Sereno, Thomas John, Jr.; Shaw, Peter Robert; Shaw, Timothy Jean; Shearer, Peter Marston; Shiller, Alan Mark; Shure, Loren; Silver, Paul Gordon; Smith, Deborah Kay; Spratt, Ray Steven; Stark, Philip Bradford; Steinberg, Spencer Martin; Stout, Paul Michael; Suman, Daniel Oscar; Sverdrup, Keith Allen; Vidal Lorandi, Francisco Vicente; Volpe, Alan Max; Warner, Mark James; Webb, Spahr Chapman; Welhan, John Andrew; White, Thomas Ellis; Wirth, Gerald Sheldon; Wright, Elizabeth; Wyatt, Frank K.; Young, Peter David

University of California, Santa Barbara
Santa Barbara, CA 93106

135 Master's, 109 Doctoral

1960s: Aldrich, John Kenneth; Avila, Fred Angelo; Bereskin, S. Robert; Bereskin, Stanley Robert; Davis, Terry E.; Doerner, David P.; Edwards, Lloyd Norman; Frisch, Thomas; Greene, Richard Patrick; Higgins, Michael W.; McClure, Daniel Victor; Mero, William Edward; Meyer, George L.; Ramelli, William P.; Sinha, Akhaury K.

1970s: Anderson, Christian A.; Babcock, J. W.; Babcock, James B.; Barker, James Michael; Benmore, William C.; Benmore, William C.; Blackmur, Robert; Blair, Barry; Blick, Nicholas Hammond; Bohannon, Robert G.; Butler, Todd; Cadman, John Denys; Chapin, Douglas William; Chen, James H-Young; Chen, James Huei-Young; Church, Stanley Eugene; Collier, Kenneth; Comstock, S. C.; Crippen, Robert E.; Crowe, Bruce; Crowe, Bruce Mansfield; Darrow, Arthur Charles; Dillon, John T.; Donnelly, Ann Tipton; Douglas, Charlton; Duebendorfer, Ernest M.; Edwards, Lloyd Norman; Ehrenspeck, Helmut; Emerson, Nancy L.; Erickson, John William; Fischer, Joseph Fred; Fountain, John Crothers; Fountain, John F.; Gordon, Stuart; Grove, Ginny R.; Gulliver, Rachel M.; Hale, Christopher; Hastings, Douglas; Haxel, Gordon; Heiken, Grant H.; Higgins, Ralph Edward; Hinthorne, James Roscoe; Hoffman, D. Frederick; Howell, David G.; Jacobs, David C.; Janes, Stephen; Jayaprakash, Gubbi P.; Jensky, Wallace Arthur, II; Jorgenson, David Bruce; Kempner, William; Kienle, Clive Frederick, Jr.; Kline, Gary L.; Link, Martin H.; Lo, Su-Chu; Loh, Spencer E.-Y.; Lohmar, John M.; MacKinnon, Thomas C.; MacLeod, Norman Stewart; Madden, Dawn Hill; Mahmoud, Fida; Mattinson, James Meikle; McConnell, Robert; Meijer, Arend F.; Meijer, Arend F.; Miller, Norman F.; Moorman, Mary; Nelson, Anthony S.; Nichols, Gerald; Nokleberg, Warren; Nunes, Paul Donald; O'Brien, John Malcolm; Patterson, Deborah; Patterson, Roy; Pelka, Gary Jerome; Peters, Kenneth E.; Peterson, Martin Spencer; Platt, John Paul; Pollak, Robert; Pollard, Dwight D.; Powell, Thomas; Rennick, Walter Lee; Revol, Jacques; Revol, Jacques; Sage, Cynthia; Sage, Orrin G., Jr.; Sage, Orrin, Jr.; Saleeby, Jason Brian; Schmidt, Marcia; Schroeter, Charles; Shamlian, Robert; Shervais, John W.; Smale, Timothy; Sousa, W. P.; Stephey, Anne; Stuart, Charles J.; Thorson, Jon Peer; Throckmorton, Michael; Wallace, Chester A.; Welch, Alan Herbert; Williams, Christopher; Williams, Richard; Willott, John

1980s: Al-Rawahi, Khalid Hilal; Antrim, Lisa Kay; Baca, Brian R.; Barnes, David A.; Barreiro, Barbara Anne; Beebe, Ward Joseph; Beggs, John McIntyre; Bevier, Mary Lou; Bicknell, John Dee Rogers; Blom, Ronald George; Bogaert, Barbara Mary; Boggs, Russell Calvin; Boggs, Russell Calvin; Bonkowski, Michael Steven; Boyd, John Ritchie; Bronson, Beth Joy; Brown, James Oliver, Jr.; Buis-

ing, Anna Valetta; Byrd, John Odard Dutton; Caputo, Mario Vicente; Cariolou, Marios Andreou; Carlisle, Craig L.; Carter, James Neville; Castillo, David Andrew; Centanni, James Patrick; Chadwick, William Ward, Jr.; Clark, Joyce Lucas; Clark, Michael Neil; Collinson, Thomas Barnes; Coss, James R.; Crandall, Gregory John; Daily, Micheal Irvin; Davis, Thomas Lealand; Dean, Richard Lloyd, II; Dembroff, Glenn Rind; Erwin, Douglas Hamilton; Fabry, Victoria Joan; Florsheim, Joan Leslie; Gilkey, Karen Eileen; Glicken, Harry; Goldstein, Peter; Griffin, Karen M.; Grivetti, Mark Christopher; Halgedahl, Susan Louise; Harrington, Robert Joseph; Haston, Roger; Hickey, James Joseph; Hoernle, Kaj; Hornafius, John Scott; Howard, Jeffrey Lynn; Howell, Roger Lynn; James, Eric William; Jennings, Kenneth; John, Barbara Elizabeth; Kamerling, Marc; Keller, Barry Ruland; Kidder, David Lee; Kieniewicz, Paul Mary Michael; Kimbrough, David Lee; Kwon, Sung-Tack; Laws, Bruce R.; Legg, Mark Randall; Lin, Jin-Lu; Link, Paul Karl; Magee, Marian Eileen; Mahfi, Achmad; Marinai, Robert K.; Marks, Danny Gregory; May, Daniel Joseph; McCabe, Robert Joseph; McMenamin, Mark Allan Schulte; Meshkov, Alexandra; Miller, Julia Mary Gertrude; Miller, Michele Gean; Minck, Kathleen Marie; Morris, Kimberly Peck; Morris, William R.; Mortensen, James Kenneth; Moses, Clarice Gayle; Mozley, Peter Snow; Mukasa, Samuel Benjamin; Neuhart, Donna Marie; Pallister, John Stith; Patterson, Deborah Lynn; Perdue, Elizabeth Ann; Power, William Laurence; Ramirez, Vincent Rex; Reitz, Alison; Richard, Stephen Miller; Robyn, Elisa S.; Rockwell, Thomas Kent; Rust, Derek John; Sawyer, David Andrew; Schussler, Sherryl Ann; Seaver, Daniel Badger; Sedlock, Richard Louis; Sempere, Jean-Christophe; Sherrod, David R.; Sisson, Thomas Winslow; Sloan, Heather; Smith, Gregory Ogden; Sorensen, Paul Arnold; Stallard, Mary Loah; Steck, Lee Karl; Steritz, John William; Sundell, Kent Allan; Suneson, Neil Hedner; Tally, Taz; Tanaka, Kenneth Lloyd; Terres, Donald Albert; Terres, Richard Ralph; Thompson, Timothy James; Tosdal, Richard Mark; Up de Graff, Jaye Ellen; Valentine, Gregory Allen; Van Kooten, Gerald K.; Walker, Nicholas Warren; Walker, Timothy Dane; Waller, James Allen; Weaver, Erich; Weeks, Robin James; Wellington, Gerard Michael; Williams, Ian Stephen; Williamson, Bonnie Marie; Wright, James Earl; Zepeda, Ricardo L.; Ziegler, Carl Kirk

University of California, Santa Cruz
Santa Cruz, CA 95064

52 Master's, 86 Doctoral

1970s: Bagby, W. C.; Becker, Susan Ward; Beeson, Melvin Harry; Chastain, Charlette Elizabeth; Connelly, William Ronald; Coppersmith, Kevin Joseph; Enkeboll, Robert Halfdan; Farrington, Richard Lee; Fumal, Thomas Edward; Goff, Fraser Earl; Hein, James Rodney; Hill, Malcolm David; Kelsey, Harvey Marion, III; Khodair, Abdul-Wahab Abdul-Aziz; Liddicoat, Joseph Carl; Matthews, Vincent, III; Moll, Elizabeth Jean; Myers, Carl Weston, II; Noguchi, Naohiko; Parrish, Judith Totman; Pisciotto, Kenneth Anthony; Prowell, David Cureton; Raymond, Robert, Jr.; Rowland, Stephen Mark; Shearer, Clement Fletcher; Sunzeri, Christine Cresswell; Vaniman, David Timothy; Vaughan, Peter James; Wachs, Daniel; Wendte, John Curtis; Whitney, Olive Therese

1980s: Anderson, Linda Davis; Barrientos, Sergio Eduardo; Basham, Sandra Lynne; Bataille, Klaus Dieter; Beyer, Betsy Jo; Bhattacharyya, Tapas; Bliefnick, Deborah Marie; Blueford, Joyce Raia; Bogue, Scott Weatherly; Boison, Paul Joseph; Box, Stephen Edward; Breen, Nancy Ann; Brown, Ethan Douglas; Brown, Ethan Douglas; Brown, Ethan Douglas; Bruns, Terry Ronald; Bullen, Thomas Darwin; Byrne, Timothy Briggs; Childs, John Frazer; Coale, Kenneth Hamilton; Constantz, Brent Richard; Cowen, James Prather; Creasey, Carol La Vopa; Cutter, Gregory Allan; Delsemme, Jacques

Andre; Dolan, James Francis; El-Sabbagh, Dallilah; El-Shishtawy, Ahmed Moustafa; Evoy, Barbara Lynn; Foxx, Mark Steven; Fulton-Bennett, Kim Wilbur; Gibbons, Helen; Glen, Craig Richard; Globerman, Brian Rod; González-Ruiz, Jaime Rogelio; Govean, Frances Marie; Guendel, Federico David; Gunderson, Richard Paul; Hayes, Joseph Phillips; Heil, Darla Jo; Heywood, Charles Edward; Hicks, Darryl Murray; Hill, Gary William; Hollander, David Jon; Jacobvitz, Michael Alvin; Janes, Stephen Douglas; Jayko, Angela Susan; Kelleher, Patrick C.; Kondolf, George Mathias; Laband, Beth Leah; Leslie, Robin Bruce; Leslie, Teri Hall; Longiaru, Samuel Joseph; Lucas, Stephen Ernest; Lundberg, Neil Scott; McCaffrey, Robert; Mertz, Karl Anton, Jr.; Michaels, Anthony Francis; Miller, Nancy Jill; Morrice, Martin Gray; Mount, Jeffrey Frazer; Myers, Georgianna; Newmark, Robin Lee; Newton, Cathryn Ruth; Nielsen, Hans Peter; Nye, Christopher John; Orians, Kristin Jean; Paterson, Scott Robert; Phillips, Roberts Lawrence; Plumley, Peter William; Prasetyo, Hardi; Reagan, Mark Kenyon; Reid, Mark E.; Robinson, James Varney; Roeske, Sarah Melissa; Rorty, Melitta; Sample, James Clifford; Sampson, Daniel Edward; Savoy, Lauret Edith; Sawlan, Michael Gary; Schwartz, Deborah Ellen; Schwartz, Hilde Lisa; Screaton, Elizabeth J.; Sinclair, Sheryl Lynn; Smith, Randall Blain; Snavely, Parke Detweiler, III; Stanley, Richard Graham; Stolte, Christian; Stork, Allen Louis; Swanson, Mitchell Lee; Tagudin, Jill Ellen; Tait, James Fulton; Tarshis, Andrew Lorie; Thrupp, Gordon Alan; Tobisch, Mary Kathryn; Tuttle, Martitia Powell; Vrolijk, Peter John; Watts, Keith Fred; Weber, Gerald Eric; Wells, Ray Edward; Whelan, Peter Michael; Williams, Ross Wood; Witschard, Maurice G.; Wright, Audrey Anne; Wright, Robert Harvey; Wuthrich, Dennis Richard; Zhao, Xixi

California Institute of Technology
Pasadena, CA 91125

144 Master's, 390 Doctoral

1920s: Clements, Thomas; Gazin, C. Lewis; Maxson, John H.; Murphy, Franklin Mac; Nickell, Frank A.; Sandberg, Edward C.; Southwick, Thomas S.; Turner, Francis Earl

1930s: Anderson, George H.; Anderson, George H.; Bell, Frank W.; Benioff, Hugo; Bode, Francis D.; Bode, Francis D.; Bode, Francis D.; Bolles, Lawrence W.; Bonillas, Ygnacio, III; Borys, Edmund; Bryson, Robert P.; Cabeen, William Ross; Chawner, William D.; Church, Victor; Clark, Alex; Clements, Thomas; Cogen, William M.; Cogen, William M.; Cooksey, Charlton D., Jr.; Daly, John W.; Dawson, Charles A., Jr.; De Long, James Henry; Donnelly, Maurice; Donnelly, Maurice; Donnelly, Maurice; Dougherty, J. F.; Drescher, Arthur B.; Dreyer, Robert M.; Dreyer, Robert M.; Dreyer, Robert M.; Eckis, Rollin P.; Edwards, Everett C.; Edwards, Everett C.; Eichelberger, A. M., Jr.; Engel, Rene; Ericson, David B.; Evans, M. Harrison; Fielder, William Morris; Findlay, Willard A.; Gazin, C. Lewis; Gazin, C. Lewis; Harshman, Elbert N.; Hendry, N. W.; Henshaw, Paul C.; Holzman, Benjamin; Hookway, Lozell C.; Hopper, Richard H.; Hopper, Richard H.; Hoy, Robert B.; Judson, Jack F.; Kelley, Vincent C.; Kelley, Vincent C.; Kemnitzer, Luis E.; Kemnitzer, Luis E.; Krick, Irving P.; Lohman, Kenneth E.; Lohman, Stanley W.; Lupher, Ralph L.; Lupher, Ralph L.; MacLellan, Donald D.; MacLellan, Donald D.; Maxson, John H.; McNaughton, Duncan A.; Moore, Bernard N.; Moore, Bernard N.; Nickell, Frank A.; Orr, James M.; Osborn, Elburt F.; Patterson, J. Wilfred; Peterson, Raymond A.; Phleger, Fred B., Jr.; Popenoe, Willis P.; Putnam, William C.; Putnam, William C.; Pye, Willard Dickison; Rice, H. M. Anthony; Rice, H. M. Anthony; Ross, Roland Case; Schafer, Sidney; Scharf, David W.; Scherb, Ivan Victor; Schultz, John R.; Schultz, John R.; Shappel, Maple D.; Shappell, Maple P.; Sharp, Robert Phillip; Sklar, Maurice; Smith, Hampton;

Smith, Hampton; Soske, Joshua Lawrence; Stanton, W. Layton, Jr.; Sutherland, J. Clark; Taylor, George F.; Taylor, George F.; Thayer, Thomas P.; Thayer, Thomas P.; Uhrig, Leonard F.; Urick, R. J.; Webb, Robert Wallace; Webb, Robert Wallace; Webb, Robert Wallace; Wells, John Cawse; White, Walter Stanley; Wilson, H. D.; Wilson, Robert W.; Wilson, Robert W.; Wilson, Robert W.

1940s: Agnew, Haddon W.; Akman, Mustapha S.; Allen, Charles W.; Buffington, Edwin Conger; Carlson, Harry W.; Cebeci, Ahmet; Cutsforth, David H.; Daleon, Benjamin; Dana, Stephen W.; Dehlinger, Peter; Doolittle, Russell C.; Dort, Wakefield, Jr.; Edmundson, James W.; Edwards, Charles D.; Ergin, Kazim; Fillippone, Walter R.; Findlay, Willard A.; Findlay, Willard A.; Fu, Ch'eng-Yi; Fuller, Willard P., Jr.; Geldart, L. P.; Geldart, Lloyd Philip; Gould, Martin James; Greenwood, Robert; Grobecker, Alan J.; Hedden, Albert H., Jr.; Henshaw, Paul C.; Henshaw, Paul C.; Holloway, John N.; Holser, William T.; Howell, Benjamin F., Jr.; Howell, Benjamin F., Jr.; Howell, Benjamin F., Jr.; Jahns, Richard H.; Jahns, Richard Henry; Jordan, John T.; Karubian, Ruhollah Y.; Lance, John F.; Lance, John Franklin; Lance, John Franklin; Leighton, Freeman Beach; Levet, Melvin N.; Lewis, Lloyd A.; Lewis, William D.; Lieber, Paul; Macneil, Robert J.; Martin, Joseph Stewart; Martner, Samuel T.; Martner, Samuel T.; Menard, Henry W.; Milner, Darwin Quigley; Moore, Return F.; Muehlberger, William Rudolf; Nigra, John O.; Popenoe, W. P.; Pray, Lloyd C.; Quarles, Miller, Jr.; Regan, Louis J., Jr.; Regan, Louis J., Jr.; Roberts, Ellis E.; Rupnik, John J.; Shoemaker, Eugene M.; Silgado, Enrique F.; Smith, Alexander; Smith, Alexander; Smith, Clay T.; Smith, Clay T.; Smith, Clay T.; Smith, Raymond James; Thomas, Blakemore Ewing; Tovell, Walter Massey; Wahrhaftig, Clyde A.; Wallace, Robert Earl; Wallace, Robert Earl; Wallace, Robert Earl; White, Robert C.; Wilson, H. D. B.; Wilson, H. D. B.; Yungul, Sulhi; Yungul, Sulhi; Zebal, George Patterson

1950s: Alexander, Joseph B.; Allen, Clarence Roderick; Allingham, John W.; Barker, Fred; Barker, Fred; Bass, Manuel Nathan; Berman, Joseph Harold; Bhattacharya, Prabhat Kumar; Bieler, Barrie Hill; Birman, Joseph H.; Bodvarsson, Gunnar; Burnham, C. Wayne; Campbell, D. C.; Campbell, Douglas Dean; Campbell, Richard Bradford; Campbell, Richard Bradford; Clayton, Robert N.; Cook, Philip G.; Dehlinger, Peter; Dehlinger, Peter; Denson, M. Elner, Jr.; Denson, Mayette Elner, Jr.; DeWitte, Leendert; DeWitte, Leendert; DeWitte, Leendert; Eaton, Gordon Pryor; Ergin, Kazim; Foote, Royal S.; Forester, Robert Donald; Forester, Robert Donald; Grau, Gérard; Harris, Paul B.; Hoppin, Richard A.; Hoppin, Richard A.; Howes, Thomas B.; Irvine, Thomas Neil; Irwin, William Porter; Leighton, Freeman Beach; Leighton, Freeman Beach; Lewis, Arthur Edward; Lieber, Paul; Lohman, Kenneth E.; Lomnitz, Cinna; Lovering, John Francis; Mackevett, Edward M.; Marshall, Royal Richard; Meier, Mark Frederick; Mooney, Harold Morton; Muehlberger, William Rudolf; Nelson, Robert Leslie; Nelson, Robert Leslie; Norris, Donald Kring; O'Neill, Bernard J., Jr.; Otte, Carel, Jr.; Otte, Carel, Jr.; Peck, Dallas Lynn; Phipps, Rodney T.; Potter, Donald Brandreth; Pray, Lloyd C.; Pray, Lloyd C.; Quigley, Milner D.; Ray, Walter Barclay; Rigsby, George P.; Rigsby, George P.; Roberts, William B.; Roddick, James Archibald; Rogers, John James William; Rose, Arthur William; Ruckmick, John Christian; Ruiz-Elizondo, Jesus; Ruiz-Elizondo, Jesus; Sanford, Allan Robert; Savage, James C.; Shor, George G., Jr.; Shreve, Ronald Lee; Silver, Leon Theodore; Smith, George I.; Smith, George Irving; Spencer, Terry Warren; Taylor, Hugh Pettingill, Jr.; Winters, H. H.; Wright, Lauren A.; Wright, Lauren A.

1960s: Alexander, Shelton S.; Alexander, Shelton Setzer; Anderson, Don Lynn; Archambeau, Charles Bruce; Aronson, James Louis; Banks, Philip Oren;

Ben-Menahem, Ari; Benson, Carl Sidney; Biehler, Shawn; Brady, Arthur Gerald; Brinkmann, Robert Terry; Chapple, William Massee; Cisternas, Armando; Clark, George Richmond, II; Conel, James Ekstedt; Dodd, James Robert; Doe, Bruce Roger; Duke, Michael B.; Flinn, Edward Ambrose, III; Gardner, John Kelsey; Garlick, George Donald; Goetz, Alexander Franklin Herman; Grant, James Alexander; Gross, Meredith Grant, Jr.; Hare, Peter Edgar; Harkrider, David Garrison; Harper, Charles Woods, Jr.; Healy, John Helding; Henyey, Thomas Louis; Hollister, Lincoln Steffens; Hooke, Roger LeBaron; Husid, Raul; Hwang, Li-San; Johnson, Lane R.; Jory, Lisle Thomas; Kieffer, Hugh Hartman; Ko, Hon-Yim; Kovach, Robert Louis; Lanphere, Marvin Alder; Lloyd, Ronald Michael; McCord, Thomas Bard; McDowell, Stewart Douglas; McGetchin, Thomas Richard; McGinley, John Robert, Jr.; Naylor, Richard Stevens; O'Connell, Richard John; Phinney, Robert Alden; Raychaudhuri, Bimalendu; Raymond, Charles Forest; Roddy, David John; Savin, Samuel Marvin; Schwarcz, Henry Philip; Sellers, George August; Sharp, Robert Victor; Shieh, Yuch-ning; Simpson, Dale Rodekohr; Smith, Douglas; Smith, Stewart Wilson; Spetzler, Harmut A. W.; Stanton, Robert James, Jr.; Teng, Ta-Liang; Toksoz, M. Nafi; Trifunac, Mihailo Dimitrije; Tsai, Nien Chien; Watson, Kenneth; Wilson, John Coe; Wu, Francis Taming; Zartman, Robert Eugene

1970s: Alewine, Ralph Wilson, III; Anderson, James Rodney; Arabasz, Walter Joseph, Jr.; Baldridge, Warren Scott; Berrill, John Beauchamp; Bills, Bruce Gordon; Blasius, Karl Richard; Burdick, Lawrence James; Butler, Rhett Giffen; Chung, Wai-Ying; Conway, Clay Michael; Cutts, James Alfred John; Cuzzi, Jeffrey Nicholas; Davies, Geoffrey Frederick; DeNiro, M. J.; DePaolo, D. J.; Diner, David Joseph; Dobrovolskis, Anthony Robert; Dvorak, John Joseph; Dymek, R. F.; Dzurisin, D.; Everson, Joel Earl; Foley, M. G.; Forester, R. W.; Fuis, Gary Stephen; Furst, Marian Judith; Gaffney, Edward Stowell; Gancarz, Alexander John, Jr.; Geller, Robert James; Gibbons, Rex Vincent; Goldberg, Richard Henry; Goldman, Don Steven; Gromet, L. Peter; Hadley, David Milton; Hammack, Joseph Leonard, Jr.; Hanks, Thomas Colgrove; Hansen, Olav Louis; Hart, Robert Stuart; Heaton, Thomas Harrison; Hileman, James Alan; Hill, David Paul; Hinkley, Todd King; Hong, Tai Lin; Jensen, Arthur R.; Joesten, Raymond Leonard; Johnson, Carl Edward; Johnson, Torrence Vaino; Jordan, Thomas Hillman; Julian, Bruce Rene; Jungels, Pierre Henri; Kieffer, Susan Werner; Knauth, LeRoy Paul; Kosloff, Dan Douglas; Labotka, Theodore Charles; Lagus, Peter Leonard; Laird, Jo; Lambert, Steven Judson; Langston, Charles Adams; Lawrence, James Robert; Liu, Hsi-Ping; Ludwig, Kenneth Raymond; Malin, Michael Charles; Mellman, George Robert; Minster, Jean-Bernard Honore; Murray, Jay Dennis; Nicholson, Philip David; Okal, Emile Andre; Peterson, Lee Louis; Potter, Russell Marsh; Raikes, Susan Ann; Rial M., Jose Antonio; Richards, Paul Granston; Sammis, Charles George; Schloerb, Frederic Peter; Smith, Roger Stanley Uhr; Smith, Stephen Pritchard; Soderblom, Laurence Albert; Squires, Richard Lane; Stein, Seth Auram; Thatcher, Wayne Raymond; Udwadia, Firdaus Erach; Van Alstine, David Ralph; Ward, William Roger; Weiner, Stephen; Wenner, David Bruce; Whitcomb, James Hall; Wolfe, Stephen Howard; Wood, Spencer Hoffman; Wyss, Max

1980s: Aines, Roger Deane; Astiz Delgado, Luciana Maria de los Angeles; Bardet, Jean Pierre; Beaty, David Wayne; Benjamin, Timothy Miller; Boslough, Mark Bruce; Brownlie, William Robert; Brugman, Melinda Mary; Burnett, Michael Welch; Burridge, Paul Brian; Carter, Bruce Alan; Chael, Eric Paul; Champion, Duane Edwin; Chang, Shih-Bin Robin; Cipar, John Joseph; Cohn, Stephen Norfleet; Cole, David Martin; Conca, James Louis; Corbett, Edward John; Criss, Robert Everett; Davies, Simon Henry Richard; Dowling, Michael

California State University, Bakersfield

John; Ebel, John-Edward; Echelmeyer, Keith Alan; Edwards, Richard Lawrence; Eissler, Holly Kathleen; El-Aidi, Bahaa M.; Fawcett, John Alan; Fine, Gerald Jonathan; Gehrels, George Ellery; Gillespie, Alan Reed; Given, Jeffrey Wayne; Grand, Stephen Pierre; Gregory, Robert Theodore; Hearn, Thomas Martin; Hill, Robert Ian; Ho-Liu, Phyllis Hang-Yin; Hofmeister, Anne Marie; Houseworth, James Evan; Houston, Heidi Beth; Huang, Moh-Jiann; Humphreys, Eugene Drake; Hushmand, Behnam; Jacobsen, Stein Bjornar; Jayakumar, Paramsothy; Jeanloz, Raymond; Jones, John Hume; Kirk, Randolph Livingstone; Larson, Peter Brennan; Lay, Thorne; Le Bras, Ronan J.; Lepelletier, Thierry Georges; Lewis, Richard Edwin; Lin, Albert Niu; Liu, Hsui-Lin; Louie, John Nikolai; Lyzenga, Gregory Allen; Manduca, Cathryn Clement Allen; Mattson, Stephanie Margaret; McCulloch, Malcolm Thomas; McKinnon, William Beall; Moser, Michael Anthony; Ng, Chihang Amy; Ojakangas, Gregory Wayne; Paparizos, Leonidas G.; Parsons, Ian Dennis; Passey, Quinn R.; Pechmann, James Christopher; Piepgras, Donald John; Powell, Robert Edward; Psycharis, Ioannis N.; Quick, James Edward; Regan, Janice; Richards, Mark Alan; Rigden, Sally Miranda; Rudy, Donald James; Ruff, Larry John; Sams, David Bruce; Sanders, Christopher O'Neill; Schmitt, Douglas Ray; Scott, David Russell; Scott, Patricia Frances; Shaw, Henry Francis, III; Silver, Lynn Alison; Stapanian, Maritza Irene; Stewart, Gordon Selbie; Stork, Christof; Svendsen, Robert Frederick Jr.; Vasilakos, Nikolaos Petrou; Vassiliou, Marios S.; Vidale, John Emilio; Vizgirda, Joana Marija; Walck, Marianne Carol; Wallace, Terry Charles, Jr.; Weldon, Ray James, II; Wilson, John Charles; Yapp, Crayton Jeffery; Zhang, Jiajun

California State University, Bakersfield
Bakersfield, CA 93311

2 Master's

1960s: Ripple, Charles D.; Schorn, Howard E.

California State University, Chico
Chico, CA 95929

12 Master's

1950s: Kurtz, John Cornell; Robinson, Louis H.

1960s: Gould, Franklin David

1970s: Bendixen, Roald L.; Budlong, Gerald Michael; Chamberlin, Peter; Coggins, Vernon; Enderlin, Margot Helene; Hayes, Delvin Arnold; Hilton, R. P.; Percival, Tim

1980s: Danielson, Joanne

California State University, Fresno
Fresno, CA 93740

29 Master's

1950s: Ishimoto, Toshio Tom

1970s: Armstrong, Jeffrey A.; Bishop, James Corwith, Jr.; Brook, Charles A.; Cehrs, D.; Crowley, Jack Arthur; Elkins, E. D.; Fry, Steven; Ganjei, Hossein; Girty, G.; Greenamyer, Randolph; Hamilton, K.; Haskins, Donald; Kirk, John; Moody, Donald S.; Palmer, C.; Percival, Timothy Jerold; Roquemore, G.; Russell, Stephen John; Smith, Michael; Suek, D. H.; Wetzel, N.; Worthington, June; Worthington, June

1980s: Bero, David Alex; Casteel, Mitch; Cole, Robert Dennis; Herron, Calvin Robert; Poole, Thomas Craig, Jr.

California State University, Hayward
Hayward, CA 94542

23 Master's

1970s: Dresen, Michael D.

1980s: Bainer, Robert W.; Baskin, David A.; Blunt, David J.; Bowman, John K.; Carlson, Carl A.; Carson, Scott E.; Duchosal, Yves R.; Fischbein, Steven A.; Kayen, Robert E.; Kehoe, James Mark; Kintzer,

---

Frederick C.; Legler, June L.; Lull, John S.; Makdisi, Richard; Marcus, Barry I.; Mathieson, Scott Alan; Ollenburger, Ronald D.; Pavletich, Joseph P.; Potter, Kenneth L.; Short, William R., Jr.; Siegel, David; Specht, Daniel J.

California State University, Long Beach
Long Beach, CA 90840

47 Master's

1970s: Berman, Brigitte Helene; Hammond, Janet G.; Lang, Harold R.; Nunez, Luis; Winterer, Joan I.

1980s: Anders, Nuni-Lyn E. Sawyer; Barlow, Steven Gus; Beck, Brian; Bull, Louis; Castle, Margaret; Collender, Jack; El-Ansari, Ahmed; Forrest, Michael; Fox, Dennis; Fox, Terrance; Greenwood, Richard; Guerrero, Juan A.; Hollis, Thomas; Huang, Cheng-yi; Jones, Sally; Kraft, Richard P.; Kruk, Taras; La Violette, John; MacKinnon, Cinda; Maher, Alice Margaret; McCutcheon, Kirk; Montgomery, Michael; Muir, William; Nguyen, Sy; Park, Nam; Phibbs, John; Randall, George; Randell, David Howard; Reardon, Jeffry; Robertson, Marina; Schneeberger, Dale; Schwartz, Martha; Spindler, Richard; Stadum, Carol J.; Taban, Osman; Tracy, William Craig; Walls, Thomas; White, Rene; Wyman, Rosemary; Yi, Dar; Youn, Oong; Zhou, Hua-wei

California State University, Los Angeles
Los Angeles, CA 90032

10 Master's, 1 Doctoral

1930s: Hooper, Richard H.

1960s: Robinson, James Holt

1970s: Campbell, Alice M.; Cleath, Timothy S.; Conrad, Rae L.; Corson, Donald L.; Hamilton, Patrick; Lass, Garry L.; McJunkin, Richard Dean; Mirowka, Jack P.; Nason, Geoffrey W.

California State University, Northridge
Northridge, CA 91330

54 Master's

1970s: Blom, Ronald George; Reid, Stephen Anthony; Richmond, William C.; Testa, Stephen Michael; Thor, Devin; Werner, Michael Robert

1980s: Advocate, David Michael; Anderson, Thomas P.; Blake, Thomas Ford; Blundell, Michael Craig; Bouton, Katherine Alice; Chen, Xunhong; Clark, Frederick Emory; Custis, Kit H.; Fowler, Julie Ann; Furst, Bruce Wayne; Griffis, Robert Arthur; Hartshorn, David Robert; Hickey, Edward Paul; Johnson, Elizabeth A.; Kappeler, Kristine Ann; Kraemer, Susanne Margit Charlotte; Liggett, David Lee; Manz, Richard P.; Morrison, Lowell Russell; Muir, Stephen G.; Muskat, Judd; Norby, Philip Arthur; Oborne, Juli G.; Oborne, Mark Stephen; Parker, Jonathan David; Peterson, Larry Lynn; Popelar, Stanley James; Prevost, Dana Victor; Raftery, Peter John; Remsen, Walter E., Jr.; Ritterbush, Linda Anita; Roush, James Manfred; Roush, Kathleen Ann; Rudat, Juhani; Russell, Perry Wooten; Schymiczek, Herman Bodo; Shariff, Asghar Jahani; Sterling, Stephen Charles; Taylor, Gary Edward; Thomas, June Marie; Ticken, Edward John; Trembly, W. A., Jr.; Unruh, Mark Eugene; van Wagoner, Steven Lewis; Webb, James Franklin; White, David Ross; Woods, Madeline Marie; Yaldezian, John George, II

Carleton University
Ottawa, ON K1S 5B6

121 Master's, 54 Doctoral

1940s: Kueffner, Mary H. E.

1960s: Bartlett, Grant A.; Carson, David John Temple; Carter, Maurice Wylde; Chen, Tzong-Tzyy; Forgeron, Fabian David; Fox, Peter Edward; Franklin, James M.; Gabisi, Abdul H.; Hamilton, C. G.; Haynes, Simon John; Hounslow, Arthur William; Hounslow, Arthur William; Jambor, John Leslie;

Lovell, Howard Lawrence; Mackasey, William Oliver; McMillan, William John; Potter, Ralph Richard; Presant, Edward W.; Shafiqullah, Muhammad; Sopher, Stephen R.; Vernon, Peter David; Weston, Ayla

1970s: Ayer, John Albert; Bishop, A. G.; Card, J. W.; Casselman, M. J.; Cecile, Michael P.; Cecile, Michael P.; Chappell, J. F.; Cheang, K. K.; Clark, Frederick; Dass, A. S.; Davies, John Leslie; Dunning, Gregory; Elkhoraibi, M. C. E.; Ewert, W. D.; Exton, John; Findlay, A. R.; Gibson, Harold Lorne; Gill, J. W.; Grant, Earl Brian; Greenough, John David; Gunter, Avril E.; Gunter, Avril E.; Heslop, John B.; Hews, P. C. H.; Ho-Tun, Edwin; Hoy, T.; Hunter, A. Douglas; Hutcheon, Ian E.; Hutcheon, Ian E.; Jackson, K. S.; James, Clifford M.; Khan, T. R.; Kurt, Vace H.; Lalonde, Jean-Pierre; Lambert, Maurice Bernard; Lane, Larry Stephen; Larouche, Christiane; Larouche, Claude; Leatherbarrow, Robert Wesley; Lee, Chen-Wah; Macey, Gerald J.; Mann, Robin Carl; Mason, George David; McCorkell, Robert H.; Moore, R. L.; Morton, R. L.; Ostler, J.; Peeling, Gordon Roderick; Psutka, J. F.; Ricketts, B. D.; Rodrigues, Cyril Gerard I.; Sawford, Clayton; Sethuraman, K.; Stephenson, R.; Stewart, Robert; Tan, J. T.; Tanczyk, Elizabeth; Taylor, A. H.; Thompson, Peter Hamilton; Tippett, C. R.; Troyer, David Robert; van der Leeden, John; Watkins, David H.; Yuan, Peter Bee-Deh

1980s: Abercrombie, Hugh James; Adcock, Stephen William; Allen, D. M.; Ames, D. E.; Anderson, Robert Gordon; Ansell, Valerie; Aspler, Lawrence B.; Barham, Bruce A.; Bartlett, James Rodney; Brown, Don M.; Burn, Christopher Robert; Carr, Cynthia; Carr, Sharon D.; Chernis, Peter J.; Chiarenzelli, Jeffrey Robert; Chomyn, Beverley A.; Conlon, Paul Joseph; deKemp, Eric A.; Dillon-Leitch, Henry C. H.; Duvadi, Ashok K.; Fadaie, Kian; Ford, Kenneth Lloyd; Füstös, Árpád; Goetz, Peter Andrew; Goodz, Morrie D.; Gordon, Susan L.; Grapes, Kathyrn J.; Harnois, Luc; Hwu, Chen-Roon; Ikingura, Justinian Rwezaula; James, Donald T.; Johnson, Bradford J.; Johnstone, Robert M.; Karboski, Frank Adam; Kerans, Charles; Kim, Won Sa; Kim, Won Sa; Kingston, David; Knight, Ross; Lane, Larry Stephen; Leatherbarrow, Robert Wesley; Lefebvre, David Victor; Legault, Marc H.; Lockwood, Michael B.; MacQueen, J. Kenneth; MacRobbie, Paul; Manojlovic, Peter M.; McEwen, John H.; McKinstry, Brian William; Melling, David R.; Morton, Penelope; Murphy, Donald Currie; Murray, Jane; Mustard, Peter S.; Nadeau, Léopold; Neale, Kathryn L.; Nentwich, Franz W.; Nnolim, Chude Austine; Onuonga, Isaac Oriechi; Patterson, George Cameron; Pelletier, Karen; Perkins, Michael John; Pollock, S. P.; Rainbird, R. H.; Rees, Christopher John; Reichenbach, Ingrid G.; Richardson, Jean Madeline; Rodrigues, Cyril Gerard I.; Roots, Charles Frederick; Ross, Gerald Marckres; Sage, Ronald Parker; Scammell, Robert J.; Sharpe, Joanna L.; Shaw, David Andrew; Simandl, George; Skirrow, Roger G.; Smyk, Mark C.; Sykora, Jerry J.; Szymanski, Maciej Boleslaw; Tiong, Siong-Sui; Wen, Jianping; Whittaker, Peter James; Wilkinson, Stephen J.; Williams, E. Joy; Wilson, Bradley S.; Wood, John Anderson

Carnegie-Mellon University
Pittsburgh, PA 15213

1 Master's, 14 Doctoral

1960s: Brown, Ralph Edward; Yanizeski, George Michael

1970s: Boynton, William Vandegrift; Darby, W. P.; Erel, Bilgin; Gilpin, A. E.

1980s: Coronato, James Allen; Ejezie, Samuel Uchechukwu; Mihelcic, James Robert; Moskal, Thomas Eugene; Motazed, Behnam; Mui, Kwoon Chuen; Mullarkey, Peter William; Pieri, Robert Victor; Siller, Thomas Joseph

Case Western Reserve University
Cleveland, OH 44106

27 Master's, 60 Doctoral

1920s: Focke, Helen M.

1930s: Cutter, Paul Frank

1940s: Wood, Lucile H.

1960s: Bugh, James Edwin; Greiner, Gary Oliver George; Hall, Minard Lane; Kafescioglu, Ismail A.; Kafescioglu, Ismail Ali; Malone, Philip Garcin; Meyers, Darwin; Noble, James Peter Allison; Perry, Edward Adams, Jr.; Rozilo, Paul John; Saylor, Timothy E.

1970s: Barsotti, A. F.; Belinger, Eric Vance; Brooks, Irving Harvey; Burkley, Lewis A.; Calcagno, Frank, Jr.; Cook, William Riley, Jr.; Cotman, Richard Matthew; Durazzi, J. T.; Durst, T. L.; Eberl, Dennis Donald; Eslinger, Eric Vance; Fisher, John B.; Fletcher, M. R.; Franks, S. G.; Fukuda, Michael K.; Gedeon, James E.; Hackett, William Robert; Harris, A. B.; Hecht, Alan David; Hetrick, John Harold, Jr.; Hoffman, J. C.; Hoffman, Janet L.; Hoover, P. R.; Johnson, Allen Harold; Kherl, Dennis Donald; Kim, Hae Soo; Klar, G.; Lukert, Michael Thomas; MacTavish, John Nickolas; Mathews, Geoffrey William; May, Charles A.; Mertzman, Stanley Arthur, Jr.; Murray, Joseph Buford; Phillips, D. S.; Redding, Carter Eugene, Jr.; Rettke, R. C.; Rhodes, Flora Lee; Rudmann, Joseph Emmett; Steele, William Kenneth; Tausch, Earl Harry; Thompson, Graham R.; Wardlaw, B. R.; Welker, D. B.; Yeh, H.-W.

1980s: Barrera, Enriqueta; Barrera, Enriquetta; Bianchini, Gary Francis; Bill, Steven David; Burrows, Steven Mark; Capichioni, Maria Luisa; Cheng, Jauh-Tai; Dasch, L. E.; Du, Chenggui; Elliott, William Crawford; Gereby, Clarissa H.; Ghahremani, Darioush Tabrizi; Hart, William Kenneth; Inglis, J. Mark; Isiorho, Solomon Akpoghenobor; Jordan, M. R.; Kelly, Walton Ross; Khourey, Christopher J.; Kordesch, Elizabeth Gierlowski; Law, Eric W.; Lebreton, Claude Marie; Lee, Mingchou; Liu, Shu-Wang; Macky, Tarek Ahmed Aly; MacNeille, P. R.; Mausser, Herbert F.; Puccini, Piero Miguel; Walter, Robert Curtis; Woldegabriel, Giday

Catholic University of America
Washington, DC 20064

15 Master's, 7 Doctoral

1920s: Clawsey, Patrick J.; Fox, Joseph Peter; Nieset, Rev. C. F.

1930s: Ellsworth, Richard Gerald; Horton, Roger Goldsmith; Kieran, Mary; Lane, Thomas James; May, Timothy Crawford; May, Timothy Crawford; Meara, Joseph Edwin

1940s: Dempsey, William Joseph; Kemme, Joseph W.; Tighe, Edward J.; Williams, Marguerite Thomas

1950s: Osborn, Donald R., Jr.

1960s: Campbell, Louis F.; Coughlin, Mary St. Patrick

1970s: Bhambri, Inder Jit; Goad, C. C.; Tahir, Abdul Haleem

1980s: Danner, David Lee; Keene, Warren Elmer

Central Connecticut State University
New Britain, CT 06050

1 Master's

1980s: Plocharczyk, Elise J.

Central Washington University
Ellensburg, WA 98926

1 Master's

1960s: Foisy, Raymond Deane

Chadron State College
Chadron, NE 69337

1 Master's, 1 Doctoral

1970s: Fitzgibbon, James Lavern; Witzel, Frank Lewis

University of Chicago
Chicago, IL 60637

80 Master's, 392 Doctoral

1890s: Bain, H. Foster; Case, Ermine C.; Gordon, Charles Henry; Kummel, Henry Barnard

1900s: Alden, William Clinton; Atwood, Wallace Walter; Bastin, Edson Sunderland; Bolles, Myrick Nathanial; Branson, Edwin Bayer; Calhoun, Fred Harvey Hall; Capps, Stephen Reid, Jr.; Chamberlin, Rollin Thomas; Emmons, William Harvey; Fenneman, Nevin Melancthon; Garrey, George H.; Hopkins, Thomas Cramer; Light, William George; Logan, William Newton; Moodie, Roy Lee; Moore, Elwood S.; Schultz, Alfred Reginald; Stauffer, Clinton Raymond; Tight, George William; Wilder, Frank Alonzo

1910s: Blackwelder, Eliot; Boyce, Helen; Bretz, J. Harlen; Brokaw, Albert D.; Burwash, Edward M. J.; Cady, Gilbert H.; Calkins, R. D.; Carman, Joel Ernest; Chaney, Ralph Works; Cooke, Harold Caswell; Coryell, Horace Noble; Decker, Charles Elijah; Douthitt, Herman; Fuller, Margaret B.; Hance, James Harold; Hasslock, Augusta T.; Hole, Allen David; Kay, George Frederick; Knox, John Knox; Lees, James Henry; Leighton, Morris Morgan; Martin, Laura Hatch; Mather, Kirtley Fletcher; McKee, Howard Harper; Mehl, Maurice Goldsmith; Moore, Raymond C.; Quirke, Terence Thomas; Snider, Luther Crocker; Stephenson, Eugene Austin; Tarr, William Arthur; Tilton, John Littlefield; Tomlinson, Charles Weldon; Trowbridge, Arthur Carleton; Umpleby, Joseph Bertram

1920s: Anderson, Gustavus E.; Athy, Lawrence F.; Bacon, Charles Sumner, Jr.; Bailey, Reed Warner; Ball, John Rice; Behre, Charles H.; Bell, Alfred H.; Bevan, Arthur Charles; Bolyard, Garrett L.; Boos, C. M.; Boos, Maynard; Bradley, John H.; Branson, Carl C.; Brown, Ira Otho; Brown, Robert Wesley; Cox, Benjamin B.; Cressey, George Babcock; Cressey, George Babcock; Culbertson, John A.; Culver, Harold E.; Doggett, Ruth Allen; Evans, John R. C.; Fath, Arthur Earl; Feliciano, Jose Maria; Fenton, Carroll L.; Fisher, Daniel J.; Fisher, Daniel J.; Flint, Richard F.; Fryxell, Fritiof M.; Fuller, Margaret B.; Giles, Albert W.; Hance, James Harold; Jones, J. Claude; Kerr, Forrest Alexander; Koons, Frederic C.; Landon, Robert Emmanuel; Link, Theodore A.; MacClintock, Paul; MacKay, Bertram Reid; Mathews, Asa A. Lee; Matson, George Charlton; McFarlan, Arthur Crane; Meinzer, Oscar Edward; Melton, Frank Armon; Merritt, Clifford A.; Newmann, Fred Robert; Oldham, Albert; Perry, Eugene Sheridan; Pike, Ruthven W.; Renick, B. Coleman; Roark, Norris Wilson; Rothrock, Edgar Paul; Runner, Joseph J.; Sayre, Albert Nelson; Shepard, Francis P.; Stark, John Thomas; Stewart, Grace Anne; Sutton, Arle Herbert; Tansey, Vivian Ouray; Thomas, Abram O.; Toepelman, Walter Carl; Van Steeg, Karl; Ver Steeg, Carl; Vestal, Franklin E.; Weller, James Marvin; Wilson, Alice Evelyn; Wilson, Roy Arthur; Wing, Monta Eldo; Worcester, Philip G.; Workman, Lewis Edwin; Wright, John Frank; Yeaton, Walter James

1930s: Aberdeen, Esther; Adler, Joseph L.; Alvir, Antonio Delgado; Anderson, Alfred L.; Atherton, Elwood; Beck, Robert W.; Bennett, William Alfred Glenn; Bristol, Hubert M.; Caldwell, Lorin T.; Dorf, Erling; Dunn, Paul H.; Eicher, Donald B.; Enlows, Harold E.; Espenshade, Edward B.; Freeman, Bruce C.; Funkhouser, Harold J.; Gale, Arthur S., Jr.; Graham, Roy; Gries, John Paul; Gries, John Paul; Grove, Brandon H.; Grubbs, David M.; Hills, John Moore; Hoffman, Arnold Daniel; Hoffman, Melvin G.; Holmes, Terence C.; Horberg, Leland;

Hornung, Arthur G.; Hough, Jack L.; Hubbert, M. King; Imbt, William Clarence; Ireland, H. A.; Janssen, Raymond E.; Janssen, Raymond E.; Jones, Daniel John; Kirkham, Virgil R. D.; Kramer, William B., III; Krumbein, William C.; Lammers, Edward C. H.; Langton, Claude M.; Loewenstamm, Heinz Adolph; Lowell, Wayne Russell; Macknight, Franklin C.; Mather, William B.; Mayfield, Samuel M.; Olson, Everett C.; Olson, Everett C.; Payne, James Norman; Perdue, Henry Stewart; Potter, Franklin C.; Rasmussen, William Charles; Ridge, John D.; Ridge, John D.; Rigney, Harold William; Rigney, Harold William; Riley, Christopher; Rittenhouse, Gordon; Rittenhouse, Gordon; Robinson, Lewis C., III; Root, Towner B.; Rust, George W.; Scott, Harold William; Scott, Harold William; Sloss, Laurence L.; Smith, Homer J.; Stevens, Edward H.; Stovall, John Willis; Straley, Harrison W., III; Tansley, Wilfred; Tansley, Wilfred; Teis, Maurice R.; Thomson, Robert; Thurman, Franklin A.; Wadell, Hakon A.; Williams, Lou Page; Wray, Franklin C.; Wright, Paul Randall

1940s: Armstrong, H. S.; Bertholf, W. E., Jr.; Botero, Gilberto; Byrne, Frank E.; Carlton, James L.; Chao, Edward C. T.; Church, H. Victor; Cook, Kenneth L.; Cooper, Chalmer L.; Dowie, Paul G.; Dubois, Ernest Paul; Easton, William Heyden; Erdman, Oscar A.; Foster, Wilfred R.; Freeman, Louise B.; Freudenheim, Priscilla; Garcés-Gonzales, Hernán; Goldsmith, Julian R.; Gummer, Wilfred K.; Herbert, Paul, Jr.; Higazy, Riad A. M.; Higgins, James W.; Holden, Frederick T.; Hough, Jack L.; Hough, Margaret Jean; Juan, Vei Chow; Kidwell, Albert Laws; Kurk, Edwin H.; Loeblich, Alfred R.; Loeblich, Helen N. Tappan; Lowell, Wayne Russell; Luckhardt, Paul G. L.; Lundahl, Arthur Charles; Manjarres, Gilberto F.; McGrew, Paul O.; Meade, Grayson Eichelberger; Monk, George D.; Nelson, Elmer R.; Nelson, Vincent E.; Oehler, E. T.; Otto, George H.; Paba-Silva, Fernando; Payne, Thomas G.; Peterson, Victor E.; Plumley, William J.; Prince, Alan T.; Pye, Willard Dickison; Read, William; Reinhart, Roy H.; Riley, N. Allen; Roy, Sharat Kumar; Samiento-Alarcon, A.; Sandefur, Bennett T.; Siever, Raymond; Smalley, Robert G.; Spivak, Joseph; Sternberg, Charles William; Stevenson, Frank Vincent; Stocking, Hobart Ebey; Sun, Ming-Shan; Swineford, Ada; Thomas, Harold Eugene; Tippie, Frank E.; Tisdel, Fred Walter; Trow, James William; Trow, James William; Tullis, Edward Langdon; Turner, Gregory Larkin; Williams, Lou Page; Young, Ruth Hope

1950s: Adams, John A. S.; Anderson, Richard; Atlas, Leon Morris; Bader, Richard G.; Bader, Robert S.; Baskin, Yehuda; Bhattacharji, Somdev; Bloss, Fred Donald; Bokman, John W.; Bonham, Lawrence D.; Brown, Charles N.; Byers, Frank M.; Chao, George Yien-Chi; Chave, Keith Ernest; Chidester, Alfred Hermann; Clark, Lorin D.; Craig, Harmon B.; Currie, Kenneth Lyell; DeVore, George W.; Ehlers, Ernest G.; Emiliani, Cesare; Fahrig, Walter F.; Flint, Arthur; Friedman, Isidore I.; Ghose, Subrata; Gibson, Lee B.; Ginsburg, Robert N.; Greenman, Norman N.; Harrison, Phillip W.; Hoffer, Abraham; Hotton, Nicholas, III; Huff, Lyman Coleman; Jamieson, John C.; Johnson, Frederick A.; Kaarsberg, Ernest Andersen; Kahn, Steven James; Karlstrom, Thor N. V.; Knight, Charles; Konizeski, Richard L.; Koucky, Frank; Kretz, Ralph Albert; Krinsley, David; Kuellmer, Frederick J.; Lee, Hulbert A.; Leighton, Morris W.; Limper, Karl E.; Lundelius, Ernest L., Jr.; Magorian, Thomas R.; McCrossan, Robert G.; McGlynn, John C.; McGrossan, R.; Mueller, Robert Frances; Nanz, Robert Hamilton; Nickel, Ernest H.; Olsen, Edward John, Jr.; Olson, Jerry; Osterwald, Frank W.; Potter, Paul Edwin; Robert, Richard A.; Robie, Richard A.; Rubin, Meyer; Rusnak, Gene Alexander; Seltin, Richard J.; Sen, Sisir Kumar; Shaw, Denis M.; Shykind, Edwin B.; Siever, Raymond; Silverman, Sol Robert; Simmons, David; Sloan, Robert E.; Van Vlack, L. H.; Vitanage, Piyadas; Wasserburg, Gerald Joseph; Weeks, Wilford Frank;

Woo, Ching-Chang; Woodard, Henry H., Jr.; Yalkovsky, Ralph

1960s: Albright, James; Applegate, Shelton Pleasants; Barghusen, Herbert R.; Bayly, Maurice; Bennington, Kenneth Oliver; Birchfield, Gene E.; Byrne, Robert J.; Chang, Luke; Coleman, Neil Lloyd; Corlett, Mabel; Dandekar, Dattataya; DeMar, Robert Eugene; Forslev, Alfred William; Gage, Kenneth S.; Gaines, Alan M.; Gangopadhyay, Jibimatra; Giese, Graham S.; Gipson, Mack Jr.; Godfrey, John D.; Hessler, Robert R.; Hoskins, Hartley; Isaacs, Thelma; Jones, A. E. Nyema; Jones, Robert William; Kayode, Abiodum A.; Long, Robert S.; Louisnathan, S. John; McCallum, Ian S.; McLelland, James M.; Moore, Paul Brian; Moxham, Robert Lynn; Murray, Stephen P.; Nitecki, Matthew; Nordlie, Bert E.; Perotta, Anthony J.; Piwinskii, Alfred J.; Sestini, Julian; Simpson, Robert A.; Smith, James D.; Sommerfeld, Richard; Stenstrom, Richard; Thein, Myint Lwin; Williams, Garnett P.; Woodland, Bertram

1970s: Ablordeppey, Victor; Benzing, William Martin, III; Bishop, Finley C.; Chase, Robert Perkins; Cisne, John Luther; Coplen, Tyler B., II; Daly, Stephen F.; Deganello, Sergio; Duba, Alfred G.; Fugelso, Leif Eric; Halleck, Phillip Michael; Hansen, Kirk S.; Haseltion, Henry Trenholm; Hervig, Richard Lokke; Huang, Wuu-Liang; Ito, Emi; Johnson, Markes Eric; Kampf, Anthony Robert; Kranz, Peter M.; Lasker, Howard Robert; Lesht, Barry Mark; MacDaniel, R. P.; Mehrtens, Charlotte Jean; Merrill, Russell B.; Mohr, David W.; Olinger, Barton; Osman, R. W.; Park, T.; Pizzaferri, L.; Richards, R. Peter; Richter, Frank M.; Robertson, John K.; Smith, Mary Ann Hrivnak; Smyth, Joseph; Stern, Charles R.; Sulanowski, Jacek Kazimierz; Thompson, Ida Louise; Warburton, David Lewis; Weaver, Thomas Adrian; Weber, Hans-Peter

1980s: Artioli, Gilberto; Barrett, Stephen Francis; Barton, Mark David; Beckett, John Randall; Boyer, Larry Fred; Damuth, John Douglas; Geiger, Charles; Hansen, Edward Conrad, III; Hansen, Kirk S.; Harris, David Milo; Jenkins, David Mark; Karlsson, Haraldur R.; Koziel, Andrea; Leitch, Catherine A.; Mackin, James E., II; McLarnan, Timothy J.; Miller, Arnold I.; Otto, Jens; Parker, William C.; Plotnick, Roy Elliott; Prombo, Carol Ann; Raymond, Anne; Ribe, Neil Marshall; Sahagian, Dork L.; Sans, John Rudolfs; Schmachtenberg, William F.; Schwindinger, Kathleen Rose; Scotese, Christopher R.; Spieglan, Mark Joseph; Swihart, George H.; Tang, Ming; Ullman, William John; VanderWood, Timothy Blake; Zhang, Da-chun

University of Cincinnati
Cincinnati, OH 45221

283 Master's, 98 Doctoral

1910s: Braun, Emma Lucy

1920s: Lorenz, Eleanor Mary; Rogers, James Kenneth; Theis, Charles Vernon

1930s: Alexander, Walter Herbert; Blickle, Arthur H.; Brown, Carl B.; Collier, James E.; Demorest, Max Harrison; Desjardins, Louis Hosea; Edwards, Richard Archer; Felts, Wayne Moore; Fisk, Harold Norman; Flower, Rousseau Hayner; Howe, Ralph H.; Laurence, Robert Abraham; McClure, Robert I.; Meyer, Willis George; Miller, Lucile; Munyan, Arthur Claude; Munyan, Arthur Claude; Rogers, James Kenneth; Rouse, John Thomas; Rowland, Richard Atwell; Sandberg, Adolph Engelbrekt; Sandberg, Adolph Engelbrekt; Shanklin, Bob; Truesdell, Page Ernest; Wachs, W. C., Jr.

1940s: Altschuler, Zalman Samuel; Atkins, Frank Pearce, Jr.; Blickle, Arthur H.; Branch, John Russell; Connolly, F. Thomas; Dickson, Elizabeth; Doroshenko, Jerry; Fackler, William C.; Forsyth, Jane Louise; Gray, Russell Dent; Griffin, Robert Hardy; Han, Tsu-ming; Jones, Hal Joseph; Laird, Wilson Morrow; Manry, John Phillips; Metternich, Viola B.; Meyer, Willis George; Schlicten, Harold

Carl von; Sonnenberg, Frank Payler; Taylor, Jane M.; Tetrick, Paul Roderick

1950s: Abbott, Maxine Langford; Armstrong, Augustus Keathly; Bass, Irvin; Beaudry, Donald Arthur; Bell, Peter Mayo; Blane, John P.; Breene, Victor Martin; Brown, James Lee; Bruns, Richard Harte; Cahall, Leavitt P.; Chen, Ping-fan; Conatser, Willis E.; Dresser, Hugh; Eisenberg, Marvin; Gaither, Alfred; Gibson, William Carleton; Gilson, Edward S., Jr.; Hall, Leo Matthew; Hall, William B.; Harris, Steven H.; Hartsock, John Kaus; Hays, Frank Richard; Holland, Frank Delno, Jr.; Holland, Frank Richard; Horstman, Arden William; Hyde, David Edward; Keller, Daniel James; Kerr, Stuart Duff; Kingery, Thomas LeRoy; Kruse, Henry Oscar; La More, Francis Ellsworth; Lattman, Laurence Harold; Lattman, Laurence Harold; Macke, William Bernard; Malcolm, Daniel Connor; Mallin, James Wilson; Martin, Wayne Dudley; Mase, Jack Edgar; Meinert, Richard Joseph; Middendorf, Robert P.; Miller, Hayden Daniel; Myers, Addison Reid; Nuttall, Brandon Duncan; Phillips, David Lee; Pirkle, Earl Conley, Jr.; Pogue, Jesse B.; Province, Harold Edward; Rau, Jon Llewellyn; Rawlinson, George Harmon; Russell, Robert Thayer; Sappenfield, Luther Weidner; Sass, Daniel B.; Schaleman, Harry James; Schmidt, Ronald G.; Schweinfurth, Mark Fred; Schweinfurth, Stanley Paul; Seyfried, Conrad Kent; Single, Erwin Leroy; Stephens, James Gilbert; Tritschler, Charles W.; Truitt, Paul Bettman; Ulmer, Joseph Walter, Jr.; Van Fossen, John Doan; Vetorino, Robert Morris; Weaver, John Ferry; Welsh, Gerald Merritt; Wilson, William Jay; Witt, William John

1960s: Alpha, James W.; Arenas, Mario J.; Armstrong, Augustus Keathly; Bassarab, Dennis Rudyard; Bell, Bruce McConnell; Bohmer, Harold, Jr.; Bohmer, Harold, Jr.; Branstrator, Jon Wayne; Breuer, Joseph W.; Bullard, Reuben George; Bullard, Reuben George; Carpenter, Gene Charles; Carter, John Lyman; Coble, Ronald Winner; Conkin, James Elvin; Craig, Howard Reid; Drafall, Larry Edward; Eye, John David; Farmer, George Thomas, Jr.; Ford, David Wayne; French, Robert Rex; Galbraith, Robert Marshall, IV; Hall, Donald D.; Hester, Norman Curtis; Hester, Norman Curtis; Hu, Chung-Hung; Huff, Warren David; Jones, Robert Alan; Lienhart, David A.; Manus, Ronald Warren; Masters, John Michael; McClellan, William Alan; Naegele, Orville Dale; O'Donnell, Edward; O'Donnell, Edward; Osgood, Richard Grosvenor, Jr.; Parsley, Ronald L.; Parsley, Ronald Lee; Pohowsky, Robert Alexander; Pojeta, John; Pojeta, John; Pope, John Keyler; Portugal Tejada, Jorge A.; Portugal Tejada, Jorge A.; Rowan, Lawrence Calvin; Schaber, Gerald Gene; Schaber, Gerald Gene; Shaw, Frederick Carleton; Simms, Frederick Eugene, Jr.; Sims, John David; Singh, Raman J.; Starmer, Robert John; Stephens, Michael Anthony; Suarez Riglos, Mario; Swisher, Marian Margaret; Tappe, John; Tucker, Eva; Ward, Ronald Arthur; Warshauer, Steven M.; Wilson, Dail Adair

1970s: Acomb, Barry W.; Adams, Gregory T.; Amaral, Eugene Jordan; Aronoff, Steven Martin; Attoh, Kodjopa; Beem, Kenneth Alan; Bell, Bruce McConnell; Beltrame, Robert J.; BeMent, W. Owen; Benson, Donald Joe; Benson, Donald Joe; Bielak, LeRon E.; Branstrator, Jon Wayne; Bremer, Mary Lee; Broadhead, Ronald Frigon; Cordiviola, Steven; Cottrill, Nancy Elaine; Drafall, Larry Edward; Effimoff, Igol; Effler, Michael Edward; Elias, Robert Jacob; Elias, Robert Jacob; Ettensohn, Frank Robert; Ewers, Ralph; Fink, Robert Arthur; Fiorito, Thomas Francis; Fulton, Kenneth James; Fulton, Kenneth James; Gallant, William A.; Gardner, Thomas William; Garrison, Robert Kent; Goodman, Wayne Richard; Green, David J.; Gunal, Asuman; Hannan, Andrew J.; Harmon, William Lloyd; Harrison, Linda Kelley; Harrison, William Baxter, III; Hillman, Barry Arthur; Hrabar, Stephanie Vladimira; Jordan, Douglas W.; Kalaitjis, Michael G.; Kaplafka, Nancy A.; Kaplan, Allen Edward; Kepferle, Roy; Klein, Helen M.; Klekamp, C.

Thomas; Krebs, Elizabeth Ann; Laub, Richard Steven; Lee, John Clifford Hodges, III; Lenhart, Robert James; Lenhart, Robert James; Lundegard, Paul D.; Martin, Richard Vernon; Martin, Robin; McCandless, Richard Melvin; Meyers, Stephen Douglas; Nye, Osborne Barr, Jr.; Pohowsky, Robert Alexander; Prager, Gerald David; Provo, Linda Jeanne; Purcell, Charles Wilson; Reidel, Stephen Paul; Safford, Frederick Bargar; Samuels, Neil D.; Sarwar, Ghulam; Short, Michael Ray; Singh, Raman J.; Staley, Jack W.; Starmer, Robert John; Stephenson, James Todd; Sumartojo, Jojok; Tadkod, Mohamed-Ali; Taylor, Gary Allen; Teller, James Tobias; Thaeler, John D.; Twyman, James DeWitt; Twyman, Terry Rayno; Ulmschneider, Rene Joseph; Van Wie, William Arthur; Van Wie, William Arthur; VanArsdale, Roy Burbank; Warn, John Michael; Warn, John Michael; Warshauer, Steven M.; Wells, Stephen Gene; Wells, Stephen Gene; Werblow, Jack; Westrich, Henry R.; Whitsett, Charles K.; Williams, Frederick D.; Yu, Ho-Shing

1980s: Aytuna, Sezgin; Baum, Rex Lee; Baum, Rex Lee; Beaujon, James Sherman; Berger, Philip S.; Braide, Sokari Percival; Brandt, Danita Sue; Brockman, Charles Scott; Bryan, Timothy Michael; Camur, Mehmet Zeki; Caputo, Mario Vincent; Clement, Craig Robert; Covaleski, Ann M.; Cruikshank, Kenneth M.; Czoer, Kenneth E.; Davies, Russell King; DeReamer, John; Fisher, Mark P.; Flanigan, Donna M. Herring; Fluegeman, Richard Herbert; Franca, Almerio Barros; Frank, Robin; Geiger, Kenneth J.; Gilb, Scot H.; Goodrich, James Alan; Greenstein, Benjamin J.; Haneberg, William C.; Harrell, James Anthony; Haynes, John Tweedt; Hemlein, Kristin; Hendricks, Charles E., Jr.; Hoholick, D. John; Hudson, Thomas W.; Isea, Andreina; Jackson, Dana Scott; Jacques-Ayala, César; Jennette, David C.; Kareth, Paul Edward; Kazmer, Carol; Kelm, James S.; Keltch, Brian; Krumpolz, Bradley J.; Kryza, Edward A., Jr.; Lace, Penny J.; Lamborg, Amy Davison; Law, Maureen Min Wu; Letargo, Maria Rosario R.; Lewan, Michael Donald; Lion, Thomas E.; Liu, Tiebing; Liu, Yu-Lin; Loar, Steven J.; Lollis, Joan Cullen; Loos, Kenneth Dingwell; Losonsky, George; Lukasik, David M.; Lukasik, David M.; McComb, Thomas D.; Mersmann, Mark A.; Metarko, Thomas A.; Miller, Mark E.; Minnery, Gregory A.; Murdoch, Lawrence Corlies, III; Norrish, Winston A.; Okita, Patrick Masao; Olson, Clifford M.; Olson, Robert Laurence; Patterson, Neil Bruce; Petersn, Daniel Wayne; Pfaff, Virginia J.; Plaus, Peggy; Pohana, Richard Edward; Ranganathan, Vishnu; Rassman, Barbara; Richards, Kenneth A.; Riestenberg, Mary M.; Riestenberg, Mary M.; Roberts, Michael J.; Rodriguez Molina, Carlos; Roeser, Edgar Waldemar; Rolph, Alan Lindsay; Santa, Rick J.; Sarwar, Ghulam; Savage, Bruce A.; Savage, Kevin M.; Schneider, Jay A.; Schumacher, Gregory A.; Smith, James C.; Smyth, Ann Lindsay; Sobel, Lloyd S.; Sullivan, Michael P.; Sutton, Sally Jo; Sverdlove, Marc Selig; Tharp, Timothy C.; Thies, Jennifer L.; Tobin, Rick Curtis; Tobin, Rick Curtis; Trippi, Michael Herbert; Van Beuren, Victor Vignot; Wahlman, Gregory Paul; Waller, Stephen F.; Walters, Sylvia Jene; Webb, Elmer James; Weinheimer, Roger L.; Westerfield, Michael J.; Wheelock, Graham; Willard, James S.; Wilsey, William L. M.; Woodward, Stephen C.; Yerino, Lawrence N.; Zoghi, Manoochehr

City College (CUNY)
New York, NY 10031

4 Master's, 9 Doctoral

1970s: Cullen, Timothy R.; Farmer, M. W.; Hipsch, G.; Ruggiero, J. G.

1980s: Davies, Mark Allen; Eshet, Yoram; Fenster, Eugene Joel; Fu, Cheng-Ping David; Grande, Roger Lance; Lung, Richard Hai; Neff, Nancy Ann; Train, Fiona F.; Wang, Wei-Yeong

Clark University
Worcester, MA   01610

15 Master's, 28 Doctoral

1920s: Atwood, Rollin S.; Atwood, Wallace W., Jr.

1930s: Atwood, Wallace W.; Kingman, Celia Collins; Lemaire, M. E.; Quam, Louis O.; Ristow, Walter W.

1940s: Adkinson, Burton Wilbur; Clark, Genevieve; English, Van Harvey; Martin, Paul Felix; Williamson, Mary Frances

1950s: Delliquadri, Lawrence Michael; George, John Louis; Grotewold, Andreas Peter; Hokans, David Hamlin; Jackman, Albert Havens; Jeness, John Lewis; Kiewiet de Jonge, E. J. Coen; Lord, Arthur C.; Parmenter, Guy Norris; Renny, Edward; Smolski, Chester E.; Stacey, Karl; Vander, Pyl

1960s: Bariss, Nicholas; Brown, Roger James Evan; Fraser, John Keith; Looker, Robert B.; Lowe, John Carl; Thompson, Will Francis

1970s: Copes, Joe Lenon; Crawford, N. C.; Hubbard, E. L.; Marcus, A. L.; Meleen, N. H.; Renwick, W. H.; Spayne, Robert William

1980s: Bomah, Andrew Kelvinson; Ferguson, Kevin William; Molinelli, Jose Antonio; Way, Douglas Stewart; Zahavi, Abraham

Clarkson University
Potsdam, NY   13676

7 Doctoral

1970s: Burns, Robert R.; Wood, James L.

1980s: DeGregorio, Vincent B.; Nelligan, John D.; Raymond, Rhonda Karen; Scheffe, Richard Donald; Siah, Shu-Sheng Jonathan

Clemson University
Clemson, SC   29631

1 Master's, 10 Doctoral

1980s: Bailey, Alan Clarke; Crider, Steven Snowden; Dunn, David Lynn; Dunnivant, Frank Morris; Fay, Louis Elwyn; Heatley, William Robert, Jr.; Keel, Raybon Thomas; Pandit, Ashok; Rikard, Michael Wayne; Warner, Richard Charles; Zirschky, John Herbert

Colgate University
Hamilton, NY   13346

3 Master's

1930s: Berry, G. W.

1940s: Buermann, John M.

1980s: Keller, Dianne M.

University of Colorado
Boulder, CO   80309

442 Master's, 301 Doctoral

1900s: Crawford, Ralph D.; Jackson, Bethell H.; Underhill, James

1910s: Butters, Roy M.; Duncan, Q. Randolph; Dungan, Q. Randolph; Heaton, Ross Leslie; Worcester, Philip George

1920s: Boss, Reuel Lee; Collins, Melvin J.; Gibson, Russell; Gibson, Russell; Hall, Ellis A.; Hill, James Daniel; Livingstone, Jennie; Morey, Caroll A.; Murray, Albert Nelson; Nelson, Lloyd A.; Perini, Vincent C., Jr.; Sellers, J. E.; Thornbury, William David; Willson, Kenneth

1930s: Alf, Raymond M.; Bauer, C. Max; Blackmer, Joanne; Boatright, Byron B.; Coke, John McBrien; Dorrell, Carter Victor; Forward, Frederick; Graham, John R.; Hendricks, Thomas A.; Hill, E. Bratton, Jr.; Ives, Ronald Lorenz; Johnson, James F.; Johnson, Jesse Harlan; Kimball, Edgar W.; Knapp, Vernon; Koerner, Harold E.; Lester, James George; Maxwell, James M.; Murphy, Robert E.; Murray, Charles R.; Nelson, Lloyd A.; New-

bill, Thomas J., Jr.; Nygren, Walter E.; Oder, A. Louis; Osborne, Paul F.; Quam, Louis Otto; Reno, Duane Hugh; Seaman, David Martin; Stagner, Howard Ralph; Stagner, Willbur Lowell; Swem, Charles Edward; Thies, Roland Kendall; Van Beveren, Oscar F.; Van Valkenburgh, Alvin; Wahlstrom, Ernest Eugene; Waldschmidt, William Albert; Watson, Robert J.

1940s: Addy, Richard V.; Baird, Donald; Baird, Lucille Bailey; Baker, Vernon R.; Bates, Clair Ellen; Callier, Douglas Rean; Coash, John Russell; Cree, Allan; Culligan, Leland B.; Culligan, Leland B.; Curtis, Bruce F.; Davis, Ralph A.; Deere, Don Uel; Deul, M.; Dorrell, Carter Victor; Fix, Philip Forsyth; Freeman, James C.; Goldstein, August, Jr.; Greenlee, Arthur; Gross, Eugene B.; Hayford, Frank Sim; Holt, Edward L.; Hunter, Zena M.; Iglehart, Charles F.; Krill, Karl E.; Langenheim, Ralph L., Jr.; MacCornack, Richard John; Magee, John J.; McCann, Thomas P.; McKenna, James W.; McLellan, Russell R.; Neighbor, Frank; Oswald, Mary L.; Pearl, Richard M.; Peters, William C.; Rubright, Richard D.; Sample, Raymond Dewey; Sax, Norman Alfred; Smith, Albert W.; Thompson, John H.; Tollefson, Oscar W.; Toung, Kouang Shu; Wagner, Chancellor Philip; Weisberg, Maurice M.; Woodbury, Homer Olwin

1950s: Abrassart, Chester P.; Allen, Arthur Thomas, Jr.; Ashbaugh, James Graham; Beckett, Robert L.; Bird, Allan G.; Bolyard, Dudley W.; Bradley, Whitney A.; Breed, Charles E.; Broin, Irene J.; Broin, Thayne Leo; Broin, Thayne Leo; Butler, Charles R.; Butler, James Robert; Campbell, John Arthur; Cavender, Wayne S.; Chmelik, Frank B.; Cieslewicz, Walter John; Clough, George A.; Dearth, Albert E.; Duncan, Robert Louis; Dunn, Harold L., Jr.; Eicher, Don Lauren; Ellis, Charles Howard; Ettinger, Morris J.; Field, Robert Joseph; Fischer, William A.; Fowler, William A.; Fundingsland, Ernest L.; Gard, Leonard Meade, Jr.; Gates, Olcott; Gates, Olcott; Goode, Harry Donald; Hampton, O. Winston; Hampton, O. Winston; Hanshaw, Bruce Busser; Harms, John Conrad; Harris, David V.; Hatfield, Lloyd E.; Hauck, Rogers Austin; Hayes, John R.; Hendrickson, Glen; Holmes, Allen Whitney; Honea, Russel M.; Hornback, V. Quintin; Horner, Wesley Pate; Hubert, John F.; Hudson, Belva D.; Humphrey, Arthur G.; Hupp, William Ervin; Kim, Ok Joon; Kinnaman, Ross Lorrain; Kittleman, Lawrence; Larsen, Veryl E.; Laurent, J. Scott; Litsey, Linus R.; Luedke, Robert George; Mark, Anson; Mason, Thomas Paxton; Masters, Charles Day; Masters, John A.; McElroy, James Ralph; Millet, Marion T.; Moore, Samuel L.; Muenzinger, John W.; Munger, Robert D.; Murray, Harrison F.; Nottingham, Marsh Whitney; Parkinson, Lucius J., Jr.; Perry, Albert J.; Peters, William C.; Pilkington, Harold Dean; Pillmore, Charles L.; Pinckney, Darrell Mayne; Poole, Forrest Graham; Praetorius, H. W.; Quinlivan, William D.; Randall, John A.; Ranspot, Henry W.; Regout, Robertus; Richmond, Gerald M.; Ridlon, James Barr; Riley, Paul E.; Robinson, Charles Sheerwood; Rold, John W.; Rowlinson, Norman R.; Safford, Wilbur L.; Sainsbury, Cleo L.; Scott, Kenneth Eugene; Sharps, Seymour L.; Shearer, Eugene Merle; Shenkel, Claude W., Jr.; Sheridan, David S.; Shullaw, Byron L.; Siapno, William David; Skipp, Betty A.; Speno, Leo Anthony; Stafford, Wilbur L.; Stauffer, John Edward; Steinbach, Robert C.; Stevens, Calvin H.; Stough, Joan B.; Struby, William D.; Sundharovat, Swai; Thayer, James Bliss; Thompson, Craig Dickerson; Thrailkill, John Vernon; Tietz, Frederic A.; Timroth, Ron A.; Tollefson, Oscar W.; Toulmin, Priestley, III; Trask, Newell J., Jr.; Van Diver, Bradford Babbitt; Veeh, Hans Herbert; Wakefield, Lawrence Woodbury; Wanek, Leo James; Ward, Dederick C., III; Ward, Dwight Edward; Weist, William G., Jr.; Welder, George Edward; Whitebread, Donald Harvey; Whitney, Fred L.; Willhour, Robert R.; Witkind, Irving Jerome

1960s: Addison, Michael Earl; Anderson, Thomas B.; Anderson, Thomas Bertram, III; Armbrust,

George Aimé; Baars, Donald Lee; Baker, Richard Graves; Bank, Evelyn Ruth Jastram; Barnard, Frederick L.; Barosh, Patrick James; Bartleson, Bruce Landon; Berg, Thomas Miles; Berger, Wolfgang H.; Berger, Wolfgang Helmut; Boggs, Sam, Jr.; Bookstrom, Arthur Albin; Brennan, William Joseph; Brodsky, Harold; Bucknam, Robert Campbell; Budge, Wallace Don; Burrell, Stephen D.; Butcher, Robert H.; Calvert, Ronald H.; Campbell, John Arthur; Campbell, Newell Paul; Carlson, Ernest H.; Cary, Russell S., Jr.; Clark, Jerry H.; Coates, Donald Allen; Cody, Robert Dow; Conner, Jon J.; Connor, Jon James; Conroy, A. Richard; Cook, Peter John; Corbett, Marshall Keene; Curry, Robert R.; Curtin, Gary C.; Cys, John McKnight; Duffet, Walter Nelson; Duke, Walter; Eggler, David Hewitt; Eyer, Jerome Arlan; Fenske, Paul Roderick; Files, Frederic G.; Fritts, Paul Jan; Garvin, Paul Lawrence; Gawarecki, Stephen Jerome; Gibbs, James F.; Giegengack, Robert F.; Gilbert, John Robert; Griggs, David Gould; Gustafson, Donald L.; Hall, Robert Dean; Hammuda, Omar Suleiman; Harper, Melvin Louis; Hawley, Charles Caldwell; Heidrick, Tom Lee; Hepp, Mary Margaret; Holt, Henry E.; Horstman, Arden William; Hott, Albert C.; Hoyt, John Harger; Hughes, Travis Hubert; Hunter, John Henry; Irvine, Ben M.; Jacob, Arthur Frank; Keighin, Charles William; Keighin, Charles William; Kelly, James Michael; Kennedy, Vance Clifford; Kent, Harry Christison; Kittredge, Tylor F.; Kucera, Richard Edward; Lamb, George Marion; Larsen, Margaret Kreider; Larson, Edwin Erie; Laughon, Robert B.; LeMasurier, W. E.; Lewis, John Hubbard; Long, Clarence Sumner, Jr.; Lowman, Paul Daniel, Jr.; Mann, Donald M.; Matson, Neal A., Jr.; Mayer, Victor J.; McMahon, B. E.; McMahon, Beverly Edith; Merifield, Paul Milton; Merk, George P.; Millard, Richard C.; Mull, Charles G.; Murray, Frederick N.; Murray, Frederick N.; Mutschler, Felix Ernest; Newman, Karl Robert; Nordlie, Bert Edward; Nutalaya, Prinya; Nutalaya, Prinya; O'Connor, Joseph Tappan; O'Connor, Thomas E.; Page, William Delano; Pannella, Giorgio; Phipps, James B.; Poelchau, Harold S.; Power, Peter Edward; Prather, Thomas Leigh; Prather, Thomas Leigh; Raup, Omer Beaver; Riecker, Robert Edward; Rizvi, Syed Saghir Ahmed; Rowekamp, Edward Terry; Schmidt, Peter B.; Schroeder, Warren Lee; Sharp, John Van Alstyne; Sharps, Seymour Lytton; Small, John, Jr.; Smith, Richard Paul; Stephens, Jerry Clair; Stone, George Thomas; Swann, Gordon Alfred; Swett, Keene; Tharalson, Darryl Bruce; Tillman, Roderick Whitbeck; Trammell, John W.; Treckman, John F.; Ulrich, George Erwin; Waldrop, Henry A.; Ward, Albert Noll, Jr.; Williams, Wilbur S.; Wohlford, Duane Dennis; Worstell, Paula Jane; Yeend, Warren E.; Zayed, Mostafa Mahmoud

1970s: Abbott, Jeffrey Tarbell; Addy, Sunit Kumar; Agard, Sherry Sue; Alford, Donald Leslie; Anderson, L. W.; Armstrong, Richard L.; Ashton, L. W.; Bailey, T. P.; Baker, Victor Richard; Birmingham, Thomas John; Bischke, Richard Edward; Blakestad, Robert Byron, Jr.; Boberg, Walter W.; Bond, Wendell A.; Boyer, Stephen Joseph; Bradbury, J. W.; Brenckle, Paul Louis; Brill, Russell Martin; Buck, Stanley P.; Burke, Raymond Merle; Byerley, Keith Alan; Campbell, L. F., Jr.; Carrara, Paul Edward; Christopherson, Karen Rae; Clark, J. A.; Cole, J. C.; Colman, Steven Michael; Coombs, Margery Chalifoux; Coombs, Walter Preston, Jr.; Courtright, T. R.; Cowan, W. R.; Crone, A. J., Jr.; Damuth, John Erwin; Delson, Eric; Downey, Cameron Ingraham; Downs, W. F.; Duckson, D. W., Jr.; Dula, William F., Jr.; Dyke, A. S.; Ebaugh, W. F.; Edgerton, G. K.; Ellis, E. G.; Eshett, Ali; Evetts, Michael J.; Ewing, J. W.; Fahey, Barry D.; Fitch, Thomas Jelstrup; Frischknecht, Frank Conrad; Frush, Mary Penelope; Gamble, Bruce Martin; Gawne, Constance Elaine; Gawthrop, W. H.; Grey, L.; Grocock, G. R.; Gunow, Alexander James; Hammuda, Omar Suleiman; Hiss, W. L.; Hoblitt, Richard Patrick; Holcomb, D. J.; Ingraffea, A. R.; Isherwood, D.; Isherwood, W. F.; Jochim, Candace

University of Colorado at Colorado Springs

L.; Johnson, J. B.; Johnston, A. C.; Judd, J. B.; Kaback, D. S.; Katz, B. G.; Kellogg, Karl Stuart; Kizis, Joseph Anthony, Jr.; Komarkova, V.; Kridelbaugh, Stephen Joseph; LaForge, R.; LaFountain, Lester James, Jr.; LaPoint, D. J.; Lawless, Steven James; LeGendre, Gary Ralph; Leonhart, Scott W.; Lindberg, R. D.; Locke, W. W., III; Ludington, S. D.; Lueck, S. L.; Lundquist, G. M.; Mabee, S. B.; Machette, Michael N.; Mahaffy, M.-A. W.; Mahaney, William Cornelius; Manley, K.; Markos, G.; Martens, Ronald Wayne; Martin, H. L.; Mather, Terry James; Mears, Arthur Irvin; Meierding, T. C.; Miller, Clifford Daniel; Miller, G. H.; Miller, H. F., II; Mosburg, Shirley Krauthausen; Nelson, A. R.; Nesse, W. D.; Netoff, D. I.; Nowlan, G. A.; O'Neill, J. M.; Oesleby, T. W.; Okumura, Terrence A.; Ostergard, Deborah; Page, William Delano; Pendleton, J. A.; Perkins, Michael; Perkins, R. M.; Perry, R. V.; Peterson, J. E.; Peterson, J. E.; Pheasant, David Richard; Punongbayan, Raymundo Santiago; Razum, Brynne Anne; Reheis, M. J.; Rendu, Jean-Michel Marie; Retherford, Robert Morse; Reynolds, R. L.; Reynolds, Richard L.; Rottman, Marcia Louise Gaines; Sanderson, I. D.; Savino, John Michael; Seeling, Alan; Severy, Charles Lamb; Severy, Niles Ian; Sharpe, D. R.; Shelton, David Camp; Short, S. K.; Shroba, R. R.; Shropshire, Kenneth Lee; Smith, Richard P.; Solounias, N.; Stevens, James Crosby; Sture, S.; Sullivan, Michael W.; Thurman, E. M.; Topper, Ralf E.; Toth, Margo Irene; Trench, N. R.; Vinckier, Thomas Alan; Wahr, J. M.; Wallace, A. R.; Walsh, James Paul Jr.; Warrick, R. A.; Weaver, Ronald Lee; Wesner, George Mack; Wilder, D. G.; Williams, Lawrence Darryl; Wilson, Lee; Wiman, S. K.; Woodward, D. G.; Young, H. L.; Yount, James C.

1980s: Alawi, Mohamed Mohamed; Albino, Katharine Chase; Amini, Mohammad-Hassan; Askar, Hasan Ghuloom; Azevedo, Roberto Francisco; Baker, Fred G.; Baker, Stephanie Ashburn; Barber, Larry Billingsley, II; Barlow, Lisa Katharine; Barovich, Karin Marie; Batt, Richard James; Beeson, Dale Clayton; Bell, Jean Louise; Bercaw, Louise B.; Bieber, David William; Blair, Terence C.; Boler, Frances Michele; Botts, Michael Edward; Bove, Dana Joseph; Bowman, J. Roger; Bowman, J. Roger; Brigham, Julie K.; Brigham, Julie Kay; Brodsky, Nancy S.; Brodsky, Nancy S.; Budding, Karin Elisabeth; Burns, Scott Frimoth; Caine, Jennifer M.; Campbell, Gregg Tyler; Cannon, Susan Hilary; Caporuscio, Florie Andre; Chen, Jing-Wen; Clark, Peter Underwood; Clendenen, William Sterling; Coss, John Michael; Creighton, Ann; Crespi, Jean M.; Crifasi, Robert R.; Cruz Calderon, Gonzalo; Crysdale, Bonnie L.; Davis, Andrew Owen; Davis, John Wesley; Davis, Philip Thompson; Day, Harry Clinton, Jr.; Dentler, Patricia L.; Diaz, Henry Frank; Dickinson, Warren William; Dixon, John Charles; Dubiel, Russell F.; Dyni, John Richard; Eaton, Jeffrey Glenn; Eckert, Anne Douglas; Elder, William Perdue; Elias, Scott Armstrong; Esmaili, Esmail; Estey, Louis Howard; Finn, Carol Ann; Finn, Carol Ann; Fishman, Neil Steven; Forbes, Robert Lyle; Forester, Elisabeth Brouwers; Forman, Steven Lawrence; Francis, Kevin Albert; Garza, Roberto; Gautier, Donald Lee; Geary, Dana Helen; Geirsdottir, Aslaug; Gephart, John Wesley; Gerlitz, Carol Nan; Glenister, Linda Marie; Glibota, Thomas J.; Gockley, Catherine Kristin; Gross, Richard Stewart; Grout, Marilyn Ann; Guccione, Margaret Josephine Weatherhead; Gunow, Alexander James; Habermann, Ray Edward; Haller, Kathleen Monger; Hansen-Bristow, Katherine Jane; Harden, Deborah Reid; Harrington, Robert John; Harris, Timothy Donovan; Harvey, Danny James; Hawkins, Fred Frost; Hearty, Paul Joseph; Hindman, David Jerome; Holliday, Vance Terrell; Hon, Kenneth Alan; Hong, Kappyo; Horita, Masakuni; Huang, Zhongxian; Huber, Thomas Patrick; Hudson, Ann Elizabeth; Indeck, Jeff; Jackson, Michael Eldon; Janoo, Vincent Clement; Jennings, Anne Elizabeth; Johnson, Carla; Johnson, Claudia C.; Johnson, Kurt Warren; Kane, Ward Thompson;

Karimpour, Mohammad Hassan; Kihm, Allen James; Kim, Myoung Mo; Kindred, Valerie Prescott; Kiteley, Louise W.; Klinger, Lee Francis; Kost, Linda Suzanne; Krarti, Moncef; Kraus, Mary Jean; Kron, Donald Gordon; Lanham, Robert Evans; Larson, Jay Leo; Lawrence, Viki Ann; Lawrence, William W., Jr.; Laymon, Charles Alan; Leckie, Robert Mark; Lee, Charles Gordon; Leifer, John C.; Leonard, Eric Michael; Lidke, David J.; Lindberg, Ralph DeWitt; Litaor, Michael Iggy; Locke, William Willard, III; Loeffler, Bruce Marston; Lund, Karen; Madden, Cary Thomas; Mander, Marsha L. Morton; Marvil, Joshua D.; Massoudi, Nasser; McCoy, William Dennis; McGimsey, Debra Hanson; McGimsey, Robert Gamewell; Meertens, Charles Mangelaar; Meertens, Charles Mangelaar; Middleton, Michael D.; Miller, Douglas Robert; Minor, Scott Alan; Mode, William Niles; Morris, Scott Edward; Mould, John Calvin, Jr.; Mozley, Peter Snow; Mruk, Denise Helen; Muhs, Daniel Robert; Muller, David S.; Munoz Bravo, Jorge Oswaldo; Murphy, W. Dale; Muth, Lorant Andreas; Nelson, Karl Russell; Nichols, Thomas Chester, Jr.; Niebauer, Timothy Michael; Nielsen, Norman C.; Nordstog, Kim Thomas; Ontuna, Kazim Ates; Osterman, Lisa Ellen; Pane, Vincenzo; Patterson, Charles G.; Patterson, Penny Ellen; Pavlik, Hannah Flora; Phillips, Victor Duzerah, III; Podsen, Donald Wayne; Pohlman, John Carl; Reheis, Marith Cady; Remy, Robert Reginald; Rhode, David Ronald; Ritchie, James Graham; Rocken, Christian; Roecken, Christian; Rohtert, William R.; Rosenbaum, Joseph Griffin; Rueger, Bruce Francis; Saether, Ola Magne; Sayili, Alaittin; Schatz, Barry Allen; Schnapp, Michael Gordon; Schneider, Jill Leslie; Schroeder, David Alan; Sedenquist, Daniel Frederic; Selverstone, Jane Elizabeth; Shaver, Kenneth Charles; Shea, Valois Ruth; Sheu, Wa-Ye; Shin, Charng-Jeng; Silliman, Alan Holt; Skewes, Milka Alexandra; Skipp, Betty Ann Lindberg; Spirakis, Charles Stanley; Steefel, Carl Iver; Sterr, Horst Michael; Stevens, Jeffry Lowell; Stoker, Carol R.; Stollenwerk, Kenneth George; Stover, Bruce King; Stravers, Jay Anthony; Stravers, Loreen K. Stegeman; Stubbs, Gale Susan; Summer, Rebecca Mae; Swanson, Peter Lee; Swanson, Peter Lee; Swenson, Alan Lee; Szavits-Nossan, Vlasta; Tainter, Patrick A.; Taylor, Emily Mahealani; Tewksbury, Barbara Jarvis; Toth, Margo Irene; Turner-Peterson, Christine Elizabeth; Vallance, James W.; Vaughan, Patrick Robison; Villien, Alain; Von Holdt, Laura Lynn; Voorhies, Coerte Van; Wahli, Catherine; Walker, Marilyn Drew; Walters, James Ettore; Wanklyn, Robert Paul; Webb, John Charles; Weimer, Paul; Weissmann, Gary Stephen; Welty, Curtis Michael; Werner, Alan; Whipple, James Wilburn; Whitaker, Stephen Taylor; Williams, Carol Alvis; Williams, Kerstin Margareta; Williams, Kerstin Margareta; Wineteer, Craig Brian; Wong, Kong-Cheng; Zimbelman, David R.; Znidarcic, Dobroslav; Zúñiga, Francisco Ramón

University of Colorado at Colorado Springs
Colorado Springs, CO   80933

6 Master's

1980s: Felling, Richard A.; Jackson, James Milton; Nelson, Garry Richard; Stanton, Michael D.; Stanton, Peter T.; Thornton, Wendy Barclay

Colorado College
Colorado Springs, CO   80903

6 Master's

1920s: Keyte, Wilbur Ross

1930s: Bennett, Billie; Ragle, Richard Charles

1950s: Hula, Charles William

1980s: Rath, Bruce A.; Walker, Valerie A.

Colorado School of Mines
Golden, CO   80401

711 Master's, 264 Doctoral

1910s: Gow, Thomas T.

1920s: Aguerrevere, Pedro I.; Courtier, William Henry; Foulkes, Thomas G.; Johnson, Jesse Harlan; Manuel, William Asbury; Marvin, Theodore; Serviss, Fred L. F.; Thomas, George W.

1930s: Aldredge, Robert F.; Boyd, James; Boyd, James; Brewer, Quenton L.; Fenwick, Willis H.; Focken, C. M.; Gabriel, Vittaly Gavrilovich; Gabriel, Vittaly Gavrilovich; Hawkins, James E.; Herbert, Donald L.; Hyslop, Ralph Craig; Ingham, W. I.; Jameson, Maynard H.; Johnson, Frank Melvin S.; Ku, Kong-Gyiu; Loring, Ralph C.; Manhart, Thomas A.; Marr, John D.; Masse, Lucien; Mitera, Zygmut; Parker, Ben Hutchinson; Parker, Ben Hutchinson; Paterson, Raymond G.; Pirson, Sylvain J.; Pugh, William Emerson; Sanjeevareddi, Buggana S.; Somers, George Brooks; Stroud, Paul W.; Tattam, Charles M.; Wantland, Dart; Winkler, Bruno Oscar

1940s: Acharya, Ashutosh; Anderson, Thomas P.; Azim, Syed A.; Baird, John; Bediz, Pertev I.; Bench, Bernard M.; Bosco, Francis N.; Bozbag, Hamdi A.; Chandiok, Kailash Chandra; Chapman, John J.; Cutter, Russell C.; Ferris, Bernard J.; Gabelman, John W.; Gabelman, John W.; Gude, Arthur J., III; Hawkins, James E.; Hohlt, Richard B.; LeRoy, Leslie W.; Manfredini, Antonio; Masse, Lucien; McCutchen, Wilmont R.; Moody, John Drummond; Musselman, Elmer Thomas; Myers, Otto Jay; Negi, Balwant Singh; Opland, Homer N.; Robb, George L.; Robbins, Edward M., Jr.; Rue, Edward E.; Saptarshi, Vidyadhar Chintamen; Schaeffer, Stanley E.; Shaeffer, Stanley B.; Skeeters, Warren Ware; Thomas, Clark L.

1950s: Adamson, Robert Clarence; Alaric, Saw; Ali, Hamzah; Ames, Vincent Eugene; Arora, O. P.; Ballew, William Harold; Banghart, Roger Clinton; Bauer, William Henry; Beattie, Donald Andrew; Berry, J. E.; Bhutta, Mohammed Afhar; Birdsall, Edwin F.; Bombolakis, Emanuel G.; Bradley, Wayles B.; Cannaday, Francis X.; Chamney, T. P.; Clement, Jean F.; Comstock, Sherman S.; Conley, Curtis D.; Cook, Douglas R.; Corbett, John Dennen; Corn, Russell Morrison; Davidson, R. N.; Decker, Robert W.; Demaison, G.; Dias, Manuel de Bettencourt; Diker, Salahi; Domenico, Samuel N.; Erwin, John W.; Fogarty, Charles F.; Fouret, James Howard; Ganguli, Ajit Kumar; Garcia, Marcial V.; Garrett, Howard L.; Gondouin, Michel; Hagen, John Christopher; Hartman, Frederick H.; Hastings, James S.; Herrera, Amilcar Oscar; Hodder, Edwin J.; Hollingsworth, J. A. C.; Holmer, Ralph C.; Horino, Frank G.; Iradji, Amir Houshang; Johnson, Robert Alfred; Jonson, David Carl; Kanizay, Stephen P.; Kerr, Bobby G.; Kim, Ok Joon; Kline, Mortimer A., Jr.; Konishi, Kenji; Levings, William S.; Lovejoy, Earl M. P.; MacKay, Ian H.; Malek-Aslani, Morad K.; Malek-Aslani, Morad K.; Meissner, Fred F.; Miller, Joseph H.; Morehouse, George E.; Morgan, George B., Jr.; Morrison, Lawrence S.; Musgrave, Albert W.; Ogden, Lawrence; Oppel, Richard E.; Osborne, Robert Edward; Parsons, Marshall Clay; Paul, Santosh Kumar; Pickett, George R.; Pierce, Arthur P.; Quiett, Frederick T.; Qureshy, Mohammed N.; Rahman, Yousuf H.; Ramirez Berrera, Andreas; Reichert, Stanley O.; Robertson, Lloyd B.; Roy, Amalendu; Rugg, Edwin Stanton; Ruley, Eugene E.; Sears, Charles Edward, Jr.; Sinha, Bindeshwari Narain; Slebir, Edward Joseph; Sonido, Ernesto P.; Stewart, Walter Alan; Stoddard, Robert Russell; Stommel, Harrison Edfred; Stout, John L.; Strommel, H. E.; Troseth, Frank Paton; Van Arsdale, B. E., Jr.; Waterman, Herbert Douglas; Williams, Floyd James; Willis, John Bruce; Wise, Richard Atlee; Wolfe, John A.; Woodward, Thomas C.; Zoerb, Richard M.

1960s: Al-Khafaji, S. A.; Ali, Mohammad; Altan, Ozer; Applegate, James K.; Balachandran, K.; Baldwin, Harry L., Jr.; Becker, R. M.; Benyamin, Ninos B.; Bianchi, Luiz; Blood, W. Alexander; Bloom, Duane N.; Bloom, Harold; Bridwell, Richard Joseph; Brown, Delmer; Brown, Lynn A.; Bublitz, Richard F.; Buchanan, Peter H.; Butler, David; Calkin, William S.; Camacho, Ricardo; Chao, Tsu Ko; Chao, Tsu Ko; Chapin, Charles E.; Collings, Stephen P.; Coolbaugh, David F.; Covington, George H., III; Crews, George; Cronoble, James M.; De Voto, Richard H.; Del Castillo, G. Luis; Denslow, Lathrop V. B.; Diebold, Frank E.; Donovan, Peter R.; DuHamel, Jonathan E.; Dumas, Michael C.; Edwards, Jonathan, Jr.; Evans, W. D.; Gardner, Maxwell E.; Graebner, Peter; Gray, Russell L.; Grybeck, Donald; Hadsell, Frank A.; Hamzawi, Anwar T.; Harthill, Norman; Hedge, Carl E.; Henderson, Don K.; Hill, David Paul; Homuth, Emil F.; Hoover, Donald B.; Hutchinson, R. Alan; Hyun, Byung Koo; Jacobson, Jimmy J.; Johnson, Donald H.; Jones, Peter; Karig, Daniel E.; Kilpatrick, Bruce E.; Koch, Robert Winfield; Koesoemadinata, R. P.; Koksoy, Mumin; Kouther, Jameel H.; Kowalski, Michael A.; LaFehr, Thomas R.; Lauman, Gary W.; Leighton-Puga, Tomas; Leitinger, J. Hans; Lobato, Fabiano Sayao; Lozano, Efraim; Maberry, John O.; Machado, J. E.; Manning, Gerald E.; Maung, Tun U.; Mayhew, John D.; McMahon, Barry K.; McPeek, Lawrence A.; Michels, Donald E.; Michels, Donald E.; Miles, Charles H.; Morgan, Robert R.; Morris, Gary R.; Moya, Eduardo A.; Mukherjee, Nilendu S.; Munson, Robert C.; Murray, Thomas H., Jr.; Navas, Jaime; Nelson, Mark T.; Odouli, Khalil; Ong, Han-Ling; Ozcandarli, Tevfik Demir; Parker, Ben Hutchinson, Jr.; Pasquali-Zanin, Jean; Penttila, William C.; Perry, John Kent; Pierce, Walter; Pilcher, Stephen H.; Porter, Christopher; Ramarathnam, Sethurama; Reimer, Louis R.; Reynolds, Edward B.; Richards, David Barton; Romig, Phillip R., Jr.; Rouse, George E.; Sabet, Mohamed A.; Sabet, Mohamed A.; Sage, Ronald; Sawatsky, Don L.; Sawatzky, Don L.; Schwarz, Frederick P., Jr.; Seymour, David; Shaffer, Mark E.; Siems, Peter L.; Silva, Luis R.; Sinclair, William; Smookler, Stanley; Snow, Charles Bruce; Snyder, Donald D., III; Stevens, Douglas N.; Stewart, Dennis D.; Thomson, Ker C.; Thuren, John B.; Trimble, James K.; Turpin, Bernard; Valdes, Gustavo E.; Vargas D., Francisco H.; Walton, David S.; Wang, Yung-Liang; Warner, Don L.; Welch, Fred., Jr.; Welsh, Fred, Jr.; Welsh, Thomas M.; Wideman, Charles; Yacoub, Nazieh K.; Yingst, Parke O.; Young, Peter C.; Zeuss, Hilario

1970s: Abbott, David M., Jr.; Acosta, Ramon; Aguilera, Roberto; Ahmad, Mahmood U.; Ahmed, Adnan A.; Ajayi, Clement Olatunde; Alves, Carlos A.; Anderson, James P.; Antony, John J.; Aranda-Gomez, Jose; Arestad, John F.; Bacelar de Oliveira, Ruy Bruno; Bakhshandeh, Farhad; Barker, Steven A.; Barnett, Colin T.; Barreda, Willy Z. Rodriguez; Barrero, Dario; Barrows, Lawrence J.; Barrows, Lawrence J.; Beggs, George; Berghorn, Claude E.; Berman, Arthur E.; Best, James R.; Biller, Edward J.; Billingsley, Lee T.; Bird, William H.; Blair, Robert W., Jr.; Blanco, Stephen R.; Booth, Franklin O., III; Borges, Rafael E.; Brazie, Mike E.; Briceño M., Henry O.; Briscoe, Harry J., Jr.; Brown, William J.; Bruns, Dennis L.; Buchanan, Larry J.; Butler, James; Butler, Robert; Cain, Douglas L.; Candee, Christopher R.; Candito, Robert J.; Cardwell, Aubrey L.; Carloss, James C.; Carlson, Kenneth W.; Cattany, Ronald W.; Cavanaugh, Eugene T.; Chairat, Trakarn; Chilcoat, Steven R.; Clark, Betty A.; Collins, Bruce A.; Collins, Bruce A.; Connors, Roland A.; Correa, Aberbal Caetano; Cox, Kathleen L.; Crewdson, Robert A.; Crewdson, Robert A.; Crompton, James S.; Cronoble, James M.; Crous, Christiaan Mauritz; Cruson, Michael G.; Cuffney, Robert G.; Darken, William H.; De la Cruz, Luis A.; del Rozas Elqueta, Eduardo; DeRidder, Eduard; Diaz, Ricardo Navarro; Dimelow, Thomas E.; Dinkmeyer, Paul R.; Dodge, Rebecca Lee; Drobeck, Pete A.; Druecker, Michael D.; Duarte, R. Armando; Dupree, Jean A.; Ege, John R.; Eggert, Douglas J.; Eichler, Rodney J.; El-Hindi, Mohamed A.; Erickson, Richard A.; Erol, Vasfi; Evans, James L.; Fairer, George M.; Fassett, Jack W.; Fisher, James C.; Francis, Dennis C.; Fryberger, Steven G.; Furgerson, Robert Bernard; Furnaguera C., Jose A.; Gallagher, James Robert; Garcia S., Eduardo; Geyer, Richard G.; Giulj, Dominique; Glanzman, Richard K.; Gobel, Volker W.; Gokturk, Erkin; Guu, Jeng-Yih; Hadley, Linda M.; Halla, Marilyn Margaret; Hamilton, Robert D.; Haworth, Randol A.; Hebb, David T.; Hecox, Gary R.; Heiple, Linda J.; Hendricks, Michael L.; Hilterman, Fred J.; Holcombe, Lawrence J.; Huber, Gary C.; Huntley, David; Hurr, R. Theodore; Irtem, Oquz; Irtem, Oquz; Jado, Abdul-Rasof; Jansen, Walrave; Jaworski, John J.; Jenkins, Robert E., II; Jennings, Joan K.; Johnson, Stephen M.; Jordan, John M.; Kanaan, Faisel Mohamed; Kehmeier, Richard J.; Kirkman, Roy C.; Kirkwood, Steven G.; Klein, John M.; Kline, Robert J.; Knepper, Daniel H.; Kratochvil, Gary L.; Krish, Edward J.; Krug, Jack A.; Krygowski, Daniel; Krygowski, Daniel; Kumamoto, Lawrence H.; Kumamoto, Lawrence H.; Land, Cooper B.; Landress, R. A.; Langston, David J.; Lebel, Andre; Lee, Chong Y.; Lee, Chong Y.; Lee, Doo-Sung; Lewis, Linda L.; Lewis, Warren S.; Limbach, Fred W.; Lindberg, Cheryl A.; Lindstrom, Linda Jane; Lisle, Richard E.; Lopez Eyzaguirre, Carlos J.; Love, Alonza H.; Lowman, Bambi M.; Lusty, Quayle C.; MacMillan, Logan; Maione, Steven J.; Malone, S. J.; Manzolillo, Claudio D.; Marek, John M.; Marrs, Ronald W.; Martinez Muller, Remigio; Maslyn, Raymond M.; Mathews, Mark A.; McGlasson, James A.; McParland, Brian J.; Mergner, Marcia; Metzger, Chris W.; Miles, Deborah R.; Miller, James R.; Money, Nancy R.; Montazer, P. M.; Morris, Drew; Mueller, Joseph W.; Murray, J. C.; Myers, Robert C.; Nelson, Karl R.; Newell, Roger A.; Nicolais, Stephen M.; Nicolaysen, Gerald; Nolting, Richard M., III; Nwangwu, Uka; Nwangwu, Uka; Obernyer, Stanley L.; Odien, Robert J.; Oduolowu, Olusegun Akinyemi; Ofrey, Ofiafate; Olsen, Roger L.; Olson, William R.; Omari, Sid'Ali N.; Orr, Donald G.; Palencia, Cesar M.; Pansza, Arthur J., Jr.; Parra, Jorge; Parsons, Clifford C.; Paschis, James A.; Pasquali-Zanin, Jean; Paul, Alexander H.; Peel, Frederick A.; Pelizza, Mark S.; Perez, Omar; Perry, Harry A.; Peterson, Michael John; Peterson, Steven D.; Pinel, Mark J.; Poleschook, Daniel, Jr.; Porter, Karen W.; Posada, Jorge H.; Pritchard, James I.; Prost, Gary Leo; Pyrih, Roman Z.; Pyrih, Roman Z.; Rahmanian, Victor D.; Raines, Gary L.; Raines, Gary L.; Ramirez Lopez, Miguel; Ramirez Rojas, Armando J.; Ranta, Donald E.; Reeder, Robert T.; Reeve, William; Reid, Allan R.; Restrepo, Jorge J.; Ringrose, Charles D.; Rizo, Jaime A.; Robinson, James E.; Rodriguez, N. S.; Roghani, Foad; Rozelle, John W.; Ruiz de la Pena Horcasitas, Gerardo; Rutherford, David W.; Sanders, George F., Jr.; Schimschal, Ulrich; Schimschal, Ulrich; Schlaks, Francine D.; Schwochow, S. D.; Serna-Isaza, Mario J.; Shadul, Shadul A.; Shannon, James R.; Shehata, William Makram; Shipley, Webster E., III; Shoffner, David A.; Shuck, Edward L.; Silitonga, Parahum H.; Skelly, William A.; Skokan, C. K.; Skokan, Catherine A. King; Smith, Shea C.; Smith, William H.; Snyder, Linden E.; Sperandio, Robert J.; Speziale, Mario H.; Spoelhof, Robert William; Spoelhof, Robert William; Sprouse, Dan P.; Stewart, Collin L.; Stockton, Scott L.; Stoughton, Dean; Su, Fu-Chen; Suh, Jung Hee; Suryanto, Untung; Taranik, James V.; Tascin, Mustafa T.; Taufen, Paul M.; Tavora, Flavio J.; Taylor, Andrew M.; Taylor, Bruce E.; Tekiner, Yasar; Tien, Jean H.; Towle, Guy H.; Trost, Paul B.; Tsubota, K.; Turner, Dennis L.; Van der Spuy, Peter M.; Van Horn, Michael D.; Van Wyk, Stefanus J.; Waegli, Jerome A.; Watkins, Thomas A.; Watson, John E.; Weber, James R.; Weiland, Erick F.; West, Michael W.; Westfahl, Diane E.; Whan, Wen J.; Whitney, Mark S.; Williams, John T.; Williams, Steven John; Wilson, Ellery H.; Wittstrom, Martin D., Jr.; Wychgram, Daniel C.; Yang, Chieh-Hou; Young, Terence K.; Young, Terence K.; Yousefpour, M. V.; Yousefpour, Mohammed Vali; Zainy, Muhammad A.; Zogg, William D.; Zuccaro, Bruce

1980s: Abass, Hazim H.; Abdelmalik, Mohamed B. A.; Abrahao, Dirceu; Abrahao, Dirceu; Afifi, Abdulkader M.; Aikin, Andrea R.; Al-Faraj, Mohammed; Al-Fares, Mohammed H.; Al-Mashouq, Khalid; Al-Nemer, Jaafar Muhammad; Al-Shakhis, Amir A.; Al-Yacoub, Tassier; Alam, Mohamed Nour; Alatorre, Armando E.; Alazar, Tesfalul; Allen, Samuel W.; Altschuld, Kenneth R.; Amerman, Roger E.; Andersen, Henrik T.; Anderson, Brent C.; Arauz, Alejandro J.; Araya, Kidane; Audemard, Felipe; Aulia, Karsani; Bahavar, Manouchehr; Baker, David W.; Ball, Theodore T.; Bateman, Philip Walker; Bean, Susan M.; Belcher, Wayne R.; Bellatti, John T.; Benedict, Frank Christopher; Bennett, Jon Lewis; Bergeon, Thomas C.; Berkman, Frederick Eugene; BeVier, Laura M.; Bhasavanija, Khajohn; Bigarella, Laertes P.; Billings, Patty; Boden, David R.; Boden, Linda; Bond, Marc A.; Boreck, Donna L.; Borges Olivieri, Cesar A.; Botha, Willem J.; Bowman, Scott A.; Breit, George Nicholas; Breit, George Nicholas; Briceño M., Henry O.; Brinton, Lise; Brophy, James G.; Bruce, Robert MacAllister; Bryn, Sean M.; Budge, Suzanne; Bussey, Steven D.; Cable, Steven W.; Carpenter, Donald J.; Catts, John G.; Ceazan, Marnie L.; Chamberlin, Richard M.; Chang, Pingsheng; Chapin, Mark A.; Chatham, James Randall; Chen, Deng-Bo; Clark, J. Robert; Clark, Reino; Claussen, John P.; Cole, David M.; Collier, James D.; Connors, Katherine A.; Cooke, Dennis; Cooke, Dennis A.; Copper, Lon M.; Corbett, Mary Diane; Cress, Leland D.; Crisi, Peter A.; Crock, James Gerard; Cunningham, David A.; Dalke, Roger A.; Damron, Larry A.; Daniels, Robert P.; Day, Katherine Walton; de Rojas, Isabel; Dean, James Scott; Dewhurst, Warren Taylor; Dézé, Jean-Francois; Dickinson, Marc; Djuanda, Handoko; Docherty, Paul; Dodge, Rebecca Lee; Doebrich, Jeff L.; Domoracki, William Joseph; Donaldson, James C.; Donovan, Arthur Dean; Doyle, James David; Dunkhase, John A.; Durfee, Steven L.; Durrani, Javaid A.; Durrenberger, Sally; Duster, David W.; Dwyer, Ruth-Ann; Eiseman, Hope H.; Eisenmenger, Karl Kenneth; Emme, James J.; Eriksson, Carl L.; Farmer, Cathy L.; Fernandez Casals, Javier; Fillmore, Barbara J.; Fleckenstein, Martin; Franczyk, Karen J.; Franotovic, Davor; Freyr Thorarinsson; Fry, Michael F.; Gable, Douglas M.; Gaffke, Thresa M.; Garg, Nek R.; Garg, Nek R.; Garrett, Bruce T.; Geer, Kristen Anders; Geoltrain, Sebastien; Gesink, Marc L.; Gibbs, William Kirk, Jr.; Godbey, Will E.; Gonzalez, José Manuel Souto; Graaskamp, Garret W.; Graebner, Mark; Graham, Rodney W.; Grauch, V. J. S.; Graves, Ramona M.; Greubel, Scott P.; Grube, John P.; Guerrero, Benito; Guest, Peter R.; Guu, Cindy Kuei-Ding; Hall, John F., Jr.; Halliwell, Beverly Ann; Han, Soong Soo; Hansen, Brian G.; Harper, Charles Thomas; Hartner, John D.; Harwell, Jeffrey W.; Hasan, Mohammad Nurul; Hato, Masami; Heiple, Paul W.; Hendricks, Michael L.; Hendrickson, Denise M.; Herrera, Peter Ariel; Herrod, Wilson H.; Hestmark, Martin C.; Hickey, James C.; Hickey, Thomas H.; Hicks, John R.; Hill, Virginia S.; Hinchman, Steven B.; Hofstra, Albert H.; Holden, William F.; Horlacher, Craig F.; Horton, Robert A., Jr.; Howarth, Susan M. T.; Hsi, Ching-Kuo Daniel; Hudson, Mark Ransom; Hudson, Mark Ransom; Hunt, Walter; Hunter, James C.; Ibrahim, Kamil E.; Ikwuakor, Killian Chinwuba; Iroe, Hindartono D.; Jaacks, Jeffrey A.; James, Bryan A.; Janak, Peter M.; Janzon, Hans A.; Johnson, Gregory R.; Johnson, Stephen A.; Jones, Alison H.; Jorden, Thomas E.; Josten, Nicholas E.; Jurich, David M.; Kaczkowski, Peter; Kao, Jason Chin-Sen; Karnes, Kerri A.; Katahira, Yo; Kay, Bruce D.; Keller, J. David; Kelly, Anne O.; Kelly, Kevin E.; Kim, Jin-Hoo; Kim, Kun Deuk; Klein, Terry L.; Kline, John H.; Kramer, Ann M.; Krueger, Paul G.; Kruse, Curtis; Kruse, Fred A.; Kruse, Fred A.; Kubik, West T.; Kuo, Shih-Yeng; Kupecz, Julie A.; Kutrubes, Doria Lee; Lambert, David Dillon; Larson, Daniel M.;

Colorado State University

Larson, John Edgar; Laskowski, Keith A.; Lasky, Loren R.; Laurie, Robert John; Lee, Shu-Schung; Lee, Tai Sup; Lee, Yuan-cheng; Leibold, Anne M.; Leighton, Cheryl D.; Lerch, Frederick George; Lessenger, Margaret A.; Letsch, Dieter K.; Levorsen, Mark K.; Li, Yuesheng; Lin, Hung-Liang; Lippitt, Clifford R.; Lopez P., Wilfredo Armando; Lopez, David A.; Lu, Ming; Lu, Ming-Tar; Ma, Xiaochun; Madden, Dawn J.; Magnusson, Stefan G.; Malley, Michael J.; Mamah, Luke I.; Manfrino, Carrie; Margulies, Todd D.; Marrone, Frank J.; Martin, Linda G.; Martin, Marshall A.; Martinek, Brian C.; Matheson, Gordon M.; McCalpin, James; McDowell, Ronald R.; McEntee, Robert A.; Mdala, Chisengu L.; Mead, Richard H.; Meis, Philip J.; Melchior, Daniel C.; Melchior, Daniel Carl, III; Mendoza, Carlos; Meng, Haiyan; Merkel, David C.; Miles, Thomas O.; Milne, Wendy; Min, Kyoung Won; Montazer, Parviz; Morrison, Greg; Mueller, Tanya L.; Musselman, Thomas E.; Myint, Khin Maung; Nicoletis, Serge; Northrop, Harold Roy; O'Brien, Patrick W.; O'Rourke, T. J.; Oglesby, Chris A.; Olm, Mark C.; Padgett, Michael F.; Parduhn, Nancy Louise; Passalacqua, Herminio; Pawlewicz, Mark J.; Pawlowski, Robert S., Jr.; Peck, Charles W.; Penas, Carlos; Perry, Sandra Linthicum; Peters, Douglas C.; Pfeifer, Mary Catherine; Phillips, Kent D.; Pierce, Walter H.; Pillmore, Kathryn A.; Pincus, William J.; Pivonka, Lee J.; Plikk, Martin; Prost, Gary Leo; Quintus-Bosz, Robert L.; Ray, Robert R.; Rebne, Claudia A.; Record, Richard Storey; Rees, Terry F.; Reese, Ronald S.; Reeve, Donna M.; Reeves, James J.; Regueiro S., Jose; Rehn, Warren M.; Reich, Matthew A.; Rhoads, Holly; Rice, James A.; Rice, Thomas L.; Richardson, Archie M.; Richey, Scott R.; Richter, Brian E.; Riese, Arthur Carl; Rindsberg, Andrew Kinney; Rittersbacher, David J.; Roberts, Colleen T.; Rossow, Joerg; Rossow, Joerg; Rousseau, Joseph P.; Runnels, Tyson D.; Ryding, John; Sadowski, Raymond M.; Saito, Akira; Saleh, Saad T.; Saleh, Saad Turky; Samsela, John J., Jr.; Santos, Michele M.; Saunders, James Alexander; Sawyer, Michael B.; Sayre, Ted M.; Schavran, Gabrielle; Schneider, William A., Jr.; Schoendaller, Karen Sue; Schuyler-Rossie, Christine; Selby, Jonathan B.; Serenko, Thomas J.; Shank, Robert D.; Shanley, Keith W.; Shannon, James R.; Shannon, Lee T.; Shine, Brendan F.; Shuter, Kelli A.; Sinton, Peter O.; Sixta, David P.; Smaglik, Suzanne M.; Smith, David B.; Smith, Garon Corder; Smith, Kathleen S.; Smith, Larry B.; Smith, Linda F.; Smith, Patricia Gould; Smith, Rodney S.; Snyder, Gregory A.; Soeripto, Ruskamto; Solano, Marcelo; Sonnenberg, Stephen A.; Squires, Stewart G.; Stewart, Michael; Stites, Robert L.; Strack, Kurt-Martin; Sullivan, Michael Francis; Swayze, Gregg A.; Szabelak, Stanley A.; Tarache, Crisalida; Thomas, Mark H.; Tongtaow, Chalermkiat; Tower, Chris; Tschupp, Edward Walter; Tucker, James D.; Tucker, Robert E.; Unruh, Daniel M.; Upp, Charles S.; Valasek, David W.; Valusek, Jay E.; Vella, Alfred J.; Viana, R. T.; Villaescusa Cordova, Ernesto; Walker, Constance M.; Walker, Robert J.; Wanty, Richard B.; Washburne, James C.; Weaver, Stephen George; Weber, Janine F.; Weber, Richard G.; Webster, James D.; Welch, Paul M.; Wellborn, Jewel E. F.; Wessel, Gregory Ralph; Wheeler, David M.; Wiggs, Calvin R.; Wiley, Robert W.; Witter, David N.; York, Terrell M.; Yost, Lawrence W.; Young, Genevieve B. C.; Youngkin, Mark T.; Yousif, Majeed H.; Zalan, Pedro V.; Zalan, Pedro Victor; Zalan, Thomas Anthony; Zelewski, Gregg; Zuker, J. Stevens

Colorado State University
Fort Collins, CO 80523

268 Master's, 191 Doctoral

1950s: Simons, Daryl B.

1960s: Akhavi, Manouchehr Sadat; Athaullah, Muhammad; Barnett, Lloyd Oris; Beverly, Charles E.; Bibby, R.; Black, Peter Elliot; Brown, George W.; Cerrillo, Lawrence A.; Chu, Show-Chuyan; Chunkao, Kasem; Djordjevic, N.; Dourojeanni, Axel Charles; Fleming, William M.; Frank, Ernest C.; Freethey, Geoffrey W.; Ganow, Harold C.; Hawkins, Richard H.; Hawkins, Richard H.; Heede, Burchard N.; Helgesen, John O.; Holtje, R. Kenneth, Jr.; Hubbard, John Edward; Jobson, Harvey Eugene; Johnson, Kendall L.; Jones, Everett Bruce; Juneidi, Mohmoud J.; Khan, Abdur K.; Kuhn, Alan Karl; Kunkle, Samuel H.; Lee, Richard; Longenbaugh, R. A.; McComas, Murray Ratcliffe; Meiman, James R.; Mercer, Jerry Wayne; Meyers, Alan E.; Muir, Clifford Donald; Murray, David L.; Nordin, Carl F., Jr.; Ohlander, Coryell A.; Page, Oliver S.; Pyle, William D., Jr.; Qutub, Musa Y.; Reddell, Donald Lee; Richardson, Stuart; Roark, Philip W.; Romero, John C.; Sayre, William W.; Sheng, Tse Cheng; Sherman, Frank B., Jr.; Silverts, Leif E.; Spangenberg, Norma E.; Stockton, Charles W.; Stollar, Robert L.; Suryanarayana, Bhamidipaty; Talbott, Lyle W.; Tischendorf, Wilheim; Ulugur, Mustafa Elmas; Van Schaik, Jacob Cornelis; White, Donald E.; Yang, Tsung; Zoghet, Mouine Fahed; Zorich, Theodore M.

1970s: Al-Rashid, Y.; Alldredge, A. William; Arehart, Gregory B.; Barnum, Bruce E.; Bazaraa, A. S.; Bean, Daniel W.; Beathard, R. Michael; Begin, Ze've B.; Beissel, Dennis R.; Berry, Joyce Kempner; Bickford, Hugh L.; Black, Kenneth D.; Brutsaert, Willem Frans; Cacek, Terrance L.; Cameron, Douglas R.; Camp, Wayne K.; Carrigan, P. H.; Chamberlin, David C.; Chandler, R. L.; Chang, Tien-Po; Ching, Paul W.; Daly, C. J.; Dass, P.; Davis, Ronald; Dean, Robert M.; Dewitt, Henry Gray; Donnelly, Michael E.; Edgar, Dorland E.; Elliot, John G.; Fechner, Steven A.; Filson, Robert H.; Flug, M.; Fogg, James L.; Galbraith, Alan Farwell; Gardner, Thomas W.; Gerry, David L.; Ghooprasert, W.; Goeke, James W.; Gonzalez, D. D.; Gutierrez, Julian Castillo; Haddock, David R.; Hamilton, Patricia A.; Hart, Thomas C.; Hartmann, Leo A.; Haufler, J. B.; Haug, P. T.; Heinemeyer, G. R.; Helweg, O. J.; Hermelin, Michel G.; Hill, John Jerome; Hung, C.; Illangasekare, T.; Ingraham, Mark G.; Isailovic, D.; Itkowsky, Francis A.; Jackson, Timothy J.; Janssen, Robert J.; Joench-Clausen, T.; Johnson, Steven R.; Khan, A. K. M. Hamidur Rahman; Klein, Terence L.; Kluth, Mary Jo Ann Morgan; Knisel, Walter Gus, Jr.; Krasowski, Dennis J.; Krishnamurthi, N.; Lagasse, P. F.; Lane, L. J.; Lane, W. L.; Laronne, Jonathan B.; Leavesley, George Haslam; Lee, San Wei; Lin, Tzeu-Lie; Loucks, Robert R.; Mabarak, Charles D.; Macke, David L.; Mahar, James W.; Margheim, G. A.; McClure, Paul Frederick; McCrumb, Dennis R.; McKean, James A.; McLane, Charles F., III; Mejia Angel, Jose M.; Millon, Eric R.; Mosley, M. Paul; Mosley, M. Paul; Mueller, David K.; Naas, S. L.; Nadler, Carl Theodore, Jr.; Newman, James W.; Nualchawee, K.; Olson, Roger W.; Ortiz, N. V.; Osterwald, Edward J.; Park, J. K.; Parker, Randolph S.; Patton, Peter C.; Pearson, Robert L.; Phelps, Richard T.; Pillsbury, Norman H.; Price, James N.; Prommool, Suthep; Puckett, James L.; Ramirez, Octavio; Ramirez-Rivera, Jaime; Rice, Raymond Martin; Riordan, E. J.; Robinson, Sara L.; Rodriguez-Amaya, R.; Root, Ralph R.; Rovey, C. E.; Rundquist, L. A.; Sabol, George V.; Samuelson, Donald R.; Schettig, William J.; Schneider, E. J.; Scroggs, Doyle L.; Senturk, H. A.; Shafer, J. M.; Shepherd, Russell G.; Shepherd, T. A.; Sheppard, John Selwyn; Simpson, Donald H.; Smith, Craig B.; Smith, Freeman M.; Smith, Mark E.; Spane, Frank A., Jr.; Spangenberg, Norman E.; Spearnak, Mark; Spicker, Fredrick A.; Stark, Jack H.; Stelter, L. H.; Steppuhn, Harold W.; Sundeen, Kerry D.; Tai, K. C.; Tamburi, Alfred J.; Tamburi, Alfred J.; Tompkin, J. M.; Tyler, Theodore F.; Unites, Dennis F.; Van Haveren, Bruce P.; Vargas-Semprun, D.; Ward, T. J.; Weiner, Mitchell S.; Wescott, William A.; Wiegand, J. Patrick; William, Owen R.; Womack, W. Raymond; Woolsey, Thomas S.; Wu, Y. H.; Wymore, Ivan F.; Young, James Stanton; Ziemer, Robert Ruhl; Zimpfer, Gerald Lee

1980s: Abdelbary, Mohamed Rafeek; Abdulrazzak, Mohamed Jamil; Abt, Steven Roman; Adamsen, Floyd James; Al-Ghamisi, Hezam Hazzaa; Al-Muttair, Fouad Fahad; Albino, George V.; Alexander, William G.; Alexandri-Rionda, Rafael; Alhassoun, Saleh Abdullah; Anderson, Scott A.; Andrews, Sarah; Andrievich, Ellen; Ater, Patricia C.; Baggs, Charles Chaplin; Bahjat, Abdullah Mohammed; Banta, Edward R.; Barrett, Robert E.; Bartlett, Robert D.; Baumgardner, Thomas F.; Bergstrom, Frank W.; Berry, Catherine E.; Boyer, Jeffrey T.; Boyle, Ray E.; Bradley, Jeffrey Brent; Bradley, Michael T.; Bradley, Scott D.; Brazil, Larry E.; Brown, Bobby Joe; Buchanan, John Petrella; Buchanan, John Petrella; Burch, Alvin L.; Burnett, Adam W.; Caicedo, Nelson Oswaldo Luna; Campbell, Allan G.; Campbell, Donald H.; Carlson, Jon Andrew; Carter, Thomas E.; Cervantes R., J. Alfredo; Chang, Tien-Po; Chavez D., Eduardo E.; Chebaane, Mohamed; Chehata, Mondher; Choudhary, Muhammad Rafiq; Combs, Samuel Theodore; Cope, Edward L.; Craig, Gerald N., II; Craig, Steven D.; Culp, Stuart L.; Davidson, James M.; de Miranda, Antonio Nunes; Dobak, Paul J.; Dolson, John; Dwelley, Peter C.; Eccker, Sandra; Edgar, Thomas Viken; Edwards, Jeffrey S.; Eggert, Kenneth George; Erickson, James R.; Eschner, Thomas Richard; Evans, Robert George; Fernandez, Bonifacio; Finley, Jim B.; Fiuzat, Abbas-Ali; Foruria, Jon; Gander, Malcolm J.; Garbrecht, Jurgen D.; Gellis, Allen C.; Ghaheri, Abbas; Ghoneim, Ghoneim Abdel-Azim; Gilley, John Edwards; Gowen, Peter J.; Graves, Timothy; Gregory, Daniel I.; Griswold, Mark L.; Guertin, David Phillip; Gutub, Saud Abdulaziz; Hanif, Muhammad; Hann, Megan; Harlan, J. Bruce; Harris, Jane; Harvey, Michael David; Havlin, John LeRoy; Henao, Angela M.; Hepler, Jeffrey Alan; Hertel, John W.; Herzog, Dave; Heston, Deborah A.; Hilton, Joanne; Hippe, Daniel J.; Hodge, Cara Jyl; Hohman, John Craig; Holt, William K.; Hunsaker, Ernest Leon, III; Ingram, John Jeffrey; Ingwersen, James B.; Israel, Alan M.; Jackson, Mary L. W.; Jarrett, Robert David; Jayasena, H. A. Hemachandra; Johansing, Robert J.; Johnson, Douglas Edward; Johnson, Richard K.; Johnson, Robert B., Jr.; Kaldenbach, Thomas; Kaplin, John L.; Karr, Leonard J.; Kashkuli, Heydar Ali; Kasomekera, Zachary Mark; Katiyar, Vidya; Khadr, Hassan Ali Abdel-Aziz; Khan, Shakeel Ahmed; Kinney, Thomas E.; Kirkley, Melissa B.; Koch, Roy W.; Kolenbrander, Lawrence Gene; Laird, Jeffrey R.; Lange, Peter C.; Layla, Rasheed Ibrahim; Lazaro, Rogelio Cruz; Leighton, Van L.; Lidstone, Christopher D.; Loen, Jeffrey Scott; Lopez-Rendon, Jorge E.; Lou, Wellington Coimbra; Loureiro, Blanor Torres; Lovelace, Kenneth A., Jr.; Lu, Jau-Yau; Lujan Sierraalta, Carlos Alberto; Luvira, Somboon; Lyttle, Thomas; Maghsoudi, Nosratollah; Maharjan, Bhuwon D.; Maher, Brian; Mahmoodian-Shooshtari, Mohamad; Mansouri, Tareg Ahmad; Marcouiller, Barbara A.; Marks, Thomas R.; Martin, Joseph Paul; Martinson, Holly A.; Matondo, Jonathan Ihoyelo; McAndrews, Kevin P.; Meixner, Richard Eugene; Menzer, Fred J.; Merrill, William G.; Meyer, David Frederick; Mohorjy, Abdullah Mustafa; Moll, Stanton H.; Moore, Richard C.; Muhaimeed, Ahmad Saleh; Musgrave, John A.; Mussard, Donald E.; Neff, Linda M.; Nelson, James Douglas; Newbry, Brooks Walter; Nibbelink, Kenneth A.; Nogueira, Vicente de Paulo Queiroz; Nunes Correia, Francisco Carlos da Graca; Nyumbu, Inyambo Liyambila; O'Brien, Jimmy Steven; O'Brien, Michael; Osborne, Lesslie W., Jr.; Osman, Mohamed Akode; Ouchi, Shunji; Padgett, Joel P.; Paine, Alasdair D. M.; Papson, Ronald P.; Parra-Mata, Juan Jose; Patton, Jean J.; Peterson, Thomas Charles; Peyton, Robert Lee, Jr.; Phamwon, Sanguan; Phillips, Loren F.; Pierson, John R.; Pitlick, John; Pitlick, John Charles; Pranger, Harold S.; Radloff, David Lee; Raeissi-Ardakani, Ezatollah; Rais, Samira; Rathnayake, Rathnayake M. D.; Reeder, Jean Dolan; Renda, Christine A.; Restrepo Mejia, Jorge I.; Rhodes, Randolph A.; Rice, John Albert; Richter, Brian; Riordan, Carol J.; Robbins, Elizabeth A.; Rogers,

Jack A., Jr.; Roig, Lisa C.; Rosenlund, Gene C.; Ruvalcaba-Ruiz, Delfino; Ruvalcaba-Ruiz, Delfino Concepcion; Santos, Emidio Gil; Saperstone, Herb I.; Saracino, Anthony M.; Sastrowihardjo, Sutadi; Saunders, Denise Marie; Schall, James Douglas; Schimel, David Steven; Schutz, Joseph Leroy; Seanor, Clinton E.; Shaikh, Alaeddin; Sherman, Thomas Floyd; Sidharta, Ananta Sigit; Simons, Robert Keith; Smith, Henry H.; Sriburi, Thavivongse; Stegen, Ralph J.; Steininger, Roger Claude; Steppe, Michael C.; Strumpher, Phillipus J.; Stumpe, Kim Michael; Sumner, Wendolyn R.; Tabios, Guillermo Quesada, III; Thomsen, Anton Gaarde; Thurber, James E.; Tomes, Reynold Joseph; Turbak, Abdulaziz S.; Tyler, Noel; Vendetti, Michael J.; Verry, Elon Sanford; Wade, Kenneth; Wana-Etyem, Charles; Wang, Wen Chiang; Wang, Yuan Sung; Warner, James Walter; Watson, Chester Conrad; Watson, Louis Camille; Weaver, William E.; Weech, Philip S.; Weider, Mark F.; Whitely, Joy L.; Williams, Thomas R.; Wilson, Michael Phillip; Woo, Hyoseop; Wood, Maria Luisa; Woodzick, Thomas L.; Yassin, Yassin Yahya; Zazueta Ranahan, Fedro Sigmundo; Zerpa, Oswaldo; Zimpfer, Gerald Lee

Columbia University
Palisades, NY   10964

63 Master's, 162 Doctoral

1930s: Adams, George F.

1940s: Knox, Margaret S.

1950s: Giffin, Charles E.; Schull, Herman Walker, III

1960s: Abdel-Monem, Abdalla A.; Abdel-Monem, Abdalla A.; Alexandrov, Eugene Alexander; Anderson, Thomas F.; Auld, Bruce Charles; Balachandran, Nambath K.; Banghar, Amru Ram; Bentley, Robert Donald; Bernstein, Robert L.; Berry, John L.; Booy, Emmy Catherine; Boucher, Gary Wynn; Buchbinder, Goetz Gustav Rudolf; Burgess, William Joseph; Cameron, Barry Winston; Cameron, Barry Winston; Cameron, Richard William; Chiang, Yu-Jen; Chute, John Lawrence, Jr.; Clark, George S.; Cutler, John Fredrick; Damuth, John Erwin; Davidson, Maurice James; Dickson, Geoffrey Owen; Donahue, Jack David; Drew, Isabella Milling; Eastler, Thomas Edward; Edgar, Norman Terence; Eldredge, Robert Niles; Florer, Linda E.; Folger, David Winslow; Fruth, Lester Sylvester, Jr.; Fruth, Lester Sylvester, Jr.; Gaffney, Eugene Spencer; Gardner, James Vincent; Gavasci, Anna Teresa; Gordon, Arnold L.; Gornitz, Vivien Monisa; Gornitz, Vivien Monisa; Gould, Stephen Jay; Haji-Vassiliou, Andreas; Haji-Vassiliou, Andreas; Hansen, Harry J.; Hausen, Donald Martin; Hayes, Dennis Edward; Hekinian, Roger; Helwig, James Anthony; Herron, Ellen Mary; Herron, Thomas Joseph; Hollister, Charles D.; Horne, Gregory Stuart; Karp, Edwin; Khogia, Abdelhadi; Knowles, David Martin; Ku, Teh-Lung; Kutschale, Henry Walter; Ladd, John Walcott; Langer, Arthur M.; Langseth, Marcus Gerhardt, Jr.; Latham, Gary Vincent; Mayhew, Michael Allen; McConn, Virginia Barr; McDowell, Fred Wallace; McGarr, Arthur Francis; McIntyre, Andrew; Mellett, James Silvan; Miller, Henry J.; Missallati, Amin Abdulla; Mitronovas, Walter; Morris, Robert William; Morris, Robert William; Morse, Robert Harold; Nabighian, Misac N.; Nash, John Thomas; Nash, John Thomas; Neuman, Lawrence Donald; Oversby, Brian Sedgwick; Oversby, Brian Sedgwick; Oversby, Virginia M.; Page, Robert Alan, Jr.; Piermattei, Rodolfo; Pitman, Walter Clarkson, III; Powell, Benjamin Neff; Powell, Benjamin Neff; Rich, Patricia Vickers; Roellig, Harold Frederick; Rollins, Harold Bert; Rooney, Thomas Peter; Ruddiman, William Fitzhugh; Rudnick, Jon; Sarpi, Ernesto; Schmitt, Leonard Joseph, Jr.; Schneider, Eric Davis; Schwartz, Maurice Leo; Shaw, David Montgomery; Sheridan, Robert E.; Sheridan, Robert E.; Skinner, William Robert; Smith, Jerry D.; Su, Sergio S.; Surovell, Elizabeth Jean; Swainbank, Ian G.; Tan, Li-

Ping; Thomsen, Leon; Tobin, Don Graybille; Tolderlund, Douglas Stanley; Voight, Barry; Wall, Robert Ecki; Waller, Thomas Richard; Walthall, Bennie Harrell; Ward, Peter Langdon; Washburn, Robert Henry; Weiss, Richard Lawrence; Whalen, Patricia Ann; Wolgemuth, Kenneth M.; Yegulalp, Tuncel Mustafa

1970s: Aggarwal, Yash P.; Alterman, Ina Brown; Anderson, John G.; Baker, T. N.; Barazangi, Muawia; Bender, Michael L.; Berglof, William Randall; Boylan, John C.; Breeding, J. Ernest, Jr.; Brocoum, Alice V.; Brocoum, Stephan John; Buchanan, Hugh; Caldwell, Donald Martin; Chander, Ramesh; Chen, Chi-Chieu; Chen, Pei-Hsin; Chia, Yee-Ho; Connary, Stephen Dodd; Connary, Stephen Dodd; Demarest, Harold Hunt, Jr.; Donahue, Jessie Gilchrist; Drake, William Edward; Early, William P.; Eastler, Thomas Edward; Edsall, Douglass W.; Einarsson, Paul; Eittreim, Stephen Lawrence; Emerson, S.; Emry, Robert John; Fletcher, John P. B.; Floran, Robert J.; Foster, John Harold; Franck, C. J.; Gal-Chen, Tzvi; Gardner, James Vincent; Georgi, Daniel T.; Ghellali, Salem M.; Gilbert, Jean Ann; Gregersen, S.; Gumper, Frank J.; Hall, John Kendrick; Hammond, Douglas E.; Hamza, Mokhtar S. A.; Herron, Ellen Mary; Hinds, Robert Warren; Ho, Ching-Oh; Hollcombe, Troy L.; Hunt, Robert Molyneaux, Jr.; Ileri, Saldiray; Jachens, Robert C.; Jacoby, Gordon Campbell, Jr.; Johnson, Tracy L.; Kaneps, Ansis Girts; Kay, Robert Woodbury; Kelleher, John A.; Kellogg, Thomas Bartlett; Kent, Dennis V.; King, Kenneth, Jr.; Klein, Frederick W.; Krstulovic L., G.; Ladd, J. W.; Lahr, John Clark; Lahr, John Clark; Lammlein, David Raymond; Lovegreen, J. R.; Lozano, J. A.; Maasha, Ntungwa; Manson, Douglas Martin Vincent; Maurrasse, Florentin J.-M. R.; Molnar, Peter H.; Montes, Hernan A.; Murphy, Andrew J.; Neuman, Lawrence Donald; Pan, Yuh-Shyi; Payne, Lawrence H.; Perruzza, Albert; Prell, Warren L.; Rabinowitz, Philip David; Rich, Patricia Vickers; Rich, Thomas Hewitt; Richardson, Darlene S.; Robertson, James H.; Ryan, William Bradley Frear; Rynn, John M. W.; Sbar, Marc Lewis; Schneider, Eric Davis; Schull, Herman Walter, III; Shih, Chi-Yu; Simpson, Harry James, Jr.; Singh, Shri K.; Smith, Philip A.; Stoffa, P. L.; Sun, Shine-Soon; Swiderski, Helena S.; Swiderski, Helena S.; Tatham, Robert H.; Tsou, J.-L.; Van Donk, Jan; Wald, S. D.; Walker, A. T., III; Wang, Shin; Wehmiller, John F.; Wolgemuth, Kenneth M.; Won, Ihn Jae; Zahariev, G. K.

Columbia University, Teachers College
New York, NY   10027

536 Master's, 551 Doctoral

1870s: Church, John A.; Love, Edward G.; Munroe, Henry

1880s: Britton, Nathaniel L.; Cornwall, Henry B.; Eliot, Walter G.; Elliot, Arthur H.; Irving, Roland D.; Marsh, Charles W.; Munsell, Charles E.; Newberry, Spencer B.; Neyman, Percy; Northrup, John I.; Porter, John B.; Simonds, Francis M.; Wiechmann, Ferdinand G.

1890s: Irving, John D.; Jouet, Cavalier H.; Luquer, Lea M.; Matthew, William D.; Merrill, Frederick J.; Moses, Alfred J.; Pope, Frederick J.; Ries, Heinrich; Smyth, Charles H., Jr.; Struthers, Joseph; Vanderpoel, Frank; White, Theodore G.

1900s: Anderson, Gustavus E.; Brown, Thomas C.; Catherinet, Jules; Curtin, Margaret H.; Dickson, Charles W.; Fenner, Clarence N.; Finlay, George I.; Gordon, Clarence E.; Humphreys, Edwin W.; Hyde, Jesse E.; Johnson, Douglas Wilson; Johnson, Herman F.; Kellogg, Lee Olds; Kingsbury, Joseph W.; Kneisley, George W.; Kong, Shun Tet; Lamme, Maurice A.; Leavenworth, George; Lincoln, Francis C.; Lohman, Henry William; Lull, Richard S.; Morrison, Charles E.; Ogilvie, Ida H.; Pack, Fred James; Pack, Fred James; Prather, John K.; Queneau, Augustin L. J.; Rogers, Austin Flint; Shamel, Charles Harmonas; Simpson, Kenneth M.; Starr, Charles

Comfort; Stewart, Charles A.; Swartz, Charles C.; Wang, Chung Yu; Wilson, John H.; Wood, Elvira

1910s: Alling, Harold L.; Ames, Edward W.; Baker, Ethel; Barrows, Walter L.; Billingsley, Paul; Blanchard, Ralph C.; Boyle, Albert C.; Boyle, Albert C.; Bruce, Everend L.; Bruce, Everend L.; Brunton, James S. L.; Burne, Eleanor; Burr, Freeman F.; Clarke, Alexander C.; Condit, Daniel D.; Crowell, William R.; Fenner, Clarence N.; Fettke, Charles Reinhard; Fettke, Charles Reinhard; Frank, Jeannette; Franklin, Adele; Gaby, Walter E.; Gordon, Clarence E.; Hintze, Ferdinand F.; Hodge, Edwin T.; Holzwasser, Florrie; Hubbard, Bela; Jacobs, Elbridge C.; Jensen, Joseph; Jones, Edward L.; Jud, Friedolina C.; Jui, Pao Vung; Kinney, Harry D.; Kurtz, Anna E.; Kuthy, Olga; Latham, Everett B.; Lincoln, Francis C.; Lobeck, Armin K.; Long, Eleanor T.; Long, Emilie O.; Loveman, Michael H.; Merrill, Bertha; Millard, William J.; Mitchell, Graham J.; Mitchell, Graham J.; Mook, Charles C.; Moon, Evangeline A.; Morris, Frederick K.; Nagelberg, J. Leo; O'Connell, Marjorie; O'Connell, Marjorie; Reed, Norman H.; Rice, Marion; Rice, Winfield L.; Roesler, Max; Rogers, Gaillard S.; Saxe, Joseph Archibald; Semmes, Douglas R.; Smith, Warren Slocum; Smith, Warren Slocum; Spearman, Charles; Stephen, Walter Mitchell; Steubing, William C.; Stewart, Charles A.; Taylor, Henry B.; Terrill, Arthur C.; Thompson, Arthur P.; Van Tuyl, Francis M.; Wang, Yinchang T.; Welleck, Mary N.; Wood, Elvira; Wright, Frank J.; Young, I.; Ziegler, Victor; Zons, Frederick W.

1920s: Abouchard, Sylvain S.; Alling, Harold L.; Bain, George W.; Bain, George W.; Bandy, Mark Chance; Barger, Edith M.; Barton, Otis; Beckwith, Radcliffe H.; Blauvelt, Bessie; Blood, Pearl; Boggs, Samuel W.; Bohlin, Howard G.; Brown, John S.; Burgess, Frances C.; Carhart, Grace M.; Charleton, F.; Clendenin, Thomas P.; Collins, Robert F.; Dake, Charles L.; Davis, Henry G.; Davis, Thornton; Diamond, Benjamin T.; Dixon, Max Mueller; Doane, George H.; Dorr, James B.; Earle, F. M.; Fluhr, Thomas W.; Fong, Kin Lan; Furse, George Norman D.; Gallagher, Helen D.; Goodwin, Ralph T.; Greenland, Cyril Walter; Guernsey, Terrant Dickie; Gyss, Emile B.; Halter, Clarence R.; Hawkins, Glenn De Wayne; Hinn, Helen B.; Hitchcock, Margaret R.; Holzwasser, Florrie; Honess, Charles W.; Hubbard, Bela; Jones, Ray S.; Kahrs, Anna E.; Kay, George M.; Kilinski, Edward A.; Knappen, Russel S.; Knight, Samuel Howell; Li, Shih Chang; Lott, Frederick S.; Lowman, Shepard Wetmore; Manning, Paul De Vries; Maucini, Joseph J.; Meyerhoff, Howard A.; Miller, Leroy C.; Morgan, George D.; Morgan, George D.; Morrey, Margaret; Muilenburg, Garret A.; Nichols, David A.; Ohlsen, Violet E.; Perry, Vincent D.; Raisz, Erwin J.; Raisz, Erwin J.; Reed, Ethbert F.; Rossin, Edgar L.; Sharp, Henry S.; Spock, Leslie E.; Spock, Leslie E.; Stolfus, Martha A.; Strader, Harold L.; Van Esbroeck, Guillaume; Varady, Ella H.; White, Maynard P.; White, Maynard P.; Wissler, Stanley G.; Wood, Horace E., 2nd; Woodward, Harriette B.; Yih, Liang Foo

1930s: Anderson, Richard J.; Anisgard, Harry V.; Antine, Helen M.; Arnold, Herbert Julius; Aronovici, Vladimir S.; Babenroth, Donald L.; Barbour, George B.; Barkman, Leilya K.; Bates, Robert Ellery; Baumgarten, Hortense; Bell, Gordon Knox, Jr.; Billings, Gladys D.; Birdsall, John Manning; Bogert, Bernard O.; Booth, Robert T.; Brackmier, Gladys Helen; Brauneck, Dorothy A.; Braunstein, Jules; Bunker, Rachel; Burbridge, Clarence E.; Butler, John W.; Caccaviello, Vivian M.; Callaghan, Eugene; Cameron, Eugene N.; Canfield, Charles R.; Chayes, Felix; Chu, Sen; Colbert, Edwin H.; Colbert, Edwin H.; Cooper, Margaret; Cuskley, Virginia A.; Denison, Robert H.; Denison, Robert H.; Dolloff, Norman H.; Durfee, Wilda; Dusenberry, Arthur N.; Embich, John Reigle; Emendorfer, Earl; Farwell, Fred W.; Fields, Suzanne; Foote, Priscilla; Forde, Margaret E.; Fowler, Katherine Stevens; Fraser, Donald M.; Frondel, Clifford; Gianella, Vincent P.; Gordon, Sadie C.

B.; Haff, John Coles; Hagner, Arthur F.; Happ, Stafford C.; Hatfield, Willis C.; Helprin, Sydney; Hitchcock, Charles B.; Hooper, Medora L.; Howard, Arthur D.; Hurwitz, Garvin L.; Irwin, Dorothy W.; Irwin, William H.; Irwin, William Harold; Itter, Harry A.; Jenney, Charles Phillip; Jenney, Charles Phillip; Jennings, Philip H.; Jennings, Phillip H.; Johnson, Samuel C.; Jones, Murray L. G.; Jung, Dorothy Anne; Kane, Julian; Keppel, David; King, Jessie M.; Knight, Pauline U.; Konkoff, Vladimir I.; Krieger, Phillip; Krieger, Phillip; Li, Ching-Yuan; Loetterle, Gerald John; Lougee, Richard Jewett; Lowe, Kurt Emil; Macar, Paul F. J.; Mackin, J. Hoover; Mackin, J. Hoover; Macksoud, Adrienne M.; Malkin, Doris S.; Mason, Arthur E.; Merritt, Philip L.; Merritt, Philip L.; Messina, Angela R.; Meyerhoff, Howard A.; Miller, Buford Maxwell; Miller, Burford Maxwell; Miller, Ralph L.; Milne, Beulah L.; Minkofsky, Anna; Morgan, Arthur M.; Morgan, Arthur M.; Morris, Frederick K.; Mossman, Reuel Wallace; Mugler, Dorothy S.; Namowitz, Samuel N.; Osorio, Gustave A.; Packer, Ethel; Pascoe, H. L.; Pattin, Virginia L.; Peltier, Louis Cook; Pier, Katherine D.; Pollak, Miriam B.; Preuss, Charlotte; Punches, Richard K.; Rogatz, Henry; Rogatz, Henry; Rozanski, George; Rudell, Marjorie; Russo, Amelia Gloria; Salmon, Eleanor S.; Sample, Charles H.; Schlaikjer, Erich M.; Schlaikjer, Erich M.; Schmidt, Ruth A. M.; Secor, Dana M.; Segal, Lucille B.; Segal, Rose J.; Selfridge, George C.; Selfridge, George C., Jr.; Sharpe, Charles F. S.; Sharpe, Charles F. S.; Shupack, Benjamin; Shupack, Dora; Smith, Virginia H.; Sohn, Israel G.; Sperb, Grace; Stewart, Lincoln Adair; Stobbe, Helen R.; Thatcher, Russell N.; Thomas, Horace Davis; Valentine, Wilbur G.; Van Burkalow, Anastasia; Van Steeg, Karl; Vitaliano, Charles J.; Wallace, Dorothy M.; Warthin, Aldred S., Jr.; Wasson, Isabel B.; Wheeler, Girard Emory; Whiteman, Edith J.; Williams, Harriet L.; Williams, Ira A.; Williams, Robert C.; Williamson, Marjorie; Wing, Margaret M.; Wolfanger, Louis A.; Wood, Albert E.; Woodward, Herbert P.; Woodward, Herbert P.; Woolford, Jean H.; Young, Frederick P, Jr.; Zernitz, Emilie R.

1940s: Adler, Hans H.; Agatston, Robert Stephen; Ahmed, Mesbahuddin; Allen, Walter C.; Baldwin, W. Brewster; Bank, Walter; Bassett, Ann B.; Bates, Thomas F.; Bates, Thomas F.; Biemesderfer, George K.; Bispham, Robert G.; Bodenlos, Alfred J.; Bohrn, Marie T.; Booth, Verne H.; Booth, Verne H.; Brown, Jean C.; Cady, Wallace M.; Carlston, Charles W.; Chancellor, Robert E.; Chayes, Felix; Chenoweth, Philip A.; Chronic, Byron John, Jr.; Coates, Donald R.; Craig, Lawrence C.; Craig, Lawrence C.; Curtis, Lawrence W.; Dobrin, Milton B.; Dobrin, Stefanie Z.; Donaldson, Francis, Jr.; Douglas, Robert J. W.; Dykstra, Franz R.; Elias, Helen V.; Erdman, Mary Cordelia; Evans, James E. L.; Fisher, Lysabeth Ann; Fuller, James O.; Galavis, Jose A.; Garwick, Robert; Glass, Herbert D.; Grace, Katherine; Graf, Donald L.; Graf, Donald L.; Gray, Carlyle; Grossman, Irving; Grossman, William L.; Halpern, Joseph B.; Hamilton, Peggy Kay; Haselau, Olivia V.; Hewitt, William Paxton; Hintze, Lehi F.; Hole, Gilbert L.; Holland, Heinrich D.; Holmes, Ralph J.; Huffman, George Garrett; Hussain, Mehdi; James, Carolyn; Jones, Stewart M.; Kauffman, Albert J.; Keppel, David; Koons, Edwin Donaldson; Koons, Edwin Donaldson; Krieger, Medora H.; Kummel, Bernhard; Larson, Edward Richard; LeCount, Dorothy A.; Levin, S. Benedict; Li, Ching-Yuan; Lomerson, William W.; Lowe, Jurt Emil; Lowman, Shepard Wetmore; Mahard, Richard H.; Mahard, Richard H.; Main, Frederic Hall; Mallory, William L.; Mallory, William Wyman; Marsh, Marion J.; Martinez, Abraham; Miller, Maynard M.; Miller, Victor C.; Mintz, Yale; Mitcham, Thomas W.; Moorhouse, Walter W.; Northrup, John; O'Brien, James J.; Oxley, Philip; Patterson, Charles M.; Patterson, Charles M.; Pavlides, Louis; Penha, Lala; Perfetti, Jose N.; Pertusio, Serge M.; Petrick, Audrey B.; Pincus, Howard J.; Press, Frank; Press, Frank; Pretzer, Elizabeth J.; Prouty, Chilton E.; Revere, Althea; Riggs, Richard M.; Roberts, Thomas G.; Roedder, Edwin W.; Rush, Richard W.; Salmon, Eleanor S.; Sandoval, Jose C.; Schmidt, Ruth A. M.; Seymour, Catherine; Simpson, E. S.; Smith, Kenneth G.; Socolow, Arthur Abraham; Steenland, Nelson C.; Stobbe, Helen R.; Strahler, Arthur N.; Strahler, Arthur N.; Stringham, Bronson; Tan, Ek-Khoo; Taylor, Melvin Hall; Thompson, Andrew A.; Thurston, William Richardson; Tolstoy, Ivan; Trites, Albert F.; Van Burkalow, Anastasia; Vitaliano, Charles J.; Walthier, Thomas N.; Walthier, Thomas N.; Walton, Matthew S., Jr.; Wang, Kung-Ping; Webb, Gregory W.; Wells, Dana; Wertz, Jacques B.; Winters, Stephen Samuel; Witkind, Irving Jerome; Worzel, J. Lamar; Worzel, J. Lamar; Wright, P. J.; Wright, Robert J.; Wright, Robert J.; Young, Ford; Young, Frederick P., Jr.; Ysalgue, Sarah E.; Yuen, Chew F.

1950s: Abdel-Gawad, Abdel-Moneim M.; Agatston, Robert Stephen; Alexandrov, Eugene A.; Allan, Urban S., Jr.; Amendolagine, Emanuel; Anazalone, Salvatore A.; Anzalone, Salvatore A.; Arnott, Ronald James; Arnow, Theodore; Ault, Wayne U.; Ault, Wayne U.; Baldwin, W. Brewster; Barber, Irene E.; Barnes, Hubert Lloyd; Barrington, Jonathan; Barton, Paul B., Jr.; Barton, Paul B., Jr.; Bassett, Allen M.; Bassett, Allen Mordorf; Bassett, William S.; Bate, George L.; Batten, Roger L.; Batten, Roger L.; Bauman, Carl F., Jr.; Be, A. W. H.; Be, Allan W.; Beaumont, Donald F.; Beaumont, Donald F.; Belt, Charles B., Jr.; Belt, Charles Banks, Jr.; Benavides-Caceres, Victor E.; Bentley, Charles R.; Bergeron, Robert; Bethke, P. M.; Bethke, Philip Martin; Biren, Helen A.; Biren, Helen Antine; Blau, Barbara J.; Bodenlos, Alfred J.; Bodine, Marc W., Jr.; Boyd, Donald W.; Brilliant, Renee M.; Brixey, Austin Day, Jr.; Broecker, W. S.; Brophy, Gerald P.; Brophy, Gerald Patrick; Byrd, Richard E.; Byrne, John V.; Caceres, Victor B.; Carr, Donald R.; Carr, Donald R.; Chaffe, Robert Gibson; Chatfield, Ernest J.; Cheetham, Alan; Cheney, Monroe G., Jr.; Chronic, Halka; Clift, William Orrin; Coates, Donald R.; Cobb, James Curtis; Cohen, William J.; Collins, George M., Jr.; Cook, William R., Jr.; Cooper, J. F.; Corgan, James X.; Craddock, J. Campbell; Crawford, John P.; Croft, William J.; Cserna, Eugene; Cserna, Eugene George; Cserna, Gloria A.; Cummings, J. S.; Dahl, Harry M.; Dahl, Harry Martin; Dally, Jesse LeRoy; de Cserna, Zoltan; DeJoia, Frank J.; Dennis, John G.; Dennis, John G.; Dennis, John G.; Dincer, Hikmet; Doll, Charles G.; Donn, William L.; Dott, Robert H., Jr.; Douglas, Hugh; Drashevska, Lubov; Drummond, Paul Linwood; Dunne, James A.; Duschatko, Robert W.; Eckelmann, F. Donald; Eckelmann, F. Donald; Eckelmann, W. R.; Edwards, John D.; Edwards, John D.; Edwards, John L.; Elston, Wolfgang; Elston, Wolfgang; Ericksen, George Edward; Fails, Thomas G., Jr.; Feely, Herbert W.; Felber, B. E.; Finks, R. M.; Fischer, Alfred G.; Folger, David W.; Foster, Donald I.; Friedman, Gerald M.; Fyles, J. T.; Gabrielse, Hubert; Gast, Paul W.; Gast, Paul Werner; Giese, Rossman F., Jr.; Giletti, B. J.; Gilkey, Arthur K.; Gilkey, Arthur K.; Glass, Herbert D.; Gonzalez, Alberto Rex; Gray, Carlyle; Green, Howard R.; Green, Jack; Greene, John M.; Harris, Rae L., Jr.; Hawley, L. David; Haworth, Raymond Harrison; Heaslip, W. G.; Heezen, Bruce C.; Heezen, Bruce C.; Hill, Patrick Arthur; Hintze, Lehi F.; Hole, Gilbert L.; Holland, Hans J.; Holland, Heinrich D.; Holm, E. Richard; Holser, William T.; Houston, Robert Stroud, Jr.; Howard, Calhoun L. H.; Howe, Herbert J.; Hull, Joseph P. D., Jr.; Hull, Joseph P. D., Jr.; Jacob, Leonard, Jr.; Jacob, Leonard, Jr.; Jacobson, John B.; Jahn, Jeanne E.; Jicha, Henry Louis, Jr.; Jicha, Henry Louis, Jr.; Katz, Samuel; Kelley, Dana Robineau; Kellogg, Harold E.; Kelly, William Crowley; Kelly, William Crowley; Kent, P.; Kleinkopf, Merlin Dean; Kling, Stanley A.; Klovan, John E.; Knight, Augustus S., Jr.; Kopp, O. C.; Kopp, Otto C.; Kornicker, L. S.; Kotschar, Vincent F.; Kozary, Myron Theodore; Landisman, Mark G.; Lapham, D. M.; Lapham, Davis M.; Larsen, Leonard H.; Larsen, Leonard Hills; Larson, Edward Richard; Leland, Rodney C.; Leroy, Paul G.; Leroy, Paul G.; Lippitt, Louis; Long, Austin; Long, Leon E.; Long, Leon Eugene; Lovejoy, Donald W.; Lovejoy, Donald W.; Lowell, James Diller; Lowell, James Diller; Luchsinger, S. E.; Main, Frederic Hall; Malkin, Doris S.; Marshall, W. S.; Mathias, David L., Jr.; McCauley, John F.; McGuire, Odell S.; Mears, Brainerd; Melton, Mark A.; Merrit, P. C.; Miller, David W.; Miller, Edward T.; Miller, Edward T.; Miller, Leo J.; Miller, Victor C.; Mitcham, Thomas W.; Molloy, Martin W.; Nagy, Bartholomew S.; Norton, Dorita A.; Norton, James Jennings; Norton, Matthew Frank; Officer, Charles B., Jr.; Oglesby, Woodson R.; Oliver, Jack; Osmond, John C.; Oxley, Philip; Panek, Louis A.; Parker, Raymond Lawrence; Peng, Chi-Jui; Perlmutter, Nathaniel M.; Pincus, Howard J.; Ravneberg, N. M.; Regnier, Jerome Philippe Mathieu; Reiser, Wendel; Remson, Irwin; Renzetti, Phyllis A.; Reso, A.; Rigby, J. Keith; Robertson, David S.; Roedder, Edwin W.; Rohl, Arthur N.; Rush, Richard W.; Sarda, Gouri Saukar; Sawyer, J. F.; Schmidt, Ronald G.; Schmoll, H. R.; Schreiber, Hans W.; Schumm, Stanley A.; Sciacca, Thomas P., Jr.; Seagle, E. P.; Seel, Charles P.; Sharp, John V. A.; Silverman, Arnold J.; Smith, Kenneth G.; Socolow, Arthur Abraham; Staatz, Mortimer H.; Stebbins, Robert H.; Steenland, Nelson C.; Stehli, Francis G.; Stoertz, G. E.; Strongin, Oscar; Strongin, Oscar; Stubbs, James E.; Sutton, George H.; Sutton, George H.; Swanson, Vernon Emanuel; Swinchatt, Peter F. E.; Takahashi, Taro; Taubeneck, William Harris; Thurston, William Richardson; Tolstoy, Ivan; Tryon, Lansing E.; Turekian, Karl; Turekian, Karl K.; Upham, Sidney H., Jr.; Van Vloten, Roger; Vasquez, L.; Victor, Iris; Volchok, Herbert L.; Volchok, Herbert L.; Voorhees, John K.; Walton, Matthew S., Jr.; Warren, David H.; Webb, Gregory W.; Weeks, Lewis Austin; Wharton, H. M.; Wheeler, J. O.; Wheeler, John Oliver; Wilcox, Ronald E.; Williams, Floyd J.; Williams, Norman Charles; Winograd, Isaac J.; Winters, Stephen Samuel; Woolard, Louis Eugene; Woolard, Louis Eugene; Worthington, J. E.; Wright, Harold D.; Wuenschel, Paul Clarence; Yochelson, Ellis L.

1960s: Abdel-Gawad, A. M.; Adler, Hans Henry; Alper, Allen M.; Alvarez-Osejo, J. Alberto; Amy, Vincent Peter; Anestad-Fruth, Elizabeth; Atia, Abdel-Kadir Mohamed H.; Barrington, Jonathan; Bassett, William A.; Beckman, Marian C.; Bernardini, Gian P.; Bollin, Edgar Marshall; Booy, Emmy C.; Broscoe, Andy J.; Bruen, James N.; Bryers, Wesley E.; Burgess, William J.; Buseck, Peter Robert; Butler, James Robert; Call, Richard D.; Cann, Ross S.; Casella, Clarence Joseph; Catanzaro, Edward John; Celasun, Merih; Cherukupalli, Nehru E.; Ciriacks, Kenneth Wilmer; Cohen, Charles; Crawford, John P.; Crosby, Gary Wayne; Davidson, Donald M.; Davidson, Donald Miner, Jr.; Dorman, Henry J.; Douglas, Richard F.; Douglas, Richard Franklin; Dunne, James Arthur; Dureck, Joseph J.; Durek, Joseph John; Engels, Joan Carol; Erickson, Glen P.; Erickson, Glen P.; Espinosa, Alvaro F.; Ewing, M.; Fagan, John J.; Faill, Rodger T.; Faill, Rodger Tanner; Fanale, Fraser Partington; Freyman, A.; Fruth, Elisabeth Anestad; Giese, Rossman Frederick, Jr.; Gilleti, Bruno J.; Glass, Billy Price; Grim, Paul J.; Grow, John A.; Hansen, Harry J., III; Hawkins, William H.; Hays, James Douglas; Heaslip, William Graham; Heyman, Arthur Mark; Howard, Calhoun Ludlow Harper; Howard, John Hall; Howe, Herbert James; Isacks, Bryan L.; Ishihara, Shunso; Jacobs, Marian Beckmann; Jordan, William M.; Kamhi, Samuel; Karp, Edwin; Kaufman, Aaron; Keller, Allen Seely; Kerr, James William; Kimmel, Grant E.; King, Robert Nephew; Klovan, John E.; Kornicker, Louis S.; Lancaster, Mary Jane; Langer, Arthur M.; Laporte, Leo F.; Leveson, David Jeffrey; Liebermann, Robert Cooper; Liebling, Richard S.; Liebling, Richard Stephen; Lippitt, Louis; Lobanoff-Rostovsky, Nikita; Lubowe, Joan K.; MacFadyen, John A., Jr.; MacIntyre, Giles Ternan;

Major, Maurice W.; Mangus, Marlyn; Maxwell, James Christie; McThenia, Andrew W., Jr.; Megrue, George H.; Megrue, George Henry; Mihm, Richard J.; Miller, Donald Spencer; Molloy, Martin William; Moore, Willard S.; Morisawa, Marie E.; Moshiri-Yazdi, Reza; Munoz J., Nicolas G.; Munoz, J. Nicholas G.; Nace, Raymond L.; Oliver, J.; Olson, Edwin Andrew; Ostrom, John H.; Otooni, M. A.; Oxley, Philip; Peterson, Christoph R.; Pitcher, Max Grow; Pomeroy, Paul W.; Prinz, Martin; Purdy, Edward George; Riva, John F.; Rooney, Thomas P.; Russell, Dale Alan; Schmitt, Leonard J., Jr.; Schwartz, Maurice; Silverman, Arnold Joel; Simpson, Eugene S.; Smith, Muriel C.; Spencer, Edgar W.; Streeter, Sereno Stephen; Sykes, Lynn R.; Talwani, Manik; Tan, Li-Ping; Thurber, David Lawrence; Turkeli, Arif; Turner, Alan Keith Finlay; Van Cott, Harrison C.; Wampler, J. Marion; Weiss, Richard L.; Widmier, John Michael; Wiegand, Jeffrey W.; Wilcox, John T.; Wilcox, John Thomas; Wiles, William W.; Winters, Stephen S.; Wright, Frederick F.; Yasso, Warren E.; Yasso, Warren E.

1970s: Boatwright, J. L.; Bopp, R. F.; Bruhn, R. L.; Cande, S. C.; Chapman, M. E. D.; Chen, K. H.; Cho, D.; Choy, G. L.; Cochran, J. R.; Collins, W.; Cooke, D. W.; Cormier, V. F.; Dutch, S. I.; Embley, R. W.; Engelmann, G. F.; Fornari, Daniel John; Fox, Paul Jeffrey; Gordon, S. I.; Gorini, M. A.; Hesslein, R. H.; Hwang, L. F.; Kipphut, G. W.; Kligfield, R.; Kranz, R. L.; Kristoffersen, Y.; LaBrecque, J. L.; Lee, S. S.; Macfadden, B. J.; Marfurt, K. J.; Morley, J. J.; Olsen, C. R.; Perfit, M. R.; Pichulo, R. O.; Quay, P. D.; Rampino, M. R.; Rigby, J. K., Jr.; Rowlett, H. E.; Sarmiento, J. L.; Schroeder, F. W.; Torgersen, T. L.; von=Baumgaertner, I.; Ward, Peter Langdon; Zielinski, G. W.

1980s: Abbott, Dallas Helen; Adler, Dennis Marvin; Agee, Carl Bernard; Al-Harari, Zaki Y.; Allen, Richardson Beardsell; Amdurer, Michael; Andors, Allison Victor; Barghoorn, Steven Frederick; Bodine, John Howard; Bogen, Nicholas Louis; Bower, Peter Michael; Boyd, Thomas Muryl; Brown, Stephen Ray; Brown, Walter Emerson; Buhl, Peter; Cember, Richard Paul; Chen, Thomas Chui-Tung; Cifelli, Richard Lawrence; Cifuentes, Ines Lucia; Clement, Bradford Mark; Coffin, Millard Filmore, III; Coldiron, Ronn William; Coles, Kenneth Spencer; Cook, Robert Bradley; Cox, Simon Jonathan David; da Silva, Maria Augusta Martins; Dai, Tingfang; de Azevedo, Antonio Expedito Gomes; Deck, Bruce Linn; Devlin, William Joseph; Diebold, J. B.; Ellins, Katherine Kelly; Elthon, Donald Lee; Evander, Robert Lane; Farre, John Andrew; Flynn, John J.; Forsythe, Randall David; Foster, Douglas John; Fox, Christopher Gene; Frankel, Arthur David; Freeman-Lynde, Raymond Paul; Gaddis, M. Francis; Gamboa, Luiz Antonio Pierantoni; Girty, Gary H.; Goldberg, David S.; Goldstein, Steven Lloyd; Gyebi, Osei Kwabena; Hanson, Richard Eric; Hassanzadeh, Siamak; Hauksson, Egill; Hegarty, Kerry Anne; Herczeg, Andrew Leslie; Herman, Bruce Meyer; House, Leigh Scott; Huang, Jau-Inn; Hurst, Kenneth Joslin; Jacobi, R. D.; Jones, Glenn A.; Jones, Williams Maury; Kadko, David Charles; Kappel, Ellen Sue; Karner, Garry David; Khan, Mohammad Javed; Kominz, Michelle Anne; Kong, Michael; Lazarus, David B.; Leith, William Stanley; Leslie, Robin Bruce; Lewis, Stephen Dana; Madden, Victoria Hope; Marshak, Stephen; Martinez, Fernando; Martinson, Douglas George; Massa, Audrey Adams; McCann, William Richard; McNutt, Stephen Russell; Menke, William H.; Merguerian, Charles Michael; Michael, Peter John; Miller, John Donley; Mithal, Rakesh; Mix, Alan Campbell; Mori, James Jiro; Mountain, Gregory Stuart; Mrozowski, Cary Louis; Mutter, John Colin; Naini, B. R.; Nelson, Eric Paul; Neville, Colleen Ann; Newmark, Robin L.; Nicholson, Craig Claverie; Nishenko, Stuart Paul; O'Connell, Suzanne Bridget; Perez A., Omar J.; Plumb, Richard Allen; Pokras, Edward Mathew; Prothero, Donald Ross; Quittmeyer, Richard Charles; Reisberg, Laurie Ceil; Rohl, Arthur N.; Schiff, Sherry Line;

Schuster, Gerard Thomas; Sereno, Paul Callistus; Shepherd, Charles Edward; Spariosu, Dann Jack; Steckler, Michael Stuart; Stein, Carol Ann; Tan, Robert Yaunchan; Tauxe, Lisa; Taylor, Brian; ten Brink, Marilyn Rae Buchholtz; Ten Brink, Uri S.; Thorne, Julian Arthur; Toggweiler, John Robert; Wanninkhof, Richard Hendrik; Wesnousky, Steven Glenn; Wilson, Terry Jean; Winslow, Margaret Anne; Yang, Jih-Ping

Concordia University
Montreal, PQ H4B 1R6

1 Master's

1980s: Gocevski, Vladimir

University of Connecticut
Storrs, CT 06268

30 Master's, 42 Doctoral

1960s: Dorrler, Richard; Frankfort, Donald; Giddings, Marston Todd; Hubbard, Edwin L.; Khanna, Sat Dev; Osgood, John O.; Parker, Donald Andrew; Ramanjaneya, Gundarlahalli Sankarasetty; Vilinskas, Peter; Vitali, Rino

1970s: Bertoni, R. S.; Curry, Richard Porter; Danbom, S. H.; Dixon, John McConkey; Duke, John Murray; Dzis, R. J.; Gerardin, V.; Haskell, N. L.; Hunt, C. D.; Kasper, Andrew Edward, Jr.; Kastning, E. H.; Kropf, F. W.; Long, Jerome Pillow, III; Lyons, W. B.; Major, Richard Paul; Markl, R. G.; Ogushwitz, P. R.; Proctor, N. S.; Siraki, E. S.; Szabo, Nicholas; Tisoncik, D. D.; Wakeland, M. E., Jr.

1980s: Alfano, John Joseph; Allen, Stephen Jeffrey; Apotria, Theodore G.; Baillie, Priscilla Woods; Bowen, James Howland; Chichester-Constable, David John; Civco, Daniel Louis; Close, Edward Joseph; Conway, Michael Francis; Dalphin, Richard James; Doyle, Christopher Denis; Ellefsen, Karl J.; Focazio, Michael Joseph; Gill, Gary Arthur; Gurrieri, Joseph Thomas; Hankins, John B.; Herman, Gregory C.; Hesler, Donald J.; Izraeli, Ruth L.; Jeng, Yih; Jurczyk, Gayle Katherine; Kanazawich, Michael F.; Kim, Jonathan Philip; Kim, Tae Moon; Kjelleren, Gary Palmer; Lewis, Catherine Louise; Martello, Angela R.; Matson, Ernest Augustus; Musiker, Laurie B.; Pan, Jeng-Jong; Pecci, Anthony Salvatore; Ryan, Scott S.; Sakamoto-Arnold, Carole; Scott, David; Taylor, David Winship; Videlock, Shari Lynn; Washington, Paul Allyn; Wenk, Warren Joel; Wingfield, Betsey C.; Zavada, Michael Stephan

Cornell University
Ithaca, NY 14853

230 Master's, 331 Doctoral

1870s: Foote, Charles Whittlesey

1880s: Chester, Frederick Dixon; Chester, Frederick Dixon; Comstock, Theodore B.; Cushing, Henry Platt; Prosser, Charles Smith

1890s: Case, Ermine Cowles; Lockhead, William; Watson, Thomas Leonard

1900s: Butler, Bert S.; Carney, Frank; Cushing, Henry Platt; Dales, Benton; Hubbard, George David; Kyser, Kathryn J.; Martin, James O.; Matson, George C.; Maury, Carlotta J.; Prosser, Charles Smith; Reinecke, Leopold; Rich, John Lyon; Sheldon, Pearl Gertrude; Smith, Charles Edward

1910s: Davis, Norman Bruce; Galpin, Sidney Longman; Galpin, Sidney Longman; Honess, Charles William; Hopper, Walter Everett; Keele, Joseph; Long, E. Tatum; Martin, Lawrence; Mordoff, Richard Alan; Mulliner, Beulah A.; Perrine, Irving; Perrine, Irving; Rich, John Lyon; Sheldon, Pearl Gertrude; Shideler, William Henry; Somers, Ransom Evarts; Storrer, James; Teas, Livingstone Pierson; Verwiebe, Walter August; Von Engeln, Oscar Dierich; Whitney, Francis Luther; Wolcott, Henry Newton; Wong, Parkin

1920s: Arnold, Chester Arthur; Burfoot, James Dabney, Jr.; Carlson, Fred Albert; Cole, W. Storrs; Conant, Louis Cowles; Cornish, Cornelia Baker; Duncan, Georgianna Hawley; Filmer, Edwin Alfred; Forrester, James Donald; Fridley, Harry Marion; Gustafson, Axel Ferdinand; Gwynne, Charles Sumner; Hard, Edward Wilhelm; Harper, Francis; Hayes, William Harold; Hedberg, Hollis Dow; Heminway, Caroline Ella; Hodson, Floyd; Hodson, Floyd; Kinsman, Daniel Francis; Langford, George B.; Lee, Herbert V.; Martens, James H.; Martens, James H.; Megathlin, Gerrard R.; Monnett, Victor E.; Neumann, Fred Robert; Nevin, Charles M.; Nevin, Charles M.; Niimony, Kunitaro; Niinomy, Kunitaro; Palmer, Katherine V.; Papish, Jacob; Pegau, Arthur August; Phipps, William Mason; Robinson, Ernest Guy; Saint John, Ruth Nimmo; Shaub, Benjamin Martin; Shaub, Benjamin Martin; Sherrill, Richard Ellis; Smith, Richard Wellington; Smythe, Donald DeCou; Stillman, Francis Benedict; Stow, Marcellus H.; Stucky, Jasper L.; Tucker, Helen Ione; Weisbord, Norman Ed; Wheeler, Everett Pepperell; Whitney, Francis Luther

1930s: Alexander, Alexander Emil; Baker, Elizabeth; Baril, Roger Wilfrid; Berry, George Willard; Berthiaume, Sheridan Alba; Beverly, Burt, Jr.; Bissell, Bradford; Bolton, Beatrice E.; Bowen, Ward Culver; Burroughs, Wilbur Greeley; Campbell, Donald Fergus; Caster, Kenneth Edward; Caster, Kenneth Edward; Chelikowsky, Joseph R.; Chelikowsky, Joseph R.; Cole, W. Storrs; Conant, Louis Cowles; Dyson, James Lindsay; Dyson, James Lindsay; Edmundson, Raymond Smith; Ferry, Chamberlain; Forrester, James Donald; Gabriel, Alton; Gillespie, Ruth Frances; Grant, Allan Marshall; Graton, Louis Caryl; Griffis, Arthur Thomas; Hadley, Wade H.; Haseman, Joseph Fish; Herrick, Stephen Marion; Hills, Robert Chadwick; Ingham, Albert I.; Johnson, Howard F.; Jones, Verner Everett; Jones, Verner Everett; Jordan, Richard Hollister; Langton, Claude M.; LeRoy, Frank L.; Ludlum, John C.; Mayo, Evans B.; McConnell, Duncan; Megathlin, Gerrard R.; Meyers, Harold G.; Michener, Charles Edward; Morrison, Roger B.; Moss, Rycroft Gleason; Murphy, Eleanor F.; Osborne, Margaret Eleanor; Parker, John Mason, III; Parker, John Mason, III; Patnode, Homer Whitman; Pepper, James Franklin; Price, Paul Holland; Rappenecker, Caspar; Rodgers, John; Rogers, Reginald Douglas, Jr.; Rowland, Richard A.; Sadler, James William; Safonov, Anatole Ivanovitch; Schoonover, Lois Margaret; Seery, Virginia B.; Sherrill, Richard E.; Short, Allan McIlroy; Stow, Marcellus H.; Sweet, Lois Bigelow; Thomas, Dale Edmund; Trainer, David Woolsey, Jr.; Tuck, Ralph; Tucker, Helen Ione; Wagner, Norman Spencer; Watson, John Gaul; Wedel, Arthur Albert; Wells, John West; Wells, John West; Wheeler, Everett Pepperell; Wilbur, David Truxton; Wold, John Schiller

1940s: Allen, Rhesa McCoy, Jr.; Baldwin, Ewart Merlin; Banks, Harlan Parker; Beeson, Kenneth Crees; Berry, George Willard; Brown, William Randall; Cady, John Gilbert; Carr, William Lester; Chisnell, Thomas Cutter; Chu, Ching-Jui; Cline, Marlin George; Colligan, Richard Vincent; Crafts, Shirley; Durham, Forrest; Farrington, William Benford; Fox, Harold Dixon; Gealey, William Kelso; Gilmore, Raymond Maurice; Hawley, Luther David; Hunter, Frank R.; Johnsgard, Gordon Alexander; Johnson, Clayton Henry, Jr.; Ludlum, John C.; Macquown, William C.; Moore, Charles H., Jr.; Moore, John C. G.; Moore, Wayne E.; Netschert, Bruce Carlton; Newcomb, Edward L.; Nugent, Lawrence E., Jr.; Patton, William Wallace, Jr.; Pitzrick, Raymond August; Proctor, Paul Dean; Rhodes, Rene George; Ross, Mary Harvey; Schmidt, Victor Edward; Symonds, Paul S.; Tailleur, Irvin L.; Tyler, Henry Johnson; Vookerding, Clifford J.; Wintringham, Neil Andrews; Wright, James Clifton

1950s: Bailey, Roy Alden; Brent, William Bonney; Brent, William Bonney; Bryant, Jay Clark; Carlisle,

Frank Jefferson, Jr.; Chisnell, Thomas Cutter; Corbett, Marshall Keene; Davis, Clarence King; Dressner, Elliott Francis; Dyott, Mark Hamilton; Edwin, Robert Bruce; Erwin, Robert Bruce; Fernow, Leonard Reynolds; Fisher, Stanley Parkins, Jr.; Flach, Klaus Werner; Fry, Wayne L.; Gehman, Harry Merrill, Jr.; Geiger, Kenneth Warren; Goodman, Richard Edwin; Grossman, Robert Bruce; Hewitt, Philip Cooper; Hinrichs, Edgar Neal; Holmes, Stanley Winchester; Hueber, Francis Maurice; Humphreville, James A.; Isachsen, Yngvar W.; Knox, Ellis Gilbert; Konig, Ronald Howard; Konig, Ronald Howard; MacLeod, William George; Matthews, Burton Clare; McCaleb, Stanley Bert; McCracken, Ralph Joseph; McLeod, William George; Mellen, James Vedrey; Mollard, John Douglas Ashton; Moore, Wayne E.; Nichols, David Ryden; Oliver, William Albert, Jr.; Oliver, William Albert, Jr.; Perhac, Ralph Matthew; Pessagno, Emile Anthony, Jr.; Pomerening, James Albert; Price, James Kennedy; Quinn, Harold Arthur; Reilly, Edgar Milton, Jr.; Richards, Paul William; Rickard, Lawrence Vroman; Ross, Mary Harvey; Schneer, Cecil Jack; Secor, Donald Terry, Jr.; Shockey, Philip Nelson; Shumaker, Robert Clarke; Sirkin, Leslie Arthur; Squires, Donald Fleming; Starr, Robert Brewster; Stone, Donald Sherwood; Storr, John Frederick; Turner, Philip Ambrose; Walters, Charles Philip; Webb, James Edward; Williams, George Quigley; Wilson, Philo Calhoun; Wilson, Thomas Carroll; Winder, Charles Gordon; Winder, Charles Gordon; Young, Robert Spencer; Zimmerli, Edward Joseph

1960s: Acaroglu, Ertan Riza; Bemben, Stanley Michael; Berkey, Edgar; Boekenkamp, Richard Paul; Brew, Douglas Crocker; Brew, Douglas Crocker; Brunskill, Gregg John; Bullock, Peter; Carluccio, Leeds Mario; Chaturvedi, Lokeshwa Nathr N.; Clement, Stephen Caldwell; Conner, Joanne Marie; Crierson, James Douglas, Jr.; DeMent, James Alderson; Diffenbach, Robert Nevin; Dijkerman, Joost Christiaan; Donovan, John Francis; Duskin, Douglas John; Fahnestock, Robert Kendall; Fernow, Leonard Reynolds; Geiger, Kenneth Warren; Gonzales, Serge; Grasso, Thomas Xavier; Grierson, James Douglas; Gupta, Barun Kumar Sen; Harrington, Jonathan Waldo; Harrington, Jonathan Waldo; Hatheway, Richard Brackett; Hopkins, Henry Robert; Hueber, Francis Maurice; Jha, Parmeshwari Prasad; Kirchgasser, William Thomas; Kirchgasser, William Thomas; Klimowski, Richard Joseph; Kothe, Kenneth Ralph; Laub, Richard Steven; Lavkulich, Leslie Michael; Ludlam, Stuart Dietrich; Majtenyi, Steven Istvan; Martini, Jose Alberto; Matten, Lawrence Charles; McEwen, Robert Barlow; Morales, Pedro Augusto; Peralta-Cardenas, Gale; Reed, Kenneth John; Reed, Kenneth John; Romkens, Mathias Joseph Marie; Roth, James Richard; Sachs, Kelvin Norman, Jr.; Sangrey, Dwight Abram; Scherffius, William Edward; Schumaker, Robert Clarke; Shumaker, Robert Clarke; Sivarajasingham, Sivasupramaniam; Southard, Alvin Reid; Tiller, Kevin George; Torrance, James Kenneth; Verma, Rameshwar Dayal; Weeden, Harmer Allen; Wilcox, Margret Schnaitman; Witty, John Edward; Wolff, Manfred Paul; Yamani, Mohamed Abdullah Abdu; Yamani, Mohamed Abdullah Abdu; Yudhbir; Zenger, Donald Henry

1970s: Aguilar, Eduardo; Ballantyne, J. C.; Barton, William Thomas; Bell, Michael Stewart; Bilby, R. E.; Botto, R. I.; Bouchard, A. B.; Brice, William Riley; Brown, Larry Douglas; Burgess, Lawrence Charles Norman; Cabrera, John George; Caldwell, John Gerrit; Campbell, Bruce S.; Chen, Tzong-Tzyy; Citron, Gary P.; Clancy, Robert Todd; Coen, Gerald Marvin; Cole, Gregory Lawrence; Connerney, J. E. P.; Dempster, Kelly; Dickerson, Carol Adrienne; Dillon, M. J.; Fellows, Steven Neal; Forrester, John Douglas; Friedman, Daniel Bruce; Frohlich, C. A.; Goodman, Karen Jeanne; Grant, Douglas Roderick; Hague, Nancy Ellen; Harrison, Ellen Zucker; Hartman, C. M.; Hawes, R. A.; Haxby, W. F.; Hayes, Christopher George;

Hodgson, Edward Askew; Hughes, Susan Elaine McAlear; Kromah, Fodee; Ladd, John Howard; Lawson, Jeffrey Thomas; Liu, C. C.-K.; Markham, John Joseph; McAdoo, D. C.; Montgomery, Scott Lyons; Moore, G. F.; Nakiboglu, Sadik Mete; Namy, Dominique; Nekut, A. G., Jr.; Nieber, J. L.; Novak, Irwin Daniel; Parmentier, Edgar M.; Pennington, Wayne David; Pieri, D. C.; Pollard, M. W. S.; Pueschel, C. M.; Rathke, William Wilson; Reilinger, Robert Eric; Rockwell, Charles; Rodriquez, Raphael T.; Scheckler, Stephen Edward; Schoenberg, Michael; Smith, Allan Conrad, Jr.; Sondergeld, Carl H.; Sparacin, W. G.; Stephens, Christopher Douglas; Tag, Peter Harrison; Tarbox, David L.; Taylor, Frederick Wiley, Jr.; Thomas, Peter Chew; Tyson, Robert Michael; Umari, A. M. A.; Van Raij, Bernardo; Wang, Chang; Weathers, Maura Susan; Weathers, Maura Susan; Wentworth, T. R.; Wold, John Pearson; Yamada, John A.; Yassa, Guirguis Fahmy; York, James Earl, III; Younce, Gordon Baldwin; Zall, L. S.

1980s: Ahern, J. W.; Ahn, Kyu-Hong; Alter, Benjamin; Brown, Larry; Amin, Magdy Ibrahim; Angevine, Charles Leon; Arnow, Jill Ann; Bevis, Michael Graeme; Billington, Selena; Boyce, Steven Craig; Bray, Cynthia Jean; Brewer, Jonathan Andrew; Buratti, Bonnie Jean; Burton, James Hutson, III; Cardwell, Richard Kenneth; Cavallaro, Nancy; Chandlee, George Oliver; Cheng, Hung-Darh Alexander; Chinn, Douglas Samuel; Chiu, Jer-Ming; Citron, Gary P.; Clukey, Edward Charles; Conrad, Walter Karr; Cook, Frederick Ahrens; Cooper, Reid Franklin; Crespo, Esteban; D'Andrea, R. A.; De Voogd, Beatrice; Driscoll, C. T.; El-Kadi, Aly Ibrahim; Emerman, Steven Howard; Fitts, Charles R.; Flemings, Peter Barry; Forbes, Terence Robert; French, Peter Newton; Furnish, Michael David; Gallahan, David Michael; Geary, Edward Eugene; Geary, Edward Eugene; Gilpin, Lawrence Mellick; Goodrich, Cyrena Anne; Grove, Thurman Lee; Haeck, Gary Dennis; Hamburger, Michael Wile; Hamburger, Michael Wile; Harding, David John; Harris, Ruth Audrey; Hay, Bernward Josef; Heath, Jenifer Sue; Hewitt, Allan Edward; Hibbard, James Patrick; Hsu, Jeffrey Tsen-Jer; Huang, I-Chen Eugene; Huang, Yih-Ping; Hughto, Richard John; Jamnongpipatkul, Pichit; Johnson, Mark Galen; Kadinsky-Cade, Katharine A.; Keach, R. William, II; Kellogg, Louise Helen; Kenyon, Patricia May; Klemperer, Simon Louis; Kung, Samuel King-Jau; Lafe, Olurinde Ebenezer; Lee, Steven Wendell; Lenhardt, Duane Rudolph; Lennon, G. P.; Lillie, Robert James; Liu, Jue-Yu; Long, George Henry; Macedo, Jamil; Malhotra, Renu; Marthelot, Jean-Michel; Mayer, James Roger; Mbagwu, Joe Sonne Chinyere; McBride, John Henry; McCarty, Thomas Richard; McConnaughey, Paul Kevin; McDowell, William Hunter, II; Miesen, David Lee; Mummert, Mark Christopher; Murray-Rust, Douglas Hammond; Nelson, Priscilla; Neuweld, Mark Adam; Ni, James Fu; O'Brien, Thomas Francis; O'Day, Peggy Anne; Perucchio, Renato S.; Phillips, Sandra; Phillips, Wayne Jude; Renner, Rebecca; Reynolds, Jeffrey; Ricoult, Daniel Louis; Rosales, Anibal Marcia L.; Rubenstone, James L.; Sandford, William Edward; Sarewitz, Daniel R.; Schilt, Frank Steve; Schweller, William John; Serpa, Laura Fern; Sharp, James William; Shelton, Deborah H.; Shen, Pouyan; Showers, Kate Barger; Smalley, Robert, Jr.; Smith, Pamela Louise; Snyder, David Bufton; Snyder, Kent Everett; Snyder, Victor Abram; Squyres, Steven Weldon; Stanturf, John Alvin, IV; Steiner, David Robert; Stewart, James Pirtle, Jr.; Strecker, Manfred Reinhard; Svendsen, Mark T.; Taigbenu, Akpofure Efemena; Teng, William Ling; Thompson, Bruce Gregory; Timlin, Dennis James; Trautmann, Charles Home; Turner, John Patrick; Underwood, Michael Bruce; Vertucci, Frank Anthony; Vogel, Richard Mark; Weaver, Tamie Renee; Welbourn, Martha Lynne; Widom, Elisabeth; Wijeyesekera, Sunil David; Wille, Douglas Michael; Willemann, Raymond James; Wirth, Karl R.; Zecharias, Yemane Berhan; Zhu, Tianfei

Dalhousie University
Halifax, NS B3H 3J5

152 Master's, 52 Doctoral

1870s: Cochrane, N. A.

1930s: Turnbull, Lionel Graham

1950s: Berger, A. R.; Carter, A. L.; Davison, Wilbert Lloyd; Gass, Nicholas James; Gourley, Albert Carlisle; Hogg, W. A.; Hogg, William A.; Johnston, James William Derek; Peters, H. R.; Phillips, John Edward; Rowley, Eric Alfred; Shea, Frank Stoddard; Take, W. F.

1960s: Allen, Robert Gordon Hamilton; Aumento, Fabrizio; Aumento, Fabrizio; Barrett, Donald Larry; Berger, Jonathan Joseph; Blundon, Sandra J.; Buchbinder, Goetz Gustav Rudolph; Campbell, F. H. A.; Chen, Hsien-su; Cochrane, Norman; Cook, Robert H.; Creed, Robert M.; Dainty, Anton Michael; Desborough, John; Duff, Sheila Louise; Evans, Robert; Ewing, Gerald Neil; Fenwick, Donald Kenneth Bruce; Grant, Douglas R.; Harris, I. McK.; Hennigar, Terry W.; James, Noel P.; Keen, Charlotte Elizabeth; Kent, George Robert; Krauel, David Paul; Lambert, Anthony; Lawrence, D. E.; Lines, Roland Arnold Granville; Lyall, Anil K.; MacIntyre, William G.; MacNab, Ronald Finlay; Manchester, Keith Stanton; McAllister, Ronald Eric; Miller, J. A.; Milner, Michael; Mossman, David John; Nolan, Francis J.; Nyland, Edo; Palmer, William James; Parsons, Roger Clare; Pezzetta, John Mario; Ruffman, Alan S.; Silverberg, Norman; Simpson, David William; Tsernoglou, Demetrius; Turker, Yucel; Vilks, G.; Watt, Walton Delbert; Weeks, Ross M.; Weiler, Roland R.

1970s: Abdel-Aal, Osama Youssef; Aksu, Ali Engin; Alam, Mahmood; Alam, Mahmood; Barnes, Neal E.; Beaumont, Christopher; Bhattacharyya, P. J.; Blinn, L. J.; Bugden, G.; Charest, Marc H.; Chinn, Alan F.; Choo, Kangsoo; Clark, D. F.; Cok, Anthony E.; Cross, H.; Dale, Christopher Thomas; Dayal, R.; Dickie, J. R.; Drapeau, Georges; Drury, Malcolm J.; Farley, E.; Fiess, K. M.; Fricker, Aubrey; Gatien, M. G.; Graves, M. C.; Gregory, M. R.; Gustajtis, K. Andrew; Hall, Blaine R.; Hartwell, J. M.; Hatt, B. L.; Herb, G.; Horowitz, M. R.; Hume, Howard; Iqual, Javed; Johnson, James S.; Keeley, J. R.; Kepkay, Paul E.; Kublick, Ernest E.; Kubuck, E. E.; Langhus, Bruce G.; Lewis, J. F.; Liew, M. J. C.; Lyttle, Norman A.; MacCutcheon, Murray; MacGregor, D. R. C.; MacLean, J. Arthur; Mageau, Camille M.; McGraw, Patricia A.; McKenzie, C. B.; Miller, C. K.; Murray, Daniel A.; Nielsen, Erik; Palmer, A.; Parrott, Richard J. E.; Pride, C. R.; Pye, Graham D.; Railton, J. B.; Rankin, D. S.; Renwick, Gregory K.; Rice, Phillip; Rohrbacher-Carls, M. R.; Ryall, Patrick J. C.; Sadler, H. E.; Sarkar, P. K.; Scott, David B.; Secord, David J.; Sinna, Ravindra P.; Stehman, C. F.; Stewart, John McG.; Stow, Dorrik, A. V.; Stucchi, D. J.; Stukas, Vidas; Sullivan, Kathryn D.; Tee, K.-T.; Ulriksen, Carlos E.; Underwood, J. K.; von Borstal, B. E.; Watt, Peter J.; Wightman, D.; Willey, J. D.; Wilson, Robert F.

1980s: Adams, Peter J.; Akande, Samuel Olusegun; Aksu, Ali Engin; Bailey, David G.; Barrie, Charles Q.; Bird, Donna J.; Blanchard, Marie-Claude; Bonham, Oliver J. H.; Bourque, Paul Daniel; Brew, David Scott; Chatterjee, A. K.; Clarke, John Edward Hughes; Cordsen, Andreas; Cullen, John D.; Cullen, Michael Paul; DeIure, Anita Mary; Doucet, Pierre; Douma, Marten; Douma, Stephanie Leigh; Elias, Peter; Fitzgerald, Colleen E.; Fralick, Philip W.; Haines, John W.; Ham, Linda J.; Hebert, David Lawrence; Helgason, Johann; Honig, Cecily A.; Issler, Dale; Logothetis, John; McLaren, Shirley Anne; Miller, Ann-Alberta Louise; Mudie, Peta J.; Myers, Robert; Nicks, Linda P.; Noguera Urrea, Victor Hugo; O'Reilly, George A.; Parrish, Christopher C.; Peterson, Carol Audrey; Plint, Heather Elizabeth; Prime, Garth A.; Ravenhurst, Casey Edward; Richard, Linda R.; Schroder, Claudia J.; Smith, D. L.; Smith, William D.; Stam, Beert; Stea,

Rudolph R.; Stergiopoulos, Apostolo Basil; Thomas, F. C.; Todd, Brian Jeremy; Todd, Brian Jeremy; Wallace, Derek; Wightman, Daryl M.; Williamson, Marie-Claude; Williamson, Mark A.; Wolfson, Isobel K.; Yule, Alan; Zhou, Xianliang

Dartmouth College
Hanover, NH 03755

168 Master's, 41 Doctoral

1940s: Murphy, John F.; Woodard, Henry H.

1950s: Baker, Richard C.; Bates, Robert G.; Boudette, Eugene L.; Elberty, William T., Jr.; Gosman, Robert F.; Hamilton, Charles L.; LaFreniere, Gilbert F.; Lupton, D. Keith; MacKenzie, W. Bruce; Martin, David B.; McCrehan, Richard E.; McDowell, John Parmelee; Noel, James Arthur; Ragle, Richard H.; Ratte, Charles A.; Ratte, James Clifford; Rockwood, Walter G.; Snyder, George L.; Snyder, John Lemoyne; Stuart, Roy Armstrong; Taylor, Lawrence D.; Tremaine, John W.; Zenger, Donald Henry

1960s: Adams, Samuel S.; Allan, Roderick James; Armstrong, W. D.; Barnard, Walther M.; Bary, David O.; Black, Thomas John; Bothner, M. H.; Christopher, P. A.; Colburn, James A.; Eggers, Albert Allyn; Gaffney, Edward S.; Juang, Franz H. T.; Kilmer, Kilmer W.; Kneidel, Eliezer; Lange, Ian M.; Lessing, Peter; Linkletter, George O.; Linscott, Robert Orrin; Lofgren, Gary E.; Mock, Steven J.; Muench, Robin D.; Naeser, Charles W.; Newell, Wayne L.; Norwick, Stephen A.; Ormiston, Allen R.; Peck, John H.; Potter, Noel, Jr.; Riggs, Stanley; Steidtmann, James; Stockard, Donald P.; Taylor, Paul Scott; Thompson, Graham R.; Unger, John Duey; Van Ingen, Robert; Zahony, Stephen G.

1970s: Aleinikoff, John N.; Aleinikoff, John N.; Anderson, Robert C.; Barndt, Jeffrey K.; Bathurst, B. W.; Bieler, David B.; Birnie, Richard W.; Boudette, Eugene L.; Breitzman, Lynne L.; Bullen, Thomas D.; Carlson, Gerald G.; Carr, Michael J.; Carr, Michael J.; Clark, Russell Gould, Jr.; Collins, B. I.; Condit, William S.; Cox, Gordon F. N.; Cox, Gordon, F. N.; Crafford, T. C.; Dean, Bradley W.; Doherty, J. T.; Douthitt, C. B.; Duke, Edward F.; Dykstra, Jon D.; Dykstra, Jon D.; Eggers, Albert Allyn; Eng, K. J.; Englund, Evan John; Gleason, Richard J.; Green, Julian W.; Harlow, David H.; Hazlett, Richard W.; Hoisington, W. David; Hughes, John Michael; Johnson, Arthur H., Jr.; Keller, Homer M.; Koutz, Fleetwood R.; Lamb, Henry J.; Liddicoat, Joseph C.; Loucks, Thomas A.; Malinconico, Lawrence Lorenzo, Jr.; Malone, Gary B.; Mayer, Lawrence M.; Mayer, Lawrence M.; McClelland, Lindsay R.; McGirr, Robert R.; Merritt, David H.; Merry, Carolyn J.; Montgomery, Carla Westlund; Murrow, Patricia J.; Nadeau, Paul H.; Nelson, Carl E.; Nibbelink, Mark P.; Nielson, Dennis Lon; Nielson, Dennis Lon; Nielson, Dianne Ruth Gerber; Nielson, Dianne Ruth Gerber; Page, Frederick W.; Petersen, Erich Ulrech; Pipes, William Vaughn; Plumb, Richard A.; Reynolds, James Howard, III; Rose, William Ingersoll, Jr.; Schrock, Ronald L.; Snellenburg, Jonathan W.; Spydell, D. Randall; Stonehouse, James Mrcus; Summers, Donna M.; Taylor, Paul Scott; van Oss, Hendrik G.; Vitousek, Peter Morrison; Williams, Stanley Nichols

1980s: Barnett, Daniel; Benjamin, Michael T.; Benton, Lynda M.; Bray, Timothy D.; Burbank, Douglas West; Campbell, Jeffrey Erle; Campbell, Jeffrey Erle; Carr, Richard S.; Cerveny, Philip F., III; Chamberlain, C. Page; Chico, Eduardo; Connor, Charles Benjamin; Connor, Charles Benjamin; Conrad, Mark E.; Cronin, Vincent S.; DeFeo, Nancy J.; Douglass, David Neil; Douglass, David Neil; Duke, Edward F.; Duke, Genet Ide; Eusden, John Dykstra, Jr.; Francica, Joseph R.; Fuchs, Viveka; Gardiner, Janice L.; Gemmell, John Bruce; Gemmell, John Bruce; Gibson, Lisa M.; Hafner-Douglass, Katrin; Hanson, Carl R.; Helsel, Laura; Hluchy, Michele Marie; Hluchy, Michele Marie; Hughes, John Michael; Johnson, William P.; Johnsson, Mark; Johns-

son, Patricia; Kavanagh, Paul E.; Keith, Douglas W.; Kramer, Michael S.; Leschen, Melanie R.; Lynch, F. Leo, III; Malhotra, R. Veena; Malinconico, Lawrence Lorenzo, Jr.; Malinconico, MaryAnn Love; Mango, Helen N.; McKoy, Mark L.; McRae, Lee E.; Nadeau, Paul Henry; Newhall, Christopher George; Olson, Terrilyn M.; Parnell, Roderic Alan; Pogorzelski, Brett Katherine; Price, Curtis Yarnay; Prosser, Jerome T.; Pytte, Anthony Mark; Raynolds, Robert Gregory Honshu; Reynolds, James Howard, III; Roberts, Janis L.; Roggensack, Kurt; Shiekh, Khalid; Shumway, Dinah O'Sullivan; Sterne, Edward J.; Stone, Thomas A.; Sussman, David; Tabbutt, Kenneth Dean; Tellier, Kathleen E.; Tellier, Kathleen E.; Torcoletti, Paul James; Turner, Patricia Ann; Urquhart, Joanne; Walker, Jeffrey Ross; Walker, Jeffrey Ross; Williams, Stanley Nichols; Wills, Robert H.; Wilson, Frederick Henley; Zeitler, Peter Karl; Zeitler, Peter Karl

University of Delaware
Newark, DE 19716

101 Master's, 27 Doctoral

1970s: Allen, Elizabeth A.; Allen, Elizabeth A.; Baker, Ralph N.; Barlow, Rodney A.; Barone, John; Belknap, Daniel F.; Belknap, Daniel F.; Bopp, Frederick, III; Branca, John; Brickman, Eugene; Chase, Carol; Cooper, Neil F.; Corman, David; Demarest, James Monroe, II; Elliott, Glenn K.; Fithian, Patricia A.; Frank, Barbara A.; Goettle, Marjorie S.; Gohn, Gregory S.; Gohn, Gregory S.; Golovchenko, Xenia; Hager, Glenn M.; Halsey, Susan D.; Helsel, Dennis R.; Holzinger, Philip; John, Chacko J.; John, Chacko J.; Kearns, Lance E.; Kearns, Lance E.; Keenan, Everly Mary; Kent, Kathleen; Leis, Walter M.; Maurmeyer, Evelyn M.; Maurmeyer, Evelyn M.; Merrill, Laura; Metz, Rebecca W.; Minck, Robert J.; Moose, Roger David; Oostdam, Bernard Lodewijk; Pate, David L.; Porter, William M.; Richter, Alan; Rosen, Louis; Sheaffer, Sandra; Sheedy, Katherine A.; Sherman, John W.; Sherman, Sharon F.; Strom, Richard N.; Strom, Richard N.; Swetland, Paul J.; Tziavos, Christos C.; Weil, Charles B., Jr.; Zalusky, Donald W.; Zwart, Michael J.; Zwart, Peter A.

1980s: Andrews, Wayne; Atwater, Brian Franklin; Bates, Leonard Gordon, Jr.; Beasley, Elizabeth L.; Benamy, Elana; Bolakas, John Frank; Bopp, Frederick, III; Burns, Christopher; Cherry, Philip John; Chrzastowski, Michael J.; Coblentz, Alex; Collins, Daniel John; Conrad, John Adrian; Coughanowr, Christine; Crosby, James Thompson; Decker, Stephen; Demarest, James Monroe, II; Demicco, Peter M.; Dreier, Christine de Angelis; DuBois, David L.; Dunn, Pete J.; Eisner, Mark Walter; Fitzgerald, Sharon; Fletcher, Charles Henry III; Fletcher, Charles Henry, III; Fletcher, Ruth Reilly; Fordes, George D.; Foster, Christopher Allen; Gardiner, Mary Anne; Goldstein, Abram; Golike, David C.; Hansen, Kevin; Harrington, Anne; Howard, C. Scott; Hoyt, William Henry; Hulmes, Leita Jean; Jengo, John William; Johnson, Glenn Wilbur; Jones, Garry Davis; Keenan, Everly Mary; Lauffer, James Robert; Lee, Irene; Leon, Ralph Richard; Lerner, David H.; Leslie-Bole, Benjamin; Magruder, George Lloyd; Maguire, Timothy James; Maley, Kevin; Marx, Pat (Washburn); McDonald, Kathleen; Meger, Steven Anthony; Mirecki, June Elizabeth; Newsom, Steven Wayne; Peck, Timothy Joseph; Perrone, Emily F.; Pierson, Beverly A.; Rahaim, Stephen David; Ramsey, Kelvin Wheeler; Roberts, John; Shepley, Susan I.; Smith, Alison Jean; Smith, Richard V.; Sugarman, Peter J.; Toscano, Marguerite Ann; Villas, Cathleen Anna; Wheatley-Doyle, Michelle D.; Wisker, George E.; Yancheski, Tadeusz; Yang, Lien-Chu; Yang, Wei-Chong; Yau, Dah-Miin; York, Linda Louise; Zarra, Lawrence

University of Delaware, College of Marine Studies
Lewes, DE 19958

12 Master's, 16 Doctoral

1960s: Varrin, Robert Douglas

1970s: Bartlett, D. S.; Fallah-Araghi, M. H.; Guala, John Riddoch, Jr.; Lepple, F. K.; Parraras-Carayannis, G.; Suszkowski, D. J.

1980s: Abdel-Kader, Adel; Cifuentes, Luis Arturo; Cifuentes, Luis Arturo; Farah, Osman Mohamed; Fox, Lewis E.; Howell, Barbara A.; Leu, David Jack; Lord, C. J., III; Masse, Ann Katherine; Mathur, Shashi; O'Donnell, James; Ogwada, Richard Ayoro; Pape, Edwin Henry, III; Scibek, John C.; Scotto, Susan; Scrivens, Paul R.; Shapiro, Neal; Stumpf, Richard Paul; Terchunian, Aram V.; Washburne, Catharine Lorena; Wells, Ian

University of Denver
Denver, CO 80208

2 Master's, 6 Doctoral

1910s: Ziegler, Victor

1970s: Clark, Robert Owen; Paludan, C. T. N.; White, Paul Gary

1980s: Kuhaida, Andrew Jerome, Jr.; Lahlou, Mourad; Rudd, Lawrence P.; Walker, Graham Thomas

DePauw University
Greencastle, IN 46135

2 Master's, 1 Doctoral

1970s: Eder, Richard M.; Price, Katherine H.; Weathers, L. Michael

Drexel University
Philadelphia, PA 19104

8 Doctoral

1970s: Bender, Joel R.; Kovacs, Sandor; McCabe, W. M.; Metry, Amir Alfi

1980s: Deutsch, William Louis, Jr.; Kam, Moshe; McElroy, John Joseph, Jr.; Panigrahi, Bijay Kumar

Duke University
Durham, NC 27706

150 Master's, 33 Doctoral

1960s: Boutwell, Gordon Powers, Jr.; Cleary, William James, Jr.; Colinvaux, Paul Alfred; Dill, Charles Edward, Jr.; Doyle, Larry James; Estes, Ernest Lathan; Field, Michael Ehrenhart; Jorgeson, Eric Charles; Judd, James Brian; Kier, Jerry Stephen; Koerner, Robert M.; Luternauer, John Leland; Mantuani, Mark Anthony; Molnia, Bruce Franklin; Morton, Robert Whelden; Offutt, Patrick A.; Stehman, Charles F.; Terlecky, Peter Michael, Jr.

1970s: Airola, T. M.; Al-Awkati, Z. A.; Arbogast, Jeffrey Scott; Audibert, Jean M. E.; Bennetts, Kimberly Robert Winter; Berelson, William M.; Bisson, M. A.; Blackwelder, Patricia Lurie; Bornhold, Brian Douglas; Brauer, Constance J.; Brauner, J. F.; Brown, Nicholas Arthur; Budd, David A.; Bush, David M.; Chapman, Diane; Crissman, Stephen C.; DeFelice, David; Diecchio, Richard Joseph; Ditty, Patrick Scott; Duda, A. M.; Edwards, Brian Douglas; Elmore, Richard Douglas; Fritz, Steven James; Fuerst, Samuel I.; Geen, Alfred Francis; Golden, Bruce L.; Green, Marsha A.; Grossman, Zev N.; Haines, Evelyn B.; Halsey, Susan Dana; Hamilton, Jean L.; Hamilton, Paul Lawrence; Harbridge, William Frank; Harding, John D.; Harmon, Carol Jean; Harvey, T. J.; Hastings, Susan Carol; Herbert, John R.; Holdship, S. A.; Holloway, D. M.; Kamens, John S.; Keer, Frederick Rhoades; Kilham, Susan Soltau; Klasik, John Arthur; Lee, S.; Leister, G. L.; Lighty, Robin Greg; Lyon, Robert B., Jr.; May, Jeffrey A.; Melack, J. M.; Mench, Patricia Anne; Miller, William C., III; Monrad, John R.; Moslow, Thomas F.;

Mueller, James Manning; Mullins, Henry T.; Murphy, Donald J., Jr.; Murphy, Philip Joseph; Nelson, Jeffrey Carter; Oliver, Steffenie Anne; Petersen, Richard Randolph; Phelps, Daniel Craig; Piotrowski, Robert G.; Pratt, Franklin Pierce; Rooney, William Stephen, Jr.; Russell, Steven Duffy; Sarle, Laura L.; Scarlett, Ervin Wesley, Jr.; Schneider, Rita; Schold, Gary Paul; Schrader, Edward Leon, Jr.; Skerky, Barbara Blanche; Sparks, Thomas Norton; Stanczyk, Dennis T.; Stout, Paul Michael; Susman, Kenneth R.; Thornton, Scott Ellis; Tsentas, Constantine I.; Van Tassell, Jay; Wagner, Keith Brian; Watkins, Jeffery Alan; Whitcomb, Natalie Jo; Wilber, Robert Jude; Zeff, Marjorie Lee

1980s: Adams, Michael Anthony; Allen, Marian Ruth; Anderson, Virginia Ruth; Barany, Istvan, Jr.; Benes, Paula S.; Bishop, Michele Gregg; Brackett, Robert Stevens; Brauer, Julie Fay; Brown, Jennifer Ruth; Burgess, Carl Foulds; Burney, David Allen; Burns, Donna Jane; Burns, Stephen James; Byle, Chris S.; Chen, Christina M.; Chen, Yao-Tang; Clemenceau, George Robert; Conti, Edward P.; Cook, Ariadne Helen Olga; Curtis, Patchin Crandall; Davis, Thomas Weldon; Doull, Mary Elizabeth; Dunkelman, Thomas Julian; Erwin, Parrish Nesbitt, Jr.; Flannery, James William, Jr.; Green, Carolyn A.; Haberyan, Kurt August; Haddad, Geoffrey Allen; Halfman, John David; Henderson, Virginia W.; Herrick, Dean H.; Hobert, Linda A.; Hoff, Jean Louise; Hokanson, Claudia L.; Horstmann, Kent M.; Jacobson, Susan Kay; Johnson, Andrea Marie; Kacena, Jeffrey A.; Katz, Scott D.; Kiefer, Karen Bernice; Leaver, June; Leonard, Lynn Ann; Lightner, Jon T.; Lincoln, Rush B., III; Lorber, Peter Mark; Lynch, Elizabeth Linfield; Manyak, David Michael; Marlay, Lisa Emerick; Martin, Ellen Eckels; Martin, Jonathan Bowman; Murray, Louis Charles, Jr.; Naidoo, Devamonie D.; Ng'ang'a, Patrick; Nilsson, Kristen; Nolet, Gilbert J.; Patterson, Marcus Brent; Polo, Jesus Miguel; Prehmus, Cynthia Anne; Proctor, Christian Jennings; Rach, Nina Marie; Reverman, Karla M.; Reynolds, David James; Richter, Daniel deBoucherville, Jr.; Sander, Stephan; Sayed, Sayed Mourad; Schaeffer, Charlotte Louise; Schorsch, Laurie Jane; Schwab, William Charles; Scott, Deborah L.; Shorb, William Murray; Smith, Roberta Lynn; Snyder, Michael Robert; Specht, Thomas David; Stager, Jay Curt; Steed, Robert W.; Steele, George Alexander, III; Versfelt, Joseph W.; Wieg, Paul Kenneth; Winters, Alec T.; Woodruff, John M.; Zeppieri, James Benjamin

Earlham College
Richmond, IN 47374

1 Master's

1900s: Lamar, F. S.

East Carolina University
Greenville, NC 27858

59 Master's

1970s: Ayers, Mark W.; Byrum, Scott R.; Christopher, Michael T.; Coble, James F.; Dayvault, Richard D.; Duque, Thomas A.; Gall, Daniel G.; Hartness, Thomas Scott; Maddry, John W.; Mauger, Lucy L.; Moorefield, Thomas P.; Pearson, Daniel R.; Slagle, E. S.; Spruill, Richard K.

1980s: Allison, Mead A.; Auch, Timothy W.; Bedell, Theodore E., III; Benton, Stephen B.; Bergen, Christopher L.; Blount, Jonathan G.; Campbell, Steven K.; Capps, Richard C.; Corbitt, Christopher L.; Corbitt, Lisa B.; Crowson, Ronald A.; Danahy, Thomas V.; DiRenzo, Vincent N., Jr.; Eames, Gary B.; Ellington, Michael D.; Gay, Norman Kennedy; Gray, Brian Erwin; Grundy, Allen T.; Hale, Walter R.; Hardaway, C. Scott; Hartsook, Alan F.; Hines, Robert A.; Indorf, Michael S.; Jones, William E.; Katrosh, Mark Ralph; Kirkland, Michael John; Koehler, Adrienne; Lewis, Don W.; Loftin, L. K.; Lyle, Michael; Mallette, Patrick M.; Mallinson, David J.; Moore, Teresa L.; Moretz,

Leonard C.; Pearson, D. R.; Powers, Eric R.; Privette, Robert W.; Roberts, Sarah Ann; Scarborough, A. Kelly; Schiappa, Christopher; Stewart, T. Lori; Varlashkin, Charlotte M.; Waters, Virginia J.; Wedemeyer, Richard C.; Workman, Robert R., Jr.

East Tennessee State University
Johnson City, TN 37614

1 Master's

1970s: Barrett, James Edward

East Texas State University
Commerce, TX 75428

16 Master's

1960s: Alexander, Gilbert R.

1970s: Berggren, C. F.; Blankenship, O.; Christensen, Michael W.; Dickey, B. C.; Dirin, J.; Dod, B. D.; Goebel, J. E.; Griffith, Charles E.; Lanmon, L. B.; McDaniel, P. A.; McKinney, R. D.; Meeks, R. L.; Pashuck, R. J.; Roy, R. H.; Shehabi, G.

Eastern Kentucky University
Richmond, KY 40475

109 Master's

1970s: Anderson, Robert W.; Bell, Vernon Lynn; Berkheiser, Samuel W., Jr.; Bolivar, Stephen L.; Burton, Dale M.; Cantrell, Carlton L.; Cobb, James C.; Currens, James Calvin; Davis, Michael W.; Dillenberger, Douglas S.; Ellsworth, George W., Jr.; Elton, William G.; Engelhardt, Richard Lee; Faulkner, Barry M.; Fink, Susan L.; Goble, Robert S.; Kuchenbuch, Pamela A.; Kuhnhenn, Gary L.; MacGill, Peter L.; MacGill, Rotha A.; Mandell, Wayne A.; Marashi, Hamidedin; Martin, Archie H., III; May, S. Judson; McLoughlin, Thomas F.; Merritt, Roy Dale; Moore, Linda J.; Neal, Donald W.; Niskanen, Keith A.; Norris, Marc J.; Phillips, Donald T., II; Sanders, Anthony W.; Sanders, Thomas P.; Sasso, Jane A.; Scott, Lewis P., IV; Scott, Thomas M.; Sergeant, Richard E.; Slone, George T.; Stevens, Stanley S.; Szymanski, William N.; Taylor, Frederick W.; VanArsdall, David E.; Walter, John R.; Weber, Lawrence C.; Wetmore, Clinton C.; Williams, David A.

1980s: Amr, Amr M.; Anderson, Richard L.; Beckner, Jeffrey S.; Bouknight, Dan L.; Bowling, Edward C., III; Bronner, Raymond L.; Burston, Michael R.; Clodfelter, Rebecca A.; Cook, Douglas R.; Cox, J. Michael; Cross, Wayne A.; Davidson, Oscar B.; Devilbiss, Thomas S.; Dugan, Thomas E.; Duncan, Truman E., Jr.; Furr, James E.; Gauthier, Marilyn; Gilchrist, William B.; Gooding, Patrick J.; Gorman, Kelly M.; Grosse, Charles W.; Gustafson, Thomas K.; Gustafson, Timothy J.; Henning, Russell J.; Howard, Lawrence H.; Jacobs, Brent B.; Johnson, Steven T.; Ketani, Rapheal V.; Lawal, Adetunji A.; Lazarsky, Jennifer J.; Lovitt, Ronald L.; Marcelletti, Nicholas; Milici, Alfred W.; Monroe, Charles Jr.; Moody, Jack R.; Morgan, Baylus K.; Morris, Frank R., IV; Ochsenbein, C. Douglas; Papp, Alexander R.; Paul, Dennis A.; Perkinson, Mary C.; Pieracacos, Nicholas J.; Price, Garry L.; Rea, Ronald G.; Romanik, Peter B.; Ross, Bruce C.; Schroder, Richard A.; Slucher, Ernie R.; Smath, Richard A.; Songer, Nathan L.; Spurgeon, Paul A.; Stickney, James Francis; Swinford, Edward M.; Tenharmsel, Ronald L.; Tillman, Joseph W.; Traub, John H.; Tully, Deborah G.; Unuiboje, Felix E.; Vincent, Harold R.; Wilson, James K.; Wilson, R. Wayne; Wilson, Richard T., Jr.; Young, Timothy J.

Eastern Washington University
Cheney, WA 99004

99 Master's

1970s: Arnold, Harold B.; Burke, Karl D.; Du, Ming-Ho; Finn, Dennis D.; Fritz, Lloyd G.; Hall, Richard J.; Harbour, Jerry L.; Hughes, Roger D.; Jayne, Douglas I.; Johnston, James S.; Kemple, Harold F.; Kennedy, Donald B.; Lachance, David

J.; Nesbit, Lee C.; Nickmann, Rudy J.; Nijak, Walter F.; Olson, Theodore M.; Pernsteiner, Robert K.; Powers, Mark William; Russell, Edward F.; Schipper, Louis B., III; Senter, Lance E.; Snyder, Edward M.; Visger, Frank J.; Vogel, Irene D.; West, Barbara J.; West, Jerry R.; Wright, Ernest George; Yost, Carl R.

1980s: Abrams, Mark J.; Anderson, Randall L.; Ansell, Mark Willis; Asmerom, Yemane; Barnett, Douglas B.; Benham, John R.; Besse, Linda; Biglow, Crague C.; Bjornstad, Bruce N.; Calder, Craig Paul; Cammarata, Thomas Joseph; Chiang, Robert Huai; Cho, Gin; Chung, Curtis Dean; Dennison, Douglas I.; Doak, Roger W.; Doh, Seong-Jae; Donovan, Sean; Dunn, Michael D.; Ernst, David Raymond; Ernst, Robert P.; Etienne, John E.; Freudenberg, Connie M.; Fullmer, Corey Y.; Garman, James E.; Gheddida, Mehemed S.; Griffin, Mark E.; Gulick, Charles Wyckoff, III; Hahn, Raimund; Hall, Bruce S.; Harz, Mary Catherine; Henderson, David W.; Hillmann, Bob; Janzen, John H.; Kietzman, Donald R.; Kinart, Kirk P.; Lewellen, Dennis G.; Lovett, Cole K.; Lucas, Harold E.; Luker, James A., Jr.; Martin, John D.; Martin, Kyle; Martin, Walt M.; Massa, Philip Joseph; McHugh, Brian; Mihalasky, Mark John; Milliken, Mark D.; Mork, Andrew R.; Nazarian, Mohammad H.; Nordstrom, Paul M.; Orlander, Peter R.; Orlean, Howard M.; Pan, Kuan-Chou; Rau, Robert L.; Reed, Darrell; Ridgway, Eric R.; Rupp, John Andrew; Schneck, William M.; Smith, Bruce L.; Steigerwald, Celia H.; Stevens, D. Scott; Tallyn, Lee Ann K.; Turner, Lawrence D.; Vande Kamp, Brad Douglas; Wang, Hau-Ran; Wang, Kong; Werle, James L.; Wicklund, Mark A.; Wilbur, J. Scott; Wingerter, Jeffrey Hush

Ecole Polytechnique
Montreal, PQ H3C 3A7

235 Master's, 35 Doctoral

1960s: Bertrand, Claude; Boissonnault, Jean; Crepeau, Pierre M.; Durand, Marc; Girard, Paul; Hardy, Richard; Joncas, Gilles; Lavoie, Clermont; Sylla, N'Fanly; Tanguay, Marc G.

1970s: Ally-Gregoire, Raymonde; Ballivy, Gérard; Bazinet, Robert; Bazinet, Robert; Beauchemin, Yves; Bergeron, Michel; Bergeron, Robert; Campiglio, Carlo; Carignan, Jacques; Chouteau, Michel C.; Desjardins, Jean-Pierre; Do Lam Sinh; Do, Lam Sinh; Dolle, Yves; Doucet, Daniel; Filion, Gilles; Filion, Gilles; Fortin, Normand; Foscal-Mella, Gabriel; Gagnon, Denis-Claude; Gauthier, Michel; Gauthier, Michel; Gentile, Francesco; Gentile, Fransesco; Guertin, Kateri; Harvey, Y.; Harvey, Yves; Kheang, Lao; Laforte, Marc-Antoine; Laifa, Embarek; Laifa, Embarek; Lambert, Roger; Lao, Kheang; Le Du, Raymonde; Mellinger, Michel; Poulin, R.; Poulin, Richard; Provost, Gilles; Provost, Gilles; Richard, Pierre; Robert, Jean-Marc; Roy, Jean; Sabourin, Raymond; Saint-Amant, Marcel; Sampara, Michel Ulrich; Spitz, Guy; St-Hilaire, Camil; Trudel, Pierre

1980s: Abet Yao, Marcel; Arbour, Guy; Barthelemy, Ernst; Beaudoin, Alain; Belisle, Jean-Marc; Bellehumeur, Claude; Bergeron, Michel; Bernier, Louis; Birkett, Tyson C.; Blanchard, Chrystian; Bouchard, Karl; Bourgault, Gilles; Bourget, André; Carboni, Salvatore; Carignan, Jacques; Cazavant, Alain; Chainey, Michel; Chakridi, Rachid; Chaouai, Nour-Eddine; Chartrand, Francis; Chartrand, Francis; Chouteau, Michel C.; Cloutier, Marc-André; Cossette, Denis; Couture, Diane; Daigneault, Réal; de l'Etoile, Robert; De Villiers, Johanne; des Rivières, Jean; Desbarats, Alexandre; Deschamps, Fernand; Doyon, Martin; Dugas, Helene; Durand, Benoit; Dussault, Chantal; Gagnon, Yves; Gaulin, Raymond; Gaumond, André; Gauthier, Michel; Ghanem, Youcef; Giguère, Christine; Gilbert, Michel; Gilbert, Michel; Grant, Maureen; Houde, Robert; Jenkins, Cecilia L.; Jutras, Marc; Kheang, Lao; Konan, Gilbert K.; Labrecque, Paul; Lacroix,

Robert; Lacroix, Sylvain; Lacroix, Sylvain; Lambert, Marc; Lao, Kheang; Laplante, Richard; Laverdure, Louise; Lebel, Jeanne; Losier, Lisanne; Marcotte, Denis; McCann, A. James; McCann, James; Méthot, Yves; Michel, Sylvestre Georges; Moreau, Alain; Nimpagaritse, Gérard; Prud'homme, Michel; Rainville, Serge; Rainville, Serge; Robert, François; Robert, Francois; Roy, Charles; Sea, Frédéric; Trottier, Jacques; Trudel, Pierre; Verly, Georges; Vu, Xuan-Lan; Wahnon, Ethel; Waitzenegger, Bernard

Emory University
Atlanta, GA   30322

96 Master's, 5 Doctoral

1940s: Arden, Daniel Douglas, Jr.; Butler, Howard Putnam; Cofer, Harland Elbert, Jr.; Grant, Willard Huntington; Pinson, William Hamet, Jr.; Pirkle, Earl Conley, Jr.

1950s: Albritton, John Allen; Blodgett, Jack W.; Buzarde, Laverne Ernest; Callahan, James Emmett; Cappel, Howard Noble, Jr.; Clement, William Gilbert; Crawford, Thomas Jones; Cribb, Robert Eugene; Crisler, R. M.; Darling, Robert William; Dicus, Joseph Martin; Gould, Joseph Charles; Grumbles, George Robert; Guttery, Thomas H.; Hanson, Hiram Stanley; Holland, Willis A., Jr.; Hooten, John Albert; Hurst, Vernon James; Hutcheson, Lewis Bryan; Ingram, Frank Thompson; Ingram, William Franklin; Jackson, Lawson Erwin, Jr.; King, James A., V; Kirkpatrick, Samuel Roger; Lamb, George Marion; Marquis, Urban Clyde; McClain, Donald Schofield, Jr.; Mitchell, William Louis; Moore, John Byron; Moore, William Halsell, Jr.; Murphy, Robert Edward; Owen, Vaux, Jr.; Pierson, Richard Edwin; Pound, James Hannon, Jr.; Pruitt, Robert Grady, Jr.; Renshaw, Ernest Wilroy; Rogers, Wiley Wamuel; Rosenfeld, Sigmund Judith; Schepis, Eugene Louis; Sheridan, John Thomas; Smith, James William; Smith, William LaRue; Stuart, Alfred Wright; Taylor, Henry G.; Truxes, Lee Sayles; Vest, Ernest Louis, Jr.; Walter, Kenneth Gaines; Wheeler, Garland Edgar; Windham, Steve R.; Wright, David Craig

1960s: Almand, Charles William; Bowen, Boone Moss, Jr.; Durham, David Peterson; Fountain, Richard Calhoun; Gardner, Charles Hardwood; Graham, Robin Spear; Higgins, Michael Wicker; Mohr, David Wildred; Ostrander, Charles C.; Preston, Charles Dean; Reade, Ernest Herbert, Jr.; Reighard, Kenneth Frederick; Sandlin, Walter Lee, Jr.; Schultz, Roger Stephen; Spalvins, Karlis; Wright, Nancy Elin Peck

1970s: Burbanck, George Palmer; Gray, Marion Glover; Hay, J. D.; Jones, Donovan Deronda, Jr.; Jordan, Larry Eugene; Jordon, Larry Eugene; Kendall, D. R.; Lefkowitz, Paul Allen; Meadows, George Richard; Mitchell, Jeffrey Leonard; Nunan, Walter Edward; Parks, William Scott; Prowell, David Cureton; Schalles, J. F.

1980s: Convert, Jean; Dean, Lewis Shepherd; Diprime, Leonard Joseph, Jr.; Dula, Philip Charles; Gottschalk, Marlin Ralph; McCullogh, Debia Hershelle Fine; Morris, William Lee; Nixon, Roy Arthur, III; Renwick, Patricia Louise; Ritter, John Ernest; Shaw, Thomas Howard; Shegewi, Omar M.; Stieve, Alice Leutung; Vargo, Elena Fisher; Zeigler, Lynn E.

Emporia State University
Emporia, KS   66801

10 Master's, 1 Doctoral

1960s: Breemer, Dale A.

1970s: Hodison, Starlyn T.

1980s: Abdelsaheb, Ibrahim Z.; Ezerendu, Friday O.; Hedstrom, Bradley L.; Law, Michael S.; Nutter, Brian L.; Riley, John M.; Saint Clair, Gregory M.; Sinnett, Donald L.; Thornton, William S.

University of Florida
Gainesville, FL   32611

170 Master's, 91 Doctoral

1950s: Auffenberg, Walter; Brown, Eugene; Darlington, Julian Trueheart; Weigel, Robert D.

1960s: Assefa, Getaneh; Baker, Robert Allison, III; Benedict, Barry Arden; Bonner, William Paul; Boulware, Joe Wood; Conklin, Carleton Veith; Cooper, David Michael; Cowart, James Bryant; Dimmick, Charles William; Dolliver, Claire Vincent; Echols, Ronald James; Eppert, Herbert Charles; Floyd, James Gordon; Garcia Bengochea, Jose Ignacio R.; Girard, Oswald Woodrow, Jr.; Goldsmith, William Alee; Hamon, J. Hill; Higgins, Brenda Baer; Hill, Raymond Leslie; Hirschfeld, Sue Ellen; Isphording, Wayne Carter; Jenkins, David Maurice; Kashirad, Ahmad; Keller, Roland Bradford; Kontrovita, Hervin; Langford, Neal Gerald; Lester, Robert Worth; Lewis, Jackson Ellis; Lipchinsky, Zelek Lawrence; Martin, Robert Allen; Maxey, Larry R.; McClellan, Guerry Hamrick; McCoy, John J.; McCullough, Louis Marshall; Meitzke, Jane Ellen; Mitchell, Charles Leonard; Mojab, Fathollah; Moore, Raymond Kenworthy; Mott, Charles James; Novak, Irwin Daniel; Rieckin, Charles Christopher; Scolaro, Reginald Joseph; Skirvin, Raymond Taylor; Sniffen, Jane; Teleki, Paul Geza; Tessman, Norman; Thompson, Jon Louis; Vormelker, Joel David; Witmer, Richard Everett

1970s: Adams, William David; Adenle, Oladepo Adeoye; Baldwin, Dorothy Esther; Beatty, Gwendolyn Faye; Beckman, Gary Lee; Bedient, Philip Bruce; Bennett, Kathleen C.; Burnson, Terry Quentin; Buros, Oscar Krisen; Calhoun, Frank Gilbert; Campbell, Kenneth Eugene; Carriker, Neil Edward; Carter, Anna Dombrowski; Cason, James Hubert; Ceryak, Ronald Joseph; Coleman, James Mark; Cook, Gregory Allan; Copeland, Richard Evan; Crews, Patricia Ann; Cummings, George Howard, III; Cutright, Bruce Lee; Davis, Michael Paul; Dixon, Frederik Sigurd; Eger, James Douglas; Eichler, Gary Edward; Emhof, John Warren; Evans, Andrew Joseph; Feast, Charles Frederick; Fein, Michael Neal; Fenk, Edward Michael; Fisher, Darrell Reed; Frazier, Michael K.; Fuller, Wayne Ross; Gartland, Jeffrey Dale; Garvey, Michael Joseph; Gibson, Christy Rae; Gillespie, Dennis Patrick; Gilliland, Martha Winters; Glass, John Patrick; Goldstein, Susan Twyla; Gregory, Robert George; Grinnell, Philip Collins; Gunning, S. P.; Halvatzis, Gregory James; Harper, John Andrew; Hecker, Kenneth E., Jr.; Hess, Susan Lynne; Hickey, Edwin Weyman; Hoenstine, Ronald W., Jr.; Hoganson, John William; Holly, James Benjamin; Hunt, David K.; Jack, Howard Corwin; Jackson, Dale Robert; Johnson, Larry Douglas; Johnson, Richard Alan; Jones, Douglas Frank; Kirby, Edward George, III; Krantz, Gary Wayne; Labowski, James Lawrence; Lawrence, Marc Arnold; Lentell, Randall Lynn; Levy, Alexandro Gustavo; Limoges, L. D.; Liu, Jeun-Shyang; Lukas, Theodore Chris; Marcus, Steven Roy; Markun, Charles Daniel; Martin, Ronald Edward; McKinney, Curtis Ross; Meeder, John Frank; Messer, Jay James; Metrin, Deborah B.; Mims, Jimmie Floyd; Miura, Ryosuke; Moore, Mark Power; Morgan, Gary Scott; Morris, Frederick W., IV; Nettles, Norton Sandy; Nickens, Dan Alan; Nottingham, Larry Curtis; Nuckels, Clarence Edward; Ogden, Palmer Raphael, Jr.; Palacios, Alejandro; Persaud, Naraine; Pirkel, Fredric Lee; Pourzadeh-Boushehri, Jalil; Pratt, Daniel Allen; Ripy, Bruce Johnson; Rogers, Caroline Sutherland; Ruppel, Stephen C.; Russell, John Phillip; Saroop, Hayman Cecil; Sarver, Timothy John; Schmidt, Allan Thomas; Sears, Stephen O'Reilly; Segretto, Peter Ssalvatore; Siefken, David Lee; Smith, Russell Clarence; Soper, Donald Arthur; Stone, Gary Calvin; Swindler, James P.; Taylor, George Johnston; Valleau, Douglas Nelson; Vaughan, Hague Hingston; Vespucci, Paul Daniel; Walton, Todd Leon, Jr.; Wicker, Russell Alan; Williams, Kenneth

E.; Woessner, William Wendling; Zachos, Louis George

1980s: Aardema, James A.; Abbott, Tom Austin; Abernathy, Sarah Allison; Abu Bakar, Othmanbin; Angley, Joseph Timothy; Baker, Lawrence Alan; Baker, Philip Craig; Baskin, Jon Alan; Becker, Jonathan, J.; Bloom, Jonathan I.; Bodge, Kevin Robert; Boghrat, Alireza; Brenner, Mark; Brown, Mark Theodore; Bryan, Jonathan R.; Burklew, Richard Hill, Jr.; Butts, Christopher Lloyd; Byers, Gerald Eugene; Carson, Matt Wayne; Cline, Patricia V.; Diblin, Mark C.; Dierberg, Forrest Edward; Dufresne, Douglas Paul; Emslie, Steven Douglas; Gaston, Lewis Andrew; Goforth, Gary F. E.; Goode, Richard Whitfield, III; Gupta, Ramesh Chandra; Hancock, Michael Curtis; Handel (Tutt), Donna J.; Hansen, John Kenneth; Hayter, Earl Joseph; Hearn, Douglas James; Heimburg, Klaus Frederick; Hood, Larry Quentin; Hughes, Steven Allen; Hulbert, Richard Charles, Jr.; Jacome Villanueva, Enrique Osmar; Japy, Kate Elisabeth; Jarrett, Marcus Lee; Jenkins, Dwight; Jex, Garnet Wolseley; Johnson, John Clifton; Kalisz, Paul John; Kangas, Patrick Carl; Kanis, Oak K.; Koulekey, Kodjo C.; Lafrenz, W. B.; Lazo, Edward Nicholas; Lin, Chen Hsin; Lin, Chung-Po; Liu, Ko-Hui; Lytton, Rome Gaffney; Maa, Peng-Yea; Maalel, Khlifa; Mallard, Elliott A.; McKenzie, Harvey Kenneth; McKinney, Michael L.; Mitchell, Charles Clifford, Jr.; Myers, Ronald Lewis; Ng, Elliot Kin; Ntokotha, Enock Mangwiyo; Ogburn, Reuben Walter, III; Opper, Steven Carl; Ovalles, Francisco Antonio; Papadopoulos, Panayiotis Charilaou; Parchure, Trimbak Mukund; Paterek, James Robert; Pratt, Ann E.; Quinn, Michael J.; Rose, Seth E.; Ross, Mark Allen; Rushton, Betty Toombs; Sawyer, Robert Knowlton; Schult, Mark Frederick; Seereeram, Devo; Self, Gregory Alan; Sharpe, Charles Lee; Short, David George; Smathers, Nancy Preas; Smith, Matthew Clay; Smith, Michael Forrest; Sompongse, Duangporn; Sprague, Charles Warren; Stanley, Daniel R.; Stevens, Anthony John; Stockhausen, Edward Joseph; Stone, Kenneth Coy; Stross, Richard Anthony; Struthers, Robert Allen; Tabatabai, Habibollah; Tuschall, John Richard, Jr.; Wiener, Jacky M.; Yang, Hsi Chi; Yekini, Bourahm Bourahim; Zapata, Raul Emilio; Zhu, Bingfu

Florida Institute of Technology
Melbourne, FL   32901

26 Master's, 3 Doctoral

1970s: Bates, Peter Paul; Beazley, Robert W.; Browne, David Richard; Capaldo, Paul S.; Carey, Max Raymond; Cohenour, Bernard C.; Daggett, Joyce M.; Dallemagne, Pierre G.; Gallop, Roger G.; Haeger, Steven D.; Hale, David A.; Kennimer, Mary Ann Y.; Komelasky, Michael Charles; Kriegel, Robert V.; Lawson, Robert A.; Reed, Timm L.; Richardson, David Bruce; Roulo, David L.; Sramek, Steven F.; Takayanagi, Kazufumi; Tomasello, Richard S.; Tower, Deborah A.; Trees, Charles Connett; Tyler, Jeremy Guy Anthony; Woodsum, Glenn Craig; Zarkanellas, Antois J.

1980s: Metz, Simone; Pierce, Robert Remsen; Shieh, Chih-Shin

Florida State University
Tallahassee, FL   32306

213 Master's, 87 Doctoral

1950s: Andrews, Franklin; Brenneman, Lionel; Cazeau, Charles Jay; Enrich, Grover H.; Haley, Francis L.; Haley, Patrick C.; Hendry, Charles W., Jr.; Kirby, James P.; Lammers, George; Lapinski, William James; Larsen, Alfred L.; McCoy, Henry J.; McKnight, William M., Jr.; Meyers, William D.; Miller, Lawrence F.; Milton, Jesse W.; Mullins, Allen T.; Nettles, James Edward; Raasch, Albert C., Jr.; Revell, Steve; Roberts, William B.; Saffer, Parke E.; Vause, James E., Jr.; Waskom, John D.; Yon, James W., Jr.

1960s: Barackman, Milan A.; Barnum, Dean C.; Bell, David L.; Benda, William K.; Blair, Donald; Brogden, William B., Jr.; Carpenter, John Richard; Carpenter, John Richard; Chen, Chih Shan; Earley, Charles F.; Edgar, N. Terence; Edwards, Dennis S.; Emerson, John Wilford; Evans, Richard G.; Fennell, Edward L.; Finney, Vernon Lee; Fisher, Victor A.; Fleece, James B.; Garman, Ray Keith; Gleece, James B.; Goldsmith, Victor; Grant, Hogn Bruce; Gremillion, Louis Ray; Gresens, Randall Lee; Grose, Peter; Holmes, Charles Ward; Holmes, Charles Ward; Horvath, George; Huang, Ter-Chien; Huddlestun, Paul Francis; Hyne, Norman John, Jr.; Kahn, Michael I.; Kaufman, Matthew Ivan; Kirtley, David; Kofoed, John W.; Koster, Samuel; Kozo, Thomas; Kramer, Kenneth Francis; Kronfeld, Joel; Kunzler, Robert Henry; Lee, Wang Chih-ming; Malloy, John; Margolis, Stanley; Mark, Norman; Martens, Christopher; Mather, Thomas T.; Mathews, Haywood; McCutchen, William T.; Meyland, Maurice A.; Milligan, Donald B.; Mitterer, Richard Max; Noble, David Frederick; Norman, Mark; Olds, T. S.; Oman, Carl Henry; Parker, Neal M.; Paster, Theodore Phillip; Pastula, Edward J.; Pflum, Charles E.; Pilkey, Orrin H.; Pollard, Charles Oscar, Jr.; Pollard, Lin Davis; Reynolds, William Roger; Reynolds, William Roger; Richey, James M.; Ross, Landon T.; Ross, Landon T., Jr.; Rydell, Harold Stanford; Rydell, Harold Stanford; Schnable, Jon Edwin; Severance, Robert W.; Shannon, Dennis L.; Shier, Daniel Edward; Stahl, Lloyd E.; Stewart, Richard; Stone, James A.; Stonebraker, Jack Douglas; Sward, Cynthia A.; Taylor, Vernon A.; Vanstrum, Vincent B.; Vaos, Stephanos Pantelis; Vickers, Michael A.; Wallick, Edward Israel; Whisonant, Robert Clyde; Whisonant, Robert Clyde; Whitton, Elliott

1970s: Anderson, John B.; Andren, Anders Wikar; Andress, Noel Eugene; Bauer, David Thomas; Beckett, M. Patricia; Berg, W. W., Jr.; Briel, Lawrence I.; Burk, Roger; Burkett, David; Busen, Karen E.; Busen, Ken; Campbell, Kenneth M.; Chaiffetz, Michael; Chaki, Susan J.; Chapuis, Ralph A.; Ciesielski, Paul F.; Ciesielski, Paul F.; Coe, Curtis; Cohen, Lynne; Coleman, Craig J.; Colton, Richard C.; Constans, Richard E.; Cook, N. W.; Covington, Sidney L., Jr.; Cowart, James B.; Daugherty, Gregory Lynn; DeFelice, David R.; Deininger, Donald T.; Drehle, William; Dreyer, Charles; Emmerling, Michael Dean; Entsminger, Lee; Ertel, John R.; Fellers, Thomas J.; Fetter, Franklin; Filewicz, Mark V.; Galicki, Alan M.; Geitzenauer, Kurt R.; Gitschlag, Gregg; Goetschius, David W.; Goldstein, Robert Fritz; Gombos, Andrew M., Jr.; Gram, Ralph; Guy, Jerry L.; Hajash, Andrew, Jr.; Hajishafie, Manoutcher; Hammer, Jay A.; Hart, T. L.; Horvath, George J.; Huang, Ter-Chien; Hurlburt, Harley Ernest; Husain, Mohammad Asghar; Husted, John Edwin; Immel, R. L.; Immel, Robert L.; Keany, John; Kearsley, Fie; Ketchen, Harold G.; King, Keith; Kirtley, David W.; Kish, Stephen; Kraemer, Thomas F.; Kwader, Thomas; Lader, Gary R.; Lawson, Douglas R.; Lindberg, S. E.; Lisco, Neil; Ma, C. M.; Mackensie, Duncan T., III; Magley, Wayne C.; Maxon, Jonathan Rolfe; Maxwell, Robert; May, James P.; Missimer, Thomas M.; Mitchell-Tapping, Hugh J.; Mitchell-Tapping, Hugh J.; Miyajima, Melvin H.; Mohajer-Ashjai, Arsalan; Monastero, Francis C.; Montgomery, R. T.; Mooney, Thomas Rodney; Na, J.; Niedoroda, Allen W.; Ogden, George Malcolm; Osburn, William L.; Outler, Brenda; Peck, Douglas M.; Peng, C. Y.; Phillips, Stanley Michael; Raymond, William F.; Russell, Gail S.; San Juan, Francisco Claudio, Jr.; Schmidt, Walter; Schornick, James C., Jr.; Setlow, Loren W.; Shultz, David James; Silberman, Louis; Sivaraman, Tirupattur V.; Slater, David H.; Stapor, Frank W., Jr.; Stonebraker, Jack Douglas; Sutton, C.; Thompson, James D., Jr.; Turner, R. R.; Turner, Ralph R.; Vanderwood, Timothy B.; Walton, Frank Dennis; Wayne, Christopher J.; Weaver, Fred M.; Weaver, Fred M.; Weaver, Trinchitella Marianne; Weimer, Robert H.; Weinstein, Robert P.; Weintraub, Gary S.; Williams, George K.; Wind, Frank; Wind, Frank H.; Zemmels, Ivar

1980s: Anderson, Thomas; Armstrong, John; Ausburn, Kent; Ausburn, Mark P.; Bedosky, Stephen J.; Bergen, James A.; Bergen, James Alan; Bergmann, Peter C.; Bolling, Sharon J.; Byrd, James Tillman; Byrne, Christian Jean; Clark, David Raymond; Clark, Murlene Wiggs; Crane, James John; Croft, Melvin; Cummins, Laura Elaine; Cushman-Roisin, Mary; D'Andrea, Julie; Dabous, Adel Ahmed; Deetae, Suchint; Defant, Marc J.; Dobbs, Frederick Courtrite; Drummond, Mark Stephen; Ferek, Ronald John; Franks, Bernard Jeffrey; Gerami, Abbas; Goldstein, Elaine; Harwood, David M.; Hattner, John George; Hoenstine, Ronald Woodrow; Holland, Willis Algeon, Jr.; Huddlestun, Paul Francis; Hull, Robert; Hummel, Richard; Humphreys, Cynthia; Kaul, Lisa Wells; Keen, Timothy R.; Kety, Irvin; Kim, Kee Hyun; Kirkpatrick, Gerald; Klinzing, Susan L.; Knuttle, Stephen; Korosy, Marianne; Kwader, Thomas; Lang, Thomas; LeHuray, Anne P.; Leroy, Ronald; Muza, Jay P.; Neale, John; Parker, Mary; Peacock, Steve; Petty, Steven Matthew; Pontigo, Felipe A., Jr.; Reik, Barry; Roe, Kevin K.; Rupert, Frank; San Juan, Francisco Claudio, Jr.; Schade, Carleton; Schmidt, Walter; Schneider, Harvey Ira; Schramm, Linda Sue; Scott, Thomas Melvin; Sivaraman, Tirupattur V.; Socci, Anthony D.; Sol, Ayhan; Spicola, John; Vairavamurthy, Appathurai; Vaos, Stephanos Pantelis; Watkins, David Kibler; Whittington, David; Yonover, Robert

Fordham University
Bronx, NY 10458

1 Master's, 3 Doctoral

1930s: Crowley, Mark Thomas

1940s: Kovach, Edward Michael

1960s: Lugay, Josefina

1970s: Scanlon, Valerie C.

Fort Hays State University
Hays, KS 67601

32 Master's

1930s: Runyon, H. Everett

1950s: Drees, Linus S.

1970s: Bretz, Richard F.; Caprez, Lionel Preston; Drees, Robert Henry; Fisher, R. D.; Fritz, T. R.; Gnidovec, D. M.; Kolb, K. K.; Kovach, James Thomas; Niermeier, V. D.; Seyrafian, A.; Tallan, M. E.; Wallace, Kenneth C.; Zehr, Danny D.

1980s: Bermudez, Vilma Isabel Perez; Conley, Steven J.; Crooks, Deborah Marie; de Albuquerque, John Stephen; DeBoer, Daniel Alan; Desai, Pankaj; Folwarczny, Joseph James; Groneck, John E., III; Heimann, William Henry; Jepsen, Karl Oscar; LaGarry, Hannan E.; Louden, Robert James; Podorsky, Robert A.; Pomes, Michael L.; Stimac, John P.; Thompson, Nils Wilder; Williams, Jack Edward

Franklin and Marshall College
Lancaster, PA 17604-3003

18 Master's, 1 Doctoral

1960s: Cadwell, Donald Herbert; Campbell, Lyle David; DeWindt, Justus Thomas; Diehl, Paul Emmett; Gill, James Burton; Kier, Robert Spencer; Love, George Edmond Wilson, Jr.; McKelvey, Gregory Ellis; Rios, Juan H.; Ten Brink, Norman Wayne

1970s: Berg, Jonathan Henry; Brown, Dorothy Claire; Collins, Margaret R.; Gresh, Roger Theodore; Holbrook, Philip William; Lewis, Ralph S.; Miles, Christine E.; Sylvester, Steven J.; Talley, John H.

George Washington University
Washington, DC 20052

88 Master's, 57 Doctoral

1920s: Bowles, Oliver; Gidley, James William; Hansen, George H.; Holmes, Grace Bruce; Mansfield, Wendell Clay; Melcher, Charles Francis; Pohl, Erwin Robert; Stearns, H. T.

1930s: Arnaud, Elaine P.; Baulsir, George Edward; Easton, William Heyden; Edwards, Ira; Emmerich, Harry Henry; Jespersen, Anna; Jobe, William T.; Johnston, William D., Jr.; Kruger, Gustav Otto, Jr.; Manion, Ester Ann; Martin, Robert Joseph; Miller, John Charles; Mutchler, W. H.; Norris, Mary Lillian; Reavis, Betty Hill; Ridgway, Robert H.; Runner, Delmar G.; Stumm, Erwin Charles; Thom, Emma N.; Werner, Walter C.; Williams, James Stewart; Williard, John Earlton

1940s: Ahrens, Thomas Patrick; Baker, William Samuel; Barker, William Samuel; Gilkey, Earle Will; Howard, George Wilberforce; Jaster, Marion Charlotte; Moran, William Edward

1950s: Seeger, Charles Ronald

1960s: Amenta, Roddy V.; Baldauf, Phillip Dayle; Beauchamp, Robert G.; Borella, Peter Edward; Briggs, William Melrose, Jr.; Cargill, Simon M.; Collier, Frederick J.; Gassaway, John Duncan; Gassaway, John Duncan; Goett, Harry; Hoover, Earl Gerald; Jackson, Jeremy; Kouns, Charles Wilmarth; Lampiris, Nicholas; Lindskold, John Eric; Loring, Richard Blake; Miller, Eldon S.; Papaspyros, A. G.; Pasley, Douglas C.; Rabchevsky, George A.; Salahi, Dirgham Rida; Schroeder, Johannes Herbert; Stancoiff, Andrew Simeon; Stephens, George C.; Stone, Irving Charles; Stone, Irving Charles, Jr.; Vaz, Jesus Eduardo; Wasilewski, Peter Joseph

1970s: Adabi, Mohammad H.; Adenle, O. A.; Al-Temeemi, A. Y.; Alsup, Stephen Alex; Blackwelder, Blake W.; Blackwelder, Blake Winfield; Bloch, S.; Blodget, H. W.; Blodget, Herbert; Boosman, Jaap Wim; Chapelle, Francis H.; Clarke, R. S., Jr.; Dewall, A. E.; Dionne, J. V.; Ekstrom, Carol; Fagin, Stuart; Feigenson, Mark D.; Field, M. E.; Finkelman, Robert Barry; Fritz-Miller, Molly; Giles, Alice B.; Hailer, J. G.; Hazlett, Jean; Hearn, P. P.; Helmold, K.; Holecek, Thomas J.; Huffman, Arlie C.; Huffman, Arlie C., Jr.; Klavans, A. S.; Koch, C. F.; Koch, Fred; Kravitz, J. H.; La Piana, Margo K.; Martell, J.; Mielke, James Edward; Miller, C.; Miller, M. F.; Morton, Robert Wheldon; Muller, Eric Charles; Perry, Richard Baker; Peter, George; Peters, James F.; Rabchevsky, George A.; Roberts, William; Rye, Raymond T., III; Sayala, D.; Sperandio, R. J.; Thom, M.; Valentine, Page; Weems, R. E.; Weissman, M.; Wetlaufer, Pamela H.; Wright, Thomas O.; Wright, Thomas O.

1980s: Akbarpour, Abbas; August, Lisa Layne; Baedecker, Mary Jo; Carter, Virginia Perkins; Chapelle, Francis Hughes; Crowley, Sharon S.; Dionne, J. V.; Gore, Pamela Jeanne Wheeless; Johnson, James Weldon; Kelafant, Jonathan Robert; Kravitz, Joseph Henry; Kulyk, Valerie-ann; Lee, Jacqueline San Miguel; Miller, Jeffrey Peter; Monet, William Francis; Ruppert, Leslie F.; Santas, Photenos; Saylan, Serif; Scott, Edith Elizabeth Bohrer; Soller, David Rugh; Vespucci, Paul Daniel; Watters, Thomas Robert; Wilson, Frederick Albert; Wu, Arthur Han; Zilczer, Janet Ann

University of Georgia
Athens, GA 30602

183 Master's, 73 Doctoral

1960s: Ashbaugh, Alexander Cleveland; Austin, Roger Seth; Bailey, Arthur C., Jr.; Cook, Robert Bigham, Jr.; Evenden, Leonard Jesse; Fouts, James Allen; Gergel, Thomas Joseph; Gergel, Thomas Joseph; Giles, Robert Talmadge; Jinks, Douglas David; Klett, William Young; Lawton, David Edward; Levy, John Sanford; Logan, Thomas Francis, Jr.; Matthews, Vincent, III; McLemore, William Hickman; McSween, Harry Y., Jr.; Medlin, Jack

Harold; Millians, Robert Wilson; Myers, Carl Weston, II; Poole, Donald Hudson; Rihani, Rushdi F.; Simmons, William Bruce, Jr.; Tiwari, Suresh Chandra

1970s: Aniya, M.; Aranz, William B.; Arnold, Anthony J.; Austin, Roger Seth; Basan, Paul B.; Baskin, George D.; Burns, R. G.; Chalmers, A. G.; Cook, Robert Bigham, Jr.; Courtney, P. S.; Dallmeyer, Mary D. Gilmore; de Araujo Filho, Jose Oswaldo; Deery, John Richard; Doughtery, Daniel; Duever, Michael James; Dupuis, Roy H.; Edwards, James Michael; Eppihimer, Richard M.; Ferens, Mary C.; French, Leanne Sue; Glass, Douglas Edward; Greer, Sharon A.; Gunter, William L.; Hartley, Marvin Eugene, III; Hayden, R. S.; Head, C. M.; Henry, Richard Lee; Hess, James R.; Horowitz, Carol G.; Humphrey, Ronald C.; Hunt, Jesse L., Jr.; Hunter, Donald Reid; Johnson, Milford Ronald; Kaplan, David Mark; Kilbourne, Richard T.; Kilbourne, Richard T.; King, G. M.; Letzsch, W. Stephen; Libby, Stephen Charles; Luckett, Michael A.; Martin, Benjamin Frank, Jr.; Mathieu, R. J.; Maybin, Arthur H., III; McLemore, William Hickman; Miller, R. W.; Moore, Sally Pennington; Morgan, Warren P., Jr.; Moritz, Harold W.; Nash, Robert E.; Needham, Robert Edmund; O'Kelley, Robert V.; Oliveira, Jose A.; Paris, T. A.; Pferd, Jeffery William; Potluri, Ramamohan Rao; Prather, Jesse Preston; Price, Edwin Henry; Reusing, Stephen P.; Richter, Dennis Max; Rihani, Rushdi F.; Robinson, Gene D., Jr.; Robinson, Nelson M., Jr.; Rogers, Lewis F.; Roth, Janet; Rozen, Robert W.; Ruff, Barbara L.; Saunders, James Alexander; Schroder, Charles H.; Schultz, Frederick J.; Scott, Ralph Carter, Jr.; Scott, Richard M.; Shapre, Roger Dale; Sharpe, R. D.; Smith, C. L.; Temples, Tommy Joe; Thurmond, Carol J.; Trimble, Stanley Wayne; Troendle, C. A.; Webster, J. R.; Wilson, James Lee; Woolsey, James R., Jr.; Woolsey, James R., Jr.; Zarillo, Gary A.

1980s: Abrams, Charlotte E.; Ahmad, Fachri; Ainsworth, Calvin Carney; Akbari, Gholamlali Estahbanati; Al-Sanabani, Gaber Ali; Alcorn, Stephen Richard; Allison, Eric T.; Ames, Richard M.; Annis, Malcolm Paul; Auble, Gregor Thomas; Bailey, William M.; Baker, Donald John; Baldasari, Arthur; Barba Pingarron, Luis Alberto; Barker, Gregg S.; Barker, William Wayne; Beaulieu, Giselle M.; Bernier, Pierre Yves; Biggs, Thomas H.; Blaeser, Christopher; Blay, Oliver T.; Blood, Elizabeth Reid; Brazell, Thomas Nathan; Brock, John Campbell; Brock, M. Michael; Bulfinch, Douglas L.; Bynum, Fred J., Jr.; Cheang, Kok Keong; Chernow, Robert Michael; Cowart, Jack H.; Cuffney, Francie Lou Smith; Culp, Randolph Alan; Cumbest, Randolph Josh; Daniel, Barbara J.; Davidson, John W.; Davis, Gary J.; Davis, Jerry Douglas; Davis, Kenneth R.; Davison, Fred C., Jr.; Dean, Nancy E.; Delia, Ronald G.; Donahue, John C.; Dorais, Michael John; Duncan, Glen A.; Durocher, Marcel Elzear Emery; Earley, Mark A.; Efteland, Jon Norquist; Elrod, Mary Melinda; Fay, William Martin; Fitzpatrick, Kathleen; Flebbe, Patricia Ann; Foley, Francis Daniel, Jr.; Folkoff, Michael Edward; Forsthoff, Harry S.; Garcia, Joseph E.; Gibb, Dorothy Margaret; Gilliam, William W.; Gillon, Kenneth A.; Godfrey, Pamela E.; Goldstein, Stuart B.; Grinstead, Gary Patrick; Groce, John A.; Grove, John Hamman; Gruber, Paul; Guthrie, Verner Noel; Hames, Willis Emory; Hamilton, David Bennett; Hardy, Lisa Steward; Henderson, Stephen William; Hendrickson, Ole Quist, Jr.; Hoyt, Gregory Dana; Hudson, Thomas Allen; Hurst, Marc Vernon; Hyatt, Robert Allen; Idris, Faisal Mohamed; Jacobs, Elliott B.; Jones, Tracy; Jost, Hardy; Kamola, Diane L.; Kania, Henry Joseph; Kaufmann, Robert A.; Kellam, Jeffrey A.; Kim, Choon-Sik; Kline, Stephen Warren; Kremer, Thomas; Langmyer, Kathryn; Lapallo, Christopher M.; Lens, Larry F.; Lin, Thomas T. H.; Lott, Thomas L., Jr.; Lovingood, Daniel; Manning, Philip L.; Martinez, Jaime Orlando; Massad, Marilyn L.; McClain, William R.; McFadden, Stephen S.; Mercer, David Morris; Millar, William Winston; Miller, David Martin; Miller,

John William, Jr.; Morris, Michael M.; Morrison, Jean; Moskow, Michael Gideon; Murphy, Barbara Ann; Murphy, Sean C.; Murphy, Susan Hope; Nagel, Steven P.; Noel, J. R.; O'Leary, William J.; Obenshain, Karen R.; Osborn, Noel Irene; Owen, Jerry A.; Papacharalampos, Demetrios; Paulson, Gary David; Potter, Phillip M.; Rabek, Karen Elaine; Reddy, Ramesh Kumar T.; Redwine, James C.; Reid, Jeffrey Clinton; Rindsberg, Andrew K.; Roberts, Malissa A.; Scarborough, Charles T.; Shaw, Cynthia A.; Sheehan, Michael Anthony; Shellebarger, Jeffrey; Sibley, Michael J.; Smith, David M.; Smith, Jennifer Margaret; Sprague, Edward K.; Standridge, Mark C.; Steflik, Martin; Steinhilber, Patricia Mary; Sun, Chin-Hong; Tan, Kuo-Yu; Tripp, William James; Turner, William L.; van Middelaar, Wilhelmus T.; Van Nostrand, Amy K.; Wadsworth, Joseph Rogers, Jr.; Watwood, Mary Elizabeth; Wells, Douglas E.; Wheatcroft, Robert Arthur; White, Roger Lee; Wilkinson, Peter H.; Wooten, M. W.; Wooten, Richard Mark; Yu, Keun Bai; Zimmerman, Marc James; Zippi, Pierre A.

Georgia Institute of Technology
Atlanta, GA 30332

81 Master's, 32 Doctoral

1930s: Wollard, George Prior

1960s: Abdul-Latif, Numan A. R.; Neilson, Frank Murray; Schwartz, Arnold Edward; Williams, Ronald Calvin

1970s: Arnonne, Robert; Benoit, Jeffrey R.; Bhate, Uday Ramesh; Bigham, Gary Neil; Bloomer, Daniel R.; Bridges, Samuel Rutt; Broekstra, Bradley R.; Butler, James M.; Champion, J. W., Jr.; Chen, Ching-Rua; Cooke, Gary; Denman, Harry Edward, Jr.; Dooley, Robert E.; Dunbar, David M.; Dunn, Townsend H.; Faust, Nicholas L.; Fulford, James Kenny; Gevrak, Ihsan; Guinn, Stewart A.; Highsmith, Patrick B.; Holcomb, Derrold; Hsiao, Helmut Y. A.; Intraprasart, S.; Junhavat, Suphachai; Kean, Alan E.; Liscum, F.; Martin, Steven James; Mathur, Uday Prakash; Maye, Peter Robert, III; McKee, John H.; Mullins, B. M.; Munasifi, Wasim G. A.; Nance, Steven; Neiheisel, James; Obaoye, Michael Olajide; Palaniappan, E. A. C.; Patterson, Patricia L.; Piccola, Larry J.; Rahn, William R., Jr.; Rawls, W. J.; Rice, Donald L.; Rothe, George Henry, III; Saad, A. A.; Sedivy, Robert Alan; Volz, William Richard; Waslenchuk, Dennis G.; Webb, Lyndall

1980s: Alexander, C. Shafe; Allison, Jerry Dewell; Bean, David; Bergantz, George; Billington, Edward; Caines, Gary L.; Chang, Ker-Chi; Chew, Chye Heng; Chin, David Arthur; Collins, Kenneth Robert; Collins, Steve Allan; Conner, Trent; Creamer, Frederick; Cunningham, James P.; Draper, Stephen Elliot; Duckworth, Robert M.; Frazier, James Edward; Friddell, Michael S.; Hall, Anne Marie; Harsha, Senusi; Hassanipak, Ali Asghar; Herbert, James C.; Hernandez, Heroel; Hinton, Douglas; Hornbeck, David Earl; Hovland, Nancy K.; Huff, Glenn; Jessup, David; Johnson, Anthony; Johnston, Gregory Lamar; Jones, Frank Burdette; Kuang, Jian; Lee, Chang Kong; Liow, Jeih-San; Ngoddy, Adaeze; O'Nour, Ibrahim; Ogilvie, Jeffrey; Ormsby, Marka R.; Padan, Ady; Parks, William Scott, Jr.; Patterson, Patricia Lynn; Petelka, Martin Frank; Propes, Russell; Radford, Wilbur Edward; Ross, Barbara; Smith, Gordon Egbert; Springfield, Charles Winston, Jr.; Steigert, Frederick; Storti, Frank W.; Thoroman, Marilyn; Tie, An; Tschirhart, Rochie; Tzeng, Wen Shyr; Wahlig, Barry Glenn; Wallace, Blanche; Watson, Thomas; Whang, Jooho; Wilson, Jeffrey Kent; Winester, Daniel; Zakikhani, Mansour; Zelt, Karl-Heinz

Georgia State University
Atlanta, GA 30303

4 Master's, 1 Doctoral

1980s: Lawrence, Thomas Spencer; Mallary, McKenzie; Piette, Robyn A.; Sullivan, John Denis; Torres, Max Antonio

University of Guelph
Guelph, ON N1G 2W1

18 Master's, 14 Doctoral

1960s: Asamoa, Godfried Kofi

1970s: Acton, Clifford John; Chatarpaul, L.; Chege, A. M.; Cox, G. L.; Crosson, L. S.; Fox, C. A.; Hawkins, R. K.; Heath, C. G.; Hilliard, B. C.; Hons, D. B.; Jowett, A. A.; La Hay, K. L.; Laryea, K. B.; Nicholaichuk, W.; Nik Wan, N. M. B.; Patterson, G. T.; Patton, G. D.; Raad, Awni Tewfiq Saleh; Tarzi, J. G.; Venkataraman, Sundaram

1980s: Abboud, Salim A.; Bélisle, Jacqueline; Clarke, K. E.; Garrett, R.; Grinham, David F.; King, W. Allan; Loi, Kuong-Soon; McBride, Raymond Allan; Montgomery, A. N.; Reynolds, William Daniel; Robin, Michel J. L.

Harvard University
Cambridge, MA 02138

7 Master's, 539 Doctoral

1870s: Benton, Edward Raymond; Wadsworth, Marshman Edward

1880s: Jackson, Robert Tracy; Lane, Alfred Church; Penrose, Richard A. F., Jr.; Wolff, John Elliot

1890s: Collie, George Lucius; Daly, Reginald Aldworth; Foerste, August Frederic; Gulliver, Frederic Putnam; Harris, Thaddeus William; Jaggar, Thomas Augustus, Jr.; Ladd, George Edgar; Westgate, Lewis Gardner

1900s: Bell, James Mackintosh; Goldthwait, James Walter; Grabau, Amadeus William; Howe, Ernest; La Forge, Laurence; Mansfield, George R.; Smith, Phillip Sidney; Vaughan, Thomas Wayland; Wilson, Alfred W. G.; Woodman, Joseph Edmund

1910s: Barton, Donald Clinton; Field, Richard Montgomery; Foye, Wilbur Garland; Haynes, Winthrop Perrin; Lahee, Frederick H.; Locke, Augustus; McLaughlin, Donald H.; Merwin, Herbert E.; Murdoch, Joseph; Powers, Sidney; Shuler, Ellis William; Wandtke, Alfred; Wigglesworth, Edward

1920s: Billings, Marland P.; Chen, Harold Hwai; Clark, Thomas Henry; Connolly, Joseph P.; Croneis, Carey G.; Eggleston, Julius Wooster; Ettlinger, Isadore Aaron; Gibson, Russell; Gilbert, Philip Geoffrey Britton; Greig, Joseph W.; Harvey, Roger Douglas; Hinds, Norman Ethan Allen; James, Howard Turnbull; Landes, Kenneth Knight; Lusk, Randolph Gordon; McKinstry, Hugh Exton; Peattie, Roderick; Powers, Howard Adorno; Short, Maxwell Naylor; Swinnerton, Allyn Coates; Talmage, Sterling Booth; Thomson, Ellis; Vanderwilt, John W.; Willard, Bradford

1930s: Albritton, Claude Carrol; Ball, Clayton Garrett; Bandy, Mark Chance; Berman, Harry; Bowditch, Samuel I.; Buie, Bennett Francis; Burgess, Charles Harry; Campbell, Catherine Chase; Campbell, Ian; Chalmer, John Roy; Chapman, Carleton Abramson; Chapman, Randolph Wallace; Chute, Newton Earl; Cleaves, Arthur Bailey; Cornelius, Searle H., Jr.; Dane, Ernest B., Jr.; Davidson, Stanley C.; de Laguna, Wallace; Delo, David Marion; Denny, Charles Storrow; Dodge, Theodore A.; Doggett, Ruth Allen; Dunham, Kingsley Charles; Dunkle, David H.; Fairbairn, Harold Williams; Fraser, Horace John; Goldthwait, Richard Parker; Goranson, Edwin A.; Goranson, Roy Walter; Gustafson, John Kyle; Hadley, Jarvis Bardwell; Harcourt, George Alan; Harris, Reginald Wilson; Haskell, Norman Abraham; Hinchey, Norman S.; Hurlbut, Cornelius Searle, Jr.; Jenks, William Furness; Johnson, H. Norton; Jones, Stephen Barr; Kransdorff, David; Lalicker, Cecil Gordon; Leet, Lewis Don; Mason, Shirley Lowell; Miller, Franklin S.; Modell, David Isaiah; Moehlman, Robert S.; Moore, Thomas G.; Moses, John H.; Noble, James A.; Peale, Rodgers; Phleger, Fred B., Jr.; Pough, Frederick Harvey; Quinn, Alonzo Wallace; Quinn, Howard Edmond; Ray, Louis Lamy, Jr.; Richmond, Wallace Everett, Jr.; Roy, Chalmer John; Schalk, Marshall; Schmedeman, Otto C.; Sharp, Robert Phillip; Smith, Harold T. U.; Smith, J. Fred, Jr.; Stratton, Everett Franklin; Thiesmeyer, Lincoln R.; Tunell, George; Turneaure, Frederick S.; Upson, Joseph E. II; Wahlstrom, Ernest E.; Waldo, Allen Worcester; Welker, Kenneth Kramer; Wickenden, Robert T. D.; Williams, Charles R.; Wilson, Eldred Dewey; Yates, Arthur Berkeley

1940s: Beck, Carl Wellington; Blumberg, Roland Krezdorn; Brown, Ira Charles; Burrell, Herbert Cayford; Butler, Arthur Pierce, Jr.; Chace, Frederick Mason; Cooke, Hermon Richard, Jr.; Curtis, Bruce Franklin; Dowse, Alice Mary; Eric, John Howard; Fisher, Bernard; Fowler, Phillip Teague; Freedman, Jacob; Greenwood, Robert; Hack, John Tilton; Heald, Milton Tidd; Heinrich, Eberhardt William; Holmes, George William; Huffington, Roy Michael; Joralemon, Peter; Judson, S. Sheldon, Jr.; Kennedy, George Clayton; Kruger, Frederick C.; Larsen, Esper Signius, III; Lyons, John Bartholomew; Menard, Henry W., Jr.; Mitchell, Wilson Doe; Moke, Charles B.; Moore, George Emerson, Jr.; Moseley, John R.; Moss, John H.; Nichols, Robert Leslie; Oliphant, Charles W.; Pecora, William Thomas, II; Peltier, Louis Cook; Rabbitt, John Charles; Schulman, Edmund; Shaw, Alan B.; Smith, Althea Page; Sohnge, Paul Gerhard; Spackman, William Jr.; Switzer, George; Walker, Eugene H.; Wengerd, Sherman Alexander; Wheeler, Robert Reid; White, Walter Stanley; Whitmore, Frank C., Jr.; Winchell, Horace; Wolfe, Caleb Wroe; Wright, Herbert Edgar, Jr.; Young, John A., Jr.

1950s: Albee, Arden Leroy; Allen, Robert Dorchester; Baird, Donald; Bean, Robert Jay; Berman, Robert Morris; Boucot, Arthur J.; Bowman, Edgar Cornell; Boyd, Francis Raymond, Jr.; Bradbury, James Clifford; Brush, Lucien Munson, Jr.; Byerly, Perry Edward; Caldwell, Dabney Withers; Chamberlain, Joseph Annandale; Chang, Ping-Hsi; Cheriton, Camon Glenn; Cifelli, Richard; Clabaugh, Stephen Edmund; Clark, Sydney Procter, Jr.; Diment, William Horace; Drury, W. H.; Emery, David James; Eschman, Donald Frazier; Everhart, Donald Lough; Fernald, Arthur Thomas; Fisher, Irving Sanborn; Gaines, Richard Venable; Gealy, Betty Lee; Gealy, William James; Hall, Leo Matthew; Hall, Wayne Everett; Hart, Pembroke J.; Hartshorn, Joseph Harold; Hawes, Julian; Henderson, Donald Munro; Herrin, Eugene Thornton, Jr.; Hopkins, David Moody; Horen, Arthur; Howard, Peter Felix; Joyner, William Blish; Koch, George Schneider, Jr.; Lachenbruch, Arthur Herold; Lacy, Willard C.; Leahy, Richard Gordon; Leopold, Luna B.; Macdonald, Gordon James Fraser; Markham, Neville Lawrence; McCartney, William Douglas; McCartney, William Douglas; McLain, Jay Forman; Meyer, Charles; Miller, John P.; Montgomery, Arthur; Moore, John C. G.; Moritz, Carl A.; Moustafa, Youssef S.; Nobles, Laurence H.; Ohle, Ernest L.; Osberg, Philip Henry; Raup, David Malcolm; Redden, Jack Allison; Rich, Charles Clayton; Robertson, Eugene C.; Rosebsloom, Eugene Holloway, Jr.; Rosenfeld, John Lang; Said, Rushdi; Schmalz, Robert Fowler; Schmidt, Robert G.; Silman, Jack Forrest Banning; Skehan, James W.; Skinner, Brian J.; Smith, Edgar Ernest Norval; Sriramadas, Aluru; Stearns, Charles E.; Stewart, David Benjamin; Sticht, John H. S.; Stone, Solon W.; Toulmin, Priestley, III; Trainer, Frank Wilson; Traverse, Alfred F., Jr.; Tuttle, Sherwood D.; Vaughn, Peter P.; Wahrhaftig, Clyde A.; Wandke, Alfred D., Jr.; Wenden, Henry Edward; Williams, Edwin Philip; Williams, Robert Lee; Wolman,

Markley G.; Wood, Francis W.; Zeigler, John Milton; Zen, E-an; Zieglar, Donald Lowell

1960s: Adams, Samuel Sherman; Anderson, James Arthur; Anthony, John W.; Aristarain, Lorenzo Francisco; Balsley, James Robinson, Jr.; Barnes, Ivan Keiler; Bell, Peter M.; Berner, Robert Arbuckle; Blackwell, David Douglas; Bowen, Zeddie Paul; Brett, Peter R.; Bricker, Owen; Burtner, Roger L.; Butler, Patrick, Jr.; Cahoon, Bobby Glenn; Carpenter, Alden Bliss; Cox, Doak; Crosby, Percy; Decker, Edward Ronald; DeMott, Lawrence L.; Dickinson, Stanley Key, Jr.; Dixon, Helen Roberta; Drake, John Craig; Eckstrand, Olof Roger; Einaudi, Marco Tullio; Fardon, Ross Stuart Harpur; Field, William; Fitzpatrick, Michael Morson; Forman, David John F.; Fujii, Takashi; Gaposchkin, Edward Michael; Gaucher, Edwin Henri Stanislas; Grant, Raymond Wallace; Green, John Chandler; Greene, Robert C.; Guidotti, Charles Vincent; Gustafson, Lewis Brigham; Hanor, Jeffrey S.; Hanshaw, Bruce Busser; Harwood, David; Hatch, Norman Lowrie, Jr.; Hays, James F.; Helgeson, Harold Charles; Henderson, Frederick B., III; Hess, Paul Charles; Honea, Russell Morgan; Hostetler, Paul Blair; Hunt, Allen Standish; Jones, Alexander Gordon; Jopling, Alan Victor; King, Elbert Aubrey, Jr.; King, Elbert Aubry, Jr.; Klein, Cornelis, Jr.; Koch, William Jerry; Kothavala, Rustam Zal; Lamarche, Valmore C.; Langmuir, Donald; Layman, Frederic G.; Leavens, Peter B.; LeComte, Paul; Lerman, Abraham; Linn, Kurt O.; Loughridge, Michael Samuel; Lovell, Julian Patrick Bryan; Lustwig, Lawrence Kenneth; Marvin, Ursula Bailey; McCubbin, Donald Gene; Milligan, George Clinton; Milton, Daniel Jeremy; Newberg, Donald W.; Norton, Stephen Allen; Ormiston, Allen R.; Pankiwyskyj, Kost A.; Peck, Dallas Lynn; Petersen, Ulrich B.; Peterson, Melvin Norman Adolph; Phillips, John Stephen; Rahm, David Allen; Rankin, Douglas Whiting; Ray, Clayton Edward; Raymahashay, Bikash Chandra; Reed, Bruce; Reitzel, John S.; Robinson, Peter; Ross, Malcolm; Roy, Robert F.; Rumble, Douglas, III; Runnells, Donald DeMar; Sahakian, Armen Souren; Schmitt, Harrison Hagan; Schoen, Robert; Schwab, Frederic L.; Shan, Frederick C.; Shaw, Frederick C.; Sill, William Dudley; Simmons, Marvin Gene; Solomon, Peter J.; Southard, John B.; Still, Arthur Rood; Swinchatt, Jonathan Phillip; Szekely, Thomas S.; Taggart, James Nash; Thompson, Mary E.; Tillman, Chauncey Glenn; Tolbert, Gene Edward; Trask, Newell J., Jr.; Truesdell, Alfred Hemingway; Verma, Raj Kumar; Waldbaum, David R.; Walker, Laurence G.; Wang, Chi-Yuen; Warner, Jeffrey L.; Wiese, Robert George, Jr.; Wilde, Pat; Wilson, John Joseph; Wilson, Robert J.

1970s: Abbott, Richard Newton, Jr.; Andrews, Harold Edward, III; Apps, John Anthony; Arem, Joel Edward; Arzi, Avner A.; Atkinson, William W., Jr.; Austria, Benjamin Suarez; Awramik, Stanley M.; Bakker, Robert Thomas; Beck, Kevin Charles; Beger, Richard Myron; Behrensmeyer, Anna K.; Bickel, Charles Eliot; Binstock, Jutta Lore Hager; Birnie, Richard Williams; Bloomer, Gail; Brady, John Ballard; Burt, Donald McLain; Callender, Jonathan Ferris; Chipman, David Walter; Chou, Teh-An George; Cook, Lawrence Paul; Creasy, John W.; Dean, Claude S.; Dingman, Stanley Lawrence; Eaton, Andrea Drake; Farrell, Clifton William; Feiss, Paul Geoffrey; Ferry, John Mott; Fisher, Daniel Claude; Foster, Merrill W.; Gibbs, Allan Kendrick; Gillmeister, Norman Maack; Goetze, Christopher; Goodell, Philip Charles; Grew, Edward Sturgis; Griscom, Andrew; Grove, Timothy L.; Hanscom, Roger Herbert; Harper, Howard Earl, Jr.; Hazen, Robert Miller; Hedberg, Ronald M.; Hepburn, John Christopher; Hovis, Guy Leader; Jones, Carol C.; Kamilli, Robert J.; Kastner, Miriam; Kelley, Patricia Hagelin; Klein, Christopher William; Knoll, Andrew Herbert; Leo, Richard Francis; Littlejohn, John Joseph; Longhi, John A.; Lonker, Steven Wayne; Lundeen, Margaret Thompson; Lyttle, Peter T.; Ma, Che-Bao; Mao, Nai-Hsien; Maynard, James Barry; McCoy, Floyd W.,

Jr.; McSween, Harry Younger, Jr.; Menge, J. L.; Mottl, Michael James; Mutti, Laurence Joseph; Nitsan, Uzi; Ohashi, Yoshikazu; Rehmer, Judith; Rich, Robert Alan; Richardson, Catherine Kessler; Richardson, Steven McAfee; Roberts, Steven Arland; Robinson, Gilpin Rile, Jr.; Sack, Richard Olmstead; Sailor, Richard Vance; Sanford, Richard Frederick; Schindel, David Edward; Sepkoski, Joseph John, Jr.; Sprinkle, James Thomas; Stein, Carol Lynn; Stolper, Edward Manin; Stout, James Harry; Sundquist, Eric Thorsten; Thomas, Roger David Keen; Tsui, Tien Fung; Veblen, David Rodli; Verma, Pramod Kumar; von Metzsch, Ernst Hans; Wahlert, John Howard; Walker, David; Walker, W. W., Jr.; Woronow, Alexander; Wu, I. Hsiung

1980s: Aliberti, Elaine Angela; Allen, Fred Mitchell; Allmon, Warren Douglas; Arnold, Anthony Jay; Boak, Jeremy Lawrence; Brown, Michael Jonathan; Brush, Laurence Henry; Bushnell, Steven Ensign; Campbell, Andrew Robert; Candela, Philip Anthony; Chamberlain, C. Page; Cohen, Ronald Elliott; Compton, John Sternbergh; Converse, David Rhys; Cranstone, Donald Alfred; Crocetti, Charles Alfred; Davidson, John Matthew; Doblas, Miguel M.; Docka, Janet Anne; Downie, Elizabeth Anne; Duran, Alexander Paul; Ekstrom, Goran Anders; Erslev, Eric Allan; Follo, Michael Ford; Gale, Peter Edward; Geary, Dana Helen; Gilinsky, Norman Lawrence; Goldstein, Steven Joel; Goreau, Thomas Joaquin; Green, Julian Wiley; Hackbarth, Claudia Jane; Heaton, Timothy Howard; Johnson, Mary Louise; Kappelman, John Wesley, Jr.; Keto, Lisette Scott; Kornacki, Alan Stanley; Kriens, Bryan Jon; Kring, David Allen; Lenk, Cecilia; Lesher, Charles Edward; Lipschultz, Fredric; Loucks, Robert Ray; Lovison, Lucia Cecilia; Maliva, Robert George; Marin, Carlos Mariano; Meddaugh, William Scott; Metzger, Bernhard Hugo; Mitchell, Charles Emerson; Pilskalin, Cynthia Hughes; Pimentel-Klose, Mario Rafael; Pinckney, Linda Ruth; Powell, Mildred A.; Raynolds, Mary Vera; Rouhani, Shahrokh; Schiffries, Craig Mason; Schneiderman, Jill Stephanie; Siegel, Malcolm Dean; Sonder, Leslie Jean; Steim, Joseph Michael; Sues, Hans-Dieter; Tse, Simon Tak-Chan

University of Hawaii
Honolulu, HI 96822

200 Master's, 122 Doctoral

1930s: Beach, A. R.; Chinn, E. Y. H.; Jones, A. E.; Kubota, H.; Okubo, S.

1940s: Fujimoto, C. K.; Gill, W. R.; Iwashita, S.; Kanehiro, Y.; Man, K. T.

1950s: Blomberg, N. E.; Chu, A. E. C.; Chun, E. H. L.; Hagihara, H. H.; Kawano, Y.; MacCracken, W. L.; Matsusaka, Y.; Nakamura, M. T.; Redman, F. H.; Robertson, J. B.; Suzuki, C. K.; Terada, K.; Uehara, G.

1960s: Agarwal, A. S.; Ahmed, S.; Atkinson, Ian Athol Edward; Barnes, I. L.; Bathen, K. H.; Bauer, Glenn R.; Bigelow, G. E.; Briones, Angelina Mariano; Busche, F. D.; Caskey, M. C.; Chotimon, A.; Cropper, A. G.; De Silva, G. L. R.; Duennebier, F. K.; Ebersole, W. C.; Feeney, J. W.; Fernandez, N. C.; James Bruce; Funkhouser, John Gray; Gardiner, H. C.; Gazdar, M. N.; Grunwald, R. R.; Hammond, L. L.; Harvey, R. R.; Hassan, T. S.; Heald, Emerson Francis; Herlicska, E.; Houng, Kun- Huang; Hubbard, Norman Jay; Hubbard, Philip Scott; Hurd, D. C.; Hussain, Sultan M.; Hussong, D.; Ishizaki, K.; Johnson, R. H.; Juang, T. C.; Kanehiro, Y.; Kapteyn, R. J.; Kaushal, S. K.; Khan, Mohammud Attaullah; Kimura, H. S.; King, D. L.; Klim, D. G.; Kroenke, L. W.; Langford, Stephen Arthur; Malahoff, Alexander; Maske, N. D.; Mathur, Surendra Pratap; Maynard, G. L.; McCoy, Floyd; McGuire, D. M.; Mekaru, T.; Moore, Larry Joe; Motooka, P. S.; Noble, Clyde S.; Northrop, John; Oshiro, K.; Pandey, Sheo Ji; Pararas-Carayannis, G.; Pescador, P.; Phongbetchara, R.; Pottratz, S. W.; Rahman, A.; Raymundo, M. E.; Richmond, R. N.; Rixon, A. J.;

Saing, S.; Sangtian, C.; Sego, P. D.; Sharma, M. L.; Shetty, Y. G.; Shirazi, G. A.; Singh, B. R.; Stice, Gary Dennis; Taira, H.; Tamimi, Y. N.; Tenorio, P. A.; Thiagalingam, K.; Tracy, R. W.; Tsuji, G. Y.; Von Seggern, D. H.; Walker, D. A.; Walker, J. L.; Wolfe, L. A.; Yamashiro, C. H.; Yokoyama, J. S.; Zachariadis, R. G. P.

1970s: Au, G. H. C.; Bainbridge, C. W., III; Balasubramanian, V.; Beers, L. D.; Bell, J. A.; Bentley, L. R.; Bigelow, G. E.; Bonnar, R. U.; Braide, J. O.; Broyles, M.L.; Bruce, R. C.; Burke, S. K.; Burnett, W. C.; Burnett, W. C.; Buyannanonth, V.; Carter, J. A.; Christensen, N., IV; Coulbourn, W. T.; Coulbourn, W. T.; Covey, W. P., III; Craig, J. D.; Craig, J. D.; Culp, S. K.; Cutler, S.; Dale, R. H.; Daniel, T. H.; Daniel, T. H.; Davenport, R. R.; De Silva, G. L. R.; Derby, J. V.; Dollar, S. J.; Dudley, W. C., Jr.; Dugolinsky, B. K.; Easton, R. M.; Epp, David; Eshlemann, A. M.; Fein, C. D.; Ferrall, C. C., Jr.; Foreman, Jerome A.; Foreman, Jerome A.; Fraley, C. M.; Fryer, G. J.; Fryer, P.; Fugate, James K.; Fujishima, K. Y.; Gilliard, T. C.; Gonzalez, F. I., Jr.; Gooding, J. L.; Goodney, D. E.; Graham, D. G.; Gramlich, J. W.; Gribble, G. W.; Halada, R. S.; Halunen, A. J.; Hammond, D. A.; Hammond, S. R.; Hammond, S. R.; Handschumacher, David W.; Hargis, D.; Harvey, R. R.; Heutmaker, D. L.; Hollett, K. J.; Horton, K. A.; Houck, James Edward; Huber, R. D.; Hubred, G. L.; Hudnall, W. H.; Hufen, T. H.; Hurd, D. C.; Hussong, D. M.; Jellinger, M.; Kanehiro, B.; Katahara, K. W.; Kauahikaua, J.; Kaya, M. H.; Keeling, D. L.; Keeling, D. V.; Killingley, J. S.; Kinsey, D. W.; Klein, D. P.; Klein, D. P.; Knutson, D. W.; Kroenke, L. W.; Kwon, T.; Landmesser, C. W.; Langford, Stephen Arthur; LaTraille, S. L.; Lee, C. R.; Lee, J. H.; Lim, S. K.; Lineberger, P. H.; Lum, L. W. K.; Luoma, Samuel N.; Malinowski, M. J.; Mashina, K. I.; Matsui, T.; Maynard, G. L.; McMurtry, G. M.; Meyer, J.; Meylan, M. A.; Michael, M. O.; Michael, M. O.; Milholland, Phillip Delbert; Montagne, H. W.; Morgenstein, M. E.; Muller, P. M. H.; Mussels, J. H.; Nayak, U. B.; Odegard, M. E.; Oldnall, R. J.; Orwig, T. L.; Pesret, F.; Principal, P. A.; Puccetti, A.; Rahman, A.; Rai, C. B.; Rogers, G. C.; Rosendahl, B. T.; Rotert, J.; Saboski, E. M.; Sakoda, E. T.; Santo, L. T.; Sato, H. H.; Schenck, B. E.; Schroth, Charles Lorenz; Sehgal, M. M.; Seyb, S. M.; Sicks, G. C.; Sinanuwong, S.; Smith, Stephen Vaughan; Sokolowski, T. J.; Soroos, R. L.; Southworth, J. H.; Steinhilper, F. A.; Stone, C.; Suyenaga, W.; Suyenaga, W.; Tama, K.; Tanabe, M. J.; Thomas, D.; Turner, B. W.; Valencia, M. J.; Vandrevu, B. R.; Voss, J.; Voss, R. L.; Walker, D. A.; Webb, M. D.; Wheatcraft, S. W.; Wilcox, L. E.; Wilcox, L. E.; Woodruff, J. L.; Yaibuathes, N.; Yanagida, R. Y.; Yim, Patrick C. Q.; Zachariadis, R. G. P.; Zee, G. T. Y.

1980s: Agegian, Catherine Rose; Ambos, Elizabeth Luke; Anderson, Paul; Anthony, Stephen S.; Atkinson, Marlin J.; Bartlett, Wade A.; Beal, Kenton L.; Bruckenthal, Eileen A.; Buxton, Donna S.; Byers, Charles D.; Caress, Mary Elizabeth; Cessaro, Robert K.; Chase, Wilbert Gordon; Christie, David Mark; Cloutis, Edward Anthony; Cooper, Patricia Ann; Dorn, Wolfgang Ulrich; Ferrall, Charles C., Jr.; Gaddis, Lisa R.; Gaffey, Susan Jenks; Giambelluca, Thomas Warren; Hagen, Ricky A.; Haggerty, Janet Ann; Herrero-Bervera, Emilio; Hoffmann, John P.; Izuka, Scot Kiyoshi; Kennedy, Kevin; Kim, Dae Choul; Knight, Michael Don; Lauritzen, Robert A.; Lee, Patricia D.; Legowo, Eko; Lindwall, Dennis; Mallick, Subhasis; Matson, Dean W.; Milholland, E. Cheney Snow; Milholland, Phillip Delbert; Miller, Daniel J.; Miller, Mark E.; Morgan, Lisa Ann; Oberdorfer, June Ann; Ogujiofor, Ikechukwu Jonathan; Reed, Thomas B.; Rodgers, Dorothy Lynne; Roush, Ted L.; Rowland, Scott K.; Sager, William Warren; Sen, Mrinal K.; Shettigara, K. V.; Showers, William J.; Trangmar, Bruce Blair; Wang, Chung-Ho; Wedgeworth, Bruce

Steven; Wiltshire, John C.; Yan, Chun-Yeung; Zent, Aaron Patrick

University of Hawaii at Manoa
Honolulu, HI 96822

17 Master's, 2 Doctoral

1970s: Blank, Richard; Brill, R. C., Jr.; Duennebier, Frederick K.; Erlandson, D. L.; Getts, T. R.; Gibson, B. S.; Palmiter, D. B.; Watts, G. P.; Wisham, C. M., Jr.

1980s: Imada, Jewelle Akie; Kearney, Terrence J.; Lo, Kwong Fai Andrew; Mortera Gutierrez, Carlos Angel Q.; Price, Patricia E.; Pringle, Mary Katherine Williams; Rice, Craig W.; Roush, Ted L.; Tsutsui, Bruce Osamu; Zbinden, Elizabeth Anne

University of Houston
Houston, TX 77204

219 Master's, 34 Doctoral

1940s: Black, J. P.; Fisher, Juanita Prior; Miron, Sam

1950s: Allison, David Bryan; Benke, Mary Lee; Berggren, William Alfred; Bogart, Vera Jo; Branham, Thomas B.; Brooks, Paul L., Jr.; Brown, Warren L.; Caillouet, Howard J.; Chimene, Calvin Alphonse; Clardy, Arthur L.; Courtney, Cecil; Enlow, Donald Hugh; Forney, L. Bruce; Geer, Lucius C.; Gordon, George E.; Grace, Marvin; Hall, Robert G.; Harvill, Lee L.; Horn, Myron K.; Johnson, Robert Edwin; Kasim, S. A.; Kellough, Gene Ross; Kister, Tom L.; Leeds, Anna L.; Liebe, Richard Milton; Lynch, Vance M.; Mallory, James A.; McBride, William Joseph; McCarty, Dana G.; Miller, Cliff Q., Jr.; Park, Sam, III; Parker, Kenneth L.; Pulley, Thomas E.; Reiter, Jessie Oscar; Sale, Hershel E.; Shaw, Daniel B.; Walker, Thomas Wiley; Weintritt, Donald J.

1960s: Alewine, James W.; Ballard, Eva Oakley; Barnett, Richard S.; Barnhart, John T.; Beardsley, Donald W.; Bishop, Allen David, Jr.; Cameron, Kenneth L.; Cameron, Maryellen; Caskey, Thomas Lee; Chaplin, James R.; Clopine, Gordon A.; Fisher, Fred Eugene; Flory, Donald Andrew; Fowler, James W.; Frazier, Don W.; Fuex, Anthony Nichols; Gann, Donald P.; Goheen, Hunter C.; Harlan, Ronald W.; Horne, Jerry D.; Horne, Jerry D.; Howard, James F.; James, Keith H.; Jarrell, Mary Kathryn; Krowski, Stanley P.; Lewis, Douglas W.; McCracken, Willard A.; Moredock, Duane E.; Myers, Robert L.; Newby, McInnis S.; Nooner, Daryl Wilburn; Schneider, Howard John; Smith, Charles Culberson; Solliday, James Richard; Sterling, Richard P.; Sutherland, Berry; Taylor, Donald S.; Turner, Robert D.

1970s: Adedokun, Oluwatele Alabi; Altman, Thomas DeWitt; Armistead, Gary Anthony; Baker, B. A.; Banholzer, Gordon S., Jr.; Blankenship, John C., Jr.; Brautigam, Gerald L.; Brooks, Warren W.; Burnaman, M. D.; Creger, Robert B.; Crews, A. L.; Dean, J. R.; Droddy, M. J., Jr.; Forsman, Nels F.; Fruland, Ruth Marcia; Garbett, Elizabeth C.; Gaston, Wilbert P.; Grasel, Peter Corbin; Greene, G. M.; Hawkins, Gayne Patrick; Heuer, Wolfgang C.; Hodgkinson, R. J.; Ilukewitsch, Alejandro G.; Indest, Stanley J.; Johnson, Pratt H.; Jones, Irene B. Carter; King, D. T., Jr.; Kocurek, Gary; Ladle, Garth H.; Lim, Sung J.; Loep, Kenneth J.; Miller, Kenneth W.; Moysey, D. G.; Nakayama, Kazuo; Newcomb, Frederick A.; Olson, Russel D.; Parrish, John George; Patel, Arunkumar Kalidas; Petersen, Harry W.; Pettus, David S.; Piety, William Duncan; Powers, Russell S.; Robinson, B. Spence; Robinson, Bob R.; Rogers, John L.; Rosa, Andre L. R.; Schuster, Gerard Thomas; Smith, Rita Monahan; Souza, Murilo M.; Taber, Edward C., III; Thomas, J. E.; Van Nieuwenhuise, Donald S.; Vance, G. F., Jr.; Velden, Trude Vander; Wen-Jen Wu; Wood, David G.; Woronow, Alexander Nick; Zabel, Garrett Edward

1980s: Ambs, Loran D.; Ashabranner, Donald; Battié, John E.; Baysal, Edip; Blaney, Geoffrey W.; Brink, Ronald; Bruno, Lawrence; Buczynski, Chris; Carter, Kent; Cashore, Jac; Cate, Alta; Chimene, Julius, III; Chou, Lynn; Cominguez, Alberto Horacio; Cox, John; Damle, Mayurika; Davin, Christopher Gerard; Dillman, George; Duffy, Robert E., Jr.; Ebrom, Daniel; Egan, Mark S.; England, D. Kent; Eyer, Andrew; Faraguna, John; Farrow, Hillary; Fisher, David; Forel, David; Frey, Susan; Geeslin, Jill H.; Geno, Kirk R.; Gomez-Moran, Concepcion; Graber, Karen; Ha, Tiong; Harrison, Jane; Hellstern, Donald; Herlinger, David L.; Herrick, Robert; Heydinger, Andrew Gerard; Hillendbrand, Charles, III; Hindlet, Francois, J. F.; Hornbostel, Scott; Huang, Hann-Chen; Hwang, Grace; Inderwiesen, Philip Leon; Jaramillo Mejia, Jose Maria; Kammer, David; Karrenbach, Martin; Kemner, Mark; Kerr, Ronald; Kharas-Khumbata, Nazneen; Klein, John P.; Kleist, Ronald J.; Komor, Stephen Charles; Lane, Richard; Liang, Luh-Cheng; Lock, Susan; Locklin, Jo Ann C.; Love, Karen; Marhadi; Matsui, Kunio; Maxwell, Geraldine; McCormack, Michael David; McIntosh, Allen; McNew, Mark; McQuillen, Daniel; Meier, Thomas Allan; Melnyk, David H.; Meredith, John C.; Morgan, Thomas Richard; Myers, Keith; O'Brien, William; Ottmann, Jeffry D.; Owusu, Joseph Kwame; Pan, Naide; Pascual, Ruben; Peterson, Caroline; Purnell, Guy; Radack, Phyllis; Randazzo, Santi; Rao, Meera; Reyes, Carlos; Robertson, Paul; Ross, Daniel; Ross, James; Sarmiento, Raul; Savci, Gultekin; Savic, Milos; See, Thomas; Shepherd, Ashley; Siroky, Francis; Smith, Daniel; Smith, Susan; Smith, Thomas Andrew; Spetseris, Jerry; Squires, Livia J.; Stewart, Randall; Sung, Roger; Tanner, Jeffrey; Tatalovic, Radmilo; Thompson, Rodney; Tzeng, Rong-Fung; Ugaz, Oscar Guillermo; Urosevic, Milovan; Usmani, Tariq U.; Utech, Nancy; Verm, Richard Wayne; Voight, Kenneth; Wang, Peter; Wang, Sou-Yung; Williams, Adele; Wong, Daniel On-Cheong; Wu, Wen-Jen; Xia, Chunshou; Yao, Chien-Chang David; Yeung, Terence C.; Zweig, Julie

Howard University
Washington, DC 20059

1 Master's

1980s: Olowomeye, Richard Boluwaji

Humboldt State University
Arcata, CA

1 Master's

1980s: Lundstrom, Scott C.

University of Idaho
Moscow, ID 83843

456 Master's, 87 Doctoral

1920s: Anderson, Alfred Leonard; Carder, Dean Samuel; Elder, R. B.; Joyce, Edwin A.; Lokken, John Carl; Piper, Arthur Maine; Sandback, John Elmer; Shenon, Philip John; Siegfus, Stanley; Smolak, George; Sorenson, Robert Eugene; Udell, Stewart

1930s: Bryant, Boyd LaVerl; Carpenter, John Tyer; Emigh, George Donald; Faick, John N.; Hammerand, Veral Franklin; Hite, Thomas H.; McConnell, Roger Harmon; Newton, Joseph; Rasor, Charles Alfred; Tullis, Edward Langdon; Wagner, Warren R.; Willard, Max E.

1940s: Allen, Rhesa McCoy, Jr.; Bower, Guy Joseph; Browning, James S.; Carpenter, Robert D.; Finkelnburg, Oscar Carl; Holland, John Sylvester; MacKenzie, Wayne Oliver; Powers, Harold Auburn; Scheid, Vernon A.; Shefloe, Allyn Carlyle; Smedley, Jack Elwood; Thune, Howard Willis; Upson, Roberta Hastings

1950s: Alberts, Robert Kirk; Beeder, John Ralph; Bessey, Larry Eugene; Bitten, Bernard I.; Buhn, William Kenneth; Bulla, Edward William; Carlson,

John Edward; Carmichael, Virgil Wesly; Full, Roy P.; Hawley, Robert William; Hollenbaugh, Kenneth Malcolm; Holmes, David Allen; Jemmett, Joseph Paul; Kern, Billy Francis; Kinnison, Phillip Taylor; Kopp, Richard S.; Le Moine, Denis; Lee, Heungwon; Leland, George R.; Martin, Roger C.; McDonald, James Vernon; Mehelich, Miro; Melear, John David; Milner, Carlos E., Jr.; Nalwalk, Andrew Jerome; Nielsen, Merrill Longhurst; Ringe, Louis Don; Schipper, Warren Bailey; Schwarze, David Martin; Sidler, Aubrey Gene; Stinson, Melvin C.; Storey, Lester Oscar; Sturm, Frederick Henry; Sweeney, Gerald Thornton; Treves, Samuel B.

1960s: Abbott, Jesse Walter; Alief, Mohammed Hassan; Asher, Roderick R.; Baillie, William Norman; Bains, Trilochan Singh; Barber, Dean Austin; Barnes, Charles Winfred; Bolm, John Gary; Brackebusch, Fred W.; Brim, Raymundo Jose Portella; Butler, Thomas Abraham; Carmichael, Virgil Wesly; Chan, Samuel Shu Mou; Chauhan, Ehsanul Haque; Chin, Lennard Hilton; Clark, Allen LeRoy; Clark, Allen LeRoy; Clark, Sandra Helen Becker; Clark, Sandra Helen Becker; Coffin, Peter E.; Cook, John Dee; Eier, Douglas Dexter; Eng, Frank Gee; Garber, Lowell Wilbur; Gillespie, Gary L.; Glerup, Melvin Obert; Greenwood, William Rucker; Greenwood, William Rucker; Hall, Henry Thompson; Hamilton, James Allen; Hollenbaugh, Kenneth Malcolm; Holm, Richard Frank; Howard, Terry R.; Hsi, Huey-rong; Jones, Walter V.; Kunter, Richard Sain; Landreth, John Orlin; Lee, Tien-Chang; Liu, Kannson T. H.; McDole, Robert Elroy; McDole, Robert Elroy; Morrison, Donald A.; Murtaugh, John Graham; Myers, David Arthur; Neary, Thomas Hubert; Nevin, Andrew Emmet; Nord, Gordon L., Jr.; Palmer, Irven France; Paris, Gabriel; Park, Robert Gene; Pober, Patricia Taylor; Reyes, Benjamin Panganiban; Ross, Sylvia Yvonne Hall; Shah, Safdar Ali; Shah, Syed Mohammad Ibrahim; Shah, Syed Mohammad Ibrahim; Shannon, Spencer Sweet, Jr.; Smith, Clyde Louis; Soregaroli, Arthur Earl; Suhr, David Olaf; Toron, Praphat; Walker, William Peter, Jr.; Wilson, Richard Shirl

1970s: Abegglen, Donn E.; Ahamed, Aziz U.; Allman, David William; Anderson, Roy Arnold; Anderson, Roy Arnold, Jr.; Bailey, William C.; Baker, Edward Daniel; Baldwin, Joe Allen; Barrash, Warren; Baum, Lawrence Frederick; Bennett, Earl Healen, II; Bezan, Abduelhafid Mohamed; Bhatt, Bipinkumar J.; Bhatti, Nasir Ali; Bishop, Donald Thomas; Bjornsson, Bjorn Johann; Bolm, John Gary; Breeser, Patsy J.; Broili, Christopher J.; Buffa, John Warren; Caddey, Stanton William; Cannon, M. R.; Carter, Daniel Bradley; Cass, Hilton K.; Coffin, Jeffrey Hart; Cohen, Philip Leon; Conners, John Anthony; Corbet, T. F.; Crist, Elliott McDonald; Dart, Robert Henry; Davis, Jonathan O.; Davis, Jonathan O.; De Renne, Paul; De Renne, Paul; De Sonneville, Joseph Leonardus Johannes; Derkey, Pamela Dunlap; Dobey, Allen B.; Doler, Robert Earl; Driesner, Douglas Arthur; Druffel, Leroy; Duff, James Kenneth; Edelman, Delbert Wayne; Edwards, Thomas Kyle; Eliagoubi, Bahlul Ali Hameid; Eliagoubi, Bahlul Ali Hameid; Ellis, Clarence E.; Embree, Glenn F.; Fabiyi, Ekundayo E.; Fenne, Frank Karl; Fitzgerald, James Francis, Jr.; Flanders, Richard William; Fortier, David Harvey; Fryxell, Roald; Gaillot, Gary; Galbraith, James Herbert; Galbraith, James Herbert; Goldman, Dennis; Gordon, Allen Stewart; Grant, Stanley Cameron; Hansen, Henry Eugene; Hansen, Steven C.; Hashim, Hashim Mohammed; Hemud, Abdul Rahman; Herdrick, Melvin A.; Hernandez, Pedro Amando; Huppert, George N.; Ibenye, Ikechi S.; Ioannou, Christos; Johnson, James Blake; Juras, Dwight Stephen; Juras, Dwight Stephen; Kaal, Ayad Said; Kauffman, John David; Kealy, Charles D.; Keeley, Joseph Francis; Kehew, Alan E.; Kettenacker, William Charles; Khatib, Abdulhamid Ahmad; King, John G.; Kirkpatrick, Glen Edgar; Kun, Peter; Larsen, Ronald Edward; Lockard, David W.; Lynch, Maurice Butler; Mabes, Deborah Lynn; Maley, Terry Samuel; Malloy, Robert W.;

Mathewson, David C.; Mayfield, Charles F.; McClernan, Henry G.; McNary, Samuel W.; McNeill, Albert Russell; Meehan, Kenneth Tillotson; Miller, Don Adair, Jr.; Mink, Leland Leroy; Mink, Leland Leroy; Mohammad, Omar Mohammad Joudeh; Mohammad, Omar Mohammad Joudeh; Morilla, Alberto Garcia; Motzer, William Erhardt; Murray, Sharon Ann; Nahring, Eldon L.; Najjar, Ismail Muhammad; Nakai, Theresa Sigl; Newton, Garth David; Norbeck, Peter M.; Norman, Lonnie Dale; Ortman, Dale; Osiensky, James Leo; Pennifill, Roger Alan; Price, Susan Alys Medlicott; Quevedo, Ermel B.; Rahn, Jerry Everett; Ralston, Dale R.; Reece, Dennis E.; Reed, Richard W.; Roales, Paul A.; Robinette, Michael Joseph; Ross, Martin Edward; Sagstad, Steven R.; Salami, Satari Olatunde; Salman, Diab Salmon Ahmad; Seitz, Harold R.; Shadid, Omar Shakir Abdu Samara; Shea, Michael Curtis; Sheehan, Leo; Sheikh, Abdul Mannon; Shenker, Alan Edward; Shepherd, H. E.; Shively, Margaret V.; Sibbett, Bruce Scott; Singh, Harbhajan; Smith, Robert Martin; Snyder, Kenneth Dele; Standish, Richard Perkins; Steen-McIntyre, Virginia C.; Steinman, Dale Marie P.; Stiles, Craig A.; Sylvester, Kenneth Albert; Toukan, Ziad R.; Trexler, Bryson D., Jr.; Vandell, Terry Delores; Vandiver-Powell, Lorraine; Wagstaff, Donald Allan; Wallace, Richard Warren; Walsh, Thelma Helaine; Wikoff, Penny Marie; Williams, James Frank, III; Williams, James Oliver; Williams, Michael M.; Wilson, Monte Dale; Winter, Gerry Vernon; Woods, Paul Fredric; Yinger, Mark Andrew; Yoo, Kyung Hak

1980s: Aadland, Rolf Konrad; Adams, Wayne C.; Albers, Doyle Francis; Anastasi, Frank S.; Anderson, G. Witt; Apgar, Julie L.; Arrigo, John A.; Bachtel, Steven L.; Baghai, Nina Lucille; Baglio, Joseph V., Jr.; Barinaga, Charles Joe; Barrash, Warren; Barrett, Ruth Anne; Beaman, Brian Roy; Beck, Chris C.; Benson, Robert G.; Bernt, John Dodson; Bhatti, Nasir Ali; Biddle, John H.; Bittner-Gaber, Enid; Blank, Robert Raymond; Blank, Robert Raymond; Bliss, Douglas Allen; Blount, Gerald C.; Bockius, Samuel Harrison; Borovicka, Thomas G.; Boyd, Austin E., III; Bradley, Michael Dennis; Broadhead, Roxane; Brooks, Thomas David; Brown, Robert J.; Bruhl, Elliot J.; Burrell, Steve C.; Caddey, Eric Lee; Campbell, Richard B.; Campbell, William G.; Campo, Arthur M.; Canter, Karen Lyn; Carlisle, Scott P.; Carter, Cole H.; Cazes, Debra Kay; Chaney, Gregory Paul; Chavez, Joel Edmund; Cherry, Janet G.; Chiang, Hsien-Hsiang; Cleveland, Scott R.; Clough, Albert Hughes; Cochran, Bruce Duane; Cockrum, Dave A.; Collins, Harold P.; Constantopoulos, James Theodore; Courtright, Kelly Dean; Dahl, Mary Katherine; Dansart, William Joseph; Davidson, William R.; de Long, Richard F.; de Long, Richard F.; Dechert, Thomas Van; Deick, Jan F.; Derkey, Robert Erwin; Devine, Steven C.; Dieziger-Kim, Donna; Dingler, Craig Mitchell; Dolenc, Max Rudolph; Dudziak, Suzanne; Duncan, Ronald C.; Durgin, Philip Bassett; Eberle, Frederick Claude; Eckwright, Terry Alan; Erikson, Daniel L.; Ervin, Melanie K.; Felkey, Jack R.; Fischer, Howard J.; Fisher, Laura Lee; Fitzsimmons, Clifford Lynn; Foster, Scot Alan; Fox, Stephen Edward; Frankel, Paul; Gage, D. R.; Galvin, Timothy Joseph; Gehlen, William T.; Gemperle, Richard J.; Gilmore, Tyler J.; Giraud, Richard Ernest; Goldman, Dennis; Good, William K.; Grady, Michael; Graham, David L.; Greene, Earl A.; Greene, Steven E.; Greybeck, James D.; Griffing, David H.; Gross, Michael Robert; Gruenenfelder, Charles R.; Guarino, Joe C.; Hammond, Carol J.; Hansen, Dorrell Reed, III; Harris, Mary Katherine; Hartman, Mary J.; Harvey, Andrew Frank, III; Haskell, Kenneth G.; Hauntz, Charles E.; Hayden, Terry John; Hays, Ronny A.; Healy, Michael P.; Heitz, Leroy Fredrick; Hemphill, Gary Brian; Hill, Bradley M.; Hipple, Karl Walter; Holmes, Rebecca Ann; Hubbell, Joel M.; Hultman, William A.; Hunt, Joel A.; Ingraham, Peter Curwood; Inverso, George Anthony; Jacob, Thomas M.; Jampton, Kathleen M.;

Janik, Michael Garland; Jenks, Margaret Dana; Jeong, Sangman; Johnson, Gary Steven; Jokisaari, Allan O.; Jones, Allen L.; Joolazadeh, Mohammad; Jordan, David Charles; Jorstad, Robert B.; Kadoch, Teresa Lynn; Kagel, Carla Turner; Kauffman, Daniel F.; Kelsey, Richard Kelly; Kendra, William R.; Killman, Kathryn Susan; Kim, Sung; Klipfel, Paul Dexter; Krom, Thomas D.; Lafko, Eric M.; Lahabi, Ahmad-Ali; Lamb, Kevin J.; Lane, Margaret Lucille; Leck, Scott MacLeod; Lee, Harry William; Linder, Gerhard Martin; Lingren, John E.; Logan, David Craig; Luttrell, Stuart Paul; MacDonald, Elizabeth Cora; Mach, Leah E.; Maltz, Gary; Mancuso, Thomas Kaye; Mansur, Milud Abdulkrim; Martin, Lawrence James; Matzner, Robert A.; May, Thomas Patrick; Mayo, Alan Lee; McCurley, Earl B.; McIntyre, Colleen; McKiness, John Paul; Meyer, Paul Eaton; Mitchell, Victoria E.; Mohsenisaravi, Mohsen; Mok, Wai Man; Moody, Ula Laura; Moody, Ula Laura; Moore, Beth Anne; Morell, Douglas Jeffrey; Motzer, Mary Ellen Benson; Motzer, William Erhardt; Moye, Falma Jean; Nammah, Hassan Audah; Neely, Kenneth Wray; Nelson, Deborah J.; Nelson, Mark; Nemser, Katherine B.; Nielson, Eric S.; Niemi, Warren Lee; Norton, Marc A.; Osiensky, James Leo; Ott, Lawrence E.; Pancoast, Laurence Edwin; Parkinson, Craig Leonard; Parsley, G. Peter; Paul, E. Kenneth, Jr.; Pawlowski, Michael Raymond; Peale, Robert Newton; Pereus, Steven Charles; Perley, Philip Charles; Pfau, Mark A.; Plumley, Patrick S.; Pontius, Jeffrey Allan; Powell, John Edwin; Priesmeyer, Steven T.; Pusc, Steve W.; Rashrash, Salem Mohamed; Reid, Steven K.; Rigby, James Gordon; Riley, John A.; Robinson, Robert Blair; Rogers, Patrick C.; Russell, Charles W.; Sablock, Jeanette M.; Sader, Steven Alan; Scanlan, Terry M.; Schalck, Diane Kate; Schiebel, Lawrence Glenn; Sexton, William T.; Shaw, Christopher William; Shiveler, Donna Jean; Shrake, Tom; Siegmann, Sheryl A.; Simpson, Kenneth Reed; Simpson, Thomas Mason; Sims, Francis R.; Singer, Stephen H.; Singh, Harnek; Sitler, Gary Wilson; Slavik, Harold Joseph, Jr.; Smith, Donald Allen; Smitherman, James R.; Smoot, John Leach; Smykowski, Anthony Steve; Snook, James Donald; Souder, Karl Cameron; Spence, Jeffrey Gordon; Stamm, Robert G.; Stanford, Loudon Roberts; Stephens, Robert Leck; Sterling, Richard P.; Strowd, William Bruce; Stryhas, Bart Andrew; Swanson, L. Craig; Tarr, Karen M.; Taylor, Daniel T.; Thoreson, Ronald F.; Tracy, Paul W.; Truitt, Duane J.; Uhrich, William G.; Underwood, Joan Elizabeth; Vance, Randall Blaine; Venkatakrishnan, Ramesh; Wang, Ching-Pi; Wavra, Craig Scott; Waylett, Annette Shelton; Welford, Mark R.; Westfall, Jonathan E.; Whyatt, J. K.; Willoughby, Janice Kay Sowards; Willoughby, William W.; Winkelmaier, Joseph R.; Wittreich, Curtis D.; Wotruba, Patrick Roy; Wytzes, Jetze; Yake, Daniel Glen; Yeatman, Robert Andrew

Idaho State University
Pocatello, ID 83209

108 Master's, 2 Doctoral

1960s: Carlson, Roger Allan; Klauss, Thomas E.; Meyers, W. C.; Niccum, Marvin Richard; Wise, Joseph Patrick; Zilka, Nicholas T.

1970s: Allen, Russell Warren; Allexan, John Stephen; Anderson, Norman N.; Bingaman, Paul T.; Cline, K. Michael; Craig, Robert R.; Damp, Jeffery N.; Doyle, James M.; Duncan, Gary Alan; Durkee, Steven; Entzminger, David Jacob; Fuller, David Richard; Galyen, Robert L.; Giardinelli, Anthony; Gurney, James Walter; Hand, Perry A.; Hawkins, Robert B.; Henrich, William J.; Hotchkiss, Samuel A.; Imse, John P.; Jackson, Dicky Joe; Kamis, James Edward; Kern, Richard R.; Lilley, Wesley Wayne; Martin, Peter W.; McKinnie, Nancy Jayne; Middleton, Larry T.; Muller, Sean Conroy; Murk, Ronald Clarence; Perkins, Russell W.; Ridenour, James; Ruden, Stuart Michael; Shields, Richard H.; Thompson, Robert M.; White, Robert J.; Wieland, Edward Paul; Williams, R. Dave

1980s: Anderson, I. Carl; Barker, K. Scott; Bodnar, Theodor; Bogdanski, John K.; Burgel, William D.; Burton, Bradford Robert; Bush, Richard R.; Christenson, Tod D.; Clayton, Janine; Cochran, Daniel John; Conrad, Gregory Stevens; Cook, Jerry Robert; Cumming, Harry John Karns; Cunningham, Gregory D.; Danzl, Ralph; Darling, Robert S.; Dunn, Sandra Louise Dimitre; Dvoracek, Douglas; Enwall, Robert E.; Farahmand, Seyedhassan Hashemi; Ferdock, Gregory Christopher; Gallucci, Richard Nicholas; Geslin, Jeffrey K.; Goldstein, Flora J.; Halimdihardja, Piushadi; Hatfield, Harold Edmond; Hengesh, James V.; Henrich, Catherine; Hladky, Frank R.; Ishibashi, Gary Duane; Jansen, Stephen T.; Jasmer, Rodney M.; Kimball, Robert Vail; Lande, Andrew C.; LeFebre, George B.; Lindsey, Kevin A.; Litke, Richard Timothy; Luessen, Michael J.; MacGowan, Donald B.; Mahoney, J. Brian; May, Geoffrey R.; Morabbi, Mohammad; Morin-Jansen, Ann; Moulton, David Richard, Jr.; Neace, Thomas Foster; Olivier, Wendell Gregory; Palmer, Carl Riley; Palmer, Mark A.; Pogue, Kevin R.; Quinn, Laughlin C.; Renier, Joseph Maurice; Roberts, John Calvin; Ruebelmann, Kerry L.; Scheu, Steven Ray; Shoemaker, William A.; Steele, Elizabeth Anne; Stewart, Dave; Studley, Gregory Wayne; Sullivan, Brian G.; Thompson, Braden Jay; Ungate, Carol A.; Upson, Susan Adelaide; Walker, Charles Stephen; Wenger, Lloyd Miller, Jr.; Zahn, Paul D.; Zielinski, Gregory Anthony; Zimmerman, David W.

University of Illinois, Chicago
Chicago, IL 60680

81 Master's, 15 Doctoral

1910s: Brown, Robert Wesley; Clark, Clifton Wirt; Fleener, Frank Leslie; Hsu, Tsung Han; Nebel, Merle Louis; Read, Mason Kent; Ross, Clarence Samuel; Ross, Clarence Samuel; Thompson, David Grosh

1920s: Claypool, Chester Burns; Griffin, Judson Roy; Holmes, Leslie Arnold; Love, William Wray; Meyer, Alfred Herman Ludwig; Murray, Albert Nelson; Netzband, William Ferdinand; Poor, Russell Spurgeon; Poor, Russell Spurgeon; Roy, Sharat Kumar; Smith, Clyde Moffett; Toler, Henry Miles; Wagner, Oscar Emil; Waldo, Allen Worcester; Willman, Harold Bowen

1930s: Clark, Charles Roosevelt; Cohee, George Vincent; Fuller, Melville Weston; Griffin, Judson Roy; Grubb, Carl Frederich; MacVeigh, Edwin Lester; Scott, Harold William; Trefethen, Joseph Muzzy; Utterback, Donald Desmond; Wagner, Oscar Emil; Willman, Harold Bowen

1950s: Tsiza, Stephen Thomas; Weeks, Wilford Frank

1960s: Blair, Alexander Marshall; Peirce, Robert W.

1970s: Bement, W. Owen; Bogner, Jean A.; Boyle, Joseph; Bromberg, Janet; Buss, Barbara Ann; Cygan, Gary L.; Duckworth, D. L.; Duckworth, Diana; Fukui, Larry M.; Gagnard, Philip E.; Hendrix, Gary G.; Holst, Norman Benjamin, Jr.; Kampf, Anthony R.; Kennedy, S. K.; Kern, Ronald Arthur; Krivz, Andrea L.; Long, David Timothy; Lunardi, L. F.; Massion, Peter J.; McArdle, John E.; McConnell, D. R.; Nardin, Barbara A.; Pivorunas, August; Riekels, Lynda M.; Ripley, David P.; Rubin, A. E.; Skalnik, Petr; Vendl, M. A.; Visser, W.; Vredevoogd, James J.; White, J. D.; Wulkowicz, Gerald; Yeh, T. C. J.

1980s: Anderson, Marilyn Lea; Bouchard, Stephen M.; Clough, Stephen Ronald; Feng, Wei-Lin; Full, William Edward; Hong, Mingde; Lee, Jung Hoo; Leshchinsky, Dov; McClure, Dennis; Melcer, Allen; Miklius, Asta; Peters, John Fredrick; Phillips, Andrew; Quas, Marilyn; Read, Steven Edward; Roy, Stephen Donald; Rudloff, Gregory A.; Sumarac, Dragoslav; Terpstra, Paul; van der Laan, Sieger Robbert; Vogt, Eric; Wulff, Julie L.; Yeskis, Douglas Jerome; Zanoria, Elmer

University of Illinois, Urbana
Urbana, IL 61801

530 Master's, 483 Doctoral

1900s: Fox, Harry Bert; Rolfe, Martha Deette

1910s: Crooks, Harold Fordyce; Ekblaw, Walter Elmer; Ellis, Arthur Jackson; Engle, Edgar Wallace; Heitkamp, George William; Hutton, Joseph Gladden; Kennedy, Luther Eugene

1920s: Bassett, Charles Fernando; Dutton, Carl Evans; Ekblaw, George Elbert; Kennedy, Luther Eugene; Knipe, Ralph Ernest; McMackin, Samuel Carl; Utzig, Esther W.; Wildman, Ernest Atkins; Young, Jackson Smallwood

1930s: Allen, William Hammond; Bean, Beryl Kenneth; Billings, M. Hewitt; Borger, Harvey Daniel; Burns, Robert Obed; Claypool, Chester Burns; Cohee, George Vincent; Culbertson, John Archer; Dietz, Robert Sinclair; Ekblaw, Sidney Everette; Elias, Maxim Maximavich; Emery, Kenneth Orris; Franklin, Donald Wilbert; Geis, Harold Lorenz; Gutschick, Raymond Charles; Hagan, Wallace Woodrow; Hoover, William Farrin; Kilian, Harry Stephen; Larsen, Everett Christian; Larson, Carl Leonard, Jr.; Lee, Charles Denard; Lester, John Lawrence; Love, William Wray; McCabe, Louis Cordell; McCabe, Louis Cordell; McCabe, William Stokes; Newton, William Albert; Oder, Charles Rollin Lorain; Parker, Thomas Reilly; Schroth, Eugene Howard; Utterback, Donald Desmond; Williams, John Raynesford; Winkler, Virgil Dean; Wrath, William Frederick

1940s: Agnew, Allen Francis; Alexander, Joseph Watrous; Barnes, Mary Elizabeth; Bauer, Charles Bruce; Beard, Charles Noble; Brokau, Arnold Leslie; Brophy, John Allen; Cassin, Richard Joel; Clay, John Otis; Cordell, Robert James; Dietz, Robert Sinclair; Dillon, Edward Lamblin; Eddings, Arnold Lester; Ellingwood, Robert Whitcomb; Emery, Kenneth Orris; Eveland, Harmon Edwin, Jr.; Feray, Daniel E.; Finfrock, Lawrence J.; Fisher, James Harold; Fisher, Richard Forrest; Ford, Glen Melvin; Geisler, Jean Marie; Girhard, Mary Nancy; Gollnick, Robert L.; Gregg, William Nathan, Jr.; Grote, Benjamin; Gutschick, Raymond Charles; Hagan, Wallace Woodrow; Harrison, John Albert; Henton, John Melvin, Jr.; Holmes, Leslie Arnold; Honea, Robert Clair; Irish, Ernest James Wingett; Irvin, William Carl; Irwin, Melvin LeRoy; Johnson, Dorothy Bernice; Kennedy, Virgil John; Kesling, Robert Vernon; Kesling, Robert Vernon; Knodle, Robert Day; Koenig, Karl Joseph; Koenig, Karl Joseph; Kraye, Robert Frank; Lamb, Robert Reid; Lewark, James Edward; Livesay, Elizabeth Ann; Lynch, Bernard Walden; Meyer, Marvin Phillip; Miller, Don John; Morton, Robert Brading; Oesterling, William Arthur; Osment, Frank Carter; Otton, Edmond George; Pampe, William Riley; Patton, Howard Lewis; Pendleton, Margaret Meda; Reynolds, Robert Ramon; Rogers, Robert Errett; Saxby, Donald B.; Shaver, Robert Harold; Simon, Jack Aaron; Simpson, Howard Edwin, Jr.; Smith, Maurice Harold; Spotti, Adler E.; Stephens, Robert Monroe; Summerson, Charles Henry; Summerson, Charles Henry; Taylor, Warren LeRoy; Templeton, Justus Stevens; Threet, Richard Lowell; Walk, Hugh Gerard; Weinberg, Edgar Leon; White, Harold Richard; White, William Arthur; Wilson, George Miller; Winkler, Virgil D.; Ziebell, Walter Richard

1950s: Allsman, Paul Lewis; Ames, John Alfred; Amos, Dewey Harold; Amos, Dewey Harold; Andersen, Marvin John; Ault, Curtis Henry; Aye, Tin; Baird, Donald Wallace; Baker, Jack; Baldwin, Donald Carl; Ballmann, Donald Lawrence; Ballmann, Donald Lawrence; Bandy, James Chapman; Bartleson, Bruce Landon; Barton, Charles Addison; Baxter, James Watson; Benson, Richard Hall; Benson, Richard Hall; Berman, Byrd Louis; Bickford, Marion Eugene, Jr.; Bierschenk, William Henry; Biggs, Donald Lee; Bishop, Robert Eugene; Boardman, Richard Stanton; Bohor, Bruce Forbes; Borden, Eu-

gene William; Braumiller, Allen Spooner; Bredehoeft, John Dallas; Brockhouse, Robert Burton; Brophy, John Allen; Brown, George Donald, Jr.; Brown, Henry Seawell; Brown, Henry Seawell; Brownfield, Robert Lee; Burgener, John Albert; Buschbach, Thomas Charles; Buschbach, Thomas Charles; Byrne, Patrick James Sherwood; Caldwell, William Stone; Century, Jack Remo; Chamblin, William Jack; Chapman, John Judson; Christy, Robert Brandt; Clegg, Kenneth Edward; Cofer, Harland Elbert, Jr.; Collins, Barbara Jane Schenck; Collins, Lorence Gene; Collins, Lorence Gene; Conlin, Richard Renault; Corchary, George Sutter; Cramer, Howard Ross; Cropp, Frederick William III; Cropp, Frederick William, III; Cygan, Norbert Everett; Decker, Jack Minrod; Deere, Don Uel; Dellenback, Charles Richard; Dickie, George Allan; Doehler, Robert William; Doehler, Robert William; Doyle, Frank Larry; Droste, John Brown; Droste, John Brown; Dyni, John Richard; Eberly, Lyle Dean; Eccles, John Kerby; Etheredge, Forest de Royce; Eveland, Harmon Edwin, Jr.; Farrelly, Peter Joseph; Ferguson, John Alexander; Fisher, James Harold; Fisher, Robert Wilson; Foss, Ted Harry; Foster, Jack D.; Fox, Mary Ann Strouse; Fox, Robert Eugene; Frund, Eugene; Fuchs, Robert Louis; Gadd, Nelson Raymond; Garrett, James Hugh; Gilman, Richard Atwood; Gore, Dorothy Jean; Gossett, Charles Joseph; Grinnell, Robert Newell; Grossman, Stuart; Haack, Norman Erwin; Hackett, James Edward; Hallstein, William Weyrich; Hardie, Charles Henning; Harrison, Jack Edward; Hathaway, John Cummins; Hay, William Winn; Heinz, David Michael; Hopkins, M. E.; Howard, Richard Henry; Hunt, John Bancroft; Hutcheson, Donald Wade; Johns, William Davis; Johnson, Robert Britten; Jonas, Edward Charles; Jonas, Edward Charles; Karrow, Paul Frederick; Kidda, Michael Lamond; Kosanke, Robert M.; Lane, Donald Wilson; Lane, Phillip Jene; Langer, Milton Friedrich; Lennon, Russell Bert; Levish, Murray; Lewis, Joseph Thomas; Lloyd, Ronald Michael; Locker, Walter Augustine, Jr.; Lucas, Margaret Jennifer; Lucas, Margaret Jennifer; Lundwall, Walter Raymond, Jr.; Lynch, Thomas Wimp; Major, Charles Fredrick, Jr.; Mason, Arnold Caverly; McAllister, Raymond Francis, Jr.; McCormick, Wade Lowery; McDivitt, James Frederick; McGregor, Jackie Delaine; McNitt, James Raymond; Mercurio, Richard Nicholas; Metzger, William John; Meyer, Jurg Walter; Miller, William Frank; Monroe, Eugene Allen; Moody, Dwight Millington; Moore, John Ezra; Morrill, David Currier; Motts, Ward Sundt; Mueller, Joseph Charles; Muller, Ernest Hathaway; Murray, Hayden Herbert; Murray, Haydn Herbert; Myers, Robert Errol; Narain, Kedar; Neely, Florence; Nelson, Bruce Warren; Newell, Hildreth Adele; Niemann, Robert Leslie; Niyogi, Dipankar; Norman, Emmerson Kirkpatrick; Odom, Ira Edgar; Offield, Terry Watson; Ostrom, Meredith Eggers; Ostrom, Meredith Eggers; Palmer, James Edward; Parham, Walter Edward; Parker, Margaret Ann; Patterson, Ben Arnold; Patterson, Dale Duane; Patterson, Jacqueline Woodman; Patterson, Sam Hunting; Peppers, Russel Allen; Phillips, Sanford Ingels; Phillips, Scott Harlan; Pierce, Jack Warren; Plebuch, Raymond Otto; Porter, John Seaman; Powers, Richard James; Pryor, Wayne Arthur; Pullen, Milton William, Jr.; Rall, Elizabeth Pretzer; Rall, Raymond Wallace; Randall, Allan Dow; Rich, Mark; Rimsnider, Donald Orin; Rioux, Robert Lester; Rioux, Robert Lester; Roberson, Herman Ellis; Robertson, Donelson Anthony; Roddie, W. G.; Rogers, John Robert; Roth, Robert Sidney; Roth, Robert Sidney; Schultz, Leonard Gene; Schultz, Leonard Gene; Scott, Alan Johnson; Scott, Harold; Scott, Robert Brown; Searight, Thomas Kay; Sestak, Andrew Aloysius; Shannon, Ellen Carol; Shaver, Robert Harold; Shelton, John Wayne; Shelton, John Wayne; Shepps, Vincent Chester; Shepps, Vincent Chester; Shrode, Raymond Scott; Siddhanta, Sushil Kumar; Silverman, Maxwell; Sims, Dewey Leroy; Sitler, Robert Francis; Sitler, Robert Francis; Slovinsky, Raymond LeRoy; Smith, Guy William; Smith, William Horn;

Smoot, Thomas William; Sneed, Henry Eugene; Snodgrass, Donald Blaine; Snyder, Robert Dean; Sohl, Norman Frederick; Sohl, Norman Frederick; Spencer, Charles Winthrop; Sprouse, Donald West; Staffeld, Byron Clifford, Jr.; Staplin, Frank Lyons; Stevenson, Wilbur Lloyd; Stone, John Elmer; Susong, Bruce Irvin; Tharin, James Cotter; Tisza, Stephen Thomas; Titus, Frank Bethel, Jr.; Tom, Gene Francis; Tooker, Edwin Wilson; Toy, Billy Reynolds; Tranter, Charles Enoch, Jr.; Van Den Berg, Jacob; Van Horn, Clifford Layne; Vera, Elpidio de la Cruz; Vineyard, William Lawton; Voris, Richard Hensler; Wafer, James Oscar; Wahl, Floyd Michael; Wahl, Floyd Michael; Wainwright, John Ernest Nolan; Wainwright, John Ernest Nolan; Walters, Mathias Joseph; Weart, Richard Claude; Webb, David Knowlton, Jr.; Wehrenberg, John Patterson; Wehrenberg, John Patterson; Weill, Daniel Francis; Wertman, Ronald La Mar; White, William Arthur; Whiting, Lester Le Roy; Williams, Eugene Griffin; Williams, Frederick Enslow; Williamson, Lee Foster; Willis, Ronald Porter; Wilson, Gene Douglass; Wilson, Roger Lenox; Winar, Richard Marion; Winslow, John Durfee; Witherspoon, Paul Adams, Jr.; Wright, Roland Finley; Wright, Thomas Lee; Zadnik, Valentine Edward; Ziebell, Warren Gilbert; Ziemba, Eugene Anthony; Zirkle, Robert Gale

1960s: Adams, Russell Stanley, Jr.; Aiyer, Arunachalam Kulathu; Alcordo, Isabelo Suelo; Allen, Donald Bruce; Allman, David William; Amin, Mohammad; Avcin, Matthew John, Jr.; Baker, Robert Jethro, Jr.; Balbach, Margaret Kain; Baroffio, James Richard; Benak, Joseph Vincent; Berger, Richard Lee; Berger, Richard Lee; Bickford, Marion Eugene, Jr.; Bleuer, Ned Kermit; Boudreaux, Joseph Edes; Boyer, Paul Rice; Boyer, Paul Rice; Bredehoeft, John Dallas; Bromfield, Calvin Stanton; Brower, Ross Dean; Burns, Gerald Ray; Butler, Louis Winters, II.; Carl, James Dudley; Carl, James Dudley; Carr, Peter Alexander; Carss, Brian Williams; Carss, Brian Williams; Cassity, Paul Edward; Chase, Livingston; Cheng, Chaonang; Cherry, John Anthony; Christiansen, Earl Alfred; Chryssafopoulos, Hanka Wanda Sobczak; Clayton, Lee Stephen; Clemency, Charles Valentine; Colquhoun, Donald John; Coogan, Alan Hall; Coon, Richard Floyd; Cording, Edward James; Cote, William Emerson; Couch, Elton Leroy; Cygan, Norbert Everett; Davis, Richard Albert, Jr.; de Figueiredo Filho, Paulo Miranda; de Villiers, Johan Pieter Roos; DiBiagio, Elmo Lawrence; Dickson, Beryl Ann; Dixon, Gordon Elliott; Dow, Garnett McCormick; Dow, Garnett McCormick; Dudley, William Wyatt, Jr.; Dudley, William Wyatt, Jr.; Dunlap, William Howard; Dunn, Darrel Eugene; Eades, James Lynwood; Ealey, Peter John; Ealey, Peter John; El-Ashry, Mohamed Taha; El-Ashry, Mohamed Taha; Eldridge, William Frederick; Elliott, Richard Alden; Ellwood, Robert Brian; Ely, Richard Woodman; Emrich, Grover Harry; Faccioli, Ezio; Farvolden, Robert Norman; Fenner, Peter; Fenner, Peter; Ferrell, Ray Edward, Jr.; Ferrell, Ray Edward, Jr.; Finch, William Anderson, Jr.; Flaate, Kaare Sigfred; Frankenberg, Julian Myron; Frazee, Charles Joseph; Frost, Stanley Harold; Frost, Stanley Harold; Fullagar, Paul David; Furlong, Robert Burton; Furlong, Robert Burton; Gamble, James Clifton; Gartner, Stefan, Jr.; Gartner, Stefan, Jr.; Gates, Richard Holt; Gaudette, Henri Eugene; Gaudette, Henri Eugene; Gilman, Richard Atwood; Glover, Albert Douglas; Godwin, Robert Paul; Goodfield, Alan Granger; Goodfield, Alan Granger; Gordon, Joan Esther; Gorman, Donald Robert; Graf, Robert Bernard; Gray, Lewis Richard; Grice, Reginald Hugh; Gross, David Lee; Gross, David Lee; Guber, Albert Lee; Hanagan, Elizabeth Jean; Hatch, Joseph Ray; Hawley, John William; Heath, Christopher Peter Macclesfield; Heath, Christopher Peter Macclesfield; Heigold, Paul Clay; Heim, George Edward, Jr.; Henderson, Geral Vernon; Heuer, Ronald Eugene; Holder, Jon Thomas; Horne, John Corbett; Horne, John Corbett; Horton, Gary Walker; Hughes, George Muggah;

Hughes, Paul Warren; Hulse, John Arthur; Hulsey, Jess Dale; Hulsey, Jess Dale; Hussey, Arthur Mekeel, II; Inden, Richard Francis; James, Alan Thomas; Jester, Guy Earlscort; Johnson, Kenneth Sutherland; Johnson, Paul Richard; Johnson, William Hilton; Johnson, William Hilton; Karner, Frank Richard; Keller, George Henrik; Kempton, John Paul; Kennerley, John Brian; Kenney, Leland Frederick; Kiefer, John David; Kraatz, Paul; Krueger, Allen Reed; Lacey, James Edward; Lang, William Joseph; Langfelder, Leonard Jay; Leung, Samuel Seh-Shue; Leung, Samuel Seh-Shue; Liao, Kao Hsiung; Lowe, Donald Ray; Luce, Robert William; Lumsden, David Norman; Lundin, Robert Folke; Lundin, Robert Folke; Manos, Constantine Thomas; Manos, Constantine Thomas; Marks, John Wallace; Marszalek, Donald Stanley; Marszalek, Donald Stanley; McClellan, Guerry Hamrick; McComas, Murray Ratcliffe; McCullough, Patrick Terrence Peter; McGeary, David Fitz Randolph; McGinnis, Lyle David; McGregor, Jackie Delaine; McGuire, Odell S.; McLure, John William; McQuiston, Ian Brice; Meneley, William Allison; Merritt, Andrew Hutcheson; Merritt, Andrew Hutcheson; Merritt, Eleanor Walton; Mesri, Gholamreza; Metzger, Charles Frederick; Metzger, Charles Frederick; Metzger, William John; Miller, Donald George, Jr.; Misiaszek, Edward Thomas; Mitchell, Nolan William Ralph; Monroe, Eugene Allen; Moore, Duane Milton; Moore, Duane Milton; Moore, John Ezra; Moran, Stephen Royse; Moran, Stephen Royse; Morrison, Anne Kranek; Motto, Harry Lee; Mueller, Bruce Elmo; Niemann, Robert Leslie; North, William Gordon; North, William Gordon; Nunnally, Nelson Rudolph; O'Brien, Kathryn Gronberg; O'Brien, Neal Ray; O'Brien, Neal Ray; Odom, Ira Edgar; Orlopp, Donald Easton; Orlopp, Donald Easton; Page, Norman John; Palomino Cardenas, Jack Roger; Parham, Walter Edward; Parizek, Richard Rudolph; Patton, Franklin Davis; Peikert, Ernest William; Peppers, Russel Allen; Pierce, Robert William; Pierce, Robert William; Pinder, George Francis; Piskin, Rauf; Plusquellec, Paul Lloyd; Plusquellec, Paul Lloyd; Pride, Douglas Elbridge; Pulanco, Demetrio Hidalgo; Pulpan, Hans; Reinking, Robert Louis; Reinking, Robert Louis; Richardson, James Bushnell, III; Riggs, Elliott Arthur; Roche, James Edward; Roche, James Edward; Rogers, James Samuel; Rosenshein, Joseph Samuel; Roy, Donald Hilaire; Ruotsala, Albert Peter; Schaeffer, Katherine Maude Marie; Schmidt, Birger; Schnitker, Detmar Friedrich; Scott, John Stanley; Seaber, Paul Robert; Sevon, William David, III; Shea, James Herbert; Shideler, Gerald Lee; Shover, Edward Franklin; Simmonds, Robert Tobin; Simonds, Charles Henry, III; Smith, Richard Elbridge; Smith, William Calhoun; Soderman, Jarmo Georg William; Soderman, Jarmo Georg William; Soeria-Atmadja, Rubini; Somers, Lee Bert Hamill; Sorrell, Charles Arnold; Souter, James Edwin; Spanski, Gregory Thomas; Spanski, Gregory Thomas; Squier, Lyman Radley; Stephenson, David Arthur; Stieglitz, Ronald Dennis; Stone, John Elmer; Sturgul, John Roman; Tettenhorst, Rodney Tampa; Textoris, Daniel Andrew; Tharin, James Cotter; Totten, Stanley Martin; Totten, Stanley Martin; Towe, Kenneth McCarn; Trescott, Peter Chapin; Trescott, Peter Chapin; Triplehorn, Don Murray; Tubb, John Beaufort, Jr.; Tubb, John Beaufort, Jr.; Vadnais, Raymond R.; Vail, Ruth Staron; Vukovich, John William; Walpole, Robert Leonard; Wang, Fun-Den; Webb, David Knowlton, Jr.; Wedderburn, Leslie Ansel; Weiner, John Louis; Whitaker, Sidney Hopkins; Wilband, John Truax; Williams, Donald Roy; Williams, Roy Edward; Wilson, William Edward, III; Wingard, Paul Sidney; Wise, Sherwood Willing, Jr.; Withers, James Henry; Wobber, Francis John; Wolff, Roger Glen; Wolff, Roger Glen; Wright, Cynthia Ann Roseman; Wright, Cynthia Roseman; Wright, Ramil Carter; Wright, Ramil Carter; Yund, Richard Allen; Zadnik, Valentine Edward; Zeizel, Arthur John

1970s: Adachi, Kakuichiro; Aigen, A. A.; Allen, Donald Bruce; Andrews, R. W.; Avcin, Matthew J., Jr.; Ayer, Nathan John, Jr.; Bailey, J. B.; Balazs, Rodney J.; Banks, D. C.; Barnhardt, M. L.; Beaudry, Frederick H.; Benzel, William Marc; Bourque, Michael W.; Brownlee, M. E.; Buckley, Glenn Robert; Buckley, S. B.; Busch, William Henry; Byrd, William J.; Carroll, Michael Timothy; Cartwright, Keros; Castle, J. W.; Chamberlin, Thomas L.; Chamberlin, Thomas L.; Cole, Sally Ann; Coleman, D. D.; De Brito, Sergio N. A.; Der Kiureghian, Ahmen; Dowell, Thomas Perry Laning, Jr.; Dowell, Thomas Perry Laning, Jr.; Dreifuss, Sophie M.; Ettensohn, F. R.; Febres-Cordero, E. E.; Figueroa, J. L.; Follmer, Leon Robert; Forester, Elizabeth Brouwers; Forester, Richard M.; Forester, Richard M.; Fraser, Gordon Simon; Fraser, Gordon Simon; Gamble, James Clifton; Ganow, H. C.; Gerber, Murry S.; Getzen, Rufus T.; Goldman, Dennis; Gombos, Andrew Michael, Jr.; Goulter, I. C.; Graf, J. B.; Granath, James Wilton; Haldar, A.; Hamdan, Abdul-Latif; Hansmire, W. H.; Hanson, M. W.; Hatch, Joseph R.; Hebert, Glenn P.; Hill, Alan T.; Hiseler, Robert Bruce; Hooten, James E.; Hughes, Randall Edward; Hunt, S. R.; Jackson, R. G., II; Johnson, Donald Otto; Johnson, Peter Roy; Ju, Fu-Shyong; Kanji, Milton Assis; Karshenas, M.; Kemmis, Timothy J.; Kern, R. A.; Keys, John N.; Khan, M. H.; Khawlie, Mohamad R.; Khawlie, Mohamad R.; Khoury, H. N.; Khoury, M. A.; Kiefer, John David; Kim, D.; Kirkpatrick, R. James; Koelsch, T. A.; Kolata, Dennis Robert; Kuhn, Alan Karl; Kuhnhenn, G. L.; Kulla, J. B.; Lahann, R. W.; Lappin, Allen R.; Lawson, D. E.; Lee, S.; Lin, Chang-Lu; Link, R. L.; Lu, Po-Yung; Luman, D. E.; Ma, T.; Mahar, J. W.; Mahlburg, Suzanne E.; Maltman, Alexander James; Maltman, Alexander James; Martin, Michael David; Maruri, Raul D.; Mason, Robert Michael; Masters, Bruce Allen; McCullough, Patrick Terrence Peter; McCullough, Sahar A.; McKay, E. D., III; McKay, E. D., III; Mercer, James Wayne, Jr.; Miller, Donald David; Miller, James A.; Miller, William Gossett; Monsees, James Eugene; Moore, Michael C.; Mosher, S.; Nelson, Walter John, Jr.; Newberry, Ralph Jeffrey; Nieto-Pescetto, A. S.; Norby, Rodney D.; Nowak, Frank John; Nowak, Frank John; O'Rourke, T. D.; Olimpio, J. C.; Olimpio, J. C.; Olsson, William Arthur; Palladino, Donald Joseph; Palomino Cardenas, Jack Roger; Parkhurst, J. I.; Price, Annette; Price, Larry Wayne; Provo, Linda Jeanne; Ranalli, Giorgio; Rao, Prasada C.; Reinbold, Mark Lester; Rice, William David; Ringler, R. W.; Risatti, J. B., Jr.; Rogers, James E., Jr.; Rogers, James E., Jr.; Rogers, Leah Lucille; Russell, Suzanne J.; Ryan, John F.; Sanders, Richard Pat; Santogrossi, Patricia A.; Sargent, Michael L.; Scheihing, Mark H.; Schluger, Paul Randolph; Schmidt, Alan James; Schneidermann, Nahum; Schwartz, Franklin Walter; Seto, H. G.; Seyyedian-Choobi, M.; Sharp, John Malcolm, Jr.; Shayani, Sohrab; Simonds, Charles Henry; Sinclair, Brian J.; Size, William Bachtrup; Skrzyniecki, Alan Francis; Skrzyniecki, Alan Francis; Slorp, L. H.; Smallwood, Alan Robert; Smosna, Richard Allan; Smosna, Richard Allan; Spangler, D. R.; Stanker, Larry Henry; Steele, Ian McKay; Steinmetz, J. C.; Stephens, Michael P.; Stepusin, Susan M.; Stieglitz, Ronald Dennis; Stindl, Heribert; Stricker, Gary Dale; Suchomel, T. J.; Tarkoy, P. J.; Taylor, Gilbert D., Jr.; Thakur, Tukrel Radhakishin; Tills, Linda Ann; Tissue, Jeffery Stephen; Tollefson, Linda Joyce Sindelar; Trefz, Richard Joseph; Troncoso, J. H.; Ullrich, C. R.; Van Ryswyk, Roy J.; Von Rhee, Robert Weston; Walker, J. P.; Ward, James G.; Weiner, William; White, Owen Lister; Whitehead, N. H., III; Whitney, C. G.; Wickham, J. T.; Wickham, Susan Specht; Wise, Sherwood Willing, Jr.; Worsley, Thomas Raymond; Wright, William Herbert, III; Wunder, Susan Jean; Zanbak, C.

1980s: Altaner, Stephen Paul; Babb, Robert Frederick, II; Bakush, Sadeg H.; Banaee, Jila; Bauer, Robert Alan; Benzel, William Marc; Bertani, Renato

Tadeu; Bethke, Craig Martin; Bhagat, Snehal; Bieler, David Bruce; Black, Nancy R.; Boakye, Samuel Yamoah; Bowers, Glenn Lee; Cahill, Richard Allen; Cardinell, Alex Phillip; Casavant, Deborah Ilgenfritz; Cepeda Díaz, Abel Fernando; Cheng, Bih-Ling Monica; Cheng, Shui-Tuang; Chia, Yee Ping; Choi, Ling-Kit; Chu, Kai-Dee; Cisneros, Lee Anne; Cobb, James Collins; Colten, Virginia Ann; Cornelius, Jeffrey Bernard; Costanza, Suzanne Helene; Couto dos Anjos, Sylvia Maria; Daniels, Eric Joseph; Dawers, Nancye Helen; Dawson, William Craig; Demir, Ilham; Diaby, Ibrahima; Diaby, Ibrahima; Dikmen, Seyyit Umit; dos Anjos, Sylvia Maria Couto; Elnawawy, Osman Ali; Elsbree, Hope Carole; Falkenhein, Frank Ulrich Helmut; Feizna, Sadat; Feiznia, Sadat; Fleeger, Gary Mark; Foote, Gary Ray; Fox, John Martin; Franke, Milton Romeu; Fritz, Jeffrey Lynn; Fryer, Karen Helene; Grimison, Nina Louise; Grosser, Paul W.; Guensburg, Thomas Edgar; Gustafson, Craig Warren; Hackley, Keith Crowell; Hageman, Steven James; Hansel, Ardith Kay; Harris, Henry John Hayden; Hartline, Laurie Elizabeth; Haskin, Mark Allan; Hayashi, Hiroshi; Haydon, Paul Richard; Haymes, David Edward; Heinrich, Paul Victor; Hess, Barry Samuel; Holm, Paul Eric; Hong, Sung Wan; Horton, Duane Gale; Huff, Bryan Gregory; Hunt, S. R.; Hutasoit, Lambok M.; Jacobson, Russel James; Kettles, Inez M.; Kim, Kwang Jin; Klassen, Rodney Alan; Kosobud, Ann Maxine; Krueger, Scott Raymond; Kung, Shyh-Yuan; Kuo, Lung-Chuan Joseph; Kuo, Lung-Chuan Joseph; Lasemi, Y.; Lasemi, Zakaria; Laubach, Stephen Ernest; Laubach, Stephen Ernest; Lear, Paul Robert; Leary, Richard Lee; Lee, Chung I.; Lee, Han-Lin; Lee, Yong Il; Lin, Cheng-Leo George; Lipman, Eric William; Marsaglia, Kathleen Marie; Matthews, James Coert; Maycotte, Jorge I.; McCabe, Steven Lee; McEachran, David Ballard; McHone, John Frank, Jr.; Miller, Michael Vernon; Mills, Patrick Clarence; Momen, Hassan Mostafa; Moore, David Warren; Mostaghimi, Saied; Nau, James Michael; Oestrike, Richard Wilson, Jr.; Oestrike, Richard Wilson, Jr.; Okhravi, Rasool; Olsen, Mikael Per Jexen; Ou, Joyce Ling-Mei; Owen, Michael Rainey; Owen, Michael Rainey; Park, Seung Woo; Park, Young, Jr.; Pearson, Corrinne Dorset; Phienweja, Noppadol; Phillips, Bruce Edwin; Pires, Jose Antunes; Pollock, David Warren; Popp, Brian Nicholas; Popp, Brian Nicholas; Reichelderfer, Jan L.; Rich, D. W.; Rodriguez Perez, C. E.; Rojstaczer, Stuart Alan; Roscoe, Bradley Albert; Roy, William Robert; Saadeghvazari, Mohamad Ala; Sale, Michael John; Schaetzl, Randall John; Schlosser, Isaac Joseph; Schrodt, Joseph Keith; Schuster, David Conway; Shepard, John Lynn; Shroba, Cynthia Susan; Siyam, Youssuf Mustafa; Smith, Laura Ann; Smith, Lawson Mottley; Snorrason, Arni; Snyder, Edward McKinley; Stanke, Faith Alane; Stecyk, Amy N.; Stephanatos, Basilis Nikolaos; Stump, Daniel; Styles, Thomas Richard; Sweeney, J. J.; Sweet, Michael Louis; Tabor, John Raymond; Timken, Hye Kyung Cho; Treworgy, Janis Driver; Vaiden, Robert Clifford; Van Sint Jan, Michel Leopold; Vergo, Norma; Von Bergen, Donald; Von Bergen, Donald; Vonderohe, Alan Paul; Watso, David Charles; Weibel, Carl Pius; Weibel, Carl Pius; Westgate, Linda Marie; Wildanger, Edward George; Winston, Richard Baury; Wolff, Breno; Woo, Kyung Sik; Yang, Wang Hong; Yang, Wang-Hong Alex; Yong, Yan; Zerva, Aspasia

Illinois Institute of Technology
Chicago, IL   60616

1 Master's, 5 Doctoral

1960s: Hsia, Yu-ping; Wetzel, Richard Allen

1970s: Boonlayangoor, C.; Varadhi, S. N.

1980s: Gausseres, Richard Francis; Jahedi, Jamshid

Indiana University of Pennsylvania
Indiana, PA   15705

13 Master's

1960s: Liebfreid, Doris Jean; Ramsey, James E.; Wolfe, Ronald A.

1970s: Ellenberger, John L.; Feather, Ralph Merle, Jr.; Park, Edwena Kay Eger; Pina, Jon J.; Wells, James Alan; Zampogna, Ralph V.

1980s: Brink, Terry L.; Lightcap, Dixon Samuel, II; McCandless, Susan L.; Repine, Thomas Edward, Jr.

Indiana State University
Terre Haute, IN   47809

51 Master's, 21 Doctoral

1920s: Reeves, John Robert; Reeves, John Robert

1960s: Dinga, Carl F.

1970s: Ainscough, Harlen R.; Alger, L. H.; Bruner, Diane Hyslop; Dinga, Carl F.; Ehrenzeller, Jeffrey L.; Elhami, Rahmatollah; Friedmann, Anton R.; Gardner, Joseph Vincent; Garza, Roberto; Harris, John O.; Kelty, Barbara M.; Kilgore, David L.; Kind, T. C.; Love, D. L.; Maness, Lindsey V., Jr.; Mueller, Donald; Ozier, Ronald L.; Patrick, R. R.; Sakalowsky, P. P., Jr.; Shahabi, Mohammad Ali; Van Maness, L., Jr.; Weber, Neil Victor; White, David L.; Witherspoon, James Mark

1980s: Adidas, Eric O.; Al-Hurban, Adeeba; Alger, Leonard Hugh, Jr.; Badiei, Jalil; Barry, William Leo; Bayless, Michael Lynn; Beaven, Lee Wilson; Beck, Robert Lynn; Bengert, Stephen R.; Bridges, Katherine H.; Cwick, Gary J.; Duweluis, John A.; Elbert, J. A.; Elbert, Julie Ann; Faflak, Richard E.; Fitzpatrick, Michael F.; Hammen, John Leo, III; Hermance, William; Herner, Robert R.; Hoggatt, Leslie J.; Howe, Martin R.; Huang, Kai-Yi; Hyde, Richard Franklin; Kalaswad, Sanjeev; Lee, Jae K.; Levine, Norman Seth; Lulla, Kamlesh Parsram; Madison, Patrick James; Manwaring, Mark S.; Marh, Bhupinder Singh; Moore, Jamison S.; Omolo, Fenner O.; Parraga, Felipe; Pipentacos, John; Posner, Alex; Repic, Randall L.; Schinderle, Denis W.; Spraker, Larry A.; Stanley, Steven A.; Talbett, Michael Steven; Thompson, Donald Merrell; Truccano, Norman D.; Vaughn, Danny M.; Wilson, William L.; Woodson, Frederick Jennings

Indiana University, Bloomington
Bloomington, IN   47405

469 Master's, 223 Doctoral

1880s: Moore, David Ross

1890s: Call, Richard Ellsworth; Price, James Arra

1900s: Greene, Frank Cook; Mauck, Abram Vardiman; Reagan, Albert B.; Shannon, Charles William; Smith, Essie Alma; Snider, Luther Crocker; Tucker, William Motier

1910s: Bonsib, Ray Myron; Bybee, Halbert Pleasant; Bybee, Halbert Pleasant; Coryell, Horace Noble; Cumings, E. R.; Dubois, Henry Mathusalem; Galloway, Jesse James; Galloway, Jesse James; Jackson, Thomas Franklin; Jackson, Thomas Franklin; Malott, Clyde Arnett; Malott, Clyde Arnett; McEwan, Eula (Davis); McEwan, Eula (Davis); Tucker, William Motier; Vickrey, Earl Wayne; Whitmarsh, James Hardin; Wood, Harry Warren

1920s: Addington, Archie Rombaugh; Baker, Lora May; Bartle, Glenn Gardner; Cumings, Edgar Roscoe; Esarey, Ralph Emerson; Ferguson, Luther Short; Hunt, Raymond Samuel; Lee, Glen A.; Lucas, Elmer Lawrence; Malott, Burton Joseph; Martin, Viva Erma; Moore, Prentiss D.; Rawles, William Post; Roark, Louis; Schrock, Robert Rakes; Shrock, Robert Rakes; Shrock, Robert Rakes; Stockdale, Paris Buell; Stouder, Ralph Eugene; Thomas, William Avery; Whitlatch, George Isaac; Zierer, Clifford Maynard

1930s: Allen, William Odis; Tipsword, Howard Lee; Bartle, Glenn Gardner; Bates, Robert Ellery; Beard, Charles Noble; Bradfield, Herbert Henry; Bradfield, Herbert Henry; Fidlar, Marion Moore; Fix, Philip Forsyth; Freed, George Richard; Harrell, Marshall Allen; Heap, George; Huddle, John Warfield; Manning, Christine; Payne, Kenneth Armstrong; Reeves, James Elmo; Rose, John Kerr; Ross, Theodore William; Spencer, Charles Grason; Stockdale, Paris Buell; Thornbury, William David; Whitlatch, George Isaac

1940s: Bajza, Charles Carl; Bajza, Ester Ruth; Bandy, Orville Lee; Bush, James; Childs, Lewis; Dawson, Thomas Albert; Ericksen, George Edward; Fender, Hollis Blair; Fidlar, Marion Moore; Harris, Hobart Byron; Harris, John Rodefer; Heminway, Caroline Ella; Kottlowski, Frank Edward; McGrain, Preston; Parker, Raymond Lawrence; Passel, Charles Fay; Patton, John Barratt; Proctor, Paul Dean; Roberts, Carroll Norton; Schweers, Frederick Paul; Spangler, Walter Blue; Waddell, Courtney; Warner, Esther Ruth; Weidman, Robert McMaster; Winston, George Otis

1950s: Axenfeld, Sheldon; Barnes, James Virgil; Barr, John George; Barua Remy, Victor Felix; Bieberman, Robert Arthur; Biggs, Maurice Earl; Bohor, Bruce Forbes; Bowen, Richard Lee; Boyce, Malcom Walter; Boyer, Robert Ernst; Brookley, Arthur Clifford; Brueckmann, John Edward; Brunton, George Delbert; Bundy, Wayne Miley; Bundy, Wayne Miley; Cameron, Donald Kenzie; Cass, John Tufts; Chen, Pei-Yuan; Christensen, Evart Wayne; Clark, Dean Stanley; Connaughton, Charles Robert; Core, Eugene Howard; Cowen, Michael Terence; Crites, William Henry; Deane, Harold Lutz; Devening, Donald Clayton; Edwards, Benjamin; Erd, Richard Clarkson; Erickson, Wayne Albert; Fiandt, Dallas N.; Fish, Ferol Fredric; Fitch, Richard; Flanagan, William Hamilton; Fowler, Wayne Edward; Fowler, Wayne Edward; Frielinghausen, Karl William; Frugoni, James John; Galloway, Jesse J.; Gilmore, John W.; Gravenor, Conrad Percival; Greenberg, Seymour Samuel; Greenberg, Seymour Samuel; Grender, Gordon Conrad; Hale, Elbert, Lee; Harrison, Jack Lamar; Harrison, Jack Lamar; Heisterkamp, Warren Craig; Henderson, Eric P.; Henry, Gary E.; Herr, George Albert; Holloway, Perry Gregory; Holt, Olin R.; Horowitz, Alan Stanley; Hughes, John Herbert; Hutchison, Harold Christy; Hyer, Donald Eugene; Jenkins, Robert David; Johnson, Frank Marion; Jones, Thomas David; Kaska, Harold Victor; Kline, Beryl Dale; Koenig, James Bennett; Kottlowski, Frank Edward; Kugler, Harry Wesley; Larson, Richard Walter; Lebauer, Lawrence Robert; Leininger, Richard Keith; Mahorney, James Robert; McCammon, Helen Mary; McCammon, Richard Baldwin; Miesch, Alfred Thomas; Moore, George Thomas; Moore, George Thomas; Nevers, George Morrison; Newcomer, Earle Seifried; Noel, James Arthur; Ogle, Ronald Kent; Olson, Victor Emanuel; Patton, John Barratt; Peirce, Howard Wesley; Pickering, Ranard Jackson; Pinsak, Arthur Peter; Pinsak, Arthur Peter; Prinz, Martin; Pruett, Frank Donald; Puscas, George; Rago, Frank Thomas; Raymond, Paul Cletus; Renzetti, Bert Lionel; Renzetti, Bert Lionel; Reshkin, Mark; Revetta, Frank A.; Riely, Samuel Leander; Riggs, Kenton Nile; Roemermann, Donald Gregory; Rooney, Lawrence Fredrich; Rooney, Rosalia Eugenia (Rey); Rudman, Albert Julius; Rudnyansky, Albert Julius; Saenger, Robert Craig; Sargent, Robert Edward; Sayyab, Abdullah Shakir; Schemehorn, Neil R.; Scudder, Phyllis Jeanne; Sirvas, Ernesto; Six, Don Eldon; Smith, John Millard; Smith, Ned Myron; Sorgenfrei, Harold; St. Jean, Joseph; St. Jean, Joseph; Summers, William Kelly; Taylor, Ira Daniel; Taylor, Stuart Ross; Taylor, Waller Eugene; Theodosis, Steven Daniel; Treadway, Keith Richard; Tudor, Daniel Strain; Vance, Kenneth Raymond; Von Tress, David Edward; Waddell, Courtney; Wagner, George Robert; Wakeman, James Fisher; Warner, Marvin Eugene; Warren, Elbert Clay; Warren, Jack Roland; Waters, Kenneth Montelle, Jr.; Wayne,

William John; Wayne, William John; Whaley, Joseph Floyd; White, Richard LeRoy; Wier, Charles Eugene; Wier, Charles Eugene; Wirey, Gary Lee; Yurkas, George John

1960s: Balthaser, Lawrence Harold; Barrero, Dario; Beckman, Richard John; Berry, Bernard Richard; Bhattacharya, Nityananda; Bloom, James Clifford; Bork, Kennard Baker; Brown, George Donald; Bubb, John Neal; Burfeind, Walter John; Burger, Henry Robert; Butler, Roy Elbert; Campbell, Carl Earl; Capozza, Frank Cataldo; Carr, Donald Dean; Cleveland, John Herbert; Cohen, Stephen; Colville, Alan Andrew; Conley, Jack Francis; Cox, Hollace Lawton; Crane, Ronald Clinton; Cuffey, Roger James; Davis, Edward Louis; DeRudder, Ronald Dean; DeRudder, Ronald Dean; Dixon, William Gordon; Doheny, Edward John; Eckerty, Donald Gayle; Elberty, William Turner; Engelhardt, Donald Wayne; Frey, Robert Wayne; Gan, Tjiang-Liong; Gates, Gary Rickey; Gates, Gary Rickey; Girdley, William Arch; Glass, Susan Elizabeth; Goulden, Clyde Edward; Guennel, Gottfried Kurt; Hanna, William Francis; Harvey, Richard David; Hatfield, Craig Bond; Hatfield, Craig Bond; Heckard, John Martin; Heim, Herbert Carl; Hess, David Filbert; Hirschmann, Thomas Simon; Howard, James Franklin; Hrabar, Stephanie Vladimira; Jacobs, Alan Martin; Johnson, Gerald Homer; Khawaja, Ikram Ullah; Kissling, Don Lester; Klusman, Ronald William; Knudtson, Lee Gardiner; Laney, Robert L.; Lebauer, Lawrence Robert; Leckie, George Gallie; Leckie, George Gallie; Lineback, Jerry Alvin; Ludman, Allan; Luther, Lars Christian; Mathews, David Lane; Megard, Robert Ordell; Miller, Ralph Wayne; Mound, Michael Charles; Mound, Michael Charles; Mueller, Wayne Paul; Nelson, Warren Lee; Nicoll, Robert Sherburne; Orr, Robert William; Palmer, Arthur Nicholas; Parker, John Stephen; Parrish, Irwin S.; Perkins, Ronald Dee; Pirie, Robert Gordon; Pirie, Robert Gordon; Ploger, Sheila Lynn Wagner; Pratt, Alan Rogers; Renick, Howard, Jr.; Renzetti, Phyllis Jean; Reshkin, Mark; Reynolds, Douglas Wade; Richard, Benjamin Hinchcliffe; Rodriguez, Joaquin; Root, Forrest Keith; Rudman, Albert Julius; Sardi, Otto; Schaiowitz, Michael; Smith, Ned Myron; Stevenson, Ralph Girard, Jr.; Straw, William Thomas; Straw, William Thomas; Sunderman, Jack Allen; Tank, Ronald Warren; Utgaard, John Edward; Utgaard, John Edward; Van Coutren, Lewis Anderson; Wagner, Sheila L.; Wilson, Daniel Allen; Wiltse, Milton Adair, Jr.; Wood, Joseph Miller

1970s: Alamos Ovejero, Julio; Alexander, Richard Raymond; Amadi, Philip Uchenna Mbanu; Anderson, Garry Gayle; Anderson, Stanley Wayne; Anstey, Robert Leland; Archer, Allen W.; Ausich, William Irl; Ausich, William Irl; Bassett, John L.; Basu, Abhijit; Bauman, Jeanette M.; Belak, Ronald; Benham, Steven R.; Bennett, Nathan Paul; Bergeron, Marcel P.; Biggs, Maurice Earl; Binford, M. W.; Blakely, Robert Fraser; Borgerding, Janet Lee; Bottjer, David J.; Bradbury, Kenneth Rhoads; Brittain, Alan Lee; Brumbaugh, David Scott; Brunson, Karen L.; Budnick, Dorene M.; Calengas, Peter Leonard; Caserotti, Phillip Mark; Chandler, Val William; Christiansen, Jack Hilbert; Clark, David W.; Clere, David Russell; Cody, Clyde A.; Cordua, William Sinclair; Corneliussen, Eric Frantz; Crisman, T. L.; Crisp, Edward Lee; Dahl, Peter Steffen; Darnell, Nancy Rebecca; Dembicki, Harry; Desmarais, David John; Dixon, Joseph A.; Dobecki, Thomas Lee; Duigon, Mark Thomas; Duncan, Mark Stewart; Epstein, Mark L.; Fetter, Charles Willard; Forbes, Dorothy Ann; Fout, James Scott; Fowler, Michael Lee; Friberg, James Frederick; Friberg, LaVerne Marvin; Gaddah, Ali Hadi; Games, Larry Martin; Giffin, Jon W.; Glore, Charles Richard; Graham, Michael James; Haase, C. S.; Hall, Robert Dean; Hall, Stephen J.; Hamilton, Michael Miller; Hamilton, Timothy Scott; Hanley, Thomas Brainard; Harrington, Charles Dare; Hartsough, Gregory Warren; Henderson, Gerald J.; Henderson, Stephen William; Hirsch, Stuart; Hohn, Michael Edward; Hohn, Michael Edward; Holm, Melody Ruth;

Huang, Wen Yen; Huffman, Samuel Floyd; Immega, Inda Proske; Immega, Neal Terry; Jackson, Thomas Joseph; James, William Calvin; Kammer, Thomas William; Kayes, Douglas M.; Kelly, Stuart Mackenzie; King, Norman Ralph; Knapp, Ralph William; Knox, Larry William; Koehler, Steven William; Krisher, Daniel Lee; Kues, Barry Stephen; Kwon, Byung-Doo; Lake, Ellen A.; Lambert, Michael W.; Lane, Michael Arthur; Lazor, John David; Leonard, Mark Steven; Lesher, Carl Michael; Mack, Gregory Harold; Mack, Gregory Harold; Massell, Wulf Friedrich; Mazalan, Paul Alan; McLane, Michael John; Meyers, James Harlan; Millholland, Madelyn Ann; Mitchell, James Michael; Moreau, Peter Allan; Murchie, Bonnie; Okla, Saleh Mohamed; Olson, Carolyn G.; Olson, Carolyn G.; Orlich, Michael S.; Oslund, Jeffrey S.; Paine, Michael Henry; Pheifer, R. N.; Pickering, Ann; Porter, Elise White; Reynolds, John Lawrence; Rimstidt, Daniel L.; Rinaldi, Gianfrumco Giuseppe L.; Risser, Dennis W.; Rochester, Haydon; Satoskar, Vijay Vishnu; Schmidt, Christopher John; Schwartz, Robert Karl; Sexton, John Lloyd; Sheehan, Mark Charles; Sheffy, M. V.; Sheikh Ali, Khadim S.; Siemers, Charles Troy; Smith, David P.; Smith, John Livingstone; Stratton, James Forrest; Suchomel, Diane Marie; Sundeen, Daniel Alvin; Tendall, Bruce Alan; Thompson, Glenn Michael; Tilander, Nathaniel G.; Tilford, Maxwell Joseph; Tudor, Daniel Strain; Vane, Gregg Allen; Varma, Madan Mohan; Volz, Steven Alan; Wade, Jay Alan; Wahlman, Gregory Paul; Waldrip, David Bennett; Walker, Steven W. H.; Waterman, Arthur Stephen; Waters, Johnny Arlton; Waters, Johnny Arlton; Welch, James Robert; Wells, Jane Freeman; Wiechmann, Mark Jerdone; Wimberg, William B.; Wojtal, Aileen Marie; Young, Steven Wilford; Young, Steven Wilford

1980s: Abdulkareem, Talal Faisel; Abolkhair, Yahya Mohammed Sheikh; Al-Alawi, Jomaah Abd-Ulraheem Awad; Al-Aswad, Ahmad Abdullah; Al-Jassar, Tariq Jamil; Al-Jassar, Tariq Jamil; Alawi, Jomaah A.; Alger, Dean Wesley; Amadi, Philip Uchenna Mbanu; Ambers, Clifford P.; Andrews, Mark Stephen; Archer, Allen W.; Baker, Kevin L.; Ballard, William Turpin; Bandy, William F., Jr.; Bangs, Carol Lynn; Basick, James T.; Bayless, Edward Randall; Bean, Clarke Lee; Becker, Michael James; Behymer, Thomas David; Beier, Joy Ann; Beier, Joy Ann; Berkhouse, Gregory A.; Bernitz, John Alexander; Bhagavathula, Rao V.; Blickwedel, Roy; Bogardus, James W.; Bolton, David W.; Boucherle, Mary M.; Breedon, David H.; Brewster, David P.; Brockman, Allen R.; Broekstra, Scott Douglas; Bromley, Bruce Warren; Brown, Mark A.; Brown, Thomas William; Burke, David Alan; Carter, Pamela Hobart; Chang, Chia-Yu; Charles, Donald Franklin; Cocroft, John E.; Cutler, Jodi Lyn; De Celles, Peter G.; Dombrowski, Thomas; Dutta, Prodip Kumar; Ebbott, Kendrick Alan; Edkins, John E.; Fara, Daniel Ray; Farley, Martin B.; Feldman, Howard Randall; Feldman, Howard Randall; Ferderer, Robert J.; Filippini, Mark G.; Fishbaugh, David A.; Fituri, Hussein Saleh; Flores, Richard J.; Floyd, Larry Wayne; Foell, Christopher J.; Fout, James Scott; Funkhouser, Roy V.; Gibboney, Melissa J.; Giles, Billy E.; Graham, Michael James; Griffiths, Scott A.; Grove, Arlen K.; Gruver, Barbara L.; Gubala, Chad Paul; Guzman, Humberto A.; Hacker, Robert D.; Harter, Ty Andrew; Harvey, Colin Charles; Hash, Troy M.; Heberton, Richard P.; Heger, Paul A.; Hettenhausen, Roger L.; Himes, Gregory Tait; Hirt, David S.; Hoffman, Bradley C.; Hokanson, Neil B.; Hood, Lindsay Ann; Houck, Karen J.; Howard, W. Brant; Hudson, Michael R.; Isard, Scott Alan; Jackson, Stephen T.; Janssen, Janelle L.; Jenkins, Steven Drexel; Johnson, John J.; Johnson, Marcus W.; Johnson, Neal Carter; Jones, Jay H.; Jorgensen, Ronald Wilbur; Kammer, Thomas William; Kaufman, Alan Jay; Kelly, Stuart Mackenzie; King, James Michael; Kovach, Warren L.; Kron, Terryl Ray; Kuminecz, Cary Phillip; Kwolek, James Michael; Laferriere, Alan Price; Laferriere,

Alan Price; Langford, Richard P.; Laurin, Priscilla Rehnquist; Leibold, Arthur W., III; Lewis, Daniel D.; Leyden, Barbara Wilhelmina; Libra, Robert D.; Loretto, Thomas McLean; Madrid, George A.; Maliva, Robert G.; Manchester, Steven Russell; Mangold, Kent M.; Maples, Christopher Grant; Markisohn, David B.; Mathews, Jane E.; McGrath, Dennis J.; Merkl, Roland S.; Meyerholtz, Keith A.; Michaud, Denis Paul; Middleman, Bruce H.; Miller, Michael E.; Miller, Michael E.; Monson, K. David; Moore, Craig Hayden; Moran, Martin V.; Morgan, William Tony; Morganwalp, David William; Muffler, Steven A.; Nellist, William Edward; Nicol, Dorian L.; O'Connell, Anne F.; Oliver, Joseph W.; Opell, Douglas A.; Owens, Anthony D.; Patzkowsky, Mark E.; Perrin, Shannon S.; Perucca, Melissa A.; Peters, Janet; Petricca, Ann M.; Petrovski, David M.; Petzold, Daniel D.; Pfau, Gerchard Edmund; Porter, Elise White; Price, Robert C.; Pruett, Robert J.; Ray, Wendy L.; Reazin, David G.; Reifenstein, Mark; Reynolds, Douglas Wade; Ridgway, Kenneth D.; Risley, David E.; Robinson, William H.; Roy, William R.; Ryan, Ruth M.; Saines, Steven James; Salter, Timothy L.; Sarwar, A. K. M.; Saturni, Ben A.; Saunders, Margaret M.; Schaefer, John G.; Scholl, Layne A.; Schubert, Christopher E. Kohler; Schultz, Albert West; Schuyler, Jeffrey N.; Schwarzwalder, Robert Nathan, Jr.; Seelen, Mark Allan; Shannon, Kathleen Marie; Shewmaker, Sherman Nelson; Shultz, Albert W.; Sieverding, Jayne Louise; Smith, Bruce C.; Smith, Christopher R.; Smith, David L.; Smith, Stephen C.; Snyder, Kossouth; Soranno, Michael Andrew; Specht, Thomas Henry; Stuenitz, Holger; Stumpf, Gary A.; Swanson, William Alfred; Szpakowski, James; Szpakowski, Sally; Takigiku, Ray; Tanner, George F.; Taylor, Randall E.; Ten Eyck, James R.; Thomas, Andrew R.; Thomas, James M.; Thomas, Kimberly Kodidek; Thompson, Todd A.; Thompson, Todd Alan; Totten, William B.; Tweddale, John B.; Vandivier, John Carl, III; Vierma, Luis F.; Visher, Peggy M.; Warner, Scott David; Warren, Victoria L.; Webster, John Robert; Weintraub, Jill; Weiss, Garrett D.; Weiss, Paula; Yates, Martin G.; Yates, Martin G.; Yoder, Gary Eldon; Zinn, Lori A.

University of Iowa
Iowa City, IA 52242

493 Master's, 244 Doctoral

1890s: Houser, Gilbert Logan; Savage, Thomas Edmund; Thompson, George Fayette

1900s: Thomas, Abram Owen; Wilson, Malcolm E.

1910s: Cable, Emmett James; Dewey, Arthur Howard; Dick, Robert Irving; Howell, Jesse V.; Leighton, Morris Morgan; Muilenberg, Garrett Anthony; Patton, LeRoy Thompson; Shipton, Washburne D.; St. Clair, Stuart; Van Tuyl, Francis Maurice; Williams, Arthur James

1920s: Adams, John Emery; Apfel, Earl Taylor; Apfel, Earl Taylor; Barragy, Edward J.; Bay, Harry X.; Brown, Irvin Cecil; Cornwall, Dean Torrey; Dille, Glenn Scott; Dille, Glenn Scott; Fields, Harry Basil; Fillman, Louise Anna; Fillman, Louise Anna; Freie, Alvin John; Freie, Alvin John; Grawe, Oliver Rudolph; Howell, Jesse V.; Jordan, Rudolph Henry; Kay, George Marshall; King, Philip Burke; Ladd, Harry Stephen; Laudon, Lowell Robert; Littlefield, Max Sylvan; Lonsdale, John Tipton; Lugn, Alvin Leonard; Lugn, Alvin Leonard; Mortimore, Morris Edmon; Mortimore, Morris Edmon; Patton, LeRoy Thompson; Quinn, Alonzo Wallace; Raney, James H.; Rowser, Edwin M.; Schoewe, Walter Henry; Searight, Walter Vernon; Searight, Walter Vernon; Seashore, Robert Holmes; Sidwell, Raymond; Sidwell, Raymond; Stainbrook, Merrill Addison; Stainbrook, Merrill Addison; Thomas, Leonard C.; Unash, Cora Louise; Walter, Otto T.; Wentworth, Chester Keeler; Wentworth, Chester Keeler; Wilbur, Doris Marion; Williams, Arthur James; Williams, Myron T.; Woods, Earl Hazen

1930s: Atwater, Gordon I.; Baker, Roger Crane; Banks, Joseph Edwin; Bates, Robert Latimer; Bates, Robert Latimer; Bay, Harry X.; Bissell, Joseph Harold; Boyle, Huron Lee; Brady, Frank Howard; Brandenbure, F. Merrill; Briard, Vernon Eugene; Buhle, Merlyn Boyd; Cameron, Cornelia C.; Campbell, Charles Lyman; Clark, Edward Lee; Clement, George Muller; Clement, George Muller; Cline, Lewis Manning; Cline, Lewis Manning; Cooper, Byron Nelson; Cooper, Byron Nelson; Couser, Chester Wendall; Couser, Chester Wendall; Curry, H. Donald; Curtis, Dwight Kenneth; Ditsworth, Glenn William; Donnelly, Clarence William; Edmund, Rudolph William; Folk, Stewart Huntley; Frye, John Chapman; Frye, John Chapman; Furcron, Aurelius Sidney; Furnish, William Madison; Furnish, William Madison; Georgesen, Neils Christian; Gould, Donald B.; Gould, Donald B.; Grim, Ralph Early; Guest, Hardy Grady; Guest, Henry Grady; Hamilton, Robert Gilbert; Hamilton, Robert Gilbert; Johnson, Henry Luther; Jones, Victor Harlan; Knaack, Edward Leslie; Laudon, Lowell Robert; Leatsler, Maynard E.; Lees, Laurence Fitch; MacGaw, Bradford Kuhns; Miller, Paul Theodore; Miller, Paul Theodore; Moore, Carl Allphin; Nelson, Paul Hugh; Petrick, Glen; Powers, Elliot Holcomb; Powers, Elliot Holcomb; Rowser, Edwin M.; Scobey, Ellis Hurlbut; Scobey, Ellis Hurlbut; Smith, Cale Clinton; Spivey, Robert Charles; Spivey, Robert Charles; Stookey, Donald Graham; Stookey, Donald Graham; Talley, Gilbert Arthur; Tapper, Wilfred Bonno; Taylor, Earle Frederick; Taylor, Garvin Lawrence; Taylor, Garvin Lawrence; Thomas, Leonard C.; Thompson, Marcus Luther; Thompson, Marcus Luther; Vernon, Robert Orion; Vesely, Leon Robert; Whealdon, Edwin Phillips; Wilbur, Doris Marion; Wiringa, Leon Otis; Wood, John Edwin; Work, Paul Murray; Yoho, William Herbert

1940s: Adams, Clifford; Adams, Clifford; Alcock, Charles William; Amsden, Thomas William; Anderson, Franklin Joseph; Aronow, Saul; Baskette, Harry Buchanan; Beekly, Emerson Keagy; Bengtson, Carl Aners; Bennett, Robert Turner; Berg, John Robert; Berg, John Robert; Berninghausen, William Henry; Beveridge, Thomas Robinson; Beveridge, Thomas Robinson; Bissell, Joseph Harold; Boardman, Donald Chapin; Cameron, Cornelia C.; Carrier, John Baldwin; Cheetham, Robert Nelson; Condon, James Carl; Damon, Henry Gordon; Davis, Dan Arthur; Deuth, John Eakle; Downs, Robert Harold; Downs, Robert Harold; Easker, David George; Edmund, Rudolph William; Fan, Paul Hsiu-Tsu; Fan, Paul Hsui-Tsu; Graham, Jack Bennett; Graham, Jack Bennett; Gray, Warren Wilbur; Gross, Robert Erwin; Harris, Robert Alan; Harris, Stanley Edwards; Harris, Stanley Edwards; Hendricks, Herbert Edward; Hendricks, Herbert Edward; Hershelman, William Lee; Horick, Paul Joseph; Huffman, George Garrett; King, Evan Shelby, Jr.; Kos, Charles George; Larson, George Delmore; Larson, Wilbert Sanford; Lindvall, Robert Marcus; Marshall, Charles Harding; Mickelson, John Chester; Mickelson, John Chester; Milton, Arthur Alvern; Moore, Carl Allphin; Parizek, Eldon Joseph; Parizek, Eldon Joseph; Patterson, Sam Hunting; Peterson, Lorenz August; Peterson, Richard Frank; Péwé, Troy Lewis; Pierce, Guy Russell; Rau, Weldon Willis; Ringena, Delbert Bearne; Rodin, Evald Maurice; Ross, Donald C.; Schuldt, Walter Carl; Shirley, Brooke Howard; Smith, William Oliver; Summerford, H. Edgar; Swenson, Frank Albert; Swenson, Frank Albert; Travis, Fred Lee; Unklesbay, Athel Glyde; Unklesbay, Athel Glyde; Vorhis, Robert Carson; Wenberg, Edwin Hugo; White, Max Gregg, Jr.; Youngquist, Walter Lewellyn; Youngquist, Walter Lewellyn

1950s: Allison, Marvin Dale; Anderson, Gerald Kenneth; Bacho, Andrew Benjamin; Baker, Forrest Grant; Barker, Norman Kay; Beghtel, Floyd Woodrow; Block, Douglas Alfred; Bossort, Dallas Overton; Briard, Vernon Eugene; Burchfield, Gail Robert; Burgchardt, Carl Robert; Carpenter, Lee B.; Carson, John Richard; Case, Lyle Eldon;

Chenoweth, William Charles; Clark, David Leigh; Collinson, Charles William; Collinson, Charles William; Cox, Robert Gerald; Davis, Stanton Hoffman; Degenfelder, George John; Dixon, Howard Raymond; Dow, Verne Eugene; Dryden, Jacob Edward; Dunn, David Lawrence; Edmund, Richard Amos; Ethington, Raymond Lindsay; Farmer, Richard Echols; Feulner, Alvin John; Fisher, Cecil Coleman; Garner, Hessle Filmore; Garner, Hessle Filmore; Giedt, Norman Ray; Gleim, David Thomas; Gleim, David Thomas; Glenister, Brian Frederick; Gluskoter, Harold Jay; Goebel, Edwin DeWayne; Gooding, Ansel Miller; Gooding, Ansel Miller; Graham, Charles Edward; Gustafson, Velman Oscar; Hansman, Robert Herbert; Hansman, Robert Herbert; Harrington, Frederick Irving; Hayes, William Clifton, Jr.; Henderson, L. Brooke; Hoffer, Jerry Martin; Jenkinson, Lewis Frank; Jennings, Arnold Harvey; Jennings, Ted Vernon; Johnson, Carlton Robert; Johnson, Carlton Robert; Kallsen, Clarence Edward; Kastler, Neil Blair; Larimer, Ted Ray; Leighton, Donald Lewis; Lock, Frank Loren; Maher, Louis J. Jr.; Malmberg, Glenn Thomas; Mattox, Richard Benjamin; Meier, Mark Frederick; Millman, Dean Beardsley; Mooney, Albert Russell; Morgan, Charles Orville; Nave, Floyd Roger; Ostrander, Robert Earl; Palensky, John Joseph; Paslay, Jack Duane; Petrie, William Leo; Pickford, Peter John; Raitz, Charles Henry; Rau, Weldon Willis; Rexroad, Carl Buckner; Rogers, William Patrick; Ruhe, Robert Victory; Sargent, Kenneth Albert; Satterthwaite, Laurence Cyrus; Savre, Wayland Carlyle; Schroeder, Eugene Robert; Schroeder, Robert John; Skolnick, Herbert; Smart, Burton; Stenberg, Carroll Dean; Stevenson, David Lloyd; Stone, Gerald Leslie; Swanson, Cleo R.; Swanson, Roger Glenn; Sweet, Walter Clarence; Tasch, Paul; Thomas, Clifford Ward, Jr.; Traylor, Henry Grady; Treloar, Anne Marie; Warfield, Robert George; Wilson, Robert Lake; Wingert, John Richard; Zimmer, Louis George

1960s: Aaronson, Donald Bruce; Ali, Odeh Said; Ali, Odeh Said; Anderson, Henry Robert; Anderson, Wayne Irvin; Anderson, Wayne Irvin; Annambhotla, Venkata Subramanya Shastri; Armbrustmacher, Theodore Joseph; Barnett, Stockton Gordon, III; Bateman, Marcus Kelden; Beghtel, Floyd Woodrow; Beinert, Richard James; Borger, John Godfrey, II; Bowman, Jimmy Dean; Bromberger, Samuel H.; Bromberger, Samuel H.; Brown, Lewis M.; Cahill, Kevin Edwin; Campbell, Russell Boone; Church, Norman K.; Collins, Desmond Harold; Davis, Richard Arnold; Davis, Richard Arnold; Dechert, Hedy S.; Dixon, Joe Scott; Drahovzal, James Alan; Drahovzal, James Alan; Foster, Norman Holland; Goodwin, Peter Warren; Goodwin, Peter Warren; Gordon, Donivan Lewis; Guldenzopf, Emil Charles; Hart, Richard Royce; Hinman, Eugene Edward; Hodgkinson, Kenneth Allred; Holte, Karl Emrud; Hudson, Robert Frank; Jennings, Ted Vernon; Johnson, Richard Gustave; Keen, Douglass C.; Kirkpatrick, Terrence D.; Klapper, Gilbert John; Knapp, William Dale; Knorr, Jack H.; Knox, Burnal Ray; Koch, Donald L.; Kozak, Samuel Joseph; Kullman, John D.; La Valle, Placido D.; Lane, Harold Richard; Lane, Harold Richard; Lanyk, James Louis; Liebe, Richard Milton; Lohnes, R. A.; Luth, William Clair; Macomber, Richard Wiltz; Manger, Walter Leroy; McCaleb, James Abernathy; Mikesh, David L.; Mikesh, David Leonard; Milling, Marcus Eugene; Milling, Marcus Eugene; Moore, Richard Lee; Mossler, John Hamilton; Moyle, Richard W.; Nassichuk, Walter William; Nassichuk, Walter William; Nelson, Robert Stanley, Jr.; Nichols, Chester Encell; Oberg, Rolland; Oberg, Rolland; Oertel, George Frederick, II; Olson, Norman Keith; Pabst, Larry Dean; Palmquist, John Charles; Pestana, Harold Richard; Petersen, Gary Gene; Petersen, Morris Smith; Peterson, Morris Smith; Price, John E.; Prusok, Ridi Albin; Randau, Paul Clemens; Rohr, Gene M.; Rose, Jeannette Noel; Sanders, Robert Bruce; Sargent, Kenneth Albert; Schroeder, Richard John; Sellers, David Henry Aikins; Sholes, Mark

Allen; Simmons, George Mills; Smit, David E.; Smith, Robert Kay; Spinosa, Claude; Spinosa, Claude; Squarer, David; Steiner, Richard James; Straka, Joseph John; Straka, Joseph John, II; Thompson, Thomas Luther; Uyeno, Thomas Tadashi; Uyeno, Thomas Tadashi; Valentine, Robert Miles; Vanek, James R.; Warner, Mont Marcellus; Williams, Lyman O'Dell, Jr.

1970s: Akinbola, Johnson A.; Anderson, Harold Dean; Anderson, Raymond R.; Austin, George Stephen; Baesemann, John F.; Bailey, George B.; Barrick, James Edward; Barrick, James Edward; Becker, Paul John; Beinert, Richard James; Berchenbriter, Dean Kenneth; Blakey, Ronald Clyde; Bounk, Michael Joseph; Broadhead, Thomas Webb; Burdick, Dennis W.; Burkart, Michael R.; Caldwell, James Phaon; Campbell, Russell Boone; Carnaghi, Gary Louis; Chauff, K. M.; Chrisinger, Danny L.; Churchill, R. R.; Cibula, Duane Allen; Clarke, Pamela R.; Cocke, Julius Marion; Crawford, D. Steven; Culbert, Llewellyn Borlaug; Davis, L. C.; Dilamarter, Ronald Raymond; Dow, Robert Russell; Dulian, James Joseph; Eagan, James Matthew; Eshelman, Ralph E.; Faflak, Richard E.; Fay, Leslie Porter; Fitzgerald, David J.; Frederking, Ray Lynn; Frest, Terrence James; Frye, Kenneth Lee; Gaffey, Michael J.; Gennett, Judith Ann; Gerhardt, Roger A.; Girard, William W.; Graham, Russell W.; Hall, Stephen Austin; Hallberg, G. R.; Hanson, B. V.; Harlin, J. M.; Hartwig, N. L.; Healey, John M.; Heathcote, Susan Kay Hudson; Heiland, James; Heitzman, Donald Paul; Herr, Stephen Richard; Holtzman, Allan F.; Hughes, John E.; Ingram, Gary R.; Isuk, Edet Effing; Jain, Subhash Chandra; Jenkins, John T., Jr.; Johnson, David Bruce; Johnson, David Bruce; Johnson, Martin Chester; Johnson, Paul Curtis; Johnson, Richard Gustave; Jones, James Ogden; Kettenbrink, Edwin Carl, Jr.; Kirchner, James Gary; Kittleson, Kendell Lloyd; Knowling, Richard Dean; Knox, James Clarence; Kramer, Thomas L.; Lee, Chunsun; Legg, Thomas E.; Lehman, Donald D.; Lewis, Ronald Dale; Lloyd, Ronald LaVerne; Lucas, James R.; Ludvigson, Gregory Alan; Lueck, Everett William; Luth, Iris Annette; Lyman, Robert M.; Lynn, R. J., II; Mackey, Gary W.; Manger, Walter L.; Manley, Ronald D.; Mapes, Royal H.; Matzko, John Rodney; Mayou, Taylor Vinton; McClelland, Steven W.; McMullen, L. D.; Meyer, Michael; Milne, Bonnie L.; Mitchell, John Charles; Moeng, Bruce C.; Mossler, John H.; Moultrie, William A.; Nakato, Tatsuaki; Nelson, David L.; Nelson, Robert Stanley, Jr.; Nicoll, Robert Sherburne; O'Brien, D. E.; Oertel, George Frederick, Jr.; Olyphant, G. A.; Opstad, Erik Alan; Palmer, Steven Dale; Pan, Kuo-Liang; Pavlik, Marcy Lynn; Perlmutter, Barry; Perry, Thomas C.; Person, Jennifer Ann; Priz, Paula Marie; Ravn, Robert Lee; Rebertus, Donald Gene; Reihman, Mary Ann; Ririe, George Todd; Roldan-Quintana, J.; Rouhani, M.; Rudesill, Roger C.; Sammis, Catherine G.; Sammis, Neil C.; Saunders, William Bruce; Scales, John Robert; Schedl, Andrew D.; Schutter, Stephen Richard; Schwartz, P. H.; Senich, Michael A.; Senich, Michael A.; Smith, David Ernst; Smith, Robert Kay; Specht, Ralph W.; Stepanek, John G.; Stubblefield, William Lynn; Swenson, Alan L.; Szabo, John Paul; Tappmeyer, D. M.; Taylor, Harold R.; Taylor, John Dallas; Tharalson, Darryl Bruce; Theiling, Stanley Cecil; Theiling, Stanley Cecil; Thorsteinsson, Erik; Tynan, Mark Christian; Van Zant, Kent Lee; Van Zant, Kent Lee; VerPloeg, Alan James; Vitek, John Dennis; Voldseth, Nels Edward; Ward, Jeffrey Kost; Warner, Albert J.; Watson, Michael Guy; Webster, J. M.; Wills, Donald L.; Windle, Delbert Leroy, Jr.; Witinok, Patricia Mary; Witzke, Brian J.; Wood, Robert H., II; Woods, Edmund Bert; Yaghubpur, A.; Yaghubpur, Abdolmajid; Zawistowski, Stanley J.

1980s: Aden, Leon John; Aronoff, Steven Martin; Baik, Ho Yeal; Baker, Cathy; Bardwell, Jennifer; Bils, Julie; Black, Ross Allen; Bowman, Phillip Robert; Brusseau, Mark Lewis; Calhoun, Steven H.; Carman, Mary Ruth Cote; Chao, Jih-Yuh; Chaplinsky, Peter P., Jr.; Chatelain, Edward Ellis; Christianson, Linda J.; Crowder, Rowley Keith; Cumerlato, Calvin Lee; Daut, Steven William; Day, James Edgar, II; Denesen, Stephen Louis; Dixit, Shailaja R.; Dockal, James Allan; Dogan, Ahmet Umran; Dombrowski, Anna; Edwards, Jeffrey Craig; Egner, Barbara E.; Esling, Steven Paul; Febres Cedillo, Hector Enrique; Fischer, Lorraine Eleanor; Foley, Robert LeRoy; Fouke, Bruce William; Frest, Terrence James; Funkhouser-Marolf, Myra C.; Geppert, Timothy J.; Goebel, Katherine A.; Graeff, Ronald W.; Greenberg, Helene; Gregory, Janet L.; Gross, David Thomson; Groves, John Reid; Hager, Richard Charles; Hajic, Edwin Robert; Heathcote, Richard Carl; Henderson, Barry Leon; Hesler, James Lewis; Hoeksema, Robert James; Howes, Mary Rachel; Huang, Liang-Hsiung; Huang, Long-Cheng; Huang, Zhixin; Hudak, Curtis Martin; Hudak, Curtis Martin; Hudelson, Peter Marc; Jaramillo Torres, Wilson Fabian; Jerpbak, Mark James; Karim, M. Fazle; Kidder, David Lee; Kim, Hyung Keun; Kim, Hyung Keun; Klare, Matthew William; Klug, Curtis Robert; Knight, Kimbell Lee; Kolpin, Dana Ward; Korpel, Joost Adrian; Kross, Burton Clare; Ku, Chi Young; Lardner, James Edward; Lawler, Sydney Kent; Le Seur, Linda Perkins; Lemke, Karen A.; Logel, John Duane; Ludvigson, Gregory Alan; Malinky, John Mark; Mann, Keith Olin; Martin, James Hosmer; McAdams, Mark Patrick; McHargue, Timothy Reed; McKechnie, Deborah Jean; Megivern, Katherine Jean; Megivern, Stephen James; Metzger, Ronald Allen; Mitchell, John Charles; Mohan, Madhukar; Morrissey, Arthur Michael; Morton, Marc K.; Mosconi, Carlos Eduardo; Mou, Duenchien C.; Moussavi-Harami, Reza; Nations, Brenda K.; Nelson, Murray Robert; Nielsen, Mark Andrew; Niemann, William L.; Nikolaidis, Nikolaos P.; Nollsch, David Allen; Nott, Jerry Alan; Nunn, Jerald Ralph; Osweiler, Donna Jean; Park, Inbo; Parkinson, Randall William; Parson, Charles Grady; Pavlicek, John A.; Pavlicek, Meeyoun In; Pearson, Steven Gerald; Price, Rex Clayton; Putney, Kevin Lee; Railsback, Loren Bruce; Rao, Narasinga Bandiatmakur; Ravn, Robert Lee; Reinholtz, Philip N.; Ressmeyer, Paul F.; Rhodes, Richard Sanders, II; Ririe, George Todd; Rose, Susan Humphrey; Rosenberg, Robert Steven; Routh, Darcia Layne; Rowden, Robert D.; Ruiz Calzada, Carlos Edgardo; Ryan, Michael Patrick; Sabol, Donald Edwin, Jr.; Satorius-Fox, Marsha R.; Schutter, Stephen Richard; Seigley, Lynette Sue; Siebels, Charles Joseph; Sixt, Shirley Claire Smith; Sklar, Paul Jeffrey; Soesilo, Dwisuryo Indroyono; Spasojevic, Miodrag P.; Sullivan, Amy E.; Sutton, Margot Jean; Svoboda, Joseph Otto; Swade, John W.; Swihart, George Hammond; Tang, Wen Jian; Torney, Barbara Calhoon; Tvrdik, Timothy N.; Tynan, Mark Christian; Underwood, Mark Roland; Updegraff, Richard Alan; Vadnal, John Louis; Van Nest, Julieann; Venchiarutti, Daniel A.; Waln, Kirk Alexander; Walsh, David Vernon; Wang, Tzupo; Wehmeyer, Lisa Kathryn; Wheeler, Karen Lynn; White, Christine Anne; Whitley, Donald Lee; Winfrey, Elaine Clare; Witzke, Brian J.; Woodman, Neal; Worthington, Ralph E.; Yang, Jinn-Chuang

Iowa State University of Science and Technology
Ames, IA 50011

205 Master's, 122 Doctoral

1890s: Hartman, Russell T.; Houser, Gilbert L.

1900s: William, I. A.

1910s: Howe, F. B.; Smith, John E.

1920s: Jones, Victor H.; Ladd, Harry S.; Littlefield, Max Sylvan; Webber, B. S.

1930s: Bissell, Harold Joseph; Geib, Horace Valentine; Goddard, Ira; Kinne, Raymond Charles; Miner, Neil Alden; Thompson, William Allen; Vogt, Robert R.; Wood, Lyman W.

1940s: Bissell, Harold Joseph; Cuthbert, Frederick L.; Hsui-Tsu-Fan, Paul; Lawson, Ralph W.;

McClelland, John E.; Phillips, Kenneth A.; Ruhe, Robeert Victory; Wilcox, Charles R.

1950s: Armstrong, Escar Weldon; Aye, Tin; Balster, Clifford A.; Berry, Keith David; Bisque, Ramon Edward; Bisque, Ramon Edward; Bjoraker, Robert Wayne; Carlson, Paul Richard; Dahl, A. R.; Dorheim, Fred H.; Ethington, Raymond Lindsay; Galloway, John Duncan; Giltner, Robert; Gooding, Ansel M.; Handy, Richard L.; Handy, Richard L.; Hansen, John Andrew, Jr.; Hayes, J. B.; Haynes, John Bernard; Hiltrop, Carl L. R.; Huedepohl, Earnest Brady; Larsen, William Roger; Lindholm, Gerald Franklin; Lyon, Craig Alfred; Lyons, C. A.; Nyman, Dale James; O'Sullivan, John Blandford; Payton, Charles Ellis; Posey, Donald Rue; Riggs, K. A., Jr.; Rosenfeld, George Albert; Rush, Francis Eugene; Sayyab, Abdullah Shakir; Schneider, R. C.; Stevens, Robert Louis; Stone, Gerald L.; Storm, Paul Vissing; Stump, Richard Webster; Sweet, Walter Clarence; Tench, Robert Norman; Valletta, Robert Michael; Weissmann, Robert Charles; Wickstrom, Alden Eugene; Wilcox, Ronald Erwin; Williams, W. W.; Zimmerman, H. L.

1960s: Anderson, Gary Swen; Assaf, Karen Klare; Backsen, L. B.; Binder, Frank Hewson; Carlson, K. J.; Carson, Charles Edward; Christensen, John Edward; Collins, Gary Brent; Curry, Sharon G.; Dahl, Arthur Richard; DeKoster, Gene R.; Diebold, Frank Enri; Eilers, Lawrence John; Elwell, James Halsey; Elwell, James Halsey; Faas, Richard W.; Faas, Richard William; Fenton, Thomas Eugene; Foster, John David; Frerichs, W. E.; Glenn, George Rembert; Glenn, Jeryy Lee; Guldenzopf, Emil Charles; Hall, George Frederick; Harwood, James Robert; Haskell, Norman Leif; Hilltrop, Carl Lee Roy; Huddleston, James Herbert; Isenberger, Kenyon Jay; Johnson, Gary Dean; Kent, Douglas Charles; Knochenmus, D. D.; Kovar, Erlece Paree; Lohnes, Robert Alan; Mason, James R., Jr.; McConnell, Harold Lee; Michelson, Ronald Wayne; Moore, William Joseph; Moores, Eugene Albert; Neasham, John West; Noble, Calvin Athelward; O'Sullivan, John Blandford; Ogden, Joseph Cornelius; Pedersen, D. E.; Plato, Phillip Alexander; Qutub, Musa Yacub; Reckendorf, Frank Fred; Roderick, Gilbert Leroy; Schoell, John; Sendlein, Lyle Vernon Archie; Senich, Donald; Shuman, Fred Leon, Jr.; Simon, David Eugene; Singamsetti, Surya Rao; Smith, Samuel Joseph; Staub, William Praed; Stensland, Robert Dean; Stoermer, Eugene Filmore; Stone, Randolph; Tarman, Donald; Tinoco, Fernando Heriberto; Vander Ley, J. W.; Ver Steeg, David James; Vredenburgh, Larry Dale; Wagner, James Kendall; Wallace, Charles Michael; Wallace, Richard Warren; Warner, Albert J.; Werner, Michael Askam; Wier, Donald Raymond; Youd, Thomas Leslie

1970s: Acuff, Hoyt Nealy; Akhavi, Manouchehr Sadat; Badger, William Wiley; Bainbridge, Russell Benjamin, Jr.; Basmaci, Y.; Becker, Dennis Lee; Bible, Gary G.; Bible, Gary Gill; Blair, Michael Reed; Bowen, Bruce E.; Bowen, Bruce Eugene; Bredall, S.; Brunotte, D.; Bunn, R. L.; Burch, S. L.; Burggraf, Daniel Robert, Jr.; Burggraf, Gloria Butson; Cammack, C. H.; Canfield, D. E., Jr.; Cerling, Thure Edward; Cisar, Marilyn Taggi; De Bruin, Rodney H.; Dockal, James Allan; Drennon, Clarence Bartow, III; Elleboudy, A.; Erol, O.; Frank, Hal J.; Genrich, Donald Allen; Goepfert, W. M.; Gopfert, Wolfgang Martin; Gross, Barry L.; Hamilton, Robert David; Hansen, Daniel Lloyd; Hooper, John Marten; Hutchinson, Roderick; Iles, D. L.; Johnson, Gary Dean; Johnson, Mary Lou; Jones, John R.; Khan, Muhammad Yunus; Klefstad, Gilbert Eugene; Kozimko, Leo M.; Ladd, Robert Edward; Leone, John Michael, Jr.; Lutenegger, A. J.; Mairaing, W.; Mathisen, M. E.; Morehouse, David Frank; Naylor, L. M.; Neasham, John West; Nicklin, Michael Earl; Onyeagocha, Anthony Chukwuma; Peckenpaugh, J. M.; Peterson, Ronald Milton; Ramp, E. R.; Reese, Joseph L.; Rich, Michael A.; Rossmiller, R. L.; Shaser, Joseph Leroy; Simon, D. E.; Siudyla, E. A.; Smith, D. E.; Smith,

K. E.; Smith, Mark Francis; Smith, Thomas Andrew; Spencer, John Michael; Spitz, Dan Spencer; Stangl, David William; Stark, Michael Paul; Stephens, Mark Randall; Stone, Randolph; Tuncer, E. R.; Van Driel, James Nicholas; Vreeken, Willem Jaap; Wallace, Donna; Watney, Willard Lynn; White, Howard James; Yazicigil, H.; Young, Michael Steven

1980s: Ahmed, Mushtaque; Amonette, James Edward; Ankeny, Mark Dwight; Antosch, Larry Michael; Arfa, Hossein; Arjmand, Olya; Babalola, Stephen Oladele; Bard, Gary G.; Barnum, James Bradford; Bicki, Thomas James; Birdseye, Richard Underwood; Bohlken, Bruce Arthur; Book, Patricia O'Donnell; Bruns, Joan Marie Burnet; Bruns, Joseph John; Bryant, Deborah Jean Allen; Carr, Robert Sidney; Carson, Dana Woodruff; Chadima, Sarah Anne; Chamberlain, Rick Earl; Christman, Joseph Robert; Chung, Sang-Ok; Cloud, Thomas Arthur; Cole, Melanie Ruth Will; Collins, Mary Elizabeth; Dawson, Malcolm Robert, II; Dawson, Malcom Robert, II; DeDecker, Kenneth Arnold; Doden, Arnold Gabriel; El-Ghoul, Arebi B.; Elliot, William John; Escalante, Margarito Coballes; Feibel, Craig Stratton; Filkins, Jeffrey Elliott; Finley, Mark Edward; Fox, Darwin Eugene; Galloway, Kathleen Elise; Giraud, Joel Robert; Golchin, Jahanshir; Green, Timothy Myron; Hashem, Fadel Musa; Hayes, David Thomas; Hill, Robert Lee; Hobson, Winston Edward; James, Harry Rudolph; Jensen, Joseph Matthew; Jones, Henry D.; Kipp, James Alden; Knightly, John Paul; Kocken, Roger James; Koellner, Mark S.; Kramer, Matthew Joseph; Kutz, Keith Brian; Kvale, Cindy Marie; Kvale, Erik Peter; Kvale, Erik Peter; Kwun, Soon-Kuk; Lamb, Robert Odell; Lin, Tso-Wang; Luwe, Randall Scott; Manahl, Kenneth A.; Mason, Edward William; Massoudi, Heidargholi; Masterson, Tina K.; Mathisen, Mark Evan; Monnens, Lee Edwin; Nassar, Ibrahim Nassar; Noggle, Karen Sue; Okereke, Victor Onuzurike Irokanulo; Olander, Jon David; Osolin, Robert Lyle; Park, Seung O.; Peck, Curtis Allen; Pederson, Roger Lynn; Pedrick, Jane Nuli; Petershagen, John Haynes; Pierson, Milton Lee; Pitt, John Michael; Postlethwaite, Clay Edward; Postlethwaite, Clay Edward; Potter, Kenneth Neil; Potter, Lee Shefte; Reppe, Calvin Clark; Roglans-Ribas, Jordi; Ryan, Timothy Harold; Saleh, Hamed Hussein; Salih, Hammed Mohammad; Schilling, Keith Edwin; Shahghasemi, Ebrahim; Shaw, Carl G.; Slaughter, Cecil Bryan; Soliman, Hosny El-Desouky Ahmed; Thieben, Scott E.; Thompson, Gregory L.; Tse, Eric Wai Keung; Tuan-Sarif, Tuan Besar Bin; Uhlir, David Mason; Wang, Jia Shung; Weed, Daniel Del; White, Howard James; Wilke, Kurtis Merle; Wolka, Kevin K.; Wonder, James David; Wysocki, Donald John; Yang, Jane-Fu Jeff; Yang, Pe-Shen

The Johns Hopkins University
Baltimore, MD 21218

23 Master's, 355 Doctoral

1880s: Bayley, W. S.; Haworth, Erasmus; Hobbs, William Henry; Lawson, Andrew C.

1890s: Abbe, Cleveland; Bagg, Rufus Mather; Bascom, Florence; Beyer, Samuel W.; Bibbins, Arthur; Cragin, Francis W.; Gane, Henry Stewart; Glenn, L. C.; Grant, Ulysses S.; Grimsley, George P.; Keyes, C. R.; King, F. P.; Leonard, A. G.; Mathews, Edward B.; O'Harra, Cleophas C.; Shattuck, G. B.; Smith, George O.; Spencer, Arthur C.

1900s: Bonsteel, Jay A.; Grasty, John S.; Johannsen, Albert; Martin, George C.; Maynard, T. P.; Miller, Ben L.; Miller, William J.; O'Hern, D. W.; Prouty, W. F.; Richardson, G. B.; Rowe, R. B.; Rutledge, J. J.; Singewald, Joseph T., Jr.; Stephenson, Lloyd W.; Swartz, C. K.; Twitchell, M. W.

1910s: Bassler, Harvey; Clark, Burton W.; Cooke, Charles W.; Dorsey, George E.; Gardner, Julia Anna; Goldman, Marcus I.; Hopkins, Oliver B.; Hull, Joseph P. D.; Hunter, John Frederick; Insley,

Herbert; Lee, Willis Thomas; Little, Hower P.; Merite, John B.; Mertie, John B., Jr.; Overbeck, Robert M.; Price, W. A., Jr.; Reeside, J. B. Jr.; Reeves, Frank; Roberts, Joseph; Sears, Julian D.; Thom, William T., Jr.; Wallis, B. Franklin; Williams, Richard C.; Woodring, Wendell P.

1920s: Anderson, J. D.; Berry, E. Willard; Brown, Roland Wilbur; Collins, R. E. L.; Dobbin, C. E.; Douglas, J. G.; Eby, James B.; Fisher, Lloyd Wellington; Hoffmeister, J. E.; Hoffmeister, W. S.; Jones, Walter B.; Kellogg, Frederic H.; Kellum, Lewis B.; Knechtel, M. M.; Knetchtel, M. M.; Manger, George Edward; Milton, Charles; Ohrenschall, Robert D.; Roberts, Joseph K.; Sandidge, John R.; Secrist, Mark H.; Singewald, Quenton D.; Smith, Laurence Lowe; Smith, Walter R.; Spieker, Edmund M.; Stewart, Ralph B.; Swartz, Frank M.; Swartz, Joel H.; Watson, Edward H.

1930s: Berry, Charles T.; Bowles, Edgar; Broedel, Carl Huntington; Carter, Charles William; Cohen, C. J.; Corbin, M. W.; Dickey, Parke Atherton; Dryden, A. L., Jr.; Gildersleeve, Benjamin; Gillette, Tracy; Hanley, J. B.; Hershey, Howard Garland; Hill, Malcom W., Jr.; Hughes, Richard V.; Johnson, George Duncan; Kellogg, Frederic H.; Lochman, Christina; Marshall, John; Maurice, Charles S.; Philbrick, Shailer Shaw; Pike, W. S., Jr.; Stocking, Hobert Ebey; Tong, James A.; Wade, Franklin Alton; Waters, Arnold E., Jr.; Wolford, John J.; Woodberry, Marjorie

1940s: Agron, Sam L.; Aitken, Janet M.; Aitken, Janet M.; Broughton, John Gerard; Cobban, William Aubrey; Fellows, Robert E.; Fernandez-Concha, Jaime; Foose, Richard M.; Gair, Jacob E.; Gault, Hugh Richard; Graham, John W.; Guild, Phillip W.; Hamilton, Edward A.; Hass, W. H.; Joesting, Henry R.; Lange, Marie L.; Merrill, Robert J.; Neuman, Robert B.; Palmer, Richard B.; Scharon, Harry Leroy; Scheid, Vernon Edward; Shifflett, Frances Elaine; Slaughter, Turbit Henry; Sowers, George M.; Stefansson, Karl; Stephenson, Robert C.; Wagner, Warren R.; Walters, Robert F.; Warner, Lawrence Allen

1950s: Afshar, Freydown A.; Allard, Gilles O.; Andrews, L. A.; Appleman, Daniel E.; Barnes, Harley, Jr.; Black, Robert Foster; Choquette, Philip W.; Ernst, Wallace Gary; Evitt, W. E., II; Grant, Willard Huntington; Halferdahl, Laurence B.; Hearn, Bernard Carter, Jr.; Herrmann, Leo Anthony; Herz, Norman; Hopson, Clifford Andrae; Hurst, Vernon James; Jensen, Fred S.; Juhle, R. Werner; Lintz, Joseph, Jr.; Long, Miner B.; Long, Miner Barton; Moore, James G.; Nickelsen, Richard P.; Pelletier, Bernard R.; Pipiringos, George Nicholas; Poborski, Stanislaw Jozef; Power, W. Robert, Jr.; Ray, Richard G.; Reed, John Calvin; Ryan, John Donald; Sando, William J.; Sauve, Pierre; Schlanger, Seymour O.; Schlee, John Stevens; Scotford, David M.; Stevens, George R.; Sutton, Robert George; Thomas, Byron K.; Trexler, David William; Weaver, Kenneth Newcomer; Whitaker, John Carroll; Yeakel, Lloyd Stanley

1960s: Adams, Robert W.; Bromery, Randolph W.; Chown, Edward H. M.; Clifton, H. Edward; Donaldson, J. Allen; Fairley, William Merle; Fisher, George W.; Fiske, Richard Sewell; French, Bevan M.; Glaser, John Donald; Grolier, Maurice Jean; Hadley, Donald Gene; Hardie, Lawrence Alexander; Huebner, J. Stephen; Hunter, Ralph E.; Jones, Blair F.; Kujawa, Frank Benedict; Lindholm, Roy Charles; Lindsey, David Allen; Lindsley, Donald H.; McBride, Earle F.; McDowell, John P.; McIver, Norman L.; McKay, James Hughes, Jr.; Moody, David Wright; Munoz, James Loomis; Newbury, Robert William; Prostka, Harold J.; Reeburgh, William Scott; Rieder, Milan; Rutherford, Malcolm John; Schmincke, Hans-Ulrich; Schubel, Jerry Robert; Sheppard, Richard Abner; Skippen, George Barber; Southwick, David Leroy; Swanson, Donald Alan; Turnock, Allan Charles; Wickham,

John S.; Wise, William Stewart; Wright, Thomas Llewellyn

1970s: Bailey, R. A.; Blumberg, A. F.; Boicourt, W. C.; Boland, John J.; Bosch, Herman F.; Boyer, S. E.; Braun, D. D.; Bray, John Thomas; Brenchley, Gayle Anne; Callahan, Jeffrey Edwin; Carter, Charles Henry; Chou, I-M.; Cleaves, Emery Taylor; Costa, John Emil; Feazel, C. T.; Foster, A. B.; Foster, C. T., Jr.; Frantz, John Duncan; Froomer, N. L.; Garrett, Peter; Gault, H. R.; Geiser, Peter A.; Gunter, William Daniel; Gupta, Avijit; Han, G. C.; Harrison, Stanley Cooper; Henderson, John B.; Hirsch, R. M.; Hoersch, A. L.; Hoffman, Paul F.; Hohl, Arthur Henry; Holdren, G. R.; Howard, Alan Dighton; Kaiser, William Richard; Makurath, J. H.; Maos, Jacob O.; Matisoff, G.; McComas, C. H., III; Mitra, G.; Mitra, S.; Morse, D. G.; Newell, Wayne Linwood; O'Connor, Bruce James; Olsen, Sakiko Nakaya; Owens, W. B.; Pavich, M.; Potter, K. W.; Reinhardt, Juergen; Smith, Derald Glen; Smith, Terence Robert; Smoot, Joseph P.; Stone, B. D.; Swan, F. H.; Taft, J. L., III; Tourek, Thomas James; Troup, Bruce Neil; Wanless, Harold Rogers; Weaver, G. D.; Williams, Richard John; Wilson, Robert E.; Wood, J. R.

1980s: Allen, Julia Coan; Alley, William McKinley; Anastasio, David John; Atkin, Steven Allen; Banerjee, Bakul; Barrett, Mary Louise; Bauer, Bernard Oswald; Bergantz, George Walter; Brophy, James Gerald; Buss, L. W.; Cashman, Katharine Venable; Chuang, W. S.; Coleman, Derrick Job; Cutler, Jonathan Mitchell; Davis, Frank Willard; DeFries, Ruth Sarah; Demicco, Robert Victor; Didwall, Edna Mary; Diegel, Fredric A.; Domagalski, Joseph Leo; Edmonds, Robert Lee; Fournelle, John Harold; Gilotti, Jane A.; Gislason, Sigurdur Reynir; Goldhammer, Robert Kent; Goswami, Dulal Chandra; Graham, Donald Steven; Grant, Gordon Elliot; Gunnarsson, Bjorn; Haley, J. Christopher; Harris, Mark Thomas; Hershler, Robert; Hertz, Terrance L.; Hinojosa, Juan Homero; Ilton, Eugene Saul; Jackson, Marie Dolores; Jacobson, Robert B.; Jephcoat, Andrew Philip; Kandasamy, Kumarasamy; Karabinos, Paul M. S.; Kat, Pieter W.; Knopman, Debra S.; Kovach, Linda Anne; Lidgard, Scott Harrison; Lintner, Stephen Francis; Lowenstein, Tim K.; Massare, Judy Ann; McKinney, D. Brooks; Miller, Andrew J.; Mitchell, Raymond Weatheral, III; Morris, Stephen John Samuel; Myers, J. D.; Neuzil, Christopher Eugene; Nguyen, Chau Trung; Peterson, Tony Douglas; Pine, F. W.; Rylaarsdam, Katharine Worcester; Sambol, Melvin; Scatena, Frederick N.; Schlinger, Charles Martin; Schmidt, John Christian, III; Seidell, Barbara Castens; Shapiro, Alan M.; Shore, Michael James; Shyu, Jinn-Hwa; Signor, Philip White, III; Simonson, Bruce Miller; Singer, Harvey A.; Spencer, Ronald James; Spratt, Deborah Anne; Steiner, Roland Christian; Steneck, Robert Steven; Summers, Robert Michael; van Valkenburgh, Blaire; Weissman, Arthur Bruce; Wetmore, Karen Louise; Whitcomb, John Byington; Whitney, John William; Wilderman, Candie Caplan; Williams, Sherilyn Coretta; Wilson, Edith Newton; Wilson, Glenn Alan; Winn, William M.; Wojtal, Steven F.; Wojtal, Steven Francis; Woodward, Nicholas Brugger; Yang, Xiangning

University of Kansas
Lawrence, KS 66045

433 Master's, 172 Doctoral

1880s: Haworth, Erasmus

1890s: Beede, Joshua W.; Grover, Charles H.

1900s: Bedell, Frank G.; Bedell, Frank G.; Eyerly, Terma LeClerc; Kuchs, Oscar M.; Logan, Spencer R.; Mayberry, J. W.; Shaler, Millard K.; Tobey, William H.

1910s: Belchic, George; Culbertson, Alex E.; Ellis, H. A.; Foster, William H.; Hixson, Arthur Warren; Hoffman, R. N.; Miller, Forrest J.; Miller, Ray-

mond F.; Roberts, Clay; Shotts, T. W.; Smith, Lewis B.

1920s: Abernathy, George Elmer; Benson, Dale L.; Charles, Homer H.; Croneis, Carey G.; Delo, David Marion; Henbest, Lloyd G.; Hipp, Thomas; Hoffman, Olive L.; Jewett, John Mark; Knight, James Brookes; McGee, Dean; Tester, Allen Crawford; Wells, Dana; Wing, Monta Eldo

1930s: Abernathy, George Elmer; Courtier, William H.; Hoover, William Farrin; Lane, Joseph H., Jr.; Lumb, Wallace E.; McLaughlin, Thad G.; Mettner, Francis E.; Mitchell, Robert C.; Moreman, Walter L.; Morrow, A. L.; Moss, Rycroft Gleason; Newell, Norman Dennis; Patterson, Joseph M.; Whitla, Raymond E.; Wismer, Raymond J.

1940s: Amstutz, Platte T., Jr.; Bramlette, William Allen; Chronic, Byron John, Jr.; Crain, Hugh F.; Dubins, M. Ira; Fairchild, Paul W.; Goodrich, Willard Dale; Grey, Charles Edwin; Jeffords, Russell M.; Jeffords, Russell M.; Jewett, John Mark; Kaiser, Charles Philip; Kaiser, Charles Philip; Kleihege, Bernard W.; Lemmons, Jacob E.; Leonard, John R.; Lewis, Paul J.; McBee, William, Jr.; McGregor, Duncan J.; Purrington, Wealthy; Redman, Kenneth G.; Requist, Norris N.; Spreng, Alfred Carl; Stoneburner, Roger W. Q.; Swain, Frederick Morrill, Jr.; Wallace, Maurice H.; Williams, Charles Coburn; Zeller, Edward J.

1950s: Adams, Donald J.; Angino, Ernest Edward; Arper, William Burnside, Jr.; Asquith, Donald O.; Ball, Mahlon Marsh; Ball, Stanton Mock; Barney, Emmet C.; Bates, Wayne E.; Beck, Henry Voorhees; Beu, Robert D.; Bigelow, Nelson, Jr.; Bishop, Samuel Wills; Boker, Thomas A.; Brown, William G.; Brown, William Lindop; Byers, Philip C.; Carlson, William A.; Chakravorty, Sailendra K.; Chapman, John S.; Collins, Donald N.; Conkin, Barbara Moyer; Conkin, James E.; Cooley, Douglas R.; Davis, Darrell E.; Davis, Stanley N.; Dietsch, Harold A.; Dodge, Harry W., Jr.; Douglass, Myrl Robert; Duane, David Bierlein; DuBar, Jules R.; Dufford, Alvin E.; Eastwood, William P.; Fisher, William L.; Foster, Glen Lloyd; Franks, Paul C.; Galbreath, Edwin C.; Grinnell, Robert S., Jr.; Gutentag, Edwin D.; Gwinn, Billy W.; Hager, Glenn G.; Hambleton, William W.; Harbaugh, John W.; Hattin, Donald Edward; Hattin, Donald Edward; Hawryszko, Julian W.; Haynes, Edward H.; Hickox, John E.; Hilpman, Paul Lorenz; Hockens, Sidney N.; Hodson, Warren G.; Holland, B. D.; Howard, Leonard W.; Howe, Wallace Brady; Hughes, Owen L.; Hughes, Owen L.; Ives, William, Jr.; Johnston, Paul L.; Kelly, John M.; King, Ralph H.; Klapper, Gilbert J.; Klein, George deVries; Koenig, John Waldo; Lamb, Ralph C., Jr.; Lamerson, Paul R.; Lane, Norman Gary; Lane, Norman Gary; Laughlin, Dwight J.; Laukel, Quinn C.; Lins, Thomas Wesley; Mack, Leslie E.; Mack, Leslie E.; Mann, C. John; Martin, Charles Arthur, Jr.; McCrae, Robert O.; McCullough, Douglas L.; McLaren, Donald B.; McManus, Dean Alvis; McManus, Dean Alvis; McMillan, Neil John; McNellis, Jesse M.; Mehl, Robert L.; Mendoza, Herbert A.; Merriam, Daniel Francis; Mettler, Don E.; Miller, Halsey W., Jr.; Miller, Robert R.; Muehlberger, Eugene Bruce; Ni Ta Pe; Nicholas, Richard Ludlam; Norris, Robert Peter; O'Connor, Howard G.; Odem, Wilbert I.; Owen, Donald Edward; Padgham, John B.; Parkhurst, Robert W.; Pearn, William Charles; Peltier, Edward J.; Perkins, Hamilton C.; Procter, Richard Malcolm; Purzer, James J.; Reese, Dale O.; Reynolds, J. Rex; Richards, H. Glenn; Ronca, Luciano Bruno; Ryther, Thomas E.; Sanders, Donald T.; Schmidt, Harold A.; Schulte, George S.; Self, Edward Moss; Sensintaffar, Jack L.; Siegel, Frederic H.; Siler, Jerry C.; Sloanaker, Charles Jasper; Squires, Donald Fleming; Squires, Jean M.; Stocker, George Robert; Tartamella, Natale John; Thorsteinsson, Raymond; Underwood, Prescott, Jr.; Van Sant, Jan Franklin; Wahrhaftig, Leon; Walter, Lawrence E., Jr.; Webster, Gary D.; Wilbert, Louis Joseph, Jr.; Wilson, James A.; Wilson, John Coe; Winchell, Richard L.;

Wood, Roger L.; Yochelson, Ellis Leon; Zajic, William E.; Zinser, Robert W.

1960s: Abdel Wahid, Ibrahim; Adams, Larry W.; Ahmed, Syed Sirtajuddin; Angino, Ernest Edward; Badon, Calvin L.; Baldwin, Arthur D., Jr.; Ball, Mahlon Marsh; Ball, Stanton Mock; Bebout, Don Gray; Bondurant, Charles E.; Bower, Richard R.; Brady, Lawrence Lee; Brosseau, Hubert N.; Brown, Dwight Alan; Brown, Sally Sue Liggett; Brown, Thomas D.; Buchwald, Caryl Edward; Campbell, Charles L.; Canard, Carlos; Church, Stanley Eugene; Coleman, George L., II; Conrad, Omar G.; Cramer, John A., Jr.; Crawford, William A.; Cridland, Arthur Albert; Crow, Billy B.; Currey, Donald Rusk; Dixon, Val R.; Duane, David Bierlein; Edwards, George H., III; Ehret, Albert L., Jr.; Emery, Philip A.; Empie, Joel S.; Fay, Robert Oran; Fishburn, Maurice D.; Fisher, William L.; Foster, Norman Holland; Franks, Paul C.; Frost, Jackie Glenn; Garber, Murray S.; Gautier, Theodore Gary; Gerhard, Lee C.; Gerhard, Lee C.; Gerhard, Roberta G.; Goebel, Edwin D.; Gogel, Anthony J.; Gordon, W. Richard; Green, Loring K.; Grossman, Stuart; Habib, Daniel; Haglund, Wayne Milton; Hall, Billy P.; Harris, Leaman D.; Hatcher, David A.; Hathaway, Lawrence Robbins; Hays, James K.; Hedberg, Ronald M.; Hicks, Reginald V.; Hill, Walter E., Jr.; Hilpman, Paul Lorenz; Ho, Tong-yun; Hodgden, H. Jerry; Hollweg, William A.; Horner, William J.; Howard, James D.; Ibrahim, Abd El Wahid; Jacques, Theodore Emil; Jamkhindikar, Suresh M.; Johns, Wendell S.; Johnson, Lynn A.; Jungmann, William L.; Kaesler, Robert LeRoy; Kaesler, Robert LeRoy; Kahle, Charles Franz; Kelly, Thomas E.; Kinell, Carl B., III; Krause, Hans H.; Lefebvre, Richard H.; Lillegraven, Jason Arthur; Lineback, Jerry A.; Lins, Thomas Wesley; Lippert, Rudolph H.; Lucken, John E.; Lynch, John Douglas; MacDonald, Harold Carleton; MacDonald, Harold Carleton; Maddocks, Rosalie F.; Maddocks, Rosalie F.; Martin, Ronald B.; McCoy, Roger Michael; McCrone, Alistair William; McElroy, Marcus Nelson; McKellar, Tommy R.; McKinney, R. E.; McMurray, Keith S.; Meek, Richard M.; Merriam, Daniel Francis; Metcalf, Artie Lou; Meyer, Richard Fastabend; Michelson, J. E.; Miller, Don E.; Mills, John Peter; Monger, James W. H.; Morris, David Albert; Morris, David Albert; Mose, Douglas George; Naff, John Davis; Ojala, Gary L.; Osborn, James B.; Owen, Donald Edward; Painter, Alice; Pearn, William Charles; Peterson, James Leonard; Pierce, Jack Warren; Pooser, William Kenneth; Qandil, Yacoub A.; Radke, Frank, Jr.; Reams, Max W.; Roberson, Michel I.; Rochna, David A.; Rodriguez, Luis R.; Ronca, Luciano Bruno; Ross, David A.; Roy, Malcom Bernard; Sackett, Duane H.; Sandlin, Larry F.; Saueracker, Paul Robert; Scafe, Donald W.; Schabilion, Jeffry Tod; Schleh, Edward E.; Schroeder, Marvin L.; Schuman, Richard L.; Scott, Robert W.; Seevers, William J.; Segal, Ronald Henry; Siegel, Frederic Richard; Slewitzke, Edward B.; Smedley, Gary Lee; Smith, Michael A.; Snyder, James D.; Sorauf, James Edward; Spencer, Randall Scott; Spencer, Randall Scott; Steffen, Robert W.; Steinker, Don C.; Stelljes, Von D.; Stone, Wilbur Alan; Tatro, James O.; Thomas, Barbara R.; Thomas, John Jenks; Thompson, Thomas L.; Tien, Pei-Lin; Turner, Brian Buddington; Twell, Charles F.; Van Sant, Jan Franklin; Vaz, Jesus Eduardo; Vincent, Douglas Anderson; Wahlstedt, Warren J.; Wainwright, Kenneth E.; Walton, Robert G.; Welch, Robert Gerald; West, Ronald R.; Williams, Roger Bennett; Wynne, Milo E.; Zandell, Charles H.

1970s: Aber, James S.; Aber, James S.; Ashton, Jean Hadley; Baxendale, R. W.; Beaumont, Edward A.; Bennett, Debra Kim; Berry, Archie William, Jr.; Brady, Lawrence Lee; Brondos, Michael David; Burnett, Neill C.; Cadot, H. Meade, Jr.; Cadot, H. Meade, Jr.; Campbell, J. B., Jr.; Caspall, Frederick Charles; Chang, Yi-Maw; Cihlar, Josef; Crick, Rex L.; Cudzilo, Thomas Frederick; Cudzilo, Thomas Frederick; Dart, Stephen W., Jr.; Degner,

Dennis A.; DeNooyer, LeRoy L.; DuBois, Susan Morrison; Einsohn, Sudi D.; El Zouki, Ashour Y.; Elks, John E.; Elliott, Robert G.; Engleman, Mary; Evans, Robert; Farmer, Jack D.; Fisher, William L.; Funk, James M.; Gautier, Theodore Gary; Ghuma, Mohamed A.; Gilmore, John B.; Gould, George F.; Haack, Richard C.; Haggiagi, Musa A.; Hakes, William G.; Hakes, William G.; Hammond, Roger Darril; Harrison, J. A.; Harrower, Karen L.; Henderson, Floyd Merl; Holdoway, Katrine A.; Holdoway, Katrine A.; Honderich, Jeff P.; Hopkins, Robert T.; Hulen, Paul Leon; Hulse, William J.; Jarjur, Salah Z.; Jefferis, Lee H.; Johnson, Donald Lee; Jordan, Jeffrey M.; Kimbrough, C. W.; Kirk, J. Norman; Knapp, Roy Marvin; Knepper, Daniel H., Jr.; Knoll, Kenneth Mark; Koepnick, Richard B.; Koepnick, Richard B.; Krause, Frederico F.; Laney, Randy T.; Lavin, Stephen J.; Lentell, Thomas L.; Lewis, Anthony J.; Lilley, Wesley Wayne; Lister, K. H.; Long, David Timothy; Lucente, Michael Eugene; MacFarlane, P. Allen; Maerz, R. H., Jr.; Mason, Robert M.; Mathews, Wilbert L.; McBride, David James; McBride, David James; McCauley, James R.; McCauley, James R.; McClure, John W.; McGee, Joseph William; McNeely, John B.; Mills, John Peter; Morain, Stanley Alan; Morency, Maurice L.; Mose, Douglas George; Muhs, Daniel R.; Nazer, Naji M.; Odom, Arthur LeRoy; Paul, R. W.; Penley, Gary N.; Perkins, T. W.; Perrin, Nancy Ann; Peterson, R. M.; Pollard, William D.; Pregill, Gregory K.; Rao, M. Kesava; Rasmussen, Donald L.; Rees, Margaret N.; Schneider, Harvey I.; Scully, Elizabeth Mary; Seyrafian, Ali; Shortridge, Barbara Gimla; Shuster, Robert Duncan; Sides, James Ronald; Simms, John J.; Songsirikul, Benja; Soule, Mary Alice; Stanley, G. D., Jr.; Steinke, Theodore R.; Stewart, Gary Franklin; Tarkington, Daniel K.; Taylor, Ronald Shearer; Teifke, Robert H.; Tien, Pei-Lin; Traylor, Charles Tim; Turner, Mortimer Darling; Van Dyke, R. J.; Vera, Ramon H.; Von Bitter, Peter H.; Waller, Thomas H.; Wells, Anke Marie Neumann; Williams, Michael E.; Williams, Michael E.; Wing, Richard Sherman; Woolsey, Leonard L.; Yukler, M. A.; Yukler, M. A.

1980s: Abdullah, Talat Y.; Arriola Torres, Alfredo; Attwood, Paul J.; Ball, David S.; Basocak, Cihat; Bennett, Debra Kim; Berg, James A.; Bitter, Mark R.; Blades-Zeller, Elizabeth L.; Blum, Cynthia Elizabeth; Bowring, Samuel Anthony; Branham, Keith L.; Brondos, Michael David; Butler, David Ray; Caldwell, Craig D.; Caruso, Nancy E.; Chesser, Kevin C.; Chorn, John Douglas; Conroy, Elizabeth A.; Crook, Maurice Clifford; Cummings, David O.; Daly, Eleanor; DuBois, James; DuBois, Martin; Eger, Martha J. Erickson; Enciso, Gonzalo; Evans, David G.; Ezekwe, John Nnaemeka; Fisher, William Lee; Foster, David Wayne; Foster, David Wayne; Frailey, Carl David; Franz, Richard H.; Geil, Sharon Anne; Gilmartin, Patricia Purcell; Gilson, Mary M.; Grannis, Jonathon L.; Gray, J. E.; Guvenir, Ibrahim M.; Hahn, Roger K.; Hentz, Tucker Fox; Holien, Christopher W.; Hoppe, Wendel J.; Hoppie, James R.; Izuka, Scot K.; Killen, David B.; Kopaska-Merkel, David Crispin; Kopsick, Deborah A.; Kritikos, William Paul; La Fon, Neal A.; Lam, Chi-Kin; Lampe, Leslie Kent; Lilo, Yehuda; Lin, Zsay-Shing; Livingston, Neal D.; Lui, Chung-Yao; Maness, Timothy R.; May, Michael T.; McKibben, Mark Eugene; Merchant, James William, Jr.; Meyer, Michael T.; Mueller, Raymond George; Nelson, Bruce K.; Nelson, Gerald E.; Nusbaum, Robert L.; Olea, Ricardo Antonio; Palmer, Beth Ann; Patton, Jeffrey Connor; Pereira da Cunha, Roberto; Perry, Michael; Premo, Wayne R.; Prochnow, Suzanne; Rees, Margaret N.; Richards, Barry Charles; Richardson, Clinton Preston; Roark, Clayton R.; Rofheart, Douglas H.; Rutan, Debra; Salter, Timothy L.; Schweitzer, Peter Neil; Shuster, Robert Duncan; Sims, Richard Carlton; Spray, Karen L.; Stander, Thomas; Stanley, Barbara (Vis); Stavnes, Sandra A.; Stewart, Joe Dean; Tao, Shu; Torres-Robles, Rafael; Treadway, Jeffrey A.; Tucker, Glenn Jefferson; Vargas, John F.; Vogl,

Eric G.; Vorwald, Gary R.; Wallace, Kirk D.; Wallace, Ronald James; Walters, Robert L.; Watney, Willard Lynn; Willard, Jane M.; Yarlot, Mark; Zell, Mary G.; Zetterlund, Dale; Zhou, Di

Kansas State University
Manhattan, KS 66506

166 Master's, 5 Doctoral

1930s: Walters, Charles Phillip

1940s: Beck, Henry Voorhees; Coombs, Vincent B.; Harned, C. Hal; Johnson, Wendell B.; Lill, Gordon G.; Matthews, Claude W.; McMillen, Hugh O.; Mudge, Melville R.; Neff, Arthur W.; Tye, Rennie V.

1950s: Archer, Rex Donald; Asmussen, Loris E.; Baehr, William M.; Barrett, William J.; Bergman, Denzil W.; Beshears, Glenn T.; Biegler, Norman W.; Bonchonsky, Andrew P.; Bridge, Thomas E.; Briggeman, Homer W.; Brown, Alton R.; Bruton, Roger L.; Bryson, William R.; Carr, Donald D.; Clark, William K.; Crumpton, Carl F.; Davis, Michael E.; Drake, Larson Y.; Eastty, Frederick D., Jr.; Gassaway, Mack A., III; Geil, Donald D.; Hargadine, Gerald D.; Hartig, Robert L.; Herman, Charles W.; Holcombe, Walter B.; Hooker, Richard A.; Howard, James R.; Koons, Donald L.; Kotoyantz, Alexander A.; Logsdon, Truman F.; Markley, Lewis C.; Matthews, Jerry L.; McPherron, Donald S.; Mendenhall, Richard A.; Merryman, Raleigh J.; Metz, Harold L.; Metz, Jerry P.; Muehlhauser, Helmut C.; Myers, Ronald E.; Nelson, Paul D.; Nicholas, Raymond H.; Olson, Dale R.; Pan, Chih-Wei; Parish, Kenneth L.; Ratcliff, Gene A.; Ricci, Armando T.; Rieb, Sidney L.; Rohrbough, Claude A.; Sandlin, Gary Stuart; Seiler, Charles D.; Shapley, Robert A.; Sleeman, Lyle H., Jr.; Smith, James T.; Soucek, Charles H.; Strunk, Paul M.; Swanson, Wallace A.; Swett, Earl R., Jr.; Taylor, Wallace K.; Tibbetts, Benton L.; Tucker, Norman Alvi; Twiss, Page Charles; Waterman, Willis D.; Watkins, Kenneth N.; Welch, Vorrin J.; Wells, John D.; Wilbur, Robert Olas; Wilson, Frank W.

1960s: Ansari, Noorul Wase; Ardell, Robert J.; Baysinger, Billy Lynn; Booth, Arthur L.; Brown, Lawrence Edward; Butler, John B.; Dowell, Albert Roger; Dowling, Paul L., Jr.; Dulekoz, Erhan; Eastwood, Raymond Lester; Gregory, John L.; Grossnickle, William E.; Hansen, Thomas J.; Hinshaw, Gaylord C.; Holmes, John Ferrell; Huber, Darrell Dean; Husain, Athar; Hylton, Gary K.; Little, John Marshall; Little, John Marshall; Maderak, Marion L.; McDermott, Vincent J.; McQuillan, Michael W.; Moulthrop, James S.; Pattengill, Maurice Glenn; Pearson, Robert Stanley; Renfro, Arthur R.; Rosa, Felipe; Russell, John L.; Sander, Edgar Anthony; Singh, Gambhir; Sloan, Kenneth W.; Smith, Jerry P.; Snow, Dale L.; Snyder, Donald L.; Sternin, Jay E.; Stewart, John W.; Stindl, Heribert; Strong, Richard M.; Suess, Erwin; Veatch, Maurice D.; Warren, Kenneth M., Jr.; Wiman, W. David

1970s: Anderson, C. E.; Arnold, B.; Bell, T. C.; Black, Tyrone James; Cline, Royce L.; Dunlap, Lloyd E.; Dyer, Roger Gregory; Gilliland, W. J.; Griffin, J. R., Jr.; Gundrum, Lois Elizabeth; Hall, Robert Arthur; Hansen, Terry Jay; Harris, R. J.; Jeppesen, Jon A.; Johnson, Martin S.; Kilbane, N. A.; Koch, Richard J.; Konig, Michael; Lee, Moon Joo; Lorenzen, D. J.; Mathias, Kenneth E.; McClain, T. J.; McMullen, Terrence Leigh; Methot, Robert Leo; Miesse, J. V.; Pearce, R. W.; Reimer, L. J.; Roden, M. K.; Sawin, Robert Scott; Schmidt, Winfried; Scott, D. R.; Smith, Barbara J.; Steeples, Donald Wallace; Switek, John; Voran, Roxie Lynn; Voss, James D.; Woods, Michael J.; Yarrow, G. R.; Yeh, Long-Tsu

1980s: Bisby, Curtis G.; Falatah, Abdulrazag Mohammed; Faulkender, DeWayne J.; Ibrahim, Ismail K.; Reinecke, Kurt M.; Reitz, Bruce K.; Richter, Joseph Gustav; Schumacher, John G.; Spaid-Reitz,

Malia K.; Stell, Mary Jane Armitage; Stone, James M.

Kent State University, Kent
Kent, OH 44242

213 Master's, 4 Doctoral

1960s: Barclay, Craig C.; Metzler, Jean M.; Mrakovich, John Vincent; Stone, William J.

1970s: Artzner, D. G.; Austin, John Michael; Beikirch, Dale W.; Brown, Paul Henry; Bucher, Edward J.; Buller, Robert David; Butz, Todd Randall; Carden, John R.; Carothers, Thomas Arthur; Cooper, Milton L.; Couri, Clay C.; Decatur, Stephen Henley; Dellechaie, Frank; Dribus, John R.; Frost, Kenneth Robert; Gallagher, Gerald L.; Gardner, George Dennis; Garlauskas, Algirdas Benedict; Gawell, Mark J.; Gay, Frank Thomas; Grant, Timothy C.; Hahn, Kenneth R.; Hill, James Gregory; Hollenbaugh, Donald William; Honeycutt, Floyd Mitchell; Howell, Lamar Allan; Iivari, Thomas A.; Judy, J. R.; Kay, W. T.; Kell, Scott Randolph; Kingsbury, Richard Howard, Jr.; Kinney, Frederick Dawless; Knudson, Thomas D.; Kraemer, Curtis Allen; Krist, Hazel Fagley; Lahola, Irene; Malcuit, Robert J.; Manzer, Gerald K., Jr.; Maxey, Marilyn Helen; Meldgin, Neil J.; Mezga, Lance Joseph; Misko, T.; Mitchell, George Scott; Mychkovsky, George; Myers, Robert G.; Nacht, Steve Jerry; Nelson, Gordon C.; Oneacre, John William; Oros, Robert; Palubniak, Daniel S.; Perez, Emmanuel DeJesus; Perez, Stephanie; Phuphatana, Amorn; Robinson, George M.; Rogers, Dennis J.; Romito, Anthony A.; Ross, Martin E.; Schiering, Mark Harrison; Schlorholtz, Michael William; Schultz, Mark Gulliver; Schultz, Thomas Allan; Shotwell, L. Brad; Sigmund, James Martin; Simon, Peter R.; Smith, Harris Theodore; Spicer, Stanley R.; Stanley, Richard J.; Stollar, Richard Lloyd; Suphasin, Chai; Szczepanowski, Stanley Peter; Trudick, Lee S.; Uthe, Richard E.; Van Buskirk, Donald Robert; Vinopal, Robert J.; Wachter, Jack P.; Wallace, James; Wallace, Ronald Louis; Warrner, Charles Joseph; Wells, Terry L.; Wittine, Arthur H.; Wittman, Glenn Howard

1980s: Allen, Timothy J.; Anderson, Jane L.; Babcock, Loren Edward; Barber, William Bruce; Barker, Gary Wayne; Bendula, Richard A.; Berk, Jeffrey A.; Biros, Daniel J.; Bjerstedt, Thomas W.; Blackman, John Tristan; Brock, Dennis; Burns, David Bruce; Burrows, Vernon C.; Cannon Suva, Melinda S.; Cirbus, Lisa; Cook, Brad D.; Cox, Jason Charles; Culek, Thomas E.; Dalrymple, Margaret Ann; Deering, Mark F.; Dice, Mark A.; Dick, Jeffrey C.; Dwyer, Thomas Edward; Emadian, Nazila; Factor, David F.; Ferber, Charles Thomas; Ferry, Robert Allen; Figuli, Samuel P.; Finta, Susan F.; Fisher, Daniel S.; Folger, Charles Lee, Jr.; Freeman, Mimi J.; Fryman, Mark David; Grasso, Anthony Louis; Hanna, Ronald E.; Hannibal, Joseph Timothy; Hanson, James Phillip; Harker, David E.; Hau, Joseph A.; Heaton, Kevin P.; Hendricks, Robert Craig; Hilton, Don A.; Hose, David R.; Hoyt, Brian R.; Hunt, Patricia Kelly; Illes, Robert John; Johnston, John B.; Jones, Mark Lewis; Kammer, Heidi W.; Kerschner, David R.; King, Timothy Allen; Kormendy, Kenneth J.; Kraig, Scott A.; Krulik, Joseph W.; Lanigan, John Carroll, Jr.; Lopez, Carlos Alberto; MacDonald, Alistair P. T.; Maki, Mark Urho; Malik, Roberto F.; Maurath, Garry; McCauslin, S. E.; McCoy, David L.; McPherson, Constance Barbara; McQuown, M. Scott; Middleton, Dennis L.; Mignery, Thomas J.; Mohr, Eileen T.; Moore, Rosalie Carol; Natali, Patricia March; Nelson, Bradford E.; Norris, Cynthia R.; Novak, Stephanie Anne; Osten, Mark Allen; Park, Lawrence A.; Parsons, Kenneth E.; Patzke, Jeffrey A.; Penso, Sharon Marie Hirt; Petersen, Mark A.; Plevniak, John E.; Radon, Stanley F.; Rau, Theresa; Raymondi, Michael Joseph; Reeve, Richard L.; Richards, Susan Staben; Rintala, William E., Jr.; Roberts, Barry L.; Robertson, William L.; Roeper, Timothy R.; Ruof, Mark Anthony; Sanders, Laura Lourdes; Santini, Ronald J.; Schmidt, Mark F.;

Schmidt, Mark Thomas; Schmidt, Martin Leo; Schwimmer, Barbara A.; Sergoulopoulos, Alexandros; Sinclair, Richard H.; Singer, Karen M.; Singer, Michael P.; Skopec, Robert A.; Spurney, John C.; Stanley, Thomas M.; Stillings, Lisa L.; Stroebel, Kenneth H.; Stukel, Donald Joseph, II; Sypniewski, Bruce F.; Taft, A. G.; Taylor, Karen S.; Teepen, Kristina L.; Thomas, Andrew Russell; Thomas, John B.; Tshudy, Dale; Tucker, Annette B.; Tuzinski, Patrick A.; Underberg, Gregory L.; Urian, Brett A.; Vogel, Karen L.; Wagner, Joseph J.; Walters, Gerard Michael; Weber, Mitch W.; Wehn, David C.; Weidner, William E.; Weisgarber, Sherry Lee; Wickstrom, Lawrence H.; Wieckowski, Miriam Anna; Williams, F. W.; Wilson, John T.; Wilson, Margaret T.; Worstall, Robert Stewart; Yannacci, Dawna S.; Young, Brian T.; Young, Lawrence Edward

University of Kentucky
Lexington, KY 40506

270 Master's, 36 Doctoral

1890s: Brock, Lafayette Breckinridge; Downing, G.; Reynolds, Nelly A.

1900s: Averitt, S. D.; Marshall, Albert Ross

1920s: Beebe, Morris Wilson; Glenn, Howard E.; Perry, Eugene Sheridan; Pirtle, George William

1930s: Averitt, Paul; Bach, William Earl; Barton, Anna Louise; Haag, William George, Jr.; Hicklin, Richard Stuart; Hirsch, Jack; Marvin, Mary Lewis; Miles, Phil Middleton; Mills, Joseph Henry; Parker, Herbert F. R.; Purnell, James A.; Sandefur, Bennett Toy; Trautman, Ray Love; Welch, Robert Newman; Wesley, George Rutherford; Wilder, Newell Morris; Young, David Marion

1940s: Bruce, Clemont Hughes; Hamilton, Daniel Kirk; Lewis, James Otis; Stokley, John Allen; Straughan, George M.; Sunderman, Harvey Cofer; Thomas, R. N.; Young, James Lewis, Jr.

1950s: Brown, Alfred Louis; Conyers, William Patrick; Dukes, Bill Jady; Eldridge, Charles A.; Elmore, Robert Thompson; Flege, Robert Frederick, Jr.; Ford, Louis McKee; Ford, Russell James; Freeman, Ralph Neptune; Fugate, George W., Jr.; Gibson, Joseph Gallagher; Greenfield, Roy Emmett; Hamilton, Carter John; Hough, James Emerson; Jackson, William Ernest; Johnson, James F. H.; King, Lowell Franklin; King, William Roy, Jr.; Kramsky, Melvin Bernard; Laughlin, George Ray; McCord, Wallace R.; McCreary, Gary B.; Murphy, James Howard; Nosow, Edmund; Patterson, Logan Reid; Perkins, Jerome Hunt; Ponsetto, Louis R.; Ringo, William Pryor, Jr.; Rubarts, William Eugene; Rutledge, Henry Mitchell; Stanonis, Francis L.; Stoeckinger, William T.; Thomas, William Andrew; Walker, Frank Haff; Wood, Edward Boyne; Wood, Leonard E.

1960s: Allen, Corbett U., Jr.; Baedecker, Philip A.; Byrne, Richard Michael; Carrington, Thomas J.; Clarke, Murray K.; Dohm, Francis Paul; Felty, Kent K.; Fullerton, Donald S.; Hall, James Monroe, III; Haney, Donald C.; Hazel, James W.; Helton, Walter L.; Hine, George T.; Holbrook, Charles E.; Huffman, Jerald Dwight; Hurd, Robert James; Jacobs, Charles M.; Lewis, Wardell Lavon; Lieberman, Kenneth Warren; McKown, David Melvin; Miller, Delmon W.; Moser, Paul H.; Plummer, Leonard Niel, Jr.; Rebagay, Teofila Velasco; Reeves, Thomas Leslie; Rice, Glenn S.; Smith, Kenneth G.; Smith, Melvin Owen; Stafford, Thomas F., Jr.; Stallard, Paul C.; Sumartojo, Jojok; Tanner, James Thomas; Taylor, Ronald S.; Turner, William B.; Whaley, Peter Walter

1970s: Acquaviva, Daniel Joseph; Adams, Halbert Eden; Ammerman, Michael Lee; Beiter, David P.; Beniwal, R.; Blancher, Donald W.; Bland, Alan Edward; Bondurant, William Stewart; Breeding, N. Kelly; Carey, Daniel Irvin; Carney, Kevin Michael; Coskren, Thomas Dennis; Covert, Linda Lee; Crisp, Edward Lee; Davis, Philip A., Jr.; Deister, Walter

J., Jr.; Dever, Garland Ray., Jr.; Diamond, William Patrick; Eicher, Michael Lee; Elizalde, Leonardo; Ellis, John Hazle; Etter, Evelyn Mary; Graham, George Martin; Greenburg, Jeffrey King; Griswold, Thomas Baldwin; Griswold, Thomas Baldwin; Hawke, Bernard Ray; Hazel, James W.; Hetterman, John Leslie; Hopkins, John Walter; Huang, Scott Lin; Johnson, John Thomas; Kaldy, Windsor John; Kearby, James Kimbro; Lenhart, Stephen Wayne; Lieber, Robert Barry; Lin, T. C.; Mackey, Ronald Taylor; Markowitz, Gerald; McCann, Michael R.; McCulloch, Charles Malcolm; Miller, Michael L.; Navarro, Enrique Farran; Navarro, Enrique Farran; Nicholson, Frank Herbert; Norton, Hiram A., Jr.; Okolo, Stephen Anago; Pavona, Kennon Vincent; Pierson, Bernard J.; Ping, Russell Gordon; Powers, Stephen John; Price, Peter Elliot; Raab, Enrique Pedro Gentzsch; Rankin, James Scott; Reeves, Robert G.; Rettew, David Mark; Richers, David Matthew; Robl, Thomas L.; Robl, Thomas L.; Rosa, Eugene; Schwendeman, J. F.; Short, Michael Ray; Showalter, Donald Lee; Soderberg, Roger Kenneth; Spies, William Andrew; Stollenwerk, Kenneth G.; Swager, Dennis Ray; Thurman, Charles P.; Urbani, Franco; Urbani, Franco; Wachs, Thomas C.; Wixted, James Bernard

1980s: Allex, Mark; Allsop, Charles Mark; Amig, Bruce Clement; Amster, Andrew L.; Assad, Jamal M.; Baird, Robert Alan; Baker, Christopher Thomas; Barron, Lance S.; Bayan, Mohammad Reza; Baynard, David Nicoll; Bertsch, Paul Michael; Birch, Michael Joseph; Bolton, James Christopher; Brice, Donald A.; Brock, Martha Morgan; Brumfield, Kevin E.; Bruno, Anthony C.; Byrd, Phillip E.; Campagna, David John; Cheng, Chunyuen Raymond; Chestnut, Donald R.; Chestnut, Donald Rader, Jr.; Coates, Anna L.; Coskren, Thomas Dennis; Couch, Amber W.; Couch, David L.; Coyle, James L.; Currie, Michael Thomas; Dillman, Scott Brian; Dimmock, Pamela E.; DuBois, Sarah Barton; Elam, Timothy D.; Esterle, Joan Sharon; Fenske, John M., Jr.; Fierro, Pedro; Foley, William Clark; Fouts, John Douglas; Frankie, Kathleen Adams; Galceran, Carlos Manuel, Jr.; Gavett, Kerry Lea; Geller, Kris L.; George, John Samuel; Gordon, Larry; Gouzie, Douglas R.; Graese, Anne M.; Greb, Stephen Francis; Grossnickle, Effie Ann; Grow, Jeffrey S.; Hale, Alma Phillips; Hetherington, Peter Alan; Higgins, Lee; Hook, Robert Warren; Hopper, William M.; Horrell, Mark Alan; Huggins, Camillus B.; Jacobs, Gary Wayne; Jacobs, Gary Wayne; Johnson, Robert; Jones, David Kerrell; Kells, Bruce Lynn; Kemp, Robert E.; King, Brian Charles; Klapheke, Jeffrey G.; Kohles, Kevin Michael; Konkler, Jonathan L.; Lacazette, Alfred Julian; Lekhakul, Somjintana; Lenhart, Stephen W.; Lewellen, Dennis Gilbert; Lovins, Eric E.; Maddox, Eric; Manger, Katherine Chang; Manger, Phillip Herrmann; Mathis, Harry Leon, Jr.; Matthews, Neffra Alice; Maynor, Gregory Keith; Mensah, Winterford W.; Metcalf, Rodney Virgil; Metzner, David Craig; Miller, Timothy Robert; Moore, Timothy Allen; Muthig, Michael Gregory; Muthig, Paul Joseph; Neavel, Kenneth Edward; Neeley, Don Hitt; Netzband, Michael K.; Neuder, Gary Leslie; Nicholson, Toni Jost; Nickias, Peter N.; Nuckols, John Robert; Nunez del Arco, Eugenio; O'Hare, Andrew T.; Pashin, Jacob Charles; Pear, James Lewis; Portig, Elisabeth R.; Preziosi, Gregory Joseph; Raione, Richard Paul; Rezayat, Mohsen; Richers, David Matthew; Rogers, Timothy Joseph; Roulston, John S.; Rymer, Rodney Keith; Samarasinghe, Ananda Mahinda; Santos, Vanessa Anne; Scanlon, Bridget R.; Schaub, William J.; Schreiber, Anne Marie; Seale, Gary L.; Sharangpani, Shirish C.; Shepard, J. Scott; Shy, Timothy Laurence; Smith, Constant Ann; Spalding, Thomas D.; Spangler, Lawrence E.; Steinemann, Christopher F.; Sullivan, Stephen Bradley; Trimble, David Charlton; Turner, Joseph Gresham; Vance, Robert Kelly; Ward, Andrew David; Warwick, Peter Delawet; Watson, Kenneth Wayne; Waymire, Kelly Sue; Webb, James Sutton; Webster, Thomas

Craig; Wilkins, Jerry I.; Wonderley, Patricia Faith; Woock, Robert David; Zekulin, Alexander Darius

Lafayette College
Easton, PA   18042

1 Master's

1910s: Smith, Laurence Lowe

Lakehead University
Thunder Bay, ON   P7B 5E1

20 Master's

1980s: Arne, D. C.; Berger, Ben R.; Brown, G. H.; Cheadle, S.; Devaney, J. R.; El Tawashi, A. M. H.; Jackson, P. A.; Jago, B. C.; Jennings, E. A.; Kennedy, M. C.; Laderoute, D.; Lehto, Douglas Andrew Warren; Lukosius-Sanders, J.; McKay, D. B.; Poulsen, Knud Howard; Sarvas, P.; Schnieders, B. R.; Steel, O. J.; Tabrez, A.; Zayachkivsky, B.

Laurentian University, Sudbury
Sudbury, ON   P3E 2C6

16 Master's

1970s: Born, Peter; Campbell, G. J.; Everitt, Richard; Grant, R. W. E.; Innes, D. G.; Nikolic, Slobodan; Osa'-Idahosa, A.; Robertson, D.; Soucie, Gordon E.

1980s: Brunton, Frank R.; Dewing, Keith; Foley, Steven L.; Jin, Jisuo; Jiricka, D. E.; Kwok, Kai Ming; Ma Xueping

Universite Laval
Ste.-Foy, PQ   G1K 7P4

151 Master's, 45 Doctoral

1940s: Béland, Jacques Robert; Blois, Roland de; Gadd, Nelson Raymond; Grenier, Paul Emile; Maurice, Ovide Dollard; Melihercsik, Stephen J.; Taylor, Edward Drummond; Tremblay, Léo-Paul

1950s: Benoit, Fernand Wilbrod; Bérard, Jean; Berard, Jean; Bergeron, Robert; Bienvenu, Léo R.; Blais, Roger A.; Ezeani, Chuba; Gélinas, Léopold; Gelinas, S.; Grenier, Paul Emile; Grondin, Gaston Guy; Hannah, G. J. Raymond; Hannah, G. J. Raymond; Laporte, Jean; Laurin, Andre; Laurin, André Fredéric Joseph; Lyall, H. Bruce; Marleau, Raymond Alban; Melihercsik, Stephen J.; Miller, Robert J. M.; Morin, Marcel; Morin, Marcel; Nunes, Arturo de F.; Pérusse, Jacques; Sabourin, Robert Joseph Edmond; Sabourin, Robert Joseph Edmond; Smith, J. R.; Tessier, G. Robert

1960s: Arbour, Roger; Cimon, Jules; Depatie, Jean; Duquette, Gilles; Gélinas, Léopold; Hashimoto, Tsutomu; Kiss, Leslie; Lakatos, Stephen; Lamarche, Robert Y.; Lamarche, Robert Y.; Orajaka, Stephen; Robert, Jean Louis; Sears, Peter J.; St. Julien, Pierre; Williams, David

1970s: Biron, Serge; Bussières, Louise; Charbonneau, Jean-Marc; Descarreaux, Jean; Duberger, Reynald; Lamothe, Daniel; Laroche, Paul; Leroux, Jean Pierre; Rouleau, A.;

1980s: Asselin, Esther; Audet-Lapointe, Martine; Beaudoin, Georges; Benn, Keith; Beullac, Raymond; Brazeau, André; Caron, Alain; Champagne, Michel; Choquette, Marc; Cockburn, Daniel; Cousineau, Pierre A.; d'Astous, Jacques; De Broucker, Gilles; Demers, Denis; Desrochers, Andre; Dion, Claude; Doré, Guy; Drolet, Michel; Drouin, Ruth; Dubé, Benoît; Fortin, Danielle; Fournier, Benoit; Fréchette, Ghislain; Gahé, Emile; Gaudreau, Roch; Girard, Marie-Josée; Gosselin, Charles; Hébert, Réjean; Hebert, Yves; Labbé, Jean-Yves; Laforest, André; Laliberté, Jean-Yves; Lapointe, Daniel; Lavoie, Denis; Lavoie, Denis; Leboeuf, Denis; Lefebvre, Yues; Lessard, Denis; Martel, Richard; Morin, Claude; Padilla, Francisco; Padilla, Francisco; Pelletier, Lise; Pinsonnault, François; Plante, Langis; Raymond, Lynda; Robitaille, René; Rousseau, Normand; Savard, Martine; Slivitsky, Anne; Soucy, René; Therrien, Pierre; Thibault,

Yves; Togola, N'Golo; Vallières, André; Wilhelmy, Jean-Francois

Lehigh University
Bethlehem, PA   18015

158 Master's, 55 Doctoral

1900s: Evans, Morris De B.; McCaskey, Hiram Dryer; Penneypacker, N. R.

1910s: Bartlett, Ralph L.; Fraim, Parke Benjamin

1920s: Burke, James M.; Chamberlin, Dale S.; Lawall, Charles E.; Talmage, Sterling Booth

1930s: Buie, Bennett Frank; Getz, Albert Julius; Myers, Philip B.

1940s: Agocs, William B.; Dellwig, Louis Field; Hersey, John Brackett; Mills, John Ross; Petersen, Richard G.; Ryan, John Donald; Skinner, Walter Swart; Tooker, Edwin Wilson; Warmkessel, Carl Andrew

1950s: Alter, B. E. K.; Bellerjeau, Orwyn Tilton; Bergenback, Richard E.; Bowman, Delbert A.; Burns, Robert Earle; Chow, Minchen Ming; Ern, Ernest Henry, Jr.; Fox, G. R.; Goodwin, Bruce Kesseli; Graham, Theodore Kenne; Johnsen, John Herbert; Kell, James Alexander; Layman, Frederick G.; Lein, Carl A.; Lessentine, Ross Henry; Long, Morris Andrew; Macfadyen, John A., Jr.; McCallum, John S.; McGuchen, John G.; Rader, Herbert L.; Ray, Satyabrata; Read, Edward Wade; Stevenson, Robert Evans; Stingelin, Ronald W.; Taylor, Russell N.; Trexler, John Peter; Warmkessel, Carl Andrew

1960s: Arce, Carlos; Biggs, Robert Bruce; Craig, James Roland; Eby, George Nelson; Kaufmann, Karl W., Jr.; Land, Lynton Stuart; Larimer, John William; Mackenzie, Frederick Theodore; Mentzer, Thomas Cartwright; Moose, Louis; Myers, Paul Benton, Jr.; Neumann, Andrew Conrad; Rhindress, Richard C.; Roland, George Warren; Sherwood, William Cullen; Strong, David F.; Suess, Erwin; Taylor, Lawrence A.; Vincent, Robert J.; Virgin, William Wallace, Jr.

1970s: Allan, John R.; Arcaro, Nick P.; Bass, Jay David; Basu, Debabrata; Bauer, Jon F.; Berglund, Pete; Borry, Barrett E.; Cameron, Aubrey T., Jr.; Carius, Terry L.; Carvalho, Antone V., III; Citrone, Jeffrey; Connelly, John R.; Domino, Grant A.; Dunleavy, Jeffrey M.; Fergus, John Howard, Jr.; Fleming, Robert S., Jr.; Force, Eric Ronald; Force, Lucy McCartan; Ganis, George R.; Gould, David R.; Halladay, Christopher R.; Hardiman, Andrew L.; Huff, David W.; Hyman, David; Kaplan, Sanford; Kastelic, Robert L., Jr.; Kelley, Joseph T.; Kluger, Karen Lee; Koelmel, Mark H.; Kran, Neil; Lash, G. G.; Levy, Joel B.; Luther, Frank R.; Martin, Harry; Meglio, Joanne Teresa; Mehta, Sudhir; Meyerson, Arthur Lee; Morzenti, Stephen P.; Nyobe, Jean Blaise; O'Mahoney, Laurence; Pasquini, Thomas; Pietrobon, Vincent J., Jr.; Popper, George H. P.; Raring, Andrew Michael; Rugh, Alex L.; Sama-Nupa-Win, Bancha; Sassen, Roger; Sassen, Roger; Schultz, Lane D.; Schultz, Lane D.; Snitbhan, N.; Squiller, Samuel F.; Stephens, George Christopher; Stewart, Robert A.; Stokke, Per R.; Suwanasing, Akanit; Swanson, Mark Thomas; Usselman, Thomas Michael; Veitch, John D.; Wegner, Robert C.; Westerman, David Scott; Wigley, William C.; Winn, Robert; Yaniga, Paul M.; Yuan, Jennwei; Zamel, Abdulla Z. Al

1980s: Alevizos, Anastasios; Andres, Alan Scott; Anson, Gwendolyn L.; Awak, Collins T.; Begley, Alisa L.; Benimoff, Alan Irwin; Benoit, Paul Harland; Blanchard, Allan Marc; Bloomfield, James Miller; Bond, Robert M.; Brugger, Keith A.; Bufo, David; Carney, Keith F.; Carpenter, Martha Alice; Cauller, Stephen J.; Chapin, Daivd A.; Clay, Julia S.; Clinch, J. Michael; Cole, David Andrew; Collins, David G.; Cotter, James F. P.; Cotter, James F. P.; Crouse, George W.; Deamer, Gay A.; Declercq, Eric P.; Evans, Jeffrey Clinton; Fielding, Stanley J.; Gallagher, Robert Anthony; Gawarecki, Susan L.;

Gibson, Robert G.; Glassburn, Tracy; Griesemer, Jeffrey Crane; Hall, Frank Reginald; Hall, Mary Jo; Healy, Mary Jo; Holliday, Valerie E.; Keating, Robert William; Kelley, Alice A. Repsher; Kelley, Joseph Timothy; Kim, Adeline; Krafft, Alison D.; Lagas, Philip Joseph; Lash, Gary George; Marshall, Robert C.; Meglis, Andrew J.; Mengel, Martin; Mikroudis, George Konstantinos; Monteverde, Donald H.; Novillo, Mary Muscarella; Nyobe, Jean Blaise; O'Neill, Robert L.; O'Toole, Patrick Brian; Pearce, Suzanne M.; Pruitt, Maria Pankos; Redmond, Roy J.; Ridge, John C.; Ritger, Scott D.; Rozov, Wendy Cara; Scamman, Robert L.; Schroeder, Tom Scot; Seibel, Geoffrey C.; Serfes, Michael Edward; Sharga, Paul J.; Sine, Franklin Arthur, Jr.; Sirois, Brenda; Siwiec, Steven F.; Snyder, Geoffrey William; Stamatakos, John A.; Sudano, Peter L.; Taylor, Theodore Warren; Valentino, Albert J.; Vetter, James R., Jr.; Vittorio, Louis F.; Volpi, Mary; Wagner, Annette; Walker, Alfred Thomas, III; Weakliem, John Herbert; Witte, Ron W.; Zigmont, James H.

Loma Linda University
Riverside, CA   92515

13 Master's, 4 Doctoral

1970s: DeBord, Philip L.; Fisk, L. H.; Hodges, L. T.

1980s: Barnett, Stephen F.; Britton, Douglas R.; Clausen, Benjamin L.; Cushman, Robert A., Jr.; Hanson, David L.; Hornbacher, Dwight; Jensen, Karen Grace; Jensen, Roy E.; Nick, Kevin E.; Rasmussen, Michael G.; Sandefur, Craig A.; Schremp, Lee A.; Vyhmeister, Gerald Erwin; Wareham, Stephen I.

Long Island University, C. W. Post Campus
Brookville, NY   11548

2 Master's

1960s: Penn, Sheldon H.

1970s: Bruderer, Barry E.

Louisiana State University
Baton Rouge, LA   70803

255 Master's, 191 Doctoral

1920s: Alexander, William Albert; Harris, Reginald Wilson; Tatum, Emmett Perry, Jr.

1930s: Blouin, Cecil F.; Broussard, Marion U.; Chambers, Jack; Chawner, William D.; Dohm, Christian F.; Dunbar, Clarence P.; Frink, John W.; Garret, Julius B., Jr.; Gooch, Dee David; Hough, Leo Willard; Huner, John, Jr.; Hussey, Keith Morgan; Lea, Joseph W.; Martin, James L., Jr.; Mayer, Maurice J., Jr.; McDougall, John E.; McGuirt, James H.; McGuirt, James H.; Mincher, Albert Russel; Monsour, Eli Thomas; Monsour, Emil; Moresi, Cyril Killian; Murray, Grover E., Jr.; Postell, William Dosite; Pyeatt, Lloyd M.; Roberts, Marion S.; Rukas, Justin M.; Shively, C.; Simpson, Jack Ezelle; Smith, Frederick E.; Taylor, Ralph E.; Wallace, William E., Jr.; Warner, Ambrose Deidriche

1940s: Ashby, William H.; Aycock, Lester C.; Bailey, Robert; Barry, John O'Keefe; Belchic, George, Jr.; Bernard, Hugh Allen; Blake, Daniel B.; Bloomer, Philip A., Jr.; Cameron, Harriet V.; Glockzin, Albert Richard; Goldstein, August, Jr.; Halsey, Ramond E.; Hecker, Edward N.; Holland, Wilbur Charles; Hussey, Keith Morgan; Kaplan, Lazard Harold; Kolb, Charles R.; Lamont, Norman; LeBlanc, Rufus J.; LeRoy, Tom E.; Martin, James L., Jr.; McDonald, Stanley M.; McLaughlin, Kenneth Phelps; McLean, James D.; Meagher, David Pope; Monsour, Edward; Murray, Grover E., Jr.; Oakes, Ramsey L.; Osanik, Alec N.; Pope, David E.; Smith, Denver Jeter; Smith, Robert Hendell; Tator, Benjamin Almon; Tator, Benjamin Almon; Thomas, Emil Paul; Varvaro, Gasper Gus; Vernon, Robert Orion; Wallace, William E., Jr.; Wasem, Richard; Weingeist, Leo; Wiengeist, Leo; Wilbert,

Louis Joseph, Jr.; Woodward, Truman P.; Wooley, William Leeman

1950s: Andersen, Harold V.; Autin, Leonard J.; Bernard, Hugh Allen; Boyd, Donald Ray; Buis, Otto J.; Butler, Elizabeth M.; Chandler, Charles; Cockerham, Kirby L., Jr.; Contreras, Carlos E.; Cox, Charles L., Jr.; Crawford, Frank C.; Delaney, Patrick; Dinnean, Robert F.; Forde, Robert H.; Grigg, Robert P., Jr.; Hanai, Tetsuro; Joerger, Arthur P.; Jones, Douglas; Jones, Paul H.; Kimmel, Marion L.; Krinitzsky, Ellis Louis; Lawrence, Robert M.; Levert, Charles F., Jr.; Marianos, Andrew; Martinez, Joseph Didier; Mixon, Robert B.; Morgan, James P.; O'Shields, Richard L.; Olive, Wilds W.; Puri, Harbans S.; Rizvi, S. Ali Ibne Hamid; Sexton, James V.; Sloane, Bryan Jennings; Smith, Charles I.; Smithwick, Jack Allison; Sternberg, Hilgard O'Reilly; Stricklin, Fred Lee, Jr.; Sun, Ming-Shan; Thorsen, Carl Elmer; Thorsen, Carl P. E.; Tonti, Edmond Charles; Treadwell, Robert Cuthrell; Treadwell, Robert Cuthrell; Van Lopik, Jack Richard; Varvaro, Gasper Gus; Waldron, Robert P.; Wang, Kia-Kang; Warren, Albert David; Welder, Frank A.; Zimmerman, Thomas J.; Zingula, Richard P.

1960s: Anderson, Don Randolph; Artusy, Raymond Longino; Bergeron, William Joseph; Blackman, Abner; Blount, Donald Neal; Bolander, Richard John; Bonis, Samuel B.; Boutte, Andre L.; Breaux, James E.; Brooks, Robert Alexander; Carpenter, Glenn F.; Castle, Richard A.; Cavaroc, Carolyn Wynn; Cavaroc, Victor Viosca, Jr.; Cavaroc, Victor Viosca, Jr.; Coleman, James Malcolm; Coleman, James Malcom; Collins, Robert Joseph, Sr.; Corgan, James Xavier; Davies, David K.; Davis, Carol Waite; Deboo, Phili B.; Dixon, Louis H.; Dixon, Mark A.; Dolan, Robert; Drouant, Ronald George; Dukes, George Houston, Jr.; Durham, Cordelia Louise; Ehrlich, Robert; Ehrlich, Robert; Elliott, Herbert A., Jr.; Emmer, Rodney E.; Erdogan, Solmaz Z.; Esker, George C., III; Ethridge, Frank G.; Facundus, Michael R.; Ferguson, Hershal C., Jr.; Flores, Romeo Marzo; Franzmann, Frederick K.; Gagliano, Sherwood Moneer; Gimbrede, Louis de A.; Glawe, Lloyd Neil; Hamil, Martha M.; Hazel, Joseph Ernest; Henderson, Barry Keith; Henderson, Barry Keith; Hobday, David K.; Howard, C. Edward; Jones, Charles L.; Jones, Hershel Leonard; Jones, Paul Hastings; Kelly, George A., Jr.; King, Robert E.; Kolb, Charles Rudolph; Krutak, Paul Russell; Krutak, Paul Russell; Kury, Theodore William; Kwon, Hyuck Jae; Lee, Keenan; Majewske, Otto P.; Marcantel, Emily L.; Marcantel, Jonathan; McArthur, David Samuel; McCloy, James Murl; McGowan, Clyde Ronald; McKee, James W.; McKee, James Walker; McNiel, Norman; McRee, David E.; Merrill, Glen K.; Mixon, Robert Burnley; Morales, Frias, Gustavo; Morales-Frias, Gustavo Adolfo; Mullins, Robert L.; Mumma, Martin Oale; Nikravesh, Rashel; O'Niell, Charles A., III; Ogier, Stephen Hahn; Ouellette, Dorice J.; Parsons, Brian E.; Paulson, Oscar Lawrence, Jr.; Psuty, Norbert Phillip; Raphael, Constantine Nicholas; Reiser, Samuel G.; Rios, Michael; Roberts, Harry Heil; Roberts, Harry Heil; Rosen, Norman Charles; Rucker, James B.; Sachdev, Sham Lal; Sandberg, Philip A.; Saucier, Roger Thomas; Shaw, Nancy Sue; Shaw, Nolan Gail; Shayes, F. P.; Siesser, William G.; Smith, Charles G., Jr.; Snead, Rodman Eldredge; Stephens, Raymond Weathers, Jr.; Stevens, Waldo Eugene; Thom, Bruce Graham; Van Beek, Johannes Laurens; Wall, James Roy; Wappler, John H.; Warter, Janet Lee Kirchner; Watson, Stuart T.; Webb, James W.; Weidie, Alfred E., Jr.; Wermund, Edmund G., Jr.; Whaley, Peter Walter; Wiedie, Alfred E., Jr.; Worley, George T.; Zimmerman, Ronald K.; Zimmerman, Ronald K.

1970s: Abington, Oscar D.; Ahmad, N.; Albach, Douglas C.; Allen, Stephen H.; Ambuehl, Alan W.; Arguello, Ottoniel; Arndorfer, David James; Badon, Calvin Lee; Bailey, Janet E.; Baker, Alfred H., Jr.; Banas, P. J.; Becher, Jack W.; Blondeau, Kenneth M.; Boellstorff, John David; Boleneus, David E.; Bonnet, Daniel J.; Bordine, Burton William; Brannon, James M.; Brooks, Robert Alexander; Brown, James Lee, Jr.; Carney, John L.; Chakrabarti, Chinmoy; Choung, Haeung; Christopher, Raymond Anthony; Clear, Harry C.; Coates, Eugene Joseph; Collins, Earl M.; Conklin, Jack S.; Cramer, George H., II; Cratsley, D. W.; Crout, R. L.; Crow, Sidney Alfred, Jr.; Crowe, Clifford T.; Cruz-Orozco, R.; Cruz-Orozco, Rodolfo; Darrell, James Harris, II; Delcourt, Hazel Roach; Delcourt, Paul Allen; Devine, Stanley Bevan; Dinnean, Robert F.; Dobbins, Ralph J.; Doiron, Linda; Donellan, Monica Sue; Drennan, William T., III; Earle, D. W., Jr.; Fields, Noland Embry, Jr.; Forth, David R.; Gatenby, Glen Michael; Goodarzi, Nasrin K.; Gradijan, Stephen J.; Hall, John W.; Hambrick, Gordon A., III; Handford, Charles R.; Hanna, John Clark; Harper, Howard E., Jr.; Harper, J. R.; Harper, J. R.; Harper, Wallace F.; Harris, M. K.; Harrison, Frank W., III; Harrison, W. E.; Hart, D. L.; Hebert, Roger L.; Hernandez-Avila, Manuel L.; Hessenbruch, John M.; Hiltabrand, Robert R.; Hixon, Robert Louis; Hoffman, James H.; Hunerman, Aybars E.; Inden, Richard F.; Jackson, Marie C.; Jendrzejewski, John P.; Jervey, Macomb T.; Johnston, S. W.; Jones, Daphne L.; Josey, William L.; Kaye, John Morgan; Kessinger, Walter Paul, Jr.; Khan, Rashid Ali; Kirst, Timothy L.; Klasik, John A.; Kress, Margaret R.; Kubera, Paula A.; Kumar, Madhurendu Bhushan; LeBlanc, Robert C.; Lindstedt, D. M.; LoPiccolo, Robert D.; Markham, Thomas A.; McManis, K. L.; Meaney, William R.; Milan, C. S.; Monte, J. A.; Morgan, David J.; Mumme, Stephen T.; Naymik, Thomas G.; Neal, Ronald E.; Nemeth, Donald F.; O'Neil, Thomas J.; Oivanki, Stephen M.; Peake, J. S.; Pederson, David E.; Petta, Timothy J.; Pierce, Robert W.; Price, C. Allen; Ray, Pulak K.; Reese, Robert J.; Reitz, Donald D.; Rexroad, Richard L.; Rives, John S.; Robichaud, Stephen R.; Ross, William C.; Saxena, Ram S.; Schewe, John H.; Seyfried, William E., Jr.; Shanks, Wayne C., III; Short, Andrew Damien; Simmons, Weldon A., Jr.; Small, Benjamin A., III; Smith, Carl C.; Spurr, Malcolm R.; Steineck, Paul Lewis; Stoessell, Ronald K.; Sturm, David H.; Sutherland, John W., Jr.; Taylor, Gilbert D.; Teleki, Paul Geza; Thibodaux, Bernadette L.; Tubman, Michael W.; Van Beek, Johannes Laurens; Vincent, Frank S.; Waddell, Evans; Walker, Charles William; Walter, Lynn M.; Weisblatt, Edward A.; Wells, J. T.; Wells, Richard F.; Wesson, John Nolan; Wicker, Karen M.; Williams, Charles D.; Wright, Lynn Donelson

1980s: Angelich, Michael Terry; Aubert, Winton G.; Bair, John H.; Baker, Nancy Tucker; Baykal, Gokhan I.; Beckman, Scott Warren; Bender, Russell Berryman, Jr.; Brannon, James Milton; Brock, Frank C., Jr.; Caffrey, Jane Marie; Cassell, David Terrance; Cavanaugh, James F.; Colten, Virginia Ann; Croft, Wayne S.; Dahmani, Mohamed Amine; DiMarco, Michael J.; Dinnel, Scott Page; Dinnel, Scott Page; Dozier, Malcolm David; Feijtel, Tom Cornelis Jan; Fithian, Patricia Ann; Funderburg, Eddie Ray; Goh, Yong Soon; Hartnell, Jill Ann; Hatton, Richard S.; Hughes, Robert Hilton; Johns, Hilary Desmond; Kahn, Jacob Henry; Kemp, George Paul; Kim, Soon Tae; Kim, Yeadong; Kiousis, Panagiotis Demetrios; Kosters, Elisabeth Catharina; Krstanovic, Predrag Felix; Lowry, Philip; Machan-Castillo, Maria Luisa; Majumdar, Arunaditya; Mastin, Gary Arthur; Miller, Carolyn Ann; Mohammadi, Mohammad; Moore, Philip Alderson, Jr.; Moshier, Stephen Oakley; Nocita, Bruce William; Pamukcu, Sibel; Ramsey, Albert Frank; Ranganathan, Vishnu; Roy, Amitava; Saller, Arthur Henry; Schrodt, Augusta Kay; Seguinot-Barbosa, Jose; Sklar, Fred Hal; Snedden, John William; Standhardt, Barbara R.; Suter, John Robert; Thibodeaux, Jerry Lee; Tye, Robert S.; van Heerden, Ivor Llewellyn; van Heerden, Ivor Llewellyn; Walker, Nan Delene; Wrenn, John Harry; Yuan, Peter B.

Louisiana Tech University
Ruston, LA   71272

22 Master's, 3 Doctoral

1960s: Box, Jerry W.; Holder, Robert E.; Jones, Garnet W., Jr.; Meade, Carroll Wade; Miller, James Andrew; Poole, Russel Wayne; Taylor, Bobbey Ben; Whitfield, Merrick S., Jr.; Wray, Cloyd Field

1970s: Dupre, David Carl; Espeseth, Robert Lynn; Franz, Richard Lewis; Hinthong, Chaiyan; Ketchum, Robert L.; Nakhinbodee, Veerasak; Oudomugsorn, Prakal; Russel, Ronald P.; Russell, Ronald Paul; Santos, John Joseph, Jr.; Taylor, Richard David; Williamson, Robert L.

1980s: Frizzell, Larry Glen; Hejazi, Assadollah; Poffenberger, Michael Robert; Siddiqi, Khalid Omar

University of Louisville
Louisville, KY   40292

7 Doctoral

1960s: Noland, Anne Vinson

1970s: Hill, F. C.; Horowitz, A.; Leuthart, C. A.; Riehl, A. M.; Tittlebaum, M. E.

1980s: Hill, Paul Lester, Jr.

Loyola University
Chicago, IL   60611

1 Master's

1980s: Rimkus, Wayne Vincent

University of Maine
Orono, ME   04573

80 Master's, 2 Doctoral

1940s: Donohue, John Joseph

1950s: Coney, Peter J.; Fairley, William Merle; Farley, William H.; Forsyth, William T.; Frank, Glenn W.; Groselle, Francis X.; Kyte, Harold F.; Linger, Robert E.; Wing, Lawrence Alvin; Zink, Robert Miller

1960s: Bonnett, Richard Brian; Calkin, William S.; Cavalero, Richard; Erinakes, Dennis C.; Goodspeed, Robert Marshall; Hennings, Richard Armond; Jacobson, Marvin LeRoy; Mickelson, David M.; Rhoades, David Alan; Stoeser, Douglas; Supkow, Donald; Tays, Gerald; Williamson, Terrence C.

1970s: Abbott, Richard N., Jr.; Attig, J. W., Jr.; Baker, Robert W.; Ballaron, Paula Balcom; Bradstreet, Theodore E.; Caruso, Louis J.; Davis, P. T.; Dignes, T. W.; Heinonen, Charles E.; Karlen, Wibjorn; Kenoyer, Galen; Kite, James Steven; Lingle, C. S.; Liotta, Frederick P.; McKeon, John B.; McSwiggin, P. L.; Olson, Roger K.; Pollock, Stephen G.; Sasseville, D. R.; Simpson, R. W., Jr.; Surgenor, J. W.; Thompson, Stewart N.; Van Beever, Hank G.

1980s: Berry, Henry N., IV; Biederman, John L.; Birnie, Robert I.; Brewer, George F.; Burgess, Margaret V.; Carlson, David John; Cotter, Mark Patrick; Crossen, Kristine J.; Dagel, Mark A.; Hay, Bradley W. B.; Hyland, Mark R.; Johnston, Steven E.; Kahl, Jeffrey S.; Knight, James A.; Lepage, Carolyn A.; Lindstrom, Dean R.; Lowell, Thomas V.; Miller, Sarah B.; Moyse, David Wayne; Mulry, Christopher J.; Pelto, Mauri S.; Pfeffer, Tad; Pinette, Steven R.; Prentice, Michael L.; Rappaport, Paul Aaron; Sorenson, Eric R.; Stone, Christopher Talbott; Teichmann, Friedrich; Tepper, Dorothy H.; Virta, Robert Lee; Wall, Ellen R.; Walsh, J. Andrew; Weed, Rebecca; West, David P., Jr.; Williams, John S.

University of Manitoba
Winnipeg, MB   R3T 2

275 Master's, 51 Doctoral

1910s: Hanson, George

1920s: Baker, W. F.; Birse, Donald John; Brownell, George McLeod; Childerhose, Allan Jerome; Fraser, Horace John; Leith, Edward Issac; Maynard, James E.; McCartney, Garnet Chester; Merritt, Clifford A.; Quinn, Reay Pullar; Spratt, Joseph Grant; Thompson, Lucas G.; Ward, G.; Yarwood, Walter S.

1930s: Horwood, H. C.; Lane, Harry Campbell; Prest, Victor Kent; Ruttan, George Douglas; Smith, F. Gordon; Spivack, Joe; Webb, John Benwell

1940s: Arnott, Ronald James; Colcleugh, V. D.; Davies, J. F.; Johnson, H. A. C.; Morgan, John Harold; Oliver, Thomas Albert; Spector, I. H.

1950s: Allen, Clifford Marsden; Anderson, Donald Thomas; Anderson, Andrew D.; Bannatyne, Barry B.; Binda, Louis S.; Bramadat, Kelvin; Butler, Roy Leslie; Caldwell, Charles Keith; Calich, Rade; Childs, Gerald Dewitt; Clark, Patricia J.; Davies, John Clifford; Dornian, Nicholas; Emslie, Ronald Frank; Fletcher, G. L.; Hunter, Hugh E.; Irvine, T. N.; Irvine, Thomas Neil; Kilburn, Lionel Clarence; Knutson, Robert A.; Koop, W. J.; Lowther, Jack; Macauley, George; McCabe, Hugh Ross; Michalkow, Albert; Moorhouse, Michael David; Organ, David William; Osborne, Thomas Cramer; Papirchuk, W.; Proctor, R. M.; Proctor, Richard M.; Root, Samuel; Shepherd, Jackson Howard; Stott, Donald Franklin; Troop, Andrew John; Turnock, Allan Charles; Venour, E. R.; Vernour, E. R.

1960s: Anderson, Donald Thomas; Andrews, Peter W.; Bailes, Alan Harvey; Bari, Shah Fazoul; Barker, G. S.; Barrett, Vernon A.; Bell, Kenneth William; Bristol, Calvert C.; Bristol, Calvert C.; Brown, Robert James; Buckingham, James Gordon; Callender, Edward; Clarke, Peter J.; Coats, Colin J. A.; Cranstone, Donald A.; Davies, John C.; Dwibedi, K.; Freund, Harold H.; Gait, Robert I.; Gait, Robert I.; Genik, Gerard Julian; Gilliland, J. A.; Gurbuz, Behic; Haugh, Ian; Hodgkinson, John Morris; James, Michael N. G.; Jardine, Donald Edwin; Kilburn, Lionel Clarence; King, Kenneth R.; Kornik, Leslie J.; Kuryliw, Chester J.; Lambo, W. A.; Lapointe, Guy; Laznicka, Peter; MacPherson, Robert A.; McDonald, John A.; McKennitt, D. B.; McPherson, Robert A.; Oddy, Richard William; Paulus, George Edmund; Pearse, Gary H.; Pollock, Gerald D.; Pollock, Gerald D.; Putt, David J.; Quaraishi, A. A.; Rector, Roger J.; Riley, C. J.; Rousell, D. H.; Rousell, Donald H.; Saunderson, Carol Patricia; Schmidt, H. G.; Scoates, Reginald Francis Jon; Sherwood, Herbert Gordon; Sherwood, Herbert Gordon; Silversides, David A.; Smith, Donald Leigh; Solohub, J. T.; Trembath, Lowell Thomas; Tweedy, Norma A.; Vagt, G. Oliver; Wicks, F. J.; Zakus, Paul D.; Zwanzig, H.

1970s: Amukan, Samuel E.; Anderson, Neil L.; Anderson, Richard Kime; Andrews, Douglas James; Baer, David Walter; Bailes, Alan H.; Bailes, Richard James; Bakhtiari, Raiani Hamid; Baldwin, David Arthur; Bamburak, James David; Bates, Allan Clifford; Beakhouse, Gary Philip; Beswick, Barry Thomas; Blewett, Kenneth W.; Bond, William Douglas; Boyce, Barry James Simon; Bridge, Dane Alexander; Buck, Peter Stanley; Budrevics, Valdis; Butrenchuk, Stephen; Cameron, Peter Forrest; Campbell, Frederick H. A.; Campbell, L. B.; Campbell, Susan Wendy; Cerna, Ivanka; Chagarlamudi, Pakiriah; Cheung, Sha-Pak; Chute, Michael Earl; Clister, William Eugene; Coles, Richard Leslie; Coles, Richard Leslie; Corkery, Maurice Timothy; Cowan, John R.; Cranstone, John R.; Delancey, Peter R.; Desmarais, Ralph J.; Dorn, Thomas Franz; Eilers, R. G.; Elphick, Patricia Margaret; Elphick, Stephen Conrad; Enns, Steve Gerhard; Fenton, Mark Macdonald; Friesen, George Henry; Frohlinger, Thomas Gordon; Giesbrecht, Karl Otto; Goff, Kenneth J.; Green, Nathan Louis; Grice, Joel Denison; Grice, Joel Denison; Grove, G. D.; Gumbo, F. J.; Hajnal, Zoltan; Hargreaves, Roy; Homeniuk, Leonard Anthony; Hooper-Reid, N. M.; Hopkins, L. A.; Humiski, Robert Nicholas; Hutnik, F. T. A.; Ingram, Ruth M.; Islam, S. M. Nazrul;

Jackson, Michael Ralph; James, David Richard; Jones, Robert Douglas; Josse, Genic Raymond; Juhas, Allan Paul; Keatinge, Penelope Rosanna Gann; Kohut, Alan Peter; Kor, Philip S. G.; Kushnir, Donald William; Lalor, Jim; Lamb, Craig Forbes; Last, William Michael; Laznicka, Peter; Lebel, John Laurence; Lenton, Paul G.; Lustig, Gary Norman; Macek, Josef Jan; Mackie, B. W.; Manchuk, Barry; Marr, John; Masson, Arthur Guy; Matthews, Robert Norman; Maynard, J. E.; McGinn, R. A.; McPherson, Robert A.; Menzies, Douglas G.; Morin, James Arthur; Morrice, Martin Gray; Muhammad, Mir Jan; Mwanang'onze, E. H. B.; Mwanang'onze, E. H. B.; Mysyk, Walter K.; Obinna, F. C.; Pedora, John Michael; Penner, Alvin Paul; Posehn, Gary Arnold; Provins, Dean Allen; Pryslak, Anthony Paul; Raudsepp, Mati; Render, Francis William; Rinaldi, Romano; Ritchie, Paul Michael; Santos, Rebecca R.; Schwartz, Franklin Walter; Scoates, Reginald Francis Jon; Seccombe, Philip Kenneth; Shemeliuk, Edward Michael; Shemeliuk, Virginia A. B.; Simpson, Frederick Muir; Singhroy, V. H.; Smith, Scott Raymond; Solonyka, Edward Richard; Sopuck, Vladimir Joseph; Standing, Keith F.; Staubo, John Peder; Steeves, Michael Albert; Stephenson, John Francis; Taylor, J. R.; Trueman, David Lawrence; Wallace, Garnet Cecil Grant; Warwick, W. F.; Ziehlke, Daniel V.

1980s: Ahmed, M.; Al-Taweel, Bashir Hashim; Alley, Hugh A.; Anderson, Alan; Bald, Roberta C.; Baldwin, David Arthur; Barrett, Kent R.; Brown, Bruce A.; Car, Dwayne Peter; Carswell, Allan; Chackowsky, L. E.; Chagarlamudi, Pakiraiah; Christianson, Carlyle Bruce; Coniglio, Mario; Conley, Glenn; de Landro, Wanda-Lee; Duke, Norman Albert; Duncan, E. J. Scott; Eby, Raymond K.; Egan, Martin D.; Ercit, Timothy Scott; Ferreira, K. J.; Ferreira, William S.; Findlay, Donald J.; Gaboury, Bernard E.; Glenday, Keith Stuart; Goad, Bruce E.; Gordanier, Wayne Derek; Groat, Lee Andrew; Hanneson, Donna L.; Hanneson, James Edward; Hansen, Beverly J.; Hillary, Elizabeth M.; Hinds, Ronald C.; Khan, Sheraz M.; Knapp, Caroline J.; Knight, David Cooper; Kovac, L. J.; Last, William Michael; Lau, Meng Hoo Sebastian; Leggett, Sidney R.; Loiselle, James Richard; Mare, Marius P. H.; Marius, Maré; Maxwell, Brian; McAuly, Roger J.; McGowan, E. B.; McGregor, Catherine R.; Mehner, David T.; Meintzer, Robert Ells; Messfin, Derbew; Metcalfe, Paul; Misra, Kiran Shanker; Noble, Ian A.; Obi, Adeniyi Olubunmi; Olson, P. E.; Owusu, John; Rahman, Mohammed Golzar; Raudsepp, Mati; Schweyen, Timothy; Sepehr, Keyvan; Servos, Mark Roy; Sheppard, Stephen Charles; Sibiya, Victor; Strobel, Guye; Tang, Alan M.; Tang, Roger; Thomas, M. W.; Thorleifson, L. Harvey; Trembath, Gerald; Trueman, David Lawrence; Ucakuwun, Elias Kerukaba; Ushah, Abdurrazag; Wise, Michael Anthony

Marshall University
Huntington, WV 25755

2 Master's, 1 Doctoral

1970s: Durfee, Daniel

1980s: Calandra, John D.; Chappell, George A., Sr.

University of Maryland
College Park, MD 20742

11 Master's, 58 Doctoral

1960s: Harmison, Lowell Thomas; Larson, Jerome Valjean

1970s: Ailin-Pyzik, I. B.; Buhl, P. H.; Caponi, E. A.; Clark, Pamela Elizabeth; Healy, R. P.; Horzempa, L. M.; Jackson, T. J.; Kessel, Richard H.; Krickenberger, K. R.; Law, S. L.; O'Brien, Dennis J.; Pyzik, A. J.; Rubincam, David Parry; Simon, F. O.; VanderBrug, G. J.; Weaver, J. F.

1980s: Al-Sanad, Hasan Abdul Aziz; Amer, Mohamed Ibrahim Mohamed; Amini, Farshad; Athanas, Louis Chris; Barnard, Pamela Bright;

Beauchamp, Robert George; Bell, Douglas Alan; Bell, Kenneth Robert; Bell, Stephen Craig; Buckingham, William Forrest; Burke, Todd M.; Burris, Robert Leroy; Christy, Michael Scott; Chung, Donald Ta-Lung; Clegg, Robert Henry; Darmody, Robert George; El-Damak, Reda Abdu El-Hay Mohamed Ali; El-Hemry, Ismail Ibrahim Mohamed; Fellows, Jack D.; Fendinger, Nicholas Joseph; Finkelman, Robert Barry; Finnegan, David Lawrence; Fitzgerald, John Joseph; Galarraga, Federico Antonio; Groves, James Robert; Hall, Dorothy Kay; Helwa, Mohamed Fawzy; Hwang, Daekyoo; Jordan, Stephen James; Kotra, Ramakrishna; Libert, John M.; Lofti, Hani Abdel-Latif; Matthias, Cheryl Louise; Miller, Richard E.; Morgan, John Milton, III; Mostafa, Ayman A.; Ossi, John C.; Pagoaga, Mary Katherine; Paik, Sun Mok; Phelan, Janet Meredith; Powell, Robert Leslie; Shea, Damian; Sinex, Scott Alden; Sircar, Jayanta Kumar; Stone, Craig A.; Symborski, Mark Andrew; Tacker, Robert Christopher; Tawfiq, Kamal Sulaiman; Wagner, Daniel P.; Walters, Clifford Carol; Wijayaratne, Rammali Devlina

University of Massachusetts
Amherst, MA 01003

221 Master's, 84 Doctoral

1930s: Stevens, Nelson Pierce

1940s: Fisher, Joseph O.

1950s: Ames, Herbert Tate; Andersen, Allen E., Jr.; Berry, W. F.; Donaldson, A. C.; Dutcher, Russel R.; Erickson, Norman K.; Fisher, John; Hulsman, Robert B.; Kelley, Dana Robineau; LeBlanc, Arthur E.; Malloy, R. E.; Messinger, Curtis; Nicholeris, N.; Parks, Donald A.; Percy, Cynthia W.; Rotan, R. A.; Sarris, Nelson James; Saulnier, Henry Siddell; Shea, James F., Jr.; Tetreault, Andre R.; Tynan, Eugene Joseph; Waloweek, W.

1960s: Bazakas, Peter C.; Beer, Lawrence P.; Carranza, Carlos; DaBoll, Joan; DeWyk, Bruce H.; Felsher, Murray; Fitzgerald, Kim; Frye, Charles I.; Gaffney, Joseph Walter; Gonthier, Joseph Bernard; Groat, Charles George; Guthrie, Daniel Albert; Hagar, David Jon; Halpin, David Lawrence; Hinthorne, James Roscoe; Knapp, George Leroy; Lloyd, Orville Bruce; Makower, Jordan; Matz, David B.; McCormick, C. L.; Merrill, Robert D.; Peper, John Dunkak; Pestrong, Raymond; Pike, Thomas Mace; Popper, George H. P.; Ruhle, James L.; Shroder, John F., Jr.; Sommers, David Arthur; Walker, Robert F.; Wessel, James McCandless; Wessell, J. M.; Zimmerman, Herman B.

1970s: Abele, Ralph Warren, Jr.; Ahmad, F. I.; Alavi, M.; Anan, Fayez Shaban; April, Richard H.; Ashenden, David D.; Ashley, Gail M.; Ashwal, Lewis D.; Ashwall, L.; Baillieul, Thomas A.; Berg, Jonathan Henry; Bucci, S. A.; Bwerinofa, Obadiah K.; Caggiano, Joseph Anthony, Jr.; Callahan, Richard; Campbell, Kerry Jacquith; Chandler, William E.; Clebnik, S. M.; Dana, Richard H., Jr.; Davies, Hope M.; deLorraine, William F.; Dowdall, Wayne Larry; Durgin, Philip Bassett; Eskenasy, Diane M.; Fairbairn, Patrick W.; Farrell, Stewart C.; Farrell, Stewart C.; Feuer, S. M.; Field, Michael Timberlake; Fitzpatrick, John Cole; Franz, Arthur J.; Gilchrist, James Michael; Goldsmith, Victor; Goldstein, Arthur Gilbert; Golombek, Matthew Philip; Grady, Stephen J.; Graves, Catherine A.; Gunter, Karl D.; Gustav, Spence H.; Gustavson, Thomas Carl; Guthrie, James O.; Hall, David Joseph; Hall, David Joseph; Hamilton, James A.; Handy, Walter A.; Hartwell, Alan D.; Heeley, Richard William; Hendrix, W. G.; Hine, Albert C.; Hobbs, Carl H., III; Hollands, Garrett G.; Huntington, J. Craig; Inners, J. D.; Jackson, Richard A.; Jezek, P. A.; Kaczorowski, Raymond T.; Kalamarides, Ruth I.; Kelly, William M.; Kelly, William Morgan; Keyes, Scott Wellington; Kick, John Frederick; Kinney, Vincent Lewis; Kroll, Richard Lawrence; Ku, W.-C.; Laird, H. Scott; Larsen, Frederick Duane; Leftwich, John Thomas; Lodge, Lynn; Lund, Kendall G.; Moser, John

Archer; Mulholland, James Willard; Newton, R. M.; Nilsson, Harold Daniel; Nilsson, Harold Daniel; O'Brien, Dennis Craig; Onasch, Charles M.; Piepul, R. G.; Pinet, Paul Raymond; Potter, Donald Brandreth, Jr.; Principe, Paul A.; Pywell, H. R., III; Raber, Ellen; Ranson, William Albrecht; Reed, Alan A.; Retelle, Michael James; Rhodes, E. G.; Rosen, Peter S.; Saines, Marvin; Sancton, Joel A.; Shirey, S. B.; Smith, Philip Gerard; Snow, P. D.; Soloyanis, Susan Constance; Stoeck, Penelope L.; Stromquist, Albert W., Jr.; Suchecki, Robert K.; Symmes, K. H.; Tracy, Robert J.; Trudeau, P. N.; Tucker, R. D.; Walsh, J. E.; Weddle, Thomas K.; Wessel, James McCandless; White, Christopher Minot; Wiener, Richard Witt; Williams, Gerald; Williams, Larry Dean; Wolff, Robert A.

1980s: Alford, David Dorman; Allison, John P.; Allison, Merle Lee; Autio, Laurie Knapp; Bailey, Janet Allen; Ball, Stephen; Barrows, Peter Scott; Batchelder, Gail; Birney, Carol C.; Blum, Brian Allen; Boudreau, Lawrence Joseph; Brandon, James P.; Burr, Jonathan L.; Caffall, Nancy M.; Cambareri, Thomas Christian; Clayton, Timothy J.; Day-Lewis, Robert Ernest; Denatale, Douglas Robert; Dermer, Michele Suzy; Drake, Natalie E. R.; Eberly, Paul O.; Fayer, Michael James; Field, Stephen W.; Field, Stephen Walter; Filipov, Allan James; Flanagan, James Joseph; Forlenza, Michael Francis; Francis, Robert A.; Fulton, Christopher Robert; Gilbert, John Robert, Jr.; Gilchrist, Carol Mary; Goldstein, Arthur Gilbert; Golombek, Matthew Philip; Good, John Conrad; Gray, Floyd; Greene, David Carl; Hall, James Creevey; Haniman, Kurt Christopher; Harrington, James R.; Hatfield, Kirk; Hayes, Martha Anne; Hicks, Jason F.; Hills, Doris Volz; Hodgkins, Catherine E.; Hollocher, Kurt T.; Hollocher, Kurt Thomas; Hozik, Michael Jacob; Huntington, Hope Davies; Hyde, Matthew G.; Jackman, Anthony Edwin; Jackson, Richard A.; Jacobson, Peter R.; Jasaitis, Richard A., Jr.; Jennings, A. A.; Jewell, T. K.; Johnson, Cheryl Elaine; Joneja, Danielle C.; Josephson, John J.; Kalamarides, Ruth I.; Kalamarides-Berg, Ruth Irene; Kolker, Allan; Kuo, Ching-Liang; Leavell, Daniel Nelson; LeFond, Joanne M.; Leonard, Wendy C.; Leung, Jana C.; Liias, Raimo Arnold; Limentani, Giselle Beth; Lin, Hsiuan; Llewellyn-Smith, Timothy M.; Luckey, Frederick E.; Maalouf, George Y.; Maczuga, David E.; Mazzone, Peter; McCrory, Thomas Alan; McElroy, Thomas Alan; McIlvride, William Allen; McMahon, Brendan Michael; Meriney, Paul E.; Mertz, Karl Anton; Michener, Stuart Reid; Mitchell, Christopher B.; Morgan, Cedwyn; Morton, Peter S.; Nicholson, Suzanne Warner; Norman-Gregory, Gillian Margaret; O'Brien, Keith M.; O'Laskey, Robert H.; Obi, Curtis Mitsuru; Ogawa, Hisashi; Ollila, Paul William; Orrell, S. Andrew; Oshchudlak, Martha E.; Owens, Brent Edward; Perkins, Russell Edward; Perkins, Warren W.; Peters, Norman Edward; Peterson, Virginia L.; Pferd, Jeffrey William; Pohanka, Susan J.; Potter, Donald Brandreth, Jr.; Retelle, Michael James; Rogers, Frederick S.; Roll, Margaret A.; Schlain, Mildred Rachel; Schumacher, John Charles; Schwalbaum, William Jesse; Serreze, Mark C.; Shannon, Margarita C.; Shearer, Charles Kenneth; Simon, Steven Bruce; Slosek, Jean; Soukup, James J.; Stam, Marianne; Steenstrup, Susan Jeanette; Stekl, Peter J.; Stopen, Lynne E.; Storms, Erik; Taterka, Bruce D.; Thomas, George McConnell; Thompson, Peter James; Tirsch, Franklin Steven; Toft, Paul Bernard; Tollo, Richard Paul; Tompkins, Linda Anne; Truettner, Laura Elizabeth; Valdés, José de Jesús; Valentine, Michael James; Walen, Sarah Kimball; Wall, William Patrick; Walsh, Daniel Charles; Warner, Philip Edmund; Weeraratne, Saroj Premasiri; Weingarten, Baruch; Weiss, Alan E.; Weisse, Patricia A.; Wenz, Kenneth P., Jr.; Werkheiser, William H.; Wiener, Richard Witt; Zielinski, Gregory A.; Zimmermann, Regula Dorothea

Massachusetts Institute of Technology
Cambridge, MA   02139
331 Master's, 453 Doctoral

1900s: Phalen, W. C.; Trueman, Joseph Douglas

1910s: Allan, J. A.; Clapp, C. H.; Dolmage, Victor; Mackenzie, John David; Means, Alan Hay; Powers, Sidney; Schofield, Stuart James; Whitehead, W. L.

1920s: Abbott, A. C.; Benedict, Platt C.; Boydell, H. C.; Buerger, Martin J.; Buerger, Martin J.; Burbank, W. S.; Burton, W. D.; Callahan, W. H.; Creveling, J. G.; Davies, H. F.; Davy, W. M.; Dennen, William Llewellyn; Flaherty, G. F.; Gillson, J. L.; Gillson, J. L.; Gledhill, T. L.; Gunning, H. C.; Gunning, H. C.; Hanson, George; Hodges, P. A.; Hurst, M. E.; Lary, H. N.; McClelland, H. W.; McKinstry, Hugh Exton; Meng, H. M.; Morse, W. C.; Muller, Charles Julian; Muller, Charles Julian; Newhouse, W. H.; Newhouse, W. H.; Papenfus, E. B.; Putnam, P. C.; Rexford, E. P.; Smitheringale, W. V.; Waldschmidt, William A.; Williams, W. H.

1930s: Beaton, N. S.; Brown, L. S.; Buerger, Newton W.; Butler, Robert D.; Carman, K. W.; Chase, H. D.; Dumbros, Nicholas; Fisher, J. D., Jr.; Flaherty, G. F.; Foster, F. L.; Frondel, Clifford; Furnival, G. M.; Gold, K. M.; Graves, W. H., Jr.; Gussow, W. C.; Herpers, Henry F.; Holden, M. A.; Horwood, C. H.; Ilsley, Ralph; Jordan, Louise; Jordan, Louise; Kania, Joseph Ernest Anthony; Keith, M. L.; Lopez, V. M.; Lopez, V. M.; Lord, C. S.; Lukesh, J. S.; McMurry, H. V.; Mencher, Ely; Parker, F. L.; Raymond, L. C.; Rove, Olaf N.; Shimer, J. A.; Stevenson, J. S.; Stoiber, R. E.; Twinem, J. C.; Wilson, Earl O.; Zuloaga, Guillermo

1940s: Allan, J. D.; Auger, P. E.; Beers, R. F.; Bell, K. G.; Bell, K. G.; Berger, Louis; Brawner, J. D., Jr.; Bray, Joseph M.; Buerger, Newton W.; Campbell, Neil; Cutten, William; Dennen, W. H.; Faul, Henry; Frederickson, A. F.; Frueh, A. J., Jr.; Frueh, A. J., Jr.; Goodman, C.; Hamburger, G. E.; Hawkes, H. E., Jr.; Hogg, Nelson; Houssiere, L. I.; Hurley, Patrick M.; Jalichandra, Nithipatana; Keiser, Edward P.; Kirman, Z. M.; Klein, G. E.; Mead, Judson; Mills, J. W.; Moss, John H.; Neterwala, M. P.; Nogami, H. H.; Parrish, William; Pollock, J. P.; Sagoci, H. F.; Sargent, T. H. E.; Seibert, W. E., Jr.; Shimer, J. A.; Stewart, Robert W.; Stoll, W. C.; Stoll, W. C.; Thompson, George A., Jr.; Tripp, Russell Maurice; Tuttle, Orville Frank; Vincent, Kenneth C.; Walker, Edward Bullock, III; Walton, Paul Talmadge; Washken, Edward; Yoder, H. S., Jr.

1950s: Allan, J.; Anderson, D. H.; Azaroff, L. F.; Backus, M. M.; Bell, Christopher K.; Bowman, R.; Brace, William F.; Breger, Irving A.; Briscoe, Howard; Bullwinkel, Henry J.; Canney, Frank C.; Cloke, P. L.; Cormier, Randall F.; Coryell, Lawrence Ritchie Brooke; Decker, Robert W.; Dewart, Gilbert; Duffin, W. J.; Edie, Ralph William; Farrington, William B.; Galvin, Cyril J.; Gilbert, J. F., Jr.; Gore, Roger C.; Gowen, Walter K.; Gower, John A.; Grine, Donald; Grine, Donald R.; Hagen, John C.; Hallof, Phillip G.; Haq, K. E.; Hendricks, Walter J.; Herzog, Leonard F., II; Hill, Claude P. T.; Hilton, E. R.; Holyk, Walter; Huppi, R. G.; Jacobson, Herbert S.; James, Allan H.; Jensen, Mead Leroy; Johnston, W. G.; Kaminsky, P. D.; Kelley, D. G.; Kelley, Danford Greenfield; Kermabon, A. J.; King, C. W.; King, Lewis W.; Kolb, J. D.; Kranck, Svante Hakan; Leavitt, William Z.; Leonard, R. B.; Lopez, Linares; Marshall, Donald J.; McAllister, John J.; McIntire, William L.; Murray, Bruce; Murray, Bruce; Ness, Norman F.; Neves, A. S.; Niizeki, N.; Oca, G. R.; Phinney, Robert A.; Phinney, William C.; Phinney, William C.; Pinson, William Hamet, Jr.; Podolsky, T.; Posen, Harold; Powell, R. M.; Puech, J. F. C. G.; Richardson, Paul W.; Robinson, Enders; Rolnick, L. S.; Rose, Edwin R.; Ross, Virginia F.; Sage, Nathaniel McLean, Jr.; Sage, Nathaniel McLean, Jr.;

1960s: Adger, John B., Jr.; Anderson, Philip J.; Arkani Hamed, Jafar Gholi; Beall, George H.; Beamish, Peter C.; Beger, Richard M.; Beiser, Erna; Bence, Alfred E.; Blackburn, William H.; Bless, Stephen J.; Bolka, David F.; Bombolakis, Emanuel G.; Bowker, David E.; Brookins, D.G.; Brownlow, Arthur H.; Bruns, Terry Ronald; Burnham, Charles W.; Byerlee, James D.; Byrd, William; Camfield, Paul Adrian; Cantwell, Thomas; Carlisle, Donald Hugh; Carter, Richard Michael; Cid-Dresdner, Hilda; Clark, Robert C., Jr.; Clearbout, Jon; Crocket, James H.; Cudjoe, John E.; Dakin, Francis; Devorkin, Donald B.; Dingler, John R.; Dollase, Wayne A.; Dos Santos, Edson R.; Downs, James P.; Drapeaus, Georges; Drinker, Philip A.; Dulaney, Ernest N.; Eckhardt, Donald H.; Everett, James Edward; Fahlquist, Davis A.; Faramarzpour, Faramarz; Faure, Gunter; Folinsbee, Robert Allin; Frasier, Clint Wellington; Galbraith, James Nelson, Jr.; Galvin, Cyril J.; Garaycochea-Wittke, Isabel; Ginsburg, Merrill Stuart; Goldstein, Myron; Green, Edward J.; Greenewalt, David; Greenfield, Roy J.; Halverson, Ward D.; Hamilton-Smith, Terence; Harita, Yoichi; Harita, Yoichi; Harper, Charles W., Jr.; Hart, Stanley R.; Hauck, Anthony M.; Heath, Edward W.; Heath, Marla M.; Heath, Stanley A.; Horodyski, Robert Joseph; Horowicz, Leon; Hwang, Jae-Young; Jackson, David Diether; Johnson, David A.; Jokela, Arthur; Kehrer, Harold Henry; Kelley, Arthur M.; Kenyon, Kern; Krotser, Donald J.; Kuenzler, Howard W.; Lewis, Lloyd F.; Madden, Theodore R.; Maehl, Richard H.; Magnell, Bruce A.; Martin, Randolph J., III; Mcnutt, Robert H.; Miller, Elliot White; Miller, Richard C.; Moon, Warren D.; Moore, John M.; Nourbehecht, Bijan; Nur, Amos Michael; Pan, Cheh; Paul, Donald Lee; Paulding, Bartlett W.; Paulson, Edwin G.; Payson, Harold, Jr.; Peacor, Donald R.; Peacor, Donald R.; Perry, Eugene Carleton, Jr.; Philpotts, John Aldwyn; Phipps, Donald; Posadas, Veronica Gomez de; Powell, James L.; Prewitt, Charles T.; Prewitt, Charles T.; Querol, Sune Francisco; Quesada, Antonio E.; Quigley, Robert M.; Recks, Elizabeth Helen; Redden, Martha J.; Reesman, Richard H.; Ritter, Charles J.; Roe, Glenn D.; Russell, James D.; Saint-Amant, Marcel Michel Yvon; Sanford, Thomas B.; Saul, John M.; Sax, Robert L.; Schilling, Jean-Guy E.; Schneider, William A.; Schnetzler, Charles C.; Scholz, Christopher H.; Schroeder, Gerald; Shaw, Donald H.; Shields, Robert McC.; Sill, William Robert; Sleep, Norman Harvey; Smitheringale, William G.; Spilhaus, Athelstan F.; Spirn, Regin V.; Spooner, Charles M.; St. Amant, Marcel M. Y.; Swift, Charles M., Jr.; Sykes, Lynn R.; Taxer, Karlheinz J.; Thompson, Keith F.; Towell, David Garrett; Trojer, Felix J.; Trojer, Felix J.; Tsai, Yi-Ben; Unger, John D.; Waldbaum, David R.; Walters, Lester J., Jr.; Wandzilak, Michael; Ware, Jerry Allen; Warren, Bruce A.; Webster, Thomas F.; White, James A. L.; Whitney, James A.; Whitney, Philip R.; Wing, Charles G.; Wittels, Razel A.; Wones, David R.; Woodside, John M.; Wuensch, Bernhardt John; Wunsch, Carl I.; Yearsley, John R.; Yules, John A.

1970s: Abu-Eid, Ratab Muhmood; Agunloye, Alfred Olusegun; Andersen, Kristine Louise; Anderson, Kenneth Robert; Anderson, Kenneth Robert; Andresen, Brian Dean; Angoran, Yed Esaie; Arney, Barbara Holota; Austin, James Albert, Jr.; Bacon,

Sammel, Edward A.; Saull, Vincent A.; Seraphim, Robert H.; Shaw, William S.; Short, Nicholas N.; Shumway, George Alfred, Jr.; Simpson, M. S., Jr.; Smith, George F., Jr.; Smith, M. K.; Southwick, P.; Southwick, Peter Frederick; Southwick, Stanley Harpham; Stacy, Maurice Cyrus; Stewart, Robert W.; Tang, Alice C.; Tenny, Ralph Emil; Thompson, J. B., Jr.; Thompson, William B.; Tooley, Richard D.; Towse, Donald Frederick; Treitel, S.; Treitel, Sven; Tupper, William M.; Van Lewen, Melvin C.; Vozoff, K.; Wadsworth, Donald V.; Walsh, W. P.; Webber, George R.; Welby, Charles W.; Wetzel, John Hall; White, James A. L.; Whiting, Francis B.; Wittels, Mark C.; Wylie, R. W.; Young, Edward J. A.; Young, Edward J. A.; Zoltai, Tibor Z.

Michael Putnam; Barker, Terrance Gordon; Batzle, Michael Lee; Ben-Avraham, Zvi; Berman, Joel; Besancon, James Robert; Bird, George Peter; Blake, Brenda Jean; Boguchwal, Lawrence Allen; Boore, David M.; Bornhold, Brian Douglas; Bouchon, Michel Paul; Boyle, Edward Allen; Briggs, Peter Laurence; Brown, Raymon Lee; Buma, Grant; Burr, Norman Charles; Burroughs, Richard Hansford; Carroll, Beverly Mildred; Chandler, John Frederick; Chapman, Clark Russell; Chapman, Edward Dewey; Charles, Robert W.; Chen, Wang-Ping; Cheng, Chuen Hon Arthur; Chernosky, Joseph V.; Chouet, Bernard A.; Chouet, Bernard Alfred; Combs, James B.; Consolmagno, Guy Joseph; Cooper, Herman William; Copeland, Richard A.; Corea, William Charles; Costello, Warren Russell; Cox, Larry Paul; Das, Shamita; Davis, Karleen Ethel; Davis, Robert Alvin; Dengler, Alfred Theodore; Detrick, Robert Sherman; Durham, William Bryan; Duschenes, Jeremy David; Eichelberger, John; Ellsworth, William Leslie; England, Anthony W.; Erez, Jonathan; Evans, James Brian; Fehler, Michael; Feves, Michael Lawrence; Fitterman, David Vincent; Fitzgerald, William Francis; Flagg, Charles Noel; Folinsbee, Robert Allin; Forsyth, Donald William; Foster, Stephen Eric; Francis, Donald Michael; Fryer, Brian J.; Gaffey, Michael James; Galen, William Mamoru; Galson, Daniel Allen; Gardner, Wilford Dana; Gates, Todd M.; Goettel, Kenneth Alfred; Goettel, Kenneth Alfred; Goettel, Mary Jane Westervelt; Goins, Neal Rodney; Hadley, Kate Hill; Hao, Wei Min; Hawley, Nathan; Hazen, Robert M.; Heinze, William D.; Hellinger, Steven J.; Hon, Rudolph; Huested, Sarah S.; Huggins, Francis Edward; Huguenin, Robert Louis; Hull, Marylee Witner; Humphris, Susan Elizabeth; Johnson, Carl Edward; Johnston, David Hervey; Johnston, Janet Catherine Pruszenski; Kahn, Gail Anne Heffner; Kasameyer, Paul William; Knight, Curtis Alan; Konwar, Lohit Narayan; Kreimendahl, Frank Alan; Kuster, Guy Thierry; Laine, Edward Paul; Larner, Kenneth Lee; Leavy, Donald; Leavy, Donald Lucien; Lebofsky, Larry Allen; Lee, Wook Bae; Libicki, Charles Melvin; Loiselle, Marc Charles; Lopez Escobar, Leopoldo; Lopez-Escobar, Leopoldo; Louden, Keith Edward; Macdonald, Kenneth Craig; MacIlvaine, Joseph Chad; Mackintosh, Michael Edward; Madariaga Meza, Raul I.; Maderazzo, Marc Matthew; Martin, Randolph J., III; Mendiguren, Jorge Andres; Menke, William Henry; Miller, Stephen P.; Montgomery, Carla Paige Westlund; Moo, Charles Anthony; Morrow, Carolyn Alexandria; Murray, James Wray; Nautiyal, Chandra Mohan; Ng, Albert Tung-Yiu; Niazy, Adnan Mohammed; Nikhanj, Yashvir A.; Nolet, Daniel Arthur; Noyes, Harold James; Obata, Masaaki; Olszewski, William John; Ostro, Steven Jeffrey; Parkin, Kathleen Marie; Patton, Howard John; Pavlin, Gregory B.; Peirce, John Wentworth; Pelke, Paul A.; Pieters, Carle Ellen; Pieters, Carle Ellen; Pines, Philip Jacques; Poehls, Kenneth Allen; Ranganayaki, Rambabu Pothireddy; Reasenberg, Paul Allen; Reed, David Allen; Reid, John B., Jr.; Reisz, A. Colbert; Richardson, Randall Miller; Robertson, Douglas Scott; Robin, Pierre-Yves François; Roy, David C.; Roy, Stephen Donald; Sayer, Suzanne; Schaller, Phillip H.; Schatz, John F.; Scheimer, James Francis; Schutts, Larry Davis; Schwartz, Kenneth Bruce; Schwenn, Mary Bernadette; Scott, Robert Earl; Scranton, Mary Isabelle; Sen Gupta, Mrinal Kanti; Sen Gupta, Mrinal Kanti; Settle, Mark F.; Shlien, Seymour; Siegfried, Robert Wayne; Simon, Henry Francis; Skibo, Donald Nicholas; Sleep, Norman Harvey; Smith, Albert Turner; Smith, Albert Turner; Sneeringer, Margaret Riggs; Sneeringer, Margaret Riggs; Solomon, Sean C.; Sprunt, Eve S.; Sprunt, Hugh Hamilton; Stesky, Robert Michael; Suen, Chi-Yeung John; Sung, Chien-Min; Takeuchi, Shozaburo; Thonis, Michael; Todd, Terrence P.; Tucholke, Brian Edward; Ulrich, Gilbert Wayne; Vagt, William Arthur; Vilas, Faith; Waldrop, Ann Lyneve Chapman; Wang, Chi-fung; Wang, Herbert Fan; Ward, Ronald W.; Warsi, Waris Ejaz Khan; Watson, Edward Bruce; Weidenschilling, Stuart John; Weidner, Donald

James; Whipple, Earle Raymond; Williams, Anthony Brackett; Williams, David Lee; Willis, Mark Elliott; Wolf, Eric R.; Wolfe, Jack C.; Wong, George T. F.; Wong, George Tin Fuk; Young, Robert Alexander; Young, Robert Alexander; Yulke, Sandra Gay; Zandt, George; Zarrow, Lorraine; Zielinski, Robert A.; Zimmerman, M. B.

1980s: Anderson, Robert Frederick; Asper, Vernon L.; Axen, Gary James; Baker, Michael Baldwin; Baranowski, Jean M.; Bartley, John Michael; Bennett, Brian R.; Bennett, Sara L.; Bergman, Eric Allen; Beydoun, Wafik Bulind; Beydoun, Wafik Bulind; Blackman, Donna Kay; Blumberg, George Micah Connor; Bohacs, Kevin Michael; Braatz, Barbara Vanston; Bratt, Steven Richard; Brownawell, Bruce J.; Bryan, Carol J.; Buck, Walter Roger; Cameron, Christopher Scott; Cao, Tianqing; Cappallo, Roger James; Caristan, Yves Denis; Casas, Federico Pardo; Celerier, Bernard; Chen, Chu-Yung; Choi, Jin Beom; Clark, Roger Nelson; Collier, Robert William; Comer, Robert Pfahler; Commerford, Janine; Corea, William Charles; Coyner, Karl B.; Crowley, Peter Duncan; Cunningham, Paul S.; Cuvelier, Gaëtan Jean Francois Joseph; Danna, James G.; Davis, Daniel Michael; de Baar, Hein J. W.; Dyar, Melinda Darby; Ebinger, Cynthia Joan; Ebinger, Cynthia Joan; Elkins, Linda Tarbox; Fitzgerald, Michael Gerard; Freedman, Adam Paul; Fry, Virginia Ann; Gerlach, David Christian; Godkin, Carl B.; Goldberg, David Samuel; Goldfarb, Marjorie Styrt; Gonguet, Christophe; Goodwin, Jeffrey Thomas; Goreau, Peter David Efran; Graham, David William; Gray, Dale Franklin; Grazer, Robert Anthony; Gulen, Levent; Gurriet, Philippe Charles; Guth, Peter Lorentz; Haines, Harvey Hartman; Hall, Jennifer Lynn; Hardin, Ernest L.; Harpin, Raymond Joseph; Henderson, Jeremy Robert; Herring, Thomas Abram; Hess, Thomas E.; Hickey, Rosemary Louise; Hickmott, Donald Degarmo; Hodges, Kip Vernon; Huang, Paul Yi-Fa; Hubbard, Mary Syndonia; Hudson, Andrew G.; Ikeda, Keiichiro; Ikeda, Keiichiro; Jasper, John Paul; Jaupart, Claude; Jones, Craig Howard; Jones, Lucille Merrill; Judge, Anne Victoria; Keho, Timothy H.; Kennedy, Allen Ken; Kinzler, Rosamond Joyce; Klepacki, David Walter; Kuhnle, Roger Alan; Lambie, John M.; Larrère, Marc H.; LaTorraca, Gerald Alan; Leinbach, Alan Edward; Libes, Susan M.; Lineman, David J.; Little, Sarah Alden; Lo, Tien When; Lo, Tien When; Lupo, Mark Joseph; Lyon-Caen, Hélène; Martin, William R.; Matarese, Joseph Richard; McNichol, Ann P.; Mellen, Michael H.; Mendelson, J. D.; Miller, Kenneth George; Morgan, Frank Dale Oliver; Morris, Julie Dianne; Muller, James Louis; Nabelek, John L.; Nathman, Douglas Robert; Nelson, Michael Roy; Novich, Bruce Eric; Nowack, Robert L.; Nozette, Stewart David; Okubo, Paul G.; Olgaard, David LeClair; Orange, Daniel Lewis; Park, Stephen Keith; Paternoster, Benoit J.; Pegram, William Joseph; Pfirman, Stephanie Louise; Phillips, William Scott; Pierce, Stephen Davis; Pulli, Jay J.; Ray, Glenn Lamar; Reid, Mary Ruth; Richardson, Stephen Hilary; Roden, Michael Frank; Roecker, Steven William; Rohr, Kristin Marie Michener; Rooze, Tom W.; Rosa, Joao Willy Correa; Royden, Leigh Handy; Ruppel, Carolyn Denise; Saltzer, Sarah Dawn; Sawyer, Dale Stewart; Schaefer, Martha Williams; Schneider, Marie Diane; Selverstone, Jane Elizabeth; Shakal, Anthony Frank; Sharry, John; Shaughnessy, Anna Catarina; Shedlock, Kaye M.; Sheffels, Barbara Moths; Shen, Glen T.; Sherman, David Michael; Shih, John Shai-Fu; Smith, Eric R.; Singer, Robert Bennett; Smith, Abigail Marion; Sneeringer, Mark Albert; Solberg, Teresa Christine; Speer, Kevin George; Spencer, Jon Eric; Spivack, Arthur J.; Stallard, Robert Forster; Sternlof, Kurt Richard; Stewart, Robert R.; Stock, Joann Miriam; Stock, Joann Miriam; Stockman, Harlan Wheelock; Suarez, Gerardo; Suárez, Gerardo; Swift, Stephen Atherton; Taras, Brian Daniel; Taylor, Steven Renold; Thurber, Clifford H.; Tilke, Peter Gerhard; Tréhu, Anne Martine; Tubman, Kenneth M.; Von Damm, Karen Louise;

Vrolijk, Peter John; Wagner, Richard A., Jr.; Walker, James Douglas; Walker, James Douglas; Watkins, Guyton Hampton; Watkins, Guyton Hampton, Jr.; Weisenstein, Debra Kay; Wernicke, Brian Philip; Wilcock, Peter Richard; Wingo, James Raymond; Wissler, Thomas Martin; Wong, Teng-Fong; Wu, Ru-Shan; Yang, Mai; Yomogida, Kiyoshi; Zhang, Peizhen; Zindler, Gregory Alan

McGill University
Montreal, PQ H3A 2A7

336 Master's, 263 Doctoral

1900s: McIntosh, D. S.; Strangways, H. F.

1910s: Wilson, Morley Evans

1920s: Bostock, Hugh Samuel; Bray, A. C.; Davidson, S. C.; Ellis, D. H.; Gerson, H. S.; Hopper, R. V.; James, William Fleming

1930s: Brown, Robert A.; Buckland, Francis Channing; Buckland, Francis Channing; Burton, F. R.; Byers, Alfred R.; Cleveland, Courtney E.; Davis, C. W.; Denis, Bertrand T.; Gray, Richard H.; Grimes-Graeme, Rhoderick C. H.; Grimes-Graeme, Rhoderick C. H.; Halet, Robert Alfred Frans; Halet, Robert Alfred Frans; Hall, J. D.; Harris, J. J.; Hart, E. A.; Hutt, G. M.; Keating, B. J.; Lowther, George K.; MacDonald, M. V.; Malouf, Stanley E.; Moss, Albert E.; Neeland, W. D.; Okulitch, Vladimir; Price, P.; Riddell, J. E.; Riordon, Peter H.; Robinson, R. F.; Robinson, William G.; Schindler, Norman R.; Schindler, Norman R.; Schlemm, L. G. W.; Selmser, C. B.; Shaw, G.; Trenholm, L. S.; Williamson, J. R.; Wilson, Norman L.; Wilson, Norman L.; Wykes, E. R.

1940s: Antrobus, Edmund S. A.; Baird, David McCurdy; Black, James M.; Black, Philip T.; Bray, R. C. E.; Brossard, L.; Cameron, H. L.; Castle, R. O.; Christie, Archibald; Cleveland, Courtney E.; Cote, Pierre E.; Cunnington, F. A.; Denton, W. E.; Douglas, J. M.; Dufresne, Cyrille; Eade, Kenneth Edgar; Eakins, Peter R.; Fortier, Yves O.; Gerryts, Egbert; Gilbert, Joseph E. J.; Gilbert, Joseph E. J.; Gillies, N. B.; Gray, Richard H.; Harding, S. R. L.; Hill, L. S.; Howells, William C.; Imbault, Paul E.; Jooste, Rene F.; Kirkland, Robert W.; L'Esperance, Robert L.; Lee, Burdett W.; Malouf, Stanley E.; Mauffette, P.; McAllister, Arnold L.; McDougall, David J.; McPherson, William John; Milner, Robert L.; Moss, Albert E.; Mulligan, Robert; Neilson, James M.; Robinson, H. R.; Robinson, William G.; Simpson, David Hope; Sinclair, George W.; Stalker, Archibald M.; Thorson, E. F.; Tiphane, M.; Veilleux, B. M.; Wahl, William G.; Whiting, Francis B.; Yu, P. L.

1950s: Anderson, F. D.; Anderson, F. D.; Antrobus, Edmund S. A.; Assad, Robert Joseph; Assad, Robert Joseph; Averill, E. L.; Averill, E. L.; Avison, A. T.; Bahyrcz, G. S.; Bassett, Henry G.; Beall, George H.; Bell, K. C.; Benoit, Fernand Wilbrod; Benson, David G.; Berrange, Jevan Pierre; Black, Ernest D.; Black, Philip T.; Blake, Donald A. W.; Brown, Norman Elwood; Brummer, Johannes J.; Buck, W. K.; Buckley, R. A.; Burley, Brian John; Byrne, A. W.; Cameron, R. A.; Carboneau, Come; Carbonneau, C.; Carriere, G. E.; Carter, George F. E.; Carter, George F. E.; Clark, Lloyd A.; Cooper, Gerald E.; Cornwall, F. W.; Cornwall, F. W.; Cumberlidge, John T.; De Romer, Henry Severyn; Dean, Ronald S.; Deland, Andre Normand; Dubuc, F.; Dufresne, Cyrille; Dugas, Jean; Eade, Kenneth Edgar; Eadie, Dorothy Ann; Eakins, Peter R.; Emo, Wallace B.; Emo, Wallace B.; Engineer, B. B.; Ferguson, John; Findlay, David Christopher; Freeman, Peter Verner; Freeman, Peter Verner; Freeman, Peter Verner; Fullerton, H. D.; Gerryts, Egbert; Gillet, Lawrence Britton; Gleeson, Christopher F.; Godard, J. D.; Gorman, William Alan; Gorman, William Alan; Grady, J. C.; Grant, I. C.; Hawkins, W. M.; Henderson, Gerald G. L.; Hoffmann, H. J.; Hogan, Howard R.; Hogan, Howard R.; Hogarth, Donald

David; Hogg, William A.; Holmes, Stanley W.; Horscroft, F. D. M.; Horsecroft, F. D.; Husain, B. R.; Imbault, Paul E.; Irwin, Arthur B.; Jackson, G. D.; James, William; Jardine, W. G.; Jeffery, W. G.; Jeffrey, W. G.; Jenkins, John Trevor; Kirkland, Robert W.; Klugman, Michael Anthony; Klugman, Michael Anthony; Kranck, Svante Hakan; L'Esperance, Robert; Larochelle, André; Laurin, André Frédéric Joseph; Lee, Burdett W.; Leeson, J. I.; Leuner, W. R.; Lunde, Magnus; Lyall, H. Bruce; Macdougall, J. F.; MacDougall, J. F.; MacGregor, A. Roy; Macgregor, A. Roy; Machamer, J. F.; Macintosh, James Alexander; MacLaren, Alexander S.; Maclaren, Alexander S.; Maclean, Donald Wardrope; Mann, E. L.; Mannard, George William; Marieau, Raymond Alban; Marler, P.; Mattinson, Cyril R.; Mattinson, Cyril R.; McAllister, Arnold L.; McCuaig, James A.; McCuaig, James A.; McDougall, David J.; McDougall, J. F.; McDougall, J. F.; McPhee, D. S.; Meikle, Brian Keith Michael; Meikle, Brian Keith Michael; Moore, T.; Moore, Thomas Howard; Moore, Thomas Howard; Morris, P. G.; Morrison, Euen Ritchie; Morrow, Harold Francis; Morse, Stearns A.; Mueller, G. V.; Mulligan, Robert; Mumtazaddin, M.; Mumtazuddin, M.; Murray, L. G.; Nelson, Samuel J.; Oja, Reino Verner; Owens, Owen E.; Owens, Owen E.; Petruk, William; Pollock, Donald William Thomas; Pollock, Donald William Thomas; Prusti, Bansi D.; Rejhon, G.; Rejon, G.; Relly, B. H.; Riddell, John Evans; Riley, G. C.; Riley, G. C.; Riordon, Peter H.; Robinson, J. E.; Sater, G. S.; Schmidt, Richard Carl; Schmidt, Richard Carl; Shields, R. C.; Sikka, Desh B.; Simpson, David H.; Simpson, David Hope; Sims, W. A.; Skinner, Ralph; Slaght, W. H.; Slipp, R. M.; Slipp, R. M.; Slipp, R. M.; Spat, A. G.; Spino, D.; Stalker, Archibald M.; Stevenson, Ira M.; Stevenson, Ira M.; Stocken, C. G.; Syme, A. M.; Taylor, F. C.; Taylor, F. C.; Taylor, F. C.; Taylor, W. L. W.; Thompson, Harold Reid; Tiphane, M.; Tremblay, Mosseau; Tuffy, F.; Usher, John Leslie; Vollo, N. B.; Wolofsky, Leib; Wolofsky, Leib; Wolofsky, Leib; Woolverton, Ralph S.; Woolverton, Ralph S.; Woolverton, Ralph S.; Young, William L.; Zwartendyk, Jan

1960s: Barton, Jackson M.; Berry, J. R. Malcolm; Blecha, Matthew; Blecha, Matthew; Bliss, Neil Welbourne; Brett, B.D.; Butler, R. B.; Cabri, Louis Jean Pierre; Carlson, Ernest H.; Chagnon, Jean Yves; Chagnon, Jean Yves; Coates, D. F.; Coates, Maurice J.; Currie, Lester J. E.; David, Peter Pascal; David, Peter Pascal; Davies, Raymond; Dean, Ronald S.; Dhar, B. B.; Doig, R.; Dorr, Andre; Erdosh, George; Erdosh, George; Fou, Joseph T. K.; Gandhi, Sunilkumar S.; Gibbs, Graham W.; Golightly, John Paul; Gray, Norman Henry; Greig, Stanley; Gwyn, Quintin H. J.; Hansuld, J. A.; Hansuld, John A.; Hashimoto, T.; Hodgson, Christopher J.; Hodgson, Christopher J.; Hoen, E. W.; Hoffmann, H. J.; Hosein, I.; Hubert, C. M.; Jackson, G. D.; Katz, M.; Keeler, Charles Martyn; Klein, Cornelius; Lajoie, J. R.; LaSalle, P.; Lee, Sang Man; Lesperance, Pierre J.; Lewis, Douglas Windsor; Lickus, Robert John; Longe, Robert Vernon; Macintyre, Ian G.; MacKean, B. E.; MacLean, Wallace H.; Mahtab, A. M.; Mallick, K. A.; Mannard, George William; Mason, G. David; Mc-Donald, D. G.; McKyes, Shirley Edward; Mehrotra, P. N.; Morrison, P. F.; Morse, Stearns Anthony; Nakashiro, Masaykui; Newham, W. D. N.; Papezik, V. S.; Pattison, Edward Foyer; Pendala, Krishnamurthy; Peredery, Walter V.; Petryk, Allen Alexander; Philpotts, A. R.; Philpotts, J. A.; Pike, Dale P.; Pouliot, G.; Preto, Victor A. G.; Rajasekaran, Konnur C.; Raudsepp, John J.; Raychaudhuri, Sunilkumar; Redpath, Bruce Beckwith; Relly, B. H.; Roberts, Robert G.; Roscoe, William Edwin; Roth, Horst; Sakrison, Herbert C.; Salman, T.; Sangster, D. F.; Scott, Susan A.; Seguin, Maurice; Seguin, Maurice; Shah, Dasharathlal K.; Sharma, Bijon; Shewman, Robert W.; Shih, T-M.; Sikka, Deshbandhu; Sizgoric, Martha; Skinner, Ralph; Smajovic, I.; Soles, James A.;

Spence, John Alen; Sygusch, Jurgen E.; Tait, Sandra Elizabeth; Tan, Francis C.; Tempelman-Kluit, Dirk J.; Thorniley, B. K.; Underhill, Douglas H.; Upitis, Uldis; Urquhart, Glen; Van Loan, Paul Rose; Vaughan, W. S.; Vincent, J. S.; Williams, Frederick M. G.; Wolhuter, Louis E.; Yu, Y-S

1970s: Allard, M.; Arcuri, J.; Arnold, K. C.; Asad, Syed Ali; Barron, Lawrence Murray; Barton, Erika S.; Barton, Jackson M.; Beaton, William Douglas; Bliss, Neil Welbourne; Braithwaite, R. J.; Breakey, A. R.; Breakey, E. C.; Brinkman, D. B.; Brookes, Ian Alfred; Chen, D. D.-S.; Cheung, C. H.; Chorlton, Lesley B.; Chough, S. K.; Clack, W. J. F.; Coppold, Murray; Duke, John M.; Easdon, Michael M.; Eivemark, Michael M.; El-Sabh, Mohammed Ibrahim; Fowler, Anthony David; Fox, Joseph S.; Frenkel, Oded J.; Frith, R. Anthony; Frith, R. Anthony; Gauthier, C.; Gilmour, Ralph G.; Girard, Paul; Goldie, Raymond J.; Gray, James T.; Gray, Norman Henry; Gregory, A.; Grove, E. W.; Hardy, L.; Haslam, Christopher R. S.; Hauseux, M. A.; Hebil, Keith Edmund; Henriquez, F. J.; Hoag, R. B., Jr.; Hodgins, Larry E.; Hoffman, E. L.; Hopkins, John Charles; Horsky, S.; James, Noel P.; Kapp, Ulla; Kobluk, David R.; Kumarapeli, P. S.; Lambert, R.; Lapierre, D. P.; Lee, Hyun-Ha; Levoie, Clermont; Maccarone, Umberto; MacGeachy, J. K.; MacGeachy, J. K.; MacGeehan, P. J.; Makovicky, Emil; Marchand, Michael; Marsden, M.; McGillivray, James G.; Medford, Gary A.; Mercure, S.; Milner, M. W.; Moody, J. B.; Morton, Ronald; Paul-Douglas, Gabrielle; Pearce, Andrew John; Pierce, Andrew; Pressburger, Alexander; Prochnau, John F.; Reisz, R.; Roscoe, William Edwin; Salloum, John Duane; Schultheiss, Norbert H.; Scott, William James; Siemiatkowska, Krystyna; Silvestri, V.; Slankis, John Aris; Smith, Eric C.; Soonawala, N. M.; Srivastava, Prem N.; Stamatelopoulou-Seymour, Karen Catherine; Tabba, M. M.; Tam, S. S.; Taylor, Colin Hubert; Teskey, D. J.; Thomas, R.; Trudeau, M.; Trzcienski, Walter Edward Jr.; Umar, P. A.; Walls, R. A.; Watson-White, M.; Waychison, M.; Webb, Anthony J.; Wong, Poh-Poh; Young, Frederick Griffin; Yu, Thiann-Ruey; Zeman, A. J.

1980s: Abdel Warith, Mostafa Mohamed; Abdel-Rahman, Abdel-Fattah Mostafa; Abdon, Abdol-Reza; Abid, Iftikhar A.; Acker, Kelly L.; Aftabi, Alijan; Aftabi, Alijan; Al-Geroushi, Rajab A.; Al-ammawi Alsayed, Alsayed Mouhamed; Bail, Pierre; Baranyi, Elizabeth; Beaudry, Charles; Bedard, Jean H. J.; Bonavia, Franco Ferdinand; Boone, Erica Rechnitzer; Bouchard, Michel A.; Breteler, Ronald Johannes; Cattalani, Sergio; Chow, Andre M. C.; Courchesne, François; Currie, Philip John; Dalton, Edward; Duba, Daria; Elmonayeri, Diaa Salah; Evans, Robert Douglas; Forest, Richard C.; Fowler, Anthony David; Fransham, Peter Bleadon; García-Banda, Rosalba; Gartner, John F.; Gebert, James; Gent, Malcolm Richard; Hardardottir, Vigdis; Helie, Robert G.; Higgins, Michael Denis; Hill, Robert E.; Hughson, Robert Carl; Hwang, Chung-Yung; Islam, Shafiul; Kaylor, Donald Charles; Keating, Pierre Benjamin; Kim, Chun-Soo; LaFleche, Paul Thomas; Lalonde, Andre E.; Larsson, Sven Y.; Le Gallais, Christopher J.; Leonard, Richard; Linnen, Robert; Liu, Mian; Lucotte, Marc Michel; Machel, Hans-Gerhard; Mah, Anmarie Janice; Majdi, Abbas; Mattes, B. W.; Meyers, R. E.; Millette, Jacques Armand; Mohamed, Abdel-Mohsen Onsy; Momayezzadeh, Mohammed; Morogan, Viorica; Murphy, James Brendan; Nast, Heidi J.; Nurse, Leonard Alfred; Ogunyomi, Olugbenga; Olson, Karin Elizabeth; Ozoray, Judit; Paquette, Jeanne; Pasitschniak, Anna; Payette, Christine; Petzold, Donald Emil; Pilkington, Mark; Polan, Kevin Patrick; Prabhu, Mohan Keshav; Prescott, John Whitman; Rashid-Noah, Augustine Bundu; Rigoti, Augustinho; Roy, Jean L.; Sawiuk, Myron J.; Schandl, Eva S.; Simonetti, Antonio; Smith, Gary Parker; Smith, Philip Alson; Stamatelopoulou-Seymour, Karen Catherine; Stevenson, Ross Kelley; Tasse, Normand; Teitz, Martin W.; Tella, Subhas; Tessier, Gérard;

Todoeschuck, John Peter; Tyraskis, Panagiotis A.; van Stempvoort, Dale; Veldhuyzen, Hendrik; Wallace, Graeme M. B.; Wellstead, Carl Frederick; Wright, Richard Kyle; Yang, Wen-Cai; Zhong, Shaojun

McMaster University
Hamilton, ON   L8S 4M1

166 Master's, 98 Doctoral

1910s: Stewart, I. E.

1920s: Hainstock, H. N.; Macpherson, A. P.

1930s: Bowler, E. L.; Breakey, B.; Chalmers, John A.; Perdue, H. S.; Scott, Henry Kenneth

1950s: Bourne, Donald Alleyne; Deuters, Barrie Eugene; Filby, Royston Herbert; Freeman, Edward Bicknell; Gietz, Otto; Gittins, John; Harrison, William Donald; Hurd, Donald W.; Lapkowsky, Walter W.; MacLeod, John; Maxwell, J. A.; Moxham, Robert Lynn; Nickel, Ernest Henry; Pamenter, Charles Beverly; Pearson, G. Raymond; Pelletier, Bernard R.; Reeves, John Edward; Siroonian, Harold Ara; Sturgeon, Ernest Sidney; Sutterlin, Peter G.; Webber, George Roger; Weber, Jon N. E.

1960s: Anderson, Peter Ascroft MacKenzie; Anderson, Peter Ascroft MacKenzie; Bray, R. G.; Bugry, Raymond; Candy, Graham John; Chesworth, Ward; Chiang, Ming Chen; Church, Barry Neil; Chyi, Lindgren Lin; Cruft, Edgar Frank; Edgar, Alan Douglas; Getty, Theodore Alexander; Goruk, Gerald L.; Grootenboer, Johan; Haughton, David Roderick; Henderson, John Russell; Hsieh, Shuang-Shii; Hsu, Mao-Yang; Hurst, Donald Lindsay; James, Richard Stephen; Jordan, Frank William; Keays, Reid Roderick; Kretschmar, Ulrich Horst; Kudo, Albert Masakiyo; Kwak, Teunis Adrianus Pieter; Lee, Pei Jen; Lin, Szu-Bin; Lusk, John; MacDougall, J. Douglas; MacRae, Neil Donald; MacRae, Neil Donald; Martindale, Robert David; Martini, Irendo Peter; Mason, Ian MacLean; Mitchell, Roger Howard; Onions, Diane; Parkash, Barham; Payne, John Garfield; Prasad, Nirankar; Reilly, George Alexander; Sheppard, Simon Mark Foster; Simony, Philip Steven; Skippen, George Barber; Smoor, Peter Bernard; Spaven, Harvey Robert; Teeter, James Wallis; Troup, Arthur George; Van de Kamp, Peter Cornelius; Van Peteghem, James Karl; Vemuri, Ramesam; Verma, Harish Mitter; Watkinson, David Hugh; Watts, Terrance Roger; Williams, Harold H.

1970s: Alcock, Fred G.; Barker, James Franklin; Bhattacharjee, Shyama Bijoy; Birk, W. Dieter; Booty, William Gordon; Bray, R. G.; Breaks, Frederick W.; Burnie, S. W.; Campbell, J. A.; Cant, Douglas J.; Cant, Douglas J.; Chen, Shu-Meei; Chyi, Lindgren Lin; Conroy, Nels; Costello, Warren Russell; Cowan, Patricia; Coward, Julian Michael Henry; Craig, Harold Douglas; Cuddy, Robert Graham; Dalrymple, Robert Walker; Davies, Ian Charles; Dostal, Jaroslav; Douglas, G. B.; Eynon, George; Fish, Johnnie Edward; Fung, Patrick Chuen-Fai; Gartzos, Eutheme G.; Gascoyne, Melvyn; Gibbins, Walter A.; Gonzalez-Bonorino, Gustavo; Gonzalez-Bonorino, Gustavo; Gower, Charles Frederick; Griep, Jacobus L.; Hall, Russell Lindsay; Hamblin, Anthony P.; Harmon, Karen A.; Harmon, Russell Scott; Hawthorne, Frank C.; Hein, Frances J.; Hein, Frances J.; Helsen, Jan Nicolaas Walter; Helsen, Jan Nicolaas Walter; Herdman, David J.; Hiscott, Richard Nicholas; Houston, William Norman; Hsu, Mao-Yang; Hyde, Richard Stuart; Jeffries, Dean Stuart; Jennings, David S.; Johnson, Barry Allen; Kim, Ki-Tae; Knight, R. John; Kobluk, David Ronald; Kuo, Hsiao-Yu; Kuo, Hsiao-Yu; Kwong, Yan-Tat John; Lambiase, Joseph J.; Lin, Szu-Bin; Longstaffe, Frederick John; Marchand, Michael; Marttila, Raymond K.; McMaster, Glenn Edward; Mercer, William; Mudroch, Alena; Mummery, Robert Craig; Munro, D. S.; Nair, G. P.; Olson, Eric Robert; Pacesova, Magdalena G.; Pemberton, Stuart George; Pemberton, Stuart George; Pett, John Woodfull; Rector, Roger Joseph;

Schindler, John Norman; Semkin, Raymond Garry; Sivenas, Prokopios; Skipper, Keith; Smith, Duncan Ross; Stewart, W. Douglas; Strong, Percy George; Teal, Philip Rae; Teal, Suzanne Elizabeth; Teitsma, A.; Teruta, Yuko; Thompson, Peter; Tihor, Sharon Louise; Turner, Colin C.; Underhill, Douglas Henry; Van den Berg van Saparoea, C. M. G.; Verma, Harish Mitter; Vicencio, Raul; Wallace, Peter Ian; Ward, Peter Douglas; Warry, Norman David; Westerman, Christopher John; Whittaker, Peter J.; Witteman, John P.; Wojdak, P. J.; Wolff, John Marvin; Yeo, Ross K.

1980s: Adediran, Sulleiman Adebayo; Aitken, Alec Edison; Bartlett, Jeremy John; Beakhouse, Gary Philip; Beeden, David Robert; Bergeron, Mario; Bergman, Katherine Mary; Bergman, Katherine Mary; Bezys, Ruth Krista Angela; Blackwell, Bonnie; Blum, Norbert; Booty, William Gordon; Bourgoin, Bernard Patrick; Buhay, William M.; Cheel, Richard James; Collins, David J.; Connare, Kevin M.; Cortes, Jorge N.; Davies, Stephen; Downing, Karen Pamela; Duke, William Lewis; Evans, Noreen J.; Ferguson, G. Scott; Franklyn, Michael T. (Bone); Fuelen, Frank; Fyon, John Andrew; Fyon, John Andrew; Ghazban, Fereydoun; Hassan, Ishmael; Hassan, Ishmael; Heaman, Larry M.; Heaman, Larry Michael; Huhn, Frank Jones; Hurley, Teresa Dawn; Karakostanoglou, Iakovos; Keith, Donald Alexander Walter; Kusmirski, Richard Taddeusz Michael; Latham, Alfred G.; Lavigne, Maurice Jean, Jr.; Leckie, Dale Allen; Leggitt, Shelley Maureen; MacKinnon, Paula; MacRae, William Edgar; Marshall, Michael Cameron; Massey, Nicholas William David; Mayer, Tatiana; McInnes, Brent Ian Alexander; McLean, D. J.; McRoberts, Gordon D.; Miller, Thomas Edward; Mohamad, Daud Bin; Moreno-Hentz, Pedro E.; Moritz, Robert Peter; Nadon, Gregory Crispian; O'Donnell, Lynn Louise; Oshin, Igbekele Oyeyemi; Panko, Andrew William; Pattison, Simon Alan James; Power, Bruce Andrew; Pozzobon, Joseph G.; Prasad, Mithilesh Nandan; Prevec, Stephen Anthony; Rice, M. Craig; Rice, Randolph James; Rosenthal, Lorne Richard Phillip; Ross, David Ian; Rutka, Margaret A.; Sherriff, Barbara Lucy; Smith, Paul Laurence; Taylor, David R.; Taylor, Ian Edward; Teeter, Kathy; Thompson, Danny Lee; Turner, Laurie Jeanne; Turner, Laurie Jeanne; Vilks, Peter; Vilks, Peter; Yonge, Charles J.; Zymela, Steve

Memorial University of Newfoundland
St. John's, NF A1B 3X5

152 Master's, 44 Doctoral

1950s: Williams, Harold

1960s: Barning, Kwasi; Colman-Sadd, Stephen Peter; Cooper, Gordon Evans; Dawson, James M.; DeZoysa, Terence Henry; Fong, Christopher Chung-Kuen; Gale, George Henry; Gibbons, Rex Vincent; Greene, Bryan A.; King, Arthur F.; Lilly, Hugh Dalrymple; McKillop, John H.; Misra, Shiva Balak; Mullins, John; Murthy, Gummuluru S.; Nautiyal, Avinash Chandra; Pandit, Bhaskar Iqbal; Pearce, George William; Singh, Chatra Ket; Somayajula, Chavali Rama; Stevens, Robert Keith; Utting, John

1970s: Alley, Douglas Wayne; Annan, Alexander Peter; Baker, Donald Frederick; Blackwood, Reginald Frank; Brown, Peter Alan; Brown, Peter Alan; Calcutt, Michael; Clark, Anthony Miles Stapleton; Clark, Anthony Miles Stapleton; Coates, Howard James; Colman-Sadd, Stephen Peter; Comeau, Reg L.; Dal Bello, Anthony Eugene; Dayal, Umesh; Dean, Paul L.; DeGrace, John Russell; Dickson, William Lawson; Eyles, Nicholas; Fleming, John M.; Haardeng-Pedersen, G. P.; Hawkins, David Wilfred; Hibbard, James Patrick; Hsu, Eugene Ying-Chih; Hughes, Steven; Hunter, David Roy; Hussey, Eric Maurice; Iams, William James; Jaayasinghe, Nimal Ranjith; Jamieson, Rebecca Anne; Jayasinghe, Nimal Ranjith; Jesseau, Conrad Wayne; Kean, Baxter Frederick; Keats, Harvey Franklin; Knight, Ian; Koh, In Seok; Kristjansson,

Leo Geir; Levesque, Rene Joseph; Maher, John Bernard; Malpas, John Graham; Malpas, John Graham; Marten, Brian Ernest; Marten, Brian Ernest; McArthur, John Gilbert; McCann, Allan Mervyn; McGonigal, Michael Henry; Miller, Hugh G.; Minatidis, Demitris George; Nixon, Graham Tom; Norman, Ryburn E.; Nowlan, Godfrey Shackleton; O'Brien, Felicity Heather Claire; O'Brien, Sean James; O'Driscoll, Cyril Francis; Patzold, Raymund Rainer; Payne, John Garnett; Pratt, Brian Richard; Ranger, Michael Joseph; Rao, Kunduri Viswa Sundara; Ryan, Arthur Bruce; Sayeed, Usman Ahmed; Shaikh, Zafar Mohammed; Shearer, James Moxley; Smyth, Walter Ronald; Swinden, Harold Scott; Taylor, Sidney William; Taylor, William R.; Teng, Hau Chong; Thurlow, John Geoffrey; Todoeschuck, John Peter; Tuach, John; Tucker, Christopher M.; Upadhyay, Hansa Datt; Upadhyay, Hansa Datt; Uzuakpunwa, Anene Benedict; Waring, Ronald Anthony; Weerasinghe, Asoka; Weir, Harvey C.; Whalen, Joseph Bruce

1980s: Ahmed, Farazi Kamaluddin; Anglin, Carolyn Diane; Atkinson, Lee Chaflin; Barrette, Paul Dominique; Bell, Trevor J.; Belland, Rene J.; Boland, David Craig; Botsford, Jack William; Boyce, William Douglas; Brown, Dennis Lewis; Cameron, Kevin J.; Chaoka, Thebeyame R.; Chorlton, Lesley Bronwyn; Chow, Nancy; Coniglio, Mario; Desrochers, Andre; Desrochers, Andre; Dewey, Christopher Paul; Dix, George Roger; Douglas, John Leslie; Dunning, Gregory Ralph; Dyer, Alison K.; Easton, Robert Michael; Elias, Peter; Evans, D. J. A.; Fang, Changle; Finn, Gregory Clement; Foley, Stephen Francis; Gagnon, Robert E.; Gall, Quentin; Gardiner, Scott; Gillespie, Randall Thomas; Godfrey, Stephen C.; Gower, David Patrick; Greenough, John David; Haywick, Douglas Wayne; Higgins, Neville Charles; Hildebrand, Robert S.; Hudson, Karen A.; Johnson, David Ian; Kay, Elizabeth Alexandra; Kennedy, Denis Patrick Stephen; Kerr, Andrew; Kilfoil, Gerald Joseph; Knapp, Douglas Alan; Knight, Ian; Kodybka, Richard Joseph; Langdon, George Stanley; Langille, Andrew Benjamin; Lorenz, Brenna Ellen; MacDougall, Craig S.; Martineau, Yvon Arthur; Mellars, Gillian; Mengel, Flemming Cai; Mohanty, Priya Ranjan; Moreton, Christopher; Morgan, David; Mosher, David Cole; Myrow, Paul Michael; North, Jon W.; O'Neill, Patrick P.; Owen, John Victor; Pal, Badal Kanti; Parsons, Marion Grace; Peavy, Samuel Thomas; Pickett, Jacob Wayne; Pohler, Suzanne Margarete Luise; Prasad, Jagat Nandan; Prasad, Jagat Nandan; Pulchan, Kalidas; Quinn, Louise A.; Relf, Carolyn Diane; Reusch, Douglas N.; Robinson, James William; Ryley, Charles Christopher; Sandeman, Hamish A. I.; Saunders, Cynthia Margaret; Schillereff, Herbert Scott; Smith, Simon Andrew; Solomon, Steven M.; Stander, Edward; Stewart, Peter William; Stouge, Svend Sandbergh; Swinden, Harold Scott; Talkington, Raymond Willis; Thurlow, John Geoffrey; Tuach, John; van Nostrand, Timothy Stuart; Ware, Michael James; Watson, Simon Timothy; Weaver, Faye Janet; White, Charlotte Anne; Wilton, Derek Harold Clement; Woodworth-Lynas, Christopher M. T.; Wynne, Paula Jane

Memphis State University
Memphis, TN 38152

68 Master's, 1 Doctoral

1970s: Abner, H.; Battle, J. M., Jr.; Burkholder, James Franklin; Burshears, C. A.; Chimahusky, J. S.; Cowell, R. C.; Crockett, J. A.; Crowder, W. T., Jr.; Etienne, David; Gustavson, J. B.; Hares, E. M.; Haw, Tong Chee; Hildebrand, S. L.; Jecha, T. E.; Kalk, Thomas Raymond; Klazynski, Ralph J.; Ledbetter, Michael T.; Mayo, G. D., Jr.; McCullough, J. D., Jr.; McDowell, Stephen; Moore, Gary Lance; Moore, P. N.; Proffitt, J. R., III; Puchstein, Richard L.; Rowland, Bret; Sasser, Walter B., III; Schanck, J. W.; Smith, George Taylor; Spencer, Sherman Glenn; Stanley, George D.,

Jr.; Taylor, James C. Milton; Templeton, Terry R.; Thompson, Glenn Michael

1980s: Anderson, Susan Jean; Benecke, Daniel M.; Bizzio, Renato R.; Bob, Matthew Regis; Cash, Leon D.; Chamberlain, Barry N.; Daigle, Deborah M.; English, Jordan W.; Galicki, Stanley J.; Gerlock, Jeffrey Lee; Gracey, George Dennis, III; Gregory, Phillip Glyde; Hill, Robert E.; Horowitz, Warren Lee; Hughes, Conway Todd, III; Jeffers, William Larry; Jones, Byron K.; Kane, David George; Kershaw, David M.; Mann, Richard A.; Nava, Susan Jane; Norman, Charles Darrel; Reed, Phillip Lewis; Reid, J. Barry; Robbs, Edward E.; Roberts, Jeffrey B.; Ross, William R.; Shareghi, Ehsan A.; Smythe, Frank Ward, Jr.; Stoddard, Paul V.; Stone, Denise M.; Thomas, Lee A.; White, Maurice Douglas; Wilkinson, Rex; Williams, David M.; Williams, Wayne K.

University of Miami
Coral Gables, FL 33149

75 Master's, 59 Doctoral

1960s: Alexander, James Edward; Andrews, James Einar; Bock, Wayne Dean; Broida, Saul; Clausner, Edward, Jr.; Delnore, Victor E.; Fink, Loyd Kenneth, Jr.; Finlen, James Rendell; Johnson, Robert Frederick; Kaighin, Hall Young; Kline, Gheretein; Kohout, Francis A.; Lee, Chun Chi; Lidz, Louis; Milliman, John Douglas; O'Brien, Edward J., III; Rosholt, John Nicholas; Sidjabat, Mulia M.; Smith, John Alan; Supko, Peter Richard; Szabo, Barney Julius; Traganza, Eugene Dewees; Wagstaff, Ronald A.; Wanless, Harold

1970s: Almasi, Mohammad; Barron, Eric James; Behensky, J. F., Jr.; Bohlke, Brenda; Bohlke, John Karl; Brooks, D. A.; Bybell, Laurel Mary; Chermak, Andrew; Crevello, Paul; Dalziel, Mary Catherine; Dravis, Jeffrey; Emmet, Robert Temple; Fine, R. A.; Gifford, John A.; Glaccum, Robert; Gomberg, David; Gomberg, David Norman; Harlem, Peter Wayne; Harris, Karen; Harris, Paul M.; Idris, Faisal; Johnson, Donald Ray; Katz, Barry; Kirst, Paul; Kraemer, Thomas F.; Leung, W. H.; Long, Robert Bryan; McGregor, Bonnie Ann; Paull, Charles; Ramirez, Edward; Richardson, E. S.; Schubert, Carl Eric; Steinmetz, John; Supko, Peter Richard; Wang, D.-P.; Warzeski, E. Robert

1980s: Almasi, Mohammad; Almasi, Mohammad Naghash; Alt, Jeffrey C.; Alt, Jeffrey C.; Babashoff, George, Jr.; Baddour, Frederick R.; Badiey, Mohsen; Barron, Eric James; Barros, Jose Antonio; Beach, David Kent; Berler, Daniel H.; Blank, Marsha Ann; Bullwinkel, Paul E.; Burton, Elizabeth Ann; Carle, Henry Mark; Cartwright, Richard; Child, Charles Joseph; Chung, Gong Soo; Cofer-Shabica, Nancy B.; Cooper, William James; Corso, William; Craig, Genevieve Susan; Dawans, Jean-Michel L.; de Andrade Nery Leao, Zelinda Margarida; de Figueiredo, Alberto Garcia, Jr.; Dedick, Eugene; Dominguez, Jose Maria Landim; Douglas, Nancy Browning; Droxler, Andre Willy; Eberhart, Ginger Lea; Evans, Charles; Evans, Charles Carroll; Figueiredo, Alberto; Gerhardt, Daniel Joseph; Guzikowski, Michael Vincent; Harrison, Steven Adam; Hayling, Kjell; Hayling, Kjell Lennart; Huang, Qilin; Introne, Douglas Stuart; Jabali, Habib Hilmi; Johnson, Christopher A.; Kurkjy, Karen Anne; Leao, Zelinda; Lindh, Thomas Bertil; Meeder, John Frank; Miskell, Kimberlee Jeanne; Mucci, Alfonso; Nelsen, Terry Allen; Nelson, Terry A.; Parkinson, Randall W.; Perlmutter, Martin A.; Petuch, Edward James; Pierson, Bernard J.; Prager, Ellen Joyce; Prasad, Schindra; Reid, Ruth Pamela; Rine, James Marshall; Robbins, Lisa Louise; Rossinsky, Victor; Saltzman, Eric; Shinn, Richard; Sloan, James; Soldate, Albert Mills, Jr.; Spencer, Mary Jo; Thurmond, Valerie; Vahrenkamp, Volker C.; Van Valin, Reed; Walter, Lynn Marie; Waltz, Michael David; Westphall, Michael; Whitman, Jill M.; Williams, Stuart Charles; Wittpenn, Nancy Ann

Miami University (Ohio)
Oxford, OH 45056

336 Master's, 27 Doctoral

1930s: Strete, Ralph F.

1940s: Batchelor, James W.; Boies, Robert B.; Brand, John P.; Campbell, Richard N.; Currie, John Morgan; Flaschen, Steward S.; Haverfield, John Joseph; Morris, Robert Wynn; Morrow, David L.; Smith, James Mitchell; Yang, King-Chih

1950s: Balseiro, Lina M.; Bergman, Sheldon Cornelius; Bergmann, Robert J.; Bersticker, Albert C.; Bishop, William F.; Bock, Charles Mitchell; Butler, Robert E.; Carlson, Robert John; Carlyle, George Alva; Church, Richard R.; Cody, Martha P.; Davies, Rhys J.; De Wys, Egbert Christiaan; Dickas, Albert Binkley; Eiffert, James Howard; Fetzer, Kenneth Rolland; Garman, George Walter, Jr.; Garrison, Gene; Gerrard, Thomas A.; Gillespie, Walter Lee; Glaeser, John Douglas; Godowic, Paul Francis; Gotautas, Vito A.; Harris, Ann G.; Harris, Clyde E., Jr.; Hatch, Gregory C.; Hazel, Harold F.; Healy, James S.; Helman, Ronald Paul; Herbert, Frank J.; Holladay, Curtis O.; Holmes, Robert F., Jr.; Homeister, Owen E.; Horak, Ralph L.; Kahle, Charles F.; Kanizay, Stephen P.; Kelsey, Martin C., Jr.; Kneller, William Arthur; Knupke, James Albert; Kolb, John Edward; Larson, Kenneth A.; Lind, Carl R.; Lindner, Robert Frederick; Luke, Gene Edward; Lutz, Delbert Henry; Magaw, Mary C.; Magbee, Byron D.; Mattox, Richard B.; McDaniel, Willard R.; McDonald, Ralph L.; Mushake, William I., Jr.; Ohotnicky, Raymond E.; Parker, Harold Barnes; Parkinson, Charles R.; Pauly, Harold Porter; Peterson, Gerald Edwin; Pinter, Thomas J.; Reisland, Jack N.; Roddy, David John; Runge, Erwin John, Jr.; Schatz, Frank Lee; Scheufler, Lowell W.; Schneider, Michael C.; Shade, Harry D.; Shafor, Kenneth W.; Skvarla, John J.; Smallwood, James C.; Smoot, Thomas W.; Sorrell, Charles A.; Stanley, James Theodore; Stearns, Donald L.; Stewart, Douglas Bruce; Storch, Robert H.; Sutton, Eral Maurice; Tanksley, David Arthur; Thomas, Hugo F.; Thompson, William Joseph; Thoms, John A.; Thrasher, Ronald E.; Toeppe, Victor Francis; Versfelt, Porter LaRoy, Jr.; Vian, Richard W.; Warner, Robert Adolph; White, James Robert; Wingard, Paul Sidney

1960s: Armbrustmacher, Theodore J.; Aufmuth, Raymond E.; Baker, Bernard Boyd; Baker, Bruce H.; Baptista, Braulio M.; Bayha, David C.; Behling, Robert E.; Bowles, Frederick A.; Brace, Benjamin R.; Brownstein, Jack M.; Bunch, Theodore Eugene; Burkett, William C.; Butler, John Charles; Butler, John Charles; Camp, Thomas M.; Cannon, William Francis; Coppinger, Walter W.; Distler, George E.; Dolgoff, Anatole; Farber, Steve L.; Freas, Robert C.; Geisler, Thomas A.; Giles, David Lee; Goldstein, Fredric R.; Gonzales, Serge; Grubbs, Larry Stanley; Grygo, Roland; Heele, Gordon L.; Henniger, Bernard Robert; Hipple, Dennis L.; Hoekstra, Karl E.; Houpt, John Ronald; Kotila, David Arthur; Kuder, Harry Bruce; Kuryvial, Robert J.; Lambert, Robert A.; Liston, Thomas C.; Luhn, Judith K.; Lyon, Stephen R.; McCormick, Charles Larry; Miska, William S.; Negas, Taki; Negas, Taki; Nietert, Thomas Christian; Noble, John H.; O'Brien, Dennis C.; Passero, Richard N.; Rassam, Ghassan Noel; Raymond, William H.; Rosholt, John Nicholas, Jr.; Russell, Robert O.; Saines, Marvin; Sardi, Otto; Semler, Charles E., Jr.; Shade, John W.; Shilts, William W.; Slitor, Truman Wentworth; Spoley, Robert J.; Swinehart, Thomas W.; Taggart, Joseph Edgar, Jr.; Taylor, Stanley D.; Thomas, John B.; Thompson, Bruce A.; Thompson, Dale Richard; Thompson, Esther H.; Tucker, Daniel R.; Tucker, Melinda R.; Van Hart, Dirk; Vandersluis, George D.; Weidner, Jerry R.; Wetzel, John M.; Worcester, Peter A.

1970s: Atalan, Namik K.; Beskid, Nicholas J., Jr.; Blackhall, Raymond N.; Blackman, Myron James; Brown, Stephen W.; Camp, Victor E.; Carpenter, David W., Jr.; Chapman, Curtis R.; Chappell, D.

F.; Chimney, Peter J.; Coppinger, Walter W.; Cortellini, Edmund A.; Crawford, Charles Mark; Crowell, Catherine Shafer; Crowell, Douglas L.; Curtis, John B., Jr.; Davis, Katherine Renee; Davis, Philip A., Jr.; Distler, George Edward; Duggan, Peter M.; Egler, Alan P.; Fischer, Howard J.; Fisher, R. Stephen; Fitzpatrick, Kenneth Thomas; Fluegeman, Richard H., Jr.; Fortuna, Raymond; Frey, Robert Charles; Gardinier, Clayton Frank; Glaser, Ann H.; Gopinath, Tumkur R.; Guccione, Margaret J.; Harris, Frank W.; Hassell, Donald R.; Hay, Helen B.; Hillman, C. Thomas; Hoda, Syed Nurul; Holmes, Michael E.; Ikramuddin, Mohammed; Jenkins, Gale F.; Jones, Thomas Z.; Kaldon, Richard C.; Kallio, Thomas A.; Keogh, Richard J.; Korzeb, Stanley L.; Kowalczyk, Frank J.; Kramer, Terry M.; Laskowski, Thomas Edward; Lee, Gerald B.; Lorenz, Douglas M.; Lovrak, Steven R.; Ludlum, Nathaniel Burroughs, Jr.; Lynn, Jeffrey S.; Maurer, Robert J.; Meyers, David C.; Miller, James E.; Molling, Philip Andrew; Moore, Kim; Morell, Douglas J.; Neale, Patrick S.; O'Brien, John M.; Oldfield, James H.; Parr, David F.; Price, William H.; Renken, Robert A.; Rich, Douglas H.; Rike, William M.; Riley, Jill K.; Rothman, Edward M.; Ruez, Paul H.; Sachdev, Suresh C.; Scanlon, David S.; Schafer, Robert W.; Scholl, Carol J.; Shirley, John E.; Shuff, Sheldon G.; Silver, Ronald; Sinex, Scott A.; Slack, John F.; Smith, Edgar M.; Soller, David; Spinnler, Gerard Eugene; Spitzer, Jeanette; Spraitzar, Ronald F.; Stumpenhaus, Cathie L.; Sun, Kwang-Hua David; Sunwall, Mark T.; Timbel, Ned R.; Tucker, Daniel R.; Tyson, R. Michael; Wagner, Richard A.; Walia, Daman S.; Walia, Daman S.; Watson, Robert L.; White, David L.; Williams, Jeffrey R.; Winkelbauer, Howard M.; Worcester, Peter A.; Wright, Frank M.

1980s: Adenuga, Oladipo S.; Andersen, C. Brannon; Beckerman, Joseph H.; Benton, Douglas Chamberlin; Betz, Christopher E.; Bliss, Franklin E.; Bliss, Miranda C. Hoch; Borchelt, Thomas R.; Brockmann, Mark E.; Chalokwu, Christopher Iloba; Christian, Barbara S.; Close, Jay C.; Colvin, Michael Dale; Connair, Dennis P.; Cooper, Jeanne L.; Daugherty, David R.; Dickhaut, Lisa A.; Donn, Thomas Frank; Dulin, Lise A.; Eaton, Matthew Richard; Fay, Donald A.; Fleischmann, Karl H.; Fraley, Peter Allen; Franzi, David A.; Frey, Robert Charles; Galya, Thomas Andrew; Gardinier, Clayton Frank; Gargi, Satya Parkash; Giuseffi, David Francis; Gospodarec, Judith A.; Hacker, David B.; Hartwick, Wayde M.; Hay, Helen B.; Hinkley, Everett A.; Hinterlong, Gregory Dale; Hoff, John Anderson; Hunter, Gerhart Eugene; Kenter, Richard J.; Knecht, Matthew D.; Kochan, Mark; Laskowski, Thomas Edward; Legge, Paul William; Lehmann, David F.; Lepak, Robert James; Lierman, Robert Thomas; Lyke, Frederick P.; MacNaughton, David R.; Maher, Thomas M.; Martin, Anthony J.; Marzano, Michael S.; Mavris, George; Mazzone, Peter; McCartney, Richard F.; Mellott, Mark G.; Mitock, Joanne R.; Murchison, David K.; Newdale, Karen Marie; Olson, David Peck; Opalka, Richard B.; Orndorff, Harold Anton; Paxson, Kevin B.; Paxton, Stanley Turner; Pignolet, Susanne; Pritchard, George Flory; Rambo, Daniel J.; Ratliff, Jeffrey Allan; Reddin, Nancy Jean; Reid, Brian C.; Rendina, Michael A.; Rotondi, Paul L.; Sasala, Connie S.; Schmitt, Clifford T.; Schneider, Nicholas P.; Schoenborn, William Anthony; Serna, Carlos J.; Smith, Barry Samuel; Starkey, Sarah J.; Staursky, Geoffrey N.; Stinnett, James William, Jr.; Stoffer, Philip Ward; Sutherland, Susan M.; Taliaferro, Lindsay C., III; Thomas, R. Scott; Thompson, John W., III; Thomsen, Mark Andrew; Vasilkovs, Irene N.; Voner, Frederick Ronald; Voner, Frederick Ronald; Weimer, Douglas James; Whitaker, Kent Y.; White, Charlene K.; Wilder, Graham; Winningham, Bruce R.; Woodruff, Michael S.; Yau, Yu-Chyi Lancy; Zhong, William J. S.

University of Michigan
Ann Arbor, MI 48109

615 Master's, 401 Doctoral

1880s: Holmes, Mary Emilie

1900s: Scott, Irving Day; Sherzer, William Hittell;

1910s: Cook, Charles Wilford; Cook, Charles Wilford; Hunt, Walter Frederick; Staples, Lloyd William

1920s: Ayers, Vincent Leonard; Belknap, Ralph Leroy; Bergquist, Stanard Gustaf; Bullard, Fred Mason; Bush, Louise Altha; Clark, Robert Watson; Deiss, Charles Fred, Jr.; Gould, Laurence McKinley; Haydon, Osborne; Holden, Edward Fuller; Hussey, Russell Claudius; Jones, Leland Willard; Lougee, Richard Jewett; MacLachlan, Donald Claude; Peck, Albert Becker; Pepper, James Franklin; Ramsdell, Lewis Stephen; Senstius, Maurits Wilhelm; Slawson, Chester Baker; Walcott, Albert J.; Wolcott, Albert

1930s: Bassett, Charles Fernando; Bell, William Charles; Bergquist, Stanard Gustaf; Berner, Ruth Eva; Brill, Kenneth Gray; Brill, Kenneth Gray; Campbell, Charles Duncan; Carlson, William S.; Chapman, Donald Harding; Chen, Ly Kwong; Davis, Charles Moler; Dickey, Robert McCullough; Donner, Henry Frederick; Eddy, Gerald Ernest; Ehlers, George Marion; Faust, George Tobias; Glendinning, Robert Morton; Goddard, Edwin Newell; Guthe, Otto Emmor; Hard, Edward Wihelm; Hatch, Robert Alchin; Heller, Frederick Klach; Imlay, Ralph Willard; Imlay, Ralph Willard; Jones, Theodore Henry; Kendall, Henry Madison; Kline, Virginia Harriet; Kline, Virginia Harriet; Knapp, Thomas Stevens; Long, Persis Marian; MacLachlan, Donald Claude; Maebius, Jed Barnes; Marshall, Earl Elmore; McNair, Andrew Hamilton, Jr.; Miner, Ernest Lavon; Moffett, James William; Morse, Margaret Louise; Myers, William Marsh; Newcombe, Robert John Burgoyne; Radabaugh, Robert Eugene; Rickett, R. L.; Rigg, Robert Mader; Rockwood, Dwight Nelson; Stanley, George Mahon; Stanley, George Mahon; Stearns, Margaret Dorothy; Stearns, Margaret Dorothy; Stewart, Duncan; Wilkerson, Albert Samuel; Woodruff, John Grunt

1940s: Adam, William Louis; Albaugh, Frederick W.; Bachrach, Ruth Esther; Bastanchury, Ruth Frances; Baykal, Orhan; Bayless, John Clinton; Beard, Thomas; Becker, Robert William; Beeker, Ralph Edward; Bejnar, Waldemere; Benner, Richard Walter; Bergren, Arthur Learoyde, Jr.; Bowers, Gerald Frank; Bradley, Daniel Albert; Brant, Russell Alan; Buckwalter, Tracy Vere; Bull, Stratton Hemptead; Bussey, Floyd Robert; Calver, James Lewis; Childs, Orlo Eckersley; Christman, Robert Adam; Cook, Janet Arlene; Cooley, Gerald Allison; Cummins, Dean Lewis; Daviess, S. N.; Davis, Ann H.; Davis, Robert Irving; DePree, Lynn Julius; Dorr, John Adam, Jr.; Drake, William Robert; Drexler, James Michael; Elmer, Nixon; Enyert, Richard Lyle; Evans, Oren Frank; Foster, Helen Laura; Foster, Helen Laura; Galbraith, Lyman Edgar; Garvey, Phillip J.; Gillespie, W. A.; Gorton, Kenneth Arnold; Gorton, Kenneth Arnold; Gray, Alice Viola King; Gray, Henry Hamilton; Griffitts, Wallace Rush; Hatch, Robert Alchin; Hayes, John Jesse; Hazelworth, John Beemon; Heyman, Louis; Honkala, Frederick Saul; Hornbaker, Allison Lynn; Humphrey, William Elliot; Humphrey, William Elliot; Hutchinson, Robert Maskiell; Jobin, Daniel Alfred; Keenmon, Kendall Andrews; Kellum, L. B.; Kildal, Edwin; Krusekopf, Henry Herman, Jr.; Krusekopf, Lily Marie Carter; Kupsch, Walter Oscar; Kyselka, Will; LaRocque, Joseph Alfred Aurele; LaRocque, Joseph Alfred Aurele; Lemish, John; Levinson, Alfred Abraham; Lipp, Edward George; Loeser, Cornelius James; Mannion, Lawrence Edward; Marsik, Dolores Dorothy; McAlpin, Archie Justus; McIntosh, Joseph Arthur; McUsic, James Michael; Mignery, Florence Perkins; Mok, Chee Sun; Monnett, Victor Brown;

Morris, Mary Walker; Newcomb, Ester Hollis; Norton, Margaret Marilyn; Notley, Donald Frances; O'Connor, John Edward; O'Halloran, Daniel John; Owens, L. D.; Peterson, John Robert; Planck, Robert F.; Plank, Robert Forrest; Radabaugh, Robert Eugene; Richards, Arthur; Richards, Edith Jean; Saunders, Jack McLeod; Scholten, Robert; Seglund, James Arnold; Siegler, Violet Bernice; Smith, William Theodore; Snively, Norman Ray; Swann, David Henry; Swinney, C. M.; Tague, Glenn Charles; Tarbell, Eleanor; Tharp, Marie; Thibault, Newman William; Tweto, Ogden Linne; Van Dyke, Lindell Howard; Vaughn, William James, Jr.; Wallace, Stewart Raynor; Wharton, Mary E.; Wheeler, Walter H.; Wilson, John Andrew; Wright, Dorothy Alden Davis; Wyman, Anne F.; Wyman, R.

1950s: Amoruso, John Joseph; Archbold, Norbert Lee; Aughenbaugh, Nolan Blaine; Austin, Ward Hunting; Baetcke, Gustav Berndt; Baker, Walter Edwin; Barnes, John McGregor, Jr.; Beauclair, William Alfred; Becker, Herman Frederick; Behrens, Earl William; Berner, Robert Arbuckle; Bever, James Edward; Blanchard, Frank Nelson; Boeckerman, Ruth Bastanchury; Bottoms, Kenneth P.; Boudouris, James; Boydston, Donald; Boyer, Robert Eernst; Brabb, Earl Edward; Bradley, Daniel Albert; Brasher, George Kirt; Breitenwischer, Robert; Bruder, Karl Fritz; Burge, Furman Horace, Jr.; Burgess, Curtis William, Jr.; Burnel, Ralph Sherman; Chapman, William Brewer; Charlesworth, Lloyd James, Jr.; Choate, Raoul; Chrow, James Kenneth; Clements, Donald Harry; Copeland, Murray John; Copeland, Murray John; Corbett, Robert Guy; Corey, Allen Frank; Coupal, Frank Edward; Craig, Bruce Gordon; Craig, Bruce Gordon; Crick, Richard Wayne; Cropper, Wallace John; Curtis, Alan Deane; Dahlem, Robert Dale; Davis, George Hardy, III; Davis, Robert Irving; Dellwig, Louis Field; Denning, Reynolds McConnell; Dibble, Edwin Thompson; Dixon, William Hyatt; Dorr, John Adam, Jr.; Dott, Robert Henry, Jr.; Duane, John William; Dudar, John Steven; Duggan, William LeRoy; Easton, William Wonch; Eggleston, Richard Elton; Enyert, Richard Lyle; Erickson, Lance; Evans, Stewart Thompson; Ewing, John Maclyn; Fauser, Walter Bernard, Jr.; Fellows, Larry Dean; Fenske, Paul Roderick; Fitzpatrick, Robert Charles; Focht, John Doster; Forsythe, James Thorp; Frarey, Murray James; Frarey, Murray James; Freeman, Leroy Bradford; French, Alice Elizabeth; Frew, William Michner; Gacek, Walter Frank; Galas, Christodoulos Alexander; Gallaher, John Taylor; Geyer, Alan Raymond; Giardini, Armando Alfonzo; Goodfriend, Cecil Thomas; Grayson, John Francis; Grieve, Robert Oliver; Griffitts, Wallace Rush; Haas, James J.; Hall, Sylvia Duncan; Hamberg, Lawrence Roger; Hamblin, William Kenneth; Hamil, Brenton M.; Hardy, David Graham; Harker, Peter; Harley, William Frank; Hayes, John Jesse; Hazimah, Ibrahim Anis; Heany, Franklin Maurice; Heim, George Edward, Jr.; Henderson, Janet Elizabeth Rieder; Hentz, Max Ferdinand; Herndon, Chesley Coleman, Jr.; Herndon, Thomas; Hewitt, Charles Hayden; Hilles, Robert; Hilmy, Mohamed Ezzeldin; Hody, Harold Martin; Hoheisel, Charles Richard; Hoke, John Humphrey; Holmes, Douglas Allen; Hoover, Linn; Horvath, Allan Leo; Hosmer, Henry Liggett; Hosmer, Henry Liggett; Howell, Jay Lee; Hoyt, John Harger; Hulstrand, Richard F.; Hultman, John Richard; Hume, James David; Hume, James David; Hunt, Allen Standish; Hunt, Hubert Bush; Jacobson, Russell Carl; Jacques, Richard Dewey; Jarre, Guntram A.; Jefferson, Clinton Frank; Johnson, Alvin Charles, Jr.; Jones, Robert Joseph; Jurko, Robert Clarence; Kaarsberg, J. T.; Kalafatis, Christodoulos; Kasabach, Haig Frederick; Kauffman, Erle Galen; Kaufmann, Charles H., Jr.; Kavary, Emabeddin; Keeler, John White; Keenmon, Kendall Andrews; Kelly, John Joseph; Kerr, Joe Harriss; Kersting, Cecil Carl; Kier, Porter Martin; Kilgore, John Elija; Kinder, Anthony Stanley; Klosterman, Gregory Elmer; Knaffle, Leonard Ludwig; Kohn, Jack Arnold; Kramer, James Rich-

ard; Kramer, James Richard; Kunkle, George Robert; Kupsch, Walter Oscar; Kyle, Douglas Haig; Lammers, Leo Joseph; Leeder, Robert W.; LeMay, William Joseph; Lemish, John; Leney, George Willard; Lesperance, Pierre Jacques; Levan, Donald Clement; Levandowski, Donald William; Levinson, Alfred Abraham; Liddicoat, William Keith; Linsley, Robert Martin; Lootens, Douglas Joseph; Lowther, John Stewart; Lucas, Peter Thomas; Lundy, Curtis Lee; Lusk, Loren Douglas; MacDowell, John Fraser; Malin, William John; Mandarino, Joseph Anthony; Manulik, Alexander John; Masten, Douglas Everett; Masterson, James Arthur; May, Paul Russell; May, Paul Russell; McCammon, Richard Baldwin; McClurg, James Edson; McCulloch, David Sears; McGregor, Duncan J.; McLaren, Digby Johns; McLean, W. F.; McMillan, Gordon Warner; Meisler, Harold; Melhorn, Wilton Newton; Melrose, Thomas Graham; Merchant, John Stines; Mitchell, Richard Scott; Moore, E. James; Moore, Reginald George; Musselman, George Hayes; Myers, Arthur John; Newman, Frederick George; Newman, Karl Robert; Parker, John Williams; Perkins, Bobby Frank; Pessl, Fred, Jr.; Peterson, Rex Marion; Pi-Sunyer, James; Poindexter, Edward Haviland; Pope, John Keyler; Powell, Louise M.; Preble, Harold Douglas; Pusey, Richard Downing; Raup, Robert Bruce; Read, Robert Olcott; Rector, Glasco Windrom; Rector, Willis Edward; Reeder, William Glase; Reid, John Reynolds, Jr.; Reinke, Charles Austin, Jr.; Reiter, Frederick Howard; Richter, James B.; Rigg, Charles Gordon; Robinson, Edwin Simons; Robinson, Paul T.; Rogers, James Palmer; Rogers, Kenneth Joseph; Roof, Raymond Bradley, Jr.; Rose, Hugh; Rose, John Kerr; Ross, Alex R.; Ross, Alex R.; Ross, Richard Bush; Sable, Edward George; Salotti, Charles A.; Santos, Elmer S.; Satin, Lowell Robert; Scheiern, Milton Ralph; Scholten, Robert; Schultz, Arthur H.; Scott, Richard A.; Signor, Carl Wilson, Jr.; Smith, Alan Barrett; Smith, Jane Elizabeth Inch; Smith, Russell; Soronen, George Charles; Spaulding, Walter Miles; Spies, David C.; St. John, Jack W.; Stephens, J. J.; Stephens, John James, III; Stewart, John Rolland; Stoever, Edward Carl, Jr.; Stoever, Edward Carl, Jr.; Strausberg, Sanford Irvin; Sunderman, Jack; Sutton, Walter John; Swartz, Daniel Herbert; Sweet, John M.; Tabor, Norman Richard; Tatlock, Derek Bruce; Taylor, Denver Walter; Tillman, Chauncey; Tillman, John Robert; Tinker, Clarence N.; Trudell, Laurence G.; Truettner, Walter James, Jr.; Van Eck, Orville J.; Vollendorf, William Charles; Wagner, Philip Lee; Walker, John Weldon; Wallace, Stewart Raynor; Watkins, Jackie Lloyd; Weiss, Martin; Weiss, Martin; Wendt, Roy L.; Wheeler, Charles Thomas, Jr.; White, James Edward Moseby; Willard, Gates; Willis, David Edwin; Wilson, Herschell Thomas; Winslow, Donald Clarence; Wojciechowski, Walter Anthony; Wolfe, Joseph Andrew; Woodhams, Richard L.; Woodhull, Patricia; Wright, Jean Davies; Wulf, George Richard; Zaborniak, Helen Mary

1960s: Ait-Laoussine, Nordine; Allen, C. U.; Archbold, Norbert Lee; Archibald, Gary Mervyn; Baker, Richard Calvin; Bakr, Muhammed Abu; Bensinger, Herbert Schatzlein; Bjork, Philip Reese; Blackwell, Richard Joseph; Blair, Robert G.; Blanchard, Frank Nelson; Blasdel, Eugene Sherwood; Bloor, David Trent; Boardman, Shelby Jett; Boneham, Roger Frederick; Book, James Burgess, IV; Boucher, Michael Lee; Brown, Alexander Cyril; Brown, Alexander Cyril; Brown, Judith Barbara Moody; Buchi, Sylvia Duncan Hall; Buckwalter, Tracy Vere; Bufe, Charles Glenn; Burford, Arthur Edgar; Campbell, Kenneth Eugene, Jr.; Canstein, Ruth Marilyn; Chapman, William Frank; Charleston-Aviles, Santiago; Charlesworth, Lloyd James, Jr.; Cline, Denzel Riste; Conrad, Malcolm Alvin; Corbett, Robert Guy; Corey, Ronald Stewart; Corliss, Bruce Clyde; Courtis, David Michael; Coveney, Raymond Martin; Crafts, Frederick S.; Crawford, James J., Jr.; Cuthbert, Margaret Elizabeth; Cvancara, Alan Milton; Dahlem, David Harrison; Danforth, Isabel Levin; Darby, David Grant; Darby,

David Grant; DeLong, James Edward, Jr.; Denison, David Floyd; Denison, David Floyd; DeYoung, John Hulbert, Jr.; Dole, Robert Malcolm, Jr.; Drexler, Christopher William; Driscoll, Egbert Gotzian, Jr.; Drnevich, Vincent Paul; Dudar, John Steven; Ehman, Donald Allen; Fagerstrom, John Alfred; Fan, William Reun-Sen; Farrand, William Richard; Fast, Susan Elaine Jeffries; Fichter, Lynn Stanton; Foit, Franklin Frederick, Jr.; Foit, Franklin Frederick, Jr.; Foreman, Neil; Freed, Robert Lowell; Freed, Robert Lowell; Freed, Robert Lowell; French, William Edwin; French, William Edwin; Frezon, Sherwood E.; Fritts, Crawford Ellsworth; Gallo, Benedict James; Gernant, Robert Everett; Gernant, Robert Everett; Goel, Subhash Chandra; Goldberg, Paul; Greenwood, Richard John; Gross, Eugene Bischoff; Guyton, J. Stephen; Hall, Donald D.; Henes, Walter E.; Herman, Theodore Coxon; Hileman, Mary Esther; Hixon, Sumner Best; Hutchinson, Thomas Weston; Hutton, John R.; Jhaveri, Dilip Purshottamdas; Johnson, Alvin Charles, Jr.; Kapp, Ronald Ormond; Kauffman, Erle Galen; Kneller, William Arthur; Komar, Paul Douglas; Kramberger, John Joseph; Ku, Henry Fu Heng; Kunkle, George Robert; Kurtz, Timothy David; Lang, Thomas Pursell; Langenbahn, William Edward; Langway, Chester Charles, Jr.; Lasca, Norman Paul; Lasca, Norman Paul, Jr.; Leary, Richard Lee; Legault, Raymond Z.; Linsley, Robert Martin; Lizotte, Marcel Romuald; Lund, Richard; Lynde, Harold William, Jr.; MacLean, William Finley; MacNish, Robert Dick; MacNish, Robert Dick; McCulloch, David Sears; Melik, James Charles; Miller, Barry Bennett; Miller, Barry Bennett; Miller, Charles Nash, Jr.; Miller, Lee Durward; Mintz, Leigh Wayne; Mook, Anita Louise; Moore, Dwight Garrison; Moore, Reginald George; Moser, Frank; Moyer, Paul Tyson, Jr.; Murphy, Daniel Lawson; Myers, Paul Edward; Nussmann, David George; Nussmann, David George; Ogden, Maynard Blair; Parsons, Myles Lyle; Paulson, Gerald Raymond; Pentsill, Benjamin Kobina; Perhac, Ralph Matthew; Perry, William James, Jr.; Peterson, Rex Marion; Pezzetta, John Mario; Pike, Jane Ellen Nielson; Pike, Richard Joseph, Jr.; Ploch, Richard Allen; Pollack, Henry Nathan; Quon, Shi Haung; Reid, John Reynolds, Jr.; Reiter, Leon; Reuss, Robert Lester; Rhodes, James A.; Robertson, James Magruder; Rosen, Sherman Jay; Ross, Gerhard John; Rouse, Roland Carl; Rubel, Daniel Nicholas; Ryan, Richard C.; Sable, Edward George; Salotti, Charles Anthony; Sanvordenker, Viola Chang; Schultz, Gerald Edward; Sehnke, Errol Douglas; Semken, Holmes Alford, Jr.; Shappirio, Joel Rez; Sharma, Ghanshyam Datta; Shubak, Kenneth Arnold; Skeels, Margaret Anne; Smith, Charles Isaac; Smith, Gerald Ray; Smith, Raymond Newton; Smith, Raymond Newton; Smith, Russell; Sokolsky, George E.; Somers, Lee Bert Hamill; Spearing, Darwin Robert; Stacy, Howard Elwell; Steidtmann, James Richard; Steller, David DeLong; Stevens, James Bowie; Takagi, Robert Shigern; Thomas, John Alroy; Tillman, John Robert; Tischler, Herbert; Trexler, John Peter; Turpening, Roger Munson; Turpening, Roger Munson; Tyler, John Howard; Vian, Richard Wright; Volckmann, Richard Peter; Volckmann, Richard Peter; Wagner, William Philip; Wagner, William Phillip; Webb, William Martin; Weiss, Benjamin; Well, Ralph Gordon; Williams, John Stuart; Williams, Richard S., Sr.; Williams, Richard Sugden, Jr.; Williams, Thomas Clifford; Willis, David Edwin; Wilson, Richard Leland; Wilson, Richard Leland; Woodburne, Michael Osgood; Wright, Robert Paul; Zakrzewski, Richard Jerome; Zakrzewski, Richard Jerome

1970s: Abdul-Razzaq, Sabeekah; Abdul-Razzaq, Sabeekah; Abraham, Mazeeh Younis; Abraham, Nazeen Younis; Afifi, Sherif El-sayed Ahmed; Akersten, William Andrew; Al-Abdul-Razzaq, Sabeekah K.; Al-Muneef, Nasser Saad; Alexander, Donald Henry; Anderson, Donald G.; Ardrey, Robert Holt; Aufdemberge, Theodore Paul; Bennet, Robert Edwin, Jr.; Boardman, Shelby Jett; Bohlen,

Steven Ralph; Bohlen, Steven Ralph; Boris, C. M.; Bourbonniere, R. A.; Bowman, John Randall; Brandon, Dale Edward; Brett, C. E.; Brosnahan, David Ramsey; Brown, Kenneth Alan; Brown, Philip Edward; Budros, Ronald Charles; Burgess, Diane Eleanor; Burgis, Winifred Ann; Burgis, Winifred Ann; Chapman, David Spencer; Chapman, William Frank; Charleston-Aviles, Santiago; Chase, Terry Lee; Chon, C. S.; Collier, William Wayne; Coveney, Raymond Martin, Jr.; Crook, Wilson Walter, III; Cross, Timothy Aureal; Crough, Sherman Thomas; Culp, B. R.; Davis, George Herbert; Dell, Carol Irene Green; Deutsch, Harvey A.; Devore, Cynthia Helen; DeWitt, David Bruce; Dixon, Kenneth Randall; Dooley, Peter Comstock; Druce, Edric Charles; Duque, Pablo; Engel, Ruth Flora; Eshelman, Ralph Ellsworth; Estelle, Duane Kendall; Evenson, Edward Bernard; Fichter, Lynn Stanton; Filipek, L. H.; Finiol, Gary Walter; Fluharty, D. L.; Fountain, David Michael; French, Rowland Barnes; Garske, David Herman; Geissman, John William; Gilbertson, Roger Lee; Gill, Dan; Glass, Steven Wilbur; Goldberg, Paul S.; Goldman, Gary C.; Gomez Reggio, Jose de Jesus; Goodwin, C. W.; Greene, John Frederick; Grubbs, Kenneth Lee; Grummon, Mark Longden; Hall, Stephen Austin; Harrington, Mark Terrell; Harris, Jeanne Elizabeth; Hellinger, Terry Scott; Henry, Stephen George; Hileman, Mary Esther; Hill, James David; Huh, John Mun Suk; Jackson, John G.; Jackson, Philip Larkin; Jesse, Judith Mary; Johansen, K. A.; Johnson, D. M.; Johnson, Robert Lane; Joity, John Frank; Jones, Frances Gwynn; Jones, Meridee; Jurdy, Donna Marie; Katsikas, C. A.; Kimmel, Peter Gerrit; Kimmel, Peter Gerrit; Koch, William Frederick, II; Koenings, J. P.; Kulvanich, Sermsakdi; Lackey, Laurence Evan; Landing, Edward William; Landing, Edward William; Landwehr, J. M.; Lanney, Nicholas Anthony; Lasheen, M. R. M. W.; Lattanzi, Robert David; Levy, Michael Arnold; Liddell, William David; Liou, C. P.; Lu, C. L.; Lung, W.-S.; MacKenzie, Wallace Bruce; Malila, William A.; Maresca, J. W., Jr.; Marshall, Ernest Willard; Mauk, Frederick John; Mauk, Frederick John; McIntosh, George Clay; Meloy, David Urey; Metzger, Fredrick William; Meyer, Franz Oswald; Meyer, Franz Oswald; Mortensen, Martin Eckert; Mullin, Rosemary Patricia; Murphy, David Hazlett; Nash, David Byer; Nash, David Byer; Nesbitt, Bruce Edward; Nesbitt, Bruce Edward; Netkowski, Thomas Frank; Newman, Daniel Benjamin; O'Brien, Lawrence Edward; Oldroyd, John David; Pattridge, Katherine Amanda; Perkins, Dexter, III; Perkins, Dexter, III; Pierce, Douglas Stanley; Pillars, William Wynn; Price, Floyd Ray; Rajagopal, Rangaswamy; Reilly, George Alexander; Reiter, Leon; Reuss, Robert Lester; Rial M., Jose Antonio; Ritter, Charles John; Robertson, C. K.; Robertson, James Magruder; Rose, K. D.; Rossmann, Ronald; Rouse, Roland Carl; Rust, James Edward; Salmon, Bette Christine; Sarsfield, L. J.; Scherzer, Howard Jay; Schieck, David Ernest; Schreiber, Richard Lee; Seibel, Erwin; Shaw, Brian Robert; Shedlock, Robert John; Simmons, William Bruce Jr.; Smith, Lawrence Ralph; Sobol, Joseph Walter; Sommarstrom, S. J.; Somogyi, F.; Spencer, Barry Craig; Stott, Laurence Richard; Struhsaker, James Frederick; Swirydczuk, Krystyna; Tallamraju, R. K. M. R.; Thies, Barry Peter; Titman, G. D.; Tremper, Lauren Roy; Vallejo, L. E.; Valley, John William; Van Dellen, Kenneth J.; Van Den Berg, Alexander Nicolaas; Vincent, Robert Keller; Vitorello, Icaro; Vitorello, Icaro; Wadsworth, Joseph Rogers; Wagner, Richard Joseph; Watson, Donald Whitman; Watson, James Knox; Watts, D. R.; Williams, Barbara Jean Radovich; Williams, Daryll Wayne; Winter, Gary Allan; Wood, Gordon Daniel, II; Wright, Robert Paul; Zajac, Ihor Stephan

1980s: Ahn, Jung-Ho; Ahn, Jung-Ho; Alexander, Amanda Joyce; Alexander, Donald Henry; Allard, Margaret Jane; Aminian, Khashayar; Anders, Mark Hill; Anderson, Max Leroy; Anovitz, Lawrence M.; Anovitz, Lawrence Michael; Athanasopoulos,

George Andreas; Bajwa, Rajinder Singh; Baker, Colin Woods; Ballard, Martha M.; Ballard, Sanford; Ballard, Sanford, III; Bartels, William Stephen; Bartels, William Stephen; Beck, Susan Lynn; Bickart, K. Jeffrey; Blake, David Frederick; Blake, David Frederick; Bolsenga, Stanley Joseph; Bosscher, Peter Jay; Bowen, Timothy Dana; Braunsdorf, Neil Robert; Brown, Peter McKay; Brown, Peter McKay; Brown, Philip Edward; Budai, Joyce Margaret; Carlson, Sandra Jean; Carlson, Sandra Jean; Carroll, Alan Robert; Cercone, Karen Rose; Chapra, Steven Christopher; Chen, Joy C.; Chester, Frederick Michael; Christensen, Douglas H.; Conley, Daniel Joseph; Cowan, C. A.; Craddock, John Paul; Craddock, John Paul; Czerniakowski, Lana Ann; Davis, Rhonda Lynn; Derman, A. Sami; Drexler, Christopher William; Duston, Nina Marie; Eastman, Daniel Brian; Edwards, R. Lawrence; Elder, Ruth Lucinda; Eldredge, Sarah; Elmore, Richard Douglas; Essene, Eric J.; Essere, Eric J.; Farr, John Vail; Finkel, Elizabeth A.; Frank, Marc Hilary; Fusilier, Wallace Eaton; Futyma, Richard Paul; Gaskill, Daniel Wills; Geissman, John William; Gesink, Joel A.; Given, Robert Kevin; Given, Robert Kevin; Gle, Dennis Ray; Glover, Rebecca Marie; Gonzalez, Luis A.; Gough, Lia Ann Fong; Grant, Ulyses Simpson, III; Gunnell, Gregg Frederick; Haynes, Frederick Mitchell; Heikes, Brian Glenn; Henry, Steven George; Herman, John D.; Hinkel, Kenneth Mark; Hoffman, Karen Sue; Hurley, Neil Francis; Husby-Coupland, Karen Joanne; Huspeni, Jeffrey Ralph; Isaacs, Andrew Mansfield; Ivy, Logan Dudley; Jackson, Michael James; Jackson, Michael James; Janecek, Thomas Raphael; Johnson, Craig Alden; Johnson, Rex J. E.; Johnson, Rex J. E.; Kaiser, Charles John; Kappmeyer, Janet Carol; Katili, Amanda Niode; Keeler, Gerald Joseph; Kesler, Stephen E.; Khilar, Kartic Chandra; Kilsdonk, Bill; Kopania, Andrew A.; Krakker, Linda A.; Krause, David James; Krause, David Wilfred; Kreutzberger, Melanie E.; Krezoski, John Roman; Kureth, Charles L., Jr.; Lee, Jung Hoo; Leenheer, Mary Janeth; Livnat, Alexander; Long, Keith Richard; Loureiro, Celso de Oliveira; Lynnes, Christopher Scott; Lynnes, Christopher Scott; Lyon, John Grimson; Marcotty, Louise-Annette; McCabe, Chad Law; McCabe, Charles Law; McIntosh, George Clay; Miller, Carol Pomering; Miller, Cass Timothy; Moecher, David Paul; Moldovanyi, Eva Paulette; Morley, David Patterson; Musgrove, Frank William; Nelson, Frederick Edward; Nishioka, Gail Keiko; Page, G. W., III; Perigo, Russell Edward; Petersen, Erich Ulrich; Postawko, Susan Elaine; Presnell, Ricardo Davis; Quan, Richard A.; Rautman, Alison Eunice; Ray, Richard Paul; Richardson, Stephen Vance; Ruiz, Joaquin; Scavia, D.; Schedl, Andrew David; Schenk, Christopher Joseph; Schwab, Anne Marie; Schwaller, Mathew K.; Schwartz, Susan Ynid; Schwartz, Susan Ynid; Scott, George Norman; Selvius, Douglas Brian; Sengupta, Arijeet; Sharp, Zachary David; Shuchman, Robert A.; Smith, Scot Earle; Smith, Ted J.; Smith, Ted J.; Stearns, Carola Hill; Stearns, Carola Hill; Stein, William Earl, Jr.; Stepanek, Bryan Earl; Suo, Lisheng; Sutton, Stephen T.; Swirydczuk, Krystyna; Tabachnick, Rachel; Taylor, Carl H.; Taylor, Mark; Tinker, Scott W.; Torres-Roldan, Victor; Treiman, Allan Harvey; Tsao, Tze-Tzong; Udegbunam, Emmanuel Onyekwelu; Vahrenkamp, Volker Christian; Valley, John Williams; Walsh, James F.; Welc-LePain, Joan L.; Wells, Neil Andrew; Wicke, Heather Dawn; Wight, David Clayton; Winkler, Dale Alvin; Wisniowiecki, Michael James; Wu, Cho-Sen; Wu, Shi-ming; Yau, Yu-Chyi Lancy; Young, Christopher J.; Zempolich, William G.

Michigan State University
East Lansing, MI 48824

380 Master's, 182 Doctoral

1930s: Gibbs, Clifford J.

1940s: Davies, William Edward; Erickson, Ralph LeRoy; Faul, Henry; Henry, Waldo Henry; Hobbs, Robert Adelbert; Hofstra, Warren Elwin; Jewell, William Franklin; Kilbourne, Deane Earle; King, Victor Hugo; McCallum, Marjorie Louise; Schmitt, George Theodore; Smith, George Wendell; Stewart, David Perry; Thornburn, Thomas Hampton; Verhoeven, Cornelius Simon

1950s: Allen, Bonnie L.; Bernardon, Milo A.; Bradford, William Fay; Brigham, Robert John; Brooke, Gerald L.; Campau, Donald Edmund; Christofferson, Keith A.; Cooke, Laurence S., Jr.; Crane, Marilyn Joyce; Curren, Robert F.; DeMarte, Domenic L.; DeVries, Neal H.; Dewey, David E.; Dice, Bruce Burton; Din, Min; Egleston, David Lewis; El-Khalidi, Hatem Hussein; Elman, Stanley Harold; Engel, Theodore, Jr.; Gallagher, Alton V.; Goodrich, Donald Larry; Grabau, Warren Edward; Hagni, Richard Davis; Hefner, John Hardin; Hoffman, George Albert; Hogberg, Rudolph K.; Holmquest, Harold John, Jr.; Holway, William; Hornstein, Owen Merle; Hudson, Richard James; Husband, Philip M.; Jackson, Robert Paul; Jodry, Richard Louis; Johnson, Kenneth G.; Kohl, Karl W.; Kropschot, Robert E.; Kuehner, Irvin Verne; Long, Joseph Bacon; Long, Richard Arthur; Melhorn, Wilton Newton; Mitten, Hugh T.; Moore, Roger Kent; Naegeli, Faith I.; Neal, James T.; Nutter, Neill Hodges; O'Connell, James F.; O'Hara, Norbert Wilhelm; Oden, Arlo Leigh; Paige, David Stanley; Patrick, Joseph L.; Perry, Raymond Clair; Rawls, Vernon C., Jr.; Rhodehammel, Edward Charles; Riggs, Karl A., Jr.; Ross, Roy M., Jr.; Sahakian, Armen S.; Sander, John Egan; Sawtelle, E. Rossiter, Jr.; Schiller, Edward Alexander; Schmidt, Earl Albert; Slaughter, Arthur Edwin; Smale, Gordon R.; Smith, Henry Carl, Jr.; Solberg, Roger A.; Star, Clarence; Steder, Robert M.; Swanson, Roy Ivar; Tara, Muriel Elizabeth; Terwilliger, F. Wells; Thaden, Robert Emerson; Tinklepaugh, Betty M.; Tinklepaugh, Betty M.; Trethewey, Ben Clifford; Uhri, Duane C.; Utter, Gordon Stanley; Vehrs, Robert Alan; Villar, James Walter; Walker, Robert Dean; Wayman, Cooper Harry; Wild, Robert C.; Wilson, Laurence MacKenzie; Wood, Leonard Eugene; Wray, James E.; Wuckert, Arthur Emil; Young, Robert Thomas

1960s: Abu-gheida, Othman Mohammad; Adams, John Rodger; AlNouri, Ilham; Andress, Edward C.; Asseez, Laidiyu Olayinka; Asseez, Liadiyu Olayinka; Babb, Carlton Scott; Bailey, Alan; Baranyai, Paul D.; Barratt, Michael William; Benedict, Ellis Neil; Bishop, William Clifton; Blake, Daniel B.; Bloomer, Alfred Travers; Bork, Jonathan; Bradley, James W.; Brett, James Walter; Brown, Robert Ernest; Burns, James William; Buzas, Alfons; Carroll, James F.; Cascaddan, Ann E.; Chafetz, Henry S.; Chipman, David Walter; Chowdiah, Attru Mallikarjuniah; Christensen, Donald Robert; Clark, Russell Gould, Jr.; Colwell, John Allison; Connally, George Gordon; Cummings, David; Dickas, Albert Binkley; Dixon, Richard A.; Dobar, Walter I.; Douglas, Arthur Gordon; Dryden, Donald A.; Ebtehadj, Khosrow; Egan, Christopher Paul; Firouzian, Assadolah; Fisher, James C.; Freers, F. Theodore; Griffin, Villard Stuart, Jr.; Griggs, Peter Humphrey; Groughnour, Roy Robert; Gustafson, Lewis Allan; Hamil, David F.; Herbert, Thomas Allan; Heuser, Robert Frederick; Hill, Donald Gardner; Hillard, David L.; Hoffman, James Irvie, III; Hoffman, James Irvie, III; Horowitz, Martin; Juo, Anthony Shiang-Ru; Kashfi, Mansour S.; Keith, Warren E.; Kellogg, Richard L.; Kerman, Charles E.; Kirschke, William H.; Kitani, Osamu; Klasner, John S.; Kudlac, John J.; Lai, Tung-Ming; Lammons, James Monroe; Lawler, Thomas L.; LeAnderson, P. James; LeMone, David VonDenburg; Leong, Wing K.; Lifshin, Arthur; Lifshin, Arthur; Loh, Abraham Kwan-Yuen; Lowden, James E.; Majedi, M.; Mancuso, Joseph John; Manley, Thomas Richard; Mencenberg, Frederick E.; Merritt, Donald W.; Merritt, Donald W.; Meshref, Wafik M.; Meyer, Howard J.; Miller, Norton George; Miller, Roy Malcolm; Miller, William

Roger; Moore, Donald James; Murrish, Charles H.; Nelson, Clifford M.; O'Hara, Norbert Wilhelm; Ohlhber, Robert F.; Olmsted, James Frederick; Olmsted, James Frederick; Orr, William Norton; Pailoor, Govind; Peebles, Roger Waite; Pennington, Erwin K.; Puffer, John H.; Raman, Athipet Bashyam; Redmond, Brian; Redmond, Charles Edward; Reed, Robert C.; Richtmyer, Allan G.; Rosenberger, Eric J.; Roth, John N.; Schafer, John William, Jr.; Servos, Gary Gordon; Shaffer, Bernard LeRoy; Shaw, Allen Vaughan; Smith, LeRoy W.; Sorrwar, Gholam; Sparks, Dennis Michael; Steinkraus, William E.; Stelzer, William T.; Stevenson, Jack C.; Su, Hon-Hsieh; Swanston, Douglas N.; Thiruvathukal, John V.; Thompson, Gary Gene; Thompson, Richard J.; Thompson, Stephen L.; Tidwell, William D.; Timmerman, David Harold; Travis, Patricia Ann Asiala; Villar, James Walter; Waggoner, Thomas D.; Webster, Michael Stilson; Wheeler, Merlin L.; Whitaker, James T.; Whiting, William Martin; Wilson, Chester H.; Wingard, Norman Edward; Wingard, Norman Edward; Wood, Warren W.; Wood, Warren W.; Yettaw, Gordon A.; Zaitzeff, James Boris; Zaitzeff, James Boris

1970s: Al-Agidi, Waleed Khalid Hassan; Al-Khafaji, Abdul-Amir Wadi Nasif; Aldrich, Arthur G., Jr.; Alyanak, Nancy; Anderson, Carl W.; Autra, Katherine Balshaw; Autra, Marshall D.; Bacon, Douglas J.; Bailey, Alan; Balombin, Michael T.; Baranowski, James; Bates, Edward R.; Boker, Thomas Dominic Nmah; Burger, John Robert; Byerly, Gary Ray; Byerly, Gary Ray; Cahow, A. C.; Canfield, Douglas John; Carlson, Barry Albin; Carroll, John Edward; Castillon, David Alan; Catacosinos, Paul Anthony; Chaichanavong, Thira; Chambers, Richard Lee; Charlie, W. A.; Chazen, Stephen I.; Chung, Pham Kim; Cline, Jeffrey Thomas; Dali, Ayad H.; Daniels, Jeffrey J.; Dastanpour, Mohammad; Davis, Mary Walter; Davis, Phillip Burton; Delmet, Dale Aaron; Eames, Leonard E.; Eftaxiadis, Thrasos; Egan, Christopher Paul; Eicher, Constance Carolyn; Ewald, Frederick Charles; Fincham, William J.; Flis, James Edward; Folsom, Michael MacKay; Fortuna, Mark Allen; Freeman, Kevin John; Gere, Milton A., Jr.; Gies, Theodore Fredrick; Griggs, Peter Humphrey; Haimila, Norman Edward; Haji-Djafari, Sirous; Hamrick, R. J.; Hanson, Stuart; Henny, Robert Warren; Henry, Robert W.; Herbert, Thomas Allan; Hewitt, Jeanne L.; Hood, Edward John; Hyde, Michael Kevin; Ibrahim, Abdelwahid; Isham, Julian C.; Iversen, Christine M.; Johnson, Kenneth F.; Jones, Vernon K.; Kao, Ching-nan; Kellogg, Richard L.; Kempany, Ryan Glenn; Kennedy, James Walton; Khattak, Anwar S.; Kidson, Evan Joseph; Kim, Joon Yol; Kolbash, Ronald Lee; Kotsch, Richard William; Krause, David James; Kromah, Fodee; Laaksonen, Harry J.; Lasemi, Yaghoob; Lehtola, Kathleen Anne; Lentz, Rodney Ward; Lewis, Jerry D.; Li, John Chien-Chung; Lietzke, David A.; Lively, Richard S.; Lovato, Joseph; MacClure, Thomas William; Maher, John Kelly; Mahjoory, Ramez; Malcuit, Robert Joseph; Mandle, Richard J.; Mariotti, Philip Arno; McKosky, John A.; Merk, George Philip; Miller, Lewis Ruthardt; Morris, W. J.; Morrison, Stanley J.; Mrakovich, John Vincent; Natarajan, Palamadai S.; Newhart, Richard Eugene; Nortey, Peter Alphonsus; Nowak, Ronald P.; Nurmi, Roy David; Nutter, Neill Hodges; Onofryton, Jerry K.; Oray, Erdogan; Orlowski, Louis Allen; Orr, Patricia Lynne; Orzeck, John Joseph; Ossian, Clair Russell; Pachut, Joseph F., Jr.; Paris, Randy Max; Parker, Lee Ross; Pawling, John W.; Pentony, Kevin John; Petro, William L.; Phillips, Wesley M.; Picken, Cyrus Seeley, Jr.; Plopper, Christopher Stevens; Podell, Mark Edward; Poopath, Visharn; Pothacamury, Innaiah; Potter, Darrell L.; Prather, Barry W.; Preston, John Kante; Prezbindowski, Dennis; Price, Michael Lee; Reagan, Jeffrey F.; Regan, Robert David; Richardson, Nathaniel Reginal, Jr.; Richter, Howard L.; Rieck, Richard Louis; Rodriguez, Jaime Alberto; Runyon, Stephen Lane; Ryan, Kim Kathleen; Ryan, Michael Patrick; Ryder, Graham; Sander, John

Egan; Sanderson, Dewey Dennis; Schock, Michael Reed; Seavey, Donald Barker; Seyler, Douglas J.; Shah, Balkumar P.; Shah, Balkumar Prataprai; Shanabrook, David Clark; Shannon, Eugene Himie; Shaw, Richard Michael; Sonaike, Susanna Yetunde; Sorrwar, Gholam; Steinfurth, Carl; Stone, Joseph Fred; Strutz, Timothy Arthur; Swearingen, Ted L.; Syrjamaki, Robert M.; Taggart, R. E.; Tallman, Ann M.; Taylor, Thomas Raymond; Ten Have, Lewis Earl; Test, Thomas Alvin; Tillman, Stephen Edward; Tingey, John Craig; Todd, Raymond C.; Torres, Pedro Leon; Travis, Jack Watson; Tuckey, Michael Edward; Turpening, Walter Ray; Van Dam, George Henry; van Leeuwen, Wim; Verma, Ambika Prasad; Von Almen, William Frederick; Waanders, Gerald Lee; Walker, Bruce M.; Walker, Bruce Michael; Waltz, Stephen Ray; Ward, Marsha Jane; Wasuwanich, Pipob; Welsh, James Patrick Jr.; Westjohn, David B.; Wharton, Richard J.; Widmayer, Ronald Edward; Williams, Ella Ruth Tews; Wilson, Chester H.; Wilson, Thomas Vincent; Wilson, V. V.; Wood, Gordon Daniel; Wood, William; Young, T. C.; Younker, Jean Kay; Younker, Jean Lower; Younker, Leland Wilbur; Younker, Leland Wilbur; Zainuddin, Syed Mohammad; Zenone, Chester R.

1980s: Al-Moussawi, Hassan M.; Alwahhab, Riyadh Mostafa; Anderson, Lynn G.; Asrar, Ghassem; Bartlett, Timothy R.; Bartley, John William; Batterson, Ted Randall; Breithart, Mark S.; Buckler, William Roger; Bullen, Susan Brook; Bush, Charles Vincent; Bust, Vivian Kay; Buxton, Timothy Montrose; Campbell, Kevin Todd; Carlton, Ronald Ray; Carty, James Michael; Chakel, John Anthony; Coley, Michael J.; Colmenares, Omar A.; Cook, David Bruce; Crissman, Susan Elizabeth; Cunniff, Robert Thomas; de Gruyter, Philip Clarence; Dedoes, Robert E.; Delcore, Manrico; Dodge, Sheridan Lee; Dodson, Russell Leslie; Dufek, Debra Ann; Dworkin, Stephen Irving; Everse, Douglas Gene; Farmer, Randall Allen; Fay, Leslie Porter; Feng, Bing-Cheng; Fisher, Jeanne Anne; Flood, Timothy P.; Florian, Marc D.; Fox, Thomas P.; Gell, James Walter; Gephart, Carol J.; Gephart, Gregory David; Ghavidel-Syooki, Mohammed; Glandon, Robert Paul; Goitom, Tesfai; Gold, Arthur J.; Grantham, Jeremy Hummon; Gregg, Jay Mason; Gudramovics, Robert; Guentert, James S.; Gustafson, Richard Dale; Harrington, John Ausman, Jr.; Hartsell, Mickey York; Hascall, Allan P.; Hicks, David Robert; Hill, Keith Charles; Hill-Rowley, Richard; Hogarth, Craig G.; Hoin, Steven James; Host, George Edward; Jameossanaie, Abolfazl; Jank, Mary Ellen; Jessup, Donald David; Johns, Warren Harvey; Johnson, Bradley Scott; Kampmueller, Elaine; Kapaldo, David Wayne; Karteris, Michael Apostolos; Kehres, Cheryl A.; Lee, Jiunn-Fwu; Lewallen, Noble F., II; Lusch, David Paul; MacDonald, Neil William; Marsh, John W., Jr.; Mattson, Steven R.; Mattson, Steven R.; McMullen, Cindy Anne; Melia, Michael Brendan; Mescher, Paul K.; Meyer, Rudolf; Middlekauff, Bryon Douglas; Miller, Michael Allan; Moyer, Raymond B.; Musser, Kathryn Jeanne; Myers, Gary A.; Nefe, Erick C.; Newberry, James Tyler; Nieman, Timothy Lynn; Offer, Stuart Adam; Orr, G. Daniel; Paddock, David Ray; Pelowski, Sandra M.; Petrie, Mark Alan; Premo, Bette Jayne; Raab, James Michael; Rabbio, Salvatore Frank; Ragnarsdottir, Kristin Vala; Regalbuto, David Philip; Rezabek, Dale Henry; Richardson, Douglas Burton; Richey, Ronald Glenn; Ritter, Michael H.; Robords, Alan Carman; Rodabaugh, Gary Lee; Rogers, William John, Jr.; Rohr, Steven Anthony; Rose, Timothy Patrick; Rutland, Carolyn; Sack, William R.; Salvino, John Francis; Satchell, Loretta Simmons; Schock, Susan C.; Schultink, Gerhardus; Schuraytz, Benjamin Charles; Shirey, Burrell Peter; Siami, Mehdi; Silber, Jay Brian; Sims, James Thomas; Slayton, David F.; Soo, Sweanum; Sperling, Tedd F.; Starcher, Robert Warren; Sullivan, Robert Michael; Swiderski, Donald L.; Takacs, Michael James; Taylor, Allan Beowulf; Taylor, Julie Marie; Taylor, Thomas Raymond;

Thomas, Michael Robert; Tituskin, Susan E.; Tolbert, James N.; Trebing, Harry Evan; Tu, John Siuming; Tuckey, Michael Edward; Walden, Kyle Douglas; Walles, Frank E.; Watson, Bruce F.; Wee, Soo-Meen; Zalidis, George C.; Zwicker, Deborah L.

Michigan Technological University
Houghton, MI 49931

198 Master's, 13 Doctoral

1930s: Braund, Robert William; Erdahl, William Mitchell; Knaebel, Carl Henry; Leedy, Forrest Benton; Macintosh, Albert N.; Patek, John Mark; Spiroff, Kiril; Zinn, Justin

1940s: Blade, Lawrence V.; Denning, Reynolds McConnell; Livingston, Clifton Walter; Longacre, William A.; Myers, Arthur John; Porturas-Plaza, Antonio; Sermon, Thomas Croxford

1950s: Bhatt, Kireet Jivanram; Broman, William H.; Browne, James L.; Burns, Robert Donald; Campbell, Robert Emerson; Conrad, Malcolm A.; Doane, Virginia L.; Durfee, George Austin; Fotouchi, Manuchehre; Frantti, Gordon E.; Fritts, Crawford Ellsworth; Haynes, Charles W.; Herron, Thomas Joseph; Jackson, Kern Chandler; Knowles, David M.; Lee, Sang Man; Mandarino, Joseph Anthony; Margenau, Roy E., Jr.; Matthews, Wilfred J.; McChesney, Robert Douglass Ross; Orajaka, Stephen Onyebueke; Papadakis, James; Parker, Dana C.; Raza, Saiyid; Remick, Jerome H., III; Saraby, Fereydoon; Van Altena, Peter James; Williams, Sidney A.; Wilson, James G.; Wyble, D. O.

1960s: Aho, Gary D.; Beard, Richard C.; Beger, Richard M.; Brady, John M.; Brothers, Jack Anthony; Carter, Nicholas C.; Cupal, Jerry J.; Day, Damon P.; Dzierwa, David John; Finnegan, Stephen Allan; Guarnera, Bernard John; Hussin, James Joseph; Johnson, Allan Michael; Johnson, William Wallace; Juilland, Jean D.; Kantor, Joseph Alan; Koons, Gerald Jay; Kustra, Clarence Ronald; La Point, Ronald; Langill, Richard Francis; Larson, Robert James; Leonardson, Robert William; Ludeman, Frank L.; Macauley, Terrence M.; Middleton, Robert Stuart; Moulton, Gail Francis, Jr.; Nuttall, Jeffrey Clarke; Phillion, George William; Reuss, Robert James; Sigsby, Robert M.; Sterling, Donald Lowell; Taylor, G. Lynn; Tepedino, Victor; Thompson, Frederick Henry; Tobias, Theodore; Van Voorhis, Gerald D.; Wanzong, Walter F.; Woznessensky, Boris; Zaher, Mohammad Abduz

1970s: Babcock, L. L.; Barlow, Roger Brock; Bell, James M.; Bennett, George H.; Berger, Richard J.; Bingham, Michael W.; Blaisdell, George L.; Bodwell, Willard Arthur; Bowden, Douglas Richard; Carlson, Gerald G.; Chittrayanont, Sumeth; Clark, Jeffrey L.; Cole, John Frederick; Dann, Jesse C.; Drexler, John William; Dyl, Stanley J., II; Fenwick, Kenneth George; Fultz, Lawrence Anthony; Garcia, Francisco Raul; Harris, Robert D.; Hase, Harold W., Jr.; Henderson, Frank A.; Hughes, Gordon J., Jr.; Hughes, Gordon J., Jr.; Jacobsen, S. I.; Johnson, Allan Michael; Johnson, Robert L.; Johnson, William James; Kelley, Gary M.; Landress, Mark R.; Lewan, Michael Donald; Mahajan, Amrish K.; Murray, John C.; Nelson, Richard C.; Nielsen, Eric Richard; Page, Gary Walter; Phillips, Gerald R.; Pires, Fernando Roberto Mendes; Rogers, Robert K.; Rognerud, Walter N.; Scherkenbach, Daryl A.; Scofield, Nancy L. C.; Shoja-Taheri, Jafar; Sriruang, Somsakdi; Stuart, Edmund J.; Supina, Richard D.; Taylor, Gerald Lynn; Tolunay, Aykut; Trusler, James R.; Turner, Thomas R.; Villarroel, Patricio

1980s: Alao, David Afolayan; Anderson, Timothy D.; Balls, Scott Nelson; Basher-Riani, Mustafa; Baxter, David A.; Beshish, Ghaith K.; Boben, Carolyn L.; Bornhorst, Laurie E.; Brojanigo, Antonio; Browning, Timothy D.; Bumgarner, Edward L.; Capaul, William A.; Chartier, Torrie A.; Chesner, Craig Alan; Chesner, Craig Alan; Counts, C. David; Crisman, David P.; Deans, Brian D.;

Drexler, John William; Dyke, Gary A.; Ensign, Paul S.; Gertje, Henry; Glennon, Mary Ann; Godaih, Sulaiman H.; Hagley, Mark T.; Haig, Trevor D.; Hall, Charlene R.; Hepp, Eric; Hiatt, Cheryl Rae; Hoehl, Eberhard J.; Hoerger, Steven Fred; Hoerger, Steven Fred; Hoffman, Mark Allen; Hoffman, Mary Frances; Jaeger, David J.; Johnson, Daniel L.; Johnson, Rodney C.; Kent, Gretchen R.; Kitchen, Mark R.; Koivunen, Alan; Koncuk, Fatih; Kurtulus, Cengiz; Li, Huilin; MacLellan, Mary L.; McClannahan, Kevin M.; Mitchell, Robert J.; Neumann, Peter C.; Nordeng, Stephan H.; Owens, Eric O.; Paces, James Bryant; Peterson, Paula S.; Prosen, Barbara J.; Rees, Todd Howard; Repasky, Ted R.; Rossell, Dean M.; Rush, Randy J.; Schleiss, Wolfgang A.; Shelnutt, John P.; Shepeck, Anthony W.; Sikkila, Kevin M.; Smith, Casey C.; Soroka, William L.; Stadnik, Paul M.; Strauss, Robert G.; Symonds, Robert B.; Taylor, Lisa Gale; Temudom, Ladda; Van Horn, John E.; Van Roosendaal, Dan J.; Warburton, Wayne L.; Warren, Elmer John; Weeks, Victor L.; Wolfe, Steven P.; Wunderman, Richard Lloyd; Wunderman, Richard Lloyd; Zhou, Xinquan

Millersville University
Millersville, PA 17551

13 Master's, 1 Doctoral

1960s: Capen, Robert C.; Nearhoof, Elmer G.; Witmer, Daniel C.

1970s: Chase, Robert Perkins; Davis, Charles Wayne; Fish, Thomas William; Kiehl, Edwin L.; Kiley, Edward H.; Kreiger, E. William, Jr.; Kuhnert, Richard Franklin; Martin, Richard A.; Moxley, Frances M.; Powell, L. R.; Swift, Robert N.

University of Minnesota, Duluth
Duluth, MN 55812

90 Master's, 12 Doctoral

1920s: Carlson, E. N.; Quinn, Howard Edmond; Smith, C. J.

1930s: Chute, Newton E.

1950s: Gunderson, James Ronald Novotny

1960s: Bulin, George Vincent, Jr.; Jongedyk, Howard Albert; Siratovich, Edmund Norman

1970s: Alwin, B. W.; Anderson, Ronald E.; Barr, Kelton; Burnell, J. R., Jr.; Churchill, Ronald; Cooper, Roger Wayne; Curet, A. F.; des Autels, David; Elterman, Joan; Everson, C. I.; Feirn, W. C.; Gladen, L. W.; Goodner, D. C.; Kilberg, James A.; Listerud, W. H.; Magessis, Aberra; Makeig, Kathyrn; Mattis, Allan; McLimans, Roger; Moss, C. M.; Orlando, M. E.; Pope, N. M.; Prutzman, John M., Jr.; Reid, D. F.; Schandle, Thomas M.; Seeling, R. R.; Severson, M. J.; Stevenson, R. J.; Tilley-Grogan, Carol; Vinje, S. P.; Welsh, J. L.; Zarth, R. A.

1980s: Boerboom, Terrence John; Brown, Timothy Reed; Campbell, Frederick Kedney; Carlson, Diane Helen; Clark, Richard C.; Connolly, Marc Robert; Crum, James Robert; Davis, Doug; Eames, Valerie; England, Daniel L.; Erickson, Denis Roger; Ervin, Sarah Mills; Ferguson, Richard R.; Flood, Timothy P.; Frantes, James R.; Giangrande, Peter Anthony; Gross, Laura Blanche; Groves, David Alan; Halfman, Barbara Mary; Holm, Daniel Keith; Huber, James Kenneth; Hyrkas, Gerald Lee; Jirsa, Mark Alan; Johnson Rozacky, Wendy; Karabalis, Dimitris L.; Kohlmann, Nickolas Alfred John; Lannon, Patrick Michael; Leach, Elizabeth Katrin; Lehman, George Albert; Matlack, William Fuller; Motamedi, Shoaullah; Murchie, Scott Lawrence; Nebel, Mark Louis; Neumann, Scott Nelson; Norton, Arthur Randolph; O'Rourke, Elizabeth Frances; Olson, Jean Marie; Olson, Jean Marie; Osterberg, Mark Warren; Osterberg, Steven Arvid; Oulgout, Bassou; Palmer, Elizabeth Ann; Reichhoff, Colin Lee; Reichhoff, Jayne A.; Rismeyer, Neil W.; Rosen, Lawrence Collinger; Sahimi, Muhammad; Scholz, Christopher Alfred;

Spyrakos, Constantine Christoforos; Stone, David J.; Swain, Edward Balcom; Sykes, Julia Ann; Thayer, Valerie Lynn; Urban, Noel Richard; Vander Horck, Mark Patrick; Vervoort, Jeffrey D.; Wangensteen, Martin Walter; Weber, Richard Elmo; Wei, Chenkou; Wonson-Liukkonen, Barbara; Wurdinger, Stephanie; Yeomans, Bruce Wyatt

University of Minnesota, Minneapolis
Minneapolis, MN 55455

413 Master's, 325 Doctoral

1890s: Berkey, Charles Peter; Elftman, Arthur Hugo

1900s: Sorkness, H. O.; Truesdell, W. H.

1910s: Aldrich, Henry R.; Broderick, Thomas M.; Coryell, Lewis S.; Dresser, Myron A.; Foley, L. L.; Gannett, R. W.; Gruner, John W.; Hodapp, Aloys Philip; Hodge, Edwin T.; Hodgman, Robert F.; Hosted, Joseph Orrin; Hubbard, William E.; Ingersoll, Guy E.; Krey, Frank; Lang, Walter B.; Levorsen, A. I.; Mallon, Alfred E.; Nishihara, George Shikataro; Nissen, Arvid E.; Notestein, Frank B.; Quinn, Howard E.; Ravicz, Louis G.; Segall, Julius; Soper, Edgar K.; Soper, Edgar K.; Sweetman, Edwin A.; Wheeler, James D.

1920s: Adams, Frank C.; Allen, Victor Thomas; Allison, Ira S.; Armstrong, Lee C.; Bauernschmidt, August John; Bayliss, W. Z.; Bloom, John R.; Bodman, Geoffrey Baldwin; Broman, Moritz L.; Brown, William Horatio; Brownell, George McLeod; Carlson, Earl Reinhold; Chadbourn, Charles H.; Clay, J. Withers; Conhaim, Howard J.; Copeland, W. A.; Cram, Ira H.; Davidson, Donald Miner; Davidson, Donald Miner; Davies, Herman F.; Davies, Nathan C.; DeLury, Justin Sarsfield; Edwin, John; Erdmann, Charles Edgar; Erdmann, Charles Edgar; Ernster, Omer Francis; Fetzer, Wallace Gordon; Ffolliott, J. H.; Foreman, Fred; Fosness, John Leslie; Foss, Adolph L.; Friedl, Arthur John; George, William Owsley; Graeber, Clyde P.; Graham, William A. P.; Graham, William A. P.; Gray, F. Anton; Griswold, Willis R.; Gruner, John W.; Hansen, Mayer G.; Heins, Melburn E.; Hendry, Lynne D.; Henkel, Howard L.; Kamb, Hugo R.; Kegler, Vern L.; Kendall, Hugh F.; Knutson, Clarence J.; LaTendresse, Henri L.; Leonard, Raymond Jackson; Lilly, Richard J.; Lin, Sze Chen; Lovering, Thomas S.; Lovering, Thomas Seward; Lovering, Thomas Seward; Marx, Archer H.; Middleton, John L.; Nelimark, John H.; Olson, Walter Sigfrid; Patten, Richard C.; Peterson, Eunice; Peterson, Eunice; Pettijohn, Francis J.; Pixler, Everett T.; Sanders, Clarence Whitney, Jr.; Schmitt, Harrison Ashley; Schmitt, Harrison Ashley; Schwartz, George M.; Shenon, Philip J.; Strunk, William L.; Thiel, George Alfred; Tieje, Arthur Jerrold; Tollefson, Everett Harold; Tousley, Robert M.; Walsh, R. P.; Walz, Clarence M.; Ward, George William; Wells, Francis Gerritt; Wilcox, Fred H.; Wolfer, Donald H.

1930s: Adams, R. S.; Alexander, Hugh S.; Allen, Charles C.; Andrews, Thomas G.; Armstrong, Lee Charles; Ashley, Burton Edward; Backman, Olen L.; Bacon, Walter S.; Bennett, Theodore W.; Berg, Ernest Lyle; Berg, Gilman A.; Berge, Olaf T.; Bergquist, Harlan Richard; Bergquist, Harlan Richard; Bickford, Kenneth F.; Calton, Robert; Carlson, Gustaf Magnus; Crowley, Appleton J.; Dennis, Lyman Clark; Dobrick, Edward G.; Dobrick, Edward G.; Downs, George Reed; Dreveskracht, Lloyd R.; Dutton, Carl Evans; Eaton, Wentworth C.; Ericson, Warren Tongo; Fetzer, Wallace Gordon; Fischer, Donald; Fontaine, Paul Jean; Frey, Maurice Gordon; Frey, Maurice Gordon; Gardner, William Irving; Gebhardt, R. C.; Geehan, R. W.; Gibson, George Randall; Goldich, Samuel S.; Gorman, William Albert; Grogan, Robert Mann; Haley, Alva Justice; Hayes, John H.; Herness, Kermit; Hewitt, Robert Leigh; Hicks, Harold Smith; Johnson, Edward William; Johnson, Harold F.; Jolley, Ted R.; Kendall, Richard Garsed; Kinser, James Hanford; Kragness, Ned Low; Kristofferson, Ole Herman; Kruger, Frederick C.; Krum, William

Mark; Kutz, Clarence A.; Lacy, Robert J.; Lavine, Irvin; Lewis, Lloyd Allen; Liao, Yu Jen; Lindner, Joseph Leicht; Lindner, Joseph Leicht; Longley, William Warren; Lundstrom, Orville Glebe; MacLean, Willis John; Manning, John P., Jr.; Matheson, Archie Farquhar; Mathisrud, Gordon C.; Maxwell, John Crawford; McConnell, Duncan; McCorquodale, Ross J.; McMillen, Ralph E.; McMurchy, Robert Connell; Ogryzlo, Stephen Peter; Page, Lincoln R.; Page, Lincoln R.; Park, Charles Frederick, Jr.; Patterson, W. Ray; Peck, Benejhar J.; Pettijohn, Francis J.; Powell, Louis Harvey; Ranta, Reino A.; Rasmussen, Clayton R.; Ronbeck, Arthur C.; Russell, George A.; Savage, James W.; Smith, Terence A.; Sorenson, Seval C.; Stephens, Maynard Moody; Stephens, Maynard Moody; Strand, Edwin H.; Sundeen, Curtis R.; Sundeen, Stanley Wilford; Swanson, Harold A.; Swanson, Roger Warren; Taber, Arthur P.; Thomes, Margaret S.; Titcomb, Jane; Trengove, Stanley Albin; Voigt, Virginia; Wayland, Russell Gibson; Wayland, Russell Gibson; Willard, Paul D., Jr.; Wilson, Robert E.; Woodward, Warren M.; Wright, Willis I.

1940s: Allen, Charles C.; Apsouri, Constantin Nicolas; Beck, Warren R.; Bradley, Edward; Chu, H. J.; Cowie, Roger H.; de Figueiredo, Joao Neiva; Eastwood, George Edmund Peter; Feniak, Michael Walter; Feniak, Michael Walter; Feniak, Oliver William; Folinsbee, Robert E.; Folinsbee, Robert E.; Fruehling, S. W.; Fryklund, Verne C., Jr.; Gillingham, Thomas E., Jr.; Grogan, Robert Mann; Gryc, George; Hsu, Ke-Chin; Hsu, Ke-Chin; Kendall, John Manford; Kurtz, Vincent E.; Lacabanne, W. David; Lathram, Ernest H.; Manly, Robert L.; Marmaduke, Richard C.; Nelson, Clemens A.; Nelson, Clemens A.; O'Neill, Thomas Francis; Pan, Chung-Hsiang; Perry, Louis M.; Pickering, Warren; Poirier, C. A.; Rantala, Raymond Henry; Riley, Charles Marshall; Skillman, Margaret W.; Skillman, Margaret W.; Todd, James Hodkins; Tu, Kwang-Chi

1950s: Alt, David; Anderson, Edwin E.; Anderson, Gerald Edward; Anderson, Henry W., Jr.; Appledorn, Conrad R.; Barton, Robert H.; Bayer, Thomas Norton; Beckman, Charles Allan; Benson, Carl Sidney; Berg, Arthur Brede; Berg, Robert R.; Blake, Rolland Laws; Bleifuss, Rodney L.; Bolin, Edward J.; Brobst, Donald Albert; Brown, Norman James; Buchheit, Richard L.; Bulin, George Vincent, Jr.; Burr, John H., Jr.; Burwash, Ronald Allen McLean; Bury, Curtis A.; Cedarleaf, Darwin C.; Cornell, James Richard; Corwin, Gilbert; Crain, William E.; Crosby, Garth M.; Dickinson, Kendall A.; Dobbins, David A.; Dollof, John H.; Ehrlich, Walter Arnold; Ellinwood, Howard L.; Engel, Paul Louis; Erickson, Ralph LeRoy; Ernst, Wallace Gary; Flammer, Gordon H.; Fogelson, David E.; Ford, Graham Rudolph; Fraser, James Allan; Frye, John K.; Garmoe, Walter James; Gehman, Harry Merrill, Jr.; Gelineau, William J.; Gheith, Mohamed Ahmed; Grant, Richard E.; Hadley, Richard Frederick; Hansen, Donald L.; Hendrickson, Harald F.; Henrickson, Eiler Leonard; Hoeft, David Ralph; Horstman, Elwood Louis; Horstman, Elwood Louis; Humphrey, William R.; Jenness, Stuart Edward; Kaasa, Robert A.; Knox, James A.; Kohls, Donald William; Kraft, John Christian; Kraft, John Christian; Krueger, Harold W.; Langenheim, Ralph L., Jr.; Lepp, Henry; Lewis, David V.; Lucia, F. Jerry; Lund, Ernest H.; Majewski, Otto Paul; Mangen, Lawrence Raymond; Manheim, Frank; Mattson, Louis Arthur; Maxwell, John A.; McGannon, Donald E., Jr.; McGill, George Emmert; Mead, Edwin Ruthven; Meader, Robert Wooten; Menheim, Frank; Miller, Murray L.; Neilson, James M.; Nelson, Ronald G.; Nicolaou, Anthony; Palacas, James George; Palmer, Allison R.; Patten, Harvey L.; Peterman, Zell E.; Peterson, James A.; Pierce, Richard LeRoy; Quaschnick, Ralph Kohler; Quirke, Terence Thomas, Jr.; Quirke, Terence Thomas, Jr.; Riley, Charles Marshall; Roberts, David Blair; Roepke, Harlan H.; Rogers, John James William; Rozendal, Roger Anthony; Rudd, Neilson; Ruotsala, Albert

Peter; Sato, Motoaki; Schneider, Allan Frank; Sheridan, Douglas M.; Smith, Deane Kingsley, Jr.; Sturm, Edward; Taylor, Harry L.; Taylor, Richard Bartlette; Taylor, Richard Bartlette; Taylor, Richard Spence; Taylor, Richard Spence; Thomas, John A.; Thompson, Willis H., Jr.; Tyler, Stanley Roy; Watson, Richard A.; Weiss, Malcolm Pickett; Whelan, James Arthur; White, David Archer; Woncik, John; Yardley, Donald H.; Zumberge, James H.

1960s: Allen, R.; Anderson, Daniel Harvie; Anderson, John J.; Austin, George S.; Baker, Richard Graves; Barazangi, Muawia; Bastien, Thomas W.; Bay, Roger Rudolph; Bayer, Thomas Norton; Beasley, Charles Alfred; Becker, Jacques; Bell, Robert E.; Benson, Richard Norman; Bleifuss, Rodney L.; Bonnichsen, Bill; Cahoon, Elizabeth Jerabek; Callaway, Richard Joseph; Cushing, Edward John; Deischl, Dennis George; Dickinson, Kendall A.; Dolence, Jerry D.; Drake, Benjamin; Farnham, Paul Rex; Finger, Larry Wayne; Gnirk, Paul Farrell; Griffin, William Lindsay; Gunn, Donald William; Gunther, Fredrick John; Haimson, Bezalel; Hanson, Gilbert Nikolai; Hanson, Gilbert Nikolai; Hardyman, Richard F.; Himmelberg, Glen Ray; Hirekerur, Laxmikant Rangrao; Ikola, Rodney Jacob; Imam, Hassan Fahmy El-Sayed; Jahanbagloo, Iraj Cyrus; Johnson, Bernard T.; Johnson, Lane R.; Johnson, Stephen Hans; Jones, Norris William; Karklins, Olgerts L.; Karklins, Olgerts Longins; Kehle, Ralph Ottmar; Kirwin, Peter H.; Kline, Jerry Robert; Knox, William P.; Kohls, Donald William; Krech, Warren Willard; Linder, Harold W.; Loren, J. D.; Loy, William George; Loy, William George; Maher, Louis J., Jr.; Matsch, Charles L.; McAndrews, John Henry; McGannon, Donald E., Jr.; Melchior, Robert Charles; Miller, Peter L.; Miller, Thomas P.; Mitchell, Brian James; Morey, Glenn B.; Morey, Glenn Bernhardt; Muncaster, Neill K.; Nathan, Harold Decantillon; Nelson, Carlton Hans; Niehaus, James R.; Norton, Norman James; Nutter, Larry J.; Nyquist, Laurence Elwood; Papike, James Joseph; Paulsen, Gerald W.; Perumalswami, P. R.; Potter, Noel, Jr.; Radtke, Arthur S.; Rampton, Vernon Neil; Rassam, Ghassan Noel; Renner, Joel L.; Rogers, Marion Alan; Rogers, Marion Alan; Rutford, Robert Hoxie; Ryu, Jisoo; Sahni, Ashok; Schmidt, Paul Gerhard; Schultz, Gerald E.; Searle, Clark Wellington; Seddon, George; Seeland, David Arthur; Sharma, Bijon; Shoemaker, Robert Earl; Snider, Henry Irwin; Soholt, Donald Eugene; Speliotis, Dionysios Elias; Sutton, Thomas C.; Thill, Richard E.; Vaux, Walter Gregson; Veith, Karl F.; Volz, Gary Arlen; Wawersik, Wolfgang R.; Webers, Gerald F.; Webers, Gerald F.; Weiblen, Paul Willard; Wilkes, Jerry; Winter, Thomas C.

1970s: Albinson, Tawn; Assefa, Getaneh; Bagdadi, Khaled Ali; Baker, Robert W.; Black, David Charles; Braun, Gerald E.; Burch, Terrill Lee; Burton, Bruce H.; Carlson, R. E.; Carlson, Steven R.; Carlson, Thomas Warren; Convery, Michael P.; Cornet, F. H.; Crouch, Steven L.; Cummings, Michael L.; Dablow, John F.; Daemen, J. J. K.; Dahlin, Brian B.; Davidian, Beth Eileen Kramer; Davis, Philip A.; Delcourt, Hazel Marie; Delcourt, Paul A.; Driscoll, Fletcher G.; Drown, David Birke; Edens, Mark; Elkins, Steven R.; Evans, James Brian; Evensen, N. M.; Frankel, Leah Shirley; Geldon, Arthur L.; Germano, Richard Joseph; Gingrich, Dean Alan; Glaser, Paul H.; Graber, Ronald Gene; Greenhalgh, Stewart A.; Hangari, Khaled; Harland, Gregg H.; Harris, Alfred Ray; Hartman, Joseph H.; Hemingway, Bruce S.; Hess, Gordon R.; Highland, William Robert; Hollod, Gregory J.; Holst, Timothy B.; Holtzman, Richard Charles; Hudson, John A.; Huntsberger, David V., II; Jacobson, George Lloyd, Jr.; Jahn, Bor-ming; Johnson, Larry D.; Johnston, Alan Dana; Kane, Douglas Lee; Kim, Chin Man; Kowalik, Joseph, Jr.; Kraatz, Paul; Labno, Bruce A.; Landa, Edward R.; Landis, Gary Perrin; Levine, Steven L.; Li, Fu Shung; Li, Fu-Shung; Lie, G. B.; Lund, Steven Phillip; Mansilla, Enrique; Marland, Gregg Hinton; McBride, Mark Stuart; Miller, John Agnew, III; Mohr, Richard Earl; Moore, Ian D.; Mosher, Charles Clinton;

Mudrey, Michael George, Jr.; Norman, David Irwin; O'Hayre, Arthur P.; Olsen, Bruce Michael; Parker, Gary; Pastrana, Jose Manuel; Pollock, David W.; Poppe, James; Porter, Darrell D.; Ripley, Edward M.; Robertson, Eddie B.; Roegiers, Jean-Claude; Rye, Danny Michael; Sabelin, Tatiana; Saint, Prem Kishor; Sansome, Constance Jefferson; Saporito, Mark; Schey, N. D.; Schulz, Klaus Jurgen; Schulz, Klaus Jurgen; Senft, W. H., II; Shapiro, Lewis Harold; Shin, Myong Sup; Singh, Sudarshan; Smith, Douglas Lee; Sobel, Phyllis A.; Swain, Albert M.; Swain, Patricia C.; Tassos, Stavros T.; Van Eeckhout, Edward M.; Van Hessert, Christian; Venkatesan, Thandalalai R.; Viswanathan, Subramanian; Von Schonfeldt, Hilmar; Waddington, Jean C. B.; Walker, Alta; Wilson, Wendell Eugene; Winter, Thomas C.; Wolberg, Donald L.; Wolter, John A.; Wright, Richard Frederic; Wright, Sally; Young, Michael A.; Yu, Shu-Cheng

1980s: Akiyama, Juichiro; Almendinger, James Edward; Almendinger, John Curtis; Andricevic, Roko; Asgian, Margaret Isabelle; Badri, Mohammed; Bagdadi, Khaled Ali; Baker, Joel Eric; Barlaz, Dora; Barten, Paul Kevin; Bauer, Robert Louis; Beck, John Warren; Beck, Stuart Murray; Berndt, Michael Eugene; Bickel, James; Blake, Natalie Ruth; Bradof, Kristine Lynn; Brandenburg, John L.; Brasaemle, Karla Anne; Brice, William Charles; Brown, John Michael; Callahan, Gary Delmar; Capel, Paul David; Carlson, Kelley Elaine; Carlson, Thomas Warren; Cavaleri, Mark Eugene; Chen, Yohchia; Chernicoff, Stanley Edward; Churchill, Ronald Keith; Clark, James Samuel; Clausen, John Campbell; Cohen, Gary B.; Cook, Kevin V.; Curtis, Thomas Gray, Jr.; Czarnecki, John Brian; Daggett, Maxcy DeWitt, III; Dalgleish, Janet Blair; Dasgupta, Biswajit; Davis, Owen Kent; Day, Warren C.; Detournay, Emmanuel Michel; Dosso, Laure; Eames, Valerie; Earley, Drummond, III; Edwards, Harold Hermann; Eldougdoug, Abdelmonem Abdelfattah; Engstrom, Daniel Russell; Erdmann, Anne Lana; Erickson, Denis R.; Evans, James Erwin; Fedors, Randall W.; Ferderer, Robert Joel, Jr.; Ferguson, Richard R.; Ford, Mary Spencer; Forrest, Kimball; Foster, Michael; Gafni, Abraham; Gardner, David Ward; Ghosh, Santi Kumar; Goldstein, Barry Samuel; Goudreault, Paul Richard; Grande, R. Lance; Grimm, Eric Christopher; Groschen, George Earl; Grow, Sheila Roseanne; Guertin, David Phillip; Hackenmueller, Joseph M.; Haitjema, Hendrik Marten; Halfman, John D.; Hart, Roger Dale; Hartman, Joseph Herbert; Hunt, Christopher Paul; Huss, Gary Robert; Idike, Francis Igboji; Janecky, David Richard; Johannesson, Helgi; Johnston, Alan Dana; Jordan, Martha Josephine Ellis; Keen, Kerry Lee; Kim, Dong Jin; King, George Anthony; King, John William; Kuhns, Roger James; Larson, Ronald Gary; Leach, Elizabeth; Leete, Jeanette Helen; Lemos, Jose Antero Senra Vieira; Leung, Kon Lim; Lindner, Ernest Norman; Lorig, Loren Jay; Lund, Steven Phillip; Lundy, James Russell; Malterer, Thomas John; McGuiggan, Patricia Marie; Miller, James Duane, Jr.; Milske, Jodi A.; Mohring, Eric H.; Mooers, Howard DuWayne; Mosher, Charles Clinton; Moskowitz, Bruce Matthew; Murphy Rohrer, William Lyman; Nourse, Susan Marie; O'Dell, Catherine Hobbs; O'Hanley, David Sean; O'Hanley, Hilda Nevius; Okusami, Temitope Abayomi; Olson, Kurt Nathaniel; Padilla, Washington Augusto; Pizzuto, James Eugene; Plymate, Thomas George; Plytus, Michael; Radle, Nancy; Rice, William Forrester; Riley, Michael James; Robert, Pierre Camille; Roloff, Glaucio; Romano, Joseph Vincent; Rosenberry, Donald Orville; Ross, Brian; Saccocia, Peter James; Samson, Scott Douglas; Scherkenbach, Daryl Andrew; Schultz-Ela, Daniel Dennett; Scott, Charles Thomas; Seaberg, John Karl; Seewald, Jeffrey Steven; Seltzer, Geoffrey Owen; Sharp, Samuel; Shimko, Kenneth Andrew; Siegel, Donald Ira; Simmons, Stuart Frank; Simmons, Stuart Frank; Smith, Janet Yvonne; Spear, Ray William; Sprowl, Donald Richard; Ste-

venson, Robert James; Streitz, Andrew Ryan; Swanson, David Karl; Tabor, John Raymond; Thornton, Edward Clifford; Usdansky, Steven Ira; Vollmer, Frederick Wolfer; Walker, James Steven; Wang, Yongjia; Wattrus, Nigel James; Werth, Lee Forrest; Wiens, Roger Craig; Williams, Richard Trudo; Williams, Susan; Wright, Stephen F.; Wright, Stephen F.; Zaadnoordijk, Willem Jan; Zerwick, Susan; Zimmerer, Sheilah Marie

University of Mississippi
University, MS 38677

47 Master's, 6 Doctoral

1930s: Gault, Alta Ray; Shook, Ellen L.

1940s: Darwin, Helen; Jones, Margaret Grace McCorkle

1950s: Attaya, James S.; Causey, Miles Andrew; Gaudet, Philip Arthur, Jr.; Gustafson, Rayford B.; Little, Robert Lewis; Lusk, Tracy W.; Peeples, Joseph D.; Stacy, Curtis Clyde, Jr.; Turner, James

1960s: Ainsworth, B. D.; Allen, William Henry, Jr.; Bailey, John Richard; Harned, Wentworth V.; Lindsey, Reavis Hall, Jr.; Lipsey, James Allen; Lowe, Donald Wayne; McPhail, Robert Louis; Murphey, Joseph Bledsoe; Nicholas, Billy F.; Walker, Charles William; Weissinger, John Leonard

1970s: Autin, Whitney J., Jr.; Borah, D. K.; Breland, Fritz Clayton, Jr.; Broussard, Matthew C.; Dockery, David T., III; Duplantis, Merle James; Franks, Stephen G.; Gazzier, Conrad A.; Hall, Dan O.; Hawks, Paul H.; Hoffman, Bruce Frederick; Lee, Patricia Tsean-Shu; O'Donnell, Terrence T.; Pellegrin, Freddie John; Russell, Eugene; Thompson, Dan A.; Uttamo, Wutti

1980s: Adeff, Sergio Everardo; Aziz, Nadim Mahmoud; Barfus, Brian Lawrence; Bargeron, Dorothy L.; Blanton, Carol Lynn; Cooper, Daniel Howell; Latifi'naieni, Abdolhamid; O'Brien, Harry Deforest, Jr.; Roquemore, Sam Kendall; Spencer, James R.; Wilson, Andrew Moore

Mississippi State University
Mississippi State, MS 39762

111 Master's, 3 Doctoral

1920s: Needham, Claude Ervin

1930s: Neal, Henry Percy; Seiler, Frank Carl

1940s: Mapp, Marcus B.

1950s: Anthony, Clyde C., Jr.; Arledge, Edward Abner; Berryhill, Walter; Birchum, Jack Roy; Bristow, Joseph Dalton; Bryant, Luther Frye; Burchfield, George Edward; Byrd, Thomas Wayne; Carr, John William; Clarke, Harris G.; Edwards, Douglas Edwin, Jr.; Garrett, Marion E.; Gibson, James Bedford, Jr.; Goodman, Marjorie Jeanne; Greco, John Antonio; Harper, William David; Hayes, Arnell Saucier; Hayes, William Errol, Jr.; Holland, Richard Rainey; Hunt, George Lewis, Jr.; Hutto, Andrew Clifton, Jr.; Johnson, Ollie Henry, Jr.; Kaye, John Morgan; Keady, Donald M.; Marion, Cecil Price, Jr.; Martin, Gene B.; Martin, Robert E.; Myers, John D.; Parks, William Scott; Parsons, Willie Frank; Paulson, Oscar L., Jr.; Phelps, William Eugene; Phillips, Ned H.; Pittman, Franklin T.; Pruden, Jimmy Lee; Reed, Jack Morce; Reynolds, John Maurice; Russell, Ernest Everett; Smith, James August; Stegall, Melton J.; Terry, Ira James; Upshaw, Charles Francis; Upshaw, Laurel P.; Wall, James Ray; Whatley, Arthur F.; Wright, Jesse F.

1960s: Brooks, Frank Leroy; Carmichael, Vernon Owen; Carson, Thomas Gordon; Christian, James Terry; Coleman, John M.; Dinkins, Theo H., Jr.; Freeman, M. Lawrence; Gee, Wing Lin; Greeley, Ronald; Haddox, Jimmy V.; Kirkland, John K., Jr.; Mattox, Robert M.; McGee, John B.; Newell, James H.; O'Quinn, Edgar Byron; Parnell, Robert H.; Savage, George Richard; Scott, Harold B.; Scott, Owen A.; Simpson, Lloyd William, Jr.; Stowers, Douglas

I.; Thomas, Carol Varner; Thomas, Warren Baxter; Torries, Thomas F.; Vodrazka, Walter C.; Williams, Albert Earl

1970s: Coleman, James Lawrence, Jr.; Gunter, Charles Phillip; Johnson, Frank Ernest; Kuo, Frank Fu-Kwei; Mitlin, Lucille List; Newman, Stephen Miller; Pandya, Dinesh N.; Risatti, James B.; Shukla, Narendra R.

1980s: Abu-Agwa, Fawzy El-Shazly; Aide, Michael Thomas; Aksoy, M. Zihni; Arikan, Ender; Barcellona, Bruce; Carlson, Alane R.; Clark, Daryl Darnell; Clark, Michael S.; Cook, Philip Ray; Devery, Hugh Blase; Eren, Ahmet Aytac; Garbisch, Jon Ootek; Heller, Noah Russell; Lacko, Peter J.; Lloyd, Ruth E.; Nelson, John Wayne; Olive, Robert Southerland; Pace, Forrest Wilson, Jr.; Pody, Robert Dale; Puckett, Terry Markham; Rodrigues, Francisco Soland de Oliveira; Rogers, David M.; Salomon, Ralph A.; Statom, Richard A.; Stowers, Robert Earl, II; Tabora, Oscar; Taylor, Richard H.; Vogel, Peter Nicholas; Yip, Freddy F.

University of Missouri, Columbia
Columbia, MO 65211

558 Master's, 97 Doctoral

1910s: Bratton, Samuel T.; Connelly, Joseph Peter; Kelley, Ward Wesley; Longwell, Chester Ray; Markham, E. O.; McCoy, A. W.; Owen, Edgar Wesley; Scott, Halley Mering; Thomas, Lewis F.; Wilson, Walter Byron

1920s: Andrews, D. A.; Bailey, Lester; Bailey, Willard Francis; Branson, Carl Colton; Bumgardner, Louis Samuel; Case, Leslie C.; Clark, Joseph Marsh; Craig, Thomas Council; de Irisarri, A. M.; Decker, LaVerne; Glassman, Donald; Glines, Aubrey Leon; Graham, Ida Ellen; Gray, Shapleigh Gardom; Hall, Roy Homes; Keller, Walter David; Keyte, I. Allen; Koester, Edward Albert; Maddox, Gerald Caton; Mathias, Henry Edwin; McQueen, Henry S.; Miller, Arthur K.; Moore, Gilbert Parvin; Nelson, Thomas Mason; Peck, Raymond E.; Price, Arthur S.; Rutledge, Richard Boyden; Rutledge, Richard Boyden; Shayes, Frederick Pine; Swartzlow, Carl Robert; Ware, John McKee; Whorton, Chester Deward; Williams, James Steele; Williams, James Steele

1930s: Babcock, Dorothy Fern; Bailey, Willard Francis; Branson, Carl Colton; Bruner, Frank Henry; Bryan, Joseph Jefferson; Buffum, James Ted; Burnley, Gertrude I.; Clark, Robert Scarth; Cline, Wilford La Verne; Conselman, Frank Buckley; Cooke, Strathmore Ridley Barnott; Cotey, Bradford James; Davies, James Dudley; Davis, Dorothy Taylor; Doyle, Robert; Drake, Robert Tucker; Ellison, Samuel P.; Engel, Albert Edward John; Farmer, Russell; Fletcher, Herbert C.; Gallagher, Robert Taylor; Gallemore, Roy Thornhill; Gardner, Jack Winston; Gault, Hugh Richard; Gleason, Charles D.; Griley, Horace Longin; Gunnell, Francis Hawkes; Hackett, Robert S.; Hammond, Weldon W.; Johnson, Clayton Henry, Jr.; Keller, Walter David; Kraus, Paul S.; Langendoerfer, Martha F.; LaRoge, Clifford Thomas; Lix, Henry W.; Lutz, James F.; Moore, George Emerson, Jr.; Morey, Phillip Stockton; Murdock, James Neil; Oliver, Donald McCreery; Olson, Walter Sigfrid; Owen, John Wallace; Peck, Raymond E.; Peery, Trusten Edwin; Randall, Duane Chilton; Rhodes, Mary Louise; Roberts, John F.; Robinson, Harry R.; Roy, Chalmer John; Schwartzlow, Carl R.; Swartzlow, Carl Robert; Thackrey, Edmund Lee; Trowbridge, Raymond Maxwell; Trowbridge, Raymond Maxwell; Wallace, Cloyd Russell; Walter, Henry Glenn; Warner, William Crim; Wilson, Woodrow Pitkin; Wood, Hiram Budd

1940s: Allen, Henry Whitney; Allen, William Burrows; Aylor, Richard Burns; Barclay, Joseph Ellis; Barnes, James Virgil; Barrett, Henry Haldred; Beavers, Alvin H.; Becker, Leroy Everett; Bender, Hallock John; Clark, Edward Lee; Colson, William

Edward; Cordell, Robert James; Eisner, Stephan Max Leopold; Ellis, Bernett Eston; Ellison, Evard Pitts; Ellison, Samuel P.; Farmer, Paul; Ferrell, Max Everett; Frossard, Robert Louis; Gatchell, John H.; Griggs, Roy Lee; Hardy, Roy Paul; Haseman, Joseph Fish; Heller, Robert Leo; Hembree, Max Reed; Honkala, Frederick Saul; Hopf, Robert W.; Howe, Wallace Brady; Jones, Joseph Maxfield; Keathley, Frances Kathleen Stephens; Keenan, James Edward; Light, Aaron Mitchell; Littlefield, Romaine Faye; Looney, Hugh Marvin; McLaughlin, Kenneth P.; Meyer, Jane Doris E.; O'Byrne, Thomas James; Peck, Joseph Howard, Jr.; Peery, Trusten Edwin; Pennington, Jack, Jr.; Peterson, Jahn Jean; Prossard, R. L.; Prunty, Merle C., Jr.; Quigley, Claud Merle, Jr.; Quinn, William H.; Reker, Carl Caspar; Ryan, William Alexander, Jr.; Sanders, John Warren, Jr.; Smith, William Calhoun; Stephens, Hal Grant; Strothmann, Frederick Henry; Taylor, Louis; Thomas, Leo Almor; Thomas, Leo Almor; Thomas, Murrell Dee; Waldram, Robert James; Westcott, James Franklin; Williams, Philip Anthony; Wolf, Michael Walter; Yeh, Chih-Cheng

1950s: Adams, James Warren; Agee, George Ray; Allen, Billy Dean; Barkdull, James Edwin; Biggs, Donald Lee; Bintasan, L.; Boyd, Richard G.; Bridges, William Clayton; Brown, James Harrison, Jr.; Brown, William Lee; Burgess, Jack Donald; Burst, John Frederick; Burton, Guy C., Jr.; Caneer, William T.; Carini, George Francis; Carver, Robert E.; Cochran, Wendell; Cockran, Wendell A., Jr.; Connelly, James Leslie; Danser, James Weart; Daughdrill, William E.; Diem, Robert Denton; Ditzell, Leon Sebastian; Doermus, Eugene Henry; Doremus, Eugene Henry; Dorman, James H.; Dorman, James Hubert; Duewel, Dennis Brandon; Dwight, Marvin Linn; Ehlmann, Arthur, Jr.; Ellis, Jessie Bird S.; Fairchild, Raymond Eugene; Forbes, Gerald Eugene; Foxworth, Richard Dear; Fraunfelter, George Henry; Galegor, William Baker; Garvin, Donald S.; Gentile, Richard J.; Goeger, Donald Ernest; Goodrich, Edward Arrott; Gord, Clarence E.; Granata, Walter Harold, Jr.; Gregory, Billy Warren; Hahn, Glenn Walter; Haller, Charles Regis; Hambleton, Thomas; Hamilton, Richard C.; Hampstead, Howard A.; Hanson, Robert F.; Harlan, John Lee; Hatcher, Roy Alvin, Jr.; Heller, Robert Leo; Hendren, John Blair; Herbst, Emmett Lee; Herrell, George Leonard; Hoag, William Myrl; Hoare, Richard David; Hoare, Richard David; Holland, Frank Deleno, Jr.; Hooper, William F.; Hover, Frank Bryan; Johnson, Cordell M.; Jones, Daniel Hubbard; Kebert, Fay Dean; Klausing, Robert L.; Kretsch, Donald Lee; Larsen, Kenneth G.; Laswell, Troy James; Leone, Raymond John; Levin, Harold Leonard; Markward, Ellen L.; Marshall, John Harris; Marvin, Leslie Kenneth; Mason, Robert Clifton; McBride, Earle F.; McCarty, Tedford A.; McMillen, Dan E., Jr.; McMillen, Dan E., Jr.; Miller, Mary Helen Alexander; Miller, Roger Glenn; Murphy, Daniel Lawson; Nash, Victor; Niewoehner, Walter B.; Noble, James Eugene; Ojakangas, Dennis R.; Ortiz, Gustavus A.; Ott, Henry Louis; Ott, Henry Louis; Otten, William John; Patterson, Elmer Davisson; Perry, Harry Mcaughton; Perry, Norton; Perry, Norton R.; Planalp, Roger Newton; Ponder, Herman; Potts, Ray Horton; Powers, Richard Blake; Prevey, John Leo; Pulliam, James Millard; Ragan, Wendell J.; Rapp, David William; Rayl, Robert Lee; Reinertsen, David Louis; Rexroad, Carl Buckner; Robbins, Carl Richard; Roehrs, Robert C.; Rush, Thomas Dudley; Schindler, Jack Frederic; Schmaltz, Lloyd John; Schmaltz, Lloyd John; Schmieg, Robert Eugene; Searight, Thomas Kay; Sheth, Pranlal Girdharlal; Shikoh, Mirza M.; Sinclair, Calvin R.; Slaughter, Maynard; Slaughter, Maynard; Smith, Robert Ryland; Smith, William Thomas; Smoot, Virginia Ellen B.; Soderstrom, Glen S.; Spitznas, Roger L.; Spotts, John Hugh; Strassberg, Morton D.; Street, Billy Andres; Swift, William Arnold; Sylvester, Robert Kilbur; Taber, Robert William; Taylor, James Rulie; Taylor, Roy Owen; Thomasson, Maurice R.; Ting, Chuen Pu; Todd, Robert G.; Tucker, Rodney John; Tuthill,

Schuyler K.; Twenter, Floyd Robert; Upshaw, Charles Francis; Van Lieu, Junius A.; Von Almen, William F.; Wall, John Hallet; Wall, John Hallett; Ware, Thomas III; Williams, James Hadley; Woodruff, Edwin C.; Woods, Everett Kenneth; Wright, Leo Milfred; Zalusky, Donald W.; Zeidner, Martin Aaron; Zimmerman, Tom Van

1960s: Andresen, Marvin John; Baker, Dewey Allen; Balgord, William D.; Birkhead, Paul Kenneth; Bishop, Richard S.; Brahana, John Van; Brown, Dwight Delon; Buder, Theodore A.; Canis, Wayne Francis; Canis, Wayne Francis; Carini, George Francis; Carver, Robert Elliott; Chesney, Claybourne; Cole, David Lee; Courdin, James L.; Craig, William Warren; David, Jimmy Leon; De Camara, Richard P.; de Moraes, Joao A. P.; Deike, George Herman; Doty, Robert W.; Echols, John Bowlus; Eyer, Jerome Arlan; Foster, Robert Lutz; Foster, Robert Lutz; Fraunfelter, George Henry; Funk, John L., III; Golden, Jerry B.; Goldman, David John; Granata, Harold Peter; Hagni, Richard Davis; Harris, Frank Gaines, III; Hasseltine, George H.; Hatheway, Richard B.; Hazel, Joe Ernest; Heflin, Larry Holden; Hight, Robert, Jr.; Holcombe, Troy Leon; Holifield, Billy Ray; Honea, Elmont G.; Hopkins, Don Eugene; Jackson, Togwell Alexander; Jacobson, Roger Leif; Kavary, Emadeddin; Kennedy, Henry David; Kinerney, Eugene James; Kinsley, Gerald W.; Kling, Donald Lee; Knapp, William Dale; Lange, Stephen Stanley; Leimer, Harold Wayne; Lohrengel, Carl F.; Looff, Karl M.; Lynch, Matthew J.; Mallory, Bob Franklin; Merkle, Arthur Beiser; Miller, Jim Patrick; Miller, William Donald; Minke, Joseph Garrett; Minor, John A.; Moraes, Joao A. P.; Morales, Gustavo Adolfo; Mumma, Martin Dale; Neal, William Joseph; Neal, William Joseph; Nold, John Lloyd; Ojakangas, Richard W.; Otto, David Arthur; Parrish, David; Patrick, David Maxwell; Pauken, Robert J., Jr.; Payton, Charles Ellis; Pearl, Richard Howard; Reed, Byram E., Jr.; Reesman, Arthur Lee; Robbins, William H.; Robertson, Charles E.; Rucker, James Bivin; Saum, Nicholas Mather; Schneider, Raymond H.; Scrivner, Clarence Leland; Shaffer, Bernard Leroy; Sleeman, Lyle Herman, Jr.; Smith, Arthur Edward; Snowden, Jesse Otho; Snowden, Jesse Otho, Jr.; Sommer, Nicholas A.; Sullivan, John S., Jr.; Tennissen, Anthony Cornelius; Thomas, Hugo Frederick; Tlapek, John William; Van de Graaff, Fredric Ray; Vineyard, Jerry Daniel; Walker, Russell Allen; Weiser, Robert Neal; Whitehead, Jack Waters

1970s: Abel, Kathleen D.; Adshead, John Douglas; Alford, Robert L.; Almon, W. R.; Banet, A. C., Jr.; Barnes, Jerry D.; Bauer, Robert L.; Baumann, D. K.; Bergfelder, W.; Berkley, John L.; Bez, Lauri; Brahana, John Van; Brand, U.; Brenner, Robert Lawrence; Butera, Joseph G.; Chowdhury, Yusuf; Clark, Christopher N.; Clayton, Jerry L.; Combs, M. J.; Dando, Mark; Darr, James M.; De Moraes, Joao A.; DeVine, Carolyn S.; Diefendorf, Andrew F.; Dreiss, Shirley Jean; Drummond, Keith F.; Dyess, James N.; Feder, Gerald Leon; Felton, Richard M.; Ferber, Daniel; Foreman, Terry Lee; Founie, Alan; Fox, Laura; Frank, James R.; Fullerton, Marilynn; Garbarini, J. Michael; Gebhard, Paul; George, M.; Glassinger, Craig L.; Goldstein, B. A.; Goydan, Paul Alexander; Grannemann, N. G.; Grethen, Bruce L.; Hall, Morris D.; Hamil, Martha M.; Hart, Richard M.; Hebberger, J. J., Jr.; Hesemann, Thomas; Hilliard, Henry D.; Huang, Wen Hsing; Hunter, B. E.; Jackimovicz, Joseph James; Jackson, Kenneth J.; Jayyusi, G. S.; Jones, R. T.; Karlo, John F.; Kastler, Everett J.; Kemmer, David Andrew; Kennedy, D. J.; Kennedy, John David; Knirk, Ernest P.; Knox, Larry M.; Kussow, Roger Glenn; Lambert, Marshall Brice; Leach, David L.; Leach, David Lamar, Jr.; Lipke, Audrey C.; Lischer, Lowell K.; Lyle, John; Madalosso, Antonio; Mansker, William L.; Markwell, Kenneth E.; Martin, Michael W.; McGee, Kenneth A.; McHorgue, Timothy Reed; McManus, Jeffrey; McQuillan, Kirk A.; Miles, R. C.; Mitchell, Gary Clark; Moffett, Tola B.; Moore, Arthur Howard; Murray,

James P.; Nelson, Cheryl Ann; Nelson, R.; Oglesby, T. W.; Petersen, M. E.; Peterson, Mark E.; Petrus, Richard T.; Potter, Benjamin A.; Potter, C. W.; Pritchard, Charles L.; Quearry, M. W.; Raymond, Scott; Reinert, S. L.; Repetski, J. E.; Repetski, John E.; Rice, Roger F.; Roper, Edith D.; Ryland, Stephen Lane; Saum, Nicholas Mather; Scambilis, N. A.; Schumacher, Dietmar; Shepherd, Rodney D.; Shore, Rochelle C.; Siebert, R. M.; Siebert, Robert M.; Sisk, Steve W.; Sommer, Nicholas Anthony; Stevens, Neil E.; Stoufer, R. N.; Tacker, Allen B.; Tennyson, L. C.; Tharpe, L. W.; Trout, Michael Lynn; Ulmo, George J., Jr.; Vessell, Richard; Walter, Gary R.; Ward, Paul T.; Waring, Juliana; Warmbrodt, R. E.; Werner, William G.; Williams, Donald Lee; Williamson, Eddie A.; Woolverton, D. G.; Yunker, Gerald G.; Zatezalo, M. P.; Zeiner, Thomas C.; Zick, A. D.; Zumwalt, Robert Wayne

1980s: Al-Basso, Khaled M. S.; Aloui, Tahar; Aucremann, Leslie J.; Aucremann, Leslie J.; Babaei, Abdolali; Bajsarowicz, Caroline J.; Bauer, Dennis A.; Bernthal, Mike; Bidwell, Matthew E.; Breshears, Terry L.; Broberg, Steven Kent; Broom, B. B.; Brown, Stephen H.; Carroll, Cynthia J.; Chapman, Kenneth R.; Claassen, Daniel R.; Dacre, George J.; Dew, Mary McClure; Doesburg, James M.; Downs, Dana V.; Draney, David; Drew, Thomas A.; Elliott, Kenneth L.; Engel, Kevin; Esslinger, Brad A.; Fagerlin, Stanley Charles; Fairhurst, William; Falteisek, Jan D.; Focht, Thomas; Fridley, Mark S.; Fulton, David A.; Fulton, Sara M. Mauldin; Garney, Ronald T.; Ghidey, Fessehaie; Gorday, Lee; Greenberg, Dallas W.; Guthrie, John M.; Haensel, Jose Mariano; Hart, William D.; Hathon, Eric Gene; Heet, Steve; Hicks, Brian Douglas; Hoeffner, Steven Lewis; Horton, Robert B.; Hotrabhavananda, Tachpong; Houser, Kenneth L.; Hsueh, Chao-min; Hutchison, Hillary W.; Hutson, Frederick John; Kalish, Robert S.; Kertis, Carla A.; Khafagi, Om Mohamed Ahmed; King, David Thompson, Jr.; Kortenhof, Michael H.; Kuhn, Mark A.; Lent, Mary C.; Lewis, Jean; Lofstrom, Dotty Mae; Lozano-Chavez, Guillermo; Matlack, Keith S.; Matteo, Brett; Matthews, Steven N.; McDonald, Kirk W.; McKallip, Thomas E.; Medary, Tom A.; Mindheim, Bruce K.; Moore, Thomas R.; Morreale, Steve; Morse, Robert K.; Murphy, Russell King; Murphy, Timothy B.; Newton, Morgan Roe; Norville, Charles R.; Nugent, R. Michael; Older, Kathy; Pallesen, Thomas J.; Papusch, Richard G.; Parman, Lynn; Patterson, Daniel J.; Perry, Stanley James; Phasukyud, Prapon; Phillips, Stephen T.; Poole, Leslie Ann; Quinn, Harold Edward, III; Raker, Gary; Reader, John Malcolm; Rothbard, David Rod; Satterfield, Joe; Schmitz, Larry; Schwandt, Craig Stuart; Scott, Kriston H.; Seifert, Gregory G.; Settles, Patricia Leigh; Slagley, Scott A.; Smart, Miles Millard, III; Smith, Terrance Lee; Sporer, Peggy; Stackelberg, Paul E.; Stallman, Georgia; Stewart, Craig; Stewart, John M.; Strong, Robert H.; Sturgess, Steven W.; Taylor, John Felix; Tedesco, Lawrence; Thomas, Mark A.; Turner, Jeffrey S.; Van Horn, Stephen R.; Weaverling, Paul Harrison; Wheatley, Todd L.; White, James D. L.; Williams, Eula C.; Williams, Stephen N.; Yanoski, Mark A.; Yesberger, William Lloyd, Jr.; Young, David Lucius; Zeller, Craig G.; Zhao, Naiyu

University of Missouri, Kansas City
Kansas City, MO 64110

12 Master's

1940s: Shurnas, Marshall Kenneth

1980s: Allen, Ashley V.; Blasch, Sheila R.; Campbell, Bruce G.; Costello, J. Patrick; Crow, Roxane J.; Hoyt, Albert J.; Martin, Sharon P.; Ragan, Virginia M.; Rudy, Richard J.; Spencer, Charles G.; Verhulst, Galen G.

University of Missouri, Rolla
Rolla, MO 65401

309 Master's, 109 Doctoral

1870s: Greason, Arthur

1900s: Fay, Albert Hill; Moore, Stanley Ralston

1910s: Dean, Reginald S.; McNutt, V. H.

1920s: Abernathy, George Elmer; Black, Clarence J.; Bowles, J. H.; Bowles, John Hyer; Bridge, J.; Burg, Robert Stanley; Chavez, Raul; Cole, Virgil Bedford; Cordry, Cletus D.; de Cousser, Kurt H.; Dolman, Phil B.; Eulich, Artileus V.; Hatmaker, Paul C.; Hopkins, James; Ingerson, M. J.; Ingerson, M. J.; Irwin, Joseph S.; Lloyd, S. H.; McCartney, William H.; Miller, Edwin L.; Miller, John Charles; Millikan, Carl E.; Murphy, Thomas C.; Netzband, William F.; Quillam, William E.; Schaeffer, Willard A.; Tedrow, Harvey Louis; Wilson, Joseph M.; Zoller, Lawrence J.

1930s: Clair, J. R.; Cullison, J. S.; Dresbach, C. H.; Farrar, Willard; Grohskopf, John G.; Householer, Earl R.; Kay, W. W.; Lynch, Shirley Alfred; Martyn, Phillip F.; Prouty, Chilton E.; Reid, Joseph Hugh; Zvanut, Frank Joseph

1940s: Chaney, J. B.; Graves, Roy William; Hayes, William Clifton; Haynes, W. C.; James, Jack Alexander; Mabrey, P. R.; McLean, D. C.; Nackowski, Matthew Peter; Pomerene, J. B.; Rasor, J. P.; Wightman, R. H.

1950s: Anderson, D. K.; Beatty, William A.; Boeckman, G. O.; Butterfield, Gale Eugene; Chico, Raymundo L.; Christiansen, C. R.; Clark, J. W.; Clarke, Peter John; Colson, C. M.; Cotter, R. D.; Crockett, N. E.; Cronk, R. J.; Deaver, B. G.; Doe, B. R.; Dokozoglu, Hilmi; Drake, A. A.; Dukozoglu, Hilmi; Dunscombe, Thomas D.; Emery, J. A.; Evans, Lanny L.; French, G. B.; Hyatt, E. P.; James, Jack Alexander; Jeffries, Norman W.; Jeffries, Norman William; Kurtz, Peter; Ligasacchi, Attilio; Ligasacchi, Giovanna R.; Martin, J. A.; May, J. E.; Middour, E. S.; Miller, D. N.; Miller, H. N.; Mueller, H. E.; Nackowski, Matthew Peter; Orlansky, Ralph; Perry, Bobbie L.; Plunkett, J. D.; Sarapuu, Erich; Schafer, R. P.; Stevens, Robert Paul; Taylor, L. B.; Van Duym, Dirk Peter; Waheb, M. A.; Weixelman, Wesley D.; Yorston, H. J.; Zarzavatjian, Papken A.; Zimmerman, Richard Albert

1960s: Abul-Husn, Adnan Asid; Aguilar, Oscar; Al-Hashimi, A. R. K.; Al-Omari, Farouk Sunallah; Al-Shaieb, Zuhair; Aldrich, Charles Allen; Anderson, John H.; Andrews, Harold Edward; Babu, Peethambaram; Bachinski, Donald John; Basson, Philip Walter; Bearce, Denny N.; Brown, Donald L.; Casquino Rey, Walter T.; Chamon, Nagib; Chan, Samuel S. M.; Chan, Siew Hung; Chen, Roland Lee-Ping; Chen, Tu Kao; Cheng, Shih-Cheang; Choudary, Narendra; Chun, Joong Hee; Crouch, James H.; Daneshy, Abbas Ali; Dar, Ikram-ul-Hag; Dayley, Richard D.; Deatherage, Jas. H.; del Prete, Anthony; Desai, Arvind A.; Dinkel, Ted Richard; El Baz, Farouk; El Baz, Farouk el Sayed; El-Etr, Hasan A.; El-Etr, Hasan A.; Fernandez, Henry E.; Fisher, Henry Hugh; Fore, James Gary; Frey, Richard Paul; Frouzan, Faramarz; Gallagher, John; Gandhi, Rajni K.; Ganthavee, Somkiat; Gentile, Richard Joseph; Germundson, Robert Kenneth; Ghole, Jagannath Rao; Goldstone, Martin; Gopal, Vijender Nath; Greeley, Michael Nolan; Greeley, Ronald; Gupta, Sujoy; Han, Daesuk; Hansen, Spenst Mitchell; Haycocks, Christopher; Hedden, William J.; Hornsey, Edward Eugene; Horton, Wayne C.; Jasim, Rafid A. H.; Kabeiseman, William Joseph; Kane, William Theodore; Kantor, Tedral; Karwoski, William James; Kasapoglu, Kadri E.; Kehrman, Robert F.; Kisvarsanyi, Eva B.; Kisvarsanyi, Geza; Knewtson, Steve; Kohl, Barry; Konya, Calvin Joseph; Kraus, Gregory Paul; Krause, Jerome Bernard; Lamber, C. Kurt; Lassley, Richard Harold; Lee, Larry; Mantei, Erwin J.; Mantei, Erwin J.; Martin, Richard A.; Masiello,

Remo Antonio; Miller, John; Mirbaba, Mehdi H.; Misra, Krishna Kant; Mullins, L. E.; Naiknimbalkar, Narendra M.; Namdarian, Faridoon Ardeshire; Narupon, Anant S.; Nawrocki, Michael Andrew; Nichols, Renny Roger; Nute, Alton John; Nute, Alton John; O'Leary, Dennis; Oscar, Aguilar M.; Owens, Willard G.; Palmer, James Edward; Park, Won Choon; Peacock, David N.; Piskin, Kemal; Pointer, Gary Neal; Potamianos, Socrates N.; Quan, Choon Kooi; Reddy, Varakantham S.; Reesman, Arthur Lee; Reijenstein d'Acierno, Carlos Enrique; Reinhard, Mahlon J. A., III; Robinson, John Charles; Robison, James B.; Saad, Afif Hani; Saadallah, Adnan A.; Schmitz, Richard Joseph; Schot, Eric H.; Sheth, Kertikant R.; Shrivastava, Jai Nandan; Smith, Cole L.; Smith, Frederick J.; Stainbrook, Don J.; Sunda, Laxman Singh; Thieme, Martin Alan; Thomas, Faiz N.; Thomas, John Neil; Thompson, Stanley D.; Thornton, Robert C.; Tibbs, Nicholas Howard; Trapp, John Siegfried; Troell, Arthur R.; Tumialan, Pedro H.; Van Besien, Alphonse Camille; Van Vesien, Alphonse C.; Vander Schaaff, Bertis J. III; Wicklein, Phillip; Zambrano, Elias; Zenor, John Julian; Zevallos, Raul A.; Zimmerman, Richard Albert

1970s: Aboud, Jesus M.; Aboud, Nelson; Achmad, Grufron; Adam, M. E.; Afzali, Behzad; Akbar, Ali Mohammed; Al-Omari, Farouk Sunallah; Al-Shaieb, Zuhair; Alexander, Emmit Calvin, Jr.; Al-usow, Edward W.; Attiga, M. A.; Bhatia, D. M. S.; Bippus, William John; Bird, Melvin Leroy; Borahay, Abd El Aziz El Hady A.; Borahay, Abd El-Aziz El-Hady Ahmad; Bown, John S.; Bradley, M. F.; Butherus, D. L.; Butz, Todd R.; Chareonsri, Prachon; Chen, Li-King; Chen, Shih-Tsu; Chollett, D.; Choudary, Narendra Y. B.; Chun, Joong Hee; Clark, Dean Stanley; Coen, Larry P.; Collins, Susan J. B.; Collins, Terry Moore; Collins, William E.; Conroy, Peter J.; Corwine, John W.; Craig, Ted William; Creveling, J. B.; Davis, Jerry Lee; Deming, H. Michael; Doraibabu, Peethambaram; Dutta, Virendra Kumar; Fielding, D. H.; Fletcher, Charles S.; Francisco, German; Frank, Gregory Bryan; Gann, D. E.; Garrison, Edwin J.; Gastreich, K. D.; Guharoy, Prasanta Kumar; Haines, Forest E.; Head, W. J.; Hedden, W. J.; Higgins, Jerry Don; Honapour, Mehdi; Hudson, D. D.; Ilavia, Piloo Eruchshaw; Jessey, D. R.; Kalia, Hemendra Nath; Kettenbrink, Edwin Carl, Jr.; King, M. J.; Ko, Kyung Chul; Koederitz, Leonard Frederick; Kreider, John E.; Kunkel, Richard B.; Lance, R. J.; Lane, Charles A.; Lexa, David J.; Long, L. L.; Manusmare, Purushottam; Meinecke, L., III; Meisenheimer, James Kenneth; Melton, Douglas C.; Mercer, Barry P.; Merchant, Abdul Rashid; Misra, Brij Raj; Naiknimbalker, N. M.; Najjor, Abdullatif; Narendra, Choudary Y. B.; Nichols, Chester E.; Nuelle, Laurence M.; O'Brien, William P.; Ozkaya, Ismail; Parikh, Upendra J.; Park, Duk-Won; Patel, Dinesh; Patrick, W. C.; Payken, C. L.; Peacock, David Nuse; Peters, John F.; Posey, Harry Howard; Rath, David L.; Rauch, Peter C.; Reagan, R. L.; Riggs, C. O.; Salem, Mostafa; Salem, Mostafa Juma; Schaefer, M. G.; Serviss, Curtis Raymond, Jr.; Silverman, Alan N.; Singh, Krishna Kumar; Sinha, Bhudeo Narayan; Smith, Gregory Paul; Smith, N. S.; Smith, R. H.; Stewart, David Mark; Stinchcomb, Bruce L.; Stovall, Robert M.; Su, Chen-Bin; Sullivan, J. S.; Taleb, T.; Thacker, Joseph L., Jr.; Thompson, David; Tibbs, Nicholas Howard; Till, Henry Anthony; Tolley, M. A.; Toweh, Solomon H.; Trancynger, Thomas C.; Trapp, John Siegfried; Ucar, R.; Von Demfange, W. C., Jr.; Walker, Wayne T., Jr.; Wedge, William Keith; Wedge, William Keith; Wei, Chi-Sheng; Wessel, Gregory R.; Williams, A. L.; Williams, James H.; Wilson, Tommie Claud; Wongsawat, S.

1980s: Algan, Ugur; Birgisson, Gunnar Ingi; Bond, William Howard; Brandon, Steven Howard; Brown, Vernon Max; Carpenter, Gregory Wallace; Cole, Jimmy Dale; Cooper, Michelle; Duckett, Kathleen Carey; Elifrits, Charles Dale; Erten, Zeynep Mujde; Fass, Fred W. R.; Fleck, Kenneth Stewart; Flori, Ralph Emil, Jr.; Foster, Douglas John; Fulton,

Dwight David; Ganjidoost, Hossein; Garrido, Robert William; Gavin, Brian; Geronsin, Rolin Lee; Harbaugh, Michael William; Hatfield, Stanley Christopher; Higgins, Jerry Don; Horrall, Kenneth Bruce; Huang, Scott Lin; Jessey, David Ray; Johnson, William McDaniel; Joseph, Allan Jeffrey Anthony; Khandoker, Jalal Uddin; Lee, Young-Hoon; Liang, Huh-Yuan; Lin, Wuu-Jyh; Lux, Jeffrey John; Mallery, Linda Leigh; McCallister, Phyllis Grace; Mugel, Douglas N.; Netzler, Bruce William; Nusbaum, Robert L.; Pignolet-Brandom, Susanne; Puri, Vijay Kumar; Reed, James Edward; Richardson, David Newton; Rickman, Douglas L.; Robold, Edward Lynn; Roca-Ramisa, Luis; Sacre, Jeffrey Allen; Simon, Andrew Dorsey; Sinha, Bhudeo Narayan; Stevens, Marian Merrill; Tangchawal, Sanga; Taylor, Donald Richard; Vidrine, Dana Marie; Woolsey, Leonard Lee; Yang, Tsung-Wen; Zoukaghe, Mimoun

University of Montana
Missoula, MT 59812

271 Master's, 48 Doctoral

1890s: Douglass, Earl

1910s: Wilson, Roy Arthur

1920s: Rowe, Royle Carlton

1930s: Bell, William Charles; Blackstone, Donald LeRoy, Jr.; Clapp, Michael M.; Denson, Norman Maclaren; Duncan, Donald Cave; Duncan, Helen M.; McNair, Andrew Hamilton, Jr.; Tweto, Ogden Linne

1950s: Achauer, Charles Woodrow; Anderson, Roy Ernest; Clawson, Paul Norman; Eakins, Gilbert Royal; Eisenbeis, H. Richard; Groff, Sidney Lavern; Hutchison, David Malcolm; Leischner, Lyle Myron; Long, William A.; McGuire, Robert Hillary, Jr.; Montgomery, Joel Kenneth; Pilkey, Orrin H.; Sieja, Donald Michael; Smallwood, Kenneth Keith; Stone, Jerome; Sweeney, George Le Jeune; Toler, Larry Gene; Woodward, Lee Albert

1960s: Bentzin, David Allan; Berg, Richard Blake; Bhatt, Bharat K.; Brenner, Robert Lawrence; Chase, Ronald Buell; Chase, Ronald Buell; Cremer, Edward A., III; Decker, Gary L.; Ege, John Rodda; Fisher, Victor Arthur; Fox, Richard Dale; Gibbs, Frank Kendall; Gilmour, Ernest Henry; Gilmour, Ernest Henry; Goers, John William; Hall, Frank Washington, II; Hall, Frank Washington, II; Harris, William Langseth; Hintzman, Davis Eugene; Hood, William Calvin; Illich, Harold Aallen; Jansons, Uldis; Jerome, Norbert Hugh; Johnson, Durwood Milton; Kelly, James A.; Kuenzi, Wilbur David; Kuenzi, Wilbur David; Langfield, Peter Michael; Latuszynski, Felix Victor; Manghnani, Murli Hukumal; Matson, Robert Ernest; Maxwell, Dwight Thomas; McKay, Rodney H.; Mowatt, Thomas Charles; Nold, John L.; O'Connor, Michael Peter; Oltz, Donald F., Jr.; Pevear, David R.; Pevear, David R.; Rasmussen, Donald Linden; Riel, Stanley Joseph; Riggs, Stanley Robert; Schell, Elmer Morris; Shoemaker, Robert Earl; Stuart, Charles Juhami; Thiel, Paul Thomas; Velde, Bruce Dietrich; White, Brian George

1970s: Ahlstrand, Dennis Carl; Aklstrand, Dennis Carl; Badley, Ruth Hall; Bateridge, Thomas Earl; Baty, Joseph Bruce; Beall, Joseph John; Benoit, Walter Richard; Bielak, James Walter; Bleiwas, Donald I.; Bodholt, Frederick Brunson; Bratney, William A.; Breuninger, Ray Hubert; Brittenham, Marvin Del; Bryan, Thomas Scott; Burnside, Michael James; Calbeck, James Morgan; Castle, Bruce; Cavanaugh, Patrick Charles; Chambers, Richard Lee; Cheney, John Thomas; Ciliberti, Vito A., Jr.; Cleary, John Gladden; Cole, George Paul; Collins, Benjamin I.; Cox, Bruce Ellis; Custer, Stephan G.; Derkey, Robert Erwin; Desmarais, Neal Raymond; Desormier, William Leo; Douglas, Jesse King; Egan, Roger Thad; Ehinger, Robert Ferris; Enterline, Theodore R.; Ferguson, James Ardon; Feucht, Alex T.; Flood, Raymond Edward, Jr.; Forster, John R.; Geldon, Arthur L.; Gensamer, Alan

Richard; Gordon, Elizabeth A.; Grimestad, Garry R.; Haartz, Eric R.; Harris, Richard Huntington; Harris, William Langseth; Hawe, Robert Glen; Hawley, Katherine Taft; Hecht, Paul David; Hoffman, Dale Sheridan; Huebschman, Richard Patrick; Iagmin, Paul Jean; Jens, John Christian; Johnson, Bruce R.; La Tour, Timothy Earle; Lankston, Robert Wayne; LaPoint, Dennis John; LaPointe, Daphne D.; Laudon, John Lowell; Lelek, Jeffrey John; Lemoine, Stephen R.; Lisle, Thomas Edwin; Marin Rivera, Pedro A.; Marsh, Phyllis Scudder; McClellan, Thomas Stewart; McKee, James M.; Mejstrick, Peter Francis; Monroe, James Stewart; Nichols, Ralph; Nolan, K. Michael; Norwick, Stephen Allan; Nowell, William Benjamin; Otto, Bruce Richard; Pearson, Monte Laurence; Petkewich, Richard Mathew; Pincomb, Arthur Chesney; Piombino, Joseph J.; Presley, Mark Whitehead; Rushin, Carol Jo; Shurr, George W.; Smith, Donald Laurence; Spindel, Sylvia Frances; Strickler, William John; Struhsacker, Debra Winter; Tysdal, Russell Gene; Van der Poel, Washinton I., III; Wackwitz, Linda K.; Walker, Thomas Franklin; Wheeler, Robert J.; White, Brian George; Whiting-McBride, Celia Kathleen; Williams, Larry D.; Williams, Roderick David; Williams, Thomas Roy; Willis, Gerald F.; Winegar, Robert Charles; Wiswall, Charles Gilbert; Wiswall, Charles Gilbert; Wold, Ronald Odin; Woods, Michael J.

1980s: Achuff, Jonathan M.; Ackman, Brad C.; Al-Khirbash, Salah; Alleman, David G.; Angeloni, Linda Marie; Axelrod, Russell B.; Ballard, James H.; Barkmann, Peter E.; Barlow, David A.; Barrett, Philip M.; Beyer, William C.; Bloomfield, Susan L.; Brandon, William Campbell; Briar, David W.; Brook, Edward J.; Brooks, Rebekah; Burnham, Robert Lawrence; Butler, Barbara A.; Byer, Gregory B.; Cantwell, Jonathan; Carlson, Garry J.; Carstensen, Andrew B.; Carter, Bruce Applegate; Cartier, Kenn D. W.; Clark, Kenneth W.; Clement, William P.; Connor, Cathy Lynn; Cossaboom, C. Carey; Crabtree, David Rockwell; Cronin, Christopher; Dalby, Charles E.; Davis, John Steven; Dea, Peter A.; Detra, Earl H.; Dickman, Lynn R.; Dolberg, David Michael; Dunlap, Dennis Gordon; Eaton, Gary D.; Edmond, Carolyn Lorraine; Eduardo, Benjamin E.; Embry, Paige A.; Emmart, Laura A.; Erler, Elise L.; Finstick, Sue Ann; Foggin, G. Thomas, III; Foland, Sara S.; Folger, Peter F.; Foster, David Allen; Foster, Fess; French, Larry B.; Fritz, William Jon; Garcia, Daniel; Gary, Steven D.; Geiger, Beth Carol; Gierke, William Gordon; Godlewski, David W.; Goodge, John William; Grotzinger, John P.; Hansen, Vicki L.; Harris, David William; Harrison, Sylvia L.; Heise, Bruce A.; Herberger, Don; Herndon, Stephen D.; Hoffman, Jonathan D.; Holloway, Carleen D.; Hudak, George; Jennings, Scott; Johnson, Larry M.; Johnson, Milo J.; Johnson, Peter P.; Joyce, Michael J.; Kendrick, George C.; King, David Alexander; Kirchner, Gail L.; Knadle, Marcia E.; Kogan, Jerry; Kremer, Marguerite C.; Kruger, John M.; Kuhn, Jeffrey A.; Kuhn, Paul W.; LaPasha, Constantine Anthony; Larson, Jeffery E.; Lazuk, Raymond; Leppert, Dave Eric; Lippert, James Brent; Liptak, Alan Robert; Lister, James C.; Lofgren, Donald L.; Lonn, Jeff; Luthy, Stephen T.; Mauk, Jeffrey L.; McDonough, Daniel T.; McGrane, Daniel J.; McGroder, Michael F.; McLeod, Paul J.; Moore, Richard F.; Murray, Christopher J.; Myers, Sharon A.; Peery, William M.; Pottinger, Michael H.; Quattlebaum, David M.; Rankin, Peter Watson; Rhodes, Brady P.; Ripley, Anneliese A.; Roberts, Sheila M.; Roemmel, Janet Sue; Rubin, Charles M.; Runkel, Anthony Charles; Schaefer, Michael J.; Schissel, Donald J.; Sengebush, Robert M.; Shore, Lawrence R.; Sikkink, Pamela Gayla Lindell; Simpson, Stephen J.; Sixt, Karen C.; Slover, Susan M.; Smith, Kevin; Snyder, Hollice Andrew; Sorensen, Gary Frank; Stickney, Michael C.; Stoffel, Keith L.; Stradley, Ann Chalmers; Thomas, Gerald M.; Thomas, Michael B.; Thomas, Robert Curtiss; Thompson, William Ross; Udaloy, Anne Greenough; Uthman, William; Vuke, Susan M.;

Waisman, Dave; Walker, Susan Claire; Watson, Ian A.; Watson, S. Michelle; Weeks, Gary C.; Weiss, Christopher Paul; Whalen, Michael T.; Wilson, Michael L.; Wogsland, Karen L.; Woods, Marvin O.; Young, Maria Leigh; Zehner, Richard E.; Zieg, Gerald A.

Montana College of Mineral Science & Technology
Butte, MT 59701

120 Master's

1930s: Blixt, John Elmer; Corry, Andrew Vincent; Elliott, Harold Charles; Fritzsche, Hans; Grassmück, Gerhard; Hamblin, Ralph Hugh; Kuechler, Adolph Harmon; McMillan, Donald Theodore; Powe, George Robert; Rabbitt, John Charles; Sahinen Uuno, Mathias; Soal, Norman; Warde, John Maxwell; Wendell, Clarence Adami; White, William A.; Wilson, Arthur Oliver; Zeihen, Lester Gregory

1940s: Brinker, Willard Franklin; Fitzgerald, Wilfred Harold; Fowells, Joseph Edward; Gallant, Raymond Bockles; Goudarzi, Hossein; Jones, Roy Meyrick Price; Mitchell, William, Jr.; Nelson, Harry Eugene; Puumala, Paava Pellervo; Ramboseck, August F.; Reyner, Millard Lester; Richards, John Charles; Roe, Joseph Thomas; Stejer, Francis Adrien, Jr.; Stout, Koehler Sheridan; Thurlow, Ernest Emmanuel

1950s: Archibald, John C., Jr.; Au-Ngoc-Ho; Banta, Howard E.; Clement, James Hallowell; Crowley, Francis Allen; Dahlem, David Harrison; Eyde, Theodore Henrik; Guttormsen, Paul Andrew, Jr.; Hallock, Allan Richard; Hilpert, Frederick Martin; Johns, Willis Merle; Marvin, Richard F.; Miller, Richard Nelson; Paine, William Rhodes; Pepper, Miles Warren; Regnier, Jerome Philippe Mathieu; Stanley, Kirk William; Williams, Higbee George; Win, Maung Soe

1960s: Birkholz, Donald O.; Chelini, J. M.; DenHartog, Stephen Ludwig; Fox, Kenneth Francis; Gillette, Christopher B.; Hruska, Donald C.; Ingersoll, Robert George, Jr.; Lindquist, Alec E.; McClernan, Henry G.; McPherson, Roger Ian; Phelps, George B.; Pulju, Hugo James; Ramseier, Frederic Neil; Wendel, Clifford Arthur; Winters, Allen S.; Wolfe, David F.; Young, Francis Millard

1970s: Abdul-Malik, Muhammed M.; Blumer, John W.; Clendenin, Charles William, Jr.; Dahl, Gardar G., Jr.; Dobb, David E.; Forrest, Richard A.; Jan, Ming-Ju; Jenke, Dennis R.; Johnson, Edward A.; Lockrem, Larry L.; Minervini, George B.; Mungi, Julio C. Castañada; Obolewicz, David; Pederson, Robert J.; Robson, James M.; Snyder, Robert D., Jr.; Storm, Linda M.; Suydam, John R.; Whitehead, Mark L.; Whitlock, James D.

1980s: Ahmed, Shemsudin; Anderson, Garry E.; Christensen, Kim C.; Cordalis, Charles; Dahy, James P.; Dresser, Douglas W.; Ferguson, Scott D.; Gogas, John G.; Halvorson, James Walter; Heald, B. Patrick; Holmes, Margaret A.; Ikeda, Margaret; Jakubiak, Annette Leisner; Kleinschmidt, John C.; Lupindu, Kandidus P.; Miller, Brent L.; Newcomer, Darrell R.; Ott, Lawrence E.; Palke, Dale Robert; Pasecznyk, Michael J.; Patton, Thomas W.; Payne, Richard Allan; Peterson, Janet L.; Peterson, Mark P.; Robocker, J. E.; Schofield, James Dean; Semmens, Dave; Shaller, Philip J.; Truckle, Daniel M.; Weekes, David C.; Wilde, Edith M.

Montana State University
Bozeman, MT 59717

98 Master's, 10 Doctoral

1950s: Tompson, Willard D.

1960s: Andretta, Daniel B.; Basler, Albert L.; Bluemle, John P.; Boyd, Donald William; Bush, John Harold, Jr.; Chalmers, Ann L. Stradley; Glancy, Patrick A.; Kavanagh Yllarramendi, John A.; Love, Charles M.; Mifflin, Martin D.; Ropes, Leverett H.; Shelden, Arthur William; Spahn, Ronald A., Jr.;

Todd, Stanley G.; Tysdal, Russell Gene; Van Voast, Wayne Adams; Weber, William Mark

1970s: Aram, Richard Bruce; Bailey, Robert L.; Bonnet, Adrienne Thornley; Feichtinger, Sylvia H.; Galloway, Michael J.; Goolsby, Jimmy Earl; Grabb, Robert F.; Gyan-Gorski, S. S.; Harp, Edwin L.; Kaczmarek, Michael B.; Kehew, Alan E.; Leeson, Bruce Frank; Lunceford, Robert A.; Mackin, Richard L.; McPartland, John T.; Montagne, Clifford; Montagne, Clifford; O'Haire, Daniel P.; Petticord, David V.; Roper, Michael William; Schneider, Gary B.; Schrunk, Verne K.; Shaver, Kenneth C.; Sollid, Sherman A.; St. Lawrence, William; Struhsacker, Eric M.; Tilley, Craig W.; Tonnsen, John J.; Walsh, Thelma Helaine; Wash, Thelma H.; Weinheimer, Gerald Joseph; Whalen, Stephen C.; Widmayer, Margaret A.

1980s: Anderson, Dale Lewis; Baken, Jeffrey Frank; Barrick, Paula Jean; Bauman, Bruce John; Bearzi, James Paul; Bibler, Carol Jean; Black, Geoffrey Alan; Black, Janette Louise Young; Bowen, David Wayne; Braun, Roger Elmer; Bugosh, Nicholas; Callmeyer, Thomas J.; Clark, Michael Lee; Cole, Marshall Morris; Craiglow, Carol Jean; Davis, Thomas Edward; Dent, Jimmie Duane; Drain, Vance Keith; Fox, Norman Albert; Fryxell, Jenny Christine; Garrett, Paul Allen; Gavin, William Morris Bauer; Grady, Thomas Richard; Griffith, Earl Francis; Guthrie, Gary Eich; Harlan, Stephen Scott; Hartsog, William Smith; Hayes, Graham Stephen; Hazen, David Ralph; Heins, Vasco Ann M.; Hughes, Gary Claude; Ihle, Bethany A.; Johnson, Ronald Frederick; Kinley, Teresa May; Knight, Jonathan Charles; Mannick, Matthew Lee; May, Karen Anne; Meyer, Grant Arnold; Miller, Erick W. B.; Moore, Bonnie K.; Olson, Timothy John; Personius, Stephen Francis; Richmond, Douglas P.; Rose, Charles Cleland; Ruth, John Helms; Salt, Kenneth Julian; Singdahlsen, Donald Scott; Stanley, Alice Roberta; Stimson, James Roy; Stine, Alan D.; Suydam, James David; Thurston, Peter Bouck; Trombetta, Michael J.; Tsai, Kuang-Jung; Vandervoort, Dirk Sheridan; Wyatt, Glen Milton; Yang, Jae Eui

Montclair State College
Upper Montclair, NJ 07043

16 Master's

1970s: Baker, Gerald Lawrence; Johnson, J. Kent; Wagenhoffer, Albert J.

1980s: Anagnostos, Nicholas; Cameron, Diana; Cardona, Alberto; Devries, Donald Charles; Gaito, Richard A.; Godfrey, Patricia Kathryn; Jefopoulos, Timothy; Refolo, Perry J.; Rizzo, Jean G.; Storm, Evelyn V.; Titus, Russell Gerard; Toder, Daniel R.; Volkert, Richard Allen

Universite de Montreal
Montreal, PQ H3C 3J7

162 Master's, 27 Doctoral

1960s: Bourque, Pierre-André; Descarreaux, Jean; Dumesnil, Jean-Claude; Hocq, Michel; Jobin, Claude; Robillard, Jacques; Roy, Denis; Woussen, Gérard

1970s: Barondeau, Bernard; Barondeau, Bernard; Barraud, Claude; Barraud, Claude; Barthelemy, Ernst Nels; Beaulieu, Jean; Beaupre, Michel; Beriault, Andre; Beriault, André; Bertrand, Rudolf; Bertrand, Rudolf; Bouchard, Michel; Bouchard, Michel; Bouillon, Jean Jacques; Bourque, Pierre Andre; Brisebois, Daniel; Carignan, Jacques; Caty, Jean Louis; Caty, Jean Louis; Chagnon, Andre; Chauvin, Luc; Chevalier, Jean; Coulomb, Jean-Jacques; Ducrot, Claude; Gentile, Francesco; Gentile, Francesco; Granger, Bernard; Granger, Bernard; Guilbault, Jean-Pierre; Heroux, Yvon; Heroux, Yvon; Hocq, Michel; Janes, Donald A.; Janes, Donald A.; Kerba, Mona; Lebuis, Jacques; Leonard, Marc-Andre; Letendre, Jacques; Letendre, Jacques; Machado Fernandes, Nuno; Martineau, Gismond; Mathey, Bernard; Nadeau, Andre;

Nantel, Suzanne; Payette, Francine; Rocheleau, Michel; Rocheleau, Michel; Shalaby, Hany; Tasse, Normand; Trudel, Pierre; Vallieres, Andre; Verpaelst, Pierre; Woussen, Gérard

1980s: Arpin, Marc; Auclair, François; Babineau, Jacques; Belisle, Jean-Marc; Birkett, Tyson C.; Boily, Michel; Boily, Michel; Cadieux, Bernard; Chabot, Nathalie; Chainey, Daniel; Charbonneau, Robert; Charland, Anne; Chartrand, Francis; Cossette, Danielle; Deragon, Robert; Desbiens, Sylvain; Deschamps, Fernand, Jr.; Gauthier, Gilles; Gauthier, Louise; Gauthier, Lysanne; Goyette, André; Hamel, Cathy; Harvey, Yves; Hilali, Atika; Indares, Aphrodite; Kirkwood, Donna; LaFleche, Marc; Lapointe, Martine; Malo, Michel; Marcotte, Claude; Nimpagaritse, Gérard; Perreault, Serge; Seuthé, Chantal; Shalaby, Hany; Simard, Alain

Mount Allison University
Sackville, NB

1 Master's

1950s: Murray, D. W.

Mount Holyoke College
South Hadley, MA 01075

3 Master's

1940s: Albrecht, Jean Irene; Rogers, Betty Ross

1960s: Crampton, Janet Wert

Murray State University
Murray, KY 42071

13 Master's

1980s: Baumgarten, Diane M.; Bower, Scott M.; Elder, Barbara L.; Gesch, D.; Hynes, P.; Kreighbaum, D.; Major, Jeffrey D.; Meier, P.; Pearson, R.; Rushing, V.; Schmidt, Lane T.; Shelby, Lynn; Spillman, T.

Naval Postgraduate School
Monterey, CA 93943

11 Master's

1960s: Davis, J. H.

1970s: Anderson, Richard G.; Bodie, Jeffrey G.; Frydenlund, David Dexter; Hamlin, J. S.; Keith, John L.; Maratos, Alexander; McCord, William K.; Sheridan, John Joseph; Souto, Antonio Pedro Dias; Vieira, Mario E. C.

University of Nebraska, Lincoln
Lincoln, NE 68508

317 Master's, 67 Doctoral

1900s: Bengston, Nels A.; Condra, George E.; Fisher, Cassius Asa; Gould, Charles Newton; Gould, Charles Newton; Knight, Wilbur Clinton; Pepperberg, Roy O.; Rowe, Jesse Perry; Rowe, Jesse Perry; Schramm, Eck Frank; Storrs, Lucius S.; Woodruff, Elmer G.

1910s: Borrowman, George; Burnett, Jerome B.; Moore, Calvin Turner; Pool, Raymond John; Whitford, Arley C.

1920s: Whyman, L. O.

1930s: Bennett, Robert B.; Busby, Clarence Edward; Clark, Clare M.; Hayes, Frank A.; Hewitt, L. W.; Hornaday, Albert C.; Johnson, William R.; Kraemer, John L.; Link, John T.; Loetterle, Gerald John; Lueninghoener, Gilbert Carl; Lukert, Louis Henry; Maher, John Charles; Meade, Grayson Eichelberger; Mechling, George William; Mills, Lloyd Clarence; Reed, Eugene Clifton; Scherer, Oliver J.; Schultz, Charles Bertrand; Smedley, Harold Orian; Stout, Thompson Mylan; Upp, Jerry E.; Upson, M. E.; Wainner, Kenneth Fred; Waite, Herbert A.

1940s: Amato, Francis L.; Bell, Frank James; Brier, Ervin Ernest; Burkholder, Paul; Buthman, B. David; Crosbie, James Morton; Elias, Gregory Konrad; Frankforter, Weldon Deloss; Fuenning,

Paul; Gordon, Ellis Davis; Griffith, James Hendrie; Hauptman, Charles McNerney; Heidtbrink, Werner Henry, Jr.; Hendy, William James; Honkala, Adolf Uno; Horney, William Rolland; Hubka, James Lewis, Jr.; Kersey, James Doyle; Lorenz, Howard Wilhelm; Lueninghoener, Gilbert Carl; Morris, Donald Arthur; Robbins, Harold Weston; Rogers, Wilbur Frank; Schultz, Charles Bertrand; Stacy, Howard Elwell; Stoesz, LeRoy Warren; Tychsen, Paul Charles; Wahl, Carl Christian; Wright, Jerome J.

1950s: Anderson, Arthur Erick; Ansari, Homayon Jaberi; Backlund, Alvin Lorenzo, Jr.; Beardsley, Henry S., Jr.; Bender, Marvin J.; Biesiot, Peter Gerard; Braden, William Joseph; Brandorff, William A.; Brown, Bahngrell Walter; Brown, Bahngrell Walter; Burchett, Raymond Richard; Busby, John Cifford; Cambridge, Thomas Ross; Carson, Charles Edward; Castellano, Rocco Horatio; Christensen, Richard M.; Chuman, Richard Wayne; Copeland, Jerry Harold; Cox, Williard E.; Dietrich, Ernest S.; Docekal, Jerry; Douglass, Raymond C.; Downey, Marlan Wayne; Dreeszen, Vincent Harold; Driscoll, Egbert G., Jr.; Erwin, Eugene; Folsom, Jerry Robert; Frankel, Larry; Graham, Graydon Elliott; Green, William; Harden, Rollin Wayne; Harding, Kenneth Stanley; Harvey, Cyril Hingston, II; Holmberg, Russell L.; Horton, Marvin Dean; Howe, John Alfred; Hunter, William J.; Juilfs, John D.; Karabatsos, George Tom; Knoop, John William; Kowalski, John Francis; Lampshire, Wayne Gilbert; Leonard, Benjamin Franklin; Lorenz, Donald Paul; Lucas, Robert Charles; Lugn, Richard Victor; Malin, Eugene R.; Marvin, Raymond Glenn; McCrone, Alistair William; Mendenhall, Gerald Vernon; Mendenhall, Maurice Elvin; Moore, Vinton Aubrey; Nekritz, Richard; Nelson, Robert Harry; Newville, Harold Lee; Nielsen, Mitchell Frederic; Nygreen, Paul Wallace; O'Donnell, Paul J., Jr.; Ogbukagu, Ikechukwu Nwafo; Peckham, Alan Embree; Rasmussen, Noel Fredrich; Reider, Eugene R.; Robinson, Robert Bradley; Robinson, Vincent A.; Rollins, John Flett; Ruede, George M.; Ryan, Robert; Sabatka, Edward Frank; Sahl, Howard Leroy; Sakowski, Henry A.; Sauer, Frank Joseph; Schooler, Owen Edwin; Schrott, Robert Otto; Schulte, John Joseph; Seff, Philip; Slama, Don C.; Stacy, Robert R.; Stephenson, Larry Gene; Stoley, Aaron Kenneth; Svoboda, Richard Frank; Swanson, Richard Wayne; Swenson, Donald Bruce; Toohey, Loren Milton; Truxell, Robert Eugene; Tychsen, Paul Charles; Vondra, Carl Frank; Wakely, William James; Watson, Stuart Tucker; Weidler, Mark E.; Yetter, Riley Glen

1960s: Aadland, Arne Johannes; Aadland, Rolf; Abed, Fawzi M. A. H.; Alker, Julius; Allinton, John Richard; Avers, Darrell D.; Babcock, Gary B.; Bamford, John Ross; Bartel, Douglas J.; Beckmann, Douglas D.; Bernasek, Rodney A.; Boellstorff, John David; Buffington, John William; Carlson, Marvin Paul; Carlson, Marvin Paul; Castellano, Rocco Horatio; Chaffin, Herbert Scott, Jr.; Christensen, Richard Martin; DeGraw, Harold M.; Derieg, George William; Dideriksen, Carrell J.; Diffendal, Robert Francis, Jr.; Dishman, Bill D.; Estasen, Elena; Girardot, Stephen Lee; Goll, Carroll Leon; Gundersen, Wayne Campbell; Harper, John LeRoy; Harvey, Cyril Hingston, II; Hillerud, John Martin; Howe, John Alfred; Irwin, Don Dennis; Jakway, G. E.; Jones, Larry LeRoy; Kent, Douglas Charles; Khan, Muhammad Aslam; Kreycick, Karen A.; Luebke, Laurence Orville; Machovec, Marvin Anthony; Martin, Larry D.; Mayo, Curtis Ray; McCormick, Charles D.; McLellan, Robert Charles; Miller, Laurence S.; Minshew, Velon Haywook, Jr.; Mizula, Joseph William; Morgan, Stanley Sherwood; Pampe, William Riley; Pampe, William Riley; Pollack, Henry N.; Ready, Jeffery A.; Rollins, John Flett; Roper, Paul James; Schrott, Robert Otto; Schuett, Edwin Clarence, Jr.; Singler, Charles R.; Singler, Charles Richard; Snyder, Barry L.; Stoffey, Philip Stephen; Svendsen, Augie Eugene; Tohill, Bruce Owen; Vondra, Carl Frank; Walker, William B.; Washburn, Robert Henry; Weakly, Edward Cletus; Wehrman, Ken C.; Wellman, Samuel Sidney; Whitcomb, Charles W.; Yang, Tsu-Hsi

1970s: Atkinson, Jon Charles; Avery, C.; Baird, Gordon C.; Bart, Henry A.; Blodgett, Robert Hugh; Bowe, Richard J.; Boyden, Ernest Duree; Breyer, John Albert; Busch, John Daniel; Busch, Karl M.; Campion, Kirt Michael; Corner, Richard George; Diffendal, Robert F., Jr.; Docekal, Jerry; Dugan, J. T.; Edwards, Paul D.; Evander, Robert L.; Eversoll, Duane A.; Gerken, Antony N.; Giardino, J. R.; Goodenkauf, O.; Griesemer, Allan David; Grosskopf, F. W.; Guzel, Nuri; Hillerud, John Martin; Hintlian, Raymond Arthur; Horsburgh, Martha Sennett; Hosek, Ronald Joseph; Irons, L. A.; Jacobs, Stephen Emanual; Jensen, Wayne Gale; Karl, Herman A.; Kirumakki, Nagaraja Subraya; Korth, W. W.; Larson, David Roy; Lindsay, Robert R.; Maroney, D.; Messin, Gerard M. L.; Miller, Daniel D.; Moeglin, T. D.; Moeglin, Thomas Dean; Moore, Vinton Aubry; Murphy, Catherine Marie; Nash, K. G.; Nicholson, Jane E.; Pabian, Roger Karr; Palmer, S. E.; Phillips, R. L.; Pilger, Rex H., Jr.; Piskin, Rauf; Reider, Richard Gary; Resser, Kurt Douglas; Rickles, Sue E.; Rinehart, Jon; Rowell, Bruce Fenton; Rowell, Bruce Fenton; Ruffin, Isiah Washington; Russell, John Lysle; Sayeed, Usman Ahmed; Sirkin, Gerald L.; Spellman, James Wheeler; Stauffer, Truman Parker, Sr.; Stilwell, D. P.; Suhm, Raymond Walter; Swinehart, J. B., II; Turner, Mary Ann; Watts, Stephen H.; Wellstead, Carl F.; Westgate, J. W.; Whitney, John W.; Wiig, Stephen Victor; Wilson, Samuel M.; Zarins, Andrejs

1980s: Al-Abed, Souhail Radhi Ali; Al-Khirbash, Salah A.; Alabi, Kolade Ebenezer; Anderson, Randall L.; Arbab, Mahmood; Barton, James Wesley; Baumer, Otto Weichsel; Bloom, Margaret Louise Haroldson; Bolitho, Mason R.; Brookner, Paul L.; Bryda, Anthony P.; Chu, Tyan-Ming; Davis, Ralph K.; Donofrio, Chris Joseph; Druliner, Allan Douglas; Ehrman, Richard L.; Elliott, Roy William; Fiorillo, Anthony R.; Folkman, David N.; Fowler, Charles Sidney; Frankforter, Matthew J.; Fulton, John W.; German, Kenneth E., Jr.; Ghazi, Ali Mohamad; Ghohestani-Bojd, Hamid; Ginsberg, Marilyn H.; Gless, Ingrid Maria Verstraeten; Gowen, Linn H.; Gross, Jonathan A.; Hanna, Abdulaziz Yalda; Hedin, Rae Ann; Hiskey, Robert Marshall; Holly, Dean E.; Holterhoff, Peter F.; Huang, Steve Kuo-Yi; Joeckel, Robert Matthew; Johnson, Jeffrey S.; Kanna, Sanousi S.; Katz, Gary Joshua; Kintner, Henry B.; Kitchen, Lisa; Krueger, James P.; Kuntz, Gregory Brent; Kuzila, Mark Steven; Leite, Michael B.; Low, Dennis James; Mays, Major Dewayne; McCarty, Kevin L.; McCool, Kevin E.; Moran, Michael James; Murphy, Kathleen; Murray, Gene L.; O'Connor, Thomas Mark; Page, Eric J.; Parrott, Jack D.; Pipes, Jeffrey W.; Pluim, Scott B.; Rankis, Linda Victoria; Rowan, Charles David V.; Rowan, M. Elizabeth Anderson; Schwegal, Steven R.; Silvers, Eric Richard; Simpson, Edward L.; Steele, Anthony D.; Stracher, Glenn B.; Swisher, Carl Celso, III; Tabidian, Mohamad Ali; Uhl, David A.; Uzochukwu, Godfrey A.; van Noort, Peter John; Vanderhill, James B.; Walz, David M.; Watts, Linda Jean; Wetzstein, Eric E.; Williams, Richard H.; Wolfram, Katherine; Woodcock, Deborah; Wright, David Brian; Wright, Janet Decker; Wu, Guoping

University of Nevada
Reno, NV 89557

339 Master's, 36 Doctoral

1920s: Gianella, Vincent P.

1950s: Collagan, Robert Bruce; Divens, Donald F.; Fulton, Fred J.; Godwin, Larry H.; Hand, David; Lundby, William; Middlebrook, John; Young, Peter Frederick; Zones, Christe P.

1960s: Abdullah, S. K. M.; Armstrong, Charles F.; Back, William; Behnke, Jerold J.; Booth, G. Martin, III; Brennan, Peter Anderson; Cartwright, Keros; Carver, Gary Alen; Chipp, Eddie Ray;

Clark, Wesley Inman; Cline, Robert Bruce; Cordova, Tommy; Davis, Terry E.; DeWyk, Bruce H.; Domenico, Patrick Anthony; Fabbi, Brent P.; Garside, Larry J.; Gedney, Larry D.; Gentry, Donald William; Glenn, Robert Jerrell; Green, William Randolph; Greensfelder, Roger W.; Greenslade, William Murray; Hoffman, Frederic; Hughes, Jerry L.; Jennings, Olin R.; Jucevic, Edward Paul; Larsen, William Robert; Lee, Chiekyo; McGillis, John L.; McLelland, Douglas; Mifflin, Martin David; Mindling, Anthony Leo; Morrison, Roger Barron; Onuschak, Emil; Osborne, David H.; Rai, Vijai Narain; Ranta, Donald Eli; Scales, Bert; Stathis, George John; Taylor, Michael Francis; Tingley, Joseph V.; Tippett, Michael Charles; VanWormer, James D.; White, Craig Kenneth; Wilson, George M.; Zeizel, Eugene Paul

1970s: Abrams, Gerard Joseph; Alkaseh, Ahmidi Ali; Barnes, James Irvin; Basinski, Paul; Bateman, Richard L.; Batra, Ravi; Benito, Hugo Oscar; Bird, John Wilbur; Blomquist, John T.; Bonné, Jochanan; Borbas, Steve; Bradhurst, Stephen Thomas; Bryan, Dennis Paul; Buff, Paul Jeffrey; Butler, Robert Scott; Carnahan, Chalon L.; Carraher-Muto, Ruth; Carroll, James C.; Castor, Stephen Baird; Charlton, Douglas W.; Chen, Hsiu-hsiung; Cinque, Mark J.; Cochran, Gilbert Francis; Cornwall, Diane Elinor; Cutler, Sandra Kay; Dabbagh, Ali A.; Davis, Robert Martin; Decker, Donald James; Dircksen, Paul Eric; Dixon, James B.; Dixon, Richard Lee; Drowley, David; Drumheller, Richard E.; Dunn, Mildred Carneal; Duren, Fred Kenneth, Jr.; Fairfax, Vella L.; Farsad, Ebrahim; Federici, James M.; Fink, Richard Christopher; Fisher, John Bailey; Garbrecht, David A.; Gibbons, James Arthur; Grant, Terry A.; Gupta, Ashok K.; Gustafson, Fridolf V.; Hamlin, Scott Norman; Hansen, Donald S.; Hardyman, R. F.; Hart, Daniel Douglas; Haworth, William D.; Hiner, John Edward; Hoffman, Lee Ellis; Howard, Lauran L.; Howe, Daniel Marshall; Hudson, Donald M.; Hull, Donald Albert; Hutton, Robert A.; Ivosevic, Stanley Wayne; Jager, Douglas John; Johnson, Dane S.; Johnson, Robert Crandall; Jones, Richard Burton; Khanna, Satish Kumar; Kilbreath, Steven Perry; Kirkham, Robert Marshall; Kronberg, Merrily; Kurtak, Joseph M.; Laule, Susan W.; Luethe, Ronald D.; Mahin, Donald Alan; Malone, Stephen D.; McCleary, Jefferson Rand; McGarvie, Scott Douglas; McKinney, Roy Franklin; Miller, Donn William; Mizell, Nancy Brent Hunt; Mizell, Stephen A.; Mock, Ralph G.; Nadolski, John A.; Naff, Richard Louis; Navoy, Anthony S.; Nelson, Steven Wayne; Nimsic, Thomas L.; Nowak, Gregory; Oberlander, Phil Louis; Oldham, Richard Lewis; Olson, George Harrison; Orsen, David A.; Passmore, Gary William; Pautsch, Richard Joseph; Pease, Robert Charles; Peterson, David Michael; Phariss, Edward Irvin; Pidcoe, William W., Jr.; Ponsler, Harley E.; Powers, Sandra L.; Priest, George R.; Priestley, Keith F.; Puchlik, Kenneth Phillip; Quade, Jack Gehring; Rai, Vijai Narain; Richardson, Larie Kenneth; Richins, William D.; Rogers, David K.; Rudy, Samuel; Saulnier, George J., Jr.; Savage, William Underwood; Schaff, Schuyler C.; Scheibach, Robert Bruce; Soeller, Stephen Anton; Sonderman, Frank James; Spencer, Ronald J.; Stevens, David Lee; Tafuri, William Joseph; Trivedi, Nikhilesh Chandrakant; Troutman, Thomas William; Trudeau, Douglas A.; Waggoner, Raymond Russell; Wallace, Andy B.; Ward, Timothy James; Watson, Phillip Charles; Wells, John H.; Westhoff, David Edward; Westphal, Jerome Anthony; Wickwire, Joy McIntosh; Wilson, Geoffrey Evans; Zellmer, John Theodore

1980s: Adams, Opal Fay; Alvarez-Ayesta, Jose Alfredo; Amateis, Larry Joe; Amin, Isam Eldin; Arghin, Salem Saleh; Banks, Eric W.; Bard, Thomas R.; Barry, Jeffrey M.; Benham, Julia Anne; Blair, Michael L.; Bohm, Burkhard W.; Boone, Richard L.; Bosma-Douglas, Julia; Boughton, Carol Jean; Bratberg, David; Briner, William D.; Broadbent, Robert C.; Bronson, Brent R.; Bruce, James L.; Bruce, Lorraine; Bruha, Douglas James; Bryce, Robert William; Buck, Caroline Jo; Bugenig, Dale

C.; Burbey, Thomas J.; Carlton, Stephen M.; Cave, Deborah L.; Chavez, David Eliseo; Clayton, Carol Aurelia; Cohen, Donald K.; Collins, Amy Hutsinpiller; Collord, E. J.; Connell, Larry E.; Cox, John Waldron; Crone, Walter Richard; Dale, Michael W.; Day, Garrett Arthur; Desrochers, Gary J.; Dowden, John E.; Duttweiler, Karen A.; Elliott, Charles S., Jr.; Emme, David H.; Erikson, Susan J.; Evashko, Anna Helen; Fanning, David James; Feeney, Thomas Aquinas; Feldman, Sandra C.; Fezie, Glenn Stephen; Findley, David Paul; Finn, Dale Robert; Fischer, Jeffrey M.; Flint, David C.; Foster, Joseph M.; Fox, Forrest L.; Franzone, Joseph G.; Frick, Elizabeth A.; Fricke, Rodney A.; Frost, Karl Albert; Fuller, Lynn Roy; Geasan, Dennis L.; Ghasemi, Amir Mohammad Soltani; Gibson, Peter Craig; Goldfarb, Richard Jeffrey; Graney, Joseph Robert; Hadiaris, Amy K.; Hallee, Mark C.; Harpel, Greg; Hayes, Garry Fallis; Heggeness, John O.; Henne, Mark Siegfried; Herman, Marc Edward; Heuberger, Mark Oscar; Hudson, Donald McComb; Inghram, Brent J.; Ingraham, Neil L.; Ismail, Azmi; Jackson, Mac Roy, Jr.; Jackson, Philip Richard; Johnson, Cady Leonard; Johnson, Edwin Lionel; Jones, Steven K.; Juncal, Russell Wright; Karst, Gary B.; Kautsky, Mark; Kearl, Peter M.; Kemp, Wayne Russell; Killeen, Katheryn Marie; Kirk, Marcia W. Olson; Kirk, Stephen T.; Kirkham, Richard A.; Klemme, Michael; Klimberg, David M.; Kneiblher, Carolyn Ruth; Koltermann, Christine Rinzel; Koltermann, Howard H.; Krank, Kenneth D.; Krause, Alan Joel; Krause, Kerwin J.; Kwa, Boo Leong; Lagoni, Jack R.; Laird, Brien A.; Leatham, Stacey; Lide, Chester Scott; Linderfelt, William R.; Lopes, Thomas J.; Lugaski, Thomas P.; Lyles, Bradley F.; Markos, Andrew George; McBeth, Paul Edward, Jr.; McCormack, John Kevin; McCulla, Michael; McDaniel, Scott Byron; McFarlane, Deborah Nyal; McFarlane, Michael James; Metcalf, Linda Anne; Mitchell, Burke M.; Molinari, Mark Philip; Moss, Kenneth Lee; Muto, Paul; Nichol, Michael R.; Nichol, Michael John; Nitchman, Steve P.; Nork, Diane M.; Nosker, Richard Ernest; Nosker, Sue Anderson; O'Malley, Peg Ann; Odt, David Albert; Oliveira, Jose Auto Lancaster; Ornstein, Peter; Osborne, Marla A.; Page, Tench C.; Panian, Thomas F.; Park, Steven Lynn; Parr, Andrew Joseph; Perry, Richard Michael; Pittman, Kate L.; Pottorff, Edward J.; Price, Donald R.; Prokop, Christopher Jon; Pulido, Oscar H.; Pullman, Steven A.; Purington, Paul Richard; Putney, Thomas R.; Raker, Sarah L.; Ralston, Edward Charles; Ramelli, Alan Ray; Renken, Paul; Rennie, Douglas Paul; Rhodes, Jonathan J.; Robbins, Charles Henry; Roelofs, Nicolas Henry; Ronkos, Charles Joseph; Roquemore, Glenn Raymond; Roth, James G.; Royse, Susan E.; Schroth, Brian K.; Scott, Troy Calvin; Seidl, Richard F.; Sharp, Gregory A.; Shields, Hilbert Nathaniel; Shump, Kenneth W.; Sirles, Phil C.; Smith, Corilss M., Jr.; Smith, Michael Roy; Spatz, David Moore; Sprecher, Terry Ann; Strawson, Frederick MacLeod, Jr.; Strobel, Robert J.; Svoboda, Mark Scott; Szecsody, James Edward; Taylor, Joseph K.; Taylor, Kendrick Cashman; Thiesse, Mark F.; Thomas, Trevor James; Tingley, Icyl Cathryn; Truschel, Anthony D.; Varnum, Nick C.; Walker, Nancy Denning; Walker, Patrick M.; Weaver, Sarah C.; Weaver, Stephen George; Weiss, Steven I.; Welch, Alan Herbert; Wendell, Daniel E.; Whitney, Robert A.; Wigglesworth, John Bradley; Wolverson, Nancy Jean; Zelinsky, Anne E.; Zellmer, John Theodore; Zinner, Ronald Eric

1990s: Ross, Wyn Charles

University of Nevada, Las Vegas
Las Vegas, NV 89154

18 Master's

1980s: Blegen, Ronald Paul; Crow, Henry Clay, III; Feuerbach, Daniel Lee; Goings, David Bruce; Hayworth, Joel Stacey; Hurst, Thomas L.; Mills, James Gordon, Jr.; Myers, Ingrid A.; Naumann, Terry Richard; Noack, Richard Eric; Panttaja,

Susan Kay; Parolini, Joseph R.; Rice, Jonathan Aaron; Russell, Charles Eugene; Scott, Allan J.; Sewall, Angela Jean; Thomas, Edward F.; Timm, John Jay

University of Nevada - Mackay School of Mines
Reno, NV   89557

50 Master's, 11 Doctoral

1910s: Hart, James J. P.

1920s: Romig, Woodfred Edward; Smith, Cassius Crowell

1930s: Ericson, Norman John; Henricksen, Raymond Milton

1940s: Davis, Dudley L.; Hannifan, Martin K.; Humphrey, Fred L.; Overton, Theodore D.

1950s: Brooks, Howard; Byrkit, James W.; Carlson, John E.; Dempsey, Earle V.; Ehrlinger, Henry Phillip, III; Flangas, William G.; Harris, Charles M.; Kesterke, Donald G.; Kral, Victor E.; Martin, Conrad; Newman, William J.; Olds, Edward B.; Ramsdell, Jack D.; Riva, John F.; Scott, James B.; Stephens, Robert W.; Swain, Robert L.; Tokoro, Atsuo; Wheeler, Alfred H.

1960s: Behnke, Jerold J.; Bonham, Harold F.; Bortz, Louis C.; Crain, John Russell; Eisinger, V. John; Harsh, John Franklin; Hartzog, Laurence David; Herber, Lawrence J.; Horton, Robert Carlton; Lohr, Lewis Stillman; Lyons, Mark S.; Sales, John Keith; Schryver, Robert F.; Schuyler, Donald Richard, II; Van Gilder, Kerry L.

1970s: Allen, Merrill Peter; Bateman, Richard L.; Blackburn, Wilbert Howard; Collins, Daniel E.; Cunningham, Alfred B.; Dinger, James Sheldon; Dunn, Anthony Charles; Edwards, John; Holmes, Paul J.; Houng-Ming, Joung; Johnson, James Mark; Lai, Eong-Lip; Malghan, Subhaschandra G.; Mitchell, A. Wallace; Mohler, Amy Szumigala; Smith, Thomas Edward; Spane, Frank A., Jr.

1980s: Tsui, Ping-Sheng

University of New Brunswick
Fredericton, NB   E3B 5A3

118 Master's, 44 Doctoral

1920s: Good, George A.

1930s: Brown, Douglas F.; Freeze, Arthur C.

1950s: Baldwin, Andrew B.; Benson, David G.; Brittain, William H.; Brown, Albert Anthony; Church, Joseph F.; Cummings, Leslie M.; Hachey, Philip Osmund; Hale, William Ernest; Johnston, J. F.; Langmaid, Kenneth K.; Magnusson, Donald Harry; Patterson, John Murray; Potter, Ralph Richard; Rutledge, Donald W.; Scott, Fenton J.; Sharpe, John I.; Smith, John C.; Stewart, Keith John; Sund, J. Olaf; Tupper, William M.

1960s: Ali, Sayed I.; Bennett, Gerald; Clark, George Sydney; Dastidar, Priyabrata Ghosh; Davies, John Leslie; Fraser, Douglas Culton; Globensky, Yvon Raoul; Globensky, Yvon Raoul; Grant, Alan Carson; Gunter, William Daniel; Hamilton, John Bonar; Harris, Frederick Robson; Helmstaedt, Herwart; Ikramuddin Ali, Syed; Jones, Robert Alan; Laughlin, C. H.; Lawson, David E.; Leavitt, Gene Millidge; Lee, David T. C.; McIlwaine, William Hardy; McNutt, James R. A.; Melvin, Robert L.; Mersereau, Terence Gerard; Nash, Walter A.; Ruitenberg, Arie Anne; Sevillano, Arturo Cabreros; Simpson, Peter Robert; van de Poll, Henk Wouter; Wilband, John Truax; Wilbrand, John T.

1970s: Austria, V. B., Jr.; Ball, F. D.; Bamwoya, J. J.; Barnett, D. E.; Bhatia, Dil Mohan Singh; Butt, Khurshid Alam; Chapman, R. P.; Cherry, M. E.; Chork, C. Y.; Crosby, R. M.; Donohoe, H. V., Jr.; Dryer, S.; Elhadi, N. D. A.; Erdogan, B.; Felder, F.; Fyffe, Leslie Robert; Gagnon, P.; Galanos, D. A. M.; Gandhi, S. M.; Garnett, John Arthur; Gemmell, D. E.; Goodfellow, W. D.; Gupta, V. K.; Howells, K. D. M.; Inasi, James C.;

Korpijaakko, Martti Jaakko; Kuan, Soong; Lahti, Howard Reino; Lahti, Howard Reino; Lajtai, N. V.; Lee, Hee Jin; Luff, W. M.; McBride, D. E.; McHaro, B. A.; McLeod, Malcolm John; Naing, W.; Nassar, M. M. A.; O'Brien, B. H.; Owsiacki, Leonede; Panayiotou, A.; Pilch, Peter George Henry; Porter; Pringle, G. J.; Punwasee, J. D. N.; Pwa, Aung; Rencz, A. N.; Richards, Nancy A.; Roulston, B. V.; Saif, S. I.; Stirling, John Alexander Robert; Subhas, Tella; Szabo, N. L.; Tejirian, Haig G.; Thomson, D. B.; Uthe, R. E.; Villard, D. J.; Wahl, John Lesslie; Wardle, R. J.; Webb, G. R.; Whitehead, Robert Edgar; Yoon, Tai Nam

1980s: Alison, Jamie Richard; Antonuk, Caroline-Nathalie; Burton, Donald MacLaren; Carter, David C.; Chandra, Jagdesh James; Chen, Yong-Qi; d'Orsay, Albert Murray; Dickson, William Lawson; Elliott, Colleen Georgia; Emory-Moore, Margot; Fay, Vincent Kevin; Fillion, Denis; Foley, Patricia Louise; Friske, Peter Wilhelm Bruno; Gao, Ruixiang; Hassan, Hassan-Hashim; Juras, Stephen Joseph; Kambampati, Mohan V.; Kettles, Karen R.; Keys, David Gerald; Lee, Don-Jin; Lee, Yuk Cheung; Leger, Albert Joseph; Lentz, David Richard; Lutes, Glenn Gordon; MacLellan, H. Elizabeth; McCann, Tommy; McDonald, Dean William Arthur; Mertikas, Stilianos P.; Miller, John Albert; Mugridge, Samantha-Jane; Nelson, Grant E.; Pagiatakis, Spiros Demitris; Parker, John Stephen Dawson; Pope, Christiana Sheldon; Rankin, Leslie Darrell; Rudnick, Barbara J.; Schneider, Dieter; Simpson, Evanna Lois; Smee, Barry Warren; Steeves, Robin Roy; Stupak, William A.; Tanoli, Saifullah Khan; Van Der Pluijm, Bernardus Adrianus; van Staal, Cees R.; Watson, Gordon Peter; Webb, Timothy C.; Young, Graham Arthur

University of New Hampshire
Durham, NH   03824

85 Master's, 12 Doctoral

1930s: Johnson, Ruth Helen

1960s: Drooker, Penelope B.

1970s: Armstrong, Peter B.; Bjerklie, David M.; Blomshield, Richard J.; Briggs, Peter B.; Bursaw, Richard B.; Dowse, Mary E.; Flight, Wilson R.; Gerla, Philip J.; Gundlach, Erich R.; Hanson, Gary M.; Haug, Frederick W.; Hensley, Carol; Herlihy, Daniel M.; Hill, David B.; Jackson, James R.; Johng, DuSik; Keene, Howard W.; Kolbe, E. R.; Malkoski, Mark; Mills, Thomas; Moore, Richard; Nelson, Eric G.; Nevins, Judith B.; Shanley, Gerard E.; Sharp, John; Shearer, Charles; Swenson, Erick M.; Talkington, Raymond; Thompson, Cheryl I.; Thornton, Jeffrey A.; Tollo, Richard; Trask, Richard P.; Tugal, Halil; White, Steven C.; Wilson, Kevin M.

1980s: Allen, Boyd, III; Bither, Katherine M.; Boxwell, Mimi A.; Britton, Kathy Booth; Brooks, John A.; Burack, Anna Camille; Campisano, Cynthia Dane; Carnese, Michael J.; Carrigan, John A.; Cheatham, Michael M.; Cheatham, Terri L.; Chormann, Frederick H., Jr.; Collins, Donald W.; Earl, Forrest C.; Eby, Richard Kerr; Eusden, J. Dykstra; Fagan, Timothy Jay; Falotico, Robert J., Jr.; Fanelli, Eileen M.; Fischer, Conrad G.; Foster, Jeffrey S.; Hancox, Gregory A.; Hassinger, Jon Miller; Hayward, Jennifer A.; Hedberg, Kim E.; Hines, Mark Edward; Howard, Carolyn Kheboian; Hurd, Michael L.; Jahrling, Chris E.; Kelly, William J., Jr.; Kim, Jonathan P.; Koch, Thomas J.; Larson, Albert O.; Leavitt, Karen M.; Lewis, Dion A.; Meese, Debra A.; Morency, Robert E., Jr.; Orem, William Henry, V; Piette, Marjorie A.; Regan, Terence R.; Rickerich, Steven F.; Roe, Gene Vincent; Rust, Lee D.; Sakakeeny, Stephen Anthony; Schmidt, Edward John; Schrager, Gene; Seely, Diana M.; Shevenell, Thomas Cortland; Shope, Steven B.; Simmons, J. A. Kent; Simmons, James Andre Kent; Simmons, Kent J. A.; Smith, Gordon McNeal, IV; Spader, David H.; Templeton, George Daniel, III; Tinkham, Daniel J.; Trombley, Thomas

J.; Wathen, John B.; Whitaker, Laura Rothenberg; Wood, Eric S.

University of New Mexico
Albuquerque, NM   87131

344 Master's, 73 Doctoral

1900s: Ross, Edmund

1930s: Bisbee, Wallace A.; Harrington, Eldred R.; McGuinness, Charles L.; Vann, Richard Pickard

1940s: Beaumont, Edward C.; DeCoster, George L.; Harrison, Earl P.; Hendren, Celia Faith; Johnson, Ross B.; Murphey, Leslie V.; Silver, Caswell; Silver, Leon Theodore

1950s: Ash, Henry O.; Baltz, Elmer Harold, Jr.; Bogart, Lowell Eldon; Borton, Robert L.; Bruns, John J.; Brunton, George Delbert; Bushnell, Hugh Pearce; Caldwell, John William; Cargo, David N.; Carten, Thomas L.; Chenoweth, William Lyman; de Cserna, Zoltan; Del Mar, Robert; Deurmyer, James Justin; Disbrow, Alan Eastman; Dixon, George H.; Emerick, William L.; Emmanuel, Robert J.; Fallis, John F., Jr.; Finney, Joseph J.; Fitter, Francis L.; Galloway, Sherman Elsworth; Gaskill, David L.; Givens, David B.; Goldsmith, Louis H.; Gratton, Patrick John Francis; Harbour, Jerry; Hayes, Philip T.; Hill, John Davis; Homme, Frank C.; Hutson, Osler C.; Jacobs, Richard C.; Johnston, Herbert C., Jr.; Jordan, Jack Gerald; Kirkland, Douglas Wright; Lookingbill, John L.; Lovejoy, Bill P.; Lustig, Lawrence Kenneth; Lyons, Thomas R.; Maxwell, Charles H.; McRae, Otis M.; Mills, Donald E.; Mohar, John, Jr.; Mutschler, Felix Ernest; Noble, Edwin A.; O'Sullivan, Robert Brett; Parry, Marshall Eugene; Perkins, Ronald Dee; Peterson, John W.; Rawson, Donald Eugene; Reynolds, Charles B.; Sears, Richard Sherwood; Sharp, Kenneth Denver; Smith, James Allen; Soister, Paul E.; Stevenson, Ralph G., Jr.; Swift, Ellsworth Rowley; Szabo, Ernest; Thompson, Sam, III; Toomey, Donald Francis; Ugrinic, George M.

1960s: Anderson, Jerome E.; Ash, Sidney R.; Atkinson, William W., Jr.; Baltz, Elmer Harold, Jr.; Bandoian, Charles Asa; Black, Bruce Allen; Blagbrough, John Wilkinson; Bowers, William E.; Bradbury, John Platt; Brady, John Francis, Jr.; Brown, Bert N.; Brown, William T., Jr.; Burgoyne, Alfred Alexander; Burton, Robert Clyde; Campbell, Jock Albert; Caprio, Eugene R.; Carter, James Franklin; Catacosinos, Paul A.; Chao, Pao-Chin; Clark, Kenneth Frederick; Clark, Kenneth Frederick; Clemons, Russell E.; Coney, Peter James; Cordell, Lindreth E.; Dean, Walter Edward, Jr.; Dean, Walter Edward, Jr.; Dilworth, Ottis L.; Dodge, Charles Fremont, III; Edmonds, Robert J.; Emerson, John W.; Farkas, Steven Eugene; Feinberg, Herbert; Fitch, David C.; Furlow, James Warren; Giles, David Lee; Gonzalez, Ralph Alan; Goolsby, Robert Stark; Gustafson, William G.; Haines, Richard Arthur; Harris, Arthur Horne; Hatchell, William O'Donnell; Headley, Klyne; Helms, Phyllis Borden; Hinds, Jim S.; Huber, James R.; Irwin, Charles Dennis; Johnson, Eldred; Kaplan, Seymour Fred; Kasiraj, Iyadurai; Kirkland, Douglas Wright; Kirkland, Peggy L.; Krimsky, Glenn A.; Lambert, Paul Wayne; Lambert, Paul Wayne; Lease, Robin Clair; Lisenbee, Alvis Lee; Lodewick, Richard J.; M'Gonigle, John William; Master, Pilsum Phiroze; McCleary, John T.; McLeroy, Donald F.; McSwain, James L.; Melvin, James W.; Melvin, Norman Wayne; Merkle, Arthur B.; Moore, Dwight G., Jr.; Morgan, J. R.; O'Neill, John Michael; Pastuszak, Robert A.; Petersen, John W.; Phillips, Charles Heulan; Pradham, Bi-swa M.; Rasho, John L.; Reddy, George R.; Rimal, Durga Nath; Ryberg, George Ernest; Saucier, Alva Eugene; Schowalter, Timothy T.; Seager, William R.; Shomaker, John Wayne; Simms, Richard W.; Singh, Yogendra L.; Smith, Eugene Irwin; Snider, Henry Irwin; Snyder, Don Otis; Squires, Richard Lane; Stukey, Arthur Herbert, Jr.; Sturdevant, James A.; Szabo, Ernest; Thompson, Tommy B.; Thompson, Tommy B.; Titus, Frank Bethel, Jr.;

Wagner, Lawrence Henry; Webster, David Alexander; Weilbacher, Carol A.; Werrell, William Lewis; Yale, Fred Roger

1970s: Affholter, Kathleen Ann; Aldrich, Merritt J.; Andersen, Richard L.; Anderson, John B.; Anderson, Orin J.; Arendt, Ward W.; Aubele, Jayne Christine; Bachhuber, Frederick W.; Baer, Roger Lawrence; Bangert, James C.; Berkley, John Lee; Billo, Saleh Mohammad; Black, Bruce Allen; Bolivar, Stephen L.; Bornhorst, Theodore Joseph; Brandwein, Sidney S.; Bregman, Martin Louis; Broomfield, Robin E.; Brown, H. Gassaway, IV; Broxton, David E.; Burroughs, Richard L.; Busche, Frederick D.; Cagle, Fred R., Jr.; Causey, James D.; Cook, Clarence W.; Corbitt, Lonnie L.; Crumpler, Larry S.; Deal, Edmond Graham; Dickson, John Richard; Dillon, John F.; Dorn, Geoffrey Alan; DuChene, Harvey R.; Durham, Jon A.; Edwards, Duncan L.; Enz, Robert David; Eppler, Dean B.; Erb, Edward Edeburn, Jr.; Feldman, Sandra C.; Flesch, Gary A.; Fodor, Ronald V.; Fullas, George H.; Gail, George Joseph; Gibson, Gail G.; Gooding, James Leslie; Goodknight, Craig S.; Gorham, Timothy W.; Graeber, Edward John; Grainger, James R.; Green, Jonathan A.; Hill, Carol A.; Hirt, Warren; Hlava, Paul Frank; Hock, Philip F., Jr.; Hoge, Harry Porter; Huss, Gary R.; Huzarski, Jan Ralph; Jagnow, David Henry; Kaharoeddin, Francis Amrisar; Kasten, James A.; Kaufman, William H.; Kessler, L. Gifford, II; Krohn, Douglas H.; Lambert, Raymond S.; Lawrence, John R.; Lee, Moon Joo; Lopez, David A.; Love, David Waxham; Love, Linda Lou A.; Lux, Gayle E.; Maldonado, Florian; Mannhard, Gregory W.; Mansker, William L.; Marple, M. L.; Martin, James Lee; Martinez, Ruben; Martinez, Ruben; Mason, James Trimble; McAnulty, William Noel, Jr.; Moore, Richard B.; Mukhopadhyay, Bimal; Nelsen, Craig J.; Northrop, Harold R.; Nuhfer, Edward B.; Perkins, Michael; Peterson, Stephen L.; Planner, Harry N.; Planner, Harry N.; Potter, Jared Michael; Purdue, Gary Lynn; Reed, Richard K.; Register, Joseph K., Jr.; Register, Marcia E.; Rhodes, Rodney Charles; Ricciardi, Karen; Riese, Walter Charles; Riesmeyer, William Duncan; Ross, Joseph Ray; Ruetschilling, Richard L.; Ryan, Edmond P.; Sayala, Dasharatham; Schumacher, Otto L.; Shaffer, William Leroy; Sibray, Steven Sherman; Sigleo, Anne M. C.; Skeryanc, Anthony J.; Slack, Paul B.; Smith, Christian S.; Smith, Eugene Irwin; Smith, Gregory Warren; Snyder, Don Otis; Soule, James McGovern; Spradlin, Ernest J.; Steinborn, Terry L.; Sullivan, Catherine E.; Swenson, David R.; Timmer, Robert Scott; Turner, William Morrow; Warren, Richard G.; Weinberg, David M.; Widdicombe, Roberta E.; Wilson, John E.; Woltz, David; Wortman, Richard A.; Zilinski, Robert E., Jr.; Zimmerman, Charles J.

1980s: Abashian, Mark S.; Abitz, Richard; Ander, Mark Embree; Atwood, Glen William; Banowsky, Bill Raymond; Barker, Steven E.; Bauer, Paul Winston; Beard, Robert D.; Bornhorst, Theodore Joseph; Brothers, Sara C.; Bryan, Charles R.; Bullard, Thomas Fitts; Cavin, William J.; Codding, David B.; Colvard, Elizabeth Monroe; Condit, Christopher Dana; Connolly, James R.; Cowen, Rachel; Coxe, Berton Woodward; Criswell, C. William; Croose, Daivd; Della Valle, Richard Saverio; Dickinson, Tamara L.; Dickinson, Tamara Lynn; Duggan, Toni J.; Emanuel, Karl M.; Erskine, Daniel W.; Farris, Stephen Robert; Ford, Richard Lee, III; Fulp, Michael S.; Gerety, Michael Thomas; Gilbeau, Kevin P.; Glenn, Rosemary Thompson; Grimm, Joel Patrick; Guilbeau, Kevin P.; Gutiérrez, Alberto Alejandro; Harvey, Bruce A.; Hester, Patricia M.; Hicks, Randall Thackery; Holbrook, John M.; Holcombe, Horace Truman; Hudson, Jeanna Sue; Hultgren, Michael Charles; Hutchinson, Peter John; Jercinovic, Devon Eldridge; Jercinovic, Michael J.; Jones, David Pierce; Karas, Paul A.; Kasten, Terri Ann; Kelly, John C.; Kelson, Keith Irvin; Kepes, Gerald Joseph; Kisucky, Michael J.; Krier, Donathon James; Lambert, Ellen E.; Leopoldt, Winfried; Leyenberger, Terry; Logsdon, Mark J.;

Longden, Markham R.; Longmire, Patrick; Looff, Kurt M.; Love, David Waxham; Lowy, Robert Michael; Lozinsky, Richard Peter; Maggiore, Peter; Mahoney, Maureen; Matheney, Ronald K.; May, S. Judson; Maynard, Stephen R.; McCarty, Rose Mary; McCormick, Kelli A.; McCormick, Tamsin Cordner; McCraw, David Jackson; McKinley, James P.; McKinley, Susan G.; Menges, Christopher M.; Merker, Robert Randall; Merrick, Margaret Anne; Miller, Jerry Russell; Murphy, Mark Thomas; Nimick, Karol Gillespie; North, Robert; Olsen, Clayton E.; Parchman, Mark Alan; Persico, John L.; Peterson, John L.; Picha, Mark Gregory; Pickle, John D.; Place, Jeannie Theresa; Plummer, David A.; Recca, Steven I.; Reiter, David Ernest; Renshaw, James L.; Riese, Walter Charles; Ristorcelli, Steven Joseph; Ritter, John B.; Rubin, Alan Edward; Sares, Steven William; Sarkar, Gautam Prasad; Schultz, Jerald David; Seaman, Sheila J.; Singer, Bradley Sherwood; Smith, Douglas Michael; Smith, Frank C.; Smith, Larry Noel; Smith, Lawrence Noel; Smith, Roger F.; Stein, Harlan L.; Steinpress, Martin Garth; Taylor, Keith R.; Trumbull, Robert Bruce; Vazzana, Michael Eugene; Vogler, Herbert A., III; Ward, David Barry; Weber, Jo Ann T.; Wentworth, Susan Jane; Wesling, John R.; White, William Dennis; Williams, Michael L.; Williamson, Thomas E.; Woodard, Thomas William

New Mexico Institute of Mining and Technology
Socorro, NM 87801

259 Master's, 47 Doctoral

1930s: Harley, George Townsend; Kelly, John Martin; Riddell, Walter; Stonerook, William H.; Terry, Owen W.

1940s: Gray, Ralph L.; Walter, Rudolph J.

1950s: Anderson, Richard C.; Berkley, Richard J.; Carlin, Joseph T.; Chen, Kung-Yung; Doyle, James C.; Frische, Richard H.; Gentile, Anthony L.; Hambleton, Arthur W.; Hand, John E.; Johnson, Dean; Johnson, James T.; Johnson, John Burlin, Jr.; Kalish, Phillip; Kintzinger, Paul Raymond; LaVergne, Michel; Lombardi, Oreste W.; McCaslin, John; Schilling, John Harold; Sykes, Robert E.; Williams, Duane H.; Winkler, Hartmut A.

1960s: Berg, Eric L.; Biles, Norman; Bingler, Edward C.; Brimhall, Ronald M.; Butler, Patrick, Jr.; Carapetian, Ara G.; Carman, John H.; Chen, Tse Pu; Clark, Richard D.; Columbus, Nathan; DeBrine, Bruce E.; Deju, Raul A.; Escalera, Saul J.; Evans, George Carman; Geddes, Richard W.; Ghazarian, Ghazar Boulos; Grace, Ken A.; Haederle, Wolfgagng F.; Halepaska, John C.; Hassan, Ahmad Amin Abdel Khalek; Herber, Lawrence J.; Hillard, Patrick D.; Howard, Edward Viet; Ibrahim, Abou-Bakr K.; Keeney, Joseph W.; Kershner, James D.; Kopicki, Robert J.; Latham, Don Jay; Long, Leland Timothy; Lowell, Gary Richard; Mallon, Kenneth M.; Marino, Miguel A.; McCrory, Roy; Olsen, Royce W.; Papadopulos, Istavros S.; Poe, Thomas I., III; Rabinowitz, D. Daniel; Ramananantoandro, Ramanantsoa; Ratcliff, Marvin W.; Reddy, Vavula Srinivas; Rejas, Angel; Renault, Jaques R.; Riese, Ronald W.; Roman, Ronald J.; Saad, Kamal Farid; Saleem, Zubair A.; Sheffer, Herman Weaver; Siddiqui, Wasit Ahmed; Soto-Vargas, M. Fernando; Spiegel, Zane; Stacy, Ann L.; Sullivan, William C.; Thayer, Donald D.; Tobey, Eugene Francis; Uy, Dominador C.; Visocky, Adrian P.; Wedekind, Frank E.; Williams, Dennis E.; Williams, Dennis E.; Wyckoff, Barkley Sudduth

1970s: Allmedinger, Roger J.; Allmendinger, Roger J.; Alptekin, Omer; Beaver, Donald W.; Beers, Charles A.; Binsariti, Abdalla; Birsoy, Rezan; Birsoy, Yuksel K.; Birsoy, Yuksel K.; Blodgett, Daniel D.; Bloom, Mark S.; Bolton, W. R.; Bonem, Rena Mae; Brown, Daniel R.; Brown, David McKendree; Bruning, James Earl; Buapang, Somkid; Cappa, James A.; Caravella, Frank; Cepeda, Joseph Cherubini; Chamberlin, Richard Martin; Charukalas, Banhan; Charukalas, Benja-

min; Cookro, T. M.; Croxell, Thomas Ray; Dee, Mark Philip; Duffy, Christopher J.; Durtsche, J. S.; Edwards, C. L.; Edwards, C. L.; Escorce, Eufredo B.; Ewing, Thomas; Faith, Stuart E.; Farquhar, Paul Thomas; Fischer, Jeffrey Allan; Fisher, Walter William; Fleischhauer, Henry Louis, Jr.; Flores, W. Adán Emigdio Z.; Flores, W. Adán Emigdio Z.; Gromer, James; Hartman, Harold Joseph, Jr.; Hassen-Bey, Tarak Mustafa; Haubold, Reiner G.; Hayslip, David L.; Hines, Stephen Anthony; Hoffman, George L.; Hook, Stephen Charles; Hook, Stephen Charles; Iovenitti, J.; Jacobson, Rudolph Harry; Jaramillo, L.; Jaworski, Michael John; Jenkins, David A.; Jones, Duke Forrest; Kandemir, Burhaneddin Hamit; Kapta, Mohammed Shafi; Koehn, Henry Hans; Koehn, Marsha A.; Lewchalermvong, Chettavat; McLafferty, Susan W.; Mercado, Abraham; Morley, Raymond L.; Morrison, Ronald C.; Mott, Richard P., Jr.; Niesen, Preston L.; Nogueira, Alexandrino Cosme; Oralratmanee, Komol; Park, David Eugene; Pawlowicz, Richard Melvin; Rabinowitz, D. Daniel; Rickman, Douglas Lee; Rinehart, Eric John; Rodphothong, Somphong; Roffman, Haia; Sakdejayont, Kiet; Sampattavanija, Suvit; Schiffman, Robert A.; Schnake, Carol J.; Schnake, David W.; Shearer, Charles Raymond; Shelburne, Kevin Lee; Shuleski, Paul J.; Siemers, William Terry; Simon, Donald Bruce; Singh, Surendra; Smith, Roger Dayton; Sonderegger, John L., II; Strachan, Donald G.; Suthakorn, Phairat; Swenson, David Howard; Taber, James Tobert; Thienprasert, Ammuayachai; Toppozada, Tousson Mohamed Roushdy; Umshler, Dennis B.; Vichit, Pongsak; Ward, R. M.; Way, Shao Chih; Wilkinson, William Holbrook, Jr.; Williams, Charles E.; Wongwiwat, Kraiwut; Woodward, Thomas Michael; Yousef, Ali Abdullah; Yuras, Walter; Zelinski, William P.

1980s: Abramson, Beth S.; Alford, Dean E.; Allen, Philip; Allen, Reid; Andres, Robert J.; Apodaca, Lori E.; Ardito, Cynthia Paula; Arkell, Brian W.; Baker, Bruce W.; Barker, David L.; Barrie, Donald Show; Batory, Bruce L.; Bauch, John H. A.; Bauer, Paul Winston; Behr, Christina B.; Bernhardt, Carl A.; Bickford, David A.; Bigelow, Eric A.; Bijak, Martin Kenneth; Boadi, Issac Opoku; Bobrow, Danny J.; Boryta, Mark D.; Bowie, Mark R.; Bowling, G. Patrick; Brouillard, Lee A.; Brown, Karen B.; Bruneau, Jeffrey Allan; Carmichael, Alan Barnett; Carpenter, Philip John; Clarkson, Gerry W.; Coffin, Greg C.; Colpitts, Robert Moore, Jr.; Cook, Kevin H.; Copeland, Peter; Cox, Eugene W.; Craig, Alan S.; Crow, Henry Clay; De Melas, John P.; Duffy, Christopher J.; Dunbar, Nelia W.; Eaton, Larry G.; Eggleston, Ted Leonard; Eggleston, Ted Leonard; Fagrelius, Kurt H.; Falkowski, Stephen Kenneth; Ferguson, Charles A.; Gabelman, Joan L.; Gerwe, Jeffrey E.; Gibbons, Thomas Lynn; Gibson, Thomas R.; Gil, April V.; Gramont, Bertrand; Guilinger, David R.; Hammond, Charles M.; Haney, Joseph M.; Harrison, Richard W.; Harrover, Robin D.; Hemingway, Mark P.; Hermann, Michael; Hill, Steve R.; Hunt, Adrian P.; Jenkins, John E.; Jochems, Theodore Paul; Kedzie, Laura L.; Klich, Ingrid; Krukowski, Stanley T.; Lanzirotti, Antonio; Laroche, T. Matthew; Laskin, John; Leavy, Brian David; Lemley, Kenneth Ray; Linden, Ronald M.; Little, Gregory E.; Lowe, Roger S.; Lowey, Robert Francis; Lozinsky, Richard Peter; Martell, Charles; Maulsby, Joe; McCrink, Marie Taaffe; McKallip, Curtis Jr.; McLemore, Virginia T.; Meeker, Kimberly A.; Minier, Jeffrie D.; Moore, James Allan; Nguene, Francois Roger; Noll, Mark R.; Noll, Philip D., Jr.; Perry, Patricia Lynn; Post, Tim E.; Purson, John D.; Raby, Andrew G.; Reed, James Robert; Reimers, Richard F.; Rice, James E.; Robinson, Richard W.; Robinson-Cook, Sylveen E.; Rundell, Bruce M.; Schafer-Perini, Annette L.; Sexsmith, Suzanne L.; Shellhorn, Mark A.; Sivakumar, Ramamurthy; Smith, Robert W.; Smith, Stewart; Spahr, Cynthia L.; Specter, Robert Michael; Spell, Terry Lee; Thacker, Mark Sloan; Wallin, E. Timothy; Ward, Theresia A.; Weber, L.

James; Wilkening, Lee; Wright-Grassham, Anne C.; Young, John D.; Zody, Steven P.

New Mexico State University, Las Cruces
Las Cruces, NM 88003

41 Master's, 13 Doctoral

1960s: Charles, Michel; Dominick, Wayne Paul; Freeman, Charles Edward, Jr.; Taylor, Andrew M.

1970s: Brown, Lionel F.; Daggett, Paul H.; Hills, R. G.

1980s: Al-Dabagh, Abdulsattar Rashied; Al-Dabbagh, Ahmed Assim; Avery, Donald C.; Beebe, Matthew A.; Beyer, Joan Brown; Bowman, Robert Stephen; Brown, Glen Arthur; Coons, Lawrence M.; Cravens, Daniel Lester; Daggett, Paul Henry; Dicey, Timothy R.; Donnan, Gary Thomas; Elkins, Ned Zane; Galemore, Joe A.; Grigsby, Jeffrey D.; Gross, James T.; Hooker, Andrew T.; Hyde, Tinka G.; Ilchik, Susan Emilie; Kaczmarek, Edward L., Jr.; Kalesky, John F.; Kies, Bouziane; Kolins, Warren B.; Kuo, Rong-Heng; Lemley, Irene Savanyo; Lohse, Richard L.; Lopez, Stephen G.; Lynch, Hugh David, Jr.; Mack, Pamela Diane Carpenter; Marvin, Peter R.; Mayer, Anton Bernard; Morse, Earl L.; Newcomer, Robert W.; Rasmussen, Keith A.; Rupert, Michael G.; Rybarczyk, Sandra M.; Sanders, Peter A.; Schaal, William Conrad; Serrag, Salaheddin Ali; Smith, James Richard; Smyth, Andrew H.; Snyder, Jay; Stageman, J. Christopher; Taylor, David W. A., Jr.; Walker, James D.; Williamson, Anne L.; Wilson, Gregory C.

University of New Orleans
New Orleans, LA 70148

139 Master's, 2 Doctoral

1970s: Barr, Richard Kevin; Barton, Eric Watson; Batzner, Jay C.; Bechtel, Michael Joseph; Best, Bryan M.; Brown, William Donald; Burns, Thomas Daniel; Carrol, Michael J.; Chou, Shuh-Dar Frank; Crecca, Arthur Joseph; Douglas, Chris W.; Draper, Louise Pierce; Hafner, Robert Otto; Hart, Jeanne C.; Lanier, Hershel Dale; Lanigan, Dennis Michael; Lassus, Roy E., Jr.; Lazzeri, Joel Joseph; Lemastus, Steven Wayne; Marr, John Donald; Marshall, Robert Harden; McDonald, Kent Charles; Ramirez, Guillermo; Ranson, William Albrecht; Scheldt, John Christian; Sgouras, John D.; Smith, Donnie Fay; Sticker, Edwin E.; Thornhill, Ronald Roger; Vinet, Marshall Justin; Wickstrom, Charles W.

1980s: Alford, Elizabeth V.; Altobelli, Randall Wilson; Aston, Robert Earl, Jr.; Avenius, Christopher Gerald; Aycox, Tracie Leigh; Baldwin, John Schuyler; Bankley, Erik Stefan; Barnes, Philips Jeffrey; Bathke, Susan Ann; Bilinski, Peter Walter; Billingsley, Arthur Lee; Blickwede, Jon Frederic; Bordelon, Irion, Jr.; Boronow, Thomas Carlton; Bothner, Bryan R.; Bourgeois, Jason; Brewster, Renee Harrison; Cagle, David Anthony; Caluda, Carol Ann; Cancienne, Gary Peter; Caravella, Joseph Christopher; Cochran, Walter House; Collier, Robert L.; Crocker, Jonathan A.; Dahl, William Martin; Dando, Kathryn Hickman; Danielson, Daryl Arthur, Jr.; Danielson, Stephen Eric; Deliz, Michael John; Ford, Bruce Hicks; Forsthoff, Gary M.; Fortunato, Kathleen Susan; Free, Brickey Rae; Frischhertz, Robert P.; Gaudet, Donald Joseph, Jr.; Gratton, Sara M.; Green, James A.; Gregory, Janice Lynne; Hamlin, Tracy L.; Hanan, Mark Allen; Hankinson, Peter Kent; Hansen, Stephanie Thomas; Harris, Nancy Jensen; Hernandez, Gilbert Xavier; Hill, John A.; Hogan, Patrick Joseph; Horner, Greg James; Horner, Rick Oliver; Jee, Jonathan Lucas; Johanson, David Bryan; Johnson, Anthony Gerard; Julius, Jonathan Fred; Kindred, Fred R.; Koenigsberg, Andrew M.; Krause, Karen Webber; Kuhn, Douglas Eugene; Lahiere, Leon; Lavoie, Dawn L.; Lee, Maxie Turner; Legendre, Kerry John; Leger, William R.; Levy, Stephen E.; Love, Timothy Christopher; Lustick, Charles Francis; Martin, Frank J.; Martinez, Paul Edwin; McCleery, Raymond Scott; McCoy, Melinda Delle; McDevitt,

Marybeth C.; McDonnell, Sheila Louise; McHugh, Alice Ellen; Meyer, Martha Grose; Moise, Theodore, Jr.; Moline, Myrna Marie; Moschella, Victor Charles; Murphy, Edward Joseph; Newby, Ray A.; Parker, John Arthur; Pinney, Reese Bruner; Powell, Lynn Gladieux; Prather, Bradford E.; Richards, James Anthony; Ritchie, Eric Lee; Rodriquez, Carlos Jose; Roesler, Toby Albert; Rog, Andre Mark; Rutherford, Gary J.; Ryan, William P., Jr.; Salisbury, Beverly J.; Sharpe, Karen Broderick; Shinn, Rory K.; Showalter, James Aswell; Skidmore, Charles M.; Smith, Charles Randy; Snavely, Richard K.; Soileau, Lyndon Sewell; Spencer, Jeffrey Allen; St. Romain, Samuel Joseph; Stilwell, Randy; Sura, Michael Anthony; Sura, Suzanne Hopkins; Thompson, Daniel Lee; Van Sant, Mary Jane; Wallace, Kevin John; Wayne, David Matthew; Whiddon, Deborah Justice; Witherspoon, Robert R.; Woods, Henry Harper; Woods, Marion Marshall; Young, Carl Michael

New York University
New York, NY 10003

93 Master's, 53 Doctoral

1920s: Barwick, Arthur R.; Barwick, Arthur Richardson; Behre, Charles H., Jr.

1930s: Bohlin, Howard G.; Butler, Bertram Theodore; Ellis, Brooks Fleming; O'Connell, Daniel Triggott; Thompson, Henry Dewey

1940s: Appleby, Alfred Noel

1950s: Baskerville, Charles Alexander; Behm, Juan Joaquin; Biel, Ralph; Blaik, Maurice; Brenner, Gilbert J.; Brock, Jerome A.; Clemency, Charles; Cliadakis, William C.; Cousminer, Harold Leopold; Dieterle, Gifford Aas; Dolgoff, Abraham; Feigl, Frederick J.; Fink, Sidney; Fournier, Jorge C.; Garet, Gerald H.; Geraghty, James J.; Germeroth, Robert; Grekulinski, Edmund F.; Guarraia, Ernest; Healy, Henry G.; Hemer, Darwin O.; Hughes, David D.; Hunt, Lynn Bogue; Isaacs, Thelma; Kratchman, Jack; Laufer, Arthur R.; Leffingwell, Harry A.; Myerson, Bertram L.; Niever, Emanuel J.; Pemsler, Paul; Polugar, Morton; Popper, Robert J.; Puig, Joseph Albert; Revel, Humbert S.; Robinson, Vincent H.; Ronai, Lili E.; Ronai, Peter; Rudin, Cyril; Sachs, Jules; Savage, E. Lynn; Schull, H. W.; Shishkevish, Leo; Sikka, Desh B.; Sperrazza, Joseph Thomas; Sperrazza, Josephine; Sulek, John A.; Talmadge, Thomas White; Tilson, Seymour; Ward, Richard F.; Wasserman, Gilbert; Weiss, Lawrence; Westerholm, Allan Sixten; Zui, Yuval

1960s: Atwater, Marshall A.; Baer, Norbert Sebastian; Bartlett, Grant Aulden; Baskerville, Charles Alexander; Blank, Efrom; Bollwinkel, Donald; Busenberg, Eurybiades; Chang, David; Chapietta, Richard; Charmatz, Richard; Charmatz, Richard; Cousminer, Harold Leopold; Danker, Jeanne A.; Dimopoullos, Thomas J.; Fischer, Hans J. E.; Gorycki, Michael A.; Hebard, Edgar B.; Hoiles, Edwin K.; Jaffe, Samuel; Johnson, George; Kaplin, Stephen, Jr.; Kimyai, Abbas; Koening, Martin; LeGeros, Racquel Zapanta; Loring, Arthur P.; Lubke, E. Ronald; Luce, Philip G.; MacKenzie, Michael E.; Meyers, Bernard M.; Modzeleski, Vincent E.; Molitor, William C.; Newman, Walter S.; Nicholas, John; Nichols, Douglas J.; Potter, Delbert E.; Quick, Allen N.; Russo, Grace-Louise M.; Saia, Robert; Schafer, Charles T.; Schafer, Charles Thomas; Schuberth, Christopher John; Segeler, Marie-Louise; Shishkevish, Leo J.; Sirkin, Leslie A.; Smith, Foster D., Jr.; Steineck, Paul; Tuozzolo, Peter A.; Van der Leeden, Fritz; Weiss, Dennis; Wietrzychowski, Joseph

1970s: Belling, A. J.; Bennett, B. G.; Bishop, Joseph Michael; Burckle, Lloyd H.; Cohen, Lawrence Kenneth; Costa-Ribeiro, Carlos Antonio De Leers; Georgens, Robert E.; Griscom, Clement A.; Hairr, L. M.; Herrera, Luis Enrique; Heusser, Linda E.; Florer, Howard, K. M.; Klonowski, John E.; Mackler, Anne; Mogolesko, Fred J.; Overland,

James Edward; Rachele, L. D.; Schaffel, Simon; Selywn, Stephen; Weiss, Dennis

1980s: Balter, Howard; Gaffin, Stuart Roger; Hazen, Robert Edward; Heusser, Calvin J.; Ibe, Ralph Anthony; Lei, Wayne; Linsalata, Paul; Loeb, Robert Eli; Lowney, Karen Anne; Lynn, Leslie Michael; Pavlakis, Parissis P.; Peteet, Dorothy Marie; Steiner, Werner; Volk, Tyler

University of North Carolina, Chapel Hill
Chapel Hill, NC 27599

302 Master's, 141 Doctoral

1900s: Allen, Risden Tyler

1910s: Brownson, Allyn R.; Cobb, William Battle; Davis, Martin Jones; Dobbins, Charles Nelson; Goldston, Walter L., Jr.; Lambert, Henry D.; Randolph, Eldred O.; Ray, Hubert Roy

1920s: Amick, Harold Clyde; Babb, Josiah Smith; Boyce, Henry Spurgeon; Bryson, Herman Jennings; Butt, William Horace; Cobb, William Battle; Davis, Harry Towles; Johnston, Claud Stuart; Lee, Samuel Bayard; Lineberry, R. A.; Lohr, Burgin Edison; MacCarthy, Gerald R.; MacCarthy, Gerald Raleigh; Marsh, Herman Earl; Martin, Irving Lee; Miller, Clarence Edmund; Miller, James Bennett; Powell, Thomas Edward, Jr.; Sasscer, Reverdy G.; Shearer, Ralph Duward; Slavens, Margaret Dever; Stuckey, Jasper L.; Thompson, Henry Travis; Walker, Carl Hampton; Webster, Maude Martha

1930s: Edwards, Richard Archer; Greene, Naomi Esther; Holland, William Thompson; Hornbeck, Ross Wright; Kesler, Thomas Lingle; Laird, Wilson Morrow; LeBaron, Philip Mallory; Maurice, Charles Stewart; Norburn, Martha Elizabeth; Straley, Harrison Wilson, III; Vitz, Howard Engeler; Watkins, Joe Henry; White, William Alexander; White, William Alexander

1940s: Berryhill, Henry Lee, Jr.; Berryhill, Louise Russell; Bloomer, Robert Oliver; Clair, Joseph Robinson, Jr.; Clark, O. S.; Hamilton, Daniel Kirk; Harrington, John Wilbur; Harrington, John Wilbur; Jeter, Douglas DeL.; Krinitzsky, Ellis Louis; Martin, Romeo Jarrett; McCampbell, John Caldwell; Steel, Warren George; Turner, W. N.; Walls, James Gray; Watkins, Joe Henry

1950s: Ballard, James Alan; Batten, Roland Wesley; Berry, Edward Clark; Bowman, Frank Otto, Jr.; Bright, Mont Jackson, Jr.; Brown, Charles Quentin; Charles, William Curtis; Clarke, Thomas Graham; Councill, Richard Jefferson; Dawson, Robert William; Ebert, Charles H. V.; Fleming, Robert Eugene, Jr.; Goedicke, Thomas R.; Gooch, Edwin Octavius; Griffin, George Melvin, Jr.; Heron, Stephen Duncan, Jr.; Hooks, William Gary; Johnson, Fritz Kreisler; Kirstein, Dewey S., Jr.; Madison, James Ambrose; Murchison, Roderick Goldston, Jr.; Powers, Maurice Cary; Powers, Maurice Cary; Reves, William Dickenson, Jr.; Simons, Merton Eugene; Sinha, Evelyn Zepel; Skean, Donald Minter; Smith, Ernest Marshall, Jr.; Tingle, Woodrow Wilson; Zablocki, Frank Stefan

1960s: Allen, David William; Allen, Gary Curtiss; Bartlett, Charles Samuel, Jr.; Benson, Paul Harrison, III; Benson, Paul Harrison, III; Bird, Samuel Oscar, II; Birkhead, Paul Kenneth; Brett, Charles Everett; Brett, Charles Everett; Burt, Edward Ramsey, III; Cabaup, Joseph John; Callahan, John Edward; Cazeau, Charles Jay; Centini, Barry Austin; Chalcraft, Richard George; Copeland, Charles Wesley, Jr.; Cunliffe, James Edwin; Curran, Harold Allen; Curran, Harold Allen; Dobbins, David Ashmun; Duke, James Alan; Eckhoff, Oscar Bradley; Edwards, Robert Wheless; Fallaw, Wallace Craft; Fallaw, Wallace Craft; Fisher, John Joseph; Fisher, John Joseph; Fleisher, Penrod Jay; Glover, Dale Prince; Guidroz, Ralph Robert; Guy, Samuel Cole; Hayes, Lawrence Douglas; Helwig, Jo Wilson; Hermes, Oscar Don; Hermes, Oscar Don; Hooks, William Gary; Jagannadham, Gollakota; Justus, Philip Stanley; Lemmon, Robert Edgar; Mallory,

William Mason; May, James P.; McKinney, Frank Kenneth; McKinney, Marjorie Jackson; Michalek, Daniel D.; Morgan, Benjamin Arthur, III; Pels, Robert John; Phillips, Edward Lindsey, Jr.; Pickett, T. E.; Pickett, Thomas Ernest; Price, Vaneaton, Jr.; Price, Vaneaton, Jr.; Randazzo, Anthony F.; Randazzo, Anthony F.; Rogers, Wiley Samuel, III; Schnitker, Detmar Friedrich; Scrudato, Ronald John; Singh, Harinder, Jr.; Snipes, David Strange; Snyder, John Frank; Staheli, Albert C.; Swe, Win; Swift, Donald Josiah Palmer; Syder, J. F.; Thayer, Paul Arthur; Wagener, Henry Dickerson; Walker, Kenneth Russell; Weigand, Peter Woolson; Woodas, Nicholas A.

1970s: Aghajanian, John Gregory; Anderson, Gary Dale; Avary, Katherine Lee; Bailey, Richard Hendricks; Bailey, Richard Hendricks; Banks, Roland Stewart; Baum, Gerald Robert; Baum, Gerald Robert; Bernstein, Howard Alan; Best, David Malcolm; Best, David Malcolm; Beyer, Paul Joseph; Black, William W.; Bland, Alan Edward; Boardman, M. R.; Bond, Paulette Alice; Brewer, W. S., Jr.; Byrd, William John; Cannon, Robert P.; Cavanaugh, Thom; Cavanaugh, Thomas Daniel; Chalcraft, Richard George; Clark, Tony Franklin; Constantino-Herrera, Sergio E.; Cooley, Tillman Webb, Jr.; Coron, Cynthia R.; Custer, E. S., Jr.; Dabbagh, Abdullah E.; Daniel, Charles Camp, III; Dischinger, James B., Jr.; Dittmar, Edward I.; Donaldson, Marybeth; Drez, Paul E.; Dunning, Jeremy D.; Edzwald, James Kenneth; Eichenberger, Nancy L.; Eralp, Atal Enerjin; Estes, Ernest Lathan, III; Falls, Darryl Lee; Ferguson, John F.; Filer, Jonathan K.; Frazier, William James; Fritz, Steven J.; Galipeau, Joan Mary; Goff, William T.; Greenberg, Jeffrey K.; Guarin, Gilberto; Hansen, Michael Wayne; Harper, Stephen Brewer; Harris, W. Burleigh; Hauck, Steven A.; Hayes, Michael John; Horton, James Wright, Jr.; Horton, James Wright, Jr.; Howell, David Ernest; Hu, Tse-chuang; Huang, Y. S.; Huf, William Langley; Huggett, Thomas K.; Hundley, Emily M.; Hurley, B. W.; Ibrahim, Yarub Khalid; Indorf, Christopher P.; Jackson, Robert Eugene, Jr.; Jagannadham, Gollakota; Justus, Philip Stanley; Katuna, Michael P.; Kirchgessner, David Arthur; Kline, Charles C.; Klinge, David Michael; Kurtz, Jeffrey Paul; LeFurgey, Edoris Ann; Lemmon, Robert Edgar; Lewis, Stanley Royce; Lewis, Wardell L.; Lindley, Julia Ione; Lozano, Hernando; Lyday, Travis Quinton; McGhee, George Rufus, Jr.; McHone, James Gregory; McKinney, Frank Kenneth; McNelis, David N.; Meyer, Dann; Miller, James A.; Mills, Hugh H., Jr.; Moller, Stuart A.; Moore, David Warren; Morrow, Hyland B.; Muangnoicharoen, Nopadon; Mulholland, P. L.; Mullins, H. T.; Murphy, C. J., III; Nagy, Richard Michael; Newton, Cathryn Ruth; Nielsen, Kent Christopher; Odom, Arthur Leroy; Otte, Lee J.; Park, Byong Kwon; Petree, David Hoke, Jr.; Phelps, Gertrude Gunia; Pirkle, William Arthur; Pirkle, William Arthur; Pratt, Lisa M.; Rappaport, Stephen M.; Rodriguez, Rafael W.; Roper, Paul James; Rush, James D.; Sando, Thomas W.; Schumacher, James David; Shew, Roger D.; Slaymaker, Susan Clark; Smeds, Russell Clarence; Staheli, Albert Clifford; Steele, Kenneth Franklin; Stein, Holly Jean; Stirewalt, Gerry Lewis; Stock, Carl W.; Sullivan, James G.; Thomas, Peter C.; Tillman, Peter Douglas; Trygstad, Joyce C.; Upchurch, Michael Lee; Wagener, Henry Dickerson; Walls, Richard A.; Wanger, Johnny P.; Waskom, John Dennis; Weigand, Peter Woolson; Whisnant, Jack Summey; Whitehead, Neil Harwood, III; Wilkinson, Sarah E.; Wilson, Augustus O'Hara, Jr.; Wooden, Joseph Lovell

1980s: Acker, Louis L.; Aelion, Claire Marjorie; Albert, Daniel Bruce; Amarasiriwardena, Dulasiri Dayananda; Bailar, Elizabeth F.; Ballard, J. A.; Barifaijo, Erasmus; Barrell, Sharon; Beckman, Steven C.; Belgea, Paul; Beuthin, John D.; Braun-Adams, Karla A.; Bulley, Enid Joan; Bullock, Walter Richard; Burns, Stephen James; Cabe, Suellen; Cabe, Suellen; Callahan, Elaine J.; Chacko, Thomas; Chanton, Jeffrey Paul; Chuilli, Allan T.;

Crill, Patrick Michael; Custer, Edward Scheid, Jr.; Dabbagh, Mohammad Eesa; Dabbagh, Mohammad Eesa; Davis, Ronnie McConnell; Diecchio, Richard Joseph; Diehl, Katharine Benkelman; Dover, Robert Allen; Druham, Robert M.; Dwiggins, George Albert; El Makhrouf, Ali A.; Eligman, Don; English, Brian L.; Erlich, Robert N.; Esawi, E. K.; Folsom, Cynthia Elizabeth; Ford, Margaret Meisburger; Freeman, John H.; Fryxell, Joan Esther; Gelinas, Robert L.; Giacomini, David; Gibson, Joan Reynolds; Glass, B. D.; Gloeckler, Emily Frances; Gregory, Joel P.; Griffin, William Timothy; Gruen, Mary Abbott; Gulley, Gerald Lee, Jr.; Hageman, Mark Robert; Hall, Jack Charles; Hodges, Rex Alan; Holland, Ann Elizabeth; Hughes, Elizabeth; Hughes, William Theodore; Hunt, Emily Lee; Hurst, Stephen D.; Jenkins, Joe Earl; Jolley, Richard Michael; Jones, Shannon Elizabeth; Kaczor, Laurel; Keoughan, Kathleen M.; King, Sharon Lynne; Kish, Stephen Alexander; Klump, Jeffrey Val; Kronenfeld, Kathi R.; Kurtz, Jeffrey Paul; Lewis, Sharon Elizabeth; Lindley, Julia Ione; Lineberger, David Howard, Jr.; Loomis, Dana Paul; Lyke, William LeRoy; Madzsar, Elizabeth Marie; Martin, Mark W.; McCarthy, Susan Mary; McDonnell, Daniel E.; McGill, Kathryn A.; McKee, L. H.; McNeese, Linda Roberts; Meltzer, Anne S.; Meyer, Scott C.; Mies, Jonathan Wheaton; Miller, Michael Byron; Mims, Charles Van Horn; Mitchell, Bruce T.; Monrad, John Raymond; Mueller, Joseph Fred, Jr.; Murray, Margaretha E.; Norwood, Daniel Lee; Nunan, Adrienne Nichola; Nunan, Walter Edward; Otte, Lee James; Parker, Ronald Alvin; Parks, John T.; Posey, Harry H.; Powell, Richard Justin; Prufert, Leslie E.; Read, Barry Steven; Rollins, Francis O'Rourke; Rossbach, Thomas J.; Russell, Merlin D.; Sanner, Wayne K.; Sansone, Francis Joseph; Sexauer, Mae Lynn; Shell, Jesse Allen; Shomo, Sarah J.; Simons, Jeffrey H.; Smelik, Eugene Alan; Solomons, Eugenie; Sorauf, Christine M.; Spruill, Richard Kent; Stagg, Julie Wenger; Stein, Holly Jayne; Steltenpohl, Mark Gregory; Strobel, John Stuart; Stroh, Patricia Tucker; Suayah, Ismail B.; Sullivan, Eileen M.; Supplee, Jeffrey A.; Taylor, Carl Alvin, Jr.; Tiety, P. J.; Tietz, Paul G.; Tobiassen, Richard Torre; Turner, Ryan David; Unger, Henry E.; Ussler, William, III; Ussler, William, III; van Camp, Scott Gregory; VanGundy, Robert D.; Wagstaff, Melvin D., Jr.; Waltman, M. R.; Weiland, Thomas Joseph; Weiland, Thomas Joseph; White, Elijah, Jr.; Whitehead, Ann Waybright; Whitehurst, Thomas M.; Whiting, Brian M.; Wilber, Robert Jude; Wild, Thomas; Wildman, Sally Ann; Williams, Nancy Susan; Witner, Thomas W.; Wood, Laura Fain; Wright, Sarah D.; Yeilding, Cindy Ann; Zastrow, M. E.; Zimmerman, Neil Jay; Zmoda, Andrew J.; Zylstra, Elise

North Carolina State University
Raleigh, NC 27695

141 Master's, 72 Doctoral

1930s: George, D'arcy Roscoe

1940s: Goedicke, Thomas Robert; Miller, Edwin Lawrence, Jr.

1950s: Bowerman, James Nelson; Farquhar, Clyde Randolph; Fortson, Charles Wellborn, Jr.; Hope, Robert C.; Houston, Robert Stroud, Jr.; Howard, Clarence Edward; Nixon, Edward Calvert

1960s: Carter, Berkeley Roger; Centini, Barry Austin; Collins, James Finnbarr; Cook, John Thomas; Craig, R. M.; Custer, Richard Lewis Payzant; Dickey, Jerry Bland; Dolman, Jan Dirk; Eaddy, Donald Workman; Gamble, Erling Edward; Goss, Don Woodson; Grannell, Dana Bradford; Gupton, Charles Pernell; Jarrett, J. T.; Leonard, Ralph Avery; Losche, Craig Kendall; Malcolm, Ronald Lee; Miller, James Anderson; Mills, Herbert Cornell; Nettleton, Wiley Dennis; Patterson, Orus Fuquay, III; Philbrick, Charles Russell; Poplin, Jack Kenneth; Ray, James Allen; Schauble, Carl Eugene; Smith, Ronald Ellis; Soileau, John Millard; Stafford, Donald Bennett; Ung, M. K.; Upchurch, Clyde Neil; Wilkins, Richard Llewellyn

1970s: Abdelzahir, Mohmed Abdelzahir; Anderson, Michael B.; Babiker, Hashim Musa, II; Barberio, Stephen John; Basaran, A. K. T.; Bennett, Earl Healen, II; Birdseye, Richard Underwood; Bliven, F. L., Jr.; Bowman, John Thomas; Boyd, Harry William; Brown, James Eugene; Burnette, John Paul, III; Carpenter, Perry Albert, III; Carrilho, Cid; Chang, K.-R.; Cox, Gary Wriston; Dodge, William Stuart; Eiumnoh, A.; El-Samani, Karimeldin Z.; Fernandez, Louis Osvaldo; Gayer, Martin Jerome; Green, George Bruton, Jr.; Gryta, Jeffrey John; Harding, Sherie Cerise; Harrison, W. G.; Harvey, Bruce Warren; Harvey, Constance St. Clair; Holman, Robert E., III; Julian, Edward L.; Julian, Louise Chandler; Kautzman, Robert R.; Kays, B. L.; Lewis, Sharon E.; Liggon, George Herbert; Lugo, Hector Manuel; Mann, William Rhodes; Marcuson, William Frederick, III; Marley, Walter Ellis; Masterson, Robert P., Jr.; McDaniel, Ronald Dean; McDonald, Robert Lacy, Jr.; Mohamed, Magzoub Ahmed; Nivargikar, Vasantrao R.; Osman, Abdelaziz A.; Philen, O. D., Jr.; Radwan, A. M.; Rochester, Eugene Wallace, Jr.; Rosenboom, Arthur Kenneth; Shiver, Richard Steven; Singletary, Henry McLean; Slowey, Austin Henry; Smith, Bill Ross; Stanley, Larry Gerald; Stone, William Leroy; Teseneer, Ronald Lee; Tilden, Jean Ellen; Trexler, Bryson Douglas, Jr.; Weant, George Edward, III; Westbrook, Stephen Henry; Wilson, Joseph Raymond; Wilson, Thomas Virgil

1980s: Abu-Jaber, Nizar Shabib; Alegre, Julio Cesar; Alvarado, Alfredo; Angle, David G.; Argenbright, Dean Nelson; Bellis, Brian James; Bellis, Caroline Johnson; Blake, David Edward; Bolich, Richard E.; Boltin, William Randolph; Bowden, William Breckenridge; Brown, John Michael; Cambareri, Greg; Canavello, Douglas A.; Clark, Lindsey D.; Cookey, Melanie; Dadgari, Farzad; Delorey, Catherine Marie; Denton, Harry Paul; Dineen, Michael J.; Ditbanjong, Sandusit; Dolfi, Robert Michael; Doucette, William Henry, Jr.; Edelman, William Dennis; El-Idrissi, Mirghani Elsayed; Elbashir, Mohamed M. Elhassan; Fenaish, Taher Ali; Fontes, Mauricio Paulo Ferreira; France, Noelle A.; Ganesan, Sudalaimuthu; Gaylor, Robert Marshall; Gent, James Albert, Jr.; Gilbert, Lewis Edward; Godfrey, Joseph E., Jr.; Graham, Robert C.; Heartz, William Thomas; Helou, Amin Habib; Hicks, Henry Thomas, Jr.; Hoag, Robert Eugene; Hunsberger, Gloria Grace; Ibanga, Iniobong Jimmy; Idris, Eltahir Osman; Jacobs, Robert Sanger; Jernigan, Bruce Lee; Johnson, Germaine P.; Kite, Lucille Eggborn; Kuehl, Steven Alan; Layas, Fathi Mohamed; Lee, Sang-Ho; Lent, Robert M.; Lentz, Leonard James; Liddle, Susan Krongold; Lins, Ibere Delmar Gondim; Lynn, LaRee, Jr.; Lyon, Henry Wortham; McDade, Joel A.; McKee, Brent Andrew; McKinney, Richard Bowen; Meade, James Sherwood; Miller, Richard Lincoln; Mohammad, Fawzi Said; Moniz, Antonio Carlos; Moore, Billy R.; Moye, Robert Josephus, Jr.; Muldoon, William James; Nelsen, John Edward, Jr.; Nelson, Thomas M.; Niederreither, Michael Scott; Oates, Kenneth Michael; Otts, Charlotte; Pait, Eugene D.; Pal, Sakti K.; Palczuk, Nicholas C., Jr.; Parsons, John Edward; Payne, Robert A.; Powers, Jonathan Andrew; Purisinsit, Pitsamai; Putcha, Sastry Purnanjaneya; Rebertus, Russell A.; Reed, James Patrick; Reynolds, Robin Raible; Rice, Thomas John, Jr.; Robbins, Kevin Douglas; Rudek, Evelyn Anne; Seyedghasemipour, Seyedjavad; Shumac, Karen May; Siedlecki, Mary; Simmerman, Graham Hanson, Jr.; Smith, Bradley Wayne; Smith, Sandra Leslie Rhyne; Solomah, Ahmed Gabr; Son, Kang-Hyee; Southhard, Randal Jay; Spanjers, Raymond Peter; Stephens, Edward Harrison; Strum, Stuart; Subagio, Hardjosubroto; Tang, Pei-Tau; Tew, Katherine Hine; Toth, John C.; Ubiera Castro, Antonio Amilcar; van der Meyden, Hendrik Jan; van Es, Harold Mathijs; Vetter, Scott Keith; Warwick, Peter Delawet; Whipkey, Charles Evans; Wilson, Thomas M.; Womack, Bernard Anderson;

University of North Dakota

Wylie, Albert Sidney, Jr.; Yang, Wan-Fa; Zickus, Thomas Arunas

University of North Dakota
Grand Forks, ND 58202

178 Master's, 49 Doctoral

1910s: Quirke, Terence Thomas

1930s: Alpha, Andrew G.; Lohn, Cecil O.; Peterson, Hjalmer V.; West, Philip W.

1940s: Thorsteinsson, Thorsteinn

1950s: Crosby, Spurgeon C., II; Cvancara, Alan M.; Erickson, Roland I.; Hall, Gary Owen; Hansen, Dan E.; Haraldson, Harald C.; Meldahl, Elmer G.; Petter, Charles K., Jr.; Russell, Alice E.; Schmitz, Emmett Richard; Wilson, Everett E.

1960s: Bakken, Wallace E.; Ballard, Frederick V.; Block, Douglas A.; Bonneville, John W.; Callender, Edward; Callender, Edward; Carlson, Clarence G.; Chmelik, James; Clark, Michael B.; Clayton, Lee; Crawford, Jack W.; Delimata, John J.; Dow, Wallace G.; Erickson, John Mark; Faigle, George A.; Feldmann, Rodney M.; Feldmann, Rodney M.; Friestad, Harlan K.; Frye, Charles Isaac; Gustavson, Thomas C.; Hamilton, Thomas M.; Harrison, Samuel Sterrett; Harrison, Samuel Sterrett; Kresl, Ronald J.; Kume, Jack; Madenwald, Kent A.; Marafi, Hussein; McCollum, Morris J.; Merritt, James C.; Morgan, Douglass H.; Nielsen, Dennis Niels; Pernichele, Albert D.; Reishus, Mark; Reith, Howard Cartnick; Royse, Chester Franklin, Jr.; Rude, LaVerne C.; Salisbury, Richard A.; Sanjines, Raul; Sherrod, Neil R.; Sigsby, Robert J.; Tinker, John R., Jr.; Tuthill, Samuel James; Tuthill, Samuel James; Vig, Reuben J.; Walker, Thomas F.; Williams, Barrett J.; Willson, Robert G.; Ziebarth, Harold Clarence

1970s: Anderson, Curtis A.; Arndt, Michael B.; Bickley, William B., Jr.; Bickley, William B., Jr.; Bjorlie, Peter F.; Bluemle, John P.; Bluemle, Mary E.; Brekke, David W.; Brinster, Kenneth F.; Butler, Raymond Darrell; Camara, Michael; Caramanica, Frank Phillip; Carroll, Kipp W.; Cherven, Victor B.; Cook, C. W.; Deal, Dwight Edward; Degenstein, Joel A.; Delimata, John J.; Erickson, John Mark; Erickson, Kirth; Fashbaugh, Earl F.; Fenner, William E.; Fenner, William E.; Fulton, Clark; Furman, Marvin J.; Grenda, James C.; Groenewold, Gerald Henry; Groenewold, Gerald Henry; Groenewold, Joanne Van Ornum; Hagmaier, Jonathan Ladd; Hamilton, Thomas M.; Harris, Kenneth L.; Harris, Kenneth L.; Heck, Thomas J.; Hemish, LeRoy A.; Hickey, William K.; Himebaugh, John P.; Hobbs, Howard; Hobbs, Howard C.; Honeyman, Leslie R.; Johnson, Kent A.; Johnson, Robert Post; Kornbrath, Richard W.; Kulland, Roy E.; Lee, David Robert; Malick, Kenneth C.; Malmquist, Kevin Lee; Meyer, Gary N.; Moore, Richard B.; Morin, Kevin; Nielsen, Dennis Niels; Novak, Robert M.; Okland, Howard E.; Okland, Linda E.; Pederson, Darryll T.; Perkins, Roderick L.; Pilatzke, Richard H.; Ramsey, Bruce L.; Ray, John T.; Reede, Roger John; Richardson, Rondald K.; Sackreiter, Donald K.; Scattolini, Richard; Scattolini, Richard; Schulte, Frank J.; Schulte, Frank J.; Scott, Mary Woods; Smyers, Larry F.; Starks, T. L.; Steiner, Mark A.; Stone, William J.; Tinker, John R., Jr.; Ulmer, James H.; Van Alstine, James Bruce; Walsh, Robert G., Sr.; Wehrfritz, Barbara D.; Wosick, Frederick D.; Ziebarth, Harold Clarence

1980s: Anderson, Douglas B.; Anderson, Garth S.; Bailey, Palmer K.; Baumann, Rodney M.; Beal, William A.; Beaver, Frank W., Jr.; Beaver, Frank W., Jr.; Boettger, William M.; Bulger, Paul R.; Butler, Raymond Darrell; Catt, Diane M.; Chipera, Steve J.; Dumonceaux, Gayle M.; Durall, Rebecca L.; Falcone, Sharon K.; Farris, Robert A.; Ferguson, Lori J.; Fischer, David W.; Forsman, Nels Frank; Giddings, Steven D.; Halle, Richard E.; Halvorson, Don Llewellyn; Hayes, Michael D.; Henke, Kevin R.; Hoganson, John W.; Huang, Yue-Chain; Huber, Timothy P.; Iverson, Louis Robert; Jenner, Gordon A., Jr.; Kelley, Lynn Irvin; Kenaley, Douglas Scott; Kleesattel, David R.; Langtry, Tina M.; Larsen, Richard A.; Lindholm, Rosanne M.; Lobdell, Frederick K.; Lobdell, Frederick K.; LoBue, Charles L.; Loeffler, Peter T.; Logan, Katherine J.; Lord, Mark Leavitt; Lord, Mark Leavitt; Luther, Kathryn C.; Luther, Mark R.; Maletzke, Jeffrey D.; Mayer, Gale G.; Millsop, Mark D.; Murphy, Edward C.; Nesemeier, Bradley D.; O'Toole, Frederick S.; Obelenus, Thomas J.; Parsons, Michael W.; Perkins, Rodney K.; Perrin, Nancy A.; Pound, Wayne R.; Prichard, Gordon H.; Quinn, Christopher F.; Reiskind, Jeremy; Robinson, Scott E.; Ronnei, David M.; Roob, Christine K.; Sandberg, Brian S.; Schnacke, Arthur W., Jr.; Schwartz, Dirk Anson; Seidel, Robert Eugene; Steadman, Edward N.; Stephens, Randall A.; Sturm, Stephen D.; Thompson, Stephen C.; Thrasher, Lawrence C.; Van Alstine, James Bruce; Wallick, Brian P.; Wartman, Brad L.; Waters, Douglas L.; Webster, Rick L.; White, Stanley F.; Wilkinson, Michael; Williams, David L.; Winbourn, Gary D.; Winczewski, Laramie Martin; Young, Daniel R.; Zimmer-Dauphinee, Susan A.; Zygarlicke, Christopher J.

North Dakota State University
Fargo, ND 58105

2 Doctoral

1980s: Skarie, Richard Luther; Wolf, James Kurt

University of North Texas
Denton, TX 76203

1 Master's, 2 Doctoral

1980s: Bates, Tim Frank; Stuart, Tom Jeffrey; Zgambo, Thomas Patrick

Northeast Louisiana University
Monroe, LA 71209

94 Master's, 1 Doctoral

1960s: Birdwell, Maurice Nixon; Cook, William T.; Geissler, Edwin L.; Harper, Edward S, III; Heard, Edward T.; Holaway, Rose Mary White; McGowen, Clyde B.; Newton, Albert N.; Nolan, Donny Ray; Stover, Stewart L.

1970s: Abbott, William Harold, Jr.; Baria, Lawrence Robert; Bender, Russell B., Jr.; Bergeron, William M.; Breard, Sylvester Q.; Burton, Jacqueline C.; Carson, David B.; Dick, Jay D.; Divine, Douglas Wayne; Doss, Daniel L.; Earl, Michael W.; Edwards, Gerald; Farmer, Ronald B.; Fiddler, Linda Carol; Fluker, James C., III; Galya, Thomas A.; Gann, Delbert Eugene, Jr.; Ghazizadeh, Mahmood; Hall, Johnny L.; Harrison, Peter F.; Harrison, Susan B.; Jehn, Paul J.; Kavanaugh, Ernest G.; Kucsma, Paul J.; Kuentag, Chumpon; Kumanchan, Prasert; Lasuzzo, Anthony; Mabibi, Mohammad J.; Maxwell, Garry S.; Maxwell, Glenn B.; McIntyre, Kenneth E., Jr.; Meyer, Thomas; Mills, John Robert; Mohayej, Zeinalabedin; Nichols, George H.; Ragland, Kenneth E.; Reagan, Wiley S.; Schell, Roy T., III; Sheppard, Don G.; Smith, March E.; Smith, Thomas G.; Sobba, Donald H.; Steed, Michael O.; Stringer, Gary L.; Trudnak, Gerald S.; Venso, Nolan J.; White, Joe R., Jr.; Wooley, Jerry D.; Young, Jay D.; Young, Robert C.

1980s: Alexander, Nancy Jo; Alston, Mickey Wayne; Amerson, Michael Daniel; Barnett, Stanley David; Boyd, Oliver Ray; Breen, David John; Brewer, Bobby Lee; Carter, Jesse Louis, Jr.; Doney, Kim Elizabeth; Fontana, Ronald Victor; Hippensteel, David Lee; Hopper, Jack T., Jr.; Kelly, Glen Eric; Kimball, Colin Edward; Latson, Rebecca Lynn; Leonard, Clifford, Jr.; Loggins, Susan Karleane Jones; Mathews, James Clay; McDonald, David Wilson; McPhearson, Ronald Dean; Mittler, Richard Wayne; Morris, Bradley Allen; Putimanitpong, Supalak; Ratchford, Timothy Daniel; Reid-Green, John Douglas; Reising, Gayle Angela; Shulkaew, Pitak; Simmons, Marlys Gail; Slapp, Kevin P.; Smith, Mark Newton; Smith, Ricky; Townsend,

Roger Neal; Whitmore, Randall Paul; Wilson, Guy D.; Wiygul, Gary J.

Northeastern Illinois University
Chicago, IL 60625

20 Master's, 1 Doctoral

1970s: Affolter, R. H.; Ahuja, H. S.; Bankole, B. O.; Barnett, T. R.; Berlin, L. A.; Maillis, A.; Mishkin, L. A.; Ricou, Michel L.; Sipiera, P. P., Jr.; Sotonoff, M. L.; Stromdahl, A. W.; Thompson, P. C.; Thomsen, K. O.; Thudium, C. L.

1980s: Bader, Roger H.; Chalokwu, Christopher I.; Kay, Mary Ann; Kuecher, Gerald Joseph; Peczkis, Jan; Pranschke, Frank A.; Sapper, Samuel E.

Northeastern University
Boston, MA 02115

1 Master's

1970s: Silvia, M. T.

Northern Arizona University
Flagstaff, AZ 86011

161 Master's, 4 Doctoral

1960s: Scholtz, Judith Fessenden; Thompson, John Robert., Jr.

1970s: Abendshein, Mark; Alexander, William J.; Auld, Thomas W.; Benfer, Jon Alan; Billingsley, George H., Jr.; Boyce, Joseph Michael; Bradshaw, Emily C.; Briscoe, Melanie; Broomhall, Robert W.; Burgert, Barrett L.; Celestian, Susan Myers; Cheeseman, Raymond J.; Condit, Christopher D.; Corken, R. James; Dalton, Russell O., Jr.; Daneker, Thomas M.; Emmons, Patrick Jay; Flinn, Douglas Lowell; Gambell, Neil Austin; Gebel, Dana Carl; Geesaman, Richard Carl; Gilmont, Norman L.; Gomez, Ernest; Harrison, Gary Clyde; Hendricks, John D.; Henkle, William R., Jr.; Hereford, Richard; Hewett, Robert Lewis; Himanga, James Carlo; Hughes, Scott Stevens; Jenness, M. I.; Kent, W. Norman; Klockenbrink, Thomas L.; Kluth, Charles F.; Knox, Debra; Koval, David B.; Kovas, Edward J.; Lane, Charles L.; Langman, James W., Jr.; Lewis, Norman M., Jr.; Light, Thomas D.; Lufholm, Peter Henry; Marshall, Donald R.; Martinsen, Randi S.; Matthews, John Joseph; McCabe, Kirk; McCain, Ronald Gordon; McDonald, Sandra D.; Murray, Kent Stephen; Peacock, Edward W.; Pettengill, James G.; Pope, Robert William; Pugmire, Ralph U.; Reed, V. Stephen; Reid, Reginald E.; Rogers, Thornwell; Sanders, David E.; Schmidt, Gregory Thomas; Scott, Kenneth Charles; Scott, Phyllis Wilk; Smith, Jeffrey William; Staab, Robert F.; Stevenson, Gene M.; Turner, Christine E.; Voorhees, Brent J.; Walters, Kenneth Allan

1980s: Altany, Robert M.; Alvis, Mark R.; am Ende, Barbara Ann; Baughman, Richard Lee; Bayne, Bradley J.; Benz, Sandra; Brady, Timothy Brian; Brathovde, James Edgar; Bremner Cramer, Jane Alison; Burns, Beverly Ann; Caputo, Mario Vincent; Carr, David Alan; Carr, James E.; Cassell, David Terrance; Chapman, Mary G.; Cheevers, Craig W.; Christensen, Paul K.; Cloud, Robert A.; Cluer, Brian L.; Cunion, Edward Joseph, Jr.; Czaplewski, Nicholas Jay; Darrach, Mark Edward; Davey, John Raymond; Day, James E., II; Dennis, Michael D.; Dew, Elizabeth Ann; DeWitt, Ronald H.; Donchin, Jason H.; Doty, Robert L.; Duffield, James A.; Edwards, David P.; Espegren, William A.; Farrar, Christopher D.; Frank, Andrew Jay; Gallaher, David W.; Glotfelty, Marvin Frank; Gonzales, David Alan; Gordon, Michael; Granata, James Samuel, II; Gray, David S.; Gubitosa, Richard; Gustason, Edmund R.; Hall-Burr, Marty Joanne; Haney, Eileen M.; Hardy, James A., Jr.; Haslett, James M.; Henningsgaard, Jeffrey J.; Hobbs, William H.; Hopkins, Ralph L.; Horstman, Kevin C.; Hughbanks, Julia A.; Ismail, Razimah; Jensen, Tim R.; Johns, Michael E.; Kirkland, James I.; Krokosz, Michael; Lewis, Patrick R.; Lieblang, Sean;

Lockrem, Timothy Mark; Luttrell, Patty Rubick; Martin, Daryl Lynn; McAllen, William R.; McEwen, Alfred S.; Muehlberger, Eric William; Nealey, Lorenza David; Noll, John H.; Ohlman, James R.; Olesen, James; Ordonez, Steve; Pendergrass, T. Michael; Peterson, Richard Robert; Pierson, Lowell Craig; Porter, Michael L.; Pregger, Brian H.; Puls, David Donald; Richards, Alexander Moreno; Roller, Julie Ann; Sargent, Colleen G.; Schmid, Karl J.; Sherlock, Sean M.; Shirley, Dennis H.; Skelly, Michael F.; Smith, Mark A.; Sutphin, Hoyt Baldwin; Sydow, Marc Wolfgang; Tinl, Teresa J.; Trujillo, Alan P.; Tucker, Mary L.; Vance, Richard; Vonderharr, Jerry; Vrba, Sheryl L.; Waters, Jeffrey Phillip; Weisman, Melanie Custer; Weiss, Gayle C.; Werme, Douglas R.; Wilder, William; Winkler, Fred E.

University of Northern Colorado
Greeley, CO 80639

3 Master's, 3 Doctoral

1970s: Augustin, B. D.; Howell, Jack W.; Huppert, G. N.; Shannon, W. J.

1980s: Callahan, Debra; Schofield, Donald W.

Northern Illinois University
De Kalb, IL 60115

192 Master's, 4 Doctoral

1950s: Bowden, Kenneth Lester; Deisner, Richard Herbert; Lounsbury, John Thomas

1960s: Burkart, Michael R.; Comella, Joseph Robert; Cwiak, Ronald Alvin; Davis, Carl G.; Dilamarter, Ronald R.; Englishman, Doanld Ellsworth; Gilberto, Richard Joseph; Gooden, Don C.; Hesler, James L.; Horall, Kenneth Bruce; Johnson, Donald Otto; Kirby, Emery, Jr.; Kveton, Edward J.; Lebo, Marvel Hope; Lukert, Michael T.; Mudrey, Michael George, Jr.; Palm, Richard Steven; Rubner, David Paul; Russell, Rick Harold; Sanders, Richard Pat; Simonis, Edvardas Karolis; Size, William Bachtrup; Solyom, Val; Sowayan, Abdulrahman M.; Tabor, Richard L.; Vail, Ronald Grant; Warner, Paul Freeman; Wenzel, Robert John; Zacate, Michael Everett

1970s: Abraham, Dwaine G.; Aleman, Antenor M.; Allen, Brian L.; Allen, Brian L.; Ankenbauer, Gilbert A., Jr.; Arola, John L.; Bakos, Nancy A.; Baynas, Christopher H.; Behrens, Gene K.; Brady, Howard Thomas; Carlsen, Greg M.; Carlson, David Roy; Catlin, James E.; Clark, Clifford Charles; Crown, Robert W.; Davis, Harold G.; Di Paolo, William Dominic; Distefano, Mark; Dittman, Fred Melvin, Jr.; Dugan, Joseph P., Jr.; Dyman, Thaddeus Stanley; Eble, Edward; Ericksen, Rick L.; Fedor, Dennis George; Flurkey, Andrew J.; Ford, William Harry; Gensmer, Richard P.; Georgiou, John C.; Ginsel, Marvin G.; Graff, Paul; Hamilton, Charles T.; Harrison, William James; Hartzell, Stephan P.; Hiatt, John Ludlow; Hirst, Brian; Huffman, Kenneth Jay; Hughes, Dolores M.; Hunter, William Clay; Jackson, Jeffrey K.; Jensen, Thomas E.; Kenney, Robert John; Kiester, Scott A.; Kilbourne, John Lyle; Koch, Michael Robert; Kohsmann, James J.; Kramer, David J.; Lahti, Victor R.; Lassin, Richard J.; Leedy, John B.; Levay, Joseph; Levy, Thomas M.; Lucey, Keith J.; Majewski, David G.; Maroney, Michael H.; McBroom, Mark N.; McCarthy, John Patrick; McLeod, John David; Metz, Robert Louis; Metzger, Robert; Meyer, David F.; Montgomery, Gerald Edward; Morgan, Brenda E.; Nienkerk, Monte M.; Niewold, Cary L.; Norman, Lonnie Dale; Oberts, Gary Leonard; Olson, Donald L.; Olson, Robert K.; Osby, Donald R.; Oskvarek, Jerome David; Paarlberg, Norman; Palmer, Paul W.; Pederson, Dale Russell; Peters, Walter G.; Peterson, Kent A.; Placher, George A.; Pokorny, Harvey Dreifuss; Pottorf, Robert J.; Proctor, Kenneth E.; Raidl, Robert F.; Read, David L.; Realini, Michael J.; Rongitsch, Brian A.; Schafersman, Jacqueline Sue; Schroeder, David Alan; Suda, Robert U.; Swulius, Thomas; Terrell, Bruce C.; Timm, Ronald W.; Tschopp, D. G.; Vendl, Law-

rence J.; Voss, Patrick Charles; Wallace, Robert J.; Wamser, Robert Charles; Ward, Barbara L.; Wegrzyn, Richard S.; Wheeler, Robert B.; Willand, T. N.; Wolf, Michael Gene; Wolffing, Craig L.; Wrenn, John Harry Wycoff; Yagishita, Koji

1980s: Andrews, William J.; Bahr, Charles H.; Brossman, James J.; Burdelik, William Joseph; Caithamer, Celine E.; Cartwright, George C.; Clark, Donald L.; Cowan, Ellen Anne; Cowan, Ellen Anne; D'Agostino, Anthony; de Prado, Connie A.; Docka, Janet Anne; Engelhardt, Nancy L.; Fasnacht, Timothy Lee; Fengler, Timothy A.; Fleming, Alfred John; Frederick, Daniel; Goldman, Barbara Ellen; Grundl, Timothy J.; Hank, Robert Allen; Heiny, Janet S.; Hemzacek, Jean Marie; Joslin, Peter Schuyler; Kamm, John L.; Kapchinske, John M.; Kiester, Jeffrey A.; Klewin, Kenneth Wade; Klewin, Kenneth Wade; Larson, Timothy Howe; Leckie, R. Mark; Liu, Shih-Tseng; Lytwyn, John N.; Mackiewicz, Nancy E.; Malander, Mark William; Marek, Norman J.; Marko, Joel; Meltz, Robert E.; Millen, Timothy M.; Olson, Christopher J.; Olson, David N.; Pencak, Michael Stanley; Power, Kathleen M.; Saric, James A.; Schenk, Paul M.; Schenning, James W.; Schuh, Mary Louise; Sticha, Jill Marie; Styzen, Michael J.; Taylor, Sheryl M.; Theis, William P.; Tripp, Angela M.; Tuftee, Kelly Krenz; Ueng, Charles Wen-Long; Vagt, Peter John; Vagt, Peter John; Van de Voorde, Barbara Wiley; Volkert, David G.; Wallace, Debbie L.; Werbach, David; White, Richard J.; Wilson, David D.; Yonk, Allen K.

Northwestern State University
Natchitoches, LA 71457

6 Master's

1970s: Jordan, Patrick J. W.; McKnight, Randy Henry; Pippen, Fred M., Jr.; Ryals, Gary N.; Stewart, Harry Edward; Stiles, Walter W.

Northwestern University
Evanston, IL 60201

189 Master's, 178 Doctoral

1890s: Quereau, Edmund C.

1900s: Burchard, Ernest F.; Higgins, Daniel F., Jr.

1910s: Ball, John R.; Cady, Gilbert H.; Cline, Justis H.; Merritt, John Wesley; Scott, Horace A.; Troxell, Edward L.

1920s: Apple, Olive F.; Berry, Hally L.; Currier, Louis W.; Devou, Marie L.; Erb, Elizabeth; Gillson, Joseph L.; Lamey, Carl Arthur; Plummer, Helen S.; Post, Paul T.; Potter, Franklin C.; Powers, William Edwards; Ridgway, Lucille; Seager, Oramel Ainsworth; Stark, John Thomas; Trainer, David Woolsey

1930s: Aberdeen, Esther J.; Ball, John R.; Banfield, Armine F.; Barnes, Farrell Francis; Barnes, Harley, Jr.; Belyea, Helen R.; Beutner, Edward L.; Butler, John W., Jr.; Cady, Francis H.; Cady, Wallace M.; Cannon, Ralph S.; Cort, John J., Jr.; Crosby, Eleanor J.; Dapples, Edward C., Jr.; Foose, Richard M.; Garrels, Robert M.; Graham, Joseph J.; Harris, David V.; Henderson, Charles F.; Hoagland, Alan D.; Howland, Arthur L.; Jackson, Alvin M.; Jahns, Richard H.; Jones, Alice J.; Klaer, Fred H.; Lamey, Carl Arthur; Lucy, Harold P.; Maxwell, Ross A.; Miller, Howard W.; Morse, Margaret L.; Needham, Claude Ervin; Osborn, Elburt F.; Palmer, Arthur H., Jr.; Rainwater, Edward H.; Roe, Walter B.; Runne, Marjorie E.; Russell, Robert T.; Schultz, John R.; Schwade, Irving T.; Scott, Erwin Ralph; Sleight, Vergil G.; Taylor, David O.; Thayer, Thomas P.; Todd, Jean P.; Van Alstine, Ralph Erkstine

1940s: Banfield, Armine F.; Bieber, Charles L.; Brown, Glen F.; Brown, Glen Francis; Cushman, Robert V.; Devlin, Frank; Dow, Donald H.; Dowling, Helen E.; Flint, Delos E.; Garrels, Robert M.; Gram, Oscar E.; Hall, Robert B.; Hambleton, William W.; Hay, Richard Leroy; Hinrichs, Frederick W.; Jones, Charles L.; Kent, Deane F.; Kozary,

Myron Theodore; Krimmel, Carl P.; Norton, James J.; Pierce, Guy Russell; Quigley, Darwin M.; Robertson, Cameron P.; Rominger, Joseph F.; Savage, Carleton N.; Sharpe, Lois Kremer; Smith, Edgar E. N.; Staatz, Mortimer H.; Stark, Jessie B.; Stewart, Robert H.; Theodosis, Steven Daniel; Turpin, Evelyn M.; Warn, G. Frederick; Williams, William P.; Wilpot, Ralph H.

1950s: Andrichuk, John Michael; Bailey, James Stuart; Baillie, Andrew Dollar; Basham, William L.; Bird, Melvin Leroy; Blythe, Jack Gordon; Bourn, Oscar B.; Bowin, Carl Otto; Brooks, James E.; Burgess, Lawrence C. N.; Burk, Cornelius Franklin, Jr.; Castano, John R.; Christensen, D. F.; Conti, Mario A.; Cramer, Howard Ross; Dewees, Allen H.; Dodd, Philip H.; Dorman, Henry J.; Erickson, Robert H.; Forgotson, James Morris, Jr.; Forgotson, James Morris, Jr.; Francis, David Roy; Glaister, Roland P.; Goodell, Horace Grant; Goodell, Horace Grant; Gutstadt, Alan Morton; Hansen, Alan R.; Havers, Murray Hall; Hemley, John Julian; Herbaly, Elmer L.; House, Richard D.; Huber, Norman K.; Huber, Norman K.; Karlstrom, Adabell; Karstrom, Adabell; Kauffman, Marvin Earl; Kilburn, Chabot; King, James F.; Kistler, James O.; Kraetsch, Ralph B., Jr.; Kroon, Harris M.; Lewis, Donald W.; Libby, Willard G.; Macomber, Bruce E.; Macomber, Richard; Martin, Leonard John; McCabe, Hugh R.; Miles, Roy G.; Mitchum, Robert Mitchell, Jr.; Nagel, Fritz G.; Peterson, Marvin L.; Peterson, Melvin N. A.; Pitt, William D.; Probst, David Arthur; Rominger, Joseph F.; Rudd, Robert Dean; Schmitt, George Theodore; Snyder, John LeMoyne; Stearns, Richard Gordon; Strand, Rudolph G.; Sutterlin, Peter George; Thomas, Harry George; Vail, Peter Robbins; Vail, Peter Robbins; Visher, Glenn S.; Whitford, Stanley D.; Wilson, John R.; Yeakel, Lloyd S.

1960s: Adeyeri, Joseph Babalola; Anderson, Franz E.; Atchison, Michael E.; Ayrton, William G.; Ayrton, William Grey; Baker, Wallace Hayward; Banaszak, Konrad J.; Barnes, Robert P.; Cain, James A.; Cain, James Allan; Chen, Chin Shan; Churkin, Michael, Jr.; Coates, Mary H.; Cox, Walter M.; Emslie, Ronald Frank; Felber, Bernard E.; Fischer, Donald E.; Fowler, Donald R.; Fox, William Templeton; Fox, William Templeton; Franklin, Arley Graves; Fulton, Robert John; Gibbins, Walter A.; Gibbs, Ronald E.; Glaeser, J. Douglas; Graham, David W.; Graham, David Wilson; Hempkins, William B.; Henderson, John R.; Ho, Michael Man-Kai; Hobson, Richard D.; Hobson, Richard David; Hughes, John Derek; Ingels, Jerome J. C.; James, Clarence Hubert Cavendish; James, William R.; James, William R.; Jones, Thomas A.; Jones, Thomas Allen; Kuntz, Mel A.; Lafon, Guy Michel; Lefebvre, Richard Harold; Lewis, Laurence A.; Link, Arthur J.; Link, Arthur Jurgen; Mahmoud, Idris Ahmed; Matuszak, David R.; McCabe, Hugh R.; Miesch, Alfred Thomas; Nigrini, Andrew; Nigrini, Andrew; Nugent, Robert Charles; Ogren, David Ernest; Perelman, David S.; Riehle, James Donald; Roy, Kenneth James; Sangree, John Brewster, Jr.; Scherer, Wolfgang; Shannon, John P.; Shannon, John Philip, Jr.; Shurr, George W.; Sims, John David; Smith, Peter H.; Smith, Peter Henderson; Smith, Stephen; Steinmetz, Richard; Stuart, William D.; Thein, Maung; Thein, Maung; Thomas, John J.; Thorstenson, Donald Carl; Threinen, David T.; Upchurch, Sam B.; Vacher, Henry Leonard; Visher, Glenn S.; Wadsworth, William B.; Wadsworth, William Bingham; Whitney, Harold Tichenor, Jr.; Wilson, Michael D.; Wilson, Michael David; Winters, Harold Abraham; Zelasko, Joseph Simon

1970s: Abdelhamid, M. S.; Allen, Rodger F.; Ansal, A. M.; Attoh, Kodjopa; Badiozamani, Khosrow; Banaszak, K. J.; Beane, Richard E.; Blazquez, R.; Brown, Thomas Howard; Casteleiro, M.; Castillo Ron, Enrique; Chung, R. M.; Cogbill, A. H., Jr.; Cuellar, V.; Dodge, C. H.; Egli, P.; El-Moursi, H. E.-D. H.; Elzaroughi, A. A.; Fujita, K.; Fujita, Kazuya; Gardner, Weston C.; Geddes, A.

J. S.; Giger, M. W.; Holtz, Robert Dean, II; Jin, J. S.; Lantzy, R. J.; Larue, D. K.; Larue, D. K.; Leeper, Robert; Link, D. A.; Lo, Hoom-bin; Lorenz, Douglas McNeil; Mac Nabb, Bert E.; MacMillan, John R.; MacNabb, Bert E.; Matthews, Martin David; Mock, S. J.; Monte, J. L.; Moore, Timothy Joseph; Oldow, J. S.; Perlman, Vicky A.; Plummer, Leonard N.; Riehle, James Donald; Ristvet, B. L.; Rogers, Jerry Rowland; Rosenfarb, J. L.; Salem, A. M.; Scherer, Wolfgang; Semrau Lago, R.; Sener, C.; Snell, N. S.; Snyder, Jeremy; Spirakis, Charles Stanley; Stablein, Newton Kingman; Stoffyn, M. A.; Stuart, William D.; Subramanian, Vaidyanatha; Sulima, J. H.; Sulima, John H.; Toth, D. J.; Turgut, Suleyman; Upchurch, Sam Bayliss; Vacher, Henry Leonard; Wetterauer, R. H.; Wetterauer, Richard H.; White, A. F.; Wolery, T. J.

1980s: Adams, Gerald Edwin, Jr.; Aimone, Catherine Taylor; Arnseth, Richard Wayne; Babaie, Hassan Ali; Bina, Craig Richard; Bischoff, William David; Brown, Irving Foster; Chang, Eric Yea-Yuan; Chang, Hui Sing; Chang, Hung Kiang; Chou, Lei; Chu, Shi-Chih; DeMets, Dennis Charles; Elison, Mark W.; Engeln, Joseph Francis; Frizado, Joseph Pacheco; Fulthorpe, Craig Stephen; Heck, Frederick Richard; Henderson, Laurel Jean; Hill, James Martin; Horii, Hideyuki; Hull, Amy Berg; Hutson, Robert William; Iwakuma, Tetsuo; Jansma, Pamela Elizabeth; Joyce, James; Kim, Jin-Keun; Koutsibelas, Dimitrios A.; Kuo, Mao-Kuen; Labuz, Joseph F.; Langan, Robert Thomas; Lundgren, Paul Randall; Muszynski, Isabelle; Nunn, Jeffrey Allen; O'Connor, Kevin Myles; Pigott, John Dowling; Rowshandel, Badiollah; Russell, Branch James; Schoonmaker, Jane E.; Stahl, Stephen David; Sung, Jen-Chun; Torrini, Rudolph Edward, Jr.; Vanko, David Alan; Wiens, Douglas Alvin; Yen, Hsiang-Jen

University of Notre Dame
Notre Dame, IN 46556

1 Master's, 2 Doctoral

1970s: Bierman, Victor Joseph, Jr.; McCabe, Peter Joseph

1980s: Kiphart, Kerry

Oberlin College
Oberlin, OH 44074

5 Master's, 1 Doctoral

1930s: Goldstone, Selma L.; Thomsen, Harry L.; Wenberg, Edwin Hugo

1940s: Bunch, Rosella L.; Keeler, Jane V.; Laswell, Troy J.

Ohio State University
Columbus, OH 43210

447 Master's, 313 Doctoral

1890s: Bownocker, John A.; Grimsley, George P.

1900s: Lamb, George F.; Morse, William C.; Stauffer, Clinton Raymond

1910s: Baumiller, George N.; Lamborn, Helen Morningstar; Mark, Clara G.; Mix, Sidney E.; O'Rourke, Edward Joseph; Turkopp, John

1920s: Bognar, Edwin J.; Conrey, Guy Woolard; Dunn, Paul H.; Emery, Alden H.; Fisher, Mildred; Fisk, Henry Grunsky; Foster, Wilder D.; Meyers, Theodore Ralph; Mitchell, Robert H.; Morgan, Richard; Moses, Clarence F.; Schaefer, Jacob Edward; Schillhahn, Ernest O.; White, George Willard; Wolford, John J.

1930s: Baker, Merle V.; Bernhagen, Ralph John; Blair, Helen Mae; Bond, Ralph Hurd; Busch, Danial Adolph; Busch, Daniel Adolph; Bush, Daniel A.; Chappars, Michael S.; Cummins, James Walter; Eberele, Robert Francis; Griggs, David T.; Hendrix, William Edwin; Hobbs, Susan Smith; Kelley, Joseph A.; Kelly, Joseph Allen; Klepser, Harry John;

Ludena, S. E.; Melvin, John Harper; Mohr, Elizabeth B.; Nesbitt, Robert H.; Phelps, Willard B.; Priddy, Richard Randall; Priddy, Richard Randall; Rogers, Maynard; Schoenlaub, Robert A.; Schoff, Stuart L.; Schoff, Stuart L.; Shaffer, Paul Raymond; Snow, Roland B.; Stephenson, Edgar L.; Sturgeon, Myron T.; Sturgeon, Myron T.; White, George Willard

1940s: Babisak, Julius; Bean, Robert Taylor; Bonar, Chester Milton; Bowen, Charles Henry; Brown, Donald Marvin; Fagadau, Sanford Payne; Faulk, Niles; Fisher, Mildred; Flint, Norman Keith; Flint, Norman Keith; Gilliland, William Nathan; Hardy, Clyde Thomas; Hardy, Clyde Thomas; Hunt, Robert Elton; Hutton, Charles Wetherill; Johnson, Mike Sam; Karhi, Louis; Kate, Frederick H.; Katherman, Vance Edward; Martin, G. C.; Maxey, Julian S.; Merrill, William Meredith; Rehn, Edgar Ernest; Reutinger, Charles Anton; Shaffer, Paul Raymond; Sheriff, Robert Edward; Smith, William Henking; Swick, Leo Emmett; Taylor, Dorothy Ann; Washburn, George Robert; Weiss, Eriol Joseph; Wilson, Mark Dale; Zeller, Howard Davis

1950s: Alexander, Robert John; Alpdogan, Sami; Arkle, Thomas, Jr.; Aydin, A. F.; Bachman, Mattias Edgar; Banfield, Oscar M.; Baroffio, James Richard; Bartlett, John David; Baughman, Russell Leroy; Bayley, Richard William; Bayley, Richard William; Beard, Thomas Noble; Bell, Gerald Laverne; Blackee, Benson D.; Blake, Oliver Duncan; Blakney, William Gilbert G.; Boble, John D.; Bowen, Anita Schenck; Bowen, Charles Henry; Bowman, Richard Spencer; Burkard, Richard Killiam; Campbell, Lois Jeannette; Case, James Boyce; Case, James Boyce; Christopher, James Ellis; Christopher, James Ellis; Clark, Armin Lee; Cockfield, James E.; Conley, James Franklin; Cooper, John Edmond, Jr.; Cornejo, John; Crombie, Richard Howard; Defeu, Edwin Leroy; Dowdy, James Marshall; Edwards, William Russell; Erickson, Thomas David; Ewing, Clair Eugene; Fagadau, Sanford Payne; Farnsworth, Don Willard; Farrand, William Richard; Fograscher, Arthur Carl; Forsyth, Jane Louise; Foster, John Webster; Franklin, George Joseph; Frazier, Noah Arthur; Friedman, Samuel Arthur; Froelich, Albert Joseph; Fulweiler, Robert Edward; Fyles, John Gladstone; Gill, James Rodger; Gliozzi, James; Goodman, Jerome; Gordon, David Walker; Gray, Henry Hamilton; Gregory, James Finley; Hall, John Frederick; Hilty, Robert Emil; Hohler, James Joseph; Hopkins, Roy Marshall; Horowitz, Alan Stanley; Hsu, Kenneth Jinghwa; Humphris, Curtis Carlyle; Hunt, Robert Elton; Jackson, Robert Reed; Jessup, Donald Edward; Jones, Robert Lewis; Jury, Harold Louie; Kantrowitz, Irwin Howard; Katich, Philip Joseph; Kaula, William M.; Kempton, John Paul; Khin, Maung Aung; King, Edward Larnard; Kleinhampl, Frank Joseph; Knipling, Louis Henry; Konecny, Gottfried; Kucera, Richard Edward; Kutlu, Nurettin; Laubach, James Taylor; Lautenschlager, Herman Kenneth; Lauzon, E. P.; Lee, Kwang-Yuan; Lee, Kwang-Yuan; Lehman, Russell John; Lehner, Robert Eugene; Leutze, Willard Parker; Lossman, Edward A.; Macomber, Mark M.; Mahaffey, Jack L.; Mahoney, William Clement; Marple, Mildred Fisher; Mase, Russell Edwin; McGookey, Donald Paul; Merchant, Dean Charles; Merrill, William M.; Metter, Raymond Earl; Metter, Raymond Earl; Miller, Paul Melby; Montero, Felipe J.; Moore, Clarence Victor; Mourad, Asa George; Mowery, Dale Harris; Muessig, Siegfried Joseph; Multer, Harold Gray; Norling, Donald Leonard; Norman, Carl Edgar; O'Brien, Leslie J.; Ockert, Donn Lee; Osborne, Wiley Wilson; Pashley, Emil Frederick, Jr.; Pedry, John Joseph; Pendell, Ray; Pennell, Ray, Jr.; Peterson, William R.; Phalakarakula, Charas; Prinz, William Charles; Pulse, Richard Reid; Reynolds, Martin Bruce; Richards, Gene Edward; Rodriguez, Joaquin; Root, Samuel I.; Roseboom, Eugene Holloway, Jr.; Rouse, Glen E.; Ryland, Robert R.; Savoy, Donald DeCoursey; Schapiro, Norman; Schuh, Henry Allen; Schuster, Robert Lee; Scott, David Kendall;

Sedam, Alan Charles; Seik, Lawrence Michael; Sharp, Everett Ray; Sheriff, Robert Edward; Smyth, Pauline; Struble, Richard Allen; Swick, Norman Eugene; Szmuc, Eugene Joseph; Szmuc, Eugene Joseph; Teflian, Samuel; Teichmann, Warren James; Thompson, Gerald Leon; Tinker, Wesley R.; Treves, Samuel Blain; Tucker, Leroy Maddy; Turco, Caroline Ann; Turpin, Robert David; Uotila, Urho Antti Kalevi; Veis, George; Vogel, James William; Warner, Earl, Jr.; Weiss, Richard Marion; Wilkie, Lorna Christine; Williamson, Richard Edward; Winslow, Marcia Ring; Young, Robert Glenn; Zimmerman, James Arthur

1960s: Abby, Darwin G.; Adler, Ron K. H.; Adutwum, John B.; Ali, Syed A.; Allen, Robert S.; Anderson, George B.; Anderton, Peter Wightman; Aukeman, Frederick Neil; Barnes, Gordon L.; Barnett, Stockton Gordon, III; Barrett, Peter John; Batsche, Ralph W.; Beard, William Clarence; Bennett, Joseph E.; Bennett, Truman W.; Benson, Anthony Lane; Bergford, Paul M.; Birle, John David; Blake, Jerry Wayne; Blake, Weston, Jr.; Boster, Ronald S.; Bradley, James E.; Brahma, Chandra Sekhar; Brecher, Henry H.; Brown, William W.; Burger, William Hunt; Butterman, William Charles; Butterman, William Charles; Calkin, Parker Emerson; Cameron, Richard Leo; Cameron, Richard Leo; Campbell, Andrew Clare; Carnein, Carl Robert; Carpenter, William H.; Caswell, William Bradford, Jr.; Chaisrakeo, Meechai; Chaudhuri, Sanbhudas; Chopra, Kailash C.; Clowers, Stanley; Crane, Robert Lee; Crane, Robert Lee; Cronk, Caspar; Cunningham, Leslie L.; Daugherty, Kenneth Ivan; Davis, James William; de Jong, Sybren Hendrik; Decker, Billy Louis; DeLoach, William, Jr.; Devereaux, Alfred B., Jr.; Dewart, Gilbert; Diaz-Garzon, Alfonso E.; Dickman, Everitt W.; Dietz, Earl Daniel; Dow, John Wilson; Dragg, James L.; Drake, Lon David; Dunn, James V.; Durupinar, Ahmet T.; Eastin, Rene; Emrick, Harry W.; Everette, Kaye Ronald; Ewing, Don R.; Fenton, Michael Dwight; Fishel, Nathan; Fitzpatrick, Kenneth; Ford, John Philip; Foreman, Dennis Walden, Jr.; Franklin, George Joseph; Frechette, Andre B.; Furry, Robert Edward; Gentile, Anthony L.; Ghosh, Sanjib Kumar; Gibson, Gail G.; Gielisse, Peter Jacob; Goll, Robert Miles; Goode, Sterling D.; Groening, Donald I.; Hamilton, Wayne Lee; Hancock, Kenneth J.; Hart, William E.; Haselton, George Montgomery; Hays, James Douglas; Heilbronner, Heinrich K.; Heiniger, Keith D.; Herat, Samson T.; Hernandez, Cristy R.; Hernandez, Federico; Herring, John Charles; Hicks, Forrest L.; Hill, Richard Lee; Hoekstra, Karl E.; Holdsworth, Gerald; Holdsworth, Gerald; Holway, Orlando, III; Horvath, Allan Leo; Janssens, Adriaan; Janssens, Adriaan; Janssens, Arie; Johnson, Gerald; Jonah, Maxwell V.; Jones, Lois Marilyn; Kalb, George William; Karren, J. R.; Kaufmann, Robert F.; Kelly, Gary G.; King, Robert B.; Kinnan, Joseph E.; Kinzelman, David J.; Kivioja, Lassi Antti; Kohut, Joseph James; Koopman, Donald Edward; Kovach, Jack; Krushensky, Richard Dean; Kryger, Adolph H.; Kurtis, Mehmet S.; Kvaale, Sigurd O.; Lager, James Lee; Laubscher, Alan L.; Lee, Bennon; Lee, Shuh-Chai; Lewis, Thomas Leonard; Limerick, C. J., Jr.; Lin, Hsi-Che; Linclon, James Bruce; Lindsay, John Francis; Long, William E.; Long, William Ellis; Lyon, Stephen Reed; Macomber, Mark M.; Madkour, Mohamed F.; Madkour, Mohamed Fathi; Madole, Richard Frank; Madole, Richard Frank; Mahoney, William Clement; Marangunic, Cedomir Damianovic; Martin, James P.; Martucci, Louis; Massmann, Thomas A.; Mayer, Richard R.; Mayhew, George Herbert; Mayhew, George Herbert; McClish, Richard F.; McCormick, George Robert; McCormick, George Robert; McCoy, Robert L.; McKenzie, Garry Donald; Miller, Frederick Powell; Minshew, Velon H., Jr.; Mirsky, Arthur; Mitchell, James; Mitchell, Michael M.; Montero, Felipe J.; Mueller, Ivan Istvan; Nave, Floyd Roger; Needham, Paul E.; Negron, Jenaro R.; Negus, Kenneth D.; Nicholson, Roy J.; Nikravesh, Rashel; Norman, Carl Edgar; Novotny,

Robert F.; Obenson, Gabriel Francis; Orlin, Hyman; Osborn, Roger T.; Osborne, Robert Howard; Pawlowicz, Edmund Frank; Peterson, Donald Neil; Pope, Allen J.; Potter, Robert K.; Powell, William I., Jr.; Ramey, Everett H.; Ramsey, Nancy Jo; Rapp, Richard H.; Rapp, Richard Henry; Reeber, Robert Richard; Reese, Thomas J.; Reeves, Donald W.; Rhoades, Rendell; Richards, James K.; Richardson, Donald A.; Rios, Julio C.; Robson, Walter M.; Rockett, Thomas John; Rockie, John D.; Rosen, Norman C.; Roy, Edward C.; Rudin-Rodriguez, Fernando M.; Russell, Edgar Ernest; Rust, Claude Charles; Rutledge, Elliott Moye; Ryan, Roger M.; Ryan, Wallace; Sastrosoedirdjo, Djoko W.; Saunders, David W.; Scatterday, James Ware; Schopf, Thomas Joseph Morton; Semler, Charles Edward, Jr.; Shallom, Lizzie J.; Sharni, Dan; Shaw, Jimmy E.; Sheatsley, Larry Lee; Sheldon, Lyndon L.; Shultz, Charles High; Skoch, Edwin James; Sloan, John F.; Smith, Geoffrey Wayne; Smith, Merlin C., Jr.; Snowden, John M.; Soliman, Afifi H.; Solter, Donald D.; Sparling, Dale Richard; Sprinsky, William Harold; St. Clair, James H.; Stibbe, Ehud; Stohl, Frances Virginia; Struble, Richard Allen; Stukhart, George, Jr.; Taylor, Lawrence Dow; Teller, James T.; Textoris, Daniel Andrew; Therkelsen, Edward R.; Thomas, Gilbert Edward; Thomason, Thomas J.; Tobin, Don Grayville; Tomajczyk, Charles F.; Upperco, Jesse R.; Vamosi, Sendor; Vatis, Martin D.; Versic, Ronald James; Wallace, Ronald G.; Wallace, Ronald Gary; Weissman, Simha; Wharton, Ralph E.; Williams, Nelson Noel; Williams, Robert L.; Winchell, Robert Eugene, Jr.; Wintz, Edward K.; Wintz, Edward K.; Wolfe, Edward Winslow; Yurdakul, Ali R.; Zahn, Jack C.

1970s: Ager, T. A.; Ahern, Judson Lewis; Alexander, Earl Betson, Jr.; Ali, E. M.; Allong, Albert Francis; Alzaydi, A. A.; Arur, Manohar; Ashworth, E. T.; Ayeni, O. O.; Baird, Mary Rebecca; Baker, D. J.; Baranovic, Michael Joseph; Barbis, Frederic C.; Baxter, Sonny; Baybrook, Thomas G.; Behling, Robert Edward; Bickel, Edwin David; Birsa, D. S.; Blackman, M. J.; Boger, J. B.; Boger, P. D.; Bole, Clifton Eugene; Bonnett, Richard Brian; Bossler, John David; Botoman, George G.; Bowman, John Randall; Bruce, L. G.; Camp, Mark J.; Campion, K. M.; Careaga, Richard Oliver; Carnein, C. R.; Carnes, J. B.; Carnes, Susan Fraker; Carwile, Roy H.; Chang, N.-Y.; Chapel, J. D.; Clark, Douglas B.; Clifford, Michael James; Cocker, Mark David; Cocker, Mark David; Cooper, B. J.; Couch, Robert F.; Cox, Herbert M.; Cox, Herbert Michael; Cox, Jeffrey Martin; Croft, John S.; Curl, J. E.; Cushman, Solomon Frederick; Dawson, Robert Scott; de Gruyter, D. A.; DeHaas, Ronald J.; Derksen, S. J.; Derksen, S. J.; DesCamps, Julius Robert; Devereaux, Alfred Boyce; Eastin, Rene; Elghazali, M. S.; Emrick, Harry William; Epstein, Anita Fishman; Epstein, Jack Burton; Erb, Edward Edeburn, Jr.; Erickson, Thomas David; Fasola, A.; Fetzer, Joseph A.; Finney, S. C.; Foley, Duncan; Foley, Duncan; Frederick, V. R., Jr.; Frye, Mark W.; Gable, Kristine M.; Ghist, J. M.; Gonterman, J. Ronald; Gopalapillai, Sivasithamparam; Greaney, Peter H.; Gunner, John Duncan; Haden, J. M.; Hariharan, Ganesan; Hasenmueller, Walter; Hasenohr, E. J.; Hayes, Larry Ross; Heatwole, L. C.; Hellert, John R.; Henning, Roger John; Hines, J. M.; Hofmann, Douglas A.; Holterhoff, F. K.; Hower, James Clyde; Hoyer, M. C.; Hsu, J. R.; Ikpeama, Mmajuogu Onyelankea U.; Jacobson, Stephen Richard; Jakob, P. G.; Kalinowski, Donald D.; Kaufman, John Warren; Khedr, S. A.; Knipling, Louis Henry, Jr.; Kopacz, M. A.; Kovach, Jack; Kumar, M.; Kuryvial, Robert J.; Larson, G. J.; Larson, Grahame J.; Lawrey, J. D.; Lemire, Jerome A.; Loon, J. C.; Ludlum, Gloria King; Lumsden, Jesse Beadles, III; Madeley, Hulon Matthews; Madley, Hulon Matthews; Marcantel, E. L.; Marcantel, Jonathan Benning; Marcos, Zilmar Ziller; Marino, Rrobert John; Martin, David Lichty; Mayewski, Paul Andrew; McCullough, L. A.; McKeon, J. B.; McSaveney, E. C.; McSaveney, Eileen Craven;

McSaveney, M. J.; Merrill, William Meredith; Mickelson, David Melvin; Mikan, Frank M.; Miller, M. A.; Miller, P.; Moos, Milton; Murtaugh, J. G.; Myers, C. P.; Nagaraja, Hebbur Narasimhamurthy; Nardone, Craig D.; Naymik, T. G.; Nixon, Kenneth Ray; Obenson, Gabriel Francis També; Ogden, Julius Sterly; Olds, Michael Warren; Orheim, Olav; Owen, Lawrence Barry; Pace, Karen Klusmeyer; Palombo, D. A.; Parkinson, Robert J.; Pegram, W. J.; Philip, Aldwyn Thomas; Pinker, Robert; Quinn, M. J.; Quinn, Michael J.; Rampal, K. K.; Ray, Phillip T.; Reed, George Bruce; Reilly, James Patrick; Rhodes, M. L.; Richard, J. K.; Roscoe, Michael; Rosengreen, Theodore Ernest; Schultz, Thomas R.; Schwietering, Joseph Francis; Selby, Andrew C.; Shaffer, N. R.; Shanklin, Robert Elstone; Shatzer, D. C.; Shearer, Gerald Brian; Simpson, C. Leon; Simpson, Leon; Smith, Glen N.; Smith, Timothy Ellis; Snee, Lawrence Warren; Soler, T.; Stewart, Rae Alden; Stockey, R. A.; Studlick, J. R. J.; Stump, E.; Swank, Willard J.; Thompson, E. M.; Thompson, Ellen M.; Thompson, L. G.; Thompson, Leon Garfield; Thompson, Lonnie G.; Thompson, William E.; Thompson, Woodrow Burr; Timson, Glenn H.; Trautman, T. A.; Tremba, Edward Louis; Trent, Donald Eugene; Tuller, Jack N.; Turinetti, James D.; Uhrin, David C.; Ukayli, Mustafa Ahmad; Uttley, J. S.; Van Horn, Robert Gary; Von Horn, Robert; Votaw, Robert B.; Walton, Wayne J. A., Jr.; Warlow, Joseph Charles; Watts, Doyle Robin; Weatherington, Julie B.; Weinle, Arthur R.; Whillans, I. M.; Williams, Ernest B.; Wootton, C. F.; Yang, Houng-Yi

1980s: Ahmed, Ahmed El-Sayed; Amba, Etim Anwanna; Anton, Ann; Archinal, Brent Allen; Baker, Edward Michael; Baraka, Moustafa Ahmed; Bauer, Jeffrey A.; Bock, Yehuda; Chang, Ting-Chieh; Cichan, Michael Anthony; Cruz, Jaime Yap; de Castro, Celso Filho; de Vries, Thomas John; Dedes, George C.; Goodwin, Robert Glenn; Greenfeld, Joshua Shlomo; Haghighi, Rahim Ghorbanzadeh; Hall, Jack Charles; Hansen, Michael Christian; Harwood, David Michael; Hasenohr, Edward Joseph; Huber, Brian Thomas; Iz, Huseyin Baki; Jaynes, William Frederick; Jekeli, Christopher; Katsambalos, Kostas Evangelos; Kim, Yoo Bong; Kleffner, Mark Alan; Krieg, Lenny Albert; Kulatilake, Pinnaduwa Howa S. W.; Leatham, W. Britt; Lee, In Mo; Libicki, Charles Melvin; Lucius, Jeffrey E.; Lux, Daniel R.; Macellari, Carlos Enrique; Madani, Mostafa Seyed; Mainville, Andre; Mast, V. A.; Mensing, Teresa Marie; Milbert, Dennis Gerard; Morra, Matthew John; Mossaad, Mostafa El-Sayed; Moussa, Osama Moursy; Myers, Jed Anthony; Norton, Lloyd Darrell; Nosseir, Mostafa Kamel; Nyerges, Timothy Lee; Oliver, George Rick; Pacht, Jory Allen; Palais, Julie Michelle; Patias, Petros Georgios; Pavlis, Erricos C.; Pichtel, John Robert; Pigg, Kathleen Belle; Potter, John Claude; Powell, Ross David; Priovolos, George Jim; Ramirez P., Raul; Ransom, Michel Doyle; Reddi, Lakshmi Narayana; Roy, Bimal Chandra; Saif, Hakeem Thamir; Schwans, Peter; Shaw, David William; Smoot, Edith L.; Snee, Lawrence Warren; Thapa, Khagendra; Valizaden-Alavi, Hedayatollah; Vavra, Charles Lee; Williams, Richard Lynn; Wyers, Gerard Paul

Ohio University, Athens
Athens, OH   45701

183 Master's, 4 Doctoral

1960s: Baker, Herbert Arney; Bjurstrom, Stanley Theodore; Bush, Edward Calvin; Colson, Calvin Thomas; Current, George Thomas; Fanaff, Allan S.; Fassett, Bernard Donald; Haines, Forest E., Jr.; Herdendorf, Charles Edward, III; Hoge, Harry Porter; Hylbert, David Kent; Kellenberger, Jack Eugene; Kozusko, Raymond George; Neff, Jerry W.; Nelson, Ronald Harry; Oinonen, Russell Lee; Owens, Gordon L.; Picking, Larry Webb; Pinney, Robert I.; Smith, Bradley Earl; Stonestreet, Albert Lee; Taggart, Ralph E.; Thompson, Loren Edward, Jr.; Turrill, Sheldon Lee

1970s: Al-Hajji, Yacoub Y.; Al-Sarawi, Mohammad; Alfano, Joseph Michael; Baker, George Oliver; Bakush, Sadeg Hasan; Barnhart, Stephen F.; Barnhouse, John Douglas, Jr.; Birkhimer, Cheryl Patricia; Blackie, Gary William; Bobba, Arabinda Ghosh; Bossong, Clifford Robert; Burling, Robert Jeffrey; Butler, Mark L.; Canter, Neil W.; Carlson, Thomas R.; Cemen, Ibrahim; Cole, Mark R.; Cook, Robert A.; Couchot, Michael Lee; Daftary, Homayoun; Dogan, Ahmet U.; Dogan, Meral; Doughri, Abdoolrhman K.; Eddib, Ali Ahmed; Elbakhbkhi, Mohamed Abolgasen; Faulk, Kenneth L.; Ford, Theodore Lester; Frey, David A.; Hanes, S. D.; Hansen, Michael Christian; Hartke, Edwin Joseph; Hedges, Robert Bruce; Huang, Wu-Shung; Johnson, Richard C.; Kantner, David Arthur; Kantner, Lynn M.; Khalaf, Mukhtar Hammali; Knappe, Roy, Jr.; Kufs, Charles T., Jr.; Kutz, William J.; Lant, Kevin J.; Lee, Chun-sun; Lipp, Russell L.; Lovrak, Peter William; Lowe, E. Charles, Jr.; Mahoney, John; Manoogian, Peter R.; Megerisi, Mohamed Fadlalla; Merschat, Walter R.; Mic'dleman, Allen; Minnick, Edward; Msek, Salahaddin Akif; Munne, Aarne Iivari; Nelligan, Frederick M.; Ortega, Jose F.; Parsley, Robert M.; Ricketts, Edward W.; Rieser, Robert B.; Roberts, Charles L.; Sage, Orrin G., Jr.; Sainey, Timothy J.; Salem-Mehemed, Salem S.; Schlaefer, Jill T.; Schlueter, James C.; Seay, John G., Jr.; Shaath, Samir Khali; Shanmugam, Ganapathy; Shields, Kermit E.; Smith, Norman Walter; Soukup, William G.; Sowers, John William; Stemen, Kim S.; Turco, Kevin; Vargas, Jose Eusebio; Walker, John Anthony, Jr.; Wanyeki, Simon; Webb, Douglas R.; Whaley, Keith Ray; Windle, Delbert Leroy, Jr.; Yacoub, U. Al-Hajji

1980s: Abufila, Taher M.; Adams, Vincent; Ahmed, Abdelazim I.; Ahsan, Abdus Salam; Al-Amri, M.; Al-Bassam, Abdulaziz M.; Al-Shammari, Lateef T.; Anderson, Patricia Ann; Atha, Thomas M.; Backush, Ibrahim M.; Bayless, Richard C.; Bebel, Dennis; Beljin, Milovan; Beljin, Milovan Slavko; Bentz, Mark G.; Beraithen, Mohammed I.; Bernitsas, Nikolaos; Boardman, Darwin Rice, II; Borch, Mary Ann; Boston, William Bryan; Campbell, Michael D.; Carpenter, Carey C.; Chaboudy, Louis R., Jr.; Dunn, Paul M.; Fadhli, Fathi Ali; Finneran, Joseph M.; Garrett, George R.; Ghul, Sharef; Greenlee, Mac; Gregorowicz, Timothy Joseph; Gresko, Mark J.; Hammuda, Khalifa Salem; Handwerk, Roger H.; Hubbard, Frank Steven; Hughart, Joseph L.; Hussein, Adel M.; Jackson, Patrick A.; Jong, Ron S.; Khalil, Kabiru; Larson, Jill Marie; Madi, Lutfi Ali; Malinky, John M.; Martin, James A., Jr.; Massey-Norton, John T.; Mele, Thomas Anthony; Mickle, James Earl; Muza, Richard E.; N'Guessan, Yao; Nealon, Dennis J.; O'Brien, Ann Marie; Omar, Mazmumah Mamudah; Painter, Brian D.; Park, Choon-Byong; Paska, Michael A.; Peterson, Joseph D.; Petty, Rebecca J.; Quinn, Kenneth J.; Raleigh, Robert Eugene, Jr.; Rashrash, Salem M.; Reddy, Chemicala Janardhan; Risner, Jeffrey Keith; Schroeder, Kim Erik; Seaman, Jane Marie; Sekel, David M.; Shupe, Mark G.; Sims, Michael S.; Smith, Donald Eugene; Starn, Jon Jeffrey; Stellman, Terry Allen; Stephens, Harold Criss; Susko, John M.; Tokhais, Ali; Tomastik, Thomas E.; Vogel, Donald A.; Wang, Shih-Hsien; Warner, James Brian; Werle, Kevin J.; Wilson, Russell L.; Young, Michael H.; Youshah, Bashir M.; Zaleha, Michael James; Zatezalo, Jo Lynn; Zubari, Waleed Khalil

University of Oklahoma
Norman, OK   73019

929 Master's, 164 Doctoral

1900s: Kirk, Charles Townsend

1910s: Aurin, Fritz; Buttram, George Franklin; Cullen, John; MacKay, Hugh James; Monnett, Victor E.; Trout, Laurence Emory; Vanderpool, Harold Claude; Waite, Verdi V.

1920s: Bell, Hillis F.; Brown, Levi S.; Bullard, Fred M.; Butcher, Seldon D.; Clifton, Roland LeRoy; Cloud, Wilbur Frank; Cooper, Chalmer Lewis; Denison, Albert Rodger; English, Leon E.; Gahring, W. Ross; Harris, Forest Klaire; Hayes, John F.; Hedrick, O. F.; Henderson, George G.; Ireland, H. Andrew; Jones, Boone; Jones, Robert Lee; McCollough, Edward Heron; Meland, Norman; Millison, Clark Drury; Moore, Raymond A.; Mulky, Francis P.; Oakes, Malcolm C.; Pool, R. Harold; Pratt, Ernest S.; Radler, Dollie; Redfield, John Stowe; Sampson, Edward W.; Six, Ray L.; Smith, Gerald Nelson; Swiger, Rual Bower; Tharp, Paul A.; Tillotson, Harold Harman; Vanderpool, Harold Claude; Vanzant, James Harvey; Waters, James A.; Watkins, William A.; Weinzierl, John F.; Wright, Andrew Clemmons

1930s: Anderson, Rudolph F.; Armor, Mildred Virginia; Awbrey, Elizabeth; Ballard, William Norval; Coil, Fay; Constant, Warren LeRoy; Cooksey, Calvin Leavelle; Daugherty, Clarence Gordon, Jr.; Davis, John Roland; Easton, Harry Draper, Jr.; Edmiston, Eudora Fern; Eley, Hugh Moore; Foster, Paul Woodward; Frost, Victor Le Roy; Garrison, Martyna; Gillin, John A.; Grubbs, David M.; Ham, William Eugene; Hamner, Edward J.; Harris, Claude Milner; Hickcox, Charles Atwood; Hollingsworth, Richard Vincen; Husband, Edna Maurine; James, Bela Louis; Jones, Daniel John; Kimmel, Garman O.; Lalicker, Cecil Gordon; Loeblich, Alfred Richard, Jr.; Lucas, Elmer Lawrence; Maxey, James Roy; Maxwell, Ross A.; Meents, Richard O.; Miller, Edward Buford; Muir, James Lawrence; Park, Lee Brown; Patterson, Luther Edwin, Jr.; Perkinson, Floyd; Peters, Herbert N., Jr.; Posey, Ellen; Robertson, Leo L.; Savage, Donald Elvin; Spencer, Maria Frances; Strain, William Samuel; Tappan, Helen Nina; Travis, Abe; Uhl, Ben Forrest; Vieaux, Don George; Wallace, Pollack Austin; Wethington, William Orville; Wharton, Jay Bigelow, Jr.

1940s: Akers, Wilburn Holt; Allen, John O.; Anderson, Kenneth Clyde; Arper, William Burnside, Jr.; Austin, Robert Burton; Baumeister, Dorothy; Billingsley, Harold Ray; Branson, John Wallace; Burditt, Marvin Reece; Cantrell, Peggy Francis Parthenia; Carver, George Evans, Jr.; Chandler, William Alton; Crumpley, Bobby Kelly; Dewey, Robert Flanders; Dudley, Raymond Wesley; Farmilo, Alfred William; Fisher, Stanley Perkins; Fox, Frederick Glenn; Gifford, John Dempster; Giles, Alfred E.; Hadler, Harry George; Harrison, Edward Vernon; Heilborn, George; Hendrick, Thomas K.; Hoyle, Lorraine E.; Hudson, Roy Browning; Huie, James Powell; Ingram, Roy Lee; Jackson, Neil A.; Jacobsen, Eloise Tittle; Jacobsen, Lynn; Kassander, Arno Richard, Jr.; Kuhleman, Milton Henry; Laird, Joe Alex; Langston, Wann, Jr.; Latta, Lee Allen; Liddell, Jessie Kelsey; Lintz, Joseph, Jr.; MacGregor, James Donald; Martin, Monty Gene; Mayes, John Wilmot; McAnulty, William Noel; McCollum, Jack H.; McGregor, James D.; McNutt, Gordon Russell; Mohler, Charles Edwin; Neustadt, Walter, Jr.; Pate, James Durwood; Perry, Bernard James; Polk, Thomas Robb; Record, Walter Ross, Jr.; Renfroe, Charles Albert; Ries, Edward Richard; Rogers, William Donald; Sarles, John E.; Schacht, David Waldron; Schoonover, Floyd Eldon; Scull, Berton James; Skolnick, Herbert; Stevens, William Walter, Jr.; Stewart, Francis Jr.; Stratton, J. Lynn; Swesnik, Robert Malcolm; Vestal, Jack Herring; Walker, Keith F.; Walper, Jack Louis; Webb, John Hanor; Westervelt, Mary Lynn; Westmoreland, Harry; Williams, Harold L.; Wilson, George William; Wonfor, John Stephen

1950s: Abels, Thomas Allen; Akmal, Mohammed Gawid; Alexander, Richard Dolphin; Alexander, Russell James; Allen, Albert E., Jr.; Andrews, Ralf E., Jr.; Armstrong, Bobby D.; Arnold, Billy M., Sr.; Atkinson, Walter Edward; Baker, David Alan; Baker, Frank Elmore; Ballard, William Wayne; Barby, Boardman Gene; Barker, James Charles; Barker, Robert E.; Beckwith, Clyde Grosvenor, Jr.;

Bell, Robert Joe; Bell, Walton; Bellizzia, Alirio Antonio; Bellizzia, Cecilia Martin; Benoit, Edward L.; Bercutt, Henry; Berry, Dean Harold; Beveridge, Richard Clark; Billings, Roger Lewis; Blakeley, David C.; Blanchard, Kenneth Stephen; Bleil, C. E.; Blumenthal, Morris B.; Blythe, Jack Gordon; Boeckman, Charles H.; Bohart, Philip H., Jr.; Bohnsack, Richard Lee; Boler, Milton E.; Bollman, James Franklin; Bowen, John E.; Bowles, Jack Paul Fletcher, Jr.; Bowman, Eugene A.; Bracken, Barth W.; Bradshaw, Robert Donald; Branson, Robert Burns; Brauer, Clemens P.; Braun, Jordan C.; Browder, George Thomas; Bryan, Robert Calvin; Bryant, David Gerald; Bullock, Jack M.; Burke, Jenie Lee, III; Burkett, Gerald G.; Butler, David Ray; Cade, Cassius A., III; Campbell, David Gwynne; Capps, William M.; Carl, Joseph Buford; Carnahan, George Gilbert; Carter, James A., Jr.; Cary, Logan W.; Caylor, Floyd Martin; Caylor, James Warren; Champlin, Stephen C.; Chandler, Philip Prescott; Chase, Gerald Warren; Cheatham, Bruce Ned; Chrisman, Louis Paul; Christian, Harry E.; Claxton, Charles Dale; Cohoon, Richard Roy; Cole, James Morgan; Cole, John Albert; Cole, Mary Jane; Coleman, Walter F.; Collins, R. L.; Connell, James Frederick Louis; Connell, James Frederick Louis; Cowan, Jack Vincent; Cronenwett, Charles Emanuel; Cullins, Henry Long, Jr.; Culp, Eugene Forrest; Cumella, Ronald; Dana, George F.; Dannenberg, Roy Berry; Darnell, Richard Douglas; Daw, Robert Norman; DeGraffenreid, Norman Bruce; DeJong, Gerard; DeLay, John Milton; Denison, Rodger E.; Despot, Camille C.; Dietrich, Ray Francis, Jr.; Dillé, Alan Charles Francis; Dillon, George R.; Disney, Ralph Willard; Dobervich, George; Donovan, Joe D.; Douglass, H. Marvin; Druitt, Charles Edward; Duck, James H.; Dunham, Robert Jacob; Eckert, Thomas Joseph; Edwards, John D.; Faucette, James Robert; Fenoglio, Anthony F.; Fisher, Henry Coleman, Jr.; Fleming, Ray Edward; Ford, William Jack; Fox, Charles S.; Furlow, Bruce; Gallaspy, Irvin Lee; Gardner, William Edgar; Gearhart, Harry L.; Gelphman, Norman Ray; Gibbons, Kenneth E.; Gibbs, Harry Daniel; Gillert, Martin Peter; Gillum, Cecil Conrad; Godfrey, Jack Martin; Gold, Irwin B.; Govett, Raymond Weston; Govett, Raymond Weston; Graves, John Milton; Green, Jack Harlan; Gregware, William; Grieg, Paul Bennett, Jr.; Grieg, Paul Bennett, Jr.; Griffith, Alan Fraser; Grimes, Wayne Harlan; Gruman, William Paul; Gullatt, Ennis Murray, Jr.; Gunter, Craig E.; Hamberger, Kimball Lee; Hansen, James C.; Hansen, Robert F., Jr.; Harris, Donald G.; Hassinger, Russell Neal; Hayes, Lyman Neal; Heinzelmann, Gerald Mathias, Jr.; Hill, Frank Eugene; Hruby, Alexander Joseph; Hughes, Richard David; Hull, Paul W.; Hurt, Thomas Wayne; Hvolboll, Victor T.; Hyde, Jimmie Collins; Jobe, Billye Irene; Jobe, Thomas C.; Johnson, Eben Lennart; Johnson, Hamilton McKee; Johnson, Robert Kern; Johnson, Roderick H., Jr.; Johnston, Robert C.; Jones, Cecil L., Jr.; Jones, Jackson G.; Kellett, Charles Richard; Kerstetter, Frank Linwood, Jr.; Kimberlein, Za Grant, Jr.; Kinard, John C.; Kirk, Myrl Stuart, Jr.; Kozak, Frank Daniel; Krueger, Robert Carl; Land, Cooper B., Jr.; Lang, Robert Campbell, III; LaPorte, William D.; LaRue, John W.; Lauderback, Ralph Lewis; Lee, Harry C.; Leitner, Donald G.; Lively, John Robert; Lohman, Clarence, Jr.; Lontos, Jimmy T.; Lord, Robert L.; Luff, Glen Charles; MacQueen, Peter A.; Mahoney, Carroll F.; Manhoff, Charles N., Jr.; Mann, Wallace; Masters, Kenneth E.; Matesich, Charles O.; Matthews, Roy Edgar; Matuszak, David Robert; Maughan, John Bohan; May, Milton E.; McBryde, Thomas J.; McCall, Robert R.; McCollough, Edward L.; McCulloch, John Snyder; McDade, Laddie Burl; McDaniel, Gary A.; McDuffie, Roger H.; McKenny, Jere Wesley; McKinley, Glenn Ernest; McKinley, Myron Earnest; McKinney, James S.; McMurtry, Robert Paul; McNulty, Charles Lee, Jr.; Meek, Robert A.; Meltzer, Bernard David; Meyer, Richard Burt; Miller, Buster W.; Miller, James H.; Mills, Earl Lee; Mohn, Jean Doris; Mondy, Holland H.; Monk, Wilfred Jerale; Moore, Leslie Ray; Mor-

gan, Bill R.; Morgan, James Leland; Morgan, James Thornton; Murphey, Clifford W.; Nakayama, Eugene; Namazie, Mizra Hussain Ali; Nance, Richard Leon; Nelms, Jerry L.; Noll, Charles Richard, Jr.; Nolte, Clifton Jerry; Norton, David Lee; Oakes, David Thomas; Ottenstein, Robert Paul; Oxley, Marvin L.; Page, Kenneth G.; Paine, Jack W.; Pate, Joe Henry; Pietschker, Harold L.; Pollack, Jerome Marvin; Pollack, Jerome Marvin; Powell, Boyd DeWitt Hartley, Jr.; Powell, Clarence Cave; Pownell, Leland D.; Prestridge, Jefferson D.; Querry, Jamie L.; Ramay, Charles Lee; Rambo, George Daniel; Rath, Otto; Reed, Billy Kirk; Reeves, Corwin C.; Renfro, Kenneth McDonald; Ries, Edward Richard; Roach, Carl H.; Robertson, Billy Gene; Rockwell, Charles; Roe, Newton Charles; Rotter, Harold A.; Rowell, Thomas David, III; Rowland, Tom Lee; Russell, Dearl T.; Russell, Orville Ray; Sartain, Maxwell Roland; Sartin, John Philip; Saylor, Weldon Wayne; Schmalz, J. P., Jr.; Schulze, Jack D.; Scott, George L., Jr.; Scull, Berton James; Sears, Joseph McHutchon; Seely, Donald Randolph; Shankle, John Dyer, III; Shannon, Patrick Joseph; Shaw, Richard Frank, Jr.; Siemens, Allen G.; Simpson, I. D., Jr.; Skinner, Hubert C.; Skinner, Hubert C.; Slocum, R. C.; Smith, Avery Edward; Smith, Earl Winston; Smith, Edward Thornton, Jr.; Snodgrass, Elvis Dean; Sonnamaker, Charles P.; Soule, Kenneth Dana; Sparks, Billy Joe; Spear, John H.; Speer, John H.; Spooner, Harry V., Jr.; Spradlin, Charles Buckner; Stafford, Lester Earl; Stall, Albert M.; Stark, Charles Edwin, Jr.; Stephens, Edward Vernon; Stevenson, Raymond H.; Stine, Joseph G.; Stringer, C. Pleas, Jr.; Stringer, Richard S.; Talbott, William Charles; Talley, James Bishop; Tanner, William Francis, Jr.; Taylor, Jack Allen; Taylor, Robert Clark; Tettleton, Burvon B.; Thornton, Wayne D.; Tillman, Jack Louis; Tirey, Homer Luvois, Jr.; Tolgay, Mitat Yumnu; Umpleby, Stuart Standish; Vanderpool, Robert E.; Veal, Harry Kaufman; Ventress, William Pynchon Stewart; Vosburg, David Lee; Waddell, Dwight Ernest; Wahl, Harry Albert; Wallace, Don Lee; Walters, Donald Lee; Ward, Fred Darrell; Ware, Herbert Earl, Jr.; Watson, John; Weaver, Oscar Dee, Jr.; Webb, Frank S.; Webb, Philip K.; West, Alvin E.; White, John M., Jr.; Whitesides, Virgil Stuart, Jr.; Wilkinson, Thomas Allen; Williams, Billye Roan; Williams, Vernon Leslie; Willis, Paul Dewey; Wilmott, Charles L., Jr.; Withrow, Philip Charles; Wood, Mary Connor; Worden, John A.

1960s: Ahdoot, Hooshang; Ahmeduddin, Mir; Al-Khersan, Hashim Fadil; Albano, Lorenzo Luis; Alberstadt, Leonard Philip; Alexander, Wayland B.; Alfonsi, Pedro P.; Arro, Eric; Ausburn, Brian Edwin; Babcock, Robert Earl; Baharaloui, Abdolhossein; Barrett, Lynn Wandell, II; Barthelman, William Bruce; Bedwell, John Lewis; Bellis, William Henry; Berg, Orville Roger; Bergeson, Jerry R.; Berryhill, Richard A.; Bircum, Joe Michael; Blackwood, Charles F.; Blair, Arthur John, II; Blau, Peter E.; Blazenko, Eugene J.; Bond, Thomas Alden; Bond, Thomas Alden; Boone, Richard Lee; Bordeau, Kenneth Vernon; Bordeau, Kenneth Vernon; Bower, Richard Raymond; Bowers, John Richard; Bowlby, David C.; Bozovich, Slobodan; Bradley, Wilbur C.; Bradshaw, Donald Dean; Brookby, Harry England; Bross, Gerald L.; Brown, Robert Ludger; Bucke, David Perry, Jr.; Bucke, David Perry, Jr.; Burrough, Herman C.; Cannon, Philip Jan; Cassidy, Martin M.; Clare, Patrick Henry; Clarke, Robert Travis; Clarke, Robert Travis; Clayton, John Mason; Clements, Kenneth Paul; Cocke, Julius Marion; Cole, Joseph Glenn; Cole, Joseph Glenn; Cook, Harwin T.; Copley, Albert J.; Cronoble, William R.; Culp, Chesley Key, Jr.; Currier, John D., Jr.; Cutolo-Lozano, Francisco José; Dalton, Dale V.; Dalton, Richard Clyde; Davis, Phillip Nixon; Davis, Phillip Nixon; Dempsey, James Edward; Dempsey, James Edward; Disney, Ralph Willard; Dolly, Edward Dawson; Dolly, Edward Dawson; Duarte, Andrew Henry; Duffield, John Burton, Jr.; Durham, Charles Albert, Jr.; Edwards, William Arthur; Ellis, Charles

Allen; Ellzey, Robert T., Jr.; England, Richard L.; Ervin, James Kirk; Esfandiari, Bijan; Esfandiari, Bijan; Everett, Ardell Gordon; Fambrough, James Warren; Ferguson, David Bryan; Flood, James Ray, Jr.; Flores, Jorge G.; Fox, Walter F., Jr.; Frech, Richard Eugene; Fronjosa, Ernesto; Gafford, Edward Leighman, Jr.; Gamero, Gonzalo A.; Gamero, Maria Lourdes; Ganser, Robert W.; Garner, Gary Lee; George, John H.; Gibbons, Kenneth Edward; Gibbs, James A.; Gibson, Lee B.; Gilbert, M. Charles; Giles, Albert H.; Glaser, Gerald Clement; Glenn, David Hendrix; Gonzalez-P., Gustavo C.; Gordji, Nasser; Greene, William Mordock; Greer, Jerry Kenneth; Guest, Michael E.; Hamilton, William, Jr.; Hamric, Burt Ervin; Hamric, Burt Ervin; Hancock, James Martin, Jr.; Hanke, Harold Wayne; Hare, Ben Dean; Harris, Reginald Wilson, Jr.; Harrison, William Earl; Harvey, Ralph Leon; Haugh, Bruce Nisson; Hedlund, Richard Warren; Hedlund, Richard Warren; Helander, Donald Peter; Hellman, John Dale; Hessa, Samuel Lyndon; Higgins, Maurice J.; Hilchie, Douglas Walter; Hiss, William Louis; Hoffman, Edward Arthur, Jr.; Howery, Sherrill D.; Huang, Ying-Yan; Hunt, David Gardiner; Hurley, Patrick J.; Hutcheson, Harvie Leon, Jr.; Iranpanah, Assad; Iranpanah, Touran Soltanzadeh; Jeary, Gene L.; Jeffries, Edwin Lee; Johnson, Henry Derr, Jr.; Johnson, Kenneth S.; Jones, Eugene Laverne; Jones, Hershel Leonard; Karns, Anthony Wesley Warren; Kerns, Raymond LeRoy, Jr.; Kesebir, Mehmet; Khaiwka, Moayrad Hamid; Kidder, Gerald; Killian, Anna Mae; Kitchen, Earl William; Koutahi, Mohammed John; Krivanek, Connie Mac; Kunz, Howard E.; Kurash, George E., Jr.; Kurtz, Vincent Ellsworth; Langston, Jackson Maurice; Logan, John Merle; Logan, John Merle; Lovett, Frank D.; Lowe, Kenneth Lance; Madeley, Hulon M.; Mairs, Tom; Manley, Frederick Harrison, Jr.; Marchetti, John William, Jr.; Markas, John Mitchell; Marsh, Philip Wienecke; Matalucci, Rudolph Vincent; McDaniel, George O., Jr.; McElroy, Marcus Nelson; Meinert, Joseph G.; Mogharabi, Ataolah; Mogharabi, Ataolah; Morgan, Bill Eugene; Morgan, Bill Eugene; Morgan, Ralph Archie; Morris, Robert Jones; Musgrove, Carl D.; Nalewaik, Gerald Guy; Neff, Everett Richard; Nichols, Clayton Ralph; O'Brien, Brian E.; Olcay, Kaya Yilmay; Olson, Lawrence John, Sr.; Ospovat, Alexander Meier; Peace, H. W., II; Pereira-Soarez, Orlando; Piatt, Larry L.; Porter, John Robert, Jr.; Presley, Olan Dee; Prewit, Billie Neil; Purcell, Thomas E.; Pybas, Gerald Wayne; Rambo, Charles Edward; Rashid, Muhammad Abdur; Ray, George Dale; Redman, Robert H.; Reese, Donald Leon; Rhoads, Ray William; Richter, Robert W.; Riley, Alton O'Neil; Robb, Marion Glenn; Robertson, John Louis; Rotan, Pat Malone; Roundtree, Robert L.; Rowett, Charles Llewellyn, Jr.; Ruffin, James H.; Saint Clair, Charles Spencer; Salas, Guillermo Armando; Sanchez, Victor M.; Sanders, Robert Bruce; Sargent, Kenneth Aaron; Schramm, Martin William, Jr.; Scofield, Nancy Lou; Seely, Donald Randolph; Self, Robert Patrick; Shaarawy, Mostafa A. Razik; Sharma, Shatish K.; Sheikh, Abdul Razzak; Slate, Houston Leale; Smith, Alvin H.; Sorrel, Frank D.; Spencer, Alexander B.; Stanbrow, Gregory E., Jr.; Starke, John Metcalf, Jr.; Stewart, Gary F.; Stith, David Allen; Strong, Daniel McSpadden; Taylor, Raymond John, Jr.; Thatcher, Robert James; Thomas, Laurence E.; Todd, Harry Wayne; Trapnell, Don Edward; Trekell, Rex Elroy; Tsou, Po; Tuttle, Jesse L., Jr.; Tynan, Eugene Joseph; Urban, James Bartel; Urban, James Bartel; Urban, Logan L.; Vanbuskirk, John Reed; Vosburg, David Lee; Waddell, Dwight Ernest; Wahl, Harry Albert; Waldroop, William W.; Walker, George Ernest; Wall, Leeman Jack; Warren, Tom Hillary; Wasteneys, Richard Alan; Watkins, Henry Vaughan, Jr.; Wiggins, Virgil D.; Withers, Ronald Carlton; Withrow, Jon R.; Wolfson, Michael Stephen; Wong, Her Yue; Wong, Her Yue; Woodson, John Pierce; Wu, Dah Cheng; Wu, Dah Cheng; Young, Carl R.; Young, Leonard Maurice; Young, Robert Thomas

1970s: Agbe-Davies, Victor F.; Ahmed, Anees Uddin; Albano, Michael Anthony; Anderson, Roger N.; Anessi, Thomas Joseph; Anthony, James Michael; Archinal, Bruce Edward; Baker, Randall Keith; Bartley, John William; Basham, William Lassiter; Baumgartner, Eric Paul; Benenati, Francis E.; Benne, Robert R.; Bentley, R. H.; Benton, John William; Bonem, Rena Mae; Booth, Sherry Linette; Boras, Jaime Buitrago; Borras, Jaime Buitrago; Bradshaw, David Curtis; Breeze, Arthur F.; Bridges, Kenneth Francis; Brown, Vernon Max; Campbell, Robert A.; Caprara, John Robert; Carter, Darryl Wayne; Chapman, Jimmy Lee; Charles, Raymond Grover; Chen, James Chian-Tung; Childers, David Wayne; Clupper, David Richie; Craney, Dana Leon; Crews, Gary Alan; Cromwell, David Williams; Curran, Claude Warren; DeLuca, Frederick Peter; Diba, Mahmoud Hossein; Dionisio, Leonard C., Jr.; Donica, David R.; Donofrio, Richard R.; Dougherty, Dortha Lea; Doughty, Roger Keith; Dunagan, Joseph F., Jr.; Echols, Betty Joan; Emmendorfer, Alan Paul; Feenstra, Roger Ernest; Feinstein, Shimon; Fleury, Mark Gerald Roland; Flud, Lowell Randle; Frasca, J. W.; Gentry, Robert W.; Ghazal, Ralphael Louis; Gonzalez, Adelso Vera; Goodbread, Drew Robert; Goodwin, Elisabeth Rayner Krause; Green, Edwin Thomas; Grisso, Julie Martin; Guilarte, Fernando Anibal; Haas, Eugene Anthony; Hare, Ben Dean; Harrell, James Anthony; Harris, Reginald Wilson, Jr.; Harris, Sherod A.; Hart, Suchit Suthirachartkul; Hart, Thomas Allen; Hawk, David Harold; Henry, Thomas W.; Henry, Thomas Wood; Holtzman, Alan McKim, Jr.; Hsue, Tien Shaing; Isokrari, Ombo Ferguson; Isom, John William; Jackson, Andrew Carlton; Jha, K.; Jones, Bradley Blake; Jones, Richard Lewis; Jones, Richard Lewis; Kaup, Carl B., III; Koff, Leonid Roland; Kotila, David Arthur; Kousparis, Dimitrios; Krancer, Anthony Edward; Kumar, Subodh; Kunzer, Alexander Hourwich; Leagault, Jocelyne Andree; Lee, Sheng-Shyong; Legault, Jocelyne Andree; Lewis, Charles Downing, Jr.; Lewis, Fletcher Sherwood; Li, Ching-Chang E.; Lockwood, Mary Glenn; Lockwood, Richard Patrick; London, William W.; Luke, Robert Franklin; Lyon, Garth Monk; Main, Linda Darlene; Mannhard, Gregory William; Manz, Ronald E.; McQuillan, Michael William; Meglen, Joseph Francis; Merritt, Michael Louis; Michaelis, Sara E.; Miller, Ernest George; Miller, Thomas N.; Mistretta, Suzanne Barrerre; Morrison, Garrett Louis; Moussavi-Harami, Reza; Nelsen, J. J.; Nichols, Clayton Ralph; Nichols, Roland Franklin; Nicks, A. D.; Orgren, April Hoefner; Orgren, Mark David; Pate, James D., Jr.; Patrick, David Maxwell; Paz, Jose Gabriel; Pearson, Daniel L.; Petzel, Gerald J.; Pruatt, Martin Aaron; Pulling, David Michael; Randolph, Ellis Edwin; Rensick, David Gene; Richardson, Jimmie Larry; Rios, Nelson Guillermo; Roach, W. R.; Robinson, Andres J., Jr.; Roegner, Harold F., Jr.; Rowland, Tommy Lee; Ryan, Patrick Joseph; Saether, Ola Magne; Santiago, Donald Jose; Sargent, Kenneth Aaron; Sawyerr, Olumuyiwa Akinnade; Schultz, Douglas J.; Scott, Jerry Douglas; Sibley, Duncan Fawcett; Simpson, Larry Clark; Skinner, John Russo; Sriisraporn, Somchai; Stein, Ronald John; Stone, William Burgess, Jr.; Tapp, Gayle Standridge; Tapp, James Bryan; Thomas, John Byron; Tiab, Djebbar; Totten, Matthew Wayne; Toussaint, C. R.; Trexler, James Hugh, Jr.; Türkarslan, Muharrem; Tway, Linda Elaine; Waters, Roger Kenneth, III; Watson, Jerry Palmer; West, Ronald Robert; Westerdahl, Howard Ellsworth; Wiltse, Elliott Woodrow; Wingate, Frederick Huston; Woller, Kevin Lowell; Yim, T. B.; Yu, Jen Haur; Zabawa, Pamela Jean; Zeliff, Clifford W.; Zimbrick, Grant David

1980s: Abu-Rizaiza, Omar Seraj; Alam, Muhammad Waqi Ul; Alarcon, Alcocer Carolos Felipe; Allam, Awad Moustafa; Anno, Phil Dean; Aquilar, Jannette; Austin, Michael Neal; Axtmann, Tyrrell Charles; Bakel, Allen J.; Ballard, David Wayne; Banta, John Elliott; Barrett, Christopher Mathias;

Bartlett, Elizabeth Hancock; Bastidas, Ramon O.; Bauernfeind, Paul Edward; Becker, David G.; Benthien, Ross Howard; Bernero, Clare Ann; Bieneman, Paul Martin; Boatright, Daniel Thomas; Bonilla Franco, Jose Vicente; Brady, Karen S.; Brown, Charles Michael; Brown, Darren Leo; Brownlee, Diane Elizabeth; Bryan, Lynn Claire; Burdick, Donald G.; Butler, Thomas Allen; Butler, Thomas Harry; Callender, Alistaire Blyden Bruce; Carter, Michael Howard; Caspar, Barry Christman; Chaney, Mark Anthony; Cheng, Mo Chun; Chin, S-Len Richard; Clark, Jon Peter; Clopine, William Walter; Cochran, Karen Anne; Cole, Tony; Coughlin, Matthew Kent; Crawford, Lisa Doris; Curiale, Joseph Anthony; Dennen, Mark M.; Diez de Medina, Diana Magdalena; Dihrberg, Edward Ernest; Dikeou, Panayes John; Ditmars, Richard C.; Drexler, Timothy John; Dunn, William John; Duruewuru, Anthony U.; El-Bokle, Farouk Mohamed; Fabian, Robert S.; Farabee, Michael Jay; Fitter, Jeffrey L.; Forsyth, Prentice Mark; Fruit, David J.; Galvez Sinibaldi, Alfredo Salvador; Gasteiger, Carla Maria; Gazi, Md Nazmul Hossain; Geurin, Stanley Paul; Goldhammer, Robert Kent; Goodman, Kathleen Stack; Graetzer, Miguel K.; Grayson, Robert Calvin, Jr.; Groves, John R.; Grubbs, Robert Kent; Habibafshar, Azar; Hajali, Paris Andraos; Hayden, Joseph M.; Herd, Leslie Lee; Hidore, John Warren; Hillman, Daniel Marc Jan; Howe, James Robert; Huff, Donald Ross; Inyang, Aniefiok David; Iqbal, Ghulam M. M.; Johnson, Douglas Wade; Jones, Peter John; Kang, Joo-Myung; Khan, Mohammed Gulnawaz; Kilgore, Brian Douglas; Kim, Ki Young; Kim, Ki Young; Knox, Robert Charles; Kuhlman, Steven Larry; Larson, Dana Christine; Laughlin, Jefferson Edwin; Lee, Sa Ba; Lewis, Charles Downing, Jr.; Liesch, Aaron Robert; Lin Lewis, Sue Jane; Lin, Ching-Weei; Lin, Li-Hua; Linscott, Jeffrey Parrish; Loucks, Virginia L.; Luker, Richard Stephen; Macleod, Norman Scott; Maley, Michael Paul; Manesh, Abdulkarim Nick; Mazariegos Alfaro, Ruben Alberto; McCollum, Robert Andrew; McConnell, Cary Lewis; McCormick, Dennis Joseph; McGee, David Thomas; McLean, Thomas Richard; Medhani, Rezene Gurmu; Meek, Frederick Barber, III; Metcalf, William James, III; Michael, Gerald Eric; Milavec, Gary John; Morgan, George Beers, VI; Morgan, George Beers, VI; Mrkvicka, Steven Robert; Murgatroyd, Carolyn Drake; Murray, Donald James; Nagengast, Timothy John; Nageotte, Alton Lee; Najjar-Bawab, M. Mummtaz; Nakornthap, Kurujit; Neese, Douglas G.; Nelson, Michael Ray; Nixon, Gail Alice; Numbere, Daopu Thompson; O'Donnell, Michael Raymond; O'Neal, Marianne Victoria; Ofoh, Ebere Paulinus; Okoye, Christian Udokwu; Osanloo, Gholi Morteza; Oung, Jung Nan; Peck, Craig Jonathan; Perez, Jorge M.; Perrot, Jeannine A.; Perry, Christopher L.; Pilgrim, Alan Thomas; Pino, Henry; Prucha, Christopher P.; Quillin, Michael Edward; Radi, Mahmoud Diab; Rafalska-Bloch, Janina; Randall, Bruce Loyal; Raymer, John Herbert; Reddy, Raja Palpunuri; Reis, James Martin; Rezigh, A. A.; Rippee, David Scott; Roberts, Stephen M.; Robison, Vaughn David; Rosewitz, Lura Ellen; Roshong, Carolyn Grace; Ruffel, Alice Veronica; Saisasong, Atapon; Sawyer, Kenneth Charles, III; Schenewerk, Philip Andrew; Scott, Thomas Dwayne, Jr.; Sediqi, Atiqullah; Shinol, John Henry; Shirley, Steven Hayden; Shoup, Robert Charles; Shrestha, Rajendra K.; Smith, Steven Don; Spaulding, Karen Lee; Spooner, Jill A.; Steffens, Gary Scott; Stever, Richard Clay; Stewart, Gary C.; Sullivan, Karen Louise; Tahiri, Mohamed; Tapp, James Bryan; Tennant, Steven Hunter; Tenney, Christopher M.; Thomson, William A., III; Todd-Brown, William Edward, Jr.; Trumbly, Nancy Irene; Tsay, Siuh-Chun; Tsiris, Vassilios L.; Turmelle, Thomas Jeffrey; Tway, Linda Elaine; Ubani, Ephraim Agbawo; Van Keuren, Lewis Karl, III; Venkatesh, Eswarahalli S.; Vernon, James Hayes; Walters, Jon K.; Webb, Gregory Edward; Westerman, Julius D.; Whitney, Dan D.; Whitney, Sam Weslie; Wijeyawickrema, Chandrasi; Williams, Daniel Bernhard; Winters, Jay

Oklahoma State University

Arthur; Yang, Jun-Yang Chen; Yeh, Yaw-Huei; Yu, John Pingshun; Zdzinski, Alexander Jules

Oklahoma State University
Stillwater, OK   74078

176 Master's, 46 Doctoral

1950s: deGruchy, James H. B.

1960s: Henderson, William Garth; LeFevre, Elbert Walter, Jr.; Monahan, Edward James; Ozkol, Sedat; Stahnke, Clyde Raymond; Vowell, Bobby Gene

1970s: Abbott, Marvin Milton; Adams, Scott Randall; Al-Sumait, Abdulaziz Jasem; Alipouraghtapeh, Samad; Armstrong, Jim Richard; Astarita, Arthur Michael; Azimi, Esmaeil; Blair, John Anthony; Buckner, Duane Herbert; Burman, Howard Richard, Jr.; Candler, Charlotte Evans; Case, Harvey Lee, III; Catalano, Lee Edward; Cheung, Paul Kwon-Shun; Chinsomboon, Vichol; Cohoon, Richard Roy; Cook, Gregory Lee; Cox, Roy Edwin; Davidson, Joe Dwain; Dawud, Awni Yaqub; Eginton, Charles William; Fawzy, Aly Mahmoud; Ferguson, Jerry Duane; Ford, Gary Wayne; Fowler, Jack; Franks, James Lee; Fritz, Richard Dale; Garden, Arthur John; Gregg, Jay Mason; Hansen, Charles Allan; Hanson, Richard Eric; Heine, Richard Ralph; Hollrah, Terry Lewis; Holtzclaw, Mark John; Ireland, Jarrette Lynn; Karvelot, Michael D.; Keasler, Walter Robin; Kemmerly, Phillip Randall; Kochick, James P.; Kranak, Peter Val; Kriengsiri, Pirote; Kwang, John Ako; Lane, Steven Dale; Levings, Gary Wayne; Loo, Walter Wei-To; McGuire, Michael James; Meyer, Gary Dean; Mileff, Robert John; Morganelli, Daniel; Morris, David Gordon; Morrison, Charles Michael; Namminga, Harold Eugene; Naney, James Wesley; Nayyeri, Cyrus; Ngah, Khalid Bin; Noble, Raymond Lee; O'Bannon, Charles Edward; Olmsted, Richard Warren; Perkins, Robert Allen; Petry, Thomas Merton; Pulfrey, Robert John; Rahimi, Hassan; Reed, Thomas Willis; Robbins, Gerald Duane, Jr.; Ross, John Sawyer; Roy, Emery Bernard; Sheridan, John Francis; Shipley, Raymond Dale; Silka, Lyle Ramsay; Silker, Ted H.; Srinivasan, Venkatraman; Terrell, Don Michael; Towns, Danny Joe; Townsend, Frank Charles; Trent, William Richard; Verish, Nicholas Paul; Watts, Kenneth Robert; Wennagel, Dale Anderson; Wepfer, A. J.; White, Stephen Joseph; Williams, Charles Enyart; Wilson, Todd Montgomery

1980s: Al-Momani, Ayman Hassan; Alberta, Patricia Lynne; Allen, Roy Frank; Almoghrabi, Hamzah Abdulgader; Ammentorp, Alan David; Ausmus, Judith Erlene; Babaei, Abdolali; Back, David Bishop; Balke, Scott Carter; Bat, David Thomas; Beardall, Geoffrey Bonser, Jr.; Beauchamp, Weldon Harold; Beausoleil, Yvan Joseph; Bengtson, Richard Lee; Bentkowski, James Edward; Birdwell, Bobby Thomas; Bissell, Clinton Randall; Biyikoglu, Yusuf; Black, Grant Eugene; Bowker, Kent Alan; Bramlett, Richard Randall; Bridges, Steven Dwayne; Cairns, Janet Lorraine; Campbell, Nancy E.; Chang, Chi-Chung; Clark, James Alan; Cockrell, Dale Reed; Collins, Kathryn Hope; Ditzell, Curtis Leon; Duckwitz, George Herman; Duckwitz, Lester Dean; Dulaney, Brenda Sue; Edwards, Dwayne Ray; El-Hassanin, Adel Saad; Elrod, Dennis Dean; Ferraro, Thomas Edward; Fields, Perry Merle, III; Fies, Michael Wayne; Fleming, William Jeffrey; French, Tracy Alan; Fritts, John Raymond; Gentile, Leo Frederick; Glass, Jon Lawrence; Godard, Stephen Thomas; Gong, Jingyao; Green, James Edward Peter; Hagen, David J.; Haiduk, John Paul; Hall, Robert Lynn; Hanlon, Edward A., Jr.; Hemann, Mark Richard; Hemyari, Parichehr; Hester, William Christopher; Hooker, Ellen Ostroff; Hoque, Mohammed Mozzammel; Hossain, Syed Abul; Hoyle, Blythe Lynn; Jobe, Tracy Hutch; Johnson, Christopher Lee; Keely, Joseph Francis, Jr.; Ketcher, Austin; Kidwai, Mohammed Ali; King, Gordon Patrick; Kuykendall, Michael Douglas; Lee, Bryan Edward; Lilburn, Ralph Anthony; Lojek, Carole Ann; Lyons, Timothy Donald; Manni, Frederica M.; Markert, John Conrad; Mason, Eric Paul;

McBride, Katherine Kretow; McConnell, David Alan; McKenna, Robert Daniel; Medlock, Patrick Lee; Medlyn, Gary Wayne; Menke, Kathleen Patricia; Michlik, David Michael; Miller, Andrew Michael; Miller, Jeffery Allen; Munsil, John Michael; Nwaogazie, Ifeanyi Lawrence; O'Conner, Franklin Austin; Ohland, Grant Lawrence; Padgett, Philip C. O.; Patterson, James William, Jr.; Paukstaitis, Eric John; Pooler, Michael Lee; Poore, Clark Alan; Pybas, Kevin M.; Quinn, Kenneth Elmus; Race, Charles Dana; Race, Kelley Ann Clinton; Rafalowski, Mary Beth; Ragland, Deborah Ann; Reinsch, Thomas Glynn; Richardson, Jennifer Lynn; Robberson, Thomas Allen; Robertson, Kenneth Scott; Rountree, John H.; Schachter, Stanley Mark; Schipper, Mark Raymond; Seale, John David; Simmons, Gregory R.; Smith, Cindy Lynn; South, Mark Veeder; Swartz, Daniel Herbert, III; Sweet, Rebecca Gail; Tate, William Lewis; Teague, Richard Darnell; Tee, Deebari Porobe; Tindall, Terry Allen; Torabian, Ali; Tortorelli, Robert Louis; Tsegay, Tekleab; Udayashankar, K. V.; Vaden, David W.; Wade, Bruce Jerome; Walker, Lawrence Price; Walker, Patricia Ellen Grove; Walker, Robert Keith; Walker, Willis Lavern; Walter, R. Bryan; Wang, Wei-Ming; Wanty, Duane Allen; Way, Helen Sue Kincaid; Weber, Scott James; Wheatcraft, Suzanne Bragg; White, Harold O., Jr.; Yelken, Douglas Lynn; You, Yet-Cheng; Younger, Paul Lawrence; Yuan, Pao-Chiang

Old Dominion University
Norfolk, VA   23529

51 Master's, 10 Doctoral

1960s: Bunch, James W.

1970s: Boehmer, Walter Richard; Boone, Charles Glenn; Brush, E. R.; Campbell, D. E.; Cunningham, Richard Carson, Jr.; Domurat, G. W.; Feuillet, J. P.; Firek, F.; Fitchko, R. M.; Granat, M. A.; Hecker, Stanley; Holliday, Barry W.; Katsaounis, A.; McGrath, Dennis G.; McHone, John F., Jr.; Melchor, James R.; Needham, Bruce Harry; Nivens, William; Riggenbach, D. K.; Sanford, Robert Bailey, Jr.; Saumsiegle, W. J.; Sears, Philip C.; Weinman, Zvi H.; Wells, John Thomas; Whitlock, C. H., III

1980s: Anderson, Robert C.; Babuin, Michael L.; Barringer, Richard Alan; Berger, Thomas J.; Byrnes, Mark Richard; Collins, Eric S.; Council, Edward Augustus, III; Cross, Joseph William; Davison, Gordon E.; Emry, Janet Salyer; Evans, Allen E., Jr.; Fraser, William Brian; Garrett, Jim R.; Gary, Anthony C.; Goshorn, JoAnn H.; Jasper, Alan K.; Kang, Hyo Jin; Kunzinger, Frederick William, Jr.; Landrum, James Hanford; Lane, Daniel Stephen; Lundberg, Dennis LeRoy; Magee, Alfred W., III; Manfredi-Mathews, Terri; McDaniel, Robert C.; McMillan, T. Britt; Miller, Jack Edward; Orndorff, Randall C.; Oyler, Daniel Leland; Perillo, Gerardo Miguel Eduardo; Strauss, Ruth Ann; Takayanagi, Kazufumi; Talay, Theodore A.; Todd, James Forrest; Velinsky, David Jay; Weyenberg, Lynn Ellen

University of Oregon
Eugene, OR   97403

275 Master's, 71 Doctoral

1920s: Callaghan, Eugene; Campbell, Ian; Cox, Edwin Payne; Fraser, Donald M.; Hendon, Bryon; Lupher, Ralph Leonard; Powers, Howard Adorno; Ramsey, Margaret; Schenck, Hubert Gregory; Souza, Manuel Edward; Stovall, James Curl; Tuck, Ralph; Zimmerman, Don Z.

1930s: Allen, John Eliot; Barnes, Farrell Francis; Bogue, Richard G.; Butler, John W., Jr.; Fisk, Harold Norman; Handley, Howard W.; Marlatte, Charles Raymond; Schenk, Edward T.; Sheets, M. Meredith; Smith, Helen V.; Spreen, Christian August; Stafford, Howard Straus; Wilkinson, William Donald

1940s: Greenup, Wilbur; Isotoff, Andrei; Johns, William Roy; Roberts, Albert Eugene

1950s: Appling, Richard N.; Ashwill, Walter R.; Bales, William E.; Barlow, James L.; Beeson, John H.; Berry, Norman J.; Bowen, Richard Gordon; Bray, Richard A.; Bristow, Milton M.; Brown, Robert D.; Calkins, James A.; Carlat, James Eugene; Cheesman, William C.; Corcoran, Raymond Ervin; Cummings, Jan A.; Dale, Robert H.; Daugherty, Lloyd F.; Desonie, Dana; Doak, Robert Alvin, Jr.; Feely, Herbert W.; Feichtinger, John Rudolph; Fryberger, John S.; Gray, Wilfred Lee; Halstead, Perry Neil; Hausen, Donald M.; Humphrey, Thomas M., Jr.; James, Ellen Louise; Jolly, James H.; Jolly, Janice L.; Kennedy, Joseph Max; Ladwig, Lewis R.; Lewis, Richard Quintin; Macpherson, Bruce A.; Mathias, Donald Ernest; Merewether, Edward Allen; Meyers, Joseph Duncan; Mobley, Bruce Justin; Napper, Jack E.; Nelson, Adrian Marian; Peterson, Norman Vernon; Porter, Philip Weldon; Pritchett, Frank Ide; Privrasky, Norman Calvin; Ramp, Lenin; Richardson, Hibbard Ellsworth; Schnaible, Dean R.; Spiller, Jason; Thomas, George Martz; Thompson, Calvin John; Westhusing, James K.; Wilkerson, William Louis; Williams, Rodney King; Wolff, Ernest Nichols

1960s: Anderson, Robert Warner; Armentrout, John Myers; Bateman, Richard L.; Beeson, Marvin H.; Beeson, Melvin H.; Bezzerides, Theodore L.; Born, Stephen M.; Bradshaw, Herbert E.; Bruemmer, Jerry L.; Burns, Lary Kent; Champ, John G.; Clark, Terry E.; Crowley, Karl C.; Curry, Donald Lee; Derksen, Charlotte Meynink; Dodds, R. Kenneth; Ehlen, Judy; Elphic, Lance; Engelhardt, Claus L.; Fairchild, Roy W.; Farooqui, Saleem M.; Fifer, Henry Clay; Fouch, Thomas Dee; Gakle, Arthur Frederick; Gardner, Douglas Hansen; Girard, William W.; Godchaux, Martha Miller; Goebel, Leo R.; Green, Arthur R.; Gregory, Cecilia Dolores; Haddock, Gerald Hugh; Hagood, Allen Roland; Ham, Herbert Hoover; Harper, Kennard R.; Hauck, Samuel M.; Heinrich, M. Allen; Helming, Bob Hager; Hess, Paul Dennis; Hickman, Carole Jean Stentz; Hicks, Donald L.; Higgs, Nelson B.; Hill, John Jerome; Hixson, Harry C.; Ikeagwuani, Frederick Duaka; Jan, M. Qasim; Johnson, Arvid M.; Johnson, Wallace Ray; Kim, Chong Kwan; Kittleman, Laurence Roy, Jr.; Kleck, Wallace D.; Klohn, Melvin Larry; Lawrence, John K.; Lent, Robert Louis; Lydon, Philip Andrew; Maddox, Terrance; Magoon, Leslie B., III; McMurray, Jay Maurice; Millhollen, Gary Lloyd; Morrison, Robert Fairchild; Mossel, Leroy Gene; Nelson, Eric Bruce; Neugebauer, Henry Edwin Otto; Nowak, Michael; Patterson, Peter Vosper; Payton, Clifford Charles; Peterson, Benjamin Leland; Phillips, Robert Lawrence; Pigg, John H., Jr.; Ramer, Alan Rutledge; Redmond, John Lynn; Russell, Robert Guy; Ryden, Bonnie L. Stepp; Schetter, William Cameron; Shaw, John Holmes; Smith, David Lawrence; Stadler, Carl Albert; Stearns, Charles Edward; Stewart, Richard Lee; Switek, Michael John; Thompson, Richard Lee; Trigger, James Kendall; Weeden, Dennis Alvin; Wolff, Ernest Nichols

1970s: Alemi, Mohammad A.; Baitis, Hartmut W.; Barnes, Calvin Glenn; Beyer, Robert Lee; Bodvarsson, Gudrun M.; Bow, Craig Sherwood; Bowman, Anthony Frank; Brown, Bruce A.; Brownfield, Michael E.; Carlson, Craig Iver; Cole, D. N.; Conley, Steven Meril; Cordell, Donald Allen; Cornell, Josiah H., III; Dalheim, Peggy Ann; Dickinson, Roger G.; Donato, Mary Margaret; Donley, Michael William; Drake, Michael Julian; Edwards, Alan Frances; Endrodi, Sandra Monroe; Faulhaber, John Jacob; Ferns, M. L.; Ferrate-Felice, L. A.; Forbes, Charles Frank; Frye, Wayne Herschel; Gandera, William Edward; Gentile, John Richard; Gest, Donald Evan; Gettings, Mark Edward; Hammitt, Ray Wesley; Hanson, Kathryn Lee; Haq, Zubair Noorul; Hawley, John Edward; Henage, Lyle Frederick; Henage, Lyle Frederick; Hickcox, David Hunter; Jerome, Dominique Yves; Johannesen, Nils Poorbaugh; Jones, Charles Alan; Judkins, Thomas W.; Kim, Chong Kwan; Kozarek, Robert James; Krans, Ainslie Earl Browen; Kugler, Ralph Leonard; Leeman, William Prescott; Lewis, Fletcher

Sherwood; Lienkaemper, George William; Lindstrom, David John; Lindstrom, Marilyn Martin; Lissner, Frederick Gordon; Maynard, LeRoy Carson; McKay, Gordon Alan; Miles, Gregory Allen; Miles, Gregory Allen; Moklestad, T. Charles; Munts, Steven Rowe; Murphy, William Marshall; Naslund, Howard Richard; Olsen, Robert Roger; Perkins, James Morgan; Quine, Richard Lyle; Ritchey, Joseph L.; Robertson, Richard Douglas; Robyn, Thomas Lynn; Rottmann, Carmen Juanita Farr; Rud, John Orlin; Ryberg, Paul Thomas; Seeley, William Oran; Shuford, Marlene Eloise; Sinton, John M.; Smith, Harry Dean; Soper, Elmer Gail; Steinborn, Terry Laurence; Stembridge, James Edward, Jr.; Stoeser, Douglas Benjamin; Storm, Allen Bruce; Sutton, Kenneth George; Swanson, Frederick John; Sweet, Randolph; Tobey, Eugene Francis; Travis, Paul Leonard, Jr.; Uppuluri, Venkata Rao; Van Deusen, John Ernest, III; Wang, Wha-ching; Wells, Ray Edward; Wickstrom, Conrad Eugene; Wise, Michael Terence

1980s: Ahmad, Raisuddin; Al-Mudaiheem, Khalid Nasser; Allen, Charlotte Mary; Aranda-Gomez, Jose Jorge; Avramenko, Walter; Back, Judith Mae; Bailey, Lee Eldon; Barkeley, Susan Jeanette; Barnes, Calvin Glenn; Barnes, Melanie Ames Weed; Bestland, Erick Anthony; Boudreau, Alan Ernest; Brandon, Alan D.; Brikowski, Tom Harry; Burton, William Chapin; Carey, James William; Carpenter, Paul Kenneth; Chambers, John Mark; Christensen, Ralph Warren; Clark, James Gregory; Clingman, William Warren; Colbath, George Kent; Colbath, Sharon Larson; Cook, Sterling S.; Cooney, Thomas Francis; Cullen, Andrew; Davie, Ellen Ingraham, II; Dorais, Michael John; Feakes, Carolyn Ruth; Feiereisen, Joseph John; Flaherty, Gerard Martin; Geist, Dennis James; Goldberg, Steven Amiel; Goodfellow, Robert W.; Gray, Thomas Eastman; Grover, Timothy Warren; Grover, Timothy Warren; Hagen, Randall Alan; Hering, Carl William; Hirschmann, Marc M.; Hobbet, Randall Douglas; Hoover, James David; Houghton, Marcia Lea; Hunt, Paul Thomas; Jeddeloh, George; Johnson, Kathleen Esther; Katsura, Kurt Toshiro; King, Ellen Jean; Lang, Helen Marie; Leavitt, James Douglas; Lewison, Maureen Ann; Lieberman, Joshua Elliot; Linder, Robert Andrew; Lu, Changsheng; Mac Caskie, Dennis Raymond; McChesney, Stephen Michael; McKeown, Rosalyn Rae; Miller, Paul R.; Monroe, Sheila A.; Morris, Sandra Lee; Naslund, Howard Richard; Obermiller, Walter A.; Peretsman, Gail Sue; Petersen, Scott Walter; Pratt, Jennifer Adams; Prueher, Elizabeth M.; Rawson, Shirley Ann; Ritchie, Beatrice; Rogers, John Hiram, Jr.; Ruendal, Aime Pamela; Russell, William John; Schieber, Juergen; Seymour, Richard Scott; Sidder, Gary Brian; Smith, Grant Sackett; Soja, Constance Meredith; Sorensen, Mark Randall; Spycher, Nicolas François; Starr, James Patrick; Stimson, Eric Jordan; Tyburczy, James Albert; Urquhart, Scott Allen; Vaskey, Gordon Thomas; Vicenzi, Edward; von Bargen, Nikolaus; White, Craig McKibben; Wilgus, Cheryl Kathleen; Wilson, Nathaniel Carl; Woodland, Alan Butler; Wright, Judith; Wright-Clark, Judith

Oregon Graduate Institute of Science and Technology
Beaverton, OR 97006

1 Master's, 2 Doctoral

1980s: DeCesar, Richard T.; Johnson, Richard Lee, Jr.; Rosen, Michael Elliott

Oregon State University
Corvallis, OR 97331

401 Master's, 215 Doctoral

1930s: Felts, Wayne M.; Gonzales, B. Norman; Harris, Quinton P., Jr.; McKitrick, William Ernest; Mundorff, Maurice J.; O'Neill, Thomas Francis; Stokesbury, Walter Allen

1940s: Bandy, Orville Lee; Bowman, Flora Jean; Coleman, Robert Griffin; Dobell, Joseph Porter;

Dole, Hollis M.; Harper, Herbert E.; Hutchinson, Murl W.; Jones, Stewart M.; Leever, William H.; Lowry, Wallace D.; Mote, Richard H.; Mundorff, Norman L.; Teir, Lennart T.

1950s: Bartley, Ronald Clark; Bedford, John William; Bowers, Howard E.; Brandenburg, Norman R.; Brogan, John P.; Conrad, Clarence F.; Dawson, John W.; Deacon, Robert J.; Du Bar, Jules R.; Gray, Allan W.; Hampton, Eugene R.; Harms, James E.; Heacock, Robert L.; Herron, John E.; Hogenson, Glenmore Melvin; Howard, Conrad B.; Irish, Robert J.; Jones, Robert W.; McIntyre, Loren B.; Moore, James F., Jr.; Nesbit, Robert A.; Ogren, David Ernest; Pilcher, Stephen H.; Schlicker, Herbert G.; Shelton-V, Bert J.; Smith, Raymond I.; Snook, James R.; Swarbrick, James C.; Taubeneck, William Harris; Waisgerber, William

1960s: Albin, Arthur G.; Alexander, John B.; Banks, Ernest Robey; Beal, Miah Allan; Beasley, Thomas M.; Blanton, Jackson O.; Boettcher, Richard Scott; Bourke, Robert H.; Bradford, Wesley L.; Bubb, John Neal; Buffo, Lynn Karen; Bushnell, David; Carlson, Paul Roland; Carlton, Richard W.; Carnahan, Gary L.; Carstea, Dumitru Dumitru; Chambers, David Marshall; Chiburis, Edward Frank; Christie, Harold Hans; Cissell, Milton Charles; Collins, Curtis A.; Connors, Donald Nason; Couch, Richard William; Culberson, Charles H.; Cutshall, Norman Hollis; Duedall, Iver Warren; Duncan, John Russell, Jr.; Ellison, Bruce Edward; Emilia, David Arthur; Ensminger, Henry R.; Erickson, Barrett H.; Fankhauser, Robert E.; Favorite, Felix; Forth, Michael; Frederick, Lawrence Churchill; Gallagher, John Neil; George, Gene Richard; Glasheen, Richard Michael; Glenn, Jerry Lee; Greene, Frank F.; Griggs, Gary Bruce; Hanson, Peter James; Hartman, Donald A.; Hill, David R.; Hobbs, Billy B.; Hunger, Arthur A.; Hutt, Jeremy R.; Ingham, Merton Charles; Jarman, Gary D.; Jennings, C. David; Jennings, Charles David; Johnson, George D.; Johnson, Vernon Gene; Jones, Mark M.; Kester, Dana Ray; Krammes, Jay Samuel; Kulm, LaVerne Duane; Lane, Robert Kenneth; Lauer, Timothy Campbell; Laun, Philip Royal; Lee, Kuo-heng; Lembach, Dixie Jane; Long, Leland Timothy; Lukanuski, James N.; Mackay, Angus James; Maloney, Neil J.; Maloney, Neil Joseph; Manske, Douglas Charles; Manske, Douglas Charles; Mathews, Frank Samuel; Matson, Adrian L.; McKnight, William Ross; Mesecar, Roderick Smit; Muntzert, James K.; Nelson, Carlton Hans; North, William Benjamin; Odegard, Mark E.; Olson, Boyd E.; Paeth, Robert Carl; Papageorge, George Elefterios; Patterson, Robert L.; Ray, Jimmie Dell; Renfro, William C.; Rinehart, Verrill Joanne; Rosé, Robert Rowland; Runge, Erwin John, Jr.; Russell, Kenneth Lloyd; Sarmah, Suryya Kanta; Schatz, Clifford Eugene; Shih, Keh-gong; Singleton, Paul C.; Skorpen, Allan J.; Skov, Niels Aage; Skov, Niels Aage; Slater, Mical N.; Souders, Robert H.; Stevenson, Merritt Raymond; Stump, Arthur Darrell; Taylor, Edward M.; Templeton, Bonnie Carolyn; Thiruvathukal, John Varkey; Thronton, Edward Bennett; Trembly, Lynn Dale; Trembly, Lynn Dale; Vallier, Tracy Lowell; Vossler, Donald A.; Westermann, Dale Thomas; Wetherell, Clyde E.; Whitcomb, James; White, Willis Harkness; White, Willis, H.; Wood, Allan D.; Yao, Neng-chun

1970s: Abuzkhar, Ahmed A.; Adotevi-Akue, George Modesto; Aguilar-Tuñón, Nicholás A.; Allen, David William; Allmaras, J. M.; Alvarez-Borrego, Saúl; Anderson, David Lawrence; Atlas, Elliot L.; Atzet, T.; Axelsen, Claus; Barday, Robert J.; Barnes, Thomas John; Baumgartner, T. R.; Bée, Michel; Boler, Frances Michele; Borchardt, Glenn Arnold; Bourke, Robert Hathaway; Bowman, Kenneth Charles, Jr.; Brewster, Nancy A.; Briceño-Guarupe, Luis Alberto; Brown, Randall B.; Bruce, Wayne Royal; Bryant, George T.; Burnham, Rollins; Calderón Riveroll, Gustavo; Carlton, Richard W.; Carter, Jack M.; Carter, James W.; Chintakovid, Vanit; Clark, James G.; Cohen, Yuval; Coperude, Shane Patrick; Coryell, George Fossas;

Cox, Gregory M.; Cressy, Frank B., Jr.; Culberson, Charles Henry; Cunningham, Cynthia Taylor; Dauphin, Joseph Paul; Davenport, Ronald Edmond; De Keyser, Thomas Lee; Dingus, Delmar D.; Dinkelman, Menno Gustaaf; Doak, William H.; Dotter, Jay Albert; Dowding, Lynn G.; Drake, Edwin A.; Dudas, Marvin Joseph; Dyhrman, R. F.; Eggers, Dwight E.; Eggers, Dwight Edward; Eklund, W. A.; Elliott, Monty Arthur; Ellis, David Burl; Enfield, David B.; Evans, David William; Evans, David William; Fiske, Douglas A.; Flegal, A. R., Jr.; Flory, Richard A.; Franklin, Wesley Earlynne; Franklin, William Talbert; French, William Stanley; Frey, B. E.; Gaughan, Michael K.; Gaughan, Michael K.; Gillespie, Clinton D.; Goebel, Vaughn; Goodwin, Clinton J.; Graham, Rhea L.; Graichen, Ronald E.; Greenman, Celia; Gumma, William H.; Gunther, Frederick J.; Hakes, Peter O.; Hammermeister, Dale P.; Hammitt, Jay W.; Hanson, Jonathan M.; Hanson, William B.; Harle, David Sig; Harlett, John Charles; Harris, Billy L.; Harrold, Jerry; Hassanzadah, Siamak; Heggen, R. J.; Hemstrom, Miles A.; Henricksen, Thomas A.; Hewitt, Samuel L.; Hodges, Wade Allan; Hodler, Thomas W.; Hudson, Jon P.; Huehn, Bruce; Huggins, Jonathan Wayne; Isaacson, Laurie Brown; Isaacson, Peter Edwin; Jarman, Clara Birchak; Jenne, David Allen; Jeter, Hewitt Webb; Johnson, Floyd R.; Johnson, Kenneth S.; Johnson, Stephen H.; Johnson, Vernon Gene; Jones, Michael B.; Jones, Paul R., III; Jones, Paul R., III; Kachelmeyer, John Michael; Karlin, Robert; Keene, Donald F.; Kendall, George W.; Kendrick, John W.; Keser, Judith; Kim, So Gu; King, John R.; Kitchen, J. C.; Klanderman, David S.; Koch, William F., II; Kowsmann, Renato Oscar; Larsen, David P.; Larson, Douglas William; Leinen, Margaret; Lenaers, W. Michael; Lidstrom, John Walter, Jr.; Lillie, R. J.; Lizarraga, Arciniéga Jose R.; Long, Roney C.; Lopez, Carlos; Lowell, Robert Paul; Lu, Richard Shih-Ming; Lu, Richard Shih-Ming; Lyle, Mitchell; Lynn, Walter S.; MacFarlane, William T.; Malfait, Bruce Terry; Martell, C. Michael; Masias Echegaray, Juan A.; McKinney, Barbara A.; McKnight, Brian Keith; Miller, Martin C.; Milnes, Peter Treadwell; Molina-Cruz, Adolfo; Molina-Cruz, Adolfo; Moran, John L.; Muehlberg, Gary E.; Muellenhoff, William P.; Mullen, E. D.; Murray, Clyde L.; Nachman, Daniel A.; Naidu, Janakiram Ramaswamy; Neel, Robert H.; Nelsen, Terry Allen; Nelson, Michael Peter; Ness, Gordon Everett; Niebuhr, Walter W.; Olson, Gregory A.; Olson, James P.; Oser, Robert K.; Owen, Philip C.; Packard, John A., Jr.; Paeth, Robert Carl; Parker, Donald James; Peargin, Thomas R.; Penoyer, Peter E.; Pequegnat, J. E.; Peters, James F.; Peterson, Robert E.; Peterson, Robert E.; Phipps, James Benjamin; Pisias, N. G.; Pitts, Gerald Stephen; Plawman, Thomas Leon; Pope, Stephen Van Wyck; Porter, Richard W.; Potter, Eric C.; Prince, R. A.; Pungrassami, Thongchai; Rabii, H. A.; Rai, Dhanpat; Rea, Campbell C.; Rea, David K.; Reckendorf, Frank Fred; Ridlon, James Barr; Rinne, Richard W.; Roach, David M.; Rohr, David M.; Rohr, David. M.; Rojanasoonthon, Santhad; Rollins, Anthony; Rooth, Guy Harlan; Rosato, V. J.; Roush, Robert C.; Rowe, Winthrop A.; Sancetta, Constance A.; Schalla, Robert Allen; Schaubs, Michael Paul; Scheidegger, Kenneth Fred; Scheidt, Ronald C.; Schriener, Alexander, Jr.; Schweller, William J.; Selk, Bruce W.; Sharp, George Carter, Jr.; Shorey, Edwin J.; Shyu, Chuen T.; Simmons, Michael L.; Skurla, Steven J.; Smith, Roy E.; Smith, Roy Edward; Smith, Thomas N.; Spigai, Joseph John; Spycher, G.; Stakes, Debra S.; Stensland, Donald E.; Stickney, Roger B.; Storch, Sara Glen Power; Sturdavant, Charles D.; Su, Chong-G.; Swift, Stephen A.; Tang, Rex Wai Yuen; Taskey, Ronald D.; Terich, Thomas Anthony; Thrasher, Glenn P.; Tolle, Timothy V.; Tolson, Patrick M.; Toth, John R.; Tower, Dennis B.; Trauba, W. C.; Trojan, William R.; Tucker, Elizabeth R.; Utterback, W. C.; Vail, Scott G.; Van Atta, Robert Otis; Victor, Linda; Wakeham, Susan Elizabeth; Wallace, William John, Jr.; Walsh, Stephen J.; Weinkauf, Ronald

Albert; Wendell, William G.; Wenkam, Chiye; White, Richard E.; Whitsett, Robert M.; Wise, Diane; Woodcock, S. F.; Wracher, David A.; Wrolstad, Keith H.; Yee, Carlton S.; Zaneveld, Jacques Ronald Victor; Zdanowicz, Ted. A.

1980s: Amegee, Kodjo Yahwondu; Avera, William Edgar; Axelsson, Gudni; Axelsson, Gudni; Azim, Bilqees; Baba, Jumpei; Baba, Jumpei; Badayos, Rodrigo Briones; Bailey, Mark H.; Baker, Dan M.; Baker, Linda J.; Barrow-Hulbert, Sarah A.; Bartlett, Mark William; Baumgartner, Timothy Robert; Bée, Michel; Bingert, Neil J.; Blodgett, Robert B.; Bonelli, Douglas T.; Boss, Theodore Robert; Braman, David E.; Brandsdóttir, Bryndis; Briskey, Joseph A., Jr.; Brown, James O.; Brunson, Burlie Allen; Busch, William Henry; Butler, James Hall; Calhoun, Jeannette A.; Cannon, Debra May; Casaceli, Robert J.; Chauvot, Isabelle P.; Chesley, John Theodore; Clark, Charles W.; Clayton, James Lindou; Clemens, Karen E.; Connard, Gerald George; Conrey, Richard M.; Cook, Jeffrey A.; Cooper, David Michael; Cowell, Peter F.; Cranswick, Mark S.; Cullough, Dale Alan; Dahm, Clifford Neal; Davies-Colley, Robert James; De Den, F. Michael; Dippon, Duane Roy; Donegan, David P.; Doucette, John; Douglas, Clyde Lee, Jr.; Drake, Ellen Tan; Dubendorff, Bruce H.; Duroy, Yannick; Ehret, Gayle Ann; Elfrink, Neil; Elrick, Maya; Emerick, Christina Marie; Fahlstrom, Beverly E.; Federman, Alan Neil; Fifarek, Richard H.; Finney, Bruce Preston; Fischer, Kathleen Mary Brigid; Fok-Pun, Luis; Foote, Robert W.; Freeman, Lawrence K.; Ganoe, Steven J.; Garrow, Holly C.; Glasmann, Joseph Reed; Goalan, Jeffrey S.; Goetze, Brigitte Ricarda; Goodhue, William V., Jr.; Goodman, Dean; Graham, John Paul; Grigsby, F. Bryan; Hamilton, William L.; Han, Myung Woo; Hanson, David W.; Hayman, Glenn A.; Hays, Patricia E.; Heironimus, Thurman L.; Hicks, Bryan A.; Higinbotham, Larry R.; Hildreth, Gail Darice; Hill, Brittain Eames; Hillesland, Larry L.; Hoffman, Sarah Elizabeth; Holliday, Joseph; Hook, Richard; Hoover, Amy L.; Horning, Thomas S.; Howd, Peter A.; Huftile, Gary J.; Hughes, Scott Stevens; Huppunen, JoAnne Louise; Ibach, Darrell Henry; Ibach, Lynne E. Johnson; Jackson, Patrick Allan; Jackson, William Longstreth; Jakes, Mary Clare; Jarvis, Linda Jane; Jaumé, Steven C.; Jay, Jeremy Barth; Karlin, Robert; Kaufmann, Philip Robert; Keats, Donna G.; Key, Colin F.; Kim, Tae In; Klecker, Richard A.; Krissek, Lawrence Alan; Kuiper, John L.; Lane, Jeffrey W.; Leathers, Michael R.; Lee, Jaw-Fang; Lenhard, Robert James; Li, Zhenlin; Lilley, Marvin Douglas; Lipka, Joseph T., II; Little, Stephen W.; Loken, Trygve; Loubere, Paul Walter; Mack, Gregory Stebbins; Madin, Ian P.; Mahood, Richard O.; Marston, Richard Alan; Matherne, Anne Marie; McCarter, Paul; McCulla, Michael S.; McDonald, William P.; McDougall, James William; McDougall, James William; McDowell, Theodore R.; McHugh, Margaret H.; McLain, William Henry; Mikulic, Donald George; Mitchell, Terry Edward; Moser, John Christian; Mullen, Ellen D.; Murphy, Kim Marie; Murphy, Thomas M.; Murray, David W.; Murray, David William; Naim, Shamim; Nellis, Marvin Duane; Nelson, David Ezra; Nelson, Dennis O.; Ness, Gordon Everett; Nevins, Barbara B.; Oakley, Stewart M.; Olbinski, James S.; Olson, Daniel J.; Omana, Miguel Angel Alvarado; Palmer-Rosenberg, Paul S.; Parsons, Michael Raymond; Pearson, Monte L.; Pendergast, Margaret A.; Pennock, Edward S.; Peterson, Carolyn Pugh; Peterson, Curt Daniel; Piepgras, Donald J.; Pikul, Joseph Lawrence, Jr.; Potter, Alfred Warren; Poujol, Michel; Powell, Heidi Sara; Power, Sara Glen; Price, Georgina; Priest, George R.; Prihar, Douglas W.; Reimers, Clare Elizabeth; Richards, Matthew E.; Ricks, Cynthia L.; Sadler, R. Kumbe; Sánchez Zamora, Osvaldo; Sanchez Zamora, Oswaldo; Sans, Roger Stephen; Sansome, Constance Jefferson; Sarewitz, Daniel R.; Schneider, Richard C.; Schuette, Gretchen; Schultz, Karin L.; Schulz, Michael Gerhard; Seedorf, Douglas Christopher; Shepard, P. J.; Sholin, Michael Hugh; Sidder, Gray Brian;

Smith, Gary Allen; Solano-Borrego, Ariel E.; Solano-Borrego, Ariel Enrique; Spitz, Herbert M.; Stitt, Leonard Timothy; Swanson, Sherman Roger; Thomason, Robert Edward; Thormahlen, David J.; Thornburg, Todd Mark; Thornburg, Todd Mark; Thrall, F.G.; Van Heeswijk, Marijke; Veen, Cynthia A.; Verplanck, Emily Pierce; Verplanck, Philip L.; Villamayor, Faustino Paysan; Visconti, Robert Vincent; von Breymann, Marta T.; Wallace, Alan Ryon; Walsh, Ian David; Weidenheim, Jan Peter; Weliky, Karen; Whitaker, John Henry; White, David Dean; Wozniak, Karl C.; Yogodzinski, Gene M.

University of Ottawa
Ottawa, ON   K1N 6N5

52 Master's, 34 Doctoral

1960s: Barnes, C. R.; Broad, D.; Coakley, J.; Cote, Philip Richard; Giguere, J.; Legault, J.; Mayr, U.; Miall, Andrew; Sikander, A. H.; Steele, M.; Suparman, Agus; Tauchid, Mohamad; Tuke, M. F.

1970s: Bissonnette, R.; Brand, Uwe; Carrara, Alberto; Cass, John I.; Cheel, Richard James; Dixon, J.; Durocher, M. E.; Francoeur, D.; Gibling, Martin R.; Hartree, Ron; Jen, L.-S.; Jones, B.; Kamineni, D. C.; Koster, E. H.; Krausz, K.; Lafond, J. M.; Lafontaine, Michel Albert Georges; Lapointe, P.; Nixon, C. M.; Rambaldi, E. R.; Rao Divi, Sri Ramachandra; Rey, N. A. C.; Rivers, C.; Robinson, S. D.; Romanelli, R.; Savelle, J. M.; Sempels, Jean-Marie; Singh, Sudesh Kumar; Sozanski, A. G.; Walker, W.; Waslenchuk, D. G.; Wong, A. S.

1980s: Al-Aasm, Ihsan Shakir; Best, M. A.; Bottomley, Dennis James; Burbidge, Geoffrey Harrison; Campbell, Ian D.; Culshaw, Nicholas G.; D'Iorio, Marc A.; Digel, Mark Richard; Diles, Shawn James; Dillon-Leitch, Henry C. H.; Fabbri, A. G.; Graf, Gary C.; Grunsky, Eric Christopher; Habib, M. K.; Holroyd, Michael Thomas; Hurdle, E. J.; Kerr, Daniel Ernest; Kilias, Stephanos; Lafleur, Pierre Jean; Legun, Andrew S.; Majid, Abdul Hamid; Masson, Arthur Guy; McQuade, B. N.; Muir, Iain D.; Muir, Iain D.; Naldrett, Dana L.; Naldrett, Dana L.; Narbonne, G. M.; Nazli, Kazim; Packard, Jeffrey J.; Paktunc, A. Dogan; Poey, Jean-Luc; Reny, J. J. P. Gilles; Roach, Dan; Salas, Carlos J.; Thomas, Peter B.; Wadleigh, Moire Anne; Wright, Daniel Frederick; Zaitlin, Brian A.

Pacific Lutheran University
Tacoma, WA

1 Doctoral

1980s: Conley, R. H.

University of Pennsylvania
Philadelphia, PA   19104

14 Master's, 40 Doctoral

1890s: Brown, Amos Peaslee; Ehrenfeld, Frederick

1900s: Travis, Charles; Wherry, Edgar Theodore

1910s: Lefferts, Walter

1920s: Oldach, Frederick Maier; Storm, Paul Jennings; Wanner, Henry Eckert

1930s: Berman, Joseph; Howard, Edgar B.; Postel, Albert Williams; Postel, Albert Williams; Storm, Paul Jennings

1940s: Chaffe, Robert Gibson

1960s: Kovisars, Leons; Laub, Mary G.; Ludman, Allan; Petersen, N. F.; Smith, R. S.; Way, J. H., Jr.

1970s: Button, R. M.; Drost, B. W.; Friel, J. J.; Grauch, Richard Irons; Heffner, J. D.; Lou, Y.-S.; Lutz, T. M.; Pasternack, E. S.; Ralph, Elizabeth Kennedy; Reimer, G. Michael; Schluger, Paul Randolph; Serencsits, Colleen McCabe; Socratous, G.; Stewart, Michael K.; Thomas, Herman Hoit; Thomson, Margaret C.; Weingarten, Baruch; Zimmermann, R. A.

1980s: Andersen, Sarah Beale; Friedland, Andrew Jay; Handy Barringer, Julia L.; Hegemann, David Alan; Huang, Shu-Li; Kaplan, Louis Arnold; Lonergan, Stephen Colnon; Lu Huan-Zhang; Mersky, Ronald Lee; Omar, Gomaa Ibrahim; Pardi, Richard Raymond; Scheihing, Mark Henry; Srogi, Elizabeth Lee Ann; Turner, Robert Spilman; Weishampel, David Bruce; Wnuk, Christopher

Pennsylvania State University, University Park
University Park, PA   16802

651 Master's, 537 Doctoral

1890s: Atkinson, Elizabeth Allen

1910s: Cathcart, Stanley Holman; Caudill, Samuel Jefferson; Smith, Lloyd Beecher

1920s: Ayers, Vincent Leonard; Fisher, Lloyd Wellington; Graeber, Charles Karsner; Steele, Frederick Abbott; Taylor, Thomas Garrett

1930s: Barnes, Kenneth Burton; Jones, J. Robert; Levine, Joseph Samuel; Lewis, James Albert; Rosenkrans, Robert Russell; Swain, Frederick Morrill, Jr.; Whitmore, Frank Clifford, Jr.; Young, George Husband

1940s: Adams, Linn Frank; Bacon, Lloyal Orrin; Bye, Doris Lippincott; Calhoun, John C., Jr.; Doan, David Bentley; Ferm, John Charles; Haworth, Charles C.; Kennedy, Vance Clifford; Landis, Sam Wallace; Levine, Joseph Samuel; Mangus, Marvin Dale; Miranda Barbosa, Aluzio Liciniode; Mitchell, Lane; Pelto, Chester Robert; Ping, Kuo; Rees, Rhys Willis; Richardson, Eugene Stanley, Jr.; Roy, Della Martin; Siefert, August Carl; Tait, Donald Burkholder; Tuttle, Orville Frank; Zeller, Robert Allen, Jr.

1950s: Aleshin, Eugene; Andrews, Alday Bishop; Andrews, Alday Bishop; Ansari, A. M. Azheruddin; Bailey, Ralph Fraser; Berg, Joseph Wilbur, Jr.; Berg, Joseph Wilbur, Jr.; Biederman, Edwin Williams, Jr.; Bieler, Barrie Hill; Bressler, Calder Tupper; Brune, William Arthur; Buckner, Dean Alan; Budenstein, David; Cadigan, Robert Allen; Cameron, Alexander Rankin; Carpenter, David W.; Cervik, Joseph; Channabasappa, Kenkere C.; Chase, Leonard Richard; Chen, Wen-Lan; Cobb, Robert Eugene; Cook, Donald Jean; Cook, John Call; Crowley, Michael Summers; Curray, Joseph Ross; Dachille, Frank; DeVries, Robert Charles; Dolsen, Charles Philip; Donaldson, Alan Chase; Duecker, John Cecil; Duey, Herbert David; Emerson, Donald Orville; Emerson, Donald Orville; Emery, John Rathbone; Ferm, John Charles; Fetterman, James William; Flaschen, Steward Samuel; Folk, Robert Louis; Folk, Robert Louis; Gee, Kenneth Homer; Gerhard, Jacob Esterly; Glasser, Frederick Paul; Good, Richard Standish; Grace, John Dale; Gross, C. M.; Gross, Geraldo Wolfgang; Haines, David Vincent; Hambleton, Harvey Jay; Haney, Warren Dale; Hasson, Mohey El-Din M. T.; Hill, Vincent G.; Hobson, John Peter; Holmes, Charles Robert; Hough, Van Ness Dearborn; Huber, Robert Evans; Hubert, John F.; Hulbe, Christoph W. H.; Hutta, Joseph John; Illsley, Charles Truman; Insley, Robert Hiteshew; Jacobsen, Lynn; Jarmell, Solomon; Johnson, Gordon Harene; Kahn, James Steven; Kaukonen, Everett Konstantine; Keller, George Vernon; Keller, George Vernon; Klemic, Harry; Klingsberg, Cyrus; Landy, Richard Allen; Lane, Maurice Vincent; Leonard, Arnold David; Licastro, Pasquale Hallison; Licastro, Pasquale Hallison; Luedemann, Lois Ann Weiser; Luft, Stanley Jeremie; Macauley, George Raymond, Jr.; MacChesney, John Burnette; Majumdar, Amalendu J.; Mathur, Surendra Pratap; McConnell, Elliott Bonnell, Jr.; McGlade, William George; McIntyre, Donald David; Mellon, George Barry; Metz, Clyde Thomas; Montgomery, Hugh Brinton; Muan, Arnulf Ingau; Mumpton, Frederick Albert; Mumpton, Frederick Albert; Nagy, Bartholomew Stephen; Navias, Robert Alexander; Neavel, Richard Charles; Nelson, Bruce Warren; O'Neil, Robert Lester; Outlaw, Donald Elmer; Palacas, James G.; Percival, Stephen F.; Phillips, Bert;

Putman, George Wendell; Riffelmacher, Wallace Edwin; Roberts, John Lenox; Rones, Morris; Rosenfeld, Melvin Arthur; Rosenfeld, Melvin Arthur; Roy, Della Martin; Ruiz-Menacho, Carmen Maria; Russell, Edmund Louis, Jr.; Ryan, John Arthur; Saha, Prasenjit K.; Sand, Leonard Bertram; Schneider, Allan Frank; Shadle, Harry Wallace; Shafer, Elena Camilli; Shulhof, William Peter; Siegrist, Henry Galt, Jr.; Silverman, Eugene Norton; Smith, Chester Martin, Jr.; Stanley, Edward Alexander; Stanonis, Frank L.; Steinmetz, Richard; Stengle, Eugene Henry; Strahl, Erwin O.; Strickler, Donald Ward; Sutton, Willard Holmes; Swineford, Ada; Tasch, Paul; Terriere, Robert Theodore; Thamm, John Kenneth; Turley, Mitchell Reed; Van Harryok, Harry Jerrold; Van Hook, Harry Jerrold; Vozoff, Keeva; Warshaw, Charlotte Marsh; Warshaw, Israel; Watson, Robert Joseph; Weaver, Charles Edward; Weaver, Charles Edwin; Weitz, John Hills; White, Eugene Wilbert; Wigginton, William Barclay; Williams, Eugene Griffin; Wintermute, Thomas Judson; Wyble, Donald O.

1960s: Ackermann, Hans D.; Adler, Alan A.; Andrews, Peter William; Badger, William Barton; Balgord, William Dwyer; Barker, M. S.; Barker, Richard M.; Barks, Ronald E.; Barnard, Walther M.; Barry, Thomas Leo; Bates, Carl Hobart; Bauer, John W.; Bayer, James Lawrence; Benedict, Louis G.; Bennett, Gordon D.; Bennion, Douglas Wilford; Bergenback, Richard Edward; Berman, Roslyn; Berry, William Francis; Beutner, Edward Chandler; Biggers, James Virgil; Binder, Charles Regis; Boettcher, Arthur Lee; Boettcher, Arthur Lee; Brenner, Gilbert Jay; Brinkley, Charles Alexander; Bubeck, Robert Clayton; Butler, Phillip Edward; Calkins, James A.; Cameron, Alexander Rankin; Cano, Octavio; Carman, John Homer; Caruccio, Frank Thomas; Caruccio, Frank Thomas; Caslavsky, Jaroslav Ladislav; Cassidy, William Arthur; Chao, Tze; Cheng, Yung-Yu; Ciciarelli, John Anthony; Clark, George Michael; Clark, John Harris; Cline, George Douglas; Coatney, Richard Lee; Cochran, John A.; Cohen, Arthur David; Cohen, Howard Melvin; Coleman, K. Fred; Cook, Donald Jean; Cooley, Richard Lewis; Crawford, Samuel W.; Crelling, John Crawford; Dahl, Hilbert Douglas; Dahlberg, Eric Charles; Dahlberg, Eric Charles; Datta, Ranajit Kumar; Davis, Alan; Deike, George Herman, III; Deines, Peter; Deines, Peter; Deuser, Werner George; Deuser, Werner George; Dolliver, Claire Vincent; Douglass, Peter Mack; Drew, Lawrence James; Drew, Lawrence James; Duffy, Leo Joseph; Dunn, Peter Ayres; Durisek, E. Jane; Dutcher, Russell Richardson; Eichler, Roland; Elbel, William P.; Ellison, Robert Lee; Erickson, Edwin Sylvester, Jr.; Falla, William S., Jr.; Fang, Jen-Ho; Farlekas, George M.; Fauth, John Louis; Fauth, John Louis; Fetterman, James William; Fish, Ferol Fredric, Jr.; Fisher, James Russell; Flock, William Merle; Flueckinger, Linda Ann; Franz, Gilbert Wayne; Frederiksen, Norman Oliver; Frye, John Keith; Fudali, Robert F.; Gard, Theodore Max; Gardner, Leonard Robert; Gardner, Leonard Robert; Gedde, Roger W.; Gehris, Clarence Winfred; Gerencher, Joseph James; Ghaffer-Adly, Rahmat; Gibbs, Gerald V.; Gillis, John William; Graham, Earl Kendall, Jr.; Grender, Gordon Conrad; Grisafe, David Anthony; Grisafe, David Anthony; Groth, Linda Williamson; Groth, Peter K. H.; Grutzeck, Michael William; Gucwa, John Henry; Haas, John Lewis; Habib, Daniel; Haefner, Richard Charles; Hait, Mortimer Hall, Jr.; Halbig, Joseph B.; Halbig, Joseph B.; Hand, Bryce Moyer; Hanson, Henry William Andrew, III; Hanson, Henry William Andrew, III; Harris, DeVerle Porter; Hawkins, Daniel Ballou; Hea, James Paul; Hedberg, William H.; Hinckley, David Narwyn; Horowitz, Daniel Henry; Huh, Oscar Karl, Jr.; Huh, Oscar Karl, Jr.; Hutta, Joseph John; Ikawa, Haruyoshi; Irving, Stephen Myles; Jobling, John Lloyd; Johnson, Arvid Mauritz; Johnson, Gerald Glenn, Jr.; Johnson, Philip W.; Keim, James Will; Kelley, William N., Jr.; Kilinc, Ishak Attila; Kilinc, Ishak Attila; Kim, Kee Hyong; Kimura, Shigeyuki; Kissin, Stephen

A.; Knowles, Raymond Robert; Knowles, Raymond Robert; Koesters, Baerbel; Konikow, Leonard Franklin; Kreidler, Eric Russell; Kunasz, Ihor Andrew; Landis, Charles A., Jr.; Landon, Ronald Arthur; Landy, Richard Allen; Lane, Burke E.; Lavery, Norman Garnsey; Lavin, Peter M.; LeBlanc, Gabriel; Lees, John Allen; Lenker, Earle Scott; Lifschutz, Arthur Paul; Lin, Jin-Long; Liu, Hok-Shing; Lucchitta, Baerbel Koesters; Lucchitta, Ivo; Lundquist, Gary M.; Luth, William Clair; M'Gonigle, John William; Machamer, Jerome Frank; MacKenzie, George Donald; Maneval, David R.; Mansfield, S. P.; Martin, Robert F. C.; Mathers, Lewis John; Matzke, Richard H.; McCarl, Henry Newton; McCarl, Henry Newton; McCauley, James Weymann; McCauley, Ronald Arthur; McEachern, Slater E., Jr.; McKague, Herbert Lawrence; McKinstry, Herbert Alden; Meagher, Edward Patrick; Medlin, Jack Harold; Merdler, Stephen C.; Merkel, Richard H.; Merrin, Seymour; Meyer, Harvey John; Middleton, Bruce Donald; Miller, Harold Ellis, Jr.; Modarresi, Hassan Ghavami; Modarrsei, Hassan Ghavami; Moench, Allen F.; Monmonier, Mark Stephen; Moore, Raymond Kenworthy; Morris, Arthur Edward; Nafziger, Ralph Hamilton; Nakamura, Yosio; Neavel, Richard Charles; Nickey, David Allen; Novak, Gary A.; Okuma, Angelo Frederick; Ondrick, Charles William; Ondrick, Charles William; Ostrander, William J.; Ougland, Ronald M.; Pachman, Jerrold Marvin; Perry, Frederick Welford; Pfluke, John Henry; Pillay, K. K. Sivasankara; Piwinski, Alf J.; Popovich, Daniel Eugene; Poth, Charles Warner; Powdysocki, Melvin Henri; Presnall, Dean Carl; Putman, George Wendell; Ragone, Stephen Edward; Rahn, Perry Hendricks; Ramspott, Lawrence D.; Rankin, William E.; Rao, B. V. Parameswara; Rapp, George Robert, Jr.; Ratcliffe, Nicholas Morley; Redmond, John Charles; Reidenouer, David Raymond; Riegel, Walter Leonard; Roberts, Michael Taylor; Robertson, P. Blyth; Roeder, Peter L.; Romberger, Samuel Bergstresser; Rosenberg, Philip E.; Ross, Howard Persing; Ross, Howard Persing; Rothman, Robert L.; Ryder, Robert Thomas; Savanick, George Adrian; Schlaudt, Charles McCammon; Schleicher, David Lawrence; Scott, Steven Donald; Segovia Nerhot, Antonio Valentin; Shade, John William; Shank, John C.; Shulhof, William Peter; Siddiqui, Shamsul Hasan; Siegrist, Henry Galt, Jr.; Simonberg, Elliott Mark; Simons, Philip Yale; Singer, Donald Allen; Siskind, David Eugene; Smith, Chester Martin, Jr.; Smith, Jan G.; Smith, Richard Elbridge; Smith, Ronald D.; Smith, William Gill; Sohon, Robert S.; Sommer, Sheldon Emanuel; Speidel, David Harold; Spelman, Allen Rathjen; Spengler, Charles Joseph; Staley, Walter Goodwin, Jr.; Stanley, Edward Alexander; Stingelin, Ronald Werner; Tan, Francis C.; Taylor, Allan Maurice; Taylor, P. T.; Taylor, Robert Wesley; Thompson, Richard Rogers; Thompson, Thomas Dick, III; Thompson, Thomas Dick, III; Ting, Francis Ta-Chuan; Trafton, Burke O.; Tressler, Richard Ernest; Trotter, Charles Leonard; Turner, William Morrow; Tuthill, Rosalind L.; Ujueta, Guillermo; Ulmer, Gene Carlton; Vaughn, David E. W.; Viletto, John, Jr.; Walter, Louis Simon; Wang, Yuan; Warshaw, Israel; Watkinson, David Hugh; Weidner, Jerry Raymond; Wells, James Aertsen; Wentworth, Sally Ann; White, Eugene Wilbert; White, James R.; White, William Blaine; Williams, Richard Sugden, Jr.; Williams, Robert Bruce; Witte, Herman C.; Wood, George V.; Yiu, Shih-Kao; Young, Davis A.

1970s: Aaron, John Marshal, III; Abriel, William Lee; Adams, Bruce E., Jr.; Aitken, Frances Kenneth; Aivano, John Peter; Aktan, M. Tunc; Al-Temeeni, Ali Yousuf; Alvarado, Rodrigo B.; Amell, Alexander Renton; Ananaba, Simon Enyinnah; Apgar, Michael A.; Applin, Kenneth Richard; Arnstein Breuer, Roberto John; Austin, Steven Arthur; Bahr, John Robert; Bair, Edwin Scott; Baker, James John; Baker, Paul Arthur; Barnett, Donald W.; Basile, Laura Lorraine; Bebout, John Wardell; Bechtel, William Lott; Bell, Alan George Ridley;

Bender, John F.; Berger, Roger John; Bhardwag, M. C.; Biadgelgne, Abraham; Bifano, Francis Vincent; Bish, David Lee; Block, Fred; Bolt, Charles T.; Bragonier, William Atwood; Brant, Lynn Alvin; Brown, Charles Edward; Brown, Charles Edward; Brown, Robert Lewis; Bucek, Milena F.; Cameron, John Ian; Canich, M. R.; Carmony, John Rodman; Carroll, Richard M.; Carter, Brian John; Casadevall, Tom; Casadevall, Tom; Casagrande, Daniel Joseph; Chang, Tien-Show; Chavez-Martinez, Marlo Luis; Chelius, Carl Robert; Chewning, John R.; Christen, Randolph Frederick; Ciciarelli, John Anthony; Clark, Connie M.; Cole, David Robert; Collins, Henry B.; Colman, Steven Michael; Corbett, Edward Sisk; Cornet, Walter Bruce; Craig, Richard Gary; Craig, Richard Gary; Crelling, John Crawford; Crerar, David Alexander; Crock, James Gerard; Crouse, Harry Lynn; Cybriwsky, Zenon Alexander; D'Andrea, Ralph F.; Davis, Nicholas Falconer; Davis, Thomas Mooney; Dein, James Lindall; Delfel, Deborah Lynn; DeRose, William R.; Diehl, Paul Emmett; DiGiacomo, Harry Joseph, Jr.; Dobrzykowski, David B.; Downs, William Frederick; Drean, Thomas Alen; Drexler, Wallace William; Dunay, Robert Edmund; Duncan, Douglas Wells; Ebaugh, Walter Fielding; Eby, James Robert; Edmunds, Frederick Robin; Engelder, P. Richard; Engelder, P. Richard; Erikson, Robert Lawrence; Essenfeld, Martin; Everly, Robert A.; Exarchos, Constantine Christos; Famy, Syed Mohamad; Faust, Charles Russell; Fisher, Wilson, Jr.; Flanigan, Kenneth R.; Fleer, Varda Nanette; Flueckinger, Linda Ann; Flynn, Ronald Thomas; Foerster, Bernhard; Fonda, Shirley Smith; Frisillo, Albert Lawrence; Fry, Harold Chester, Jr.; Fuchs, William Arthur; Furst, George Arrowsmith; Furukawa, Toshiharu; Gaither, Bruce Edward; Galiette, Stephen Joseph; Gander, Craig Robert; Gang, Michael W.; Garihan, Anne Burroughs Lutz; Garihan, Anne Burroughs Lutz; Garihan, John Michael; Gerhart, James M.; Giddings, Marston Todd, Jr.; Gigl, Paul Donald; Giordano, Thomas Henry; Gleason, Patrick James; Gold, D. P.; Gong, Henry; Goscinski, John S.; Gough, William R.; Griffiths, Sally A.; Grutzeck, Michael William; Guswa, John Henry; Gutzler, Robert Quenton; Guzofsky, David Paul; Haefner, Richard Charles; Haggerty, Janet Ann; Hang, Pham Thi; Harmon, Russell S.; Hart, Margaret Lynn; Hawk, Joseph H.; Hayba, Daniel Owen; Haygood, Christine Cricket; Helz, George Rudolph; Helz, Rosalind Tuthill; Herrick, David C.; Hess, John Warren; Hill, Donald W.; Holbrook, Philip William; Holloway, John Requa; Holmes, Thomas Connor; Horn, Richard A.; Horne, Mary E.; Houck, Richard Thomas; Houseknecht, David Wayne; Hower, James Clyde; Hsi, Ching-Kuo Daniel; Hsieh, Shuang-shi; Hsu, Fu-tzu; Huang, Chi-I; Hunt, John A.; Hunter, Philip M.; Imbalzano, John Francis; Isaacs, Charles Manning, Jr.; Isaak, Donald G.; Iversen, Gary M.; Jacobson, Roger Leif; James, Laurence Pierson; Jobling, John Lloyd; Johnson, C. Branning; Jones, Kathleen L.; Justice, Mahlon Gilbert; Justice, Pamela Rose; Kasapoglu, Kadri Ercin; Khattab, Khattab Mansour M.; Kim, Ran Young; King, Kenneth Ross; Kishbaugh, James Wilbur; Kobelski, Bruce Joseph; Konikow, Leonard Franklin; Korner, Lisa Ann; Kosich, Deborah Frances; Kowalik, William Stephen; Kowatch, John S.; Kraft, Gordon D.; Krajewski, Stephen A.; Krajewski, Stephen A.; Kramer, John Howard; Kreiger, Edgar William, Jr.; Krohn, Melvyn Dennis; Krothe, Noel Calvin; Krothe, Noel Calvin; Krupka, Kenneth Michael; Kuehn, Kenneth William; Kunasz, Ihor Andrew; Kunze, Adolf Wilhelm Gerhard; Kusiak, John Robert; Kuzio, Michael Kay; Labovitz, Mark Larry; Labovitz, Mark Larry; Langton, Christine A.; Lanning, Robert Maye; Leap, Darrell Ivan; Lewis, Alvin; Libby, Stephen Charles; Lipin, Bruce Reed; Lisenbee, Alvis Lee; Lithgow, Enrique W.; Loudin, Michael George; Love, John Eric; Lowenhaupt, Douglas E.; Lowright, Richard Henry; Luster, Gordon Ray; MacKallor, Jules A.; Mahar, Ddennis Lee; Martin, C. J.; Martin, Mary Margaret; Mattison, George David; Maud, Randall Lee; May, Fred E.; Mazurak, Robert E.; McCowan,

Douglass William; McIntosh, William; McLimans, Roger Kenneth; McNeal, James Marr; Meiser, E. William, Jr.; Meiser, Edward William, Jr.; Menzie, W. David, II; Menzie, W. David, II; Merkel, Richard H.; Michlik, Rudolph R.; Miller, Charles Elden, Jr.; Miller, Floyd E., Jr.; Miller, Robert Norman; Millhollen, Gary Lloyd; Mitchell, Gareth D.; Modreski, Peter John; Modreski, Peter John; Moore, Joseph Neal; Moore, Joseph Neal; Moorshead, Frank Arthur; Morris, James Robert; Mozumdar, Bijoy Kumar; Mundi, Emmanuel Kengnjisu; Mundi, Emmanuel Kengnjisu; Murowchick, James Bernard; Mysen, Bjoorn Olav; Naert, Karl Achiel; Naert, Karl Achiel; Negri, Daniel R.; Newton, Carl Adams; Newton, Geoffrey Bruce; Nichols, Douglas James; O'Brien, Philip Joseph; O'Leary, Dennis William; Okuma, Angelo Frederick; Oldham, David Wayne; Onasch, Charles Martin; Onasch, Christine Condon; Otton, James Keith; Ozsvath, David Lynn; Palmer, Carl David; Parrish, Jay Bennett; Pennington, Dennis Ira; Petrus, Carolyn Ann; Pirc, Simon; Pirkle, Frederic Lee; Pisutha-Arnond, Visut; Platco, Nicholas L., Jr.; Potter, Robert William, II; Powdysocki, Melvin Henri; Quaah, Amos Ofori; Rahmanian, Victor David K.; Rasmussen, Karen Hasine; Rauch, Henry William; Reardon, Eric John; Resley, William E.; Reyes-Navarro, Jaime; Rich, Frederick James; Riggins, Earl Michael; Rimstidt, James Donald; Ripley, Edward Michael; Rising, Brandt Albert; Roberts, James Morgan, Sr.; Roberts, Michael Taylor; Rothman, Robert L.; Russ, David Perry; Ruths, Mark Allen; Ryan, Michael P.; Samuelson, Alan Conrad; Scanlin, Michael A.; Schasse, Henry William; Scheetz, Barry Earl; Scheetz, Barry Earl; Schmiermund, Ronald Lee; Schreifels, Walter Arthur; Schubert, Jeffrey Paul; Schultz, Thomas R.; Sears, Stephen O'reilly; Sgambat, Jeffrey Peter; Shapiro, Earl A.; Shaub, Francis Jean; Sherburne, Roger Wayne; Shettel, Don Landis, Jr.; Shettel, Don Landis, Jr.; Shore, Michael J.; Shuart, Susan Rae; Shuster, Evan Thomas; Siegal, Barry Steven; Siegal, Barry Steven; Siegel, Donald I.; Simonsen, August Henry; Singer, Donald Allen; Siskind, David Eugene; Slaughter, John; Slingerland, Rudy Lynn; Slingerland, Rudy Lynn; Smith, Cameron Outcalt; Smith, Robert Charles, II; Solomon, George Cleve; Soroka, Leonard Gregory; Sorrentino, Anthony Vincent; Spence, William John; Spiller, Reginal Wayne; Stepp, Jesse Carl; Stewart, Dion Carlyle; Stewart, Nanna Beth Bolling; Stohl, Francis Virginia; Stolar, John, Jr.; Stottmann, Walter; Stoyer, Charles Hayes; Sturdevant, James Anton; Suarez, Donald Louis; Sukhajintanakan, Warawan; Sundheimer, Glenn Robert; Sweigard, Richard Joseph; Tamm, Lucille C.; Taylor, Robert Warren; Terzakis, George N.; Teufel, Michael Richard; Theokritoff, Sergius; Tobias, Steven Martin; Tole, Peter Mwakio; Tregaskis, Scott W.; Troutt, William Richard; Turnbull, Lawrence Stur Levant, Jr.; Verbeek, Earl Raymond; Verbeek, Karen Jane Wenrich; Verbeek, Karen Jane Wenrich; Villaume, James F.; Volk, Karen Wagner; Volk, Karen Wagner; Vonarx, Clifford E.; Voultsos, Mark; Waddel, Claudia True; Waddell, Richard Kent, Jr.; Walawender, Michael John; Walsh, Deborah Anne; Wan, Hsien-Ming; Warg, Jamison B.; Warner, David John; Watson, Alicia Tyler; Weed, Charles Edward; Weinman, Barry L.; Weirauch, Douglas Allan, Jr.; Welch, Jane Marie; Welch, Jane Marie; Wen, W.; Wendlandt, Richard Frederick; Werner, Matthew Lambert III; Westlund, Carlyle W.; Whelan, Joseph F.; White, Elizabeth Loczi; Whitney, John Wilber; Whittemore, Donald Osgood; Windom, Kenneth Earl; Wittur, Glen Eric; Wright, John Clinton, Jr.; Yang, Ker-Chi; Yeh, Yeong Tein; Yu, Shu-cheng; Zeiss, Harvey S.

1980s: Allshouse, Sharon Dale; Almashoor, Syed Sheikh; Alves, David J.; Anderson, Christine Alexis; Applin, Kenneth Richard; Arens, Nan Crystal; Arnold, Walter Allen; Arthur, Randolph Clyde; Ayers, John C.; Azuola Valls, Hannia; Baag, Chang-Eob; Bachman, Leon Joseph; Badie, Ahmad; Bair, Edwin Scott; Baker, Don Read; Banwell, Gail Marie; Barbaro, Ralph Wesley; Barker, Jeffrey

Scott; Barker, Jeffrey Scott; Barry, John P.; Baumgardt, Douglas Reid; Baxter, James Edward; Beegle, Douglas Brian; Bell, Christy Anne; Bianchi-Mosquera, Gino Cesar; Bilgesu, Huseyin Ilkin; Blackey, Mark E.; Blackmer, Gale Corless; Blakeney, Beverly A.; Bluth, Gregg J.; Bodnar, Robert John; Booth, Colin John; Borkowski, Annette Hottman; Bourcier, William Louis; Brachman, Steven H.; Bradbury, John; Bradbury, John William; Brant, Lynn Alvin; Breen, Kevin John; Brickey, David Wayne; Burch, Charles Ivan; Burger, Roy W.; Buss, David Roger; Cameron, Peter John; Campbell, Bruce Samuel; Carter, Brian John; Casey, William Howard; Cemen, Ibrahim; Chacko, Thomas; Chaffin, David Leland; Choi, Duck Keun; Christy, Joseph J.; Clouser, Robert H.; Cohen, Robert Mark; Cole, David Robert; Connolly, James Alexander Denis; Connors, Kathryn Francis; Cox, Robert Sayre; Cravotta, Charles Angelo, III; Cronce, Richard Charles; Crum, Steven V.; Curatolo, Joel Charles; Cygan, Randall Timothy; Cygan, Randall Timothy; Dade, William Brian; Davidheiser, Carolyn Elizabeth; Davies, Thomas Daniel; Davis, Willard Frew; Dermengian, John Michael; DeWitt, Ed Howard; Dhaliwal, Hardave; Dodd, Kurt A.; Dresel, P. Evan; Drummond, Segal Edward, Jr.; DuBois, Dean Paul; Duffield, Glenn M.; Duffy, William Joseph; Dymond, Randel Leo; Eary, L. Edmond, III; Eary, L. Edmond, III; Ediger, Volkan S.; Eggert, Roderick Glenn; Ehleiter, John Edward; Eldridge, Charles Stewart; Eldridge, Charles Stewart; Ersavci, M. Nedim; Erwin, Leslie Eugene; Evans, Karl Vierling; Ewart, James Alfred, Jr.; Faria Santos, Claudio A. F.; Faria Santos, Claudio A. F.; Farinelli, Joseph Augustine; Farley, Martin Birtell; Fausey, William Herbert; Felch, Roger N.; Filley, Thomas Howard; Fleer, Varda Nanette; Franco, Marcia Clara; Francois, Darryl K.; Fregeau, Elizabeth J.; Fulcher, Richard Alfred, Jr.; Galadanchi, Habeeb I.; Gammons, Christopher Hall; Garabedian, Stephen P.; Garmezy, Lawrence; Garmezy, Lawrence; Gayes, Paul Thomas; Geer, Kristen Anders; Gerald, Rosemary Elaine; Gercek, Hasan; Gerencher, Joseph James, Jr.; Gerety, Kathleen Mary; Giammarco, Joseph H.; Gize, Andrew Paul; Glazier, Robert M., Jr.; Glick, David C.; Goerold, William Thomas; Goerold, William Thomas; Goss, Brian Glen; Grenot, Charles H.; Gryta, Jeffrey J.; Guebert, Michael Dean; Hansen, Peter Michael; Hanson, Clifford Gail; Hare, Paul William; Harouaka, Abdallah; Haupt, Robert William; Hawman, Robert Barrett; Heninger, Steven G.; Henke, Jeffrey R.; Henry, Hollis Earl; Henson, Ivan Hendrix; Herbert, Louis; Herece, Erdal Ibrahim; Herman, Janet Suzanne; Hewitt, Marshall Cooper; Heydari Laibidi, Ezatoliah; Hilbert, Eric George; Ho, Phyllis Hang-Yin; Hong, Eason; Hoover, David Samuel; Hoover, Michael Thomas; Hornberger, Roger J.; Houck, Richard Thomas; Howe, Stephen Sherwood; Hu, Nien-Tsu Alfred; Hugo, William D.; Hull, Laurence Charles; Hutter, Adam Richard; Jabro, Jalal David; Jacobs, Gary Kermit; Jarrin, Keith Manuel; Jaynes, Dan Brian; Jefferis, Robert Gilpin; Johnson, Norma Grace; Johnston, David Earle; Jorgensen, David Wayne; Joyce, David Brian; Justice, Mahlon Gilbert, Jr.; Justice, Pamela Rose Eckberg; Kaiser, Charles John; Kaktins, Teresa L.; Karasevich, Ellen Lee Richter; Karasevich, Lawrence Paul; Kimler, Scott Thomas; Klins, Mark Albert; Kovarik, Mary Beth; Krupka, Kenneth Michael; Kubilius, Walter Paris; Kuehn, Carl Anton; Kuehn, Deborah Wilbur; Kuehn, Deborah Wilbur; Kuehn, Kenneth William; Lamberson, Michelle Noreen; Lambert, Joseph Michael, Jr.; Lee, Barbara Jeanette; Lee, Jia Ju; Lemieux, Corinne Renee; Lenaugh, Thomas C.; Lester, Barry Henry; Levine, Elissa Robin; Levine, Jeffrey Ross; Levine, Jeffrey Ross; Lew, Laurence Reed; Lew, Lawrence Reed; Li, Todd Ming Chun; Liaw, Zen-Sen; Lin, Rui; Litwin, Ronald James; Litwin, Ronald James; Loule, Jean-Pierre; Low, Steven P.; Lund, Karen Ivy; MacLeod, Anne Jacquelyn; Major, Jon Joseph; Mamulas, Ned; Mao Chen Ge; Mark, Christopher; Mason, Scott Edward; Mathews, Melissa J.; Matters, Seth Eugene; Mat-

thews, Leo Gerard; Maud, Randall Lee; Mazid, Mohamad; Mazza, Thomas A.; McAuliffe, James Michael; McBean, Alan Johnston, II; McBrinn, Geraldine E.; McCandless, David Oliver; McCollough, William F.; McKibben, Michael Andersen; McNally, Joseph T.; Meen, James Kenneth; Meglis, Irene Llewellyn; Merkel, Gregory Albert; Merzbacher, Celia Irene; Merzbacher, Celia Irene; Michrina, Barry Paul; Mikucki, Edward J.; Miller, Cynthia Kay; Mills, Alison M.; Mirkin, Adam Nicholas; Mosconi, Deborah Anne; Mowrey, Gary Lee; Muncill, Gregory Ernest; Munly, Walter C.; Murowchick, James Bernard; Nachlas, Jesse; Narbut, Susan Margaret; Nathman, Neal J.; Nekvasil, Hanna; Nelson, Carolynn; Ness, Mark William; Newman, David Alan; Ng, Carolyn Yee-Han; Nichols, Christine C.; Niemann, Nancy L.; O'Neill, Dennis Charles; Olafsson, Magnus; Orkan, Nebil I.; Ossman, Robert M.; Ousey, John Russell, Jr.; Pabalan, Roberto Tuason; Parrish, Jay Bennett; Pavelka, Anne; Pavlin, Gregory Byron; Paxton, Stanley Turner; Phelps, Lee Barry; Pinta, James, Jr.; Pisutha-Arnond, Visut; Pottorf, Robert John; Prave, Anthony Robert; Prave, Anthony Robert; Pretorius, Eugene B.; Prosser, Rex Michael; Ramirez-Rojas, Armando Jose; Ramsey, Robert M.; Ravenhurst, Casey Edward; Rhodes, John Arthur; Rice, Benjamin John; Rimmer, Susan Margaret; Risser, Jeffrey Allen; Robbins, Eleanora Iberall; Roebuck, Sheila Joan; Root, Robert William, Jr.; Rosencrans, Richard D.; Ruder, Michal Ellen; Rumbaugh, James Orville, III; Russell, Suzanne Jeannette; Ryan, Nancy Joan; Sanford, Ward Earl; Sartell, Jonathan Floyd; Sasowsky, Ira D.; Schmidt, Bennetta Lee; Schneider, Mark Edward; Schramke, Janet Ann; Schroeder, Mark E.; Schuyler, Andrew; Schuyler, Sharon Stowe; Schwab, Jean Ann; Schwessinger, William T.; Scott, Robert Karl; Seme-Abomo, Richard; Senftle, Joseph Thomas; Sentner, David A.; Shenberger, David M.; Shoemaker, James Scovell; Shope, Steven Michael; Shuman, Christopher A.; Sienko, Dennis Alan; Smith, Arthur Tremaine; Smith, Arthur Tremaine; Smith, Michael William; Snow, Robin Scott; Snyder, Randy William; Sopkin, Sandra Meryl; Stamm, John Francis; Star, Ira; Steele, Sarah Ellen; Stephens, George; Sterner, Steven Michael; Stewart, Dion Carlyle; Stockar, David V.; Stout, Scott Alan; Stout, Scott Alan; Streb, Lawrence Lambert; Sung, Wonmo; Swanson, Karen Anne; Sweigard, Richard Joseph; Taioli, Fabio; Tang, Su; Tanner, Stephen Bruce; Taylor, Terry Dean; Tchinda, Fidele; Tietbohl, Douglass Ralph; Tole, Mwakio Peter; Tormey, Brian B.; Touysinhthiphonexay, Kimball C. N.; Touysinhthiphonexay, Kimball Cooke Nettleton; Touysinhthiphonexay, Yen; Trocki, Linda Katherine; Turner, Kenneth Harold; Varchol, Douglas J.; Vogfjord, Kristin S.; Voight, Donald Edward; von Seggern, David Henry; Wagner, Gregory S.; Waltman, William John; Wardrop, Richard T.; Warwick, John Jules; Waters, Roger Kenneth; Waters, Susan Alice; Watson, Robert William; Weedman, Suzanne Dallas; Weedman, Suzanne Dallas; Weir, L. Alison; Weir, Robert H., Jr.; Werner, Sarah R.; Wertz, William Earl; Wesolowski, David; White, Daniel J.; Williams, David E.; Williams, John Herbert; Wu, Sheng Tung; Yan, B.; Yeakel, Jesse David; Young, Kirby David; Zellmer, Lauren Ann; Zimmerman, Laurie S.; Zipf, R. Karl, Jr.; Zolensky, Michael Ewing; Zolensky, Michael Ewing

University of Pittsburgh
Pittsburgh, PA   15260

213 Master's, 109 Doctoral

1910s: Eaton, Harry Nelson

1920s: Brewer, Charles; Hayasaka, Ichiro; Herrick, Stephen Marion; Huber, Theodore A.; Kammer, Glenn D.; Kohler, Frederick William; McCobb, Harry W.; Merriman, Ray Warren; Millward, William; Zimmerman, Charles C.

1930s: Aitken, William Ernest; Benthack, Louis; Billings, M. Hewitt; Branckstone, Hugh R.; Brooks, Betty Watt; Brower, Gilbert K.; Burke,

John James; Charles, John Roy; Clark, John; Colley, Ralph S.; Dana, Frederick F.; Dragusanu, Juliu Basile; Dresbach, Charles Howard; Eller, Eugene R.; Gealy, Wendell Baum; Gibbs, Harley S.; Hamilton, James E.; Harris, Sidney L.; Holland, Wilbur Charles; James, Jay R.; Leighton, Helen Elizabeth; Martinoff, Alexander D.; Meyers, Clay Kenton; Myers, Clay Kenton; Parris, Frank G.; Pirson, Sylvain J.; Ross, Ralph B.; Rutherford, Homer Morgan; Rynearson, Sylvester; Schindler, John Henry; Sidhu, Surain Singh; Wilson, Thomas Carroll

1940s: Batuk, Hamzer; Baykal, Turan S.; Cahn, Joe Harold; Christ, Charles Milton; Danforth, Carroll F.; Dickey, Leonard Claude; Ebright, John Richard; Goth, Joseph Herman, Jr.; Graham, Gordon Marion; Hardman, Charles F.; Hoffacker, Benjamin Franklin, Jr.; Huffman, Kenneth Paul; Hull, Thomas Edward; Johnson, Wayne S.; Lanning, David Roy; Latta, William Love, Jr.; Layfield, Moody E.; Miklausen, Anthony J.; Miles, Daniel John; Moushegian, Richard; Murcy, Richard James; Murdy, Richard James; Newmeyer, Amel James, Jr.; Popovich, Michael Joseph; Power, Harry H.; Powers, E. Lloyd; Probst, David Arthur; Roeder, Joseph Harrison; Stann, Leon Kruk; Theiss, Mary Elizabeth; Webb, Edwin James; Wigbels, Frank B.; Wikander, Frederick Gerdes

1950s: Aarons, Irwin Isaac; Ackenheil, Alfred Curtis; Adams, Henry James; Angerman, Thomas Westley; Bender, Martin S.; Boyd, John A.; Boyer, James Francis, Jr.; Carr, George T.; Darakos, William Efstratos; Demshar, Ludwig Stanley; Doney, Hugh Holt; Edmunds, William Edward; Engman, Harry Arthur; Ewing, David Jay; Fleck, William Pyle; Frantz, Wendelin Robert; Frazier, Samuel Bowman; French, Gordon; Gazdik, Gertrude Christie; Ingram, Robert James; Jones, Earl Verner; Kafka, Frederick Thomas; Kardos, William Gustave; King, William Lyle; Kuhn, Thomas Alfred; Latshaw, Warren Leroy; Lusk, Edwin Wallace; Macpherson, Louis Alan; Martin, John Raymond; Matter, Conrad F.; McCollough, William Matthew; McWilliams, George Robert; Mershon, Robert E.; Messineo, Anthony Vincent; Miller, Kenneth Joseph; Moser, Erwin Leroy, Jr.; Ozol, Michael A.; Pachman, Jerrold Marvin; Payne, Franklin Russell, Jr.; Radisi, John; Raine, John Wesley, III; Rieg, Louis Eugene; Schramm, Martin William, Jr.; Sheafer, William Lesley, II; Sitler, Guy F.; Tallon, Walter Adam; Thomas, John Moore; Timko, Donald Joseph; Tindell, William Norman; Trump, George William; Waag, Charles J.; Walker, Philip Caleb; Wayman, Cooper Harry; Weis, George Franklin; Werner, James Edward; Wing, Robert Claude; Worley, John Cochran

1960s: Alvi, Javaid; Balsinger, Daniel Francis; Barsdate, Robert John; Bunch, Theodore Eugene; Carmichael, Robert Stewart; Carmichael, Robert Stewart; Chen, Kuang-Chian; Darby, Dennis Arnold; de la Torre Robles, Jorge; Dougherty, M. T.; Frantz, Wendelin Robert; Gavenda, Alan Paul; Groff, Donald William; Grubbs, Donald Keeble; Gupta, Alok Krishna; Hoque, Mominul; Johnson, William W.; Lacey, James Edward; Leeper, Wayne S.; Lowrie, William; McLean, Willis John; McQuillian, Tom Alan; Nalwalk, Andrew Jerome; Park, Frederick; Rao, Chelluri; Reid, Archibald McMillan; Rolin, John; Salver, Henry Arthur; Seeger, Charles Ronald; Slaughter, Maynard; Smith, Randall William; Smith, Randall William; Tarr, Arthur Charles; Wasilewski, Peter; Wendelin, R. Franz

1970s: Akpati, Benjamin Nwaka; Bannister, E. N.; Berger, Z.; Borawski, Teddy W., Jr.; Cain, Bruce; Carothers, Marshall C.; Chen, C. Y.; Chou, Chen-Lin; Cisowski, Stanley M.; Clauter, Dean A.; Cobucci, Dolores Ann; Cunningham, Thomas; Day, Ronald; De Gasparis, Aurelio Alfonso Amedeo; Dickson, Peter A.; Dodson, Richard E.; El-Harram, F. A.; Elias, M. R.; Erdlac, Richard John, Jr.; Fierstein, John; Haarr, Doris T.; Hamel, James Victor; Harper, John A.; Hassan, Farkhonda; Humbertson, P. G.; Iannacchione, Anthony; Janezic, Gary;

Kean, William Francis, Jr.; Kilburg, James A.; Kim, Ann Gallagher; Kudlac, John Joseph; Laosebikan, Samuel Cyebanji; Lee, Kiehwa; Lee, Richard Kenneth; Lindahl, David; Mahadev, P. D.; Makar, Laila; Maynard, B. R.; Milewski, R. G.; Mizumura, K.; Nelson, Robert; Norton, Charles Warren; Novak, S. M. d'O.; Paruso, Debbie; Payne, Michael A.; Pimentel, Nelly R.; Planinsek, Frances; Pollak, Henry; Price, Michael Louis; Puglio, Donald; Rigotti, Peter A.; Robison, Mary; Rosenfeld, Charles Louis; Rosenquest, Darl; Sandstrom, Melissa; Sarg, Frederick; Shaak, Graig Dennis; Shalaby, Mohamed A. E. A.; Shulik, Stephen J.; Solanki, Jawahirlal J.; Thornburg, Robert; Tzeng, Shih-Ying; Watson, Thomas; Wells, Eddie N.; Winters, Dermont; Witkowski, Robert; Wu, YeeMing Timothy; Yarussi, Michael

1980s: Adams, William Russell, Jr.; Adekeye, Jacob Ishola Dele; Akgunduz, Recep; Al-Qayim, Basim A.; Anderson, John Robert, II; Archer, Hugh Victor; Arikan, Fehmi; Basilone, Tim; Behum, Paul; Brezinski, David Kevin; Buis, Patricia Frances; Busch, Richard Munroe; Ceci, Vincent; Chapman, Bradley; Chiang, Liann; Chiou, Jyh-Dong; Cohen, Karen Kluger; Corona, Francesco; D'Urso, Gary John; Duck, John; Duerring, Nancy; El Emam, Mohamed; Etzler, Paul; Glohi, Boblai; Gutowski, Vincent Peter; Hayward, William; Heinecke, Thomas A.; Hill, Jeffrey; Hsiung, Ennchi David; Jones, J. Richard; Jorstad, Tom; Kaplan, Sanford Sandy; Kersting, Joseph Jeffrey; Knapik, James; Korth, William Willard; Kuntz, Timothy; Kurshin, James; Larbah, Mohamed Ali; LaSota, Kenneth Alan; Lee, Der-Shing; Lin, Gwo-Fong; Lin, Roscow Ching-Hsing; Lin, Wei-Hsiung; Luce, Robert James; Maala, Mohamed; Madar, James Michael; Maniar, Papu Dayalal; Mario, Annette; Marrs, Thomas; McCullough, John; Meehan, Michael; Molinda, Gregory; Murin, Timothy; New, Robert A.; O'Connor, Joyce; O'Neil, Caron; Olaniyan, Olufemi; Olaniyan, Olufemi; Packariyangkun, Adisorn; Panian, John; Partlow, Deborah Paruso; Piccoli, Philip; Pontoriero, Pasquale; Presley, Susan; Rava, Barry; Robertson, David John; Robinson, James; Rodgers, Michael Robert; Rojas-Gonzalez, Luis Fernando; Sabin, Andrew; Snyder, Frank S.; Stephens, William; Thomson, Robert D.; Tisin, Abdulmehdi Bektash; Toprak, Selami; Tucker, Mark; Vento, Frank; Vento, Frank John; Wagner, Jeffrey Karl; Wall, David Joseph; Wells, Karen; Welsh, Robert; Yeager, Richard Neil; Yu, Jianxin; Yuan, Ding-Wen; Zell, Paul

Polytechnic University
Brooklyn, NY   11201

1 Master's, 9 Doctoral

1960s: Karafiath, Leslie L.; Levy, Roy

1970s: Bagchi, S.

1980s: Burger, Theodore Bernhard; Fagan, George Lawrence, Jr.; Gupta, Ram Swaroop; Khondker, Sufian A.; Lauria, Jeffrey M.; Reilly, Thomas Eugene; Zavesky, Richard Roy

Pomona College
Claremont, CA   91711

37 Master's, 2 Doctoral

1920s: Hill, Mason L.

1930s: Bellemin, George Jean; Hill, H. Stanton

1940s: Olmstead, Franklin H.; Shelton, John S.; Stark, Howard Everett

1950s: Baird, Alex K.; Burnham, Willis Lee; Clark, George Hollinger; Colburn, Ivan Paul; Dolton, Gordon Lee; Dudley, P. H., Jr.; English, H. Duncan; Forman, John Alexander; Gray, C. B.; Gray, Clifton H.; Heath, Edward G.; Hilton, George S.; Hoskins, Cortez William; Kundert, Charles J.; Owens, George V.; Pampeyan, Earl H.; Price, Maurice Carlton; Raleigh, Cecil Baring; Richmond, James Frank; Smith, Patsy J.; Taylor, James Carlton; Tay-

lor, James Carlton; Van West, Olaf; Vedder, John G.; Weldon, John B., Jr.; Yerkes, Robert F.

1960s: Buchholtz, Herbert F.; Doehring, Donald O.; Larsen, Norman R.; MacColl, Robert S.; Madlem, Kathleen W.; Welday, Edward E.; Williams, James D.

Portland State University
Portland, OR   97207

88 Master's, 6 Doctoral

1970s: Anderson, James Lee; Avolio, G. W.; Baker, Charles Allen; Black, Gerald Lee; Callender, A. D., Jr.; Carter, L. M.; Chitwood, Lawrence Allan; Clayton, C. M.; Fiksdal, Allen James; Gaston, L. R.; Istas, L. S.; Jackson, James Streshley; Jackson, Ronald Laverne; Kent, Mavis Hensley; Kent, Richard C.; Koler, Thomas Edward; Lentz, R. T.; Lofgren, David Carl; Ludowise, Harry; Mathiot, R. K.; Matty, David Joseph; Neal, K. G.; Pedersen, Steven; Perttu, R. K.; Redfern, R. A.; Schmela, Ronald J.; Sidle, William C.; Smith, Paul L.; Tawfik, F. M.; Taylor, D. G.; Timm, Susan; Veen, Cynthia Ann; Wood, J. D.

1980s: Al-Azzaby, Fathi Ayoub; Al-Eisa, Abdul-Rahman Mohammed; Berri, Dulcy Annette; Bounds, Jon Dudley; Brown, Edward Charles; Budai, Christine M.; Burck, Martin Stuart; Byrnes, M. E.; Cameron, Kenneth Allan; Carlin, Rachel Ann; Cole, David Lee; Cunderla, Brent Joseph; d'Agnese, Susanne L.; Davis, Steven Allen; Doerr, John Timothy; Eberhardt, Ellen; Evans, Carol Susan; Gabor, Reka Katalin; Gannett, Marshall W.; Gannon, Brian Lee; Gullixson, Carl Fredrick; Haas, Nina; Hagood, Michael Curtis; Hoffman, Charles William; Jackson, Michael Keith; Jaffer, Rebecca K.; Johnson, Michael J.; Kadri, Moinoddin Murtuzamiya; Kelty, Kevin Blair; Ketrenos, Nancy Tompkins; LaViolette, Paul Alex; Lytle, Charles Russell; Marty, Richard Charles; Maywood, Paul S.; McClincy, Matthew John; McGowan, K. I.; Mtundu, Nangantani Davies Godfrey; Nazy, David John; Orzol, Leonard Lee; Perttu, Janice C.; Pfaff, Virginia Josette; Polivka, David R.; Pollock, J. Michael; Rankin, David Karl; Raymond, Richard Brian; Roché, Richard Louis; Shaw, Neil B.; Simkover, Elizabeth Gail; Stine, C. M.; Swanson, Rodney Duane; Timmons, Dale M.; Titus, Willard Sidney, III; Tolan, Terry Leo; Townley, Paul Joseph; Trone, Paul Max; Visconty, Greg; Vogt, Beverly Frobenius; Whitson, David Neale; Wilkening, Richard Matthew; Winters, Warren Jon; Woller, Neil M.

Princeton University
Princeton, NJ   08544

7 Master's, 419 Doctoral

1910s: Buddington, Arthur Francis; Cockfield, William Egbert; Dale, Nelson C.; Giraud, A. P.; Hayes, Albert Orion; Mack, Edward, Jr.; Martin, James Cook

1920s: Agar, William MacDonough; Alexander, Charles Ivan.; Bannerman, Harold MacCall; Bridge, Josiah; Buffam, Basil Scott Whyte; Cairnes, Clive Elmore; Davis, Newton Fraser Gordon; Evans, Charles Sparling; Fulle, Richard M.; Gill, James E.; Honess, A. P.; Howell, Benjamin Franklin; James, William Fleming; Jewell, Willard Brownell; Kelly, W. A.; Mawdsley, James Buckland; Norman, George W. H.; Pegrum, Reginald Herbert; Perry, Elwyn Lionel; Sampson, Edward; Walker, John Fortune; Wanless, Harold Rollin

1930s: Beaven, Arthur P.; Betz, Frederick, Jr.; Blackstone, Donald LeRoy, Jr.; Cannon, Ralph S.; Clark, John; Cooper, John Roberts; Demorest, Max H.; Doten, Robert K.; Eardley, A. J.; Espenshade, Gilbert H.; Fanshawe, John Richardson; Fischer, Richard Phillip; Foley, Frank C.; Fox, Stephen Knowlton, Jr.; Gillanders, Earle Burdette; Haycock, Maurice Hall; Hess, Harry Hammond; Heyl, George Richard; Holland, Stuart Sowden; Howland, Arthur Lloyd; Jepsen, Glenn Lowell; Jolliffe, Alfred W.; Kidd, Desmond F.; Kindle, Cecil Haldane; Landes, Robert William; Lang, Arthur Hamilton; Lucke, John Becker; MacNeil, Donald Jonathan; McGerrigle, Harold W.; Parsons, Willard Hall; Peoples, Joe Webb; Pierce, William G.; Reed, John Calvin; Retty, Joseph Arlington; Ridland, G. Carmen; Rosenkrans, Robert Russell; Rouse, John Thomas; Russell, Loris Shano; Sanford, John Theron; Skeels, Dorr Covell; Smiser, Jerome Standley; Snelgrove, Alfred Kitchener; Stumm, Erwin Charles; Toulmin, Lyman D., Jr.; Vhay, John S.; Whitcomb, Lawrence; White, Donald E.; Wilson, Charles William, Jr.; Wilson, John Tuzo; Wishart, James Scotland; Woollard, George Prior

1940s: Christiansen, Francis Wyman; Cornwall, Henry Rowland; Dengo, Gabriel; Denson, Norman Maclaren; Engel, Albert Edward John; Freeze, Arthur C.; Greig, Edmund W.; Heroy, William B., Jr.; Hotz, Preston Enslow; James, Harold Lloyd; Klemme, Hugh Douglas; Kriz, Stanislav Jaroslav; Leech, Geoffrey Bosdin; Lozo, Frank Edgar, Jr.; Lynott, William John; MacDonald, Roderick Dickson; MacLean, Hugh James; Mason, John Frederick; Maxwell, John Crawford; Meier, Dudley R.; Patmore, William Henry; Phair, George; Roots, Ernest Frederick; Stephenson, Hubert Kirk; Stokes, William Lee; Tanner, Joseph Jarratt; Van Alstine, Ralph Erkstine; Van Houten, Franklyn Bosworth; Waage, Karl M.; Watson, Kenneth DePencer; Whitmore, Duncan Richard Elmer; Wolfe, Peter Edward

1950s: Alexander, Roger Gordon, Jr.; Anderman, George Gibbs; Arnold, Ralph Gunther; Badgley, Peter Coles; Baker, Donald Roy; Bartholome, Paul Marie; Bass, Manuel Nathan; Bassett, Henry Gordon; Beland, Jacques; Best, Raymond Victor; Braddock, William A.; Brown, A. S.; Brown, Charles William; Bushman, James Richard; Campbell, Finley Alexander; Christman, Robert Adam; Coleman, Leslie Charles; Crowl, George Henry; Curry, William Hirst, III; Dahlstrom, Clinton Dennis Augustine; Deffeyes, Kenneth Stover; Donnelly, Thomas Wallace; Fuller, Arthur Orpen; Garbarini, George Stephen; Geyer, Richard Adam; Hargraves, Robert Bero; Hay, Richard Leroy; Henderson, Gerald Gordon Lewis; Heyl, Allen Van, Jr.; Hriskevich, Michael Edward; Kalliokoski, Jorma Osmo Kalervo; Kavanagh, Paul Michael; Konigsmark, Theodore Albert; Lenz, Alfred Carl; Leonard, Benjamin F., III; Lougheed, Milford Seymour; Mackenzie, David Brindley; MacLachlan, James Crawford; Mann, John Allen; Mattson, Peter Humphrey; Maxey, George Burke; McGill, George Emmert; McMannis, William Junior; Merrill, John R.; Miller, Roswell, III; Moberly, Ralph Moon, Jr.; Morris, William Joseph; Newmarch, Charles Bell; Nicolaysen, Louis Otto; Ohlen, Henry R.; Olsson, Richard Keith; Poole, William Hope; Poulter, Glenn Joseph; Price, Raymond Alex; Prucha, John James; Reesor, John Elgin; Richardson, Eugene Stanley, Jr.; Robinson, Malcolm Campbell; Shagam, Reginald; Shoemaker, Eugene M.; Simons, Elwyn Laverne; Sims, Paul Kibler; Skidmore, Wilfred Brian; Slodowski, Thomas Raymond; Smith, J. R.; Smith, James Robert; Smith, Raymond James; Souther, Jack Gordon; Stewart, John Conyngham; Stott, Donald Franklin; Stuart, Roy Armstrong; Subranmaniam, Anantharama Parameswara; Sutherland-Brown, Atholl; Tonking, William Harry; Toohey, Loren Milton; Van Siclen, Dewitt Clinton; Varrin, Robert Douglas; Verrall, Peter; Widmer, Kemble; Wise, Donald Underkofler; Young, John Cannon

1960s: Allen, Jack Christopher, Jr.; Alvarez, Walter; Anderson, Alfred Titus, Jr.; Ayres, Lorne Dale; Bamber, Edward Wayne; Barker, Daniel Stephen; Barnes, William Charles; Bell, John Sebastian; Bell, Richard Thomas; Bergh, Hugh W.; Bierwagen, Elmer Emanuel; Birch, Francis Sylvanus; Bowin, Carl Otto; Bukry, John, David; Burk, Creighton A.; Card, Kenneth Darius Huycke; Chase, Richard Lionel St. Lucien; Childers, Milton Orville; Cotter, Edward Joseph; Csejtey, Bela, Jr.; Cunningham-

Dunlop, Peter K.; Davis, Brian Thomas Canning; De, Aniruddha; Dickey, John Sloan, Jr.; Dodd, Robert Taylor, Jr.; Dougan, Thomas William, Jr.; Drever, James Irving; Eaton, Jerome F.; Eisbacher, Gerhard Heinz; Flemal, Ronald Charles; Garrison, Robert Edward; Gibson, Thomas George; Gillett, Lawrence B.; Glover, Lynn, III; Gordon, Terence Michael; Greenwood, Hugh John; Gwinn, Vinton E.; Handfield, Robert C.; Hedberg, James D.; Helsey, Charles Everett; Hickey, Leo Joseph; Hinners, Noel William; Kauffman, Marvin Earl; Langford, Fred Frazer; Lawrence, David Reed; Lockwood, John Paul; Lumbers, Sydney Blake; Luttrell, Eric Martin; MacDonald, William David; MacGregor, Ian D.; MacQueen, Roger Webb; Marshall, Alan E.; Melson, William Gerald; Menendez, Alfredo; Metz, Harold L.; Monty, Claude Leopold Victor; Moores, Eldridge Morton, III; Morgan, Benjamin A., III; Muffler, Leroy John Patrick; Murray, James Wolfe; Mutch, Thomas A.; Nagle, Frederick, Jr.; Nolf, Bruce Owen; Ohmoto, Hiroshi; Otalora, Guillermo; Oxburgh, Ernest Ronald; Palmer, Donald Frank; Palmer, Henry Currie; Pessagno, Emile Anthony, Jr.; Phillips, Joseph Daniel; Piburn, Michael D.; Picard, Meredith Dana; Pinckney, Darrell Mayne; Ritter, Dale Franklin; Russell, Kenneth Lloyd; Rye, Robert Orph; Sawkins, Frederick J.; Seiders, Victor Mann; Simkin, Thomas Edward; Smith, Alan Gilbert; Soydemir, Cetin; Taylor, Gordon Cosmos; Temple, Peter G.; Thrailkill, John Vernon; Vreeland, John Howard; Wellman, Samuel S.; Whetten, John T.; Williamson, Alexander M.; Yates, Michael Timothy; Zimmerman, Jay, Jr.

1970s: Anderson, Timothy Allan; Arthur, Michael A.; Ashwal, Lewis D.; Baba, Nobuyoshi; Berry, Robert Chapman; Burmester, Russell Frederick; Burruss, Robert Carlton; Cathles, Lawrence M., III; Chang, Ki Hong; Cherkauer, Douglas Stuart; Duguid, James O.; Foose, Michael Peter; Frazer, Laurie Neil; Friedlander, Susan J.; Fullerton, David S.; Grambling, Jeffrey A.; Grandstaff, David Eugene; Harlow, George Eugene; Hey, Richard N.; Howe, Milton W.; James, Harold Edward, Jr.; Karanjac, Jasminko B.; Kenah, Christopher; Kimberley, Michael M.; Kirschvink, Joseph Lynn; Kleinmann, Robert L. P.; Lappin, Allen Ralph; Lehner, Florain K.; Lewis, Roger James Gollan; Loomis, Timothy Patrick; Malin, Peter Eric; Maresch, Walter Victor; Moore, J. Casey; Muessig, Karl Walter; Mulcahy, Marjorie; Murray, Cecil George; Openshaw, Ronald E.; Patterson, Ronald James; Percival, Stephen F., Jr.; Perkins, William Enfield; Powell, Christine A.; Prieto-Portar, Luis A.; Roberts, James C.; Rotunno, Richard; Roy, Denis W.; Rupke, Nicolaas Adrianus; Safai, Nader M.; Scholle, Peter Allen; Schopf, Paul S.; Semtner, Albert J., Jr.; Siegel, John Alan; Skerlec, Grant M.; Smith, Martin L.; Steinthorsson, Sigurdur; Strelitz, Richard A.; Suarez, Max J.; Travers, William Brailsford; Twiss, Robert J.; Vierbuchen, Richard C., Jr.; Wallace, Robert James; Ward, Steven N.; Wheeler, Russell L.; Wier, Stuart Kirkland; Woodhead, James A.; Woodsworth, Glenn James; Yuretich, Richard Francis

1980s: Abriola, Linda Marie; Ahlfeld, David Philip; Allen, Myron Bartlett, III; Atobrah, Kobina; Bathurst, Bruce Warren; Bergman, Steven Clark; Bhattacharyya, Debaprasad; Borgia, Andrea; Bowman, Kenneth Paul; Brantley, Susan Louise; Brocher, Thomas Mark; Busby-Spera, Cathy Jeanne; Chirlin, Gary Richard; Chyi, Michael So; Clark, Stephen Christopher Lane; Covey, Michael Conrad; Crisp, David; Crisp, Joy Anne; Crowley, Kevin David; Davis, James Peter; Dougherty, David Emery; Douglas, Bruce James; Elgamal, Ahmed-Waeil Metwalli; Ellis, Glenn W.; Evans, John Richard; Feigenson, Mark Daniel; Griffin, Thomas T.; Hasson, Phyllis Fairbanks; Hawman, Robert Barrett; Hay, Randall Stuart; Hebson, Charles S.; Heestand, Richard Lee; Hennet, Remy Jean-Claude; Henson, Ivan Hendrix; Herbert, Timothy D.; Hill, Mary Catherine; Hill, Mary Louise; Husch, Jonathan Mark; Jones, Douglas Stephen; Kabala, Zbigniew Jan; Kellogg, James Nelson;

Kleinspehn, Karen Lee; Krystinik, Lee Franklin; Lawson, Charles Alden; Lorenz, John Clay; MacPherson, Glenn Joseph; Maest, Ann Sharon; Maze, William Bronson; McLaughlin, Dennis B.; Means, Jeffrey Lynn; Medwedeff, Donald Arthur; Murnane, Richard James; Namson, Jay Steven; Onstott, Tullis Cullen; Page, Roger Henry; Palmer, Amanda Ann; Pan, Gee-Shang; Phipps, Stephen Paul; Powers, Dennis Wayne; Pratt, Lisa Mary; Purucker, Michael E.; Roggenthen, William Michael; Rossbacher, Lisa Ann; Shapiro, Allen Marc; Shudofsky, Gordon N.; Sisson, Virginia Baker; Snow Boles, Jennifer Lee; Stowell, Harold Hilton; Susak, Nicholas John; Vink, Gregory Evans; Wanamaker, Barbara Jo; Wang, Simon Yaou-Dong; Williams, Loretta Ann; Wood, Scott Alan; Yang, Mary Mei-ling; Zelt, Frederick Bruce; Zhao, Wu-Ling

University of Puget Sound
Tacoma, WA 98416

4 Master's, 1 Doctoral

1940s: Hedges, Joseph W.

1950s: Dinsmore, James P.

1960s: McNeely, Warren L.

1970s: Konicek, Daniela L.; Stricklin, Claude R.

Purdue University
West Lafayette, IN 47907

206 Master's, 173 Doctoral

1930s: Bennett, W. R.

1940s: Alvarado, Pacifico M.; Dolch, William L.; Hill, Frederick B.; McCullough, Charles R.; Metcalf, Charles T.; Mollard, John D.; Parvis, Merle; Pollard, William S., Jr.; Yang, Shih Te

1950s: Altschaeffl, Adolph G.; Bronson, Roy DeBolt; Clark, Robert E.; Fears, Fulton K.; Herrin, Moreland; Howe, Robert Hsi Lin; Larew, Hiram G.; Leighty, Robert D.; Lennartz, Carl R.; Litt, Donald D.; Ludwick, Jimmy Donald; McGregor, Duncan D.; Moore, Bruce H.; Moultrop, Kendall; Nishimura, Katsuyoshi; Phlainen, John A.; Ramiah, B. K.; Shepard, James R.; Shurig, Donald G.; Stephens, Jack E.; Stylianopoulos, L. C.; Venters, Edwards; Yeh, Pai-Tao

1960s: Alpay, Okan; Andersland, Orlando B.; Aughenbaugh, Nolan B.; Avedissian, Yegishe Murad; Baladi, George Youssef; Baron, William; Barr, David John; Beals, Harold Oliver; Bjelke, William; Brown, Cyril Benjamin; Brumund, William Frank; Burns, Allan Fielding; Chang, Ting Pao; Dash, Umakant; Dolch, William L.; Girault, Pablo; Jaeger, Ralph Roger; James, Wesley P.; Jorgensen, Per; Ledoux, Robert Louis; Lewis, Glenn Charles; McGammon, Norman R.; Mishu, Louis Petrous; Mishu, Louis Petrous; Rockaway, John Dobbling, Jr.; Russell, Donald Arthur; Schiff, Anshel Judd; Schuster, Robert L.; Simon, Andrew L.; Sooky, Attilla A.; Sridharan, Asuri; Sweet, Arnold Lawrence; Tanguay, Marc Gilles; Terrel, Ronald L.; Van Heerden, Willem Maartens; Wu, I-Pai; Zachary, Alvin Leslie

1970s: Abeyesekera, R. A.; Adams, Jerald M.; Ahbe, J. B.; Antai, A. E.; Aragon, R. A.; Aten, Robert Eugene; Athanasiou-Grivas, D.; Baker, Mark Richard; Bannister, Timothy Allen; Barnard, R. S.; Bartolucci-Castedo, Luis A.; Bartolucci-Castedo, Luis A.; Bhasin, R. N.; Bishop, W. E.; Black, Paul R.; Boctor, Nabil Zaki; Borenstein, Herbert; Bowman, Paul L.; Brand, S. R.; Brodie, Gregory A.; Chandler, V. W.; Coffman, Daniel M.; Cortes Lombana, Abdon; Coyle, Lynn A.; Davis, S. E.; Devenny, D. W.; Dowell, Knneth E.; Edgar, D. E.; Espindola, Juan M.; Essed, A. S.; Fralick, T. N.; Frater, J. B.; Gardner, Michael C.; Hanten, J. B.; Hasan, S. E.; Haws, W. J.; Hossain, Aolad; Hume, Howard Robertson; James, Dennis R.; Johansen, Nils I.; Johnson, Nancy Perry; Johnson, Steven A.; Kalkani, E. C.; Keller, Edward Anthony; King,

Clifford L.; Kornegay, Francis Clyde; Koski, John S.; Lachmar, Thomas E.; Landau, H. G., Jr.; Landwer, William R.; Lester, Mark; Lizak, John B., Jr.; Lo, Kwok-wai Kenneth; Lomax, Francis Earl; Lowrey, William Stephen; Luca, Anthony J.; Lund, Lanny Jack; Maarouf, Abdelrahman; Magnus, Keith R.; Mazzella, Frederick E.; McGrew, Gloria; Meadows, Guy Allen; Murray, William L.; Noel, Stephen D.; Okagbue, Celestine Obialo; Otto, Ellen E.; Paulet, Manuel R.; Perry, A. O.; Peyton, T. O.; Plenge, Gustavo Carlos; Powell, R. L.; Price, F. T.; Ramirez, Abelardo Luis; Richardson, G. T.; Richardson, Nathaniel R., Jr.; Roberts, William Stephan; Romani-Cardenas, J. A. Fredy; Romano, R. R.; Saltzman, Uzi; Sinnock, Scott; Stillman, Neil Warren; Stohr, Christopher J.; Tsai, Helen M. H.; van Scoyoc, G. E.; van Zyl, D. J. A.; Von Bargen, David J.; Von Bargen, David J.; von Frese, Ralph Robert Benedict; Warncke, Darryl Dean; Witczak, Matthew Walter; Woodfill, Robert Dean; Woodring, S. M.; Worland, Vincent Peter; Young, Gregory B.; Zentani, A. S.

1980s: Accame, Guillermo M.; Adams, Scot Crawford; Al-Daghastani, Nabil Subhi; Alarcon-Guzman, Adolfo; Alfaro, Luis Domingo; Allotta, Thomas Lawrence; Ankeny, Lee Andrew; Awan, Muhammed Amjad; Baffaut, Claire; Baker, Charles Hays; Baldwin, James L.; Baldwin, Richard Taylor; Ballotti, Dean M.; Basharkhah, Mohammad Ali; Baud, Richie Darren; Bauer, Linda Rose; Bidin, Abdul-Aziz; Bowman, Michael J.; Brewster, Christine; Brumbaugh, Mark Virgil; Bryant, Ray Baldwin; Burke, Christopher Brian; Campbell, Michael James; Cassol, Elemar Antonino; Cervantes, Jose Enrique; Chan, Chien-Lu; Chang, Cheng-Jung; Chang, Fi-John; Chang, Ming-Fang; Chen, Jin-Song; Chen, Rong-Her Jimmy; Chen, Tai-Shan; Chiang, Chao-Sheng; Chu, Jia-Bao; Coffin, D. Todd; Cogo, Neroli Pedro; Crosby, George; Cull, Frances Anne; Curi, Nilton; Curry, Ben Brandon; Darrag, Ahmad Amr; Daudt, Carl Ransford; de Rezende, Servulo Batista; DeGraff, Alejandra Escobar; DeGraff, James Michael; Dextraze, Brenda Lynn; Dillaha, Theo Alvin, III; Dombrouski, Richard Paul, Jr.; Dorich, Roderick Alan; Doyle, Polly Ann; Ebel, Denton Seybold; Elbring, Gregory Jay; Elkin, Robert Rich; Engel, Bernard Allen; Fan, Jen-Chen; Fauria, Thomas; Faz, Jorge J.; Fein, Matthew; Fernandez, Ramon Norberto; Feth, Elle; Fox, Adam Jeffrey; Franti, Thomas George; Frey, Leo Joseph, III; Ganju, Ashutosh M.; Gee, Lauren Louise; Gelberg, Russ; German, Robert Allen; Giglierano, James D.; Ginn, Timothy Rollins; Goin, John Samuel, Jr.; Green, William; Greengold, Gerald; Griffin, Mitchell Lee; Haji, Mustapha; Harwood, Christine Lee; Haselow, John Stevens; Head, Roger Wayne; Hill, Roy Louis; Hodgson, Philip Richard; Holail, Hanafy Mahmoud; Holdrege, Thomas J.; Hoover, Julie Ann; Hsieh, Chang-hsin; Huang, An-Bin; Huang, Chen Tair; Hume, Howard Robertson; Hummeldorf, Raymond George; Humphrey, Dana Norman; Hunt, Paula J.; Hussein, Mohammad Hasan; Huston, Ted Jay; Hwang, Jiann-Yang; Irish, Neil Frederick; Jardine, William George; Jean, Jiin-Shuh; Jones, Paul; Jordan, Paul J.; Juang, Charng-Hsein; Kaczaral, Patrick Walter; Kairo, Suzanne; Kamel, Asad; Kaplan, Paul Garry; Karr, Michael Charles; Kenaga, Steven Gerald; Kochanski, Mark Alan; Krishna, Paul P., Jr.; Kucek, Leo; Kuo, Kung Chia; L'Heureux, David Maurice; Leosewski, John Fitzgerald; Lewis, Jonathan C.; Lewis, Richard Dale; Liang, Yueh; Lin, Ping-Sien; Lingner, David William; Lippus, Craig Stephen; Lipten, Eric Jack Henry; Lo, Yung-Kwong Terence; Loiacono, Nancy; Longacre, Mark; Losee, Bruce Anthony; Loughnane, Brian Keith; Lovely, Daniel Arthur; Lutter, William John; Lyslo, Jeffery Allen; Mao, Liang-Tsi; Marciano, Eugene; Martindale, Kevin; Masood, Hamid; Masood, Hamid; McClain, Shannon; McGinnis, John Patrick; McKee, Larry Douglas; McPhee, James P.; Meade, Ronald Bartholomew; Meyer, William Vincent; Miller, David; Mitchell, David L.; Mitchell, Jay Preston; Mizuno, Eiji; Morneau, Richard A.;

Neal, Charles William; Neibling, William-Howard; Nwabuokei, Samuel Onyeabor; O'Leary, Michael J.; Okagbue, Celestine Obialo; Okonkwo, Ignatius Okechukwu; Okonkwo, Ignatius Okechukwu; Okonny, Isaac Peri; Oliphant, Joseph Lawrence; Oschman, Kurt Patrick; Padres, Fidel Calimoso, Jr.; Padmanabhan G.; Palmieri, Francesco; Paredes, Carmela Hernandez; Parrott, Mark H.; Paul, Rick Lee; Pavey, Richard R.; Peregrine, Keith; Perez-Ramirez, Gerardo Antonio; Perry, Christopher; Piegat, James Jan; Piegat, James Jan; Plappert, John Wesley; Pohlmann, Karl; Poljak, Marijan; Poole, Vickie Lynn; Prapaharan, Sinnadurai; Raabe, Kenneth Charles; Randall, Karl Gordon; Ravat, Dhananjay Narendra; Read, John Russell Lee; Reagan, Mary; Reed, Jon Edward; Reed, Timothy; Ridgway, Jeffrey; Roberts, Paula; Salcedo, Marco Antonio; Saleeb, Atef Fatthy; Salleh, Mustapha Haji Mohd; Santana, Derli Prudente; Schroeder, William Floyd; Schultz, Richard Allen; Shakoor, Abdul; Shakshuki, Mokhtar; Sharma, Sunil; Sivakugan, Nagaratnam; Smith, James George; Smith, Randall; Sparlin, Mark Alan; Starich, Patrick; Stilwell, Jeffrey Darl; Strayer, Geoffrey Ben; Sudar, Susan A.; Szymanski, Daniel; Teme, So-Ngo Clifford; Tomczyk, Ted; Tomlinson, Robin Gayle; Travers, Mark Aaron; Ullmann-Beck, Gary; Verkouteren, R. Michael; Verner, Frederick Carr; von Frese, Ralph Robert Benedikt; Watts, Chester Frederick; Weinreb, Gary; Weishar, Lee; Welch, John Charles; Whitacre, Thomas James; White, Thomas Edward; Wilcox, Douglas Abel; Wiley, Michael T.; Yanez Pintado, Galo; Zellouf, Khemissi; Zhang, Zhenzhong

Universite du Quebec a Chicoutimi
Chicoutimi, PQ G7H 2B1

91 Master's

1970s: Bélanger, Jacques; Bélanger, Jean; Boudreault, Alain P.; Gauthier, André; Hébert, Claude; Kouassi, Frih; Maillet, James; Wagner, Wayne R.

1980s: Archer, Paul; Bédard, L. Paul; Bergeron, Alain; Bouchard, Gilles; Bureau, Serge; Côté, Denis; Cousineau, Pierre; Couture, Jean-François; Crevier, Michel; Dagenais, Solange; Dembele, Yahaya; Hervet, Michel; Hildebrand, Kanzira; Jourdain, Vincent; Lacoste, Pierre; Lange-Brard, Françoise; Lapointe, Bernard; Leduc, Maxime; Marquis, Robert; Martin, Etienne; Otis, Marlène; Ouellet, Eric; Ouellet, Rodrigue; Owen, Victor J.; Pearson, Vital; Piché, Mathieu; Pilote, Pierre; Poitras, Alain; Racicot, Denis; Sanschagrin, Yves; Sansfacon, Robert; Shareck, Andre; Simoneau, Pierre; Tait, Larry; Tremblay, André; Tremblay, François; Tremblay, Guy; Tremblay, Michel; Trudeau, Yvon

Universite du Quebec a Montreal
Montreal, PQ H3C 3P8

27 Master's

1980s: Boisvert, Denis; Chilakos, Peter; Corbeil, Paul; de Corta, Hugues; Decroix, Dominique; Desbiens, Harold; Desmarais, Luc; Gagnon, Pierre; Godue, Robert; La Haye, Jean; Lapierre, Guy; Perras, Danielle; Podkhlebnik, Yvette; Poirier, Ghislain

Universite du Quebec a Rimouski
Rimouski, PQ G5L 3A1

6 Master's

1980s: Laroche, Bernard; Ouellet, Guy; Pelletier, Marc

Queen's University
Kingston, ON K7L 3N6

326 Master's, 141 Doctoral

1910s: Williams, Thomas Bowerman

1930s: Atchison, D. W.; Beaton, N. S.; Beavan, A. P.; Bridger, J. R.; Cormie, J. M.; Frasner, N. H. C.; Furnival, G. M.; Gummer, W. K.; Gussow, W.

C.; Harcourt, G. A.; Harcourt, G. A.; Harding, W. D.; Henderson, J. F.; Hewitt, R. L.; Horwood, H. C.; Jewitt, Walter; Johnston, A. W.; Jolliffe, F. T.; Keith, M. L.; Keys, M. R.; Kindle, E. D.; Macdonald, R. D.; McGill, W. J.; Runnalls, N. D.; Russell, G. A.; Sutton, W. R.; Turner, J. D.; Wilson, B. T.; Zurbrigg, H. F.

1940s: Abraham, E. M.; Aitkens, D. F.; Allen, J. D.; Ames, H. G.; Brown, I. C.; Burr, S. V.; Campbell, E. E.; Carmichael, A. D.; Colgrove, G. L.; Cornford, E. H. G.; Cousineau, Yvon; Dix, N. D.; Evans, J. E. L.; Fawley, A. P.; Forrest, R. M.; Ginn, A.; Graham, A. R.; Haffner, B. K.; Harrison, J. M.; Harrison, J. M.; Haw, V. A.; Hoiles, R. G.; Hriskevich, M. E.; Leech, G. B.; MacKay, D. G.; Marshall, H. I.; Martison, N. W.; McGlynn, J. C.; McTaggart, K. C.; Merrill, R. J.; Morrow, H. F.; Nesbitt, B. I.; Parsons, W. H.; Quinn, H. A.; Robinson, M. C.; Robinson, S. C.; Robson, G. M.; Roscoe, S. M.; Rose, E. R.; Rose, K. C.; Stanton, M. S.; Whitmore, Duncan; Wright, G. M.; Yardley, D. H.

1950s: Allard, Gilles; Appleyard, Edward Clair; Baragar, W. R. A.; Burns, C. A.; Campbell, F. A.; Carlson, H. D.; Christie, N. T.; Clarke, Peter Johnston; Clarke, R. J.; Coates, Colin; Coats, Colin John Alastair; Coleman, L. C.; Evoy, E. F.; Gill, J. C.; Ginn, Robert M.; Graham, A. D.; Grant, James Alexander; Groeneveld Meijer, Willem Otto Jan; Gross, G. A.; Hale, W. E.; Hale, W. E.; Halferdahl, Laurence Bowes; Hewlett, C. G.; Hogg, G. M.; Huston, W. J.; James, D. H.; Jeffs, Donald N.; Jones, R. E.; Kirkland, S. J. T.; Kirkland, S. J. T.; Kretz, R. A.; Langford, Fred F.; LeComte, Paul; Lewis, C. L.; Maycock, Ian D.; Nichol, Ian; Pearson, George Raymond; Pearson, Walter John; Philips, K. A.; Pienaar, Petrus J.; Pienaar, Petrus J.; Podolsky, T. M.; Richter, D. H.; Rimsaite, Yadvyga; Rose, E. R.; Rose, E. R.; Rowland, J. F.; Sadler, J. F.; Sauve, Pierre; Schwellnus, Jurgen Erdmann Gotthilf; Schwellnus, Jurgen Erdmann Gotthilf; Scott, Barry; Silman, J. F. B.; Speers, E. C.; Speers, E. C.; Sproule, W. R.; Stubbins, J. B.; Traill, R. J.; Traill, R. J.; Walker, J. W. R.; Weeks, L. J.; Wegenast, W. G.; Wright, Charles Malcolm; Wynne-Edwards, Hugh Robert; Wynne-Edwards, Hugh Robert

1960s: Allan, James Frederick; Barua, Mridul C.; Beecham, Arthur W.; Berkhout, Aart Wouter Jan; Bielenstein, Hans U.; Brown, Donald D.; Byers, Peter N.; Campbell, Frederick E.; Cargill, Donald G.; Carswell, Henry Thomas; Cochrane, Donald R.; Colwell, John Allison; Cook, Donald G.; Cook, Robert B.; Cowan, Michael F.; Ermanovics, Ingomar Frank; Evans, Anthony Meredith; Findlay, David Christopher; Fletcher, Christopher John Nield; Fox, Peter Edward; Froese, Edgar; Gupta, V. K.; Harju, Hendric O.; Hay, Peter W.; Heidecker, Eric Joseph; Heidecker, Eric Joseph; Hill, Robin E. T.; Jacoby, Russel S.; Jen, Lo-Sun; Kingston, Paul W. E.; Krogh, Thomas Edward; Krogh, Thomas Edward; Laasko, Raymond Kalervo; Lowes, Brian Edward; MacGregor, Ian Duncan; Meloche, Marvin J.; Mothersill, John S.; Muraro, Theodore W.; Naldrett, Anthony J.; Naldrett, Anthony James; Oja, Reino Verner; Park, Frederick B.; Pearce, Thomas H.; Peterson, Nathan N.; Quist, Lawrence G.; Radcliffe, Dennis; Reinhardt, Edward Wade; Riley, R. A.; Robertson, James Alexander; Robertson, James Alexander; Rodriguez, Simon E.; Sawford, Edward C.; Sharma, Kamal N. M.; Smith, Arthur Young; Smith, Morland E.; Thorpe, Ralph Irving; Trembath, Lowell Thomas; Vasquez, Julio C.; Venkitasubramanyan, Calicut S.; Westervelt, Ralph Donaldson; Yorath, Christopher J.

1970s: Abbott, James Grant; Allen, John Murray; Allen, John Murray; Arancibia Ramos, Olga Nanet; Archibald, Douglas Arthur; Armstrong, Robert Clarke; Armstrong, Robert Clarke; Bailey, David Gerard; Barendregt, R. W.; Barongo, Justus Obiko; Benvenuto, Gary Louis; Binney, William Paul; Birkett, Tyson Clifford; Blain, Christopher F.; Bogle, Edward Warren; Bond, Ivor John; Bourne,

James H.; Bourne, James Hillary; Bovell, George R. L.; Bowlby, J. R.; Brown, Anton; Caelles, Juan Carlos; Caley, W. F.; Callahan, John Edward; Clark, Thomas; Closs, Lloyd Graham; Coker, William Bernard; Collins, Jon A.; Collins, Jon Alexander; Comba, C. D. A.; Cooper, Herman William; Cooper, Murray F. J.; Cordsen, Andreas; Dabitzias, Spyros Georgiou; Daly, Alan Ronald; Davenport, Peter H.; Dawes, Bruce Cameron; Dick, Lawrence Allan; Dykes, Shaun Methuen; Erdmer, Philippe; Fong, David G.; Foo, Wayne Kim; Foster, John Robert; Frape, Shaun Keith; Frape, Shaun Keith; Fuh, Tsu-Min; Gardner, Douglas A. C.; Geldsetzer, Helmut; Glover, J. Keith; Goble, Ronald James; Goldie, Raymond James; Gordey, Steven P.; Gorman, Barry E.; Goulet, Normand; Grant, Alan H.; Gunton, John Eric; Hamdan, Abdul R. A.; Haughton, David R.; Haynes, S. J.; Henry, Brian J.; Henry, Joseph B.; Holmes, Gary S.; Hoy, Trygve; Hudson, Geofrey Robert; Jackson, Robert George; Johnston, Laura M.; Johnston, Laura M.; Kalogeropoulos, Stavros Ilia; Karim, S. Abdul; Karvinen, William Oliver; Kehlenbeck, Manfred M.; Klassen, Rodney Alan; Kluyver, Huybert M.; Knight, Rosemary; Krause, Jerome B.; Larsen, Chris Robert; Lavin, Owen Patrick; LeAnderson, Paul James; Lefebure, David V.; Lortie, Ralph Burton; Lydon, John W.; McBride, Derek E.; McBride, Sandra L.; McBride, Sandra L.; McConnell, John Wilson; McDougall, Gillian Frances Ellen; McIlreath, Ian A.; Molinsky, Linda; Morse, R. H.; Muir, Thomas L.; Murray, David Lloyd; Newson, Ralph; Nikols, Carol A.; Pearson, William Norman; Peatfield, Giles Russum; Percival, John Allan; Picklyk, Donald D.; Pirie, James; Potter, Elizabeth Anne; Poulton, T. P.; Quirt, G. Stewart; Raeside, Robert Pollock; Riverin, Gerald; Roberts, Andrew Clifford; Robinson, George W.; Roddick, J. C.; Roddick, Susan L.; Rosenstein, E. S.; Sangster, A. L.; Sargent, M. W.; Schink, Ernest Allen; Schulze, Daniel J.; Scott, J. D.; Scott, Waldemar F.; Sears, James Walter; Slessor, David K.; Smith, Roger W.; Sopuck, Vladimir Joseph; Speirs, David; Stephens, L. E.; Sturman, Bozidar D.; Theis, Nicholas James; Theis, Nicholas James; Thompson, Robert I.; Thornber, Carl Richard; Tosdal, Richard Mark; Travers, Ian C.; Tremblay, Pierrette; Trueman, Edward Albert George; Vandine, Douglas F.; Wallach, Joseph Leonard; Walton, Godfrey John; Wares, Roy M.; Way, Dana Clark; West, James M.; Williams, D. R.; Williams-Jones, Anthony Eric; Wilson, Bruce Craig; Woods, Dennis V.; Xenophontos, Costas; Young, Harvey Ray; Zaw, Khin; Zentilli, Marcos; Zwanzig, Herman V.

1980s: Addo-Abedi, Frederick Yaw; Allard, Michel; Allen Dick, Beryl J.; Allen, Mary E. Theobald; Allen, Michael Steven; Allen, Rodney L.; Amor, Stephen Donald; Anderson, Paul Gordon; Archibald, Douglas Arthur; Ashton, Kenneth Earl; Bailey, Gordon C.; Bardoux, Marc-Victor; Barnhill, Susan Jean; Berger, Paula Marie; Bloom, Lynda B.; Bogle, Edward Warren; Bostock, Michael Gerhard; Burk, Raymond Ronald; Carbotte, Suzanne Marie; Chapman, R. S. G.; Charusiri, Boonsiri; Christie, Brian James; Clark, Megan Elizabeth; Cohen, David Ronald; Colton, Ilsley Daniel; Connelly, James N.; Corriveau, Louise; Creasy, David Edward Jack; de Souza, Euler Magno; Devlin, John Frederick; Dey, Sarmistha; Dick, Lawrence Allan; Dixit, Sushil Sharan; Doggett, Michael David; Edwards, Thomas W. D.; El-Hakim, Ahmed Zaki; Erdmer, Philippe; Evans, Daniel Frederick; Evenchick, Carol Anne; Fermor, Peter Robin; France, Lynne June; Garcia, Esmeralda; George, Hubert; Gerasimoff, Michael D.; Grimes, Douglas James; Guindon, David Leslie; Hall, Douglas Charles; Hamilton, John Vernon; Hanish, Mark Burton; Harms, Tekla A.; Harris, Sandra; Heather, Kevin B.; Heinrich, Silvia Maria; Hollingshead, Stephen C.; Jackson, Steven Leonard; Jamieson, Heather Edith; Johnson, Priscilla L.; Journeay, John Murray; King, Janet Elizabeth; King, Janet Elizabeth; Kontak, Daniel Joseph; Krentz, Daniel Hugh; Le, Duc; Leclair, Alain Daniel; Leclair, Alain Dan-

iel; LeGresley, Eric M.; Luk, Grace King Yan; Lyon, Kenneth E.; MacFarlane, Darryl B.; Mathieson, Gillian Ann; Mathieson, Neil Alexander; Mazurski, Marcia Ann; McGaughey, W. John; McLeod, Robert Andy; McMechan, Margaret Evaline; McMechan, Robert Douglas; Medwedeff, Donald Arthur; Miller, Iori P.; Morasse, Suzanne; Moreton, E. Peter; Morritt, Robin Frederick Charles; Neuman, Cheryl Lynn; Notley, Keith Roger; Nozdryn-Plotinicki, Michael John; Palma, Vicente Vladimir; Pearson, William Norman; Pennock, Daniel John; Percival, John Allan; Pirie, Ian David; Piroshco, Darwin W.; Poulsen, Knud Howard; Reichhard-Barends, Enrique; Richards, Peter Alan Leslie; Rigg, David Michael; Robinson, R. L.; Sayao, Otavio de Sampaio Ferraz Jardim; Scowen, P. A. H.; Seal, Robert R., II; Shelp, Gene Sidney; Simpson, David Gordon; Smol, John Paul; Spooner, Ian Stewart; St-Onge, Marc Robert; Staargaard, Christiaan Frederik; Sterenberg, Velma Zwaantje; Stone, David Michael Raymond; Sveinsdottir, Edda Lilja; Swinamer, Ralph Terrance; Taylor, Sharon I.; Tippett, Clinton Raymond; Trudu, Alfonso Giacomo; Tsikos, George; Tucker, Bruce C.; Verleun, Leo Johannes; Watkins, John Joseph; Webb, David Ralph; Wells, Gary Steven; Whang, Chen-Wen; Whyte, James Bernard; Wilson, Ann Catherine; Zaitlin, Brian Allen; Zolnai, Andrew S.; Zweng, Paul L.

Queens College (CUNY)
Flushing, NY   11367

74 Master's

1960s: Espejo, Anibal C.; Sulenski, Robert J.

1970s: Arbucci, R. P.; Aurisano, R.; Benda, M. N.; Bennett, G. V.; Bower, P. M.; Cavallero, Lillian; Cinquemani, L. J.; Dadourian, P.; Della Valle, R. S.; Fenster, D. F.; Filep, E. J.; Fredericks, Carol M.; Goddard, John G.; Goldberg, M. A.; Gonzalez, E., Jr.; Hertling, M. M.; Jantzen, R. E.; Kaiteris, P.; Kalisky, Maurice; Katuna, M. P.; Kobre, N. A.; Koszalka, E. J.; Krauser, R. F.; Levy, A. S.; Markl, Rudi G.; McNeill, R. J.; Mesticky, L. J.; Munsart, C. A.; Murray, E. H., Jr.; O'Hara, P.; Perissoratis, C.; Peterson, J. J.; Plichta, C.; Prince, Arthur M.; Roche, Michael B.; Rokach, Allen; Roux, P. H.; Sambol, M.; Sblendorio Levy, J. S.; Schneck, M. C.; Schreifels, W. A.; Schwartz, David P.; Scofield, Jane Ann Rutledge; Seymour, B. O.; Sondergeld, Carl H.; Spanglet, M.; Sverdlove, Mark Selig; Thies, K. J.; Vargas, A.; Witrock, R. B.; Wolterding, D.

1980s: Apostolou, Charalampos; Boudreau, M.; Bromble, Sandra L.; Buis, Patricia; Casson, Robert N.; Clappin, Philip F.; Filomena, Joseph James; Flax, Philip; Fogarty, Mark; Grey, Carllett; Helman, Marc; Klimentidis, R.; Knapp, Susan; Kocher, Frederick; Roth, Mark; Sachi-Kocher, Afsar; Sassos, Michael; Swanson, Andrew; Tavolaro, John F.; Wehrle, Mary E.; Zotto, Maria

University of Regina
Regina, SK   S4S 0A2

19 Master's

1970s: Dwairi, Ibrahim; Garven, Audrey Curry; Hulbert, Larry John; Roberts, Keith

1980s: Abraham, Andrew Peter; Adamson, David William; Akhurst, Maxine Carole; Crabtree, Harry Thomas; Eriyagama, Sarath Chandra; Haidl, Frances Margaret; Leibel, Robert John; MacEachern, James Anthony; Potter, Dean Edward George; Rees, Christopher John; Stasiuk, Lavern D.; Thomas, David James; Walker, Nola Constance; Walters, Kenneth Lamont; Walters, Stephen

Rensselaer Polytechnic Institute
Troy, NY   12180

124 Master's, 79 Doctoral

1950s: Bird, John M.; Hutchings, Roy Theodore, Jr.; Laskowski, Edward A.

1960s: Ahrens, Thomas J.; Allen, George P.; Baker, Seymour R.; Baum, Bruce R.; Bird, John Malcolm; Bishko, Donald; Borst, Roger Lee; Brost, Roger L.; Carragan, William Dillard; Chen, Albert T. F.; Cutcliffe, William E.; Dillon, William P.; DiPiazza, Nicholas John; Elam, Jack Gordon; Foord, Eugene E.; Gavish, Eliezer Kneidel; Geiger, Earl G., Jr.; Gerhard, F. Bruce, Jr.; Gevirtz, Joel L.; Gevirtz, Joel L.; Gilmore, Robert Snee; Haglund, David S.; Han, Dongyup; Herman, Howard R.; Hudec, Peter P.; Hudec, Peter Paul; Johnson, Kenneth G.; Kaufman, John W.; Kolesar, Peter Thomas, Jr.; LaBrake, Richard F.; LaFleur, Robert George; Lahoud, Joseph A.; McKinney, Thomas Francis; Mercado, Edward John; Metz, Robert; Mohammed, Mahdi; Murphy, Kenneth Robert; O'Brien, P. J.; Ozol, Michael Arvid; Rasmussen, John A.; Rebull, Peter Mario; Schock, Robert Norman; Schwartz, Michael; Schwarzer, Rudolph R., Jr.; Schwarzer, Rudolph Reynolds; Schwarzer, Theresa F.; Schwarzer, Theresa Frances; Schwarzer, Theresa Frances; Whipple, Janice M.; Woeber, A. Frederick; Yatsevitch, Yuri

1970s: Agostino, Patrick Noel; Albanese, James R.; Ali, Syed A.; Anderson, George A., III; Arnon, Boas; Baker, Seymour R.; Bentley (Pyzanowski), Barbara; Bonner, Brian P.; Borak, Barry; Bosworth, William P.; Boyd, Kenneth A.; Buttner, Peter J. R.; Buyce, M. Raymond; Carlson, Gorden Anders, Jr.; Collins, Catherine M.; Conway, Stephen W.; Cullen, James J., IV; Dahl, John; deCaprariis, Pascal Peter; Del Prete, Anthony; DeSimone, David J.; Dineen, R. J.; Down-Logan, Kathleen; Epstein, Samuel A.; Fisher, Louis A.; Floess, C. H. L.; Frank, William M.; Garber, Raymond Alan; Goter, Edwin R., Jr.; Haimes, Robert; Hanson, E.; Harris, Roger L.; Hatt, Timothy A.; Helsinger, Marc H.; Helsinger, Marc H.; Hill, Constance H.; Humphreys, Matthew; Jones, Thomas W.; Keith, Brian D.; Kimmel, Tammy; Klara, Eugene W.; Kochem, Edward J.; Kontis, Angelo L.; Kott, Michael J.; Kramers, John William; Krecow, Frank C.; Kuslansky, Gerald H.; Lakatos, Stephen; Leahy, P. Patrick; Lindemann, Richard H.; Mathis, Robert L.; Mazzullo, Salvatore J.; Middleton, Jack A.; Moore, David John; Mylroie, J. E.; Nemetz, Arthur C.; Nurmi, Roy D.; Ogidan, Richardson D.; Ordan, Laura D.; Owen, Roy W.; Paik, Y. S.; Porter, Leonard A.; Pratte, J. Frances; Puppolo, David G.; Radke, Bruce M.; Ricart y Menendez, Fernando Osacar; Ross, Alan; Rubin, David M.; Schrank, Joseph A.; Schreiber, B. Charlotte; Seitz, John N.; Singer, Jeff; Smith, Jeffry A.; Sneh, Amihai; Sneh, Amihai; Strife, Stuart C.; Toub, Joyce Silverstein; Treesh, Michael Irvin; Turner, Donald A.; Waldman, Estella; Way, John H., Jr.; Wroblewski, Frank G.; Yeates, David G.

1980s: Bellemore, Barbara A.; Bigman, Nathan; Bin Mohamad, Ramli; Bond, William Earl; Cannizzaro, Carl R.; Capobianco, Christopher; Castro, Gerardo; Cattafe, Joseph S., Jr.; Cousens, Ellis E.; Crist, W. Konrad; Curl, Mary W.; Davis, Jeffery A.; DeSimone, David J.; Dolfi, Ronald U.; Donelick, Raymond A.; Donelick, Raymond Allen; Duskin, Priscilla; Dyvik, Rune; El-Far, Sherif Ahmed Kamal; Finley, Sharon G.; Galli, Kenneth G.; Garber, Raymond Alan; Goldmintz, Amy Jo; Greiner, Gerald F.; Harrop-Williams, Kingsley Ormonde; Hennessey, Russell B.; Holden, Peter Newhall; Howland, Jonathan Dean; Ianniello, Michael L.; Itell, Karyn Marie; Jamsheed, Behsheed; Johnson, Robert; Jurewicz, Amy Jo Goldmintz; Jurewicz, Stephen Richard; Jurewicz, Stephen Richard; Klein, Roger W.; Koral, Hayrettin; Kulansky, Gerald Harry; Lawler, Kevin P.; Lindemann, Richard Henry; Maiurano, Karen; May, William P., Jr.; Maynard, Donald; Mazzo, Carl R.; Nasim, Mushtaq Ahmad; Petrakis, Emmanuel; Rapp, Robert Paul; Redmond, Brian Thomas; Regan, Peter T.; Ruzyla, Kenneth; Salerno, Catherine M.; Shukla, Vijai; Sinclair, Horace A.; Souflis, Constantinos I.; Sternbach, Charles A.; Sternbach, Charles Alan; Sternbach, Linda Raine; Stuart, Charles K.; Tierney, Michael T.; Urschel, Stephen F.; Vasquez-Herrera, Andres R.; Vicente, Ernesto Edgardo; Voorhees,

David H.; Vucetic, Mladen; Whitbeck, Luanne F.; Wood, Lawrence A.; Zagorski, Theodore W.

University of Rhode Island
Kingston, RI 02881

92 Master's, 108 Doctoral

1960s: Baseler, Thomas W.; Bean, Daniel Joseph; Christopher, Raymond A.; Cornell, William C.; Dillon, William Patrick; Fain, Gilbert; Garrison, Louis Eldred; Giese, Graham Sherwood; Hindle, Robinson Joseph; McGregor, Bonnie A.; Savard, Wilfred L.; Schwartz, James

1970s: Albert, Robert L.; Amerigian, C.; Ballard, R. D.; Barrett, Jonathan R.; Barrett, Karen W.; Beale, Richard A.; Belmont, Ronald A.; Betzer, Peter R.; Boehm, P. D.; Brunner, C. A.; Byrne, R. H.; Christofferson, Eric; Coddington, Wayne J.; Collins, B. P.; Corliss, B. H.; Corrigan, Donald; Daub, Gerald J.; Demars, K. R.; Diesl, Warren F.; Dignes, Thomas W.; Donlon, Thomas J.; Dow, Roberta L.; Ellwood, B. B.; Erchul, Ronald Anton; Fanning, Kent Abram; Fillon, Richard Henry; Fisk, M. R.; Froelich, P. N., Jr.; Gaines, A. G., Jr.; Gallagher, James; Garner, John C., Jr.; Gautie, Stephen C.; Goetz, Michael J.; Goodale, Jonathan L.; Graham, W. F.; Hagstrom, Earl L.; Hess, K. W.; Hogan, Colleen A.; Jendrzejewski, John P.; Johnson, David Leslie; Kahn, N. M.; Keany, J.; Keigwin, L. D., Jr.; Kern, Christian A.; Kerr, R. A.; Klinkhammer, G. P.; Lackoff, Martin Robert; Lambiase, Joseph J., Jr.; Leavy, Brian D.; Ledbetter, M. T.; Leinen, Margaret Sandra; Lorens, R. B.; Lowry, Bruce E.; McClennen, Charles Eliot; McGrail, David W.; McGrail, David W.; Miller, Gerard Roland, Jr.; Milligan, S. D.; Milne, Peter C.; Murphy, John F.; Myers, Allen Cowles; O'Connor, Thomas Patrick; Pierce, Richard H., Jr.; Pierce, Thomas A.; Pinet, Paul Raymond; Piotrowicz, S. R.; Pisias, N. G.; Pope, David M.; Pope, Joan; Regan, Donald R.; Rittschof, William F.; Robb, James M.; Samson, Michael R.; Schultz, D. M.; Schwab, William C.; Shipman, Wayne D.; Simpson, Elizabeth J.; Stolzman, Robert A.; Thompson, Gary L.; Thunell, R. C.; Unni, C. K.; Urish, D. W.; Van Vleet, E. S.; Wade, M. J.; Wade, T. L.; Wallace, G. T., Jr.; Walsh, P. R.; Weisberg, R. H.; White, W. M.; Williams, D. F.; Zarillo, Gary A.

1980s: Adelman, David; Aldinger, Paul Bruce; Allard, David; Amankwah, Samuel Asare; Baldwin, Kenneth Charles; Bein, Margaret G.; Blais, Alan G.; Bouse, Robin; Boyle, Steven T.; Call, Terry D.; Carey, Steven Norman; Chin, Yu-Tung; Cornell, Winton C.; Cornell, Winton Charles; Curlin, Kimberly B.; Danforth, William W.; Dauphin, Joseph Paul; Devine, Joseph Driscoll, III; Doh, Seong-Jae; Douglas, Gregory Scott; Dunn, Dean Alan; Dunne, Lorie A.; Evans, Robert James; Friedrich, Nancy E.; Garber, Jonathan Hunt; Gibeaut, James C.; Glenn, Craig Richard; Grant, John A., III; Hamidzada, Nasir; Hanson, Alfred Kenneth, Jr.; Hanson, Lawrence G.; Hatton, Josephine; Hodell, David Arnold; Hong, Huasheng; Hughes, William D.; Huizenga, Douglas Lee; Hutchinson, Deborah Ruth; Johari, Akbar; Jones, David E.; Jordan, Bradley C.; King, D. Whitney; Kocis, Diane E.; Kowalski, Richard G.; Lai, David Yuekchung; Lee, Richard A.; Liang, Duohaw; Loutit, Tom Stuart; Madsen, John Alfred; Maguire, Thomas F.; Maring, Hal Barton; Maris, Cynthia Robin Parmalee; McCarty, Harry Brinton, Jr.; McClory, Joseph Patrick; McGinn, Stephen R.; McGregor, Michael Andrew; Meyer, Peter Sheafe; Mills, Gary Lawrence; Morin, Roger Henri; Mosher, Byard William; Muerdter, David Robert; Muller, François; Noble, Marlene Ann; Oakman, Marial; Owen, David J.; Pac, Timothy J.; Parke, Craig D.; Patterson, Thomas Lee; Pearson, Carl A.; Peters, Colen R.; Pickart, David; Powell, Harriet; Pratson, Lincoln; Pszenny, Alexander A. P.; Requejo, Adolfo G.; Richardson, Mark; Riebessel, Ulf; Riegler, Paul William; Romine, Karen Kay; Rosenberg, Murray J.; Rudnick, David Thornton; Sanders, David P.; Savarese, Joseph G.; Schultz,

Norbert; Severson, Roger H.; Sowers, Todd; Sundvik, Michael Todd; Swanson, J. Craig; Symes, James Leo, III; Szak, Caroline; Thompson, John; Thumtrakul, Wilaiwan; Twichell, David Cushman, Jr.; Viekman, Bruce; Vogel, James William; Wei, Kuo-Yen; Weisel, Clifford Paul; Zachos, James C.

Rice University
Houston, TX 77251

196 Master's, 178 Doctoral

1940s: Macgregor, J. E.

1950s: Adams, Henry Clay, Jr.; Dawson, Ross Elmo, Jr.; Fessenden, Franklin W.; Griffith, Lawrence S.; Hammill, Gilmore Semmes, IV; Harkrider, David Garrison; Head, William Burres; Huff, William Jennings; Lawhorn, Thomas Warren; Longshore, John David; McEwen, Michael C.; Miller, Dale Everett; Pliler, Richard; Pliler, Richard; Powell, William Frank; Powell, William Frank; Richardson, Keith Allan; Schutz, Donald F.; Strong, Cyrus; Whitfield, John M.

1960s: Aguilera, Raymundo; Ahr, Wayne Merrill; Allen, Gary Curtis; Almy, Charles Coit, Jr.; Almy, Charles Coit, Jr.; Alvarado, Salvador; Aspiroz, Rogilio; Balacek, Kenneth Joseph; Banks, Thomas H.; Barks, Ronald E.; Beck, Barry Frederick; Behrens, Earl William; Billings, Gale Killmer; Billings, Gale Killmer; Brann, Bethany Celia; Brimhall, Willis Hone; Burkart, Burke; Camargo, Antonio; Carter, James Lee; Chang, Andre Chi-Chao; Chen, Ju-Chin; Chen, Ju-Chin; Cochran, Michael David; Coker, C. Eugene; Cook, Beverly Kay Gatlin; Cram, Ira H.; Crane, David Clinton; da Silva, Zenaide Carvalho Goncales; Dalrymple, Don Wayne; Deininger, Robert W.; Dolgoff, Abraham; Ebanks, William James, Jr.; Ebanks, William James, Jr.; Evans, John Keith; Failing, Martha S.; Foss, Ted Harry; Friedman, Melvin; Gibbon, Donald Leroy; Graf, Claus Heinrich; Graf, Claus Heinrich; Griffin, George Martin; Hammill, Gilmore Semmes, IV; Harriss, Robert Curtis; Harriss, Robert Curtis; Hartung, Jack Burdair; Heckel, Philip Henry; Hernandez, Salvador; High, Lee Rawdon, Jr.; Himes, David Madero; Horn, Myron Kay; Huff, William Jennings; Jacka, Alonzo David; Jarvis, Harry Aydelotte, Jr.; Kline, Mary-Cornelia; Krawiec, Wesley; Krog, Marilyn K.; Kruger, William Charles, Jr.; Lane, Donald Wilson; Leeman, William P.; Lidiak, Edward George; Lidiak, Edward George; Livingston, John Lee; Livingston, John Lee; Longshore, John David; Longshore, Judith Clark; Madrigal, Luis; Mahdavi, Azizeh; Martin, Edgar Keith; Matthews, Robley Knight; Matthews, Robley Knight; McEwen, Michael C.; McKay, David Stewart; McKay, Gordon Alan; McKay, Sheila Mahan; McKniff, Joseph Michael; Minear, John Wesley; Olson, Henry David; Palafox, Hector; Pan, Poh-Hsi; Park, David Eugene; Pelton, Peter John; Ponce de Leon, Jose G.; Pusey, Walter Carroll, III; Ragland, Paul C.; Ragland, Paul C.; Reeves, Thomas Kenneth, Jr.; Rehkemper, Leonard James; Renard, Vincent Paul Augustine; Reso, Anthony; Riber, Joshua I.; Rice, Raymond H.; Richardson, Keith Allan; Schubert, Carlos; Schubert, Carlos; Scott, Martha Lyles Richter; Scott, Robert Blackburn, III; Stow, Stephen Harrington; Stow, Stephen Harrington; Sutter, John Frederick; Taylor, G. Jeffrey; Tebbutt, Gordon Edward; Teeter, James Wallace; Terrell, John H.; Toomey, Donald Francis; Troell, Arthur Richard, Jr.; Viveros, José G.; Wang, Kung Teh; Wantland, Kenneth Franklin; Wantland, Kenneth Franklin; Wilson, Raymond C., Jr.; Winters, Martha Diane; Woeber, Arthur Frederick; Wu, Changsheng; Wu, Changsheng

1970s: Abbott, Earl William; Abbott, Earl William; Baldwin, Otha Don; Barretto, Paulo Marcos de Campos; Barretto, Paulo Marcos de Campos; Bauer, Mary Alvina; Beck, Barry Frederick; Biddle, Kevin Thomas; Biddle, Kevin Thomas; Bosc, Eric Antoine; Boyer, Jannette Elaine; Boyer, Paul Slayton; Brady, Michael John; Buckley, Christo-

pher Paul; Butler, D. M.; Cameron, Christopher Scott; Campbell, Michael David; Capers, William A., Jr.; Carr, Michael David; Carr, Michael David; Cashby, Susan Margaret; Cole, Mary Lou; Cook, Beverly Kay Gatlin; Couples, Gary Douglas; Donaldson, John William; Duex, Timothy W.; Dumas, David Byron; Dunne, George Charles; Dunning, Charles Preston; Edwards, James Michael; Ekdale, Allan Anton; Ekdale, Allan Anton; Ekdale, Susan Faust; Fainstein, Roberto; Ferreira, Justo Camejo; Forrest, Joseph Turner; Freeland, George Lockwood; Friesen, Larry Jay; Fuex, Anthony Nichols; Gans, William Thomas; Georges, Danae; Ghosh, Protip Kumar; Ghuma, Mohamed Ali; Gust, David Allen; Hakkinen, Joseph William; Hamilton, John Richard; Hickcox, Alice Ellen; Hickcox, Charles Woodbridge; Hohlt, Richard B.; Houston, Mark Harig; Hueni, Camille D.; James, Alan Thomas; James, Gerard W.; Jaramillo, Jose Maria; Jenks, Susan Elizabeth; Johnson, Edward Allison; Johnson, Linda Ann; Johnson, Michael Lee; Jordan, Clifton F., Jr.; Jordan, Clifton F., Jr.; Jordan, Jimmie Lynn; Khoja, Elhadi Razzagh; Kronfeld, Joel; Kruger, Steven T.; Kunze, Florence Mollie; Kurtz, Dennis Darl; Lambert, William Robert; Lundell, Leslie Lee; Lux, Daniel R.; Mackay, Thomas Stephen; Manker, John Phillip; Manzer, Geraald K., Jr.; Marcucci, Ettore; Marcucci, Ettore; McCrevey, John Alfred; McHuron, Eric Jay; McMillen, Kenneth James; McMillen, Kenneth James; Meyers, William John; Miller, Elizabeth Louise; Miller, Elizabeth Louise; Moed, Barbara A.; Mueller, Paul Allen; Murray, Marc Michael; Nagy, Richard Michael; Nelson, Eric Paul; Novitsky-Evans, Joyce Marie; Novitsky-Evans, Joyce Marie; O'Rourke, Terence Lee; Olson, Robert Wendell; Parrish, David Keith; Pearson, Daniel Bester, III; Phelps, David William; Phelps, David William; Pratt, David E.; Radovich, Barbara Jean; Reynolds, Richard Alan; Reynolds, Richard Alan; Sanchez-Barreda, Luis Antonio; Santamaria, Francisco Jose; Schupbach, Martin Albert; Self, Robert P.; Sharief, Farooq Abdulsattar M.; Sharief, Farooq Abdulsattar M.; Shipley, Thomas Howard; Sloss, Peter William; Snowdon, Lloyd R.; Spaw, Joan Mussleer; Spaw, Joan Mussler; Spaw, Richard Hoencke; Stalmach, Daniel Miles; Standlee, Larry Aaven; Standlee, Larry Aven; Stanley, Richard Graham; Sutter, John Frederick; Talukdar, Suhas Chandra; Taylor, G. Jeffrey; Thornton, Edward Clifford; Tula, Alex; Tull, James Franklin; Tyler, David Lynn; Urish-McLatchie, Carol Lynn; Van Wagoner, John Charles; Van Wagoner, John Charles; Walker, Alta S.; Walraven, David; Wang, Chen Yu; Wang, William Cruse; Warford, Andrew Craig; Wegner, Robert Carl; Weisenberg, Charles William; Weisenberg, Charles William; Wheeler, Richard Brian; Wilson, Patricia McDowell; Wilson, William M.; Winchester, Paul Drake; Winchester, Paul Drake; Yau, Mary Wing-Chi

1980s: Adams, Roy Donald; Ageli, Hadi S.; Ander, Holly Dockery; Andrews, Barbara Ann; Archer, Robert Edward; Baegi, Mohamed Bashir; Balshaw-Biddle, Katherine M.; Beck, Candyce Lynne; Bowen, Corey Scott; Brake, Christopher French; Brown, Lauren Shelley; Budai, Joyce Margaret; Carson, Thomas L.; Chacko, Soman; Chen, Shiahn-Jauh; Chiang, Chen Yu; Cleveland, Michael N.; Cole Hoerster, Mary Lou; Cooper, J. Calvin; Coward, Robert Irvin; Cuddihee, John Lee; Dahl, Jeremy Eliot; Dockery, Holly A.; Domack, Eugene Walter; Domack, Eugene Walter; Dotson, Kirk Wayne; Dravis, Jeffrey James; Driskill, Lorinda Elizabeth; Fisco, Mary Pamala Polite; Gammill, Laura May; Gelber, Arthur Winston; Gerlach, David C.; Gibson, Bruce Sanderson; Gorody, Anthony Wagner; Gottschalk, Richard Robert, Jr.; Grabowski, George Joseph, Jr.; Griffith, Thomas W.; Harlan, Janis G.; Hawk, Jody M.; Herron, Margaret J.; Honjo, Norio; Jeffers, John Douglas; Kennedy, Douglas S.; Kennedy, Jerry Wilson; Killen, Rosemary Margaret; Koehler, Robert Paul; Kunze, Florence Mollie; Kurtz, Dennis Darl; Lamb, William Marion; Lee, Michael Donald; Leventer, Amy

Ruth; Lewis, Dana Lyn; Liu, Chingju; Lord, Jacques Passerat; MacDonald, Scott Edward; Matthews, William K., III; Matty, David Joseph; Matty, Jane Miller; Mauch, Elizabeth Ann; May, Jeffrey Allyn; Meinwald, Javan N.; Metcalfe, Cynthia Watson; Milam, Robert Wilson; Milliken, Jeffrey V.; Minnis, Steffi Ann; Mirkin, Andrew S.; Mora, Claudia Ines; Myers, Nathan Cebren; Nelson, Carl Owen; Norman, Marc Douglas; Oddo, John Edward; Ostos-Rosales, Marino; Pearson, Daniel Bester, III; Pereira, Enio Bueno; Perez Guzman, Ana Maria; Phelps, James Carl; Putzig, Nathaniel E.; Roqueplo-Brouillet, C.; Schafersman, Steven Dale; Schmidt, William Jay; Schmidt, William Jay; Seidensticker, C. Michael; Singer, Jill Karen; Singer, Jill Karen; Smith, Diane Ruth; Smith, Diane Ruth; Smith, Michael John; Watkins, Elizabeth Ann; Webber, Karen Louise; Weidemann, Donna Elizabeth; Weinheimer, Amy L.; Wenger, Lloyd Miller, Jr.; West, Howard Bruce; Wielchowsky, Charles Carl; Wigley, Cynthia R.; Wigley, Cynthia R.; Wolfteich, Carl Martin; Wong, Peter Kin; Wright, Robyn; Wright, Robyn; Yeo, Ross Kenneth

University of Rochester
Rochester, NY 14627

177 Master's, 42 Doctoral

1900s: Chadwick, George H.; Heaton, Charles David; Sarle, Clifton J.

1910s: Giles, Albert William; Sinclair, Joseph H.

1920s: Alling, Merle K.; Dunbar, Elizabeth Urquhart; Edwards, Ira; Valentine, Wilbur G.; Wishart, James Scotland

1930s: Agey, Charles; Broughton, John Gerard; Chasey, Kenneth LeMay; Dings, McClelland Griffith; Dodge, Nelson B.; Dollen, Bernard Halloran; Gillette, Tracy; Grossman, William Lewis; Jensen, David Edward; Karleskind, Lorene Cora; Kremer, Lois Antoinette; Payne, Thomas Gibson; Reed, Charles M.; Simmons, Benjamin Titus; Smith, Marguerite Adelle; Suess, Irma L.; Walters, Robert Fred; Wray, Irene

1940s: Anderson, Keith Elliott; Baird, David McCurdy; Bush, Alfred Lerner; Chamberlain, John David; Davis, Ethel M.; Fay, Robert Lawrence; Fellows, Robert Ellsworth; Gros, Frederick Christian; Hooper, Jane; Hoyt, Virginia; Lowry, Wallace Dean; MacQuown, William Charles; Platt, Robert M.; Williams, George Quigley; Withington, Charles Francis; Wray, Charles F.

1950s: Anderson, Walter A.; Ashmead, Lawrence Peel; Campbell, A. Richard; Cohen, Philip; Connally, Gordon; Dyer, Charles; Fisher, Donald W.; Freeman, Harvey Albert; Fulreader, Rufus Everett, Jr.; Hamlen, Dale Alexander; Harris, Ellwood Glendenning; Hoskins, Donald Martin; Howd, Frank H.; Kennedy, Edward; Lewis, Thomas L.; Manley, Frederick; Moyer, Paul T., Jr.; Nagell, Raymond Harris; Neel, Robert; Paulus, Fred; Pefley, David R.; Rickard, Lawrence Vroman; Sachs, Kelvin Norman; Sass, Daniel Benjamin; Seaber, Paul R.; Smith, Marvin Lyle; Stover, Lewis E.; Sutton, Robert George

1960s: Campbell, Josephine Kay; Coch, Nicholas K.; Egemeier, Stephen Jay; Eld, Terry Johnson Hammeken; Fletcher, Frank William; Gordon, Andrew Hunt; Grosvenor, Florence Anne; Guu, Jengyih; Hammond, Douglas E.; Hoffman, John W.; Humes, Elmer C., Jr.; Huzzen, Carl Stewart; Jung, George; Karmen, Andrew A.; Killip, Colbeth; Krawiec, Wesley; Krueger, William Charles; Liu, John Lin-gun; Lucier, Wallace Anthony; Mao, Ho-Kwang; Mao, Ho-Kwang; Nugent, Robert Charles; Oyer, W. Brian; Peper, John Dunkak; Pratt, Howard Riley; Pratt, Howard Riley; Rukavina, Norman Andrew; Silberman, Miler Louis; Sommers, David Arthur; Terlecky, Peter Michael Jr.; Twigg, Daniel Bruce; White, Stanton M.; Wolff, Manfred Paul; Woodrow, Donald Lawrence; Woodrow, Donald Lawrence

1970s: Baird, Gordon Cardwell; Batt, Paul; Brill, Edward Tobias; Bubeck, Robert Clayton; Chamberlain, John Andrew, Jr.; Chandlee, George O.; Chang, Yung-Kang Mark; Chapman, Ralph Ebener; Cottrell, John; Cottrell, John Francis; Crick, Rex Edward; Cutler, Alan Hughes; Deck, Bruce L.; Derstler, Kraig Lawrence; Dockstader, David R.; Dockstader, David Roy; Dygert, Harold Paul, III; Ebblin, Claude Paul; Gartland, Eugene F.; German, Rebecca; Graus, Richard Raphael; Gray, Lee Malcolm; Halka, Jeffrey P.; Hill, Jennie Lu Hill; Hrubec, John Anthony; Jelacic, Allan Joseph; Ju, Chi-Rei; Kinsland, Gary Lynn; Kinsland, Gary Lynn; Lasker, Howard Robert; Lidgard, Scott; Liu, Lin Gun; Lockner, David Avery; Lux, Richard Alan; McGhee, George Rufus, Jr.; Meguid, Fayek; Merrill, Leo; Mihalyi, Dale L.; Mihalyi, Dale Lynn; Ming, Li-chung; Ming, Li-chung; Ming, Pam P. Chou; Muller, Otto Helmuth; Muller, Otto Helmuth; Muller, Otto Helmuth; Nelson, Ann B.; Papantonis, Dimosthenis; Plotnick, Roy E.; Ramsayer, George Ralph; Ramsayer, George Ralph; Remz, Stuart R.; Revetta, Frank Alexander; Roe, Leon Mergeson, II; Roe, Leon Mergeson, II; Rubin, David M.; Schwartz, Dale Ann; Selleck, Bruce Warren; Selleck, Bruce Warren; Sharry, John; Signor, Philip W., III; Silberman, Miles Louis; Smith, Charles Andrew Francis, III; Tang, Hwei-Feng Nadine; Thurrell, Robert F., III; Urban, Thomas Charles; Weaver, J. Scott; Weaver, John Scott; Wilburn, David R.

1980s: Adams, John Edward; Brunelle, Thomas M.; Callaway, Jack M.; Chellman, Kathryn King; Conard, Nicholas John; Davis, Richard Laurence; Dick, Vincent B.; Domenick, Michael A.; Druke, Carmen B.; Ehrets, James Russell; Faggart, Billy E.; Giffuni, Genaro F.; Gill, Ivan P.; Gray, Lee Malcolm; Hamell, Richard David; Hertel, Fritz; Hull, Joseph Michael; Khandaker, Nazrul Islam; Kramer, Matthew Joseph; Lin, Bea-Yeh; LoDuca, Steven T.; Lumino, Karen Marie; Miller, Keith Brady; Moritz, Gail; Ongley, Jennifer S.; Parsons, Karla Moreau; Pivnik, David; Protzman, Gretchen Marie; Rasmussen, Kenneth A.; Rosen, Michael Robert; Rubury, Eric Alan; Savarese, Michael L.; Scherzer, Jolie L.; Sharp, Gerald L.; Silverman, Stephen J.; Snow, Jonathan E.; Speyer, Patricia M.; Speyer, Stephen E.; Speyer, Stephen Eric; Srimal, Neptune; Srivastava, Praveen; Talpey, James G.; Teng, Ray Tsao Dah; Zadins, Zintars Z.

Rose-Hulman Institute of Technology
Terre Haute, IN 47803

1 Doctoral

1900s: Andrews, C. B. M.

Rutgers, The State University, New Brunswick
New Brunswick, NJ 08903

135 Master's, 121 Doctoral

1930s: Greacen, Katherine Fielding; Johnson, Andrew Leigh

1940s: Davidson, Edward Sheldon; Keller, Fred, Jr.; Phair, George; Ramsdell, Robert C.; Waltman, Reid Martin

1950s: Adams, John Kendal; Aukland, Merrill Forrest; Brownson, Ernest Maitland; Cuppels, Norman Paul; Dix, Fred Andrew, Jr.; Donohue, John Joseph; Edsall, Thomas D., III; Farley, William Horace; Friedman, Melvin; Gill, Harold Edward; Gordon, Lawrence; Horton, Ernest Henderson; Jones, Albert Vincent; Kline, James E.; Krebs, Robert Dixon; Light, Mitchell Arron; Lynd, Langtry E.; McCormack, Robert Keith; McMaster, Robert Luscher; Meditz, Richard Donald; Minard, James Pierson; Morris, Robert Hall; Muskatt, Herman Solomon; Mutch, Thomas Andrew; Nichols, Paul Harry; Nichols, Paul Harry; Nine, Ogden Wells, Jr.; Olsson, Richard K.; Page, Richard Adams; Pryor, Wayne Arthur; Schaffel, Simon; Schlanger, Seymour Oscar; Scudder, Ronald Jay; Smith, William Lee; Sturm, Edward; Thomson, Alan Frank;

Vecchioli, John; Voshinin, Natalie; Whelan, Thomas Joseph, Jr.

1960s: Allen, Walter Carl; Baker, Ralph N.; Berdanier, Charles Reese, Jr.; Block, Fred; Bowman, James Floyd, II; Chapman, Diana Ferguson; Chapman, Diana Ferguson; Davids, Robert Norman; De Ratmiroff, Gregor N.; Drobnyk, John Wendel; El-Khayal, Abd El-Malik Abd Allah; Enright, Richard Louis, Jr.; Goodspeed, Robert Marshall; Haagensen, Robert B.; Hammell, Laurence; Harmon, Kathryn Parker; Hellerman, Joan; Iden, Lee J.; Isphording, Wayne Carter; James, Arthur Darryl; Johnson, Eben Lennart; Jordan, Carl Frederick; Lipman, Leonard H., II; Mac Donald, Robert B.; MacNamara, Edlen E.; Macomber, Bruce Edkins; Manspeizer, Warren; McDonald, Robert B.; Nogan, Donald Stanley; Oman, Carl Henry; Parrillo, Daniel G.; Rajagopalan, Natasayyer; Rao, Shankaranarayana R. N.; Savage, E. Lynn; Schreiber, B. Charlotte; Servilla, Thomas; Sparks, William E.; Spence, William Henry; Spink, Walter John; Spink, Walter John; Ugolini, Fiorenzo C.; Wieluns, Robert

1970s: Ahmed, Riaz; Allen, James R.; Allen, John Francis, Jr.; Althoff, Penelope L.; Anderson, Peter Wilfred William; Au, Wing-Cheong; Aydin, Fehmi Numan; Bambrick, James, Jr.; Bandoian, Charles Asa; Berthoud, Charles E., Jr.; Black, William J.; Cirello, John; Crandall, Thomas M.; Crossan, Arthur Brook, III; Cummings, Warren LeRoy; Cunliffe, James E.; Dahlgren, Paul B.; Dunning, Jeremy David; Eilenberg, Sarah; English, John Richard; Fairbrothers, Gregg E.; Feldman, Howard Ross; Franceschini, Timothy; Gillings, O. J.; Goldenberg, Jeffrey E.; Goldstein, Fredric Robert; Grasso, Santo Vincent; Harper, David Paul; Hirsch, Alfred Martin; Holzer, Robert Albert; Houlik, Charles William; Houlik, Charles William, Jr.; Hunt, Margo Elaine; Jogan, Brenda M. H.; Kennish, Michael Joseph; Kennish, Michael Joseph; Kerr, James McKinnon, Jr.; Kewer, Robert Parker; Klonsky, Louis Farrell; Koch, Robert Clement; Krall, Donald Bowman; Kraybill, Richard Lancaster; Kulpecz, Alexander A.; Lan, Ching-Ying; Martino, Ronald Layton; Massa, Vito, Jr.; Mattis, Allen Francis; Maxey, Lawrence R.; Mayfield, Darrell G.; McWhorter, James G.; Muller, Frederick Lorenz, Jr.; Nemickas, Bronius; Nieswand, George Heinz; Nordstrom, Karl Fredrik; Novak, Robert James; O'Grady, Martin Dempsey; Onega, Lawrence Kerokadho; Pendleton, Martha Warren; Petters, Sunday W.; Pollock, Stephen Garrett; Pontier, Nancy Kilbridge; Rafaelis, Maris; Raman, Swaminathan Venkat; Rehm, John M., Jr.; Rhett, Douglas William; Robertson, Bruce Edward; Robertson, David Kenneth; Samsel, W. A.; Sawhill, Gary S.; Shelton, Theodore Brian; Sibley, Duncan Fawcett; Slaughter, John; Smith, Ethan Timothy; Strojan, C. L.; Thompson, Donald Eugene; Thompson, Peter Robert; Thurlow, Ernest Huntington; Turner, Ronald Fredric; Ulrich, Barbara Carol; Walters, James Carter; Wang, Ming Kuang; Werner, Eberhard Wolfgang; Yersak, Thomas E.; Youssefnia, Iradj; Yuan, Wen Lin

1980s: Agosto, William N.; Anderson, Richard Mark; Arnold, Lynne J.; Aurisano, Richard Warren; Bam, Swagat Arvind; Boerner, Ralph E. J.; Buteux, Christopher Blaine; Chang, Pyoung Wuck; Charletta, Anthony Charles; Cipolletti, Debbie L.; Cipolletti, Robert M.; Clausen, John Eric; Colucci, Michael T.; David, Pierre; Delu, Jacqueline; Dimmick, Ross Alan; Dombroski, Daniel R., Jr.; Ekwurzel, Brenda; Ferland, Marie Ann; Field, Mary Leslie; Foote, Mary Ann; Frasco, Barry Ri▮ard; Gastrich, Mary Downes; Gesumaria, R▮ Hugh; Gillespie, Thomas D.; Gorycki, Michae▮ thony; Haag, Gary H.; Habrukowich, Ri▮ George; Harriott, Theresa A.; Hart, William ▮ Hedin, Robert Stewart; Hughes, Theresa M.; H▮ inson, Wayne Robert; Jannik, Nancy Olga; ▮ Tzay-Rong T.; Jesinkey, Christopher; Ka▮ Margeret E.; Korfiatis, George Panayiotis; ▮ Edward Charles; Kruger, Anne Longsworth; ▮ Chenfang; Malla, Prakash Babu; Martino, Rona▮

Layton; McCluskey, James M.; McLaughlin, Franklin B., III; McNevin, Thomas F.; Melillo, Allan Joseph; Melillo, Allan Joseph; Milionis, Peter Nicholas; Mitchell, James Porter; Moser, Fredrika C.; Motta, Christopher J.; Nakashima, Lindsay D.; Nwaochei, Ben Nnaemeka; Nyong, Eyo Etim; Nyong, Eyo Etim; Osamor, Chukwuka Azubuike; Phillips, Jonathan David; Radomsky, Patrick M.; Reimer, Gerda Elise; Reynolds, William Jerome; Rowan, Andrew Thomas; Sacco, Paul Augustus; Sahagian, Dork; Schofield, Richard Edward; Schreiber, Berta L.; Schuyler, T. Kent; Sehayek, Lily; Starcher, Robert Warren; Sugihara, Teruo; Szuwalski, Daniel Robert; Toskos, Theodoros; Trela, John Joseph; Ungrady, Timothy Edward; Vassallo, Carol Frances; von Schondorf, Amy; Walker, James Allen; Walker, James Allen; Warne, Andrew G.; Wu, Jy-Shing; Zeff, Marjorie Lee; Ziegenfus, Robert Charles; Zimmer, Bonnie Jeanne

Rutgers, The State University, Newark
Newark, NJ 07102

58 Master's

1970s: Blauvelt, Robert P.; Charletta, Anthony Charles; Cichetti, Maureen J.; Coco, Arlene M.; Cohen, Robert Stuart; D'Angelo, Louellen; Fontaine, David Alex; Grosso, Stephen T.; Hurtubise, Donlon O.; Lechler, Paul; Meekins, Keith Leroy; Meyer, Gary Peter; O'Bryan, James William; O'Neill, Aloysius Joseph; Orlowski, Wayne C.; Peters, Joseph John; Petersen, Edward Arnt; Sciarrillo, Joanne R.; Stellas, Michael James

1980s: Abrams, Nelson Jay; Asemota, Isaac; Barkemeyer, Eric; Bauer, Janet; Bello, Donald M.; Bolen, Michael M.; Burak, Roman W.; Caamano, Edward; Cirilli, Jerome P.; de Mauret, Kevin; De Rose, Nicholas; Ehlenberger, Robert Gordon; Geiger, Fredric J.; Germine, Mark; Gomes, Patricia M.; Helbig, Steffan Reed; Houlday, Mark; King, John J.; Kirby, Mark William; Klas, Mieczyslawa; Kortis, Phillip C.; Koto, Robert Y.; Koutsomitis, Dimitrios; Krone, Steven; Lucas, Mark; Maresca, Gerard P.; McClellan, Bruce S.; McGowan, Michael; Mrotek, Kathryn Anastasia; Nelridge, Richard Alan; Patel, Bharat A.; Pestana, Edith M.; Posnick, Allan Edward; Renzulli, Michael J.; Ruisch, Edeltraud; Seborowski, K. Damian; Spohn, Thomas; Torlucci, Joseph, Jr.; Wong, Margaret S.

University of San Diego
San Diego, CA 92110

21 Master's

1960s: Brogan, George E.; Dreessen, Richard S.; Fife, Donald L.; Flynn, Clinton J.; Maytum, James R.; McGee, David C.; Moscoso, Belisario A.; Nordstrom, Charles E.; Reed, Robert G.; Schroeder, James E.

1970s: Bell, Robert E.; Copenhaver, George C.; Itson, Sonja P.; Kopel, Jerry H.; Law, Benny E.; McEldowney, Roland C.; Mickey, Michael B.; Shawa, Monzer S.; Slyker, Robert G.; Smyers, Norman B.; Waldbaum, Raymond

San Diego State University
San Diego, CA 92182

245 Master's

1950s: Darnell, William I.

1960s: Boyce, Robert E.; Butler, Louis Winters; DeLisle, Mark James; Holden, John Clinton; Liska, Robert D.; Phillips, Edward Hayden; Thompson, Donald E.

1970s: Adams, Mark Allen; Allinger, Richard Jack; Allison, M. L.; Arleth, K. F.; Ashley, Randal Jack; Barthelmy, D. A.; Bender, Gretchen L.; Berggreen, R. G.; Bowersox, J. R.; Boyd, R. W.; Briedis, N. A.; Brown, Lawrence Gregory; Buck, D. J.; Callaway, T. M.; Chanpong, R. R.; Consort, James Jeremiah; Crane, D. J.; Dawson, M. K.; Fink, J. W.; Fink, K. A.; Fleming, Sheryl Denise; Fourt, Robert; Gassaway, J. S.; Gera, A. V.; Gibbs, Alan

D.; Henry, M. J.; Hoffman, D. R.; Howard, R. P.; Hoyt, D. H.; Hurst, C. H.; Jones, Steven Dennis; Lamb, T. N.; Lillis, P. G.; Lower, S. R.; Mandel, D. J., Jr.; Mattox, W. A.; McCarthy, R. J.; Mendeck, M. F.; Millikan, Gregory Robert; Moore, Thomas E.; Nicholson, G. E.; Nocita, B. W.; Nuccio, R. M.; Peart, J. E.; Peterson, John E.; Peterson, John Ellis; Pierce, S. E.; Pischke, Gary Michael; Ravenscroft, Arthur William; Riley, C. O.; Robinson, J. W.; Ruisaard, Chris Ivan; Sawicki, David A.; Schatzinger, R. A.; Schile, C. A.; Schile, Charles A.; Seekins, William C.; Sheely, Milton Jerome, Jr.; Sherrer, P. L.; Smith, Michael L.; Sturz, A. A.; Sundberg, Frederick Allen; Sweeney, J. A.; Troughton, G. H.; Waggoner, James Allen; Ware, Don Westmont; Williams, Timothy Anderson; Woidneck, Robert Keith; Woolley, J. J.

1980s: Academia, Imelda Garcia; Adams, Mark Alan; Alba, Christopher Anthony; Anderson, Paul Victor; Arneill, Lynn; Ashton, Donald Alan; Bachman, Wayne R.; Balch, Duane Clark; Baltz, Rachel May; Bartling, William Allen; Bass-Laszlo, Sarah Luann; Battelle, Sarah Jane; Bell, Patricia J.; Belyea, Richard R.; Bement, Kenneth Arthur; Blain, Paul Guy; Blundell, Lane Cameron; Brake, James Frank; Bryant, Brian Alan; Buch, I. Philip; Butterworth, Joseph E.; Callian, James Thomas; Cameron, Gregory Joseph; Campbell, Michael John; Carroll, Forrest Arthur; Clinkenbeard, John Patrick; Costello, Stephen Charles; Craig, Christy; Crocker, James R.; Cunningham, Arnold Bryce; Curiel-Mitchell, Helen; Dahm, Jerry B.; Deen, Patricia Ann; Demaree, Randall Gene; der Sarkissian, Volga; Detterman, Mark E.; Dockum, Mark Steven; Eastman, Benjamin Gordon; Edelman, Steven Harold; Edwards, Jeffrey Craig; Ehleringer, Bruce Ernest; Eisenberg, Leonard I.; Elder, Dorian Lizabeth; Ellis, Margaret Jane; Emerson, William Stewart; English, Douglas John; Erskine, Bradley G.; Farquhar, Scot Paul; Fenton, Scott Bruce; Feragen, Edward Sebastian; Furu, Edward James; Galvan, Geoffrey Scott; Gaona, Michael Thomas; Germinario, Mark Philip; Gester, Kenneth Clark; Gjerde, Michael Wolf; Gladden, Scott Charles; Gray, Lynn D.; Griffith, Roger Clinton; Groffie, Frank Johannes; Gross, Warren William; Grotts, Tim Douglas; Gunn, Susan Helen; Gustafson, Edward Paul; Hankins, David D.; Harrington, John Mark; Harvey, Timothy William; Hatch, Michael E.; Haug, Guido Alfredo; Haworth, Roger Alan; Heaton, Kevin Michael; Hempy, Daniel Willett; Herbert, Elizabeth Lee; Hillemeyer, Frank Lloyd; Hintzmann, Kathleen Joyce; Hoobs, John H.; Hoppler, Harl; Hughes, John Patrick, Jr.; Irwin, Randal L.; Jacobson, Gary Louis; Jansen, Lawrence T.; Jarzabek, Dianne Palmer; Johnson, Keith Eric; Kerr, Dennis R.; Kies, Ronald Paul; Kimzey, Jo Ann; Kofron, Ronald J.; Komjima, Russell Kei; Kraft, Michael Thomas; Kwasnica, Edward Anthony; Lafferty, Mark Robert; Leier-Engelhardt, Paula Jean; Leveille, Gregory Paul; Locke, Glenn Leslie; Logan, Robert Ellis, III; Lorenson, Thomas D.; Lorkowski, Robert Michael; Lothringer, Carl J.; Lowe, Gary Duane; Lyle, John Hyer, III; Mace, Neal Wayne; Mark, David L.; McCormick, William Vincent, III; McDonough, Scott David; McGuire, Michael Dale; McNaboe, Gerald Joseph; Medina, Julian Michael; Miele, Martin J.; Miller, Cydney Michele; Miller, Victor Van; Mills, William Glenn; Mishler, Harry Michael; Mitchell, Peter S.; Mossman, Brian John; Mueller, Karl Jules; Muncill, Gregory Ernest; Murphy, David Andrew; Natenstedt, Christopher J.; Nigro, Danny Michael; Oliver, Deborah M.; Paden, Edward A.; Pappajohn, Steven; Pardini, Charles Holliger; Phillips, James Richard; Pinault, C. Thomas; Pincus, Scott D.; Prall, John Russell; Price, Carol A.; Pridmore, Cynthia Lee; Raede, Deborah Lynn; Reaber, Douglas W.; Reddig, Ransom P.; Reed, James Bradly; Richardson, Steven Michael; Ross, Christopher George Arthur; Roszkowski, George Antoni; Ryder, Albert; Sangines, Eugenia Maria; Sankey, Scott Norman; Schmalfuss, Bradford Roger; Schug, David Lynn; Seutter, Andrew Edward, III; Shaw, Martha Jane;

Smith, Craig Anthony; Stang, Peter M.; Steer, Bradley Laurance; Strand, Carl Ludvig; Struck, Rodney Grant; Swenson, Guy Andrew, III; Taylor, Gregson William; Teel, Derrick Brehm; Tirrell, Ann Louise; Trumbly, Philip Nelson; Tucker, Robert Scott; Turner, Joseph Brian; Veseth, Michael K.; Vidal, Francisco Suarez; Wallace, Robert Duncan; Weaver, Benjamin Franklin; Wells, Donald Loren; Wernicke, Rolf Stephan; Weslow, Vanessa Maria; Wiedlin, Matthew Paul; Williams, D. Sam; Williams, John Thomas; Willian, Mark A.; Young, Jay Marc; Zlotnik, Elias

San Fernando Valley State University
Sepulveda, CA 91343

3 Master's

1960s: Bailey, Robert G.; Bell, Brian H.; MacDonald, Angus A.

San Francisco State University
San Francisco, CA 94132

8 Master's

1960s: Stromberg, Peter A.

1970s: Coffelt, Richard M.; Cosmos, George J.; Dow, Gerald R.; Hendrick, John W.; Sheridan, James T.; Whelan, James P.; Williamson, James A.

San Jose State University
San Jose, CA 95192

174 Master's

1960s: Anderson, Gery F.; Berkland, James; Blanchard, Maxwell B.; Buckley, Christopher P.; Conomos, John T.; Cotton, William R.; Gram, Ralph; Johnson, Edward A.; Lewis, Robert E.; Pinkerton, James B.; Quinterno, Paula J.; Raymond, Loren A.; Reese, James; Robbins, Stephen L.; Smith, Eugene L.; Wolf, Stephen C.

1970s: Albert, Nairn R. D.; Armin, Richard A.; Austin, Steven Arthur; Bargar, Keith E.; Bartsch-Winkler, Susan; Batchelder, John N.; Bauer, Paul G.; Bennett, Michael J.; Bennett, Reb. E.; Brown, James Peter, Jr.; Brown, Janet L.; Calk, Lewis Clifton; Carothers, William W.; Carter, C. H.; Conley, David E.; Cooper, Alan; Cress, Leland D.; Cunningham, Gary G.; Davis, Alice S.; DeCoster, Judith M.; Dittmer, Eric Rheydt; Dockter, Roger D.; Dunne, George Charles; Elayer, Robert W.; Endo, Elliot Toru; Frame, Philip A.; Greene, H. Gary; Haltenhoff, Frederick W.; Hill, James M.; Holden, Kenneth D.; Hummert, B. A.; Husk, Robert H.; Keith, William J.; Kelley, J. S.; Kyte, Frank T.; LeCompte, James R.; Leopold, Lawrence C.; Levine, Saul R.; Littlefield, Robert G.; Locke, James Leroy; Macy, Jonathan S.; Mankinen, Edward A.; McLaughlin, Robert J.; Metz, Jenny; Mitchell, Martha Jeanne; Moffitt, John; Nakata, John K.; O'Kane, John Anthony, Jr.; Olsen, Robert C.; Osbun, Erik; Pearl, James E.; Peterson, David; Porter, Edward J.; Randall, Richard G.; Reid, George O.; Ridley, Albert Paul; Rodeick, Craig A.; Ross, Bruce E.; Rymer, Michael J.; Seidelman, Paul J.; Sheriff, Akbar; Sheth, Madhusudan; Silva, L. R.; Simoni, Tully R., Jr.; Simpson, Garey L.; Solomon, Barry J.; Sorensen, Martin L.; Spencer, Clyde H., Jr.; Stuart, James Edward; Sumsion, R. S.; Todd, William C.; Torres, Adrian R.; Valdes, Fernando, Jr.; Wagner, D. B.; Wagner, David L.; Whiteley, Karen R.; Younse, Gary A.; Ziemianski, Wayne P.

1980s: Allaway, William; Anderson, Peter; Bader, Jeffrey W.; Bork, Kenneth R.; Borum, Jeffrey; Bottaro, Joseph L.; Braun, Don; Chin, John; Chinburg, Susan; Christie, Kyle; Clynne, Michael; Coyle, John M.; Curtis, George; Diggles, Michael; DiLeonardo, Christopher G.; Doukas, Michael; Drinkwater, James; Eberz, Noel; Egbert, Robert; Fedewa, William; Flora, Larry; Fraticelli, Luis; Garlow, Richard A.; Gartner, Anne E.; Glick, Linda Lee; Golia, Ralph; Hardin, Jack; Harding, Michael J.; Heath, Kathryn Carol; Hedel, Charles; Helm, Ronnie L.; Hopkins, Willard N.; Hovland, David;

Howe, Robert; Humphreys, Richard; Johannesen, Dann; Kaplan, Terry; Klingman, Darrell S.; Klise, David; Lico, Michael; Luken, Michael; Magginetti, Robert T.; Martindale, Steven; McCoy, Gail; McCusker, Robert; Moore, David; Moore, Steven; Nagel, David; Nelson, James; Oberlindacher, H. P.; Orlando, Robert C.; Orris, Greta J.; Ostendorf, Paul; Perkins, James A.; Ponce, David; Powell, John R.; Raviola, F. P.; Roberts, Eve; Schwartz, David; Sklenar, Scott; Springer, James E.; Stotz, Tina M.; Strandberg, Carl H.; Tomson, Janice; Torresan, Michael E.; Treat, Cheryl Lee; Trimble, Deborah; Van Buren, Mark; Vanderhurst, William Lee; Vercoutere, Thomas; Villalobos, Hector; Wade, W. Michael; Ward, William O.; Wilbur, Lyman; Williams, Kathleen Marie; Williams, Ronald O.; Winsor, Henry; Witham, Roger; Wolfe, Mitchell Dean; Woodward, Philip V.

University of Saskatchewan
Saskatoon, SK   S7N 0W0

179 Master's, 49 Doctoral

1940s: Aston, H. F.; Dahlstrom, C. D. A.; McClelland, John Edward A.

1950s: Almond, P.; Ambler, J. S.; Berg, Clifden A.; Boyke, Waldimer Paul; Christiansen, Earl Alfred; Clark, Lloyd A.; Cumming, George Leslie; Dahlstrom, Clinton D. A.; Eckstrand, Olof Roger; Edwards, Robert Garry; Froese, Edgar; Hovdebo, H. R.; Kent, Donald M.; Kermeen, James Seton; Kirkland, S. J. T.; Kirkland, Samuel John Thomas; Magdich, F. S.; McCamis, John Graham; McMillan, N. J.; McPherson, William J.; Meneley, Robert Allison; Mullock, J. E.; Pearson, Walter John; Petruk, W.; Powley, D.; Pyke, Murray W.; Rotherham, D. C.; Sawatsky, L. H.; Shackleton, James Stephen; Shklanka, Roman; Vigrass, Laurence William; Waddell, William Henry; Williams, F. J.

1960s: Acar, Kazim Zafer; Berven, Robert James; Bostock, Charles Alexander; Brandt, John Lawrence; Chernoff, C. N.; Clark, Alan Raymond; Copper, Paul; Delorme, Denis Larry; Dixon, Owen Arnold; Dyck, John Henry; Evans, John Keith; Eweida, Ahmed Mahmoud Farag; Faulkner, Edward Leslie; Faulkner, Edward Leslie; Ferris, Clinton S., Jr.; Finlayson, George Barry; Fischbuch, Norman Robert; Forsythe, Leander Harold; Friesen, Menno; Frison, Eugene Hubert John; Gantela, Christopher; Gaskarth, Joseph William; Gendzwill, Don John; Goldak, George Robert; Goldak, George Robert; Guliov, Paul; Hajnal, Zoltan; Hamilton, Wylie Norman; Hogg, William Andrew; Holter, Milton Edward; Hoque, Monirul; Huffman, David Patrick; Jordan, David Lohman; Kaufmann, William Lawrence Martin; King, Herman Leo; King, Roger Hatton; Klassen, Rudolph Waldemar; Lahey, Barry Armstrong Lloyd; Lissey, Allan; Lissey, Allan; Malik, Om Parkash; McIntosh, Ronald Alexander; Meneley, W. A.; Miedema, Oene; Mukherjee, Amar Chandra; North, Beatrice Ruth; Park, Jack Melvin; Parry, John Powell; Parsons, Myles Lyle; Pawliw, Paul Andrew; Perkins, George David; Petryk, Allen Alexander; Pfeffer, Beverley James; Pissarides, Andreas Savva; Pocock, Yvonne Patricia; Popowich, James Leslie; Pyke, Dale Randolph; Rainsberry, Lois Eileen; Rask, Dale Hugo; Reinhardt, Edward Wade; Roed, Murray A.; Rogan, Allan Douglas; Rutherford, Malcolm John; Sangameshwar, Salem Ramachandra Rao; Scott, Blaine Pierce; Sharp, James Wilfrid George; Sibul, Ulo; Statham, Kenneth Francis; Streeton, Dwight Harold; Sykes, Donald Windsor; Truscott, Marilyn Gail; Vonhof, Jan Albert; Vonhof, Jan Albert; Ward, William James; Watson, Douglas William; Wild, Jack; Wohlberg, Elwood Geron; Wyder, John Ernest; Wyder, John Ernest; Yont, D. R.

1970s: Ahuja, Suraj Prakash; Ashton, Kenneth Earl; Bamford, Thomas Sayers; Bradford, Martin Ronald; Brooke, Margaret Martha; Burnett, Andrew Isaac; Fensome, Robert Allan; Fillo, Wayne Joseph; Fumerton, Stewart Lloyd; Garg, Own Prakash; Har-

ker, Stuart David; Harker, Stuart David; Harper, Charles Thomas; Jensen, Larry Sigfred; Jonescu, M. E.; Kennedy, David Scott; Kim, Haeyoun; Koo, Jahak; Laplante, Bernard Eric; Lee, Derek Gordon; Lomenda, Melvin George; Luciuk, Gerald Michael; Macfarlane, N. D.; MacFarlane, Neil Daniel; Malik, Om Parkash; Martel, Yvon; McClure, James Edward; McKellar, Ronald Lawrence; McLean, James Ross; McNeil, D. H.; Misko, Ronald Michael; Morgan, R. E.; Morin, James A.; Mukherjee, Amar Chandra; Nautiyal, Avinash Chandra; Oro, Fe Haresco; Pearson, John G.; Pobran, Vernon Stephen; Reinson, Gerald Edward; Robertson, Ray Henry; Sangameshwar, Salem Ramachandra Rao; Seo, Haeyoun; Sereda, Ivan Theodore; Skwara, Theresa; Souster, W. E.; Streeton, Eric Grant; Sutherland, Garry Neil; Syme, Eric Charles; Truscott, Marilyn Gail; Watkins, Russell Allen; Wiley, William Eldon; Wilson, Malcolm Alan; Woolf, Theresa Skwara; Wright, Charles Edward

1980s: Atkin, Kenneth Thomas James; Broughan, Fionnuala M.; Burianyk, Michael J. A.; Campbell, Janet E.; Chouinard, Paul Norman; Cluff, Garnet Robert; Congram, Angela M.; Coyne, Carmel Anne; Dasog, Ghulappa; Fay, Ignatius Charles; Featherstone, Raymond Paul; Fensome, Robert Allan; Fowler, Sharon Patricia; Halabura, Stephen Philip; Haltie, Ian Edward; Harrison, Shane M.; Hejazi, Sayyed Hossein; Huebner, Mark; Jensen, Larry Sigfred; Jin, Jisuo; Juma, Noorallah Gulamhusein; Kleiman, Laura Elena; Koziol, Brenda L.; Lomas, Margaret Kathleen; Longiaru, Samual Joseph; MacDonald, Colin Campbell; Martz, Lawrence Wilfred; Mathison, Joseph Edward; Maynard, Douglas George; McMonagle, Anne Linette; McNamara, Susan Joan; Millard, MacDonald John; Penner, Lynden A.; Powers, Laura J.; Prugger, Arnfinn F.; Qayyun, M. Abdul; Reilkoff, Brian Rory; Robertson, Benjamin Telfer; Saggar, Surinder Kumar; Santos, Mauro Carneiro; Schreiner, Bryan T.; Scott, Douglas Lindsay; Tortosa, Delio J.; Van de Reep, Thomas W.; Wall, Duncan Arthur; Wallis, Paul Francis; Wang, Min-Chao; Wheatley, Kenneth Lewis; Wheeler, James William; Wilks, Maureen E.; Wilson, James Steven; Wilson, Malcolm Alan; Wilson, Mark Robert; Wilson, Nancy L.; Wing, Samuel James Courtney; Wittrup, Mark B.; Yee, Eugene Chan; Zebarth, Bernard John

Simon Fraser University
Burnaby, BC   V5A 1S6

1 Master's, 7 Doctoral

1970s: Mark, D. M.; Nanson, G. C.

1980s: Krhoda, George Okoye; Walker, Ian Richard; Warner, Barry G.; Wetzel, Wayne Allen; Williams, Harry F. L.; Yarnal, Brenton Murray

Slippery Rock University
Slippery Rock, PA   16057

6 Master's

1960s: Smith, Ian S.

1970s: Dicks, Alice L.; Glenn, Cheryl A. Dudo; Opitz, Dale A.; Sanborn, John D.; Sims, Richard M.

Smith College
Northampton, MA   01063

45 Master's

1900s: Heine, Aida A.

1910s: Gregory, Elizabeth; Perkins, Elizabeth Gregory

1920s: Anthony, Helen V.; Burgess, Anne E.; Ferris, Phebe; Hobbet, Anna; Hubbell, Marion; Kingsley, Louise; Merchant, Dorothy; Nelson, Rosie; Pond, Adela

1930s: Church, Mary S.; Harper, Margaret Francis; Lochman, Christina; Matteson, Jane S.; Olmstead, Elizabeth Warren; Schmedtje, Lucille C.; Wolcott, Helen

1940s: Barker, Rachel M.; Bogert, Elizabeth A.; Caldwell, Eleanor; Charles, Maureen E.; Flahive, Mary E.; Hoyt, Virginia; Powell, Wyveta; Starquist, Virginia L.; Tuttle, Frances; Wallace, Jane House

1950s: Clark, Sylvia Robb; Compton, Nancy H.; Geisse, Elaine; LaLonde Schake, Celia May; Reed, Mary Catherine; Robinson, Rosalind; Rotan, Cleone M.; Schenck, Barbara Jane; Volz, Marilyn; Weissenborn, Helen Frances; Werner, Marian Adair

1960s: Banka, Eleanor C.; Gallena, Jane; Ingraham, Jean; Prajmovsky, Atalia; Simpson, Jo-anna R.

University of South Carolina
Columbia, SC   29208

184 Master's, 125 Doctoral

1920s: Barker, M. E.

1930s: Smith, Laurence Lynwood

1940s: Cole, James Henderson; Eargle, Dolan Hoye; Newcome, Roy, Jr.; Stewart, Otis Floyd; Taylor, Vernon A.

1950s: Butler, Kim Robert; Heron, Stephen Duncan, Jr.; Johnson, Ragnar Edwin, Jr.; Neiheisel, James; Pooser, William K.; Pooser, William K.; Smith, LeBrun N.

1960s: Bernat, Phoebe E.; Bright, William E.; Brown, Timothy S.; Bundy, Jerry Lowell; Burnett, Thomas L.; Clark, William D.; Drummond, Kenneth McCoy; Duncan, Donald A.; Evans, Ian; Getzen, Rufus D., Jr.; Kiff, I. T.; Lehocky, Alan John; McClure, William C.; McKenzie, John C.; Morrison, Garrett L.; Paradeses, William D.; Phyfer, Daniel Wade; Privett, Donald R.; Ridgeway, David C.; Riley, Louise Anderson; Tewhey, John; Thompson, George M.

1970s: Abbott, William Harold, Jr.; Aburawi, Ramadan; Alcorn, Steve; Amick, David Carroll; Andrew, James Alexander; Andrews, Edwin Eads, III; Archie, Andrea; Astwood, Phillip M.; Astwood, Phillip M.; Baganz, Bruce P.; Baganz, Bruce P.; Baldwin, Jeffrey; Barrell, Steve; Barwis, John H.; Bascle, Barbara; Bascle, Robert; Benomran, Omran; Bifani, Ronald; Blackwelder, Patricia L.; Blount, Ann E.; Boast, John; Boothroyd, Jon C.; Boyd, Harry William; Brown, Jeffrey; Brown, Jeffrey; Brown, P. J.; Brown, Paul Jeffrey; Brown, Roy Harold; Burgess, Christopher; Cable, Gregory; Campbell, Lyle D.; Campbell, Sarah L. C.; Canfield, Joyce; Carlson, Gregory D.; Carpenter, W. David, Jr.; Chiang, J. H.; Cho, Y.-Y.; Cleary, William J.; Clemens, Robert; Comer, C. Drew; Corvinus, Dorothy Anne; Cossey, Steve; Cossey, Steve P. J.; Costello, Oliver P., Jr.; Cummins, Gloria; Czyscinski, Kenneth; Czyscinski, Kenneth; Davies, John; Dennis, Alan; Drew, Alice J. R.; Dunn, Jeff L.; Eason, James; El Fazzani, Ashour; Ellis, Paul; Elmoudi, Salem M.; Elsinger, Robert John; Eppler, Duane; Ernissee, John J.; Ernissee, John J.; Fate, Thomas; Fico, Cary; Findlay, Marsha G.; Finkelstein, Kenneth; Finley, Robert J.; Fischer, Ian A.; FitzGerald, Duncan M.; Fleischmann, Marianne Lynn; Fletcher, Jonathan; Frank, Barbara Joyce; Galloway, Malcolm Charles Bell Bradsworth; Geidel, Gwendelyn; Gettys, William R.; Gimmel, J. C.; Gonsiewski, James; Grace, James; Griffen, Villard; Grinnell, Daniel Voorhis; Grothaus, Brian; Grothaus, Brian T.; Gundlach, Erich R.; Hanselman, David Henry; Harding, Andrew George; Harding, Andrew George; Harrigan, Joseph A.; Harris, Thomas G.; Herting, David Allen; Hine, Albert C., III; Hohos, Edward F.; Hope, Robert C.; Hornig, Carl A.; Howell, David; Hubbard, D. K.; Hubbard, Dennis K.; Huckabay, William; Hudson, Carolyn Brauer; Huguley, Robert William; Hulse, Robert C.; Humphries, Stanley M.; Jacobeen, Frank; Jacobeen, Frank H., Jr.; Johnson, Michael; Johnson, Thomas F.; Jones, Thomas; Kaczorowski, R. T.; Kana, Timothy W.; Kana, Timothy W.; Karasek, Richard; Keeling, Theodore; Kheoruenromne, Irb; Knoth, Jeff; Lee, Christopher W.; Levey, R. A.; Lorenz, John; Maher, Harmon Droge, Jr.; Martin, David L.;

Mathew, David; Mazzullo, James; McLaren, Patrick; Mellegard, Andy; Melton, Robert A.; Metzgar, Craig R.; Michel, Jaqueline M.; Millwood, Lynn; Mitchell, John C.; Molnia, Bruce Franklin; Montenyohl, Victor I.; Montenyohl, Victor I.; Moody, Marjorie; Morabit, Almoundir; Mullin, Peter R.; Mundy, Brian Roy; Neauhauser, Kenneth; Neauhauser, Kenneth; Nelson, Douglas DeWayne; Neuhauser, K. R.; Neuhauser, Kenneth Reed; Nystrom, Paul G.; Owens, Edward H.; Padgett, Guy; Padgett, Guy V., Jr.; Payne, Myron William; Payne, R. R.; Payne, Robert Ridley; Pedlow, George W., III; Peeples, Vernon; Pelletier, Michael; Perlman, Stephen H.; Porter, Gerald; Przygocki, Robert; Radford, Robert M.; Ressetar, Robert M.; Ressetar, Robert M.; Reynolds, David E.; Richardson, Ralph O., Jr.; Roberts, Michael Anderson, Jr.; Robinson, Michael; Rowland, David; Ruby, Christopher; Said, Faraj; Sauber, Jeanne; Scheffler, Peter K.; Schmidt, Gerry Lee; Schmidt, Gerry Lee; Schmitt, Thomas J.; Schultz, Douglas J.; Settlemyre, Julius L., III; Settlemyre, Julius L., III; Sewell, John Michael; Shearer, David; Simpson, Michael; Smith, George E., III; Smithwick, Maureen; Staub, James Rodney; Stephens, Daniel Guy; Stephens, Daniel Guy; Stevenson, Donald A.; Sullivan, A. M.; Taleb, Mohammed; Turner, William; Van Nieuwenhuise, Donald S.; Van Nieuwenhuise, Robert; Visvanathan, Thellur Rangswamy; Vos, Richard G.; Ward, Larry Guy; Ward, Larry Guy; Ward, Lauck; Waring, C. Joseph; Weisenfluh, Gerald Alan; Witkus, M. A.; Woollen, Ian D.; Woollen, Ian D.; Yarus, Jeffrey M.; Yarus, Jeffrey M.; Zabawa, C. F.

1980s: Abul-Nasr, Radwan Abdel-Aziz; Abuzied, Hassan T. H.; Al-Sarawi, Mohammad A.; Alsharhan, Abdulrahman Sultan; Bedford, Betty; Bollinger, Marsha Spencer; Bou-Rabee, Firyal Ahmed; Bramlett, Kenneth; Burt, Ronald Allen; Buswell, Karyn; Butts, JoLynn; Cable, Mark Stephan; Cao, Song; Cavanaugh, Michael Dennis; Chapnick, Susan D.; Chen, Hsien Su; Chin, Maureen; Christensen, Eric Joseph; Coffield, Dana Quentin; Conley, James Franklin; Corvinus, Dorothy Anne; Crabtree, Sterling James, Jr.; Dooley, Robert Ervin; Duc, Aileen Wojtal; Eggers, Margaret Royall; El Shazly, Hanssan; Elsinger, Robert John; Ember, Leon M.; Eppler, Duane Thomas; Full, William Edward; Gaitanaros, Alexandros P.; Gary, Anthony Cavedo; Gawarecki, Susan L.; Geidel, Gwendelyn; Goodman, Emery; Gumati, Yousef Daw; Healy-Williams, Nancy; Hermanrud, Christian; Heyn, Teunis; Hodge, John A.; Hooper, Robert James; Horkowitz, Kathleen O'Neill; Howell, David J.; Howell, Michael Wade; Hurdle, David; Jenkins, Janice A.; Karaske, Richard Mark; Kennedy, Stephen Kenneth; Killeen, Terrence P.; Kornder, Steven Charles; Krantz, David Eugene; Leventer, Amy; Levey, Raymond Allen; Mazzullo, James Michael; McCreesh, Catherine A.; Michel, Jacqueline M.; Moslow, Thomas Francis; Mou, Ching-Hua; Murday, Mayloganaden; Nakayama, Kazuo; Nardi, Guiseppi; Nassr, Mohammad Nashaat Gad; Pai, Miao-Li M.; Palmer, Andrew James Malcolm; Pantano, John James; Perry, Stephen Kenneth; Ramsey, Elijah William, III; Riester, Debra; Ruby, Christopher Houston; Sadd, James Lester; Sexton, Walter Jerome; Shipp, Craig; Smith, Marian McNally; Smith, William Algene, Jr.; Staub, James Rodney; Stephen, Michael Frederick; Ward, Lauck Walton; Weisenfluh, Gerald Alan; Yuan, Li-Ping

University of South Dakota
Vermillion, SD 57069

42 Master's

1950s: Albert, Tom; Baird, James Kaye; Collins, Sam G.; Douglas, Charles H.; French, Tipperton J.; Jorgensen, Donald Gene; Lutzen, Edwin Earl; Pettyjohn, Wayne A.; Schoon, Robert A.; Sevon, William David; Skogstrom, H. Clifford; Steece, Fred V.; Taft, William H.; Tipton, Merlin J.

1960s: Bruce, Richard L.; Christensen, Cleo M.; Fournier, Rene E.; Fox, James E.; Hammill, Robert W.; Hedges, Lynn S.; Hoff, Jerald Herbert; Johnson, Curtis Leonard; Lange, Alan Ulrich; Mickel, Charles Joseph; Mickel, Edward G.; Nelson, Michael E.; Walker, Ian Robert; Wong, H. Donald

1970s: Arndt, B. Michael; Barari, Rachel A.; Baron, Donald M.; Beaver, George; Beffort, Joseph D.; Buehrer, David W.; Cramer, Ronald Thomas; Helgerson, Ronald N.; Kahil, Alain; Sarnecki, Joseph Charles; Schroeder, Wayne E.; Stach, Robert L.; Stockdale, Richard G.

1980s: Carter, Kristine D.

South Dakota School of Mines & Technology
Rapid City, SD 57701

274 Master's, 24 Doctoral

1940s: Janosky, R. A.; Onoda, Kiyoka

1950s: Ackerman, Walter; Ackroyd, Earl Arthur; Andersen, D.; Arcilise, Casper; Bagan, Richard John; Ballou, William D.; Berg, D. A.; Brennan, Daniel J.; Brooks, James Rolland; Cetrone, Ronald; Connor, Mike; Coulson, Francis M.; Dyer, Charles F.; Flinn, Donald J.; Glerup, Melvin O.; Hummel, Charles L.; Johnson, Charles; Johnson, Charles Frederick; Kane, Byron; Kane, Byron L.; Kelly, Herbert A.; Kepferle, Roy Clark; King, Robert L.; Knapp, Crawford; Lane, Robert W.; Mancuso, James Dominic; Martin, Harold; Nicknish, John; Papcke, David E.; Paschal, Lawrence W., Jr.; Potter, Lloyd D.; Potter, Lloyd Dean; Ritter, John R.; Rosenberg, Louis J.; Seefeldt, David R.; Shortridge, Charles Glen; Smith, Joseph Blake; Sottek, T. C.; Speice, Charles; Spring, S.; Stearns, David W.; Steege, Lauren; Sturgeon, David A.; Vigoren, LaVerne; Wilson, John M.; Wolff, Roger G.; Wulf, George Richard

1960s: Alsayegh, Abdul Hadi Y.; Bartels, Richard L.; Bertrand, Walley E.; Bishop, Gale A.; Bjork, Philip R.; Black, Douglas F. B.; Bryant, Laurie J.; Crooks, Thomas J.; Ellis, M. J.; Evans, John P.; Fisher, John K.; Getz, Roger C.; Grissom, Holly D.; Hanson, C. Bruce; Harsh, John F.; Johnson, Gary D.; Kernaghan, James S.; Knight, F. J.; Kulik, Joseph W.; Lillegraven, Jason A.; Luza, Kenneth Vincent; Niven, David W.; Parris, David C.; Patraw, James M.; Pearson, David Victor; Roadifer, Jack E.; Solmonson, Donald W.; Stinnett, Landy A.; Thurn, Richard L.; Walawender, Michael J.; Whippo, Robert; Wilde, Roger D.; Wilshusen, John P.

1970s: Abuannaja, Abdullah S.; Aho, John E.; Alkazmi, Rajab Abdussalam; Anna, Lawrence O.; Aparicio, Agustin; Atkinson, Ross David; Baghanem, Ali M.; Bahabri, Mohammad Sultan; Bahadoran, Behzad; Bahadur, Sher; Baker, Richard K.; Barnum, Dean Charles; Beck, James Aubrey, Jr.; Bell, Richard Arthur; Bidgood, Thomas Warren; Bidgood, Thomas Warren; Briggs, John Peter; Campbell, Frank W.; Carda, Dan D.; Ching, Paul D.; Chonglakmani, Chongpan; Dana, John Kenneth; Davis, Arden D.; Dheeradilok, Phisit; Dolan, John D.; Elwood, Michael Warren; Engstrom, William Scott; Farkas, Frank S.; Fayyaz, Mohammad S.; Gendi, Mohamed H.; Gerlach, Paul Joseph; Gersic, Joseph; Gilani, Mohammad Ali Sadighi; Gjere, Robert Allen; Greenwald, Michael T.; Grimes, John H.; Grunwald, Ross Richard; Hadji-Sabbagh, Mehdi; Heidt, J. Harmon; Hendrick, Steven J.; Higgins, Donald W.; Ingle, Steven Carl; Isarangkoon, Piphop; Isarankura, Somsak; Iskander, Atef Fanzi; Iyer, L. Srinivasa; Johnson, Dennis V.; Johnson, Thomas Lee; Jumnongthai, Junya; Jumnongthai, Manit; Kar, Lakshmidhar; Keene, Jack R.; Kihm, Allen James; Kim, Jong Dae; Kokcharoensup, Wichai; Larsen, Richard K.; Larson, Jay L.; Loskot, Carole Lynn; Luza, Kenneth Vincent; Maranate, Srisopa; Martin, James Edward; Matthews, Curtis B., III; McMillan, Richard C.; Mellon, Steven Allen; Miller, Richard H.; Morea, Michael Frank; Najjar-Bawab, Moumtaz

1960s: Mohammad; Nakanart, Araya; Nixon, Roy Arthur, III; Owen, Sandra Joan; Pakkong, Mongkol; Parkhill, Thomas A.; Payne, Curtis M.; Punatar, Gajendra Kesshavlal; Rapp, John S.; Rawlins, David M.; Reitenbach, Jacob Andrew; Rice, Lee R.; Rockey, David L.; Roth, John E.; Samai, Mahdi; Sateesha, Malalur K.; Schneider, Gary Bradley; Shaddrick, David R.; Siok, William J.; Sofranoff, Stephanie E.; Steffen, Lyle J.; Szigeti, George Joseph; Tyler, Ronald D.; Usiriprisan, Chamroon; Vandike, James E.; Ward, William D., Jr.; Welch, Carl Martin; White, Jon M.; White, Jon M.; Willibey, Tom Dean; Woodhouse, Bruce Alan; Yancey, Clyde L.; Zakir, Fawaz Abdul Rahman

1980s: Ainsworth, Michael R.; Alkhazmi, Rajab Abdussalam; Bajabaa, Saleh A. S.; Bangsund, William J.; Beck, James Aubrey, Jr.; Bergeron, Brian P.; Best, William Allen; Blake, Bonnie Janine; Bond, William D.; Borden, Richard K.; Brook, Hilary James; Bush, James Gilbert; Buswell, Michael Douglas; Campbell, Thomas J.; Chaille, John Lee; Charoen-Pakdi, Dawaduen; Christiansen, William D.; Cleath, Richard Allen; Coker, Diane; Dadoly, John Peter; Dahlstrom, David James; Daly, Cathryn Hayes; Daly, William E.; Dandavati, Kumar S.; Davis, Arden D.; DeTample, Craig; Dorsett, Russell K.; Durkin, Thomas V.; Eberlin, John E.; Effinger, James A.; El-Ghoul, Muhktar Taher; Elliot, R. John; Elwood, Sompis Chuntamee; Erb, David; Everson, Douglas D.; Faircloth, Susan Lynne; Fantone, Kenneth Scott; French, Gregory McNaughton; Fricke, John N.; Fry, Joyce Ann; Gaines, Roberta Kay Sampson; Galbreath, Kevin C.; Gasser, Michael M.; Gates, William C. B.; Ghassemi, Ahmad; Gillespie, Janice M.; Goodrum, Christopher K.; Gosselin, David Charles; Hafi, Zuhair; Hall, Rowland L.; Hamyouni, Ezzidin Ahmed; Hangari, Khaled M.; Hendrickson, Brent R.; Holzheimer, Joanne M.; Huq, Syed Y.; Jenkins, Creties D., Jr.; Jensen, Martin; Johnson, Kathryn Olive; Jolliff, Bradley L.; Jolliff, Bradley L.; Kazdal, Recep A.; Khan, Mumtaz Ahmed; Kleiter, Kathryn Jean; Knell, Gregory W.; Knirsch, Karen; Koopersmith, Craig Allen; Krahulec, Kenneth A.; Kremin-Smith, Denise J.; Kroeger, Timothy J.; Kuhl, Tim O.; Kyllonen, David P.; Lofholm, Stephen T.; Machacha, Tafilani P.; Macleod, Roderick J.; Mannai, Mohamed A.; Marin, Jon Randal; Meier, Laurence F.; Modisi, Motsoptse P.; Motes, Arthur Glenn, III; Musa, Nagieb S.; Mutlu, Hamlin; Namoglu, A. Coskun; Norby, John W.; Ostrander, Gregg; Ouyang, Shoung; Pekas, Bradley S.; Peter, Kathy Dyer; Peterman, Bruce D.; Pinsof, John David; Pish, Timothy A.; Refai, Refai Taher; Robertson, Dennis; Roca, R. Luis; Rueb, Ronald A.; Sabel, Joseph M.; Sabtan, Abdullah A.; Schreuder, Kenneth M.; Schubbe, Dennis Lee; Settle, Alberta L.; Sharata, Salem Muftah; Simon, Steven B.; Slifko-Welch, Christine M.; Spilde, Michael N.; Stearns, Bruce G.; Stokowski, Steven J., Jr.; Sussman, Jeffery A.; Tabrum, Alan Robert; Talbot, W. Robert; Tapper, Charles Joseph; Tuysuz, Necati; Van Stone, Larry J.; Vogt, Timothy J.; Weissenborn, Paul R.; Whitmore, Janet Lynn; Wuolo, Ray Wilbert; Zakir, Fawaz Abdul Rahman; Zigich, Daniel K.

University of South Florida, St. Petersburg
St. Petersburg, FL 33701

40 Master's, 5 Doctoral

1970s: Barnard, Leo Allen; Bolger, George Walton; Burton, Michael D.; Chiou, Wen-An; Conner, W. G.; Fruland, Robert M.; Gartner, Jeffrey; Goetz, Carole L.; Gray, James R.; Guerin, William F.; Hayward, Gary Lewis; Jones, Sandra Lynn; McConnell, C. L.; Neurauter, Thomas William; Rogers, Scott W.; Schlemmer, Frederick C.

1980s: Acker, James Gardner; Breland, Jabe A., II; Brooks, Gregg R.; Brooks, Gregg R.; Burke, Roger Allen, Jr.; Cable, Mark Stephan; Cantrell, Jason Kirk; Cantrell, Kirk Jason; Costello, David K.; Evans, Mark W.; Hebert, Jean A.; Hutton, Joan G.; Joyce, Rosanne M.; Kump, Lee R.; Martin, David

University of South Florida, Tampa

W.; Matteucci, Thomas D.; McNeillie, Jennifer I.; Osking, Erick B.; Parker, Douglar M.; Pauly, George; Schroeder, Paul A.; Steward, Robert G.; Szydlik, Stephen J.; Thompson, Shannon Wesley; Torres, Linda M.; Walker, Steven T.; Wall, Frederick M.; Willis, John W.; Young, Richard Wescoe

University of South Florida, Tampa
Tampa, FL 33620

100 Master's, 1 Doctoral

1960s: Harrison, William Baxter; Manker, John P.; Steingraber, Walter A.; Witous, John M.

1970s: Bell, Leslie; Birdsall, Barton C.; Brame, Jeffrey W.; Brown, David P.; Brown, Michael P.; Coleman, Neil M.; Conrad, Eric Hale; Dalton, Matthew G.; Fretwell, Judy D.; Gallivan, Lyle Bradshaw; Goetz, James E.; Grace, Scott R.; Gurr, Theodore M.; Heimmer, Donald; Huang, Hui-Lun; Huse, Scott M.; Hutchinson, Craig Brandt; Jones, Phyllis Case; Kaufman, Ronald S.; Kiang, Wen Chao; Lawrence, Fred W.; Lehman, Linda L.; Lynch-Blosse, Michael A.; Mericle, James E.; O'Neill, Charles W.; Popek, John P.; Razem, Allan C.; Rea, Raymond A.; Reel, David A.; Rosen, Douglas S.; Ross, Frederick W.; Rust, Aaron B.; Tedrick, Patricia Ann; Tompkins, Robert E.; Vogler, David L.; Wolansky, Richard M.; Wright, Alexandra P.

1980s: Aisner, Jonathan Alan; Al-Wail, Tahir A.; Alamri, Abdullah M.; Andronaco, Margaret; Barton, Churchill J.; Bengtsson, Terrance; Bland, Michael J.; Bloomberg, Diane; Bretnall, Robert Edward, Jr.; Chacartegui, Fernando J.; Crowe, Douglas E.; Culbreth, Mark A.; D'Aluisio-Guerrieri, Gary M.; DeHaven, Eric C.; Dehen, Timothy; Estes, Carol; Gay, Michael Charles; Goodson, Robert H.; Gregory, John S.; Hagemeyer, R. Todd; Jewell, Pliny; Jones, Gregg W.; Kick, Robert M.; Kim, Jonathan J.; Knowles, Stephen C.; Kuhn, Bernard J.; Kwiatowski, Peter; Lawrence, Thomas A.; Layton, Michael C.; Lee, Donald J.; Littlefield, James R.; Lizanec, Theodore J.; McFadden, Maureen; Moore, David L.; Nuckels, Mark Gordon; O'Connor, Lynn D.; Pearce, Edward Wayne; Putzier, Paul; Restrepo, Juan F.; Rodiguez, Eileen; Rodriguez, Maria C.; Schrader, David L.; Shaw, Jonathan E.; Simmon, Douglas E.; Smith, David A.; Smith, David S.; Solebello, Louis P., Jr.; Spechler, Rick M.; Starks, Michael J.; Stebnisky, Richard J.; Stodghill, Allan M.; Stump, James Duffield; Sussko, Roger J.; Trommer, Jeff; Tyson, Rogert G., Jr.; Weber, Kenneth A.; Williams, Michael J.; Williams, Sheila R.; Wood, John W.; Woods, Terri Lee

Southeast Missouri State University
Cape Girardeau, MO 63701

1 Master's

1980s: Scott, Margaret Ann

University of Southern California
Los Angeles, CA 90089

326 Master's, 151 Doctoral

1920s: Farrand, William Hoffman; Moyer, Dorothy A.

1930s: Artusy, Ray; Barnes, Sydney U.; Baudino, Frank Joseph; Brockhouse, Thomas E.; Crouch, Smith A.; Driver, Herschel L.; Ferguson, Glenn C.; Harding, Maynard W.; Hazzard, John C.; Kimm, Diamond; Mason, John Frederick; Merenbach, Simon Eugene; Mitchell, Mark W.; Moore, Helen Louise; Oakeshott, Gordon Blaisdell; Osterholt, William Russell B.; Parsons, Robeert L.; Perry, Esther p.; Robertson, George K.; Sheldon, D. H.; Shelton, Dean H.; Spencer, Frank Darwyn; Stewart, Katherine C.; Stewart, Roscoe E.; Stolz, Harry P.; White, Ella Marie; Winckel, E. E.

1940s: Bettinger, Charles Edward; Carsola, Alfred James; Chambers, E. F.; Crouch, Robert Wheeler; Dana, Stephen W.; Fernandes, Robert J.; Foster, Joseph F.; Hamilton, Warren Bell; Jeffreys, Stanley R.; Mann, J. F.; Nelson, A. Graham; O'Bert, Lawrence Kay; Pfaffman, George A.; Ruhlman, Fred Lee; Sinclair, John Taylor, Jr.; Trostel, Everett G.; Wade, Franklin Russell; Wright, Lauren A.

1950s: Arnal, Robert E.; Babcock, Burt A.; Baldwin, E. Joan; Barnard, Ralph M.; Byrne, John Vincent; Carriel, James Turner; Ceylan, Rasit; Chilingar, George V.; Clary, Michael R.; Conrey, Bert Louis; David, Richard S.; Davis, Richard Spencer; Dill, Robert Floyd; Ehring, Theodore W.; Evans, Barry Louis; Evans, James R.; Eymann, James L.; Fissell, Donald Evan; Gaal, Robert; Goodwill, David; Gorsline, Donn Sherrin; Gorsline, Donn Sherrin; Gould, Howard Ross; Green, Keith W.; Hannah, William G.; Haskell, Barry S.; Holwerda, James Gerhardus; Holwerda, James Gerhardus; Holzman, Johnston Earl; Huckaba, William Arden; Jones, Benjamin F.; Keesling, Stuart Allan; Knight, Raymond Louis; Laurie, Archibald M.; Long, Robert Edwin; Lownes, Richard E.; Mann, John Francis, Jr.; Marlette, John William; Martin, Lewis; McGlasson, Robert H.; McNaughton, Duncan Anderson; Mead, Richard George; Merriam, Patricia J.; Moore, David G.; Patchick, Paul F.; Pierce, Richard L.; Pipkin, Bernard W.; Polski, William; Ragan, Donal M.; Reiter, Martin; Resig, Johanna Martha; Riccio, Joseph F.; Rich, Mark; Richards, Carrol A.; Roth, Eldon Sherwood; Scholl, David William; Schupp, Robert Donald; Seastrom, Wesley C.; Shuler, Edward Hooper; Simpson, Altus L.; Slosson, James E.; Slosson, James Edward; Staub, Harrison L.; Stevenson, Robert E.; Stone, Richard O'Neil; Stone, Richard O'Neill; Terry, Richard Dean; Tripp, Eugene C.; Uchupi, Elazar; Waller, Harry O'Neal; Walrond, Henry; Watkins, James G.; White, William Robert; Wozab, David Hyrum; Zalesny, Emil R.

1960s: Anderson, Gordon John; Anderson, Richard E.; Balkwill, Hugh R.; Ball, Alexander Ross; Barca, Richard Albert; Bartholomew, Mervin J.; Beer, Robert M.; Bell, David I.; Bell, Stuart A.; Bradford, William T.; Brown, James Alexander, Jr.; Bunnell, Victoria D.; Butram, Glen N.; Buttram, Glen Neil; Byer, John W.; Carson, Matthew V., III; Cook, David Olney; Cooper, William Clinton; Curtis, Charles M.; Dobbs, Phillip Hale; Echols, Ronald James; Felix, David W.; Findlay, William F.; Fisk, Edward P.; Fleury, Bruce; Fowler, Gerald Allan; Frerichs, William E.; Gaal, Robert Arthur Paul; Garrett, Donald Maurice; Hackett, Barbara E.; Haga, Hideyo; Hand, Bryce M.; Harman, Robert Allison; Headlee, Larry A.; Heiner, Williams R.; Heintz, Louis O.; Horn, Ancel Dan; Hunt, Gail S.; Hyne, Norman John, Jr.; Inderbitzen, Anton L.; Ingle, James C., Jr.; Ingle, James Chesney, Jr.; Kaplan, Isaac Raymond; Karim, Mostafa Fahmy; Keene, Arthur G.; Kesse, Godfried Opang; Kheradpir, Ahmad; Kilian, Henry Martin; Knox, Robert E.; Kolpack, Ronald Lloyd; Kolpack, Ronald Lloyd; Lane, Bernard Owen; Leslie, Robert James; Leslie, Robert James; Lessard, Robert Henry; Lidz, Louis; Loop, Taylor V.; Malloy, Richard James; Manera, Thomas E.; Martin, Bruce Delwyn; McCurdy, Robert; Means, Kendall D.; Medall, Sheldon E.; Merrill, John D.; Merselis, William B.; Moll, Robert F.; Morley, Earl R., Jr.; Nahama, Rodney; Neblett, Sidney S.; Nemeth, Donald F.; Oates, N. D.; Palmer, Harold D.; Palmer, Harold Dean; Pratt, Willis Layton, Jr.; Preston, Michael M.; Purcell, Francis A., Jr.; Quinn, Harry M.; Reade, Harold Leslie, Jr.; Reike, Herman H., III; Riccio, Joseph Frank; Rieke, Herman Henry, III; Rodolfo, Kelvin Schmidt; Rodolfo, Kelvin Schmidt; Rodrigue, Raymond Fredrick; Roth, Eldon Sherwood; Rottweiler, Kurt A.; Scheliga, John Thomas, Jr.; Schiffman, Arnold; Scott, Norman Jackson, Jr.; Seiple, Willard R.; Seward, Allan E.; Sherman, Douglas B.; Slater, Richard A.; Smith, Duane D.; Southwick, Robert S.; Specht, Glenwood W.; Stapleton, Richard Pierce; Stevens, Calvin Howes; Szatai, John Endre; Tamura, Allen Y.; Taweel, Michael Elias, Jr.; Terry, Richard Dean; Thompson, James H.; Trexler, Dennis Thomas; Turner, Neil L.; Uchupi, Elazar; Valencia, Shirley M.; Vernon, James Wesley; Ward, Robert Alan; Warnke, Detlef Andaeas; Weismeyer, Albert L., Jr.; Werner, Sanford L.; West, Philip J.; Wilcoxon, James A.; Wildharber, Jimmie L.; Willis, Joseph P., Jr.; Wimberley, C. Stanley; Wright, Frederick Fenning; Yelverton, Charles A.

1970s: Abrams, Michael Allan; Adams, Paul Michael; Adcock, T. D.; Ando, Clifford Joseph; Barnes, Peter William; Barton, C. L.; Beckett, K. L.; Blake, G. H.; Bloom, Laurie; Booth, J. S.; Booth, James S.; Borella, P. E.; Brenninkmeyer, Benno Max S.J.; Buffington, E. C.; Buika, James Alexander; Calzia, James P.; Carter, Louis D.; Cashman, Patricia Hughes; Cerri, S. T.; Chandler, P. B.; Clancy, J.; Colazas, Zenophon C.; Coleman, Teresa Ann; Combellick, Rodney Alan; Cotton, Mary Lou; Crawford, Harry Michael; Cross, T. A.; Davis, R. A., Jr.; Day, Peter Conrad; Denere, Thomas Ashley; Dokka, R. K.; Doyle, Larry James; Drake, David E.; Edwards, Brian Douglas; Evans, Karl Vierling; Evenson, W. A.; Fernandez, Manuel Nicolas; Fischer, Peter J.; Fishel, Ken W.; Fleischer, Peter; Fleisher, R. L.; Fortsch, David E.; Fulwider, Roy Wesley; Gatto, Lawrence W.; Gibson, James M.; Goldstein, Alan S.; Goodrich, J. A.; Gourley, James W., III; Grabyan, R. J.; Grant, David J.; Gundlach, David Lou; Hackett, John P., Jr.; Haner, Barbara E.; Hartman, B. A.; Hawkins, H. G.; Heitman, Hal Louis; Hess, Frank D.; Hess, Frank Devereaux; Johnson, Bruce Alan; Jones, J. A.; Junge, W. R.; Kahn, M. I.; Karl, H. A.; Kimmel, Margaret A.; Knaup, William Wade; Knauss, K. G.; Koehnken, P. J.; Kovaltchouk, A. G.; Krohn, J. P.; Ledingham, G. W., Jr.; Lee, M. Y.; Lee, Tien-Chang; Leneman, M.; LeRoy, S. D., Jr.; Limerick, Samuel Hazzard; Lingrey, Steven Howard; Little, Richard Douglas; Lobo, C. F.; Lobo, Cyril Francis; Loring, Anne K.; Lowe, B. V.; Lu, J. C.-S.; Maley, Richard P.; Malouta, D. N.; Marian, Melinda Lee; Marincovich, Louie N.; Marincovich, Louie Nick, Jr.; McDougall, Donald S.; McDougall, Dristin Ann; McDougall, K. A.; Meyer, W. C.; Molnar, James Stephen; Morin, Ronald W.; Mulhern, M. E.; Murray, S. M.; Murray, Steven M.; Murray, Steven M.; Nardin, T. R.; Nelson, Diane Marie; Nett, Allan R.; Niemitz, J. W.; Oh, Jin S.; Oki, Malemi; Paluzzi, Peter Ronald; Pao, Gloria Ai-yi; Pasho, David W.; Pavlak, Stephen John; Pilger, Rex Herbert, Jr.; Ploessel, Michael R.; Podruski, James Allan; Prensky, S. E.; Presch, William Frederick, Jr.; Real, Charles R.; Rice, R. M.; Rieke, Herman Henry, III; Robertson, J. O., Jr.; Robinson, B. A.; Robinson, James Parker; Rohatgi, N. K.; Roig, J. H.; Rosenbauer, R. J.; Rosenthal, Robert John; Rosentraub, M. S.; Savula, N. A.; Schmitt, V. L.; Seekins, L. C.; Seyfried, W. E., Jr.; Shackelford, T. J.; Shaefer, George; Shanks, W. C., III; Shiller, Gerald I.; Slade, R. C.; Smith, David Robertson; Spear, Steven G.; Sprague, Douglas W.; Stephen, Michael F.; Stumpf, H. G.; Suchsland, Reinhard John; Theyer, Fritz; Tung, J. P.-Y.; Vargo, Jan M.; Vaughan, J. L., Jr.; Vincent, Edith S.; Vonder Haar, Stephen P.; Vonder Haar, Stephen P.; Walch, C. A.; Wall, Linda Sue; Wilcoxon, J. A.; Willis, Donald Kenyon; Wilson, Jeffrey Leigh; Wilson, Jerry C., Jr.; Wilson, John Thomas; Woodruff, Fay; Yen, Zora Meei-meei; Yu, Y. K.; Zrupko, M. M.

1980s: Abdassah, Doddy; Ahlschwede, Kelly; Amabeoku, Maclean Oluka; Amaefule, Jude Ogbonnah; Anderson, James Lee; Ayer, Robert Mitchell; Azari, Mehdi; Banerdt, William Bruce; Berelson, William Max; Biegel, Ronald L.; Birch, Douglass Wanell; Blake, Gregory Howard; Broadhead, Sean; Chan, Mankin Kenneth; Chang, Syhhong; Chao, Chung-Huei; Chiu, Hung-Chie; Cusimans, George; Dahlen, Margaret; Dokka, Roy Karl; Dominguez-Vargas, Guillermo Cruz; Donoghue, Joseph Francis; Droser, Mary Louise; El-Arabi, Mahgiub Ali; Faulkner, John; Feng, Chi-Chin; Fitzgerald, Mary; Gehrels, George Ellery; Ghassemi, Farhad; Grossman, Ethan Lloyd; Hammond, Janet Louise Griswold; Haner, Barbara Elizabeth; Hartman, Blayne Alan; Hauge, Thomas

Armitage; Hazlett, Richard W.; Hoisch, Thomas David; Huh, Chih-An; Jalali-Yazdi, Younes; Jenkins, Jean; Jordanovski, Ljupco R.; Kojic, Slobodan B.; Kottmeier, Steven Thornton; Kwan, Jonathan Tak Pui; Larrain, Alberto P.; Lerch, Christopher; LeRoy, Samuel David, Jr.; Li, Yong-Gang; Link, Martin Hans; Marshall, Timothy Robert; Moeen-Vaziri, Nasser; Moslem, Kaazem; Nahhas, Tariq Mohammed; Nardin, Thomas Richard; Omoregie, Osazuwa Sunday; Orrell, Suzanne; Owusu, Lawrence A.; Pariente, Vita; Partowidagdo, Widjajono; Pepe, Philip John; Pike, Stephen Joseph; Piper, Kenneth Allen; Plescia, Jeffrey B.; Powers, James; Reynolds, Suzanne; Roberts, Lillian; Sacker, Joshua; Sadeghi, Mohammad-Ali; Savrda, Charles Edward; Sharma, Mukul M.; Smith, Linda; Stewart, George; Tang, James I. S.; Taylor, Read; Teng, Louis Suh-Yui; Thornton, Scott Ellis; White, Brian Nelson; Yasuda, Memorie; Yin, An; Zahary, Robert Gene; Zekri, Abdurrazzag Yusef

Southern Connecticut State College
New Haven, CT 06515

1 Master's

1960s: Mongillo, Joseph C.

Southern Illinois University, Carbondale
Carbondale, IL 62901

246 Master's, 16 Doctoral

1960s: Baesemann, John Frederick; Carter, Neal Allen; Cerven, James F.; Corbitt, Lonnie Leroy; Desborough, George Albert; Dial, Don C.; Engstrom, James Charles; Epie, Ebenezer E. E.; Fabry, Fredric Carl; Faust, Robert J.; Fenzel, Frank Walker; Flowers, Glen Dwight; Forrest, Ronald J.; Gauss, Joseph Charles; Givens, Terry J.; Grenda, James C.; Grimmer, John C.; Guass, Joseph Charles; Hall, John Whitling; Heivilin, Fred G.; Kehlenbach, Richard W.; Kerns, Raymond LeRoy, Jr.; Khawaja, Ikram Ullah; Klimstra, Richard Kent; Knight, Lawrence W.; Kolesar, John Charles; Krey, Frank; Kubicek, Leonard; Lemmon, Robert David; Marko, Paul Joseph; McCormick, Louis M.; Meacham, James F.; Morgan, John Harrison; Nance, Roger B.; Norton, James Austin; Olsson, William A.; Pickard, Frank Robert; Porter, Joseph A.; Reinbold, Grover; Robinson, Paul David; Satterfield, Ira Robert; Smunt, Frank Michael; Stevenson, Robert Louis; Strean, Bernard Max, Jr.; Suhm, Raymond Walter; Taylor, Robert F.; Tucker, R. Lee; Warthen, Robert Carl; Wetendorf, Fred H.; Zehner, Harold H.

1970s: Adair, Marcia B.; Adams, Scott A.; Allison, Dennis A.; Anderson, Gerald D.; Anderson, Gregg R.; Andrews, Joe A.; Bailey, Loren T.; Bluhm, Christopher T.; Blumthal, James E.; Bohnert, John E.; Boyer, David Layne; Chang, Susie Yung; Clark, Redmond R.; Cole, Gary A.; Contessa, Joseph V.; Cornish, Bruce E.; Czimer, Marilyn A.; Daniel, Marshall Edward, IV; Dyroff, Terry L.; Econ, George D.; Eggert, Donald L.; Evansin, David Paul; Fowler, Scott K.; Gastaldo, Robert A.; Gognat, Timothy A.; Gopinath, Tumkur Raja Rao; Guensburg, Thomas Edward; Haberfeld, Joeseph L.; Harper, Denver; Haynes, Ronnie J.; Healey, Neil D.; Henderson, Donpaul; Hoda, Syed Nurul; Hood, Sandra Diane; Houseknecht, David W.; Jasieniecki, Michael Simon; Johnson, Verner C.; Kaegi, Dennis D.; Keck, David Alan; Kochel, Robert Craig; Koeninger, C. Allan; Krantz, Dennis V.; Krivanek, Kenneth R.; Lambert, Stephen W.; Land, Gary F.; Langrand, Edgar L.; Leming, Stephen L.; Levine, Charles R.; Lewis, Bernard A.; Lingamallu, Surya Narayana; Little, Maynard N.; Malkames, Judith Ann; Maniocha, Michael L.; McClain, Linda K.; Mercier, Michael J.; Murrie, Gary Wayne; Nagai, Richard Brian; Neumann, William Henry; Ochs, Allan Michael; Panno, Samuel V.; Pinchock, John; Popp, John T.; Rahsman, Robert G., II; Randall, John W.; Reif, Henry Ernest, Jr.; Revell, Stephen; Rosso, Weymar Allen; Roush, Thomas L.; Ryerson,

Charles Curtis; Shelton, Marlyn Lyle; Shomali, Bahman Saghatchian; Shukis, Paul S.; Slechta, John J.; Sliva, Thomas W.; Smith, Michael J.; Somasekhara, Kananur V.; Stanonis, Gregory L.; Steidl, Peter F.; Sumner, Richard Lee; Thomas, Jimmy N.; Titus, Charles A. O.; Wade, Edward J.; Waller, Louis Raymond; Ward, JoAnn; Wescott, William A.; White, Martha; Wildeman, Joseph W.

1980s: Abegg, Frederick E.; Adams, Keith; Adem, Adem Osman; Allen, Robert Stanton; Averill, Sally Ann; Babcock, Douglas Lee; Baumann, Dean R.; Becker, John E.; Bensley, David F.; Bird, Shane R.; Black, Kenneth C.; Bohm, Steven M.; Boisture, Timothy A.; Bosse, Mark K.; Bradburn, Frederick R.; Buchanan, Douglas Mitchell; Buelter, Donald Paul; Burchfield, Margaret R.; Burd, James R.; Burdette, David James; Burk, Mitchell Keith; Cardott, Brian Joseph; Cascia, Malvin Charles; Childs, Michael S.; Childs, Susan Marie Stell; Chroback, David Allen; Chruscicki, Jean B.; Cleaveland, Thomas H.; Close, Jay Charles; Cohen, Mitchell L.; Cole, Charles Andrew; Czechowski, Douglas A.; de Cort, Thierry Michael; DeJarnette, Mark Lynn; Delph, Bryan Clifford; Devera, Joseph A.; Distefano, Lee; Dittrich, Harold Steven; Douglass, Donald P.; Dunbar, Gordon Douglas; Eisner, Michael H.; Ellison, Robert J.; English, Robert D.; Ford, James Timothy; Fowler, Kathleen Anne; Fraser, Douglas R.; George, Joseph Peter, Jr.; Germanoski, Dru; Ghaznavi, Muhammad Ishaq; Graham, Richard C.; Gould Ronald Wayne; Graves, Barbara Jean; Greif, M. Andrew; Guzan, Michael John; Hardesty, Alan F.; Harris, Clayton D.; Hayes, Benjamin R.; Henderson, Elizabeth Darrow; Hilmoe, Cynthia; Honeycutt, Thomas K.; Hoving, Sheryl J.; Howes, Susan Dawn; Hugentobler, Michael Ned; Hughes, William Brian; Infanger, Michael F.; Jaeger, Paul; Joliat, Steven A.; Kelly, Martin Henry; Kiouses, Stephan; Kistler, Barbara R.; Kravits, Christopher M.; Landis, Charles R.; Lange, Rolf V.; Lannon, Mary Susan; Laughland, Matthew M.; Loomis, Edward Charles; Mansholt, Michael Scott; Mattern, Joel K.; McCarn, Steve T.; McKay, Richard H.; Myers, James M.; O'Connell, Dennis B.; Oertel, Allen O.; Oliver, Lynne; Padgett, Jeffrey Thomas; Pasley, Mark A.; Potochnik, Mark; Reilly, Stephen Moran; Riley, Greg; Robinson, Bret A.; Rogers, Philip R.; Roman, Juan A. Deliz; Roths, Pamela J.; Runyon, Cassandra J.; Salaymeh, Talab A.; Saul, Michael T.; Schneider, Kari J.; Shebl, Mamdouh Abdel-Aal; Shoemaker, Craig Alan; Simmons, David W.; Sirota, Thomas; Smith, Cathlee; Sonnefield, Robert D.; Stafford, Mark R.; Stanley, Roderick G.; Stedman, Thomas Gentry; Strunk, Kevin Lee; Stutzman, Paul E.; Tanner, William Roger; Tedesco, Steve A.; Teerman, Stanley C.; Terranova, Thomas F.; Van Biersel, Thomas P. V.; Verseput, Timothy Dean; Vessal, Ali; Vice, Mari Ann; Von Stemle, Steven; Walker, Monte Eugene; Weaver, Sidney Mark; Weber, John C.; Werner, Alan; Wilson, Barry James; Wilson, Gregory C.; Wissinger, Diane E.; Zaengle, Donald G.

Southern Illinois University, Edwardsville
Edwardsville, IL 62026

1 Master's

1960s: Hosking, Peter Leighton

Southern Methodist University
Dallas, TX 75275

137 Master's, 46 Doctoral

1940s: Austin, Earl Bowen, Jr.; Bauchman, John Allen; Brown, Owen Cleveland, Jr.; Goldberg, Jerald Melvin; Johnson, Wylie Bruce; Monroe, John Napier; Moss, Muriel Ellen; Steinhoff, Raymond Okley

1950s: Aschenbrenner, Bert Claus; Becker, Joseph Henry; Beddoes, Leslie R., Jr.; Bryan, Tolbert Wilson; Budd, Harrell J., Jr.; Clement, Joseph Frederick; Clement, Mark Anthony, Jr.; Dodge, Charles Fremont; Dunn, David E.; Fellows, Ralph Harold, Jr.; Free, Dwight Allen, Jr.; Godfrey, Charlie

Brown; Guerrero, Richard Gonzales; Hall, George Waverly Briggs, Jr.; Herrin, Eugene Thornton, Jr.; Hightower, Charles Henry, Jr.; Hull, Louis Vincent; Ingels, Jerome J. C.; Irvin, Hollie F., Jr.; Jeffers, Joseph William; Johnson, Charles Craig; Laramore, Baylis Harriss; Laughbaum, Lloyd Ronald; McJunkin, Herbert Henry, Jr.; Munchrath, Marvin A.; Neece, Neal, Jr.; Overymyer, Dale Owen; Peabody, William Wirt; Perkins, Bobby Frank; Pitkin, James A.; Reaser, Donald Frederick; Reed, Louis Calvin; Reid, William Thomas; Riley, James Lemuel; Roberts, Carl Nelson; Robertson, Herbert Chapman, Jr.; Santillan, Hector Manuel; Schell, William Willkomm; Shaw, Nolan Gail; Sholl, Vinton Hubbard; Simmons, Marvin Gene; Taggart, James Nash; Towles, Henry Clay, Jr.; Turner, William Louis; Watkins, Jackie Lloyd; Wayland, John Rex; Wiggins, Peter Nelson, III; Williams, Thomas Ellis; Williamson, Paul Bain; Winn, Vernard

1960s: Bilelo, Maria A. Marques; Bretsky, Peter William; Briggs, Garrett; Dooley, Duane; Driscoll, William J.; Erikson, Erik Harold, Jr.; Fellowes, Terrence Leigh; Jacobson, John Martin; Kimball, Newton Scott, Jr.; Laury, Robert Lee; Lobdell, John Little; Mack, Harry; Meier, Robert William; Naeser, Charles Wilbur; Poche, David John; Poort, Jon Michael; Seiner, Maureen Bernice; Seymour, Frank F.; Sorrells, Gordon Guthrey; Spall, Henry Roger; Steiner, Maureen Bernice; Steward, Hugh Leighton; Swanberg, Chandler Alfred; Thurmond, John Tydings; Thurmond, John Tydings; Vormelker, R. S.

1970s: Alexander, Nancy Sue; Allen, Peter Martin; Brott, Charles Arthur; Chu, Jiaw; Davis, Joseph Redmond; Dees, Jerry Lee; Der, Zoltan Andrew; Feizpour, Ali A.; Fix, James Edward, Sr.; Gant, Orland James, Jr.; Gillette, David D.; Goforth, Thomas Tucker; Gose, Wulf Achim; Gosnold, William David, Jr.; Haas, Herbert; Hassan, Afifa Afifi; Herrin, Toni Elizabeth; Jin, Doo Jung; Johnson, Gary Dee; Kast, Joe Alex; Kehler, Philip Leroy; Kohl, Karen Brummett; Leeper, Robert H., Jr.; Lezak, Jennifer Linn; Liaw, Liang-Chi; Malarcher, Falvey L.; Martin, Sheila Shinn; Masse, Robert Patrick; Mazzella, Frederick E.; McKinley, George Alvin; Meyer, Robert Lee; Michael, Fouad Yousry; Mitchell, Brian James; Moore, Leonard Vanard; Newkirk, Steven Ross; Novak, James Michael; Osten, Lawrence William; Patzewitsch, Wwndy W.; Petefish, David Michael; Prigmore, Susan Marcheta; Prigmore, Susan Marcheta; Pronold, Thomas George; Pruitt, Jacqueline Davis; Rafipour, Bijan; Rainey, Mary Tindal; Reinke, Robert Edward; Sartin, Austin Albert, Jr.; Smart, Eugene; Sorrells, Gordon Guthrey; Steele, John Lisle; Steele, John Lisle; Swanberg, Chandler Alfred; Talley, Kieth L.; Taylor, Henry Clyde; Veith, Karl Fredrick; Waugh, John Russell, II; Willimon, Edward Lloyd; Young, Ching Ju Jennifer

1980s: Alden, Kathleen A.; Becker, David Joseph; Brown, Carolyn Ruth; Brown, Charles Douglas; Cerny, John H.; Covington, Morton Douglas; Daniyan, Muhammad Abdullahi; Dickerson, Robert Paul; Dutrow, Barbara Lee; Dutrow, Barbara Lee; Elghadamsi, Fawzi E.; Ferguson, John Franklin; Flynn, Elizabeth Chittenden; Gagnier, Michelle Annette; Golden, Paul W.; Grant, Lori T.; Hagedon, Dan Newman; Johnson, Kathleen Elizabeth; Kasulis, Paul Francis; Kelley, Shari Anne; Kempton, Pamela Dara; Kempton, Pamela Dara; Kinsel, Erick Paul; MacLeod, N.; Marquis, Samuel Austin, Jr.; McMillan, Nancy Jeanne Stoll; Murry, Phillip Anthony; Pattey, Phillip D.; Phillips, Robert Laurence; Rafipour, Bijan; Reese, Nathan Mark; Rehder, Timothy Ray; Reid, Fredrick Samuel, Jr.; Shore, Patrick John; Smith, Dana N.; Sweetkind, Donald Steven; Viglino, Janet Atkinson; von der Hoya, H. Austin, II; Woolridge, Bruce Alan; Ziagos, John Peter

University of Southern Mississippi
Hattiesburg, MS 39406

12 Master's, 4 Doctoral

1960s: Gilliland, William A.

1970s: Aultman, W. L.; Chen, Tzann-Hwang; Cook, P. L., Jr.; Denehie, R. B.; Gilliland, W.; Havard, Deborah Ann; Johnson, J. J.; Johnson, Stephen A.; McCarty, J. E.; Phillips, John Asa; Ray, T. M.; Reel, Ted Wesley; Tarbutton, R. J.; Warnock, Frank B.

1980s: May, James H.

Southwest Missouri State University
Springfield, MO 65804

23 Master's

1980s: Alioha, Iheakachuku George; Barner, Wendell; Bush, Bruce Allen; Coonrod, David L.; Emrie, Gail Estelle; Entrup, Karen; Harpine, Joseph E.; Helm, Sue; Hipple, Robert; Hosey, Lisa Elaine; Hough, Ronald David; Kintner, Stephen S.; Krizanich, Gary W.; Morris, Kent D.; Neill, Jerry R.; Pumphrey, Phillip L.; Sappington, Eric Jon; Schloss, Jeffrey A.; Steinkamp, Allen L.; Stettes, Steven S.; Sutton, Victoria A.; Varner, Vicki; Vujnich, Joseph William

University of Southwestern Louisiana
Lafayette, LA 70504

75 Master's

1960s: Brown, Billy H.; Christensen, Arthur Francis; Ferguson, Floyd Jay; Labat, Clevland; Lovelace, Bobby G.

1970s: Bergeron, Dalton J.; Bradley, Christopher H.; Buck, Don E., Jr.; Duhon, Michael P.; Dungan, James R.; Finley, William R.; Friedel, George F.; Martin, Jack Philip, Jr.; McFarlain, Tommy; Mehenni, Mourad; Tabbi-Anneni, Abdelhafid; Trepagnier, Albert James; Weber, Anthony J.; Zuraff, Steven J.

1980s: Barrilleaux, Janell; Bell, Mary K.; Bell, Steven R.; Bowdon, M. M.; Caffery, Stephen; Callihan, John Brent; Cancienne, Joseph A.; Darling, Bruce K.; Despot, Martin; Desselle, Bruce A.; Ferber, Robin J.; Fowler, Chris K.; Gamble, James A.; Gautreaux, John W.; Green, Stephen H.; Guilbeau, Ellis R.; Hall, Cassandra; Hearn, Frank; Hearne, James H.; Heron, Stephen D., III; Hosseini, Masood S.; Humphris, David J.; Judice, Philip C.; Kurth, Randall J.; Lalonde, Kayron F.; Landry, Richard G.; Lautier, Jeffrey C.; Lazarus, Norman H.; Louque, Roland J.; Lyons, William S.; McCurdy, Maureen; Moore, Dan; Mulsow, Miriam H.; Mulsow, Randall R.; Praditan, Surawit; Rahmatian, Mansour; Rex-Pelkey, Ilene; Rutherford, Mark S.; Salahuddin, Qazi; Saul, William L.; Saxton, Deborah C.; Schramm, William H.; Sherrill, John F., III; Smith, Gilbert B.; Strickland, Matthew O.; Swick, Steven E.; Thorkelson, John M., Jr.; Troyanowski, Larry D.; Truax, Stephen, III; Vermillion, Peter A.; Verrastro, Robert T.; Vesey, Brian K.; Vesey, Jamsie Roberts; Zanghi, Elizabeth M.; Zekmi, Nadir; Zeosky, Joseph E.

St. Francis Xavier University
Antigonish, NS B2G 1C0

1 Master's

1960s: Doucet, J. A.

St. Louis University
St. Louis, MO 63103

125 Master's, 85 Doctoral

1920s: Dillon, Vincent Francis; Repetti, William C.

1930s: Blum, Victor Joseph; Bradford, Donald Comnick; Dahm, Cornelius G.; Dahm, Cornelius G.; Dahm, Cornelius G.; Heinrich, Ross Raymond; Hodgson, Ernest A.; Kremer, Benedict Peter; Miller, Carl John; O'Donnell, George Anthony; Oefelein, Rosalie Teresa; Ramirez, Jesus Emillio;

Ramirez, Jesus Emillio; Reynolds, Thomas Emmett; Robertson, Florence; Westland, Anthony James

1940s: Birkenhauer, Henry Francis; Birkenhauer, Henry Francis; Blum, Victor Joseph; Frank, Albert Joseph; Frank, Albert Joseph; Jennemann, Vincent Francis; Kisslinger, Carl; Robertson, Florence; Roemer, Cletus D.; Walter, Edward Joseph; Walter, Edward Joseph

1950s: Adams, William Mansfield; Adams, William Mansfield; Baker, Ray Gordon; Bingham, Henry Todd; Brayer, Roger C.; Brereton, Roy G.; Brereton, Roy George; Burrows, Lees Joslyn, Jr.; Chandiok, Kailash Chandra; Chang, Feng-Keng; Cherry, Jesse Theodore, Jr.; Connor, James J., Jr.; Douthit, Thomas D. Nathan; Erwin, James Walter; Fava, James Archie; Ford, Ralph Joseph; Fox, James Henry; Fox, James Henry; Gabriel, Walter J.; Ganguli, Debkumar; Garland, George David; Ginzel, Edwin Charles; Haines, Carroll Eugene; Hanson, Roy E.; Hasenpflug, Harry John, Jr.; Heelan, Patrick Aidan; Hoffschwelle, John William; Hofmann, Renner Bergene; Holmes, Charles R.; Honnell, Pierre Marcel; Italia, Santo; Jones, George P., Jr.; Kailasam, Lakshmi Narayan; Kisslinger, Carl; Knapp, Gregory Anthony; Larochelle, Andre; Leaf, Howard Westley; Mateker, Emil Joseph, Jr.; Miatech, Gerald James; Miller, Henry Joseph; Mondschein, Herman F.; Moore, Fred Edward; Neal, Donald Arthur; Nuttli, Otto William; Nuttli, Otto William; Opp, Albert G.; Oriard, Lewis L.; Pantall, Jack Travis, Jr.; Schaefer, George J.; Schwendinger, William W.; Villars, Paul Emile; Volk, Joseph Anthony; Walsh, Daniel Hallaron; Walsh, Daniel Hallaron; Webb, John Purcell; Whipple, Arthur Paul; Whitmore, John Day; Wiggins, John Henry, Jr.; Young, Durward Dudley

1960s: Al-Khafaji, Saadi Abbas; Al-Sinawi, Sahil Abdulla; Aparicio, Miguel Pablo; Bahjat, Dhari Saaid; Bahjat, Dhari Saaid; Batllo-Ortiz, Josep; Bollinger, Gilbert Arthur; Boudreau, Francis C.; Carr, Theodore George; Castano, Juan Carlos; Chandra, Umesh; Cherry, Jesse Theodore, Jr.; Davis, James W.; Dowling, John Joseph; Engdahl, Eric Robert; Espinosa, Alvaro Felipe; Farouq, Fadullah Meer; Farrell, Thomas G.; Fernandez, Luis Maria; Fernandez, Luis Maria; Furumoto, Augustine Sadamu; Ganse, Robert Anthony; Goodwin, Michael Lawrence; Goodwin, Michael Lawrence; Gupta, Indra Narayan; Gupta, Indra Narayan; Hannon, Willard James, Jr.; Hansink, James D.; Heigold, Paul Clay; Ibrahim, Abou-Bakr Khalil; Khattri, Kailash Nath; Khattri, Kailash Nath; Kim, Won Ho; King, Harry J.; Mateker, Emil Joseph, Jr.; McEvilly, Thomas Vincent; Messmer, William J.; Necioglu, Altan; Papazachos, Basil C.; Pfluke, John Henry; Stewart, Samuel Woods; Udias, Agustin Vallina; Voss, James T.; Wagner, Richard J.

1970s: Bennett, Theron Joseph; Bhattacharya, Bireswar; Canas, Jose Antonio; Cheng, Shiang-ho; Correig, Antoni M.; Dasgupta, Shivaji N.; Fischer, Gerard William; Ganse, Robert Anthony; Goel, Shailendra Kumar; Grover, Anil; Guha, Shyamal Kanti; Guja, Nasser Hossain; Habib, Nashat M.; Hashim, Braik M.; Herrmann, Robert B.; Huaco, Daniel; Kane, Martin Francis; Khan, Abdul Qadir; Kim, So Gu; Kramer, Mark Thomas; Mezcua, Julio; Mualchin, Lalliana; Necioglu, Altan; Powell, James Adrian; Rodriguez E. Rene; Rogers, Albert Mitchell, Jr.; Sayman, Ali; Schaefer, Stephen Felix; Stirling, William Alex; Street, Ronald Leon; Syed, Atiq Ahmad; Tuksal, Ilker; Wagner, Donald E.; Wen, Huei-Yuin; Yacoub, Naziek K.; Yamamoto, Jaime; Yasar, Tuncay; Yasar, Tuncay; Yu, Guey-Kuen

1980s: Aiinehsazian; Canas, Jose Antonio; Chan, Wing-Wah Winston; Chan, Wing-Wah Winston; Chen, Jing-Jong; Cheng, Chiung-Chuan; Chiou, Shyh-Jeng; Chulick, John Alexander; da Boa Hora, Marco Polo Pereira; Dablain, Mark Albert; Dwyer, John Joseph; Goertz, Michael Joseph; Gordon, David Walker; Himes, Larry Douglas; Hwang, Horng-Jye; John, Viera; Kohsmann, James Joseph; Leite, Lourenildo Williame Barbosa; Leu, Peih-Lin;

Leu, Peih-Lin; Lydon, Michael Thomas; Mandal, Batakrishna; Mindevalli, Oznur; Mokhtar, Talal Ali; Nguyen, Bao Van; Nguyen, Bao Van; Nicholson, Craig; Perry, Robert Gayle; Raoof-Malayeri, Mehdi; Reidy, Denis; Russell, David Ray; Saikia, Chandan Kumar; Shieh, Chiou-Fen; Shin, Tzay-Chyn Tony; Singh, Sudarshan; Wang, Chien-Ying; Wang, Huei-Yuin Wen; Woods, Mark Thomas; Zollweg, James Edward

Stanford University
Stanford, CA 94305

710 Master's, 844 Doctoral

1890s: Ashley, George H.; Ashley, George H.; Drake, Noah F.; Drake, Noah F.; Hopkins, Thomas Cramer; Means, John H.; Newsom, John Flesher

1900s: Arnold, Ralph; Arnold, Ralph; Crandall, Roderic; Haehl, Harry L.; Knecht, Carl Emil; Kramm, Hugo E.; Newsom, John Flesher; Shedd, Solon S.; Smith, Warren DuPre

1910s: Beal, Carl H.; Chapin, Theodore; Clark, William O.; Crandall, Hector; Davis, Charles H.; Garfias, Valentine Richard; Guild, Frank Nelson; Hawley, Henry J.; Hook, Joseph S.; Jenkins, Olaf P.; Lee, Wah Seyle; Loel, Wayne F.; Nelson, Wilbur A.; Prescott, Basil; Ray, James C.; Shedd, Solon S.; Templeton, Eugene C.; Vickery, Frederick P.; Waring, Clarence A.

1920s: Adams, Sidney F.; Addison, Carl C.; Baddley, Elmer R.; Briggs, Otis E.; Buss, Fred Earle; Carson, Carlton M.; Cartwright, Lon D., Jr.; Christensen, Andrew L.; Church, Clifford C.; Corby, Grant W.; Crawford, Arthur L.; Crickmay, Colin H.; Donnay, Joseph D. H.; Ekblaw, George Elbert; Evans, David L.; Ewing, Sydney C.; Faustino, Leopoldo Aldo; Faustino, Leopoldo Aldo; Gifford, Charles D.; Herold, Stanley C.; Hertlein, Leo George; Hertlein, Leo George; Hillis, Donuil M.; Hoots, Harold W.; Howe, Henry V.; Hughes, Donald Dudley; Janish, Jeanne R.; Janish, Jeanne Russell; Jordan, Eric K.; Kerr, Paul F.; Kildale, Malcolm B.; Kleinpell, Robert Minssen; L'Egraye, Michael P. H.; Lambert, Gerald S.; Lynn, Harold F.; Mackenzie, Andrew N.; Mathews, Asa A. Lee; Mayo, Evans B.; Mitchell, Harold G.; Muller, Siemon W.; Mulryan, Henry; Palmer, Robert H.; Pike, Ruthven W.; Powers, Delmer Lance; Ray, James C.; Reagan, Albert B.; Reed, Ralph D.; Reinhart, Phillip Wingate; Richardson, Remond W.; Sparks, Dale D.; Stipp, Thomas F.; Suffel, George G.; Touwaide, Marcel E.; Vickery, Frederick P.; Von Estoff, Fritz E.; Wiedey, Lionel William; Wiedey, Lionel William; Willis, Robin; Willis, Robin

1930s: Adams, Bradford C.; Anderson, Frank Marion; Campbell, Charles D.; Ellsworth, Elmer W.; Field, Ross; Frizzel, Donald L.; Gale, Hoyt R.; Galliher, Edgar W.; Galliher, Edgar W.; Grant, Ulysses S.; Grant, Ulysses S., IV; Green, David Ely; Greninger, Alden Buchannon; Hake, Benjamin F.; Hedberg, Hollis D.; Jenkins, Olaf P.; Johnston, Francis N.; Kartchner, Wayne E.; Keenan, Marvin F.; Kildale, Malcolm B.; Kleinpell, Robert M.; Kornfeld, Moses M.; Krauskopf, Konrad Bates; Landwehr, Walter R.; Lemmon, Dwight M.; Lemmon, Dwight M.; Martin, Lois Ticknor; McAllister, James R.; McMasters, John H.; Melcon, Zenas K.; Muller, Siemon W.; Page, Ben M.; Page, Ben Markham; Penn, William Y.; Poland, Joseph F.; Putlitz, Fritz H.; Putnam, William C.; Ransome, Alfred L.; Reinhart, Phillip Wingate; Richards, George L.; Smith, Merritt B.; Smith, Norman H.; Snedden, Loring B.; Staples, Lloyd W.; Taber, Edward Carroll; Taylor, William Harlan; Thompson, Warren O.; Todd, Wallace; Tolman, Frank Bronson; Verhoogen, Jean; Wheeler, Harry E.; Wheeler, Harry E.; Wilson, Robert Rogers; Woodward, Albert F.

1940s: Agnew, Allen Francis; Amer, Hamed H.; Bailey, Edgar H.; Bradbury, Albert E.; Campbell, Charles Virgil; Carpenter, Robert H.; Carpenter,

Robert H.; Cassell, John K.; Childs, Theodore S.; Compton, Robert R.; Cook, Rufus E.; Crume, Robert W.; Curran, John F.; Drummond, C. Hanford; Fortier, Yves O.; Foxhall, Harold B.; Freeman, John C.; Fulton, Robert B., III; Funkhouser, Lawrence W.; Girard, Cecil M.; Green, Charles Frederic; Grimm, Kenneth E.; Haffner, Robert Louis; Heikkila, Henry H.; Humphrey, Frederic Gavin; Kelley, Frederic R.; Klein, Ira; Kroger, Robert S.; Lombardi, Leonard Volk; Macleod, George M.; Marks, Jay G.; McCroden, Thomas J.; McKinlay, Phillip France; McPherson, William John; Miller, Gardner Burnham; Miranda, Leandro J.; Moses, Selma; Nauss, Arthur W.; Nicol, David; Nicol, David; O'Malley, Frank Ward; Paguirigan, Francisco; Pantin, Jose H.; Pattison, Halka M.; Pekkan, Ahmet; Penney, Dolores Jeanne; Pierce, Jack William; Roberts, Ellis E.; Robins, Alfred Raymond; Sinnott, Allen; Stopper, Robert Francis; Swinney, Chauncey Melvin; Thompson, George A., Jr.; Thorup, Richard Russell; Troster, John Gooch; Walker, George W.; Wasson, Edward Bassett; Watson, Elizabeth Anne; Zimmerman, John, Jr.

1950s: Abell, Philip Webster; Abell, Philip Webster; Adams, John Bright; Albers, John Patrick; Albers, John Patrick; Albrizzio, Carlos; Ancieta, Hugo Alfonso; Anderson, Arthur Taylor; Atchley, Frank William; Bailly, Paul Alain; Baker, Arthur, III; Barnola, Alberto; Barnola, Alberto; Barrow, Thomas Davies; Berry, Frederick Almet Fulghum; Beveridge, Alexander James; Blaustein, Morton Katz; Brabb, Earl Edward; Bradley, William Crane; Brooke, Robert Clymer, Jr.; Brooke, Robert Clymer, Jr.; Brooke, Robert Clymer, Jr.; Brown, Eugene; Brown, Willis Reider; Burchfiel, Burrell Clark; Burke, Harold W.; Burtner, Roger Lee; Burtner, Roger Lee; Cady, James Richard; Campbell, Charles Virgil; Christian, Louis; Classen, James Stark; Coleman, Robert Griffin; Cox, Dennis Purver; Creveling, Louis; Crosby, Donald Gladstone, Jr.; Crozier, Robert N.; Cummings, Jon Clark; Davis, Charles George; Demirmen, Ferruh; Dickinson, William Richard; Dickinson, William Richard; Donath, Fred Arthur; Dort, Wakefield, Jr.; Douglass, Raymond Charles; Eckert, William Frederic, Jr.; Edgell, Henry Stewart; Elbishlawi, Mohamed Husni Morad; Esser, Robert Worth; Fish, John L.; Flint, William B.; Flint, William B.; Forkgen, Peter Edward; Gomes, Joao Bosco Ponciano; Gomes, Joao Bosco Ponciano; Gonzales, Arsenio Geronimo; Griggs, Allan B.; Grose, Lucius Trowbridge; Gulbrandsen, Robert Allen; Hall, Clarence Albert, Jr.; Hamilton, Edwin Lee; Hash, Bender; Hassan, Ahmad; Henricksen, Donald Anton; Hilton, Allan Decou; Hilton, Allan Decou; Hoskins, Cortez William; Hughes, Thomas Hastings; Hurley, Neal Lilburn; Jenkins, Elmer Leroy; Johnson, Grace Phillips, II; Jones, D. L.; Jones, David L.; Kanaya, Taro; Key, Carlos Eduardo; Kistler, R. B.; Kuo, John Tsung Fen; Langan, Leon Verdin, Jr.; Lawton, John Edward; Lee, Donald Edward; Lee, James William; Leon, Hernan Jose; Lombardi, Leonard Volk; Lounsbury, Richard William; Lund, Lamar; Lund, Lamar; MacDiarmid, Roy Angus; Mack, John Erick, Jr.; Mahrholz, Wolfgang Werner Ekkehardt; Mandra, York Tooree; Manning, John Draige; Marks, Jay G.; Marsh, Owen Thayer; McAllister, James F.; McCollom, Robert Lucien, Jr.; McGirk, Lon S., Jr.; McKee, Elliott B., Jr.; Melendres, Mariano M., Jr.; Meyerhoff, Arthur A.; Miller, Charles Parker; Miller, Richard D.; Miller, Robert S.; Moreno Agreda, Francisco; Moreno Agreda, Francisco; Mundt, Philip Amos; Nagell, Raymond Harris; Nagell, Raymond Harris; Osten, Erimar Alfred von der; Paredes, Manuel; Patton, William Wallace, Jr.; Payne, Anthony L.; Persons, Philip; Péwé, Troy Lewis; Philips, Richard P.; Plantevin, Jean Paul; Richmond, James F.; Richmond, James Frank; Robinson, Russell Dow; Rosa, Hermenegildo; Rosa, Hermenegildo; Roscoe, Stuart M.; Salvador, Amos; Sanborn, Albert F.; Schmidt, Otto Mackenty; Shawe, Daniel Reeves; Sheldon, Richard Porter; Shimazaki, Yoshihiko; Silberling, Norman John;

Simons, Frank S.; Sisler, John Joseph; Soliman, Soliman Mahmoud; Spotts, John Hugh; Stelck, Charles R.; Stone, John Grover, II; Strom, Robert Gregson; Sulaym-an, Sulaym-an Mahm-ud; Sutton, J. S.; Taft, William Harrison; Taft, William Harrison; Tamesis, Emmanuel Valerio; Taylor, Samuel Guy, Jr.; Touring, Roscoe Manville; Travers, William Brailsford; Van Denburgh, Alber Stevens; Verastegui-Mackee, Pedro; VerPlanck, William E., Jr.; von den Osten, Erimar Alfred; Wagner, Frances Joan Estelle; Waldron, John Francis; Walters, Richard Francis; Waterman, Glenn C.; Weber, Paul Wesley; Weimer, Robert Jay; Wheeler, Charles Brown; Williams, Thomas J.; Willis, David Grinnell; Wilson, Richard Fairfield; Wise, William S.; Wrucke, Chester T.; Zevallos-Herrera, Francisco J.

1960s: Akhtar, Salim; Anderson, Robert Neil; Anderson, Roger Yates; Anttonen, Gary Jacob; Anttonen, Gary Jacob; Arguelles, Victor; Armstrong, Frank Clarkson; Ashley, Roger Parkmand; Aso, Kazuo; Baldwin, Arthur Dwight, Jr.; Ballew, Gary I.; Bansbach, Louis Philip, III; Bayoumi, Abdel Rehim Imam; Beall, Arthur Oren, Jr.; Beck, Myrl Emil; Benavides, Victor; Bezara, Miguel; Birkeland, Peter Wessel; Bjorck, Frederick Richard; Blacet, Philip Merrell; Blake, Milton Clark, Jr.; Bonilla, Manuel George; Brabb, Earl Edward; Brant, David Mann; Breiner, Sheldon; Breitenbach, Eugene Allen; Brew, David Alan; Brito, Ignacio Machado; Brown, Eugene H.; Brown, Walter William; Bryant, Donald Glassell; Bull, William Benham; Burch, Stephen Howell; Burford, Robert Oliver; Burke, Dennis Bernan; Burt, Robert John; Buss, Walter Richard; Cabral de Farias, Luiz Carlos; Cady, Gilbert Victor; Campbell, Robert W.; Cartaya, Rafael A.; Carter, Claire; Christiansen, Robert Lorenz; Chuber, Stewart; Clark, Bruce Robert; Clark, Howard Charles, Jr.; Clark, Joseph Clyde; Clark, Malcolm Mallory; Clement, William Glenn; Coan, Eugene Victor; Colburn, Ivan P.; Collinson, James Waller; Converse, Glenn Leland; Corvalan, Jose Idamor; Crosson, Robert Scott; Cummings, Jon Clark; Czamanske, Gerald Kent; Dale, Ruth; Darling, Richard Graydon; Davidson, Lynn Blair; Del Solar, Carlos W.; Demirmen, Ferruh; Demirmen, Ferruh; Dodge, Franklin Charles Walter; Doll, Barry; Douze, Eduard Jan; Dowty, Eric; Duffield, Wendell Arthur; Earlougher, Robert Charles, Jr.; Edwards, Kenneth Lang; Elboushi, Ismail Mudathir; Enciso, Salvador; Enos, Paul Portenier; Ersoy, Demir; Euribe Dulanto, Alejandro; Fairborn, John William; Farquhar, Roger P.; Fernandez, Carlos E.; Ferretti A., Jorge; Fleck, Robert Joseph; Fowles, George Richard; Frost, John Elliot; Gale, Robert Earle; Gay, Sylvester Parker; Gimlett, James Irwin; Glover, Peter; Goodnow, Warren Hastings; Grantz, Arthur; Gregson, Victor Gregory, Jr.; Greve, Gordon Madsen; Griffin, William Lindsay; Guzman, Armando; Hall, Francis Ramey; Hallberg, Jack Arthur; Hamilton, Douglas Holmes; Harwood, Robert James; Hay, Peter William; Hay, William Winn; Hedberg, James Dow; Heinrichs, Donald Frederick; Henderson, Frederick Bradley, III; Herkenhoff, Earl Frederic; Hickman, Robert G.; Hodges, Carroll Ann; Holstrom, Geoffrey Burwell; Hornberger, George Milton; Howard, James Campbell; Howard, James Hatten, III; Hoyt, Philip Munro; Hunkins, Kenneth Leland; James, David Evan; James, David Evan; James, Laurence Pierson; James, Odette Francine Bricmont; James, Odette Francine Bricmont; James, Odette Francine Bricmont; Jarvis, Daniel; Kesler, Stephen Edward; Khan, Suhail; King, Paul Hamilton; Korpi, Glen Kaye; Kuntz, Mel Anton; Kvenvolden, Keith Arthur; LaFehr, Thomas Robert; Lawrence, Robert Dale; Lazier, Bruce Earl; Lee, Keenan; LeMasurier, Wesley Ernest; Leo, Gerhard William; Lewis, Richard Wheatley, Jr.; Linstedt, Kermit Daniel; Lipman, Peter W.; Lobo Guerrero U., Alberto; Lobo Guerrero U., Alberto; Lofgren, Gary Ernest; Loomis, Alden Albert; Lowe, Donald R.; Luce, Robert William; MacDiarmid, Roy A.; Machado Brito, Ignacio Autrliano; Mahai, Thomas; Malfait, Bruce; Mann-

ion, Lawrence E.; Martin, Lewis; Martin, Robert Francois Churchill; Martinez D., Ignacio; Martinez, Luis; Martinez, Maximo; Martinez, Maximo; McFarlane, Robert Craig; McLeroy, Donald Frazier; Meade, Robert Herber, Jr.; Mederos H., Alfredo; Meister, Laurent Justin; Menzie, Thomas Eugene; Miller, Fred Key; Millitante, Priscilla Juan; Moisseeff, Alexis Nicolas; Montgomery, Edward Sharar; Moore, Henry John, II; Moore, Richard Thomas; Moore, Richard Thomas; Morey, Booker Williams; Morris, Elliot C.; Munroe, Robert John; Mursky, Gregory; Naimi, Ali Ibrahim; Narasimhan, Tyagarajan; Nazikoglu, Zekai; Neel, Thomas Howard; Negev, Moshe; Nikias, Peter A.; Noble, Donald Charles; Normark, William Raymond; Ojakangas, Dennis Roger; Ojakangas, Richard Wayne; Olander, Harvey Chester, Jr.; Paredes, Manuel; Park, Allan Morey; Park, Frederick Blair; Pestrong, Raymond; Petersen, Carl Frank; Peterson, Donald William; Peterson, Frank Lynn; Peterson, Fred; Pickering, Ranard Jackson; Playford, Phillip Elliott; Pollard, David Dierker; Pratt, Walden Penfield; Puffer, John Harold; Putnam, Bruce McCormick; Radtke, Arthur Sears; Redmond, John Lynn; Redmond, John Lynn; Reeves, Elaine Louise; Reeves, Robert Grier; Rich, Ernest I.; Richmond, William O.; Rietman, Jan David; Riley, Francis Stevenson; Robinson, Russell; Rofe, Rafael; Rowland, David Andrew; Saad, Afif Hani; Sacris, Eduardo Milan; Sadighi, Soleman; Saidi, Ali Mohammad; Sainsbury, Cleo Ladell; Salehi, Iraj A.; Sandvik, Peter Olaf; Santos-Ynigo, Luis Marcial; Scarpelli, Wilson; Schilling, Frederick A., Jr.; Scholl, David William; Schwarzman, Elisabeth C.; Secor, Donald Terry, Jr.; Sedore, Jacquelin; Seibert, Barre Alan; Seyfert, Carl Keenan, Jr.; Sheridan, Michael Francis; Shive, Peter Northrop; Shklanka, Roman; Sides, James Wesley; Sieck, Herman C.; Simonds, Charles Henry; Sims, Samuel John; Sinnokrot, Ali Amin; Sisler, John Joseph; Smith, David Dwyer; Smith, James Gordon, II; Smith, Lee Anderson; Smith, Ross Wilbert; Smith, Rossman William; Smith, Thomas Edward; Snetsinger, Kenneth George; Snetsinger, Kenneth George; Sokol, Daniel; Speed, Robert Clarke; Spieker, Andrew Maute; Spresser, Ralph G.; Stauffer, Karl Walter; Stauffer, Peter Hermann; Steele, Timothy Doak; Stewart, John Harris; Stuart, David J.; Stuart-Alexander, Desiree Elizabeth; Swe, Win; Taft, William Harrison; Tamesis, Emmanuel Valerio; Taylor, Charles Mosser; Taylor, James Grover V.; Taylor, Patrick Timothy; Terry, Judith Shoemaker; Terry, Judith Shoemaker; Thomas, Gordon Wallace; Thompson, Thomas Luman; Tieh, Thomas Ta-pin; Tipton, Ann; Tuman, Vladimir S.; Turk, Leland Jan; Veatch, Maurice Deyo; Verdejo, Cecilia; Vigrass, Laurence William; Vincelette, Richard Roy; Vogel, John David; Wahl, Ronald Richard; Wallace, William; Waltz, James Patterson, II; Warren, John Stanley; Wattenbarger, Robert Allen; Wentworth, Carl Merrick, Jr.; Wesson, Robert L.; Whiting, Jerry Max; Wiebe, Robert Alan; Wiedmann, John Philip; Willden, Charles Ronald; Wobus, Reinhard Arthur; Wodzicki, Antoni; Wood, Milton Darroll; Wrucke, Chester Theodore, Jr.; Yates, Robert Giertz; Yeh, William Wen-Gong; Young, Chapman, III; Young, Chapman, III; Young, Roger A.; Zantop, Half Al; Zohdy, Adel Abd El-Rahman

1970s: Abry, Claude Georges; Acunzo, Antonio Carlos; Aguado, E. A.; Ahmed, Ajaz; Aitken, B. G.; Albayrak, A. Feridun; Alley, William McKinley; Allmendinger, R. W.; Anderson, E. R.; Anderson, Robert Stewart; Angulo, Raul E.; Anttonen, Gary Jacob; Arditty, Patricia C.; Arihara, Norio; Aruna, Muhammadu; Atwater, B. F.; Avakian, Robert W.; Avotins, Peter; Awald, John Theodore; Aydin, A.; Azih, O. D.; Bahia-Guimaraes, Paulo Fernando; Bailey, G. B.; Baker, P.; Bamford, Robert Wendell; Barker, Benjamin Joseph; Bartow, J. A.; Basse, R. A.; Bassett, R. L.; Bateson, J. T.; Bauder, J. M.; Beaulieu, John David; Bela, Jim; Bell, M. L.; Bergquist, J. R.; Berlanga, J. M.; Berry, Anne L.; Beyer, Larry Albert; Biancardi,

John M.; Bilodeau, W. L.; Bilzi, Paul; Bishop, R. S.; Bittencourt-Netto, Otto; Blaisdell, Thomas; Blakely, Richard J.; Blencoe, James G.; Bookstrom, A. A.; Bostick, Neely Hickman; Boyd, Harry R.; Braga, B. P. F., Jr.; Brock, Kenneth Jack; Brunengo, M. J.; Burg, John Parker; Burke, Dennis Bernan; Bush, William Robert, Jr.; Butler, Robert F.; Cady, John W.; Caixeiro, E.; Callard, J. G.; Campbell, Alan Neil; Canales-L., L.; Carroon, C.; Carson, William Pierce; Casey, Tom Ann L.; Casse, F. J.; Castillo S., Jesus M.; Cavounidis, S.; Charbeneau, Randall J.; Chen, Hsiu-Kuo; Chiang, J. C.; Chicoine, Stephen Duane; Chin, Ti-hau; Chipping, David Hugh; Chisholm, Duncan M.; Christensen, O. D.; Cinco Ley, Heber; Clark, S. R.; Clarke, Denis Edmund; Cleveland, Michael N.; Cobb, William Marshall; Cogan, J.; Collier, M. P.; Connor, C. L.; Cooper, A. K.; Cooper, M. R.; Correa, A. C.; Courtillot, Vincent E.; Cowan, Darrel Sidney; Cramer, C. H.; Crawford, Michael F.; Crichlow, Henry Brent; Crough, S. T.; Cunningham, Charles Godvin, Jr.; Daniel, R. G.; Davies, Hugh Lucius; Davis, Brent E.; Davis, James A.; Day, Theodore James; de Almeida, Erasto Boretti; de Almeida, Erasto Boretti; De Eston, S. M.; De Nault, Kenneth James; de Vasconcelos, Jose Aluizio; Delaney, Paul Theodore; Denby, Gordon Morrison; Denham, Charles R.; Denlinger, R. P.; Dent, Brian Edward; DeVilbiss, J. W.; Dickert, Paul Fisher; Doherty, S. M.; Dohrenwend, J. C.; Dohrenwend, J. C.; Donnelly, M. F.; Dowsett, Frederick Richard, Jr.; du Bray, Edward Arthur; Dunbar, W. S.; Duncan, Robert A.; Dupre, W. R.; Eaton, Marilyn Keller; Ebert, Janet; Echeverria R., L. M.; Egemeier, Stephen Jay; Eichelberger, John Charles; Eidlin, Michael B.; El-Hawat, Ahmed Saleh; El-Naggar, Mohamed Mamdouh Abdalla; Ellen, Stephenson Davis; Elliot, Terence Martin; Elliott, James Edward; Ellsworth, William L.; Ercan, Ahmet; Ervine, Warren Basil; Estevez L., R. J.; Estrada M., Armando; Evarts, R. C.; Ewing, Rodney Charles; Faggioli, Justin M.; Falade, Gabriel Kayode; Fandriana, Lilian; Faure, François M.; Fenn, Philip Michael; Ferraz, Celso Pinto; Ferry, John Mott; Finch, Christian C.; Fink, J. H.; Fink, Kendrick Claude; Finley, J. A.; Finnemore, Erhardt John; Fisher, Michael A.; Fitch, Frank Williams, III; Fleming, Robert William; Fliegner, J. F.; Foord, E. E.; Fowler, David; Fox, Kenneth Francis, Jr.; Frizzell, V. A., Jr.; Fuenzalida, Ricardo H.; Gaisie, J. S.; Gilbert, Wyatt Graves; Gillam, Mary L.; Gilman, J. A.; Glenn, Lawrence Edward; Godfrey, R. J.; Gokce, Ali Onder; Gordon, R. G.; Graham, S. A.; Gray, W. C.; Greene, David Terrell; Greene, H. G.; Gringarten, Alain Charles; Guppy, K. H.; Gutmann, James Trafton; Haddad, A.; Hampton, Monty Allen; Harden, Deborah R.; Harris, Michael; Harris, N. B.; Hartley, P. D.; Hatfield, M. A.; Heffern, Edward L.; Henneberger, Roger; Hickman, C. J.; Hillhouse, J. W.; Hoexter, D. F.; Holcombe, Rodney John; Holzer, Thomas Lequear; Holzhausen, G. R.; Honea, E. G.; Hotson, Crispian John; Houle, Julie; Howland, Mark Douglas; Hsu, Kuan-Hsiung; Hudson, T. L.; Huffington, Terry L.; Huston, John; Ikoku, Chinyere Ukeagumo; Inderbitzen, Anton Louis; Ingersoll, Raymond Vail; Iverson, Richard Matthew; Jackson, Lionel E.; Jin, Doo Jung; John, David Allen; Johnson, Ansel Grieg; Johnson, M. T.; Jordan, T. E.; Jusbasche, Joachim Michael; Keefer, D. K.; Keene, D. G.; Keller, Chester K.; Keller, G.; Keskinen, M. J.; Kiremidjian, Anne Setian; Knott, S. A.; Knuepfer, P. L.; Kodama, K. P.; Korringa, Marjorie Kitchel Whallon; Koski, R. A.; Kosro, P. M.; Krishnan, N. Gopala; Kroft, D. J.; Krumhansl, James Lee; Kucuk, Fikri; Kulkarni, Srikant N.; Kuo, Ming-Ching T.; Lamon, Kathryn Ann; Landers, Thomas E.; Langer, Laura L.; Laraya, Rogelio Gotardo; Layman, E. B.; Le Clerk, R. V., II; Lemonnier, Thierry R. L.; Loayza, Oscar; Long, P. E.; Lu, Lee; Lufkin, John Laidley; Lundeen, Lloyd John; Mahrer, K. D.; Maillet, Jean; Mana, A. I.; Mannon, Leslie Susan; Mansfield, Charles Frederic; Mao, Ming-Ling; Mark, Robert Kent; Marks, N. S.; Marlow, Michael Stewart; Marsh, S. E.; Marshall, Claude Monte, Jr.; Marshall, S.; Maruo, Jiro; Marval, Francisco Rafael; Matti, J. C.; Mavko, G. M.; Mavor, Matthew John; May, Daniel J.; May, R. J.; McCardle, Michael F.; McCracken, Willard Alton; McHugh, S. L.; McLean, Dewey Max; McLeroy, Carol Ann Chmura; McMasters, Catherine R.; McMillan, D. K.; McTigue, D. F.; Miller, Thomas Patrick; Mills, Bradford Alan; Missallati, Amin A.; Mitchell, James F.; Mitchell, Martha Jeanne; Mitchell, Peter Ashley; Moore, D. E.; Moore, William Joseph; Morales C., Enrique; Morgan, David S.; Morgan, T. H.; Morrell, R. P.; Morris, Mark W.; Morrow, William Bruce; Mortgat, Christian Pierre; Moughamian, J. M.; Müller-Henneberg, Matthias; Murphy, R. J.; Mustart, David Alexander; Naney, M. T.; Ndombi, J. M.; Neill, William Marshall; Nelsen, K. P.; Newberry, R. J. J.; Newell, R. A.; Niay, Robert A.; Nichols, Kathryn Marion; Nicholson, T. J.; Niemi, Leslie Owen; Nieto-Antunez, Antonio; Nordstrom, D. K.; Numbere, D. T.; Nutt, C. J.; Oberste-Lehn, D.; Oberste-Lehn, Deane; Oberste-Lehn, Deane; Okandan, Ender; Okandan, Ender; Oliver, Adolph A., III; Ortiz Vertiz, Salvador; Osaimi, Aayed Eid; Ostby, M.; Pantin, Ronald C.; Parker, H. M.; Parker, T. S.; Parker, Timothy Scott; Payne, William Downes; Peabody, Carey Evans; Peng, Syh-Deng; Pereira, Helton; Pering, Katherine Lundstrom; Phillips, J. D.; Piaggio, Arthur Donald; Pike, Jane E. N.; Pikul, Mary Frances; Pineo, Charles C.; Pisciotto, Kenneth A.; Plafker, George; Platt, J. B.; Plongeron, A.; Poincloux, P. A.; Ponti, Daniel J.; Prelat, Alfredo Eduardo; Price, James Edward; Prisbrey, Keith A.; Querol-Sune, Francisco; Raghavan, Rajagopal; Randolph, John; Raynolds, R. G. H.; Reches, Z.; Recny, C. J.; Riley, D. C.; Robinson, Russell, Jr.; Rock, K, N.; Rodine, J. D.; Romero, E.; Rosenbaum, James; Ross, James J.; Ruetz, Joseph William; Rytuba, J. J.; Sahin, E.; Salas, Guillermo Armando; Salem, Bruce B.; Salvador, Phillip; Sanyal, Subir Kumar; Satman, Abdurrahman; Savage, Godfrey Hamilton; Sbeta, Ali M.; Schiffman, P.; Schlesinger, Benjamin; Schmidt, Jeanine Marie; Schnapp, Madeline; Schultz, P. S.; Schweickert, Richard Allan; Scofield, D. H.; Scorer, John D.; Selleck, William Lewis; Sengul, Mustafa; Shanahan, Edward; Shanahan, Peter; Shanley, F. E.; Shi, Chung-Shin; Shinohara, Kiyoshi; Sieh, K. E.; Simpson, R. W., Jr.; Sitar, N.; Sitar, N.; Slack, J. F.; Slagle, Letha P.; Smith, David Duane; Snoke, Arthur Wilmot; Snyder, W. S.; Soliman, Mohamed; Sonnevil, Ronald Alan; Soto, A. E.; Speer, Michael Carr; Spencer, J. W., Jr.; Spreiter, Terry Anne; Sprunt, E. C. S.; Squyres, John Benjamin; Steele, W. C.; Steeples, D. W.; Steiner, Jeffrey Carl; Stevenson, Andrew J.; Stewart, Richard John; Stierman, D. J.; Stratton, Alan Jerome; Strobel, Calvin Jerome; Stuckless, John Shearing; Suczek, C. A.; Sumner, John R.; Sutch, Patricia Leigh; Swanson, Samuel Edward; Talwani, Pradeep; Tan, D. Y.; Tariq, Syed Mohammad; Taylor, B.; Taylor, M.; Tekneci, Zeki; Tinsley, J. C.; Todd, Victoria Roy; Townsend, T. E.; Trautmann, Charles H.; Treiman, Allan Harvey; Tripathi, Vijay Shankar; Trummel, John E.; Trummel, John E.; Turnbull, R. W.; Turner, Barbara Lee; Ukaji, K.; Utine, Mehmet T.; Uzoigewe, Andrew Chukudebelu; Van Hecke, Michael Clement; Vanderpool, N. L.; Velho, Luis Rousset; Vennum, Walter Robert; Vikre, P. G.; Von Dohlen, Edward Lee; Vonder Linden, Karl; Wachter, Bruce George; Walters, Mark A.; Warner, Richard Dudley; Weinbrandt, Richard Mickey; Weiss, Richard B.; Weissenburger, Ken William; Wesson, Robert Laughlin; Whitney, James Arthur; Widhelm, Sally; Williams, John Wharton; Williams, Richard E., Jr.; Williams, Stephen Loring; Williamson, J. W.; Winkler, K. W.; Witte, Duncan M.; Wright, James L.; Wright, R. M.; Yadon, Douglas Mark; Yilmaz, O.; Yu, J. P.; Yuan, Georgia; Yuen, Dexter L.; Yun, S.; Zaghi, Nourollah; Zengeni, Teddy Godfrey; Zoback, M. D.; Zoback, M. L. C.

1980s: Abadie, Victor Hugo, III; Abbaszadeh-Dehghani, Maghsood; Aboim-Costa, Carlos Alfredo Ferreira; Abracosa, Ramon Panoy; Ahmed, Gulfaraz; Al-Khalifah, Abdul-Jaleel Abdullah; Al-Yahya, Kamal Mansour; Al-Yousef, Hasan Yousef; Alabert, François Georges; Albert, Nairn Randolph; Allain, Olivier; Ambrason, Ellen P.; Anagnos, Thalia; Araktingi, Udo Gaetan; Atwood, Dorothy Fisher; Avon, Lizanne; Azevedo, Joao Jose Rio Tinto de; Bachus, Robert Charles; Bahr, Jean Marie; Bahr, Jean Marie; Barrow, Kenneth Thomson; Barton, Colleen A.; Bartos, Paul Joseph; Bate, Matthew Adam; Beaupre, Gary Scott; Beery, John Arlington; Belitz, Kenneth; Benavides Alfaro, Jorge Daniel; Bendimerad, Mohamed Fouad; Benoit, Jean; Bent, James VanEtten; Bernstein, Lawrence Richard; Bettini, Cláudio; Bint, Anthony Neil; Bishop, Barbara Parks; Black, John Ernest; Black, Thomas Cummins; Bodden, Wilfred Rupert, III; Boden, David Rendall; Boehm, Mark Charles Francis; Boissonnade, Auguste Claude; Bond, Linda Darlene; Booth, Derek Blake; Borja, Ronaldo Israel; Bouett, Lawrence W.; Bourbie, Thierry; Brandriss, Mark Elliott; Brigham, Robert Hoover; Brush, Randal Moorman; Burns, Laurel E.; Bush, Robert Nelson; Butler, James Johnson, Jr.; Butler, James Johnson, Jr.; Butler, James Johnson, Jr.; Butler, Paula Jean; Buxton, Bruce Edward; Carlson, Christine; Carten, Richard Bell; Carter, Jack Bryan; Catchings, Rufus Douglas; Cederberg, Gail Anne; Chameau, Jean-Lou; Cherven, Victor Bruce; Chiang, Wei-Ling; Cho, Moonsup; Cindrich, Richard B.; Clark, Douglas H.; Clayton, Robert Webster; Cole, Jeffrey E.; Contant, Cheryl Katherine; Cooley, Scott Alfred; Corley, Daniel I.; Crawford, Michael Francis; Da Prat, Giovanni C.; Danskin, Wesley Robert; Davies, Peter Bowen; Davies, Tarin Smith; Davis, James Edward; de Jong, Bernardus Hermenigildus W. S.; Decker, John Evans, Jr.; DeHerrera, Milton Augusto; Delaney, Paul Theodore; Demitrack, Anne; Demitrack, Anne; Deruyck, Bruno Guy; Desbarats, Alexander Jean; Deutsch, Clayton Vernon; DeVilbiss Munoz, J. W.; Dibble, Walter Earl, Jr.; Dilles, John Hook; Diner, Yehuda A.; Dobson, David Charles; Dobson, Patrick Foley; Dobson, Patrick Foley; Donahoe, Rona Jean; Donato, Mary Margaret; Douglas, Ian Hedberg; Doyen, Philippe Marie; Dreiss, S. J.; Dyer, James Russell; Eaby, Jacqueline S.; Eastman, H. S.; Economides, Michael J.; El-Shazly, Aley El-Din Khaled; Elvidge, Christopher David; Engebretson, David C.; Erickson, Laurie Lynn; Erikson, Johan P.; Essaid, Hedeff Izzudeen; Essaid, Hedeff Izzudeen; Evans, Barbara; Fassihi, Mohammad Reza; Ferriz-Dominguez, Horacio Gerardo; Finke, Eberhard A. W.; Finno, Richard Joseph; Fletcher, Darby Ian; Fligelman, Haim; Fossum, Martin Peter; Fowler, Paul J.; Fowler, William Lane; Fox, Charles E.; Frei, Leah Shimonah; Freyberg, David Lewis; Fridrich, Christopher J.; Fridrich, Christopher John; Frost, Thomas Philip; Galton, Julie Hope; Gambill, David Thomas; Gan, Qigao; Gans, Phillip Bruce; Gans, Phillip Bruce; Gavigan, Catherine Louise; Geist, Eric; Gerard, Matthew G.; Gilardi, John R.; Gillespie, Blake W.; Gobran, Brian David; Goldstein, Bruce Leon; Gomez Hernandez, Jose Jaime; Gonzalez-Serrano, Alfonso; Goodman, David Karns; Gorelick, Steven Marc; Gortner, Catherine Willis; Granados, Eduardo; Green, Kenneth A.; Green, Robert Otis; Greensfelder, Roger Weir; Greenwald, Robert M.; Grier, Susan Patricia; Gruenenfelder, Jane B.; Grunder, Anita Lizzie; Guertin, Kateri Valerie; Guillemette, Renald Norman; Hadidi-Tamjed, Hassan; Hadj Hamou, Taric Aly; Hagstrum, Jonathan Tryon; Hale, Ira David; Hall, Nelson Timothy; Hamilton, Douglas Holmes; Han, De-hua; Hansen, Susan Sharp; Harbaugh, Dwight Warvelle; Harbert, William Perry; Harlan, William Stephen; Harris, N. B.; Harun, Happy; Harvey, Ronald William; Haven, Elizabeth Lorraine; Hayes, Julie Allison; Hayes, Timothy S.; Hayes, Timothy Scott; Heinzler, C. Thomas; Helenes-Escamilla, Javier; Helenes-Escamilla, Javier; Helmold, Kenneth Paul; Hess, Gordon Russell; Hess, Kathryn Marie; Hill, Lawson Bruce; Hinckley, Bern Schmehl; Hitzman, Murray Walter; Hochella, Michael Frederick, Jr.; Holbrook, W. Steven; Holcomb, Robin Terry; Honeyman, Bruce Donald; Hoshi, Kazuyoshi; Hreggvidsdottir,

Halldora; Hsieh, Hsii-Sheng; Hubbard, Richard Jon; Huedepohl, Anita; Husson, Didier Emmanuel; Ingebritsen, Steven Eric; Ingebritsen, Steven Eric; Isaacs, C. M.; Isaaks, Edward Harold; Iverson, Richard Matthew; Jacobs, Allan Samuel; Jensen, Clair Lynn; Jibson, Randall Wade; John, David Allen; Johns, Robert Anthony; Johnson, Clark Montgomery; Johnson, Clark Montgomery; Johnson, Diane Louise; Johnson, Stephen Edward; Johnston, Paul Roche; Jones, Terry Dean; Jong, Hsing-Lian; Jong, Hsing-Lian; Kanter, Lisa Ruth; Karish, Charles R.; Kark, Margaret Jeanne; Karl, Susan Margaret; Kasali, Gyimah; Kasper, David Conlin; Kawakatsu, Hitoshi; Keefer, Keith Douglas; Khemici, Omar; Kim, Won Hyung; Kiser, Nancy Louise; Kjartansson, E.; Knight, Rosemary Jane; Knott, Stephanie Ann; Koch, Franklyn Gordon; Kokinos, John Peter; Komor, Paul Stuart; Kostov, Clement; Kowalik, William Stephen; Kroeger, Glenn Charles; Kuespert, Jonathan Godard; Kulachol, Konthi; Lagoe, Martin Brooks; Lamarre, Michele; Larkin, Brett James; Lashkari-Irvani, Bahman; Laudati, Robert P.; Lawton, Timothy Frost; Layer, Paul William; Lefkoff, Lawrence Jeffrey; Levander, Alan R.; Levin, Stewart Arthur; Li, Yianping; Li, Zhiming; Lichtman, Grant S.; Lindh, A. G.; Lion, Leonard W.; Lippincott, Diane Kay; Little, Timothy Alden; Little, Timothy Alden; Lubetkin, Lester Kenneth Cantelow; Lucas-Clark, Joyce Emily; Lucking, John Chase; Lynn, H. B.; Lynn, W. S.; Macias-Chapa, Luis; Madrid, Raul John Jose; Magill, James Robert; Manning, Craig Edward; Marcou, John Andrew; Marion, Roy Clarence; Martel, Stephen Joseph; Martel, Stephen Joseph; Martinez, Paul Anthony; Martinez, Rodolfo Ignacio; Mastin, Larry Garver; Mastin, Larry Garver; Mathieson, Elizabeth Lincoln; Mavko, B. B.; McCann, Martin William, Jr.; McCarthy, Jill; McCloy, Cecelia; McCrory, Patricia Alison; McGeary, Susan Emily; McGrew, Allen J.; McGuire, Anne Vaughan; McGuire, Douglas Joseph; McKay, Wayne Irving; McKeown, David Alexander; Meeks, Yvonne Joyce; Meinert, Lawrence David; Mendoza, Jorge Segundo; Menell, Richard; Mensah-Dwumah, Francis Kwabena; Menzies, Anthony J.; Merritts, Dorothy J.; Metz, Jenny; Michael, Andrew Jay; Milam, Robert Wilson; Millage, Andra H.; Millage, Clayton Dodge; Miller, Eric G.; Miller, Greta E.; Miller, Mark A.; Miller, Mary Meghan; Miller, Nevin Lane; Miller-Hoare, Martha Lynn; Mishra, Srikanta; Miyazaki, Yoshinori; Moll-Stalcup, Elizabeth Jean; Moore, Thomas Edward; Moos, Daniel; Mora, Peter Ronald; Morley, Laurence Charles; Morris, Sandra Plumlee; Morritt, Robin Frederick Charles; Mortimer, Nicholas; Morton, Janet Lee; Mueller, Charles Scott; Murphy, Donald Currie; Murphy, William Francis, III; Newberry, Rainer Jerome Joachim; Niemi, Tina Marie; Noblett, J. B.; Novak, Steven William; Nuckolls, Helen Marie; O'Brien, William Jay; O'Brient, James David; O'Rourke, John T.; Okaya, David Akiharu; Olson, Hilary Clement; Olson, Steven Frederick; Omre, Karl Henning; Ortiz, Keith; Ortiz-Ramirez, Jaime; Osiecki, Richard Alan; Ottolini, Richard Albert; Pauschke, Joy Marie; Pedrosa, Oswaldo Antunes, Jr.; Pence, Jennifer Joan; Peterson, Eric Thomas; Pittinger, Lyndon Frank; Pohl, Demetrius Christmus; Poland, Joseph Fairfield; Ponader, Carl Wilson; Ponader, Heather Boek; Ponti, Daniel John; Pope, Kevin Odell; Pope, Kevin Odell; Potter, Jared Michael; Prescott, William Herbert; Ramos-Martinez, Luis; Rappeport, Melvyn Lewis; Reeder, John William; Reichard, Eric George; Reichard, Eric George; Riley, Thomas Andrew; Roberts, Dar Alexander; Rodgers, David Walter; Rodriguez de la Garza, Fernando Javier; Rojstaczer, Stuart Alan; Ronen, Joshua Mordechai; Rose, Nicholas Martin; Rossi, Mario Eduardo; Roth, Richard A.; Rothman, Daniel Harris; Rowles, Lisa Dianne; Rubin, Allan Mattathias; Rubinstein, Jacobo; Saller, Arthur Henry; Sander, Mark VanDyke; Sander, Mark VanDyke; Scheidle, Diana Lynn; Schmidt, Ehud Jeruham; Schmidt, Jeanine Marie; Schofield, Neil Aubrey; Schoof, Craig Crandall; Schult, Frederick Roy; Schweig, Eugene Sidney, III; Schweig, Eugene

Sidney, III; Scott, Norman; Sedlock, Richard Louis; Seeburger, Donald Alan; Seedorff, Charles Eric; Seedorff, Charles Eric; Segall, Paul; Semprini, Lewis; Shafii Rad, Nader; Shaver, Stephen Allen; Sheridan, Sally Wright; Sherr, Margot S.; Shigley, James Edwin; Sirey, Cordelia R.; Solow, Andrew R.; Sperber, Christine Martina; Spieth, Mary Ann; Srivastava, Rae Mohan; Stauber, D. A.; Stearns, Stanley W.; Stefani, Joseph Paul; Stein, Jeffrey Allen; Stein, Ross Simmon; Stern, Laura A.; Stone, Paul; Stonestrom, David Arthur; Sufi, Arshad Hussain; Sullivan, Jeffery Alan; Suro Perez, Vinicio; Swanger, Henry Jay; Sweeney, Brian Philip; Sword, Charles Hege, Jr.; Tarduno, John Anthony; Temeng, Kwaku Ofori; Tetzlaff, Daniel Matias; Thomas, Susan; Thorson, Jeffrey R.; Tiliouine, Boualem; Toldi, John L.; Tolson, Ralph Bradley; Tosaya, Carol Ann; Townsend, Timothy Elwood; Tripathi, Vijay Shankar; Turner, Robert John Whitlock; Turner, Robert John Whitlock; Turrin, Brent David; Tütüncü, Azra Nur; Ueng, Wen-Long Charles; Upp, Robert Rexford; Valocchi, Albert Joseph; Wagner, Brian Jeffrey; Wagner, Brian Jeffrey; Walbridge, Stephen Rorick; Walder, Joseph Scott; Walder, Joseph Scott; Wallmann, Peter Caswell; Walls, Joel Dan; Walter, John William; Walter, John William; Wang, Shi-Chen; Ward, Richard Brendan; Webb, Robert Howard; Weber, Linda; Weeks, John David; Weissenberger, Ken William; Wells, Liss Eleanor; Wharton, David Ian; Whelan, Hugh Thomas More; White, Christopher D.; White, Christopher David; Whiteford, William B.; Wiley, Thomas James; Williams, Richard Llewellyn; Wilson, Douglas Slade; Wolters, Bernd; Wust, Stephen Louis; Yale, David Paul; Yale, Leslie Berlincourt; Zamora Guerrero, David Hipolito; Zhang, Zhimeng; Zientek, Michael Leslie; Zucca, John Justin

State University of New York, College of Environmental Science and Forestry
Syracuse, NY 13210

1 Master's, 6 Doctoral

1970s: Skaller, P. M. G.

1980s: Caslick, James Frederick; David, Mark Barnett; Huggins, Andrew; Owe, Manfred; Schwengber, Janet Ruth; Zipperer, Wayne C.

Stephen F. Austin State University
Nacogdoches, TX 75962

71 Master's, 1 Doctoral

1970s: Bonora, P. F.; Irwin, J. E.

1980s: Abernethy, Robert Morris; Adamick, John Alton; Alford, Gary W.; Arfele, Anthony Thomas, Jr.; Barrett, Mary L.; Belk, Jerrel Keith; Bennett, Marvin Edward, III; Berger, Kent H.; Biehle, Alfred A.; Blount, Scott Brian; Boland, Gary D.; Brown, Robert Bruce; Byram, Kelly Gene; Callender, William Russell; Cason, Russell R.; Chadwick, Michael L.; Cheatham, Thomas L.; Corley, Robert Andrew; Davidson, Robert H.; Davis, Robin Alane; Dear, Timothy B.; Denny, James H., Jr.; Dewitt, Nancy; Dotsey, Pete; Dye, Jane Elizabeth; Ehrhart, Thomas F.; Eubanks, Darrell Lynn; Farnham, Jack D.; Fertitta, Jay C.; Finneran, Jane Beeman; Fuchs, James W.; Fuhr, Joseph M.; Garza, Abato John; Geffert, Michael A.; Gonzales, Benjamin Ray; Green, Mary L.; Greer, Catherine Bowman; Hagar, Richard Allen; Hayes, Joseph J.; Hitchcock, Kenneth Brent; Hultman, John R., Jr.; Jolly, Glenn Douglass; Kalbacher, Karl F.; Klein, Jack Jay, Jr.; Magouirk, Deborah A.; Mahbubullah, A. K. M.; Miller, Michael Carl; Mohn, Kenneth William; Mosconi, Louis S.; O'Neal, Jill Evans; O'Sullivan, William H., Jr.; Oliver, Ricky Dean; Parmley, William P.; Pate, Christopher Scott; Queen, Elizabeth Bolton; Reiner, Steven Roy; Rushing, Emmett O., III; Russell, Billy Joe, Jr.; Santamaria, Stephen V.; Schiltz, Debra Lynn; Slaughter, Thad A.; Smith, Charles David; Smith, Elizabeth J.; Sprague, Gloria Davis; Stella, Mark Phillip; Strange, William D., Jr.; Sutley, William Christopher; Toelle, Brian Ed-

ward; Wise, Karen Ann Winterhoff; Wylie, James Louis

Stevens Institute of Technology
Hoboken, NJ 07030

1 Master's

1980s: Chang, Ja-Shian

Sul Ross State University
Alpine, TX 79832

19 Master's

1980s: Bloom, M. A.; Calkins, Cary Peter Howard; Ebisch, James F.; Fallon, James H.; Gadou, Georges S.; McDonough, William F.; Mohammed, K.; Murley, William H.; O'Connor, M. J.; Owen, Dennis Ray; Reck, Donald F., Jr.; Reeves, Keith D.; Rudine, S. F.; Rudnick, Roberta L.; Scott, A. R.; Simpkins, Tim H.; Tellez, J. R.; Urbanczyk, Kevin M.; Wilcox, Robert

SUNY, College at Cortland
Cortland, NY 13045

1 Master's

1960s: Kirkland, James T.

SUNY, College at Fredonia
Fredonia, NY 14063

33 Master's, 1 Doctoral

1970s: Bolle, Doris J.; Bugliosi, Edward F.; Burrier, Dale; Buxton, Herbert T.; Clute, Peter R.; Gagliardi, James S.; Harriger, Ted; Jagoda, John L.; Keller, Stephen M.; Kronman, George; Lewis, Richard; Messinger, Donald J.; Nairn, James P.; Rhinehart, Julie; Walton, John C.; Wilson, Michael P.; Winter, Richard

1980s: Bennett, Bruce A.; Coulter, Lynne E.; Domagala, Mark A.; Johnson, Thomas M.; Jordan, Thomas E.; Lilga, Mary Colburn; Manings, Gordon C.; McCarthy, Brian P.; Metzger, Stacy Lynn; Meyer, Walter F.; Rollins, David D.; San Vicente, Napoleon Otero; Schumacher, Mark J.; Seyler, Beverly; Tomik, John C.; Weekes, Alan F.; Wysocki, Martin

SUNY, College at New Paltz
New Paltz, NY 12561

1 Master's, 1 Doctoral

1970s: Muller, Peter Dale

1980s: Ferng, Yue-Lang

SUNY, College at Oneonta
Oneonta, NY 13820

15 Master's

1970s: Byrne, James Richard; Coskey, Robert J.; DeMatties, Theodore A., Jr.; Gieschen, Paul Allen; Hornbeck, James M.; Melia, Michael B.; Palmer, Margaret V.; Risley, Lawrence J.; Witherbee, Kermit G.

1980s: Clikeman, Paul W.; Costolnick, David E.; Emhof, Stewart A.; Mozer, Robert J.; Stevens, Gordon M.; Walker, John

SUNY at Albany
Albany, NY 12222

57 Master's, 22 Doctoral

1960s: Lang, Dorothy M.

1970s: Bostwick, T. R.; Chisholm, Sallie W.; Gregg, William J.; Gregg, William J.; Grippi, Jack; Hoffman, Mark A.; Hoyt, William H.; Jacobi, Louise Delano; Karson, Jeffrey A.; Karson, Jeffrey A.; Lee, Victor J. B.; Malcolm, Frieda L.; Nelson, K. D.; Nisbet, Bruce W.; O'Connell, Suzanne; Peddada, Anantaramam; Scanlon, Kathryn M.; Sengor, Ali Mehmet Celal; Shibata, Tsugio; White, Carla A.; Yue, G. K.-L.

1980s: Ach, Jay A.; Aparisi, Michelle; Baldwin, Suzanne Louise; Baldwin, Suzanne Louise; Be, Kenneth; Blake, Robert Whitney; Bobyarchick, Andy Russell; Bosworth, William Paul; Bradley, Dwight C.; Bradley, Lauren Magin; Casey, John Francis; Cooper, J. Calvin; Cushing, Grant W.; Dyer, Julie M.; Gallo, David G.; Gultekin, Savci; Hall, Peter C.; Harris, Janet M.; Hempton, Mark Robert; Hoak, Thomas E.; Hofmann, Peter Michael; Idleman, Bruce D.; Idleman, Katrina A. J.; Jessell, Mark Walter; Jones, F. Ross; Klimetz, Michael P.; Kozinski, Jane; Kusky, Timothy M.; Livaccari, Richard F.; Loureiro, Daniel; Mahon, Keith I.; Mann, William Paul; Mihalich, John P.; Moody, Richard H., Jr.; Mora, Jorge; Mudambi, Anand Rajagopal; Ohr, Matthias; Park-Jones, Rosann; Pindell, James Lawrence; Roma-Hernandez, Mauricio; Rosencrantz, E. J.; Rowley, David B.; Rowley, David Ballantyne; Sengör, Ali Mehmet Celâl; Steinhardt, Christoph K.; Stella, Pamela J.; Stroup, Janet B.; Sullivan, Jerry W.; Swanson, Mark Thomas; Tanski, Stephen A.; Thiessen, Richard Leigh; Vollmer, Frederick W.; Washington, Paul A.; Will, Thomas Michael; Wolf, Douglas A.; Xia, Zong-Guo; Young, James R.

SUNY at Binghamton
Binghamton, NY 13901

130 Master's, 43 Doctoral

1960s: Caramanica, Frank P.; Childs, Philip; Flint, Jean Jacques; Harrison, John E.; Hekinian, Roger; Jindrich, Vladimir; Kaplan, Harvey I.; Kraemer, Philip G.; Patchen, Douglas G.; Yamanka, T.

1970s: Ash, D. W.; Basan, Paul B.; Blatter, C. L.; Bottjer, David J.; Cadwell, Donald Herbert; Campbell, R. C.; Cipar, John J.; Clark, M. E.; Cook, J. R.; Diemer, J. A.; Doolan, Barry Lee; Dunn, Martha Jean; Dupler, Philip C.; Farrar, S. S.; Fernalld, Thomas; Fessenden, R.; Flint, Jean-Jacques; Flynn, T.; Gerety, Kathleen M.; Ghosh, S. K.; Gilje, Stephen Arne; Ginsburg, Marilyn; Grinnell, Robert Stone, Jr.; Hinaman, Gary; Howard, James Jennings; Jindrich, Vladimir; Jolly, Wayne Travis; Katock, Robert; Kirkland, James Totten; Kowall, S. J.; La Flure, Ernest; Landon, Susan; Langille, Gerald Burton; Lawrence, D. P.; Leitzke, Peter Andrew; Leventhal, B. A.; Li, Pun-yuk Daniel; Liebes, Eric; Lu, C.-P.; Lu, Chih-Ping; Marble, J. P.; Mayer, Peter W.; Mendoza Sanchez, V.; Mendoza, Vicente; Moshier, S.; Nelson, Bruce R.; Neubeck, William Sidney; Newcomb, W. E.; Newton, Robert; Nielsen, Peter; Ott, Kyle R.; Palmer, L. M.; Parodi, Margaret; Pfann, H. D.; Podoff, Nedda; Polasek, John; Rideg, Peter; Ruzyla, Kenneth; Schwartz, D. P.; Simmons, Dale L.; Spevack, B. Z.; Stock, Carl; Thomas, David J.; Tompkins, K. E.; Tyler, Ronald Douglas; Visco, C.; Ward, J. A.; Wilkens, R. H.; Wu, Wen-Jen

1980s: Altman, Amy Bentley; Ammon, Charles James; Argast, Scott Frederick; Argast, Scott Frederick; Barbaro, Jeffrey Ralph; Bembia, Paul J.; Bennett, Sean Joseph; Bittner, M.; Bozza, A. W.; Braddock, J. A.; Brandon, David Earl; Brauer, D. F.; Browne, Kathleen M.; Campisi, Joseph S., Jr.; Caprio, R. C., Jr.; Casas, Enrique; Cassa, Mary Rose; Clark, Roger Alan; Collamer, Stephen Vaughn; Corbo, Salvatore; Coughlin, Robert Michael; Craft, James Homer; Curran, Donald Walter; Davis, Mary; Dawe, Steven E.; Dibb, Jack; Dibb, Jack Eaton, Jr.; Diemer, John Andrew; Dingman, Lorraine Elizabeth; Dominic, D. F.; Droser, Mary Louise; Ebert, James Roger; Ervin, E. M.; Fillo, E. J.; Fisher, Robert E.; Geller, Bruce Alan; Gillespie, Robert Howard, Jr.; Ginzler, S. L.; Gordon, Elizabeth Adams; Grebe, Steven Walter; Greene, Kimberly Richmond; Grigor, Lynne Jones; Gubitosa, Matthew; Halperin, Alan D.; Hauser, K.; Helm, Kenneth Richard; Huppert, Lawrence Norman; Hutchings, Lawrence John; Ioannidou, Eleni I.; Johnson, Eric Lee; Johnson, Kent Raymond; Jones, J. G.; Kossoff, Martin Jay; Lenney, Thomas William; Lifrieri, L.; Mangino, Stephen George; McCollum, L. B.; Michael, Peter Robert; Miller,

Keith Brady; Minero, Charles John; Murdoch, P. S.; Natishan, Joseph John; Nelson, John R.; Nickelsen, B. H.; Nyman, Matthew William; Obert, G. E.; Ozsvath, David Lynn; Pagano, Timothy Samuel; Penzo, Michael Anthony; Phelan, Kevin; Pickens, Kathleen L.; Pratt, Sandra; Prehoda, W. P.; Preisig, Joseph Richard Mark; Randazzo, Peter Joseph; Raybuck, M. S.; Rayome Goldblatt, Rosann E.; Rosenfeld, Joshua Henry; Scanlon, G. A.; Spitz, A. H.; Stancel, Steven George; Steinberg, D. J.; Stone, Timothy Storer; Tchombe, Laurence Puande; Terry, David Brian; Treadwell, Carol Jane; Turner, J. M.; Wang, Jeen-Hwa; Warzeski, Edward Robert; Whitlow, Sallie Ida; Wiles, Greg; Wilhelm, Philip Arthur, Jr.; Willis, Brian James; Wygant, G. T.

SUNY at Buffalo
Buffalo, NY 14222

202 Master's, 42 Doctoral

1930s: Cuthbert, Frederick L.; Wedow, Helmuth Jr.

1940s: Buehler, Edward J.; Dwornik, Edward J.; Fisher, Donald W.; Hoffman, Carlyle; Owens, James P.; Sargent, John D.; Tesmer, Irving Howard; Wolkodoff, Vladimir E.

1950s: Blackmon, Paul D.; Blackmon, Paul D.; Dwornik, Stephen E.; Dzimian, Raymond; Dzimian, Raymond; Jaffe, Gilbert; Kammer, Charles; Kammer, Charles; King, John S.; King, John S.; Kopf, Rudolph; Kopf, Rudolph; Poth, Charles Warner; Schnabel, Robert W.; Schnabel, Robert W.; Wiesnet, Donald R.

1960s: Abel, Vernon G.; Apmann, Robert Proctor; Bartolomucci, Henry A.; Beck, Lawrence D.; Blumreich, William, III; Bobola, John M.; Boehme, Richard William; Devon, John W.; Feder, Allen M.; Gahnoog, Abdillahi; Galas, F. Brian; Geitzenauer, Kurt R.; Grzybek, Paul Stanley; Ho, Diana Yunn; Izard, John Emmette; Janowsky, Ronald E.; Kaldor, Michael; Kirst, Paul William; Kirst, Paul William; Lumsden, David N.; Luther, Frank R.; Michalski, Paul J.; Muscallo, David; Nork, William Edward; Opera, John L.; Owens, James P.; Peterson, Robert W.; Rettke, Robert Clark; Rosenshein, Arthur N.; Schwimmer, David R.; Scott, Billy; Sweeney, John Francis, Jr.; Symecko, Ronald Edward; Taublieb, Edward J.; Walko, George R.; Wolfe, Robert Willard

1970s: Abbey, Dale A.; Barnes, John H.; Beck, William W., Jr.; Bendig, Daniel J.; Berg, S. A.; Biggi, Robert J.; Brennan, Sandra F.; Brett, C. E.; Busenberg, Eurybiades; Carmichael, Thomas J.; Cavanaugh, M. D.; Champion, Duane E.; Chevillon, Chas. Victor; Christ, Janice M.; Cottrell, Daniel J.; Datta, Pabindranath; Delisle, Georg; Dentan, Catherine M.; Dimitriadis, B. D.; Ellis, James M.; Fickett, Paul V.; Foster, Brayton P.; Foster, Paul W.; Freedenberg, H.; Friedman, Joan; Greiner, Gretchen G.; Harbin, Gary M.; Hoar, S. S. L.; Hoffman, S. A.; Johnson, R. C.; Jones, Richard Edwin; Jurdy, Donna M.; Kaplowitz, Phyllis S.; Karlo, J. F. M.; Kelley, L. M.; Kirchgessner, David Arthur; Kirk, Allan Robert; Klouda, G. A.; Koscielniak, Daniel E.; Leedy, Willard Page; Lemmon, Harry W.; Lenhardt, Duane R.; Mallick, Brian Charles; McMullen, Richard J., Jr.; Meisen, Daniel S.; Mercer, Mark F.; Muck, Karl L.; Murphy, James D.; Murphy, James W.; Mysore, R. K.; Nemcek, David Francis; Oaksford, Edward T.; Ostrye, T. F.; Palmer, Eugene C.; Pan, W. S.; Patterson, J. F.; Petraske, A. K.; Pollastro, R. M.; Poppendeck, Mark C.; Proctor, B. L.; Pryor, M.; Ring, M. J.; Scott, Graham Howard; Seiler, Robert C.; Snyder, Robert H.; Spear, D. B.; Spear, D. B.; Stoiber, G. A.; Subbarayudu, G. V.; Sweeney, John Francis, Jr.; Thompson, D. A.; Uchida, A.; Underwood, W. D.; Watson, William W.; Weir, G. M.; Wilson, Neil T.; Wolfe, R. W.; Wolkdoff, Vladimir E.; Womer, M. B.; Yoo, T.-S.

1980s: Al-Habash, Muyyed; Aloysius, David L.; Arnold, Eve Maureen; Awosika, Olakunle A.; Bar-

num, Harry P.; Bastedo, Jerold C.; Baumgras, Lynne M.; Becker, Dale A.; Borkland, Jay A.; Boyd, John; Brady, Brian K.; Bretches, John E.; Bruen, Michael P.; Burkett, Gerald R.; Burns, Phillip E.; Bush, Mark M.; Campbell, George E.; Chatterjee, Subir K.; Chen, Kuang-Hsiang; Costanzo, Patricia M.; Costanzo, Patricia Marie Vogt; Crook, Maurice C.; Dargush, Gary Franklyn; Datondji, Apollinaire; De Rito, Robert F.; De Vincenzo, Theresa E.; Del Signore, Alan G.; Diringer, Michael F.; Dudak, Richard M.; Dull-Coleman, Michele M.; Eckert, Raymond A., Jr.; Economou, Harris; Ellis, James Manning; Ellis, Lynn Doyle; Erb, Denise; Fahey, Timothy J.; Fargo, Thomas R.; Fortune, Kim M.; Frappa, Richard H.; Fromm, Kurt A.; Geier, Richard J.; Goldman, Daniel; Halter, Eric Francis; Haworth, Leah A.; Hendry, Robert D.; Herron, Michael Myrl; Herron, Susan Lynne; Hetherly, David Christopher; Hilfiker, Kenneth G.; Hsu, Tzu-Li; Jowett, Richard A.; Karim, Usama Farhan; Klein, Charles A.; Klima, Walter Francis, Jr.; Kloc, Gerald J.; Kommeth, Bryan M.; Koslosky, Robert A.; Kumbhojkar, Arvind Sadashiv; Lamb, Beth; Lin, Feng-Chih; Lorenz, Brenna E.; Lowell, Thomas Vinal; Maiga, Bokary S.; Marcus, David W.; Marsh, Leeda Elizabeth; Martin, Kathleen M.; May, Glenn M.; Mazierski, Paul J.; Molnar, Paul S.; Norris, Janice A.; Oleynek, Fred J.; Panus, Elizabeth Alice; Pérez, Libardo Aquiles; Peterson, Michael Paul; Petsrillo, Ira; Phillips, Robert Arthur; Przybyl, Bruce J.; Ranger, Julie Ann; Ritter, R. M.; Roy, Andre Gerald; Sen, Rajan; Shelefka, Michael A.; Sher, Mohammad Tahir; Simmons, Ardyth M.; Stanton, Michael; Stewart, Scott; Stoops, Sheryl Lynn; Studley, Kermit; Talkiewicz, Joseph M.; Tsai, Chong-Shien; Tsai, Meng-Chin; Uutala, Allen J.; Vassallo, Kathryn L.; Waddell-Sheets, Carol; Wasowski, Janusz J.; Wasowski, Janusz Josef; Wieczorek, William Frederick; Williams, Terry Lynn; Yaghmour, Farouk Abdul-Khaleg; Yousif, Nesreen Bashir; Zawacki, Stephen John; Zinter, Glenn G.

SUNY at Stony Brook
Stony Brook, NY 11794

40 Master's, 62 Doctoral

1960s: Gatehouse, Colin G.; Hallford, C. M.; Kavanaugh, James; Moore, W. S.

1970s: Arth, Joseph George, Jr.; Arth, Joseph George, Jr.; Ayuso, R. A.; Brande, S.; Cheng, Albert; Chytalo, Karen Nadia; Cook, J. G.; Dallmeyer, Ray David; Delano, J. W.; Eby, David Eugene; Fisher, Nicholas Seth; Floran, R. J.; George, R. P., Jr.; Halley, Robert Bruce; Hauser, E. E.; Heyse, J. V.; Hirschberg, David Jacob; Jayaraman, Ramurthy; King, H. E., Jr.; Levien, L.; Lohmann, K. C.; Malhotra, Ramesh; Mazzullo, L. J.; Mercier, J. C. C.; Miyazaki, J. M.; Muzyka, L. J.; Norris, T. L.; Queen, J. M.; Rajamani, V.; Reichlin, R. L.; Rice, S. B.; Richard, G. A.; Schaefer, P. J.; Schwimmer, David Richard; Sevian, Walter Andrew; Simmons, E. C.; Simmons, K. R.; Snellenburg, J. W.; Spiller, J.; Vaughan, F. R.; Vaughan, M. T.; Zaikowski, A.

1980s: Andersen, David John; Au, Andrew Yu-Chung; Banner, Jay Lawrence; Bass, Jay David; Bayri, Halis Muhtesem; Bender, John Francis; Bianchetti, Susan Fullam; Bolsover, Leslie Ruth; Burton, Benjamin Paul; Chu, Gordon Robert; Daniels, Lawrence David; Davidson, Paula Marie; Eby, Robert George; Ehrhart, Louis Edward, III; Epler, Nathan Andrew; Evans, Owen Cope; Fuhrman, Miriam Lea; Gasparik, Tibor; Gayes, Paul Thomas; Gillett, Stephen Lee; Kafka, A. L.; Kandelin, John Jacob; Karpin, Timothy Lee; King, David Edward; Ko, Jaidong; Kress, Victor Charles, II; Krogstad, Eirik Jens; Langmuir, Charles Herbert, II; Leighton, Carl Winslow; Liu, James Tsu-chien; Liu, Xing; Marintsch, Edward Joseph; Nabelek, Peter Igor; Novelli, Paul C.; Pucci, Amleto Arthur, Jr.; Roethel, Frank Joseph; Ross, Jeffrey Allen; Russ, Carol Alice; Sarwar, Golam; Schwartz, Kenneth Bruce; Seal, Thomas Lee; Shirey, Steven Bottome; Slater,

Jennifer Margaret; Smith, Francis deSales; Spencer, Khalil Joseph; Swanson, Donald Keith; Tanski, Joseph James; Terracciano, Stephen Alan; Um, Junho; Vickery, Ann Marie; Vocke, Robert Donald, Jr.; Walker, Richard John; Wechsler, Barry Andrew; Wolosz, Thomas Henry Matthew; Zeitlin, Michael J.; Zhang, Jiaxiang

Syracuse University
Syracuse, NY 13244

202 Master's, 85 Doctoral

1880s: Rice, H. J.

1890s: Schneider, P. F.

1900s: Clark, Burton W.; Proudy, William F.

1910s: Brainerd, A. E.; Camp, S. H.; Church, Earl; Collister, Morton C.; Conway, Ernest F.; Eldredge, Frank E.; Holcomb, Samuel; Holmes, Harvey N.; Jones, Daniel J.; Perry, Clinton W.; Turner, Homer G.

1920s: Bolton, Beatrice; Brainerd, William F.; Cabeen, Charles K.; Chisholm, David B.; Gwynne, Charles Sumner; Holmes, Chauncey DeP.; Huck, Florence; Ploger, Louis W.

1930s: Apsouri, Constantin N.; Currier, Louis W.; Goldrich, Samuel S.; Hershelman, William Lee; Hooker, Marjorie M.; Kaiser, Edward P.; Kaiser, Russell F.; Klepser, Harry John; Koch, Gustave H.; Miner, Neil Alden; Mozola, Andrew J.; Mulholland, Malcolm M.; Rowell, Eleanor M.; Saint Clair, Donald W.; Stewart, Glen William; Thibault, Newman William

1940s: Avenius, Rodney; Black, Robert F.; Digman, Ralph E.; Griswold, Russell E.; Hopkins, William H.; Huang, Wei Ta; Jones, James R.; Jones, Richard L.; Ketterer, Walter P.; Klemme, Daniel N.; LeMar, Harold K.; Maslowski, Edith; McNulty, Charles L., Jr.; Newell, John G.; Pike, Stewart J.; Scobey, Warren Barrett; Stone, Solon W.; Weart, Richard Claude; Weigle, James; Weiser, Jeanne

1950s: Ackerbloom, Donald R.; Allen, James M.; Allenson, Sherman; Anastasio, John; Apfel, John B.; Benz, Robert; Blagbrough, John; Blij, Harm Jan de; Boosman, Jaap W.; Brown, Richard L., Jr.; Buzzalini, Arnold; De Bruin, James H.; De Groff, Edward; Devaul, Robert W.; Difford, Winthrop Cecil; Dunkerley, Robert; Durham, Forrest; Eidie, Harold D., Jr.; Elston, Donald P.; Etson, Neil R.; Faltyn, Norbert E.; Fielding, Howard; Fix, Carolyn E.; Furbush, Malcolm; Gantnier, Robert; Gilg, Joseph G.; Glenn, Sidney; Harrington, Robert B.; Hasser, Edward G.; Higgins, Grove L., Jr.; Horowitz, Seymour; Howe, Jerry R.; Kaiser, Russell Florentine; La Grange, John; Langley, Edward J.; Leachtenauer, Jon C.; Leutze, Willard; Lichtler, William F.; Mack, Seymour; Mack, Seymour; Marks, Milton R.; Meaker, Harold N.; Mozola, Andrew J.; Multer, Harold Gray; Murphy, Allen Emerson; Myers, Paul E.; Newman, Walter S.; Novotny, Robert F.; Pees, Samuel T.; Phillips, John Stephen; Reid, John; Rezak, Richard; Ross, Stewart H.; Sachs, Peter L.; Sanders, Robert; Simmonds, Robert Tobin; Smith, Bennett Lawrence; Statkewicz, Edmund; Stevens, Samuel S.; Stewart, David Perry; Struthers, Parke H., Jr.; Tennissen, Anthony; Tesmer, Irving Howard; Van Tyne, Arthur M.; Werner, Harry Jay; Wetterhall, Walter S.; White, Sidney E.; Wolle, Peter; Woodmansee, Helen; Woodmansee, Walter

1960s: Agne, Russell Maynard; Babcock, Elkanah Andrew; Buchwald, Caryl Edward; Caggiano, Joseph A., Jr.; Cannon, William Francis; Clement, Sara J.; Coon, Richard F.; Craft, Jesse Leo, Jr.; Domenico, Patrick A.; Fernandez, Louis Anthony; Finley, Robert; Heyburn, Malcolm M.; Kaktins, Uldis; Kastner, Sidney Oscar; Krall, Donald Bowen; Lessing, Peter; Letteney, Cole DeWitt; Lewis, John Richard; Morgenstein, Maury; Muskatt, Herman S.; Nichols, Lee C.; Nichols, William David; Oberlander, Theodore Marvin; Parkinson, Robert W.; Piper, David Zink; Rhodes,

Richard L., Jr.; Sakkaf, Ali; Savage, William Z.; Snow, Phillip D.; Street, James S.; Street, James Stewart; Sutherland, Jeffrey Clark; Tomikel, John; Veinus, Julia; Vick, Alphonso Roscoe; Wallach, Joseph; Watson, Ralph Mayhew, Jr.

1970s: Barber, B. G.; Bartberger, C. E.; Bellotti, M. J.; Blanton, T. L., III; Bornemann, E.; Bourke, John Francis; Burroughs, William Alfred; Chambers, T. M.; Cohn, B. P.; Cohn, B. P.; Collyer, P. L.; Cucci, M. A.; Donahue, John Joseph, Jr.; Dugolinsky, B. K.; Effler, S. W.; Eggleston, Jane; Eppler, Duane T.; Foland, Richard L.; Forster, Stephen W.; Frank, Charles Otis; Froehlich, David J.; Gardner, Peter M.; Garshasb, Masoud; Genes, Andrew Nicholas; Geraghty, E. P.; Ghosh, S. K.; Gibbons, J. F.; Gilligan, Eileen Dombroski; Glennie, James Stanley; Guzowski, R. V.; Hanley, J. T.; Harrison, John Edward; Heath, D. E.; Heffner, T. A.; Jordan, R. J.; Keith, Brian D.; Kelley, G. C.; Koller, G. R.; Koller, G. R.; Krishnanath, Raghava; Kroll, Richard L.; Leetaru, H.; Levendosky, W. T.; Millendorf, S. A.; Miller, Jesse W.; Mulhern, K.; Newell, W. A.; Newman, William Alexander; Patchen, Douglas Gene; Plopper, Christopher Steven; Posamentier, H. W.; Proett, B. A.; Rhodes, Dallas D.; Ribeiro, Julio C.; Roberts, R. B.; Rubins, Charles Curtis; Salomon, N. L.; Serra, S.; Shannon, E. H. M.; Shaw, B. R.; Shilts, William Weimer; Singh, Vijay; Srivastava, G. H.; Tabesh, Elahe; Thomson, James Alan; Thrivikramaji, K. P.; Tillman, J. Edward; Trask, C. B.; Vehrs, T. I.; Vere, Victor Kurt; Willette, P. D.; Willette, P. D.; Wissig, George Conrad, Jr.; Woodward, C. W. D.; Woodward, Charles W. D.; Wright, Frank Myron, III; Yamamoto, S.; Yamamoto, S.

1980s: Abdulla, Khalifa Ahmed; Algeria, Steven Alan; Baker, Sherry Lynn; Banikowski, Jeffrey E.; Beinkafner, Katherine Jorgensen; Berndt, Marian Patricia; Browne, Bryant Alan; Burroughs, William Alfred; Cataldo, Robert Mario; Cheng, Chang-Chi; de Lacroix, Pierre; Dix, George Roger; Douglas, Debra Rena; Duchossois, George Earl; Foresti, Robert J.; Franz, Kristen Elizabeth; Franzi, David Alan; Gachowski, Christin Marianne; Gardulski, Anne Frances; Gilligan, Eileen Dombroski; Goodmen, William Walter; Gould, Gerald; Haddad, Marwan Najeh; Harth, Peter Marc; Ladzekpo, Doe Henry; Lawrence, Gregory Brad; Lowey, James M.; Maloney, William Vincent; Marvinney, Robert George; Metzger, Ellen Pletcher; Metzger, Ellen Pletcher; Oleck, Robert Francis, Jr.; Ridge, John Charles; Schafran, Gary Charles; Schwab, Joseph Patrick; Singh, Vijay; Sklenar, Walter Martin; Stehm, Mark; Szustakowski, Robert James; Taylor, Wanda Jean; Wang, Jason; Wilson, Michael Peter; Yarka, Paul James

Technical University of Nova Scotia
Halifax, NS B3J 2X4

3 Master's

1950s: Warren, K. R.

1970s: Atkinson, K. D.; Davis, J. D.

Temple University
Philadelphia, PA 19122

8 Master's

1960s: Flory, Richard A.

1970s: De Santis, J. E.; Duchaine, R. P.; Frischmann, Peter S.; Johnson, R. C.; Pazdersky, G. J.; Sutphen, C. F.; Tearpock, Daniel John

University of Tennessee, Knoxville
Knoxville, TN 37996

360 Master's, 78 Doctoral

1920s: Moneymaker, Beren C.

1930s: Ayres, Emma; Hyde, Victor Albert; Martin, G. C.; Turner, William Newton; Walls, James G.; West, Walter Scott

1940s: Allen, Arthur Thomas, Jr.; Bingham, Edgar; Burchfield, William W., Jr.; Cagle, Joseph W., Jr.; Harvey, Edward J.; Irwin, David; Jones, Reece A.; Lane, Charles F.; Lessig, Joseph Watson; Maher, Stuart W.; Moore, James H.; Parhial, Leimo I.; Poole, Joe Lester; Ricketts, James Edward; Rittgers, Fred Henry; Sawyer, Noah Gus, Jr.; Sniegocki, Richard Ted; Spangler, John Franklin; Swingle, George D.; Teng, Hai-Chuan

1950s: Atik, Ertugrul A.; Boyd, Gilbert H.; Brown, Charles K.; Bumgarner, James G.; Byerly, Don Wayne; Carlisle, Joseph T.; Cathey, Joseph B., Jr.; Causey, Marion E., Jr.; Cofield, William H.; Collins, Stephen E.; Cummings, David; Cundiff, Jerry Allen; Dail, John Hugh; Dail, Rhea A.; Davis, David Chandler; Elder, Ben Frank; Fagerstrom, John Alfred; Finlayson, Carroll P.; Gibbs, Gerald V.; Gilbert, Leonard C.; Goldhaber, Martin M.; Greene, Adrian Vance; Greene, Robert C.; Harding, James L.; Harris, Lawrence A.; Harvell, George R., Jr.; Hathaway, Donald Joseph; Hawkins, John O.; Heller, John Lowell; Hewitt, Phillip Cooper; Hill, William T.; Hoover, Karl Victor; Jackson, Paul; Kemp, Malcolm W.; Kemp, Peter Evans; Kemp, Thomas Earl; Leak, Robert E.; Lineberger, Ralph D.; Lomenick, Thomas F.; Lounsbury, Richard E.; Maclay, Robert Weaver; McCallum, Malcolm E.; McLaughlin, Robert Everett; McMaster, William M.; Mebane, R. Alan; Milici, Robert Calvin; Miller, Glen A.; Nelson, Arthur E.; Phillips, Harris Edwards; Rackley, Ruffin I.; Raymond, Richard H.; Shekarchi, Ebraham; Sherkarchi, Ebraham; Smith, Ollie L., Jr.; Steiner, Robert L.; Steuerwald, John B.; Stickney, Webster Fairbanks; Tiedemann, Herbert Allen; Tucker, Charlie Alexander, Jr.; Walter, Louis S.; White, Lloyd Arthur; Witzel, William Thomas

1960s: Aven, Russell Edward; Avignone, Joseph; Barlow, James Arthur; Bearce, Denny Neil; Benson, Lawrence I.; Biery, Jerry N.; Blythe, Ernest W., Jr.; Bolt, Larry R.; Byerly, Don Wayne; Coker, Alfred Eugene; Cox, Norman J.; Darrell, James H., Jr.; Emerson, Matthew S.; Feder, Gerald; Fetters, Robert Thomas, Jr.; Fields, Noland E., Jr.; Frank, William; Freeman, Timothy F.; French, Vernon Edwin; Gianotti, Frank B., III; Gibbs, Clare H.; Hagegeorge, Charles G.; Hajosy, Roger Alan; Haney, Donald Clay; Harper, Delbert D.; Hasson, Kenneth Owen; Hatcher, Robert Dean, Jr.; Havryluk, Ihor; Hayfield, George H.; Helton, Walter Lee; Henderson, Arnold R.; Hetrick, John; Hofstetter, Oscar Bernard, III; Hooker, Andrew M.; Hulme, James A.; Jeran, Paul William; Johnson, Robert C.; Jones, C. Keith; Jones, Clarke; Jones, Michael; Kasey, Arthur R., III; King, John S.; Kohland, William Francis; Krotzer, Chris J.; Little, Robert Lewis; Lomenick, Thomas Fletcher; Mann, Charles F.; Manus, Ronald W.; Marie, James R.; Martin, Ray G.; McConnell, Robert; McKinney, Thomas F.; McReynolds, J., Jr.; Merschat, Carl; Milici, Robert Calvin; Miller, John David; Miller, Robert C.; Moehl, William R.; Moore, James L.; Murray, Joseph Buford; Nadeau, Joseph; Nalewaik, Gerald Guy; Noonan, Albert F.; Oliver, Harold L.; Owen, Lawrence B.; Owens, Michael; Palmer, Raleigh A.; Parsons, Barbara Mae; Philley, John C.; Pickering, Samuel Marion; Porter, Samuel G.; Potosky, Robert; Privett, Donald Ray; Pruitt, Glenn N.; Pugh, Lewis E.; Quick, Kurt; Ratliff, Larry Eugene; Rife, David Leroy; Ripley, William F.; Robinson, Gene D., Jr.; Russell, Ernest Everett; Russell, Timothy Gray; Saltman, David; Saltzman, D.; Serim, Hakki Erdem; Shows, Thaddeus N.; Skinner, Roland B.; Smith, James Wiliam; Spigai, Joseph J.; Sykes, Charles Ronald; Tarkoy, Peter J.; Troensegaard, Kingdon W., II; Tung, Ping Ya James; Turner, Irving L.; Valachi, Laszlo Zoltan; Vest, William C.; Wiener, Leonard S.; Wilson, Robert Lake; Wolfe, James Alvis; Wood, Barry R.; Wood, John A.; Worsley, Thomas; Yarbrough, Ronald Edward; Zimmer, James A.

1970s: Ahler, Bruce Allen; Asher, Bruce Robert; Barrett, Harold Elliott, Jr.; Bartlett, Charles Samuel,

University of Texas, Arlington

Jr.; Bayer, Robert J.; Belvin, William Mark; Blythe, Ernest W., Jr.; Bogucki, Donald Joseph; Bohanan, Earl Roger, Jr.; Bowlin, Benjamin; Brera, A. M.; Brower, John Charles; Bryan, Benjamin K., Jr.; Burton, Jacqueline C.; Churnet, Habte Giorgis; Clark, Armin Lee; Claytor, Gale Catherine; Cobb, LaVerne Burkhart; Collins, Steven Lee; Combs, Douglas W.; Crosby, E. C.; Davis, Richard D.; DeGroodt, James H., Jr.; Dregne, James Michael; dWest, Joseph E.; Ebers, Michael L.; Escobar, Ricardo Reyes; Fallis, Susan Mary; Ferrigno, Kenneth F.; Fletcher, Clark S.; Franks, Christopher D.; Garrett, Theodore Watrous, Jr.; Gentry, Stephen Swift; Hasson, Kenneth Owen; Hersch, James Barry; Hill, William T.; Hopkins, Richard A.; Horton, Robert A., Jr.; Hsu, Vindell; Hu, Hsien-Neng; Hylbert, David Kent; Jacobs, Alan Korach; Johnson, Verner Carl; Kashfi, Mansour S.; Kennedy, James A.; Ketelle, Richard H.; Koch, Carl Allinger; Kyle, James Richard; LaFollette, Stephen G.; Larsen, Roland M.; LeRoux, Gay Breton, III; Livingston, John E., Jr.; Lowe, Nathan Ted; Lown, David J.; Maitland, Michael R.; Masuoka, Edward Jay; McWilliams, Edward B.; Metcalfe, Susan Judd; Minkin, Steven C.; Moore, Harry Leander, III; Moore, Nelson Kinzly; Nabelek, Peter Igor; Norman, Marc D.; Ossi, Edward John, III; Ott, D. W.; Pack, Donald David; Parker, William Charles; Patel, Kishore N.; Payne, William W., III; Penley, H. Michael; Philley, John Calvin; Poole, James Leroy; Price, Charles Errol; Ratliff, Larry Eugene; Rishel, John Curtis; Roberts, Thomas Adolph; Rogers, W. J.; Roper, Douglas C.; Ruppel, Stephen C.; Samman, Nabil Fahmi; Sandrock, George Stephen; Schneider, Raymond H.; Schrader, Edward J., Jr.; Shanmugam, Ganapathy; Sickafoose, Donald Kim; Siribhakdi, Kanchit; Sitterly, Preston; Slusarski, Mark Leo; Smith, Melvin Owen; Smith, Richard J.; Snyder, John Frank; Stanin, Frederick Theodore; Stephenson, John P.; Sutton, Thomas Culver; Thompson, Candace M.; Thurmer, G. Sidney; Tieman, David J.; Tisdale, Ronald Marion; Tung, H.-S.; Upham, Gregory A.; Via, William Noel; Wallace, James Ray; West, Joseph Edward; West, Joseph Edward; Whelan, Charlene J.; White, John Fullington; Wiethe, John David; Wilson, James Robert; Wilson, Steven M.; Yarbrough, Ronald Edward; Yust, William W.; Zuberi, Zaheer H.

1980s: Absher, Bobby Steven; Achaibar, Jaikisan; Agee, Jeffrey J.; Allen, Milton; Allen, William Turner; Allison, Michael C.; Asreen, Robert C., Jr.; Berg, James Donald; Bienkowski, Lisa Sophia; Borowski, Walter S.; Breland, Fritz Clayton, Jr.; Brewer, Roger Clay; Brewster, Charles L.; Brite, S. E. A.; Cain, Parham Mikell; Cantrell, Dave Lee; Capaccioli, Deborah Ann; Cathey, William B.; Clark, Stephen Rex; Coffey, Michael Lynn; Colson, Russell Owen; Conte, Jonathan A.; Conway, Charles Daniel; Crafts, Anne S.; Crattie, Thomas Bradford; Crider, Don; Cristil, Anita Ione; Cronin, Thomas Paul; Cudzill, Mary R.; Deetz, Stephan F.; Diehl, Wesley K.; Dobson, Mary Lynn; Dorsey, Anne E.; Duddy, Mark Morgan; Durazzo, Aldo; Easthouse, Kurt Allen; Ferguson, Tony Lee; Fischer, Mark W.; Fronabarger, Allen Kem; Fronabarger, Allen Kem; Garrison, Judy West; Ghazizadeh, Mahmood; Gibson, Michael Allen; Gilbert, Oscar Edward, Jr.; Glazzard, Charles F.; Goswami, Ram Kishore; Gratz, Jeffrey F.; Green, Thomas Kent; Hammer, Richard D.; Hammond, Patrick Allen; Harden, James Thomas; Harlow, George Edward, Jr.; Harris, Charles Steven; Heine, Christian J.; Helms, Thomas S.; Henrot, Jacqueline Francoise; Hoffman, Frank Owen, Jr.; Howze, Bryn David; Jackson, David Ernest; Jernigan, Dana Gregory; Johnson, Alan Roy; Johnson, Robert Eric; Kath, Randal Lee; King, Wendell Christopher; Kissling, Randall Douglas; Kittelson, Roger; Knapp, Steven A.; Kozar, Michael Glenn; Kung, Hsiang-Te; Larabee, Peter A.; Lawson, John Sheldon; Lewis, Jonathan C.; Lutz, Charles Talbott; Marks, Gregory Thomas; Masuoka, Penny McFarlan; Matlock, Joseph Franklin; Mayfield, Michael Wells; McClellan, Elizabeth A.; McComb,

Ronald; McDonald, David C.; McElhaney, Matthew Stuart; McGill, Mary Margaret; McGinn, Carl Wilson; McReynolds, Joseph A.; Mitchell, Michael M.; Monger, Hugh Curtis; Moss, Thomas Allen; Mulligan, Patrick John; Neff, Nancy E.; Norton, Willard Eugene; Novick, Jonathan S.; Nwadialo, Bernard-Shaw Emeje; Olsen, Barbara A.; Paul, J. Bryan; Poppelreiter, Barbara Savage; Pryce-Harvey, Jacqueline Simone; Reese, Stuart O.; Reid, Sarah R.; Robert, Lance Christian; Rose, Raymond R.; Royall, P. Daniel; Runyon, Gary A.; Scales, Anthony Scott; Schnoebelen, Douglas J.; Schoner, Amy Elizabeth; Schumann, Paul L.; Shafer, Daivd Scott; Sherrill, Timothy Wilson; Simmons, William Alexander; Simonson, John C. B.; Skelly, Raymond Lee; Sledz, James John; Sledz, Janine Gajda; Smith, Everett Newman; Sneyd, Deana S.; Steinberg, Roger T.; Stone, Charles David; Strange, Elizabeth Allison; Tarabzouni, Mohamed Ahmed; Thompson, James R.; Triegel, Elly Kirsten; Turnmire, J. B.; van Gelder, Susan M.; Vazin, Hassan; Walters, Randy R.; Weber, Lawrence James, Jr.; Wedekind, James E.; Weiss, Eric A.; Weitz, John H., Jr.; Wilkins, Gary; Wise, Thomas W.; Witherspoon, William Dale; Wrightson, Walter, Jr.

University of Texas, Arlington
Arlington, TX   76019

145 Master's, 1 Doctoral

1970s: Abedin, K. Z.; Bacon, Randall W.; Black, Paula Jo; Blauser, William H.; Botros, Effat S.; Bowers, Roger Lee; Box, Michael R.; Brezina, James Lewis; Buehrle, Paul Michael; Champlin, Maurice Anthony; Corley, J. B.; Dawson, Mary L.; Dawson, William Craig; Fee, David Wayne; Files, Nelson; Foley, William James; Fowler, Todd A.; Harrington, Charles Eston, Jr.; Hart, Alan W.; Hatley, Michael D.; Hendrickson, Walter John; Hively, Roger E.; Hughston, Mark D.; Hursky, M. J., Jr.; Ice, Robert G.; Johnson, Ronald O.; Jumper, Robert S.; Kotila, Nancy Lee; Kovschak, Anthony Andrew, Jr.; Laali, Hooman; Lindsey, Douglas Dewitt; Litke, Gene Richard; Lovick, G. P.; Magee, Robert Wright; Maluf, Fred W.; Marcus, Donald L.; McGibbon, Douglas H.; McMullin, W. Dennis; Middleton, Kenneth Douglas; Mitchell, Gary C.; Mockbee, Gael A.; Moor, Ann Lynnette; Murlin, Jack Ronald; Nelson, Ralph L.; Ramsey, Jacqueline M.; Richardson, Jimmy David; Root, Stephen Allen; Ross, Mark A.; Sandlin, Gary L.; Sheu, D. D.; Skiles, David Glenn; Smith, Michael William; Snyder, J. M.; Sweezy, John L.; Taylor, Ricky Joe; Umphress, A. M.; Vaughan, Michael J.; Watson, William Gorom; Webster, Robert E.; Wells, David Rolfe; Wiedmann, Sebastian Paul; Wilson, George Newton, Jr.; Yang, Shyue-rong; Zahedi, Jafar

1980s: Adams, Craig W.; Balsley, Steven Devry; Baltensperger, Paul; Barranco, Frank Thomas, Jr.; Barrett, Elizabeth E.; Beeson, David L.; Bergan, Gail Renae; Blair, Terence C.; Bowles, Kelly L.; Broome, James Richard; Carr, Mary Margaret; Cheek, Catherine A.; Cox, Martin L.; Crabaugh, Jeff Patrick; Cree, Susan Bentley; Darwin, Robert Louis; Deaton, B. C.; Dellinger, Philip B.; Demases, Tamrara; Duncker, Katherine Elizabeth; Estes, Larry D.; Evans, W. Scot; Faul, Cydney L.; Ford, John W.; Galbiati, Larry Dale; Godfrey, Warren C.; Hansen, Diana Kay Thomas; Hathaway, Wayne L.; Hensleigh, Diane E.; Holland, Donna J. Little; Hoskins, Benjamin Wayne; Hough, Alan N.; Hudson, Richard M.; Humphreys, Curtis H.; Indest, Daniel J.; Kircher, Dorcas Elizabeth; Kitz, Mary Beth; Knight, Michael T.; Kuentz, David C.; Lowry-Chaplin, Barbara L.; Mann, Keith Olin; Martin, Ben Stephen; McFarland, Veronica T.; Mechler, Lina S.; Meyer, Kevin S.; Miller, Bruce Calvin; Miller, Duane Jay; Moravec, David Morgan, James Cyrus, III; Neybert, Daniel Steven; Ottensman, Vicki Vieroski; Papenguth, Hans William; Parker, James M.; Piñero, Joanne Louise; Purcell, David Richard James; Raabe, Bruce A.; Ray, Bradley Stephen; Reynolds, Stephen Kempster;

Roberts, Larry E.; Robinson, William Conrad; Romanak, Martin; Ronalder, Nina Lynn Walker; Roush, Tod Wayne; Sachs, Scott Donald; Schafer, Daniel B.; Shapiro, Bruce E.; Shelley, Geoffrey K.; Sheu, Nien-Jen Wang; Steele, David R.; Stockton, Marjorie Moore; Talbot, James Paul; Trevino, Ramon H., III; Turbeville, Bruce N.; van der Loop-Avery, Mary Louise; vonGonten, Glenn; Waite, Lowell E.; Waresback, Damon B.; Wilson, Mark A.; Winslow, Michael L.; Wyszynski, Joseph; Younger, Michael Alan; Zemboski, Steven S.

University of Texas, Austin
Austin, TX   78712

1128 Master's, 358 Doctoral

1890s: Hill, Benjamin Felix

1900s: Whitten, Harriet Virginia

1910s: Kniker, Hedwig Thusnelda

1920s: Arick, Millard Boston; Barrow, Leonidas Theodore; Brill, Virgil August; Buford, Selwyn Oliver; Cannon, Robert Lee; Christner, James Blaine; Cuyler, Robert Hamilton; Damon, Henry Gordon; Deen, Arthur Harwood; Green, Guy Emmett; Hancock, William Tarrant, Jr.; King, John Joseph; McCarter, William Blair; Milton, William Billingslea, Jr.; Richey, Oleta May; Ries, Minette Lillian; Tyson, Alfred Knox; Winkler, Hans

1930s: Allen, Stanley Randolph; Archer, Katherine; Bivens, Wilmer E., Jr.; Bramlette, William Allen; Breedlove, Robert Leeroy; Broughton, Martin Napoleon; Canaan, Morris; Cartwright, Weldon Emerson; Cole, Charles Taylor; Conway, Edward Spurgeon; Cook, Carroll Edwin; Cooper, Robert Peyton; Cox, William Edgerton; Cronin, Kenneth Stewart; Cumley, Russell Walters; Cuyler, Robert Hamilton; Dalton, Mary Chalk; Davis, Flavy Eugene; Durham, Charles Albert; Eifler, Gus Kearney, Jr.; Fletcher, Claude Osborne; Fouts, John Martin, Jr.; Frazell, William Davis; Frost, Jay Miles, III; Gardner, Frank Johnson; Hancock, James Martin; Hatfield, Arlo Clark; Hornberger, Joseph, Jr.; Horne, Stewart Walsh; Ikins, William Clyde; Konz, Leo Wilford; McCallum, Henry DeRosset; McClung, Esther Carroll; McCollum, Audrey Britton; McFarland, LaRue Buzan; McGowan, Francis Herbert; McNutt, Gordon Russell; Meadows, James Lawson; Moorhead, Johnny Bob; Nickell, Clarence Oliver; Parker, Travis Jay; Patton, Jacob Luther; Pilcher, Benjamin Luther, Jr.; Reedy, Milton Frank, Jr.; Sandifer, Donald Ford; Sargent, Elwood Cather; Seals, Wilburn Hale; Shelby, Thomas Hall, Jr.; Sparenberg, George Russell; Stafford, Gerald Maner; Stiles, Aden Edmund; Teagle, John; Wendler, Arno Paul; Wendler, Arno Paul; Wheeler, Joseph Bowen; Whitney, Marion Isabelle; Whitney, Marion Isabelle; Wilson, Forest Wayne; Woods, Raymond Douglas; Yates, Harvey Emmons

1940s: Anderson, Irvin J.; Banks, Luis Maria; Barrow, Thomas Davies; Bloodworth, Billy Lloyd; Bloomer, Richard Rodier; Brown, Jack Ralston; Bryant, James Elwood; Carter, Robert Daniel; Clabaugh, Stephen Edmund; Cockrum, Amil Blake, Jr.; Cocovinis, Dimitri Basil; Craddock, William Percival; Crawford, Gayle Posey; Culbertson, Thomas Milton; Darling, R. M.; de Mohrenschildt, George Sergius; DeLancey, Charles, Jr.; Dixon, Louis Helprein; Dodson, Edward Auld; Ellsworth, Ralph Irving; Fox, Hewitt Bates; Fuller, Warren Philips; Gardner, Edgar Jackson; Gardner, Frank Johnson; Gee, David Easton; George, Clement Enos, III; Gipson, William Earl; Gould, John David; Graves, Roy William; Guess, Roy Hayes, Jr.; Halbouty, James Jubron; Harvard, Charles Gentry; Hays, Norbert Alan; Head, Thomas Franklin; Headington, Clare Wesley; Hendricks, Charles Leo; Henry, John Francis; Holland, Daniel Edward; Howard, Jesse James; Hughes, Richard John, Jr.; Hunter, Jack; Ikins, William Clyde; Izgi, Mehlika Fahri; Jackson, James Roy, Jr.; Jacobson, Jule Marion; Jager, Eric Howard; Kennedy, Edward Reynolds, Jr.; Keyser, Joseph Edward; Kirk, Bruce

Glenn; Koenig, Joseph Baldwin; Kucukcetin, Adnan Mehmet; Langford, Eldon Woodrow; Langford, Othell Franklin; Lewis, Jean; Lieb, Carl Varney; Lind, C. M.; Lyth, Ambrose Lee; Major, Millard Holland; Marquez, Gustavo Enrique; Mathis, Robert Warren; McCammon, John Henry, Jr.; McCampbell, William Gibson, Jr.; McCracken, Weaver H., Jr.; McFarlan, Edward, Jr.; McKinlay, Ralph Harold; Means, John Albert; Means, John Brittian, Jr.; Moon, Charles Gardley; Musselman, George Abraham; Nicholson, John Hirston; Nobles, Melvin A.; Outlaw, Donald Elmer; Patterson, Archibald Balfour, III; Patton, E. C.; Payne, Billie Rex; Peterson, Hazel Agnes; Petty, John Kirkpatrick; Petty, Van Alvin, Jr.; Pflucker, Eduardo Cabieses; Plummer, Roger Sherman, Jr.; Pruitt, Earl Joseph, Jr.; Redfield, Robert Crim; Remick, David Brear; Richardson, Raymond Moseley; Rodgers, Jack Pinknea; Schmidt, Timothy Germer; Scholl, Milton Richard, Jr.; Sebring, Louie, Jr.; Sheldon, William Knowles; Skelly, Lawrence; Slingluff, Frank Peter; Smith, Joe Earl; Souaya, Fernand Joseph; Spindler, William M., III; Stapp, Wilford Lee; Starnes, Jasper Leon; Stern, Thomas Whital; Tanaka, Harry Harumi; Tariki, Abdulla Homoud; Throop, Robert Neblett; Timm, Bert Clifford; Treybig, Lucille Evelyn; Waddell, Richard Kent; Wadsworth, Albert Hodges, Jr.; Walton, Wahnes; Ward, J. Harold Edgar; Weaver, Oscar David, Jr.; Webb, Sam Nail; Weeks, Albert William; Wiley, Samuel Rogers; Wilie, Enid Evelyn; Williams, Harry Franklin; Wilson, James Lee; Zapp, Alfred Dexter

1950s: Adams, George Baxter; Adams, Gordon Edward; Adams, James Bethel, Jr.; Alexander, Robert Harwood; Allday, Edwin; Allen, Martin; Allen, Robert; Amsbury, David Leonard; Anderson, Arthur Edward; Arrington, Robert Newton; Ashworth, Edwin Thomas; Atchison, Dick Eric; Baker, Gus Bowman; Balke, Bennie Kuno; Bannahan, Annabelle Richardson; Bapuji, Soli Jehangir; Barnhill, William Burrough; Bauchman, James Bell; Bay, Thomas A., Jr.; Bennett, Richard Edwin; Bilbrey, Don Gene; Bills, Terry Vance, Jr.; Bix, Cecil Charles; Blackwell, Thomas Sanford; Blankenship, Asa Lee, Jr.; Blankenship, William Dave; Blatt, Harvey; Bogardus, Egbert Hall; Bookout, John Frank, Jr.; Booth, Charles Clinton; Bostwick, Douglas Leland; Boyle, Walter Victor; Braithwaite, Philip; Brand, John Paul; Bridges, Luther Wadsworth, II; Bridges, William Elmer; Brogdon, Dewey Robert; Bronaugh, Richmond Lee; Brown, Noel King, Jr.; Brown, Thomas Edwards; Brown, Wilton J.; Brundrett, Jesse Lee; Brunson, Wallace Edward; Bullard, Fredda Jean; Buongiorno, Benny; Byrd, William Martin; Callender, Dean Lynn; Campbell, Richard Allan; Carleton, Alfred Townes, Jr.; Carlisle, Joel Christie; Carter, Lee Steven; Cartwright, Jack Cleveland; Cassell, Dwight Eugene; Chapman, Ruthven Hoyt; Chatham, Ernest Walter, Jr.; Chin, Wai Suey; Clark, Joseph Clyde; Clutterbuck, Donald Booth; Cochrum, Arthur Leroy; Cocke, Robert Robinson, III; Colton, Clark Roper; Colton, Earl Glenn, Jr.; Cooley, Beaumont Brewer, Jr.; Cotera, Augustus S., Jr.; Craig, Dexter Hildreth; Cross, J. H.; Crutcher, Thomas Dent; Dasch, Ernest Julius, Jr.; Daugherty, Franklin Wallace; Davis, Morgan Jefferson, Jr.; DeCook, Kenneth James; Dehlinger, Martin Emery; Denson, John Lane, III; Dietrich, John William; Dincel, Mehmet Bedi; Dixon, William Ronald; Doyle, Robert Emmett, Jr.; Drake, Dennis Adolph; Duchin, Ralph Charles; Dzilsky, Thomas Edward; Early, Alberta Jenne Kunz; Echols, Betty Joan; Edson, Dwight James, Jr.; Elliott, Arthur Beverly, Jr.; Ely, Lael Marguerite; Ferguson, John David; Ferguson, Walter Keene Linscott; Ferrell, Alton Durane; Finley, Judge Dinsmore; Fitzpatrick, Jack Cleo; Frantzen, Danie Ray; Freeman, Worth Merle; Fulgham, Henry Leroy; Fuqua, Frank Jones; Gaines, Robert Byron, Jr.; Gatlin, Leroy; Giannone, Ralph John; Giddens, Leslie Wylie, Jr.; Gillerman, Elliot; Gimbrede, Louis de Agramonte; Girard, Roselle Margaret; Gonulden, Parisa; Gordon, James

Eddie; Grant, Richard Evans; Gray, Donald McLeod; Green, Thomas Edgar, Jr.; Green, Willard Russell; Greenfield, Leslie Lohr; Gurel, Mehmet; Haeggni, Walter Tiffany; Halamicek, William Arnold, Jr.; Hall, Ward Lee; Harpster, Robert Eugene; Harrington, David Haymond; Harris, John Richard; Harrison, Hubert James; Hartwig, Albert Ernest, Jr.; Harwell, George Mathis, Jr.; Hay-Roe, Hugh; Hay-Roe, Hugh; Hayes, James Frederick; Hewitt, Edward Ringwood, II; Hightower, Maxwell Lee; Hixon, Sumner Best; Holasek, Raymond Joseph; Holt, Charles Lee Roy, Jr.; Howell, Richard Shelby, Jr.; Hughston, Edward Wallace; Humble, Emmett Arl; Hutchinson, Robert Maskiell; Hutchison, J. L., Jr.; James, O. L.; Jansen, Gerhard Cyril Julius; Janszen, Milton Hugo; Jaroska, Robert Stanley; Jenkins, Evan Cramer; Jenkins, William Adrian; Johnston, Carl Hewitt; Jones, Hal Joseph; Justen, John Joseph; Kelly, John L.; Kent, Leon Alfred; King, Victor LeRoy, Jr.; Knabe, Robert George; Komie, Earl Esar; Krause, Erwin Koerps; Kurie, Andrew Edmunds; Lampert, Leon Max; Lee, Herbert Louis, Jr.; Lehkemper, Leonard James; Leve, Gilbert Warren; Levin, Max; Levin, Samuel; Lohse, Edgar Alan; Major, Rufus Orville; Mankin, Charles John; Mankin, Charles John; Mann, Hugh Thomas; Marks, Edward; Marr, Ronald James; Mason, Curtis Calvin; Mayo, Robert Truitt; McAnulty, William Noel; McCandless, Garrett Clair, Jr.; McCarthy, Jeremiah Francis; McFall, Clinton Carew; McGee, Edward Franklin; McGrew, Bill Judson; McIntire, William Leigh; McKinney, Robert Geers; McReynolds, J. Carroll; Mear, Charles Eugene; Miller, Daniel Newton, Jr.; Miller, John Collins; Miller, Wayne Davis; Milner, Charles Porter; Moon, Charles Gardley; Moore, Clyde Herbert, Jr.; Moran, Sidney Stuart; Morris, Charles Brady; Motsch, Aaron Sherrill; Moyer, Grant Luke; Murrah, William Eugene; Newberry, William Bohning; Nichols, John Conner; Nienaber, James H.; Nogues, DeWitt Collier; Noyes, Alvin Peter, Jr.; O'Brien, Bob Randolph; Oden, Josh Winters; Owen, Donald Eugene; Parker, Travis Jay; Peterson, James Eugene; Pettigrew, Robert William; Pickens, William Robert III; Porter, Charles Earnest; Porter, Robert Bowden; Quinn, James Harrison; Rehkemper, Leonard James; Renaud, Charles Benham; Reynolds, William Francis; Richardson, Everett Ellsworth; Ridley, Wade Clark; Ripple, Alfred Louis; Rix, Cecil Charles; Rix, Cecil Charles; Rizvi, Saiyed Mohammed Naseer; Roberson, Herman Ellis; Robert, Jean-Paul; Robertson, Roland Secrest; Rose, Peter Robert; Rothschild, Donald Isador; Roux, Wilfred Francois, Jr.; Rowe, Andrew Jackson; Russell, Jimmie Norton; Rutledge, Floyd Wayne; Sahanhaya, Sait; Schneider, Thomas; Schnurr, Paul Eugene; Schulenberg, John Theodore; Sealy, George, Jr.; Selim, Mohammed Abdel-Moniem; Sewell, Charles Robertson; Shambaugh, John Scott; Sharp, William Wheeler; Sims, Samuel John; Sirrine, George Keith; Slocum, Gilbert; Smith, Harry Lee; Smith, Joseph Thurston, Jr.; Smitherman, Eugene Alston; Sneed, Edmund David; Snider, John Luther; Spice, John Overstreet; Staplin, Frank Lyons; Stead, Frederick Lee; Stein, Walter William, Jr.; Stengl, Gerald Edward; Stevens, James Crosby; Stinson, William Dank; Strong, Walter Morrill; Swadley, W. C.; Taylor, Dennis Ritch; Thames, Clement Beal, Jr.; Tipton, William Everett; Todd, Thomas Waterman; Travis, Everett Joyce; Trice, Edwin Leslie, Jr.; Tunnell, Felix Maxwell; Twining, John Theodore; Twiss, Page Charles; Tydlaska, LeRoy Jerome; Underwood, James Ross, Jr.; Vest, Harry Arthur; Wade, Don Earl; Walker, Joe Dudgeon, Jr.; Walls, Billy; Walper, Jack Louis; Walper, Jack Louis; Walter, Joseph Charles, Jr.; Ward, Daniel Lee; Ward, William Cruse; Watson, Fred Somervill, Jr.; Webernick, Nelson Ellsworth; Weston, Ray Franklin; Wheeler, Joseph Orby; White, J. C.; Wightman, Robert Bradford; Wills, Bill Frank; Wilson, Duncan Campbell Ogden, Jr.; Wilson, John Ewing; Wilson, Wilbur Dean; Wimberley, C. Stanley; Winston, Donald, II; Winter, Claud Victor; Wollman, Constance Elizabeth; Woodward, John Eylar;

Woodward, Thomas Canby; Woodyard, Kenneth Eugene; Woollett, LeRoy Andrew; Wright, John Buel; Zabriskie, Walter Edward; Zaman, Mohammad Qamar; Zimmerman, James Blaisdell; Zinn, Robert Leonard

1960s: Abbott, Patrick Leon; Akersten, William Andrew; Al-Khersan, Hashim Fadil; Alt, David Dolton; Anan, Fayez Shaban; Anderson, Jay Earl, Jr.; Anderson, Jay Earl, Jr.; Anderson, John Jerome; Anderson, Thomas Howard; Anderson, Thomas Howard; Andrews, Peter Bruce; Ardila, Luis Ernesto; Asbury, Larry Marshall; Atwill, Edward Robert, IV; Avadisian, Antoine Mehran; Balkwill, Hugh Robert; Ballard, William Wayne; Bell, James John; Bence, Alfred Edward; Bhatrakarn, Tanakarn; Bingler, Edward Charles; Bishop, Bobby Arnold; Bjorklund, Thomas Keith; Bloomquist, Marvin Gaines; Blount, Donald Neal; Boyd, Alston; Boyer, Bruce W.; Bradshaw, Lael Marguerite Ely; Bridge, Thomas E.; Bridges, Luther Wadsworth, II; Brown, Thomas Edwards; Bryant, Vaughn Motley, Jr.; Burkart, Burke; Burmester, Russell Frederick; Burnett, Harold Morris; Burnitt, Seth Charles; Calder, John Archer; Campbell, Donald Harvey; Carew, James Leslie; Carrasco-Velazquez, Baldomero; Chen, Pei-Yuan; Clabaugh, Patricia Sutton; Clanton, Uel S., Jr.; Clanton, Uel S., Jr.; Clemons, Russell Edward; Cook, Lawrence Paul; Cooper, John Doyne; Cordoba-Mendez, Diego Arturo; Cotera, Augustus S., Jr.; Craig, William Warren; Crawley, Richard Alvin; Daugherty, Franklin Wallace; Davis, George H.; Davis, James Harrison; Davis, Richard Albert, Jr.; Davis, William Edwin, Jr.; Defandorf, May; Dekker, Frederik Ernst; DeLong, Stephen Edwin; Denison, Rodger Espy; Desai, Chandrakant S.; Dickerson, Eddie Joe; Dietrich, John William; Dill, George Meyer; Dobkins, James E., Jr.; Doney, Hugh Holt; Dunaway, William Edmond; Dunn, David Evan; Ebanks, Gerald Keith; Esmail, Omar Jubran; Evans, James Parham, III; Everett, Ardell Gordon; Everett, John Raymond; Fakundiny, Robert Harry; Frank, Ruben Milton; Freeman, Paul Swift; Freeman, Thomas Jewell, Jr.; Frishman, Steven Arthur; Fuentes, Ruderico Procopio; Galloway, William Edmond; Garner, Norman Earl; Gates, Cameron Herschel; Gayle, Henry Boyes; Gieger, Ronald Maney; Girijavallabhan, Chiyyarath V.; Goforth, Tommy Tucker; Gonzalez, Jose Grover Percy; Groshong, Richard Hughes, Jr.; Gross, Robert Olvin; Gumert, William Richard; Haeggni, Walter Tiffany; Hamilton, Samuel Clinton; Hamman, Henry Royden; Hammond, Weldon Woolf, Jr.; Harris, John Michael; Harris, William Howard; Harvill, Martin Lavell; Hayes, Miles Oren; Heiken, Grant Harvey; Hempkins, William Brent; Holden, William Robert; Hoover, Richard Alan; Hopkins, Edgar Member; Horn, David Russell; Hoskin, Charles Morris; Houser, John Foster; Iranpanah, Assad; Jackson, James Robert; Jones, Darrell King; Kessler, L. Gifford, 2nd; Kimberly, John Eli; King, Elbert Aubrey, Jr.; King, Harvey Dennis; Kudo, Akira; Kuich, Nicholas Franklin; Lattimore, Robert Kehoe; Laux, John Peter, III; Lindemann, William Lee; Lindholm, Roy Charles; Longacre, Susan Ann Burton; Longgood, Theodore Edward, Jr.; Lytton, Robert Leonard; Macedo-Raa, Albino Reynaldo; Martin, Kenneth Glenn; Mayfield, Jack Hastings, Jr.; McAnulty, William Noel, Jr.; McGehee, Richard Vernon; McGowen, Joseph Hobbs; McKinney, W. N., Jr.; McKnight, John Forrest; McKnight, John Forrest; McLamore, Roy Travis; McQueen, Jereld Edward; Merrill, Glen Kenton; Messina, Mario Leo; Meyer, Joachim Dietrich; Milton, Arnold Powell; Molnar, Ralph E.; Moore, Clyde Herbert, Jr.; Morelock, Jack; Myers, Ralph Lawrence, II; Namy, Jerome Nicholas; Newton, Robert Stirling; O'Sullivan, William Joseph; Orr, Harold D.; Orr, Robert William; Oweis, Issa Sebeitan; Parker, John William; Parks, Peter; Patton, Thomas Hudson, Jr.; Patton, Thomas Hudson, Jr.; Pearson, Frederick Joseph, Jr.; Pearson, Frederick Joseph, Jr.; Podio-Lucioni, Augusto; Porsch, Herman W., Jr.; Powell, James Daniel; Purushothaman, Krishnier;

Ragsdale, James Allan; Ramsey, John William, Jr.; Reid, William McCormick; Richey, Charles Irwin; Rightmire, Craig Turner; Ritchie, Alexander Webb; Robison, Richard Ashby; Rodgers, Benjamin Kirby; Rogers, Charles William; Rogers, James Edwin; Rogers, Lowell Thompson; Rogers, Margaret Anne Christie; Rose, Peter Robert; Rowley, Peter DeWitt; Russell, Richard Verner; Schake, Wayne Eugene; Schlaudt, Charles McCammon; Schwarzbach, Theodore Jeremiah; Seewald, Clyde Ray; Semken, Holmes Alfred, Jr.; Shafiq, Moayad Abdulla; Shelby, Cader Alverd; Shull, Roger Don; Sims, William Eldon; Sipperly, David William; Smith, Robert Earl; Spencer, Alexander Burke; Spiegelberg, Frederick, III; St. John, Billy Eugene; St. John, Billy Eugene; Stevens, James Bowie; Stitt, James Harry; Stitt, James Harry; Strain, William Samuel; Sultan, Ghazi Hashim; Terriere, Robert Theodore; Thomas, George Ligon; Tikrity, Sammi Sherif; Tucker, Delos Raymond; Underwood, James Ross, Jr.; Urbanec, Don Alan; Valencia, Mark John; Waitt, Mary Cooper; Walker, George Pinckney, III; Walton, Anthony Warrick; Warner, Ralph Hartwin; Watkins, Joel Smith, Jr.; Watson, Richard L.; Weber, Gerald Eric; Wershaw, Robert Lawrence; White, John Weldon; White, Rex Harding, Jr.; Wilbert, William Pope; Wiley, Michael Alan; Wilson, William Feathergail; Winston, Donald, II; Winter, Johannes Antonius Franciscus; Wise, James Charlton; Wolleben, James Anthony; Wood, John William; Workman, Charles Edwin; Workman, William Edward; Yager, Milan King; Yeager, John Conner; Youash, Younathan Yousif; Youash, Younathan Yousif; Youash, Younathan Yousif; Young, Leonard Maurice; Zachry, Doy Lawrence, Jr.; Zamora, Lucas Guillermo

1970s: Abbott, Patrick Leon; Achalabhuti, Charan; Agagu, Olusegun K.; Al-Hinai, Khalifa; Albert, Donald G.; Amdurer, Michael; Anderson, William B.; Anderton, Arlo Jo Payne; Anepohl, Jane K.; Argenal, R.; Baker, Robert Allison, III; Barrett, Michael E.; Barton, Gerald Stanley; Bastug, Mustafa Cengiz; Becker, Bruce D.; Beckham, Elizabeth; Begle, Elsie A.; Belcher, Robert C.; Belforte, Alberto Santiago; Berg, Edgar Lowndes; Berumen, Manuel, Jr.; Birsa, David S.; Bishop, Gale A.; Bloxsom, Walter Eden; Bockoven, Neil T.; Boehl, J. E.; Boggs, Ann S.; Bommer, P. M.; Boone, John L.; Bosch, Silverio C.; Boyce, Robert L.; Bozanich, Richard G.; Brewton, Joseph Lawrence; Broadhead, T. W.; Browning, Lawrence A.; Bumgardner, John E.; Burt, Edward Ramsey, III; Busbey, Arthur B., III; Butterworth, Ronald Arthur; Byrne, James R.; Cadwgan, Richard Morgan; Caffey, Kyle C.; Campbell, Archibald R., III; Carew, James Leslie; Caskey, Deborah Jane; Caughey, Charles A.; Cepeda, Joseph C.; Chafetz, Henry Simon; Chan, King Nam; Chiquito, Freddy Jesus; Clark, Thomas Phillips; Cleaves, Arthur W., II; Cleaves, Arthur Wordsworth, II; Colchin, Michael P.; Comer, John Bennett; Connors, Harry E., III; Cooper, John Doyne; Cornish, Frank G.; Crawley, Richard Alvin; Cuellar-Chavez, R.; Daghlian, C. P.; Davis, Louis Lloyd, Jr.; DeLong, Stephen Edwin; Dorfman, M. H.; Doyle, James D.; Dresser, Anita E.; Droddy, M. Jackson, Jr.; Dupré, William Roark; Ece, Omer Isik; Edwards, John Emerson; Eisenbraun, Paul H.; Elder, Ruth L.; Elliott, Thomas L.; Erxleben, Albert Walter; Etter, Stephen D.; Evans, Daniel S.; Everett, John Raymond; Evertson, D. W.; Fakundiny, Robert Harry; Felsher, Murray; Ferrusquia-V., Ismael; Finch, Richard Carrington; Fiore, Richard N.; Fish, Johnnie Edward; Fok, Henry W.; Fortier, J. Daniel; Fredrikson, Goran; Fredrikson, Goran; French, Lawrence Nelson; Fritz, Deborah M.; Funk, Alan C.; Galloway, William Edmond; Gallup, Marc R.; Garcia-Solorzano, Roberto; Garner, L. Edwin; Garrison, James R., Jr.; Gearing, P. J.; Goetz, Lisa K.; Gomez-Masso, A. J.; Gorski, Daniel Everett; Goter, Edwin Robert, Jr.; Govin, Charles T., Jr.; Graham, Russell W.; Greenberg, Redge L.; Gries, John Charles; Gries, Ruth Roberta Rice; Grimshaw, Thomas W.; Grimshaw, Thomas Walter; Groat, Charles George; Gucwa, Paul Ramon; Gucwa, Paul Ramon; Guendel-Umana, Federico D.; Guevara-Sanchez, Edgar Humberto; Guevara-Sanchez, Edgar Humberto; Gunn, Vincent C.; Guo, H. Y.; Gustafson, Eric P.; Hall, William Douglas; Harwood, Peggy J.; Haulenbeek, Roderick Beazley; Haynes, Cynthia L.; Hedges, J.; Heil, Richard John; Henry, Christopher Duval; Henry, Christopher Duval; Hinote, Russell E.; Hodges, Floyd N.; Holland, Walter Fox, Jr.; Hoyle, Blythe L.; Hulbert, Richard Charles, Jr.; Hulke, Steven D.; Hunter, William C.; Hyun, I.; Isokrari, O. F.; Jeng, W.-L.; Jogi, P. N.; Johnson, Bruce D.; Johnston, John E.; Jolly, Wayne Travis; Jordan, Michael Andrew; Jordan, Michael Andrew; Jungyusuk, Nikom; Katz, Steven G.; Keizer, Richard Paul; Keller, Peter Charles; Keller, Peter Charles; Kerr, Ralph S.; Kier, Robert Spencer; Kirschner, Carolyn Elisabeth; Kleist, John Raymond; Kolvoord, R. W.; Kreitler, Charles W.; Kreitler, Charles W.; Laird, Charles Elbert, Jr.; Landers, Ronald Alfred; Laudon, Robert C.; Lawson, Douglas A.; Lehman, David H.; Lentz, Robert C.; Leonard, Raymond C.; Levich, Robert A.; Levy, Susan S.; Lewis, Paul S.; Lichaa, Pierre Michel; Lindquist, Sandra J.; Long, John M.; Longman, Mark W.; Loocke, Jack E.; Looney, R. Michael; Loucks, Robert E.; Luttrell, Pamela E.; Luzardo, Manuel A.; Lyons, James I., Jr.; Manley, Walker D.; Martell, Hildebrando Jose; Martinez-Garcia, Enrique; Matos, Jose Francisco; McBryde, J. C.; McCarley, Lon Allen; McCulley, Bryan L.; McCulloh, Richard P.; McMahon, David A., Jr.; Mead, James Glen; Megaw, Peter Kenneth McNeill; Mejia, Daisy; Merrill, Robert D.; Merrill, Robert David; Milliken, Kitty Lou; Mochizuki, S.; Moran, Robert E.; Morrow, David Watts; Morrow, David Watts; Morton, John Phillip; Moseley, Marianne G.; Mueller, Harry W., III; Munson, Michael G.; Murphy, T. Dennis; Mutis-Duplat, Emilio; Nemeth, Kenneth E.; Netto, A. Sergio; Newcomb, John Hartnell; Nordquist, Ronald W.; Ogley, David S.; Oliver, William Benjamin, IV; Olivier, Jacques M.; Orchard, David Merle; Ossian, Clair Russell; Parker, Donnie F., Jr.; Parker, Donnie F., Jr.; Parrish, Walter C.; Pattarozzi, Michael; Patton, Peter C.; Pedone, Vicki A.; Person, C. P.; Petering, George Wilfred; Peterson, Shirley J.; Phongprayoon, Pongsak; Piazza, Idelso Antonio; Pigott, John D.; Plamondon, Michael P.; Poth, Stephen; Quinlan, James F.; Raney, Jay A.; Reaser, Donald R.; Reeve, Scott Cleveland; Reid, Jeffrey Clinton; Reid, William McCormick; Richmann, Debra L.; Ricoy, Jose Ulises; Rios, R. A.; Ritchie, Alexander Webb; Roden, Michael F.; Roepke, Harlan Hugh; Rogers, William Brokaw; Rudolph, Kurt W.; Rutland, Carolyn; Salem, Mohamed Rafik Ibrahim; Salyapongse, Sirot; Sanchez, B. V.; Schiebout, Judith Ann; Schiebout, Judith Ann; Schwarz, Mary E. Bowers; Seagle, S. M.; Sears, R. Bonner; Seekatz, Jeffrey G.; Seni, Steven John; Seo, Jung Hoon; Sepassi, Bahman; Sever, Julia Rebecca; Shankar, Nilakantan Jothi; Shaw, Stephen Lynn; Shepherd, Russell G.; Shih, Tai-Chang; Sholes, Mark A.; Siegmann, James M.; Sivaborvorn, Vichai; Skolasky, Robert A.; Smith, Gary E.; Smith, Michael A.; Smith, Richard M.; Solis-Iriarte, Raul Fernando; Sorenson, Raymond P.; St. Clair, Ann Elizabeth; Stanton, George D.; Strange, Nettie S.; Sullins, Charles Jefferson; Summer, Rebecca M.; Swanson, Eric Rice; Swanson, Eric Rice; Testarmata, Margaret M.; Thompson, M. Gary; Tipple, Gregory L.; Tondu, R. Joe.; Trask, Charles Brian; Turner, Neil Lee; Tyner, Grace Nell; Valastro, Salvatore, Jr.; Valente, Jose T.; Van Allen, Bruce R.; Waddell, Richard K., Jr.; Waechter, Noel B.; Wahl, David E., Jr.; Waisley, Sandra L.; Waitt, Richard Brown, Jr.; Walton, Anthony Warrick; Walz, David Henry; Warning, Karl R.; Watson, Richard L.; Weise, Bonnie Renee; Wenzel, Roger A.; Wiley, Michael Alan; Wilkinson, Bruce Harvey; Williamson, Charles R.; Williamson, Turner Franklin; Winker, Charles David; Wood, Raymond A.; Woodman, James T.; Woodruff, Charles Marsh, Jr.; Worrall, Dan M.; Zuniga Izaguirre, M. A.

1980s: Acurero Salas, Luis Armando; Adamek, Scott Harper; Adilman, Daivd; Agra, Jefferson de Mello; Aguirre-Diaz, Gerardo de Jesus; Allie, Adrienne Dee; Alsop, Janice L.; Ambrose, William Anthony; Anderson, James Howard; Anderson, Richard Garland; Anderson, Richard Kent; Angstadt, David Moris; Atzmon, Gil; Ayers, Walter Barton, Jr.; Babalola, Olufemi O.; Badachhape, Abhaya Ramachandra; Bailey, Jonas William; Ballinger, Philip; Barcelo-Duarte, Jaime; Barratt, John C.; Barrie, Charles Prescott; Barron, Barbara Rae; Barron, Terry Jay; Bartlett, Peter McIntyre; Bath, William W.; Bay, Annell Russell; Bebout, Gray Edward; Beike, Dieter; Bentley, Michael Emmons; Berge, Timothy Bryan Swearingen; Bernal, Juan Bautista; Berryhill, Alan Walter; Bertagne, Allen John; Birmingham, Scott Daniel; Black, Curtis Wendell; Blanchard, Paul Edward; Blum, Michael David; Bobeck, Patricia Ann; Bockoven, Frances Dart; Bockoven, Neil Thomas; Bodner, Daniel Paul; Boeker, Ralph; Bollinger, Becky; Bond, Steven Craig; Bowland, Christopher Lee; Boyd, Felicia Michelle; Boyles, Joseph Michael; Bracken, Bryan Reed; Bradford, Cynthia A.; Bramson, Emil; Braschayko, Thomas; Bristol, David Arthur, Jr.; Broderick, Gregory Philip; Budd, David A.; Bunker, Russell Craig; Burbach, George VanNess; Burbach, George VanNess; Burks, Rachel Jane; Burks, Rachel Jane; Cagle, Clinton D.; Calaway, Edward Lee; Camargo, Jorge M. T.; Capo, Rosemary Clare; Caran, Samuel Christopher; Carballo, Jose Domingo, Jr.; Carlson, Steven Michael; Carr, David L.; Carter, Karen Eileen; Casey, James Michael; Cast, Martha E.; Castagna, John Patrick; Cather, Steven Martin; Cather, Steven Martin; Catto, Antonio J.; Caughey, Michael Eugene; Cazier, Edward Coin, III; Cervantes, Michael Arthur; Chandler, Mark Arnold; Chang, Jui-Yuan; Chapin, Robert Ira; Chapin, Thomas Scott; Chapman, Jeannette Burgen; Chen, Huei-Tsyr Jeremy; Cheng, Minkang; Chieruzzi, Gianni Oswaldo; Childs, Constance Smythe; Chornesky, Elizabeth Ann; Chuchla, Richard Julian; Chun, Insik; Clark, John Thaddeus; Cobb, Robert Charles; Coffman, Paul Eugene, Jr.; Coley, Katharine Lancaster; Collins, Ann M.; Coltrin, Donald George, Jr.; Conlon, Sean Thomas; Connally, Thomas Chambless, Jr.; Connolly, John Patrick; Conover, William V.; Conti, Robert D.; Copeland, William Barton; Cornelius, Reinold R.; Corrigan, Jeffrey Delon; Corso, William P.; Cowan, Kenneth Lee; Coxe, Cynthia Louise; Craig, Lisa Ellen; Cumela, Stephen Paul; Curchin, John Montgomery; Curry, David James; Curtis, Rene Virginia; Daniel, David Edwin, Jr.; Dauzacker, Modesto Victor; Davidsen, Erik Kennedy; Davies, Kyle Linton; Davis, Joseph Redmon; Davis, Scott Daniel; de Figueiredo, Antonio Manuel Ferreira; de Souza, Jairo Marcondes; de Zoeten, Ruurdjan; DeCamp, Dodd Werner; DeMis, William Dermot; Dempsey, John; Dennis, Norman Dale, Jr.; Devine, Paul Ellis; Dingus, William Frederick; Dobbs, Steven Lawrence; Donnelly, Andrew Charles Alexander; Dreier, RaNaye Beth; Duex, Timothy William; Dumitru, Trevor Alan; Dunbar, John Andrew, Jr.; Duncan, Edward A.; Duncan, Mary Anne; Dutton, Alan Robert; Dutton, Shirley Peterson; Ebeniro, Joseph Onukansi; Ebeniro, Joseph Onukansi; El Jard, Mustapha R.; Elder, Susan Rachel; Elliott, Laura Ann; Ellis, Patricia Mench; Emmet, Peter Anthony; Entzeroth, Lee Catherine; Erdlac, Richard John, Jr.; Eschner, Terence Brent; Evans, Carol Anne; Faecke, David Charles; Fagin, Stuart William; Farr, Mark Randall; Farrand, Richard Brownlow; Farrelly, John James; Farrens, Christine M.; Faust, Michael Jess; Figueiredo, Antonio M.; Finn, Christopher Jude; Fisher, Robert Stephen; Flanigan, T. Edward, III; Flores, Victor; Flores-Espinoza, Emilio; Fly, Sterling Harper, III; Fogg, Graham Edwin, Jr.; Fortier, Alfred Joseph, III; Fox, Michael; Frazier, Melvin; Fredericks, Paul Edward, Jr.; Gabay, Steven Howard; Gahagan, Lisa Marie; Garcia Delgado, Victor; Gaskell, Barbara Ann; Gates, Bruce Cameron; Gee, Carole Terry; Germiat, Steven John; Ghazi, Samir Abd-el-Rahman; Giltner, John Patrick; Goggin, David Jon;

Gold, Paul B.; Gorham, Scott Brady; Graber, Ellen Ruth; Granata, George Edward; Grant, David Edward; Gray, Gary George; Greenburg, Joseph Gary; Greene, Jeremy Theodore; Gregory, James L.; Groh, Douglas; Gu, Hongren; Guimarães, Paulo de Tarso Martins; Gutierrez, Gay Nell; Guzman-Speziale, Marco; Haldorsen, Helge Hove; Hall, Michael Scott; Hallam, Susan Lee; Hamlin, Herbert Scott; Hammond, Weldon Woolf, Jr.; Han, Jong Hwan; Hardwick, James Fredrick, Jr.; Harris, Susan Frye; Harris, Therese; Harwood, Roderick James; Havholm, Karen Gene; Helper, Mark Alan; Hensarling, Larry Reid, Jr.; Herber, Jon Philip; Herrington, Karen Laverne; Herwig, Jonathan Charles; Heubeck, Christoph Egbert; Hiebert, Franz Kunkel; Hinnov, Linda A.; Hoar, Richard James; Hoel, Holly D.; Honda, Hiromi Rigakushi; Hong, Chong-Huey; Houle, Julie A.; Houston, Betty Green; Hovorka, Susan Davis; Howe, Roger; Huerta, Raul; Huffington, Terry Lynn; Hummel, Gary Alan; Hurry, Debra Jean; Huston, Daniel Cliff; Ichara, Mark Josiah; Ide, Susan; Immitt, James Peter; Ingram, Gregory D.; Jacobs, James Alan; Jeng, Shian-Woei; Johansen, Steven John; Johns, David Ainslie; Johns, Ronald Alan; Johnson, Larry C.; Jones, Jon Rex, Jr.; Kabir, Muhammad Ismat; Kariyawasam, Hettigamage Cyril; Kastning, Ernst H., Jr.; Kauschinger, Joseph Lewis; Kautz, Steven Arthur; Kempter, Kirt Anton; Kim, Woo Han; Knight, Julia Baret; Kochel, Robert Craig; Kolb, Richard Alan; Kraft, Jennifer Lucille; Kugler, Ralph Leonard; LaFave, John Irwin; Laguros, George Andrew; Lall, Upmanu; Lanan, Holly Kay; Larkin, Randall George; Lawton, Jeffery L.; Layman, Thomas Bruce; Leary, David Austin; Leason, Jonathan Oren; Lee, Tung-Yi; Lehman, Thomas Mark; Lehman, Thomas Mark; Leininger, Roland L.; Lewis, Ronald Dale; Lin, Eugene Ching-Tsao; Lin, Meei-Ling Teresa; Lin, Tung-Hung Thomas; Logan, William Stevenson, IV; Long, James Howard; Lopez, Cynthia M.; Lundegard, Paul David; Luneau, Barbara Ann; Machenberg, Marcie Debra; Mack, Lawrence Edward; Macko, Stephen Alexander; MacPherson, Gwendolyn Lee; Mahler, Julianne Phyllis; Mann, Steven D.; Mashburn, Leslie Edwin; Masterson, Wilmer Dallam, IV; Mather, Gordon Scott; Mayes, Catherine Lynn; McCartney, Merle G.; McCrary, Megan Marie; McDermott, Robert W.; McDermott, Robert Wayne; McDonald, Susan; McDowell, Kenneth Otto; McElroy, John; McGookey, Douglas A.; McGraw, Maryann Margaret McDonough; McIntyre, John Francis, III; McLaren, James Peter; McMahon, Peter B.; McMurry, Jude B.; McNeish, Jerry A.; McNulty, Edmund Gregory; Meador, Karen Jean; Melius, Douglas James; Meneses-Rocha, Javier de Jesus; Merritt, Linda Carol; Millberry, Kimberlee Whitney; Miller, James K.; Milliken, Kitty Lou; Miser, Donald Evans; Mitchell, Jeffrey Todd; Mok, Young Jin; Moor, Amanda; Morton, John Phillip; Moustafa, Adel Ramadan; Murray, Robert Cozzens; Musgrove, Lee Ann; Naiman, Ellen Rose; Nam-Koong, Wan; Nance, Hardie Seay, III; Nazarian, Soheil; Nelis, Mary Karen; Nelson, Katherine Helen; Nelson, Robert H., Jr.; Nepomuceno, Francisco Filho; Neuberger, Daniel John; Newman, Jerry Savrda; Ni, Sheng-Huoo; Nielson, Jamie Adler; Noe, David Charles; Northam, Mark Alexander; Oakes, Chandler A.; Ohkuma, Hiroshi; Okada, Airton Hiroshi; Okoye, David Mobike; Onstott, Gregory Erle; Orr, Cynthia Dolores; Orr, Elizabeth Decker; Padilla y Sanchez, Ricardo Jose; Paige, Richard E.; Palmer, David Paul; Palmer, Jeffrey John; Parsley, Matthew Jay; Patterson, Joseph E.; Payne, Janie Hopkins; Payne, John Beckwith; Peterson, Christine Mary; Pew, Elliott; Pfeiffer, Deborah Susan; Phair, Ronald Leslie; Pisasale, Eugene T.; Pol, James Campalans; Pollman, Keith S.; Prezbindowski, Dennis Robert; Price, Vicky Irene; Prieto Cedraro, Rodulfo; Pursell, Victoria Jane; Pyle, Phillip F.; Rainwater, Kenneth Alvis; Ramage, Joseph Robert; Ramirez Serafinoff, Rafael Esteban de la Cruz; Rangel, Hamilton Duncan; Reck, Brian Harrison; Remondi, Benjamin William; Renkin, Miriam L.; Reynolds, Shawn Arvin; Richter, Bernd Chris-

tian; Rivera, Jorge Enrique Lugo; Robertson, Stephen Wood; Rosborough, George Walton; Rosenthal, David Bruce; Rothwell, Sally Ann; Rubin, Jeffrey Neil; Ruggiero, Robert Winslow; Runkel, Anthony Charles; Sadd, James Lester; Sagasta, Paul Frederick; Sanchez-Barreda, Luis Antonio; Sánchez-Salinero, Ignacio; Sander, Paul Martin; Sarzenski, Darci José; Satterfield, Will McSwain; Sauve, Jeffrey Allen; Savinelli, Peter; Schatzinger, Richard Allen; Schneyer, Joel David; Sedlacek, Wanda Jane; Self, Daniel Eugene; Senger, Rainer Klaus; Serlin, Bruce Steven; Shackelford, Charles Duane; Sheu, Jiun-Chyuan; Shorey, Mark David; Shum, Che-Kwan; Sicking, Charles John; Sikora, Paul J.; Simmons, James Layton, Jr.; Slator, Dorothy Stevenson; Smith, Brian Alan; Smith, Nathaniel Greene; Smits, James Robert; Soar, Linda Katherine; Solis-Iriarte, Raul Fernando; Souza, Jairo M.; Speer, Stephen William; Spencer, Alice Whitham; Spinler, Paul; Spradlin, Scott Dunbar; Stancliffe, Richard John; Standen, Allan Richard; Stark, Tracy Joseph; Stimac, James Alan; Storrs, Glenn William; Suchecki, Robert Kenneth; Suddhiprakarn, Chairat; Sullivan, Joseph Edward; Sullivan, Keith Barry; Sullivan, Michael Parnell; Suneson, Mark A.; Suter, John Robert; Sutton, Stanley Matthew, Jr.; Taha, Rozlan Mohammad; Tauvers, Peter Rolfs; Taylor, Alisa J.; Taylor, Steven; Tebedge, Sleshi; Tebedge, Sleshi; Thomas, Kimberly Jaye; Thompson, Diana M.; Thompson, Keith Goodwin; Thompson, Susan Lewis; Thurwachter, Jeffrey E.; Tiezzi, Pamela Anne; Todd, Charles Payson; Townsend, Margaret Anne; Travis, Deborah Sue; Treadgold, Galen E.; Triana, Rebecca; Tsai, Ching-Chang James; Tyner, Grace Nell; Ulrich, Suzanne Danner; Unver, Olcay Ismail Hakki; Utseth, Rolf Halvor; Van Dalen, Stephen Craig; Van der Ven, Paulus Hendrikus; Van Saun, Richard; Vanderhill, James Burke; Vasconcelos, Paulo M.; VerHoeve, Mark W.; Verross, Victoria Ann; Visser, Alex Theo; Vogel, Kenneth Daniel; Vogt, Jay Nathan; Walsh, Mark Patrick; Walters, Diana; Walters, Robert Derek; Walton, Anne Helene; Wanakule, Nisai; Wang, Ben; Wark, David Austin; Weatherill, Philip Mathew; Weiner, Stephen Paul; Westgate, James William; Wiggin, Roger Clay; Wiggins, William David, III; Wilkerson, Amy; Williams, Jefferson Boone; Wilson, Clayton Hill; Winans, Melissa Constance; Winkler, Dale Alvin; Wirojanagud, Prakob; Wirojanagud, Wanpen; Witebsky, Susan Nadine; Wittke, James Henry; Wolff, Martin; Wong, Henry Kwok-Hin; Wood, Becky Leigh; Woods, Arnold Martin; Worayingyong, Kaweepoj; Woronick, Robert Eugene; Worrall, John Griggs, III; Worrel, Elizabeth Ann; Wright, Stephen S.; Yeh, Hund-Der; Yilmaz, Pinar Oya; Yong, Kingston Cheng Wu; Young, Susan L.; Zapata, Vito Joseph; Zemlicka, George

Texas A&I University
Kingsville, TX 78363

5 Master's

1980s: Basham, Hal J.; Duerr, Michael David; Pitakpaivan, Kasana; Scott, Andrew William; Soonthornsaratul, Chekchanok

Texas A&M University
College Station, TX 77843

626 Master's, 287 Doctoral

1930s: Baughn, Milton H.; Bevan, Stewart; Elms, Morris A.; Gulmon, Gordon W.; Halbouty, Michel T.; McAdams, Frederick W.; Mueller, Frederick W.

1940s: Beckman, Michael W.; Graham, Daniel W.; Parmelee, E. Bruce; Schoenfeld, Perry C.; Smith, Edward James, Jr.; Tisdale, Ernest Edward

1950s: Alexander, William L.; Ammer, Bobby R.; Baker, Eldon R.; Balderas, Jack Moreno, Jr.; Bates, Charles C.; Blackstone, James P.; Blankenship, Joseph Croxton; Blumberg, Randolph; Bryant, George T.; Carter, Thomas R.; Coughran, Theodore; Creagor, Joe Scott; Creagor, Joe Scott; Dannemiller, George D.; Dunlap, John Bettes; Duvall, Victor M.; Egar, Joseph M.; Fallis, Jasper N.; Fos-

ter, R. Leon; Fritz, Joseph F.; Fuller, Robert L.; Goodwyn, James T., Jr.; Goolsby, Jay Lee; Gray, Eddie V.; Grote, Fred Rankin; Grubbs, Edward L.; Harwood, William E.; Henry, Vernon J., Jr.; Hope, Alvin C., Jr.; Kelly, Thomas Eugene; Marland, Frederick Charles; Marshall, Hollis D.; McAllister, Raymond Francis, Jr.; McDowell, Alfred N.; McGrath, Bernard D.; Miller, George H.; Morris, Gerald B.; Morris, Thomas J.; Mosteller, Stanley A.; Mounce, Donald D.; Napp, Donald E.; Neumann, Conrad Andrew; Noble, Robert A.; Parke, Robert Preston; Pedrotti, Daniel A.; Perry, Richard B.; Peterson, Don H.; Peterson, Thomas L.; Polk, Ted P.; Rogers, Luther Franklin, Jr.; Rokke, Stephen R.; Rokke, Stephen Richard; Rolf, E. Gerald; Scaife, Norman Caldwell; Seward, Clay L.; Seward, Clay L.; Shenton, Edward; Sliger, Kenneth Leon; Sweet, William Edward, Jr.; Thompson, Warren Charles; Vinson, George Larry; Walton, William L.; Wauters, John Ferdinand; Wilson, Edmon D.; Wilson, Guilford, James, Jr.; Woolsey, Issac W.

1960s: Ahr, Wayne M.; Airhart, Tom Patterson; Andrews, Charles Hubert; Atwell, Buddy H.; Baie, Lyle Frederick; Barfield, Billy Joe; Bhattacharyya, Tapan Kanti; Bowles, Frederick Albert; Brewster, James B.; Brown, Clifford L., III; Busby, Roswell F.; Campbell, Donald Harvey; Cebulski, Donald E.; Cernock, Paul John; Chan, Paul Chi-Keung; Chauvin, Aaron L.; Chiburis, Edward F.; Chowdhury, Dipak Kumar; Cook, Billy C.; Crocker, Marvin Carey, Jr.; Crocker, Marvin Carey, Jr.; Daugherty, Thomas D.; Davis, Donald Ray; Davis, Kenneth E.; DeWitt, Gary R.; Dixon, Bryan W.; Dunlap, Wayne Alan; Eaves, Glenn P.; Edwards, Goldsborough Serpell; Elsik, William Clinton; Elsik, William Clinton; Eusufzai, Hossain Sekandar H. Khan; Fisher, Neil E.; Frazier, David E.; Fredericks, Alan D.; Gibson, Roy B., Jr.; Gilmore, Walter E.; Gontko, Robert N.; Gordon, Patrick T.; Graczyk, Edward J., Jr.; Greenwood, Bobby M.; Hagerty, Roayl Moncrief; Hampton, Loyd Donald; Harding, James Lombard; Harlan, Ronald Wade; Harrell, Glenn C.; Hayes, David Wayne; Henderson, Garry Couch; Henderson, Garry Couch; Henry, Vernon J., Jr.; Herring, Maxwell, Jr.; Holt, Jack Haston; Hooks, James E.; Howle, Arlen G., Jr.; Janakiramaiah, Bollapragada; Jennings, Albert Ray; Jennings, Albert Ray; Jones, Billy Ray; Jurik, Paul P.; Kelly, F. Randolph; Kmiecik, Jerome Gregory; Kmiecik, Jerome Gregory; Knibbe, Willem Gerard Johan; Lepley, Larry K.; Linder, Henry D.; Lyons, Charles Gene; Maggio, Carlos M.; Majlis, Muhammad Ali Kahn; Malone, Carl Hubert, Sr.; Mangum, Charles R.; Maxwell, John R.; McBrayer, Michael A.; Miller, David I.; Miller, Robert E.; Molinari, Robert L.; Moore, George E.; Moore, Richard W.; Moore, Walter Richard; Moriton, William Thomas; Morris, Gerald Brooks; Morton, William R.; Murdock, Don M.; Mutis-Duplat, Emilio; Neathery, Orphie, III; Noakes, John Edward; Park, Edward C., Jr.; Pearring, Jerome Richard; Pitzer, Carroll D.; Poobrasert, Suparb; Pool, Alexander S.; Porter, Charles O.; Pyle, Thomas Edward; Raba, Carl Franz, Jr.; Roberson, Lindon B.; Rowe, Gilbert T.; Scafe, Donald William; Sealy, Brian E.; Siegert, Rudolf B.; Simpson, Jimmie D.; Slowey, James Frank, Jr.; Smith, James B.; Smith, John C.; Smith, Lester B., Jr.; Snead, Robert G.; Sommer, Sheldon; Song, Byong-Mu; Steinhoff, Raymond Okley; Sveter, Owen D.; Swoboda, Allen Ray; Swolfs, Henri Samuel; Trenchard, Walter H.; Trivedi, Harshadrai P.; Twell, Buddy H.; Vastan, Andrew Charles; Von Schwind, Joseph J.; Wadhwa, Nand P.; Walsh, Don; Waters, Ronald Hobart; Watson, Jerry A.; Wert, Richard T.; White, Dixon N.; Williams, Joseph D.; Yanex, Amade; Yanez-Correa, Amado; Yungul, Sulhi H.

1970s: Adams, D. G.; Al-Layla, Mohamad Tayeb Hussain; Ali, M. A.; Allen, Woods Wilkinson, Jr.; Altman, L. W.; Anderson, Don Randolph; Andrews, Robert Sanborn; Anspach, David Harold; Antoine, John W.; Appelbaum, Bruce S.; Archer, Paul Lawrence; Armstrong, Jenifer Ann; Baie, Lyle Frederick; Barber, David Williams; Barber, David

Williams; Barnes, Charles; Barnes, William Charles, III; Bassin, N. J.; Bateman, Barry Lynn; Baumgardner, Robert Welcome, Jr.; Bennett, R. H.; Bernard, B. B.; Bhambhani, Deepak J.; Bishop, Mary Augusta; Blackwell, Michael Lloyd; Blanton, Thomas Lindsay, III; Blanz, R. E.; Boenig, Charles Martin; Boone, Peter Augustine; Bowman, Linda Gail; Brewer, R. L.; Brock, William G.; Brooks, J. M.; Brooks, James Mark; Burbach, S. P.; Burnett, Thomas Lawrence, Jr.; Burns, William A.; Butler, Kenneth Bryan; Byun, B. S.; Byun, B. S.; Calmes, Grady Allen; Calogero, Frank; Castello, R. R.; Castleberry, Joe P., II; Cernock, Paul John; Chandler, Gary Wayne; Charles, Robert John; Chatham, Randall James; Chen, M. P.; Chmelik, Frank Bernard; Chung, H. M.; Clary, James Heath; Conrad, Robert E., II; Cook, Robert Annan; Cook, Robert Annan; Cool, T. E.; Cool, Thomas Edward; Cordero Ardila, V. F.; Cordero, Vladimir; Corry, C. E.; Coulter, R. E.; Coulthard, Dale Eugene; Coyne, John C.; Cunningham, S. E.; Custodi, George L.; Dengo, Carlos Arturo; DePaul, Gilbert John; Destefano, Mark; Distefano, M. P.; Dobson, Benjamin Mark; Drew, Fred Prescott; Dunphy, J. L.; Dyke, Lawrence Dana; Dyke, Lawrence Dana; Edwards, C. M.; Edwards, Goldsborough S.; Engelder, James Terry; Ethridge, Frank Gulde; Evans, Ian; Fall, Steven A.; Farrar, P. D.; Feely, Richard Alan; Feely, Richard Alan; Ferebee, T. W., Jr.; Ferebee, T. W., Jr.; Findley, Richard Lee; Fishman, Paul Harold; Fitzgerald, Duncan Martin; Font, Robert Geoseph; Foss, Deane Campbell; Frank, Donald James; Fugitt, David Spencer; Gallagher, John Joseph, Jr.; George, L.; George, Larry; Giammona, C. P., Jr.; Gilbert, Pat Kader; Graybeal, G. E.; Green, Jimmie Logan; Hajash, Andrew, Jr.; Hall, Gary L.; Hall, Gary L.; Harber, D. L.; Harker, George R.; Harris, John Elliott; Heinze, Daniel William; Heinze, W. D.; Heinze, W. D.; Helwick, S. J., Jr.; Helwick, S. J., Jr.; Henry, M.; Herndon, James M.; Hill, Gerhard William, Jr.; Hill, J. M.; Hinson, C. A.; Hluchanek, James Andrew; Ho, W. K.; Hottman, W. E.; Hottman, W. E.; Hugman, Robert H. H., III; Humbertz, Jon M.; Humston, John; Iwamura, S.; Iwasaki, Takeshi; James, Bela Michael; Jamison, William Richard; Jensen, P. A.; Johnson, Charles M.; Johnson, Charles T. L.; Johnson, Kenneth Walter; Johnston, Beatrice Bryant; Jones, L. D.; Jones, Leo David; Kan, David Lanrong; Kan, David Lan-rong; Katherman, C. E.; Keady, Donald Myron; Kincaid, George Preston, Jr.; Kuzela, Robert Christian; Lamping, Neal Edward; Larberg, Gregory Martin; Leblanc, Rufus Joseph, Jr.; Ledger, Ernest Broughton, Jr.; Lee, S. J.; Levitan, Mark Leslie; Lin, Joseph Tien-Chin; Lindquist, P.; Ludwig, Claudia Petra; Makoju, C. A.; Mancini, Ernest Anthony; Mareschal, J.-C.; Mareschal, Marianne; Marshall, William Dustin; Marti, J.; Mason, Curtis; Mathews, Thomas Delbert; McHam, Robert M.; McKee, Kathryn Merkle; McKee, T. R.; Miller, Michael Eugene; Mills, Earl R.; Min, K. D.; Minter, Larry Lane; Mitchell, Michael Harold; Mitchell, Thomas M.; Mitchell, Thomas M.; Moore, Walter Richard; Morris, Joe Lockhardt, Jr.; Morse, James Donald; Nelms, Katherine Currier; Nelson, Pennelope Conover; Nelson, Ronald Alan; Nelson, Ronald Alan; Nichols, Thomas Chester, Jr.; Nufer, Janet Ann; Nuzzo, M. L.; Olling, C. R.; Ongley, Lois K.; Parashivamurthy, Agasanapura Subbanna; Parker, R. A.; Parks, Louis Steven; Parks, Pamela Hennis; Pattison, Linda; Payne, Myron William; Peoples, M. W.; Pluenneke, Judith Louise; Pollard, Richard Mark; Powell, Raina Rae; Prasse, Eric Martin; Preston, Ralph J.; Pugh, C. E.; Pyle, C. A.; Pyle, Thomas Edward; Rangel, Jorge Enrique; Rannefeld, James W.; Rasor, R. W.; Rawlings, Gary Don; Reid, D. F.; Reimers, David D.; Roemer, Lamar B.; Roemer, Lamar B.; Rohani-Najafabadi, Behzad; Ruckman, David W.; Salter, P. F.; Sauer, T. C., Jr.; Savage, William Z.; Schneider, William Joseph; Schumacher, Madelyn; Sealy, James E., Jr.; Seay, Christopher S.; Serra, Sandro; Shephard, L. E.; Shimamoto, Toshihiko; Shirley, Richard H., Jr.; Shoemaker, Phillip Wayne; Shokes, R. F.; Sidner, B. R.; Sidner, Bruce Robert; Silver, Wendy Ilene; Smith, L. B., Jr.; Snedden, John William; Soemarso, C.; Sonnad, Jagadeesh Ramana; Sonnenberg, Stephen Arnold; Spalding, Roy Follansbee; Staff, George McDonald; Stearns, David W.; Stewart, Robert Donald; Stinson, James E., II; Stoudt, David L.; Sungy, Eugene D.; Swanson, Steven Brian; Sweet, William Edward, Jr.; Swolfs, Henri Samuel; Tam, Kwok F.; Tanenbaum, Ronald Joel; Tarantolo, P. J., Jr.; Tatum, T. E., Jr.; Tedford, Frederick J.; Temple, D. M. G.; Temple, Vernon J., Jr.; Teufel, Lawrence William; Teufel, Lawrence William; Tiezzi, Lawrence James; Tinkel, Anthony Robert; Tirey, Martha Margaret; Trabant, P. K.; Trabant, Peter K.; Trefry, J. H., III; Tresslar, R. C.; Tuefel, Lawrence William; Turner, James R.; Vaughan, R. J.; Vaughn, Patty Hollyfield; Vincent, Jerry William; Vyas, Y. K.; Ward, Suzanne; Waring, Juliana; Warring, Juliana; Watson, Joseph Quealy; Weinberg, David Michael; Weinmeister, Marcus Paul; Weiss, W. G., Jr.; Weynand, G. W.; Whalin, Robert Warren; Wilson, Raymond Carl Jr.; Wong, William Wai-Lun; Woodward, R. J.; Yang, S. J.; Yang, S. J.; Yen, Tzuhua Edward; Zupan, Alan-Jon Wellward

1980s: Abbott, Caroline L.; Abder-Ruhman, Mohammed; Adam, Adam Ibrahim; Adlis, David Scott; Ahlrichs, John Sigurd; Aldrich, Jeffrey B.; Alexander, Steven C.; Almalik, Mansour Saleh; Archer, Jerry Alan; Armstrong, Scott C.; Arroyo, Patricio Goyes; Ayan, Cosan; Baker, Joel F.; Bandy, William Lee; Barbosa Levy, Miguel Rudy; Barnard, Leo Allen; Barrell, Kirk Arthur; Bartlett, Wendy Louise; Barton, Robert A.; Bates, Charles; Bauer, Stephen Joseph; Bauer, Stephen Joseph; Bausch, Walter Charles; Beard, Les Paul; Becker, Joseph; Benavidez, Alberto; Benavidez, Alberto; Bicknell, James Scott; Billingsley, Lee Travis; Black, Cynthia E.; Bloom, Mark Alan; Blount, William; Bomber, Brenda Jean; Bond, Grego Benton; Bonds, James A.; Borbely, Evelyn Susanna; Borkowski, Richard M.; Bott, Winston; Bowers, Keith Douglas; Brabston, William Newell; Brandstrom, Gary Wayne; Brasher, James Everett; Brillinger, Allan; Brotherton, Mark Allison; Buck, Arvo Viktor; Bui, Elisabeth Nathalie; Bukowski, Charles T., Jr.; Bullion, David Nelson; Burch, Gary K.; Burkett, Patti Jo; Butler, Dwain Kent; Calvert, Craig Steven; Canova, Judy Lynn; Caram, Hector L.; Carriere, Patrick Edwige; Casarta, Lawrence Joseph; Cason, Cynthia Lynn; Cato, Kerry D.; Cauffman, Toya L.; Cavanaugh, Martin James; Cecil, Thomas Martin; Cefola, David Paul; Celaya, Michael Augustine; Chang, Heu-Cheng; Chen, Chia-Shyun; Chester, Frederick M.; Chester, Judith; Chiou, Wen-An; Chow, Jinder; Clark, Virginia Ann; Clowe, Celia A.; Coffman, Bryan Keith; Coffman, Jeffery Dale; Cole, Jay Timothy; Cole, Kathleen Patricia Hicks; Cole, William Fletcher; Conner, Steven P.; Connolly, William Marc; Conover, Dale; Conrad, Curtis Paul; Corbett, Kevin Patrick; Coryell, Jeffrey J.; Couples, Gary Douglas; Covington, Daniel Joseph; Covington, Thomas; Cregg, Allen Kent; Crisp, Jeffery; Cronin, Vincent Sean; Cross, Bradley D.; Cross, Scott Lewis; Cummins, Robert Hays; Cunningham, David; Curtis, Robert; Dally, David J.; Daniel, Joseph Hawkins; Davies, David John; Dean, Christopher William; Dedominic, Joseph Robert; Dengo, Carlos Arturo; Diebold, Michael Patrick; Dillon, William Gregory; Donahue, John Michael; Dougless, Thomas C.; Dowling, Sharron Lea; Dramis, Louis Albert; Dransfield, Betsy J.; Drees, Larry Richard; Dula, William Frederick, Jr.; Duncan, Mardon; Duree, Dana K.; Eady, Craig; Ebeling, Lynn Louis; Eckert, James Olin, Jr.; Edwards, David Arthur; Eicher, Randall Neal; Evans, James P.; Evans, James Paul; Fahlquist, Lynne S.; Faucette, Robert Christian; Feeley, Mary Hart; Feeley, Mary Hart; Fenner, Frederick Donald; Ferry, James Gerard; Feucht, Lynn Janet; Ford, Patrick; Fortner, David William; Fouret, Kent L.; Fowler, Rhonda M.; Frand, David M.; Frank, David M.; Franklin, Stanley P.; Freitag, Helen Clare; Frossard, Michael Louis; Fryer, Alan Ernest; Gaidusek, Barbara Ursel Marie; Gatto, Henrietta; Gazonas, George Aristotle; Gazonas, George Aristotle; George, Lawrence; George, Peter G.; Gibson, John L.; Godwin, David; Goldburg, Barbara L.; Gore, Larry D.; Gormican, Sheila Catherine; Goulet, William H.; Gowan, Samuel Ward; Gowan, Samuel Ward; Gray, Timothy J.; Green, Deborah; Gregg, Jack H.; Groschel, Henrike; Gruebel, Marilyn May; Habeck, Mark Fredrick; Hahn, Robert; Haines, John Beverly, Jr.; Hall, Steven D.; Halper, Fern Beth; Hand, Linda Mimura; Hanger, Rex Alan; Hankins, Donald Wayne; Hansen, Francis Dale; Harder, Paul Henry, II; Harris, William Maurice; Harris, William Maurice, Jr.; Harriz, J. Kimberly; Harville, Donald G.; Hastings, John O.; Hastings, Thomas Worcester; Hays, Phillip Dean; Hedgcoxe, Reiffery H.; Helsley, Robert; Henke, Kim Ann; Hennier, Jeffrey H.; Hennings, Peter Hill; Henry, Mitchell Earl; Hiemann, Mary Helen; Higgs, Nigel Gordon; Hinz, David W.; Hojnacki, Robert Stephen; Holley, Carolayne Elizabeth; Hoover, Caroline; Hopkins, Kenneth W.; Hopkins, Theodor William; Horne, Doyle Jackson; Hsu, Tung-Wen; Hudder, Karen A.; Hufford, Walter R.; Hull, Harris Benjamin; Huntsman, Brent S.; Hyman, Marian; Irwin, Frank Albert; Isenhower, Daniel Bruce; Jaganathan, James; Jeffrey, Alan William Adams; Jenkins, John Stacy; Jennings, Robert H.; Jiang, Ming-Jung; Johns, Mark William; Johnson, Charles G.; Johnson, Mark; Jones, James Winston; Jones, Jayne Ann; Joyce, John Edward; Juddo, Edward Paul; Jung, Woo-Yeol; Kashatus, Gerard Paul; Keaton, Jeffrey Ray; Keeney-Kennicutt, Wendy Lisabeth; Kelley, Van; Kennedy, James Lawrence, III; Kennedy, Noel Lynne; Kennicutt, Mahlon Charles, II; Key, Robert Marion; Kientop, Gregory Allen; Kim, Eul Soo; Kinnebrew, Quin; Kirkland, Brend L.; Klass, Marcia Jean; Knapp, Steven; Kornicker, William Alan; Kraig, David Harry; Kranz, Dwight Stanley; Ku, Kelly T.; Kuh, Hsien-Chien; Kuo, Tsai-Bao; Kuzior, Jerry L.; Lambert, Rebecca Bailey; Lantz, James R.; Lavenue, Arthur Marsh; Ledger, Ernest Broughton, Jr.; Lee, Chao-Shing; Lee, Suk Jin; Leethem, John T.; Leschak, Pamela; Levine, Stephen D.; Levitan, Arlette E. S.; Lin, Saulwood; Lindau, Charles Wayne; Linn, Anne; Locke, Kathleen A.; Londergan, John Thomas; Lopez, Hector S. J. L.; Lovell, Stephen Edd; Lowenstein, Glenn Robert; Lupo, Mark Joseph; Magenheimer, Stewart J.; Makarim, Chaidir Anwar; Malicse, Jose Ariel Enriquez; Mani, Philip C.; Mann, Robert Gordon; Mardon, Duncan; Mart, Joseph; May, Clifford John; May, James Herbert; Mayer, Terry Ann; Mays, Linda Lowry; Mazzullo, Elsa K.; McBride, Karen; McCallister, Dennis Lee; McCaskey, Michael D.; McClain, Anthony; McConnell, David Alan; McDonald, Thomas Joseph; McFarland, Mark Lee; McGee, Patricia; McLerran, Richard D.; Medlin, Linda Karen; Michie, Joanna; Miles, Randall Jay; Miller, Gregory Radford; Ming, Douglas Wayne; Minnery, Gregory Andrew; Mittsdarffer, Alan; Monti, Joseph; Moore, David W.; Moran, David Rick; Morey, Erol D.; Moriarty, Thomas D.; Morrison, Gregory D.; Munsey, John Sal; Murathan, Mustafa; Murdy, William; Nagy, Kathryn Louise; Narahara, Gene Masao; Neurauter, Thomas William; Nolan, Erich; O'Keefe, Arthur Francis Xavier; Olarewaju, Joseph Shola; Oldham, David Martin; Olsen, Rebecca Sarah; Omole, Olusegun; Orr, James Conrad; Palko, Gregory Jonathan; Panozzo, Renée Heilbronner; Paolini, Michael Joseph; Patterson, W. Samuel, Jr.; Patton, Thomas Lewis, III; Pepper, Gail Louise; Peterson, Mark Andrew; Petrini, Rudolf Harald Wilhelm; Phillips, Sandra; Piatt, David Allan; Pickett, Kendell; Pike, John David; Pinero, Edwin; Piper, Larry Dean; Poindexter, Marian Elizabeth; Pollock, Clifford Ralph; Ponton, James D.; Pope, Leslie Anne; Powers, Brian Kenneth; Price, Ronald Harlow; Proust, Rodrigo Diez; Prusak, Deanne; Quintana, Miguel Alfredo; Rabenhorst, Martin Capell; Rahim, Zillur; Ramanlal, Kirti Kumar; Rapport, Eric J.; Raskin, Greg Steven; Rauenzahn, Kim Ann; Ray, Earl S.; Reed, Roy Edwin, II; Reinarts, Mary Susan; Reutter, David; Rigert, James Aloysius; Riggins, Michael; Risch, David Lawrence; Ritter, Christine; Robbins, Gary Alan; Roden, Rocky Ray; Ross, Sheila Lynn; Roth, Susan Viola; Rotter, Richard Joseph; Royo, Gilberto Rafael; Rubio-Montoya, David; Russell, Karen L.; Rust, Richard Reynolds; Ryan, David; Sahl, Lauren Elizabeth; Sai, Joseph Obodai; Santi, Paul Michael; Schaftenaar, Carl Howard; Schaftenaar, Wendy Elizabeth; Scheevel, Jay R.; Schmittle, John M.; Schrull, Jeffrey Lee; Sequeira, Jose F., Jr.; Shelvey, Stephanie Anne; Shephard, Les Edward; Sheu, Der-Duen; Shih, Chung-Chi; Sims, Donald R., Jr.; Sinclair, Steven Whitney; Singleton, Scott Wayne; Smith, David A.; Smith, Gregory Alan; Smith, Richard L.; Smith, Trevor David; Sneed, David Richard; Sneider, John Scott; Sonnad, Jagadeesh Ramanna; Staff, George McDonald; Stafford, John Michael; Stearns, Steven Vincent; Steffensen, Carl K.; Straccia, Joseph Robert; Strautman, Sabina Y.; Strong, Catherine C.; Stubblefield, William Lynn; Sullivan, John J.; Szerbiak, Robert Bruce; Taylor, Elliott; Taylor, Robert Joseph; Theis, Sidney Wayne; Theiss, Richard M.; Theiss, Richard M.; Tiezzi, Lawrence James; Tinkle, Anthony Robert; Tobola, David Philip; Tompkins, Robert Eugene; Torrey, Victor Hugo, III; Travis, Lynne S.; Trojan, Michael; Tsenn, Michael C.; Tubbs, Robert E., Jr.; Tucker, James William; Ueckert, John Fant; Ukazim, Emenike Otuonyeadike; Urban, Stuart D.; van Voorhis, David; Vaught, Richmond Murphy; Volman, Kathleen Cushman; Wainright, Elizabeth Jane; Walters, Donna Lynn; Walters, James K.; Warsi, Waris Ejaz Khan; Watkins, John M.; Wei Wang; Welker, Mary; Weltz, Mark Allen; West, Larry Thomas; West, Mark Allen; Westen, Diane; Wharton, Amy Laura; White, Marjorie Ann; Whiting, Phillip Howard; Whitten, Christopher James; Whynot, John David; Wiesenburg, Denis Alan; Williams, Matt B.; Winker, Gregory James; Withers, Katrina D.; Woo, Kyung Sik; Woodhouse, Elizabeth Gail; Work, David L.; Work, Rebecca Diana; Wright, David S.; Wright, Robert John; Wu, Ru-Chuan; Wu, Ru-Chuan; Yale, Mark William; Yang, Young Kyu; Yao, Chia-Chi George; Yerima, Bernard Palmer Kfuban; Young, Stephen Robert; Zouwen, Dawn Elaine Vander; Zullig, James Joseph

University of Texas at Dallas
Richardson, TX 75083

105 Master's, 65 Doctoral

1970s: Abdul-Rahman, Mogda M. T.; Aguayo, Camargo Joaquim E.; Allen, Michael Lee; Baag, Czang-go; Carter, Paul W.; Castellanos, Mario Ruiz; Chaipayungpun, W.; Chockalingam, Solayappa; Christensen, Frank Deon; Cloud, Kelton Wayne; El Masry, Alaa Eldin Mohsen; Guerrero-Garcia, Jose Celestino; Horvath, Peter; Houde, Richard Francis; Hoyt, David Ellsworth; Jarzabek, Dave; Johnson, Douglas Martin; Jones, Rebecca Anne; Keating, Barbara Helen; Keating, Barbara Helen; Key, M. D.; Kriausakul, Nivat; Langston, Melana; Levy, Lawrence S.; Lienert, B. R.; Longoria-Trevino, Jose Francisco; Lopez, Liisa Maki; Newport, Roy Leo; Newport, Roy Leo; O'Donnell, Thomas Henry; Padovani, E. L. R.; Quillin, Robert Lynn; Reeve, Scott Cleveland; Reining, Joseph Bradley; Ruiz Castellanos, Mario; San Filipo, William Anthony; Schwartz, Daniel Evan; Scott, Gary Robert; Scott, Gary Robert; Sinton, John Blatnik; Smith, Charles Culberson; St. John, Jack W.; Steiner, Maureen B.; Taira, A.; Taylor, Henry Clyde; Vaughn, Elliott Benson; Wethington, Lynette Diane; Young, Chi Yuh; Zinz, Barry Lynn

1980s: Allen, Erlece Paree Green Kovar; Baldwin, Randall Wayne; Bandyopadhyay, Pinaki; Beach, Terry Lee; Bell, David Allan; Biggers, Barbara; Blome, Charles David; Boyd, Homer Joe; Boyer, Renee C.; Butler, Jeannette Stier; Caldwell, Kenneth Robert; Cast, Mary Elizabeth; Caughey, Michael E.; Chang, Toshi; Chang, Wen-Fong; Chen, Chen-Hong; Chen, Tzann-Hwang; Cheng, Kuei-Yu Yeh; Cheng, Kuei-Yu Yeh; Cheng, Yen-Nien; Chern, How-Hueir; Cherng, Juling-Chaun; Cloft,

Harriet S.; Correa, Victor Julio Lopez; Crump, James O., Jr.; Cunningham, Robert, Jr.; Davila-Alcocer, Victor M.; Dixon, James Robert, Jr.; Dulaney, James Patrick; Dumas, David Byron; Dzou, Iyh-Ping Leon; El Ibiary, Nabil Yakout; Elorrieta, Nimio Juvenal Tristan; Fekete, Steven Ralph; Foos, Anabelle Mary; Francki, Benjamin Joseph; Garey, Christopher Lee; Grotte, James Robert; Heidesch, Russell J.; Hersey, David Ralph; Herve, Miguel; Hilton, Glenn G.; Hong, Ming-Ren; Howell, Roy Patton, III; Hu, Liang-Zie; Hubert, Loren Matthew; Hurd, Robert Lee; Hussain, Mahbub; Idowu, Adebayo Aderemi; Ikpah, Azhinoto Ozodio; Islam, Quazi Taufiqul; Javaherian, Abdolrahim; Jimenez-Salas, Oscar Hugo; Kennedy, William David; Kim, Jung Joon; Klenk, Charlotte Dillon; Kodosky, Lawrence Gerard; Lai, Shang-Fei; Larson, Paul Andrew; Lasley, Bert A.; Lavado, Marcelo; Leonhardt, Frederick H.; Liang, Long-Cheng; Liao, Ching-Yi; Liaw, Hong-Bing; Lin, Tzeu-Lie; Liou, Jia-Shing; Liu, Teh-Ching; Lopez Correa, Victor Julio Lopez; Lopez, Liisa Maki; MacLeod, Norman; Massey, Kathleen Willis; McDonald, Cecilia Louise; McGehee, Thomas Lee; Meyer, Beatrice I.; Miranda, Roger M.; Mitchell, Charles Dale, Jr.; Montgomery, Homer Albert, Jr.; Morin, Karen Marie; Murphy, Robert Parsons; Nash, Sandra Lee; Nelson, Bruce Howard; Norman, Mark Daniel; Nutini, John, Jr.; Ottensman, Donald Clay; Parent, Allan A.; Ragland, Betty Catherine Sims; Rashak, Edward P.; Ray, Amal Kumar; Reamer, Sharon Kae; Reilly, James Francis, II; Reynolds, David Johnson; Roberts, Charles Thomas; Ross, Malcolm Ingham; Rowe, William D., Jr.; Schellhorn, Robert Wayne; Schulze, Daniel James; Scott, Thurman Eugene, Jr.; Selznick, Martin Richard; Sen, Gautam; Sleeper, James Lockert, Jr.; Southernwood, Renee; Sprague, Anthony Ross Grafton; Sun, Robert Jencheu; Thompson, Laird Berry; Tickner, Bruce; Tippit, Phyllis Russell; Tristan, Elorrieta Nimio Juvenal; Voegeli, David Afred; von Feldt, Ann Elizabeth; Wen, Jing; Whalen, Patricia Ann; Wheeler, William A.; Wherry, Stephen D.; Wilson, Douglas Hord; Yang, Chung-Tien; Yang, Qun; Yang, Shyue-Rong Vincent; Yin, David D.; Yoon, Kwi-Hyon; Zaback, Doreen Ann

University of Texas at El Paso
El Paso, TX 79968

196 Master's, 25 Doctoral

1960s: Bolich, Leonard C.; Myers, John B.

1970s: Al-Shamlin, Ali Abdula; Allouani, Rabah Nadir; Arnold, Randal Irad; Austin, C. Bradford; Barrie, F. J.; Bersch, Michael G.; Biggerstaff, Brad P.; Callahan, Chester James, Jr.; Campuzano, Jorge; Chacon, Roberto; Curtis, Robert E., Jr.; Davis, Gene H.; de Simpson, R.; Deen, Roy D.; Djeddi, Rabah; Dowdney, Jack R.; Dye, James L.; ElFoul, Djamal; Ellis, Roger David; Flih, Baghdad; Garcia, Rafael A.; Glover, Thomas J.; Gunn, Robert C. M.; Hair, Gregory L.; Hobbs, Thomas M. C.; Hoenig, Margaret A.; Hoffer, Robin L.; Hornedo, Mercedes; Huskinson, Edward July, Jr.; Kadhi, Abdul; Kasadarli, Mustapha E.; Kopp, Richard A.; Kramer, Walter V.; Larson, Marvin E.; Lattu, Andrew C.; Leggett, Bob D.; Macer, Robert J.; Martin, William R., IV; Massingill, G. L.; Messenger, Harold M., III; Millican, Richard S.; Mims, Robert Lewis, Jr.; Morris, Richard W.; Mraz, Joseph R.; Murry, Danny H.; Nelson, Martin Andrew; Nielsen, Thomas W.; Nodeland, Steven K.; Oden, James Russell; Ortiz, Terri S.; Page, Richard Owens; Petty, Andrew J., Jr.; Rabah, N. A.; Reid, Steven Graham; Rosado, Roberto Victor; Roueche, William Lee III; Setra, Abdelghani; Shaheen, Elias J.; Smartt, Richard A.; Swift, Douglas Baldi; Teal, Lewis W.; Thomson, Kenneth; Tschirhart, Sid C.; Uphoff, Thomas L.; Varnell, Ronnie J.; Verrillo, Dan E.; Wacker, Herbert James; Wagner, Harry Arthur, III; Wallace, Andy Bert; Wilcox, Ralph E., Jr.; Wilhelm, Rudolph; Willcox, R. E., Jr.; Winston,

Michael R.; Wise, Henry M.; Wollschlager, Larry R.

1980s: Abrams, Gary S.; Abugares, Youssef Issa; Abushagur, Sulaiman Ahmed; Abuzekri, Sadegh Khalifa; Adame, Javier; Aiken, Mary J.; Aiken, Olaf W.; Aluka, Maduegboaka Innocent Jude; Baker, Mark Richard; Basden, Wayne A.; Beard, Thomas Christopher; Bell, Richard C.; Belle, Eddie R.; Bodnar, Theodore; Broderick, John C.; Brown, Michael L.; Bullock, James S.; Burrows, Lloyd A., III; Campbell, Michael P.; Chavez-Quirarte, Ramon; Cofer, Richard S., III; Cornelius, Howard E.; Coughlon, John Patrick; Coultrip, Robert; Danko, Jeffrey H.; Davis, Lisa M.; Davison, Charles H., Jr.; De Angelo, Michael V.; Decker, Guy M.; Deshler, Richard M.; Dicke, Craig A.; Dupuy, John R.; Durfuee, Barbara A.; Edwards, Gerald; El-Dadah, Ghazi; Evans, Kathryn Christina; Figuers, Sands Hardin; Fisher, Marci A.; Flint, Frederick F.; Frantes, Thomas J.; Gates, Edward E.; Gilmer, Allen L.; Gomez, Filiberto P.; Gurrola, Harold; Guthrie, Robert S.; Handschy, James W.; Harden, Stephen N.; Harder, Steven Henry; Harder, Vicki M.; Harder, Vicki Marie; Harkey, Donald A.; Harvey, David B.; Hoffer, Roberta Lynne; Holt, Robert M.; Holtzclaw, Stuart R.; House, Larry A.; Jacobs, Michael A.; Kahn, Peter A.; Kiely, James M.; Knoll, Martin Albert; Kondelin, Robert J.; Kornas, Barbara Ellen; Kourse, Lauralee D.; Kruger, Joseph Michael; Kwarteng, Andrews Mensah Yaw; La Freniere, Jon Edmund; Lance, James Odell, Jr.; Lehtonen, Lee R.; Leonard, Mary L.; Lewis, Laurel M.; Lueth, Virgil Walter; Lueth, Virgil Walter; Madden, H. Douglas; Matel, Joel E.; Mauldin, Randall A.; McCabe, Marcella R.; McCutcheon, Janice A.; McCutcheon, Tim; McEvers, Lloyd K.; Miranda G., Miguel A.; Mitchell, Stephen M.; Muela, Pedro, Jr.; Newman, Brent D.; Norland, William D.; Nyunt, U.; Orajaka, Ifeanacho Paul; Osleger, David A.; Papesh, Henry; Pau, Joseph H. K.; Pearson, John W.; Pearson, Mark F.; Peters, Lisa; Phillips, Joseph D.; Phillips, Stephen E.; Pickens, Craig A.; Ponce, Benjamin F.; Powell, Darron Lee; Pyron, Arthur J.; Rahman, John L.; Raksaskulwong, Manop; Ray, David R.; Renbarger, K. Scott; Reuter, Stephen E.; Reyes C., Ignacio A.; Riess, C. Maurine; Riley, Robert; Rimando, Philip M.; Rinowski, Robert D.; Roark, Robert C.; Robinson, Bob Russell; Robinson, Rosalie M.; Roepke, Timothy J.; Russo, Joseph F.; Ryan, Elizabeth B.; Saenz, Guadalupe; Scheubel, Frank R.; Schiel, Kathryn A.; Schneider, Robert V.; Seigler, William C.; Shannon, William M.; Sheffield, Tatum M.; Shepard, Mark D.; Simpson, Ronald D.; Sinno, Yehia Ahmed; Sivils, David J.; Smith, Karl J.; Strickland, Frank G.; Suleiman, Abdunnur S.; Suleiman, Ibrahim Sharif; Taylor, Bruce; Thomann, William Frederick; Trentham, Robert Craig; Trujillo, Mario R.; Veldhuis, Jerry H.; Weise, James Richard; Wen, Cheng-Lee; Wilkerson, Gregg; Willingham, Daniel L.; Wilmar, Glenn C.; Wood, Douglas R., II; Woodrome, Larry S.; Wuellner, Dirck E.; Yousef, Ali Abdulah; Zamzow, Craig Edward

Texas Christian University
Fort Worth, TX 76129

150 Master's, 2 Doctoral

1910s: Sweeney, J. S.

1920s: Alexander, Charles Ivan; Bohart, Morris Fielding; Bowser, William Franklin; Carpenter, Margaret; Carrell, Olleon; Hill, Benjamin Harvey; Mahon, Sadie; Moore, Marcus Harvey; Moreman, Walter L.; Norris, Will Victor; Scott, Gayle; Self, Selden R.; Smiser, Jerome Standley; Stangl, Frank J., Jr.; Williams, Leonora May

1930s: Bennett, Ethel Evans; Carrell, Charles Howard; Grubbs, William Howard; Hendricks, Leo; Lozo, Frank Edgar, Jr.; Nicol, David; Smith, John Peter; Smith, Ralph Emerson; Stroud, Charles Brasher

University of Texas Health Science Center at Houston School of Public Health

1940s: Barber, Thomas David; Brooks, Jack Alexander; Jarvis, Daniel; Matthews, William Henry, III; O'Gara, William Thomas; Renfro, Millicent Aloyse; White, French Robertson, Jr.; White, James Robert

1950s: Bishop, Bobby A.; Dameron, Wyllie Frank; Glass, David Lawrence; Greenwood, Eugene; Hodgson, Walter Dale; Holland, Jasper L.; Kelley, Millard Lee; Knopp, David A.; Markovic, Francis Xaviare, Jr.; McGill, Daniel W.; Menut, D. Charles; Ohlen, Robert Henry; Parks, Oattis Elwyn; Russell, William E.; Sale, Clarence; Slocki, Stanley Francis; Smith, Lee Anderson; Walker, Jack R.; Whitaker, Doyle Gene; Wilde, Garner Lee; Wright, Alfred Edwin

1960s: Cable, Louis Walter; Carter, Paul Henry, Jr.; Couch, Elton Leroy; Ellinghausen, Robert Henry; Enis, Hunter; Gibson, Carleon, Jr.; Gilliland, John Dale; Harvard, Paul Odom; Hopper, George Steven; Kuehn, William Jackson; Lary, Brenda Brants; Perkins, Max Allen; Raish, Henry Dean Eugene; Renner, Richard Eugene; Roe, Glenn Dana; Shannon, Lynn Carlton; Shiever, John Wayne; Slaydon, Robert Earl, Jr.; Stewart, Edwin Mack; Stone, Joseph Fred; Watson, Lowell Brent; Wood, Michael Lee

1970s: Alaniz, Roberto Trevino; Ammon, Walter L.; Bagley, Roy Louis, Jr.; Barry, Dereck Michael; Benedetti, Steven Jess; Council, Konrad Koert; Council, Konrad Koert; Cunningham, K. D.; Devery, Dora Maria; Devery, Justin V.; El-Atrash, Mohamed Elmahdi; Felch, Roger N.; Flippin, Jerel Wayne; Hawkins, Connie M.; Holcomb, Richard Alfred; Horton, Frank R.; Ivey, Marvin Lee, Jr.; Ivy, David; Jackson, Dan Herman; Lazo, L. A. R.; Maerz, Richard Hugh; Manka, Leroy Louis; Mellor, Edgar I., Jr.; Mohorich, Leroy Martin; Moore, William S. DeGaspe; Ng, David Tai Wai; Patterson, William Dean, II; Pendergrass, James M.; Pense, Glenn Martin; Reilly, Maurine Brigid; Roman, Luis Alberto; Self, George William, Jr.; Steuer, Mark R.; Tleel, Jack Wadie; Vernon, Gail Franklin, Jr.; Viard, James Philip; Vicars, Robert Glenn; Ward, Christopher Allan; Western, Stephen Kent; Yu, Ho-Shing

1980s: Ali, Mohammad Sanwar; Benekas, Sandy L.; Bulling, Thomas Peter; Carlton, Keith H.; Dahl, David Alvin; Drake, Steve Allen; Fouch-Flores, Donna Lynn; Frank, Kevin James; Griffin, Andree French; Henk, Floyd (Bo) H., Jr.; Hill, William A.; Holland, Richard A.; Knights, William Jay; Koesters, Donna Baird; Land, David M.; Lindquist, Tina Walburga; Logan, Homer H.; Malek, Debra Jean; Matthews, Ross Butler; Mauch, Joseph James; Nalepa, Randolph; Nash, Tamie Rene; Neely, Laura Lea; Ness, Deborah Lee; Nolley, Janis Mergele; Osterlund, David Paul; Parris, Thomas Martin; Pershouse, Jonathan Ralph; Raschilla, Stephen Nicholas; Ross, Michael Lee; Rothammer, Christine Marie; Schildt, Timothy Allen; Shepard, Timothy Mark; Ward, Rosalyn Julia; Wiberg, Leanne; Wilhelm, Steven John

University of Texas Health Science Center at Houston School of Public Health
Houston, TX 77225

1 Master's, 1 Doctoral

1970s: Assaf, K. K.; Cechova, Irina Maclin

University of Texas of the Permian Basin
Odessa, TX 79762

17 Master's

1980s: Arama, Rachelle Brooker; Bednarski, Sheila Palmer; Davis, Danny Ray; Ellison, Charles Ralph, III; Hansen, Steven Michael; Hunter, Paul Kirk; Lambert, Patricia Frost; Lydecker, William Frederick; Martin, James Oliver; Metz, Cheryl Lynn; Partain, Bruce Robert; Setzler, Robert Eric; Sherif, Alamin Abdalla; Spencer, Jesse Garvin; Walker, Stephen Dade; WIlliams, Bobby G.; Zdinak, Andrew Patrick

Texas Tech University
Lubbock, TX 79409

317 Master's, 33 Doctoral

1930s: Abbott, Ralph E.; Cantrell, Ralph B.; Cole, Clarence A.; Fulton, G. Lyman; Gibson, Donald Thomas; Langford, Maxine; Madera, Ruford Francisco; McCullough, Edward Allen; Roach, Samuel; Rogers, Jesse Armstead; Stults, Arthur Carl; Tanner, William Francis, Jr.; Williamson, John C.

1940s: Alexander, Theodor W.; Bailey, Marshall W.; Clark, Walter Thomas, Jr.; Coon, Lester A.; Haliburton, John Leo; Hinson, H. H.; Huntington, George C.; Hurst, Ray Eugene; Perusek, Cyril J.; Rodgers, T. Deane; Sheldon, Wichita F.; Sleeper, James Lockert, Jr.; Soper, Harland; Woods, Delmer Maurice

1950s: Anderson, Gerald K.; Bailey, Paul T.; Bass, John H.; Belknap, Barton Austin; Brooks, Lon Clyde; Burress, George Thomas; Burton, Robert Clyde; Butler, Roy; Carmack, Ray P.; Chisholm, Earl J.; Clair, Virginia; Clarke, Charles Edward; Clifton, Billy Dean; Coleman, Carl R.; Cooke, Selman C.; Couch, Herbert E.; Cox, William B.; Cullinan, Thomas Anthony; Foster, James A.; Garrett, Paul Winslow, Jr.; Graves, Frank Douglas; Green, Francis Earl; Green, Francis Earl; Hasson, Richard C.; Hatley, Allen Grady, Jr.; Hawkins, Ralph D.; Head, James L.; Holt, Richard Wayne; Huzarski, Richard George; Jones, Billy R.; Kessinger, Walter Paul, Jr.; King, Charles Edward; LaPrade, Kerby Eugene; Libby, Frederick Ernest; Lokke, Donald Henry; Main, Talmage; Martin, Clyde D.; McLamore, Vernon Reid; Miller, William Donald; Morton, Maurice Warner; Munn, James Knox; Neef, George Herman; Nixon, Achilles Harry; Pittman, Gardner M.; Powell, James Daniel; Priddy, Charles Parrish; Probandt, William Taylor; Sanders, Malcolm Keith; Smith, Shelby W.; Smith, William Henry; Speed, Bert Lewis; Stennett, Albert J.; Stever, Rex Hale; Stout, Earl Douglas; Tanner, James Henry; Tonroy, Lucky Less; Tucker, Charles Odell; Vick, William Edward; Wallis, Thomas Irvin; Weldon, Charles S.; Williams, Jack Riley; Wood, John William; Young, Wilfred Ray

1960s: Allen, Ron R.; Ashour, Abdurrahim Mohammed; Baker, Lynn Edward; Barnette, Carr Howard; Bates, Thomas Robert; Beck, Ray Hall; Bitgood, Charles D.; Blazey, Philip Thomas; Blumentritt, Russell A.; Bostik, Wayne Charles; Brawley, Tommy R.; Carter, John Swain, Jr.; Cayce, William Powers, Jr.; Danbom, Stephen H.; Darden, Larry B.; De Hon, René Aurel; Dixon, John Robert; Foxworth, Wycliff Riley; Geddes, Richard W.; Greenlee, David Walden; Harper, Melvin Louis; Harrison, Stanley Cooper; Hart, William George; Himmelberg, Glen Ray; Huffman, Marion Edward; Keller, George Randy, Jr.; Kiatta, Howard William; Kothman, Winnard Sidney; LaPrade, Kerby Eugene; Leach, Jerald Wayne; Lees, William R.; LeTourneau, Nelson Joseph; Lillard, Douglas Ray; McGregor, Dan R.; McGregor, Don L.; McLean, Steven Arthur; Mount, James Russell; Normand, David Ernest; Pac, Floyd; Pendery, Eugene Christian, III; Prewitt, Ronald H.; Reed, James Courtney; Schaefer, William Alvin, Jr.; St. Germain, Louis Charles; Suggs, James De Shae; Sutcliffe, John Russell; Thomas, Carroll Morgan; Thomerson, Jamie Edward; Toney, Jimmie C.; Wilbanks, John Randall; Wilbanks, John Randall; Williams, Clifford Ralph; Williams, Karl Wendel; Williams, Karl Wendel; Winn, Robert Maurice; Yeats, Vestal Liarly; Yoder, Nelson B.

1970s: Ashour, Abdurrahim Mohammed; Ashraf, Abdul Aziz; Barone, William Edward; Bassett, Randy Lynn; Bauer, Larry Paul; Book, Gerald Wayne; Boothby, Donald Roy; Borella, L. G.; Brand, John Frederick; Buchanan, Cathy McGhee; Buchanan, John Wilson; Buika, Paul H.; Burden, Cecil Ronald; Carr, Timothy R.; Castro, Louis Reyes; Castro, Louis Reyes; Cathey, Carl A.; Clanton, Jerry S.; Coffman, Bruce P.; Compton, Joe Larry; Crawley, Mark Edward; Cronin, Thomas Crawford; D'Lugosz, Joseph J.; De Hon, René Aurel; Erpenbeck, Michael Francis; Evans, Kathryn C.; Florstedt, James Edward; Foley, Donald Charles; Franco, Lamberto-Augusto; Friess, John Paul; Galey, Jimmy L.; Goebel, Joseph Edward; Goolsby, Jimmy Earl; Gray, Terry Lee; Greer, Efford Wayne, Jr.; Grills, Richard Barbee, Jr.; Guzman, Alfredo Eduardo; Harrison, Earl Preston; Held, Harry L.; Hunter, Bruce Edward; Kasino, Raymond Edward; Keller, George Randy, Jr.; Kocurko, Michael John; Krishtalka, L.; Landreth, Robert Allen; Lee, Li-Jien; Lee, Roger William; Leiker, Loren Michael; Logan, Lloyd Eugene; Maley, Elaine Gail; Markey, Daniel Gene; Martino, David S.; McAdoo, Richard Lee; McArthur, Richard Earl; Meers, Ronald B.; Morahan, George Thomas; Muir, Nancy Jean; Mullican, Jerry W.; Normand, David Ernest; Pease, Rodney Wayne; Pierce, Harold George; Pierce, Harold George; Pinkerton, Roger Parrish; Powe, Walker H., III; Raghu, D.; Railsback, Rickard Reed; Reeves, Corwin C., Jr.; Reeves, James Ray; Russell, Lee Robin; Russell, Lee Robin; Sanford, William Casey; Schwab, Joseph Alan; Seithlheko, Edwin M.; Setoguchi, Takeshi; Setoguchi, Takeshi; Stanton, James Clifford; Stoutamire, Steve; Sturm, John J.; Sundin, Philip James; Sutton, John F.; Tully, Stephen Anthony; Tur, Stephen Martin; Vessell, Richard K.; Washburn, Judy; Webb, Chris Cynthia; Webb, Steven Ray; Wells, Randall W.; Wernlund, Russell J.; Wiginton, Randal Lynn; Winn, Robert Maurice; Womochel, Daniel Robert; Worthen, John Aldrich; Zinz, Barry Lynn; Zurinski, Stephanie Ann

1980s: Albehbehani, Abdulsamee S. K.; Alnes, Joel R.; Ateiga, Abdalla A.; Bagstad, David Peter; Baker, Lisa M.; Barnes, Nora L.; Beaver, James L.; Bernhard, Ronald P.; Boardman, James Joseph; Bohanon, James Paul; Branch, Colby Lloyd; Branch, Jill N. Haywa; Brimberry, David L.; Canfield, Beverly Anne; Chadwick, Russell John; Chen, Jinxing; Cheney, Richard Stephen; Chitale, Dattatraya V.; Chitale, Jayashree Dattatraya; Clemons, Robert Rickard; Clemons, Robert Rickard; Cornwell, Kevin J.; Cox, Ricky G.; Crider, Richard L.; Davis, Jondahl; Drake, Jerry T.; Evans, David G.; Finlay, Corey D.; Francka, Benjamin Joseph; Frelier, Andrew P.; Fu, Yun-ta; Gazdar, Muhammad Nasir; Girardot, Gerald B.; Glenn, Sidney E.; Gonzales, Eduardo; Goodwin, Peter B.; Gribble, Robert F.; Harrison, Ben S.; Hendrickson, Charles R.; Henningsen, Gary R.; Hill, Lena Elizabeth; Holtz, Kenneth Roger; Horner, Timothy C.; Howard, Kerry S.; Humphrey, John Fitzgerald; Hurley, Timothy James; Irwin, William Kenneth Arthur; Ismail, Mohamad I. B.; Jackson, William Daniel; Jarvi, Thomas Robert; Krenik, Kevin J.; Kuhn, Carlos Alfredo Clebsch; Landenberger, Daniel Ross; Lanford, Colleen Loretta; Lanter, Robert B.; Lee, Li-Jien; Lepp, Casey Louis; Malekahmadi, Fatemah; Markgraf, Philip C.; Martell, Paul W.; Mauser, John Kemmer; May, A. Brent; Mayfield, Ricky L.; McCasland, Ross Duncan; McWilliams, David Bruce; Medford, Richard M.; Meenaghan, Susan Lee; Meyer, Charles Richard; Moore, Brian Keith; Moore, Darrell C.; Moore, V. Scott; Mullican, William Franklin, III; Nossaman, Leslie Norene; Ormond, John; Panfil, Daniel John; Petraitis, Michael John; Portnoy, Michael B.; Potratz, Victoria Yeko; Price, Alan Paul; Proctor, David D.; Rahman, Ata Ur; Rahman, Ata Ur; Reid, Gary Carl; Ries, Gaila Vawn; Sanders, Richard Bryan; Sathiyakumar, Neelakandan; Schroeder, M. Richard; Shih, Xiao Rung; Sikorski, Peter Edwin; Siy, Suzan Elizabeth; Smelley, Randal Keith; Spesshardt, Scott Alan; Stillwell, Martha; Temple, James M.; Thompson, Amy Gale; Trabelsi, Ali M.; Uba, Humphreys Douglas; Vincent, Judy Ann; Wagner, Dallas M.; Waritay, Lanfia T. S.; Welsh, Joyce R.; White, Larry Michael; Wickland, David Charles; Wofford, Melissa K. Young; Yeko, John D.; Yockum, Eric T.; Young, Joe B.; Zheng, Hong; Zuravel, David Lee

University of Toledo
Toledo, OH 43606

139 Master's, 1 Doctoral

1960s: Bourne, Harold L.; Bruning, James Earl; Johnson, Russell Paul; Pawlowicz, Richard M.; Rohrbacher, Timothy J.; Rowland, Mark R.

1970s: Alther, George R.; Armour, Michael D.; Armstrong, William B.; Bienkowski, Henry G.; Boyd, Thomas M.; Buchanan, Kenneth J.; Burgett, Thomas L.; Burke, Michael R.; Buss, David R.; Camp, Mark J.; Chester, Mik E.; Clark, Karen A.; Copley, David L.; Cordas, John J.; Elliott, Brian; Everett, Edward E.; Ferguson, Pamela; Folkoff, Donald W.; Gallagher, Ronald Eric; Gindlesperger, Gary D.; Glaze, Michael V.; Hilty, Robert D.; Hopfinger, Carl; Huffman, Mark E.; Kerekgyarto, William L.; Killberg, Glen C.; King, James M.; King, Mark C.; Koechlein, Harold D.; Kowalczyk, Gary R.; Kuntz, Michael G.; Layman, John W.; Logan, Stewart Michael; Lopez, Raymond; Luppens, James Alan; Majchszak, Frank L.; Mauer, Kenneth; McAvey, Michael B.; McBride, Robert T. J.; McLin, Stephen G.; Meyers, James B.; Minning, Robert C.; Moster, Neal H.; Nations, Darrell L.; Oddo, John Edward; Olle, John Michael; Pennino, James D.; Perkins, Clarence Michael; Reimann, Martha Campbell; Rose, Edward K.; Ryder, Henry L.; Scheerens, Clark L.; Schneider, Ronald L.; Sheahan, Joseph W.; Sherif, Nasreddin; Sikora, Gerald S.; Silverman, Marc S.; Skinner, Jeffrey; Skrzyniecki, Randal G.; Swemba, Michael J.; Syvert, Raymond J.; Taylor, Larry E.; Turner, William S.; Valkenburg, Nicholas; Venturoli, Karen A.; Weis, Lawrence A.; Winegardner, Duane L.; Wu, Chia-Hsin

1980s: Baggett, Stephen Myles; Baranoski, Mark T.; Behbehani, Abdulsamee S. K.; Bertoli, Lou; Blanco, Julius M.; Blessing, Dennis R.; Brunsman, Mark J.; Chyi, Kwo-Ling; Ciner, Attila T.; Coon, Cathy A.; Cousino, Matthew A.; Cravens, Stuart James; DeMasi, Amy E.; DiCesare, Joseph A.; Dimit, Georgette; DiPlacido, Arthur J.; Dunkin, Joyce Sattler; Dunkin, Ned R., Jr.; Eller, Lynn Hansack; Ellison, Andrew Bell; Fesko, Gregory R.; Hallfrisch, Michael Paul; Herringshaw, Dennis Charles; Hj-Elias, Mohd Rohani; Hoover, John A.; Hosfeld, Richard K., Jr.; Hua, Zhang; Hunt, Timothy J.; Kessler, Kirk J.; Kihn, Gary Edward; King, Mary Anne; Kose, Celal; Kreutzfeld, James E.; Kribbs, Gary M.; Krywany, Joseph M.; LaGrange, Renee Leone; Lempke, Douglas A.; Lewis, Amy Heywood; Madigan, Terence J.; Maxwell, Gary P.; McLaughlin, Jeffrey Donald; Mohd-Nurin, Abdul Razak; Palombo, Kevin M.; Paulson, James D.; Quick, Jeffrey Charles; Reichard, James E.; Resnick, Alan J.; Rorick, Andrew Hammond; Spengler, Thomas J.; Stanley, Roy A.; Starkey, Michael J.; Stephens, Thomas C.; Stuckey, George H.; Swanson, Mark H.; Taylor, Amy E.; Tipton, Ronald M.; Tracy, Bradford M.; Tsai, Louis Loung-Yie; Watkins, Michael L.; Wavrek, David A.; Wegert, Emily Landris; Whitacre, Timothy Patrick; Wicks, John L.; Wolberg-de Venecia, Kathryn Elizabeth; Zaeff, Gene D.; Zuch, Donald M., II

University of Toronto
Toronto, ON M5S 1A1

398 Master's, 358 Doctoral

1900s: Parks, W. A.

1910s: Burwash, E. M. J.; Ellsworth, H. V.; McNairn, William H.

1920s: Bell, Archibald M.; Bell, L. V.; Carmichael, Mary F.; Dyer, William S.; Fritz, Madeleine A.; Gerrie, William; Haller, M. C.; Jones, Islwyn Wyn; Kerr-Lawson, D. E.; Maynard, J. E.; Sanderson, James O. G.; Scatterly, Jack; Warren, Percival S.

1930s: Allen, J. S.; Amrstrong, H. S.; Baker, James Morgan; Bartley, M. W.; Bell, L. V.; Berry, Leonard Gascoigne; Britton, J. W. M.; Brown, William Lester; Brown, William Lester; Caley, J. F.; Carter,

O. F.; Charlewood, G. H.; Chubb, P. A.; Corking, W. P.; Crombie, G. P.; Dadson, A. S.; Dadson, A. S.; Derby, Andrew W.; Dodson, Alexander S.; Downer, H.; Duffell, Stanley; Gardiner, McE. C.; Glauser, Alfred; Griffis, Arthur T.; Johnson, Helgi; Johnston, Wilbert E.; Jones, William A.; Laird, H. C.; Leedy, Evert John; Mackenzie, G. S.; McDonald, William L.; Meen, Victor B.; Meen, Victor B.; Moorhouse, Walter W.; Newman, William Roy; Nowlan, James P.; Perry, S. C.; Prince, Alan T.; Savage, William S.; Shaw, Ernest W.; Sprouble, John C.; Stadelman, D. K.; Teskey, Maurice Forgie; Way, Harold G.; Wright, J. D.

1940s: Barnes, Frederick Q.; Bartley, M. W.; Beland, Rene; Beland, Rene; Belding, Herbert Frederick; Berry, Leonard Gascoigne; Bolton, Thomas E.; Brown, R. A. C.; Buchanan, R. M.; Chisholm, Edward Owen; Clark, Arthur Roy; Clarke, W. J. G.; Claveau, Jacques; Claveau, Jacques; Crombie, G. P.; Currie, John B.; Dawson, Kenneth R.; Deane, Roy E.; Ferguson, Robert B.; Ferguson, Robert B.; Ferguson, Stewart Alexander; Ferguson, Stewart Alexander; Firth, D. A.; Fleming, H. W. W.; Forman, Sydney A.; Frantz, J. C.; Gilman, W. F.; Graham, Robert B.; Graham, Robert B.; Gross, William H.; Hammond, W. P.; Ingham, Walter Norman; Ingham, Walter Norman; Kaiman, S.; Keeler, Gordon T.; Keevil, A. R.; Kerr, J. W. M.; Leaming, S. F.; Liberty, Bruce Arthur; Little, Heward W.; Low, John Hay; Maurice, Ovide Dollard; McGregor, J. P.; Michener, Charles Edward; Miller, Murray L.; Milne, Ivan Herbert; Moddle, D. A.; Neal, H. E.; Nuffield, Edward Wilfrid; Parks, Thomas; Peach, Peter A.; Perry, Thomas G. J.; Pitts, P. D.; Prest, Victor Kent; Rowe, Robert Burton; Seigel, Harold O.; Senftle, Frank E.; Springer, G. D.; Sternberg, Raymond Martin; Thompson, R. M.; Tremblay, Leo-Paul; White, William Harrison

1950s: Allan, D. W.; Allen, C. M.; Allen, C. M.; Allenby, R. J.; Anderson, D. V.; Anderson, David V.; Anderson, G. M.; Arnold, Ralph Gunther; Bacon, William R.; Bain, Ian; Barnes, Frederick Q.; Beales, Francis W.; Bergey, W. R.; Betz, J. E.; Betz, J. E.; Blackadar, R. G.; Blackadar, R. G.; Blais, Roger A.; Blanchard, J. E.; Blanchard, Jonathan E.; Bolton, Thomas E.; Boyle, Robert W.; Bradshaw, B. A.; Brooker, E. J.; Brown, R. A. C.; Bruce, G. S. W.; Cameron, R. A.; Carlson, Hugh Douglas; Carlson, Hugh Douglas; Chambers, J. F.; Christie, R. L.; Christie, R. L.; Christie, Robert Loring; Christie, Robert Loring; Collins, C. B.; Colquhoun, Donald John; Copeland, J. G.; Copeland, J. G.; Copeland, M. J.; Cranswick, J. S.; Cumming, George Leslie; Cunningham, R. C.; Cunningham-Dunlop, P. K.; Currie, John B.; Das Gupta, Samir Kumar; Davies, J. F.; Dawson, Kenneth R.; Dearden, E.; Dell, Carol I.; Earley, J. W.; Eisenbrey, E. H.; Elliott, William J.; Farquhar, Ronald McCunn; Fitzpatrick, M.; Flanagan, J. F.; Forman, Sydney A.; Freberg, R. A.; Giblin, P. E.; Gillespie, W. G.; Gorman, D. H.; Gorrell, H. A.; Graham, Albert R.; Grant, F. S.; Gray, A. B.; Gross, William H.; Haper, H. G.; Harper, H. G.; Harwood, T. A.; Hester, B. W.; Hestor, Brian W.; Hilchey, Gordon; Hill, Vincent G.; Hoadley, John W.; Hodgson, John H.; Hoffer, A.; Hogarth, D. D.; Hogg, J. E.; Hogg, J. E.; Hutchison, W. W.; Innes, G. M.; Innes, Morris, J. S.; Irvine, W. T.; Jackson, William Henry; James, W.; Johns, R. W. C.; Jones, E. A. W.; Jones, R. C.; Jones, R. E.; Kidd, Donald J.; Kilgour, J. A.; Killin, Alan Ferguson; LaRocque, J. A.; Lawton, K. D.; Lemon, R. R. H.; Lemon, R. R. H.; Lemon, R. R. H.; Liberty, Bruce Arthur; Little, William Meldrum; Loudon, J. Russell; Lowre, D. A.; MacDonald, G. H.; MacDonald, Gilbert H.; Mack, Stanley Z.; Macleod, R. B.; MacPherson, H. G.; MacPherson, H. G.; Maynes, A. O.; McAlary, J. D.; McAndrew, John; McConnell, G. W.; McDermott, A. A.; McLen, P. C.; Milne, Ivan Herbert; Milne, V. G.; Mirynech, E.; Mitra, Rabindranath; Mloszewski, M. J.; Morley, Lawrence W.; Morris, H. R.; Mousuf, Abdul K.; Mutch, A. D.; Nelson, R. C.; Newton, A. C.;

Noakes, John E.; Noel, G. A.; Norris, Arnold Willy; Northwood, Thomas D.; O'Brien, P. N. S.; O'Brien, Peter N. S.; O'Flaherty, K. F.; Oldham, Charles H. G.; Ollerenshaw, N. C.; Ollerenshaw, Neil C.; Parkinson, R. N.; Parks, Thomas; Patchett, Joseph Edmund; Paterson, Norman R.; Patku, B.; Peach, Peter A.; Perrault, Guy S.; Perrault, Guy S.; Perry, Thomas G. J.; Pfeffer, Helmut W.; Phipps, C. V. G.; Phipps, C. V. G.; Pye, Edgar G.; Pye, Edgar George; Quadri, Shah M. G. J.; Quigley, R. M.; Robinson, Edwin G.; Robinson, Ernest Guy; Ross, J. S.; Russell, Richard Doncaster; Saha, A. K.; Saha, A. K.; Scott, H. S.; Serson, Paul H.; Shepherd, Norman; Shillibeer, Harry A.; Shillibeer, Harry A.; Sinclair, A. J.; Slack, Harold A.; Slack, Howard A.; Speers, E. C.; Springer, D. G. T.; Stauft, P.; Stevens, J. R.; Stonehouse, Harold B.; Sturm, J. F.; Sutherland, D. B.; Szetu, Sui-Shing; Thompson, Walter H.; Tovall, Walter Massey; Tovell, Walter Massey; Tozer, Edward T.; Van Loan, Paul Rose; Waines, H. R.; Waines, R. H.; Ward, S. H.; Weber, Wilfred W. L.; West, Gordon F.; Whitham, K.; Woo, A. A. Desmond

1960s: Aarden, Henrikus Marinus; Alcock, R. A.; Anderson, G. M.; Azzaria, L. M.; Bain, Ian; Baldwin, Andrew Bennett; Banerjee, Ajit Kumar; Bell, R. T.; Berry, Michael John; Bonham-Carter, Graeme Francis; Boorman, Roy Slater; Boyd, John Malcolm; Brooker, Edward James; Chi, Wen-Wei; Clarke, Gerald K. C.; Clarke, Gerald K. C.; Clee, T. Edward; Cooke, David Lawrence; Cooke, David Lawrence; Crerar, David Alexander; Davies, James Frederick; Dunlop, David John; Dunlop, David John; Edmond, Brian Alexander; Edmond, Katherine Louise; Edmunds, Frederick Robin Kitchener; Ervine, W. B.; Farrar, Edward; Giblin, Peter Edwin; Gibson, David Whiteoak; Gill, Frederick David; Ginn, R. M.; Graterol, Magaly; Grieve, Richard Andrew Francis; Guillet, G. R.; Gupta, Ravindra Nath; Harper, John David; Harris, Donald Clayton; Harris, Donald Clayton; Hastie, L. M.; Hayatsu, A.; Hermance, J. F.; Hore, R. C.; Hunt, Catherine Minna; Hutchison, William Watt; Jackson, Stewart Albert; Janes, Joseph Robert; Jones, A. H. M.; Katz, Michael Barry; Kendrick, Guy; Kick, J. F.; Kingwell, Lorne; Knight, C. J.; Knight, Colin Joseph; Knight, Colin Joseph; Kunar, Lloyd S. N.; Lajtai, Emery Zoltan; Lajtai, Emery Zoltan; Lal, Ravindra Kumar; Lewis, Charles Frederick Michael; Lewis, Charles Frederick Michael; Loudon, J. Russell; Loveless, A. J.; Macintyre, Robert Mitchell; MacKenzie, Warren Stuart; MacKenzie, Warren Stuart; Macqueen, R. W.; Malcolm, T. J.; Maniw, John George; McConnell, R. K., Jr.; Mensah, M. K.; Milne, Victor Gordon; Mirynech, E.; Mitchell, J. G.; Morgan, Nabil Assad; Mountjoy, E. W.; Murphy, D. K.; Neave, Kendal Gerard; Norman, Franklin John; Nwachukwu, Silas Ogo Okonkwo; Oke, William Crompton; Ollerenshaw, N. C.; Ostry, R. C.; Patchett, Joseph Edmund; Pinchin, E.; Pullen, Michael John Leslie Thomas; Purdy, John Winston; Purdy, John Winston; Reik, Gerhard Albert; Renault, Jacques Roland; Ridler, Roland Hartley; Robin, Pierre Yves; Robinson, R. E.; Rogers, David P.; Scott, Gertrude Murray; Scott, James Douglas; Scott, W. J.; Shepard, Norman; Shepherd, Norman; Sibul, U.; Siragusa, Giorgio; Spector, Allan; Symons, David Thorburn Arthur; Tesky, D. J.; Thompson, E. C.; Turner, D. J.; Turner, L. A.; Weaver, J.; Weber, N. E.; Wells, J. M.; Wenban-Smith, Alan Kenneth; White, O. L.; White, Owen L.; Williams, Harold; Williams, Richard M.; Wright, James Arthur; Wright, James Arthur

1970s: Alfonso-Roche, J.; Allis, R. G.; Allis, R. G.; Annan, A. P.; Arengi, Joseph; Arndt, N. T.; Awai-Thorne, B. V.; Bachechi, Fiorella; Bailey, M. E.; Baksi, A. K.; Barnes, Stephen John; Bau, A. F. S.; Bayer, Marvin Benno; Berger, Glenn W.; Bird, Gordon W.; Bottomley, R. J.; Buchan, K.; Buchan, K. L.; Burden, Elliott Thomas; Cermignani, C.; Clark, Thomas; Clee, T. Edward; Comparan, Jose L.; Cubins, Arnis Gunnars; Currie, A. L.; Curtis, L. W.; Dales, R. Graeme; Dampney, C. N. G.; Das

Gupta, U.; Davidson-Arnott, R.; Davies, E. H.; De la Cruz, Servando; De Rosen-Spence, A. F.; Dey-Sarkar, S.; Dey-Sarkar, S. K.; Dillon, Edward Patrick; Dillon, P. J.; Duckworth, P. B.; Dunbar, W. Scott; Duncan, P. M.; Duncan, P. M.; Fisher, D.; Fitchko, Y.; Fletcher, Ian Robert; Fletcher, Ian Robert; Freedman, B.; Ghosh, Mrinal Kanti; Goldstein, Myron A.; Gomez-Trevino, Enrique; Graterol, Victor; Green, A. H.; Greenman, Lawrence; Grieve, Richard Andrew Francis; Grunsky, Eric Christopher; Haddad, P.; Hall, Chris Michael; Halls, Henry Campbell; Hamann, Richard John; Hanes, John A.; Hanes, John A.; Hazell, S.; Hewins, Roger Herbert; Hodych, Joseph Paul; Hoffman, E. L.; Howell, E. C.; Hoyer, Peter Wlater; Huang, Y. F.; Hutchinson, Deborah Ruth; Hutchison, M. N.; Jackson, E. L.; Jarzen, David M.; Jones, Brian A.; Jowett, Edwin Craig; Kennedy, Barbara Ann; Kissin, S. A.; Koziar, A. F.; Kreczmer, Marek Jozef; Kretschmar, U.; Kunar, Lloyd S. N.; Kurtz, Ronald D.; Lajoie, Jules J.; Lamontagne, Yves; Lamontagne, Yves; Lanoix, Monique; Lavoie, Jacques S.; Lean, David Robert Samuel; Lee, H.; Lodha, Ganpat S.; Lodha, Ganpat S.; MacDonald, Alasdair James; MacKidd, David G.; Macnae, James Charles; Madrid-Gonzalez, J. A.; Mainwaring, P. R.; Mak, Eddy K. C.; McMechan, G.; McOnie, A. W.; McWilliams, Michael O.; Miller, Randy Robert; Moore, David William; Morgan, C. L.; Morgan, John; Morley, L. C.; Muir, John E.; Myers, Delbert Edward, Jr.; Narr, Wayne Mark; Neave, Kendal Gerard; Nowina-Zlotnicki, Stephan F.; Nriagu, Jerome Okonkwo; Olhoeft, G. R.; Palacky, George Joseph; Palmer, J. H. L.; Pandit, Bhaskar Iqbal; Pearce, G. W.; Peredery, Walter Volodymyr; Pesonen, Lauri Juhani; Pesonen, Lauri Juhani; Peters, Robert Henry; Pliva, Gustav L.; Prieto, Corine; Ramaekers, P. P. J.; Rannie, W. F.; Reeve, Edward John; Reik, Gerhard Albert; Rodriguez Gonzalez, Argenis; Rossiter, J. R.; Rossiter, J. R.; Runnalls, R. J.; Sabag, Shahé Fares; Sampson, Geoffrey Alexander; Saunderson, H. C.; Sbag, Shahe Fares; Scribbins, Brian Thomas; Sermer, Tamara; Sharpe, H. N.; Sharpe, Robert James; Shegelski, Roy Jan; Shegelski, Roy Jan; Silverberg, B. A.; Silverstone, Brahm S.; Simpson, James William; Siriunas, John Michael; Smith, Joyceanne; Smith, Roy George Gerhard; Sorbara, James Paul; Stesky, Robert Michael; Storer, John Edgar, III; Strung, J.; Suchit, O. H.; Szewczyk, Z. J.; Thompson, John Francis Hugh; Tome, R. F.; Waddington, Dennis Howson; Waddington, J. B.; Wallace, Henry; Watson, R. I.; Watts, Raymond Douglas; Weishampel, David Bruce; Wilson, Jeffrey Warren; Wilson, Mark Vincent Hardman; Wilson, William Hugh; Wong, Joseph; Wong, Joseph; Wright, Gordon R.; Wright, J.; Yanase, Y.

1980s: Abdel Rahman, Abdel Fattah Mostafa; Bailey, Monika Ella; Bambrick, James; Barnes, Sarah-Jane; Barnes, Stephen J.; Bawden, William Frederick; Beckett, Martyn Frank; Bending, David Alexander Glen; Boerner, David E.; Boerner, David Eugene; Booth, Geoffrey Warren; Borthwick, Alastair Andrew; Bottomley, Richard John; Brackmann, Anne Jordan; Bregman, Nina Diane; Bryndzia, L. Taras; Bryndzia, Taras Lubomyr; Bryndzia, Taras Lubomyr; Burns, James Arthur; Burrows, David Robert; Carter, Terry Robert; Cavaliere, Antonio; Chavez-Sequra, Rene Efrain; Cho, Moonsup; Ciccone, Anthony Donato; Clark, Bryan Malcolm; Coron, Cynthia Rose; Costanzo-Alvarez, Vincenzo Francesco; Crawford, Adrian Mercer; Crawford, Mark Justin; Cresswell, Richard George; de Gasparis, Silvana; del Valle, Raul; Desbiens, Rejean; Dhindsa, Rupinder Singh; Dutton, Brian Charles; Dyck, Alfred Victor; Eberth, David Anthony; Eckert, James Douglas; Eiche, Gregory; Enkin, Randolph Jonathan; Ernst, Richard Everett; Eyles, Carolyn Hope; Eyles, Carolyn Hope; Farkas, Arpad; Farr, Jocelyn Elizabeth; Farr, Mark Randall; Fasola, Armando; Flores-Luna, Carlos Francisco; Forgie, David John Leslie; Fralick, Philip William; Garland, Mary Isabelle; Geberl, Hilary Ann Plint; Gill, John F.; Gomez-Trevino, Enrique;

Gyongyossy, Zoltan; Hale, Christopher James; Halim-Dihardja, Marjammanda; Hall, Chris Michael; Hannan, Selim Sarwar; Hannington, Mark Donald; Hardy, Jenna-Lee; Harrison, John Christopher; Hart, Thomas Robert; Havas, Magda; Heider, Franz; Higuchi, Kazufumi; Hodges, William Kaufman; Hogg, Andrew Jenner Cowper; Holladay, John Scott, III; Hurley, Peter; Hyde, William Thomas; Hyodo, Hironobu; Jowett, Edwin Craig; Kalogeropoulos, Stavros I.; Ko, Jachung; Kranidiotis, Prokopis; Larocque, Cynthia; Lau, Ka Ching; Layne, Graham Donald; Lee, James Corbett; Lee, Kenneth; Li, Wan-Bing; Lightfoot, Peter Charles; Liu, Kam-Biu; Liversage, Robert Richard; Lopez-Martinez, Margarita; Lye, Joseph Alexander; MacDonald, Alasdair James; MacDonald, Glen Michael; Macnae, James Charles; Manns, Francis Tucker; Maslliwec, Anatolij; Mazza, Antonio Gennaro; McCarthy, Francine Marie Gisele; McConachy, Timothy Francis; McDonald, Hugh Gregory; McLennan, John David; McMaster, Natalie Dawn; Miller, Randy Robert; Mittler, Peter Robert; Morgan, John; Murck, Barbara Winifred; Nakamura, Eizo; Nobes, David Charles; Noble, S. R.; Noor, Iqbal; Ofeegbu, Goodluck Iroanya; Oladipo, Emmanuel Olukayode; Osadetz, Kirk Gordon; Ovenden, Lynn Elise; Peter, Jan Matthias; Qing, Hairuo; Rasmussen, Patricia Elizabeth; Renders, Peter Joseph Norbert; Richards, Jeremy Peter; Richardson, Jean Madeline; Rivard, Benoit; Robinson, Andrew G.; Rodriguez Gonzalez, Argenis; Samson, Claire; Sanborn, Mary Margaret; Sawyer, Edward William; Seymour, Kevin Lloyd; Shanks, William Scott; Sherman, Douglas Joel; Sivenas, Prokopis; Skulski, Thomas; Smith, Patrick Edmund; Smyth, William David; Solheim, Larry Peter; Spry, Paul Graeme; Stevens, Kirk; Stewart, John Patrick; Stix, John; Stone, Denver Cedrill; Stott, Gregory Myles; Tanaka, Roderick Taira; Tanoli, Saifullah Khan; Tasillo, Anne Marie; Thomas, Anne Valerie; Thompson, John Francis Hugh; Torrance, Jeffrey Gordon; Troop, Douglas Grant; Tuffnell, Pamela Anne; Twyman, James DeWitt; Twyman, Terry Rayno; Urquhart, William Edward S.; Vallee, Marc-Alex; van Bosse, Jacqueline Y.; van Kranendonk, Martin Julian; Vandamme, Luc Michel Pierre; Waheed, Abdul; Weber, Michael H.; Wee, Pamela Sui Lian; Weirich, Frank; Westrop, Stephen Richard; Westrop, Stephen Richard; Wiles, Terry David; Wolf, Detlef Karl-Heinz; Wolfgram, Peter Arthur August; Wood, Peter Colin; Wu, Patrick Pak-Cheuk; Yagishita, Koji; Young, Graham Arthur

Tufts University
Medford, MA   02155

1 Master's

1980s: Belkacemi, Smain

Tulane University
New Orleans, LA   70118

67 Master's, 26 Doctoral

1920s: Reeder, Sarah Jane

1950s: McLaughlin, Robert Everett; Rowett, Charles Llewellyn, Jr.

1960s: Alberstadt, Leonard P.; Beem, Kenneth Alan; Beyer, Robert F.; Carson, Robert James, III; Conatser, Willis Eugene; Cullinan, Thomas Anthony; Dimmick, Charles William; Eppert, Herbert Charles, Jr.; Gertman, Richard Leo; Glaser, Gerald C.; Jones, Albert M., Jr.; Kamp, Katherine Marland; L'Orsa, Anthony Theophile; Moffett, James Robert; Morrow, Harold Julin; Piaggio, Arthur D.; Pine, Clyde A.; Sandy, John Jr.; Scolaro, Reginald Joseph; Scrudato, Ronald John; Sellars, Robert T., Jr.; Sellers, Robert T., Jr.; Vokes, Emily Hoskins; Vokes, Emily Hoskins; Warren, Ray Noble; Zier, Steven Jonathan

1970s: Akers, Wilburn Holt; Beechler, Theodore W.; Boutte, Brian; Campbell, Charles Lillie; Collie, Carolyn; Cooke, James Crawford; Corona, Charles

Jude; Daughdrill, William Eugene; DeBartolo, Bruce Alan; Edson, James E., Jr.; Esteves, Ieda R. Forti; Grimwood, C.; Hopkins, Owen; Hunt, Amanda M.; John, Charles Bedford; Kontrovitz, Mervin; Mackenzie, Michael G.; Mahfoud, Robert; Martin, Bruce J.; Poag, Claude Wylie; Reimers, David D.; Rolling, David M.; Sachs, Jules Barry; Snyder, Scott William; UnKauf, John Cameron; Wakelyn, Brian D.; Wiggins, William David; Wilbert, William Pope; Williams, Cassandra; Wiltenmuth, Kathleen

1980s: Ayoub, Wafic Tawfic; Britsch, Louis D.; Butler, Denise M.; Cooke, Alison; Deremer, Lori Ann; Dominey, Jerry; Dudek, Kathleen B.; Farmer, Jeanne; Furlong, William J.; Gilbert, Lawrence William; Giosa, Thomas A.; Greenwood, E. Allen, Jr.; Halpern, Yvonne; Hegre, Jo Ann B.; Hemming, N. Gary; Herrington, Dawn; Hollander, Eileen E.; Howdeshell, Jeffrey C.; Joseph, George John; Kohl, Barry; Kordesh, Kathleen; Leshner, Orrin; Livieres, Ricardo; Miller, William Charles; Payne, Gary; Pennington, Jerry B.; Rasbury, Sidney A.; Ryan, Robert N., Jr.; Sackheim, Margo J.; Shepard, Nancy; Strachan, Betsy M.; Strider, Mark; Warren, Deborah R.; Woods, Diana M.

University of Tulsa
Tulsa, OK   74104

138 Master's, 31 Doctoral

1940s: Austin, Jesse William; Brant, Ralph A.; Goerner, Hugh H.; Hamilton, Irving B.; Piggott, Guido M., Jr.; Scruton, Philip Challacombe; Wright, Leo Milfred

1950s: Bexon, Roger; Burford, Arthur Edgar; Gore, Clayton Edwin; Hallgarth, Walter Ervin; Hansen, Beauford Victor; Harris, John F.; Hobart, Henry M., Jr.; Hyden, Harold J.; Jaske, Robert J.; Lowry, John C. A., Jr.; Malkoc, Selahaddin; Montgomery, James H.; Morrisey, Norman S.; O'Brien, William R.; Perry, Lawrence Dean; Peterson, Earl Thomas; Phillips, Jonathan Wilton; Schell, Bill Joseph; Smith, B. G.; Smith, Riley Seymour, Jr.; Waller, Michael Reginald; Wilson, Francis Smith; Winland, Hubert Dale

1960s: Akin, Ralph; Anglin, Marion Edward; Austin, C. Thomas; Bacon, Joan Irene; Berg, Orville Roger; Berry, Cameron George; Blankenship, Robert William; Brieva, Jorge A.; Buettner, Peter Erhard; Cartmill, John Craig; Connelly, Michael C.; Creel, David Versal; Cruz, Jaime A.; Desai, Kantilal Panachand; Dogan, Nevzat; Fajardo, Ivan; Fernandez, Louis Anthony; Flores, Romeo M.; Gutierrez, Francisco Javier; Hawisa, Ibrahim Sh; Hudson, Adonnis S.; Huff, Ray V.; Huval, Isaac Martin; Jam, L. Pedro; Jones, Elbert Russ; Juhan, Joe Paul; Keplinger, Henry Ferdinand; Knight, William Victor; Koinm, David N.; Landrum, Ralph Avery; Lawson, James Edward, Jr.; Liu, Hsin-Hsi; Lloyd, Robert A.; Lozano, Jose A.; McGinley, John Robert; McKeague, Gordon Clark; Meyers, William Cady; Miller, Francis Xavier; Miller, Robert E.; Phares, Rod S.; Qualls, Robert Ralph; Rathbun, Fred Charles; Saitta, Bertoni Sandro; Severson, George D.; Shriram, Calcutta R.; Shulman, Chaim; Simpson, Howard Muncie; Thompson, George David; Walsh, Marcus Whitley; Walthall, Bennie H.; Ward, Albert Noll; Zagaar, Abdussalam

1970s: Akinmade, Olufemi Ernest; Al-Laboun, Abdullah Aziz; Albarracin, Jose; Baharloui, Abdolhossein; Ben-Saleh, Faraj F.; Bierley, Robert; Caton, Paul William; Chen, Wu-Shong; Collins, A. Gene; Cooper, William Amos, Jr.; Cunningham, Russ DeWitt; Ekebafe, Samson Bandele; Ensley, Michael; Forero Esguerra, Orlando; Foresman, James B.; Freeman, William E.; Geyer, Robert Lee; Glass, Cecil Robertson; Gossett, Lloyd David; Hagen, Kurt Brian; Hayes, James Joseph; Hobson, Mary Michael; Holcomb, Haden Ray; Jennemann, Vincent Francis; Kolmer, Joseph R.; Kousparis, Dimitrios; Krumme, George W.; Laidig, Larry Wayne; Lalla, Wilson; Lawson, James Edward, Jr.; Lundy, William Leon; Lyday, John Reed; M'rah,

Mustapha; Masrous, Luis Felipe; McMahan, Gregory Lee; Meyers, William C.; Miller, Randy Vernon; Murray, William Wallace; Newcomb, Joseph Judge; Peacock, Kenneth L.; Pereira V., Jesus Orangel; Pita, Frank; Reed, Lanny Joe; Reeder, Louis Robert; Sagoe, Kweku-Mensah Olakunle; Sancar, Mustafa Sitki; Schmidt, Gene W.; Siddiqui, Sayeed Ahmed; Sommer, Michael Anthony, II; Sommer, Michael Anthony, II; Soto Ruiz, Carlos; Stanzel, Theodore Edward; Tassone, Jeffrey Allen; Torkelson, Bruce Emil; Valderrama, Rafael; Vasquez, Enrique Eduardo; Vedros, Stephen G.; Wagenhofer, Paul Joseph; Wai, U. Thit; Youn, Sung Ho; Zaaza, M. W.; Zaaza, Mahomoud Wafaie; Zwart, David

1980s: Becker, Russell Dail; Becker, Thomas Edward; Belvedere, Paul Gerard; Bennett, Curtis Owen; Black, Robert Bernard; Breckon, Curtis Eugene; Carlson, Mikel Carl; Chu, Wei Chun; Daniel, Maria M.; Ece, Omer Isik; Francis, Billy Max, II; Keys, Robert Gene; Lui, Chung-Yao; Meshri, Indurani Dayal; Messa, John Francis; Nwachukwu, Joseph Iheanacho; Ogle, Robert Allen; Ohaeri, Uche Charles; Olson, Robert Kenneth; Prijambodo, Raden; Reber, Jennifer Joy; Serra, Kelsen Valente; Shin, Chang-soo; Watt, Terry L.

Union College
Schenectady, NY 12308

6 Master's, 1 Doctoral

1930s: Benson, A. Raeburn; Parsons, William H.; Toppan, Frederick Willcox; Vaughan, Henry; Wilson, T. Yates

1940s: Stone, Donald B.

1970s: Jackson, G. E.

United States Naval Academy
Annapolis, MD 21402

79 Master's

1960s: Anderson, R. S., Jr.; Breidenstein, J. F.; Brennan, J. F.; Caster, W. A.; Davis, V. H.; Dooley, John J.; Dorman, C. E.; Eubanks, Glen E.; Gatje, P. H.; Glenn, W. H.; Griffin, P. A.; Harlett, J. C.; Harper, J. N.; Hoernemann, M. J.; Hohenstein, C. G.; Hopper, J. F.; I'Anson, Lawrence W., Jr.; Ivey, C. G., Jr.; Jaeger, J. W.; Jones, D. L.; King, J. D.; Koehr, J. E.; Labyak, P. S.; Lennox, R. J.; Meaux, R. P.; Monteath, G. M.; Moritz, C. A.; Neish, J. F.; Njus, I. J.; O'Conner, P.; Pizinger, D. D.; Raines, W. A.; Roberts, C. K.; Rohrbough, R. D.; Stevenson, C. D.; Sturr, H. D., Jr.; Wallin, S. R.; Webb, L. E.

1970s: Arfman, John Frederick, Jr.; Berg, John Stoddard; Bieda, George E.; Bollow, George Edward; Booth, Gregory Seeley; Brooks, Robert Andrew, Sr.; Brueggeman, John Lyle; Carlmark, Jon William; Carter, Lee Scott; Cepek, Robert Joseph; Colomb, Herbert Palfrey, Jr.; Cronyn, Brian Sullivan; Dias Souto, Antonio Pedro; Edelson, Stuart K., Jr.; Engel, Gregory Allen; Frederick, Margaret A.; Heck, Jerome R.; Henderson, Joseph C.; Hoag, Robert W., II; Howell, Buford Fredrick; Hunter, William Patterson; Kazanowska, Maria; Kramer, Steven Barker; Lodge, Charles D.; Malone, Michael J.; Martinek, Charles Allen; McKay, Dennis A.; Miller, Ralph Rillman, III; Morgan, John Henry, II; Peterson, John Christian; Shaar, Edwin Willis, Jr.; Simpson, John Page, III; Singler, James Charles; Smith, Dan Howard; Spikes, Clayton Henry; Voelker, George Edmund; Walsh, William Egan, Jr.; Westfahl, Richard K.; Williamson, John David; Woodson, Walter Browne, III; Zardeskas, Ralph Anthony

University of Utah
Salt Lake City, UT 84112

549 Master's, 202 Doctoral

1900s: Fox, Feramorz; Umpleby, Joseph Bertram

1910s: Nokes, Charles Mormon, Jr.; Stott, George F.; Wegg, D. S., Jr.

1920s: Clark, Albert F.; Gray, Ralph S.; Siegfried, Joshua Floyd; Siegfus, Stanley Spencer; Tanner, Vasco Myron

1930s: Bryan, G. Gregory; Byron, G. G.; Childs, Orlo E.; Christiansen, Francis Wyman; Conkhite, George; Cronkhite, George; Hodgson, Russell Beales; Kemmerer, John L.; Kemmerer, Mahlon; Marsell, Ray Everett; Paris, Oliver L.; Reiser, Allan R.; Wilcken, Phyllis D.

1940s: Anderson, Robert G.; Birch, Rondo O.; Bullock, Nedra D.; Chorney, Raymond; Edvalson, Frederick Merlin; Erickson, Max Perry; Jacobsen, Alfred Thurl; Keller, Kenneth Frank; Lambert, Hubert C.; Lofgren, Benjamin E.; Morris, Hal T.; Scott, W. F.; Sharp, Byron J.; Thomas, Wayne Barker; Walton, Paul Talmadge; Whipple, Ross

1950s: Ames, Lloyd Leroy, Jr.; Ames, Lloyd Leroy, Jr.; Amin, Surendra R.; Anderson, Barbara; Anderson, Warren LeGrande; Arnold, Dwight Ellsworth; Austin, Carl Fulton; Austin, Carl Fulton; Baker, Walker Holcombe; Barrett, David Wilburn; Bauer, Herman Louis, Jr.; Beard, John H.; Bell, Gordon Leon; Bermes, Boris John; Bowes, William A.; Brooke, John Percival; Buckner, Dean Alan; Burger, John Allan; Carlson, Allan Eugene; Cohen, Carel Lodewijk David; Cohenour, Robert Eugene; Cohenour, Robert Eugene; Coody, Gilbert L.; Cooper, Laurence C.; Dahl, Charles Laurence; Dolan, William M.; Donovan, Jack H.; Duke, David Allen; Earll, Fred Nelson; Edmisten, Neil; Egbert, Robert Lamar; Ehlmann, Arthur J.; Eskelson, Quinn Morrison; Evans, Hilton B.; Everett, Kaye R.; Frischknecht, Frank Conrad; Gardner, Weston Clive; Gauger, David Justin; Gerwels, Richard P.; Gin, Thon Too; Gin, Thon Too; Glissmeyer, Carl Howard; Green, Paul Reed; Griesbach, Frederick Richard; Groff, Sidney Lavern; Hansen, Alan Ray; Hinckley, David Newyn; Holt, Robert Eugene; Hooper, Warren G.; Jerome, Stanley Everett; Johnson, John Burlin, Jr.; Johnson, Melvin C.; Johnson, William W.; Johnston, William Percy; Keller, Allen Seely; Keller, George H.; Kim, Daniel Yon Su; King, Alan G.; King, Norman J.; Lankford, Robert Renninger; Laraway, William Harlan; Larsen, Willard N.; Larsen, Willard N.; Larson, Kenneth Williams; Laub, Donald; Liese, Homer C.; Lum, Daniel; Madsen, James Henry, Jr.; Maise, Charles Richard; Mandel, Peter, Jr.; Marquardson, Kent F.; Martin, Bruce Delwyn; McDougald, William D.; McEuen, Robert Blair; Merrell, Harvey Webb; Morris, Elliot C.; Moss, Steven A.; Mount, Donald Lee; Mumcu, Hasan H.; Murphy, William Owen, Jr.; Myers, Richard Lee; Narans, Harry Donald, Jr.; Nixon, Robert Paul; Novotny, Robert T.; O'Toole, Walter Leonard; Olsen, Donald R.; Paddock, Robert Edwards; Parry, William Thomas; Payne, Anthony; Peterson, Reed H.; Phillips, William Revell; Phillips, William Revell; Randall, Arthur G.; Regis, Andrew J.; Remington, Newell C.; Resler, Ray Chester; Rogers, Allen S.; Root, Robert Lee; Sadlick, Walter; Schick, Robert Bryant; Schreiber, Joseph Frederick, Jr.; Sharp, Byron J.; Slawson, William Francis; Slentz, Loren Williams; Smith, Glenn Scott; Smith, Theodore Lee; Stark, Norman Paul; Steuer, Fred; Stewart, Samuel Woods; Sylvester, James F.; Teichert, John A.; Viksne, Andris; Weintraub, Judy Montoya; Welsh, John Elliott; Wheelright, Mona Yvonne; Willden, Charles Ronald; Wood, William James; Young, John C.

1960s: Abou-Zied, Mohamed Saleh; Acosta, Alvaro; Adair, Donald H.; Afify, Abdel-Azeem Hendy; Allen, Jimmy E.; Alvarez, Leonardo Schultz; Amisial, Roger A. Gabriel; Anderson, Barbara Jean; Anderson, Christian Donald; Anderson, Darrell John; Anderson, Norman Roderick; Anderson, Paul Leon; Baetcke, Gustav Berndt; Balsley, John Kimball; Bardsley, Stanford Ronald; Barnett, Jack Arnold; Beer, Lawrence Peter; Belliston, William Hilton; Benvegnu, Carl Jerome; Blue, Donald M.; Bohidar, Naikananda K.; Botbol, Joseph Moses; Botbol, Joseph Moses; Bray, Robert Eldon; Brice, Glyn Alan; Bright, Robert C.; Bright, Robert C.; Brooke, John Percival; Brox, George

Stanley; Buttgereit, Charles D.; Byrd, William David, II; Cardone, Anthony Thomas; Cargo, David Niels; Chapusa, Frank Winthrope Peter; Clement, Stephen C.; Coles, Joan Link; Condie, Kent C.; Costain, John Kendall; Cox, Rulon Walter; Crosson, Robert Scott; Dalness, William Michael; Day, Blaine Spencer; Doelling, Helmut Hans; Duke, David Allen; El Mahdy, Omar Rasheed; El Shatoury, Hamad Mohamed; Epstein, Bernard; Eriksesson, Yves; Fisher, Donald Gene; Fox, Richard C.; Galli, Carlos Alberto; Galli, Carlos Alberto; Gardiner, Errol Murray; Garvin, Robert Franklin; Gates, Joseph Spencer; Ginsburg, Merrill Stuart; Gomah, Aly Hemedah; Goris, James; Grant, Sheldon Kerry; Gray, Robert Charles; Gray, Russell L.; Gray, Stephen Ralph; Groenwold, Bernard Cyrus; Gross, Larry Thomas; Hackett, Gary Kenneth; Halverson, Mark O.; Hamil, Brenton McCreary; Hansen, Leon Alden; Hardman, Elwood; Harnett, Richard Allen; Harrill, James R.; Hashad, Ahmad Hassanain; Heiner, Ted, Jr.; Hepworth, Richard Cundiff; Heylmun, Edgar Baldwin; Hoagland, James A.; Hoffman, John Paul; Holmes, Clifford Newton; Horvath, Edward Alexander; Hoskins, John Richard; Isherwood, William F.; Jeong, Bongil; Johnson, Paul Howard; Kahle, Michael Brinkman; Kayser, Robert Benham; Kent, Ray Clarke; Khan, Mohammud Attaullah; Khattab, Mohamed Mamdouh; Kolvoord, Roger Williams; Leamer, Richard James; Leonard, Fred Andrew; Liese, Homer C.; MacKenzie, Michael Vincent, Jr.; Mardirosian, Charles Azad; Marine, Ira Wendell; Matzner, Ingrid Adelheid Maria; McMurdie, Dennis Stoddard; Miller, Dale Everett; Millgate, Marvin Leroy; Mitra, Devi; Moon, Morgan Ray; Morin, Wilbur Joseph; Morin, Wilbur Joseph; Mortensen, Kay Sherman; Moussa, Mounir Tawfik; Mudgett, Philip Michael; Murany, Ernest Elmer; Murthy, Nallur Prahlada; Neff, Thomas Rodney; Nicol, Alan Boswell; Nohara, Tomohide; Norris, John W.; Oderkick, Jerry Ray; Olmore, Stephen Duane; Olsen, Donald R.; Olson, Richard Hubbell; Orlansky, Ralph; Park, Gerald; Parr, Clayton Joseph; Parry, William Thomas; Price, Charles Edgar, Jr.; Quitzau, Robert P.; Reinhart, Wilbur Allen; Ridd, Merril Kay; Rinehart, Wilburn Allan; Roberts, Philip Kenneth; Robson, Richard Michael; Rodriguez, Enrique Levy; Sadlick, Walter; Saleknejad, Hossein; Salter, Robert Joel; Sayyah, Taha Ahmed; Schaeffer, Frederick Ernst, Jr.; Schiller, Edward Alexander; Schwind, Joseph John; Seegmiller, Ben Lorin; Seegmiller, Ben Lorin; Seeland, David Arthur; Setty, M. G. Anantha Padmanabha; Shifflett, Howard Richard; Shroder, John Ford, Jr.; Smith, Hugh Preston; Smith, Robert Baer; Snow, Geoffrey Greacen; Snow, Geoffrey Greacen; Solak, Mustafa Remzi; Sontag, Richard Joseph; Stanley, Donald Alvora; Stephens, James D.; Stepp, Jesse Carl; Stifel, Peter Beekman; Stone, Dwayne David; Stouffer, Stephen Gerald; Strawn, Mary Baker; Suekawa, Harry S.; Tanis, James Iran; Temple, Dennis Charles; Thompson, C. Sheldon; Tint, Maung Thaw; Tolman, Richard Robbins; Turner, Gerry H.; Turner, Gerry H.; Viele, George Washington; Vlam, Heber Adolf; Wiley, Dennis Roy; Wilson, William Harold; Wingate, Frederick Huston; Wood, Lawrence Charles; Wright, Phillip Michael; Young, Dae Sik; Zazou, Samiha Mahoud; Zbur, Richard Thomas; Zimbeck, Donald Allen; Zimmerman, James T.

1970s: Albers, Sherly Hammond; Allen, Douglas Ray; Alley, Lonnie Bruce; Ambjah, Rachmadi; Andersen, David William; Andersen, David William; Anderson, Jerry Myron; Anderson, Paul Bradley; Baghoomian, Ovaness; Bailey, James Peter; Barnes, Marvin Peterson; Baroyant, V.; Barrett, Larry Frank; Beck, Paul J.; Beers, Armand Henry; Blackett, Robert Earl; Blakey, Ronald Clyde; Bolland, Robert Finley; Bones, Dennis George; Bowdler, Jay Laurence; Bowers, Dale; Braile, Lawrence Wendell; Bridwell, Richard Joseph; Brown, Robert Parker; Brumbaugh, William Donald; Bryant, Nancy Lee; Bryant, Peter Franklin; Bucher, Robert Louis, II; Buck, Brian Willima; Bucurel, Hildred Gail;

Butkus, Timothy Anton; Calkins, William G.; Campbell, Douglas Patrick; Campbell, Edith Ciora Allison; Campbell, Jack Albert; Carey, Mary Alice; Carter, James Allen; Case, Robert William; Caskey, Charles Frederick, Jr.; Christensen, R. J.; Christiansen, William Joseph; Cleary, Michael Duane; Cole, Rex Don; Collings, Gay Madsen; Corry, Charles Elmo; Coyuran, Vedat; Crebs, Terry Joseph; Crockett, Frederick James; Croes, Marc Kalman; Dedolph, Richard Edwin; DeWitt, Grant Whitney; Downey, Lewis Marshal; Doyuran, Vedat; Eppich, Gilbert Keith; Erler, Yusuf Ayhan; Estill, Robert Eugene; Evans, Stanley H., Jr.; Evoy, Jeffrey Allen; Faddies, Thomas Blair; Fishman, Howard Stephan; Frank, T. D.; Free, Michael Royce; Freidline, Roger Alan; Gaiser, James E.; Gallacher, Mark Hayes; Gertson, Rodney Curtis; Glenn, William Edward; Grogger, Paul Karl; Gronseth, Kenneth Allen; Guenther, Edwin Michael; Gwynn, John Wallace; Haddadin, Munir Abdullah; Halliday, Mark Everett; Hammond, David Richard; Hampton, Donald Arthur; Hamtak, Frank James; Han, Uk; Han, Uk; Harp, Edwin Lynn; Hausel, William Dan; Hawley, Bronson Waugh; Heaney, Richard John; Hedberg, Leonard L.; Hendrajaya, Lilik; Herr, Randy Gerard; Hollett, Douglas Whitlock; Howell, James Robert, III; Hurley, William Daniel, Jr.; Hutsinpiller, Amy; Inman, Joseph Robert, Jr.; Jacobs, David Cal; Jansons, Uldis; Jensen, Lynn E.; Johnson, Eric Henry; Johnson, Martin; Johnson, Martin; Jones, Bradley William; Kastrinsky, Alan Jay; Killpack, Terry Joe; Kilty, Kevin Thomas; Kimball, B. A.; King, Guy Quintin; Klein, James David; Lane, Jerry Leroy; Leefang, Willem Evert; Leeflang, W. E.; Lenzi, Gary Wilson; Lessard, Robert Henry; Liang, Dah-Ben; Lindenburg, George J.; Mainzer, George Frederick; Mann, Daven Craig; Martz, Alan Matthew; Mase, Charles W.; Mason, James Leighton; Matthews, James Emory; Maurer, Robert Eugene; Maxwell, Theodore Allen; Maxwell, Theodore Allen; May, Thomas Wayne; McCarter, Michael Kim; Michaels, Paul; Mikulich, Matthew Jonathan; Miller, Charles David; Miller, John David; Monsalve, Obdulio Alfonso; Monson, Lawrence Milton; Montgomery, Jerry R.; Morley, Lloyd Albert; Nellis, Jose Carlos; Nelson, Michael Earl; Nye, Roger K.; Oldroyd, John David; Olmore, Stephen Duane; Olson, Tes Lewis; Otis, Robert Michael; Otto, Ernest Paul; Parker, Robert Alan; Paulsen, Thomas Arne; Pavlis, Terry Lynn; Pelton, John R.; Pelton, William Harvey; Petersen, Carol Ann; Petersen, David Ward; Peterson, James B.; Podrebarac, Thomas Joseph; Price, David Tennyson; Pridmore, Donald Francie; Purvance, David Thomas; Randolph, Robert Lee; Reblin, Michael Thomas; Rees, Delbert Clyde; Rene, Raymond Morgan; Rijo, Luiz; Sadeghi, Ali Reza; Salmassy, Vladimir Baroyant; Sawyer, Robert Frank; Saxon, Fred Chalmers; Sayre, Robert Lewis; Schellinger, David Kenneth; Schilly, Michael McKernan; Schroedl, Alan Robert; Schurer, Victoria Christine; Selk, Donald Clair; Smith, Bruce Dyfrig; Smith, James Thomas; Smith, Rebecca Ann Pope; Smith, Timothy Ben; Snow, John Humphrey; Solbczyk, Stanley Michael; Stanley, William Dal; Staub, Ann Marie; Stodt, John Allan; Su, Bo-Chin; Swanson, Stephen Robert; Tew, John H.; Thangsuphanich, Ittichai; Thompson, Kenneth; Thompson, William David; Thomson, Kenneth Clair; Trevena, A. S.; Trimble, Allen Ben; Trimble, Larry Merc; Tripp, Allan C.; Turley, Charles H.; Ugland, Richard Olav; Van Deventer, Bruce Robert; VanArsdale, Roy Burbank; Vaninetti, Gerald Eugene; Varney, Peter J.; Vaughn, Rodney Lynn; Villas, Raimundo Netuno Nobre; Wang, Yun Fei; Wender, Lawrence Edwin; White, William Wesley, III; Williamson, Charles Ross; Wilson, James Robert; Wilson, Wesley Raphiel; Winkler, Gary R.; Winkler, Patrick Lynn; Wong, Ivan Gynmun; Yen, Fu-Su; Young, Dae Sik; Yusas, Michael Ray

1980s: Ackerman, Dawn Ramsey; Adams, Michael C.; Adhidjaja, Jopie Iskandar; Adhidjaja, Jopie Iskandar; Afrasiabi, Hedayat; Ahmed, Ahmed

Zakaria; Ajlani, Mohammad Ghiath; Aksell, Allan Carl; Alley, Sharon L.; Anderson, Robert; Bakhtar, Khosrow; Ballantyne, Geoffrey Hugh; Ballantyne, Judith Mary; Banks, Elizabeth Young; Barker, Craig Alan; Bartel, David Clark; Bashore, William McClellan, Jr.; Bauer, Michael Steven; Beck, Susan L.; Benz, Harley Mitchell; Benz, Harley Mitchell; Bjarnason, Ingi Thorleifur; Bodell, John Michael; Boschetto, Harold Bradley; Bowling, David Lynn; Bracken, Bryan Reed; Bradley, Michael Dennis; Brokaw, Mark Alan; Bromley, Karl Sydney; Butterworth, Nancy Ann; Bye, Bethany Ann; Cady, Candace Clark; Cameron, Robert E.; Carlston, Karen Jean; Carter, Carl Mitchell; Chen, Gianming James; Chu, Jean Juming; Clawson, Steven Ralph; Clement, Monica Diane; Congdon, Roger Duane; Connelly, Michael Peter; Cook, Stephen James; Covert, John Joseph; Crecraft, Harrison Ruffin; Curtis, Janet; da Silva, Joao Batista Correa; Davies, Stephen Farrel; Davis, Deborah Ann; Deming, David; DePangher, Michael; DePangher, Michael; DeSisto, John A.; Di Guiseppi, William Harris; Dilts, Roger David; Doser, Diane Irene; Doser, Diane Irene; Earnest, Patricia Miller; Eaton, Perry Alan; Eaton, Perry Alan; Eddington, Paul Kendall; Edquist, Ronald K.; Fassio, Thomas D.; Feibel, Craig Stratton; Fisher, Susan Richards; Flis, Marcus F.; Friz, David R.; Fuchs, William Arthur; Furlong, Kevin Patrick; Gabbert, Stephen Charles; Gallagher, Dan Jeffrey; Gallagher, Peggie R.; Gants, Donald G.; Gardner, John Darrell; Gerstner, Michael Roy; Gibler, Pamela R.; Gorman, Angela K.; Gorman, Charles M.; Green, Ronald Thomas; Griscom, Melinda; Groenewold, John Carl; Gunderson, Brian M.; Guth, Lawrence Roland; Haileab, Bereket; Hajitaheri, Jafar; Hansen, Marcia Elaine; Hansen, Ronald Lee; Haynes, Steven Anthony; Hill, Julie Anne; Hills, Scott Jean; Hinchman, Judith Anne; Hindman, James Richard; Hollis, James Richard; Holmes, Kurt Quentin; Hopkins, Debbie L.; Horine, Robert Lee; Houghton, Wendy Priestley; Howell, Jack; Huertas, Fernando; Huntoon, Jacqueline; Ingall, Ellery D.; Isby, John Scott; Jennison, Margo J.; Jewell, Paul William; Johnson, Daniel Paul; Julander, Dale Richard; Karya, Kim Aiko; Keckler, Douglas James; Keho, Timothy Henson; Kemp, William Madison; Kilty, Kevin Thomas; King, Guy Quintin; Kjos, Einar Jarle; Klein, James David; Knight, Russell Vincent; Kropp, Walter Paul; Kruer, Stacie Ann; Krumbach, Keith Ronald; LaBreque, Douglas John; Langford, Richard Parker; Larson, Mark J.; Lehman, Jay A.; Leu, Ling-Ling Lillian; Lin, Liang Ching; Lindquist, Robert C.; Lippoth, Richard Edward; Little, Thomas Marvin; Loeb, Derek T.; Lord, Gregory David; Luchetti, Cynthia A.; Lynch, John Scott; Lynch, William Charles; Maarouf, Abdelrahman Mohammad Shafik; Massoth, Terry Wayne; Matheny, James Paul; Matulevich, Myrna Rae Monk; McCandless, Tom Elden; McKee, Mary Eileen; Meyer, Brenda S.; Misra, Manoranjan; Moon, Hyunkoo; Moon, Hyunkoo; Morrison, Stan Jay; Muller, Leigh Neville; Newkirk, Deborah J.; Newman, Gregory Alex; Newman, Stephen Lars; Nielson, Russell LaRell; Nordquist, Gregg Anson; Norman, Elizabeth A. S.; Novak, Mark Thomas; Ochs, Steffen; Osakada, Did; Oviatt, Charles Gifford; Owens, Thomas Joseph; Owens, Thomas Joseph; Paul, Bradley Compton; Pavlis, Terry Lynn; Pederson, Bernhardt L.; Peindo, Jorge Fernando; Pellerin, Louise Donna; Petersen, James Frederick; Petrick, William Robert; Pfaff, Bruce Justin; Pinto-Auso, Montserrat; Planke, Sverre; Plavidal, Kay Rosalie; Price, Kevin Paul; Riess, Stephani Kay; Rogers, Robert John; Rohrs, David Tullar; Roxlo, Katherine Spencer; Sabisky, Matthew Andrew; Sack, Dorothy Irene; San Filipo, William Anthony; Sandberg, Stewart Kim; Serpa, Laura Fern; Shea, Robert Michael; Shrier, Tracy; Smith, Kelsey Anne; Sobocinski, Robert Walter, Jr.; Solomon, Douglas Kip; Stearley, Ralph Francis; Steemson, Gregory Hugh; Stevens, Mark Gerald; Stodt, John Allan; Susong, David Dunbar; Sweeney, Mary Jo; Tafuri, William Joseph; Thompson, Troy Richard; Thorbjarnardottir, Bergthora; Ting, San Chen-Shin; Tippie, Mark William;

Tokarz, Marek T.; Tomten, David Charles; Tripp, Alan Craig; Tu, Jizheng; Turner, David Raiford; Tygesen, Jeffery Dean; Uygur, Kadir; Uygur, Kadir; Van Dam, Dale A.; Viveiros, John J.; Wachtell, Douglas Lowell; Wang, Ning-Wu; Wannamaker, Philip Ein; Weis, Michelle A.; West, Richard C.; Western, Wayne H.; Whited, Joseph Michael; Willett, Sean D.; Williams, Donna Jo; Wilson, Paula Nelson; Wilson, Wesley Raphiel; Win, Pe; Winn, Peter Stewart; Wong, Ken

Utah State University
Logan, UT 84322

106 Master's, 75 Doctoral

1920s: Cooley, Lavell I.; Peterson, Harold

1930s: Peterson, Victor E.; Young, J. Llewellyn

1940s: Hanson, Alvin M.; Maxey, George Burke; Yolton, James S.

1950s: Adamson, Robert D.; Beus, Stanley S.; Ezell, Robert L.; Gardner, Walter Hale; Gelnett, Ronald H.; Haynie, Anthon V., Jr.; Prammani, Prapath; Willard, Allen Dale

1960s: Adamek, James Conrad; Adams, O. Clair; Axtell, Drew Cunningham; Bittinger, Morton Wayne; Budge, David Rush; Buenaventura, Alfredo Capistrano; Davidson, Dean Frederick; Davis, Clinton L.; Dover, R. Joseph; Eliason, James F.; Hafen, Preston L.; Hansen, Steven Charles; Hariri, Davoud; Judd, Harl Elmer; King, Harley Dee; Majumdar, Dalim Kumar; Maw, George Glayde; Milligan, James Homer; Mostafa, Abd-elmonem Sayed-ahmad; Murdock, Clair N.; Narayana, V. V. Dhruva; Neyestani, Mohammad; Nyquist, David; Rao, Manam Venkata Panduranga; Riley, John Paul; Sanders, David Thomas; Smith, Robert B.; Taylor, Michael Evan; van de Graaff, Fredric R.; VanDorston, Philip L.; Wach, Phillip Hanby; Williams, Edmund Jay

1970s: Biesinger, James C.; Bilbey, Sue Ann; Blau, Jan G.; Blood, W. H.; Brockway, C. E.; Burton, Steven Mark; Chappelle, John C.; Chatelain, Edward Ellis; Chery, D. L., Jr.; Dadkhah, Manouchehr; De Vries, George A.; DeGraff, Jerome Vernon; Dixon, L. S.; Fifield, J. S.; Finney, B. A.; Francis, George Gregory; French, Don E.; Fuller, Richard H.; Galloway, Cheryl Leora; Gardiner, Larry L.; Gray, Wayland E.; Howes, Ronald Clarence; James, William Calvin; Jones, Craig T.; Kaveh, F.; Krishna, J. Hari; Loope, Walter Lee; Malone, Ronald F.; Mayor, Jerrold N.; McClurg, Larry William; Mecham, Brent H.; Mendenhall, Arthur J.; Miller, Judith M.; Natur, F. S.; Perkins, William D.; Rauzi, Steven L.; Raymond, Larry C.; Renk, R. R.; Robertson, George C., III; Sakhan, Kousoum South; Schulingkamp, Warren John, II; Shearer, Jay Nevin; Shewman, Frederick Charles; Spalding, James Simon; Sprinkel, Douglas A.; Sweide, Alan P.; Thomas, William Dennis; Tulucu, K.; Twedt, Thomas J.; Van Luik, A. E. J.; Wakeley, Lillian Donley; Willard, Parry Don

1980s: Abbas, Fadhil Migbel; Al-Hassan, Sumani; Amrhein, Christopher; Aryani, Cyrus; Bergado, Dennes Taganajan; Bhasker, Rao Kidiyoor; Boss, Stephen K.; Brown, Aaron Donald; Buterbaugh, Gary Jay; Caoili, Abraham Albano; Chaiyadhuma, Wirote; Conrad, Keith T.; Crook, Stephen R.; Cundy, Terrance William; Dadkhah, Arsalan; Davis, Matthew C.; de Groot, Philip Henry; De Leon, Alfredo Aniano; de Vries, Janet L.; Deckelman, James A.; Deputy, Edward James; Fairbanks, Paul E.; Farmer, Eugene Edward; Finnie, John Irwin; Garr, John D.; Green, Douglas A.; Greenman, Elizabeth R.; Hamp, Lonn P.; Hansen, David Ernest; Hansen, Roger Dennis; Hare, E. Matthew; Hay, Howard William, Jr.; Hines, Gary Keith; Jadkowski, Mark Andrew; James, William Robert; Jones, Alice Jane; Kerr, Steven Brent; Kienast, Val A.; Kohler, James F.; Liu, Jack Shan; Liu, Win-Kay; Lowe, Michael V.; Ludvigsen, Phillip John; Maase, David Lawrence; Madabhushi, Govindachari Venkata; Mahmood, Ramzi Jamil; Malek,

Ali; McGurk, Bruce James; Morgan, Susan K.; Nelson, Craig V.; Nezafati, Hooshang; Olesen, Marc H.; Pack, Robert Taylor; Park, Kapsong; Paydar, Zahra; Peterson, Stanley Ross; Phillips, Johnnie O., Jr.; Puchy, Barbara J.; Raubvogel, David R.; Rice, John B., Jr.; Rice, Karen C.; Rich, Thomas B.; Robison, Robert M.; Rogers, Daniel T.; Rooyani, Firouz; Russell, Scott Lewis; Saleh, Ali; Samadi, Suleiman Afif; Scarbrough, Bruce E.; Schirmer, Tad William; Selby, Douglas Allen; Sepehr, Mansour; Sharp, Kevan Denton; Shearer-Fullerton, Amanda; Smith, Kent W.; Voit, Roland L.; Wang, Yunshuen; Wuerch, Helmuth Victor, III; Yassin, Adel Taha; Zakaria, Abdul Aziz; Zelazek, David Paul; Zomorodi-Ardebili, Kaveh

Vanderbilt University
Nashville, TN 37235

92 Master's, 7 Doctoral

1890s: Brown, Calvin Smith; Jones, Paul M.

1900s: Jones, Ernest Victor; Pugh, Griffith Thompson; Wood, Arthur Eugene

1910s: Blake, Vachel

1920s: Andrews, Thomas G.; Francis, Mary Lee; Lollar, Earl H.; Meacham, Reid Phillip; Peoples, Joe Webb; Pilkington, Edgar M.; Pilkinton, Edgar; Sandidge, John R.; Stovall, John Willis; Wilson, Charles William, Jr.

1930s: Allen, Harris Hughes; Born, Kendall Eugene; Burwell, Howard Beirne; Holt, Richard D.; McCampbell, John; Mistler, Alvin Jess; Ross, Robert Motague; Spain, Ernest

1940s: Alexander, Frank McEwen; Beimfohr, Oliver Wendell; Ferguson, Herman H.; Hardeman, William D.; Harris, Lloyd Addis; Moore, Elmer Glendon; Spain, David; Stearns, Richard Gordon

1950s: Barnes, Laverne Ellsworth; Barnes, Robert Howell; Colvin, John McRae; Davis, Clarence Jackson; Floyd, Bobby Joe; Hiers, Miles Terry; Ivey, John Barn; Jewell, John William; Luther, Edward Turner; Marrow, William Earl; McCary, Charles Edgar Little; Miles, Alfred; Mitchum, Robert Mitchell, Jr.; Morrow, William Earl; Oden, Thomas Ellsworth; Rascoe, Bailey; Rose, William · D., Jr.; Statler, Anthony Trabue; Tertz, James; Wertz, James Claude

1960s: Ferguson, Carl Council; Finch, Richard C.; Ganster, Maurice W., II; Hatcher, Robert; Hughes, Travis Hubert; Persson, Lars Evar; Puryear, Sam M.; Wilson, John M.; Woodruff, Charles M., Jr.; Zack, Allen Lad

1970s: Benedict, G. L., III; Berquist, Carl R., Jr.; Brooke, Jefferson Packard; Eilender, Herbert; Garrett, M. W.; Haselton, Thomas M.; Hoblitzell, Timothy A.; Lockyear, Eugene David; Mallory, M. J.; Matthews, Larry Edwin; Miller, R. A.; Moore, Nelson Kinzly; Moran, Mary Shanks; Smith, David Burl; Smith, James Ronald; Sprinkle, C. L.; Steidl, P. A.; Tsau, Jau-Ping; Wade, W. J.; White, S. L.; Woods, Ella Jean; Zuppann, C. W., Jr.; Zurawski, Ronald Philip

1980s: Brandes, William Frederick; Deibler, Deborah; Figueroa-Garcia, Eduardo Anibal; Gilmore, Herman Lee, Jr.; Hay, David Evan; Horton, Albert Bergen; Jaffe Duchmann, Peter Rudolf; Miller, James L.; Nataraj, Mysore Subbarao; Sauve, Judith Ann; Sciple, Larry; Sparkes, Ann Katherine; Stanonis, Frank L., III; Syriopoulou, Dimitra

University of Vermont
Burlington, VT 05405

62 Master's, 2 Doctoral

1910s: Atwood, Alfred R.

1930s: Bailey, Clarence G.; Carleton, Natalie E.

1950s: Migliore, John J., Jr.

1960s: Blakeman, William B.; Clement, Richard F.; Cline, Lawrence B.; Dinger, James S.; England,

Evan J.; Fillon, Richard H.; Jenks, Maurice; Johnson, David G.; Kasvinsky, J. Robert; Kodl, Edward; Millett, John; Ogden, Duncan G.

1970s: Acomb, T. J.; Agnew, P. C.; April, Richard H.; Arndt, Richard E.; Aubrey, W. M., III; Becker, L. R.; Bottner, R. W.; Caldwell, Katherine G.; Chase, Jack S.; Closs, L. Graham; Corneille, E. S., Jr.; Detenbeck, J. C.; Johnson, Philip H.; Knapp, D. A.; Kolar, B. W.; Malter, John A.; Marcotte, R. A.; McHone, J. G.; Mullen, John C.; Noyes, J. E.; Sarkesian, Arthur C.; Sherman, John W.; Slavin, E. J.; Stone, Byron D.; Thompson, P. J.; Thompson, Roger B., Jr.; Thompson, Woodrow B.; Thresher, John E., Jr.; Townsend, Peter H.; Turner, P. J.; Villamil, R. J., Jr.; Waite, Burt A.

1980s: Butler, Robert Grant, Jr.; Christe, Geoff; DelloRusso, Vincent; DiPietro, Joseph Anthony, III; Dowling, William M.; Haydock, Samuel Rotch; Jammallo, Joseph M.; Lapp, Eric Tod; Leonard, Katherine Esther; MacLean, David Alexander; Myrow, Paul Michael; O'Loughlin, Sharon Beth; Parker, Ronald Lewis Michael; Pyke, Anne Rutherford; Rathburn, Anthony Earl; Whittemore, Arthur Snow, III

University of Victoria
Victoria, BC V8W 2Y2

10 Doctoral

1970s: Dawson, Trevor William; Nienaber, Wilfred; Ogunade, Samuel Olumuyiwa; Thomson, David James; Wuorinen, Vilho

1980s: Brown, Sharon-Dale; Heard, Garry John; Hu, Wenbao; Schwarz, Ursula Agnes Maria; Stooke, Philip John

University of Virginia
Charlottesville, VA 22903

172 Master's, 31 Doctoral

1890s: Peebles, John Kevan

1900s: Lambeth, William Alexander

1910s: Taber, Stephen

1920s: Bass, Charles E.; Burfoot, James D., Jr.; Cardwell, Dudley H.; Furcron, Aurelius Sidney; Lewis, Mordecai, II; Longdale, John Tipton; Oder, Charles Rollin Lorain; Robeson, John Maxwell, Jr.

1930s: Anderegg, Fred; Bloomer, Robert Oliver; Bowman, Paul W.; Brown, William Randall; Cocke, Elton Cromwell; Dennis, Wilbert C.; Edmundson, Raymond Smith; Gildersleeve, Benjamin; Hazard, Allan W.; Kearfoot, Carl; Maddex, Robert M.; Manning, Leslie D.; McGavock, Cecil B., Jr.; Moore, Charles H., Jr.; Moore, Fred H.; Rickard, Hilton L.; Strange, Louis C.; Sutherland, Mortimer Y., Jr.; Twardy, Stanley A.; Van Ward, Roland; Watkins, Irvine Cabell

1940s: Barnes, Charles Wynn; Bates, John Davis; Bloomer, Richard Rodier; Brantly, John E., Jr.; Caliga, Charles F.; Clough, William Allen; Gooch, Edwin O.; Husted, John E.; Lockwood, Charles W., Jr.; Meador, John P.; Milner, Carlos E., Jr.; Overstreet, William C.; Parrott, Emory W.; Peterson, Warren S.; Tarleton, William Addison; Trainer, Frank Wilson

1950s: Applegate, Shelton Pleasants; Beard, Donald Chamberlin; Brantley, Mims McGehee; Browning, William Fleming; Burns, James Richard; Caskie, Robert A.; Caskie, Robert Alden; Cooke, Horace B., Jr.; Cordova, Robert Murray; Crist, Claude Walker, Jr.; Cross, Whitman, II; Eades, James Lynwood; Flewellen, Barbour H.; Giannini, William Fenwick; Harnsberger, Wilbur Trout, Jr.; Hay, Nicholas R. T.; Hopkins, Henry Robert; Lowdon, Jack; Mack, Tinsley; Moore, Rossie E., Jr.; Patterson, Joseph Gilbert; Peare, Robert Kunkel; Prutzman, William James; Ramsey, Elmer W.; Rector, William Kenna; Revilla, Charles E.; Rothenberger, Jay Anderson; Rowan, Lawrence Calvin; Schultz, William R.; Sherwood, William Cullen; Smith,

Robert Hamilton; Tazelaar, James Fulton; Thompson, Thomas Marvin; Vernon, Roger Clay; Young, Robert Spencer

1960s: Atakol, Kenan; Barnes, Robert Clay; Campbell, Frank Howard, III; Cole, John M.; Delaney, Holly Johanna; Delaney, John Rutledge; Farmer, George Thomas; Fink, Dwayne Harold; Fitzgerald, Francis Bell, III; Gatlin, Garnett Auman; Geitgey, Ronald Paul; Gosnell, Gary Johnston; Griffin, Villard Stuart, Jr.; Grubbs, Donald Keeble; Haglund, David Seymour; Hayes, Pamela Dee; Henika, William Sinclair; Katz, Arthur S.; Land, Ralph Joseph; McGavock, Edwin Harris; Plaster, Rodger W.; Rader, Eugene Kenton; Roberts, Clarence Everett; Rodgers, William Howard; Ryan, John Joseph, Jr.; Spangler, Daniel Patrick; Spiker, Carlisle Titus, Jr.; Wigley, Perry Braswell, Jr.; Wood, Robert Staples; Workman, William Edward

1970s: Anders, Fred John; Barnett, Thomas MacDonough; Bolyard, Thomas Harner; Chisolm, Stoney P.; Clerman, Robert Joseph; Davis, Andrew; Diaz, R. J.; Drifmeyer, Jeffrey Eugene; Dunford-Jackson, Carey Stanly; Elmer, Deborah Ann; Embree, John Marvin; Fausak, Leland Edward; Felder, Wilson Norfleat; Felder, Wilson Norfleat; Fritz, Barrett Robert; Glassen, Robert Carl; Glassey, Richard; Gleason, Mark Lawrence; Gulbrandsen, Leif Fontaine; Hewitt, Clark S.; Hipskind, Roderick Stephen; Hoffman, Carlton Scott; Hopkins, Edgar Member; Hubbard, David Adam, Jr.; Huber, William Gregor; Keene, William Charles; Kerby, Ernest Gordon; Knowlton, Sandra; Kuhlthau, Richard Harold; Leatherman, Stephen Parker; Lederman, T. C.; McCollister, Linda Suzanne; McCuskey, Sue Ann; Miller, Debra Janel; Murray, William Gerard; Peck, Gregory Erman; Poche, David John; Prichett, Wilson, III; Resio, Donald Thomas; Rude, Lawrence Culver; Sallenger, A. H., Jr.; Smith, Thomas Joseph, III; Stauble, Donald Keith; Stottlemyer, Laura Lee; Tufts, Susan; Van Horn, William Lewis; Vincent, Charles; Vincent, Charles Linwood; Vogel, Richard Mark; Watson, Jeter Marvin; Webb, John W.; Wilson, N. F.; Yewisiak, Paul P.; Zuzo, P. L.

1980s: Barnard, William Rives; Bell, Pamela Elizabeth; Benelmouffok, Djamel E.; Burns, Douglas A.; Buttleman, Kim Parker; Carpenter, John Martin; Clapp, Roger Burnham; Dagenhart, Thomas Vernon, Jr.; Garman, Phyllis Metrolis; Glaspey, Robin Gail; Goodwin, Steven Dale; Hamroush, Hany Ahmed; Herlihy, Alan Tate; Ista, Jane Pohtilla; Jacobs, Karen Stine; Kenworthy, W. Judson; Lent, Stephanie Jean; LoCastro, Richard Peter; May, Suzette Kimball; McIntire, Pamela Ellen; Meintzer, Robert Ells; Moses, Carl Owen; Moses, Carl Owen; Nash, Barry Stuart; Pien, Natalie Chen-Hsi; Price, Rene Marie; Rafalko, Leonard Gervus; Reiter, Michael Anthony; Schmidt-Fonseca, Susan; Selman, Michael Lamar; Sharrett, Janice Beechwood; Tisdale, Todd Street; Walton, John Calvin; Wassel, Raymond Anthony; Weaver, Timothy Otis; Wolock, David Michael

Virginia Polytechnic Institute and State University
Blacksburg, VA 24061

265 Master's, 142 Doctoral

1930s: Barlow, Wallace Dudley; Johnson, James Howe; Kessler, Jane; Moomaw, Benjamin Franklin; Sears, Charles Edward, Jr.; Smerchanski, Mark Gerald; Waesche, Hugh Henry

1940s: Morgan, Cecil G.; Warringer, Ben, IV

1950s: Brown, Charles Quentin; Chauvin, E. Noel; Chauvin, Edward N.; Chen, Ping-Fan; Ciaramilla, Philip Stephen; Diggs, William Edward; Edwards, Jonathan; Fara, Mark; Fitzgerald, Haile Vandenburgh, Jr.; Gilbert, Ray Clark; Glover, Lynn, III; Hergenroder, John D.; Hobbs, Charles Roderick Bruce; Hobbs, Charles Roderick Bruce; Marshall, Frederick C.; McCutcheon, Fletcher Snead; Meyertons, Carl Theile; Meyertons, Carl Theile; Nichol, Robert F.; Phillips, Howard Cottrell; Sabol,

Joseph W.; Schaff, Herbert Linwood, Jr.; Shanholtz, Wendell H.; Shufflebarger, Thomas Edwin, Jr.; Stevens, David W.; Webb, Fred; Williams, George K.

1960s: Aiken, Lewis J.; Amato, Roger V.; Anderson, E. R.; Aronson, David; Aronson, David Allen; Baillio, Robert H.; Barnhisel, Richard Irven; Bauerlein, Henry J.; Bowen, David G.; Bregman, Martin L.; Brown, Gordon Edgar, Jr.; Carrington, Thomas Jack; Cashion, W. W.; Derby, James R.; Derby, James Richard; Eckroade, William M.; Esmer, Erkan; Eubank, R. T.; Farnham, Paul Rex; Fiedler, Forest J.; Flock, William Merle; Francis, Robert E. L., Jr.; French, B. E.; Hale, Robin C.; Hall, Monte R.; Hamilton, Charles L.; Hardy, Henry Reginald, Jr.; Hazlett, William Henry, Jr.; Hergenroder, John David; Hillhouse, Douglas Neil; Jones, Norris William; Koch, Ellis; Kreglo, James R.; Lee, Fitzhugh T.; Leonard, Robert Benjamin, Jr.; Marland, Frederick Charles; McDowell, Robert C.; McDowell, Robert Carter; McTague, Stephen B., Jr.; Ming, James; Ming, James D.; Moon, William A.; Moore, Donald P.; Murphy, D. J.; Ouchark, William F.; Overshine, Alexander T.; Reiter, Marshall Allan; Riecken, Charles Christopher; Ritter, George S.; Ross, Arthur Henry, Jr.; Sanderson, Robert Michael; Schmoker, James William; Schwind, R. F.; Spencer, S. M.; Thomas, William A.; Tyler, John H.; Via, Edwin K.; Waller, James O.; Webb, Fred, Jr.; Whitman, Harry M.; Wigley, Perry Braswell, Jr.; Worthington, David W.

1970s: Ayers, Robert L.; Bartholomew, Mervin Jerome; Barton, M. D.; Bell, Raymond T.; Benson, D. G., Jr.; Blancher, Donald W., Jr.; Bobyarchick, A. R.; Bourland, William C.; Briggs, David F.; Broughton, Paul L.; Broughton, Paul L.; Brown, Gordon Edgar, Jr.; Cameron, Kenneth L.; Cameron, Maryellen; Campbell, J. K.; Clark, Horace B., III; Cohen, J. P.; Cooper, Brian J.; Cox, W. E.; Davis, Robert G.; Davison, William D., Jr.; deRosset, W. H. M.; Dumper, Thomas A.; Dumper, Thomas Apted; Edsall, Robert W.; Ford, Leonard N.; Francis, Carl Arthur; Frieders, T. Y.; Gambill, John A.; Glass, Frank R.; Goodman, David K.; Griffen, D. T.; Griffen, Dana T.; Griffiths, Donald Ward; Grover, George A.; Hall, S. T.; Hanan, B. B.; Hayes, Arthur Wesley; Hayman, James W.; Helfrich, Charles T.; Helsel, D. R.; Henry, Donald Kenneth; Heyman, Lou; Higgins, John Britt; Higgins, John Britt; Hight, David H.; Hochella, M. F., Jr.; Hopper, Margaret G.; Hougland, E. S.; Howell, L. W., Jr.; Judah, Othman Mohammed; Karpa, John B.; Kettren, Lee P.; Kolich, Thomas M.; Kreisa, Ronald D.; Lager, George A.; Lampiris, N.; Langer, C. J.; Leible, K. A.; Lindbloom, Joseph T.; Lumpkin, G. R.; Mahgerefteh, Khosrow; May, F. E.; McConnell, Keith I.; Miller, J. W., Jr.; Mitchell, Judson T.; Novak, Gary Alan; Novak, Stephen W.; Pfeil, R. W., Jr.; Phillips, Michael W.; Plants, H. F.; Poland, F. B.; Popp, R. K.; Popp, Robert K.; Prunier, A. R., Jr.; Rendon-Herrero, Oswald; Rhoads, G. H., Jr.; Riddle, J. M.; Robinson, Keith; Sandhu, I. S.; Schlenker, J. L.; Schultz, Arthur Philip; Selkregg, Kevin R.; Slocomb, J. P.; Sodbinow, E. S.; Speer, J. A.; Speer, John A.; Staten, Walter T.; Stiegler, James Harold; Stubbs, J. L., Jr.; Suter, David R.; Truman, W. E., III; Tso, J. L.; Van der Hoeven, G. A.; Vliek, P. J.; Vossler, Donald Alan; Waller, J. O.; Weand, B. L.; Weems, Robert E.; Wells, P. D.; Whitehurst, B. B.; Whitman, C. M.; Whitney, B. L.; Wiggins, Lovell B.; Williams, Richard T., II; Willoughby, R. H.; Wilson, Deborah Crotty; Wilson, John M.; Witmer, Roger J.; Wolfe, H. E.; Wright, James E.

1980s: Achtermann, Roger D.; Adamson, Richard Floyd; Affholter, Kathleen Ann; Agioutantis, Zacharias George; Ayuso, Robert Armando; Bajak, Doris M.; Bartelmehs, Kurt Lane; Belcher, Steven W.; Bova, John A.; Boyd, Thomas M.; Brennan, Jeanne L.; Brooker, Donald Duane; Brown, K. Elizabeth; Bryan, Robert A.; Caless, Jonathan R.; Chakoumakos, Bryan Charles; Chapman, Raymond Scott; Chaudet, Roy Edward; Chen, Victor J.; Chermak, John Alan; Chyi, Kwo-Ling; Cochrane,

Judith Christian; Connell, Joseph Francis, Jr.; Cooper, Brian Jay; Crowell, Mark; Cumbest, Randolph J.; d'Angelo, Richard M.; Dalton, David C.; Davis, Laura E.; Davison, Frederick Corbet, Jr.; Dawson, James W.; Deck, Linda Theresa; Degnan, Keith Terence; DeLuca, James L.; Dorobek, Steven Louis; Dove, Patricia Martin; Downs, James Winston; Downs, James Winston; Dysart, Paul Stephen; Dytrych, William Joseph; Ecevitoglu, Berkan Galip; Ecevitoglu, Berkan Galip; Edmonds, William Joseph; Ehlers, Ernest George, Jr.; Eid, Walid Khaled; Eitani, Ibrahim Mustafa; Evans, Nicholas Hartford; Everett, Charles Jay; Faris, Craig Duncan; Firth, John Victor; Fitzpatrick, Thomas Frank; Foley, Jeffrey Arthur; Foley, Nora Katherine; Francis, Carl Arthur; Gagen, Patrick Michael; Gates, Alexander E.; Geisinger, Karen Leslie; Ghafory-Ashtiany, Mohsen; Gibson, Richard G., III; Gibson, Richard G., III; Grabowski, Richard J.; Greiner, Daniel Joseph; Gresko, Mark Joseph; Groen, John Corwyn; Grotzinger, John Peter; Grover, George Adelbert, Jr.; Gunter, Mickey; Gunter, Mickey E.; Guy, Russell E.; Hanan, Barry Benton; Hanan, Barry Benton; Harris, Charles William; Harris, Charles William; Harris, Willie Garner; Hazneci, T. Hakan; Herman, Julie D.; Hickman, Gary Thomas; Hodges, Steven Clarke; Hogan, John P.; Holland, Dwight Allen; Hopper, M. G.; Houser, Brenda; Huggins, Michael James; Hund, Erik A.; Iwabuchi, Jotaro; Johnson, Neil Evan; Johnson, Neil Evan; Kaszuba, John Paul; Keith, Laura A.; Keller, Mary Ruth; Knight, Cheryl L. Erickson; Koerschner, William F., III; Kool, Jan Bart Jacobus; Krail, Paul Michael; Kreisa, Ronald Dean; Laczniak, Randell J.; Laughlin, Kenneth J.; Lemine, James; Lesser, Richard Peter; Lessley, John C.; Levy, David J.; Li, Gordon Chi-Kwong; Li, Zhongxue; Lightner, John Gwin, III; Lindsay, Curtis George; Loferski, Patricia J.; Luongo, Ronald F.; Magette, William Lawson; Mandeville, Charles W.; Mangan, Margaret; Marangakis, Andrew; Maxson, Anne E.; Mayu, Philippe Henri; McCarron, Kathryn R.; McClung, Wilson S.; McHugh, Mary Lopina; McKeever, Lauren Joann; McKinnon, William Beall; McManus, Kathleen M.; Miller, Elizabeth V.; Miller, Mark L.; Miller, Steven B.; Miller, William Paul; Monz, David J.; Morris, Mark Steven; Moses, Michael J.; Mullenax, Arthur Craig; Munsey, Jeffrey W.; Mussman, William J.; Nanda, Atul; Needham, Daniel L.; Nelson, Anthony; Newton, Maury Claiborne, III; Niemann, James Cottier; Parker, John Charles; Partin, Elizabeth; Patterson, Judith Gay; Peterson, Ronald Charles; Pettingill, Henry S.; Pratt, Thomas Lee; Pratt, Thomas Lee; Pyrak, Laura J.; Randall, A. Henry, III; Rebbert, Carolyn Rose; Reilly, Joseph Michael; Rogers, Melissa J. B.; Rounds, Thomas Richard, Jr.; Russell, Laura M.; Sage, Janet D.; Scambos, Theodore A.; Schaefer, Vernon Ray; Schilizzi, Paul P. G.; Schorr, Gregory Thomas; Schultz, Arthur Philip; Schweitzer, Janet; Scott, Stephen M.; Seaton, William Joseph; Sheng, Jopan; Shirasuna, Takeshi; Sibol, Matthew Steven; Simmons, Noel G.; Simpson, Edward Leonard; Singh, Yash Pal; Siriwardane, Hema Jayalath; Soegaard, J. K.; Soegaard, Kristian; Solie, Diana Nelson; Southerland, Elizabeth; Spears, David B.; Springer, Dale Ann; Stanley, Charles Bernard; Stark, Timothy D.; Stovall, Robert L.; Su, Shu-Chun; Swanson, Donald K.; Tamburro, Edie T.; Taylor, David Wyatt Aiken, Jr.; Teague, Alan Gaither; Tiyamani, Chanchai; Todd, Eric Donald; Trumburro, Edie; Tso, Jonathan Lee; Turek, Jeffery Lee; Viret, Marc; Walsh, Carol Ann; Weary, David John; Webster, Stephen Leroy; Wehr, Frederick Lewis, II; Weisenburger, Kenneth William; Wharton, Robert Andrew, Jr.; Wiersma, Cynthia Leigh; Wigington, Parker Jamison, Jr.; Wilson, Michael Alan; Witmer, Roger J.; Wong, Sam J.; Wu, Wei; Zentmeyer, Jan Penn; Zhou, Yingxin

Virginia State University
Petersburg, VA 23803

44 Master's, 1 Doctoral

1960s: Barham, Samuel D., III; Bohon, William Oliver, Jr.; Ditty, William E.; Dunbar, Cecil, Jr.; Henry, C. Wayne; Johnson, Ernest; Kennedy, Charles E., Jr.; Major, Virginia L.; Targgart, F. Arthur; Trail, Robert Bruce; Whiita, Richard A.

1970s: Admiraal, Peter; Andrisin, Mary E.; Bartelle, John Clemente; Brown, Carroll Parker; Chance, John Matthew; Craft, Helen Faith; Dorn, Mary S.; Foreman, Willie Earl; Fountain, Aubroy W., II; Hallock, Waite D.; Hofler, Vivian Estelle; Holland, Thelma H.; Lanham, James H.; Lauden, Edward J., Jr.; Manning, Retta A.; Mattison, Willie W.; Merriweather, Annie Pearl; Mintz, Milton; Mushinsky, Edward Stephen; Ramsey, Christina U.; Robbins, Peter; Scheele, R. A.; Shipp, Thomas C.; Sweetland, Theodore Wessley; Walker, Lydia P.; Wildy, Vernon L.; Williams, Dennis S.; Williams, Mary Louise; Wood, J. H.; Wright, Reginald D.; Yonce, Joseph B.

1980s: De Kimpe, Nancy; Sheffey, Renata; Watterson, Karen

Wake Forest University
Winston-Salem, NC 27109

1 Master's

1970s: Kincaid, D. T.

University of Washington
Seattle, WA 98195

523 Master's, 387 Doctoral

1910s: Fischer, Arthur Homer; Jillson, Willard Rouse; Packard, Earl Leroy; Powell, Edward Reed; Rhodenbaugh, Edward Franklin; Whittier, William Harrison

1920s: Ash, Simon Harry; Bauer, Hubert A.; Clifton, Clarence Cathcart; Curtis, Carl Edward; Etherington, Thomas John; Fuller, Richard Eugene; Glover, Sheldon Latta; Goodner, Ernest Francis; McAneny, John Maurice; McClellan, R. D.; McKnight, Edwin Thor; McLellan, Roy D.; McLellan, Roy Davison; McLeod, Arthur A.; Nicholson, Walter Allen; Nightingale, William Thomas; Roth, Robert Ingersoll; Tegland, Nellie May; Ward, Alfred H.; Waters, Aaron C.; Weymouth, Andrew Allen; Wilson, Leslie Edward

1930s: Berkelhamer, Louis Harry; Blanchard, Melbourne Kenneth; Brandt, William Otis; Bravinder, Kenneth Mason; Chappell, Walter Miller; Chappell, Walter Miller; Cline, Robert William; Coats, Robert Roy; Coombs, Howard Abbott; Coombs, Howard Abbott; Cooms, Howard Abbott; Couch, Albert Harris; Dammann, Arthur; Fuller, Richard Eugene; Goin, Fred L.; Granger, Arthur E.; Hansen, Henry Paul; Hurst, Thomas Leonard; Key, John Ambrose; Kotschevar, D. D.; Kravik, Gerald Enestvedt; Mitchell, Harold Delong; Norbisrath, Nans; Page, George Ava; Pask, Joseph Adam; Raine, Frank Frederick; Reed, Robert Marion; Roberts, Ralph Jackson; Russell, Mary Ellen; Southard, Lloyd Colman; Tennant, Harold Ellsworth; Todd, Margaret Ruth; Tsai, Cheng Yun; Zvanut, Frank Joseph

1940s: Armstrong, Frank C.; Beck, George Frederick; Bell, John William; Brennan, Charles Victor; Burcham, Donald Preston; Campbell, Robert Arthur; Clauson, Victor; Cook, Earl Ferguson; Creager, Marcus Orange; Danner, Wilbert Roosevelt; Davies, Ben; Davis, Franklin Theodore; Eyerly, George Brown; Farrar, Robert Lynn, Jr.; Fitzsimmons, J. Paul; Fulmer, Charles Virgil; Gibson, John Frank; Goring, Arthur William; Granquist, Donald Paul; Griffin, Bert Eldon; Griffith, Robert Fiske; Grose, Lucius Towbridge; Jones, Robert Sprague; Jonte, John Haworth; Ketzlach, Norman; Laval, William Norris; Luppold, John Hugh; McMichael, Lawrence Bradley; Morrow, John George; Oliver, Earl Davis; Parker, G. G.;

Reinertsen, Robert Wessley; Sanderman, L. A.; Seitz, James F.; Swift, Roy Erwin; Tully, John P.; Turner, Thomas Edward; Tyrrell, Miles Edward; Utterback, C. L.; Weintraub, David Leon; Whittemore, Osgood James, Jr.; Wilson, Raymond Edgar

1950s: Abbott, Agatin Townsend; Adams, John Bright; Alexander, Frank; Allison, Richard Chase; Alto, Bruno Raymond; Anderson, Norman Roderick; Andrews, Leslie W.; Babcock, Harold Earl; Babley, Ralph E.; Bagley, Ralph Eugene; Bayne, George Wallace; Becraft, George Earle; Bengston, Kermit Bernard; Bengtson, Kermit Bernard; Bethel, Horace Lloyd; Blank, Horace Richard, Jr.; Blank, Horace Richard, Jr.; Bond, John Gilbert; Bradley, John S.; Brooks, James Elwood; Bryant, Bruce Hazelton; Budinger, Thomas F.; Bush, James; Carroll, Neil Patrick; Chow, Tsaijwa J.; Cook, Earl Ferguson; Coxon, Donald Allan; Danner, Wilbert Roosevelt; Dixon, Roy Wilbur; Dobell, Joseph Porter; Dore, James Ernest; Drugg, Warren Sowle; DuBois, Robert Lee; East, Edwin; Ellingson, Jack Anton; Ellis, Ross Courtland; Fisher, Richard Virgil; Forbes, Robert Briedwell; Ford, Arthur Barnes; Ford, Arthur Barnes; Foster, Robert John; Foster, Robert John; Fritz, William Harold; Galster, Richard William; Gnagy, Jean; Goldsmith, Richard; Gould, Ramon John; Grant, Alan Robert; Gray, Irving Raymond; Harlow, George; Harvey, Joseph L.; Heywood, William Walter; Heywood, William Walter; Horsfall, John Clayton; Jones, Robert William; Jones, Robert William; Kremer, Dale Ernest; Lang, Andrew J.; Laval, William Norris; Loney, Robert Ahlberg; Marshall, Nissim Joseph; Maurer, Donald Leo; McCarthy, William R.; Miller, Gerald Matthew; Moore, James G.; Morrison, Melvin E.; Mumby, Joyce I.; Nayudu, Y. Rammohanroy; Nelson, Afred M.; Nelson, James Warren; Nelson, Robert B.; Nelson, Robert B.; Nelson, Willis H.; Newman, Thomas Stell; Oles, Keith Floyd; Paige, Russell; Pelton, Harold A.; Phetteplace, Thurston Mason; Pitard, Alden McLellan; Pratt, Richard Murray; Pratt, Richard Murry; Reichert, William H.; Reid, Rolland R.; Reid, Rolland R.; Reim, Kenneth Maurice; Riley, Roger Ray; Robertson, Forbes Smith; Roddick, James Archibald; Ryason, Daniel John; Schmidt, Dwight Lyman; Scott, Willard Frank; Shapiro, Howard E.; Shuck, Gordon R.; Smedes, Harry Wynn; Snelson, Sigmund; Snelson, Sigmund; Sorensen, Arthur; Steele, Grant; Stout, Martin Lindy; Stout, Martin Lindy; Subbarao, Eleswarapu Chinna; Swanson, Earl H.; Tabor, Rowland Whitney; Thoms, Richard Edwin; Threet, Richard Lowell; Vance, Joseph Alan; Waldichuk, Michael; Wang, Feng-Hui; Wennekens, Marcel Pat; Williams, Paul Lincoln; Willis, Clifford Leon; Yeats, Robert S.; Yeats, Robert Sheppard; Yeats, Robert Sheppard, Jr.

1960s: Aagaard, Knut; Adams, Charles E.; Adams, John Bright; Adams, Robert William; Anderson, Charles Alfred; Anderson, Franz Elmer; Anderson, Norman K.; Anikouchine, William Alexander; Anikouchine, William Alexander; Avent, Jon Carlton; Avent, Jon Carlton; Ballard, Ronald Lee; Barakos, Peter A.; Barkley, Richard A.; Barrus, Robert Bruce; Baum, Lawrence Frederick; Bayley, Emery Perham, Jr.; Biederman, Donald D.; Bond, John Gilbert; Brayton, Darryl M.; Breitsprecher, Charles Hepner; Brundage, Walter L., Jr.; Buddemeier, Robert Worth; Burns, Robert Earle; Campbell, Kenneth Vincent; Carlson, Roseann J.; Cebull, Stanley Edward; Chen, John Teh-Jen; Chen, Shih-Fang; Coachman, Lawrence K.; Conomos, Tasso John; Conway, Richard Dean; Crawford, William James Page; Curran, Theodore Allan; Dana, Robert W.; Dechert, Curt Peter; Dechert, Curt Peter; Diery, Hassan Deeb; Dover, James Herbert; Dover, James Herbert; Easterbrook, Donald James; Enbysk, Betty Joyce Blomgren; Erikson, Erik Harold, Jr.; Falck, Arnold; Franklin, Wesley Earlynne; Frisken, Jim Gilbert; Fritz, William Harold; Fullam, Timothy Jewell; Fullam, Timothy Jewell; Fyock, Tad L.; Geldsetzer, Helmut; Gilbert, Wyatt G.; Giovanella, Carlo; Grant, Alan Robert; Green, Richard; Grill,

Edwin Vatro; Gualtieri, J. L.; Hamilton, Thomas Dudley; Hamlin, William Henry; Hammond, Paul Ellsworth; Hansen, Donald V.; Harris, Constance; Hawkins, James Wilbur, Jr.; Heath, Michael Thomas; Hibbard, Malcolm; Hibbard, Malcolm; Hobsen, Louis Arthur; Holm, Richard F.; Holmgren, Dennis Arthur; Holmgren, Dennis Arthur; Hopkins, Kenneth Donald; Howard, David Ayers; Huttrer, Gerald; Hyde, Jack Herbert; Jones, Michael B.; Jung, Jim Grant; Kepper, Jack C.; Kepper, John Charles; Kirkpatrick, Doug; Knoll, Kenneth Mark; Koch, Allan James; Lange, Ian Muirhead; Lawrence, David Parker; Lee, Wilfred K.; Libby, Willard Gurnea; Lindquist, John Warren; Macalpine, Steven; Mari, David Lee; Martinson, Arthur David; McClellan, William Alan; McLean, Hugh; McWilliams, Robert G.; McWilliams, Robert Gene; Menzer, Frederick John, Jr.; Merrill, Douglas E.; Miller, Clifford Daniel; Miller, Gerald Matthew; Miller, Michael Schas; Milliman, John Douglas; Minning, Gretchen V.; Monney, Neil Thomas; Moore, James Leslie; Mullineaux, Donald Ray; Naugler, Frederick Paist; Oliphant, Jerrelyn; Palmer, Leonard Arthur; Patton, Thomas Charles; Peterson, David Holmen; Peterson, Frederick F.; Peterson, Gary Lee; Peterson, Gary Lee; Plummer, Charles Carlton; Plummer, Charles Carlton; Ptacek, Anton Donald; Ptacek, Anton Donald; Race, Ronald Williams; Ragan, Donal MacKenzie; Raish, Dan; Rector, Richard James; Reid, Allan Robert; Robinson, Carl Francis; Rosenberg, Ernest A.; Rosengreen, Theodore Ernest; Rosenmeier, Frederich Joseph; Royse, Chester F., Jr.; Schleh, Edward Eugene; Schmidt, Dwight Lyman; Shaw, George Hamill, III; Sherman, Don Kerry; Shideler, James Henry, Jr.; Silver, Burr Arthur; Smith, Donald Leith; Snook, James Ronald; Sternberg, Richard Walter; Sternberg, Richard Walter; Steuber, Alan M.; Strain, Lamar Asal, Jr.; Strong, Ceylon Perseus, Jr.; Strong, Ceylon Perseus, Jr.; Stull, Robert John; Stull, Robert John; Tabor, Rowland Whitney; Thorman, Charles Hadley; Thorman, Charles Hadley; Thorndale, C. William; Van Diver, Bradford Babbitt; Vredenbrugh, Larry Dale; Wadekamper, Donald; Wallin, Charles Stanton; Warren, Roy Kenneth; White, Stanton Morse; Wiebe, Robert Alan; Williams, Paul Lincoln; Woodward, Lee Albert; Young, Ronald Earl

1970s: Abiodun, Adigun A.; Adams, Nigel Bruce; Aiken, Carlos Lynn Virgil; Anderson, E. H.; Anderson, James J.; Ansfield, Valentine Joseph; Armentrout, John M.; Ashleman, James C.; Ashman, S. H.; Askren, David R.; Atkin, Steven Allen; Babcock, Randall Scott; Baier, Roger W.; Baker, Edward Thomas, Jr.; Balistrieri, Laurie S.; Barnard, William Dana; Barnes, Robert Stith; Barnosky, Cathy Lynn; Barrus, Robert Bruce; Battis, James Craig; Beget, James Earl; Bernath, Hans Jakob; Bierley, Janice; Bindschadler, R. A.; Birch, Peter Barrett; Birch, Peter Barrett; Bishop, Thomas Norton; Blank, Richard G.; Boak, Jeremy Lawrence; Bockheim, James Gregory; Bor, Sheng-Sheang; Bothner, Michael Henry; Braile, Lawrence Wendell; Brooks, William Earl, Jr.; Bryant, Vicki Yolanda; Buckovic, William Alan; Burbank, Douglas West; Burk, Robert L.; Burk, Robert L.; Burnet, Frederick William; Bush, Asahel; Buza, John W.; Campbell, Kenneth Vincent; Carlson, Richard L.; Carlson, Richard L.; Carmack, Eddy; Carson, Bobb; Carson, Robert James, III; Carter, Phillip K.; Carver, Gary Alen; Cashman, Susan Moran; Cashman, Susan Moran; Clayton, Daniel Noble; Codispoti, Louis A.; Colbeck, Samuel C.; Cole, Mark Rolland; Conway, Richard Dean; Cranston, Raymond Earle; Crawford, William James Page; Crecelius, Eric A.; Cullen, Janet M.; Davis, Curtiss Owen; Davis, Earl Edwin; Deliman, Daryl G.; Dethier, David Putnam; Dethier, David Putnam; Dietrich, William Eric; Downing, John Peabody; Dungan, Michael Allen; Dungan, Michael Allen; Durgin, Dana C.; Edmondson, Samuel A.; Fahey, Patrick Louis; Fairchild, Lee Hamlin; Filson, Robert Harold; Fluorie, Eric Juan de Dios; Fountain, David Michael; Foxall, William; Frost, Bryce Ron-

ald; Frost, Bryce Ronald; Fukuta, Nobuhiko; Fuste, Luis Alberto; Getsinger, Jennifer Suzanne; Glassley, William Edward; Glassley, William Edward; Goetsch, Sherree Ann; Gonen, Behram; Goullaud, Lee H.; Greene, Glen Stonefield; Greer, C. E.; Gustafson, Eric Paul; Hanson, Larry Gene; Harrison, Paul James; Harrison, Peter; Hartley, Alan H.; Hartley, Susan; Hartline, B. K.; Hartman, Donald Albert; Heath, Michael Thomas; Hedderly-Smith, David Arthur; Henshaw, Paul Carrington, Jr.; Henshaw, Paul Carrington, Jr.; Herd, Darrell G.; Herd, Darrell G.; Hersch, John Timothy; Hibbert, Dennis Mark; Hirsch, Robert M.; Hitzman, Murray W.; Hodge, Steven McNiven; Hoffman, David Gordon; Holmes, Mark Lawrence; Hopkins, Kenneth Donald; Hopkins, Thomas; Horn, Toya D.; Horton, Marc Allan; Houghton, Jonathan Parks; Hubert, Kathleen Ann; Hyde, Jack Herbert; Ishibashi, Isao; Jayaraman, K. N.; Johnson, Harlan Paul; Johnson, Paula Ann; Johnson, Samuel Yorks; Johnston, David A.; Johnston, David A.; Jones, Kathleen Ferris; Kachel, David G.; Kazzaz, H. H.; Kearnes, James K.; Keller, Barry; Khalid, R.; Kilpatrick, John Thomas, III; Kinder, T. H.; Klosterman, Keith Edward; Knapp, John Stafford, Jr.; Knebel, Harley John; Koch, Allan James; Kroft, David Jeffrey; Langbein, J. O.; Laravie, Joseph A.; Larson, A. G.; Leo, Sandra Rose; Levi, S.; Levine, Paul Elliot; Lieberman, S. H.; Likarish, Daniel Matthew; Lin, Jia-Wen; Ling, S. C.; Linkletter, George Onderdonk, II; Lovseth, Timothy Peter; Lowes, Brian Edward; Lupe, Robert Douglas; Madej, Mary Ann; Maknoon, Reza; Manfrida, Jerry Lynn; Mari, David Lee; Markham, Deborah Kesselring; Marshall, Paul Wellington; Marshall, Philip Schuyler; Martin, James Edward; Mayers, Ian Richard; McCarthy, Conrad Joseph; McClain, J. S.; McClung, David M.; McCollom, Robert Lloyd; McDougall, Kristin Ann; McKinnie, Diana B.; McLean, Hugh; McLean, Stephen Russell; McMillen, Daniel David; Meeder, C. A.; Merrill, Milford S., Jr.; Metcalf, Richard Carl; Miers, John Harlow; Miller, Robert Bruce; Mills, Hugh Harrison; Mogk, David William; Mojica, Iran H.; Moore, Stephen C.; Moore, Stephen Carlisle; Mountain, David; Muench, Robin; Mulcahey, Michael Thomas; Nelson, Jon Sherwood; Nelson, Robert Edward; Newbauer, Thomas Raymond; Newton, John LeBaron; Nichols, Bruce MacKenzie; Nimick, David Acheson; Nittrouer, Charles Albert; Nittrouer, Charles Albert; O'Clair, C. E., Jr.; Olmstead, Dennis L.; Onyeagocha, Anthony Chukwuma; Page, Richard James; Page, Richard James; Patton, Thomas Charles; Pavish, Marie; Pearson, W. C.; Peters, David Cornelius; Peterson, Michael L.; Pierson, Thomas Charles; Pierson, Thomas Charles; Pilson, Michael L.; Pongsapich, Wasant; Pongsapich, Wasant; Porter, Lee; Priestley, Keith F.; Pytlak, Shirley Ruth; Radrikrjengkrai, P.; Raedeke, Linda Dismore; Ramalingaswamy, V. M.; Ramananantoandro, Ramanantsoa; Rasrikriengkrai, Piyamit; Rice, Jack Morris; Rice, Jack Morris; Riedell, Karl Brock; Rigby, F. A.; Rivera, Louis George; Rogers, William Patrick; Ross, William Michael; Rothe, George Henry, III; Salisbury, Matthew Harold; Salisbury, Matthew Harold; Scott, William E.; Shaw, George Hamill, III; Shaw, John Damon; Shelley, Joanne Ross; Silling, Rose Mary; Silverberg, Norman; Simon, Ruth B.; Simpson, David Paul; Skinner, Robert G.; Skinner, Robert G.; Slater, L. E.; Smethie, W. M., Jr.; Snydsman, W. E.; Stengel, K. C.; Stensrud, Howard Lewis; Stottlemyre, James Arthur; Stroh, James M.; Stroh, James Michael; Svensson, Cynthia T.; Swain, Walter C.; Sylwester, Richard L.; Tallyn, Robert Bernard; Tang, C. H.-W.; Tatman, James B.; Ten Brink, Norman Wayne; Tennyson, Marilyn Elizabeth; Tennyson, Marilyn Elizabeth; Thompson, Joyce Ann; Thorndike, Alan; Thorndycraft, R. B.; Thorson, Robert Mark; Tien, Yu Bun; Trammell, John W.; Tsuchiya, C.; Tubbs, Donald W.; Tubbs, Donald Willis; Tucker, Glennda B.; Turner, E. J. J.; Twiss, Elizabeth S.; Vance, Dana Joslyn; Waitt, Richard Brown Jr.; Wakeham, Stuart Glenwood; Wald, Alan R.; Wallace, Wesley K.; Walters, Roy

A.; Wang, Hau-Ran; Ward, A. Wesley, Jr.; Ward, Alexander Wesley, Jr.; Ward, Peter Douglas; Weaver, Craig Steven; Weber, William Mark; Westbrook, Marston, Jr.; Wheeler, Gregory R.; Wheeler, Gregory R.; Whitney, John Wallis; Wildrick, Linton Leigh; Wiley, Bruce Henry; Williams, Van Slyck; Williams, Van Slyck; Williamson, Michael; Wingert, Everett Arvin; Winter, John Keith; Winter, John Keith; Wissmann, Gerd; Wu, J. C.; Yang, Albert In Che; Yeh, Betty; Yett, Jan Reynolds; Zirino, Albert Rocco; Zurcher, Hannes

1980s: Adjali, Salim; Alexander, Robert Houston; Anderson, Glenn Richard; Anderson, Paul Ralph; Anderson, Robert Stewart; Arnold, Scott; Asnake, Mesfin; Atallah, Raja Hanna; Aubry, Brain Francis; Baker, Glen E.; Bakken, Barbara M.; Balise, Michael John; Balzarini, Maria Anne; Bame, Dorthe Ann; Barker, Sally; Barnosky, Anthony David; Barnosky, Anthony David; Barnosky, Cathy Whitlock; Beget, James Earl; Benda, Lee E.; Bennett, Joseph Thomas; Bethel, John Patterson; Blechschmidt, Gretchen Louise; Boness, David Arno; Booth, Derek Blake; Borgeld, Jeffrey Calvert; Borns, David James; Boudreau, Alan Ernest; Bowers, Fred Howard; Brandon, Mark Thomas; Brandon, Mark Thomas; Brooks, William Earl, Jr.; Burnham, Robyn Jeanette; Burnham, Robyn Jeanette; Bushara, Mohammed N.; Bussod, Gilles Yves Albert; Butler, B. F.; Butler, Brian Faraday; Calacal, Elias L.; Carroll, Paul Richard; Cheng, Wen-Lon; Christensen, Thomas Hoejlund; Clark, Robert Charles, Jr.; Clarke, Anthony David; Clayton, Geoffrey Alden; Colligan, Thomas Henry; Collins, Brian David; Cool, Colin Anthony; Cosens, Barbara Anne; Criscenti, Louise Jacqueline; Croll, Timothy Caryl; Curtiss, Brian; Davenport, Thomas Edward; Denny, Stuart; Desmarais, Neal Raymond; Dietrich, William Eric; Downing, John Peabody; Dube, Thomas Eugene; Dunwiddie, Peter William; Egbert, Gary David; El-Moslimany, Ann Paxton; Endo, Elliot Toru; Ertel, John Richard; Evans, Diane Louise; Evans, James Erwin; Fairchild, Lee Hamlin; Fang, Yung-Show; Farr, Thomas Galen; Ferguson, Sue Ann; Folami, Samuel Lekan; Forbes, Jeffrey; Frank, David Gerard; Frederick, Jan Elizabeth; Furlong, Edward Thomas; Gager, Barry Robert; Gelfenbaum, Guy Richard; Ghazali, Fouad Muhammed; Gibson, Kenneth Mark; Gilbert, Deborah; Gitlin, Ellen C.; Gitlin, Ellen C.; Glover, D. W.; Graumlich, Lisa; Gudmundsson, Olafur; Haase, P. C.; Hanks, Catherine Leigh; Hanley, Christine Naomi; Haugerud, Ralph Albert; Hauptman, Julie L.; Hendrickson, M. A.; Herlihy, D. R.; Hesser, Duane Harvey; Hetherington, Martha J.; Hofstetter, Abraham; Hofstetter, Abraham; Hoover, Karin A.; Horng, Fu-Wen; Humphrey, Neil Frank; Humphrey, Neil Frank; Istas, Laurence Stewart; Jacobs, Lucinda Ann; Jahnke, Richard Alan; Jay, David Alan; Jóhannesson, Tómas; Johnson, David; Johnson, James Elmer; Johnson, Samuel Yorks; Jung, Heeok; Kachel, Nancy Brandeberry; Kaneda, Ben Keith; Karsten, Jill Leslie; Karsten, Jill Leslie; Kaufman, Darrel Scott; Kelemen, Peter Boushall; Kelley, Deborah E.; Kelley, Stephen M.; Kennard, Paul M.; King, Stagg Lipscomb; Knapp, John Stafford, Jr.; Kolle, Jack John; Ku, Tsu-Wei; Kuivila, Kathryn Marie; Laetz, Thomas J.; Lanning, Eric N.; Lapp, David B.; Lazoff, Steven Barry; Lea, Peter Donald; Leaver, Donald S.; Leaver, Donald S.; Lee, Chong Do; Lee, Lyndon Charles; Lees, Jonathan; Leipertz, Steven Lee; Leithold, Elana Lynn; Levine, Carol Alice; Lewis, Reed Stone; Leytham, Keith Malcolm; Lieberman, Joshua Elliot; Liu, Tze-Kung; Lundquist, Susan Marya; Ma, Li; MacQueen, Jeffrey Donald; Magloughlin, J. F.; Magnusson, Magnus Mar; Mann, Daniel Hamilton; Margolis, Jacob; Marrett, David Joseph; Martin, David; Martin, David Carl; Mathez, Edmond Albigese; May, Dann Joseph; McCarthy, Conrad Joseph; McClain, Kevin John; McClurg, Dai C.; McConnaughey, Ted Alan; McCorkle, Daniel Charles; McCrea, Maureen; McGroder, Michael F.; Mertes, Leal Anne Kerry; Michaelson, Caryl Ann; Miller, Robert Bruce; Mogk, David William;

Moll, Nancy Eileen; Moon, Thomas Scott; Moran, Jean Elizabeth; Muir, Mark P.; Nelson, Carol Jeanne; Nelson, Jonathan Mark; Nelson, Robert Edward; O'Keefe, Jane Frances; O'Malley, Robert Thomas; Odom, Robert Irving, Jr.; Ogren, John Addison; Oliver, Dean Stuart; Oliver, L. A.; Orr, Kristin Elizabeth; Oshinowo, Babatunde Oluwasegun; Owen, Claudia; Paine, Jeffrey G.; Pan, Yii-Wen; Patrick, Brian Ellsworth; Pavlis, Gary Lee; Peckham, David Arthur; Peckman, D. A.; Pfeffer, William Tad; Pollock, Stephen Matthew; Potter, Christopher John; Prahl, Fredrick George; Quintana, C. W.; Raedeke, Linda Dismore; Ralston, June Kathleen; Reid, Leslie Margaret; Rhodes, Brady P.; Richey, Joanna Sloane; Robertson, Christopher Alan; Rohay, Alan Charles; Rovetta, Mark Rino; Rusmore, Margaret Elizabeth; Russell, Charles William; Sanchez, Arthur Ledda; Sawlan, Jeffrey J.; Schreiber, Sue Anne; Schultz, Adam; Schwitter, Michael; Shi, Nungjane Carl; Shipley, Susan; Simonson, Eric Robb; Smith, Guy Michael; Smith, Jeremy Torquil; Spencer, Patrick Kevin; Spicer, R. C.; Squires, G. H.; Stoddard, Andrew; Stottlemyre, James Arthur; Taber, Joseph John, Jr.; Tepper, Jeffrey Hamilton; Thompson, William Hayes; Till, Alison Berna; Tivey, Maurice Anthony; Trexler, James Hugh, Jr.; Tuthill, Jonathan Dale; Vanderwal, K. S.; Vita, Charles Ludwig; Vitayasupakorn, Vichai; Walker, Ann Leslie; Wallace, Wesley K.; White, P. J.; Wiberg, Patricia Louise; Wigmosta, Mark Steven; Wilkens, Roy Henry; Wolde-Medhin, Bekele; Wong, Paul Kwok-Ting; Wood, David Mahlon; Wood, Karrie Champneys; Xu, Song; Yelin, T. S.; Zabowski, Darlene; Zemansky, Gilbert Marek; Zervas, Chris Eugene; Zervas, Chris Eugene; Zitek, W. O.; Zurcher, Hannes George

Washington & Lee University
Lexington, VA 24450

1 Master's

1900s: White, Americus Frederic

Washington State University
Pullman, WA 99164

284 Master's, 92 Doctoral

1890s: Schraubstadtler, R. T.

1920s: Barnes, Virgil Everett; Bennett, William Alfred Glenn; Cooper, Herschel H.; Denman, Cedric Eugene; Stewart, Charles A.; Treasher, Ray C.

1930s: Anderson, Roy Arnold; Baldwin, Ewart Merlin; Dennis, Terence Edwin; Fowler, Claude Stewart; Hougland, Everett; Maxfield, Ray A.; Melrose, John Walter; Newcomb, Reuben Clair; Search, Marshall Allen; Thomson, John Pretiss; Twiss, Stuart Nelson; Warren, Walter Cyrus; Wilson, James R.

1940s: Crosby, James Winfeld, III; Graham, Charles Edward; Hewett, Henry B.; Huntting, Marshall Tower; New, William Randal; Schroeder, Melvin Carroll; Stevenson, Robert Evans; Strand, Jesse Richard; Tuttle, Sherwood Dodge; Valentine, Grant M.; Zimmer, Paul William

1950s: Becraft, George Earle; Bennington, Kenneth Oliver; Conybeare, Charles Eric Bruce; Curtiss, Robert Eugene; Eccles, John Kerby; Enbysk, Betty Joyce Blomgren; Ferrians, Oscar J.; Haddock, Gerald Hugh; Hinman, Eugene Edward; Howd, Frank Hawver; Jizba, Zdenek Vaclav; Keppler, Belva Hudson; Kleweno, Walter P.; Koskinen, Victor K.; Kraszewski, Stefan; Lotspeich, Frederick B.; Mason, George William; McCreery, Robert Atkeson; Merriam, Robert Willis; Mueller, Paul M.; Mullineaux, Donald R.; Oles, Keith Floyd; Ore, Henry Thomas; Person, Donald W.; Peterson, Dallas Odell; Peterson, Donald W.; Post, Edwin Vaulton, Jr.; Reed, Bruce Loring; Rieger, Samuel; Schroeder, Melvin Carroll; Simmons, George Clarke; Stark, William James; Terry, Orlyn Lee; Thomas, Gerald William; Tipper, Howard Watson; Watson, Donald Whitman; Wilson, Philo Calhoun

1960s: Barker, Peter Arnold; Brackett, Michael Howard; Cavin, Richard E.; Clark, John H.; Cunningham, Robert Lester; Ellersick, Donald K.; Ellingson, Jack Anton; Eyrich, Henry Theodore; Fleisher, Penrod Jay; Foutz, Dell Riggs; Franklin, Jerry Forest; Girdley, William Arch; Gunn, Donald William; Hansen, Peter A.; Hansen, Peter Allen; Hoffer, Jerry Martin; Kamin, Thomas C.; Lewis, Standley Eugene; Lin, Chang-Lu; Lyons, David James; Mattraw, Harold Claude, Jr.; McIntyre, David Harry; McKague, H. Lawrence; Nordby, George Roy; Osborne, Robert Howard; Purves, William John; Ringe, Louis Don; Ross, Theodore William; Ruzicka, Joseph Frederick; Schroeder, Thomas Francis; Shumway, Ramon Dwight; Souders, Robert Patton; Steen, Virginia Carol; Stephenson, David A.; Stephenson, Gordon R.; Taylor, Edward Morgan; Thorsen, Gerald Wayne; Tozer, Warren Wilson; Van Ryswyk, Albert Leonard; Westfall, Dwayne Gene

1970s: Alwin, John Arnold; Amara, Mark Steven; Ayers, Jerry Floyd; Banning, Davey Lee; Bard, Catherine Sundstrom; Beka, Francis Thomas; Blinman, Eric; Bressler, Jason Robert; Broch, Michael John; Brown, Jeffrey C.; Bush, John Harold, Jr.; Camp, Victor Eric; Carr, Gerald L.; Cochran, Bruce Duane; Deaver, Franklin Kennedy, Jr.; Dombrowski, John; Donnelly, Brian James; Ekambaram, Vanavan; Fenton, Robert Leo; Fewkes, Ronald Hubert; Foley, Lucy Loughlin; Foster, Allan Royal; Gentry, Herman Raymond; Green, William Randolph; Griffis, Robert John; Griffis, Robert John; Hammatt, Hallett H.; Holden, Gregory Spry; Ichimura, Vernon T.; Janbaz, John Elisha, Jr.; Kiesler, James Peter; Kleck, Wallace Dean; Knutsen, Gale Curtis; Ko, Chong An; Kolva, David Allen; Korzendorfer, David Paul; LeBret, George Curtis; Lee, John Scott; Leonhardy, Frank Clinton; Lindberg, Jonathon W.; Marshall, Alan Gould; Mattigod, S. V.; Mattson, John Lyle; McDonald, Robert Joseph; Mellott, James Charles; Mercier, John Michael; Metz, Michael C.; Moody, Ula Laura; Morganti, John Michael; Morton, Jack Andrew; Moser, Kenneth Robert; Nadeau, Joseph Edward; Neitzel, Thomas William; Nordstrom, Harold Edward; Orazulike, Donatus Maduka; Petrone, Anthony; Phillips, William Morton; Rafuse, Bruce Elwood; Ream, Lanny Ray; Reidel, Stephen Paul; Reinhart, William Robert; Routson, Ronald Chester; Sanford, Steven Ray; Shovic, Henry Folke; Shubat, Michael Andrew; Siems, Barbara Ann; Summers, Karen Varley; Sylvester, Kevin John; Taylor, Terry Lee; Thole, Ronald Henry; Todd, Stanley Glenn; Tucker, Robert William; Wotruba, Nancy Jane

1980s: Adedotun, Adekoya Adedayo; Adekoya, Adeodotun Adedayo; Aimo, Nino J.; Akers, Richard H.; Alexander, James Iwan David; Ayers, Jerry Floyd; Baichtal, James Fay; Bailey, James Stuart; Bailey, Michael Mathewson; Barsotti, A. T.; Beane, John E.; Beane, John Edward; Beaver, Dennis Earl; Beck, John Walter; Beka, Francis Thomas; Biggane, John Howard; Bloomer, Gail Elizabeth; Borg, Heinz; Brainard, Ray Carter; Branham, Alan David; Briggs, Thomas D.; Bristow, Keith Leslie; Brooks, Jeffrey W.; Bush, Thomas A.; Caffrey, Gregory Michael; Calagari, Ali Ashghar; Carlson, Diane Helen; Carsten, Forrest Paul; Cheshier, Roby Albert; Cochran, Michael Patrick; Conklin, Bonnie Jean; Cook, Kerry Brian; Cotton, William Robert, Jr.; Cozza, Leonard Martin; Dainty, Norman Dale; Davis, Larry Eugene; Davis, Larry Eugene; Davoren, Anthony; di Bona, Pietro Alphonse; Dingee, Brad E.; Djuth, Gerald Joseph; Doughty, John Daniel; Drobny, Gerald Francis; Duncan, Gregory Wade; Dyman, Thaddeus Stanley, Jr.; Eby, Elaine T.; Egemeier, Robert Jack; Eliason, Jay R.; Ellis, Michael Alexander; Ellison, Patrick James; Engh, Kenneth R.; Engle, M. S.; Eves, Robert Leo; Fraser, Gregory Thomas; Gamache, Mark Thomas; Gavenda, Robert Thomas; Ginther, Paul G.; Goodman, Howard Mark; Grant, Philip Robert; Grieco, Robert Anthony; Groffman, Louis H.; Guess, Sam C.; Hafley, Daniel James; Hagan, Teresa A.;

Hanbury, Jonathan B.; Hanbury, Patricia Melling; Harris, Elaine; Haviland, Terrance; Hawksworth, Mark A.; Hickey, Michael Glenn; Hogge, Curt Edward; Holder, Grace Amelia McCarley; Holder, Robert Wade; Hollabaugh, Curtis Lee; Hooper, Robert Louis; Hooper, Robert Louis; Houseman, Michel Dirk; Hunt, Sharon Barbara; Hunter, Craig Russell; Hurley, Bruce William; Ichimura, Vernon T.; Ingemansen, Dean Brian; Jerde, Eric A.; Jirik, Richard Steven; Johnson, Clarence Richard; Johnson, Diane Marilyn; Johnson, Lisa Kaye; Johnson, Peter Eric; Juul, Steve Thorvald Julius; Kabir, Jobaid; Kohn, Sara E.; Kuhns, Mary Jo Pankratz; Kuhns, Roger James; La Salata, Frank Vincent Michael; Lance, Donald M., Jr.; Landle, George Louis; Lane, Diane Estelle; Laskowski, Erich Richard; Laudon, Julie Ann; Laudon, Katherine Jean; Lawton, Evert Carl; Letzsch, W. Stephen; Lim, Jose Bernardo R.; Lindsey, Kevin A.; Link, J. E.; Liszewski, Michael Joseph; Livesay, D. M.; Lofland, Darlane Kathryn; Ludwig, S. L.; Lundquist, John H.; Lynch, Joseph Vincent Gregory; Mack, Christopher Brown; Mackie, Thomas L.; Mahamah, Dintie Shaibu; Martin, Barton Sawyer; Mauritsen, Mark Vernon; Mauritsen, Mark Vernon; McDonald, Eric V.; McFarland, William Douglas; McGary, Etta Gaynell; Meints, Joyce P.; Melling, Patricia Hanbury; Meyer, Michael R.; Minkel, Donald H.; Mohl, Gregory Blaine; Moser, Joseph Arthur; Nack, Nissa Louise; Neal, William Scott; Nelstead, Kevin Torval; Nguyen, Son Ngoc; Nieman, William George; O'Keefe, Monica Elizabeth; O'Malley, David P.; Omer, Mekki A.; Orazulike, Donatus Maduka; Pacherneggg, Sheila M.; Palmer, Curtis Allyn; Peoples, Darrell D.; Petersen, Kenneth Lee; Peterson, D. A.; Pierson, Frederick Barker, Jr.; Pitz, Charles Forrest; Poeter, Eileen; Porter, Lee; Pottmeyer, J. A.; Price, Edwin Henry; Rahimi, Saeed; Read, John J.; Reed, John Daniel; Reef, John W.; Rhoades, Matthew J.; Richman, Lance Ramon; Rieken, E. R.; Robbins, Donald A.; Roberts, James W.; Sass, Bruce M.; Saur, William F.; Scheffler, Joanna March; Schriber, Craig Norman; Scrivner, James V.; Siebenmann, Kathleen F.; Singler, Caroline Susan; Smith, Roger Norman; Smith, Tad Monnett; Starlin, Leigh Ann; Starr, Mark Andrew Michael; Strong, Despina; Stryhas, Bart Andrew; Swan, Victor LaMarr; Swanson, Eric Craig; Taylor, Stephen Bernard; Taylor, Teresa Ann; Van Berkel, Gary Joseph; Van Klavaren, Richard William; Van Liew, Michael Wayne; Vance, David B.; Vedagiri, Velpari; Volk, J. A.; Waggoner, Gail Louise; Waldo, Jaunell Jean; Walkey, Clifton; Weberg, Erik D.; Wheeler, Mark Thomas; White, Randal Ocee; Wickwire, Daniel William; Widness, Scott E.; Wiedenhoeft, G. R.; Wigand, Peter Ernest; Wilson, Joseph Raymond; Wong, Jade Starr; Wood, Thomas R.; Yates, Douglas Morris; Yates, Eugene Adams, III; Youngs, Steven Wilcox; Zachara, John M.

Washington University
St. Louis, MO 63130

291 Master's, 92 Doctoral

1870s: Barrou, S. S.; Gibson, Victor Rutledge; Patrick, W. F.; Wilson, N. R.

1880s: Boyle, John, Jr.; Comstock, W. O.; Conzelman, William E.; Emmons, N. H.; Gazzam, J. P.; Gluck, Leo; Hodges, A. B. W.; Hutchinson, R. B.; Jackson, Edward F.; Kinealy, James R.; Lebens, E. H.; McCulloch, Richard; Monell, Joseph; Mudd, S. W.; Nicholson, Frank; Robertson, J. D.; Rombauer, Alfred B.; Rombauer, Alfred B.; Sauer, J. H.; Zukoski, C. F.; Zukoski, E. L.

1890s: Francis, Alfred; McCulloch, Richard; Pope, John L., Jr.; Tuttle, Arthur L.; Zelle, William C.

1920s: Cozzens, Arthur B.; Foster, Vellora Meek; Geddes, Francis N.; Grawe, Oliver R.; Hartnagel, Florence A.; Hinchey, Norman; Knapp, Esther Laura; Krenning, Erna Louise; Markley, Joseph Hooker; Mason, Charles Clifford; Maxwell, Riley G.; Meleen, Elmer E.; Radsbaugh, John Wesley;

Russell, Eugene Merle; Shaw, Earl Bennett; Stringfield, Victor Timothy; Studt, Charles W.; Thatcher, Richard Whitfield; Thompson, John Peters

1930s: Barr, Joe William; Brightman, George Forsha; Cozzens, Arthur B.; Crosby, Mary Francis; Crow, Louis Milton; Denham, Richard L.; Dickey, Robert I.; Dings, McClelland G.; Draper, Richard Brandt; Draper, Richard Brandt; Gillerman, Elliott; Gollhofer, Rolla Linz; Graves, Howard B.; Graves, Howard B.; Gsell, Ronald N.; Gunnell, Emory Mitchell; Jenner, John Slaten; Johnson, Hugh Nelson; Koch, Heinrich Louis; Magness, Catherine Virginia; Marshall, Willis Woodbury; Mayer, Harold Melvin; Meyer, Charles; Ostrander, Alan; Pough, Frederick Harvey; Ray, Louis Lamy; Rebholz, Irma; Robertson, Percival; Rowe, Ruth; Straube, Elsie Joan; Sutton, Donald Grant; Walka, Joseph August; Wallace, Alan Joseph; Wellman, Dean Castor; Wilgus, Wallace LaFetra; Wing, Robert Busch; Wood, Mabel Vivian; Yenne, Keith Austin; Zeip, Vera Lydia

1940s: Aschemeyer, Esther Louise; Blake, Mabel Louise; Bolger, Robert Courtney; Bonham, Lawrence Cook; Bousfield, John Channing; Campbell, Arthur Byron; Croninger, Adele Bullen; Fary, Raymond Wolcott, Jr.; Forrestal, Geraldine; Good, John Maxwell; Gruner, Thayer Meredith; Isachsen, Yngvar William; Jenke, Arthur Louis; Kidwell, Albert Laws; Levinson, Stuart Alan; Narten, Parry Foote, Jr.; Ohle, Ernest L.; Pendexter, Charles; Rezak, Richard; Robertson, Forbes Smith; Robitshek, Melvin F.; Stickel, John Frederick, Jr.; Walker, Thomas Henry; Werner, Harry Jay; Wesley, Richard Hal; Wilson, Edward Norman; Wraight, Joseph

1950s: Algermissen, Sylvester Theodore; Algermissen, Sylvester Theodore; Anderson, Burton R.; Baerns, Rudolph; Baldwin, William Felbert Jackson, Jr.; Bauman, Paul Thomas; Baumgartner, Warren Francis; Beach, Paul Ronald; Bedford, John Phillips; Benner, Velma; Berry, Richard Warren; Bethke, William Martin; Black, Rudolph Allan; Bonham, Lawrence Cook; Bronder, Joseph Bertram; Brown, Lynn A.; Collins, George Alexander; Danielson, Richard Earl; Davies, William James; Dillingham, Hervie, Jr.; Doman, Robert Charles; Dornbach, John Ellis; Ellis, Thomas Morgan; Etter, John; Faber, James Warren; Felix, Charles Jeffrey; Fink, Don Roger; Fink, Helen Binkley; Flege, Robert Frederick, Jr.; Fox, Bruce Wendell; Gerlach, George Smith; Gibson, Lee Boring; Glenn, Richard Allen; Gouty, John J.; Graf, Robert Bernard; Gregson, Victor Gregory, Jr.; Griffin, John Joseph; Haack, David Arno; Hail, William James, Jr.; Hammerquist, Donald William; Haughey, William Henry; Hays, Walter Wesley; Hill, Bernard Louis; Hill, Wayne Noe; Holke, Kenneth Arthur; Hower, John, Jr.; Hower, John, Jr.; Hunt, Mahlon Seymour; Jacobson, Rollyn Philip; Johnson, Hugh Nelson; Kays, Marvin Allan; Krummel, William J., Jr.; Legate, Carl Eugene; Legate, Carl Eugene; Levin, Harold Leonard; Levinson, Stuart Alan; McFarland, John Barnett; Meidav, Tsvi; Mercado, Edward John; Mueller, Carl Vincent; Mundt, Philip Amos; Munsey, Gorson Cloyd, Jr.; Nasmith, Hugh Wallis; Pappas, Thomas; Patterson, William A.; Pendexter, Charles; Quist, Earl Francis; Rechtien, Richard Douglas; Reynolds, Merrill Johnson; Reynolds, Robert Coltart, Jr.; Sacket, William Malcolm; Schaeffer, Frederick Ernest, Jr.; Schaeffer, Katherine Maude Marie; Schneider, Stephen Jay; Schroeder, Thomas Francis; Schroeder, Thomas Francis; Short, Nicholas Martin; Smith, Mahlon; Stuart, John William; Tettenhorst, Rodney Tampa; Thomas, Charles Ward, Jr.; Thomas, Chester Ward, Jr.; Trapp, Henry, Jr.; Truscott, Frederick Wilson, Jr.; Veesaert, Marlin Joseph; Wagner, Frederick John, Jr.; Weathers, Gerald; Wermeyer, Raymond Alfred; Wuestner, Charles E. R.

1960s: Agashe, Shripad Narayanrao; Allen, Harry; Anderson, Roy Ernest; Baag, Czang-Go; Baker, John Hudson; Ballard, Geoffrey Edwin Hall; Bal-

lon, Wilfredo; Beckham, Wallace Edgar, Jr.; Beckham, Wallace Edgar, Jr.; Berry, Richard Warren; Beyer, Harry; Bissada, Kadry Kaddis; Bissada, Kadry Kaddis; Bissada, Mona; Blanchard, Richard Lee; Boyer, Luc; Brennan, William Joseph; Budke, Earl Hugo, Jr.; Calvert, Ronald Harold; Chang, Yi-Maw; Cheng, Mary Mei-Ling Huang; Collins, George Alexander; Danusawad, Thawisak; Ehrlich, Marvin Irwin; Ellis, Thomas Morgan; Ervin, Clarence P.; Esker, George Cornelius, III; Fink, Richard P.; Foweraker, John Charles; Frey, Richard Paul; Gaskill, James R.; Golden, Julia; Gravely, Marion Shelor; Hallquist, John Berger; Hanss, Robert Edward; Hayes, Miles Oren; Hays, Walter Wesley; Hsu, I-Chi; Joerger, Arthur Peter; Jud, William F.; Kays, Marvin Allan; Kienzle, John Kenneth; Ku, Chao-cheng; Kuest, Louis John, Jr.; Laing, Eoghan MacRuaraidh; Ling, Hsin-Yi; Macke, John Edward; Madison, James Ambrose; Meidav, Tsvi; Moll, William Francis, Jr.; Nieto-Pescetto, Alberto Santiago; Paravincini, P. Guido; Peralta, Tobias Requejo; Phelan, Michael Joseph; Phelan, Michael Joseph; Phillips, Tommy Lee; Reams, Max W.; Rechtian, Richard Douglas; Samanez, Wilfredo Ballon; Segar, Robert L.; Sen Gupta, Pradip Kumar; Sendlein, Lyle Vernon Archie; Sonido, Ernesto P.; Staley, Walter Goodwin, Jr.; Staub, William Praed; Stinchcomb, Bruce Leonard; Stueber, Alan Michael; Sullivan, Dan Allen, Jr.; Tikrity, Sammi Sherif; Turner, Warren H.; Wegweiser, Arthur Ervin; West, Terry Ronald; Williams, Frank Brierley; Yen, Luis; Young, Rechard Andrew

1970s: Almon, William Robert; Barsky, C. K.; Blaxland, Alan; Blouse, R. S.; Botts, Michael Edward; Bragg, Susan Lynn; Breese, T. E.; Bridgett, L. S.; Chauff, K. M.; Collins, H. L.; Crossey, Laura Jones; Davis, Richard Laurence; Depke, T. J.; Early, Thomas Oren; Fan, Jieun-jeou; Fix, M. F.; Hongnusonthi, A-ngoon; Hsu, I-Chi; Hughes, Edward Stewart; Jubb, T. M.; Katz, Steven George; Lee, Larry Jack; Lee, Steven Wendell; Lewis, Sally Beth; Lin, Jia-wen; Lo, Howard Hunghsin; Mason, Chester Bowden; McQueen, Donald James; Morgan, C. J.; Ouellette, Robert; Payne, Kathryn Lynn; Potts, Mark John; Price, L. G.; Rafle, M. A.; Rafle, M. A.; Saeger, William Eldon; Sallak, Sulieman; Scarato, Robert Anthony; Scharnberger, Charles Kirby; Scott, William E.; Sherwood, Ronald W.; Shifflett, Howard Richard; Shimoyama, Akira; Shourd, Melvin Lee; Sibley, S. F.; Skow, Donald Lester; Snavely, D. S.; Sun, Stanley S. S.; Szekeley, Francisco L.; Thompson, Donald John

1980s: Abrajano, Teofilo Aniag, Jr.; Becker, Richard Charles; Bernatowicz, Thomas James; Bloch, John Daniel; Brannon, Joyce C.; Burton, Elizabeth Ann; Caffee, Marc William; Castillo, Paterno Reyes; Dale-Bannister, Mary Ann; de Chazal, Suzanne Marie; Decker, Deborah Ann; Dromgoole, Edward Lee; Dunn, Lisa Gay; Eddy, Michele Sharon; Edwards, Margaret Helen; Eilerts, Toni Lynn; Fetyani, Ahmad Ali; Gardner, James Edward; Green, Glen Martin; Guinness, Edward Albert, Jr.; Heatherington, Ann Louise; Jacobberger, Patricia Ann; Kennedy, Burton Mack; Koberle, A.; Kramer, Frank Edward; Kremser, Daniel Timothy; Lee, Bor-Jen; Leff, Craig Ernest; Lewis, Robert Harry; McKeegan, Kevin Daniel; Meng, Yanxi; Michel, Sandra Jo Hagni; Mueller, Steven Wayne; O'Leary, Ellen Frances; O'Reilly, Thomas Clark; Peck, Nathan Russell; Petroy, David Edward; Presley, Marsha Ann; Ragan, Jerry Michael; Roca, Henri Joseph, III; Rogers, Mark Arleigh; Salpas, Peter Andrew; Schenk, Paul Michael; Seely, Mark Richard; Shoberg, Thomas Gilford; Simon, John Frederick; Smith, Terri Lynn; Strebeck, John William; Sturchio, Neil Colrick; Sultan, Mohamed Ibrahim; Sutton, Stephen Roy; Sylvester, Paul Joseph; Villeneuve, Michael E.; Viscio, Paul James

University of Waterloo
Waterloo, ON N2L 3G1

228 Master's, 66 Doctoral

1960s: Srinivasan, Vajapeyam S.

Wayne State University

1970s: Ackermann, D.; Akiti, T. T.; Anderson, Thane Wesley; Atobrah, K.; Baechler, F. E.; Baker, C. L.; Barker, J. F.; Barnett, Peter James; Baron, J.; Beland, A.; Benoit, E. G.; Betcher, R. N.; Bottomley, D. J.; Carnevali, J.; Ceroici, W. J.; Chorley, D. W.; Cooper, A. J.; Dakin, R. A.; Davison, C. C.; Day, M. J.; Dredge, L. A.; Egboka, Boniface Chukwura Ezeanyaoho; Eyisi, A. E. O.; Forster, C. B.; Gevaert, D. M.; Grisak, G. E.; Guiton, R. S.; Gupta, Santosh Kumar; Hall, B. V.; Harrington, R.; Harvey, E. T.; Hendry, Michael James; Hilton, S. W.; Hims, A. G.; Hipel, K. W.; Hoffman, Douglas Weir; Isherwood, A.; Jackson, D. A.; Jackson, R. E.; Johnston, H. M.; Kennedy, K. G.; Kewen, T. J.; Killey, R. W. D.; Lebedin, J.; Lee, P. F. Y.; Liard, A. C.; Logan, L.; Logan, L. A.; McCracken, A. D.; McNaughton, D. C. M.; Memarian, H.; Michel, F. A.; Mirza, K.; Mitchell, S. L.; Mohsen, M. F. N.; Morris, J. H.; Munro, I.; Novakovic, B.; Nowicki, V.; Nowlan, G. S.; Panko, A. W.; Poulin, M.; Pucovsky, G. M.; Rehm, B. W.; Rendigs, R.; Reynolds, W. D.; Robertson, W. D.; Schwert, D. P.; Scott, W. D.; Sharan, S. K.; Simard, G.; Sinclair, R. D.; Sklash, M. G.; Sklash, M. G.; Smith, D. P.; Smith, J. B.; Soares, E. F.; Soyupak, S.; Stiebel, W. H.; Stott, G. A.; Thomas, B. L.; Tong, Y.; Tremblay, P.-R.; Troup, W. R.; Vandor, M.; Verge, M. J.; Victor, R.; Wallis, P. M.; Welhan, J. A.; Woldetensae, H.

1980s: Abbott, D. E.; Abdul, Abdul Shaheed; Adegoke, Oluwole Johnson; Akindunni, Festus Funso Folorunso; Allsop, Heather Allyne; Armstrong, D. K.; Ash, P. O.; Bajc, A. F.; Barbash, J. E.; Barnett, Peter James; Basharmal, M.; Ben-Miloud, K.; Berry-Spark, K. L.; Berwanger, D. J.; Blackmer, Andrew John; Blackport, Raymond J.; Blair, R. D.; Blowes, D. B.; Bolha, J., Jr.; Bradshaw, K. L.; Broughton, D. W.; Bruce, J. R. G.; Bunner, W. D.; Burnett, R. D.; Buszka, Paul Mark; Calder, Lynn M.; Callander, P. F.; Carson, David Marshall; Chatzis, Ioannis; Churcher, P. L.; Clark, Ian D.; Clark, P. U.; Clark, W. S.; Coakley, John Phillip; Cosgrave, T. M.; Dance, John Thomas; Daniels, Shirland Augustus; Daus, A. D., III; Dearlove, J. P. L.; Desaulniers, Donald Edouard; Desaulniers, Donald Edouard; Dickin, A. C.; Dickinson, Robert Rowland; Dollar, P. S.; Dubord, M. P.; Dubrovsky, Neil Michael; Duffield, Susan Linda; Durrant, Richard Lee; Easton, J. A.; Edwards, Thomas Wellington Deavitt; Egboka, Boniface Chukwura Ezeanyaoha; Ehret, K. S.; Elliott, S. R.; Feenstra, Stanley; Finamore, P. F.; Fitzgerald, William Donald; Fordham, C. J.; Francis, R. M.; Franklin, Steven Eric; Frost, L. H.; Germain, M. S. D.; Germain, M. S. D.; Glynn, Pierre David; Godwin, A. G.; Gomer, M. D.; Goodings, C. R.; Goodwin, M. J.; Harding, W. R.; Harris, R. D.; Harrison, E. M.; Healing, David William; Hendry, Michael James; Hewetson, J. P.; Hodges, D. J.; Hokkanen, Gary Elmer; Jagannath, R.; Jayatilaka, Chandrika Jayakanthi DeSilva; Jones, M. G.; Jones, Norman Kenneth; Kantzas, Apostolos; Kay, B. G.; Keller, C. K.; Keller, Chester Kent; Kerr-Lawson, L.; King, K. S.; Kornelsen, P. J.; LeBlanc, H. G.; Legall, Franklyn David; Leith, Rory Marshall Montgomery; Lesack, K. A.; Lindsay, Louise Elizabeth; Lockwood, G. C.; Loftsson, M.; Lolcama, J. L.; Lotimer, A. R.; Love, D. A.; MacDonald, I. M.; MacFarlane, D. S.; MacQuarrie, K. T. B.; Maerz, N. H.; McKee, J. A.; McLachlin, D.; McLaren, R. G.; McLeod, Norman Stuart; McMurray, M.; Michel, Frederick A.; Mielke, R.; Miles, C. M.; Miller, D. J.; Miller, Randall Francis; Miller, Randall Francis; Milton, Gwendolyn Margaret; Mkumba, J. T. K.; Moddle, P. M.; Molson, J. W. H.; Montgomery, Keith; Moore, M. B.; Moore, Ruth Albertine; Morin, Kevin Andrew; Mozeto, Antonio A.; Myrand, D.; Nadon, R.; Nicholson, Ronald Vincent; Nicholson, Ronald Vincent; Novakowski, Kentner Stephen; Nwankor, G. I.; Nwankwor, Godwin Ifedilichukwu; O'Brien, C.; O'Shea, K. J.; Omara-Ojungu, Peter Hastings; Ophori, Duke Urhobo; Or-

tega Guerrero, A.; Page, C. E.; Pair, D.; Palmer, Carl David; Patch, W. R.; Patrick, G. C.; Peacock, Simon M.; Pearson, Elizabeth M.; Pehme, Peeter Enn; Petrie, J. M.; Pickens, John Franklin; Pilny, J.; Piotrowski, J.; Proulx, I.; Ptacek, C. J.; Quinn, O. P.; Qureshi, Riffat Mahmood; Qureshi, Riffat Mahmood; Rannie, E. H.; Raven, K. G.; Reading, David John Richard; Reesor, S. N.; Reimer, James Denis; Roberts, Alice Lynn; Robertson, E.; Rodvang, S. J.; Ross, L. C.; Rouleau, Alain; Rudolph, D. L.; Ruland, W. W.; Ryan, M. C.; Sado, Edward Vincent; Sauchyn, David John; Scheier, Nicholas William; Schellenberg, S. L.; Schmidtke, Klaus-Dieter; Scott, A.; Serrano, Sergio Enrique; Sinha, M. N.; Slaine, D. D.; Smith, P. M.; Smyth, D. J. A.; Speranza, Angelo; St-Arnaud, L.; Starr, Robert Charles; Starr, Robert Charles; Stipp, S. L.; Sudicky, Edward Allan; Sweeney, S.; Taruvinga, Peter Rangarirai; Taylor, Brian Burke; Tessman, J. S.; Thompstone, Robert Marshal; Thomson, G.; Trudell, Mark Russell; Tworo, A.; Unrau, J. D.; Usher, S. J.; van Everdingen, David Allard; Van Walsum, N.; Veska, Eric; Vorauer, A. G.; Walker, Douglas J.; Webb, R. J.; Weitzman, M. J.; Wexler, E. J.; Winters, Steven L.; Zapico, N. M.; Zilans, A.

Wayne State University
Detroit, MI 48202

114 Master's, 8 Doctoral

1940s: Kelley, Robert W.; Scopel, Louis Joseph; Simons, Merton Eugene

1950s: Berner, Paul C.; Brady, Michael B.; Bryden, Elmer Louis; Chen, Chin; Chyenoweth, Clyde E.; Coley, Tyrol B.; Drabkowski, Robert S.; Elkington, Robert B.; Gartig, Derry G.; Krushensky, Richard D.; Manley, Thomas R.; Marlian, Myron G.; Marschner, Arthur W.; Morden, Audley D.; Paul, Harriet E.; Roulidis, Christos Z.; Rowe, Dean E.; Rubel, Daniel Nicholas; Scheufler, John H.; Sparling, Dale R.; Steiner, Waldon W.; Svetlich, William G.; Wooten, Maria Jane

1960s: Adams, Adam R.; Banar, Frank J.; Boneham, Roger F.; Bruehl, Donald H.; Chang, Hsing Chi; Chen, Pei-Hsin; Fassett, James E.; Fiesinger, Donald W.; Fisher, Frederick S.; Friend, Joseph E.; Gosser, Charles F.; Imam, Ali; Kirchner, James G.; Mariotti, Philip A.; Newton, Geoffrey B.; Poprik, Lee Albert; Potocki, Sigmund R.; Rihani, Rushdi Freih; Sachdev, Suresh C.; Shelden, Francis D.; Starr, Stephen G.; Stricker, Gary D.; Van Wyckhouse, Roger J.; Walston, Gerald M.; Young, Malcolm G.

1970s: Abo-Elela, R. M.; Balakrishna, Thirumale; Bohman, R. P.; Buchanan, J. J.; Buckley, Glenn R.; Chaivre, Kenneth R.; Chivalak, S.; De La Pena, Edward C.; Doherty, D. J.; Dombkowski, Francis; Dziewa, T. J.; Florek, Robert J.; Gill, Gerald M.; Greimel, Thomas C.; Hammer, W. R.; Hoda, Badrul; Konkel, David C.; Lee, Theodore David; Lilienthal, Richard; Marchel, Ronald Joseph; Miller, Charles N.; Mitchell, Steven W.; Morse, Edwin W.; Nankervis, Jeffrey Chambers; Olszewski, Gregory P.; Papadopoulos, Haralambos I.; Pitcher, Jacob J.; Pitzak, A. N.; Rulli, Vernon G.; Sarniak, Terry Michael; Sayek, T. F.; Shaver, R. B.; St. Aubin, Thomas E.; Sulanowski, Jacek S. K.; Swanson, Ruth A.; Venkateswaran, Ravi T.; Wang, Yun Chung; Ward, William Paul; Warner, Ronald L.; Weltin, Timothy P.; Wilson, Steven E.

1980s: Absi, George A.; Ague, Jay J.; Barcas, Kestutis; Bata, Mazin Y.; Bean, Lawrence Edward; Beever, Frances Kay; Bowers, Timothy L.; Burns, Richard L., Jr.; De Fauw, Sherri Lynn; Dorr, Lawrence L.; Dutton-Melendy, Victoria L.; Gron, Shelley A.; Hinchman, Nancy; Kanat, Leslie Howard; Kulpanowski, Stephen E.; La Fortune, Irene A.; Lawrence, Christopher H.; Moffett, Joanne Lynn; Nasr, Athanacios Nabeh; Powell, Michael A.; Radlick, Thomas; Ratliff, Robert A.; Strybel, Daniel Z.; Thornton, Patricia Ann; Troschinetz, John; Tyler, J. Gary; VanDorpe, Paul E.; Walniuk, Daria M.; Wilczynski, Michael V.; Zhao, Zhong Yan

Wesleyan University
Middletown, CT 06457

28 Master's

1960s: Arsalan, Ahmad; Chang, Chung Chin; Inthuputi, Boonmai; Japakasetr, Thawat; Wotorson, Cletus S.

1970s: Farrar, Michael J.; Graf, Alexander N.; Handman, Elinor H.; Kim, Sang Wook; McDonald, Nicholas Grant; Naylor, Rrichard G.; O'Bara, Jeffrey Brian; Potisat, Somsak; Rastegar, Iraj; Snider, Frederic G.; Thomas, Robert D., Jr.

1980s: Altamura, Robert James; Bean, Merit W., Jr.; Clifton, Amy E.; Gruntmeyer, Paul Alexander; Hickenlooper, John W., Jr.; LeTourneau, Peter Mark; Lodise, Lisa; Metcalf, Thomas P.; Paijitprapapon, Vivat; Pecora, William C.; Scatena, Frederick N.; Toney, Jennifer Diana

West Texas State University
Canyon, TX 79016

56 Master's

1970s: Allison, Michael Duane; Brown, Gilbert Daleth; Claughton, James L.; Cramer, Scott L.; Deaver, Boyd Edwin; Gage, John E.; Gibson, C. R.; Gilbert, Jerry L.; Hamilton, David P.; Hood, H. C.; Hull, M. J.; Hung, C. C.; Jeffery, D. A.; Lynn, Alvin R.; Massingill, Gary L.; Miller, H. T.; Morrison, Ernest Robert; Shadix, Shirley J.; Smith, Robert E.; Speer, Roberta D.; Stevens, Richard K.; Stout, Jerry Dale; Taylor, S. J.

1980s: Arzaghi, Mohamad Mehdi; Ashraf, Saied; Babb, Janet L.; Bartolino, James R.; Bertl, Jeffery D.; Bourbon, William Bruce; Boyd, Thomas Lee; Brewster, Arthur V.; Cameron, Donald Edward, Jr.; Caudle, Karen L.; Cleaver, Jeffrey S.; Cleland, Jane M.; Donaldson, William Allen, III; Drake, John Franklin; Ellinger, Scott T.; Hardisty, Russell D.; Hardy, Hughey E.; Haroon, Mohammed A.; Jaafar, Idris Bin; Kantaatmadja, Budi P.; McAlister, Randall Lee; McBride, Donald David; Mignardot, Eddie Roy; Munson, Timothy Wayne; Pertl, David Joseph; Rapstine, Inge Frances; Richards, Trenton Hubert; Ritter, Steven Paul; Rivas, Charlie, Jr.; Scrimshire, Elven Rick; Shearer, Carroll Dean; Sparling, Kirk Darren; Wilson, Gregory Allen

West Virginia University
Morgantown, WV 26506

163 Master's, 68 Doctoral

1920s: Price, Paul Holland

1930s: Heck, Edward T.; Law, Lewis B.; Nutent, Lawrence E., Jr.

1940s: Conrad, George J.; Difford, Winthrop C.; Dotson, James M.; Heck, Edward T.; Martin, Wayne Dudley; Murphy, Allen Emerson; Tucker, Charles E.; Webster, Russell N.; Withers, James Gordon, Jr.; Yost, Coyd Bickley, Jr.

1950s: Anderegg, Charles; Arnold, Eldon Drewes; Bain, George L.; Bentzel, Ruby H.; Brown, Paul M.; Cardea, Harry S.; Collins, Horace R.; Dean, John R.; Denton, George H.; Dyar, Robert Francis; Flowers, Russel R.; Fonner, Robert F.; Garnar, Thomas E., Jr.; Gray, Ralph Joseph; Gwinn, James E.; Hickman, Paul R.; Kanes, William Henry; Kapnicky, George; Latimer, Ira Sanders, Jr.; Lin, Francis Chien-Ming; McAndrews, Harry; McColloch, Samuel; McCullough, Edgar J.; Moebs, Noel N.; Norwood, Edward M., Jr.; Plants, Kenneth D.; Renton, John Johnston; Robinson, George Calver; Salgado, Peter G.; Schapiro, Norman; Schemel, Mart P.; Shockey, Philip Nelson; Stenger, William J.; Swales, William E.; Thomson, Alan Frank; Twigg, Robert W.; Warman, James Clark; Watkins, William Merle, II; Watts, Royal J.; Wentworth, David W.; Wilcox, Floyd B.; Wilmoth, Benton M.; Wray, John Lee

1960s: Barlow, James Arthur; Bluman, Dean Edward; Bowers, Richard F.; Callahan, Robert L.;

Carpenter, David; Cecil, Charles Blaine; Cecil, Charles Blaine; Cheek, Robert B.; Clendening, John A.; Collin, Martin L.; Conti, Louis Joseph, Jr.; Dean, Stuart Linden; Dean, Stuart Linden; Duncan, William M.; Dyckes, Jan Allan; Eddy, Greg Edward; Fasching, George E.; Fox, John Thomas; Gooch, James L.; Growitz, Douglas; Gwilliam, William; Hahman, W. Richard; Hall, Richard P.; Harris, William; Heffner, Larry B.; Hennen, Gary James; Hoover, John; Hughart, Richard D.; Hutchison, David Malcolm; Kanes, William H.; Kilcommins, John Peter; Kirr, James N.; Kulander, Byron Rodney; Kulander, Byron Rodney; Larese, Richard E.; Leonard, Arnold David; Luzier, James E.; Malone, Rodney D.; Martin, Richard Harold; Martin, Richard Harold; Matthews, Martin David; McPherson, Ronald Bruce; Mefford, James; Mennen, Gary; Merritt, Gary L.; Miller, Don R.; Milner, William Collier; Minke, Joseph Garrett; Mitchell, Francis J.; Morton, Robert A.; Nimickas, Bronius; Nock, Harvey; Nordeck, Robert E.; Nuhfer, Edward; Page, Roland C.; Perkins, Robert Lee; Perkins, Robert Lee; Renton, John Johnson; Russ, David P.; Ryan, William M.; Shafer, Harry E., Jr.; Simms, Frederick Eugene; Stilson, Willam P.; Streib, Donald L.; Thomas, Lloyd; Travis, Jack Watson; West, Cliff Merrell, Jr.; Yedlosky, Robert Joseph; Youse, Arthur

1970s: Akers, D. J.; Arnold, George Edward; Berger, P. S.; Carpenter, T. W.; Cheema, M. R.; Clendening, John Albert; Clendining, John Albert; Dixon, J. M.; Drake, Gerald M.; Eddy, Greg Edward; Estep, Patricia Anne; Glagola, Peter A.; Grady, William C.; Hall, M. M.; Hamilton, Walter M.; Harr, J. L.; Harris, Paul M.; Henderson, Charles D.; Henniger, Bernard Robert; Hidalgo, Robert Valeriano; Hollingsworth, Terry J.; Kemerer, Thomas F.; Kerns, Earl; Kimutis, Robert A.; King, Gary L.; Kirr, James Neil; Komar, C. A.; Lacaze, John A., Jr.; Larese, R. E.; Lo, Hoom-bin; Martin, Ray Earl; Meghji, M. H.; Miller, M.; Moore, S. B.; Morton, Robert; Moyer, Carol B.; Neal, Donald Wade; Negus-de Wys, J.; O'Neil, J. R.; Ogden, A. E.; Parker, W. W.; Powell, L. R.; Presley, M. W.; Rao, S. K.; Reger, J. P.; Selby, Curt McKee; Sites, R.; Sites, R. S.; Streib, Donald Lamar; Strickland, L. D.; Tepperman, Mark; Teti, M. J.; Trumbo, D. B.; Williamson, Norman L.; Wilson, Peter G.

1980s: Allen, Joan Park; Amanat, Jamshid; André, Richard A.; Auxt, Tara Lou; Beras, Manuel E.; Bird, Debra L.; Bjerstedt, Thomas William; Boswell, Ray Marcellus; Boswell, Ray, Jr.; Boyer, Douglas Gene; Calvert, Gary C.; Carney, Cindy Kay; Carter, Burchard D., III; Chen, Ran-Jay; Chobthum, Worapot; Conrad, James Matthew; Dixon, Denise Yvonne; Dominic, David Francis; Dowse, Mary Elizabeth; Dunn, Gilbert Riley; Eble, Cortland F.; Ferrill, David Alexander; Gerritsen, Steven Scott; Hall, Pamela S.; He, Guoqi; Heller, Sara Anne; Ho, Sheng-Zong John; Ivahnenko, Tamara I.; Kelley, Peter Alexander; Khan, Subhotosh; King, Hobart Morse, II; Kirwan, Laura; Kokli, Kewal Krishan; Lin, Po-Ming; Massaquoi, Joseph George Momodu; Ogbonlowo, David Babajide; Pogue, Peggy Todd; Reddy, Nagendra Peesary; Rohaus, Donna Marie; Smith, Diane M.; Su, Wen Huane; Swales, David L.; Tang, David Hsin-Ying; Tsai, Louis Loung-Yie; Vinopal, Robert James; Walker, April Rubens; Wilson, Thomas Hornor; Wrightstone, Gregory; Yeatts, Daniel Solomon; Zurbuch, Jeffrey S.

Western Carolina University
Cullowhee, NC

1 Master's

1970s: Maas, R. P.

Western Connecticut State University
Danbury, CT 06810

4 Master's

1980s: Hall, Louis S., III; Nazar, Edward; Obeda, Barbara A.; Smoliga, John A.

Western Michigan University
Kalamazoo, MI 49008

82 Master's, 4 Doctoral

1960s: Lemerand, Martin

1970s: Al-Jallal, Ibrahim Abdulla M.; Clark, Stacy Lon; Cookman, Charles Willard; Coons, William Ellsworth, III; Fingleton, Walter George; Fowler, John Henry; Frykberg, W. R.; Fuller, Jonathan A.; Gebben, Dennis J.; Ghahremani, Darioush Tabrizi; Heinsius, John Walter; Johanns, Williams Mathias; Kerhin, Randall Thomas; Kern, Ernest Lee; Kimmel, Richard Elmer; LoPiccolo, Robert David; Lovan, Norman Alan; Malanchak, John E.; Moss, Michael James; Murray, Eugene J.; Naeve, Valarie Adrienne; Peterson, James Carl; Tripp, Steven Edward

1980s: Apak, Sukru Nail; Arruda, Edward Charles; Barnett, James Matthew; Barnick, Sandra K.; Bartel, James Robert; Barton, Gary James; Brown, Jeffrey Scott; Carey, Neal J.; Cookman, Richard G.; Dennis, Scott Timothy; Dexter, James J.; Farnsworth, James W.; Fenner, Linda B.; Fici, Huseyin; Fronczek, Daniel V.; Fulker, Katharine D.; Fults, Michelle Ellen; Gallagher, Michael G.; Garrett, Elizabeth M.; Ghatge, Suhas Laxman; Hahnenberg, James J.; Hall, David W.; Hanna, Thomas Murray; Henderson, William Charles; Hermann, Jon Michael; Horton, Paul; Howell, David Adams; Klanke, John Emil; Koehler, Janet; Leja, Stanislaw, Jr.; Leonard, Barbara June; Lindsay, David Walter; Little, Ann Carol; Luby, Thomas Patrick; Mahan, Thomas Kent; Martin, Jeffrey R.; Mayotte, Timothy; McBride, Barry Christopher; Meisel, Kent E.; Montgomery, Eric Lee; Niewendorp, Clark Alan; Norman, William Robert; Park, Scott Gregory; Pogoncheff, Nicholas C.; Porcher, Eric N.; Prochaska, Kevin M.; Reichert, Randall Lee; Rodwan, John Charles; Rudder, James, Jr.; Samuelson, Kiff James; Sheedlo, Mark Kenneth; Spruit, Jeffry Dean; Stefaniak, Gary John; Streeter, Michael Edward; Suh, Mancheol; Sullivan, Kevin James; Talanda, Jean; Varga, Lisa Louise; Warwick, David B.; Wigger, Stephen Thomas; Williams, William Thomas; Wireman, Michael

University of Western Ontario
London, ON N6A 5B7

199 Master's, 153 Doctoral

1940s: McGill, William Peter

1950s: Cook, R. J. B.; Crowe, C.; Ellis, Robert M.; Folinsbee, J. C.; Forrester, Macquorn Rankine; Knox, Keith Sifton; Mason, Clive S.; McClure, M. E.; McConnel, Denis B.; Muir, W.; Sandomirsky, Peter; Sawyer, J. B. Paul; Surkan, Alvin J.; Tanner, James G.; Zelonka, Frederick A.

1960s: Anglin, F. M.; Armstrong, Calvert William; Bertrand, Claude E.; Blackburn, Charles; Buckley, Dale E.; Casshyap, Satyendra M.; Chandler, Frederick William; Chandra, B.; Cowan, John Christopher; Davis, Harry Osmond; Denholm, J. G.; Ermanovics, Ingomar Frank; Ferrigno, Kenneth Francis; Graham, Robert Alexander Fergus; Gunn, Christopher Bruce; Hammond, Barry M.; Harron, Gerald Allan; Hunter, James A. M.; Irwin, Gerald J.; Johnson, Alan Eugene; Jones, John Frederick; Killeen, Patrick G.; Law, F. E. M.; Lee, Pei Jen; Leslie, John A.; Lilley, F. E. M.; Mason, David; McFadden, C. P.; McGrath, Peter H.; McMillan, Ronald Hugh; Mereu, Robert F.; Milne, W. G.; Mukherji, Kalyan Kumar; Murty, Rama C.; Mustonen, E. D.; Pearce, Thomas Hulme; Piotrowski, Joseph Martin; Platt, Richard Garth; Rance, Hugh; Reid, Archibald M.; Risk, Michael J.; Rukavina, Norman Andrew; Sass, John H.; Scott,

Steven Donald; Skibo, D. N.; Slankis, J. A.; Sood, Manmohan Kumar; Sood, Manmohan Kumar; Stevens, Anne E.; Thapar, Mangat R.; Vagners, Uldis Janis

1970s: Aaquist, B. E.; Abu Bakar, M. Y.; Adams, G. W.; Akande, S. O.; Andrews, Anthony James; Atkinson, Dorothy; Banerjee, Syamadas; Barnett, D. M.; Barnett, E. S.; Barnett, R. L.; Berti, Albert A.; Bouley, Bruce Albert; Brar, N. S.; Brewster, G. R.; Brigham, Robert John; Brown, Donald Dawson; Brown, J. R.; Brule, D. G.; Bryant, J. G.; Calhoun, Thomas Addison, II; Carson, J. M.; Carvalho, I. G.; Chewaka, S.; Chiang, Kam Kuen; Coish, R. A.; Colvine, A. C.; Comline, Stuart Robert; Conaway, J. G.; Conaway, J. G.; Craft, J. L.; Darbha, D. M.; De Saboia, L. A.; Dilabio, R. N.; Duke, N. A.; Edwards, W. A. D.; Etiebet, Donatus O.; Falls, Robert Meredith; Feenstra, B. H.; Fenton, M. M.; Ferguson, L. J.; Fisher, David Frederick; Fodemesi, Stephen Paul; Franklin, James McWillie; Gale, John Edward; Giles, P. S.; Gishler, C. A.; Graham, Robert Alexander Fergus; Grant, J. A.; Gurkan, T. H.; Gwyn, Hugh; Hakim, H. D. M.; Hamza, Valiya Mannathal; Harley, D. N.; Harrison, C. J.; Hearty, David Joseph; Hill, J. M.; Hinckley, T. K.; Hinzer, J. C.; Hughes, Marie Jeanne; Hunter, James A. M.; Jefferson, Charles Wilson; Jenner, G. A.; Jobidon, G.; Johnson, Alan Eugene; Johnson, Wayne Lawrence; Johnston, Maureen Dawne; Judge, Alan Stephen; Kacira, Niyazi; Katigema, F. D.; Kerswill, J. A.; Killeen, Patrick G.; Kreutzwiser, R. D.; Kronberg, B. I.; Kyle, J. R.; Lamarre, Albert Leroy; Lamarre, R. A.; LaTour, T. E.; Law, K. T.; Lewis, T. J.; Long, Darrel Graham Francis; Long, Darrel Graham Francis; Ludvigsen, Rolf; Ludvigsen, Rolf; Lyons, J. A.; MacIntyre, Donald George; MacIntyre, Donald George; Maher, R. V.; Marshall, Paul Arthur; Marzouki, F. M. H.; Matthews, L.; Maxwell, Frank Kristian; May, Ronald William; McLennan, S. M.; McMillan, Ronald Hugh; Meade, H. D.; Merz, B.; Miller, A. R.; Misra, Kula Chandra; Mitchell, C. E.; Nawab, Z. A. H.; Nguyen, B. T.; Nogami, T.; Ntiamoah-Adjaquah, R. J.; O'Donnell, Neil Dennis; Olson, Reginald Arthur; Parker Plaut, Lynda Marjorie; Parry, Steven Elliott; Parviainen, Esko Atso Uolevi; Patel, Jayantilal P.; Perlikos, Panayotis; Perry, David G.; Perry, David G.; Pinch, James Jeffrey; Plaut, M. G.; Radain, A. A. M.; Ram, A.; Riccio, Luca Michelangelo; Riccio, Luca Michelangelo; Robinson, P. C.; Rochette, Francois Jules; Rockingham, Christopher John; Schroeter, Thomas Gordon; Seara, J. L.; Shen, P.-Y.; Stanton, Roberta Anne; Stene, L. P.; Stevens, R. K.; Thomas, A.; Timco, G. W. J.; Timco, G. W. J.; Vagners, Uldis Janis; Van Hees, Edmond Harry Peter; Venkateswaran, G. P.; Waboso, Chijoke Ezekiel; Welch, David Michael; White, Joseph Clancy; Wigington, Richard James Stephen; Winfield, W. D. B.; Wood, John; Yilmaz, H.; Yuen, C. M.; Yuen, D. T.-C.; Zodrow, Erwin Lorenz

1980s: Al-Mooji, Y.; Aravena, R. O.; Arseneau, Gildar Joseph; Baerg, James R.; Banting, Douglas Ralph; Barriga, Fernando Jose Arraiano de Sousa; Bateman, Philip Walker; Beaudoin, Alwynne Bowyer; Becker, Dennis Eldon; Bell, Thomas Howard; Bernstein, Lawrence Mark; Bowles, Jane Margaret; Bradbrook, Christopher James; Broster, Bruce Elwood; Campbell, Robert Anderson; Carter, Alan Scott; Chance, Patrick Neville; Cheadle, Burns Alexander; Clarke, Donald Shane; Costa, Umberto Raimundo; Cox, Thomas Philip; de Figueiredo, Mario Cesar Heredia; de Freitas, Timothy A.; Delaney, Gary Donald; Delitala, Frank Antony; Diamond, Fiona M.; Dunn, Anthony Price; Ebinger, Elizabeth Jane; Edwards, Garth Richard; El-Hifnawy, Laila Mahmud; El-Sharnouby, Bahaa el Ahmed; Eliopoulos, Demetrios George; Epstein, Rachel Sophia; Feuer, Wendy Jo; Finn, Gregory C.; Ford, R. Craig; Gaba, Robert G.; Galley, Alan George; Geddes, Robert Stewart; Goad, Robin E.; Gorman, Barry Edward; Greenwood, Richard Charles; Guthrie, Alan Edgar; Hall, Richard Drummond; Hart, Brian R.; Henderson, Grant Stephen;

Hicock, Stephen Robert; Hill, John David; Holmes, Peter Winchester; Imasuen, Okpeseyi Isaac S.; Innis, John William; Jackson, Andrew Rupert Needham; Jefferson, Charles Wilson; Jiménez-Mosquera, Carlos José; Jonasson, Ralph George; Jones, Ian Frederick; Junnila, Randy Michael; Kearney, Michael Sean; Kennedy, Lawrence Patrick; King, James Gagwane; Kishida, Augusto; Klinkenberg, Brian; Kuehner, Scott Milton; Ladeira, Eduardo Antonio; Lamothe, Michel; Larson, John Edgar; Lee, Young-Nam; Lobato, Lydia Maria; Lougheed, Peter John; MacDonald, Andrew Harrington; Mann, Henrietta; Marcotte, David L.; Margeson, G. Bradford; Mark, Frazer John; Maxwell, Frank Kristian; McCaig, Andrew Malcolm; McCracken, Alexander Duncan; McGill, Glen C.; McNeil, Andrew Malcolm; Melchin, Michael Jerome; Melchin, Michael Jerome; Meloche, John Dennis; Morrison, Gregg William; Munha, Jose Manuel Urbano; Murray, Faye Helen; Mwenifumbo, Campbell Jonathan; Ng, Robert Man Chiu; Nielsen, Soren Bom; Nilson, Ariplinio Antonio; Norris, Malcolm Stewart; Nwankwor, G. I.; Ogawa, Toyokazu; Ojo, Samuel Bakare; Olorunfemi, Biodun Elijah Nathaniel; Osborne, Margery D.; Parbery, David; Parent, Michel; Parkin, Gary William; Powell, Michael A.; Prosh, Eric C.; Ravenhurst, William Richard; Rimkus, Arvid J.; Robinson, Donald James; Rowell, William Frank; Rye, Kenneth Alan; Sauer, R. Tayler; Scratch, Richard Boyd; Secco, Richard Andrew; Setterfield, Thomas Neal; Sheta, Mohamed Aly; Shortt, Trevor Alan Leslie; Shotyk, William; Simigian, Sandra Lynn; Smith, Alan Robert; Soderman, Kristopher Lorne; Somers, George Henry; Stewart, Robert Arthur; Stone, William Edward; Studemeister, Paul Alexander; Sutcliffe, Richard Harry; Sutton, P. A.; Talman, Stephen James; Tampoe, Tara J.; Taylor, Alan Bruce; Thomson, Margaret Lee; Threlkeld, William Earl; Thurston, Phillips Cole; Titaro, Dino; Tronnes, Reidar Gjermund; Valenca, Joel Gomes; Valliant, Robert Irwin; Van Der Flier, Eileen; Vibetti, Ndoba Joseph; Waboso, Chijoke Ezekiel; Wai, Raymond Sheung-Che; Walker, Stephen David; Watson, Gordon Peter; Woeller, R. M.; Wu, Jianjun; Wu, Tsai-Way; Yanful, Ernest Kwesi; Yeo, Gary Matthew

Western Oregon State College
Monmouth, OR 97361

1 Master's

1970s: Jones, W. M., Jr.

Western Washington University
Bellingham, WA 98225

147 Master's

1970s: Almy, Robert, III; Bernardi, Mitchell L.; Beske, Suzanne J.; Burke, Raymond; Burr, Cynthia D.; Cockerham, Robert S.; Crandall, Robert; Deeter, Jerald D.; Diehl, James; Ellis, Steven D.; Falley, R. Thomas; Gill, Roger; Gusey, Daryl L.; Hartwell, James N.; Heller, Paul; Hepp, Michael Arthur; Horton, Duane; Jones, Garry; Kelly, Joseph Michael; Keuler, Ralph F.; Lingley, William S.; McKeever, Douglas; Morrison, Michael L.; Morrison, Scott; Mustoe, George Edward; Niski, James T.; Noson, Linda Jeanne; Othberg, Kurt L.; Petrie, Gregg M.; Phillabaum, Stephen D.; Schmidt, Stephen L.; Scott, David B.; Sheriff, Steven D.; Siegfried, Robert T.; Smith, Mackey; Sondergaard, Jon N.; Soule, Ralph P.; Spasari, John V.; Stavert, Larry W.; Thomas, Erich; Videgar, Frank D.; Vonheeder, Ellis R.; Walen, Michael B.; Whitney, C. Gene; Williams, V. Eileen; Wilson, Daniel L.; Wright, W. L.; Zamboras, Robert L.

1980s: Adams, Benjamin Nickolas; Anderson, James M.; Anderson, Kurt Soe; Ashworth, Kathryn King; Batchelor, Carl F.; Bates, Roger G.; Bazard, David R.; Bigelow, Phillip Kenneth; Birk, Robert H.; Blackwell, David L.; Blankenship, Dana G.; Bream, Susan Elaine; Bubnick, Steven C.; Carkin, Brad A.; Carrick, Stanley J.; Christenson, Bruce W.;

Christenson, Lief G.; Chrzastowski, Michael J.; Craig, Douglas E.; Cruver, Jack Richard; Cruver, Susan Kinder; Dodd, Stanton P.; Dussell, Eric; Eccleston, Chuck; Einarsen, Jon M.; Faxon, Michael F.; Foley, Lucy L.; Franklin, Russell J.; Frasse, Frederic I.; Fuller, Steven R.; Furlong, Paul O.; Gallagher, Michael Patrick; García, Alfredo R.; Globerman, Brian R.; Goldstrand, Patrick M.; Graham, Don; Granirer, Julian L.; Hadley, David G.; Harp, Brad D.; Harrison, William J.; Hatfield, David M., Jr.; Haugerud, Ralph A.; Heliker, C. Christina; Hess, Jeffrey D.; Hyde, Bert Q.; Jacobsen, Edmund E.; Jacobson, David T.; James, Eric William; Jewett, Peter D.; Jones, Jeffrey T.; Karachewski, John A.; LaManna, John M.; Leiggi, Peter A.; Lindsay, Charles S.; Liszak, Jerry Lee; Logan, Robert L.; Mahala, James; Maloy, John A.; Marcott, Keith; Marcus, Kim L.; Melim, Leslie A.; Mesoloras, Nancy; Monroe, Scott, C.; Morgan, Scott R.; Moyer, Ronald D.; Pincha, Pamela M.; Pine, Keith A.; Plumley, Peter W.; Purdy, Joel W.; Rady, Paul M.; Rauch, William E.; Redfield, Thomas F.; Reinink-Smith, Linda M.; Reiswig, Kenneth N.; Reller, Gregory Joseph; Robert, Ray, Jr.; Roland, John L.; Schmierer, Kurt E.; Schwarz, Charles G.; Schwimmer, Peter M.; Sevigny, James H.; Shultz, Julianna M.; Silverberg, David Scott; Skalbeck, John D.; Smith, Moira T.; Spencer, Patrick K.; Stauss, Lynne D.; Street-Martin, Leah V.; Strickler, David L.; Syverson, Tim L.; Taggart, Bruce E.; Thompson, John N.; Turner, Robert J.; Waldron, Richard L.; Wallace, R. Scott; Wice, Richard B.; Winters, Mark B.; Yandell, Lon R.; Ziegler, Charles B.

Wichita State University
Wichita, KS 67208

143 Master's

1930s: Vickery, Ward Rollin

1950s: Noah, Calvin G.; Wynn, Lester L.

1960s: Calvin, Don G.; Cooper, Warren W.; Dalrymple, Don W.; Ehm, Arlen E.; Gafford, Edward Leighman; Gill, Hugh W., Jr.; Hostetler, James M.; Irion, Ronald L.; Kidson, Evan Joseph; Knighton, Philip M.; Lammons, James Monroe; Leach, Carl L.; Malone, Donald J.; Miesner, James F.; Myers, Dennis E.; Myers, James Edward; Robbins, James L.; Schierling, Eldon J.; Schumaker, Robert D.; Sphar, Joe D.; Stude, Jerry R.; Trapp, Harold R.; Underwood, James O.; Ward, John Robert; Welsh, James P.; Williams, Charles Dudley; Young, Charles Robert

1970s: Al-Arabi, Nizar A.; Andersen, Rodney E.; Arnett, Bruce A.; Avis, Loren E.; Behairy, Abdel-Kader Ali; Ben Omran, Abdelmoneim; Brainard, Richard H.; Carlton, Dennis R.; Clark, S. B.; Crouch, Michael Leonard; Darrow, Douglas D.; Dorsey, Michael T.; Erlenwein, Susan D.; Fudge, Melvin Ray; Fullerton, Larry Bryant; Graff, Harry J.; Hart, Philip; Hutter, Terry J.; Jackson, William; Kellogg, Donald Walter; King, Galen E., Jr.; Korphage, M. L.; Manes, Monna Lea; McGuire, Emily; Nichols, Charles W.; Oliver, Lonnie G.; Pawel, David T.; Petta, Timothy Joseph; Reynolds, Daniel M.; Richardson, Larry J.; Shumard, C. Brent; Stobbe, William Joseph; Tetrick, Martha Jane; Verhulst, Albert T.; West, Eldon S.; Woods, Robert S.

1980s: Aikins, Charles W., III; Almouslli, Mohamad O.; Armbruster, John David; Baab, Patricia; Bagheri, Saeed; Baker, Joseph M.; Bamberger, Mark J.; Bazrafshan, Khosrow; Bober, Danny R.; Bou-Rabee, Firyal; Bradshaw, Donald T.; Butler, Kim R.; Callen, Jon M.; Callewaert, David L.; Coskun, Sefer B.; Covey, Curtis E.; Cullen, Terry R.; Davison, J. Lynne; Dietterich, Robert J.; Dubiskas, Richard A.; Dunaway, Sabrina G.; Dwivedi, Rajeev Lochan; El-Hussain, Issa Watban; Emery, Martin; Funk, Thomas J.; Gagne, Michael P.; Gardner, Joseph R.; Gentet, Robert Eugene; Griffith, Gary Lee; Grommesh, Mark W.; Hendrix, Bill; Henning, Leo G.; Hylton, Alisa K.; Jackman, Toni Kay; Jenkins,

R. V.; Jewett, David G.; Kambampati, Mohan V.; Kissick, Brian J.; Knight, Cole D.; Kopper, Randal W.; Lamoreaux, Scott B.; Lane, Michael E., III; Laney, Patrick T.; Linehan, John M.; Matzen, Thomas A.; Murray, David H., Jr.; Nchako, Felix N.; Nordstrom, John Eric; O'Neal, Marc A.; Palal, Vistasp R.; Patterson, Joel M.; Peeler, James A.; Pike, Stanley F.; Priest, Tim; Quinn, Terrence Michael; Reynolds, Thomas B.; Rosowitz, Donald W.; Salisbury, James P.; Scheffe, Gregory L.; Schultz, Richard B.; Sengupta, Somnath; Seward, Paul A.; Shonfelt, John P.; Siemens, Michael A.; Silfer, Jeffrey A.; Sorensen, Charles Elliott; Stoneburner, Richard Kelty; Swilley, Gerald Kirk; Textoris, Steven D.; Travis, Steven L.; Trotter, Sherry F.; Van Buskirk, Steven C.; Wagner, Karma L.; White, Lee H.; Wolf, Gerard V.; Zimmerman, Peter J.; Zink, Larry A.

College of William and Mary
Williamsburg, VA 23185

39 Master's, 29 Doctoral

1960s: Boon, John D., 3rd; Calder, Dale R.; Dillon, William A.; Grant, George C.; Kerwin, James A.; Kraeuter, John N.; Lawler, Adrian R.; McCain, John C.; O'Brien, Michael; Stauffer, Thomas B.; Stone, Richard B.; Tuck, D. Richard, Jr.; Warinner, J. Ernest; Young, David K.

1970s: Able, Kenneth; Boon, John Daniel, III; Boule, Mark E.; Bullock, Paul A.; Carron, Michael Joseph; D'Amico, Angela; DeAlteris, Joseph T.; Gibson, Victoria R.; Hershner, Carlton H., Jr.; Huggett, R. J.; Ingram, Carey; Jones, J. Claiborne; Lake, C. A.; Lake, Carol A.; Lake, J. L.; Lake, James L.; Lawler, Adrian R.; Manzi, John Joseph; Markle, Douglas F.; Markle, Douglas Frank; Orzech, Mary Ann Terese; Pickett, Robert Lee; Quensen, John F., III; Rackley, David Holland; Richardson, William S.; Rosen, P. S.; Rosen, P. S.; Stanley, Everett Michael; Thompson, Alyce D.; Vassaro, John Joseph; Weishar, L. L.; Windsor, J. G., Jr.; Windsor, John Golay, Jr.

1980s: Berquist, Carl Richard, Jr.; Croonenberghs, Robert Emile; de Alteris, Joseph Thomas; Espourteille, François A.; Fedosh, Michael Stephen; Finkelstein, Kenneth; Frisch, Adam Arthur; Green, Malcolm Omand; Kim, Chang Shik; Ledwin, Jane M.; Lukin, Craig G.; Milligan, Donald Bristowe; Mitchell, Martin Lane; Moustafa, Mary Sue Jablonsky; Peebles, Pamela C.; Rizzo, William Martin; Savage, Rebbeca Jo; Stauffer, Thomas Bennett; Unger, Michael Allen; Wattayakorn, Gullaya; Weston, D. P.

University of Windsor
Windsor, ON N9B 3P4

74 Master's, 4 Doctoral

1970s: Fedikow, Mark A. F.; Hannoura, A. A.; Heine, Thomas Hermann; Johnson, Michael David; Kapnistos, Minas Michael; Lantos, Etienne Alexandre; Lee, Tai Y.; Londry, John W.; Low, Barry M.; Mabbula, Sunand Shadrach; Miller, Paul M.; Mitchell, K. Daniel; Oworu, Oyewola Oyeniyi; Phelps, Robert Karl; Quirt, David Hulse; Ridley, Kevin J. D.; Rogers, Christopher J.; Stoakes, Franklin A.; Talerico, Frank; Vergos, Spiros; Wheeler, Cary F. R.; Writt, Robert Joseph

1980s: Aduamoah-Larbi, Joseph; Ali, Jaafar; Annesley, Irvine R.; Attanayake, Premadasa M.; Atuanya, Udemezue Obidigwe; Bacopoulos, Ioannis; Balsdon, Jason Thomas; Brathwaite, Samuel L. A.; Caldwell, Gary; Carson, Thomas M.; Cheung, Paul Chi-Tak; Choudhry, Abdul G.; Connolly, Garrett Morgan; Coyle, David Alexander; Deklerk, Robert Peter; Dessouki, Abdelrahim Khalil Mohamed; Dey, Sudhindra Nath; Dodt, Matthew Edward; Dunsmore, Dennis Joseph; Galinski, Christine Helen; Gossiaux, Barbara Marie; Grant, Brian D.; Gunter, W. Richard; Jiwani, Riyazali N.; Kamal-Aldin, Saad; Karanja, Samuel W.; Katham, Abd Al-Wahab N.; Klein, Kenneth Paul; Kolasa,

William B.; Larbi, Emmanuel Yaw; Marentette, Kris Allen; Mason, Sharon A.; McNaughton, Kenneth C.; Mihoren, Jerry John; Mwangi, Martin Peter; Nantais, Philip Thomas; Nyagah, Kivuti; O'Connell, Shaun C.; Obradovic, Milan Mitch; Orpwood, Timothy Gordon; Osmani, Ikramuddin Ahmad; Peck, David C.; Quick, Arthur William; Ravina, Amnon Nathan; Rigbey, Stephen J.; Robinson, Richard N.; Rogers, Cassandra T.; Singh, Maghar; Smith, Patrick E.; Taylor, Simon D.; Timmins, Edwin Allen; Vandall, Thomas Andrew; Walley, David Stephen; Wilson, Bruce A.; Yanful, Ernest K.; Zytner, Richard Gustav

University of Wisconsin-Green Bay
Green Bay, WI 54302

4 Master's

1980s: Allen, Paula E.; Johnson, Scot B.; Johnston, Donald J.; Marquardt, Tezz C.

University of Wisconsin-Madison
Madison, WI 53706

750 Master's, 606 Doctoral

1880s: Van Hise, Charles R.

1890s: Buckley, Ernest R.; Parmley, Walter C.; Sessinghaus, Gustavus; Steere, Eugene A.; Thompson, James R.; Van Hise, Charles R.; Weidman, Samuel

1900s: Bowen, Charles F.; Harder, Edmund Cecil; Laney, Francis B.; Leith, Charles K.; Smith, Warren DuPree; Steidtmann, Edward; Thwaites, Fredrik T.; Wegemann, Carroll H.; Zapffe, Carl

1910s: Ball, Sydney H.; Bean, Ernest F.; Becker, Max A.; Broderick, Thomas M.; Collins, William H.; Corbett, Clifton S.; Cox, Guy H.; Culver, Harold E.; Dake, Charles L.; Davis, Melvin K.; Edson, Fanny C.; Edwards, Merwin G.; Ellis, Robert W.; Hall, Durand A.; Harder, Edmund Cecil; Harvie, Robert; Holden, Roy J.; Hotchkiss, William Otis; Kirch, Annie B.; Kirk, Charles Townsend; Knappen, Russell S.; Longyear, Robert D.; McConnell, Wallace R.; Merritt, John Wesley; Paine, Francis W.; Pearsall, William G.; Schneider, Hyrum; Schwartz, George M.; Shearer, Harold K.; Steidtmann, Edward; Tanton, Thomas L.; Tomlinson, Charles W.; Trueman, Joseph Douglas; Uber, Harvey A.; Uglow, William L.; Uglow, William L.; Williams, Frank E.

1920s: Anderson, Wells Foster; Baker, Glenn J.; Brown, Earl D.; Brown, Ralph H.; Buckstaff, Sherwood; Burgy, Jacob H.; Carlson, Charles G.; Chang, Chen Ping; Chu, Ting Oo; Dickinson, Clyde G.; Diebel, Lyndall J.; Doering, Effie A.; Doerr, John E., Jr.; Durand, Loyal, Jr.; Edwards, Everett C.; Emmons, Richard C.; Frey, John W.; Fritzsche, Kurt W.; Greer, William L. C.; Hansell, James Myron; Harbaugh, Marion D.; Hart, Lyman H.; Hawley, James E.; Hollister, Donald E.; Hsi Chou T'an; Hsieh, Chia Yung; James, Henry F.; James, Howard T.; Jones, Russell H. B.; Jure, Albert E.; Just, Evan; Knight, Garold L.; Koschmann, Albert H.; Leith, Andrew; Loft, Genivera L.; Lund, Richard Jacob; Mead, Warren J.; Mueller, Amy F.; Murphy, Raymond E.; Neidy, Carrie L.; Ockerman, John William; Osgood, Wayland; Rand, Wendell Phillips; Rove, Olaf N.; Rutherford, Ralph L.; Schneider, Hyrum; Schubring, Selma L.; Scott, William A.; Sheldon, Estelle L.; Shoemaker, Helen E.; Squires, Henry Dayton; Stearn, Noel Hudson; Strachan, Clyde G.; Swanson, Clarence Otto; Tester, Allen Crawford; Tillman, Arthur G.; Timothy, Mary; Tyler, Stanley A.; Voskuil, Walter H.; Voskuil, Walter H.; Wanenmacher, Joseph M.; Weeks, Albert W.; Weeks, Herbert J.; Whitbeck, Florence; Williams, Frank E.; Williams, Thomas Bowerman; Wilson, Gilbert; Wood, Ella L.

1930s: Aldrich, Henry R.; Allen, Alice S.; Anderson, Wells Foster; Aszklar, Stanley Joseph; Atwater, Gordon I.; Bays, Carl A.; Bell, Archibald M.; Bethune, Pierre F.; Bradley, Charles C.; Brock,

Byron B.; Chapman, Winifred M.; Dapples, Edward Charles; Dickey, Robert McCullough; Dodge, Theodore A.; Draper, Mary B.; Drindak, Joseph T.; Edward, Albert; Ellsworth, Elmer W.; Field, George W.; Filaseta, Leonard; Fritts, Charles C., Jr.; Gillies, Donald F.; Greer, William L. C.; Hage, Conrad O.; Hansell, James Myron; Harding, William Duffield; Havard, John F. R.; Hedley, Mathew S.; Hedley, Mathew S.; Henderson, James F.; Hill, Mason L.; Hunzicker, Ashley A.; Jeffries, Charles D.; Jenks, William F.; Jure, Albert E.; Karges, Burton E.; Kindle, Edward Darwin; Leahey, Alfred; Leith, Andrew; Lund, Richard Jacob; Marsden, Ralph Walter; Marsden, Ralph Walter; Mayer, Edward A.; McCartney, Garnet Chester; McCormick, Robert B.; McCormick, Robert B.; McKelvey, Vincent Ellis; Meek, Ward B.; Mitchell, Wilson Doe; Ostrander, Allen R.; Rand, Wendell Phillips; Salton, George H.; Schmedeman, Otto C.; Stockwell, Clifford Howard; Thomson, James E.; Tiemann, Theodore D.; Trefethen, Helen B.; Trefethen, Joseph M.; Turk, Lon B.; Tyler, Stanley A.; Volk, Norman J.; Wanenmacher, Joseph M.; Wang, Chu C.; Wilgus, Wallace LaFetra; Zinn, Justin

1940s: Anderson, Thornton Earl; Bailey, Sturges Williams; Blexrud, Owen Hefte; Bullock, Kenneth C.; Burma, Benjamin H.; Chapman, Wilson A.; Clark, Robey Harned; Colgrove, Gordon L.; Crump, Robert M.; Deal, Clyde Stanley; Erickson, Robert C.; Feray, Daniel E.; Frederickson, Edward Arthur; Gates, Robert M.; Hanners, Albert James; Hanson, Alvin M.; Heuer, Edward; Hole, Francis Doan; Ingram, Roy Lee; Jacobs, Cyril; Johnson, David Perrin; Lyons, Erwin John; Marais, Jacobus Jan; McKelvey, Vincent E.; McMinn, Paul Meloy; Meek, Ward B.; Raasch, Gilbert O.; Renfro, Harold Bell; Schiesser, Clarence Frederick; Snyder, Frank George; Starke, George Wesley; Turner, Daniel Stoughton; Warren, Virginia Ada; Wilcox, Ray E.; Young, Keith Preston

1950s: Adair, Donald L.; Adams, Budd Berwyn; Ahlen, Jack L.; Amante, A.; Ammentorp, Willis Fay; Amundson, Burton; Anderson, Gerald J.; Anderson, Robert John; Andrews, George W.; Andrews, George William; Aronow, Saul; Babcock, Russel C.; Bailey, Henry H.; Bakker, Daniel; Beaver, Harold H.; Bebout, Don Gray; Becker, Ervin S.; Behrendt, John Charles; Bell, Robert A.; Benson, James C.; Bergeron, Thomas Joseph; Bergstrom, Robert E.; Bergstrom, Robert Edward; Beringer, Robert O.; Best, Edward W.; Bird, Samuel Oscar; Black, William A.; Boardman, Donald C.; Boebel, Richard Wallin; Boeck, Robert V.; Bonini, William Emory; Borst, Roger Lee; Bostwick, David Arthur; Bradley, Charles C.; Brown, Bruce E.; Brown, Leonard F., Jr.; Brown, Leonard Franklin, Jr.; Bryan, Wilfred Bottrill, Jr.; Bryan, Wilfred Bottrill, Jr.; Burns, Robert Parker; Carlisle, Donald; Chadwick, Robert Aull; Chapman, Rodger Hale; Claus, Richard John; Cumming, Leslie M.; Davis, James Frazier; Devries, David A.; Digert, Frederick E.; Dixon, James W., Jr.; Dixon, Joe Boris; Dollison, Roberts S.; Drescher, William J.; Drew, C. Wallace; Dyer, Henry Bennett; Emerson, Mark E.; Engels, Gary G.; Englund, Kenneth John; Ericson, Donald Martin; Fetzner, Richard W.; Flaten, Luvern L.; Ford, Robert B.; Fox, William Blake; Frakas, Steven Eugene; Fraser, J. A.; Fratt, Walter James; Freas, Donald Hayes; Freas, Donald Hayes; Gehrig, John Leonard; Gibson, Thomas G.; Goodwin, Alan M.; Goodwin, Alan M.; Gravenor, Conrad Percival; Gray, Robert Hugh; Green, Lewis H.; Green, Lewis H.; Gregory, Alan Frank; Grey, Charles E.; Gross, Gordon Arnold; Guilbert, John M.; Haas, Bruno J.; Hackett, James E.; Hall, Hubert H.; Hammes, Richard Robert; Hanson, George F.; Harbaugh, John W.; Harding, Norman C.; Harris, Rodger S.; Hartman, James A.; Hartman, James Austin; Harvey, John Frank; Harvey, John Frank; Hase, Donald H.; Hase, Donald H.; Hayward, Oliver Thomas; Heck, William J.; Hedlund, David Carl; Hembre, Donald R.; Hendrix, Thomas Eugene; Hewitt, Donald F.; Hewlett, Cecil George;

Howell, John Edward; Hutchinson, Richard William; Hutchinson, Richard William; Hutchinson, Robert D.; Jacka, Alonzo David; Jackson, Kerne Chandler; Jizba, Zdenek V.; Johansson, Folke Carl, Jr.; Johnson, Noye Monroe; Johnson, Robert William, Jr.; Jones, John Brett; Junemann, Paul Martin; Karl, Robert Otto; Ketner, Keith Brindley; Kilps, James R.; King, William Edward; Kingston, Dave Russell; Klinger, Frederick Lindsley; Koenen, Kenneth H.; Kuenzi, Laurence M.; Kux, Otto; LaBerge, Gene L.; Larson, Lawrence Tilford; Larson, Thomas C.; Laudon, Richard B.; Laudon, Richard Baker; Lawson, Ralph W.; Lehmann, Elroy P.; Lehmann, Elroy P.; Lindgren, Donald W.; Link, Peter Karl; Lorenzen, Robert M.; Lowther, Harold C.; Mancuso, Joseph John; Mann, Virgil I.; Marrall, Gerald E.; Martin, Charles Wellington; Matuszczak, Roger A.; McCauley, Victor T.; Mengel, Joseph Torbitt, Jr.; Meyer, Robert Paul; Moore, Walter Leroy; Moore, Walter Leroy; Moretti, Frank Joseph; Moretti, Frank Joseph; Morgridge, Dean L.; Moshiri-Yazdi, Reza; Mossman, Malcolm H.; Mueller, Robert F.; Murray, Raymond C.; Murray, Raymond C.; Nelson, Henry F.; Neville, William D.; Nordeng, Stephan C.; Nordeng, Stephen Carl; O'Rourke, Joseph Edward; Oetking, Philip F.; Omernik, John Beebe; Oppel, Theodore Wells; Osmond, John K.; Osmond, John K.; Paape, Donald W.; Packard, Frank A.; Palen, Frank S.; Parchman, William; Parker, Calvin A.; Parker, Calvin Alfred; Parker, Pierce D.; Parks, James Marshall; Patrick, T. O. H.; Paull, Richard Allen; Paull, Richard Allen; Pemberton, Roger H.; Pitrat, Charles W.; Pitrat, Charles William; Pitt, William D.; Pollack, Sidney Solomon; Randall, Gaither M.; Reyer, Robert Winslow; Ribbe, Paul Hubert; Riggs, Elliot Arthur; Risley, George A.; Roehl, Perry O.; Rose, John Creighton; Ross, Wayne A.; Rowe, Robert Burton; Sanderson, George Albert, Jr.; Scheerer, Paul Ervin; Schmidt, Richard Arthur; Schmidt, Robert G.; Schoenike, Howard; Scott, Gerald Lee; Scott, James W.; Severson, John Louis; Shelburne, Orville B.; Shelburne, Orville Berlin, Jr.; Snodgrass, Thomas W.; Sorauf, James Edward; Sorem, Ronald Keith; Spreng, Alfred Carl; Stephenson, Thomas Edwin; Sumner, John Stewart; Sundelius, Harold Wesley; Sunderman, Harvey Cofer; Suttner, Lee Joseph; Swingle, George D.; Tabbert, Robert Leland; Tank, Ronald Warren; Thede, Ray John; Thomasson, Maurice Ray; Tobison, Norman Murray; Tupas, Mateo H.; Twenhofel, William S.; Verville, George Julius; Vitcenda, John Frederick; Walch, Andrew F.; Walker, Robert Edwin; Walker, Theodore Roscoe; Warner, Maurice Armond; Weege, Randall J.; Weis, Paul L.; Weiss, Edwin M.; Wesh, Richard Adams; Wing, Richard Sherman; Woldenburg, M. J.; Woodward, Harold Walter; Wornardt, Walter William, Jr.; Wray, John Lee; Yehle, Lynn A.; Younger, James Allen; Zeller, Doris Eulalia Nadine; Zeller, Edward J.; Zimmerman, Donald A.

1960s: Aalto, Kenneth Rolf; Acharya, Hemendra Kumar; Akers, Ronald H.; Akers, Ronald Hugh; Al-Rawi, Amin Hamad; Allong, Albert F.; Alsobrook, Albert Francis; Andrew, John Alexander; Ansfield, Valentine J.; Anzoleaga, Rodolfo; Asquith, George B.; Asquith, George B.; Asthana, Virendra; Asthana, Virendra; Atkinson, Robert F.; Babcock, Jack Arthur; Babcock, Laurel Clarke; Bachhuber, Frederick Willard; Bain, Roger John; Bancroft, Genevieve R.; Barnes, Charles Winfred; Barness, Donald Lawrence; Baver, Leonard D., Jr.; Bayrock, Luboslaw Antin; Beaver, Albert John; Behken, Fred Henry; Behrendt, John Charles; Benedict, James B., Jr.; Bennett, Hugh Frederick; Bentley, Eugene Macke, III; Berge, William Victor; Berkson, Jonathan Milton; Berry, Donald W.; Bhatt, Jagdish J.; Birch, Francis S.; Bird, Kenneth John; Bird, Kenneth John; Black, Bruce Allen; Bock, Wayne D.; Bostock, Hewitt Hamilton; Brauner, Michael R.; Brezonik, Patrick Lee; Briggs, Garrett; Brooks, Elwood Ralph; Brower, James Clinton; Brown, Bruce Elliot; Brown, Jim McCaslin; Burt, William D.; Cahoon, Bobby G.; Carlson, Barry; Carpenter, Robert H.; Carpenter, Robert Heron; Casas, Jamie Lopez; Cassie, Robert MacGregor; Cenedella, Louise G.; Charlton, David S.; Chernosky, Joseph V.; Christensen, Nikolas I.; Christensen, Nikolas Ivan; Cinnamon, Charles Gerald; Cinnamon, Charles Gerald; Ciolkosz, Edward John; Cleveland, John Herbert; Coetzee, Gerrard Louis; Cohen, Theodore Jerome; Collins, Kenneth Alan; Coons, Richard L.; Craig, Douglas Bennell; Danley, William M.; Davis, James Frazier; Davis, James Howell; Delfino, Joseph John; Denechaud, E. Barton; Dennison, John Manley; Desborough, George Albert; Dewees, Donald J.; Dollase, Wayne A.; Dollinger, Gerald Lee; Doman, Robert Charles; Dorman, Leroy Myron; Dowling, Forrest Leroy; Dunn, David Lawrence; Eggleton, Richard Anthony; Erickson, Alvin J.; Everts, C. H.; Evoy, Ernest Franklin; Fanning, Delvin Seymour; Feldhausen, Peter Homer; Fellows, Larry Dean; Fiero, G. William, Jr.; Finney, Joseph Jessel; Frederickson, Norman Oliver; Froming, George T.; Furer, Lloyd Carroll; Gerlach, Terrence M.; Gilbert, John M.; Gilbertson, Roger Lee; Giovinetto, Mario Bartolomé; Gore, Dorothy Jean; Greisemer, Allen D.; Hackbarth, Douglas A.; Halpern, Martin; Halpern, Martin; Hamilton, Stanley Kerry; Hamilton, Stanley Kerry; Hamilton, Thomas D.; Hammes, Richard Robert; Hanna, Augustine Booya; Hart, Orville Dorwin; Hart, Orville Dorwin; Hashimoto, Isao; Hawks, Graham Parker; Hayes, John Bernard; Hedberg, William Hollis; Hendrix, Thomas Eugene; Higgins, David Thomas; Hill, John Gilmore; Hill, John Gilmore; Hite, David Marcel; Hite, David Marcel; Hodges, Carroll Ann; Howard, John K.; Howe, Dennis M.; Howe, Robert Crombie; Howe, Robert Crombie; Jackson, Togwell A.; Jenkins, Robert Allen; Jiracek, George; Johansson, Folke Carl, Jr.; Johansson, Margaret V.; Johnson, Kent Erwin; Johnson, Kent Erwin; Johnson, Noye Monroe; Jones, James Irvin; Jones, James Irwin; Jordan, William Malcolm; Kaddou, Nadheema Salih; Kaiser, William R.; Ketner, Keith Brindley; Kirkham, Rodney Victor; Kissling, Don L.; Klinger, Frederick Lindsley; Koch, John Gerhard; Koch, John Gerhard; Korompai, Americo E.; Kunishi, Harry Mikio; LaBerge, Gene L.; LaFountain, Lester J.; Larson, Lawrence Tilford; Laudon, Robert C.; Laudon, Thomas Stanzel; Laury, Robert Lee; Lenzer, Richard Charles; Lind, Aulis Olaf; Lindorff, David E.; Link, Peter Karl; Lister, Gordon Frank; Lister, Judith Smith; Lockwood, Richard P.; Lohr, Jerrold R.; Luchterhand, Dennis; Luttrell, Eric M.; Lynts, George Willard; Lynts, George Willard; Mack, John Wesley, Jr.; Macurda, Donald Bradford, Jr.; Mann, Christian John; Mansfield, Charles Frederic; Martin, Charles Wellington; Martin, Dewayne C. H.; McCamy, Keith; McDaniel, Alice; McDonald, John Angus; McEvoy, Thomas D.; Melby, John Harold; Mengel, Joseph Torbitt, Jr.; Milfred, Clarence James; Miller, James Fredrick; Morris, Robert Clarence; Morris, Robert Clarence; Morrison, Bradford Crary; Mosher, Loren Cameron; Mosher, Loren Cameron; Mouat, Malcolm M.; Nilsen, Tor Helge; Nilsen, Tor Helge; Nriagu, Jerome O.; Nybakken, Bette Helene Halvorsen; Oakes, Edward L.; Ocola, Leonidas; Olcott, Perry Gail; Ostenso, Ned Allen; Owen, Donald Eugene; Padgham, William Albert; Palmquist, Robert Clarence; Park, Richard Avery, IV; Park, Richard Avery, IV; Parker, Michael; Parker, Pierce Dow; Patenaude, Robert W.; Patenaude, Robert William; Piette, Carl R.; Pinney, Robert Ivan; Pooley, Robert Neville; Price, Myron W.; Pride, Douglas E.; Ranney, Richard Willard; Rasmussen, Gerald E.; Rawson, Richard Ray; Richgels, Henry J.; Ridler, Roland Hartley; Rightmire, George Philip; Robinson, Edwin Simons; Rollins, Harold B.; Roshardt, Mary Ann; Rospenda, Robert E.; Roth, Charles Barron; Ruedisili, Lon Chester; Ruedisili, Lon Chester; Sabbagh, Suzanne Kathleen Boram; Sahu, Basanta Kumar; Saidji, Mohamed; Salem, Mohammad Zarif; Salstrom, Philip L.; Saluja, Sundar Singh; Sanker-Narayan, P.; Schenk, Paul Edward; Schenk, Paul Edward; Schmidt, Richard Arthur; Scholten, Arnold Gerhard; Scholz, Sally A.; Schumacher,

Dietmar; Schwab, Frederick L.; Schweger, Charles E.; Scott, Gerald Lee; Scott, Kevin McMillan; Scott, Sally C.; Seifert, Karl Earl; Shea, James H.; Shideler, Gerald Lee; Smith, Donald L.; Smith, Judith Marilyn; Smith, Sharon Lynne; Sneider, Robert Morton; Stanley, Kenneth Oliver; Stapor, Francis W., Jr.; Stark, Philip Herald; Stark, Philip Herald; Steinhart, John Shannon; Steuerwald, Bradley A.; Stoll, Sarah Johanna; Sumner, Roger Dean; Sundeen, Stanley Paul; Sundeen, Stanley Paul; Suttner, Lee Joseph; Tank, Nihat; Thomas, Everett R.; Thorpe, Ralph Irving; Threadgold, Ian Malcolm; Tillman, Roderick Whitbeck; Turner, John Charles; Twomey, Arthur Allen; Usbug, Enis; Van der Plank, Adrian; Van Rensburg, Willem Cornelius Janse; Vogel, Thomas A.; Vogel, Thomas Adolph; Vogt, Peter Richard; Vogt, Peter Richard; Waller, Thomas R.; Walters, Lawrence Albert; Warren, Thomas Ernest; Weart, Wendell Duane; Weis, Leonard Walter; Whelan, Peter M.; Widmier, John Michael; Wold, Richard John; Wolff, Roger Gene; Woodbury, Jerry L.; Wright, Charles Malcolm; Yeend, Warren Ernest

1970s: Aalto, Kenneth Rolf; Acomb, Lawrence Joseph; Adams, R. L.; Ahrnsbrak, William Frederick; Al-Mishwt, A. T.; Al-Mishwt, Ali Theyab; Ali, Hassan Mohamed; Allen, Ralph Orville, Jr.; Anderson, J. L.; Anderson, J. L.; Anderson, J. Lawford; Anderson, Lance Christopher; Anderson, Richard Lee; Andrew, John Alexander; Andrews, Charles Bryce; Andrews, Charles Bryce; Anstett, Terrance F.; Anzoleaga, Rodolfo; Ayres, Dean Esmond; Babcock, Jack Arthur; Babcock, Laurel Clarke; Baker, Harold Wellington, Jr.; Bartlein, P. J.; Baxter, Franklin Paul; Beeghly, Sallie; Behnken, Fred Henry; Beitzel, John Edward; Berge, Charles William; Berkson, Jonathan Milton; Bethke, Karl J.; Billingsley, Randal Lee; Black, Frederick Michael; Bleuer, Ned Kermit; Blount, Alice McDaniel; Bond, Gerard Clark; Born, Stephen Michael; Brewer, Wayne Martin; Brooks, A. James; Bultman, Thomas R.; Burgener, J. D., IV; Burrell, Jennifer Ann; Bussa, Kathleen Louise; Byers, R. A.; Campbell, John S.; Carpenter, S. R.; Castle, James W.; Chamberlain, Charles Kent; Chapman, Stanley Lane; Charpentier, Ronald Russell; Cheney, J. T.; Cichowicz, Nancy Lee; Clough, John W.; Clough, John W.; Cluff, Robert Murri; Colman, J. A.; Conatore, Paul D.; Cooper, Roger Wayne; Cota, T. F.; Cowen, William F.; Crawford, George Allan; Crawford, K. A.; Cruickshank, M. J.; Cullers, Robert Lee; Cummings, Michael Levi; Daneshvar, Mohammad Reza; Daneshvar, Mohammad Reza; Darby, Dennis Arnold; Das, Braja M.; Davidson, E. L.; Davis, A. M.; Dawson, James Clifford; DeKeyser, Thomas Lee; Delgado, D. J.; Denne, Jane Elizabeth; Dhowian, A. W.; Dirlam, D. M.; Doe, Thomas William; Dolcater, David Lee; Dorman, Leroy Myron; Drew, Patricia; Driese, Steven George; Dubois, Roger Normand; Dudley, John G.; Dumoulin, J. A.; Eheart, J. W.; Eisen, Craig E.; El-Attar, Hatim Abdelwahab Ahmed; Elger, Jerry Bruce; Emanuel, Richard Paul; Ervin, Clarence Patrick; Evans, Bruce William; Everts, Craig Hamilton; Fagerlin, Stanley Charles; Felmlee, Judith K.; Fricke, Carl A. P.; Friedman, R. M.; Frodesen, Eric Wells; Fuh, G. F.; Furgason, David C.; Gamber, James H.; Gavlin, Suzanne; Gettrust, J. F.; Glover, E. D.; Goodwin, Robert Glenn; Greischar, Larry L.; Grether, William John; Griffith, Christine Marie; Guggenheim, S. J.; Habermann, Gail M.; Hackbarth, Douglas A.; Hall, Carol Anne; Hall, Stephen Harvey, Jr.; Hansen, William Bradley; Hardin, Nancy S.; Hauser, Ernest Clinton; Hayatdavoudi, A.; Helmke, Philip August; Henderson, James Henry; Hennings, Ronald George; Henry, D. M.; Henry, Darrell James; Henry, Diana Louise; Henry, William Edward; Henry, William Edward; Hershberg, Edward Leonard; Hesler, Roy Earl; Heuer, Edward; Hickman, A. Elizabeth Wenger; Hickman, Robert Gunn; Hickman, Robert Gunn; Hoffman, Thomas Frank; Holdren, G. R., Jr.; Holland, Peter T.; Hopper, Sheridan Eileen; Huang, Kung; Huber, M. E.; Hurley, N. F.; Illfelder, H.

M. J.; Ishaq, A. M.; James, Johnny; Jezek, Kenneth Charles; Johnson, Michael G.; Johnson, Stuart Donald; Johnson, Thomas Mark; Johnson, W. C.; Johnston, S. C.; Jones, David Gordon; Joy, James Anthony; Kaczmarowski, J. H., Jr.; Kan, Tze-Kong; Kan, Tze-Kong; Karnauskas, Robert James; Kaufmann, Robert Frank; Kay, P. A.; Kelly, J. M.; Ketelle, Martha J.; Kim, K.; Kimerling, A. J.; Kirchner, Joseph F.; Kitchell, Jennifer A.; Kleist, John Raymond; Koehler, R. P.; Koellner, Susan Elaine; Kohler, Martha Hansen; Komatar, Frank Donald; Koo, Joseph Lok-shan; Koons, R. D.; Korotev, R. L.; Kosiewicz, Stanley Timothy; Kososki, Bruce Alan; Koss, George Michael; Kowallis, Bart Joseph; Kronig, Donald M.; La Pointe, Paul Reggie; Lagoe, M. B.; Lahr, Melvin M.; Langran, K. J.; Larson, David Warren; Larson, John A.; Larson, John A.; Lehmann, Patrick Jon; Lenzer, Richard Charles; Leong, Wing Kwong; Lewis, Brian Thomas Robert; Lonker, S. W.; Loveless, Janet Kay; Maass, Randall Steven; MacDonald, R. H.; Mackey, Scudder Draper; Madison, Frederick William; Mangham, J. R.; Markart, K. D.; Massie, L. R.; Matsch, Charles Leo; May, H. M.; McCarthy, Michael Martin; McCarthy, Thomas Richard; McCartney, M. Carol L.; McCartney, M. Carol L.; McCord, Daniel Lee; McGovney, J. E. E.; Melcon, P. Z.; Melenberg, R. R.; Metcalfe, A. P.; Michelson, Peter C.; Miller, David Wayne; Miller, James Frederick; Mills, B. A.; Milner, Sam; Minicucci, David Andrew; Mitchell, Phillip Dwight; Mode, William Niles; Mokma, Delbert E.; Montgomery, William Willson; Moody, William Clyde, Jr.; Mooney, W. D.; Moore, Patricia D.; Mooring, Carol Elizabeth; Morgan, C. L.; Morgan, K. M.; Morgan, Kirk A.; Morgan, William Andrew; Munter, James Arnold; Musolf, Gene Emil; Myers, Wallace Darwin, II; Myles, James Robert; Nebrija, E. L.; Nedland, Daniel E.; Neese, D. G.; Nelson, Alan Robert; Newell, K. D.; Niem, Alan Randolph; O'Neill, B. J.; Ocola, Leonidas; Ojanuza, Abayomi G.; Olson, Dan E.; Onesti, Lawrence Joseph; Owen, R. M.; Parker, Dale Edward; Pennington, W. D.; Peterson, Gilbert M.; Peterson, M. L.; Phillips, Thomas L.; Porter, J.; Possin, Boyd Nelson; Rautman, Christopher A.; Rautman, Christopher Arthur; Riehl, William George; Rinaldo-Lee, Marjory Beach; Robertson, J. D.; Robertson, James Douglas; Roder, Dennis Lee; Roeloffs, Evelyn Anne; Root, Ruth Eva; Sagher, A.; Sandness, Gerald Allyn; Sarg, J. F.; Schoenwald, Carolyn Paulette; Schwartz, Karen Ann; Setlock, G. H.; Sherwood, Elizabeth Schneider; Sherwood, Kirk W.; Sherwood, Kirk W.; Shukla, Surendra Shanker; Sikes, C. S.; Simpkins, W. W.; Sinclair, Patricia Drew; Sinclair, William D.; Sinclair, William David; Small, Thomas Wayne; Smith, Christy Harvey L.; Solien, M. A.; Solien, Mark Aldon; Sorenson, Curtis James; Souto-Maior Filho, Joel; Sridharan, Nagalaxmi; Stark, James Roland; Stasko, Lawrence E.; Sternberg, Ben K.; Sternberg, Ben K.; Sterrett, Robert John; Stetz, Donna Jane; Stewart, John Arden; Stewart, M. T.; Stewart, Mark Thurston; Stolzenberg, J. E.; Stone, Ralph Arthur; Stuhr, Steven Walter; Tabet, David Elias; Tanck, Glen Steven; Tharp, Thomas M.; Theyab, A.; Thiede, D. S.; Thiede, D. S.; Tolman, A. L.; Tong, Chiun Shing; Tsao, A. C.; Tyler, D. A.; Umhoefer, Paul John; Warner, Maurice Lee; Weaver, Robert Michael; Wegner, Warren W.; Wendland, Wayne M.; Wendte, John C.; White, Craig M.; Whiting, L. R.; Whitman, Rick R.; Whittecar, George Richard, Jr.; Whittecar, George Richard, Jr.; Wiese, Larry Bruce; Winn, R. D., Jr.; Woessner, William W.; Wopat, Michael A.; Wynn, S. L.; Young, C. T.; Yurewicz, D. A.; Zaporozec, A.; Zipp, Joel Frederick

1980s: Abdel Rahman, Mostafa A.; Abel, Cole D.; Adeniyi, Joshua O.; Ahmed, Gaafar Abbashar; Alexanian, Daniel Albert; Ali, Hassan Mohamed; Alley, Richard Blaine; Anderson, Cynthia S.; Anderson, David Lee; Andrews, Mary Catherine; Anstett, Terrance F.; Apel, Robert A.; Aster, Richard C.; Atkins, Elizabeth Dale; Attig, John William, Jr.;

Avasthi, Jitendra Mohan; Bahmanyar, Gholam Hossein; Bakheit, Abdalla Kodi; Barghouthi, Amjad Fawzi; Beauheim, Richard Louis; Bengtson, Mark Eric; Berndt, Michael Eugene; Bey, Ahmad; Bickford, Barbara J.; Bjornerud, Marcia G.; Bjornerud, Marcia G.; Blanchard, Margaret C.; Blankenship, Donald D.; Bohling, Geoffrey C.; Boone, William J.; Bourgeois, Joanne; Boyd, Robin Francis; Bradbury, Kenneth Rhoads; Brailey, David Elton; Brasino, John Sheldon; Bray, Alexander G.; Brewer, Wayne Martin; Brownell, James R.; Buchheim, Martin Paul; Bultman, Mark William; Bultz, Deanna Jean; Candelaria, Magell Phillip; Canfield, Howard Evan; Carey, Stephen Paul; Carlson, Steven Ray; Carr, Timothy Robert; Catchings, Rufus; Cavanaugh, Lorraine Marie Monnier; Chan, Marjorie Ann; Charlton, David Samuel; Charpentier, Ronald Russell; Cheng, Amy I-Mei; Cheng, Amy I-Mei; Chern, Laura Allison; Cho, Taechin F.; Clark, Douglas Robison; Cochrane, Peter John; Coffman, James F.; Connell, Douglas Edward; Connelly, Johnston P., II; Cotkin, Spencer Jerome; Cotkin, Spencer Jerome; Craig, Daniel J.; Crawford, George Allan; Cutler, Mark A.; Davis, Hugh R.; Decker, Paul Lloyd; Decker, Paul Lloyd; Dell'Agnese, Daniel James; Doe, Thomas William; Doll, William Eugene; Doll, William Eugene; Dong, Allen; Driese, Steven George; Dunn, Steven Robert; Durand, Thomas Jean-Paul; Eckstein, Barbara Ann; Einberger, Carl M.; Elsenheimer, Donald William; Esser, Kjell Bjorgen; Esu, Esu Obukho; Fastovsky, David Eliot; Fathulla, Riyadh Najeeb; Fausett, Robert Julian; Faustini, John M.; Feinstein, Daniel T.; Fekete, Thomas E.; Ferreira, Maria da Graca de Vasconcelos Xavier; Fielder, Gordon W., III; Filut, Marlene; Fleming, Anthony H.; Francek, Mark A.; Franseen, Evan K.; Frolking, Tod Alexander; Gajewski, Konrad J.; Gant, Jonathan L.; Gardner, David A.; Gassett, Roger; Geiger, Charles Arthur; Gilbert, Mark W.; Goldstein, Robert Howell; Goldstein, Robert Howell; Goodroad, Lewis Leonard; Gorczyca, Nancy Elizabeth; Greger, Joel G.; Greischar, Lawrence Lee; Gresham, Cyane W.; Grimm, Kurt Andrew; Haddox, C. A.; Hampton, Bret D.; Handley, Bruce; Harnish, David Emmanuel; Harris, Mark T.; Hartz, Kenneth Eugene; Hatleberg, Eric Warner; Hauser, Ernest Clinton; Hazelwood, Anna Marie; Hendrix, Marc S.; Henry, Darrell James; Herr, Paul Edward; Hieshima, Glenn B.; Hoaglund, John Robert, III; Hogan, Gregory G.; Hogler, Jennifer Alice; Hsu, Rongshin; Hunt, Randall James; Hurley, James Patrick; Imbrigiotta, Thomas Edward; James, L. Allan; Janowiak, Matthew James; Jezek, Kenneth Charles; Johnson, Mark Dale; Johnson, Mark Dale; Johnson, William Jeffery; Johnston, Carol Arlene; Judziewicz, Emmet Joseph; Juster, Thomas C.; Kalinec, James A.; Kalman, Linda Susan; Keables, Michael John; Keith, Jeffrey Davis; Keith, Jeffrey Davis; Kendy, Eloise; Kenoyer, Galen John; Khatri-Chhetri, Tej Bahadur; Kimball, Clark Gregory; Kimball, Karen Lee; Kirkby, Kent Charles; Kite, James Steven; Kladivko, Eileen Joyce; Knurr, Rick Allen; Kocurek, Gary Alexander; Konopka, Edith Hoffman; Kowallis, Bart Joseph; Krabbenhoft, David Perry; Krabbenhoft, David Perry; Kratz, Timothy Kellogg; LaKind, Judy Sue; Lamb, William Marion; Lampert, Jordan Keith; LaPointe, Paul Reggie; Lathrop, Richard Gilbert, Jr.; Lee, Kyoo-seock; Lehrmann, Daniel J.; Lentini, Michael Robert; Liao, Jih-Sheng; Lim, Chin Huat; Lingle, Craig Stanley; Long, John Douglas; Luetgert, James Howard; Luh, Gary Gwo-Fea; Lundblad, Steven Paul; MacDonald, Ruth Heather; Mace, Thomas Hooker; Machesky, Michael Lawrence; MacKinney, John; Maclean, Ann Louise; Mahdyiar, Mehrdad; Maher, Harmon Droge; Maher, Kevin A.; Maher, Linda M.; Mako, David Alan; Mankiewicz, Carol; Manser, Richard J.; Marin, Luis Ernesto; Marquard, Randall Steven; May, David William; McDowell, Patricia Frances; McKellar, Barbara J.; Meek, Reed Harold; Mennicke, Christine M.; Meyer, Thomas Scott; Mochtar, Indrasurya Budisatria; Mochtar, Noor Endah; Modene, Janet S.; Moecher, David Paul; Monahan, Robert H.; Mora, Claudia Ines;

Morris, Thomas Henry; Morris, Thomas Henry; Morrison, Jean; Mortenson, John J.; Motan, Eyup Sabri; Muldoon, Maureen A.; Muniz, Paul Francisco; Murosko, John E.; Myers, Lesley Louise; Myers, Mark D.; Nania, Jay C.; Nauta, Robert; Need, Edward Adams; Nelson, DeVon O.; Nunn, Susan Christopher; Nyquist, Jonathan Eugene; Okwueze, Emeka Emmanuel; Olson, Dan E.; Owens, Stephen M.; Pak, Dorothy Kim; Park, Jong-Sim; Parker, Albert John; Pauli, David Allen; Paull, Rachel Krebs; Pennequin, Didier Franz Edgar; Peters, Christopher Scott; Peterson, Daniel Eric; Peterson, Gilbert Moseley; Phillips, Therese C.; Pitt, Robert Ervin; Porter, Michael Lowry; Quirk, Bruce Kenneth; Rayne, Todd William; Rector, Sharon; Reese, Joseph F.; Reinthal, Carol Ann Armstrong; Reinthal, William Arthur; Richter, Karen June; Riciputi, Lee Remo; Ritter, Scott Myers; Ritzwoller, Michael Herman; Rodenbeck, Sue Anita; Roeloffs, Evelyn Anne; Rogers, Gary D.; Rooney, Sean T.; Rosenblum, Mark B.; Rossen, Christine; Rothschild, Edward Robert; Rule, Audrey Catherine; Savage, Martha Kane; Savage, Martha Kane; Schierow, Linda-Jo; Schneider, John Frederick; Schneider, John Frederick; Schwartz, Arthur H.; Scott, William P.; Secord, Theresa Karen; Shafer, Martin Merrill; Shrestha, Ramesh Lal; Skibicky, Taras V.; Smith, Marie Theresa; Socha, Betty Jean; Spatz, Jeffrey Michael; Spatz, Paige Herzon; Stam, Alan C.; Stanford, Scott Daniel; Stanforth, Robert Rhodes; Stenzel, Sheila R.; Sterrett, Robert John; Stoertz, Mary W.; Striegl, Robert G.; Sun, Albert Yen; Sun, Albert Yen; Sutherland, Jane Louise; Swanson, Teresa H.; Syverson, Kent Maurice; Talbot, Robert Walter; Taylor, Kendrick C., Jr.; Toha, Franciscus Xaverius; Toran, Laura Ellen; Valdes, Carlos M.; van Hissenhoven, René; Voight, David Scott; Warner, Frederic Kent; Watson, Vicki Jean; Weaver, Tamie R.; Welkie, Carol Jean Jigliotti; Welsh, James Lowell; Wetherbee, Paul K.; Whiting, James Freeman; Wilcox, Lee Warren; Wills, Christopher J.; Winfree, Keith Evan; Winkler, Marjorie Green; Wong, Albert H.; Wood, Cynthia; Yelderman, Joe C., Jr.; Yoon, Kern Shin; Zeltner, Walter Anthony; Zhao, Zhongliang; Zheng, Chunmiao; Zheng, Hong; Zierenberg, Robert A.; Zolidis, Nancy Ritter

University of Wisconsin-Milwaukee
Milwaukee, WI 53201

264 Master's, 17 Doctoral

1880s: Bascom, Florence

1910s: Bassett, Herbert; Blanchard, William O.; Rehfuss, Isidore L.

1920s: Ashton, Bessie L.; Barnwell, George F.; Blanchard, William O.; Bostock, Hugh Samuel; Corbett, Clifton S.; Fowler, Katharine S.; Rettger, Robert E.; Rudolph, Joseph; Wilkerson, Albert S.

1930s: Barnes, Virgil Everett; Pentland, Arthur G.; Rawles, William Post

1960s: Bischke, Richard E.; Comer, John Bennett; Distelhorst, Carl A. R.; Kocurko, John M.; Lohr, Jerrold R.; Olup, Bernard J., Jr.; Reeve, Edward John

1970s: Alderks, David F.; Anderson, James W.; Andrews, Richard Duane, Jr.; Baier, David; Ballou, S. W.; Bartz, Gerald L.; Baumann, Warren A.; Behling, Richard W.; Bollman, D. D.; Bollmann, Dennis Dean; Bruning, Curtis J.; Brzozowy, Carl P.; Burns, Scott D.; Chien, Tammy Chin-Hsia; Cohen, Joseph Charles, Jr.; Curth, Patrick J.; Cutler, Robert M.; Davies, Scott L.; Davis, Craig B.; Deverse, George D.; Eliseuson, Thomas G.; Erwin, Charles R., Jr.; Evenson, Edward B.; Fleming, Alfred J.; Fortune, Gladys M.; Frankovic, Edward A.; Gajkowski, Wynn A.; Gartmann, Charles W.; Gartmann-Siemann, Susan; George, Gary D.; Ghosh, Swapan K.; Gibson, Wayne Ross; Gooday, Andrew J.; Greathead, Colin; Gruber, David P.; Gruetzmacher, Jeff C.; Hall, Daniel W.; Hall, George Ian; Hampton, George R.; Hill, John R.;

Holloway, D. C.; Hursey, Michael J.; Jaworski, Bill L.; Johnson, John F.; Kattman, Robert J.; Kelley, Barbara Ann; Klauk, Robert H.; Klebold, Thomas E.; Kumkum, Ray; Larson, Thomas A.; Larson, William C.; Lawton, Dennis R.; Leonard, Patricia J.; Lukowicz, Leo Joseph; Maercklein, Douglas R.; Marshall, Timothy B.; Martinson, Timothy P.; Meddaugh, William Scott; Melnychenko, Paul; Mendoza, Carlos; Miron, Mark J.; Moretti, George, Jr.; Mulica, Walter S.; Murphy, Peter J.; Nemchak, Frank M.; Neuschafer, Gregory F.; O'Connell, Michael R.; Ostenso, Nile; Overton, Deborah J.; Oxley, David R., Jr.; Pacquett, Arthur Leon, Jr.; Paull, Rachel Kay; Peltonen, Dean R.; Pinta, James, Jr.; Piotrusczewicz, Michael; Poetzl, Kenneth G.; Power, Paul C., Jr.; Pruit, John Dave; Raveck, Karen L.; Ravenscroft, John H.; Ray, K. B.; Reddy, Kevin M.; Revock, K. L.; Richter, Michael A.; Roberts, Jeannette E.; Rogers, Robert B.; Rose, James P.; Rothwell, Bret; Saltzer, Charles E.; Schmitt, Dennis A.; Schramm, Donna J.; Schriver, George H.; Shakal, A. F.; Shefchik, William T.; Shunik, Thomas W.; Sivon, Paul A.; Smith, Jan H.; Socci, Anthony D.; Steckley, Robert Cecil; Steffens, Thomas J.; Stoelting, P. K.; Strasen, James L.; Strickland, Douglas K.; Stubenrauch, Alan L.; Sylvester, Kathleen M.; Tarazona, Carlos; Tatar, Philip J.; Thompson, Mark E.; Turner, Donald H.; Venditti, Anthony R.; Venzke, Carl Peter; Voight, David J.; Volkmann, Robert G.; Weihaupt, John George, Jr.; Weirauch, Douglas A., Jr.; Wierman, Douglas A.; Wilder, Steven V.; Wimmer, Joseph L.; Winkle, Candace J.; Winston, John G.; Wipf, Robert A.; Witkowski, Marlene Ann; Wolbrink, Mark A.; Wolosin, Carl A.; Woodzick, Thomas L.; Yellin, Samuel; Yokley, John W.; Yuen, Cheong-Yip; Zimdars, Marjorie Ann; Zinkgraf, Joel P.

1980s: Albrecht, Richard K.; Aluka, Innocent J.; Anderson, Amy Louise; Anderson, Robert H.; Bailey, William; Barreto, Laura L.; Bartling, Brian T.; Benante, Joanne M.; Biller, Martin J., III; Bishea, Douglas M.; Bleem, Jeanice C.; Borquaye, Samuel A.; Borucki, Mark K.; Brennan, Brian Daniel; Brinkmann, Robert; Brukardt, Susan A.; Brusky, Eugene S.; Bues, Diane Jean; Bugs, Donna M.; Burke, Collette Dick; Byers, Jay Morgan; Carman, Erick P.; Carney, Michael Joseph; Carpenter, Roger M.; Coholich, Philip A.; Colville, Valerie R.; Cook, Timothy Ralph; Coorough, Patricia Jo; Cruciani, Cynthia L. W.; de Kruyter, Mark; Deering, Lynn Greiner; Doyle, Alison Beth; Doyle, Kevin Michael; Eckert, Jeffrey C.; Eisert, Janet Lynn; Emerick, John A.; Ethetton, Laura Kay Herrick; Ethetton, Lee Wayne; Fenelon, Bernard G.; Fenelon, Joseph Martin; Frechette, William G.; Friedel, Michael J.; Fritz, Dale A.; Gagnon, William P.; Gnabasik, Barbara J.; Goodell, Michael W.; Goodrich, Ross Edgell; Gross, Oliver; Grueter, Joyce C.; Hackenberg, Robert L.; Hagermann, Steffen Gerd; Hampton, Mark W.; Hassler, Michael H.; Henckel, Elaine M.; Hense, Robert R.; Hinkley, Bruce F.; Holmes, David Brian; Honma, Shigeo D. E.; Hopper, John Wallace; Jansen, John Richard; Jerskey, Richard Garrard; Johnson, James Kenneth; Jorsch, James J.; Kane, Kevin J.; Kostenko, James J.; Kraemer, Bradley Robert; Krolow, Mark R.; Krumenacher, Mark J.; Kumbalek, Steven Charles; Labandeira, Conrad Christopher; Ladwig, Kenneth J.; Lambert, Stephen P.; Lascelles, Peter A.; Lawton, Kevin W.; Leipzig, Martin R.; Leischer, Clayton Carter; LeNoble, Michael J.; Liaskos, Dimitrios Anastasios; LoDuca, Steven T.; Lyon, Robert P.; Martin, Steven Lee; Marz, Penny A.; McBride, John M.; McKereghan, Peter Fleming; McLinn, Gene; Melka, Timothy M.; Mercer, David A.; Milender, Kenneth Westcott; Miller, Roger Arthur; Mize, Thomas R.; Moll, J. Gregory; Mullen, Christopher Edward; Mullen, Donna Marie; Neumann, Lynda L.; Nibbe, Rod K.; Nieland, Connie Lynn; O'Bright, David Edward; Palispis, Jaime Rafael; Pandolfi, John M.; Paul, Duane G.; Peters, Joseph A.; Rosauer, Mark Steven; Rosen, Carol J.; Rovey, Charles W.; Schallhorn, Janis K.; Schiefelbein, Debra Ruth Jessica; Schlipp, Wayne Richard;

Schwetz, Diane L.; Sclosser, Thomas N.; Shen, Elizabeth Jean; Smith, Larry; Spoerl, Carol Lynn; Stanton, Linda Janice; Sunde, Robert Lynn; Swingen, Regina Anne; Synowiec, Karen A.; Tedrahn, David C.; Thompson, Debora B.; Vinson, Thomas Edward; Voltz, Charles Frederick; Weyer, Laura L.; White, Richard J.; Winkler, Ron; Winsor, Mark F.; Wolske, Roxanne L.; Zager, John P.; Zakrzewski, Allan G.; Zvibleman, Barry

Woods Hole Oceanographic Institution
Woods Hole, MA 02543

41 Doctoral

1970s: Austin, James Albert, Jr.; Burroughs, R. H., III; Cacchione, David A.; Erez, Jonathan; Erickson, Albert J.; Flood, Roger Donald; Forsyth, Donald William; Gardner, Wilford D.; Macdonald, Kenneth Craig; Poehls, Kenneth Allen; Shor, Alexander N.; Tapscott, Christopher Robert

1980s: Asper, Vernon L.; Benoit, Gaboury; Bremer, Mary Lee; Chave, Alan Dana; Collins, John Anthony; Cornuelle, Bruce Douglas; Crowe, John; de Baar, Hein J. W.; Delaney, Margaret Lois; Driscoll, Mavis Lynn; Fitgerald, Michael G.; Glenn, Scott Michael; Goud, Margaret Redding; Green, Kenneth Edward; Hay, Bernward Josef; Hotchkiss, Frances Luellen Stephenson; Jemsek, John P.; Kaminski, Michael Anthony; Kurz, Mark David; Miller, Kenneth George; Paola, Christopher; Pfirman, Stephanie Louise; Repeta, Daniel James; Richardson, Mary Josephine; Robinson, Elizabeth M.; Speer, Paul Edward; Takahashi, Kozo; Toomey, Douglas Ray; Trehu, Anne Martine

Worcester Polytechnic Institute
Worcester, MA 01609

1 Master's

1980s: Gifford, Gregory Paul

Wright State University
Dayton, OH 45435

192 Master's

1970s: Bennett, Timothy John; Bitar, Richard F.; Bolze, Claude E.; Brennan, Terrance Patrick; Burke-Griffin, Barbara Mary; Cheng, Song-Lin; Cobb, Craig Carroll; Contrino, Charles Thomas; Cornyn, Michael Robert; De Vito, Steven A.; Eaton, Richard G.; Eichen, David; Elliott, Edward S.; Flaugher, David Michael; Fredericks, Kenneth J.; Frost, Jack Philip; Gephart, Roy E.; Harrison, George B.; Heidorn, Marjorie Arline; Henry, George, Jr.; Herbert, Joseph J.; Hinkley, Carole M.; Holsinger, S. L.; Hoover, Steven Patrick; Hull, Dennis H.; Huntsman, Brent Elliot; Idowu, Ayorinde O.; Jehn, James Lawrence; Kastritis, George John; King, Alan D.; Kompanik, Gary Steven; Koschal, Gerald J.; Kulibert, Richard James; Lawrence, Paul; Lillie, John T.; Lubner, Katherine E.; Massar, Bruce Allen; Maxfield, William Kinsey; Mercer, Richard B.; Miller, Wesley L.; Neese, Michael Charles; Neville, Allen Sneed; O'Sullivan, Terence Patrick; Parker, Boyd Kent; Pearson, Jerome; Pender, Jeffrey Thomas; Persons, Jeffrey L.; Riggsbee, Wade H.; Rose, Gregory Lloyd; Rossi, Dennis A.; Sain, Herman Andrew; Schuller, Rudolph M.; Selden, Robert W.; Siegal, Sherman M.; Steele, John Davis; Taylor, Pamela Sue; Thoburn, Thomas C.; Toomey, William J.; Viste, Daniel Ralph; Walker, Jerry C.; Wilk, Charles Kenneth

1980s: Abernathy, Gary Lance; Acharya, Arvind B.; Allred, Ronald Dean; Altic, Mark Alan; Armstrong, Ernest Elwood; Armstrong, William M.; Baldwin, Thomas Ashley; Biaglow, Joseph Anthony; Blyskun, George James; Bolla, William Owen; Brooks, Mark W.; Brown, W. R.; Bunk, Gregory L.; Burdick, Joseph Matthew; Burgdorf, Gregory John; Burkhart, Patrick A.; Chamberlain, John Mark; Christie, Michael Alexander; Cooper, Budoin-Brutus J.; Criswell, James Richard; Culver, Stephen Eric; Czekalski, Steve James; Dailey, Dale V.; Daugherty, Colleen Mae; Daugherty, Helen Ro-

berta; Doyle, Marianne C.; Egan, David E.; Fago, Thomas Arthur; Fairbank, Philip Keith; Farrell, Michael Thomas; Faw, Dorothea M.; Faw, Jeffrey W.; Fenno, Dan P.; Ferry, Joe P.; Gitelson, Geoffrey A.; Grodi, Ernest D.; Hager, James Marion; Hammond, Scott Alan; Hawkins, James Gregory, Jr.; Herin, James Christopher; Hewlett, James Scott; Hillman, Douglas L.; Hockensmith, Brenda Louise; Holdeman, Timothy G.; Hook, Thomas E.; Hoose, Lori A.; Hoose, Randolph Henry; Howell, Mark Joseph; Hudak, Paul F.; Jehn, Theresa C.; Jones, Bruce Charles; Jordan, John Edgar, Jr.; Kastenhuber, Lynn Edward; Kelly, Patrick Vizard; Kendrick, Michael Brian; Kidd, Gerald Daniel; King, Scott E.; Kirchoff-Stein, Kimberly Susan; Kratky, Mark Anthony; Krupa, John; Lanchman, Gregory Joseph; LePain, David Lloyd; Lewis, Richard Timothy; Lindecke, Joseph Werner; Lockman, Dalton; Macarevich, Roger L.; Martt, Shirley Lee; Maxis, Ike; Middlebrooks, Peter Kendrick; Mika, James E.; Mitchell, James K.; Monks, Edwin Tod; Monks, Katherine Schauder; Mullett, Douglas J.; Mumpower, Douglas S.; Naas, Bruce Edward; Nevero, Ann Bernadette; Nimri, Faris Tawfiq; Onyia, Ernest Chijioke; Ossege, John; Ossinger, Richard A.; Paquette, Douglas Edward; Parks, Sandra Moffett; Pavlik, John D.; Peterson, Lance Eric; Plomer, James R.; Portman, Mark E.; Price, Susan Gay; Priore, William J.; Randall, William Arthur, Jr.; Ratliff, Terry Wayne; Rich, Gretchen Remington; Rickertsen, Mark Andrew; Ricketts, Brian M.; Riggi, Salvador Anthony, Jr.; Riggle, Mark Robbins; Ritzi, Robert William, Jr.; Sauls, Brian D.; Schackne, Michael L.; Schiferle, Jane Carol; Schrantz, Jonathan K.; Scott, Diane M.; Scott, Earl Harold; Simmons, Thomas Paul; Simon, Robert B.; Smith, Bruce Edward; Smith, William Howard; Smyth, John Thomas; Sofranko, Ronald A.; Sontag, Karen D.; Stanley, Jennifer Sue; Stover, Mark K.; Sweazy, Christopher L.; Thibodeau-Jordan, Dawne Marie; Thum, James Arthur; Tochtenhagen, Mark S.; Venys, James Joseph; Vermeulen, Mark V.; Vetter, Mark; von Maluski, Barbara Janine; von Shaffer, Ronald; Walker, James K.; Waren, Kirk Bernon; Wiedman, Lawrence Alan; Wiley, Kenneth George; Williams, Dwight Drue; Willie, Kelly Delon; Wilson, Forrest Raymond; Wittoesch, David F.; Woods, Barry Bradford; Zuberi, Shafiq Ahmed

University of Wyoming
Laramie, WY 82071

581 Master's, 118 Doctoral

1890s: Taylor, J. W.

1920s: Spalding, Robert W.; Thomas, Horace Davis

1930s: Baldwin, H. T.; Bradford, C. E.; Brown, Orman Presley; Buehner, J. H.; Dockery, W. Lyle; Ferren, Jack E.; Giddings, Harrison J.; Hallock, Donald H. V.; Hand, H. D.; Harrison, J. W.; Heathman, J. H.; Isberg, J. T.; Jenkins, Page T.; Johnston, J. C.; Konkel, Phillip; Loeffler, Richard J.; Love, John David; Martin, A. B.; Mort, F. P.; Nace, Raymond L.; Neely, Joseph; Peterson, Arthur F.; Shoemaker, Richard Walter; Starkey, Caldwell; Umbach, Paul Henry; Wolf, G. H.

1940s: Albanese, John P.; Ashley, William H.; Berryman, R. J.; Bertagnolli, Alex J., Jr.; Brady, R. T.; Carlson, Charles Edward; Del Monte, Lois; Dengo, Gabriel; Diemer, Raymond A.; Dunbar, R. O.; Edwards, Acus Rex; Gooldy, Penn Lawrence; Gray, L. O., Jr.; Hammond, Charles R.; Haun, John D.; Heisey, Edmund L.; Houlette, Kenneth N.; Kivi, W. J.; Knight, Wilbur H.; Lawson, D. E.; Lynn, John R.; Maravich, M. D.; McCurdy, Harland R.; McKay, E. J.; Osterwald, Doris B.; Osterwald, Frank W.; Partridge, Lloyd R.; Pipiringos, George Nicholas; Ritzma, Howard R.; Salisbury, Gerald P.; Sears, William Arthur, Jr.; Sims, Frank Chanberg; Umbach, Elmer Dean; Walker, Cecil Lester; Weimer, Robert Jay; Young, K. P.

1950s: Allspach, H. G.; Anderson, V. H.; Andrau, W. E.; Ary, M. D.; Badgley, Edmund Kirk, Jr.; Barlow, James A., Jr.; Barlow, James A., Jr.;

Barnes, W. C.; Barnwell, William W.; Basko, Donald B.; Bauer, E. J.; Beck, Fredrick M.; Bell, W. G.; Bergstrom, John R.; Bergstrom, John R.; Berman, Jack E.; Berry, Richard M.; Biggs, Charles A.; Bogrett, J. W.; Bozanic, Dan; Brooks, B. G.; Brown, Prescott L.; Buffett, R. N.; Burk, Creighton Alvin; Cardinal, D. F.; Carey, Byrl D., Jr.; Carey, Byrl D., Jr.; Carpenter, Leo C.; Catanzaro, E. J.; Childers, M. O.; Cleven, G. W.; Cooper, Herschel T.; Currey, D. R.; Davis, R. W.; de la Montagne, John; de la Montagne, John; Deardorff, D. L.; Del Mauro, Gene Louis; DeLand, C. R.; Dresser, H. W.; Drwenski, Vernon R.; Dunnewald, J. B.; Durkee, Edward F.; Ebbett, Ballard; Elias, D. W.; Epstein, J. B.; Espach, R. H., Jr.; Espenschied, E. K.; Faulkner, Glen L.; Fiero, G. William, Jr.; Finnell, Tommy L.; Flanagan, Philip E.; French, J. J.; Frey, D. M.; Fulkerson, Donald H.; Gilliland, J. D.; Gillum, J. P.; Gilman, M. N.; Gist, J. G.; Grace, Robert M.; Grant, S. C.; Groth, F. A.; Hale, Lyle A.; Hanagan, E. J.; Hanson, Bernold M.; Harston, Lee W.; Hinds, George W.; Hoodmaker, F. C.; Hubbell, Roger G.; Hummel, J. M.; Hunter, James M.; Hunter, LaVerne D.; Jenkins, Carl E.; Johnston, R. H.; Keefer, W. R.; Keefer, W. R.; Keller, Marvin A.; Koenig, Afton A., Jr.; Larsen, John H.; Long, E. G.; Long, J. S., Jr.; Lyon, Denton Lloyd; MacClintock, C.; Manion, R. E.; Maret, R. E.; Matus, I.; McCoy, M. R.; McCue, J. J.; McDavid, J. D.; McGookey, D. P.; McGraw, R. B., Jr.; McGrew, L. W.; Merritt, Zen S.; Mescher, Paul A.; Michalek, D. D.; Millice, Roy; Minick, J. V.; Mitchell, S. D.; Mogensen, P.; Moore, Fred Edward; Morgan, Joseph K.; Morisawa, Marie; Murphy, Richard W.; Murray, F. E.; Myers, W. G.; Mytton, James W.; Osmond, John C., Jr.; Oster, L. D.; Patel, C. B.; Patterson, Kenneth D.; Pederson, Selmer Lane; Phillips, D. P.; Powell, Joe Douglas; Rachou, John F.; Ramsey, Rodney Dean; Richardson, Albert L.; Riedl, G. W.; Riva, Joseph Peter, Jr.; Robinson, James Richard; Roehler, H. W.; Roth, Kingsley William; Royse, F., Jr.; Ruppel, Edward Thompson; Sacrison, W. R.; Schoen, R.; Schwarberg, T. M.; Schwarbert, T. M., Jr.; Severn, W. P.; Sheffer, Bernard Douglas; Shipp, B. G.; Shlemon, R. J.; Short, B. L.; Smith, J. R.; Smithson, Scott B.; Spelman, A. R.; Stephen, J. N.; Stephens, E., Jr.; Swain, B. W.; Taucher, Leonard Max; Thompson, Raymond M.; Trotter, John Francis; Troyer, Max L.; Tudor, Matthew Sanford; Unfer, Louis, Jr.; Walsh, M. H.; Walters, R. F.; Weichman, B. E.; Welsh, John Elliot; White, Vincent Lee; Willis, Richard Porter; Wills, J. G.; Wilson, Jacqueline B.; Wilson, Richard W.; Wilson, William Harold; Workum, R. H.; Wroble, John Lee; York, H. F.; Zakis, W. N.; Zell, R.; Zoble, Jerry E.

1960s: Anderson, James E.; Baker, Gordon K.; Barwin, J. R.; Bay, Kirby Whitmarsh; Bell, Lyndon H.; Bishop, D. T.; Boles, James R.; Bothner, W. A., Jr.; Bragdon, Frederick F.; Breckenridge, Roy Melvin; Bullock, James M.; Burritt, C. C.; Castro, E. J.; Chadeayne, Dennis; Clausen, Eric N.; Cody, R. D.; Conley, C. D.; Cramer, L. W.; Davidson, Peter S.; Davis, James R.; Davis, James R.; Davis, John C.; Davis, John C.; Deal, D. E.; DeNault, Kenneth, Jr.; Dickey, M. L.; Dimitroff, Pencho B.; Doehring, Donald O.; Dreier, John E.; Dunrud, C. R.; Eaton, George M.; Ebens, Richard J.; Ebens, Richard J.; Ellerby, R. S.; Fenton, Michael Dwight; Ferris, Clinton S., Jr.; Fields, Edward D.; Fikkan, Philip R.; Fisher, Frederick S.; Froidevaux, Claude M.; Fruchey, R. A.; Furer, L. C.; Gardner, Henry J.; Gliozzi, James; Gloor, Edward Alfred; Good, L. W.; Granata, Walter Harold, Jr.; Graveson, David H., Jr.; Gries, John C.; Guyton, J. W.; Guyton, J. W.; Hall, William B.; Hamilton, H. E.; Hanson, Bruce Vernor; Hauf, C. B.; Heydenburg, Richard J.; Hodge, Dennis S.; Hodge, Dennis S.; Husain, F.; Jarre, G. A.; Johnson, F. O.; Kelley, James C.; King, J. R.; King, John S.; Kisling, D. C.; Kiver, Eugene P.; Lackey, Larry L.; Landau, David; Lawrence, J. C.; Litchford R. F., Jr.; Manion, L. J.;

Mariner, Robert Howard; Max, Michael D.; McCallum, M. E.; McConnell, M. D.; McGrew, Alan R.; Merry, Ray D.; Mills, Rodger K.; Nicoll, G. A.; Noble, E. A.; Orback, C. J.; Ore, H. T.; Pedersen, Kenneth; Pledge, Neville Stewart; Prochaska, E. J.; Risley, R. G. Jr.; Robertson, Richard A.; Ruehr, B. B.; Saulnier, George J.; Schoenfeld, Mark Jean; Scott, R. W.; Sikich, S. W.; Smith, Bruce D.; Sorenson, G. E., Jr.; Stensrud, Howard Lewis; Stuart, William J., Jr.; Suydam, R. B.; Swetnam, Monte N.; Talbot, Curtis L.; Tebbutt, Gordon E.; Toogood, David J.; Toots, Heinrich A.; Toots, Heinrich A.; Ulteig, J. R.; Voorhees, Gerald E.; Voorhies, Michael R.; White, Ronald K.; Wied, Otto J.; Wiegman, Ronald W.; Wilkinson, Bruce H.; Wood, C. B.; Woodfill, Robert D.; Worl, Ronald G.; Worl, Ronald G.

1970s: Adams, Penny R.; Ahlbrandt, Thomas Stuart; Andersson, Kent Albert; Atherton, Charles C.; Baker, K. H.; Balcells, Roberto; Baldwin, E. A.; Banks, Carlie Elisabeth; Barton, Raymond; Bekkar, Hamed; Benaissa, Saddok; Benniran, M. M.; Bergstresser, Thomas James; Berryman, William M.; Bickford, Fred E.; Black, Paul R.; Blackstone, Robert E.; Boodoo, W.; Bown, T. M.; Breckenridge, Roy M.; Brooks, Kenneth J.; Buchheim, H. P.; Carlisle, W. Joseph; Cassiliano, M. L.; Coalson, Edward B.; Cook, F. A.; Cook, W. R., III; Coolidge, C. N.; Copeland, David Ashley; Davis, B. M.; Deiss, A. P.; Diehl, J. F.; Diehl, S. B.; Dring, Nancy Beth; Duhling, William H.; Dunn, Thomas Lowell; Earle, Janet L.; Fiki, M. H.; Fox, James E.; Frederickson, A. F.; Froman, N. L.; Galey, John T.; Gaskill, Charles H., Jr.; Goodwin, C. N.; Goodwin, Jonathan H.; Graff, P. J.; Gwinner, Don; Hager, Michael Warring; Hallin, James S.; Hand, Forrest E., Jr.; Hanley, J. H.; Haywood, Harry C.; Heasler, Henry Peter; Heiman, Mary E.; Heiman, Mary E.; Herrick, Rodney C.; Herrmann, Raymond; Hoffman, Monty E.; Holden, G. S.; Huang, Chi-I.; Hughes, Mark A.; Huycke, David T.; Irwin, Barbara R.; Isaacson, Laurie Brown; Jensen, Stephen D.; Kaabar, Salah M.; Kaminsky, Barney; Karasa, N. L.; Karlstrom, Karl Edward; Keefer, C. M.; Kirn, Douglas J.; Kistner, Frank B.; Knapp, R. R.; Koenig, Robert L.; Kolm, K. E.; Kolm, K. E.; Kornegay, G. L.; Kraus, Mary J.; Kron, Donald Gordon; Krosky, S.; Lageson, David Rodney; Lanthier, L. R.; Levinson, Richard A.; Lundell, L. L.; Lundy, D. A.; Mackenzie, Robert John; Madson, M.; Mankiewicz, David; Mariner, Robert Howard; Master, Timothy D.; McCullar, Dan Brett; McFaul, Michael; Miller, William R.; Miller, William R.; Moncure, George Kinser; Morel, J. D.; Mott, L. V.; Murphy, Donald James; Murphy, J. D.; Myers, Jonathan; Oakes, Edward H.; Ochoa, Rafael Eugenio; Ogden, P. R., Jr.; Olander, Paul A., Jr.; Oliver, Kenneth L.; Oviatt, Charles G.; Pacht, J. A.; Pearson, Eugene Favre; Pekarek, Alfred H.; Perez, K. R.; Pokras, Edward M.; Price, Chadderdong; Prichinello, Katherine A.; Pruss, E. F.; Puwakool, Suchit; Ramberg, E. M.; Repetto, F. L.; Ridgley, Jennie L.; Ridgley, Neill H.; Sahai, S. K.; Sands, C. D.; Schmitt, James G.; Schock, William Wallace, Jr.; Schuster, J. E.; Sears, James W.; Sever, C. K.; Sherer, Richard Lowell; Shero, B. R.; Simnacher, Faroy; Smith, Cole L.; Sowers, Nancy R.; Sukup, J. W.; Sylvester, George H.; Taylor, D. J.; Taylor, M. W.; Thompson, Keith S.; Tombaugh, Karen; Van Ingen, L. B., III; Vargas, H. Rodrigo; Vietti, Barbara Tomes; Vietti, John S.; Vinton, R. P.; Waldon, Colin E.; Wells, R. A.; Westervelt, T. N.; Wheelock, Thomas G. B.; Whitaker, Richard M.; Wilson, B. D.; Winterfeld, Gustav F.; Wolfbauer, Claudia A.; Worrall, D. M.; Wu, H. C.; Yatkola, Daniel Arthur; Zoerner, Frederick P.

1980s: Adler, Kevin R.; Andersson, Knut Albert; Antweiler, Ronald Chisholm; Babits, Steven J.; Balsam, Robert C., Jr.; Barrett, Ramsay A.; Barron, Andrew Morrow; Batt, Richard J.; Beatty, Charles A.; Beiswenger, Jane Miller; Benedict, Jonathan F.; Berg, Richard M., Jr.; Berg, William R.; Bergstresser, Thomas James; Bernaski, Greg E.;

Bierei, Mark Alan; Blair, Kevin P.; Bland, Douglas M.; Blevens, Dale M., Jr.; Blum, Alex E.; Blundell, J. Stuart; Bochensky, Paul; Bottjer, Richard J.; Branch, Charles N.; Breithaupt, Brent H.; Brown, Alison Y.; Bucher, Gerald Joseph; Buelow, Kenneth L.; Burch, Ellen; Burke, Margaret M.; Chambers, Henry Peyton; Chronic, Lucy M.; Clarey, Timothy L.; Conard, JoAnn B.; Constenius, Kurt N.; Copland, John Robin; Coughlan, James Patrick; Cozzens, James Robert; Crossey, Laura J.; Cutler, Elizabeth Reinen; Daley, Roberta L.; Dam, William L.; Davidson, Jeanne R.; Deemer, Sharon J.; deJarnett, Jeffrey G.; Denis, John R.; Desmond, Robert J., Jr.; Doelger, Nancy Micklich; Donovan, Robert C.; Doremus, Dale M.; Dudley, Julia Lynn; Duebendorfer, Ernest Martin; Easley, Dale H.; Eaton, Jeffrey G.; Edman, Janell Diane; Evans, Christine Victoria; Fannin, Timothy Edward; Fischer, Karin J.; Flurkey, Andrew James; Foote, Martin William; Fuller, Brian N.; Fuller, Brian N.; Gaylord, David Russell; Gaylord, David Russell; Gentry, Dianna J.; Gilmer, Douglas R.; Gjelsteen, Thor W.; Gorham, Julie M.; Greer, Phillip L.; Gubbels, Timothy Louis; Hagen, Earl Sven; Halcomb, Robert Allan; Hanson, Douglas Wade; Harder, Steven; Harding, Matthew B.; Harris, R. E.; Hartman, Jane E. Z.; Heasler, Henry Peter; Heyman, Oscar Glenn; Hoare, Brian Stuart; Hunter, Robert Bruce; Hurcomb, Douglas R.; Hurich, Charles A.; Hurich, Charles A.; Hurlow, Hugh A.; Hurst, Donald J.; Iltis, Steven T.; Iverson, William Paul; Jackson, Kenneth E.; Jaeger, Kenneth B.; Jarvis, W. Todd; Jaworowski, Cheryl C.; Jett, Guy A.; Johnson, Roy A.; Jung, Kwang Seop; Kablanow, Raynold I., II; Kablanow, Raynold Irvin, II; Karlstrom, Karl Edward; Key, Scott C.; Kimball, Briant A.; King, Jonathan K.; Kirschner, William A.; Kling, John F.; Koesterer, Mary Ellen; Kolb, Grant; Kopriva, Suzanne J.; Kratochvil, Anthony L.; Kulas-Adler, Helen A.; Lageson, David Rodney; Langstaff, George D.; Lee, Teh-Quei; LeFebre, George Bradburn; LeFebre, Valerie S.; Lehrer, Mark G.; Liebes, Eric; Londe, Michael David; Loope, David Bittle; Loughlin, William Dornan; Love, Frank R.; Lowry, Anthony R.; Manning, Leslie Kay; Marcus, Michael Dean; Marks, Janet E.; Marshall, David M.; Mason, Glenn Michael; Mason, John M.; McElhaney, David A.; McGee, Linda C.; Medlin, W. Eric; Mettes, Kim J.; Meyer, John E.; Miao, Desui; Middleton, Larry Thomas; Miller, Stanley Mark; Minnich, Gene W.; Moncure, George Kinser; Moran, Mary E.; Mueller, Robert Emerson; Murray, William B.; Myers, Jonathan; Nice, David E.; Nissen, Thomas C.; Noffsinger, Kent Eugene; Novakovich, Bruce D.; Nyblade, Andrew A.; Oliver, Robert L.; Parker, Steven E.; Patchen, Allan D.; Paylor, Earnest D., II; Pierson, William R.; Pujol, Jose M.; Ratterman, Nancy G.; Renner, James M.; Richter, Henry Robert, Jr.; Rochette, Elizabeth A.; Rowe, David W.; Rudkin, G. Thomas; Ryan, David C.; Schmidt, Thomas G.; Schmitt, James G.; Schock, William Wallace, Jr.; Serbeck, John W.; Shawesh, Othman Mohamed; Sheriff, Steven D.; Shuster, Mark William; Sipe, Dwight Randy; Sippel, Katharine N.; Skinner, Orion L.; Smith, Anne Lauren; Spencer, Sue Ann; Sperr, Jay T.; Starkey, Kimberly J.; Starkey, Robert James; Steele, Kenneth Kane; Stock, Mark D.; Sundell, Kent A.; Swauger, David A.; Swift, Peter N.; Taucher, Paul J.; Taucher, Susan E. Gray; Treloar, Nathan A.; Tromp, Paul L.; Valasek, Paul A.; Valenti, Gerard L.; Walker, Danny Norbert; Wallem, Daniel B.; Ware, Douglas C.; Way, Shao-Chih; Weichman, David A.; West, Christine M.; Williams, Jonathan D.; Williams, Martin C.; Williams, William C.; Wilson, Sharon Lee; Winslow, Nancy S.; Winterfeld, Gustav F.; Wolberg, Peter W.; Woods, Thomas F.; Yin, Peigui; Yingling, Virginia Leigh; Yonkee, W. Adolph; Yose, Lyndon A.; Yule, J. Douglas; Zawislak, Ronald Lynn

Yale University
New Haven, CT   06511

23 Master's, 365 Doctoral

1860s: Nelson, Edward T.; Rice, William N.

1870s: Dana, Edward S.

1880s: Barbour, Erwin Hinckley; Beecher, Charles E.; Grinnell, George B.; Hovey, Edmund O.; Williston, Samuel W.

1890s: Eaton, George F.; Farrington, Oliver C.; Girty, George H.; Gregory, Herbert E.; Kindle, Edward M.; Pratt, Joseph H.; Warren, Charles Hyde

1900s: Barrell, Joseph; Bowman, Isaiah; Cleland, Herdman F.; Cumings, Edgar R.; Forbes, Edwin H.; Ford, William E.; Harvey, Ruth S.; Huntington, Ellsworth; Lane, Francis B.; Loughlin, Gerald F.; Noble, Levi F.; Pogue, Joseph E.; Raymond, Percy E.; Robinson, Henry H.; Sarle, Clifton J.; Savage, Thomas Edmund; Sellards, Elias H.; Snelling, Walter O.; Stanley, Frederick C.; Talbot, Mignon; Ward, Freeman; Weller, Stuart; Wieland, George R.; Young, George A.

1910s: Alcock, Frederick J.; Bateman, Alan M.; Bissell, Malcolm H.; Crawford, Ralph D.; DeLorme, Donaldson C.; Drysdale, Charles W.; Dunbar, Carl O.; Emery, Wilson B.; Gabriel, Ralph Henry; Grout, Frank F.; Lauer, Arnold W.; McLean, F. H.; McLearn, Frank H.; O'Niell, John J.; Perkins, Edward P.; Reber, Louis E.; Reeds, Chester A.; Reinecke, Leopold; Robinson, Clair W.; Robinson, Wilber I.; Roesler, Max; Rose, Bruce; Stewart, James S.; Thornton, William Mynn, Jr.; Thorpe, Malcolm R.; Troxell, Edward L.; Twenhofel, William H.; Williams, Merton Y.; Wright, William J.

1920s: Armstrong, Paul F.; Bell, Walter A.; Bissell, Malcolm H.; Bradley, Wilmot H.; Bryan, Kirk; Chang, Peter Yun Tsin; Cooper, Gustav Arthur; Ferguson, Henry G.; Gilluly, James; Glock, Waldo S.; Goodridge, Randolph; Hare, Joseph E.; Hewett, Donnel F.; Hickok, William O.; Horton, Leo V.; Hume, George S.; King, Philip Burke; King, Robert E.; Lasky, Samuel G.; Longwell, Chester Ray; McCann, William Sidney; Nolan, Thomas B.; Northrop, Stuart; Osborne, Freleigh Fitz; Palmer, Harold S.; Russell, William L.; Russell, William L.; Schairer, John F.; Schmidt, Bruno M.; Simpson, George Gaylord; Stone, John B.; Tolman, Carl; Tolman, Carl; Wickwire, Grant T.; Young, Addison

1930s: Ambrose, John W.; Baker, Arthur A.; Barksdale, Julian Deverau; Bateman, John D.; Bramlette, Milton N.; Brockunier, Sawyer R.; Crickmay, Geoffrey W.; Dane, Carle H.; Elias, Maxim K.; Fleming, William L.; Gallagher, David; Hickok, William O.; Holmes, Chauncey DeP.; Ingerson, Fred Earl; Knight, James Brook; Koerner, Harold E.; Krynine, Paul D.; Lewis, George E.; Love, John David; Miller, Arthur K.; Newell, Norman D.; Perkins, William A.; Riley, Leonard B.; Robertson, George McAfee; Sanford, Wendell Glenn; Seager, George F.; Silk, Ernest S.; Smith, Ward C.; Van Gilder, Harold R.; Warren, Charles R.; Waters, Aaron C.; Yang, Tsun-Yi

1940s: Amsden, Thomas W.; Beach, Hugh H.; Berdan, Jean M.; Broderick, Alan T.; Brown, Randall E.; Cloud, Preston E.; Cobb, Edward H.; Cullison, J. S.; Eifler, Gus K., Jr.; Hobbs, Samuel W.; McTaggart, Kenneth C.; Mikami, Harry M.; Morrow, Aubrey Lyndon; Ordway, Richard J.; Oriel, Steven S.; Rice, William A.; Roberts, Ralph Jackson; Rodgers, John; Ross, Reuben J., Jr.; Sullivan, John W.; Troelsen, Johannes C.; Washburn, Albert L.; Wilson, James Lee

1950s: Allen, James Michael; Benson, William E. B.; Berry, William B. N.; Bick, Kenneth Fletcher; Binney, Edwin, Jr.; Bloom, Arthur L.; Boone, Gary McGregor; Buehler, Edward J.; Burger, John Allan; Bushnell, Kent Orpha; Carroll, Gerald V.; Clarke, James Wood; Coash, John Russell; Coulter, Henry W., Jr.; Crandell, Dwight R.; Davis, Stanley N.;

Yeshiva University

Dechow, Ernest W. C.; Deland, André Normand; Dietrich, Richard V.; Drewes, Harald D.; Dutro, John Thomas, Jr.; Eicher, Don Lauren; Elson, John Albert; Flawn, Peter Tyrell; Friedman, Jule Daniel; Friedman, Jule Daniel; Gealy, John R.; Greiner, Hugo R.; Ham, William Eugene; Hays, W. H.; Hays, William Henry; Heimlich, Richard A.; Hickox, Charles Frederick, Jr.; Hodgson, Robert Arnold; Ignatius, Heikki Gustaf; Imbrie, John; Jenness, Stuart Edward; Kinney, Douglas M.; Klepper, Montis Ruhl; Kupfer, Donald H.; Lesure, Frank G.; Livingston, Daniel A.; Lowry, Elizabeth J.; Lundgren, Lawrence Williams; Masursky, Harold; McFall, Clinton Carew; Murthy, Varanasi Rama; Neale, Ernest R. W.; Norford, Brian Seeley; Odum, Howard Thomas; Ogden, James Gordon, III; Orville, Philip Moore; Prinz, William Charles; Rau, Jon Llewelyn; Ross, Charles A.; Sabins, Floyd F., Jr.; Salisbury, John William, Jr.; Sanders, John Essington; Sclar, Charles B.; Simpson, Howard E., Jr.; Stearn, Colin W.; Strobell, John Dixon; Stugard, Frederick, Jr.; Thornton, Charles P.; Tracey, Joshua I., Jr.; Tyrrell, Willis Woodbury, Jr.; Weitz, Joseph L.; Wheeler, Walter H.; Wright, Grant MacL.

1960s: Ames, Roger Lyman; Armstrong, Richard Lee; Bambach, Richard Karl; Belt, Edward Scudder; Benson, Gilbert Thomas; Bernold, Stanley; Berry, Richard Harry; Biscaye, Pierre Eginton; Bretsky, Peter William, Jr.; Bretsky, Sara Stewart; Brown, Seward Ralph; Brueckner, Hannes Kurt; Burchfiel, Burrell Clark; Buzas, Martin Alexander; Carr, Michael H.; Cheney, Eric S.; Coch, Nicholas Kyros; Crowley, William Patrick; Dasch, E. Julius, Jr.; Denton, George Henry; Dieterich, James Herbert; Doyle, Roger Whitney Stevens; Dunham, Robert J.; Ellis, Charles W.; Enos, Paul Portenier; Giegengack, Robert, Jr.; Hansen, Edward Carlton; Hills, Francis Allan; Howard, Keith Arthur; Jenkins, Farish Alston, Jr.; Klein, George DeVries;

Masters, Charles Day; McAlester, Arcie Lee; McDonald, Barrie Clifton; Mello, James F.; Moore, George W.; Myer, George Henry; Oaks, Robert Quincy, Jr.; Offield, Terry Watson; Perry, Kenneth, Jr.; Platt, Lucian; Porter, Stephen Cummings; Powers, R. W.; Radinsky, Leonard B.; Robinson, Peter; Rona, Peter Arnold; Scott, William Henry; Speden, Ian Gordon; Stanley, Rolfe S.; Stanley, Steven Mitchell; Suppe, John Edward; Tilling, Robert Ingersoll; Vidale, Rosemary Jacobson; Walker, Kenneth Russell; Weitz, Joseph Leonard; Wellman, Thomas Robert; Williams, Thomas Ellis; Wolfe, William John

1970s: Allcock, J. B.; Aller, R. C.; Bachinski, Donald John; Bachinski, Sharon L. W.; Barabas, A. H.; Barry, J. C.; Benninger, L. K.; Bertine, Kathe Karlyn; Bokuniewicz, H. J.; Bossy, K. V. H.; Brakel, W. H.; Brass, Garret William; Bultman, T. R.; Byers, Charles, Wesley, II.; Carter, J. G.; Chai, B. H.-T.; Cochran, J. K.; Davis, A. M.; DeMaster, D. J.; Dick, H. J.; Dodge, R. E.; Dodson, Peter; Durden, Christopher J.; Ellis, D. E.; Gamble, R. P.; Gingerich, Philip Dean; Graf, J. L., Jr.; Grossman, Lawrence; Grover, John Emerson; Hansen, T. A.; Heathcote, I. W.; Hewitt, David A.; Hotchkiss, Frederick; Jablonski, D. I.; Keir, R. S.; Levinton, Jeffrey S.; Lewis, D. M.; McCaffrey, R. J.; Miller, S. D.; Morse, John Wilbur; Perkins, Philip Laurence; Perry, William J., IV.; Petrovic, Radomir; Porter, James W.; Rackoff, J. S.; Roberts, David; Rosenfeld, J. K.; Ryer, T. A.; Schamel, Steven; Seidemann, D. E.; Stewart, Alastair James; Stocker, Richard Louis; Tattersall, Ian; Thayer, Charles Walter; Williams, N.

1980s: Badgley, Catherine Elizabeth; Barton, Christopher Cramer; Boudreau, Bernard Paul; Brandt Velbel, Danita; Bulau, James Ronald; Canfield, Donald Eugene; Collins, Laurel Smith; Cuomo, Marie Carmela; Dion, Eric Paul; Dowd, John F.;

Fracasso, Michael Anthony; Gaudreau, Denise Claire; Graustein, William Chandler; Jo, Bong Gon; Keller, Frederick Brian; Key, Marcus M., Jr.; Kidwell, Susan Marie; Krill, Allan George; Landman, Neil H.; Lawrence, David Trowbridge; Losh, Steven Lawrence; Lucas, Spencer George; McKinney, Michael Lyle; Meinke, Deborah Kay; Monaghan, Marc Courtney; Nolen-Hoeksema, Richard Clarence; Ohlhorst, Sharon Lee; Olsen, Paul Eric; Padian, Kevin; Pasteris, Jill Dill; Peck, Lindamae; Reaves, Christopher Madison; Schoch, Robert Milton; Shelton, Kevin Louis; Steele-Petrovich, Helen Miriam; Stewart, Lisa Maureen; Storrs, Glenn William; Sverjensky, Dimitri Alexander; Swapp, Susan Mathilda; Tucker, Robert David; Velbel, Michael Anthony; Waldron, Kim Ann; Walter, Laurie Rianne; Westrich, Joseph Theodore; Wing, Scott Louis; Wood, Timothy Eldridge; Woodwell, Grant R.; Wright, Ellen Margrethe Marie Krogh

Yeshiva University
New York, NY   10033-3299

1 Doctoral

1970s: Teller, Jacob Abe

York University
North York, ON   M3J 1P3

7 Doctoral

1970s: Linton, J. A.; Merriam, J. B.; Pendrel, J. V.

1980s: Bardecki, Michal James; Petrachenko, William Terry; Roberts, Arthur Cecil Batt; Stergiopoulos, Stergios

THESIS SUBJECT DISTRIBUTION BY DECADE AND BY INSTITUTION

Areal Geology, General

4,836 Master's, 1,313 Doctoral

Theses per decade

1880s:	1 Master's, 6 Doctoral
1890s:	11 Master's, 15 Doctoral
1900s:	19 Master's, 21 Doctoral
1910s:	55 Master's, 43 Doctoral
1920s:	125 Master's, 76 Doctoral
1930s:	244 Master's, 137 Doctoral
1940s:	453 Master's, 104 Doctoral
1950s:	1,533 Master's, 341 Doctoral
1960s:	1,065 Master's, 352 Doctoral
1970s:	771 Master's, 159 Doctoral
1980s:	557 Master's, 59 Doctoral

Theses by Institution

Acadia University   8 M
University of Akron   13 M
University of Alabama   13 M
University of Alaska, Fairbanks   15 M
University of Alberta   14 M, 5 D
University of Arizona   114 M, 29 D
Arizona State University   16 M, 1 D
University of Arkansas, Fayetteville   57 M
Auburn University   3 M
Baylor University   9 M
Boise State University   1 M
Boston College   1 M, 1 D
Boston University   2 M, 13 D
Bowling Green State University   6 M
Brigham Young University   83 M
University of British Columbia   53 M, 13 D
Brock University   3 M
Brown University   7 M, 2 D
Bryn Mawr College   11 M, 4 D
University of Calgary   5 M, 3 D
University of California, Berkeley   103 M, 62 D
University of California, Davis   12 M, 4 D
University of California, Los Angeles   167 M, 25 D
University of California, Riverside   9 M, 2 D
University of California, San Diego   9 M, 2 D
University of California, Santa Barbara   8 M, 8 D
University of California, Santa Cruz   1 M, 3 D
California Institute of Technology   55 M, 24 D
California State University, Chico   3 M
California State University, Fresno   1 M
California State University, Hayward   1 M
California State University, Long Beach   1 M

California State University, Los Angeles   3 M
California State University, Northridge   3 M
Carleton University   6 M, 3 D
Case Western Reserve University   1 M, 1 D
Catholic University of America   2 M
Chadron State College   1 M
University of Chicago   10 M, 48 D
University of Cincinnati   9 M, 7 D
City College (CUNY)   1 M
Clark University   1 D
University of Colorado   96 M, 31 D
University of Colorado at Colorado Springs   1 M
Colorado College   1 M
Colorado School of Mines   118 M, 25 D
Colorado State University   14 M
Columbia University   3 M, 4 D
Columbia University, Teachers College   51 M, 48 D
University of Connecticut   3 M
Cornell University   16 M, 17 D
Dalhousie University   8 M
Dartmouth College   10 M, 4 D
DePauw University   1 M
Duke University   1 M
East Carolina University   5 M
East Tennessee State University   1 M
East Texas State University   1 M
Eastern Kentucky University   1 M
Eastern Washington University   11 M
Ecole Polytechnique   1 M
Emory University   21 M
University of Florida   7 M
Florida State University   4 M
Fort Hays State University   3 M
Franklin and Marshall College   4 M
George Washington University   12 M, 1 D
University of Georgia   44 M, 1 D
Harvard University   1 M, 73 D
University of Hawaii   5 M, 2 D
University of Houston   6 M
University of Idaho   76 M, 6 D
Idaho State University   23 M
University of Illinois, Chicago   4 M, 1 D
University of Illinois, Urbana   25 M, 19 D
Indiana University of Pennsylvania   2 M
Indiana State University   1 M
Indiana University, Bloomington   38 M, 11 D
University of Iowa   118 M, 16 D

Iowa State University of Science and Technology   15 M, 1 D
The Johns Hopkins University   3 M, 28 D
University of Kansas   48 M, 8 D
Kansas State University   17 M
Kent State University, Kent   6 M
University of Kentucky   26 M
Lafayette College   1 M
Lakehead University   1 M
Universite Laval   5 M, 15 D
Lehigh University   11 M, 5 D
Loma Linda University   1 M
Louisiana State University   14 M, 13 D
University of Maine   8 M
University of Manitoba   14 M, 3 D
University of Massachusetts   28 M, 6 D
Massachusetts Institute of Technology   11 M, 17 D
McGill University   34 M, 47 D
McMaster University   5 M
Memorial University of Newfoundland   22 M, 3 D
Memphis State University   1 M
University of Miami   1 M, 1 D
Miami University (Ohio)   30 M, 1 D
University of Michigan   94 M, 43 D
Michigan State University   11 M
Michigan Technological University   9 M
Millersville University   1 M
University of Minnesota, Duluth   8 M
University of Minnesota, Minneapolis   30 M, 23 D
University of Mississippi   3 M
Mississippi State University   16 M
University of Missouri, Columbia   106 M, 6 D
University of Missouri, Kansas City   1 M
University of Missouri, Rolla   23 M
University of Montana   29 M, 2 D
Montana College of Mineral Science & Technology   23 M
Montana State University   15 M
Universite de Montreal   1 M, 2 D
University of Nebraska, Lincoln   22 M, 4 D
University of Nevada   34 M, 1 D
University of Nevada, Las Vegas   2 M
University of Nevada - Mackay School of Mines   8 M, 1 D
University of New Brunswick   5 M
University of New Hampshire   7 M
University of New Mexico   92 M, 8 D
New Mexico Institute of Mining and Technology   33 M, 1 D
New Mexico State University, Las Cruces   2 M

University of New Orleans   10 M
New York University   9 M, 1 D
University of North Carolina, Chapel
    Hill   34 M, 8 D
North Carolina State University   18 M
University of North Dakota   12 M, 2 D
Northeast Louisiana University   8 M, 1 D
Northern Arizona University   34 M
Northern Illinois University   13 M
Northwestern University   22 M, 4 D
Ohio State University   61 M, 28 D
Ohio University, Athens   16 M
University of Oklahoma   235 M, 10 D
Oklahoma State University   6 M
University of Oregon   115 M, 5 D
Oregon State University   118 M, 4 D
University of Ottawa   1 M
University of Pennsylvania   3 D
Pennsylvania State University, University
    Park   24 M, 12 D
University of Pittsburgh   14 M, 3 D
Pomona College   21 M
Portland State University   14 M
Princeton University   118 D
University of Puget Sound   1 M
Purdue University   4 M, 3 D
Universite du Quebec a Chicoutimi   1 M
Universite du Quebec a Montreal   1 M
Queen's University   21 M, 5 D
Queens College (CUNY)   5 M
University of Regina   4 M
Rensselaer Polytechnic Institute   4 M, 3 D
Rice University   7 M, 17 D
University of Rochester   19 M, 2 D
Rose-Hulman Institute of Technology   1 M
Rutgers, The State University, New
    Brunswick   8 M, 1 D
University of San Diego   10 M
San Diego State University   27 M
San Jose State University   34 M
University of Saskatchewan   4 M, 3 D
Smith College   5 M
University of South Carolina   22 M, 3 D
University of South Dakota   10 M
South Dakota School of Mines &
    Technology   32 M, 1 D
University of South Florida, Tampa   2 M
University of Southern California   74 M, 4 D
Southern Illinois University,
    Carbondale   19 M
Southern Methodist University   30 M
Southwest Missouri State University   1 M
University of Southwestern Louisiana   5 M
Stanford University   95 M, 45 D
Stephen F. Austin State University   2 M
Sul Ross State University   4 M
SUNY, College at New Paltz   1 D
SUNY at Albany   6 M, 2 D
SUNY at Binghamton   3 M, 5 D
SUNY at Buffalo   7 M
SUNY at Stony Brook   1 D
Syracuse University   28 M, 8 D
University of Tennessee, Knoxville   78 M, 4 D
University of Texas, Arlington   12 M
University of Texas, Austin   214 M, 45 D
Texas A&M University   58 M, 1 D
University of Texas at Dallas   3 M
University of Texas at El Paso   33 M, 2 D
Texas Christian University   10 M
University of Texas of the Permian
    Basin   7 M
Texas Tech University   12 M, 2 D
University of Toledo   1 M
University of Toronto   15 M, 11 D
Tulane University   6 M, 3 D
University of Tulsa   6 M
Union College   2 M
University of Utah   64 M, 20 D
Utah State University   15 M
Vanderbilt University   21 M, 1 D
University of Vermont   11 M
University of Virginia   28 M, 1 D
Virginia Polytechnic Institute and State
    University   32 M, 5 D
Virginia State University   15 M

University of Washington   80 M, 48 D
Washington State University   27 M
Washington University   39 M, 1 D
Wayne State University   14 M
Wesleyan University   3 M
West Texas State University   4 M
West Virginia University   31 M, 1 D
Western Michigan University   5 M, 1 D
University of Western Ontario   3 M
Western Washington University   10 M
Wichita State University   8 M
University of Windsor   2 M
University of Wisconsin-Madison   51 M, 16 D
University of Wisconsin-Milwaukee   24 M, 1 D
Wright State University   1 M
University of Wyoming   154 M, 7 D
Yale University   2 M, 80 D

Areal Geology, Maps and Charts

9 Master's, 20 Doctoral

Theses per decade
1940s: 1 Master's
1950s: 2 Master's
1960s: 1 Master's
1970s: 4 Master's, 14 Doctoral
1980s: 1 Master's, 6 Doctoral

Theses by Institution
University of Arizona   1 M
City College (CUNY)   1 M
University of Colorado   1 M, 1 D
Colorado State University   1 D
Dartmouth College   1 D
University of Idaho   1 D
Iowa State University of Science and
    Technology   1 D
University of Kansas   5 D
University of Manitoba   1 D
University of Maryland   1 D
University of Michigan   1 M
University of Minnesota, Minneapolis   1 D
University of Missouri, Rolla   1 M
Northwestern State University   1 M
Ohio State University   4 D
University of Oklahoma   1 D
SUNY at Buffalo   1 D
Syracuse University   1 D
University of Texas, Austin   1 M
University of Virginia   1 M
University of Wyoming   1 M

Economic Geology, Energy Sources

1,355 Master's, 284 Doctoral

Theses per decade
1870s: 1 Doctoral
1880s: 1 Master's, 1 Doctoral
1890s: 1 Master's, 2 Doctoral
1900s: 5 Master's, 2 Doctoral
1910s: 17 Master's, 5 Doctoral
1920s: 70 Master's, 19 Doctoral
1930s: 69 Master's, 20 Doctoral
1940s: 83 Master's, 7 Doctoral
1950s: 183 Master's, 21 Doctoral
1960s: 87 Master's, 30 Doctoral
1970s: 236 Master's, 55 Doctoral
1980s: 603 Master's, 121 Doctoral

Theses by Institution
University of Akron   4 M, 1 D
University of Alabama   9 M
University of Alaska, Fairbanks   3 M
University of Alberta   17 M, 4 D
University of Arizona   4 M, 4 D
Arizona State University   3 M
University of Arkansas, Fayetteville   17 M
Auburn University   1 M
Ball State University   3 M
Bates College   1 M
Baylor University   25 M
Boston College   1 M
Bowling Green State University   2 M, 1 D
Brigham Young University   5 M, 1 D
University of British Columbia   5 M
University of Calgary   10 M, 2 D

University of California, Berkeley   10 D
University of California, Davis   2 M
University of California, Los Angeles   4 M, 4 D
University of California, Riverside   10 M, 2 D
University of California, Santa Barbara   1 M
California Institute of Technology   1 M, 3 D
California State University, Long Beach   4 M
California State University, Los
    Angeles   2 M
California State University, Northridge   3 M
Carleton University   2 M
Case Western Reserve University   1 M, 1 D
Catholic University of America   3 M, 1 D
University of Chicago   4 M, 4 D
University of Cincinnati   8 M
Clark University   1 D
Clarkson University   1 D
University of Colorado   21 M, 2 D
Colorado College   2 M
Colorado School of Mines   57 M, 21 D
Colorado State University   2 M
Columbia University   1 M
Columbia University, Teachers College   7 M, 8 D
Cornell University   1 M, 1 D
Dalhousie University   3 M, 1 D
Dartmouth College   2 M
East Carolina University   1 M
Eastern Kentucky University   6 M
École Polytechnique   2 M
Emory University   1 M
Emporia State University   1 M
University of Florida   4 M
Florida State University   3 M
Fort Hays State University   2 M
George Washington University   1 M, 1 D
University of Georgia   2 M
Harvard University   2 D
University of Hawaii   2 M
University of Houston   14 M
University of Idaho   7 M, 3 D
Idaho State University   3 M
University of Illinois, Chicago   6 M, 1 D
University of Illinois, Urbana   14 M, 8 D
Indiana University of Pennsylvania   2 M
Indiana State University   6 M, 1 D
Indiana University, Bloomington   30 M, 1 D
University of Iowa   5 M, 4 D
Iowa State University of Science and
    Technology   5 M, 2 D
The Johns Hopkins University   1 M, 5 D
University of Kansas   17 M, 1 D
Kansas State University   15 M
Kent State University, Kent   15 M
University of Kentucky   22 M, 1 D
Laurentian University, Sudbury   1 M
Lehigh University   1 M
Loma Linda University   1 M
Louisiana State University   13 M, 1 D
Louisiana Tech University   8 M, 1 D
University of Manitoba   4 M
University of Maryland   2 M, 1 D
University of Massachusetts   1 M
Massachusetts Institute of Technology   3 M, 7 D
McGill University   4 M, 3 D
McMaster University   3 M
Memorial University of Newfoundland   2 M
Memphis State University   1 M
Miami University (Ohio)   4 M
University of Michigan   19 M, 6 D
Michigan State University   17 M, 2 D
Michigan Technological University   4 M
University of Minnesota, Duluth   1 M
University of Minnesota, Minneapolis   9 M, 2 D
University of Mississippi   1 M
Mississippi State University   6 M
University of Missouri, Columbia   16 M, 1 D
University of Missouri, Kansas City   1 M
University of Missouri, Rolla   21 M, 4 D
University of Montana   4 M
Montana College of Mineral Science &
    Technology   7 M
Montana State University   4 M
Mount Allison University   1 M
Murray State University   1 M
University of Nebraska, Lincoln   7 M, 1 D

University of Nevada   4 M, 1 D
University of New Brunswick   1 M
University of New Mexico   8 M
New Mexico Institute of Mining and
   Technology   4 M, 1 D
New Mexico State University, Las Cruces   1 M
University of New Orleans   9 M
New York University   1 M
University of North Carolina, Chapel
   Hill   2 M
North Carolina State University   4 M, 1 D
University of North Dakota   8 M
Northeast Louisiana University   8 M
Northeastern Illinois University   2 M
Northern Arizona University   2 M
Northern Illinois University   1 M
Northwestern State University   1 M
Northwestern University   3 M
Ohio State University   5 M, 1 D
Ohio University, Athens   10 M
University of Oklahoma   106 M, 5 D
Oklahoma State University   23 M
University of Oregon   2 M
Oregon State University   5 M, 2 D
Pennsylvania State University, University
   Park   21 M, 20 D
University of Pittsburgh   33 M
Portland State University   4 M
Princeton University   4 D
Purdue University   8 M, 2 D
Queen's University   2 M
Queens College (CUNY)   1 M
University of Regina   5 M
Rensselaer Polytechnic Institute   1 M
University of Rhode Island   3 M, 3 D
Rice University   2 M, 2 D
University of Rochester   3 M
San Diego State University   4 M
San Jose State University   1 M
University of Saskatchewan   6 M
University of South Carolina   2 M, 11 D
University of South Dakota   1 M
South Dakota School of Mines &
   Technology   11 M
University of Southern California   9 M, 6 D
Southern Illinois University,
   Carbondale   17 M, 1 D
Southern Methodist University   4 M
University of Southwestern Louisiana   26 M
St. Louis University   1 M
Stanford University   55 M, 33 D
Stephen F. Austin State University   10 M
SUNY at Buffalo   6 M
Syracuse University   4 M, 1 D
University of Tennessee, Knoxville   2 M
University of Texas, Arlington   2 M
University of Texas, Austin   55 M, 14 D
Texas A&I University   3 M
Texas A&M University   29 M, 7 D
University of Texas at Dallas   6 M, 2 D
University of Texas at El Paso   10 M, 1 D
Texas Christian University   6 M
Texas Tech University   15 M, 2 D
University of Toledo   9 M
University of Toronto   3 M, 5 D
Tulane University   7 M
University of Tulsa   11 M
University of Utah   18 M, 4 D
Utah State University   3 M
Vanderbilt University   3 M
University of Virginia   2 M
Virginia Polytechnic Institute and State
   University   1 M
Virginia State University   1 M
University of Washington   9 M
Washington State University   5 M, 1 D
Washington University   5 M, 3 D
University of Waterloo   4 M
Wayne State University   3 M
West Texas State University   6 M
West Virginia University   14 M, 3 D
Western Michigan University   1 M
University of Western Ontario   2 M
Western Washington University   1 M
Wichita State University   25 M

University of Windsor   3 M
University of Wisconsin-Madison   10 M, 8 D
University of Wisconsin-Milwaukee   3 M
Wright State University   10 M
University of Wyoming   9 M, 5 D
Yale University   3 M, 3 D

Economic Geology, General, Mining Geology

600 Master's, 224 Doctoral

Theses per decade
1880s: 6 Master's
1890s: 2 Master's, 2 Doctoral
1900s: 2 Master's
1910s: 10 Master's, 5 Doctoral
1920s: 29 Master's, 18 Doctoral
1930s: 45 Master's, 24 Doctoral
1940s: 45 Master's, 21 Doctoral
1950s: 97 Master's, 30 Doctoral
1960s: 86 Master's, 56 Doctoral
1970s: 137 Master's, 41 Doctoral
1980s: 141 Master's, 27 Doctoral

Theses by Institution
University of Alabama   2 M
University of Alaska, Fairbanks   11 M
University of Alberta   4 M, 1 D
American University   1 M
University of Arizona   48 M, 12 D
Arizona State University   1 M
University of Arkansas, Fayetteville   7 M
Boston University   1 D
Bowling Green State University   3 M
Brigham Young University   7 M
University of British Columbia   21 M, 4 D
Brock University   1 M
Brown University   1 M
University of Calgary   3 M
University of California, Berkeley   9 M, 8 D
University of California, Davis   1 M
University of California, Los Angeles   5 M, 1 D
University of California, Riverside   1 D
University of California, Santa Barbara   1 M
California Institute of Technology   2 M, 4 D
Carleton University   2 M, 1 D
Catholic University of America   2 M
University of Chicago   7 D
University of Cincinnati   2 M
University of Colorado   9 M, 6 D
Colorado School of Mines   26 M, 9 D
Colorado State University   4 M, 2 D
Columbia University   2 M, 2 D
Columbia University, Teachers College   9 M, 14 D
Cornell University   5 M, 6 D
Dalhousie University   1 M, 1 D
University of Delaware, College of Marine
   Studies   1 M
Eastern Washington University   1 M
Ecole Polytechnique   9 M, 3 D
Florida State University   2 M
George Washington University   2 M
University of Georgia   3 M
Harvard University   11 D
University of Hawaii   1 M
University of Idaho   37 M, 2 D
Idaho State University   2 M
University of Illinois, Chicago   1 D
University of Illinois, Urbana   1 M, 2 D
Indiana University, Bloomington   4 M, 6 D
University of Iowa   2 M, 1 D
Iowa State University of Science and
   Technology   1 M
The Johns Hopkins University   2 D
University of Kansas   1 M
Kansas State University   1 D
Kent State University, Kent   1 M
University of Kentucky   3 M
University of Manitoba   6 M, 2 D
Massachusetts Institute of Technology   6 M, 4 D
McGill University   7 M, 9 D
McMaster University   1 M, 1 D
Memorial University of Newfoundland   1 M
Miami University (Ohio)   2 M
University of Michigan   6 M, 5 D
Michigan State University   1 M

Michigan Technological University   3 M
University of Minnesota, Duluth   2 M
University of Minnesota, Minneapolis   21 M, 3 D
University of Mississippi   1 M
University of Missouri, Columbia   2 M
University of Missouri, Rolla   11 M, 3 D
University of Montana   3 M
Montana College of Mineral Science &
   Technology   13 M
Montana State University   1 M
Universite de Montreal   2 M
University of Nebraska, Lincoln   1 M
University of Nevada   15 M
University of Nevada, Las Vegas   1 M
University of Nevada - Mackay School of
   Mines   6 M
University of New Brunswick   1 D
University of New Mexico   11 M
New Mexico Institute of Mining and
   Technology   6 M, 1 D
New Mexico State University, Las Cruces   1 M
University of New Orleans   2 M
New York University   1 M
University of North Carolina, Chapel
   Hill   4 M
North Carolina State University   2 M
Northern Arizona University   1 M
Northern Illinois University   1 M
Northwestern University   4 M, 1 D
Ohio State University   3 M, 1 D
Ohio University, Athens   1 M
University of Oklahoma   4 M, 1 D
Oklahoma State University   1 M
University of Oregon   1 M
Oregon State University   6 M
University of Ottawa   1 D
Pennsylvania State University, University
   Park   10 M, 11 D
University of Pittsburgh   1 M
Portland State University   1 M
Princeton University   15 D
Purdue University   1 M
Universite du Quebec a Chicoutimi   1 M
Queen's University   17 M, 5 D
Queens College (CUNY)   1 M
Rutgers, The State University, New
   Brunswick   1 D
San Jose State University   2 M
University of Saskatchewan   6 M
South Dakota School of Mines &
   Technology   2 M
University of Southern California   2 M
Southern Illinois University,
   Carbondale   1 M
Stanford University   16 M, 16 D
SUNY at Buffalo   3 M
Syracuse University   2 M
University of Tennessee, Knoxville   4 M
University of Texas, Austin   13 M, 2 D
University of Texas at Dallas   2 M
University of Texas at El Paso   9 M
Texas Christian University   1 M
Texas Tech University   1 M, 1 D
University of Toronto   9 M, 4 D
Tulane University   1 M
University of Tulsa   1 M
United States Naval Academy   1 M
University of Utah   14 M, 5 D
Utah State University   2 M
University of Virginia   3 M
Virginia Polytechnic Institute and State
   University   2 M, 1 D
Virginia State University   1 M
University of Washington   11 M, 3 D
Washington State University   5 M
Washington University   9 M
West Texas State University   1 M
West Virginia University   1 M, 1 D
University of Western Ontario   5 M
Western Washington University   1 M
University of Wisconsin-Madison   7 M, 11 D
University of Wisconsin-Milwaukee   2 M, 1 D
Wright State University   1 M
University of Wyoming   5 M
Yale University   1 M, 5 D

Economic Geology, Metals

2,397 Master's, 988 Doctoral

Theses per decade

1870s: 5 Master's, 1 Doctoral
1880s: 11 Master's, 2 Doctoral
1890s: 3 Master's, 5 Doctoral
1900s: 15 Master's, 12 Doctoral
1910s: 42 Master's, 33 Doctoral
1920s: 84 Master's, 46 Doctoral
1930s: 115 Master's, 77 Doctoral
1940s: 119 Master's, 57 Doctoral
1950s: 226 Master's, 122 Doctoral
1960s: 235 Master's, 135 Doctoral
1970s: 560 Master's, 229 Doctoral
1980s: 981 Master's, 269 Doctoral

Theses by Institution

Acadia University   4 M
University of Akron   1 M
University of Alabama   3 M
University of Alaska, Fairbanks   14 M
University of Alberta   42 M, 10 D
American University   2 M
University of Arizona   113 M, 34 D
Arizona State University   7 M
University of Arkansas, Fayetteville   14 M
Auburn University   4 M
Boston University   2 M, 4 D
Bowling Green State University   14 M
Brigham Young University   6 M, 1 D
University of British Columbia   62 M, 19 D
Brock University   3 M
Brooklyn College (CUNY)   3 M
Brown University   1 M
Bryn Mawr College   2 M
University of Calgary   11 M
University of California, Berkeley   13 M, 34 D
University of California, Davis   4 M
University of California, Los Angeles   14 M, 4 D
University of California, Riverside   5 M, 3 D
University of California, San Diego   2 M, 1 D
University of California, Santa Barbara   5 M, 1 D
California Institute of Technology   8 M, 15 D
California State University, Fresno   2 M
California State University, Hayward   3 M
California State University, Northridge   1 M
Carleton University   29 M, 11 D
Case Western Reserve University   1 M, 2 D
Catholic University of America   1 M
University of Chicago   1 M, 18 D
University of Cincinnati   10 M, 2 D
City College (CUNY)   1 M
University of Colorado   31 M, 5 D
Colorado School of Mines   71 M, 33 D
Colorado State University   25 M, 3 D
Columbia University   3 M, 13 D
Columbia University, Teachers College   76 M, 52 D
Cornell University   16 M, 16 D
Dalhousie University   10 M, 2 D
Dartmouth College   16 M, 4 D
University of Delaware   1 M, 1 D
Duke University   1 D
East Carolina University   1 M
Eastern Washington University   31 M
Ecole Polytechnique   44 M, 7 D
University of Florida   1 M
Florida State University   3 M, 3 D
George Washington University   3 M, 1 D
University of Georgia   6 M, 5 D
Harvard University   65 M
University of Idaho   44 M, 9 D
Idaho State University   5 M
University of Illinois, Chicago   1 M, 1 D
University of Illinois, Urbana   6 M, 6 D
Indiana State University   1 D
Indiana University, Bloomington   12 M, 7 D
University of Iowa   12 M, 4 D
Iowa State University of Science and Technology   4 M
The Johns Hopkins University   2 M, 11 D
University of Kansas   8 M, 2 D
Kansas State University   1 M
Kent State University, Kent   8 M, 1 D

University of Kentucky   5 M, 1 D
Lakehead University   1 M
Laurentian University, Sudbury   3 M
Universite Laval   8 M, 2 D
Lehigh University   14 M, 3 D
Louisiana State University   1 M
Louisiana Tech University   2 M
University of Manitoba   38 M, 9 D
University of Maryland   1 M, 2 D
University of Massachusetts   3 M, 1 D
Massachusetts Institute of Technology   25 M, 25 D
McGill University   54 M, 29 D
McMaster University   16 M, 5 D
Memorial University of Newfoundland   17 M, 5 D
Memphis State University   2 M
Miami University (Ohio)   5 M
University of Michigan   37 M, 22 D
Michigan State University   10 M
Michigan Technological University   30 M, 2 D
University of Minnesota, Duluth   15 M
University of Minnesota, Minneapolis   54 M, 25 D
Mississippi State University   1 M
University of Missouri, Columbia   10 M, 4 D
University of Missouri, Kansas City   1 M
University of Missouri, Rolla   44 M, 12 D
University of Montana   28 M, 1 D
Montana College of Mineral Science & Technology   25 M
Montana State University   4 M
Montclair State College   1 M
Universite de Montreal   4 M, 2 D
University of Nebraska, Lincoln   1 M, 1 D
University of Nevada   73 M, 4 D
University of Nevada, Las Vegas   1 M
University of Nevada - Mackay School of Mines   13 M
University of New Brunswick   25 M, 15 D
University of New Hampshire   1 M
University of New Mexico   9 M, 5 D
New Mexico Institute of Mining and Technology   37 M, 4 D
New Mexico State University, Las Cruces   1 M
University of New Orleans   1 M
New York University   1 D
University of North Carolina, Chapel Hill   11 M, 3 D
North Carolina State University   12 M, 1 D
University of North Dakota   3 M
Northeast Louisiana University   2 M
Northeastern Illinois University   1 M
Northern Arizona University   2 M
Northern Illinois University   4 M
Northwestern University   9 M, 6 D
Ohio State University   71 M, 2 D
Ohio University, Athens   3 M
University of Oklahoma   1 M, 2 D
Oklahoma State University   15 M
Old Dominion University   1 M
University of Oregon   15 M, 3 D
Oregon State University   15 M, 8 D
University of Ottawa   6 M, 2 D
University of Pennsylvania   3 D
Pennsylvania State University, University Park   41 M, 29 D
University of Pittsburgh   3 M, 1 D
Portland State University   5 M
Princeton University   25 D
Purdue University   4 M, 3 D
Universite du Quebec a Chicoutimi   12 M
Universite du Quebec a Montreal   3 M
Queen's University   106 M, 27 D
Queens College (CUNY)   5 M
University of Regina   3 M
Rensselaer Polytechnic Institute   1 M
University of Rhode Island   2 M
Rice University   2 M, 1 D
University of Rochester   6 M, 1 D
Rutgers, The State University, New Brunswick   4 M
Rutgers, The State University, Newark   1 M
University of San Diego   1 M
San Diego State University   9 M
University of Saskatchewan   10 M, 3 D
Smith College   2 M

South Dakota School of Mines & Technology   24 M, 2 D
University of South Florida, Tampa   1 M
University of Southern California   3 M, 1 D
Southern Illinois University, Carbondale   7 M
Stanford University   58 M, 81 D
Stephen F. Austin State University   2 M
Sul Ross State University   1 M
SUNY, College at Oneonta   1 M
SUNY at Albany   1 M
SUNY at Binghamton   1 M
SUNY at Stony Brook   1 M
Syracuse University   4 M
University of Tennessee, Knoxville   21 M, 2 D
University of Texas, Arlington   1 M
University of Texas, Austin   37 M, 3 D
Texas A&M University   8 M, 2 D
University of Texas at Dallas   4 M, 4 D
University of Texas at El Paso   11 M, 7 D
Texas Christian University   2 M
Texas Tech University   5 M
University of Toledo   2 M
University of Toronto   78 M, 44 D
Tulane University   1 M
University of Tulsa   1 M
University of Utah   34 M, 17 D
Utah State University   1 M
Vanderbilt University   1 M
University of Vermont   2 M
University of Virginia   7 M, 2 D
Virginia Polytechnic Institute and State University   15 M, 1 D
University of Washington   30 M, 4 D
Washington State University   29 M, 7 D
Washington University   28 M, 3 D
University of Waterloo   12 M
Wesleyan University   3 M
West Texas State University   1 M
Western Connecticut State University   1 M
Western Michigan University   2 M
University of Western Ontario   59 M, 27 D
Western Washington University   12 M
Wichita State University   3 M
University of Windsor   8 M
University of Wisconsin-Madison   41 M, 39 D
University of Wisconsin-Milwaukee   5 M, 1 D
Wright State University   4 M
University of Wyoming   24 M, 3 D
Yale University   5 M, 26 D

Economic Geology, Nonmetals

467 Master's, 147 Doctoral

Theses per decade

1870s: 1 Doctoral
1880s: 2 Doctoral
1890s: 4 Doctoral
1900s: 3 Master's
1910s: 5 Master's, 6 Doctoral
1920s: 24 Master's, 10 Doctoral
1930s: 52 Master's, 9 Doctoral
1940s: 43 Master's, 10 Doctoral
1950s: 81 Master's, 45 Doctoral
1960s: 61 Master's, 30 Doctoral
1970s: 89 Master's, 15 Doctoral
1980s: 109 Master's, 15 Doctoral

Theses by Institution

Acadia University   2 M
University of Alabama   3 M
University of Alaska, Fairbanks   1 M
University of Alberta   2 M
University of Arizona   10 M, 3 D
University of Arkansas, Fayetteville   2 M
Boston University   1 M
Bowling Green State University   1 M
Brigham Young University   4 M
University of British Columbia   1 M
Brown University   1 M
University of California, Berkeley   1 M, 2 D
University of California, Davis   1 D
University of California, Los Angeles   1 M, 1 D
University of California, Riverside   1 D
University of California, San Diego   1 D

California Institute of Technology   4 M, 4 D
Carleton University   1 M
Catholic University of America   1 M
University of Chicago   2 M, 4 D
University of Cincinnati   1 M, 1 D
University of Colorado   5 M, 2 D
Colorado School of Mines   6 M, 2 D
Colorado State University   3 M
Columbia University   2 D
Columbia University, Teachers College   12 M, 15 D
University of Connecticut   1 M
Cornell University   7 M, 7 D
Dalhousie University   4 M
Dartmouth College   3 M
Duke University   1 M
East Carolina University   2 M
Eastern Kentucky University   1 M
Eastern Washington University   4 M
Ecole Polytechnique   1 M
Emory University   1 M
University of Florida   3 M
Florida State University   3 M, 3 D
George Washington University   1 M, 1 D
University of Georgia   3 M, 2 D
Georgia Institute of Technology   3 M, 1 D
Georgia State University   1 M
Harvard University   6 D
University of Houston   3 M
University of Idaho   17 M, 1 D
Idaho State University   1 M
University of Illinois, Chicago   1 D
University of Illinois, Urbana   4 M, 6 D
Indiana University, Bloomington   9 M, 5 D
University of Iowa   4 M, 1 D
Iowa State University of Science and Technology   4 M, 1 D
The Johns Hopkins University   4 D
University of Kansas   7 M, 1 D
Kansas State University   2 M
Kent State University, Kent   6 M
University of Kentucky   4 M
Universite Laval   1 M
Lehigh University   4 M
University of Manitoba   1 M
University of Massachusetts   1 M
Massachusetts Institute of Technology   1 M, 2 D
McGill University   9 M, 5 D
Memorial University of Newfoundland   1 M
Miami University (Ohio)   2 M
University of Michigan   4 M, 8 D
Michigan State University   2 M, 2 D
Michigan Technological University   1 M
University of Minnesota, Minneapolis   7 M, 2 D
University of Mississippi   2 M
Mississippi State University   1 M
University of Missouri, Columbia   7 M, 2 D
University of Missouri, Rolla   12 M, 3 D
University of Montana   2 M, 1 D
Montana College of Mineral Science & Technology   4 M
Montana State University   1 M
Montclair State College   1 M
Murray State University   1 M
University of Nebraska, Lincoln   2 M
University of Nevada   4 M
University of Nevada - Mackay School of Mines   2 M
University of New Brunswick   2 M
University of New Mexico   2 M
New Mexico Institute of Mining and Technology   3 M
New York University   2 M
University of North Carolina, Chapel Hill   5 M, 2 D
North Carolina State University   7 M
Northeast Louisiana University   2 M
Northern Arizona University   2 M
Northern Illinois University   1 M
Northwestern University   3 M
Ohio State University   2 M, 1 D
Ohio University, Athens   1 M
University of Oklahoma   5 M, 2 D
Oklahoma State University   1 M
University of Oregon   2 M

Oregon State University   2 M, 1 D
University of Ottawa   1 M
Pennsylvania State University, University Park   7 M, 7 D
Princeton University   2 D
Purdue University   2 M
Universite du Quebec a Chicoutimi   1 M
Queen's University   3 M, 2 D
Queens College (CUNY)   2 M
Rensselaer Polytechnic Institute   1 D
University of Rhode Island   1 D
Rice University   1 M
Rutgers, The State University, New Brunswick   3 M
Rutgers, The State University, Newark   6 M
San Diego State University   1 M
San Jose State University   1 M
University of Saskatchewan   3 M
University of South Carolina   2 M
South Dakota School of Mines & Technology   5 M
University of Southern California   1 M, 1 D
Southern Illinois University, Carbondale   6 M
University of Southern Mississippi   1 M
Southwest Missouri State University   1 M
St. Louis University   1 M
Stanford University   7 M, 5 D
Stephen F. Austin State University   1 M
Sul Ross State University   1 M
SUNY at Binghamton   2 M
SUNY at Stony Brook   1 D
Syracuse University   4 M
University of Tennessee, Knoxville   8 M
University of Texas, Austin   6 M
Texas A&M University   4 M
University of Texas at El Paso   4 M
Texas Tech University   3 M
University of Toledo   3 M
University of Toronto   5 M, 7 D
Tulane University   2 M, 1 D
University of Tulsa   1 M
University of Utah   8 M
Utah State University   2 M
University of Virginia   3 M
Virginia Polytechnic Institute and State University   4 M
Virginia State University   4 M
University of Washington   37 M, 2 D
Washington State University   5 M
Washington University   1 M
University of Waterloo   1 M
West Texas State University   3 M
West Virginia University   1 M
University of Western Ontario   1 M
University of Windsor   2 M
University of Wisconsin-Madison   8 M, 5 D
University of Wisconsin-Milwaukee   1 M
University of Wyoming   8 M
Yale University   2 D

Engineering Geology

1,273 Master's, 1,613 Doctoral

Theses per decade
1880s: 2 Master's
1890s: 1 Master's
1900s: 4 Master's, 1 Doctoral
1910s: 5 Master's, 1 Doctoral
1920s: 23 Master's, 2 Doctoral
1930s: 19 Master's, 12 Doctoral
1940s: 34 Master's, 9 Doctoral
1950s: 101 Master's, 22 Doctoral
1960s: 169 Master's, 216 Doctoral
1970s: 411 Master's, 451 Doctoral
1980s: 503 Master's, 899 Doctoral

Theses by Institution
Acadia University   1 M
Adelphi University   1 M
University of Akron   4 D
University of Alabama   1 M, 4 D
University of Alaska, Fairbanks   8 M, 2 D
University of Alberta   5 M, 8 D
American University   1 D

University of Arizona   40 M, 43 D
Arizona State University   6 M, 6 D
University of Arkansas, Fayetteville   1 M, 1 D
Auburn University   1 M
Baylor University   4 M
Boise State University   1 M
Boston College   1 M
Boston University   3 M
Bowling Green State University   2 M
Brigham Young University   1 D
University of British Columbia   7 M, 15 D
Brooklyn College (CUNY)   3 M
Brown University   2 M, 3 D
University of Calgary   6 M, 2 D
University of California, Berkeley   9 M, 144 D
University of California, Davis   3 M, 20 D
University of California, Irvine   1 D
University of California, Los Angeles   5 M, 26 D
University of California, Riverside   2 M, 1 D
University of California, San Diego   5 D
University of California, Santa Barbara   1 M, 2 D
University of California, Santa Cruz   5 M, 2 D
California Institute of Technology   1 M, 23 D
California State University, Chico   1 M
California State University, Hayward   3 M
California State University, Long Beach   3 M
Carleton University   1 M, 4 D
Carnegie-Mellon University   11 D
Case Western Reserve University   4 D
Catholic University of America   1 M, 3 D
University of Chicago   2 M, 4 D
University of Cincinnati   13 M, 5 D
City College (CUNY)   1 M, 1 D
Clark University   1 D
Clarkson University   4 D
Clemson University   2 D
University of Colorado   9 M, 29 D
Colorado School of Mines   47 M, 13 D
Colorado State University   16 M, 41 D
Columbia University   4 D
Columbia University, Teachers College   11 M, 15 D
Concordia University   1 D
University of Connecticut   5 D
Cornell University   8 M, 23 D
Dartmouth College   1 M, 2 D
University of Delaware   1 M, 2 D
University of Delaware, College of Marine Studies   1 M, 1 D
Drexel University   4 D
Duke University   1 M, 8 D
Eastern Kentucky University   5 M
Ecole Polytechnique   9 M
Emory University   1 M
Emporia State University   1 M
University of Florida   6 M, 15 D
Florida Institute of Technology   1 M
George Washington University   2 M, 4 D
University of Georgia   2 M, 3 D
Georgia Institute of Technology   3 M, 7 D
Harvard University   5 D
University of Hawaii   6 M, 4 D
University of Houston   4 M, 6 D
University of Idaho   52 M, 10 D
Idaho State University   2 M
University of Illinois, Chicago   3 M, 4 D
University of Illinois, Urbana   19 M, 81 D
Illinois Institute of Technology   4 D
Indiana University of Pennsylvania   1 M
Indiana State University   2 M, 1 D
Indiana University, Bloomington   14 M, 2 D
University of Iowa   2 M, 11 D
Iowa State University of Science and Technology   16 M, 21 D
The Johns Hopkins University   4 D
University of Kansas   4 M, 6 D
Kansas State University   2 M
Kent State University, Kent   21 M, 1 D
University of Kentucky   4 M, 3 D
Universite Laval   6 M, 1 D
Lehigh University   3 M, 3 D
Louisiana State University   5 M, 10 D
Louisiana Tech University   1 D
University of Maine   1 M
University of Manitoba   1 M, 2 D

University of Maryland   9 D
University of Massachusetts   1 M, 6 D
Massachusetts Institute of Technology   14 M, 14 D
McGill University   5 M, 17 D
McMaster University   2 D
Memorial University of Newfoundland   1 D
Memphis State University   5 M
University of Miami   1 D
Miami University (Ohio)   4 M
University of Michigan   10 M, 23 D
Michigan State University   2 M, 22 D
Michigan Technological University   26 M, 1 D
University of Minnesota, Duluth   1 M, 5 D
University of Minnesota, Minneapolis   5 M, 29 D
University of Mississippi   1 M, 1 D
Mississippi State University   2 M
University of Missouri, Columbia   7 M, 3 D
University of Missouri, Kansas City   2 M
University of Missouri, Rolla   37 M, 31 D
University of Montana   1 M, 1 D
Montana College of Mineral Science & Technology   1 M
Montana State University   4 M, 3 D
Montclair State College   1 M
Universite de Montreal   1 M
Murray State University   1 M
University of Nebraska, Lincoln   2 D
University of Nevada   23 M
University of Nevada - Mackay School of Mines   5 M
University of New Brunswick   1 M, 2 D
University of New Hampshire   1 M, 3 D
University of New Mexico   3 M, 2 D
New Mexico Institute of Mining and Technology   14 M, 2 D
New Mexico State University, Las Cruces   2 D
University of New Orleans   2 M
New York University   2 M, 2 D
University of North Carolina, Chapel Hill   3 M, 2 D
North Carolina State University   4 M, 12 D
University of North Dakota   2 M
Northeast Louisiana University   2 M
Northern Arizona University   2 M
Northern Illinois University   3 M
Northwestern University   1 M, 40 D
Oberlin College   1 M
Ohio State University   18 M, 16 D
Ohio University, Athens   12 M
University of Oklahoma   23 M, 28 D
Oklahoma State University   15 D
Old Dominion University   3 M
University of Oregon   1 M, 1 D
Oregon State University   7 M, 6 D
University of Pennsylvania   1 D
Pennsylvania State University, University Park   14 M, 30 D
University of Pittsburgh   14 M, 13 D
Polytechnic University   4 D
Pomona College   1 M
Portland State University   5 M
Princeton University   7 D
Purdue University   38 M, 61 D
Universite du Quebec a Montreal   1 M
Queen's University   7 M, 7 D
Rensselaer Polytechnic Institute   7 M, 21 D
University of Rhode Island   6 M, 3 D
Rutgers, The State University, New Brunswick   7 M, 6 D
Rutgers, The State University, Newark   1 M
San Diego State University   3 M
San Fernando Valley State University   1 M
San Francisco State University   1 M
San Jose State University   14 M
University of Saskatchewan   5 M
University of South Carolina   12 M, 6 D
University of South Dakota   1 M
South Dakota School of Mines & Technology   15 M, 6 D
University of South Florida, St. Petersburg   1 M
University of South Florida, Tampa   4 M
Southeast Missouri State University   1 M
University of Southern California   22 M, 17 D

Southern Illinois University, Carbondale   6 M
Southern Methodist University   1 D
University of Southern Mississippi   2 M
Southwest Missouri State University   5 M
St. Louis University   7 M, 4 D
Stanford University   76 M, 105 D
Stevens Institute of Technology   1 D
SUNY, College at Oneonta   1 M
SUNY at Binghamton   5 M
SUNY at Buffalo   3 M, 11 D
Syracuse University   3 D
University of Tennessee, Knoxville   4 M, 4 D
University of Texas, Arlington   1 D
University of Texas, Austin   18 M, 35 D
Texas A&M University   42 M, 53 D
University of Texas at El Paso   1 M
Texas Christian University   3 M
Texas Tech University   4 M, 1 D
University of Toledo   7 M
University of Toronto   5 M, 14 D
Tufts University   1 D
Tulane University   3 M, 3 D
University of Tulsa   6 M, 7 D
United States Naval Academy   8 M
University of Utah   24 M, 20 D
Utah State University   2 M, 15 D
Vanderbilt University   1 M, 1 D
University of Vermont   2 M
University of Virginia   11 M, 7 D
Virginia Polytechnic Institute and State University   8 M, 25 D
University of Washington   14 M, 23 D
Washington State University   8 M, 3 D
Washington University   13 M, 2 D
University of Waterloo   18 M, 11 D
Wayne State University   2 M, 5 D
Wesleyan University   1 M
West Virginia University   7 M, 18 D
Western Connecticut State University   1 M
Western Michigan University   3 M
University of Western Ontario   2 M, 13 D
Western Washington University   4 M
Wichita State University   3 M
College of William and Mary   1 M
University of Windsor   8 M, 2 D
University of Wisconsin-Green Bay   1 M
University of Wisconsin-Madison   13 M, 26 D
University of Wisconsin-Milwaukee   30 M, 2 D
Worcester Polytechnic Institute   1 D
Wright State University   7 M
University of Wyoming   1 M, 2 D

Environmental Geology

698 Master's, 746 Doctoral

Theses per decade
190s: 1 Doctoral
1910s: 2 Master's
1920s: 4 Master's, 2 Doctoral
1930s: 3 Master's, 1 Doctoral
1940s: 4 Master's, 1 Doctoral
1950s: 17 Master's, 4 Doctoral
1960s: 20 Master's, 9 Doctoral
1970s: 327 Master's, 336 Doctoral
1980s: 320 Master's, 392 Doctoral
1990s: 1 Master's

Theses by Institution
Acadia University   1 M
University of Akron   10 M
University of Alabama   1 D
University of Alaska, Fairbanks   1 M, 1 D
University of Alberta   1 M, 1 D
University of Arizona   13 M, 19 D
Arizona State University   15 M, 4 D
University of Arkansas, Fayetteville   4 M, 3 D
Auburn University   1 D
Ball State University   1 M, 2 D
Baylor University   8 M
Boston University   9 M, 3 D
Bowling Green State University   5 M
Brigham Young University   1 M
University of British Columbia   5 M, 6 D
Brooklyn College (CUNY)   2 M

University of California, Berkeley   10 D
University of California, Davis   6 M, 12 D
University of California, Irvine   5 D
University of California, Los Angeles   8 M, 32 D
University of California, Riverside   2 M, 5 D
University of California, San Diego   4 D
University of California, Santa Cruz   1 M, 2 D
California Institute of Technology   4 D
California State University, Hayward   1 M
California State University, Long Beach   2 M
Carleton University   1 M
Carnegie-Mellon University   2 D
Case Western Reserve University   4 M
Catholic University of America   1 D
University of Cincinnati   2 M, 3 D
Clark University   2 D
Clarkson University   1 D
Clemson University   4 D
University of Colorado   4 M, 11 D
Colorado School of Mines   11 M, 2 D
Colorado State University   18 M, 24 D
Columbia University, Teachers College   1 M, 6 D
University of Connecticut   1 M, 4 D
Cornell University   8 D
Dalhousie University   7 M
Dartmouth College   1 M, 1 D
University of Delaware   5 M, 1 D
University of Delaware, College of Marine Studies   1 M, 5 D
University of Denver   2 D
DePauw University   1 M
Drexel University   2 D
Duke University   2 M, 3 D
East Texas State University   2 M
Eastern Kentucky University   1 M
Emory University   1 M, 2 D
Emporia State University   1 M
University of Florida   5 M, 15 D
Florida Institute of Technology   7 M, 1 D
Florida State University   1 M, 5 D
Franklin and Marshall College   1 M
George Washington University   4 M, 2 D
University of Georgia   3 M, 9 D
Georgia Institute of Technology   4 D
University of Guelph   1 M, 2 D
Harvard University   2 D
University of Hawaii   1 M, 3 D
University of Hawaii at Manoa   1 M
University of Houston   1 M
Howard University   1 D
University of Idaho   24 M, 6 D
University of Illinois, Urbana   2 M, 6 D
Illinois Institute of Technology   1 D
Indiana University of Pennsylvania   1 M
Indiana State University   6 M, 3 D
Indiana University, Bloomington   11 M, 6 D
University of Iowa   2 M, 4 D
Iowa State University of Science and Technology   5 M, 5 D
The Johns Hopkins University   7 D
University of Kansas   5 M, 5 D
Kansas State University   1 M
Kent State University, Kent   9 M
University of Kentucky   3 M, 1 D
Universite Laval   1 M
Lehigh University   1 M
Louisiana State University   4 M, 8 D
University of Louisville   5 D
University of Maine   4 M
University of Manitoba   1 D
University of Maryland   8 D
University of Massachusetts   1 M, 8 D
Massachusetts Institute of Technology   3 D
McGill University   3 D
McMaster University   6 M, 5 D
University of Miami   2 M
Miami University (Ohio)   2 M
University of Michigan   2 M, 24 D
Michigan State University   3 M, 16 D
Michigan Technological University   1 M
University of Minnesota, Duluth   1 D
University of Minnesota, Minneapolis   7 M, 14 D
Mississippi State University   4 M
University of Missouri, Columbia   5 M, 4 D
University of Missouri, Kansas City   1 M

University of Missouri, Rolla   6 M, 2 D
University of Montana   9 M, 2 D
Montana College of Mineral Science & Technology   1 M
Montana State University   7 M, 3 D
Montclair State College   1 M
Naval Postgraduate School   1 M
University of Nebraska, Lincoln   1 M, 1 D
University of Nevada   10 M
University of Nevada - Mackay School of Mines   1 M, 3 D
University of New Brunswick   1 D
University of New Hampshire   7 M, 3 D
University of New Mexico   5 M, 1 D
University of New Orleans   1 M, 1 D
New York University   1 M, 8 D
University of North Carolina, Chapel Hill   4 M, 4 D
North Carolina State University   3 M, 5 D
University of North Dakota   6 M, 2 D
Northeast Louisiana University   1 M
Northeastern Illinois University   5 M
Northern Arizona University   1 M, 2 D
University of Northern Colorado   3 D
Northern Illinois University   3 M
Northwestern University   1 M, 1 D
Ohio State University   8 M, 3 D
Ohio University, Athens   5 M, 1 D
University of Oklahoma   1 M, 13 D
Oklahoma State University   7 M, 8 D
Old Dominion University   2 M, 2 D
University of Oregon   3 D
Oregon Graduate Institute of Science and Technology   1 D
Oregon State University   2 M, 21 D
University of Pennsylvania   4 D
Pennsylvania State University, University Park   20 M, 10 D
University of Pittsburgh   1 M, 4 D
Polytechnic University   2 D
Portland State University   4 M, 1 D
Princeton University   1 M, 4 D
Purdue University   7 M, 12 D
Queen's University   6 M, 3 D
Queens College (CUNY)   2 M
Rensselaer Polytechnic Institute   1 M, 2 D
University of Rhode Island   3 M, 7 D
Rice University   3 M, 1 D
University of Rochester   1 M, 1 D
Rutgers, The State University, New Brunswick   2 M, 21 D
Rutgers, The State University, Newark   4 M
San Diego State University   7 M
San Francisco State University   2 M
San Jose State University   2 M
University of Saskatchewan   2 M, 1 D
University of South Carolina   3 M, 5 D
University of South Dakota   1 M
South Dakota School of Mines & Technology   3 M
University of South Florida, St. Petersburg   2 M
University of South Florida, Tampa   6 M
University of Southern California   1 M, 7 D
Southern Illinois University, Carbondale   6 M, 5 D
Southern Methodist University   1 D
University of Southern Mississippi   1 M, 1 D
Southwest Missouri State University   12 M
University of Southwestern Louisiana   1 M
Stanford University   20 M, 8 D
State University of New York, College of Environmental Science and Forestry   3 D
SUNY, College at Fredonia   4 M
SUNY, College at New Paltz   1 D
SUNY, College at Oneonta   1 M
SUNY at Albany   1 D
SUNY at Binghamton   9 M, 1 D
SUNY at Buffalo   4 D
SUNY at Stony Brook   2 M, 2 D
Syracuse University   2 M, 5 D
University of Tennessee, Knoxville   10 M, 10 D
University of Texas, Arlington   2 M
University of Texas, Austin   10 M, 10 D

Texas A&M University   14 M, 16 D
University of Texas at Dallas   5 D
Texas Tech University   1 M
University of Toledo   4 M
University of Toronto   7 D
Tulane University   1 D
University of Tulsa   1 M
University of Utah   3 D
Utah State University   6 D
Vanderbilt University   2 M, 1 D
University of Vermont   3 M
University of Victoria   2 D
University of Virginia   15 M, 4 D
Virginia Polytechnic Institute and State University   1 M, 10 D
Virginia State University   2 M
University of Washington   7 M, 17 D
Washington State University   6 M, 4 D
Washington University   3 M, 3 D
University of Waterloo   28 M, 9 D
Wayne State University   2 M
Wesleyan University   1 M
West Texas State University   2 M
West Virginia University   4 M
Western Carolina University   1 M
Western Connecticut State University   2 M
Western Michigan University   7 M, 1 D
University of Western Ontario   3 D
Western Oregon State College   1 M
Western Washington University   2 M
College of William and Mary   5 M, 7 D
University of Windsor   3 M, 1 D
University of Wisconsin-Madison   2 M, 24 D
University of Wisconsin-Milwaukee   9 M, 1 D
Wright State University   6 M
University of Wyoming   2 M, 3 D
Yale University   2 D
York University   1 D

Extraterrestrial Geology

111 Master's, 163 Doctoral

Theses per decade
  1950s: 1 Doctoral
  1960s: 14 Master's, 10 Doctoral
  1970s: 58 Master's, 91 Doctoral
  1980s: 39 Master's, 61 Doctoral

Theses by Institution
  University of Arizona   4 M, 11 D
  Arizona State University   7 M, 7 D
  University of Arkansas, Fayetteville   1 M
  Boston College   1 M
  Boston University   1 D
  University of British Columbia   2 M
  Brooklyn College (CUNY)   2 M
  Brown University   1 M, 12 D
  University of California, Berkeley   3 D
  University of California, Los Angeles   3 M, 16 D
  University of California, San Diego   3 D
  University of California, Santa Barbara   1 M, 1 D
  California Institute of Technology   26 D
  University of Chicago   1 D
  University of Colorado   1 M, 2 D
  Colorado State University   1 M
  Columbia University   1 M, 1 D
  University of Connecticut   1 M
  Cornell University   6 D
  Dartmouth College   1 M
  University of Delaware   2 M
  Emporia State University   1 M
  George Washington University   1 M, 1 D
  Harvard University   2 D
  University of Hawaii   2 M
  University of Hawaii at Manoa   1 D
  University of Houston   8 M
  University of Idaho   1 M
  Indiana University, Bloomington   1 M, 1 D
  University of Kentucky   1 M
  University of Maryland   3 D
  University of Massachusetts   4 M, 1 D
  Massachusetts Institute of Technology   12 M, 18 D
  University of Michigan   2 M, 2 D
  Michigan State University   2 M, 1 D
  University of Minnesota, Minneapolis   2 D

University of Missouri, Columbia   1 D
University of Missouri, Rolla   1 D
University of New Mexico   4 M, 3 D
University of North Carolina, Chapel Hill   1 M
Northern Arizona University   1 M
Northern Illinois University   2 M
Northwestern State University   1 M
Ohio State University   1 D
University of Oklahoma   1 M
Old Dominion University   1 M
University of Oregon   1 D
Pennsylvania State University, University Park   2 M, 3 D
University of Pittsburgh   2 M, 2 D
Princeton University   2 D
Purdue University   1 M
Rensselaer Polytechnic Institute   1 M
University of Rhode Island   2 M
Rice University   1 M, 5 D
San Jose State University   2 M
University of South Carolina   1 M, 2 D
South Dakota School of Mines & Technology   1 M, 1 D
Stanford University   2 M, 2 D
SUNY at Stony Brook   1 M, 3 D
University of Tennessee, Knoxville   6 M
University of Texas, Austin   1 M
Texas A&M University   2 M
University of Texas at Dallas   2 D
Texas Tech University   1 D
University of Toronto   1 D
University of Utah   2 M, 2 D
University of Victoria   1 D
University of Washington   1 M, 3 D
Washington University   7 M, 3 D
University of Wisconsin-Madison   1 D
University of Wyoming   1 M

Geochemistry

1,462 Master's, 1,443 Doctoral

Theses per decade
  1880s: 1 Doctoral
  1890s: 3 Doctoral
  1900s: 2 Doctoral
  1910s: 3 Master's, 3 Doctoral
  1920s: 4 Master's, 7 Doctoral
  1930s: 29 Master's, 14 Doctoral
  1940s: 28 Master's, 18 Doctoral
  1950s: 90 Master's, 78 Doctoral
  1960s: 194 Master's, 291 Doctoral
  1970s: 618 Master's, 505 Doctoral
  1980s: 496 Master's, 521 Doctoral

Theses by Institution
  Acadia University   1 M
  University of Akron   5 M
  University of Alabama   3 M
  University of Alaska, Fairbanks   5 M, 7 D
  University of Alberta   17 M, 11 D
  American University   2 M, 1 D
  University of Arizona   18 M, 16 D
  Arizona State University   22 M, 13 D
  University of Arkansas, Fayetteville   8 M, 11 D
  Baylor University   2 M, 1 D
  Boston University   2 M, 6 D
  Bowling Green State University   7 M
  Brigham Young University   1 M, 2 D
  University of British Columbia   26 M, 23 D
  Brock University   4 M
  Brooklyn College (CUNY)   4 M
  Brown University   3 M, 11 D
  Bryn Mawr College   3 M
  University of Calgary   2 D
  University of California, Berkeley   5 M, 31 D
  University of California, Davis   1 M, 12 D
  University of California, Irvine   2 D
  University of California, Los Angeles   10 M, 35 D
  University of California, Riverside   10 M, 12 D
  University of California, San Diego   55 D
  University of California, Santa Barbara   7 M, 8 D
  University of California, Santa Cruz   2 M, 6 D
  California Institute of Technology   33 D
  California State University, Chico   1 M

California State University, Fresno   1 M
California State University, Hayward   1 M
California State University, Long Beach   3 M
California State University, Los
   Angeles   1 M
California State University, Northridge   3 M
Carleton University   5 M, 3 D
Case Western Reserve University   1 M, 8 D
University of Chicago   1 M, 23 D
University of Cincinnati   5 M, 1 D
City College (CUNY)   1 D
Colgate University   1 M
University of Colorado   7 M, 15 D
Colorado School of Mines   22 M, 12 D
Colorado State University   3 M, 3 D
Columbia University   7 M, 12 D
Columbia University, Teachers College   10 M,
   31 D
University of Connecticut   2 M, 4 D
Cornell University   4 M, 9 D
Dalhousie University   9 M, 7 D
Dartmouth College   20 M, 2 D
University of Delaware   2 M, 1 D
University of Delaware, College of Marine
   Studies   2 M, 2 D
Duke University   3 M, 3 D
East Carolina University   3 M
Eastern Kentucky University   1 M
Eastern Washington University   1 M
Ecole Polytechnique   5 M
Emory University   1 D
University of Florida   13 M, 6 D
Florida Institute of Technology   2 M, 1 D
Florida State University   16 M, 11 D
Fordham University   2 D
Franklin and Marshall College   2 M
George Washington University   2 M, 4 D
University of Georgia   10 M, 11 D
Georgia Institute of Technology   11 M, 3 D
University of Guelph   1 M, 1 D
Harvard University   20 D
University of Hawaii   26 M, 20 D
University of Hawaii at Manoa   2 M
University of Houston   13 M, 4 D
University of Idaho   10 M, 7 D
Idaho State University   1 M
University of Illinois, Chicago   5 M
University of Illinois, Urbana   12 M, 32 D
Illinois Institute of Technology   1 D
Indiana University, Bloomington   14 M, 14 D
University of Iowa   3 M
Iowa State University of Science and
   Technology   14 M, 7 D
The Johns Hopkins University   1 M, 13 D
University of Kansas   3 M, 4 D
Kansas State University   5 M
Kent State University, Kent   7 M
University of Kentucky   24 M, 10 D
Lakehead University   4 M
Laurentian University, Sudbury   2 M
Universite Laval   3 M
Lehigh University   6 M, 5 D
Louisiana State University   11 M, 3 D
Loyola University   1 D
University of Maine   2 M, 1 D
University of Manitoba   10 M, 1 D
University of Maryland   2 M, 13 D
University of Massachusetts   5 M, 5 D
Massachusetts Institute of Technology   20 M, 68 D
McGill University   7 M, 12 D
McMaster University   29 M, 20 D
Memorial University of Newfoundland   3 M, 1 D
Memphis State University   1 M
University of Miami   10 M, 9 D
Miami University (Ohio)   20 M, 3 D
University of Michigan   14 M, 18 D
Michigan State University   24 M, 6 D
Michigan Technological University   10 M, 3 D
University of Minnesota, Minneapolis   22 M, 21 D
University of Mississippi   1 M
University of Missouri, Columbia   13 M, 3 D
University of Missouri, Kansas City   2 M
University of Missouri, Rolla   14 M, 6 D
University of Montana   3 M, 2 D

Montana College of Mineral Science &
   Technology   5 M
Universite de Montreal   10 M, 1 D
University of Nebraska, Lincoln   3 M, 1 D
University of Nevada   9 M, 1 D
University of Nevada - Mackay School of
   Mines   4 M
University of New Brunswick   8 M, 1 D
University of New Hampshire   13 M, 3 D
University of New Mexico   7 M, 6 D
New Mexico Institute of Mining and
   Technology   23 M, 4 D
New Mexico State University, Las Cruces   2 D
University of New Orleans   7 M
New York University   2 M, 4 D
University of North Carolina, Chapel
   Hill   16 M, 15 D
North Carolina State University   5 M, 5 D
University of North Dakota   4 M
University of North Texas   1 D
Northeastern Illinois University   1 M
Northern Illinois University   11 M
Northwestern University   1 M, 13 D
University of Notre Dame   2 D
Ohio State University   15 M, 23 D
Ohio University, Athens   2 M
University of Oklahoma   13 M, 9 D
Oklahoma State University   3 M
Old Dominion University   3 M, 2 D
University of Oregon   6 M, 10 D
Oregon State University   21 M, 16 D
University of Ottawa   7 M, 3 D
University of Pennsylvania   6 D
Pennsylvania State University, University
   Park   58 M, 81 D
University of Pittsburgh   8 M, 4 D
Pomona College   1 M
Portland State University   4 M
Princeton University   9 D
Purdue University   6 M, 8 D
Universite du Quebec a Chicoutimi   9 M
Universite du Quebec a Montreal   1 M
Queen's University   12 M, 8 D
Queens College (CUNY)   4 M
University of Regina   2 M
Rensselaer Polytechnic Institute   4 M, 8 D
University of Rhode Island   1 M, 35 D
Rice University   19 M, 18 D
University of Rochester   14 M
Rutgers, The State University, New
   Brunswick   3 M, 3 D
Rutgers, The State University, Newark   8 M
San Diego State University   3 M
San Jose State University   4 M
University of Saskatchewan   4 M, 2 D
University of South Carolina   8 M, 6 D
University of South Dakota   1 M
South Dakota School of Mines &
   Technology   5 M, 6 D
University of South Florida, St.
   Petersburg   7 M, 3 D
University of South Florida, Tampa   6 M
University of Southern California   5 M, 13 D
Southern Illinois University,
   Carbondale   10 M
Southern Methodist University   3 M, 3 D
University of Southern Mississippi   1 M
University of Southwestern Louisiana   1 M
Stanford University   11 M, 34 D
State University of New York, College of Environ-
   mental Science
   and Forestry   1 D
Stephen F. Austin State University   1 M
SUNY at Albany   1 M, 1 D
SUNY at Binghamton   5 M, 1 D
SUNY at Buffalo   11 M, 3 D
SUNY at Stony Brook   3 M, 11 D
Syracuse University   3 M, 2 D
University of Tennessee, Knoxville   23 M, 3 D
University of Texas, Arlington   6 M
University of Texas, Austin   10 M, 18 D
Texas A&M University   17 M, 32 D
University of Texas at Dallas   4 M, 2 D
University of Texas at El Paso   5 M
Texas Christian University   3 M

University of Texas of the Permian
   Basin   1 M
Texas Tech University   5 M
University of Toledo   6 M
University of Toronto   32 M, 17 D
University of Tulsa   9 M, 4 D
United States Naval Academy   1 M
University of Utah   24 M, 9 D
Utah State University   1 M, 6 D
Vanderbilt University   3 M
University of Virginia   8 M, 3 D
Virginia Polytechnic Institute and State
   University   11 M, 4 D
University of Washington   13 M, 36 D
Washington & Lee University   1 D
Washington State University   4 M, 6 D
Washington University   5 M, 10 D
University of Waterloo   12 M, 8 D
Wayne State University   9 M
West Virginia University   8 M, 4 D
Western Michigan University   2 M
University of Western Ontario   11 M, 10 D
Western Washington University   4 M
Wichita State University   1 M
College of William and Mary   5 M, 2 D
University of Windsor   6 M
University of Wisconsin-Madison   11 M, 38 D
University of Wisconsin-Milwaukee   8 M
Woods Hole Oceanographic Institution   6 D
Wright State University   12 M
University of Wyoming   6 M, 5 D
Yale University   23 D

Geochronology
282 Master's, 194 Doctoral

Theses per decade
   1920s: 1 Master's
   1930s: 2 Doctoral
   1940s: 1 Doctoral
   1950s: 23 Master's, 21 Doctoral
   1960s: 49 Master's, 48 Doctoral
   1970s: 92 Master's, 63 Doctoral
   1980s: 117 Master's, 59 Doctoral

Theses by Institution
   University of Alaska, Fairbanks   2 M
   University of Alberta   9 M, 3 D
   University of Arizona   8 M, 4 D
   University of Arkansas, Fayetteville   1 D
   Boston University   1 M
   Brigham Young University   1 M
   University of British Columbia   5 M, 3 D
   Brooklyn College (CUNY)   3 M
   Brown University   8 M
   University of Calgary   1 M
   University of California, Berkeley   2 M, 8 D
   University of California, Los Angeles   2 M, 6 D
   University of California, Riverside   1 M
   University of California, San Diego   1 M, 4 D
   University of California, Santa Barbara   9 D
   University of California, Santa Cruz   1 D
   California Institute of Technology   7 D
   California State University, Hayward   1 M
   California State University, Los
      Angeles   1 M
   Carleton University   3 M, 1 D
   Case Western Reserve University   3 M, 3 D
   University of Chicago   1 M
   University of Cincinnati   1 M
   University of Colorado   4 M
   Colorado School of Mines   2 M
   Columbia University   1 M, 4 D
   Columbia University, Teachers College   8 M, 10 D
   Dalhousie University   5 M, 2 D
   Dartmouth College   7 M, 2 D
   University of Delaware   1 M
   East Carolina University   1 M
   Florida State University   7 M, 3 D
   University of Georgia   1 M
   Georgia Institute of Technology   5 M
   University of Hawaii   3 M, 5 D
   University of Houston   1 M
   University of Idaho   1 D
   Indiana State University   1 M

University of Iowa   1 M
University of Kansas   15 M, 2 D
Kansas State University   1 M, 1 D
Louisiana State University   2 D
University of Maine   2 M
University of Manitoba   6 M
Massachusetts Institute of Technology   7 M, 17 D
McGill University   4 M, 2 D
McMaster University   2 M, 5 D
Memorial University of Newfoundland   1 M
Memphis State University   1 M
University of Miami   1 M, 1 D
Miami University (Ohio)   4 M, 3 D
Michigan State University   2 M
University of Minnesota, Minneapolis   2 M, 5 D
University of Missouri, Columbia   1 D
University of Montana   2 M
Montana State University   2 M
Universite de Montreal   2 M
University of Nevada   1 M
University of New Hampshire   2 M
University of New Mexico   2 M, 1 D
New Mexico Institute of Mining and
   Technology   1 M, 2 D
University of North Carolina, Chapel
   Hill   7 M, 3 D
North Carolina State University   1 M
Northern Illinois University   4 M
Ohio State University   5 M, 3 D
University of Oklahoma   1 D
University of Oregon   2 M, 1 D
Oregon State University   2 M
University of Pennsylvania   1 M, 3 D
Pennsylvania State University, University
   Park   3 M, 3 D
University of Pittsburgh   2 M, 1 D
Portland State University   3 M
Princeton University   2 D
Purdue University   1 D
Universite du Quebec a Montreal   1 M
Queen's University   7 M, 4 D
Queens College (CUNY)   2 M
Rensselaer Polytechnic Institute   5 M, 1 D
University of Rhode Island   1 D
Rice University   6 M, 4 D
San Diego State University   4 M
San Jose State University   3 M
University of Saskatchewan   1 D
University of Southern California   1 M
Southern Methodist University   2 M, 3 D
Stanford University   1 M, 1 D
SUNY at Albany   1 M
SUNY at Buffalo   1 M, 1 D
SUNY at Stony Brook   4 M
University of Tennessee, Knoxville   1 M
University of Texas, Austin   9 M, 4 D
Texas A&M University   2 D
University of Texas at Dallas   5 M, 3 D
University of Texas at El Paso   1 D
Texas Tech University   1 M
University of Toronto   10 M, 18 D
University of Tulsa   1 M
University of Utah   3 M, 1 D
Virginia Polytechnic Institute and State
   University   4 M
University of Washington   1 D
Washington State University   1 M
Washington University   2 M, 4 D
Wesleyan University   1 M
Western Michigan University   4 M
University of Western Ontario   1 M, 1 D
Western Washington University   1 M
University of Windsor   3 M
University of Wisconsin-Madison   2 M, 3 D
University of Wisconsin-Milwaukee   4 M
Yale University   2 D

Geophysics, Applied

1,677 Master's, 746 Doctoral

Theses per decade
1900s: 1 Master's
1910s: 2 Master's
1920s: 4 Master's
1930s: 28 Master's, 5 Doctoral
1940s: 28 Master's, 19 Doctoral
1950s: 181 Master's, 67 Doctoral
1960s: 393 Master's, 130 Doctoral
1970s: 569 Master's, 263 Doctoral
1980s: 468 Master's, 262 Doctoral

Theses by Institution
University of Akron   5 M
University of Alabama   1 M
University of Alaska, Fairbanks   7 M, 1 D
University of Alberta   16 M, 15 D
American University   2 M
University of Arizona   28 M, 8 D
Arizona State University   1 D
University of Arkansas, Fayetteville   10 M
Baylor University   3 M
Boston College   15 M
Boston University   2 M, 1 D
Bowling Green State University   12 M
Brigham Young University   1 M
University of British Columbia   32 M, 17 D
Brooklyn College (CUNY)   1 M
Brown University   1 M, 2 D
University of Calgary   14 M, 1 D
University of California, Berkeley   10 M, 32 D
University of California, Davis   1 M
University of California, Los Angeles   4 M, 6 D
University of California, Riverside   16 M, 3 D
University of California, San Diego   2 M, 17 D
University of California, Santa Barbara   1 M, 1 D
California Institute of Technology   10 M, 13 D
California State University, Hayward   1 M
California State University, Long Beach   9 M
Carleton University   1 M
Catholic University of America   1 D
University of Chicago   2 D
University of Cincinnati   3 M
University of Colorado   6 M, 7 D
Colorado School of Mines   122 M, 61 D
Colorado State University   9 M
Columbia University   3 M, 13 D
Columbia University, Teachers College   15 M,
   22 D
University of Connecticut   3 D
Cornell University   5 M, 7 D
Dalhousie University   12 M, 1 D
Dartmouth College   10 M, 2 D
University of Delaware   3 M
University of Denver   1 D
East Texas State University   1 M
Eastern Kentucky University   1 M
Eastern Washington University   2 M
Ecole Polytechnique   16 M
Emory University   1 M
University of Florida   8 M, 1 D
Florida Institute of Technology   1 M
Florida State University   5 M
George Washington University   2 M, 4 D
Georgia Institute of Technology   10 M, 1 D
Harvard University   15 D
University of Hawaii   16 M, 9 D
University of Hawaii at Manoa   2 M
University of Houston   35 M, 16 D
University of Idaho   1 M
Idaho State University   4 M
University of Illinois, Urbana   5 M, 2 D
Indiana State University   2 M
Indiana University, Bloomington   25 M, 9 D
University of Iowa   14 M, 1 D
Iowa State University of Science and
   Technology   5 M, 1 D
The Johns Hopkins University   3 D
University of Kansas   14 M, 4 D
Kansas State University   5 M
Kent State University, Kent   4 M
University of Kentucky   2 M, 1 D
Lehigh University   5 M, 2 D
Louisiana State University   3 M, 2 D
University of Manitoba   15 M, 2 D
Marshall University   1 M
University of Maryland   2 D
University of Massachusetts   3 M
Massachusetts Institute of Technology   31 M, 29 D
McGill University   9 M, 10 D
Memorial University of Newfoundland   6 M, 1 D

Memphis State University   3 M
University of Miami   1 M, 2 D
Miami University (Ohio)   4 M
University of Michigan   17 M, 6 D
Michigan State University   18 M, 11 D
Michigan Technological University   22 M
University of Minnesota, Minneapolis   21 M, 4 D
University of Missouri, Columbia   2 M, 2 D
University of Missouri, Rolla   14 M, 7 D
University of Montana   1 M
Montana College of Mineral Science &
   Technology   5 M
Montana State University   1 M
Montclair State College   1 M
Universite de Montreal   2 M
Murray State University   1 M
Naval Postgraduate School   2 M
University of Nevada   2 M
University of New Brunswick   3 M, 7 D
University of New Hampshire   1 M
University of New Mexico   5 M, 1 D
New Mexico Institute of Mining and
   Technology   17 M, 4 D
New Mexico State University, Las Cruces   6 M
University of New Orleans   4 M
New York University   6 M
University of North Carolina, Chapel
   Hill   9 M, 3 D
North Carolina State University   4 M, 1 D
University of North Dakota   2 M, 1 D
Northeastern University   1 D
Northern Arizona University   9 M
Northern Illinois University   8 M
Northwestern University   1 M
Ohio State University   94 M, 34 D
Ohio University, Athens   6 M
University of Oklahoma   19 M, 3 D
Oklahoma State University   1 D
Old Dominion University   1 M, 1 D
University of Oregon   3 M, 1 D
Oregon State University   12 M, 9 D
University of Ottawa   2 D
Pennsylvania State University, University
   Park   47 M, 7 D
University of Pittsburgh   4 M, 1 D
Portland State University   1 M
Princeton University   4 D
University of Puget Sound   3 M
Purdue University   31 M, 3 D
Queen's University   9 M, 1 D
Rensselaer Polytechnic Institute   2 M
University of Rhode Island   6 M, 5 D
Rice University   8 M, 2 D
University of Rochester   2 M, 2 D
San Diego State University   6 M
San Jose State University   1 M
University of Saskatchewan   7 M, 4 D
Slippery Rock University   1 M
University of South Carolina   3 M, 1 D
South Dakota School of Mines &
   Technology   4 M
University of South Florida, Tampa   2 M
University of Southern California   6 M, 3 D
Southern Illinois University,
   Carbondale   5 M
Southern Methodist University   10 M, 5 D
University of Southwestern Louisiana   1 M
St. Louis University   30 M, 8 D
Stanford University   36 M, 44 D
Sul Ross State University   1 M
SUNY, College at Fredonia   2 M
SUNY at Binghamton   1 M
SUNY at Buffalo   9 M, 2 D
SUNY at Stony Brook   1 M, 1 D
Syracuse University   2 M
Technical University of Nova Scotia   1 M
University of Tennessee, Knoxville   5 M, 2 D
University of Texas, Austin   27 M, 18 D
Texas A&M University   36 M, 14 D
University of Texas at Dallas   9 M, 8 D
University of Texas at El Paso   16 M, 3 D
Texas Christian University   2 M
University of Texas of the Permian
   Basin   3 M
Texas Tech University   5 M

Geophysics, General

University of Toledo   1 M
University of Toronto   37 M, 47 D
Tulane University   1 M, 1 D
University of Tulsa   8 M, 6 D
United States Naval Academy   9 M
University of Utah   86 M, 30 D
Vanderbilt University   2 M
University of Vermont   1 M
University of Victoria   5 D
University of Virginia   1 M
Virginia Polytechnic Institute and State
   University   14 M, 6 D
Virginia State University   1 M
University of Washington   15 M, 13 D
Washington State University   4 M, 1 D
Washington University   27 M, 2 D
University of Waterloo   4 M
Wesleyan University   2 M
West Texas State University   1 M
West Virginia University   3 M
Western Michigan University   2 M
University of Western Ontario   27 M, 7 D
Western Washington University   1 M
Wichita State University   3 M
University of Windsor   1 M
University of Wisconsin-Madison   30 M, 28 D
University of Wisconsin-Milwaukee   14 M
Woods Hole Oceanographic Institution   1 D
Wright State University   43 M
University of Wyoming   14 M, 4 D
Yale University   1 M, 1 D
York University   2 D

Geophysics, General

119 Master's, 241 Doctoral

Theses per decade
1920s: 1 Master's
1930s: 3 Master's
1940s: 1 Doctoral
1950s: 8 Master's, 7 Doctoral
1960s: 23 Master's, 37 Doctoral
1970s: 53 Master's, 118 Doctoral
1980s: 31 Master's, 78 Doctoral

Theses by Institution
University of Alberta   4 M, 3 D
American University   1 M
University of Arizona   1 D
Arizona State University   2 D
Ball State University   1 M
University of British Columbia   2 M, 8 D
Brown University   2 M, 1 D
University of Calgary   1 M
University of California, Berkeley   18 D
University of California, Davis   1 D
University of California, Los Angeles   2 M, 14 D
University of California, San Diego   8 D
University of California, Santa Barbara   1 M, 2 D
University of California, Santa Cruz   1 D
California Institute of Technology   10 D
Case Western Reserve University   1 M
University of Chicago   3 D
University of Colorado   1 M, 4 D
Colorado School of Mines   4 M, 1 D
Columbia University   7 D
Columbia University, Teachers College   7 D
Cornell University   1 M, 6 D
Dalhousie University   2 M, 2 D
Dartmouth College   1 M
Ecole Polytechnique   1 D
Florida Institute of Technology   1 M
Florida State University   3 M, 1 D
George Washington University   1 M
University of Georgia   2 M
Harvard University   3 D
University of Hawaii   2 M, 1 D
University of Illinois, Chicago   1 M
University of Illinois, Urbana   5 D
Indiana University, Bloomington   1 M
University of Iowa   1 M
The Johns Hopkins University   2 D
Lehigh University   1 M, 2 D
University of Manitoba   1 M
Massachusetts Institute of Technology   6 M, 9 D

McGill University   1 M, 1 D
Memorial University of Newfoundland   3 M, 1 D
University of Miami   1 M
Miami University (Ohio)   1 M
University of Michigan   4 M
Michigan State University   1 M
Michigan Technological University   2 M
University of Minnesota, Duluth   1 M
University of Minnesota, Minneapolis   2 M, 6 D
New Mexico Institute of Mining and
   Technology   1 M
New York University   1 M
Northwestern University   2 D
Ohio State University   7 D
University of Oklahoma   8 M, 1 D
University of Oregon   2 M, 1 D
Oregon State University   2 D
Pennsylvania State University, University
   Park   16 M, 7 D
University of Pittsburgh   1 M, 8 D
Princeton University   4 D
Purdue University   1 M, 2 D
Queens College (CUNY)   1 M
Rensselaer Polytechnic Institute   1 D
University of Rhode Island   2 D
University of Rochester   6 M, 4 D
University of South Carolina   1 D
University of Southern California   1 D
Southern Methodist University   1 M, 1 D
St. Louis University   1 D
Stanford University   8 D
SUNY at Stony Brook   7 D
University of Texas, Austin   1 M, 2 D
Texas A&M University   2 M, 6 D
University of Texas at Dallas   2 D
University of Toronto   3 M, 6 D
Tulane University   1 D
University of Tulsa   1 M
University of Utah   2 M, 2 D
University of Victoria   1 D
Virginia Polytechnic Institute and State
   University   2 M
University of Washington   4 M, 6 D
Washington University   1 M, 2 D
West Virginia University   1 D
University of Western Ontario   3 M, 8 D
University of Windsor   1 D
University of Wisconsin-Madison   1 M, 3 D
University of Wisconsin-Milwaukee   2 M
University of Wyoming   2 M, 2 D
Yale University   2 D
Yeshiva University   1 D
York University   2 D

Geophysics, Seismology

617 Master's, 600 Doctoral

Theses per decade
1920s: 2 Master's, 2 Doctoral
1930s: 12 Master's, 13 Doctoral
1940s: 13 Master's, 10 Doctoral
1950s: 42 Master's, 58 Doctoral
1960s: 101 Master's, 104 Doctoral
1970s: 215 Master's, 214 Doctoral
1980s: 232 Master's, 199 Doctoral

Theses by Institution
University of Akron   1 M
University of Alaska, Fairbanks   8 M
University of Alberta   5 M, 7 D
American University   1 M
University of Arizona   1 M
Arizona State University   2 M
University of Arkansas, Fayetteville   4 M
Boston College   12 M
Boston University   3 M, 1 D
Bowling Green State University   3 M
University of British Columbia   12 M, 6 D
Brown University   4 M, 1 D
University of Calgary   1 M, 1 D
University of California, Berkeley   3 M, 54 D
University of California, Davis   2 M, 1 D
University of California, Los Angeles   3 M, 12 D
University of California, Riverside   6 M
University of California, San Diego   22 D

University of California, Santa Barbara   4 M, 1 D
University of California, Santa Cruz   4 M, 4 D
California Institute of Technology   6 M, 72 D
California State University, Long Beach   1 M
Case Western Reserve University   1 M, 1 D
Central Connecticut State University   1 M
University of Chicago   1 M, 2 D
University of Colorado   6 M, 9 D
Colorado School of Mines   13 M, 15 D
Colorado State University   1 D
Columbia University   6 M, 16 D
Columbia University, Teachers College   3 M, 26 D
University of Connecticut   4 D
Cornell University   4 M, 9 D
Dalhousie University   2 M
Dartmouth College   2 M, 1 D
Emporia State University   1 M
Georgia Institute of Technology   24 M, 4 D
Harvard University   1 M, 9 D
University of Hawaii   7 M, 5 D
University of Hawaii at Manoa   1 M
University of Houston   2 M, 1 D
University of Idaho   1 M
University of Illinois, Urbana   1 M, 1 D
Indiana University, Bloomington   7 M, 4 D
The Johns Hopkins University   2 D
University of Kansas   6 M
Kansas State University   1 M
University of Kentucky   3 M
Universite Laval   2 M
University of Manitoba   3 M
Massachusetts Institute of Technology   29 M, 47 D
McGill University   2 M
Memphis State University   2 M
University of Michigan   14 M, 9 D
Michigan State University   10 M, 1 D
Michigan Technological University   1 M
University of Minnesota, Duluth   1 D
University of Minnesota, Minneapolis   3 M, 4 D
University of Missouri, Rolla   6 M, 5 D
Montana College of Mineral Science &
   Technology   1 M
Montana State University   1 M
University of Nebraska, Lincoln   1 D
University of Nevada   9 M, 2 D
University of Nevada - Mackay School of
   Mines   1 M
University of New Brunswick   1 D
New Mexico Institute of Mining and
   Technology   17 M, 2 D
New Mexico State University, Las Cruces   1 D
University of New Orleans   1 M
New York University   1 M
University of North Carolina, Chapel
   Hill   4 M, 1 D
North Carolina State University   1 M
Northern Arizona University   1 M
Northern Illinois University   1 M
Northwestern University   2 D
University of Oklahoma   4 M
University of Oregon   1 D
Oregon State University   8 M, 7 D
Pennsylvania State University, University
   Park   39 M, 21 D
University of Pittsburgh   4 M, 3 D
Polytechnic University   1 D
Portland State University   1 M
Princeton University   10 D
Purdue University   6 M, 3 D
Queen's University   1 M
Rensselaer Polytechnic Institute   2 M
University of Rhode Island   1 M
Rice University   4 M, 2 D
Rutgers, The State University, New
   Brunswick   1 M
Rutgers, The State University, Newark   1 M
San Diego State University   1 M
San Jose State University   1 M
University of Saskatchewan   1 M
Slippery Rock University   1 M
University of South Carolina   5 M, 2 D
University of Southern California   3 M, 5 D
Southern Illinois University,
   Carbondale   2 M
Southern Methodist University   8 M, 10 D

St. Louis University   59 M, 54 D
Stanford University   13 M, 38 D
SUNY at Binghamton   7 M, 4 D
SUNY at Stony Brook   1 M
University of Texas, Austin   10 M, 6 D
Texas A&M University   7 M, 6 D
University of Texas at Dallas   7 M, 6 D
University of Texas at El Paso   1 M
Texas Tech University   12 M, 1 D
University of Toronto   8 M, 7 D
University of Tulsa   4 M, 3 D
United States Naval Academy   3 M
University of Utah   36 M, 5 D
Vanderbilt University   1 M
Virginia Polytechnic Institute and State
   University   13 M, 3 D
University of Washington   22 M, 14 D
Washington State University   1 M
Washington University   1 D
University of Waterloo   1 M
University of Western Ontario   3 M, 6 D
University of Wisconsin-Madison   8 M, 10 D
University of Wisconsin-Milwaukee   13 M
University of Wyoming   3 M, 1 D
York University   1 D

Geophysics, Solid-Earth

501 Master's, 554 Doctoral

Theses per decade
   1870s: 1 Doctoral
   1910s: 1 Master's
   1920s: 1 Master's
   1930s: 1 Master's, 2 Doctoral
   1940s: 3 Master's, 5 Doctoral
   1950s: 5 Master's, 12 Doctoral
   1960s: 57 Master's, 59 Doctoral
   1970s: 155 Master's, 193 Doctoral
   1980s: 277 Master's, 282 Doctoral

Theses by Institution
   University of Alaska, Fairbanks   5 M, 2 D
   University of Alberta   10 M, 7 D
   University of Arizona   4 M, 4 D
   Arizona State University   2 M
   University of Arkansas, Fayetteville   1 M
   Boston College   4 M
   Boston University   2 M
   Brigham Young University   2 M
   University of British Columbia   10 M, 7 D
   Brown University   9 D
   Bryn Mawr College   2 M
   University of Calgary   3 M
   University of California, Berkeley   2 M, 18 D
   University of California, Davis   2 M, 3 D
   University of California, Los Angeles   3 M, 20 D
   University of California, Riverside   3 M, 2 D
   University of California, San Diego   30 D
   University of California, Santa Barbara   7 M, 5 D
   University of California, Santa Cruz   3 M, 10 D
   California Institute of Technology   1 M, 24 D
   Carleton University   5 M, 2 D
   Case Western Reserve University   1 D
   University of Chicago   8 D
   University of Colorado   5 M, 5 D
   Colorado School of Mines   3 M, 6 D
   Columbia University   1 M, 8 D
   Columbia University, Teachers College   35 D
   University of Connecticut   1 M
   Cornell University   8 M, 21 D
   Dalhousie University   12 M, 4 D
   Dartmouth College   3 M, 3 D
   University of Delaware   1 M
   Duke University   8 M
   Eastern Washington University   2 M
   Ecole Polytechnique   1 M
   University of Florida   4 M
   George Washington University   1 D
   University of Georgia   1 M, 1 D
   Georgia Institute of Technology   8 M, 1 D
   Harvard University   3 D
   University of Hawaii   15 M, 12 D
   University of Hawaii at Manoa   3 M
   University of Houston   3 M
   University of Illinois, Chicago   1 M

University of Illinois, Urbana   2 M, 1 D
Indiana University, Bloomington   3 M, 1 D
University of Iowa   5 M
The Johns Hopkins University   8 D
University of Kansas   1 M
Kent State University, Kent   3 M
University of Kentucky   5 M
Lakehead University   1 M
Lehigh University   1 M
Louisiana State University   1 M
University of Manitoba   6 M, 2 D
University of Massachusetts   1 D
Massachusetts Institute of Technology   19 M, 41 D
McGill University   1 M, 2 D
Memorial University of Newfoundland   3 M, 2 D
University of Miami   10 M, 1 D
University of Michigan   5 M, 7 D
Michigan State University   2 M
Michigan Technological University   2 M, 1 D
University of Minnesota, Duluth   1 M
University of Minnesota, Minneapolis   1 M
University of Missouri, Columbia   1 M
University of Montana   2 M
University of Nevada   3 M, 1 D
University of New Brunswick   1 D
University of New Mexico   1 D
New Mexico Institute of Mining and
   Technology   4 M, 3 D
New Mexico State University, Las Cruces   1 M,
   3 D
New York University   1 M, 1 D
University of North Carolina, Chapel
   Hill   3 M, 3 D
Northeastern Illinois University   1 M
Northern Arizona University   1 M
Northern Illinois University   6 M
Northwestern University   1 M, 8 D
Ohio State University   5 M, 6 D
University of Oklahoma   5 M
Oregon State University   19 M, 11 D
Pennsylvania State University, University
   Park   18 M, 6 D
University of Pittsburgh   3 D
Portland State University   1 M
Princeton University   15 D
Purdue University   20 M, 3 D
Universite du Quebec a Montreal   1 M
Queen's University   4 M, 1 D
Queens College (CUNY)   1 M
Rensselaer Polytechnic Institute   1 M
University of Rhode Island   1 M, 3 D
Rice University   6 M, 8 D
University of Rochester   3 D
Rutgers, The State University, New
   Brunswick   3 M
Rutgers, The State University, Newark   1 M
San Diego State University   6 M
University of Saskatchewan   2 M
University of South Carolina   2 M, 2 D
South Dakota School of Mines &
   Technology   1 M
University of South Florida, Tampa   4 M
University of Southern California   2 M, 3 D
Southern Illinois University,
   Carbondale   1 M
Southern Methodist University   5 M, 4 D
St. Louis University   6 M, 12 D
Stanford University   7 M, 36 D
SUNY at Albany   4 M, 2 D
SUNY at Binghamton   3 M
SUNY at Buffalo   6 M, 2 D
SUNY at Stony Brook   1 M, 3 D
Syracuse University   1 M
University of Texas, Austin   15 M, 6 D
Texas A&M University   8 M, 6 D
University of Texas at Dallas   6 M, 1 D
University of Texas at El Paso   8 M, 3 D
Texas Christian University   1 M
Texas Tech University   3 M, 2 D
University of Toronto   9 M, 8 D
Tulane University   1 M
University of Tulsa   2 M, 1 D
Union College   1 D
University of Utah   10 M, 8 D

Virginia Polytechnic Institute and State
   University   6 M, 2 D
Virginia State University   1 M
University of Washington   14 M, 15 D
Washington State University   1 D
Washington University   3 M, 2 D
University of Western Ontario   4 M, 4 D
Western Washington University   4 M
University of Wisconsin-Madison   6 M, 11 D
University of Wisconsin-Milwaukee   2 M
Woods Hole Oceanographic Institution   13 D
University of Wyoming   14 M, 4 D
Yale University   3 D

Hydrogeology and Hydrology

2,104 Master's, 1,066 Doctoral

Theses per decade
   1880s: 2 Doctoral
   1900s: 2 Doctoral
   1910s: 5 Master's, 1 Doctoral
   1920s: 8 Master's, 7 Doctoral
   1930s: 32 Master's, 8 Doctoral
   1940s: 20 Master's, 4 Doctoral
   1950s: 122 Master's, 30 Doctoral
   1960s: 257 Master's, 131 Doctoral
   1970s: 647 Master's, 335 Doctoral
   1980s: 1,009 Master's, 546 Doctoral

Theses by Institution
   University of Akron   16 M
   University of Alabama   7 M, 2 D
   University of Alaska, Fairbanks   8 M, 4 D
   University of Alberta   10 M, 5 D
   American University   1 M, 3 D
   University of Arizona   103 M, 71 D
   Arizona State University   7 M, 4 D
   University of Arkansas, Fayetteville   31 M, 5 D
   Auburn University   1 D
   Baylor University   10 M
   Boston College   2 M
   Boston University   25 M, 6 D
   Bowling Green State University   1 M
   Brigham Young University   11 M, 1 D
   University of British Columbia   10 M, 7 D
   Brock University   1 M
   Brown University   1 M
   Bryn Mawr College   1 D
   University of Calgary   1 M
   University of California, Berkeley   6 M, 25 D
   University of California, Davis   2 M, 20 D
   University of California, Irvine   1 D
   University of California, Los Angeles   12 M, 17 D
   University of California, Riverside   7 M, 2 D
   University of California, San Diego   2 D
   University of California, Santa Barbara   2 M, 3 D
   University of California, Santa Cruz   5 M, 1 D
   California Institute of Technology   4 M, 2 D
   California State University, Chico   1 M
   California State University, Fresno   3 M
   California State University, Hayward   1 M
   California State University, Los
      Angeles   2 M, 1 D
   California State University, Northridge   2 M
   Carleton University   1 M, 1 D
   Case Western Reserve University   3 D
   University of Chicago   5 D
   University of Cincinnati   6 M, 1 D
   City College (CUNY)   1 D
   Clark University   1 D
   Clarkson University   1 D
   Clemson University   5 D
   University of Colorado   9 M, 10 D
   University of Colorado at Colorado
      Springs   1 M
   Colorado School of Mines   25 M, 3 D
   Colorado State University   70 M, 87 D
   Columbia University, Teachers College   7 M, 10 D
   University of Connecticut   6 M, 5 D
   Cornell University   2 M, 21 D
   Dalhousie University   3 M
   Dartmouth College   7 M
   University of Delaware   9 M, 1 D
   University of Delaware, College of Marine
      Studies   1 D

Thesis Subject Distribution

Marine Geology and Oceanography

Drexel University   1 D
Duke University   1 M, 4 D
East Carolina University   1 M
East Texas State University   1 M
Eastern Kentucky University   5 M
Eastern Washington University   2 M
Ecole Polytechnique   2 M
Emory University   1 M, 1 D
University of Florida   12 M, 16 D
Florida Institute of Technology   6 M
Florida State University   17 M, 8 D
Fort Hays State University   2 M
George Washington University   1 M, 4 D
University of Georgia   4 M, 7 D
Georgia Institute of Technology   2 M, 7 D
University of Guelph   1 M, 2 D
Harvard University   9 D
University of Hawaii   13 M, 5 D
University of Hawaii at Manoa   1 M
University of Houston   2 M
University of Idaho   70 M, 13 D
Idaho State University   1 M, 1 D
University of Illinois, Chicago   6 M
University of Illinois, Urbana   14 M, 31 D
Indiana University of Pennsylvania   1 M
Indiana State University   7 M, 1 D
Indiana University, Bloomington   22 M, 9 D
University of Iowa   23 M, 8 D
Iowa State University of Science and
   Technology   11 M, 18 D
The Johns Hopkins University   1 M, 8 D
University of Kansas   17 M, 7 D
Kansas State University   14 M, 1 D
Kent State University, Kent   41 M
University of Kentucky   13 M, 4 D
Universite Laval   8 M, 1 D
Lehigh University   4 M
Louisiana State University   7 M, 10 D
Louisiana Tech University   3 M
University of Maine   5 M
University of Manitoba   10 M, 1 D
University of Maryland   1 M, 7 D
University of Massachusetts   29 M, 8 D
Massachusetts Institute of Technology   7 M, 1 D
McGill University   4 M, 9 D
McMaster University   8 M, 4 D
Memorial University of Newfoundland   1 M, 1 D
University of Miami   1 M, 1 D
Miami University (Ohio)   5 M
University of Michigan   9 M, 10 D
Michigan State University   18 M, 13 D
Michigan Technological University   7 M
University of Minnesota, Duluth   8 M, 1 D
University of Minnesota, Minneapolis   22 M, 17 D
University of Mississippi   1 M, 3 D
Mississippi State University   2 M
University of Missouri, Columbia   27 M, 4 D
University of Missouri, Kansas City   1 M
University of Missouri, Rolla   19 M, 5 D
University of Montana   12 M, 4 D
Montana College of Mineral Science &
   Technology   4 M
Montana State University   8 M, 2 D
Montclair State College   1 M
Murray State University   1 M
Naval Postgraduate School   1 M
University of Nebraska, Lincoln   35 M, 8 D
University of Nevada   80 M, 15 D
University of Nevada, Las Vegas   6 M
University of Nevada - Mackay School of
   Mines   1 M, 5 D
University of New Brunswick   1 M
University of New Hampshire   20 M, 1 D
University of New Mexico   6 M
New Mexico Institute of Mining and
   Technology   26 M, 10 D
New Mexico State University, Las Cruces   6 M,
   3 D
University of New Orleans   2 M
New York University   3 M
University of North Carolina, Chapel
   Hill   5 M, 2 D
North Carolina State University   11 M, 6 D
University of North Dakota   13 M, 5 D
University of North Texas   1 D

Northeast Louisiana University   2 M
Northeastern Illinois University   1 M
Northern Arizona University   12 M
Northern Illinois University   20 M, 2 D
Northwestern State University   1 M
Northwestern University   1 M, 7 D
Ohio State University   9 M, 10 D
Ohio University, Athens   34 M
University of Oklahoma   7 M, 8 D
Oklahoma State University   33 M, 10 D
Old Dominion University   2 M
University of Oregon   5 M, 2 D
Oregon Graduate Institute of Science and
   Technology   2 D
Oregon State University   4 M, 14 D
University of Pennsylvania   1 M, 2 D
Pennsylvania State University, University
   Park   36 M, 28 D
University of Pittsburgh   5 M, 6 D
Polytechnic University   3 D
Pomona College   1 M
Portland State University   1 M, 3 D
Princeton University   1 M, 13 D
Purdue University   10 M, 13 D
Universite du Quebec a Montreal   2 M
Universite du Quebec a Rimouski   1 M
Queen's University   2 M, 2 D
Queens College (CUNY)   2 M
Rensselaer Polytechnic Institute   3 M, 2 D
University of Rhode Island   10 M, 3 D
Rice University   1 M, 3 D
University of Rochester   1 D
Rutgers, The State University, New
   Brunswick   8 M, 8 D
Rutgers, The State University, Newark   2 M
San Diego State University   24 M
San Jose State University   2 M
University of Saskatchewan   10 M, 1 D
Simon Fraser University   1 D
Slippery Rock University   1 M
University of South Carolina   4 M, 2 D
University of South Dakota   4 M
South Dakota School of Mines &
   Technology   30 M, 2 D
University of South Florida, St.
   Petersburg   1 M
University of South Florida, Tampa   23 M
University of Southern California   14 M, 3 D
Southern Illinois University,
   Carbondale   13 M, 1 D
Southwest Missouri State University   4 M
University of Southwestern Louisiana   1 M
Stanford University   31 M, 45 D
State University of New York, College of Environ-
   mental Science
   and Forestry   2 D
Stephen F. Austin State University   3 M
SUNY, College at Fredonia   5 M
SUNY, College at Oneonta   3 M
SUNY at Binghamton   13 M
SUNY at Buffalo   3 M, 1 D
SUNY at Stony Brook   2 M, 1 D
Syracuse University   14 M, 4 D
University of Tennessee, Knoxville   5 M, 6 D
University of Texas, Arlington   1 M
University of Texas, Austin   37 M, 19 D
Texas A&M University   26 M, 13 D
University of Texas at Dallas   2 D
University of Texas at El Paso   4 M
Texas Christian University   2 M
University of Texas Health Science Center at Hous-
   ton School of
   Public Health   2 D
Texas Tech University   12 M, 2 D
University of Toledo   32 M
University of Toronto   3 D
Tulane University   1 M
University of Tulsa   5 M
University of Utah   18 M, 6 D
Utah State University   3 M, 36 D
Vanderbilt University   4 M, 3 D
University of Vermont   3 M, 1 D
University of Victoria   1 D
University of Virginia   12 M, 2 D

Virginia Polytechnic Institute and State
   University   5 M, 11 D
Virginia State University   1 M
University of Washington   14 M, 14 D
Washington State University   24 M, 7 D
Washington University   5 M
University of Waterloo   88 M, 20 D
Wayne State University   1 M
Wesleyan University   1 M
West Texas State University   1 M
West Virginia University   3 M, 4 D
Western Michigan University   10 M, 1 D
University of Western Ontario   4 M, 2 D
Western Washington University   6 M
Wichita State University   1 M
College of William and Mary   4 M
University of Windsor   8 M
University of Wisconsin-Madison   41 M, 33 D
University of Wisconsin-Milwaukee   35 M, 3 D
Woods Hole Oceanographic Institution   1 D
Wright State University   42 M
University of Wyoming   21 M, 3 D
Yale University   3 D

Marine Geology and Oceanography

1,006 Master's, 610 Doctoral

Theses per decade
   1920s: 5 Master's
   1930s: 8 Master's, 5 Doctoral
   1940s: 14 Master's, 6 Doctoral
   1950s: 45 Master's, 38 Doctoral
   1960s: 209 Master's, 137 Doctoral
   1970s: 387 Master's, 237 Doctoral
   1980s: 337 Master's, 186 Doctoral

Theses by Institution
   Acadia University   2 M
   Adelphi University   1 M
   University of Akron   1 M
   University of Alabama   2 M
   University of Alaska, Fairbanks   2 M, 4 D
   American University   2 M
   University of Arizona   3 M, 1 D
   Baylor University   1 M
   Boston University   6 M, 2 D
   Bowling Green State University   4 M
   University of British Columbia   9 M, 10 D
   Brooklyn College (CUNY)   6 M
   Brown University   4 M, 3 D
   Bryn Mawr College   1 D
   University of Calgary   2 D
   University of California, Berkeley   4 M, 4 D
   University of California, Davis   1 M, 2 D
   University of California, Los Angeles   7 M, 18 D
   University of California, Riverside   1 M
   University of California, San Diego   3 M, 47 D
   University of California, Santa Barbara   2 M, 2 D
   University of California, Santa Cruz   4 M, 5 D
   California State University, Hayward   2 M
   California State University, Northridge   9 M
   Carleton University   2 M, 1 D
   Case Western Reserve University   2 D
   University of Chicago   2 M, 6 D
   University of Cincinnati   3 M, 1 D
   Clark University   1 D
   University of Colorado   1 M, 1 D
   Colorado School of Mines   4 M, 2 D
   Colorado State University   1 M
   Columbia University   13 M, 19 D
   Columbia University, Teachers College   10 M,
      15 D
   University of Connecticut   6 D
   Cornell University   4 D
   Dalhousie University   18 M, 9 D
   Dartmouth College   3 M
   University of Delaware   10 M, 1 D
   University of Delaware, College of Marine
      Studies   5 M, 6 D
   Duke University   48 M, 3 D
   East Carolina University   6 M
   Emory University   2 M
   University of Florida   8 M, 4 D
   Florida Institute of Technology   4 M, 1 D
   Florida State University   31 M, 12 D

George Washington University 3 M, 6 D
University of Georgia 7 M, 1 D
Georgia Institute of Technology 5 M
University of Guelph 1 M
Harvard University 7 D
University of Hawaii 26 M, 11 D
University of Hawaii at Manoa 1 M
University of Houston 6 M
University of Illinois, Chicago 3 M
University of Illinois, Urbana 12 M, 13 D
Indiana University, Bloomington 1 M, 3 D
University of Iowa 2 M, 1 D
Iowa State University of Science and
   Technology 2 M
The Johns Hopkins University 1 M, 11 D
University of Kansas 3 M, 1 D
Lehigh University 12 M, 5 D
Long Island University, C. W. Post
   Campus 1 M
Louisiana State University 17 M, 9 D
University of Maine 1 M
University of Manitoba 1 M
University of Maryland 1 D
University of Massachusetts 6 M, 1 D
Massachusetts Institute of Technology 20 M, 20 D
McGill University 5 M, 4 D
McMaster University 2 M, 1 D
Memorial University of Newfoundland 1 M
Memphis State University 4 M
University of Miami 25 M, 24 D
Miami University (Ohio) 2 M
University of Michigan 4 M, 4 D
Michigan State University 3 M, 2 D
Millersville University 4 M
University of Minnesota, Minneapolis 4 M, 2 D
University of Mississippi 1 M
Mississippi State University 3 M
University of Missouri, Columbia 8 M, 1 D
University of Missouri, Rolla 1 D
University of Montana 1 M
Murray State University 1 M
Naval Postgraduate School 3 M
University of New Brunswick 3 M
University of New Hampshire 7 M, 2 D
University of New Orleans 11 M
New York University 6 M, 3 D
University of North Carolina, Chapel
   Hill 10 M, 6 D
North Carolina State University 6 M, 3 D
Northern Illinois University 5 M
Northwestern University 2 M, 3 D
Ohio State University 5 M, 1 D
Ohio University, Athens 1 M
University of Oklahoma 10 M
Old Dominion University 15 M, 4 D
University of Oregon 2 M
Oregon State University 45 M, 47 D
Pennsylvania State University, University
   Park 5 M, 3 D
University of Pittsburgh 1 D
Portland State University 2 M
Princeton University 5 D
Purdue University 1 M, 1 D
Universite du Quebec a Rimouski 2 M
Queen's University 4 M, 2 D
Queens College (CUNY) 6 M
Rensselaer Polytechnic Institute 6 M, 4 D
University of Rhode Island 8 M, 16 D
Rice University 21 M, 9 D
University of Rochester 1 M
Rutgers, The State University, New
   Brunswick 6 M, 3 D
Rutgers, The State University, Newark 2 M
San Diego State University 9 M
San Francisco State University 1 M
San Jose State University 13 M
University of Saskatchewan 1 M
University of South Carolina 10 M, 8 D
University of South Florida, St.
   Petersburg 20 M, 1 D
University of South Florida, Tampa 14 M
University of Southern California 46 M, 27 D
Southern Methodist University 1 M
St. Louis University 1 D
Stanford University 6 M, 6 D

Stephen F. Austin State University 2 M
SUNY, College at Oneonta 1 M
SUNY at Albany 2 M
SUNY at Binghamton 4 M, 1 D
SUNY at Buffalo 4 M
SUNY at Stony Brook 1 M, 1 D
Syracuse University 3 M, 1 D
University of Texas, Arlington 5 M
University of Texas, Austin 32 M, 12 D
Texas A&M University 71 M, 36 D
Texas Christian University 2 M
University of Texas of the Permian
   Basin 1 M
Texas Tech University 3 M
University of Toledo 2 M
University of Toronto 3 M, 5 D
Tulane University 2 D
United States Naval Academy 33 M
University of Utah 2 M
Utah State University 2 M
University of Virginia 1 D
Virginia Polytechnic Institute and State
   University 2 M
University of Washington 22 M, 34 D
Washington State University 3 M, 1 D
Washington University 2 M, 1 D
University of Waterloo 1 M
Wayne State University 1 M
Wesleyan University 1 M
Western Michigan University 2 M
University of Western Ontario 2 M
Western Washington University 3 M
Wichita State University 9 M
College of William and Mary 7 M, 7 D
University of Wisconsin-Madison 11 M, 5 D
University of Wisconsin-Milwaukee 4 M
Woods Hole Oceanographic Institution 18 D
University of Wyoming 1 M
Yale University 5 D

Mineralogy and Crystallography

849 Master's, 803 Doctoral

Theses per decade

1880s:	1 Master's, 3 Doctoral
1890s:	1 Master's, 9 Doctoral
1900s:	5 Master's, 7 Doctoral
1910s:	11 Master's, 5 Doctoral
1920s:	25 Master's, 21 Doctoral
1930s:	60 Master's, 37 Doctoral
1940s:	60 Master's, 28 Doctoral
1950s:	127 Master's, 114 Doctoral
1960s:	201 Master's, 241 Doctoral
1970s:	254 Master's, 214 Doctoral
1980s:	104 Master's, 124 Doctoral

Theses by Institution

University of Akron 1 M
University of Alaska, Fairbanks 3 M
University of Alberta 3 M
Alfred University 2 D
American University 4 M
University of Arizona 18 M, 4 D
Arizona State University 6 M, 7 D
University of Arkansas, Fayetteville 1 M
Baylor University 1 D
Boston University 7 M, 3 D
Bowling Green State University 5 M
Brigham Young University 3 M
University of British Columbia 9 M, 4 D
Brooklyn College (CUNY) 8 M
Brown University 6 M, 13 D
Bryn Mawr College 4 M, 3 D
University of Calgary 1 M
University of California, Berkeley 14 M, 23 D
University of California, Davis 5 D
University of California, Los Angeles 6 M, 15 D
University of California, Riverside 1 M, 3 D
University of California, San Diego 1 D
University of California, Santa Barbara 2 M, 1 D
California Institute of Technology 1 M, 9 D
Carleton University 6 M, 3 D
Carnegie-Mellon University 1 D
Case Western Reserve University 6 D
Catholic University of America 1 M

University of Chicago 1 M, 43 D
University of Cincinnati 9 M, 5 D
University of Colorado 3 M, 3 D
Colorado School of Mines 4 M, 5 D
Columbia University 2 M, 2 D
Columbia University, Teachers College 44 M,
   27 D
University of Connecticut 1 D
Cornell University 7 M, 8 D
Dalhousie University 3 M, 1 D
Dartmouth College 9 M, 1 D
University of Delaware 1 M, 2 D
Drexel University 1 D
Ecole Polytechnique 6 M, 1 D
Emory University 5 M
University of Florida 7 M
Florida State University 3 M, 4 D
George Washington University 2 M, 2 D
University of Georgia 1 D
Georgia Institute of Technology 1 D
University of Guelph 1 M, 1 D
Harvard University 1 M, 60 D
University of Hawaii 1 M, 3 D
University of Houston 3 M
University of Idaho 2 M
University of Illinois, Chicago 6 M
University of Illinois, Urbana 7 M, 26 D
Indiana State University 1 M
Indiana University, Bloomington 5 M, 5 D
University of Iowa 2 M, 2 D
Iowa State University of Science and
   Technology 3 M, 3 D
The Johns Hopkins University 2 M, 8 D
University of Kansas 4 M
Kansas State University 1 M
Kent State University, Kent 2 M
University of Kentucky 5 M
Universite Laval 4 M
Lehigh University 6 M, 3 D
Louisiana State University 1 M
University of Maine 2 M
University of Manitoba 16 M, 6 D
University of Maryland 2 D
University of Massachusetts 2 M
Massachusetts Institute of Technology 29 M, 45 D
McGill University 20 M, 13 D
McMaster University 10 M, 7 D
Memorial University of Newfoundland 1 M
University of Miami 1 M
Miami University (Ohio) 22 M, 6 D
University of Michigan 27 M, 29 D
Michigan State University 13 M, 5 D
Michigan Technological University 7 M
University of Minnesota, Minneapolis 26 M, 26 D
University of Missouri, Columbia 15 M, 3 D
University of Missouri, Rolla 2 M, 4 D
University of Montana 1 M
Montana College of Mineral Science &
   Technology 2 M
Montclair State College 2 M
Mount Holyoke College 1 M
University of Nebraska, Lincoln 2 M
University of Nevada 3 M, 1 D
University of Nevada - Mackay School of
   Mines 5 M, 1 D
University of New Brunswick 4 M, 1 D
University of New Mexico 15 M, 3 D
New Mexico Institute of Mining and
   Technology 4 M, 1 D
University of New Orleans 1 M
New York University 2 M, 2 D
University of North Carolina, Chapel
   Hill 4 M, 3 D
North Carolina State University 3 D
University of North Dakota 1 M
University of North Texas 1 D
Northeast Louisiana University 1 M
Northern Illinois University 3 M
Northwestern University 5 M, 4 D
Ohio State University 5 M, 20 D
Ohio University, Athens 1 M
University of Oklahoma 4 M, 6 D
Oklahoma State University 1 D
University of Oregon 2 M, 1 D
Oregon State University 1 M, 1 D

University of Ottawa 2 M, 1 D
University of Pennsylvania 5 D
Pennsylvania State University, University Park 34 M, 65 D
University of Pittsburgh 4 M, 11 D
Princeton University 13 D
Purdue University 3 M, 5 D
Queen's University 12 M, 8 D
Queens College (CUNY) 2 M
Rensselaer Polytechnic Institute 3 M, 1 D
Rice University 3 M, 1 D
University of Rochester 9 M, 1 D
Rutgers, The State University, New Brunswick 3 M, 7 D
Rutgers, The State University, Newark 1 M
University of Saskatchewan 6 M, 2 D
University of South Carolina 2 M
South Dakota School of Mines & Technology 4 M
University of South Florida, Tampa 2 M
Southern Illinois University, Carbondale 8 M
Southern Methodist University 1 D
Stanford University 8 M, 23 D
SUNY at Binghamton 1 M, 1 D
SUNY at Buffalo 9 M, 6 D
SUNY at Stony Brook 2 M, 8 D
Syracuse University 7 M, 2 D
University of Tennessee, Knoxville 5 M, 1 D
University of Texas, Arlington 1 M
University of Texas, Austin 7 M, 4 D
Texas A&M University 4 M, 2 D
University of Texas at El Paso 1 M
Texas Christian University 3 M, 1 D
Texas Tech University 3 M, 1 D
University of Toledo 6 M
University of Toronto 28 M, 27 D
Union College 1 M
University of Utah 10 M, 11 D
Vanderbilt University 1 M
University of Vermont 1 M
University of Virginia 9 M, 1 D
Virginia Polytechnic Institute and State University 32 M, 22 D
University of Washington 9 M, 2 D
Washington State University 2 M, 3 D
Washington University 10 M, 6 D
University of Waterloo 1 M
Wayne State University 3 M
West Virginia University 4 M
Western Michigan University 1 M
University of Western Ontario 1 M, 1 D
Wichita State University 1 M
University of Wisconsin-Madison 38 M, 26 D
University of Wisconsin-Milwaukee 2 M, 1 D
Wright State University 3 M
University of Wyoming 2 M, 3 D
Yale University 12 D

Miscellaneous and Mathematical Geology

81 Master's, 45 Doctoral

Theses per decade
1900s: 1 Doctoral
1920s: 3 Master's, 1 Doctoral
1930s: 3 Master's
1940s: 1 Doctoral
1950s: 9 Master's, 1 Doctoral
1960s: 22 Master's, 12 Doctoral
1970s: 31 Master's, 23 Doctoral
1980s: 13 Master's, 6 Doctoral

Theses by Institution
University of Alberta 1 M, 2 D
Baylor University 1 M
Bowling Green State University 1 M
Brooklyn College (CUNY) 1 M
University of California, Los Angeles 1 D
University of California, Santa Barbara 1 D
California State University, Chico 2 M
Chadron State College 1 M
University of Cincinnati 1 M
University of Colorado 1 D
Columbia University 1 M

Columbia University, Teachers College 2 M, 2 D
Cornell University 1 M
East Texas State University 1 M
Ecole Polytechnique 1 M, 1 D
University of Florida 2 M
George Washington University 2 M
University of Georgia 1 D
Harvard University 1 D
University of Hawaii at Manoa 1 M
University of Houston 1 M
University of Idaho 1 M, 1 D
University of Illinois, Urbana 1 M
University of Iowa 1 M
Iowa State University of Science and Technology 1 M, 2 D
Massachusetts Institute of Technology 1 D
McMaster University 1 D
University of Michigan 3 M
Michigan State University 1 D
Millersville University 4 M
University of Mississippi 1 D
University of Missouri, Rolla 1 M
Montana State University 1 M
University of Nevada 1 M
University of Nevada - Mackay School of Mines 1 M
University of New Brunswick 1 M
University of North Dakota 3 D
Northeastern Illinois University 1 M
Northern Illinois University 3 M
Northwestern University 1 M, 1 D
Ohio State University 6 M, 6 D
University of Oklahoma 1 D
Princeton University 1 D
Purdue University 2 M
Rensselaer Polytechnic Institute 1 M
University of Rhode Island 1 M
University of Rochester 1 M
San Jose State University 1 M
University of South Carolina 5 D
University of South Dakota 1 M
South Dakota School of Mines & Technology 1 M
University of Southern California 1 M, 1 D
Southern Illinois University, Carbondale 1 D
Southern Methodist University 1 M, 1 D
Stanford University 7 M, 2 D
University of Texas at Dallas 1 M
Texas Tech University 1 M
University of Toledo 1 M
University of Toronto 1 M
Tulane University 1 M
University of Utah 3 M
University of Virginia 2 M
Virginia State University 1 M
University of Washington 2 M, 2 D
University of Waterloo 1 M
Wichita State University 1 M
University of Wisconsin-Madison 3 D
University of Wisconsin-Milwaukee 1 M
University of Wyoming 1 D

Paleontology, General

158 Master's, 74 Doctoral

Theses per decade
1860s: 1 Doctoral
1900s: 1 Doctoral
1910s: 3 Master's
1920s: 4 Master's, 2 Doctoral
1930s: 11 Master's, 5 Doctoral
1940s: 4 Master's, 1 Doctoral
1950s: 29 Master's, 2 Doctoral
1960s: 50 Master's, 18 Doctoral
1970s: 47 Master's, 35 Doctoral
1980s: 10 Master's, 9 Doctoral

Theses by Institution
University of Alabama 1 M
University of Alberta 4 M
University of Arizona 4 M, 1 D
Arizona State University 1 M
University of Arkansas, Fayetteville 2 M
Baylor University 2 M

Boston University 3 M, 1 D
Brigham Young University 2 M, 1 D
University of British Columbia 2 M
Brooklyn College (CUNY) 2 M
Brown University 2 D
University of Calgary 1 M
University of California, Berkeley 4 M, 1 D
University of California, Davis 2 M, 1 D
University of California, Los Angeles 4 M, 2 D
University of California, Riverside 1 M
University of California, San Diego 1 D
University of California, Santa Barbara 2 D
University of Chicago 1 M, 1 D
University of Cincinnati 1 M
University of Colorado 2 M, 1 D
Columbia University 1 M
Columbia University, Teachers College 1 M, 1 D
Cornell University 1 M, 2 D
Emory University 1 M
University of Florida 2 M
Florida Institute of Technology 1 M
Florida State University 1 M, 1 D
Fordham University 1 D
George Washington University 1 M, 1 D
University of Georgia 1 M
Harvard University 2 D
University of Houston 1 M
University of Illinois, Chicago 1 D
University of Illinois, Urbana 5 M, 3 D
Indiana University, Bloomington 2 M, 1 D
University of Iowa 6 M, 4 D
Iowa State University of Science and Technology 2 M
The Johns Hopkins University 1 D
University of Kansas 1 M, 2 D
Kansas State University 1 M
University of Kentucky 1 M
Louisiana State University 1 M, 1 D
McGill University 1 M
McMaster University 1 M
Memorial University of Newfoundland 2 M
Miami University (Ohio) 1 M
University of Michigan 1 M, 2 D
Michigan State University 2 M, 2 D
University of Minnesota, Minneapolis 1 D
University of Missouri, Columbia 11 M, 3 D
University of Missouri, Rolla 1 M
University of Nebraska, Lincoln 3 M
University of New Mexico 2 M
New York University 1 M
University of North Carolina, Chapel Hill 1 M
University of North Dakota 1 M
Northeast Louisiana University 2 M
Northern Illinois University 2 M
Northwestern University 1 M
Ohio State University 4 M, 4 D
Ohio University, Athens 1 M
University of Oklahoma 3 M, 3 D
University of Oregon 1 M, 1 D
Pennsylvania State University, University Park 6 M
Princeton University 1 D
Queens College (CUNY) 1 M
University of Rhode Island 1 M
University of Rochester 5 M
Rutgers, The State University, New Brunswick 1 D
University of San Diego 1 M
San Jose State University 1 M
South Dakota School of Mines & Technology 1 M
University of Southern California 1 M
Southern Illinois University, Carbondale 1 M
Southern Methodist University 1 M
St. Louis University 1 M
Stanford University 1 D
SUNY at Binghamton 1 D
University of Tennessee, Knoxville 1 M
University of Texas, Austin 4 M, 1 D
Texas Tech University 5 M
Tulane University 1 D
University of Utah 1 D

Virginia Polytechnic Institute and State
University   1 M, 3 D
Washington State University   2 M
Washington University   2 M
University of Waterloo   2 M
West Texas State University   1 M
Western Michigan University   1 M
University of Western Ontario   1 M, 1 D
Wichita State University   1 M
College of William and Mary   1 M, 1 D
University of Wisconsin-Madison   6 M, 7 D
University of Wyoming   1 M, 1 D
Yale University   2 D

Paleontology, Invertebrate

1,151 Master's, 606 Doctoral

Theses per decade
1880s: 4 Doctoral
1890s: 2 Master's, 3 Doctoral
1900s: 10 Master's, 7 Doctoral
1910s: 11 Master's, 9 Doctoral
1920s: 36 Master's, 22 Doctoral
1930s: 81 Master's, 31 Doctoral
1940s: 64 Master's, 19 Doctoral
1950s: 208 Master's, 67 Doctoral
1960s: 264 Master's, 158 Doctoral
1970s: 315 Master's, 204 Doctoral
1980s: 160 Master's, 82 Doctoral

Theses by Institution
Acadia University   4 M
University of Akron   2 M
University of Alabama   5 M
University of Alaska, Fairbanks   2 M
University of Alberta   13 M
American University   1 M
University of Arizona   1 M
Arizona State University   2 M, 1 D
University of Arkansas, Fayetteville   1 M
Auburn University   2 M
Ball State University   2 M
Baylor University   3 M
Boston College   2 M, 1 D
Boston University   1 M
Bowling Green State University   23 M
Brigham Young University   7 M, 2 D
University of British Columbia   7 M, 2 D
Brooklyn College (CUNY)   7 M
Brown University   5 M, 3 D
Bryn Mawr College   2 M, 1 D
University of Calgary   3 M, 1 D
University of California, Berkeley   41 M, 30 D
University of California, Davis   12 M, 11 D
University of California, Los Angeles   7 M, 11 D
University of California, Riverside   1 M
University of California, San Diego   1 M, 5 D
University of California, Santa Barbara   3 D
University of California, Santa Cruz   3 D
California Institute of Technology   1 M, 7 D
California State University, Northridge   1 M
Carleton University   3 M
Case Western Reserve University   1 M, 6 D
Central Washington University   1 M
University of Chicago   3 M, 16 D
University of Cincinnati   26 M, 20 D
University of Colorado   5 M, 3 D
Colorado School of Mines   1 M
Colorado State University   1 M
Columbia University   4 M, 10 D
Columbia University, Teachers College   40 M, 17 D
Cornell University   10 M, 15 D
Dalhousie University   6 M, 1 D
University of Delaware   2 M
Duke University   8 M, 1 D
East Carolina University   3 M
Eastern Washington University   2 M
Emory University   3 M
University of Florida   9 M, 2 D
Florida Institute of Technology   1 M
Florida State University   17 M, 2 D
George Washington University   8 M, 3 D
University of Georgia   5 M, 2 D
Harvard University   27 D

University of Hawaii   3 M, 2 D
University of Houston   13 M
University of Idaho   3 M, 1 D
University of Illinois, Chicago   3 M
University of Illinois, Urbana   40 M, 25 D
Indiana State University   1 M
Indiana University, Bloomington   21 M, 21 D
University of Iowa   25 M, 31 D
Iowa State University of Science and
Technology   3 M, 2 D
The Johns Hopkins University   1 M, 27 D
University of Kansas   48 M, 17 D
Kansas State University   3 M
Kent State University, Kent   14 M
University of Kentucky   3 M
Laurentian University, Sudbury   2 M
Lehigh University   1 M
Louisiana State University   30 M, 16 D
Louisiana Tech University   3 M
University of Louisville   1 D
University of Maine   1 M
University of Massachusetts   3 M
Massachusetts Institute of Technology   5 M, 2 D
McGill University   6 M, 2 D
McMaster University   4 M, 5 D
Memorial University of Newfoundland   2 M, 1 D
Memphis State University   1 M
University of Miami   5 M, 3 D
Miami University (Ohio)   16 M, 2 D
University of Michigan   31 M, 25 D
Michigan State University   16 M, 3 D
Michigan Technological University   1 M
University of Minnesota, Minneapolis   4 M, 6 D
University of Mississippi   1 M
Mississippi State University   7 M
University of Missouri, Columbia   22 M, 5 D
University of Missouri, Rolla   12 M, 2 D
University of Montana   4 M
Montana College of Mineral Science &
Technology   1 M
Montclair State College   2 M
Universite de Montreal   6 M, 1 D
University of Nebraska, Lincoln   11 M, 1 D
University of Nevada   2 M, 1 D
University of New Brunswick   3 M
University of New Mexico   2 M, 1 D
New Mexico Institute of Mining and
Technology   1 D
University of New Orleans   3 M
New York University   18 M, 6 D
University of North Carolina, Chapel
Hill   11 M, 7 D
University of North Dakota   6 M, 8 D
Northeast Louisiana University   3 M
Northern Arizona University   2 M
Northern Illinois University   4 M
Northwestern University   3 M, 1 D
Ohio State University   13 M, 7 D
Ohio University, Athens   5 M
University of Oklahoma   12 M, 3 D
Oklahoma State University   1 M
Old Dominion University   3 M
University of Oregon   5 M
Oregon State University   5 M, 6 D
University of Ottawa   1 M
Pennsylvania State University, University
Park   17 M, 9 D
University of Pittsburgh   6 M, 2 D
Pomona College   1 M
Princeton University   11 D
Purdue University   2 M
Queen's University   2 D
Queens College (CUNY)   9 M
Rensselaer Polytechnic Institute   1 M, 1 D
University of Rhode Island   3 M, 1 D
Rice University   12 M, 7 D
University of Rochester   9 M, 6 D
Rutgers, The State University, New
Brunswick   5 M, 4 D
Rutgers, The State University, Newark   2 M
University of San Diego   1 M
San Diego State University   3 M
San Jose State University   2 M
University of Saskatchewan   6 M, 3 D
Smith College   7 M

University of South Carolina   4 M, 3 D
University of South Dakota   1 M
University of South Florida, St.
Petersburg   1 D
University of South Florida, Tampa   1 M
University of Southern California   29 M, 4 D
Southern Illinois University,
Carbondale   10 M
Southern Methodist University   8 M, 1 D
St. Louis University   1 M
Stanford University   15 M, 18 D
Stephen F. Austin State University   4 M
SUNY, College at Oneonta   1 M
SUNY at Binghamton   4 M, 2 D
SUNY at Buffalo   7 M
SUNY at Stony Brook   2 M, 3 D
Syracuse University   1 M, 1 D
University of Tennessee, Knoxville   4 M
University of Texas, Arlington   5 M
University of Texas, Austin   23 M, 8 D
Texas A&I University   1 M
Texas A&M University   6 M, 4 D
University of Texas at Dallas   2 M, 4 D
University of Texas at El Paso   1 M
Texas Christian University   7 M
Texas Tech University   6 M
University of Toledo   1 M
University of Toronto   7 M, 3 D
Tulane University   4 M, 7 D
United States Naval Academy   1 M
University of Utah   3 M
Utah State University   2 M
Vanderbilt University   1 M
University of Virginia   6 M
Virginia Polytechnic Institute and State
University   5 M
Virginia State University   4 M
University of Washington   10 M, 2 D
Washington State University   4 M, 1 D
Washington University   17 M, 2 D
University of Waterloo   2 M, 1 D
Wayne State University   8 M
West Virginia University   1 M
University of Western Ontario   4 M, 2 D
Western Washington University   1 M
Wichita State University   2 M
College of William and Mary   3 M, 1 D
University of Wisconsin-Green Bay   1 M
University of Wisconsin-Madison   20 M, 16 D
University of Wisconsin-Milwaukee   6 M
Wright State University   1 M
University of Wyoming   9 M, 2 D
Yale University   1 M, 26 D

Paleontology, Paleobotany

132 Master's, 240 Doctoral

Theses per decade
1880s: 1 Doctoral
1890s: 1 Doctoral
1910s: 3 Master's, 2 Doctoral
1920s: 6 Master's, 3 Doctoral
1930s: 5 Master's, 6 Doctoral
1940s: 3 Master's, 2 Doctoral
1950s: 13 Master's, 12 Doctoral
1960s: 37 Master's, 67 Doctoral
1970s: 42 Master's, 107 Doctoral
1980s: 23 Master's, 39 Doctoral

Theses by Institution
University of Akron   1 M
University of Alberta   4 M, 5 D
Andrews University   1 M
University of Arizona   4 M, 8 D
Arizona State University   2 D
University of Arkansas, Fayetteville   1 M
Ball State University   2 D
Bowling Green State University   1 M
Brigham Young University   1 D
University of British Columbia   2 M, 3 D
Brown University   2 D
University of Calgary   2 M, 1 D
University of California, Berkeley   7 M, 6 D
University of California, Davis   1 M, 1 D
University of California, Los Angeles   7 D

University of California, Riverside   1 D
University of California, Santa Barbara   2 M, 1 D
California Institute of Technology   1 D
California State University,
   Bakersfield   1 M
University of Chicago   1 M, 1 D
University of Cincinnati   1 M, 3 D
City College (CUNY)   2 D
University of Colorado   1 M, 2 D
Columbia University, Teachers College   2 M, 1 D
University of Connecticut   4 D
Cornell University   2 M, 10 D
University of Delaware   1 D
Duke University   2 M, 5 D
East Texas State University   1 M
University of Florida   1 M, 1 D
Florida State University   1 M, 2 D
Fordham University   1 D
George Washington University   1 D
University of Georgia   1 M
Harvard University   3 D
University of Hawaii   1 D
University of Houston   1 D
University of Idaho   2 M
University of Illinois, Urbana   5 M, 9 D
Indiana University, Bloomington   1 M, 8 D
University of Iowa   2 D
Iowa State University of Science and
   Technology   1 M
The Johns Hopkins University   3 D
University of Kansas   3 D
Kansas State University   2 M
Kent State University, Kent   1 M
Lehigh University   1 M
Loma Linda University   2 D
Louisiana State University   1 M, 6 D
University of Maine   1 M
University of Manitoba   1 D
University of Massachusetts   3 M, 1 D
Massachusetts Institute of Technology   1 M
Memorial University of Newfoundland   1 M
Miami University (Ohio)   1 D
University of Michigan   6 D
Michigan State University   15 D
University of Minnesota, Duluth   1 M
University of Minnesota, Minneapolis   7 D
Mississippi State University   1 M
University of Missouri, Columbia   4 M, 3 D
University of Missouri, Rolla   1 D
Naval Postgraduate School   1 M
University of Nebraska, Lincoln   2 M
New Mexico State University, Las Cruces   1 D
New York University   1 M, 7 D
University of North Carolina, Chapel
   Hill   1 M, 3 D
North Carolina State University   1 D
Northeast Louisiana University   2 M
University of Notre Dame   1 D
Ohio State University   7 D
Ohio University, Athens   2 D
University of Oklahoma   7 M, 9 D
University of Oregon   3 M
Oregon State University   1 D
University of Pennsylvania   1 D
Pennsylvania State University, University
   Park   2 M, 4 D
University of Pittsburgh   2 D
Princeton University   1 D
Purdue University   1 D
Queens College (CUNY)   2 M
Rensselaer Polytechnic Institute   1 D
University of Rhode Island   5 M, 2 D
Rutgers, The State University, New
   Brunswick   3 D
University of Saskatchewan   1 M, 3 D
University of South Carolina   1 M, 1 D
South Dakota School of Mines &
   Technology   1 M
University of Southern California   1 M, 1 D
Southern Illinois University,
   Carbondale   2 M, 2 D
Southern Methodist University   1 M
Stanford University   2 M, 6 D
Stephen F. Austin State University   1 M
SUNY at Albany   1 D

SUNY at Binghamton   1 D
SUNY at Stony Brook   1 D
University of Tennessee, Knoxville   2 M, 3 D
University of Texas, Austin   2 M, 4 D
Texas A&M University   2 M, 1 D
Texas Christian University   1 M
Texas Tech University   2 M
University of Toronto   1 M, 1 D
Tulane University   1 M, 1 D
University of Tulsa   1 M
University of Utah   2 M, 3 D
Vanderbilt University   1 D
University of Virginia   2 M
Virginia Polytechnic Institute and State
   University   1 M, 2 D
Wake Forest University   1 D
University of Washington   3 M, 3 D
Washington University   1 M, 4 D
University of Waterloo   1 M
West Virginia University   3 M
University of Western Ontario   1 M
Wichita State University   1 M
College of William and Mary   1 D
University of Wisconsin-Madison   3 M, 3 D
Yale University   1 D

Paleontology, Vertebrate

219 Master's, 253 Doctoral

Theses per decade
   1880s: 2 Doctoral
   1890s: 1 Doctoral
   1900s: 1 Master's, 5 Doctoral
   1910s: 5 Doctoral
   1920s: 6 Master's, 4 Doctoral
   1930s: 18 Master's, 16 Doctoral
   1940s: 13 Master's, 14 Doctoral
   1950s: 13 Master's, 20 Doctoral
   1960s: 41 Master's, 45 Doctoral
   1970s: 76 Master's, 80 Doctoral
   1980s: 51 Master's, 61 Doctoral

Theses by Institution
   University of Alabama   1 M
   University of Alberta   4 M, 1 D
   University of Arizona   8 M, 6 D
   University of Arkansas, Fayetteville   1 M
   Baylor University   1 M
   Boston University   1 M, 2 D
   Bowling Green State University   4 M
   Brigham Young University   2 M
   University of Calgary   1 M
   University of California, Berkeley   34 M, 43 D
   University of California, Los Angeles   7 D
   University of California, Riverside   4 M, 2 D
   California Institute of Technology   8 M, 7 D
   University of Chicago   4 M, 20 D
   City College (CUNY)   2 D
   University of Colorado   3 M, 7 D
   Columbia University   1 M, 7 D
   Columbia University, Teachers College   4 M, 15 D
   Cornell University   2 D
   University of Delaware   1 M
   University of Florida   3 M, 9 D
   Florida State University   1 M
   Fort Hays State University   4 M
   Franklin and Marshall College   1 M
   George Washington University   1 M, 2 D
   University of Georgia   2 M
   Harvard University   11 D
   University of Houston   2 M
   Idaho State University   1 M
   University of Illinois, Chicago   1 M
   University of Iowa   3 M
   The Johns Hopkins University   2 D
   University of Kansas   2 M, 9 D
   Kent State University, Kent   1 M
   University of Kentucky   1 D
   Loma Linda University   1 M
   Louisiana State University   1 M, 1 D
   University of Massachusetts   3 M, 1 D
   McGill University   4 D
   Miami University (Ohio)   2 M
   University of Michigan   12 M, 11 D
   Michigan State University   3 M, 2 D

University of Minnesota, Duluth   1 M
University of Minnesota, Minneapolis   3 M, 3 D
University of Missouri, Columbia   2 M
University of Missouri, Rolla   1 M
University of Montana   3 M, 1 D
University of Nebraska, Lincoln   14 M, 4 D
University of New Mexico   1 M, 1 D
New York University   2 D
Northeast Louisiana University   2 M
Northern Arizona University   1 D
Ohio State University   1 D
University of Oklahoma   4 M, 1 D
University of Pennsylvania   1 D
University of Pittsburgh   2 M, 1 D
Princeton University   2 D
University of Rochester   4 M
San Diego State University   1 M
University of Saskatchewan   1 M
Smith College   1 M
South Dakota School of Mines &
   Technology   18 M
University of Southern California   1 M, 3 D
Southern Illinois University,
   Carbondale   1 D
Southern Methodist University   2 M, 4 D
University of Southern Mississippi   2 D
St. Louis University   1 D
University of Texas, Austin   13 M, 6 D
University of Texas at El Paso   2 M
Texas Tech University   2 M, 3 D
University of Toronto   2 M, 5 D
University of Washington   3 M, 2 D
Wayne State University   3 D
Wesleyan University   1 M
West Texas State University   2 M
College of William and Mary   2 M, 2 D
University of Wisconsin-Madison   1 D
University of Wyoming   3 M, 2 D
Yale University   26 D

Petrology, Igneous and Metamorphic

3,326 Master's, 1,569 Doctoral

Theses per decade
   1870s: 2 Doctoral
   1880s: 4 Master's, 3 Doctoral
   1890s: 1 Master's, 18 Doctoral
   1900s: 12 Master's, 4 Doctoral
   1910s: 14 Master's, 13 Doctoral
   1920s: 65 Master's, 20 Doctoral
   1930s: 116 Master's, 67 Doctoral
   1940s: 134 Master's, 48 Doctoral
   1950s: 330 Master's, 144 Doctoral
   1960s: 510 Master's, 346 Doctoral
   1970s: 1,085 Master's, 482 Doctoral
   1980s: 1,055 Master's, 422 Doctoral

Theses by Institution
   Acadia University   10 M
   Adelphi University   1 M
   University of Akron   11 M
   University of Alabama   4 M
   University of Alaska, Fairbanks   14 M, 4 D
   University of Alberta   20 M, 10 D
   American University   1 M
   University of Arizona   48 M, 28 D
   Arizona State University   34 M, 15 D
   University of Arkansas, Fayetteville   11 M, 6 D
   Auburn University   4 M
   Baylor University   1 M
   Boise State University   2 M
   Boston College   1 D
   Boston University   9 M, 7 D
   Bowling Green State University   18 M
   Brigham Young University   21 M, 1 D
   University of British Columbia   58 M, 10 D
   Brock University   5 M
   Brooklyn College (CUNY)   20 M
   Brown University   19 M, 12 D
   Bryn Mawr College   24 M, 7 D
   University of Calgary   14 M, 5 D
   University of California, Berkeley   28 M, 54 D
   University of California, Davis   24 M, 6 D
   University of California, Los Angeles   21 M, 43 D
   University of California, Riverside   7 M, 2 D

University of California, San Diego   1 M, 14 D
University of California, Santa Barbara   18 M, 14 D
University of California, Santa Cruz   5 M, 13 D
California Institute of Technology   12 M, 35 D
California State University, Fresno   6 M
California State University, Hayward   4 M
California State University, Northridge   6 M
Carleton University   15 M, 11 D
Case Western Reserve University   4 M, 6 D
University of Chicago   1 M, 42 D
University of Cincinnati   25 M, 8 D
University of Colorado   22 M, 15 D
Colorado College   1 M
Colorado School of Mines   26 M, 10 D
Colorado State University   31 M, 1 D
Columbia University   3 M, 3 D
Columbia University, Teachers College   44 M, 33 D
University of Connecticut   2 M
Cornell University   24 M, 13 D
Dalhousie University   17 M, 4 D
Dartmouth College   33 M, 6 D
University of Delaware   9 M, 1 D
Duke University   5 M
East Carolina University   5 M
Eastern Kentucky University   5 M
Eastern Washington University   12 M
Ecole Polytechnique   16 M, 5 D
Emory University   9 M
University of Florida   3 M
Florida State University   3 M, 5 D
Fort Hays State University   1 M
Franklin and Marshall College   2 M
George Washington University   8 M, 2 D
University of Georgia   20 M, 6 D
Georgia Institute of Technology   2 M, 1 D
Georgia State University   2 M
Harvard University   2 M, 71 D
University of Hawaii   13 M, 7 D
University of Hawaii at Manoa   4 M
University of Houston   14 M, 3 D
University of Idaho   23 M, 9 D
Idaho State University   14 M
University of Illinois, Chicago   8 M, 3 D
University of Illinois, Urbana   34 M, 24 D
Indiana University, Bloomington   24 M, 13 D
University of Iowa   21 M, 5 D
Iowa State University of Science and Technology   10 M, 2 D
The Johns Hopkins University   4 M, 53 D
University of Kansas   14 M, 7 D
Kansas State University   12 M
Kent State University, Kent   31 M
University of Kentucky   12 M, 6 D
Lakehead University   7 M
Laurentian University, Sudbury   5 M
Universite Laval   17 M, 5 D
Lehigh University   12 M, 9 D
Louisiana State University   6 M, 3 D
University of Maine   9 M
University of Manitoba   52 M, 10 D
Marshall University   1 M
University of Maryland   4 D
University of Massachusetts   30 M, 13 D
Massachusetts Institute of Technology   32 M, 28 D
McGill University   79 M, 33 D
McMaster University   25 M, 13 D
Memorial University of Newfoundland   27 M, 9 D
Memphis State University   12 M
University of Miami   1 M, 1 D
Miami University (Ohio)   31 M, 3 D
University of Michigan   23 M, 13 D
Michigan State University   37 M, 16 D
Michigan Technological University   22 M, 2 D
Millersville University   1 M, 1 D
University of Minnesota, Duluth   22 M
University of Minnesota, Minneapolis   52 M, 33 D
University of Mississippi   2 M
University of Missouri, Columbia   19 M
University of Missouri, Kansas City   1 M
University of Missouri, Rolla   16 M, 2 D
University of Montana   57 M, 13 D
Montana College of Mineral Science & Technology   7 M

Montana State University   12 M
Montclair State College   3 M
Universite de Montreal   15 M, 1 D
University of Nebraska, Lincoln   9 M, 1 D
University of Nevada   15 M, 2 D
University of Nevada - Mackay School of Mines   1 M
University of New Brunswick   25 M, 3 D
University of New Hampshire   9 M
University of New Mexico   55 M, 13 D
New Mexico Institute of Mining and Technology   18 M, 4 D
New Mexico State University, Las Cruces   2 M
University of New Orleans   6 M
New York University   4 M
University of North Carolina, Chapel Hill   38 M, 22 D
North Carolina State University   19 M
University of North Dakota   15 M, 2 D
Northeast Louisiana University   5 M
Northeastern Illinois University   4 M
Northern Arizona University   5 M
University of Northern Colorado   1 M
Northern Illinois University   25 M, 1 D
Northwestern University   32 M, 14 D
Ohio State University   16 M, 10 D
Ohio University, Athens   3 M
University of Oklahoma   26 M, 2 D
Oklahoma State University   3 M
University of Oregon   52 M, 21 D
Oregon State University   10 M, 10 D
University of Ottawa   7 M, 5 D
Pacific Lutheran University   1 M
University of Pennsylvania   1 M, 4 D
Pennsylvania State University, University Park   45 M, 43 D
University of Pittsburgh   5 M, 9 D
Pomona College   3 M
Portland State University   8 M
Princeton University   48 D
Purdue University   4 M, 11 D
Universite du Quebec a Chicoutimi   14 M
Queen's University   60 M, 25 D
Queens College (CUNY)   3 M
University of Regina   2 M
Rensselaer Polytechnic Institute   12 M, 2 D
University of Rhode Island   12 M, 4 D
Rice University   26 M, 25 D
University of Rochester   18 M, 4 D
Rutgers, The State University, New Brunswick   19 M, 9 D
Rutgers, The State University, Newark   7 M
University of San Diego   2 M
San Diego State University   21 M
San Francisco State University   1 M
San Jose State University   19 M
University of Saskatchewan   18 M, 5 D
Slippery Rock University   1 M
Smith College   12 M
University of South Carolina   17 M, 2 D
University of South Dakota   1 M
South Dakota School of Mines & Technology   26 M, 1 D
University of South Florida, Tampa   2 M, 1 D
University of Southern California   7 M, 3 D
Southern Illinois University, Carbondale   9 M
Southern Methodist University   9 M, 4 D
University of Southern Mississippi   2 M
St. Louis University   4 M, 1 D
Stanford University   42 M, 80 D
Stephen F. Austin State University   1 M
Sul Ross State University   3 M
SUNY, College at Oneonta   1 M
SUNY at Albany   12 M, 4 D
SUNY at Binghamton   13 M, 6 D
SUNY at Buffalo   39 M, 2 D
SUNY at Stony Brook   9 M, 7 D
Syracuse University   16 M, 11 D
Technical University of Nova Scotia   2 M
Temple University   3 M
University of Tennessee, Knoxville   31 M, 3 D
University of Texas, Arlington   16 M
University of Texas, Austin   54 M, 19 D
Texas A&M University   10 M, 2 D

University of Texas at Dallas   3 M, 5 D
University of Texas at El Paso   21 M, 3 D
Texas Christian University   8 M
Texas Tech University   16 M
University of Toledo   7 M
University of Toronto   54 M, 46 D
Tulane University   6 M
University of Tulsa   3 M
University of Utah   40 M, 11 D
Utah State University   14 M
Vanderbilt University   4 M
University of Vermont   10 M
University of Virginia   10 M
Virginia Polytechnic Institute and State University   24 M, 4 D
University of Washington   59 M, 41 D
Washington State University   31 M, 11 D
Washington University   24 M, 12 D
University of Waterloo   5 M
Wayne State University   15 M
Wesleyan University   8 M
West Texas State University   7 M
West Virginia University   6 M, 2 D
Western Michigan University   4 M
University of Western Ontario   31 M, 23 D
Western Washington University   20 M
Wichita State University   2 M
University of Windsor   5 M
University of Wisconsin-Madison   79 M, 47 D
University of Wisconsin-Milwaukee   17 M, 2 D
Wright State University   6 M
University of Wyoming   31 M, 12 D
Yale University   2 M, 31 D

Petrology, Sedimentary

4,329 Master's, 1,201 Doctoral

Theses per decade

1890s: 3 Master's, 1 Doctoral
1900s: 7 Master's, 4 Doctoral
1910s: 8 Master's, 5 Doctoral
1920s: 50 Master's, 12 Doctoral
1930s: 102 Master's, 22 Doctoral
1940s: 124 Master's, 15 Doctoral
1950s: 471 Master's, 103 Doctoral
1960s: 660 Master's, 309 Doctoral
1970s: 1,273 Master's, 420 Doctoral
1980s: 1,630 Master's, 310 Doctoral

Theses by Institution

Acadia University   2 M
University of Akron   23 M, 1 D
University of Alabama   25 M, 1 D
University of Alaska, Fairbanks   9 M, 1 D
University of Alberta   39 M, 4 D
American University   4 M
University of Arizona   34 M, 9 D
Arizona State University   16 M
University of Arkansas, Fayetteville   42 M, 1 D
Auburn University   1 M
Ball State University   4 M
Baylor University   14 M
Boston College   2 M
Boston University   6 M
Bowling Green State University   45 M
Brigham Young University   24 M, 1 D
University of British Columbia   8 M, 5 D
Brock University   3 M
Brooklyn College (CUNY)   6 M
Brown University   9 M, 12 D
Bryn Mawr College   8 M, 1 D
University of Calgary   52 M, 5 D
University of California, Berkeley   16 M, 13 D
University of California, Davis   9 M, 2 D
University of California, Los Angeles   19 M, 19 D
University of California, Riverside   14 M, 6 D
University of California, San Diego   1 M, 10 D
University of California, Santa Barbara   14 M, 7 D
University of California, Santa Cruz   3 M, 6 D
California Institute of Technology   2 M, 2 D
California State University, Bakersfield   1 D
California State University, Chico   1 M
California State University, Fresno   3 M
California State University, Long Beach   1 M

California State University, Northridge   8 M
Carleton University   9 M, 3 D
Carnegie-Mellon University   1 D
Case Western Reserve University   3 M, 7 D
University of Chicago   12 M, 24 D
University of Cincinnati   53 M, 17 D
City College (CUNY)   1 D
Colgate University   2 M
University of Colorado   47 M, 15 D
Colorado School of Mines   24 M, 6 D
Colorado State University   22 M, 3 D
Columbia University   3 M, 7 D
Columbia University, Teachers College   19 M, 12 D
University of Connecticut   1 M
Cornell University   11 M, 9 D
Dalhousie University   8 M, 2 D
Dartmouth College   10 M, 4 D
University of Delaware   8 M, 2 D
Duke University   25 M
Earlham College   1 M
East Carolina University   12 M
East Texas State University   2 M
Eastern Kentucky University   20 M
Eastern Washington University   10 M
Ecole Polytechnique   2 M
Emory University   2 M
Emporia State University   1 M
University of Florida   27 M, 1 D
Florida Institute of Technology   1 M
Florida State University   30 M, 3 D
Fort Hays State University   8 M
Franklin and Marshall College   2 M
George Washington University   11 M, 4 D
University of Georgia   21 M, 5 D
Georgia Institute of Technology   6 M, 2 D
Georgia State University   1 M
Harvard University   1 M, 27 D
University of Hawaii   5 M, 2 D
University of Houston   29 M
University of Idaho   17 M, 3 D
Idaho State University   8 M
University of Illinois, Chicago   5 M
University of Illinois, Urbana   102 M, 67 D
Indiana University of Pennsylvania   2 M
Indiana State University   2 M
Indiana University, Bloomington   68 M, 29 D
University of Iowa   53 M, 29 D
Iowa State University of Science and Technology   37 M, 12 D
The Johns Hopkins University   3 M, 31 D
University of Kansas   56 M, 19 D
Kansas State University   32 M
Kent State University, Kent   6 M, 1 D
University of Kentucky   44 M, 2 D
Lakehead University   1 M
Universite Laval   10 M
Lehigh University   23 M, 4 D
Loma Linda University   2 M, 1 D
Long Island University, C. W. Post Campus   1 M
Louisiana State University   43 M, 27 D
Louisiana Tech University   4 M
University of Maine   4 M
University of Manitoba   29 M, 1 D
University of Maryland   3 M, 2 D
University of Massachusetts   24 M, 5 D
Massachusetts Institute of Technology   13 M, 12 D
McGill University   33 M, 13 D
McMaster University   24 M, 16 D
Memorial University of Newfoundland   8 M, 3 D
Memphis State University   23 M
University of Miami   7 M, 5 D
Miami University (Ohio)   65 M, 2 D
University of Michigan   42 M, 14 D
Michigan State University   51 M, 12 D
Michigan Technological University   18 M, 1 D
Millersville University   1 M
University of Minnesota, Duluth   6 M
University of Minnesota, Minneapolis   30 M, 8 D
University of Mississippi   12 M
Mississippi State University   14 M
University of Missouri, Columbia   110 M, 13 D
University of Missouri, Rolla   25 M, 4 D
University of Montana   25 M, 7 D

Montana College of Mineral Science & Technology   4 M
Montana State University   7 M
Montclair State College   1 M
Universite de Montreal   11 M, 1 D
University of Nebraska, Lincoln   37 M, 5 D
University of Nevada   9 M, 1 D
University of Nevada, Las Vegas   4 M
University of Nevada - Mackay School of Mines   2 M
University of New Brunswick   7 M
University of New Hampshire   3 M
University of New Mexico   22 M, 4 D
New Mexico Institute of Mining and Technology   18 M, 2 D
New Mexico State University, Las Cruces   5 M
University of New Orleans   28 M
New York University   2 M, 1 D
University of North Carolina, Chapel Hill   32 M, 14 D
North Carolina State University   16 M, 1 D
University of North Dakota   33 M, 5 D
Northeast Louisiana University   30 M
Northeastern Illinois University   3 M
Northern Arizona University   21 M
University of Northern Colorado   1 M
Northern Illinois University   18 M
Northwestern State University   1 M
Northwestern University   21 M, 13 D
Oberlin College   2 M
Ohio State University   35 M, 12 D
Ohio University, Athens   22 M, 1 D
University of Oklahoma   90 M, 22 D
Oklahoma State University   46 M
Old Dominion University   6 M
University of Oregon   14 M, 4 D
Oregon State University   13 M, 4 D
University of Ottawa   9 M, 7 D
University of Pennsylvania   3 M, 1 D
Pennsylvania State University, University Park   62 M, 48 D
University of Pittsburgh   44 M, 8 D
Pomona College   3 M
Portland State University   1 M
Princeton University   18 D
Purdue University   10 M, 5 D
Queen's University   11 M, 8 D
Queens College (CUNY)   8 M
University of Regina   2 M
Rensselaer Polytechnic Institute   35 M, 23 D
University of Rhode Island   5 M, 4 D
Rice University   20 M, 30 D
University of Rochester   21 M, 7 D
Rutgers, The State University, New Brunswick   12 M, 8 D
University of San Diego   1 M
San Diego State University   39 M
San Jose State University   15 M
University of Saskatchewan   13 M, 1 D
Slippery Rock University   1 M
Smith College   1 M
University of South Carolina   33 M, 25 D
University of South Dakota   2 M
South Dakota School of Mines & Technology   25 M, 2 D
University of South Florida, St. Petersburg   4 M
University of South Florida, Tampa   15 M
University of Southern California   34 M, 12 D
Southern Connecticut State College   1 M
Southern Illinois University, Carbondale   59 M
Southern Methodist University   18 M
University of Southern Mississippi   3 M, 1 D
University of Southwestern Louisiana   16 M
St. Louis University   3 M
Stanford University   35 M, 35 D
Stephen F. Austin State University   30 M
Sul Ross State University   4 M
SUNY, College at Fredonia   6 M
SUNY, College at Oneonta   2 M
SUNY at Albany   3 M
SUNY at Binghamton   23 M, 9 D
SUNY at Buffalo   22 M, 1 D
SUNY at Stony Brook   6 M, 4 D

Syracuse University   27 M, 5 D
Temple University   2 M
University of Tennessee, Knoxville   42 M, 7 D
University of Texas, Arlington   34 M
University of Texas, Austin   153 M, 42 D
Texas A&I University   1 M
Texas A&M University   89 M, 21 D
University of Texas at Dallas   14 M, 4 D
University of Texas at El Paso   12 M
Texas Christian University   23 M, 1 D
Texas Tech University   111 M, 7 D
University of Toledo   29 M
University of Toronto   20 M, 9 D
Tulane University   10 M
University of Tulsa   25 M, 6 D
United States Naval Academy   12 M
University of Utah   37 M, 7 D
Utah State University   20 M, 1 D
Vanderbilt University   18 M
University of Vermont   11 M
University of Virginia   5 M, 3 D
Virginia Polytechnic Institute and State University   20 M, 8 D
Virginia State University   2 M
University of Washington   23 M, 8 D
Washington State University   22 M, 4 D
Washington University   25 M, 6 D
University of Waterloo   11 M, 1 D
Wayne State University   22 M
Wesleyan University   1 M
West Texas State University   11 M
West Virginia University   27 M, 16 D
Western Michigan University   12 M
University of Western Ontario   7 M, 8 D
Western Washington University   18 M
Wichita State University   28 M
College of William and Mary   4 M, 3 D
University of Windsor   5 M
University of Wisconsin-Madison   79 M, 57 D
University of Wisconsin-Milwaukee   18 M
Wright State University   17 M
University of Wyoming   56 M, 13 D
Yale University   11 D

Stratigraphy, Historical Geology, Paleoecology

6,613 Master's, 2,092 Doctoral

Theses per decade
1860s: 1 Doctoral
1880s: 2 Master's, 1 Doctoral
1890s: 6 Master's, 18 Doctoral
1900s: 16 Master's, 20 Doctoral
1910s: 48 Master's, 37 Doctoral
1920s: 174 Master's, 65 Doctoral
1930s: 330 Master's, 148 Doctoral
1940s: 384 Master's, 119 Doctoral
1950s: 1,397 Master's, 342 Doctoral
1960s: 1,055 Master's, 527 Doctoral
1970s: 1,342 Master's, 395 Doctoral
1980s: 1,853 Master's, 419 Doctoral

Theses by Institution
Acadia University   17 M
University of Akron   12 M
University of Alabama   31 M, 2 D
University of Alaska, Fairbanks   18 M, 5 D
University of Alberta   77 M, 13 D
American University   1 M
Amherst College   1 M
Andrews University   1 M
University of Arizona   92 M, 34 D
Arizona State University   13 M, 4 D
University of Arkansas, Fayetteville   82 M
Auburn University   9 M
Ball State University   4 M
Baylor University   56 M
Boise State University   1 M
Boston College   1 M
Boston University   12 M, 6 D
Bowling Green State University   32 M
Brigham Young University   82 M, 6 D
University of British Columbia   36 M, 11 D
Brock University   5 M
Brooklyn College (CUNY)   5 M
Brown University   9 M, 10 D

Bryn Mawr College   6 M, 7 D
University of Calgary   28 M, 14 D
University of California, Berkeley   90 M, 87 D
University of California, Davis   18 M, 8 D
University of California, Los Angeles   45 M, 29 D
University of California, Riverside   30 M, 7 D
University of California, San Diego   3 M, 3 D
University of California, Santa Barbara   34 M,
   23 D
University of California, Santa Cruz   3 M, 9 D
California Institute of Technology   13 M, 26 D
California State University, Fresno   9 M
California State University, Hayward   2 M
California State University, Long Beach   14 M
California State University, Northridge   11 M
Carleton University   16 M, 5 D
Case Western Reserve University   3 M, 5 D
University of Chicago   20 M, 61 D
University of Cincinnati   64 M, 14 D
City College (CUNY)   1 D
University of Colorado   64 M, 42 D
University of Colorado at Colorado
   Springs   1 M
Colorado College   2 M
Colorado School of Mines   90 M, 19 D
Colorado State University   3 M
Columbia University   3 M, 11 D
Columbia University, Teachers College   95 M,
   69 D
University of Connecticut   1 M, 2 D
Cornell University   43 M, 28 D
Dalhousie University   3 M, 2 D
Dartmouth College   18 M, 1 D
University of Delaware   16 M, 2 D
Duke University   14 M
East Carolina University   11 M
East Texas State University   4 M
Eastern Kentucky University   47 M
Eastern Washington University   10 M
Ecole Polytechnique   1 M
Emory University   34 M
Emporia State University   1 M
University of Florida   19 M, 1 D
Florida State University   40 M, 14 D
Fort Hays State University   9 M
Franklin and Marshall College   4 M
George Washington University   12 M, 6 D
University of Georgia   20 M, 1 D
Georgia Institute of Technology   1 M
Harvard University   49 D
University of Hawaii   7 M, 4 D
University of Hawaii at Manoa   1 M
University of Houston   35 M, 1 D
University of Idaho   31 M, 6 D
Idaho State University   28 M, 1 D
University of Illinois, Chicago   17 M, 2 D
University of Illinois, Urbana   163 M, 52 D
Indiana University of Pennsylvania   4 M
Indiana State University   4 M
Indiana University, Bloomington   102 M, 33 D
University of Iowa   122 M, 68 D
Iowa State University of Science and
   Technology   35 M, 6 D
The Johns Hopkins University   1 M, 68 D
University of Kansas   106 M, 41 D
Kansas State University   25 M
Kent State University, Kent   25 M, 1 D
University of Kentucky   60 M, 1 D
Lakehead University   2 M
Laurentian University, Sudbury   2 M
Universite Laval   12 M, 4 D
Lehigh University   17 M, 8 D
Loma Linda University   5 M, 2 D
Louisiana State University   58 M, 34 D
Louisiana Tech University   1 M
University of Maine   10 M
University of Manitoba   28 M, 2 D
University of Massachusetts   25 M, 4 D
Massachusetts Institute of Technology   14 M, 12 D
McGill University   28 M, 18 D
McMaster University   16 M, 3 D
Memorial University of Newfoundland   29 M, 8 D
Memphis State University   6 M
University of Miami   2 M, 5 D
Miami University (Ohio)   67 M, 3 D

University of Michigan   124 M, 59 D
Michigan State University   83 M, 20 D
Michigan Technological University   8 M, 1 D
University of Minnesota, Duluth   9 M, 1 D
University of Minnesota, Minneapolis   52 M, 28 D
University of Mississippi   8 M
Mississippi State University   46 M
University of Missouri, Columbia   128 M, 23 D
University of Missouri, Kansas City   1 M
University of Missouri, Rolla   20 M, 9 D
University of Montana   39 M, 11 D
Montana College of Mineral Science &
   Technology   10 M
Montana State University   8 M
Montclair State College   2 M
Universite de Montreal   10 M, 4 D
University of Nebraska, Lincoln   127 M, 12 D
University of Nevada   14 M, 1 D
University of Nevada, Las Vegas   2 M
University of New Brunswick   16 M, 3 D
University of New Hampshire   2 M
University of New Mexico   51 M, 14 D
New Mexico Institute of Mining and
   Technology   24 M, 3 D
New Mexico State University, Las Cruces   10 M
University of New Orleans   44 M
New York University   19 M, 3 D
University of North Carolina, Chapel
   Hill   46 M, 19 D
North Carolina State University   7 M
University of North Dakota   35 M, 9 D
Northeast Louisiana University   19 M
Northern Arizona University   50 M
Northern Illinois University   29 M
Northwestern University   55 M, 35 D
Oberlin College   3 M
Ohio State University   69 M, 40 D
Ohio University, Athens   36 M
University of Oklahoma   259 M, 21 D
Oklahoma State University   24 M, 1 D
Old Dominion University   2 M
University of Oregon   30 M, 6 D
Oregon State University   67 M, 10 D
University of Ottawa   7 M, 10 D
University of Pennsylvania   3 M, 3 D
Pennsylvania State University, University
   Park   47 M, 31 D
University of Pittsburgh   40 M, 14 D
Pomona College   1 M, 2 D
Portland State University   25 M, 1 D
Princeton University   1 M, 51 D
University of Puget Sound   1 M
Purdue University   4 M
Universite du Quebec a Chicoutimi   6 M
Queen's University   18 M, 8 D
Queens College (CUNY)   8 M
University of Regina   1 M
Rensselaer Polytechnic Institute   15 M, 1 D
University of Rhode Island   6 M, 11 D
Rice University   18 M, 19 D
University of Rochester   35 M, 7 D
Rutgers, The State University, New
   Brunswick   29 M, 25 D
Rutgers, The State University, Newark   10 M
University of San Diego   5 M
San Diego State University   45 M
San Francisco State University   1 M
San Jose State University   21 M
University of Saskatchewan   56 M, 8 D
Smith College   9 M
University of South Carolina   26 M, 14 D
University of South Dakota   12 M
South Dakota School of Mines &
   Technology   34 M, 1 D
University of South Florida, St.
   Petersburg   4 M
University of South Florida, Tampa   4 M
University of Southern California   36 M, 15 D
Southern Illinois University,
   Carbondale   20 M
Southern Methodist University   24 M, 5 D
University of Southern Mississippi   1 M
University of Southwestern Louisiana   15 M
St. Francis Xavier University   1 M
St. Louis University   8 M

Stanford University   104 M, 75 D
Stephen F. Austin State University   10 M
Sul Ross State University   4 M
SUNY, College at Fredonia   8 M
SUNY, College at Oneonta   1 M
SUNY at Albany   7 M
SUNY at Binghamton   13 M, 2 D
SUNY at Buffalo   31 M, 1 D
SUNY at Stony Brook   1 M, 3 D
Syracuse University   30 M, 14 D
Temple University   2 M
University of Tennessee, Knoxville   61 M, 16 D
University of Texas, Arlington   48 M
University of Texas, Austin   230 M, 44 D
Texas A&M University   116 M, 9 D
University of Texas at Dallas   24 M, 14 D
University of Texas at El Paso   40 M, 3 D
Texas Christian University   61 M
University of Texas of the Permian
   Basin   4 M
Texas Tech University   61 M, 5 D
University of Toledo   12 M
University of Toronto   32 M, 36 D
Tulane University   16 M, 3 D
University of Tulsa   46 M, 1 D
Union College   1 M
University of Utah   55 M, 18 D
Utah State University   18 M
Vanderbilt University   20 M
University of Vermont   4 M
University of Virginia   19 M, 1 D
Virginia Polytechnic Institute and State
   University   30 M, 14 D
Virginia State University   2 M
University of Washington   49 M, 35 D
Washington State University   43 M, 15 D
Washington University   35 M, 11 D
University of Waterloo   8 M, 4 D
Wayne State University   22 M
Wesleyan University   2 M
West Texas State University   13 M
West Virginia University   33 M, 10 D
Western Michigan University   6 M
University of Western Ontario   14 M, 9 D
Western Washington University   31 M
Wichita State University   41 M
College of William and Mary   1 M
University of Windsor   14 M
University of Wisconsin-Madison   171 M, 80 D
University of Wisconsin-Milwaukee   34 M
Woods Hole Oceanographic Institution   2 D
Wright State University   8 M
University of Wyoming   130 M, 22 D
Yale University   2 M, 62 D

Structural Geology

2,361 Master's, 924 Doctoral

Theses per decade
1890s: 2 Master's, 1 Doctoral
1900s: 1 Master's, 1 Doctoral
1910s: 10 Master's, 5 Doctoral
1920s: 30 Master's, 5 Doctoral
1930s: 63 Master's, 46 Doctoral
1940s: 80 Master's, 38 Doctoral
1950s: 236 Master's, 82 Doctoral
1960s: 277 Master's, 163 Doctoral
1970s: 606 Master's, 234 Doctoral
1980s: 1,056 Master's, 346 Doctoral

Theses by Institution
Acadia University   5 M
University of Akron   4 M
University of Alabama   6 M
University of Alaska, Fairbanks   2 M, 1 D
University of Alberta   12 M, 2 D
American University   2 M
University of Arizona   57 M, 30 D
Arizona State University   6 M, 1 D
University of Arkansas, Fayetteville   16 M
Auburn University   3 M
Ball State University   2 M
Baylor University   7 M
Boise State University   1 M
Boston College   5 M

Boston University   2 M, 1 D
Bowling Green State University   12 M
Brigham Young University   17 M, 1 D
University of British Columbia   22 M, 8 D
Brock University   2 M
Brooklyn College (CUNY)   3 M
Brown University   8 M, 9 D
Bryn Mawr College   5 M, 3 D
University of Calgary   23 M, 2 D
University of California, Berkeley   7 M, 26 D
University of California, Davis   6 M, 4 D
University of California, Los Angeles   17 M, 23 D
University of California, Riverside   10 M, 2 D
University of California, San Diego   1 M, 7 D
University of California, Santa Barbara   17 M, 10 D
University of California, Santa Cruz   5 M, 17 D
California Institute of Technology   6 M, 20 D
California State University, Fresno   4 M
California State University, Hayward   1 M
California State University, Long Beach   9 M
California State University, Los Angeles   1 M
California State University, Northridge   3 M
Carleton University   10 M, 4 D
Case Western Reserve University   1 M
University of Chicago   2 M, 22 D
University of Cincinnati   11 M, 6 D
University of Colorado   33 M, 14 D
Colorado School of Mines   22 M, 12 D
Colorado State University   2 M
Columbia University   1 M, 8 D
Columbia University, Teachers College   16 M, 22 D
University of Connecticut   4 M, 3 D
Cornell University   27 M, 26 D
Dalhousie University   5 M, 2 D
Dartmouth College   6 M, 1 D
University of Delaware   3 M
University of Denver   1 D
Duke University   6 M
East Carolina University   1 M
Eastern Kentucky University   15 M
Eastern Washington University   2 M
Ecole Polytechnique   2 M
Emory University   7 M
Emporia State University   1 M
University of Florida   4 M
Florida State University   2 M, 2 D
Fort Hays State University   1 M
Franklin and Marshall College   2 M
George Washington University   2 M, 1 D
University of Georgia   4 M
Georgia Institute of Technology   1 M
Harvard University   29 D
University of Hawaii   1 M, 2 D
University of Houston   11 M, 1 D
University of Idaho   9 M, 5 D
Idaho State University   8 M
University of Illinois, Chicago   3 M
University of Illinois, Urbana   15 M, 9 D
Indiana State University   4 M, 3 D
Indiana University, Bloomington   20 M, 7 D
University of Iowa   26 M, 6 D
Iowa State University of Science and Technology   8 M, 1 D
The Johns Hopkins University   26 D
University of Kansas   21 M, 7 D
Kansas State University   13 M
University of Kentucky   21 M, 1 D
Lakehead University   2 M
Laurentian University, Sudbury   1 M
Universite Laval   6 M, 3 D
Lehigh University   15 M, 2 D
Louisiana State University   14 M, 7 D
University of Maine   1 M
University of Manitoba   19 M, 1 D
University of Maryland   1 D
University of Massachusetts   25 M, 9 D
Massachusetts Institute of Technology   21 M, 27 D
McGill University   13 M, 7 D
McMaster University   6 M, 3 D
Memorial University of Newfoundland   13 M, 6 D
Memphis State University   2 M
University of Miami   1 M

Miami University (Ohio)   24 M, 2 D
University of Michigan   34 M, 10 D
Michigan State University   26 M, 3 D
Michigan Technological University   16 M, 1 D
Millersville University   2 M
University of Minnesota, Duluth   5 M, 1 D
University of Minnesota, Minneapolis   12 M, 11 D
University of Mississippi   2 M
Mississippi State University   1 M
University of Missouri, Columbia   28 M, 4 D
University of Missouri, Rolla   13 M, 4 D
University of Montana   35 M, 2 D
Montana College of Mineral Science & Technology   7 M
Montana State University   7 M
Universite de Montreal   11 M, 2 D
Mount Holyoke College   1 M
Murray State University   3 M
Naval Postgraduate School   1 M
University of Nebraska, Lincoln   13 M, 2 D
University of Nevada   14 M, 4 D
University of Nevada, Las Vegas   1 M
University of Nevada - Mackay School of Mines   1 D
University of New Brunswick   9 M, 7 D
University of New Hampshire   2 M
University of New Mexico   16 M, 3 D
New Mexico Institute of Mining and Technology   7 M
New Mexico State University, Las Cruces   5 M
University of New Orleans   5 M
New York University   4 M, 1 D
University of North Carolina, Chapel Hill   22 M, 9 D
North Carolina State University   11 M
University of North Dakota   6 M
Northeast Louisiana University   4 M
Northern Arizona University   12 M
Northern Illinois University   9 M
Northwestern University   11 M, 12 D
Ohio State University   14 M, 3 D
Ohio University, Athens   14 M
University of Oklahoma   63 M, 6 D
Oklahoma State University   6 M
Old Dominion University   2 M
University of Oregon   6 M, 1 D
Oregon State University   28 M, 2 D
University of Ottawa   1 M, 4 D
University of Pennsylvania   1 M
Pennsylvania State University, University Park   45 M, 19 D
University of Pittsburgh   15 M, 2 D
Pomona College   3 M
Portland State University   7 M
Princeton University   3 M, 25 D
Purdue University   13 M, 7 D
Universite du Quebec a Chicoutimi   3 M
Queen's University   20 M, 18 D
Queens College (CUNY)   3 M
Rensselaer Polytechnic Institute   5 M, 2 D
University of Rhode Island   4 M, 2 D
Rice University   22 M, 13 D
University of Rochester   10 M, 2 D
Rutgers, The State University, New Brunswick   9 M
Rutgers, The State University, Newark   3 M
San Diego State University   25 M
San Jose State University   15 M
University of Saskatchewan   8 M, 1 D
Smith College   1 M
University of South Carolina   9 M, 10 D
South Dakota School of Mines & Technology   10 M, 4 D
University of South Florida, Tampa   3 M
University of Southern California   8 M, 13 D
Southern Illinois University, Carbondale   14 M
Southern Methodist University   4 M, 2 D
University of Southern Mississippi   1 M
University of Southwestern Louisiana   3 M
St. Louis University   3 M, 1 D
Stanford University   34 M, 57 D
Stephen F. Austin State University   3 M
Sul Ross State University   1 M
SUNY, College at Fredonia   1 M

SUNY at Albany   20 M, 11 D
SUNY at Binghamton   5 M, 1 D
SUNY at Buffalo   10 M, 1 D
SUNY at Stony Brook   1 M, 2 D
Syracuse University   28 M, 5 D
Temple University   1 M
University of Tennessee, Knoxville   34 M, 7 D
University of Texas, Arlington   8 M
University of Texas, Austin   101 M, 20 D
Texas A&M University   59 M, 27 D
University of Texas at Dallas   13 M
University of Texas at El Paso   15 M, 2 D
Texas Christian University   11 M
University of Texas of the Permian Basin   1 M
Texas Tech University   12 M, 1 D
University of Toledo   10 M
University of Toronto   19 M, 12 D
Tulane University   1 M
University of Tulsa   4 M, 2 D
University of Utah   32 M, 7 D
Utah State University   16 M
Vanderbilt University   5 M
University of Vermont   9 M
University of Virginia   8 M
Virginia Polytechnic Institute and State University   28 M, 10 D
Virginia State University   2 M
University of Washington   28 M, 16 D
Washington State University   35 M, 7 D
Washington University   11 M, 4 D
University of Waterloo   5 M
Wayne State University   11 M
Wesleyan University   2 M
West Texas State University   2 M
West Virginia University   11 M, 4 D
Western Michigan University   8 M
University of Western Ontario   3 M, 7 D
Western Washington University   13 M
Wichita State University   10 M
University of Windsor   2 M
University of Wisconsin-Madison   43 M, 24 D
University of Wisconsin-Milwaukee   10 M
Wright State University   21 M
University of Wyoming   67 M, 9 D
Yale University   3 M, 20 D

Surficial Geology, Geomorphology

1,060 Master's, 574 Doctoral

Theses per decade
1880s: 1 Doctoral
1890s: 2 Master's, 4 Doctoral
1900s: 6 Master's, 6 Doctoral
1910s: 18 Master's, 5 Doctoral
1920s: 47 Master's, 12 Doctoral
1930s: 59 Master's, 22 Doctoral
1940s: 47 Master's, 12 Doctoral
1950s: 88 Master's, 47 Doctoral
1960s: 165 Master's, 130 Doctoral
1970s: 346 Master's, 184 Doctoral
1980s: 282 Master's, 151 Doctoral

Theses by Institution
University of Akron   8 M
University of Alabama   3 M
University of Alaska, Fairbanks   3 M
University of Alberta   5 M, 3 D
American University   2 M
University of Arizona   12 M, 4 D
Arizona State University   3 M, 1 D
University of Arkansas, Fayetteville   1 M
Ball State University   1 M
Baylor University   16 M
Boston College   1 M
Boston University   6 M, 4 D
Bowling Green State University   8 M
Brigham Young University   5 M
University of British Columbia   4 M, 6 D
Brock University   1 M
Brooklyn College (CUNY)   1 M
Brown University   3 M, 2 D
University of Calgary   3 M
University of California, Berkeley   9 M, 10 D
University of California, Davis   2 M, 2 D

University of California, Los Angeles   21 M, 28 D
University of California, San Diego   3 M, 2 D
University of California, Santa Barbara   5 M, 2 D
University of California, Santa Cruz   5 M, 1 D
California Institute of Technology   2 M, 4 D
California State University, Chico   3 M
Case Western Reserve University   1 D
Catholic University of America   1 M, 1 D
University of Chicago   7 M, 8 D
University of Cincinnati   19 M, 1 D
Clark University   5 M, 16 D
University of Colorado   11 M, 5 D
University of Colorado at Colorado
    Springs   2 M
Colorado School of Mines   4 M, 3 D
Colorado State University   32 M, 15 D
Columbia University   2 M, 3 D
Columbia University, Teachers College   37 M,
    21 D
University of Connecticut   6 M
Cornell University   8 M, 8 D
Dalhousie University   1 D
Dartmouth College   2 M, 2 D
University of Delaware   8 M, 1 D
University of Delaware, College of Marine
    Studies   1 M
University of Denver   1 M, 3 D
DePauw University   1 M
Duke University   7 M
East Carolina University   2 M
East Texas State University   1 M
Eastern Washington University   3 M
Emory University   4 M, 1 D
University of Florida   7 M
Florida Institute of Technology   1 M
Florida State University   9 M, 6 D
Franklin and Marshall College   1 M
George Washington University   3 M
University of Georgia   7 M, 4 D
University of Guelph   3 M
Harvard University   9 D
University of Hawaii   1 D
University of Houston   3 M
University of Idaho   6 M, 1 D
Idaho State University   2 M
University of Illinois, Chicago   3 M
University of Illinois, Urbana   15 M, 16 D
Indiana State University   12 M, 8 D
Indiana University, Bloomington   11 M, 7 D
University of Iowa   8 M, 14 D
Iowa State University of Science and
    Technology   11 M, 5 D
The Johns Hopkins University   1 M, 21 D
University of Kansas   17 M, 9 D
Kansas State University   4 M
Kent State University, Kent   6 M
University of Kentucky   7 M
Universite Laval   1 M
Lehigh University   5 M
Loma Linda University   1 M
Louisiana State University   13 M, 27 D
University of Maine   7 M
University of Manitoba   4 M, 1 D
University of Maryland   1 D
University of Massachusetts   13 M, 4 D
Massachusetts Institute of Technology   1 M
McGill University   4 M, 8 D
McMaster University   2 M, 2 D
Memorial University of Newfoundland   1 D
Memphis State University   2 M
Miami University (Ohio)   1 M
University of Michigan   9 M, 17 D
Michigan State University   4 M, 2 D
Michigan Technological University   1 D
University of Minnesota, Minneapolis   8 M, 10 D
University of Mississippi   4 M
Mississippi State University   4 M
University of Missouri, Columbia   7 M, 2 D
University of Missouri, Rolla   3 M, 2 D
University of Montana   1 M
Montana State University   6 M
Universite de Montreal   3 M
Murray State University   2 M
Naval Postgraduate School   2 M
University of Nebraska, Lincoln   8 M, 5 D

University of Nevada   4 M
University of New Brunswick   1 M, 1 D
University of New Hampshire   6 M
University of New Mexico   13 M, 4 D
New Mexico State University, Las Cruces   1 D
New York University   2 M, 3 D
University of North Carolina, Chapel
    Hill   9 M, 6 D
North Carolina State University   6 M, 2 D
University of North Dakota   8 M, 1 D
Northern Arizona University   4 M
Northern Illinois University   7 M, 1 D
Northwestern State University   1 M
Northwestern University   3 M, 2 D
Ohio State University   10 M, 10 D
Ohio University, Athens   6 M
University of Oklahoma   8 M, 3 D
Oklahoma State University   3 M
Old Dominion University   2 M, 1 D
University of Oregon   5 M, 4 D
Oregon State University   5 M, 7 D
University of Ottawa   1 M
University of Pennsylvania   1 M, 1 D
Pennsylvania State University, University
    Park   17 M, 17 D
University of Pittsburgh   2 M, 5 D
Pomona College   2 M
Portland State University   1 M
Princeton University   6 D
Purdue University   10 M, 10 D
Queen's University   1 D
Rensselaer Polytechnic Institute   2 M, 2 D
University of Rhode Island   8 M
Rice University   4 M, 3 D
University of Rochester   6 M, 1 D
Rutgers, The State University, New
    Brunswick   3 M, 7 D
Rutgers, The State University, Newark   3 M
San Diego State University   1 M
San Fernando Valley State University   2 M
San Francisco State University   1 M
San Jose State University   8 M
University of Saskatchewan   1 M, 2 D
Simon Fraser University   5 D
Smith College   5 M
University of South Carolina   9 M, 6 D
University of South Dakota   1 M
South Dakota School of Mines &
    Technology   10 M
University of South Florida, Tampa   7 M
University of Southern California   8 M, 3 D
Southern Illinois University,
    Carbondale   19 M, 3 D
Southern Illinois University,
    Edwardsville   1 D
Southern Methodist University   1 M
University of Southwestern Louisiana   1 M
St. Louis University   1 D
Stanford University   17 M, 16 D
State University of New York, College of Environ-
    mental Science
    and Forestry   1 D
Stephen F. Austin State University   2 M
SUNY, College at Cortland   1 M
SUNY, College at Fredonia   7 M
SUNY, College at Oneonta   1 M
SUNY at Binghamton   9 M, 3 D
SUNY at Buffalo   15 M, 4 D
SUNY at Stony Brook   1 M, 1 D
Syracuse University   11 M, 7 D
University of Tennessee, Knoxville   7 M, 2 D
University of Texas, Austin   26 M, 10 D
Texas A&M University   8 M, 3 D
University of Texas at Dallas   1 M, 1 D
University of Texas at El Paso   1 M
Texas Christian University   3 M
Texas Tech University   9 M, 1 D
University of Toledo   2 M
University of Toronto   2 M, 2 D
Tulane University   1 M, 1 D
University of Tulsa   2 M
Union College   2 M
United States Naval Academy   2 M
University of Utah   11 M, 4 D
Utah State University   3 M, 1 D

Vanderbilt University   3 M
University of Vermont   2 M
University of Virginia   12 M, 2 D
Virginia Polytechnic Institute and State
    University   2 M, 4 D
Virginia State University   6 M
University of Washington   13 M, 7 D
Washington State University   4 M, 1 D
Washington University   7 M, 3 D
University of Waterloo   5 M, 5 D
West Texas State University   2 M
West Virginia University   3 M, 2 D
Western Michigan University   7 M
University of Western Ontario   2 M, 3 D
Western Washington University   6 M
Wichita State University   1 M
College of William and Mary   5 M, 2 D
University of Windsor   1 M
University of Wisconsin-Green Bay   1 M
University of Wisconsin-Madison   25 M, 15 D
University of Wisconsin-Milwaukee   7 M, 2 D
Wright State University   7 M
University of Wyoming   8 M, 6 D
Yale University   2 D

Surficial Geology, Quaternary Geology

1,422 Master's, 872 Doctoral

Theses per decade
  1870s: 2 Doctoral
  1890s: 2 Master's, 3 Doctoral
  1900s: 6 Master's, 7 Doctoral
  1910s: 15 Master's, 9 Doctoral
  1920s: 21 Master's, 8 Doctoral
  1930s: 43 Master's, 19 Doctoral
  1940s: 44 Master's, 18 Doctoral
  1950s: 111 Master's, 62 Doctoral
  1960s: 160 Master's, 128 Doctoral
  1970s: 444 Master's, 302 Doctoral
  1980s: 575 Master's, 314 Doctoral

Theses by Institution
  Acadia University   4 M
  University of Akron   24 M
  University of Alabama   3 M, 1 D
  University of Alaska, Fairbanks   14 M, 1 D
  University of Alberta   15 M, 10 D
  University of Arizona   19 M, 25 D
  Arizona State University   6 M, 2 D
  University of Arkansas, Fayetteville   2 M
  Ball State University   3 M
  Boston College   2 M
  Boston University   6 M, 1 D
  Bowling Green State University   13 M
  Brigham Young University   7 M
  University of British Columbia   10 M, 7 D
  Brock University   8 M
  Brooklyn College (CUNY)   1 M
  Brown University   3 M, 13 D
  Bryn Mawr College   1 D
  University of Calgary   28 M, 4 D
  University of California, Berkeley   8 M, 17 D
  University of California, Davis   2 M, 9 D
  University of California, Los Angeles   12 M, 12 D
  University of California, Riverside   1 M
  University of California, San Diego   1 M, 5 D
  University of California, Santa Barbara   3 M, 1 D
  University of California, Santa Cruz   2 M, 3 D
  California Institute of Technology   7 M, 13 D
  California State University, Hayward   2 M
  California State University, Northridge   4 M
  Carleton University   4 M
  Case Western Reserve University   1 M, 3 D
  Catholic University of America   1 M
  University of Chicago   4 M, 18 D
  University of Cincinnati   9 M, 3 D
  Clark University   11 M, 3 D
  University of Colorado   32 M, 45 D
  University of Colorado at Colorado
      Springs   1 M
  Colorado School of Mines   5 M, 3 D
  Colorado State University   3 M
  Columbia University   3 M, 4 D
  Columbia University, Teachers College   9 M, 13 D
  University of Connecticut   2 M

Cornell University   11 M, 10 D
Dalhousie University   13 M, 11 D
Dartmouth College   3 M, 2 D
University of Delaware   19 M, 10 D
University of Delaware, College of Marine
   Studies   1 M
Duke University   17 M, 6 D
East Carolina University   5 M
East Texas State University   2 M
Eastern Kentucky University   1 M
Eastern Washington University   6 M
Emory University   1 M
Emporia State University   3 M
University of Florida   11 M, 3 D
Florida State University   12 M, 1 D
Fort Hays State University   2 M
George Washington University   2 M, 3 D
University of Georgia   14 M, 6 D
Georgia State University   1 M
Harvard University   19 M
University of Hawaii   5 M, 4 D
University of Houston   9 M
Humboldt State University   1 M
University of Idaho   12 M, 5 D
Idaho State University   5 M
University of Illinois, Chicago   5 M
University of Illinois, Urbana   26 M, 29 D
Indiana State University   2 M, 2 D
Indiana University, Bloomington   16 M, 16 D
University of Iowa   33 M, 28 D
Iowa State University of Science and
   Technology   11 M, 13 D
The Johns Hopkins University   7 D
University of Kansas   11 M, 9 D
Kansas State University   6 M
Kent State University, Kent   6 M
Lakehead University   1 M
Universite Laval   1 M
Lehigh University   15 M, 3 D
Loma Linda University   1 M
Louisiana State University   10 M, 8 D
Louisiana Tech University   1 M
University of Louisville   1 D
University of Maine   21 M, 1 D
University of Manitoba   6 M, 2 D
Marshall University   1 M
University of Massachusetts   11 M, 10 D
Massachusetts Institute of Technology   4 M, 2 D
McGill University   5 M, 5 D
McMaster University   5 M, 4 D
Memorial University of Newfoundland   8 M, 1 D
Memphis State University   3 M
University of Miami   6 M, 5 D
Miami University (Ohio)   22 M
University of Michigan   15 M, 25 D
Michigan State University   19 M, 12 D
Michigan Technological University   5 M
University of Minnesota, Duluth   10 M, 1 D
University of Minnesota, Minneapolis   14 M, 26 D
University of Mississippi   4 M
Mississippi State University   2 M
University of Missouri, Columbia   6 M, 4 D
University of Missouri, Rolla   1 M, 1 D
University of Montana   9 M, 1 D
Montana State University   9 M
Universite de Montreal   4 M
Mount Holyoke College   1 M
Murray State University   1 M
University of Nebraska, Lincoln   18 M, 7 D
University of Nevada   6 M, 1 D
University of Nevada, Las Vegas   1 M
University of New Brunswick   2 M
University of New Hampshire   4 M
University of New Mexico   11 M, 2 D
New Mexico Institute of Mining and
   Technology   3 M
University of New Orleans   3 M
New York University   3 M, 8 D
University of North Carolina, Chapel
   Hill   11 M, 3 D
North Carolina State University   3 M
University of North Dakota   24 M, 10 D
Northeast Louisiana University   1 M
Northeastern Illinois University   2 M
Northern Illinois University   9 M

Northwestern University   5 M, 7 D
Ohio State University   36 M, 41 D
Ohio University, Athens   5 M
University of Oklahoma   12 M, 1 D
Oklahoma State University   3 M
Old Dominion University   6 M
University of Oregon   3 M, 1 D
Oregon State University   5 M, 11 D
University of Ottawa   6 M
University of Pennsylvania   3 M, 2 D
Pennsylvania State University, University
   Park   9 M, 13 D
University of Pittsburgh   3 M, 5 D
Portland State University   1 M
Princeton University   4 D
Purdue University   10 M, 1 D
Universite du Quebec a Montreal   4 M
Queen's University   2 M, 3 D
Queens College (CUNY)   6 M
Rensselaer Polytechnic Institute   12 M, 3 D
University of Rhode Island   4 M, 5 D
Rice University   10 M, 5 D
University of Rochester   7 M
Rutgers, The State University, New
   Brunswick   7 M, 4 D
Rutgers, The State University, Newark   5 M
San Diego State University   6 M
San Francisco State University   1 M
San Jose State University   8 M
University of Saskatchewan   7 M, 1 D
Simon Fraser University   2 D
Slippery Rock University   1 M
Smith College   2 M
University of South Carolina   7 M, 10 D
University of South Dakota   5 M
South Dakota School of Mines &
   Technology   3 M
University of South Florida, St.
   Petersburg   1 M
University of South Florida, Tampa   4 M
University of Southern California   7 M, 4 D
Southern Illinois University,
   Carbondale   12 M
Southern Methodist University   2 M
University of Southwestern Louisiana   4 M
Stanford University   11 M, 20 D
SUNY, College at Fredonia   1 M
SUNY, College at Oneonta   2 M
SUNY at Binghamton   8 M, 5 D
SUNY at Buffalo   16 M, 2 D
SUNY at Stony Brook   2 M, 1 D
Syracuse University   14 M, 15 D
University of Tennessee, Knoxville   5 M
University of Texas, Arlington   3 M
University of Texas, Austin   16 M, 6 D
Texas A&M University   8 M, 8 D
University of Texas at Dallas   1 M, 1 D
University of Texas at El Paso   1 M
Texas Christian University   2 M
Texas Tech University   3 M, 3 D
University of Toledo   3 M
University of Toronto   13 M, 11 D
Tulane University   4 M, 1 D
University of Tulsa   1 M
University of Utah   10 M, 6 D
Utah State University   3 M
Vanderbilt University   1 M, 1 D
University of Vermont   4 M
University of Virginia   1 M, 4 D
Virginia Polytechnic Institute and State
   University   1 D
Virginia State University   1 M
University of Washington   30 M, 27 D
Washington State University   15 M, 11 D
Washington University   6 M, 4 D
University of Waterloo   16 M, 7 D
Wayne State University   1 M
Wesleyan University   1 M
West Virginia University   3 M
Western Michigan University   4 M, 1 D
University of Western Ontario   6 M, 13 D
Western Washington University   9 M
Wichita State University   1 M
College of William and Mary   1 M, 3 D
University of Windsor   3 M

University of Wisconsin-Green Bay   1 M
University of Wisconsin-Madison   41 M, 31 D
University of Wisconsin-Milwaukee   14 M, 2 D
Wright State University   3 M
University of Wyoming   9 M, 5 D
Yale University   2 M, 14 D
York University   1 D

Surficial Geology, Soils

252 Master's, 548 Doctoral

Theses per decade
1900s: 3 Master's, 1 Doctoral
1910s: 11 Master's
1920s: 9 Master's, 6 Doctoral
1930s: 8 Master's, 9 Doctoral
1940s: 13 Master's, 10 Doctoral
1950s: 41 Master's, 14 Doctoral
1960s: 39 Master's, 110 Doctoral
1970s: 82 Master's, 90 Doctoral
1980s: 46 Master's, 308 Doctoral

Theses by Institution
University of Alabama   1 M
University of Alaska, Fairbanks   1 M, 1 D
University of Alberta   4 M
American University   2 M, 1 D
University of Arizona   11 M, 14 D
University of Arkansas, Fayetteville   1 D
Auburn University   2 D
Baylor University   1 M
Boston College   1 M
Brigham Young University   1 D
University of British Columbia   4 M, 10 D
Brock University   1 M
University of Calgary   2 M
University of California, Berkeley   7 D
University of California, Davis   15 D
University of California, Los Angeles   1 M, 1 D
University of California, Riverside   5 M, 37 D
University of California, Santa Cruz   3 M
California State University, Fresno   1 M
Case Western Reserve University   1 M
Catholic University of America   2 M
University of Chicago   1 D
University of Cincinnati   1 M
Clark University   1 D
Colgate University   1 M
University of Colorado   2 M, 8 D
Colorado School of Mines   3 M
Colorado State University   7 M, 11 D
Columbia University, Teachers College   3 M, 2 D
Cornell University   8 M, 39 D
Dartmouth College   2 D
University of Delaware, College of Marine
   Studies   1 D
Ecole Polytechnique   1 M
Emory University   1 M
University of Florida   3 M, 18 D
George Washington University   1 M, 1 D
University of Georgia   7 D
University of Guelph   10 M, 8 D
University of Hawaii   40 M, 19 D
University of Hawaii at Manoa   1 D
University of Houston   1 M
University of Idaho   10 M, 3 D
University of Illinois, Urbana   1 M, 10 D
Indiana State University   1 D
Indiana University, Bloomington   5 M, 2 D
University of Iowa   3 D
Iowa State University of Science and
   Technology   1 M, 18 D
The Johns Hopkins University   1 M, 1 D
University of Kansas   1 M, 4 D
Kansas State University   3 M, 2 D
University of Kentucky   2 M, 4 D
Universite Laval   1 M
Louisiana State University   1 M, 3 D
Louisiana Tech University   1 M
University of Maine   1 M
University of Manitoba   3 M, 3 D
University of Maryland   1 M, 2 D
Massachusetts Institute of Technology   1 D
McGill University   2 D
McMaster University   1 M

Miami University (Ohio)   1 M
University of Michigan   1 M, 3 D
Michigan State University   13 D
University of Minnesota, Minneapolis   1 M, 7 D
University of Mississippi   1 D
Mississippi State University   1 M, 3 D
University of Missouri, Columbia   1 M, 5 D
University of Missouri, Kansas City   1 M
University of Missouri, Rolla   3 M
Montana State University   2 D
Murray State University   1 M
University of Nebraska, Lincoln   1 M, 11 D
University of Nevada   4 M
University of New Brunswick   1 M
New Mexico Institute of Mining and
   Technology   1 D
New Mexico State University, Las Cruces   1 D
University of North Carolina, Chapel
   Hill   9 M, 3 D
North Carolina State University   2 M, 31 D
North Dakota State University   2 D
University of Northern Colorado   1 D
Ohio State University   2 M, 12 D

University of Oklahoma   1 D
Oklahoma State University   11 D
Old Dominion University   2 M
Oregon State University   2 M, 14 D
Pennsylvania State University, University
   Park   5 M, 10 D
Purdue University   7 M, 19 D
Queen's University   2 M, 1 D
Rutgers, The State University, New
   Brunswick   2 M, 11 D
Rutgers, The State University, Newark   1 M
University of Saskatchewan   2 M, 7 D
University of South Carolina   2 M
University of South Dakota   1 M
University of Southern California   1 D
Southern Illinois University,
   Carbondale   1 D
Southern Methodist University   1 M
St. Louis University   2 M
Stanford University   3 D
Syracuse University   1 D
University of Tennessee, Knoxville   8 D
University of Texas, Arlington   1 M

University of Texas, Austin   3 M
Texas A&M University   1 M, 15 D
Texas Tech University   4 M
University of Toledo   2 M
University of Toronto   1 M, 2 D
United States Naval Academy   1 M
University of Utah   1 M, 2 D
Utah State University   9 D
University of Virginia   6 M, 1 D
Virginia Polytechnic Institute and State
   University   2 M, 6 D
Virginia State University   1 M
University of Washington   1 M, 7 D
Washington State University   3 M, 8 D
Washington University   3 M, 1 D
University of Waterloo   1 M, 1 D
West Virginia University   2 D
University of Western Ontario   2 M, 5 D
Wichita State University   1 M
University of Wisconsin-Madison   3 M, 34 D
University of Wyoming   2 D
Yale University   1 M, 1 D